THIRD EDITION

GENETICS

ANALYSIS & PRINCIPLES

Multiple Choice Quiz
(See related pages)

Results Reporter

Out of 25 questions, you answered 3 correctly, for a final grade of 12%.

3 correct (12%)
4 incorrect (16%)
18 unanswered (72%)

Your Results:

The correct answer for each question is indicated by a ☺.

1 CORRECT The nucleotide change ________ is an example of a transversion, while the nucleotide change ________ is an example of a transition.

A) A→G; C→G
B) C→G; A→G
C) T→C; A→G
D) C→T; G→A
E) G→C; C→G

2 CORRECT Which type of mutation is most likely to revert?

A) deletion
B) translocation
C) inversion
D) transposition
E) transition

3 CORRECT Which type of mutation is least likely to revert?

A) deletion
B) translocation
C) inversion
D) transposition
E) transition

4 INCORRECT The hydrolysis of an -NH2 group from a base is called ________ by causing ________:

A) deamination; transversions

Routing Information

Date: Tue Sep 09 09:13:24 CDT 2003
My name:
Email these results to:
Email address: Format:
Me: Text
My Instructor: Text
My TA: Text
Other: Text
E-Mail The Results

Test Yourself

Take a quiz at the Genetics Online Learning Center to gauge your mastery of chapter content. Each chapter quiz is specifically constructed to test your comprehension of key concepts. Immediate feedback on your responses explains why an answer is correct or incorrect. You can even e-mail your quiz results to your professor!

Tweaking the Experiment

For the in-depth experiments that are found in each chapter, a self-help quiz is found. These questions ask you to predict how changes in the experimental parameters may affect the outcome of the experiment.

Interactive Activities

Fun and exciting learning experiences await you at the Genetics Online Learning Center! Each chapter offers a series of interactive crossword puzzles, vocabulary flashcards, and other engaging activities designed to reinforce learning.

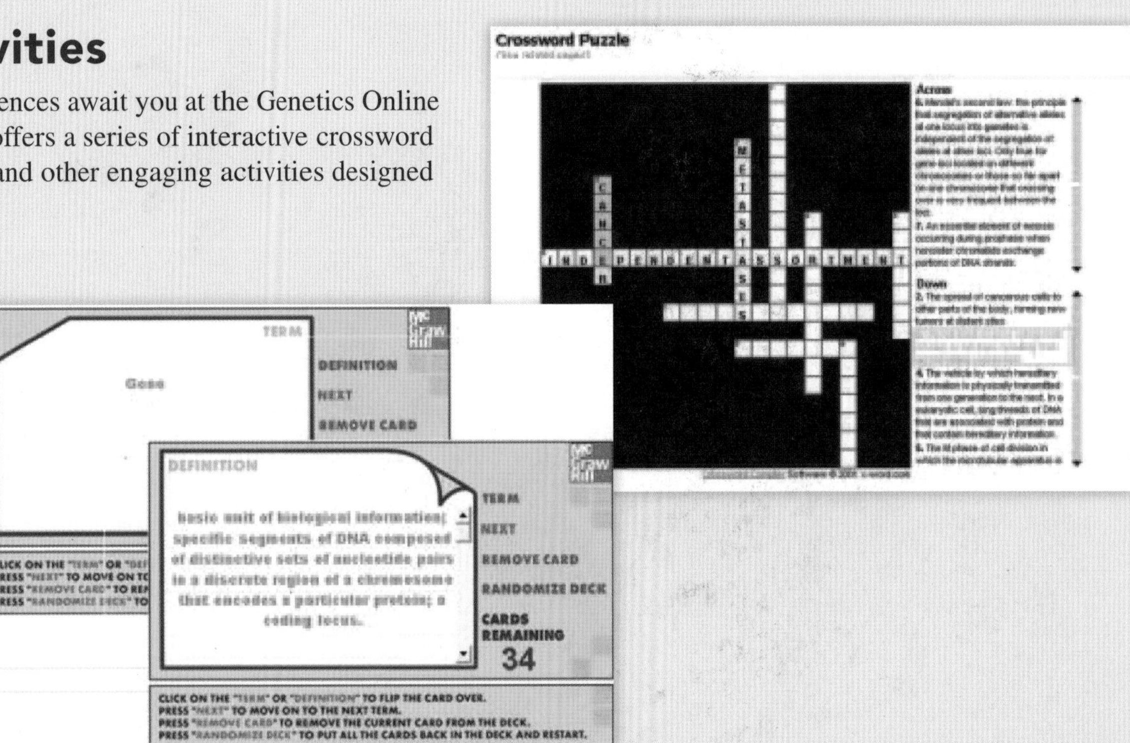
Crossword Puzzle
TERM
Gene
DEFINITION
NEXT
REMOVE CARD
DEFINITION
basic unit of biological information; specific segments of DNA composed of distinctive sets of nucleotide pairs in a discrete region of a chromosome that encodes a particular protein; a coding locus.
TERM
NEXT
REMOVE CARD
RANDOMIZE DECK
CARDS REMAINING
34
CLICK ON THE "TERM" OR "DEFINITION" TO FLIP THE CARD OVER.
PRESS "NEXT" TO MOVE ON TO THE NEXT TERM.
PRESS "REMOVE CARD" TO REMOVE THE CURRENT CARD FROM THE DECK.
PRESS "RANDOMIZE DECK" TO PUT ALL THE CARDS BACK IN THE DECK AND RESTART.

:: ROBERT J. BROOKER

UNIVERSITY OF MINNESOTA–MINNEAPOLIS

third edition

GENETICS

ANALYSIS & PRINCIPLES

Boston Burr Ridge, IL Dubuque, IA New York San Francisco St. Louis
Bangkok Bogotá Caracas Kuala Lumpur Lisbon London Madrid Mexico City
Milan Montreal New Delhi Santiago Seoul Singapore Sydney Taipei Toronto

GENETICS: ANALYSIS & PRINCIPLES, THIRD EDITION

 This book is printed on recycled, acid-free paper containing 10% postconsumer waste.

1 2 3 4 5 6 7 8 9 0 DOW/DOW 0 9 8

ISBN 978–0–07–299278–6
MHID 0–07–299278–6

Publisher: *Janice Roerig-Blong*
Executive Editor: *Patrick E. Reidy*
Senior Developmental Editor: *Lisa A. Bruflodt*
Marketing Manager: *Barbara Owca*
Senior Project Manager: *Jayne Klein*
Senior Production Supervisor: *Laura Fuller*
Senior Media Project Manager: *Jodi Banowetz*
Designer: *John Joran*
Interior Designer: *Rokusek Design, Inc.*
Cover Illustration: *Imagineering Media Services Inc.*
Senior Photo Research Coordinator: *John C. Leland*
Photo Research: *Pronk&Associates*
Supplement Producer: *Mary Jane Lampe*
Compositor: *Lachina Publishing Services*
Typeface: *10/12 Minion*
Printer: *R. R. Donnelley Williard, OH*

The credits section for this book begins on page 823 and is considered an extension of the copyright page.

Library of Congress Cataloging-in-Publication Data

Brooker, Robert J.
Genetics : analysis & principles / Robert J. Brooker. — 3rd ed.
p. cm.
Includes bibliographical references and index.
ISBN 978–0–07–299278–6 — ISBN 0–07–299278–6 (hard copy : alk. paper) 1. Genetics. I. Title.
QH430.B766 2009
576.5--dc22

2007038523

www.mhhe.com

ABOUT THE AUTHOR

Robert J. Brooker is a Professor in the Department of Genetics, Cell Biology, and Development at the University of Minnesota–Minneapolis. He received his B.A. in Biology from Wittenberg University in 1978 and his Ph.D. in Genetics from Yale University in 1983. At Harvard, he conducted post-doctoral studies on the lactose permease, which is the product of the *lacY* gene of the *lac* operon. He continues his work on transporters at the University of Minnesota. Dr. Brooker's laboratory primarily investigates the structure, function, and regulation of iron transporters found in bacteria and *C. elegans.* At the University of Minnesota, he teaches undergraduate courses in biology, genetics, and cell biology.

DEDICATION

To my wife, Deborah, and our children,
Daniel, Nathan, and Sarah

BRIEF CONTENTS

PREFACE

In the third edition of *Genetics: Analysis & Principles,* the content has been updated to reflect current trends in the field. In addition, the presentation of the content has been improved in a way that fosters active learning. As an author, researcher, and teacher, I want a textbook that gets students actively involved in learning genetics. To achieve this goal, I have worked with a talented team of editors, illustrators, and media specialists who have helped me to make the third edition of *Genetics: Analysis & Principles* a fun learning tool. The features that we feel are most appealing to students, and which have been added to or improved upon in the third edition, are the following.

- **Interactive exercises** Education specialists have crafted interactive exercises in which the students can make their own choices in problem-solving activities, and predict what the outcomes will be. These exercises often focus on inheritance patterns and human genetic diseases. (For example, see Chapters 4 and 22.)
- **Animations** Our media specialists have created over 50 animations for a variety of genetic processes. These animations were made specifically for this textbook, and use the art from the textbook. The animations literally make many of the figures in the textbook "come to life."
- **Experiments** As in the previous editions, each chapter (beginning with Chapter 2) incorporates one or two experiments that are presented according to the scientific method. These experiments are not "boxed off" from the rest of the chapter. Rather, they are integrated within the chapters and flow with the rest of the text. As you are reading the experiments, you will simultaneously explore the scientific method and the genetic principles that have been discovered using this approach. For students, I hope this textbook will help you to see the fundamental connection between scientific analysis and principles. For both students and instructors, I expect that this strategy will make genetics much more fun to explore.
- **Art** The art has been further refined for clarity and completeness. This makes it easier and more fun for students to study the illustrations without having to go back and forth between the art and the text.
- **Engaging text** A strong effort has been made in the third edition to pepper the text with questions. Sometimes these are questions that scientists considered when they were conducting their research. Sometimes they are questions that the students might ask themselves when they are learning about genetics.

Overall, an effective textbook needs to accomplish three goals. First, it needs to provide comprehensive, accurate, and up-to-date content in its field. Second, it needs to provide students with an exposure to the techniques and skills that are needed for them to become successful in that field. And finally, it should inspire students so they want to pursue that field as a career. The hard work that has gone into the third edition of *Genetics: Analysis & Principles* has been aimed at achieving all three of these goals.

HOW WE EVALUATED YOUR NEEDS

ORGANIZATION

In surveying many genetics instructors, it became apparent that most people fall into two camps: **Mendel first** versus **Molecular first.** I have taught genetics both ways. As a teaching tool, this textbook has been written with these different teaching strategies in mind. The organization and content lend themselves to various teaching formats.

Chapters 2 through 8 are largely inheritance chapters, while Chapters 24 through 26 examine population and quantitative genetics. The bulk of the molecular genetics is found in Chapters 9 through 23, although I have tried to weave a fair amount of molecular genetics into Chapters 2 through 8 as well. The information in

Chapters 9 through 23 *does not assume* that a student has already covered Chapters 2 through 8. Actually, each chapter is written with the perspective that instructors may want to vary the order of their chapters to fit their students' needs.

For those who like to discuss inheritance patterns first, a common strategy would be to cover Chapters 1 through 8 first, and then possibly 24 through 26. (However, many instructors like to cover quantitative and population genetics at the end. Either way works fine.) The more molecular and technical aspects of genetics would then be covered in Chapters 9 through 23. Alternatively, if you like the "Molecular first" approach, you would probably cover Chapter 1, then skip to Chapters 9 through 23, then return to Chapters 2 through 8, and then cover Chapters 24 through 26 at the end of the course. This textbook was written in such a way that either strategy works just fine.

Finally, since genetics is such a broad discipline ranging from the molecular to the populational levels, many instructors have told us that it is a challenge for students to see both "the forest and the trees." It is commonly mentioned that students often have trouble connecting the concepts they have learned in molecular genetics with the traits that occur at the level of a whole organism (i.e., What does transcription have to do with blue eyes?). To try to make this connection more meaningful, certain figure legends in each chapter, designated **Genes→ Traits,** remind students that molecular and cellular phenomena ultimately lead to the traits that are observed in each species (for example, see Figure 4.20).

ACCURACY

Both the publisher and I acknowledge the fact that inaccuracies can be a source of frustration for both the instructor and students. Therefore, throughout the writing and production of this textbook we have worked very hard to catch and correct errors during each phase of development and production.

Each chapter has been reviewed by a minimum of ten people. At least eight of these people were faculty members who teach the course and/or conduct research in genetics. In addition, two development editors have gone through the material, at least twice, to check for accuracy in art, and consistency between the text and art. We also had a team of students work through all the problem sets and one development editor also checked them. The author personally checked every question and answer when the chapters were completed.

PEDAGOGY

Based on our discussions with instructors from many institutions, some common goals have emerged. Instructors want a broad textbook that clearly explains concepts in a way that is interesting, accurate, concise, and up-to-date. Likewise, most instructors want students to understand the experimentation that revealed these genetic concepts. In this textbook, concepts and experimentation are woven together to provide a story that enables students to learn the important genetic concepts that they will need in their future careers, and also to be able to explain the types of experiments that allowed researchers to derive such concepts. The end-of-chapter problem sets are categorized according to their main focus, either conceptual or experimental, although some problems contain a little of both. The problems are meant to strengthen students' abilities in a wide variety of ways.

- By bolstering their understanding of genetic principles.
- By enabling students to apply genetic concepts to new situations.
- By analyzing scientific data.
- By organizing their thoughts regarding a genetic topic.
- By improving their writing skills.

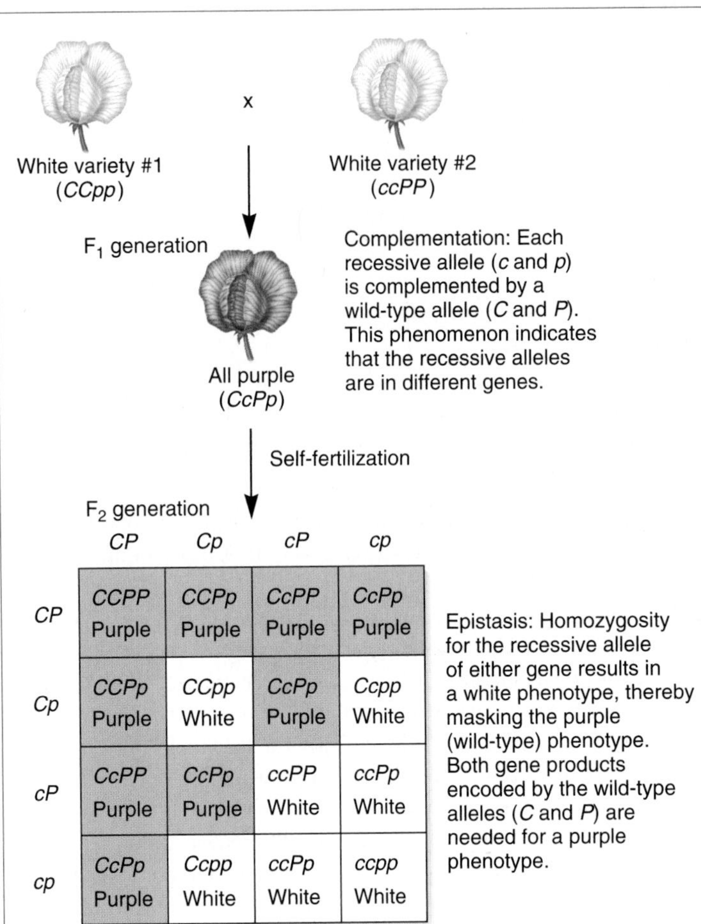

FIGURE 4.20 A cross between two different white varieties of the sweet pea.

Genes →Traits The color of the sweet pea flower is controlled by two genes, which are epistatic to each other and show complementation. Each gene is necessary for the production of an enzyme required for pigment synthesis. The recessive allele of either gene encodes a defective enzyme. If an individual is homozygous recessive for either of the two genes, the purple pigment cannot be synthesized. This results in a white phenotype.

ILLUSTRATIONS

In surveying students whom I teach, I often hear it said that most of their learning comes from studying the figures. Likewise, instructors frequently use the illustrations from a textbook as a central teaching tool. For these reasons, the greatest amount of effort in improving the third edition has gone into the illustrations. The illustrations have been refined with four goals in mind:

1. **Completeness** For most figures, it should be possible to understand an experiment or genetic concept by looking at the illustration alone. Students have complained that it is difficult to understand the content of an illustration if they have to keep switching back and forth between the figure and text. In cases where an illustration shows the steps in a scientific process, the steps are described in brief statements that allow the students to understand the whole process (for example, see Figure 12.10). Likewise, such illustrations should make it easier for instructors to explain these processes in the classroom.
2. **Clarity** The figures have been extensively reviewed by students and instructors. This has helped us to avoid drawing things that may be confusing or unclear. I hope that no one looks at an element in any figure and wonders, "What is that thing?" Aside from being unmistakably drawn, all new elements within each figure are clearly labeled.
3. **Consistency** Before we began to draw the figures for the second edition, we generated a style sheet that contained recurring elements that are found in many places in the textbook. Examples include the DNA double helix, DNA polymerase, and fruit flies. We agreed upon the best way(s) to draw these elements and also what colors they should be. Therefore, as students and instructors progress through this textbook, they become accustomed to the way things should look.
4. **Realism** An important goal of this and previous editions is to make each figure as realistic as possible. When drawing macroscopic elements (for example, fruit flies, pea plants), the illustrations are based on real images, not on cartoonlike simplifications. Our most challenging goal, and one that we feel has been achieved most successfully, is the realism of our molecular drawings. Whenever possible, we have tried to depict molecular elements according to their actual structures, if such structures are known. For example, the ways we have drawn RNA polymerase, DNA polymerase, DNA helicase, and ribosomes are based on their crystal structures. When a student sees a figure in this textbook that illustrates an event in transcription, RNA polymerase is depicted in a way that is as realistic as possible (for example, see Figure 12.10).

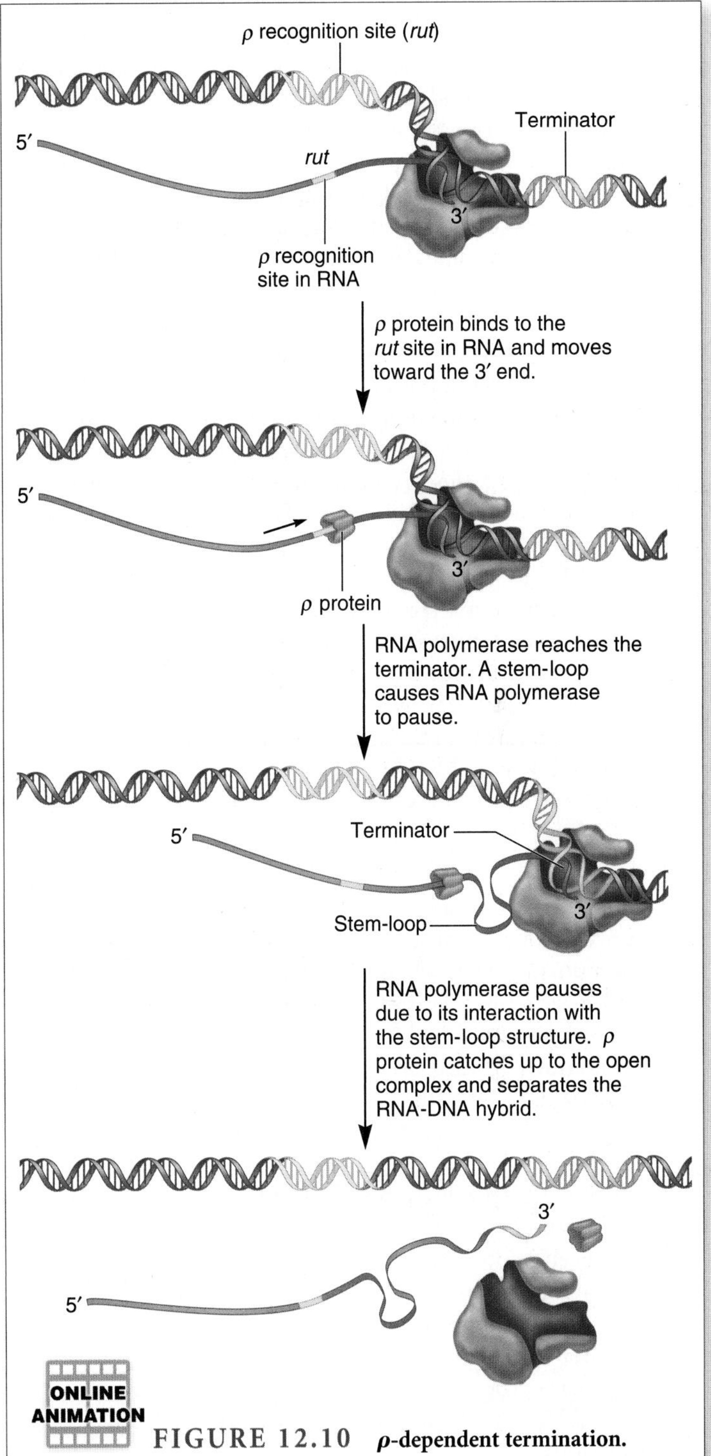
ρ recognition site (rut)
Terminator
5′
rut
3′
ρ recognition site in RNA
ρ protein binds to the rut site in RNA and moves toward the 3′ end.
5′
3′
ρ protein
RNA polymerase reaches the terminator. A stem-loop causes RNA polymerase to pause.
5′
Terminator
Stem-loop
3′
RNA polymerase pauses due to its interaction with the stem-loop structure. ρ protein catches up to the open complex and separates the RNA-DNA hybrid.
3′
5′
ONLINE ANIMATION

FIGURE 12.10 **ρ-dependent termination.**

WRITING STYLE

Motivation in learning often stems from enjoyment. If someone enjoys what they're reading, they are more likely to spend longer amounts of time with it and focus their attention more crisply. The writing style of this book is meant to be interesting, down to earth, and easy to follow. Each section of every chapter begins with an overview of the contents of that section, usually with a table or figure that summarizes the broad points. The section then examines how those broad points were discovered experimentally, as well as explaining many of the finer scientific details. Important terms are introduced in a boldface font. These terms are also found in the glossary.

There are various ways to make a genetics book interesting and inspiring. The subject matter itself is pretty amazing, so it's not difficult to build on that. In addition to describing the concepts and experiments in ways that motivate students, it is important to draw upon examples that bring the concepts to life. In a genetics book, many of these examples come from the medical realm. This textbook contains lots of examples of human diseases that exemplify some of the underlying principles of genetics. Students often say that they remember certain genetic concepts because they remember how defects in certain genes can cause disease. For example, defects in DNA repair genes cause a higher predisposition to develop cancer. In addition, I have tried to be evenhanded in providing examples from the microbial and plant world. Finally, students are often interested in applications of genetics that impact their everyday lives. Because we hear about genetics in the news on a daily basis, it's inspiring for students to learn the underlying basis for such technologies. Chapters 18 to 21 are devoted to genetic technologies, and applications of these and other technologies are found throughout this textbook. By the end of their genetics course, students should come away with a greater appreciation for the impact of genetics in their lives.

EXAMPLES OF SIGNIFICANT CONTENT CHANGES IN THE THIRD EDITION

- **Interactive Exercises** Many interactive exercises are available for students to understand inheritance patterns and human genetic diseases. These are indicated with an icon next to the relevant figures. They are largely found in Chapters 2, 4, 5, 7, and 22.
- **Animations** Nearly every chapter has one or more custom animations to help students understand genetic processes. These are also indicated by an icon next to the relevant figure.
- **Chapter 3 (Reproduction and Chromosome Transmission)** The illustrations and micrographs of mitosis and meiosis have been improved.
- **Chapter 4 (Extensions of Mendelian Inheritance)** The topic of suppressor mutations has been expanded.
- **Chapter 8 (Variation in Chromosome Structure and Number)** The technique of comparative genomic hybridization has been added, and included as a Feature Experiment.
- **Chapter 10 (Chromosome Organization and Molecular Structure)** An illustration and expanded discussion has been added for SMC proteins, which are involved with chromosome condensation and sister chromatid cohesion. Also, the histone code hypothesis is discussed and illustrated.
- **Chapter 11 (DNA Replication)** Illustrations in this chapter now depict the step-by-step formation of Okazaki fragments (for example, see Figure 11.10).
- **Chapter 13 (Translation of mRNA)** This chapter has undergone some reorganization to bring the basic concepts of translation to the first section.
- **Chapter 15 (Gene Regulation in Eukaryotes)** The topic of RNA interference has been expanded, including a Feature Experiment of the work of Fire and Mellow.
- **Chapter 16 (Gene Mutation and DNA Repair)** To complement some new material in Chapter 4, several examples of suppressor mutations are described. In addition, homologous recombination repair and nonhomologous end joining repair are illustrated and discussed within the context of repairing double-stranded DNA breaks.
- **Chapter 19 (Biotechnology)** Some new topics in biotechnology have been added including GloFish® and Bt corn.
- **Chapter 20 (Genomics I: Analysis of DNA)** The approach of shotgun DNA sequencing is greatly expanded along with complementary illustrations.
- **Chapter 21 (Genomics II: Functional Genomics, Proteomics, and Bioinformatics)** The method of chromatin immunoprecipitation (ChIP) has been added.
- **Chapter 22 (Medical Genetics and Cancer)** Several new topics have been added including the use of molecular markers in the mapping of human genetic diseases, and the molecular profiling of cancer cells using DNA microarrays.
- **Chapter 24 (Population Genetics)** This chapter contains an expanded discussion of natural selection with graphical representations of the various types. An entire new section on Sources of Genetic Variation has been added so students can appreciate how/why genetic variation is so prevalent.
- **Chapter 26 (Evolutionary Genetics)** The cladistic approach has been greatly expanded. In addition, an entire section on Evolutionary Developmental Biology (Evo-Devo) has been added.

SUGGESTIONS WELCOME!

It seems very appropriate to use the word evolution to describe the continued development of this textbook. I welcome any and all comments. The refinement of any science textbook requires input from instructors and their students. These include comments regarding writing, illustrations, supplements, factual content, and topics that may need greater or less emphasis. You are invited to contact me at:

Dr. Rob Brooker
Dept. of Genetics, Cell Biology, and Development
University of Minnesota
6-160 Jackson Hall
321 Church St.
Minneapolis, MN 55455
brook005@umn.edu

TEACHING AND LEARNING SUPPLEMENTS

McGraw-Hill offers various tools and technology products to support the third edition of *Genetics: Analysis & Principles.*

FOR THE INSTRUCTOR:

Online Learning Center (www.mhhe.com/brookergenetics3e)
The text specific website offers an extensive array of learning and teaching tools. In addition to student resources, the site also includes an instructor's manual, test bank, and lecture outlines.

Presentation Center
Complete set of electronic book images and assets for instructors
Build instructional materials wherever, whenever, and however you want! Accessed from your textbook's Online Learning Center, Presentation Center is an online digital library containing photos, artwork, animations, and other media types that can be used to create customized lectures, visually enhanced tests and quizzes, compelling course websites, or attractive printed support materials. All assets are copyrighted by McGraw-Hill Higher Education, but can be used by instructors for classroom purposes. The visual resources in this collection include:

- **Art** Full-color digital files of all illustrations in the book can be readily incorporated into lecture presentations, exams, or custom-made classroom materials. In addition, all files are pre-inserted into PowerPoint slides for ease of lecture preparation.
- **Photos** The photo collection contains digital files of photographs from the text, which can be reproduced for multiple classroom uses.
- **Tables** Every table that appears in the text has been saved in electronic form for use in classroom presentations and/or quizzes.
- **Animations** Numerous full-color animations illustrating important processes are also provided. Harness the visual impact of concepts in motion by importing these files into classroom presentations or online course materials.

Also residing on your textbook's Online Learning Center are:

- **PowerPoint Lecture Outlines** Ready-made presentations that combine art and lecture notes are provided for each chapter of the text.
- **PowerPoint Slides** For instructors who prefer to create their lectures from scratch, all illustrations, photos, and tables are pre-inserted by chapter into blank PowerPoint slides.

FOR THE STUDENT:

Student Study Guide/Solutions Manual
The solutions to the end-of-chapter problems and questions will aid the students in developing their problem-solving skills by providing the steps for each solution. The Study Guide follows the order of sections and subsections in the textbook and summarizes the main points in the text, figures, and tables. It also contains concept-building exercises, self-help quizzes, and practice exams.

Online Learning Center (www.mhhe.com/brookergenetics3e)
The text specific website offers an extensive array of learning tools, including a variety of quizzes for each chapter, interactive genetics problems, animations, and more.

ACKNOWLEDGMENTS

The production of a textbook is truly a collaborative effort, and I am greatly indebted to a variety of people. The first, second, and third editions went through multiple rounds of rigorous revisions that involved the input of faculty, students, editors, educational specialists, and media specialists. Their collective contributions are reflected in the final outcome.

Let me begin by acknowledging the many people at McGraw-Hill whose efforts are amazing. My highest praise goes to Lisa Bruflodt (Senior Developmental Editor), who managed and scheduled nearly every aspect of this project. I simply don't understand how one person can be so well organized and manage to stay calm at the same time. I also would like to thank Patrick Reidy (Executive Editor) for his patience in overseeing this project. He has the unenviable job of managing the budget for the book and that is not an easy task. I'm sure that he had some lively discussions with Janice Roerig-Blong (Publisher) during the course of this project. Other people at McGraw-Hill have played key roles in producing an actual book and the supplements that go along with it. In particular, Jayne Klein (Project Manager) has done a superb job of managing the components that need to be assembled to produce a book, along with Laura Fuller (Production Supervisor). I would also like to thank John Leland (Photo Research Coordinator), who acted as an interface between me and the photo company. In addition, my gratitude goes to John Joran (Designer), who provided much input into the internal design of the book as well as creating an awesome cover. Finally, I would like to thank Barb Owca (Marketing Manager), whose major efforts begin when the third edition comes out!

With regard to the content of the book, two editors worked closely with me in developing a book that is clear, consistent, and easy for students to follow. Deborah Brooker (Art/Text Coordinating Editor) analyzed all of the chapters in the textbook and made improvements with regard to art and text coordination. With great care, she examined the content in the text and determined if it matched what is found in the illustrations; she also checked every illustration for clarity and completeness. The high quality and pedagogy of the art are largely due to her efforts. She also scrutinized the text for clarity and logic. In addition, Joni Fraser (Freelance Development Editor) analyzed each chapter for its overall content and made many valuable suggestions for improvements. She did a great job of sharpening the writing. I am particularly grateful for her meticulous analysis of the problem sets. In her spare time, I think she may have worked all of the problems. In addition to Deb and Joni, I would also like to thank Jane DeShaw (Freelance Copy Editor) for making grammatical improvements throughout the text and art.

I would also like to extend a special thanks to everyone at Imagineering, the art house that produced the illustrations in the second and third editions. Though we initially considered several art houses, when we saw how Imagineering could draw RNA polymerase, it was clear who the winner was. Their ability to draw three-dimensional molecular structures is outstanding. I am grateful to the many artists and contact people at Imagineering who have played important roles in developing the art for the second and third editions. Likewise, the people at Pronk & Associates have done a great job of locating many of the photographs that have been used in the third edition. Also, Lachina Publishing Services worked with great care in the paging of the book, making sure that the figures and relevant text are as close to each other as possible.

Finally, I want to thank the many scientists who reviewed the chapters of this textbook. Their broad insights and constructive suggestions were an important factor that shaped its final content and organization. I am truly grateful for their time and effort.

REVIEWERS

Shivanthi Anandan, *Drexel University*
Alan G. Atherly, *Iowa State University*
William Baird, *Clemson University*
Gail S. Begley, *Northeastern University*
Michael Benedik, *University of Houston*
Kelly Bidle, *Rider University*
Andrew J. Bohonak, *San Diego State University*
Mark A. Brick, *Colorado State University*
James J. Campanella, *Montclair State University*
Arden Campbell, *Iowa State University*
Gerard Campbell, *University of Pittsburgh*
Diane A. Caporale, *University of Wisconsin–Stevens Point*
J. Aaron Cassill, *University of Texas at San Antonio*
Richard W. Cheney, Jr., *Christopher Newport University*
Joseph P. Chinnici, *Virginia Commonwealth University*
Brian Condie, *University of Georgia*
Gregory Copenhaver, *University of North Carolina*
Stephen J. D'Surney, *University of Mississippi*
Sumana Datta, *Texas A&M University*
Jeff DeJong, *University of Texas at Dallas*
Robert S. Dotson, *Tulane University*
Johnny El-Rady, *University of South Florida*
Susan Elrod, *California Polytechnic State University*
Bert Ely, *University of South Carolina*
Les Erickson, *Salisbury University*
Michael L. Foster, *Eastern Kentucky University*
Robert G. Fowler, *San Jose State University*
Richard F. Gaber, *Northwestern University*
Sudhindra R. Gadagkar, *University of Dayton*
Matthew R. Gilg, *University of North Florida*
Jack R. Girton, *Iowa State University*
Michael A. Goldman, *San Francisco State University*
Elliott S. Goldstein, *Arizona State University*
Nels H. Granholm, *South Dakota State University*
Alison Hill, *Duke University*
Ralph Hillman, *Temple University*
David C. Hinkle, *University of Rochester*
J. Spencer Johnston, *Texas A&M University*
Gregg Jongeward, *University of the Pacific*
Christopher Kvaal, *St. Cloud State University*
Arlene T. Larson, *University of Colorado at Denver*
Elena Levine Keeling, *California Polytechnic State University*
Rebecca Lyczak, *Ursinus College*
Jocelyn Malamy, *University of Chicago*
Philip M. Mathis, *Middle Tennessee State University*
Terry R. McGuire, *Rutgers University*
Mark E. Meade, *Jacksonville State University*
Roderick M. Morgan, *Grand Valley State University*
Robert Moss, *Wofford College*
Jacqueline Peltier Horn, *Houston Baptist University*
David K. Peyton, *Morehead State University*
Chara J. Ragland, *Texas A&M University*
David H. Reed, *University of Mississippi*
Linda L. Restifo, *University of Arizona*
Inder Saxena, *University of Texas at Austin*
Deemah N. Schirf, *University of Texas at San Antonio*
Mark Schlueter, *Xavier University of Louisiana*
Ekaterina G. Sedia, *The Richard Stockton College of New Jersey*
Jeff Sekelsky, *University of North Carolina at Chapel Hill*
Richard N. Sherwin, *University of Pittsburgh*
Joan Smith Sonneborn, *University of Wyoming*
Adriana Stoica, *Georgetown University*
Martin Tracey, *Florida International University*
Carol Trent, *Western Washington University*
Sarah Ward, *Colorado State University*
Dan E. Wells, *University of Houston*
Douglas Wendell, *Oakland University*
Frederick Whipple, *California State University–Fullerton*
Mark Winey, *University of Colorado–Boulder*
David Wofford, *University of Florida*
Daniel Wang, *University of Miami*

ACCURACY CHECKERS

Ruth Ballard, *CSU–Sacramento*
Michel Baudry, *USC*
Michael Benedik, *University of Houston*
Carey Booth, *Portland State University*
Matt Bozovsky, *Loyola University*

Joel Chandlee, *University of Rhode Island*
Henry Chang, *Purdue University*
Karen Cichy, *Michigan State University*
Brian Condie, *University of Georgia*
Erin Cram, *Northeastern University*
Jeff DeJong, *University of Texas at Dallas*
Rodney Dyer, *Virginia Commonwealth University*
Kevin Edwards, *Illinois State University*
Aboubaker Elkharroubi, *Johns Hopkins University*
Cedric Feschotte, *UT–Arlington*
David Foltz, *Louisiana State University*
Michael Foster, *Eastern Kentucky University*
Thomas Fowler, *Southern Illinois University*
Anne Galbraith, *UW–Lacrosse*
Jack Girton, *Iowa State University*
Elliott Goldstein, *ASU*
Christine Gray, *University of Puget Sound*
Tara Harmer Luke, *The Richard Stockton College of NJ*
Marilyn Hart, *Mankato State University*
Daryl Henderson, *Stony Brooker University*
Adam Hrincevich, *LSU*
Michael Ibba, *Ohio State University*
Margaret Jefferson, *CSU–LA*
David Kass, *Eastern Michigan University*
Cindy Malone, *CSU–Northridge*
John Mishler, *Delaware Valley College*
James Morris, *Clemson University*
Sang-Chul Nam, *Baylor University*
Daniel Odom, *California State University–Northridge*
Michael Polymenis, *Texas A&M University*
David Reed, *University of Mississippi*
Laurie Russell, *St. Louis University*
Mark Seeger, *Ohio State University*
Theresa Spradling, *University of Northern Iowa*
JD Swanson, *University of Central Arkansas*
James Thompson, *University of Oklahoma*
John Tomkiel, *U of NC–Greensboro*
Carol Trent, *Western Washington University*
Fyodor Urnov, *University of California–Berkeley*
Harald Vaessin, *Ohio State University*
Alan Waldman, *University of South Carolina*
Daniel Wang, *University of Miami*
Sarah Ward, *Colorado State University*
Matt White, *Ohio University*
Dwayne Wise, *Mississippi State University*
Yang Yen, *South Dakota State University*
Malcolm Zellars, *Georgia State University*
Jianzhi Zhang, *University of Michigan*

A Visual Guide to
GENETICS: ANALYSIS & PRINCIPLES

ROBERT J. BROOKER
UNIVERSITY OF MINNESOTA–MINNEAPOLIS

Brooker's Genetics *brings key concepts to life with its unique style of illustration.*

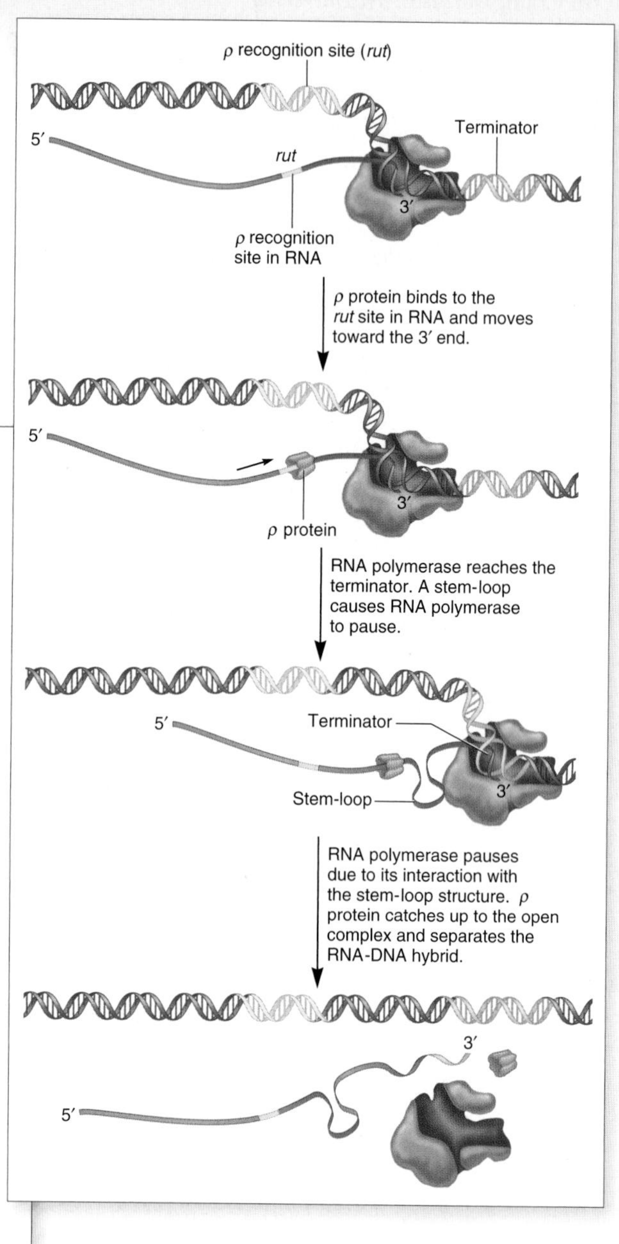

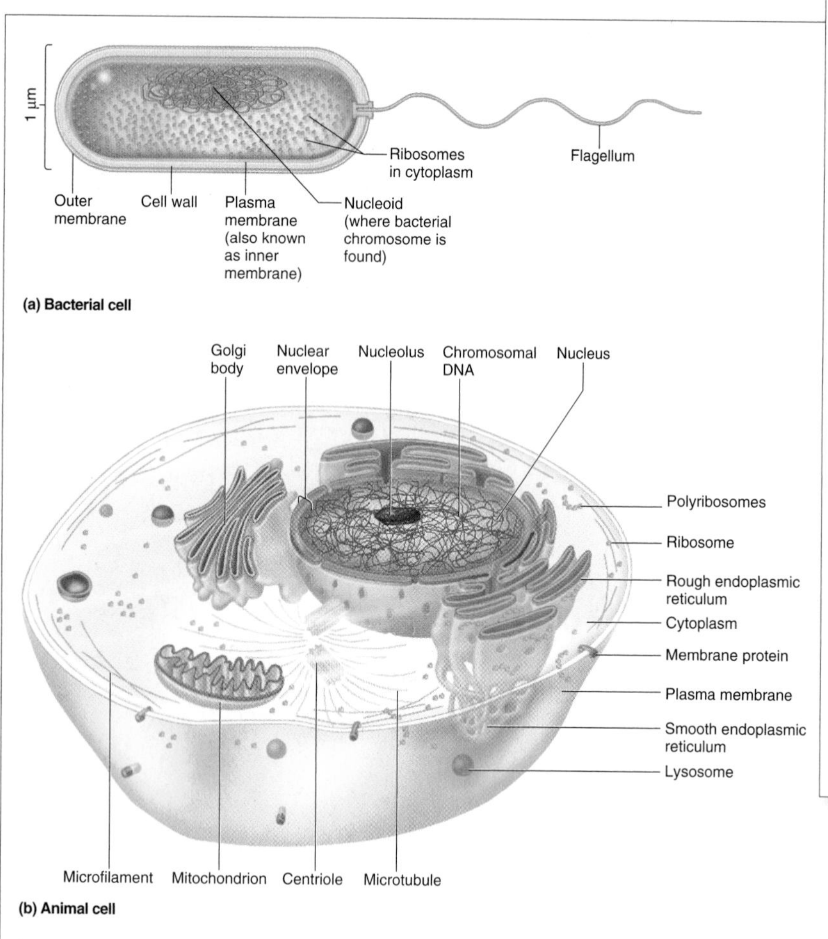

The digitally rendered images have a vivid three-dimensional look that will stimulate a student's interest and enthusiasm.

Each figure is carefully designed to follow closely with the text material.

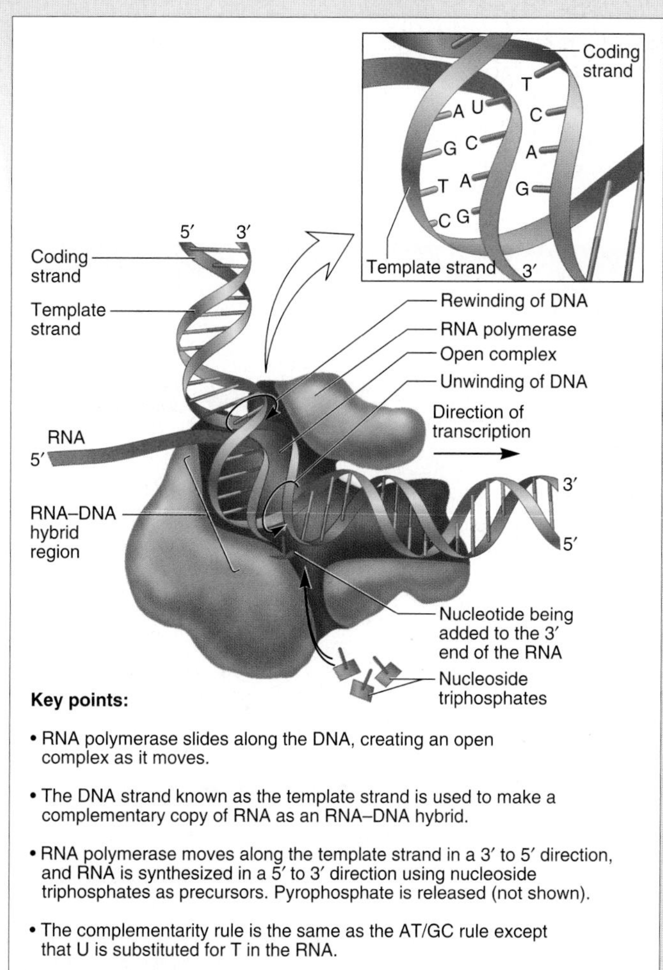
Coding strand
T
C
A
G
A U
G C
T A
C G
Template strand
3′
5′
3′
Coding strand
Template strand
Rewinding of DNA
RNA polymerase
Open complex
Unwinding of DNA
Direction of transcription
RNA
5′
3′
5′
RNA–DNA hybrid region
Nucleotide being added to the 3′ end of the RNA
Nucleoside triphosphates
Key points:
• RNA polymerase slides along the DNA, creating an open complex as it moves.
• The DNA strand known as the template strand is used to make a complementary copy of RNA as an RNA–DNA hybrid.
• RNA polymerase moves along the template strand in a 3′ to 5′ direction, and RNA is synthesized in a 5′ to 3′ direction using nucleoside triphosphates as precursors. Pyrophosphate is released (not shown).
• The complementarity rule is the same as the AT/GC rule except that U is substituted for T in the RNA.

DNA strands separate at origin, creating 2 replication forks.

Origin of replication

Replication forks

Primers are needed to initiate DNA synthesis. The synthesis of the leading strand begins in the direction of the replication fork. The first Okazaki fragment of the lagging strand is made in the opposite direction.

Leading strand

Primer

Direction of replication fork

Primer

First Okazaki fragment of the lagging strand

The leading strand elongates, and a second Okazaki fragment is made.

Second Okazaki fragment

First Okazaki fragment

The leading strand continues to elongate. A third Okazaki fragment is made, and the first and second are connected together.

Third Okazaki fragment

First and second Okazaki fragments have been connected to each other.

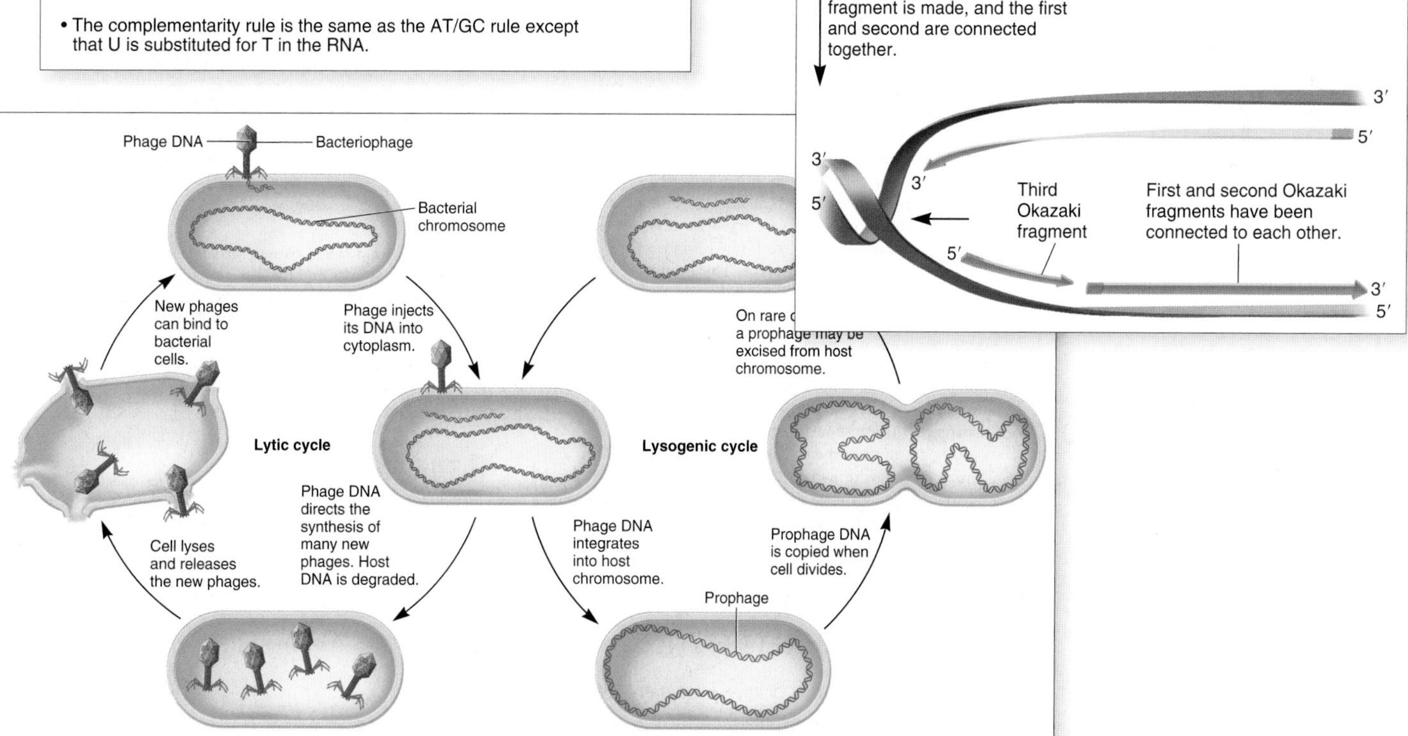
Phage DNA
Bacteriophage
Bacterial chromosome
New phages can bind to bacterial cells.
Phage injects its DNA into cytoplasm.
On rare occasions, a prophage may be excised from host chromosome.
Lytic cycle
Lysogenic cycle
Cell lyses and releases the new phages.
Phage DNA directs the synthesis of many new phages. Host DNA is degraded.
Phage DNA integrates into host chromosome.
Prophage DNA is copied when cell divides.
Prophage

Every illustration was drawn with four goals in mind: completeness, clarity, consistency, and realism.

Each chapter (beginning with Chapter 2) incorporates one or two experiments that are presented according to the scientific method. These experiments are integrated within the chapters and flow with the rest of the textbook. As you read the experiments, you will simultaneously explore the scientific method and the genetic principles learned from this approach.

STEP 1: BACKGROUND OBSERVATIONS

Each experiment begins with a description of the information that led researchers to study an experimental problem. Detailed information about the researchers and the experimental challenges they faced help students to understand actual research.

EXPERIMENT 5A

Creighton and McClintock Showed That Crossing Over Produced New Combinations of Alleles and Resulted in the Exchange of Segments Between Homologous Chromosomes

As we have seen, Morgan's studies were consistent with the hypothesis that crossing over occurs between homologous chromosomes to produce new combinations of alleles. To obtain direct evidence that crossing over can result in genetic recombination, Harriet Creighton and Barbara McClintock used an interesting strategy involving parallel observations. In studies conducted in 1931, they first made crosses involving two linked genes to produce parental and recombinant offspring. Second, they used a microscope to view the structures of the chromosomes in the parents and in the offspring. Because the parental chromosomes had some unusual structural features, they could microscopically distinguish the two homologous chromosomes within a pair. As we will see, this enabled them to correlate the occurrence of recombinant offspring with microscopically observable exchanges in segments of homologous chromosomes.

Creighton and McClintock focused much of their attention on the pattern of inheritance of traits in corn. This species has 10 different chromosomes per set, which are named chromosome 1, chromosome 2, chromosome 3, and so on. In previous cytological examinations of corn chromosomes, some strains were found to have an unusual chromosome 9 with a darkly staining knob at one end. In addition, McClintock identified an abnormal version of chromosome 9 that also had an extra piece of chromosome 8 attached at the other end (**Figure 5.5a**). This chromosomal rearrangement is called a translocation.

Creighton and McClintock insightfully realized that this abnormal chromosome could be used to demonstrate that two homologous chromosomes physically exchange segments as a result of crossing over. They knew that a gene was located near the knobbed end of chromosome 9 that provided color to corn kernels. This gene existed in two alleles, the dominant allele *C* (colored) and the recessive allele *c* (colorless). A second gene, located near the translocated piece from chromosome 8, affected the texture of the kernel endosperm. The dominant allele *Wx* caused starchy endosperm, and the recessive *wx* allele caused waxy endosperm. Creighton and McClintock reasoned that a crossover involving a normal chromosome 9 and a knobbed/translocated chromosome 9 would produce a chromosome that had either a knob or a translocation, but not both. These two types of chromosomes would be distinctly different from either of the parental chromosomes (**Figure 5.5b**).

As shown in the experiment of **Figure 5.6**, Creighton and McClintock began with a corn strain that carried an abnormal chromosome that had a knob at one end and a translocation at the other. Genotypically, this chromosome was *C wx*. The cytologically normal chromosome in this strain was *c Wx*. This corn plant, termed parent A, had the genotype *Cc Wxwx*. It was crossed to a strain called parent B that carried two cytologically normal chromosomes and had the genotype *cc Wxwx*.

They then observed the kernels in two ways. First, they examined the phenotypes of the kernels to see if they were colored or colorless, and starchy or waxy. Second, the chromosomes in each kernel were examined under a microscope to determine their cytological appearance. Altogether, they observed a total of 25 kernels (see data of Figure 5.6).

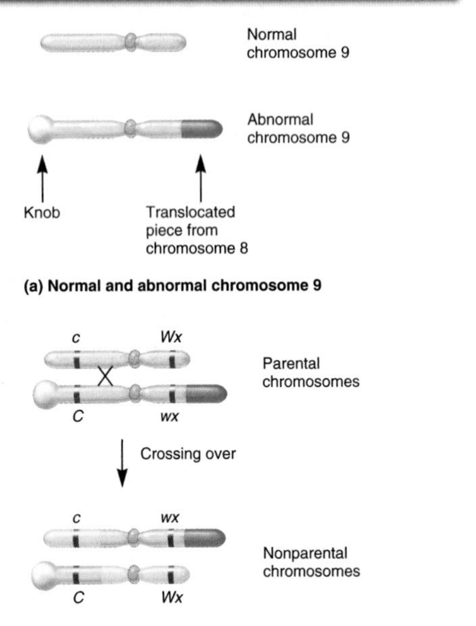

FIGURE 5.5 Crossing over between a normal and abnormal chromosome 9 in corn. **(a)** A normal chromosome 9 in corn is compared to an abnormal chromosome 9 that contains a knob at one end and a translocation at the opposite end. **(b)** A crossover produces a chromosome that contains only a knob at one end and another chromosome that contains only a translocation at the other end.

THE HYPOTHESIS

Offspring with nonparental phenotypes are the product of a crossover. This crossover should create nonparental chromosomes via an exchange of chromosomal segments between homologous chromosomes.

STEP 2: HYPOTHESIS

The student is given a statement describing the possible explanation for the observed phenomenon that will be tested. The hypothesis section reinforces the scientific method and allows students to experience the process for themselves.

■ **TESTING THE HYPOTHESIS — FIGURE 5.6 Experimental correlation between genetic recombination and crossing over.**

Starting materials: Two different strains of corn. One strain, referred to as parent A, had an abnormal chromosome 9 (knobbed/translocation) with a dominant *C* allele and a recessive *wx* allele. It also had a cytologically normal copy of chromosome 9 that carried the recessive *c* allele and the dominant *Wx* allele. Its genotype was *Cc Wxwx*. The other strain (referred to as parent B) had two normal versions of chromosome 9. The genotype of this strain was *cc Wxwx*.

	Experimental level	Conceptual level
1. Cross the two strains described. The tassel is the pollen-bearing structure, and the silk (equivalent to the stigma and style) is connected to the ovary. After fertilization, the ovary will develop into an ear of corn.		
2. Observe the kernels from this cross.		Each kernel is a separate seed that has inherited a set of chromosomes from each parent.

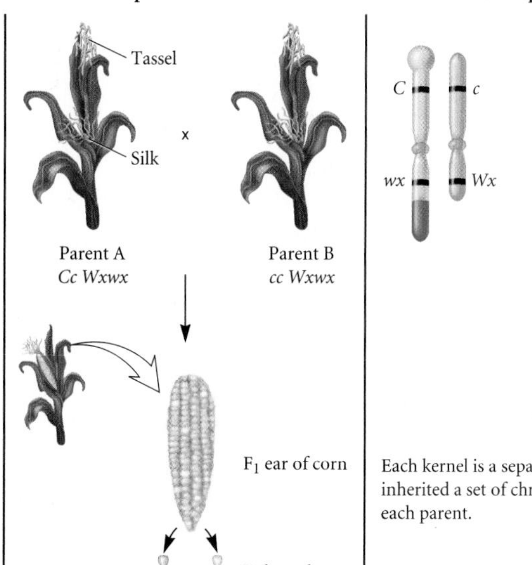

STEP 3: TESTING THE HYPOTHESIS

This section illustrates the experimental process, including the actual steps followed by scientists to test their hypothesis. Science comes alive for students with this detailed look at experimentation.

STEP 4: THE DATA

Actual data from the original research paper help students understand how actual research results are reported. Each experiment's results are discussed in the context of the larger genetic principle to help students understand the implications and importance of the research.

■ **THE DATA**

Phenotype of F_1 Kernel	*Number of Kernels Analyzed*	*Cytological Appearance of a Chromosome in F_1 Offspring**		*Did a Crossover Occur During Gamete Formation in Parent A?*
Colored/waxy	3	Knobbed/translocation C wx	Normal c wx	o
Colorless/starchy	11	Knobless/normal c Wx	Normal c Wx or c wx	No
Colorless/starchy	4	Knobless/translocation c wx	Normal c Wx	Yes
Colorless/waxy	2	Knobless/translocation c wx	Normal c wx	Yes
Colored/starchy	5	Knobbed/normal C Wx	Normal c Wx or c wx	Yes
Total	25			

*In this table, the chromosome on the left was inherited from parent A, and the chromosome on the right was inherited from parent B.

STEP 5: INTERPRETING THE DATA

This discussion, which examines whether the experimental data supported or disproved the hypothesis, gives students an appreciation for scientific interpretation.

■ **INTERPRETING THE DATA**

By combining the gametes in a Punnett square, the following types of offspring can be produced:

Parent B

♀ \ ♂	*c Wx*	*c wx*	
C wx	*Cc Wxwx* Colored, starchy	*Cc wxwx* Colored, waxy	Nonrecombinant
c Wx	*cc WxWx* Colorless, starchy	*cc Wxwx* Colorless, starchy	Nonrecombinant

nt A

Parent A	*Parent B*
C wx (nonrecombinant)	*c Wx*
c Wx (nonrecombinant)	*c wx*
C Wx (recombinant)	
c wx (recombinant)	

As seen in the Punnett square, two of the phenotypic cate ries, colored, starchy (*Cc Wxwx* or *Cc WxWx*) and colorless, star (*cc WxWx* or *cc Wxwx*), were ambiguous because they could a from a nonrecombinant and from a recombinant gamete. In ot words, these phenotypes could be produced whether or not rec bination occurred in parent A. Therefore, let's focus on the unambiguous phenotypic categories: colored, waxy (*Cc wxwx*) colorless, waxy (*cc wxwx*). The colored, waxy phenotype could h

These problems are crafted to aid students in developing a wide range of skills. They also develop a student's cognitive, writing, analytical, computational, and collaborative abilities.

CONCEPTUAL QUESTIONS

Test the understanding of basic genetic principles. The student is given many questions with a wide range of difficulty. Some require critical thinking skills, and some require the student to write coherent essay questions.

Conceptual Questions

C1. What is the meaning of the term genetic material?

C2. After the DNA from type IIIS bacteria is exposed to type IIR bacteria, list all of the steps that you think must occur for the bacteria to start making a type IIIS capsule.

C3. Look up the meaning of the word transformation in a dictionary and explain whether it is an appropriate word to describe the transfer of genetic material from one organism to another.

C4. What are the building blocks of a nucleotide? With regard to the 5' and 3' positions on a sugar molecule, how are nucleotides linked together to form a strand of DNA?

C5. Draw the structure of guanine, guanosine, and deoxyguanosine triphosphate.

C6. Draw the structure of a phosphodiester linkage.

C7. Describe how bases interact with each other in the double helix. This discussion should address the issues of complementarity, hydrogen bonding, and base stacking.

C8. If one DNA strand is 5'–GGCATTACACTAGGCCT–3' what is the sequence of the complementary strand?

C9. What is meant by the term DNA sequence?

C10. Make a side-by-side drawing of two DNA helices, one with 10 base pairs (bp) per 360° turn and the other with 15 bp per 360° turn.

C11. Discuss the differences in the structural features of A DNA, B DNA, and Z DNA.

C12. What parts of a nucleotide (namely, phosphate, sugar, and/or bases) occupy the major and minor grooves of double-stranded DNA, and what parts are found in the DNA backbone? If a DNA-binding protein does not recognize a specific nucleotide sequence, do you expect that it recognizes the major groove, the minor groove, or the DNA backbone? Explain.

C1

C1

C1

C16. Compare the structural features of a double-stranded RNA structure with those of a DNA double helix.

C17. Which of the following DNA double helices would be more difficult to separate into single-stranded molecules by treatment with heat, which breaks hydrogen bonds?

A. GGCGTACCAGCGCAT

CCGCATGGTCGCGTA

B. ATACGATTTACGAGA

TATGCTAAATGCTCT

Explain your choice.

C18. What structural feature allows DNA to store information?

C19. Discuss the structural significance of complementarity in DNA and in RNA.

C20. An organism has a G + C content of 64% in its DNA. What are the percentages of A, T, G, and C?

C21. Let's suppose you have recently identified an organism that was scraped from an asteroid that hit the earth. (Fortunately, no one was injured.) When you analyze this organism, you discover that its DNA is a triple helix, composed of six different nucleotides: A, T, G, C, X, and Y. You measure the chemical composition of the bases and find the following amounts of these six bases: A = 24%, T = 23%, G = 11%, C = 12%, X = 21%, Y = 9%. What rules would you propose govern triplex DNA formation in this organism? Note: There is more than one possibility.

C22. Upon further analysis of the DNA described in conceptual question C21, you discover that the triplex DNA in this alien organism is composed of a double helix, with the third helix wound within the major groove (just like the DNA in Figure 9.20). How would you propose that this DNA is able to replicate itself? In your answer, be specific about the base pairing rules within

EXPERIMENTAL QUESTIONS

Test the ability to analyze data, design experiments, or appreciate the relevance of experimental techniques.

Experimental Questions

E1. Genetic material acts as a blueprint for an organism's traits. Explain how the experiments of Griffith indicated that genetic material was being transferred to the type IIR bacteria.

E2. With regard to the experiment described in Figure 9.3, answer the following:

A. List several possible reasons why only a small percentage of the type IIR bacteria was converted to type IIIS.

B. Explain why an antibody must be used to remove the bacteria that are not transformed. What would the results look like, in all five cases, if the antibody/centrifugation step had not been included in the experimental procedure?

C. The DNA extract was treated with DNase, RNase, or protease. Why was this done? (In other words, what were the researchers trying to demonstrate?)

E3. An interesting trait that some bacteria exhibit is resistance to killing by antibiotics. For example, certain strains of bacteria are resistant to tetracycline, whereas other strains are sensitive to tetracycline. Describe an experiment you would carry out to demonstrate that tetracycline resistance is an inherited trait encoded by the DNA of the resistant strain.

E4. With regard to the experiment of Figure 9.6, answer the following:

A. Provide possible explanations why some of the DNA is in the supernatant.

B. Plot the results if the radioactivity in the pellet, rather than in the supernatant, had been measured.

C. Why were ^{32}P and ^{35}S chosen as radioisotopes to label the phages?

D. List possible reasons why less than 100% of the phage protein was removed from the bacterial cells during the shearing process.

E5. Does the experiment of Figure 9.6 rule out the possibility that RNA is the genetic material of T2 phage? Explain your answer. If it does not, could you modify the approach of Hershey and Chase to show that it is DNA and not RNA that is the genetic material of T2 bacteriophage? Note: It is possible to specifically label DNA or RNA by providing bacteria with radiolabeled thymine or uracil, respectively.

E6. In Chapter 9, we considered two experiments—one by Avery, MacLeod, and McCarty and the second by Hershey and Chase—that indicated DNA is the genetic material. Discuss the strengths

E7.

E8.

answer the following:

A. What is the purpose of paper chromatography?

prior to its exposure to the plant tissue?

STUDENT DISCUSSION/ COLLARBORATION QUESTIONS

Encourage students to consider broad concepts and practical problems. Some questions require a substantial amount of computational activities, which can be worked on as a group.

Questions for Student Discussion/Collaboration

1. Try to propose structures for a genetic material that are substantially different from the double helix. Remember that the genetic material must have a way to store information and a way to be faithfully replicated.

2. How might you provide evidence that DNA is the genetic material in mice?

Note: All answers appear at the website for this textbook; the answers to even-numbered questions are in the back of the textbook.

www.mhhe.com/brookergenetics3e

Visit the Online Learning Center for practice tests, answer keys, and other learning aids for this chapter. Enhance your understanding of genetics with our interactive exercises, quizzes, animations, and much more.

OVERVIEW OF GENETICS

1

Hardly a week goes by without a major news story involving a genetic breakthrough. The increasing pace of genetic discoveries has become staggering. The Human Genome Project is a case in point. This project began in the United States in 1990, when the National Institutes of Health and the Department of Energy joined forces with international partners to decipher the massive amount of information contained in our **genome**—the DNA found within all of our chromosomes (**Figure 1.1**). Working collectively, a large group of scientists from around the world has produced a detailed series of maps that help geneticists navigate through human DNA. Remarkably, in only a decade, they determined the DNA sequence (read in the bases of A, T, G, and C) covering over 90% of the human genome. The first draft of this sequence, published in 2001, is nearly 3 billion nucleotide base pairs in length. The completed sequence, published in 2003, has an accuracy greater than 99.99%; fewer than one mistake was made in every 10,000 base pairs!

Studying the human genome allows us to explore fundamental details about ourselves at the molecular level. The results of the project are expected to shed considerable light on basic questions, like how many genes we have, how genes direct the activities of living cells, how species evolve, how single cells develop into complex tissues, and how defective genes cause disease. Furthermore, such understanding may lend itself to improvements in modern medicine by leading to better diagnoses of diseases and the development of new medicines to treat them.

As scientists have attempted to unravel the mysteries within our genes, this journey has involved the invention of many new technologies. This textbook emphasizes a large number of these modern approaches. Paradoxically, one could argue that these advancements may have a greater impact on our lives than the original discoveries upon which they were based. For example, the technology of gene cloning originally arose from discoveries concerning bacterial enzymes that "cut and paste" DNA fragments together. These scientific discoveries have made it possible to create medicines that would otherwise be difficult or impossible to make. An example is human recombinant insulin, termed *Humulin*, which is synthesized in a strain of *Escherichia coli* bacteria that has been genetically altered by the addition of the gene for human insulin. The bacteria are grown in a laboratory and make large amounts of human insulin. As discussed in Chapter 19, this insulin is purified and administered to many people with insulin-dependent diabetes.

CHAPTER OUTLINE

Copycat, the first cloned pet. In 2002, the cat shown here, called Copycat, was produced by cloning, a procedure described in Chapter 19.

PART I

Introduction

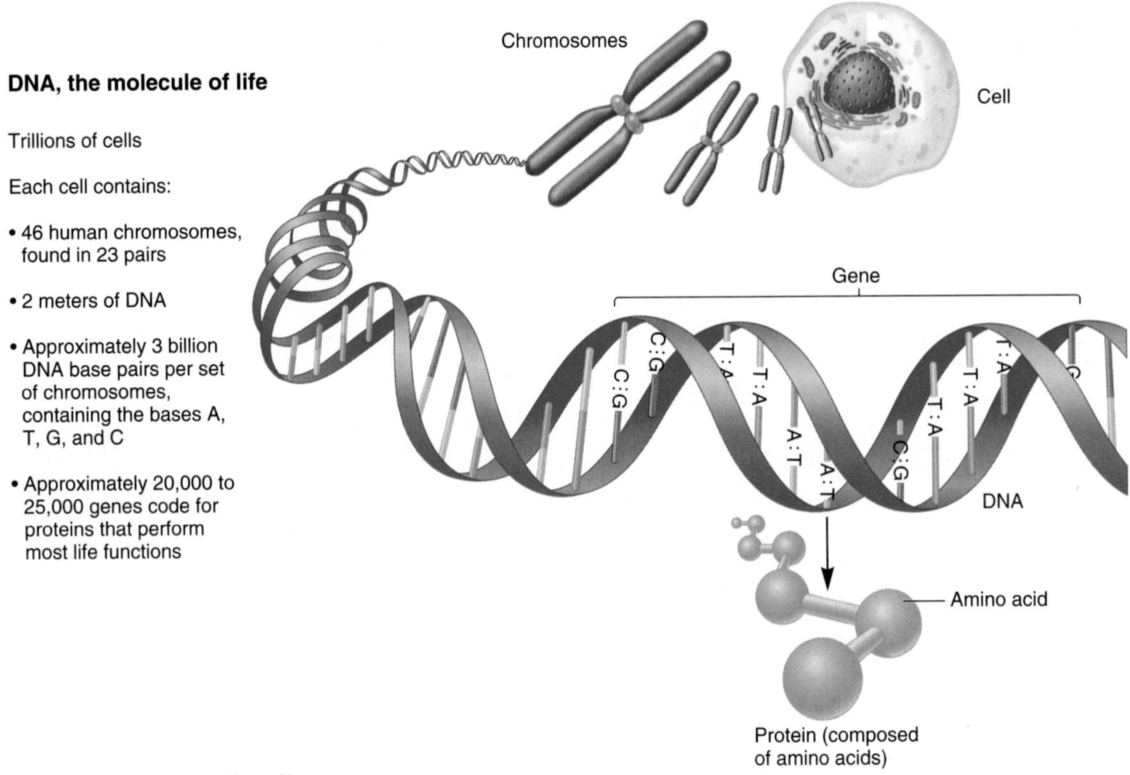

(a) The genetic composition of humans

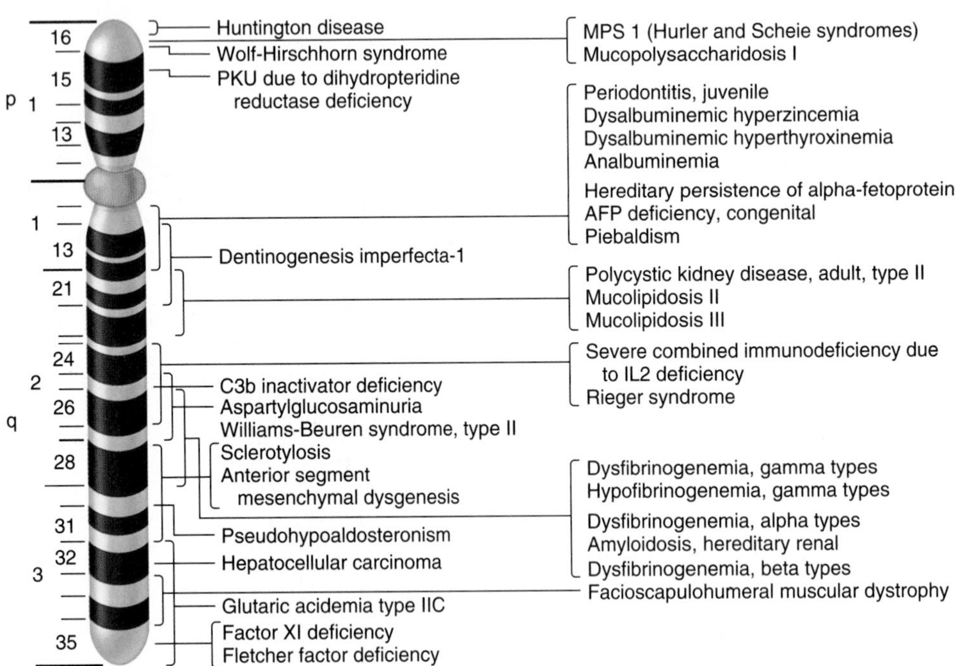

(b) Genes on one human chromosome that are associated with disease when mutant

FIGURE 1.1 **The Human Genome Project.** (**a**) The human genome is a complete set of human chromosomes. People have two sets of chromosomes, one from each parent. Collectively, each set of chromosomes is composed of a DNA sequence that is approximately 3 billion nucleotide base pairs long. Estimates suggest that each set contains about 20,000 to 25,000 different genes. This figure emphasizes the DNA found in the cell nucleus. Humans also have a small amount of DNA in their mitochondria, which has also been sequenced. (**b**) An important outcome of this work is the identification of genes that contribute to human diseases. This illustration depicts a map of a few genes that are located on human chromosome 4. When these genes carry certain rare mutations, they can cause the diseases designated in this figure.

New genetic technologies are often met with skepticism and sometimes even with disdain. An example would be DNA fingerprinting, a molecular method to identify an individual based on a DNA sample (see Chapter 24). Though this technology is now relatively common in the area of forensic science, it was not always universally accepted. High-profile crime cases in the news cause us to realize that not everyone believes in DNA fingerprinting, in spite of its extraordinary ability to uniquely identify individuals. A second controversial example would be mammalian cloning. In 1997, Ian Wilmut and his colleagues created clones of sheep, using mammary cells from an adult animal (**Figure 1.2**). More recently, such cloning has been achieved in several mammalian species, including sheep, cows, mice, goats, pigs, and cats. In 2002, the first pet was cloned, a cat named Copycat (see photo at the beginning of the chapter). The cloning of mammals provides the potential for many practical applications. With regard to livestock, cloning would enable farmers to use cells from their best individuals to create genetically homogeneous herds. This could be advantageous in terms of agricultural yield, although such a genetically homogeneous herd may be more susceptible to certain diseases. However, people have become greatly concerned with the possibility of human cloning. This prospect has raised serious ethical questions. Within the past few years, legislative bills have been introduced that involve bans on human cloning.

Finally, genetic technologies provide the means to modify the traits of animals and plants in ways that would have been unimaginable just a few decades ago. **Figure 1.3a** illustrates a bizarre example in which scientists introduced a gene from jellyfish into mice. As you may know, certain species of jellyfish emit a "green glow." These jellyfish have a gene that encodes a bioluminescent protein called green fluorescent protein (GFP). When exposed to blue or ultraviolet (UV) light, the protein emits a striking green-colored light. Scientists were able to clone the *GFP* gene from a sample of jellyfish cells and then introduce this gene into laboratory mice. The GFP protein is made throughout the cells of their bodies. As a result, their skin, eyes, and organs give off an eerie green glow when exposed to UV light. Only their fur does not glow.

FIGURE 1.2 **The cloning of a mammal.** The lamb on the left is Dolly, the first mammal to be cloned. She was cloned from the cells of a Finn Dorset (a white-faced sheep). The sheep on the right is Dolly's surrogate mother, a Blackface ewe. A description of how Dolly was created is presented in Chapter 19.

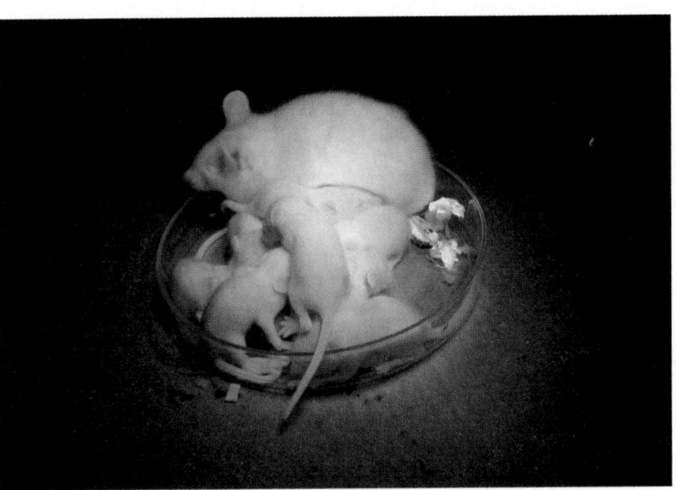

(a) GFP expressed in mice

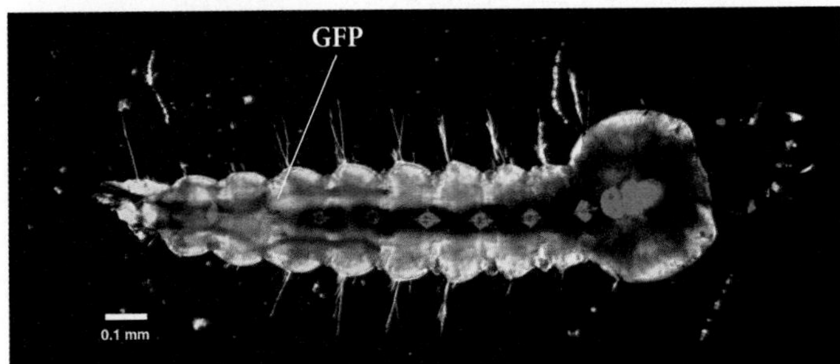

(b) GFP expressed in the gonads of a male mosquito

FIGURE 1.3 **The introduction of a jellyfish gene into laboratory mice and mosquitoes.** **(a)** A gene that naturally occurs in the jellyfish encodes a protein called green fluorescent protein (GFP). The GFP gene was cloned and introduced into mice. When these mice are exposed to UV light, GFP emits a bright green color. These mice glow green, just like jellyfish! **(b)** GFP was introduced next to a gene sequence that causes the expression of GFP only in the gonads of male mosquitoes. This allows researchers to identify and sort males from females.

GFP allows researchers to identify particular proteins in cells or specific body parts. For example, Andrea Crisanti and colleagues have altered mosquitoes to express a green fluorescent protein only in the gonads of males (**Figure 1.3b**). This enables the researchers to identify and sort males and females. Why is this useful? The ability to rapidly sort mosquitoes makes it possible to produce populations of sterile males and then release the sterile males without the risk of releasing additional females. The release of sterile males may be effective at controlling mosquito populations because females only breed once before they die. Mating with a sterile male will prevent a female from producing offspring.

Overall, as we move forward in the twenty-first century, the excitement level in the field of genetics is high, perhaps higher than it has ever been. Nevertheless, the excitement generated by new genetic knowledge and technologies will also create many ethical and societal challenges. In this chapter, we will begin with an overview of genetics and then explore the various fields of genetics and their experimental approaches.

1.1 THE RELATIONSHIP BETWEEN GENES AND TRAITS

Genetics is the branch of biology that deals with heredity and variation. It stands as the unifying discipline in biology by allowing us to understand how life can exist at all levels of complexity, ranging from the molecular to the population level. Genetic variation is the root of the natural diversity that we observe among members of the same species as well as among different species.

Genetics is centered on the study of genes. A gene is classically defined as a unit of heredity, but such a vague definition does not do justice to the exciting characteristics of genes as intricate molecular units that manifest themselves as potent contributors to cell structure and function. At the molecular level, a **gene** is a segment of DNA that produces a functional product. The functional product of most genes is a polypeptide, which is a linear sequence of amino acids that folds into units that constitute proteins. In addition, genes are commonly described according to the way they affect **traits,** which are the characteristics of an organism. In humans, for example, we speak of traits such as eye color, hair texture, and height. The ongoing theme of this textbook is the relationship between genes and traits. As an organism grows and develops, its collection of genes provides a blueprint that determines its characteristics.

In this section of Chapter 1, we will examine the general features of life, beginning with the molecular level and ending with populations of organisms. As will become apparent, genetics is the common thread that explains the existence of life and its continuity from generation to generation. For most students, this chapter should serve as a cohesive review of topics they learned in other introductory courses such as General Biology. Even so, it is usually helpful to see the "big picture" of genetics before delving into the finer details that are covered in Chapters 2 through 26.

Living Cells Are Composed of Biochemicals

To fully understand the relationship between genes and traits, we need to begin with an examination of the composition of living organisms. Every cell is constructed from intricately organized chemical substances. Small organic molecules such as glucose and amino acids are produced from the linkage of atoms via chemical bonds. The chemical properties of organic molecules are essential for cell vitality in two key ways. First, the breakage of chemical bonds during the degradation of small molecules provides energy to drive cellular processes. A second important function of these small organic molecules is their role as the building blocks for the synthesis of larger molecules. Four important categories of larger cellular molecules are **nucleic acids** (i.e., DNA and RNA), **proteins, carbohydrates,** and **lipids.** Three of these—nucleic acids, proteins, and carbohydrates—form **macromolecules** that are composed of many repeating units of smaller building blocks. Proteins, RNA, and carbohydrates can be made from hundreds or even thousands of repeating building blocks. DNA is the largest macromolecule found in living cells. A single DNA molecule can be composed of a linear sequence of hundreds of millions of nucleotides!

The formation of cellular structures relies on the interactions of molecules and macromolecules. For example, nucleotides are the building blocks of DNA, which is a constituent of cellular chromosomes (**Figure 1.4**). In addition, the DNA is associated with a myriad of proteins that provide organization to the structure of chromosomes. Within a eukaryotic cell, the chromosomes are contained in a compartment called the cell nucleus. The nucleus is bounded by a membrane composed of lipids and proteins that shields the chromosomes from the rest of the cell. The organization of chromosomes within a cell nucleus protects the chromosomes from mechanical breakage and provides a single compartment for genetic activities such as gene transcription. As a general theme, the formation of large cellular structures arises from interactions among different molecules and macromolecules. These cellular structures, in turn, are organized to make a complete living cell.

Each Cell Contains Many Different Proteins That Determine Cellular Structure and Function

To a great extent, the characteristics of a cell depend on the types of proteins that it makes. As we will learn throughout this textbook, proteins are the "workhorses" of all living cells. The range of functions among different types of proteins is truly remarkable. Some proteins help determine the shape and structure of a given cell. For example, the protein known as tubulin can assemble into large structures known as microtubules, which provide the cell with internal structure and organization. Other proteins are inserted into cell membranes and aid in the transport of ions and small molecules across the membrane. Proteins may also function as biological motors. An interesting case is the protein known as myosin, which is involved in the contractile properties of muscle cells. Within multicellular organisms, certain proteins also function in cell-to-cell recognition and signaling. For example, hormones such as insulin are secreted by endocrine cells and

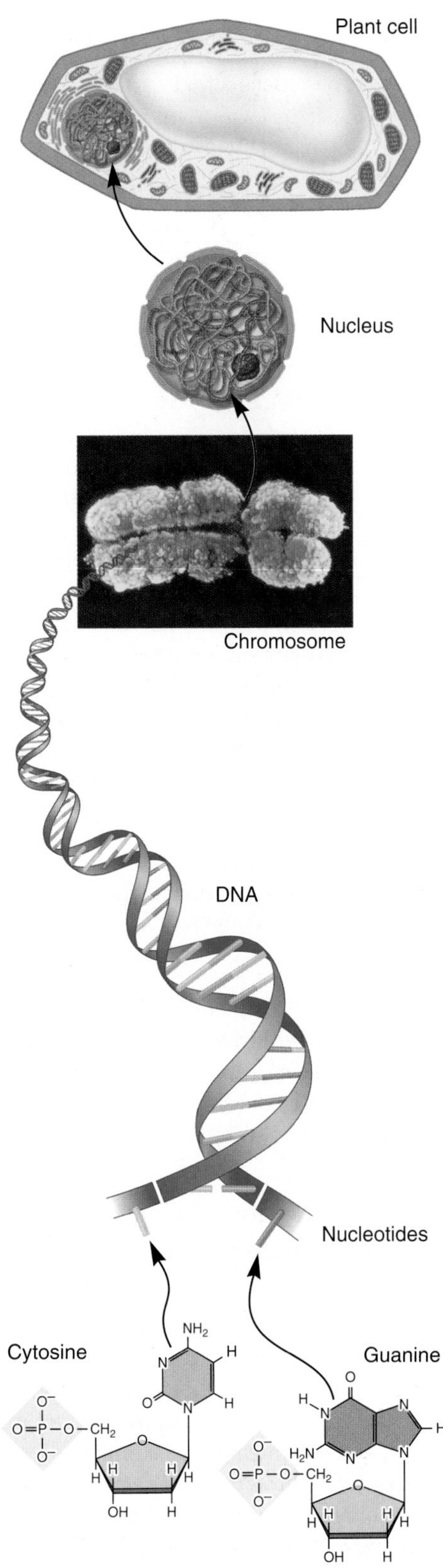

FIGURE 1.4 **Chemical composition of living cells.** Cellular structures are constructed from smaller building blocks. In this example, DNA is formed from the linkage of nucleotides to produce a very long macromolecule. The DNA associates with proteins to form a chromosome. The chromosomes are located within a membrane-bounded organelle called the nucleus, which, along with many different types of organelles, is found within a complete cell.

bind to the insulin receptor protein found within the plasma membrane of target cells.

Enzymes, which accelerate chemical reactions, are a particularly important category of proteins. Some enzymes play a role in the breakdown of molecules or macromolecules into smaller units. These are known as catabolic enzymes and are important in the utilization of energy. Alternatively, anabolic enzymes and accessory proteins function in the synthesis of molecules and macromolecules throughout the cell. The construction of a cell greatly depends on its proteins involved in anabolism because these are required to synthesize all cellular macromolecules.

Molecular biologists have come to realize that the functions of proteins underlie the cellular characteristics of every organism. At the molecular level, proteins can be viewed as the active participants in the enterprise of life.

DNA Stores the Information for Protein Synthesis

The genetic material is composed of a substance called **deoxyribonucleic acid,** abbreviated **DNA.** The DNA stores the information needed for the synthesis of all cellular proteins. In other words, the main function of the genetic blueprint is to code for the production of cellular proteins in the correct cell, at the proper time, and in suitable amounts. This is an extremely complicated task because living cells make thousands of different proteins. Genetic analyses have shown that a typical bacterium can make a few thousand different proteins, and estimates among higher eukaryotes range in the tens of thousands.

DNA's ability to store information is based on its molecular structure. DNA is composed of a linear sequence of **nucleotides.** Each nucleotide contains one nitrogen-containing base, either adenine (A), thymine (T), guanine (G), or cytosine (C). The linear order of these bases along a DNA molecule contains information similar to the way that groups of letters of the alphabet represent words. For example, the "meaning" of the sequence of bases ATGGGCCTTAGC differs from that of TTTAAGCTTGCC. DNA sequences within most genes contain the information to direct the order of amino acids within polypeptides according to the **genetic code.** In the code, a three-base sequence specifies one particular **amino acid** among the 20 possible choices. One or more polypeptides form a functional protein. In this way, the DNA can store the information to specify the proteins made by an organism.

DNA Sequence	Amino Acid Sequence
ATG GGC CTT AGC	METHIONINE GLYCINE LEUCINE SERINE
TTT AAG CTT GCC	PHENYLALANINE LYSINE LEUCINE ALANINE

In living cells, the DNA is found within large structures known as **chromosomes.** Figure 1.5 is a photograph of the 46 chromosomes contained in a cell from a human male. The DNA of an average human chromosome is an extraordinarily long, linear, double-stranded structure that contains well over a hundred

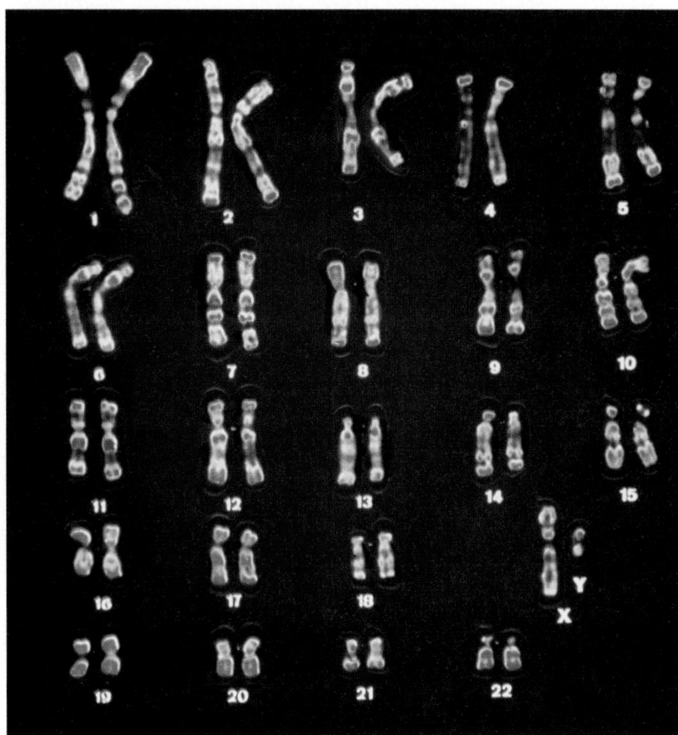

FIGURE 1.5 **A micrograph of the 46 chromosomes found in a cell from a human male.**

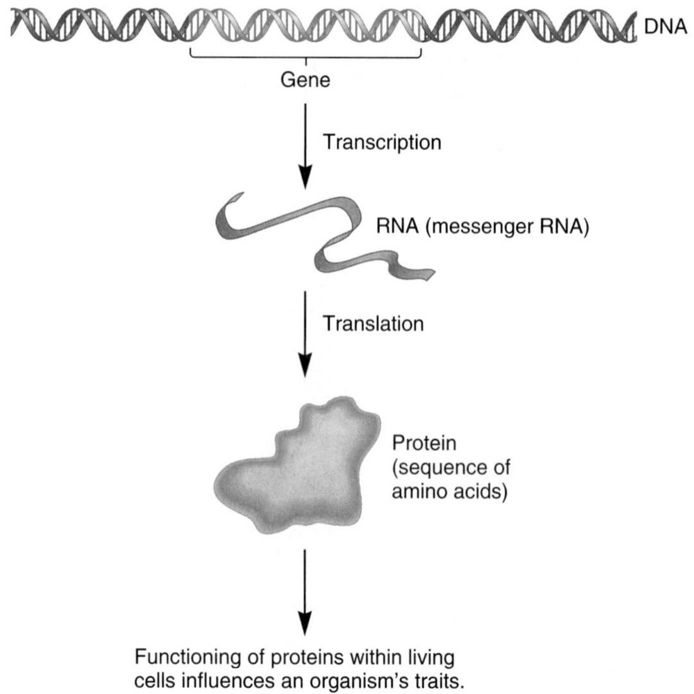

ONLINE ANIMATION

FIGURE 1.6 **Gene expression at the molecular level.** The expression of a gene is a multistep process. During transcription, one of the DNA strands is used as a template to make an RNA strand. During translation, the RNA strand is used to specify the sequence of amino acids within a polypeptide. One or more polypeptides produce a protein that functions within the cell, thereby influencing an organism's traits.

million nucleotides. Along the immense length of a chromosome, the genetic information is parceled into functional units known as genes. An average-sized human chromosome is expected to contain about 1,000 different genes.

The Information in DNA Is Accessed During the Process of Gene Expression

To synthesize its proteins, a cell must be able to access the information that is stored within its DNA. The process of using a gene sequence to affect the characteristics of cells and organisms is referred to as **gene expression.** At the molecular level, the information within genes is accessed in a stepwise process. In the first step, known as **transcription,** the DNA sequence within a gene is copied into a nucleotide sequence of **ribonucleic acid (RNA).** Most RNAs contain the information for the synthesis of a particular polypeptide. This type of RNA is called **messenger RNA (mRNA).** For polypeptide synthesis to occur, the sequence of nucleotides transcribed in an mRNA must be **translated** (using the genetic code) into the amino acid sequence of a polypeptide (**Figure 1.6**). After a polypeptide is made, it folds into a three-dimensional structure. As mentioned, a protein is a functional unit. Some proteins are composed of a single polypeptide, and other proteins consist of two or more polypeptides.

Gene expression results in the production of proteins with specific structures and functions. The unique relationship between gene sequences and protein structures is of paramount importance because the distinctive structure of each protein determines its function within a living cell or organism. Mediated by the process of gene expression, therefore, the sequence of nucleotides in DNA stores the information required for synthesizing proteins with specific structures and functions.

The Molecular Expression of Genes Within Cells Leads to an Organism's Traits

A trait is any characteristic that an organism displays. In genetics, we often focus our attention on **morphological traits** that affect the appearance of an organism. The color of a flower and the height of a pea plant are morphological traits. Geneticists frequently study these types of traits because they are easy to evaluate. For example, an experimenter can simply look at a plant and tell if it has red or white flowers. However, not all traits are morphological. **Physiological traits** affect the ability of an organism to function. For example, the rate at which a bacterium metabolizes a sugar such as lactose is a physiological trait. Like morphological traits, physiological traits are controlled, in part, by the expression of genes. **Behavioral traits** also affect the ways that an organism responds to its environment. An example would be the mating calls of bird species. In animals, the nervous system plays a key role in governing such traits.

A difficult, yet very exciting, aspect of genetics is that our observations and theories span four levels of biological organization: molecules, cells, organisms, and populations. This can make it difficult to appreciate the relationship between genes and traits. To understand this connection, we need to relate the following phenomena:

1. Genes are expressed at the **molecular level.** In other words, gene transcription and translation lead to the production of a particular protein, which is a molecular process.
2. Proteins often function at the **cellular level.** The function of a protein within a cell will affect the structure and workings of that cell.
3. An organism's traits are determined by the characteristics of its cells. We do not have microscopic vision, yet when we view morphological traits, we are really observing the properties of an individual's cells. For example, a red flower has its color because the flower cells make a red pigment. The trait of red flower color is an observation at the **organism level.** Yet the trait is rooted in the molecular characteristics of the organism's cells.
4. A **species** is a group of organisms that maintains a distinctive set of attributes in nature. The occurrence of a trait within a species is an observation at the **population level.** Along with learning how a trait occurs, we also want to understand why a trait becomes prevalent in a particular species. In many cases, researchers discover that a trait predominates within a population because it promotes the reproductive success of the members of the population. This leads to the evolution of beneficial traits.

As a schematic example to illustrate the four levels of genetics, **Figure 1.7** shows the trait of pigmentation in butterflies. One is light-colored and the other is very dark. Let's consider how we can explain this trait at the molecular, cellular, organism, and population levels.

At the molecular level, we need to understand the nature of the gene or genes that govern this trait. As shown in Figure 1.7a, a gene, which we will call the pigmentation gene, is responsible for the amount of pigment that is produced. The pigmentation gene can exist in two different forms called **alleles.** In this example, one allele confers a dark pigmentation and one causes a light pigmentation. Each of these alleles encodes a protein that functions as a pigment-synthesizing enzyme. However, the DNA sequences of the two alleles differ slightly from each other. This difference in the DNA sequence leads to a variation in the structure and function of the respective pigmentation enzymes.

At the cellular level (Figure 1.7b), the functional differences between the pigmentation enzymes affect the amount of pigment that is produced. The allele causing dark pigmentation, which is shown on the left, encodes a protein that functions very well. Therefore, when this gene is expressed in the cells of the wings, a large amount of pigment is made. By comparison, the allele causing light pigmentation encodes an enzyme that functions poorly. Therefore, when this allele is the only pigmentation gene that is expressed, little pigment is made.

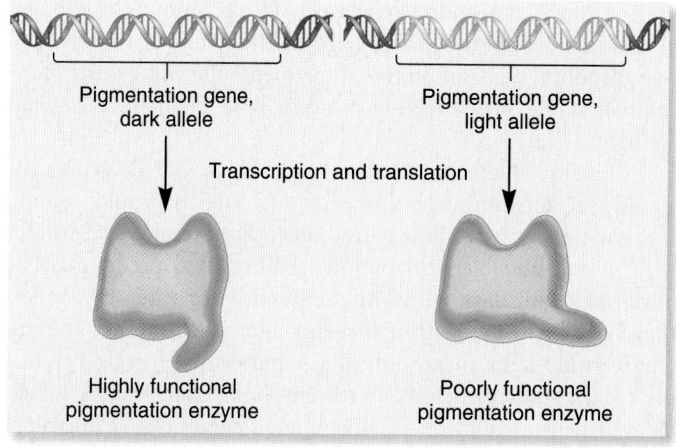

(a) Molecular level

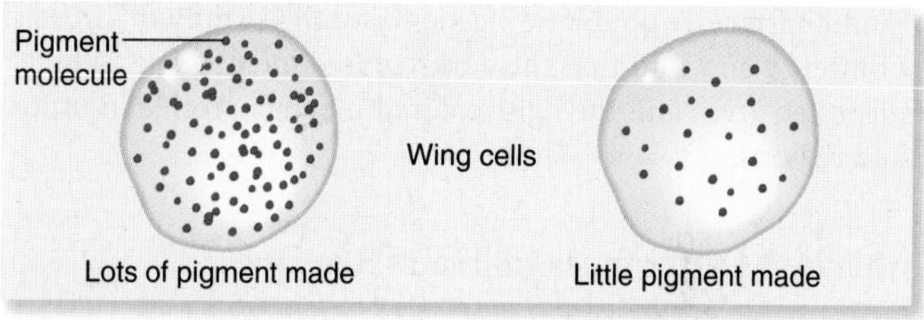

(b) Cellular level

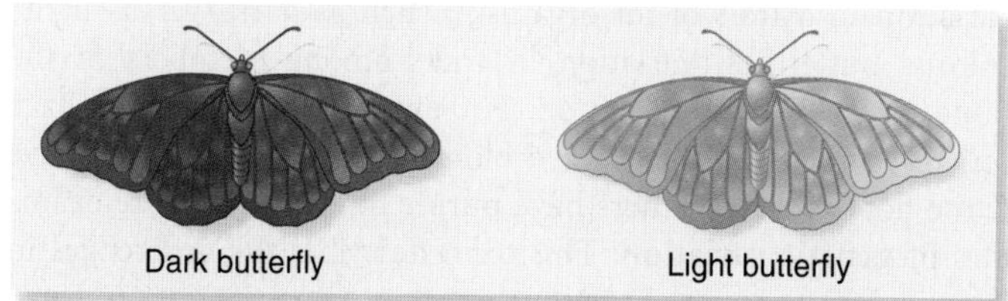

(c) Organism level

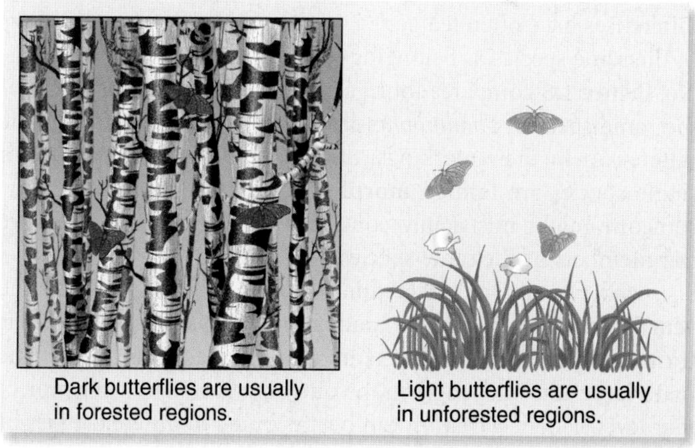

(d) Population level

FIGURE 1.7 **The relationship between genes and traits at the (a) molecular, (b) cellular, (c) organism, and (d) population levels.**

At the organism level (Figure 1.7c), the amount of pigment in the wing cells governs the color of the wings. If the pigment cells produce high amounts of pigment, the wings are dark-colored; if the pigment cells produce little pigment, the wings are light.

Finally, at the population level (Figure 1.7d), geneticists would like to know why a species of butterfly would contain some members with dark wings and other members with light wings. One possible explanation is differential predation. The butterflies with dark wings might avoid being eaten by birds if they happen to live within the dim light of a forest. The dark wings would help to camouflage the butterfly if it were perched on a dark surface such as a tree trunk. In contrast, the lightly colored wings would be an advantage if the butterfly inhabited a brightly lit meadow. Under these conditions, a bird may be less likely to notice a light-colored butterfly that is perched on a sunlit surface. A population geneticist might study this species of butterfly and find that the dark-colored members usually live in forested areas and the light-colored members reside in unforested regions.

Inherited Differences in Traits Are Due to Genetic Variation

In Figure 1.7, we considered how gene expression could lead to variation in a trait of an organism, such as dark- versus light-colored butterflies. Variation in traits among members of the same species is very common. For example, some people have brown hair, while others have blond hair; some petunias have white flowers, while others have purple flowers. These are examples of **genetic variation.** This term describes the differences in inherited traits among individuals within a population.

In large populations that occupy a wide geographic range, genetic variation can be quite striking. In fact, morphological differences have often led geneticists to misidentify two members of the same species as belonging to separate species. As an example, **Figure 1.8** compares four garter snakes that are members of the same species, *Thamnophis ordinoides.* They display dramatic differences in their markings. Such contrasting forms within a single species are termed **morphs.** You can easily imagine how someone might mistakenly conclude that these four snakes are not members of the same species.

FIGURE 1.8 **Four garter snakes showing different morphs within a single species.**

Changes in the nucleotide sequence of DNA underlie the genetic variation that we see among individuals. Throughout this textbook, we will routinely examine how variation in the genetic material results in changes in the outcome of traits. At the molecular level, genetic variation can be attributed to different types of modifications.

1. Small or large differences can occur within gene sequences. These are called **gene mutations.** This type of variation, which produces two or more alleles of the same gene, was previously described in Figure 1.7. In many cases, gene mutations will alter the expression or function of the protein that the gene specifies.
2. Major alterations can also occur in the structure of a chromosome. A large segment of a chromosome can be lost, rearranged, or reattached to another chromosome.
3. Variation may also occur in the total number of chromosomes. In some cases, an organism may inherit one too many or one too few chromosomes. In other cases, it may inherit an extra set of chromosomes.

Variations within the sequences of genes are a common source of genetic variation among members of the same species. In humans, familiar examples of variation involve genes for eye color, hair texture, and skin pigmentation. Chromosome variation—a change in chromosome structure and/or number—is also found, but this type of change is often detrimental. Many human genetic disorders are the result of chromosomal alterations. The most common example is Down syndrome, which is due to the presence of an extra chromosome (**Figure 1.9a**). By comparison, chromosome variation in plants is common and often can lead to strains of plants with superior characteristics, such as increased resistance to disease. Plant breeders have frequently exploited this observation. Cultivated varieties of wheat, for example, have many more chromosomes than the wild species (**Figure 1.9b**).

Traits Are Governed by Genes and by the Environment

In our discussion thus far, we have considered the role that genes play in the outcome of traits. Another critical factor is the **environment**—the surroundings in which an organism exists. A variety of factors in an organism's environment profoundly affect its morphological and physiological features. For example, a person's diet greatly influences many traits such as height, weight, and even intelligence. Likewise, the amount of sunlight a plant receives affects its growth rate and the color of its flowers. The

(a) (b)

FIGURE 1.9 **Examples of chromosome variation.** **(a)** A person with Down syndrome competing in the Special Olympics. This person has 47 chromosomes rather than the common number of 46, because she has an extra copy of chromosome 21. **(b)** A wheat plant. Bread wheat is derived from the contributions of three related species with two sets of chromosomes each, producing an organism with six sets of chromosomes.

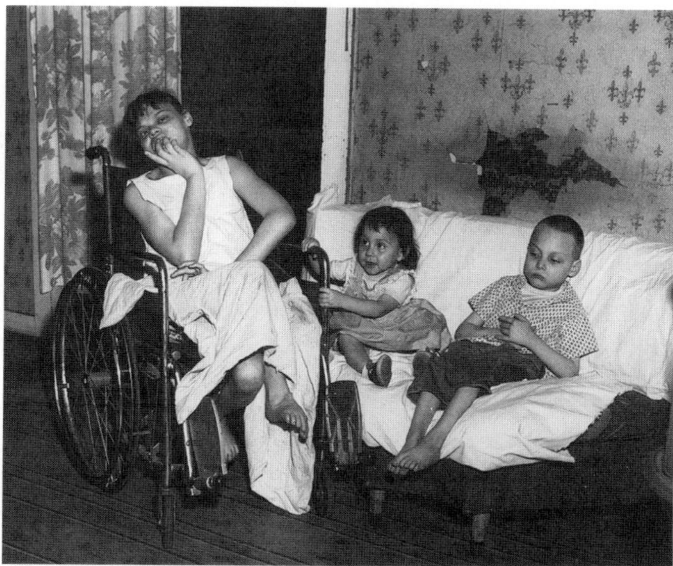

FIGURE 1.10 **Environmental influence on the outcome of PKU within a single family.** All three children pictured here have inherited the alleles that cause PKU. The child in the middle was raised on a phenylalanine-free diet and developed normally. The other two children were born before the benefits of a phenylalanine-free diet were known and were raised on diets that contained phenylalanine. Therefore, they manifest a variety of symptoms, including mental retardation. People born today with this disorder are usually diagnosed when infants. (*Photo from the March of Dimes Birth Defects Foundation.*)

term **norm of reaction** refers to the effects of environmental variation on an individual's traits.

External influences may dictate the way that genetic variation is manifested in an individual. An interesting example is the human genetic disease **phenylketonuria (PKU).** Humans possess a gene that encodes an enzyme known as phenylalanine hydroxylase. Most people have two functional copies of this gene. People with one or two functional copies of the gene can eat foods containing the amino acid phenylalanine and metabolize it properly.

A rare variation in the sequence of the phenylalanine hydroxylase gene results in a nonfunctional version of this protein. Individuals with two copies of this rare, inactive allele cannot metabolize phenylalanine. This occurs in about 1 in 8,000 births among Caucasians in the United States. When given a standard diet containing phenylalanine, individuals with this disorder are unable to break down this amino acid. Phenylalanine accumulates and is converted into phenylketones, which are detected in the urine. PKU individuals manifest a variety of detrimental traits, including mental retardation, underdeveloped teeth, and foul-smelling urine. In contrast, when PKU individuals are identified at birth and raised on a restricted diet that is low in phenylalanine, they develop normally (**Figure 1.10**). Fortunately, through routine newborn screening, most affected babies in the United States are now diagnosed and treated early. PKU provides a dramatic example of how the environment and an individual's genes can interact to influence the traits of the organism.

During Reproduction, Genes Are Passed from Parent to Offspring

Now that we have considered how genes and the environment govern the outcome of traits, we can turn to the issue of inheritance. A centrally important matter in genetics is the manner in which traits are passed from parents to offspring. The foundation for our understanding of inheritance came from the studies of Gregor Mendel in the nineteenth century. His work revealed that genetic determinants, which we now call genes, are passed from parent to offspring as discrete units. We can predict the outcome of genetic crosses based on Mendel's laws of inheritance.

The inheritance patterns identified by Mendel can be explained by the existence of chromosomes and their behavior during cell division. As in Mendel's pea plants, sexually reproducing species are commonly **diploid.** This means they contain two copies of each chromosome, one from each parent. The two

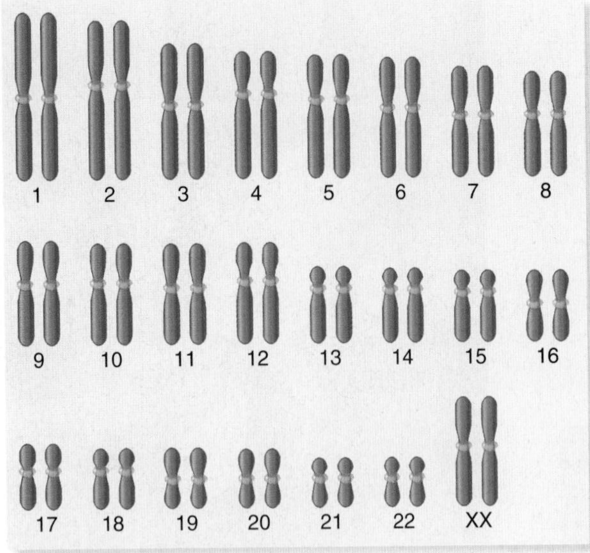

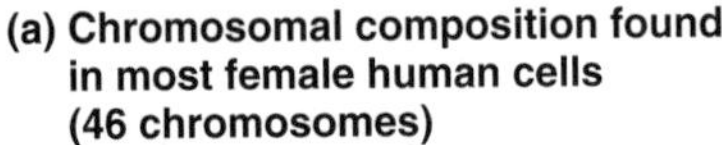
(a) Chromosomal composition found in most female human cells (46 chromosomes)

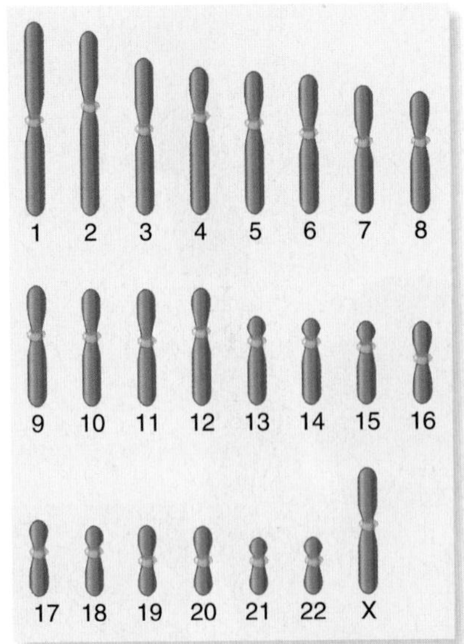

(b) Chromosomal composition found in a human gamete (23 chromosomes)

FIGURE 1.11 **The complement of human chromosomes in somatic cells and gametes. (a)** A schematic drawing of the 46 chromosomes of a human. With the exception of the sex chromosomes, these are always found in homologous pairs. **(b)** The chromosomal composition of a gamete, which contains only 23 chromosomes, one from each pair. This gamete contains an X chromosome. Half of the gametes from human males would contain a Y chromosome instead of the X chromosome.

copies are called **homologues** of each other. Because genes are located within chromosomes, diploid organisms have two copies of most genes. Humans, for example, have 46 chromosomes, which are found in homologous pairs (**Figure 1.11a**). With the exception of the sex chromosomes (namely, X and Y), each homologous pair contains the same kinds of genes. For example, both copies of human chromosome 12 carry the gene that encodes phenylalanine hydroxylase, which was discussed previously. Therefore, an individual has two copies of this gene. The two copies may or may not be identical alleles.

Most cells of the human body that are not directly involved in sexual reproduction contain 46 chromosomes. These cells are called **somatic cells.** In contrast, the **gametes**—sperm and egg cells—contain half that number and are termed **haploid** (**Figure 1.11b**). The union of gametes during fertilization restores the diploid number of chromosomes. The primary advantage of sexual reproduction is that it enhances genetic variation. For example, a tall person with blue eyes and a short person with brown eyes may have short offspring with blue eyes or tall offspring with brown eyes. Therefore, sexual reproduction can result in new combinations of two or more traits that differ from those of either parent.

The Genetic Composition of a Species Evolves over the Course of Many Generations

As we have just seen, sexual reproduction has the potential to enhance genetic variation. This can be an advantage for a population of individuals as they struggle to survive and compete within their natural environment. The term **biological evolution** refers to the phenomenon that the genetic makeup of a population can change over the course of many generations.

As suggested by Charles Darwin, the members of a species are in competition with each other for essential resources. Random genetic changes (i.e., mutations) occasionally occur within an individual's genes, and sometimes these changes lead to a modification of traits that promote reproductive success. For example, over the course of many generations, random gene mutations have lengthened the neck of the giraffe, enabling it to feed on leaves that are high in the trees. When a mutation creates a new allele that is beneficial, the allele may become prevalent in future generations because the individuals carrying the allele are more likely to reproduce and pass the beneficial allele to their offspring. This process is known as **natural selection.** In this way, a species becomes better adapted to its environment.

Over a long period of time, the accumulation of many genetic changes leads to rather striking modifications in a species' characteristics. As an example, **Figure 1.12** depicts the evolution of the modern-day horse. A variety of morphological changes occurred, including an increase in size, fewer toes, and modified jaw structure.

1.2 FIELDS OF GENETICS

Genetics is a broad discipline encompassing molecular, cellular, organism, and population biology. Many scientists who are interested in genetics have been trained in supporting disciplines such as biochemistry, biophysics, cell biology, mathematics, microbi-

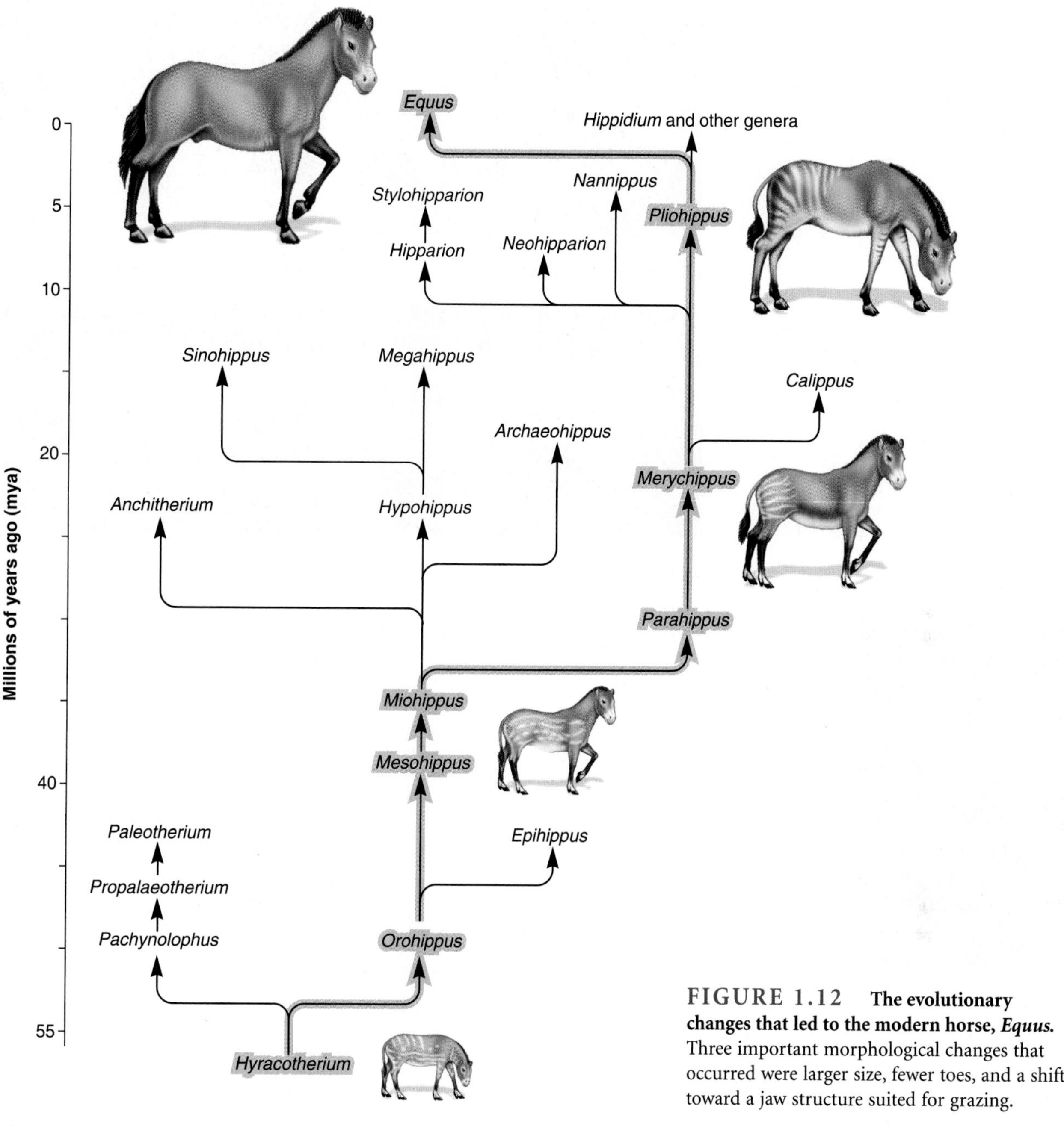

FIGURE 1.12 The evolutionary changes that led to the modern horse, *Equus*. Three important morphological changes that occurred were larger size, fewer toes, and a shift toward a jaw structure suited for grazing.

ology, population biology, ecology, agriculture, and medicine. Experimentally, geneticists often focus their efforts on **model organisms**—organisms studied by many different researchers so they can compare their results and determine scientific principles that apply more broadly to other species. Figure 1.13 shows some common examples, including *Escherichia coli* (a bacterium), *Saccharomyces cerevisiae* (a yeast), *Drosophila melanogaster* (fruit fly), *Caenorhabditis elegans* (a nematode worm), *Danio rerio* (zebrafish), *Mus musculus* (mouse), and *Arabidopsis thaliana* (a flowering plant). By limiting their work to a few such model organisms, researchers can more easily unravel the genetic composition of a given species. Furthermore, the genes found in model organisms often function in a similar way to those found in humans.

The study of genetics has been traditionally divided into three areas: transmission, molecular, and population genetics, although overlap is found among these three fields. In this section, we will examine the general questions that scientists in these areas are attempting to answer.

Transmission Genetics Explores the Inheritance Patterns of Traits as They Are Passed from Parents to Offspring

A scientist working in the field of transmission genetics examines the relationship between the transmission of genes from parent to offspring and the outcome of the offspring's traits. For example,

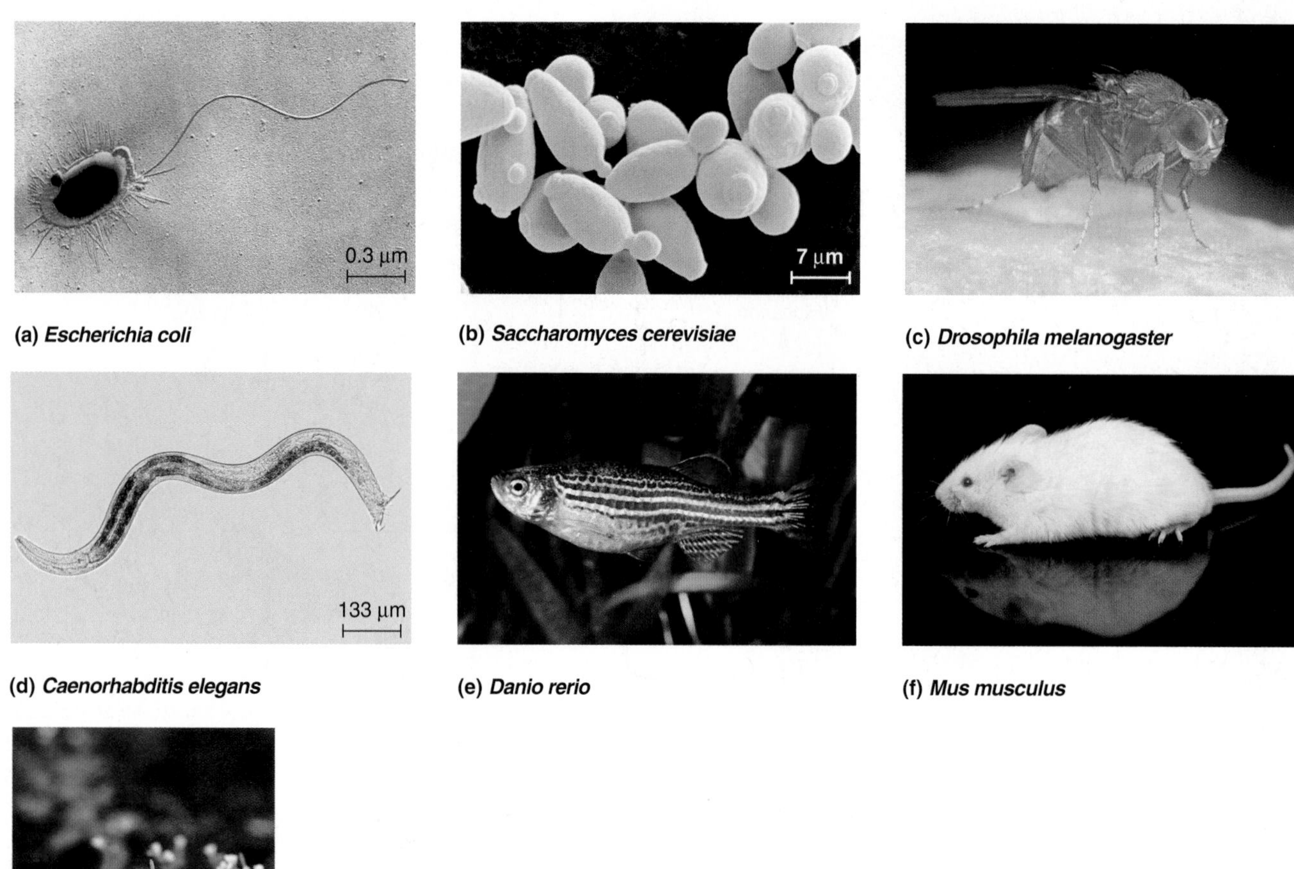

(a) *Escherichia coli*

(b) *Saccharomyces cerevisiae*

(c) *Drosophila melanogaster*

(d) *Caenorhabditis elegans*

(e) *Danio rerio*

(f) *Mus musculus*

(g) *Arabidopsis thaliana*

FIGURE 1.13 **Examples of model organisms studied by geneticists.** **(a)** *Escherichia coli* (a bacterium), **(b)** *Saccharomyces cerevisiae* (a yeast), **(c)** *Drosophila melanogaster* (fruit fly), **(d)** *Caenorhabditis elegans* (a nematode worm), **(e)** *Danio rerio* (zebrafish), **(f)** *Mus musculus* (mouse), and **(g)** *Arabidopsis thaliana* (a flowering plant).

how can two brown-eyed parents produce a blue-eyed child? Or why do tall parents tend to produce tall children, but not always? Our modern understanding of transmission genetics began with the studies of Gregor Mendel. His work provided the conceptual framework for transmission genetics. In particular, he originated the idea that genetic determinants, which we now call genes, are passed as discrete units from parents to offspring via sperm and egg cells. Since these pioneering studies of the 1860s, our knowledge of genetic transmission has greatly increased. Many patterns of genetic transmission are more complex than the simple Mendelian patterns that are described in Chapter 2. The additional complexities of transmission genetics are examined in Chapters 3 through 8.

Experimentally, the fundamental approach of a transmission geneticist is the **genetic cross.** A genetic cross involves breedding two selected individuals and the subsequent analysis of their offspring in an attempt to understand how traits are passed from parents to offspring. In the case of experimental organisms, the researcher chooses two parents with particular traits and then categorizes the offspring according to the traits they possess. In many cases, this analysis is quantitative in nature. For example, an experimenter may cross two tall pea plants and obtain 100 offspring that fall into two categories: 75 tall and 25 dwarf. As we will see in Chapter 2, the ratio of tall and dwarf offspring provides important information concerning the inheritance pattern of this trait.

Throughout Chapters 2 to 8, we will learn how researchers seek to answer many fundamental questions concerning the passage of traits from parents to offspring. Some of these questions are as follows:

What are the common patterns of inheritance for genes?
Chapters 2–4

When two or more genes are located on the same chromosome, how does this affect the pattern of inheritance? **Chapters 5, 6**

Are there unusual patterns of inheritance that cannot be explained by the simple transmission of genes located on chromosomes in the cell nucleus? **Chapter 7**

How do variations in chromosome structure or chromosome number occur, and how are they transmitted from parents to offspring? **Chapter 8**

Molecular Genetics Is Focused on a Biochemical Understanding of the Hereditary Material

The goal of molecular genetics, as the name of the field implies, is to understand how the genetic material works at the molecular level. In other words, molecular geneticists want to understand the molecular features of DNA and how these features underlie the expression of genes. The experiments of molecular geneticists are usually conducted within the confines of a laboratory. Their efforts frequently progress to a detailed analysis of DNA, RNA, and/or protein, using a variety of techniques that are described throughout Parts III, IV, and V of this book.

Molecular geneticists often study mutant genes that have abnormal function. This is called a **genetic approach** to the study of a research question. In many cases, researchers analyze the effects of gene mutations that eliminate the function of a gene. This type of mutation is called a **loss-of-function mutation,** and the resulting gene is called a **loss-of-function allele.** By studying the effects of such mutations, the role of the functional, nonmutant gene is often revealed. For example, let's suppose that a particular plant species produces purple flowers. If a loss-of-function mutation within a given gene causes a plant to produce white flowers, one would suspect the role of the functional gene involves the production of purple pigmentation.

Studies within molecular genetics interface with other disciplines such as biochemistry, biophysics, and cell biology. In addition, advances within molecular genetics have shed considerable light on the areas of transmission and population genetics. Our quest to understand molecular genetics has spawned a variety of modern molecular technologies and computer-based approaches. Furthurmore, discoveries within molecular genetics have had widespread applications in agriculture, medicine, and biotechnology.

Some general questions within the field of molecular genetics are the following:

What are the molecular structures of DNA and RNA? **Chapters 9, 18**

What is the composition and conformation of chromosomes? **Chapters 10, 20**

How is the genetic material copied? **Chapter 11**

How are genes expressed at the molecular level? **Chapters 12, 13, 18, 19, 21**

How is gene expression regulated so that it occurs under the appropriate conditions and in the appropriate cell type? **Chapters 14, 15, 18, 23**

What is the molecular nature of mutations? How are mutations repaired? **Chapter 16**

How does the genetic material become rearranged at the molecular level? **Chapter 17**

What is the underlying relationship between genes and genetic diseases? **Chapter 22**

How do genes govern the development of multicellular organisms? **Chapter 23**

Population Genetics Is Concerned with Genetic Variation and Its Role in Evolution

The foundations of population genetics arose during the first few decades of the twentieth century. Although many scientists of this era did not accept the findings of Mendel and/or Darwin, the theories of population genetics provided a compelling way to connect the two viewpoints. Mendel's work and that of many succeeding geneticists gave insight into the nature of genes and how they are transmitted from parents to offspring. The work of Darwin provided a natural explanation for the various types of characteristics observed among the members of a species. To relate these two phenomena, population geneticists have developed mathematical theories to explain the prevalence of certain forms of genes within populations of individuals. The work of population geneticists helps us understand how the forces of nature have produced and favored the existence of individuals that carry particular genes.

Population geneticists are particularly interested in genetic variation and how that variation is related to an organism's environment. In this field, the prevalence of alleles within a population is of central importance. Some general questions in population genetics are the following:

Why are two or more different alleles of a gene maintained in a population? **Chapter 24**

What factors alter the prevalence of alleles within a population? **Chapter 24**

What are the contributions of genetics and environment in the outcome of a trait? **Chapter 25**

How do genetics and the environment influence quantitative traits, such as size and weight? **Chapter 25**

What factors have the most impact on the process of evolution? **Chapter 26**

How does evolution occur at the molecular level? **Chapter 26**

Genetics Is an Experimental Science

Science is a way of knowing about our natural world. The science of genetics allows us to understand how the expression of our genes produces the traits that we possess. Researchers typically follow two general types of scientific approaches—hypothesis testing and discovery-based science. In **hypothesis testing,** also called the **scientific method,** scientists follow a series of steps to reach verifiable conclusions about the world in which we live. Although scientists arrive at their theories in different ways, the scientific method provides a way to validate (or invalidate) a particular hypothesis. Alternatively, research may also involve the collection of data without a preconceived hypothesis. For example, researchers might analyze the genes found in cancer cells to identify those genes that have become mutant. In this case, the scientists may not have a hypothesis about which particular genes may be involved. The collection and analysis of data without the need for a preconceived hypothesis is called **discovery-based science** or, simply, discovery science.

In traditional science textbooks, the emphasis often lies on the product of science. Namely, many textbooks are aimed primarily at teaching the student about the observations that scientists have made and the theories that they have proposed to explain these observations. Along the way, the student is provided with many bits and pieces of experimental techniques and data. Likewise, this textbook also provides you with many observations and theories. However, it attempts to go one step further. Each of the following chapters contains one or two experiments that have been "dissected" into five individual components to help you to understand the entire scientific process:

1. Background information is provided so that you may appreciate what previous observations were known prior to conducting the experiment.
2. Most experiments involve hypothesis testing. In those cases, the figure states the hypothesis that the scientists were trying to test. In other words, what scientific question was the researcher trying to answer?
3. Next, the figure follows the experimental steps the scientist took to test the hypothesis. The steps necessary to carry out the experiment are listed in the order they were conducted. The figure contains two parallel illustrations labeled Experimental Level and Conceptual Level. The illustration shown in the Experimental Level helps you to understand the techniques that were followed. The Conceptual Level helps you to understand what is actually happening at each step in the procedure.
4. The raw data for each experiment are then presented.
5. Last, an interpretation of the data is offered within the text.

The rationale behind this approach is that it enables you to see the experimental process from beginning to end. Hopefully, you will find this a more interesting and rewarding way to learn about genetics. As you read through the chapters, the experiments will help you to see the relationship between science and scientific theories.

As a student of genetics, you will be given the opportunity to involve your mind in the experimental process. As you are reading an experiment, you may find yourself thinking about different approaches and alternative hypotheses. Different people can view the same data and arrive at very different conclusions. As you progress through the experiments in this book, you will enjoy genetics far more if you try to develop your own skills at formulating hypotheses, designing experiments, and interpreting data. Also, some of the questions in the problem sets are aimed at refining these skills.

Finally, it is worthwhile to point out that science is a social discipline. As you develop your skills at scrutinizing experiments, it is fun to discuss your ideas with other people, including fellow students and faculty members. Keep in mind that you do not need to "know all the answers" before you enter into a scientific discussion. Instead, it is more rewarding to view science as an ongoing and never-ending dialogue.

PROBLEM SETS & INSIGHTS

Solved Problems

S1. A human gene called the *CF* gene (for cystic fibrosis) encodes a protein that functions in the transport of chloride ions across the cell membrane. Most people have two copies of a functional *CF* gene and do not have cystic fibrosis. However, a mutant version of the cystic fibrosis gene is found in some people. If a person has two mutant copies of the gene, he or she develops the disease known as cystic fibrosis. Are the following examples a description of genetics at the molecular, cellular, organism, or population level?

A. People with cystic fibrosis have lung problems due to a buildup of mucus in their lungs.

B. The mutant *CF* gene encodes a defective chloride transporter.

C. A defect in the chloride transporter causes a salt imbalance in lung cells.

D. Scientists have wondered why the mutant cystic fibrosis gene is relatively common. In fact, it is the most common mutant gene that causes a severe disease in Caucasians. Usually, mutant genes that cause severe diseases are relatively rare. One possible explanation why the *CF* gene is so common is that people who have one copy of the functional *CF* gene and one copy of the mutant gene may be more resistant to diarrheal diseases such as cholera. Therefore, even though individuals with two mutant copies are very sick, people with one mutant copy and one functional copy might have a survival advantage over people with two functional copies of the gene.

Answer:

A. Organism. This is a description of a trait at the level of an entire individual.

B. Molecular. This is a description of a gene and the protein it encodes.

C. Cellular. This is a description of how protein function affects the cell.

D. Population. This is a possible explanation why two versions of the gene occur within a population.

S2. Explain the relationship between the following pairs of terms:

A. RNA and DNA

B. RNA and transcription

C. Gene expression and trait

D. Mutation and allele

Answer:

A. DNA is the genetic material. DNA is used to make RNA. RNA is then used to specify a sequence of amino acids within a polypeptide.

B. Transcription is a process in which RNA is made using DNA as a template.

C. Genes are expressed at the molecular level to produce functional proteins. The functioning of proteins within living cells ultimately affects an organism's traits.

D. Alleles are alternative forms of the same gene. For example, a particular human gene affects eye color. The gene can exist as a blue allele or a brown allele. The difference between these two alleles is caused by a mutation. Perhaps the brown allele was the first eye color allele in the human population. Within some ancestral person, however, a mutation may have occurred in the eye color gene that converted the brown allele to the blue allele. Now the human population has both the brown allele and the blue allele.

S3. How are genes passed from generation to generation?

Answer: When a diploid individual makes haploid cells for sexual reproduction, the cells contain half the number of chromosomes. When two haploid cells (e.g., sperm and egg) combine with each other, a zygote is formed that begins the life of a new individual. This zygote has inherited half of its chromosomes and, therefore, half of its genes from each parent. This is how genes are passed from parents to offspring.

Conceptual Questions

C1. Pick any example of a genetic technology and describe how it has directly impacted your life.

C2. At the molecular level, what is a gene? Where are genes located?

C3. Most genes encode proteins. Explain how the structure and function of proteins produce an organism's traits.

C4. Briefly explain how gene expression occurs at the molecular level.

C5. A human gene called the β-globin gene encodes a polypeptide that functions as a subunit of the protein known as hemoglobin. Hemoglobin is found within red blood cells; it carries oxygen. In human populations, the β-globin gene can be found as the common allele called the Hb^A allele, but it can also be found as the rare Hb^S allele. Individuals who have two copies of the Hb^S allele have the disease called sickle-cell disease. Are the following examples a description of genetics at the molecular, cellular, organism, or population level?

A. The Hb^S allele encodes a polypeptide that functions slightly differently from the polypeptide encoded by the Hb^A allele.

B. If an individual has two copies of the Hb^S allele, that person's red blood cells form a sickle shape.

C. Individuals who have two copies of the Hb^A allele do not have sickle-cell disease, but they are not resistant to malaria. People who have one Hb^A allele and one Hb^S allele do not have sickle-cell disease, and they are resistant to malaria. People who have two copies of the Hb^S allele have sickle-cell anemia, and this disease may significantly shorten their lives.

D. Individuals with sickle-cell disease have anemia because their red blood cells are easily destroyed by the body.

C6. What is meant by the term genetic variation? Give two examples of genetic variation not discussed in Chapter 1. What causes genetic variation at the molecular level?

C7. What is the cause of Down syndrome?

C8. Your textbook describes how the trait of phenylketonuria (PKU) is greatly influenced by the environment. Pick a trait in your favorite plant and explain how genetics and environment may play important roles.

C9. What is meant by the term diploid? Which cells of the human body are diploid, and which cells are not?

C10. What is a DNA sequence?

C11. What is the genetic code?

C12. Explain the relationships between the following pairs of genetic terms:

A. Gene and trait

B. Gene and chromosome

C. Allele and gene

D. DNA sequence and amino acid sequence

C13. With regard to biological evolution, which of the following statements is not correct? Explain why.

A. During its lifetime, an animal evolves to become better adapted to its environment.

B. The process of biological evolution has produced species that are better adapted to their environments.

C. When an animal is better adapted to its environment, the process of natural selection makes it more likely for that animal to reproduce.

C14. What are the primary interests of researchers working in the following fields of genetics?

A. Transmission genetics

B. Molecular genetics

C. Population genetics

Experimental Questions

E1. What is a genetic cross?

E2. The technique known as DNA sequencing (described in Chapter 18) enables researchers to determine the DNA sequence of genes. Would this technique be used primarily by transmission geneticists, molecular geneticists, or population geneticists?

E3. Figure 1.5 shows a micrograph of chromosomes from a normal human cell. If you performed this type of experiment using cells from a person with Down syndrome, what would you expect to see?

E4. Many organisms are studied by geneticists. Of the following species, do you think it would be more likely for them to be studied by a transmission geneticist, a molecular geneticist, or a population geneticist? Explain your answer. Note: More than one answer may be possible.

A. Dogs

B. *E. coli*

C. Fruit flies

D. Leopards

E. Corn

E5. Pick any trait you like in any species of wild plant or animal. The trait must somehow vary among different members of the species. For example, some butterflies have dark wings and others have light wings (see Figure 1.7).

A. Discuss all of the background information that you already have (from personal observations) regarding this trait.

B. Propose a hypothesis that would explain the genetic variation within the species. For example, in the case of the butterflies, your hypothesis might be that the dark butterflies survive better in dark forests, while the light butterflies survive better in lighter fields.

C. Describe the experimental steps you would follow to test your hypothesis.

D. Describe the possible data you might collect.

E. Interpret your data.

Note: When picking a trait to answer this question, do not pick the trait of wing color in butterflies.

Note: All answers appear at the website for this textbook; the answers to even-numbered questions are in the back of the textbook.

www.mhhe.com/brookergenetics3e

Visit the Online Learning Center for practice tests, answer keys, and other learning aids for this chapter. Enhance your understanding of genetics with our interactive exercises, quizzes, animations, and much more.

MENDELIAN INHERITANCE

2

An appreciation for the concept of heredity can be traced far back in human history. Hippocrates, a famous Greek physician, was the first person to provide an explanation for hereditary traits (ca. 400 B.C.E.). He suggested that "seeds" are produced by all parts of the body, which are then collected and transmitted to the offspring at the time of conception. Furthermore, he hypothesized that these seeds cause certain traits of the offspring to resemble those of the parents. This idea, known as **pangenesis,** was the first attempt to explain the transmission of hereditary traits from generation to generation.

For the next 2,000 years, the ideas of Hippocrates were accepted by some and rejected by many. After the invention of the microscope in the late seventeenth century, some people observed sperm and thought they could see a tiny creature inside, which they termed a homunculus (little man). This homunculus was hypothesized to be a miniature human waiting to develop within the womb of its mother. Those who held that thought, known as spermists, suggested that only the father was responsible for creating future generations and that any resemblance between mother and offspring was due to influences "within the womb." During the same time, an opposite school of thought also developed. According to the ovists, the egg was solely responsible for human characteristics. The only role of the sperm was to stimulate the egg onto its path of development. Of course, neither of these ideas was correct.

The first systematic studies of genetic crosses were carried out by Joseph Kölreuter from 1761 to 1766. In crosses between different strains of tobacco plants, he found that the offspring were usually intermediate in appearance between the two parents. This led Kölreuter to conclude that both parents make equal genetic contributions to their offspring. Furthermore, his observations were consistent with **blending inheritance.** According to this view, the factors that dictate hereditary traits can blend together from generation to generation. The blended traits would then be passed to the next generation. The popular view before the 1860s, which combined the notions of pangenesis and blending inheritance, was that hereditary traits were rather malleable and could change and blend over the course of one or two generations. However, the pioneering work of Gregor Mendel would prove instrumental in refuting this viewpoint.

In Chapter 2, we will first examine the outcome of Mendel's crosses in pea plants. We begin our inquiry into genetics here because the inheritance patterns observed in peas are fundamentally related to inheritance patterns found in other eukaryotic species such as humans, mice, fruit flies, and corn. We will

CHAPTER OUTLINE

The garden pea, studied by Mendel.

PART II
Patterns of Inheritance

discover how Mendel's insights into the patterns of inheritance in pea plants revealed some simple rules that govern the process of inheritance. In Chapters 3 through 8, we will explore more complex patterns of inheritance and also consider the role that the chromosomes play as the carriers of the genetic material.

In the second part of this chapter, we will become familiar with general concepts in probability and statistics. How are statistical methods useful? First, probability calculations allow us to predict the outcomes of simple genetic crosses, as well as the outcomes of more complicated crosses described in later chapters. In addition, we will learn how to use statistics to test the validity of genetic hypotheses that attempt to explain the inheritance patterns of traits.

2.1 MENDEL'S LAWS OF INHERITANCE

Gregor Johann Mendel, born in 1822, is now remembered as the father of genetics (**Figure 2.1**). He grew up on a small farm in Hynčice (formerly Heinzendorf) in northern Moravia, which was then a part of Austria and is now a part of the Czech Republic. As a young boy, he worked with his father grafting trees to improve the family orchard. Undoubtedly, his success at grafting taught him that precision and attention to detail are important elements of success. These qualities would later be important in his experiments as an adult scientist. Instead of farming, however, Mendel was accepted into the Augustinian monastery of St. Thomas, completed his studies for the priesthood, and was ordained in 1847. Soon after becoming a priest, Mendel worked for a short time as a substitute teacher. To continue that role, he needed to obtain a teaching license from the government. Surprisingly, he failed the licensing exam due to poor answers in the areas of physics and natural history. Therefore, Mendel then enrolled at the University of Vienna to expand his knowledge in these two areas. Mendel's training in physics and mathematics taught him to perceive the world as an orderly place, governed by natural laws. In his studies, Mendel learned that these natural laws could be stated as simple mathematical relationships.

In 1856, Mendel began his historic studies on pea plants. For eight years, he grew and crossed thousands of pea plants on a small 115- by 23-foot plot. He kept meticulously accurate records that included quantitative data concerning the outcome of his crosses. He published his work, entitled "Experiments on Plant Hybrids," in 1866. This paper was largely ignored by scientists at that time, possibly because of its title or because it was published in a rather obscure journal (*The Proceedings of the Brünn Society of Natural History*). Another reason his work went unrecognized could be tied to a lack of understanding of chromosomes and their transmission, a topic we will discuss in Chapter 3. Nevertheless, Mendel's groundbreaking work allowed him to propose the natural laws that now provide a framework for our understanding of genetics.

Prior to his death in 1884, Mendel reflected, "My scientific work has brought me a great deal of satisfaction and I am convinced that it will be appreciated before long by the whole world." Sixteen years later, in 1900, the work of Mendel was independently rediscovered by three biologists with an interest in plant genetics: Hugo de Vries of Holland, Carl Correns of Germany, and Erich von Tschermak of Austria. Within a few years, the impact of Mendel's studies was felt around the world. In this section, we will examine Mendel's experiments and consider their monumental significance in the field of genetics.

FIGURE 2.1 **Gregor Johann Mendel, the father of genetics.**

Mendel Chose Pea Plants as His Experimental Organism

Mendel's study of genetics grew out of his interest in ornamental flowers. Prior to his work with pea plants, many plant breeders had conducted experiments aimed at obtaining flowers with new varieties of colors. When two distinct individuals with different characteristics are mated or **crossed** to each other, this is called a **hybridization** experiment, and the offspring are referred to as **hybrids.** For example, a hybridization experiment could involve a cross between a purple-flowered plant and a white-flowered plant. Mendel was particularly intrigued, in such experiments, by the consistency with which offspring of subsequent generations showed characteristics of one or the other parent. His intellectual foundation in physics and the natural sciences led him

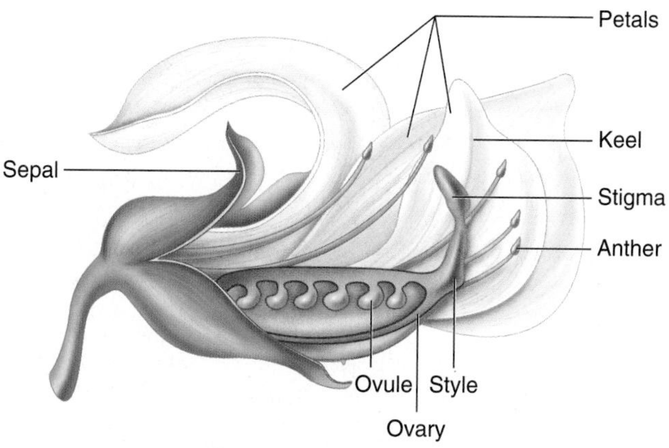

(a) Structure of a pea flower

(b) A flowering pea plant

Pollen grain
Stigma
Pollen tube
Two sperm (each $1n$)
Style
Ovary
Central cell with 2 polar nuclei (each $1n$)
Ovule (containing embryo sac)
Egg ($1n$)
Micropyle

Pollen tube grows into micropyle.
One sperm unites with the egg, and the other sperm unites with the 2 polar nuclei.

Endosperm nucleus ($3n$)
Zygote ($2n$)

(c) Pollination and fertilization in angiosperms

FIGURE 2.2 **Flower structure and pollination in pea plants. (a)** The pea flower can produce both pollen and egg cells. The pollen grains are produced within the anthers, and the egg cells are produced within the ovules that are contained within the ovary. A modified petal called a keel encloses the anthers and ovaries. **(b)** Photograph of a flowering pea plant. **(c)** A pollen grain must first land on the stigma. After this occurs, the pollen sends out a long tube through which two sperm cells travel toward an ovule to reach an egg cell. The fusion between a sperm and an egg cell results in fertilization and creates a zygote. A second sperm fuses with a central cell containing two polar nuclei to create the endosperm. The endosperm provides a nutritive material for the developing embryo.

to consider that this regularity might be rooted in natural laws that could be expressed mathematically. To uncover these laws, he realized that he would need to carry out quantitative experiments in which the numbers of offspring carrying certain traits were carefully recorded and analyzed.

Mendel chose the garden pea, *Pisum sativum*, to investigate the natural laws that govern plant hybrids. The morphological features of this plant are shown in **Figure 2.2a** and **b**. Several properties of this species were particularly advantageous for studying plant hybridization. First, the species was available in several varieties that had decisively different physical characteristics. Many strains of the garden pea were available that varied in the appearance of their height, flowers, seeds, and pods.

A second important issue is the ease of making crosses. In flowering plants, reproduction occurs by a pollination event (**Figure 2.2c**). Male gametes (**sperm**) are produced within **pollen grains** formed in the **anthers,** while the female gametes (**eggs**) are contained within **ovules** that form in the **ovaries.** For fertilization to occur, a pollen grain lands on the **stigma,** which stimulates the growth of a pollen tube. This enables sperm cells to enter the stigma and migrate toward an ovule. Fertilization occurs when a sperm enters the micropyle, an opening in the ovule wall, and fuses with an egg cell. The term **gamete** is used to describe haploid reproductive cells that can unite to form a zygote. It should be emphasized, however, that the process that produces gametes in animals is quite different from the way that gametes are produced in plants and fungi. These processes are described in greater detail in Chapter 3.

In some experiments, Mendel wanted to carry out **self-fertilization,** which means that the pollen and egg are derived from the same plant. In peas, a modified petal known as the keel covers the reproductive structures of the plant. Because of this covering, pea plants naturally reproduce by self-fertilization. In fact, pollination occurs even before the flower opens. In other experiments, however, Mendel wanted to make crosses between different plants. How did he accomplish this goal? Fortunately, pea plants contain relatively large flowers that are easy to manipulate, making it possible to make crosses between two particular plants and study their outcomes. This process, known as **cross-fertilization,** requires that the pollen from one plant be placed on the stigma of another plant. This procedure is shown in Figure 2.3. Mendel was able to pry open immature flowers and remove the anthers before they produced pollen. Therefore, these flowers could not self-fertilize. He would then obtain pollen from another plant by gently touching its mature anthers with a paintbrush. Mendel applied this pollen to the stigma of the flower that already had its anthers removed. In this way, he was able to cross-fertilize his pea plants and thereby obtain any type of hybrid he wanted.

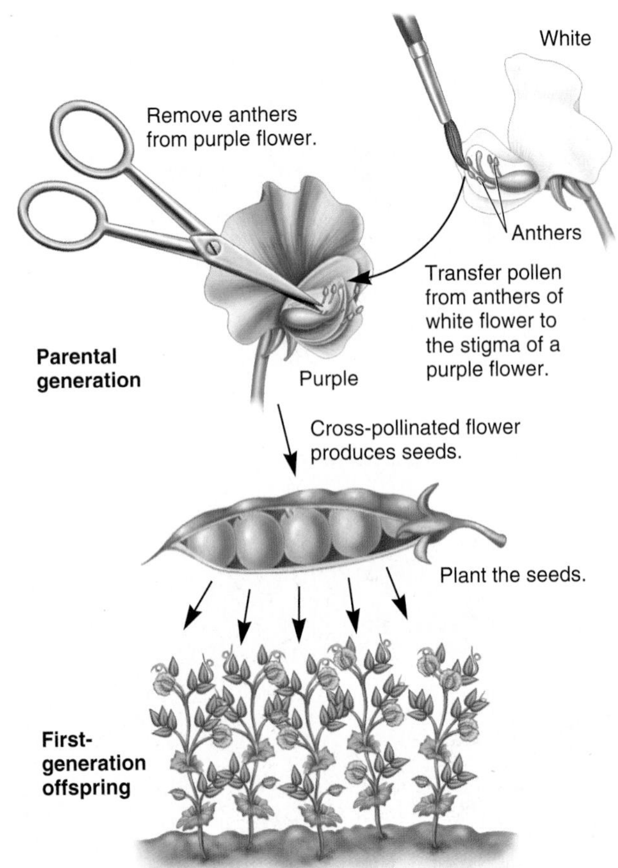

FIGURE 2.3 How Mendel cross-fertilized two different pea plants. This illustration depicts a cross between a plant with purple flowers and another plant with white flowers. The offspring from this cross are the result of pollination of the purple flower using pollen from a white flower.

Mendel Studied Seven Traits That Bred True

When he initiated his studies, Mendel obtained several varieties of peas that were considered to be distinct. These plants were different with regard to many morphological characteristics. Such characteristics of an organism are called characters, or **traits.** Over the course of two years, Mendel tested the strains to determine if their characteristics bred true. This means that a trait did not vary in appearance from generation to generation. For example, if the seeds from a pea plant were yellow, the next generation would also produce yellow seeds. Likewise, if these offspring were allowed to self-fertilize, all of their offspring would also produce yellow seeds, and so on. A variety that continues to produce the same characteristic after several generations of self-fertilization is called a **true-breeding line,** or **strain.**

Mendel next concentrated his efforts on the analysis of characteristics that were clearly distinguishable between different true-breeding lines. Figure 2.4 illustrates the seven traits that Mendel eventually chose to follow in his breeding experiments. All seven were found in two variants. The term **variant** refers to a particular trait that may be found in two or more versions within a single species. For example, one trait he followed was height, which was found in two variants—tall and dwarf plants. Mendel studied this trait by crossing the variants to each other. A cross in which an experimenter is observing only one trait is called a **single-factor cross,** also called a **monohybrid cross.** When the two parents are different variants for a given trait, this type of cross produces single-trait hybrids, also known as **monohybrids.**

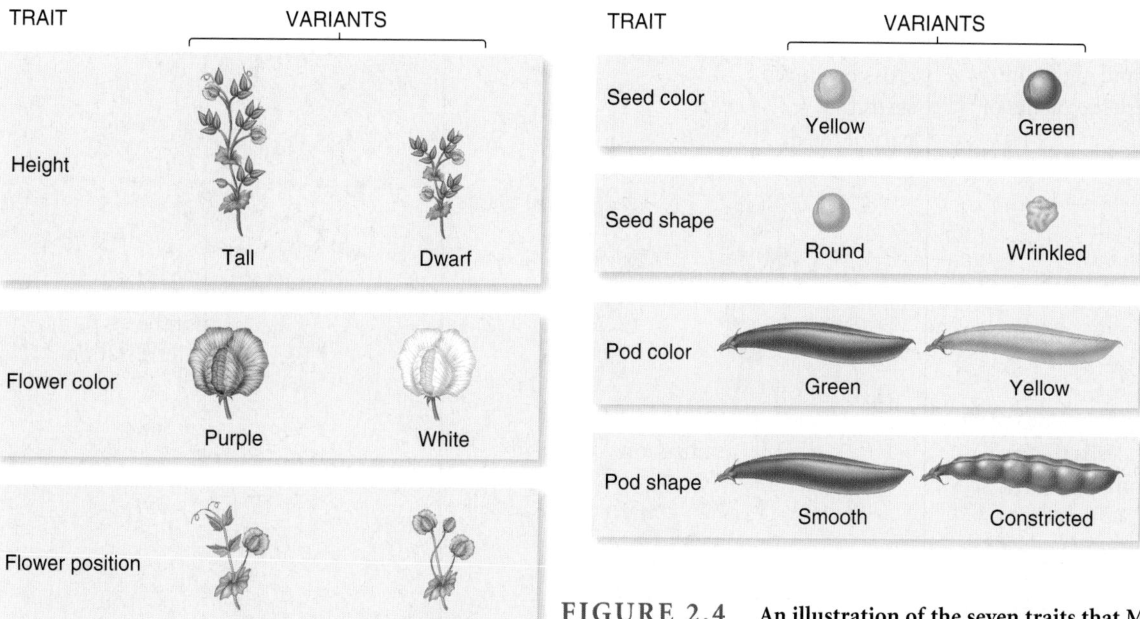

FIGURE 2.4 **An illustration of the seven traits that Mendel studied.** Each trait was found as two variants that were decisively different from each other.

EXPERIMENT 2A

Mendel Followed the Outcome of a Single Trait for Two Generations

Prior to conducting his studies, Mendel did not already have a hypothesis to explain the formation of hybrids. However, his educational background caused him to realize that a quantitative analysis of crosses may uncover mathematical relationships that would otherwise be mysterious. His experiments were designed to determine the relationships that govern hereditary traits. This rationale is called an **empirical approach.** Laws that are deduced from an empirical approach are known as empirical laws.

Mendel's experimental procedure is shown in **Figure 2.5.** He began with true-breeding plants that differed with regard to a single trait. These are termed the **parental generation,** or **P generation.** When the true-breeding parents were crossed to each other, this is called a P cross, and the offspring constitute the **F_1 generation,** for first filial generation. As seen in the data, all plants of the F_1 generation showed the phenotype of one parent but not the other. This prompted Mendel to follow the transmission of this trait for one additional generation. To do so, the plants of the F_1 generation were allowed to self-fertilize to produce a second generation called the **F_2 generation,** for second filial generation.

THE GOAL

Mendel speculated that the inheritance pattern for a single trait may follow quantitative natural laws. The goal of this experiment was to uncover such laws.

ACHIEVING THE GOAL — FIGURE 2.5 Mendel's analysis of single-factor crosses.

Starting material: Mendel began his experiments with true-breeding pea plants that varied with regard to only one of seven different traits (see Figure 2.4).

1. For each of seven traits, Mendel cross-fertilized two different true-breeding lines. Keep in mind that each cross involved two plants that differed in regard to only one of the seven traits studied. The illustration at the right shows one cross between a tall and dwarf plant. This is called a P (parental) cross.

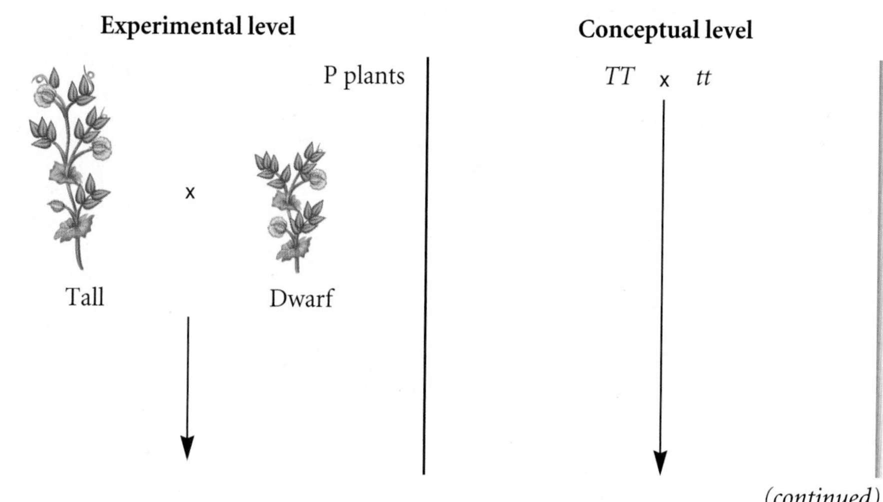

(continued)

2. Collect many seeds. The following spring, plant the seeds and allow the plants to grow. These are the plants of the F_1 generation.

3. Allow the F_1 generation plants to self-fertilize. This produces seeds that are part of the F_2 generation.

4. Collect the seeds and plant them the following spring to obtain the F_2 generation plants.

5. Analyze the characteristics found in each generation.

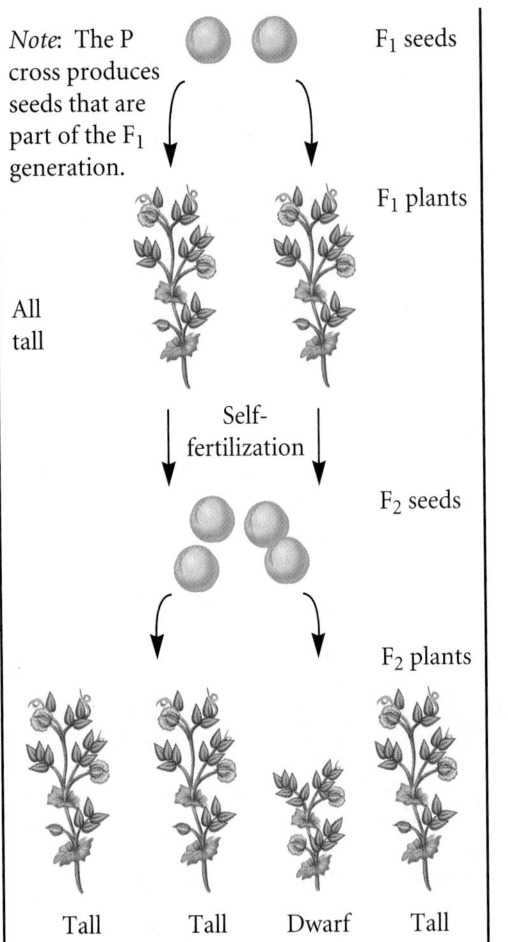

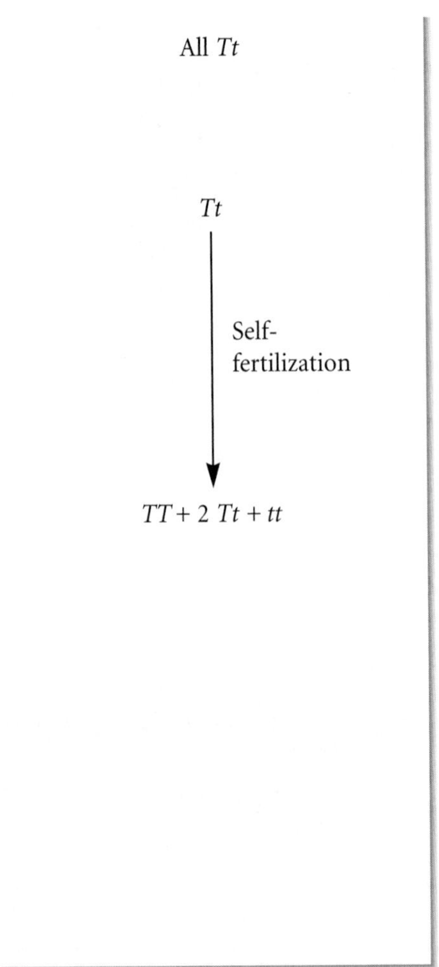

THE DATA

P Cross	*F_1 Generation*	*F_2 Generation*	*Ratio*
Tall × dwarf stem	All tall	787 tall, 277 dwarf	2.84:1
Purple × white flowers	All purple	705 purple, 224 white	3.15:1
Axial × terminal flowers	All axial	651 axial, 207 terminal	3.14:1
Yellow × green seeds	All yellow	6,022 yellow, 2,001 green	3.01:1
Round × wrinkled seeds	All round	5,474 round, 1,850 wrinkled	2.96:1
Green × yellow pods	All green	428 green, 152 yellow	2.82:1
Smooth × constricted pods	All smooth	882 smooth, 299 constricted	2.95:1
Total	All dominant	14,949 dominant, 5,010 recessive	2.98:1

INTERPRETING THE DATA

The data shown in Figure 2.5 are the results of producing an F_1 generation via cross-fertilization, and an F_2 generation via self-fertilization of the F_1 monohybrids. A quantitative analysis of these data allowed Mendel to propose three important ideas:

1. Mendel's data argued strongly against a blending mechanism of heredity. In all seven cases, the F_1 generation displayed characteristics that were distinctly like one of the two parents rather than traits intermediate in character. Using genetic terms that Mendel originated and are still used today, his first proposal was that the variant for one trait is **dominant** over another variant. For example, the variant of green pods is dominant to that of yellow pods. The term **recessive** is used to describe a variant that is masked by the presence of a dominant trait but reappears in subsequent generations. Yellow pods and dwarf stems are examples of recessive variants. They can also be referred to as recessive traits.
2. When a true-breeding plant with a dominant trait was crossed to a true-breeding plant with a recessive trait, the dominant trait was always observed in the F_1 generation. In the F_2 generation, some offspring displayed the dominant characteristic, while a smaller proportion showed the recessive trait. However, none of the offspring exhibited intermediate traits. How did Mendel explain this observation? Because the recessive trait appeared in the F_2 generation, he formed a second proposal—the genetic determinants of traits are passed along as "unit factors" from generation to generation. His data were consistent with a **particulate theory of inheritance,** in which the genes that govern traits are inherited as discrete units that remain

unchanged as they are passed from parent to offspring. Mendel called them unit factors, but we now call them genes.

3. A third important interpretation of Mendel's data is related to the proportions of offspring. When Mendel compared the numbers of dominant and recessive offspring in the F_2 generation, he noticed a recurring pattern. Within experimental variation, he always observed approximately a 3:1 ratio between the dominant trait and the recessive trait. Mendel was the first scientist to apply this type of quantitative analysis in a biological experiment. As described next, this quantitative approach allowed him to make a third proposal—genes **segregate** from each other during the process that gives rise to gametes.

A self-help quiz involving this experiment can be found at the Online Learning Center.

The 3:1 Phenotypic Ratio That Mendel Observed Is Consistent with the Segregation of Alleles, Now Known as Mendel's Law of Segregation

Mendel's research was aimed at understanding the laws that govern the inheritance of traits. At that time, scientists did not understand the molecular composition of the genetic material or its mode of transmission during gamete formation and fertilization. We now know that the genetic material is composed of deoxyribonucleic acid (DNA), a component of chromosomes. Each chromosome contains hundreds or thousands of shorter segments that function as genes—a term that was originally coined by the Danish botanist Wilhelm Johannsen in 1909. A **gene** is defined as a "unit of heredity" that may influence the outcome of an organism's traits. Each of the seven traits that Mendel studied is influenced by a different gene.

Most eukaryotic species, such as pea plants and humans, have their genetic material organized into pairs of chromosomes. For this reason, eukaryotes have two copies of most genes. These copies may be the same or they may differ. The term **allele** refers to different versions of the same gene. With this modern knowledge, the results shown in Figure 2.5 are consistent with the idea that each parent transmits only one copy of each gene (i.e., one allele) to each offspring. **Mendel's law of segregation** states that:

The two copies of a gene segregate (or separate) from each other during transmission from parent to offspring.

Therefore, only one copy of each gene is found in a gamete. At fertilization, two gametes combine randomly, potentially producing different allelic combinations.

Let's use Mendel's cross of tall and dwarf pea plants to illustrate how alleles are passed from parents to offspring (**Figure 2.6**). The letters *T* and *t* are used to represent the alleles of the gene that determines plant height. By convention, the uppercase letter represents the dominant allele (*T* for tall height, in this case), and the recessive allele is represented by the same letter in lowercase (*t*, for dwarf height). For the P cross, both parents are true-breeding plants. Therefore, we know each has identical copies of the height gene. When an individual possesses two identical copies of a gene, the individual is said to be **homozygous** with respect to that gene. (The prefix *homo* means like, and the suffix *zygo* means pair.) In the P cross, the tall plant is homozygous for the tall allele *T*, while the dwarf plant is homozygous for the dwarf allele *t*. The term **genotype** refers to the genetic composition of an individual. *TT* and *tt* are the genotypes of the P generation in this experiment. The term **phenotype** refers to an observable characteristic of an organism. In the P generation, half of the plants are phenotypically tall and half are dwarf.

In contrast, the F_1 generation is **heterozygous,** with the genotype *Tt*, because every individual carries one copy of the tall allele and one copy of the dwarf allele. A heterozygous individual carries different alleles of a gene. (The prefix *hetero* means different.) Although these plants are heterozygous, their phenotypes are tall because they have a copy of the dominant tall allele.

The law of segregation predicts that the phenotypes of the F_2 generation will be tall and dwarf in a ratio of 3:1 (see Figure 2.6). Because the parents of the F_2 generation are heterozygous, each parent can transmit either a *T* allele or a *t* allele to a particular offspring, but not both, because each gamete carries only one of the two alleles. Therefore, *TT*, *Tt*, and *tt* are the possible genotypes of the F_2 generation (note that the genotype *Tt* is the same as *tT*). By randomly combining these alleles, the genotypes are produced in a 1:2:1 ratio. Because *TT* and *Tt* both produce tall phenotypes, a 3:1 phenotypic ratio is observed in the F_2 generation.

A Punnett Square Can Be Used to Predict the Outcome of Crosses

An easy way to predict the outcome of simple genetic crosses is to use a **Punnett square,** a method originally proposed by Reginald Punnett. To construct a Punnett square, you must know the genotypes of the parents. With this information, the Punnett square enables you to predict the types of offspring the parents are expected to produce and in what proportions. We will follow a step-by-step description of the Punnett square approach using a cross of heterozygous tall plants as an example.

Step 1. Write down the genotypes of both parents. In this example, a heterozygous tall plant is crossed to another heterozygous tall plant. The plant providing the pollen is considered the male parent and the plant providing the eggs, the female parent.

Male parent: *Tt*

Female parent: *Tt*

Step 2. *Write down the possible gametes that each parent can make.* Remember that the law of segregation tells us that a gamete can contain only one copy of each gene.

Male gametes: *T* or *t*

Female gametes: *T* or *t*

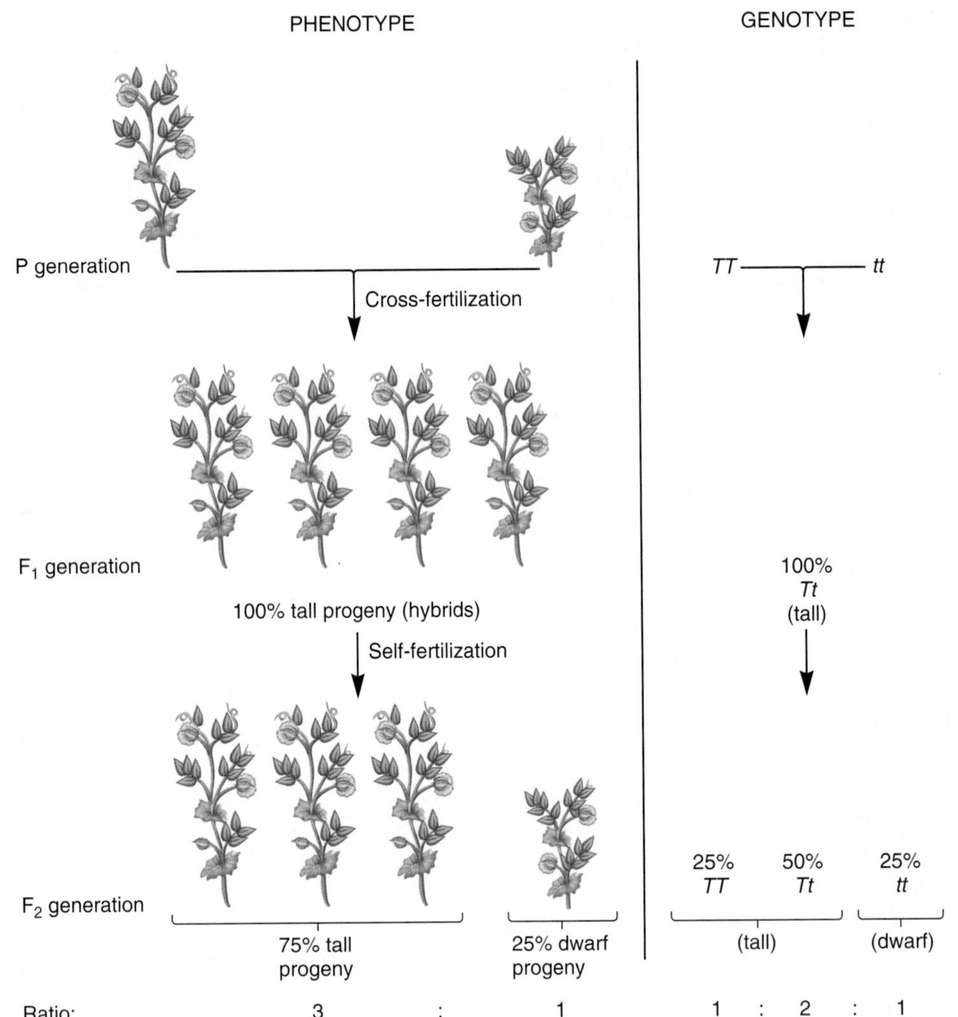

INTERACTIVE EXERCISE

FIGURE 2.6 **Mendel's law of segregation.** This illustration shows a cross between a true-breeding tall plant and a true-breeding dwarf plant and the subsequent segregation of the tall (T) and dwarf (t) alleles in the F_1 and F_2 generations.

Step 3. *Create an empty Punnett square.* In the examples shown in this textbook, the number of columns equals the number of male gametes, and the number of rows equals the number of female gametes. Our example has two rows and two columns. Place the male gametes across the top of the Punnett square and the female gametes along the side.

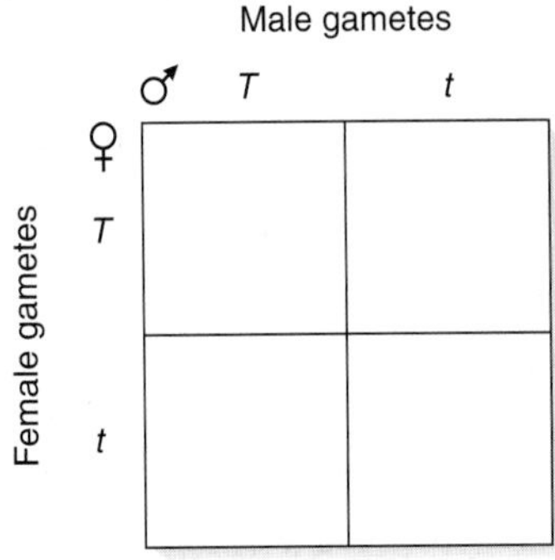

Step 4. *Fill in the possible genotypes of the offspring by combining the alleles of the gametes in the empty boxes.*

Male gametes

♂ / ♀ (Female gametes)	T	t
T	TT	Tt
t	Tt	tt

Step 5. *Determine the relative proportions of genotypes and phenotypes of the offspring.* The genotypes are obtained directly from the Punnett square. They are contained

within the boxes that have been filled in. In this example, the genotypes are *TT*, *Tt*, and *tt* in a 1:2:1 ratio. To determine the phenotypes, you must know the dominant/recessive relationship between the alleles. For plant height, we know that *T* (tall) is dominant to *t* (dwarf). The genotypes *TT* and *Tt* are tall, whereas the genotype *tt* is dwarf. Therefore, our Punnett square shows us that the ratio of phenotypes is 3:1, or 3 tall plants : 1 dwarf plant. Additional problems of this type are provided in the Solved Problems at the end of this chapter.

EXPERIMENT 2B

Mendel Also Analyzed Crosses Involving Two Different Traits

Though his experiments described in Figure 2.5 revealed important ideas regarding hereditary laws, Mendel realized that additional insights might be uncovered if he conducted more complicated experiments. In particular, he conducted crosses in which he simultaneously investigated the pattern of inheritance for two different traits. In other words, he carried out **two-factor crosses,** also called **dihybrid crosses,** in which he followed the inheritance of two different traits within the same groups of individuals. For example, let's consider an experiment in which one of the traits was seed shape, found in round or wrinkled variants; the second trait was seed color, which existed as yellow and green variants. In this dihybrid cross, Mendel followed the inheritance pattern for both traits simultaneously.

What results are possible from a dihybrid cross? One possibility is that the genetic determinants for two different traits are always linked to each other and inherited as a single unit (**Figure 2.7a**). If this were the case, the F_1 offspring could produce only two types of gametes, *RY* and *ry*. A second possibility is that they are not linked and can assort themselves independently into haploid gametes (**Figure 2.7b**). According to independent assortment, an F_1 offspring could produce four types of gametes, *RY*, *Ry*, *rY*, and *ry*. Keep in mind that the results of Figure 2.5 have already shown us that a gamete carries only one allele for each gene.

The experimental protocol of one of Mendel's two-factor crosses is shown in **Figure 2.8**. He began with two different strains of true-breeding pea plants that were different with regard to two traits. In this example, one plant was produced from seeds that were round and yellow; the other plant from seeds that were wrinkled and green. When these plants were crossed, the seeds, which contain the plant embryo, are considered part of the F_1 generation. As expected, the data revealed that the F_1 seeds displayed a phenotype of round and yellow. This was observed because round and yellow are dominant traits. It is the F_2 generation that supports the independent assortment model and refutes the linkage model.

THE HYPOTHESES

The inheritance pattern for two different traits follows one or more quantitative natural laws. Two possible hypotheses are described in Figure 2.7.

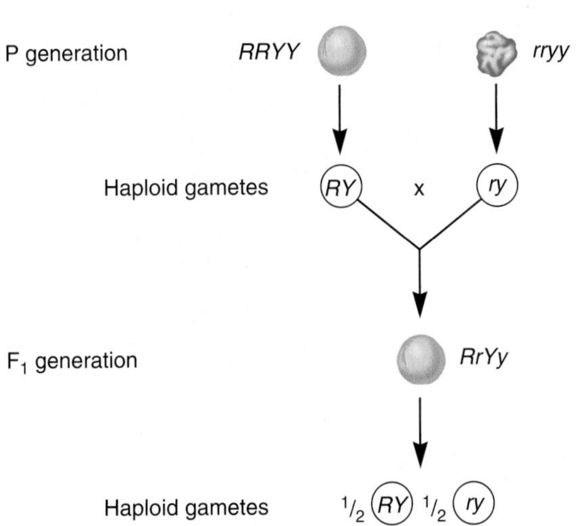

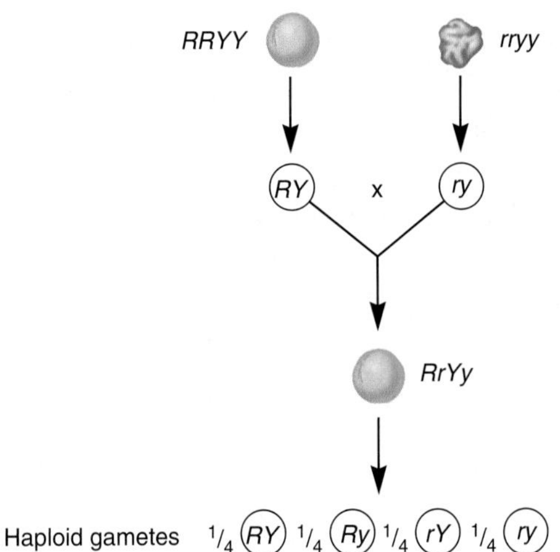

FIGURE 2.7 **Two hypotheses to explain how two different genes assort during gamete formation.** (**a**) According to the linked hypothesis, the two genes always stay associated with each other. (**b**) In contrast, the independent assortment hypothesis proposes that the two different genes randomly segregate into haploid cells.

TESTING THE HYPOTHESES — FIGURE 2.8 Mendel's analysis of two-factor crosses.

Starting material: In this experiment, Mendel began with two types of true-breeding pea plants that were different with regard to two traits. One plant had round, yellow seeds (*RRYY*); the other plant had wrinkled, green seeds (*rryy*).

1. Cross the two true-breeding plants to each other. This produces F_1 generation seeds.

2. Collect many seeds and record their phenotype.

3. F_1 seeds are planted and grown, and the F_1 plants are allowed to self-fertilize. This produces seeds that are part of the F_2 generation.

4. Analyze the characteristics found in the F_2 generation seeds.

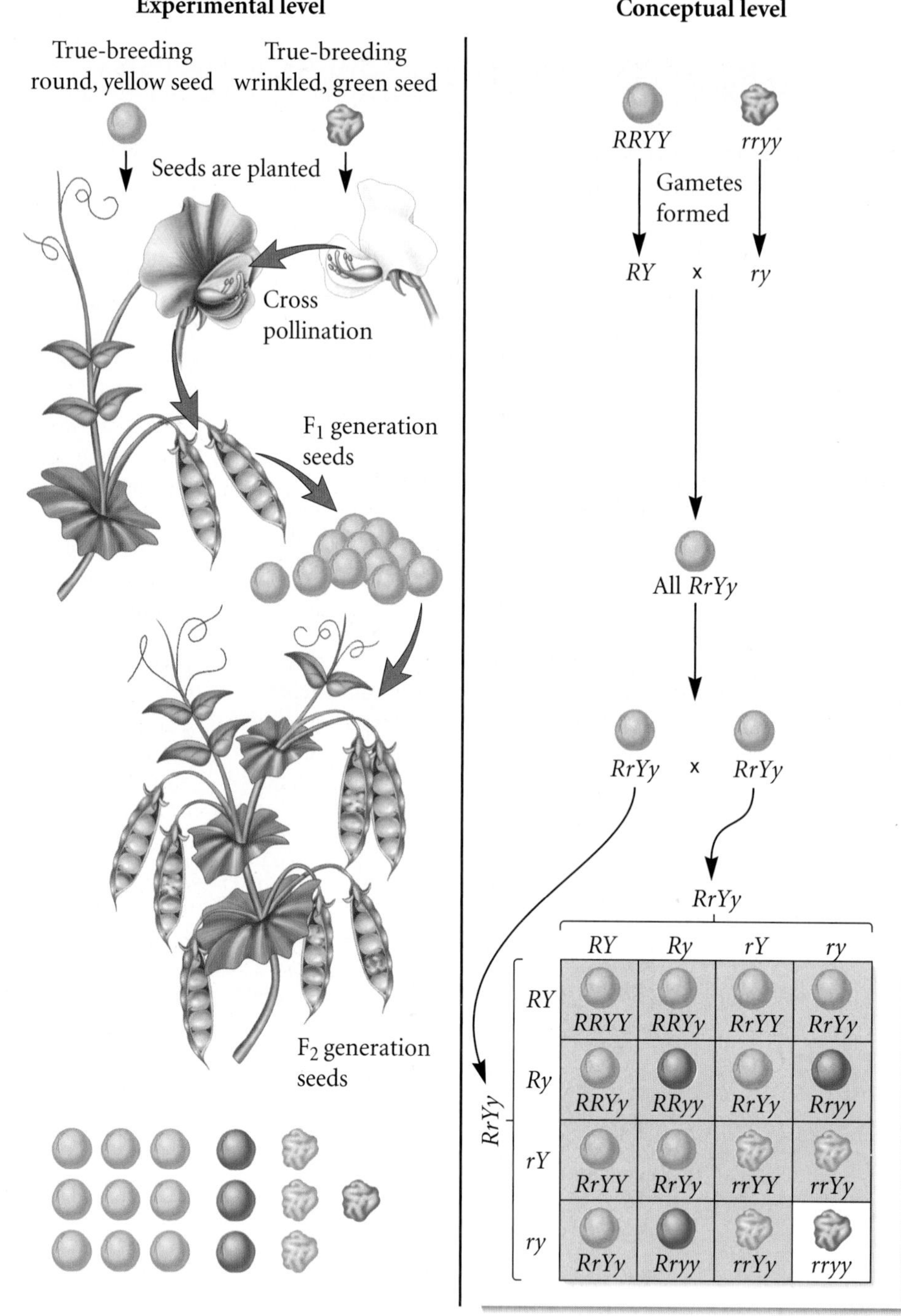

THE DATA

P cross	*F_1 generation*	*F_2 generation*
Round, yellow × wrinkled, green seeds	All round, yellow	315 round, yellow seeds 108 round, green seeds 101 wrinkled, yellow seeds 32 wrinkled, green seeds

INTERPRETING THE DATA

The F_2 generation had seeds that were round and green and seeds that were wrinkled and yellow. These two categories of F_2 seeds are called **nonparentals** because these combinations of traits were not found in the true-breeding plants of the parental generation. The occurrence of nonparental variants contradicts the linkage model. According to the linkage model, the *R* and *Y* variants should be linked together and so should the *r* and *y* variants.

If this were the case, the F_1 plants could produce gametes that are only *RY* or *ry*. These would combine to produce *RRYY* (round, yellow), *RrYy* (round, yellow), or *rryy* (wrinkled, green) in a 1:2:1 ratio. Nonparental seeds could not be produced. However, Mendel did not obtain this result. Instead, he observed a phenotypic ratio of 9:3:3:1 in the F_2 generation.

Mendel's results from many dihybrid experiments rejected the hypothesis of linked assortment and, instead, supported the hypothesis that different traits assort themselves independently during reproduction. Using the modern notion of genes, **Mendel's law of independent assortment** states:

> *Two different genes will randomly assort their alleles during the formation of haploid cells.*

In other words, the allele for one gene will be found within a resulting gamete independently of whether the allele for a different gene is found in the same gamete. Using the example given in Figure 2.8, the round and wrinkled alleles will be assorted into haploid gametes independently of the yellow and green alleles. Therefore, a heterozygous *RrYy* parent can produce four different gametes—*RY*, *Ry*, *rY*, and *ry*—in equal proportions.

In an F_1 self-fertilization experiment, any two gametes can combine randomly during fertilization. This allows for 4^2, or 16, possible offspring, although some offspring will be genetically identical to each other. As shown in **Figure 2.9**, these 16 possible combinations result in seeds with the following phenotypes: 9 round, yellow; 3 round, green; 3 wrinkled, yellow; and 1 wrinkled, green. This 9:3:3:1 ratio is the expected outcome when a dihybrid is allowed to self-fertilize. Mendel was clever enough to realize that the data for his dihybrid experiments were close to a 9:3:3:1 ratio. In Figure 2.8, for example, his F_1 generation produced F_2 seeds with the following characteristics: 315 round, yellow seeds; 108 round, green seeds; 101 wrinkled, yellow seeds; and 32 wrinkled, green seeds. If we divide each of these numbers by 32 (the number of plants with wrinkled, green seeds), the phenotypic ratio of the F_2 generation is 9.8:3.2:3.4:1.0. Within experimental error, Mendel's data approximated the predicted 9:3:3:1 ratio for the F_2 generation.

The law of independent assortment held true for all seven traits that Mendel studied in pea plants. However, in other cases, the inheritance pattern of two different genes is consistent with the linkage model described earlier in Figure 2.7a. In Chapter 5, we will examine the inheritance of genes that are linked to each other because they are physically within the same chromosome. As we will see, linked genes do not assort independently.

An important consequence of the law of independent assortment is that a single individual can produce a vast array of genetically different gametes. As mentioned in Chapter 1, diploid species have pairs of homologous chromosomes, which may differ with respect to the alleles they carry. When an offspring receives a combination of alleles that differs from those in the parental generation, this phenomenon is termed **genetic recombination.** One mechanism that accounts for genetic recombination is independent assortment. A second mechanism, discussed in Chapter 5, is crossing over, which can reassort alleles that happen to be linked along the same chromosome.

The phenomenon of independent assortment is rooted in the random pattern by which the homologues assort themselves

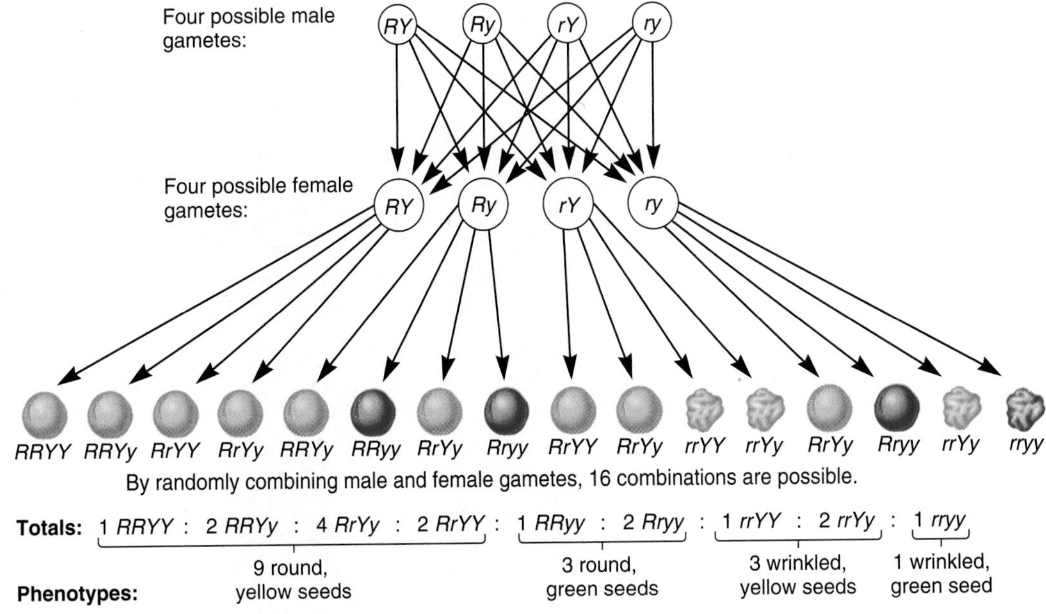

INTERACTIVE EXERCISE

FIGURE 2.9 Mendel's law of independent assortment.

Genes→Traits The cross is between two parents that are heterozygous for seed shape and seed color (*RrYy* × *RrYy*). Four types of male gametes are possible: *RY, Ry, rY,* and *ry*. Likewise, four types of female gametes are possible: *RY, Ry, rY,* and *ry*. These four types of gametes are the result of the independent assortment of the seed shape and seed color alleles relative to each other. During fertilization, any one of the four types of male gametes can combine with any one of the four types of female gametes. This results in 16 types of offspring, each one containing two copies of the seed shape gene and two copies of the seed color gene.

during the process of meiosis, a topic addressed in Chapter 3. If a species contains a large number of homologous chromosomes, this creates the potential for an enormous amount of genetic diversity. For example, human cells contain 23 pairs of homologous chromosomes. These pairs can randomly assort into gametes during meiosis. The number of different gametes an individual can make equals 2^n, where n is the number of pairs of chromosomes. Therefore, humans can make 2^{23}, or over 8 million, possible gametes, due to independent assortment. The capacity to make so many genetically different gametes enables a species to produce individuals with many different combinations of traits. This allows environmental forces to select for those combinations of traits that favor reproductive success.

A self-help quiz involving this experiment can be found at the Online Learning Center.

A Punnett Square Can Also Be Used to Solve Independent Assortment Problems

As already depicted in Figure 2.8, we can use a Punnett square to predict the outcome of crosses involving two or more genes that assort independently. Let's see how such a Punnett square is made by considering a cross between two plants that are heterozygous for height and seed color (**Figure 2.10**). This cross is *TtYy* × *TtYy*. When we construct a Punnett square for this cross, we must keep in mind that each gamete has a single allele for each of two genes. In this example, the four possible gametes from each parent are

TY, *Ty*, *tY*, and *ty*

In this dihybrid experiment, we need to make a Punnett square containing 16 boxes. The phenotypes of the resulting offspring are predicted to occur in a ratio of 9:3:3:1.

In crosses involving three or more genes, the construction of a single large Punnett square to predict the outcome of crosses becomes very unwieldy. For example, in a trihybrid cross between two pea plants that are *Tt Rr Yy*, each parent can make 2^3, or 8, possible gametes. Therefore, the Punnett square must contain $8 \times 8 = 64$ boxes. As a more reasonable alternative, we can consider each gene separately and then algebraically combine them by multiplying together the expected outcomes for each gene. Two such methods, termed the **multiplication method** and the **forked-line method,** are shown in solved problem S3 at the end of this chapter.

Independent assortment is also revealed by a **dihybrid testcross.** In this type of experiment, dihybrid individuals are mated to individuals that are doubly homozygous recessive for the two traits. For example, individuals with a *TtYy* genotype could be crossed to *ttyy* plants. As shown here, independent assortment would predict a 1:1:1:1 ratio among the resulting offspring.

	TY	*Ty*	*tY*	*ty*
ty	*TtYy* Tall, yellow	*Ttyy* Tall, green	*ttYy* Dwarf, yellow	*ttyy* Dwarf, green

Cross: *TtYy* x *TtYy*

♀ \ ♂	*TY*	*Ty*	*tY*	*ty*
TY	*TTYY* Tall, yellow	*TTYy* Tall, yellow	*TtYY* Tall, yellow	*TtYy* Tall, yellow
Ty	*TTYy* Tall, yellow	*TTyy* Tall, green	*TtYy* Tall, yellow	*Ttyy* Tall, green
tY	*TtYY* Tall, yellow	*TtYy* Tall, yellow	*ttYY* Dwarf, yellow	*ttYy* Dwarf, yellow
ty	*TtYy* Tall, yellow	*Ttyy* Tall, green	*ttYy* Dwarf, yellow	*ttyy* Dwarf, green

Genotypes: 1 *TTYY* : 2 *TTYy* : 4 *TtYy* : 2 *TtYY* : 1 *TTyy* : 2 *Ttyy* : 1 *ttYY* : 2 *ttYy* : 1 *ttyy*

Phenotypes: 9 tall plants with yellow seeds; 3 tall plants with green seeds; 3 dwarf plants with yellow seeds; 1 dwarf plant with green seeds

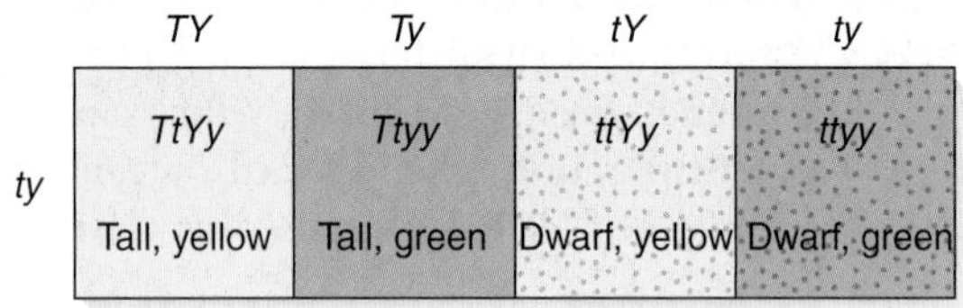

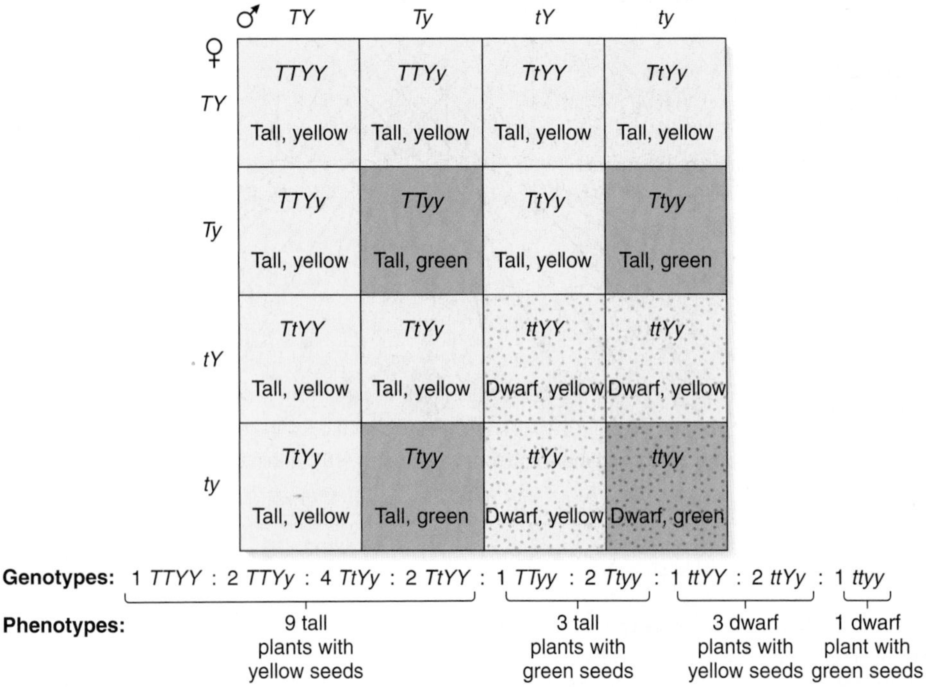

FIGURE 2.10 **A Punnett square for a dihybrid cross.** The Punnett square shown here involves a cross between two pea plants that are heterozygous for height and seed color. The cross is *TtYy* × *TtYy*.

Modern Geneticists Are Often Interested in the Relationship Between the Molecular Expression of Genes and the Outcome of Traits

Mendel's work with pea plants was critically important because his laws of inheritance pertain to all eukaryotic organisms, such as fruit flies, corn, roundworms, mice, and humans, that transmit their genes through sexual reproduction. During the past several decades, many researchers have focused their attention on the relationship between the phenotypic appearance of traits and the molecular expression of genes. This theme will recur throughout the textbook (and we will draw attention to it by designating certain figure legends with a "Genes → Traits" label). As mentioned in Chapter 1, most genes encode proteins that function within living cells. The specific function of individual proteins affects the outcome of an individual's traits. A genetic approach can help us understand the relationship between a protein's function and its effect on phenotype. Most commonly, a geneticist will try to identify an individual that has a defective copy of a gene to see how that will affect the phenotype of the organism. These defective genes are called **loss-of-function alleles,** and they provide geneticists with a great amount of information. Unknowingly, Gregor Mendel had studied seven loss-of-function alleles among his strains of plants. The recessive characteristics in his pea plants were due to genes that had been rendered defective by a mutation. Such alleles are often inherited in a recessive manner, though this is not always the case.

How are loss-of-function alleles informative? In many cases, such alleles provide critical clues concerning the purpose of the protein's function within the organism. For example, we expect the gene affecting flower color (purple versus white) to encode a protein that is necessary for pigment production. This protein may function as an enzyme that is necessary for the synthesis of purple pigment. Furthermore, a reasonable guess is that the white allele is a loss-of-function allele that is unable to express this protein and therefore cannot make the purple pigment. To confirm this idea, a biochemist could analyze the petals from purple and white flowers and try to identify the protein that is defective or missing in the white petals but functionally active in the purple ones. The identification and characterization of this protein would provide a molecular explanation for this phenotypic characteristic.

Pedigree Analysis Can Be Used to Follow the Mendelian Inheritance of Traits in Humans

Before we end our discussion of simple Mendelian traits, let's address the question of how we can analyze inheritance patterns among humans. In his experiments, Mendel selectively made crosses and then analyzed a large number of offspring. When studying human traits, however, researchers cannot control parental crosses. Instead, they must rely on the information that is contained within family trees. This type of approach, known as a **pedigree analysis,** is aimed at determining the type of inheritance pattern that a gene will follow. Although this method may be less definitive than the results described in Mendel's experiments, a pedigree analysis can often provide important clues concerning the pattern of inheritance of traits within human families. An expanded discussion of human pedigrees is provided in Chapter 22, which concerns the inheritance patterns of many different human diseases.

In order to discuss the applications of pedigree analyses, we need to understand the organization and symbols of a pedigree (**Figure 2.11**). The oldest generation is at the top of the pedigree, and the most recent generation is at the bottom. Vertical

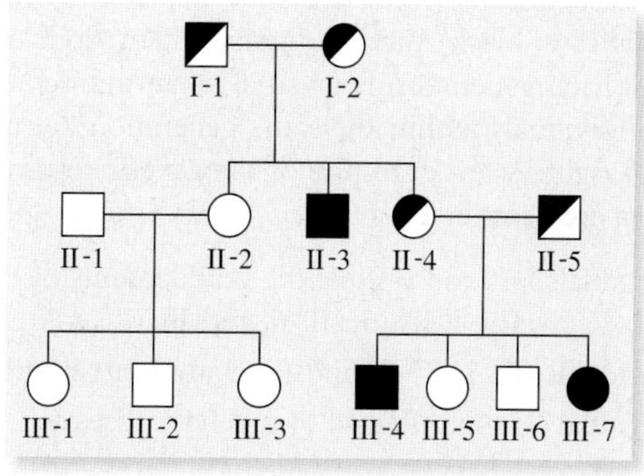

(a) Human pedigree showing cystic fibrosis

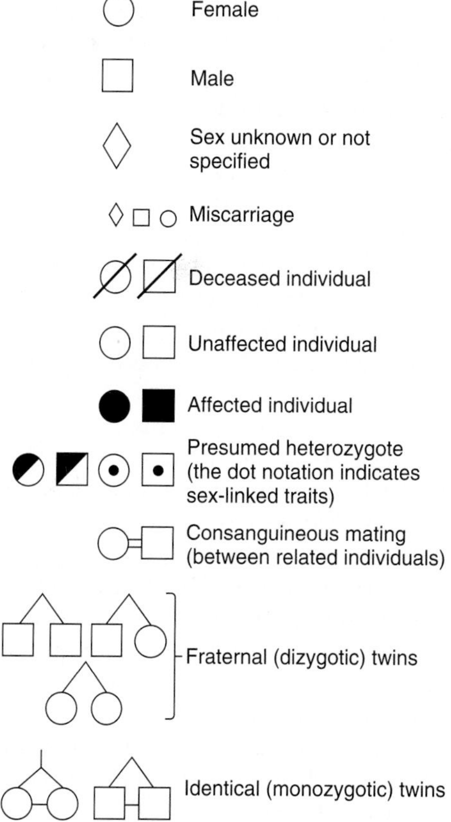

(b) Symbols used in a human pedigree

FIGURE 2.11 **Pedigree analysis.** (a) A family pedigree in which some of the members are affected with cystic fibrosis. Individuals I-1, I-2, II-4, and II-5 are depicted as presumed heterozygotes because they produce affected offspring. (b) The symbols used in a pedigree analysis. Note: In pedigrees shown in this textbook, such as those found in the problem sets, the heterozygotes are not shown as half-filled symbols. Most pedigrees throughout the book show individuals' phenotypes—open symbols are unaffected individuals and filled (closed) symbols are affected individuals.

lines connect each succeeding generation. A man (square) and woman (circle) who produce one or more offspring are directly connected by a horizontal line. A vertical line connects parents with their offspring. If parents produce two or more offspring, the group of siblings (brothers and sisters) is denoted by two or more individuals projecting from the same horizontal line.

When a pedigree involves the transmission of a human trait or disease, affected individuals are depicted by filled symbols (in this case, black) that distinguish them from unaffected individuals. Each generation is given a roman numeral designation, and individuals within the same generation are numbered from left to right. A few examples of the genetic relationships in Figure 2.11a are described here:

Individuals I-1 and I-2 are the grandparents of III-1, III-2, III-3, III-4, III-5, III-6, and III-7

Individuals III-1, III-2, and III-3 are brother and sisters

Individual III-4 is affected by a genetic disease

The symbols shown in Figure 2.11 depict certain individuals, such as I-1, I-2, II-4, and II-5, as presumed heterozygotes because they are unaffected with a disease but produce homozygous offspring that are affected with a recessive genetic disease. However, in many pedigrees, such as those found in the problem sets at the end of the chapter, the inheritance pattern may not be known, so the symbols reflect only phenotypes. In most pedigrees, affected individuals are shown with closed symbols, and unaffected individuals, including those that might be heterozygous for a recessive disease, are depicted with open symbols.

Pedigree analysis is commonly used to determine the inheritance pattern of human genetic diseases. Human geneticists are routinely interested in knowing whether a genetic disease is inherited as a recessive or dominant trait. One way to discern the dominant/recessive relationship between two alleles is by a pedigree analysis. Genes that play a role in disease may exist as a normal allele or a mutant allele that causes disease symptoms. If the disease follows a simple Mendelian pattern of inheritance and is caused by a recessive allele, an individual must inherit two copies of the mutant allele to exhibit the disease. Therefore, a recessive pattern of inheritance makes two important predictions. First, two heterozygous normal individuals will, on average, have 1/4 of their offspring affected. Second, all offspring of two affected individuals will be affected. Alternatively, a dominant trait predicts that affected individuals will have inherited the gene from at least one affected parent (unless a new mutation has occurred during gamete formation).

The pedigree in Figure 2.11a concerns a human genetic disease known as cystic fibrosis (CF). Among Caucasians, approximately 3% of the population are heterozygous carriers of this recessive allele. In homozygotes, the disease symptoms include abnormalities of the pancreas, intestine, sweat glands, and lungs. These abnormalities are caused by an imbalance of ions across the plasma membrane. In the lungs, this leads to a buildup of thick, sticky mucus. Respiratory problems may lead to early death, although modern treatments have greatly increased the life span of CF patients. In the late 1980s, the gene for CF was identified. The CF gene encodes a protein called the cystic fibrosis transmembrane conductance regulator (CFTR). This protein regulates the ion balance across the cell membrane in tissues of the pancreas, intestine, sweat glands, and lungs. The mutant allele causing CF alters the encoded CFTR protein. The altered CFTR protein is not correctly inserted into the plasma membrane, resulting in a decreased function that causes the ionic imbalance. As seen in the pedigree, the pattern of affected and unaffected individuals is consistent with a recessive mode of inheritance. Two unaffected individuals can produce an affected offspring. Although not shown in this pedigree, a recessive mode of inheritance is also characterized by the observation that two affected individuals will produce 100% affected offspring. However, for human genetic diseases that limit survival and/or fertility, there may never be cases where two affected individuals produce offspring.

2.2 PROBABILITY AND STATISTICS

A powerful application of Mendel's work is that the laws of inheritance can be used to predict the outcome of genetic crosses. In agriculture, for example, plant and animal breeders are concerned with the types of offspring their crosses will produce. This information is used to produce commercially important crops and livestock. In addition, people are often interested in predicting the characteristics of the children they may have. This may be particularly important to individuals who carry alleles that cause inherited diseases. Of course, we cannot see into the future and definitively predict what will happen. Nevertheless, genetic counselors can help couples to predict the likelihood of having an affected child. This probability is one factor that may influence a couple's decision whether to have children.

In this section, we will see how probability calculations are used in genetic problems to predict the outcome of crosses. To compute probability, we will use three mathematical operations known as the sum rule, the product rule, and the binomial expansion equation. These methods allow us to determine the probability that a cross between two individuals will produce a particular outcome. To apply these operations, we must have some knowledge regarding the genotypes of the parents and the pattern of inheritance of a given trait.

Probability calculations can also be used in hypothesis testing. In many situations, a researcher would like to discern the genotypes and patterns of inheritance for traits that are not yet understood. A traditional approach to this problem is to conduct crosses and then analyze their outcomes. The proportions of offspring may provide important clues that allow the experimenter to propose a hypothesis, based on the quantitative laws of inheritance, that explains the transmission of the trait from parent to offspring. Statistical methods, such as the chi square test, can then be used to evaluate how well the observed data from crosses fit the expected data. We will end this chapter with an example that applies the chi square test to a genetic cross.

Probability Is the Likelihood That an Event Will Occur

The chance that an event will occur in the future is called the event's **probability.** For example, if you flip a coin, the probability is 0.50, or 50%, that the head side will be showing when the coin lands. Probability depends on the number of possible outcomes. In this case, two possible outcomes (heads or tails) are equally likely. This allows us to predict a 50% chance that a coin flip will produce heads. The general formula for probability (P) is

$$\text{Probability} = \frac{\text{Number of times an event occurs}}{\text{Total number of events}}$$

$$P_{\text{heads}} = 1 \text{ heads} / (1 \text{ heads} + 1 \text{ tails}) = 1/2 = 50\%$$

In genetic problems, we are often interested in the probability that a particular type of offspring will be produced. Recall that when two heterozygous tall pea plants (*Tt*) are crossed, the phenotypic ratio of the offspring is 3 tall:1 dwarf. This information can be used to calculate the probability for either type of offspring.

$$\text{Probability} = \frac{\text{Number of individuals with a given phenotype}}{\text{Total number of individuals}}$$

$$P_{\text{tall}} = 3 \text{ tall} / (3 \text{ tall} + 1 \text{ dwarf}) = 3/4 = 75\%$$

$$P_{\text{dwarf}} = 1 \text{ dwarf} / (3 \text{ tall} + 1 \text{ dwarf}) = 1/4 = 25\%$$

The probability is 75% of obtaining a tall plant and 25% of obtaining a dwarf plant. When we add together the probabilities of all the possible outcomes (tall and dwarf), we should get a sum of 100% (here, 75% + 25% = 100%).

A probability calculation allows us to predict the likelihood that an event will occur in the future. The accuracy of this prediction, however, depends to a great extent on the size of the sample. For example, if we toss a coin six times, our probability prediction would suggest that 50% of the time we should get heads (i.e., three heads and three tails). In this small sample size, however, we would not be too surprised if we came up with four heads and two tails. Each time we toss a coin, there is a random chance that it will be heads or tails. The deviation between the observed and expected outcomes is called the **random sampling error.** In a small sample, the error between the predicted percentage of heads and the actual percentage observed may be quite large. By comparison, if we flipped a coin 1,000 times, the percentage of heads would be fairly close to the predicted 50% value. In a larger sample, we expect the random sampling error to be a much smaller percentage.

The Sum Rule Can Be Used to Predict the Occurrence of Mutually Exclusive Events

Now that we have an understanding of probability, we can see how mathematical operations using probability values allow us to predict the outcome of genetic crosses. Our first genetic problem involves the use of the **sum rule,** which states that *the probability that one of two or more mutually exclusive events will occur is equal to the sum of the individual probabilities of the events.* As an example, let's consider a cross between two mice that are both heterozygous for genes affecting the ears and tail. One gene can be found as an allele designated *de*, which is a recessive allele that causes droopy ears; the normal allele is *De*. An allele of a second gene causes a crinkly tail. This crinkly tail allele (*ct*) is recessive to the normal allele (*Ct*). If a cross is made between two heterozygous mice (*Dede Ctct*), the predicted ratio of offspring is 9 with normal ears and normal tails, 3 with normal ears and crinkly tails, 3 with droopy ears and normal tails, and 1 with droopy ears and a crinkly tail. These four phenotypes are mutually exclusive. For example, a mouse with droopy ears and a normal tail cannot have normal ears and a crinkly tail.

The sum rule allows us to determine the probability that we will obtain any one of two or more different types of offspring. For example, in a cross between two heterozygotes (*Dede Ctct* × *Dede Ctct*), we can ask the following question: What is the probability that an offspring will have normal ears and a normal tail or have droopy ears and a crinkly tail? In other words, if we closed our eyes and picked an offspring out of a litter from this cross, what are the chances that we would be holding a mouse that has normal ears and a normal tail or a mouse with droopy ears and a crinkly tail? In this case, the investigator wants to predict whether one of two mutually exclusive events will occur. A strategy for solving such genetic problems using the sum rule is described here.

The Cross: *Dede Ctct* × *Dede Ctct*

The Question: What is the probability that an offspring will have normal ears and a normal tail or have droopy ears and a crinkly tail?

Step 1. *Calculate the individual probabilities of each phenotype.* This can be accomplished using a Punnett square.

The probability of normal ears and a normal tail is $9/(9 + 3 + 3 + 1) = 9/16$

The probability of droopy ears and a crinkly tail is $1/(9 + 3 + 3 + 1) = 1/16$

Step 2. *Add together the individual probabilities.*

$$9/16 + 1/16 = 10/16$$

This means that 10/16 is the probability that an offspring will have either normal ears and a normal tail or droopy ears and a crinkly tail. We can convert 10/16 to 0.625, which means that 62.5% of the offspring are predicted to have normal ears and a normal tail or droopy ears and a crinkly tail.

The Product Rule Can Be Used to Predict the Probability of Independent Events

We can use probability to make predictions regarding the likelihood of two or more independent outcomes from a genetic cross. When we say that events are independent, we mean that the occurrence of one event does not affect the probability of another event. As an example, let's consider a rare, recessive human trait known as congenital analgesia. Persons with this trait can distinguish between sharp and dull, and hot and cold, but do not perceive extremes of sensation as being painful. The first case of congenital analgesia, described in 1932, was a man who made his living entertaining the public as a "human pincushion."

For a phenotypically unaffected couple, each being heterozygous for the recessive allele causing congenital analgesia, we can ask the question, What is the probability that the couple's first three offspring will have congenital analgesia? To answer this question, the **product rule** is used. According to this rule, *the probability that two or more independent events will occur is equal to the product of their individual probabilities.* A strategy for solving this type of problem is shown here.

The Cross: $Pp \times Pp$ (where P is the common allele and p is the recessive congenital analgesia allele)

The Question: What is the probability that the couple's first three offspring will have congenital analgesia?

Step 1. *Calculate the individual probability of this phenotype.* As described previously, this is accomplished using a Punnett square.

The probability of an affected offspring is 1/4 (25%).

Step 2. *Multiply the individual probabilities.* In this case, we are asking about the first three offspring, and so we multiply 1/4 three times.

$$1/4 \times 1/4 \times 1/4 = 1/64 = 0.016$$

Thus, the probability that the first three offspring will have this trait is 0.016. In other words, we predict that 1.6% of the time the first three offspring of a couple, each heterozygous for the recessive allele, will all have congenital analgesia. In this example, the phenotypes of the first, second, and third offspring are independent events. In this case, the phenotype of the first offspring does not have an effect on the phenotype of the second or third offspring.

In the problem described here, we have used the product rule to determine the probability that the first three offspring will all have the same phenotype (congenital analgesia). We can also apply the rule to predict the probability of a sequence of events that involves combinations of different offspring. For example, consider the question, What is the probability that the first offspring will be unaffected, the second offspring will have congenital analgesia, and the third offspring will be unaffected? Again, to solve this problem, begin by calculating the individual probability of each phenotype.

Unaffected = 3/4

Congenital analgesia = 1/4

The probability that these three phenotypes will occur in this specified order is

$$3/4 \times 1/4 \times 3/4 = 9/64 = 0.14, \text{ or } 14\%$$

In other words, this sequence of events is expected to occur only 14% of the time.

The product rule can also be used to predict the outcome of a cross involving two or more genes. Let's suppose an individual with the genotype *Aa Bb CC* was crossed to an individual with the genotype *Aa bb Cc*. We could ask the question, What is the probability that an offspring will have the genotype *AA bb Cc*? If the three genes independently assort, the probability of inheriting alleles for each gene is independent of the other two genes. Therefore, we can separately calculate the probability of the desired outcome for each gene.

Cross: *Aa Bb CC* × *Aa bb Cc*

Probability that an offspring will be *AA* = 1/4, or 0.25

Probability that an offspring will be *bb* = 1/2, or 0.5

Probability that an offspring will be *Cc* = 1/2, or 0.5

We can use the product rule to determine the probability that an offspring will be *AA bb Cc*;

$$P = (0.25)(0.5)(0.5) = 0.0625, \text{ or } 6.25\%$$

The Binomial Expansion Equation Can Be Used to Predict the Probability of an Unordered Combination of Events

A third predictive problem in genetics is to determine the probability that a certain proportion of offspring will be produced with particular characteristics; here they can be produced in an unspecified order. For example, we can consider a group of children produced by two heterozygous brown-eyed (*Bb*) individuals. We can ask the question, What is the probability that two out of five children will have blue eyes?

In this case, we are not concerned with the order in which the offspring are born. Instead, we are only concerned with the final numbers of blue-eyed and brown-eyed offspring. One possible outcome would be the following: firstborn child with blue eyes, second child with blue eyes, and then the next three with brown eyes. Another possible outcome could be firstborn child with brown eyes, second with blue eyes, third with brown eyes, fourth with blue eyes, and fifth with brown eyes. Both of these scenarios would result in two offspring with blue eyes and three with brown eyes. In fact, several other ways to have such a family could occur.

To solve this type of question, the **binomial expansion equation** can be used. This equation represents all of the possibilities for a given set of unordered events.

$$P = \frac{n!}{x!(n-x)!} p^x q^{n-x}$$

where

P = the probability that the unordered outcome will occur

n = total number of events

x = number of events in one category (e.g., blue eyes)

p = individual probability of x

q = individual probability of the other category (e.g., brown eyes)

Note: In this case, $p + q = 1$.

The symbol ! denotes a factorial. $n!$ is the product of all integers from n down to 1. For example, $4! = 4 \times 3 \times 2 \times 1 = 24$. An exception is 0!, which equals 1.

The use of the binomial expansion equation is described next.

The Cross: $Bb \times Bb$

The Question: What is the probability that two out of five offspring will have blue eyes?

Step 1. *Calculate the individual probabilities of the blue-eye and brown-eye phenotypes.* If we constructed a Punnett square, we would find the probability of blue eyes is 1/4 and the probability of brown eyes is 3/4:

$$p = 1/4$$

$$q = 3/4$$

Step 2. *Determine the number of events in category* x *(in this case, blue eyes) versus the total number of events.* In this example, the number of events in category x is two blue-eyed children among a total number of five.

$$x = 2$$

$$n = 5$$

Step 3. *Substitute the values for* p, q, x, *and* n *in the binomial expansion equation.*

$$P = \frac{n!}{x!(n-x)!} p^x q^{n-x}$$

$$P = \frac{5!}{2!(5-2)!} (1/4)^2 (3/4)^{5-2}$$

$$P = \frac{5 \times 4 \times 3 \times 2 \times 1}{(2 \times 1)(3 \times 2 \times 1)} (1/16)(27/64)$$

$$P = 0.26 = 26\%$$

Thus, the probability is 0.26 that two out of five offspring will have blue eyes. In other words, 26% of the time we expect a $Bb \times Bb$ cross yielding five offspring to contain two blue-eyed children and three brown-eyed children.

In solved problem S7 at the end of this chapter, we consider an expanded version of this approach that uses a **multinomial expansion equation.** This equation is needed to solve unordered genetic problems that involve three or more phenotypic categories.

The Chi Square Test Can Be Used to Test the Validity of a Genetic Hypothesis

We now look at a different issue in genetic problems, namely **hypothesis testing.** Our goal here is to determine if the data from genetic crosses are consistent with a particular pattern of inheritance. For example, a geneticist may study the inheritance of body color and wing shape in fruit flies over the course of two generations. The following question may be asked about the F_2 generation: Do the observed numbers of offspring agree with the predicted numbers based on Mendel's laws of segregation and independent assortment? As we will see in Chapters 3 through 8, not all traits follow a simple Mendelian pattern of inheritance. Some genes do not segregate and independently assort themselves the same way that Mendel's seven traits did in pea plants.

To distinguish between inheritance patterns that obey Mendel's laws versus those that do not, a conventional strategy is to make crosses and then quantitatively analyze the offspring. Based on the observed outcome, an experimenter may make a tentative hypothesis. For example, it may seem that the data are obeying Mendel's laws. Hypothesis testing provides an objective, statistical method to evaluate whether or not the observed data really agree with the hypothesis. In other words, we use statistical methods to determine whether the data that have been gathered from crosses are consistent with predictions based on quantitative laws of inheritance.

The rationale behind a statistical approach is to evaluate the **goodness of fit** between the observed data and the data that are predicted from a hypothesis. This is sometimes called a **null hypothesis** because it assumes there is no real difference between the observed and expected values. Any actual differences that occur are presumed to be due to random sampling error. If the observed and predicted data are very similar, we can conclude that the hypothesis is consistent with the observed outcome. In this case, it is reasonable to accept the hypothesis. However, it should be emphasized that this does not prove a hypothesis is correct. Statistical methods can never prove a hypothesis is correct. They can provide insight as to whether or not the observed data seem reasonably consistent with the hypothesis. Alternative hypotheses, perhaps even ones that the experimenter has failed to realize, may also be consistent with the data. In some cases, statistical methods may reveal a poor fit between hypothesis and data. In other words, a high deviation would be found between the observed and expected values. If this occurs, the hypothesis is rejected. Hopefully, the experimenter can subsequently propose an alternative hypothesis that has a reasonable fit with the data.

One commonly used statistical method to determine goodness of fit is the **chi square test** (often written χ^2). We can use the chi square test to analyze population data in which the members of the population fall into different categories. This is the kind of data we have when we evaluate the outcome of genetic crosses, because these usually produce a population of offspring that differ with regard to phenotypes. The general formula for the chi square test is

$$\chi^2 = \sum \frac{(O-E)^2}{E}$$

where

O = observed data in each category

E = expected data in each category based on the experimenter's hypothesis

Σ means to sum this calculation for each category. For example, if the population data fell into two categories, the chi square calculation would be

$$\chi^2 = \frac{(O_1 - E_1)^2}{E_1} + \frac{(O_2 - E_2)^2}{E_2}$$

We can use the chi square test to determine if a genetic hypothesis is consistent with the observed outcome of a genetic cross. The strategy described next provides a step-by-step outline for applying the chi square testing method. In this problem, the experimenter wants to determine if a dihybrid cross is obeying the laws of Mendel. The experimental organism is *Drosophila melanogaster* (the common fruit fly), and the two traits affect wing shape and body color. Straight wing shape and curved wing shape are designated by c^+ and c, respectively; gray body color and ebony body color are designated by e^+ and e, respectively. Note: In certain species, such as *Drosophila melanogaster*, the convention is to designate the common (wild-type) allele with a plus sign. Recessive mutant alleles are designated with lowercase letters and dominant mutant alleles with capital letters.

The Cross: A true-breeding fly with straight wings and a gray body ($c^+c^+e^+e^+$) is crossed to a true-breeding fly with curved wings and an ebony body ($ccee$). The flies of the F_1 generation are then allowed to mate with each other to produce an F_2 generation.

The Outcome:

F_1 generation:	All offspring have straight wings and gray bodies
F_2 generation:	193 straight wings, gray bodies
	69 straight wings, ebony bodies
	64 curved wings, gray bodies
	26 curved wings, ebony bodies
Total:	352

Step 1. *Propose a hypothesis that allows us to calculate the expected values based on Mendel's laws.* The F_1 generation suggests that the trait of straight wings is dominant to curved wings and gray body coloration is dominant to ebony. Looking at the F_2 generation, it appears that offspring are following a 9:3:3:1 ratio. If so, this is consistent with an independent assortment of the two traits.

Based on these observations, the hypothesis is:

Straight (c^+) is dominant to curved (c), and gray (e^+) is dominant to ebony (e). The two traits segregate and assort independently from generation to generation.

Step 2. *Based on the hypothesis, calculate the expected values of the four phenotypes.* We first need to calculate the individual probabilities of the four phenotypes. According to our hypothesis, there should be a 9:3:3:1 ratio in the F_2 generation. Therefore, the expected probabilities are:

9/16 = straight wings, gray bodies

3/16 = straight wings, ebony bodies

3/16 = curved wings, gray bodies

1/16 = curved wings, ebony bodies

The observed F_2 generation contained a total of 352 individuals. Our next step is to calculate the expected numbers of each type of offspring when the total equals 352. This can be accomplished by multiplying each individual probability by 352.

9/16 × 352 = 198 (expected number with straight wings, gray bodies)

3/16 × 352 = 66 (expected number with straight wings, ebony bodies)

3/16 × 352 = 66 (expected number with curved wings, gray bodies)

1/16 × 352 = 22 (expected number with curved wings, ebony bodies)

Step 3. *Apply the chi square formula, using the data for the expected values that have been calculated in step 2.* In this case, the data include four categories, and thus the sum has four terms.

$$\chi^2 = \frac{(O_1 - E_1)^2}{E_1} + \frac{(O_2 - E_2)^2}{E_2} + \frac{(O_3 - E_3)^2}{E_3} + \frac{(O_4 - E_4)^2}{E_4}$$

$$\chi^2 = \frac{(193 - 198)^2}{198} + \frac{(69 - 66)^2}{66} + \frac{(64 - 66)^2}{66} + \frac{(26 - 22)^2}{22}$$

$$\chi^2 = 0.13 + 0.14 + 0.06 + 0.73 = 1.06$$

Step 4. *Interpret the calculated chi square value. This is done using a chi square table.*

Before interpreting the chi square value we have obtained, we must understand how to use **Table 2.1**. The probabilities, called ***P* values,** listed in the chi square table allow us to determine the likelihood that the amount of variation indicated by a given chi square value is due to random chance alone, based on a particular hypothesis. For example, let's consider a value—0.00393—listed in row 1. (The meaning of the rows will be explained shortly.) Chi square values that are equal to or greater than 0.00393 are expected to occur 95% of the time when a hypothesis is correct. In other words, 95 out of 100 times we would expect that random chance alone would produce a deviation between the experimental data and hypothesized model that is equal to or greater than 0.00393. A low chi square value indicates a high probability that the observed deviations could be due to random chance alone. By comparison, chi square values that are equal to or greater than 3.841 are expected to occur less than 5% of the time due to random sampling error. If a high chi square value is obtained, an experimenter becomes suspicious that the high deviations have occurred because the hypothesis is incorrect. A common convention is to reject the null hypothesis if the chi square value results in a probability that is less than 0.05—less than 5%.

In our problem involving flies with straight or curved wings and gray or ebony bodies, we have calculated a chi square value of 1.06. Before we can determine the probability that this deviation would have occurred as a matter of random chance, we must first determine the degrees of freedom (*df*) in this experiment. The **degrees of freedom** is a measure of the number of categories that are independent of each other. When phenotype categories are derived from a Punnett square, it is typically $n - 1$, where n equals the total number of categories. In the preceding problem, $n = 4$ (the categories are the phenotypes: straight wings and gray body; straight wings and ebony body; curved wings and gray body; and curved wings and

TABLE 2.1

Chi Square Values and Probability

Degrees of Freedom	$P = 0.99$	0.95	0.80	0.50	0.20	0.05	0.01
						Null Hypothesis rejected	
1.	0.000157	0.00393	0.0642	0.455	1.642	3.841	6.635
2.	0.020	0.103	0.446	1.386	3.219	5.991	9.210
3.	0.115	0.352	1.005	2.366	4.642	7.815	11.345
4.	0.297	0.711	1.649	3.357	5.989	9.488	13.277
5.	0.554	1.145	2.343	4.351	7.289	11.070	15.086
6.	0.872	1.635	3.070	5.348	8.558	12.592	16.812
7.	1.239	2.167	3.822	6.346	9.803	14.067	18.475
8.	1.646	2.733	4.594	7.344	11.030	15.507	20.090
9.	2.088	3.325	5.380	8.343	12.242	16.919	21.666
10.	2.558	3.940	6.179	9.342	13.442	18.307	23.209
15.	5.229	7.261	10.307	14.339	19.311	24.996	30.578
20.	8.260	10.851	14.578	19.337	25.038	31.410	37.566
25.	11.524	14.611	18.940	24.337	30.675	37.652	44.314
30.	14.953	18.493	23.364	29.336	36.250	43.773	50.892

From Fisher, R. A., and Yates, F. (1943) *Statistical Tables for Biological, Agricultural, and Medical Research.* Oliver and Boyd, London.

ebony body); thus, the degrees of freedom equals 3.* We now have sufficient information to interpret our chi square value of 1.06.

With $df = 3$, the chi square value of 1.06 we have obtained is slightly greater than 1.005, which gives a P value of 0.80, or 80%. What does this P value mean? If the hypothesis is correct, chi square values equal to or greater than 1.005 are expected to occur 80% of the time based on random chance alone. To reject the hypothesis, the chi square would have to be greater than 7.815. Because it was actually far less than this value, we are inclined to accept that the null hypothesis is correct.

We must keep in mind that the chi square test does not prove a hypothesis is correct. It is a statistical method for evaluating whether or not the data and hypothesis have a good fit.

CONCEPTUAL SUMMARY

The work of Gregor Mendel has led to great insight into the transmission of traits from parent to offspring. First, his work showed us that the determinants of traits are discrete units that are passed via gametes from generation to generation. This particulate theory of inheritance refuted previous ideas such as blending inheritance and pangenesis. In addition, Mendel deduced two fundamental laws that pertain to the transmission of many traits in eukaryotic organisms. The law of segregation tells us that two variants for a single trait will segregate during their passage to offspring. This means that the two alleles for a particular gene will segregate from each other during the formation of haploid cells, so a gamete will contain only one copy of an allele. Mendel's law of independent assortment says that two different genes assort independently of each other during the formation of haploid cells.

An understanding of genetic inheritance allows us to make predictions about future offspring and also to test the validity of genetic hypotheses. Probability is the likelihood that an event will occur in the future. By knowing the genotypes of parents, we can determine the probability of obtaining an offspring with a particular trait. Furthermore, mathematical operations can be used to predict combinations of offspring that may, or may not, occur in a specified order. The sum rule is used in genetic problems that involve mutually exclusive events; the product rule is used to analyze independent events; and the binomial expansion equation is used to predict unordered events.

In other genetic problems, the goal is to deduce the pattern of inheritance from parents to offspring. This requires an analysis of genetic crosses in order to propose a genetic hypothesis that is consistent with the observed data. A chi square test is used to evaluate whether or not the genetic hypothesis predicts values that have a reasonable agreement with the observed data. In this approach, the amount of deviation between the observed and

* If our hypothesis already assumed that the law of segregation is obeyed, the degrees of freedom would be one (see Chapter 5).

expected values is used to determine the **goodness of fit** between the data and hypothesis. If the deviation is so large that it would be expected to occur less than 5% of the time as a result of random sampling error, then it is appropriate to reject the hypothesis. If, on the other hand, the chi square value is low, the null hypothesis is accepted.

In Chapter 2, a foundation has been provided on which to base our understanding of simple inheritance patterns. The two laws that we examined, segregation and independent assortment, provide us with insight into the transmission of traits from parent to offspring. While they form a cornerstone for our understanding of genetics, we will see in later chapters that not all genes obey these laws. In Chapter 3, we will take a close look at the composition and behavior of chromosomes, particularly during meiosis. As we will see, the segregation and assortment of chromosomes provides a cellular explanation for Mendel's laws.

EXPERIMENTAL SUMMARY

In Chapter 2, we examined the research studies of Gregor Mendel. By making single-factor crosses and examining the traits of offspring for two generations, Mendel found that the determinants of traits do not blend. Rather, they are discrete units that are passed along unaltered from generation to generation. He quantitatively analyzed the outcome of his self-fertilization experiments among monohybrids and obtained a 3:1 phenotypic ratio for the F_2 generation. This observation allowed him to deduce the law of segregation.

In dihybrid experiments, Mendel followed the inheritance pattern of two genes relative to each other. His data supported the idea that two different genes assort independently of each other during the formation of haploid cells, producing a 9:3:3:1 phenotypic ratio in the F_2 generation. The law of independent assortment held true for all seven traits that Mendel studied. However, as we will see in Chapter 5, some genes follow a linked pattern of inheritance.

PROBLEM SETS & INSIGHTS

Solved Problems

S1. A heterozygous pea plant that is tall with yellow seeds, *TtYy*, is allowed to self-fertilize. What is the probability that an offspring will be either tall with yellow seeds, tall with green seeds, or dwarf with yellow seeds?

Answer: This problem involves three mutually exclusive events, and so we use the sum rule to solve it. First, we must calculate the individual probabilities for the three phenotypes. The outcome of the cross can be determined using a Punnett square.

Cross: *TtYy* x *TtYy*

♀ \ ♂	*TY*	*Ty*	*tY*	*ty*
TY	*TTYY* Tall, yellow	*TTYy* Tall, yellow	*TtYY* Tall, yellow	*TtYy* Tall, yellow
Ty	*TTYy* Tall, yellow	*TTyy* Tall, green	*TtYy* Tall, yellow	*Ttyy* Tall, green
tY	*TtYY* Tall, yellow	*TtYy* Tall, yellow	*ttYY* Dwarf, yellow	*ttYy* Dwarf, yellow
ty	*TtYy* Tall, yellow	*Ttyy* Tall, green	*ttYy* Dwarf, yellow	*ttyy* Dwarf, green

$$P_{\text{tall with yellow seeds}} = 9/(9 + 3 + 3 + 1) = 9/16$$

$$P_{\text{tall with green seeds}} = 3/(9 + 3 + 3 + 1) = 3/16$$

$$P_{\text{dwarf with yellow seeds}} = 3/(9 + 3 + 3 + 1) = 3/16$$

Sum rule: $9/16 + 3/16 + 3/16 = 15/16 = 0.94 = 94\%$

We expect to get one of these three phenotypes 15/16, or 94%, of the time.

S2. As described in Chapter 2, a human disease known as cystic fibrosis is inherited as a recessive trait. Two unaffected individuals have a first child with the disease. What is the probability that their next two children will not have the disease?

Answer: An unaffected couple has already produced an affected child. To be affected, the child must be homozygous for the disease allele and thus has inherited one copy from each parent. Therefore, because the parents are unaffected with the disease, we know that both of them must be heterozygous carriers for the recessive disease-causing allele. With this information, we can calculate the probability that they will produce an unaffected offspring. Using a Punnett square, this couple should produce a ratio of 3 unaffected : 1 affected offspring.

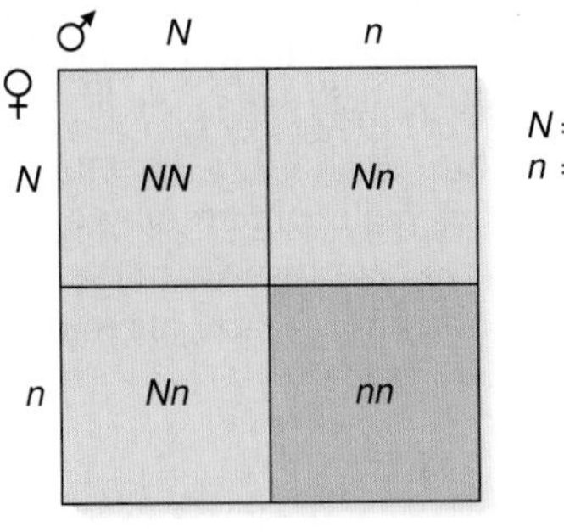

N = common allele
n = cystic fibrosis allele

The probability of a single unaffected offspring is

$$P_{\text{unaffected}} = 3/(3 + 1) = 3/4$$

To obtain the probability of getting two unaffected offspring in a row (i.e., in a specified order), we must apply the product rule.

$$3/4 \times 3/4 = 9/16 = 0.56 = 56\%$$

The chance that their next two children will be unaffected is 56%.

S3. A cross was made between two heterozygous pea plants, *TtYy* × *TtYy*. The following Punnett square was constructed:

♀ \ ♂	*TT*	*Tt*	*Tt*	*tt*
YY	*TTYY*	*TtYY*	*TtYY*	*ttYY*
Yy	*TTYy*	*TtYy*	*TtYy*	*ttYy*
Yy	*TTYy*	*TtYy*	*TtYy*	*ttYy*
yy	*TTyy*	*Ttyy*	*Ttyy*	*ttyy*

Phenotypic ratio:

9 tall, yellow seeds : 3 tall, green seeds : 3 dwarf, yellow seeds : 1 dwarf, green seed

What is wrong with this Punnett square?

Answer: The outside of the Punnett square is supposed to contain the possible types of gametes. A gamete should contain one copy of each type of gene. Instead, the outside of this Punnett square contains two copies of one gene and zero copies of the other gene. The outcome happens to be correct (i.e., it yields a 9:3:3:1 ratio), but this is only a coincidence. The outside of the Punnett square must contain one copy of each type of gene. In this example, the correct possible types of gametes are *TY*, *Ty*, *tY*, and *ty* for each parent.

S4. A pea plant is heterozygous for three genes (*Tt Rr Yy*), where *T* = tall, *t* = dwarf, *R* = round seeds, *r* = wrinkled seeds, *Y* = yellow seeds, and *y* = green seeds. If this plant is self-fertilized, what are the predicted phenotypes of the offspring, and what fraction of the offspring will occur in each category?

Answer: You could solve this problem by constructing a large Punnett square and filling in the boxes. However, in this case, eight different male gametes and eight different female gametes are possible: *TRY*, *TRy*, *TrY*, *tRY*, *trY*, *Try*, *tRy*, and *try*. It would become rather tiresome to construct and fill in this Punnett square, which would contain 64 boxes. As an alternative, we can consider each gene separately and then algebraically combine them by multiplying together the expected phenotypic outcomes for each gene. In the cross *Tt Rr Yy* × *Tt Rr Yy*, the following Punnett squares can be made for each gene:

♀ \ ♂	*T*	*t*
T	*TT* Tall	*Tt* Tall
t	*Tt* Tall	*tt* Dwarf

3 tall : 1 dwarf

♀ \ ♂	*R*	*r*
R	*RR* Round	*Rr* Round
r	*Rr* Round	*rr* Wrinkled

3 round : 1 wrinkled

♀ \ ♂	*Y*	*y*
Y	*YY* Yellow	*Yy* Yellow
y	*Yy* Yellow	*yy* Green

3 yellow : 1 green

Instead of constructing a large, 64-box Punnett square, we can use two similar ways to determine the phenotypic outcome of this trihybrid cross. In the **multiplication method,** we can simply multiply these three combinations together:

(3 tall +1 dwarf)(3 round + 1 wrinkled)(3 yellow + 1 green)

This multiplication operation can be done in a stepwise manner. First, multiply (3 tall + 1 dwarf) by (3 round + 1 wrinkled).

(3 tall + 1 dwarf)(3 round + 1 wrinkled) = 9 tall, round + 3 tall, wrinkled + 3 dwarf, round, + 1 dwarf, wrinkled

Next, multiply this product by (3 yellow + 1 green).

(9 tall, round + 3 tall, wrinkled + 3 dwarf, round + 1 dwarf, wrinkled)(3 yellow + 1 green) = 27 tall, round, yellow + 9 tall, round, green + 9 tall, wrinkled, yellow + 3 tall, wrinkled, green + 9 dwarf, round, yellow + 3 dwarf, round, green + 3 dwarf, wrinkled, yellow + 1 dwarf, wrinkled, green

Even though the multiplication steps are also somewhat tedious, this approach is much easier than making a Punnett square with 64 boxes, filling them in, deducing each phenotype, and then adding them up!

A second approach that is analogous to the multiplication method is the **forked-line method.** In this case, the genetic proportions are determined by multiplying together the probabilities of each phenotype.

Tall or dwarf	Round or wrinkled	Yellow or green	Observed product	Phenotype
$^3/_4$ tall	$^3/_4$ round	$^3/_4$ yellow	$(^3/_4)(^3/_4)(^3/_4)$ =	$^{27}/_{64}$ tall, round, yellow
		$^1/_4$ green	$(^3/_4)(^3/_4)(^1/_4)$ =	$^9/_{64}$ tall, round, green
	$^1/_4$ wrinkled	$^3/_4$ yellow	$(^3/_4)(^1/_4)(^3/_4)$ =	$^9/_{64}$ tall, wrinkled, yellow
		$^1/_4$ green	$(^3/_4)(^1/_4)(^1/_4)$ =	$^3/_{64}$ tall, wrinkled, green
$^1/_4$ dwarf	$^3/_4$ round	$^3/_4$ yellow	$(^1/_4)(^3/_4)(^3/_4)$ =	$^9/_{64}$ dwarf, round, yellow
		$^1/_4$ green	$(^1/_4)(^3/_4)(^1/_4)$ =	$^3/_{64}$ dwarf, round, green
	$^1/_4$ wrinkled	$^3/_4$ yellow	$(^1/_4)(^1/_4)(^3/_4)$ =	$^3/_{64}$ dwarf, wrinkled, yellow
		$^1/_4$ green	$(^1/_4)(^1/_4)(^1/_4)$ =	$^1/_{64}$ dwarf, wrinkled, green

S5. For an individual expressing a dominant trait, how can you tell if it is a heterozygote or a homozygote?

Answer: One way is to conduct a testcross with an individual that expresses the recessive version of the same trait. If the individual is heterozygous, half of the offspring will show the recessive trait, whereas if the individual is homozygous, none of the offspring will express the recessive trait.

Dd × *dd*	or	*DD* × *dd*
↓		↓
1*Dd* (dominant trait)		All *Dd* (dominant trait)
1 *dd* (recessive trait)		

Another way to determine heterozygosity involves a more careful examination of the individual at the cellular or molecular level. At the cellular level, the heterozygote may not look exactly like the homozygote. This phenomenon is described in Chapter 4. Also, gene cloning methods described in Chapter 18 can be used to distinguish between heterozygotes and homozygotes.

S6. In dogs, black fur color is dominant to white. Two heterozygous black dogs are mated. What would be the probability of the following combinations of offspring?

A. A litter of six pups, four with black fur and two with white fur.

B. A litter of six pups, the firstborn with white fur, and among the remaining five pups, two with white fur and three with black fur.

C. A first litter of six pups, four with black fur and two with white fur, and then a second litter of seven pups, five with black fur and two with white fur.

D. A first litter of five pups, four with black fur and one with white fur, and then a second litter of seven pups in which the firstborn is homozygous, the second born is black, and the remaining five pups are three black and two white.

Answer:

A. Because this is an unordered combination of events, we use the binomial expansion equation, where $n = 6$, $x = 4$, $p = 0.75$ (probability of black), and $q = 0.25$ (probability of white).

The answer is 0.297, or 29.7%, of the time.

B. We use the product rule because the order is specified. The first pup is white and then the remaining five are born later. We also need to use the binomial expansion equation to determine the probability of the remaining five pups.

(probability of a white pup)(binomial expansion for the remaining five pups)

The probability of the white pup is 0.25. In the binomial expansion equation, $n = 5$, $x = 2$, $p = 0.25$, and $q = 0.75$.

The answer is 0.066, or 6.6%, of the time.

C. The order of the two litters is specified, so we need to use the product rule. We multiply the probability of the first litter times the probability of the second litter. We need to use the binomial expansion equation for each litter.

(binomial expansion of the first litter)(binomial expansion of the second litter)

For the first litter, $n = 6$, $x = 4$, $p = 0.75$, $q = 0.25$. For the second litter, $n = 7$, $x = 5$, $p = 0.75$, $q = 0.25$.

The answer is 0.092, or 9.2%, of the time.

D. The order of the litters is specified, so we need to use the product rule to multiply the probability of the first litter times the probability of the second litter. We use the binomial expansion equation to determine the probability of the first litter. The probability of the second litter is a little more complicated. The firstborn is homozygous. There are two mutually exclusive ways to be homozygous, *BB* and *bb*. We use the sum rule to determine the probability of the first pup, which equals $0.25 + 0.25 = 0.5$. The probability of the second pup is 0.75, and we use the binomial expansion equation to determine the probability of the remaining pups.

(binomial expansion of first litter)([0.5][0.75][binomial expansion of second litter])

For the first litter, $n = 5$, $x = 4$, $p = 0.75$, $q = 0.25$. For the last five pups in the second litter, $n = 5$, $x = 3$, $p = 0.75$, $q = 0.25$.

The answer is 0.039, or 3.9%, of the time.

S7. In Chapter 2, the binomial expansion equation was used in situations where only two phenotypic outcomes are possible. When more than two outcomes are possible, we use a **multinomial**

expansion equation to solve a problem involving an unordered number of events. A general expression for this equation is

$$P = \frac{n!}{a!b!c!\ldots} p^a q^b r^c \ldots$$

where P = the probability that the unordered number of events will occur.

$$n = \text{total number of events}$$

$$a + b + c + \ldots = n$$

$$p + q + r + \ldots = 1$$

(p is the likelihood of a, q is the likelihood of b, r is the likelihood of c, and so on)

The multinomial expansion equation can be useful in many genetic problems where more than two combinations of offspring are possible. For example, this formula can be used to solve problems involving an unordered sequence of events in a dihybrid experiment. This approach is illustrated next.

A cross is made between two heterozygous tall plants with axial flowers (*TtAa*), where tall is dominant to dwarf and axial is dominant to terminal flowers. What is the probability that a group of five offspring will be composed of two tall plants with axial flowers, one tall plant with terminal flowers, one dwarf plant with axial flowers, and one dwarf plant with terminal flowers?

Answer:

Step 1. *Calculate the individual probabilities of each phenotype.* This can be accomplished using a Punnett square.

The phenotypic ratios are 9 tall with axial flowers, 3 tall with terminal flowers, 3 dwarf with axial flowers, and 1 dwarf with terminal flowers.

The probability of a tall plant with axial flowers is $9/(9 + 3 + 3 + 1) = 9/16$.

The probability of a tall plant with terminal flowers is $3/(9 + 3 + 3 + 1) = 3/16$.

The probability of a dwarf plant with axial flowers is $3/(9 + 3 + 3 + 1) = 3/16$.

The probability of a dwarf plant with terminal flowers is $1/(9 + 3 + 3 + 1) = 1/16$.

$$p = 9/16$$

$$q = 3/16$$

$$r = 3/16$$

$$s = 1/16$$

Step 2. *Determine the number of each type of event versus the total number of events.*

$$n = 5$$

$$a = 2$$

$$b = 1$$

$$c = 1$$

$$d = 1$$

Step 3. *Substitute the values in the multinomial expansion equation.*

$$P = \frac{n!}{a!b!c!d!} p^a q^b r^c s^d$$

$$P = \frac{5!}{2!1!1!1!} (9/16)^2 (3/16)^1 (3/16)^1 (1/16)^1$$

$$P = 0.04 = 4\%$$

This means that 4% of the time we would expect to obtain five offspring with the phenotypes described in the question.

Conceptual Questions

C1. Why did Mendel's work refute the idea of blending inheritance?

C2. What is the difference between cross-fertilization and self-fertilization?

C3. Describe the difference between genotype and phenotype. Give three examples. Is it possible for two individuals to have the same phenotype but different genotypes?

C4. With regard to genotypes, what is a true-breeding organism?

C5. How can you determine whether an organism is heterozygous or homozygous for a dominant trait?

C6. In your own words, describe what Mendel's law of segregation means. Do not use the word segregation in your answer.

C7. Based on genes in pea plants that we have considered in this chapter, which statement(s) is not correct?

A. The gene causing tall plants is an allele of the gene causing dwarf plants.

B. The gene causing tall plants is an allele of the gene causing purple flowers.

C. The alleles causing tall plants and purple flowers are dominant.

C8. In a cross between a heterozygous tall pea plant and a dwarf plant, predict the ratios of the offspring's genotypes and phenotypes.

C9. Do you know the genotype of an individual with a recessive trait and/or a dominant trait? Explain your answer.

C10. A cross is made between a pea plant that has constricted pods (a recessive trait; smooth is dominant) and is heterozygous for seed color (yellow is dominant to green) and a plant that is heterozygous for both pod texture and seed color. Construct a Punnett square that depicts this cross. What are the predicted outcomes of genotypes and phenotypes of the offspring?

C11. A pea plant that is heterozygous with regard to seed color (yellow is dominant to green) is allowed to self-fertilize. What are the predicted outcomes of genotypes and phenotypes of the offspring?

C12. Describe the significance of nonparentals with regard to the law of independent assortment. In other words, explain how the appearance of nonparentals refutes a linkage hypothesis.

C13. For the following pedigrees, describe what you think is the most likely inheritance pattern (dominant versus recessive). Explain your reasoning. Filled (black) symbols indicate affected individuals.

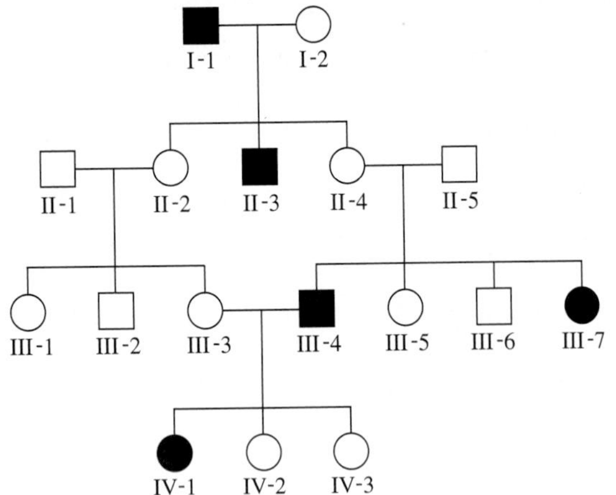

(a)

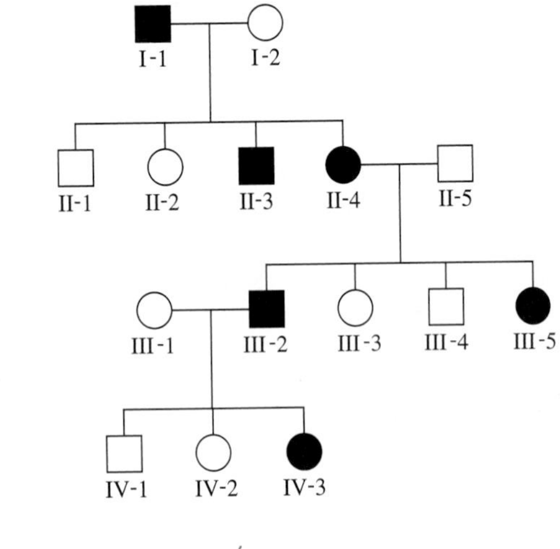

(b)

C14. Ectrodactyly, also known as "lobster claw syndrome," is a recessive disorder in humans. If a phenotypically unaffected couple produces an affected offspring, what are the following probabilities?

A. Both parents are heterozygotes.

B. An offspring is a heterozygote.

C. The next three offspring will be phenotypically unaffected.

D. Any two out of the next three offspring will be phenotypically unaffected.

C15. Identical twins are produced from the same sperm and egg (which splits after the first mitotic division), whereas fraternal twins are produced from separate sperm and separate egg cells. If two parents with brown eyes (a dominant trait) produce one twin boy with blue eyes, what are the following probabilities?

A. If the other twin is identical, he will have blue eyes.

B. If the other twin is fraternal, he/she will have blue eyes.

C. If the other twin is fraternal, he/she will transmit the blue eye allele to his/her offspring.

D. The parents are both heterozygotes.

C16. In cocker spaniels, solid coat color is dominant over spotted coat color. If two heterozygous dogs were crossed to each other, what would be the probability of the following combinations of offspring?

A. A litter of five pups, four with solid fur and one with spotted fur.

B. A first litter of six pups, four with solid fur and two with spotted fur, and then a second litter of five pups, all with solid fur.

C. A first litter of five pups, the firstborn with solid fur, and then among the next four, three with solid fur and one with spotted fur, and then a second litter of seven pups in which the firstborn is spotted, the second born is spotted, and the remaining five are composed of four solid and one spotted animal.

D. A litter of six pups, the firstborn with solid fur, the second born spotted, and among the remaining four pups, two with spotted fur and two with solid fur.

C17. A cross was made between a white male dog and two different black females. The first female gave birth to eight black pups, while the second female gave birth to four white and three black pups. What are the likely genotypes of the male parent and the two female parents? Explain whether you are uncertain about any of the genotypes.

C18. In humans, the allele for brown eye color (*B*) is dominant to blue eye color (*b*). If two heterozygous parents produce children, what are the following probabilities?

A. The first two children have blue eyes.

B. A total of four children, two with blue eyes and the other two with brown eyes.

C. The first child has blue eyes, and the next two have brown eyes.

C19. Albinism, a condition characterized by a partial or total lack of skin pigment, is a recessive human trait. If a phenotypically unaffected couple produced an albino child, what is the probability that their next child will be albino?

C20. A true-breeding tall plant was crossed to a dwarf plant. Tallness is a dominant trait. The F_1 individuals were allowed to self-fertilize. What are the following probabilities for the F_2 generation?

A. The first plant is dwarf.

B. The first plant is dwarf or tall.

C. The first three plants are tall.

D. For any seven plants, three are tall and four are dwarf.

E. The first plant is tall, and then among the next four, two are tall and the other two are dwarf.

C21. For pea plants with the following genotypes, list the possible gametes that the plant can make:

A. *TT Yy Rr*

B. *Tt YY rr*

C. *Tt Yy Rr*

D. *tt Yy rr*

C22. An individual has the genotype *Aa Bb Cc* and makes an abnormal gamete with the genotype *Aa Bc*. Does this gamete violate the law of independent assortment and/or the law of segregation? Explain your answer.

C23. Maple syrup urine disease is a disease found in humans in which the body is unable to metabolize the amino acids leucine, isoleucine, and valine. One of the symptoms is that the urine smells like maple syrup. An unaffected couple produced six children in the following order: unaffected daughter, affected daughter, unaffected son, unaffected son, affected son, and unaffected son. The youngest unaffected son marries an unaffected woman and has three children in the following order: affected daughter, unaffected daughter, and unaffected son. Draw a pedigree that describes this family. What type of inheritance (dominant or recessive) would you propose to explain maple syrup urine disease?

C24. Marfan syndrome is a rare inherited human disorder characterized by unusually long limbs and digits plus defects in the heart (especially the aorta) and the eyes. Following is a pedigree for this disorder. Affected individuals are shown with filled (black) symbols. What type of inheritance pattern do you think is the most likely?

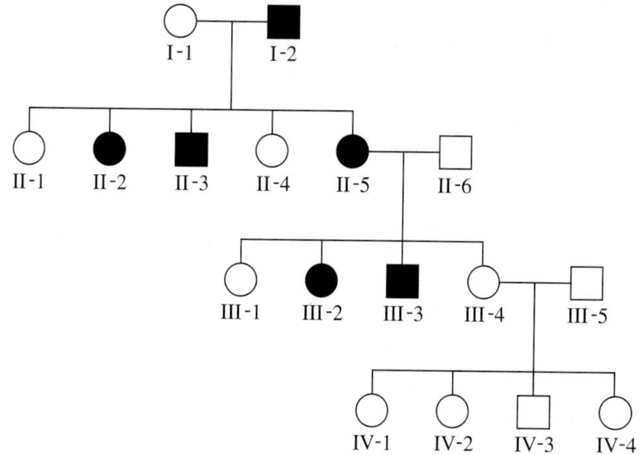

C25. A true-breeding pea plant with round and green seeds was crossed to a true-breeding plant with wrinkled and yellow seeds. Round and yellow seeds are the dominant traits. The F_1 plants were allowed to self-fertilize. What are the following probabilities for the F_2 generation?

A. An F_2 plant with wrinkled, yellow seeds.

B. Three out of three F_2 plants with round, yellow seeds.

C. Five F_2 plants in the following order: two have round, yellow seeds; one has round, green seeds; and two have wrinkled, green seeds.

D. An F_2 plant will not have round, yellow seeds.

C26. A true-breeding tall pea plant was crossed to a true-breeding dwarf plant. What is the probability that an F_1 individual will be true-breeding? What is the probability that an F_1 individual will be a true-breeding tall plant?

C27. What are the expected phenotypic ratios from the following cross: $Tt\ Rr\ yy\ Aa \times Tt\ rr\ YY\ Aa$, where T = tall, t = dwarf, R = round, r = wrinkled, Y = yellow, y = green, A = axial, a = terminal; T, R, Y, and A are dominant alleles. Note: See solved problem S3 for help in answering this problem.

C28. When an abnormal organism contains three copies of a gene (instead of the normal number of two copies), the alleles for the gene usually segregate so that a gamete will contain one or two copies of the gene. Let's suppose that an abnormal pea plant has three copies of the height gene. Its genotype is TTt. The plant is also heterozygous for the seed color gene, Yy. How many types of gametes can this plant make, and in what proportions? (Assume that it is equally likely that a gamete will contain one or two copies of the height gene.)

C29. Honeybees are unusual in that male bees (drones) have only one copy of each gene, while female bees have two copies of their genes. That is because drones develop from eggs that have not been fertilized by sperm cells. In bees, the trait of long wings is dominant over short wings, and the trait of black eyes is dominant over white eyes. If a drone with short wings and black eyes was mated to a queen bee that is heterozygous for both genes, what are the predicted genotypes and phenotypes of male and female offspring? What are the phenotypic ratios if we assume an equal number of male and female offspring?

C30. A pea plant that is dwarf with green, wrinkled seeds was crossed to a true-breeding plant that is tall with yellow, round seeds. The F_1 generation was allowed to self-fertilize. What types of gametes, and in what proportions, would the F_1 generation make? What would be the ratios of genotypes and phenotypes of the F_2 generation?

C31. A true-breeding plant with round and green seeds was crossed to a true-breeding plant with wrinkled and yellow seeds. The F_1 plants were allowed to self-fertilize. What is the probability of obtaining the following plants in the F_2 generation: two that have round, yellow seeds; one with round, green seeds; and two with wrinkled, green seeds? (Note: See solved problem S7 for help.)

C32. Wooly hair is a rare dominant trait found in people of Scandinavian descent in which the hair resembles the wool of a sheep. A male with wooly hair, who has a mother with straight hair, moves to an island that is inhabited by people who are not of Scandinavian descent. Assuming that he never leaves the island and that no other Scandinavians immigrate to the island, what is the probability that a great-grandchild of this male will have wooly hair? (Hint: You may want to draw a pedigree to help you figure this out.) If this wooly-haired male has eight great-grandchildren, what is the probability that one out of eight will have wooly hair?

C33. Huntington disease is a rare dominant trait that causes neurodegeneration later in life. A man in his thirties, who already has three children, discovers that his mother has Huntington disease though his father is unaffected. What are the following probabilities?

A. That the man in his thirties will develop Huntington disease.

B. That his first child will develop Huntington disease.

C. That one out of three of his children will develop Huntington disease.

C34. A woman with achondroplasia (a dominant form of dwarfism) and a phenotypically unaffected man have seven children, all of whom have achondroplasia. What is the probability of producing such a family if this woman is a heterozygote? What is the probability that the woman is a heterozygote if her eighth child does not have this disorder?

Experimental Questions

E1. Describe three advantages of using pea plants as an experimental organism.

E2. Explain the technical differences between a cross-fertilization experiment versus a self-fertilization experiment.

E3. How long did it take Mendel to complete the experiment in Figure 2.5?

E4. For all seven traits described in the data of Figure 2.5, Mendel allowed the F_2 plants to self-fertilize. He found that when F_2 plants with recessive traits were crossed to each other, they always bred true. However, when F_2 plants with dominant traits were crossed, some bred true while others did not. A summary of Mendel's results is shown here.

The Ratio of True-Breeding and Non-True-Breeding Parents of the F_2 Generation

F_2 Parents	True-Breeding	Non-True-Breeding	Ratio
Round	193	372	1:1.93
Yellow	166	353	1:2.13
Gray	36	64	1:1.78
Smooth	29	71	1:2.45
Green	40	60	1:1.5
Axial	33	67	1:2.08
Tall	28	72	1:2.57
TOTAL:	525	1,059	1:2.02

When considering the data in this table, keep in mind that it describes the characteristics of the F_2 generation parents that had displayed a dominant phenotype. These data were deduced by analyzing the outcome of the F_3 generation. Based on Mendel's laws, explain the 1:2 ratio obtained in these data.

E5. From the point of view of crosses and data collection, what are the experimental differences between a monohybrid and a dihybrid experiment?

E6. As in many animals, albino coat color is a recessive trait in guinea pigs. Researchers removed the ovaries from an albino female guinea pig and then transplanted ovaries from a true-breeding black guinea pig. They then mated this albino female (with the transplanted ovaries) to an albino male. The albino female produced three offspring. What were their coat colors? Explain the results.

E7. The fungus *Melampsora lini* causes a disease known as flax rust. Different strains of *M. lini* cause varying degrees of the rust disease. Conversely, different strains of flax are resistant or sensitive to the various varieties of rust. The Bombay variety of flax is resistant to *M. lini*-strain 22 but sensitive to *M. lini*-strain 24. A strain of flax called 770B is just the opposite; it is resistant to strain 24 but sensitive to strain 22. When 770B was crossed to Bombay, all the F_1 individuals were resistant to both strain 22 and strain 24. When F_1 individuals were self-fertilized, the following data were obtained:

43 resistant to strain 22 but sensitive to strain 24

9 sensitive to strain 22 and strain 24

32 sensitive to strain 22 but resistant to strain 24

110 resistant to strain 22 and strain 24

Explain the inheritance pattern for flax resistance and sensitivity to *M. lini* strains.

E8. For Mendel's data shown in Figure 2.8, conduct a chi square analysis to determine if the data agree with Mendel's law of independent assortment.

E9. Would it be possible to deduce the law of independent assortment from a monohybrid experiment? Explain your answer.

E10. In fruit flies, curved wings are recessive to straight wings, and ebony body is recessive to gray body. A cross was made between true-breeding flies with curved wings and gray bodies to flies with straight wings and ebony bodies. The F_1 offspring were then mated to flies with curved wings and ebony bodies to produce an F_2 generation.

A. Diagram the genotypes of this cross, starting with the parental generation and ending with the F_2 generation.

B. What are the predicted phenotypic ratios of the F_2 generation?

C. Let's suppose the following data were obtained for the F_2 generation:

114 curved wings, ebony body

105 curved wings, gray body

111 straight wings, gray body

114 straight wings, ebony body

Conduct a chi square analysis to determine if the experimental data are consistent with the expected outcome based on Mendel's laws.

E11. A recessive allele in mice results in an abnormally long neck. Sometimes, during early embryonic development, the abnormal neck causes the embryo to die. An experimenter began with a population of true-breeding normal mice and true-breeding mice with long necks. Crosses were made between these two populations to produce an F_1 generation of mice with normal necks. The F_1 mice were then mated to each other to obtain an F_2 generation. For the mice that were born alive, the following data were obtained:

522 mice with normal necks

62 mice with long necks

What percentage of homozygous mice (that would have had long necks if they had survived) died during embryonic development?

E12. The data in Figure 2.5 show the results of the F_2 generation for seven of Mendel's crosses. Conduct a chi square analysis to determine if these data are consistent with the law of segregation.

E13. Let's suppose you conducted an experiment involving genetic crosses and calculated a chi square value of 1.005. There were four categories of offspring (i.e., the degrees of freedom equaled 3). Explain what the 1.005 value means. Your answer should include the phrase "80% of the time."

E14. A tall pea plant with axial flowers was crossed to a dwarf plant with terminal flowers. Tall plants and axial flowers are dominant traits. The following offspring were obtained: 27 tall, axial flowers; 23 tall, terminal flowers; 28 dwarf, axial flowers; and 25 dwarf, terminal flowers. What are the genotypes of the parents?

E15. A cross was made between two strains of plants that are agriculturally important. One strain was disease resistant but herbicide sensitive; the other strain was disease sensitive but herbicide resistant. A plant breeder crossed the two plants and then allowed the F_1 generation to self-fertilize. The following data were obtained:

F_1 generation: All offspring are disease sensitive and herbicide resistant

F_2 generation:
157 disease sensitive, herbicide resistant
57 disease sensitive, herbicide sensitive
54 disease resistant, herbicide resistant
20 disease resistant, herbicide sensitive

Total: 288

Formulate a hypothesis that you think is consistent with the observed data. Test the goodness of fit between the data and your hypothesis using a chi square test. Explain what the chi square results mean.

E16. A cross was made between a plant that has blue flowers and purple seeds to a plant with white flowers and green seeds. The following data were obtained:

F_1 generation: All offspring have blue flowers with purple seeds

F_2 generation:
103 blue flowers, purple seeds
49 blue flowers, green seeds
44 white flowers, purple seeds
104 white flowers, green seeds

Total: 300

Start with the hypothesis that blue flowers and purple seeds are dominant traits and that the two genes assort independently. Calculate a chi square value. What does this value mean with regard to your hypothesis? If you decide to reject your hypothesis, which aspect of the hypothesis do you think is incorrect (i.e., blue flowers and purple seeds are dominant traits, or the idea that the two genes assort independently)?

Questions for Student Discussion/Collaboration

1. Consider the following cross in pea plants: *Tt Rr yy Aa* × *Tt rr Yy Aa*, where T = tall, t = dwarf, R = round, r = wrinkled, Y = yellow, y = green, A = axial, a = terminal. What is the expected phenotypic outcome of this cross? Have one group of students solve this problem by making one big Punnett square, and have another group solve it by making four single-gene Punnett squares and using the product rule. Time each other to see who gets done first.
2. A cross was made between two pea plants, *TtAa* and *Ttaa*, where T = tall, t = dwarf, A = axial, and a = terminal. What is the probability that the first three offspring will be tall with axial flowers or dwarf with terminal flowers and the fourth offspring will be tall with axial flowers. Discuss what operation(s) (e.g., sum rule, product rule, and/or binomial expansion equation) you used to solve them and in what order they were used.
3. Consider the following tetrahybrid cross: *Tt Rr yy Aa* × *Tt RR Yy aa*, where T = tall, t = dwarf, R = round, r = wrinkled, Y = yellow, y = green, A = axial, a = terminal. What is the probability that the first three plants will have round seeds? What is the easiest way to solve this problem?

Note: All answers appear at the website for this textbook; the answers to even-numbered questions are in the back of the textbook.

www.mhhe.com/brookergenetics3e

Visit the Online Learning Center for practice tests, answer keys, and other learning aids for this chapter. Enhance your understanding of genetics with our interactive exercises, quizzes, animations, and much more.

3 REPRODUCTION AND CHROMOSOME TRANSMISSION

CHAPTER OUTLINE

Chromosome sorting during cell division. When eukaryotic cells divide, they replicate and sort their chromosomes so that each cell receives the correct amount.

In Chapter 2, we considered some patterns of inheritance that explain the passage of traits from parent to offspring. In this chapter, we will survey reproduction at the cellular level and pay close attention to the inheritance of chromosomes. An examination of chromosomes at the microscopic level provides us with insights into understanding the inheritance patterns of traits. To appreciate this relationship, we will first consider how cells distribute their chromosomes during the process of cell division. We will see that in bacteria and most unicellular eukaryotes, simple cell division provides a way to reproduce asexually. Then we will explore a form of cell division called meiosis, which produces cells with half the number of chromosomes. By closely examining this process, we will see how the transmission of chromosomes accounts for the inheritance patterns that were observed by Mendel.

3.1 GENERAL FEATURES OF CHROMOSOMES

The **chromosomes** are structures within living cells that contain the genetic material. Genes are physically located within chromosomes. Biochemically, chromosomes contain a very long segment of DNA, which is the genetic material, and proteins, which are bound to the DNA and provide it with an organized structure. In eukaryotic cells, this complex between DNA and proteins is called **chromatin.** In this chapter, we will focus on the cellular mechanics of chromosome transmission to better understand the patterns of gene transmission that we considered in Chapter 2. In particular, we will examine how chromosomes are copied and sorted into newly made cells. In later chapters, particularly Chapters 8, 10, and 11, we will examine the molecular features of chromosomes in greater detail.

Before we begin a description of chromosome transmission, we need to consider the distinctive cellular differences between bacterial and eukaryotic species. Bacteria and archaea are referred to as **prokaryotes,** from the Greek meaning prenucleus, because their chromosomes are not contained within a separate nucleus of the cell. Prokaryotes usually have a single type of circular chromosome in a region of the cytoplasm called the **nucleoid** (**Figure 3.1a**). The cytoplasm is enclosed by a plasma membrane that regulates the uptake of nutrients and the excretion of waste products. Outside the plasma membrane is found a rigid cell wall that protects the cell from breakage. Certain species of bacteria also have an outer membrane located beyond the cell wall.

Eukaryotes, from the Greek meaning true nucleus, include some simple species, such as single-celled protists and some fungi (such as yeast), and more complex multicellular species, such as plants, animals, and other fungi. The cells of eukaryotic species have internal membranes that enclose highly specialized compartments (**Figure 3.1b**). These compartments form membrane-bounded

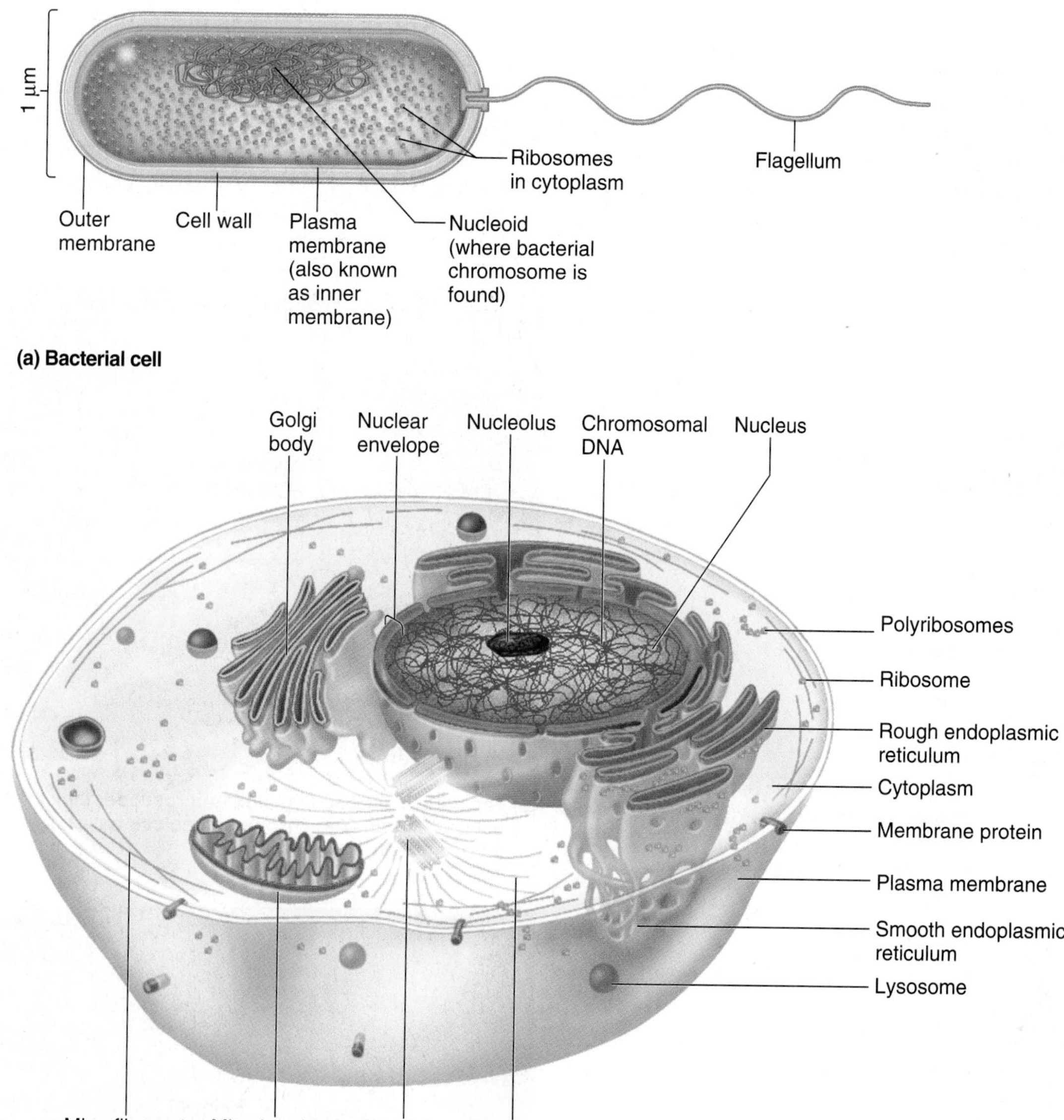

FIGURE 3.1 **The basic organization of cells.** (**a**) A bacterial cell. The example shown here is typical of a bacterium such as *Escherichia coli,* which has an outer membrane. (**b**) A eukaryotic cell. The example shown here is typical of an animal cell.

organelles with specific functions. For example, the lysosomes play a role in the degradation of macromolecules. The endoplasmic reticulum and Golgi body play a role in protein modification and trafficking. A particularly conspicuous organelle is the **nucleus,** which is bounded by two membranes that constitute the nuclear envelope. Most of the genetic material is found within chromosomes that are located in the nucleus. In addition to the nucleus, certain organelles in eukaryotic cells contain a small amount of their own DNA. These include the mitochondrion, which plays a role in ATP synthesis, and, in plant cells, the chloroplast, which plays a role in photosynthesis. The DNA found in these organelles is referred to as extranuclear or extrachromosomal DNA to distinguish it from the DNA that is found in the cell nucleus. We will examine the role of mitochondrial and chloroplast DNA in Chapter 7.

In this section, we will focus on the composition of chromosomes found in the nucleus of eukaryotic cells. As you will learn, eukaryotic species contain genetic material that comes in sets of linear chromosomes.

Eukaryotic Chromosomes Are Examined Cytologically to Yield a Karyotype

Insights into inheritance patterns have been gained by observing the behavior of chromosomes under the microscope. **Cytogenetics** is the field of genetics that involves the microscopic examination of chromosomes. The most basic observation that a **cytogeneticist** can make is to examine the chromosomal composition of a particular cell. For eukaryotic species, this is usually accomplished by observing the chromosomes as they are found in actively dividing cells. When a cell is preparing to divide, the chromosomes become more tightly coiled, which shortens them and thereby increases their diameter. The consequence of this shortening is that distinctive shapes and numbers of chromosomes become visible with a light microscope. Each species has a particular chromosome composition. For example, most human cells contain 23 pairs of chromosomes, for a total of 46. On rare occasions, some individuals may inherit an abnormal number of chromosomes or a chromosome with an abnormal structure. Such abnormalities can often be detected by a microscopic examination of the chromosomes within actively dividing cells. In addition, a cytogeneticist may examine chromosomes as a way to distinguish between two closely related species.

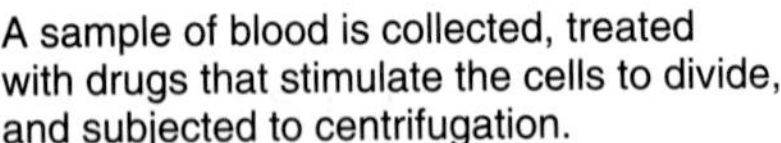

A sample of blood is collected, treated with drugs that stimulate the cells to divide, and subjected to centrifugation.

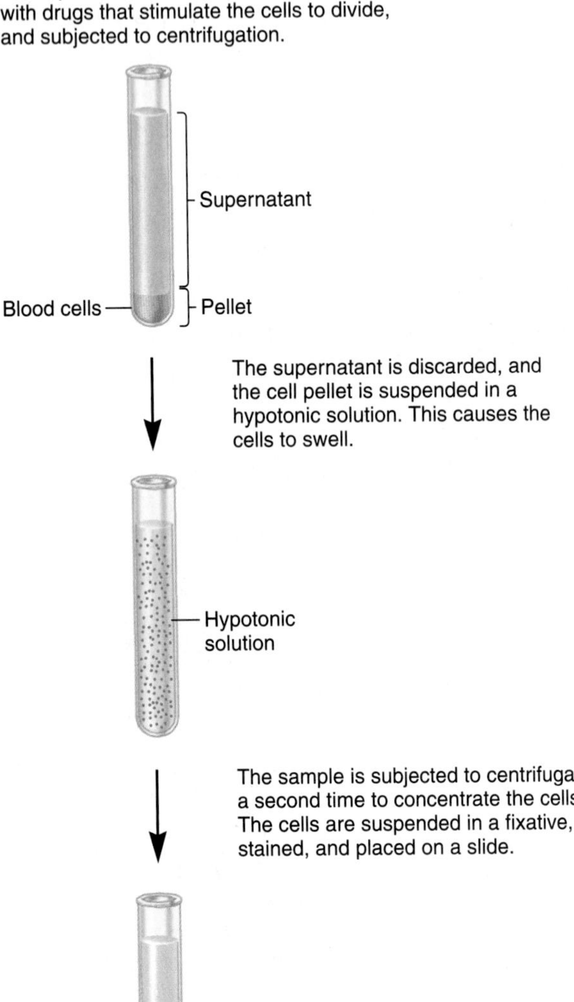

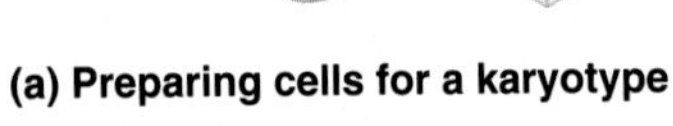

(a) Preparing cells for a karyotype

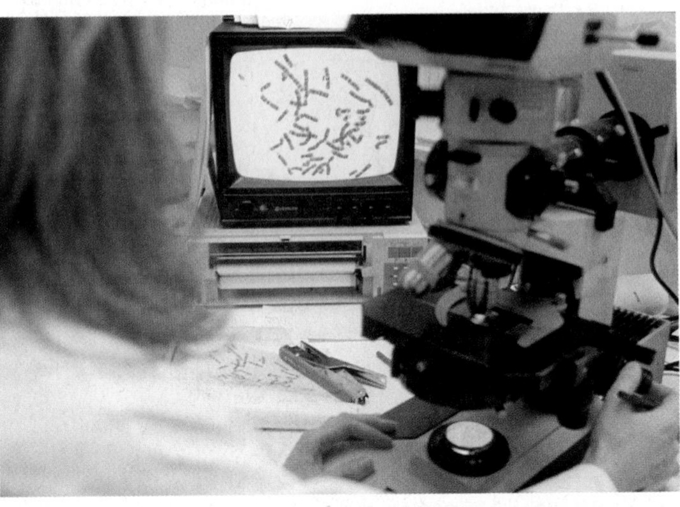

(b) The slide is viewed by a light microscope; the sample is seen on a video screen. The chromosomes can be arranged electronically on the screen.

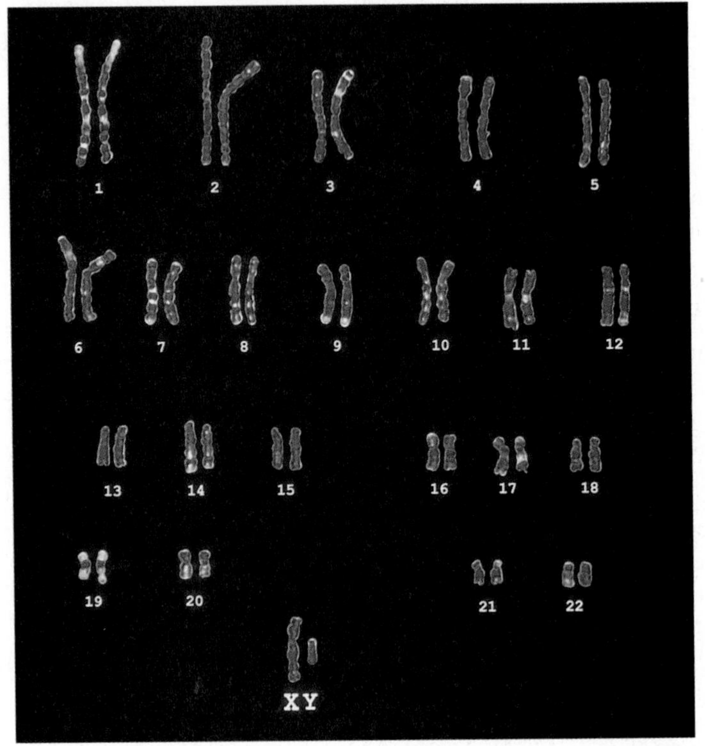

(c) For a diploid human cell, two complete sets of chromosomes from a single cell constitute a karyotype of that cell.

FIGURE 3.2 The procedure for making a human karyotype.

Figure 3.2a shows the general procedure for preparing human chromosomes to be viewed by microscopy. In this example, the cells were obtained from a sample of human blood; more specifically, the chromosomes within lymphocytes (a type of white blood cell) were examined. Blood cells are a type of **somatic cell.** This term refers to any cell of the body that is not a gamete or a precursor to a gamete. The **gametes** (sperm and egg cells or their precursors) are also called **germ cells.**

After the blood cells have been removed from the body, they are treated with drugs that stimulate them to divide. As shown in Figure 3.2a, these actively dividing cells are centrifuged to concentrate them. The concentrated preparation is then mixed with a hypotonic solution that makes the cells swell. This swelling causes the chromosomes to spread out within the cell and thereby makes it easier to see each individual chromosome. Next, the cells are treated with a fixative that chemically freezes them so that the chromosomes will no longer move around. The cells are then treated with a chemical dye that binds to the chromosomes and stains them. As discussed in greater detail in Chapter 8, this gives chromosomes a distinctive banding pattern that greatly enhances their visualization and ability to be uniquely identified (also see Figure 8.1c, d). The cells are then placed on a slide and viewed with a light microscope.

In a cytogenetics laboratory, the microscopes are equipped with a camera that can photograph the chromosomes. In recent years, advances in technology have allowed cytogeneticists to view microscopic images on a computer screen (**Figure 3.2b**). On the computer screen, the chromosomes can be organized in a standard way, usually from largest to smallest. As seen in **Figure 3.2c**, the human chromosomes have been lined up, and a number is given to designate each type of chromosome. An exception would be the sex chromosomes, which are designated with the letters X and Y. A photographic representation of the chromosomes within a cell is called a **karyotype.** A karyotype reveals how many chromosomes are found within an actively dividing somatic cell.

Eukaryotic Chromosomes Are Inherited in Sets

Most eukaryotic species are **diploid,** which means that each type of chromosome is a member of a pair. In diploid species, the somatic cells have two sets of chromosomes. For example, most human cells have 46 chromosomes—two sets of 23 each. Other diploid species, however, can have different numbers of chromosomes. For example, the dog has 39 chromosomes per set (78 total), the fruit fly has 4 chromosomes per set (8 total), and the tomato has 12 per set (24 total).

When a species is diploid, the members of a pair of chromosomes are called **homologues;** each type of chromosome is found in a homologous pair. As shown in Figure 3.2c, for example, a somatic cell has two copies of chromosome 1, two copies of chromosome 2, and so forth. Within each pair, the chromosome on the left is a homologue to the one on the right, and vice versa. The two chromosomes in a homologous pair are nearly identical in size, have the same banding pattern, and contain a similar composition of genetic material. If a particular gene is found on one copy of a chromosome, it is also found on the other homologue. However, the two homologues may carry different alleles of a given gene. As an example, let's consider a gene in humans, called *OCA2*, which is one of a few different genes that affect eye color. The *OCA2* gene is located on chromosome 15 and comes in variants that result in brown, green, or blue eyes. In a person with brown eyes, one copy of chromosome 15 might carry a dominant brown allele, while its homologue could carry a recessive blue allele.

At the molecular level, how similar are homologous chromosomes? The answer is that the sequence of bases of one homologue would usually differ by less than 1% compared to the sequence of the other homologue. For example, the DNA sequence of chromosome 1 that you inherited from your mother would be greater than 99% identical to the sequence of chromosome 1 that you inherited from your father. Nevertheless, it should be emphasized that the sequences are not identical. The slight differences in DNA sequences provide the allelic differences in genes. Again, if we use the eye color gene as an example, a slight difference in DNA sequence distinguishes the brown, green, and blue alleles. It should also be noted that the striking similarities between homologous chromosomes do not apply to the pair of sex chromosomes—X and Y. These chromosomes differ in size and genetic composition. Certain genes that are found on the X chromosome are not found on the Y chromosome, and vice versa. The X and Y chromosomes are not considered homologous chromosomes though they do have short regions of homology.

Figure 3.3 considers two homologous chromosomes that are labeled with three different genes. An individual carrying these two chromosomes would be homozygous for the dominant allele of gene *A*. The individual would be heterozygous, *Bb*, for the second gene. For the third gene, the individual is homozygous for a recessive allele, *c*. The physical location of a gene is called its **locus** (plural: **loci**). As seen in Figure 3.3, for example, the locus of gene *C* is toward one end of this chromosome, while the locus of gene *B* is more in the middle.

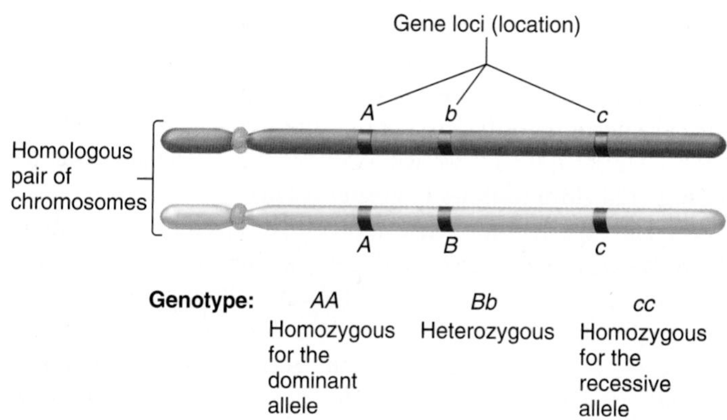

FIGURE 3.3 A comparison of homologous chromosomes. Each pair of homologous chromosomes carries the same types of genes, but, as shown here, the alleles may or may not be different.

3.2 CELL DIVISION

Now that we have an appreciation for the chromosomal composition of living cells, we can consider how chromosomes are copied and transmitted when cells divide. One purpose of cell division is **asexual reproduction.** In this process, a preexisting cell divides to produce two new cells. By convention, the original cell is usually called the mother cell, and the new cells are the two daughter cells. When species are unicellular, the mother cell is judged to be one individual, and the two daughter cells are two new separate organisms. Asexual reproduction is how bacterial cells proliferate. In addition, certain unicellular eukaryotes, such as the amoeba and baker's yeast (*Saccharomyces cerevisiae*), can reproduce asexually.

A second important reason for cell division is multicellularity. Species such as plants, animals, most fungi, and certain protists are derived from a single cell that has undergone repeated cellular divisions. Humans, for example, begin as a single fertilized egg; repeated cellular divisions produce an adult with several trillion cells. The precise transmission of chromosomes during every cell division is critical in order for all the cells of the body to receive the correct amount of genetic material.

In this section, we will consider how the process of cell division requires the duplication, organization, and distribution of the chromosomes. In bacteria, which have a single circular chromosome, the division process is relatively simple. Prior to cell division, bacteria duplicate their circular chromosome; they then distribute a copy into each of the two daughter cells. This process, known as binary fission, is described first. Eukaryotes have multiple numbers of chromosomes that occur as sets. Compared to bacteria, this added complexity requires a more complicated sorting process to ensure that each newly made cell receives the correct number and types of chromosomes. As described later in this section, a mechanism known as mitosis entails the organization and distribution of eukaryotic chromosomes during cell division.

Bacteria Reproduce Asexually by Binary Fission

As discussed earlier (see Figure 3.1a), bacterial species are typically unicellular, although individual bacteria may associate with each other to form pairs, chains, or clumps. Unlike eukaryotes, which have their chromosomes in a separate nucleus, the circular chromosomes of bacteria are in direct contact with the cytoplasm. In Chapter 10, we will consider the molecular structure of bacterial chromosomes in greater detail.

The capacity of bacteria to divide is really quite astounding. Some species, such as *Escherichia coli*, a common bacterium of the intestine, can divide every 20 to 30 minutes. Prior to cell division, bacterial cells copy, or replicate, their chromosomal DNA. This produces two identical copies of the genetic material, as shown at the top of **Figure 3.4**. Following DNA replication, a bacterial cell divides into two daughter cells by a process known as **binary fission.** During this event, the two daughter cells become separated from each other by the formation of a septum. As seen in the figure, each cell receives a copy of the chromosomal genetic material. Except when rare mutations occur, the daughter cells are usually genetically identical because they contain exact copies of the genetic material from the mother cell.

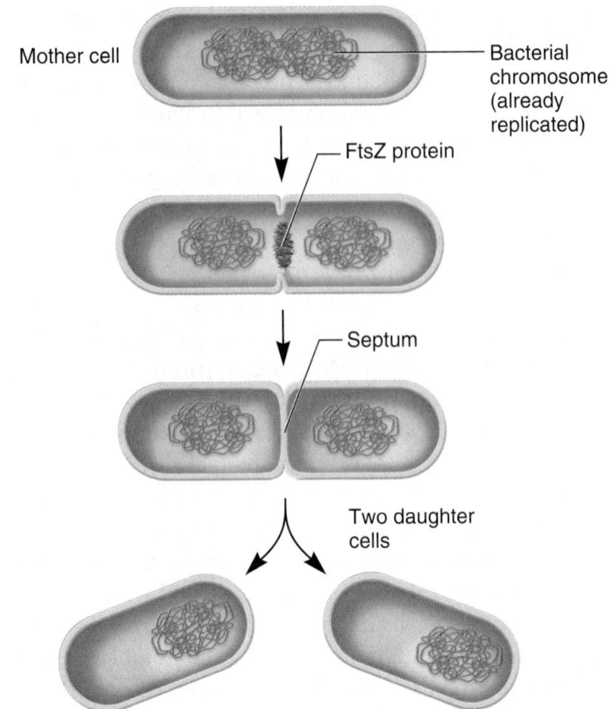

FIGURE 3.4 **Binary fission: the process by which bacterial cells divide.** Prior to division, the chromosome replicates to produce two identical copies. These two copies segregate from each other, with one copy going to each daughter cell.

Recent evidence has shown that bacterial species produce a protein called FtsZ, which is important in cell division. This protein assembles into a ring at the future site of the septum. FtsZ is thought to be the first protein to move to this division site, and recruits other proteins that produce a new cell wall between the daughter cells. FtsZ is evolutionarily related to a eukaryotic protein called tubulin. As discussed later in this chapter, tubulin is the main component of microtubules, which play a key role in chromosome sorting in eukaryotes. Both FtsZ and tubulin form structures that provide cells with organization and play key roles in cell division.

Binary fission is an asexual form of reproduction because it does not involve genetic contributions from two different gametes. On occasion, bacteria can exchange small pieces of genetic material with each other. We will consider some interesting mechanisms of genetic exchange in Chapter 6.

Eukaryotic Cells Progress Through a Cell Cycle to Produce Genetically Identical Daughter Cells

The common outcome of eukaryotic cell division is to produce two daughter cells that have the same number and types of chromosomes as the original mother cell. This requires a replica-

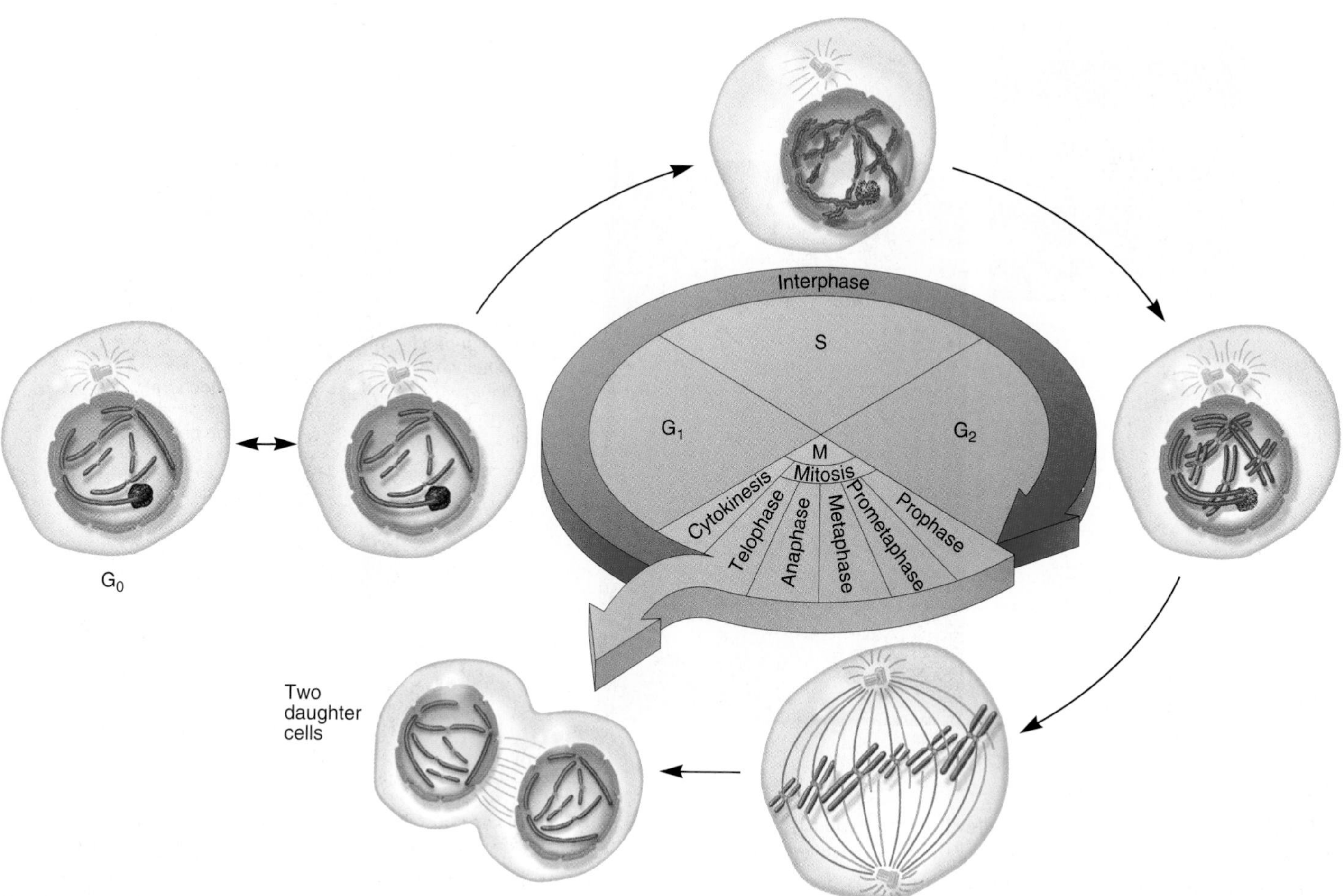

FIGURE 3.5 **The eukaryotic cell cycle.** Dividing cells progress through a series of phases, denoted G_1, S, G_2, and M phases (mitosis). This diagram shows the progression of a cell through mitosis to produce two daughter cells. The original diploid cell had three pairs of chromosomes, for a total of six individual chromosomes. During S phase, these have replicated to yield 12 chromatids found in six pairs of sister chromatids. After mitosis is complete, each of the two daughter cells contains six individual chromosomes, just like the mother cell. Note: The chromosomes in G_0, G_1, S, and G_2 phases are not condensed. In this drawing, they are shown partially condensed so they can be easily counted.

tion and division process that is more complicated than simple binary fission. Eukaryotic cells that are destined to divide progress through a series of phases known as the **cell cycle** (**Figure 3.5**). These phases are G for gap, S for synthesis (of the genetic material), and M for mitosis. There are two G phases, G_1 and G_2. The term "gap" originally described the gaps between S phase and mitosis in which it was not microscopically apparent that significant changes were occurring in the cell. However, we now know that both gap phases are critical periods in the cell cycle that involve many molecular changes. In actively dividing cells, the G_1, S, and G_2 phases are collectively known as **interphase.** In addition, cells may remain permanently, or for long periods of time, in a phase of the cell cycle called G_0. A cell in the G_0 phase is either temporarily not progressing through the cell cycle or, in the case of terminally differentiated cells, such as most nerve cells in an adult mammal, will never divide again.

During the G_1 phase, a cell may prepare to divide. Depending on the cell type and the conditions that it encounters, a cell in the G_1 phase may accumulate molecular changes (e.g., synthesis of proteins) that cause it to progress through the rest of the cell cycle. When this occurs, cell biologists say that a cell has reached a **restriction point** and is committed on a pathway that leads to cell division. Once past the restriction point, the cell will then advance to the S phase, during which the chromosomes are replicated. After replication, the two copies are called **chromatids.** They are joined to each other at a region of DNA called the **centromere** to form a unit known as a pair of **sister chromatids** (**Figure 3.6**). The **kinetochore** is a group of proteins that are bound to the centromere. These proteins help to hold the sister chromatids together and also play a role in chromosome sorting, as discussed later. When S phase is completed, a cell actually has twice as many chromatids compared to the number of chromosomes in the G_1 phase. For example, a human cell in the G_1 phase has 46 distinct chromosomes, whereas in G_2, it would have 46 pairs of sister chromatids, for a total of 92 chromatids. The term chromosome—meaning colored body—can be a bit confusing because it originally meant a distinct structure that is observable with the microscope. Therefore, the term

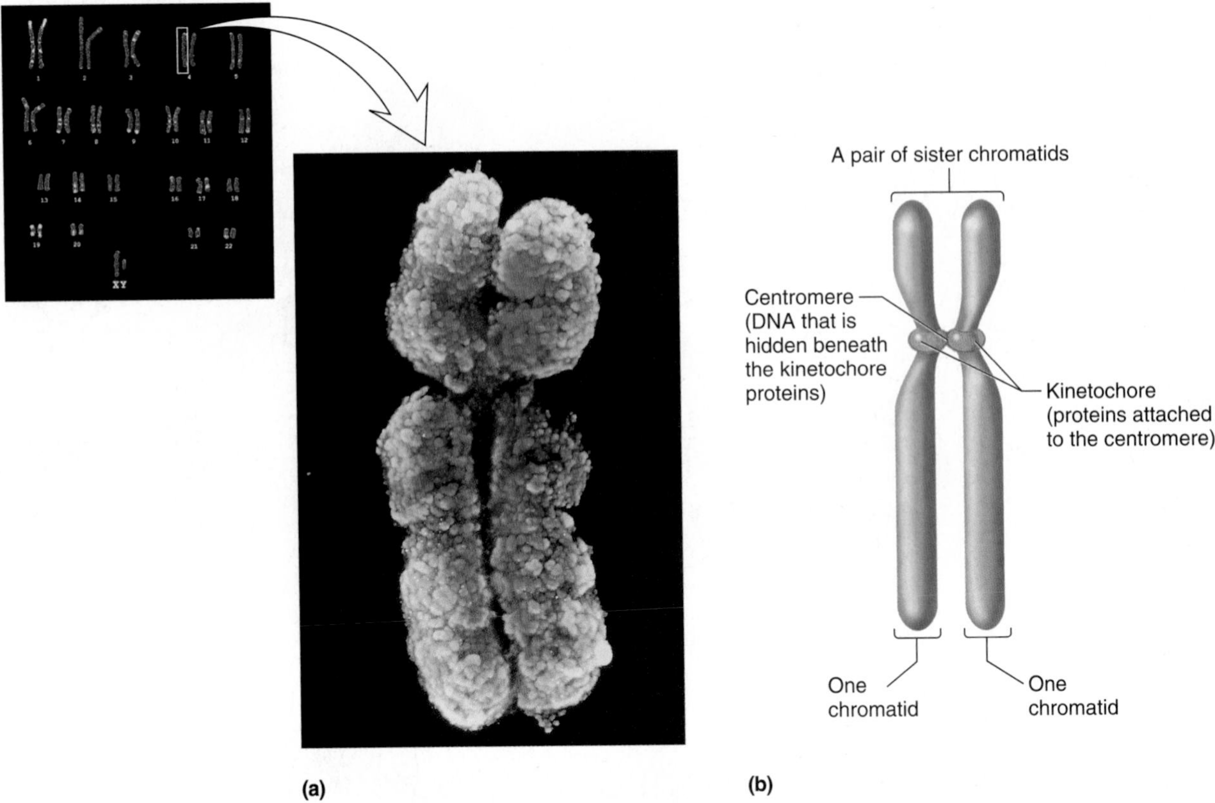

FIGURE 3.6 **Chromosomes following DNA replication.** **(a)** A photomicrograph of a chromosome in a form called a pair of sister chromatids. This chromosome is in the metaphase stage of mitosis, which is described later in the chapter. Note: Each of the 46 chromosomes that are viewed in a human karyotype is actually a pair of sister chromatids. See the white rectangular box in the inset. **(b)** A schematic drawing of sister chromatids. This structure has two chromatids that lie side by side. As seen here, each chromatid is a distinct unit. The two chromatids are held together by kinetochore proteins that bind to each other and to the centromeres of each chromatid.

chromosome can refer to either a pair of sister chromatids during the G_2 and early stages of M phase or to the structures that are observed at the end of M phase and those during G_1 that contain the equivalent of one chromatid (refer back to Figure 3.5).

During the G_2 phase, the cell accumulates the materials that are necessary for nuclear and cell division. It then progresses into the M phase of the cell cycle, when **mitosis** occurs. The primary purpose of mitosis is to distribute the replicated chromosomes, dividing one cell nucleus into two nuclei, so that each daughter cell receives the same complement of chromosomes. For example, a human cell in the G_2 phase has 92 chromatids, which are found in 46 pairs. During mitosis, these pairs of chromatids are separated and sorted so that each daughter cell receives 46 chromosomes.

Mitosis was first observed microscopically in the 1870s by the German biologist Walter Flemming, who coined the term mitosis (from the Greek *mitos*, thread). He studied the dividing epithelial cells of salamander larvae and noticed that chromosomes were constructed of two parallel "threads." These threads separated and moved apart, one going to each of the two daughter nuclei. By this mechanism, Flemming pointed out that the two daughter cells received an identical group of threads, of quantity comparable to the number of threads in the parent cell.

The Mitotic Spindle Apparatus Organizes and Sorts Eukaryotic Chromosomes

Before we discuss the events of mitosis, let's first consider the structure of the **mitotic spindle apparatus** (also known simply as the **mitotic spindle**), which is involved in the organization and sorting of chromosomes (**Figure 3.7**). In animal and plant cells, the mitotic spindle is formed from two structures called the **centrosomes.** Each centrosome is located at a **spindle pole**. In animal cells, a pair of **centrioles** at right angles to each other is found within each centrosome. However, centrioles are not found in many other eukaryotic species, including plants, and are not required for spindle formation. Each centrosome organizes the construction of structures called microtubules. These microtubules are produced from the rapid polymerization of tubulin proteins. The mitotic spindle has three types of microtubules. The aster microtubules emanate outward from the centrosome toward the plasma membrane. They are important for the positioning of the spindle apparatus within the cell and later in the process of cell division. The polar microtubules project toward the region where the chromosomes will be found during mitosis—the region between the two spindle poles. Polar microtubules

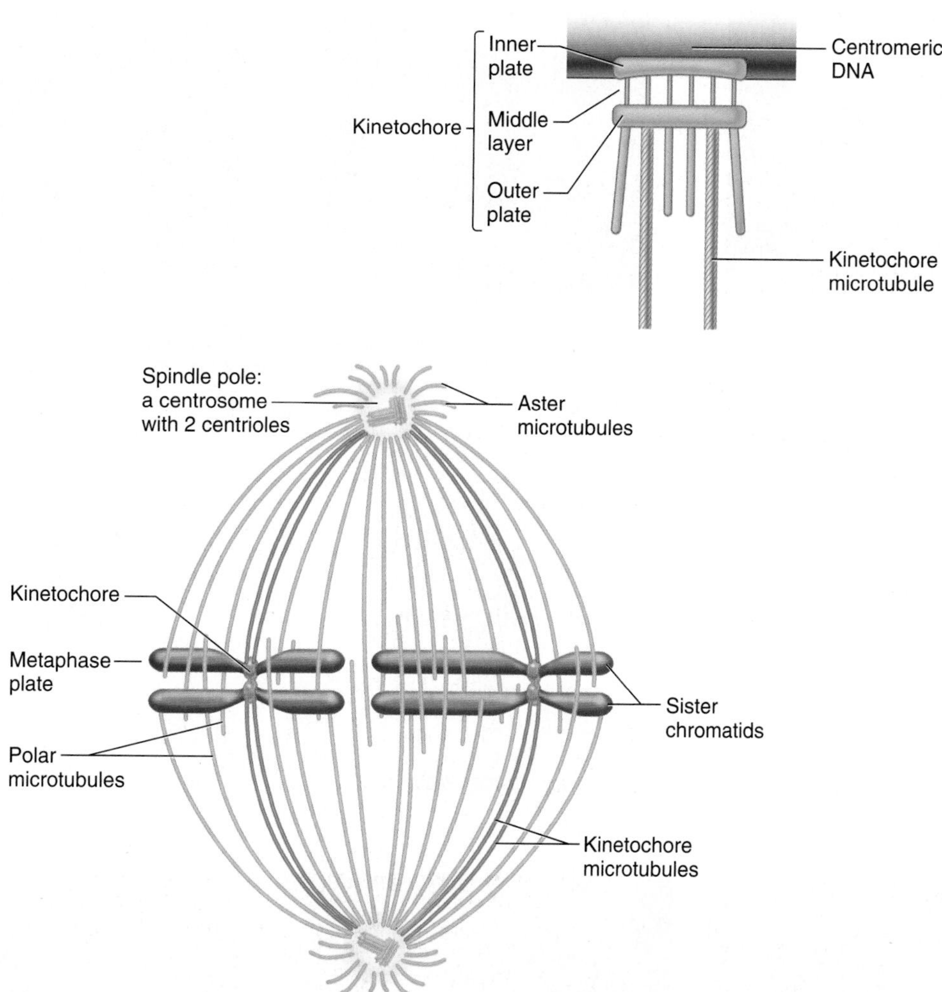

FIGURE 3.7 **The structure of the mitotic spindle.** A single centrosome duplicates during S phase and the two centrosomes separate at the beginning of M phase. The mitotic spindle is formed from microtubules that are rooted in the centrosomes. Each centrosome is located at a spindle pole. The aster microtubules emanate away from the region between the poles. They help to position the spindle within the cell and are used as reference points for cell division. The polar microtubules project into the region between the two poles; they play a role in pole separation. The kinetochore microtubules are attached to the kinetochore of sister chromatids. As seen in the inset, the kinetochore is composed of a group of proteins that form three layers: the inner plate, which recognizes the centromere; the outer plate, which recognizes a kinetochore microtubule; and the middle layer, which connects the inner and outer plates.

that overlap with each other play a role in the separation of the two poles. They help to "push" the poles away from each other. Finally, the kinetochore microtubules have attachments to the kinetochore, which is a complex of proteins that is bound to the centromere of individual chromosomes. As seen in the inset to Figure 3.7, the kinetochore proteins form three layers. The proteins of the inner plate make direct contact with the centromeric DNA, while the outer plate contacts the kinetochore microtubules. The role of the middle layer is to connect these two regions.

The mitotic spindle allows cells to organize and separate chromosomes so that each daughter cell receives the same complement of chromosomes. This sorting process, known as mitosis, is described next.

The Transmission of Chromosomes During the Division of Eukaryotic Cells Requires a Process Known as Mitosis

In **Figure 3.8**, the process of mitosis is shown for a diploid animal cell. In the simplified diagrams shown along the bottom of this figure, the original mother cell contains six chromosomes; it is diploid ($2n$) and contains three chromosomes per set ($n = 3$). One set is shown in blue, and the homologous set is shown in red. As discussed next, mitosis is subdivided into phases known as prophase, prometaphase, metaphase, anaphase, and telophase.

Prophase Prior to mitosis, the cells are in interphase, during which the chromosomes are **decondensed**—less tightly compacted—and

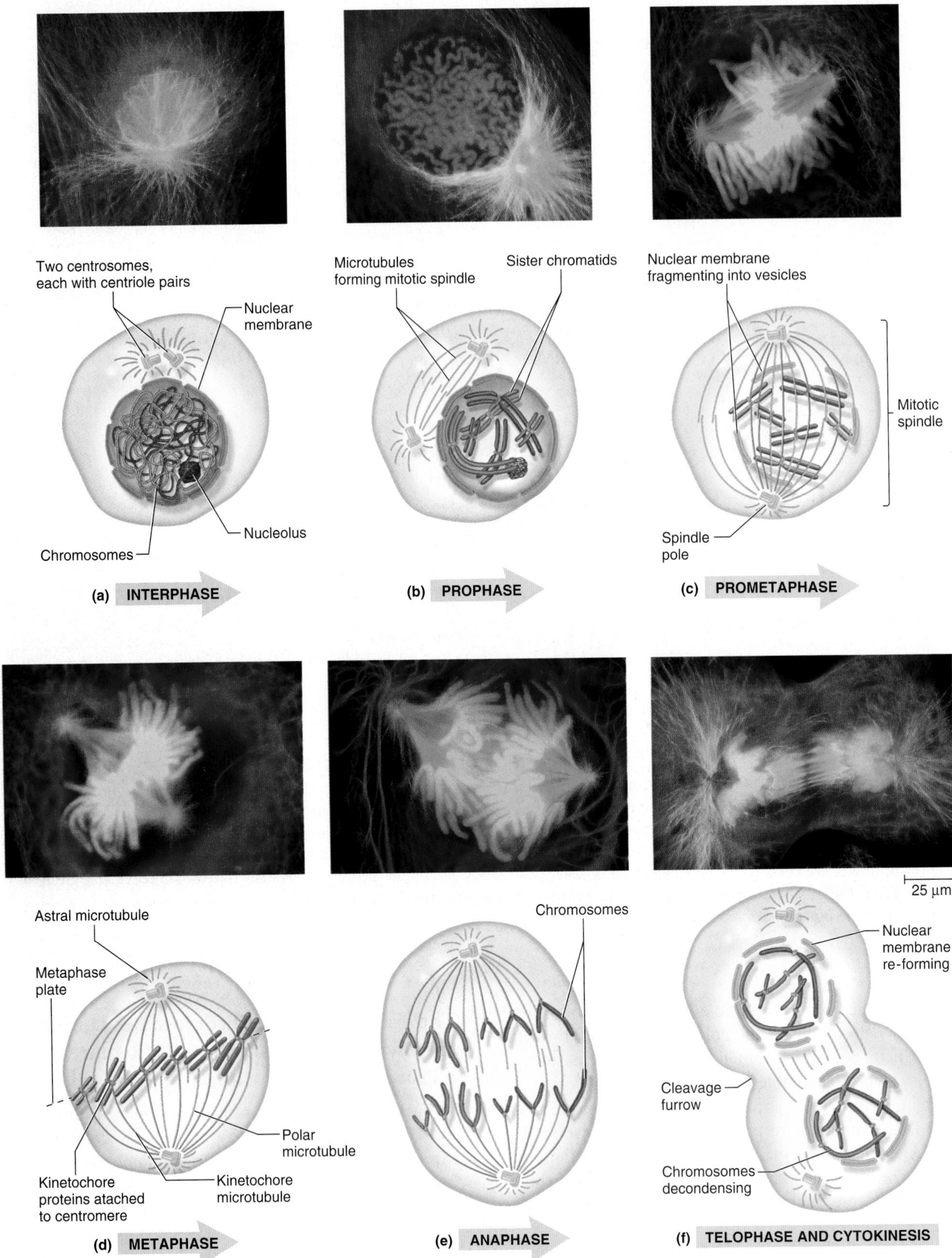

ONLINE ANIMATION

FIGURE 3.8 **The process of mitosis in an animal cell.** The top panels illustrate cells of a fish embryo progressing through mitosis. The bottom panels are schematic drawings that emphasize the sorting and separation of the chromosomes. In this case, the original diploid cell had six chromosomes (three in each set). At the start of mitosis, these have already replicated into 12 chromatids. The final result is two daughter cells each containing six chromosomes.

found in the nucleus (**Figure 3.8a**). At the start of mitosis, in **prophase,** the chromosomes have already replicated to produce 12 chromatids, joined as six pairs of sister chromatids (**Figure 3.8b**). As prophase proceeds, the nuclear membrane begins to dissociate into small vesicles. At the same time, the chromatids **condense** into more compact structures that are readily visible by light microscopy. The mitotic spindle also begins to form and the nucleolus disappears.

Prometaphase As mitosis progresses from prophase to prometaphase, the centrosomes move to opposite ends of the cell and demarcate two spindle poles, one within each of the future daughter cells. Once the nuclear membrane has dissociated into vesicles, the spindle fibers can interact with the sister chromatids. This interaction occurs in a phase of mitosis called **prometaphase** (**Figure 3.8c**). How do sister chromatids become attached to the spindle? Initially, kinetochore microtubules are rapidly formed and can be seen growing out from the two poles. As it grows, if the end of a kinetochore microtubule happens to make contact with a kinetochore, its end is said to be "captured" and remains firmly attached to the kinetochore. This random process is how sister chromatids become attached to kinetochore microtubules. Alternatively, if the end of a kinetochore microtubule does not collide with a kinetochore, the microtubule eventually depolymerizes and retracts to the centrosome. As the end of prometaphase nears, the kinetochore on a pair of sister chromatids is attached to kinetochore microtubules from opposite poles. As these events are occurring, the sister chromatids are seen to undergo jerky movements as they are tugged, back and forth, between the two poles. By the end of prometaphase, the mitotic spindle is completely formed.

Metaphase Eventually, the pairs of sister chromatids align themselves along a plane called the **metaphase plate.** As shown in **Figure 3.8d**, when this alignment is complete, the cell is in **metaphase** of mitosis. At this point, each pair of chromatids is attached to both poles by kinetochore microtubules. The pairs of sister chromatids have become organized into a single row along the metaphase plate. When this organizational process is finished, the chromatids can be equally distributed into two daughter cells.

Anaphase The next step in the division process occurs during **anaphase** (**Figure 3.8e**). At this stage, the connection that is responsible for holding the pairs of chromatids together is broken. (We will examine the process of sister chromatid cohesion and separation in more detail in Chapter 10.) Each chromatid, now an individual chromosome, is linked to only one of the two poles. As anaphase proceeds, the chromosomes move toward the pole to which they are attached. This involves a shortening of the kinetochore microtubules. In addition, the two poles themselves move farther apart due to the elongation of the polar microtubules, which slide in opposite directions due to the actions of motor proteins.

Telophase During **telophase,** the chromosomes reach their respective poles and decondense. The nuclear membrane now re-forms to produce two separate nuclei. In **Figure 3.8f,** this has produced two nuclei that contain six chromosomes each. The nucleoli will also reappear.

Cytokinesis In most cases, mitosis is quickly followed by **cytokinesis,** in which the two nuclei are segregated into separate daughter cells. Likewise, cytokinesis also segregates cell organelles such as mitochondria and chloroplasts into daughter cells. In animal cells, cytokinesis begins shortly after anaphase. A contractile ring, comprised of myosin motor proteins and actin filaments, assembles adjacent to the plasma membrane. Myosin hydrolyzes ATP, which shortens the ring and thereby constricts the plasma membrane to form a **cleavage furrow** that ingresses, or moves inwards (**Figure 3.9a**). Ingression continues until a midbody structure is formed that physically pinches one cell into two.

In plants, the two daughter cells are separated by the formation of a **cell plate** (**Figure 3.9b**). At the end of anaphase, Golgi-derived vesicles carrying cell wall materials are transported

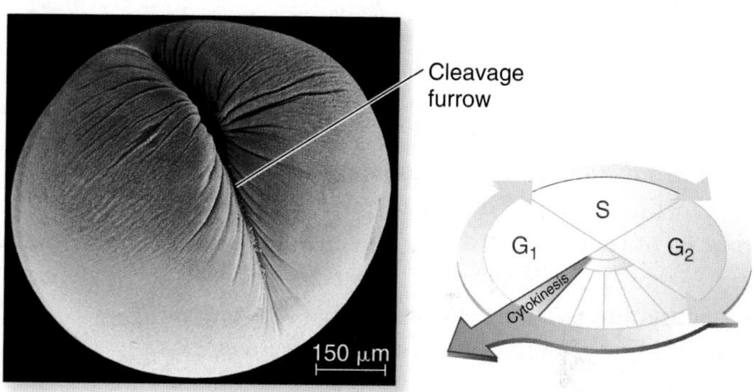

(a) Cleavage of an animal cell

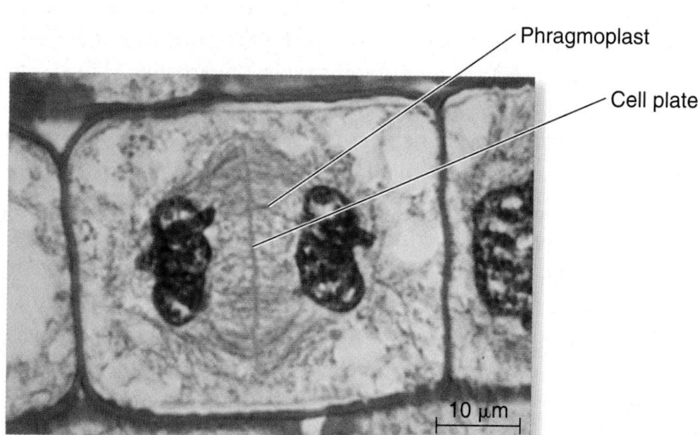

(b) Formation of a cell plate in a plant cell

FIGURE 3.9 **Cytokinesis in an animal and plant cell.** (**a**) In an animal cell, cytokinesis involves the formation of a cleavage furrow. (**b**) In a plant cell, cytokinesis occurs via the formation of a cell plate between the two daughter cells.

to the equator of a dividing cell. These vesicles are directed to their location via the phragmoplast, which is composed of parallel aligned microtubules and actin filaments that serve as tracks for vesicle movement. The fusion of these vesicles gives rise to the cell plate, which is a membrane-bound compartment. The cell plate begins in the middle of the cell and expands until it attaches to the mother cell wall. Once this attachment has taken place, the cell plate undergoes a process of maturation and eventually separates the mother cell into two daughter cells.

Outcome of mitotic cell division Mitosis and cytokinesis ultimately produce two daughter cells having the same number of chromosomes as the mother cell. Barring rare mutations, the two daughter cells are genetically identical to each other and to the mother cell from which they were derived. Thus, the critical consequence of this sorting process is to ensure genetic consistency from one somatic cell to the next. The development of multicellularity relies on the repeated process of mitosis and cytokinesis. For diploid organisms that are multicellular, most of the somatic cells are diploid and genetically identical to each other.

3.3 SEXUAL REPRODUCTION

In Section 3.2, we considered how a cell divides to produce two new cells with identical complements of genetic material. Now we will turn our attention to sexual reproduction, a common way for eukaryotic organisms to produce offspring. During sexual reproduction, gametes are made that contain half the amount of genetic material. These gametes fuse with each other in the process of fertilization to begin the life of a new organism. Gametes are highly specialized cells that are designed to locate each other and provide nutrients to the developing embryo. The process whereby gametes form is called **gametogenesis.**

Some simple eukaryotic species are **isogamous,** which means that the gametes are morphologically similar. Examples of isogamous organisms include many species of fungi and algae. Most eukaryotic species, however, are **heterogamous**—they produce two morphologically different types of gametes. Male gametes, or **sperm cells,** are relatively small and usually travel far distances to reach the female gamete. The mobility of the male gamete is an important characteristic, making it likely that it will come in close proximity to the female gamete. The sperm of most animal species contain a single flagellum that enables them to swim. The sperm of ferns and nonvascular plants such as bryophytes may have multiple flagella. In flowering plants, however, the sperm are contained within pollen grains. Pollen is a small mobile structure that can be carried by the wind or on the feet or hairs of insects. In flowering plants, sperm are delivered to egg cells via pollen tubes. Compared to sperm cells, the female gamete, known as the **egg cell,** or **ovum,** is usually very large and nonmotile. In animal species, the egg stores a large amount of nutrients that will be available to the growing embryo.

Gametes are typically **haploid,** which means they contain half the number of chromosomes as diploid cells. Compared to a diploid cell, a haploid gamete contains a single set of chromosomes. In this case, the haploid gametes are $1n$, while the diploid cells are $2n$. For example, a diploid human cell contains 46 chromosomes, but a gamete (sperm or egg cell) contains only 23 chromosomes.

During the process known as **meiosis** (from the Greek meaning less), haploid cells are produced from a cell that was originally diploid. For this to occur, the chromosomes must be correctly sorted and distributed in a way that reduces the chromosome number to half its original value. In the case of humans, for example, each gamete must receive half the total number of chromosomes, but not just any 23 chromosomes will do. A gamete must receive one chromosome from each of the 23 pairs. In this section, we will examine the cellular events of gamete development in animal and plant species and how the stages of meiosis lead to the formation of cells with a haploid complement of chromosomes.

Meiosis Produces Cells That Are Haploid

The process of meiosis bears striking similarities to mitosis. Like mitosis, meiosis begins after a cell has progressed through the G_1, S, and G_2 phases of the cell cycle. However, meiosis involves two successive divisions rather than one (as in mitosis). Prior to meiosis, the chromosomes are replicated in S phase to produce pairs of sister chromatids. This single replication event is then followed by two sequential cell divisions called meiosis I and II. As in mitosis, each of these is subdivided into prophase, prometaphase, metaphase, anaphase, and telophase.

Figure 3.10 emphasizes some of the important events that occur during prophase of meiosis I, which is further subdivided into periods known as leptotena, zygotena, pachytena, diplotena, and diakinesis. During the **leptotene** stage, the replicated chromosomes begin to condense and become visible with a light microscope. In Figure 3.10 (leptotene and zygotene stages), each structure is actually a pair of chromatids. Unlike mitosis, the **zygotene** stage of prophase of meiosis I involves a recognition process known as **synapsis,** in which the homologous chromosomes recognize each other and then align themselves along their entire lengths. The associated chromatids are known as **bivalents.** Each bivalent contains two pairs of sister chromatids, or a total of four chromatids.

In most eukaryotic species, a **synaptonemal complex** is formed between the homologous chromosomes. As shown in **Figure 3.11**, this complex is composed of parallel lateral elements, which are bound to the chromosomal DNA, and a central element, which promotes the binding of the lateral elements to each other via transverse filaments. The synaptonemal complex may not be required for the pairing of homologous chromosomes, because some species, such as *Aspergillus nidulans* and *Schizosaccharomyces pombe,* completely lack such a complex, yet their chromosomes synapse correctly. At present, the precise role of the synaptonemal complex is not clearly understood, and it remains the subject of intense research. It may play more than one role. First, although it may not be required for synapsis, the synaptonemal complex could help to maintain homologous pairing in situations where the normal process has failed. Second, the complex may play a role in meiotic chromosome structure. And

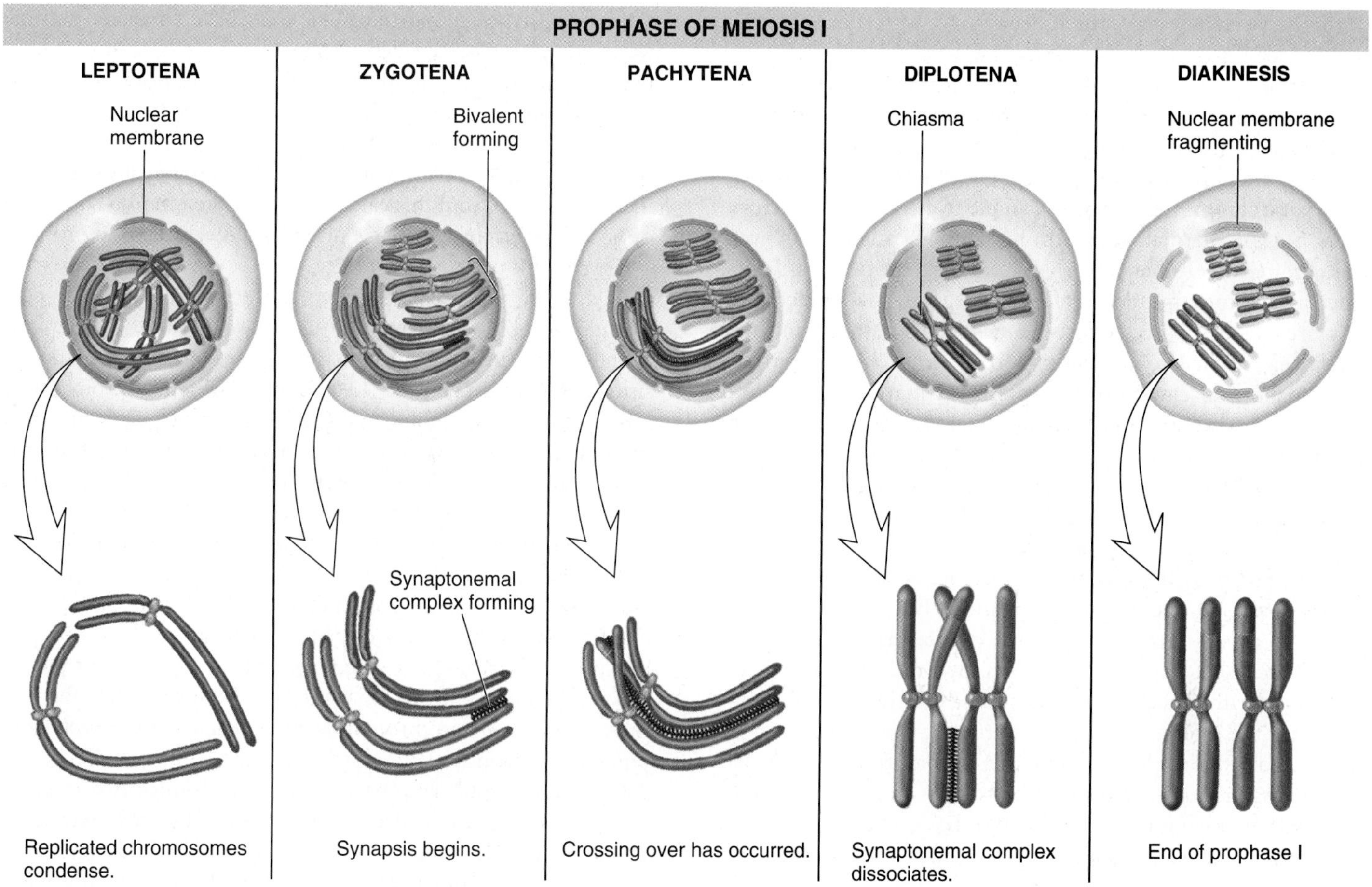

PROPHASE OF MEIOSIS I
LEPTOTENA
ZYGOTENA
PACHYTENA
DIPLOTENA
DIAKINESIS
Nuclear membrane
Bivalent forming
Chiasma
Nuclear membrane fragmenting
Synaptonemal complex forming
Replicated chromosomes condense.
Synapsis begins.
Crossing over has occurred.
Synaptonemal complex dissociates.
End of prophase I

FIGURE 3.10 **The events that occur during prophase of meiosis I.**

Synaptonemal complex
Lateral element
Central element
Chromatid
Transverse filament
(a)
(b)

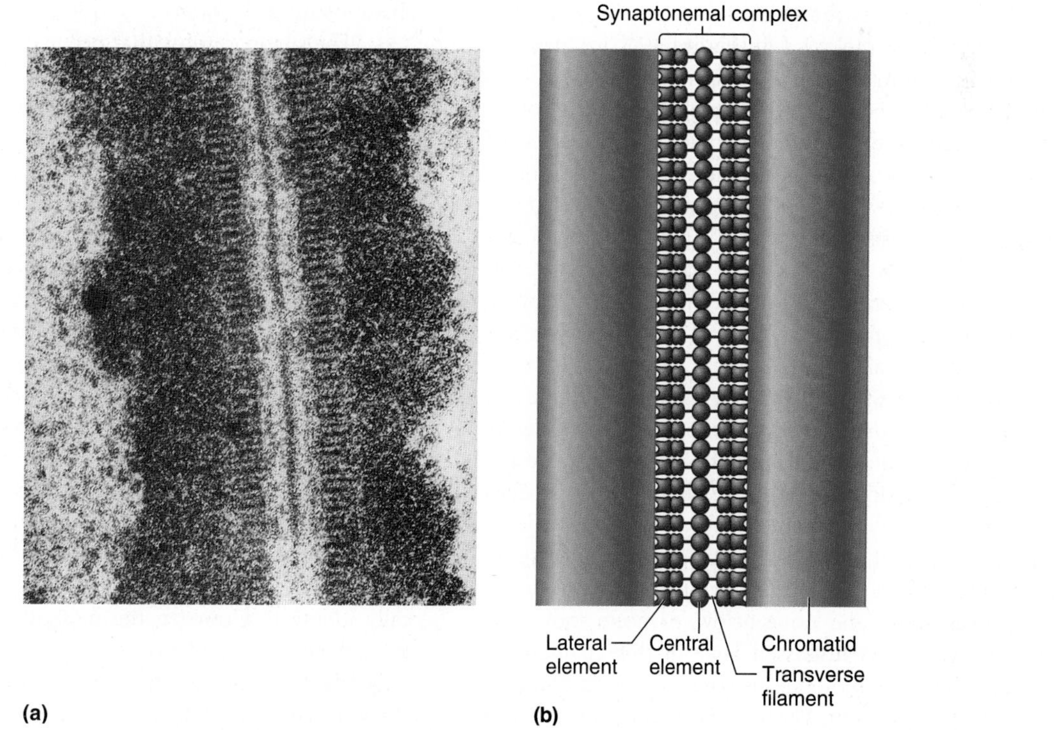

FIGURE 3.11 **The synaptonemal complex formed during prophase of meiosis I.** (**a**) Micrograph of a synaptonemal complex. (**b**) Lateral elements are bound to the chromosomal DNA of homologous chromatids. A central element provides a link between the lateral elements via transverse filaments.

third, the synaptonemal complex may serve to regulate the process of crossing over, which is described next.

Prior to the **pachytene** stage, when synapsis is complete, an event known as **crossing over** usually occurs. Crossing over involves a physical exchange of chromosome pieces. Depending on the size of the chromosome and the species, an average eukaryotic chromosome incurs a couple to a couple dozen crossovers. During spermatogenesis in humans, for example, an average chromosome undergoes slightly more than two crossovers, whereas chromosomes in certain plant species may undergo 20 or more crossovers. Recent research has shown that crossing over is critical for the proper segregation of chromosomes. In fact, abnormalities in chromosome segregation may be related to a defect in crossing over. In a high percentage of people with Down syndrome, in which an individual has three copies of chromosome 21 instead of two, research has shown that the presence of the extra chromosome is associated with a lack of crossing over between homologous chromosomes.

In Figure 3.10, crossing over has occurred at a single site between two of the larger chromatids. The connection that results from crossing over is called a **chiasma** (plural: **chiasmata**), because it physically resembles the Greek letter chi, χ. We will consider the genetic consequences of crossing over in Chapter 5 and the molecular process of crossing over in Chapter 17. By the end of the **diplotene** stage, the synaptonemal complex has largely disappeared. The bivalent pulls apart slightly, and microscopically it becomes easier to see that it is actually composed of four chromatids. A bivalent is also called a **tetrad** (from the prefix *tetra-*, meaning four) because it is composed of four chromatids. In the last stage of prophase of meiosis I, **diakinesis,** the synaptonemal complex completely disappears.

Figure 3.10 emphasizes the pairing and crossing over that occurs during prophase of meiosis I. In **Figure 3.12**, we turn our attention to the remaining events in meiosis. In prometaphase of meiosis I, the spindle apparatus is complete, and the chromatids are attached via kinetochore microtubules. At metaphase of meiosis I, the bivalents are organized along the metaphase plate. However, their pattern of alignment is strikingly different from that observed during mitosis (refer back to Figure 3.8d). Before we consider the rest of meiosis I, a particularly critical feature for you to appreciate is how the bivalents are aligned along the metaphase plate. In particular, the alignment of pairs of sister chromatids is in a double row rather than a single row as in mitosis. Furthermore, the arrangement of sister chromatids within this double row is random with regard to the blue and red homologues. In Figure 3.12, one of the blue homologues is above the metaphase plate and the other two are below, while one of the red homologues is below the metaphase plate and other two are above.

In an organism that produces many gametes, meiosis in other cells could produce a different arrangement of homologues—three blues above and none below, or none above and three below, and so on. As discussed later in this chapter, the random arrangement of homologues is consistent with Mendel's law of independent assortment. Because most eukaryotic species have several chromosomes per set, the sister chromatids can be randomly aligned along the metaphase plate in many possible ways. For example, consider humans that have 23 chromosomes per set. The possible number of different, random alignments equals 2^n, where n equals the number of chromosomes per set. Thus, in humans, this would equal 2^{23}, or over 8 million, possibilities! Because the homologues are genetically similar but not identical, we see from this calculation that the random alignment of homologous chromosomes provides a mechanism to promote a vast amount of genetic diversity.

In addition to the random arrangement of homologues within a double row, a second distinctive feature of metaphase of meiosis I is the attachment of kinetochore microtubules to the sister chromatids (**Figure 3.13**). One pair of sister chromatids is linked to one of the poles, and the homologous pair is linked to the opposite pole. This arrangement is quite different from the kinetochore attachment sites during mitosis in which a pair of sister chromatids is linked to both poles (see Figure 3.8).

During anaphase of meiosis I, the two pairs of sister chromatids within a bivalent separate from each other (see Figure 3.12). However, the connection that holds sister chromatids together does not break. Instead, each joined pair of chromatids migrates to one pole, and the homologous pair of chromatids moves to the opposite pole. Finally, at telophase of meiosis I, the sister chromatids have reached their respective poles, and decondensation occurs in many, but not all, species. The nuclear membrane may re-form to produce two separate nuclei. The end result of meiosis I is two cells, each with three pairs of sister chromatids. It is thus a reduction division. The original diploid cell had its chromosomes in homologous pairs, while the two cells produced at the end of meiosis I are considered to be haploid; they do not have pairs of homologous chromosomes.

The sorting events that occur during meiosis II are similar to those that occur during mitosis, but the starting point is different. For a diploid organism with six chromosomes, mitosis begins with 12 chromatids that are joined as six pairs of sister chromatids (refer back to Figure 3.8). By comparison, the two cells that begin meiosis II each have six chromatids that are joined as three pairs of sister chromatids. Otherwise, the steps that occur during prophase, prometaphase, metaphase, anaphase, and telophase of meiosis II are analogous to a mitotic division.

If we compare the outcome of meiosis (see Figure 3.12) to that of mitosis (see Figure 3.8), the results are quite different. (A comparison is also made in solved problem S3 at the end of this chapter.) In these examples, mitosis produced two diploid daughter cells with six chromosomes each, whereas meiosis produced four haploid daughter cells with three chromosomes each. In other words, meiosis has halved the number of chromosomes per cell. With regard to alleles, the results of mitosis and meiosis are also different. The daughter cells produced by mitosis are genetically identical. However, the haploid cells produced by meiosis are not genetically identical to each other because they contain only one homologous chromosome from each pair. Later, we will consider how the gametes may differ in the alleles that they carry on their homologous chromosomes.

MEIOSIS I

Centrosomes with centrioles

Mitotic spindle

Sister chromatids

Bivalent

Synapsis of homologous chromatids and crossing over

Nuclear membrane fragmenting

EARLY PROPHASE

LATE PROPHASE

PROMETAPHASE

Metaphase plate

Cleavage furrow

METAPHASE

ANAPHASE

TELOPHASE AND CYTOKINESIS

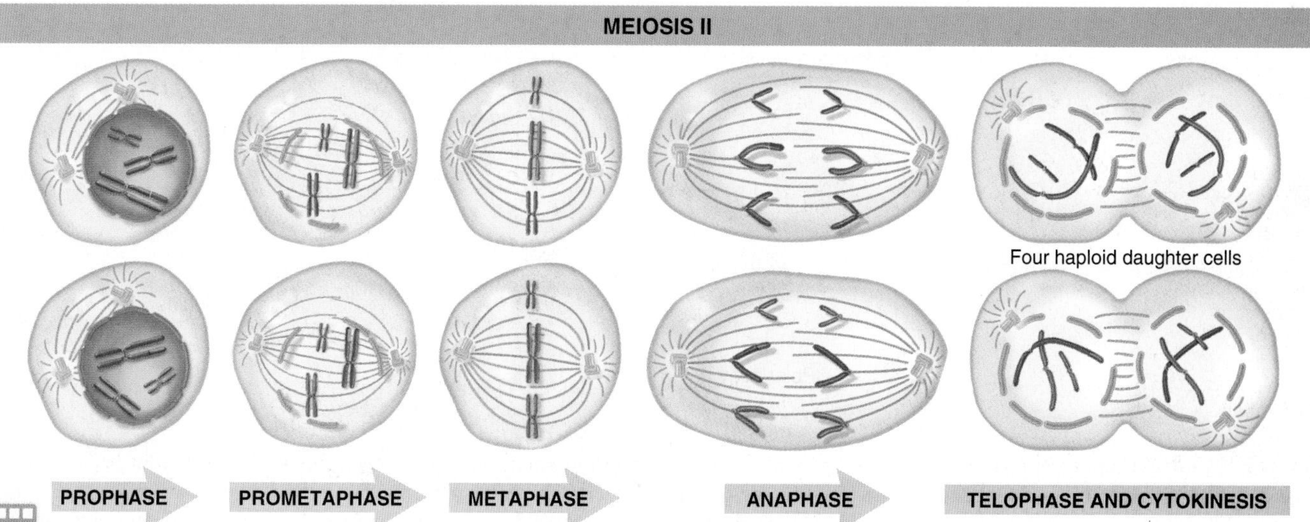

ONLINE ANIMATION

FIGURE 3.12 **The stages of meiosis in an animal cell.** See text for details.

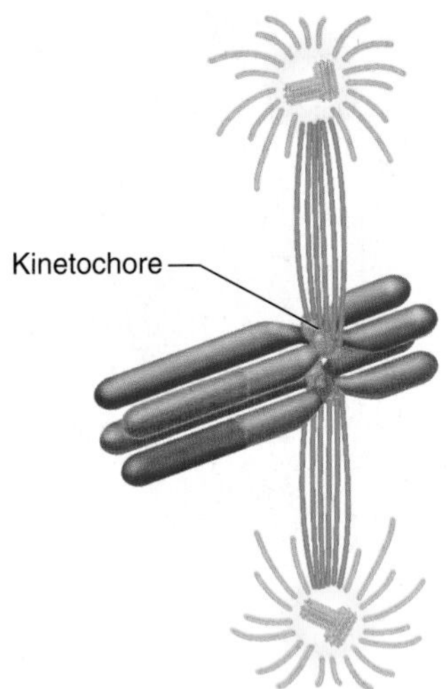

FIGURE 3.13 **Attachment of the kinetochore microtubules to replicated chromosomes during meiosis.** The kinetochore microtubules from a given pole are attached to one pair of chromatids in a bivalent, but not both. Therefore, each pair of sister chromatids is attached to only one pole.

In Animals, Spermatogenesis Produces Four Haploid Sperm Cells and Oogenesis Produces a Single Haploid Egg Cell

In male animals, **spermatogenesis,** the production of sperm, occurs within glands known as the testes. The testes contain spermatogonial cells that divide by mitosis to produce two cells. One of these remains a spermatogonial cell, and the other cell becomes a primary spermatocyte. As shown in **Figure 3.14a**, the spermatocyte progresses through meiosis I and meiosis II to produce four haploid cells, which are known as spermatids. These cells then mature into sperm cells. The structure of a sperm cell includes a long flagellum and a head. The head of the sperm contains little more than a haploid nucleus and an organelle at its tip, known as an acrosome. The acrosome contains digestive enzymes that are released when a sperm meets an egg cell. These enzymes enable the sperm to penetrate the outer protective layers of the egg and gain entry into the egg cell's cytosol. In animal species without a mating season, sperm production is a continuous process in mature males. A mature human male, for example, produces several hundred million sperm each day.

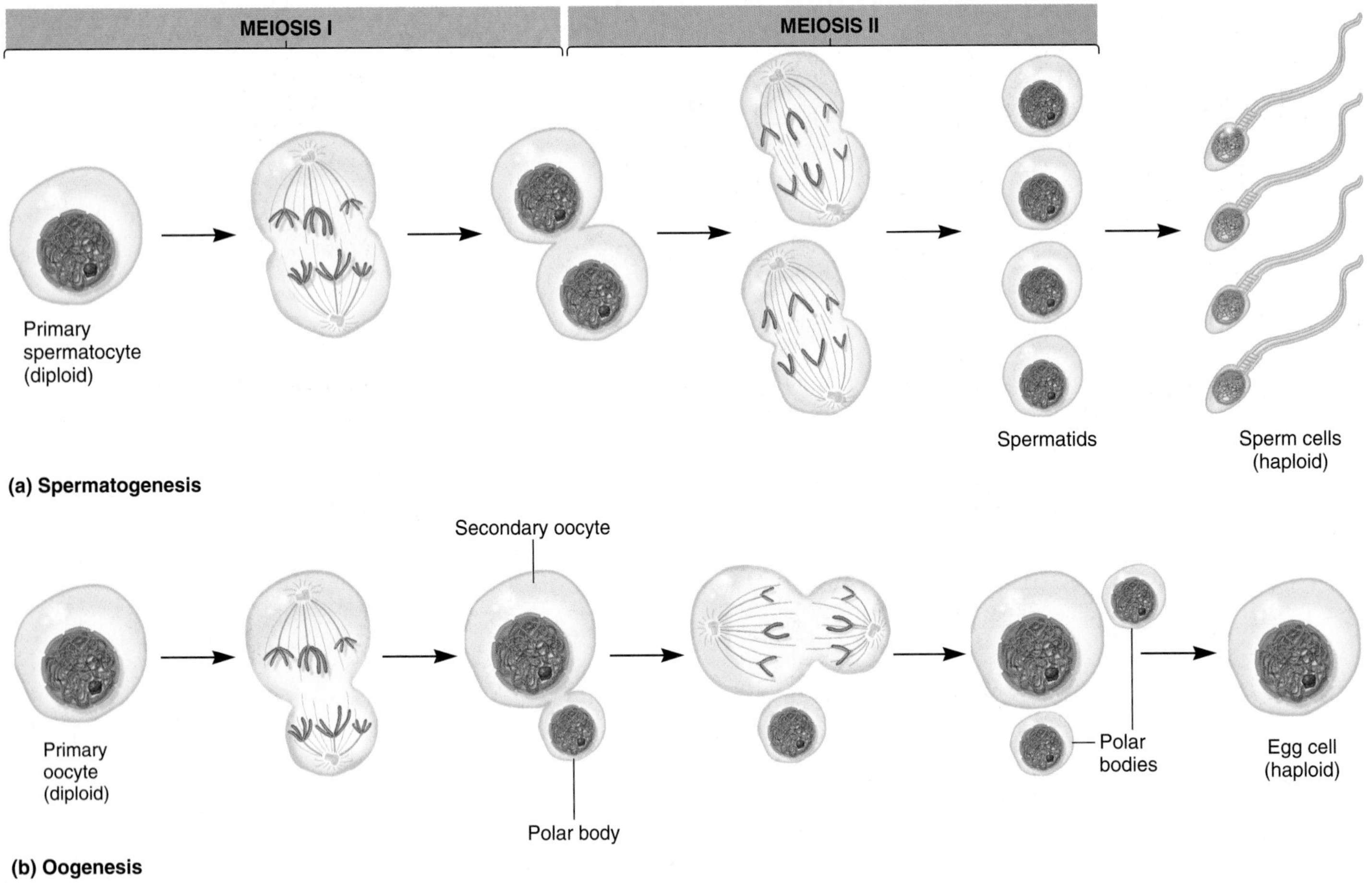

FIGURE 3.14 **Gametogenesis in animals.** (**a**) Spermatogenesis. A diploid spermatocyte undergoes meiosis to produce four haploid (n) spermatids. These differentiate during spermatogenesis to become mature sperm. (**b**) Oogenesis. A diploid oocyte undergoes meiosis to produce one haploid egg cell and two or three polar bodies. For some species, the first polar body divides, while in other species, it does not. Because of asymmetric cytokinesis, the amount of cytoplasm the egg receives is maximized. The polar bodies degenerate.

In female animals, **oogenesis,** the production of egg cells, occurs within specialized diploid cells of the ovary known as oogonia. Quite early in the development of the ovary, the oogonia initiate meiosis to produce primary oocytes. For example, in humans, approximately 1 million primary oocytes per ovary are produced before birth. These primary oocytes are arrested—enter a dormant phase—at prophase of meiosis I, remaining at this stage until the female becomes sexually mature. Beginning at this stage, primary oocytes are periodically activated to progress through the remaining stages of oocyte development.

During oocyte maturation, meiosis produces only one cell that is destined to become an egg, as opposed to the four gametes produced from each primary spermatocyte during spermatogenesis. How does this occur? As shown in **Figure 3.14b**, the first meiotic division is asymmetric and produces a secondary oocyte and a much smaller cell, known as a polar body. Most of the cytoplasm is retained by the secondary oocyte and very little by the polar body, allowing the oocyte to become a larger cell with more stored nutrients. The secondary oocyte then begins meiosis II. In mammals, the secondary oocyte is released from the ovary—an event called ovulation—and travels down the oviduct toward the uterus. During this journey, if a sperm cell penetrates the secondary oocyte, it is stimulated to complete meiosis II; the secondary oocyte produces a haploid egg and a second polar body. The haploid egg and sperm nuclei then unite to create the diploid nucleus of a new individual.

Plant Species Alternate Between Haploid (Gametophyte) and Diploid (Sporophyte) Generations

Most species of animals are diploid, and their haploid gametes are considered to be a specialized type of cell. By comparison, the life cycles of plant species alternate between haploid and diploid generations. The haploid generation is called the **gametophyte,** whereas the diploid generation is called the **sporophyte.** Meiosis produces haploid cells called spores, which divide by mitosis to produce the gametophyte. In simpler plants, such as mosses, a haploid spore can produce a large multicellular gametophyte by repeated mitoses and cellular divisions. In flowering plants, however, spores develop into gametophytes that contain only a few cells. In this case, the organism that we think of as a "plant" is the sporophyte, while the gametophyte is very inconspicuous. In fact, the gametophytes of many plant species are small structures produced within the much larger sporophyte. Certain cells within the haploid gametophytes then become specialized as haploid gametes.

Figure 3.15 provides an overview of gametophyte development and gametogenesis in flowering plants. Meiosis occurs within two different structures of the sporophyte: the anthers and the ovaries, which produce male and female gametophytes, respectively. This diagram depicts a flower from an angiosperm, which is a plant that produces seeds within an ovary.

In the anther, diploid cells called microsporocytes undergo meiosis to produce four haploid microspores. These separate into individual microspores. In many angiosperms, each microspore undergoes mitosis to produce a two-celled structure containing one tube cell and one generative cell, both of which are haploid. This structure differentiates into a **pollen grain,** which is the male gametophyte with a thick cell wall. Later, the generative cell undergoes

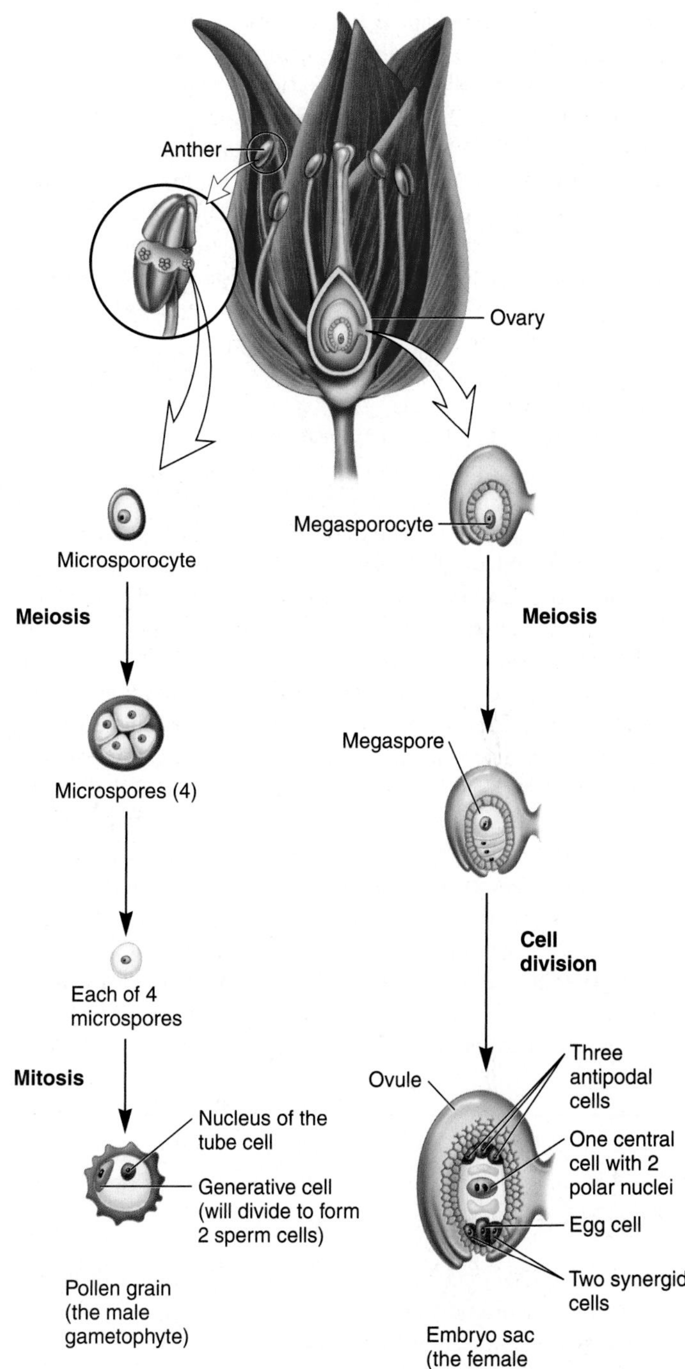

FIGURE 3.15 **The formation of male and female gametes by the gametophytes of angiosperms.**

mitosis to produce two haploid sperm cells. In most plant species, this mitosis occurs only if the pollen grain germinates—lands on a stigma and forms a pollen tube (refer back to Figure 2.2c).

By comparison, female gametophytes are produced within ovules found in the plant ovaries. A cell known as a megasporocyte undergoes meiosis to produce four haploid megaspores. Three of the four megaspores degenerate. The remaining haploid megaspore then undergoes three successive mitotic divisions accompanied by asymmetric cytokinesis to produce seven individual cells—one egg, two synergids, three antipodals, and one central cell. This seven-celled structure, also known as the **embryo sac,** is the mature female gametophyte. Each embryo sac is contained within an ovule.

For fertilization to occur, specialized cells within the male and female gametophytes must meet. The steps of plant fertilization were described in Chapter 2. To begin this process, a pollen grain lands on a stigma (refer back to Figure 2.2c). This stimulates the tube cell to sprout a tube that grows through the style and eventually makes contact with an ovule. As this is occurring, the generative cell undergoes mitosis to produce two haploid sperm cells. The sperm cells migrate through the pollen tube and eventually reach the ovule. One of the sperm enters the central cell, which contains the two polar nuclei. This creates a cell that is triploid ($3n$). This cell divides mitotically to produce **endosperm,** which acts as a food-storing tissue. The other sperm enters the egg cell. The egg and sperm nuclei fuse to create a diploid cell, the zygote, which becomes a plant embryo. Therefore, fertilization in flowering plants is actually a double fertilization. The result is that the endosperm, which uses a large amount of plant resources, will develop only when an egg cell has been fertilized. After fertilization is complete, the ovule develops into a seed, and the surrounding ovary develops into the fruit, which encloses one or more seeds.

3.4 THE CHROMOSOME THEORY OF INHERITANCE AND SEX CHROMOSOMES

Thus far, we have considered how chromosomes are transmitted during cell division and gamete formation. In this section, we will first examine how chromosomal transmission is related to the patterns of inheritance observed by Mendel. This relationship, known as the chromosome theory of inheritance, was a major breakthrough in our understanding of genetics because it established the framework for understanding how chromosomes carry and transmit the genetic determinants that govern the outcome of traits. This theory dramatically unfolded as a result of three lines of scientific inquiry (**Table 3.1**). One avenue concerned Mendel's breeding studies, in which he analyzed the transmission of traits from parent to offspring. A second line of inquiry involved the material basis for heredity. A Swiss botanist, Carl Nägeli, and a German biologist, August Weismann, championed the idea that a substance found in living cells is responsible for the transmission of traits from parents to offspring. Nägeli also suggested that both parents contribute equal amounts of this substance to their offspring. Several scientists, including Oscar Hertwig, Eduard Strasburger, and Walter Flemming, conducted studies suggesting that the chromosomes are the carriers of the genetic material. We now know the DNA within the chromosomes is the genetic material.

TABLE 3.1

Chronology for the Development and Proof of the Chromosome Theory of Inheritance

1866	Gregor Mendel: analyzed the transmission of traits from parents to offspring and showed that it follows a pattern of segregation and independent assortment.
1876–77	Oscar Hertwig and Hermann Fol: observed that the nucleus of the sperm enters the egg during animal cell fertilization.
1877	Eduard Strasburger: observed that the sperm nucleus of plants (and no detectable cytoplasm) enters the egg during plant fertilization.
1878	Walter Flemming: described mitosis in careful detail.
1883	Carl Nägeli and August Weismann: proposed the existence of a genetic material, which Nägeli called idioplasm and Weismann called germ plasm.
1883	Wilhelm Roux: proposed that the most important event of mitosis is the equal partitioning of "nuclear qualities" to the daughter cells.
1883	Edouard van Beneden: showed that gametes contain half the number of chromosomes and that fertilization restores the normal diploid number.
1884–85	Hertwig, Strasburger, and Weismann: proposed that chromosomes are carriers of the genetic material.
1889	Theodore Boveri: showed that enucleated sea urchin eggs that are fertilized by sperm from a different species develop into larva that have characteristics that coincide with the sperm's species.
1900	Hugo de Vries, Carl Correns, and Erich von Tschermak: rediscovered Mendel's work.
1901	Thomas Montgomery: determined that maternal and paternal chromosomes pair with each other during meiosis.
1901	C. E. McClung: discovered that sex determination in insects is related to differences in chromosome composition.
1902	Boveri: showed that when sea urchin eggs were fertilized by two sperm, the abnormal development of the embryo was related to an abnormal number of chromosomes.
1903	Walter Sutton: showed that even though the chromosomes seem to disappear during interphase, they do not actually disintegrate. Instead, he argued that chromosomes must retain their continuity and individuality from one cell division to the next.
1902–3	Boveri and Sutton: independently proposed tenets of the chromosome theory of inheritance. Some historians primarily credit this theory to Sutton.
1910	Thomas Hunt Morgan: showed that a genetic trait (i.e., white-eyed phenotype in *Drosophila*) was linked to a particular chromosome.
1913	E. Eleanor Carothers: demonstrated that homologous pairs of chromosomes show independent assortment.
1916	Calvin Bridges: studied chromosomal abnormalities as a way to confirm the chromosome theory of inheritance.

For a description of these experiments, the student is encouraged to read Voeller, B. R. (1968), *The Chromosome Theory of Inheritance. Classic Papers in Development and Heredity.* New York: Appleton-Century-Crofts.

Finally, the third line of evidence involved the microscopic examination of the processes of fertilization, mitosis, and meiosis. Researchers became increasingly aware that the characteristics of organisms are rooted in the continuity of cells during the life of an organism and from one generation to the next. When the work of Mendel was rediscovered, several scientists noted striking parallels between the segregation and assortment of traits noted by Mendel and the behavior of chromosomes during meiosis. Among them were Theodore Boveri, a German biologist, and Walter Sutton at Columbia University. They independently proposed the chromosome theory of inheritance, which was a milestone in our understanding of genetics. The principles of this theory are described at the beginning of this section.

The remainder of this section focuses on sex chromosomes. The experimental connection between the chromosome theory of inheritance and sex chromosomes is profound. Even though an examination of meiosis provided compelling evidence that Mendel's laws could be explained by chromosome sorting, researchers still needed to correlate chromosome behavior with the inheritance of particular traits. Because sex chromosomes, such as the X and Y chromosome, look very different under the microscope, and because many genes found on the X chromosome are not found on the Y chromosome, geneticists were able to correlate the inheritance of certain traits with the transmission of specific sex chromosomes. In particular, early studies identified genes on the X chromosome that govern eye color in fruit flies. The phenomenon researchers discovered is called **X-linked inheritance**. This inheritance pattern confirmed the idea that genes are found on chromosomes. In addition, X-linked inheritance showed us that not all traits follow simple Mendelian rules. In later chapters, we will examine a variety of traits that are governed by chromosomal genes yet follow inheritance patterns that are more complex than those observed by Mendel.

The Chromosome Theory of Inheritance Relates the Behavior of Chromosomes to the Mendelian Inheritance of Traits

According to the **chromosome theory of inheritance,** the inheritance patterns of traits can be explained by the transmission patterns of chromosomes during gametogenesis and fertilization. This theory is based on a few fundamental principles.

1. Chromosomes contain the genetic material that is transmitted from parent to offspring and from cell to cell.
2. Chromosomes are replicated and passed along, generation after generation, from parent to offspring. They are also passed from cell to cell during the multicellular development of an organism. Each type of chromosome retains its individuality during cell division and gamete formation.
3. The nuclei of most eukaryotic cells contain chromosomes that are found in homologous pairs—they are diploid. One member of each pair is inherited from the mother, the other from the father. At meiosis, one of the two members of each pair segregates into one daughter nucleus, and the homologue segregates into the other daughter nucleus. Gametes contain one set of chromosomes—they are haploid.
4. During the formation of haploid cells, different types of (nonhomologous) chromosomes segregate independently of each other.
5. Each parent contributes one set of chromosomes to its offspring. The maternal and paternal sets of homologous chromosomes are functionally equivalent; each set carries a full complement of genes.

The chromosome theory of inheritance allows us to see the relationship between Mendel's laws and chromosomal transmission. As shown in **Figure 3.16**, Mendel's law of segregation can be

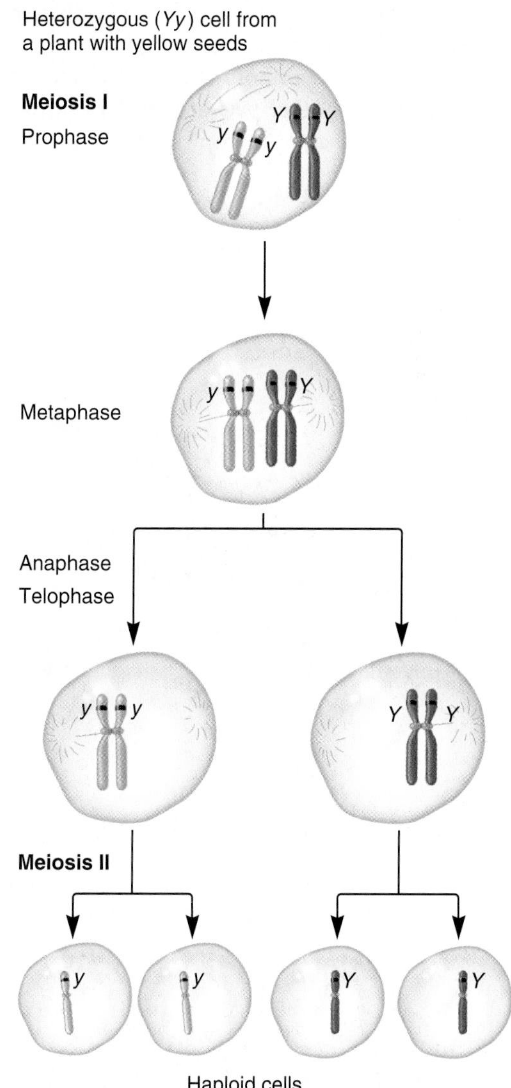

FIGURE 3.16 **Mendel's law of segregation can be explained by the segregation of homologues during meiosis.** The two copies of a gene are contained on homologous chromosomes. In this example using pea seed color, the two alleles are *Y* (yellow) and *y* (green). During meiosis, the homologous chromosomes segregate from each other, leading to segregation of the two alleles into separate gametes.

Genes→Traits The gene for seed color exists in two alleles, *Y* (yellow) and *y* (green). During meiosis, the homologous chromosomes that carry these alleles segregate from each other. The resulting cells receive the *Y* or *y* allele but not both. When two gametes unite during fertilization, the alleles that they carry determine the traits of the resulting offspring.

explained by the homologous pairing and segregation of chromosomes during meiosis. This figure depicts the behavior of a pair of homologous chromosomes that carry a gene for seed color. One of the chromosomes carries a dominant allele that confers yellow seed color, while the homologous chromosome carries a recessive allele that confers green color. A heterozygous individual would pass only one of these alleles to each offspring. In other words, a gamete may contain the yellow allele or the green allele but not both. Because homologous chromosomes segregate from each other, a gamete will contain only one copy of each type of chromosome.

How is the law of independent assortment explained by the behavior of chromosomes? **Figure 3.17** considers the segregation of two types of chromosomes, each carrying a different gene. One pair of chromosomes carries the gene for seed color: the yellow (*Y*) allele is on one chromosome, and the green (*y*) allele is on the homologue. The other pair of (smaller) chromosomes carries the gene for seed shape: one copy has the round (*R*) allele, and the homologue carries the wrinkled (*r*) allele. At metaphase of meiosis I, the different types of chromosomes have randomly aligned along the metaphase plate. As shown in Figure 3.17, this can occur in more than one way. On the left, the *R* allele has sorted with the *y* allele, while the *r* allele has sorted with the *Y* allele. On the right, the opposite situation has occurred. Therefore, the random alignment of chromatid pairs during meiosis I can lead to an independent assortment of genes that are found on nonhomologous chromosomes. As we will see in Chapter 5, this law is violated if two different genes are located close to one another on the same chromosome.

Sex Differences Often Correlate with the Presence of Sex Chromosomes

According to the chromosome theory of inheritance, chromosomes carry the genes that determine an organism's traits and are

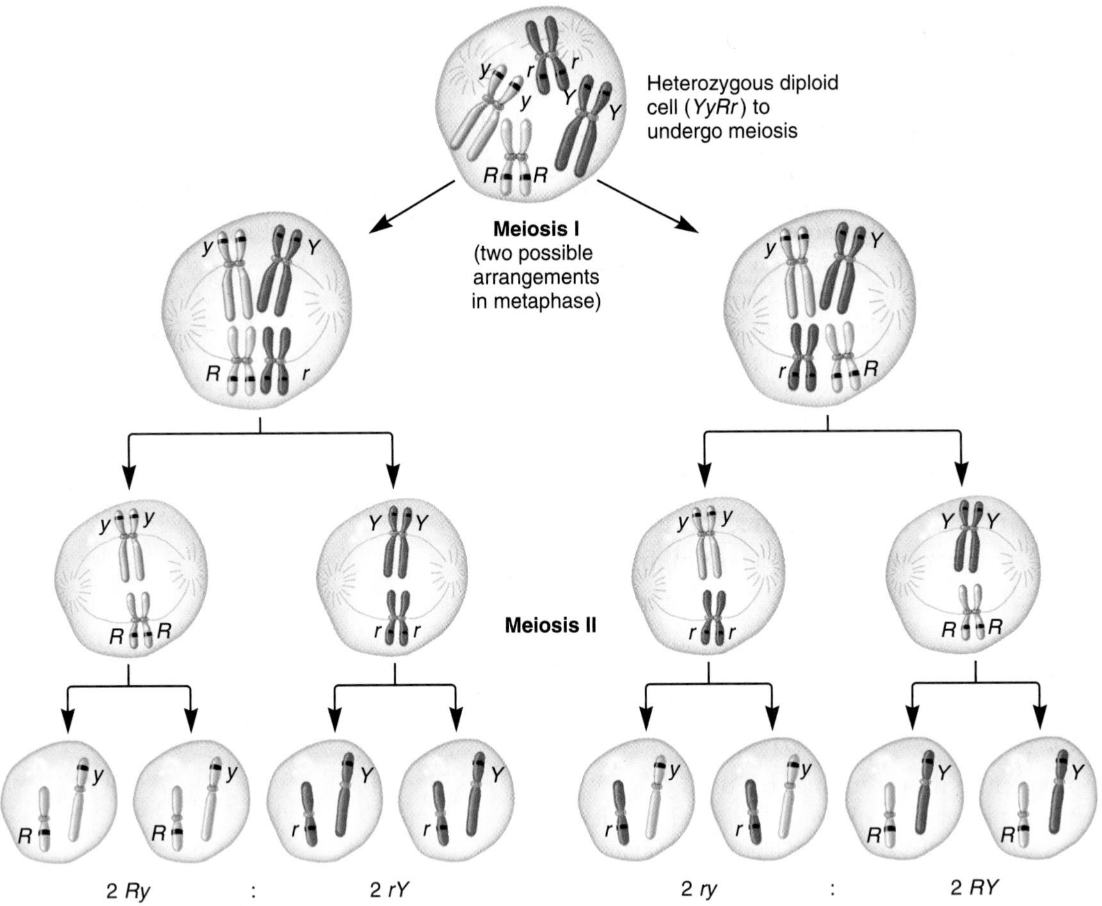

FIGURE 3.17 Mendel's law of independent assortment can be explained by the random alignment of bivalents during metaphase of meiosis I. This figure shows the assortment of two genes located on two different chromosomes, using pea seed color and shape as an example (*YyRr*). During metaphase of meiosis I, different possible arrangements of the homologues within bivalents can lead to different combinations of the alleles in the resulting gametes. For example, on the left, the dominant *R* allele has sorted with the recessive y allele; on the right, the dominant *R* allele has sorted with the dominant *Y* allele.

Genes→Traits Most species have several different chromosomes that carry many different genes. In this example, the gene for seed color exists in two alleles, *Y* (yellow) and *y* (green), and the gene for seed shape is found as *R* (round) and *r* (wrinkled) alleles. The two genes are found on different (nonhomologous) chromosomes. During meiosis, the homologous chromosomes that carry these alleles segregate from each other. In addition, the chromosomes carrying the *Y* or *y* alleles will independently assort from the chromosomes carrying the *R* or *r* alleles. As shown here, this provides a reassortment of alleles, potentially creating combinations of alleles that are different from the parental combinations. When two gametes unite during fertilization, the alleles they carry affect the traits of the resulting offspring.

the basis of Mendel's law of segregation and independent assortment. Some early evidence supporting this theory involved the determination of sex. Many species are divided into male and female sexes. In 1901, C. E. McClung, who studied fruit flies, was the first to suggest that male and female sexes are due to the inheritance of particular chromosomes. Since McClung's initial observations, we now know that a pair of chromosomes, called the **sex chromosomes,** determines sex in many different species. Some examples are described in **Figure 3.18.**

In the X-Y system of sex determination, which operates in mammals, the male contains one X chromosome and one Y chromosome, whereas the female contains two X chromosomes (Figure 3.18a). In this case, the male is called the **heterogametic sex.** Two types of sperm are produced: one that carries only the X chromosome, and another type that carries the Y. In contrast, the female is the **homogametic sex** because all eggs carry a single X chromosome. The 46 chromosomes carried by humans consist of 1 pair of sex chromosomes and 22 pairs of **autosomes**—chromosomes that are not sex chromosomes. In the human male, each of the four sperm produced during gametogenesis contains 23 chromosomes. Two sperm contain an X chromosome, and the other two have a Y chromosome. The sex of the offspring is determined by whether the sperm that fertilizes the egg carries an X or a Y chromosome.

What causes an offspring to develop into a male or female? One possibility is that two X chromosomes are required for female development. A second possibility is that the Y chromosome promotes male development. In the case of mammals, the second possibility is correct. This is known from the analysis of rare individuals who carry chromosomal abnormalities. For example, mistakes that occasionally occur during meiosis may produce an individual who carries two X chromosomes and one Y chromosome. Such an individual develops into a male.

Other mechanisms of sex determination include the X-0, Z-W, and haplo-diploid systems. The X-0 system of sex determination operates in many insects (Figure 3.18b). In such species, the male has one sex chromosome (the X) and is designated X0, while the female has a pair (two Xs). In other insect species, such as *Drosophila melanogaster*, the male is XY. For both types of insect species (i.e., X0 or XY males, and XX females), the ratio between X chromosomes and the number of autosomal sets determines sex. If a fly has one X chromosome and is diploid for the autosomes ($2n$), the ratio is 1/2, or 0.5. This fly will become a male even if it does not receive a Y chromosome. In contrast to mammals, the Y chromosome in the X-0 system does not determine maleness. If a fly receives two X chromosomes and is diploid, the ratio is 2/2, or 1.0, and the fly becomes a female.

For the Z-W system, which determines sex in birds and some fish, the male is ZZ and the female is ZW (Figure 3.18c). The letters Z and W are used to distinguish these types of sex chromosomes from those found in the X-Y pattern of sex determination of other species. In the Z-W system, the male is the homogametic sex, and the female is heterogametic.

Another interesting mechanism of sex determination, known as the haplo-diploid system, is found in bees (Figure 3.18d). The male bee, called the drone, is produced from unfertilized haploid eggs. Female bees, both worker bees and queen bees, are produced from fertilized eggs and therefore are diploid.

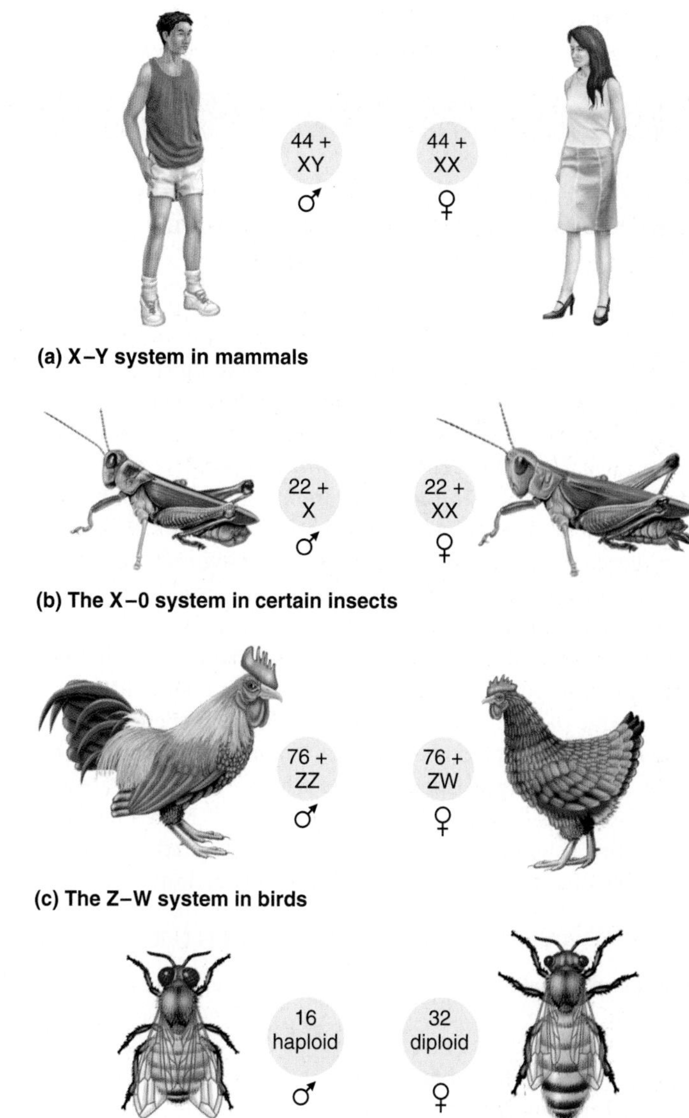

FIGURE 3.18 **Different mechanisms of sex determination in animals.** See text for a description.

Genes→Traits Certain genes that are found on the sex chromosomes play a key role in the development of sex (male versus female). For example, in mammals, genes on the Y chromosome initiate male development. In the X-0 system, the ratio of X chromosomes to the sets of autosomes plays a key role in governing the pathway of development toward male or female.

The chromosomal basis for sex determination is rooted in the location of particular genes on the sex chromosomes. The molecular basis for sex determination is described in Chapter 23.

Although sex in many species of animals is determined by chromosomes, other mechanisms are also known. In certain reptiles and fish, sex is controlled by environmental factors such as temperature. For example, in the American alligator (*Alligator mississippiensis*), temperature controls sex development. When fertilized eggs of this alligator are incubated at 33°C, nearly 100%

of them produce male individuals. When the eggs are incubated at a temperature a few degrees below 33°C, they produce nearly all females, while at a temperature a few degrees above 33°C, they produce 95% females.

EXPERIMENT 3A

Morgan's Experiments Showed a Connection Between a Genetic Trait and the Inheritance of a Sex Chromosome in *Drosophila*

In the early 1900s, Thomas Hunt Morgan carried out a particularly influential study that confirmed the chromosome theory of inheritance. Morgan was trained as an embryologist, and much of his early research involved descriptive and experimental work in that field. He was particularly interested in ways that organisms change. He wrote, "The most distinctive problem of zoological work is the change in form that animals undergo, both in the course of their development from the egg (embryology) and in their development in time (evolution)." Throughout his life, he usually had dozens of different experiments going on simultaneously, many of them unrelated to each other. He jokingly said there are three kinds of experiments—those that are foolish, those that are damn foolish, and those that are worse than that!

In one of his most famous studies, Morgan engaged one of his graduate students to rear the fruit fly *Drosophila melanogaster* in the dark, hoping to produce flies whose eyes would atrophy from disuse and disappear in future generations. Even after many consecutive generations, however, the flies appeared to have no noticeable changes despite repeated attempts at inducing mutations by treatments with agents such as X-rays and radium. After two years, Morgan finally obtained an interesting result when a true-breeding line of *Drosophila* produced a male fruit fly with white eyes rather than the normal red eyes. Because this had been a true-breeding line of flies, this white-eyed male must have arisen from a new mutation that converted a red-eye allele (denoted w^+) into a white-eye allele (denoted w). Morgan is said to have carried this fly home with him in a jar, put it by his bedside at night while he slept, and then taken it back to the laboratory during the day.

Much like Mendel, Morgan studied the inheritance of this white-eye trait by making crosses and quantitatively analyzing their outcome. In the experiment described in **Figure 3.19**, he began with his white-eyed male and crossed it to a true-breeding red-eyed female. All of the F_1 offspring had red eyes, indicating that red is dominant to white. The F_1 offspring were then mated to each other to obtain an F_2 generation.

THE GOAL

This is an example of discovery-based science rather than hypothesis testing. In this case, a quantitative analysis of genetic crosses may reveal the inheritance pattern for the white-eye allele.

ACHIEVING THE GOAL — FIGURE 3.19 **Inheritance pattern of an X-linked trait in fruit flies.**

Starting material: A true-breeding line of red-eyed fruit flies plus one white-eyed male fly that was discovered in the culture.

1. Cross the white-eyed male to a true-breeding red-eyed female.
2. Record the results of the F_1 generation. This involves noting the eye color and sexes of many offspring.
3. Cross F_1 offspring with each other to obtain F_2 offspring. Also record the eye color and sex of the F_2 offspring.

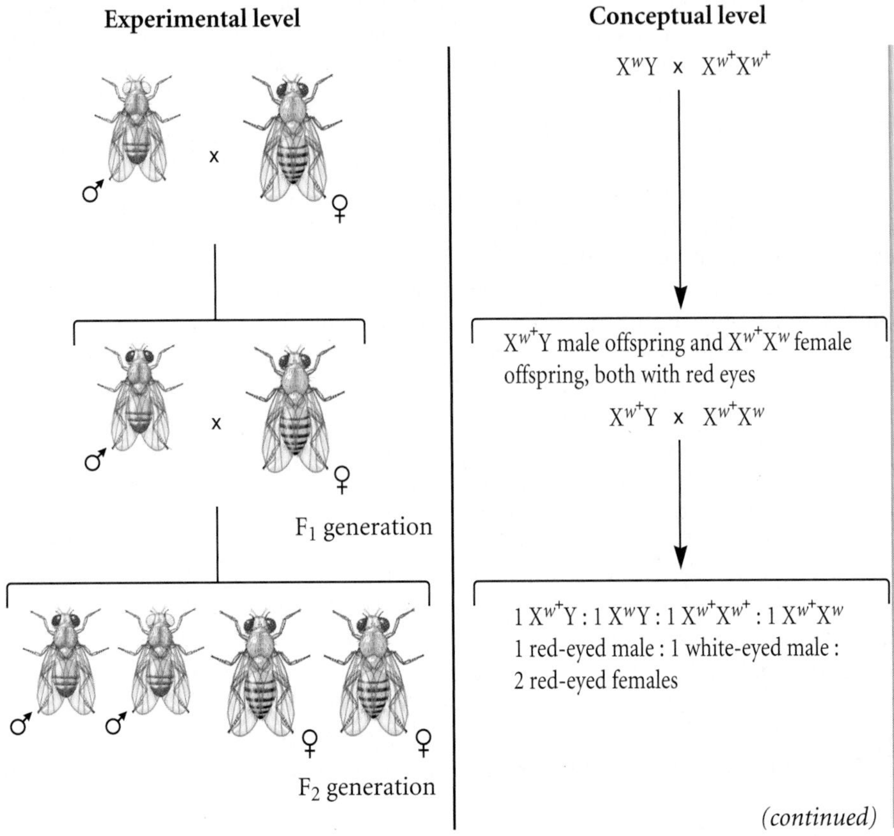

(continued)

4. In a separate experiment, perform a testcross between a white-eyed male and a red-eyed female from the F_1 generation. Record the results.

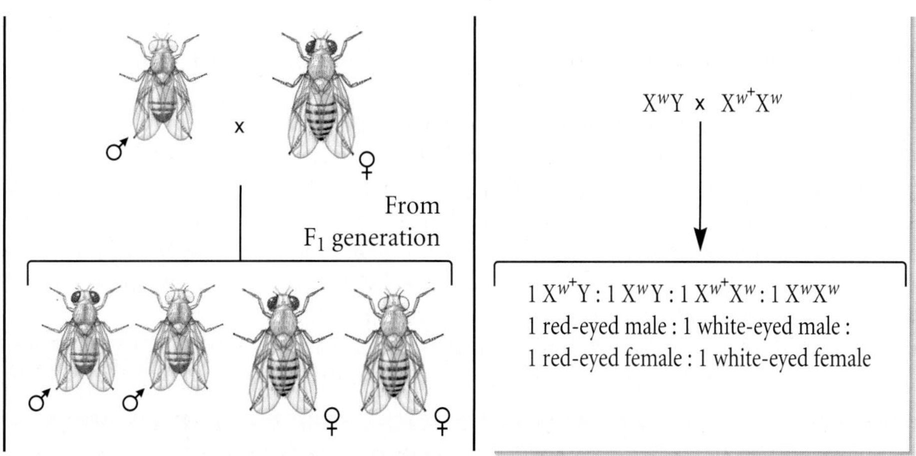

THE DATA

Cross	*Results*	
Original white-eyed male to red-eyed female	F_1 generation:	All red-eyed flies
F_1 male to F_1 female	F_2 generation:	2,459 red-eyed females 1,011 red-eyed males 0 white-eyed females 782 white-eyed males
White-eyed male to F_1 female	Testcross:	129 red-eyed females 132 red-eyed males 88 white-eyed females 86 white-eyed males

INTERPRETING THE DATA

As seen in the data table, the F_2 generation consisted of 2,459 red-eyed females, 1,011 red-eyed males, and 782 white-eyed males. Most notably, no white-eyed female offspring were observed in the F_2 generation. These results suggested that the pattern of transmission from parent to offspring depends on the sex of the offspring and on the alleles that they carry. As shown in the Punnett square here, the data are consistent with the idea that the eye color alleles are located on the X chromosome:

F_1 male is $X^{w^+}Y$
F_1 female is $X^{w^+}X^{w}$

Male gametes

Female gametes ♀ \ ♂	X^{w^+}	Y
X^{w^+}	$X^{w^+}X^{w^+}$ Red, female	$X^{w^+}Y$ Red, male
X^{w}	$X^{w^+}X^{w}$ Red, female	$X^{w}Y$ White, male

The Punnett square predicts that the F_2 generation will not have any white-eyed females. This prediction was confirmed experimentally. These results indicated that the eye color alleles are located on the X chromosome. Genes that are physically located within the X chromosome are called **X-linked genes** or **X-linked alleles.** However, it should also be pointed out that the experimental ratio in the F_2 generation of red eyes to white eyes is (2,459 + 1,011):782, which equals 4.4:1. This ratio deviates significantly from the predicted ratio of 3:1. How can this discrepancy be explained? Later work revealed that the lower-than-expected number of white-eyed flies is due to their decreased survival rate.

Morgan also conducted a **testcross** (step 4, Figure 3.19) in which an individual with a dominant phenotype and unknown genotype is crossed to an individual with a recessive phenotype. In this case, he mated an F_1 red-eyed female to a white-eyed male. This cross produced red-eyed males and females, and white-eyed males and females, in approximately equal numbers. The testcross data are also consistent with an X-linked pattern of inheritance. As shown in the following Punnett square, the testcross predicts a 1:1:1:1 ratio:

Testcross:
Male is $X^{w}Y$
F_1 female is $X^{w^+}X^{w}$

Male gametes

Female gametes ♀ \ ♂	X^{w}	Y
X^{w^+}	$X^{w^+}X^{w}$ Red, female	$X^{w^+}Y$ Red, male
X^{w}	$X^{w}X^{w}$ White, female	$X^{w}Y$ White, male

The observed data are 129:132:88:86, which is a ratio of 1.5:1.5:1:1. Again, the lower-than-expected numbers of white-eyed males and females can be explained by a lower survival rate for white-eyed flies. In his own interpretation, Morgan concluded that R (red eye color) and X (a sex factor that is present in two copies in the female) are combined and have never existed apart. In other words, this gene for eye color is on the X chromosome. Morgan was the first geneticist to receive the Nobel Prize.

Calvin Bridges, a graduate student in the laboratory of Morgan, also examined the transmission of X-linked traits. Bridges conducted hundreds of crosses involving several different types of X-linked alleles. In his crosses, he occasionally obtained offspring that had unexpected phenotypes and abnormalities in sex chromosome composition. In all cases, the parallel between the cytological presence of sex chromosome abnormalities and the occurrence of unexpected traits confirmed the idea that the sex chromosomes carry X-linked genes. Together, the work of Morgan and Bridges provided an impressive body of evidence confirming the idea that traits following an X-linked pattern of inheritance are governed by genes that are physically located on the X chromosome. Bridges wrote, "There can be no doubt that the complete parallelism between the unique behavior of chromosomes and the behavior of sex-linked genes and sex in this case means that the sex-linked genes are located in and borne by the X chromosomes." An example of Bridges's work is described in solved problem S5 at the end of this chapter.

A self-help quiz involving this experiment can be found at the Online Learning Center.

CONCEPTUAL SUMMARY

In Chapter 3, we have considered the process of reproduction at the cellular level. During binary fission in bacteria and mitotic cell division in eukaryotes, a mother cell divides to produce two daughter cells that are genetically identical to each other. In bacteria, which contain a single circular chromosome, binary fission provides a way to reproduce asexually. Eukaryotic species, which contain many linear chromosomes, must sort their chromosomes during mitotic cell division. This ensures that the chromosomes are correctly distributed to the daughter cells. In simple eukaryotes, mitotic divisions can produce new unicellular offspring; it is a form of asexual reproduction. Otherwise, the primary reason that eukaryotic species undergo mitotic cell divisions is to produce organisms that are multicellular. Diploid eukaryotes develop from a single fertilized egg cell that undergoes repeated mitotic cell divisions to produce an organism containing many cells. Most of the cells of the body are genetically identical to each other.

Sexual reproduction requires the formation of gametes, such as sperm cells or egg cells, that contain half the genetic material. For diploid organisms that have two sets of chromosomes ($2n$), the process of meiosis results in cells that are haploid, which have one set ($1n$). Meiosis requires two successive divisions. Meiosis I involves a pairing between homologous chromosomes that does not occur during mitosis. This pairing is necessary to ensure that the resultant cells will contain one complete set of chromosomes. Meiosis II is very similar to a mitotic division, except that the cells entering meiosis II begin with half the number of chromatids as compared to cells entering mitosis.

In animals, meiosis occurs in conjunction with the cellular specialization of gametes. During spermatogenesis, meiosis results in the production of four haploid sperm cells from one primary spermatocyte. In comparison, oogenesis produces a single haploid egg from each primary oocyte. Fertilization in animals involves the union between a single sperm and an egg cell. In plants, sexual reproduction involves an alternation between diploid organisms (sporophytes) and haploid organisms (gametophytes). However, the gametophytes of more complex plants contain only a few cells, which are primarily involved in producing gametes. In flowering plants, sexual reproduction is a double fertilization event in which one sperm fertilizes the egg and a second sperm unites with the central cell to produce the endosperm. The proliferation of endosperm provides nutrients to the plant embryo.

A cellular understanding of mitosis, meiosis, and fertilization allows us to appreciate how the transmission of traits is due to the passage of chromosomes. Genes, which govern the outcome of traits, are physically located within the structure of chromosomes. The chromosome theory of inheritance, proposed by Boveri and Sutton, describes how the transmission of chromosomes accounts for Mendel's laws of segregation and independent assortment. Later, researchers were able to correlate the inheritance of particular traits with the transmission of specific sex chromosomes. In particular, early studies identified genes on the X chromosome that govern eye color in fruit flies. This phenomenon, called X-linked inheritance, confirmed the idea that chromosomes carry genes that govern the outcome of traits. In some species, chromosomes also play a role in sex determination. Particular chromosomes, termed sex chromosomes, determine sex in many species.

EXPERIMENTAL SUMMARY

A central method that has been used to understand chromosomal transmission is light microscopy. Cytogeneticists can view the chromosomes in actively dividing cells under a microscope. This has allowed researchers to study the behavior of chromosomes during mitosis and meiosis. In the late nineteenth and early twentieth century, microscopic examinations of chromosomes yielded observations consistent with the idea that the chromosomes are the carriers of genes. This idea, along with Mendel's experiments in pea plants, led to the formulation of the chromosome theory of inheritance. Many subsequent experiments have confirmed the validity of this theory. In particular, investigations involving the inheritance of the X chromosome were instrumental in establishing that the inheritance of certain traits paralleled the inheritance of the X chromosome. Morgan was the first scientist to show that the inheritance of a recessive X-linked trait was consistent with the respective gene's location on the X chromosome. Since that time, the chromosomal locations of many genes have been firmly established.

PROBLEM SETS & INSIGHTS

Solved Problems

S1. A diploid cell has eight chromosomes, four per set. For the following diagram, in what phase of mitosis, meiosis I or meiosis II, is this cell?

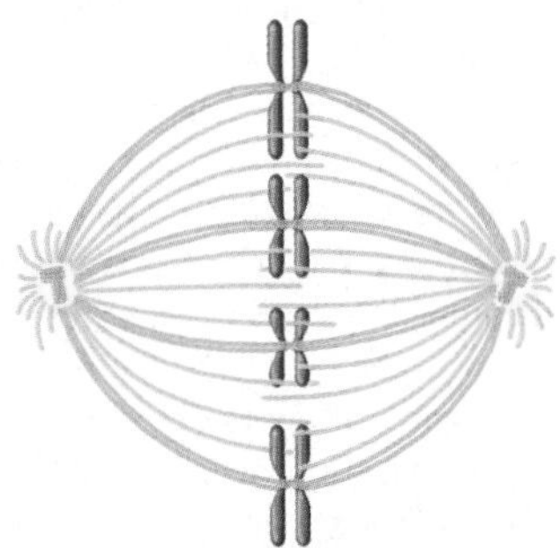

Answer: The cell is in metaphase of meiosis II. You can tell because the chromosomes are lined up in a single row along the metaphase plate, and the cell has only four pairs of sister chromatids. If it were mitosis, the cell would have eight pairs of sister chromatids.

S2. An unaffected woman (i.e., without disease symptoms) who is heterozygous for the X-linked allele causing Duchenne muscular dystrophy has children with a man with a normal allele. What are the probabilities of the following combinations of offspring?

A. An unaffected son

B. An unaffected son or daughter

C. A family of three children, all of whom are affected

Answer: The first thing we must do is construct a Punnett square to determine the outcome of the cross. *N* represents the normal allele, *n* the recessive allele causing Duchenne muscular dystrophy. The mother is heterozygous, and the father has the normal allele.

Male gametes

Female gametes ♀ \ ♂	X^N	Y
X^N	X^NX^N	X^NY
X^n	X^NX^n	X^nY

Phenotype ratio is

2 normal daughters :
1 normal son :
1 affected son

A. There are four possible children, one of whom is an unaffected son. Therefore, the probability of an unaffected son is 1/4.

B. Use the sum rule: $1/4 + 1/2 = 3/4$.

C. You could use the product rule because there would be three offspring in a row with the disorder: $(1/4)(1/4)(1/4) = 1/64 = 0.016 = 1.6\%$.

S3. What are the major differences between prophase, metaphase, and anaphase when comparing mitosis, meiosis I, and meiosis II?

Answer: The table summarizes key differences.

A Comparison of Mitosis, Meiosis I, and Meiosis II

Phase	Event	Mitosis	Meiosis I	Meiosis II
Prophase	Synapsis:	No	Yes	No
Prophase	Crossing over:	Rarely	Commonly	Rarely
Metaphase	Alignment along the metaphase plate:	Sister chromatids	Bivalents	Sister chromatids
Anaphase	Separation of:	Sister chromatids	Bivalents	Sister chromatids

S4. Among different plant species, both male and female gametophytes can be produced by single individuals or by separate sexes. In some species, such as the garden pea, a single individual can produce both male and female gametophytes. Fertilization takes place via self-fertilization or cross-fertilization. A plant species that has a single type of flower producing both pollen and eggs is termed a monoclinous plant. In other plant species, two different types of flowers produce either pollen or eggs. When both flower types are on a single individual, such a species is termed monoecious. It is most common for the "male flowers" to be produced near the top of the plant and the "female flowers" toward the bottom. Though less common, some species of plants are dioecious. For dioecious species, one individual makes either male flowers or female flowers, but not both.

Based on your personal observations of plants, try to give examples of monoclinous, monoecious, and dioecious plants. What would be the advantages and disadvantages of each?

Answer: Monoclinous plants—pea plant, tulip, and roses. The same flower produces pollen on the anthers and egg cells within the ovary.

Monoecious plants—corn and pine trees. In corn, the tassels are the male flowers and the ears result from fertilization within the female flowers. In pine trees, pollen is produced in cones near the top of the tree, and eggs cells are found in larger cones nearer the bottom.

Dioecious plants—holly and ginkgo trees. Certain individuals produce only pollen, while others produce only eggs.

An advantage of being monoclinous or monoecious is that fertilization is relatively easy because the pollen and egg cells are produced on the same individual. This is particularly true for monoclinous plants. The proximity of the pollen to the egg cells makes it more likely for self-fertilization to occur. This is advantageous if the plant population is relatively sparse. On the other hand, a dioecious species can reproduce only via cross-fertilization. The advantage of cross-fertilization is that it enhances genetic variation. Over the long run, this can be an advantage because cross-fertilization is more likely to produce a varied population of individuals, some of which may possess combinations of traits that promote survival.

S5. To test the chromosome theory of inheritance, Calvin Bridges made crosses involving the inheritance of X-linked traits. One of his experiments concerned two different X-linked genes affecting eye color and wing size. For the eye color gene, the red-eye allele

(w^+) is dominant to the white-eye allele (w). A second X-linked trait is wing size; the allele called miniature is recessive to the normal allele. In this case, m represents the miniature allele and m^+ the normal allele. A male fly carrying a miniature allele on its single X chromosome has small (miniature) wings. A female must be homozygous, mm, in order to have miniature wings.

Bridges made a cross between $X^{w,m^+} X^{w,m^+}$ female flies (white eyes and normal wings) to $X^{w^+,m}$ Y male flies (red eyes and miniature wings). He then examined the eyes, wings, and sexes of thousands of offspring. Most of the offspring were females with red eyes and normal wings, and males with white eyes and normal wings. On rare occasions (approximately 1 out of 1,700 flies), however, he also obtained female offspring with white eyes or males with red eyes. He also noted the wing shape in these flies and then cytologically examined their chromosome composition using a microscope. The following results were obtained:

Offspring	Eye Color	Wing Size	Sex Chromosomes
Expected females	Red	Normal	XX
Expected males	White	Normal	XY
Unexpected females (rare)	White	Normal	XXY
Unexpected males (rare)	Red	Miniature	X0

Data from: Bridges, C. B. "Non-Disjunction as Proof of the Chromosome Theory of Heredity," *Genetics 1*, 1–52, 107–63.

Explain these data.

Answer: Remember that in fruit flies, the number of X chromosomes (not the presence of the Y chromosome) determines sex. As seen in the data, the flies with unexpected phenotypes were abnormal in their sex chromosome composition. The white-eyed female flies were due to the union between an abnormal XX female gamete and a normal Y male gamete. Likewise, the unexpected male offspring contained only one X chromosome and no Y. These male offspring were due to the union between an abnormal egg without any X chromosome and a normal sperm containing one X chromosome. The wing size of the unexpected males was a particularly significant result. The red-eyed males showed a miniature wing size. As noted by Bridges, this means they inherited their X chromosome from their father rather than their mother. This observation provided compelling evidence that the inheritance of the X chromosome correlates with the inheritance of particular traits.

At the time of his work, Bridges's results were particularly striking because chromosomal abnormalities had been rarely observed in *Drosophila*. Nevertheless, Bridges first predicted how chromosomal abnormalities would cause certain unexpected phenotypes, and then he actually observed the abnormal number of chromosomes using a microscope. Together, his work provided evidence confirming the idea that traits that follow an X-linked pattern of inheritance are governed by genes physically located on the X chromosome.

Conceptual Questions

C1. The process of binary fission begins with a single mother cell and ends with two daughter cells. Would you expect the mother and daughter cells to be genetically identical? Explain why or why not.

C2. What is a homologue? With regard to genes and alleles, how are homologues similar to and different from each other?

C3. What is a sister chromatid? Are sister chromatids genetically similar or identical? Explain.

C4. With regard to sister chromatids, which phase of mitosis is the organization phase, and which is the separation phase?

C5. A species is diploid containing three chromosomes per set. Draw what the chromosomes would look like in the G_1 and G_2 phases of the cell cycle.

C6. How does the attachment of kinetochore microtubules to the kinetochore differ in metaphase of meiosis I compared to metaphase of mitosis? Discuss what you think would happen if a sister chromatid was not attached to a kinetochore microtubule.

C7. For the following events, specify whether they occur during mitosis, meiosis I, and/or meiosis II:

A. Separation of conjoined chromatids within a pair of sister chromatids

B. Pairing of homologous chromosomes

C. Alignment of chromatids along the metaphase plate

D. Attachment of sister chromatids to both poles

C8. Describe the key events during meiosis that result in a 50% reduction in the amount of genetic material per cell.

C9. A cell is diploid and contains three chromosomes per set. Draw the arrangement of chromosomes during metaphase of mitosis and metaphase of meiosis I and II. In your drawing, make one set dark and the other lighter.

C10. The arrangement of homologues during metaphase of meiosis I is a random process. In your own words, explain what this means.

C11. A eukaryotic cell is diploid containing 10 chromosomes (5 in each set). For mitosis and meiosis, how many daughter cells would be produced, and how many chromosomes would each one contain?

C12. If a diploid cell contains six chromosomes (i.e., three per set), how many possible random arrangements of homologues could occur during metaphase of meiosis I?

C13. A cell has four pairs of chromosomes. Assuming that crossing over does not occur, what is the probability that a gamete will contain all of the paternal chromosomes? If n equals the number of chromosomes in a set, which of the following expressions can be used to calculate the probability that a gamete will receive all of the paternal chromosomes: $(1/2)^n$, $(1/2)^{n-1}$, or $n^{1/2}$?

C14. With regard to question C13, how would the phenomenon of crossing over affect the results? In other words, would the probability of a gamete inheriting only paternal chromosomes be higher or lower? Explain your answer.

C15. Eukaryotic cells must sort their chromosomes during mitosis so that each daughter cell receives the correct number of chromosomes. Why don't bacteria need to sort their chromosomes?

C16. Why is it necessary that the chromosomes condense during mitosis and meiosis? What do you think might happen if the chromosomes were not condensed?

C17. Nine-banded armadillos almost always give birth to four offspring that are genetically identical quadruplets. Explain how you think this happens.

C18. A diploid species contains four chromosomes per set for a total of eight chromosomes in its somatic cells. Draw the cell as it would look in late prophase of meiosis II and prophase of mitosis. Discuss how prophase of meiosis II and prophase of mitosis differ from each other, and explain how the difference originates.

C19. Explain why the products of meiosis may not be genetically identical while the products of mitosis are.

C20. The period between meiosis I and meiosis II is called interphase II. Does DNA replication take place during interphase II? Explain your answer.

C21. List several ways in which telophase appears to be the reverse of prophase and prometaphase.

C22. In corn, there are 10 chromosomes per set and the sporophyte of the species is diploid. If you performed a karyotype, what is the total number of chromosomes that you would expect to see in the following types of cells?

A. A leaf cell

B. The sperm nucleus of a pollen grain

C. An endosperm cell after fertilization

D. A root cell

C23. The arctic fox has 50 chromosomes (25 per set), and the common red fox has 38 chromosomes (19 per set). These species can interbreed to produce viable but infertile offspring. How many chromosomes would the offspring have? What problems do you think may occur during meiosis that would explain the offspring's infertility?

C24. Let's suppose that a gene affecting pigmentation is found on the X chromosome (in mammals or insects) or the Z chromosome (in birds) but not on the Y or W chromosome. It is found on an autosome in bees. This gene is found in two alleles, *D* (dark), which is dominant to *d* (light). What would be the phenotypic results of crosses between a true-breeding dark female and true-breeding light male, and the reciprocal crosses involving a true-breeding light female and true-breeding dark male, in the following species? Refer back to Figure 3.18 for the mechanism of sex determination in these species.

A. Birds

B. *Drosophila*

C. Bees

D. Humans

C25. Describe the cellular differences between male and female gametes.

C26. At puberty, the testes contain a finite number of cells and produce an enormous number of sperm cells during the life span of a male. Explain why testes do not run out of spermatogonial cells.

C27. Describe the timing of meiosis I and II during human oogenesis.

C28. Three genes (*A*, *B*, and *C*) are found on three different chromosomes. For the following diploid genotypes, describe all the possible gamete combinations.

A. *Aa Bb Cc*

B. *AA Bb CC*

C. *Aa BB Cc*

D. *Aa bb cc*

C29. A phenotypically normal woman with an abnormally long chromosome 13 (and a normal homologue of chromosome 13) marries a phenotypically normal man with an abnormally short chromosome 11 (and a normal homologue of chromosome 11). What is the probability of producing an offspring that will have both a long chromosome 13 and a short chromosome 11? If such a child is produced, what is the probability that this child would eventually pass both abnormal chromosomes to one of his/her offspring?

C30. Assuming that such a fly would be viable, what would be the sex of a fruit fly with the following chromosomal composition?

A. One X chromosome and two sets of autosomes

B. Two X chromosomes, one Y chromosome, and two sets of autosomes

C. Two X chromosomes and four sets of autosomes

D. Four X chromosomes, two Y chromosomes, and four sets of autosomes

C31. What would be the sex of a human with the following numbers of sex chromosomes?

A. XXX

B. X (also described as X0)

C. XYY

D. XXY

Experimental Questions

E1. When studying living cells in a laboratory, researchers sometimes use drugs as a way to make cells remain at a particular stage of the cell cycle. For example, aphidicolin inhibits DNA synthesis in eukaryotic cells and causes them to remain in the G_1 phase because they cannot replicate their DNA. In what phase of the cell cycle—G_1, S, G_2, prophase, metaphase, anaphase, or telophase—would you expect somatic cells to stay if the following types of drug were added?

A. A drug that inhibits microtubule formation

B. A drug that allows microtubules to form but prevents them from shortening

C. A drug that inhibits cytokinesis

D. A drug that prevents chromosomal condensation

E2. In Morgan's experiments, which result do you think is the most convincing piece of evidence pointing to X-linkage of the eye color gene? Explain your answer.

E3. In his original studies of Figure 3.19, Morgan first suggested that the original white-eyed male had two copies of the white-eye allele. In this problem, let's assume that he meant the fly was X^wY^w instead of X^wY. Are his data in Figure 3.19 consistent with this hypothesis? What crosses would need to be made to rule out the possibility that the Y chromosome carries a copy of the eye color gene?

E4. How would you set up crosses to determine if a gene was Y linked versus X linked?

E5. Occasionally during meiosis, a mistake can happen whereby a gamete may receive zero or two sex chromosomes rather than one. Calvin Bridges made a cross between white-eyed female flies and red-eyed male flies. As you would expect, most of the offspring were red-eyed females and white-eyed males. On rare occasions, however, he found a white-eyed female or a red-eyed male. These rare flies were not due to new gene mutations but instead were due to mistakes during meiosis in the parent flies. Consider the mechanism of sex determination in fruit flies and propose how this could happen. In your answer, describe the sex chromosome composition of these rare flies.

E6. Let's suppose that you have karyotyped a female fruit fly with red eyes and found that it has three X chromosomes instead of the normal two. Although you do not know its parents, you do know that this fly came from a mixed culture of flies in which some had red eyes, some had white eyes, and some had eosin eyes. Eosin is an allele of the same gene that has white and red alleles. Eosin is a pale orange color. The red allele is dominant and the white allele is recessive. The expression of the eosin allele, however, depends on the number of copies of the allele. When females have two copies of this allele, they have eosin eyes. When females are heterozygous for the eosin allele and white allele, they have light-eosin eyes. When females are heterozygous for the red allele and the eosin allele, they have red eyes. Males that have a single copy of eosin allele have light-eosin eyes.

You cross this female with a white-eyed male and count the number of offspring. You may assume that this unusual female makes half of its gametes with one X chromosome and half of its gametes with two X chromosomes. The following results were obtained:

	Females*	Males
Red eyes	50	11
White eyes	0	0
Eosin	20	0
Light-eosin	21	20

*A female offspring can be XXX, XX, or XXY.

Explain the 3:1 ratio between female and male offspring. What was the genotype of the original mother, which had red eyes and three X chromosomes? Construct a Punnett square that is consistent with these data.

E7. With regard to thickness and length, what do you think the chromosomes would look like if you microscopically examined them during interphase? How would that compare to their appearance during metaphase?

E8. White-eyed flies have a lower survival rate compared to red-eyed flies. Based on the data in Figure 3.19, what percentage of white-eyed flies survived compared to red-eyed flies, assuming 100% survival of red-eyed flies?

E9. A rare form of dwarfism that also included hearing loss was found to run in a particular family. It is inherited in a dominant manner. It was discovered that an affected individual had one normal copy of chromosome 15 and one abnormal copy of chromosome 15 that was unusually long. How would you determine if the unusually long chromosome 15 was causing this disorder?

E10. Discuss why crosses (i.e., the experiments of Mendel) and the microscopic observations of chromosomes during mitosis and meiosis were both needed to deduce the chromosome theory of inheritance.

E11. A cross was made between female flies with white eyes and miniature wings (both X-linked recessive traits) to male flies with red eyes and normal wings. On rare occasions, female offspring were produced with white eyes. If we assume these females are due to errors in meiosis, what would be the most likely chromosomal composition of such flies? What would be their wing shape?

E12. Experimentally, how do you think researchers were able to determine that the Y chromosome causes maleness in mammals, whereas the ratio of X chromosomes to the sets of autosomes causes sex determination in fruit flies?

Questions for Student Discussion/Collaboration

1. In Figure 3.19, Morgan obtained a white-eyed male fly in a population containing many red-eyed flies that he thought were true-breeding. As mentioned in the experiment, he crossed this fly with several red-eyed sisters, and all the offspring had red eyes. But actually this is not quite true. Morgan observed 1,237 red-eyed flies and 3 white-eyed males. Provide two or more explanations why he obtained 3 white-eyed males in the F_1 generation.

2. A diploid eukaryotic cell has 10 chromosomes (5 per set). As a group, take turns having one student draw the cell as it would look during a phase of mitosis, meiosis I, or meiosis II; then have the other students guess which phase it is.

3. Discuss the principles of the chromosome theory of inheritance. Which principles were deduced via light microscopy, and which were deduced from crosses? What modern techniques could be used to support the chromosome theory of inheritance?

Note: All answers appear at the website for this textbook; the answers to even-numbered questions are in the back of the textbook.

www.mhhe.com/brookergenetics3e

Visit the Online Learning Center for practice tests, answer keys, and other learning aids for this chapter. Enhance your understanding of genetics with our interactive exercises, quizzes, animations, and much more.

EXTENSIONS OF MENDELIAN INHERITANCE

4

CHAPTER OUTLINE

The term **Mendelian inheritance** describes inheritance patterns that obey two laws: the law of segregation and the law of independent assortment. Until now, we have mainly considered traits that are affected by a single gene that is found in two different alleles. In these cases, one allele is dominant over the other. This type of inheritance is sometimes called **simple Mendelian inheritance** because the observed ratios in the offspring readily obey Mendel's laws. For example, when two different true-breeding pea plants are crossed (e.g., tall and dwarf) and the F_1 generation is allowed to self-fertilize, the F_2 generation shows a 3:1 phenotypic ratio of tall and dwarf offspring.

In Chapter 4, we will extend our understanding of Mendelian inheritance by examining the transmission patterns for several traits that do not display a simple dominant/recessive relationship. Geneticists have discovered an amazing diversity of mechanisms by which alleles affect the outcome of traits. Many alleles don't produce the ratios of offspring that are expected from a simple Mendelian relationship. This does not mean that Mendel was wrong. Rather, the inheritance patterns of many traits are more complex and interesting than he had realized. In this chapter, we will examine how the outcome of a trait may be influenced by a variety of factors such as the level of protein expression, the sex of the individual, the presence of multiple alleles of a given gene, and environmental effects. We will also explore how two different genes can contribute to the outcome of a single trait. Later, in Chapters 5 and 7, we will examine eukaryotic inheritance patterns that actually violate the laws of segregation and independent assortment.

Inheritance patterns and alleles. In the petunia, multiple alleles can result in flowers with several different colors, such as the three shown here.

4.1 INHERITANCE PATTERNS OF SINGLE GENES

We begin Chapter 4 with the further exploration of traits that are influenced by a single gene. **Table 4.1** describes the general features of several types of Mendelian inheritance patterns that have been observed by researchers. These various patterns occur because the outcome of a trait may be governed by two or more alleles in many different ways. In this section, we will examine these patterns with two goals in mind. First, we want to understand how the molecular expression of genes can account for an individual's phenotype. In other words, we will explore the underlying relationship between molecular genetics—the expression of genes to produce functional proteins—and the traits of individuals that inherit the genes. Our second goal concerns the outcome of crosses. Many of the

TABLE 4.1

Types of Mendelian Inheritance Patterns Involving Single Genes

Type	Description
Simple Mendelian	**Inheritance:** This term is commonly applied to the inheritance of alleles that obey Mendel's laws and follow a strict dominant/recessive relationship. In Chapter 4, we will see that some genes can be found in three or more alleles, making the relationship more complex. **Molecular:** 50% of the protein encoded by two copies of the dominant (functional) allele is sufficient to produce the dominant trait.
Incomplete dominance	**Inheritance:** This pattern occurs when the heterozygote has a phenotype that is intermediate between either corresponding homozygote. For example, a cross between homozygous red-flowered and homozygous white-flowered parents will have heterozygous offspring with pink flowers. **Molecular:** 50% of the protein encoded by two copies of the functional allele is not sufficient to produce the same trait as the homozygote making 100%.
Incomplete penetrance	**Inheritance:** This pattern occurs when a dominant phenotype is not expressed even though an individual carries a dominant allele. An example is an individual who carries the polydactyly allele but has a normal number of fingers and toes. **Molecular:** Even though a dominant gene may be present, the protein encoded by the gene may not exert its effects. This can be due to environmental influences or due to other genes that may encode proteins that counteract the effects of the protein encoded by the dominant allele.
Overdominance	**Inheritance:** This pattern occurs when the heterozygote has a trait that is more beneficial than either homozygote. **Molecular:** Three common ways that heterozygotes gain benefits: (1) Their cells may have increased resistance to infection by microorganisms; (2) they may produce more forms of protein dimers, with enhanced function; or (3) they may produce proteins that function under a wider range of conditions.
Codominance	**Inheritance:** This pattern occurs when the heterozygote expresses both alleles simultaneously. For example, in blood typing, an individual carrying the *A* and *B* alleles will have an AB blood type. **Molecular:** The codominant alleles encode proteins that function slightly differently from each other, and the function of each protein in the heterozygote affects the phenotype uniquely.
X linked	**Inheritance:** This pattern involves the inheritance of genes that are located on the X chromosome. In mammals and fruit flies, males are hemizygous for X-linked genes, while females have two copies. **Molecular:** If a pair of X-linked alleles shows a simple dominant/recessive relationship, 50% of the protein encoded by two copies of the dominant allele is sufficient to produce the dominant trait (in the female).
Sex-influenced inheritance	**Inheritance:** This pattern refers to the impact of sex on the phenotype of the individual. Some alleles are recessive in one sex and dominant in the opposite sex. An example would be pattern baldness in humans. **Molecular:** Sex hormones may regulate the molecular expression of genes. This can have an impact on the phenotypic effects of alleles.
Sex-limited inheritance	**Inheritance:** This refers to traits that occur in only one of the two sexes. An example would be breast development in mammals. **Molecular:** Sex hormones may regulate the molecular expression of genes. This can have an impact on the phenotypic effects of alleles. In this case, sex hormones that are primarily produced in only one sex are essential to produce a particular phenotype.
Lethal alleles	**Inheritance:** An allele that has the potential of causing the death of an organism. **Molecular:** Lethal alleles are most commonly loss-of-function alleles that encode proteins that are necessary for survival. In rare cases, the allele may be due to a mutation in a nonessential gene that changes a protein to function with abnormal and detrimental consequences.

inheritance patterns described in Table 4.1 do not produce a 3:1 phenotypic ratio when two heterozygotes produce offspring. In this section, we consider how allelic interactions produce ratios that differ from a simple Mendelian pattern. However, as our starting point, we will begin by reconsidering a simple dominant/recessive relationship from a molecular perspective.

Recessive Alleles Often Cause a Reduction in the Amount or Function of the Encoded Proteins

For any given gene, geneticists refer to prevalent alleles in a natural population as **wild-type alleles.** In large populations, more than one wild-type allele may occur—a phenomenon known as **genetic polymorphism.** For example, **Figure 4.1** illustrates a striking example of polymorphism in the elderflower orchid, *Dactylorhiza sambucina.* Throughout the range of this species in Europe, both yellow- and red-flowered individuals are prevalent. Both colors are considered wild type. At the molecular level, a wild-type allele typically encodes a protein that is made in the proper amount and functions normally. As discussed in Chapter 24, wild-type alleles tend to promote the reproductive success of organisms in their native environments.

In addition, random mutations occur in populations and alter preexisting alleles. Geneticists sometimes refer to these kinds of alleles as **mutant alleles** to distinguish them from the more common wild-type alleles. Because random mutations are more likely to disrupt gene function, mutant alleles are often defective in their ability to express a functional protein. Such mutant alleles tend to be rare in natural populations. They are typically, but not always, inherited in a recessive fashion.

Among Mendel's seven traits discussed in Chapter 2, the wild-type alleles are tall plants, purple flowers, axial flowers, yellow seeds, round seeds, green pods, and smooth pods (refer back to Figure 2.4). The mutant alleles are dwarf plants, white flowers, terminal flowers, green seeds, wrinkled seeds, yellow pods, and constricted pods. You may have already noticed that the seven wild-type alleles are dominant over the seven mutant alleles. Likewise, red eyes and normal wings are examples of wild-type alleles in *Drosophila*, and white eyes and miniature wings are recessive mutant alleles.

The idea that recessive alleles usually cause a substantial decrease in the expression of a functional protein is supported by the analysis of many human genetic diseases. Keep in mind that a genetic disease is usually caused by a mutant allele. **Table 4.2** lists several examples of human genetic diseases in which the recessive allele fails to produce a specific cellular protein in its active form. In many cases, molecular techniques have enabled researchers to clone these genes and determine the differences between the wild-type and mutant alleles. They have found that the recessive allele usually contains a mutation that causes a defect in the synthesis of a fully functional protein.

To understand why many defective mutant alleles are inherited recessively, we need to take a quantitative look at protein function. With the exception of sex-linked genes, diploid individuals have two copies of every gene. In a simple dominant/recessive relationship, the recessive allele does not affect the phenotype of the heterozygote. In other words, a single copy of the dominant allele is sufficient to mask the effects of the recessive allele. If the recessive allele cannot produce a functional protein, how do we explain the wild-type phenotype of the heterozygote? As described in **Figure 4.2**, a common explanation is that 50% of the functional protein is adequate to provide the wild-type phenotype. In this example, the *PP* homozygote and *Pp* heterozygote

FIGURE 4.1 **An example of genetic polymorphism.** Both yellow and red flowers are common in natural populations of the elderflower orchid, *Dactylorhiza sambucina*, and both are considered wild type.

Dominant (functional) allele: *P* (purple)
Recessive (defective) allele: *p* (white)

Genotype	*PP*	*Pp*	*pp*
Amount of functional protein P	100%	50%	0%

Phenotype	Purple	Purple	White
Simple dominant/recessive relationship			

FIGURE 4.2 **A comparison of protein levels among homozygous and heterozygous genotypes *PP*, *Pp*, and *pp*.**

Genes → Traits In a simple dominant/recessive relationship, 50% of the protein encoded by two copies of the dominant allele is sufficient to produce the wild-type phenotype, in this case, purple flowers. A complete lack of the functional protein results in white flowers.

TABLE **4.2**

Examples of Recessive Human Diseases

Disease	Protein That Is Produced by the Normal Gene*	Description
Phenylketonuria	Phenylalanine hydroxylase	Inability to metabolize phenylalanine. The disease can be prevented by following a phenylalanine-free diet. If the diet is not followed early in life, it can lead to severe mental retardation and physical degeneration.
Albinism	Tyrosinase	Lack of pigmentation in the skin, eyes, and hair.
Tay-Sachs disease	Hexosaminidase A	Defect in lipid metabolism. Leads to paralysis, blindness, and early death.
Sandhoff disease	Hexosaminidase B	Defect in lipid metabolism. Muscle weakness in infancy, early blindness, and progressive mental and motor deterioration.
Cystic fibrosis	Chloride transporter	Inability to regulate ion balance across epithelial cells. Leads to production of thick lung mucus and chronic lung infections.
Lesch-Nyhan syndrome	Hypoxanthine-guanine phosphoribosyl transferase	Inability to metabolize purines, which are bases found in DNA and RNA. Leads to self-mutilation behavior, poor motor skills, and usually mental impairment and kidney failure.

*Individuals who exhibit the disease are either homozygous for a recessive allele or hemizygous (for X-linked genes in human males). The disease symptoms result from a defect in the amount or function of the normal protein.

each make sufficient functional protein to yield purple flowers. This means that the homozygous individual makes twice as much of the wild-type protein than it really needs to produce purple flowers. Therefore, if the amount is reduced to 50%, as in the heterozygote, the individual still has plenty of this protein to accomplish whatever cellular function it performs. The phenomenon that "50% of the normal protein is enough" is fairly common among many genes.

A second possible explanation for other genes is that the heterozygote actually produces more than 50% of the functional protein. Due to the phenomenon of gene regulation, the expression of the normal gene may be increased or "up regulated" in the heterozygote to compensate for the lack of function of the defective allele. The topic of gene regulation is discussed in Chapters 14 and 15.

Incomplete Dominance Occurs When Two Alleles Produce an Intermediate Phenotype

Although many alleles display a simple dominant/recessive relationship, geneticists have also identified some cases in which a heterozygote exhibits **incomplete dominance**—a condition in which the phenotype is intermediate between the corresponding homozygous individuals. In 1905, the German botanist Carl Correns first observed this phenomenon in the four-o'clock (*Mirabilis jalapa*), involving flower color. **Figure 4.3** describes Correns's experiment, in which a homozygous red-flowered four-o'clock plant was crossed to a homozygous white-flowered plant. The wild-type allele for red flower color is designated C^R and the white allele is C^W. As shown here, the offspring had pink flowers. If these F_1 offspring were allowed to self-fertilize, the F_2 generation consisted of 1/4 red-flowered plants, 1/2 pink-flowered plants, and 1/4 white-flowered plants. The pink plants in the F_2 generation were heterozygotes with an intermediate phenotype. As noted in the Punnett square in Figure 4.3, the F_2 generation displayed a 1:2:1 phenotypic ratio, which is different from the 3:1 ratio observed for simple Mendelian inheritance.

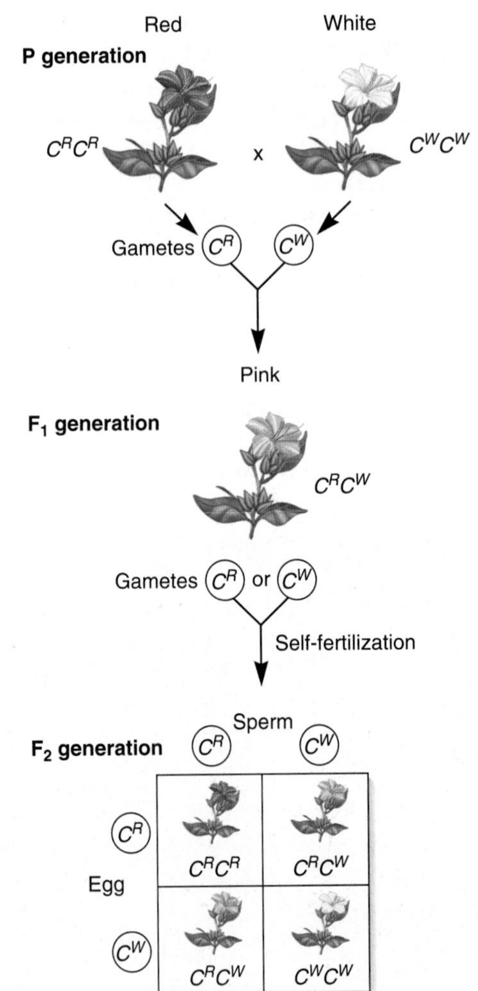

INTERACTIVE EXERCISE

FIGURE 4.3 Incomplete dominance in the four-o'clock plant, *Mirabilis jalapa*.

Genes → Traits When two different homozygotes (C^RC^R and C^WC^W) are crossed, the resulting heterozygote, C^RC^W, has an intermediate phenotype of pink flowers. In this case, 50% of the functional protein encoded by the C^R allele is not sufficient to produce a red phenotype.

In Figure 4.3, incomplete dominance resulted in a heterozygote with an intermediate phenotype. At the molecular level, the allele that causes a white phenotype is expected to result in a lack of a functional protein required for pigmentation. Depending on the effects of gene regulation, the heterozygotes may produce only 50% of the normal protein, but this amount is not sufficient to create the same phenotype as the C^RC^R homozygote, which may make twice as much of this protein. In this example, a reasonable explanation is that 50% of the functional protein cannot accomplish the same level of pigment synthesis that 100% of the protein can.

Finally, our opinion of whether a trait is dominant or incompletely dominant may depend on how closely we examine the trait in the individual. The more closely we look, the more likely we are to discover that the heterozygote is not quite the same as the wild-type homozygote. For example, Mendel studied the characteristic of pea seed shape and visually concluded that the *RR* and *Rr* genotypes produced round seeds and the *rr* genotype produced wrinkled seeds. The peculiar morphology of the wrinkled seed is caused by a large decrease in the amount of starch deposition in the seed due to a defective *r* allele. Since the time of Mendel's work, other scientists have dissected round and wrinkled seeds and examined their contents under the microscope. They have found that round seeds from heterozygotes actually contain an intermediate number of starch grains compared to seeds from the corresponding homozygotes (**Figure 4.4**). Within the seed, an intermediate amount of the functional protein is not enough to produce as many starch grains as in the homozygote carrying two copies of the *R* allele. Nevertheless, at the level of our unaided eyes, heterozygotes produce seeds that appear to be round. With regard to phenotypes, the *R* allele is dominant to the *r* allele at the level of visual examination, but the *R* and *r* alleles show incomplete dominance at the level of starch biosynthesis.

terminal flowers, green seeds, wrinkled seeds, yellow pods, and constricted pods. You may have already noticed that the seven wild-type alleles are dominant over the seven mutant alleles. Likewise, red eyes and normal wings are examples of wild-type alleles in *Drosophila*, and white eyes and miniature wings are recessive mutant alleles.

The idea that recessive alleles usually cause a substantial decrease in the expression of a functional protein is supported by the analysis of many human genetic diseases. Keep in mind that a genetic disease is usually caused by a mutant allele. **Table 4.2** lists several examples of human genetic diseases in which the recessive allele fails to produce a specific cellular protein in its active form. In many cases, molecular techniques have enabled researchers to clone these genes and determine the differences between the wild-type and mutant alleles. They have found that the recessive allele usually contains a mutation that causes a defect in the synthesis of a fully functional protein.

To understand why many defective mutant alleles are inherited recessively, we need to take a quantitative look at protein function. With the exception of sex-linked genes, diploid individuals have two copies of every gene. In a simple dominant/recessive relationship, the recessive allele does not affect the phenotype of the heterozygote. In other words, a single copy of the dominant allele is sufficient to mask the effects of the recessive allele. If the recessive allele cannot produce a functional protein, how do we explain the wild-type phenotype of the heterozygote? As described in **Figure 4.2**, a common explanation is that 50% of the functional protein is adequate to provide the wild-type phenotype. In this example, the *PP* homozygote and *Pp* heterozygote

FIGURE 4.1 An example of genetic polymorphism. Both yellow and red flowers are common in natural populations of the elderflower orchid, *Dactylorhiza sambucina*, and both are considered wild type.

Dominant (functional) allele: *P* (purple)
Recessive (defective) allele: *p* (white)

Genotype	*PP*	*Pp*	*pp*
Amount of functional protein P	100%	50%	0%
Phenotype	Purple	Purple	White
Simple dominant/recessive relationship			

FIGURE 4.2 A comparison of protein levels among homozygous and heterozygous genotypes *PP*, *Pp*, and *pp*.
Genes → Traits In a simple dominant/recessive relationship, 50% of the protein encoded by two copies of the dominant allele is sufficient to produce the wild-type phenotype, in this case, purple flowers. A complete lack of the functional protein results in white flowers.

TABLE **4.2**

Examples of Recessive Human Diseases

Disease	Protein That Is Produced by the Normal Gene*	Description
Phenylketonuria	Phenylalanine hydroxylase	Inability to metabolize phenylalanine. The disease can be prevented by following a phenylalanine-free diet. If the diet is not followed early in life, it can lead to severe mental retardation and physical degeneration.
Albinism	Tyrosinase	Lack of pigmentation in the skin, eyes, and hair.
Tay-Sachs disease	Hexosaminidase A	Defect in lipid metabolism. Leads to paralysis, blindness, and early death.
Sandhoff disease	Hexosaminidase B	Defect in lipid metabolism. Muscle weakness in infancy, early blindness, and progressive mental and motor deterioration.
Cystic fibrosis	Chloride transporter	Inability to regulate ion balance across epithelial cells. Leads to production of thick lung mucus and chronic lung infections.
Lesch-Nyhan syndrome	Hypoxanthine-guanine phosphoribosyl transferase	Inability to metabolize purines, which are bases found in DNA and RNA. Leads to self-mutilation behavior, poor motor skills, and usually mental impairment and kidney failure.

*Individuals who exhibit the disease are either homozygous for a recessive allele or hemizygous (for X-linked genes in human males). The disease symptoms result from a defect in the amount or function of the normal protein.

each make sufficient functional protein to yield purple flowers. This means that the homozygous individual makes twice as much of the wild-type protein than it really needs to produce purple flowers. Therefore, if the amount is reduced to 50%, as in the heterozygote, the individual still has plenty of this protein to accomplish whatever cellular function it performs. The phenomenon that "50% of the normal protein is enough" is fairly common among many genes.

A second possible explanation for other genes is that the heterozygote actually produces more than 50% of the functional protein. Due to the phenomenon of gene regulation, the expression of the normal gene may be increased or "up regulated" in the heterozygote to compensate for the lack of function of the defective allele. The topic of gene regulation is discussed in Chapters 14 and 15.

Incomplete Dominance Occurs When Two Alleles Produce an Intermediate Phenotype

Although many alleles display a simple dominant/recessive relationship, geneticists have also identified some cases in which a heterozygote exhibits **incomplete dominance**—a condition in which the phenotype is intermediate between the corresponding homozygous individuals. In 1905, the German botanist Carl Correns first observed this phenomenon in the four-o'clock (*Mirabilis jalapa*), involving flower color. **Figure 4.3** describes Correns's experiment, in which a homozygous red-flowered four-o'clock plant was crossed to a homozygous white-flowered plant. The wild-type allele for red flower color is designated C^R and the white allele is C^W. As shown here, the offspring had pink flowers. If these F_1 offspring were allowed to self-fertilize, the F_2 generation consisted of 1/4 red-flowered plants, 1/2 pink-flowered plants, and 1/4 white-flowered plants. The pink plants in the F_2 generation were heterozygotes with an intermediate phenotype. As noted in the Punnett square in Figure 4.3, the F_2 generation displayed a 1:2:1 phenotypic ratio, which is different from the 3:1 ratio observed for simple Mendelian inheritance.

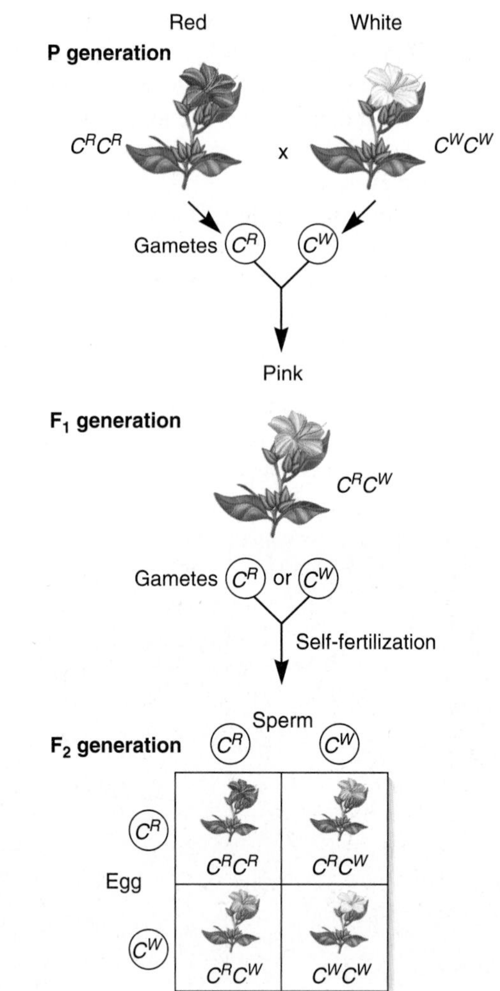

INTERACTIVE EXERCISE **FIGURE 4.3 Incomplete dominance in the four-o'clock plant, *Mirabilis jalapa*.**

Genes →Traits When two different homozygotes (C^RC^R and C^WC^W) are crossed, the resulting heterozygote, C^RC^W, has an intermediate phenotype of pink flowers. In this case, 50% of the functional protein encoded by the C^R allele is not sufficient to produce a red phenotype.

In Figure 4.3, incomplete dominance resulted in a heterozygote with an intermediate phenotype. At the molecular level, the allele that causes a white phenotype is expected to result in a lack of a functional protein required for pigmentation. Depending on the effects of gene regulation, the heterozygotes may produce only 50% of the normal protein, but this amount is not sufficient to create the same phenotype as the C^RC^R homozygote, which may make twice as much of this protein. In this example, a reasonable explanation is that 50% of the functional protein cannot accomplish the same level of pigment synthesis that 100% of the protein can.

Finally, our opinion of whether a trait is dominant or incompletely dominant may depend on how closely we examine the trait in the individual. The more closely we look, the more likely we are to discover that the heterozygote is not quite the same as the wild-type homozygote. For example, Mendel studied the characteristic of pea seed shape and visually concluded that the *RR* and *Rr* genotypes produced round seeds and the *rr* genotype produced wrinkled seeds. The peculiar morphology of the wrinkled seed is caused by a large decrease in the amount of starch deposition in the seed due to a defective *r* allele. Since the time of Mendel's work, other scientists have dissected round and wrinkled seeds and examined their contents under the microscope. They have found that round seeds from heterozygotes actually contain an intermediate number of starch grains compared to seeds from the corresponding homozygotes (**Figure 4.4**). Within the seed, an intermediate amount of the functional protein is not enough to produce as many starch grains as in the homozygote carrying two copies of the *R* allele. Nevertheless, at the level of our unaided eyes, heterozygotes produce seeds that appear to be round. With regard to phenotypes, the *R* allele is dominant to the *r* allele at the level of visual examination, but the *R* and *r* alleles show incomplete dominance at the level of starch biosynthesis.

Dominant (functional) allele: *R* (round)
Recessive (defective) allele: *r* (wrinkled)

Genotype	*RR*	*Rr*	*rr*
Amount of functional (starch-producing) protein	100%	50%	0%

Phenotype	Round	Round	Wrinkled
With unaided eye (simple dominant/ recessive relationship)			
With microscope (incomplete dominance)			

FIGURE 4.4 A comparison of phenotype at the macroscopic and microscopic levels.

Genes → Traits This illustration shows the effects of a heterozygote having only 50% of the functional protein needed for starch production. This seed appears to be as round as those of the homozygote carrying the *R* allele, but when examined microscopically, it has produced only half the amount of starch.

Traits May Skip a Generation Due to Incomplete Penetrance and Vary in Their Expressivity

As we have seen, dominant alleles are expected to influence the outcome of a trait when they are present in heterozygotes. Occasionally, however, this may not occur. **Figure 4.5a** illustrates a human pedigree for a dominant trait known as polydactyly. This trait causes the affected individual to have additional fingers and/or toes (**Figure 4.5b**). Polydactyly is due to an autosomal dominant allele—the allele is found in a gene located on an autosome (not a sex chromosome) and a single copy of this allele is sufficient to cause this condition. Sometimes, however, individuals carry the dominant allele but do not exhibit the trait. In Figure 4.5a, individual III-2 has inherited the polydactyly allele from his mother and passed the allele to a daughter and son. However, individual III-2 does not actually exhibit the trait himself, even though he is a heterozygote. This phenomenon, called **incomplete penetrance,** indicates that a dominant allele does not always "penetrate" into the phenotype of the individual. The measure of penetrance is described at the populational level. For example, if 60% of the heterozygotes carrying a dominant allele exhibit the trait, we would say that this trait is 60% penetrant. At the individual level, the trait is either present or not.

Another term used to describe the outcome of traits is the degree to which the trait is expressed, or its **expressivity.** In the case of polydactyly, the number of extra digits can vary. For example, one individual may have an extra toe on only one foot, whereas a second individual may have extra digits on both the hands and feet. Using genetic terminology, a person with several extra digits would have high expressivity of this trait, while a person with a single extra digit would have low expressivity. The extreme case, when a heterozygote does not express the trait at all, is incomplete penetrance.

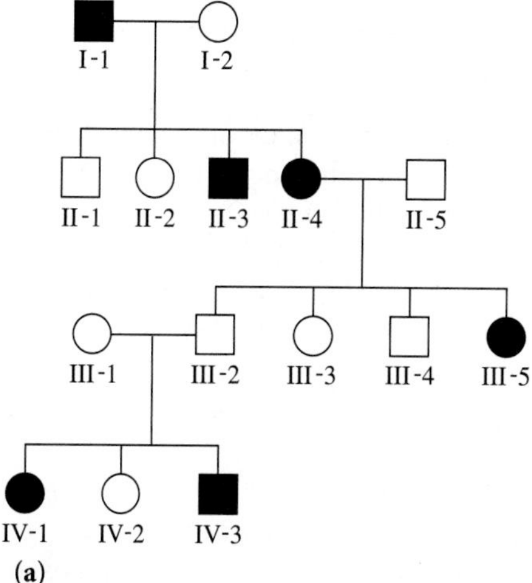

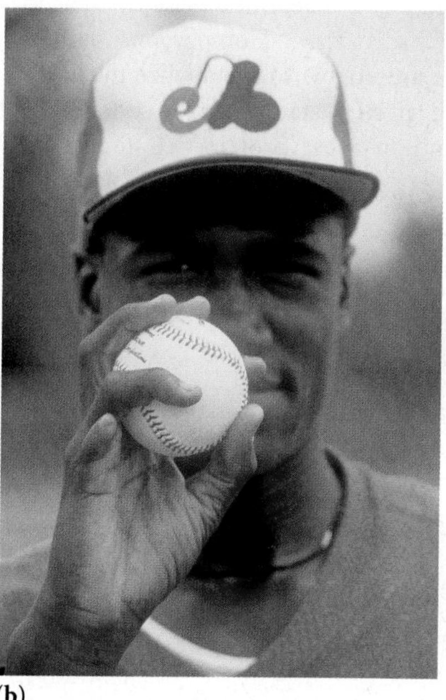

FIGURE 4.5 Polydactyly, a dominant trait that shows incomplete penetrance. **(a)** A family pedigree. Affected individuals are shown in black. Notice that offspring IV-1 and IV-3 have inherited the trait from a parent, III-2, who is heterozygous but does not exhibit polydactyly. **(b)** Antonio Alfonseca, a baseball player with polydactyly. His extra finger does not give him an advantage when pitching because it is small and does not touch the ball.

How do we explain incomplete penetrance and variable expressivity? While the answer may not always be understood, the range of phenotypes is often due to environmental influences and/or due to effects of a modifier gene in which one gene modifies the phenotypic effects of another gene. We will consider the issue of the environment next. The effects of modifier genes will be discussed later in the chapter.

The Outcome of Traits Is Influenced by the Environment

Throughout this book, our study of genetics tends to focus on the roles of genes in the outcome of traits. In addition to genetics, environmental conditions have a great impact on the phenotype of the individual. For example, the arctic fox (*Alopex lagopus*) goes through two color phases. During the cold winter, the arctic fox is primarily white, while in the warmer summer, it is mostly brown (**Figure 4.6a**). As discussed later, such temperature-sensitive alleles affecting fur color are found among many species of mammals.

A dramatic example of the relationship between environment and phenotype can be seen in the human genetic disease known as phenylketonuria (PKU). This autosomal recessive disease is caused by a defect in a gene that encodes the enzyme phenylalanine hydroxylase. Homozygous individuals with this defective allele are unable to metabolize the amino acid phenylalanine. When given a standard diet containing phenylalanine, which is found in most protein-rich foods, PKU individuals manifest a variety of detrimental traits including mental retardation, underdeveloped teeth, and foul-smelling urine. In contrast, when PKU individuals are diagnosed early and follow a restricted diet free of phenylalanine, they develop normally (**Figure 4.6b**). Since the 1960s, testing methods have been developed that can determine if an individual is lacking the phenylalanine hydroxylase enzyme. These tests permit the identification of infants who have PKU. Their diets can then be modified before the harmful effects of phenylalanine ingestion have occurred. As a result of government legislation, more than 90% of infants born in the United States are now tested for PKU. This test both prevents a great deal of human suffering and is cost-effective. In the United States, the annual cost of PKU testing is estimated to be a few million dollars, whereas the cost of treating severely retarded individuals with the disease would be hundreds of millions of dollars.

The examples of the arctic fox and PKU represent dramatic effects of very different environmental conditions. When considering the environment, geneticists often examine a range of conditions, rather than simply observing phenotypes under two different conditions. The term **norm of reaction** refers to the effects of environmental variation on a phenotype. Specifically, it is the

(a) Arctic fox in winter and summer

(b) Healthy person with PKU

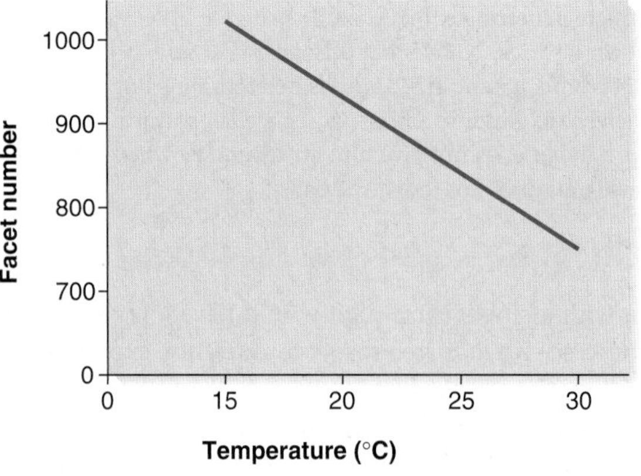

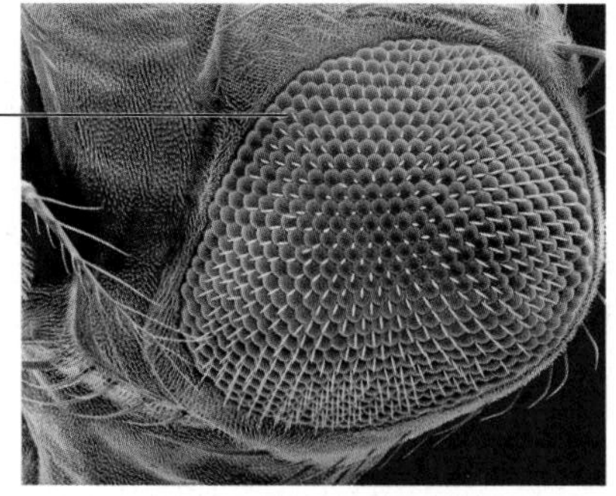

(c) Norm of reaction

FIGURE 4.6 **Variation in the expression of traits due to environmental effects.** (**a**) The arctic fox in the winter and summer. (**b**) A person with PKU who has followed a restricted diet and developed normally. (**c**) Norm of reaction. In this experiment, fertilized eggs from a population of genetically identical *Drosophila melanogaster* were allowed to develop into adult flies at different environmental temperatures. This graph shows the relationship between temperature (an environmental factor) and facet number in the eyes of the resulting adult flies. The micrograph shows an eye of *D. melanogaster.*

phenotypic range seen in individuals with a particular genotype. To evaluate the norm of reaction, researchers study members of true-breeding strains that have the same genotypes and subject them to different environmental conditions. As an example, let's consider facet number in the eyes of fruit flies, *Drosophila melanogaster.* This species has compound eyes composed of many individual facets. **Figure 4.6c** shows the norm of reaction for facet number in genetically identical fruit flies that developed at different temperatures. As shown in the figure, the facet number varies with changes in temperature. At a higher temperature (30°C), the facet number is approximately 750, while at a lower temperature (15°C), it is over 1,000.

Overdominance Occurs When Heterozygotes Have Superior Traits

As we have just seen, the environment plays a key role in the outcome of traits. For certain genes, heterozygotes may display characteristics that are more beneficial for their survival in a particular environment. Such heterozygotes may be more likely to survive and reproduce. For example, a heterozygote may be larger, disease resistant, or better able to withstand harsh environmental conditions. The phenomenon in which a heterozygote has greater reproductive success compared to either of the corresponding homozygotes is called **overdominance** or **heterozygote advantage.**

A well-documented example involves a human allele that causes sickle-cell disease in homozygous individuals. This disease is an autosomal recessive disorder in which the affected individual produces an altered form of the protein hemoglobin, which carries oxygen within red blood cells. Most people carry the Hb^A allele and make hemoglobin A. Individuals affected with sickle-cell anemia are homozygous for the Hb^S allele and produce only hemoglobin S. This causes their red blood cells to deform into a sickle shape under conditions of low oxygen concentration (**Figure 4.7a, b**). The sickling phenomenon causes the life span of these cells to be greatly shortened to only a few weeks compared with a normal span of four months, and therefore, anemia results. In addition, abnormal sickled cells can become clogged in the capillaries throughout the body, leading to localized areas of oxygen depletion. Such an event, called a crisis, causes pain and sometimes tissue and organ damage. For these reasons, the homozygous Hb^SHb^S individual usually has a shortened life span compared to an individual producing hemoglobin A.

In spite of the harmful consequences to homozygotes, the sickle-cell allele has been found at a fairly high frequency among human populations that are exposed to malaria. The protozoan genus that causes malaria, *Plasmodium,* spends part of its life cycle within the *Anopheles* mosquito and another part within the red blood cells of humans who have been bitten by an infected mosquito. However, red blood cells of heterozygotes, Hb^AHb^S, are likely to rupture when infected by this parasite, thereby preventing the parasite from propagating. People who are heterozygous have better resistance to malaria compared to Hb^AHb^A homozygotes, while not incurring the ill effects of sickle-cell disease. Therefore, even though the homozygous Hb^SHb^S condition is detrimental, the greater survival of the heterozygote has selected for the presence of the Hb^S allele within populations where malaria is prevalent. When viewing survival in such a region, overdominance explains the prevalence of the sickle-cell allele. In Chapter 24, we will consider the role that natural selection plays in maintaining alleles that are beneficial to the heterozygote but harmful to the homozygote.

Figure 4.7c illustrates the predicted outcome when two heterozygotes have children. In this example, 1/4 of the offspring are Hb^AHb^A (unaffected, not malaria resistant), 1/2 are Hb^AHb^S

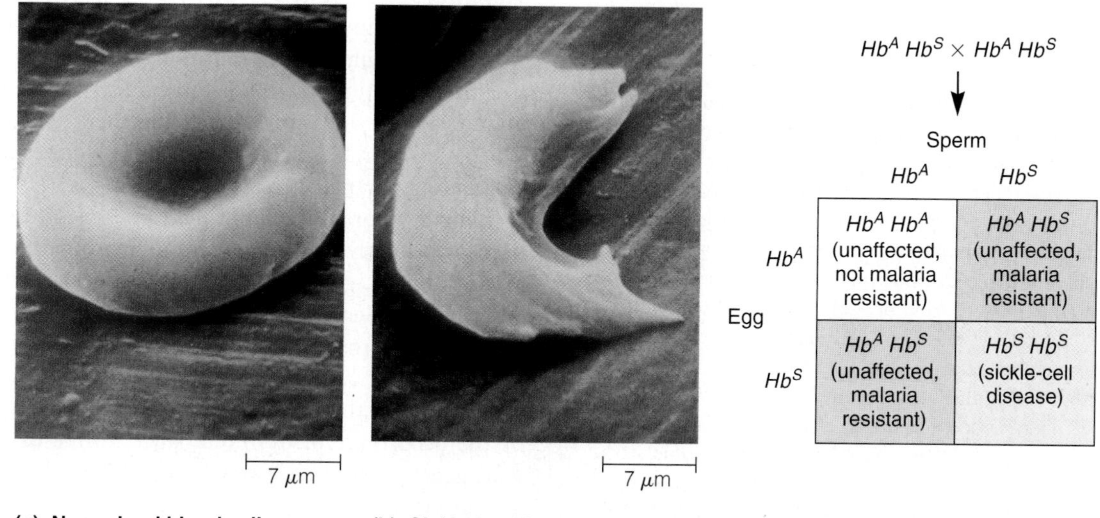

(a) Normal red blood cell **(b) Sickled red blood cell** **(c) Example of sickle-cell inheritance pattern**

FIGURE 4.7 **Inheritance of sickle-cell disease.** A comparison of (**a**) normal red blood cells and (**b**) those from a person with sickle-cell disease. (**c**) The outcome of a cross between two heterozygous individuals.

(unaffected, malaria resistant) and 1/4 are Hb^SHb^S (sickle-cell disease). This 1:2:1 ratio deviates from a simple Mendelian 3:1 phenotypic ratio.

Overdominance is usually due to two alleles that produce proteins with slightly different amino acid sequences. How can we explain the observation that two protein variants in the heterozygote produce a more favorable phenotype? There are three common explanations. In the case of sickle-cell disease, the phenotype is related to the infectivity of *Plasmodium* (**Figure 4.8a**). In the heterozygote, the infectious agent is less likely to propagate within red blood cells. Interestingly, researchers have speculated that other alleles in humans may confer disease resistance in the heterozygous condition but are detrimental in the homozygous state. These include PKU, in which the heterozygous fetus may be resistant to miscarriage caused by a fungal toxin, and Tay-Sachs disease, in which the heterozygote may be resistant to tuberculosis.

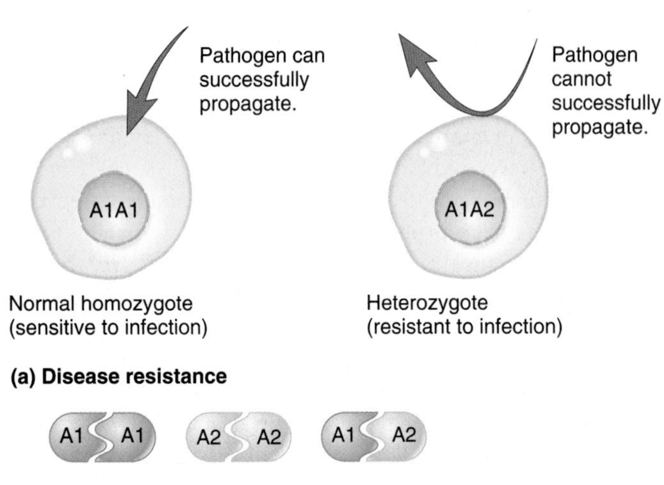

(a) Disease resistance

A1 A1 A2 A2 A1 A2

(b) Homodimer formation

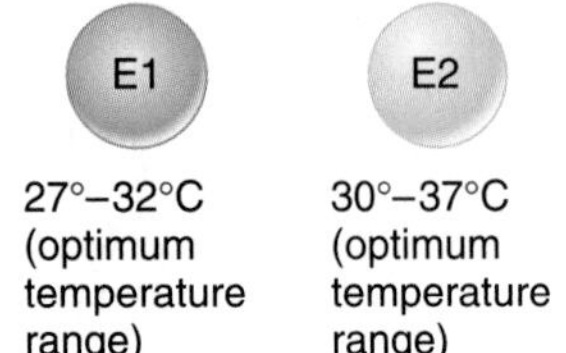

E2
30°–37°C (optimum temperature range)

(c) Variation in functional activity

FIGURE 4.8 Three possible explanations for overdominance at the molecular level. (**a**) The successful infection of cells by certain microorganisms depends on the function of particular cellular proteins. In this example, functional differences between A1A1 and A1A2 proteins affect the ability of a pathogen to propagate in the cells. (**b**) Some proteins function as homodimers. In this example, a gene exists in two alleles designated *A1* and *A2*, which encode polypeptides also designated A1 and A2. The homozygotes that are *A1A1* or *A2A2* will make homodimers that are A1A1 and A2A2, respectively. The *A1A2* heterozygote can make A1A1 and A2A2 and can also make A1A2 homodimers, which may have better functional activity. (**c**) In this example, a gene exists in two alleles designated *E1* and *E2*. The *E1* allele encodes an enzyme that functions well in the temperature range of 27° to 32°C. *E2* encodes an enzyme that functions in the range of 30° to 37°C. A heterozygote, *E1E2*, would produce both enzymes and have a broader temperature range (i.e., 27° to 37°C) in which the enzyme would function.

A second way to explain overdominance is related to the subunit composition of proteins. In some cases, proteins function as a complex of multiple subunits; each subunit is composed of one polypeptide. A protein composed of two subunits is called a dimer. When both subunits are encoded by the same gene, the protein is a homodimer. The prefix homo- means that the subunits come from the same type of gene although the gene may exist in different alleles. **Figure 4.8b** considers a situation in which a gene exists in two alleles that encode polypeptides designated A1 and A2. Homozygous individuals can produce only A1A1 or A2A2 homodimers, whereas a heterozygote can also produce an A1A2 homodimer. Thus, heterozygotes can produce three forms of the homodimer, homozygotes only one. For some proteins, A1A2 homodimers may have better functional activity because they are more stable or able to function under a wider range of conditions. The greater activity of the homodimer protein may be the underlying reason why a heterozygote has characteristics superior to either homozygote.

A third molecular explanation of overdominance is that the proteins encoded by each allele exhibit differences in their functional activity. For example, suppose that a gene encodes a metabolic enzyme that can be found in two forms (corresponding to the two alleles), one that functions better at a lower temperature and the other that functions optimally at a higher temperature (**Figure 4.8c**). The heterozygote, which makes a mixture of both enzymes, may be at an advantage under a wider temperature range than either of the corresponding homozygotes.

Before ending this topic, let's also compare overdominance with a related phenomenon. Among plant and animal breeders, a common mating strategy is to begin with two different highly inbred strains and cross them together to produce hybrids. When the hybrids display traits that are superior to both corresponding parental strains, the outcome is known as **heterosis,** or **hybrid vigor.** Within the field of agriculture, heterosis has been particularly valuable in improving quantitative traits such as size, weight, growth rate, and disease resistance. It was first described by George Shull in crosses involving different strains of corn. Heterosis differs from overdominance, because the hybrid may be heterozygous for many genes, not just a single gene. Some of the beneficial effects of heterosis may be caused by the occurrence of overdominance in one or more heterozygous genes. However, as we will see in Chapter 25, heterosis can also result from the masking of deleterious recessive alleles that tend to accumulate in highly inbred domesticated strains.

Many Genes Exist as Three or More Different Alleles

Thus far, we have considered examples in which a gene exists in two different alleles. As researchers have probed genes at the molecular level within natural populations of organisms, they have discovered that most genes exist in **multiple alleles.** Within a population, genes are typically found in three or more alleles.

An interesting example of multiple alleles involves coat color in rabbits. Figure 4.9 illustrates the relationship between genotype and phenotype for a combination of four different alleles, which are designated *C* (full coat color), c^{ch} (chinchilla pattern of coat color), c^h (himalayan pattern of coat color), and *c* (albino). In this case, the gene encodes an enzyme called tyrosinase, which is the first enzyme in a metabolic pathway that leads to the synthesis of melanin from the amino acid tyrosine. This pathway results in the formation of two forms of melanin. Eumelanin, a black pigment, is made first, and then phaeomelanin, an orange/yellow pigment, is made from eumelanin. Alleles of other genes can also influence the relative amounts of eumelanin and phaeomelanin.

Differences in the various alleles are related to the function of tyrosinase. The *C* allele encodes a fully functional tyrosinase that allows the synthesis of both eumelanin and phaeomelanin, resulting in a full brown coat color. The *C* allele is dominant to the other three alleles. The chinchilla allele (c^{ch}) is a partial defect in tyrosinase that leads to a slight reduction in black pigment and a greatly diminished amount of orange/yellow pigment, which makes the animal look more gray. The albino allele, designated *c*, is a complete loss of tyrosinase, resulting in white color. The himalayan pattern of coat color, determined by the c^h allele, is an example of a **temperature-sensitive allele.** The mutation in this gene has caused a change in the structure of tyrosinase, so it only works enzymatically at low temperature. Because of this property, the enzyme functions only in cooler regions of the body, primarily the tail, the paws, and the tips of the nose and ears. As shown in Figure 4.10, similar types of temperature-sensitive alleles have been found in other species of domestic animals, such as the Siamese cat.

(a) Full coat color *CC*, Cc^{ch}, Cc^h, or *Cc*.

(b) Chinchilla coat color $c^{ch}c^{ch}$, $c^{ch}c^h$, or $c^{ch}c$.

(c) Himalayan coat color c^hc^h or c^hc.

(d) Albino coat color *cc*.

FIGURE 4.9 **The relationship between genotype and phenotype in rabbit coat color.**

Alleles of the ABO Blood Group Can Be Dominant, Recessive, or Codominant

The ABO group of antigens, which determine blood type in humans, is another example of multiple alleles and illustrates yet another allelic relationship called codominance. To understand this concept, we first need to examine the molecular characteristics of human blood types. The plasma membranes of red blood cells have groups of interconnected sugars—carbohydrate

FIGURE 4.10 **The expression of a temperature-sensitive conditional allele produces a Siamese pattern of coat color.**

Genes → Traits The allele affecting fur pigmentation encodes a pigment-producing protein that functions only at lower temperatures. For this reason, the dark fur is produced only in the cooler parts of the animal, including the tips of the ears, nose, paws, and tail.

trees—that act as surface antigens (**Figure 4.11a**). Antigens are molecular structures that are recognized by antibodies produced by the immune system. On red blood cells, three different types of surface antigens, known as A, B, and O, may be found.

The synthesis of these surface antigens is controlled by three alleles, designated I^A, I^B, and *i*, respectively. The *i* allele is recessive to both I^A and I^B. A person who is homozygous *ii* will have type O blood. A homozygous I^AI^A or heterozygous I^Ai individual will have type A blood. The red blood cells of this individual will contain the surface antigen known as A. Similarly, a homozygous I^BI^B or heterozygous I^Bi individual will produce the surface antigen B. As Figure 4.11a indicates, surface antigens A and B have significantly different molecular structures. A person who is I^AI^B will have the blood type AB and express both surface antigens A and B. The phenomenon in which two alleles are both expressed in the heterozygous individual is called **codominance.** In this case, the I^A and I^B alleles are codominant to each other.

As an example of the inheritance of blood type, let's consider the possible offspring between two parents who are I^Ai and I^Bi (**Figure 4.11b**). The I^Ai parent makes I^A and *i* gametes, while the I^Bi parent makes I^B and *i* gametes. These combine to produce I^AI^B, I^Ai, I^Bi, and *ii* offspring in a 1:1:1:1 ratio. The resulting blood types are AB, A, B, and O, respectively.

Biochemists have analyzed the carbohydrate trees produced on the surfaces of cells of differing blood types. In type O, the tree is smaller than type A or type B because a sugar has not been attached to a specific site on the tree. This idea is schematically shown in Figure 4.11a. How do we explain this difference at the molecular level? The gene that determines ABO blood type encodes an enzyme called glycosyl transferase that attaches a sugar to the carbohydrate tree. The *i* allele carries a mutation that renders this enzyme inactive, which prevents the attachment of an additional sugar. By comparison, the two types of glycosyl transferase encoded by the I^A and I^B alleles have different structures in their active sites. The active site is the part of the protein that recognizes the sugar molecule that will be attached to the carbohydrate tree. The glycosyl transferase encoded by the I^A allele recognizes UDP-GalNAc and attaches GalNAc

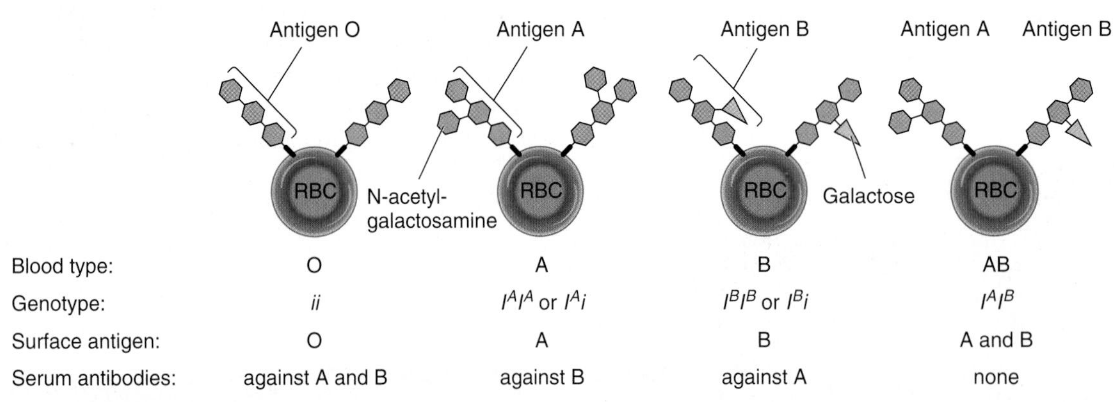

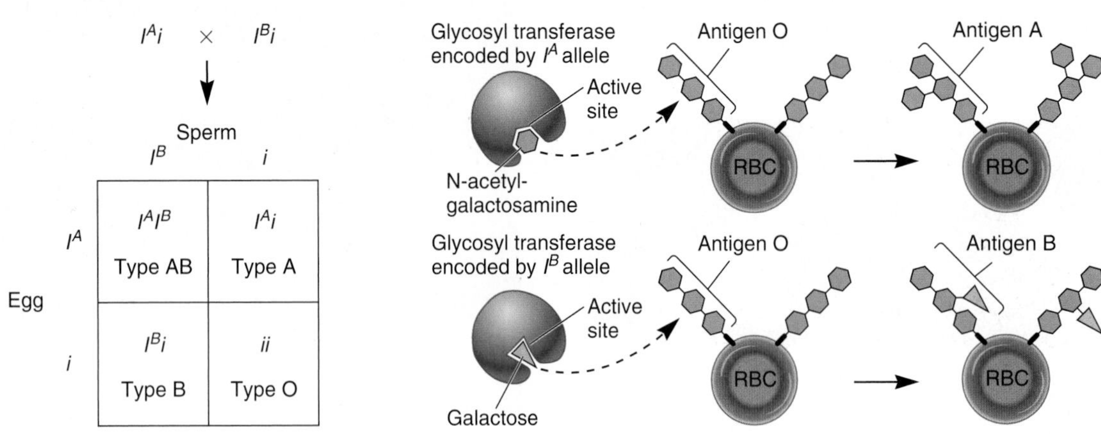

FIGURE 4.11 **ABO blood type.** (**a**) A schematic representation of blood type at the cellular level. (**b**) The predicted offspring from parents who are I^Ai and I^Bi. (**c**) The glycosyl transferase encoded by the I^A and I^B alleles recognizes different sugars due to changes in its active site. The *i* allele results in a nonfunctional enzyme.

(i.e., N-acetylgalactosamine) to the carbohydrate tree (**Figure 4.11c**). N-acetylgalactosamine is symbolized as a green hexagon. This creates the structure of surface antigen A. In contrast, the glycosyl transferase encoded by the I^B allele recognizes UDP-galactose and attaches galactose to the carbohydrate tree. Galactose is symbolized as an orange triangle. This creates the molecular structure of surface antigen B. A person with type AB blood makes both types of enzymes and thereby has a tree with both types of sugar attached.

A small difference in the structure of the carbohydrate tree, namely, an N-acetylgalactosamine in antigen A versus galactose in antigen B, explains why the two antigens are different from each other at the molecular level. These differences enable them to be recognized by different antibodies. A person who has blood type A makes antibodies to blood type B (refer back to Figure 4.11a). The antibodies against blood type B require a galactose in the carbohydrate tree for their proper recognition. Their antibodies will not recognize and destroy their own blood cells, but they will recognize and destroy the blood cells from a type B person.

With this in mind, let's consider why blood typing is essential for safe blood transfusions. The donor's blood must be an appropriate match with the recipient's blood. A person with type O blood naturally produces antibodies against both A and B antigens. If a person with type O blood is given type A, type B, or type AB blood, the antibodies in the recipient will react with the donated blood cells and cause them to agglutinate (clump together). This is a life-threatening situation that causes the blood vessels to clog. Other incompatible combinations include a type A person receiving type B or type AB blood, and a type B person receiving type A or type AB blood. Because individuals with type AB blood do not produce antibodies to either A or B antigens, they can receive any type of blood and are known as universal recipients. By comparison, type O persons are universal donors because their blood can be given to type O, A, B, and AB people.

The Inheritance Pattern of X-Linked Genes Can Be Revealed by Reciprocal Crosses

Let's now turn our attention to inheritance patterns of single genes in which the sexes of the parents and offspring play a critical role. As discussed in Chapter 3, many species have males and females that differ in their sex chromosome composition. In mammals, for example, females are XX and males are XY. In such species, certain traits are governed by genes that are located on a sex chromosome. For these traits, the outcome of crosses depends on the genotypes and sexes of the parents and offspring.

As an example, let's consider a human disease known as Duchenne muscular dystrophy (DMD), which was first described by the French neurologist Guillaume Duchenne in the 1860s. Affected individuals show signs of muscle weakness as early as age three. The disease gradually weakens the skeletal muscles and eventually affects the heart and breathing muscles. Survival is rare beyond the early 30s. The gene for DMD, found on the X chromosome, encodes a protein called dystrophin that is required inside muscle cells for structural support. Dystrophin is thought to strengthen muscle cells by anchoring elements of the internal cytoskeleton to the plasma membrane. Without it, the plasma membrane becomes permeable and may rupture.

Duchenne muscular dystrophy is inherited in an **X-linked recessive** pattern—the allele causing the disease is recessive and located on the X chromosome. In the pedigree shown in **Figure 4.12**, several males are affected by this disorder, as indicated by filled squares. The mothers of these males are presumed heterozygotes for this X-linked recessive allele. This recessive disorder is very rare among females because daughters would have to inherit a copy of the mutant allele from their mother and a copy from an affected father.

X-linked muscular dystrophy has also been found in certain breeds of dogs such as golden retrievers (**Figure 4.13a**). Like humans, the mutation occurs in the dystrophin gene, and the symptoms include severe weakness and muscle atrophy that begin at about six to eight weeks of age. Many dogs that inherit this disorder die within the first year of life, though some can live three to five years and reproduce.

Figure 4.13b (left side) considers a cross between an unaffected female dog with two copies of the wild-type gene and a male dog with muscular dystrophy that carries the mutant allele. When setting up a Punnett square involving X-linked traits, we must consider the alleles on the X chromosome as well as the fact that males may transmit a Y chromosome instead of the X chromosome. The male makes two types of gametes, one that carries the X chromosome and one that carries the Y. The Punnett square must also include the Y chromosome even though this chromosome does not carry any X-linked genes. The X chromosomes from the female and male are designated with their corresponding alleles. When the Punnett square is filled in, it predicts the X-linked genotypes and sexes of the offspring. As seen on the left side of Figure 4.13b, none of the offspring from this cross are affected with the disorder, although all female offspring are carriers.

The right side of Figure 4.13b shows a **reciprocal cross**—a second cross in which the sexes and phenotypes are reversed. In this case, an affected female animal is crossed to an unaffected

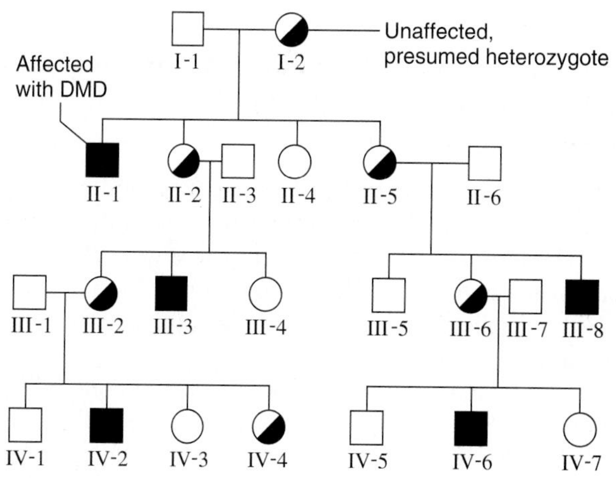

FIGURE 4.12 A human pedigree for Duchenne muscular dystrophy, an X-linked recessive trait. Affected individuals are shown with filled symbols. Females who are unaffected with the disease, but are heterozygous carriers, are shown with half-filled symbols.

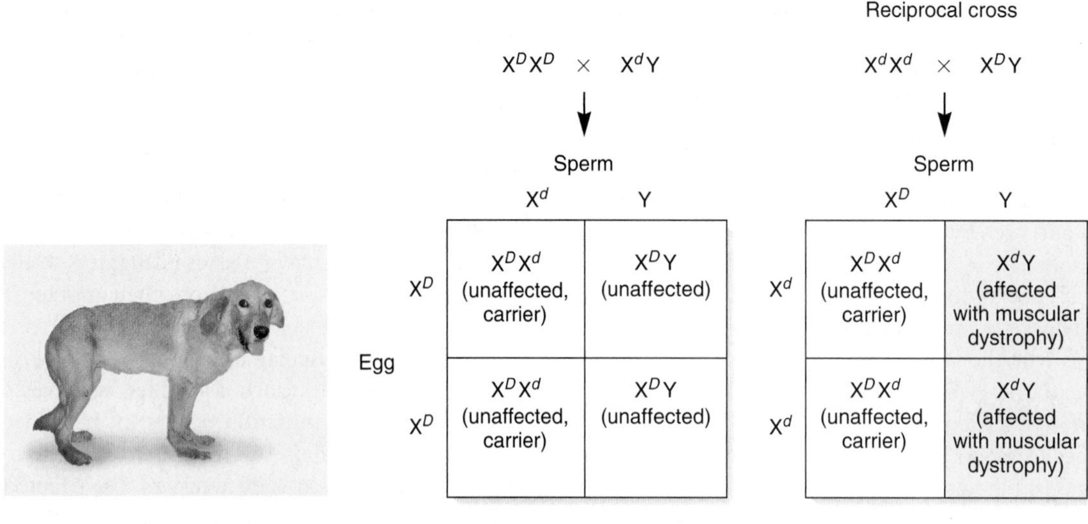

FIGURE 4.13 **X-linked muscular dystrophy in dogs.** (a) The golden retriever shown here is a male that has the disease. (b) The left side shows a cross between an unaffected female and an affected male. The right shows a reciprocal cross between an affected female and an unaffected male. *D* represents the normal allele for the dystrophin gene, and *d* is the mutant allele that causes a defect in dystrophin function.

male. This cross produces female offspring that are carriers and all male offspring will be affected with muscular dystrophy.

When comparing the two Punnett squares, the outcome of the reciprocal cross yielded different results. This is expected of X-linked genes, because the male transmits the gene only to female offspring, while the female transmits an X chromosome to both male and female offspring. Because the male parent does not transmit the X chromosome to his sons, he does not contribute to their X-linked phenotypes. This explains why X-linked traits do not behave equally in reciprocal crosses. Experimentally, the observation that reciprocal crosses do not yield the same results is an important clue that a trait may be X linked.

Genes Located on Mammalian Sex Chromosomes Can Be Transmitted in an X-Linked, a Y-Linked, or a Pseudoautosomal Pattern

Our discussion of sex chromosomes has concentrated on genes that are located on the X chromosome but not on the Y chromosome. The term **sex-linked gene** refers to a gene that is found on one of the two types of sex chromosomes but not on both. Hundreds of X-linked genes have been identified in humans and other mammals.

The inheritance pattern of X-linked genes shows certain distinctive features, such as males transmit X-linked genes only to their daughters, and sons receive their X-linked genes from their mothers. The term **hemizygous** is used to describe the single copy of an X-linked gene in the male. A male mammal is said to be hemizygous for X-linked genes. Because males of certain species, such as humans, have a single copy of the X chromosome, another distinctive feature of X-linked inheritance is that males are more likely to be affected by rare, recessive X-linked disorders.

By comparison, relatively few genes are located only on the Y chromosome. These few genes are called **holandric genes.** An example of a holandric gene is the *Sry* gene found in mammals. Its expression is necessary for proper male development. A Y-linked inheritance pattern is very distinctive—the gene is transmitted only from fathers to sons.

Besides sex-linked genes, the X and Y chromosomes also contain short regions of homology where the X and Y chromosomes carry the same genes. In addition to several smaller regions, the human sex chromosomes have three homologous regions (**Figure 4.14**). These regions, which are evolutionarily related, promote the necessary pairing of the X and Y chromosomes that occurs during meiosis I of spermatogenesis. Relatively

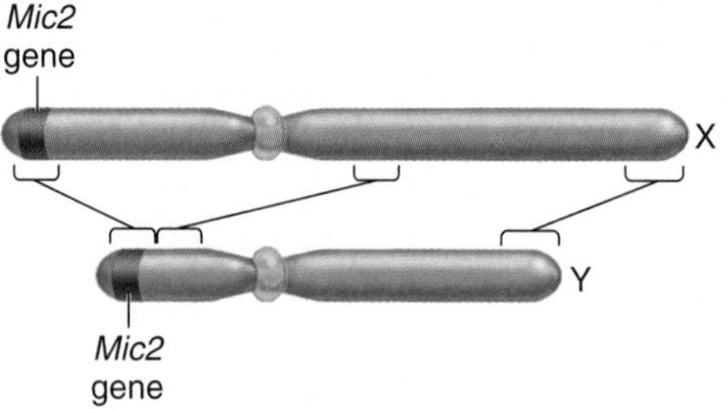

FIGURE 4.14 **A comparison of the homologous and nonhomologous regions of the X and Y chromosome in humans.** The brackets show three regions of homology between the X and Y chromosome. A few pseudoautosomal genes, such as *Mic2*, are found on both the X and Y chromosomes in these small regions of homology. Many X-linked genes are found in the nonhomologous regions of the X chromosome. A few Y-linked genes are found in the nonhomologous regions of the Y chromosome.

few genes are located in these homologous regions. One example is a human gene called *Mic2*, which encodes a cell surface antigen. The *Mic2* gene is found on both the X and Y chromosomes. It follows a pattern of inheritance called **pseudoautosomal inheritance.** The term pseudoautosomal refers to the idea that the inheritance pattern of the *Mic2* gene is the same as the inheritance pattern of a gene located on an autosome even though the *Mic2* gene is actually located on the sex chromosomes. As in autosomal inheritance, males have two copies of pseudoautosomally inherited genes, and they can transmit the genes to both daughters and sons.

Some Traits Are Influenced by the Sex of the Individual

As we have just seen, the transmission pattern of sex-linked genes depends on the sex of the parents and offspring. Sex can influence traits in other ways as well. The term **sex-influenced inheritance** refers to the phenomenon in which an allele is dominant in one sex but recessive in the opposite sex. Therefore, sex influence is a phenomenon of heterozygotes. Sex-influenced inheritance should not be confused with sex-linked inheritance. The genes that govern sex-influenced traits are usually autosomal, not on the X or Y chromosome.

In humans, the common form of pattern baldness provides an example of sex-influenced inheritance. As shown in Figure 4.15, the balding pattern is characterized by hair loss on the front and top of the head but not on the sides. This type of pattern baldness is inherited as an autosomal trait. (A common misconception is that this gene is X linked.) When a male is heterozygous for the baldness allele, he will become bald.

Genotype	Phenotype	
	Males	Females
BB	Bald	Bald
Bb	Bald	Nonbald
bb	Nonbald	Nonbald

In contrast, a heterozygous female will not be bald. Women who are homozygous for the baldness allele will develop the trait, but it is usually characterized by a significant thinning of the hair that occurs relatively late in life.

The sex-influenced nature of pattern baldness is related to the production of the male sex hormone testosterone. The gene that affects pattern baldness encodes an enzyme called 5-α-reductase, which converts testosterone to 5-α-dihydrotestosterone (DHT). DHT binds to cellular receptors and affects the expression of many genes, including those in the cells of the scalp. The allele that causes pattern baldness results in an overexpression of this enzyme. Because mature males normally make more testosterone than females, this allele has a greater phenotypic impact in males. However, a rare tumor of the adrenal gland can cause the secretion of abnormally large amounts of testosterone in females. If this occurs in a woman who is heterozygous *Bb*, she will become bald. If the tumor is removed surgically, her hair will return to its normal condition.

The autosomal nature of pattern baldness has been revealed by the analysis of many human pedigrees. An example is shown in Figure 4.16a. A bald male may inherit the bald allele from either parent, and thus a striking observation is that bald fathers can pass this trait to their sons. This could not occur if the trait was X linked, because fathers do not transmit an X chromosome to their sons. The analyses of many human pedigrees have shown that bald fathers, on average, have at least 50% bald sons. They are expected to produce an even higher percentage of bald male offspring if they are homozygous for the bald allele and/or the mother also carries one or two copies of the bald allele. For example, a heterozygous bald male and heterozygous (nonbald) female will produce 75% bald sons, while a homozygous bald male or homozygous bald female will produce all bald sons.

Figure 4.16b shows the predicted offspring if two heterozygotes produce offspring. In this Punnett square, the phenotypes are designated for both sons and daughters. *BB* offspring are bald, while *bb* offspring are nonbald. *Bb* offspring are bald if they are sons and nonbald if they are daughters. The predicted

(a) John Adams (father)

(b) John Quincy Adams (son)

(c) Charles Francis Adams (grandson)

(d) Henry Adams (great-grandson)

FIGURE 4.15 Pattern baldness in the Adams family line.

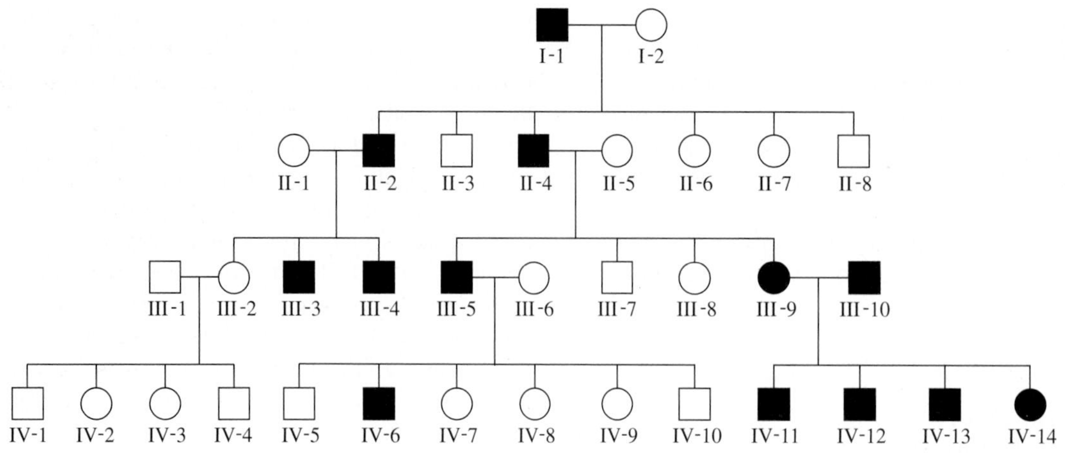

(a) A pedigree for human pattern baldness

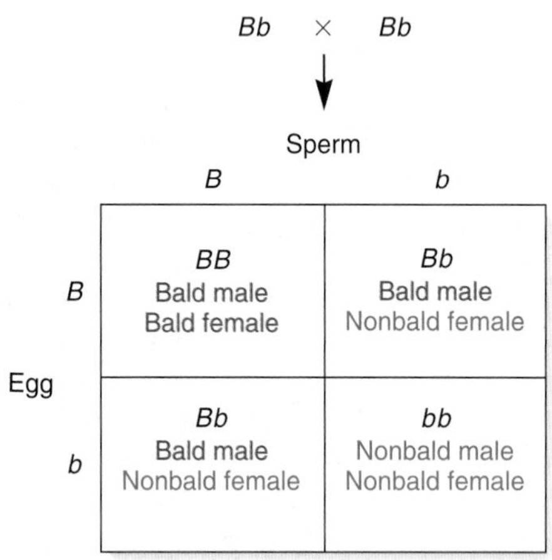

(b) Example of an inheritance pattern involving baldness

INTERACTIVE EXERCISE

FIGURE 4.16 Inheritance of pattern baldness, a sex-influenced trait involving an autosomal gene. (**a**) A family pedigree. Bald individuals are shown in black. (**b**) The predicted offspring from two heterozygous parents.

genotypic ratios from this cross would be 1 *BB* bald son : 1 *BB* bald daughter : 2 *Bb* bald sons : 2 *Bb* nonbald daughters : 1 *bb* nonbald son : 1 *bb* nonbald daughter. The predicted phenotypic ratios would be 3 bald sons : 1 bald daughter : 3 nonbald daughters : 1 nonbald son. The ratio of bald to nonbald offspring is 4:4, which is the same as 1:1.

Another example in which sex affects an organism's phenotype is provided by **sex-limited inheritance,** in which the trait occurs in only one of the two sexes. In humans, for example, breast development is a trait that is normally limited to females, whereas beard growth is limited to males. Among many types of birds, the male of the species has more ornate plumage than the female. As shown in **Figure 4.17**, roosters have a larger comb and wattles and longer neck and tail feathers than do hens. These are sex-limited features that may be found in roosters but never in normal hens. However, some varieties of chickens have males that are hen-feathered. In these species, hen-feathering is controlled by a dominant allele (*H*) that is expressed both in males and females, while cock-feathering is controlled by a recessive allele (*h*) that is expressed only in males. These alleles are located on an autosome; they are not on a sex chromosome.

Genotype	Phenotype	
	Females	Males
hh	Hen-feathered	Cock-feathered
Hh	Hen-feathered	Hen-feathered
HH	Hen-feathered	Hen-feathered

Like baldness in humans, hen-feathering depends on the production of sex hormones. If a newly hatched female with an *hh* genotype has her single ovary removed surgically, she will develop cock-feathering and look indistinguishable from a male.

Mutations That Cause a Loss of Function in an Essential Gene Result in a Lethal Phenotype

Let's now turn our attention to alleles that have the most detrimental effect on phenotype—those that result in death. An allele that has the potential to cause the death of an organism is called a **lethal allele.** These are usually inherited in a recessive manner. When the absence of a specific protein results in a lethal phenotype, the gene that encodes the protein is considered an **essential**

(a) Hen

(b) Rooster

FIGURE 4.17 **Differences in the feathering pattern in male and female chickens, an example of sex-limited inheritance.**

gene for survival. Though it varies according to species, researchers estimate that approximately 1/3 of all genes are essential genes. By comparison, **nonessential genes** are not absolutely required for survival, although they are likely to be beneficial to the organism. A loss-of-function mutation in a nonessential gene will not usually cause death. On rare occasions, however, a nonessential gene may acquire a mutation that causes the gene product to be abnormally expressed in a way that may interfere with normal cell function and lead to a lethal phenotype. Therefore, not all lethal mutations occur in essential genes, although the great majority do.

Many lethal alleles prevent cell division and thereby kill an organism at a very early stage. Others, however, may only exert their effects later in life, or under certain environmental conditions. For example, a human genetic disease known as Huntington disease is caused by a dominant allele. The disease is characterized by a progressive degeneration of the nervous system, dementia, and early death. The age when these symptoms appear, or the **age of onset,** is usually between 30 and 50.

Other lethal alleles may kill an organism only when certain environmental conditions prevail. Such **conditional lethal alleles** have been extensively studied in experimental organisms. For example, some conditional lethals will kill an organism only in a particular temperature range. These alleles, called **temperature-sensitive (ts) lethal alleles,** have been observed in many organisms, including *Drosophila.* A ts lethal allele may be fatal for a developing larva at a high temperature (30°C), but the larva will survive if grown at a lower temperature (22°C). Temperature-sensitive lethal alleles are typically caused by mutations that alter the structure of the encoded protein so it does not function correctly at the nonpermissive temperature or becomes unfolded and is rapidly degraded. Conditional lethal alleles may also be identified when an individual is exposed to a particular agent in the environment. For example, people with a defect in the gene that encodes the enzyme glucose-6-phosphate dehydrogenase (G6PD) have a negative reaction to the ingestion of fava beans. This can lead to an acute hemolytic syndrome with 10% mortality if not treated properly.

Finally, it is surprising that certain lethal alleles act only in some individuals. These are called **semilethal alleles.** Of course, any particular individual cannot be semidead. However, within a population, a semilethal allele will kill some individuals but not all of them. The reasons for semilethality are not always understood, but environmental conditions and the actions of other genes within the organism may help to prevent the detrimental effects of certain semilethal alleles.

In some cases, a lethal allele may produce ratios that seemingly deviate from Mendelian ratios. An example is an allele in a breed of cats known as Manx, which originated on the Isle of Man. (**Figure 4.18a**). The Manx cat carries a dominant mutation that affects the spine. This mutation shortens the tail, resulting in a range of tail lengths from normal to tailless. When two Manx cats are crossed to each other, the ratio of offspring is 1 normal : 2 Manx. How do we explain the 1:2 ratio? The answer is that about 1/4 of the offspring die during early embryonic development (**Figure 4.18b**). Therefore, in the homozygous condition, this dominant allele is lethal.

(a) A Manx cat

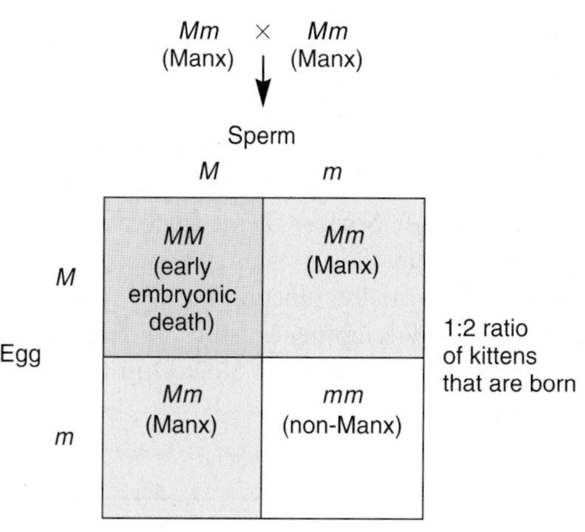

(b) Example of a Manx inheritance pattern

FIGURE 4.18 **The Manx cat, which carries a lethal allele.** (**a**) Photo of a Manx cat, which typically has a shortened tail. (**b**) Outcome of a cross between two Manx cats. Animals that are homozygous for the dominant Manx allele (*M*) die during early embryonic development.

Single Genes Have Pleiotropic Effects

Before ending our discussion of single-gene inheritance patterns, let's take a broader look at how a single gene may affect phenotype. Although we tend to discuss genes within the context of how they influence a single trait, most genes actually have multiple effects throughout a cell and/or throughout a multicellular organism. The multiple effects of a single gene on the phenotype of an organism is called **pleiotropy.** Pleiotropy occurs for several reasons, including the following:

1. The expression of a single gene can affect cell function in more than one way. For example, a defect in a microtubule protein may affect cell division and cell movement.
2. A gene may be expressed in different cell types in a multicellular organism.
3. A gene may be expressed at different stages of development.

In all or nearly all cases, the expression of a gene is pleiotropic with regard to the characteristics of an organism. The expression of any given gene influences the expression of many other genes in the genome, and vice versa. Pleiotropy is revealed when researchers study the effects of gene mutations. As an example of a pleiotropic mutation, let's consider cystic fibrosis, which is a recessive human disorder. In the late 1980s, the gene for cystic fibrosis was identified. It encodes a protein called the cystic fibrosis transmembrane conductance regulator (CFTR), which regulates ionic balance by allowing the transport of chloride ions (Cl^-) across epithelial-cell membranes. The mutation that causes cystic fibrosis diminishes the function of this Cl^- transporter, affecting several parts of the body in different ways. Because the movement of Cl^- affects water transport across membranes, the most severe symptom of cystic fibrosis is thick mucus in the lungs that occurs because of a water imbalance. In sweat glands, the normal Cl^- transporter has the function of recycling salt out of the glands and back into the skin before it can be lost to the outside world. Persons with cystic fibrosis have excessively salty sweat due to their inability to recycle salt back into their skin cells—a common test for cystic fibrosis is measurement of salt on the skin. Another effect is seen in the reproductive system of males who are homozygous for the cystic fibrosis allele. Males with cystic fibrosis may be infertile because the vas deferens, the tubules that transport sperm from the testes, may be absent or undeveloped. Presumably, a normally functioning Cl^- transporter is needed for the proper development of the vas deferens in the embryo. Taken together, we can see that a defect in CFTR has multiple effects throughout the body.

4.2 GENE INTERACTIONS

In Section 4.1, we considered the effects of a single gene on the outcome of a trait. This approach helps us to understand the various ways that alleles can influence traits. Researchers often examine the effects of a single gene on the outcome of a single trait as a way to simplify the genetic analysis. For example, Mendel studied one gene that affected the height of pea plants—tall versus dwarf alleles. Actually, many other genes in pea plants also affect height, but Mendel did not happen to study variants in those other height genes. How then did Mendel study the affects of a single gene? The answer lies in the genotypes of his strains. Although many genes affect the height of pea plants, Mendel chose true-breeding strains that differed with regard to only one of those genes. As a hypothetical example, let's suppose that pea plants have 10 genes affecting height, which we will call *K, L, M, N, O, P, Q, R, S,* and *T*. The genotypes of two hypothetical strains of pea plants may be

Tall strain:	*KK LL MM NN OO PP QQ RR SS TT*
Dwarf strain:	*KK LL MM NN OO PP QQ RR SS tt*

In this example, the alleles affecting height may differ at only a single gene. One strain is *TT* and the other is *tt*, and this accounts for the difference in their height. If we make crosses between these tall and dwarf strains, the genotypes of the F_2 offspring will differ with regard to only one gene; the other nine genes will be identical in all of them. This approach allows a researcher to study the effects of a single gene even though many genes may affect a single trait.

Researchers now appreciate that essentially all traits are affected by the contributions of many genes. Morphological features such as height, weight, growth rate, and pigmentation are all affected by the expression of many different genes in combination with environmental factors. In this section, we will further our understanding of genetics by considering how the allelic variants of two different genes affect a single trait. This phenomenon is known as a **gene interaction.** Table 4.3 considers several

TABLE 4.3

Types of Mendelian Inheritance Patterns Involving Two Genes

Type	Description
Epistasis	An inheritance pattern in which the alleles of one gene mask the phenotypic effects of the alleles of a different gene.
Complementation	A phenomenon in which two different parents that express the same or similar recessive phenotypes produce offspring with a wild-type phenotype.
Modifying genes	A phenomenon in which the allele of one gene modifies the phenotypic outcome of the alleles of a different gene.
Gene redundancy	A pattern in which the loss of function in a single gene has no phenotypic effect, but the loss of function of two genes has an effect. Functionality of only one of the two genes is necessary for a normal phenotype; the genes are functionally redundant.
Intergenic suppressors	An inheritance pattern in which the phenotypic effects of one mutation are reversed by a suppressor mutation in another gene.

examples in which two different genes interact to influence the outcome of particular traits. In this section, we will examine these examples in greater detail.

A Cross Involving a Two-Gene Interaction Can Produce Four Distinct Phenotypes

The first case of two different genes interacting to affect a single trait was discovered by William Bateson and Reginald Punnett in 1906 while they were investigating the inheritance of comb morphology in chickens. Several common varieties of chicken possess combs with different morphologies, as illustrated in **Figure 4.19a**. In their studies, Bateson and Punnett crossed a Wyandotte breed having a rose comb to a Brahma having a pea comb. All F_1 offspring had a walnut comb.

When these F_1 offspring were mated to each other, the F_2 generation consisted of chickens with four types of combs in the following phenotypic ratio: 9 walnut : 3 rose : 3 pea : 1 single comb. As we have seen in Chapter 2, a 9:3:3:1 ratio is obtained in the F_2 generation when the F_1 generation is heterozygous for two different genes and these genes assort independently. However, an important difference here is that we have four distinct categories of a single trait. Based on the 9:3:3:1 ratio, Bateson and Punnett reasoned that a single trait (comb morphology) was determined by two different genes.

R (rose comb) is dominant to *r*

P (pea comb) is dominant to *p*

R and *P* (walnut comb) are codominant

rrpp produces a single comb

As shown in the Punnett square of **Figure 4.19b**, each of the genes can exist in two alleles, and the two genes show independent assortment.

When the alleles of one gene mask the phenotypic effects of the alleles of another gene, the phenomenon is called **epistasis.** Geneticists consider epistasis relative to a particular phenotype. For example, let's consider walnut comb as our reference phenotype. Relative to a walnut comb, some geneticists may view *rr* and *pp* to be epistatic to this phenotype. In other words, both *rr* and *pp* mask a walnut comb. This is an example of **recessive epistasis** because an individual must be homozygous for either recessive allele to mask the walnut comb phenotype. As shown in the Punnett square of Figure 4.19b, four distinct phenotypes are produced.

A Cross Involving a Two-Gene Interaction Can Produce Two Distinct Phenotypes Due to Epistasis

Bateson and Punnett also discovered an unexpected gene interaction when studying crosses involving the sweet pea, *Lathyrus odoratus*. The wild sweet pea has purple flowers. However, they obtained several true-breeding mutant varieties with white flowers. Not surprisingly, when they crossed a true-breeding purple-flowered plant to a true-breeding white-flowered plant, the F_1 generation contained all purple-flowered plants and the F_2 generation (produced by self-fertilization of the F_1 generation) consisted of purple- and white-flowered plants in a 3:1 ratio.

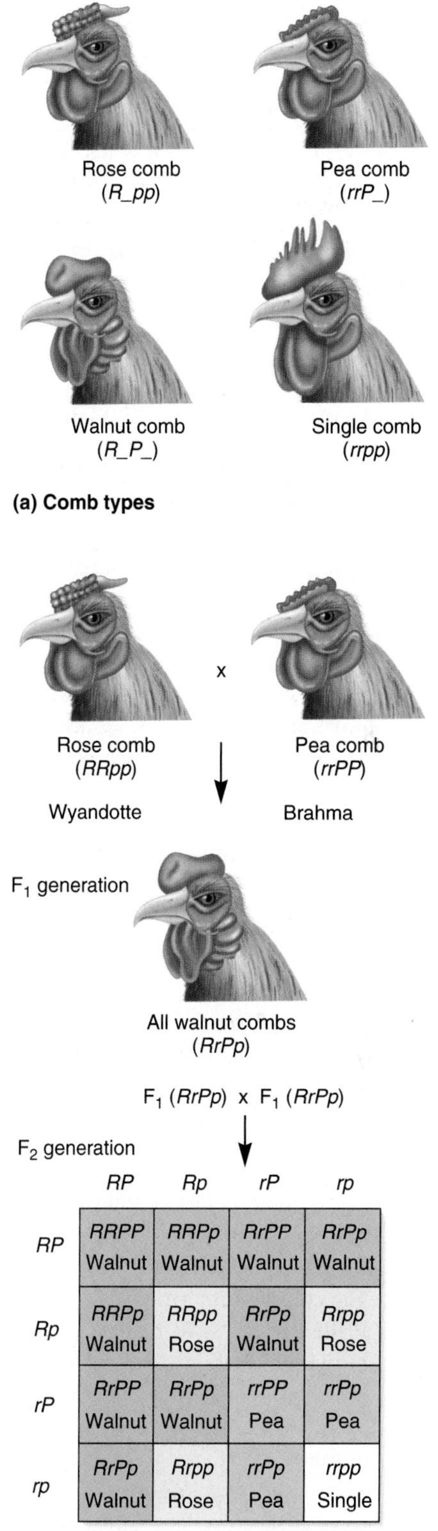

FIGURE 4.19 **Inheritance of comb morphology in chickens.** This trait is influenced by two different genes, which can each exist in two alleles. **(a)** Four phenotypic outcomes are possible. The underline symbol indicates the allele could be either dominant or recessive. **(b)** The crosses of Bateson and Punnett examined the interaction of the two genes.

A surprising result came in an experiment where they crossed two different varieties of white-flowered plants (**Figure 4.20**). All of the F_1 generation plants had purple flowers! Bateson and Punnett then allowed the F_1 offspring to self-fertilize. The F_2 generation resulted in purple and white flowers in a ratio of 9 purple : 7 white. From this result, Bateson and Punnett deduced that two different genes were involved, with the following relationship:

C (one purple-color-producing) allele is dominant to c (white)

P (another purple-color-producing) allele is dominant to *p* (white)

cc or *pp* masks the *P* or *C* alleles, producing white color

If possible, geneticists use the wild-type phenotype as their reference phenotype when describing an epistatic interaction. In this case, purple flowers are wild type. Homozygosity for the white allele of one gene masks the expression of the purple-producing allele of another gene. In other words, the *cc* genotype is epistatic to a purple phenotype, and the *pp* genotype is also epistatic to a purple phenotype. At the level of genotypes, *cc* is epistatic to *PP* or *Pp*, and *pp* is epistatic to *CC* or *Cc*. This is another example of recessive epistasis. As seen in Figure 4.20, this epistatic interaction produces only two phenotypes—purple or white flowers—in a 9:7 ratio.

Epistatis often occurs because two (or more) different proteins participate in a common function. For example, two or more proteins may be part of an enzymatic pathway leading to the formation of a single product. To illustrate this idea, let's consider the formation of a purple pigment in the sweet pea.

$$\text{Colorless precursor} \xrightarrow{\text{Enzyme C}} \text{Colorless intermediate} \xrightarrow{\text{Enzyme P}} \textbf{Purple pigment}$$

In this example, a colorless precursor molecule must be acted on by two different enzymes to produce the purple pigment. Gene *C* encodes a functional protein called enzyme C, which converts the colorless precursor into a colorless intermediate. Two copies of the recessive allele (*cc*) result in a lack of production of this enzyme in the homozygote. Gene *P* encodes a functional enzyme P, which converts the colorless intermediate into the purple pigment. Like the *c* allele, the recessive *p* allele encodes a defective enzyme P. If an individual is homozygous for either recessive allele (*cc* or *pp*), it will not make any functional enzyme C or enzyme P, respectively. When one of these enzymes is missing, purple pigment cannot be made, and the flowers remain white.

The parental cross shown in Figure 4.20 illustrates another genetic phenomenon called **complementation.** This term refers to the production of offspring with a wild-type phenotype from parents that both display the same or similar recessive phenotype. In this case, purple-flowered F_1 offspring were obtained from two white-flowered parents. Complementation typically occurs because the recessive phenotype in the parents is due to homozgyosity at two different genes. In our sweet pea example, one parent is *CCpp* and the other is *ccPP*. In the F_1 offspring, the *C* and *P* alleles, which are wild-type and dominant, complement

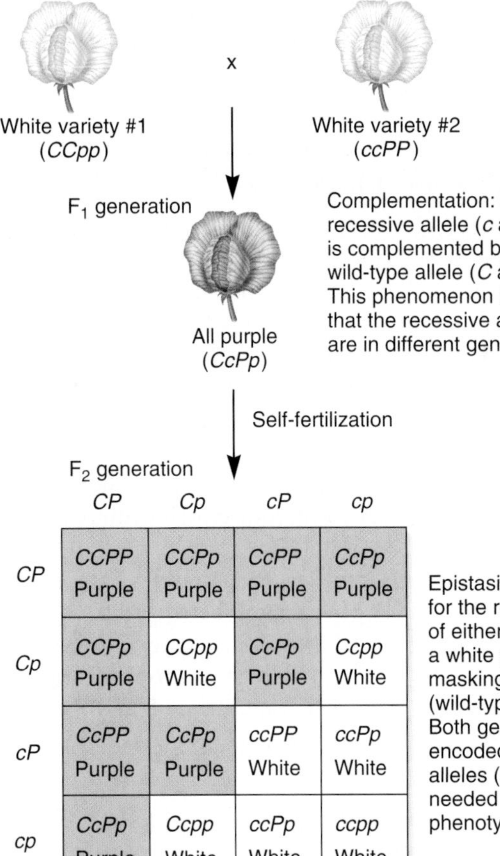

INTERACTIVE EXERCISE

FIGURE 4.20 **A cross between two different white varieties of the sweet pea.**

Genes →Traits The color of the sweet pea flower is controlled by two genes, which are epistatic to each other and show complementation. Each gene is necessary for the production of an enzyme required for pigment synthesis. The recessive allele of either gene encodes a defective enzyme. If an individual is homozygous recessive for either of the two genes, the purple pigment cannot be synthesized. This results in a white phenotype.

the *c* and *p* alleles, which are recessive. The offspring needs one wild-type allele of both genes to display the wild-type phenotype. Why is complementation an important experimental observation? When geneticists observe complementation in a genetic cross, the results suggest that the recessive phenotype in the two parent strains is caused by mutant alleles in two different genes.

A Cross Involving a Two-Gene Interaction Can Produce Three Distinct Phenotypes Due to Epistasis

Thus far, we have observed two epistatic interactions, one producing four phenotypes and the other producing only two. Coat color in rodents provides an example that produces three phenotypes. If a true-breeding black rat is crossed to a true-breeding albino rat, the result is a rat with agouti coat color. Animals with agouti coat color have black pigmentation at the tips of each hair that changes to orange pigmentation near the root. If two agouti animals of the F_1 generation are crossed to each other, they produce agouti, black, and albino offspring in a 9:3:4 ratio (**Figure 4.21**).

How do we explain this ratio? This cross involves two genes that are called *A* (for agouti) and *C* (for colored). The dominant *A* allele of the agouti gene encodes a protein that regulates hair color such that the pigmentation shifts from black (eumelanin) at the tips to orange (phaeomelanin) near the roots. The recessive allele, *a*, inhibits the shift to orange pigmentation and thereby results in black pigment production throughout the entire hair, when an animal is *aa*. As with rabbits, the colored gene encodes tyrosinase, which is needed for the first step in melanin synthesis. The *C* allele allows pigmentation to occur, while the *c* allele causes the loss of tyrosinase function. The *C* allele is dominant to the *c* allele; *cc* homozygotes are albino and have white coat color.

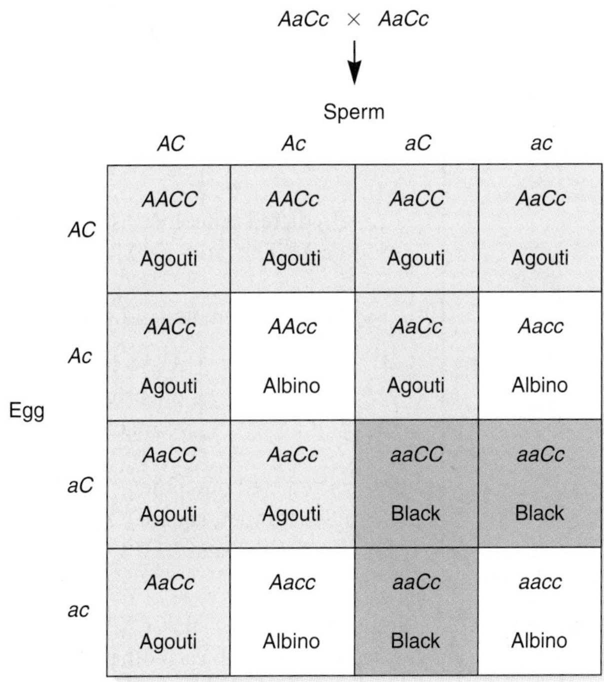

FIGURE 4.21 **Inheritance pattern of coat color in rats involving a gene interaction between the agouti gene (*A* or *a*) and the colored gene (*C* or *c*).**

As shown in Figure 4.21, the F_1 offspring are heterozygous for the two genes. In this case, *C* is dominant to *c*, and *A* is dominant to *a*. If an animal has at least one copy of both dominant alleles, the result is agouti coat color. Let's consider agouti as our reference phenotype. In the F_2 generation, if an animal has a dominant *C* allele but is *aa* homozygous, it will develop a black coat. The four cases of albino animals are all *cc* homozygous. This occurs even when an animal carries the dominant *A* allele. The *a* allele is epistatic to *C* and causes a black coat color in the *aa* homozygote, while *c* is epistatic to *A* and results in white coat color in the *cc* homozygote. This is a third example of recessive epistasis. At the molecular level, *aa* is masking the orange pigmentation, and *cc* is masking any pigmentation. Unlike the sweet pea, in which loss-of-function alleles had the same effect (white flowers), the effects of these loss-of-function alleles are not the same; *aa* yields black fur, and *cc* produces white fur.

Though the molecular effects of the *aa* genotype can be viewed as epistasis because the orange pigmentation is actually masked, some geneticists may not view this effect as epistasis but instead would call it a **gene modifier effect**—the alleles of one gene modify the phenotypic effect of the alleles of a different gene. From this alternative viewpoint, the pigmentation is not totally masked, but instead the agouti color is modified to black. As described next, geneticists may classify some alleles as gene modifiers.

EXPERIMENT 4A

Bridges Observed an 8:4:3:1 Ratio Because the Cream-Eye Gene Can Modify the X-Linked Eosin Allele but Not the Red or White Alleles

As we have seen, geneticists view epistasis as a situation in which the alleles of a given gene mask the phenotypic effects of the alleles of another gene. In some cases, however, two genes may interact to influence a particular phenotype, but the interaction of particular alleles seems to modify the phenotype, not mask it.

Calvin Bridges discovered an early example in which one gene modifies the phenotypic effects of an X-linked eye color gene in *Drosophila*. As discussed in Chapter 3, the X-linked red allele (w^+) is dominant to the white allele (*w*). Besides these two alleles, Thomas Hunt Morgan and Calvin Bridges found another allele of this gene that they called eosin (*w-e*), which results in eyes that are a pale orange color. The red allele is dominant to the eosin allele. In addition, the expression of the eosin allele depends on the number of copies of the allele. When females have two copies of this allele, they have eosin eyes. When females are heterozygous for the eosin allele and white allele, they have light-eosin eyes. Within true-breeding cultures of flies with eosin eyes, he occasionally found a fly that had a noticeably different eye color. In particular, he identified a rare fly with cream-colored eyes. Bridges reasoned that this new eye color could be explained in two different ways. One possibility is that the cream-colored phenotype could be the result of a new mutation that changed the eosin allele into a cream allele. A second possibility is that a different gene may have incurred a mutation that modified the phenotypic expression of the eosin allele. This second possibility is an example of a gene interaction. To distinguish between these two possibilities, he carried out the crosses described in **Figure 4.22**. He crossed males with cream-colored eyes to wild-type females and then allowed the F_1 generation flies, which all had red eyes, to mate with each other. As shown in the data, all F_2 females had red eyes, while males had red eyes, eosin eyes, or cream eyes.

THE HYPOTHESES

Cream-colored eyes in fruit flies are due to the effect of an allele that is in the same gene as the eosin allele or in a second gene that modifies the expression of the eosin allele.

TESTING THE HYPOTHESES — FIGURE 4.22 A gene interaction between the cream allele and eosin allele.

Starting material: From a culture of flies with eosin eyes, Bridges obtained a fly with cream-colored eyes and used it to produce a true-breeding culture of flies with cream-colored eyes. The allele was called *cream a* (c^a).

	Experimental level	Conceptual level
1. Cross males with cream-colored eyes to wild-type females.		
2. Observe the F_1 offspring and then allow the offspring to mate with each other.		
3. Observe and record the eye color and sex of the F_2 generation.		

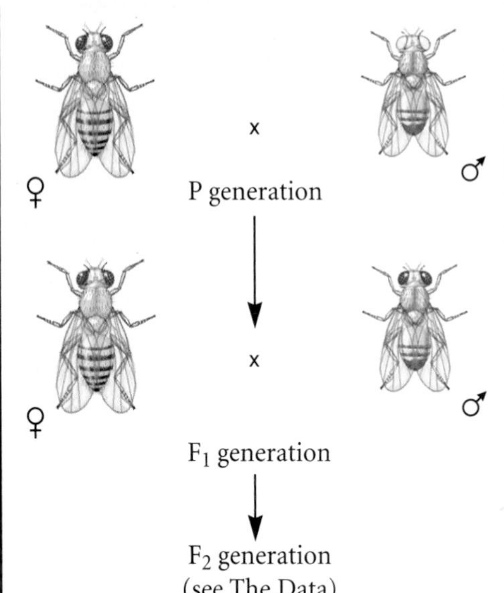

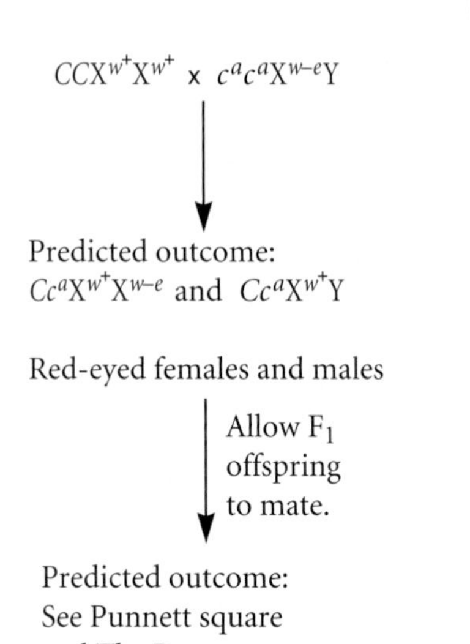

$CCX^{w^+}X^{w^+}$ x $c^ac^aX^{w-e}Y$
Predicted outcome:
$Cc^aX^{w^+}X^{w-e}$ and $Cc^aX^{w^+}Y$
Red-eyed females and males
Allow F_1 offspring to mate.
Predicted outcome:
See Punnett square and The Data.

THE DATA

Cross	*Outcome*
P cross:	
Cream-eyed male × wild-type female	F_1: All red eyes
F_1 cross:	
F_1 brother × F_1 sister	F_2: 104 females with red eyes
	47 males with red eyes
	44 males with eosin eyes
	14 males with cream eyes

INTERPRETING THE DATA

To interpret these data, keep in mind that Bridges already knew that the eosin allele is X linked. However, he did not know whether the cream allele was in the same gene as the eosin allele, in a different gene on the X chromosome, or on an autosome. The F_2 generation indicates that the cream allele is not in the same gene as the eosin allele. If the cream allele was in the same gene as the eosin allele, none of the F_2 males would have had eosin eyes; there would have been a 1:1 ratio of red-eyed males and cream-eyed males in the F_2 generation. This result was not obtained. Instead, the actual results are consistent with the idea that the male flies of the parental generation possessed both the eosin and cream alleles. Therefore, Bridges concluded that the cream allele was an allele of a different gene.

One possibility is that the cream allele is an autosomal recessive allele. If so, we can let *C* represent the dominant allele (which does not modify the eosin phenotype) and c^a represent the cream allele that modifies the eosin color to cream. We already know that the eosin allele is X linked and recessive to the red allele. The parental cross is expected to produce all red-eyed F_1 flies in which the males are $Cc^aX^{w+}Y$ and the females are $Cc^aX^{w+}X^{w-e}$. When these F_1 offspring are allowed to mate with each other, the Punnett square shown here would predict the following outcome:

$$Cc^aX^{w^+}X^{w-e} \times Cc^aX^{w^+}Y$$

Sperm

♀ Eggs \ ♂ Sperm	CX^{w^+}	CY	$c^aX^{w^+}$	c^aY
CX^{w^+}	$CCX^{w^+}X^{w^+}$	$CCX^{w^+}Y$	$Cc^aX^{w^+}X^{w^+}$	$Cc^aX^{w^+}Y$
CX^{w-e}	$CCX^{w^+}X^{w-e}$	$CCX^{w-e}Y$	$Cc^aX^{w^+}X^{w-e}$	$Cc^aX^{w-e}Y$
$c^aX^{w^+}$	$Cc^aX^{w^+}X^{w^+}$	$Cc^aX^{w^+}Y$	$c^ac^aX^{w^+}X^{w^+}$	$c^ac^aX^{w^+}Y$
c^aX^{w-e}	$Cc^aX^{w^+}X^{w-e}$	$Cc^aX^{w-e}Y$	$c^ac^aX^{w^+}X^{w-e}$	$c^ac^aX^{w-e}Y$

Outcome:

1 $CCX^{w^+}X^{w^+}$: 1 $CCX^{w^+}X^{w-e}$: 2 $Cc^aX^{w^+}X^{w^+}$: 2 $Cc^aX^{w^+}X^{w-e}$: 1 $c^ac^aX^{w^+}X^{w^+}$: 1 $c^ac^aX^{w^+}X^{w-e}$ = **8 red-eyed females**

1 $CCX^{w^+}Y$: 2 $Cc^aX^{w^+}Y$: 1 $c^ac^aX^{w^+}Y$ = **4 red-eyed males**

1 $CCX^{w-e}Y$: 2 $Cc^aX^{w-e}Y$ = **3 light eosin-eyed males**

1 $c^ac^aX^{w-e}Y$ = **1 cream-eyed male**

This phenotypic outcome proposes that the specific modifier allele, c^a, can modify the phenotype of the eosin allele but not the red-eye allele. The eosin allele can be modified only when the c^a allele is homozygous. The predicted 8:4:3:1 ratio agrees reasonably well with Bridges's data.

A self-help quiz involving this experiment can be found at the Online Learning Center.

Due to Gene Redundancy, Loss-of-Function Alleles May Have No Effect on Phenotype

During the past several decades, researchers have discovered new kinds of gene interactions by studying model organisms such as *Escherichia coli* (a bacterium), *Saccharomyces cerevisiae* (baker's yeast), *Arabidopsis thaliana* (a model plant), *Drosophila melanogaster* (fruit fly), *Caenorhabditis elegans* (a nematode worm), and *Mus musculus* (the laboratory mouse). The isolation of mutants that alter the phenotypes of these organisms has become a powerful way to investigate gene function and has provided ways for researchers to identify new kinds of gene interactions. With the advent of modern molecular techniques (described in Chapters 16, 18, and 19), a common approach for investigating gene function is to intentionally produce loss-of-function alleles in a gene of interest. When a geneticist abolishes gene function by creating an organism that is homozygous for a loss-of-function allele, the resulting organism is said to have undergone a **gene knockout.**

Why are gene knockouts useful? The primary reason for making a gene knockout is to understand how a gene affects the structure and function of cells or the phenotypes of organisms. For example, if a researcher knocked out a particular gene in a mouse and the resulting animal was unable to hear, the researcher would suspect that the role of the functional gene is to promote the formation of ear structures that are vital for hearing.

Interestingly, by studying many gene knockouts in a variety of experimental organisms, geneticists have discovered that many knockouts have no obvious effect on phenotype at the cellular level or the level of discernable traits. To explore gene function further, researchers may make two or more gene knockouts in the same organism. In some cases, gene knockouts in two different genes produce a phenotypic change even though the single knockouts have no effect (**Figure 4.23**). Geneticists may attribute this change to **gene redundancy**—the phenomenon that one gene can compensate for the loss of function of another gene.

Gene redundancy may be due to different underlying causes. One common reason is gene duplication. Certain genes have been duplicated during evolution, so a species may contain two or more copies of similar genes. These copies, which are not identical due to the accumulation of random changes during evolution, are called **paralogs.** When one gene is missing, a paralog may be able to carry out the missing function. For example, genes *A* and *B* in Figure 4.23 could be paralogs of each other. Alternatively, gene redundancy may involve proteins that are involved in a common cellular function. When one of the proteins is missing due to a gene knockout, the function of another protein may be increased to compensate for the missing protein and thereby overcome the defect.

Let's explore the consequences of gene redundancy in a genetic cross. George Shull conducted one of the first studies

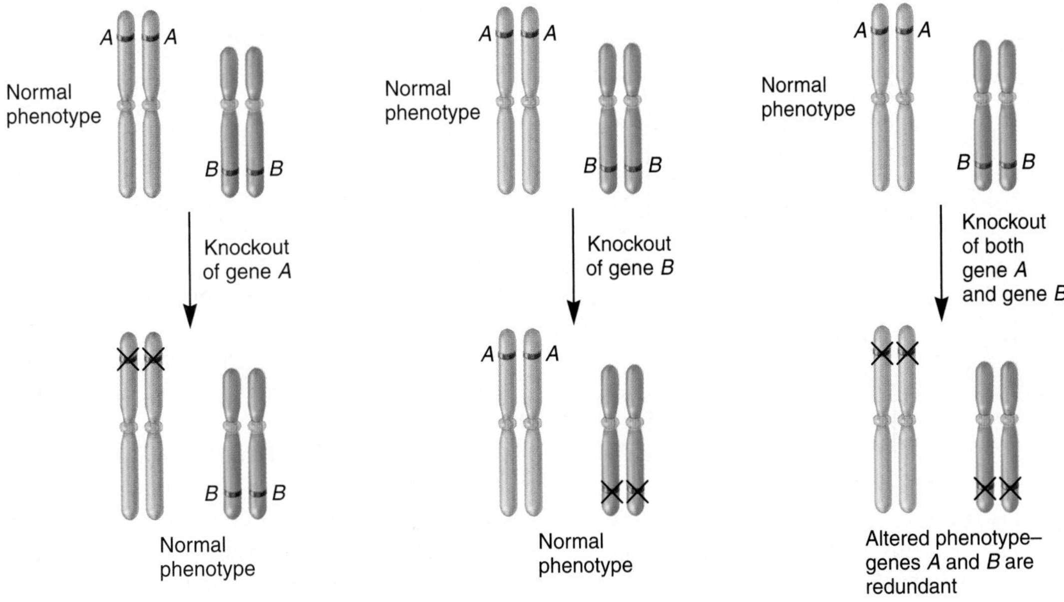

FIGURE 4.23 **A molecular explanation for gene redundancy.** To have a normal phenotype, an organism must have a functional copy of gene *A* or gene *B*, but not both. If both gene *A* and gene *B* are knocked out, an altered phenotype occurs.

that illustrated the phenomenon of gene redundancy. His work involved a weed known as shepherd's purse, a member of the mustard family. The trait he followed was the shape of the seed capsule, which is commonly triangular (**Figure 4.24**). Strains producing smaller ovate capsules are due to loss-of-function alleles in two different genes (*ttvv*). The ovate strain is an example of a double gene knockout. When Shull crossed a true-breeding plant with triangular capsules to a plant having ovate capsules, the F_1 generation all had triangular capsules. When the F_1 plants were self-fertilized, a surprising result came in the F_2 generation. Shull observed a 15:1 ratio of plants having triangular capsules to ovate capsules. The result can be explained by gene redundancy. Having one functional copy of either gene (*T* or *V*) is sufficient to produce the triangular phenotype. *T* and *V* are redundant because only one of them is necessary for a triangular shape. When the functions of both genes are knocked out, as in the *ttvv* homozygote, the capsule becomes smaller and ovate.

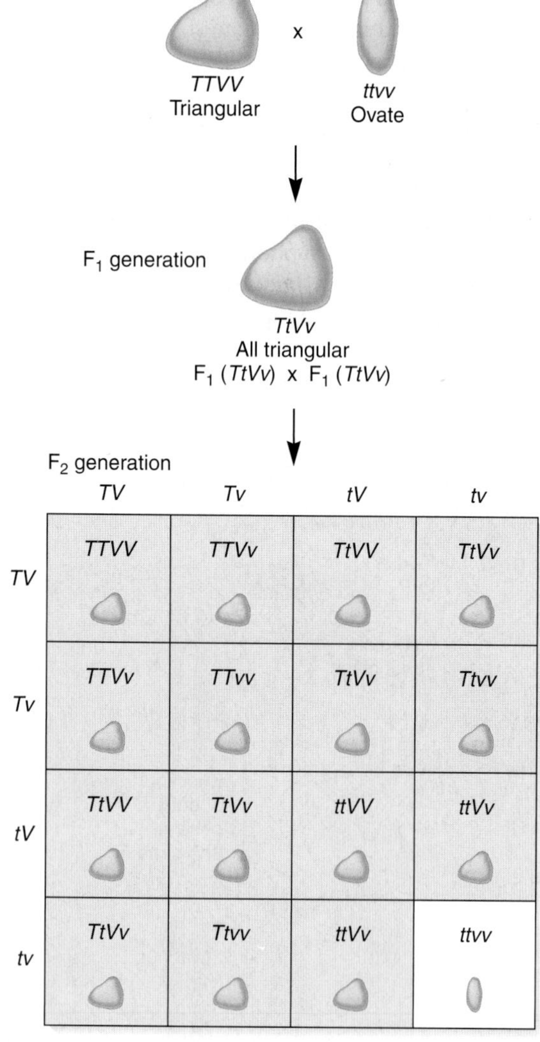

FIGURE 4.24 Inheritance of capsule shape in shepherd's purse, an example of gene redundancy. In this case, triangular shape requires a dominant allele in one of two genes, but not both. The *T* and *V* alleles are redundant.

The Phenotypic Effects of a Mutation Can Be Reversed by a Suppressor Mutation

When studying an experimental organism, a common approach to gain a deeper understanding of gene interaction is the isolation of a **suppressor mutation**—a second mutation that reverses the phenotypic effects of a first mutation. When a suppressor mutation is in a different gene than the first mutation, it is called an **intergenic suppressor** or **extragenic suppressor.**

What type of information might a researcher gain from the analysis of intergenic suppressor mutants? Usually, the primary goal is to identify proteins that participate in a common cellular process that ultimately affects the traits of an organism. In *Drosophila*, several different proteins work together in a signaling pathway that determines whether or not certain parts of the body contain sensory cells, such as those that make up mechanosensory bristles. Researchers have isolated dominant mutants that result in flies with fewer bristles. The mutated gene was named *Hairless* to reflect this phenotype. In this case, the wild-type allele is designated *h*, and the dominant mutant is *H*. After the *Hairless* mutant was obtained, researchers then isolated mutants that suppressed the hairless phenotype. Such suppressor mutants, which are in a different gene, produced flies that have a wild-type number of bristles. These mutants, which are also dominant, are in a gene that was named *Suppressor of Hairless*. The wild-type allele is designated *soh*, and the dominant mutant allele is *SoH*.

How do we explain the effects of these mutations at the molecular level? Let's first consider the functions of the proteins encoded by the normal (wild-type) genes (**Figure 4.25**). The role of the SoH protein, encoded by the *soh* allele of the *Suppressor of Hairless* gene, is to prevent the formation of sensory structures such as bristles in regions of the body where they should not be made. The Hairless protein is made in regions of the body where bristles should form, and binds to the SoH protein and inhibits its function. When the Hairless protein is properly expressed on the surface of the fly, as in an *hh* homozygote, bristles will form there.

Now let's consider the effects of a single mutation in the *Hairless* gene. In a heterozygote carrying the dominant allele (*H*), only half the amount of functional Hairless protein is made. This is not enough to inhibit all of the SoH proteins that are made. Therefore, the uninhibited SoH proteins prevent bristle formation and result in a hairless (bristleless) phenotype.

What happens in the double mutant? The suppressor mutation eliminates one of the two functional *soh* alleles. The double mutant expresses only one functional *h* allele and one functional *soh* allele. In the double mutant, the reduced amount of Hairless protein is able to inhibit the reduced amount of the SoH protein. Therefore, the ability of the SoH proteins to prevent bristle formation is stopped. Bristles form in the double heterozygote.

The analysis of a mutant and its suppressor often provides key information that two proteins participate in a common function. In some cases, the analysis reveals that two proteins physically interact with each other. As we have just seen, this type of interaction occurs between the Hairless and SoH proteins. Alternatively, two distinct proteins encoded by different genes may

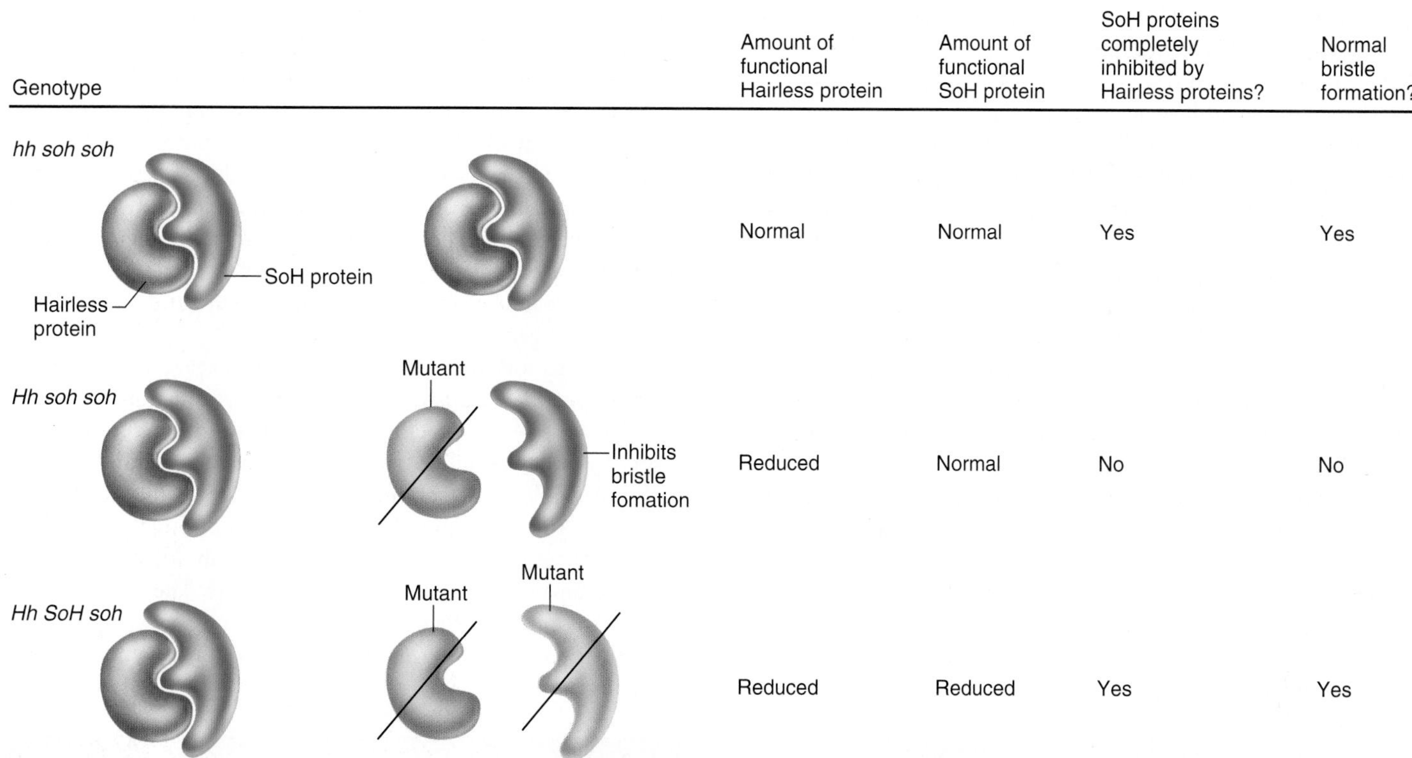

Genotype
Amount of functional Hairless protein
Amount of functional SoH protein
SoH proteins completely inhibited by Hairless proteins?
Normal bristle formation?
hh soh soh
SoH protein
Hairless protein
Normal
Normal
Yes
Yes
Hh soh soh
Mutant
Inhibits bristle fomation
Reduced
Normal
No
No
Hh SoH soh
Mutant
Mutant
Reduced
Reduced
Yes
Yes

FIGURE 4.25 **An example of a gene interaction involving an intergenic suppressor.** The *Hairless* mutation, which produces the dominant *H* allele, results in flies with fewer bristles. A dominant suppressor mutation in a second gene restores bristle formation. This dominant allele is designated *SoH.* Examination of the interactions between the mutant and its suppressor reveals that the Hairless and SoH proteins physically interact with each other to determine whether bristles are formed.

participate in a common function, but do not directly interact with each other. For example, two enzymes may be involved in a biochemical pathway that leads to the synthesis of an amino acid. A mutation that greatly decreases the amount of one enzyme may limit the ability of an organism to make the amino acid. If this occurs, the amino acid would have to be supplied to the organism for it to survive. A suppressor mutation could increase the function of another enzyme in the pathway and thereby restore the ability of the organism to make adequate amounts of the amino acid. Such a suppressor would alleviate the need for the organism to have the amino acid supplemented in its diet.

Other suppressors exert their effects by altering the amount of protein encoded by a mutant gene. For example, a mutation may decrease the functional activity of a protein that is needed for sugar metabolism. An organism harboring such a mutation may not be able to metabolize the sugar at a sufficient rate for growth or survival. A suppressor mutation in a different gene could alter genetic regulatory proteins and thereby increase the amount of the protein encoded by the mutant gene. (The proteins involved in gene regulation are described in Chapters 14 and 15.) This suppressor mutation would increase the amount of the defective protein and thereby result in a faster rate of sugar metabolism.

CONCEPTUAL SUMMARY

In Chapter 4, we have considered many examples that extend our understanding of Mendelian inheritance. We began with a consideration of the inheritance patterns involving single genes. Recessive alleles are often due to mutations that abolish gene function. In many simple dominant/recessive relationships, 50% of the normal protein is sufficient to yield a dominant phenotype. By comparison, incomplete dominance occurs when the heterozygote has an intermediate phenotype. In this case, 50% of the encoded protein does not produce a wild-type phenotype. In the pattern called incomplete penetrance, even though the dominant allele is present, the expected phenotype does not result. In other cases, the outcome depends on the degree to which the trait is expressed, or its expressivity. The outcome of traits is also due to environmental influences and/or modifier genes. The term norm of reaction refers to the effects of environmental variation on a phenotype. For certain genes, heterozygotes may display characteristics that are more beneficial for their survival in a particular environment, a phenomenon called overdominance. Overdominance may result in the prevalence of alleles, such as the sickle-cell allele, that are detrimental in the homozygous condition.

With regard to single genes, we also considered that, in natural populations, multiple alleles exist for most genes. Examples include coat color in rabbits and human blood types. The ABO blood types illustrate an inheritance pattern in which two alleles are distinctly expressed in the heterozygote, a phenomenon called codominance. We also examined how sex impacts the expression and inheritance patterns of single genes. Some genes are sex linked, meaning they are found on a sex chromosome. In mammals, X-linked genes are fairly common and produce inheritance patterns in which a reciprocal cross does not produce the same result. In sex-influenced inheritance, the phenotype of the heterozygote depends on its sex. A more extreme form of sex influence involves sex-limited traits, which are found in only one sex. Finally, we considered lethal alleles that cause the death of an organism. These are often loss-of-function alleles in essential genes. Such alleles can produce unusual ratios in a genetic cross.

We have also explored cases in which two different genes affect the outcome of a single trait, a concept known as a gene interaction. Epistasis is an inheritance pattern in which the alleles of one gene mask the phenotypic effects of the alleles of a different gene. Examples include comb shape in chickens, flower color in sweet peas, and coat color in rodents. Complementation is a phenomenon in which two different parents, both expressing the same or similar recessive phenotypes, produce offspring with a wild-type phenotype. In the example we explored, two different white strains of sweet peas, when crossed to each other, produced purple-flowered offspring. Modifying genes, such as the cream allele in *Drosophila*, alter the phenotypic outcome of the alleles of a different gene.

We also considered gene interactions that are often observed in experimental organisms. When the loss of function in a single gene has no phenotypic effect, but the loss of function of two or more genes has an effect, the genes are termed redundant with each other. Intergenic suppressors are mutations that reverse the effects of a first mutation. From the analysis of mutants and their intergenic suppressors, researchers often can identify proteins that participate in a common cellular process. The proteins may, but do not always, interact directly. Other intergenic suppressors exert their effects by altering the amount of protein encoded by a mutant gene.

EXPERIMENTAL SUMMARY

To understand how inheritance may deviate from a simple Mendelian pattern, either a genetic or molecular approach can be taken. In a genetic approach, a researcher makes crosses and then analyzes the phenotypes of the offspring. In Chapter 4, we have seen that the experimenter must weigh many factors when designing a cross and analyzing its outcome. These include the dominant/recessive relationships of alleles (namely, dominant, recessive, incompletely dominant, overdominant, or codominant), the presence of multiple alleles in a population, the relationship between the sex of the offspring and their phenotype, and the possibility of lethal alleles. In addition, complicating factors such as incomplete penetrance, the influence of the environment, and gene interactions must be considered. Based on these factors, a researcher can construct a Punnett square to see if a pattern of inheritance fits the available data obtained from crosses. Researchers may also seek to isolate particular types of mutants such as gene knockouts and suppressor mutants.

Another level of understanding in Mendelian inheritance is to appreciate how the molecular expression of genes underlies their impact on the phenotype of an organism. Using modern molecular tools that are described later in this textbook, researchers can quantitatively compare the level of gene expression between wild-type and mutant alleles. This helps us to understand how the amount of gene expression is correlated with the phenotype of the organism. Also, when gene interactions occur, an investigator may want to understand, at the molecular level, how the gene products participate in a common cellular function such as an enzymatic pathway. Therefore, the molecular investigation of gene interactions frequently involves research collaborations between geneticists, cell biologists, and biochemists.

PROBLEM SETS & INSIGHTS

Solved Problems

S1. In humans, why are X-linked recessive traits more likely to occur in males compared to females?

Answer: Because a male is hemizygous for X-linked traits, the phenotypic expression of X-linked traits depends on only a single copy of the gene. When a male inherits a recessive X-linked allele, he will automatically exhibit the trait because he does not have another copy of the gene on the corresponding Y chromosome. This phenomenon is particularly relevant to the inheritance of recessive X-linked alleles that cause human disease. (Some examples will be described in Chapter 22.)

S2. In Ayrshire cattle, the spotting pattern of the animals can be either red and white or mahogany and white. The mahogany and white pattern is caused by the allele *M*. The red and white phenotype is controlled by the allele *m*. When mahogany and white animals are mated to red and white animals, the following results are obtained:

Genotype	Phenotype	
	Females	**Males**
MM	Mahogany and white	Mahogany and white
Mm	Red and white	Mahogany and white
mm	Red and white	Red and white

Explain the pattern of inheritance.

Answer: The inheritance pattern for this trait is sex-influenced inheritance. The *M* allele is dominant in males but recessive in females, while the *m* allele is dominant in females but recessive in males.

S3. For the following pedigree involving a single gene causing an inherited disease, indicate which modes of inheritance are not possible. (Affected individuals are shown as filled symbols.)

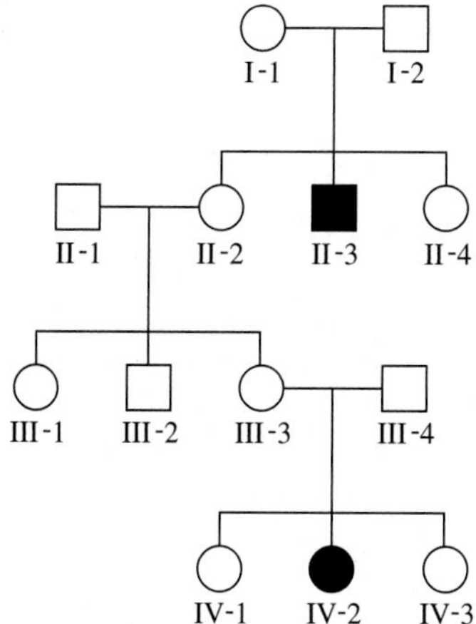

A. Recessive

B. Dominant

C. X linked, recessive

D. Sex influenced, dominant in females

E. Sex limited, recessive in females

Answer:

A. It could be recessive.

B. It is probably not dominant unless it is incompletely penetrant.

C. It could not be X-linked recessive because individual IV-2 does not have an affected father.

D. It could not be sex influenced, dominant in females because individual II-3 (who would have to be homozygous) has an unaffected mother (who would have to be heterozygous and affected).

E. It is not sex limited because individual II-3 is an affected male and IV-2 is an affected female.

S4. Red-green color blindness is inherited as a recessive X-linked trait. What are the following probabilities?

A. A woman with phenotypically normal parents and a color-blind brother will have a color-blind son. Assume that she has no previous children.

B. The next child of a phenotypically normal woman, who has already had one color-blind son, will be a color-blind son.

C. The next child of a phenotypically normal woman, who has already had one color-blind son, and who is married to a color-blind man, will have a color-blind daughter.

Answer:

A. The woman's mother must have been a heterozygote. So there is a 50% chance that the woman is a carrier. If she has children, 1/4 (i.e., 25%) will be affected sons if she is a carrier. However, there is only a 50% chance that she is a carrier. We multiply 50% times 25%, which equals $0.5 \times 0.25 = 0.125$, or a 12.5% chance.

B. If she already had a color-blind son, then we know she must be a carrier, so the chance is 25%.

C. The woman is heterozygous and her husband is hemizygous for the color-blind allele. This couple will produce 1/4 offspring that are color-blind daughters. The rest are 1/4 carrier daughters, 1/4 normal sons, and 1/4 color-blind sons. Answer is 25%.

S5. Pattern baldness is an example of a sex-influenced trait that is dominant in males and recessive in females. A couple, neither of whom is bald, produced a bald son. What are the genotypes of the parents?

Answer: Because the father is not bald, we know he must be homozygous, *bb*. Otherwise, he would be bald. A female who is not bald can be either *Bb* or *bb*. Because she has produced a bald son, we know that she must be *Bb* in order to pass the *B* allele to her son.

S6. Two pink-flowered four-o'clocks were crossed to each other. What are the following probabilities for the offspring?

A. A red-flowered plant.

B. The first three plants examined will be white.

C. A plant will be either white or pink.

D. A group of six plants contain one pink, two whites, and three reds.

Answer: The first thing we need to do is construct a Punnett square to determine the individual probabilities for each type of offspring.

Because flower color is incompletely dominant, the cross is $Rr \times Rr$.

♀ \ ♂	*R*	*r*
R	*RR* Red	*Rr* Pink
r	*Rr* Pink	*rr* White

The phenotypic ratio is 1 red : 2 pink : 1 white. In other words, 1/4 are expected to be red, 1/2 pink, and 1/4 white.

A. The probability of a red-flowered plant is 1/4, which equals 25%.

B. Use the product rule.

$1/4 \times 1/4 \times 1/4 = 1/64 = 1.6\%$

C. Use the sum rule because these are mutually exclusive events. A given plant cannot be both white and pink.

$1/4 + 1/2 = 3/4 = 75\%$

D. Use the multinomial expansion equation. See solved problem S7 in Chapter 2 for an explanation of the multinomial expansion equation. In this case, three phenotypes are possible.

$$P = \frac{n!}{a!b!c!} p^a q^b r^c$$

where

n = total number of offspring = 6
a = number of reds = 3
p = probability of reds = (1/4)
b = number of pinks = 1
q = probability of pink = (1/2)
c = number of whites = 2
r = probability of whites = (1/4)

If we substitute these values into the equation,

$$P = \frac{6!}{3!1!2!}(1/4)^3(1/2)^1(1/4)^2$$

$$P = 0.029 = 2.9\%$$

This means that 2.9% of the time we would expect to obtain six plants, three with red flowers, one with pink flowers, and two with white flowers.

Conceptual Questions

C1. Describe the differences among dominance, incomplete dominance, codominance, and overdominance.

C2. Discuss the differences among sex-influenced, sex-limited, and sex-linked inheritance. Describe examples.

C3. What is meant by a gene interaction? How can a gene interaction be explained at the molecular level?

C4. Let's suppose a recessive allele encodes a completely defective protein. If the functional allele is dominant, what does that tell us about the amount of the functional protein that is sufficient to cause the phenotype? What if the allele shows incomplete dominance?

C5. A nectarine is a peach without the fuzz. The difference is controlled by a single gene that is found in two alleles, *D* and *d*. At the molecular level, would it make more sense to you that the nectarine is homozygous for a recessive allele or that the peach is homozygous for the recessive allele? Explain your reasoning.

C6. An allele in *Drosophila* produces a "star-eye" trait in the heterozygous individual. However, the star-eye allele is lethal in homozygotes. What would be the ratio and phenotypes of surviving flies if star-eyed flies were crossed to each other?

C7. A seed dealer wants to sell four-o'clock seeds that will produce only red, white, or pink flowers. Explain how this should be done.

C8. The serum from one individual (let's call this person individual 1) is known to agglutinate the red blood cells from a second individual (individual 2). List the pairwise combinations of possible genotypes that individuals 1 and 2 could be. If individual 1 is the parent of individual 2, what are his/her possible genotypes?

C9. Which blood phenotypes (A, B, AB, and/or O) provide an unambiguous genotype? Is it possible for a couple to produce a family of children with all four blood types? If so, what would the genotypes of the parents have to be?

C10. A woman with type B blood has a child with type O blood. What are the possible genotypes and blood types of the father?

C11. A type A woman is the daughter of a type O father and type A mother. If she has children with a type AB man, what are the following probabilities?

A. A type AB child

B. A type O child

C. The first three children with type AB blood

D. A family containing two children with type B blood and one child with type AB

C12. In Shorthorn cattle, coat color is controlled by a single gene that can exist as a red allele (R) or white allele (r). The heterozygotes (Rr) have a color called roan that looks less red than the RR homozygotes. However, when examined carefully, the roan phenotype in cattle is actually due to a mixture of completely red hairs and completely white hairs. Should this be called incomplete dominance, codominance, or something else? Explain your reasoning.

C13. In chickens, the Leghorn variety has white feathers due to a dominant allele. Silkies have white feathers due to a recessive allele in a second (different) gene. If a true-breeding white Leghorn is crossed to a true-breeding white Silkie, what is the expected phenotype of the F_1 generation? If members of the F_1 generation are mated to each other, what is the expected phenotypic outcome of the F_2 generation? Assume the chickens in the parental generation are homozygous for the white allele at one gene and homozygous for the brown allele at the other gene. In subsequent generations, nonwhite birds will be brown.

C14. Propose the most likely mode of inheritance (autosomal dominant, autosomal recessive, or X-linked recessive) for the following pedigrees. Affected individuals are shown with filled (black) symbols.

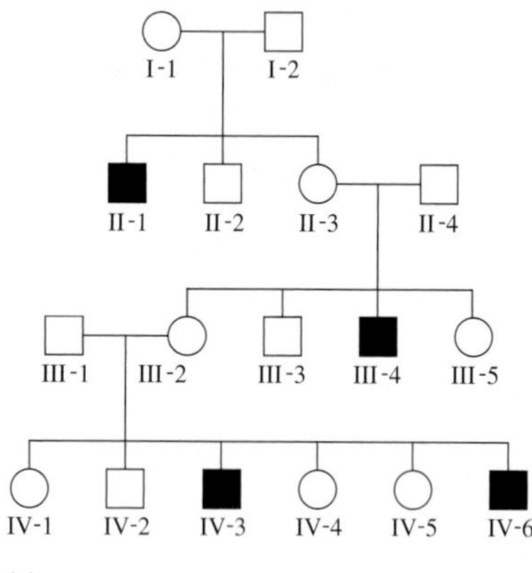

(a)

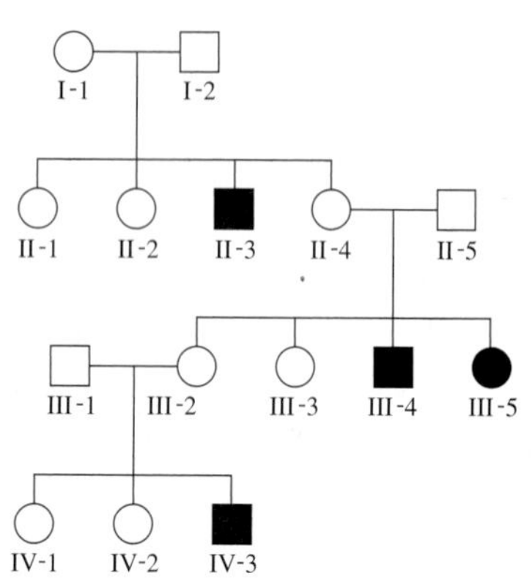

(b)

C15. A human disease known as vitamin D–resistant rickets is inherited as an X-linked dominant trait. If a male with the disease produces children with a female who does not have the disease, what is the expected ratio of affected and unaffected offspring?

C16. Hemophilia is a X-linked recessive trait in humans. If a heterozygous woman has children with a an unaffected man, what are the odds of the following combinations of children?

A. An affected son

B. Four unaffected offspring in a row

C. An unaffected daughter or son

D. Two out of five offspring that are affected

C17. Explain whether the following events could or could not happen.

A. A male has the same X chromosome as his paternal grandfather.

B. A male has the same X chromosome as his maternal grandfather.

C. A female has the same X chromosome as her maternal grandmother.

D. A female has the same X chromosome as her maternal great-great-grandmother.

C18. Incontinentia pigmenti is a rare, X-linked dominant disorder in humans characterized by swirls of pigment in the skin. If an affected female, who had an unaffected father, has children with an unaffected male, what would be the predicted ratios of affected and unaffected sons and daughters?

C19. With regard to pattern baldness in humans (a sex-influenced trait), a woman who is not bald and whose mother is bald has children with a bald man whose father is not bald. What are their probabilities of having the following types of families?

A. Their first child will not become bald.

B. Their first child will be a male who will not become bald.

C. Their first three children will be females who are not bald.

C20. In rabbits, the color of body fat is controlled by a single gene with two alleles, designated *Y* and *y*. The outcome of this trait is affected by the diet of the rabbit. When raised on a standard vegetarian diet, the dominant *Y* allele confers white body fat, and the *y* allele confers yellow body fat. However, when raised on a xanthophyll-free diet, the homozygote *yy* animal has white body fat. If a heterozygous animal is crossed to a rabbit with yellow body fat, what are the proportions of offspring with white and yellow body fat when raised on a standard vegetarian diet? How do the proportions change if the offspring are raised on a xanthophyll-free diet?

C21. A Siamese cat that spends most of its time outside was accidentally injured in a trap and required several stitches in its right front paw. The veterinarian had to shave the fur from the paw and leg, which originally had rather dark fur. Later, when the fur grew back, it was much lighter than the fur on the other three legs. Do you think this injury occurred in the hot summer or cold winter? Explain your answer.

C22. A true-breeding male fly with eosin eyes is crossed to a white-eyed female that is heterozygous for the wild-type (*C*) and cream alleles (c^a). What are the expected proportions of their offspring?

C23. In Chapter 4, we considered the trait of hen- versus cock-feathering. Starting with two heterozygous fowl that are hen-feathered, explain how you would obtain a true-breeding line that always produced cock-feathered males.

C24. In the pedigree shown here for a trait determined by a single gene (affected individuals are shown in black), state whether it would be possible for the trait to be inherited in each of the following ways:

A. Recessive

B. X-linked recessive

C. Dominant, complete penetrance

D. Sex influenced, dominant in males

E. Sex limited

F. Dominant, incomplete penetrance

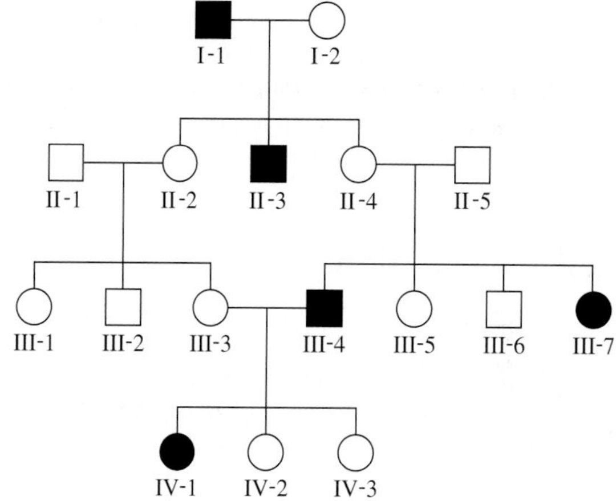

C25. The pedigree shown here also concerns a trait determined by a single gene (affected individuals are shown in black). Which of the following patterns of inheritance are possible?

A. Recessive

B. X-linked recessive

C. Dominant

D. Sex influenced, recessive in males

E. Sex limited

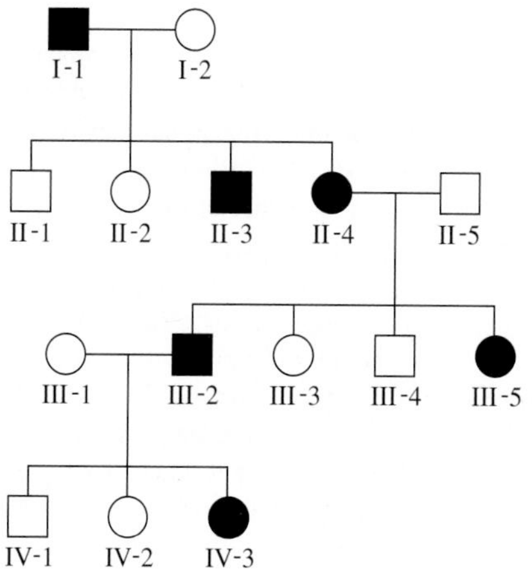

C26. Let's suppose you have pedigree data from thousands of different families involving a particular genetic disease. How would you decide whether the disease is inherited as a recessive trait as opposed to one that is dominant with incomplete penetrance?

C27. Compare phenotypes at the molecular, cellular, and organism levels for individuals who are homozygous for the hemoglobin allele, Hb^AHb^A, and the sickle-cell allele, Hb^SHb^S.

C28. A very rare dominant allele that causes the little finger to be crooked has a penetrance value of 80%. In other words, 80% of heterozygotes carrying the allele will have a crooked little finger. If a homozygous unaffected person has children with a heterozygote carrying this mutant allele, what is the probability that an offspring will have little fingers that are crooked?

C29. A sex-influenced trait in humans is one that affects the length of the index finger. A "short" allele is dominant in males and recessive in females. Heterozygous males have an index finger that is significantly shorter than the ring finger. The gene affecting index finger length is located on an autosome. A woman with short index fingers has children with a man who has normal index fingers. They produce five children in the following order: female, male, male, female, male. The oldest female offspring marries a man with normal fingers and then has one daughter. The youngest male among the five children marries a woman with short index fingers, and then they have two sons. Draw the pedigree for this family. Indicate the phenotypes of every individual (filled symbols for individuals with short index fingers and open symbols for individuals with normal index fingers).

C30. In horses, there are three coat-color patterns termed cremello (beige), chestnut (brown), and palomino (golden with light mane and tail). If two palomino horses are mated, they produce about 1/4 cremello, 1/4 chestnut, and 1/2 palomino offspring. In contrast, cremello horses and chestnut horses breed true. (In other words, two cremello horses will produce only cremello offspring and two chestnut horses will produce only chestnut offspring.) Explain this pattern of inheritance.

C31. Briefly describe three explanations for how a suppressor mutation exerts its effects at the molecular level.

Experimental Questions

E1. Mexican hairless dogs have little hair and few teeth. When a Mexican hairless is mated to another breed of dog, about half the puppies are hairless. When two Mexican hairless dogs are mated to each other, about 1/3 of the surviving puppies have hair, and about 2/3 of the surviving puppies are hairless. However, about two out of eight puppies from this type of cross are born grossly deformed and do not survive. Explain this pattern of inheritance.

E2. In chickens, some varieties have feathered shanks (legs), while others do not. In a cross between a Black Langhans (feathered shanks) and Buff Rocks (unfeathered shanks), the shanks of the F_1 generation are all feathered. When the F_1 generation is crossed, the F_2 generation contains chickens with feathered shanks to unfeathered shanks in a ratio of 15:1. Suggest an explanation for this result.

E3. In sheep, the formation of horns is a sex-influenced trait; the allele that results in horns is dominant in males and recessive in females. Females must be homozygous for the horned allele to have horns. A horned ram was crossed to a polled (unhorned) ewe, and the first offspring they produced was a horned ewe. What are the genotypes of the parents?

E4. A particular breed of dog can have long hair or short hair. When true-breeding long-haired animals were crossed to true-breeding short-haired animals, the offspring all had long hair. The F_2 generation produced a 3:1 ratio of long- to short-haired offspring. A second trait involves the texture of the hair. The two variants are wiry hair and straight hair. F_1 offspring from a cross of these two varieties all had wiry hair, and F_2 offspring showed a 3:1 ratio of wiry- to straight-haired puppies. Recently, a breeder of the short-, wiry-haired dogs found a female puppy that was albino. Similarly, another breeder of the long-, straight-haired dogs found a male puppy that was albino. Because the albino trait is always due to a recessive allele, the two breeders got together and mated the two dogs. Surprisingly, all the puppies in the litter had black hair. How would you explain this result?

E5. In the clover butterfly, males are always yellow, but females can be yellow or white. In females, white is a dominant allele. Two yellow butterflies were crossed to yield an F_1 generation consisting of 50% yellow males, 25% yellow females, and 25% white females. Describe how this trait is inherited and the genotypes of the parents.

E6. The *Mic2* gene in humans is present on both the X and Y chromosome. Let's suppose the *Mic2* gene exists in a dominant *Mic2* allele, which results in normal surface antigen, and a recessive *mic2* allele, which results in defective surface antigen production. Using molecular techniques, it is possible to identify homozygous and heterozygous individuals. By following the transmission of the *Mic2* and *mic2* alleles in a large human pedigree, would it be possible to distinguish between pseudoautosomal inheritance and autosomal inheritance? Explain your answer.

E7. Duroc Jersey pigs are typically red, but a sandy variation is also seen. When two different varieties of true-breeding sandy pigs were crossed to each other, they produced F_1 offspring that were red. When these F_1 offspring were crossed to each other, they produced red, sandy, and white pigs in a 9:6:1 ratio. Explain this pattern of inheritance.

E8. As discussed in Chapter 4, comb morphology in chickens is governed by a gene interaction. Two walnut comb chickens were crossed to each other. They produced only walnut comb and rose comb offspring, in a ratio of 3:1. What are the genotypes of the parents?

E9. In certain species of summer squash, fruit color is determined by two interacting genes. A dominant allele, *W*, determines white color, and a recessive allele allows the fruit to be colored. In a homozygous *ww* individual, a second gene determines fruit color: *G* (green) is dominant to *g* (yellow). A white squash and a yellow squash were crossed, and the F_1 generation yielded approximately 50% white fruit and 50% green fruit. What are the genotypes of the parents?

E10. Certain species of summer squash can exist in long, spherical, or disk shapes. When a true-breeding long-shaped strain was crossed to a true-breeding disk-shaped strain, all of the F_1 offspring were disk-shaped. When the F_1 offspring were allowed to self-fertilize, the F_2 generation consisted of a ratio of 9 disk-shaped : 6 round-shaped : 1 long-shaped. Assuming the shape of summer squash is governed by two different genes, with each gene existing in two alleles, propose a mechanism to account for this 9:6:1 ratio.

E11. In a species of plant, two genes control flower color. The red allele (*R*) is dominant to the white allele (*r*); the color-producing allele (*C*) is dominant to the non-color-producing allele (*c*). You suspect that either an *rr* homozygote or a *cc* homozygote will produce white flowers. In other words, *rr* is epistatic to *C*, and *cc* is epistatic to *R*. To test your hypothesis, you allowed heterozygous plants (*RrCc*) to self-fertilize and counted the offspring. You obtained the following data: 201 plants with red flowers and 144 with white flowers. Conduct a chi square analysis to see if your observed data are consistent with your hypothesis.

E12. In *Drosophila*, red eyes is the wild-type phenotype. There are several different genes (with each gene existing in two or more alleles) that affect eye color. One allele causes purple eyes, and a different allele causes sepia eyes. Both of these alleles are recessive compared to red eye color. When flies with purple eyes were crossed to flies with sepia eyes, all of the F_1 offspring had red eyes. When the F_1 offspring were allowed to mate with each other, the following data were obtained:

146 purple eyes

151 sepia eyes

50 purplish sepia eyes

444 red eyes

Explain this pattern of inheritance. Conduct a chi square analysis to see if the experimental data fit with your hypothesis.

E13. As mentioned in Experimental question E12, red eyes is the wild-type phenotype in *Drosophila*, and there are several different genes (with each gene existing in two or more alleles) that affect eye color. One allele causes purple eyes, and a different allele causes vermilion eyes. The purple and vermilion alleles are recessive compared to red eye color. The following crosses were made, and the following data were obtained:

Cross 1: Males with vermillion eyes x females with purple eyes

354 offspring, all with red eyes

Cross 2: Males with purple eyes x females with vermillion eyes

212 male offspring with vermillion eyes

221 female offspring with red eyes

Explain the pattern of inheritance based on these results. What additional crosses might you make to confirm your hypothesis?

E14. Let's suppose that you were looking through a vial of fruit flies in your laboratory and noticed a male fly that has pink eyes. What crosses would you make to determine if the pink allele is an X-linked gene? What crosses would you make to determine if the pink allele is an allele of the same X-linked gene that has white and eosin alleles? Note: The white and eosin alleles are discussed in Figure 4.22.

E15. When examining a human pedigree, what features do you look for to distinguish between X-linked recessive inheritance versus autosomal recessive inheritance? How would you distinguish X-linked dominant inheritance from autosomal dominant inheritance in a human pedigree?

Questions for Student Discussion/Collaboration

1. Let's suppose a gene exists as a functional wild-type allele and a nonfunctional mutant allele. At the organism level, the wild-type allele is dominant. In a heterozygote, discuss whether dominance occurs at the cellular or molecular level. Discuss examples in which the issue of dominance depends on the level of examination.
2. A true-breeding rooster with a rose comb, feathered shanks, and cock-feathering was crossed to a hen that is true-breeding for pea comb and unfeathered shanks but is heterozygous for hen-feathering. If you assume these genes can assort independently, what is the expected outcome of the F_1 generation?
3. In oats, the color of the chaff is determined by a two-gene interaction. When a true-breeding black plant was crossed to a true-breeding white plant, the F_1 generation was composed of all black plants. When the F_1 offspring were crossed to each other, the ratio produced was 12 black : 3 gray : 1 white. First, construct a Punnett square that accounts for this pattern of inheritance. Which genotypes produce the gray phenotype? Second, at the level of protein function, how would you explain this type of inheritance?

Note: All answers appear at the website for this textbook; the answers to even-numbered questions are in the back of the textbook.

www.mhhe.com/brookergenetics3e

Visit the Online Learning Center for practice tests, answer keys, and other learning aids for this chapter. Enhance your understanding of genetics with our interactive exercises, quizzes, animations, and much more.

5 LINKAGE AND GENETIC MAPPING IN EUKARYOTES

CHAPTER OUTLINE

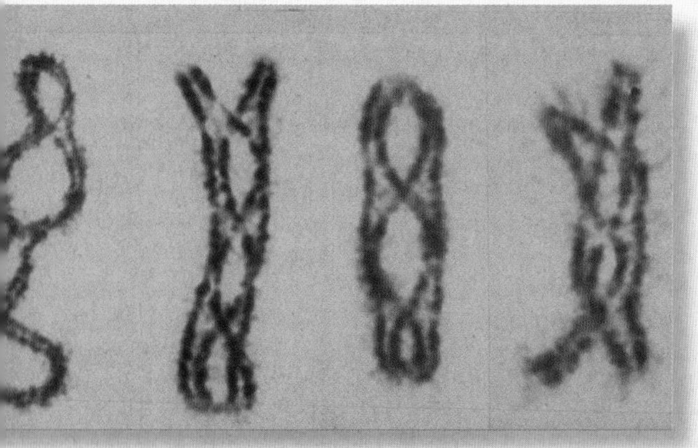

Crossing over during meiosis. This event provides a way to reassort the alleles of genes that are located on the same chromosome.

In Chapter 2, we were introduced to Mendel's laws of inheritance. According to these principles, we expect that two different genes will segregate and independently assort themselves during the process that creates gametes. After Mendel's work was rediscovered at the turn of the twentieth century, chromosomes were identified as the cellular structures that carry genes. The chromosome theory of inheritance explained how the transmission of chromosomes is responsible for the passage of genes from parents to offspring.

When geneticists first realized that chromosomes contain the genetic material, they began to suspect that a conflict might sometimes occur between the law of independent assortment of genes and the behavior of chromosomes during meiosis. In particular, geneticists assumed that each species of organism must contain thousands of different genes, yet cytological studies revealed that most species have at most a few dozen chromosomes. Therefore, it seemed likely, and turned out to be true, that each chromosome would carry many hundreds or even thousands of different genes. The transmission of genes located close to each other on the same chromosome violates the law of independent assortment.

In this chapter, we will consider the pattern of inheritance that occurs when different genes are situated on the same chromosome. In addition, we will briefly explore how the data from genetic crosses are used to construct a **genetic map**—a diagram that describes the order of genes along a chromosome. Newer strategies for gene mapping are described in Chapter 20. However, an understanding of traditional mapping studies, as described in this chapter, will strengthen our appreciation for these newer molecular approaches. More importantly, traditional mapping studies further illustrate how the location of two or more genes on the same chromosome can affect the transmission patterns from parents to offspring.

5.1 LINKAGE AND CROSSING OVER

In eukaryotic species, each linear chromosome contains a very long segment of DNA. A chromosome contains many individual functional units—called genes—that influence an organism's traits. A typical chromosome is expected to contain many hundreds or perhaps a few thousand different genes. The term **linkage** has two related meanings. First, linkage refers to the phenomenon that two or more genes may be located on the same chromosome. The genes are physically linked to each other, because each eukaryotic chromosome contains a single, continuous, linear molecule of DNA. Second, genes that are close together on the same chromosome tend to be transmitted as a unit. This second meaning indicates that linkage has an influence on inheritance patterns.

Chromosomes are sometimes called **linkage groups,** because a chromosome contains a group of genes that are physically linked together. In each species, the number of linkage groups equals the number of chromosome types. For example, human somatic cells have 46 chromosomes, which are composed of 22 types of autosomes that come in pairs plus one pair of sex chromosomes, the X and Y. Therefore, humans have 22 autosomal linkage groups, an X chromosome linkage group, and a Y chromosome linkage group.

Geneticists are often interested in the transmission of two or more traits in a genetic cross. When a geneticist follows the variants of two different traits in a cross, this is called a **dihybrid cross;** when three traits are followed, it is a **trihybrid cross;** and so on. The outcome of a dihybrid or trihybrid cross depends on whether or not the genes are linked to each other along the same chromosome. In this section, we will examine how linkage affects the transmission patterns of two or more traits.

Crossing Over May Produce Recombinant Genotypes

Even though the alleles for different genes may be linked along the same chromosome, the linkage can be altered during meiosis. In diploid eukaryotic species, homologous chromosomes can exchange pieces with each other, a phenomenon called **crossing over.** This event occurs during prophase of meiosis I. As discussed in Chapter 3, the replicated chromosomes, known as sister chromatids, associate with the homologous sister chromatids to form a structure known as a **bivalent.** A bivalent is composed of two pairs of sister chromatids. In prophase of meiosis I, a sister chromatid of one pair commonly crosses over with a sister chromatid from the homologous pair.

Figure 5.1 considers meiosis when two genes are linked on the same chromosome. One of the parental chromosomes carries the *A* and *B* alleles, while the homologue carries the *a* and *b* alleles. In Figure 5.1a, no crossing over has occurred. Therefore, the resulting haploid cells contain the same combination of alleles as the original chromosomes. In this case, two haploid cells carry the dominant *A* and *B* alleles, and the other two carry the recessive *a* and *b* alleles. The arrangement of linked alleles has not been altered.

In contrast, Figure 5.1b illustrates what can happen when crossing over occurs. Two of the haploid cells contain combinations of alleles, namely *A* and *b* or *a* and *B*, which differ from those in the original chromosomes. In these two cells, the grouping of linked alleles has been changed. An event such as this, leading to a new combination of alleles, is known as **genetic recombination.** The haploid cells carrying the *A* and *b*, or the *a* and *B*, alleles are called **nonparental** or **recombinant cells.** Likewise, if such haploid cells were gametes that participated in fertilization, the resulting offspring are called nonparental or recombinant offspring. These offspring can display combinations of traits that are different from those of either parent. In contrast, offspring

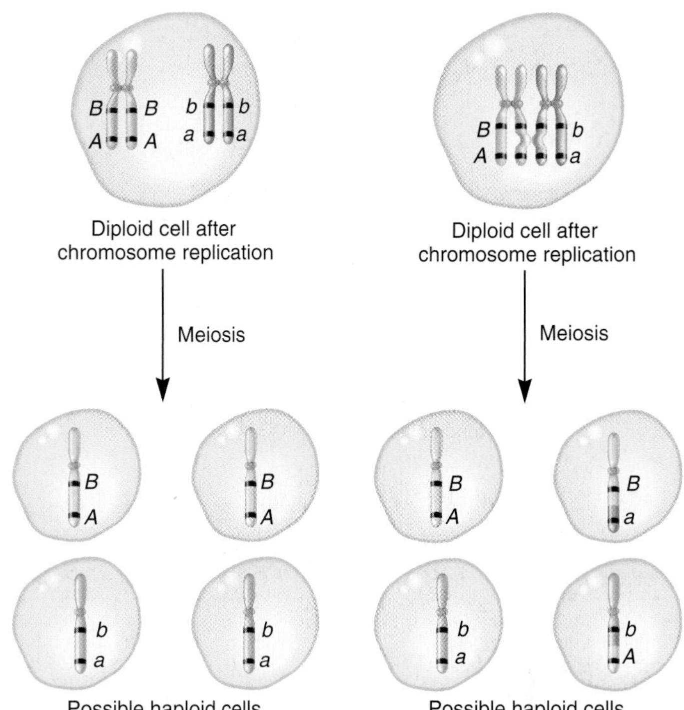

FIGURE 5.1 **Consequences of crossing over during meiosis.** (a) In the absence of crossing over, the *A* and *B* alleles and the *a* and *b* alleles are maintained in the same arrangement found in the parental chromosomes. (**b**) Crossing over has occurred in the region between the two genes, creating two nonparental haploid cells with a new combination of alleles.

that have inherited the same combination of alleles that are found in the chromosomes of their parents are known as **parental** or **nonrecombinant** offspring.

In this section, we will consider how crossing over affects the pattern of inheritance for genes linked on the same chromosome. In Chapter 17, we will consider the molecular events that cause crossing over to occur.

Bateson and Punnett Discovered Two Traits That Did Not Assort Independently

An early study indicating that some traits may not assort independently was carried out by William Bateson and Reginald Punnett in 1905. According to Mendel's law of independent assortment, a dihybrid cross between two individuals, heterozygous for two genes, should yield a 9:3:3:1 phenotypic ratio among the offspring. However, a surprising result occurred when Bateson and Punnett conducted a cross in the sweet pea involving two different traits—flower color and pollen shape.

As seen in **Figure 5.2**, they began by crossing a true-breeding strain with purple flowers (*PP*) and long pollen (*LL*) to

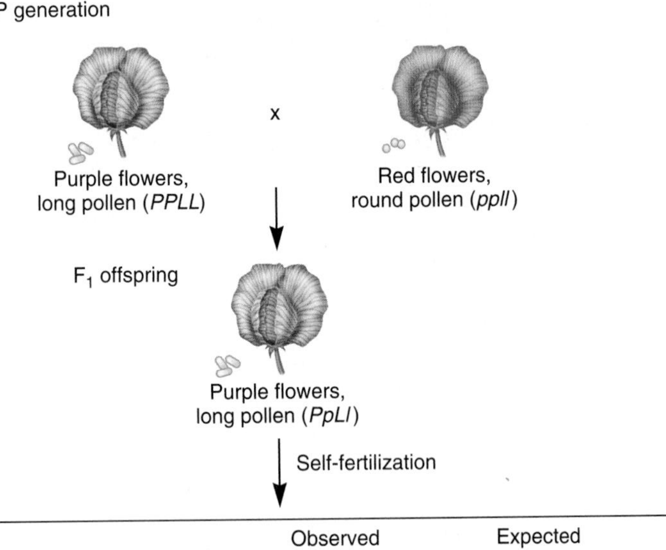

F_2 offspring	Observed number	Ratio	Expected number	Ratio
Purple flowers, long pollen	296	15.6	240	9
Purple flowers, round pollen	19	1.0	80	3
Red flowers, long pollen	27	1.4	80	3
Red flowers, round pollen	85	4.5	27	1

FIGURE 5.2 An experiment of Bateson and Punnett with sweet peas, showing that independent assortment does not always occur. Note: The expected numbers are rounded to the nearest whole number.

Genes →Traits Two genes that govern flower color and pollen shape are found on the same chromosome. Therefore, the offspring tend to inherit the parental combinations of alleles (*PL* or *pl*). Due to occasional crossing over, a lower percentage of offspring inherit nonparental combinations of alleles (*Pl* or *pL*).

a strain with red flowers (*pp*) and round pollen (*ll*). This yielded an F_1 generation of plants that all had purple flowers and long pollen (*PpLl*). An unexpected result came from the F_2 generation. Even though the F_2 generation had four different phenotypic categories, the observed numbers of offspring did not conform to a 9:3:3:1 ratio. Bateson and Punnett found that the F_2 generation had a much greater proportion of the two phenotypes found in the parental generation—purple flowers with long pollen and red flowers with round pollen. Therefore, they suggested that the transmission of these two traits from the parental generation to the F_2 generation was somehow coupled and not easily assorted in an independent manner. However, Bateson and Punnett did not realize that this coupling was due to the linkage of the flower color gene and the pollen shape gene on the same chromosome.

Morgan Provided Evidence for the Linkage of X-Linked Genes and Proposed That Crossing Over Between X Chromosomes Can Occur

The first direct evidence that different genes are physically located on the same chromosome came from the studies of Thomas Hunt Morgan in 1911, who investigated the inheritance pattern of different traits that had been shown to follow an X-linked pattern of inheritance. **Figure 5.3** illustrates an experiment involving three traits that Morgan studied. His parental crosses were wild-type male fruit flies mated to females that had yellow bodies (*yy*), white eyes (*ww*), and miniature wings (*mm*). The wild-type alleles for these three genes are designated y^+ (gray body), w^+ (red eyes), and m^+ (long wings). As expected, the phenotypes of the F_1 generation were wild-type females, and males with yellow bodies, white eyes, and miniature wings. The linkage of these genes was revealed when the F_1 flies were mated to each other and the F_2 generation examined.

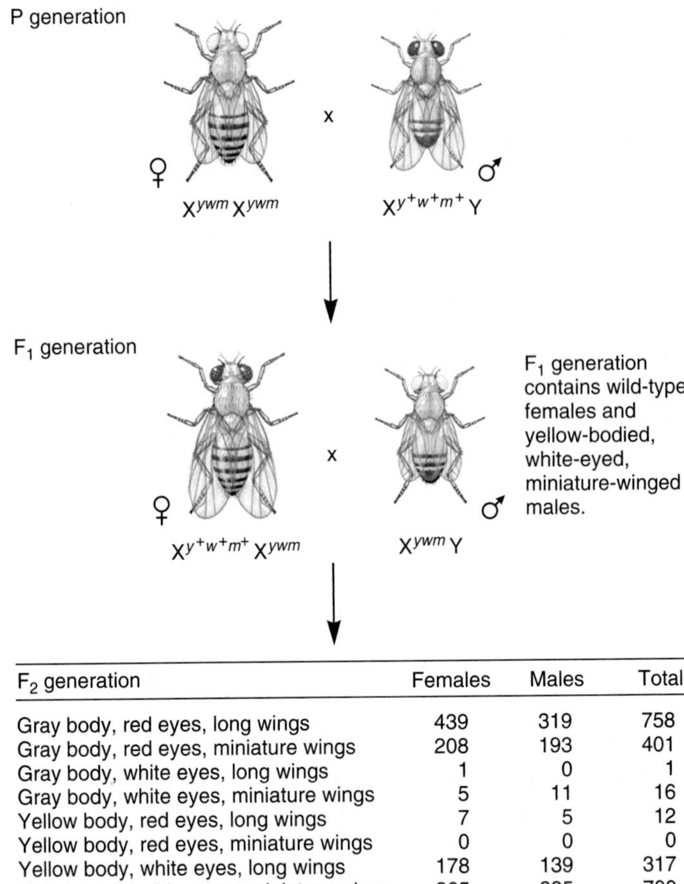

F_2 generation	Females	Males	Total
Gray body, red eyes, long wings	439	319	758
Gray body, red eyes, miniature wings	208	193	401
Gray body, white eyes, long wings	1	0	1
Gray body, white eyes, miniature wings	5	11	16
Yellow body, red eyes, long wings	7	5	12
Yellow body, red eyes, miniature wings	0	0	0
Yellow body, white eyes, long wings	178	139	317
Yellow body, white eyes, miniature wings	365	335	700

FIGURE 5.3 Morgan's trihybrid cross involving three X-linked traits in *Drosophila*.

Genes →Traits Three genes that govern body color, eye color, and wing length are all found on the X chromosome. Therefore, the offspring tend to inherit the parental combinations of alleles (y^+ w^+ m^+ or *y w m*). Figure 5.4 explains how single and double crossovers can create nonparental combinations of alleles.

Instead of equal proportions of the eight possible phenotypes, Morgan observed a much higher proportion of the combinations of traits found in the parental generation. He observed 758 flies with gray bodies, red eyes, and long wings, and 700 flies with yellow bodies, white eyes, and miniature wings. The combination of gray body, red eyes, and long wings was found in the males of the parental generation, while the combination of yellow body, white eyes, and miniature wings was the same as the females of the parental generation. Morgan's explanation for this higher proportion of parental combinations was that all three genes are located on the X chromosome and, therefore, tend to be transmitted together as a unit.

However, to fully account for the data shown in Figure 5.3, Morgan needed to interpret two other key observations. First, he needed to explain why a significant proportion of the F_2 generation had nonparental combinations of alleles. Along with the two parental phenotypes, five other phenotypic combinations appeared that were not found in the parental generation. Second, he needed to explain why a quantitative difference was observed between nonparental combinations involving body color and eye color versus eye color and wing length. This quantitative difference is revealed by reorganizing the data from Morgan's cross by pairs of genes.

Gray body, red eyes	1,159	
Yellow body, white eyes	1,017	
Gray body, white eyes	17	Nonparental offspring
Yellow body, red eyes	12	
Total	2,205	

Red eyes, long wings	770	
White eyes, miniature wings	716	
Red eyes, miniature wings	401	Nonparental offspring
White eyes, long wings	318	
Total	2,205	

Morgan found a substantial difference between the numbers of nonparental offspring when pairs of genes were considered separately. Nonparental combinations involving only eye color and wing length were fairly common—401 + 318 nonparental offspring. In sharp contrast, nonparental combinations for body color and eye color were quite rare—17 + 12 nonparental offspring.

How did Morgan explain these data? He considered the studies conducted in 1909 of the Belgian cytologist F. A. Janssens, who observed chiasmata under the microscope and proposed that crossing over involves a physical exchange between homologous chromosomes. Morgan shrewdly realized that crossing over between homologous X chromosomes was consistent with his data. He assumed that crossing over did not occur between the X and Y chromosome and that these three genes are not found on the Y chromosome. With these ideas in mind, he made three important hypotheses to explain his results:

1. The genes for body color, eye color, and wing length are all located on the same chromosome, namely, the X chromosome. Therefore, the alleles for all three traits are most likely to be inherited together.
2. Due to crossing over, the homologous X chromosomes (in the female) can exchange pieces of chromosomes and create new (nonparental) combinations of alleles.
3. The likelihood of crossing over depends on the distance between two genes. If two genes are far apart from each other, crossing over is more likely to occur between them.

With these ideas in mind, Figure 5.4 illustrates the possible events that occurred in the F_1 female flies of Morgan's experiment. One of the X chromosomes carried all three dominant alleles, while the other had all three recessive alleles. During oogenesis in the F_1 female flies, crossing over may or may not have occurred in this region of the X chromosome. If no crossing over occurred, the parental phenotypes were produced in the F_2 offspring. Alternatively, a crossover sometimes occurred between the eye color gene and the wing length gene to produce nonparental offspring with gray bodies, red eyes, and miniature wings or yellow bodies, white eyes, and long wings. According to Morgan's proposal, such an event is fairly likely because these two genes are far apart from each other on the X chromosome. In contrast, he proposed that the body color and eye color genes are very close together, which makes crossing over between them an unlikely event. Nevertheless, it occasionally occurred, yielding offspring with gray bodies, white eyes, and miniature wings, or with yellow bodies, red eyes, and long wings. Finally, it was also possible for two homologous chromosomes to cross over twice. This double crossover is very unlikely. Among the 2,205 offspring Morgan examined, he found only one fly with a gray body, white eyes, and long wings that could be explained by this phenomenon.

A Chi Square Analysis Can Be Used to Distinguish Between Linkage and Independent Assortment

Now that we have an appreciation for linkage and the production of recombinant offspring, let's consider how an experimenter can objectively decide whether two genes are linked or assort independently. In Chapter 2, we used chi square analysis to evaluate the goodness of fit between a genetic hypothesis and observed experimental data. This method can similarly be employed to determine if the outcome of a dihybrid cross is consistent with linkage or independent assortment.

To conduct a chi square analysis, we must first propose a hypothesis. In a dihybrid cross, the standard hypothesis is that the two genes are not linked. This hypothesis is chosen even if the observed data suggest linkage, because an independent assortment hypothesis allows us to calculate the expected number of offspring based on the genotypes of the parents and the law of independent assortment. In contrast, for two linked genes that have not been previously mapped, we cannot calculate the expected number of offspring from a genetic cross because we do not know how likely it is for a crossover to occur between the two genes. Without expected numbers of recombinant and parental offspring, we cannot conduct a chi square test. Therefore, we begin with the hypothesis that the genes are not linked. Recall from Chapter 2 that the hypothesis we are testing is called a **null hypothesis,** because it assumes there is no real difference between the observed and expected values. The goal is to determine whether or not the data fit the hypothesis. If the chi square value is low and we cannot reject the null hypothesis, we infer that the genes assort independently. On the other hand, if the chi square value is so high that our hypothesis is rejected, we will accept the alternative hypothesis, namely, that the genes are linked.

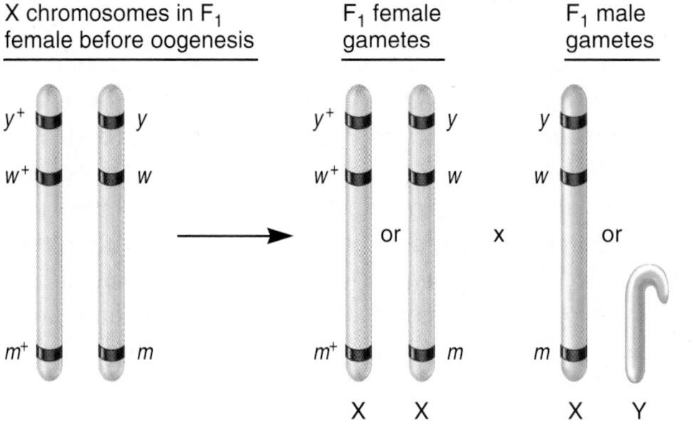

(a) No crossing over between y^+ and m^+ (most likely)

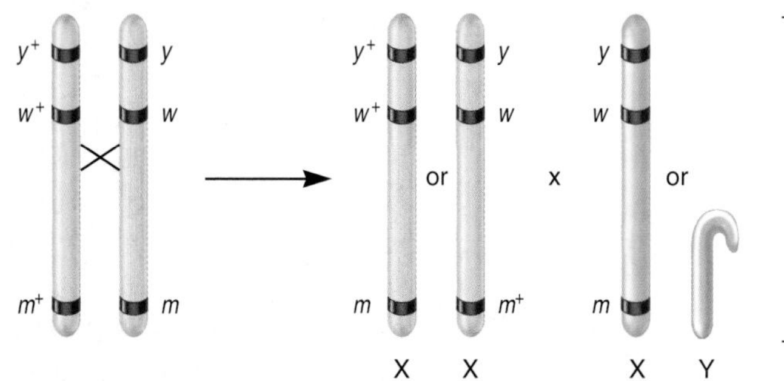

(b) Crossing over between w^+ and m^+ (fairly likely)

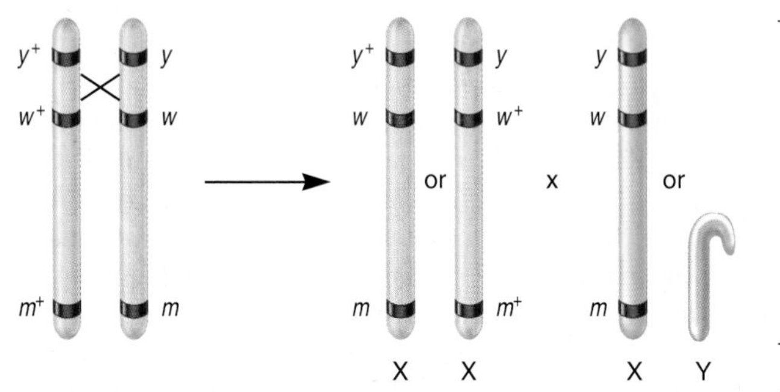

(c) Crossing over between y^+ and w^+ (unlikely)

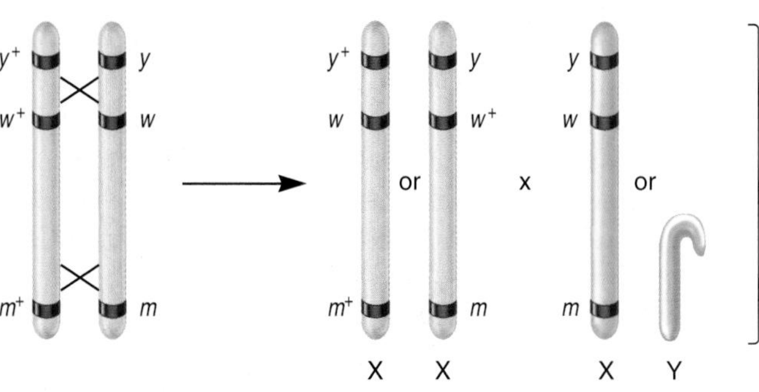

(d) Double crossing over (very unlikely)

FIGURE 5.4 The likelihood of crossing over provides an explanation of Morgan's trihybrid cross. (**a**) In the absence of crossing over, the parental chromosomes carry the $y^+w^+m^+$ or $y\ w\ m$ alleles. Alternatively, a crossover may occur between homologous X chromosomes in the female parent during oogenesis. As described in Chapter 3, crossing over actually occurs at the bivalent stage, but for simplicity, this figure shows only two X chromosomes (one of each homologue) rather than four chromatids, which would occur during the bivalent stage of meiosis. In part (**b**), a crossover has occurred between the eye color and wing length genes to produce recombinant chromosomes that carry the y^+w^+m or $y\ w\ m^+$ alleles. Because the eye color and wing length genes are relatively far apart, such recombinant offspring are common. (**c**) A crossover occurs between the body color and eye color genes to produce recombinant chromosomes that contain the $y^+w\ m$ or $y\ w^+\ m^+$ alleles. Because these two genes are very close together, these two types of recombinant offspring are fairly uncommon. (**d**) Finally, a double crossover could produce chromosomes carrying the $y^+w\ m^+$ or $y\ w^+\ m$ alleles. This occurs very rarely. Note: In this experiment, crossing over occurred between homologous X chromosomes in the female during oogenesis. A peculiar characteristic of *Drosophila* is that crossing over does not occur during spermatogenesis. In many other animal species, homologous recombination commonly occurs during spermatogenesis. Also note that this figure shows only a portion of the X chromosome. A map of the entire X chromosome is shown in Figure 5.8.

Of course, a statistical analysis cannot prove that a hypothesis is true. If the chi square value is high, we accept the linkage hypothesis because we are assuming that only two explanations for a genetic outcome are possible: the genes are either linked or not linked. However, if other factors affect the outcome of the cross, such as a decreased viability of particular phenotypes, these may result in large deviations between the observed and expected values and cause us to reject the independent assortment hypothesis even though it may be correct.

Let's consider Morgan's data concerning body color and eye color (refer back to page 103). This cross produced the following offspring: 1,159 gray body, red eyes; 1,017 yellow body, white eyes; 17 gray body, white eyes; and 12 yellow body, red eyes. However, when a heterozygous female ($X^{y^+w^+}X^{yw}$) is crossed to a hemizygous male ($X^{yw}Y$), the laws of segregation and independent assortment predict the following outcome:

F1 male gametes

F1 female gametes ♀ / ♂	X^{yw}	Y
$X^{y^+w^+}$	$X^{y^+w^+}X^{yw}$ Gray body, red eyes	$X^{y^+w^+}Y$ Gray body, red eyes
X^{y^+w}	$X^{y^+w}X^{yw}$ Gray body, white eyes	$X^{y^+w}Y$ Gray body, white eyes
X^{yw^+}	$X^{yw^+}X^{yw}$ Yellow body, red eyes	$X^{yw^+}Y$ Yellow body, red eyes
X^{yw}	$X^{yw}X^{yw}$ Yellow body, white eyes	$X^{yw}Y$ Yellow body, white eyes

Mendel's laws predict a 1:1:1:1 ratio among the four phenotypes. The observed data obviously seem to conflict with this expected outcome. Nevertheless, we stick to the strategy just discussed. We begin with the hypothesis that the two genes are not linked, and then we conduct a chi square analysis to see if the data fit this hypothesis. If the data do not fit, we will reject the idea that the genes assort independently and conclude the genes are linked.

A step-by-step outline for applying the chi square test to distinguish between linkage and independent assortment is described next.

Step 1. *Propose a hypothesis.* Even though the observed data appear inconsistent with this hypothesis, we propose that the two genes for eye color and body color obey Mendel's law of independent assortment. This hypothesis allows us to calculate expected values. Because the data seem to conflict with this hypothesis, we actually anticipate that the chi square analysis will allow us to reject the independent assortment hypothesis in favor of a linkage hypothesis. We are also assuming the alleles follow the law of segregation, and the four phenotypes are equally viable.

Step 2. *Based on the hypothesis, calculate the expected values of each of the four phenotypes.* Each phenotype has an equal probability of occurring (see the Punnett square given previously). Therefore, the probability of each phenotype is 1/4. The observed F_2 generation had a total of 2,205 individuals. Our next step is to calculate the expected numbers of offspring with each phenotype when the total equals 2,205; 1/4 of the offspring should be each of the four phenotypes:

$1/4 \times 2{,}205 = 551$ (expected number of each phenotype, rounded to the nearest whole number)

Step 3. *Apply the chi square formula, using the data for the observed values (O) and the expected values (E) that have been calculated in step 2.* In this case, the data consist of four phenotypes.

$$\chi^2 = \frac{(O_1 - E_1)^2}{E_1} + \frac{(O_2 - E_2)^2}{E_2} + \frac{(O_3 - E_3)^2}{E_3} + \frac{(O_4 - E_4)^2}{E_4}$$

$$\chi^2 = \frac{(1{,}159 - 551)^2}{551} + \frac{(17 - 551)^2}{551} + \frac{(12 - 551)^2}{551} + \frac{(1{,}017 - 551)^2}{551}$$

$$\chi^2 = 670.9 + 517.5 + 527.3 + 394.1 = 2{,}109.8$$

Step 4. *Interpret the calculated chi square value.* This is done with a chi square table, as discussed in Chapter 2. The four phenotypes are based on the law of segregation and the law of independent assortment. By itself, the law of independent assortment predicts only two categories, recombinant and nonrecombinant. Therefore, based on a hypothesis of independent assortment, the degrees of freedom equals $n-1$, which is $2-1$, or 1.

The calculated chi square value is enormous! This means that the deviation between observed and expected values is very large. With one degree of freedom, such a large deviation is expected to occur by chance alone less than 1% of the time (see Table 2.1). Therefore, we reject the hypothesis that the two genes assort independently. As an alternative, we accept the hypothesis that the genes are linked.

EXPERIMENT 5A

Creighton and McClintock Showed That Crossing Over Produced New Combinations of Alleles and Resulted in the Exchange of Segments Between Homologous Chromosomes

As we have seen, Morgan's studies were consistent with the hypothesis that crossing over occurs between homologous chromosomes to produce new combinations of alleles. To obtain direct evidence that crossing over can result in genetic recombination, Harriet Creighton and Barbara McClintock used an interesting strategy involving parallel observations. In studies conducted in 1931, they first made crosses involving two linked genes to produce parental and recombinant offspring. Second, they used a microscope to view the structures of the chromosomes in the parents and in the offspring. Because the parental chromosomes had some unusual structural features, they could microscopically distinguish the two homologous chromosomes within a pair. As we will see, this enabled them to correlate the occurrence of recombinant offspring with microscopically observable exchanges in segments of homologous chromosomes.

Creighton and McClintock focused much of their attention on the pattern of inheritance of traits in corn. This species has 10 different chromosomes per set, which are named chromosome 1, chromosome 2, chromosome 3, and so on. In previous cytological examinations of corn chromosomes, some strains were found to have an unusual chromosome 9 with a darkly staining knob at one end. In addition, McClintock identified an abnormal version of chromosome 9 that also had an extra piece of chromosome 8 attached at the other end (**Figure 5.5a**). This chromosomal rearrangement is called a translocation.

Creighton and McClintock insightfully realized that this abnormal chromosome could be used to demonstrate that two homologous chromosomes physically exchange segments as a result of crossing over. They knew that a gene was located near the knobbed end of chromosome 9 that provided color to corn kernels. This gene existed in two alleles, the dominant allele *C* (colored) and the recessive allele *c* (colorless). A second gene, located near the translocated piece from chromosome 8, affected the texture of the kernel endosperm. The dominant allele *Wx* caused starchy endosperm, and the recessive *wx* allele caused waxy endosperm. Creighton and McClintock reasoned that a crossover involving a normal chromosome 9 and a knobbed/translocated chromosome 9 would produce a chromosome that had either a knob or a translocation, but not both. These two types of chromosomes would be distinctly different from either of the parental chromosomes (**Figure 5.5b**).

As shown in the experiment of **Figure 5.6**, Creighton and McClintock began with a corn strain that carried an abnormal chromosome that had a knob at one end and a translocation at the other. Genotypically, this chromosome was *C wx*. The cytologically normal chromosome in this strain was *c Wx*. This corn plant, termed parent A, had the genotype *Cc Wxwx*. It was

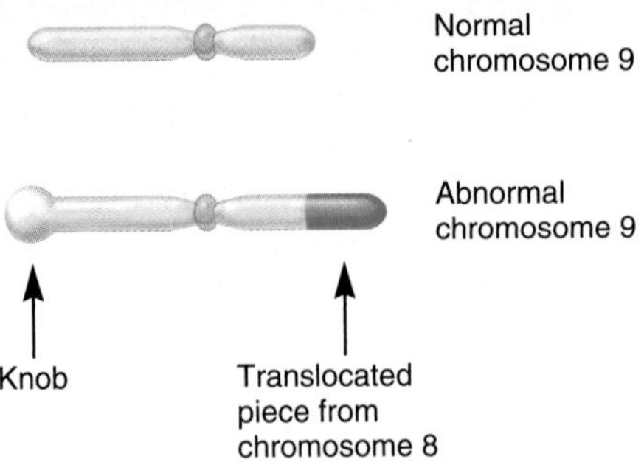

(a) Normal and abnormal chromosome 9

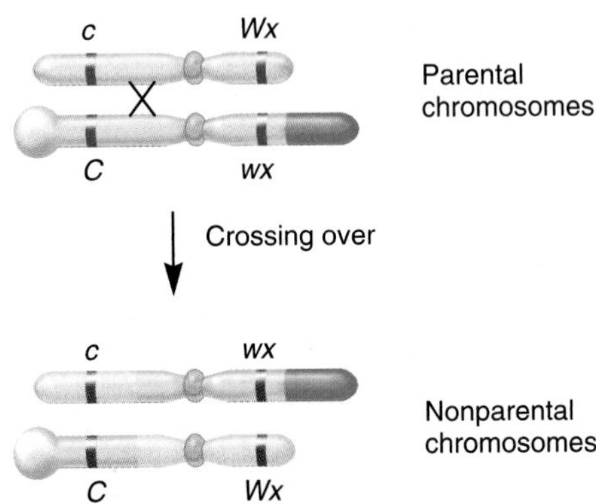

(b) Crossing over between normal and abnormal chromosome 9

FIGURE 5.5 Crossing over between a normal and abnormal chromosome 9 in corn. **(a)** A normal chromosome 9 in corn is compared to an abnormal chromosome 9 that contains a knob at one end and a translocation at the opposite end. **(b)** A crossover produces a chromosome that contains only a knob at one end and another chromosome that contains only a translocation at the other end.

crossed to a strain called parent B that carried two cytologically normal chromosomes and had the genotype *cc Wxwx*.

They then observed the kernels in two ways. First, they examined the phenotypes of the kernels to see if they were colored or colorless, and starchy or waxy. Second, the chromosomes in each kernel were examined under a microscope to determine their cytological appearance. Altogether, they observed a total of 25 kernels (see data of Figure 5.6).

THE HYPOTHESIS

Offspring with nonparental phenotypes are the product of a crossover. This crossover should create nonparental chromosomes via an exchange of chromosomal segments between homologous chromosomes.

TESTING THE HYPOTHESIS — FIGURE 5.6 Experimental correlation between genetic recombination and crossing over.

Starting materials: Two different strains of corn. One strain, referred to as parent A, had an abnormal chromosome 9 (knobbed/translocation) with a dominant *C* allele and a recessive *wx* allele. It also had a cytologically normal copy of chromosome 9 that carried the recessive *c* allele and the dominant *Wx* allele. Its genotype was *Cc Wxwx*. The other strain (referred to as parent B) had two normal versions of chromosome 9. The genotype of this strain was *cc Wxwx*.

1. Cross the two strains described. The tassel is the pollen-bearing structure, and the silk (equivalent to the stigma and style) is connected to the ovary. After fertilization, the ovary will develop into an ear of corn.

2. Observe the kernels from this cross.

3. Microscopically examine chromosome 9 in the kernels.

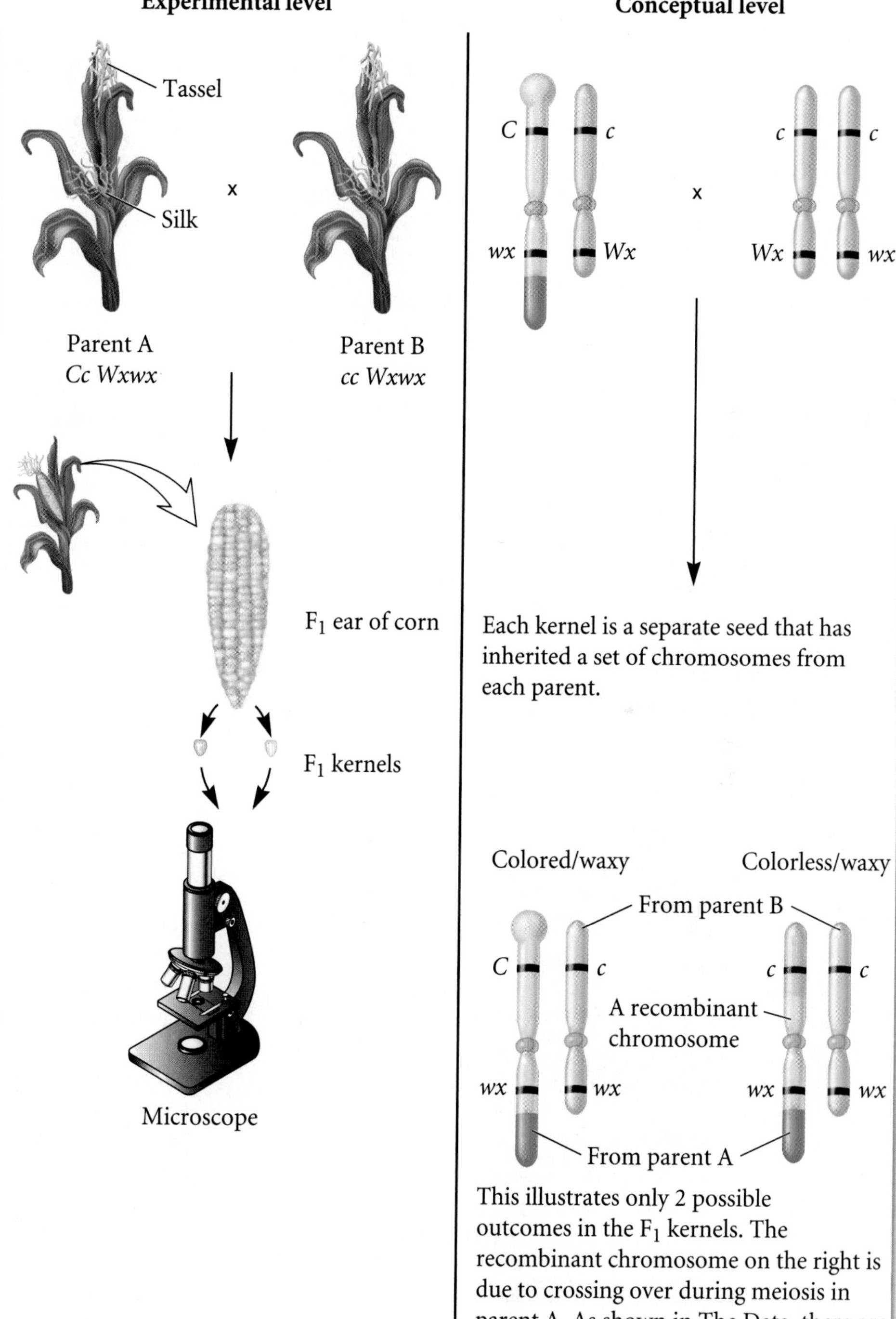

THE DATA

Phenotype of F_1 Kernel	*Number of Kernels Analyzed*	*Cytological Appearance of a Chromosome in F_1 Offspring**		*Did a Crossover Occur During Gamete Formation in Parent A?*
Colored/waxy	3	Knobbed/translocation *C* *wx*	Normal *c* *wx*	No
Colorless/starchy	11	Knobless/normal *c* *Wx*	Normal *c* or *Wx* *c* *wx*	No
Colorless/starchy	4	Knobless/translocation *c* *wx*	Normal *c* *Wx*	Yes
Colorless/waxy	2	Knobless/translocation *c* *wx*	Normal *c* *wx*	Yes
Colored/starchy	5	Knobbed/normal *C* *Wx*	Normal *c* or *Wx* *c* *wx*	Yes
Total	25			

*In this table, the chromosome on the left was inherited from parent A, and the chromosome on the right was inherited from parent B.

INTERPRETING THE DATA

By combining the gametes in a Punnett square, the following types of offspring can be produced:

Parent B

Parent A ♀ \ ♂	*c Wx*	*c wx*	
C wx	*Cc Wxwx* Colored, starchy	*Cc wxwx* Colored, waxy	Nonrecombinant
c Wx	*cc WxWx* Colorless, starchy	*cc Wxwx* Colorless, starchy	Nonrecombinant
C Wx	*Cc WxWx* Colored, starchy	*Cc Wxwx* Colored, starchy	Recombinant
c wx	*cc Wxwx* Colorless, starchy	*cc wxwx* Colorless, waxy	Recombinant

In this experiment, the researchers were interested in whether or not crossing over had occurred in parent A, which was heterozygous for both genes. This parent could produce four types of gametes, while parent B could produce only two types.

Parent A	*Parent B*
C wx (nonrecombinant)	*c Wx*
c Wx (nonrecombinant)	*c wx*
C Wx (recombinant)	
c wx (recombinant)	

As seen in the Punnett square, two of the phenotypic categories, colored, starchy (*Cc Wxwx* or *Cc WxWx*) and colorless, starchy (*cc WxWx* or *cc Wxwx*), were ambiguous because they could arise from a nonrecombinant and from a recombinant gamete. In other words, these phenotypes could be produced whether or not recombination occurred in parent A. Therefore, let's focus on the two unambiguous phenotypic categories: colored, waxy (*Cc wxwx*) and colorless, waxy (*cc wxwx*). The colored, waxy phenotype could happen only if recombination did not occur in parent A and if parent A passed the knobbed, translocated chromosome to its offspring. As shown in the data, three kernels were obtained with this phenotype, and all of them had the knobbed, translocated chromosome. By comparison, the colorless, waxy phenotype could be obtained only if genetic recombination occurred in parent A and this parent passed a chromosome 9 that had a translocation but was knobless. Two kernels were obtained with this phenotype, and both of them had the expected chromosome that had a translocation but was knobless. Taken together, these results showed a perfect correlation between genetic recombination of alleles and the cytological presence of a chromosome displaying a genetic exchange of chromosomal pieces from parent A.

Overall, the observations described in this experiment were consistent with the idea that a crossover occurred in the region between the *C* and *wx* genes that involved an exchange of segments between two homologous chromosomes. As stated by Creighton and McClintock, "Pairing chromosomes, heteromorphic in two regions, have been shown to exchange parts at the same time they exchange genes assigned to these regions." These results supported the view that genetic recombination involves a physical exchange between homologous chromosomes. This microscopic evidence helped to convince geneticists that recombinant offspring arise from the physical exchange of segments of homologous chromosomes. As shown in solved problem S4 at the end of this chapter, an experiment by Curt Stern was also consistent with the conclusion that crossing over between homologous chromosomes accounts for the formation of offspring with recombinant phenotypes.

A self-help quiz involving this experiment can be found at the Online Learning Center.

Crossing Over Occasionally Occurs During Mitosis

In multicellular organisms, the union of egg and sperm is followed by many cellular divisions, which occur in conjunction with mitotic divisions of the cell nuclei. As discussed in Chapter 3, mitosis normally does not involve the homologous pairing of chromosomes to form a bivalent. Therefore, crossing over during mitosis is expected to occur much less frequently than during meiosis. Nevertheless, it does happen on rare occasions. Mitotic crossing over may produce a pair of recombinant chromosomes that have a new combination of alleles, an event known as **mitotic recombination.** If it occurs during an early stage of embryonic development, the daughter cells containing the recombinant chromosomes continue to divide many times to produce a patch of tissue in the adult. This may result in a portion of tissue with characteristics different from those of the rest of the organism.

In 1936, Curt Stern identified unusual patches on the bodies of certain *Drosophila* strains. He was working with strains carrying X-linked alleles affecting body color and bristle morphology (**Figure 5.7**). A recessive allele confers yellow body color (*y*), and another recessive allele causes shorter body bristles that look singed (*sn*). The corresponding wild-type alleles result in gray body color (y^+) and long bristles (sn^+). Females that are $y^+y\ sn^+sn$ are expected to have gray body color and long bristles. This was generally the case. However, when Stern carefully observed the bodies of these female flies under a low-power microscope, he occasionally noticed places in which two adjacent regions were different from the rest of the body—a twin spot. He concluded that twin spotting was too frequent to be explained by the random positioning of two independent single spots that happened to occur close together. How then did Stern explain the phenomenon of twin spotting? He proposed that twin spots are due to a single mitotic recombination within one cell during embryonic development.

As shown in Figure 5.7, the X chromosomes of the fertilized egg are $y^+\ sn$ and $y\ sn^+$. During development, a rare crossover can occur during mitosis to produce two adjacent daughter cells that are $y^+y^+\ snsn$ and $yy\ sn^+sn^+$. As embryonic development proceeds, the cell on the left will continue to divide to produce many cells, eventually producing a patch on the body that has gray color with singed bristles. The daughter cell next to it will produce a patch of yellow body color with long bristles. These two adjacent patches—a twin spot—will be surrounded by cells that are $y^+y\ sn^+sn$ and have gray color and long bristles. Twin spots provide evidence that mitotic recombination occasionally occurs.

5.2 GENETIC MAPPING IN PLANTS AND ANIMALS

The purpose of **genetic mapping,** also known as gene mapping or chromosome mapping, is to determine the linear order and distance of separation among genes that are linked to each other along the same chromosome. **Figure 5.8** illustrates a simplified genetic map of *Drosophila melanogaster* depicting the locations of many different genes along the individual chromosomes. As shown here, each gene has its own unique **locus**—the site where the gene is found within a particular chromosome. For example, the gene designated *brown eyes* (*bw*), which affects eye color, is located near one end of chromosome 2. The gene designated *black body* (*b*), which affects body color, is found near the middle of the same chromosome.

Why is genetic mapping useful? First, it allows geneticists to understand the overall complexity and genetic organization of a particular species. The genetic map of a species portrays the underlying basis for the inherited traits that an organism displays. In some cases, the known locus of a gene within a genetic map can help molecular geneticists to clone that gene and thereby obtain greater information about its molecular features. In addition, genetic maps are useful from an evolutionary point of view. A comparison of the genetic maps for different species can improve our understanding of the evolutionary relationships among those species.

Along with these scientific uses, genetic maps have many practical benefits. For example, many human genes that play a role in human disease have been genetically mapped. This information can be used to diagnose and perhaps someday treat inherited human diseases. It can also help genetic counselors predict the likelihood that a couple will produce children with certain inherited diseases. In addition, genetic maps are gaining increasing importance in agriculture. A genetic map can provide plant and animal breeders with helpful information for improving agriculturally important strains through selective breeding programs.

In this section, we will examine traditional genetic mapping techniques that involve an analysis of crosses of individuals that are

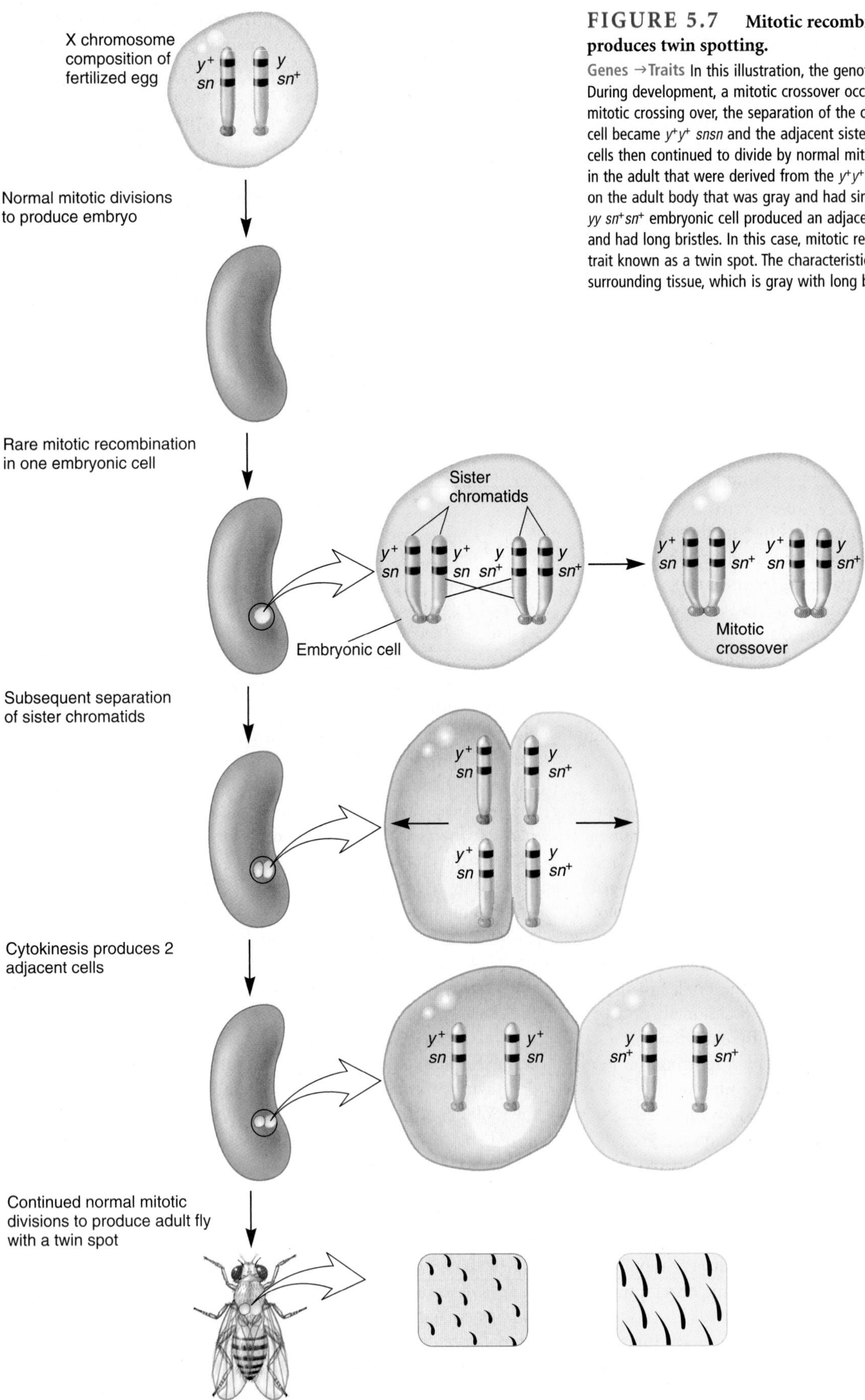

FIGURE 5.7 **Mitotic recombination in *Drosophila* that produces twin spotting.**

Genes →Traits In this illustration, the genotype of the fertilized egg was $y^+y\ sn^+sn$. During development, a mitotic crossover occurred in a single embryonic cell. After mitotic crossing over, the separation of the chromatids occurred, so one embryonic cell became $y^+y^+\ snsn$ and the adjacent sister cell became $yy\ sn^+sn^+$. The embryonic cells then continued to divide by normal mitosis to produce an adult fly. The cells in the adult that were derived from the $y^+y^+\ snsn$ embryonic cell produced a spot on the adult body that was gray and had singed bristles. The cells derived from the $yy\ sn^+sn^+$ embryonic cell produced an adjacent spot on the body that was yellow and had long bristles. In this case, mitotic recombination produced an unusual trait known as a twin spot. The characteristics of this twin spot differ from the surrounding tissue, which is gray with long bristles.

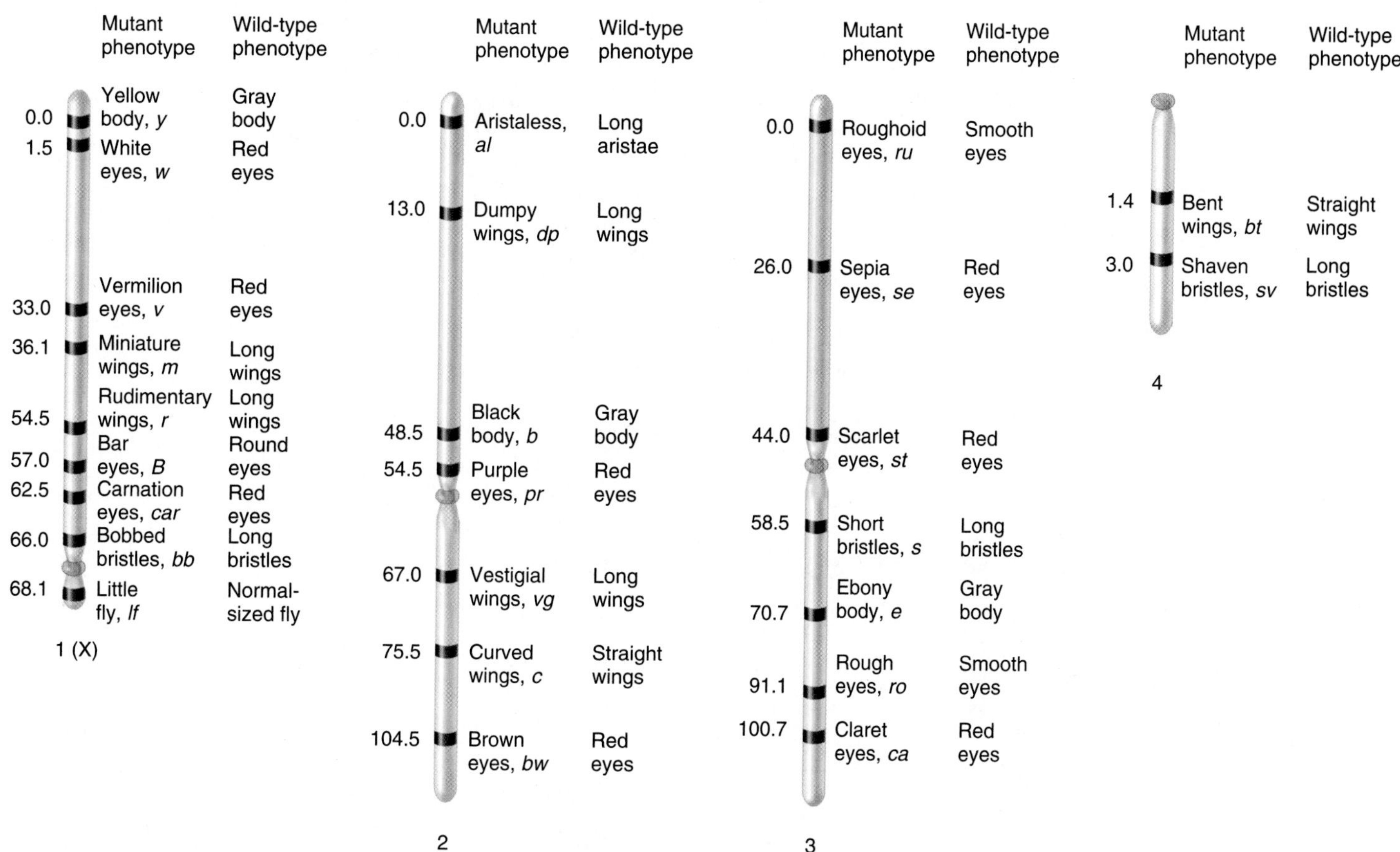

FIGURE 5.8 **A simplified genetic linkage map of *Drosophila melanogaster*.** This simplified map illustrates a few of the many thousands of genes that have been identified in this organism.

heterozygous for two or more genes. The frequency of nonparental offspring due to crossing over provides a way to deduce the linear order of genes along a chromosome. As depicted in Figure 5.8, this linear arrangement of genes is known as a **genetic linkage map.** This approach has been useful for analyzing organisms that are easily crossed and produce a large number of offspring in a short period of time. Genetic linkage maps have been constructed for several plant species and certain species of animals, such as *Drosophila*. For many organisms, however, traditional mapping approaches are difficult due to long generation times or the inability to carry out experimental crosses (as in humans). Fortunately, many alternative methods of gene mapping have been developed to replace the need to carry out crosses. As described in Chapter 20, molecular approaches are increasingly used to map genes.

The Frequency of Recombination Between Two Genes Can Be Correlated with Their Map Distance Along a Chromosome

Genetic mapping allows us to estimate the relative distances between linked genes based on the likelihood that a crossover will occur between them. If two genes are very close together on the same chromosome, a crossover is unlikely to begin in the region between them. However, if two genes are very far apart, a crossover is more likely to be initiated in this region and thereby recombine the alleles of the two genes. Experimentally, the basis for genetic mapping is that the percentage of recombinant offspring is correlated with the distance between two genes. If two genes are far apart, many recombinant offspring will be produced. However, if two genes are close together, very few recombinant offspring will be observed.

To interpret a genetic mapping experiment, the experimenter must know if the characteristics of an offspring are due to crossing over during meiosis in a parent. This is accomplished by conducting a **testcross.** Most testcrosses are between an individual that is heterozygous for two or more genes and an individual that is recessive and homozygous for the same genes. The goal of the testcross is to determine if recombination has occurred during meiosis in the heterozygous parent. Thus, genetic mapping is based on the level of recombination that occurs in just one parent—the heterozygote. In a testcross, new combinations of alleles cannot occur in the gametes of the other parent, which is homozygous for these genes.

Figure 5.9 illustrates how a testcross provides an experimental strategy to distinguish between recombinant and nonrecombinant offspring. This cross concerns two linked genes affecting bristle length and body color in fruit flies. The recessive alleles

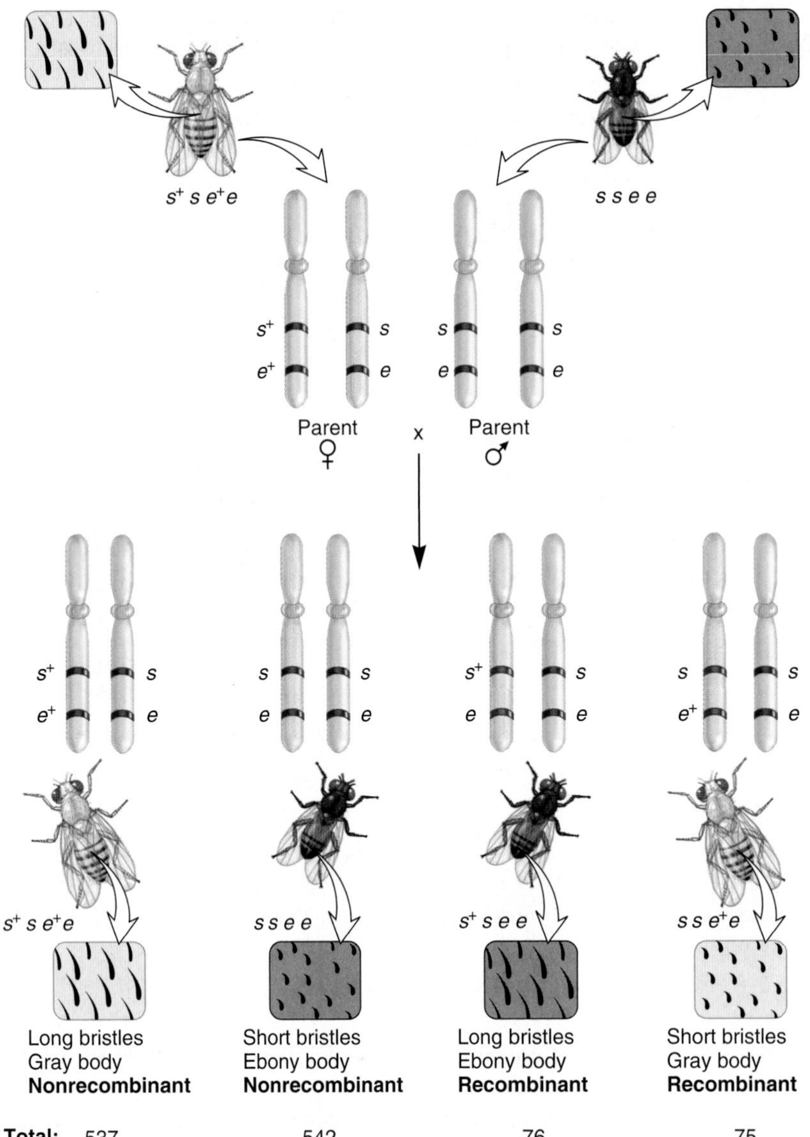

FIGURE 5.9 **Use of a testcross to distinguish between recombinant and nonrecombinant offspring.** The cross involves one *Drosophila* parent that is homozygous recessive for short bristles (*ss*) and ebony body (*ee*), and one parent heterozygous for both genes ($s^+s\ e^+e$). (Note: *Drosophila* geneticists normally designate the short allele as *ss* and a homozygous fly with short bristles as *ssss.* In the text, the allele causing short bristles is designated with a single *s* to avoid confusion between the allele designation and the genotype of the fly.)

are *s* (short bristles) and *e* (ebony body), and the dominant (wild-type) alleles are s^+ (long bristles) and e^+ (gray body). One parent displays both recessive traits. Therefore, we know this parent is homozygous for the recessive alleles of the two genes (*ss ee*). The other parent is heterozygous for the linked genes affecting bristle length and body color. This parent was produced from a cross involving a true-breeding wild-type fly and a true-breeding fly with short bristles and an ebony body. Therefore, in this heterozygous parent, we know that the *s* and *e* alleles are located on one chromosome and the corresponding s^+ and e^+ alleles are located on the homologous chromosome.

Now let's take a look at the four possible types of offspring these parents can produce. The offspring's phenotypes are long bristles, gray body; short bristles, ebony body; long bristles, ebony body; and short bristles, gray body. All four types of offspring have inherited a chromosome carrying the *s* and *e* alleles from their homozygous parent (shown on the right in each pair). Focus your attention on the other chromosome. The offspring with long bristles and gray bodies have inherited a chromosome carrying the s^+ and e^+ alleles from the heterozygous parent. This chromosome is not the product of a crossover. The offspring with short bristles and ebony bodies have also inherited a chromosome carrying the *s* and *e* alleles from the heterozygous parent. Again, this chromosome is not the product of a crossover.

The other two types of offspring, however, can be produced only if crossing over has occurred in the region between these two genes. Those with long bristles and ebony bodies or short bristles and gray bodies have inherited a chromosome that is the product of a crossover during meiosis in the heterozygous parent. As noted in Figure 5.9, the recombinant offspring are fewer in number than are the nonrecombinant offspring.

The frequency of recombination can be used as an estimate of the physical distance between two genes on the same chromosome. The **map distance** is defined as the number of recombinant offspring divided by the total number of offspring, multiplied by 100. We can calculate the map distance between these two genes using this formula:

$$\text{Map distance} = \frac{\text{Number of recombinant offspring}}{\text{Total number of offspring}} \times 100$$

$$= \frac{76 + 75}{537 + 542 + 76 + 75} \times 100$$

$$= 12.3 \text{ map units}$$

The units of distance are called **map units (mu)** or sometimes **centiMorgans (cM)** in honor of Thomas Hunt Morgan. One map unit is equivalent to a 1% frequency of recombination. In this example, we would conclude that the *s* and *e* alleles are 12.3 map units apart from each other along the same chromosome.

EXPERIMENT 5B

Alfred Sturtevant Used the Frequency of Crossing Over in Dihybrid Crosses to Produce the First Genetic Map

In 1911, the first individual to construct a (very small) genetic map was Alfred Sturtevant, an undergraduate who spent time in the laboratory of Thomas Hunt Morgan. Sturtevant wrote: "In conversation with Morgan . . . I suddenly realized that the variations in the strength of linkage, already attributed by Morgan to differences in the spatial separation of the genes, offered the possibility of determining sequences [of different genes] in the linear dimension of a chromosome. I went home and spent most of the night (to the neglect of my undergraduate homework) in producing the first chromosome map, which included the sex-linked genes, *y*, *w*, *v*, *m*, and *r*, in the order and approximately the relative spacing that they still appear on the standard maps."

In the experiment of **Figure 5.10**, Sturtevant considered the outcome of crosses involving six different mutant alleles that altered the phenotype of flies. All of these alleles were known to be recessive and X linked. They are *y* (yellow body color), *w* (white eye color), *w-e* (eosin eye color), *v* (vermilion eye color), *m* (miniature wings), and *r* (rudimentary wings). The *w* and *w-e* alleles are alleles of the same gene. In contrast, the *v* allele (vermilion eye color) is an allele of a different gene that also affects eye color. The two alleles that affect wing length, *m* and *r*, are also in different genes. Therefore, Sturtevant studied the inheritance of six recessive alleles, but since *w* and *w-e* are alleles of the same gene, his genetic map contained only five genes. The corresponding wild-type alleles are y^+ (gray body), w^+ (red eyes), v^+ (red eyes), m^+ (long wings), and r^+ (long wings).

THE HYPOTHESIS

When genes are located on the same chromosome, the distance between the genes can be estimated from the proportion of recombinant offspring. This provides a way to map the order of genes along a chromosome.

TESTING THE HYPOTHESIS — FIGURE 5.10 The first genetic mapping experiment.

Starting materials: Sturtevant began with several different strains of *Drosophila* that contained the six alleles already described.

1. Cross a female that is heterozygous for two different genes to a male that is hemizygous recessive for the same two genes. In this example, cross a female that is $X^{y^+w^+}X^{yw}$ to a male that is $X^{yw}Y$.

   This strategy was employed for many dihybrid combinations of the six alleles already described.

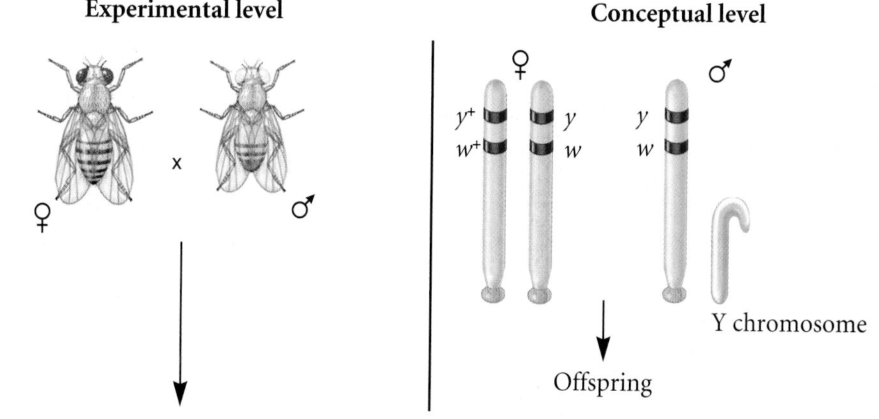

2. Observe the outcome of the crosses.

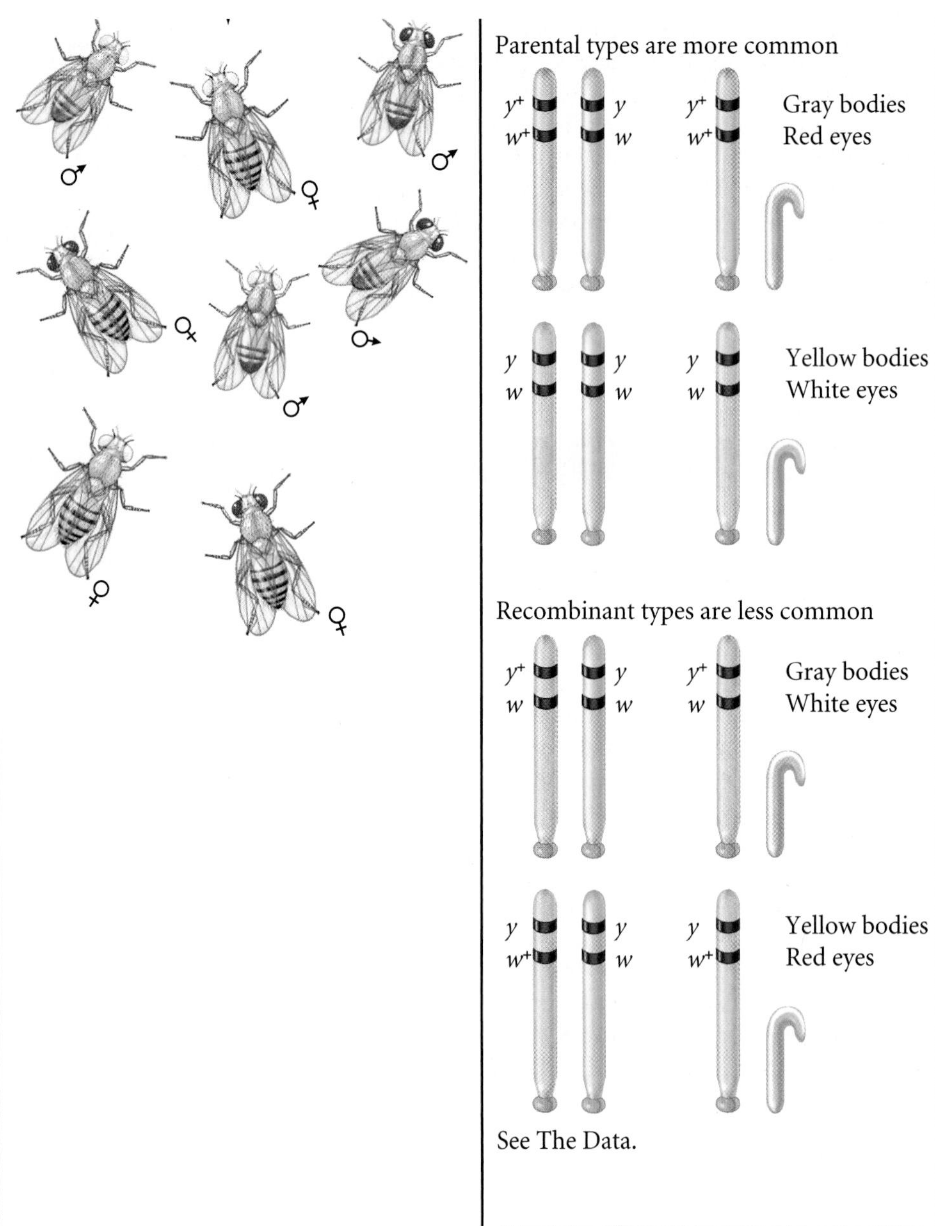

3. Calculate the percentages of offspring that are the result of crossing over (number of nonparental/total).

THE DATA

Alleles Concerned	*Number Recombinant/Total Number*	*Percent Recombinant Offspring*
y and *w*/*w-e*	214/21,736	1.0
y and *v*	1,464/4,551	32.2
y and *m*	115/324	35.5
y and *r*	260/693	37.5
w/*w-e* and *v*	471/1,584	29.7
w/*w-e* and *m*	2,062/6,116	33.7
w/*w-e* and *r*	406/898	45.2
v and *m*	17/573	3.0
v and *r*	109/405	26.9

INTERPRETING THE DATA

As shown in Figure 5.10, Sturtevant made pairwise testcrosses and then counted the number of offspring in the four phenotypic categories. Two of the categories were nonrecombinant and two were recombinant, requiring a crossover between the X chromosomes in the female heterozygote. Let's begin by contrasting the results between particular pairs of genes, shown in the data. In some dihybrid crosses, the percentage of nonparental offspring was rather low. For example, dihybrid crosses involving the *y* allele and the *w* or *w-e* allele yielded 1% recombinant offspring. This result suggested that these two genes are very close together. By comparison, other dihybrid crosses showed a higher percentage of nonparental offspring. For example, crosses involving the

v and r alleles produced 26.9% recombinant offspring. These two genes are expected to be farther apart.

To construct his map, Sturtevant began with the assumption that the map distances would be more accurate between genes that are closely linked. Therefore, his map is based on the distance between y and w (1.0), w and v (29.7), v and m (3.0), and v and r (26.9). He also considered other features of the data to deduce the order of the genes. For example, the percentage of crossovers between w and m was 33.7. The percentage of crossovers between w and v was 29.7, suggesting that v is between w and m, but closer to m. The proximity of v and m is confirmed by the low percentage of crossovers between v and m (3.0). Sturtevant collectively considered the data and proposed the genetic map shown here.

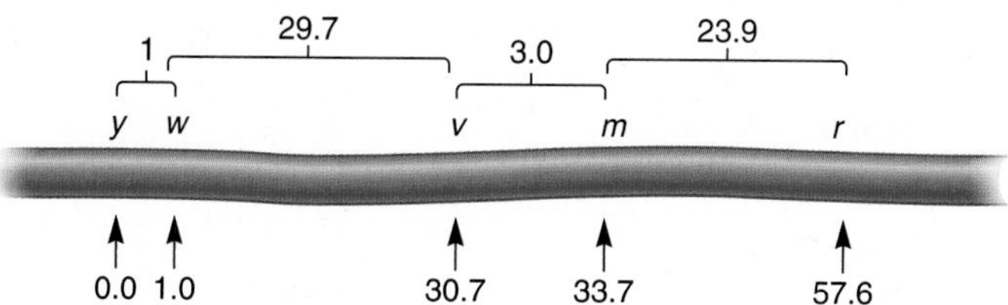

In this genetic map, Sturtevant began at the y allele and mapped the genes from left to right. For example, the y and v alleles are 30.7 mu apart, and the v and m alleles are 3.0 mu apart. This study by Sturtevant was a major breakthrough, because it showed how to map the locations of genes along chromosomes by making the appropriate crosses.

If you look carefully at Sturtevant's data, you will notice a few observations that do not agree very well with his genetic map. For example, the percentage of recombinant offspring for the y and r dihybrid cross was 37.5 (but the map distance is 57.6), and the crossover percentage between w and r was 45.2 (but the map distance is 56.6). As the percentage of recombinant offspring approaches a value of 50%, this value becomes a progressively more inaccurate measure of actual map distance (**Figure 5.11**). What is the basis for this inaccuracy? When the distance between two genes is large, the likelihood of multiple crossovers in the region between them causes the observed number of recombinant offspring to underestimate this distance. In addition, multiple crossovers set a quantitative limit on the relationship between map distance and the percentage of recombinant offspring. Even though two different genes can be on the same chromosome and more than 50 mu apart, a testcross is expected to yield a maximum of only 50% recombinant offspring. (This idea is also discussed in solved problem S5 at the end of this chapter.) When two different genes are more than 50 mu apart, they follow the law of independent assortment in a testcross.

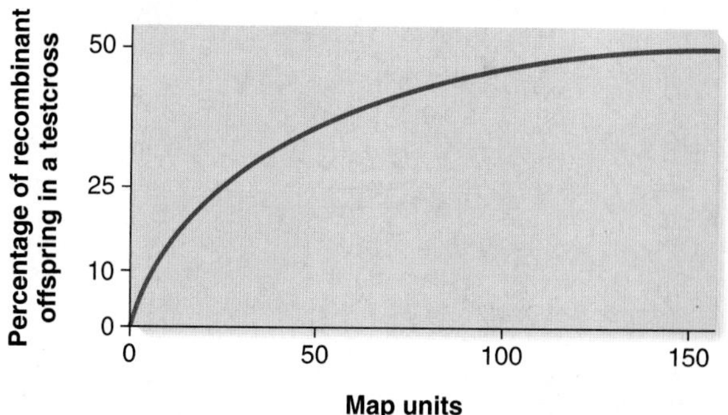

FIGURE 5.11 Relationship between the percentage of recombinant offspring in a testcross and the actual map distance between genes. The y-axis depicts the percentage of recombinant offspring that would be observed in a dihybrid testcross. The actual map distance, shown on the x-axis, is calculated by analyzing the percentages of recombinant offspring from a series of many dihybrid crosses involving closely linked genes. Even though two genes may be more than 50 map units apart, the percentage of recombinant offspring will not exceed 50%.

A self-help quiz involving this experiment can be found at the Online Learning Center.

Trihybrid Crosses Can Be Used to Determine the Order and Distance Between Linked Genes

Thus far, we have considered the construction of genetic maps using dihybrid testcrosses to compute map distance. The data from trihybrid crosses can yield additional information about map distance and gene order. In a trihybrid cross, the experimenter crosses two individuals that differ in three traits. The following experiment outlines a common strategy for using trihybrid crosses to map genes. In this experiment, the parental generation consists of fruit flies that differ in body color, eye color, and wing shape. We must begin with true-breeding lines so that we know which alleles are initially linked to each other on the same chromosome. In this example, all the dominant alleles are linked on the same chromosome.

Step 1. *Cross two true-breeding strains that differ with regard to three alleles.* In this example, we will cross a fly that has a black body (*bb*), purple eyes (*prpr*), and vestigial wings (*vgvg*) to a homozygous wild-type fly with a gray body (b^+b^+), red eyes (pr^+pr^+), and long wings (vg^+vg^+):

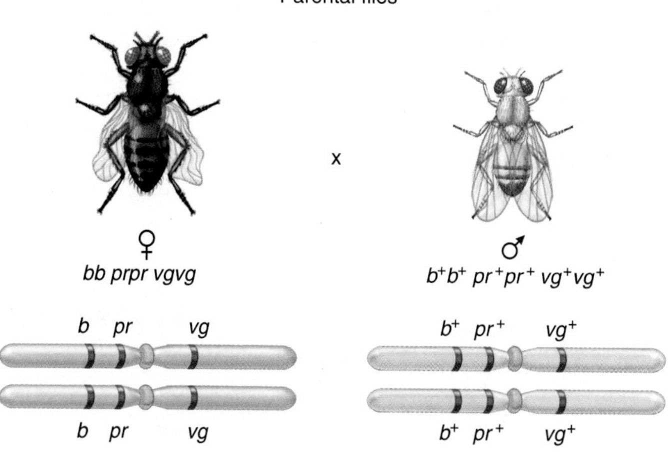

The goal in this step is to obtain F_1 individuals that are heterozygous for all three genes. In the F_1 heterozygotes, all dominant alleles are located on one chromosome, and all recessive alleles are on the other homologous chromosome.

Step 2. *Perform a testcross by mating F_1 female heterozygotes to male flies that are homozygous recessive for all three alleles (bb prpr vgvg).*

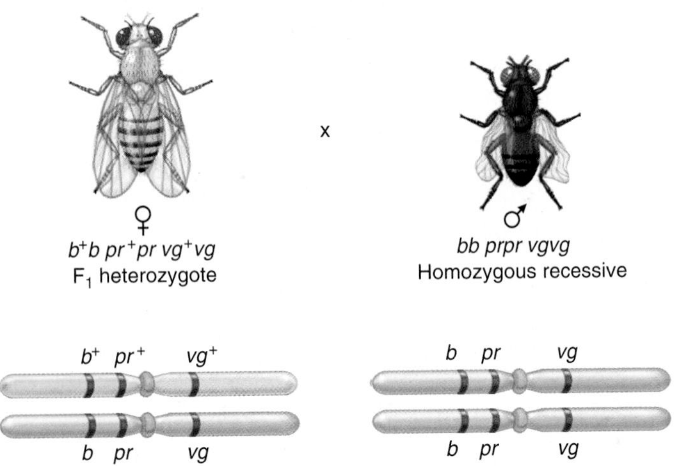

During gametogenesis in the heterozygous female F_1 flies, crossovers may produce new combinations of the three alleles.

Step 3. *Collect data for the F_2 generation.* As shown in **Table 5.1**, eight phenotypic combinations are possible. An analysis of the F_2 generation flies allows us to map these three genes. Because the three genes exist as two alleles each, we have $2^3 = 8$ possible combinations of offspring. If these alleles assorted independently, all eight combinations would occur in equal proportions. However, we see that the proportions of the eight phenotypes are far from equal.

The genotypes of the parental generation correspond to the phenotypes gray body, red eyes, and long wings, and black body, purple eyes, and vestigial wings. In crosses involving linked genes, the parental phenotypes occur most frequently in the offspring. The remaining six phenotypes are due to crossing over.

The double crossover is always expected to be the least frequent category of offspring. Two of the phenotypes—gray body, purple eyes, and long wings, and black body, red eyes, and vestigial wings—arose from a double crossover between two pairs of genes. Also, the combination of traits in the double crossover tells us which gene is in the middle. When a chromatid undergoes a double crossover, the gene in the middle becomes separated from the other two genes at either end.

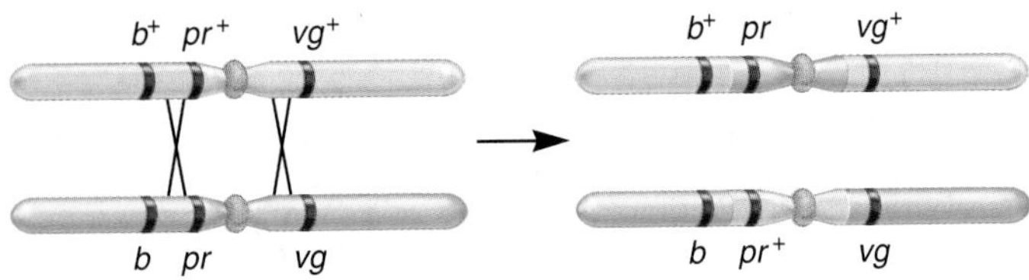

TABLE 5.1

Data from a Trihybrid Cross (see step 2)

Phenotype	Number of Observed Offspring (males and females)	Chromosome Inherited from F_1 Female
Gray body, red eyes, long wings	411	b^+ pr^+ vg^+
Gray body, red eyes, vestigial wings	61	b^+ pr^+ vg
Gray body, purple eyes, long wings	2	b^+ pr vg^+
Gray body, purple eyes, vestigial wings	30	b^+ pr vg
Black body, red eyes, long wings	28	b pr^+ vg^+
Black body, red eyes, vestigial wings	1	b pr^+ vg
Black body, purple eyes, long wings	60	b pr vg^+
Black body, purple eyes, vestigial wings	412	b pr vg

In the double crossover categories, the recessive purple-eye allele is separated from the other two recessive alleles. When mated to a homozygous recessive fly in the testcross, this yields flies with gray bodies, purple eyes, and long wings, or with black bodies, red eyes, and vestigial wings. This observation indicates that the gene for eye color lies between the genes for body color and wing shape.

Step 4. *Calculate the map distance between pairs of genes.* To do this, we need to understand which gene combinations are recombinant and which are nonrecombinant. The recombinant offspring are due to crossing over in the heterozygous female parent. If you look back at step 2, you can see the arrangement of alleles in the heterozygous female parent in the absence of crossing over. Let's consider this arrangement with regard to gene pairs:

b^+ is linked to pr^+, and *b* is linked to *pr*

pr^+ is linked to vg^+, and *pr* is linked to *vg*

b^+ is linked to vg^+, and *b* is linked to *vg*

With regard to body color and eye color, the recombinant offspring have gray bodies and purple eyes (2 + 30) or black bodies and red eyes (28 + 1). As shown along the right side of Table 5.1, these offspring were produced by crossovers in the female parents. The total number of these recombinant offspring is 61. The map distance between the body color and eye color genes is thus

$$\text{Map distance} = \frac{61}{944 + 61} \times 100 = 6.1 \text{ mu}$$

With regard to eye color and wing shape, the recombinant offspring have red eyes and vestigial wings (61 + 1) or purple eyes and long wings (2 + 60). The total number is 124. The map distance between the eye color and wing shape genes is

$$\text{Map distance} = \frac{124}{881 + 124} \times 100 = 12.3 \text{ mu}$$

With regard to body color and wing shape, the recombinant offspring have gray bodies and vestigial wings (61 + 30) or black bodies and long wings (28 + 60). The total number is 179. The map distance between the body color and wing shape genes is

$$\text{Map distance} = \frac{179}{826 + 179} \times 100 = 17.8 \text{ mu}$$

Step 5. *Construct the map.* Based on the map unit calculation, the body color (*b*) and wing shape (*vg*) genes are farthest apart. The eye color gene (*pr*) must lie in the middle. As mentioned earlier, this order of genes is also confirmed by the pattern of traits found in the double crossovers. To construct the map, we use the distances between the genes that are closest together.

6.1 12.3
b *pr* *vg*

In our example, we have placed the body color gene first and the wing shape gene last. The data also are consistent with a map in which the wing shape gene comes first and the body color gene comes last. In detailed genetic maps, the locations of genes are mapped relative to the centromere.

You may have noticed that our calculations underestimate the distance between the body color and wing shape genes. We obtained a value of 17.8 map units even though the distance seems to be 18.4 map units when we add together the distance between body color and eye color genes (6.1 mu) and the distance between eye color and wing shape genes (12.3 mu). What accounts for this discrepancy? The answer is double crossovers. If you look at the data in Table 5.1, the offspring with gray bodies, purple eyes, and long wings or with black bodies, red eyes, and vestigial wings are due to a double crossover. From a phenotypic perspective, these offspring are not recombinant with regard to the body color and wing shape alleles. Even so, we know that they arose from a double crossover between these two genes. Therefore, we should consider these crossovers when calculating the distance between the body color and wing shape genes. In this case, three offspring (2 + 1) were due to double crossovers. Because they are double crossovers, we multiply two times the number of double crossovers (2 + 1) and add this number to our previous value of recombinant offspring:

$$\text{Map distance} = \frac{179 + 2(2 + 1)}{826 + 179} \times 100 = 18.4 \text{ mu}$$

Interference Can Influence the Number of Double Crossovers That Occur in a Short Region

In Chapter 2, we considered the product rule to determine the probability that two independent events will both occur. The product rule allows us to predict the expected likelihood of a double crossover provided we know the individual probabilities of each single crossover. Let's reconsider the data of the trihybrid testcross just described to see if the frequency of double crossovers is what we would expect based on the product rule. If each crossover is an independent event, we can multiply the likelihood of a single crossover between *b* and *pr* (0.061) times the likelihood of a single crossover between *pr* and *vg* (0.123). The product rule predicts

Expected likelihood of a double crossover =
$0.061 \times 0.123 = 0.0075 = 0.75\%$

Based on a total of 1,005 offspring produced:

Expected number of offspring due to a double crossover =
$1{,}005 \times 0.0075 = 7.5$

In other words, we would expect about seven or eight offspring to be produced as a result of a double crossover. The observed number of offspring was only three (namely, two with gray bodies, purple eyes, and long wings, and one with a black body, red eyes, and vestigial wings). What accounts for the lower number? This lower-than-expected value is probably not due to random sampling error. Instead, the likely cause is a common genetic phenomenon known as **positive interference,** in which the occurrence of a crossover in one region of a chromosome decreases the probability that a second crossover will occur nearby. In other words, the first crossover interferes with the ability to form a second crossover in the immediate vicinity. To provide interference with a quantitative value, we first calculate the coefficient of coincidence (C), which is the ratio of the observed number of double crossovers to the expected number.

$$C = \frac{\text{Observed number of double crossovers}}{\text{Expected number of double crossovers}}$$

Interference (I) is expressed as

$$I = 1 - C$$

For the data of the trihybrid testcross, the observed number of crossovers is 3 and the expected number is 7.5, so the coefficient of coincidence equals 3/7.5 = 0.40. In other words, only 40% of the expected number of double crossovers were actually observed. The value for interference equals 1 − 0.4 = 0.60, or 60%. This means that 60% of the expected number of crossovers did not occur. Because I has a positive value, this is called positive interference. Rarely, the outcome of a testcross yields a negative value for interference. A negative interference value suggests that a first crossover enhanced the rate of a second crossover in a nearby region. Although the molecular mechanisms that cause interference are not entirely understood, in most organisms the number of crossovers is regulated so that very few occur per chromosome. The reasons for positive and negative interference will require further research.

5.3 GENETIC MAPPING IN HAPLOID EUKARYOTES

Before ending our discussion of genetic mapping, let's consider some pioneering studies that involved the genetic mapping of haploid organisms. You may find it surprising that certain species of simple eukaryotes, particularly unicellular algae and fungi, which spend part of their life cycle in the haploid state, have also been used in genetic mapping studies. The sac fungi, called ascomycetes, have been particularly useful to geneticists because of their unique style of sexual reproduction. In fact, much of our earliest understanding of genetic recombination came from the genetic analyses of fungi.

Fungi may be unicellular or multicellular organisms. Fungal cells are typically haploid (1*n*) and can reproduce asexually. In addition, fungi can also reproduce sexually by the fusion of two haploid cells to create a diploid zygote (2*n*) (**Figure 5.12**). The diploid zygote can then proceed through meiosis to produce four haploid cells, which are called **spores.** This group of four spores is known as a **tetrad** (not to be confused with a tetrad of four sister chromatids). In some species, meiosis is followed by a mitotic division to produce eight spores, known as an **octad.** In *Ascomycete* fungi, the cells of a tetrad or octad are contained within a sac known as an **ascus** (plural: *asci*). In other words, the products of a single meiotic division are contained within one sac. This key feature is useful to geneticists, and it dramatically differs from sexual reproduction in animals and plants. For example, in animals, oogenesis produces a single functional egg, and spermatogenesis occurs in the testes, where the resulting sperm become mixed with millions of other sperm.

Using a microscope, researchers can dissect asci and study the traits of each haploid spore. In this way, these organisms offer a unique opportunity for geneticists to identify and study all of the cells that are derived from a single meiotic division. In this section, we will consider how the analysis of asci can be used to map genes in fungi.

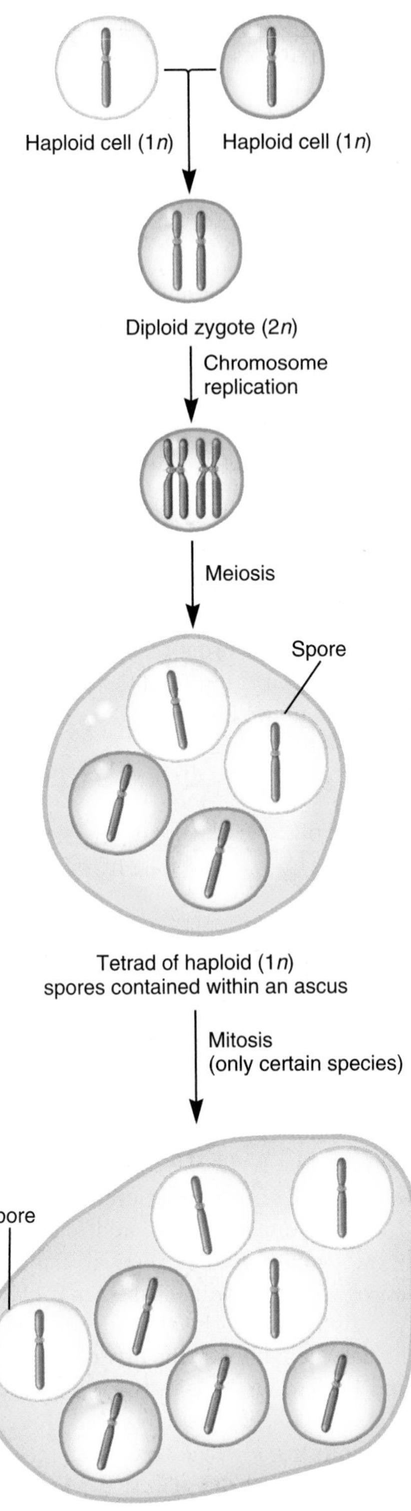

FIGURE 5.12 Sexual reproduction in ascomycetes. For simplicity, this diagram shows each haploid cell as having only one chromosome per haploid set. However, fungal species actually contain several chromosomes per haploid set.

Ordered Tetrad Analysis Can Be Used to Map the Distance Between a Gene and the Centromere

The arrangement of spores within an ascus varies from species to species (**Figure 5.13a**). In some cases, the ascus provides enough space for the tetrads or octads of spores to randomly mix together. This creates an **unordered tetrad** or **octad.** These occur in fungal species such as *Saccharomyces cerevisiae* and *Aspergillus nidulans* and also in certain unicellular algae (*Chlamydomonas reinhardtii*). By comparison, other species of fungi produce a very tight ascus that prevents spores from randomly moving around, which results in an **ordered tetrad** or **octad. Figure 5.13b** illustrates how an ordered octad is formed in *Neurospora crassa.* In this example, spores that carry the *A* allele have orange pigmentation, while spores having the *a* (albino) allele are white.

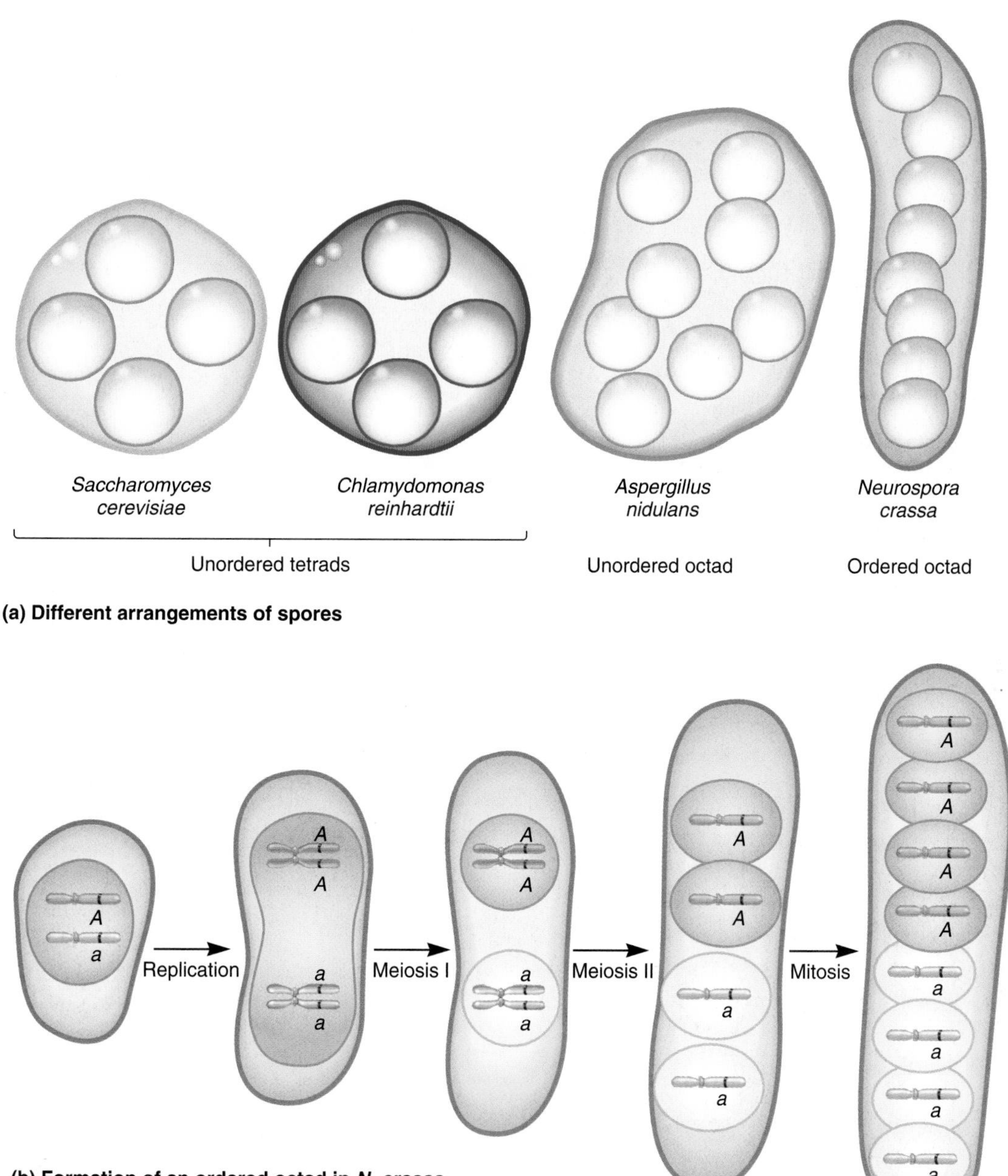

FIGURE 5.13 **Arrangement of spores within asci of different species.** **(a)** *Saccharomyces cerevisiae* and *Chlamydomonas reinhardtii* (an alga) produce unordered tetrads, *Aspergillus nidulans* produces an unordered octad, and *Neurospora crassa* produces an ordered octad. **(b)** Ordered octads are produced in *N. crassa* by meiosis and mitosis in such a way that the eight resulting cells are arranged linearly.

A key feature of ordered tetrads or octads is that the position and order of spores within the ascus reflects their relationship to each other as they were produced by meiosis and mitosis. This idea is schematically shown in Figure 5.13b. After the original diploid cell has undergone chromosome replication, the first meiotic division produces two cells that are arranged next to each other within the sac. The second meiotic division then produces four cells that are also arranged in a row. Due to the tight enclosure of the sac around the cells, each pair of daughter cells is forced to lie next to each other in a linear fashion. Likewise, when these four cells divide by mitosis, each pair of daughter cells is located next to each other.

In species that make ordered tetrads or octads, experimenters can determine the genotypes of the spores within the asci and map the distance between a single gene and the centromere. Because the location of the centromere can be seen under the microscope, the mapping of a gene relative to the centromere provides a way to correlate a gene's location with the cytological characteristics of a chromosome. This approach has been extensively exploited in *N. crassa.*

Figure 5.14 compares the arrangement of cells within a *Neurospora* ascus depending on whether or not a crossover has occurred between two homologues that differ at a gene with

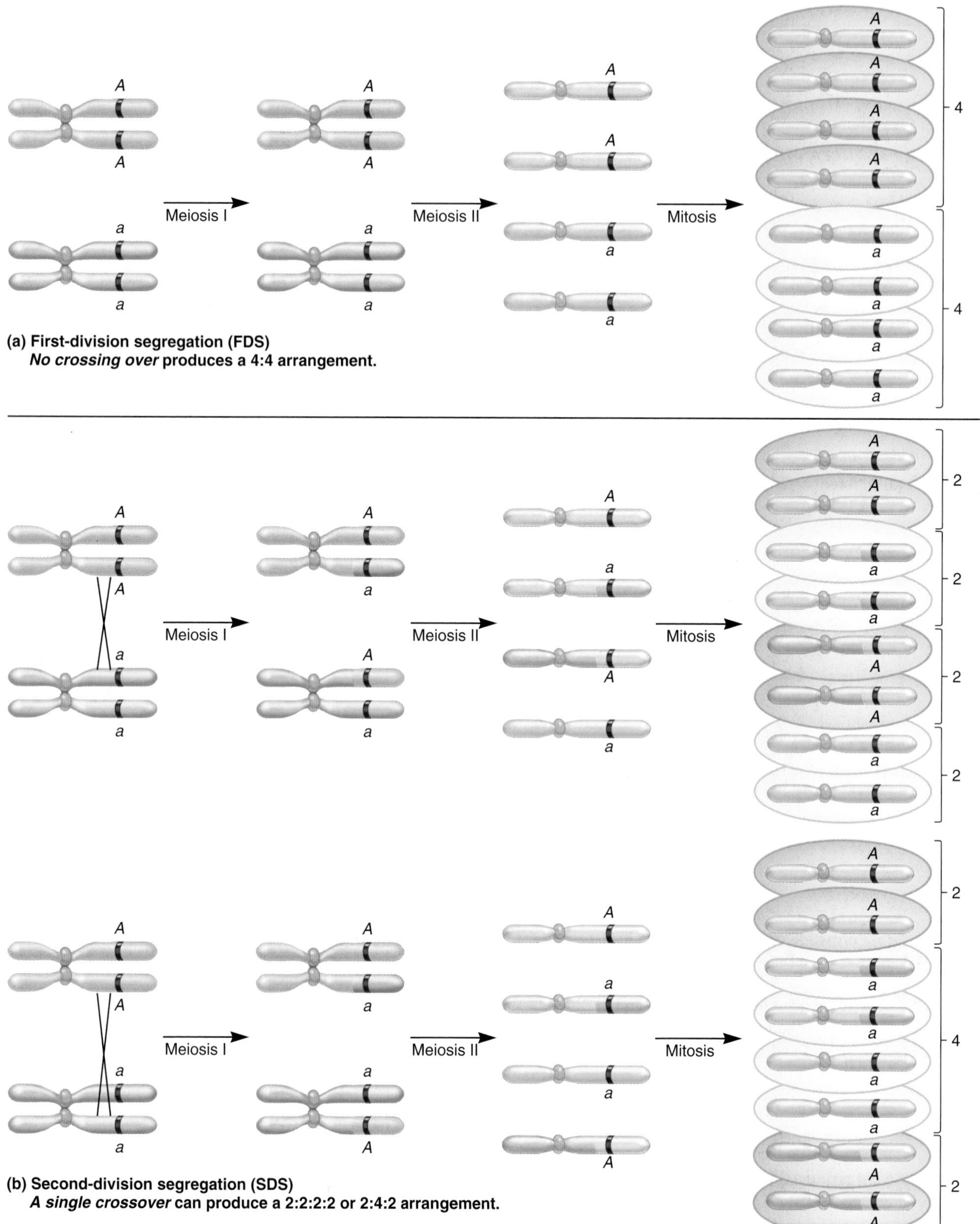

FIGURE 5.14 A comparison of the arrangement of cells within an ordered octad, depending on whether or not crossing over has occurred. (**a**) If no crossing over has occurred, the octad will have a 4:4 arrangement of spores known as an FDS or M1 pattern. (**b**) If a crossover has occurred between the centromere and the gene of interest, a 2:2:2:2 or 2:4:2 pattern, known as an SDS or M2 pattern, is observed.

alleles *A* (orange pigmentation) and *a* (albino, which results in a white phenotype). In Figure 5.14a, a crossover has not occurred, so the octad contains a linear arrangement of four haploid cells carrying the *A* allele, which are adjacent to four haploid cells that contain the *a* allele. This 4:4 arrangement of spores within the ascus is called **first-division segregation (FDS)** or an M1 pattern. It is called a first-division segregation pattern because the *A* and *a* alleles have segregated from each other after the first meiotic division.

In contrast, as shown in Figure 5.14b, if a crossover occurs between the centromere and the gene of interest, the ordered octad will deviate from the 4:4 pattern. Depending on the relative locations of the two chromatids that participated in the crossover, the ascus will contain a 2:2:2:2 or 2:4:2 pattern. These patterns are called **second-division segregation (SDS)** or M2 patterns. In this case, the *A* and *a* alleles do not segregate until the second meiotic division is completed.

Because a pattern of second-division segregation is a result of crossing over, the percentage of SDS asci can be used to calculate the map distance between the centromere and the gene of interest. To understand why this is possible, let's consider the relationship between a crossover site and the centromere. As shown in **Figure 5.15**, a crossover will separate a gene from its original centromere only if it begins in the region between the centromere and that gene. Therefore, the chances of getting a 2:2:2:2 or 2:4:2 pattern depend on the distance between the gene of interest and the centromere.

To determine the map distance between the centromere and a gene, the experimenter must count the number of SDS asci and the total number of asci. In SDS asci, only half of the spores are actually the product of a crossover. Therefore, the map distance is calculated as

$$\text{Map distance} = \frac{(1/2)\,(\text{Number of SDS asci})}{\text{Total number of asci}} \times 100$$

Unordered Tetrad Analysis Can Be Used to Map Genes in Dihybrid Crosses

Unordered tetrads contain a group of spores that are the product of meiosis and randomly arranged in an ascus. An experimenter can conduct a dihybrid cross, remove the spores from each ascus, and determine the phenotypes of the spores. This analysis can determine if two genes are linked or assort independently. If two genes are linked, a tetrad analysis can also be used to compute map distance.

Figure 5.16 illustrates the possible outcomes starting with two haploid yeast strains. One strain carries the wild-type alleles ura^+ and arg^+, which are required for uracil and arginine biosynthesis, respectively. The other strain has defective alleles *ura-2* and *arg-3*; these result in yeast strains that require uracil and arginine in the growth medium. A diploid zygote with the genotype ura^+ *ura-2* arg^+ *arg-3* was produced from the fusion of haploid cells from these two strains. The diploid cell then proceeds through meiosis to produce four haploid cells. After the completion of meiosis, three distinct types of tetrads could be produced. One possibility is that the tetrad will contain four spores with the parental combinations of alleles. This ascus is said to have the **parental ditype (PD).** Alternatively, an ascus may have two parental cells and two nonparental cells, which is called a **tetratype (T).** Finally, an ascus with a **nonparental ditype (NPD)** contains four cells with nonparental genotypes.

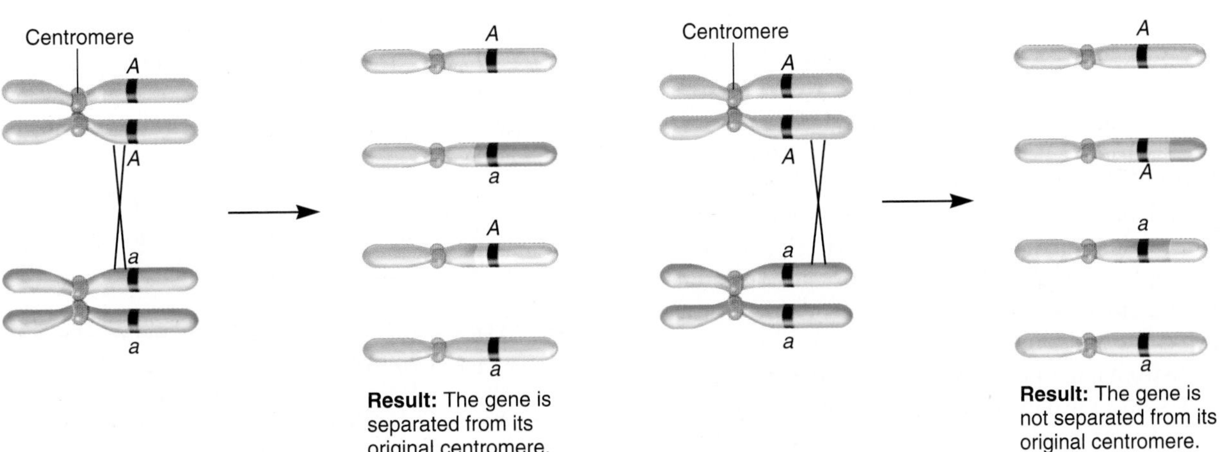
Centromere
A
A
a
a
A
a
A
a
Result: The gene is separated from its original centromere.
(a) Crossover begins between centromere and gene of interest.
Centromere
A
A
a
a
A
A
a
a
Result: The gene is not separated from its original centromere.
(b) Crossover does not begin between centromere and gene of interest.

FIGURE 5.15 **The relationship between a crossover site and the separation of an allele from its original centromere.** **(a)** If a crossover initially forms between the centromere and the gene of interest, the gene will be separated from its original centromere. **(b)** If a crossover initiates outside this region, the gene remains attached to its original centromere.

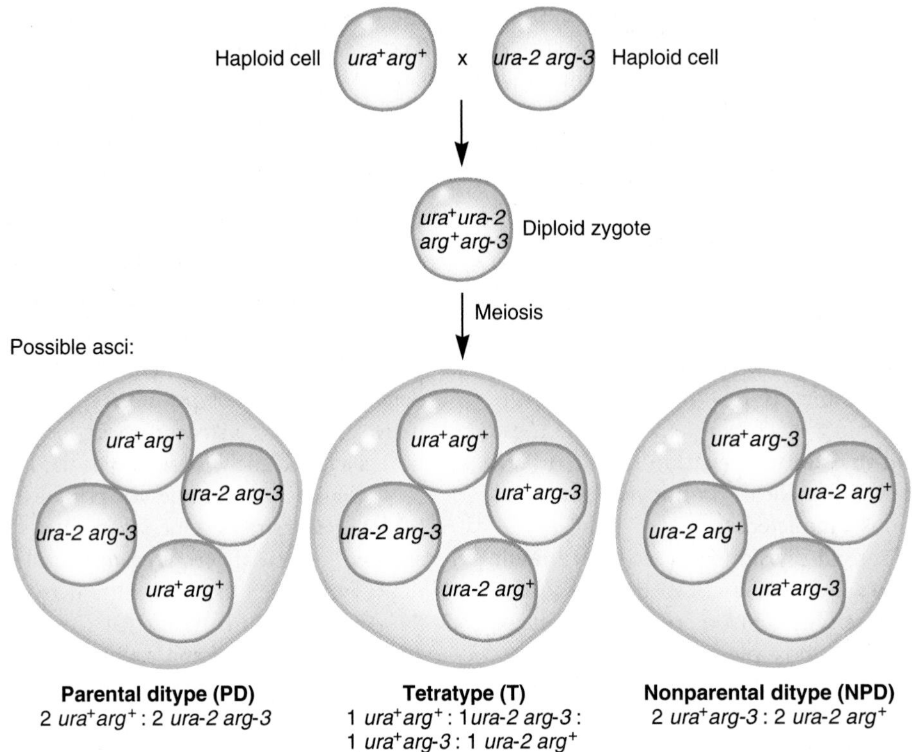

INTERACTIVE EXERCISE

FIGURE 5.16 The assortment of two genes in an unordered tetrad. If the tetrad contains 100% parental cells, this ascus has the parental ditype (PD). If it contains 50% parental and 50% recombinant cells, it is a tetratype (T). Finally, an ascus with 100% recombinant cells is called a nonparental ditype (NPD). This figure does not illustrate the chromosomal locations of the alleles. In this type of experiment, the goal is to determine whether the two genes are linked on the same chromosome.

When two genes assort independently, the number of asci having a parental ditype is expected to equal the number having a nonparental ditype, thus yielding 50% recombinant spores. For linked genes, **Figure 5.17** illustrates the relationship between crossing over and the type of ascus that will result. If no crossing over occurs in the region between the two genes, the parental ditype will be created (Figure 5.17a). A single crossover event will produce a tetratype (Figure 5.17b). Double crossovers can yield a parental ditype, tetratype, or nonparental ditype, depending on the combination of chromatids that are involved (Figure 5.17c). A nonparental ditype is produced when a double crossover involves all four chromatids. A tetratype will result from a three-chromatid crossover. Finally, a double crossover between the same two chromatids will produce the parental ditype.

The data from a tetrad analysis can be used to calculate the map distance between two linked genes. As in conventional mapping, the map distance is calculated as the percentage of offspring that carry recombinant chromosomes. As mentioned, a tetratype contains 50% recombinant chromosomes; a nonparental ditype, 100%. Therefore, the map distance is computed as

$$\text{Map distance} = \frac{\text{NPD} + (1/2)(\text{T})}{\text{total number of asci}} \times 100$$

Over short map distances, this calculation provides a fairly reliable measure of distance. However, it does not adequately account for double crossovers. When two genes are far apart on the same chromosome, the calculated map distance using this equation underestimates the actual map distance due to double crossovers. Fortunately, a particular strength of tetrad analysis is that we can derive another equation that accounts for double crossovers and thereby provides a more accurate value for map distance. To begin this derivation, let's consider a more precise way to calculate map distance.

$$\text{Map distance} = \frac{\text{Single crossover tetrads} + (2)(\text{Double crossover tetrads})}{\text{Total number of asci}} \times 0.5 \times 100$$

This equation includes the number of single and double crossovers in the computation of map distance. The total number of crossovers equals the number of single crossovers plus two times the number of double crossovers. Overall, the tetrads that contain single and double crossovers also contain 50% nonrecombinant chromosomes. To calculate map distance, therefore, we divide the total number of crossovers by the total number of asci and multiply by 0.5 and 100.

To be useful, we need to relate this equation to the number of parental ditypes, nonparental ditypes, and tetratypes that are obtained by experimentation. To derive this relationship, we must consider the types of tetrads that are produced from no crossing over, a single crossover, and double crossovers. To do so, let's take

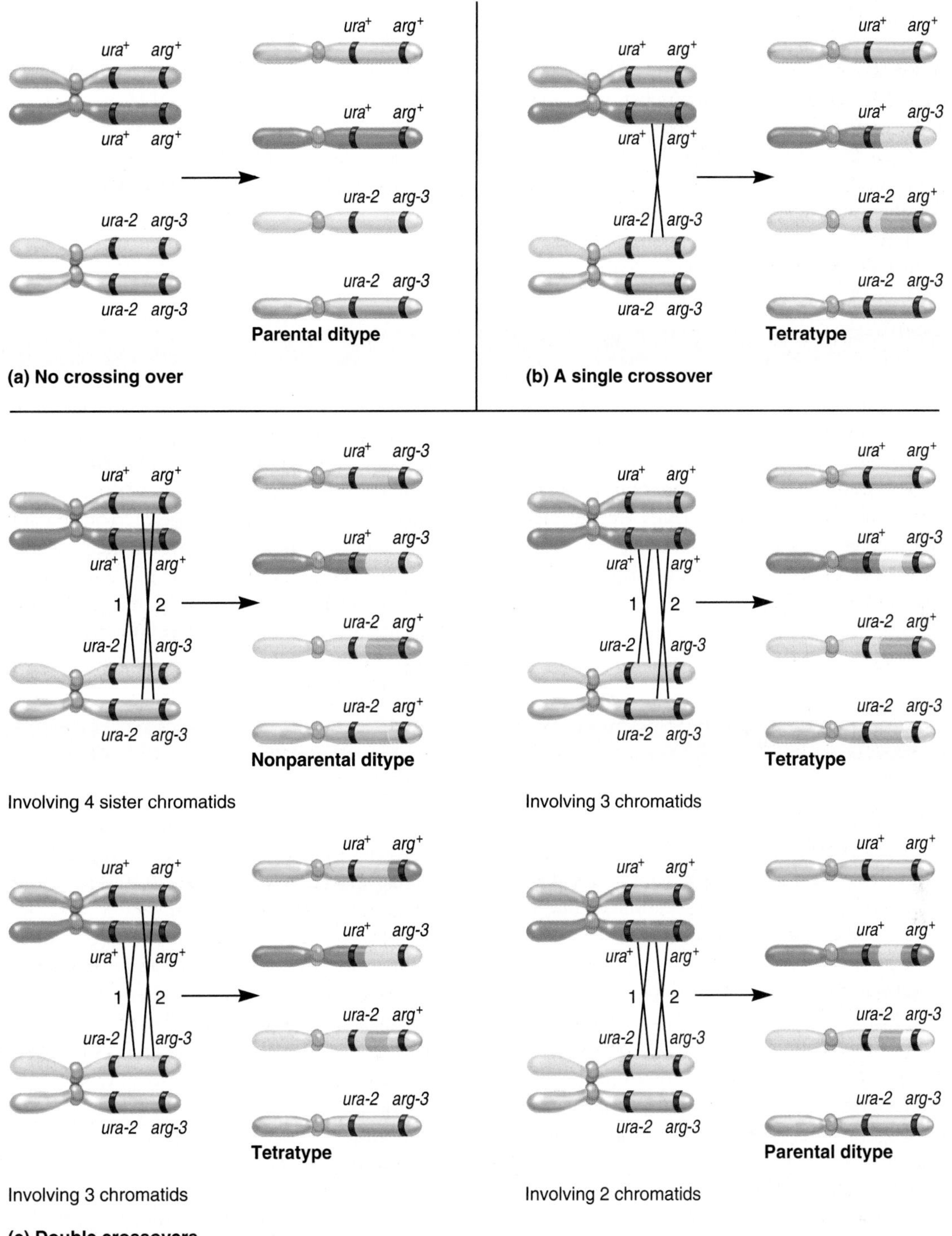

FIGURE 5.17 **Relationship between crossing over and the production of the parental ditype, tetratype, and nonparental ditype for two linked genes.** In the case of double crossovers, this figure shows the outcome in which the crossover on the left occurs first.

another look at Figure 5.17. As shown there, the parental ditype and tetratype are ambiguous. The parental ditype can be derived from no crossovers or a double crossover; the tetratype can be derived from a single crossover or a double crossover. However, the nonparental ditype is unambiguous, because it can be produced only from a double crossover. We can use this observation as a way to determine the actual number of single and double crossovers. As seen in Figure 5.17, 1/4 of all the double crossovers are nonparental ditypes. Therefore, the total number of double crossovers equals four times the number of nonparental ditypes.

Next, we need to know the number of single crossovers. A single crossover will yield a tetratype, but double crossovers can also yield a tetratype. Therefore, the total number of tetratypes overestimates the true number of single crossovers. Fortunately, we can compensate for this overestimation. Because two types of tetratypes are due to a double crossover, the actual number of tetratypes arising from a double crossover should equal 2NPD. Therefore, the true number of single crossovers is calculated as T−2NPD.

Now we have accurate measures of both single and double crossovers. The number of single crossovers equals T−2NPD, and the number of double crossovers equals 4NPD. We can substitute these values into our previous equation.

$$\text{Map distance} = \frac{(\text{T} - 2\text{NPD}) + (2)(4\text{NPD})}{\text{Total number of asci}} \times 0.5 \times 100$$

$$= \frac{\text{T} + 6\text{NPD}}{\text{Total number of asci}} \times 0.5 \times 100$$

This equation provides a more accurate measure of map distance because it considers both single and double crossovers.

CONCEPTUAL SUMMARY

Linkage refers to the phenomenon that many different genes may be located on the same chromosome. Chromosomes are sometimes called linkage groups because they contain a group of linked genes. Linkage affects the pattern of inheritance, because closely linked genes do not assort independently during meiosis. This produces a greater percentage of offspring that display parental phenotypes. Nevertheless, nonparental offspring can be produced as a result of crossing over.

The likelihood of crossing over depends on the distance between two genes. If two genes are far apart from each other on the same chromosome, it is more likely that crossing over will occur between them. Therefore, when two genes are widely separated, a substantial percentage of recombinant offspring will be obtained from a testcross. However, the percentage of recombinant offspring cannot exceed a value of 50%, even when two genes are more than 50 map units apart on the same chromosome. The relationship between the percentage of recombinant offspring and the linear distance between genes is the basis for genetic mapping.

EXPERIMENTAL SUMMARY

Experimentally, the phenomenon of linkage was deduced from genetic crosses. Bateson and Punnett were the first scientists to notice that certain genes do not assort independently. Morgan conducted crosses involving X-linked traits in fruit flies and correctly proposed that linkage is due to the location of particular genes on the same chromosome. He also hypothesized that recombinant phenotypes occur because of crossing over during meiosis. Morgan realized that the likelihood of crossing over depends on the distance between two genes. The hypothesis that crossing over can result in genetic recombination was confirmed cytologically by the studies of Creighton and McClintock, which showed that the production of recombinant offspring correlates with the physical exchange of material between chromosomes.

Genetic mapping is the determination of gene order and distance along chromosomes. In this chapter, we have considered how testcrosses are conducted as a method to map genes. Sturtevant was the first person to understand that the percentage of recombinant offspring in a testcross could be used as a measure of the relative distance between two genes. Map distance is computed as the number of recombinant offspring divided by the total number of offspring multiplied by 100. This approach can be readily applied to map genes using dihybrid and trihybrid testcrosses. Genetic mapping is most accurate when map distances are calculated between closely linked genes. As the map distance approaches 50 map units (mu) and above, the percentage of observed recombinant offspring is not a reliable measure of map distance. In Chapter 20, several molecular methods of genetic mapping are described.

The chapter ended with a discussion of gene mapping methods in fungi. A group of fungi known as the ascomycetes have been extensively used in genetic studies, because all products of a single meiosis are enclosed within an ascus. For fungi such as *Neurospora* that make an ordered octad, the spores are arranged in a manner that reflects their relationship to each other during meiosis and mitosis. Ordered octads can be analyzed to map the location of a single gene relative to the centromere. Fungal species and unicellular algae that produce unordered asci have also been used in mapping studies. In yeast, for example, dihybrid crosses are made, and the distance between the two genes can be computed by determining the proportions of parental ditypes, tetratypes, and nonparental ditypes.

In Chapter 6, we will consider the linkage of genes within bacterial chromosomes and bacteriophages. Although bacteria normally reproduce asexually, they still can transfer genetic material by various mechanisms. As we will see, these mechanisms also provide a way to map genes along the bacterial chromosome.

PROBLEM SETS & INSIGHTS

Solved Problems

S1. In the garden pea, orange pods (*orp*) are recessive to green pods (*Orp*), and sensitivity to pea mosaic virus (*mo*) is recessive to resistance to the virus (*Mo*). A plant with orange pods and sensitivity to the virus was crossed to a true-breeding plant with green pods and resistance to the virus. The F_1 plants were then testcrossed to plants with orange pods and sensitivity to the virus. The following results were obtained:

160 orange pods, virus sensitive
165 green pods, virus resistant
36 orange pods, virus resistant
39 green pods, virus sensitive
400 total

A. Conduct a chi square analysis to see if these genes are linked.

B. If they are linked, calculate the map distance between the two genes.

Answer:

A. Chi square analysis.

1. Our hypothesis is that the genes are not linked.
2. Calculate the predicted number of offspring based on the hypothesis. The testcross is

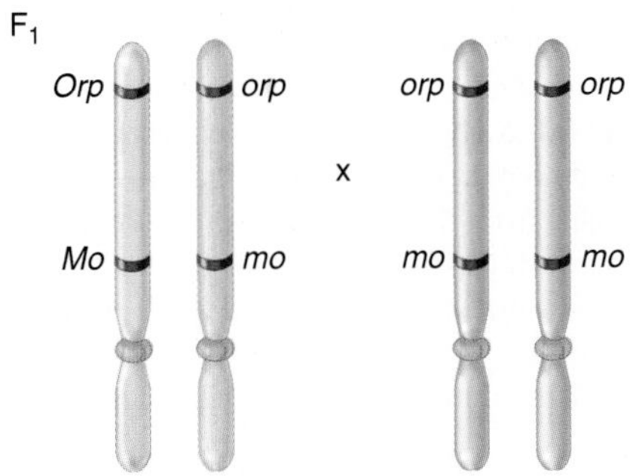

The predicted outcome of this cross under our hypothesis is a 1:1:1:1 ratio of plants with the four possible phenotypes. In other words, 1/4 should have the phenotype orange pods, virus sensitive; 1/4 should have green pods, virus resistant; 1/4 should have orange pods, virus resistant; and 1/4 should have green pods, virus sensitive. Because a total of 400 offspring were produced, our hypothesis predicts 100 offspring in each category.

3. Calculate the chi square.

$$\chi^2 = \frac{(O_1 - E_1)^2}{E_1} + \frac{(O_2 - E_2)^2}{E_2} + \frac{(O_3 - E_3)^2}{E_3} + \frac{(O_4 - E_4)^2}{E_4}$$

$$\chi^2 = \frac{(160 - 100)^2}{100} + \frac{(165 - 100)^2}{100} + \frac{(36 - 100)^2}{100} + \frac{(39 - 100)^2}{100}$$

$$\chi^2 = 36 + 42.3 + 41 + 37.2 = 156.5$$

4. Interpret the chi square value. The calculated chi square value is quite large. This indicates that the deviation between observed and expected values is very high. For one degree of freedom in Table 2.1, such a large deviation is expected to occur by chance alone less than 1% of the time. Therefore, we reject the hypothesis that the genes assort independently. As an alternative, we may infer that the two genes are linked.

B. Calculate the map distance.

$$\text{Map distance} = \frac{\text{Number of nonparental offspring}}{\text{Total number of offspring}} \times 100$$

$$= \frac{36 + 39}{36 + 39 + 160 + 165} \times 100$$

$$= 18.8 \text{ mu}$$

The genes are approximately 18.8 mu apart.

S2. Two recessive disorders in mice—droopy ears and flaky tail—are caused by genes that are located 6 mu apart on chromosome 3. A true-breeding mouse with normal ears (*De*) and a flaky tail (*ft*) was crossed to a true-breeding mouse with droopy ears (*de*) and a normal tail (*Ft*). The F_1 offspring were then crossed to mice with droopy ears and flaky tails. If this testcross produced 100 offspring, what is the expected outcome?

Answer: The testcross is

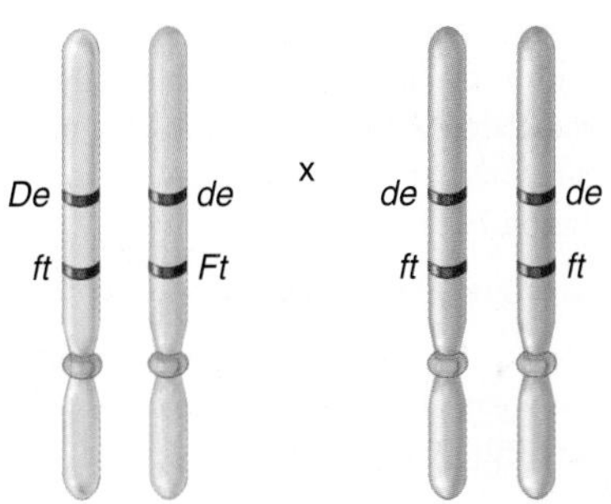

The parental offspring are

Dede ftft Normal ears, flaky tail
dede Ftft Droopy ears, normal tail

The recombinant offspring are

dede ftft Droopy ears, flaky tail
Dede Ftft Normal ears, normal tail

Because the two genes are located 6 mu apart on the same chromosome, 6% of the offspring will be recombinants. Therefore, the expected outcome for 100 offspring is

3 droopy ears, flaky tail
3 normal ears, normal tail
47 normal ears, flaky tail
47 droopy ears, normal tail

S3. The following X-linked recessive traits are found in fruit flies: vermilion eyes are recessive to red eyes, miniature wings are recessive to long wings, and sable body is recessive to gray body. A cross was made between wild-type males with red eyes, long wings, and gray bodies to females with vermilion eyes, miniature wings, and sable bodies. The heterozygous females from this cross, which had red eyes, long wings, and gray bodies, were then crossed to males with vermilion eyes, miniature wings, and sable bodies. The following outcome was obtained:

Males and Females

1,320 vermilion eyes, miniature wings, sable body

1,346 red eyes, long wings, gray body

102 vermilion eyes, miniature wings, gray body

90 red eyes, long wings, sable body

42 vermilion eyes, long wings, gray body

48 red eyes, miniature wings, sable body

2 vermilion eyes, long wings, sable body

1 red eyes, miniature wings, gray body

A. Calculate the map distance between the three genes.

B. Is positive interference occurring?

Answer:

A. The first step is to determine the order of the three genes. We can do this by evaluating the pattern of inheritance in the double crossovers. The double crossover group occurs with the lowest frequency. Thus, the double crossovers are vermilion eyes, long wings, and sable body, and red eyes, miniature wings, and gray body. Compared to the parental combinations of alleles (vermilion eyes, miniature wings, sable body and red eyes, long wings, gray body), the gene for wing length has been reassorted. Two flies have long wings associated with vermilion eyes and sable body, and one fly has miniature wings associated with red eyes and gray body. Taken together, these results indicate that the wing length gene is found in between the eye color and body color genes.

Eye color——wing length——body color

We now calculate the distance between eye color and wing length, and between wing length and body color. To do this, we consider the data according to gene pairs:

vermilion eyes, miniature wings = 1,320 + 102 = 1,422

red eyes, long wings = 1,346 + 90 = 1,436

vermilion eyes, long wings = 42 + 2 = 44

red eyes, miniature wings = 48 + 1 = 49

The recombinants are vermilion eyes, long wings and red eyes, miniature wings. The map distance between these two genes is

$(44 + 49)/(1{,}422 + 1{,}436 + 44 + 49) \times 100 = 3.2$ mu

Likewise, the other gene pair is wing length and body color.

miniature wings, sable body = 1,320 + 48 = 1,368

long wings, gray body = 1,346 + 42 = 1,388

miniature wings, gray body = 102 + 1 = 103

long wings, sable body = 90 + 2 = 92

The recombinants are miniature wings, gray body and long wings, sable body. The map distance between these two genes is

$(103 + 92)/(1{,}368 + 1{,}388 + 103 + 92) \times 100 = 6.6$ mu

With these data, we can produce the following genetic map:

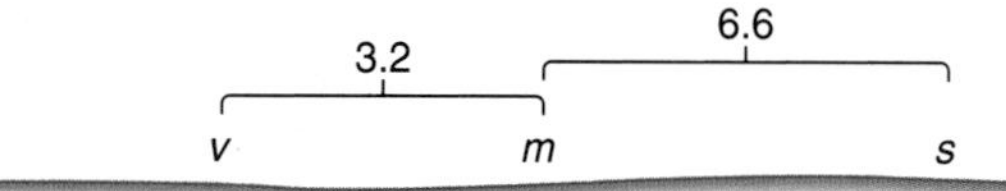

B. To calculate the interference value, we must first calculate the coefficient of coincidence.

$$C = \frac{\text{Observed number of double crossovers}}{\text{Expected number of double crossovers}}$$

Based on our calculation of map distances in part A, the percentage of single crossovers equals 3.2% (0.032) and 6.6% (0.066). The expected number of double crossovers equals 0.032×0.066, which is 0.002, or 0.2%. A total of 2,951 offspring were produced. If we multiply $2{,}951 \times 0.002$, we get 5.9, which is the expected number of double crossovers. The observed number was 3. Therefore,

$C = 3/5.9 = 0.51$

$I = 1 - C = 1 - 0.51 = 0.49$

In other words, approximately 49% of the expected double crossovers did not occur due to interference.

S4. Around the same time as the study of Creighton and McClintock, described in Figure 5.6, Curt Stern conducted similar experiments with *Drosophila*. He had strains of flies with microscopically detectable abnormalities in the X chromosome. In one case, the X chromosome was shorter than normal due to a deletion at one end. In another case, the X chromosome was longer than normal because an extra piece of the Y chromosome was attached at the other end of the X chromosome, where the centromere is located. He had female flies that had both abnormal chromosomes. On the short X chromosome, a recessive allele (*car*) was located that results in carnation-colored eyes, and a dominant allele (*B*) that causes bar-shaped eyes was also found on this chromosome. On the long X chromosome were located the wild-type alleles for these two genes (designated car^+ and B^+), which confer red eyes and round eyes, respectively. Stern realized that a crossover between the two X chromosomes in such female flies would result in recombinant chromosomes that would be cytologically distinguishable from the parental chromosomes. If a crossover occurred between the *B* and *car* genes on the X chromosome, this is expected to produce a normal-sized X chromosome and an abnormal chromosome with a deletion at one end and an extra piece of the Y chromosome at the other end.

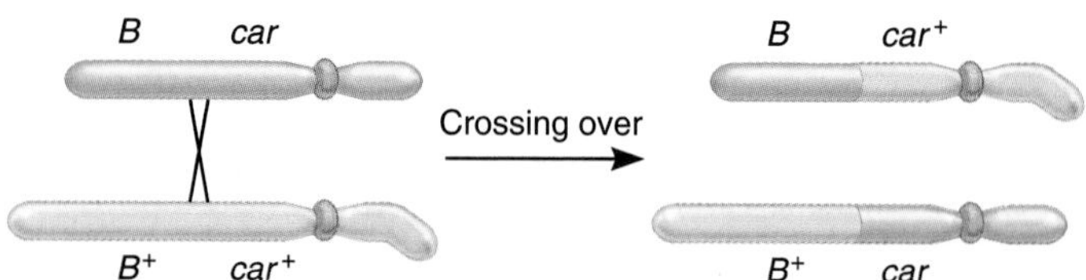

Stern crossed these female flies to male flies that had a normal-length X chromosome with the *car* allele and the allele for round eyes (*car* B^+). Using a microscope, he could discriminate between the morphologies of parental chromosomes—like those contained within the original parental flies—and recombinant chromosomes that may be found in the offspring. What would be the predicted phenotypes and chromosome characteristics in the offspring if crossing over did or did not occur between the X chromosomes in the female flies of this cross?

Answer: To demonstrate that genetic recombination is due to crossing over, Stern needed to correlate recombinant phenotypes (due to genetic recombination) with the inheritance of recombinant chromosomes (due to crossing over). Because he knew the arrangement of alleles in the female flies, he could predict the phenotypes of parental and nonparental offspring. The male flies could contribute the *car* and B^+ alleles (on a cytologically normal X chromosome) or contribute a Y chromosome. In the absence of crossing over, the female flies could contribute a short X chromosome with the *car* and *B* alleles or a long X chromosome with the car^+ and B^+ alleles. If crossing over occurred in the region between these two genes, the female flies would contribute recombinant X chromosomes. One possible recombinant X chromosome would be normal-sized and carry the *car* and B^+ alleles, while the other recombinant X chromosome would be deleted at one end with a piece of the Y chromosome at the other end and carry the car^+ and *B* alleles. When combined with an X or Y chromosome from the males, the parental offspring would have carnation, bar eyes or wild-type eyes; the nonparental offspring would have carnation, round eyes or red, bar eyes.

Male gametes

♀ Female gametes \ ♂	$carB^+$	Y	Phenotype	X chromosome from female
carB	*carB* $carB^+$	*carB* Y	Carnation, bar eyes	Short X chromosome
car^+B^+	car^+B^+ $carB^+$	car^+B^+ Y	Red, round eyes	Long X chromosome with a piece of Y
$carB^+$	$carB^+$ $carB^+$	$carB^+$ Y	Carnation, round eyes	Normal-sized X chromosome
car^+B	car^+B $carB^+$	car^+B Y	Red, bar eyes	Short X chromosome with a piece of Y

The results shown in the Punnett square are the actual results that Stern observed. His interpretation was that crossing over between homologous chromosomes—in this case, the X chromosome—accounts for the formation of offspring with recombinant phenotypes.

S5. Researchers have discovered a limit to the relationship between map distance and the percentage of recombinant offspring. Even though two genes on the same chromosome may be much more than 50 mu apart, we do not expect to obtain greater than 50% recombinant offspring in a testcross. You may be wondering why this is so. The answer lies in the pattern of multiple crossovers. At the pachytene stage of meiosis, a single crossover in the region between two genes will produce only 50% recombinant chromosomes (see Figure 5.1b). Therefore, to exceed a 50% recombinant level, it would seem necessary to have multiple crossovers within the tetrad.

Let's suppose that two genes are far apart on the same chromosome. A testcross is made between a heterozygous individual, *AaBb*, and a homozygous individual, *aabb*. In the heterozygous individual, the dominant alleles (*A* and *B*) are linked on the same chromosome, and the recessive alleles (*a* and *b*) are linked on the same chromosome. Draw out all of the possible double crossovers (between two, three, or four chromatids) and determine the average number of recombinant offspring, assuming an equal probability of all of the double crossover possibilities.

Answer: A double crossover between the two genes could involve two chromatids, three chromatids, or four chromatids. The possibilities for all types of double crossovers are shown here:

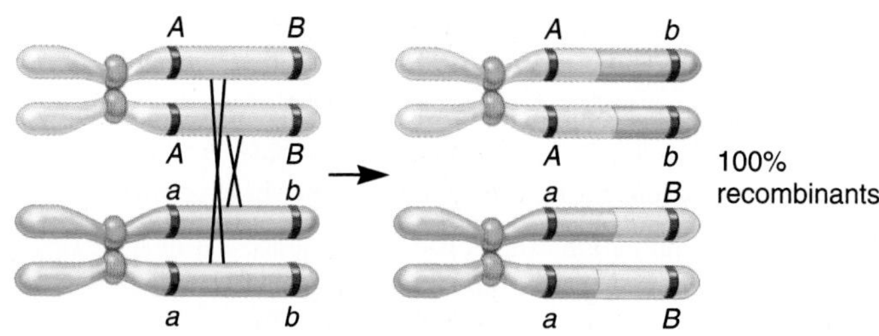

Double crossover (involving 4 chromatids)

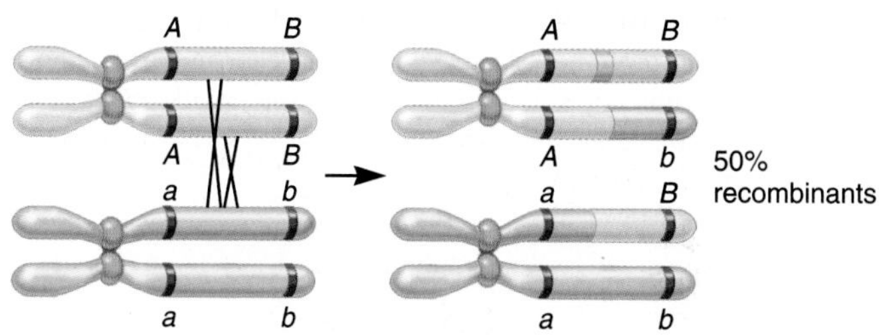

Double crossover (involving 3 chromatids)

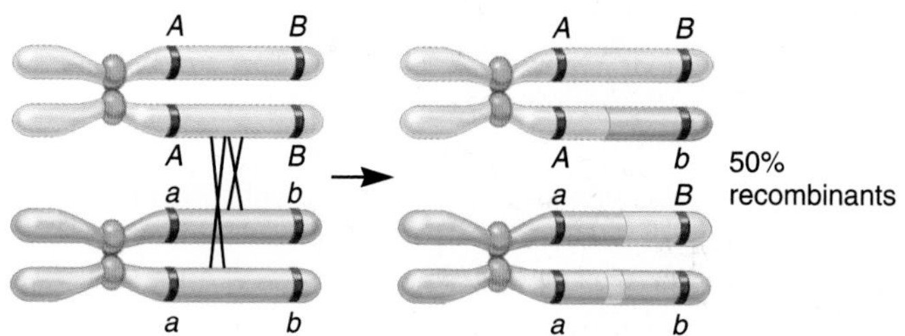

Double crossover (involving 3 chromatids)

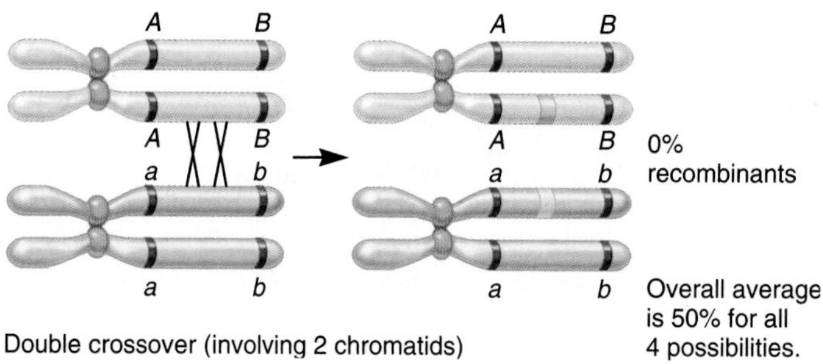

Double crossover (involving 2 chromatids)

This drawing considers the situation where two crossovers are expected to occur in the region between the two genes. Because the tetrad is composed of two pairs of homologues, there are several possible ways that a double crossover could occur between homologues. In this illustration, the crossover on the right has occurred first. Because all of these double crossing over events are equally probable, we take the average of them to determine the maximum recombination frequency. This average equals 50%.

Conceptual Questions

C1. What is the difference in meaning between the terms genetic recombination and crossing over?

C2. When applying a chi square approach in a linkage problem, explain why an independent assortment hypothesis is used.

C3. What is mitotic recombination? A heterozygous individual (*Bb*) with brown eyes has one eye with a small patch of blue. Provide two or more explanations for how the blue patch may have occurred.

C4. Mitotic recombination can occasionally produce a twin spot. Let's suppose an animal species can be heterozygous for two genes that govern fur color and length: One gene affects pigmentation, with dark pigmentation (*A*) dominant to albino (*a*); the other gene affects hair length, with long hair (*L*) dominant to short hair (*l*). The two genes are linked on the same chromosome. Let's assume an animal is *AaLl*; *A* is linked to *l*, and *a* is linked to *L*. Draw the chromosomes labeled with these alleles, and explain how mitotic recombination could produce a twin spot with one spot having albino pigmentation and long fur, the other having dark pigmentation and short fur.

C5. A crossover has occurred in the bivalent shown here.

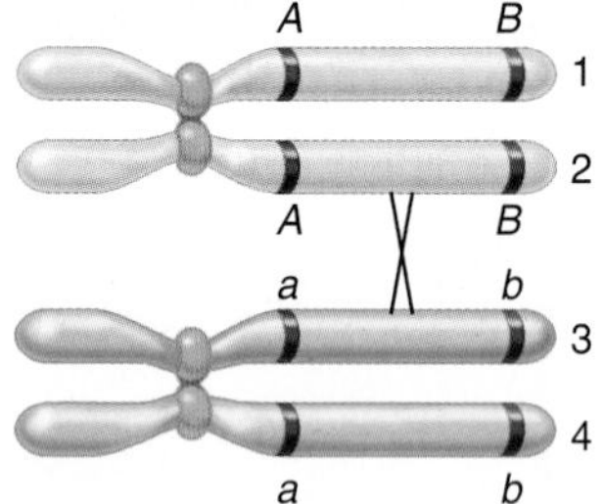

If a second crossover occurs in the same region between these two genes, which two chromatids would be involved in order to produce the following outcomes?

A. 100% recombinants

B. 0% recombinants

C. 50% recombinants

C6. A crossover has occurred in the bivalent shown here.

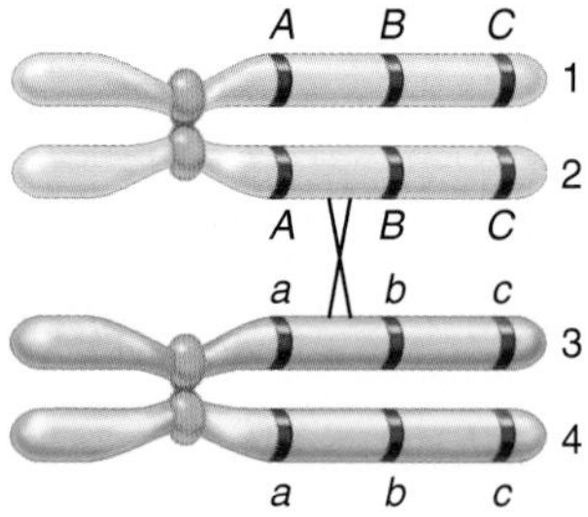

What would be the outcome of this single crossover event? If a second crossover occurs somewhere between *A* and *C*, explain which two chromatids it would involve and where it would occur (i.e., between which two genes) to produce the types of chromosomes shown here:

A. *A B C*, *A b C*, *a B c*, and *a b c*

B. *A b c*, *A b c*, *a B C*, and *a B C*

C. *A B c*, *A b c*, *a B C*, and *a b C*

D. *A B C*, *A B C*, *a b c*, and *a b c*

C7. A diploid organism has a total of 14 chromosomes and about 20,000 genes per haploid genome. Approximately how many genes are in each linkage group?

C8. If you try to throw a basketball into a basket, the likelihood of succeeding depends on the size of the basket. It is more likely that you will get the ball into the basket if the basket is bigger. In your own words, explain how this analogy also applies to the idea that the likelihood of crossing over is greater when two genes are far apart compared to when they are close together.

C9. By conducting testcrosses, researchers have found that the sweet pea has seven linkage groups. How many chromosomes would you expect to find in leaf cells?

C10. In humans, a rare dominant disorder known as nail-patella syndrome causes abnormalities in the fingernails, toenails, and kneecaps. Researchers have examined family pedigrees with regard to this disorder and, within the same pedigree, also examined the individuals with regard to their blood types. (A description of blood genotypes is found in Chapter 4.) In the following pedigree, individuals affected with nail-patella disorder are shown with filled symbols. The genotype of each individual with regard to their ABO blood type is also shown. Does this pedigree suggest any linkage between the gene that causes nail-patella syndrome and the gene that causes blood type?

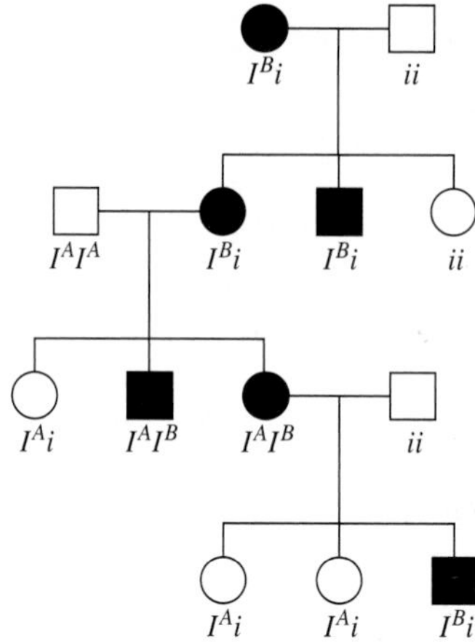

C11. When true-breeding mice with brown fur and short tails (*BBtt*) were crossed to true-breeding mice with white fur and long tails (*bbTT*), all F_1 offspring had brown fur and long tails. The F_1 offspring were crossed to mice with white fur and short tails. What are the possible phenotypes of the F_2 offspring? Which F_2 offspring are recombinant, and which are nonrecombinant? What will be the ratios of the F_2 offspring if independent assortment is taking place? How will the ratios be affected by linkage?

C12. Though we often think of genes in terms of the phenotypes they produce (e.g., curly leaves, flaky tail, brown eyes), the molecular function of most genes is to encode proteins. Many cellular proteins function as enzymes. The table that follows describes the map distances between six different genes that encode six different enzymes: *Ada*, adenosine deaminase; *Hao-1*, hydroxyacid oxidase-1; *Hdc*, histidine decarboxylase; *Odc-2*, ornithine decarboxylase-2; *Sdh-1*, sorbitol dehydrogenase-1; and *Ass-1*, arginosuccinate synthetase-1.

Map distances between two genes:

	Ada	*Hao-1*	*Hdc*	*Odc-2*	*Sdh-1*	*Ass-1*
Ada		14		8	28	
Hao-1	14		9		14	
Hdc		9		15	5	
Odc-2	8		15			63
Sdh-1	28	14	5			43
Ass-1				63	43	

Construct a genetic map that describes the locations of all six genes.

C13. If the likelihood of a single crossover in a particular chromosomal region is 10%, what is the theoretical likelihood of a double or triple crossover in that same region? How would positive interference affect these theoretical values?

C14. Except for fungi that form asci, in most dihybrid crosses involving linked genes, we cannot tell if a double crossover between the two genes has occurred because the offspring will inherit the parental combination of alleles. How does the inability to detect double crossovers affect the calculation of map distance? Is map distance underestimated or overestimated because of our inability to detect double crossovers? Explain your answer.

C15. Researchers have discovered that certain regions of chromosomes are much more likely to cross over compared to other regions. We might call such a region a "hot spot" for crossing over. Let's suppose that two genes, gene *A* and gene *B*, are 5,000,000 base pairs apart on the same chromosome. Genes *A* and *B* are in a hot spot for crossing over. Two other genes, let's call them gene *C* and gene *D*, are also 5,000,000 base pairs apart but are not in a hot spot for recombination. If we conducted dihybrid crosses to compute the map distance between genes *A* and *B*, and other dihybrid crosses to compute the map distance between genes *C* and *D*, would the map distances be the same between *A* and *B* compared to *C* and *D*? Explain.

C16. Describe the unique features of ascomycetes that lend themselves to genetic analysis.

C17. In fungi, what is the difference between a tetrad and an octad? What cellular process occurs in an octad that does not occur in a tetrad?

C18. Explain the difference between an unordered versus an ordered octad.

C19. In *Neurospora*, a cross is made between a wild-type and an albino mutant strain, which produce orange and white spores, respectively. Draw two different ways that an octad might look if it was displaying second-division segregation.

C20. One gene in *Neurospora*, let's call it gene *A*, is located close to a centromere, while a second gene, let's call it gene *B*, is located more toward the end of the chromosome. Would the percentage of octads exhibiting first-division segregation be higher with respect to gene *A* or gene *B*? Explain your answer.

Experimental Questions (Includes Most Mapping Questions)

E1. Figure 5.2 shows the first experimental results that indicated linkage between two different genes. Conduct a chi square analysis to confirm that the genes are really linked and the data could not be explained by independent assortment.

E2. In the experiment of Figure 5.6, the researchers followed the inheritance pattern of chromosomes that were abnormal at both ends to correlate genetic recombination with the physical exchange of chromosome pieces. Is it necessary to use a chromosome that is abnormal at both ends, or could the researchers have used a parental strain with two abnormal versions of chromosome 9, one with a knob at one end and its homologue with a translocation at the other end?

E3. The experiment of Figure 5.6 is not like a standard testcross, because neither parent is homozygous recessive for both genes. If you were going to carry out this same kind of experiment to verify that crossing over can explain the recombination of alleles of different genes, how would you modify this experiment to make it a standard testcross? For both parents, you should designate which alleles are found on an abnormal chromosome (i.e., knobbed, translocation chromosome 9) and which alleles are found on normal chromosomes.

E4. How would you determine that genes in mammals are located on the Y chromosome linkage group? Is it possible to conduct crosses (let's say in mice) to map the distances between genes along the Y chromosome? Explain.

E5. Explain the rationale behind a testcross. Is it necessary for one of the parents to be homozygous recessive for the genes of interest? In the heterozygous parent of a testcross, must all of the dominant alleles be linked on the same chromosome and all of the recessive alleles be linked on the homologue?

E6. In your own words, explain why a testcross cannot produce more than 50% recombinant offspring. When a testcross does produce 50% recombinant offspring, what do these results mean?

E7. Explain why the percentage of recombinant offspring in a testcross is a more accurate measure of map distance when two genes are close together. When two genes are far apart, is the percentage of recombinant offspring an underestimate or overestimate of the actual map distance?

E8. If two genes are more than 50 mu apart, how would you ever be able to show experimentally that they are located on the same chromosome?

E9. In Morgan's trihybrid testcross of Figure 5.3, he realized that crossing over was more frequent between the eye color and wing length genes than between the body color and eye color genes. Explain how he determined this.

E10. In the experiment of Figure 5.10, list the gene pairs from the particular dihybrid crosses that Sturtevant used to construct his genetic map.

E11. In the tomato, red fruit (*R*) is dominant over yellow fruit (*r*), and yellow flowers (*Wf*) are dominant over white flowers (*wf*). A cross was made between true-breeding plants with red fruit and yellow flowers, and plants with yellow fruit and white flowers. The F_1 generation plants were then crossed to plants with yellow fruit and white flowers. The following results were obtained:

333 red fruit, yellow flowers

64 red fruit, white flowers

58 yellow fruit, yellow flowers

350 yellow fruit, white flowers

Calculate the map distance between the two genes.

E12. Two genes are located on the same chromosome and are known to be 12 mu apart. An *AABB* individual was crossed to an *aabb* individual to produce *AaBb* offspring. The *AaBb* offspring were then crossed to *aabb* individuals.

A. If this cross produces 1,000 offspring, what are the predicted numbers of offspring with each of the four genotypes: *AaBb*, *Aabb*, *aaBb*, and *aabb*?

B. What would be the predicted numbers of offspring with these four genotypes if the parental generation had been *AAbb* and *aaBB* instead of *AABB* and *aabb*?

E13. Two genes, designated *A* and *B*, are located 10 mu from each other. A third gene, designated *C*, is located 15 mu from *B* and 5 mu from *A*. A parental generation consisting of *AA bb CC* and *aa BB cc* individuals were crossed to each other. The F_1 heterozygotes were then testcrossed to *aa bb cc* individuals. If we assume no double crossovers occur in this region, what percentage of offspring would you expect with the following genotypes?

A. *Aa Bb Cc*

B. *aa Bb Cc*

C. *Aa bb cc*

E14. Two genes in tomatoes are 61 mu apart; normal fruit (*F*) is dominant to fasciated fruit (*f*), and normal numbers of leaves (*Lf*) is dominant to leafy (*lf*). A true-breeding plant with normal leaves and fruit was crossed to a leafy plant with fasciated fruit. The F_1 offspring were then crossed to leafy plants with fasciated fruit. If this cross produced 600 offspring, what are the expected numbers of plants in each of the four possible categories: normal leaves, normal fruit; normal leaves, fasciated fruit; leafy, normal fruit; and leafy, fasciated fruit?

E15. In the tomato, three genes are linked on the same chromosome. Tall is dominant to dwarf, skin that is smooth is dominant to skin that is peachy, and fruit with a normal tomato shape is dominant to oblate shape. A plant that is true-breeding for the dominant traits was crossed to a dwarf plant with peachy skin and oblate fruit. The F_1 plants were then testcrossed to dwarf plants with peachy skin and oblate fruit. The following results were obtained:

151 tall, smooth, normal

33 tall, smooth, oblate

11 tall, peach, oblate

2 tall, peach, normal

155 dwarf, peach, oblate

29 dwarf, peach, normal

12 dwarf, smooth, normal

0 dwarf, smooth, oblate

Construct a genetic map that describes the order of these three genes and the distances between them.

E16. A trait in garden peas involves the curling of leaves. A dihybrid cross was made involving a plant with yellow pods and curling leaves to a wild-type plant with green pods and normal leaves. All F_1 offspring had green pods and normal leaves. The F_1 plants were then crossed to plants with yellow pods and curling leaves. The following results were obtained:

117 green pods, normal leaves

115 yellow pods, curling leaves

78 green pods, curling leaves

80 yellow pods, normal leaves

A. Conduct a chi square analysis to determine if these two genes are linked.

B. If they are linked, calculate the map distance between the two genes. How accurate do you think this distance is?

E17. In mice, the gene that encodes the enzyme inosine triphosphatase is 12 mu from the gene that encodes the enzyme ornithine decarboxylase. Let's suppose you have identified a strain of mice homozygous for a defective inosine triphosphatase gene that does not produce any of this enzyme and is also homozygous for a defective ornithine decarboxylase gene. In other words, this strain of mice cannot make either enzyme. You crossed this homozygous recessive strain to a normal strain of mice to produce heterozygotes. The heterozygotes were then backcrossed to the strain that cannot produce either enzyme. What is the probability of obtaining a mouse that cannot make either enzyme?

E18. In the garden pea, several different genes affect pod characteristics. A gene affecting pod color (green is dominant to yellow) is approximately 7 mu away from a gene affecting pod width (wide is dominant to narrow). Both genes are located on chromosome 5. A third gene, located on chromosome 4, affects pod length (long is dominant to short). A true-breeding wild-type plant (green, wide, long pods) was crossed to a plant with yellow, narrow, short pods. The F_1 offspring were then testcrossed to plants with yellow, narrow, short pods. If the testcross produced 800 offspring, what are the expected numbers of the eight possible phenotypic combinations?

E19. A sex-influenced trait is dominant in males and causes bushy tails. The same trait is recessive in females and results in a normal tail. Fur color is not sex influenced. Yellow fur is dominant to white fur. A true-breeding female with a bushy tail and yellow fur was crossed to a white male without a bushy tail. The F_1 females were then crossed to white males without bushy tails. The following results were obtained:

Males	Females
28 normal tails, yellow	102 normal tails, yellow
72 normal tails, white	96 normal tails, white
68 bushy tails, yellow	0 bushy tails, yellow
29 bushy tails, white	0 bushy tails, white

A. Conduct a chi square analysis to determine if these two genes are linked.

B. If the genes are linked, calculate the map distance between them. Explain which data you used in your calculation.

E20. Three recessive traits in garden pea plants are as follows: yellow pods are recessive to green pods, bluish green seedlings are recessive to green seedlings, creeper (a plant that cannot stand up) is recessive to normal. A true-breeding normal plant with green pods and green seedlings was crossed to a creeper with yellow pods and bluish green seedlings. The F_1 plants were then crossed to creepers with yellow pods and bluish green seedlings. The following results were obtained:

2,059 green pods, green seedlings, normal

151 green pods, green seedlings, creeper

281 green pods, bluish green seedlings, normal

15 green pods, bluish green seedlings, creeper

2,041 yellow pods, bluish green seedlings, creeper

157 yellow pods, bluish green seedlings, normal

282 yellow pods, green seedlings, creeper

11 yellow pods, green seedlings, normal

Construct a genetic map that describes the map distance between these three genes.

E21. In mice, a trait called snubnose is recessive to a wild-type nose, a trait called pintail is dominant to a normal tail, and a trait called jerker (a defect in motor skills) is recessive to a normal gait. Jerker mice with a snubnose and pintail were crossed to normal mice, and then the F_1 mice were crossed to jerker mice that have a snubnose and normal tail. The outcome of this cross was as follows:

560 jerker, snubnose, pintail

548 normal gait, normal nose, normal tail

102 jerker, snubnose, normal tail

104 normal gait, normal nose, pintail

77 jerker, normal nose, normal tail

71 normal gait, snubnose, pintail

11 jerker, normal nose, pintail

9 normal gait, snubnose, normal tail

Construct a genetic map that describes the order and distance between these genes.

E22. In *Drosophila*, an allele causing vestigial wings is 12.5 mu away from another gene that causes purple eyes. A third gene that affects body color has an allele that causes black body color. This third gene is 18.5 mu away from the vestigial wings gene and 6 mu away from the gene causing purple eyes. The alleles causing vestigial wings, purple eyes, and black body are all recessive. The dominant (wild-type) traits are long wings, red eyes, and gray body. A researcher crossed wild-type flies to flies with vestigial wings, purple eyes, and black bodies. All F_1 flies were wild type. F_1 female flies were then crossed to male flies with vestigial wings, purple eyes, and black bodies. If 1,000 offspring were observed, what are the expected numbers of the following types of flies?

Long wings, red eyes, gray body

Long wings, purple eyes, gray body

Long wings, red eyes, black body

Long wings, purple eyes, black body

Short wings, red eyes, gray body

Short wings, purple eyes, gray body

Short wings, red eyes, black body

Short wings, purple eyes, black body

Which kinds of flies can be produced only by a double crossover event?

E23. Three autosomal genes are linked along the same chromosome. The distance between gene *A* and *B* is 7 mu, the distance between *B* and *C* is 11 mu, and the distance between *A* and *C* is 4 mu. An individual who is *AA bb CC* was crossed to an individual who is *aa BB cc* to produce heterozygous F_1 offspring. The F_1 offspring were then crossed to homozygous *aa bb cc* individuals to produce F_2 offspring.

A. Draw the arrangement of alleles on the chromosomes in the parents and in the F_1 offspring.

B. Where would a crossover have to occur to produce an F_2 offspring that was heterozygous for all three genes?

C. If we assume that no double crossovers occur in this region, what percentage of F_2 offspring is likely to be homozygous for all three genes?

E24. Let's suppose that two different X-linked genes exist in mice, designated with the letters *N* and *L*. Gene *N* exists in a dominant, normal allele and in a recessive allele, *n*, that is lethal. Similarly, gene *L* exists in a dominant, normal allele and in a recessive allele, *l*, that is lethal. Heterozygous females are normal, while males that carry either recessive allele are born dead. Explain whether or not it would be possible to map the distance between these two genes by making crosses and analyzing the number of living and dead offspring. You may assume that you have strains of mice in which females are heterozygous for one or both genes.

E25. The alleles *his-5* and *lys-1*, found in baker's yeast, result in cells that require histidine and lysine for growth, respectively. A cross was made between two haploid yeast strains that are *his-5 lys-1* and his^+ lys^+. From the analysis of 818 individual tetrads, the following numbers of tetrads were obtained:

2 spores: *his-5* lys^+ + 2 spores: his^+ *lys-1* = 4

2 spores: *his-5 lys-1* + 2 spores: his^+ lys^+ = 502

1 spore: *his-5 lys-1* + 1 spore: *his-5* lys^+ + 1 spore: his^+ *lys-1* + 1 spore: his^+ lys^+ = 312

A. Compute the map distance between these two genes using the method of calculation that considers double crossovers and the one that does not. Which method gives a higher value? Explain why.

B. What is the frequency of single crossovers between these two genes?

C. Based on your answer to part B, how many NPDs are expected from this cross? Explain your answer. Is positive interference occurring?

E26. On chromosome 4 in *Neurospora*, the allele *pyr-1* results in a pyrimidine requirement for growth. A cross was made between a *pyr-1* and a pyr^+ (wild-type) strain, and the following results were obtained:

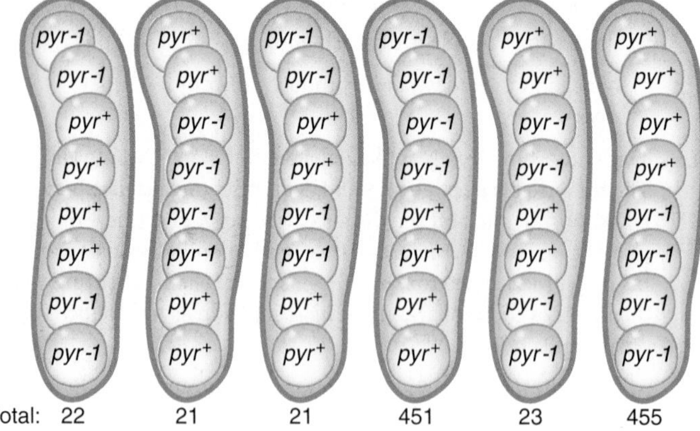

What is the distance between the *pyr-1* gene and the centromere?

E27. On chromosome 3 in *Neurospora*, the *pro-1* gene is located approximately 9.8 mu from the centromere. Let's suppose a cross was made between a *pro-1* and a pro^+ strain and 1,000 asci were analyzed.

A. What are the six types of asci that can be produced?

B. What are the expected numbers of each type of ascus?

Questions for Student Discussion/Collaboration

1. In mice, a dominant gene that causes a short tail is located on chromosome 2. On chromosome 3, a recessive gene causing droopy ears is 6 mu away from another recessive gene that causes a flaky tail. A recessive gene that causes a jerker (uncoordinated) phenotype is located on chromosome 4. A jerker mouse with droopy ears and a short, flaky tail was crossed to a normal mouse. All F_1 generation mice were phenotypically normal, except they had short tails. These F_1 mice were then testcrossed to jerker mice with droopy ears and long, flaky tails. If this cross produced 400 offspring, what would be the proportions of the 16 possible phenotypic categories?

2. In Chapter 3, we discussed the idea that the X and Y chromosomes have a few genes in common. These genes are inherited in a pseudoautosomal pattern. With this phenomenon in mind, discuss whether or not the X and Y chromosomes are really distinct linkage groups.

3. Mendel studied seven traits in pea plants, and the garden pea happens to have seven different chromosomes. It has been pointed out that Mendel was very lucky not to have conducted crosses involving two traits governed by genes that are closely linked on the same chromosome because the results would have confounded his theory of independent assortment. It has even been suggested that Mendel may not have published data involving traits that were linked! An article by Stig Blixt ("Why Didn't Gregor Mendel Find Linkage?" *Nature 256*:206, 1975) considers this issue. Look up this article and discuss why Mendel did not find linkage.

Note: All answers appear at the website for this textbook; the answers to even-numbered questions are in the back of the textbook.

www.mhhe.com/brookergenetics3e

Visit the Online Learning Center for practice tests, answer keys, and other learning aids for this chapter. Enhance your understanding of genetics with our interactive exercises, quizzes, animations, and much more.

GENETIC TRANSFER AND MAPPING IN BACTERIA AND BACTERIOPHAGES

6

CHAPTER OUTLINE

One reason researchers are so interested in bacteria and viruses is related to their impact on health. Infectious diseases caused by these agents are a leading cause of human death, accounting for a quarter to a third of deaths worldwide. The spread of infectious diseases results from human behavior, and in recent times has been accelerated by increased trade and travel, and the inappropriate use of antibiotic drugs. Although the incidence of fatal infectious diseases in the United States is relatively low compared to the worldwide average, an alarming increase in more deadly strains of bacteria and viruses has occurred over the past few decades. Since 1980, the number of deaths in the United States due to infectious diseases has approximately doubled.

Thus far, our attention in Part II of this textbook has focused on genetic analyses of eukaryotic species such as fungi, plants, and animals. As we have seen, these organisms are amenable to genetic studies for two reasons. First, allelic differences, such as white versus red eyes in *Drosophila* and tall versus dwarf pea plants, provide readily discernible traits among different individuals. Second, because most eukaryotic species reproduce sexually, crosses can be made, and the pattern of transmission of traits from parent to offspring can be analyzed. The ability to follow allelic differences in a genetic cross is a basic tool in the genetic examination of eukaryotic species.

In Chapter 6, we turn our attention to the genetic analysis of bacteria. Like their eukaryotic counterparts, bacteria often possess allelic differences that affect their cellular traits. Common allelic variations among bacteria that are readily discernible involve traits such as sensitivity to antibiotics and differences in their nutrient requirements for growth. In these cases, the allelic differences are between different strains of bacteria, because any given bacterium is usually haploid for a particular gene. In fact, the haploid nature of bacteria is one advantage that makes it easier to identify mutations that produce phenotypes such as altered nutritional requirements. Loss-of-function mutations, which are often recessive in diploid eukaryotes, are not masked by dominant alleles in haploid species. Throughout this chapter, we will consider interesting experiments that examine bacterial strains with allelic differences.

Compared to eukaryotes, another striking difference in prokaryotic species is their mode of reproduction. Because bacteria reproduce asexually, researchers do not use crosses in the genetic analysis of bacterial species. Instead, they rely on a similar mechanism, called genetic transfer, in which a segment of bacterial DNA is transferred from one bacterium to another. In the first part of this chapter,

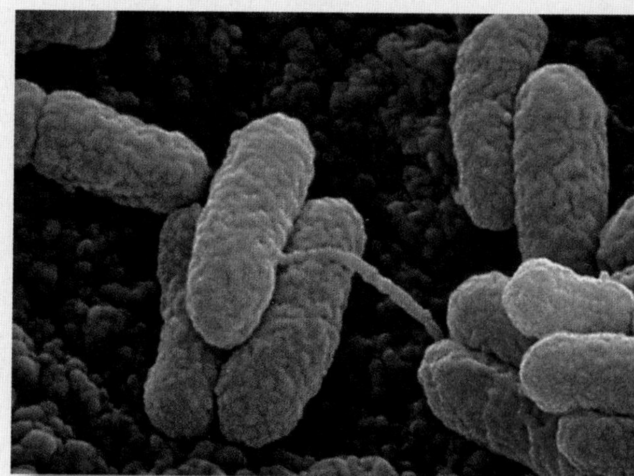

Conjugating bacteria. The bacteria shown here are transferring genetic material by a process called conjugation.

we will explore the different routes of genetic transfer. We will see how researchers have used genetic transfer to map the locations of genes along the chromosome of many bacterial species.

In the second part of this chapter, we will examine **bacteriophages** (also known as **phages**), which are viruses that infect bacteria. Bacteriophages contain their own genetic material that governs the traits of the phage. As we will see, the genetic analysis of phages can yield a highly detailed genetic map of a short region of DNA. These types of analyses have provided researchers with insights regarding the structure and function of genes.

6.1 GENETIC TRANSFER AND MAPPING IN BACTERIA

Genetic transfer is a process by which one bacterium transfers genetic material to another bacterium. Why is genetic transfer an advantage? Like sexual reproduction in eukaryotes, genetic transfer in bacteria is thought to enhance the genetic diversity of bacterial species. For example, a bacterial cell carrying a gene that provides antibiotic resistance may transfer this gene to another bacterial cell, allowing that bacterial cell to survive exposure to the antibiotic.

Bacteria can naturally transfer genetic material in three ways (**Table 6.1**). The first route, known as **conjugation,** involves a direct physical interaction between two bacterial cells. One bacte-

TABLE 6.1

Three Mechanisms of Genetic Transfer Found in Bacteria

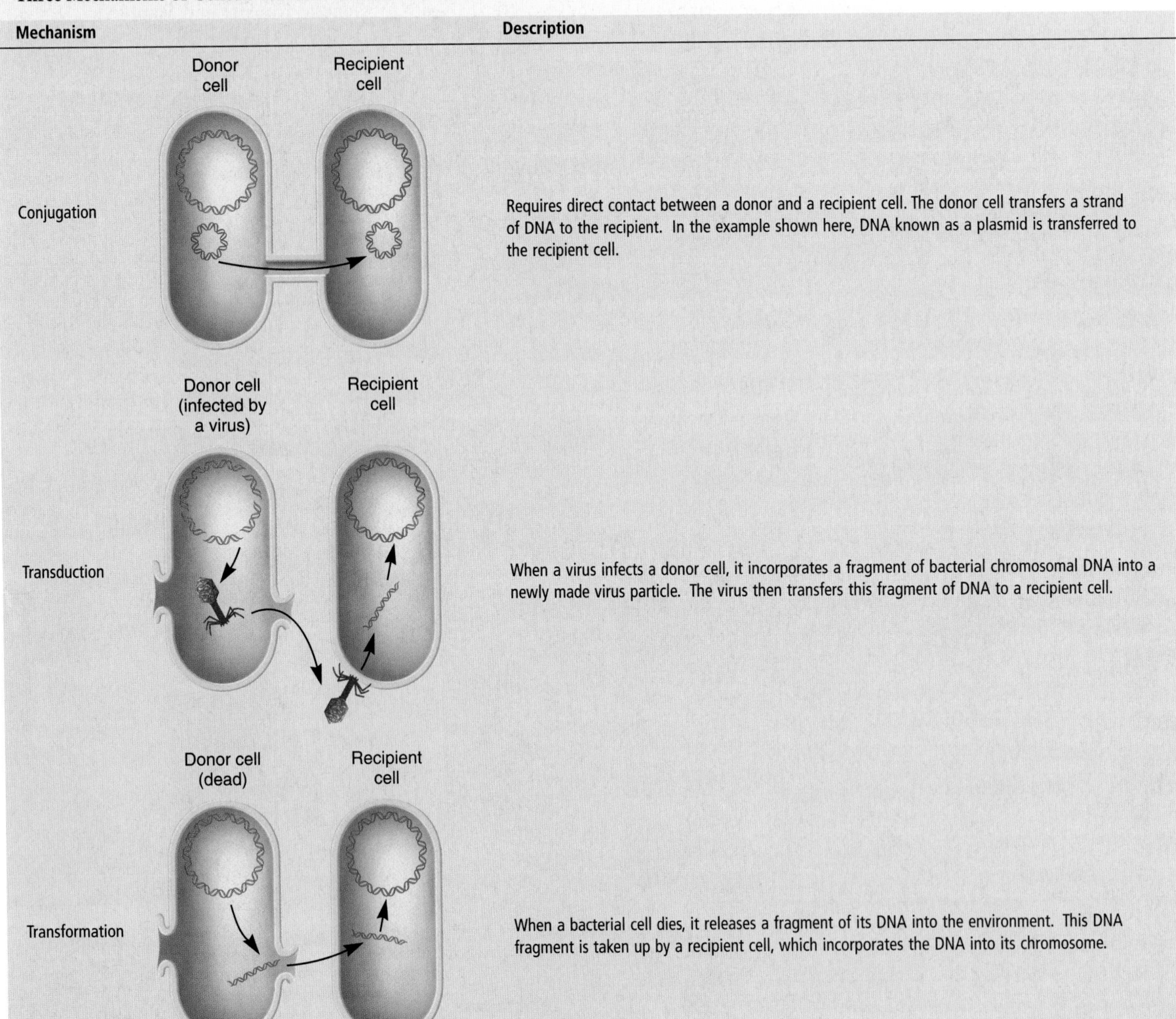

Mechanism	Description
Conjugation	Requires direct contact between a donor and a recipient cell. The donor cell transfers a strand of DNA to the recipient. In the example shown here, DNA known as a plasmid is transferred to the recipient cell.
Transduction	When a virus infects a donor cell, it incorporates a fragment of bacterial chromosomal DNA into a newly made virus particle. The virus then transfers this fragment of DNA to a recipient cell.
Transformation	When a bacterial cell dies, it releases a fragment of its DNA into the environment. This DNA fragment is taken up by a recipient cell, which incorporates the DNA into its chromosome.

rium acts as a donor and transfers genetic material to a recipient cell. A second means of transfer is called **transduction.** This occurs when a virus infects a bacterium and then transfers bacterial genetic material from that bacterium to another. The last mode of genetic transfer is **transformation.** In this case, genetic material is released into the environment when a bacterial cell dies. This material then binds to a living bacterial cell, which can take it up. These three mechanisms of genetic transfer have been extensively investigated in research laboratories, and their molecular mechanisms continue to be studied with great interest. In this section, we will examine these three systems of genetic transfer in greater detail.

We will also learn how genetic transfer between bacterial cells has provided unique ways to accurately map bacterial genes. The mapping methods described in this chapter have been largely replaced by molecular approaches described in Chapter 20. Even so, the mapping of bacterial genes serves to illuminate the mechanisms by which genes are transferred between bacterial cells and also helps us to appreciate the strategies of newer mapping approaches.

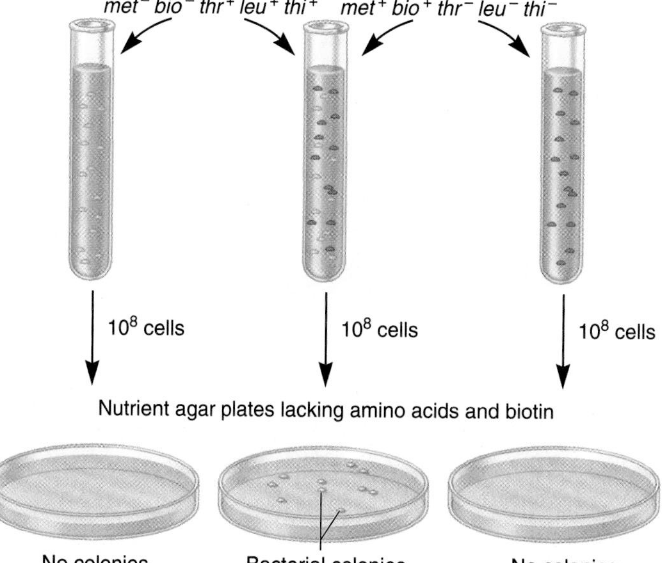

FIGURE 6.1 Experiment of Lederberg and Tatum demonstrating genetic transfer during conjugation in *E. coli.* When plated on growth media lacking amino acids, biotin, and thiamine, the *met*$^-$ *bio*$^-$ *thr*$^+$ *leu*$^+$ *thi*$^+$ or *met*$^+$ *bio*$^+$ *thr*$^-$ *leu*$^-$ *thi*$^-$ strains were unable to grow. However, if they were mixed together and then plated, some colonies were observed. These colonies were due to the transfer of genetic material between these two strains by conjugation. Note: In bacteria, it is common to give genes a three-letter name (shown in italics) that is related to the function of the gene. A plus superscript ($^+$) indicates a functional gene, and a minus superscript ($^-$) indicates a mutation that has caused the gene or gene product to be inactive. In some cases, several genes have related functions. These may have the same three-letter name followed by different capital letters. For example, different genes involved with leucine biosynthesis may be called *leuA*, *leuB*, *leuC*, and so on. In the experiment described in Figure 6.1, the genes involved in leucine biosynthesis had not been distinguished, so the gene involved is simply referred to as *leu*$^+$ (for a functional gene) and *leu*$^-$ (for a nonfunctional gene).

Bacteria Can Transfer Genetic Material During Conjugation

The natural ability of some bacteria to transfer genetic material between each other was first recognized by Joshua Lederberg and Edward Tatum in 1946. They were studying strains of *Escherichia coli* that had different nutritional requirements for growth. A strain that cannot synthesize a particular nutrient and needs that nutrient supplemented in its growth medium is called an **auxotroph.** For example, a strain that cannot make the amino acid methionine would need this amino acid in its growth medium and would be called a methionine auxotroph. By comparison, a strain that could make this amino acid would be termed a methionine prototroph. A **prototroph** does not need the nutrient in its growth medium.

The experiment in **Figure 6.1** considers one strain, designated *met*$^-$ *bio*$^-$ *thr*$^+$ *leu*$^+$ *thi*$^+$, which required one amino acid, methionine (met), and one vitamin, biotin (bio), in order to grow. This strain did not require the amino acids threonine (thr) or leucine (leu), or the vitamin thiamine (thi) for growth. Another strain, designated *met*$^+$ *bio*$^+$ *thr*$^-$ *leu*$^-$ *thi*$^-$, had just the opposite requirements. It was an auxotroph for threonine, leucine, and thiamine, but a prototroph for methionine and biotin. These differences in nutritional requirements correspond to variations in the genetic material of the two strains. The first strain had two defective genes encoding enzymes necessary for methionine and biotin synthesis. The second strain contained three defective genes required to make threonine, leucine, and thiamine.

Figure 6.1 compares the results when the two strains were mixed together and when they were not mixed. Without mixing, about 100 million (10^8) *met*$^-$ *bio*$^-$ *thr*$^+$ *leu*$^+$ *thi*$^+$ cells were applied to plates on a growth medium lacking amino acids, biotin, and thiamine; no colonies were observed to grow. This result is expected because the media did not contain methionine or biotin. Likewise, when 10^8 *met*$^+$ *bio*$^+$ *thr*$^-$ *leu*$^-$ *thi*$^-$ cells were plated, no colonies were observed because threonine, leucine, and thiamine were missing from this growth medium. However, when the two strains were mixed together and then 10^8 cells plated, approximately 10 bacterial colonies formed. Because growth occurred, the genotype of the cells within these colonies must have been *met*$^+$ *bio*$^+$ *thr*$^+$ *leu*$^+$ *thi*$^+$. How could this genotype occur? Lederberg and Tatum hypothesized that some genetic material was transferred between the two strains. One possibility is that the genetic material providing the ability to synthesize methionine and biotin (*met*$^+$ *bio*$^+$) was transferred to the *met*$^-$ *bio*$^-$ *thr*$^+$ *leu*$^+$ *thi*$^+$ strain. Alternatively, the ability to synthesize threonine, leucine, and thiamine (*thr*$^+$ *leu*$^+$ *thi*$^+$) may have been transferred to the *met*$^+$ *bio*$^+$ *thr*$^-$ *leu*$^-$ *thi*$^-$ cells. The results of this experiment did not distinguish between these two possibilities.

In 1950, Bernard Davis conducted experiments showing that two strains of bacteria must make physical contact with each other to transfer genetic material. The apparatus he used, known as a U-tube, is shown in **Figure 6.2.** At the bottom of the U-tube is a filter with pores small enough to allow the passage of genetic material (i.e., DNA molecules) but too small to permit the passage of bacterial cells. On one side of the filter, Davis added a bacterial strain with a certain combination of nutritional requirements (the *met*$^-$ *bio*$^-$ *thr*$^+$ *leu*$^+$ *thi*$^+$ strain). On the other side, he added a different bacterial strain (the *met*$^+$ *bio*$^+$ *thr*$^-$ *leu*$^-$ *thi*$^-$ strain). The application of alternating pressure and suction

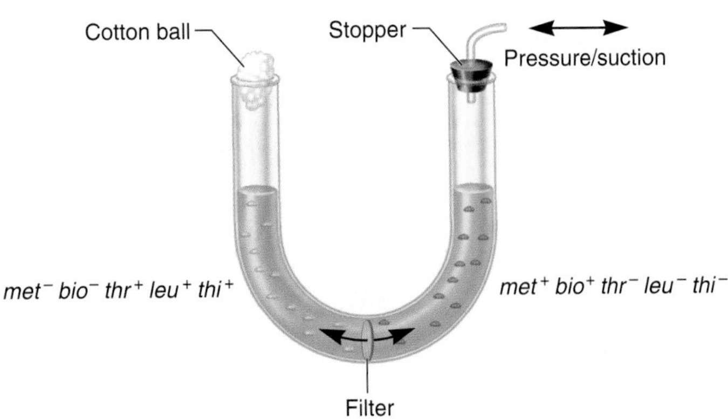

FIGURE 6.2 A U-tube apparatus like that used by Bernard Davis. The fluid in the tube is forced through the filter by alternating suction and pressure. However, the pores in the filter are too small for the passage of bacteria.

promoted the movement of liquid through the filter. Because the bacteria were too large to pass through the pores, the movement of liquid did not allow the two types of bacterial strains to mix with each other. However, any genetic material that was released from a bacterium could pass through the filter.

After incubation in a U-tube, bacteria from either side of the tube were placed on media that could select for the growth of cells that were $met^+ \ bio^+ \ thr^+ \ leu^+ \ thi^+$. These selective media lacked methionine, biotin, threonine, leucine, and thiamine but contained all other nutrients essential for growth. In this case, no bacterial colonies grew on the plates. The experiment showed that, without physical contact, the two bacterial strains did not transfer genetic material to one another.

The term conjugation is now used to describe the natural process of genetic transfer between bacterial cells that requires direct cell-to-cell contact. Many, but not all, species of bacteria can conjugate. Working independently, Joshua and Esther Lederberg, William Hayes, and Luca Cavalli-Sforza discovered in the early 1950s that only certain bacterial strains can act as donors of genetic material. For example, only about 5% of natural isolates of *E. coli* can act as donor strains. Research studies showed that a strain incapable of acting as a donor could subsequently be converted to a donor strain after being mixed with another donor strain. Hayes correctly proposed that donor strains contain a fertility factor that can be transferred to conjugation-defective strains to make them conjugation proficient.

We now know that certain donor strains of *E. coli* contain a small circular segment of genetic material known as an **F factor** (for fertility factor) in addition to their circular chromosome. Strains of *E. coli* that contain an F factor are designated F^+, while strains without F factors are termed F^-. In recent years, the molecular details of the conjugation process have been extensively studied. Though the mechanisms vary somewhat from one bacterial species to another, some general themes have emerged. F factors carry several genes that are required for conjugation to occur. For example, **Figure 6.3** shows the arrangement of genes on the F factor found in certain strains of *E. coli.* The functions of the proteins encoded by these genes are needed to transfer a strand of DNA from the donor cell to a recipient cell.

Figure 6.4a describes the molecular events that occur during conjugation in *E. coli.* Contact between donor and recipient cells is a key step that initiates the conjugation process. **Sex pili** (singular: **pilus**) are made by F^+ strains (**Figure 6.4b**). The gene encoding the pilin protein (*traA*) is located on the F factor. The pili act as attachment sites that promote the binding of bacteria to each other. In this way, an F^+ strain makes physical contact with an F^- strain. In certain species, such as *E. coli,* long pili project from F^+ cells and attempt to make contact with nearby F^- cells.

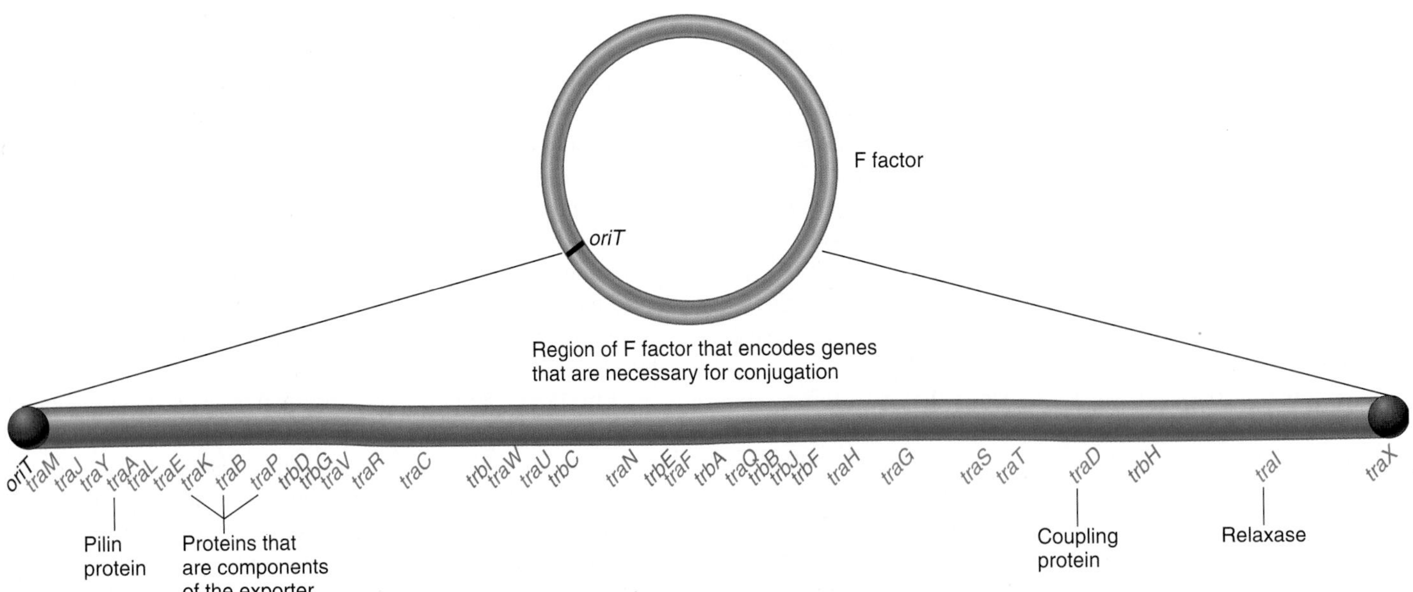

FIGURE 6.3 Genes on the F factor that play a role during conjugation. A region of the F factor contains a few dozen genes that play a role in the conjugative process. Because they play a role in the transfer of DNA from donor to recipient cell, the genes are designated with the three-letter names of *tra* or *trb*, followed by a capital letter. The *tra* genes are shown in red, and the *trb* genes are shown in blue. The functions of a few examples are indicated. The origin of transfer is designated *oriT.*

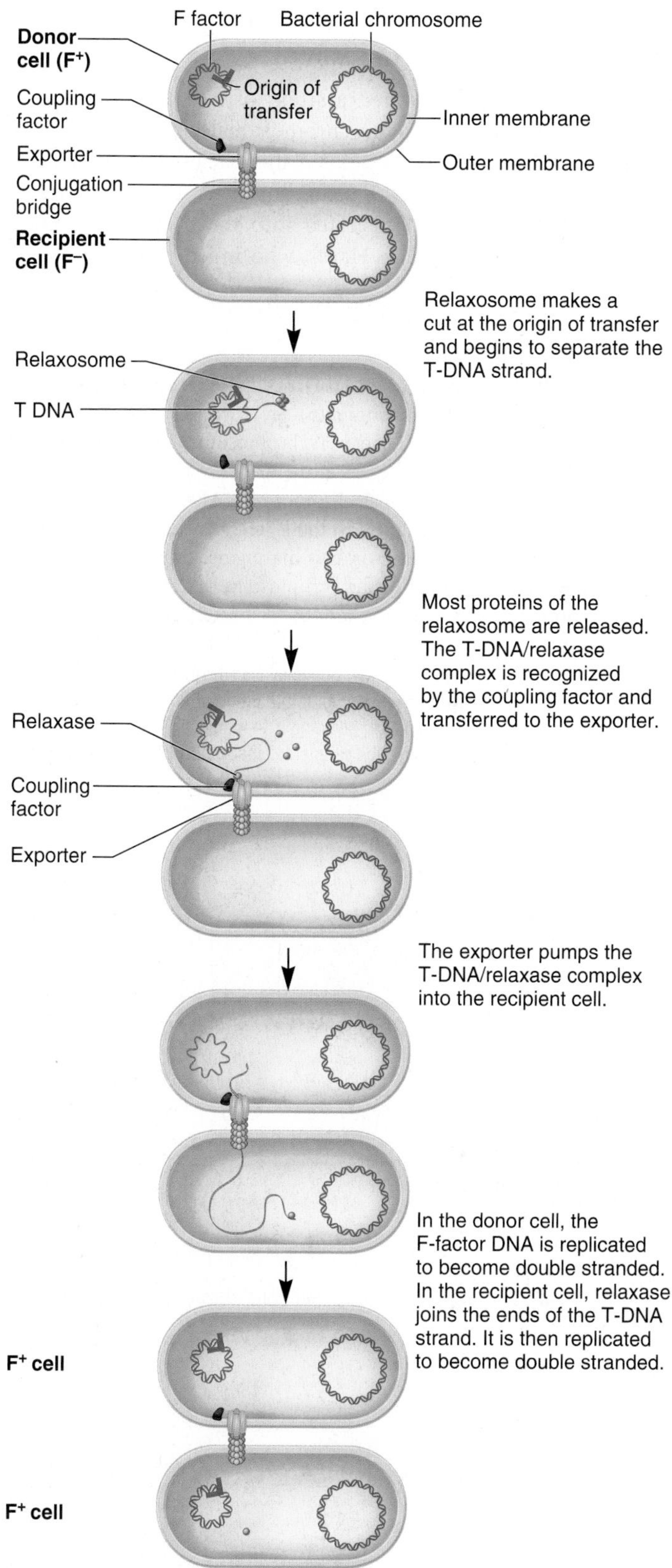

Donor cell (F⁺)
F factor
Bacterial chromosome
Coupling factor
Origin of transfer
Inner membrane
Exporter
Outer membrane
Conjugation bridge
Recipient cell (F⁻)
Relaxosome makes a cut at the origin of transfer and begins to separate the T-DNA strand.
Relaxosome
T DNA
Most proteins of the relaxosome are released. The T-DNA/relaxase complex is recognized by the coupling factor and transferred to the exporter.
Relaxase
Coupling factor
Exporter
The exporter pumps the T-DNA/relaxase complex into the recipient cell.
In the donor cell, the F-factor DNA is replicated to become double stranded. In the recipient cell, relaxase joins the ends of the T-DNA strand. It is then replicated to become double stranded.
F⁺ cell
F⁺ cell

(a) Transfer of an F factor via conjugation

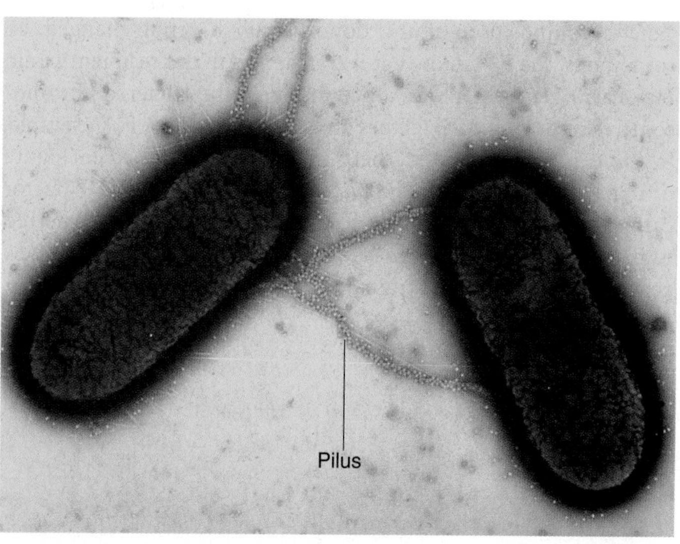

Pilus

(b) Conjugating *E. coli*

ONLINE ANIMATION

FIGURE 6.4 The transfer of an F factor during bacterial conjugation. (**a**) The mechanism of transfer. The end result is that both cells have an F factor. (**b**) Two *E. coli* cells in the act of conjugation. The cell on the left is F^+, and the one on the right is F^-. The two cells make contact with each other via sex pili that are made by the F^+ cell.

Once contact is made, the pili shorten and thereby draw the donor and recipient cells closer together. A **conjugation bridge** is then formed between the two cells, which provides a passageway for DNA transfer.

The successful contact between a donor and recipient cell stimulates the donor cell to begin the transfer process. Genes within the F factor encode a protein complex called the **relaxosome.** This complex first recognizes a DNA sequence in the F factor known as the **origin of transfer.** Upon recognition, the site is cut in one DNA strand. This cut strand of DNA is called **T DNA** because it is the strand that will be transferred to the recipient cell. The relaxosome also catalyzes the separation of the T-DNA strand from its complementary strand. As the DNA strands separate, most of the proteins within the relaxosome are released, but one protein, called relaxase, remains bound to the end of the T DNA. The complex between the T DNA and relaxase is called a **nucleoprotein** because it contains both nucleic acid (DNA) and protein (relaxase).

The next phase of conjugation involves the export of the nucleoprotein complex from the donor cell to the recipient cell. To begin this process, the T-DNA/relaxase complex is recognized by a coupling factor that promotes the entry of the nucleoprotein into the exporter, a complex of proteins that spans both inner and outer membranes of the donor cell. In bacterial species, this complex is formed from 10 to 15 different proteins that are encoded by genes within the F factor.

Once the T-DNA/relaxase complex is pumped out of the donor cell, it travels through the conjugation bridge and then into the recipient cell. As shown in Figure 6.4a, the other strand of the F factor DNA remains in the donor cell, where DNA replication restores the F factor DNA to its original double-stranded condition. After the recipient cell receives a single strand of the F factor DNA, relaxase catalyzes the joining of the ends of the linear DNA molecule to form a circular molecule. This single-stranded DNA is replicated in the recipient cell to become double stranded.

The result of conjugation is that the recipient cell has acquired an F factor, converting it from an F^- to an F^+ cell. The genetic composition of the donor strain has not changed. In some cases, the F^+ factor that is transferred to the recipient strain may carry genes that were once found in the bacterial chromosome. These types of F factors are called **F' factors.** For example, in the experiment of Lederberg and Tatum described in Figure 6.1, an F' factor carrying the *bio*$^+$ and *met*$^+$ genes or an F' factor carrying the *thr*$^+$, *leu*$^+$, and *thi*$^+$ genes may have been transferred to the recipient strain. Therefore, conjugation may introduce new genes into the recipient strain and thereby alter its genotype.

Hfr Strains Contain an F Factor Integrated into the Bacterial Chromosome

Luca Cavalli-Sforza discovered a strain of *E. coli* that was very efficient at transferring many chromosomal genes to recipient F^- strains. Cavalli-Sforza designated this bacterial strain an ***Hfr* strain** (for high frequency of recombination). In an *Hfr* strain, an F factor has become integrated into the bacterial chromosome (**Figure 6.5a**). William Hayes, who independently isolated

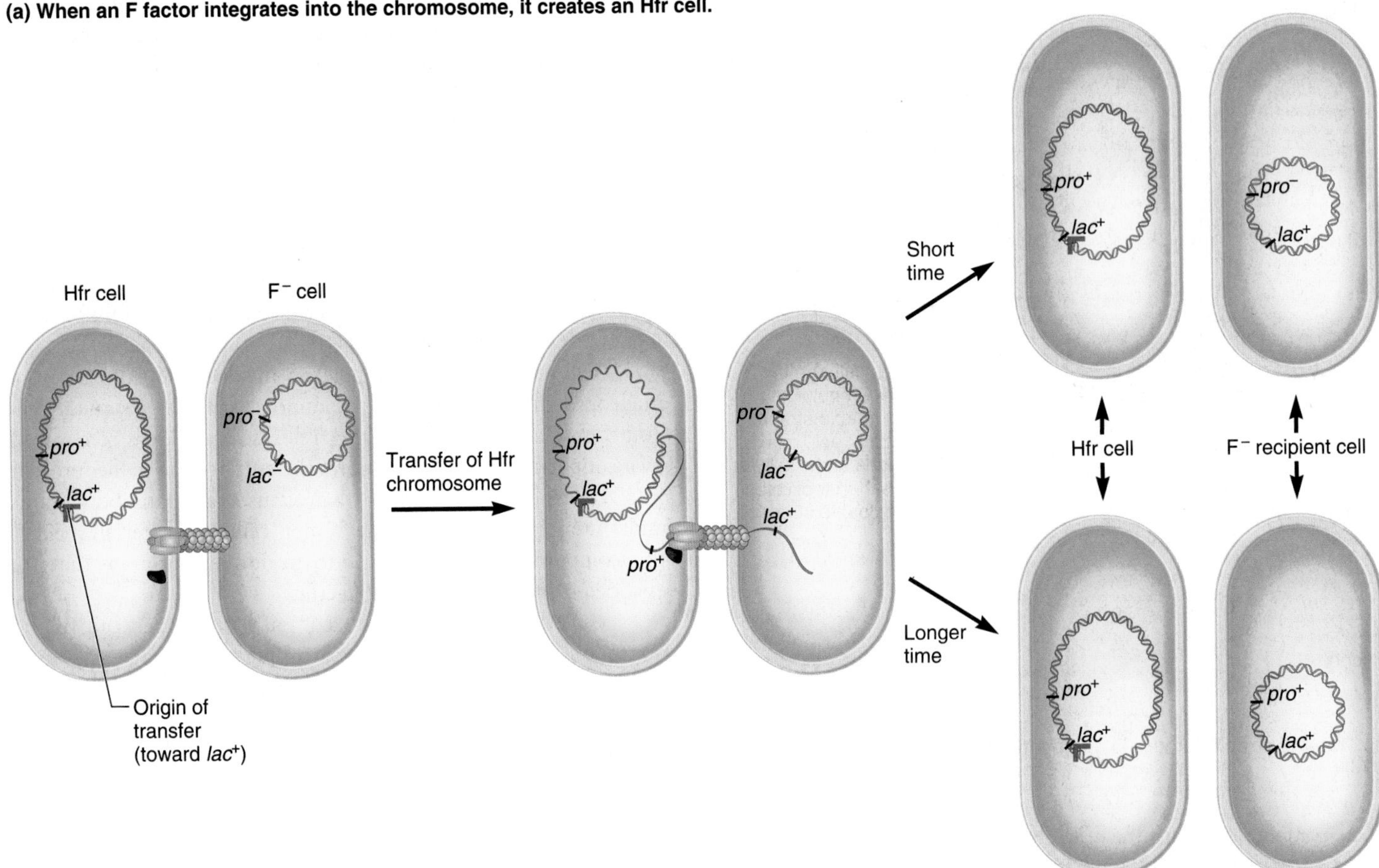

FIGURE 6.5 Formation of an Hfr cell and its ability to transfer portions of the *E. coli* chromosome. (a) An Hfr cell is created when an F factor integrates into the bacterial chromosome. **(b)** The transfer of the bacterial chromosome begins at the origin of transfer and then proceeds around the circular chromosome. After a segment of chromosome has been transferred to the F^- recipient cell, it recombines with the recipient cell's chromosome. If mating occurs for a brief period, only a short segment of the chromosome is transferred. If mating is prolonged, a longer segment of the bacterial chromosome is transferred.

Genes → Traits The F^- recipient cell was originally *lac*$^-$ (unable to metabolize lactose) and *pro*$^-$ (unable to synthesize proline). If mating occurred for a short period of time, the recipient cell acquired *lac*$^+$, allowing it to metabolize lactose. If mating occurred for a longer period of time, the recipient cell also acquired *pro*$^+$, enabling it to synthesize proline.

another *Hfr* strain, demonstrated that conjugation between an *Hfr* strain and an F^- strain involves the transfer of a portion of the bacterial chromosome from the *Hfr* strain to the F^- cell (**Figure 6.5b**). The origin of transfer within the integrated F factor determines the starting point and direction of this transfer process. One of the DNA strands is cut at the origin of transfer. This cut, or nicked, site is the starting point at which the Hfr chromosome will enter the F^- recipient cell. From this starting point, a strand of the DNA of the Hfr chromosome begins to enter the F^- cell in a linear manner. The transfer process occurs in conjunction with chromosomal replication, so the Hfr cell retains its original chromosomal composition. About 1.5 to 2 hours are required for the entire Hfr chromosome to pass into the F^- cell. Because most matings do not last that long, usually only a portion of the Hfr chromosome is transmitted to the F^- cell.

Once inside the F^- cell, the chromosomal material from the Hfr cell can swap, or recombine, with the homologous region of the recipient cell's chromosome. (Chapter 17 describes the process of homologous recombination.) How does this process affect the recipient cell? As illustrated in Figure 6.5b, this recombination may provide the recipient cell with a new combination of alleles. In this example, the recipient strain was originally lac^- (unable to metabolize lactose) and pro^- (unable to synthesize proline). If mating occurred for a short time, the recipient cell received a short segment of chromosomal DNA from the donor. In this case, the recipient cell has become lac^+ but remains pro^-. If the mating is prolonged, the recipient cell will receive a longer segment of chromosomal DNA from the donor. After a longer mating, the recipient becomes lac^+ and pro^+. As shown in Figure 6.5b, an important feature of Hfr mating is that the bacterial chromosome is transferred linearly to the recipient strain. In this example, lac^+ is always transferred first, and pro^+ is transferred later.

In any particular *Hfr* strain, the origin of transfer has a specific orientation that promotes either a counterclockwise or clockwise transfer of genes. Also, among different *Hfr* strains, the origin of transfer may be located in different regions of the chromosome. Therefore, the order of gene transfer depends on the location and orientation of the origin of transfer. For example, another *Hfr* strain could have its origin of transfer next to pro^+ and transfer pro^+ first and then lac^+.

EXPERIMENT 6A

Conjugation Experiments Can Map Genes Along the *E. coli* Chromosome

The first genetic mapping experiments in bacteria were carried out by Elie Wollman and François Jacob in the 1950s. At the time of their studies, not much information was known about the organization of bacterial genes along the chromosome. A few key advances made Wollman and Jacob realize that the process of genetic transfer could be used to map the order of genes in *E. coli.* First, the discovery of conjugation by Lederberg and Tatum and the identification of *Hfr* strains by Cavalli-Sforza and Hayes made it clear that bacteria can transfer genes from donor to recipient cells in a linear fashion. In addition, Wollman and Jacob were aware of previous microbiological studies concerning bacteriophages—viruses that bind to *E. coli* cells and subsequently infect them. These studies showed that bacteriophages can be sheared from the surface of *E. coli* cells if they are spun in a blender. In this treatment, the bacteriophages are detached from the surface of the bacterial cells, but the bacteria themselves remain healthy and viable. Wollman and Jacob reasoned that a blender treatment could also be used to separate bacterial cells that were in the act of conjugation without killing them. This technique is known as an **interrupted mating.**

The rationale behind Wollman and Jacob's mapping strategy is that the time it takes for genes to enter a donor cell is directly related to their order along the bacterial chromosome. They hypothesized that the chromosome of the donor strain in an Hfr mating is transferred in a linear manner to the recipient strain. If so, the order of genes along the chromosome can be deduced by determining the time it takes various genes to enter the recipient strain. Assuming the Hfr chromosome is transferred linearly, they realized that interruptions of mating at different times would lead to various lengths of the Hfr chromosome being transferred to the F^- recipient cell. If two bacterial cells had mated for a short period of time, only a small segment of the Hfr chromosome would be transferred to the recipient bacterium. However, if the bacterial cells were allowed to mate for a longer period of time before being interrupted, a longer segment of the Hfr chromosome could be transferred (see Figure 6.5b). By determining which genes were transferred during short matings and which required longer mating times, Wollman and Jacob were able to deduce the order of particular genes along the *E. coli* chromosome.

As shown in the experiment of **Figure 6.6**, Wollman and Jacob began with two *E. coli* strains. The donor (*Hfr*) strain had the following genetic composition:

- thr^+: able to synthesize threonine, an essential amino acid for growth
- leu^+: able to synthesize leucine, an essential amino acid for growth
- azi^s: sensitive to killing by azide (a toxic chemical)
- ton^s: sensitive to infection by bacteriophage T1. (As discussed later, when bacteriophages infect bacteria, they may cause lysis, which results in plaque formation.)
- lac^+: able to metabolize lactose and use it for growth
- gal^+: able to metabolize galactose and use it for growth
- str^s: sensitive to killing by streptomycin (an antibiotic)

The recipient (F^-) strain had the opposite genotype: thr^- leu^- azi^r ton^r lac^- gal^- str^r (r = resistant). Before the experiment, Wollman and Jacob already knew the thr^+ gene was transferred first, followed by the leu^+ gene, and both were transferred relatively soon (5 to 10 minutes) after mating. Their main goal in this experiment was to determine the times at which the other genes (azi^s ton^s lac^+ gal^+) were transferred to the recipient strain. The transfer of the str^s gene was not examined because streptomycin was used to kill the donor strain following conjugation.

Before discussing the conclusions of this experiment, let's consider how Wollman and Jacob monitored gene transfer. To determine if particular genes had been transferred after mating, they took the mated cells and first plated them on growth media that lacked threonine (thr) and leucine (leu) but contained streptomycin (str). On these plates, the original donor and recipient strains could not grow because the donor strain was streptomycin sensitive and the recipient strain required threonine and leucine. However, mated cells in which the donor had transferred chromosomal DNA carrying the thr^+ and leu^+ genes to the recipient cell would be able to grow.

To determine the order of gene transfer of the azi^s, ton^s, lac^+, and gal^+ genes, Wollman and Jacob picked colonies from the first plates and restreaked them on media that contained azide or bacteriophage T1 or on media that contained lactose or galactose as the sole source of energy for growth. The plates were incubated overnight to observe the formation of visible bacterial growth. Whether or not the bacteria could grow depended on their genotypes. For example, a cell that is azi^s cannot grow on media containing azide, and a cell that is lac^- cannot grow on media containing lactose as the carbon source for growth. By comparison, a cell that is azi^r and lac^+ can grow on both types of media.

THE GOAL (DISCOVERY-BASED SCIENCE)

The chromosome of the donor strain in an Hfr mating is transferred in a linear manner to the recipient strain. The order of genes along the chromosome can be deduced by determining the time various genes take to enter the recipient strain.

ACHIEVING THE GOAL — FIGURE 6.6 **The use of conjugation to map the order of genes along the *E. coli* chromosome.**

Starting materials: The two *E. coli* strains already described, one *Hfr* strain (thr^+ leu^+ azi^s ton^s lac^+ gal^+ str^s) and one F^- (thr^- leu^- azi^r ton^r lac^- gal^- str^r).

1. Mix together a large number of Hfr donor and F⁻ recipient cells.

2. After different periods of time, take a sample of cells and interrupt conjugation in a blender.

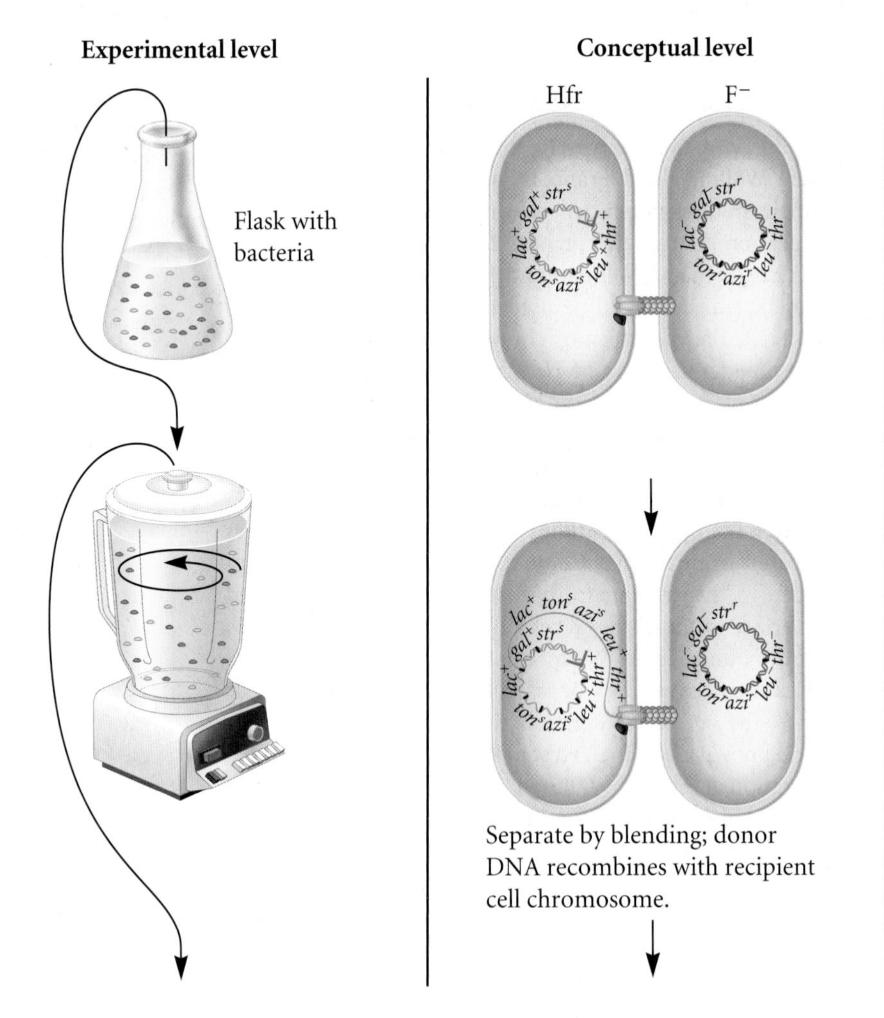

3. Plate the cells on growth media lacking threonine and leucine but containing streptomycin. Note: The general methods for growing bacteria in a laboratory are described in the Appendix.

4. Pick each surviving colony, which would have to be thr^+ leu^+ str^r, and test to see if it is sensitive to killing by azide, sensitive to infection by T1 bacteriophage, and able to metabolize lactose or galactose.

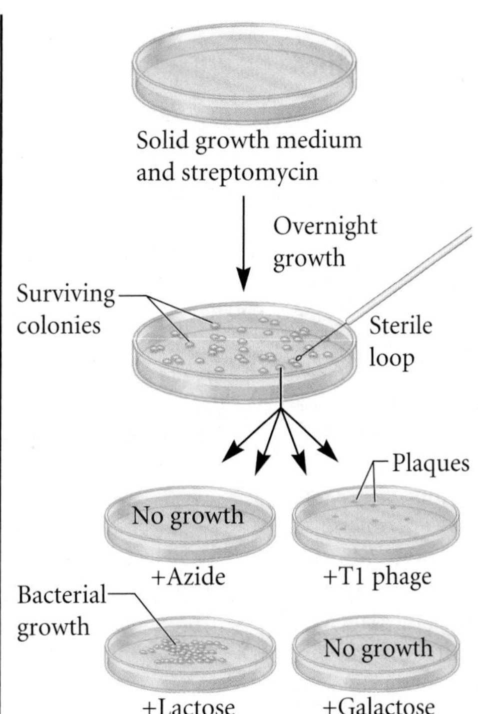

In this conceptual example, the cells have been incubated about 20 minutes.

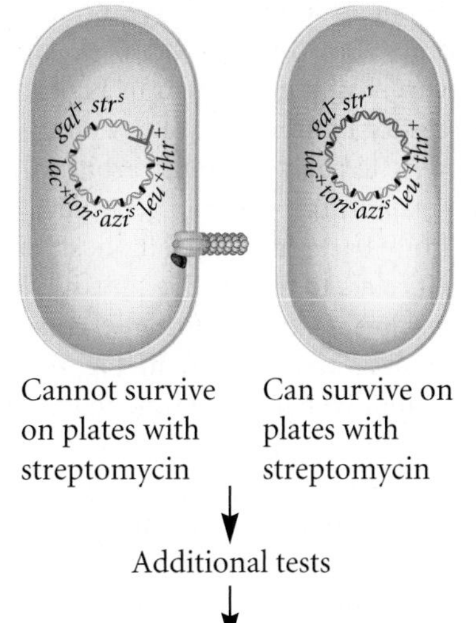

The conclusion is that the colony that was picked contained cells with a genotype of thr^+ leu^+ azi^s ton^s lac^+ gal^- str^r.

THE DATA

Minutes That Bacterial Cells Were Allowed to Mate Before Blender Treatment	*Percent of Surviving Bacterial Colonies with the Following Genotypes:* thr^+ leu^+	azi^s	ton^s	lac^+	gal^+
5	—*	—	—	—	—
10	100	12	3	0	0
15	100	70	31	0	0
20	100	88	71	12	0
25	100	92	80	28	0.6
30	100	90	75	36	5
40	100	90	75	38	20
50	100	91	78	42	27
60	100	91	78	42	27

*There were no surviving colonies within the first 5 minutes of mating.

INTERPRETING THE DATA

Now let's discuss the data shown in Figure 6.6. After the first plating, all survivors would be cells in which the thr^+ and leu^+ alleles had been transferred to the F^- recipient strain, which was already streptomycin resistant. As seen in the data, 5 minutes was not sufficient time to transfer the thr^+ and leu^+ alleles because no surviving colonies were observed. After 10 minutes or longer, however, surviving bacterial colonies with the thr^+ leu^+ genotype were obtained. To determine the order of the remaining genes (azi^s, ton^s, lac^+, and gal^+), each surviving colony was tested to see if it was sensitive to killing by azide, sensitive to infection by T1 bacteriophage, able to use lactose for growth, or able to use galactose for growth. The likelihood of surviving colonies depended on whether the azi^s, ton^s, lac^+, and gal^+ genes were close to the origin of transfer or farther away. For example, when cells were allowed to mate for 25 minutes, 80% carried the ton^s gene, whereas only 0.6% carried the gal^+ gene. These results indicate that the ton^s gene is closer to the origin of transfer compared to the gal^+ gene. When comparing the data in Figure 6.6, a consistent pattern emerged. The gene that conferred sensitivity to azide (azi^s) was transferred first, followed by ton^s, lac^+, and finally, gal^+. From these data, as well as those from other experiments, Wollman and Jacob constructed a genetic map that described the order of these genes along the *E. coli* chromosome.

This work provided the first method for bacterial geneticists to map the order of genes along the bacterial chromosome. Throughout the course of their studies, Wollman and Jacob identified several different *Hfr* strains in which the origin of transfer had been integrated at different places along the bacterial chromosome. When they compared the order of genes among different *Hfr* strains, their results were consistent with the idea that the *E. coli* chromosome is circular (see solved problem S2).

A self-help quiz involving this experiment can be found at the Online Learning Center.

A Genetic Map of the *E. coli* Chromosome Has Been Obtained from Many Conjugation Studies

Conjugation experiments have been used to map more than 1,000 genes along the circular *E. coli* chromosome. A map of the *E. coli* chromosome is shown in **Figure 6.7**. This simplified map shows the locations of only a few dozen genes. Because the chromosome is circular, we must arbitrarily assign a starting point on the map, in this case the gene *thrA*. Researchers scale genetic maps from bacterial conjugation studies in units of **minutes.** This unit refers to the relative time it takes for genes to first enter an F^- recipient strain during a conjugation experiment. The *E. coli* genetic map shown in Figure 6.7 is 100 minutes long, which is approximately the time that it takes to transfer the complete chromosome during an Hfr mating.

The distance between two genes is determined by comparing their times of entry during a conjugation experiment. As shown in **Figure 6.8**, the time of entry is found by conducting mating experiments at different time intervals before interruption. We compute the time of entry by extrapolating the data back to the x-axis. In this experiment, the time of entry of the *lacZ* gene was approximately 16 minutes, and that of the *galE* gene was 25 minutes. Therefore, these two genes are approximately 9 minutes apart from each other along the *E. coli* chromosome.

Let's look back at Figure 6.7 and consider where the origin of transfer must have been located in the donor strain that was used in the experiment of Figure 6.8. The *lacZ* gene is located at 7 minutes on the chromosome map, and the *galE* is found at 16 minutes. For the donor strain used in Figure 6.8, we can deduce that the origin of transfer was located at approximately 91 minutes on the chromosome map and transferred DNA in the clockwise direction. If we assume the origin was located here, it would take about 16 minutes to transfer *lacZ* and about 25 minutes to transfer *galE*.

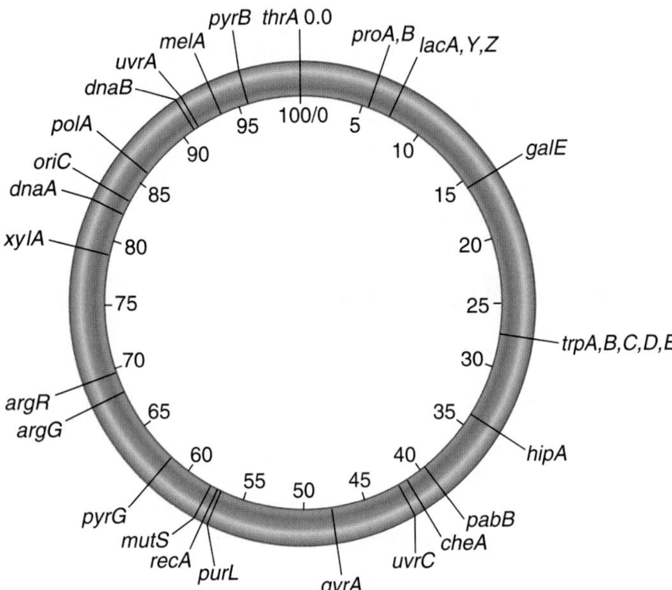

FIGURE 6.7 A simplified genetic map of the *E. coli* chromosome indicating the positions of several genes. *E. coli* has a circular chromosome with about 4,200 different genes. This map shows the locations of several of them. The map is scaled in units of minutes, and it proceeds in a clockwise direction. The starting point on the map is the gene *thrA*.

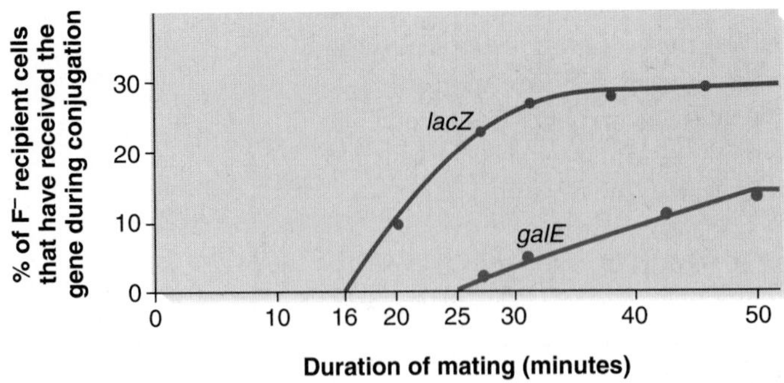

FIGURE 6.8 Time course of an interrupted *E. coli* conjugation experiment. By extrapolating the data back to the origin, the approximate time of entry of the *lacZ* gene is found to be 16 minutes; that of the *galE* gene, 25 minutes. Therefore, the distance between these two genes is 9 minutes.

Bacteria May Contain Different Types of Plasmids

Thus far, we have considered F factors, which are one type of DNA that can exist independently of the chromosomal DNA. The more general term for this structure is **plasmid.** Most known plasmids are circular, although some are linear. Plasmids occur naturally in many strains of bacteria and in a few types of eukaryotic cells such as yeast. The smallest plasmids consist of just a few thousand base pairs (bp) and carry only a gene or two; the largest are in the range of 100,000 to 500,000 bp and carry several dozen or even hundreds of genes. Some plasmids, such as F factors, can integrate into the chromosome. These are also called **episomes.**

A plasmid has its own origin of replication that allows it to be replicated independently of the bacterial chromosome. The DNA sequence of the origin of replication influences how many copies of the plasmid are found within a cell. Some origins are said to be very strong because they result in many copies of the plasmid, perhaps as many as 100 per cell. Other origins of replication have sequences that are described as much weaker, in that the number of copies created is relatively low, such as one or two per cell.

Why do bacteria have plasmids? Plasmids are not usually necessary for bacterial survival. However, in many cases, certain genes within a plasmid provide some type of growth advantage to the cell. By studying plasmids in many different species, researchers have discovered that most plasmids fall into five different categories:

1. Fertility plasmids, also known as F factors, allow bacteria to mate with each other.
2. Resistance plasmids, also known as R factors, contain genes that confer resistance against antibiotics and other types of toxins.
3. Degradative plasmids carry genes that enable the bacterium to digest and utilize an unusual substance. For example, a degradative plasmid may carry genes that allow a bacterium to digest an organic solvent such as toluene.

4. Col-plasmids contain genes that encode colicines, which are proteins that kill other bacteria.
5. Virulence plasmids carry genes that turn a bacterium into a pathogenic strain.

Bacteriophages Transfer Genetic Material from One Bacterial Cell to Another Via Transduction

We now turn to a second method of genetic transfer, one that involves bacteriophages. Before we discuss the ability of bacteriophages to transfer genetic material between bacterial cells, let's consider some general features of a phage's life cycle. Bacteriophages are composed of genetic material that is surrounded by a protein coat. As shown in **Figure 6.9**, bacteriophages bind to the surface of a bacterium and inject their genetic material into the bacterial cytoplasm. At this point, depending on the specific type of virus and its growth conditions, a phage may follow a lytic cycle or a lysogenic cycle. During the **lytic cycle,** the bacteriophage directs the synthesis of many copies of the phage genetic material and coat proteins (see Figure 6.9, left side). These components then assemble to make new phages. When synthesis and assembly are completed, the bacterial host cell is lysed (broken apart), releasing the newly made phages into the environment. **Virulent phages** follow only a lytic cycle and thus infection results in the death of the host cell.

In other cases, a bacteriophage infects a bacterium and follows the lysogenic cycle (see Figure 6.9, right side). During the **lysogenic cycle,** most types of phages integrate their genetic material into the chromosome of the bacterium. This integrated phage DNA is known as a **prophage.** A prophage can exist in a dormant state for a long time during which no new bacteriophages are made. When a bacterium containing a lysogenic prophage divides to produce two daughter cells, the prophage's genetic material is copied along with the bacterial chromosome. Therefore, both daughter cells will inherit the prophage. At some later time, a prophage may become activated to excise itself from the bacterial chromosome and enter the lytic cycle. When this happens, it will promote the synthesis of new phages and eventually lyse the host cell. A bacteriophage that usually exists in the lysogenic life cycle is called a **temperate phage.** Under most conditions, temperate phages do not produce new phages and will not kill the host bacterial cell.

With a general understanding of bacteriophage life cycles, we may now examine the ability of phages to transfer genetic material between bacteria. This process is called transduction. Examples of phages that can transfer bacterial chromosomal DNA from one bacterium to another are the P22 and P1 phages, which infect the bacterial species *Salmonella typhimurium* and *E. coli*, respectively. The P22 and P1 phages can follow either the lytic or lysogenic cycle. In the lytic cycle, when the phage infects the bacterial cell, the bacterial chromosome becomes fragmented into small pieces

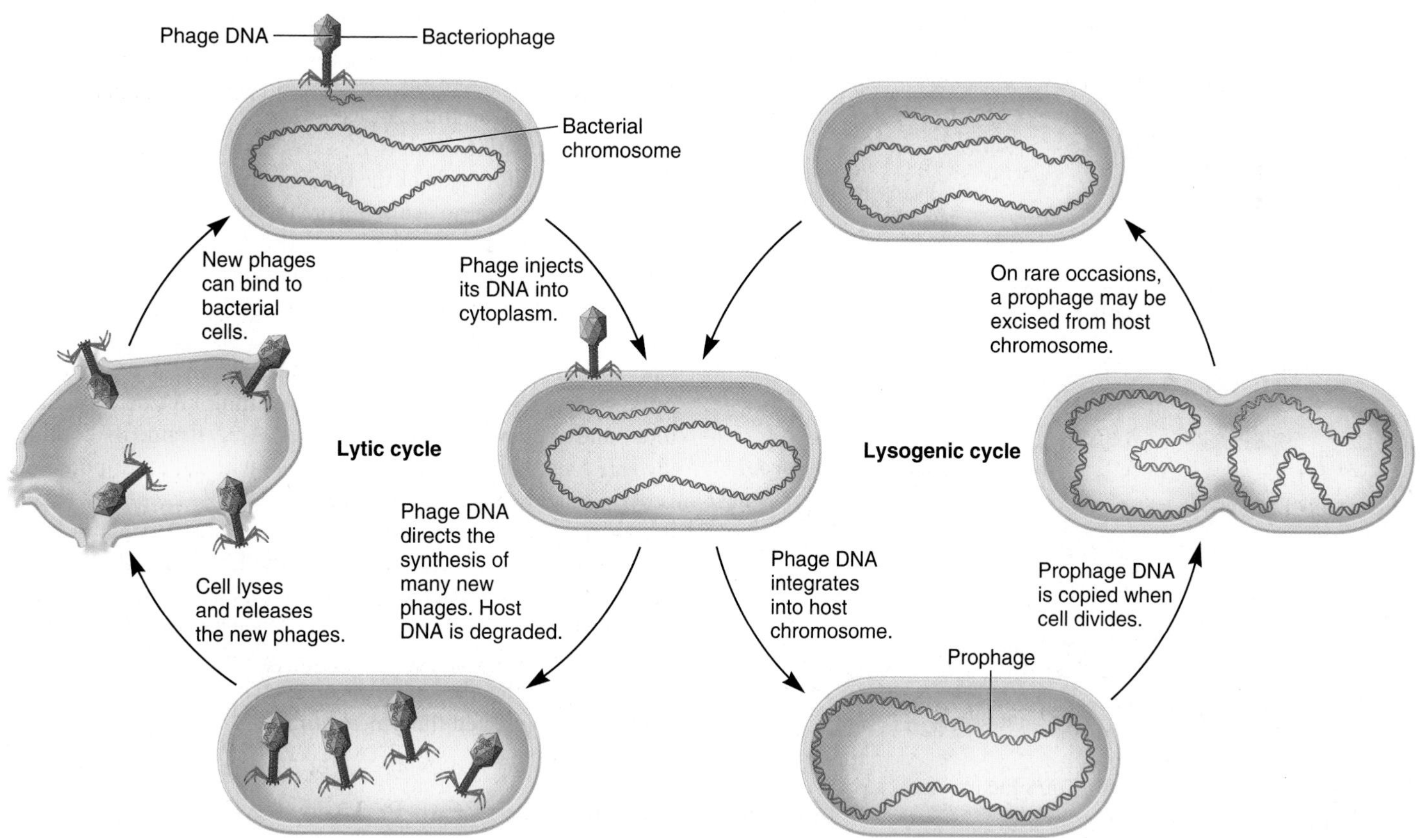
Phage DNA
Bacteriophage
Bacterial chromosome
New phages can bind to bacterial cells.
Phage injects its DNA into cytoplasm.
On rare occasions, a prophage may be excised from host chromosome.
Lytic cycle
Lysogenic cycle
Cell lyses and releases the new phages.
Phage DNA directs the synthesis of many new phages. Host DNA is degraded.
Phage DNA integrates into host chromosome.
Prophage DNA is copied when cell divides.
Prophage

FIGURE 6.9 **The lytic and lysogenic life cycles of certain bacteriophages.** Some bacteriophages, such as temperate phages, can follow both cycles. Other phages, known as virulent phages, can follow only a lytic cycle.

of DNA. The phage DNA directs the synthesis of more phage DNA and proteins, which then assemble to make new phages (**Figure 6.10**).

How does a bacteriophage transfer bacterial chromosomal genes from one cell to another? Occasionally, a mistake can happen in which a piece of bacterial DNA assembles with phage proteins. This creates a phage that contains bacterial chromosomal DNA. When phage synthesis is completed, the bacterial cell is lysed and releases the newly made phage into the environment.

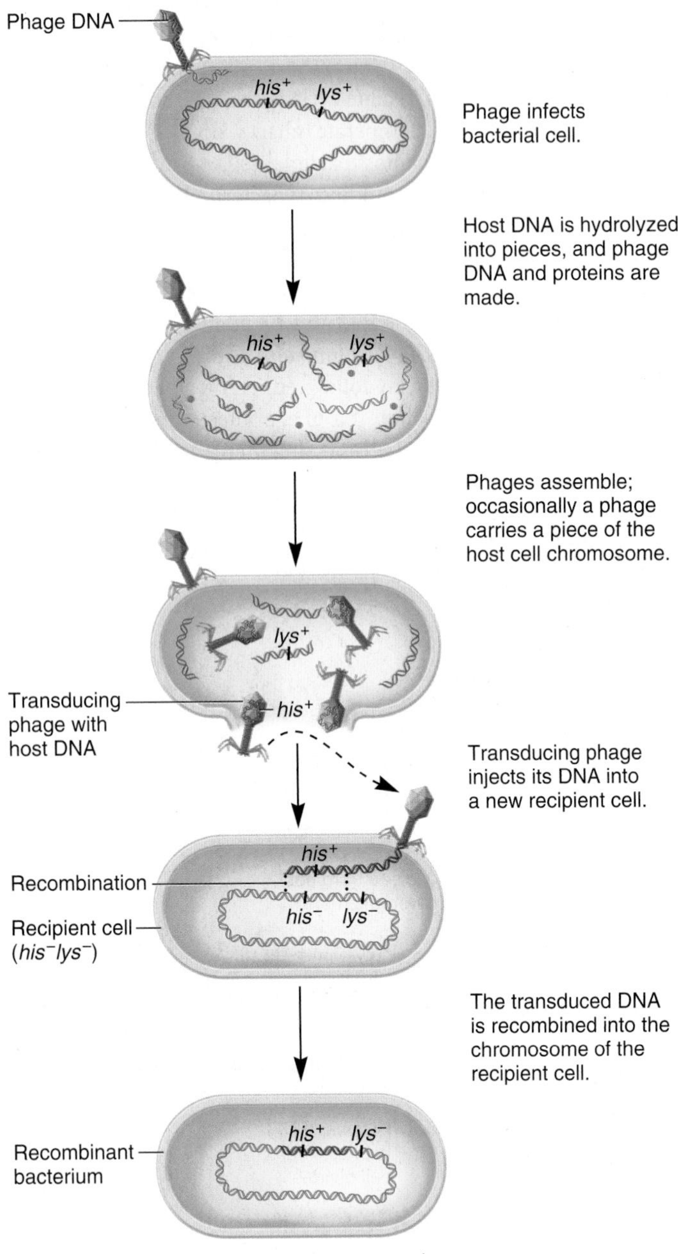
Phage DNA
his+
lys+
Phage infects bacterial cell.
Host DNA is hydrolyzed into pieces, and phage DNA and proteins are made.
his+
lys+
Phages assemble; occasionally a phage carries a piece of the host cell chromosome.
lys+
Transducing phage with host DNA
his+
Transducing phage injects its DNA into a new recipient cell.
his+
Recombination
Recipient cell (his−lys−)
his− lys−
The transduced DNA is recombined into the chromosome of the recipient cell.
his+ lys−
Recombinant bacterium
The recombinant bacterium's genotype has changed from his−lys− to his+lys−.

ONLINE ANIMATION

FIGURE 6.10 Transduction in bacteria.

Genes → Traits The transducing phage introduced DNA into a new recipient cell that was originally *his*$^-$ *lys*$^-$ (unable to synthesize histidine and lysine). During transduction, it received a segment of bacterial chromosomal DNA that carried *his*$^+$. Following recombination, the recipient cell's genotype was changed to *his*$^+$, so it now could synthesize histidine.

Following release, this abnormal phage can bind to a living bacterial cell and inject its genetic material into the bacterium. The DNA fragment, which was derived from the chromosomal DNA of the first bacterium, can then recombine with the recipient cell's bacterial chromosome. In this case, the recipient bacterium has been changed from a cell that was *his*$^-$ *lys*$^-$ (unable to synthesize histidine and lysine) to a cell that is *his*$^+$ *lys*$^-$ (able to synthesize histidine but unable to synthesize lysine). In the example shown in Figure 6.10, any piece of the bacterial chromosomal DNA can be incorporated into the phage. This type of transduction is called **generalized transduction.** By comparison, some phages carry out specialized transduction in which only particular bacterial genes are transferred to recipient cells (see solved problem S5 at the end of the chapter).

Transduction was first discovered in 1952 by Joshua Lederberg and Norton Zinder, using an experimental strategy similar to that depicted in Figure 6.1. They mixed together two strains of the bacterium *S. typhimurium*. One strain, designated LA-22, was *phe*$^-$ *trp*$^-$ *met*$^+$ *his*$^+$. This strain was unable to synthesize phenylalanine or tryptophan but was able to synthesize methionine and histidine. The other strain, LA-2, was *phe*$^+$ *trp*$^+$ *met*$^-$ *his*$^-$. It was able to synthesize phenylalanine and tryptophan but not methionine or histidine. When a mixture of these cells was placed on plates with growth media lacking these four amino acids, approximately one cell in 100,000 was observed to grow. The genotype of the surviving bacterial cells must have been *phe*$^+$ *trp*$^+$ *met*$^+$ *his*$^+$. Therefore, Lederberg and Zinder concluded that genetic material had been transferred between the two strains.

A novel result occurred when Lederberg and Zinder repeated this experiment using a U-tube apparatus, as previously shown in Figure 6.2. They placed the LA-22 strain (*phe*$^-$ *trp*$^-$ *met*$^+$ *his*$^+$) on one side of the filter and LA-2 (*phe*$^+$ *trp*$^+$ *met*$^-$ *his*$^-$) on the other. After an incubation period of several minutes, they removed samples from either side of the tube and plated the cells on media lacking the four amino acids. Surprisingly, they obtained colonies from the side of the tube that contained LA-22 but not from the side that contained LA-2. From these results, they concluded that some filterable agent was being transferred from LA-2 to LA-22 that converted LA-22 to a *phe*$^+$ *trp*$^+$ *met*$^+$ *his*$^+$ genotype. In other words, a filterable agent carrying the *phe*$^+$ and *trp*$^+$ genes was produced on the side of the tube containing LA-2, traversed the filter, and then was taken up by the LA-22 strain. By conducting this type of experiment using filters with different pore sizes, Lederberg and Zinder found that the filterable agent was slightly less than 0.1 μm in diameter, a size much smaller than a bacterium. They correctly concluded that the filterable agent in these experiments was a bacteriophage. In this case, the LA-2 strain contained a prophage, such as P22. On occasion, the prophage switched to the lytic cycle and packaged a segment of bacterial DNA carrying the *phe*$^+$ and *trp*$^+$ genes. This phage then traversed the filter and injected the genes into LA-22.

Cotransduction Can Be Used to Map Genes That Are Within 2 Minutes of Each Other

Can transduction be used to map the distance between bacterial genes? The answer is yes, but only if the genes are relatively close together. During transduction, P1 phages cannot package pieces that are greater than 2 to 2.5% of the entire *E. coli* chromosome, and P22 phages cannot package pieces that are greater than 1% of the length of the *S. typhimurium* chromosome. If two genes are close together along the chromosome, a bacteriophage may package a single piece of the chromosome that carries both genes and transfer that piece to another bacterium. This phenomenon is called **cotransduction.** Thus, the likelihood that two genes will be cotransduced depends on how close together they lie. If two genes are far apart along a bacterial chromosome, they will never be cotransduced because the bacteriophage cannot physically package a DNA fragment that is larger than 1 to 2.5% of the bacterial chromosome. In genetic mapping studies, cotransduction is used to determine the order and distance between genes that lie fairly close to each other.

To map genes using cotransduction, a researcher selects for the transduction of one gene and then monitors whether or not a second gene is cotransduced along with it. As an example, let's consider a donor strain of *E. coli* that is arg^+ met^+ str^s (able to synthesize arginine and methionine but sensitive to killing by streptomycin) and a recipient strain that is arg^- met^- str^r (**Figure 6.11**). The donor strain is infected with phage P1. Some of the *E. coli* cells are lysed by P1, and this P1 lysate is mixed with the recipient cells. After allowing sufficient time for transduction, the recipient cells are plated on a growth medium that contains arginine and streptomycin but not methionine. Therefore, these plates select for the growth of cells in which the met^+ gene has been transferred to the recipient strain, but they do not select for the growth of cells in which the arg^+ gene has been transferred, because the growth media are supplied with arginine.

Nevertheless, the arg^+ gene may have been cotransduced with the met^+ gene if the two genes are close together. To determine this, a sample of cells from each bacterial colony can be picked up with a wire loop and restreaked on media that lack both amino acids. If the colony can grow, the cells must have also obtained the arg^+ gene during transduction. In other words, cotransduction of both the arg^+ and met^+ genes has occurred. Alternatively, if the restreaked colony does not grow, it must have received only the met^+ gene during transduction. Data from this type of experiment are shown at the bottom of Figure 6.11. These data indicate a cotransduction frequency of 21/50 = 0.42 or 42%.

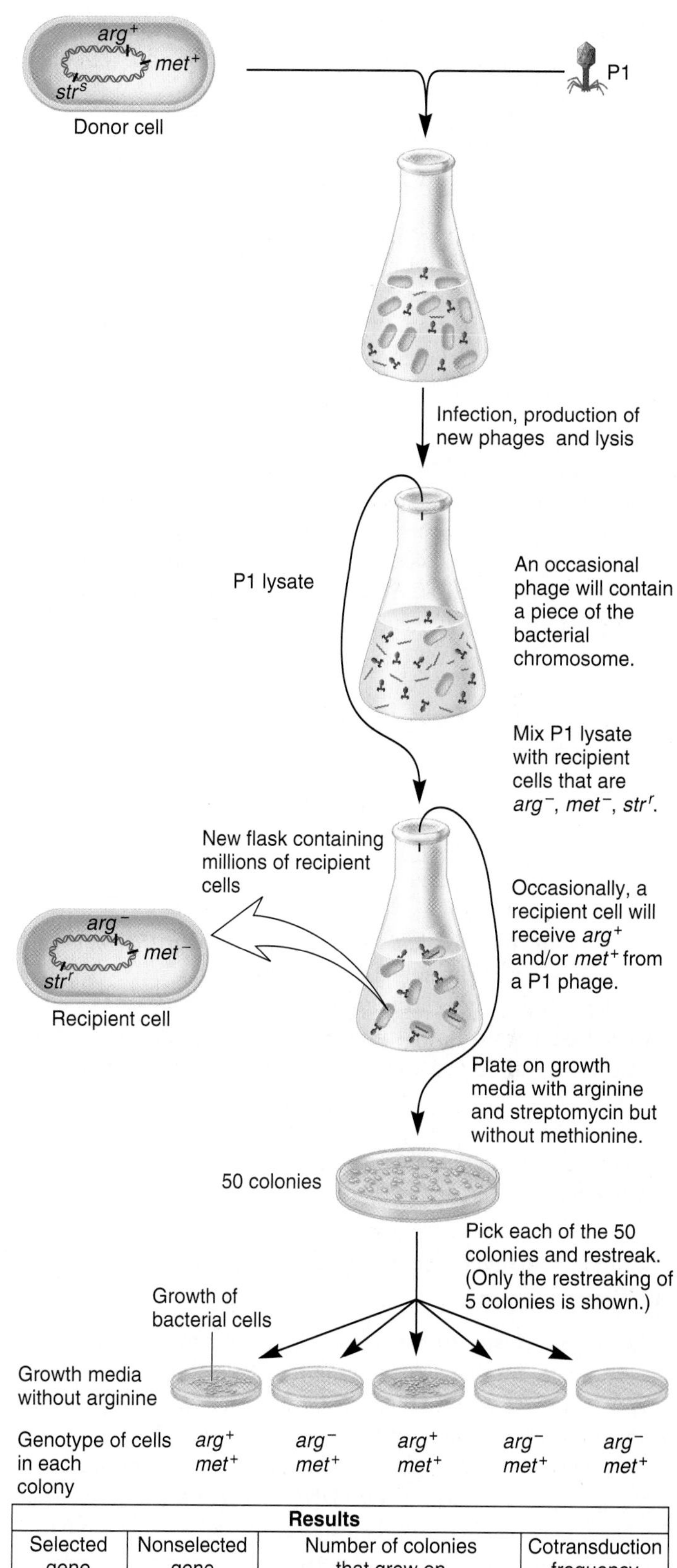

Results				
Selected gene	Nonselected gene	Number of colonies that grew on media + arginine	Number of colonies that grew on media − arginine	Cotransduction frequency
met^+	arg^+	50	21	0.42

FIGURE 6.11 **The steps in a cotransduction experiment.** The donor strain, which is arg^+ met^+ str^s (able to synthesize arginine and methionine but sensitive to streptomycin), is infected with phage P1. Some of the cells are lysed by P1, and this P1 lysate is mixed with cells of the recipient strain, which are arg^- met^- str^r. P1 phages in this lysate may carry fragments of the donor cell's chromosome, and the P1 phage may inject that DNA into the recipient cells. To identify recipient cells that have received the met^+ gene from the donor strain, the recipient cells are plated on a growth medium that contains arginine and streptomycin but not methionine. To determine if the arg^+ gene has also been cotransduced, cells from each bacterial colony (on the plates containing arginine and streptomycin) are lifted with a sterile wire loop and streaked on media that lack both amino acids. If the cells can grow, cotransduction of the arg^+ and met^+ genes has occurred. Alternatively, if the restreaked colony does not grow, it must have received only the met^+ gene during transduction.

In 1966, Tai Te Wu derived a mathematical expression that relates cotransduction frequency with map distance obtained from conjugation experiments. This equation is

$$\text{Cotransduction frequency} = (1 - d/L)^3$$

where

d = distance between two genes in minutes
L = the size of the chromosomal pieces (in minutes) that the phage carries during transduction (For P1 transduction, this size is approximately 2% of the bacterial chromosome, which equals about 2 minutes.)

This equation assumes the bacteriophage randomly packages pieces of the bacterial chromosome that are similar in size. Depending on the type of phage used in a transduction experiment, this assumption may not always be valid. Nevertheless, this equation has been fairly reliable in computing map distance for P1 transduction experiments in *E. coli.* We can use this equation to estimate the distance between the two genes described in Figure 6.11.

$$0.42 = (1 - d/2)^3$$

$$(1 - d/2) = \sqrt[3]{0.42}$$

$$1 - d/2 = 0.75$$

$$d/2 = 0.25$$

$$d = 0.5 \text{ minutes}$$

This equation tells us that the distance between the met^+ and arg^+ genes is approximately 0.5 minutes.

Historically, genetic mapping strategies in bacteria often involved data from both conjugation and transduction experiments. Conjugation has been used to determine the relative order and distance of genes, particularly those that are far apart along the chromosome. In comparison, transduction experiments can provide very accurate mapping data for genes that are fairly close together.

Bacteria Can Also Transfer Genetic Material by Transformation

A third mechanism for the transfer of genetic material from one bacterium to another is known as transformation. This process was first discovered by Frederick Griffith in 1928 while working with strains of *Streptococcus pneumoniae* (formerly known as *Diplococcus pneumoniae* or pneumococcus). During transformation, a living bacterial cell will take up DNA that is released from a dead bacterium. This DNA may then recombine into the living bacterium's chromosome, producing a bacterium with genetic material that it has received from the dead bacterium. (This experiment is discussed in detail in Chapter 9.) Transformation may be either a natural process that has evolved in certain bacteria, in which case it is called **natural transformation,** or an artificial process in which the bacterial cells are forced to take up DNA, an experimental approach termed **artificial transformation.** For example, a technique known as electroporation, in which an electric current causes the uptake of DNA, is used by researchers to promote the transport of DNA into a bacterial cell.

Since the initial studies of Griffith, we have learned a great deal about the events that occur in natural transformation. This form of genetic transfer has been reported in a wide variety of bacterial species. Bacterial cells that are able to take up DNA are known as **competent cells.** Those that can take up DNA naturally carry genes that encode proteins called **competence factors.** These proteins facilitate the binding of DNA fragments to the cell surface, the uptake of DNA into the cytoplasm, and its subsequent incorporation into the bacterial chromosome. Temperature, ionic conditions, and nutrient availability can affect whether or not a bacterium will be competent to take up genetic material from its environment. These conditions influence the expression of competence genes.

In recent years, geneticists have unraveled some of the steps that occur when competent bacterial cells are transformed by genetic material in their environment. **Figure 6.12** describes the steps of transformation. First, a large fragment of genetic material binds to the surface of the bacterial cell. Competent cells express DNA receptors that promote such binding. Before entering the cell, however, this large piece of chromosomal DNA must be cut into smaller fragments. This cutting is accomplished by an extracellular bacterial enzyme known as an endonuclease, which makes occasional random cuts in the long piece of chromosomal DNA. At this stage, the DNA fragments are composed of double-stranded DNA.

In the next step, the DNA fragment begins its entry into the bacterial cytoplasm. For this to occur, the double-stranded DNA interacts with proteins in the bacterial membrane. One of the DNA strands is degraded, and the other strand enters the bacterial cytoplasm via an uptake system, which is structurally similar to the one described for conjugation (as shown earlier in Figure 6.4a) but is involved with DNA uptake rather than export.

To be stably inherited, the DNA strand must be incorporated into the bacterial chromosome. If the DNA strand has a sequence that is similar to a region of DNA in the bacterial chromosome, the DNA may be incorporated into the chromosome by a process known as **homologous recombination,** discussed in detail in Chapter 17. For this to occur, the single-stranded DNA aligns itself with the homologous location on the bacterial chromosome. In the example shown in Figure 6.12, the foreign DNA carries a functional lys^+ gene that aligns itself with a nonfunctional (mutant) lys^- gene already present within the bacterial chromosome. The foreign DNA then recombines with one of the strands in the bacterial chromosome of the competent cell. In other words, the foreign DNA replaces one of the chromosomal strands of DNA, which is subsequently degraded. During homologous recombination, alignment of the lys^- and the lys^+ alleles results in a region of DNA called a **heteroduplex** that contains one or more base sequence mismatches. However, the heteroduplex exists only temporarily. DNA repair enzymes in the recipient cell recognize the heteroduplex and repair it. In this case, the heteroduplex has been repaired by eliminating the mutation that caused the lys^- genotype, thereby creating a lys^+ gene. In this example, the recipient cell has been transformed from a lys^- strain to a lys^+ strain. Alternatively, a DNA fragment that has entered a

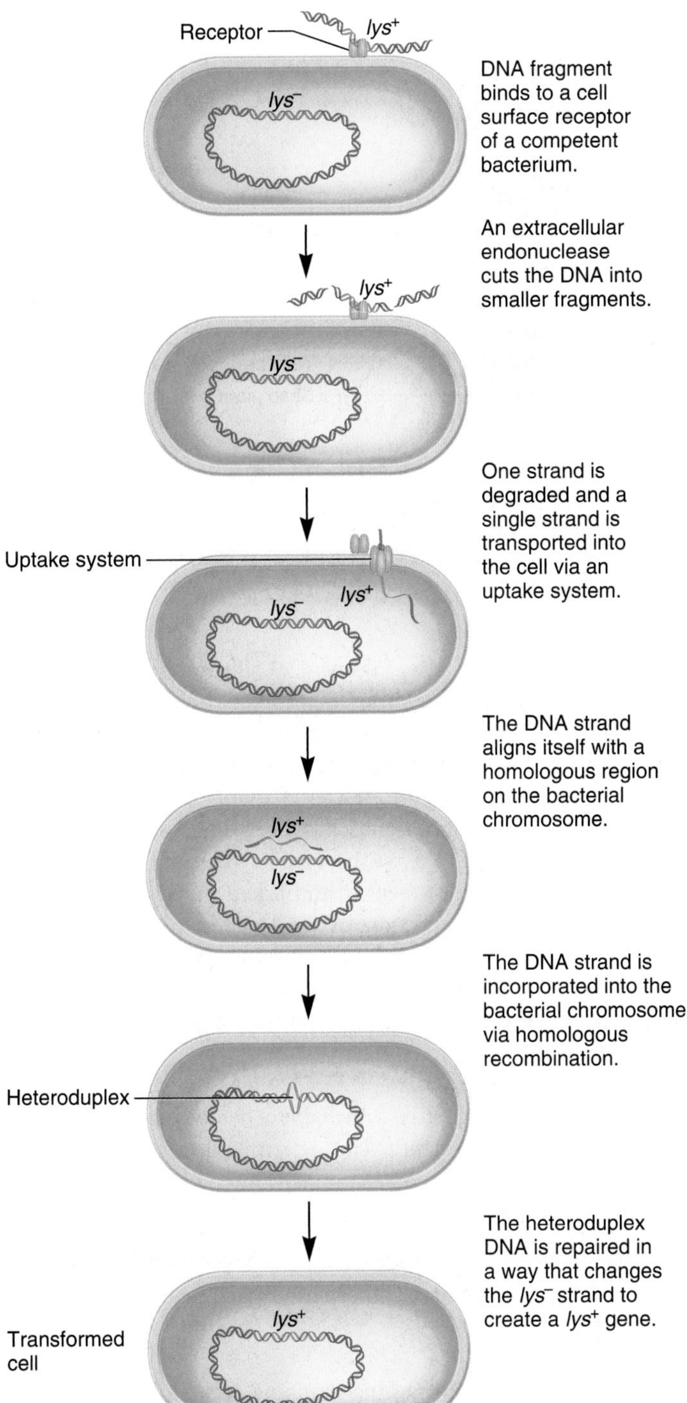

Receptor
lys+
lys−
DNA fragment binds to a cell surface receptor of a competent bacterium.
An extracellular endonuclease cuts the DNA into smaller fragments.
One strand is degraded and a single strand is transported into the cell via an uptake system.
Uptake system
The DNA strand aligns itself with a homologous region on the bacterial chromosome.
The DNA strand is incorporated into the bacterial chromosome via homologous recombination.
Heteroduplex
The heteroduplex DNA is repaired in a way that changes the lys− strand to create a lys+ gene.
Transformed cell

FIGURE 6.12 The steps of bacterial transformation. In this example, a fragment of DNA carrying a *lys*$^+$ gene enters the competent cell and recombines with the chromosome, transforming the bacterium from *lys*$^-$ to *lys*$^+$.

Genes → Traits Bacterial transformation can also lead to new traits for the recipient cell. The recipient cell was *lys*$^-$ (unable to synthesize the amino acid lysine). Following transformation, it became *lys*$^+$. This result would transform the recipient bacterial cell into a cell that could synthesize lysine and grow on media that lacked this amino acid. Before transformation, the recipient *lys*$^-$ cell would not have been able to grow on media lacking lysine.

cell may not be homologous to any genes that are already found in the bacterial chromosome. In this case, the DNA strand may be incorporated at a random site in the chromosome. This process is known as **nonhomologous** or **illegitimate recombination.**

Some bacteria preferentially take up DNA fragments from other bacteria of the same species or closely related species. How does this occur? Recent research has shown that the mechanism can vary among different species. In *Streptococcus pneumoniae,* the cells secrete a short peptide called the **competence-stimulating peptide (CSP).** When many *S. pneumoniae* cells are in the vicinity of one another, the concentration of CSP becomes high, and this stimulates the cells, via a cell-signaling pathway, to express the competence proteins that are needed for the uptake of DNA and its incorporation in the chromosome. Because competence requires a high external concentration of CSP, *S. pneuomoniae* cells are more likely to take up DNA from nearby *S. pneumoniae* cells that have died and released their DNA into the environment.

Other bacterial species promote the uptake of DNA among members of their own species via **DNA uptake signal sequences,** which are 9 or 10 bp long. In the human pathogens *Neisseria meningitidis* (a causative agent of meningitis), *N. gonorrhoeae* (a causative agent of gonorrhea), and *Haemophilus influenzae* (a causative agent of ear, sinus, and respiratory infections), these sequences are found at many locations within their respective genomes. For example, *H. influenzae* contains approximately 1,500 copies of the sequence 5'-AAGTGCGGT-3' in its genome, while *N. meningitidis* contains about 1,900 copies of the sequence 5'-GCCGTCTGAA-3'. DNA fragments that contain their own uptake signal sequence are preferentially taken up by these species compared to other DNA fragments. For example, *H. influenzae* is much more likely to take up a DNA fragment with the sequence 5'-AAGTGCGGT-3'. For this reason, transformation is more likely to involve DNA uptake between members of the same species.

Transformation has also been used to map many bacterial genes, using methods similar to the cotransduction experiments described earlier. If two genes are close together, the **cotransformation** frequency is expected to be high, whereas genes that are far apart will have a cotransformation frequency that is very low or even zero. Like cotransduction, genetic mapping via cotransformation is used only to map genes that are relatively close together.

Horizontal Gene Transfer Is the Transfer of Genes Between Different Species

The transmission of genes from mother cell to daughter cell or from parent to offspring is called **vertical gene transfer.** In comparison, a gene transferred between two different species is termed **horizontal gene transfer.** The three mechanisms of genetic transfer that we have considered—conjugation, transduction, and transformation—are important mechanisms for horizontal gene transfer among bacterial species. When analyzing the genomes of bacterial species, researchers have discovered that a sizable fraction of their genes are derived from horizontal gene transfer. For example, over the past 100 million years, *E. coli* and *Salmonella typhimurium* have acquired roughly 17% of their genes via horizontal gene transfer.

The types of genes that bacteria acquire via horizontal gene transfer are quite varied, though they commonly involve functions that are readily acted upon by natural selection. These include genes that confer antibiotic resistance, the ability to degrade toxic compounds, and pathogenicity. Geneticists have suggested that much of the speciation that has occurred in prokaryotic species is the result of horizontal gene transfer. In many cases, the acquisition of new genes allows a novel survival strategy that has led to the formation of a new species. These processes are considered in detail in Chapters 24 and 26.

The medical relevance of horizontal gene transfer is quite profound. Antibiotics are commonly prescribed to treat many bacterial illnesses, including infections of the respiratory tract, urinary tract, skin, ears, and eyes. In addition, antibiotics are used in agriculture as a supplement in animal feed and to control certain bacterial diseases of high-value fruits and vegetables. Unfortunately, however, the widespread and uncontrolled use of antibiotics has promoted the prevalence of antibiotic-resistant strains of bacteria. This phenomenon, termed **acquired antibiotic resistance,** may occur via genetic alterations in the bacteria's own genome or by the horizontal transfer of resistance genes from a resistant strain to a sensitive strain. Resistant strains carry genes that counteract the effects of antibiotics. Such resistance genes encode proteins that either break down the drug, pump the drug out of the cell, or prevent the drug from inhibiting cellular processes.

Bacterial resistance to antibiotics in community-acquired respiratory tract infections, such as pneumonia, as well as other medical illnesses, is a serious problem, and it is increasing in prevalence worldwide at an alarming rate. As often mentioned in the news media, antibiotic resistance has increased dramatically over the past few decades, and resistance has been reported in almost all species of bacteria. In many countries, for example, penicillin resistance in *Streptococcus pneumoniae* is found in over 50% of all strains, with resistance to other drugs rising as well. Likewise, the antibiotic-resistance problem in hospitals continues to worsen. Resistant strains of *Klebsiella pneumoniae* and *Enterococcus* are significant causes of morbidity and mortality among critically ill patients in intensive care units. Treating infections caused by these pathogens poses increasingly difficult therapeutic dilemmas.

6.2 INTRAGENIC MAPPING IN BACTERIOPHAGES

Let's now turn our attention to **viruses**—small particles that have genetic material and can propagate with the aid of a host cell. Biologists do not consider viruses to be living entities because they rely on a host cell for their existence and proliferation. Nevertheless, we can think of viruses as having traits because they have unique biological structures and functions. Each type of virus has its own genetic material, which contains many genes. In this section, we will focus our attention on a bacteriophage called T4. Its genetic material contains several dozen different genes encoding proteins that carry out a variety of functions. For example, some of the genes encode proteins needed for the synthesis of new viruses and the lysis of the host cell. Other genes encode the viral coat proteins that are found in the head, shaft, base plate, and tail fibers. Figure 6.13 illustrates the structure of T4. As seen here, five different proteins bind to each other to form a tail fiber. The expression of T4 genes to make these proteins provides the bacteriophage with the trait of having tail fibers, enabling it to attach to the surface of a bacterium.

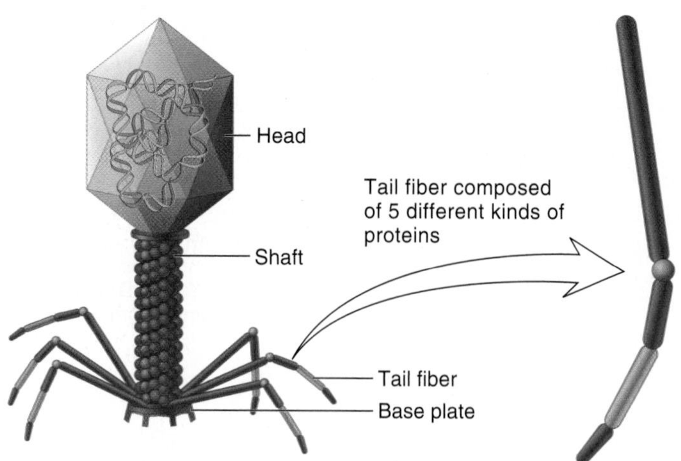

FIGURE 6.13 Structure of the T4 virus.

Genes → Traits The inset to this figure shows the five different proteins, encoded by five different genes, that make up a tail fiber. The expression of these genes provides the bacteriophage with the trait of having tail fibers, enabling it to attach itself to the surface of a bacterium.

The study of viral genes has been instrumental in our basic understanding of how the genetic material works. During the 1950s, Seymour Benzer embarked on a 10-year study that focused on the function of viral genes in the T4 bacteriophage. In this section, we will examine some of his pivotal results. We will also explore how he conducted a detailed type of genetic mapping known as **intragenic mapping** or **fine structure mapping.**

Earlier in this chapter and in Chapter 5, we explored intergenic mapping, the goal of which is to determine the distance between two different genes. The determination of the distance between genes *A* and *B* is an example of intergenic mapping:

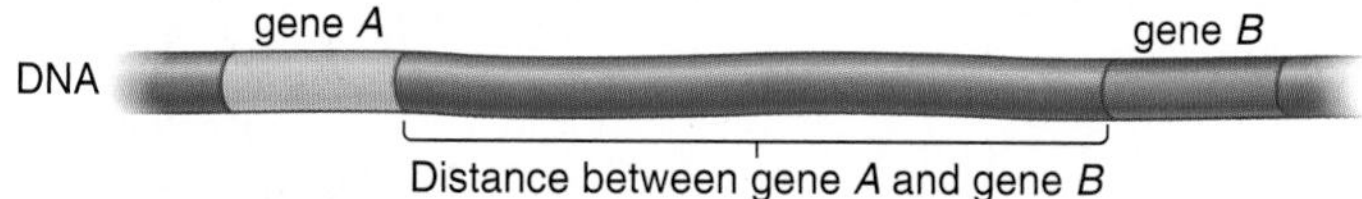

Intergenic mapping

In comparison, intragenic mapping seeks to establish distances between two or more mutations within the same gene. For example, in a population of viruses, gene *C* may exist as two different mutant alleles: One allele may be due to a mutation near the beginning of the gene; the second, to a mutation near the end:

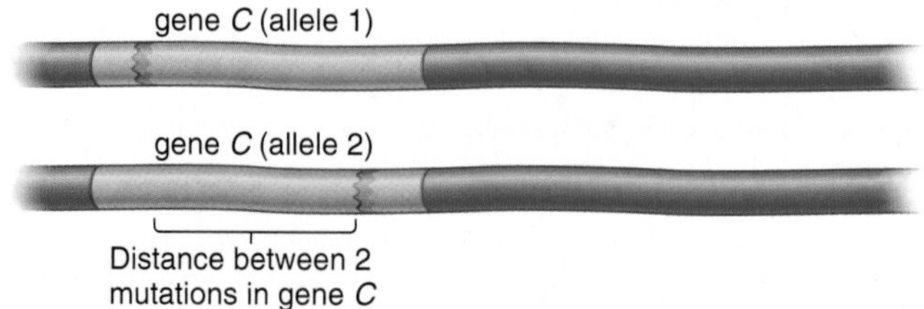

Intragenic mapping (also known as fine structure mapping)

In this section, we will explore the pioneering studies that led to the development of intragenic mapping and advanced our knowledge of gene function. Benzer's results showed that, rather than being an indivisible particle, a gene must be composed of a large structure that can be subdivided during intragenic crossing over.

Mutations in Viral Genes Can Alter Plaque Morphology

As they progress through the lytic cycle, bacteriophages ultimately produce new phages, which are released when the bacterial cell lyses (refer back to Figure 6.9, left side). In the laboratory, researchers can visually observe the consequences of bacterial cell lysis in the following way (**Figure 6.14**). A sample of bacterial cells and lytic bacteriophages are mixed together and then poured onto petri plates containing nutrient agar for bacterial cell growth. Bacterial cells that are not infected by a bacteriophage will rapidly grow and divide to produce a "lawn" of bacteria. This lawn of bacteria is opaque—you cannot see through it to the underlying agar. In the experiment shown in Figure 6.14, 11 bacterial cells have been infected by bacteriophages and these infected cells are found at random locations in the lawn of uninfected bacteria. The infected cells will lyse and release newly made bacteriophages. These bacteriophages will then infect the nearby bacteria within the lawn. These cells eventually lyse and also release newly made phages. Over time, these repeated cycles of infection and lysis will produce an observable clear area, or **plaque,** where the bacteria have been lysed around the original site where a phage infected a bacterial cell.

Now that we have an appreciation for the composition of a viral plaque, let's consider how the characteristics of a plaque can be viewed as a trait of a bacteriophage. As we have seen, the genetic analysis of any organism requires strains with allelic differences. Because bacteriophages can be visualized only with an electron microscope, it would be rather difficult for geneticists to analyze mutations that affect phage morphology. However, some mutations in the bacteriophage's genetic material can alter the ability of the phage to cause plaque formation. Therefore, we can view the morphology of plaques as a trait of the bacteriophage. Because plaques are visible with the unaided eye, mutations affecting this trait lend themselves to a much easier genetic analysis. An example is a rapid-lysis mutant of bacteriophage T4, which tends to form unusually large plaques (**Figure 6.15**). The plaques are large because the mutant phages lyse the bacterial cells more rapidly than do the wild-type phages. Rapid-lysis mutants form large, clearly defined plaques, as opposed to wild-type bacteriophages, which produce smaller, fuzzy-edged plaques. Mutations in different bacteriophage genes can produce a rapid-lysis phenotype.

Benzer studied one category of T4 phage mutants, designated *rII* (*r* stands for rapid lysis). In the bacterial strain called *E. coli* B, *rII* phages produce abnormally large plaques. Nevertheless, *E. coli* B strains produce low yields of *rII* phages because the *rII* phages lyse the bacterial cells so quickly they do not have sufficient time to produce many new phages. To help study this phage, Benzer wanted to obtain large quantities of it. Therefore, to improve his yield of *rII* phage, he decided to test its yield in other bacterial strains.

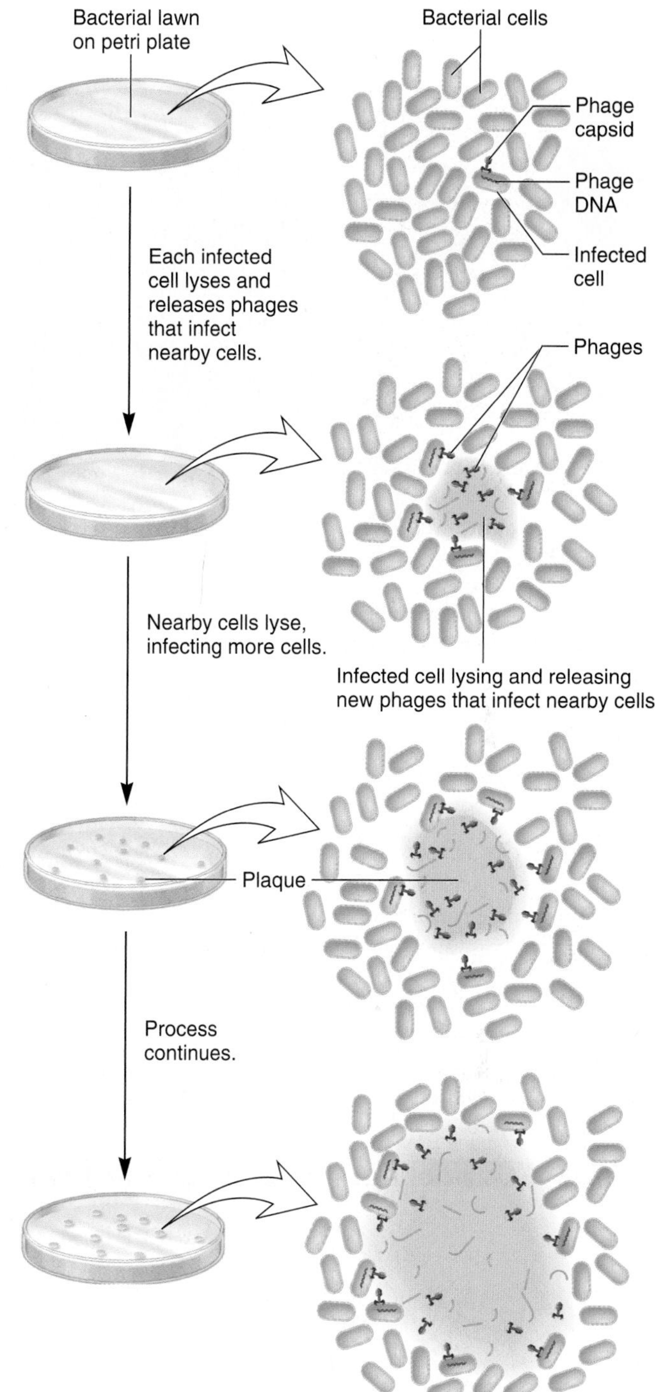

FIGURE 6.14 The formation of phage plaques on a lawn of bacteria in the laboratory. In this experiment, bacterial cells were mixed with a small number of lytic bacteriophages. In this particular example, eleven bacterial cells were initially infected by phages. The cells were poured onto petri plates containing nutrient agar for bacterial cell growth. Bacterial cells will rapidly grow and divide to produce an opaque lawn of densely packed bacteria. The 11 infected cells will lyse and release newly made bacteriophages. These bacteriophages will then infect the nearby bacteria within the lawn. Likewise, these newly infected cells will lyse and release new phages. By this repeated process, the area of cell lysis creates a clear zone known as a viral plaque.

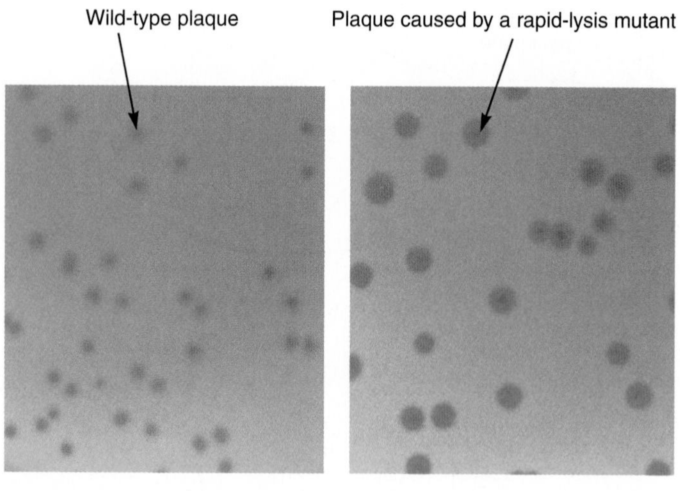

FIGURE 6.15 A comparison of plaques produced by the wild-type T4 bacteriophage and a rapid-lysis mutant.

Genes → Traits **(a)** These plaques were caused by the infection and lysis of *E. coli* cells by the wild-type T4 phage. **(b)** A mutation in a phage gene, called a rapid-lysis mutation, caused the phage to lyse *E. coli* cells more quickly. A phage carrying a rapid-lysis mutant allele yields much larger plaques with clearly defined edges.

On the day Benzer decided to do this, he happened to be teaching a phage genetics class. For that class, he was growing two *E. coli* strains designated *E. coli* K12S and *E. coli* K12(λ). He was growing these two strains to teach his class about the lysogenic cycle. *E. coli* K12(λ) has DNA from another phage, called lambda, integrated into its chromosome, whereas *E. coli* K12S does not. To see if the use of these strains might improve phage yield, *E. coli* B, *E. coli* K12S, and *E. coli* K12(λ) were infected with the *rII* and wild-type T4 phage strains. As expected, the wild-type phage could infect all three bacterial strains. However, the *rII* mutant strains behaved quite differently. In *E. coli* B, the *rII* strains produced large plaques that had poor yields of bacteriophage. In *E. coli* K12S, the *rII* mutants produced normal plaques that gave good yields of phage. Surprisingly, in *E. coli* K12(λ), the *rII* mutants were unable to produce plaques at all, for reasons that were not understood. Nevertheless, as we will see later, this fortuitous observation was a critical feature that allowed intragenic mapping in this bacteriophage.

A Complementation Test Can Reveal If Mutations Are in the Same Gene or in Different Genes

In his experiments, Benzer was interested in a single trait, namely, the ability to form plaques. He had isolated many *rII* mutant strains that could form large plaques in *E. coli* B but could not produce plaques in *E. coli* K12(λ). To attempt gene mapping, he needed to know if the various *rII* mutations were in the same gene or if they involved mutations in different genes. To accomplish this, he conducted a **complementation test.** In this type of approach, the goal is to determine if two different mutations that affect the same trait are in the same gene or in two different genes (also see Figure 4.20).

The possible outcomes of complementation tests involving mutations that affect plaque formation are shown in **Figure 6.16.** This example involves four different *rII* mutations in T4

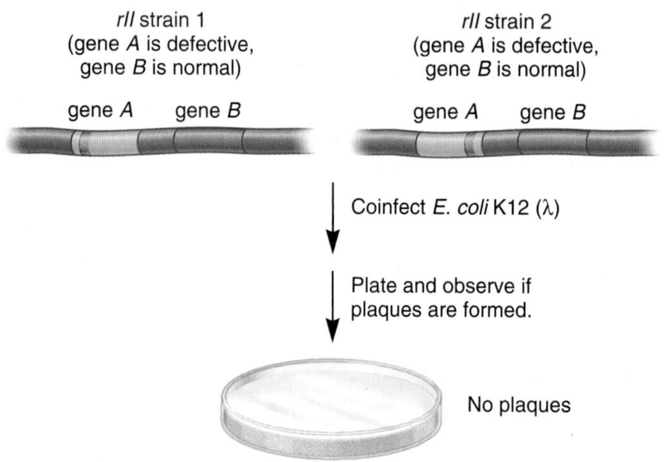

No complementation occurs, because the coinfected cell is unable to make the normal product of gene *A*. The coinfected cell will not produce viral particles, thus no bacterial cell lysis and no plaque formation.

(a) Noncomplementation: The phage mutations are in the same gene.

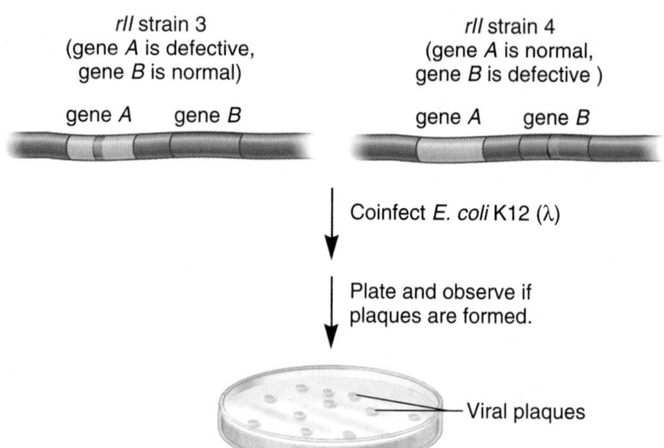

Complementation occurs, because the coinfected cell is able to make normal products of gene *A* and gene *B*. The coinfected bacterial cell will produce viral particles that lyse the cell, resulting in the appearance of clear plaques.

(b) Complementation: The phage mutations are in different genes.

FIGURE 6.16 A comparison of noncomplementation and complementation. Four different T4 phage strains (designated 1 through 4) that carry *rII* mutations were coinfected into *E. coli* K12(λ). **(a)** If two *rII* phage strains possess mutations in the same gene, noncomplementation will occur. **(b)** If the *rII* mutations are in different genes (such as gene *A* and gene *B*), a coinfected cell will have two mutant genes but also two wild-type genes. Doubly infected cells with a wild-type copy of each gene can produce new phages and form plaques. This result is called complementation because the defective genes in each *rII* strain are complemented by the corresponding wild-type genes.

bacteriophage, designated strains 1 through 4, that prevent plaque formation in *E. coli* K12(λ). To conduct this complementation experiment, bacterial cells were coinfected with two different strains of T4 phage. Two distinct outcomes are possible. In Figure 6.16a, the two *rII* phage strains possess deleterious mutations in the same gene (gene *A*). Because they cannot make a wild-type gene *A* product when coinfected into an *E. coli* K12(λ) cell, plaques will not form. This phenomenon is called **noncomplementation.**

Alternatively, if each *rII* mutation is in a different phage gene (e.g., gene *A* and gene *B*), a bacterial cell that is coinfected by both types of phages will have two mutant genes as well as two wild-type genes (Figure 6.16b). If the mutant phage genes behave in a recessive fashion, the doubly infected cell will have a wild-type phenotype. Why does this phenotype occur? The coinfected cells will produce normal proteins that are encoded by the wild-type versions of both genes *A* and *B*. For this reason, coinfected cells will be lysed in the same manner as if infected by the wild-type strain. Therefore, this coinfection should be able to produce plaques in *E. coli* K12(λ). This result is called **complementation** because the defective genes in each *rII* strain are complemented by the corresponding wild-type genes. It should be noted that, for a variety of reasons, intergenic complementation may not always work. One possibility is that a mutation may behave in a dominant fashion. In addition, mutations that affect regulatory genetic regions rather than the protein-coding region may not show complementation.

By carefully considering the pattern of complementation and noncomplementation, Benzer found that the *rII* mutations occurred in two different genes, which were termed *rIIA* and *rIIB*. The identification of two distinct genes affecting plaque formation was a necessary step that preceded his intragenic mapping analysis, which is described next. Benzer coined the term **cistron** to refer to the smallest genetic unit that gives a negative complementation test. In other words, if two mutations occur within the same cistron, they cannot complement each other. Since these studies, researchers have learned that a cistron is equivalent to a gene. In recent decades, the term gene has gained wide popularity while the term cistron is not commonly used. However, the term polycistronic is still used to describe bacterial mRNAs that carry two or more gene sequences, as described in Chapter 14.

Intragenic Maps Were Constructed Using Data from a Recombinational Analysis of Mutants Within the *rII* Region

As we have seen in Figure 6.16, the ability of strains with mutations in two different genes to produce viral plaques upon coinfection is due to complementation. Noncomplementation occurs when two different strains have mutations in the same gene. However, at an extremely low rate, two noncomplementing strains of viruses can produce an occasional viral plaque if intragenic recombination has taken place. For example, **Figure 6.17** shows a coinfection experiment between two phage strains that both contain *rII* mutations in gene *A*. These mutations are located at different places within the same gene. On rare occasions, a crossover may occur in the very short region between each mutation. This crossover produces a double mutant gene *A* and a wild-type gene *A*. Because this event has produced a wild-type gene *A*, the function of the protein encoded by gene *A* will be restored. Therefore, new phages can be made in *E. coli* K12(λ), resulting in the formation of viral plaques.

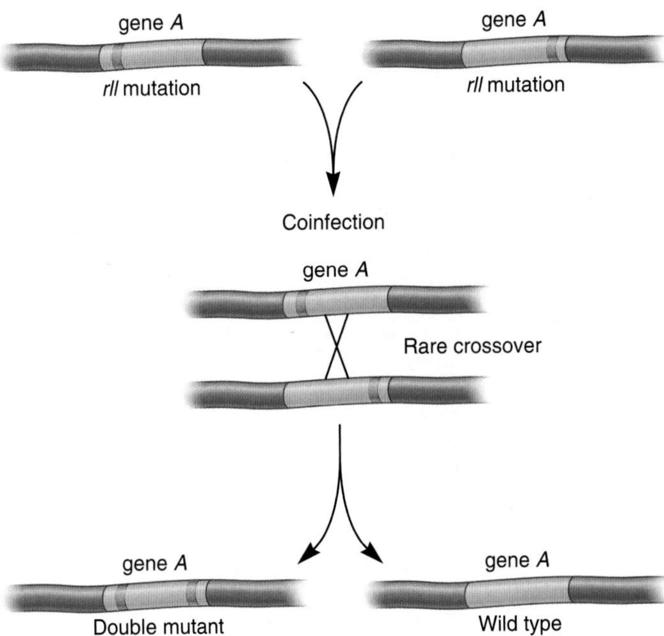

FIGURE 6.17 Intragenic recombination. Following coinfection, a rare crossover has occurred between the sites of the two mutations. This produces a wild-type phage with no mutations and a double mutant phage with both mutations.

Figure 6.18 describes the general strategy for intragenic mapping of *rII* phage mutations. Bacteriophages from two different noncomplementing *rII* phage mutants (here, *r103* and *r104*) were mixed together in equal numbers and then infected into *E. coli* B. In this strain, the *rII* mutants grew and propagated. Recall from Figure 6.17, when two different mutants coinfect the same cell, intragenic recombination can occur, producing wild-type phages and double mutant phages. However, these intragenic recombinants were produced at a very low rate. Following coinfection and lysis of *E. coli* B, a new population of phages was isolated. This population was expected to contain predominantly nonrecombinant phages. However, due to intragenic recombination, it should also contain a very low percentage of wild-type phages and double mutant phages (refer back to Figure 6.17).

How could Benzer determine the number of rare phages that were produced by intragenic recombination? The key approach is that *rII* mutant phages cannot grow in *E. coli* K12(λ). Following coinfection, he took this new population of phages and used some of them to infect *E. coli* B and some to infect *E. coli* K12(λ). After plating, the *E. coli* B infection was used to determine the total number of phages, because *rII* mutants as well as wild-type phages can produce plaques in this strain. The overwhelming majority of these phages were expected to be nonrecombinant phages. The *E. coli* K12(λ) infection was used to determine the number of rare intragenic recombinants that produce wild-type phages.

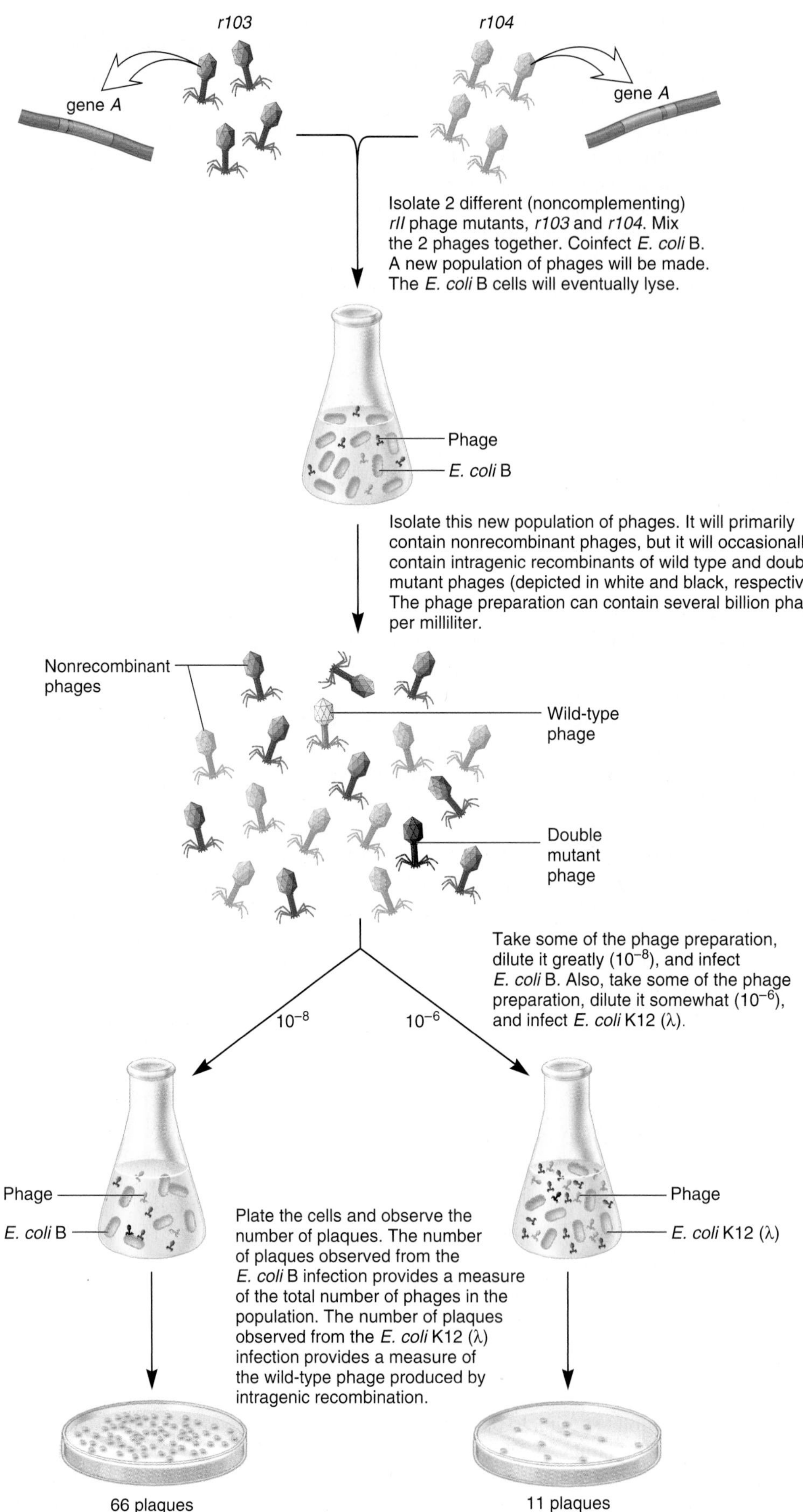
r103
r104
gene A
gene A
Isolate 2 different (noncomplementing) rII phage mutants, r103 and r104. Mix the 2 phages together. Coinfect E. coli B. A new population of phages will be made. The E. coli B cells will eventually lyse.
Phage
E. coli B
Isolate this new population of phages. It will primarily contain nonrecombinant phages, but it will occasionally contain intragenic recombinants of wild type and double mutant phages (depicted in white and black, respectively). The phage preparation can contain several billion phages per milliliter.
Nonrecombinant phages
Wild-type phage
Double mutant phage
Take some of the phage preparation, dilute it greatly (10^{-8}), and infect E. coli B. Also, take some of the phage preparation, dilute it somewhat (10^{-6}), and infect E. coli K12 (λ).
10^{-8}
10^{-6}
Phage
E. coli B
Phage
E. coli K12 (λ)
Plate the cells and observe the number of plaques. The number of plaques observed from the E. coli B infection provides a measure of the total number of phages in the population. The number of plaques observed from the E. coli K12 (λ) infection provides a measure of the wild-type phage produced by intragenic recombination.
66 plaques
11 plaques

FIGURE 6.18 Benzer's method of intragenic mapping in the *rII* region.

Figure 6.18 illustrates the great advantage of this experimental system in detecting a low percentage of recombinants. In the laboratory, phage preparations containing several billion phages per milliliter are readily made. Among billions of phages, a low percentage (e.g., 1 in every 1,000) may be wild-type phages arising from intragenic recombination. The wild-type recombinants can produce plaques in *E. coli* K12(λ), whereas the *rII* mutant strains cannot. In other words, only the tiny fraction of wild-type recombinants would produce plaques in *E. coli* K12(λ).

The frequency of recombinant phages can be determined by comparing the number of wild-type phages, produced by intragenic recombination, and the total number of phages. As shown in Figure 6.18, the total number of phages can be deduced from the number of plaques obtained from the infection of *E. coli* B. In this experiment, the phage preparation was diluted by 10^8 (1:100,000,000), and 1 ml was used to infect *E. coli* B. Because this plate produced 66 plaques, the total number of phages in the original preparation was $66 \times 10^8 = 6.6 \times 10^9$, or 6.6 billion phages per milliliter. By comparison, the phage preparation used to infect *E. coli* K12(λ) was diluted by only 10^6 (1 : 1,000,000). This plate produced 11 plaques. Therefore, the number of wild-type phages was 11×10^6, which equals 11 million wild-type phages per milliliter.

As we have already seen in Chapter 5, genetic mapping distance is computed by dividing the number of recombinants by the total population (nonrecombinants and recombinants) times 100. In this experiment, intragenic recombination produced an equal number of two types of recombinants: wild-type phages and double mutant phages. Only the wild-type phages are detected in the infection of *E. coli* K12(λ). Therefore, to obtain the total number of recombinants, the number of wild-type phages must be multiplied by two. With all this information, we can compute the frequency of recombinants using the experimental approach described in Figure 6.18.

$$\text{Frequency of recombinants} = \frac{2\left[\text{Wild-type plaques obtained in } E.\ coli \text{ K12}(\lambda)\right]}{\text{Total number of plaques obtained in } E.\ coli \text{ B}}$$

We can use this equation to calculate the frequency of recombinants obtained in the experiment described in Figure 6.18.

$$\text{Frequency of recombinants} = \frac{2(11 \times 10^6)}{6.6 \times 10^9}$$

$$= 3.3 \times 10^{-3} = 0.0033$$

In this example, approximately 3.3 recombinants were produced per 1,000 phages.

The frequency of recombinants provides a measure of map distance. In eukaryotic mapping studies, we compute the map distance by multiplying the frequency of recombinants by 100 to give a value in map units (also known as centiMorgans). Similarly, in these experiments, the frequency of recombinants can provide a measure of map distance along the bacteriophage DNA. In this case, the map distance is between two mutations within the same gene. Like intergenic mapping, the frequency of intragenic recombinants is correlated with the distance between the two mutations; the farther apart they are, the higher the frequency of recombinants. If two mutations happen to be located at exactly the same site within a gene, coinfection would not be able to produce any wild-type recombinants, and so the map distance would be zero. These are known as **homoallelic** mutations.

Deletion Mapping Can Be Used to Localize Many *rII* Mutations to Specific Regions in the *rIIA* or *rIIB* Genes

Now that we have seen the general approach to intragenic mapping, let's consider a method to efficiently map hundreds of *rII* mutations within the two genes designated *rIIA* and *rIIB*. As you may have realized, the coinfection experiments described in Figure 6.18 are quite similar to Sturtevant's strategy of making dihybrid crosses to map genes along the X chromosome of *Drosophila* (refer back to Figure 5.10). Similarly, Benzer wanted to coinfect different *rII* mutants in order to map the sites of the mutations within the *rIIA* and *rIIB* genes. During the course of his work, he obtained hundreds of different *rII* mutant strains that he wanted to map. However, making all the pairwise combinations would have been an overwhelming task. Instead, Benzer used an approach known as **deletion mapping** as a first step in localizing his *rII* mutations to a fairly short region within gene *A* or gene *B*.

Figure 6.19 describes the general strategy used in deletion mapping. This approach is easier to understand if we use an example. Let's suppose that the goal is to know the approximate location of an *rII* mutation, such as *r103*. To do so, *E. coli* K12(λ) would be coinfected with *r103* and a deletion strain. Each deletion strain is a T4 bacteriophage that is missing a known segment of the *rIIA* and/or *rIIB* gene. If the deleted region includes the same region that contains the *r103* mutation, a coinfection cannot produce intragenic wild-type recombinants. Therefore, plaques will not be formed. However, if a deletion strain recombines with *r103* to produce a wild-type phage, the deleted region does not overlap with the *r103* mutation.

In the example shown in Figure 6.19, the *r103* strain produced wild-type recombinants when coinfected with deletion strains *PB242*, *A105*, and *638*. However, coinfection of *r103* with *PT1*, *J3*, *1241*, and *1272* did not produce intragenic wild-type recombinants. Because coinfection with *PB242* produced recombinants and *PT1* did not, the *r103* mutation must be located in the region that is missing in *PT1* but not missing in *PB242*. As shown at the bottom of Figure 6.19, this region is called A4 (the A refers to the *rIIA* gene). In other words, the *r103* mutation is located somewhere within the A4 region, but not in the other six regions (A1, A2, A3, A5, A6, and B).

As described in Figure 6.19, this first step in the deletion mapping strategy localized an *rII* mutation to one of seven regions; six of these were in *rIIA* and one was in *rIIB*. Other deletion strains were used to eventually localize each *rII* mutation to one of 47 short regions; 36 were in *rIIA*, 11 in *rIIB*. At this point, pairwise coinfections were made between mutant strains that had been localized to the same region by deletion mapping. For

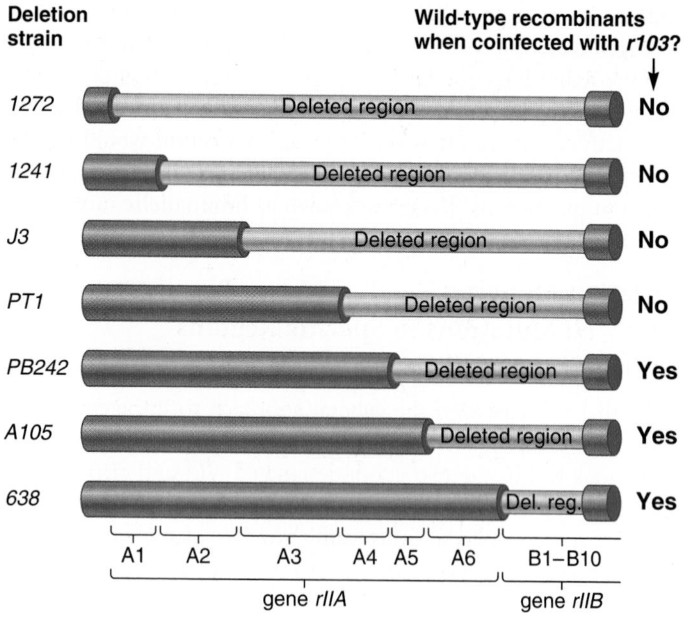

FIGURE 6.19 **The use of deletion strains to localize *rII* mutants to short regions within the *rIIA* or *rIIB* gene.** The deleted regions are shown in gray.

example, 24 mutations were deletion-mapped to a region called A5d. Pairwise coinfection experiments were conducted among this group of 24 mutants to precisely map their locations relative to each other in the A5d region. Similarly, all mutants in each of the 46 other groups were mapped by pairwise coinfections. In this way, a fine structure map was constructed depicting the locations of hundreds of different *rII* mutations (**Figure 6.20**). As seen in this figure, certain locations contained a relatively high number of mutations compared to other sites. These were termed **hot spots** for mutation.

Intragenic Mapping Experiments Provided Insight into the Relationship Between Traits and Molecular Genetics

Intragenic mapping studies were a pivotal achievement in our early understanding of gene structure. Since the time of Mendel, geneticists had considered a gene to be the smallest unit of heredity, which provided an organism with its inherited traits. In the late 1950s, however, the molecular nature of the gene was not understood. Because it is a unit of heredity, some scientists envisioned a gene as being a particle-like entity that could not be further subdivided into additional parts. However, intragenic mapping studies revealed, convincingly, that this is not the case. These studies

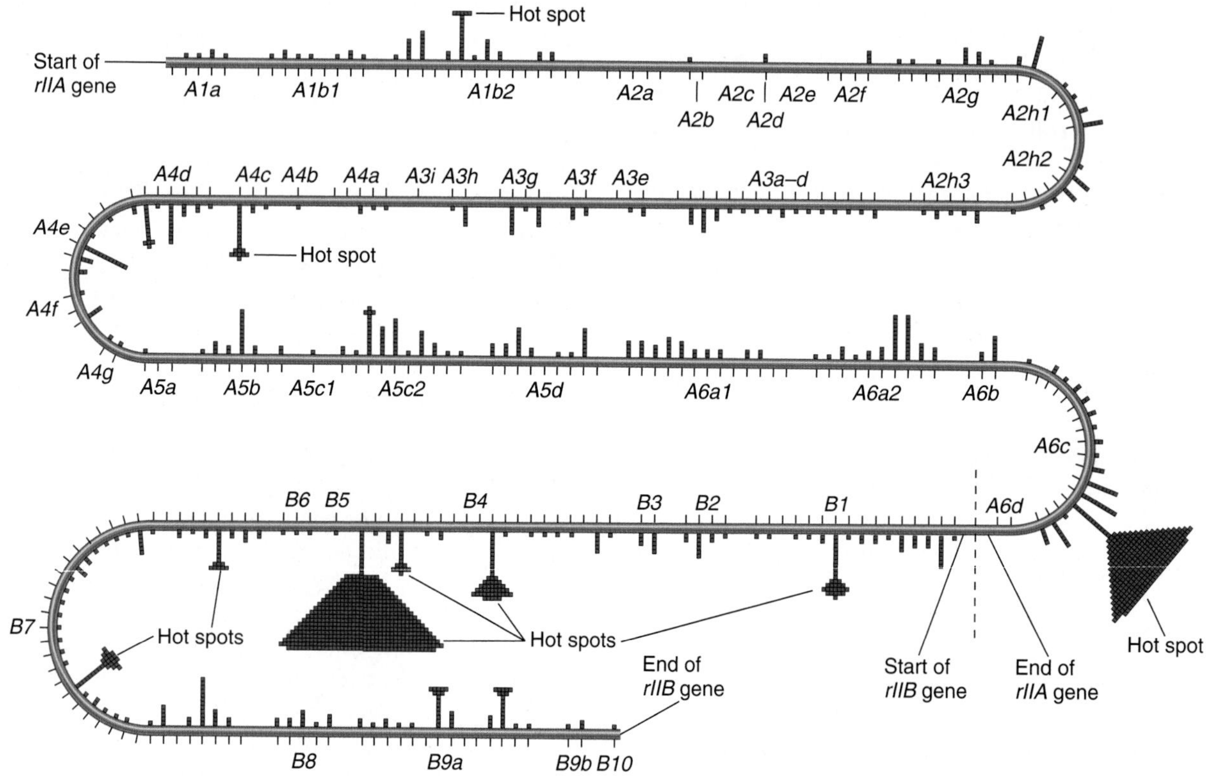

FIGURE 6.20 **The outcome of intragenic mapping of many *rII* mutations.** The blue line represents the linear sequence of the *rIIA* and *rIIB* genes, which are found within the T4 phage's genetic material. Each small purple box attached to the blue line symbolizes a mutation that was mapped by intragenic mapping. Among hundreds of independent mutant phages, several mutations sometimes mapped to the same site. In this figure, mutations at the same site form columns of boxes. Hot spots contain a large number of mutations and are represented as a group of boxes attached to a column of boxes. A hot spot contains many mutations at the same site within the *rIIA* or *rIIB* gene.

showed that mutations can occur at many different sites within a single gene. Furthermore, intragenic crossing over could recombine these mutations, resulting in wild-type genes. Therefore, rather than being an indivisible particle, a gene must be composed of a large structure that can be subdivided during crossing over.

Benzer's results were published in the late 1950s and early 1960s, not long after the physical structure of DNA had been elucidated by Watson and Crick. We now know that a gene is a segment of DNA that is composed of smaller building blocks called nucleotides. A typical gene is a linear sequence of several hundred to many thousand base pairs. As the genetic map of Figure 6.20 indicates, mutations can occur at many sites along the linear structure of a gene; intragenic crossing over can recombine mutations that are located at different sites within the same gene.

CONCEPTUAL SUMMARY

This chapter has been concerned with genetic transfer and mapping in bacteria. Three mechanisms are known for the transfer of genetic material from one bacterial cell to another: conjugation, transduction, and transformation. During conjugation, bacteria make direct physical contact with each other. Donor cells can be either F^+, F', or Hfr cells. An F^+ cell contains a circular piece of genetic material, a plasmid called an F factor that is transferred to the recipient (F^-) cell during mating. An F' factor is an F factor that also carries genes derived from the bacterial chromosome. An Hfr cell (for high frequency of recombination) has an integrated F factor and can transfer chromosomal DNA to a recipient cell. A second route of genetic transfer is transduction. In this case, a bacteriophage accidentally packages a portion of the bacterial chromosome, which is then transferred to another bacterium upon infection. Finally, a third mechanism of genetic transfer is transformation. For this to occur, a dead bacterial cell must release its genetic material into the environment. A living bacterial cell in a competent state subsequently imports the DNA, and then recombination causes the imported DNA to replace segments of genetic material along the bacterial chromosome. Overall, these three mechanisms promote gene transfer within bacterial species and also allow horizontal gene transfer among different bacterial species. From a medical perspective, this phenomenon has led to a dramatic rise in the prevalence of strains that are resistant to antibiotics.

Chapter 6 concluded with a description of intragenic mapping studies in T4 bacteriophages. Benzer constructed a fine structure genetic map of two bacteriophage genes, *rIIA* and *rIIB*. In the early 1960s, his fine structure map provided important insights into the molecular nature of the gene. It revealed that a gene is not an indivisible unit. Rather, his results indicated that mutations can occur at many sites along the linear structure of a gene and that intragenic crossing over can recombine mutations located at different sites within the same gene. These results are consistent with our current knowledge of the molecular structure of genes.

EXPERIMENTAL SUMMARY

Research studies of genetic transfer in bacteria have provided a unique experimental strategy for mapping the linear order and relative locations of bacterial genes. Conjugation has been used to map the locations of many genes along the bacterial chromosome. In this approach, map distances are determined by the number of minutes it takes for a gene to enter a recipient cell during conjugation. In addition, cotransduction and cotransformation experiments have been commonly used to accurately map bacterial genes that are relatively close to each other on the chromosome.

As mentioned in the Conceptual Summary, intragenic mapping studies of T4 bacteriophages have provided great insight into the molecular nature of genes. Benzer constructed a fine structure genetic map of two bacteriophage genes, *rIIA* and *rIIB*. This was accomplished using a coinfection strategy, in which the very low percentage of intragenic recombinants yielding wild-type phages could be identified by their unique ability to infect a strain of bacteria known as *E. coli* K12(λ). This approach illustrates the power of phage genetics compared to the analysis of offspring in eukaryotic species. The selective inability of *rII* mutants to lyse *E. coli* K12(λ) allowed detection of the very few intragenic recombinants that produced wild-type phages that could lyse *E. coli* K12(λ). In this way, the very short distance that occurs between mutations within a single gene could be determined.

PROBLEM SETS & INSIGHTS

Solved Problems

S1. In *E. coli*, the gene *bioD*$^+$ encodes an enzyme involved in biotin synthesis, and *galK*$^+$ encodes an enzyme involved in galactose utilization. An *E. coli* strain that contained wild-type versions of both genes was infected with P1, and then a P1 lysate was obtained. This lysate was used to transduce (infect) a strain that was *bioD*$^-$ and *galK*$^-$. The cells were plated on media containing galactose as the sole carbon source for growth to select for transduction of the *galK*$^+$ gene. These media also were supplemented with biotin. The colonies were then restreaked on media that lacked biotin to see if the *bioD*$^+$ gene had been cotransduced. The following results were obtained:

Selected Gene	Non-selected Gene	Number of Colonies That Grew On: Galactose + Biotin	Galactose − Biotin	Cotransduction Frequency
galK$^+$	*bioD*$^+$	80	10	0.125

How far apart are these two genes?

Answer: We can use the cotransduction frequency to calculate the distance between the two genes (in minutes) using the equation

$$\text{Cotransduction frequency} = (1 - d/2)^3$$

$$0.125 = (1 - d/2)^3$$

$$1 - d/2 = \sqrt[3]{0.125}$$

$$1 - d/2 = 0.5$$

$$d/2 = 1 - 0.5$$

$$d = 1.0 \text{ minute}$$

The two genes are approximately 1 minute apart on the *E. coli* chromosome.

S2. By conducting mating experiments between a single *Hfr* strain and a recipient strain, Wollman and Jacob mapped the order of many bacterial genes. Throughout the course of their studies, they identified several different *Hfr* strains in which the F factor DNA had been integrated at different places along the bacterial chromosome. A sample of their experimental results is shown in the following table:

Hfr strain	Origin	Order of Transfer of Several Different Bacterial Genes: First								Last
H	O	*thr*	*leu*	*azi*	*ton*	*pro*	*lac*	*gal*	*str*	*met*
1	O	*leu*	*thr*	*met*	*str*	*gal*	*lac*	*pro*	*ton*	*azi*
2	O	*pro*	*ton*	*azi*	*leu*	*thr*	*met*	*str*	*gal*	*lac*
3	O	*lac*	*pro*	*ton*	*azi*	*leu*	*thr*	*met*	*str*	*gal*
4	O	*met*	*str*	*gal*	*lac*	*pro*	*ton*	*azi*	*leu*	*thr*
5	O	*met*	*thr*	*leu*	*azi*	*ton*	*pro*	*lac*	*gal*	*str*
6	O	*met*	*thr*	*leu*	*azi*	*ton*	*pro*	*lac*	*gal*	*str*
7	O	*ton*	*azi*	*leu*	*thr*	*met*	*str*	*gal*	*lac*	*pro*

A. Explain how these results are consistent with the idea that the bacterial chromosome is circular.

B. Draw a map that shows the order of genes and the locations of the origins of transfer among these different *Hfr* strains.

Answer:

A. In comparing the data among different *Hfr* strains, the order of the nine genes was always the same or the reverse of the same order. For example, *HfrH* and *Hfr1* transfer the same genes but their orders are reversed relative to each other. In addition, the *Hfr* strains showed an overlapping pattern of transfer with regard to the origin. For example, *Hfr1* and *Hfr2* had the same order of genes, but *Hfr1* began with *leu* and ended with *azi*, while *Hfr2* began with *pro* and ended with *lac*. From these findings, Wollman and Jacob concluded that the segment of DNA that was the origin of transfer had been inserted at different points within a circular *E. coli* chromosome in different *Hfr* strains. They also concluded that the origin can be inserted in either orientation, so the direction of gene transfer can be clockwise or counterclockwise around the circular bacterial chromosome.

B. A genetic map consistent with these results is shown here.

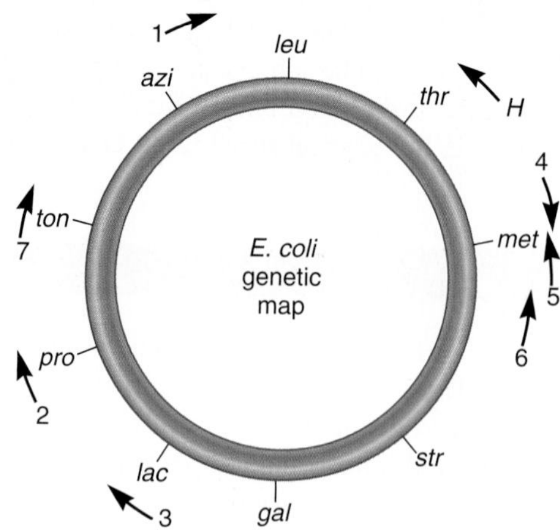

S3. An *Hfr* strain that is *leuA*$^+$ and *thiL*$^+$ was mated to a strain that is *leuA*$^-$ and *thiL*$^-$. In the data points shown here, the mating was interrupted and the percentage of recombinants for each gene was determined by streaking on media that lacked either leucine or thiamine. The results are shown.

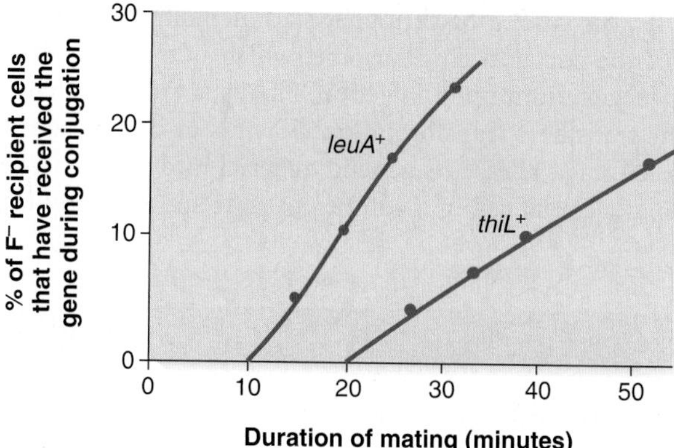

What is the map distance (in minutes) between these two genes?

Answer: This problem is solved by extrapolating the data points to the x-axis to determine the time of entry. For *leuA*$^+$, they extrapolate back to 10 minutes. For *thiL*$^+$, they extrapolate back to 20 minutes. Therefore, the distance between the two genes is approximately 10 minutes.

S4. Genetic transfer via transformation can also be used to map genes along the bacterial chromosome. In this approach, fragments of chromosomal DNA are isolated from one bacterial strain and used to transform another strain. The experimenter examines the transformed bacteria to see if they have incorporated two or more different genes. For example, the DNA may be isolated from a donor *E. coli* bacterium that has functional copies of the *araB* and *leuD* genes. Let's call these genes *araB*$^+$ and *leuD*$^+$ to indicate the genes are functional. These two genes are required for arabinose metabolism and leucine synthesis, respectively. To map the distance between these two genes via transformation, a recipient bacterium would be used that is *araB*$^-$ and *leuD*$^-$. Following transformation, the recipient bacterium may become *araB*$^+$ and *leuD*$^+$. This phenomenon is called cotransformation because two genes from the donor bacterium have been transferred to the recipient via transformation. In this type of experiment, the recipient cell is exposed to a fairly low concentration of donor DNA, making it unlikely that the recipient bacterium will take up more than one fragment of DNA. Therefore, under these conditions, cotransformation is likely only when two genes are fairly close together and are found on one fragment of DNA.

In a cotransformation experiment, a researcher has isolated DNA from an $araB^+$ and $leuD^+$ donor strain. This DNA was transformed into a recipient strain that was $araB^-$ and $leuD^-$. Following transformation, the cells were plated on media containing arabinose and leucine. On this media, only bacteria that are $araB^+$ can grow. The bacteria can be either $leuD^+$ or $leuD^-$ because leucine is provided in the media. Colonies, which grew on this media, were then restreaked on media that contained arabinose but lacked leucine. Only $araB^+$ and $leuD^+$ cells could grow on these secondary plates. Following this protocol, a researcher obtained the following results:

Number of colonies growing on arabinose plus leucine media: 57

Number of colonies that grew when restreaked on arabinose media without leucine: 42

What is the map distance between these two genes? Note: This problem can be solved using the strategy of a cotransduction experiment except that the researcher must determine the average size of DNA fragments that are taken up by the bacterial cells. This would correspond to the value of *L* in a cotransduction experiment.

Answer: As mentioned, the basic principle of gene mapping via cotransformation is identical to the method of gene mapping via cotransduction described in this chapter. One way to calculate the map distance is to use the same equation that we used for cotransduction data, except that we substitute cotransformation frequency for cotransduction frequency.

$$\text{Cotransformation frequency} = (1 - d/L)^3$$

(Note: Cotransformation is not quite as accurate as cotransduction because the sizes of chromosomal pieces tend to vary significantly from experiment to experiment, so the value of *L* is not quite as reliable. Nevertheless, cotransformation has been used extensively to map the order and distance between closely linked genes along the bacterial chromosome.)

The researcher needs to experimentally determine the value of *L* by running the DNA on a gel and estimating the average size of the DNA fragments. Let's assume they are about 2% of the bacterial chromosome, which, for *E. coli*, would be about 80,000 base pairs in length. So *L* equals 2 minutes, which is the same as 2%.

$$\text{Cotransformation frequency} = (1 - d/\text{L})^3$$

$$42/57 = (1 - d/2)^3$$

$$d = 0.2 \text{ minutes}$$

The distance between *araB* and *leuD* is approximately 0.2 minutes.

S5. In our discussion of transduction via P1 or P22, the life cycle of the bacteriophage sometimes resulted in the packaging of many different pieces of the bacterial chromosome. For other bacteriophages, however, transduction may involve the transfer of only a few specific genes from the donor cell to the recipient. This phenomenon is known as specialized transduction. The key event that causes specialized transduction to occur is that the lysogenic phase of the phage life cycle involves the integration of the viral DNA at a single specific site within the bacterial chromosome. The transduction of particular bacterial genes involves an abnormal excision of the phage DNA from this site within the chromosome that would carry adjacent bacterial genes. For example, a bacteriophage called lambda (λ) that infects *E. coli* specifically integrates between two genes designated gal^+ and bio^+ (required for galactose utilization and biotin synthesis, respectively). Either of these genes could be packaged into the phage if an abnormal excision event occurred. How would specialized transduction be different from generalized transduction?

Answer: Generalized transduction can involve the transfer of any bacterial gene, while specialized transduction can transfer only genes that are adjacent to the site where the phage integrates. As mentioned, a bacteriophage that infects *E. coli* cells, known as lambda (λ), provides a well-studied example of specialized transduction. In the case of phage lambda, the lysogenic life cycle results in the integration of the phage DNA at a site that is called the attachment site (described further in Chapter 17). The attachment site is located between two bacterial genes, gal^+ and bio^+. An *E. coli* strain that is lysogenic for phage lambda will have the lambda DNA integrated between these two bacterial genes. On occasion, the phage may enter the lytic cycle and excise its DNA from the bacterial chromosome. When this occurs normally, the phage excises its entire viral DNA from the bacterial chromosome. The excised phage DNA is then replicated and becomes packaged into newly made phages. However, an abnormal excision does occur at a low rate (i.e., about one in a million). In this abnormal event, the phage DNA is excised in such a way that an adjacent bacterial gene is included and some of the phage DNA is not included in the final product. For example, the abnormal excision may yield a fragment of DNA that includes the gal^+ gene and some of the lambda DNA but is missing part of the lambda DNA. If this DNA fragment is packaged into a virus, it is called a defective phage because it is missing some of the phage DNA. If it carries the gal^+ gene, it is designated λ*dgal* (the letter *d* designates a defective phage). Alternatively, an abnormal excision may carry the bio^+ gene. This phage is designated λ*dbio*. Defective lambda phages can then transduce the gal^+ or bio^+ genes to other *E. coli* cells.

Conceptual Questions

C1. The terms conjugation, transduction, and transformation are used to describe three different natural forms of genetic transfer between bacterial cells. Briefly discuss the similarities and differences between these processes.

C2. Conjugation is sometimes called "bacterial mating." Is it a form of sexual reproduction? Explain.

C3. If you mix together an equal number of F^+ and F^- cells, how would you expect the proportions to change over time? In other words, do you expect an increase in the relative proportions of F^+ or of F^- cells? Explain your answer.

C4. What is the difference between an F^+ and an *Hfr* strain? Which type of strain do you expect to transfer many bacterial genes to recipient cells?

C5. What is the role of the origin of transfer during F^+- and Hfr-mediated conjugation? What is the significance of the direction of transfer in Hfr-mediated conjugation?

C6. What is the role of sex pili during conjugation?

C7. Think about the structure and transmission of F factors and discuss how you think F factors may have originated.

C8. Each species of bacteria has its own distinctive cell surface. The characteristics of the cell surface play an important role in processes such as conjugation and transduction. For example, certain strains of *E. coli* have pili on their cell surface. These pili enable *E. coli* to mate with other *E. coli*, and the pili also enable certain bacteriophages (such as M13) to bind to the surface of *E. coli* and gain entry into the cytoplasm. With these ideas in mind, explain which forms of genetic transfer (i.e., conjugation, transduction, and transformation) are more likely to occur between different species of bacteria. Discuss some of the potential consequences of interspecies genetic transfer.

C9. Briefly describe the lytic and lysogenic cycles of bacteriophages. In your answer, explain what a prophage is.

C10. What is cotransduction? What determines the likelihood that two genes will be cotransduced?

C11. When bacteriophage P1 causes *E. coli* to lyse, the resulting material is called a P1 lysate. What type of genetic material would be found in most of the P1 phages in the lysate? What kind of genetic material would occasionally be found within a P1 phage?

C12. As described in Figure 6.10, host DNA is hydrolyzed into small pieces, which are occasionally assembled with phage proteins, creating a phage with bacterial chromosomal DNA. If the breakage of the chromosomal DNA is not random (i.e., it is more likely to break at certain spots compared to other spots), how might nonrandom breakage affect cotransduction frequency?

C13. Describe the steps that occur during bacterial transformation. What is a competent cell? What factors may determine whether a cell will be competent?

C14. Which bacterial genetic transfer process does not require recombination with the bacterial chromosome?

C15. Researchers who study the molecular mechanism of transformation have identified many proteins in bacteria that function in the uptake of DNA from the environment and its recombination into the host cell's chromosome. This means that bacteria have evolved molecular mechanisms for the purpose of transformation by extracellular DNA. Of what advantage(s) would it be for a bacterium to import DNA from the environment and/or incorporate it into its chromosome?

C16. Antibiotics such as tetracycline, streptomycin, and bacitracin are small organic molecules that are synthesized by particular species of bacteria. Microbiologists have hypothesized that the reason why certain bacteria make antibiotics is to kill other species that occupy the same environment. Bacteria that produce an antibiotic may be able to kill competing species. This provides more resources for the antibiotic-producing bacteria. In addition, bacteria that have the genes necessary for antibiotic biosynthesis contain genes that confer resistance to the same antibiotic. For example, tetracycline is made by the soil bacterium *Streptomyces aureofaciens*. Besides the genes that are needed to make tetracycline, *S. aureofaciens* also contains genes that confer tetracycline resistance; otherwise it would kill itself when it makes tetracycline. In recent years, however, many other species of bacteria that do not synthesize tetracycline have acquired the genes that confer tetracycline resistance. For example, certain strains of *E. coli* carry tetracycline-resistance genes, even though *E. coli* does not synthesize tetracycline. When these genes are analyzed at the molecular level, it has been found that they are evolutionarily related to the genes in *S. aureofaciens*. This observation indicates that the genes from *S. aureofaciens* have been transferred to *E. coli*.

A. What form of genetic transfer (i.e., conjugation, transduction, or transformation) would be the most likely mechanism of interspecies gene transfer?

B. Because *S. aureofaciens* is a nonpathogenic soil bacterium and *E. coli* is an enteric bacterium, do you think it was direct gene transfer, or do you think it may have occurred in multiple steps (i.e., from *S. aureofaciens* to other bacterial species and then to *E. coli*)?

C. How could the widespread use of antibiotics to treat diseases have contributed to the proliferation of many bacterial species that are resistant to antibiotics?

C17. What does the term complementation mean? If two different mutations that produce the same phenotype can complement each other, what can you conclude about the locations of each mutation?

C18. Intragenic mapping is sometimes called interallelic mapping. Explain why the two terms mean the same thing. In your own words, explain what an intragenic map is.

C19. As discussed in Chapter 12, genes are composed of a sequence of nucleotides. A typical gene in a bacteriophage is a few hundred or a few thousand nucleotides in length. If two different strains of bacteriophage T4 have a mutation in the *rIIA* gene that gives a rapid-lysis phenotype, yet they never produce wild-type phages by intragenic recombination when they are coinfected into *E. coli* B, what would you conclude about the locations of the mutations in the two different T4 strains?

Experimental Questions

E1. In the experiment of Figure 6.1, a $met^- \ bio^- \ thr^+ \ leu^+ \ thi^+$ cell could become $met^+ \ bio^+ \ thr^+ \ leu^+ \ thi^+$ by a (rare) double mutation that converts the $met^- \ bio^-$ genetic material into $met^+ \ bio^+$. Likewise, a $met^+ \ bio^+ \ thr^- \ leu^- \ thi^-$ cell could become $met^+ \ bio^+ \ thr^+ \ leu^+ \ thi^+$ by three mutations that convert the $thr^- \ leu^- \ thi^-$ genetic material into $thr^+ \ leu^+ \ thi^+$. From the results of Figure 6.1, how do you know that the occurrence of 10 $met^+ \ bio^+ \ thr^+ \ leu^+ \ thi^+$ colonies is not due to these types of rare double or triple mutations?

E2. In the experiment of Figure 6.1, Lederberg and Tatum could not discern whether $met^+ \ bio^+$ genetic material was transferred to the $met^- \ bio^- \ thr^+ \ leu^+ \ thi^+$ strain or if $thr^+ \ leu^+ \ thi^+$ genetic material was transferred to the $met^+ \ bio^+ \ thr^- \ leu^- \ thi^-$ strain. Let's suppose that one strain is streptomycin resistant (say, $met^+ \ bio^+ \ thr^- \ leu^- \ thi^-$) while the other strain is sensitive to streptomycin. Describe an experiment that could determine whether the $met^+ \ bio^+$ genetic material was transferred to the $met^- \ bio^- \ thr^+ \ leu^+ \ thi^+$ strain or if the $thr^+ \ leu^+ \ thi^+$ genetic material was transferred to the $met^+ \ bio^+ \ thr^- \ leu^- \ thi^-$ strain.

E3. Explain how a U-tube apparatus can distinguish between genetic transfer involving conjugation and genetic transfer involving transduction. Do you think a U-tube could be used to distinguish between transduction and transformation?

E4. What is an interrupted mating experiment? What type of experimental information can be obtained from this type of study? Why is it necessary to interrupt mating?

E5. In a conjugation experiment, what is meant by the time of entry? How is the time of entry determined experimentally?

E6. In your laboratory, you have an F^- strain of *E. coli* that is resistant to streptomycin and is unable to metabolize lactose, but it can metabolize glucose. Therefore, this strain can grow on media that contain glucose and streptomycin, but it cannot grow on media containing lactose. A researcher has sent you two *E. coli* strains in two separate tubes. One strain, let's call it strain *A*, has an F factor that carries the genes that are required for lactose metabolism. On its chromosome, it also has the genes that are required for glucose metabolism. However, it is sensitive to streptomycin. This strain can grow on media containing lactose or glucose, but it cannot grow if streptomycin is added to the media. The second strain, let's call it strain *B*, is an F^- strain. On its chromosome, it has the genes that are required for lactose and glucose metabolism. Strain *B* is also sensitive to streptomycin. Unfortunately, when strains *A* and *B* were sent to you, the labels had fallen off the tubes. Describe how you could determine which tubes contain strain *A* and strain *B*.

E7. As mentioned in solved problem S2, origins of transfer can be located in many different locations, and their direction of transfer can be clockwise or counterclockwise. Let's suppose a researcher mated six different *Hfr* strains that were thr^+ leu^+ ton^s str^r azi^s lac^+ gal^+ pro^+ met^+ to an F^- strain that was thr^- leu^- ton^r str^s azi^r lac^- gal^- pro^- met^-, and obtained the following results:

Strain	Order of Gene Transfer
1	ton^s azi^s leu^+ thr^+ met^+ str^r gal^+ lac^+ pro^+
2	leu^+ azi^s ton^s pro^+ lac^+ gal^+ str^r met^+ thr^+
3	lac^+ gal^+ str^r met^+ thr^+ leu^+ azi^s ton^s pro^+
4	leu^+ thr^+ met^+ str^r gal^+ lac^+ pro^+ ton^s azi^s
5	ton^s pro^+ lac^+ gal^+ str^r met^+ thr^+ leu^+ azi^s
6	met^+ str^r gal^+ lac^+ pro^+ ton^s azi^s leu^+ thr^+

Draw a circular map of the *E. coli* chromosome and describe the locations and orientations of the origins of transfer in these six *Hfr* strains.

E8. An *Hfr* strain that is $hisE^+$ and $pheA^+$ was mated to a strain that is $hisE^-$ and $pheA^-$. The mating was interrupted and the percentage of recombinants for each gene was determined by streaking on media that lacked either histidine or phenylalanine. The following results were obtained:

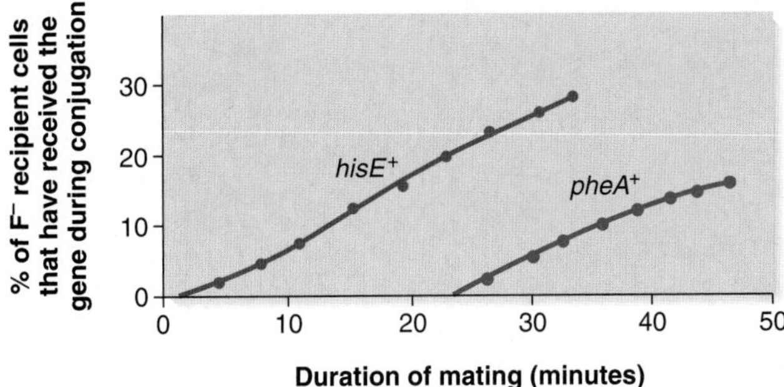

A. Determine the map distance (in minutes) between these two genes.

B. In a previous experiment, it was found that *hisE* is 4 minutes away from the gene *pabB*. *PheA* was shown to be 17 minutes from this gene. Draw a genetic map describing the locations of all three genes.

E9. Acridine orange is a chemical that inhibits the replication of F factor DNA but does not affect the replication of chromosomal DNA, even if the chromosomal DNA contains an Hfr. Let's suppose that you have an *E. coli* strain that is unable to metabolize lactose and has an F factor that carries a streptomycin-resistant gene. You also have an F^- strain of *E. coli* that is sensitive to streptomycin and has the genes that allow the bacterium to metabolize lactose. This second strain can grow on lactose-containing media. How would you generate an *Hfr* strain that is resistant to streptomycin and can metabolize lactose? (Hint: F factors occasionally integrate into the chromosome to become *Hfr* strains, and occasionally *Hfr* strains excise their DNA from the chromosome to become F^+ strains that carry an F' factor.)

E10. In a P1 transduction experiment, the P1 lysate contains phages that carry pieces of the host chromosomal DNA, but the lysate also contains broken pieces of chromosomal DNA (see Figure 6.10). If a P1 lysate is used to transfer chromosomal DNA to another bacterium, how could you show experimentally that the recombinant bacterium has been transduced (i.e., taken up a P1 phage with a piece of chromosomal DNA inside) versus transformed (i.e., taken up a piece of chromosomal DNA that is not within a P1 phage coat)?

E11. Could you devise an experimental strategy to get P1 phage to transduce the entire lambda genome from one strain of bacterium to another strain? (Note: The general features of phage lambda's life cycle are described in Chapter 14.) Phage lambda has a genome size of 48,502 nucleotides (about 1% of the size of the *E. coli* chromosome) and can follow the lytic or lysogenic life cycle. Growth of *E. coli* on minimal growth media favors the lysogenic life cycle, whereas growth on rich media and/or UV light promotes the lytic cycle.

E12. Let's suppose a new strain of P1 has been identified that packages larger pieces of the *E. coli* chromosome. This new P1 strain packages pieces of the *E. coli* chromosome that are 5 minutes long. If two genes are 0.7 minutes apart along the *E. coli* chromosome, what would be the cotransduction frequency using a normal strain of P1 and using this new strain of P1 that packages large pieces? What would be the experimental advantage of using this new P1 strain?

E13. If two bacterial genes are 0.6 minutes apart on the bacterial chromosome, what frequency of cotransductants would you expect to observe in a P1 transduction experiment?

E14. In an experiment involving P1 transduction, the cotransduction frequency was 0.53. How far apart are the two genes?

E15. In a cotransduction experiment, the transfer of one gene is selected for and the presence of the second gene is then determined. If 0 out of 1,000 P1 transductants that carry the first gene also carry the second gene, what would you conclude about the minimum distance between the two genes?

E16. In a cotransformation experiment (see solved problem S4), DNA was isolated from a donor strain that was $proA^+$ and $strC^+$ and sensitive to tetracycline. (The *proA* and *strC* genes confer the ability to synthesize proline and confer streptomycin resistance, respectively.) A recipient strain is *proA*$^-$ and *strC*$^-$ and is resistant to tetracycline. After transformation, the bacteria were first streaked on media containing proline, streptomycin, and tetracycline. Colonies were then restreaked on media containing streptomycin and tetracycline. (Note: Both types of media had carbon and nitrogen sources for growth.) The following results were obtained:

70 colonies grew on media containing proline, streptomycin, and tetracycline, while only 2 of these 70 colonies grew when restreaked on media containing streptomycin and tetracycline but lacking proline.

A. If we assume the average size of the DNA fragments is 2 minutes, how far apart are these two genes?

B. What would you expect the cotransformation frequency to be if the average size of the DNA fragments was 4 minutes and the two genes are 1.4 minutes apart?

E17. If you took a pipette tip and removed a phage plaque from a petri plate, what would it contain?

E18. As shown in Figure 6.16, phages with *rII* mutations cannot produce plaques in *E. coli* K12(λ), but wild-type phages can. From an experimental point of view, explain why this observation is so significant.

E19. In the experimental strategy described in Figure 6.18, explain why it was necessary to dilute the phage preparation used to infect *E. coli* B so much more than the phage preparation used to infect *E. coli* K12(λ).

E20. Here are data from several complementation experiments, involving rapid-lysis mutations in genes *rIIA* and *rIIB*. The strain designated *L51* is known to have a mutation in *rIIB*.

Phage Mixture	Complementation
L91 and L65	No
L65 and L62	No
L33 and L47	Yes
L40 and L51	No
L47 and L92	No
L51 and L47	Yes
L51 and L92	Yes
L33 and L40	No
L91 and L92	Yes
L91 and L33	No

List which groups of mutations are in the *rIIA* gene and which groups are in the *rIIB* gene.

E21. A researcher has several different strains of T4 phage with single mutations in the same gene. In these strains, the mutations render the phage temperature sensitive. This means that temperature-sensitive phages can propagate when the bacterium (*E. coli*) is grown at 32°C but cannot propagate themselves when *E. coli* is grown at 37°C. Think about Benzer's strategy for intragenic mapping and propose an experimental strategy to map the temperature-sensitive mutations.

E22. Explain how Benzer's results indicated that a gene is not an indivisible unit.

E23. Explain why deletion mapping was used as a step in the intragenic mapping of *rII* mutations.

Questions for Student Discussion/Collaboration

1. Discuss the advantages of the genetic analysis of bacteria and bacteriophages. Make a list of the types of allelic differences among bacteria and phages that are suitable for genetic analyses.
2. Complementation occurs when two defective alleles in two different genes are found within the same organism and produce a normal phenotype. What other examples of complementation have we encountered in previous chapters of this textbook?

Note: All answers appear at the website for this textbook; the answers to even-numbered questions are in the back of the textbook.

www.mhhe.com/brookergenetics3e

Visit the Online Learning Center for practice tests, answer keys, and other learning aids for this chapter. Enhance your understanding of genetics with our interactive exercises, quizzes, animations, and much more.

NON-MENDELIAN INHERITANCE

7

CHAPTER OUTLINE

Mendelian inheritance patterns involve genes that directly influence the outcome of an offspring's traits and obey Mendel's laws. To predict phenotype, we must consider several factors. These include the dominant/recessive relationship of alleles, gene interactions that may affect the expression of a single trait, and the roles that sex and the environment play in influencing the individual's phenotype. Once these factors are understood, we can predict the phenotypes of offspring from their genotypes.

Most genes in eukaryotic species follow a Mendelian pattern of inheritance. However, many genes do not. For example, in Chapter 5 we saw how genes that are closely linked did not obey Mendel's law of independent assortment. One could argue that such inheritance patterns are also non-Mendelian. In this chapter, we will examine several additional and even bizarre types of inheritance patterns that deviate from a Mendelian pattern. In the first two sections of this chapter, we will consider two important examples of non-Mendelian inheritance called the maternal effect and epigenetic inheritance. Even though these inheritance patterns involve genes on chromosomes within the cell nucleus, the genotype of the offspring does not directly govern their phenotype in ways predicted by Mendel. We will see how the timing of gene expression and gene inactivation can cause a non-Mendelian pattern of inheritance.

In the third section, we will examine deviations from Mendelian inheritance that arise because some genetic material is not located in the cell nucleus. Certain cellular organelles, such as mitochondria and chloroplasts, contain their own genetic material. We will survey the inheritance of organellar genes and a few other examples in which traits are influenced by genetic material that exists outside of the cell nucleus.

Shell coiling in the water snail, *Lymnaea peregra*. In this species, some snails coil to the left, while others coil to the right. This is due to an inheritance pattern called the maternal effect.

7.1 MATERNAL EFFECT

We will begin by considering genes that have a **maternal effect.** This term refers to an inheritance pattern for certain **nuclear genes**—genes located on chromosomes that are found in the cell nucleus—in which the genotype of the mother directly determines the phenotype of her offspring. Surprisingly, for maternal effect genes, the genotypes of the father and offspring themselves do not affect the phenotype of the offspring. We will see that this phenomenon is explained by the accumulation of gene products that the mother provides to her developing eggs.

The Genotype of the Mother Determines the Phenotype of the Offspring for Maternal Effect Genes

The first example of a maternal effect gene was studied in the 1920s by A. E. Boycott and involved morphological features of the water snail, *Lymnaea peregra*. In this species, the shell and internal organs can be arranged in either a right-handed (dextral) or left-handed (sinistral) direction. The dextral orientation is more common and is dominant to the sinistral orientation. Figure 7.1 describes the results of a genetic analysis carried out by Boycott. In this experiment, he began with two different true-breeding strains of snails with either a dextral or sinistral morphology. Many combinations of crosses produced results that could not be explained by a Mendelian pattern of inheritance. When a dextral female (*DD*) was crossed to a sinistral male (*dd*), all F_1 offspring were dextral. However, in the **reciprocal cross,** where a sinistral female (*dd*) was crossed to a dextral male (*DD*), all F_1 offspring were sinistral. Taken together, these results contradict a Mendelian pattern of inheritance.

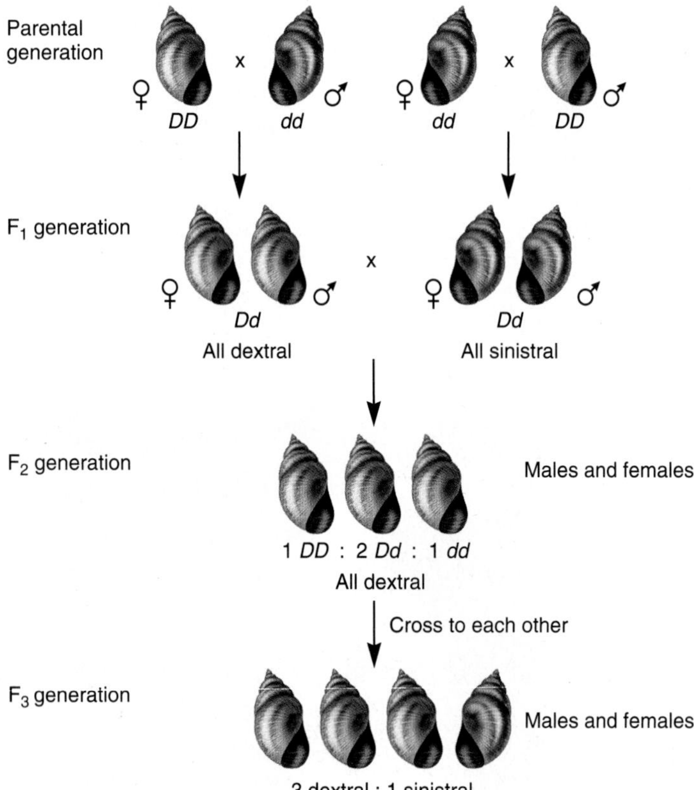

INTERACTIVE EXERCISE

FIGURE 7.1 Experiment showing the inheritance pattern of snail coiling. In this experiment, *D* (dextral) is dominant to *d* (sinistral). The genotype of the mother determines the phenotype of the offspring. This phenomenon is known as the maternal effect. In this case, a *DD* or *Dd* mother produces dextral offspring, and a *dd* mother produces sinistral offspring. The genotypes of the father and offspring do not affect the offspring's phenotype.

How can we explain the unusual results obtained in Figure 7.1? Alfred Sturtevant proposed the idea that snail coiling is due to a maternal effect gene that exists as a dextral (*D*) or sinistral (*d*) allele. His conclusions were drawn from the inheritance patterns of the F_2 and F_3 generations. In this experiment, the genotype of the F_1 generation is expected to be heterozygous (*Dd*). When these F_1 individuals were crossed to each other, a genotypic ratio of 1 *DD* : 2 *Dd* : 1 *dd* is predicted for the F_2 generation. Because the *D* allele is dominant to the *d* allele, a 3:1 phenotypic ratio of dextral to sinistral snails should be produced according to a Mendelian pattern of inheritance. Instead of this predicted phenotypic ratio, however, the F_2 generation was composed of all dextral snails. This incongruity with Mendelian inheritance is due to the maternal effect. The phenotype of the offspring depended solely on the genotype of the mother. The F_1 mothers were *Dd.* The *D* allele in the mothers is dominant to the *d* allele and caused the offspring to be dextral, even if the offspring's genotype was *dd.* When the members of the F_2 generation were crossed, the F_3 generation exhibited a 3:1 ratio of dextral to sinistral snails. This ratio corresponds to the genotypes of the F_2 females, which were the mothers of the F_3 generation. The ratio of F_2 females was 1 *DD* : 2 *Dd* : 1 *dd.* The *DD* and *Dd* females produced dextral offspring, whereas the *dd* females produced sinistral offspring. This explains the 3:1 ratio of dextral and sinistral offspring in the F_3 generation.

Female Gametes Receive Gene Products from the Mother That Affect Early Developmental Stages of the Embryo

At the molecular and cellular level, the non-Mendelian inheritance pattern of maternal effect genes can be explained by the process of oogenesis in female animals (**Figure 7.2a**). As an animal oocyte (egg) matures, many surrounding maternal cells called nurse cells provide the egg with nutrients and other materials. In Figure 7.2a, a female is heterozygous for the snail-coiling maternal effect gene, with the alleles designated *D* and *d.* Depending on the outcome of meiosis, the haploid egg may receive the *D* allele or the *d* allele, but not both. The surrounding nurse cells, however, produce both *D* and *d* gene products (mRNA and proteins). These gene products are then transported into the egg. As shown here, the egg has received both the *D* allele gene products and the *d* allele gene products. These gene products persist for a significant time after the egg has been fertilized and begins its embryonic development. In this way, the gene products of the nurse cells, which reflect the genotype of the mother, influence the early developmental stages of the embryo.

Now that we have an understanding of the relationship between oogenesis and maternal effect genes, let's reconsider the topic of snail coiling. As shown in **Figure 7.2b**, a female snail that is *DD* will transmit only the *D* gene products to the egg. During the early stages of embryonic development, these gene products will cause the egg cleavage to occur in a way that promotes a right-handed body plan. A heterozygous female will transmit both *D* and *d* gene products. Because the *D* allele is dominant, the maternal effect will also cause a right-handed body plan.

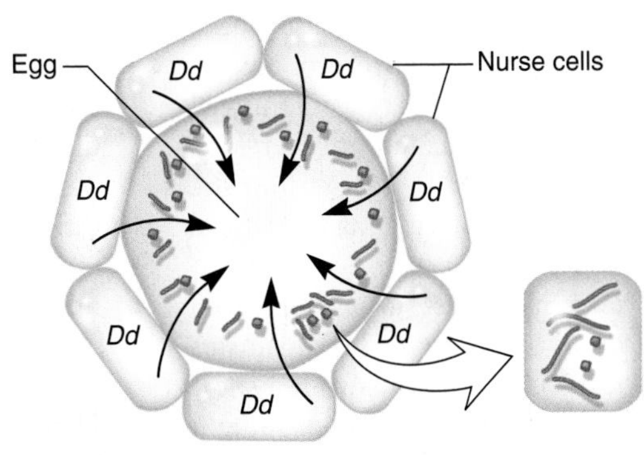

Egg
Nurse cells
Dd
Dd
Dd
Dd
Dd
Dd
Dd
The nurse cells express mRNA and/or protein from genes of the d allele (red) and the D allele (green) and transfer those products to the egg.

(a) Transfer of gene products from nurse cells to egg

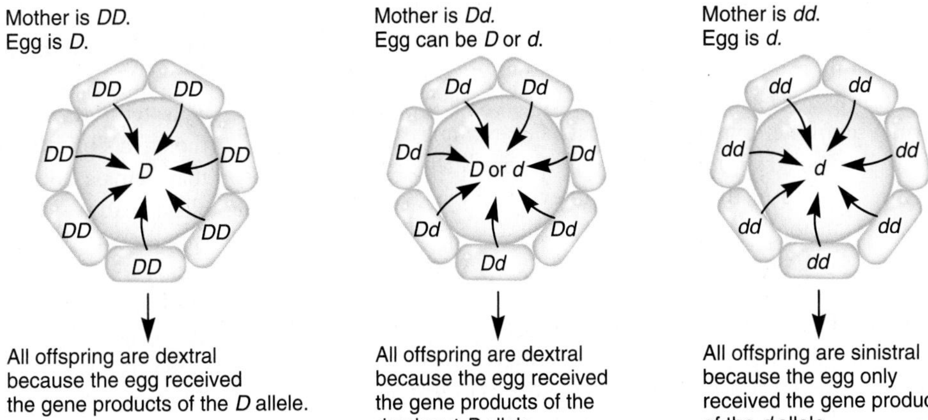

Mother is DD.
Egg is D.
DD
D
All offspring are dextral because the egg received the gene products of the D allele.
Mother is Dd.
Egg can be D or d.
Dd
D or d
All offspring are dextral because the egg received the gene products of the dominant D allele.
Mother is dd.
Egg is d.
dd
d
All offspring are sinistral because the egg only received the gene products of the d allele.

(b) Maternal effect in snail coiling

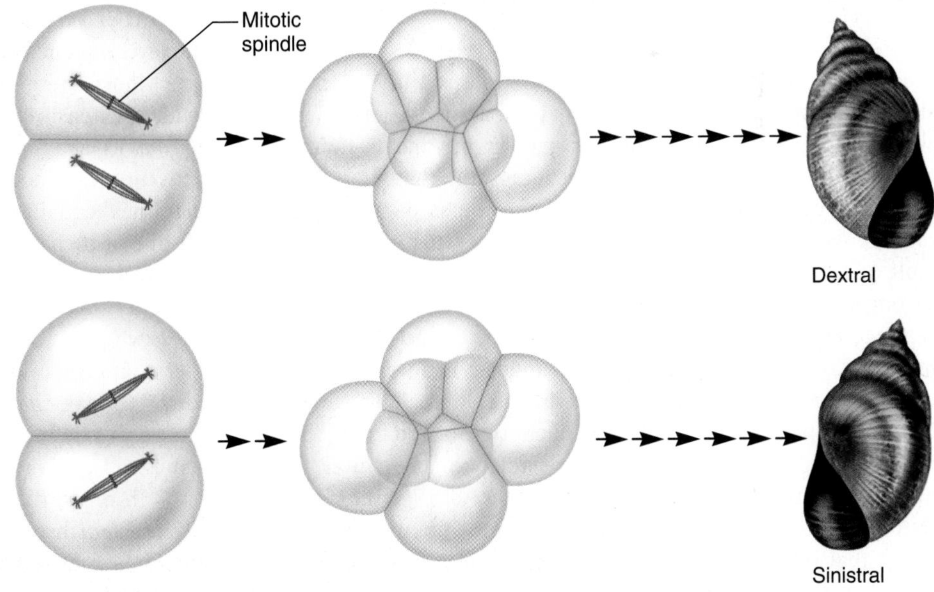

Mitotic spindle
Dextral
Sinistral

(c) An explanation of coiling direction at the cellular level

FIGURE 7.2 The mechanism of maternal effect in snail coiling. (**a**) Transfer of gene products from nurse cells to an egg. The nurse cells are heterozygous (*Dd*). Both the *D* and *d* alleles are activated in the nurse cells to produce *D* and *d* gene products (mRNA and/or proteins). These products are transported into the cytoplasm of the egg, where they accumulate to significant amounts. (**b**) Explanation of the maternal effect in snail coiling. (**c**) The direction of snail coiling is determined by differences in the cleavage planes during early embryonic development.

Genes → Traits If the nurse cells are *DD* or *Dd*, they will transfer the *D* gene product to the egg and thereby cause the resulting offspring to be dextral. If the nurse cells are *dd*, only the *d* gene product will be transferred to the egg, so the resulting offspring will be sinistral.

Finally, a *dd* mother will contribute only *d* gene products that promote a left-handed body plan, even if the egg is fertilized by a sperm carrying a *D* allele. The sperm's genotype is irrelevant, because the expression of the sperm's gene would occur too late. The origin of dextral and sinistral coiling can be traced to the orientation of the mitotic spindle at the two- to four-cell stage of embryonic development. The dextral and sinistral snails develop as mirror images of each other (**Figure 7.2c**).

Since these initial studies, researchers have found that maternal effect genes encode proteins that are important in the early steps of embryogenesis. The accumulation of maternal gene products in the egg allows embryogenesis to proceed quickly after fertilization. Maternal effect genes often play a role in cell division, cleavage pattern, and body axis orientation. Therefore, defective alleles in maternal effect genes tend to have a dramatic effect on the phenotype of the individual, altering major features of morphology, often with dire consequences.

Our understanding of maternal effect genes has been greatly aided by their identification in experimental organisms such as *Drosophila melanogaster.* In such organisms with a short generation time, geneticists have successfully searched for mutant alleles that prevent the normal process of embryonic development. In *Drosophila,* geneticists have identified several maternal effect genes with profound effects on the early stages of development. The pattern of development of a *Drosophila* embryo occurs along axes, such as the anteroposterior axis and the dorsoventral axis. The proper development of each axis requires a distinct set of maternal gene products. For example, the maternal effect gene called *bicoid* produces a gene product that accumulates in a region of the egg that will eventually become anterior structures in the developing embryo. Mutant alleles of maternal effect genes often lead to abnormalities in the anteroposterior or the dorsoventral pattern of development. More recently, several maternal effect genes have been identified in mice and humans that are required for proper embryonic development. Chapter 23 examines the relationships among the actions of several maternal effect genes during embryonic development.

7.2 EPIGENETIC INHERITANCE

As we have just seen, events during oogenesis can cause the inheritance pattern of traits to deviate from a Mendelian pattern. Likewise, **epigenetic inheritance** is a pattern in which a modification occurs to a nuclear gene or chromosome that alters gene expression, but is not permanent over the course of many generations. As we will see, epigenetic inheritance patterns are the result of DNA and chromosomal modifications that occur during oogenesis, spermatogenesis, or early stages of embryogenesis. Once they are initiated during these early stages, epigenetic changes alter the expression of particular genes in a way that may be fixed during an individual's lifetime. Therefore, epigenetic changes can permanently affect the phenotype of the individual. However, epigenetic modifications are not permanent over the course of many generations, and they do not change the actual DNA sequence. For example, a gene may undergo an epigenetic change that inactivates it for the lifetime of an individual. However, when this individual makes gametes, the gene may become activated and remain operative during the lifetime of an offspring who inherits the active gene.

In this section, we will examine two examples of epigenetic inheritance called dosage compensation and genomic imprinting. The effect of dosage compensation is to offset differences in the number of sex chromosomes. One of the sex chromosomes is altered, with the result that males and females have similar levels of gene expression, even though they do not possess the same complement of sex chromosomes. In mammals, dosage compensation is initiated during the early stages of embryonic development. By comparison, genomic imprinting happens prior to fertilization; it involves a change in a single gene or chromosome during gamete formation. Depending on whether the modification occurs during spermatogenesis or oogenesis, imprinting governs whether an offspring expresses a gene that has been inherited from its mother or father.

Dosage Compensation Is Necessary to Ensure Genetic Equality Between the Sexes

Dosage compensation refers to the phenomenon that the level of expression of many genes on the sex chromosomes (such as the X chromosome) is similar in both sexes even though males and females have a different complement of sex chromosomes. This term was coined in 1932 by Hermann Muller to explain the effects of eye color mutations in *Drosophila.* Muller observed that female flies homozygous for certain X-linked eye color alleles had a similar phenotype to hemizygous males. He noted that an X-linked gene conferring an apricot eye color produces a very similar phenotype in homozygous females and hemizygous males. In contrast, a female that has one copy of the apricot allele and a deletion of the apricot gene on the other X chromosome has eyes of paler color. Therefore, one copy of the allele in the female is not equivalent to one copy of the allele in the male. Instead, two copies of the allele in the female produce a phenotype that is similar to that produced by one copy in the male. In other words, the difference in gene dosage—two copies in females versus one copy in males—is being compensated at the level of gene expression.

Since these initial studies, dosage compensation has been studied extensively in mammals, *Drosophila,* and *Caenorhabditis elegans* (a nematode). Depending on the species, dosage compensation occurs via different mechanisms (**Table 7.1**). Female mammals equalize the expression of X-linked genes by turning off one of their two X chromosomes. This process is known as **X inactivation.** In *Drosophila,* the male accomplishes dosage compensation by doubling the expression of most X-linked genes. In *C. elegans,* the XX animal is a hermaphrodite that produces both sperm and egg cells, while an animal carrying a single X chromosome is a male that produces only sperm. The XX hermaphrodite diminishes the expression of X-linked genes on both X chromosomes to approximately 50% of that in the male.

In birds, the Z chromosome is a large chromosome, usually the fourth or fifth largest, which contains almost all of the known sex-linked genes. The W chromosome is generally a much smaller microchromosome containing a high proportion of repeat sequence DNA that does not encode genes. Males are ZZ, and females are ZW. Several years ago, researchers studied the level of expression of a Z-linked gene that encodes an enzyme called aconitase. They discovered that males express twice as much aconitase as females do. These results suggested that dosage compensation does not occur in birds. However, more recently the expression of nine Z-linked genes was examined in chickens, and at least six of the genes showed expression levels that were

TABLE 7.1

Mechanisms of Dosage Compensation Among Different Species

Species	Sex Chromosomes in: Females	Males	Mechanism of Compensation
Placental mammals	XX	XY	One of the X chromosomes in the somatic cells of females is inactivated. In certain species, the paternal X chromosome is inactivated, while in other species, such as humans, either the maternal or paternal X chromosome is randomly inactivated throughout the somatic cells of females.
Marsupial mammals	XX	XY	The paternally derived X chromosome is inactivated in the somatic cells of females.
Drosophila melanogaster	XX	XY	The level of expression of genes on the X chromosome in males is increased 2-fold.
Caenorhabditis elegans	XX*	X0	The level of expression of genes on both X chromosomes in hermaphrodites is decreased to 50% levels compared to males.

*In *C. elegans*, an XX individual is a hermaphrodite, not a female.

similar in males and females. The interpretation of these recent results is that dosage compensation occurs in birds, but perhaps not for every gene. The molecular mechanism of dosage compensation is not well understood in birds. Recent evidence by Heather McQueen and colleagues suggests that dosage compensation in birds may be achieved by the modification of histone proteins along the single Z chromosome in the female. Such a modification could provide dosage compensation by increasing the expression of Z-linked genes twofold.

Dosage Compensation Occurs in Female Mammals by the Inactivation of One X Chromosome

In 1961, Mary Lyon proposed that dosage compensation in mammals occurs by the inactivation of a single X chromosome in females. Liane Russell also proposed the same theory around the same time. This proposal brought together two lines of study. The first type of evidence came from cytological studies. In 1949, Murray Barr and Ewart Bertram identified a highly condensed structure in the interphase nuclei of somatic cells in female cats that was not found in male cats. This structure became known as the **Barr body** (Figure 7.3a). In 1960, Susumu Ohno correctly proposed that the Barr body is a highly condensed X chromosome.

In addition to this cytological evidence, Lyon was also familiar with mammalian examples in which the coat color had a variegated pattern. Figure 7.3b illustrates a calico cat, which is a female that is heterozygous for an X-linked gene that can occur as an orange or a black allele. (The white underside is due to a dominant allele in a different gene.) The orange and black patches are randomly distributed in different female individuals. The calico pattern does not occur in male cats, but similar kinds of mosaic patterns have been identified in the female mouse. Lyon suggested that both the Barr body and the calico pattern are the result of X inactivation in the cells of female mammals.

The mechanism of X inactivation, also known as the **Lyon hypothesis,** is schematically illustrated in Figure 7.4. This example involves a white and black variegated coat color found in certain strains of mice. As shown here, a female mouse has inherited an X chromosome from its mother that carries an allele conferring white coat color (X^b). The X chromosome from its father carries a black coat color allele (X^B). How can X inactivation explain a variegated coat pattern? Initially, both X chromosomes are active. However, at an early stage of embryonic development, one of the two X chromosomes is randomly inactivated in each somatic cell and becomes a Barr body. For example, one embryonic cell may have the X^B chromosome inactivated. As

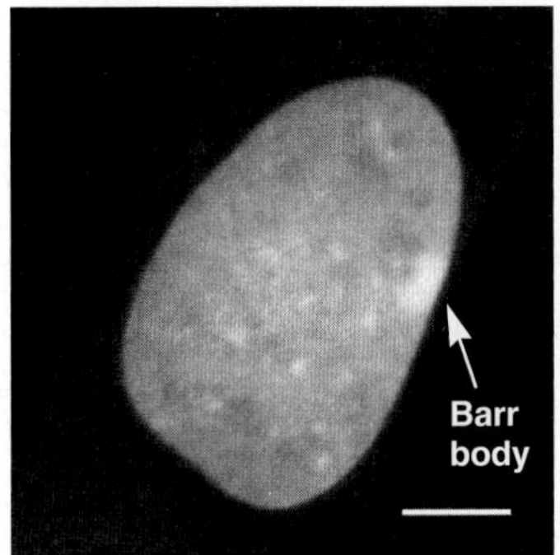

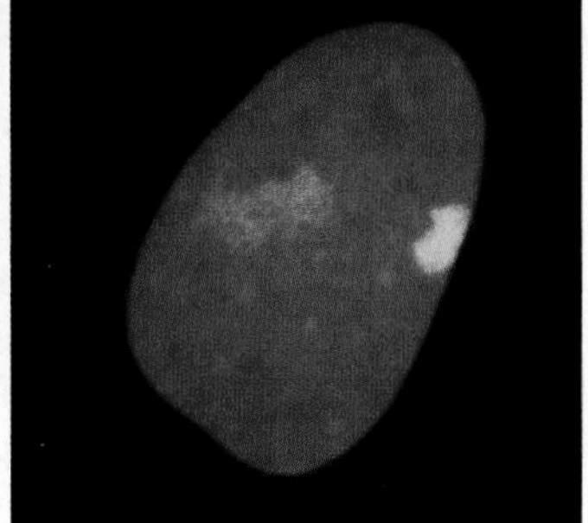

(a) Nucleus with a Barr body

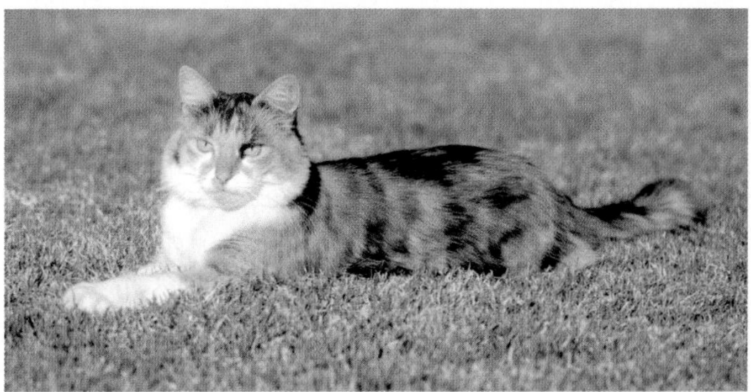

(b) A calico cat

INTERACTIVE EXERCISE

FIGURE 7.3 **X chromosome inactivation in female mammals.** (a) The left micrograph shows the Barr body on the periphery of a human nucleus after staining with a DNA-specific dye. Because it is compact, the Barr body is the most brightly staining. The white scale bar is 5 μm. The right micrograph shows the same nucleus using a yellow fluorescent probe that recognizes the X chromosome. The Barr body (inactive X) is more compact compared to the active X chromosome, which is to the left of the Barr body. **(b)** The fur pattern of a calico cat.

Genes → Traits The pattern of black and orange fur on this cat is due to random X inactivation during embryonic development. The orange patches of fur are due to the inactivation of the X chromosome that carries a black allele; the black patches are due to the inactivation of the X chromosome that carries the orange allele. In general, only heterozygous female cats can be calico. A rare exception would be a male cat (XXY) that has an abnormal composition of sex chromosomes.

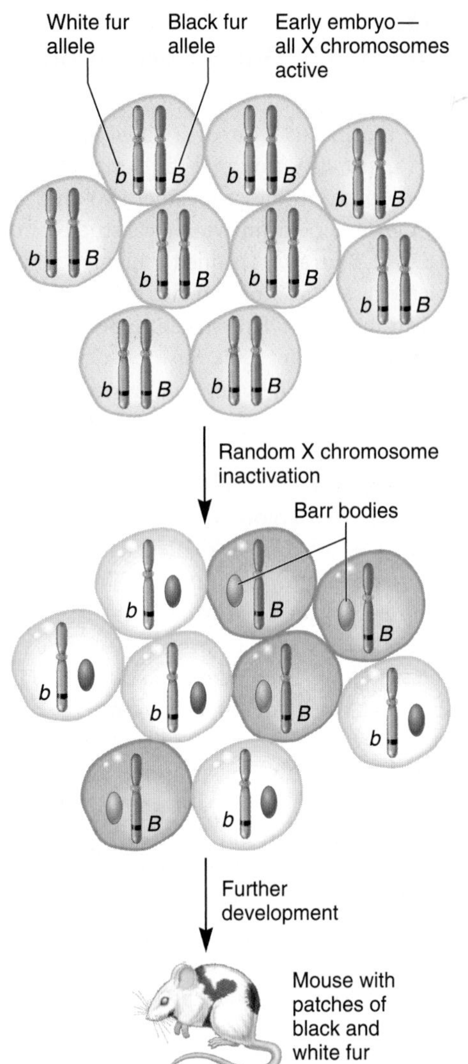

the embryo continues to grow and mature, this embryonic cell will divide and may eventually give rise to billions of cells in the adult animal. The epithelial (skin) cells that are derived from this embryonic cell will produce a patch of white fur because the X^B chromosome has been permanently inactivated. Alternatively, another embryonic cell may have the other X chromosome inactivated (i.e., X^b). The epithelial cells derived from this embryonic cell will produce a patch of black fur. Because the primary event of X inactivation is a random process that occurs at an early stage of development, the result is an animal with some patches of white fur and other patches of black fur. This is the basis of the variegated phenotype.

During inactivation, the chromosomal DNA becomes highly compacted in a Barr body, so most of the genes on the inactivated X chromosome cannot be expressed. When cell division occurs and the inactivated X chromosome is replicated, both copies remain highly compacted and inactive. Likewise, during subsequent cell divisions, X inactivation is passed along to all future somatic cells.

FIGURE 7.4 The mechanism of X chromosome inactivation.

Genes → Traits The top of this figure represents a mass of several cells that compose the early embryo. Initially, both X chromosomes are active. At an early stage of embryonic development, random inactivation of one X chromosome occurs in each cell. This inactivation pattern is maintained as the embryo matures into an adult.

EXPERIMENT 7A

In Adult Female Mammals, One X Chromosome Has Been Permanently Inactivated

According to the Lyon hypothesis, each somatic cell of female mammals will express the genes on one of the X chromosomes, but not both. If an adult female is heterozygous for an X-linked gene, only one of two alleles will be expressed in any given cell. In 1963, Ronald Davidson, Harold Nitowsky, and Barton Childs set out to test the Lyon hypothesis at the cellular level. To do so, they analyzed the expression of a human X-linked gene that encodes an enzyme involved with sugar metabolism known as glucose-6-phosphate dehydrogenase (G-6-PD).

Prior to the Lyon hypothesis, biochemists had found that individuals vary with regard to the G-6-PD enzyme. This variation can be detected when the enzyme is subjected to gel electrophoresis (see the Appendix for a description of gel electrophoresis). One *G-6-PD* allele encodes a G-6-PD enzyme that migrates very quickly during gel electrophoresis (the "fast" enzyme), whereas another *G-6-PD* allele produces an enzyme that migrates more slowly (the "slow" enzyme). As shown in **Figure 7.5**, a sample of cells from heterozygous adult females produces both types of enzymes, whereas hemizygous males produce either the fast or slow type. The difference in migration between the fast and slow G-6-PD enzymes is due to minor differences in the structures of these enzymes. These minor differences do not significantly affect G-6-PD function, but they do enable geneticists to distinguish the proteins encoded by the two X-linked alleles.

As shown in **Figure 7.6**, Davidson, Nitowsky, and Childs tested the Lyon hypothesis using cell culturing techniques. They removed small samples of epithelial cells from a heterozygous female and grew them in the laboratory. When combined together, these samples contained a mixture of both types of

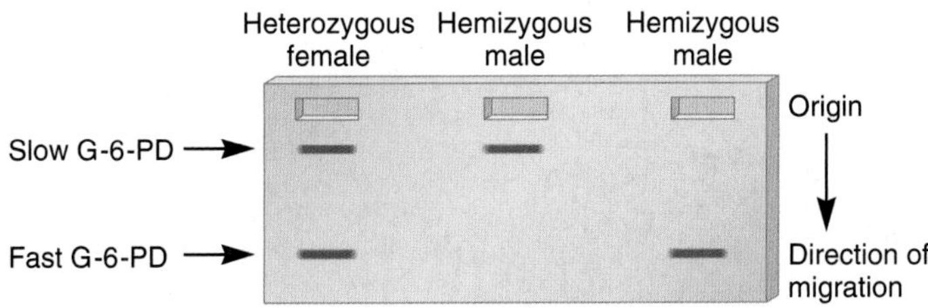

FIGURE 7.5 Mobility of G-6-PD protein on gels. *G-6-PD* can exist as a fast allele that encodes a protein that migrates more quickly to the bottom of the gel and a slow allele that migrates more slowly. The protein encoded by the fast allele is seen more toward the bottom of the gel.

enzymes because the adult cells were derived from many different embryonic cells, some that had the slow allele inactivated and some that had the fast allele inactivated. In the experiment of Figure 7.6, these cells were sparsely plated onto solid growth media. After several days, each cell grew and divided to produce a colony, also called a **clone** of cells. All cells within a colony were derived from a single cell. The researchers reasoned that all cells within a single clone would express only one of the two *G-6-PD* alleles if the Lyon hypothesis was correct. Nine colonies were grown in liquid cultures, and then the cells were lysed to release the G-6-PD proteins inside of them. The proteins were then subjected to SDS gel electrophoresis.

THE HYPOTHESIS

According to the Lyon hypothesis, an adult female who is heterozygous for the fast and slow *G-6-PD* alleles should express only one of the two alleles in any particular somatic cell and its descendants, but not both.

TESTING THE HYPOTHESIS – FIGURE 7.6 Evidence that adult female mammals contain one X chromosome that has been permanently inactivated.

Starting material: Small skin samples taken from a woman who was heterozygous for the fast and slow alleles of *G-6-PD*.

1. Mince the tissue to separate the individual cells.

2. Grow the cells in a liquid growth medium and then plate (sparsely) onto solid growth medium. The cells then divide to form a clone of many cells.

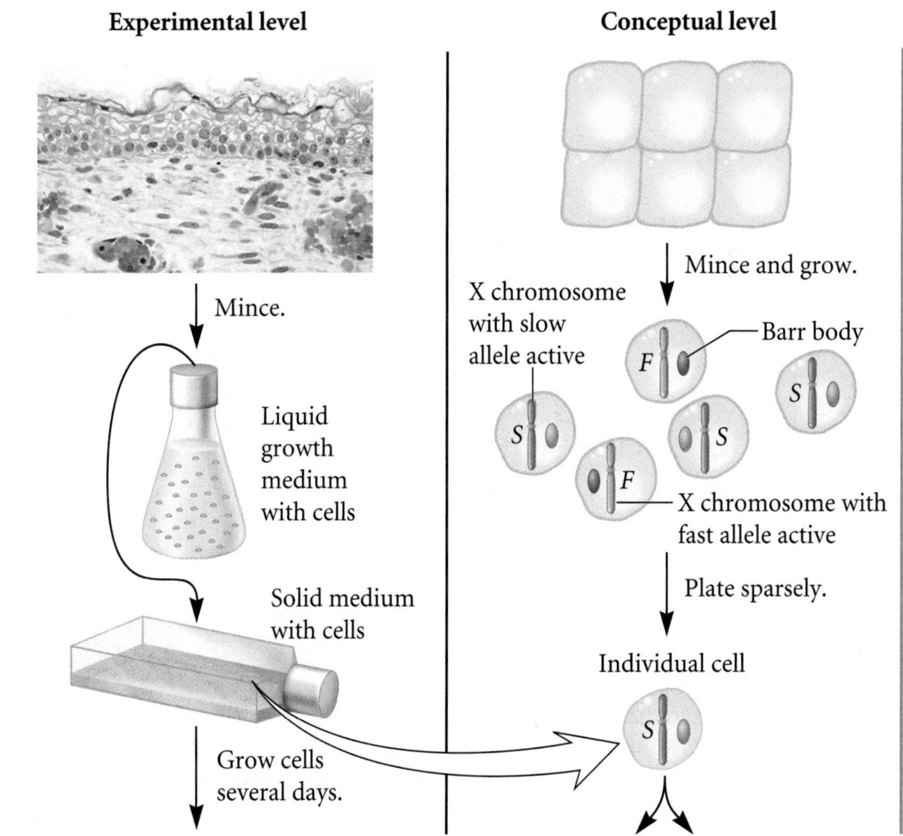

(continued)

3. Take nine isolated clones and grow in liquid cultures. (Only three are shown here.)

4. Take cells from the liquid cultures, lyse cells to obtain proteins, and subject to gel electrophoresis. (This technique is described in the Appendix.)

   Note: As a control, lyse cells from step 1, and subject the proteins to gel electrophoresis. This control sample is not from a clone. It is a mixture of cells derived from a woman's skin sample.

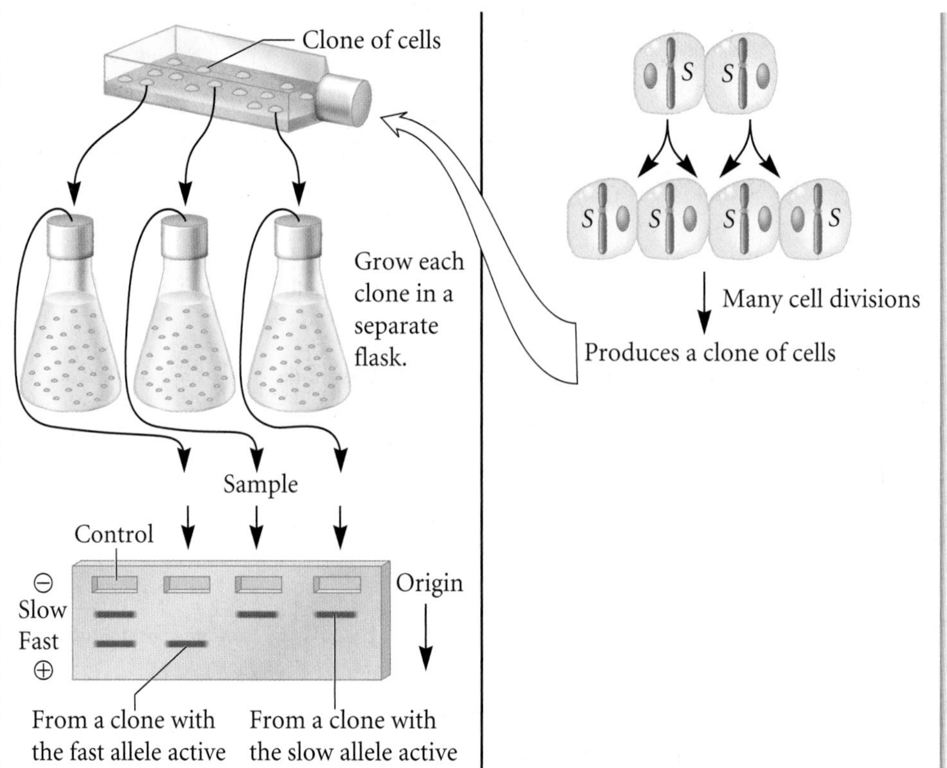

THE DATA

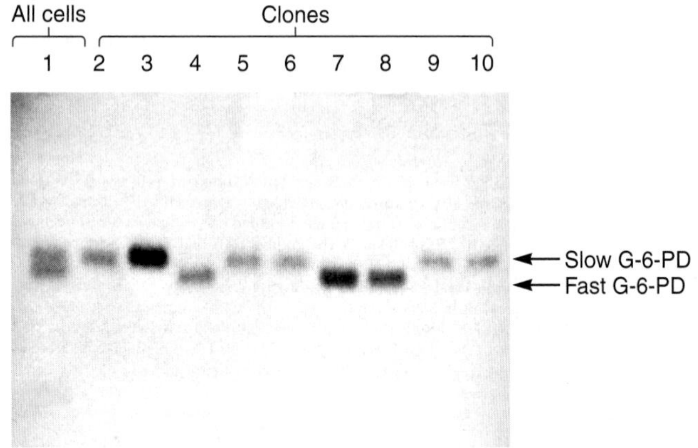

INTERPRETING THE DATA

In the data shown in Figure 7.6, the control (lane 1) was a protein sample obtained from a mixture of epithelial cells from a heterozygous woman who produced both types of G-6-PD enzymes. Bands corresponding to the fast and slow enzymes were observed in this lane. As described in steps 2 to 4, this mixture of epithelial cells was also used to generate nine clones. The proteins obtained from these clones are shown in lanes 2 to 10. Each clone was a population of cells independently derived from a single epithelial cell. Because the epithelial cells were obtained from an adult female, the Lyon hypothesis predicts that each epithelial cell would already have one of its X chromosomes permanently inactivated and would pass this trait to its progeny cells. For example, suppose that an epithelial cell had inactivated the X chromosome that encoded the fast G-6-PD. If this cell was allowed to form a clone of cells on a plate, all cells in this clonal population would be expected to have the same X chromosome inactivated—the X chromosome encoding the fast G-6-PD. Therefore, this clone of cells should express only the slow G-6-PD. As shown in the data, all nine clones expressed either the fast or slow G-6-PD protein, but not both. These results are consistent with the hypothesis that X inactivation has already occurred in any given epithelial cell and that this pattern of inactivation is passed to all of its progeny cells.

A self-help quiz involving this experiment can be found at the Online Learning Center.

X Inactivation in Mammals Depends on the X-Inactivation Center and the *Xist* Gene

Since the Lyon hypothesis was confirmed, the genetic control of X inactivation has been investigated further by several laboratories. Research has shown that mammalian cells possess the ability to count their X chromosomes and allow only one of them to remain active. How was this determined? A key observation came from comparisons of the chromosome composition of people who have been born with normal or abnormal numbers of sex chromosomes.

Phenotype	Chromosome Composition	Number of X Chromosomes	Number of Barr Bodies
Normal female	XX	2	1
Normal male	XY	1	0
Turner syndrome (female)	X0	1	0
Triple X syndrome (female)	XXX	3	2
Klinefelter syndrome (male)	XXY	2	1

In normal females, two X chromosomes are counted and one is inactivated, while in males, one X chromosome is counted and none inactivated. If the number of X chromosomes exceeds two, as in triple X syndrome, additional X chromosomes are converted to Barr bodies.

Although the genetic control of inactivation is not entirely understood at the molecular level, a short region on the X chromosome called the **X-inactivation center (Xic)** is known to play a critical role (**Figure 7.7**). Eeva Therman and Klaus Patau identified the Xic from its key role in X inactivation. The counting of human X chromosomes is accomplished by counting the number of Xics. A Xic must be found on an X chromosome for inactivation to occur. Therman and Patau discovered that if one of the two X chromosomes in a female is missing its Xic due to a chromosome mutation, a cell counts only one Xic and X inactivation does not occur. Having two active X chromosomes is a lethal condition for a human female embryo.

Let's consider how the molecular expression of certain genes controls X inactivation. The expression of a specific gene within the Xic is required for the compaction of the X chromosome into a Barr body. This gene, discovered in 1991, is named *Xist* (for X-inactive specific transcript). The *Xist* gene on the inactivated X chromosome is active, which is unusual because most other genes on the inactivated X chromosome are silenced. The *Xist* gene product is an RNA molecule that does not encode a protein. Instead, the role of the *Xist* RNA is to coat the X chromosome and inactivate it. After coating, other proteins associate with the *Xist* RNA and promote chromosomal compaction into a Barr body.

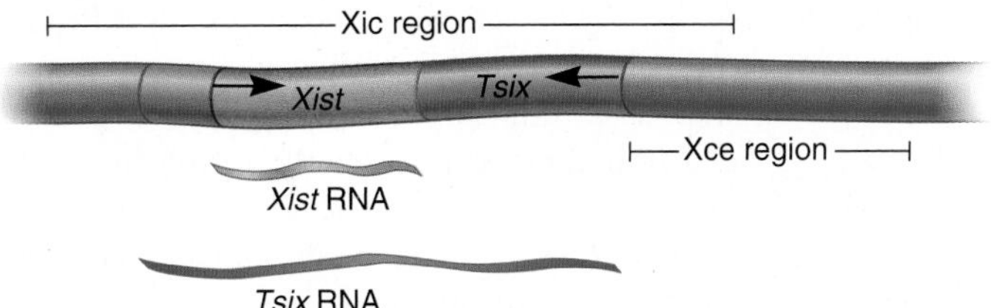

FIGURE 7.7 The X-inactivation center (Xic) of the X chromosome. The *Xist* gene is transcribed into RNA from the inactive X chromosome but not from the active X chromosome. This *Xist* RNA binds to the inactive X chromosome and recruits proteins that promote its compaction. The precise location of the Xce is not known, but it may be within the Xic or adjacent to it. Xce may regulate the transcription of the *Xist* and/or *Tsix* genes and thereby influence the choice of the X chromosome that remains active.

Another gene found within the Xic, designated *Tsix*, also plays a role in X inactivation. As shown in Figure 7.7, the *Xist* and *Tsix* genes are overlapping and transcribed in opposite directions. (The name *Tsix* is *Xist* spelled backwards). The role of the *Tsix* gene is not completely understood, but it appears to decrease the expression of the *Xist* gene. On the inactive X chromosome, the *Xist* gene is expressed, and the *Tsix* gene is not. The opposite situation occurs on the active X chromosome—*Tsix* is expressed, and *Xist* is not. The choice of which X chromosome to inactivate involves a complex interplay between the expression of the *Xist* and *Tsix* genes. Researchers have studied heterozygous females that carry a normal *Tsix* gene on one X chromosome and a defective, mutant *Tsix* gene on the other. The X chromosome carrying the mutant *Tsix* gene is preferentially inactivated.

Another region termed the **X chromosomal controlling element (Xce)** also affects the choice of the X chromosome to be inactivated. Genetic variation occurs in the Xce. An X chromosome that carries a strong Xce is more likely to remain active than an X chromosome that carries a weak Xce, thereby leading to skewed (nonrandom) X inactivation. As shown in Figure 7.7, the Xce is very close to the end of the Xic and may even lie within the Xic. Although the mechanism by which the Xce exerts its effects are not well understood, some researchers speculate that Xce serves as a binding site for proteins that regulate the expression of genes in the Xic, such as *Xist* or *Tsix*. Genetic variation in Xce that enhances *Xist* expression would tend to promote Barr body formation, whereas Xce variation that enhances *Tsix* expression would tend to prevent X inactivation.

The process of X inactivation can be divided into three phases: initiation, spreading, and maintenance (**Figure 7.8**). During initiation, one of the X chromosomes is targeted to remain active, and the other is chosen to be inactivated. During the spreading phase, the chosen X chromosome is inactivated. This spreading requires the expression of the *Xist* gene. The *Xist* RNA coats the inactivated X chromosome and recruits proteins that promote compaction. This compaction occurs via the modification of histone proteins, which are described in Chapter 10. The spreading phase is so named because inactivation begins near the X-inactivation center and spreads in both directions along the X chromosome. Both initiation and spreading occur during embryonic development. Subsequently, the inactivated X chromosome is maintained as such during future cell divisions. When a cell divides, the Barr body is replicated, and both copies remain compacted.

Some genes on the inactivated X chromosome are expressed in the somatic cells of adult female mammals. These genes are said to escape the effects of X inactivation. As mentioned, *Xist* is an example of a gene that is expressed from the highly condensed Barr body. In humans, up to a quarter of X-linked genes may escape inactivation to some degree. Many of these genes occur in clusters. Among these are the pseudoautosomal genes found on the X and Y chromosomes in the regions of homology described in Chapter 4. Dosage compensation is not necessary for X-linked pseudoautosomal genes because they are located on both the X

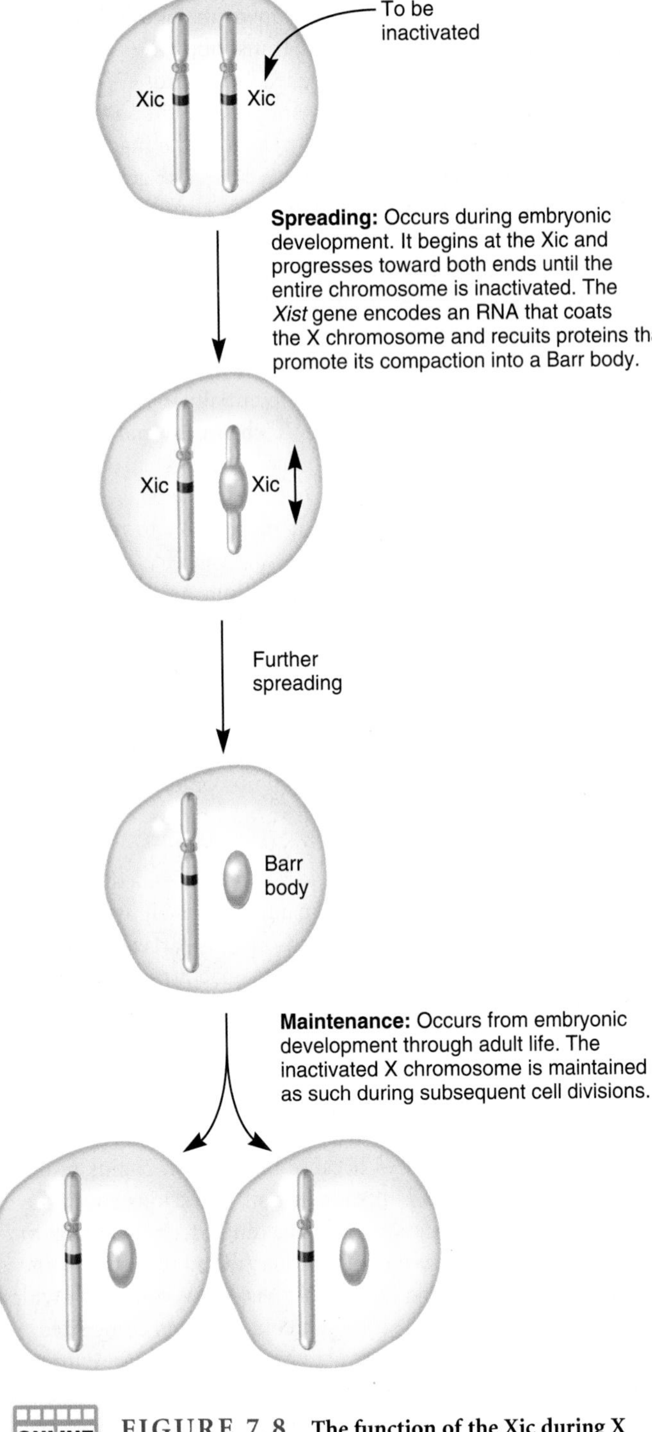

Initiation: Occurs during embryonic development. The number of X-inactivation centers (Xics) are counted and one of the X chromosomes remains active and the other is targeted for inactivation.
To be inactivated
Xic
Xic
Spreading: Occurs during embryonic development. It begins at the Xic and progresses toward both ends until the entire chromosome is inactivated. The Xist gene encodes an RNA that coats the X chromosome and recuits proteins that promote its compaction into a Barr body.
Xic
Xic
Further spreading
Barr body
Maintenance: Occurs from embryonic development through adult life. The inactivated X chromosome is maintained as such during subsequent cell divisions.

ONLINE ANIMATION **FIGURE 7.8 The function of the Xic during X chromosome inactivation.**

and Y chromosomes. How are genes on the Barr body expressed? While the mechanism is not understood, perhaps these genes are found in localized regions where the chromatin is less tightly packed and able to be transcribed.

The Expression of an Imprinted Gene Depends on the Sex of the Parent from Which the Gene Was Inherited

As we have just seen, dosage compensation changes the level of expression of many genes located on the X chromosome. We now turn to another epigenetic phenomenon known as imprinting. The term imprinting implies a type of marking process that has a memory. For example, newly hatched birds identify marks on their parents, which allows them to distinguish their parents from other individuals. The term **genomic imprinting** refers to an analogous situation in which a segment of DNA is marked, and that mark is retained and recognized throughout the life of the organism inheriting the marked DNA. The phenotypes caused by imprinted genes follow a non-Mendelian pattern of inheritance because the marking process causes the offspring to distinguish between maternally and paternally inherited alleles. Depending on how the genes are marked, the offspring expresses only one of the two alleles. This phenomenon is termed **monoallelic expression.**

To understand genomic imprinting, let's consider a specific example. In the mouse, a gene designated *Igf2* encodes a protein growth hormone called insulin-like growth factor 2. Imprinting occurs in a way that results in the expression of the paternal *Igf2* allele but not the maternal allele. The paternal allele is transcribed into RNA, while the maternal allele is transcriptionally silent. With regard to phenotype, a functional *Igf2* gene is necessary for normal size. A loss-of-function allele of this gene, designated *Igf2*$^-$, is defective in the synthesis of a functional Igf2 protein. This may cause a mouse to be a dwarf, but the dwarfism depends on whether the mutant allele is inherited from the male or female parent, as shown in **Figure 7.9**. On the left side, an offspring has inherited the *Igf2* allele from its father and the *Igf2*$^-$ allele from its mother. Due to imprinting, only the *Igf2* allele is expressed in the offspring. Therefore, this mouse grows to a normal size. Alternatively, in the reciprocal cross on the right side, an individual has inherited the *Igf2*$^-$ allele from its father and the *Igf2* allele from its mother. In this case, the *Igf2* allele is not expressed. In this mouse, the *Igf2*$^-$ allele would be transcribed into mRNA, but the mutation renders the Igf2 protein defective. Therefore, the offspring on the right has a dwarf phenotype. As shown here, both offspring have the same genotype; they are heterozygous for the *Igf2* alleles (i.e., *Igf2 Igf2*$^-$). They are phenotypically different, however, because only the paternally inherited allele is expressed.

At the cellular level, imprinting is an epigenetic process that can be divided into three stages: (1) the establishment of the imprint during gametogenesis, (2) the maintenance of the imprint during embryogenesis and in adult somatic cells, and (3) the erasure and reestablishment of the imprint in the germ cells. These stages are described in **Figure 7.10**, which shows the imprinting of the *Igf2* gene. The two mice shown here have inherited the *Igf2* allele from their father and the *Igf2*$^-$ allele from their mother. Due to imprinting, both mice express the *Igf2* allele in their somatic cells, and the pattern of imprinting is maintained in the somatic cells throughout development. In the germ cells (i.e., sperm and eggs), the imprint is erased; it will be reestablished according to the sex of the animal. The female mouse on the left will transmit only transcriptionally inactive

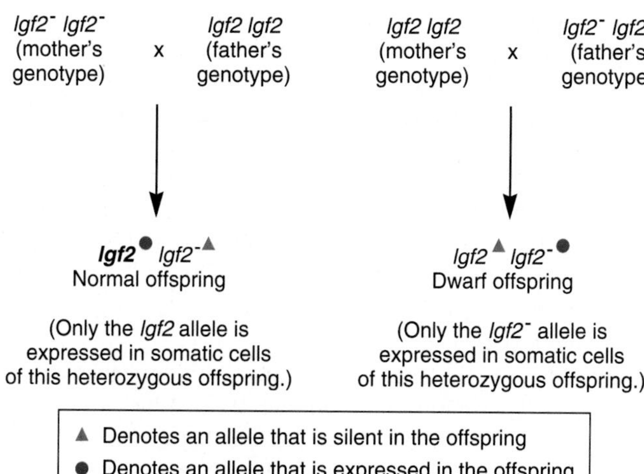
Igf2⁻ Igf2⁻ (mother's genotype)
x
Igf2 Igf2 (father's genotype)
Igf2 Igf2 (mother's genotype)
x
Igf2⁻ Igf2⁻ (father's genotype)
Igf2 Igf2⁻
Normal offspring
Igf2 Igf2⁻
Dwarf offspring
(Only the Igf2 allele is expressed in somatic cells of this heterozygous offspring.)
(Only the Igf2⁻ allele is expressed in somatic cells of this heterozygous offspring.)
▲ Denotes an allele that is silent in the offspring
● Denotes an allele that is expressed in the offspring

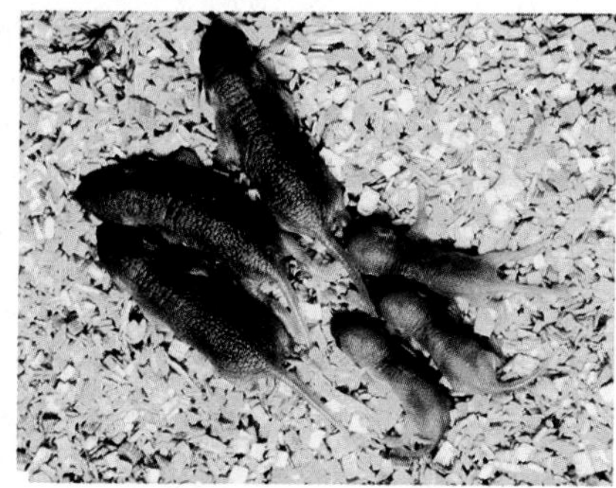

INTERACTIVE EXERCISE

FIGURE 7.9 An example of genomic imprinting in the mouse. In the cross on the left, a homozygous male with the normal *Igf2* allele is crossed to a homozygous female carrying a defective allele, designated *Igf2*⁻. An offspring is heterozygous and normal because the paternal allele is active. In the reciprocal cross on the right, a homozygous male carrying the defective allele is crossed to a homozygous normal female. In this case, the offspring is heterozygous and dwarf. This is because the paternal allele is defective due to mutation and the maternal allele is not expressed. The photograph shows normal-size (left) and dwarf littermates (right) derived from a cross between a wild-type female and a heterozygous male carrying a loss-of-function *Igf2* allele (courtesy of A. Efstratiadis). The loss-of-function allele was created using gene knockout methods described in Chapter 19 (see Figure 19.7).

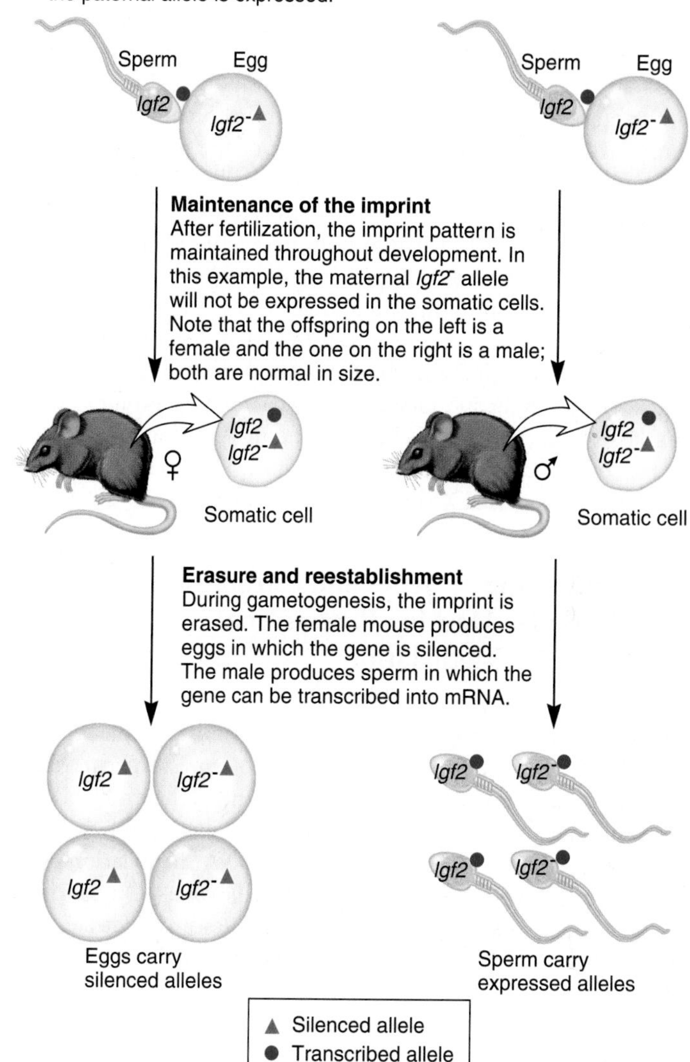
Establishment of the imprint
In this example, imprinting occurs during gametogenesis in the Igf2 gene, which exists in the Igf2 allele from the male and the Igf2⁻ allele from the female. This imprinting occurs so that only the paternal allele is expressed.
Sperm
Egg
Igf2
Igf2⁻
Sperm
Egg
Igf2
Igf2⁻
Maintenance of the imprint
After fertilization, the imprint pattern is maintained throughout development. In this example, the maternal Igf2⁻ allele will not be expressed in the somatic cells. Note that the offspring on the left is a female and the one on the right is a male; both are normal in size.
Igf2
Igf2⁻
♀
Somatic cell
Igf2
Igf2⁻
♂
Somatic cell
Erasure and reestablishment
During gametogenesis, the imprint is erased. The female mouse produces eggs in which the gene is silenced. The male produces sperm in which the gene can be transcribed into mRNA.
Igf2
Igf2⁻
Igf2
Igf2⁻
Igf2
Igf2⁻
Igf2
Igf2⁻
Eggs carry silenced alleles
Sperm carry expressed alleles
▲ Silenced allele
● Transcribed allele

FIGURE 7.10 Genomic imprinting during gametogenesis. This example involves a mouse gene *Igf2*, which is found in two alleles designated *Igf2* and *Igf2*⁻. The left side shows a female mouse that was produced from a sperm carrying the *Igf2* allele and an egg carrying the *Igf2*⁻ allele. In the somatic cells of this female animal, the *Igf2* allele is active. However, when this female produces eggs, both alleles are transcriptionally inactive when they are transmitted to offspring. The right side of this figure shows a male mouse that was also produced from a sperm carrying the *Igf2* allele and an egg carrying the *Igf2*⁻ allele. In the somatic cells of this male animal, the *Igf2* allele is active. However, the sperm from this male will contain either a functionally active *Igf2* allele or a functionally defective *Igf2*⁻ allele. Note: The *Igf2*⁻ allele transmitted by the sperm can be expressed into mRNA in the somatic cells of offspring, but the expression will not produce a functional Igf2 protein due to the mutation that inhibits the protein's function.

alleles to her offspring. The male mouse on the right will transmit transcriptionally active alleles. However, because this male is a heterozygote, it will transmit either a functionally active *Igf2* allele or a functionally defective mutant allele (*Igf2*⁻). An *Igf2*⁻ allele, which is inherited from a male mouse, can be expressed into mRNA (i.e., it is transcriptionally active), but it will not produce a functional Igf2 protein due to the deleterious mutation that created the *Igf2*⁻ allele; a dwarf phenotype will result.

As seen in Figure 7.10, genomic imprinting is permanent in the somatic cells of an animal, but the marking of alleles can be altered from generation to generation. For example, the female mouse on the left possesses an active copy of the *Igf2* allele, but any allele this female transmits to its offspring will be transcriptionally inactive.

Genomic imprinting occurs in several species, including numerous insects, plants, and mammals. Imprinting may involve a single gene, a part of a chromosome, an entire chromosome, or even all of the chromosomes from one parent. Helen Crouse discovered the first example of imprinting, which involved an entire chromosome in the housefly, *Sciara coprophila*. In this species, the fly normally inherits three sex chromosomes, rather than two as in most other species. One X chromosome is inherited from the female, and two are inherited from the male. In male flies, both paternal X chromosomes are lost from somatic cells during embryogenesis. In female flies, only one of the paternal X chromosomes is lost. In both sexes, the maternally inherited X chromosome is never lost. These results indicate that the maternal X chromosome is marked to promote its retention or paternal X chromosomes are marked to promote their loss.

Genomic imprinting can also be involved in the process of X inactivation, described previously. In certain species, imprinting plays a role in the choice of the X chromosome that will be inactivated. For example, in marsupials, the paternal X chromosome is marked so that it is the X chromosome that is always inactivated in the somatic cells of females. In marsupials, X inactivation is not random; the maternal X chromosome is always active.

The Imprinting of Genes and Chromosomes Is a Molecular Marking Process That Involves DNA Methylation

As we have seen, genomic imprinting must involve a marking process. A particular gene or chromosome must be marked dissimilarly during spermatogenesis versus oogenesis. After fertilization takes place, this differential marking affects the expression of particular genes. What is the molecular explanation for genomic imprinting? As discussed in Chapter 15, **DNA methylation**—the attachment of methyl groups onto a cytosine base—is a common way that eukaryotic genes may be regulated. Research indicates that genomic imprinting involves a **differentially methylated region (DMR)** that is located near the imprinted gene. Depending on the particular gene, the DMR is methylated in the egg or the sperm, but not both. The DMR contains binding sites for one or more proteins that regulate the transcription of a nearby gene.

For most imprinted genes, methylation at a DMR results in an inhibition of gene expression. Methylation could enhance the binding of proteins that inhibit transcription and/or inhibit the binding of proteins that enhance transcription. (The relationship between methylation and gene expression is described in Chapter 15.) For this reason, imprinting is usually described as a marking process that silences gene expression by preventing transcription. However, this is not always the case. Two imprinted human genes, *H19* and *Igf2*, provide an interesting example. These two genes lie close to each other on human chromosome 11 and appear to be controlled by the same DMR, which is a 2,000-base pair region that lies upstream of the *H19* gene (**Figure 7.11a**). It contains binding sites for proteins that regulate the transcription of the *H19* or *Igf2* genes. This DMR is highly methylated on the paternally inherited chromosome. Methylation of the DMR silences the *H19* gene because a specific protein or group of proteins bind to the DMR only after it has been methylated, and protein binding inhibits the transcription of the *H19* gene.

The same DMR that controls *H19* also controls *Igf2*, but in a different way. The unmethylated DMR inherited from a female binds a protein that causes an alteration in the structure of chromatin over a fairly long distance that includes the *Igf2* gene. This alteration in structure blocks the ability of transcriptional activator proteins to bind to the *Igf2* gene. If these activator proteins are unable to bind to the *Igf2* gene, transcription is silenced. In contrast, the protein that changes chromatin structure does not bind to a methylated DMR. Therefore, the *Igf2* gene that is inherited from a male is recognized by transcriptional activator proteins and transcribed into mRNA.

Now that we have an understanding of how methylation may affect gene transcription, let's consider the methylation process during the life of an animal. A key issue is that certain imprinted genes are differentially methylated during oogenesis compared to spermatogenesis (**Figure 7.11b**). In this example, a female and male offspring have inherited a methylated DMR from their father and a nonmethylated DMR from their mother. If this was the DMR next to the *H19* gene, the somatic cells of the offspring would express only the maternal *H19* allele. This pattern of imprinting is maintained in the somatic cells of both offspring. However, when the female offspring makes gametes, the imprinting is erased during early oogenesis, so the female will pass an unmethylated DMR to its offspring. In the male, the imprinting is also erased during early spermatogenesis, but then *de novo* (new) methylation occurs in both DMRs. In the case of the *H19* gene, the result of DMR methylation is that the male transmits inactive alleles of this gene.

Genomic imprinting is a fairly new and exciting area of research. Imprinting has been identified in many mammalian genes (**Table 7.2**). In some cases, the female alleles are transcriptionally active in the somatic cells of offspring, whereas in other cases, the male alleles are active. The biological significance of imprinting is still a matter of much speculation. Several hypotheses have been proposed to explain the potential benefits of genomic imprinting. One example, described by David Haig, involves differences in female versus male reproductive patterns in

TABLE 7.2

Examples of Mammalian Genes and Inherited Human Diseases That Involve Imprinted Genes*

Gene	Allele Expressed	Function
WT1	Maternal	Wilms tumor–suppressor gene; suppresses cell growth
INS	Paternal	Insulin; hormone involved in cell growth and metabolism
Igf2	Paternal	Insulin-like growth factor II; similar to insulin
Igf2R	Maternal	Receptor for insulin-like growth factor II
H19	Maternal	Unknown
SNRPN	Paternal	Splicing factor
Gabrb	Maternal	Neurotransmitter receptor

*Researchers estimate that approximately 1–2% of human genes are subjected to genomic imprinting, but fewer than 100 have actually been demonstrated to be imprinted.

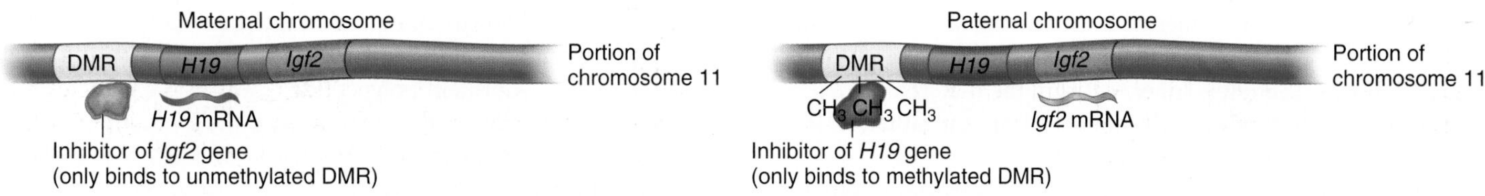

(a) Actions of regulatory proteins that bind to an unmethylated DMR or to a methylated DMR

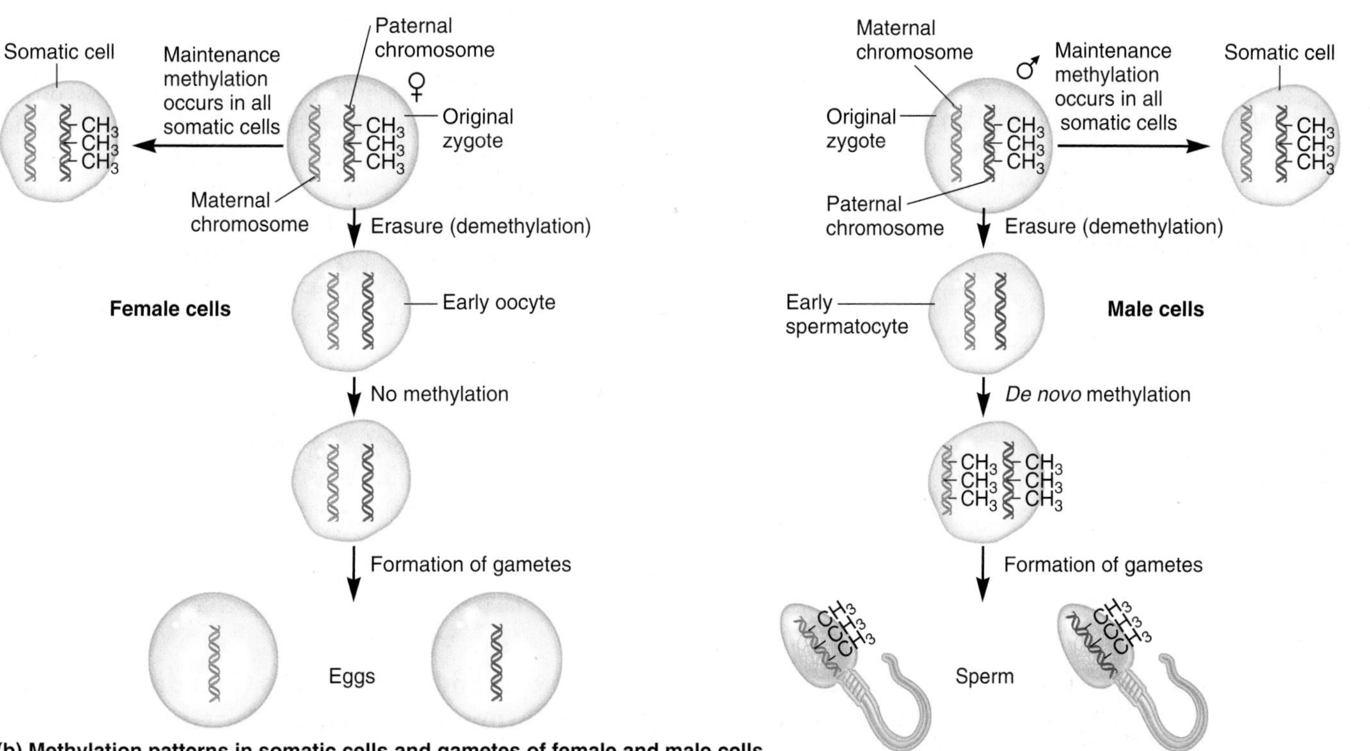

(b) Methylation patterns in somatic cells and gametes of female and male cells

FIGURE 7.11 **DNA methylation in the imprinting process.** **(a)** The effects of methylation on gene transcription. A differentially methylated region (DMR) is located next to two genes called *H19* and *Igf2*. This DMR is methylated on the paternally inherited chromosome. When it is methylated, it binds a protein that inhibits the *H19* gene. Therefore, the paternally inherited *H19* allele is silenced. In contrast, when the DMR is not methylated, it binds a different protein that inhibits the *Igf2* gene. In this way, the maternally inherited *Igf2* allele is silenced. **(b)** The pattern of DMR methylation during the life of an animal. In this example, a male and a female offspring have inherited a methylated DMR and nonmethylated DMR from their father and mother, respectively. Maintenance methylation retains the imprinting in somatic cells during embryogenesis and in adulthood. Demethylation occurs in cells that are destined to become gametes. In this example, *de novo* methylation occurs only in cells that are destined to become sperm. Haploid male gametes transmit a methylated DMR, whereas haploid female gametes transmit an unmethylated DMR.

mammals. As discussed in Chapter 24, natural selection will favor types of genetic variation that confer a survival advantage. The likelihood that favorable variation will be passed to offspring may differ between the sexes. In many mammalian species, females may mate with multiple males, perhaps generating embryos in the same uterus fathered by different males. For males, silencing genes that inhibit embryonic growth would be an advantage. The embryos of males that silence such genes would grow faster than other embryos in the same uterus, making it more likely for the males to pass their genes to future generations. For females, however, rapid growth of embryos might be a disadvantage because it could drain too many resources from the mother. According to this scenario, the mother would silence genes that cause rapid embryonic growth. From the mother's perspective, she would give all of her offspring an equal chance of survival without sapping her own strength. This would make it more likely for the female to pass her genes to future generations. The Haig hypothesis seems to be consistent with the imprinting of several mammalian genes that are involved with growth, such as *Igf2*. Females silence this growth-enhancing gene, whereas males do not. However, several imprinted genes do not seem to play a role in embryonic development. Therefore, an understanding of the biological role(s) of genomic imprinting will require further investigation.

Imprinting plays a role in the inheritance of certain human diseases such as Prader-Willi syndrome (PWS) and Angelman syndrome (AS). PWS is characterized by reduced motor function, obesity, and mental deficiencies. AS patients are thin and hyperactive, have unusual seizures and repetitive symmetrical muscle movements, and exhibit mental deficiencies. Most commonly, both PWS and AS involve a small deletion in human chromosome 15. If this deletion is inherited from the maternal parent, it leads to Angelman syndrome; if inherited from the father, it leads to Prader-Willi syndrome (**Figure 7.12**).

Researchers have discovered that this region contains closely linked but distinct genes that are maternally or paternally imprinted. AS results from the lack of expression of a single gene (*UBE3A*) that codes for a protein called E6-AP, which functions to transfer small ubiquitin molecules to certain proteins to target their degradation. Both copies of this gene are active in many of the body's tissues. In the brain, however, only the copy inherited from a person's mother (the maternal copy) is active. The paternal allele of *UBE3A* is silenced. Therefore, if the maternal allele is deleted, as in the left side of Figure 7.12, the individual will develop AS because he or she will not have an active copy of the *UBE3A* gene.

The gene(s) responsible for PWS has not been definitively determined, although five imprinted genes in this region of chromosome 15 are known. One possible candidate involved in PWS is a gene designated *SNRPN*. The gene product is part of a small nuclear ribonucleoprotein polypeptide N, which is a complex that controls RNA splicing and is necessary for the synthesis of critical proteins in the brain. The maternal allele of *SNRPN* is silenced, and only the paternal copy is active.

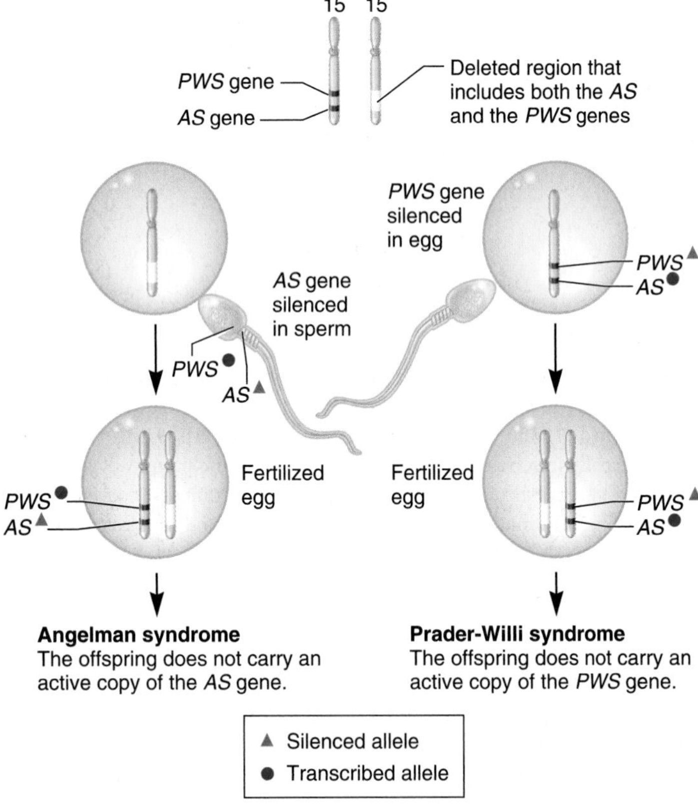

FIGURE 7.12 **The role of imprinting in the development of Angelman and Prader-Willi syndromes.**

Genes → Traits A small region on chromosome 15 contains two different genes designated the *AS* gene and *PWS* gene in this figure. If a chromosome 15 deletion is inherited from the maternal parent, Angelman syndrome occurs because the offspring does not inherit an active copy of the *AS* gene (left). Alternatively, the chromosome 15 deletion may be inherited from the male parent, leading to Prader-Willi syndrome. The phenotype of this syndrome occurs because the offspring does not inherit an active copy of the *PWS* gene (right).

7.3 EXTRANUCLEAR INHERITANCE

Thus far, we have considered several types of non-Mendelian inheritance patterns. These include maternal effect genes, dosage compensation, and genomic imprinting. All of these inheritance patterns involve genes found on chromosomes in the cell nucleus. Another cause of non-Mendelian inheritance patterns concerns genes that are not located in the cell nucleus. In eukaryotic species, the most biologically important example of extranuclear inheritance involves genetic material in cellular organelles. In addition to the cell nucleus, the mitochondria and chloroplasts contain their own genetic material. Because these organelles are found within the cytoplasm of the cells, the inheritance of organellar genetic material is called **extranuclear inheritance** (the prefix *extra-* means outside of) or **cytoplasmic inheritance.** In this section, we will examine the genetic composition of mitochondria and chloroplasts and explore the pattern of transmission of these organelles from parent to offspring. We will also consider a few other examples of inheritance patterns that cannot be explained by the transmission of nuclear genes.

Mitochondria and Chloroplasts Contain Circular Chromosomes with Many Genes

In 1951, Y. Chiba was the first to suggest that chloroplasts contain their own DNA. He based his conclusion on the staining properties of a DNA-specific dye known as Feulgen. Researchers later developed techniques to purify organellar DNA. In addition, electron microscopy studies provided interesting insights into the organization and composition of mitochondrial and chloroplast chromosomes. More recently, the advent of molecular genetic techniques in the 1970s and 1980s has allowed researchers to determine the genome sequences of organellar DNAs. From these types of studies, the chromosomes of mitochondria and chloroplasts were found to resemble smaller versions of bacterial chromosomes.

The genetic material of mitochondria and chloroplasts is located inside the organelle in a region known as the **nucleoid** (**Figure 7.13**). The genome is a single circular chromosome (composed of double-stranded DNA), although a nucleoid contains several copies of this chromosome. In addition, a mitochondrion or chloroplast often has more than one nucleoid. In mice, for example, each mitochondrion has one to three nucleoids, with each nucleoid containing two to six copies of the circular mitochondrial genome. However, this number is variable and depends on the type of cell and the stage of development. In comparison, the chloroplasts of algae and higher plants tend to have more nucleoids per organelle. **Table 7.3** describes the genetic composition of mitochondria and chloroplasts for a few selected species.

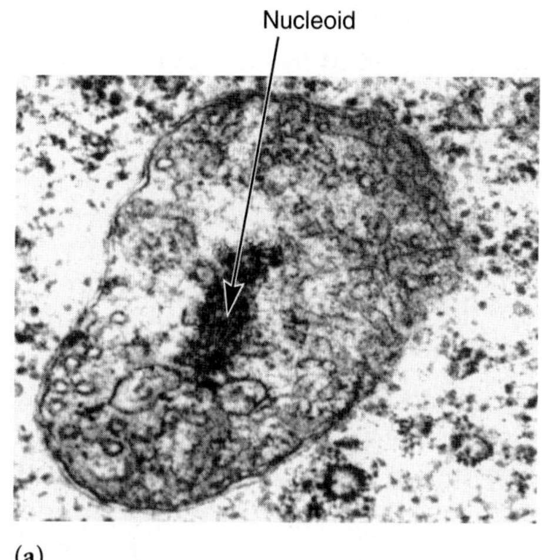

(a)

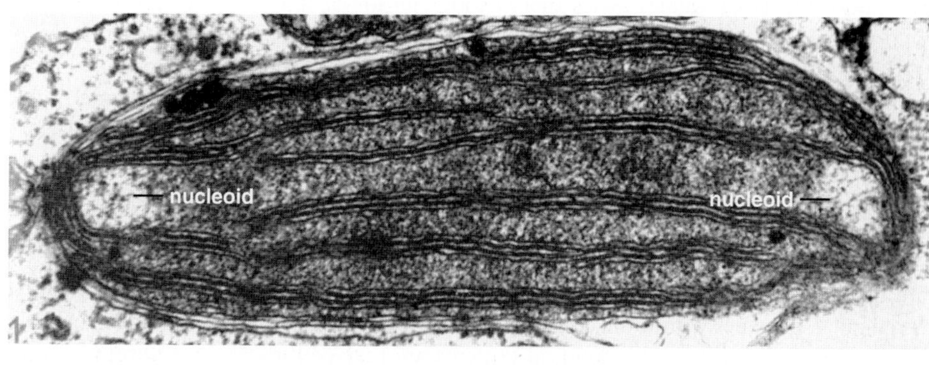

(b)

FIGURE 7.13 Nucleoids within (a) mitochondria and (b) chloroplasts. The mitochondrial and chloroplast chromosomes are found within the nucleoid region of the organelle.

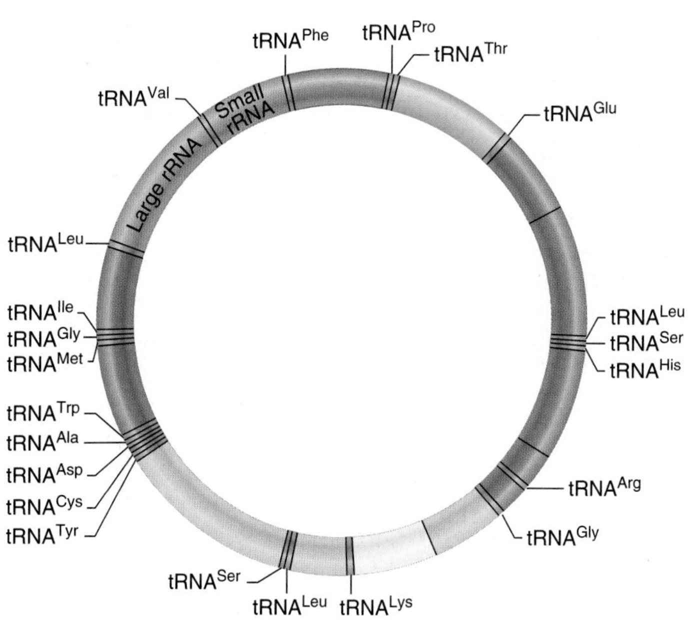

FIGURE 7.14 A genetic map of human mitochondrial DNA (mtDNA). This diagram illustrates the locations of many genes along the circular mitochondrial chromosome. The genes shown in red encode transfer RNAs. For example, tRNA^{Arg} encodes the tRNA that carries arginine. The genes that encode ribosomal RNA are shown in light brown. The remaining genes encode proteins that function within the mitochondrion. The mitochondrial genomes from numerous species have been determined.

TABLE 7.3

Genetic Composition of Mitochondria and Chloroplasts

Species	Organelle	Nucleoids per Organelle	Total Number of Chromosomes per Organelle
Tetrahymena	Mitochondrion	1	6–8
Mouse	Mitochondrion	1–3	5–6
Chlamydomonas	Chloroplast	5–6	~80
Euglena	Chloroplast	20–34	100–300
Higher plants	Chloroplast	12–25	~60

Data from: Gillham, N. W. (1994). *Organelle Genes and Genomes.* Oxford University Press, New York.

Besides variation in copy number, the sizes of mitochondrial and chloroplast genomes also vary greatly among different species. For example, a 400-fold variation is found in the sizes of mitochondrial chromosomes. In general, the mitochondrial genomes of animal species tend to be fairly small; those of fungi and protists are intermediate in size; and those of plant cells tend to be fairly large. Among algae and plants, substantial variation is also found in the sizes of chloroplast chromosomes.

Figure 7.14 illustrates a map of human **mitochondrial DNA (mtDNA).** Each copy of the mitochondrial chromosome consists of a circular DNA molecule that is only 17,000 base pairs (bp) in length. This size is less than 1% of a typical bacterial chromosome. The human mtDNA carries relatively few genes. Thirteen genes encode proteins that function within the mitochondrion. In addition, mtDNA carries genes that encode ribosomal RNA and transfer RNA. These rRNAs and tRNAs are necessary for the synthesis of the 13 proteins that are encoded by the mtDNA. The primary role of mitochondria is to provide cells with the bulk of their adenosine triphosphate (ATP), which is used as an energy source to drive cellular reactions. These 13 proteins function in a process known as oxidative phosphorylation, which enables the mitochondria to synthesize ATP. However, mitochondria require many additional proteins to carry out oxidative phosphorylation and other mitochondrial functions. Most mitochondrial proteins are encoded by genes within the cell nucleus. When these nuclear genes are expressed, the mitochondrial proteins are first synthesized outside the mitochondria in the cytosol of the cell and then transported into the mitochondria.

Chloroplast genomes tend to be larger than mitochondrial genomes, and they have a correspondingly greater number of genes. A typical chloroplast genome is approximately 100,000 to 200,000 bp in length, which is about 10 times larger than the mitochondrial genome of animal cells. **Figure 7.15** shows the **chloroplast DNA (cpDNA)** of the tobacco plant, which is a circular DNA molecule that contains 156,000 bp of DNA and carries between 110 and 120 different genes. These genes encode ribosomal RNAs, transfer RNAs, and many proteins required for photosynthesis. As with mitochondria, many chloroplast proteins are encoded by genes found in the plant cell nucleus. These proteins contain chloroplast-targeting signals that direct them into the chloroplasts.

Extranuclear Inheritance Produces Non-Mendelian Results in Reciprocal Crosses

In diploid eukaryotic species, most genes within the nucleus obey a Mendelian pattern of inheritance because the homologous pairs of chromosomes segregate during gamete formation. Except for sex-linked traits, offspring inherit one copy of each gene from both the maternal and paternal parents. The sorting of chromosomes during meiosis explains the inheritance patterns of nuclear genes. By comparison, the inheritance of extranuclear genetic material does not display a Mendelian pattern. Mitochondria and chloroplasts are not sorted during meiosis and therefore do not segregate into gametes in the same way as nuclear chromosomes.

In 1909, Carl Correns discovered a trait that showed a non-Mendelian pattern of inheritance involving pigmentation

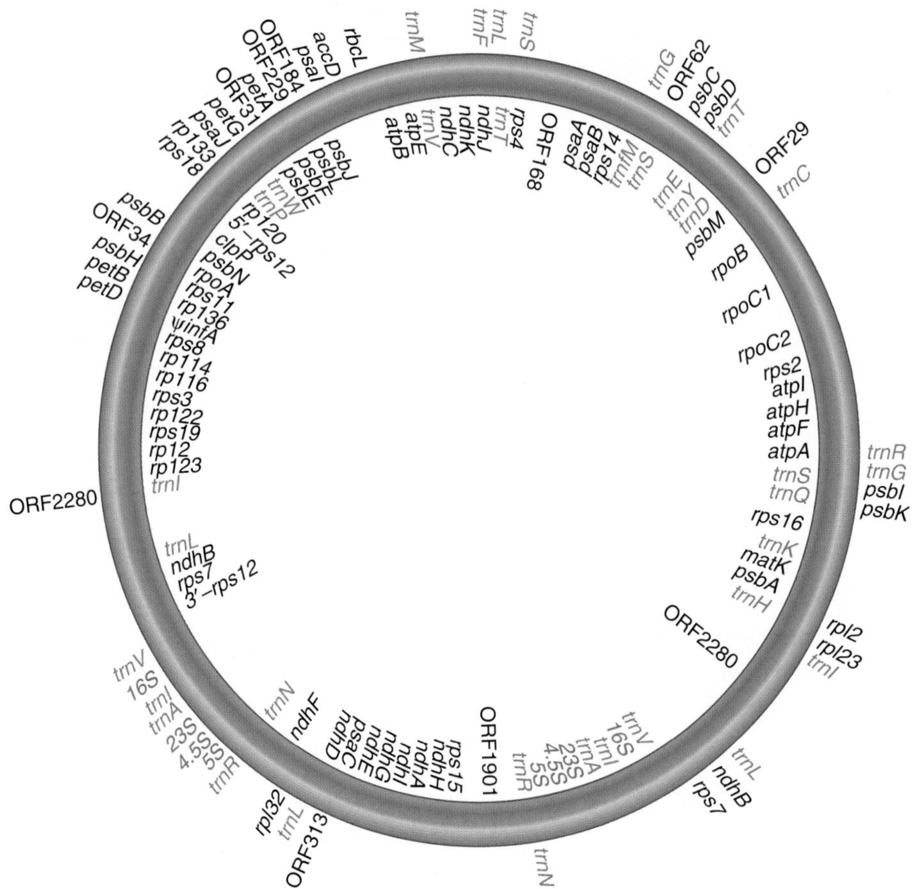

FIGURE 7.15 **A genetic map of tobacco chloroplast DNA (cpDNA).** This diagram illustrates the locations of many genes along the circular chloroplast chromosome. The gene names shown in blue encode transfer RNAs. The genes that encode ribosomal RNA are shown in red. The remaining genes shown in black encode polypeptides that function within the chloroplast. The genes designated ORF (open reading frame) encode polypeptides with unknown functions.

in *Mirabilis jalapa* (the four-o'clock plant). Leaves can be green, white, or variegated with both green and white sectors. Correns demonstrated that the pigmentation of the offspring depended solely on the maternal parent (**Figure 7.16**). If the female parent had white pigmentation, all offspring had white leaves. Similarly, if the female was green, all offspring were green. When the female was variegated, the offspring could be green, white, or variegated.

The pattern of inheritance observed by Correns is a type of extrachromosomal inheritance called **maternal inheritance** (not to be confused with maternal effect). Chloroplasts are a type of plastid that makes chlorophyll, a green photosynthetic pigment. Maternal inheritance occurs because the chloroplasts are inherited only through the cytoplasm of the egg. The pollen grains of *M. jalapa* do not transmit chloroplasts to the offspring.

The phenotypes of leaves can be explained by the types of chloroplasts within the leaf cells. The green phenotype, which is the wild-type condition, is due to the presence of normal chloroplasts that make green pigment. By comparison, the white phenotype is due to a mutation in a gene within the chloroplast DNA that diminishes the synthesis of green pigment. A cell may contain both types of chloroplasts, a condition known as **heteroplasmy.** A leaf cell containing both types of chloroplasts is green because the normal chloroplasts produce green pigment.

How does a variegated phenotype occur? **Figure 7.17** considers the leaf of a plant that began from a fertilized egg that contained both types of chloroplasts (i.e., a heteroplasmic cell). As a plant grows, the two types of chloroplasts are irregularly distributed to daughter cells. On occasion, a cell may receive only the chloroplasts that have a defect in making green pigment. Such a cell continues to divide and produce a sector of the plant that is entirely white. In this way, the variegated phenotype is created. Similarly, if we consider the results of Figure 7.16, a female parent that is variegated may transmit green, white, or a mixture of these types of chloroplasts to the egg cell and thereby produce green, white, or variegated offspring, respectively.

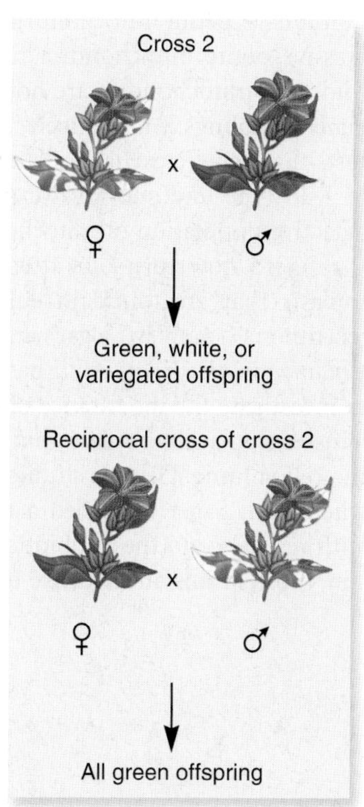

INTERACTIVE EXERCISE

FIGURE 7.16 Maternal inheritance in the four-o'clock plant, *Mirabilis jalapa*. The reciprocal crosses of four-o'clock plants by Carl Correns consisted of a pair of crosses between white-leaved and green-leaved plants, and a pair of crosses between variegated-leaved and green-leaved plants.

Genes → Traits In this example, the white phenotype is due to chloroplasts that carry a mutant allele that diminishes green pigmentation. The variegated phenotype is due to a mixture of chloroplasts, some of which carry the normal (green) allele and some of which carry the white allele. In the crosses shown here, the parent providing the eggs determines the phenotypes of the offspring. This is due to maternal inheritance. The egg contains the chloroplasts that are inherited by the offspring. (Note: The defective chloroplasts that give rise to white sectors are not completely defective in chlorophyll synthesis. Therefore, entirely white plants can survive, though they are smaller than green or variegated plants.)

Studies in Yeast and *Chlamydomonas* Provided Genetic Evidence for Extranuclear Inheritance of Mitochondria and Chloroplasts

The research of Correns and others indicated that some traits, such as leaf pigmentation, are inherited in a non-Mendelian manner. However, such studies did not definitively determine that

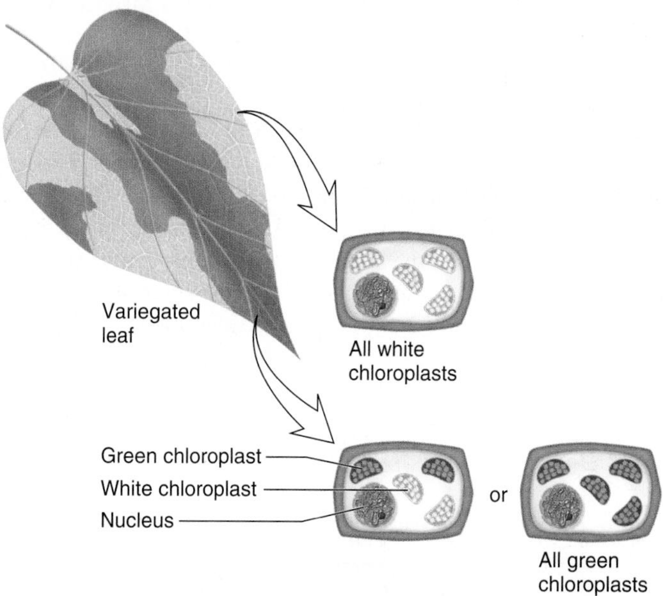

FIGURE 7.17 A cellular explanation of the variegated phenotype in *Mirabilis jalapa*. This plant inherited two types of chloroplasts—those that can produce green pigment and those that are defective. As the plant grows, the two types of chloroplasts are irregularly distributed to daughter cells. On occasion, a cell may receive only the chloroplasts that are defective at making green pigment. Such a cell continues to divide and produces a sector of the plant that is entirely white. Cells that contain both types of chloroplasts or cells that contain only green chloroplasts would produce green tissue, which may be adjacent to a sector of white tissue. This is the basis for the variegated phenotype of the leaves.

maternal inheritance is due to genetic material within organelles. Further progress in the investigation of extranuclear inheritance was provided by detailed genetic analyses of eukaryotic microorganisms, such as yeast and algae, by isolating and characterizing mutant phenotypes that specifically affected the chloroplasts or mitochondria.

During the 1940s and 1950s, yeasts and molds became model eukaryotic organisms for investigating the inheritance of mitochondria. Because mitochondria produce energy for cells in the form of ATP, mutations that yield defective mitochondria are expected to make cells grow much more slowly. Boris Ephrussi and his colleagues identified mutations in *Saccharomyces cerevisiae* that had such a phenotype. These mutants were called **petites** to describe their formation of small colonies on agar plates as opposed to wild-type strains that formed larger colonies. Biochemical and physiological evidence indicated that petite mutants had defective mitochondria. The researchers found that petite mutants could not grow when the cells only had an energy source requiring the metabolic activity of mitochondria, but could form small colonies when grown on sugars metabolized by the glycolytic pathway, which occurs outside the mitochondria.

Because yeast exist in two mating types, designated a and α, Ephrussi was able to mate a wild-type strain to his petite mutants. Genetic analyses showed that petite mutants were inherited in different ways. When a wild-type strain was crossed to a segregational petite mutant, he obtained a ratio of 2 wild-type cells to 2 petite cells (**Figure 7.18a**). This result is consistent with a Mendelian pattern of inheritance (see the discussion of tetrad analysis in Chapter 5). Therefore, segregational petite mutations cause defects in genes located in the cell nucleus. These genes encode proteins necessary for mitochondrial function. Such proteins are synthesized in the cytosol and are then taken up by the mitochondria, where they perform their functions. Segregational petites get their name because they segregate in a Mendelian manner during meiosis.

By comparison, the second category of petite mutants, known as vegetative petite mutants, did not segregate in a Mendelian manner (**Figure 7.18b**). Ephrussi identified two types of vegetative petites, called neutral petites and suppressive petites. In a cross between a wild-type strain and a neutral petite, all four haploid daughter cells were wild type. This type of inheritance contradicts the normal 2:2 ratio expected for the segregation of Mendelian traits. In comparison, a cross between a wild-type strain and a suppressive petite usually yielded all petite colonies. Thus, both types of vegetative petites are defective in mitochondrial function and show a non-Mendelian pattern of inheritance. These results occurred because vegetative petites carry mutations in the mitochondrial genome itself.

Since these initial studies, researchers have found that neutral petites lack most of their mitochondrial DNA, whereas suppressive petites usually lack small segments of the mitochondrial genetic material. When two yeast cells are mated, the daughter cells inherit mitochondria from both parents. For example, in a cross between a wild-type and a neutral petite strain, the daughter cells inherit both types of mitochondria. Because wild-type mitochondria are inherited, the cells display a normal phenotype. The inheritance pattern of suppressive petites is more difficult to explain because the daughter cells inherit both normal and suppressive petite mitochondria. One possibility is that the suppressive petite mitochondria replicate more rapidly so that the wild-type mitochondria are not maintained in the cytoplasm for many doublings. Alternatively, some experimental evidence suggests that genetic exchanges between the mitochondrial genomes of wild-type and suppressive petites may ultimately produce a defective population of mitochondria.

Let's now turn our attention to the inheritance of chloroplasts that are found in eukaryotic species capable of photosynthesis (namely, algae and plants). The unicellular alga *Chlamydomonas reinhardtii* has been used as a model organism to investigate the inheritance of chloroplasts. This organism contains a single chloroplast that occupies approximately 40% of the cell volume. Genetic studies of chloroplast inheritance began when Ruth Sager identified a mutant strain of *Chlamydomonas* with resistance to the antibiotic streptomycin (sm^r). By comparison, most strains are sensitive to killing by streptomycin (sm^s).

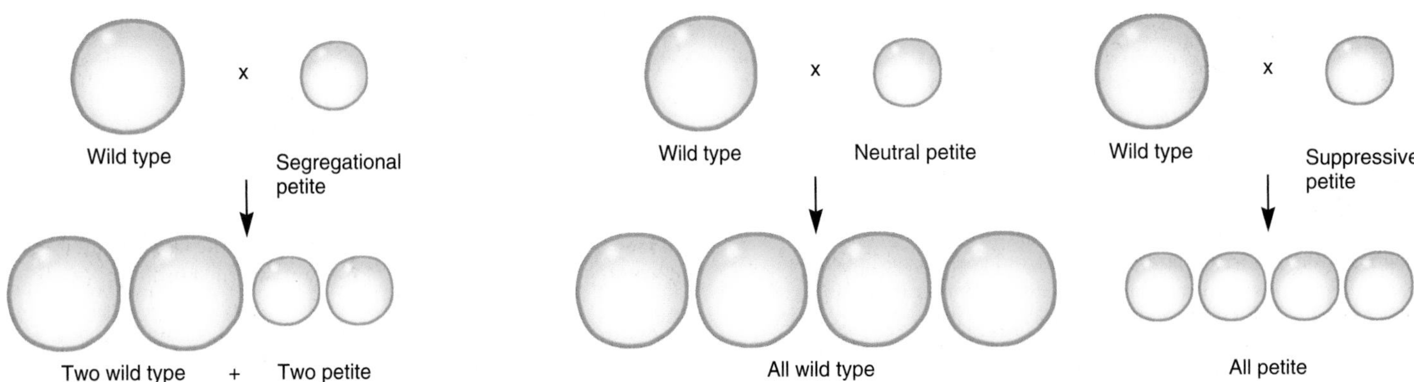

FIGURE 7.18 **Transmission of the petite trait in *Saccharomyces cerevisiae*.** (**a**) A wild-type strain crossed to a segregational petite. (**b**) A wild-type strain crossed to a neutral vegetative petite and to a suppressive vegetative petite.

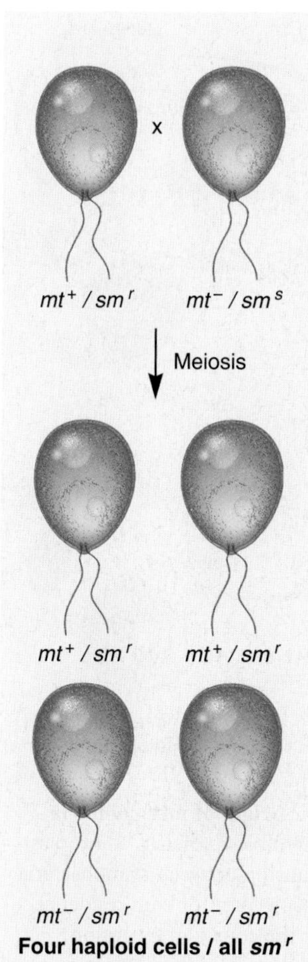

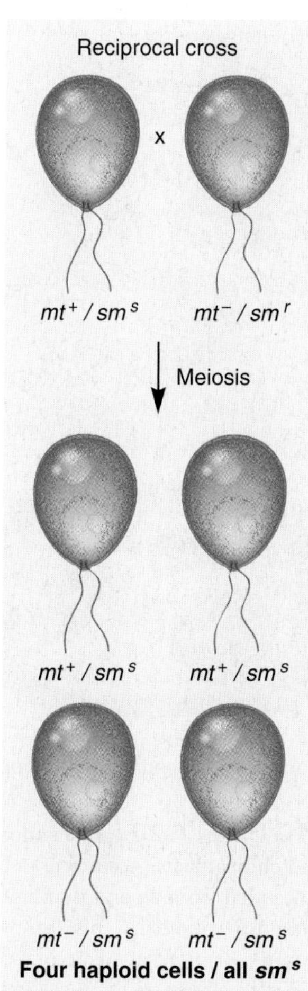

FIGURE 7.19 **Chloroplast inheritance in *Chlamydomonas*.** Mt^+ and mt^- indicate the two mating types of the organism. Sm^r indicates streptomycin resistance, whereas sm^s indicates sensitivity to this antibiotic.

Sager conducted crosses to determine the inheritance pattern of the sm^r gene. During mating, two haploid cells unite to form a diploid cell, which then undergoes meiosis to form four haploid cells. Like yeast, *Chlamydomonas* is an organism that can be found in two mating types, in this case, designated mt^+ and mt^-. Mating type is due to nuclear inheritance and segregates in a 1:1 manner. By comparison, Sager and her colleagues discovered that the sm^r gene was inherited from the mt^+ parent but not from the mt^- parent (**Figure 7.19**). Therefore, this sm^r gene was not inherited in a Mendelian manner. This pattern occurred because only the mt^+ parent transmits chloroplasts to daughter cells and the sm^r gene is found in the chloroplast genome.

The Pattern of Inheritance of Mitochondria and Chloroplasts Varies Among Different Species

The inheritance of traits via genetic material within mitochondria and chloroplasts is now a well-established phenomenon that geneticists have investigated in many different species. In **heterogamous** species, two kinds of gametes are made. The female gamete tends to be large and provides most of the cytoplasm to the zygote, whereas the male gamete is small and often provides little more than a nucleus. Therefore, mitochondria and chloroplasts are most often inherited from the maternal parent. However, this is not always the case. **Table 7.4** describes the inheritance patterns of mitochondria and chloroplasts in several selected species.

In species where maternal inheritance is generally observed, the paternal parent may occasionally provide mitochondria via the sperm. This phenomenon, called **paternal leakage,** occurs in many species that primarily exhibit maternal inheritance of their organelles. In the mouse, for example, approximately one to four paternal mitochondria are inherited for every 100,000 maternal mitochondria per generation of offspring. Most offspring do not inherit any paternal mitochondria, but a rare individual may inherit a mitochondrion from the sperm.

A Few Rare Human Diseases Are Caused by Mitochondrial Mutations

As noted previously, the human mitochondrial genome has 13 genes that encode polypeptides necessary for the synthesis of ATP. In addition, the mtDNA has genes that encode ribosomal RNA and transfer RNA molecules. Human mtDNA is maternally inherited because it is transmitted from mother to offspring via the cytoplasm of the egg. Therefore, the transmission of human mitochondrial diseases follows a strict maternal inheritance pattern.

Table 7.5 describes several mitochondrial diseases that have been discovered in humans and are caused by mutations in mitochondrial genes. These are usually chronic degenerative disorders that affect cells requiring a high level of ATP, such as nerve and muscle cells. For example, Leber's hereditary optic neuropathy (LHON) affects the optic nerve and may lead to the progressive

TABLE **7.4**

Transmission of Organelles Among Different Species

Species	Organelle	Transmission
Mammals	Mitochondria	Maternal inheritance
S. cerevisiae	Mitochondria	Biparental inheritance
Molds	Mitochondria	Usually maternal inheritance; paternal inheritance has been found in the genus *Allomyces*
Chlamydomonas	Mitochondria	Inherited from the parent with the mt^- mating type
Chlamydomonas	Chloroplasts	Inherited from the parent with the mt^+ mating type
Plants		
Angiosperms	Mitochondria and chloroplasts	Often maternal inheritance, although biparental inheritance is found among some species
Gymnosperms	Mitochondria and chloroplasts	Usually paternal inheritance

TABLE 7.5

Examples of Human Mitochondrial Diseases

Disease	Mitochondrial Gene Mutated
Leber's hereditary optic neuropathy	A mutation in one of several mitochondrial genes that encode respiratory chain proteins: *ND1, ND2, CO1, ND4, ND5, ND6,* and *cytb*
Neurogenic muscle weakness	A mutation in the *ATPase6* gene that encodes a subunit of the mitochondrial ATP-synthetase, which is required for ATP synthesis
Mitochondrial encephalomyopathy, lactic acidosis, and strokelike episodes	A mutation in genes that encode tRNAs for leucine and lysine
Mitochondrial myopathy	A mutation in a gene that encodes a tRNA for leucine
Maternal myopathy and cardiomyopathy	A mutation in a gene that encodes a tRNA for leucine
Myoclonic epilepsy with ragged-red muscle fibers	A mutation in a gene that encodes a tRNA for lysine

Data from: Wallace, D. C. (1993). Mitochondrial diseases: genotype versus phenotype. *Trends Genet.* *9,* 128–33.

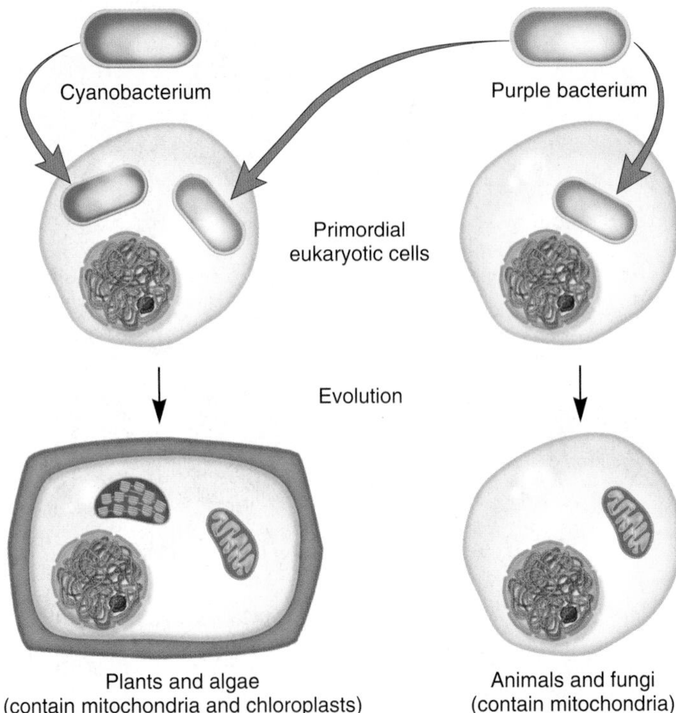

FIGURE 7.20 The endosymbiotic origin of mitochondria and chloroplasts. According to the endosymbiotic theory, chloroplasts descended from an endosymbiotic relationship between cyanobacteria and eukaryotic cells. This arose when a bacterium took up residence within a primordial eukaryotic cell. Over the course of evolution, the intracellular bacterial cell gradually changed its characteristics, eventually becoming a chloroplast. Similarly, mitochondria are derived from an endosymbiotic relationship between purple bacteria and eukaryotic cells.

loss of vision in one or both eyes. LHON can be caused by a defective mutation in one of several different mitochondrial genes. Researchers are still investigating how a defect in these mitochondrial genes produces the symptoms of this disease.

Extranuclear Genomes of Mitochondria and Chloroplasts Evolved from an Endosymbiotic Relationship

The idea that the nucleus, mitochondria, and chloroplasts contain their own separate genetic material may at first seem puzzling. Wouldn't it be simpler to have all of the genetic material in one place in the cell? The underlying reason for distinct genomes of mitochondria and chloroplasts can be traced back to their evolutionary origin, which is thought to involve a symbiotic association.

A symbiotic relationship occurs when two different species live together in a close association. The symbiont is the smaller of the two species; the host is the larger. The term **endosymbiosis** describes a symbiotic relationship in which the symbiont actually lives inside (*endo-*, inside) the host. In 1883, Andreas Schimper proposed that chloroplasts were descended from an endosymbiotic relationship between cyanobacteria and eukaryotic cells. This idea, now known as the **endosymbiosis theory,** suggested that the ancient origin of chloroplasts was initiated when a cyanobacterium took up residence within a primordial eukaryotic cell (**Figure 7.20**). Over the course of evolution, the characteristics of the intracellular bacterial cell gradually changed to those of a chloroplast. In 1922, Ivan Wallin also proposed an endosymbiotic origin for mitochondria.

In spite of these hypotheses, the question of endosymbiosis was largely ignored until researchers in the 1950s discovered that chloroplasts and mitochondria contain their own genetic material. The issue of endosymbiosis was hotly debated after Lynn Margulis published a book entitled *Origin of Eukaryotic Cells* (1970). During the 1970s and 1980s, the advent of molecular genetic techniques allowed researchers to analyze genes from chloroplasts, mitochondria, bacteria, and eukaryotic nuclear genomes. They found that genes in chloroplasts and mitochondria are very similar to bacterial genes but not as similar to those found within the nucleus of eukaryotic cells. This observation provided strong support for the endosymbiotic origin of mitochondria and chloroplasts, which is now widely accepted.

The endosymbiosis theory proposes that the relationship provided eukaryotic cells with useful cellular characteristics. Chloroplasts were derived from cyanobacteria, a bacterial species that is capable of photosynthesis. The ability to carry out photosynthesis provided algal and plant cells with the ability to use the energy from sunlight. By comparison, mitochondria are thought to have been derived from a different type of bacteria known as Gram-negative nonsulfur purple bacteria. In this case, the endosymbiotic relationship enabled eukaryotic cells to synthesize greater amounts of ATP. It is less clear how the relationship would have been beneficial to cyanobacteria or purple bacteria, though the cytosol of a eukaryotic cell may have provided a stable environment with an adequate supply of nutrients.

During the evolution of eukaryotic species, most genes that were originally found in the genome of the primordial cyanobacteria and purple bacteria have been lost or transferred from the organelles to the nucleus. The sequences of certain genes within the nucleus are consistent with their origin within an organelle. Such genes are more similar in their DNA sequence to known bacterial genes than to their eukaryotic counterparts. Therefore, researchers have concluded that these genes have been removed from the mitochondrial and chloroplast chromosomes, and relocated to the nuclear chromosomes. This has occurred many times throughout evolution, so modern mitochondria and chloroplasts have lost most of the genes that are still found in present-day purple bacteria and cyanobacteria.

Most of this gene transfer occurred early in mitochondrial and chloroplast evolution. The functional transfer of mitochondrial genes seems to have ceased in animals, but gene transfer from mitochondria and chloroplasts to the nucleus continues to occur in plants at a low rate. The molecular mechanism of gene transfer is not entirely understood, but the direction of transfer is well established. During evolution, gene transfer has occurred primarily from the organelles to the nucleus. Transfer of genes from the nucleus to the organelles has almost never occurred, although one example is known of a nuclear gene in plants that has been transferred to the mitochondrial genome. This unidirectional gene transfer from organelles to the nucleus partly explains why the organellar genomes now contain relatively few genes. In addition, gene transfer can occur between organelles. It can happen between two mitochondria, between two chloroplasts, and between a chloroplast and mitochondrion. Overall, the transfer of genetic material between the nucleus, chloroplasts, and mitochondria is an established phenomenon, although its biological benefits remain unclear.

Eukaryotic Cells Occasionally Contain Symbiotic Infective Particles

Other unusual endosymbiotic relationships have been identified in eukaryotic organisms. Several examples are known in which infectious particles establish a symbiotic relationship with their host. In some cases, research indicates that symbiotic infectious particles are bacteria that exist within the cytoplasm of eukaryotic cells. While symbiotic infectious particles are relatively uncommon, they have provided interesting and even bizarre examples of the extranuclear inheritance of traits.

In the 1940s, Tracy Sonneborn studied a trait in the protozoan *Paramecia aurelia* known as the killer trait. Killer paramecia secrete a substance called paramecin, which kills some but not all strains of paramecia. Sonneborn found that killer strains contain particles in their cytoplasm known as kappa particles. Each kappa particle is 0.4 μm long and has its own DNA. Genes within the kappa particle encode the paramecin toxin. In addition, other kappa particle genes provide the killer paramecia with resistance to paramecin.

Nonkiller paramecia are killed when mixed with killer paramecia. However, Sonnenborn found that when nonkiller paramecia were mixed with a cell extract derived from killer paramecia, the kappa particles within the extract are taken up by the nonkiller strains and convert them into killer strains. In other words, the extranuclear particle that determines the killer trait is infectious.

Infectious particles have also been identified in fruit flies. P. l'Heritier identified strains of *Drosophila melanogaster* that are highly sensitive to killing by CO_2. Reciprocal crosses between CO_2-sensitive and normal flies revealed that the trait is inherited in a non-Mendelian manner. Furthermore, cell extracts from a sensitive fly can infect a normal fly and make it sensitive to CO_2.

Another example of an infectious particle in fruit flies involves a trait known as sex ratio in which affected flies produce progenies with a large excess of females. Chana Malogolowkin and Donald Poulson discovered one strain of *Drosophila willistoni* where most of the offspring of female flies were daughters; nearly all the male offspring died. The sex ratio trait is transmitted from mother to offspring. The rare surviving males do not transmit this trait to their male or female offspring. This result indicates a maternal inheritance pattern for the sex ratio trait. The agent in the cytoplasm of female flies responsible for the sex ratio trait was later found to be a symbiotic bacterium, which was named *Spiroplasma poulsonii.* Its presence is usually lethal to males but not to females. This infective agent can be extracted from the tissues of adult females and used to infect the females of a normal strain of flies.

CONCEPTUAL SUMMARY

In this chapter, we have considered inheritance patterns that differ from Mendelian inheritance. In some cases, genes that are physically located on nuclear chromosomes fail to follow a Mendelian pattern of inheritance. For example, genes that display a maternal effect are expressed in the mother's nurse cells during oogenesis and affect the phenotype of the offspring. Epigenetic inheritance describes another class of examples in which nuclear genes do not follow a Mendelian pattern of inheritance. During dosage compensation, the level of expression of genes on the sex chromosomes is altered. In mammals, dosage compensation occurs via X inactivation. According to the Lyon hypothesis, one of the two X chromosomes in females is randomly inactivated during embryonic development to produce a transcriptionally inactive Barr body. The pattern of active and inactive X chromosomes is then inherited during later stages of development to create a female animal in which a single X chromosome is active in each cell. A second example of epigenetic inheritance is known as genomic imprinting. In this case, a gene or even an entire chromosome is marked in such a way as to be inherited in an inactive form. Imprinting can be divided into three stages: (1) the establishment of the imprint during gametogenesis, (2) the maintenance of the imprint during embryogenesis and in adult somatic cells, and (3) the erasure and reestablishment of the imprint in the germ cells. Differential DNA methylation of regions near the imprinted genes is the key event in this marking process.

Extranuclear or cytoplasmic inheritance is due to the existence of genetic material in organelles outside the nucleus. The two most important examples of extranuclear inheritance are due to genetic material within mitochondria and chloroplasts. In many cases, extranuclear genetic material follows a maternal inheritance pattern because the egg is large and, in most species, is more likely than the sperm to transmit organelles to the offspring. However, some species display biparental inheritance in which the organelle is inherited from both parents, and a few species show paternal inheritance of organelles. Some human diseases are caused by mutations in mitochondrial genes.

The extranuclear genome of mitochondria and chloroplasts is believed to be the result of the endosymbiotic origin of these organelles. According to the endosymbiosis theory, mitochondria are derived from purple bacteria, and chloroplasts are derived from cyanobacteria that were engulfed by eukaryotic cells. During evolution, many genes have been transferred from the organellar genomes to the nuclear genomes, so modern mitochondria and chloroplast genomes are small and contain relatively few genes. Other examples of extranuclear inheritance include symbiotic infective particles identified in *Paramecium aurelia* and fruit flies.

EXPERIMENTAL SUMMARY

The primary way that researchers have identified non-Mendelian patterns of inheritance is by making a series of genetic crosses and then analyzing the phenotypes of the offspring. In many cases, reciprocal crosses have yielded results that deviate from Mendelian inheritance. For genes with a maternal effect, reciprocal crosses reveal that the genotype of the mother governs the phenotype of the offspring. In genomic imprinting, reciprocal crosses indicate that one parent transmits an active form of a gene while the other parent transmits an allele that is inactive.

Similarly, the outcomes of reciprocal crosses are usually different for extrachromosomal inheritance. In most species, organelles are inherited from the maternal parent, although many exceptions are known. The inheritance of the genetic material within mitochondria and chloroplasts, first identified in studies of yeast and algae, is non-Mendelian.

Researchers have also investigated non-Mendelian inheritance at the cellular and molecular levels. The cellular explanation of maternal effect genes is that the nurse cells transfer their gene products to the developing oocyte. In Chapter 23, we will examine methods that demonstrate the accumulation of maternal effect gene products within the oocyte. Genomic imprinting is also being investigated at the molecular level. Researchers exploring the biochemistry of imprinted genes have identified DNA methylation as a key to marking. Further experimentation is needed to elucidate the mechanisms of erasure and *de novo* methylation that occur during gametogenesis.

Likewise, researchers have explored the cellular and molecular mechanisms of X inactivation. Cytological examination of female mammalian cells revealed the presence of a highly condensed X chromosome, called the Barr body. A study of the expression of an enzyme known as glucose-6-phosphate dehydrogenase (G-6-PD) at the cellular level confirmed that one X chromosome is permanently inactivated in the somatic cell lineages of females. A region on the X chromosome called Xic is required for the proper counting and inactivation of X chromosomes. Expression of the *Xist* gene located in that region is required for the X-inactivation mechanism. The embryonic expression of a gene termed *Tsix* may play a role in choosing the active X chromosome. Another region termed Xce is also important in the choice of the inactivated chromosome.

Finally, molecular tools have been used to dissect the genetic composition of organellar chromosomes. Some of these methods, which include DNA cloning and sequencing, are described in Chapter 18. These techniques have enabled researchers to identify the genes located in mitochondrial DNA (mtDNA) and chloroplast DNA (cpDNA).

PROBLEM SETS & INSIGHTS

Solved Problems

S1. Our understanding of maternal effect genes has been greatly aided by their identification in experimental organisms such as *Drosophila melanogaster* and *Caenorhabditis elegans.* In experimental organisms with a short generation time, geneticists have successfully searched for mutant alleles that prevent the normal process of embryonic development. In many cases, the offspring will die at early embryonic or larval stages. These are called maternal effect lethal alleles. How would a researcher identify a mutation that produced a recessive maternal effect lethal allele?

Answer: A maternal effect lethal allele can be identified when a phenotypically normal mother produces only offspring with gross developmental abnormalities. For example, let's call the normal allele *N* and the maternal effect lethal allele *n*. A cross between two flies that are heterozygous for a maternal effect lethal allele would produce 1/4 of the offspring with a homozygous genotype, *nn*. These flies are viable because of the maternal effect. Their mother would be *Nn* and provide the *n* egg with a sufficient amount of *N* gene product so that the *nn* flies would develop properly. However, homozygous *nn* females cannot provide their eggs with any normal gene product. Therefore, all of their offspring are abnormal and die during early stages.

S2. A maternal effect gene in *Drosophila*, called *torso*, is found as a recessive allele that prevents the correct development of anterior- and posterior-most structures. A wild-type male is crossed to a female of unknown genotype. This mating produces 100% larva that are missing their anterior- and posterior-most structures and therefore die during early development. What is the genotype and phenotype of the female fly in this cross? What are the genotypes and phenotypes of the female fly's parents?

Answer: Because this cross produces 100% abnormal offspring, the female fly must be homozygous for the abnormal *torso* allele. Even so, the female fly must be phenotypically normal in order to reproduce. This female fly had a mother that was heterozygous for a normal and abnormal *torso* allele and a father that was either heterozygous or homozygous for the abnormal *torso* allele.

$torso^+$ $torso^-$	×	$torso^+$ $torso^-$ or $torso^-$ $torso^-$
(grandmother)		(grandfather)
	↓	
	$torso^-$ $torso^-$	(mother of 100% abnormal offspring)

This female fly is phenotypically normal because its mother was heterozygous and provided the gene products of the $torso^+$ allele from the nurse cells. However, this homozygous female will produce only abnormal offspring because it cannot provide them with the normal $torso^+$ gene products.

S3. An individual with Angelman syndrome produced an offspring with Prader-Willi syndrome. Why does this occur? What are the sexes of the parent with Angelman syndrome and the offspring with Prader-Willi syndrome?

Answer: These two different syndromes are most commonly caused by a small deletion in chromosome 15. In addition, genomic imprinting plays a role because genes in this deleted region are differentially imprinted, depending on sex. If this deletion is inherited from the paternal parent, the offspring develops Prader-Willi syndrome. Therefore, in this problem, the individual with Angelman syndrome must have been a male because he produced a child with Prader-Willi syndrome. The child could be either a male or female.

S4. In yeast, a haploid petite mutant also carries a mutant gene that requires the amino acid histidine for growth. The petite his^- strain is crossed to a wild-type his^+ strain to yield the following tetrad:

2 cells: petite his^-

2 cells: petite his^+

Explain the inheritance of the petite and his^- mutations.

Answer: The his^- and his^+ alleles are segregating in a 2:2 ratio. This result indicates a nuclear pattern of inheritance. By comparison, all four cells in this tetrad have a petite phenotype. This is a suppressive petite that arises from a mitochondrial mutation.

S5. Let's suppose that you are a horticulturist who has recently identified an interesting plant with variegated leaves. How would you determine if this trait is nuclearly or cytoplasmically inherited?

Answer: Make crosses and reciprocal crosses involving normal and variegated strains. In many species, chloroplast genomes are inherited maternally, although this is not always the case. In addition, a significant percentage of paternal leakage may occur. Nevertheless, when reciprocal crosses yield different outcomes, an organellar mode of inheritance is possibly at work.

S6. A phenotype that is similar to a yeast suppressive petite was also identified in the mold *Neurospora crassa.* Mary and Herschel Mitchell identified a slow-growing mutant that they called *poky.* Unlike yeast, which are isogamous (i.e., produce one type of gamete), *Neurospora* is sexually dimorphic and produces male and female reproductive structures. When a *poky* strain of *Neurospora* was crossed to a wild-type strain, the results were different between reciprocal crosses. If a *poky* mutant was the female parent, all spores exhibited the *poky* phenotype. By comparison, if the wild-type strain was the female parent, all spores were wild type. Explain these results.

Answer: These genetic studies indicate that the *poky* mutation is maternally inherited. The cytoplasm of the female reproductive cells provides the offspring with their mitochondria. Besides these genetic studies, the Mitchells and their collaborators showed that *poky* mutants are defective in certain cytochromes, which are iron-containing proteins that are known to be located in the mitochondria.

Conceptual Questions

C1. Define the term epigenetic inheritance, and describe two examples.

C2. Describe the inheritance pattern of maternal effect genes. Explain how the maternal effect occurs at the cellular level. What are the expected functional roles of the proteins that are encoded by maternal effect genes?

C3. A maternal effect gene exists in a dominant *N* (normal) allele and a recessive *n* (abnormal) allele. What would be the ratios of genotypes and phenotypes for the offspring of the following crosses?

A. *nn* female × *NN* male

B. *NN* female × *nn* male

C. *Nn* female × *Nn* male

C4. A *Drosophila* embryo dies during early embryogenesis due to a recessive maternal effect allele called *bicoid.* The wild-type allele is designated $bicoid^+$. What are the genotypes and phenotypes of the embryo's mother and maternal grandparents?

C5. For Mendelian traits, the nuclear genotype (i.e., the alleles found on chromosomes in the cell nucleus) directly influences an offspring's traits. In contrast, for non-Mendelian inheritance patterns, the offspring's phenotype cannot be reliably predicted solely from its genotype. For the following traits, what do you need to know to predict the phenotypic outcome?

A. Dwarfism due to a mutant *Igf2* allele

B. Snail coiling direction

C. Leber's hereditary optic neuropathy

C6. Let's suppose a maternal effect gene exists as a normal dominant allele and an abnormal recessive allele. A mother who is phenotypically abnormal produces all normal offspring. Explain the genotype of the mother.

C7. Let's suppose that a gene affects the anterior morphology in house flies and is inherited as a maternal effect gene. The gene exists in a normal allele, *H*, and a recessive allele, *h*, which causes a small head. A female fly with a normal head is mated to a true-breeding male with a small head. All of the offspring have small heads. What are the genotypes of the mother and offspring? Explain your answer.

C8. Explain why maternal effect genes exert their effects during the early stages of development.

C9. As described in Chapter 19, researchers have been able to "clone" mammals by fusing a cell having a diploid nucleus (i.e., a somatic cell) with an egg that has had its (haploid) nucleus removed.

A. With regard to maternal effect genes, would the phenotype of such a cloned animal be determined by the animal that donated the egg or by the animal that donated the somatic cell? Explain.

B. Would the cloned animal inherit extranuclear traits from the animal that donated the egg or by the animal that donated the somatic cell? Explain.

C. In what ways would you expect this cloned animal to be similar to or different from the animal that donated the somatic cell? Is it fair to call such an animal a "clone" of the animal that donated the diploid nucleus?

C10. With regard to the numbers of sex chromosomes, explain why dosage compensation is necessary.

C11. What is a Barr body? How is its structure different from that of other chromosomes in the cell? How does the structure of a Barr body affect the level of X-linked gene expression?

C12. Among different species, describe three distinct strategies for accomplishing dosage compensation.

C13. Describe when X inactivation occurs and how this leads to phenotypic results at the organism level. In your answer, you should explain why X inactivation causes results such as variegated coat patterns in mammals. Why do two different calico cats have their patches of orange and black fur in different places? Explain whether or not a variegated coat pattern due to X inactivation could occur in marsupials.

C14. Describe the molecular process of X inactivation. This description should include the three phases of inactivation and the role of the Xic. Explain what happens to X chromosomes during embryogenesis, in adult somatic cells, and during oogenesis.

C15. On rare occasions, an abnormal human male is born who is somewhat feminized compared to normal males. Microscopic examination of the cells of one such individual revealed that he has a single Barr body in each cell. What is the chromosomal composition of this individual?

C16. How many Barr bodies would you expect to find in humans with the following abnormal compositions of sex chromosomes?

A. XXY

B. XYY

C. XXX

D. X0 (a person with just a single X chromosome)

C17. Certain forms of human color blindness are inherited as X-linked recessive traits. Hemizygous males are color-blind, while heterozygous females are not. However, heterozygous females sometimes have partial color blindness.

A. Discuss why heterozygous females may have partial color blindness.

B. Doctors identified an unusual case in which a heterozygous female was color-blind in her right eye but had normal color vision in her left eye. Explain how this might have occurred.

C18. A black female cat (X^BX^B) and an orange male cat (X^OY) were mated to each other and produced a male cat that was calico. Which sex chromosomes did this male offspring inherit from its mother and father? Remember that the presence of the Y chromosome determines maleness in mammals.

C19. What is the spreading stage of X inactivation? Why do you think it is called a spreading stage? Discuss the role of the *Xist* gene in the spreading stage of X inactivation.

C20. When does the erasure and reestablishment phase of genomic imprinting occur? Explain why it is necessary to erase an imprint and then reestablish it in order to always maintain imprinting from the same sex of parent.

C21. In what types of cells would you expect *de novo* methylation to occur? In what cell types would it not occur?

C22. On rare occasions, people are born with a condition known as uniparental disomy. It happens when an individual inherits both copies of a chromosome from one parent and no copies from the other parent. This occurs when two abnormal gametes happen to complement each other to produce a diploid zygote. For example, an abnormal sperm that lacks chromosome 15 could fertilize an egg that contains two copies of chromosome 15. In this situation, the individual would be said to have maternal uniparental disomy 15 because both copies of chromosome 15 were inherited from the mother. Alternatively, an abnormal sperm with two copies of chromosome 15 could fertilize an egg with no copies. This is known as paternal uniparental disomy 15. If a female is born with paternal uniparental disomy 15, would you expect her to be phenotypically normal, have Angelman syndrome (AS), or have Prader-Willi syndrome (PWS)? Explain. Would you expect her to produce normal offspring or offspring affected with AS or PWS?

C23. Genes that cause Prader-Willi syndrome and Angelman syndrome are closely linked along chromosome 15. Although people with these syndromes do not usually reproduce, let's suppose that a couple produces two children with Angelman syndrome. The oldest child (named Pat) grows up and has two children with Prader-Willi syndrome. The second child (named Robin) grows up and has one child with Angelman syndrome.

A. Are Pat and Robin's parents both normal or does one of them have Angelman or Prader-Willi syndrome? If one of them has a disorder, explain why it is the mother or the father.

B. What are the sexes of Pat and Robin? Explain.

C24. How is the process of X inactivation similar to genomic imprinting? How is it different?

C25. What is extranuclear inheritance? Describe three examples.

C26. What is a reciprocal cross? Let's suppose that a gene is found as a wild-type allele and a recessive mutant allele. What would be the expected outcomes of reciprocal crosses if a true-breeding normal individual were crossed to a true-breeding individual carrying the mutant allele? What would be the results if the gene were maternally inherited?

C27. Among different species, does extranuclear inheritance always follow a maternal inheritance pattern? Why or why not?

C28. What is the phenotype of a petite mutant? Where can a petite mutation occur: in nuclear genes, extranuclear genetic material, or both? What is the difference between a neutral and suppressive petite?

C29. Extranuclear inheritance often correlates with maternal inheritance. Even so, paternal leakage is not uncommon. What is paternal leakage? If a cross produced 200 offspring and the rate of mitochondrial paternal leakage was 3%, how many offspring would be expected to contain paternal mitochondria?

C30. Discuss the structure and organization of the mitochondrial and chloroplast genomes. How large are they, how many genes do they contain, and how many copies of the genome are found in each organelle?

C31. Explain the likely evolutionary origin of mitochondrial and chloroplast genomes. How have the sizes of the mitochondrial and chloroplast genomes changed since their origin? How has this occurred?

C32. Which of the following traits or diseases are determined by nuclear genes?

A. Snail coiling pattern

B. Prader-Willi syndrome

C. Streptomycin resistance in *Chlamydomonas*

D. Leber's hereditary optic neuropathy

C33. Acute murine leukemia virus (AMLV) causes leukemia in mice. This virus is easily passed from mother to offspring through the mother's milk. (Note: Even though newborn offspring acquire the virus, they may not develop leukemia until much later in life. Testing can determine if an animal carries the virus.) Describe how the formation of leukemia via AMLV resembles a maternal inheritance pattern. How could you determine that this form of leukemia is not caused by extranuclear inheritance?

C34. Describe how a biparental pattern of extranuclear inheritance would resemble a Mendelian pattern of inheritance for a particular gene. How would they differ?

C35. According to the endosymbiosis theory, mitochondria and chloroplasts are derived from bacteria that took up residence within eukaryotic cells. However, at one time, prior to being taken up by eukaryotic cells, these bacteria were free-living organisms. However, we cannot take a mitochondrion or chloroplast out of a living eukaryotic cell and get it to survive and replicate on its own. Why not?

Experimental Questions

E1. Figure 7.1 describes an example of a maternal effect gene. Explain how Sturtevant deduced a maternal effect gene based on the F_2 and F_3 generations.

E2. Discuss the types of experimental observations that Mary Lyon brought together in proposing her hypothesis concerning X inactivation. In your own words, explain how these observations were consistent with her hypothesis.

E3. Chapter 18 describes three blotting methods (i.e., Southern blotting, Northern blotting, and Western blotting) that are used to detect specific genes and gene products. Southern blotting detects DNA, Northern blotting detects RNA, and Western blotting detects proteins. Let's suppose that a female fruit fly is heterozygous for a maternal effect gene, which we will call gene *B*. The female is *Bb*. The normal allele, *B*, encodes a functional mRNA that is 550 nucleotides long. A recessive allele, *b*, encodes a shorter mRNA that is 375 nucleotides long. (Allele *b* is due to a deletion within this gene.) How could you use one or more of these techniques to show that nurse cells transfer gene products from gene *B* to developing eggs? You may assume that you can dissect the ovaries of fruit flies and isolate eggs separately from nurse cells. In your answer, describe your expected results.

E4. As a hypothetical example, a trait in mice results in mice with very long tails. You initially have a true-breeding strain with normal tails and a true-breeding strain with long tails. You then make the following types of crosses:

Cross 1: When true-breeding females with normal tails are crossed to true-breeding males with long tails, all F_1 offspring have long tails.

Cross 2: When true-breeding females with long tails are crossed to true-breeding males with normal tails, all F_1 offspring have normal tails.

Cross 3: When F_1 females from cross 1 are crossed to true-breeding males with normal tails, all offspring have normal tails.

Cross 4: When F_1 males from cross 1 are crossed to true-breeding females with long tails, half of the offspring have normal tails and half have long tails.

Explain the pattern of inheritance of this trait.

E5. You have a female snail that coils to the right, but you do not know its genotype. You may assume that right coiling (*D*) is dominant to left coiling (*d*). You also have male snails at your disposal of known genotype. How would you determine the genotype of this female snail? In your answer, describe your expected results depending on whether the female is *DD*, *Dd*, or *dd*.

E6. On a recent camping trip, you find one male snail on a deserted island that coils to the right. However, in this same area, you find several shells (not containing living snails) that coil to the left. Therefore, you conclude that you are not certain of the genotype of this male snail. On a different island, you find a large colony of snails of the same species. All of these snails coil to the right, and every snail shell that you find on this second island coils to the right. With regard to the maternal effect gene that determines coiling pattern, how would you determine the genotype of the male snail that you found on the deserted island? In your answer, describe your expected results.

E7. Figure 7.6 describes the results of X inactivation in mammals. If fast and slow alleles of glucose-6-phosphate dehydrogenase (*G-6-PD*) exist in other species, what would be the expected results of gel electrophoresis for a heterozygous female of the following species?

A. Marsupial

B. *Drosophila melanogaster*

C. *Caenorhabditis elegans* (Note: We are considering the hermaphrodite in *C. elegans* to be equivalent to a female.)

E8. Two male mice, which we will call male A and male B, are both phenotypically normal. Male A was from a litter that contained half phenotypically normal mice and half dwarf mice. The mother of male A was known to be homozygous for the normal *Igf2* allele. Male B was from a litter of eight mice that were all phenotypically normal. The parents of male B were a phenotypically normal male and a dwarf female. Male A and male B were put into a cage with two female mice that we will call female A and female B. Female A is dwarf, and female B is phenotypically normal. The parents of these two females were unknown, although it was known that they were from the same litter. The mice were allowed to mate with each other, and the following data were obtained:

Female A gave birth to three dwarf babies and four normal babies.

Female B gave birth to four normal babies and two dwarf babies.

Which male(s) mated with female A and female B? Explain.

E9. In the experiment of Figure 7.6, why does a clone of cells produce only one type of G-6-PD enzyme? What would you expect to happen if a clone was derived from an early embryonic cell? Why does the initial sample of tissue produce both forms of G-6-PD?

E10. Chapter 18 describes a blotting method known as Northern blotting that can be used to determine the amount of mRNA produced by a particular gene. In this method, the amount of a specific mRNA produced by cells is detected as a band on a gel. If one type of cell produces twice as much of a particular mRNA compared to another cell, the band will appear twice as dark. Also, sometimes mutations affect the length of mRNA that is transcribed from a gene. For example, a small deletion within a gene may shorten an mRNA. Northern blotting also can discern the sizes of mRNAs.

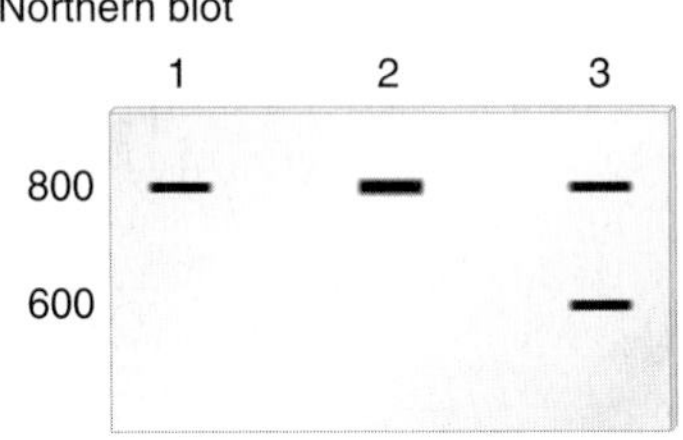

Lane 1 is a Northern blot of mRNA from cell type A that is 800 nucleotides long.

Lane 2 is a Northern blot of the same mRNA from cell type B. (Cell type B produces twice as much of this RNA compared to cell type A.)

Lane 3 shows a heterozygote in which one of the two genes has a deletion, which shortens the mRNA by 200 nucleotides.

Here is the question. Let's suppose an X-linked gene exists as two alleles, which we will call *B* and *b*. Allele *B* encodes an mRNA that is 750 nucleotides long, while allele *b* encodes a shorter mRNA that is 675 nucleotides long. Draw the expected results of a Northern blot using mRNA isolated from the same type of somatic cells taken from the following individuals:

A. First lane is mRNA from an X^bY male fruit fly.

Second lane is mRNA from an X^bX^b female fruit fly.

Third lane is mRNA from an X^BX^b female fruit fly.

B. First lane is mRNA from an X^BY male mouse.

Second lane is mRNA from an X^BX^b female mouse.*

Third lane is mRNA from an X^BX^B female mouse.*

*The sample is taken from an adult female mouse. It is not a clone of cells. It is a tissue sample, like the one described in the experiment of Figure 7.6.

C. First lane is mRNA from an X^B0 male *C. elegans.*

Second lane is mRNA from an X^BX^b hermaphrodite *C. elegans.*

Third lane is mRNA from an X^BX^B hermaphrodite *C. elegans.*

E11. A variegated trait in plants is analyzed using reciprocal crosses. The following results are obtained:

Variegated female × Normal male	Normal female × Variegated male
↓	↓
1,024 variegated + 52 normal	1,113 normal + 61 variegated

Explain this pattern of inheritance.

E12. Ruth Sager and her colleagues discovered that the mode of inheritance of streptomycin resistance in *Chlamydomonas* could be altered if the mt^+ cells were exposed to UV irradiation prior to mating. This exposure dramatically increased the frequency of biparental inheritance. What would be the expected outcome of a cross between an mt^+ sm^r and an mt^- sm^s strain in the absence of UV irradiation? How would the result differ if the mt^+ strain was exposed to UV light?

E13. Take a look at Figure 7.18 and describe how you could experimentally distinguish between yeast strains that are neutral petites versus suppressive petites.

Questions for Student Discussion/Collaboration

1. Recessive maternal effect genes are identified in flies (for example) when a phenotypically normal mother cannot produce any normal offspring. Because all the offspring are dead, this female fly cannot be used to produce a strain of heterozygous flies that could be used in future studies. How would you identify heterozygous individuals that are carrying a recessive maternal effect allele? How would you maintain this strain of flies in a laboratory over many generations?

2. What is an infective particle? Discuss the similarities and differences between infective particles and organelles such as mitochondria and chloroplasts. Do you think the existence of infective particles supports the endosymbiosis theory of the origin of mitochondria and chloroplasts?

Note: All answers appear at the website for this textbook; the answers to even-numbered questions are in the back of the textbook.

www.mhhe.com/brookergenetics3e

Visit the Online Learning Center for practice tests, answer keys, and other learning aids for this chapter. Enhance your understanding of genetics with our interactive exercises, quizzes, animations, and much more.

VARIATION IN CHROMOSOME STRUCTURE AND NUMBER

8

The term **genetic variation** refers to genetic differences among members of the same species or those between different species. Throughout Chapters 2 to 7, we have focused primarily on variation in specific genes, which is called **allelic variation.** In Chapter 8, our emphasis will shift to larger types of genetic changes that affect the structure or number of eukaryotic chromosomes. You might think small mutations that produce allelic differences would happen much more frequently than such larger changes. Surprisingly, changes in chromosome structure and number are actually fairly common. These larger alterations may affect the expression of many genes and thereby affect phenotypes. Variation in chromosome structure and number are of great importance in the field of genetics because they are critical in the evolution of new species and have widespread medical relevance. In addition, agricultural geneticists have discovered that such variation can lead to the development of new strains of crops, which may be quite profitable.

In the first section of Chapter 8, we will begin by exploring how the structure of a eukaryotic chromosome can be modified, either by altering the total amount of genetic material or by rearranging the order of genes along a chromosome. Such changes can usually be detected microscopically. The rest of the chapter is concerned with changes in the total number of chromosomes. Variation in chromosome number occurs in two subtypes: changes in the number of sets of chromosomes and changes in the numbers of individual chromosomes within a set. Natural variation in the number of sets of chromosomes is relatively common among different species, particularly in the plant kingdom. Within the same species, changes in chromosome number within a set can also occur, but such alterations are usually detrimental to the organism. We will explore how variation in chromosome number occurs and consider examples where it has significant phenotypic consequences. We will conclude by examining how changes in chromosome number can be induced through experimental treatments and how these approaches have applications in research and in agriculture.

CHAPTER OUTLINE

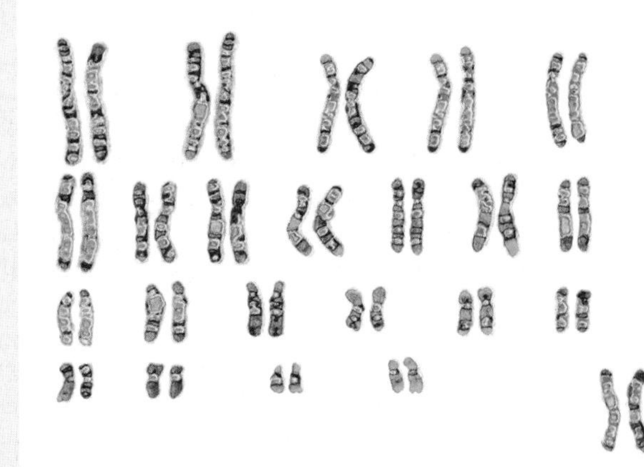

The chromosome composition of humans. Somatic cells in humans contain 46 chromosomes, which come in 23 pairs.

8.1 VARIATION IN CHROMOSOME STRUCTURE

Chromosomes in the nuclei of eukaryotic cells contain long, linear DNA molecules that carry hundreds or even thousands of genes. In this section, we will explore how the composition of a chromosome can be changed. As you will see, segments of a chromosome can be lost, duplicated, or rearranged in a new way.

We will also examine the cellular mechanisms that underlie these changes in chromosome structure. Unusual events during meiosis may affect how altered chromosomes are transmitted from parents to offspring. Also, we will consider many examples in which chromosomal alterations affect an organism's phenotypic characteristics.

Natural Variation Exists in Chromosome Structure

To appreciate changes in chromosome structure, researchers need to have a reference point for a normal set of chromosomes. To determine what the normal chromosomes of a species look like, a **cytogeneticist**—a scientist who studies chromosomes microscopically—examines the chromosomes from several members of a given species. In most cases, two phenotypically normal individuals of the same species will have the same number and types of chromosomes.

To determine the chromosomal composition of a species, the chromosomes in actively dividing cells are examined microscopically. **Figure 8.1a** shows micrographs of chromosomes from three species: a human, a fruit fly, and a corn plant. As seen here, a human has 46 chromosomes (23 pairs), a fruit fly has 8 chromosomes (4 pairs), and corn has 20 chromosomes (10 pairs). Except for the sex chromosomes, which differ between males and females, most members of the same species have very similar chromosomes. For example, the overwhelming majority of people have 46 chromosomes in their somatic cells. By comparison, the chromosomal compositions of distantly related species, such as humans and fruit flies, may be very different. A total of 46 chromosomes is normal for humans, while 8 chromosomes is the norm for fruit flies.

Cytogeneticists have various ways to classify and identify chromosomes. The three most commonly used features are location of the centromere, size, and banding patterns that are revealed when the chromosomes are treated with stains. As shown in **Figure 8.1b**, chromosomes are classified as **metacentric** (in which the centromere is near the middle), **submetacentric** (in which the centromere is slightly off center), **acrocentric** (in which the centromere is significantly off center but not at the end), and **telocentric** (in which the centromere is at one end). Because the centromere is never exactly in the center of a chromosome, each chromosome has a short arm and a long arm. The short arm is designated with the letter *p* (for the French, petite), and the long arm is designated with the letter *q*. In the case of telocentric chromosomes, the short arm may be nearly nonexistent.

Figure 8.1c shows a human karyotype. The procedure for making a karyotype was described in Chapter 3 (see Figure 3.2). A **karyotype** is a micrograph in which all of the chromosomes within a single cell have been arranged in a standard fashion. When preparing a karyotype, the chromosomes are aligned with the short arms on top and the long arms on the bottom. By convention, the chromosomes are numbered roughly according to their size, with the largest chromosomes having the smallest numbers. For example, human chromosomes 1, 2, and 3 are relatively large, whereas 21 and 22 are the two smallest. An exception to the numbering system involves the sex chromosomes, which are designated with letters (for humans, X and Y).

Because different chromosomes often have similar sizes and centromeric locations (e.g., compare human chromosomes 8, 9, and 10), geneticists must use additional methods to accurately identify each type of chromosome within a karyotype. For detailed identification, chromosomes are treated with stains to produce characteristic banding patterns. Several different staining procedures are used by cytogeneticists to identify specific chromosomes. An example is **G banding,** which is shown in Figure 8.1c. In this procedure, chromosomes are treated with mild heat or with proteolytic enzymes that partially digest chromosomal proteins. When exposed to the dye called Giemsa, named after its inventor Gustav Giemsa, some chromosomal regions bind the dye heavily and produce a dark band. In other regions, the stain hardly binds at all and a light band results. Though the mechanism of staining is not completely understood, the dark bands are thought to represent regions that are more tightly compacted. As shown in Figure 8.1c and d, the alternating pattern of G bands is a unique feature for each chromosome.

In the case of human chromosomes, approximately 300 G bands can usually be distinguished during metaphase. A larger number of G bands (in the range of 800) can be observed in prometaphase chromosomes because they are more extended than metaphase chromosomes. **Figure 8.1d** shows the conventional numbering system that is used to designate G bands along a set of human chromosomes. The left chromatid in each pair of sister chromatids shows the expected banding pattern during metaphase, while the right chromatid shows the banding pattern as it would appear during prometaphase.

Why is the banding pattern of eukaryotic chromosomes useful? First, when stained, individual chromosomes can be distinguished from each other, even if they have similar sizes and centromeric locations. For example, compare the differences in banding patterns between human chromosomes 8 and 9 (Figure 8.1d). These differences permit us to distinguish these two chromosomes even though their sizes and centromeric locations are very similar. Banding patterns are also used to detect changes in chromosome structure. As discussed next, chromosomal rearrangements or changes in the total amount of genetic material are more easily detected in banded chromosomes. Also, chromosome banding can be used to assess evolutionary relationships between species. Research studies have shown that the similarity of chromosome banding patterns is a good measure of genetic relatedness.

Changes in Chromosome Structure Include Deletions, Duplications, Inversions, and Translocations

With an understanding that chromosomes typically come in a variety of shapes and sizes, let's consider how the structures of normal chromosomes can be modified. In some cases, the total amount of genetic material within a single chromosome can be increased or decreased significantly. Alternatively, the genetic material in one or more chromosomes may be rearranged without affecting the total amount of material. As shown in **Figure 8.2**, these mutations are categorized as deletions, duplications, inversions, and translocations.

Deletions and duplications are changes in the total amount of genetic material within a single chromosome. In Figure 8.2,

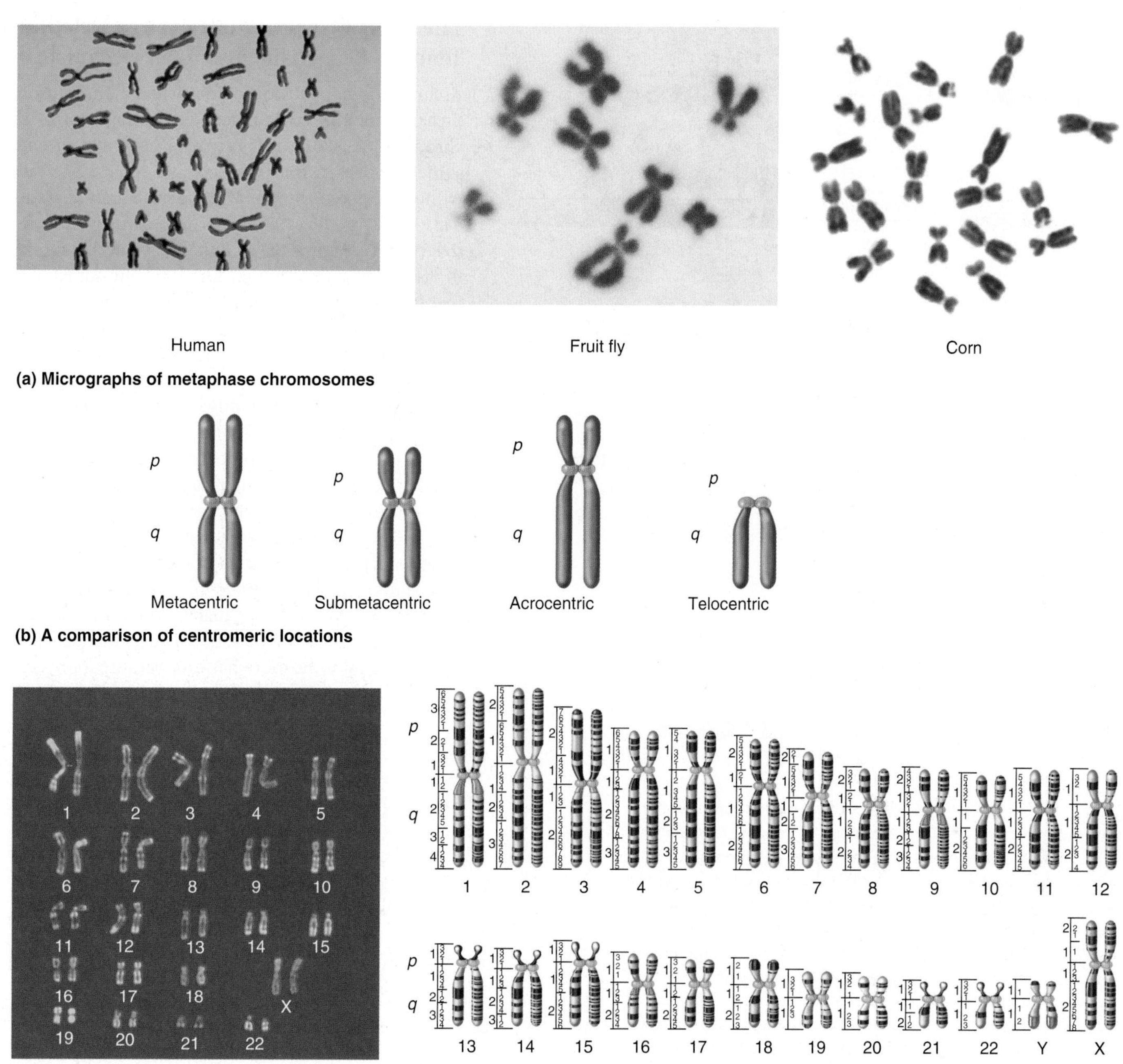

FIGURE 8.1 Features of normal chromosomes. (**a**) Micrographs of chromosomes from a human, a fruit fly, and corn. (**b**) A comparison of centromeric locations. Centromeres can be metacentric, submetacentric, acrocentric (near one end), or telocentric (at the end). (**c**) Human chromosomes that have been stained with Giemsa. (**d**) The conventional numbering of bands in Giemsa-stained human chromosomes. The numbering is divided into broad regions, which then are subdivided into smaller regions. The numbers increase as the region gets farther away from the centromere. For example, if you take a look at the left chromatid of chromosome 1, the uppermost dark band would be at a location designated p35. The banding patterns of chromatids change as the chromatids condense. The left chromatid of each pair of sister chromatids shows the banding pattern of a chromatid in metaphase, and the right side shows the banding pattern as it would appear in prometaphase. Note: In prometaphase, the chromatids are more extended than in metaphase.

human chromosomes are labeled according to their normal G banding patterns. When a **deletion** occurs, a segment of chromosomal material is missing. In other words, the affected chromosome is deficient in a significant amount of genetic material. The term **deficiency** is also used to describe a missing region of a chromosome. In contrast, a **duplication** occurs when a section of a chromosome is repeated compared to the normal parent chromosome.

Inversions and translocations are chromosomal rearrangements. An **inversion** involves a change in the direction of the genetic material along a single chromosome. For example, in Figure 8.2c, a segment of one chromosome has been inverted,

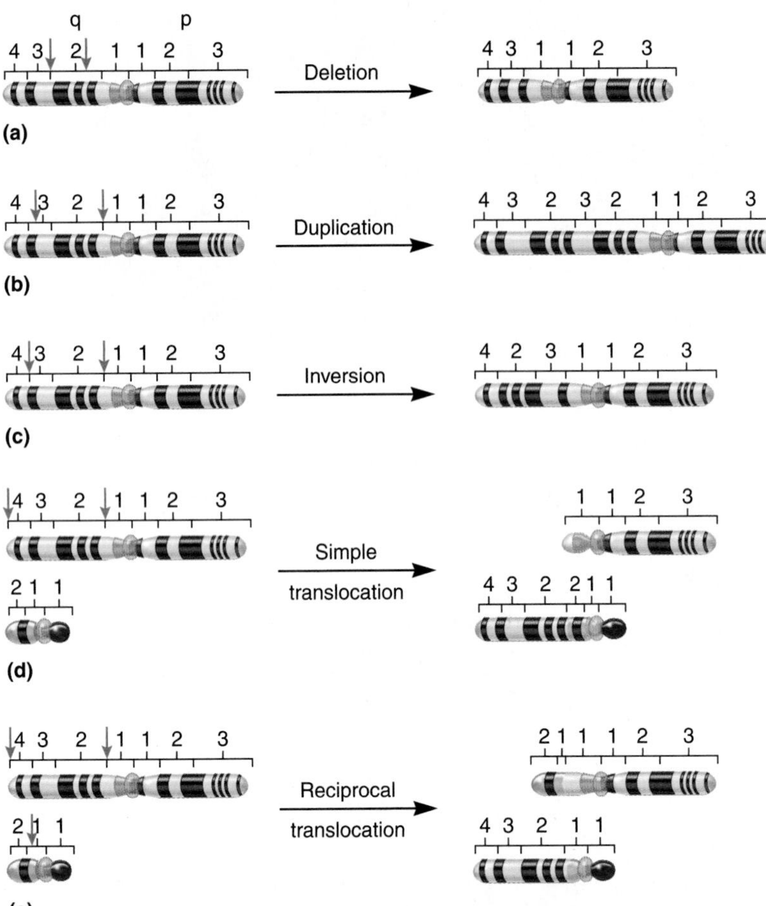

ONLINE ANIMATION **FIGURE 8.2 Types of changes in chromosome structure.** The large chromosome shown throughout is human chromosome 1. The smaller chromosome seen in (d) and (e) is human chromosome 21. **(a)** A deletion occurs that removes a large portion of the q2 region, indicated by the red arrows. **(b)** A duplication occurs that doubles the q2-q3 region. **(c)** An inversion occurs that inverts the q2-q3 region. **(d)** The q2-q4 region of chromosome 1 is translocated to chromosome 21. A region of a chromosome cannot be inserted directly to the tip of another chromosome because telomeres at the tips of chromosomes would prevent such an event. In this example, a small piece at the end of chromosome 21 would have to be removed in order for the q2-q4 region of chromosome 1 to be attached to chromosome 21. **(e)** The q2-q4 region of chromosome 1 is exchanged with most of the q1-q2 region of chromosome 21.

so the order of four G bands is opposite to that of the parent chromosome. A **translocation** occurs when one segment of a chromosome becomes attached to a different chromosome or to a different part of the same chromosome. A **simple translocation** occurs when a single piece of chromosome is attached to another chromosome. In a **reciprocal translocation,** two different types of chromosomes exchange pieces, thereby producing two abnormal chromosomes carrying translocations.

Figure 8.2 illustrates the common ways that the structure of chromosomes can be altered. Throughout the rest of this section, we will consider how these changes occur, how the changes are detected experimentally, and how they affect the phenotypes of the individuals who inherit them.

The Loss of Genetic Material in a Deletion Tends to Be Detrimental to an Organism

A chromosomal deletion occurs when a chromosome breaks in one or more places and a fragment of the chromosome is lost. In Figure 8.3a, a normal chromosome has broken into two separate pieces. The piece without the centromere will be lost and degraded. This event produces a chromosome with a **terminal deletion.** In Figure 8.3b, a chromosome has broken in two places to produce three chromosomal fragments. The central fragment is lost, and the two outer pieces reattach to each other. This process has created a chromosome with an **interstitial deletion.** Deletions can also be created when recombination takes place at incorrect locations between two homologous chromosomes. The products of this type of aberrant recombination event are one chromosome with a deletion and another chromosome with a duplication. This process is examined later in this chapter.

The phenotypic consequences of a chromosomal deletion depend on the size of the deletion and whether it includes genes or portions of genes that are vital to the development of the organism. When deletions have a phenotypic effect, they are usually detrimental. Larger deletions tend to be more harmful because more genes

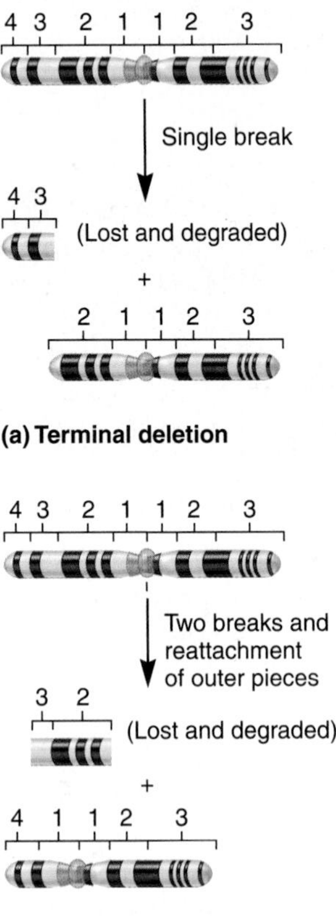

FIGURE 8.3 Production of terminal and interstitial deletions. This illustration shows the production of deletions in human chromosome 1.

are missing. Many examples are known in which deletions have significant phenotypic influences. For example, a human genetic disease known as cri-du-chat syndrome is caused by a deletion in a segment of the short arm of human chromosome 5 (**Figure 8.4a**). Individuals who carry a single copy of this abnormal chromosome along with a normal chromosome 5 display an array of abnormalities including mental deficiencies, unique facial anomalies, and an unusual catlike cry in infancy, which is the meaning of the French name for the syndrome (**Figure 8.4b**). Some other human genetic diseases, such as Angelman syndrome and Prader-Willi syndrome, described in Chapter 7, are due to a deletion in chromosome 15.

Duplications Tend to Be Less Harmful Than Deletions

Duplications result in extra genetic material. They are usually caused by abnormal events during recombination. Under normal circumstances, crossing over occurs at analogous sites between homologous chromosomes. On rare occasions, a crossover may occur at misaligned sites on the homologues (**Figure 8.5**). This results in one chromatid with an internal duplication and another chromatid with a deletion. In Figure 8.5, the chromosome with the extra genetic material carries a **gene duplication,** because the number of copies of gene *C* has been increased from one to two. In most cases, gene duplications happen as rare, sporadic events during the evolution of species. Later in this section, we will consider how multiple copies of genes can evolve into a family of genes with specialized functions.

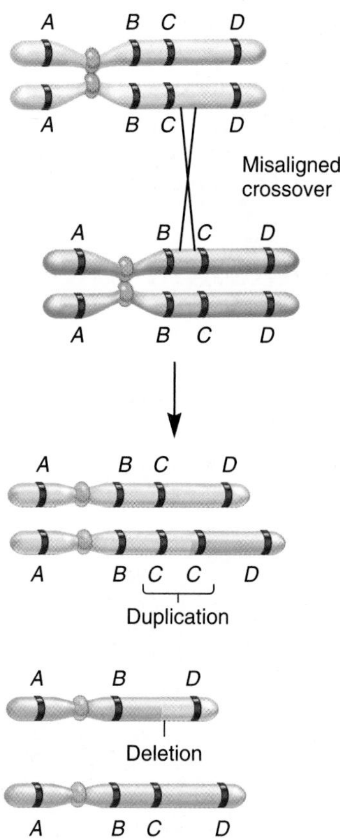

FIGURE 8.5 **Abnormal crossing over, leading to a duplication and a deletion.** A crossover has occurred at sites between genes *C* and *D* in one chromatid and between genes *B* and *C* in another chromatid. After crossing over is completed, one chromatid contains a duplication, and the other contains a deletion.

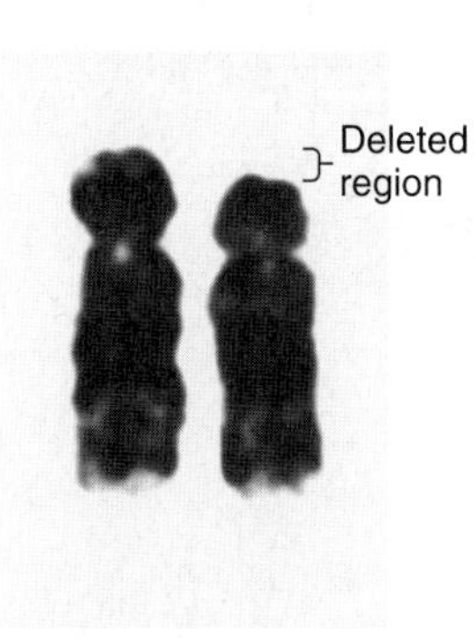

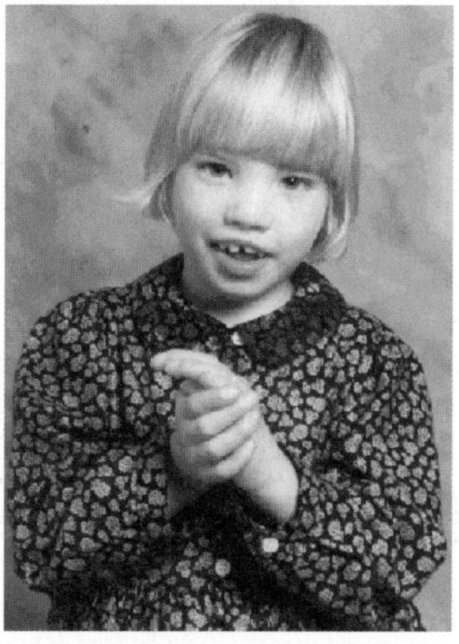

(a) Chromosome 5 **(b) A child with cri-du-chat syndrome**

FIGURE 8.4 **Cri-du-chat syndrome.** (**a**) Chromosome 5 from the karyotype of an individual with this disorder. A section of the short arm of one chromosome is missing. (**b**) An affected individual.

Genes → Traits Compared to an individual who has two copies of each gene on chromosome 5, an individual with cri-du-chat syndrome has only one copy of the genes that are located within the missing segment. This genetic imbalance (one versus two copies of many genes on chromosome 5) causes the phenotypic characteristics of this disorder, which include a catlike cry in infancy, short stature, characteristic facial anomalies (e.g., a triangular face, almond-shaped eyes, broad nasal bridge, and low-set ears), and microencephaly (a smaller than normal brain).

Like deletions, the phenotypic consequences of duplications tend to be correlated with size. Duplications are more likely to have phenotypic effects if they involve a large piece of the chromosome. In general, small duplications are less likely to have harmful effects than are deletions of comparable size. This observation suggests that having only one copy of a gene is more harmful than having three copies. In humans, relatively few well-defined syndromes are caused by small chromosomal duplications. An example is Charcot-Marie-Tooth disease (type 1A), a peripheral neuropathy characterized by numbness in the hands and feet that is caused by a small duplication on the short arm of chromosome 17.

Duplications Provide Additional Material for Gene Evolution, Sometimes Leading to the Formation of Gene Families

In contrast to the gene duplication that causes Charcot-Marie-Tooth disease, the majority of small chromosomal duplications have no phenotypic effect. Nevertheless, they are vitally important because they provide raw material for the addition of more genes into a species' chromosomes. Over the course of many generations, this can lead to the formation of a **gene family** consisting of two or more genes that are similar to each other. As shown in **Figure 8.6**, the members of a gene family are derived from the same ancestral gene. Over time, two copies of an ancestral gene can accumulate different mutations. Therefore, after many generations, the two genes will be similar but not identical. During evolution, this type of event can occur several times, creating a family of many similar genes.

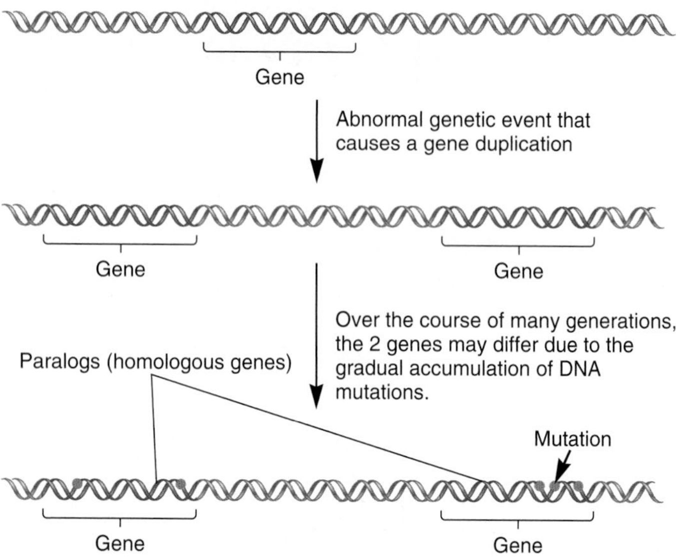

ONLINE ANIMATION **FIGURE 8.6 Gene duplication and the evolution of paralogs.** An abnormal crossover event like the one described in Figure 8.5 leads to a gene duplication. Over time, each gene accumulates different mutations.

When two or more genes are derived from a single ancestral gene, the genes are said to be **homologous.** Homologous genes within a single species are called **paralogs** and constitute a gene family. A well-studied example of a gene family is shown in **Figure 8.7**, which illustrates the evolution of the globin gene family found in humans. The globin genes encode polypeptides that are subunits of proteins that function in oxygen binding. For example, hemoglobin is a protein found in red blood cells; its function is to carry oxygen throughout the body. The globin gene family is composed of 14 paralogs that were originally derived from a single ancestral globin gene. According to an evolutionary analysis, the ancestral globin gene first duplicated about 500 million years ago and became separate genes encoding myoglobin and the hemoglobin group of genes. The primordial hemoglobin gene duplicated into an α-chain gene and a β-chain gene, which subsequently duplicated to produce several genes located on chromosomes 16 and 11, respectively. Currently, 14 globin genes are found on three different human chromosomes.

Why is it advantageous to have a family of globin genes? Although all globin polypeptides are subunits of proteins that play a role in oxygen binding, the accumulation of different mutations in the various family members has produced globins that are more specialized in their function. For example, myoglobin is better at binding and storing oxygen in muscle cells, and the hemoglobins are better at binding and transporting oxygen

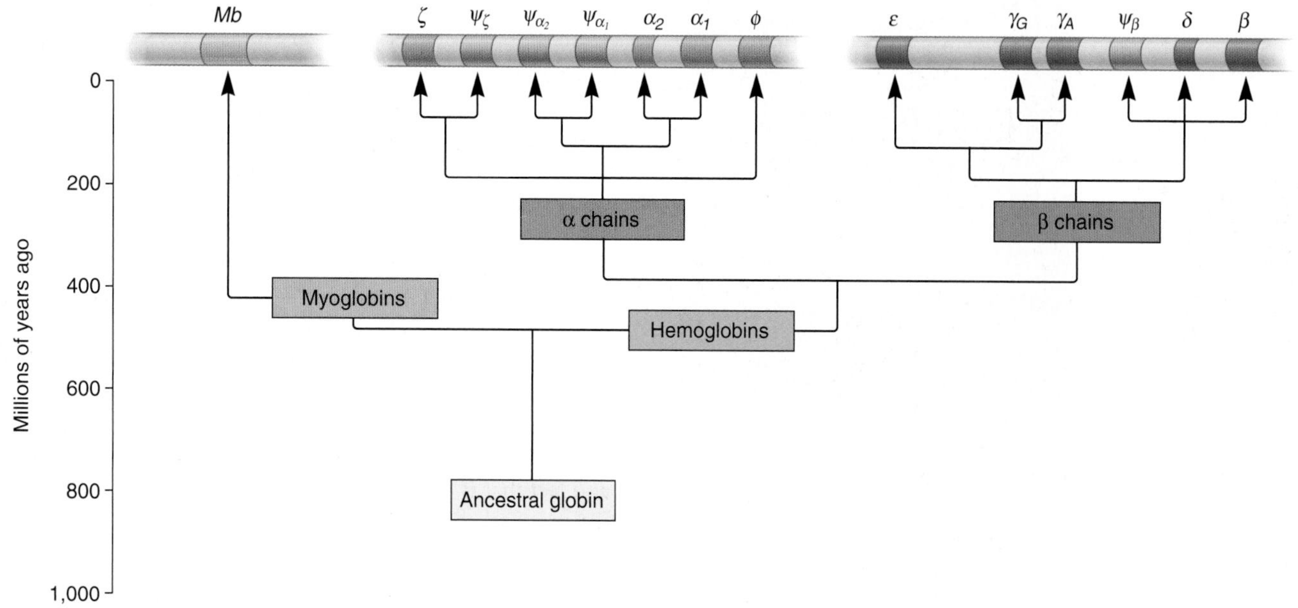

FIGURE 8.7 The evolution of the globin gene family in humans. The globin gene family evolved from a single ancestral globin gene. The first gene duplication produced two genes that accumulated mutations and became the genes encoding myoglobin (on chromosome 22) and the group of hemoglobins. The primordial hemoglobin gene then duplicated to produce several α-chain and β-chain genes, which are found on chromosomes 16 and 11, respectively. The four genes shown in gray are nonfunctional pseudogenes.

via the red blood cells. Also, different globin genes are expressed during different stages of human development. The ε- and ζ-globin genes are expressed very early in embryonic life, whereas the α-globin and γ-globin genes are expressed during the second and third trimesters of gestation. Following birth, the α-globin gene remains turned on, while the γ-globin genes are turned off and the β-globin gene is turned on. These differences in the expression of the globin genes reflect the differences in the oxygen transport needs of humans during the embryonic, fetal, and postpartum stages of life.

EXPERIMENT 8A

Comparative Genomic Hybridization Is Used to Detect Chromosome Deletions and Duplications

As we have seen, chromosome deletions and duplications can occur by a variety of mechanisms and may have an impact on the phenotypes of individuals who inherit them. One very important reason why researchers have become interested in these types of chromosomal changes is related to cancer. As discussed in Chapter 22, chromosomal deletions and duplications have been associated with many types of human cancers. Though such changes may be detectable by traditional chromosomal staining and karyotyping methods, small deletions and duplications may be difficult to detect in this manner. Fortunately, researchers have been able to develop more sensitive methods to identify changes in chromosome structure.

In 1992, Anne Kallioniemi, Daniel Pinkel, and colleagues devised a method called **comparative genomic hybridization (CGH).** This technique is largely used to determine if cancer cells have changes in chromosome structure, such as deletions or duplications. To begin this procedure, DNA is isolated from a test sample, which in this case was a sample of breast cancer cells, and also from a normal reference sample (**Figure 8.8**). The DNA from the breast cancer cells was used as a template to make green fluorescent DNA, and the DNA from normal cells was used to make red fluorescent DNA. These DNA molecules were 600 bp to 1,000 bp in length. Both types of DNA were then denatured by heat treatment. Equal amounts of the two fluorescently labeled DNA samples were mixed together and applied to normal metaphase chromosomes in which the DNA had also been denatured. Because the fluorescently labeled DNA fragments and the metaphase chromosomes had both been denatured, the fluorescently labeled DNA strands could bind to complementary regions on the metaphase chromosomes. This process is called **hybridization** because the DNA from one sample (a green or red DNA strand) forms a double-stranded region with a DNA strand from another sample (an unlabeled metaphase chromosome). Following hybridization, the metaphase chromosomes were visualized using a fluorescence microscope, and the images were analyzed by a computer that could determine the relative intensities of green and red fluorescence. What are the expected results? If a chromosomal region in the breast cancer cells and the normal cells are present in the same amount, the ratio between green and red fluorescence should be 1. If a chromosomal region is deleted in the breast cancer cell line, the ratio will be less than one, or if a region is duplicated, it will be greater than one.

THE GOAL

Deletions or duplications in cancer cells can be detected by comparing the ability of fluorescently labeled DNA from cancer cells and normal cells to bind (hybridize) to normal metaphase chromosomes.

ACHIEVING THE GOAL — FIGURE 8.8 The use of comparative genomic hybridization to detect deletions and duplications in cancer cells.

Starting materials: Breast cancer cells and normal cells.

1. Isolate DNA from human breast cancer cells and normal cells. This involved breaking open the cells and isolating the DNA by chromatography. (See Appendix for description of chromatography.)

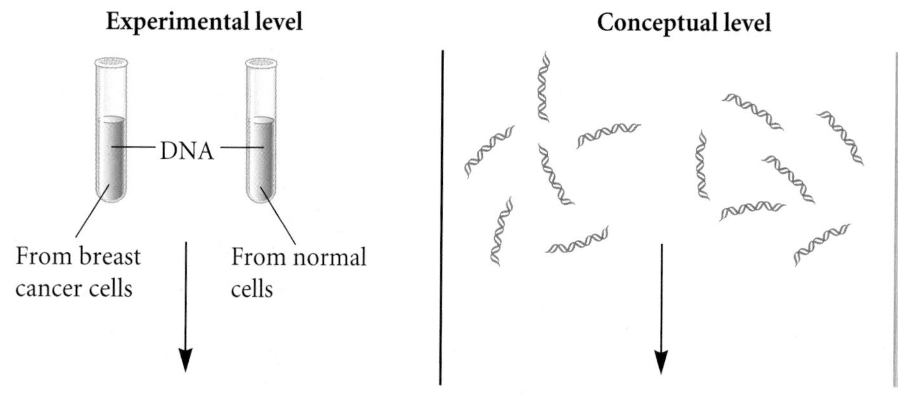

2. Label the breast cancer DNA with a green fluorescent molecule and the normal DNA with a red fluorescent molecule. This was done by using the DNA from step 1 as a template, and incorporating fluorescently labeled nucleotides into newly made DNA strands.

3. The DNA strands were then denatured by heat treatment. Mix together equal amounts of fluorescently labeled DNA and add it to a preparation of metaphase chromosomes from white blood cells. The procedure for preparing metaphase chromosomes is described in Figure 3.2. The metaphase chromosomes were also denatured.

4. Allow the fluorescently labeled DNA to hybridize to the metaphase chromosomes.

5. Visualize the chromosomes with a fluorescence microscope. Analyze the amount of green and red fluorescence along each chromosome with a computer.

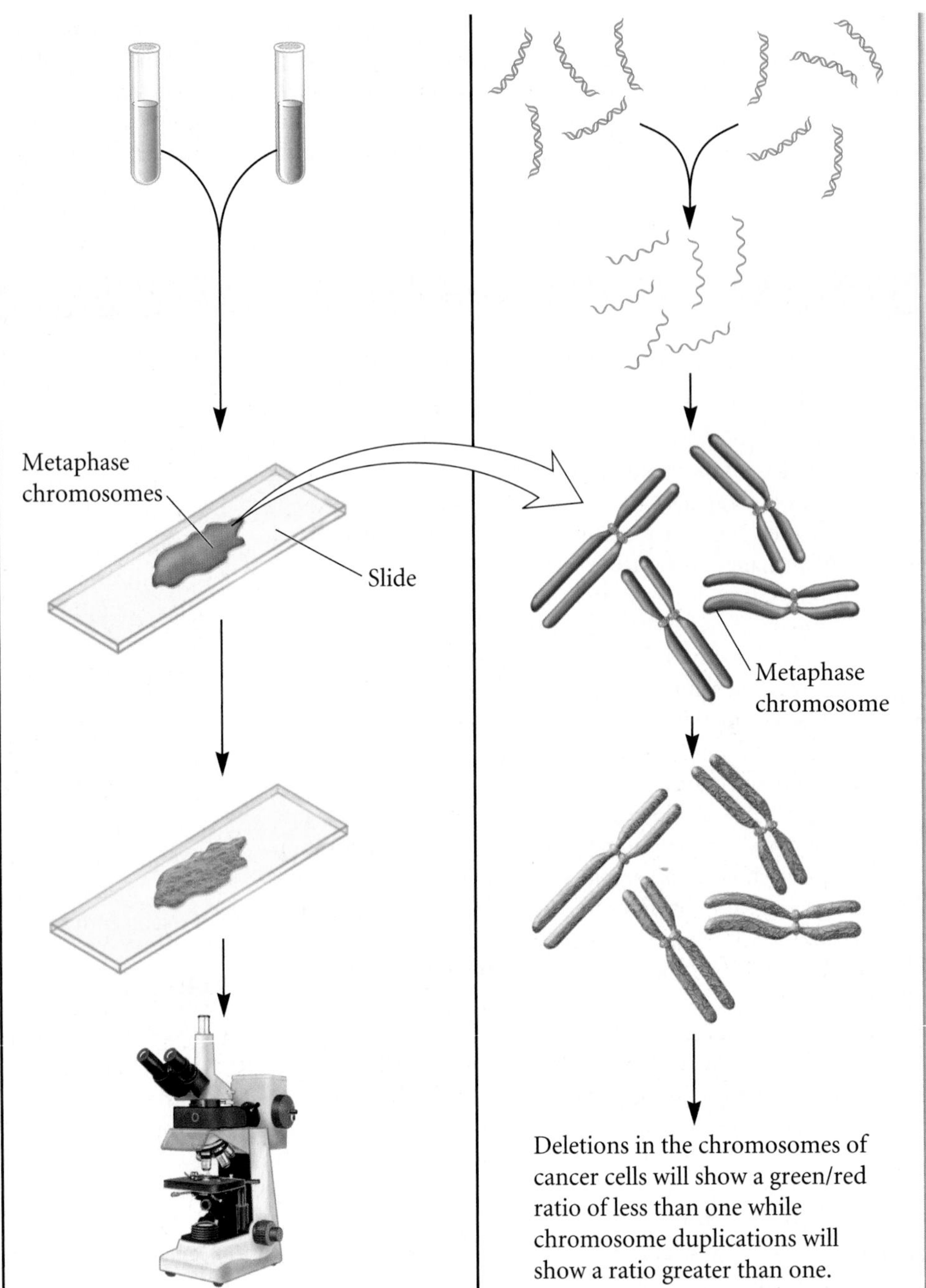
Metaphase chromosomes
Slide
Metaphase chromosome
Deletions in the chromosomes of cancer cells will show a green/red ratio of less than one while chromosome duplications will show a ratio greater than one.

THE DATA

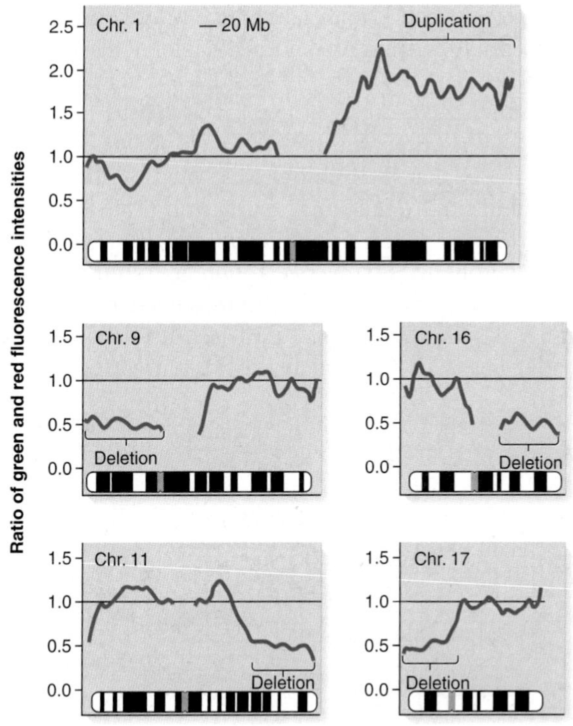

Note: Unlabeled repetitive DNA was also included in this experiment to decrease the level of nonspecific, background labeling. This repetitive DNA also prevents labeling near the centromere. As seen in the data, regions in the chromosomes where the curves are missing are due to the presense of highly repetitive sequences near the centromere.

INTERPRETING THE DATA

The data of Figure 8.8 show the ratio of green (cancer DNA) to red (normal DNA) fluorescence along five different metaphase chromosomes. Chromosome 1 shows a large duplication, as indicated by the ratio of 2. One interpretation of this observation is that both copies of chromosome 1 carry a duplication. In comparison, chromosomes 9, 11, 16, and 17 have regions with a value of 0.5. This value indicates one of the two chromosomes of these four types in the cancer cells carries a deletion, while the other chromosome of these four types does not. (If the value was zero, this would indicate both copies of a chromosome had deleted the same region.) Overall, these results illustrate how this technique can be used to map chromosomal duplications and deletions in cancer cells.

This method is named comparative genomic hybridization because a comparison is made between the ability of two DNA samples (cancer versus normal cells) to hybridize to an entire genome. In this case, the entire genome is in the form of metaphase chromosomes. As discussed in Chapter 20, the fluorescently labeled DNAs can be hybridized to a DNA microarray instead of metaphase chromosomes. This newer method, called array comparative genomic hybridization (aCGH), is gaining widespread use in the analysis of cancer cells.

A self-help quiz involving this experiment can be found at the Online Learning Center.

Inversions Often Occur Without Phenotypic Consequences

We now turn our attention to inversions, changes in chromosome structure that involve a rearrangement in the genetic material. A chromosome with an inversion contains a segment that has been flipped to the opposite direction. Geneticists classify inversions according to the location of the centromere. If the centromere lies within the inverted region of the chromosome, the inverted region is known as a **pericentric inversion** (**Figure 8.9b**). Alternatively, if the centromere is found outside the inverted region, the inverted region is called a **paracentric inversion** (**Figure 8.9c**).

When a chromosome contains an inversion, the total amount of genetic material remains the same as in a normal chromosome. Therefore, the great majority of inversions do not have any phenotypic consequences. In rare cases, however, an inversion can alter the phenotype of an individual. Whether or not this occurs is related to the boundaries of the inverted segment. When an inversion occurs, the chromosome is broken in two places, and the center piece flips around to produce the inversion. If either breakpoint occurs within a vital gene, the function of the gene is expected to be disrupted, possibly producing a phenotypic effect. For example, some people with hemophilia (type A) have inherited an X-linked inversion in which the breakpoint has inactivated the gene for factor VIII, which is a blood-clotting protein. In other cases, an inversion (or translocation) may reposition a gene on a chromosome in a way that alters its normal level of expression. This is a type of **position effect**—a change in phenotype that occurs when the position of a gene changes from one chromosomal site to a different location. This topic is also discussed in Chapter 16 (see Figures 16.2 and 16.3).

Because inversions seem like an unusual genetic phenomenon, it is perhaps surprising that they are found in human populations in significant numbers. About 2% of the human population carries inversions that are detectable with a light microscope. In most cases, these individuals are phenotypically normal and live their lives without knowing they carry an inversion. In a few cases, however, an individual with an inversion chromosome may produce offspring with phenotypic abnormalities. This event

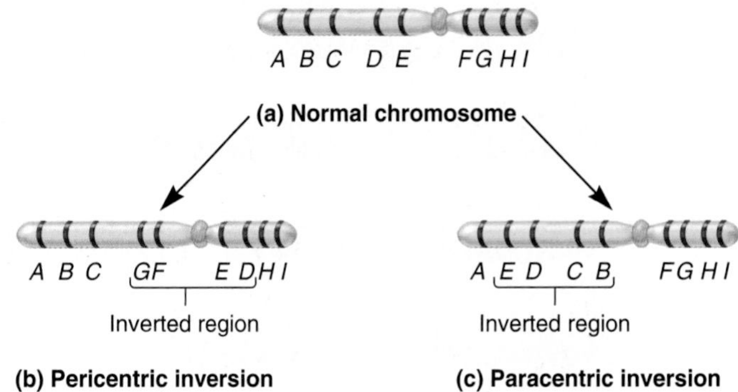

FIGURE 8.9 **Types of inversions.** (**a**) Depicts a normal chromosome with the genes ordered from *A* through *I*. A pericentric inversion (**b**) includes the centromere, whereas a paracentric inversion (**c**) does not.

may prompt a physician to request a microscopic examination of the individual's chromosomes. In this way, phenotypically normal individuals may discover they have a chromosome with an inversion. Next, we will examine why an individual carrying an inversion may produce offspring with phenotypic abnormalities.

Inversion Heterozygotes May Produce Abnormal Chromosomes Due to Crossing Over

An individual that carries one copy of a normal chromosome and one copy of an inverted chromosome is known as an **inversion heterozygote.** Such an individual, though possibly phenotypically normal, may have a high probability of producing haploid cells that are abnormal in their total genetic content. This likelihood depends on the size of the inverted segment. During meiosis, an inversion heterozygote with a fairly large inverted segment may produce a sizable fraction of abnormal haploid cells, perhaps 1/3 or even higher.

The underlying cause of abnormality is the phenomenon of crossing over within the inverted region. During meiosis I, pairs of homologous sister chromatids synapse with each other. **Figure 8.10** illustrates how this occurs in an inversion heterozygote. For the normal chromosome and inversion chromosome to synapse properly, an **inversion loop** must form to permit the homologous genes on both chromosomes to align next to each other despite the inverted sequence. If a crossover occurs within the inversion loop, highly abnormal chromosomes are produced.

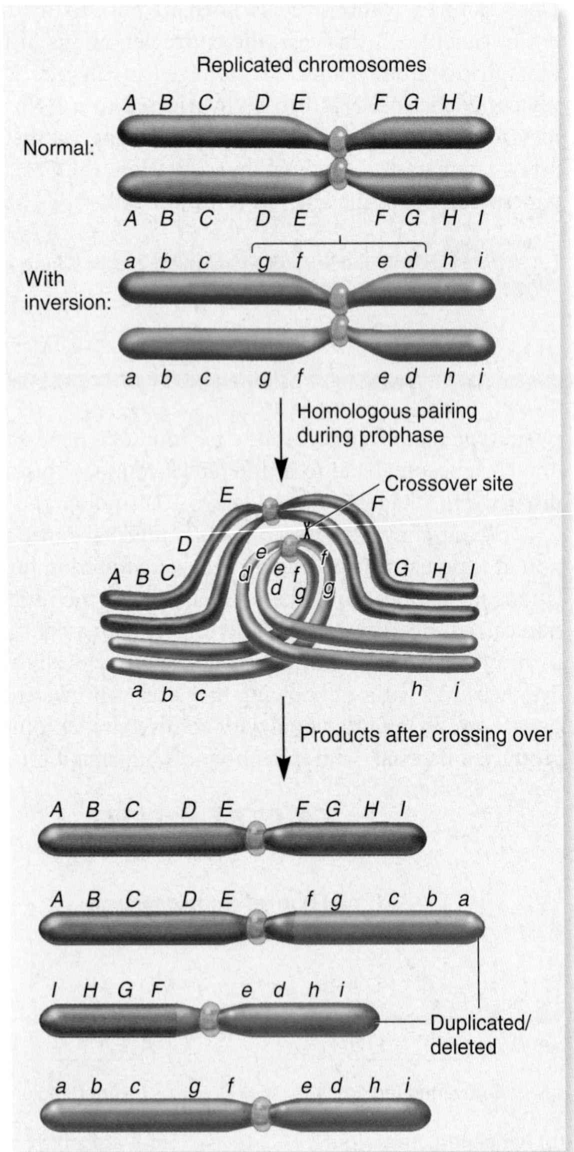

(a) Pericentric inversion

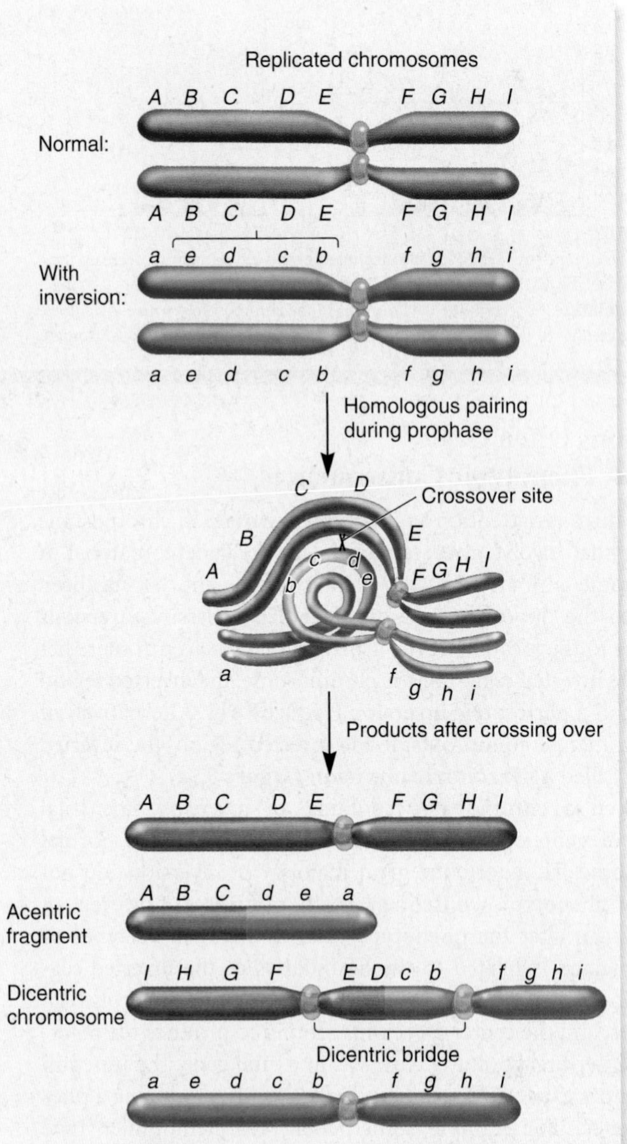

(b) Paracentric inversion

FIGURE 8.10 **The consequences of crossing over in the inversion loop.** **(a)** Crossover within a pericentric inversion. **(b)** Crossover within a paracentric inversion.

The consequences of this type of crossover depend on whether the inversion is pericentric or paracentric. Figure 8.10a describes a crossover in the inversion loop when one of the homologues has a pericentric inversion in which the centromere lies within the inverted region of the chromosome. This event consists of a single crossover that involves only two of the four sister chromatids. Following the completion of meiosis, this single crossover yields two abnormal chromosomes. Both of these abnormal chromosomes have a segment that is deleted and a different segment that is duplicated. In this example, one of the abnormal chromosomes is missing genes *H* and *I* and has an extra copy of genes *A*, *B*, and *C*. The other abnormal chromosome has the opposite situation; it is missing genes *A*, *B*, and *C* and has an extra copy of genes *H* and *I*. These abnormal chromosomes may result in gametes that are inviable. Alternatively, if these abnormal chromosomes are passed to offspring, they are likely to produce phenotypic abnormalities, depending on the amount and nature of the duplicated/deleted genetic material. A large deletion is likely to be lethal.

Figure 8.10b shows the outcome of a crossover involving a paracentric inversion in which the centromere lies outside the inverted region. This single crossover event produces a very strange outcome. One chromosome, called a **dicentric** chromosome, contains two centromeres. The region of the chromosome connecting the two centromeres is a **dicentric bridge.** The crossover also produces a piece of chromosome without any centromere—an **acentric fragment,** which will be lost and degraded in subsequent cell divisions. The dicentric chromosome is a temporary condition. If the two centromeres try to move toward opposite poles during anaphase, the dicentric bridge will be forced to break at some random location. Therefore, the net result of this crossover is to produce one normal chromosome, one chromosome with an inversion, and two chromosomes that contain deletions. These two chromosomes with deletions result from the breakage of the dicentric chromosome. They are missing the genes that were located on the acentric fragment.

Translocations Involve Exchanges Between Different Chromosomes

Another type of chromosomal rearrangement is a translocation in which a piece from one chromosome is attached to another chromosome. Eukaryotic cells have evolved telomeres, which tend to prevent translocations from occurring. As described in Chapters 10 and 11, **telomeres**—specialized repeated sequences of DNA—are found at the ends of normal chromosomes. Telomeres allow cells to distinguish where a chromosome ends and prevent the attachment of chromosomal DNA to the natural ends of a chromosome.

If cells are exposed to agents that cause chromosomes to break, the broken ends lack telomeres and are said to be reactive—a reactive end readily binds to another reactive end. If a single chromosome break occurs, DNA repair enzymes will usually recognize the two reactive ends and join them back together; the chromosome is repaired properly. However, if multiple chromosomes are broken, the reactive ends may be joined incorrectly to produce abnormal chromosomes (**Figure 8.11a**). This is one mechanism that causes reciprocal translocations to occur.

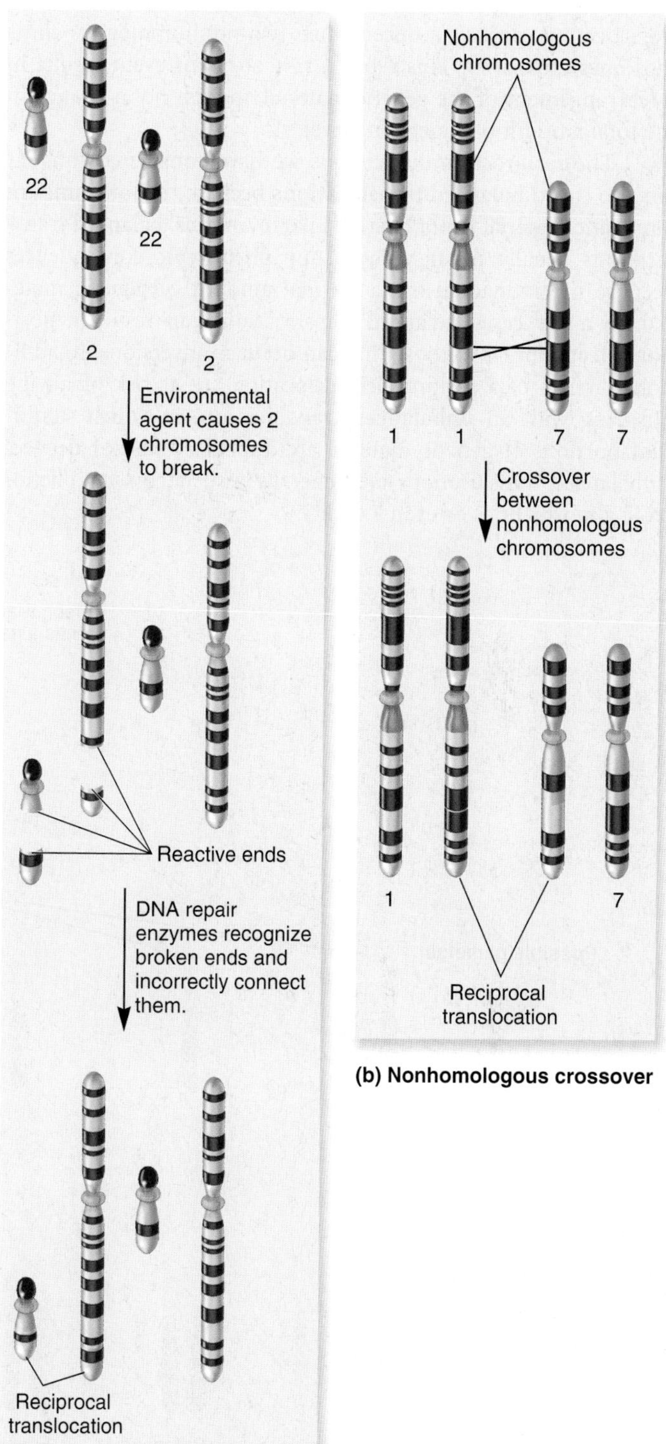

FIGURE 8.11 Two mechanisms that cause a reciprocal translocation. (**a**) When two different chromosomes break, the broken ends are recognized by DNA repair enzymes, which attempt to reattach broken ends. If two different chromosomes are broken at the same time, the incorrect ends may become attached to each other. (**b**) A nonhomologous crossover has occurred between chromosome 1 and chromosome 7. This crossover yields two chromosomes that carry translocations.

A second mechanism that can cause a translocation is an abnormal crossover. As shown in **Figure 8.11b**, a reciprocal translocation can be produced when two nonhomologous chromosomes cross over. This type of rare aberrant event results in a rearrangement of the genetic material, though not a change in the total amount of genetic material.

The reciprocal translocations we have considered thus far are also called **balanced translocations** because the total amount of genetic material is not altered. Like inversions, balanced translocations usually occur without any phenotypic consequences because the individual has a normal amount of genetic material. In a few cases, balanced translocations can result in position effects similar to those that can occur in inversions. In addition, carriers of a reciprocal translocation are at risk of having offspring with an **unbalanced translocation,** in which significant portions of genetic material are duplicated and/or deleted. Unbalanced translocations are generally associated with phenotypic abnormalities or even lethality.

Let's consider how a person with a balanced translocation may produce gametes and offspring with an unbalanced translocation. An inherited human syndrome known as familial Down syndrome provides an example. A person with a normal phenotype may have one copy of chromosome 14, one copy of chromosome 21, and one copy of a chromosome that is a fusion between chromosome 14 and 21 (**Figure 8.12a**). The individual has a normal phenotype because the total amount of genetic material is present (with the exception of the short arms of these chromosomes that do not carry vital genetic material). During meiosis, these three types of chromosomes will replicate and segregate from each other. However, because the three chromosomes cannot segregate evenly, six possible types of gametes may be produced. One gamete is normal, and one has a balanced amount of genetic material. The four gametes to the right, however, are unbalanced, either containing too much or too little material from chromosome 14 or 21. The unbalanced gametes may be inviable, of they could combine with a normal gamete. The three offspring on the right will not survive.

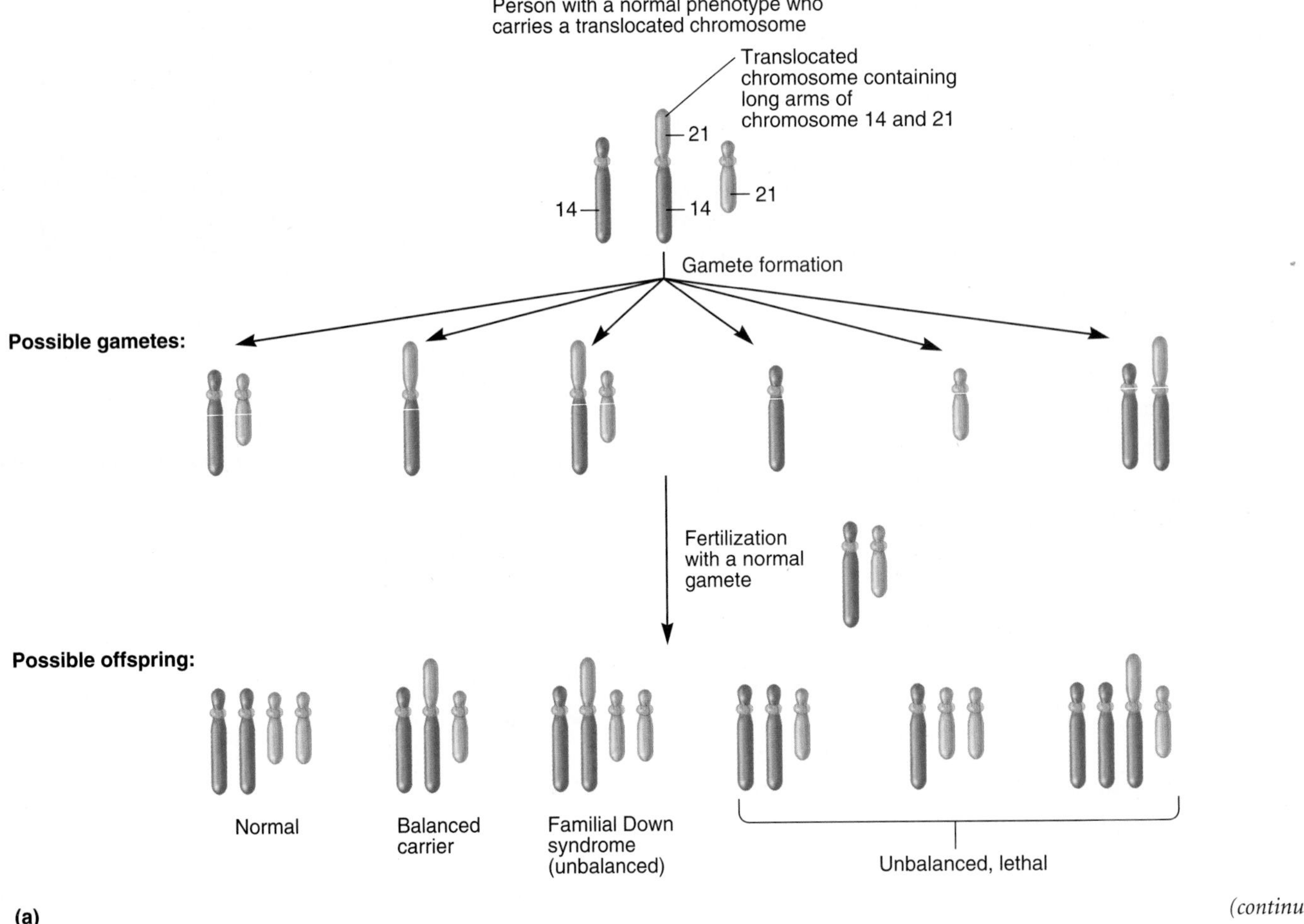

(continued)

FIGURE 8.12 **Transmission of familial Down syndrome.** (**a**) Potential transmission of familial Down syndrome. The individual with the chromosome composition shown at the top of this figure may produce a gamete carrying chromosome 21 and a fused chromosome containing the long arms of chromosomes 14 and 21. Such a gamete can give rise to an offspring with familial Down syndrome. (**b**) The karyotype of an individual with familial Down syndrome. This karyotype shows that the long arm of chromosome 21 has been translocated to chromosome 14 (see arrow). In addition, the individual also carries two normal copies of chromosome 21. (**c**) An individual with this disorder.

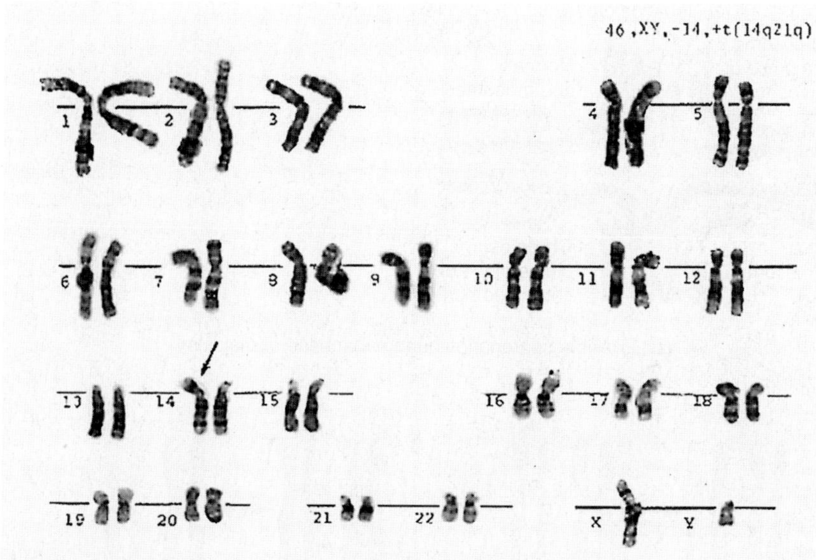

(b)

(c)

FIGURE 8.12 *(continued)*

In comparison, the unbalanced gamete that carries chromosome 21 and the fused chromosome will result in an offspring with familial Down syndrome (also see karyotype in **Figure 8.12b**). Such an offspring has three copies of the genes that are found on the long arm of chromosome 21. **Figure 8.12c** shows a person with this disorder. She has characteristics similar to those of an individual who has the more prevalent form of Down syndrome, which is due to three entire copies of chromosome 21. We will examine this common form of Down syndrome later in this chapter.

The abnormal chromosome that occurs in familial Down syndrome is an example of a **Robertsonian translocation,** named after William Robertson, who first described this type of fusion in grasshoppers. This type of translocation arises from breaks near the centromeres of two nonhomologous acrocentric chromosomes. In the example shown in Figure 8.12, the long arms of chromosomes 14 and 21 had fused, creating one large single chromosome; the two short arms are lost. This type of fusion between two nonhomologous acrocentric chromosomes is the most common type of chromosome rearrangement in humans, occurring at a frequency of approximately one in 900 live births. In humans, Robertsonian translocations involve only the acrocentric chromosomes 13, 14, 15, 21, and 22.

Individuals with Reciprocal Translocations May Produce Abnormal Gametes Due to the Segregation of Chromosomes

As we have seen, individuals who carry balanced translocations have a greater risk of producing gametes with unbalanced combinations of chromosomes. Whether or not this occurs depends on the segregation pattern during meiosis I (**Figure 8.13**). In this example, the parent carries a reciprocal translocation and is likely to be phenotypically normal. During meiosis, the homologous chromosomes attempt to synapse with each other. Because of the translocations, the pairing of homologous regions leads to the formation of an unusual structure that contains four pairs of sister chromatids (i.e., eight chromatids), termed a **translocation cross.**

To understand the segregation of translocated chromosomes, pay close attention to the centromeres, which are numbered in Figure 8.13. For these translocated chromosomes, the expected segregation pattern is governed by the centromeres. Each haploid gamete should receive one centromere located on chromosome 1 and one centromere located on chromosome 2. This can occur in two ways. One possibility is alternate segregation. As shown in Figure 8.13a, this occurs when the chromosomes diagonal to each other within the translocation cross sort into the same cell. One daughter cell receives two normal chromosomes, and the other cell gets two translocated chromosomes. Following meiosis II, four haploid cells are produced: two have normal chromosomes, and two have reciprocal (balanced) translocations.

Another possible segregation pattern is called adjacent-1 segregation (Figure 8.13b). This occurs when adjacent chromosomes (one of each type of centromere) segregate into the same cell. Following anaphase of meiosis I, each daughter cell receives one normal chromosome and one translocated chromosome. After meiosis II is completed, this produces four haploid cells, all of which are genetically unbalanced because part of one chromosome has been deleted and part of another has been duplicated. If these haploid cells give rise to gametes that unite with a normal gamete, the zygote is expected to be abnormal genetically and possibly phenotypically.

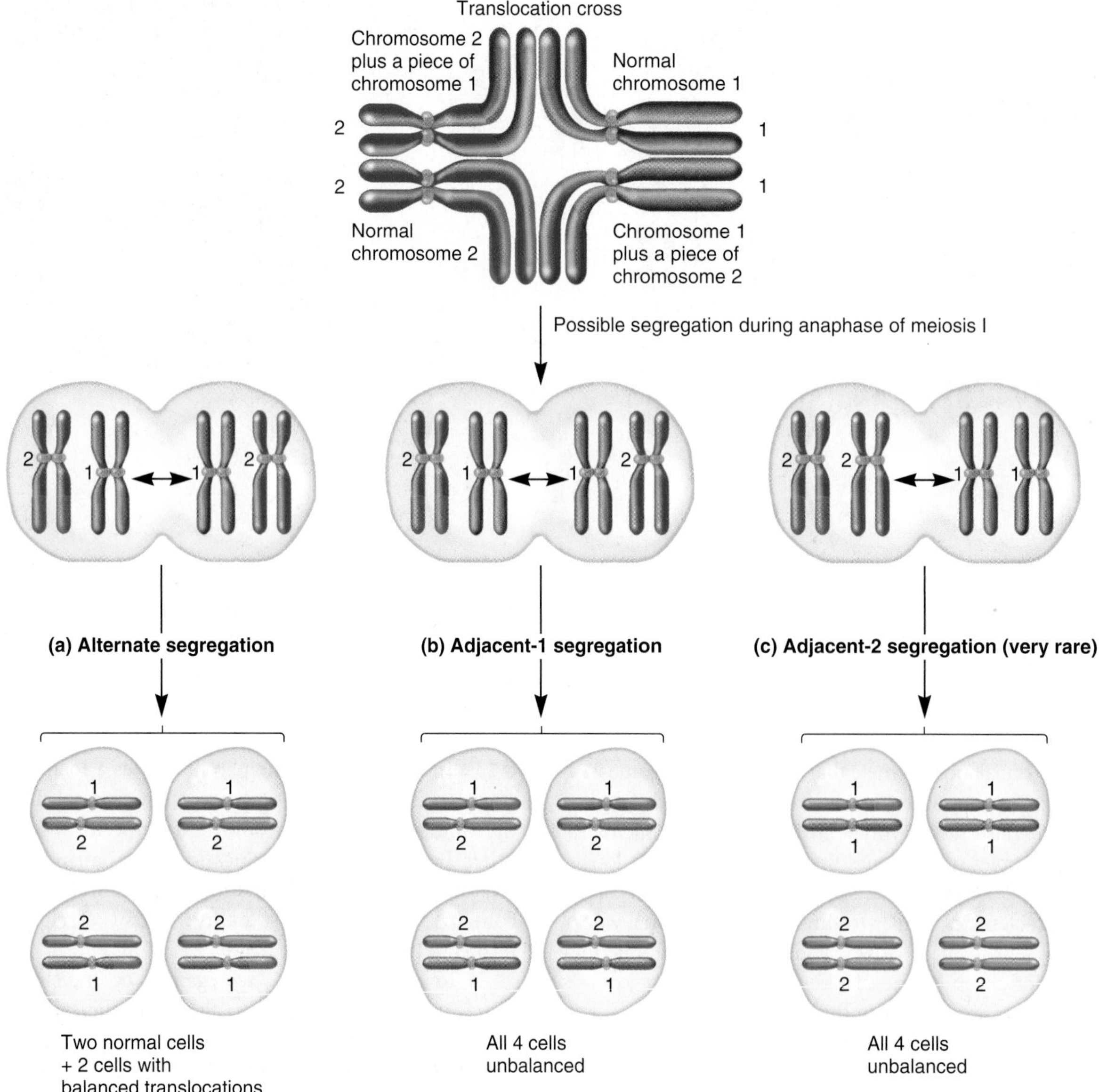

FIGURE 8.13 **Meiotic segregation of a reciprocal translocation.** Follow the numbered centromeres through each process. (**a**) Alternate segregation gives rise to balanced haploid cells, whereas (**b**) adjacent-1 and (**c**) adjacent-2 produce haploid cells with an unbalanced amount of genetic material.

On very rare occasions, adjacent-2 segregation can occur (Figure 8.13c). In this case, the centromeres do not segregate as they should. One daughter cell has received both copies of the centromere on chromosome 1; the other, both copies of the centromere on chromosome 2. This rare segregation pattern also yields four abnormal haploid cells that contain an unbalanced combination of chromosomes.

Alternate and adjacent-1 segregation patterns are the likely outcomes when an individual carries a reciprocal translocation. Depending on the sizes of the translocated segments, both types may be equally likely to occur. In many cases, the haploid cells from adjacent-1 segregation are not viable, thereby lowering the fertility of the parent. This condition is called **semisterility.**

8.2 VARIATION IN CHROMOSOME NUMBER

As we have seen in Section 8.1, chromosome structure can be altered in a variety of ways. Likewise, the total number of chromosomes can vary. Eukaryotic species typically contain several chromosomes that are inherited as one or more sets. Variations in chromosome number can be categorized in two ways: variation in the number of sets of chromosomes and variation in the number of particular chromosomes within a set.

Organisms that are **euploid** have a chromosome number that is an exact multiple of a chromosome set. In *Drosophila melanogaster*, for example, a normal individual has 8 chromosomes.

The species is diploid, having two sets of 4 chromosomes each (**Figure 8.14a**). A normal fruit fly is euploid because 8 chromosomes divided by 4 chromosomes per set equals two exact sets. On rare occasions, an abnormal fruit fly can be produced with 12 chromosomes, containing three sets of 4 chromosomes each. This alteration in euploidy produces a **triploid** fruit fly with 12 chromosomes. Such a fly is also euploid because it has exactly three sets of chromosomes. Organisms with three or more sets of chromosomes are also called **polyploid** (**Figure 8.14b**). Geneticists use the letter n to represent a set of chromosomes. A diploid organism is referred to as $2n$, a triploid organism as $3n$, a **tetraploid** organism as $4n$, and so on.

A second way in which chromosome number can vary is by **aneuploidy.** Such variation involves an alteration in the number of particular chromosomes, so the total number of chromosomes is not an exact multiple of a set. For example, an abnormal fruit fly could contain nine chromosomes instead of eight because it has three copies of chromosome 2 instead of the normal two copies (**Figure 8.14c**). Such an animal is said to have trisomy 2 or to be **trisomic.** Instead of being perfectly diploid (2n), a trisomic animal is $2n + 1$. By comparison, a fruit fly could be lacking a single chromosome, such as chromosome 1, and contain a total of seven chromosomes ($2n - 1$). This animal would be **monosomic** and be described as having monosomy 1.

In this section, we will begin by considering several examples of aneuploidy. This is generally regarded as an abnormal condition that usually has a negative impact on phenotype. We will then examine euploid variation that occurs occasionally in animals and quite frequently in plants, and consider how it affects phenotypic variation.

Aneuploidy Causes an Imbalance in Gene Expression That Is Often Detrimental to the Phenotype of the Individual

The phenotype of every eukaryotic species is influenced by thousands of different genes. In humans, for example, a single set of chromosomes contains approximately 20,000 to 25,000 different genes. To produce a phenotypically normal individual, intricate coordination has to occur in the expression of thousands of genes. In the case of humans and other diploid species, evolution has resulted in a developmental process that works correctly when somatic cells have two copies of each chromosome. In other words, when a human is diploid, the balance of gene expression among many different genes usually produces a person with a normal phenotype.

Aneuploidy commonly causes an abnormal phenotype. To understand why, let's consider the relationship between gene

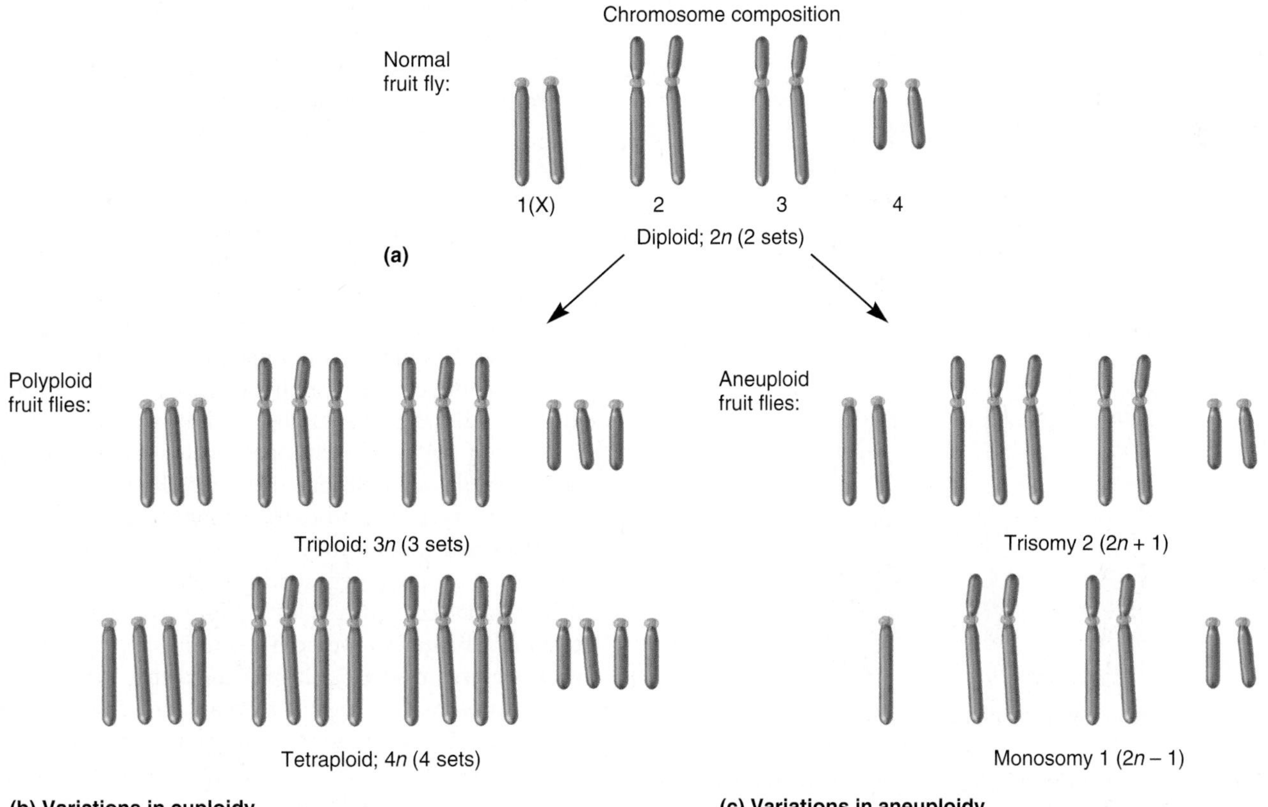

FIGURE 8.14 **Types of variation in chromosome number.** (a) Depicts the normal diploid number of chromosomes in *Drosophila.* (b) Examples of polyploidy. (c) Examples of aneuploidy.

expression and chromosome number in a species that has three pairs of chromosomes (**Figure 8.15**). The level of gene expression is influenced by the number of genes per cell. Compared to a diploid cell, if a gene is carried on a chromosome that is present in three copies instead of two, more of the gene product is typically made. For example, a gene present in three copies instead of two may produce 150% of the gene product, though that number may vary due to effects of gene regulation. Alternatively, if only one copy of that gene is present due to a missing chromosome, less of the gene product will usually be made, perhaps only 50%. Therefore, in trisomic and monosomic individuals, an imbalance occurs between the level of gene expression on the chromosomes found in pairs versus the one type that is not.

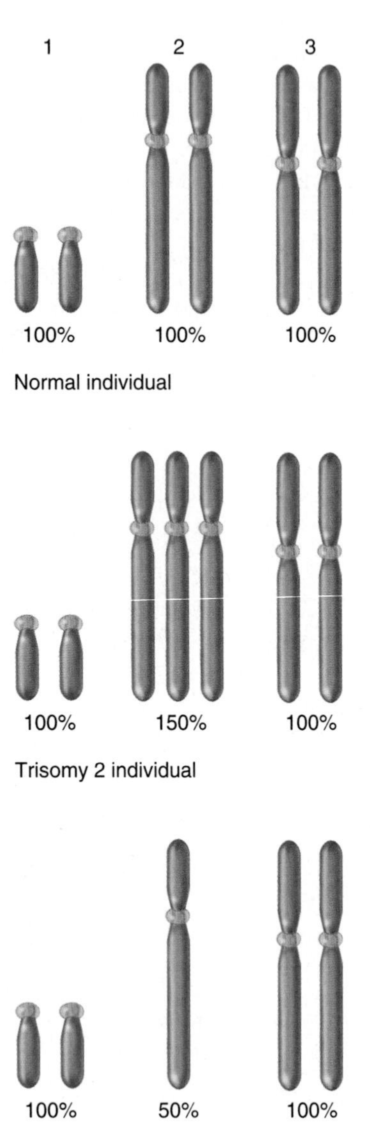

FIGURE 8.15 Imbalance of gene products in trisomic and monosomic individuals. Aneuploidy of chromosome 2 (i.e., trisomy and monosomy) leads to an imbalance in the amount of gene products from chromosome 2 compared to the amounts from chromosomes 1 and 3.

At first glance, the difference in gene expression between euploid and aneuploid individuals may not seem terribly dramatic. Keep in mind, however, that a eukaryotic chromosome carries hundreds or even thousands of different genes. Therefore, when an organism is trisomic or monosomic, many gene products will occur in excessive or deficient amounts. This imbalance among many genes appears to underlie the abnormal phenotypic effects that aneuploidy frequently causes. In most cases, these effects are detrimental and produce an individual that is less likely to survive than a euploid individual.

In the 1920s, Albert Blakeslee and his colleagues were the first to recognize the harmful effects of aneuploidy by studying Jimson weed (*Datura stramonium*). **Figure 8.16** compares normal and trisomic strains with regard to morphology of the capsule, which is the dry fruit. All 12 trisomies had capsules that were morphologically different from those of the normal diploid strain. These aneuploid plants also had many other morphologically distinguishable traits, including changes in leaf shape, size, and so forth. In many cases, the observed changes in chromosome number produced detrimental traits. For example, Blakeslee noted that the cocklebur plant (trisomy 6) is "weak and lopping with the leaves narrow and twisted."

Aneuploidy in Humans Causes Abnormal Phenotypes

A key reason why geneticists are so interested in aneuploidy is its relationship to certain inherited disorders in humans. Even though most people are born with a normal number of chromosomes (i.e., 46), alterations in chromosome number occur fairly frequently during gamete formation. About 5 to 10% of all fertilized human eggs result in an embryo with an abnormality in chromosome number! In most cases, these abnormal embryos do not develop properly and result in a spontaneous abortion very early in pregnancy. Approximately 50% of all spontaneous abortions are due to alterations in chromosome number.

In some cases, an abnormality in chromosome number produces an offspring that survives to birth or longer. Several human disorders involve abnormalities in chromosome number. The most common are trisomies of chromosomes 13, 18, or 21, and abnormalities in the number of the sex chromosomes (**Table 8.1**). Most of the known trisomies involve chromosomes that are relatively small—chromosome 13, 18, or 21—and may carry fewer genes than larger chromosomes. Trisomies of the other human autosomes and monosomies of all autosomes are presumed to produce a lethal phenotype, and many have been found in spontaneously aborted embryos and fetuses. For example, all possible human trisomies have been found in spontaneously aborted embryos except trisomy 1. It is believed that trisomy 1 is lethal at such an early stage that it prevents the successful implantation of the embryo. Variation in the number of X chromosomes, unlike that of other large chromosomes, is often nonlethal. The survival of trisomy X individuals may be explained by X inactivation, which is described in Chapter 7. In an individual with more than one X chromosome, all additional X chromosomes are converted to Barr bodies in the somatic cells of adult tissues. In an

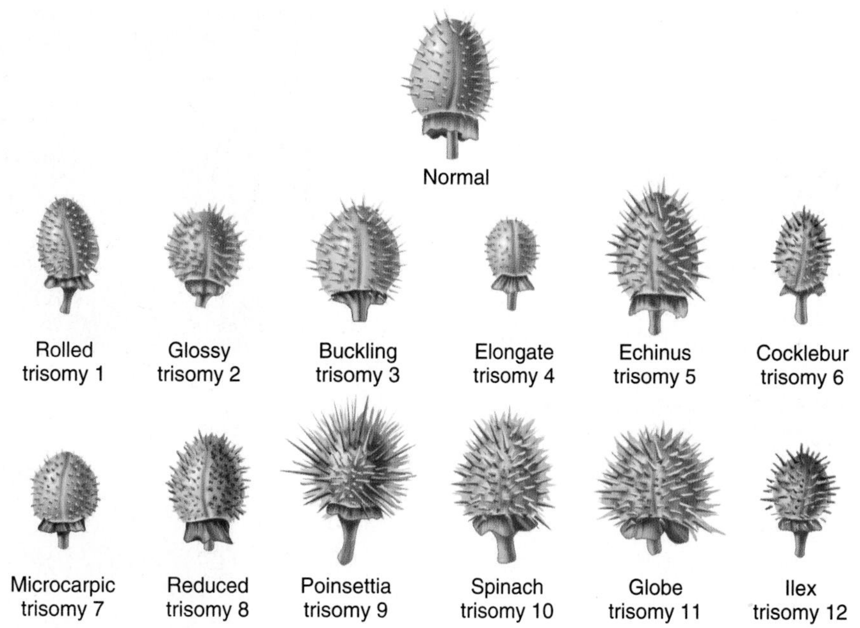

FIGURE 8.16 The effects of trisomy on the phenotype of the capsules of Jimson weed, *Datura stramonium*.
Genes → Traits As described in Figure 8.15, trisomy leads to an imbalance in the copy number of genes on the trisomic chromosomes (here, 150% of normal) versus those on the other chromosomes (100%). This drawing compares the morphology of the capsule of the *Datura* plant in a normal individual and in those carrying a trisomy of each of the 12 types of chromosomes in this species. As seen here, trisomy of each chromosome causes changes in the phenotype of the capsule that are due to changes in the balance of gene expression.

TABLE 8.1

Aneuploid Conditions in Humans

Condition	Frequency	Syndrome	Characteristics
Autosomal			
Trisomy 13	1/15,000	Patau	Mental and physical deficiencies, wide variety of defects in organs, large triangular nose, early death
Trisomy 18	1/6,000	Edward	Mental and physical deficiencies, facial abnormalities, extreme muscle tone, early death
Trisomy 21	1/800	Down	Mental deficiencies, abnormal pattern of palm creases, slanted eyes, flattened face, short stature
Sex Chromosomal			
XXY	1/1,000 (males)	Klinefelter	Sexual immaturity (no sperm), breast swelling
XYY	1/1,000 (males)	Jacobs	Tall and thin
XXX	1/1,500 (females)	Triple X	Tall and thin, menstrual irregularity
X0	1/5,000 (females)	Turner	Short stature, webbed neck, sexually undeveloped

individual with trisomy X, for example, two out of three X chromosomes are converted to inactive Barr bodies. Unlike the level of expression for autosomal genes, the normal level of expression for X-linked genes is from a single X chromosome. In other words, the correct level of mammalian gene expression results from two copies of each autosomal gene and one copy of each X-linked gene. This explains how the expression of X-linked genes in males (XY) can be maintained at the same levels as in females (XX). It may also explain why trisomy X is not a lethal condition. The phenotypic effects noted in Table 8.1 involving sex chromosomal abnormalities may be due to the expression of X-linked genes prior to embryonic X inactivation or to the expression of genes on the inactivated X chromosome. As described in Chapter 7, pseudoautosomal genes and some other genes on the inactivated X chromosome are expressed in humans. Having one or three copies of the sex chromosomes would result in an under- or overexpression of these X-linked genes, respectively.

Some human abnormalities in chromosome number are influenced by the age of the parents. Older parents are more likely to produce children with abnormalities in chromosome number. Down syndrome provides an example. The common form of this disorder is caused by the inheritance of three copies of chromosome 21. The incidence of Down syndrome rises with the age of either parent. In males, however, the rise occurs relatively late in life, usually past the age when most men have children. By comparison, the likelihood of having a child with Down syndrome

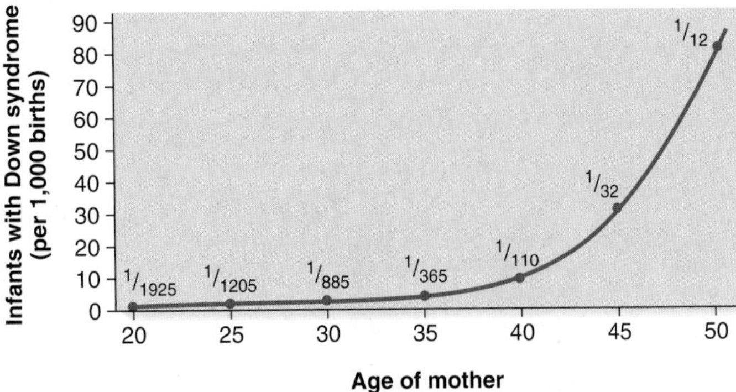

FIGURE 8.17 The incidence of Down syndrome births according to the age of the mother. The y-axis shows the number of infants born with Down syndrome per 1,000 live births, and the x-axis plots the age of the mother at the time of birth. The data points indicate the fraction of live offspring born with Down syndrome.

rises dramatically during a woman's reproductive age (**Figure 8.17**). This syndrome was first described by the English physician John Langdon Down in 1866. The association between maternal age and Down syndrome was later discovered by L. S. Penrose in 1933, even before the chromosomal basis for the disorder was identified by the French scientist Jérôme Lejeune in 1959. Down syndrome is most commonly caused by **nondisjunction,** which means that the chromosomes do not segregate properly. In this case, nondisjunction of chromosome 21 most commonly occurs during meiosis I in the oocyte.

Different hypotheses have been proposed to explain the relationship between maternal age and Down syndrome. One popular idea suggests that it may be due to the age of the oocytes. Human primary oocytes are produced within the ovary of the female fetus prior to birth and are arrested at prophase of meiosis I and remain in this stage until the time of ovulation. Therefore, as a woman ages, her primary oocytes have been in prophase I for a progressively longer period of time. This added length of time may contribute to an increased frequency of nondisjunction. About 5% of the time, Down syndrome is due to an extra paternal chromosome. Prenatal tests can determine if a fetus has Down syndrome and some other genetic abnormalities. The topic of genetic testing is discussed in Chapter 22.

(a) *Hyla chrysoscelis*

(b) *Hyla versicolor*

FIGURE 8.18 Differences in euploidy in two closely related frog species. The frog in (**a**) is diploid, whereas the frog in (**b**) is tetraploid. Both species grow to the same size.

Genes → Traits Though similar in appearance, these two species differ in their number of chromosome sets. At the level of gene expression, this observation suggests that the number of copies of each gene (two versus four) does not critically affect the phenotype of these two species.

Variations in Euploidy Occur Naturally in a Few Animal Species

We now turn our attention to changes in the number of sets of chromosomes, referred to as variations in euploidy. Most species of animals are diploid. In some cases, changes in euploidy are not well tolerated. For example, polyploidy in mammals is generally a lethal condition. However, many examples of naturally occurring variations in euploidy occur. In **haplodiploid** species, which includes many species of bees, wasps, and ants, one of the sexes is haploid, usually the male, and the other is diploid. For example, male bees, which are called drones, contain a single set of chromosomes. They are produced from unfertilized eggs. By comparison, female bees are produced from fertilized eggs and are diploid.

Many examples of vertebrate polyploid animals have been discovered. Interestingly, on several occasions, animals that are morphologically very similar to each other can be found as a diploid species as well as a separate polyploid species. This situation occurs among certain amphibians and reptiles. **Figure 8.18** shows photographs of a diploid and a tetraploid ($4n$) frog. As you can see, they look indistinguishable from each other. Their difference can be revealed only by an examination of the chromosome number in the somatic cells of the animals and by mating calls—*H. chrysoscelis* has a faster trill rate than *H. versicolor*.

Variations in Euploidy Can Occur in Certain Tissues Within an Animal

Thus far, we have considered variations in chromosome number that occur at fertilization, so all the somatic cells of an individual contain this variation. In many animals, certain tissues of the body display normal variations in the number of sets of chromosomes. Diploid animals sometimes produce tissues that are polyploid. For example, the cells of the human liver can vary to a great degree in their ploidy. Liver cells contain nuclei that can be triploid, tetraploid, and even octaploid (8*n*). The occurrence of polyploid tissues or cells in organisms that are otherwise diploid is known as **endopolyploidy.** What is the biological significance of endopolyploidy? One possibility is that the increase in chromosome number in certain cells may enhance their ability to produce specific gene products that are needed in great abundance.

An unusual example of natural variation in the ploidy of somatic cells occurs in *Drosophila* and some other insects. Within certain tissues, such as the salivary glands, the chromosomes undergo repeated rounds of chromosome replication without cellular division. For example, in the salivary gland cells of *Drosophila,* the pairs of chromosomes double approximately nine times ($2^9 = 512$). **Figure 8.19a** illustrates how repeated rounds of chromosomal replication produce a bundle of chromosomes that lie together in a parallel fashion. This bundle, termed a **polytene chromosome,** was first observed by E. G. Balbiani in 1881. Later, in the 1930s, Theophilus Painter and colleagues recognized that the size and morphology of polytene chromosomes provided geneticists with unique opportunities to study chromosome structure and gene organization.

Figure 8.19b shows a micrograph of a polytene chromosome. The structure of polytene chromosomes is different from other forms of endopolyploidy because the replicated chromosomes remain attached to each other. Prior to the formation of polytene chromosomes, *Drosophila* cells contain eight chromosomes (two sets of four chromosomes each; see Figure 8.14a). In the salivary gland cells, the homologous chromosomes synapse with each other and replicate to form a polytene structure. During this process, the four types of chromosomes aggregate to form a single structure with several polytene arms. The central point where the chromosomes aggregate is known as the **chromocenter.** Each of the four types of chromosome is attached to the chromocenter near its centromere. The X and Y and chromosome 4 are telocentric, and chromosomes 2 and 3 are metacentric. Therefore, chromosomes 2 and 3 have two arms that radiate from the chromocenter, while the X and Y and chromosome 4 have a single arm projecting from the chromocenter (**Figure 8.19c**).

Because of their considerable size, polytene chromosomes lend themselves to an easy microscopic examination. Ordinarily, we use light microscopy to visualize the highly condensed metaphase chromosomes seen during mitosis or meiosis, as in Figure 8.1a. Because polytene chromosomes are so large, we

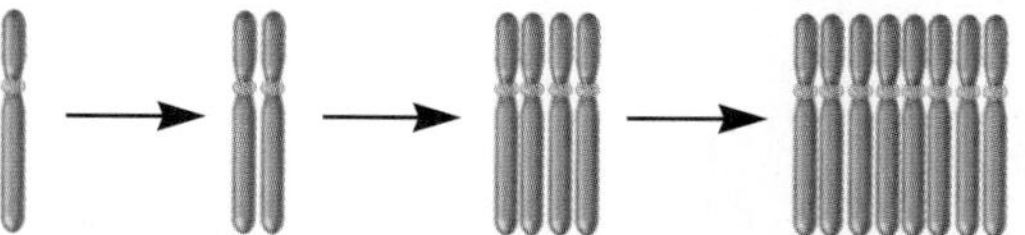

(a) Repeated chromosome replication produces polytene chromosome.

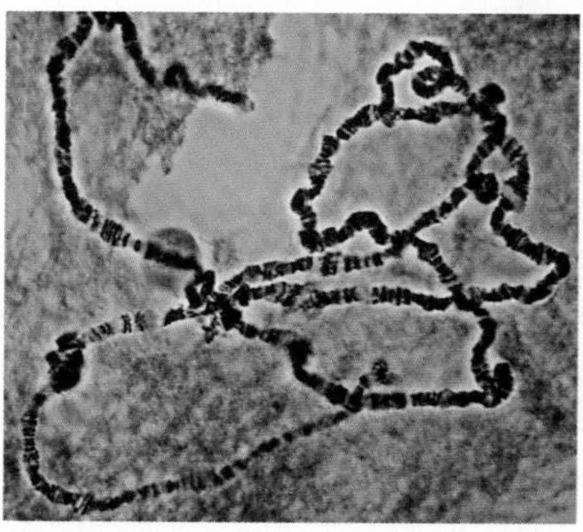

(b) A polytene chromosome

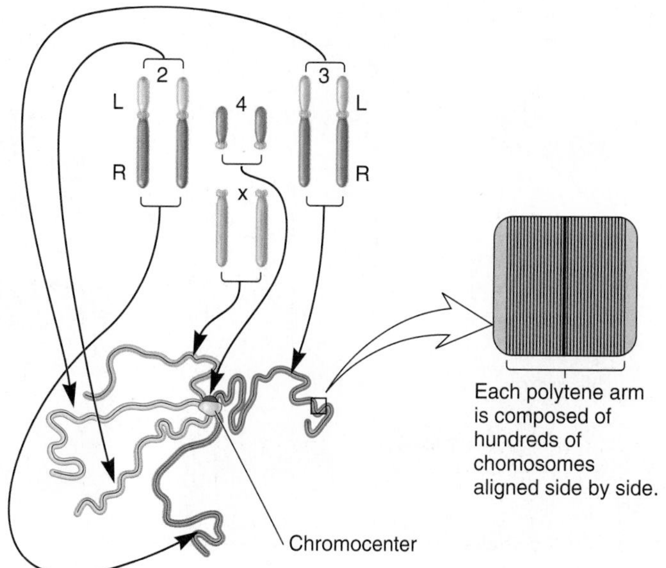

(c) Relationship between a polytene chromosome and regular *Drosophila* chromosomes

FIGURE 8.19 **Polytene chromosomes in *Drosophila.*** **(a)** A schematic illustration of the formation of polytene chromosomes. Several rounds of repeated replication without cellular division result in a bundle of sister chromatids that lie side by side. Both homologues also lie parallel to each other. This replication does not occur in highly condensed, heterochromatic DNA near the centromere. **(b)** A photograph of a polytene chromosome. **(c)** This drawing shows the relationship between the four pairs of chromosomes and the formation of a polytene chromosome in the salivary gland. The heterochromatic regions of the chromosomes aggregate at the chromocenter, and the arms of the chromosomes project outward.

can see them during interphase, when normal chromosomes are not readily visible. Remarkably, a polytene chromosome during interphase is 100 to 200 times larger than the average metaphase chromosome. As shown in Figure 8.19b, polytene chromosomes exhibit a characteristic banding pattern. Each dark band is known as a **chromomere.** The structure of the genetic material within a dark band is more compact than the interband region. More than 95% of the DNA is found within these dark bands. The banding patterns of polytene chromosomes are much more detailed than those observed in metaphase chromosomes. Cytogeneticists have identified approximately 5,000 bands along polytene chromosomes. At one time, each chromomere was thought to correspond to one gene. However, this idea was found to be incorrect because the sequencing of the entire *Drosophila* genome has revealed an approximate gene number of 14,000.

Polytene chromosomes have allowed geneticists to study the organization and functioning of interphase chromosomes in great detail. When a gene is deleted or duplicated via mutation, researchers can map the change if it results in a microscopically visible alteration in the structure of a polytene chromosome. In addition, because polytene chromosomes can be observed during interphase, the expression of particular genes in salivary cells can be correlated with changes in the compaction of certain bands in the polytene chromosome.

Variations in Euploidy Are Common in Plants

We now turn our attention to variations of euploidy that occur in plants. Compared to animals, plants more commonly exhibit polyploidy. Among ferns and flowering plants, at least 30 to 35% of species are polyploid. Polyploidy is also important in agriculture. Many of the fruits and grains we eat are produced from polyploid plants. For example, the species of wheat that we use to make bread, *Triticum aestivum*, is a hexaploid (6n) that arose from the union of diploid genomes from three closely related species (**Figure 8.20a**).

In many instances, polyploid strains of plants display outstanding agricultural characteristics. They are often larger in size and more robust. These traits are clearly advantageous in the production of food. In addition, polyploid plants tend to exhibit a greater adaptability, which allows them to withstand harsher environmental conditions. Also, polyploid ornamental plants often produce larger flowers than their diploid counterparts (**Figure 8.20b**).

Polyploid plants having an odd number of chromosome sets, such as triploids ($3n$) or pentaploids ($5n$), usually cannot reproduce. Why are they sterile? The sterility arises because they produce highly aneuploid gametes. During prophase of meiosis I, homologous pairs of sister chromatids will form bivalents. However,

(a) Cultivated wheat, a hexaploid species

Tetraploid

Diploid

(b) A comparison of diploid and tetraploid petunias

FIGURE 8.20 **Examples of polyploid plants.** (**a**) Cultivated wheat, *Triticum aestivum*, is a hexaploid. It was derived from three different diploid species of grasses that originally were found in the Middle East and were cultivated by ancient farmers in that region. (**b**) Differences in euploidy may exist in two closely related petunia species. The flower at the bottom is diploid, whereas the large one at the top is tetraploid.

Genes → Traits An increase in chromosome number from diploid to tetraploid or hexaploid affects the phenotype of the individual. In the case of many plant species, a polyploid individual is larger and more robust than its diploid counterpart. This suggests that having additional copies of each gene is somewhat better than having two copies of each gene. This phenomenon in plants is rather different from the situation in animals. Tetraploidy in animals may have little effect (as in Figure 8.18b), and it is also common for polyploidy in animals to be detrimental.

organisms with an odd number of chromosomes, such as three, will display an unequal separation of homologous chromosomes during anaphase of meiosis I (**Figure 8.21**). An odd number cannot be divided equally between two daughter cells. For each type of chromosome, a daughter cell randomly gets one or two copies. For example, one daughter cell might receive one copy of chromosome 1, two copies of chromosome 2, two copies of chromosome 3, one copy of chromosome 4, and so forth. For a triploid species containing many different chromosomes in a set, meiosis is very unlikely to produce a daughter cell that is euploid. If we assume that a daughter cell will receive either one copy or two copies of each kind of chromosome, the probability that meiosis will produce a cell that will be perfectly haploid or diploid is $(1/2)^{n-1}$, where n is the number of chromosomes in a set. As an example, in a triploid organism containing 20 chromosomes per set, the probability of producing a haploid or diploid cell is 0.000001907, or 1 in 524,288. Thus, meiosis is almost certain to produce cells that contain one copy of some chromosomes and two copies of the other chromosomes. This high probability of aneuploidy underlies the reason for triploid sterility.

Though sterility is generally a detrimental trait, it can be desirable agriculturally because it may result in a seedless fruit. For example, domestic bananas and seedless watermelons are triploid varieties. The domestic banana was originally derived from a seed-producing diploid species and has been asexually propagated by humans via cuttings. The small black spots in the center of a domestic banana are degenerate seeds. In the case of flowers, the seedless phenotype can also be beneficial. Seed producers such as Burpee have developed triploid varieties of flowering plants such as marigolds. Because the triploid marigolds are sterile and unable to set seed, more of their energy goes into flower production. According to Burpee, "They bloom and bloom, unweakened by seed bearing."

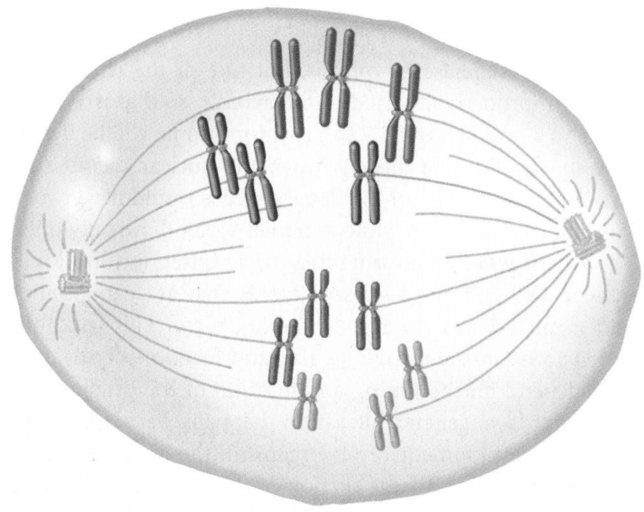

FIGURE 8.21 Schematic representation of anaphase of meiosis I in a triploid organism containing three sets of four chromosomes. In this example, the homologous chromosomes (three each) do not evenly separate during anaphase. Each cell receives one copy of some chromosomes and two copies of other chromosomes. This produces aneuploid gametes.

8.3 NATURAL AND EXPERIMENTAL WAYS TO PRODUCE VARIATIONS IN CHROMOSOME NUMBER

As we have seen, variations in chromosome number are fairly widespread and usually have a significant impact on the phenotypes of plants and animals. For these reasons, researchers have wanted to understand the cellular mechanisms that cause variations in chromosome number. In some cases, a change in chromosome number is the result of nondisjunction. The term nondisjunction refers to an event in which the chromosomes do not segregate properly. As we will see, it may be caused by an improper separation of homologous pairs in a bivalent in meiosis or a failure of the centromeres to disconnect during mitosis.

Meiotic nondisjunction can produce haploid cells that have too many or too few chromosomes. If such a cell gives rise to a gamete that fuses with a normal gamete during fertilization, the resulting offspring will have an abnormal chromosome number in all of its cells. An abnormal nondisjunction event also may occur after fertilization in one of the somatic cells of the body. This second mechanism is known as **mitotic nondisjunction.** When this occurs during embryonic stages of development, it may lead to a patch of tissue in the organism that has an altered chromosome number. A third common way in which the chromosome number of an organism can vary is by interspecies crosses. An **alloploid** organism contains sets of chromosomes from two or more different species. This term refers to the occurrence of chromosome sets (ploidy) from the genomes of different (allo) species.

In this section, we will examine these three mechanisms in greater detail. Also, in the past few decades, researchers have devised several methods to manipulate chromosome number in experimentally and agriculturally important species. We will conclude this section by exploring how the experimental manipulation of chromosome number has had an important impact on genetic research and agriculture.

Meiotic Nondisjunction Can Produce Aneuploidy or Polyploidy

Nondisjunction during meiosis can occur during anaphase of meiosis I or meiosis II. If it happens during meiosis I, an entire bivalent migrates to one pole (**Figure 8.22a**). Following the completion of meiosis, the four resulting haploid cells produced from this event are abnormal. If nondisjunction occurs during anaphase of meiosis II (**Figure 8.22b**), the net result is two abnormal

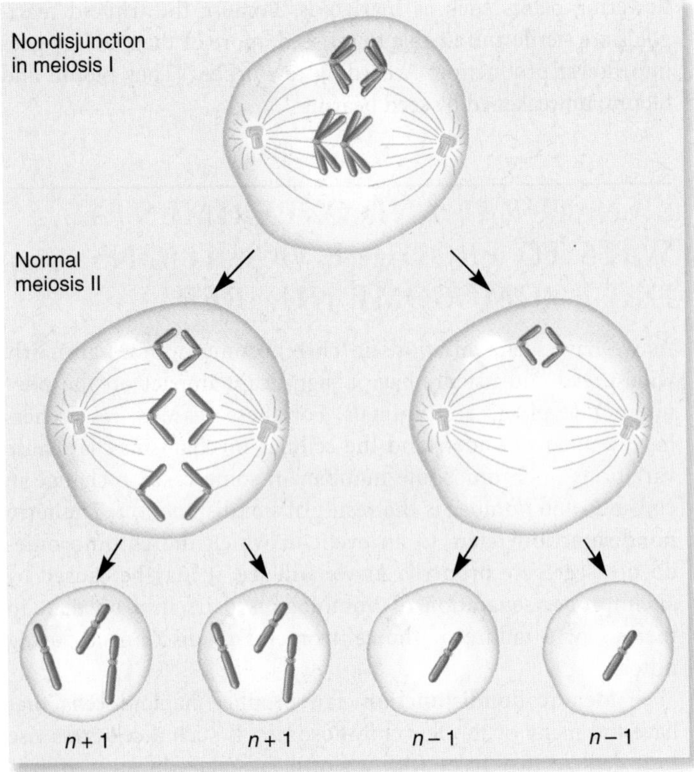

(a) **Nondisjunction in meiosis I**

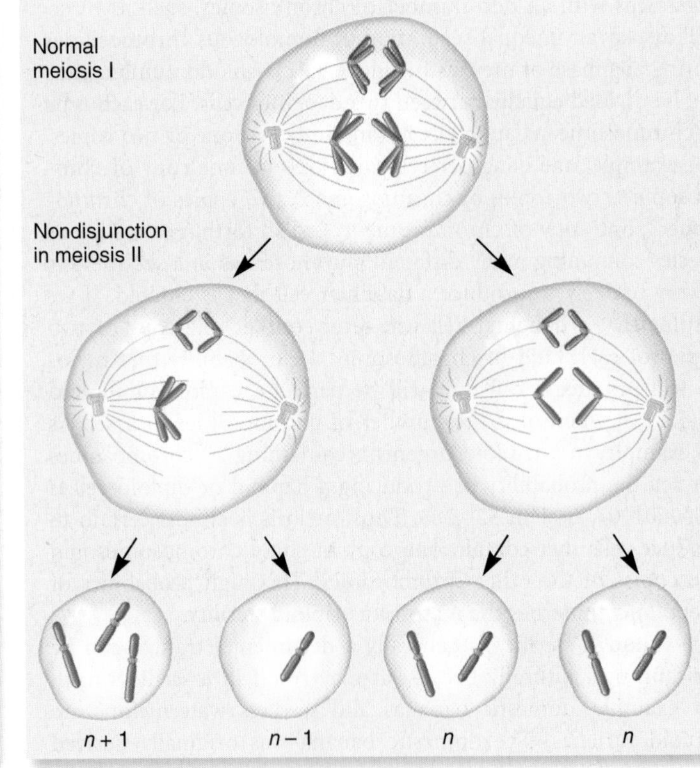

(b) **Nondisjunction in meiosis II**

FIGURE 8.22 **Nondisjunction during meiosis I and II.** The chromosomes shown in purple are behaving properly during meiosis I and II, so each cell receives one copy of this chromosome. The chromosomes shown in blue are not disjoining correctly. In (**a**), nondisjunction occurred in meiosis I, so the resulting four cells receive either two copies of the blue chromosome or zero copies. In (**b**), nondisjunction occurred during meiosis II, so one cell has two blue chromosomes and another cell has zero. The remaining two cells are normal.

and two normal haploid cells. If a gamete that is missing a chromosome is viable and participates in fertilization, the resulting offspring is monosomic for the missing chromosomes. Alternatively, if a gamete carrying an extra chromosome unites with a normal gamete, the offspring will be trisomic.

In rare cases, all of the chromosomes can undergo nondisjunction and migrate to one of the daughter cells. The net result of **complete nondisjunction** is a diploid cell and a cell without any chromosomes. While the cell without chromosomes is nonviable, the diploid cell might participate in fertilization with a normal haploid gamete to produce a triploid individual. Therefore, complete nondisjunction can produce individuals that are polyploid.

Mitotic Nondisjunction or Chromosome Loss Can Produce a Patch of Tissue with an Altered Chromosome Number

Abnormalities in chromosome number occasionally occur after fertilization takes place. In this case, the abnormal event happens during mitosis rather than meiosis. One possibility is that the sister chromatids separate improperly, so one daughter cell has three copies of that chromosome while the other daughter cell has only one (**Figure 8.23a**). Alternatively, the sister chromatids could separate during anaphase of mitosis, but one of the chromosomes could be improperly attached to the spindle, so that it would not migrate to a pole (**Figure 8.23b**). A chromosome will be degraded if it is left outside the nucleus when the nuclear membrane re-forms. In this case, one of the daughter cells would have two copies of that chromosome, while the other would have only one.

When genetic abnormalities occur after fertilization, the organism will contain a subset of cells that are genetically different from those of the rest of the organism. This condition is referred to as **mosaicism.** The size and location of the mosaic region depend on the timing and location of the original abnormal event. If a genetic alteration happens very early in the embryonic development of an organism, the abnormal cell will be the precursor for a large section of the organism. In the most extreme case, an abnormality could take place at the first mitotic division. As a bizarre example, consider a fertilized *Drosophila* egg that is XX. One of the X chromosomes may be lost during the first mitotic division, producing one daughter cell that is XX and one that is X0. Flies that are XX develop into females, and X0 flies develop into males. Therefore, in this example, one-half of the organism will become female and one-half will become male! This peculiar and rare individual is referred to as a **bilateral gynandromorph** (**Figure 8.24**).

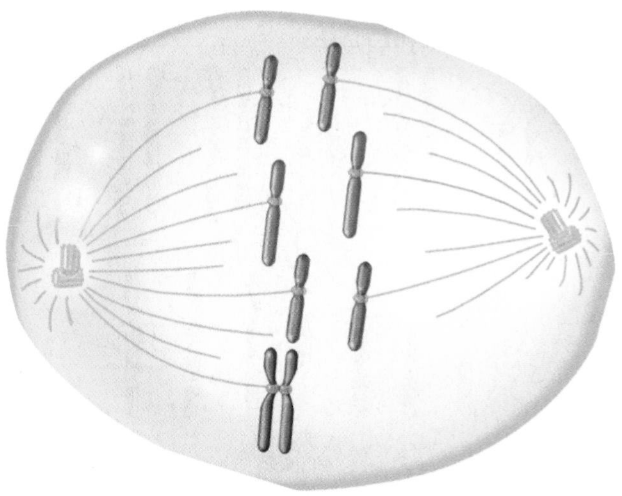

(a) Mitotic nondisjunction

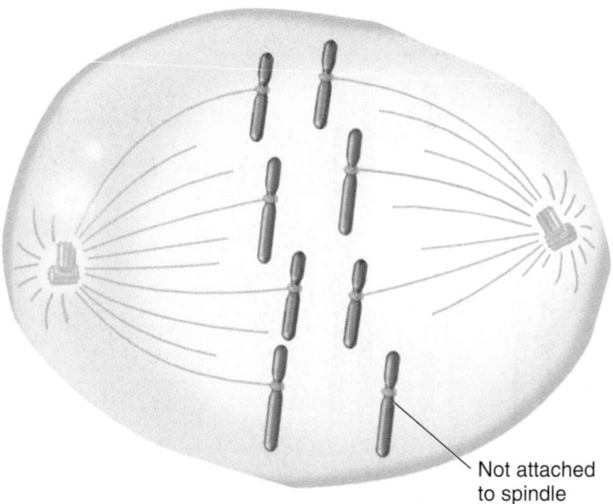

(b) Chromosome loss

FIGURE 8.23 **Nondisjunction and chromosome loss during mitosis in somatic cells.** (**a**) Mitotic nondisjunction produces a trisomic and a monosomic daughter cell. (**b**) Chromosome loss produces a normal and a monosomic daughter cell.

FIGURE 8.24 **A bilateral gynandromorph of *Drosophila melanogaster*.**

Genes → Traits In *Drosophila*, the ratio between genes on the X chromosome and genes on the autosomes determines sex. This fly began as an XX female. One X chromosome carried the recessive white-eye and miniature wing alleles, while the other X chromosome carried the wild-type alleles. The X chromosome carrying the wild-type alleles was lost from one of the cells during the first mitotic division, producing one XX cell and one X0 cell. The XX cell became the precursor for the left side of the fly, which developed as female. The X0 cell became the precursor for the other side of the fly, which developed as male with a white eye and a miniature wing.

Changes in Euploidy Can Occur by Autopolyploidy, Alloploidy, and Allopolyploidy

Different mechanisms account for changes in the number of chromosome sets among natural populations of plants and animals (**Figure 8.25**). As previously mentioned, complete nondisjunction, due to a general defect in the spindle apparatus, can produce an individual with one or more extra sets of chromosomes. This individual is known as an **autopolyploid** (Figure 8.25a). The prefix auto- (meaning self) and term polyploid (meaning many sets of chromosomes) refer to an increase in the number of chromosome sets within a single species.

A much more common mechanism for change in chromosome number, called **alloploidy,** is a result of interspecies crosses (Figure 8.25b). An alloploid that has one set of chromosomes from two different species is called an **allodiploid.** This event is most likely to occur between species that are close evolutionary relatives. For example, closely related species of grasses may interbreed to produce allodiploids. As shown in Figure 8.25c, an **allopolyploid** contains two (or more) sets of chromosomes from two (or more) species. In this case, the **allotetraploid** contains two complete sets of chromosomes from two different species, for a total of four sets. In nature, allotetraploids usually arise from allodiploids. This can occur when a somatic cell in an allodiploid undergoes complete nondisjunction to create an allotetraploid cell. In plants, such a cell can continue to grow and produce a section of the plant that is allotetraploid. If this part of the plant produced seeds, the seeds would give rise to allotetraploid offspring. Cultivated wheat (refer back to Figure 8.20a) is a plant in which two species must have interbred to create an allotetraploid, and then a third species interbred with the allotetraploid to create an allohexaploid.

Allodiploids Are Often Sterile, but Allotetraploids Are More Likely to Be Fertile

Geneticists are interested in the production of alloploids and allopolyploids as ways to generate interspecies hybrids with desirable traits. For example, if one species of grass can withstand hot temperatures and a closely related species is adapted to survive cold winters, a plant breeder may attempt to produce an interspecies hybrid that combines both qualities—good growth in the heat and survival through the winter. Such an alloploid may be desirable in climates with both hot summers and cold winters.

An important determinant of success in producing a fertile allodiploid is the degree of similarity of the different species' chromosomes. In two very closely related species, the number and types of chromosomes might be very similar. **Figure 8.26** shows a karyotype of an interspecies hybrid between the roan antelope (*Hippotragus equinus*) and the sable antelope (*Hippotragus niger*). As seen here, these two closely related species have the same number of chromosomes. The sizes and banding patterns of the chromosomes show that they correspond to one another. For example, chromosome 1 from both species is fairly large, has very similar banding patterns, and carries many of the same genes. Evolutionarily related chromosomes from two different species are called **homeologous** chromosomes (not to be confused with

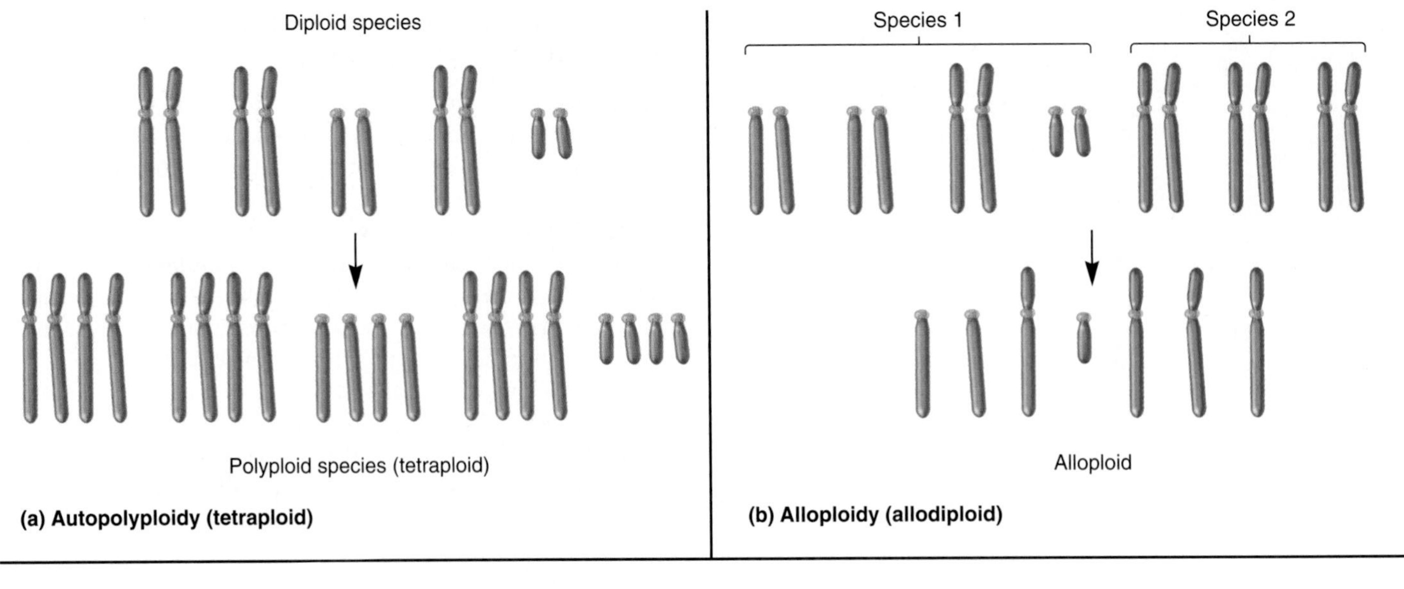

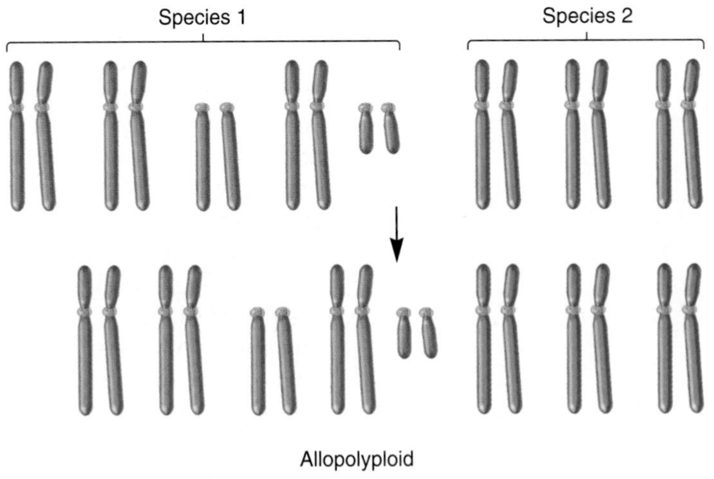

FIGURE 8.25 A comparison of autopolyploidy, alloploidy, and allopolyploidy.

homologous). This allodiploid is fertile because the homeologous chromosomes can properly synapse during meiosis to produce haploid gametes.

The critical relationship between chromosome pairing and fertility was first recognized by the Russian cytogeneticist Georgi Karpechenko in 1928. He crossed a radish (*Raphanus*) and a cabbage (*Brassica*), both of which are diploid and contain 18 chromosomes. Each of these organisms produces haploid cells containing 9 chromosomes. Therefore, the allodiploid produced from this interspecies cross contains 18 chromosomes. However, because the radish and cabbage are not closely related species, the nine *Raphanus* chromosomes are distinctly different from the nine *Brassica* chromosomes. During meiosis I, the radish and cabbage chromosomes cannot synapse with each other. This prevents the proper chromosome pairing and results in a high degree of aneuploidy (**Figure 8.27a**). Therefore, the radish/cabbage hybrid is sterile.

Among his strains of sterile alloploids, Karpechenko discovered that on rare occasions a plant would produce a viable seed. When such seeds were planted and subjected to karyotyping, the plants were found to be allotetraploids with two sets of chromosomes from each of the two species. In the example shown in **Figure 8.27b**, the radish/cabbage allotetraploid contains 36 chromosomes instead of 18. The homologous chromosomes from each of the two species can synapse properly. When anaphase of meiosis I occurs, the pairs of synapsed chromosomes

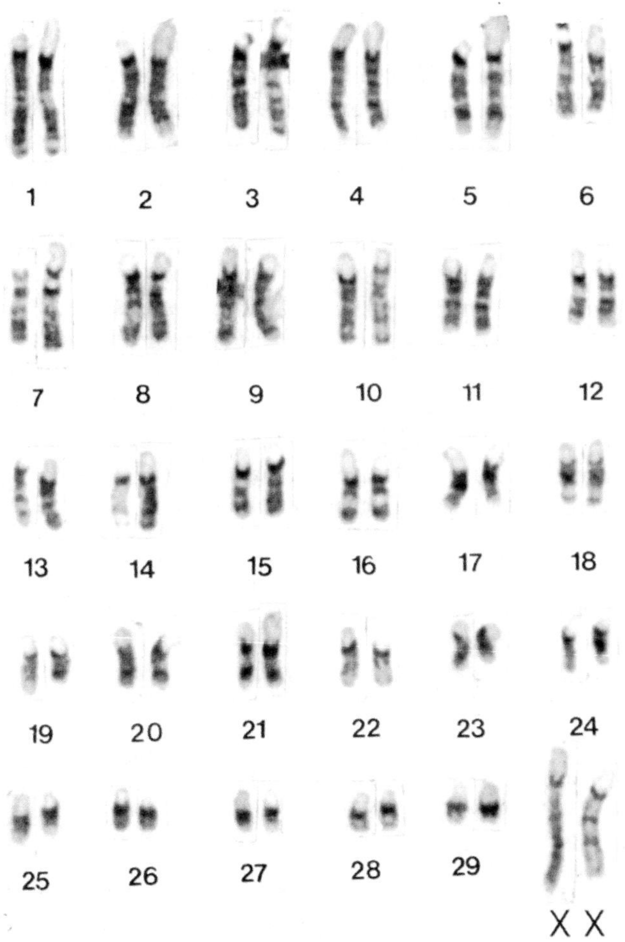

FIGURE 8.26 The karyotype of a hybrid animal produced from two closely related antelope species. In each chromosome pair in this karyotype, one chromosome was inherited from the roan antelope (*Hippotragus equinus*) and the other from the sable antelope (*Hippotragus niger*). When it is possible to distinguish the chromosomes from the two species, the roan chromosomes are shown on the left side of each pair.

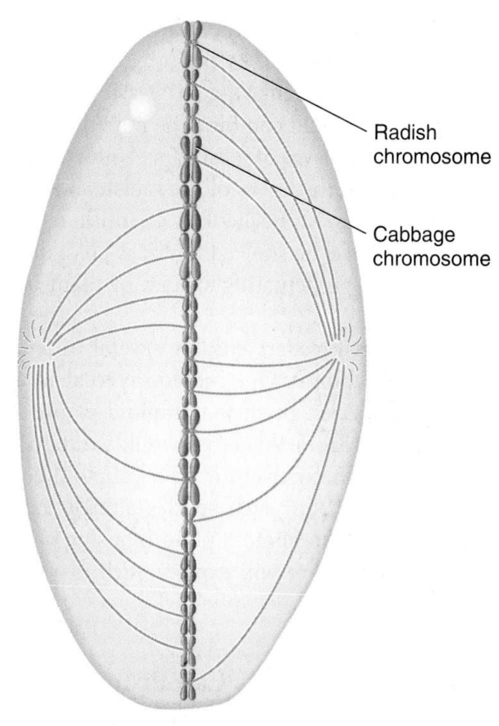

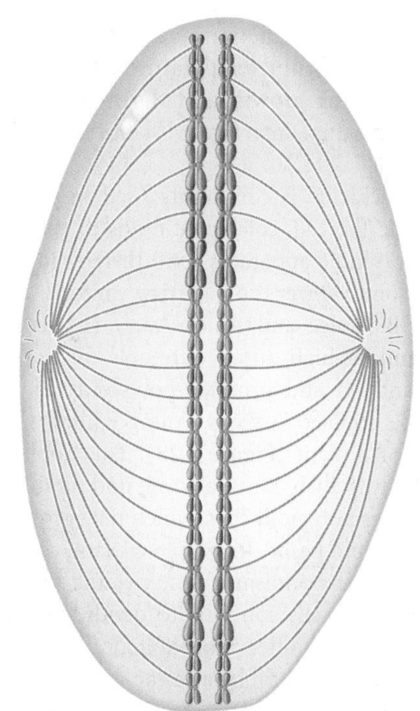

FIGURE 8.27 A comparison of metaphase I in an allodiploid and an allotetraploid. The purple chromosomes are from the radish (*Raphanus*), and the blue are from the cabbage (*Brassica*). (**a**) In the allodiploid, the radish and cabbage chromosomes have not synapsed and are randomly aligned during metaphase. (**b**) In the allotetraploid, the homologous radish and homologous cabbage chromosomes are properly aligned along the metaphase plate.

can disjoin equally to produce cells with 18 chromosomes each (a haploid set from the radish plus a haploid set from the cabbage). These cells can give rise to gametes with 18 chromosomes that can combine with each other to produce an allotetraploid containing 36 chromosomes. In this way, the allotetraploid is a fertile organism. Karpechenko's goal was to create a "vegetable for the masses" that would combine the nutritious roots of a radish with the flavorful leaves of a cabbage. Unfortunately, however, the allotetraploid had the leaves of the radish and the roots of the cabbage! Nevertheless, Karpechenko's scientific contribution was still important because he showed that it is possible to artificially produce a new self-perpetuating species of plant by creating an allotetraploid.

Modern plant breeders employ several different strategies to produce allotetraploids. One method is to start with two different tetraploid species. Because tetraploid plants make diploid gametes, the hybrid from this cross would contain a diploid set from each species. Alternatively, if an allodiploid containing one set of chromosomes from each species already exists, a second approach is to create an allotetraploid by using agents that alter chromosome number. We will explore such experimental treatments next.

Experimental Treatments Can Promote Polyploidy

Because polyploid and allopolyploid plants often exhibit desirable traits, the development of polyploids is of considerable interest among plant breeders. Experimental studies on the ability of environmental agents to promote polyploidy began in the early 1900s. Since that time, various agents have been shown to promote nondisjunction and thereby lead to polyploidy. These include abrupt temperature changes during the initial stages of seedling growth and the treatment of plants with chemical agents that interfere with the formation of the spindle apparatus.

The drug colchicine is commonly used to promote polyploidy. Once inside the cell, colchicine binds to tubulin (a protein found in the spindle apparatus) and thereby interferes with normal chromosome segregation during mitosis or meiosis. In 1937, Alfred Blakeslee and Amos Avery applied colchicine to plant tissue and, at high doses, were able to cause complete mitotic nondisjunction and produce polyploidy in plant cells. Colchicine can be applied to seeds, young embryos, or rapidly growing regions of a plant (**Figure 8.28**). This application may produce aneuploidy, which is usually an undesirable outcome, but it often produces polyploid cells, which may grow faster than the surrounding diploid tissue. In a diploid plant, colchicine may cause complete mitotic nondisjunction, yielding tetraploid ($4n$) cells. As the tetraploid cells continue to divide, they generate a portion of the plant that is often morphologically distinguishable from the remainder. For example, a tetraploid stem may have a larger diameter and produce larger leaves and flowers. Because individual plants can be propagated asexually from pieces of plant tissue (i.e., cuttings), the polyploid portion of the plant can be removed, treated with the proper growth hormones, and grown as a separate plant. Alternatively, the tetraploid region of a plant may have flowers that produce seeds. A tetraploid flower will produce diploid pollen and eggs, which combine to produce tetraploid offspring. In this way, the use of colchicine provides a straightforward method to produce polyploid strains of plants.

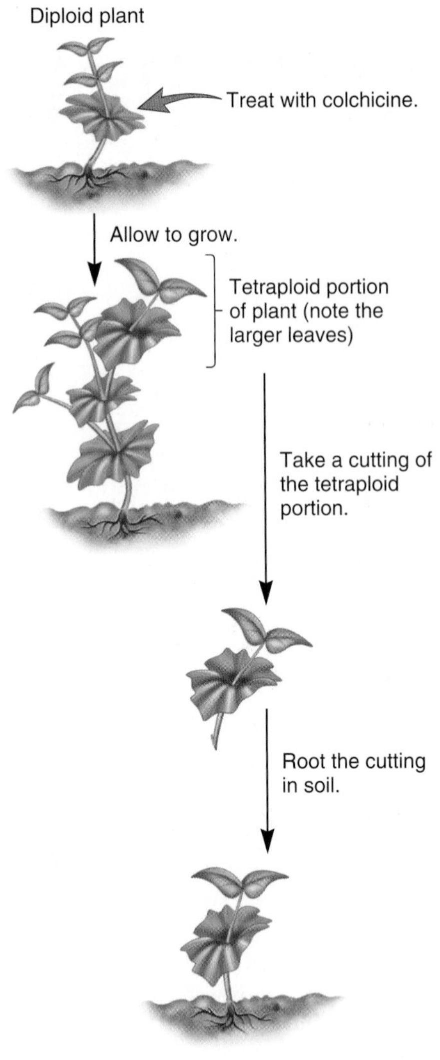

FIGURE 8.28 Use of colchicine to promote polyploidy in plants. Colchicine interferes with the mitotic spindle apparatus and promotes nondisjunction. If complete nondisjunction occurs in a diploid cell, a daughter cell will be formed that is tetraploid. Such a tetraploid cell may continue to divide and produce a segment of the plant with more robust characteristics. This segment may be cut from the rest of the plant and rooted. In this way, a tetraploid plant can be propagated.

Cell Fusion Techniques Can Be Used to Make Hybrid Plants

Thus far, we have examined several mechanisms that produce variations in chromosome number. Some of these processes occur naturally and have figured prominently in speciation and evolution. In addition, agricultural geneticists can administer treatments such as colchicine to promote nondisjunction and thereby obtain useful strains of organisms. More recently, researchers have developed cellular approaches to produce hybrids with altered chromosomal composition. As described here, these cellular approaches have important applications in research and agriculture.

In the technique known as **cell fusion,** individual cells are mixed together and made to fuse. In agriculture, cell fusion can create new strains of plants. An advantage of this approach is that researchers can artificially "cross" two species that cannot interbreed naturally. As an example, **Figure 8.29** illustrates the use of cell fusion to produce a hybrid grass. The parent cells were derived from tall fescue grass (*Festuca arundinacea*) and Italian ryegrass (*Lolium multiflorum*). Prior to fusion, the cells from these two species were treated with agents that gently digest the cell wall without rupturing the plasma membrane. A plant cell without a cell wall is called a **protoplast.** The protoplasts were mixed together and treated with agents that promote cellular fusion. Immediately after this takes place, a cell containing two separate nuclei is formed. This cell, known as a **heterokaryon,** will spontaneously go through a nuclear fusion process to produce a **hybrid cell** with a single nucleus. After nuclear fusion, the hybrid cells can be grown on laboratory media and eventually regenerate an entire plant. The allotetraploid shown in Figure 8.29 has phenotypic characteristics that are intermediate between tall fescue grass and Italian ryegrass.

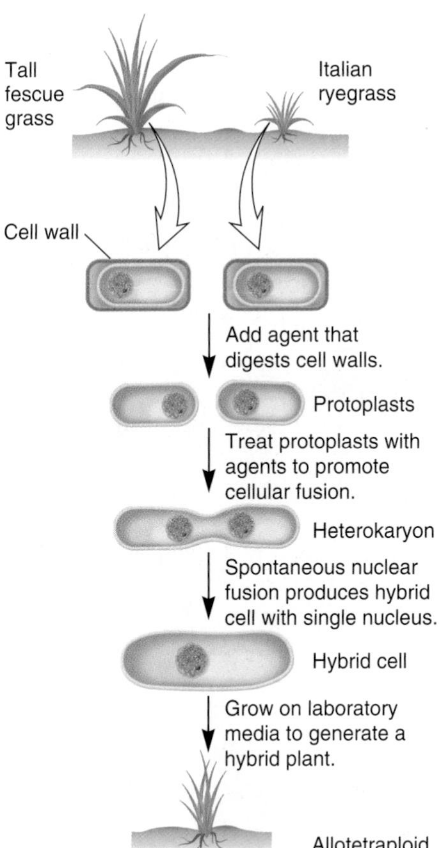

FIGURE 8.29 The technique of cell fusion. This technique is shown here with cells from tall fescue grass (*Festuca arundinacea*) and Italian ryegrass (*Lolium multiflorum*).The resulting hybrid is an allotetraploid.

Genes → Traits The allotetraploid contains two copies of genes from each parent. In this case, the allotetraploid displays characteristics that are intermediate between the tall fescue grass and the Italian ryegrass.

Monoploids Produced in Agricultural and Genetic Research Can Be Used to Make Homozygous and Hybrid Strains

A goal of some plant breeders is to have diploid strains of crop plants that are homozygous for all of their genes. One true-breeding strain can then be crossed to a different true-breeding strain to produce an F_1 hybrid that is heterozygous for many genes. Such hybrids are often more vigorous than the corresponding homozygous strains. This phenomenon, known as **hybrid vigor** or **heterosis,** is described in greater detail in Chapter 25. Seed companies often use this strategy to produce hybrid seed for many crops, such as corn and alfalfa. To achieve this goal, the companies must have homozygous parental strains that can be crossed to each other to produce the hybrid seed. One way to obtain these homozygous strains involves inbreeding over many generations. This may be accomplished after several rounds of self-fertilization. As you might imagine, this can be a rather time-consuming endeavor.

As an alternative, the production of **monoploids**—organisms that have a single set of chromosomes in their somatic cells—can be used as part of an experimental strategy to develop homozygous diploid strains of plants. Monoploids have been used to improve agricultural crops such as wheat, rice, corn, barley, and potato. In 1964, Sipra Guha and Satish Maheshwari developed a method to produce monoploid plants directly from pollen grains. **Figure 8.30** describes the experimental technique called **anther culture,** which has been extensively used to produce diploid strains of crop plants that are homozygous for all of their genes. The method involves alternation between monoploid and diploid generations. The parental plant is diploid but not homozygous for all of its genes. The anthers from this diploid plant are collected, and the haploid pollen grains are induced to begin development by a cold shock—an abrupt exposure to cold temperature. After several weeks, monoploid plantlets emerge, and these can be grown on agar media in a laboratory. However, due to the presence of deleterious alleles that are recessive, many of the pollen grains may fail to produce viable plantlets. Therefore, anther culture has been described as a "monoploid sieve" that weeds out individuals that carry deleterious recessive alleles.

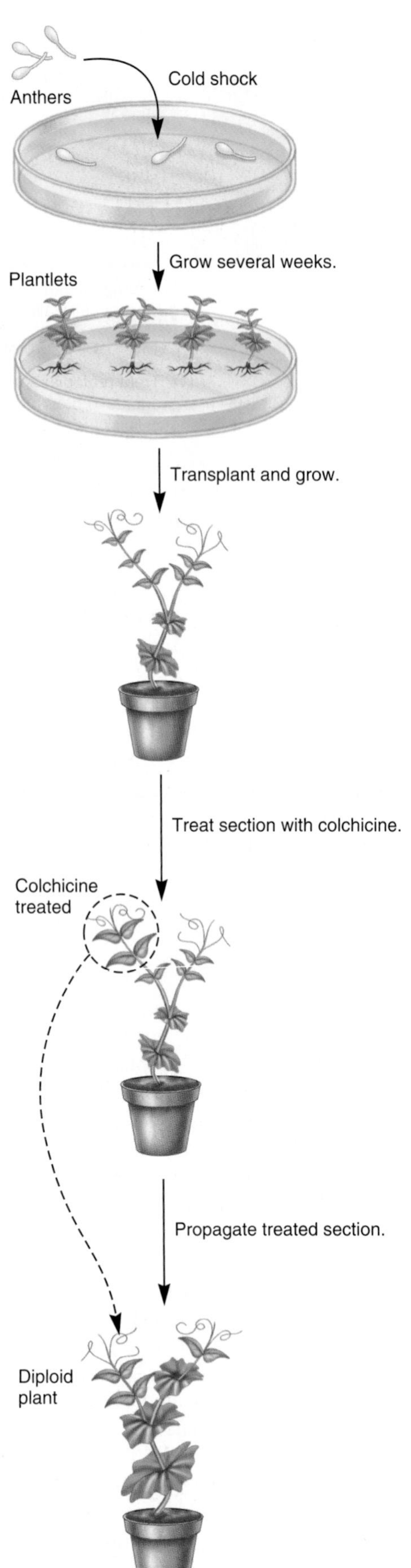

Eventually, plantlets that are healthy can be transferred to small pots. After the plants grow to a reasonable size, a section of the monoploid plant can be treated with colchicine to convert it to diploid tissue. A cutting from this diploid section can then be used to generate a separate plant. This diploid plant is homozygous for all of its genes because it was produced by the chromosomal doubling of a monoploid strain.

In certain animal species, monoploids can be produced by experimental treatments that induce the eggs to begin development without fertilization by sperm. This process is known as **parthenogenesis.** In many cases, however, the haploid zygote will develop only for a short period of time before it dies. Nevertheless, a short phase of development can be useful to research scientists. For example, the zebrafish (*Danio rerio*), a common aquarium fish, has recently gained popularity among researchers interested in vertebrate development. The haploid egg can be induced to begin development by exposure to sperm rendered biologically inactive by UV irradiation.

FIGURE 8.30 The experimental production of monoploids with anther culture. In the technique of anther culture, the anthers are collected, and the haploid pollen within them is induced to begin development by a cold shock treatment. After several weeks, monoploid plantlets emerge, and these can be grown on an agar medium. Eventually, the plantlets can be transferred to small pots. After the plantlets have grown to a reasonable size, a section of the monoploid plantlet can be treated with colchicine to convert it to diploid tissue. A cutting from this diploid section can then be used to generate a diploid plant.

CONCEPTUAL SUMMARY

We classify variations in chromosome structure as deletions (or deficiencies), duplications, inversions, and translocations. Deletions and duplications involve changes in the total amount of genetic material within a chromosome; inversions and translocations are chromosomal rearrangements. Deletions tend to be detrimental to an organism, although this depends on the size and location of the deletion. By comparison, duplications tend to be less harmful. From an evolutionary perspective, duplications provide the raw material for the addition of genes into a species' genome. This can lead to the evolution of gene families in which a group of genes encodes proteins that carry out similar yet specialized functions.

Inversions often do not affect the phenotype of the individual that carries them. They are classified according to the location of the centromere, as either pericentric (if the centromere lies within the inverted region) or paracentric (if the centromere is found outside the inverted region). A translocation occurs when one segment of a chromosome becomes attached to a different chromosome or to a different part of the same chromosome. Translocations are categorized as simple, in which a single piece of chromosome is attached to another chromosome, or reciprocal, in which two different types of chromosomes exchange pieces. In a Robertsonian translocation, the long arms of two acrocentric chromosomes fuse, creating one large single chromosome, and the two short arms are lost. In balanced translocations, there is no resulting change in the total amount of genetic material, and the resulting phenotype is usually normal.

In a few cases, inversions and balanced translocations can affect the phenotype because a breakpoint is within a vital gene or the rearrangement results in a position effect that alters gene expression or regulation. Inversions and translocations are often associated with fertility problems and a higher probability of producing abnormal offspring. Even though an individual carrying an inversion or balanced translocation may be phenotypically normal, crossing over and chromosomal segregation may lead to the production of gametes that do not have a balanced amount of genetic material. Familial Down syndrome provides an example of this phenomenon.

Chromosome number is another critical factor that determines the phenotype of an organism. Aneuploidy involves an alteration in the number of chromosomes, so the total number is not an exact multiple of a set. In aneuploidy, an individual may have an extra chromosome (trisomy) or may be missing a chromosome (monosomy); both of these are usually detrimental to the individual's phenotype. For example, various human genetic diseases, such as Down syndrome (trisomy 21), are due to irregularities in chromosome number. Likewise, in plants, aneuploidy also significantly affects the phenotype of an individual.

Organisms that are euploids have a chromosome number that is an exact multiple of a chromosome set. Variations in the number of sets of chromosomes are also common, particularly in the plant kingdom. Many of our modern crops are polyploid plants that exhibit characteristics superior to those of their diploid counterparts. Polyploids with an odd number of chromosome sets are usually sterile, and these are useful in producing seedless varieties of plants. In animals, some cells of the body may be endopolyploid—that is, they have more sets of chromosomes than other somatic cells. The polytene chromosomes found in *Drosophila* are a dramatic example of this phenomenon.

Changes in chromosome number can arise via several different mechanisms. In natural populations, the most important of these are meiotic or mitotic nondisjunction and interspecies crosses. Meiotic nondisjunction can produce cells with alterations in chromosome number, which can lead to aneuploidy or polyploidy. Either mitotic nondisjunction or chromosome loss can occur after fertilization and can lead to patches of tissue with an altered chromosomal composition, a condition termed mosaicism. Interspecies crosses can produce allodiploids, which have one set of chromosomes from each of two different species, or allopolyploids, which have two or more sets from each species. Allotetraploids, which have two complete sets of chromosomes from two different species, are more likely to be fertile because the chromosomes are able to form homologous pairs during meiosis.

EXPERIMENTAL SUMMARY

The primary experimental strategy for studying variation in chromosome structure and number is the microscopic examination of chromosomes. A karyotype is a micrograph that shows the composition of chromosomes from an actively dividing eukaryotic cell. When preparing a karyotype, the chromosomes may be stained with Giemsa dye, which gives individual chromosomes a unique G banding pattern. By observing the banding patterns and number of chromosomes under a microscope, a cytogeneticist can determine if an individual carries a change in chromosome structure or number. The technique of comparative genomic hybridization can also be used to detect changes in chromosome structure, such as deletions and duplications. This method has found wide use in the analysis of cancer cells, which often exhibit such changes.

Experimental methods that cause changes in chromosome number are also available. The drug colchicine can increase chromosome number by promoting mitotic nondisjunction. Using cell fusion techniques, researchers have artificially crossed two species that cannot interbreed naturally, creating fertile interspecies hybrids. The production of monoploid plants can be used as part of a strategy to develop homozygous diploid strains. In certain animal species, experimental treatments have induced eggs to begin development without fertilization by sperm.

PROBLEM SETS & INSIGHTS

Solved Problems

S1. Describe how a gene family is produced. Discuss the common and unique features of the family members in the globin gene family.

Answer: A gene family is produced when a single gene is copied one or more times by a gene duplication event. This duplication may occur by an abnormal (misaligned) crossover, which produces a chromosome with a deletion and another chromosome with a gene duplication.

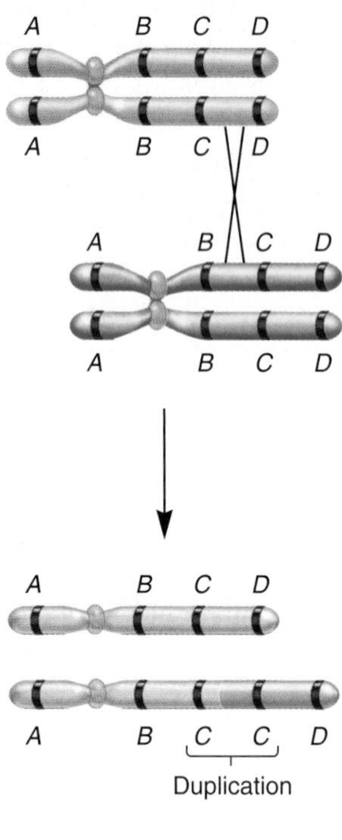

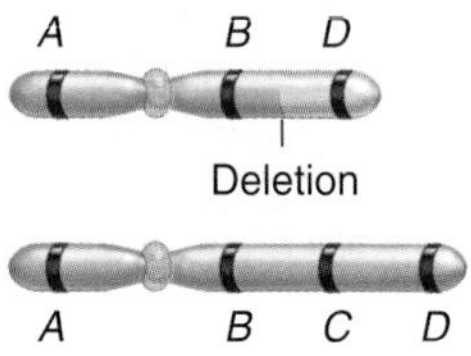

Over time, this type of duplication may occur several times to produce many copies of a particular gene. In addition, translocations may move the duplicated genes to other chromosomes, so the members of the gene family may be dispersed among several different chromosomes. Eventually, each member of a gene family will accumulate mutations, which may subtly alter their function.

All the members of the globin gene family bind oxygen. Myoglobin tends to bind it more tightly; therefore, it is good at storing oxygen. Hemoglobin binds it more loosely, so it can transport oxygen throughout the body (via red blood cells) and release it to the tissues that need oxygen. The polypeptides that form hemoglobins are predominantly expressed in red blood cells, whereas myoglobin genes are expressed in many different cell types. The expression pattern of the globin genes changes during different stages of development. The ε- and ζ-globin genes are expressed in the early embryo. They are turned off near the end of the first trimester, and then the γ-globin genes exhibit their maximal expression during the second and third trimesters of gestation. Following birth, the γ-globin genes are silenced, and the β-globin gene is expressed for the rest of a person's life. These differences in the expression of the globin genes reflect the differences in the oxygen transport needs of humans during the different stages of life. Overall, the evolution of gene families has resulted in gene products that are better suited to a particular tissue or stage of development. This has allowed a better "fine-tuning" of human traits.

S2. An inversion heterozygote has the following inverted chromosome:

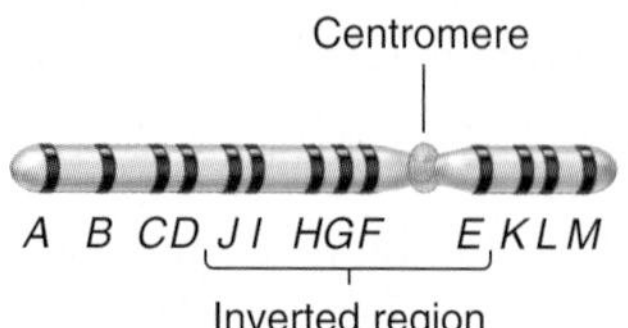

What is the result if a crossover occurs between genes *F* and *G* on one inversion and one normal chromosome?

Answer: The resulting product is four chromosomes. One chromosome is normal, one is an inversion chromosome, and two chromosomes have duplications and deletions. The two duplicated/deficient chromosomes are shown here:

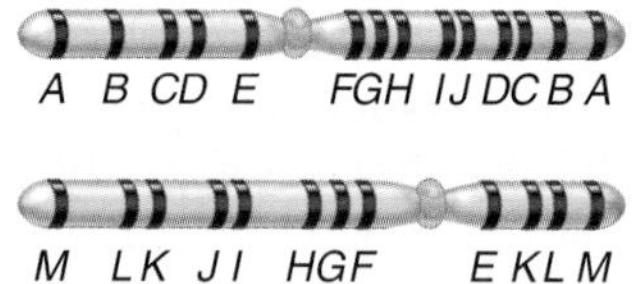

S3. In humans, the number of chromosomes per set equals 23. Even though the following conditions are lethal, what would be the total number of chromosomes for the following individuals?

A. Trisomy 22

B. Monosomy 11

C. Triploid individual

Answer:

A. 47 (the diploid number, 46, plus 1)

B. 45 (the diploid number, 46, minus 1)

C. 69 (3 times 23)

S4. A diploid species with 44 chromosomes (i.e., 22/set) is crossed to another diploid species with 38 chromosomes (i.e., 19/set). What would be the number of chromosomes in an allodiploid or allotetraploid produced from this cross? Would you expect the offspring to be sterile or fertile?

Answer: An allodiploid would have 22 + 19 = 41 chromosomes. This individual would likely be sterile, because all the chromosomes would not have homeologous partners to pair with during meiosis. This yields aneuploidy, which usually causes sterility. An allotetraploid would have 44 + 38 = 82 chromosomes. Because each chromosome would have a homologous partner, the allotetraploid would likely be fertile.

S5. **Pseudodominance** occurs when a single copy of a recessive allele is phenotypically expressed because the second copy of the gene has been deleted from the homologous chromosome; the individual is hemizygous for the recessive allele. As an example, we can consider the notch phenotype in *Drosophila*, which is an X-linked trait. Fruit flies with this condition have wings with a notched appearance at their edges. Female flies that are heterozygous for this mutation have notched wings; homozygous females and hemizygous males are unable to survive. The notched phenotype is due to a defect in a single gene called *notch* (*N*). Geneticists studying fruit flies with this phenotype have discovered that some of the mutant flies are due to a small deletion that includes the *notch* gene as well as a few genes on either side of it. Other notch mutations are due to small mutations confined within the *notch* gene itself. A genetic analysis can distinguish between notched fruit flies carrying a deletion versus those that carry a single-gene mutation. This is possible because the *notch* gene happens to be located next to the red/white eye color gene on the X chromosome. How would you distinguish between a notched phenotype due to a deletion that included the *notch* gene and the adjacent eye color gene versus a notch phenotype due to a small mutation only within the *notch* gene itself?

Answer: To determine if the notch mutation is due to a deletion, red-eyed females with the notched phenotype can be crossed to white-eyed males. Only the daughters with notched wings need to be analyzed. If they have red eyes, this means the notch mutation has not deleted the red-eye allele from the X chromosome. Alternatively, if they have white eyes, this indicates the red-eye allele has been deleted from the X chromosome that carries the notch mutation. In this case, the white-eye allele is expressed because it is present in a single copy in a female fly with two X chromosomes. This phenomenon is pseudodominance.

S6. Albert Blakeslee began using the Jimson weed (*Datura stramonium*) as an experimental organism to teach his students the laws of Mendelian inheritance. Although this plant has not gained widespread use in genetic studies, Blakeslee's work provided a convincing demonstration that changes in chromosome number have an impact on the phenotype of organisms (see Figure 8.16). Blakeslee's assistant, B. T. Avery, identified a Jimson weed mutant that he called "globe" because the capsule is more rounded than normal. In genetic crosses, he found that the globe mutant had a peculiar pattern of inheritance. The globe trait was passed to about 25% of the offspring when the globe plants were allowed to self-fertilize. Unexpectedly, about 25% of the offspring also had the globe phenotype when globe plants were pollinated by a normal plant. In contrast, when pollen from a globe plant was used to pollinate a normal plant, less than 2% of the offspring had the globe phenotype. This non-Mendelian pattern of inheritance caused Blakeslee and his colleagues to investigate the nature of this trait further. We now know the globe phenotype is due to trisomy 11. Can you explain this unusual pattern of inheritance knowing it is due to trisomy 11?

Answer: This unusual pattern of inheritance of these aneuploid strains can be explained by the viability of euploid versus aneuploid gametes and/or gametophytes. When an individual is trisomic, there is a 50% chance that an egg or sperm will inherit an extra chromosome and a 50% chance that a gamete will be normal. Blakeslee's results indicate that when a pollen grain inherited an extra copy of chromosome 11, it was almost always nonviable and unable to produce an aneuploid offspring. However, a significant percentage of aneuploid eggs were viable, so some (25%) of the offspring were aneuploid. Because an aneuploid plant should produce a 1:1 ratio between euploid and aneuploid eggs, the observation that only 25% of the offspring were aneuploid also indicates that about half of the female gametophytes or aneuploid eggs from such gametophytes were also nonviable. Overall, these results provide compelling evidence that imbalances in chromosome number can alter reproductive viability and also cause significant phenotypic consequences.

Conceptual Questions

C1. Which changes in chromosome structure cause a change in the total amount of genetic material, and which do not?

C2. Explain why small deletions and duplications are less likely to have a detrimental effect on an individual's phenotype than large ones. If a small deletion within a single chromosome happens to have a phenotypic effect, what would you conclude about the genes in this region?

C3. How does a chromosomal duplication occur?

C4. What is a gene family? How are gene families produced over time? With regard to gene function, what is the biological significance of a gene family?

C5. Following a gene duplication, two genes will accumulate different mutations, causing them to have slightly different sequences. In Figure 8.7, which pair of genes would you expect to have more similar sequences, α_1 and α_2 or ψ_{α_1} and α_2? Explain your answer.

C6. Two chromosomes have the following order of genes:

Normal: *A B C* centromere *D E F G H I*

Abnormal: *A B G F E D* centromere *C H I*

Does the abnormal chromosome have a pericentric or paracentric inversion? Draw a sketch showing how these two chromosomes would pair during prophase of meiosis I.

C7. An inversion heterozygote has the following inverted chromosome:

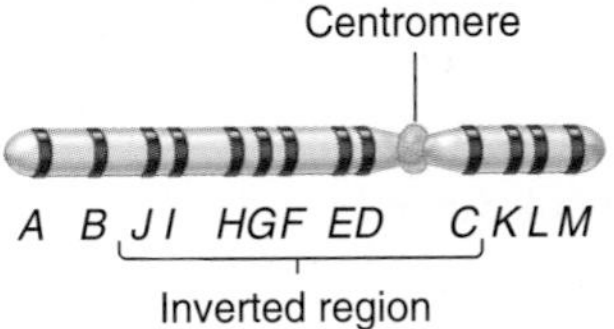

What would be the products if a crossover occurred between genes *H* and *I* on one inverted and one normal chromosome?

C8. An inversion heterozygote has the following inverted chromosome:

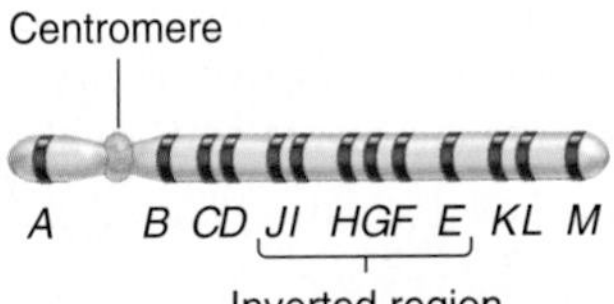

What would be the products if a crossover occurred between genes *H* and *I* on one inverted and one normal chromosome?

C9. Explain why inversions and reciprocal translocations do not usually cause a phenotypic effect. In a few cases, however, they do. Explain how.

C10. An individual has the following reciprocal translocation:

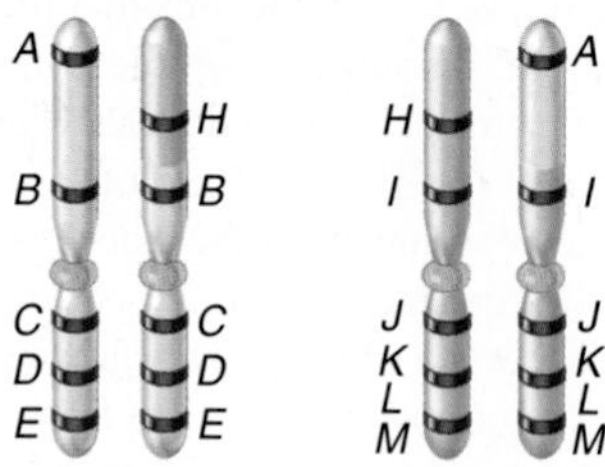

What would be the outcome of alternate and adjacent-1 segregation?

C11. A phenotypically normal individual has the following combinations of abnormal chromosomes:

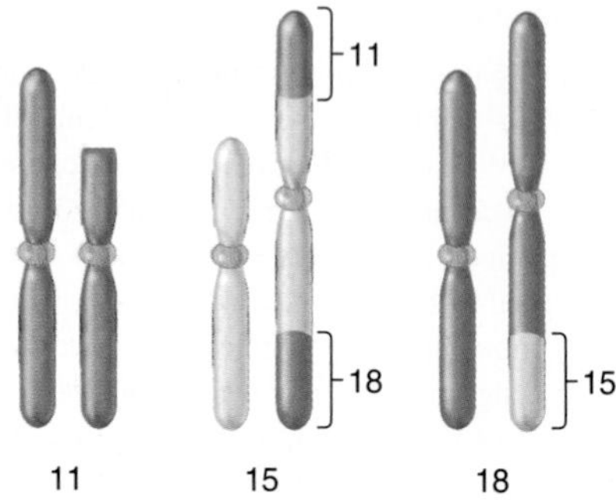

The normal chromosomes are shown on the left of each pair. Suggest a series of events (breaks, translocations, crossovers, etc.) that may have produced this combination of chromosomes.

C12. Two phenotypically normal parents produce a phenotypically abnormal child in which chromosome 5 is missing part of its long arm but has a piece of chromosome 7 attached to it. The child also has one normal copy of chromosome 5 and two normal copies of chromosome 7. With regard to chromosomes 5 and 7, what do you think are the chromosomal compositions of the parents?

C13. In the segregation of centromeres, why is adjacent-2 segregation less frequent than alternate or adjacent-1 segregation?

C14. Which of the following types of chromosomal changes would you expect to have phenotypic consequences? Explain your choices.

A. Pericentric inversion

B. Reciprocal translocation

C. Deletion

D. Unbalanced translocation

C15. Explain why a translocation cross occurs during metaphase of meiosis I when a cell contains a reciprocal translocation.

C16. A phenotypically abnormal individual has a phenotypically normal father with an inversion on one copy of chromosome 7 and a normal mother without any changes in chromosome structure. The order of genes along chromosome 7 in the father is as follows:

R T D M centromere *P U X Z C* (normal chromosome 7)

R T D U P centromere *M X Z C* (inverted chromosome 7)

The phenotypically abnormal offspring has a chromosome 7 with the following order of genes:

R T D M centromere *P U D T R*

With a sketch, explain how this chromosome was formed. In your answer, explain where the crossover occurred (i.e., between which two genes).

C17. A diploid fruit fly has eight chromosomes. How many total chromosomes would be found in the following flies?

A. Tetraploid

B. Trisomy 2

C. Monosomy 3

D. $3n$

E. $4n + 1$

C18. A person is born with one X chromosome, zero Y chromosomes, trisomy 21, and two copies of the other chromosomes. How many chromosomes does this person have altogether? Explain whether this person is euploid or aneuploid.

C19. Two phenotypically unaffected parents produce two children with familial Down syndrome. With regard to chromosomes 14 and 21, what are the chromosomal compositions of the parents?

C20. Aneuploidy is typically detrimental, whereas polyploidy is sometimes beneficial, particularly in plants. Discuss why you think this is the case.

C21. Explain how aneuploidy, deletions, and duplications cause genetic imbalances. Why do you think that deletions and monosomies are more detrimental than duplications and trisomies?

C22. Female fruit flies homozygous for the X-linked white-eye allele are crossed to males with red eyes. On very rare occasions, an offspring is a male with red eyes. Assuming these rare offspring are not due to a new mutation in one of the mother's X chromosomes that converted the white-eye allele into a red-eye allele, explain how this red-eyed male arose.

C23. A cytogeneticist has collected tissue samples from members of the same butterfly species. Some of the butterflies were located in Canada, and others were found in Mexico. Upon karyotyping, the cytogeneticist discovered that chromosome 5 of the Canadian butterflies had a large inversion compared to the Mexican butterflies. The Canadian butterflies were inversion homozygotes, whereas the Mexican butterflies had two normal copies of chromosome 5.

A. Explain whether a mating between the Canadian and Mexican butterflies would produce phenotypically normal offspring.

B. Explain whether the offspring of a cross between Canadian and Mexican butterflies would be fertile.

C24. Why do you think that human trisomies 13, 18, and 21 can survive but the other trisomies are lethal? Even though X chromosomes are large, aneuploidies of this chromosome are also tolerated. Explain why.

C25. A zookeeper has collected a male and female lizard that look like they belong to the same species. They mate with each other and produce phenotypically normal offspring. However, the offspring are sterile. Suggest one or more explanations for their sterility.

C26. What is endopolyploidy? What is its biological significance?

C27. What is mosaicism? How is it produced?

C28. Explain how polytene chromosomes of *Drosophila melanogaster* are produced and how they form a six-armed structure.

C29. Describe some of the advantages of polyploid plants. What are the consequences of having an odd number of chromosome sets?

C30. While conducting field studies on a chain of islands, you decide to karyotype two phenotypically identical groups of turtles, which are found on different islands. The turtles on one island have 24 chromosomes, while the turtles on another island have 48 chromosomes. How would you explain this observation? How do you think the turtles with 48 chromosomes came into being? If you mated the two types of turtles together, would you expect their offspring to be phenotypically normal? Would you expect them to be fertile? Explain.

C31. A diploid fruit fly has eight chromosomes. Which of the following terms should not be used to describe a fruit fly with four sets of chromosomes?

A. Polyploid

B. Aneuploid

C. Euploid

D. Tetraploid

E. $4n$

C32. Which of the following terms should not be used to describe a human with three copies of chromosome 12?

A. Polyploid

B. Triploid

C. Aneuploid

D. Euploid

E. $2n + 1$

F. Trisomy 12

C33. The kidney bean, *Phaseolus vulgaris*, is a diploid species containing a total of 22 chromosomes in somatic cells. How many possible types of trisomic individuals could be produced in this species?

C34. The karyotype of a young girl who is affected with familial Down syndrome revealed a total of 46 chromosomes. Her older brother, however, who is phenotypically unaffected, actually had 45 chromosomes. Explain how this could happen. What would you expect to be the chromosomal number in the parents of these two children?

C35. A triploid plant has 18 chromosomes (i.e., 6 chromosomes per set). If we assume a gamete has an equal probability of receiving one or two copies of each of the six types of chromosome, what are the odds of this plant producing a monoploid or a diploid gamete? What are the odds of producing an aneuploid gamete? If the plant is allowed to self-fertilize, what are the odds of producing a euploid offspring?

C36. Describe three naturally occurring ways that the chromosome number can change.

C37. Meiotic nondisjunction is much more likely than mitotic nondisjunction. Based on this observation, would you conclude that meiotic nondisjunction is usually due to nondisjunction during meiosis I or meiosis II? Explain your reasoning.

C38. A woman who is heterozygous, *Bb*, has brown eyes. *B* (brown) is a dominant allele, and *b* (blue) is recessive. In one of her eyes, however, there is a patch of blue color. Give three different explanations for how this might have occurred.

C39. What is an allodiploid? What factor determines the fertility of an allodiploid? Why are allotetraploids more likely to be fertile?

C40. What are homeologous chromosomes?

C41. Meiotic nondisjunction usually occurs during meiosis I. What is not separating properly: bivalents or sister chromatids? What is not separating properly during mitotic nondisjunction?

C42. Table 8.1 shows that Turner syndrome occurs when an individual inherits one X chromosome but lacks a second sex chromosome. Can Turner syndrome be due to nondisjunction during oogenesis, spermatogenesis, or both? If a phenotypically normal couple has a color-blind child (due to a recessive X-linked allele) with Turner syndrome, did nondisjunction occur during oogenesis or spermatogenesis in this child's parents? Explain your answer.

C43. Male honeybees, which are monoploid, produce sperm by meiosis. Explain what unusual event (compared to other animals) must occur during spermatogenesis in honeybees to produce sperm? Does this unusual event occur during meiosis I or meiosis II?

Experimental Questions

E1. What is the main goal of comparative genome hybridization? Explain how the ratio of green/red fluorescence provides information regarding chromosome structure.

E2. Let's suppose a researcher conducted comparative genomic hybridization (see Figure 8.8) and accidentally added two-fold too much (red) DNA from normal cells. What green/red ratio would you expect in a region from a chromosome from a cancer cell that carried a duplication on both chromosomal copies? What ratio would be observed for a region that was deleted on just one of the chromosomes from cancer cells?

E3. With regard to the analysis of chromosome structure, explain the experimental advantage that polytene chromosomes offer. Discuss why changes in chromosome structure are more easily detected in polytene chromosomes compared to ordinary (nonpolytene) chromosomes.

E4. Describe how colchicine can be used to alter chromosome number.

E5. Describe the steps you would take to produce a tetraploid plant that is homozygous for all of its genes.

E6. In agriculture, what is the primary purpose of anther culture?

E7. What are some experimental advantages of cell fusion techniques as opposed to interbreeding approaches?

E8. It is an exciting time to be a plant breeder because so many options are available for the development of new types of agriculturally useful plants. Let's suppose you wish to develop a seedless tomato that could grow in a very hot climate and is resistant to a viral pathogen that commonly infects tomato plants. At your disposal, you have a seed-bearing tomato strain that is heat resistant and produces great-tasting tomatoes. You also have a wild strain of tomato plants (which have lousy-tasting tomatoes) that is resistant to the viral pathogen. Suggest a series of steps you might follow to produce a great-tasting, seedless tomato that is resistant to heat and the viral pathogen.

E9. What is a G band? Discuss how G bands are useful in the analysis of chromosome structure.

E10. A female fruit fly contains one normal X chromosome and one X chromosome with a deletion. The deletion is in the middle of the X chromosome and is about 10% of the entire length of the X chromosome. If you stained and observed the chromosomes of this female fly in salivary gland cells, draw what the polytene arm of the X chromosome would look like. Explain your drawing.

E11. Describe two different experimental strategies to create an allotetraploid from two different diploid species of plants.

E12. In the procedure of anther culture (see Figure 8.30), an experimenter may begin with a diploid plant and then cold shock the pollen to get them to grow as haploid plantlets. In some cases, the pollen may come from a phenotypically vigorous plant that is heterozygous for many genes. Even so, many of the haploid plantlets appear rather weak and nonvigorous. In fact, many of them fail to grow at all. In contrast, some of the plantlets are fairly healthy. Explain why some plantlets would be weak, while others could be quite healthy.

Questions for Student Discussion/Collaboration

1. A chromosome involved in a reciprocal translocation also has an inversion. In addition, the cell also contains two normal chromosomes.

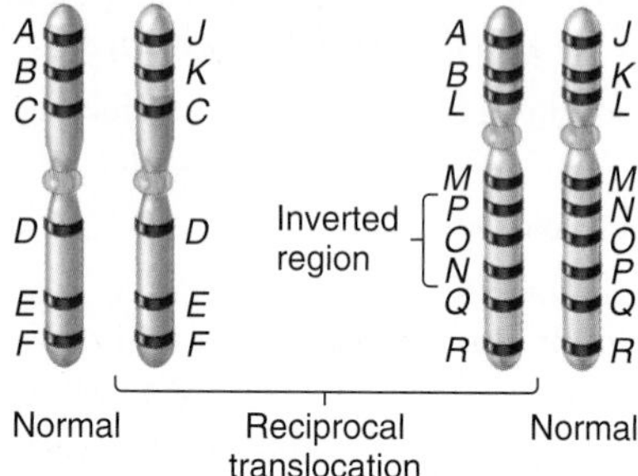

Make a drawing that shows how these chromosomes will pair during metaphase of meiosis I.

2. Besides the ones mentioned in this textbook, look for other examples of variations in euploidy. Perhaps you might look in more advanced textbooks concerning population genetics, ecology, or the like. Discuss the phenotypic consequences of these changes.

3. Cell biology textbooks often discuss cellular proteins encoded by genes that are members of a gene family. Examples of such proteins include myosins and glucose transporters. Take a look through a cell biology textbook and identify some proteins encoded by members of gene families. Discuss the importance of gene families at the cellular level.

4. Discuss how variation in chromosome number has been useful in agriculture.

Note: All answers appear at the website for this textbook; the answers to even-numbered questions are in the back of the textbook.

www.mhhe.com/brookergenetics3e

Visit the Online Learning Center for practice tests, answer keys, and other learning aids for this chapter. Enhance your understanding of genetics with our interactive exercises, quizzes, animations, and much more.

MOLECULAR STRUCTURE OF DNA AND RNA

9

CHAPTER OUTLINE

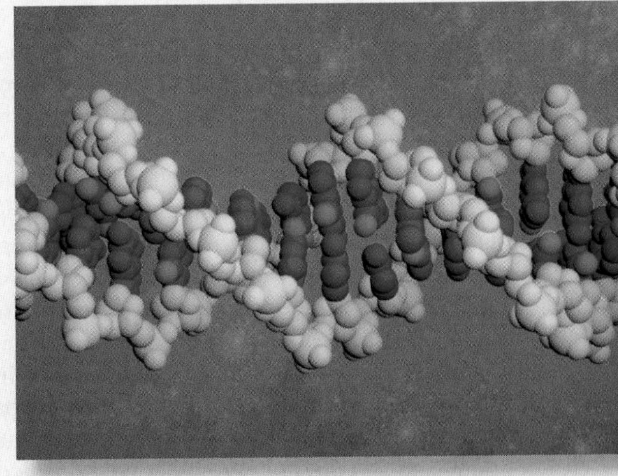

A molecular model showing the structure of the DNA double helix.

In Chapters 2 through 8, we focused on the relationship between the inheritance of genes and chromosomes, and the outcome of an organism's traits. In Chapter 9, we will shift our attention to **molecular genetics**—the study of DNA structure and function at the molecular level. An exciting goal of molecular genetics is to use our knowledge of DNA structure to understand how DNA functions as the genetic material. Using molecular techniques, researchers have determined the organization of many genes. This information, in turn, has helped us understand how the expression of such genes governs the outcome of an individual's inherited traits.

The past several decades have seen dramatic advances in techniques and approaches to investigate and even to alter the genetic material. These advances have greatly expanded our understanding of molecular genetics and also have provided key insights into the mechanisms underlying transmission and population genetics. Molecular genetic technology is also widely used in supporting disciplines such as biochemistry, cell biology, and microbiology.

To a large extent, our understanding of genetics comes from our knowledge of the molecular structure of DNA (deoxyribonucleic acid) and RNA (ribonucleic acid). In this chapter, we will begin by considering classic experiments that showed DNA is the genetic material. We will then survey the molecular features of DNA and RNA that underlie their function.

9.1 IDENTIFICATION OF DNA AS THE GENETIC MATERIAL

In his pioneering experiments, Gregor Mendel studied several different traits in pea plants. By conducting the appropriate crosses, he showed that traits are inherited as discrete units as they pass from parent to offspring. This implies that living organisms contain a genetic material that governs an individual's traits, a substance that is transferred during the process of reproduction.

To fulfill its role, the genetic material must meet several criteria.

1. **Information:** The genetic material must contain the information necessary to construct an entire organism. In other words, it must provide the blueprint to determine the inherited traits of an organism.
2. **Transmission:** During reproduction, the genetic material must be passed from parents to offspring.

PART III

Molecular Structure and Replication of the Genetic Material

3. **Replication:** Because the genetic material is passed from parents to offspring, and from mother cell to daughter cells during cell division, it must be copied.
4. **Variation:** Within any species, a significant amount of phenotypic variability occurs. For example, Mendel studied several traits in pea plants that were variable among different plants. These included height (tall versus dwarf) and seed color (yellow versus green). Therefore, the genetic material must also have variation that can account for the known phenotypic differences within each species.

Along with Mendel's work, the data of many other geneticists in the early 1900s were consistent with these four properties: information, transmission, replication, and variation. However, the experimental study of genetic crosses cannot, by itself, identify the chemical nature of the genetic material.

In the 1880s, August Weismann and Carl Nägeli championed the idea that a chemical substance within living cells is responsible for the transmission of traits from parents to offspring. The chromosome theory of inheritance was developed, and experimentation demonstrated that the chromosomes are the carriers of the genetic material (see Chapter 3). Nevertheless, the story was not complete because chromosomes contain both DNA and proteins. Also, RNA is found in the vicinity of chromosomes. Therefore, further research was needed to precisely identify the genetic material. In this section, we will examine the first experimental approaches to achieve this goal.

Experiments with Pneumococcus Suggested That DNA Is the Genetic Material

Some early work in microbiology was important in developing an experimental strategy to identify the genetic material. Frederick Griffith studied a type of bacterium known then as pneumococci and now classified as *Streptococcus pneumoniae.* Certain strains of *S. pneumoniae* secrete a polysaccharide capsule, whereas other strains do not. When streaked on petri plates containing solid growth media, capsule-secreting strains have a smooth colony morphology, whereas those strains unable to secrete a capsule have a rough appearance.

When comparing two different smooth strains of *S. pneumoniae,* researchers found that the chemical composition of their capsules can differ significantly. Using biochemical techniques, these different types of smooth strains were characterized. For example, a type II smooth strain and a type III smooth strain both make a capsule, but their capsules differ from each other biochemically. This idea is schematically shown in **Figure 9.1.** Rare mutations can occasionally convert a smooth bacterium into a rough bacterium, and vice versa. These infrequent interconversions are type-specific. For example, a type III smooth strain can mutate to become a type III rough strain. On rare occasions, a bacterium from this rough strain may mutate back to become a smooth strain. In this case, it can mutate to become only a type III smooth strain, never a type II smooth strain.

The different forms of *S. pneumoniae* also affect their virulence, or ability to cause disease. When smooth strains of *S. pneumonia* infect a mouse, the capsule allows the bacteria to escape attack by the mouse's immune system. As a result, the bacteria can grow and eventually kill the mouse. In contrast, the nonencapsulated (rough) bacteria are destroyed by the animal's immune system.

In 1928, Griffith conducted experiments that involved the injection of live and/or heat-killed bacteria into mice. He then observed whether or not the bacteria caused a lethal infection. Griffith was working with two strains of *S. pneumoniae,* a type IIIS (S for smooth) and a type IIR (R for rough). When injected into a live mouse, the type IIIS bacteria proliferated within the mouse's bloodstream and ultimately killed the mouse (**Figure 9.2a**). Following the death of the mouse, Griffith found many type IIIS bacteria within the mouse's blood. In contrast, when type IIR bacteria were injected into a mouse, the mouse lived (**Figure 9.2b**). To verify that the proliferation of the smooth bacteria was causing the death of the mouse, Griffith killed the smooth bacteria with heat treatment before injecting them into the mouse. In this case, the mouse also survived (**Figure 9.2c**).

The critical and unexpected result was obtained in the experiment outlined in **Figure 9.2d.** In this experiment, live type IIR bacteria were mixed with heat-killed type IIIS bacteria. As shown here, the mouse died. Furthermore, extracts from tissues of the dead mouse were found to contain living type IIIS bacteria! What can account for these results? Because type IIR bacteria cannot mutate to type IIIS, the interpretation of these data is that something from the dead type IIIS bacteria was transforming the

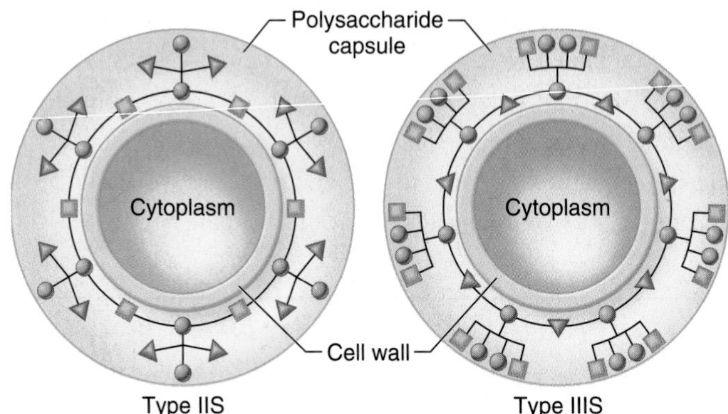

FIGURE 9.1 Schematic illustration of biochemical differences between the capsules of two strains of pneumococci. The capsule is composed of sugar units linked to form a polysaccharide material. This figure illustrates how the types of sugars (represented by circles, squares, and triangles) and their arrangement to form a polysaccharide capsule can differ among different strains. This figure is not meant to represent the actual polysaccharide capsule, which is much more complex than shown here. Instead, it emphasizes that the structures of the polysaccharide capsules are different in type IIS and type IIIS bacteria.

Genes → Traits The chemical composition of the polysaccharide capsule is governed by genes within the bacterial cell. These genes encode enzymes that synthesize the polysaccharide capsule. The genes within type IIS bacteria encode enzymes that synthesize the polysaccharide capsule depicted on the left. The genes within type IIIS bacteria encode enzymes that function somewhat differently to synthesize the capsule shown on the right.

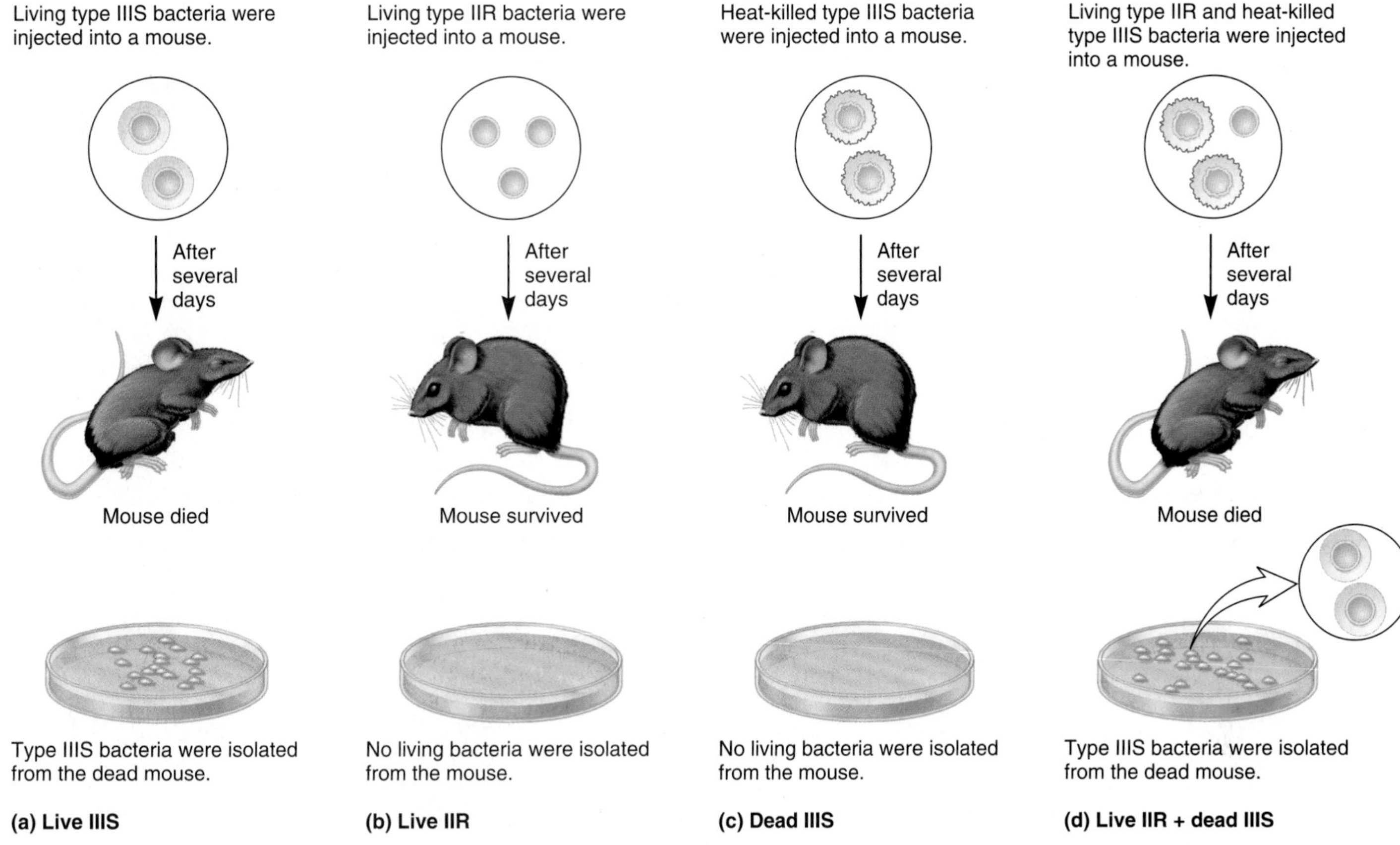

Living type IIIS bacteria were injected into a mouse.
After several days
Mouse died
Type IIIS bacteria were isolated from the dead mouse.
(a) Live IIIS
Living type IIR bacteria were injected into a mouse.
After several days
Mouse survived
No living bacteria were isolated from the mouse.
(b) Live IIR
Heat-killed type IIIS bacteria were injected into a mouse.
After several days
Mouse survived
No living bacteria were isolated from the mouse.
(c) Dead IIIS
Living type IIR and heat-killed type IIIS bacteria were injected into a mouse.
After several days
Mouse died
Type IIIS bacteria were isolated from the dead mouse.
(d) Live IIR + dead IIIS

FIGURE 9.2 **Griffith's experiments on genetic transformation in pneumococcus.**

type IIR bacteria into type IIIS. Griffith called this process **transformation,** and the unidentified substance causing this to occur was termed the transformation principle. The steps of bacterial transformation are described in Chapter 6 (see Figure 6.12).

At this point, let's look at what Griffith's observations mean in genetic terms. The transformed bacteria acquired the *information* to make a type III capsule. Among different strains, *variation* exists both in the ability to create a capsule and the type that is made. The genetic material for the type III trait must be *replicated* so that it can be *transmitted* from mother to daughter cells during cell division. Taken together, these observations are consistent with the idea that the formation of a capsule is governed by the bacteria's genetic material, meeting the four criteria described previously. Griffith's experiments showed that some genetic material from the dead bacteria had been transferred to the living bacteria and provided them with a new trait. However, Griffith did not know what the transforming substance was.

Important scientific discoveries often take place when researchers recognize that someone else's experimental observations can be used to address a particular scientific question. Oswald Avery, Colin MacLeod, and Maclyn McCarty realized that Griffith's observations could be used as part of an experimental strategy to identify the genetic material. They asked the question, What substance is being transferred from the dead type IIIS bacteria to the live type IIR? To answer this question, they incorporated additional biochemical techniques into their experimental methods.

At the time of these experiments in the 1940s, researchers already knew that DNA, RNA, proteins, and carbohydrates are major constituents of living cells. To separate these components and to determine if any of them was the genetic material, Avery, MacLeod, and McCarty used established biochemical purification procedures and prepared bacterial extracts from type IIIS strains containing each type of these molecules. After many repeated attempts with different types of extracts, they discovered that only one of the extracts, namely, the one that contained purified DNA, was able to convert the type IIR bacteria into type IIIS. As shown in **Figure 9.3**, when this extract was mixed with type IIR bacteria, some of the bacteria were converted to type IIIS. However, if no DNA extract was added, no type IIIS bacterial colonies were observed on the Petri plates.

A biochemist might point out that a DNA extract may not be 100% pure. In fact, any purified extract might contain small traces of some other substances. Therefore, one can argue that a small amount of contaminating material in the DNA extract might actually be the genetic material. The most likely contaminating substances in this case would be RNA or protein. To further verify that the DNA in the extract was indeed responsible for the transformation, Avery, MacLeod, and McCarty treated samples of

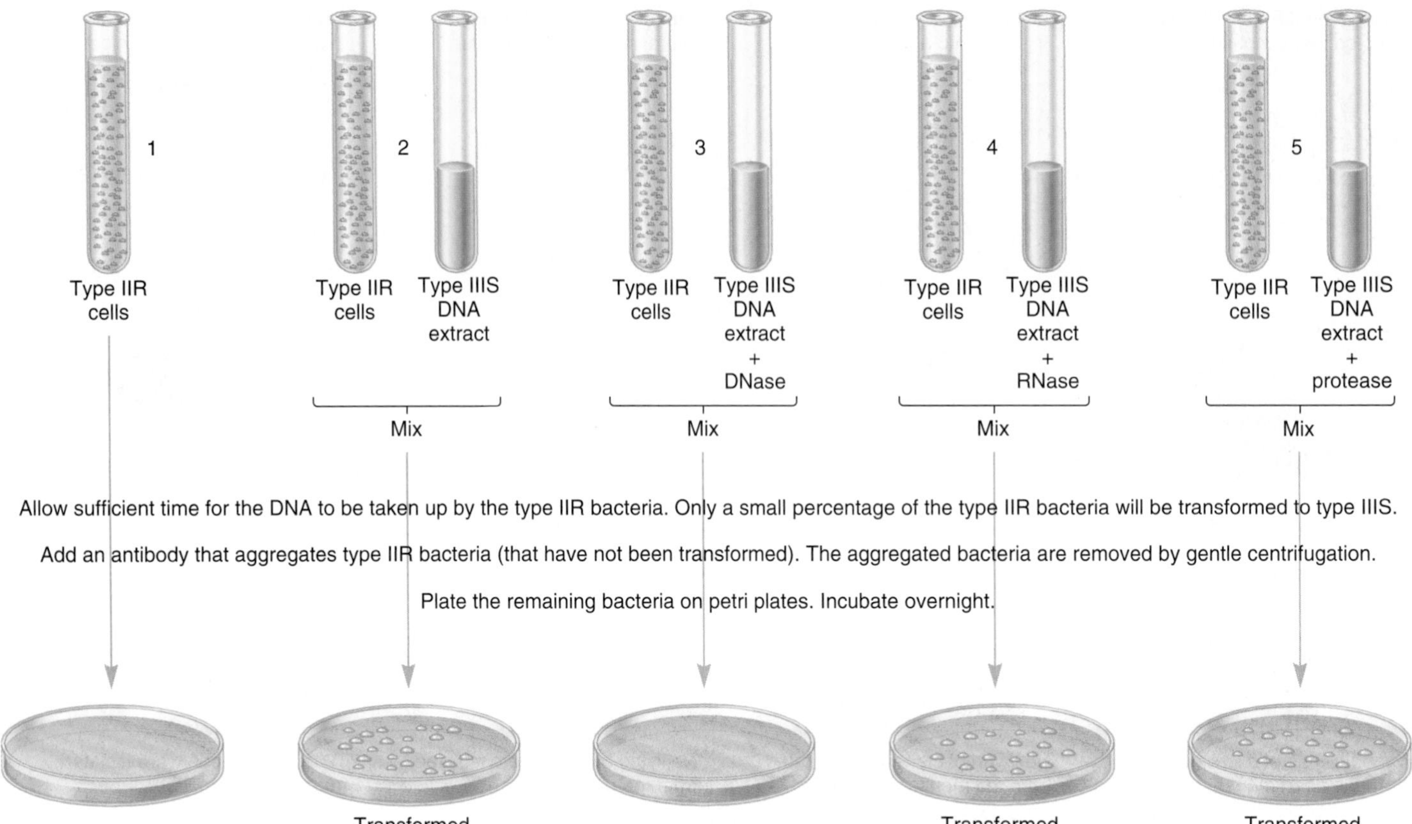

FIGURE 9.3 **Experimental protocol used by Avery, MacLeod, and McCarty to identify the transforming principle.** Samples of *S. pneumonia* cells were either not exposed to a type IIIS DNA extract (tube 1) or exposed to a type IIIS DNA extract (tubes 2–5). Tubes 3, 4, and 5 also contained DNase, RNase, or protease, respectively. After incubation, the cells were exposed to antibodies, which are molecules that can specifically recognize the molecular structure of macromolecules. In this experiment, the antibodies recognized the cell surface of type IIR bacteria and caused them to clump together. The clumped bacteria were removed by a gentle centrifugation step. Only the bacteria that were not recognized by the antibody (namely, the type IIIS bacteria) remained in the supernatant. The cells in the supernatant were plated on solid growth media. After overnight incubation, visible colonies may be observed.

the DNA extract with enzymes that digest DNA (called **DNase**), RNA (**RNase**), or protein (**protease**)(see Figure 9.3). When the DNA extracts were treated with RNase or protease, they still converted type IIR bacteria into type IIIS. These results indicated that any remaining RNA or protein in the extract was not acting as the genetic material. However, when the extract was treated with DNase, it lost its ability to convert type IIR into type IIIS bacteria. These results indicated that the degradation of the DNA in the extract by DNase prevented conversion of type IIR to type IIIS. This interpretation is consistent with the hypothesis that DNA is the genetic material. A more elegant way of saying this is that the transforming principle is DNA.

EXPERIMENT 9A

Hershey and Chase Provided Evidence That the Genetic Material Injected into the Bacterial Cytoplasm Is T2 Phage DNA

A second experimental approach indicating that DNA is the genetic material came from the studies of Alfred Hershey and Martha Chase in 1952. Their research centered on the study of a virus known as T2. This virus infects *Escherichia coli* bacterial cells and is therefore known as a **bacteriophage** or simply a **phage.** As shown in **Figure 9.4**, the external structure of the T2 phage, known as the capsid or phage coat, consists of a head, sheath, tail fibers, and base plate. Biochemically, the phage coat is composed entirely of protein, which includes several different polypeptides. DNA is found inside the head of the T2 capsid. From a molecular point of view, this virus is rather simple, because it is composed of only two types of macromolecules: DNA and proteins.

Although the viral genetic material contains the blueprint to make new viruses, a virus itself cannot synthesize new viruses. Instead, a virus must introduce its genetic material into the

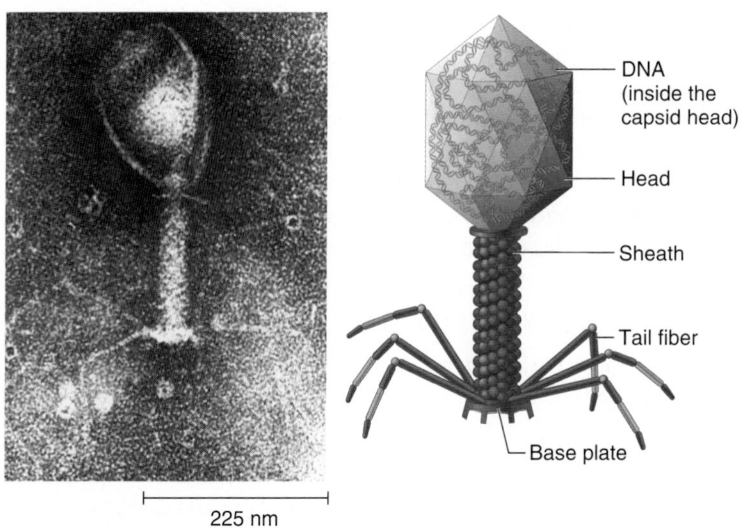

FIGURE 9.4 Structure of the T2 bacteriophage. The T2 bacteriophage is composed of a phage coat, or capsid, with genetic material inside the head of the capsid. The capsid is divided into regions called the head, sheath, tail fibers, and base plate. These components are composed of proteins. The genetic material is composed of DNA.

Genes → Traits The genetic material of a bacteriophage contains many genes, which provide the blueprint for making new viruses. When the bacteriophage injects its genetic material into a bacterium, these genes are activated and direct the host cell to make new bacteriophages, as described in Figure 9.5.

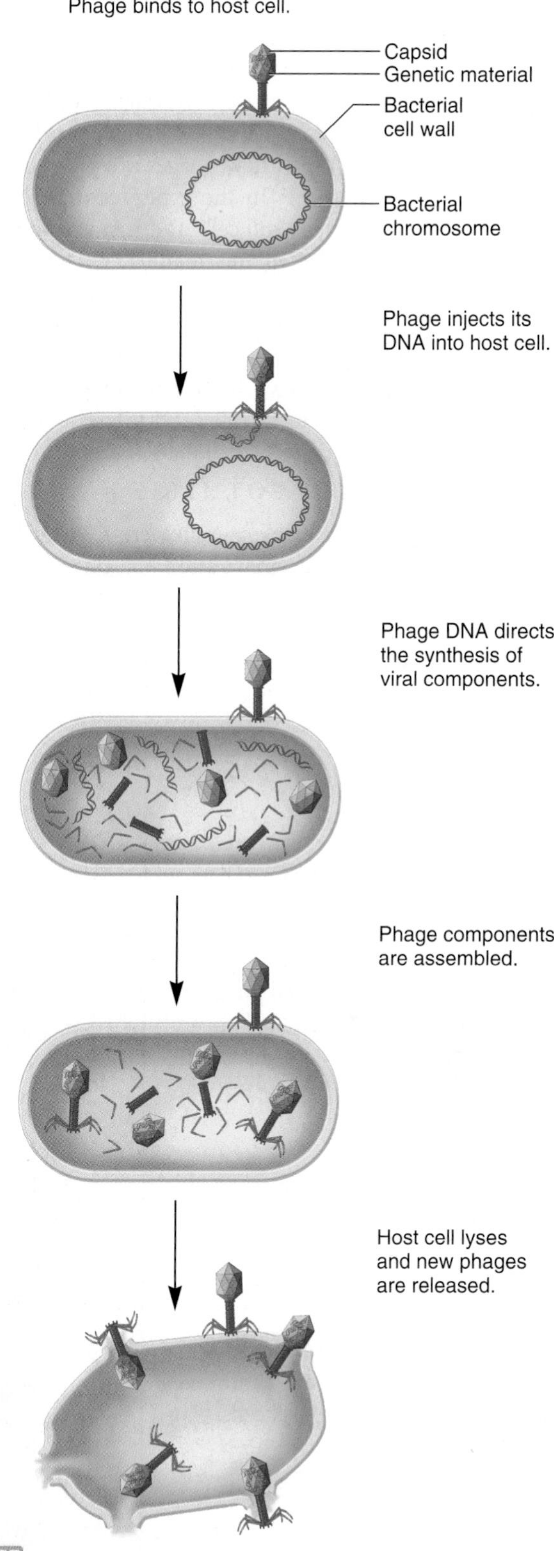

FIGURE 9.5 Life cycle of the T2 bacteriophage.

cytoplasm of a living cell. In the case of T2, this first involves the attachment of its tail fibers to the bacterial cell wall and the subsequent injection of its genetic material into the cytoplasm of the cell (**Figure 9.5**). The phage coat remains attached on the outside of the bacterium and does not enter the cell. After the entry of the viral genetic material, the bacterial cytoplasm provides all the synthetic machinery necessary to make viral proteins and DNA. The viral proteins and DNA assemble to make new viruses that are subsequently released from the cell by **lysis** (i.e., cell breakage).

To verify that DNA is the genetic material of T2, Hershey and Chase devised a method to separate the phage coat, which is attached to the outside of the bacterium, from the genetic material, which is injected into the cytoplasm. They were aware of microscopy experiments by Thomas Anderson showing that the T2 phage attaches itself to the outside of a bacterium by its tail fibers. Hershey and Chase reasoned that this is a fairly precarious attachment that could be disrupted by subjecting the bacteria to high shear forces, such as those produced in a kitchen blender. Their method was to expose bacteria to T2 phage, allowing sufficient time for the viruses to attach to bacteria and inject their genetic material. They then sheared the phage coats from the surface of the bacteria by a blender treatment. In this way, the phages' genetic material, which had been injected into the cytoplasm of the bacterial cells, could be separated from the phage coats that were sheared away.

Hershey and Chase used radioisotopes to distinguish proteins from DNA. Sulfur atoms are found in proteins but not in

DNA, whereas phosphorus atoms are found in DNA but not in phage proteins. Therefore, ^{35}S (a radioisotope of sulfur) and ^{32}P (a radioisotope of phosphorus) were used to specifically label phage proteins and DNA, respectively. Researchers can grow *E. coli* cells in media that contain ^{35}S or ^{32}P and then infect the *E. coli* cells with T2 phages. When new phages are produced, these will be labeled with ^{35}S or ^{32}P. In the experiment described in **Figure 9.6**, they began with *E. coli* cells and two preparations of T2 phage that were obtained in this manner. One preparation was labeled with ^{35}S to label the phage proteins, and the other preparation was labeled with ^{32}P to label the phage DNA. In separate flasks, each type of phage was mixed with a new sample of *E. coli* cells. The phages were given sufficient time to inject their genetic material into the bacterial cells, and then the sample was subjected to shearing force using a blender. This treatment was expected to remove the phage coat from the surface of the bacterial cell. The sample was then subjected to centrifugation at a speed that would cause the heavier bacterial cells to form a pellet at the bottom of the tube, while the light phage coats would remain in the supernatant, the liquid found above the pellet. The amount of radioactivity in the supernatant (emitted from either ^{35}S or ^{32}P) was determined using a scintillation counter.

THE HYPOTHESIS

Only the genetic material of the phage is injected into the bacterium. Isotope labeling will reveal if it is DNA or protein.

TESTING THE HYPOTHESIS — FIGURE 9.6 Evidence that DNA is the genetic material of T2 bacteriophage.

Starting materials: The starting materials were *E. coli* cells and two preparations of T2 phage. One phage preparation had phage proteins labeled with ^{35}S, and the other preparation had phage DNA labeled with ^{32}P.

1. Grow bacterial cells. Divide into two flasks.
2. Into one flask, add ^{35}S-labeled phage; in the second flask, add ^{32}P-labeled phage.
3. Allow infection to occur.
4. Agitate solutions in blenders for different lengths of time to shear the empty phages off the bacterial cells.
5. Centrifuge at 10,000 rpm.

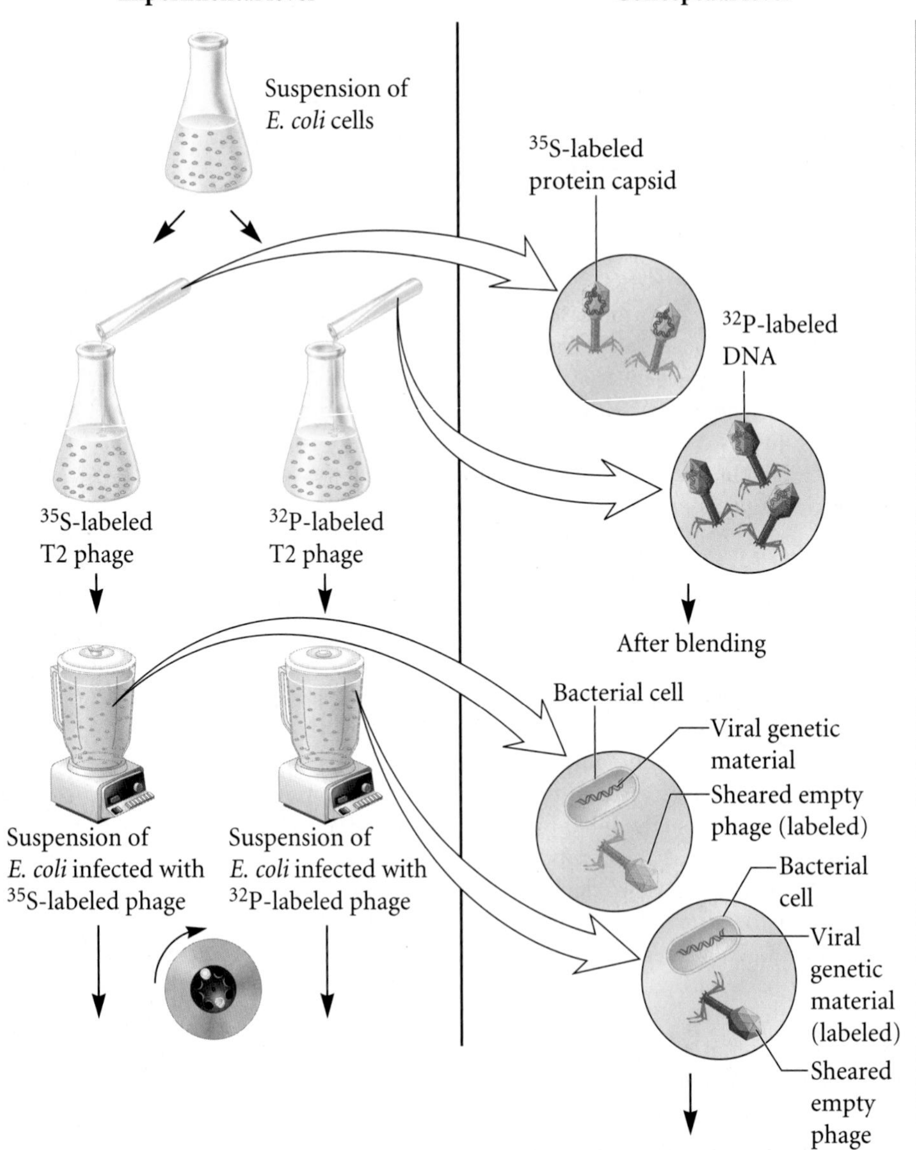

(continued)

6. The heavy bacterial cells sediment to the pellet, while the lighter phages remain in the supernatant. (See Appendix for explanation of centrifugation.)

7. Count the amount of radioisotope in the supernatant with a scintillation counter (see Appendix). Compare it with the starting amount.

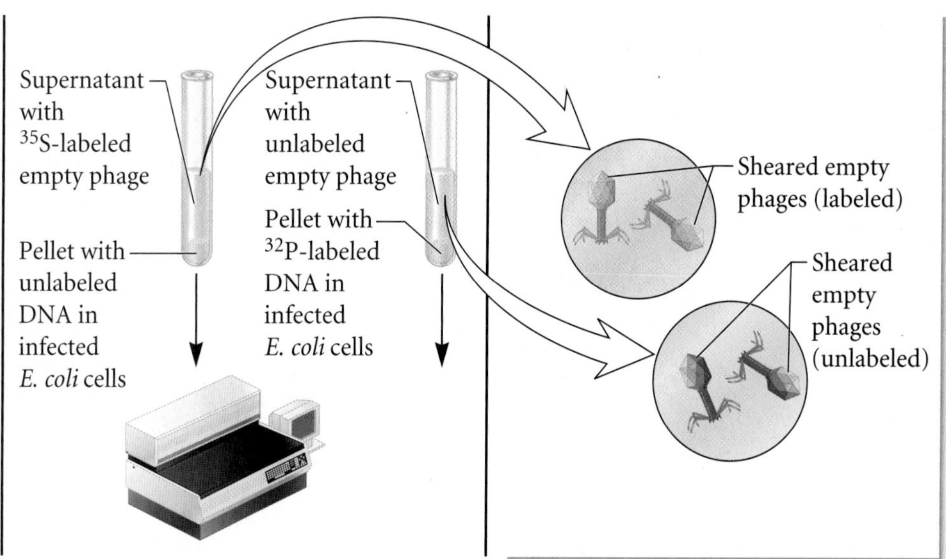

THE DATA

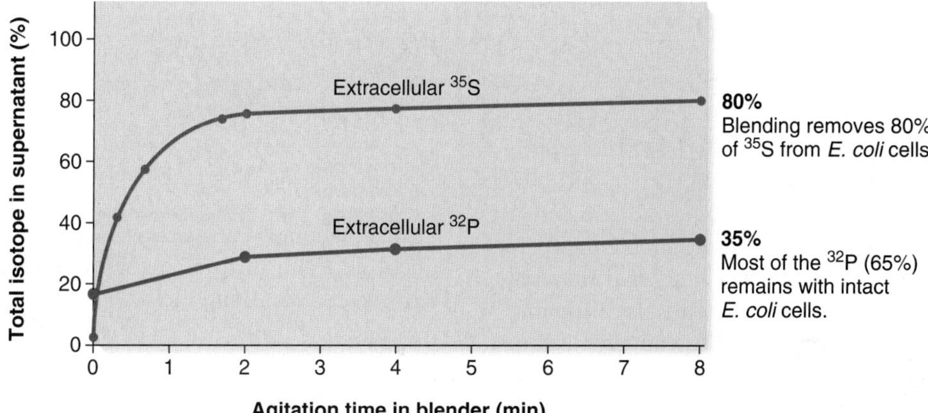

INTERPRETING THE DATA

As seen in the data, most of the ^{35}S isotope was found in the supernatant. Because the shearing force was expected to remove the phage coat, this result indicates that the empty phages contain primarily protein. By comparison, only about 35% of the ^{32}P was found in the supernatant following shearing. Therefore, most of the DNA was located within the bacterial cells in the pellet. These results are consistent with the idea that the DNA is injected into the bacterial cytoplasm during infection, which would be the expected result if DNA is the genetic material.

By themselves, the results described in Figure 9.6 were not conclusive evidence that DNA is the genetic material. For example, you may have noticed that less than 100% of the phage protein was found in the supernatant. Therefore, some of the phage protein could have been introduced into the bacterial cells (and could function as the genetic material). Nevertheless, the results of Hershey and Chase were consistent with the conclusion that the genetic material is DNA rather than protein. Overall, their studies of the T2 phage were quite influential in convincing the scientific community that DNA is the genetic material.

A self-help quiz involving this experiment can be found at the Online Learning Center.

RNA Functions as the Genetic Material in Some Viruses

We now know that bacteria, archaea, protists, fungi, plants, and animals all use DNA as their genetic material. As mentioned, viruses also have their own genetic material. Hershey and Chase concluded from their experiments that this genetic material is DNA. In the case of T2 bacteriophage, that is the correct conclusion. However, many viruses use RNA, rather than DNA, as their genetic material. In 1956, Alfred Gierer and Gerhard Schramm

TABLE 9.1

Examples of DNA- and RNA-Containing Viruses

Virus	Host	Nucleic Acid
Tomato bushy stunt virus	Tomato	RNA
Tobacco mosaic virus	Tobacco	RNA
Influenza virus	Humans	RNA
HIV	Humans	RNA
f2	*E. coli*	RNA
Qβ	*E. coli*	RNA
Cauliflower mosaic virus	Cauliflower	DNA
Herpes virus	Humans	DNA
SV40	Primates	DNA
Epstein-Barr virus	Humans	DNA
T2	*E. coli*	DNA
M13	*E. coli*	DNA

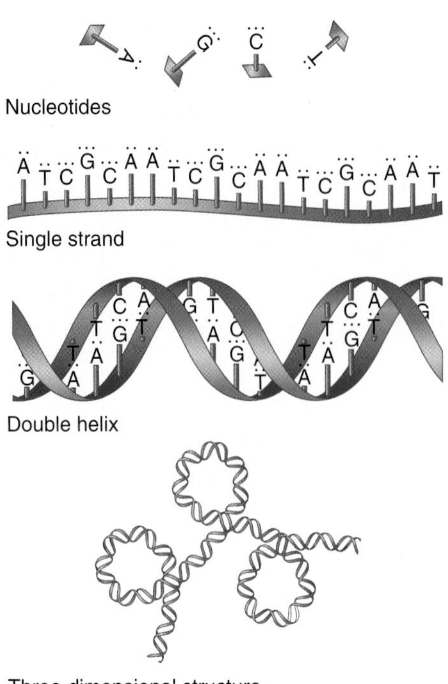

FIGURE 9.7 **Levels of nucleic acid structure.**

isolated RNA from the tobacco mosaic virus (TMV), which infects plant cells. When this purified RNA was applied to plant tissue, the plants developed the same types of lesions that occurred when they were exposed to intact tobacco mosaic viruses. Gierer and Schramm correctly concluded that the viral genome of tobacco mosaic virus is composed of RNA. Since that time, many other viruses have been found to contain RNA as their genetic material. **Table 9.1** compares the genetic compositions of several different types of viruses.

9.2 NUCLEIC ACID STRUCTURE

Geneticists, biochemists, and biophysicists have been interested in the molecular structure of nucleic acids for many decades. Both DNA and RNA are macromolecules composed of smaller building blocks. To fully appreciate their structures, we need to consider four levels of complexity (**Figure 9.7**):

1. **Nucleotides** form the repeating structural unit of nucleic acids.
2. Nucleotides are linked together in a linear manner to form a **strand** of DNA or RNA.
3. Two strands of DNA (and sometimes RNA) can interact with each other to form a **double helix.**
4. The three-dimensional structure of DNA results from the folding and bending of the double helix. Within living cells, DNA is associated with a wide variety of proteins that influence its structure. Chapter 10 examines the roles of these proteins in creating the three-dimensional structure of DNA found within chromosomes.

In this section, we will first examine the structure of individual nucleotides and then progress through the structural features of DNA and RNA. Along the way, we will also review some of the pivotal experiments that led to the discovery of the double helix.

Nucleotides Are the Building Blocks of Nucleic Acids

The **nucleotide** is the repeating structural unit of DNA and RNA. A nucleotide has three components: at least one phosphate group, a pentose sugar, and a nitrogenous base. As shown in **Figure 9.8**, nucleotides can vary with regard to the sugar and the nitrogenous base. The two types of sugars are **deoxyribose** and **ribose,** which are found in DNA and RNA, respectively. The five different bases are subdivided into two categories: the **purines** and the **pyrimidines.** The purine bases, **adenine (A)** and **guanine (G),** contain a double-ring structure; the pyrimidine bases, **thymine (T), cytosine (C),** and **uracil (U),** contain a single-ring structure. The sugar in DNA is always deoxyribose; in RNA, ribose. Also, the base thymine is not found in RNA. Rather, uracil is found in RNA instead of thymine. Adenine, guanine, and cytosine occur in both DNA and RNA. As noted in Figure 9.8, the bases and sugars have a standard numbering system. The nitrogen and carbon atoms found in the ring structure of the bases are given numbers 1 through 9 for the purines and 1 through 6 for the pyrimidines. In comparison, the five carbons found in the sugars are designated with primes, such as 1', to distinguish them from the numbers found in the bases.

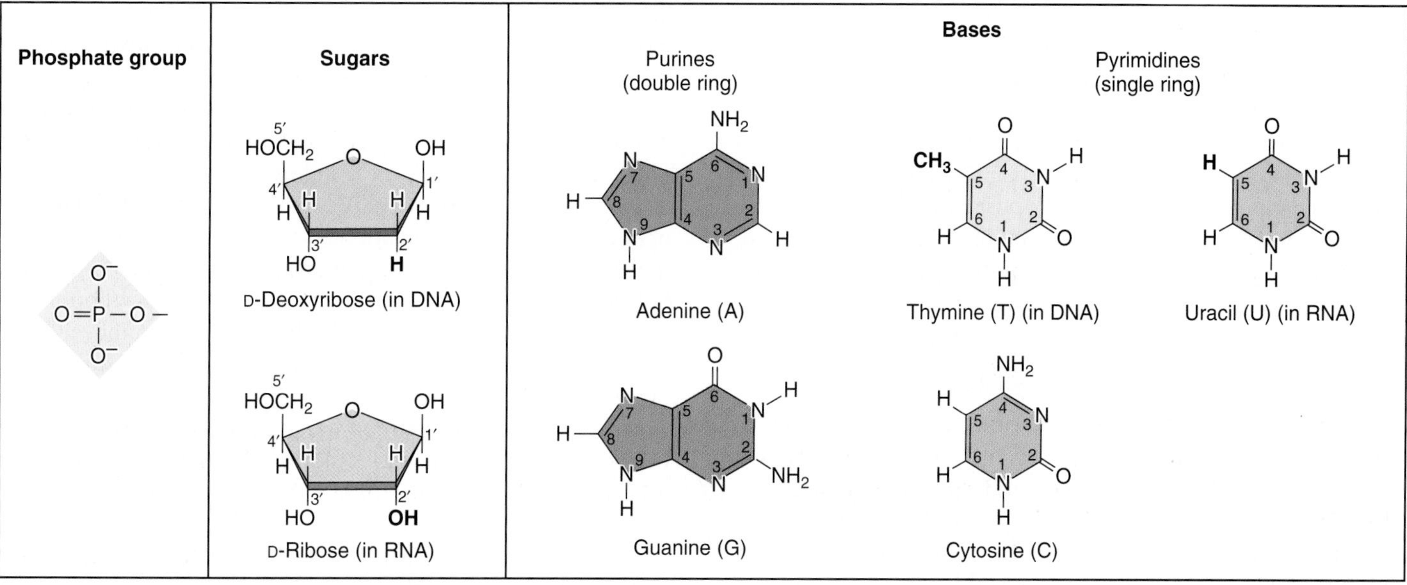

FIGURE 9.8 **The components of nucleotides.** The three building blocks of a nucleotide are one or more phosphate groups, a sugar, and a base. The bases are categorized as purines (adenine and guanine) and pyrimidines (thymine, cytosine, and uracil).

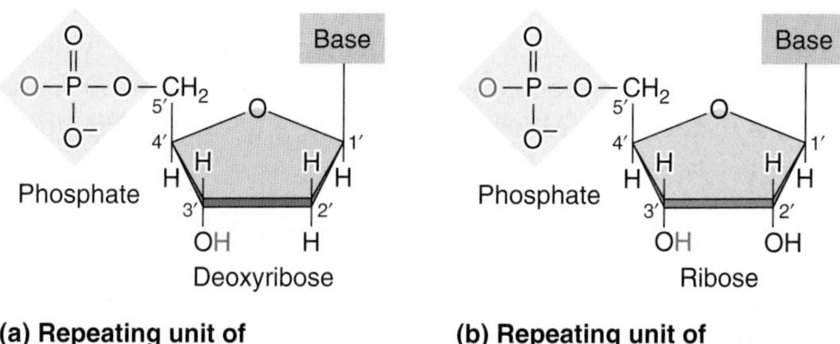

FIGURE 9.9 **The structure of nucleotides found in (a) DNA and (b) RNA.** DNA contains deoxyribose as its sugar and the bases A, T, G, and C. RNA contains ribose as its sugar and the bases A, U, G, and C. In a DNA or RNA strand, the oxygen on the 3' carbon is linked to the phosphorus atom of phosphate in the adjacent nucleotide. The two atoms (O and H) shown in red would be found within individual nucleotides but not when nucleotides are joined together to make strands of DNA and RNA.

Figure 9.9 shows the repeating unit of nucleotides found in DNA and RNA. The locations of the attachment sites of the base and phosphate to the sugar molecule are important to the nucleotide's function. In the sugar ring, carbon atoms are numbered in a clockwise direction, beginning with a carbon atom adjacent to the ring oxygen atom. The fifth carbon is outside the ring structure. In a single nucleotide, the base is always attached to the 1' carbon atom and one or more phosphate groups are attached at the 5' position. As discussed later, the –OH group attached to the 3' carbon is important in allowing nucleotides to form covalent linkages with each other.

The terminology used to describe nucleic acid units is based on three structural features: the type of base, the type of sugar, and the number of phosphate groups. When a base is attached to only a sugar, we call this pair a **nucleoside.** If adenine is attached to ribose, this nucleoside is called adenosine (**Figure 9.10**). Nucleosides containing guanine, thymine, cytosine, or uracil are called guanosine, thymidine, cytidine, and uridine, respectively. When only the bases are attached to deoxyribose, they are called deoxyadenosine,

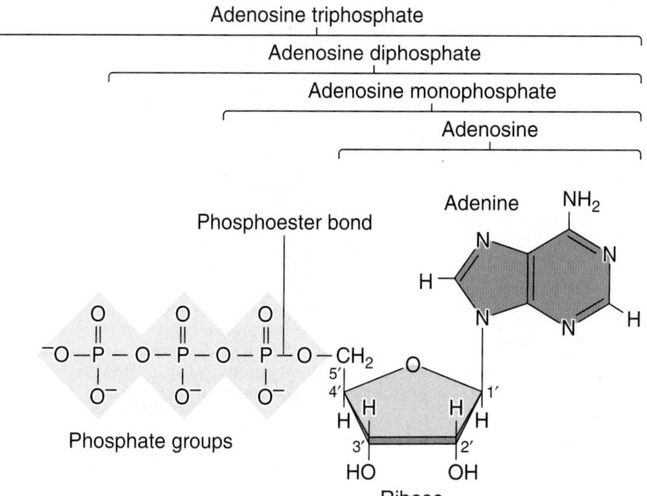

FIGURE 9.10 **A comparison between the structures of an adenine-containing nucleoside and nucleotides.**

deoxyguanosine, deoxythymidine, and deoxycytidine. The covalent attachment of one or more phosphate molecules to a nucleoside creates a nucleotide. If a nucleotide contains adenine, ribose, and one phosphate, it is adenosine monophosphate, abbreviated AMP. If a nucleotide contains adenine, ribose, and three phosphate groups, it is called adenosine triphosphate, or ATP. If it contains guanine, ribose, and three phosphate groups, it is guanosine triphosphate, or GTP. A nucleotide can also be composed of adenine, deoxyribose, and three phosphate groups. This nucleotide is deoxyadenosine triphosphate (dATP).

Nucleotides Are Linked Together to Form a Strand

A strand of DNA or RNA has nucleotides that are covalently attached to each other in a linear fashion. **Figure 9.11** depicts a short strand of DNA with four nucleotides. A few structural features are worth noting. First, the linkage involves an ester bond between a phosphate group on one nucleotide and the sugar molecule on the adjacent nucleotide. Another way of viewing this linkage is to notice that a phosphate group connects two sugar molecules. For this reason, the linkage in DNA or RNA strands is called a **phosphodiester linkage.** The phosphates and sugar molecules form the **backbone** of a DNA or RNA strand. The bases project from the backbone. The backbone is negatively charged due to a negative charge on each phosphate.

A second important structural feature is the orientation of the nucleotides. As mentioned, the carbon atoms in a sugar molecule are numbered in a particular way. A phosphodiester linkage involves a phosphate attachment to the 5' carbon in one nucleotide and to the 3' carbon in the other. In a strand, all sugar molecules are oriented in the same direction. As shown in Figure 9.11, the 5' carbons in every sugar molecule are above the 3' carbons. Therefore, a strand has a **directionality** based on the orientation of the sugar molecules within that strand. In Figure 9.11, the direction of the strand is 5' to 3' when going from top to bottom.

A critical aspect regarding DNA and RNA structure is that a strand contains a specific sequence of bases. In Figure 9.11, the sequence of bases is thymine–adenine–cytosine–guanine, abbreviated TACG. Furthermore, to show the directionality, the strand should be abbreviated 5'–TACG–3'. The nucleotides within a strand are covalently attached to each other, so the sequence of bases cannot shuffle around and become rearranged. Therefore, the sequence of bases in a DNA strand will remain the same over time, except in rare cases when mutations occur. As we will see throughout this textbook, the sequence of bases within DNA and RNA is the defining feature that allows them to carry information.

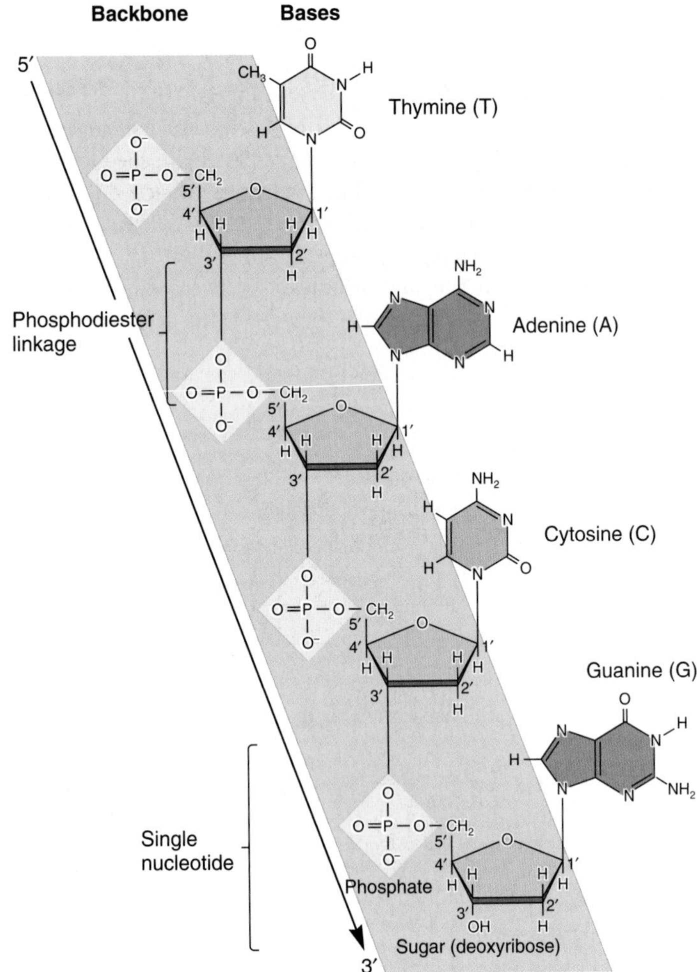

FIGURE 9.11 A short strand of DNA containing four nucleotides. Nucleotides are covalently linked together to form a strand of DNA.

A Few Key Events Led to the Discovery of the Double-Helix Structure

A major discovery in molecular genetics was made in 1953 by James Watson and Francis Crick. At that time, DNA was already known to be composed of nucleotides. However, it was not understood how the nucleotides are bonded together to form the structure of DNA. Watson and Crick committed themselves to determine the structure of DNA because they felt this knowledge was needed to understand the functioning of genes. Other researchers, such as Rosalind Franklin and Maurice Wilkins, shared this view. Before we examine the characteristics of the double helix, let's consider the events that provided the scientific framework for Watson and Crick's breakthrough.

In the early 1950s, Linus Pauling proposed that regions of proteins can fold into a secondary structure known as an α helix (**Figure 9.12a**). To elucidate this structure, Pauling built large models by linking together simple ball-and-stick units (**Figure 9.12b**). By carefully scaling the objects in his models, he could visualize if atoms fit together properly in a complicated three-dimensional structure. Is this approach still used today? The answer is "Yes," except that today researchers construct their three-dimensional models on computers. As we will see, Watson and Crick also used a ball-and-stick approach to solve the structure of the DNA double helix. Interestingly, they were well aware that Pauling might figure out the structure of DNA before they did. This provided a stimulating rivalry between the researchers.

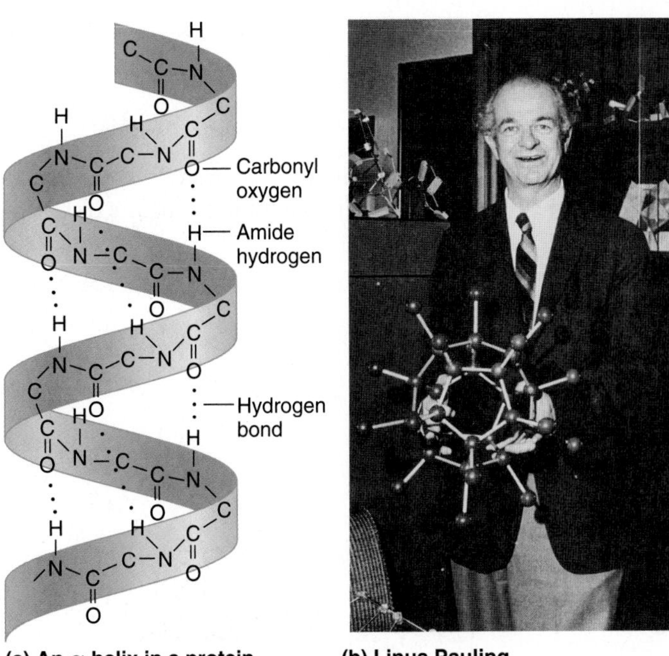

FIGURE 9.12 Linus Pauling and the α-helix protein structure. (**a**) An α helix is a secondary structure found in proteins. This structure emphasizes the polypeptide backbone (shown as a tan ribbon), which is composed of amino acids linked together in a linear fashion. Hydrogen bonding between hydrogen and oxygen atoms stabilizes the helical conformation. (**b**) Linus Pauling with a ball-and-stick model.

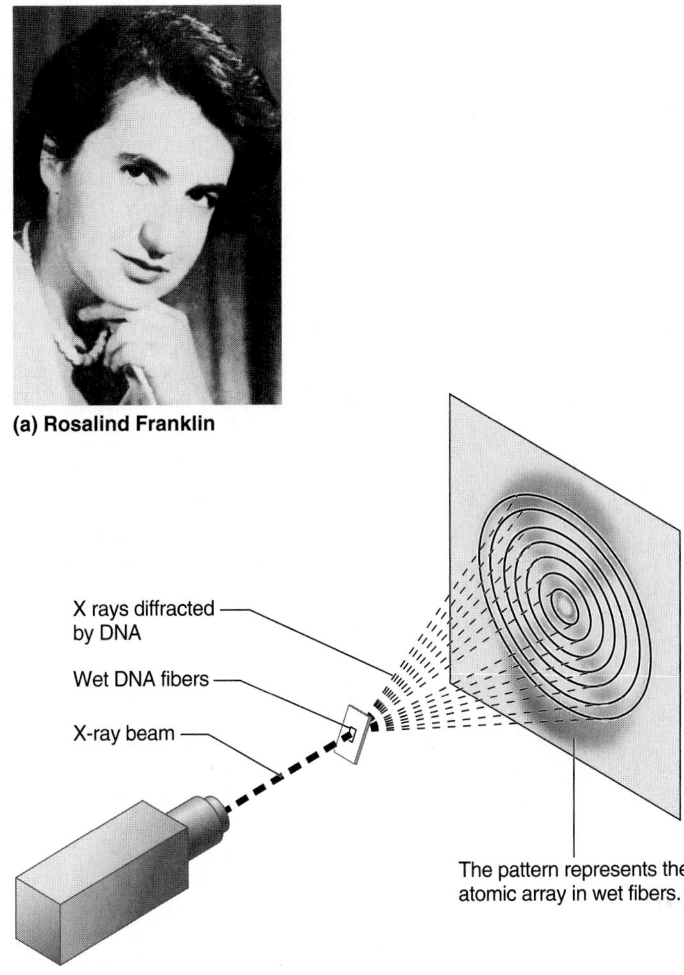

FIGURE 9.13 X-ray diffraction of DNA.

A second important development that led to the elucidation of the double helix was X-ray diffraction data. When a purified substance, such as DNA, is subjected to X-rays, it will produce a well-defined diffraction pattern if the molecule is organized into a regular structural pattern. An interpretation of the diffraction pattern (using mathematical theory) can ultimately provide information concerning the structure of the molecule. Rosalind Franklin (**Figure 9.13a**), working in the same laboratory as Maurice Wilkins, used X-ray diffraction to study wet DNA fibers. Franklin made marked advances in X-ray diffraction techniques while working with DNA. She adjusted her equipment to produce an extremely fine beam of X-rays. She extracted finer DNA fibers than ever before and arranged them in parallel bundles. Franklin also studied the fibers' reactions to humid conditions. The diffraction pattern of Franklin's DNA fibers is shown in **Figure 9.13b**. This pattern suggested several structural features of DNA. First, it was consistent with a helical structure. Second, the diameter of the helical structure was too wide to be only a single-stranded helix. Finally, the diffraction pattern indicated that the helix contains about 10 base pairs per complete turn. These observations were instrumental in solving the structure of DNA.

EXPERIMENT 9B

Chargaff Found That DNA Has a Biochemical Composition in Which the Amount of A Equals T and the Amount of G Equals C

Another piece of information that led to the discovery of the double-helix structure came from the studies of Erwin Chargaff. In the 1940s and 50s, he pioneered many of the biochemical techniques for the isolation, purification, and measurement of nucleic acids from living cells. This was not a trivial undertaking, because the biochemical composition of living cells is complex. At the time of Chargaff's work, researchers already knew that the building blocks of DNA are nucleotides containing the bases adenine, thymine, guanine, or cytosine. Chargaff analyzed the base composition of DNA, which was isolated from many different species. He expected that the results might provide important clues concerning the structure of DNA.

The experimental protocol of Chargaff is described in **Figure 9.14**. He began with various types of cells as starting material. The chromosomes were extracted from cells and then treated with protease to separate the DNA from chromosomal proteins. The DNA was then subjected to a strong acid treatment that cleaved the bonds between the sugars and bases. Therefore, the strong acid treatment would release the individual bases from the DNA strands. This mixture of bases was then subjected to paper chromatography to separate the four types. The amounts of the four bases were determined spectroscopically.

THE GOAL

An analysis of the base composition of DNA in different organisms may reveal important features about the structure of DNA.

ACHIEVING THE GOAL — FIGURE 9.14 An analysis of base composition among different DNA samples.

Starting material: The following types of cells were obtained: *Escherichia coli, Streptococcus pneumoniae* (type III), yeast, turtle red blood cells, salmon sperm cells, chicken red blood cells, and human liver cells.

1. For each type of cell, extract the chromosomal material. This can be done in a variety of ways, including the use of high salt, detergent, or mild alkali treatment. Note: The chromosomes contain both DNA and protein.

2. Remove the protein. This can be done in several ways, including treament with protease.

3. Hydrolize the DNA to release the bases from the DNA strands. A common way to do this is by strong acid treatment.

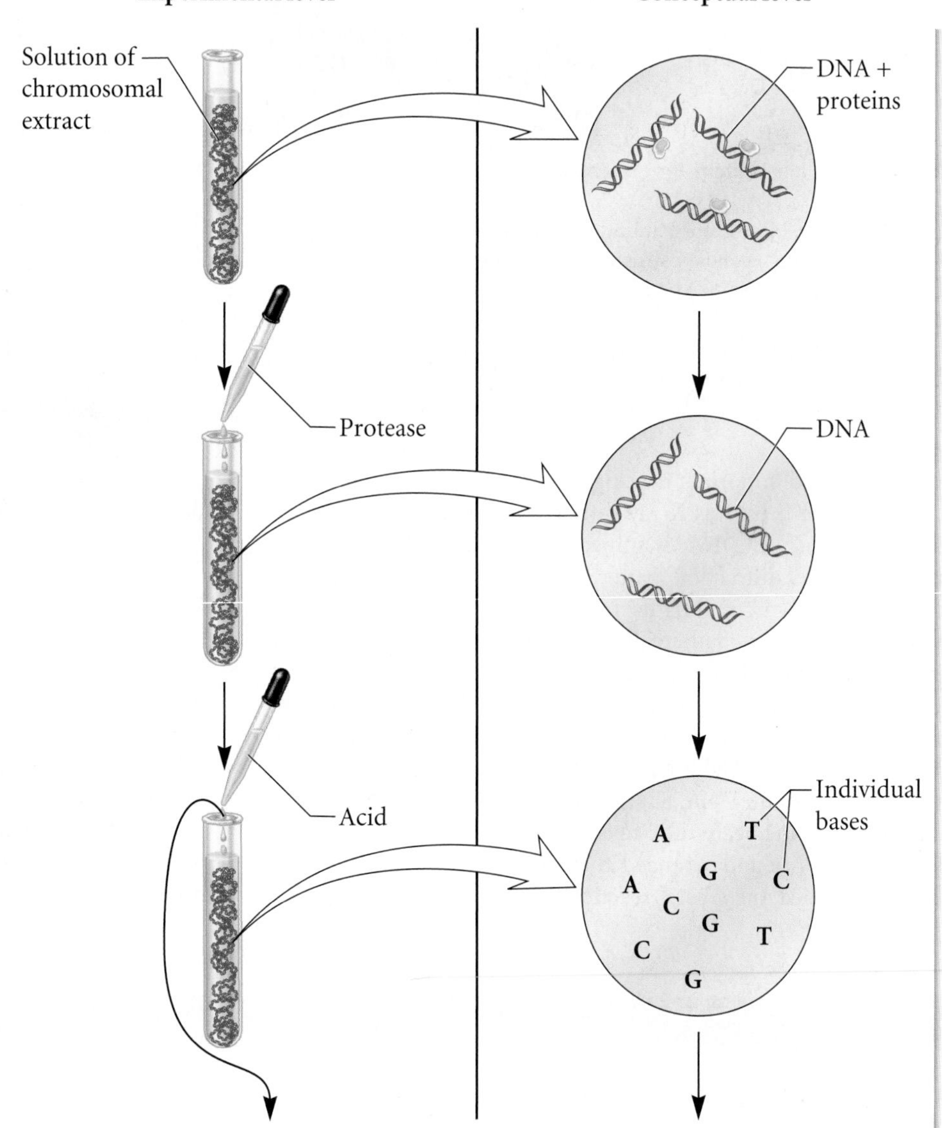

4. Separate the bases by chromatography. Paper chromatography provides an easy way to separate the four types of bases. (The technique of chromatography is described in the Appendix.)

5. Extract bands from paper into solutions and determine the amounts of each base by spectroscopy. Each base will absorb light at a particular wavelength. By examining the absorption profile of a sample of base, it is then possible to calculate the amount of the base. (Spectroscopy is described in the Appendix.)

6. Compare the base content in the DNA from different organisms.

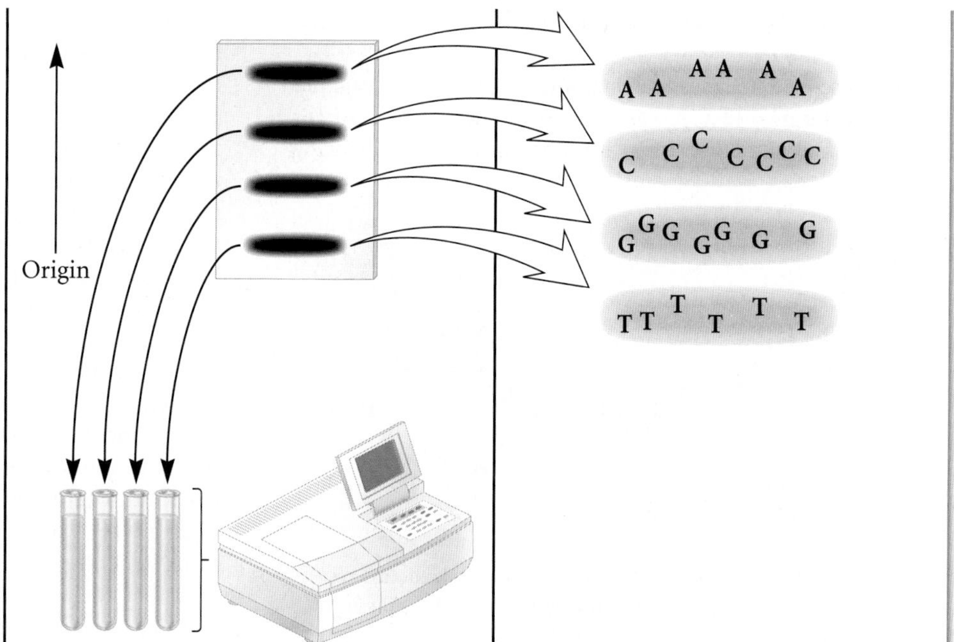

THE DATA

Base Content in the DNA from a Variety of Organisms*

	% of Bases (based on molarity)			
Organism	*Adenine*	*Thymine*	*Guanine*	*Cytosine*
Escherichia coli	26.0	23.9	24.9	25.2
Streptococcus pneumoniae (type III)	29.8	31.6	20.5	18.0
Yeast	31.7	32.6	18.3	17.4
Turtle red blood cells	28.7	27.9	22.0	21.3
Salmon sperm	29.7	29.1	20.8	20.4
Chicken red blood cells	28.0	28.4	22.0	21.6
Human liver cells	30.3	30.3	19.5	19.9

*When the base contents from different tissues within the same species were measured, similar results were obtained.

INTERPRETING THE DATA

The data shown in Figure 9.14 are only a small sampling of Chargaff's results. During the late 1940s and early 1950s, Chargaff published many papers concerned with the chemical composition of DNA from biological sources. Hundreds of measurements were made. The compelling observation was that the amount of adenine was similar to thymine, and the amount of guanine was similar to cytosine. These results were not sufficient to propose a model for the structure of DNA. However, they provided the important clue that DNA is structured so that each molecule of adenine interacts with thymine, and each molecule of guanine interacts with cytosine. A DNA structure in which A binds to T, and G to C, would explain the equal amounts of A and T, and G and C observed in Chargaff's experiments. As we will see, this observation became crucial evidence that Watson and Crick used to elucidate the structure of the double helix.

A self-help quiz involving this experiment can be found at the Online Learning Center.

Watson and Crick Deduced the Double-Helical Structure of DNA

Thus far, we have examined key pieces of information used to determine the structure of DNA. In particular, the X-ray diffraction work of Franklin suggested a helical structure composed of two or more strands with 10 bases per turn. In addition, the work of Chargaff indicated that the amount of A equals T, and the amount of G equals C. Furthermore, Watson and Crick were familiar with Pauling's success in using ball-and-stick models to deduce the secondary structure of proteins. With these key observations, they set out to solve the structure of DNA.

Watson and Crick assumed DNA is composed of nucleotides that are linked together in a linear fashion. They also assumed the chemical linkage between two nucleotides is always the same. With these ideas in mind, they tried to build ball-and-stick models that incorporated the known experimental observations. Because the diffraction pattern suggested the helix must have two (or more) strands, a critical question was, How could two strands interact? As discussed in his book *The Double Helix*, James Watson noted that in an early attempt at model building, they considered the possibility that the negatively charged phosphate groups, together with magnesium ions, were promoting an interaction between the backbones of DNA strands (**Figure**

9.15). However, more detailed diffraction data were not consistent with this model.

Because the magnesium hypothesis for DNA structure appeared to be incorrect, it was back to the drawing board (or back to the ball-and-stick units) for Watson and Crick. During this time, Rosalind Franklin had produced even clearer X-ray diffraction patterns, which provided greater detail concerning the relative locations of the bases and backbone of DNA. This was a major breakthrough that suggested a two-strand interaction that was helical. In their model building, the emphasis shifted to models containing the two backbones on the outside of the model, with the bases projecting toward each other. At first, a structure was considered in which the bases form hydrogen bonds with the identical base in the opposite strand (A to A, T to T, G to G, and C to C). However, the model building revealed that the bases could not fit together this way. The final hurdle was overcome when it was realized that the hydrogen bonding of adenine to thymine was structurally similar to that of guanine to cytosine. With an interaction between A and T and between G and C, the ball-and-stick models showed that the two strands would fit together properly. This ball-and-stick model, shown in **Figure 9.16**, was consistent with all the known data regarding DNA structure.

For their work, Watson, Crick, and Maurice Wilkins were awarded the Nobel Prize in 1962. The contribution of Rosalind Franklin to the discovery of the double helix was also critical and has been acknowledged in several books and articles. Franklin

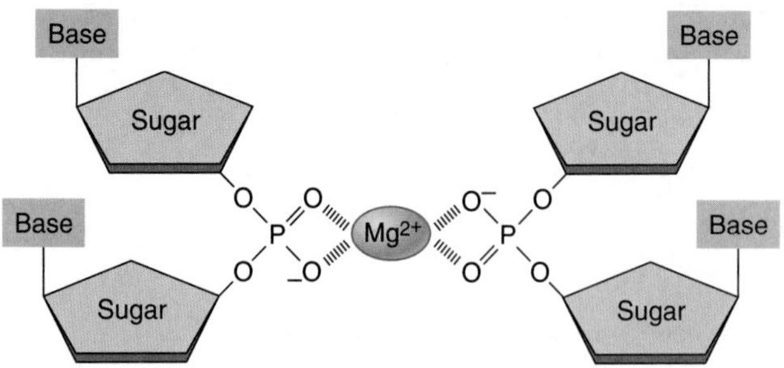

FIGURE 9.15 **An incorrect hypothesis for the structure of the DNA double helix.** This illustration shows an early hypothesis of Watson and Crick's, suggesting that two DNA strands interact by a cross-link between the negatively charged phosphate groups in the backbone and divalent Mg^{2+} cations.

(a) **Watson and Crick**

(b) **Original model of the DNA double helix**

FIGURE 9.16 **Watson and Crick and their model of the DNA double helix.** **(a)** James Watson is shown here on the left and Francis Crick on the right. **(b)** The molecular model they originally proposed for the double helix. Each strand contains a sugar-phosphate backbone. In opposite strands, A hydrogen bonds to T, and G hydrogen bonds with C.

was independently trying to solve the structure of DNA. However, Wilkins, who worked in the same laboratory, shared Franklin's X-ray data with Watson and Crick, presumably without her knowledge. This provided important information that helped them solve the structure of DNA, which was published in the journal *Nature* in April 1953. Though she was not given credit in the original publication of the double-helix structure, Franklin's key contribution became known in later years. Unfortunately, however, Rosalind Franklin died in 1958, and the Nobel Prize is not awarded posthumously.

The Molecular Structure of the DNA Double Helix Has Several Key Features

The general structural features of the double helix are shown in **Figure 9.17**. In a DNA double helix, two DNA strands are twisted together around a common axis to form a structure that resembles a circular staircase. This double-stranded structure is stabilized by **base pairs (bp)**—pairs of bases in opposite strands that are hydrogen bonded to each other. Counting the bases, if you move past 10 base pairs, you have gone 360° around the backbone. The linear distance of a complete turn is 3.4 nm; each base pair traverses 0.34 nm.

A distinguishing feature of the hydrogen bonding between base pairs is its specificity. An adenine base in one strand hydrogen bonds with a thymine base in the opposite strand, or a guanine base hydrogen bonds with a cytosine. This **AT/GC rule,** also known as Chargaff's rule, explained the earlier data of Chargaff showing that the DNA from many organisms contains equal amounts of A and T, and equal amounts of G and C (see Figure 9.14). The AT/GC rule indicates that purines (A and G) always bond with pyrimidines (T and C). This keeps the width of the double helix relatively constant. As noted in Figure 9.17, three hydrogen bonds occur between G and C but only two between A and T. For this

Key Features

- Two strands of DNA form a right-handed double helix.
- The bases in opposite strands hydrogen bond according to the AT/GC rule.
- The 2 strands are antiparallel with regard to their 5′ to 3′ directionality.
- There are ~10.0 nucleotides in each strand per complete 360° turn of the helix.

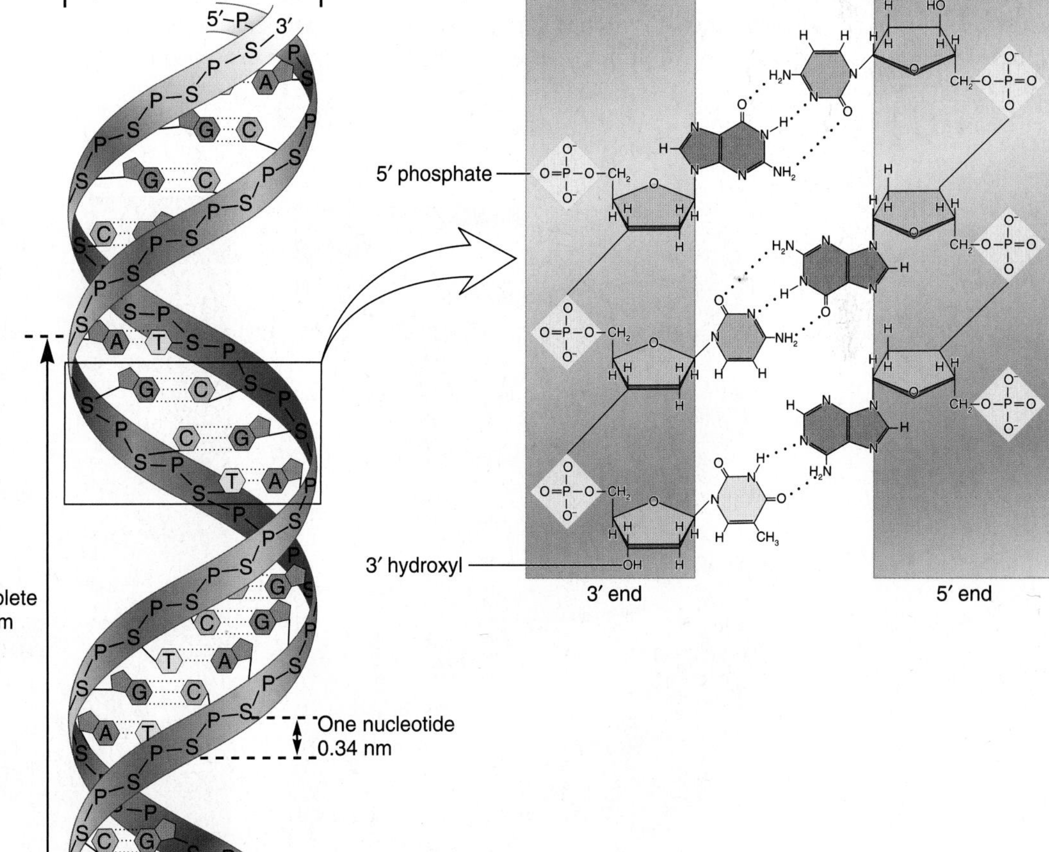

FIGURE 9.17 **Key features of the structure of the double helix.**

reason, DNA sequences that have a high proportion of G and C tend to form more stable double-stranded structures.

The AT/GC rule implies that we can predict the sequence in one DNA strand if the sequence in the opposite strand is known. For example, if one strand has the sequence of 5'–ATG-GCGGATTT–3', then the opposite strand must be 3'–TACC-GCCTAAA–5'. In genetic terms, we would say that these two sequences are **complementary** to each other or that the two sequences exhibit complementarity. In addition, you may have noticed that the sequences are labeled with 5' and 3' ends. These numbers designate the direction of the DNA backbones. The direction of DNA strands is depicted in the inset to Figure 9.17. When going from the top of this figure to the bottom, one strand is running in the 5' to 3' direction, while the other strand is 3' to 5'. This opposite orientation of the two DNA strands is referred to as an **antiparallel** arrangement. An antiparallel structure was initially proposed in the models of Watson and Crick.

Figure 9.18a is a schematic model that emphasizes certain molecular features of DNA structure. The helical structure is formed by the DNA backbone. The bases in this model are depicted as flat rectangular structures that hydrogen bond in pairs. (The hydrogen bonds are the dotted lines.) Although the bases are not actually rectangular, they do form flattened planar structures. Within DNA, the bases are oriented so that the flattened regions are facing each other, an arrangement referred to as base stacking. In other words, if you think of the bases as flat plates, these plates are stacked on top of each other in the double-stranded DNA structure. Along with hydrogen bonding, base stacking is a structural feature that stabilizes the double helix.

By convention, the direction of the DNA double helix shown in Figure 9.18a spirals in a direction that is called "right-handed." To understand this terminology, imagine that a double helix is laid on your desk; one end of the helix is close to you, and the other end is at the opposite side of the desk. As it spirals away from you, a right-handed helix turns in a clockwise direction. By comparison, a left-handed helix would spiral in a counterclockwise manner. Both strands in Figure 9.18a spiral in a right-handed direction.

Figure 9.18b is a space-filling model for DNA in which the atoms are represented by spheres. This model emphasizes the surface features of DNA. Note that the backbone—composed of sugar and phosphate groups—is on the outermost surface. In a living cell, the backbone has the most direct contact with water. In contrast, the bases are more internally located within the double-stranded structure. Biochemists use the term **grooves** to describe the indentations where the atoms of the bases are in contact with the surrounding water. As you travel around the DNA helix, the structure of DNA has two grooves: the **major groove** and the **minor groove.** As will be discussed in later chapters, certain proteins can bind within these grooves and interact with a particular sequence of bases, thereby altering DNA structure and regulating gene expression.

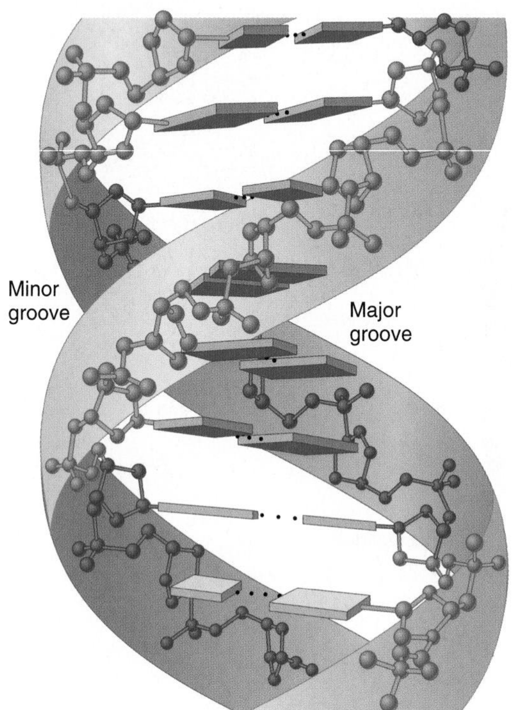

(a) Ball-and-stick model of DNA

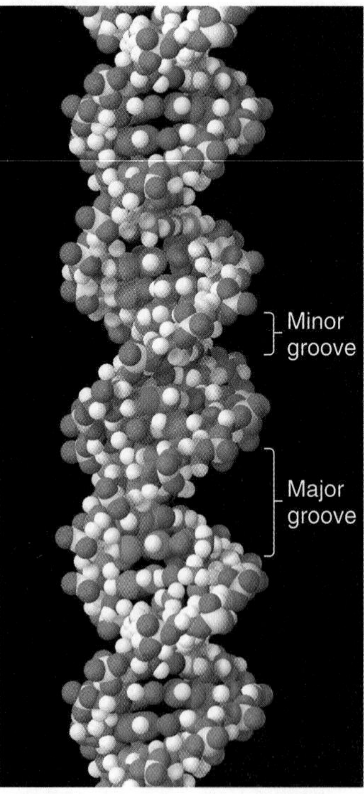

(b) Space-filling model of DNA

FIGURE 9.18 **Two models of the double helix.** (a) Ball-and-stick model of the double helix. The ribose–phosphate backbone is shown in detail, while the bases are depicted as flattened rectangles. (b) Space-filling model of the double helix.

DNA Can Form Alternative Types of Double Helices

The DNA double helix can form different types of structures. **Figure 9.19** compares the structures of **A DNA, B DNA,** and **Z DNA.** The highly detailed structures shown here were deduced by X-ray crystallography on short segments of DNA. B DNA is the predominant form of DNA found in living cells. However, under certain in vitro conditions, the two strands of DNA can twist into A DNA or Z DNA, which differ significantly from B DNA. A and B DNA are right-handed helices; Z DNA has a left-handed conformation. In addition, the helical backbone in Z

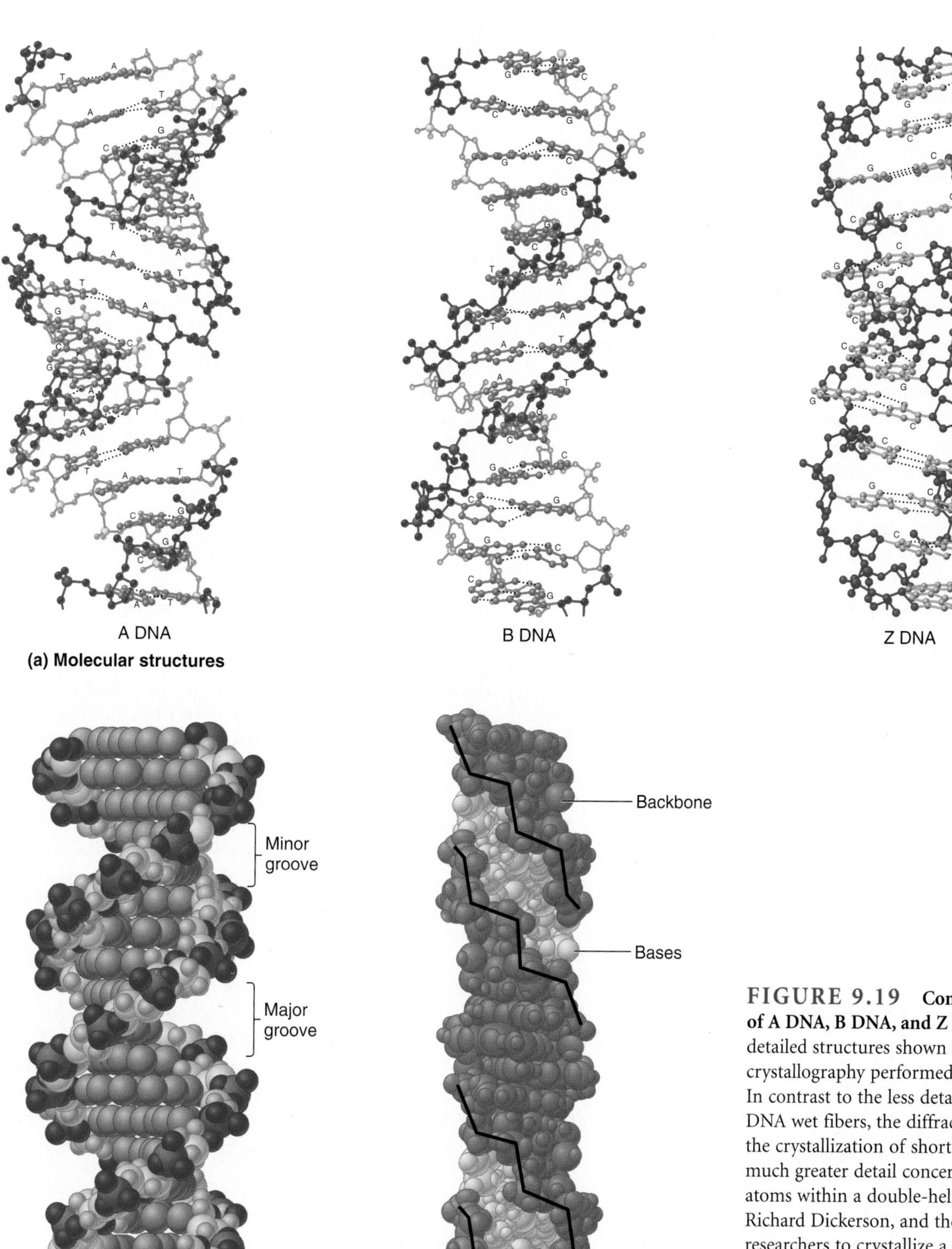

FIGURE 9.19 Comparison of the structures of A DNA, B DNA, and Z DNA. (a) The highly detailed structures shown here were deduced by X-ray crystallography performed on short segments of DNA. In contrast to the less detailed structures obtained from DNA wet fibers, the diffraction pattern obtained from the crystallization of short segments of DNA provides much greater detail concerning the exact placement of atoms within a double-helical structure. Alexander Rich, Richard Dickerson, and their colleagues were the first researchers to crystallize a short piece of DNA. (b) Space-filling models of the B-DNA and Z-DNA structures. In the case of Z DNA, the black lines connect the phosphate groups in the DNA backbone. As seen here, they travel along the backbone in a zigzag pattern.

DNA appears to zigzag slightly as it winds itself around the double-helical structure. The numbers of base pairs per 360° turn are 11.0, 10.0, and 12.0 in A, B, and Z DNA, respectively. In B DNA, the bases tend to be centrally located, and the hydrogen bonds between base pairs occur relatively perpendicular to the central axis. In contrast, the bases in A DNA and Z DNA are substantially tilted relative to the central axis.

The ability of the predominant B DNA to adopt A-DNA and Z-DNA conformations depends on certain conditions. In X-ray diffraction studies, A DNA occurs under conditions of low humidity. The ability of a double helix to adopt a Z-DNA conformation depends on various factors. At high ionic strength (i.e., high salt concentration), formation of a Z-DNA conformation is favored by a sequence of bases that alternates between purines and pyrimidines. One such sequence is

5'-GCGCGCGCG-3'

3'-CGCGCGCGC-5'

At lower ionic strength, the methylation of cytosine bases can favor Z-DNA formation. Cytosine **methylation** occurs when a cellular enzyme attaches a methyl group ($–CH_3$) to the cytosine base. In addition, negative supercoiling (a topic discussed in Chapter 10) favors the Z-DNA conformation.

What is the biological significance of A and Z DNA? Research has not found any biological role for A DNA. However, accumulating evidence suggests a possible biological role for Z DNA in the process of transcription. Recent research has identified cellular proteins that specifically recognize Z DNA. In 2005, Alexander Rich and colleagues reported that the Z-DNA binding region of one such protein played a role in regulating the transcription of particular genes. In addition, other research has suggested that Z DNA may play a role in chromosome structure by affecting the level of compaction.

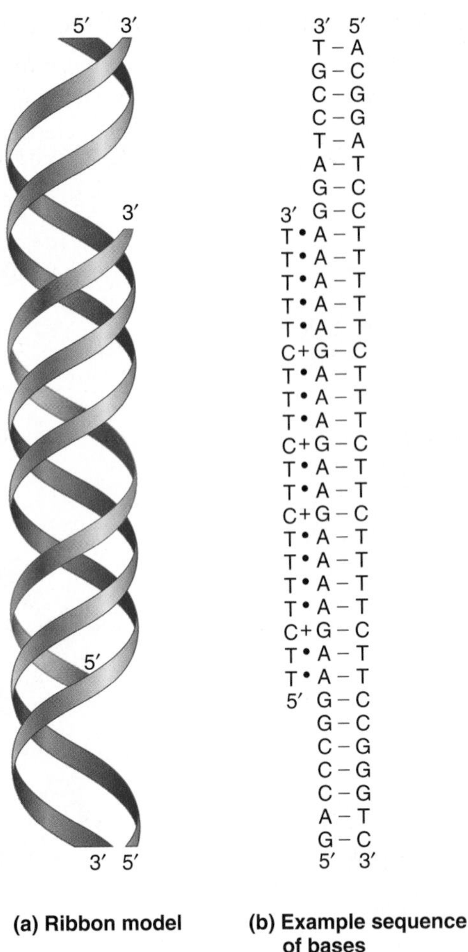

FIGURE 9.20 **The structure of triplex DNA. (a)** As seen in the ribbon model, the third, synthetic strand binds within the major groove of the double-stranded structure. **(b)** Within triplex DNA, the third strand hydrogen bonds according to the rule T to AT, and C to GC. The cytosine bases in the third strand are protonated (i.e., positively charged).

DNA Can Form a Triple Helix, Called Triplex DNA

A surprising discovery made in 1957 by Alexander Rich, David Davies, and Gary Felsenfeld was that DNA can form a triple-helical structure called **triplex DNA.** This triplex was formed in vitro using pieces of DNA that were made synthetically. Although this result was interesting, it seemed to have little, if any, biological relevance.

About 30 years later, interest in triplex DNA was renewed by the observation that triplex DNA can form in vitro by mixing natural double-stranded DNA and a third short strand that is synthetically made. The synthetic strand binds into the major groove of the naturally occurring double-stranded DNA (**Figure 9.20**). As shown here, an interesting feature of triplex DNA formation is that it is sequence specific. In other words, the synthetic third strand incorporates itself into a triple helix due to specific interactions between the synthetic DNA and the biological DNA. The pairing rules are that a thymine in the synthetic DNA will hydrogen bond at an AT pair in the biological DNA and that a cytosine in the synthetic DNA will hydrogen bond at a GC pair.

The formation of triplex DNA has been implicated in several cellular processes, including recombination, which is described in Chapter 17. In addition, researchers are interested in triplex DNA due to its potential as a tool to specifically inhibit particular genes. As shown in Figure 9.20, the synthetic DNA strand binds into the major groove according to specific base-pairing rules. Therefore, researchers can design a synthetic DNA to recognize the base sequence found in a particular gene. When the synthetic DNA binds to a gene, it inhibits transcription. In addition, the synthetic DNA can cause mutations in a gene that inactivate its function. Researchers are excited about the possibility of using such synthetic DNA to silence the expression of particular genes. For example, this approach could be used to silence genes that become overactive in cancer cells. However, future research will be needed to develop effective ways to promote the uptake of synthetic DNAs into the appropriate target cells.

The Three-Dimensional Structure of DNA Within Chromosomes Requires Additional Folding and the Association with Proteins

To fit within a living cell, the long double-helical structure of chromosomal DNA must be extensively compacted into a three-dimensional conformation. With the aid of DNA-binding proteins, such as histone proteins, the double helix becomes greatly twisted and folded. **Figure 9.21** depicts the relationship between the DNA double helix and the compaction that occurs within a eukaryotic chromosome. Chapter 10 is devoted to the topic of chromosome organization and the molecular mechanisms responsible for the packaging of genetic material in cells.

RNA Molecules Are Composed of Strands That Fold into Specific Structures

Let's now turn our attention to RNA structure, which bears many similarities to DNA structure. The structure of an RNA strand is much like a DNA strand (**Figure 9.22**). Strands of RNA are typically several hundred or several thousand nucleotides in length, which is much shorter than chromosomal DNA. When RNA is made during transcription, the DNA is used as a template to make a copy of single-stranded RNA. In most cases, only one of the two DNA strands is used as a template for RNA synthesis. Therefore, only one complementary strand of RNA is usually made. Nevertheless, relatively short sequences within one RNA molecule or between two separate RNA molecules can form double-stranded regions.

The helical structure of RNA molecules is due to the ability of complementary regions to form base pairs between A and U and between G and C. This base pairing allows short segments to form a double-stranded region. As shown in **Figure 9.23**, different types of structural patterns are possible. These include bulge loops, internal loops, multibranched junctions, and stem-loops (also called hairpins). These structures contain regions of complementarity punctuated by regions of noncomplementarity. As shown in Figure 9.23, the complementary regions are held together by connecting hydrogen bonds, while the noncomplementary regions have their bases projecting away from the double-stranded region.

Many factors contribute to the structure of RNA molecules. These include the base-paired double-stranded helices, stacking between bases, and hydrogen bonding between bases and backbone regions. In addition, interactions with ions, small molecules, and large proteins may influence RNA structure. **Figure 9.24** depicts the structure of a transfer RNA molecule known as $tRNA^{phe}$, which is a tRNA molecule that carries the amino acid phenylalanine. It was the first naturally occurring RNA to have its structure elucidated. This RNA molecule has several double-stranded and single-stranded regions. RNA double helices are antiparallel and right-handed with 11 to 12 base pairs per turn. In a living cell, the various regions of an RNA molecule fold and interact with each other to produce the three-dimensional structure.

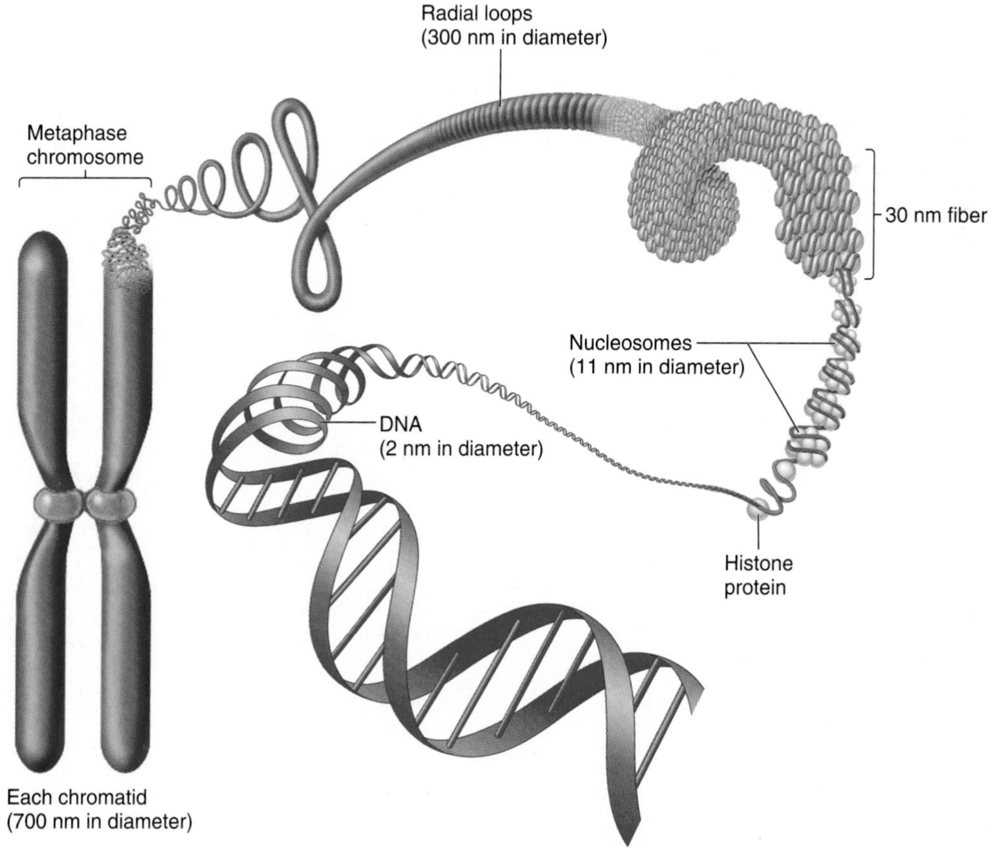

FIGURE 9.21 The steps in eukaryotic chromosomal compaction leading to the metaphase chromosome. The DNA double helix is wound around histone proteins and then is further compacted to form a highly condensed metaphase chromosome. The levels of DNA compaction will be described in greater detail in Chapter 10.

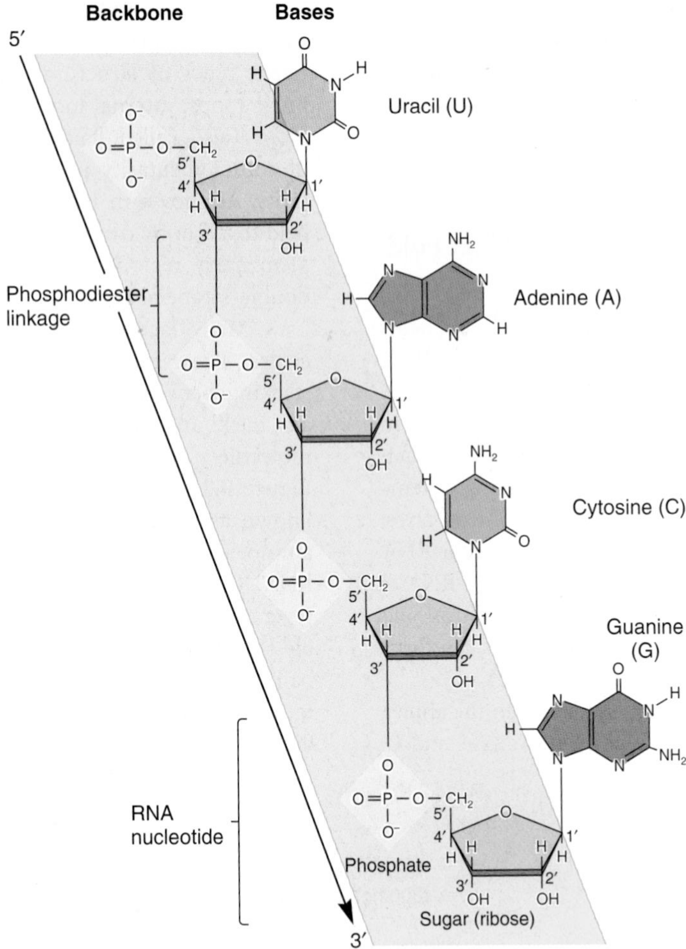

FIGURE 9.22 A strand of RNA. This structure is very similar to a DNA strand (see Figure 9.11), except that the sugar is ribose instead of deoxyribose, and uracil is substituted for thymine.

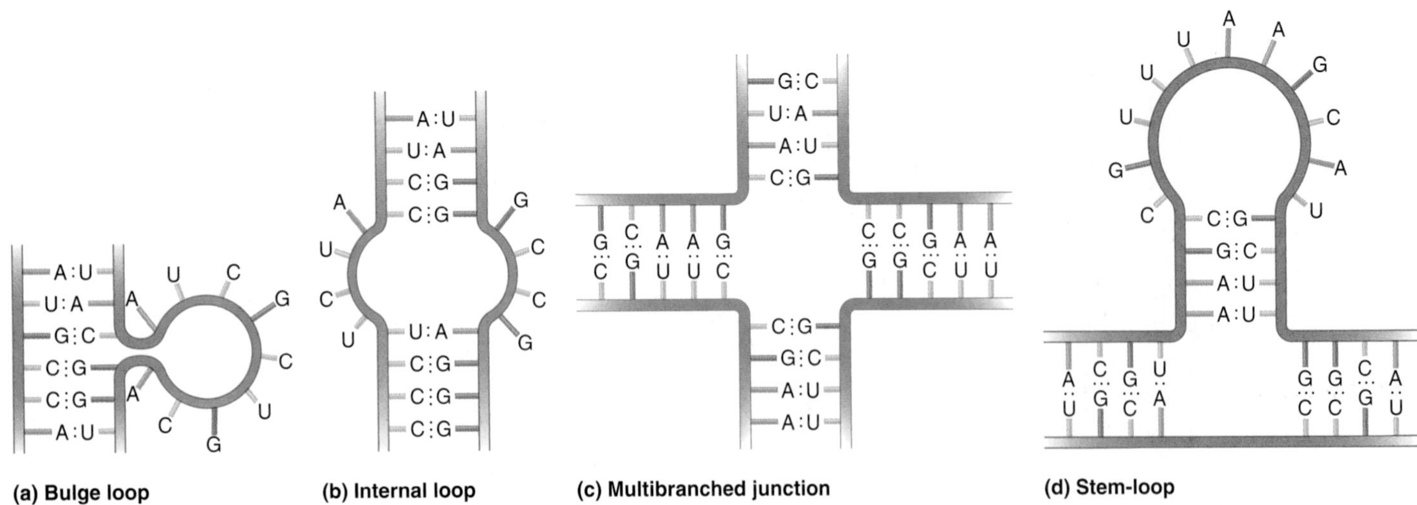

FIGURE 9.23 Possible structures of RNA molecules. The double-stranded regions are depicted by connecting hydrogen bonds. Loops are noncomplementary regions that are not hydrogen bonded with complementary bases. Double-stranded RNA structures can form within a single RNA molecule or between two separate RNA molecules.

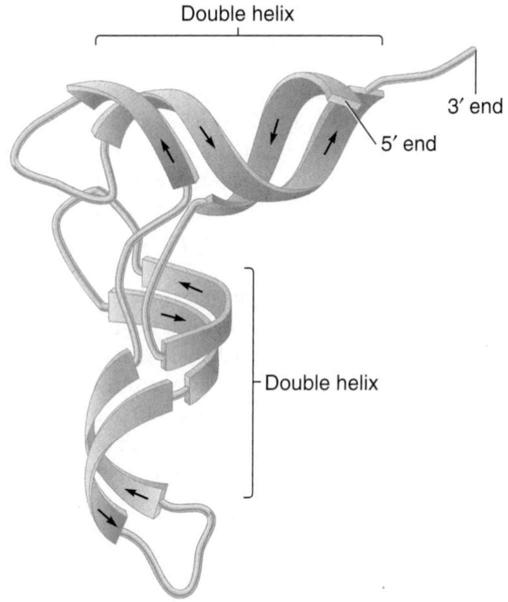

(a) Ribbon model

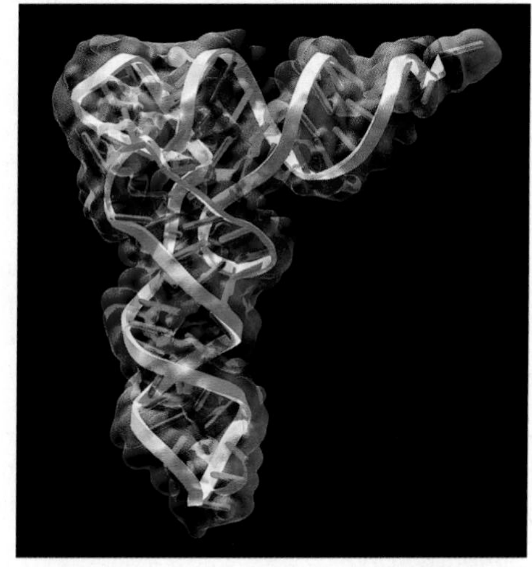

(b) Space-filling model

FIGURE 9.24 The structure of tRNA^{phe}, the transfer RNA molecule that carries phenylalanine. (a) The double-stranded regions of the molecule are shown as antiparallel ribbons. (b) A space-filling model of tRNA^{phe}.

CONCEPTUAL SUMMARY

The molecular structure of nucleic acids underlies their function. Nucleotides, which are composed of a phosphate group, pentose sugar, and nitrogenous base, form the repeating structural unit of nucleic acids. A strand of DNA or RNA contains a linear sequence of nucleotides. Strands of DNA (and occasionally RNA) can form a double helix where complementary strands interact due to hydrogen bonding between adenine and thymine (or uracil in RNA) and between guanine and cytosine. Base stacking also stabilizes a double-stranded structure. The most common form of DNA is a right-handed double helix; the backbones in the double helix are composed of sugar–phosphate linkages, with the bases projecting inward from the backbones and hydrogen bonding with each other. The two strands are antiparallel, with one strand running in the 5' to 3' direction, and the other running in the 3' to 5' direction. Three different helical conformations have been identified: A DNA, B DNA, and Z DNA. B DNA is the predominant form found in living cells, but short regions of Z DNA may play an important functional role in transcription. Within chromosomes, DNA is folded into a three-dimensional conformation with the aid of proteins. Short segments within RNA may also form double-helical structures such as bulge loops, internal loops, multibranched junctions, and stem-loops. The structure of RNA is dictated by several factors including double-helical regions, base stacking, hydrogen bonding between bases and backbone, and interactions with other molecules.

EXPERIMENTAL SUMMARY

Two different experimental approaches were used to show that DNA is the genetic material. Griffith's experiments with *S. pneumonia* showed that a "transforming principle," later identified as DNA, was capable of transferring genetic information. Avery, MacLeod, and McCarty purified DNA from type IIIS strains of *S. pneumonia* and used it to transform type IIR strains into type IIIS. Further testing showed that DNase prevented the transformation, whereas RNase and protease did not. These observations indicated that DNA is the transforming substance that provides type IIR bacteria with a new trait. Hershey and Chase showed that T2 bacteriophage injects DNA into bacterial cells, indicating that DNA is the genetic material of this virus. Taken together, the latter two studies were instrumental in convincing scientists that the genetic material is composed of DNA.

Several experimental observations enabled Watson and Crick to determine the structure of DNA. The X-ray diffraction work of Franklin indicated that DNA had a helical structure with two strands and 10 base pairs per 360° turn. Furthermore, the work of Chargaff indicated that the amount of adenine (A) equals that of thymine (T), and the amount of guanine (G) equals that of cytosine (C). With these observations, and employing Pauling's strategy of using ball-and-stick models to visualize the structure of macromolecules, Watson and Crick deduced the structure of DNA, which is a double helix in which the outer backbone is composed of antiparallel strands of alternating deoxyribose and phosphate molecules. The four types of bases are within the interior of the double helix; bases in opposite strands hydrogen bond with each other according to the AT/GC rule.

PROBLEM SETS & INSIGHTS

Solved Problems

S1. A hypothetical sequence at the beginning of an mRNA molecule is

5'–AUUUGCCCUAGCAAACGUAGCAAACG...rest of the coding sequence

Using two out of the three underlined sequences, draw two possible models for potential stem-loop structures at the 5' end of this mRNA.

Answer:

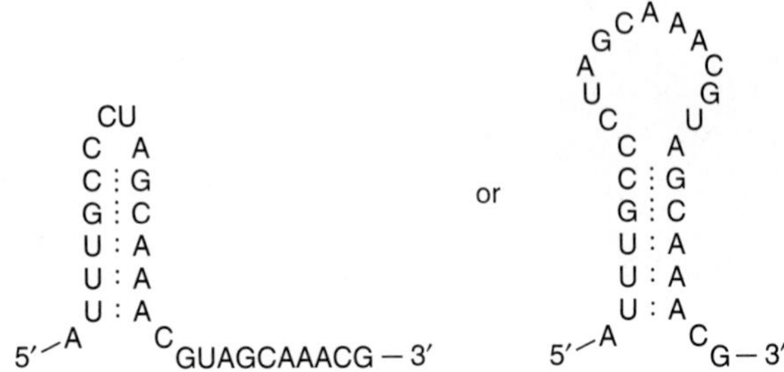

S2. Describe the previous experimental evidence that led Watson and Crick to the discovery of the DNA double helix.

Answer:

1. The chemical structure of single nucleotides was understood by the 1950s.
2. Watson and Crick assumed DNA is composed of nucleotides linked together in a linear fashion to form a strand. They also assumed the chemical linkage between two nucleotides is always the same.
3. Franklin's diffraction patterns suggested several structural features. First, it was consistent with a helical structure. Second, the diameter of the helical structure was too wide to be only a single-stranded helix. Finally, the line spacing on the diffraction pattern indicated the helix contains about 10 base pairs per complete turn.
4. In the chemical analysis of the DNA from different species, the work of Chargaff indicated the amount of adenine equaled the amount of thymine, and the amount of cytosine equaled that of guanine.
5. In the early 1950s, Linus Pauling proposed that regions of proteins can fold into a secondary structure known as an α helix. To discover this, Pauling built large models by linking together simple ball-and-stick units. In this way, he could determine if atoms fit together properly in a complicated three-dimensional structure. A similar approach was used by Watson and Crick to solve the structure of the DNA double helix.

S3. Within living cells, a myriad of different proteins play important functional roles by binding to DNA and RNA. As described throughout your textbook, the dynamic interactions between nucleic acids and proteins lie at the heart of molecular genetics. Some proteins bind to DNA (or RNA) but not in a sequence-specific manner. For example, histones are proteins important in the formation of chromosome structure. In this case, the positively charged histone proteins actually bind to the negatively charged phosphate groups in DNA. In addition, several other proteins interact with DNA but do not require a specific nucleotide sequence to carry out their function. For example, DNA polymerase, which catalyzes the synthesis of new DNA strands, does not bind to DNA in a sequence-dependent manner. By comparison, many other proteins do interact with nucleic acids in a sequence-dependent fashion. This means that a specific sequence of bases can provide a structure that will be recognized by a particular protein. Throughout the textbook, the functions of many of these proteins will be described. Some examples include transcription factors that affect the rate of transcription, proteins that bind to centromeres, and proteins that bind to origins of replication. With regard to the three-dimensional structure of DNA, where would you expect DNA-binding proteins to bind if they recognize a specific base sequence? What about DNA-binding proteins that do not recognize a base sequence?

Answer: DNA-binding proteins that recognize a base sequence must bind into a major or minor groove of the DNA, which is where the bases would be accessible to a DNA-binding protein. Most DNA-binding proteins, which recognize a base sequence, fit into the major groove. By comparison, other DNA-binding proteins, such as histones, which do not recognize a base sequence, bind to the DNA backbone.

S4. The formation of a double-stranded structure must obey the rule that adenine hydrogen bonds to thymine (or uracil) and cytosine hydrogen bonds to guanine. Based on your previous understanding of genetics (from this course or a general biology course), discuss reasons why complementarity is an important feature of DNA and RNA structure and function.

Answer: Note: Many of the topics described below are discussed in Chapters 10 through 13. One way that complementarity underlies function is that it provides the basis for the synthesis of new strands of DNA and RNA. During replication, the synthesis of the new DNA strands occurs in such a way that adenine hydrogen bonds to thymine, and cytosine hydrogen bonds to guanine. In other words, the molecular feature of a complementary double-stranded structure makes it possible to produce exact copies of DNA. Likewise, the ability to transcribe DNA into RNA is based on complementarity. During transcription, one strand of DNA is used as a template to make a complementary strand of RNA.

In addition to the synthesis of new strands of DNA and RNA, complementarity is important in other ways. As mentioned in this chapter, the folding of RNA into a particular structure is driven by the hydrogen bonding of complementary regions. This event is necessary to produce functionally active tRNA molecules. Likewise, stem-loop structures also occur in other types of RNA. For example, the rapid formation of stem-loop structures is known to occur as RNA is being transcribed and to affect the termination of transcription.

A third way that complementarity can be functionally important is that it can promote the interaction of two separate RNA molecules. During translation, codons in mRNA bind to the anticodons in tRNA (see Chapter 13). This binding is due to complementarity. For example, if a codon is 5'–AGG–3', the anticodon is 3'–UCC–5'. This type of specific interaction between codons and anticodons is an important step that enables the nucleotide sequence in mRNA to code for an amino acid sequence within a protein. In addition, many other examples of RNA-RNA interactions are known and will be described throughout this textbook.

S5. An important feature of triplex DNA formation is that it is sequence specific. The synthetic third strand incorporates itself into a triple helix, so a thymine in the synthetic DNA will bind near an AT pair in the biological DNA, and a cytosine in the synthetic DNA will bind near a GC pair. From a practical point of view, this opens the possibility of synthesizing a short strand of DNA that will form a triple helix at a particular target site. For example, if the sequence of a particular gene is known, researchers can make a synthetic piece of DNA that will form a triple helix somewhere within that gene according to the T to AT, and C to GC rule. Triplex DNA formation is known to inhibit gene transcription. In other words, when the synthetic DNA binds within the DNA of a gene, the formation of triplex DNA prevents that gene from being transcribed into RNA. Discuss how this observation might be used to combat diseases.

Answer: Triplex DNA formation opens the exciting possibility of designing synthetic pieces of DNA to inhibit the expression of particular genes. Theoretically, such a tool could be used to combat viral diseases or to inhibit the growth of cancer cells. To combat a viral disease, a synthetic DNA could be made that specifically binds to an essential viral gene and thereby prevents viral proliferation. To inhibit cancer, a synthetic DNA could be made to bind to an oncogene. (Note: As described in Chapter 22, an oncogene is a gene that promotes cancerous growth.) Inhibition of an oncogene could prevent cancer. At this point, a primary obstacle in applying this approach is devising a method of getting the synthetic DNA into living cells.

Conceptual Questions

C1. What is the meaning of the term genetic material?

C2. After the DNA from type IIIS bacteria is exposed to type IIR bacteria, list all of the steps that you think must occur for the bacteria to start making a type IIIS capsule.

C3. Look up the meaning of the word transformation in a dictionary and explain whether it is an appropriate word to describe the transfer of genetic material from one organism to another.

C4. What are the building blocks of a nucleotide? With regard to the 5' and 3' positions on a sugar molecule, how are nucleotides linked together to form a strand of DNA?

C5. Draw the structure of guanine, guanosine, and deoxyguanosine triphosphate.

C6. Draw the structure of a phosphodiester linkage.

C7. Describe how bases interact with each other in the double helix. This discussion should address the issues of complementarity, hydrogen bonding, and base stacking.

C8. If one DNA strand is 5'–GGCATTACACTAGGCCT–3' what is the sequence of the complementary strand?

C9. What is meant by the term DNA sequence?

C10. Make a side-by-side drawing of two DNA helices, one with 10 base pairs (bp) per 360° turn and the other with 15 bp per 360° turn.

C11. Discuss the differences in the structural features of A DNA, B DNA, and Z DNA.

C12. What parts of a nucleotide (namely, phosphate, sugar, and/or bases) occupy the major and minor grooves of double-stranded DNA, and what parts are found in the DNA backbone? If a DNA-binding protein does not recognize a specific nucleotide sequence, do you expect that it recognizes the major groove, the minor groove, or the DNA backbone? Explain.

C13. List the structural differences between DNA and RNA.

C14. Draw the structure of deoxyribose and number the carbon atoms. Describe the numbering of the carbon atoms in deoxyribose with regard to the directionality of a DNA strand. In a DNA double helix, what does the term antiparallel mean?

C15. Write out a sequence of an RNA molecule that could form a stem-loop with 24 nucleotides in the stem and 16 nucleotides in the loop.

C16. Compare the structural features of a double-stranded RNA structure with those of a DNA double helix.

C17. Which of the following DNA double helices would be more difficult to separate into single-stranded molecules by treatment with heat, which breaks hydrogen bonds?

A. GGCGTACCAGCGCAT

CCGCATGGTCGCGTA

B. ATACGATTTACGAGA

TATGCTAAATGCTCT

Explain your choice.

C18. What structural feature allows DNA to store information?

C19. Discuss the structural significance of complementarity in DNA and in RNA.

C20. An organism has a G + C content of 64% in its DNA. What are the percentages of A, T, G, and C?

C21. Let's suppose you have recently identified an organism that was scraped from an asteroid that hit the earth. (Fortunately, no one was injured.) When you analyze this organism, you discover that its DNA is a triple helix, composed of six different nucleotides: A, T, G, C, X, and Y. You measure the chemical composition of the bases and find the following amounts of these six bases: A = 24%, T = 23%, G = 11%, C = 12%, X = 21%, Y = 9%. What rules would you propose govern triplex DNA formation in this organism? Note: There is more than one possibility.

C22. Upon further analysis of the DNA described in conceptual question C21, you discover that the triplex DNA in this alien organism is composed of a double helix, with the third helix wound within the major groove (just like the DNA in Figure 9.20). How would you propose that this DNA is able to replicate itself? In your answer, be specific about the base pairing rules within the double helix and which part of the triplex DNA would be replicated first.

C23. A DNA-binding protein recognizes the following double-stranded sequence:

5'–GCCCGGGC–3'

3'–CGGGCCCG–5'

This type of double-stranded structure could also occur within the stem region of an RNA stem-loop molecule. Discuss the structural differences between RNA and DNA that might prevent this DNA-binding protein from recognizing a double-stranded RNA molecule.

C24. Within a protein, certain amino acids are positively charged (e.g., lysine and arginine), some are negatively charged (e.g., glutamate and aspartate), some are polar but uncharged, and some are nonpolar. If you knew that a DNA-binding protein was recognizing the DNA backbone rather than base sequences, which amino acids in the protein would be good candidates for interacting with the DNA?

C25. In what ways are the structures of an α helix in proteins and the DNA double helix similar, and in what ways are they different?

C26. A double-stranded DNA molecule contains 560 nucleotides. How many complete turns would be found in this double helix?

C27. As the minor and major grooves of the DNA wind around a DNA double helix, do they ever intersect each other, or do they always run parallel to each other?

C28. What chemical group (phosphate group, hydroxyl group, or a nitrogenous base) is found at the 3' end of a DNA strand? What group is found at the 5' end?

C29. The base composition of an RNA virus was analyzed and found to be 14.1% A, 14.0% U, 36.2% G, and 35.7% C. Would you conclude that the viral genetic material is single-stranded RNA or double-stranded RNA?

C30. The genetic material found within some viruses is single-stranded DNA. Would this genetic material contain equal amounts of A and T and equal amounts of G and C?

C31. A medium-sized human chromosome contains about 100 million bp. If the DNA were stretched out in a linear manner, how long would it be?

C32. A double-stranded DNA molecule is 1 cm long, and the percentage of adenine is 15%. How many cytosines would be found in this DNA molecule?

C33. Could single-stranded DNA form a stem-loop structure? Why or why not?

C34. As described in Chapter 15, the methylation of cytosine bases can have an important impact on gene expression. For example, the methylation of cytosines may inhibit the transcription of genes. A methylated cytosine base has the following structure:

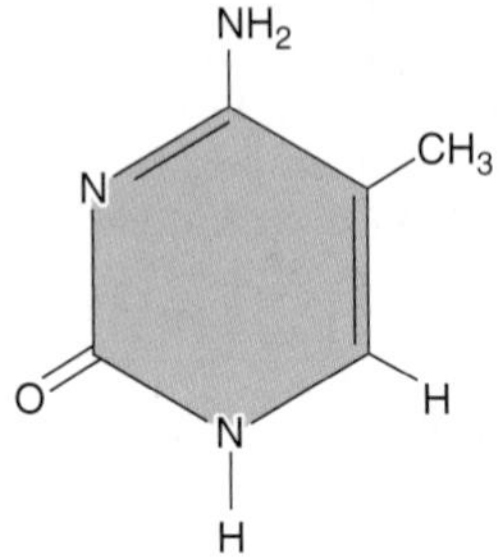

Would you expect the methylation of cytosine to affect the hydrogen bonding between cytosine and guanine in a DNA double helix? Why or why not? (Hint: See Figure 9.17 for help.) Take a look at solved problem S3 and speculate as to how methylation could affect gene expression.

C35. An RNA molecule has the following sequence:

Region 1 Region 2 Region 3
5'-CAUCCAUCCAUUCCCCAUCCGAUAAGGGGAAUGGAUCCGAAUGGAUAAC-3'

Parts of region 1 can form a stem-loop with region 2 and with region 3. Can region 1 form a stem-loop with region 2 and region 3 at the same time? Why or why not? Which stem-loop would you predict to be more stable: a region 1/region 2 interaction or a region 1/region 3 interaction? Explain your choice.

Experimental Questions

E1. Genetic material acts as a blueprint for an organism's traits. Explain how the experiments of Griffith indicated that genetic material was being transferred to the type IIR bacteria.

E2. With regard to the experiment described in Figure 9.3, answer the following:

A. List several possible reasons why only a small percentage of the type IIR bacteria was converted to type IIIS.

B. Explain why an antibody must be used to remove the bacteria that are not transformed. What would the results look like, in all five cases, if the antibody/centrifugation step had not been included in the experimental procedure?

C. The DNA extract was treated with DNase, RNase, or protease. Why was this done? (In other words, what were the researchers trying to demonstrate?)

E3. An interesting trait that some bacteria exhibit is resistance to killing by antibiotics. For example, certain strains of bacteria are resistant to tetracycline, whereas other strains are sensitive to tetracycline. Describe an experiment you would carry out to demonstrate that tetracycline resistance is an inherited trait encoded by the DNA of the resistant strain.

E4. With regard to the experiment of Figure 9.6, answer the following:

A. Provide possible explanations why some of the DNA is in the supernatant.

B. Plot the results if the radioactivity in the pellet, rather than in the supernatant, had been measured.

C. Why were ^{32}P and ^{35}S chosen as radioisotopes to label the phages?

D. List possible reasons why less than 100% of the phage protein was removed from the bacterial cells during the shearing process.

E5. Does the experiment of Figure 9.6 rule out the possibility that RNA is the genetic material of T2 phage? Explain your answer. If it does not, could you modify the approach of Hershey and Chase to show that it is DNA and not RNA that is the genetic material of T2 bacteriophage? Note: It is possible to specifically label DNA or RNA by providing bacteria with radiolabeled thymine or uracil, respectively.

E6. In Chapter 9, we considered two experiments—one by Avery, MacLeod, and McCarty and the second by Hershey and Chase—that indicated DNA is the genetic material. Discuss the strengths and weaknesses of the two approaches. Which experimental approach did you find the most convincing? Why?

E7. The type of model building used by Pauling and Watson and Crick involved the use of ball-and-stick units. Now we can do model building on a computer screen. Even though you may not be familiar with this approach, discuss potential advantages that computers might provide in molecular model building.

E8. With regard to Chargaff's experiment described in Figure 9.14, answer the following:

A. What is the purpose of paper chromatography?

B. Explain why it is necessary to remove the bases in order to determine the base composition of DNA.

C. Would Chargaff's experiments have been convincing if they had been done on only one species? Discuss.

E9. Gierer and Schramm exposed plant tissue to purified RNA from tobacco mosaic virus, and the plants developed the same types of lesions as if they were exposed to the virus itself. What would be the results if the RNA was treated with DNase, RNase, or protease prior to its exposure to the plant tissue?

Questions for Student Discussion/Collaboration

1. Try to propose structures for a genetic material that are substantially different from the double helix. Remember that the genetic material must have a way to store information and a way to be faithfully replicated.

2. How might you provide evidence that DNA is the genetic material in mice?

Note: All answers appear at the website for this textbook; the answers to even-numbered questions are in the back of the textbook.

www.mhhe.com/brookergenetics3e

Visit the Online Learning Center for practice tests, answer keys, and other learning aids for this chapter. Enhance your understanding of genetics with our interactive exercises, quizzes, animations, and much more.

10 CHROMOSOME ORGANIZATION AND MOLECULAR STRUCTURE

CHAPTER OUTLINE

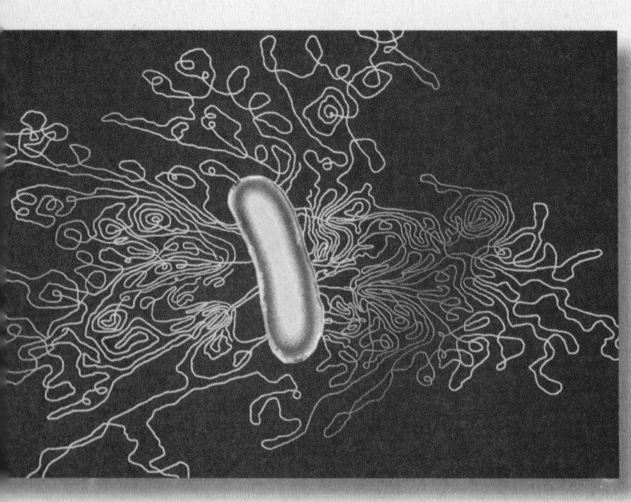

Structure of a bacterial chromosome. This is an electron micrograph of a bacterial chromosome, which has been released from a bacterial cell.

Chromosomes are the structures within cells that contain the genetic material. The term **genome** describes all the types of genetic material that an organism has. For bacteria, the genome is typically a single circular chromosome. For eukaryotes, the nuclear genome refers to one complete set of chromosomes that resides in the cell nucleus. In other words, the haploid complement of chromosomes is considered a nuclear genome. Eukaryotes have a mitochondrial genome, and plants also have a chloroplast genome. Unless otherwise noted, the term eukaryotic genome refers to the nuclear genome.

The primary function of the genetic material is to store the information needed to produce the characteristics of an organism. As we have seen in Chapter 9, the sequence of bases in a DNA molecule can store information. To fulfill their role at the molecular level, chromosomal sequences facilitate four important processes: (1) the synthesis of RNA and cellular proteins, (2) the replication of chromosomes, (3) the proper segregation of chromosomes, and (4) the compaction of chromosomes so they can fit within living cells. In this chapter, we will examine the general organization of the genetic material within viral, bacterial, and eukaryotic chromosomes. In addition, the molecular mechanisms that account for the packaging of the genetic material in viruses, bacteria, and eukaryotic cells will be described. We will begin by considering the comparatively simple genomes of viruses.

10.1 VIRAL GENOMES

Viruses are small infectious particles that contain nucleic acid as their genetic material, surrounded by a protein coat, or capsid (**Figure 10.1a**). The capsid of **bacteriophages,** which are viruses that infect bacteria, may also contain a sheath, base plate, and tail fibers (see Figure 9.4). Certain eukaryotic viruses also have an envelope consisting of a membrane embedded with spike proteins (**Figure 10.1b**). By themselves, viruses are not cellular organisms. They do not contain energy-producing enzymes, ribosomes, or cellular organelles. Instead, viruses rely on their **host cells**—the cells they infect—for replication. In general, most viruses exhibit a limited **host range,** the spectrum of host species that a virus can infect. Many viruses can infect only specific types of cells of one host species. Depending on the life cycle of the virus, the host cell may or may not be destroyed during the process of viral replication and release. In this section, we will consider the genetic composition of viruses and examine how viral genomes are packaged into virus particles.

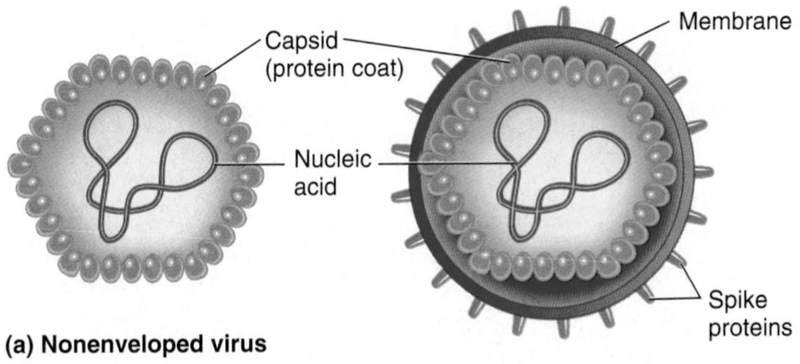

FIGURE 10.1 **General structure of viruses.** (**a**) The simplest viruses contain a nucleic acid molecule (DNA or RNA) surrounded by a capsid, or protein coat. (**b**) Other viruses also contain an envelope composed of a membrane and spike proteins. The membrane is obtained from the host cell when the virus buds through the plasma membrane.

Viral Genomes Are Relatively Small and Are Composed of DNA or RNA

A **viral genome** is the genetic material that a virus contains. The term viral chromosome is also used to describe the viral genome. Surprisingly, the nucleic acid composition of viral genomes varies markedly among different types of viruses. **Table 10.1** describes the genome characteristics of a few selected viruses. The genome can be DNA or RNA, but not both. In some cases, it is single stranded, whereas in others, it is double stranded. Depending on the type of virus, the genome can be linear or circular.

As shown in Table 10.1, viral genomes vary in size from several thousand to more than a hundred thousand nucleotides in length. For example, the genomes of some simple viruses, such as Qβ virus, are only a few thousand nucleotides in length and contain only a few genes. Other viruses, particularly those with a complex structure, have many more genes. The T even phages (T2, T4, etc.), discussed in Chapters 6 and 9, are examples of more complex viruses.

TABLE 10.1

Characteristics of Selected Viral Genomes

Virus	Host	Type of Nucleic Acid*	Size**	Number of Genes***
Parvovirus	Mammals	ssDNA	5.0	5
Fd	*E. coli*	ssDNA	6.4	10
Lambda	*E. coli*	dsDNA	48.5	71
T4	*E. coli*	dsDNA	169.0	288
Qβ	*E. coli*	ssRNA	4.2	4
TMV	Many plants	ssRNA	6.4	6
Influenza virus	Mammals	ssRNA	13.5	11
Human immuno-deficiency virus (HIV)	Primates	ssRNA	9.7	9
Herpes simplex virus, type 2 (genital herpes)	Humans	dsDNA	158.4	77

*ss refers to single stranded, and ds refers to double stranded.
**Number of thousands of nucleotides or nucleotide base pairs
***This number refers to the number of protein-encoding units. In some cases, two or more proteins are made from a single gene due to events such as protein processing.

Viral Genomes Are Packaged into the Capsid in an Assembly Process

In an infected cell, the life cycle of the virus eventually leads to the synthesis of viral nucleic acids and proteins. Newly synthesized viral chromosomes and capsid proteins must then come together and assemble to make mature virus particles. Viruses with a simple structure may self-assemble, which means that the nucleic acid and capsid proteins spontaneously bind to each other to form a mature virus. The structure of one self-assembling virus, the tobacco mosaic virus, is shown in **Figure 10.2**. As shown here, the proteins assemble around the RNA genome, which becomes trapped inside the hollow capsid. This assembly process can occur in vitro if purified capsid proteins and RNA are mixed together.

Some viruses, such as T2 bacteriophage, have more complicated structures that do not self-assemble. The correct assembly of this virus requires the help of proteins not found within the mature virus particle itself. When virus assembly requires the participation of noncapsid proteins, the process is called directed assembly, because the noncapsid proteins direct the proper assembly of the virus. What are the functions of these noncapsid proteins? Some proteins, called scaffolding proteins, catalyze the assembly process and are transiently associated with the capsid. However, as viral assembly nears completion, the scaffolding proteins are expelled from the mature virus. In addition, other noncapsid proteins act as proteases that specifically cleave viral

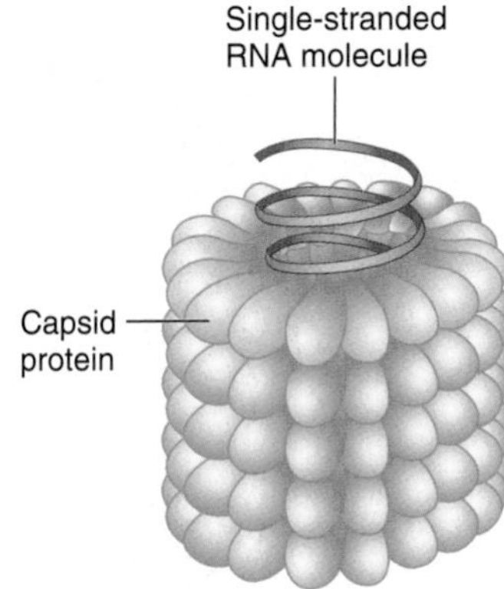

FIGURE 10.2 **Structure of the tobacco mosaic virus.** This self-assembling virus is composed of a coiled RNA molecule surrounded by 2,130 identical protein subunits. Only a portion of the tobacco mosaic virus is shown here. Several layers of proteins have been omitted from this illustration to reveal the RNA genome, which is trapped inside the protein coat.

capsid proteins. This cleavage produces a capsid protein that is somewhat smaller and able to assemble correctly. For many viruses, the cleavage of capsid proteins into smaller units is an important event that precedes viral assembly.

10.2 BACTERIAL CHROMOSOMES

Let's now turn our attention to the organization of chromosomes found in bacterial species. Inside a bacterial cell, the chromosome is highly compacted and found within a region of the cell known as the **nucleoid.** Although bacteria usually contain a single type of chromosome, more than one copy of that chromosome may be found within one bacterial cell. Depending on the growth conditions and phase of the cell cycle, bacteria may have one to four identical chromosomes per cell. In addition, the number of copies varies depending on the bacterial species. As shown in **Figure 10.3**, each chromosome occupies its own distinct nucleoid region within the cell. Unlike the eukaryotic nucleus, the bacterial nucleoid is not a separate cellular compartment bounded by a membrane. Rather, the DNA in a nucleoid is in direct contact with the cytoplasm of the cell.

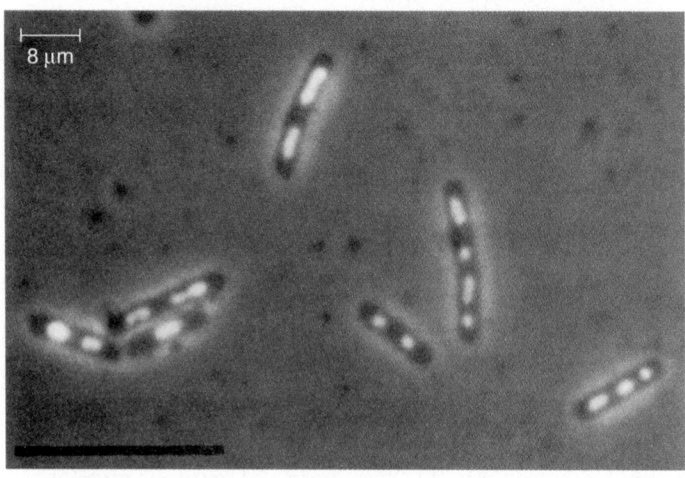

FIGURE 10.3 **The localization of nucleoids within *Bacillus subtilis* bacteria.** The nucleoids are fluorescently labeled and seen as bright, oval-shaped regions within the bacterial cytoplasm. Note that two or more nucleoids are found within each cell. Some of the cells seen here are in the process of dividing.

In this section, we will explore two important features of bacterial chromosomes. First, the organization of DNA sequences along the chromosome will be examined. Second, we will consider the mechanisms that cause the chromosome to become a compacted structure within a nucleoid of the bacterium.

Bacterial Chromosomes Contain a Few Thousand Gene Sequences Interspersed with Other Functionally Important Sequences

Bacterial chromosomal DNA is usually a circular molecule, though some bacteria have linear chromosomes. A typical chromosome is a few million base pairs (bp) in length. For example, the chromosome of one strain of *Escherichia coli* has approximately 4.6 million bp, and the *Haemophilus influenzae* chromosome has roughly 1.8 million bp. A bacterial chromosome commonly has a few thousand different genes. These genes are interspersed throughout the entire chromosome (**Figure 10.4**). **Structural genes**—nucleotide sequences that encode proteins—account for the majority of bacterial DNA. The nontranscribed regions of DNA located between adjacent genes are termed **intergenic regions.**

Other sequences in chromosomal DNA influence DNA replication, gene transcription, and chromosome structure. For

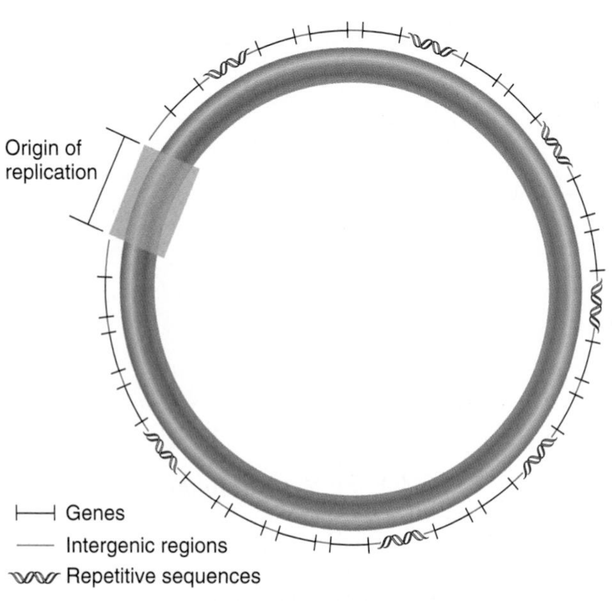

Origin of replication
Genes
Intergenic regions
Repetitive sequences
Key features:
• Most, but not all, bacterial species contain circular chromosomal DNA.
• A typical chromosome is a few million base pairs in length.
• Most bacterial species contain a single type of chromosome, but it may be present in multiple copies.
• Several thousand different genes are interspersed throughout the chromosome. The short regions between adjacent genes are called intergenic regions.
• One origin of replication is required to initiate DNA replication.
• Repetitive sequences may be interspersed throughout the chromosome.

FIGURE 10.4 **Organization of sequences in bacterial chromosomal DNA.**

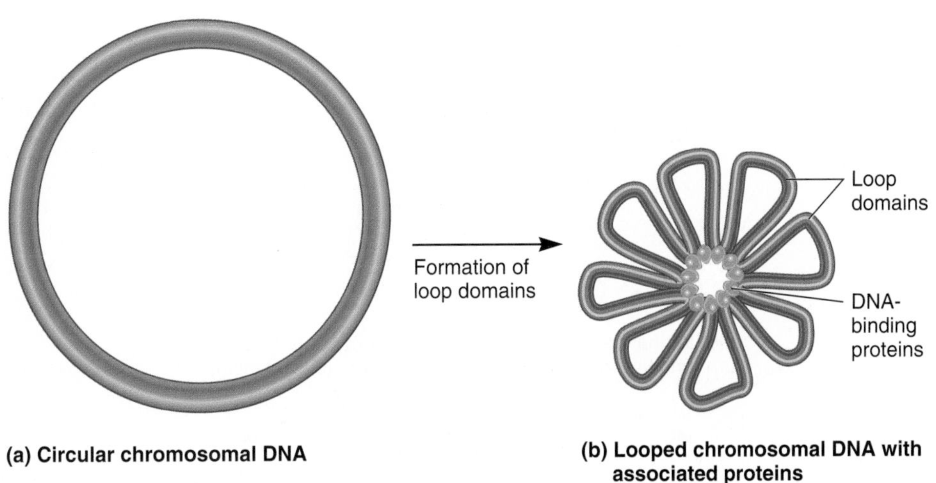

FIGURE 10.5 **The formation of loop domains within the bacterial chromosome.** To promote compaction, **(a)** the large, circular chromosomal DNA is organized into **(b)** smaller, looped chromosomal DNA with loop domains and associated proteins.

example, bacterial chromosomes have one **origin of replication,** a sequence that is a few hundred nucleotides in length. This nucleotide sequence functions as an initiation site for the assembly of several proteins required for DNA replication. Also, a variety of **repetitive sequences** have been identified in many bacterial species. These sequences are found in multiple copies and are usually interspersed within the intergenic regions throughout the bacterial chromosome. Repetitive sequences may play a role in a variety of genetic processes, including DNA folding, DNA replication, gene regulation, and genetic recombination. As discussed in Chapter 17, some repetitive sequences are transposable elements that can move throughout the genome. Figure 10.4 summarizes the key features of sequence organization within bacterial chromosomes.

The Formation of Chromosomal Loops Helps Make the Bacterial Chromosome More Compact

To fit within the bacterial cell, the chromosomal DNA must be compacted about 1,000-fold. Part of this compaction process involves the formation of **loop domains** within the bacterial chromosome (**Figure 10.5**). As its name suggests, a loop domain is a segment of chromosomal DNA folded into a structure that resembles a loop. DNA-binding proteins anchor the base of the loops in place. The number of loops varies according to the size of the bacterial chromosome and the species. In *E. coli*, a chromosome has 50 to 100 loop domains with about 40,000 to 80,000 bp of DNA in each loop. This looped structure compacts the circular chromosome about 10-fold.

DNA Supercoiling Further Compacts the Bacterial Chromosome

Because DNA is a long thin molecule, twisting forces can dramatically change its conformation. This effect is similar to twisting a rubber band. If twisted in one direction, a rubber band will eventually coil itself into a compact structure as it absorbs the energy applied by the twisting motion. Because the two strands within DNA already coil around each other, the formation of additional coils due to twisting forces is referred to as **DNA supercoiling** (**Figure 10.6**).

How do twisting forces affect DNA structure? **Figure 10.7** illustrates four possibilities. In Figure 10.7a, a double-stranded DNA molecule with five complete turns is anchored between two plates. In this hypothetical example, the ends of the DNA molecule cannot rotate freely. Both underwinding and overwinding of the DNA double helix can induce supercoiling of the helix. Because B DNA is a right-handed helix, underwinding is a left-handed twisting motion, and overwinding is a right-handed twist. Along the left side of Figure 10.7, one of the plates has been given a turn in the direction that tends to unwind the helix. As the helix absorbs this force, two things can happen. The underwinding motion can cause fewer turns (Figure 10.7b) or cause a negative supercoil to form (Figure 10.7c). On the right side of

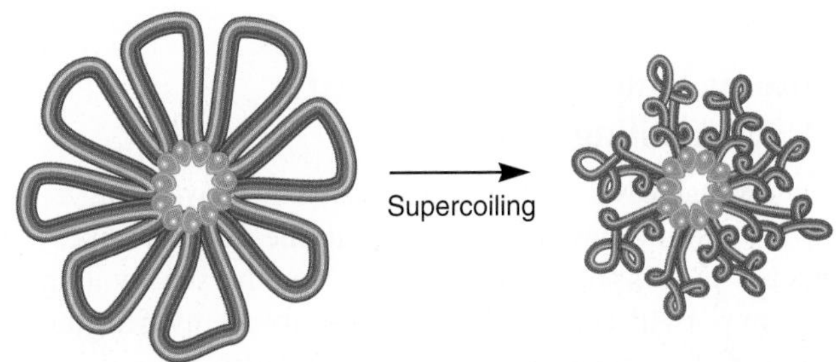

FIGURE 10.6 **DNA supercoiling leads to further compaction of the looped chromosomal DNA.** **(a)** The looped chromosomal DNA becomes much more compacted due to **(b)** supercoiling within the loops.

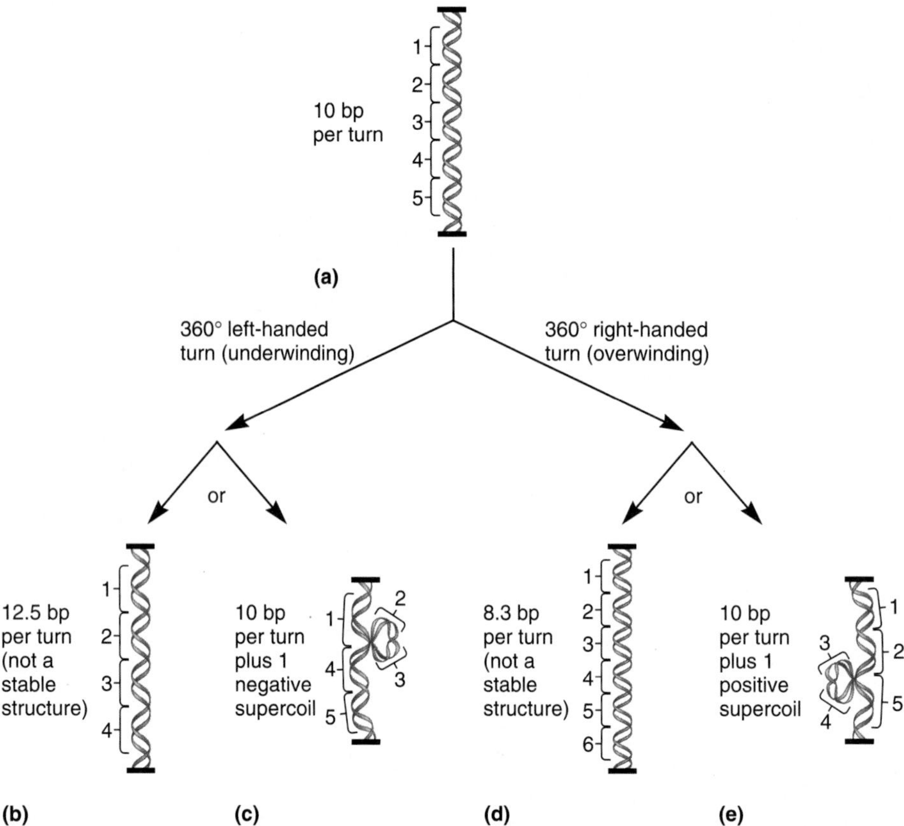

FIGURE 10.7 **Schematic representation of DNA supercoiling.** In this example, the DNA in **(a)** is anchored between two plates and given a twist as noted by the arrows. A left-handed twist (underwinding) could produce either **(b)** fewer turns or **(c)** a negative supercoil. A right-handed twist (overwinding) produces **(d)** more turns or **(e)** a positive supercoil. The structures shown in (b) and (d) are unstable.

Figure 10.7, one of the plates has been given a right-handed turn, which overwinds the double helix. This can lead to either more turns (Figure 10.7d) or the formation of a positive supercoil (Figure 10.7e). The DNA conformations shown in Figure 10.7a, c, and e differ only with regard to supercoiling. These three DNA conformations are referred to as **topoisomers** of each other. The DNA conformations shown in Figure 10.7b and d are not structurally favorable and would not occur in living cells.

Chromosome Function Is Influenced by DNA Supercoiling

The chromosomal DNA in living bacteria is negatively supercoiled. In the chromosome of *E. coli,* about one negative supercoil occurs per 40 turns of the double helix. Negative supercoiling has several important consequences. As already mentioned, the supercoiling of chromosomal DNA makes it much more compact (see Figure 10.6). Therefore, supercoiling helps to greatly decrease the size of the bacterial chromosome. In addition, negative supercoiling also affects DNA function. To understand how, remember that negative supercoiling is due to an underwinding force on the DNA. Therefore, negative supercoiling creates tension on the DNA strands that may be released by DNA strand separation (**Figure 10.8**). Although most of the chromosomal DNA is negatively supercoiled and compact, the force of negative supercoiling may promote DNA strand separation in small regions. This enhances genetic activities such as replication and transcription that require the DNA strands to be separated.

How does bacterial DNA become supercoiled? Two bacterial enzymes are primarily responsible for the level of supercoiling within living bacteria. In 1976, Martin Gellert and collaborators discovered the enzyme **DNA gyrase,** also known as topoisomerase II. This enzyme, which contains four subunits, introduces negative supercoils into DNA using energy from ATP (**Figure 10.9**) or relaxes positive supercoils when they occur. A second enzyme, **topoisomerase I,** can relax negative supercoils. The competing actions of DNA gyrase and topoisomerase I govern the overall supercoiling of the bacterial DNA.

The ability of gyrase to introduce negative supercoils into DNA is critical for bacteria to survive. For this reason, much research has been aimed at identifying drugs that will specifically block bacterial gyrase function as a way to cure or alleviate diseases caused by bacteria. Two main classes, quinolones and coumarins, inhibit gyrase and other bacterial topoisomerases and thereby block bacterial cell growth. These drugs do not inhibit eukaryotic topoisomerases, which are structurally different from

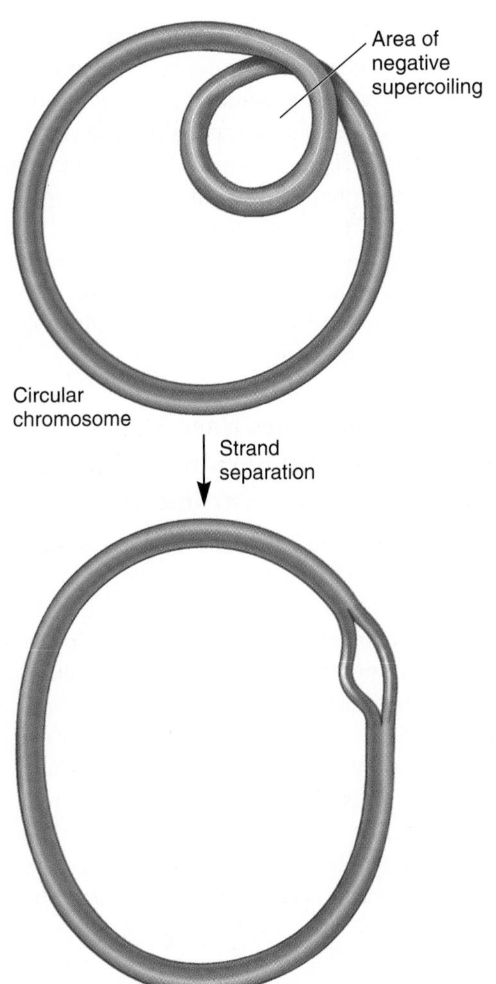

FIGURE 10.8 **Negative supercoiling promotes strand separation.**

their bacterial counterparts. This finding has been the basis for the production of many drugs with important antibacterial applications. An example is ciprofloxacin (e.g., Cipro®), which is used to treat a wide spectrum of bacterial diseases, including anthrax.

10.3 EUKARYOTIC CHROMOSOMES

Eukaryotic species have one or more sets of chromosomes; each set is composed of several different linear chromosomes (refer back to Figure 8.1). Humans, for example, have two sets of 23

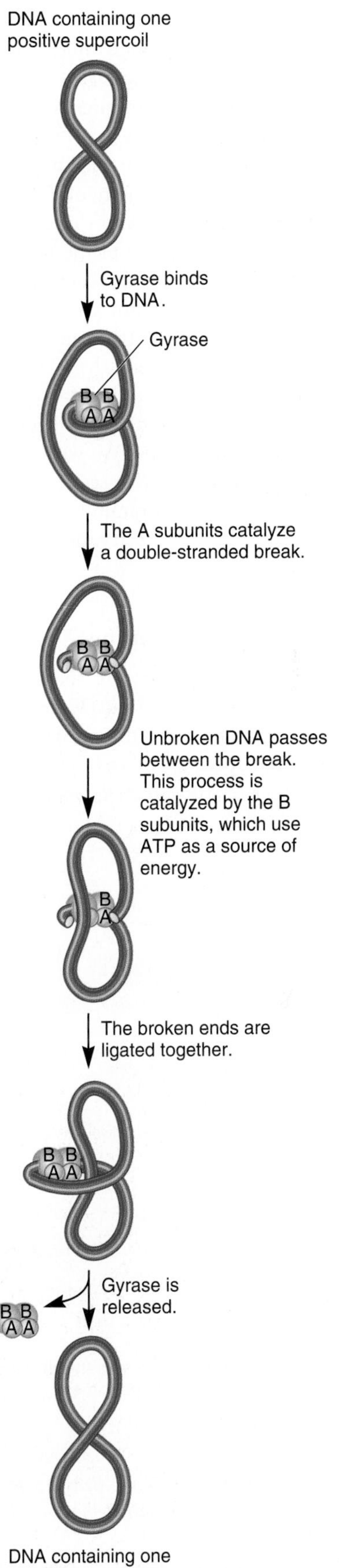

ONLINE ANIMATION

FIGURE 10.9 **The action of DNA gyrase.** DNA gyrase, also known as topoisomerase II, is composed of two A and two B subunits. The A subunits cleave the DNA. The unbroken segment of DNA then passes through the break, and the break is repaired. The B subunits capture the energy from ATP hydrolysis to catalyze this process. The result is that two negative turns have been introduced into the DNA molecule. In this example, the DNA originally contained one positive supercoil, and now it has one net negative supercoil.

chromosomes each, for a total of 46. The total amount of DNA in cells of eukaryotic species is usually much greater than that in bacterial cells. This enables eukaryotic genomes to contain many more genes than their bacterial counterparts. A distinguishing feature of eukaryotic cells is that their chromosomes are located within a separate cellular compartment known as the **nucleus.** To fit within the nucleus, the length of DNA must be compacted by a remarkable amount. As in bacterial chromosomes, this is accomplished by the binding of the DNA to many different cellular proteins. The term **chromatin** is used to describe the DNA-protein complex found within eukaryotic chromosomes. Chromatin is a dynamic structure that can change its shape and composition during the life of a cell.

In this section, we will examine the sizes of eukaryotic genomes and the organization of DNA sequences along the length of eukaryotic chromosomes. We will consider techniques to analyze the complexity of DNA sequences found in chromosomes. We will conclude by examining the mechanisms that account for the compaction of eukaryotic chromosomes during different stages of the cell cycle.

The Sizes of Eukaryotic Genomes Vary Substantially

When comparing different eukaryotic species, dramatic variation in genome size is often observed (**Figure 10.10a**; note that this is a log scale). In many cases, this variation is not related to the complexity of the species. For example, two closely related species of salamander, *Plethodon richmondi* and *Plethodon larselli*, differ considerably in genome size (**Figure 10.10b,c**). The genome of *P. larselli* is over two times larger than the genome of *P. richmondi*. However, the genome of *P. larselli* probably doesn't contain more genes. How do we explain the difference in genome size? The additional DNA in *P. larselli* is due to the accumulation of repetitive DNA sequences present in many copies. In some species, these repetitive sequences can accumulate to enormous levels. Such highly repetitive sequences do not encode proteins, and their function remains a matter of controversy and great interest. The structure and significance of repetitive DNA will be discussed later in this chapter.

Eukaryotic Chromosomes Have Many Functionally Important Regions Including Genes, Origins of Replication, Centromeres, and Telomeres

Each eukaryotic chromosome contains a long, linear DNA molecule (**Figure 10.11**). Three types of regions are required for chromosomal replication and segregation: origins of replication, centromeres, and telomeres. **Origins of replication** are chromosomal sites that are necessary to initiate DNA replication. Unlike most bacterial chromosomes, which contain only one origin of replication, eukaryotic chromosomes contain many origins, interspersed approximately every 100,000 bp apart. The function of origins of replication is discussed in greater detail in Chapter 11. **Centromeres** are DNA regions that play a role in the proper segregation of chromosomes during mitosis and meiosis. For most species, each eukaryotic chromosome contains a single centro-

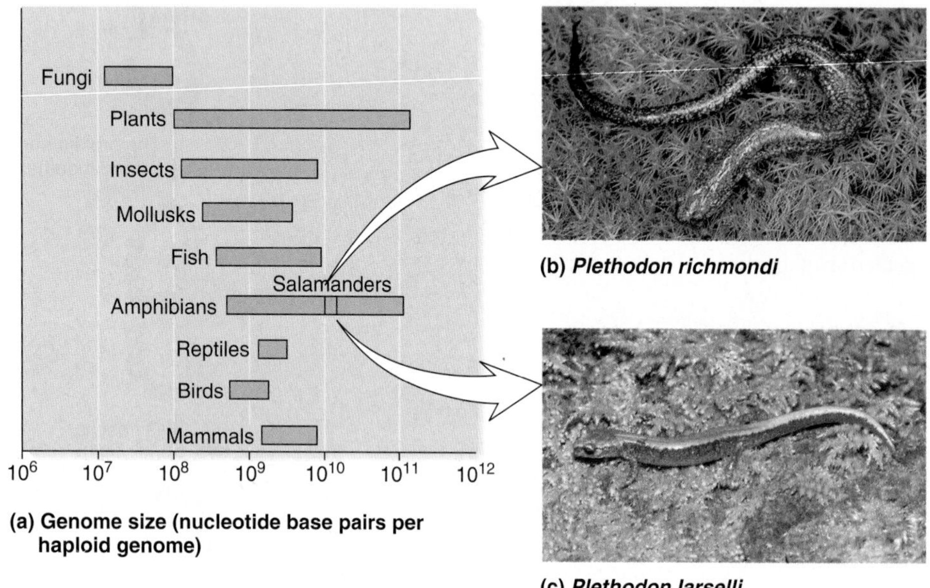

(a) Genome size (nucleotide base pairs per haploid genome)

(b) *Plethodon richmondi*

(c) *Plethodon larselli*

FIGURE 10.10 **Haploid genome sizes among groups of eukaryotic species.** (**a**) Ranges of genome sizes among different groups of eukaryotes. (**b**) A species of salamander, *Plethodon richmondi*, and (**c**) a close relative, *Plethodon larselli*. The genome of *P. larselli* is over two times as large as that of *P. richmondi*.

Genes→Traits The two species of salamander shown here have very similar traits, even though the genome of *P. larselli* is over twice as large as that of *P. richmondi*. However, the genome of *P. larselli* is not likely to contain more genes. Rather, the additional DNA is due to the accumulation of short repetitive DNA sequences that do not code for genes and are present in many copies.

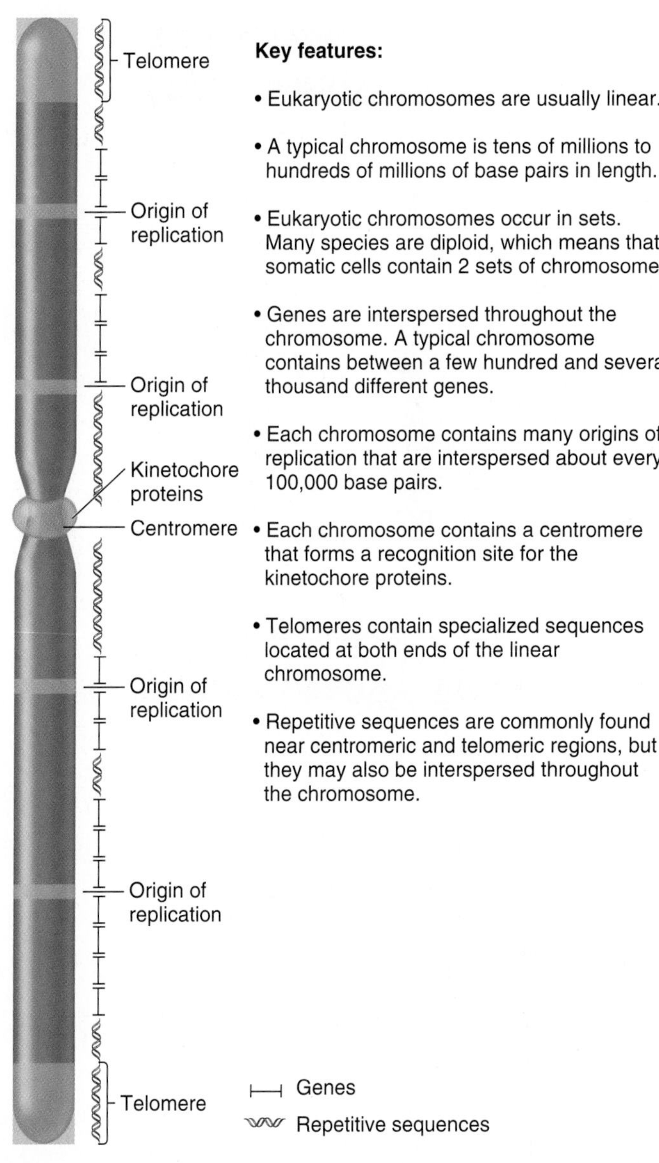

Telomere
Origin of replication
Origin of replication
Kinetochore proteins
Centromere
Origin of replication
Origin of replication
Telomere
Key features:
• Eukaryotic chromosomes are usually linear.
• A typical chromosome is tens of millions to hundreds of millions of base pairs in length.
• Eukaryotic chromosomes occur in sets. Many species are diploid, which means that somatic cells contain 2 sets of chromosomes.
• Genes are interspersed throughout the chromosome. A typical chromosome contains between a few hundred and several thousand different genes.
• Each chromosome contains many origins of replication that are interspersed about every 100,000 base pairs.
• Each chromosome contains a centromere that forms a recognition site for the kinetochore proteins.
• Telomeres contain specialized sequences located at both ends of the linear chromosome.
• Repetitive sequences are commonly found near centromeric and telomeric regions, but they may also be interspersed throughout the chromosome.
Genes
Repetitive sequences

FIGURE 10.11 **Organization of eukaryotic chromosomes.**

mere. The centromere serves as an attachment site for the **kinetochore,** a group of cellular proteins that link the centromere to the spindle apparatus during mitosis and meiosis, ensuring the proper segregation of the chromosomes to each daughter cell. Finally, at the ends of linear chromosomes are found specialized regions known as **telomeres.** Telomeres serve several important functions in the replication and stability of the chromosome. As discussed in Chapter 8, telomeres prevent chromosomes from having "sticky ends" and thereby inhibit chromosomal rearrangements such as translocations. In addition, they prevent chromosome shortening in two ways. First, the telomeres protect chromosomes from digestion via enzymes called exonucleases that recognize the ends of DNA. Secondly, as discussed in Chapter 11, an unusual form of DNA replication occurs at the telomere to ensure that eukaryotic chromosomes do not become shortened with each round of DNA replication.

Genes are located between the centromeric and telomeric regions along the entire eukaryotic chromosome. A single chromosome usually has a few hundred to several thousand different genes. The sequence of a typical eukaryotic gene is a few thousand to tens of thousands of base pairs in length. In less complex eukaryotes such as yeast, genes are relatively small and primarily contain nucleotide sequences that encode the amino acid sequences within proteins. In more complex eukaryotes such as mammals and higher plants, structural genes tend to be much longer due to the presence of **introns**—noncoding intervening sequences. Introns range in size from less than 100 bp to more than 10,000 bp. Therefore, the presence of large introns can greatly increase the lengths of eukaryotic genes.

The Genomes of Eukaryotes Contain Sequences That Are Unique, Moderately Repetitive, or Highly Repetitive

The term **sequence complexity** refers to the number of times a particular base sequence appears throughout the genome. Unique or nonrepetitive sequences are those found once or a few times within the genome. Structural genes are typically unique sequences of DNA. The vast majority of proteins in eukaryotic cells are encoded by genes present in one or a few copies. In the case of humans, unique sequences make up roughly 40% of the entire genome.

Moderately repetitive sequences are found a few hundred to several thousand times in the genome. In a few cases, moderately repetitive sequences are multiple copies of the same gene. For example, the genes that encode ribosomal RNA (rRNA) are found in many copies. Ribosomal RNA is necessary for the functioning of ribosomes. Cells need a large amount of rRNA for making ribosomes, and this is accomplished by having multiple copies of the genes that encode rRNA. Likewise, the histone genes are also found in multiple copies because a large number of histone proteins are needed for the structure of chromatin. In addition, other types of functionally important sequences can be moderately repetitive. For example, moderately repetitive sequences may play a role in the regulation of gene transcription and translation. By comparison, some moderately repetitive sequences do not play a functional role and are derived from **transposable elements**—segments of DNA that have the ability to move within a genome. This category of repetitive sequences is discussed in greater detail in Chapter 17.

Highly repetitive sequences are found tens of thousands or even millions of times throughout the genome. Each copy of a highly repetitive sequence is relatively short, ranging from a few nucleotides to several hundred in length. A widely studied example is the *Alu* family of sequences found in humans and other primates. The *Alu* sequence is approximately 300 bp long. This sequence derives its name from the observation that it contains a site for cleavage by a restriction enzyme known as *Alu*I. (The function of restriction enzymes is described in Chapter 18.) The *Alu* sequence is present in about 1,000,000 copies in the human genome. It represents about 10% of the total human DNA and occurs approximately every 5,000 to 6,000 base pairs! Evolutionary studies suggest that the *Alu* sequence arose 65 million years ago from a section of a single ancestral gene known as the *7SL RNA* gene. Since that time, this gene has become a type of transposable element called a **retroelement,** which can be transcribed into RNA, copied into DNA, and then inserted into the genome. Remarkably, over the course of 65 million years, the *Alu* sequence has been copied and inserted into the human genome to achieve the modern number of about 1,000,000 copies.

Some highly repetitive sequences, like the *Alu* family, are interspersed throughout the genome. However, other highly repetitive sequences are clustered together in a **tandem array,** also known as tandem repeats. In a tandem array, a very short nucleotide sequence is repeated many times in a row. In *Drosophila*, for example, 19% of the chromosomal DNA is highly repetitive DNA found in tandem arrays. An example is shown here.

AATATAATATAATATAATATAATATATAATAT

TTATATTATATTATATTATATTATATATTATA

In this particular tandem array, two related sequences, AATAT and AATATAT, are repeated. Tandem arrays of short sequences are commonly found in centromeric regions of chromosomes and can be quite long, sometimes more than 1,000,000 bp in length!

What is the functional significance of highly repetitive sequences? Whether highly repetitive sequences play any significant functional role is controversial. Some experiments in *Drosophila* indicate that highly repetitive sequences may be important in the proper segregation of chromosomes during meiosis. It is not yet clear if highly repetitive DNA plays the same role in other species. The sequences within highly repetitive DNA vary greatly from species to species. In fact, the amount of highly repetitive DNA can vary a great deal even among closely related species (as noted earlier in Figure 10.10).

Sequence Complexity Can Be Evaluated in a Renaturation Experiment

One approach that has proven useful in understanding genome complexity has come from renaturation studies. These kinds of experiments were first carried out by Roy Britten and David Kohne in 1968. In a renaturation study, the DNA is broken up into pieces containing several hundred base pairs. The double-stranded DNA is then denatured (separated) into single-stranded pieces by heat treatment (**Figure 10.12a**). When the temperature is lowered, the pieces of DNA that are complementary can reassociate, or renature, with each other to form double-stranded molecules.

The rate of renaturation of complementary DNA strands provides a way to distinguish between unique, moderately repetitive, and highly repetitive sequences. For a given category of DNA sequences, the renaturation rate will depend on the concentration of its complementary partner. Highly repetitive DNA sequences renature much faster because there are many copies of the complementary sequences. In contrast, unique sequences, such as those found within most genes, take longer to renature because of the added time it takes for the unique sequences to find each other.

The renaturation of two DNA strands is a bimolecular reaction that involves the collision of two complementary DNA strands. Its rate is proportional to the product of the concentrations of both strands. If C is the concentration of a single-stranded

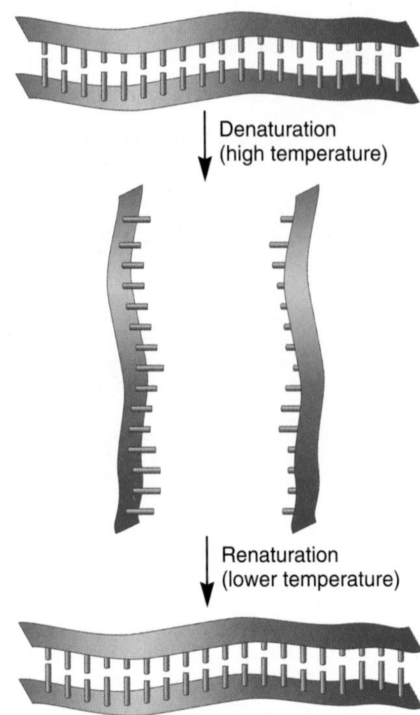

(a) Renaturation of DNA strands

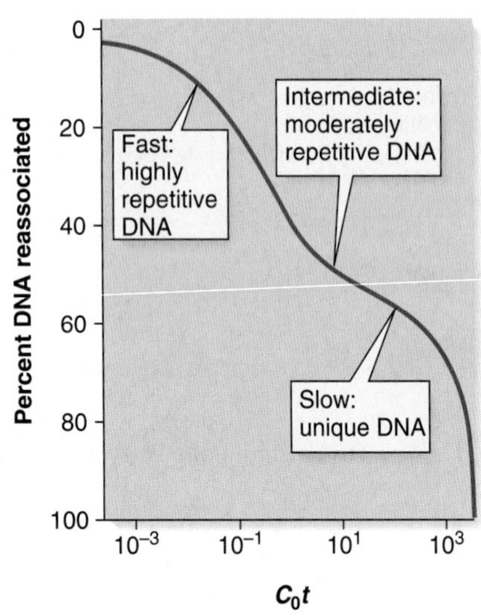

(b) Human chromosomal DNA C_0t curve

FIGURE 10.12 Renaturation and DNA sequence complexity. (**a**) Denaturation and renaturation (or reassociation) of DNA strands. (**b**) A C_0t curve for human chromosomal DNA.

DNA, then for any DNA derived from a double-stranded fragment, the concentration of one DNA strand (denoted C_1) equals the concentration of its complementary partner (denoted C_2). Letting $C = C_1 = C_2$, we see the rate of renaturation is represented by the second-order equation

$$-\,dC/dt = kC^2$$

This is called a second-order equation because the rate depends on the concentration of both reactants—C_1 and C_2. In this case, this product is simplified to C^2 because $C_1 = C_2$.

This equation says that a change in concentration of a single strand ($-\,dC$) with respect to time (dt) equals a rate constant (k) times the concentration of the single-stranded molecule squared (C^2). This equation can then be integrated to determine how the concentration of the single-stranded DNA will change from time zero to a later time.

$$\frac{C}{C_0} = \frac{1}{1 + k_2C_0t}$$

Where

C = the concentration of single-stranded DNA at a later time, t

C_0 = the concentration of single-stranded DNA at time zero

k_2 = the second-order rate constant for renaturation

In this equation, C/C_0 is the fraction of DNA still in single-stranded form after a given length of time. For example, if C/C_0 equals 0.4 after a certain period of time, 40% of the DNA is still in the single-stranded form, while 60% has renatured into the double-stranded form. A renaturation experiment can provide quantitative information about the complexity of DNA sequences within chromosomal DNA. In the experiment shown in **Figure 10.12b**, human DNA was sheared into small pieces (each about 600 bp in length), subjected to heat, and then allowed to renature at a lower temperature. The rates of renaturation for the DNA pieces can be represented in a plot of C/C_0 versus C_0t, which is referred to as a **C_0t curve** (called a Cot curve). The term Cot refers to the DNA concentration (C_0) multiplied by the incubation time (t). A fair amount of the DNA renatures very rapidly. This is the highly repetitive DNA. Some of the DNA reassociates at a moderate rate, and the remaining DNA renatures fairly slowly. From these data, the relative amounts of highly repetitive, moderately repetitive, and unique DNA sequences can be approximated. As seen in Figure 10.12b, about 40% of human DNA fragments are unique DNA sequences that renature slowly.

Eukaryotic Chromatin Must Be Compacted to Fit Within the Cell

We now turn our attention to ways that eukaryotic chromosomes are folded to fit within a living cell. A typical eukaryotic chromosome contains a single, linear double-stranded DNA molecule that may be hundreds of millions of base pairs in length. If the DNA from a single set of human chromosomes was stretched from end to end, the length would be over 1 meter! By comparison, most eukaryotic cells are only 10 to 100 μm in diameter, and the cell nucleus is only about 2 to 4 μm in diameter. Therefore, the DNA in a eukaryotic cell must be folded and packaged by a staggering amount to fit inside the nucleus.

The compaction of linear DNA within eukaryotic chromosomes is accomplished through mechanisms that involve interactions between DNA and several different proteins. In recent years, it has become increasingly clear that the proteins bound to chromosomal DNA are subject to change during the life of the cell. These changes in protein composition, in turn, affect the degree of compaction of the chromatin. Chromosomes are very dynamic structures that alternate between tight and loose compaction states in response to changes in protein composition. In the remaining parts of Chapter 10, we will focus our attention on two issues of chromosome structure. First, we will consider how chromosomes are compacted and organized during interphase—the period of the cell cycle that includes the G_1, S, and G_2 phases. Later, we will examine the additional compaction that is necessary to produce the highly condensed chromosomes found in M phase.

Linear DNA Wraps Around Histone Proteins to Form Nucleosomes, the Repeating Structural Unit of Chromatin

The repeating structural unit within eukaryotic chromatin is the **nucleosome**—a double-stranded segment of DNA wrapped around an octamer of **histone proteins** (**Figure 10.13a**). Each octamer contains eight histone subunits, two copies each of four different histone proteins. The DNA lies on the surface and makes 1.65 negative superhelical turns around the histone octamer. The amount of DNA that is required to wrap around the histone octamer is 146 or 147 bp. At its widest point, a single nucleosome is about 11 nm in diameter.

The chromatin of eukaryotic cells contains a repeating pattern in which the nucleosomes are connected by linker regions of DNA that vary in length from 20 to 100 bp, depending on the species and cell type. It has been suggested that the overall structure of connected nucleosomes resembles beads on a string. This structure shortens the length of the DNA molecule about sevenfold.

Each of the histone proteins consists of a globular domain and a flexible, charged amino terminus called an amino terminal tail. Histone proteins are very basic proteins because they contain a large number of positively charged lysine and arginine amino acids. The arginines, in particular, play a major role in binding to the DNA. Arginines within the histone proteins form electrostatic and hydrogen-bonding interactions with the phosphate groups along the DNA backbone. The octamer of histones contains two molecules each of four different histone proteins: H2A, H2B, H3, and H4. These are called the core histones. In 1997, Timothy Richmond and colleagues determined the structure of a nucleosome by X-ray crystallography (**Figure 10.13b**).

Another histone, H1, is found in most eukaryotic cells and is called the linker histone. It binds to the DNA in the linker region between nucleosomes and may help to compact adjacent nucleosomes (**Figure 10.13c**). The linker histones are less tightly bound to the DNA than are the core histones. In addition, nonhistone proteins bound to the linker region play a role in the organization and compaction of chromosomes, and their presence may affect the expression of nearby genes.

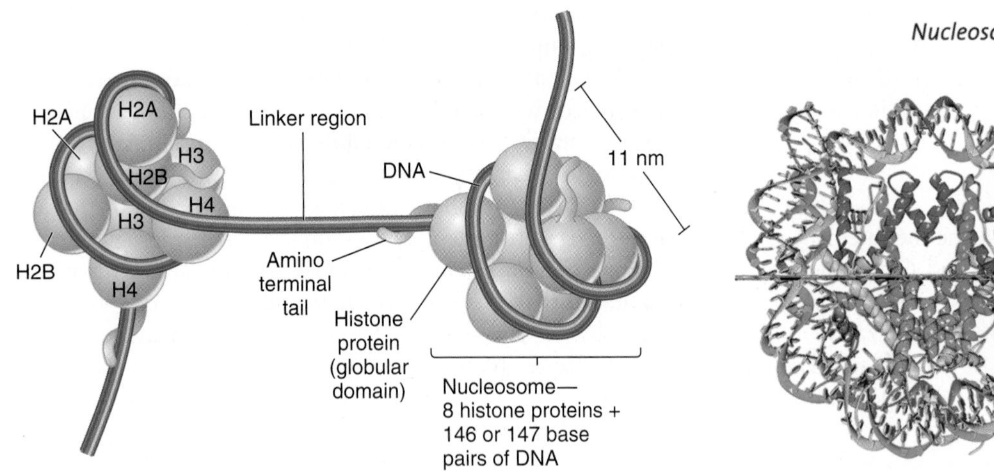

(a) Nucleosomes showing core histone proteins

(b) Molecular model for nucleosome structure

(Image courtesy of Timothy J. Richmond. Reprinted by permission from Macmillan Publishers Ltd. *Nature.* Crystal structure of the nucleosome core particle at 2.8 A resolution. Luger K., Mader, AW, Richmond, RK, Sargent, DF, Richmond, TJ. 389:6648, 251–260, 1997.)

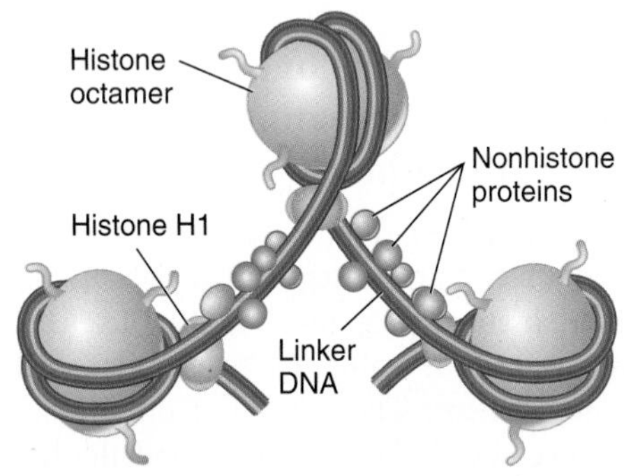

(c) Nucleosomes showing linker histones and nonhistone proteins

FIGURE 10.13 **Nucleosome structure.** (**a**) A nucleosome consists of 146 or 147 bp of DNA wrapped around an octamer of core histone proteins. (**b**) A model for the structure of a nucleosome as determined by X-ray crystallography. (**c**) The linker region of DNA connects adjacent nucleosomes. The linker histone H1 and nonhistone proteins also bind to this linker region.

EXPERIMENT 10A

The Repeating Nucleosome Structure Is Revealed by Digestion of the Linker Region

The model of nucleosome structure was originally proposed by Roger Kornberg in 1974. He based his proposal on several observations. Biochemical experiments had shown that chromatin contains a ratio of one molecule of each of the four core histones (namely, H2A, H2B, H3, and H4) per 100 bp of DNA. Approximately one H1 protein was found per 200 bp of DNA. In addition, purified core histone proteins were observed to bind to each other via specific pairwise interactions. Subsequent X-ray diffraction studies showed that chromatin is composed of a repeating pattern of smaller units. Finally, electron microscopy of chromatin fibers revealed a diameter of approximately 11 nm. Taken together, these observations led Kornberg to propose a model in which the DNA double helix is wrapped around an octamer of core histone proteins. Including the linker region, this would involve about 200 bp of DNA.

Markus Noll decided to test Kornberg's model by digesting chromatin with DNase I, an enzyme that cuts the DNA backbone, and then accurately measuring the molecular mass of the DNA fragments by gel electrophoresis. Noll assumed that the linker region of DNA will be more accessible to DNase I and, therefore, DNase I is more likely to make cuts in the linker region than in the 146-bp region that is tightly bound to the core histones. If this is correct, incubation with DNase I is expected to make cuts in the linker region and thereby produce DNA pieces approximately 200 bp in length. The size of the DNA fragments may vary somewhat because the linker region is not of constant length and because the cut within the linker region may occur at different sites.

Figure 10.14 describes the experimental protocol of Noll. He began with nuclei from rat liver cells and incubated them with low, medium, or high concentrations of DNase I. The DNA was extracted into an aqueous phase and then loaded onto an agarose gel that separated the fragments according to their molecular mass. The DNA fragments within the gel were stained with a UV-sensitive dye, ethidium bromide, which made it possible to view the DNA fragments under UV illumination.

THE HYPOTHESIS

This experiment seeks to test the beads-on-a-string model for chromatin structure. According to this model, DNase I should preferentially cut the DNA in the linker region and thereby produce DNA pieces that are about 200 bp in length.

TESTING THE HYPOTHESIS — FIGURE 10.14 DNase I cuts chromatin into repeating units containing 200 bp of DNA.

Starting material: Nuclei from rat liver cells.

1. Incubate the nuclei with low, medium, and high concentrations of DNase I. The conceptual level illustrates a low DNase I concentration.

2. Extract the DNA. This involves dissolving the nuclear membrane with detergent and extracting with the organic solvent phenol.

3. Load the DNA into a well of an agarose gel and run the gel to separate the DNA pieces according to size. On this gel, also load DNA fragments of known molecular mass (marker lane).

4. Visualize the DNA fragments by staining the DNA with ethidium bromide, a dye that binds to DNA and is fluorescent when excited by UV light.

Experimental level

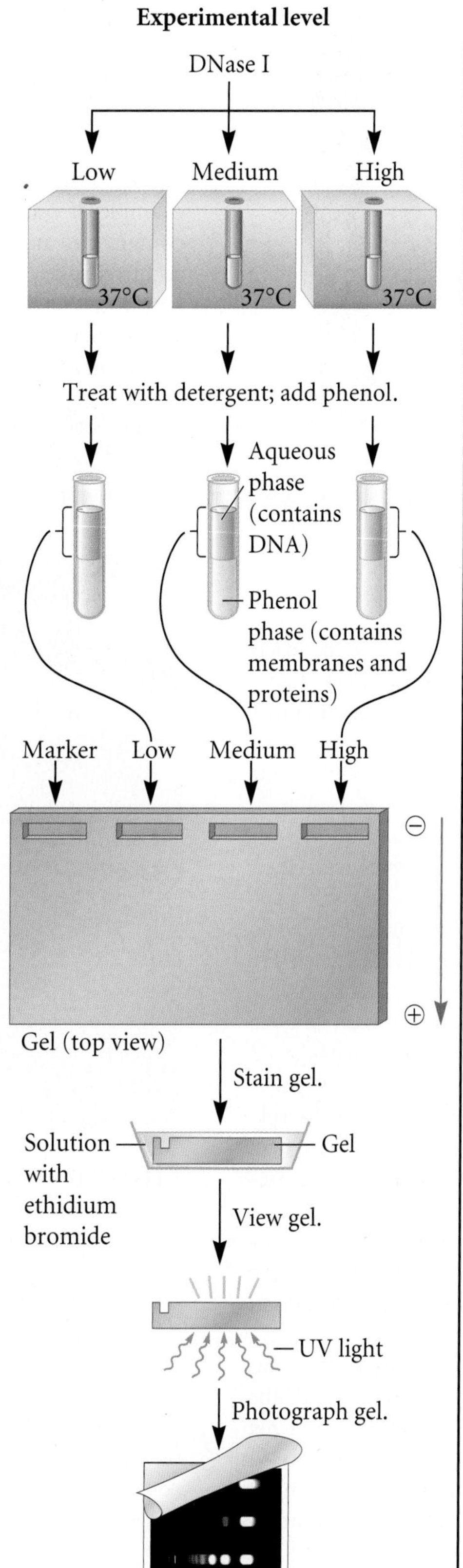

Conceptual level

Before digestion (beads on a string)

After digestion (DNA is cut in linker region)

DNA in aqueous phase

Low

⊖

⊕

THE DATA

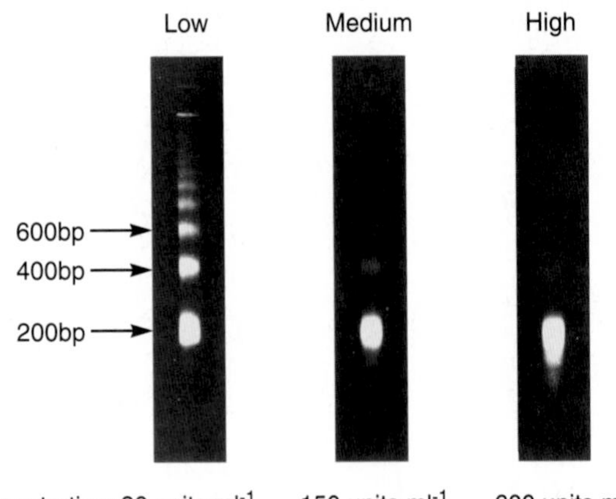

(Reprinted by permission from Macmillan Publishers Ltd. *Nature.* Subunit structure of chromatin. Markus Noll. 251:5472, 249–251. 1974.)

INTERPRETING THE DATA

As shown in the data of Figure 10.14, at high DNase I concentrations, the entire sample of chromosomal DNA was digested into fragments of approximately 200 bp in length. This result is predicted by the beads-on-a-string model. Furthermore, at lower DNase I concentrations, longer pieces were observed, and these were in multiples of 200 bp (400, 600, etc.). How do we explain these longer pieces? They occurred because occasional linker regions remained uncut at lower DNase I concentrations. For example, if one linker region was not cut, a DNA piece would contain two nucleosomes and be 400 bp in length. If two consecutive linker regions were not cut, this would produce a piece with three nucleosomes containing about 600 bp of DNA. Taken together, these results strongly supported the nucleosome model for chromatin structure.

A self-help quiz involving this experiment can be found at the Online Learning Center.

Nucleosomes Become Closely Associated to Form a 30 nm Fiber

In eukaryotic chromatin, nucleosomes associate with each other to form a more compact structure that is 30 nm in diameter. Evidence for the packaging of nucleosomes was obtained in the microscopy studies of Fritz Thoma in 1977. Chromatin samples were treated with a resin that removed histone H1, but the removal depended on the salt concentration. A moderate salt solution (100 mM NaCl) removed H1, but a solution with no added NaCl did not remove H1. These samples were then observed with an electron microscope. At moderate salt concentrations (**Figure 10.15a**), the chromatin exhibited the classic beads-on-a-string morphology. Without added NaCl (when H1 is expected to remain bound to the DNA), these "beads" associated with each other into a more compact conformation (**Figure 10.15b**). These results suggest that the nucleosomes are packaged to create a more compact unit and that H1 has a role in the packaging and compaction of nucleosomes. However, the precise role of H1 in chromatin compaction remains unclear. Recent data suggest that the core histones also play a key role in the compaction and relaxation of chromatin.

The experiment of Figure 10.15 and other experiments have established that nucleosome units are organized into a more compact structure that is 30 nm in diameter, known as the **30 nm fiber** (**Figure 10.16a**). The 30 nm fiber shortens the total length of DNA another sevenfold. The structure of the 30 nm fiber has proven difficult to determine, because the conformation of the DNA may be substantially altered when it is extracted from living cells. Most models for the 30 nm fiber fall into two main classes. The solenoid model suggests a helical structure in which contact between nucleosomes produces a symmetrically compact structure within the 30 nm fiber (**Figure 10.16b**). This type of model is still favored by some researchers in the field. However, experimental data also suggest that the 30 nm fiber may not form such a regular structure. Instead, an alternative zigzag model, advocated by Rachel Horowitz, Christopher Woodcock, and others, is based on techniques such as cryoelectron microscopy (electron microscopy at low temperature). According to the zigzag model, linker regions within the 30 nm structure are variably bent and twisted, and little face-to-face contact occurs between nucleosomes (**Figure 10.16c**). At this level of compaction, the overall picture of chromatin that emerges is an irregular, fluctuating, three-dimensional zigzag structure with stable nucleosome units

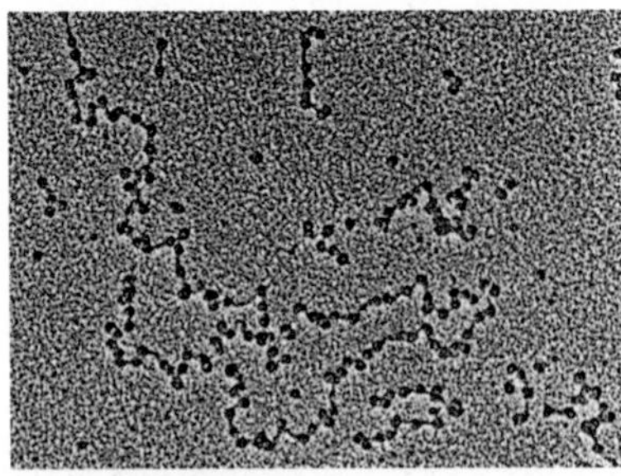

(a) H1 histone not bound—beads on a string

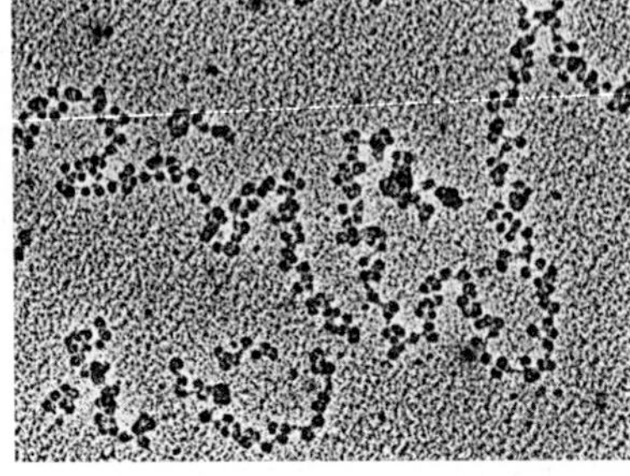

(b) H1 histone bound to linker region—nucleosomes more compact

FIGURE 10.15 The nucleosome structure of eukaryotic chromatin as viewed by electron microscopy. The chromatin in **(a)** has been treated with moderate salt concentrations to remove the linker histone H1. It exhibits the classic beads-on-a-string morphology. The chromatin in **(b)** has been incubated without added NaCl and shows a more compact morphology.

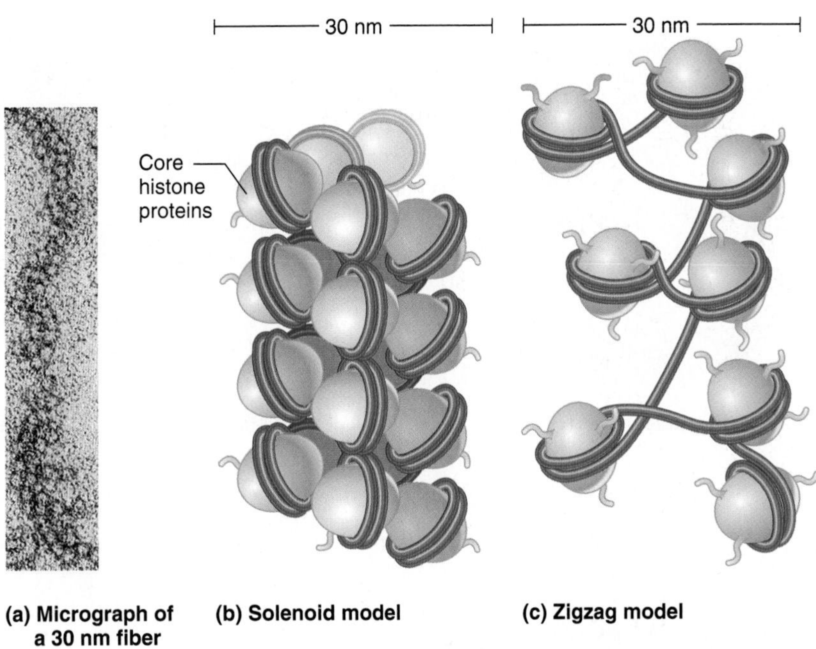

FIGURE 10.16 **The 30 nm fiber.** (**a**) A photomicrograph of the 30 nm fiber. (**b**) In the solenoid model, the nucleosomes are packed in a spiral configuration. (**c**) In the zigzag model, the linker DNA forms a more irregular structure, and less contact occurs between adjacent nucleosomes. The zigzag model is consistent with more recent data regarding chromatin conformation.

connected by deformable linker regions. In 2005, Timothy Richmond and colleagues were the first to solve the crystal structure of a segment of DNA containing multiple nucleosomes, in this case four. The structure with four nucleosomes revealed that the linker DNA zigzags back and forth between each nucleosome, a feature consistent with the zigzag model.

Chromosomes Are Further Compacted by Anchoring the 30 nm Fiber into Radial Loop Domains Along the Nuclear Matrix

Thus far, we have examined two mechanisms that compact eukaryotic DNA. These involve the wrapping of DNA within nucleosomes and the arrangement of nucleosomes to form a 30 nm fiber. Taken together, these two events shorten the DNA nearly 50-fold. A third level of compaction involves interactions between the 30 nm fibers and a filamentous network of proteins in the nucleus called the **nuclear matrix.** As shown in Figure 10.17a, the nuclear matrix consists of two parts. The **nuclear lamina** is a collection of fibers that line the inner nuclear membrane. These fibers are composed of intermediate filament proteins. The second part is an **internal nuclear matrix,** which is connected to the nuclear lamina and fills the interior of the nucleus. The internal nuclear matrix, whose structure and functional role remain controversial, is hypothesized to be an intricate fine network of irregular protein fibers plus many other proteins that bind to these fibers. Even when the chromatin is extracted from the nucleus, the internal nuclear matrix may remain intact (Figure 10.17b and c). However, the matrix should not be considered a static structure. Research indicates that the protein composition of the internal nuclear matrix is very dynamic and complex, consisting of dozens or perhaps hundreds of different proteins. The protein composition varies depending on species, cell type, and environmental conditions. This complexity has made it difficult to propose models regarding its overall organization. Further research will be necessary to understand the structure and dynamic nature of the internal nuclear matrix.

The proteins of the nuclear matrix are involved in compacting the DNA into **radial loop domains,** similar to those described for the bacterial chromosome. During interphase, chromatin is organized into loops, often 25,000 to 200,000 bp in size, which are anchored to the nuclear matrix. The chromosomal DNA of eukaryotic species contains sequences called **matrix-attachment regions (MARs)** or **scaffold-attachment regions (SARs),** which are interspersed at regular intervals throughout the genome. The MARs bind to specific proteins in the nuclear matrix, thus forming chromosomal loops (Figure 10.17d).

Why is the attachment of radial loops to the nuclear matrix important? In addition to compaction, the nuclear matrix serves to organize the chromosomes within the nucleus. Each chromosome in the cell nucleus is located in a discrete **chromosome territory.** As shown in studies by Thomas Cremer, Christoph Cremer, and others, these territories can be viewed when interphase cells are exposed to multiple fluorescent molecules that recognize

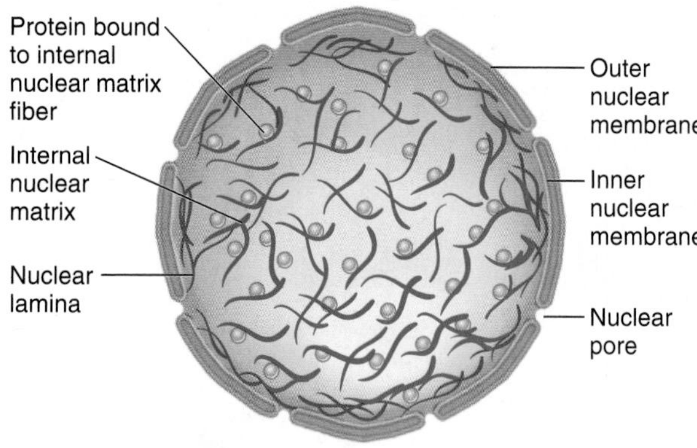

(a) Proteins that form the nuclear matrix

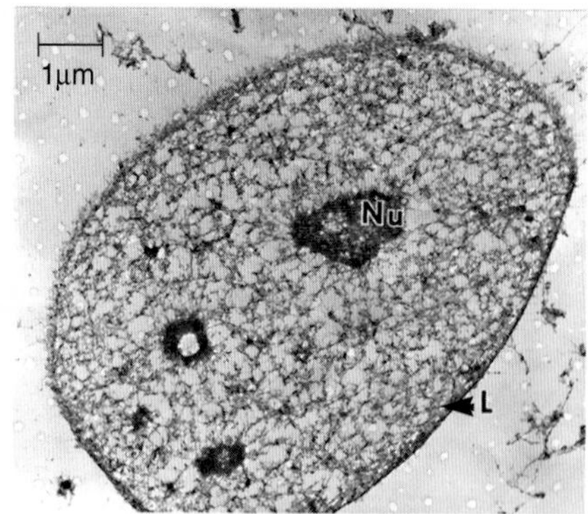

(b) Micrograph of nucleus with chromatin removed

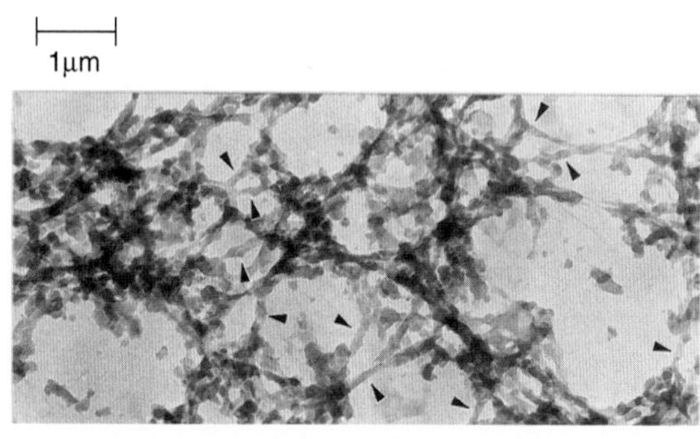

(c) Micrograph showing a close-up of nuclear matrix

(d) Radial loop bound to a nuclear matrix fiber

FIGURE 10.17 **Structure of the nuclear matrix.** (**a**) This schematic drawing shows the arrangement of the matrix within a cell nucleus. The nuclear lamina (depicted in red) is a collection of fibrous proteins that line the inner nuclear membrane. The internal nuclear matrix is composed of protein filaments (depicted in green) that are interconnected. These fibers also have many other proteins associated with them (depicted in orange). (**b**) An electron micrograph of the nuclear matrix during interphase after the chromatin has been removed. The nucleolus is labeled Nu, and the lamina is labeled L. (**c**) At higher magnification, the protein fibers are more easily seen (arrowheads point at fibers). (**d**) The matrix-attachment regions (MARs), which contain a high percentage of A and T bases, bind to the nuclear matrix and create radial loops. This causes a greater compaction of eukaryotic chromosomal DNA.

specific sequences on particular chromosomes. **Figure 10.18** illustrates an experiment in which chicken cells were exposed to a mixture of probes that recognize multiple sites along several of the larger chromosomes found in this species (*Gallus gallus*). Figure 10.18a shows the chromosomes in metaphase. The probes label each type of metaphase chromosome with a different color. Figure 10.18b shows the use of the same probes during interphase, when the chromosomes are less condensed and found in the cell nucleus. As seen here, each chromosome occupies its own distinct territory.

Before ending the topic of interphase chromosome compaction, let's consider how the compaction level of interphase chromosomes may vary. This variability can be seen with a light microscope and was first observed by the German cytologist E. Heitz in 1928. He coined the term **heterochromatin** to describe the tightly compacted regions of chromosomes. In general, these regions of the chromosome are transcriptionally inactive. By comparison, the less condensed regions, known as **euchromatin,** reflect areas that are capable of gene transcription. In euchromatin, the 30 nm fiber forms radial loop domains. In heterochromatin, these radial loop domains become compacted even further.

Figure 10.19 illustrates the distribution of euchromatin and heterochromatin in a typical eukaryotic chromosome during interphase. The chromosome contains regions of both heterochromatin and euchromatin. Heterochromatin is most abundant in the centromeric regions of the chromosomes and, to a lesser extent, in the telomeric regions. The term **constitutive heterochromatin** refers to chromosomal regions that are always heterochromatic and permanently inactive with regard to transcription. Constitutive heterochromatin usually contains highly repetitive DNA sequences, such as tandem repeats, rather than gene sequences. **Facultative heterochromatin** refers to chromatin that can occasionally interconvert between heterochromatin and euchromatin. An example of facultative heterochromatin occurs in female mammals when

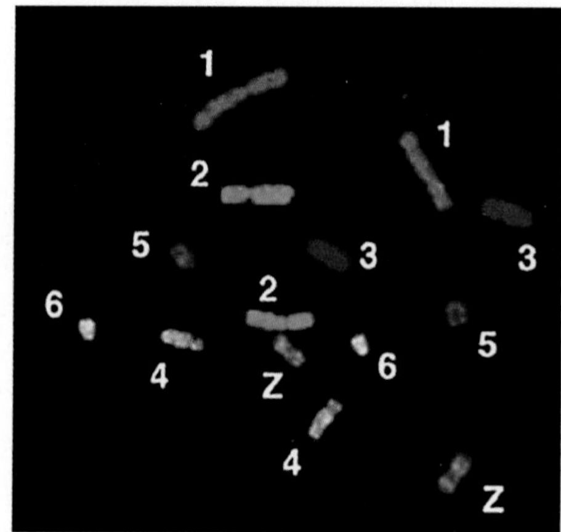

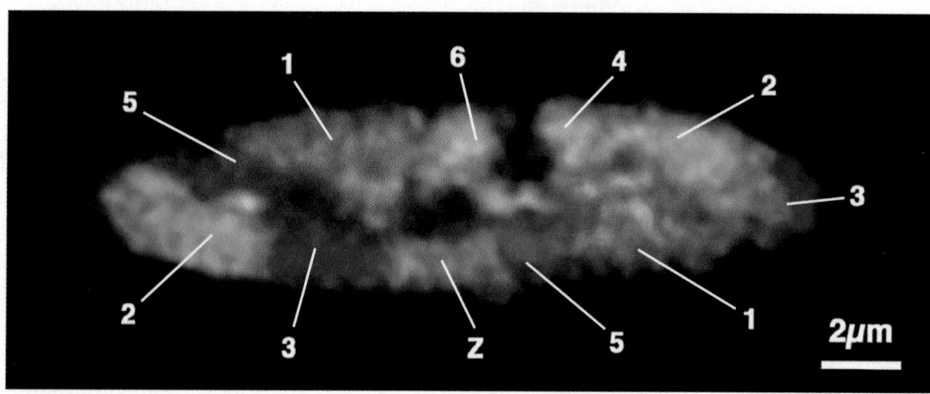

(a) Metaphase chromosomes (b) Chromosomes in the cell nucleus during interphase

FIGURE 10.18 **Chromosome territories in the cell nucleus.** (**a**) Several metaphase chromosomes from the chicken were labeled with chromosome-specific probes. Each of seven types of chicken chromosomes (i.e., 1, 2, 3, 4, 5, 6, and Z) is labeled a different color. (**b**) The same probes were used to label interphase chromosomes in the cell nucleus. Each of these chromosomes occupies its own distinct, nonoverlapping territory within the cell nucleus. (Note: Chicken cells are diploid, with two copies of each chromosome.)

(Reprinted by permission from Macmillan Publishers Ltd. *Nature Reviews/Genetics.* Chromosome territories, nuclear architecture and gene regulation in mammalian cells. Cremer, T. & Cremer, C. 2:4, 292–301, 2001.)

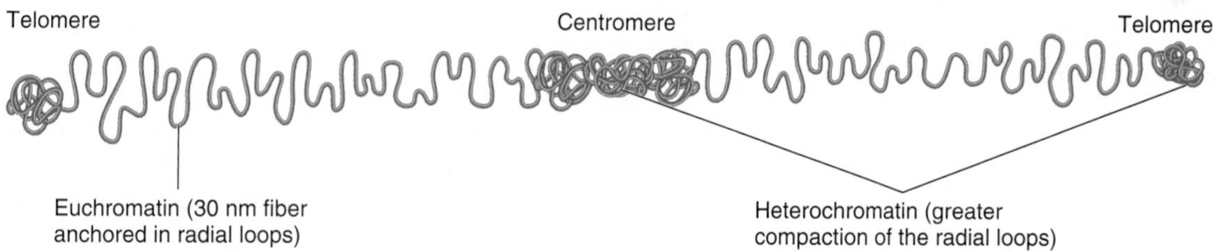

FIGURE 10.19 **Chromatin structure during interphase.** Heterochromatic regions are more highly condensed and tend to be localized in centromeric and telomeric regions.

one of the two X chromosomes is converted to a heterochromatic Barr body. As discussed in Chapter 7, most of the genes on the Barr body are transcriptionally inactive. The conversion of one X chromosome to heterochromatin occurs during embryonic development in the somatic cells of the body.

The Histone Code Controls Chromatin Compaction

Thus far, we have learned that the genomes of eukaryotic species are greatly compacted to fit inside the cell nucleus. Even euchromatin, which is looser than heterochromatin, is still compacted to such a degree that it is difficult for transcription factors and RNA polymerase to access and transcribe genes. As discussed in Chapters 12 and 15, euchromatin must be loosened up so that genes can be transcribed into RNA. Gene transcription is greatly influenced by a process known as **chromatin remodeling.** This term refers to changes in chromatin structure that regulate the ability of transcription factors to gain access to genes so they may be transcribed into RNA.

As described earlier, each of the core histone proteins consists of a globular domain and a flexible, charged amino terminus called an amino terminal tail (refer back to Figure 10.13a). The DNA wraps around the globular domains, and the amino terminal tails protrude from the chromatin. In recent years, researchers have discovered that particular amino acids in the amino terminal tails are subject to several types of covalent modifications, including acetylation, methylation, and phosphorylation. Over 50 different enzymes have been identified in mammals that selectively modify amino terminal tails. **Figure 10.20** shows examples of sites in the tails of H2A, H2B, H3, and H4 that can be modified.

How do histone modifications affect the level of chromatin compaction? First, they may directly influence interactions between nucleosomes. Second, histone modifications provide binding sites that are recognized by proteins. According to the **histone code hypothesis,** proposed by Brian Strahl, C. David Allis, and Bryan Turner in 2000, the pattern of histone modification acts much like a language or code in specifying alterations in chromatin structure. For example, one pattern might involve phosphorylation of the serine at the first position in H2A and acetylation of the fifth and eighth amino acids in H4, which are lysines. A different pattern could involve acetylation of the fifth

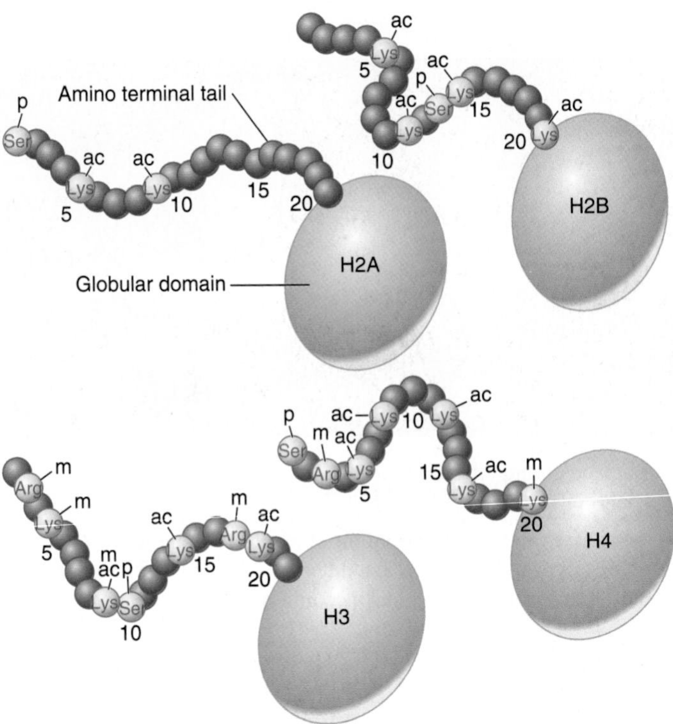

FIGURE 10.20 **Examples of covalent modifications that may occur in the amino terminal tails of core histone proteins.** The amino acids are numbered from the amino terminus. The modifications shown here are m for methylation, p for phosphorylation, and ac for acetylation. Many more modifications can occur to the amino terminal tails; the ones shown here represent common examples.

amino acid, a lysine, in H2B and methylation of the third amino acid in H4, which is an arginine.

The pattern of covalent modifications to the amino terminal tails provides binding sites for proteins that subsequently affect the degree of chromatin compaction. One pattern of histone modification may attract proteins that cause the chromatin to become even more compact, which would silence the transcription of genes in the region. A different combination of histone modifications may attract proteins, such as the chromatin remodeling enzymes discussed in Chapter 15, which serve to loosen the chromatin and thereby promote gene transcription. In this way, the histone code plays a key role in making the information within the genomes of eukaryotic species accessible. Researchers are trying to unravel which patterns of histone modifications promote compaction and which promote a loosening of chromatin structure. In other words, they are trying to decipher the effects of the covalent modifications that comprise the histone code.

Condensin and Cohesin Promote the Formation of Metaphase Chromosomes

As we have seen, several mechanisms can alter the level of chromosomal compaction. Furthermore, the degree of compaction can vary along a single eukaryotic chromosome. When cells prepare to divide, the chromosomes become even more compacted or condensed. This aids in their proper sorting during metaphase. **Figure 10.21** illustrates the levels of compaction that lead to a metaphase chromosome. During interphase, most of the chromosomal DNA is found in euchromatin, in which the 30 nm fibers form radial loop domains that are attached to a protein scaffold. The average distance that loops radiate from the protein scaffold is approximately 300 nm. This structure can be further compacted via additional folding of the radial loop domains and protein scaffold. This additional level of compaction greatly shortens the overall length of a chromosome and produces a diameter of approximately 700 nm, which is the compaction level found in heterochromatin. During interphase, most chromosomal regions are euchromatic, and some localized regions, such as those near centromeres, are heterochromatic.

As cells enter M phase, the level of compaction changes dramatically. By the end of prophase, sister chromatids are entirely heterochromatic. Two parallel chromatids have a larger diameter of approximately 1,400 nm but a much shorter length compared to interphase chromosomes. These highly condensed metaphase chromosomes undergo little gene transcription because it is difficult for transcription proteins to gain access to the compacted DNA. Therefore, most transcriptional activity ceases during M phase, although a few specific genes may be transcribed. M phase is usually a short period of the cell cycle.

In highly condensed chromosomes, such as those found in metaphase, the radial loops are highly compacted and remain anchored to a **scaffold,** which is formed from nonhistone proteins of the nuclear matrix. Experimentally, researchers can delineate the nonhistone proteins of the scaffold that hold the loops in place. **Figure 10.22a** shows a human metaphase chromosome. In this condition, the radial loops of DNA are in a very compact configuration. If this chromosome is treated with a high concentration of salt to remove both the core and linker histones, the highly compact configuration is lost, but the bottoms of the elongated loops remain attached to the scaffold composed of nonhistone proteins. In **Figure 10.22b**, an arrow points to an elongated DNA strand emanating from the darkly staining scaffold. Remarkably, the scaffold retains the shape of the original metaphase chromosome even though the DNA strands have become greatly elongated. These results illustrate that the structure of metaphase chromosomes is determined by the nuclear matrix proteins, which form the scaffold, and by the histones, which are needed to compact the radial loops.

Researchers are trying to understand the steps that lead to the formation and organization of metaphase chromosomes. During the past several years, studies in yeast and frog oocytes have been aimed at the identification of proteins that promote the conversion of interphase chromosomes into metaphase chromosomes. In yeast, mutants have been characterized that have alterations in the condensation or the segregation of chromosomes. Similarly, biochemical studies using frog oocytes resulted in the purification of protein complexes that promote chromosomal condensation or sister chromatid alignment. These two lines of independent research produced the same results. Researchers

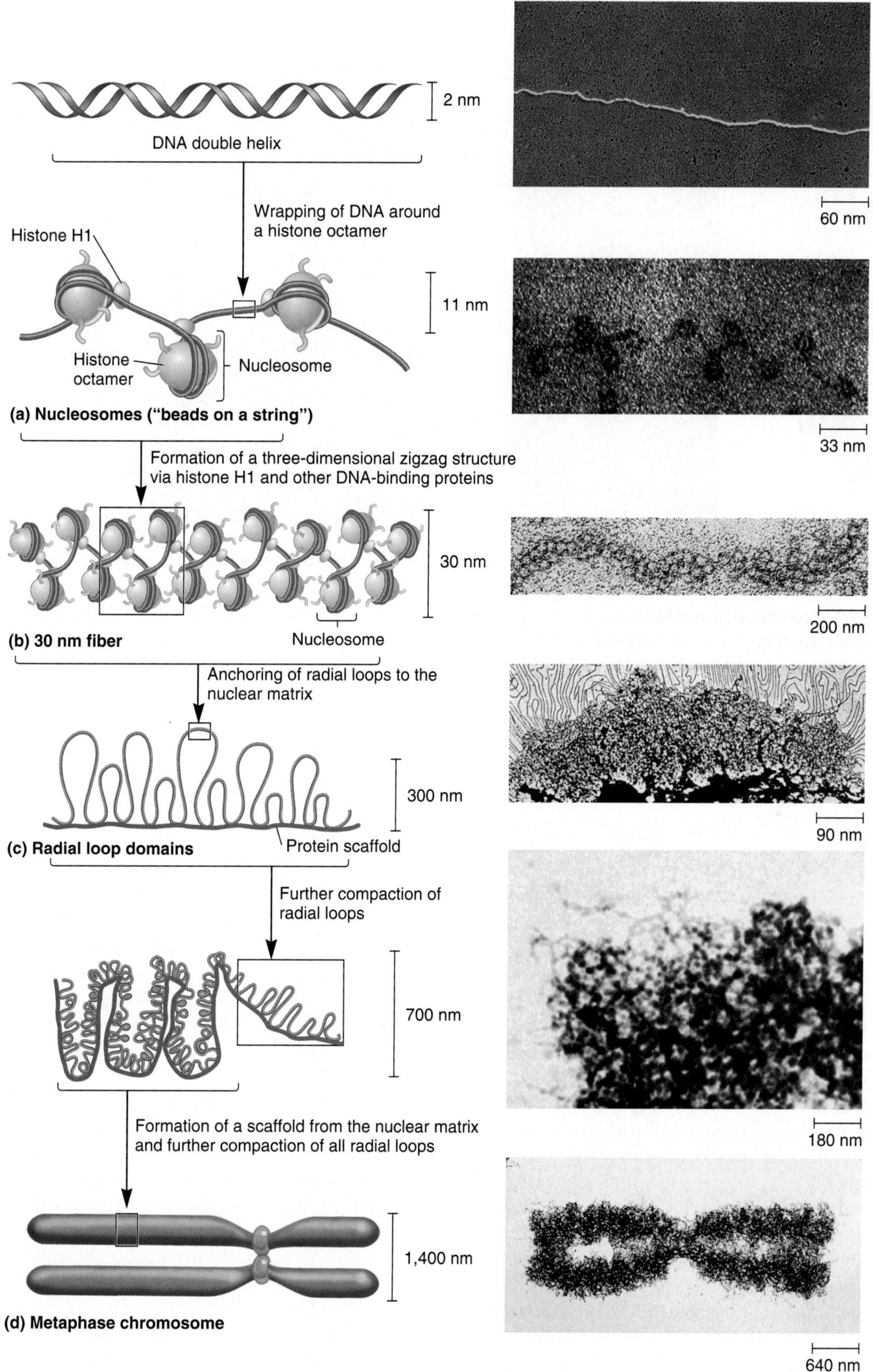

FIGURE 10.21 **The steps in eukaryotic chromosomal compaction leading to the metaphase chromosome.**

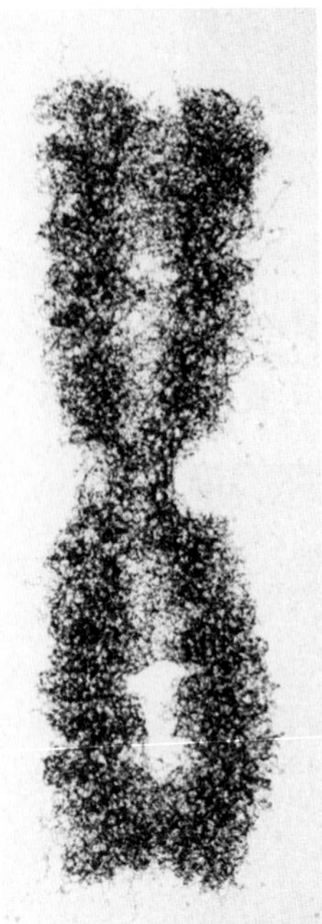

(a) Metaphase chromosome

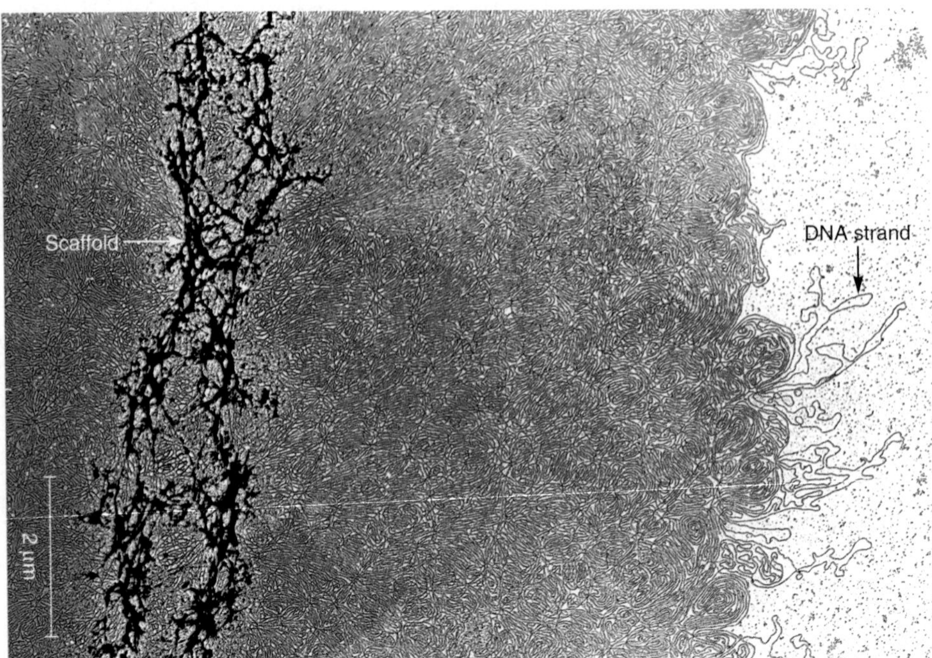

(b) Metaphase chromosome treated with high salt to remove histone proteins

FIGURE 10.22 The importance of histone proteins and scaffolding proteins in the compaction of eukaryotic chromosomes. (**a**) A metaphase chromosome. (**b**) A metaphase chromosome following treatment with high salt concentration to remove the histone proteins. The arrow on the left points to the scaffold (composed of nonhistone proteins), which anchors the bases of the radial loops. The arrow on the right points to an elongated strand of DNA.

found that cells contain two multiprotein complexes called **condensin** and **cohesin,** which play a critical role in chromosomal condensation and sister chromatid alignment, respectively.

Condensin and cohesin are two completely distinct complexes, but both contain a category of proteins called **SMC proteins.** SMC stands for structural maintenance of chromosomes. These proteins use energy from ATP to catalyze changes in chromosome structure. Together with topoisomerases, SMC proteins have been shown to promote major changes in DNA structure. An emerging theme is that SMC proteins actively fold, tether, and manipulate DNA strands. They are dimers that have a V-shaped structure. The monomers, which are connected at a hinge region, have two long coiled arms with a head region that binds ATP (**Figure 10.23**). The length of each monomer is about 50 nm, which is equivalent to approximately 150 bp of DNA.

As their names suggest, condensin and cohesin play different roles in metaphase chromosome structure. Prior to M phase, condensin is found outside the nucleus (**Figure 10.24**). However, as M phase begins, condensin is observed to coat the individual chromatids as euchromatin is converted into heterochromatin. The role of condensin in the compaction process is not well understood. Although condensin is often implicated in the process of chromosomal condensation, researchers have been able to deplete condensin from actively dividing cells, and the chromosomes are still able to condense. However, such condensed chromosomes show abnormalities in their ability to separate from each other during cell division. These results suggest that condensin is important in the proper organization of highly condensed chromosomes, such as those found during metaphase.

In comparison, the function of cohesin is to promote the binding (i.e., cohesion) between sister chromatids. After S phase and until the middle of prophase, sister chromatids remain attached to each other along their length. As shown in **Figure 10.25**, this attachment is promoted by cohesin, which is found along the entire length of each chromatid. In certain species, such as mammals, cohesins located along the chromosome arms are released during prophase. This allows the chromosome arms to separate. However, some cohesins remain attached, primarily to the centromeric regions, leaving the centromeric region as the main linkage before anaphase. At anaphase, the cohesins bound to the centromere are rapidly degraded by a protease aptly named separase, thereby allowing sister chromatid separation.

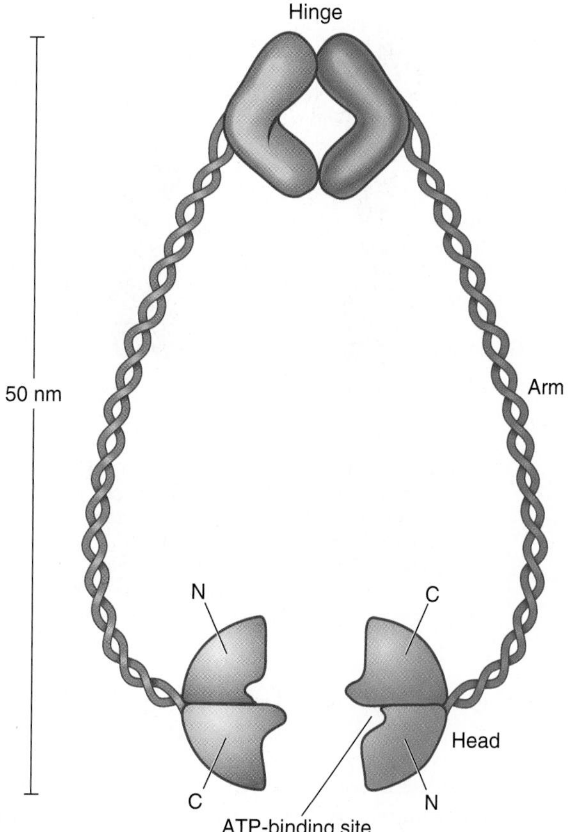

FIGURE 10.23 **The structure of SMC proteins.** This figure shows the generalized structure of SMC proteins, which are dimers consisting of hinge, arm, and head regions. The head regions bind and hydrolyze ATP. Condensin and cohesin have additional protein subunits not shown here.

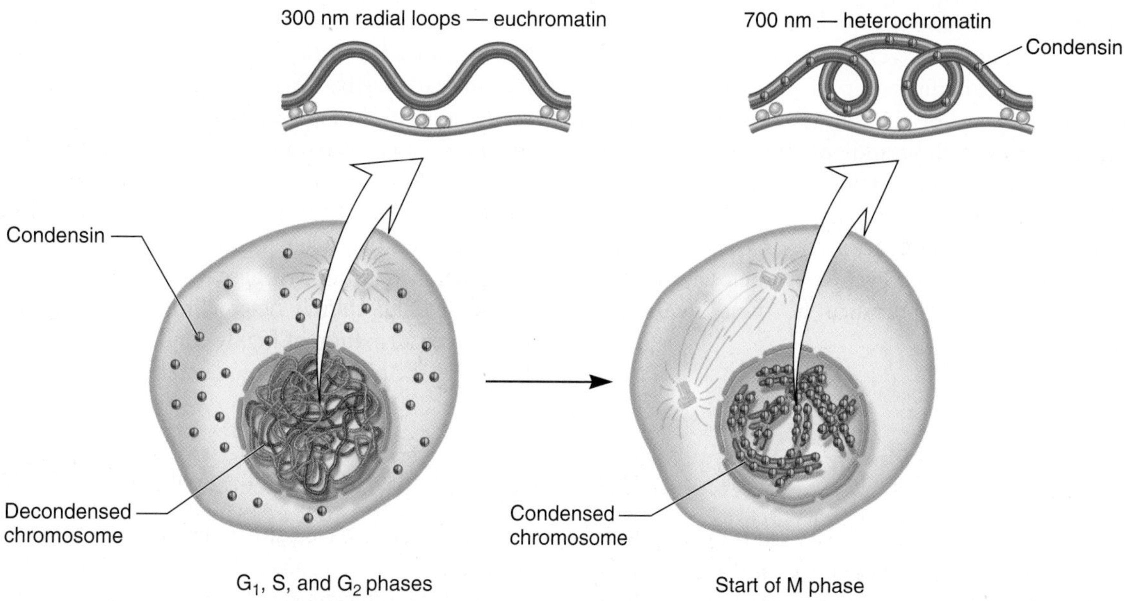

FIGURE 10.24 **The localization of condensin during interphase and the start of M phase.** During interphase (G_1, S, and G_2), most of the condensin protein is found outside the nucleus. The interphase chromosomes are largely euchromatic. At the start of M phase, condensin travels into the nucleus and binds to the chromosomes, which become heterochromatic due to a greater compaction of the radial loop domains.

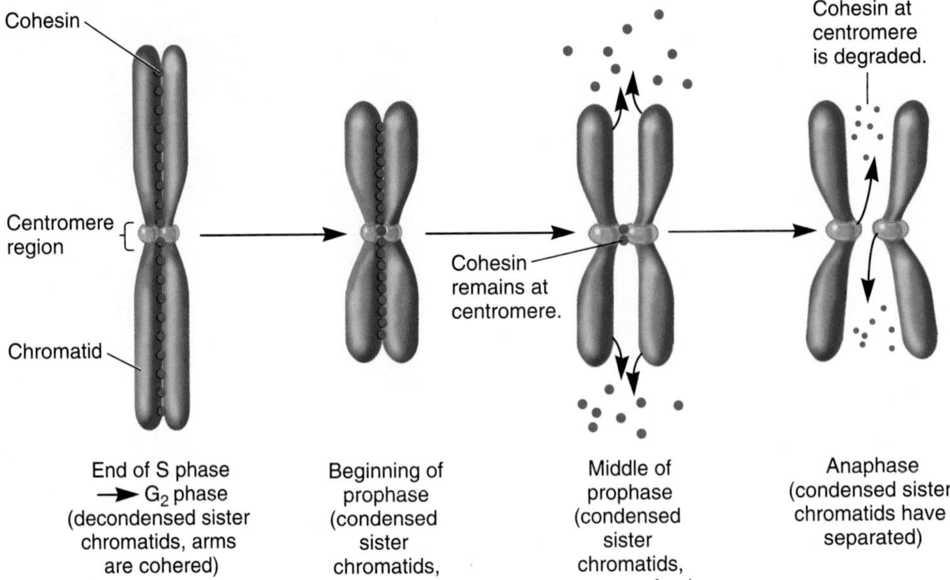

ONLINE ANIMATION **FIGURE 10.25 The alignment of sister chromatids via cohesin.** After S phase is completed, many cohesin complexes bind along each chromatid, thereby facilitating their attachment to each other. During the middle of prophase, cohesin is released from the chromosome arms, but some cohesin remains in the centromeric regions. At anaphase, the remaining cohesin complexes are rapidly degraded by separase, which promotes sister chromatid separation.

CONCEPTUAL SUMMARY

The chromosomal location and function of DNA sequences are central to our understanding of genetics. Viruses contain relatively small genomes that are composed of DNA or RNA. The viral genome is packaged into mature virus particles via self-assembly or by directed assembly. In bacteria, a few thousand different genes are interspersed throughout a circular chromosome; a single origin of replication is also present. Short repetitive sequences, which may play a role in chromosome structure and gene function, are found throughout the chromosome. By comparison, eukaryotes have many linear chromosomes. A single centromere is found on each chromosome and functions as a recognition site for the group of proteins that form the kinetochore. At the ends of the linear chromosomes are telomeres. Hundreds or even thousands of different genes are located between the centromeres and telomeres. An unusual feature of many eukaryotic species is that their DNA contains an abundance of highly repetitive sequences. The amount of highly repetitive DNA varies a great deal, even among closely related species, and the functional importance of these sequences remains unclear.

A second important issue that underlies chromosomal structure and function is the amount of compaction that occurs within chromosomal DNA. In bacteria, two main events, the formation of loop domains within the chromosome and DNA supercoiling, are responsible for the folding of the circular chromosome into a compact structure within a nucleoid. Certain DNA-binding proteins are involved in this process. In eukaryotic chromatin, the DNA is first folded into nucleosomes, which contain the DNA double helix wrapped around histone proteins. The nucleosomes are then compacted into a 30 nm fiber. The 30 nm fiber is further organized into radial loop domains, which are anchored to the nuclear matrix. This is termed euchromatin. In heterochromatin, the radial loop domains are compacted even further. During interphase, eukaryotic chromosomes are found within the cell nucleus. They consist of short, highly compacted regions that are heterochromatic and less condensed euchromatic regions. Histone modifications play a key role in controlling the level of chromatin compaction. During M phase, the chromosomes become highly condensed and entirely heterochromatic. Condensin functions in the proper formation of highly condensed chromosomes. Cohesin promotes the binding between sister chromatids.

EXPERIMENTAL SUMMARY

Our understanding of genome complexity and organization has been aided by various cytological, genetic, biochemical, and molecular techniques. In this chapter, we considered renaturation experiments as a technique to evaluate the sequence complexity of DNA. In Chapters 5 and 6, we examined ways to genetically map the locations of genes along eukaryotic and bacterial chromosomes. In Chapter 20, we will explore the use of cytological techniques such as in situ hybridization to identify particular sequences within an intact chromosome. In addition, Chapter 20 will describe molecular methods in which segments of a genome are cloned and analyzed molecularly.

The second topic in this chapter was the compaction of chromosomes to fit within living cells. This phenomenon can also be examined via several different approaches. The microscopic analysis of chromosomes provides key insight into their structure. In addition, chromatin can be analyzed biochemically by many different methods. For example, the model of repeating nucleosome structure was verified by the studies of Noll, which involved digestion of chromatin by the enzyme DNase I. More recently, the detailed structure of nucleosomes has been solved by crystallization studies. Currently, much research is aimed at identifying the proteins bound to chromosomal DNA and elucidating the roles these proteins play in determining chromatin structure.

PROBLEM SETS & INSIGHTS

Solved Problems

S1. Here is a C_0t curve for a hypothetical eukaryotic species:

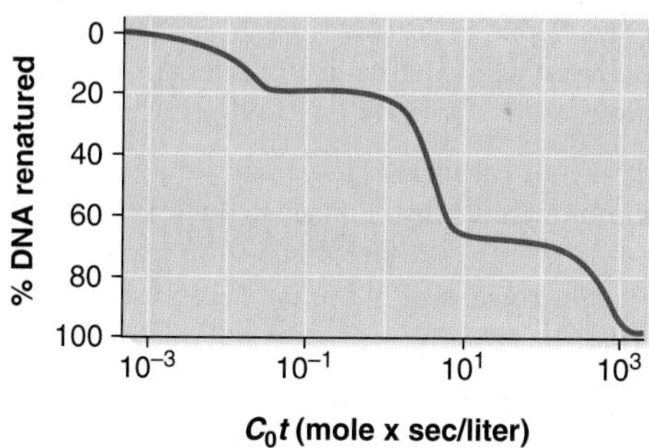

Estimate the amount of highly repetitive DNA, moderately repetitive DNA, and unique DNA.

Answer: About 20% is highly repetitive and renatures quickly, about 50% is moderately repetitive, and about 30% is unique and renatures very slowly.

S2. Let's suppose a bacterial DNA molecule is given a left-handed twist. How does this affect the structure and function of the DNA?

Answer: A left-handed twist is negative supercoiling. Negative supercoiling makes the bacterial chromosome more compact. It also promotes DNA functions that involve strand separation, including gene transcription and DNA replication.

S3. To hold bacterial DNA in a more compact configuration, specific proteins must bind to the DNA and stabilize its conformation (as shown in Figure 10.5). Several different proteins are involved in this process. These proteins have been collectively referred to as "histonelike" due to their possible functional similarity to the histone proteins found in eukaryotes. Based on your knowledge of eukaryotic histone proteins, what biochemical properties would you expect from bacterial histonelike proteins?

Answer: The histonelike proteins have the properties expected for proteins involved in DNA folding. They are all small proteins found in relative abundance within the bacterial cell. In some cases, the histonelike proteins are biochemically similar to eukaryotic histones. For example, they tend to be basic (positively charged) and bind to DNA in a non-sequence-dependent fashion. However, other proteins appear to bind to bacterial DNA at specific sites in order to promote DNA bending.

Conceptual Questions

C1. In viral replication, what is the difference between self-assembly and directed assembly?

C2. Bacterial chromosomes have one origin of replication, whereas eukaryotic chromosomes have several. Would you expect viral chromosomes to have an origin of replication? Why or why not?

C3. What is a bacterial nucleoid? With regard to cellular membranes, what is the difference between a bacterial nucleoid and a eukaryotic nucleus?

C4. In Part II of this textbook, we considered inheritance patterns for diploid eukaryotic species. Bacteria frequently contain two or more nucleoids. With regard to genes and alleles, how is a bacterium that contains two nucleoids similar to a diploid eukaryotic cell, and how is it different?

C5. Describe the two main mechanisms by which the bacterial DNA becomes compacted.

C6. As described in Chapter 9, 1 bp of DNA is approximately 0.34 nm in length. A bacterial chromosome is about 4 million bp in length and is organized into about 100 loops that are about 40,000 bp in length.

A. If it was stretched out linearly, how long (in micrometers) would one loop be?

B. If a bacterial chromosomal loop is circular, what would be its diameter? (Note: Circumference = πD, where D is the diameter of the circle.)

C. Is the diameter of the circular loop calculated in part B small enough to fit inside a bacterium? The dimensions of the bacterial cytoplasm, such as *E. coli*, are roughly 0.5 μm wide and 1.0 μm long.

C7. Why is DNA supercoiling called supercoiling rather than just coiling? Why is positive supercoiling called overwinding and negative supercoiling called underwinding? How would you define the terms positive and negative supercoiling for Z DNA (described in Chapter 9)?

C8. Coumarins and quinolones are two classes of drugs that inhibit bacterial growth by directly inhibiting DNA gyrase. Discuss two reasons why inhibiting DNA gyrase might inhibit bacterial growth.

C9. Take two pieces of string that are approximately 10 inches each, and create a double helix by wrapping the two strings around each other to make 10 complete turns. Tape one end of the strings to a table, and now twist the strings three times (360° each time) in a right-handed direction. Note: As you are looking down at the strings from above, a right-handed twist is in the clockwise direction.

A. Did the three turns create more or fewer turns in your double helix? How many turns are now in your double helix?

B. Is your double helix right-handed or left-handed? Explain your answer.

C. Did the three turns create any supercoils?

D. If you had coated your double helix with rubber cement and allowed the cement to dry before making the three additional right-handed turns, would the rubber cement make it more or less likely for the three turns to create supercoiling? Would a pair of cemented strings be more or less like a real DNA double helix compared to an uncemented pair of strings? Explain your answer.

C10. Try to explain the function of DNA gyrase with a drawing.

C11. How are two topoisomers different from each other? How are they the same?

C12. On rare occasions, a chromosome can suffer a small deletion that removes the centromere. When this occurs, the chromosome usually is not found within subsequent daughter cells. Explain why a chromosome without a centromere is not transmitted very efficiently from mother to daughter cells. (Note: If a chromosome is located outside the nucleus after telophase, it is degraded.)

C13. What is the function of a centromere? At what stage of the cell cycle would you expect the centromere to be the most important?

C14. Describe the characteristics of highly repetitive DNA.

C15. Describe the structures of a nucleosome and a 30 nm fiber.

C16. Beginning with the G_1 phase of the cell cycle, describe the level of compaction of the eukaryotic chromosome. How does the level of compaction change as the cell progresses through the cell cycle? Why is it necessary to further compact the chromatin during mitosis?

C17. If you assume the average length of linker DNA is 50 bp, approximately how many nucleosomes are found in the haploid human genome, which contains 3 billion bp?

C18. Draw the binding between the nuclear matrix and MARs.

C19. Compare heterochromatin and euchromatin. What are the differences between them?

C20. Compare the structure and cell localization of chromosomes during interphase and M phase.

C21. What types of genetic activities occur during interphase? Explain why these activities cannot occur during M phase.

C22. Let's assume the linker DNA averages 54 bp in length. How many molecules of H2A would you expect to find in a DNA sample that is 46,000 bp in length?

C23. In Figure 10.15, what are we looking at in part b? Is this an 11 nm fiber, a 30 nm fiber, or a 300 nm fiber? Does this DNA come from a cell during M phase or interphase?

C24. What are the roles of the core histone proteins compared to the role of histone H1 in the compaction of eukaryotic DNA?

C25. A typical eukaryotic chromosome found in humans contains about 100 million bp of DNA. As described in Chapter 9, 1 bp of DNA has a linear length of 0.34 nm.

A. What is the linear length of the DNA for a typical human chromosome in micrometers?

B. What is the linear length of a 30 nm fiber of a typical human chromosome?

C. Based on your calculation of part B, would a typical human chromosome fit inside the nucleus (with a diameter of 5 μm) if the 30 nm fiber were stretched out in a linear manner? If not, explain how a typical human chromosome fits inside the nucleus during interphase.

C26. Which of the following terms should not be used to describe a Barr body?

A. Chromatin

B. Euchromatin

C. Heterochromatin

D. Chromosome

E. Genome

C27. Discuss the differences in the compaction levels of metaphase chromosomes compared to interphase chromosomes. When would you expect gene transcription and DNA replication to take place, during M phase or interphase? Explain why.

C28. Explain two ways that histone modifications may affect the level of chromatin compaction.

C29. What is an SMC protein? Describe two examples.

Experimental Questions

E1. Two circular DNA molecules, which we will call molecule A and molecule B, are topoisomers of each other. When viewed under the electron microscope, molecule A appears more compact compared to molecule B. The level of gene transcription is much lower for molecule A. Which of the following three possibilities could account for these observations?

First possibility: Molecule A has 3 positive supercoils, and molecule B has 3 negative supercoils.

Second possibility: Molecule A has 4 positive supercoils, and molecule B has 1 negative supercoil.

Third possibility: Molecule A has 0 supercoils, and molecule B has 3 negative supercoils.

E2. Explain how a renaturation experiment can provide quantitative information about genome sequence complexity.

E3. In a renaturation experiment, does the copy number affect only the rate of renaturation, or does it also affect the rate of denaturation? Explain your answer.

E4. Let's suppose that you have isolated DNA from a cell and have viewed it under a microscope. It looks supercoiled. What experiment would you perform to determine if it is positively or negatively supercoiled? In your answer, describe your expected results. You may assume that you have purified topoisomerases at your disposal.

E5. We seem to know more about the structure of eukaryotic chromosomal DNA than bacterial DNA. Discuss why you think this is so, and list several experimental procedures that have yielded important information concerning the compaction of eukaryotic chromatin.

E6. An organism contains 20% highly repetitive DNA, 10% moderately repetitive DNA, and 70% unique sequences. Draw the expected C_0t curve that would be obtained from this organism.

E7. When chromatin is treated with a moderate salt concentration, the linker histone H1 is removed (see Figure 10.15a). Higher salt concentration removes the rest of the histone proteins (see Figure 10.22b). If the experiment of Figure 10.14 were carried out after the DNA was treated with moderate or high salt, what would be the expected results?

E8. Let's suppose you have isolated chromatin from some bizarre eukaryote with a linker region that is usually 300 to 350 bp in length. The nucleosome structure is the same as in other eukaryotes. If you digested this eukaryotic organism's chromatin with a high concentration of DNase I, what would be your expected results?

E9. If you were given a sample of chromosomal DNA and asked to determine if it is bacterial or eukaryotic, what experiment would you perform, and what would be your expected results?

E10. Consider how histone proteins bind to DNA and then explain why a high salt concentration can remove histones from DNA (as shown in Figure 10.22b).

E11. In Chapter 20, the technique of fluorescence *in situ* hybridization (FISH) is described. This is another method used to examine sequence complexity within a genome. In this method, a particular DNA sequence, such as a particular gene sequence, can be detected within an intact chromosome by using a DNA probe that is complementary to the sequence. For example, let's consider the β-globin gene, which is found on human chromosome 11. A probe complementary to the β-globin gene will bind to the β-globin gene and show up as a brightly colored spot on human chromosome 11. In this way, researchers can detect where the β-globin gene is located within a set of chromosomes. Because the β-globin gene is unique and because human cells are diploid (i.e., have two copies of each chromosome), a FISH experiment would show two bright spots per cell; the probe would bind to each copy of chromosome 11. What would you expect to see if you used the following types of probes?

A. A probe complementary to the *Alu*I sequence

B. A probe complementary to a tandemly repeated sequence near the centromere of the X chromosome

Questions for Student Discussion/Collaboration

1. Bacterial and eukaryotic chromosomes are very compact. Discuss the advantages and disadvantages of having a compact structure.
2. The prevalence of highly repetitive sequences seems rather strange to many geneticists. Do they seem strange to you? Why or why not? Discuss whether or not you think they have an important function.
3. Discuss and make a list of the similarities and differences between bacterial and eukaryotic chromosomes.

Note: All answers appear at the website for this textbook; the answers to even-numbered questions are in the back of the textbook.

www.mhhe.com/brookergenetics3e

Visit the Online Learning Center for practice tests, answer keys, and other learning aids for this chapter. Enhance your understanding of genetics with our interactive exercises, quizzes, animations, and much more.

11

DNA REPLICATION

CHAPTER OUTLINE

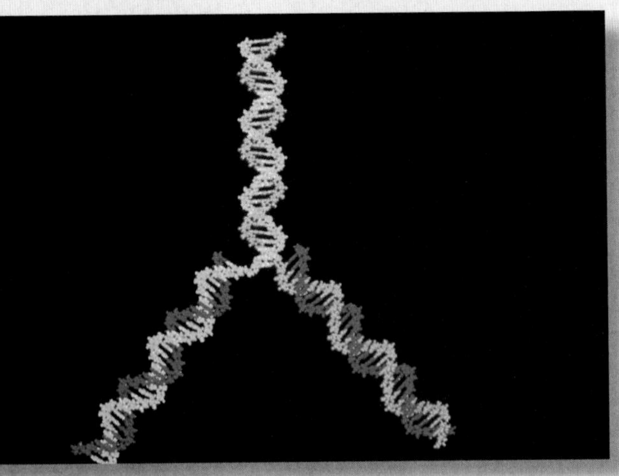

A model for DNA replication. This molecular model shows a DNA replication fork, the site where new DNA strands are made. In this model, the original DNA is yellow and blue. The newly made strands are purple.

As discussed throughout Chapters 2 to 8, genetic material is transmitted from parent to offspring and from cell to cell. For transmission to occur, the genetic material must be copied. During this process, known as **DNA replication,** the original DNA strands are used as templates for the synthesis of new DNA strands. We will begin Chapter 11 with a consideration of the structural features of the double helix that underlie the replication process. Then we will examine how chromosomes are replicated within living cells, addressing the following questions: where does DNA replication begin, how does it proceed, and where does it end? We first consider bacterial DNA replication and examine how DNA replication occurs within living cells, and then turn our attention to the unique features of the replication of eukaryotic DNA. At the molecular level, it is rather remarkable that the replication of chromosomal DNA occurs very quickly, very accurately, and at the appropriate time in the life of the cell. For this to happen, many cellular proteins play vital roles. In this chapter, we will examine the mechanism of DNA replication and consider the functions of several proteins involved in the process.

11.1 STRUCTURAL OVERVIEW OF DNA REPLICATION

Let's begin by recalling a few important structural features of the double helix from Chapter 9 because they bear directly on the replication process. The double helix is composed of two DNA strands, and the individual building blocks of each strand are nucleotides. The nucleotides contain one of four bases: adenine, thymine, guanine, or cytosine. The double-stranded structure is held together by base stacking and by hydrogen bonding between the bases in opposite strands. A critical feature of the double-helix structure is that adenine hydrogen bonds with thymine, and guanine hydrogen bonds with cytosine. This rule, known as the AT/GC rule or Chargaff's rule, is the basis for the complementarity of the base sequences in double-stranded DNA.

Another feature worth noting is that the strands within a double helix have an antiparallel alignment. This directionality is determined by the orientation of sugar molecules within the sugar–phosphate backbone. If one strand is running in the 5' to 3' direction, the complementary strand is running in the 3' to 5' direction. The issue of directionality will be important when we consider the function of the enzymes that synthesize new DNA strands. In this section, we will consider how the structure of the DNA double helix provides the basis for DNA replication.

Existing DNA Strands Act as Templates for the Synthesis of New Strands

As shown in **Figure 11.1a**, DNA replication relies on the complementarity of DNA strands according to the AT/GC rule. During the replication process, the two complementary strands of DNA come apart and serve as **template strands,** or **parental strands,** for the synthesis of two new strands of DNA. After the double helix has separated, individual nucleotides have access to the template strands. Hydrogen bonding between individual nucleotides and the template strands must obey the AT/GC rule. To complete the replication process, a covalent bond is formed between the phosphate of one nucleotide and the sugar of the previous nucleotide. The two newly made strands are referred to as the **daughter strands.** Note that the base sequences are identical in both double-stranded molecules after replication (**Figure 11.1b**). Therefore, DNA is replicated so that both copies retain the same information—the same base sequence—as the original molecule.

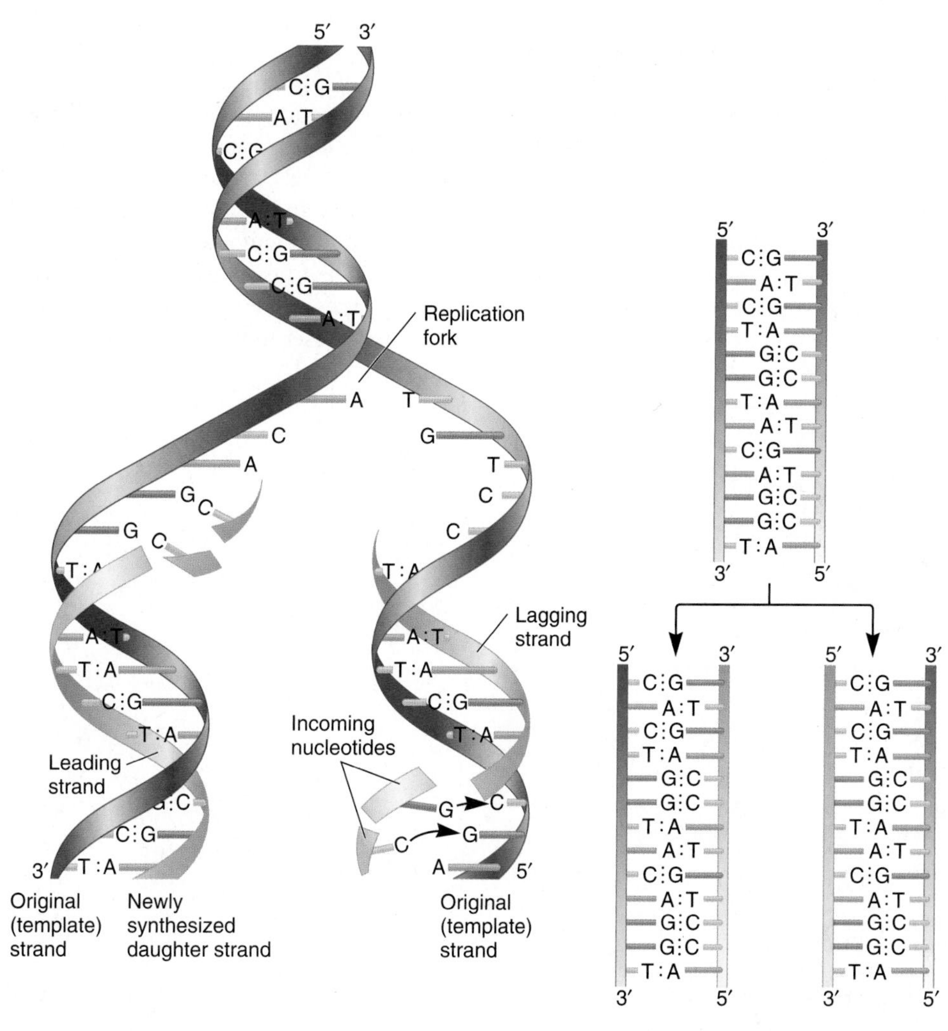

FIGURE 11.1 **The structural basis for DNA replication.** (**a**) The mechanism of DNA replication as originally proposed by Watson and Crick. As we will see, the synthesis of one newly made strand (the leading strand) occurs in the direction toward the replication fork, whereas the synthesis of the other newly made strand (the lagging strand) occurs in small segments away from the replication fork. (**b**) DNA replication produces two copies of DNA with the same sequence as the original DNA molecule.

EXPERIMENT 11A

Three Different Models Were Proposed That Described the Net Result of DNA Replication

Scientists in the late 1950s had considered three different mechanisms to explain the net result of DNA replication. These mechanisms are shown in **Figure 11.2.** The first is referred to as a **conservative model.** According to this hypothesis, both strands of parental DNA remain together following DNA replication. In this model, the original arrangement of parental strands is completely conserved, while the two newly made daughter strands also remain together following replication. The second is called a **semiconservative model.** In this mechanism, the double-stranded DNA is half conserved following the replication process. In other words, the newly made double-stranded DNA contains one parental strand and one daughter strand. The third, called the **dispersive model,** proposes that segments of parental DNA and newly made DNA are interspersed in both strands following the replication process. Only the semiconservative model shown in Figure 11.2b is actually correct.

In 1958, Matthew Meselson and Franklin Stahl devised a method to experimentally distinguish newly made daughter strands from the original parental strands. The technique they used involved heavy isotope labeling. Nitrogen, which is found within the bases of DNA, occurs in a light (^{14}N) form and a heavy (^{15}N) form. Prior to their experiment, they grew *E. coli* cells in the presence of ^{15}N for many generations. This produced a population of cells in which all of the DNA was heavy labeled. At the start of their experiment, shown in **Figure 11.3** (generation 0), they switched the bacteria to a medium that contained only ^{14}N and then collected samples of cells after various time points. Under the growth conditions they employed, 30 minutes is the time required for one doubling, or one generation time. Because the bacteria were doubling in a medium that contained only ^{14}N, all of the newly made DNA strands would be labeled with light nitrogen, while the original strands would remain in the heavy form.

Meselson and Stahl then analyzed the density of the DNA by centrifugation, using a cesium chloride (CsCl) gradient. (The procedure of gradient centrifugation is described in the Appendix.) If both DNA strands contained ^{14}N, the DNA would have a light density and sediment near the top of the tube. If one strand contained ^{14}N and the other strand contained ^{15}N, the DNA would be half-heavy and have an intermediate density. Finally, if both strands contained ^{15}N, the DNA would be heavy and would sediment closer to the bottom of the centrifuge tube.

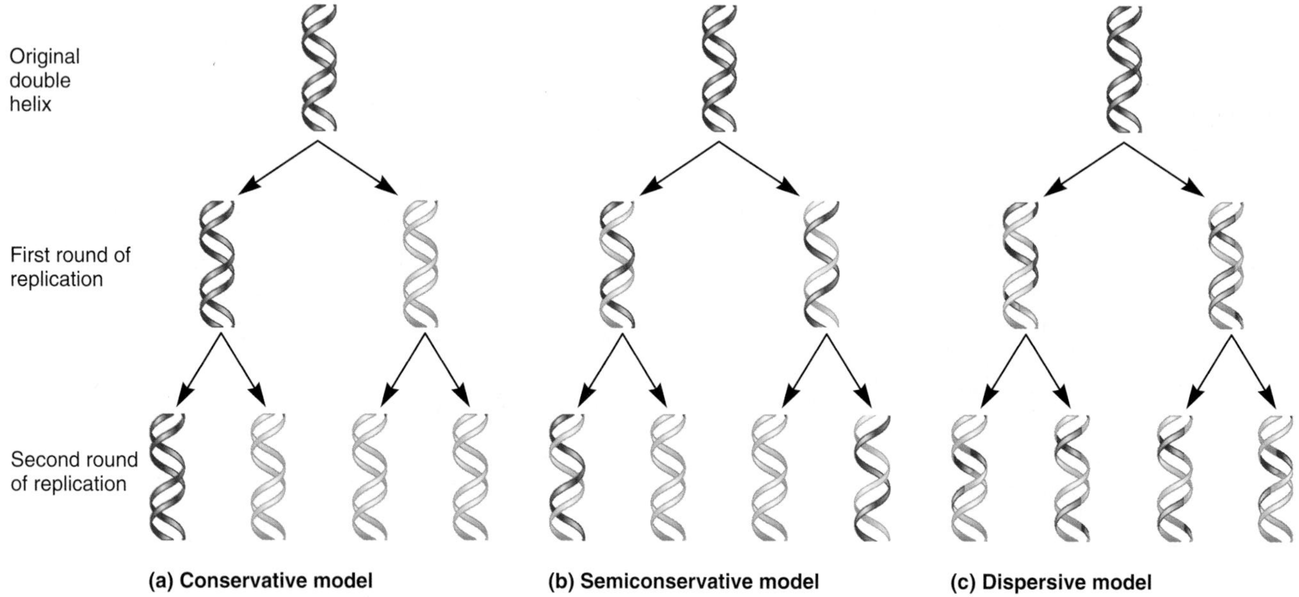

FIGURE 11.2 **Three possible models for DNA replication.** The two original parental DNA strands are shown in purple, and the newly made strands after one and two generations are shown in light blue.

THE HYPOTHESIS

Based on Watson's and Crick's ideas, the hypothesis was that DNA replication is semiconservative. Figure 11.2 also shows two alternative models.

TESTING THE HYPOTHESIS — FIGURE 11.3 Evidence that DNA replication is semiconservative.

Starting material: A strain of *E. coli* that has been grown for many generations in the presence of ^{15}N. All of the nitrogen in the DNA is labeled with ^{15}N.

1. Add an excess of ^{14}N-containing compounds to the bacterial cells so that all the newly made DNA will contain ^{14}N.
2. Incubate the cells for various lengths of time. Note: The ^{15}N-labeled DNA is shown in purple and the ^{14}N-labeled DNA is shown in blue.
3. Lyse the cells by the addition of lysozyme and detergent, which disrupt the bacterial cell wall and cell membrane, respectively.
4. Load a sample of the lysate onto a CsCl gradient. (Note: The average density of DNA is around 1.7 g/cm^3, which is well isolated from other cellular macromolecules.)
5. Centrifuge the gradients until the DNA molecules reach their equilibrium densities.
6. DNA within the gradient can be observed under a UV light.

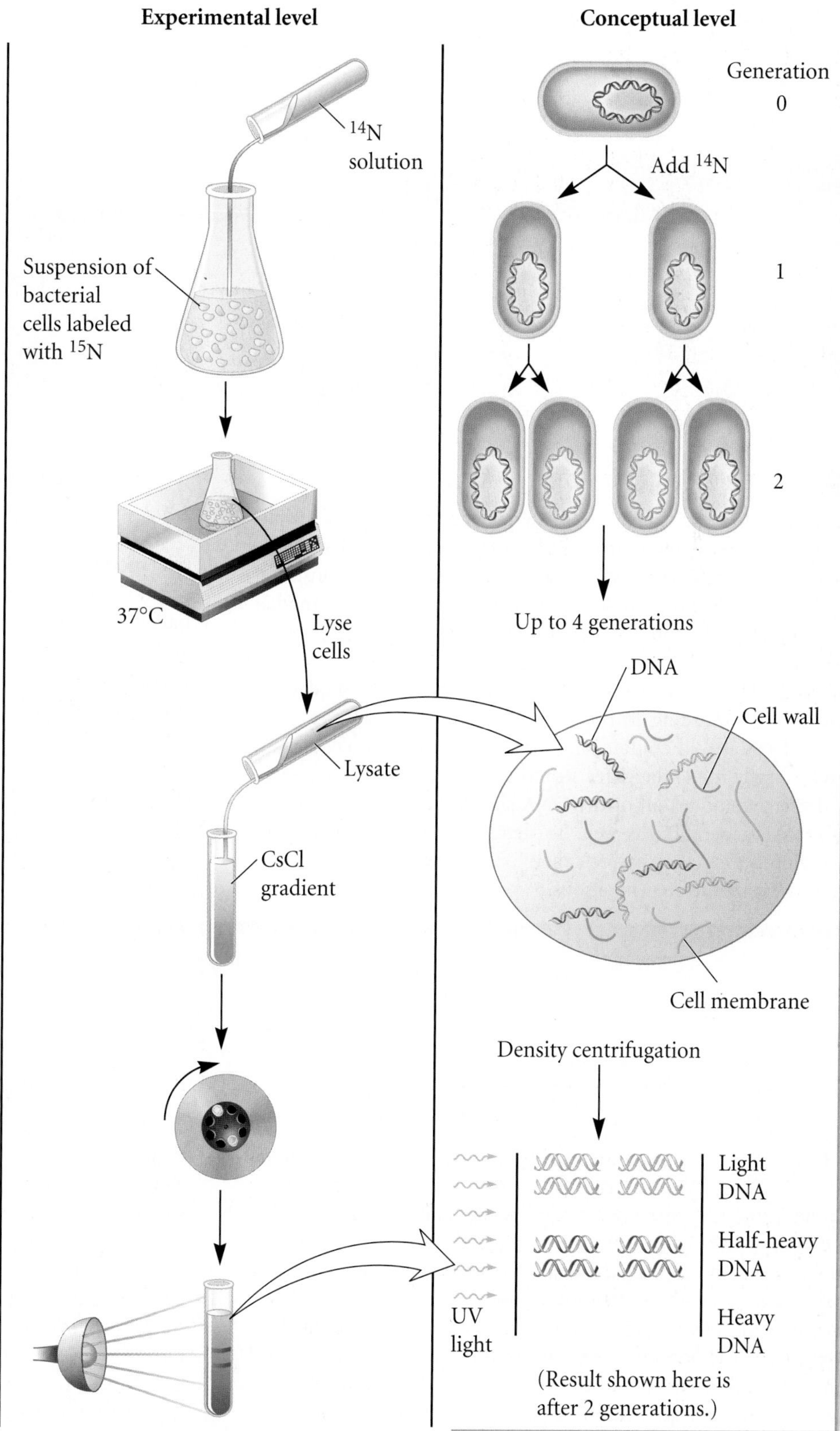

THE DATA

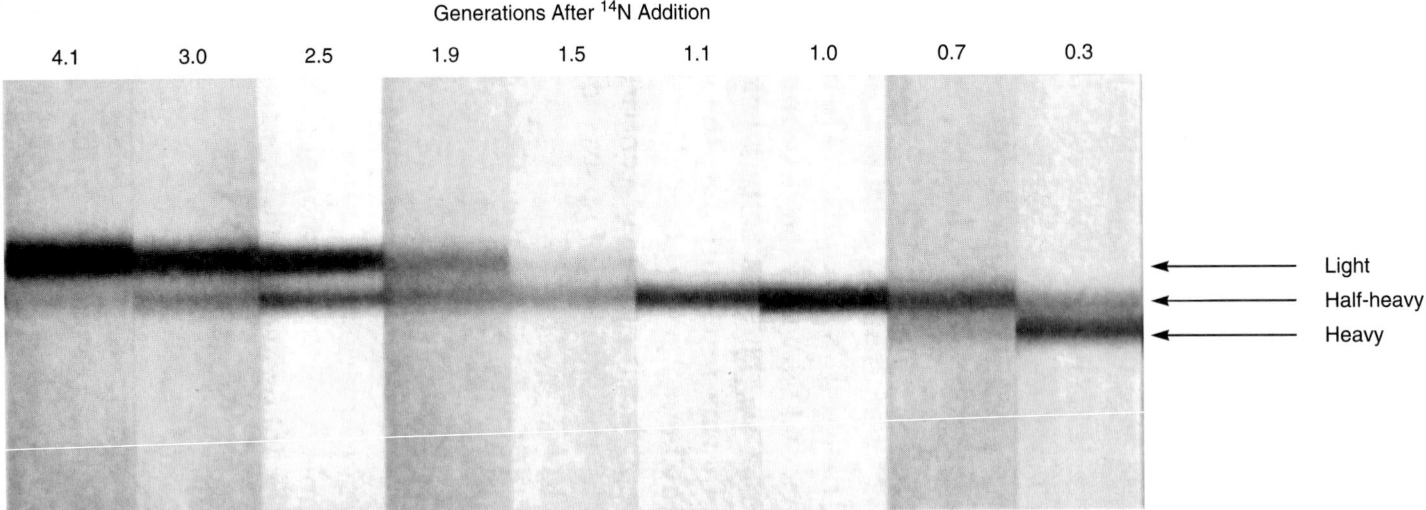

INTERPRETING THE DATA

As seen in the data of Figure 11.3, after one round of DNA replication (i.e., one generation), all of the DNA sedimented at a density that was half-heavy. Which of the three models is consistent with this result? Both the semiconservative and dispersive models are consistent. In contrast, the conservative model would predict two separate DNA types: a light type and a heavy type. Because all of the DNA had sedimented as a single band, this model was disproved. According to the semiconservative model, the replicated DNA would contain one original strand (a heavy strand) and a newly made daughter strand (a light strand). Likewise, in a dispersive model, all of the DNA should have been half-heavy after one generation as well. To determine which of these two remaining models is correct, therefore, Meselson and Stahl had to investigate future generations.

After approximately two rounds of DNA replication (i.e., 1.9 generations), a mixture of light DNA and half-heavy DNA was observed. This result was consistent with the semiconservative model of DNA replication, because some DNA molecules should contain all light DNA, while other molecules should be half-heavy (see Figure 11.2b). The dispersive model predicts that after two generations, the heavy nitrogen would be evenly dispersed among four strands, each strand containing 1/4 heavy nitrogen and 3/4 light nitrogen (see Figure 11.2c). However, this result was not obtained. Instead, the results of the Meselson and Stahl experiment provided compelling evidence in favor of only the semiconservative model for DNA replication.

A self-help quiz involving this experiment can be found at the Online Learning Center.

11.2 BACTERIAL DNA REPLICATION

Thus far, we have considered how a complementary, double-stranded structure underlies the ability of DNA to be copied. In addition, the experiments of Meselson and Stahl showed that DNA replication results in two double helices, each one containing an original parental strand and a newly made daughter strand. We will now turn our attention to how DNA replication actually occurs within living cells. Much research has focused on the bacterium *Escherichia coli.* The results of these studies have provided the foundation for our current molecular understanding of DNA replication. The replication of the bacterial chromosome is a stepwise process in which many cellular proteins participate. In this section, we will follow this process from beginning to end.

Bacterial Chromosomes Contain a Single Origin of Replication

Figure 11.4 presents an overview of the process of bacterial chromosomal replication. The site on the bacterial chromosome where DNA synthesis begins is known as the **origin of replication.** Bacterial chromosomes have a single origin of replication. The synthesis of new daughter strands is initiated within the origin and proceeds in both directions, or **bidirectionally,** around the bacterial chromosome. This means that two **replication forks** move in opposite directions outward from the origin. A replication fork is the site where the parental DNA strands have separated and new daughter strands are being made. Eventually, these replication forks meet each other on the opposite side of the bacterial chromosome to complete the replication process.

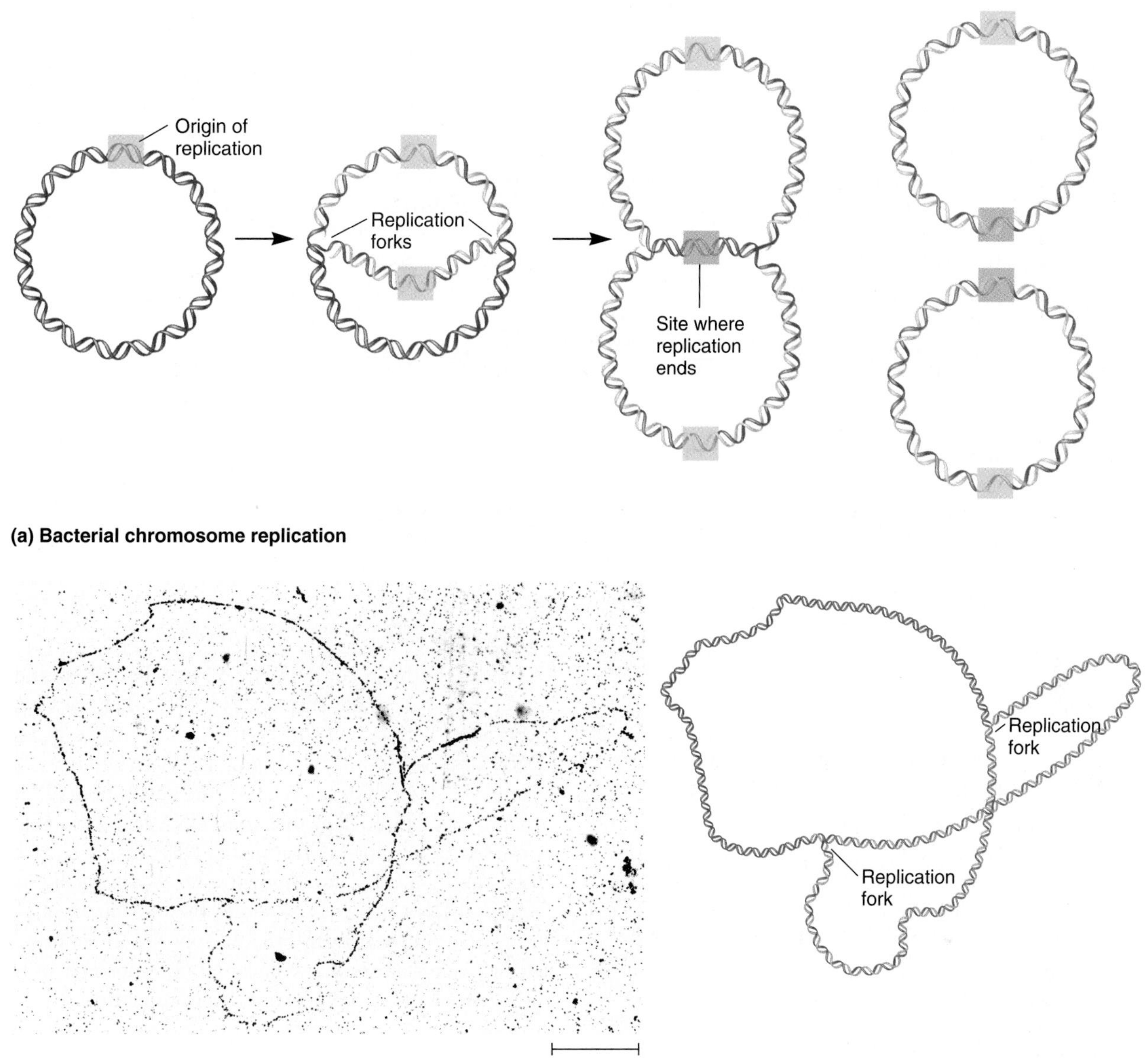

FIGURE 11.4 **The process of bacterial chromosome replication.** (**a**) An overview of the process of bacterial chromosome replication. (**b**) A replicating *E. coli* chromosome visualized by autoradiography. This chromosome was radiolabeled by growing bacterial cells in media containing radiolabeled thymidine. The diagram at the right shows the locations of the two replication forks. The chromosome is about one-third replicated. New strands are shown in blue.

Replication Is Initiated by the Binding of DnaA Protein to the Origin of Replication

Considerable research has focused on the origin of replication in *E. coli.* This origin is named *oriC* for origin of Chromosomal replication (Figure 11.5). Three types of DNA sequences are found within *oriC:* an AT-rich region, DnaA box sequences, and GATC methylation sites. The GATC methylation sites will be discussed later in this chapter when we consider the regulation of replication.

DNA replication is initiated by the binding of **DnaA proteins** to sequences within the origin known as **DnaA box sequences.** The DnaA box sequences serve as recognition sites for the binding of the DnaA proteins. When DnaA proteins are in their ATP-bound form, they bind to the five DnaA boxes in *oriC* to initiate DNA replication. DnaA proteins also bind to each other to form a complex (Figure 11.6). With the aid of other DNA-binding proteins, such as HU and IHF, this causes the DNA to bend around the complex of DnaA proteins and results in the separation of the AT-rich region. Because only two hydrogen bonds form between AT base pairs, while three hydrogen bonds occur between G and C, the DNA strands are more easily separated at an AT-rich region.

Following separation of the AT-rich region, the DnaA proteins, with the help of the DnaC protein, recruit **DNA helicase** enzymes to this site. DNA helicase is also known as DnaB protein. When a DNA helicase encounters a double-stranded region,

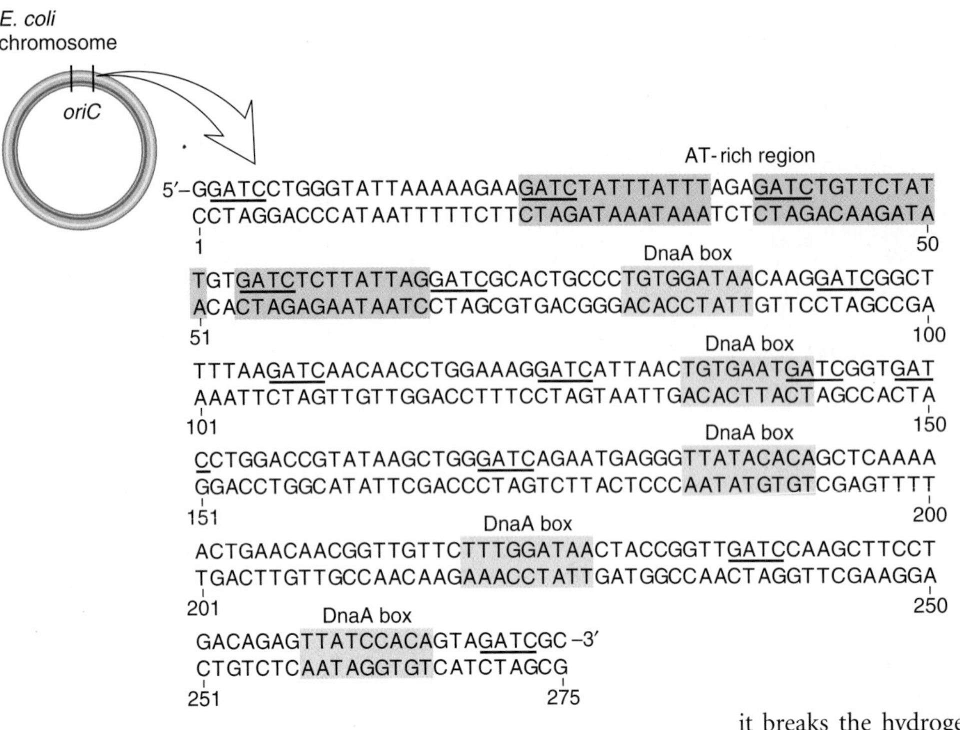

FIGURE 11.5 **The sequence of *oriC* in *E. coli.*** The AT-rich region is composed of three tandem repeats that are 13 base pairs long and highlighted in blue. The five DnaA boxes are highlighted in orange. The GATC methylation sites are underlined.

it breaks the hydrogen bonds between the two strands, thereby generating two single strands. Two DNA helicases begin strand separation within the *oriC* region and continue to separate the DNA strands beyond the origin. These enzymes use the energy from ATP hydrolysis to catalyze the separation of the double-stranded parental DNA. In *E. coli,* DNA helicases bind to single-stranded DNA and travel along the DNA in a 5' to 3' direction to keep the replication fork moving. As shown in Figure 11.6, the action of DNA helicases promotes the movement of two replication forks outward from *oriC* in opposite directions. This initiates the replication of the bacterial chromosome in both directions, an event termed **bidirectional replication.**

Several Proteins Are Required for DNA Replication at the Replication Fork

Figure 11.7 provides an overview of the molecular events that occur as one of the two replication forks moves around the bacterial chromosome, and **Table 11.1** summarizes the functions of the major proteins involved in *E. coli* DNA replication. Let's begin with strand separation. To act as a template for DNA replication, the strands of a double helix must separate. As mentioned previously, the function of DNA helicase is to break the hydrogen bonds between base pairs and thereby unwind the strands; this action generates positive supercoiling ahead of each replication

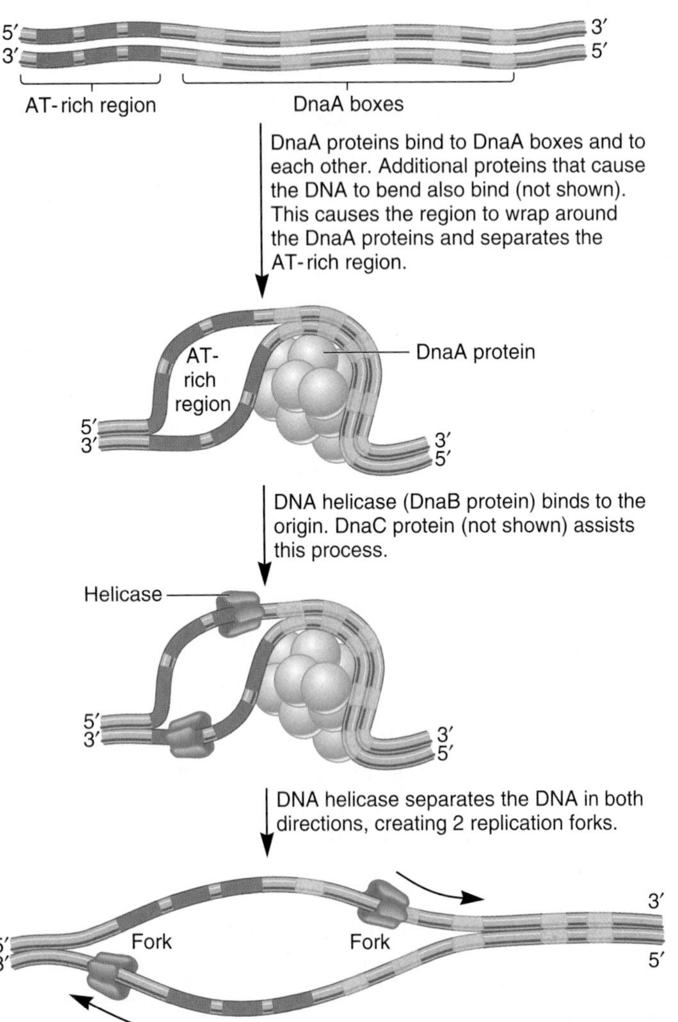

ONLINE ANIMATION

FIGURE 11.6 **The events that occur at *oriC* to initiate the DNA replication process.** To initiate DNA replication, DnaA proteins bind to the five DnaA boxes, which causes the DNA strands to separate at the AT-rich region. DnaA and DnaC proteins then recruit DNA helicase (DnaB) into this region. Each DNA helicase is composed of six subunits, which form a ring around one DNA strand and migrates in the 5' to 3' direction. As shown here, the movement of two DNA helicase proteins serves to separate the DNA strands beyond the *oriC* region.

Functions of key proteins involved with DNA replication

- DNA helicase breaks the hydrogen bonds between the DNA strands.
- Topoisomerase alleviates positive supercoiling.
- Single-strand binding proteins keep the parental strands apart.
- Primase synthesizes an RNA primer.
- DNA polymerase III synthesizes a daughter strand of DNA.
- DNA polymerase I excises the RNA primers and fills in with DNA (not shown).
- DNA ligase covalently links the Okazaki fragments together.

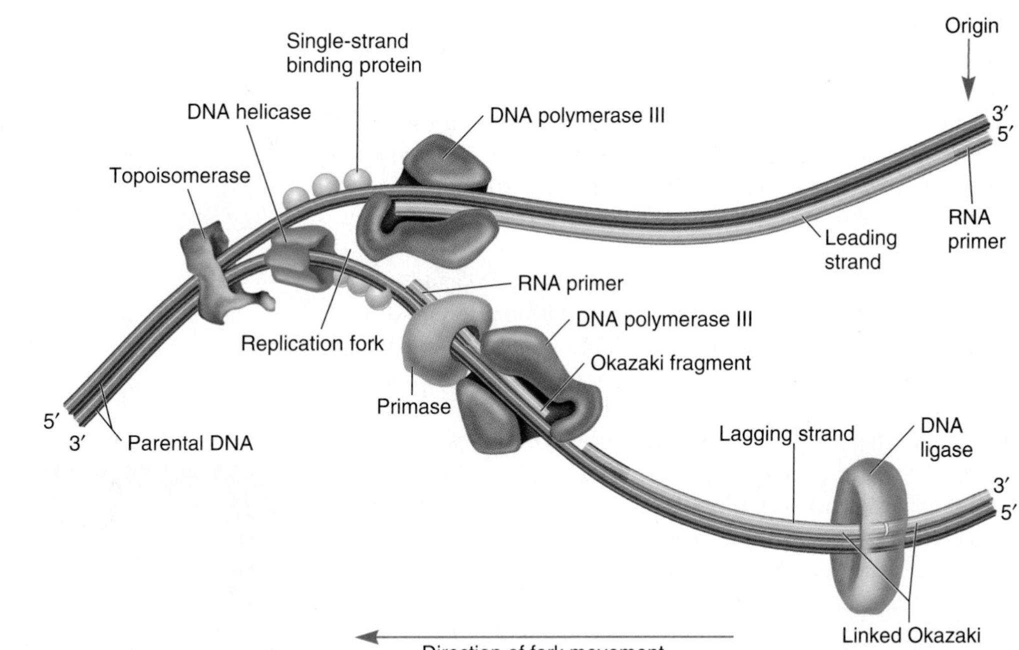

ONLINE ANIMATION

FIGURE 11.7 The proteins involved with DNA replication.

Note: The drawing of DNA polymerase III depicts the catalytic subunit that synthesizes DNA.

fork. As shown in Figure 11.7, an enzyme known as a **topoisomerase (type II),** also called **DNA gyrase,** travels in front of DNA helicase and alleviates positive supercoiling.

TABLE **11.1**

Proteins Involved in *E. coli* DNA Replication

Common Name	Function
DnaA protein	Binds to DnaA boxes within the origin to initiate DNA replication
DnaC protein	Aids DnaA in the recruitment of DNA helicase to the origin
DNA helicase (DnaB)	Separates double-stranded DNA
Topoisomerase	Removes positive supercoiling ahead of the replication fork
Single-strand binding protein	Binds to single-stranded DNA and prevents it from re-forming a double-stranded structure
Primase	Synthesizes short RNA primers
DNA polymerase III	Synthesizes DNA in the leading and lagging strands
DNA polymerase I	Removes RNA primers, fills in gaps with DNA
DNA ligase	Covalently attaches adjacent Okazaki fragments
Tus	Binds to ter sequences and prevents the advancement of the replication fork

After the two parental DNA strands have been separated and the supercoiling relaxed, they must be kept that way until the complementary daughter strands have been made. What prevents the DNA strands from coming back together? DNA replication requires **single-strand binding proteins** that bind to the strands of parental DNA and prevent them from re-forming a double helix. In this way, the bases within the parental strands are kept in an exposed condition that enables them to hydrogen bond with individual nucleotides.

The next event in DNA replication involves the synthesis of short strands of RNA (rather than DNA) called **RNA primers.** These strands of RNA are synthesized by the linkage of ribonucleotides via an enzyme known as **DNA primase,** or simply **primase.** This enzyme synthesizes short strands of RNA, typically 10 to 12 nucleotides in length. These short RNA strands start, or prime, the process of DNA replication. In the **leading strand,** a single primer is made at the origin of replication. In the **lagging strand,** multiple primers are made. As discussed later, the RNA primers are eventually removed.

A type of enzyme known as **DNA polymerase** is responsible for synthesizing the DNA of the leading and lagging strands. This enzyme catalyzes the formation of covalent bonds between adjacent nucleotides and thereby makes the new daughter strands. In *E. coli,* five distinct proteins function as DNA polymerases and are designated polymerase I, II, III, IV, and V. DNA polymerases I and III are involved in normal DNA replication, while DNA polymerases II, IV, and V play a role in DNA repair and the replication of damaged DNA.

DNA polymerase III is responsible for most of the DNA replication. It is a large enzyme consisting of 10 different subunits that play various roles in the DNA replication process (**Table 11.2**). The α subunit actually catalyzes the bond formation between adjacent nucleotides, while the remaining nine subunits fulfill other functions. The complex of all 10 subunits together is called DNA polymerase III holoenzyme. By comparison, DNA polymerase I is composed of a single subunit. Its role during DNA replication is to remove the RNA primers and fill in the vacant regions with DNA.

Though the various DNA polymerases in *E. coli* and other bacterial species vary in their subunit composition, several common structural features have emerged. The catalytic subunit of all DNA polymerases has a structure that resembles a human hand. As shown in **Figure 11.8**, the template DNA is threaded through the palm of the hand; the thumb and fingers are wrapped around the DNA. The incoming dNTPs enter the catalytic site, bind to the template strand according to the AT/GC rule, and then are covalently attached to the 3' end of the growing strand. DNA polymerase also contains a 3' exonuclease site that removes mismatched bases, as described later.

As researchers began to unravel the function of DNA polymerase, two features seemed unusual (**Figure 11.9**). DNA polymerase cannot begin DNA synthesis by linking together the first two individual nucleotides. Rather, this type of enzyme can elongate only a preexisting strand starting with an RNA primer or existing DNA strand (Figure 11.9a). A second unusual feature is the directionality of strand synthesis. DNA polymerase can attach nucleotides only in the 5' to 3' direction, not in the 3' to 5' direction (Figure 11.9b).

Due to these two unusual features, the synthesis of the leading and lagging strands shows distinctive differences (**Figure 11.10**). The synthesis of RNA primers by DNA primase allows DNA polymerase III to begin the synthesis of complementary daughter strands of DNA. DNA polymerase III catalyzes the attachment of nucleotides to the 3' end of each primer, in a 5' to 3' direction. In the leading strand, one RNA primer is made at the origin, and then DNA polymerase III can attach nucleotides in a 5' to 3' direction as it slides toward the opening of the replication fork. The synthesis of the leading strand is therefore continuous.

TABLE **11.2**

Subunit Composition of DNA Polymerase III Holoenzyme from *E. coli*

Subunit(s)	Function
α	Synthesizes DNA
ε	3' to 5' proofreading (removes mismatched nucleotides)
θ	Accessory protein that stimulates the proofreading function
β	Clamp protein, which allows DNA polymerase to slide along the DNA without falling off
τ, γ, δ, δ', ψ, and χ	Clamp loader complex, involved with helping the clamp protein bind to the DNA

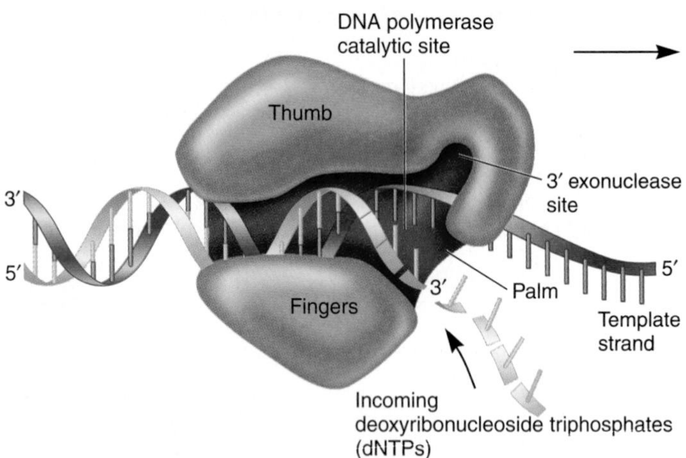

(a) Schematic side view of DNA polymerase III

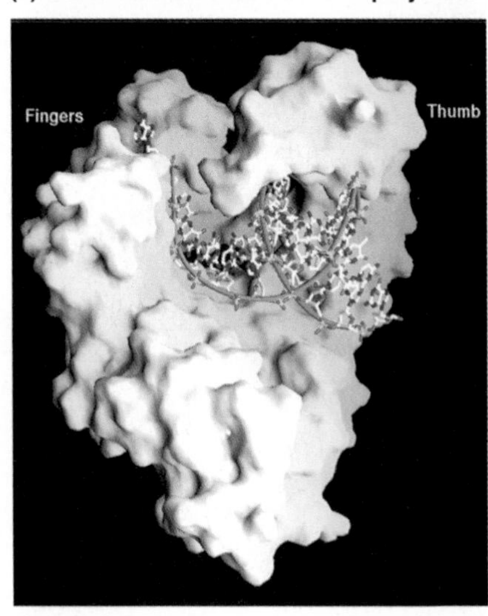

(b) Molecular model for DNA polymerase bound to DNA

(Reprinted by permission from Macmillan Publishers Ltd. *The Embo Journal.* Crystal structures of open and closed forms of binary and ternary complexes of the large fragment of *Thermus aquaticus* DNA polymerase I: structural basis for nucleotide incorporation. Ying Li et al. 17:24, 7514–7525, 1998.)

FIGURE 11.8 **The action of DNA polymerase.** (**a**) DNA polymerase slides along the template strand as it synthesizes a new strand by connecting deoxyribonucleoside triphosphates (dNTPs) in a 5' to 3' direction. The catalytic subunit of DNA polymerase resembles a hand that is wrapped around the template strand. In this regard, the movement of DNA polymerase along the template strand is similar to a hand that is sliding along a rope. (**b**) The molecular structure of DNA polymerase I from the bacterium *Thermus aquaticus.* This model shows a portion of DNA polymerase I that is bound to DNA. This molecular structure depicts a front view of DNA polymerase; part (a) is a schematic side view.

In the lagging strand, the synthesis of DNA also elongates in a 5' to 3' manner, but it does so in the direction away from the replication fork. In the lagging strand, RNA primers must repeatedly initiate the synthesis of short segments of DNA; thus, the synthesis has to be discontinuous. The length of these fragments in bacteria is typically 1,000 to 2,000 nucleotides. In eukaryotes, the fragments are shorter—100 to 200 nucleotides. Each fragment contains a short RNA primer at the 5' end, which is made by primase. The remainder of the fragment is a strand of DNA

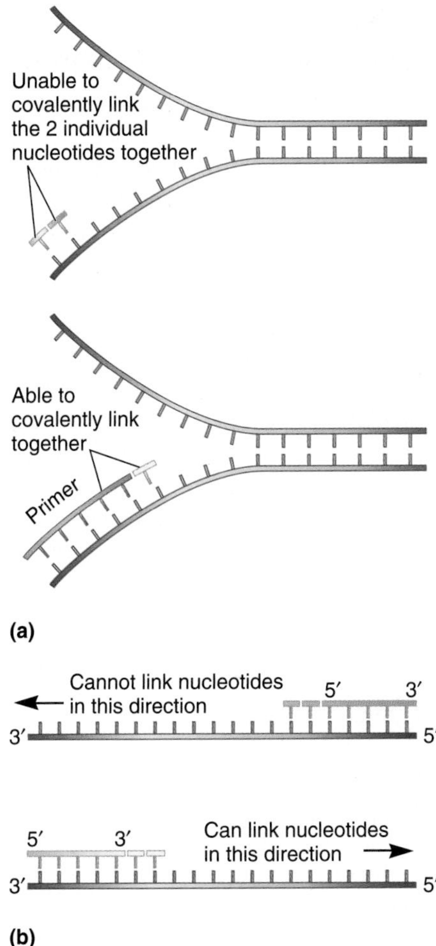

FIGURE 11.9 Unusual features of DNA polymerase function. (**a**) DNA polymerase can elongate a strand only from an RNA primer or existing DNA strand. (**b**) DNA polymerase can attach nucleotides only in a 5' to 3' direction. Note the template strand is in the opposite, 3' to 5', direction.

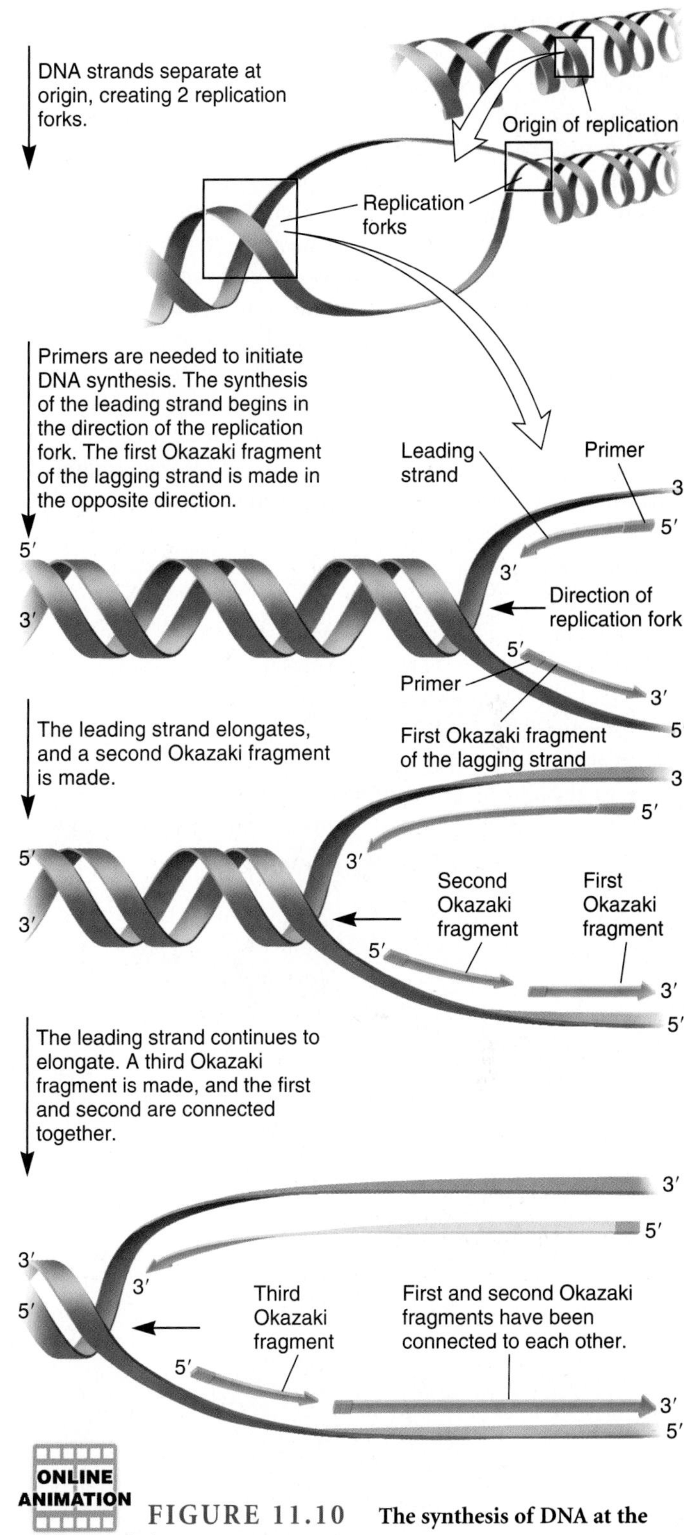

FIGURE 11.10 The synthesis of DNA at the replication fork.

made by DNA polymerase III. The DNA fragments made in this manner are known as **Okazaki fragments,** after Reiji and Tuneko Okazaki, who initially discovered them in the late 1960s.

To complete the synthesis of Okazaki fragments within the lagging strand, three additional events must occur: removal of the RNA primers, synthesis of DNA in the area where the primers have been removed, and the covalent attachment of adjacent fragments of DNA (see Figure 11.10 and refer back to Figure 11.7). In *E. coli*, the RNA primers are removed by the action of DNA polymerase I. This enzyme has a 5' to 3' exonuclease activity, which means that DNA polymerase I digests away the RNA primers in a 5' to 3' direction, leaving a vacant area. DNA polymerase I then fills in this region by synthesizing DNA there. It uses the 3' end of an adjacent Okazaki fragment as a primer. For example, in Figure 11.10, DNA polymerase I would remove the RNA primer from the first Okazaki fragment and then synthesize DNA in the vacant region by attaching nucleotides to the 3' end of the second Okazaki fragment. After the gap has been completely filled in, a covalent bond is still missing between the last nucleotide added by DNA polymerase I and the adjacent DNA strand that had been previously made by DNA polymerase III. An enzyme known as **DNA ligase** catalyzes a covalent bond between adjacent fragments to complete the replication process in the lagging strand (refer back to Figure 11.7). In *E. coli*, DNA ligase requires NAD^+ to carry out this reaction, whereas the DNA ligases found in archaea and eukaryotes require ATP.

The synthesis of the lagging strand was studied by the Okazakis using radiolabeled nucleotides. They incubated *E. coli* cells with radiolabeled thymidine for 15 seconds and then added an excess of nonlabeled thymidine. This is termed a **pulse/chase experiment** because the cells were given the radiolabeled compound for a brief period of time—a pulse—followed by an excess amount of unlabeled compound—a chase. They then isolated DNA from samples of cells at timed intervals after the pulse/chase. The DNA was denatured into single-stranded molecules, and the sizes of the radiolabeled DNA strands were determined by centrifugation. At quick time intervals, such as only a few seconds following the thymidine incubation, the fragments were found to be short, in the range of 1,000 to 2,000 nucleotides in length. At extended time intervals, the radiolabeled strands became much longer. At these later time points, the adjacent Okazaki fragments would have had enough time to link together.

DNA Polymerase III Is a Processive Enzyme That Uses Deoxyribonucleoside Triphosphates

Let's now turn our attention to other enzymatic features of DNA polymerase. As shown in **Figure 11.11**, DNA polymerases catalyze the covalent attachment between the phosphate in one nucleotide and the sugar in the previous nucleotide. The formation of this covalent (ester) bond requires an input of energy. Prior to bond formation, the nucleotide about to be attached to the growing strand is a deoxyribonucleoside triphosphate (dNTP). It contains three phosphate groups attached at the 5'–carbon atom of deoxyribose. The deoxyribonucleoside triphosphate first enters the catalytic site of DNA polymerase and binds to the template strand according to the AT/GC rule. Next, the 3'–OH group on the previous nucleotide reacts with the phosphate group adjacent to the sugar on the incoming nucleotide. The breakage of a covalent bond between two phosphates in deoxyribonucleoside triphosphate is a highly exergonic reaction that provides the energy to

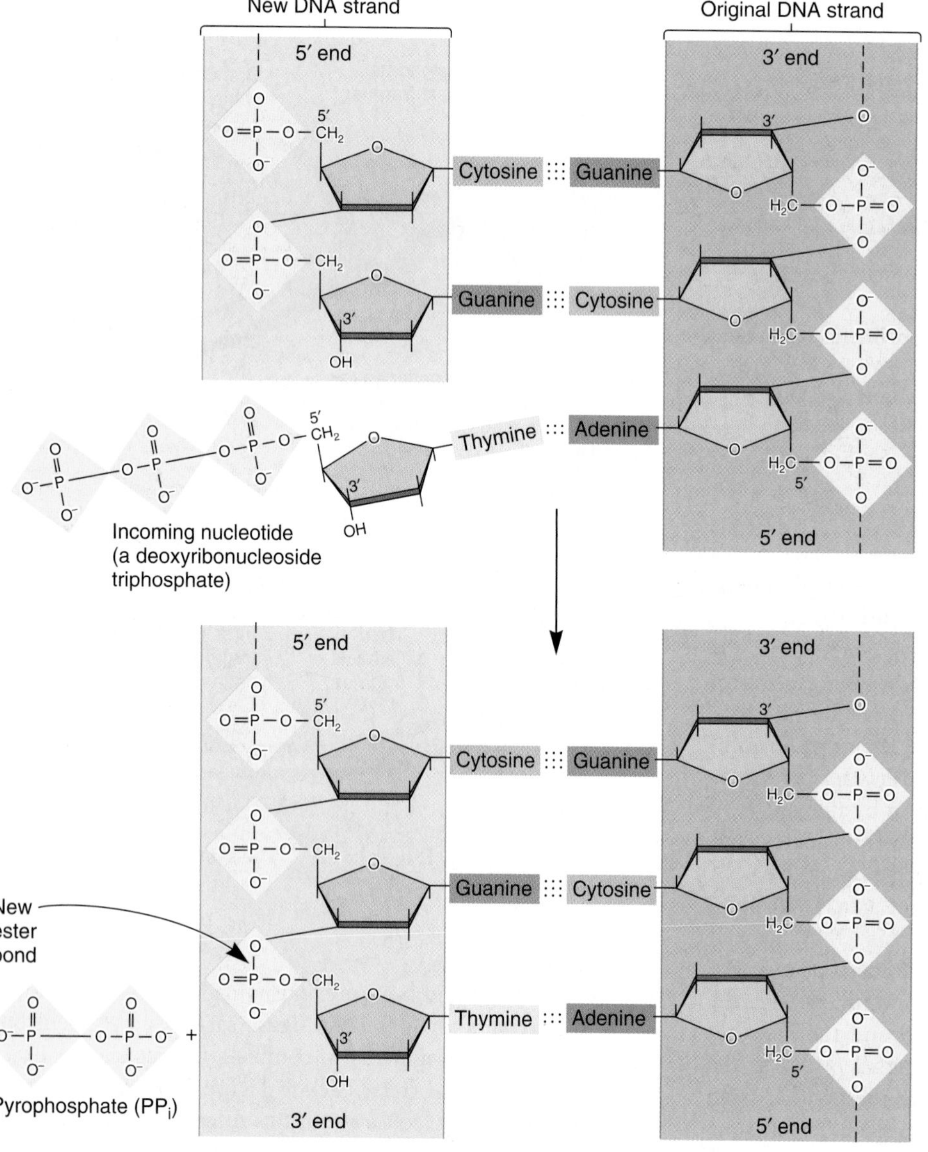

FIGURE 11.11 The enzymatic action of DNA polymerase. An incoming deoxyribouclesoside triphosphate (dNTP) is cleaved to form a nucleoside monophosphate and pyrophosphate. The energy released from this exergonic reaction allows the nucleoside monophosphate to form a covalent (ester) bond at the 3' end of the growing strand. This reaction is catalyzed by DNA polymerase. Pyrophosphate (PP$_i$) is released.

form a covalent (ester) bond between the sugar at the 3' end of the DNA strand and the phosphate of the incoming nucleotide. The formation of this covalent bond causes the newly made strand to grow in the 5' to 3' direction. As shown in Figure 11.11, pyrophosphate (PP_i) is released.

As noted in Chapter 9 (Figure 9.11), the term phosphodiester linkage (also called a phosphodiester bond) is used to describe the linkage between a phosphate and two sugar molecules. As its name implies, a phosphodiester linkage involves two ester bonds. In comparison, as a DNA strand grows, a single covalent (ester) bond is formed between adjacent nucleotides (see Figure 11.11). The other ester bond in the phosphodiester linkage—the bond between the 5'-oxygen and phosphorus—is already present in the incoming nucleotide.

DNA polymerase catalyzes the covalent attachment of nucleotides with great speed. In *E. coli*, DNA polymerase III attaches approximately 750 nucleotides per second! DNA polymerase III can catalyze the synthesis of the daughter strands so quickly because it is a **processive enzyme.** This means it does not dissociate from the growing strand after it has catalyzed the covalent joining of two nucleotides. Rather, as depicted in Figure 11.8a, it remains clamped to the DNA template strand and slides along the template as it catalyzes the synthesis of the daughter strand. The β (beta) subunit of the holoenzyme, also known as the clamp protein, promotes the association of the holoenzyme with the DNA as it glides along the template strand (refer back to Table 11.2). The β subunit forms a dimer in the shape of a ring; the hole of the ring is large enough to accommodate a double-stranded DNA molecule, and its width is about one turn of DNA. A complex of several subunits functions as a clamp loader that allows the DNA polymerase holoenzyme to initially clamp onto the DNA.

The effects of processivity are really quite remarkable. In the absence of the β subunit, DNA polymerase can synthesize DNA at a rate of approximately only 20 nucleotides per second. On average, it falls off the DNA template after about ten nucleotides have been linked together. By comparison, when the β subunit is present, as in the holoenzyme, the synthesis rate is approximately 750 nucleotides per second. In the leading strand, DNA polymerase III has been estimated to synthesize a segment of DNA that is over 500,000 nucleotides in length before it inadvertently falls off.

Replication Is Terminated When the Replication Forks Meet at the Termination Sequences

On the opposite side of the *E. coli* chromosome from *oriC* is a pair of **termination sequences** called ter sequences. A protein known as the termination utilization substance (Tus) binds to the ter sequences and stops the movement of the replication forks. As shown in **Figure 11.12**, one of the ter sequences designated T1 prevents the advancement of the fork moving left to right, but allows the movement of the other fork (see the inset to Figure 11.12). Alternatively, T2 prevents the advancement of the fork moving right to left, but allows the advancement of the other fork. In any given cell, only one ter sequence is required to stop the advancement of one replication fork, and then the other fork ends its synthesis of DNA when it reaches the halted replication fork. In other words, DNA replication ends when oppositely advancing forks meet, usually at T1 or T2. Finally, DNA ligase covalently links the two daughter strands, creating two circular, double-stranded molecules.

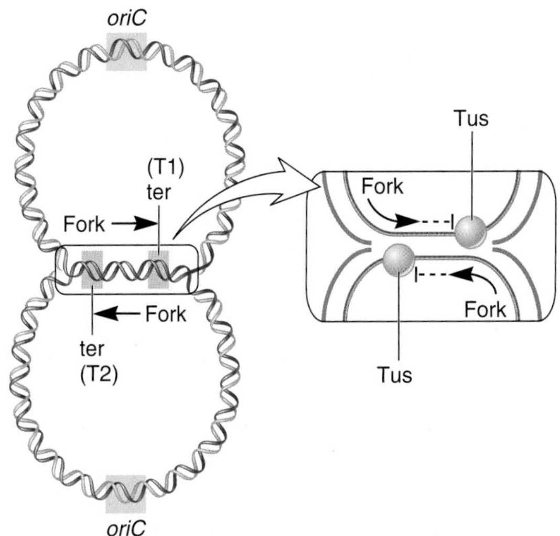

FIGURE 11.12 The termination of DNA replication. Two sites in the bacterial chromosome, shown with rectangles, are ter sequences designated T1 and T2. The T1 site prevents the further advancement of the fork moving left to right, while T2 prevents the advancement of the fork moving right to left. As shown in the inset, the binding of Tus prevents the replication forks from proceeding past the ter sequences in a particular direction.

After DNA replication is completed, one last problem may exist. DNA replication often results in two intertwined DNA molecules known as **catenanes** (**Figure 11.13**). Fortunately, catenanes are only transient structures in DNA replication. In *E. coli*, topoisomerase introduces a temporary break into the DNA strands and then rejoins them after the strands have become unlocked. This allows the catenanes to be separated into individual circular molecules.

Certain Enzymes of DNA Replication Bind to Each Other to Form a Complex

Figure 11.14 provides a more three-dimensional view of the DNA replication process. DNA helicase and primase are physically bound to each other to form a complex known as a **primosome.** This complex leads the way at the replication fork. The primosome tracks along the DNA, separating the parental strands and synthesizing RNA primers at regular intervals along the lagging strand. By acting within a complex, the actions of DNA helicase and primase can be better coordinated.

The primosome is physically associated with two DNA polymerase holoenzymes to form a **replisome.** As shown in Figure 11.14, two DNA polymerase III proteins act in concert to replicate the leading and lagging strands. The term **dimeric DNA polymerase** is used to describe two DNA polymerase holoenzymes that move as a unit toward the replication fork. For this to occur, the lagging strand is looped out with respect to the DNA polymerase that synthesizes the lagging strand. This loop allows

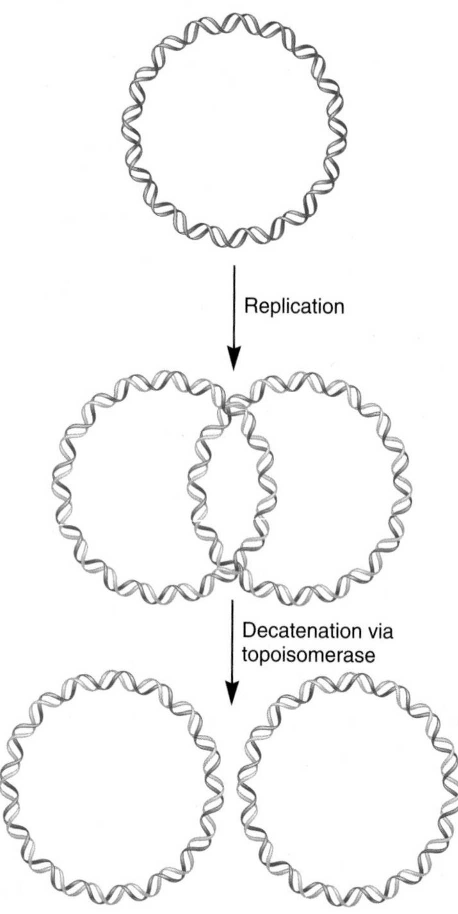

FIGURE 11.13 **Separation of catenanes.** DNA replication can result in two intertwined chromosomes called catenanes. These catenanes can be separated by the action of topoisomerase.

the lagging-strand polymerase to make DNA in a 5' to 3' direction yet move toward the opening of the replication fork. Interestingly, when this DNA polymerase reaches the end of an Okazaki fragment, it must be released from the template DNA and "hop" to the RNA primer that is closest to the fork. The clamp loader complex (see Table 11.2), which is part of DNA polymerase holoenzyme, then reloads the enzyme at the site where the next RNA primer has been made. Similarly, after primase synthesizes an RNA primer in the 5' to 3' direction, it must hop over the primer and synthesize the next primer closer to the replication fork.

The Fidelity of DNA Replication Is Ensured by Proofreading Mechanisms

With replication occurring so rapidly, one might imagine that mistakes could happen in which the wrong nucleotide could be incorporated into the growing daughter strand. Although mistakes can happen during DNA replication, they are extraordinarily rare. In the case of DNA synthesis via DNA polymerase III, only one mistake per 100 million nucleotides is made. Therefore, DNA synthesis occurs with a high degree of accuracy or **fidelity.**

Why is the fidelity so high? First, the hydrogen bonding between G and C or A and T is much more stable than between mismatched pairs. However, this stability accounts for only part of the fidelity, because mismatching due to stability considerations would account for one mistake per 1,000 nucleotides. Two additional characteristics of DNA polymerases contribute to the fidelity of DNA replication. The active site of DNA polymerase is such that it will preferentially catalyze the attachment of nucleotides when the correct bases are located in opposite strands. In other words, DNA polymerase is unlikely to catalyze bond formation between adjacent nucleotides if a mismatched base pair is formed.

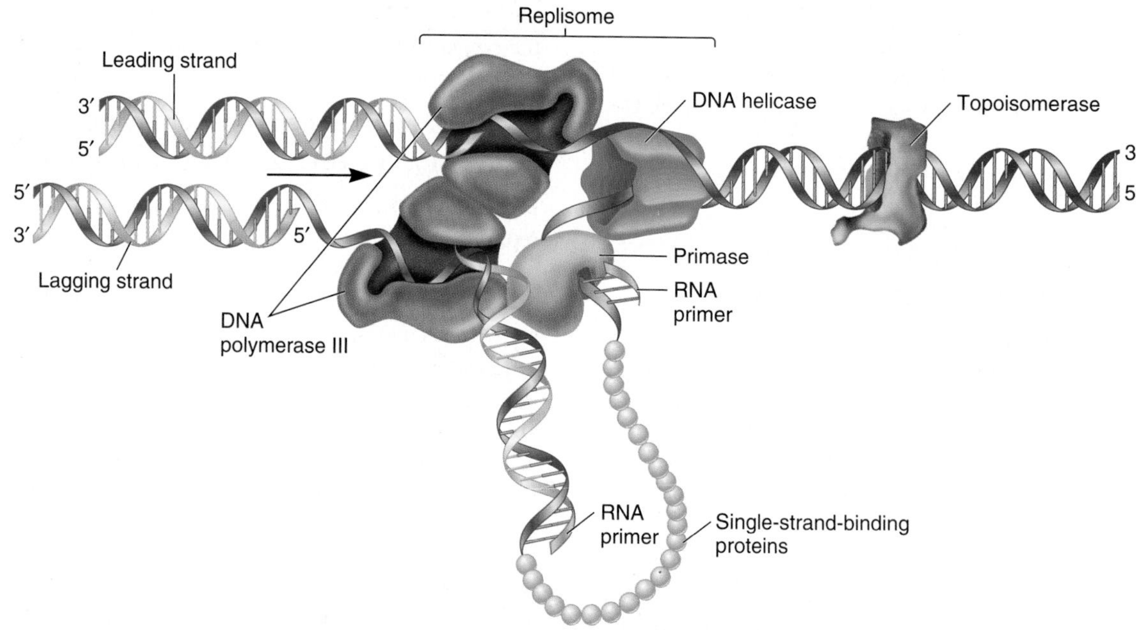

FIGURE 11.14 **A three-dimensional view of DNA replication.** DNA helicase and primase associate together to form a primosome. The primosome associates with two DNA polymerase enzymes to form a replisome.

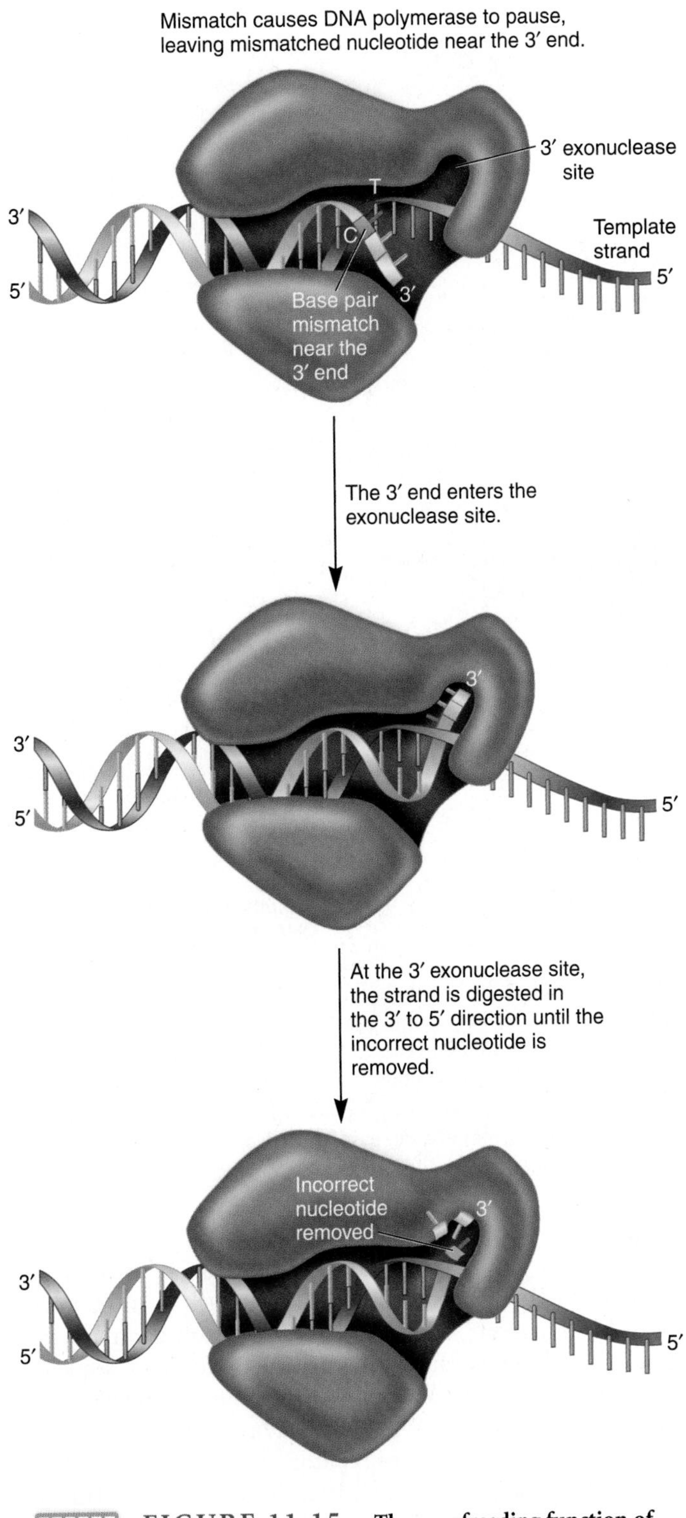

Mismatch causes DNA polymerase to pause, leaving mismatched nucleotide near the 3′ end.

3′ exonuclease site

Template strand

Base pair mismatch near the 3′ end

The 3′ end enters the exonuclease site.

At the 3′ exonuclease site, the strand is digested in the 3′ to 5′ direction until the incorrect nucleotide is removed.

Incorrect nucleotide removed

ONLINE ANIMATION

FIGURE 11.15 **The proofreading function of DNA polymerase.** When a base pair mismatch is found, the end of the newly made strand is shifted into the 3' exonuclease site. The DNA is digested in the 3' to 5' direction to release the incorrect nucleotide.

This phenomenon, called induced fit, decreases the error rate to a range of 1 in 100,000 to 1 million.

Another way that DNA polymerase decreases the error rate is by the enzymatic removal of mismatched nucleotides. As shown in Figure 11.15, DNA polymerase can identify a mismatched nucleotide and remove it from the daughter strand. This occurs by exonuclease cleavage of the bonds between adjacent nucleotides at the 3' end of the newly made strand. The ability to remove mismatched bases by this mechanism is called the **proofreading function** of DNA polymerase. Proofreading occurs by the removal of nucleotides in the 3' to 5' direction at the 3' exonuclease site. After the mismatched nucleotide is removed, DNA polymerase resumes DNA synthesis in the 5' to 3' direction.

Bacterial DNA Replication Is Coordinated with Cell Division

Bacterial cells can divide into two daughter cells at an amazing rate. Under optimal conditions, certain bacteria such as *E. coli* can divide every 20 to 30 minutes. DNA replication should take place only when a cell is about to divide. If DNA replication occurs too frequently, too many copies of the bacterial chromosome will be found in each cell. Alternatively, if DNA replication does not occur frequently enough, a daughter cell will be left without a chromosome. Therefore, cell division in bacterial cells must be coordinated with DNA replication.

Bacterial cells regulate the DNA replication process by controlling the initiation of replication at *oriC*. This control has been extensively studied in *E. coli*. In this bacterium, several different mechanisms may control DNA replication. In general, the regulation prevents the premature initiation of DNA replication at *oriC*.

After the initiation of DNA replication, DnaA protein hydrolyzes its ATP and therefore switches to an ADP-bound form. DnaA-ADP has a lower affinity for DnaA boxes and does not readily form a complex. This prevents premature initiation. In addition, the initiation of replication is controlled by the amount of the DnaA protein (Figure 11.16). As discussed previously, the DnaA protein binds to DnaA boxes within the origin of replication, causing the DNA helix to separate at the AT-rich region within the origin. To initiate DNA replication, the concentration of the DnaA protein must be high enough so that it can bind to all of the DnaA boxes and form a complex. Immediately following DNA replication, the number of DnaA boxes is double, so an insufficient amount of DnaA protein is available to initiate a second round of replication. Also, some of the DnaA protein may be rapidly degraded and some of it may be inactive because it becomes attached to other regions of chromosomal DNA and to the cell membrane during cell division. Because it takes time to accumulate newly made DnaA protein, DNA replication cannot occur until the daughter cells have had time to grow.

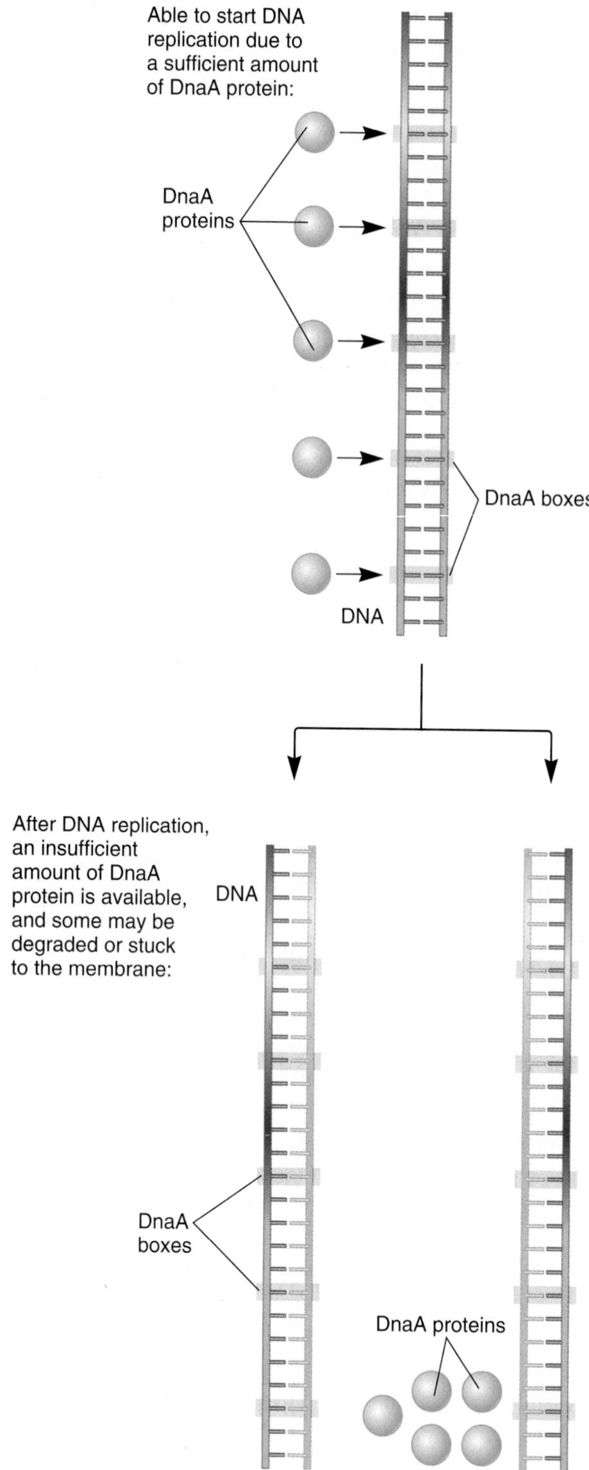

FIGURE 11.16 **The amount of DnaA protein provides a way to regulate DNA replication.** To begin replication, enough DnaA protein must be present to bind to all of the DnaA boxes. Immediately after DNA replication, insufficient DnaA protein is available to reinitiate a second (premature) round of DNA replication at the two origins of replication. This is because twice as many DnaA boxes are found after DNA replication and because some DnaA proteins may be degraded or stuck to other chromosomal sites and to the cell membrane.

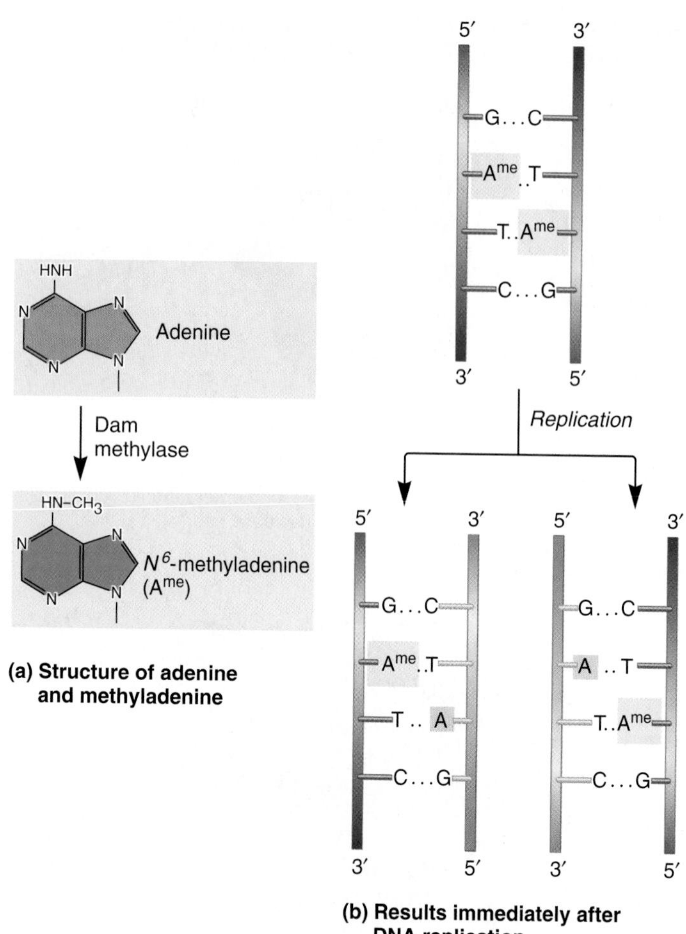

FIGURE 11.17 **Methylation of GATC sites in *oriC*.** (**a**) The action of Dam, which covalently attaches a methyl group to adenine to form methyladenine. (**b**) Prior to DNA replication, the action of Dam causes both adenines within the GATC sites to be methylated. After DNA replication, only the adenines in the original strands are methylated. Several minutes will pass before Dam will methylate these unmethylated adenines.

Another way to regulate DNA replication involves the GATC methylation sites within *oriC*. These sites can be methylated by an enzyme known as DNA adenine methyltransferase (Dam). The Dam enzyme recognizes the 5'–GATC–3' sequence, binds there, and attaches a methyl group onto the adenine base, forming methyladenine (**Figure 11.17a**). DNA methylation within *oriC* helps regulate the replication process. Prior to DNA replication, these sites are methylated in both strands. This full methylation of the 5'–GATC–3' sites facilitates the initiation of DNA replication at the origin. Following DNA replication, the newly made strands are not methylated, because adenine rather than methyladenine is incorporated into the daughter strands (**Figure 11.17b**). The initiation of DNA replication at the origin does not readily occur until after it has become fully methylated. Because it takes several minutes for Dam to methylate the 5'–GATC–3' sequences within this region, DNA replication will not occur again too quickly.

EXPERIMENT 11B

DNA Replication Can Be Studied In Vitro

Much of our understanding of bacterial DNA replication has come from thousands of experiments in which DNA replication has been studied in vitro. This approach was pioneered by Arthur Kornberg in the 1950s, who received a Nobel Prize for his efforts in 1959.

Figure 11.18 describes Kornberg's approach to monitor DNA replication in vitro. In this experiment, an extract of proteins from *E. coli* was used. Although we will not consider the procedures for purifying replication proteins, an alternative approach is to purify individual proteins from the extract and study their functions individually. In either case, the proteins would be mixed with template DNA and radiolabeled nucleotides. Kornberg correctly hypothesized that deoxyribonucleoside triphosphates (dNTPs) are the precursors for DNA synthesis. Also, he knew that deoxyribonucleoside triphosphates are soluble in an acidic solution, whereas long strands of DNA will precipitate out of solution at an acidic pH. This precipitation event provides a method to separate nucleotides—in this case, deoxyribonucleoside triphosphates—from strands of DNA. Therefore, after the proteins, template DNA, and nucleotides were incubated for a sufficient time to allow the synthesis of new strands, step 3 of this procedure involved the addition of perchloric acid. This step precipitated strands of DNA, which were then separated from the radiolabeled nucleotides via centrifugation. Newly made strands of DNA, which were radiolabeled, sediment to the pellet, while radiolabeled nucleotides that had not been incorporated into new strands remained in the supernatant.

THE HYPOTHESIS

DNA synthesis can occur in vitro if all the necessary components are present.

TESTING THE HYPOTHESIS — FIGURE 11.18 In vitro synthesis of DNA strands.

Starting material: An extract of proteins from *E. coli.*

1. Mix together the extract of *E. coli* proteins, template DNA that is not radiolabeled, and ^{32}P-radiolabeled deoxyribonucleoside triphosphates. This is expected to be a complete system that contains everything necessary for DNA synthesis. As a control, a second sample is made in which the template DNA was omitted from the mixture.

2. Incubate the mixture for 30 minutes at 37°C.

3. Add perchloric acid to precipitate DNA. (It does not precipitate free nucleotides.)

4. Centrifuge the tube.
   Note: The radiolabeled deoxyribonucleoside triphosphates that have not been incorporated into DNA will remain in the supernatant.

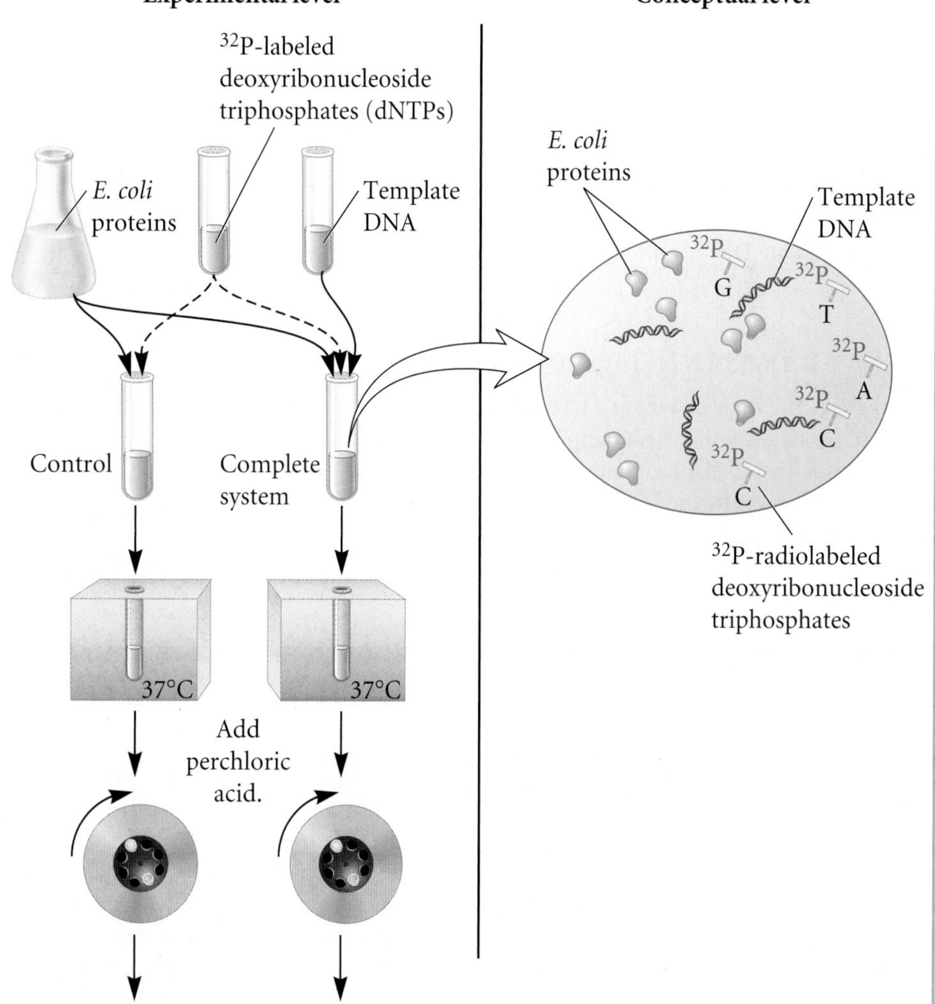

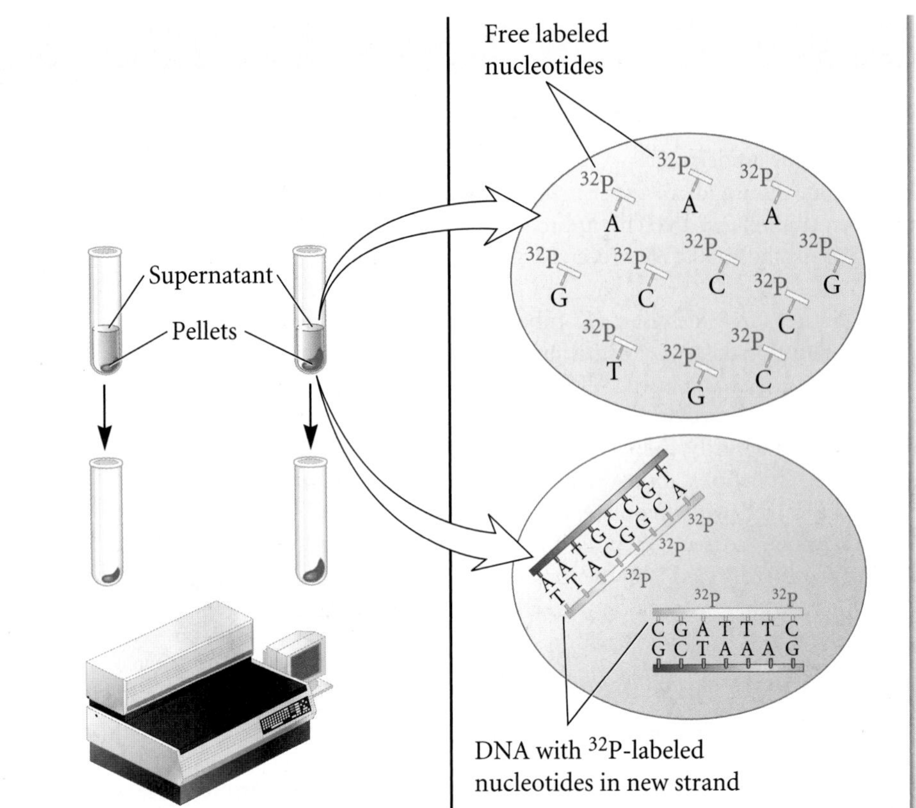

5. Collect the pellet, which contains precipitated DNA and proteins. (The control pellet is not expected to contain DNA.)

6. Count the amount of radioactivity in the pellet using a scintillation counter. (See the Appendix.)

THE DATA

Conditions	*Amount of Radiolabeled DNA**
Complete system	3,300
Control (template DNA omitted)	0

*Calculated in picomoles of ^{32}P-labeled DNA.

INTERPRETING THE DATA

As shown in the data of Figure 11.18, when the *E. coli* proteins were mixed with nonlabeled template DNA and radiolabeled deoxyribonucleoside triphosphates, an acid-precipitable, radiolabeled product was formed. This product was newly synthesized DNA strands. As a control, if nonradiolabeled template DNA was omitted from the assay, no radiolabeled DNA was made. This is the expected result, because the template DNA is necessary to make new daughter strands. Taken together, these results indicate that this technique can be used to measure the synthesis of DNA in vitro.

The in vitro approach has provided the foundation to study the replication process at the molecular level. A common experimental strategy is to purify proteins from cell extracts and to determine their roles in the replication process. In other words, purified proteins, such as those described in Table 11.1, can be mixed with nucleotides, template DNA, and other substances in a test tube to determine if the synthesis of new DNA strands occurs. This approach still continues, particularly as we try to understand the added complexities of eukaryotic DNA replication.

A self-help quiz involving this experiment can be found at the Online Learning Center.

The Isolation of Mutants Has Been Instrumental to Our Understanding of DNA Replication

In the previous experiment, we considered an experimental strategy to study DNA synthesis in vitro. In his early experiments, Arthur Kornberg used crude extracts containing *E. coli* proteins and monitored their ability to synthesize DNA. In such extracts, the predominant polymerase enzyme is DNA polymerase I. Surprisingly, its activity is so high that it is nearly impossible to detect the activities of the other DNA polymerases. For this reason, researchers in the 1950s and 1960s thought that DNA polymerase I was the only enzyme responsible for DNA replication. This situation dramatically changed as a result of mutant isolation.

In 1969, Paula DeLucia and John Cairns isolated a mutant in which DNA polymerase I lacked its 5' to 3' polymerase function but retained its 5' to 3' exonuclease function, which is needed to remove RNA primers. This mutant was identified by screening thousands of bacterial colonies that had been subjected to mutagens—agents that cause mutations. This result indicated that the DNA-synthesizing function of DNA polymerase I is not absolutely required for bacteria to replicate their DNA, because the strain harboring this mutation could grow normally. How is this possible? The researchers concluded that *E. coli* must have other DNA polymerases. Therefore, DeLucia and Cairns set out to find these seemingly elusive enzymes.

The isolation of mutants was one way that helped researchers identify additional DNA polymerase enzymes, namely DNA polymerase II and III. In addition, mutant isolation played a key role in the identification of other proteins needed to replicate the leading and lagging strands, as well as proteins that recognize the origin of replication and the ter sites. Because DNA replication is vital for cell division, most mutations that block DNA replication would be lethal to a growing population of bacterial cells. For this reason, if researchers want to identify loss-of-function mutations in vital genes, they must screen for **conditional mutants.** A type of conditional mutant is a **temperature-sensitive (ts) mutant.** In the case of a vital gene, an organism harboring a temperature-sensitive mutation can survive at the permissive temperature but not at the nonpermissive temperature. For example, a ts mutant might survive and grow at 30°C (the permissive temperature) but fail to grow at 42°C (the nonpermissive temperature). The higher temperature inactivates the function of the protein encoded by the mutant gene.

Figure 11.19 shows a general strategy for the isolation of ts mutants. Researchers expose bacterial cells to a mutagen that increases the likelihood of mutations. The mutagenized cells are plated on growth media and incubated at the permissive temperature. The colonies are then replica plated onto two plates, one incubated at the permissive temperature and one at the nonpermissive temperature. As seen here, this enables researchers to identify ts mutations that are lethal at the nonpermissive temperature.

With regard to the study of DNA replication, researchers analyzed a large number of ts mutants to discover if any of them had a defect in DNA replication. For example, one could expose a ts mutant to radiolabeled thymine (a base that is incorporated into DNA), shift to the nonpermissive temperature, and determine if a mutant strain could make radiolabeled DNA, using procedures that are similar to those described in Figure 11.18. Because *E. coli* has many vital genes not involved with DNA replication, only a small subset of ts mutants would be expected to have mutations in genes that encode proteins that are critical to the replication process. Therefore, researchers had to screen many thousands of ts mutants to identify the few involved in DNA replication. This approach is sometimes called a "brute force" genetic screen.

Table 11.3 summarizes some of the genes that were identified using this type of strategy. The genes were originally designated with the name dna, followed by a capital letter that generally refers to the order in which they were discovered. When shifted to the nonpermissive temperature, certain mutants showed a rapid arrest of DNA synthesis. These so-called rapid-stop mutations inactivated genes that encode enzymes needed for DNA replication. By comparison, other mutants were able to

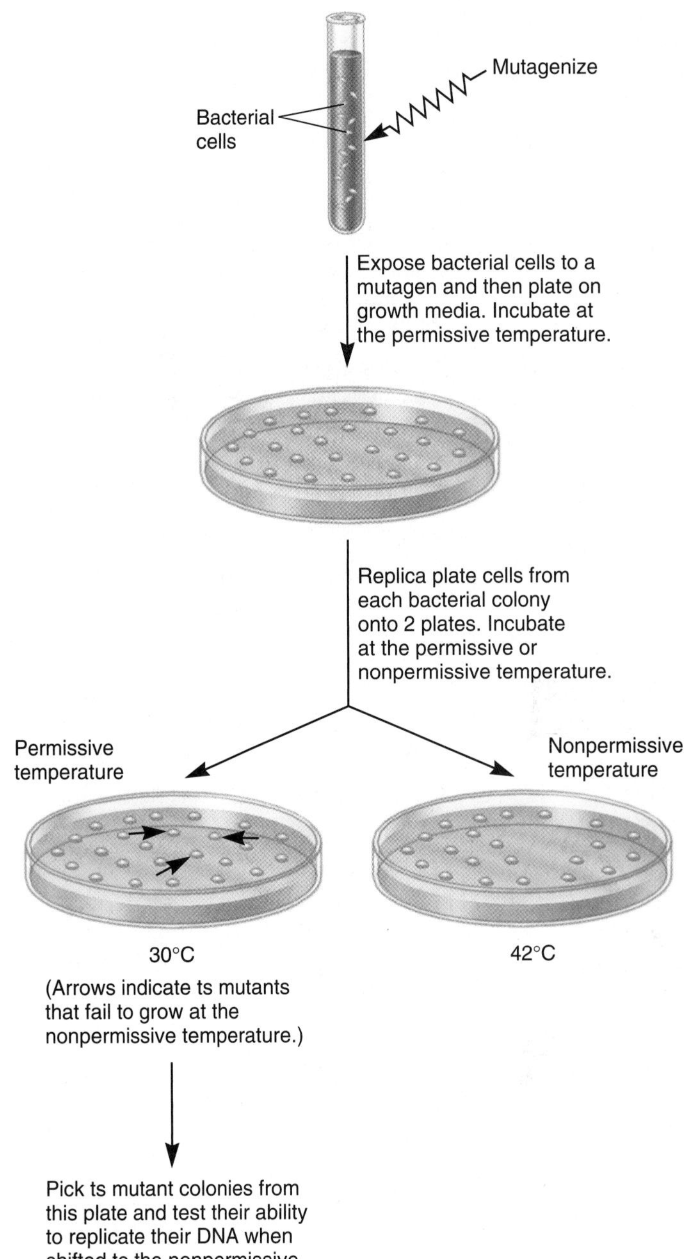

FIGURE 11.19 A strategy to identify ts mutations in vital genes. In this approach, bacteria are mutagenized, which increases the likelihood of mutation, and then grown at the permissive temperature. Colonies are then replica plated and grown at both the permissive and nonpermissive temperatures. (Note: The procedure of replica plating is shown in Chapter 16, Figure 16.7.) Ts mutants fail to grow at the nonpermissive temperature. The appropriate colonies can be picked from the plates, grown at the permissive temperature, and analyzed to see if DNA replication is altered at the nonpermissive temperature.

complete their current round of replication but could not start another round of replication. These slow-stop mutants involved genes that encode proteins needed for the initiation of replication at the origin. In later studies, the proteins encoded by these

TABLE 11.3

Examples of ts Mutants Involved in DNA Replication in *E. coli*

Gene Name	Protein Function
Rapid-Stop Mutants	
dnaE	α subunit of DNA polymerase III, synthesizes DNA
dnaX	τ subunit of DNA polymerase III, promotes the dimerization of two DNA polymerase III proteins together at the replication fork and stimulates DNA helicase
dnaN	β subunit of DNA polymerase III, functions as a clamp protein that makes DNA polymerase a processive enzyme
dnaZ	γ subunit of DNA polymerase III, helps the β subunit bind to the DNA
dnaG	Primase, needed to make RNA primers
dnaB	Helicase, needed to unwind the DNA strands during replication
Slow-Stop Mutants	
dnaA	DnaA protein that recognizes the DnaA boxes at the origin
dnaC	DnaC protein that recruits DNA helicase to the origin

genes were purified, and their functions were studied in vitro. This work contributed greatly to our modern understanding of DNA replication at the molecular level.

11.3 EUKARYOTIC DNA REPLICATION

Eukaryotic DNA replication is not as well understood as bacterial replication. Much research has been carried out on a variety of experimental organisms, particularly yeast and mammalian cells. Many of these studies have found extensive similarities between the general features of DNA replication in prokaryotes and eukaryotes. For example, DNA helicases, topoisomerases, single-strand binding proteins, primases, DNA polymerases, and DNA ligases—the types of bacterial enzymes described in Table 11.1—have also been identified in eukaryotes. Nevertheless, at the molecular level, eukaryotic DNA replication appears to be substantially more complex. These additional intricacies of eukaryotic DNA replication are related to several features of eukaryotic cells. In particular, eukaryotic cells have larger, linear chromosomes, the chromatin is tightly packed within nucleosomes, and cell cycle regulation is much more complicated. This section will emphasize some of the unique features of eukaryotic DNA replication.

Initiation Occurs at Multiple Origins of Replication on Linear Eukaryotic Chromosomes

Because eukaryotes have long, linear chromosomes, the chromosomes require multiple origins of replication so that the DNA can be replicated in a reasonable length of time. In 1968, Joel Huberman and Arthur Riggs provided evidence for multiple origins of replication by adding a radiolabeled nucleoside (^{3}H-thymidine) to a culture of actively dividing cells, followed by a chase with nonlabeled thymidine. The radiolabeled thymidine was taken up by the cells and incorporated into their newly made DNA strands for a brief period. The chromosomes were then isolated from the cells and subjected to autoradiography. As seen in **Figure 11.20**, radiolabeled segments were interspersed among nonlabeled segments. This result is consistent with the hypothesis that eukaryotic chromosomes contain multiple origins of replication.

As shown schematically in **Figure 11.21a**, DNA replication proceeds bidirectionally from many origins of replication during S phase of the cell cycle. The multiple replication forks eventually make contact with each other to complete the replication process. **Figure 11.21b** shows a region of eukaryotic DNA in the process of replication. As DNA replication radiates bidirectionally from each origin, regions are formed that contain two double helices. These regions are punctuated by other regions that have not yet replicated and consist of one double helix.

The molecular features of eukaryotic origins of replication may have some similarities to the origins found in bacteria. At the molecular level, eukaryotic origins of replication have been extensively studied in the yeast *Saccharomyces cerevisiae*. In this organism, several replication origins have been identified and sequenced. They have been named **ARS elements** (for Autonomously Replicating Sequence). ARS elements, which are about 50 bp in length, are necessary to initiate chromosome replication. ARS elements have unique features of their DNA sequences. First, they contain a higher percentage of A and T bases than the rest of the chromosomal DNA. In addition, they contain a copy of the ARS consensus sequence (ACS), ATTTAT(A or G)TTTA, along with additional elements that enhance origin function. This arrangement is similar to bacterial origins. In *S. cerevisiae*, origins of replication are determined primarily by their DNA sequences.

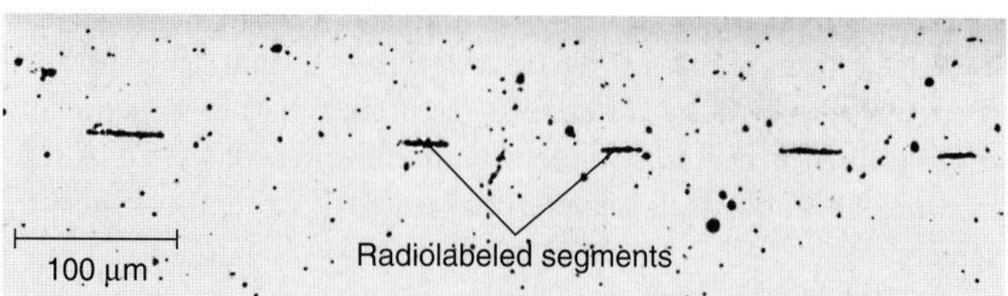

FIGURE 11.20 **Evidence for multiple origins of replication in eukaryotic chromosomes.** In this experiment, cells were given a pulse/chase of ^{3}H-thymidine and unlabeled thymidine. The chromosomes were isolated and subjected to autoradiography. In this micrograph, radiolabeled segments were interspersed among nonlabeled segments, indicating that eukaryotic chromosomes contain multiple origins of replication.

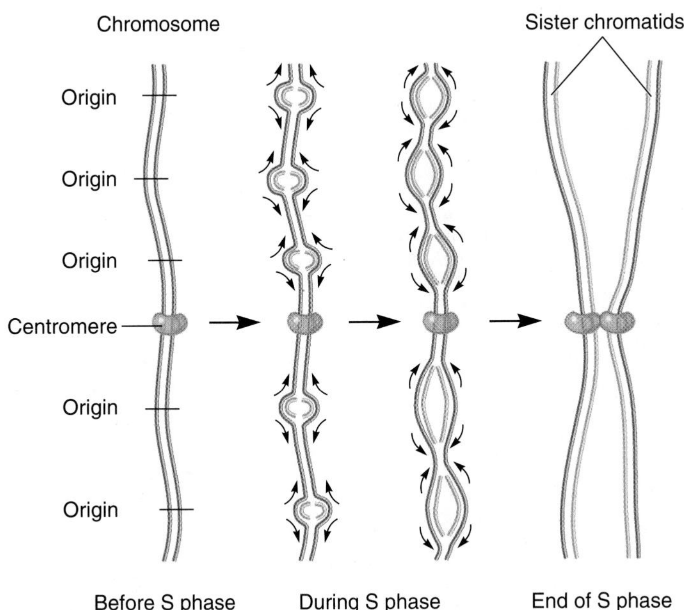

(a) DNA replication from multiple origins of replication

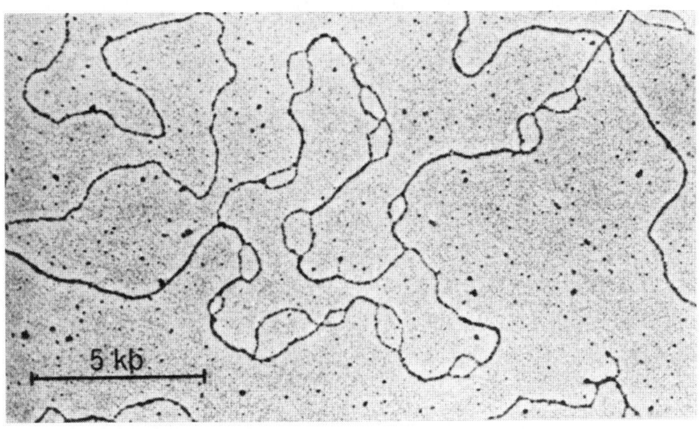

(b) A micrograph of a replicating eukaryotic chromosome

FIGURE 11.21 **The replication of eukaryotic chromosomes.** (**a**) At the beginning of synthesis (S) phase of the cell cycle, eukaryotic chromosome replication begins from multiple origins of replication. As S phase continues, the replication forks move bidirectionally to replicate the DNA. By the end of S phase, all of the replication forks have merged. The net result is two sister chromatids attached to each other at the centromere. (**b**) A micrograph of a replicating chromosome in a eukaryotic species, showing DNA replication radiating bidirectionally from each origin.

In animals, the critical features that define origins of replication are not completely understood. In some cases, origins are not determined by particular DNA sequences but instead occur at specific sites along a chromosome due to chromatin structure and protein modifications.

DNA replication in eukaryotes begins with the assembly of a **prereplication complex (preRC)** consisting of at least 14 different proteins. Part of the preRC is a group of six proteins called the **origin recognition complex (ORC)** that acts as the initiator of eukaryotic DNA replication. ORC was originally identified in yeast as a protein complex that binds directly to ARS elements. DNA replication at the origin begins with the binding of ORC, which usually occurs during G_1 phase. Other proteins of the preRC then bind, including a group of proteins called **MCM helicase.**[1] The binding of MCM helicase at the origin completes a process called **DNA replication licensing;** only those origins with MCM helicase can initiate DNA synthesis. During S phase, DNA synthesis begins when preRCs are acted upon by at least 22 additional proteins that activate MCM helicase and assemble two divergent replication forks at each replication origin. An important role of these additional proteins is to carefully regulate the initiation of DNA replication so that it happens at the correct time during the cell cycle and occurs only once during the cell cycle. The precise roles of these proteins are under active research investigation.

Eukaryotes Contain Several Different DNA Polymerases

Eukaryotes have many types of DNA polymerases. For example, mammalian cells have well over a dozen different DNA polymerases (**Table 11.4**). Four of these, designated α (alpha), δ (delta), ε (epsilon), and γ (gamma), have the primary function of replicating DNA. DNA polymerase γ functions in the mitochondria to replicate mitochondrial DNA, whereas α, δ, and ε are involved with DNA replication in the cell nucleus during S phase. In the nucleus, DNA polymerase α is the only eukaryotic polymerase

TABLE 11.4

Eukaryotic DNA Polymerases

Polymerase Types*	Function
α, δ, ε	Replication of nondamaged DNA in the cell nucleus during S phase
γ	Replication of mitochondrial DNA
η, κ, ι, ξ (lesion-replicating polymerases)	Replication of damaged DNA
α, β, δ, ε, σ, λ, μ, φ, θ, η	DNA repair or other functions†

*The designations are those of mammalian enzymes.
†Many DNA polymerases have dual functions. For example, DNA polymerases α, δ, and ε are involved in the replication of normal DNA and also play a role in DNA repair. In cells of the immune system, certain genes that encode antibodies (i.e., immunoglobulin genes) undergo a phenomenon known as hypermutation. This increases the variation in the kinds of antibodies the cells can make. Certain polymerases in this list, such as η, may play a role in hypermutation of immunoglobulin genes. DNA polymerase σ may play a role in sister chromatid cohesion, a topic discussed in Chapter 10.

[1] MCM is an acronym for minichromosome maintenance. The genes encoding MCM proteins were originally identified in mutant yeast strains that are defective in the maintenance of minichromosomes in the cell. MCM proteins have since been shown to play a role in DNA replication.

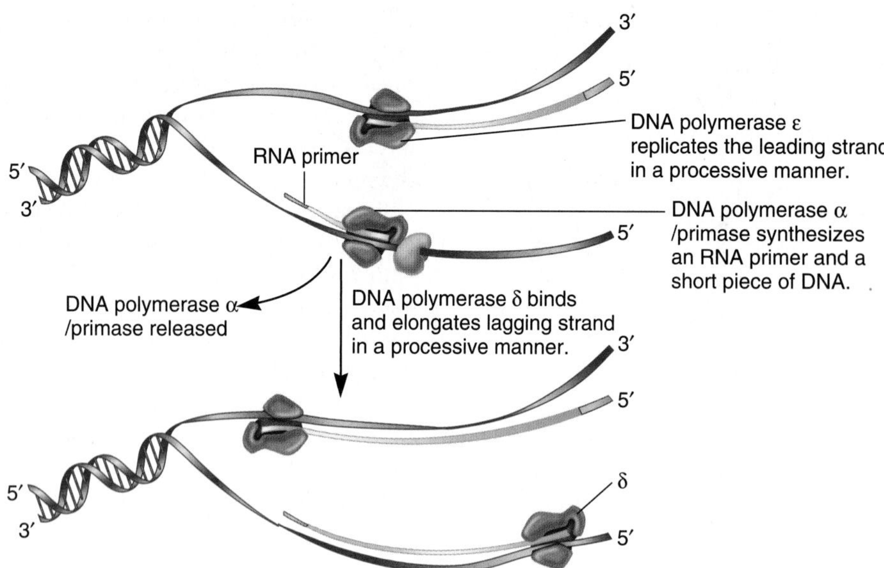

FIGURE 11.22 **An example of a DNA polymerase switch.** After DNA polymerase α/primase has made an RNA primer and short segment of DNA, this complex is released, and a processive polymerase, such as DNA polymerase δ, attaches to the template strand and rapidly synthesizes DNA. A similar mechanism could switch the α/primase complex for ε in the leading strand.

that associates with primase. The functional role of the DNA polymerase α/primase complex is to synthesize a short RNA-DNA primer of approximately 10 RNA nucleotides followed by 20 to 30 DNA nucleotides. This short RNA-DNA strand is then used by DNA polymerase δ or ε for the processive elongation of the leading and lagging strands (Figure 11.22). The exchange of DNA polymerase α for δ or ε is called a **polymerase switch.** It occurs only after the RNA-DNA primer has been made by the complex of DNA polymerase α and primase. The relative importance of DNA polymerase δ versus ε is not completely understood. Some studies have suggested that DNA polymerase ε may be involved with leading-strand synthesis, while DNA polymerase δ has a greater role in lagging-strand synthesis. However, this hypothesis will require further experimentation.

What are the functions of the other DNA polymerases? Several of them also play an important role in DNA repair, a topic that will be examined in Chapter 16. DNA polymerase β, which has been studied for several decades, is not involved in the replication of normal DNA, but plays an important role in removing incorrect bases from damaged DNA. More recently, several additional DNA polymerases have been identified. While their precise roles have not been elucidated, many of these are in a category called **lesion-replicating polymerases.** When DNA polymerase α, δ, and ε encounter abnormalities in DNA structure, such as abnormal bases or cross-links, they may be unable to replicate over the aberration. When this occurs, lesion-replicating polymerases are attracted to the damaged DNA and have special properties that enable them to synthesize a complementary strand over the abnormal region. Each type of lesion-replicating polymerase may be able to replicate over a different kind of DNA damage.

The Ends of Eukaryotic Chromosomes Are Replicated by Telomerase

Linear eukaryotic chromosomes contain **telomeres** at both ends. The term telomere refers to the complex of telomeric sequences within the DNA and the special proteins that are bound to these sequences. Telomeric sequences consist of a moderately repetitive tandem array and a 3' overhang region that is 12 to 16 nucleotides in length (Figure 11.23).

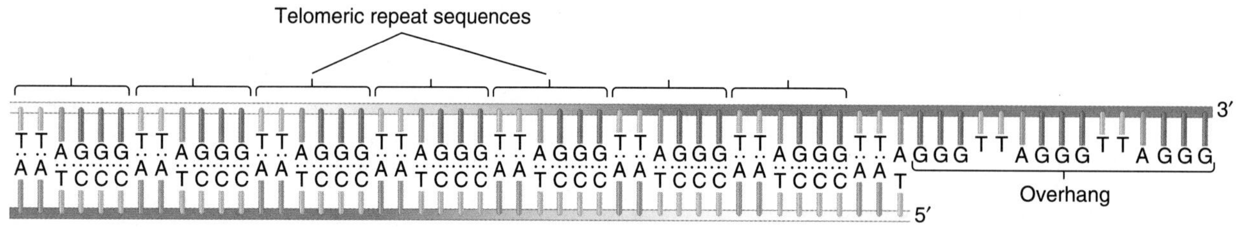

ONLINE ANIMATION FIGURE 11.23 **General structure of telomeric sequences.** The telomere DNA consists of a tandemly repeated sequence and a 12- to 16-nucleotide overhang.

The tandem array that occurs within the telomere has been studied in a wide variety of eukaryotic organisms. A common feature is that the telomeric sequence contains several guanine nucleotides and often many thymine nucleotides (Table 11.5). Depending on the species and the cell type, this sequence can be tandemly repeated up to several hundred times in the telomere region.

One reason why telomeric repeat sequences are needed is because DNA polymerase is unable to replicate the 3' ends of DNA strands. Why is DNA polymerase unable to replicate this region? The answer lies in the two unusual enzymatic features of this enzyme. As discussed previously, DNA polymerase synthesizes DNA only in a 5' to 3' direction, and it cannot link together the first two individual nucleotides; it can elongate only preexisting strands. These two features of DNA polymerase function pose a problem at the 3' ends of linear chromosomes. As shown in Figure 11.24, the 3' end of a DNA strand cannot be replicated by DNA polymerase because a primer cannot be made upstream from this point. Therefore, if this problem were not solved, the chromosome would become progressively shorter with each round of DNA replication.

To prevent the loss of genetic information due to chromosome shortening, additional DNA sequences are attached to the ends of telomeres. In 1984, Carol Greider and Elizabeth Blackburn discovered an enxyme called **telomerase** that prevents chromosome shortening. It recognizes the sequences at the ends of eukaryotic chromosomes and synthesizes additional repeats of telomeric sequences. Figure 11.25 shows the interesting mechanism

TABLE **11.5**

Telomeric Repeat Sequences Within Selected Organisms

Group	Example	Telomeric Repeat Sequence
Mammals	Humans	TTAGGG
Slime molds	*Physarum, Didymium*	TTAGGG
	Dictyostelium	$AG_{(1-8)}$
Filamentous fungi	*Neurospora*	TTAGGG
Budding yeast	*Saccharomyces cerevisiae*	$TG_{(1-3)}$
Ciliates	*Tetrahymena*	TTGGGG
	Paramecium	TTGGG(T/G)
	Euplotes	TTTTGGGG
Higher plants	*Arabidopsis*	TTTAGGG

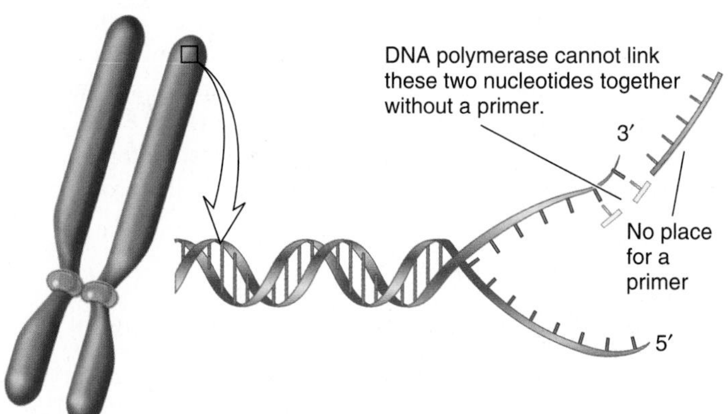

FIGURE 11.24 **The replication problem at the ends of linear chromosomes.** DNA polymerase cannot synthesize a DNA strand that is complementary to the 3' end because a primer cannot be made upstream from this site.

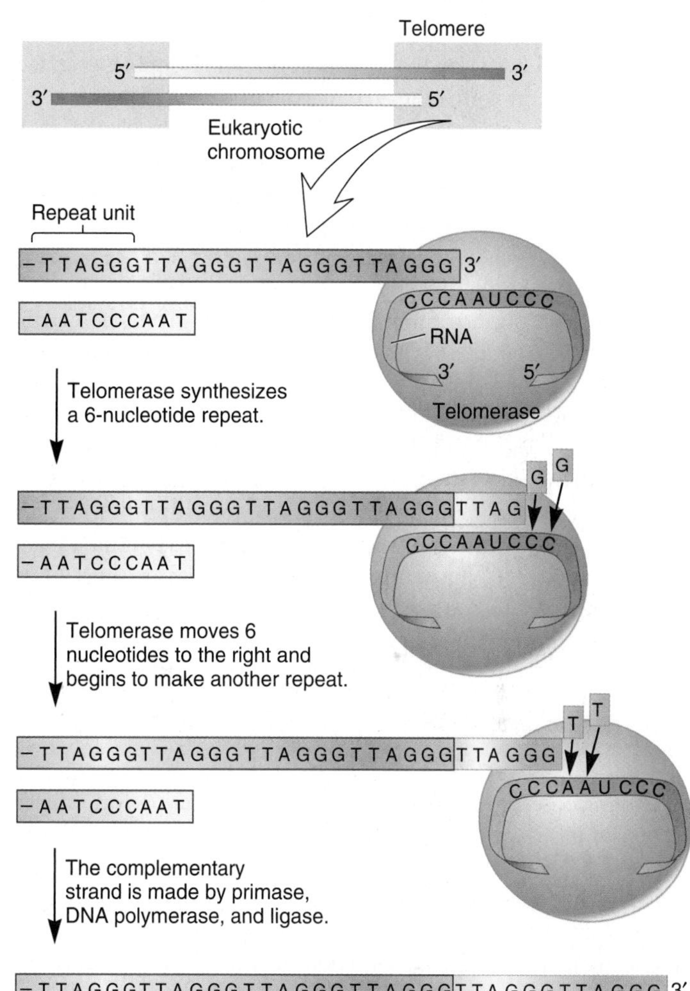

ONLINE ANIMATION

FIGURE 11.25 **The enzymatic action of telomerase.** A short, three-nucleotide segment of RNA within telomerase causes it to bind to the 3' overhang. The adjacent part of the RNA is used as a template to make a short, six-nucleotide repeat of DNA. After the repeat is made, telomerase moves six nucleotides to the right and then synthesizes another repeat. This process is repeated many times to lengthen the top strand shown in this figure. The bottom strand is made by DNA polymerase, using an RNA primer at the end of the chromosome that is complementary to the telomeric repeat sequence in the top strand. DNA polymerase fills in the region, which is sealed by ligase.

by which telomerase works. The telomerase enzyme contains both protein subunits and RNA. The RNA part of telomerase contains a sequence complementary to the DNA sequence found in the telomeric repeat. This allows telomerase to bind to the 3' overhang region of the telomere. Following binding, the RNA sequence beyond the binding site functions as a template allowing the synthesis of a six-nucleotide sequence at the end of the DNA strand. This is called polymerization, because it is analogous to the function of DNA polymerase. It is catalyzed by two identical protein subunits called telomerase reverse transcriptase (TERT). TERT's name indicates that is uses an RNA template to synthesize DNA. Following polymerization, the telomerase can then move—a process called translocation—to the new end of this DNA strand and attach another six nucleotides to the end. This binding-polymerization-translocation cycle occurs many times in a row and thereby greatly lengthens the 3' end of the DNA strand in the telomeric region. The complementary strand is then synthesized by DNA polymerase.

CONCEPTUAL SUMMARY

The structural basis for DNA replication is the double-stranded helix of DNA, in which the AT/GC rule is obeyed. This complementarity between strands allows DNA to be replicated in a semiconservative fashion. DNA replication begins at a specific site within the DNA, known as the origin of replication. Circular bacterial chromosomes have a single origin of replication. Specific proteins, such as the DnaA protein found in *E. coli*, recognize the origin and initiate the DNA replication process. The synthesis of DNA strands proceeds bidirectionally from the origin. This synthesis requires various proteins. DNA helicase breaks the hydrogen bonding between the parental strands, topoisomerase removes positive supercoils, and single-strand binding proteins hold the parental strands in a single-stranded state. Primase synthesizes RNA primers, which are necessary for DNA polymerase to elongate new daughter strands in a 5' to 3' direction. In the leading strand, synthesis of the daughter strand is continuous. In the lagging strand, the synthesis of DNA occurs in short Okazaki fragments. To complete the synthesis of Okazaki fragments within the lagging strand, the RNA primers are removed, DNA polymerase fills in the gaps, and DNA ligase covalently attaches the fragments together. DNA replication is terminated when both replication forks reach the ter sequences. If replication has resulted in two intertwined DNA molecules known as catenanes, the catenanes are separated into individual circular molecules.

Coordination of cell division and DNA replication is accomplished via the cellular regulation of replication. In bacteria, the control of DNA replication occurs at the origin. Several mechanisms can prevent premature DNA replication. Two examples are the availability of the DnaA protein and the methylation of GATC sites.

DNA replication in eukaryotes has several unique features. Eukaryotic chromosomes contain multiple origins of replication. Eukaryotes have multiple types of DNA polymerases involved in synthesizing the leading and lagging strand, DNA repair, and replicating of DNA lesions. Also, because eukaryotic chromosomes are linear, a specialized mechanism exists for the replication at the telomeres. An enzyme known as telomerase recognizes the sequences at the ends of eukaryotic chromosomes and synthesizes additional repeats of telomeric sequences. This prevents chromosome shortening with each round of DNA replication.

EXPERIMENTAL SUMMARY

The semiconservative mechanism of DNA replication was initially proposed based on the double helix structure determined by Watson and Crick. Nevertheless, researchers needed to experimentally demonstrate that this mechanism was correct. The results of the isotope labeling experiments of Meselson and Stahl were consistent with a semiconservative mode of replication in which DNA contains one parental strand and one newly synthesized or daughter strand.

Once the semiconservative model was established, researchers focused their attention on the details of DNA replication within living cells. Arthur Kornberg and his colleagues developed methods to synthesize bacterial DNA in vitro. In conjunction with the isolation of ts mutants, this allowed researchers to identify the role of individual proteins in DNA replication. Many of these components were found to be enzymes, such as DNA polymerase, primase, and DNA helicase. In eukaryotes, similar findings have been obtained, but more enzymes are involved and the sequences of the origins of replication are more complex. These added complexities have made it more difficult to analyze DNA replication in eukaryotes, although much progress has been made. As we will see in Chapter 22, the regulation of DNA replication is an important topic and an area of intense research because it may underlie the proliferation of cancer cells.

PROBLEM SETS & INSIGHTS

Solved Problems

S1. Describe three ways to account for the high fidelity of DNA replication. Discuss the quantitative contributions of each of the three ways.

Answer:

First: AT and GC pairs are preferred in the double helix structure. This provides fidelity to around one mistake per 1,000.

Second: Induced fit by DNA polymerase prevents covalent bond formation unless the proper nucleotides are in place. This increases fidelity another 100- to 1,000-fold, to about one error in 100,000 to 1 million.

Third: Exonuclease proofreading increases fidelity another 100- to 1,000-fold, to about one error per 100 million nucleotides added.

S2. What do you think would happen if the ter sequences were deleted from the bacterial DNA?

Answer: Instead of meeting at the ter sequences, the two replication forks would meet somewhere else. This would depend on how fast they were moving. For example, if one fork was advancing faster than the other fork, they would meet closer to where the slower-moving fork started. In fact, researchers have actually conducted this experiment. Interestingly, *E. coli* without the ter sequences seemed to survive just fine.

S3. Summarize the steps that occur in the process of chromosomal DNA replication in *E. coli.*

Answer:

Step 1. DnaA proteins bind to the origin of replication, resulting in the separation of the AT-rich region.

Step 2. DNA helicase breaks the hydrogen bonds between the DNA strands, topoisomerases alleviate positive supercoiling, and single-strand binding proteins hold the parental strands apart.

Step 3. Primase synthesizes one RNA primer in the leading strand and many RNA primers in the lagging strand. DNA polymerase III then synthesizes the daughter strands of DNA. In the lagging strand, many short segments of DNA (Okazaki fragments) are made. DNA polymerase I removes the RNA primers and fills in with DNA, and DNA ligase covalently links the Okazaki fragments together.

Step 4. The processes described in steps 2 and 3 continue until the two replication forks reach the ter sequences on the other side of the circular bacterial chromosome.

Step 5. Topoisomerases unravel the intertwined chromosomes, if necessary.

S4. If a strain of *E. coli* overproduced the Dam enzyme, how would that affect the DNA replication process? Would you expect such a strain to have more or less chromosomes per cell compared to a normal strain of *E. coli*? Explain why.

Answer: If a strain overproduced the Dam enzyme, DNA would replicate more rapidly. The GATC methylation sites in the origin of replication have to be fully methylated for DNA replication to occur. Immediately after DNA replication, a delay occurs before the next round of DNA replication because the two copies of newly replicated DNA are hemimethylated. A strain that overproduces Dam would rapidly convert the hemimethylated DNA into fully methylated DNA and more quickly allow the next round of DNA replication to occur. For this reason, the overproducing strain might have more copies of the *E. coli* chromosome because it would not have a long delay in DNA replication.

Conceptual Questions

C1. What are the key structural features of the DNA molecule that underlie its ability to be faithfully replicated?

C2. With regard to DNA replication, define the term bidirectional replication.

C3. Which of the following statements is not true? Explain why.

A. A DNA strand can serve as a template strand on many occasions.

B. Following semiconservative DNA replication, one strand is a newly made daughter strand and the other strand is a parental strand.

C. A DNA double helix may contain two strands of DNA that were made at the same time.

D. A DNA double helix obeys the AT/GC rule.

E. A DNA double helix could contain one strand that is 10 generations older than its complementary strand.

C4. The compound known as nitrous acid is a reactive chemical that replaces amino groups ($-NH_2$) with keto groups ($=O$). When nitrous acid reacts with the bases in DNA, it can change cytosine to uracil and change adenine to hypoxanthine. A DNA double helix has the following sequence:

```
TTGGATGCTGG
AACCTACGACC
```

A. What would be the sequence of this double helix immediately after reaction with nitrous acid? Let the letter H represent hypoxanthine and U represent uracil.

B. Let's suppose this DNA was reacted with nitrous acid. The nitrous acid was then removed, and the DNA was replicated for two generations. What would be the sequences of the DNA products after the DNA had replicated two times? Your answer should contain the sequences of four double helices. Note: During DNA replication, uracil hydrogen bonds with adenine, and hypoxanthine hydrogen bonds with cytosine.

C5. One way that bacterial cells regulate DNA replication is by GATC methylation sites within the origin of replication. Would this mechanism work if the DNA was conservatively (rather than semiconservatively) replicated?

C6. The chromosome of *E. coli* contains 4.6 million bp. How long will it take to replicate its DNA? Assuming DNA polymerase III is the primary enzyme involved and this enzyme can actively proofread during DNA synthesis, how many base pair mistakes will be made in one round of DNA replication in a bacterial population containing 1,000 bacteria?

C7. Here are two strands of DNA.

—————————————DNA polymerase–>

————————————————————————

The one on the bottom is a template strand, and the one on the top is being synthesized by DNA polymerase in the direction shown by the arrow. Label the 5' and 3' ends of the top and bottom strands.

C8. A DNA strand has the following sequence:

```
5'-GATCCCGATCCGCATACATTTACCAGATCACCACC-3'
```

In which direction would DNA polymerase slide along this strand (from left to right or from right to left)? If this strand was used as a template by DNA polymerase, what would be the sequence of the newly made strand? Indicate the 5' and 3' ends of the newly made strand.

C9. List and briefly describe the three types of sequences within bacterial origins of replication that are functionally important.

C10. As shown in Figure 11.5, five DnaA boxes are found within the origin of replication in *E. coli*. Take a look at these five sequences carefully.

A. Are the sequences of the five DnaA boxes very similar to each other? (Hint: Remember that DNA is double stranded; think about these sequences in the forward and reverse direction.)

B. What is the most common sequence for the DnaA box? In other words, what is the most common base in the first position, second position, and so on until the ninth position? The most common sequence is called the consensus sequence.

C. The *E. coli* chromosome is about 4.6 million bp long. Based on random chance, is it likely that the consensus sequence for a DnaA box occurs elsewhere in the *E. coli* chromosome? If so, why aren't there multiple origins of replication in *E. coli*?

C11. Obtain two strings of different colors (e.g., black and white) that are the same length. A length of 20 inches is sufficient. Tie a knot at one end of the black string, and tie a knot at one end of the white string. Each knot designates the 5' end of your strings. Make a double helix with your two strings. Now tape one end of the double helix to a table so that the tape is covering the knot on the black string.

A. Pretend your hand is DNA helicase and use your hand to unravel the double helix, beginning at the end that is not taped to the table. Should your hand be sliding along the white string or the black string?

B. As in Figure 11.14, imagine that your two hands together form a dimeric replicative DNA polymerase. Unravel your two strings halfway to create a replication fork. Grasp the black string with your left hand and the white string with your right hand. Your thumbs should point toward the 5' end of each string. You need to loop one of the strings so that one of the DNA polymerases can synthesize the lagging strand. With such a loop, the dimeric replicative DNA polymerase can move toward the replication fork and synthesize both DNA strands in the 5' to 3' direction. In other words, with such a loop, your two hands can touch each other with both of your thumbs pointing toward the fork. Should the black string be looped, or should the white string be looped?

C12. Sometimes DNA polymerase makes a mistake, and the wrong nucleotide is added to the growing DNA strand. With regard to pyrimidines and purines, two general types of mistakes are possible. The addition of an incorrect pyrimidine instead of the correct pyrimidine (e.g., adding cytosine where thymine should be added) is called a transition. If a pyrimidine is incorrectly added to the growing strand instead of purine (e.g., adding cytosine where an adenine should be added), this type of mistake is called a transversion. If a transition or transversion is not detected by DNA polymerase, this will create a mutation that permanently changes the DNA sequence. Though both types of mutations are rare, transition mutations are more frequent than transversion mutations. Based on your understanding of DNA replication and DNA polymerase, offer three explanations why transition mutations are more common.

C13. A short genetic sequence, which may be recognized by DNA primase, is repeated many times throughout the *E. coli* chromosome. Researchers have hypothesized that DNA primase may recognize this sequence as a site to begin the synthesis of an RNA primer for DNA replication. The *E. coli* chromosome is roughly 4.6 million bp in length. How many copies of the DNA primase recognition sequence would be necessary to replicate the entire *E. coli* chromosome?

C14. Single-strand binding proteins keep the two parental strands of DNA separated from each other until DNA polymerase has an opportunity to replicate the strands. Suggest how single-strand binding proteins keep the strands separated and yet do not impede the ability of DNA polymerase to replicate the strands.

C15. The ability of DNA polymerase to digest a DNA strand from one end is called its exonuclease activity. Exonuclease activity is used to digest RNA primers and also to proofread a newly made DNA strand. Note: DNA polymerase I does not change direction while it is removing an RNA primer and synthesizing new DNA. It does change direction during proofreading.

A. In which direction, 5' to 3' or 3' to 5' is the exonuclease activity occurring during the removal of RNA primers and during the proofreading and removal of mistakes following DNA replication?

B. Figure 11.15 shows a drawing of the 3' exonuclease site. Do you think this site would be used by DNA polymerase I to remove RNA primers? Why or why not?

C16. In the following drawing, the top strand is the template DNA, and the bottom strand shows the lagging strand prior to the action of DNA polymerase I. The lagging strand contains three Okazaki fragments. The RNA primers have not yet been removed.

The top strand is the template DNA

```
3'—————————————————————————————————————————————————————5'
5'*************________***************________*************__________3'
  RNA primer          ↑ RNA primer ↑          RNA primer
|————————————————————||——————————————————————||———————————————————————|
  Left Okazaki         Middle Okazaki          Right Okazaki
  fragment             fragment                fragment
```

A. Which Okazaki fragment was made first, the one on the left or the one on the right?

B. Which RNA primer would be the first one to be removed by DNA polymerase I, the primer on the left or the primer on the right? For this primer to be removed by DNA polymerase I and for the gap to be filled in, is it necessary for the Okazaki fragment in the middle to have already been synthesized? Explain why.

C. Let's consider how DNA ligase connects the left Okazaki fragment with the middle Okazaki fragment. After DNA polymerase I removes the middle RNA primer and fills in the gap with DNA, where does DNA ligase function? See the arrows on either side of the middle RNA primer. Is ligase needed at the left arrow, at the right arrow, or both?

D. When connecting two Okazaki fragments, DNA ligase needs to use NAD^+ or ATP as a source of energy to catalyze this reaction. Explain why DNA ligase needs another source of energy to connect two nucleotides, but DNA polymerase needs nothing more than the incoming nucleotide and the existing DNA strand. Note: You may want to refer to Figure 11.11 to answer this question.

C17. What is DNA methylation? Why is DNA in a hemimethylated condition immediately after DNA replication? What are the functional consequences of methylation in the regulation of DNA replication?

C18. Describe the three important functions of the DnaA protein.

C19. If a strain of bacteria was making too much DnaA protein, how would you expect this to affect its ability to regulate DNA replication? With regard to the number of chromosomes per cell, how might this strain differ from a normal bacterial strain?

C20. Draw a picture that illustrates how DNA helicase works.

C21. What is an Okazaki fragment? In which strand of DNA are Okazaki fragments found? Based on the properties of DNA polymerase, why is it necessary to make these fragments?

C22. Discuss the similarities and differences in the synthesis of DNA in the lagging and leading strands. What is the advantage of a primosome and a replisome as opposed to having all replication enzymes functioning independently of each other?

C23. Explain the proofreading function of DNA polymerase.

C24. What is a processive enzyme? Explain why this is an important feature of DNA polymerase.

C25. Why is it important for living organisms to regulate DNA replication?

C26. What enzymatic features of DNA polymerase prevent it from replicating one of the DNA strands at the ends of linear chromosomes? Compared to DNA polymerase, how is telomerase different in its ability to synthesize a DNA strand? What does telomerase use as its template for the synthesis of a DNA strand? How does the use of this template result in a telomere sequence that is tandemly repetitive?

C27. As shown in Figure 11.25, telomerase attaches additional DNA, six nucleotides at a time, to the ends of eukaryotic chromosomes. However, it works in only one DNA strand. Describe how the opposite strand is replicated.

C28. If a eukaryotic chromosome has 25 origins of replication, how many replication forks does it have at the beginning of DNA replication?

C29. A diagram of a linear chromosome is shown here. The end of each strand is labeled with an A, B, C, or D. Which ends could not be replicated by DNA polymerase? Why not?

```
5'-A————————————————————————————————B-3'

3'-C————————————————————————————————D-5'
```

C30. As discussed in Chapter 10, some viruses contain RNA as their genetic material. Certain RNA viruses can exist as a provirus in which the viral genetic material has been inserted into the chromosomal DNA of the host cell. For this to happen, the viral RNA must be copied into a strand of DNA. An enzyme called reverse transcriptase, encoded by the viral genome, copies the viral RNA into a complementary strand of DNA. The strand of DNA is then used as a template to make a double-stranded DNA molecule. This double-stranded DNA molecule is then inserted into the chromosomal DNA, where it may exist as a provirus for a long period of time.

A. How is the function of reverse transcriptase similar to the function of telomerase?

B. Unlike DNA polymerase, reverse transcriptase does not have a proofreading function. How might this affect the proliferation of the virus?

C31. Telomeres contain a 3' overhang region, as shown in Figure 11.23. Does telomerase require a 3' overhang to replicate the telomere region? Explain.

Experimental Questions

E1. Answer the following questions that pertain to the experiment of Figure 11.3.

A. What would be the expected results if the Meselson and Stahl experiment were carried out for four or five generations?

B. What would be the expected results of the Meselson and Stahl experiment after three generations if the mechanism of DNA replication was dispersive?

C. As shown in the data, explain why three different bands (i.e., light, half-heavy, and heavy) can be observed in the CsCl gradient.

E2. An absentminded researcher follows the steps of Figure 11.3, and when the gradient is viewed under UV light, the researcher does not see any bands at all. Which of the following mistakes could account for this observation? Explain how.

A. The researcher forgot to add ^{14}N-containing compounds.

B. The researcher forgot to add lysozyme.

C. The researcher forgot to turn on the UV lamp.

E3. Figure 11.4b shows an autoradiograph of a replicating bacterial chromosome. If you analyzed many replicating chromosomes, what types of information could you learn about the mechanism of DNA replication?

E4. The experiment of Figure 11.18 described a method for determining the amount of DNA made during replication. Let's suppose that you can purify all of the proteins required for DNA replication. You then want to "reconstitute" DNA synthesis by mixing together all of the purified components necessary to synthesize a complementary strand of DNA. If you started with single-stranded DNA as a template, what additional proteins and molecules would you have to add for DNA polymerization to occur? What additional proteins would be necessary if you started with a double-stranded DNA molecule?

E5. Using the reconstitution strategy described in experimental question E4, what components would you have to add to measure the ability of telomerase to synthesize DNA? Be specific about the type of template DNA that you would add to your mixture.

E6. As described in Figure 11.18, perchloric acid precipitates strands of DNA, but it does not precipitate free nucleotides. (Note: The term free nucleotide means nucleotides that are not connected covalently to other nucleotides.) Explain why this is a critical step in the experimental procedure. If a researcher used a different reagent that precipitated DNA strands and free nucleotides instead of using perchloric acid (which precipitates only DNA strands), how would that affect the results?

E7. Would the experiment of Figure 11.18 work if the ^{32}P-labeled nucleotides were deoxyribonucleoside monophosphates instead of deoxyribonucleoside triphosphates? Explain why or why not.

E8. To synthesize DNA in vitro, single-stranded DNA can be used as a template. As described in Figure 11.18, you also need to add DNA polymerase, dNTPs, and a primer in order to synthesize a complementary strand of DNA. The primer can be a short sequence of DNA or RNA. The primer must be complementary to the template DNA. Let's suppose a single-stranded DNA molecule is 46 nucleotides long and has the following sequence:

GCCCCGGTACCCCGTAATATACGGGACTAGGCCGGAGGTCCGGGCG

This template DNA is mixed with a primer with the sequence 5'–CGCCCGGACC–3', DNA polymerase, and dNTPs. In this case, a double-stranded DNA molecule is made. However, if the researcher substitutes a primer with the sequence 5'–CCAG-GCCCGC–3', a double-stranded DNA molecule is not made.

A. Which is the 5' end of the DNA molecule shown, the left end or the right end?

B. If you added a primer that was 10 nucleotides long and complementary to the left end of the single-stranded DNA, what would be the sequence of the primer? You should designate the 5' and 3' ends of the primer. Could this primer be used to replicate the single-stranded DNA?

E9. The technique of dideoxy sequencing of DNA is described in Chapter 18. The technique relies on the use of dideoxyribonucleotides (shown in Figures 18.14 and 18.15). A dideoxyribonucleotide has a hydrogen atom attached to the 3' carbon atom instead of an –OH group. When a dideoxyribonucleotide is incorporated into a newly made strand, the strand cannot grow any longer. Explain why.

E10. Another technique described in Chapter 18 is the polymerase chain reaction (PCR) (see Figure 18.6). This method is based on our understanding of DNA replication. In this method, a small amount of double-stranded template DNA is mixed with a high concentration of primers. Nucleotides and DNA polymerase are also added. The template DNA strands are separated by heat treatment, and when the temperature is lowered, the primers can bind to the single-stranded DNA, and then DNA polymerase replicates the DNA. This increases the amount of DNA made from the primers. This cycle of steps (i.e., heat treatment, lower temperature, allow DNA replication to occur) is repeated again and again and again. Because the cycles are repeated many times, this method is called a chain reaction. It is called a polymerase chain reaction because DNA polymerase is the enzyme needed to increase the amount of DNA with each cycle. In a PCR experiment, the template DNA is placed in a tube, and the primers, nucleotides, and DNA polymerase are added to the tube. The tube is then placed in a machine called a thermocycler, which raises and lowers the temperature. During one cycle, the temperature is raised (e.g., to 95°C) for a brief period and then lowered (e.g., to 60°C) to allow the primers to bind. The sample is incubated at the lower temperature for a few minutes to allow DNA replication to proceed. In a typical PCR experiment, the tube may be left in the thermocycler for 25 to 30 cycles. The total time for a PCR experiment is a few hours.

A. Why is DNA helicase not needed in a PCR experiment?

B. How is the sequence of each primer important in a PCR experiment? Do the two primers recognize the same strand or opposite strands?

C. The DNA polymerase used in PCR experiments is a DNA polymerase isolated from thermophilic bacteria. Why is this kind of polymerase used?

D. If a tube initially contained 10 copies of double-stranded DNA, how many copies of double-stranded DNA (in the region flanked by the two primers) would be obtained after 27 cycles?

Questions for Student Discussion/Collaboration

1. The complementarity of double-stranded DNA is the underlying reason that DNA can be faithfully copied. Propose alternative chemical structures that could be faithfully copied.

2. The technique described in Figure 11.18 makes it possible to measure DNA synthesis in vitro. Let's suppose you have purified the following enzymes: DNA polymerase, DNA helicase, ligase, primase, single-strand binding protein, and topoisomerase. You also have the following reagents available:

A. Radiolabeled nucleotides (labeled with ^{32}P, a radioisotope of phosphorus)

B. Nonlabeled double-stranded DNA

C. Nonlabeled single-stranded DNA

D. An RNA primer that binds to one end of the nonlabeled single-stranded DNA

With these reagents, how could you show that DNA helicase is necessary for strand separation and primase is necessary for the synthesis of an RNA primer? Note: In this question, think about conditions where DNA helicase or primase would be necessary to allow DNA replication and other conditions where they would be unnecessary.

3. DNA replication is fast, virtually error-free, and coordinated with cell division. Discuss which of these three features you think is the most important.

Note: All answers appear at the website for this textbook; the answers to even-numbered questions are in the back of the textbook.

www.mhhe.com/brookergenetics3e

Visit the Online Learning Center for practice tests, answer keys, and other learning aids for this chapter. Enhance your understanding of genetics with our interactive exercises, quizzes, animations, and much more.

GENE TRANSCRIPTION AND RNA MODIFICATION

12

CHAPTER OUTLINE

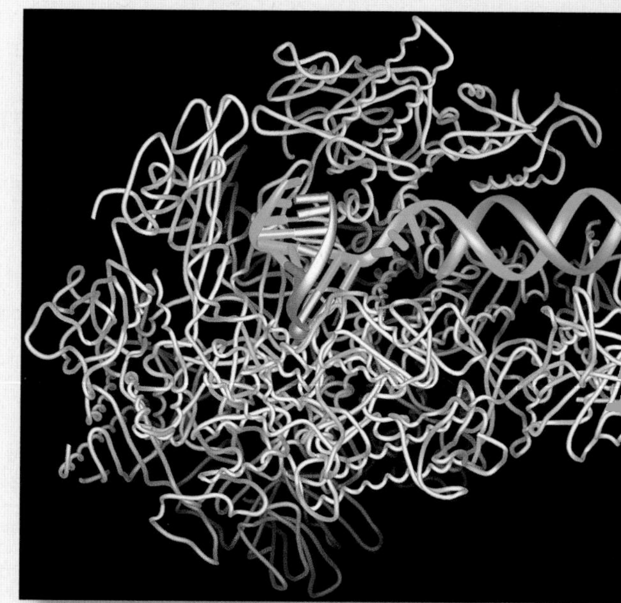

A molecular model showing the enzyme RNA polymerase in the act of sliding along the DNA and synthesizing a copy of RNA.

The function of the genetic material is that of a blueprint. It stores the information necessary to create a living organism. The information is contained within units called genes. At the molecular level, a definition of a **gene** is a segment of DNA that is used to make a functional product, either an RNA molecule or a polypeptide. How is the information within a gene accessed? The first step in this process is called **transcription,** which literally means the act or process of making a copy. In genetics, this term refers to the process of synthesizing RNA from a DNA sequence (**Figure 12.1**). The structure of DNA is not altered as a result of transcription. Rather, the DNA base sequence has only been accessed to make a copy in the form of RNA. Therefore, the same DNA can continue to store information. DNA replication, which was discussed in Chapter 11, provides a mechanism for copying that information so it can be transmitted from cell to cell and from parent to offspring.

Structural genes encode the amino acid sequence of a polypeptide. When a structural gene is transcribed, the first product is an RNA molecule known as

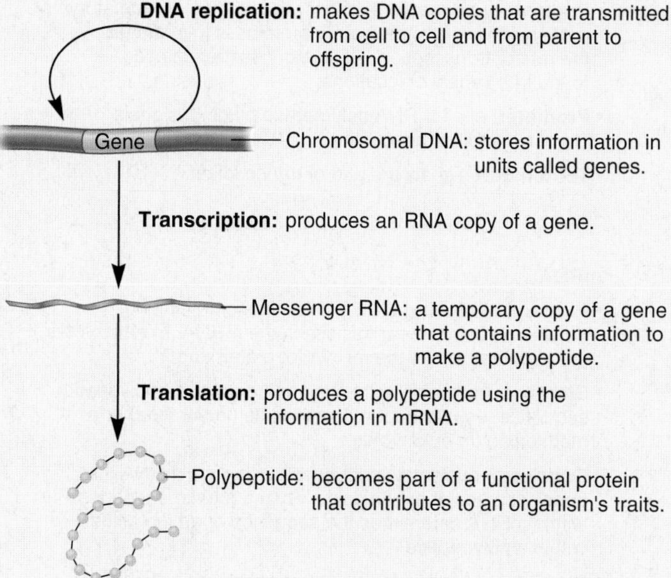

DNA replication: makes DNA copies that are transmitted from cell to cell and from parent to offspring.
Gene
Chromosomal DNA: stores information in units called genes.
Transcription: produces an RNA copy of a gene.
Messenger RNA: a temporary copy of a gene that contains information to make a polypeptide.
Translation: produces a polypeptide using the information in mRNA.
Polypeptide: becomes part of a functional protein that contributes to an organism's traits.

FIGURE 12.1 **The central dogma of genetics.** The usual flow of genetic information is from DNA to mRNA to polypeptide. Note: The direction of informational flow shown in this figure is the most common direction found in living organisms, but exceptions occur. For example, RNA viruses and certain transposable elements use an enzyme called reverse transcriptase to make a copy of DNA from RNA.

PART IV

Molecular Properties of Genes

messenger RNA (mRNA). During polypeptide synthesis—a process called **translation**—the sequence of nucleotides within the mRNA determines the sequence of amino acids in a polypeptide. One or more polypeptides then assemble into a functional protein. The synthesis of functional proteins ultimately determines an organism's traits. The model depicted in Figure 12.1, which is called the **central dogma of genetics** (also called the central dogma of molecular biology), was first enunciated by Francis Crick in 1958. It forms a cornerstone for our understanding of genetics at the molecular level. The flow of genetic information occurs from DNA to mRNA to polypeptide.

In this chapter, we begin to study the molecular steps in gene expression, with an emphasis on transcription and the modifications that may occur to an RNA transcript after it has been made. In Chapter 13, we will examine the process of translation, and Chapters 14 and 15 will focus on how the level of gene expression is regulated at the molecular level.

12.1 OVERVIEW OF TRANSCRIPTION

One key concept important in the process of transcription is that short base sequences define the beginning and ending of a gene and also play a role in regulating the level of RNA synthesis. In this section, we will begin by examining the sequences that determine where transcription starts and ends, and also briefly consider DNA sequences, called regulatory sites, that influence whether a gene is turned on or off. The functions of regulatory sites will be examined in greater detail in Chapters 14 and 15. A second important concept is the role of proteins in transcription. DNA sequences, in and of themselves, just exist. For genes to be actively transcribed, proteins must recognize particular DNA sequences and act upon them in a way that affects the transcription process. In the later part of this section, we will consider how proteins participate in the general steps of transcription, and the types of RNA transcripts that can be made.

Gene Expression Requires Base Sequences That Perform Different Functional Roles

At the molecular level, **gene expression** is the overall process by which the information within a gene is used to produce a functional product, such as a polypeptide. Along with environmental factors, the molecular expression of genes determines an organism's traits. For a gene to be expressed, a few different types of base sequences perform specific roles. **Figure 12.2** shows a common organization of base sequences needed to create a structural gene that functions in a bacterium such as *E. coli.* Each type of base sequence performs its role during a particular stage of gene expression. For example, the promoter and terminator are base sequences used during gene transcription. Specifically, the **promoter** provides a site to begin transcription, and the **terminator** specifies the end of transcription. These two sequences cause RNA synthesis to occur within a defined location. As shown in Figure 12.2, the DNA is transcribed into RNA from the end of the promoter to the termina-

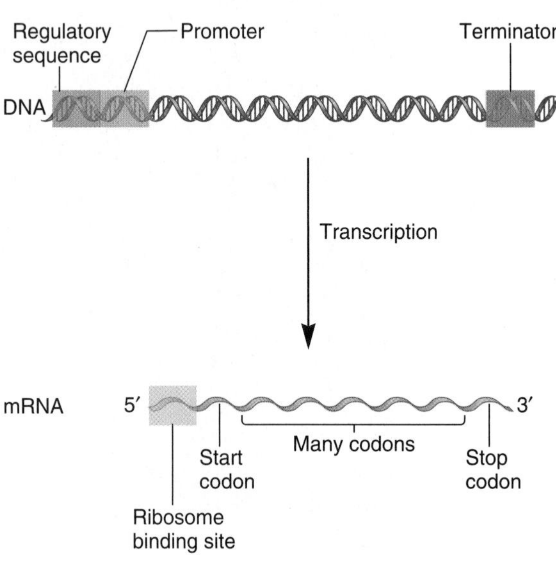

DNA:

- **Regulatory sequences:** site for the binding of regulatory proteins; the role of regulatory proteins is to influence the rate of transcription. Regulatory sequences can be found in a variety of locations.
- **Promoter:** site for RNA polymerase binding; signals the beginning of transcription.
- **Terminator:** signals the end of transcription.

mRNA:

- **Ribosomal binding site:** site for ribosome binding; translation begins near this site in the mRNA. In eukaryotes, the ribosome scans the mRNA for a start codon.
- **Start codon:** specifies the first amino acid in a polypeptide sequence, usually a formylmethionine (in bacteria) or a methionine (in eukaryotes).
- **Codons:** 3-nucleotide sequences within the mRNA that specify particular amino acids. The sequence of codons within mRNA determines the sequence of amino acids within a polypeptide.
- **Stop codon:** specifies the end of polypeptide synthesis.
- Bacterial mRNA may be polycistronic, which means it encodes two or more polypeptides.

FIGURE 12.2 **Organization of sequences of a bacterial gene and its mRNA transcript.** This figure depicts the general organization of sequences that are needed to create a functional gene that encodes an mRNA.

tor. As described later, the base sequence in the RNA transcript is complementary to the **template strand** of DNA, also called the **antisense strand.** For genes that encode proteins, the non-template DNA strand is called the **coding strand** or the **sense strand.** The sequence of nucleotides in the transcribed mRNA is the same as that found in the coding strand, except that the T's are replaced by U's.

A category of proteins called **transcription factors** recognizes base sequences in the DNA and controls transcription. Some transcription factors bind directly to the promoter and facilitate transcription. Other transcription factors recognize **regulatory sequences**—short stretches of DNA involved in the regulation of transcription. Certain transcription factors bind to such regulatory sequences and increase the rate of transcription while others inhibit transcription.

Base sequences within an mRNA are used during the translation process. In bacteria, a short sequence within the mRNA, the **ribosomal-binding site,** provides a location for the ribosome to bind and begin translation. The bacterial ribosome recognizes this site because it is complementary to a sequence in ribosomal RNA. In addition, mRNA contains a series of **codons,** read as groups of three nucleotides, which contain the information for a polypeptide's sequence. The first codon, which is very close to the ribosomal-binding site, is the **start codon.** This is followed by many more codons that dictate the sequence of amino acids within the synthesized polypeptide. Finally, a **stop codon** signals the end of translation. Chapter 13 will examine the process of translation in greater detail.

The Three Stages of Transcription Are Initiation, Elongation, and Termination

Transcription occurs in three stages: **initiation; elongation,** or synthesis of the RNA transcript; and **termination** (Figure 12.3). These steps involve protein-DNA interactions in which proteins such as **RNA polymerase,** the enzyme that synthesizes RNA, interact with DNA sequences. What causes transcription to begin? The initiation stage in the transcription process is a recognition step. The sequence of bases within the promoter region is recognized by transcription factors. The specific binding of transcription factors to the promoter sequence identifies the starting site for transcription.

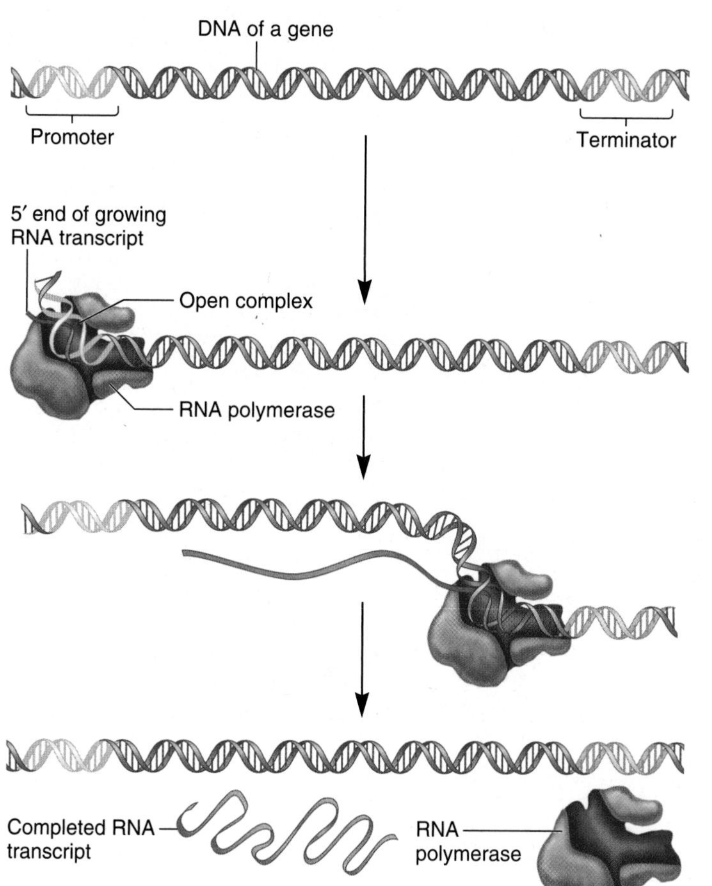

Initiation: The promoter functions as a recognition site for transcription factors (not shown). The transcription factor(s) enables RNA polymerase to bind to the promoter. Following binding, the DNA is denatured into a bubble known as the open complex.

Elongation/synthesis of the RNA transcript: RNA polymerase slides along the DNA in an open complex to synthesize RNA.

Termination: A terminator is reached that causes RNA polymerase and the RNA transcript to dissociate from the DNA.

ONLINE ANIMATION

FIGURE 12.3 Stages of transcription.

Genes→Traits The ability of genes to produce an organism's traits relies on the molecular process of gene expression. Transcription is the first step in gene expression. During transcription, the gene's sequence within the DNA is used as a template to make a complementary copy of RNA. In Chapter 13, we will examine how the sequence in mRNA is translated into a polypeptide chain. After polypeptides are made within a living cell, they fold into functional proteins that govern an organism's traits.

Transcription factors and RNA polymerase first bind to the promoter region when the DNA is in the form of a double helix. For transcription to occur, the DNA strands must be separated. This allows one of the two strands to be used as a template for the synthesis of a complementary strand of RNA. This synthesis occurs as RNA polymerase slides along the DNA, creating a small bubblelike structure known as the open promoter complex, or simply as the **open complex.** Eventually, RNA polymerase reaches a terminator, which causes both RNA polymerase and the newly made RNA transcript to dissociate from the DNA.

RNA Transcripts Have Different Functions

Once they are made, RNA transcripts play different functional roles (**Table 12.1**). Well over 90% of all genes are structural genes, which are transcribed into mRNA. For structural genes, mRNAs are made first, but the final, functional products are polypeptides that operate within proteins. The remaining types of RNAs described in Table 12.1 are never translated. The RNA transcripts from such nonstructural genes have various important cellular functions. For nonstructural genes, the functional product is the RNA. In some cases, the RNA transcript becomes part of a complex that contains both protein subunits and one or more RNA molecules. Examples of protein-RNA complexes include ribosomes, signal recognition particles, RNaseP, spliceosomes and telomerase.

TABLE 12.1

Functions of RNA Molecules

Type of RNA	Description
mRNA	Messenger RNA (mRNA) encodes the sequence of amino acids within a polypeptide. In bacteria, some mRNAs encode a single polypeptide. Other mRNAs are polycistronic—a single mRNA encodes two or more polypeptides. In most species of eukaryotes, each mRNA usually encodes a single polypeptide. However, in some species, such as *Caenorhabditis elegans* (a nematode worm), polycistronic mRNAs are relatively common.
tRNA	Transfer RNA (tRNA) is necessary for the translation of mRNA. The structure and function of transfer RNA are outlined in Chapter 13.
rRNA	Ribosomal RNA (rRNA) is necessary for the translation of mRNA. Ribosomes are composed of both rRNAs and protein subunits. The structure and function of ribosomes are examined in Chapter 13.
MicroRNA	MicroRNAs (miRNAs) are short RNA molecules that are involved in gene regulation in eukaryotes (see Chapter 15).
scRNA	Small cytoplasmic RNA (scRNA) is found in the cytoplasm of bacteria and eukaryotes. In bacteria, scRNA is needed for protein secretion. An example in eukaryotes is 7S RNA, which is necessary in the targeting of proteins to the endoplasmic reticulum. It is a component of a complex known as signal recognition particle (SRP), which is composed of 7S RNA and six different protein subunits.
RNA of RNaseP	RNaseP is an enzyme necessary in the processing of all bacterial tRNA molecules. The RNA is the catalytic component of this enzyme. RNaseP is composed of a 350- to 410-nucleotide RNA and one protein subunit.
snRNA	Small nuclear RNA (snRNA) is necessary in the splicing of eukaryotic pre-mRNA. snRNAs are components of a spliceosome, which is composed of both snRNAs and protein subunits. The structure and function of spliceosomes are examined later in this chapter.
Telomerase RNA	The enzyme telomerase, which is involved in the replication of eukaryotic telomeres, is composed of an RNA molecule and protein subunits.
snoRNA	Small nucleolar RNA (snoRNA) is necessary in the processing of eukaryotic rRNA transcripts. snoRNAs are also associated with protein subunits. In eukaryotes, snoRNAs are found in the nucleolus, where rRNA processing and ribosome assembly occur.
Viral RNAs	Some types of viruses use RNA as their genome, which is packaged within the viral capsid.

12.2 TRANSCRIPTION IN BACTERIA

Our molecular understanding of gene transcription initially came from studies involving bacteria and bacteriophages. Several early investigations focused on the production of viral RNA after bacteriophage infection. The first suggestion that RNA is derived from the transcription of DNA was made by Elliot Volkin and Lazarus Astrachan in 1956. When the researchers exposed *E. coli* cells to T2 bacteriophage, they observed that the RNA made immediately after infection had a base composition substantially different from the base composition of RNA prior to infection. Furthermore, the base composition after infection was very similar to the base composition in the T2 DNA, except that the RNA contained uracil instead of thymine. These results were consistent with the idea that the bacteriophage DNA is used as a template for the synthesis of bacteriophage RNA.

In 1960, Matthew Meselson and François Jacob found that proteins are synthesized on ribosomes. One year later, Jacob and his colleague Jacques Monod proposed that a certain type of RNA acts as a genetic messenger (from the DNA to the ribosome) to provide the information for protein synthesis. They hypothesized that this RNA, which they called messenger RNA, is transcribed from the sequence within DNA and then directs the synthesis of particular polypeptides. In the early 1960s, this proposal was remarkable, considering that it was made before the actual isolation and characterization of the mRNA molecules in vitro. In 1961, the hypothesis was confirmed by Sydney Brenner in collaboration with Jacob and Meselson. They found that when a virus infects a bacterial cell, a virus-specific RNA is made that rapidly associates with preexisting ribosomes in the cell.

Since these pioneering studies, a great deal has been learned about the molecular features of bacterial gene transcription. Much of our knowledge comes from studies in *E. coli.* In this section, we will examine the three steps in the gene transcription process as they occur in bacteria.

A Promoter Is a Short Sequence of DNA That Is Necessary to Initiate Transcription

The type of DNA sequence known as the promoter gets its name from the idea that it "promotes" gene expression. More precisely, this sequence of bases directs the exact location for the initiation of RNA transcription. Most of the promoter region is located just ahead of or upstream from the site where transcription of a gene actually begins. By convention, the bases in a promoter sequence are numbered in relation to the **transcriptional start site** (**Figure 12.4**). This site is the first base used as a template for RNA transcription and is denoted +1. The bases preceding this site are numbered in a negative direction. No base is numbered zero. Therefore, most of the promoter region is labeled with negative numbers that describe the number of bases preceding the beginning of transcription.

Although the promoter may encompass a region several dozen nucleotides in length, short **sequence elements** are particularly critical for promoter recognition. By comparing the sequence of DNA bases within many promoters, researchers have learned that certain sequences of bases are necessary to create a functional promoter. In many promoters found in *E. coli* and similar species, two sequence elements are important. These are located at approximately the –35 and –10 sites in the promoter region (Figure 12.4). The sequence at the –35 region is 5'–TTGACA–3', and the one at the –10 region is 5'–TATAAT–3'. The TATAAT sequence is sometimes called the **Pribnow box** after David Pribnow, who initially discovered it in 1975.

The sequences at the –35 and –10 sites can vary among different genes. For example, **Figure 12.5** illustrates the sequences found in several different *E. coli* promoters. The most commonly occurring bases within a sequence element form the **consensus sequence.** This sequence is efficiently recognized by proteins that

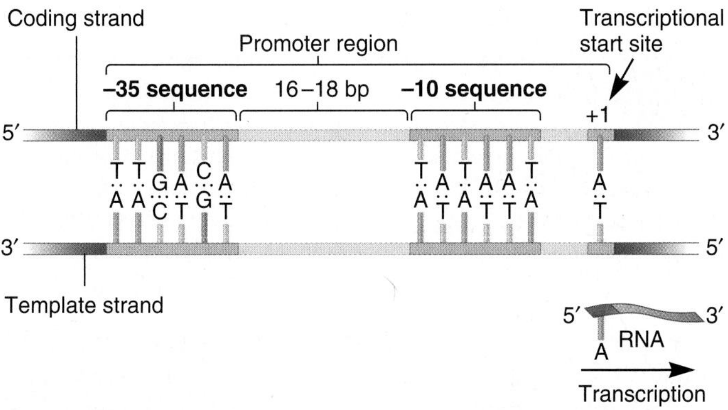

FIGURE 12.4 The conventional numbering system of promoters. The first nucleotide that acts as a template for transcription is designated +1. The numbering of nucleotides to the left of this spot is in a negative direction, while the numbering to the right is in a positive direction. For example, the nucleotide that is immediately to the left of the +1 nucleotide is numbered –1, and the nucleotide to the right of the +1 nucleotide is numbered +2. There is no zero nucleotide in this numbering system. In many bacterial promoters, sequence elements at the –35 and –10 regions play a key role in promoting transcription.

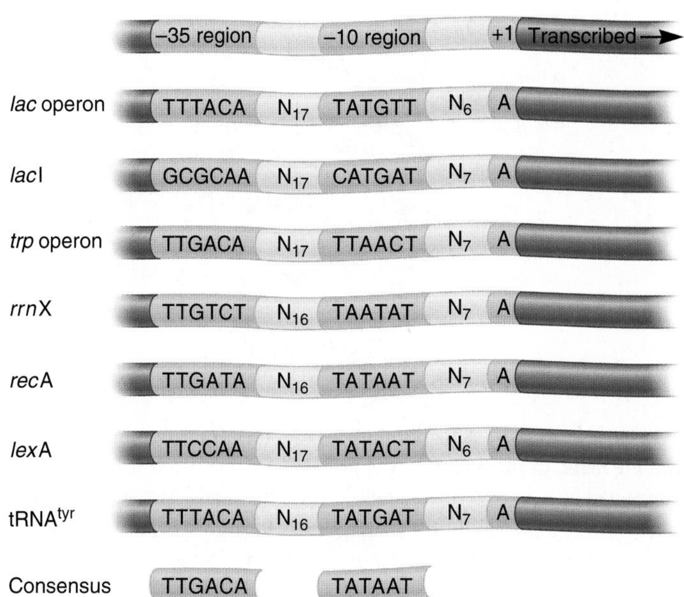

FIGURE 12.5 Examples of –35 and –10 sequences within a variety of bacterial promoters. This figure shows the –35 and –10 sequences for seven different bacterial and bacteriophage promoters. The consensus sequence is shown at the bottom. The spacer regions contain the designated number of nucleotides between the –35 and –10 region or between the –10 region and the transcriptional start site. For example, N_{17} means there are 17 nucleotides between the end of the –35 region and the beginning of the –10 region.

initiate transcription. For many bacterial genes, a strong correlation is found between the maximal rate of RNA transcription and the degree to which the –35 and –10 regions agree with their consensus sequences.

Bacterial Transcription Is Initiated When RNA Polymerase Holoenzyme Binds at a Promoter Sequence

Thus far, we have considered the DNA sequences that constitute a functional promoter. Let's now turn our attention to the proteins that recognize those sequences and carry out the transcription process. The enzyme that catalyzes the synthesis of RNA is **RNA polymerase.** In *E. coli*, the **core enzyme** is composed of five subunits, $\alpha_2\beta\beta'\omega$. The association of a sixth subunit, **sigma (σ) factor,** with the core enzyme is referred to as RNA polymerase **holoenzyme.** The different subunits within the holoenzyme play distinct functional roles. The two α subunits are important in the proper assembly of the holoenzyme and in the process of binding to DNA. The β and β' subunits are also needed for binding to the DNA and carry out the catalytic synthesis of RNA. The ω (omega) subunit is important for the proper assembly of the core enzyme. The holoenzyme is required to initiate transcription; the primary role of σ factor is to recognize the promoter. Proteins, such as σ factor, that influence the function of RNA polymerase are types of transcription factors.

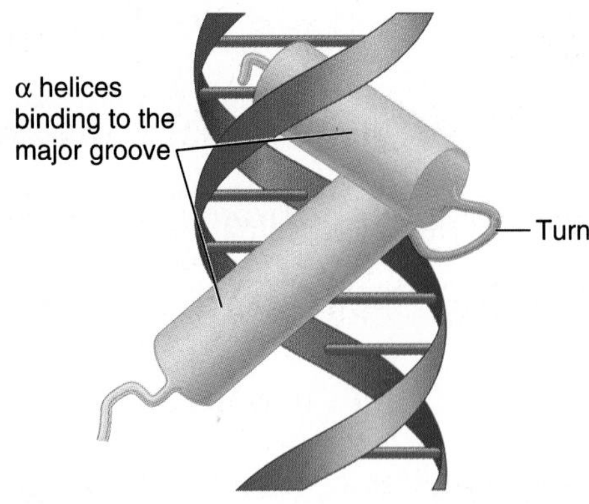

FIGURE 12.6 The binding of σ factor protein to the DNA double helix. In this example, the protein contains two α helices connected by a turn, termed a helix-turn-helix motif. Two α helices of the protein can fit within the major groove of the DNA. Amino acids within the α helices form hydrogen bonds with the bases in the DNA.

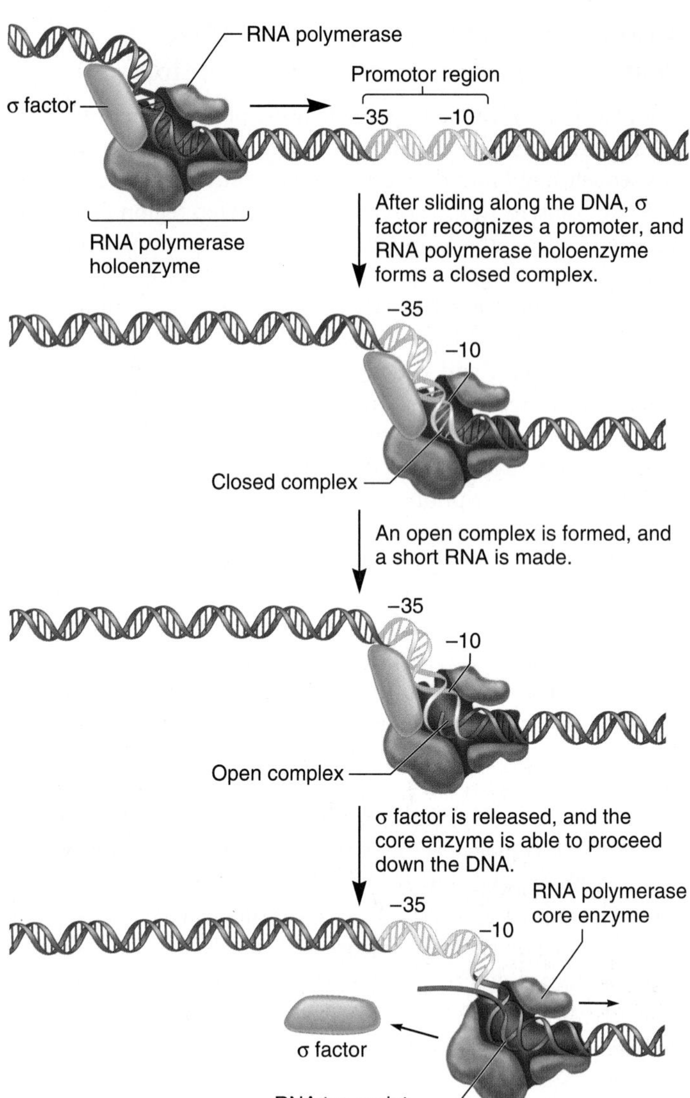

ONLINE ANIMATION **FIGURE 12.7 The initiation stage of transcription in bacteria.** The σ factor subunit of the RNA polymerase holoenzyme recognizes the –35 and –10 regions of the promoter. The DNA unwinds in the –10 region to form an open complex, and a short RNA is made. σ factor then dissociates from the holoenzyme, and the RNA polymerase core enzyme can proceed down the DNA to transcribe RNA, forming an open complex as it goes.

After RNA polymerase holoenzyme is assembled into its six subunits, it binds loosely to the DNA and then slides along the DNA, much as a train rolls down the tracks. How is a promoter identified? When the holoenzyme encounters a promoter sequence, σ factor recognizes the bases at both the –35 and –10 regions. σ factor protein contains a structure called a **helix-turn-helix motif** that can bind tightly to these regions. Alpha helices within the protein fit into the major groove of the DNA double helix and form hydrogen bonds with the bases. This phenomenon of molecular recognition is shown in Figure 12.6. Hydrogen bonding occurs between nucleotides in the –35 and –10 regions of the promoter and amino acid side chains in the helix-turn-helix structure of σ factor.

As shown in Figure 12.7, the process of transcription is initiated when σ factor within the holoenzyme has bound to the promoter region to form the **closed complex.** For transcription to begin, the double-stranded DNA must then be unwound into an open complex. This unwinding first occurs at the TATAAT sequence in the –10 region, which contains only AT base pairs, as shown in Figure 12.4. AT base pairs form only two hydrogen bonds, whereas GC pairs form three. Therefore, DNA in an AT-rich region is more easily separated because fewer hydrogen bonds must be broken. A short strand of RNA is made within the open complex, and then σ factor is released from the core enzyme. The release of σ factor marks the transition to the elongation phase of transcription. The core enzyme may now slide down the DNA to synthesize a strand of RNA.

The RNA Transcript Is Synthesized During the Elongation Stage

After the initiation stage of transcription is completed, the RNA transcript is made in the elongation stage of transcription. During the synthesis of the RNA transcript, RNA polymerase moves along the DNA, causing it to unwind (Figure 12.8). As previously mentioned, the DNA strand used as a template for RNA synthesis is called the template or antisense strand. The opposite DNA strand is the coding or sense strand; it has the same sequence as the RNA transcript except that T in the DNA corresponds to U in the RNA. Within a given gene, only the template strand is used for RNA synthesis, while the coding strand is never used. As it moves along the DNA, the open complex formed by the action of RNA polymerase is approximately 17 bp long. On average, the rate of RNA synthesis is about 43 nucleotides per second! Behind the open complex, the DNA rewinds back into a double helix.

As described in Figure 12.8, the chemistry of transcription by RNA polymerase is similar to the synthesis of DNA via DNA polymerase, which is discussed in Chapter 11. RNA polymerase always

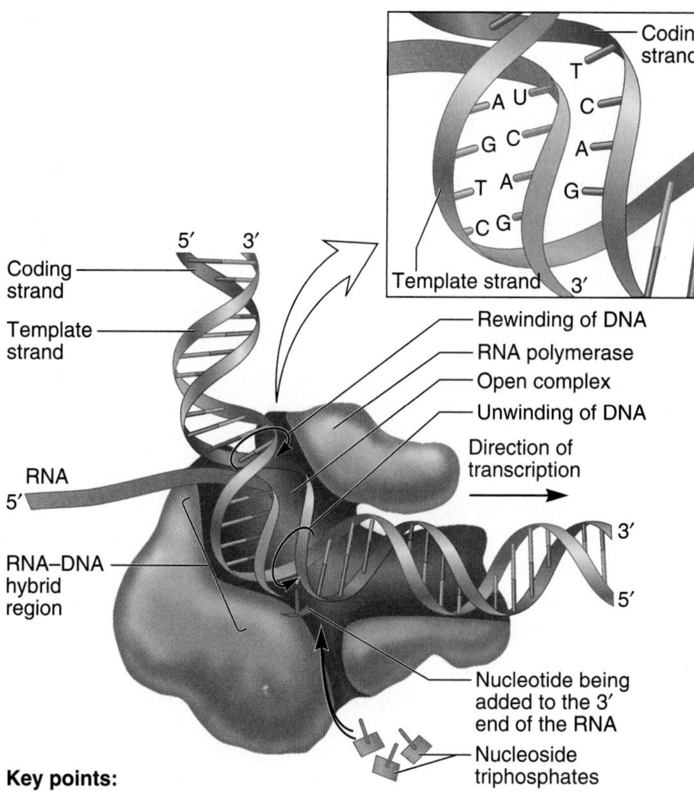
Coding strand
T
A U
C
G C
A
T A
G
C G
Template strand 3′
5′ 3′
Coding strand
Template strand
Rewinding of DNA
RNA polymerase
Open complex
Unwinding of DNA
Direction of transcription
RNA
5′
3′
5′
RNA–DNA hybrid region
Nucleotide being added to the 3′ end of the RNA
Nucleoside triphosphates
Key points:
• RNA polymerase slides along the DNA, creating an open complex as it moves.
• The DNA strand known as the template strand is used to make a complementary copy of RNA as an RNA–DNA hybrid.
• RNA polymerase moves along the template strand in a 3′ to 5′ direction, and RNA is synthesized in a 5′ to 3′ direction using nucleoside triphosphates as precursors. Pyrophosphate is released (not shown).
• The complementarity rule is the same as the AT/GC rule except that U is substituted for T in the RNA.

ONLINE ANIMATION **FIGURE 12.8** **Synthesis of the RNA transcript.**

connects nucleotides in the 5' to 3' direction. During this process, RNA polymerase catalyzes the formation of a bond between the 5' phosphate group on one nucleotide and the 3'–OH group on the previous nucleotide. The complementarity rule is similar to the AT/GC rule, except that uracil substitutes for thymine in the RNA. In other words, RNA synthesis obeys an $A_{DNA}U_{RNA}$/$T_{DNA}A_{RNA}$/$G_{DNA}C_{RNA}$/$C_{DNA}G_{RNA}$ rule.

When considering the transcription of multiple genes within a chromosome, the direction of transcription and the DNA strand used as a template can vary among different genes. Figure 12.9 shows three genes adjacent to each other within a chromosome. Genes *A* and *B* are transcribed from left to right, using the bottom DNA strand as a template. By comparison, gene *C* is transcribed from right to left and uses the top DNA strand as a template. Note that in all three cases, the template strand is read in the 3' to 5' direction, and the synthesis of the RNA transcript occurs in a 5' to 3' direction.

Transcription Is Terminated by Either an RNA-Binding Protein or an Intrinsic Terminator

The end of RNA synthesis is referred to as termination. Prior to termination, the hydrogen bonding between the DNA and RNA within the open complex is of central importance in preventing dissociation of RNA polymerase from the template strand. Termination occurs when this short RNA-DNA hybrid region is forced to separate, thereby releasing RNA polymerase as well as the newly made RNA transcript. In *E. coli*, two different mechanisms for termination have been identified. For certain genes, an RNA-binding protein known as **ρ (rho)** is responsible for terminating transcription, in a mechanism called ρ-dependent termination. For other genes, termination does not require the involvement of the ρ protein. This is referred to as ρ-independent termination.

In **ρ-dependent termination,** the termination process requires two components. First, a sequence near the 3' end of the newly made RNA, called the *rut* site for rho utilization site,

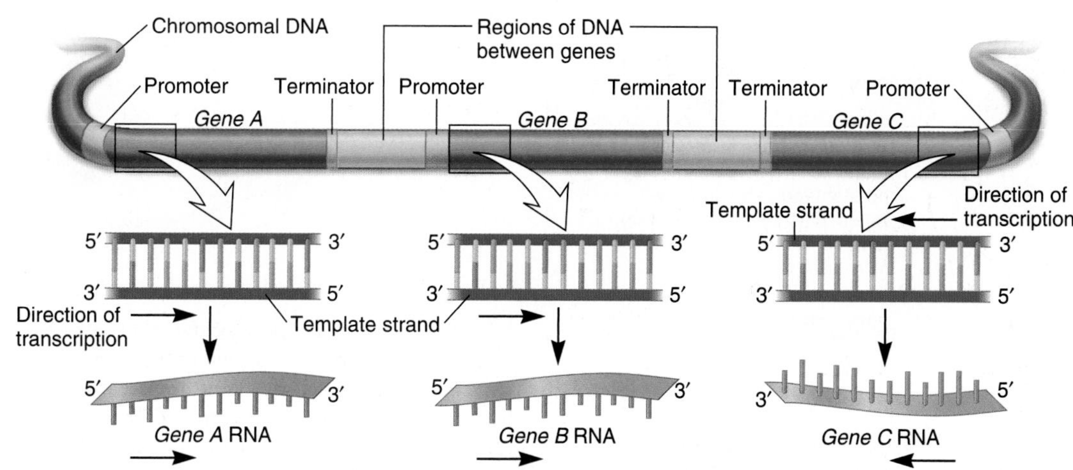
Chromosomal DNA
Regions of DNA between genes
Promoter
Terminator
Promoter
Terminator
Terminator
Promoter
Gene A
Gene B
Gene C
Template strand
Direction of transcription
5′ 3′
3′ 5′
Direction of transcription
Template strand
5′ 3′
Gene A RNA
Gene B RNA
3′ 5′
Gene C RNA

FIGURE 12.9 **The transcription of three different genes found in the same chromosome.** RNA polymerase synthesizes each RNA transcript in a 5' to 3' direction, sliding along a DNA template strand in a 3' to 5' direction. However, the use of the template strand can vary from gene to gene. For example, genes *A* and *B* use the bottom strand, while gene *C* uses the top strand.

acts as a recognition site for the binding of the ρ protein (**Figure 12.10**). How does ρ protein facilitate termination? The ρ protein functions as a helicase, an enzyme that can separate RNA-DNA hybrid regions. After the *rut* site is synthesized in the RNA, ρ protein binds to the RNA and moves in the direction of RNA polymerase. The second component of ρ-dependent termination is the site where termination actually takes place. At this terminator site, the DNA encodes an RNA sequence containing several GC base pairs that form a stem-loop structure. RNA synthesis terminates several nucleotides beyond this stem-loop structure. As discussed in Chapter 9, a stem-loop structure, also called a hairpin, can form due to complementary sequences within the RNA (refer back to Figure 9.23). This stem-loop forms almost immediately after the RNA sequence is synthesized and quickly binds to RNA polymerase. This binding results in a conformational change that causes RNA polymerase to pause in its synthesis of RNA. The pause allows ρ protein to catch up to the stem-loop, pass through it, and break the hydrogen bonds between the DNA and RNA within the open complex. When this occurs, the completed RNA strand is separated from the DNA along with RNA polymerase.

Let's now turn our attention to **ρ-independent termination,** a process that does not require the ρ protein. In this case, the terminator is composed of two adjacent nucleotide sequences that function within the RNA (**Figure 12.11**). One is a uracil-rich sequence located at the 3' end of the RNA. The second sequence is adjacent to the uracil-rich sequence and promotes the formation of a stem-loop structure. As shown in Figure 12.11, the formation of the stem-loop causes RNA polymerase to pause in its synthesis of RNA. This pausing is stabilized by other proteins that bind to RNA polymerase. For example, a protein called NusA, which is bound to RNA polymerase, promotes pausing at stem-

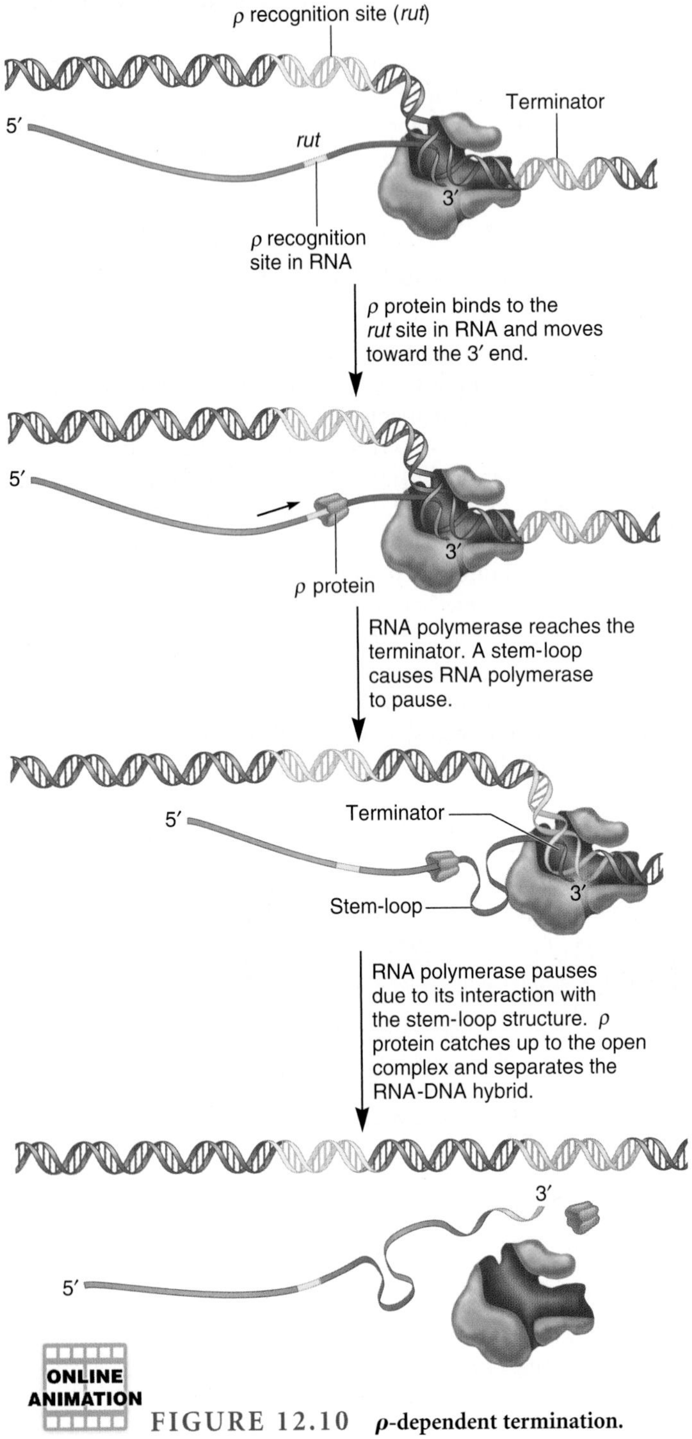

ρ recognition site (*rut*)
Terminator
5′
rut
3′
ρ recognition site in RNA
ρ protein binds to the *rut* site in RNA and moves toward the 3′ end.
5′
3′
ρ protein
RNA polymerase reaches the terminator. A stem-loop causes RNA polymerase to pause.
5′
Terminator
Stem-loop
3′
RNA polymerase pauses due to its interaction with the stem-loop structure. ρ protein catches up to the open complex and separates the RNA-DNA hybrid.
3′
5′
ONLINE ANIMATION

FIGURE 12.10 **ρ-dependent termination.**

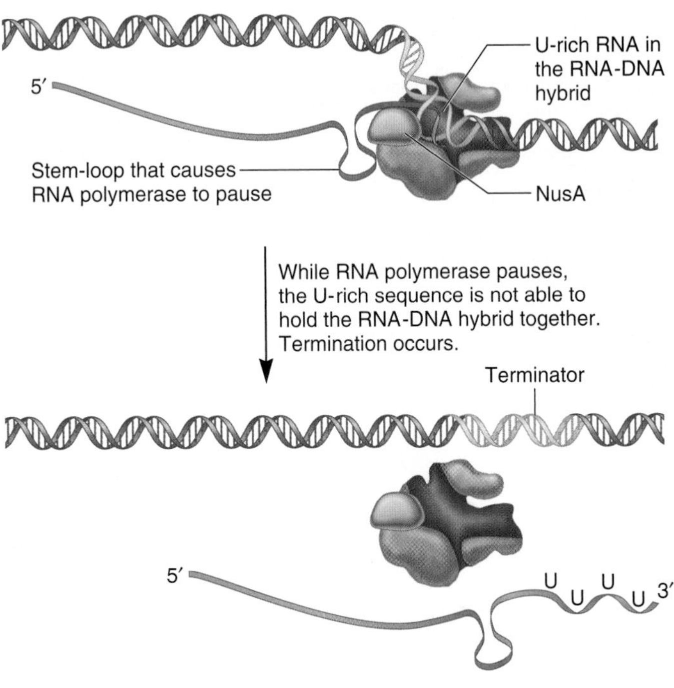

FIGURE 12.11 **ρ-independent or intrinsic termination.** When RNA polymerase reaches the end of the gene, it transcribes a uracil-rich sequence. As this uracil-rich sequence is transcribed, a stem-loop forms just upstream from the open complex. The formation of this stem-loop causes RNA polymerase to pause in its synthesis of the transcript. This pausing is stabilized by NusA, which binds near the region where RNA exits the open complex. While it is pausing, the RNA in the RNA-DNA hybrid is a uracil-rich sequence. Because hydrogen bonds between U and A are relatively weak interactions, the transcript and RNA polymerase dissociate from the DNA.

loop sequences. At the precise time RNA polymerase pauses, the uracil-rich sequence in the RNA transcript is bound to the DNA template strand. As previously mentioned, the hydrogen bonding of RNA to DNA keeps RNA polymerase clamped onto the DNA. However, the binding of this uracil-rich sequence to the DNA template strand is relatively weak, causing the RNA transcript to spontaneously dissociate from the DNA and cease further transcription. Because this process does not require a protein (the ρ protein) to physically remove the RNA transcript from the DNA, it is also referred to as **intrinsic termination.** In *E. coli,* about half of the genes show intrinsic termination, and the other half are terminated by ρ protein.

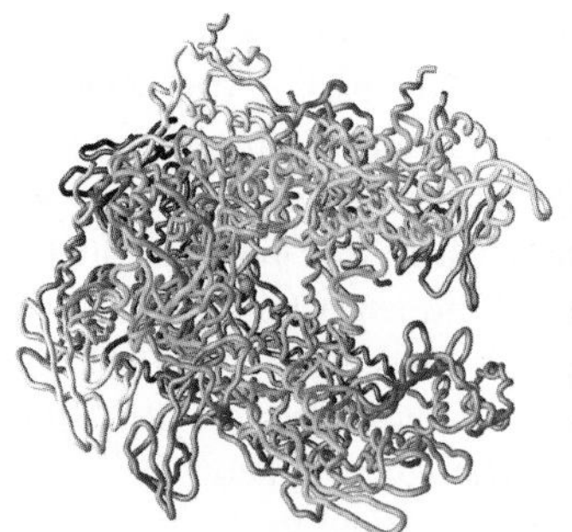

(a) Structure of a bacterial RNA polymerase

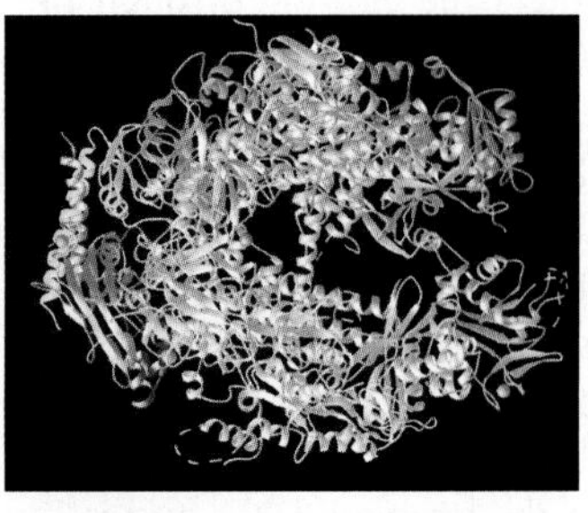

Structure of a eukaryotic RNA polymerase II (yeast)

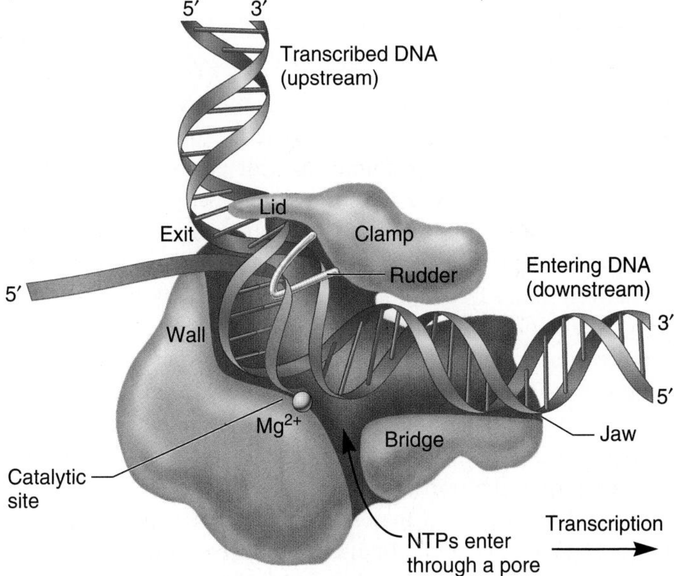

(b) Schematic structure of RNA polymerase

FIGURE 12.12 Structure and molecular function of RNA polymerase. (**a**) A comparison of the crystal structures of a bacterial RNA polymerase (left) to a eukaryotic RNA polymerase II (right). The bacterial enzyme is from *Thermus aquaticus.* The eukaryotic enzyme is from *Saccharomyces cerevisiae.* (**b**) A mechanism for transcription based on the crystal structure. In this diagram, the direction of transcription is from left to right. The double-stranded DNA enters the polymerase along a bridge surface that is between the jaw and clamp. At a region termed the wall, the RNA-DNA hybrid is forced to make a right-angle turn, which enables nucleotides to bind to the template strand. Mg^{2+} is located at the catalytic site. Nucleoside triphosphates (NTPs) enter the catalytic site via a pore region and bind to the template DNA. At the catalytic site, the nucleotides are covalently attached to the 3' end of the RNA. As RNA polymerase slides down the template, a small region of the protein termed the rudder separates the RNA-DNA hybrid. The single-stranded RNA then exits under a small lid.

12.3 TRANSCRIPTION IN EUKARYOTES

Many of the basic features of gene transcription are very similar in bacterial and eukaryotic species. Much of our understanding of transcription has come from studies in *Saccharomyces cerevisiae* (baker's yeast) and other eukaryotic species, including mammals. In general, gene transcription in eukaryotes is more complex than that of their bacterial counterparts. Eukaryotic cells are larger and contain a variety of compartments known as organelles. This added level of cellular complexity dictates that eukaryotes contain many more genes encoding cellular proteins. In addition, most eukaryotic species are multicellular, being composed of many different cell types. Multicellularity adds the requirement that genes be transcribed in the correct type of cell and during the proper stage of development. Therefore, in any given species, the transcription of the thousands of different genes that an organism possesses requires appropriate timing and coordination. In this section, we will examine features of gene transcription unique to eukaryotes. The regulation of eukaryotic gene transcription is covered in Chapter 15.

Eukaryotes Have Multiple RNA Polymerases That Are Structurally Similar to the Bacterial Enzyme

The genetic material within the nucleus of a eukaryotic cell is transcribed by three different RNA polymerase enzymes, designated RNA polymerase I, II, and III. What are the roles of these enzymes? Each of the three RNA polymerases transcribes different categories of genes. RNA polymerase I transcribes all of the genes that encode ribosomal RNA (rRNA) except for the 5S rRNA. RNA polymerase II plays a major role in cellular transcription because it transcribes all of the structural genes. It is responsible for the synthesis of all mRNA and also transcribes certain snRNA genes, which are needed for pre-mRNA splicing. RNA polymerase III transcribes all tRNA genes and the 5S rRNA gene.

All three RNA polymerases are structurally very similar and are composed of many subunits. They contain two large catalytic subunits similar to the β and β' subunits of bacterial RNA polymerase. The structures of RNA polymerase from a few different species have been determined by X-ray crystallography. A remarkable similarity exists between the bacterial enzyme and its eukaryotic counterparts. **Figure 12.12a** compares the structures of a bacterial RNA polymerase with RNA polymerase II from yeast. As seen here, both enzymes have a very similar structure. Also, it is very exciting that this structure provides a way to envision how the transcription process works. As seen in **Figure 12.12b**, DNA enters the enzyme through the jaw and lies on a surface within RNA polymerase termed the bridge. The part of the enzyme called the clamp is thought to control the movement of the DNA through RNA polymerase. A wall in the enzyme forces the RNA-DNA hybrid to make a right-angle turn. This bend facilitates the ability of nucleotides to bind to the template strand. Mg^{2+} is located at the catalytic site, which is precisely at the 3' end of the growing RNA strand. Nucleoside triphosphates (NTPs) enter the catalytic site via a pore region. The correct nucleotide binds to the template DNA and is covalently attached to the 3' end. As RNA polymerase slides down the template, a rudder, which is about 9 bp away from the 3' end of the RNA, forces the RNA-DNA hybrid apart. The single-stranded RNA then exits under a small lid.

Eukaryotic Structural Genes Have a Core Promoter and Regulatory Elements

In eukaryotes, the promoter sequence is more variable and often more complex than that found in bacteria. For structural genes, at least three features are found in most promoters: regulatory elements, a TATA box, and a transcriptional start site. Figure 12.13 shows a common pattern of sequences found within the promoters of eukaryotic structural genes. The **core promoter** is a relatively short DNA sequence that is necessary for transcription to take place. It consists of a TATAAA sequence called the **TATA box** and the **transcriptional start site,** where transcription begins. The TATA box, which is usually about 25 bp upstream from a transcriptional start site, is important in determining the precise starting point for transcription. If it is missing from the core promoter, the transcription start site point becomes undefined, and transcription may start at a variety of different locations. The core promoter, by itself, produces a low level of transcription. This is termed **basal transcription.**

Regulatory elements affect the ability of RNA polymerase to recognize the core promoter and begin the process of transcription. There are two categories of regulatory elements. Activating sequences, known as **enhancers,** are needed to stimulate transcription. In the absence of enhancer sequences, most eukaryotic genes have very low levels of basal transcription. Under certain conditions, it may also be necessary to prevent transcription of a given gene. This occurs via **silencers**—DNA sequences that inhibit transcription. As seen in Figure 12.13, a common location for regulatory elements is the –50 to –100 region. However, the locations of regulatory elements are quite variable among different eukaryotic genes. These elements can be far away from the core promoter yet exert strong effects on the ability of RNA polymerase to initiate transcription. Eukaryotic regulatory elements are recognized by transcription factors—proteins that bind to regulatory elements and influence the rate of transcription.

DNA sequences such as the TATA box, enhancers, and silencers exert their effects only over a particular gene. They are called ***cis*-acting elements.** The term *cis* comes from chemistry nomenclature meaning "next to." *Cis*-acting elements, though possibly far away from the core promoter, are always found within the same chromosome as the genes they regulate. By comparison, the regulatory transcription factors that bind to such elements are called ***trans*-acting factors** (the term *trans* means "across from"). The transcription factors that control the expression of a gene are themselves encoded by genes; regulatory genes that encode transcription factors may be far away from the genes they control. When a gene encoding a *trans*-acting factor is expressed, the transcription factor protein that is made can diffuse throughout the cell and bind to its appropriate *cis*-acting element. Let's now turn our attention to the function of such proteins.

Transcription of Eukaryotic Structural Genes Is Initiated When RNA Polymerase II and General Transcription Factors Bind to a Promoter Sequence

Thus far, we have considered the DNA sequences that play a role in the promoter region of eukaryotic structural genes. By studying transcription in a variety of eukaryotic species, researchers

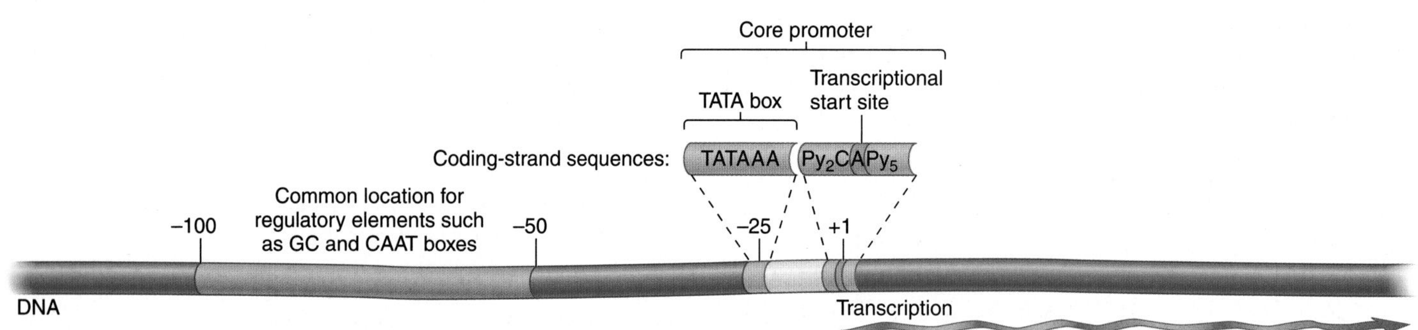

FIGURE 12.13 A common pattern found for the promoter of structural genes recognized by RNA polymerase II. The start site usually occurs at adenine; two pyrimidines (Py) and a cytosine precede this adenine, and five pyrimidines (Py; cytosine or thymine) follow it. A TATA box is approximately 25 bp upstream. However, the sequences that constitute eukaryotic promoters are quite diverse, and not all structural genes have a TATA box. Regulatory elements, such as GC or CAAT boxes, are variable in their locations but often are found in the –50 to –100 region. The core promoters for RNA polymerase I and III are quite different. A single upstream regulatory element is involved in the binding of RNA polymerase I to its promoter, while two regulatory elements called A and B boxes facilitate the binding of RNA polymerase III.

have discovered that three categories of proteins are needed for basal transcription at the core promoter: RNA polymerase II, general transcription factors, and mediator (**Table 12.2**).

Five different proteins called **general transcription factors (GTFs)** are always needed for RNA polymerase II to initiate transcription of structural genes. **Figure 12.14** describes the assembly of GTFs and RNA polymerase II at the TATA box. As shown here, a series of interactions leads to the formation of the open complex. Transcription factor IID (TFIID) first binds to the TATA box and thereby plays a critical role in the recognition of the core promoter. TFIID is composed of several subunits, including TATA-binding protein (TBP), which directly binds to the TATA box, and several other proteins called TBP-associated factors (TAFs). After TFIID binds to the TATA box, it associates with TFIIB. TFIIB promotes the binding of RNA polymerase II and TFIIF to the core promoter. Lastly, TFIIE and TFIIH bind to the complex. This completes the assembly of proteins to form a closed complex, also known as a **preinitiation complex.**

TFIIH plays a major role in the formation of the open complex. TFIIH has several subunits that perform different functions. Certain subunits act as helicases, which break the hydrogen bonding between the double-stranded DNA and thereby promote the formation of the open complex. Another subunit hydrolyzes ATP and phosphorylates a domain in RNA polymerase II known as the carboxyl terminal domain (CTD). Phosphorylation of the CTD releases the contact between RNA polymerase II and TFIIB.

TABLE **12.2**

Proteins Needed for Transcription via the Core Promoter of Eukaryotic Structural Genes

RNA polymerase II: The enzyme that catalyzes the linkage of ribonucleotides in the 5' to 3' direction, using DNA as a template. Essentially all eukaryotic RNA polymerase II proteins are composed of 12 subunits. The two largest subunits are structurally similar to the β and β' subunits found in *E. coli* RNA polymerase.

General transcription factors:

TFIID: Composed of TATA-binding protein (TBP) and other TBP-associated factors (TAFs). Recognizes the TATA box of eukaryotic structural gene promoters.

TFIIB: Binds to TFIID and then enables RNA polymerase II to bind to the core promoter. Also promotes TFIIF binding.

TFIIF: Binds to RNA polymerase II and plays a role in its ability to bind to TFIIB and the core promoter. Also plays a role in the ability of TFIIE and TFIIH to bind to RNA polymerase II.

TFIIE: Plays a role in the formation and/or the maintenance of the open complex. It may exert its effects by facilitating the binding of TFIIH to RNA polymerase II and regulating the activity of TFIIH.

TFIIH: A multisubunit protein that has multiple roles. First, certain subunits act as helicases and promote the formation of the open complex. Other subunits phosphorylate the carboxyl terminal domain (CTD) of RNA polymerase II, which releases its interaction with TFIIB and thereby allows RNA polymerase II to proceed to the elongation phase.

Mediator: A multisubunit complex that mediates the effects of regulatory transcription factors on the function of RNA polymerase II. Though mediator typically has certain core subunits, many of its subunits vary, depending on the cell type and environmental conditions. The ability of mediator to affect RNA polymerase II function is thought to occur via the CTD of RNA polymerase II. Mediator can influence the ability of TFIIH to phosphorylate CTD, and subunits within mediator itself have the ability to phosphorylate CTD. Because CTD phosphorylation is needed to release RNA polymerase II from TFIIB, mediator plays a key role in the ability of RNA polymerase II to switch from the initiation to the elongation stage of transcription.

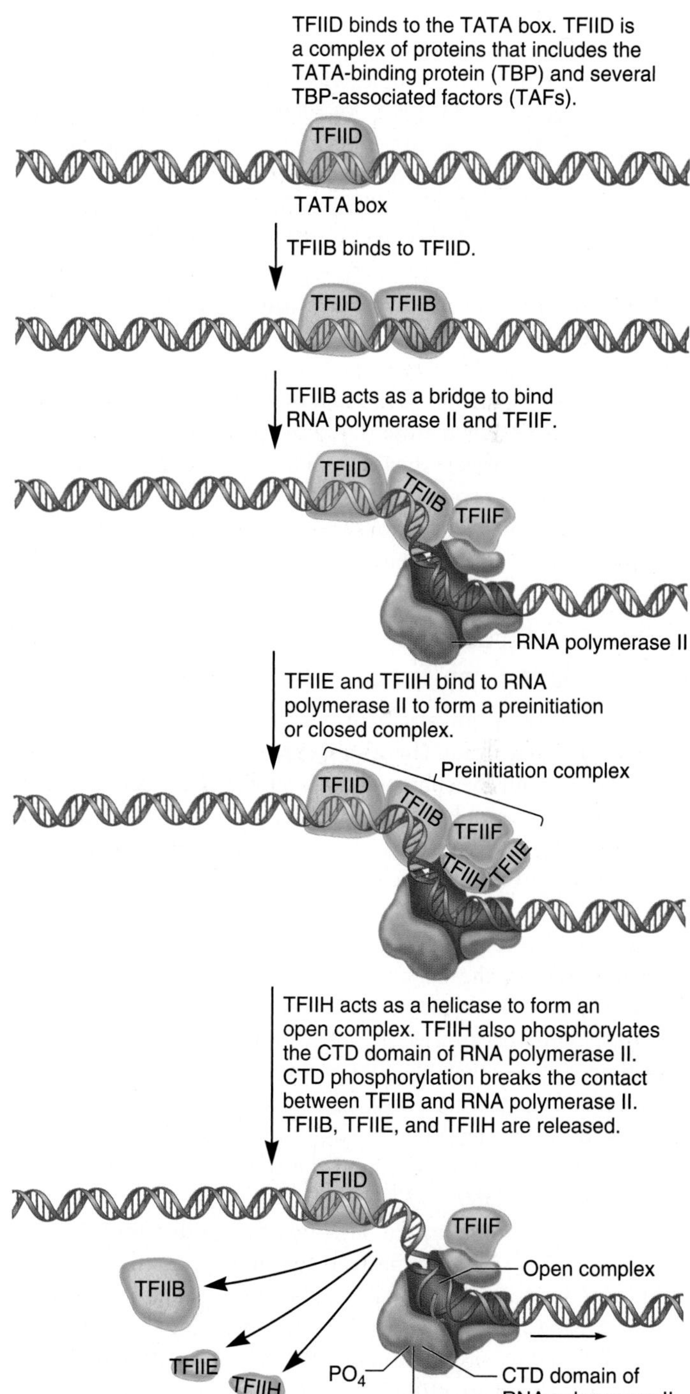

FIGURE 12.14 Steps leading to the formation of the open complex.

Next, TFIIB, TFIIE, and TFIIH dissociate, and RNA polymerase II is free to proceed to the elongation stage of transcription.

In vitro, when researchers mix together TFIID, TFIIB, TFIIF, TFIIE, TFIIH, RNA polymerase II, and a DNA sequence containing a TATA box and transcriptional start site, the DNA is transcribed into RNA. Therefore, these components are referred to as the **basal transcription apparatus.** In a living cell, however, additional components regulate transcription and allow it to proceed at a reasonable rate.

In addition to GTFs and RNA polymerase II, another component required for transcription is a large protein complex termed mediator. This complex was discovered by Roger Kornberg and colleagues in 1990. In 2006, Kornberg was awarded the Nobel Prize for his studies regarding the molecular basis of eukaryotic transcription. **Mediator** derives its name from the observation that it mediates interactions between RNA polymerase II and regulatory transcription factors that bind to enhancers or silencers. It serves as an interface between RNA polymerase II and many, diverse regulatory signals. The subunit composition of mediator is quite complex and variable. The core subunits form an elliptical-shaped complex that partially wraps around RNA polymerase II. Mediator itself may phosphorylate the CTD of RNA polymerase II and it may regulate the ability of TFIIH to phosphorylate the CTD. Therefore, it can play a pivotal role in the switch between transcriptional initiation and elongation. The function of mediator during eukaryotic gene regulation is explored in greater detail in Chapter 15.

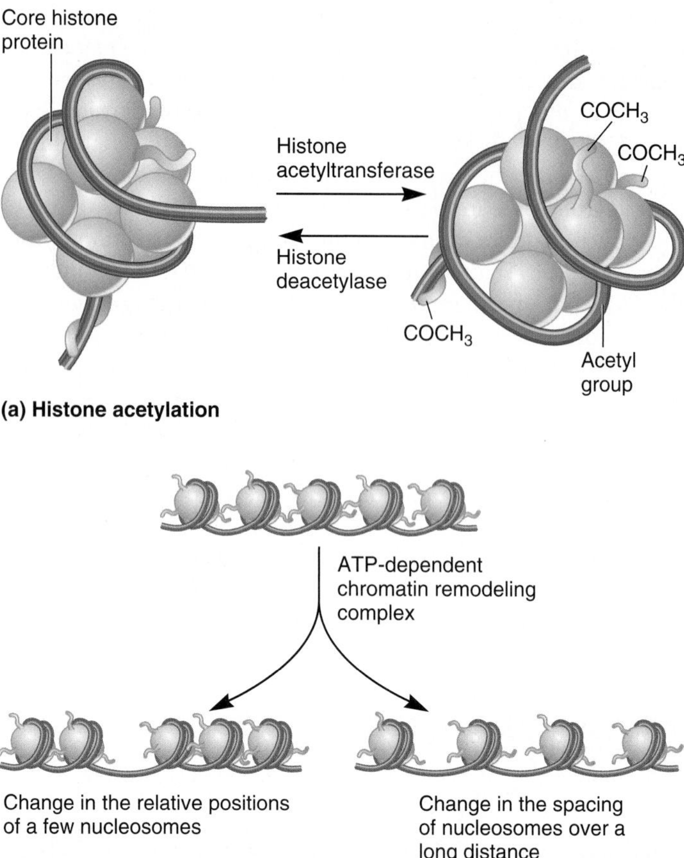

FIGURE 12.15 Mechanisms that disrupt the tight binding of DNA and histones within nucleosomes. **(a)** The covalent modification of the amino terminal tails of histones by acetylation, for example, alters their ability to bind to the DNA backbone. Histone acetyltransferase catalyzes the attachment of acetyl groups to lysine residues, thereby loosening the interactions between histones and DNA. Histone deacetylase reverses this process. **(b)** In ATP-dependent chromatin remodeling, proteins of the SWI/SNF family catalyze an ATP-dependent change in the locations of nucleosomes. This may involve a shift in nucleosomes to a new location (left) or a change in the spacing of nucleosomes over a long stretch of DNA (right). These effects may significantly alter gene transcription.

Chromatin Structure Plays a Key Role in Gene Transcription

As we have learned, eukaryotic transcription involves the binding of general transcription factors and RNA polymerase to the promoter region and the subsequent movement of RNA polymerase along the DNA double helix, allowing one strand to function as a template for transcription. The compaction of DNA to form chromatin can be an obstacle to the transcription process. During interphase, when most transcription occurs, the chromatin of eukaryotes is found in 30 nm fibers organized into radial loop domains. Within the 30 nm fiber, the DNA is wound around histone octamers to form nucleosomes (refer back to Chapter 10, Figure 10.16). The size of a histone octamer is roughly five times smaller than the complex of RNA polymerase II and GTFs. Therefore, because RNA polymerase is a very large enzyme compared to a nucleosome, the tight wrapping of DNA within a nucleosome is expected to inhibit the ability of RNA polymerase to transcribe the DNA.

How is chromatin structure loosened so that transcription can occur? Two common mechanisms alter chromatin structure. First, the amino terminal tails of the core histone proteins are covalently modified in a variety of ways, including the acetylation of lysines, methylation of lysines, and phosphorylation of serines (refer back to Chapter 10, Figure 10.20). These covalent modifications play a key role in the level of chromatin compaction. For example, positively charged lysine residues within the amino terminal tails of the core histone proteins can be acetylated by enzymes called **histone acetyltransferases (Figure 12.15a).** The attachment of the acetyl group ($-COCH_3$) eliminates the positive charge on the lysine side chain and thereby disrupts the favorable interaction between the histone protein and the negatively charged DNA backbone. This may loosen the structure of nucleosomes and the 30 nm fiber and thereby facilitate the ability of RNA polymerase to transcribe a gene. Some studies suggest that histones are completely displaced, whereas others suggest they are loosened but remain attached to the DNA. In addition, covalently modified histones can be recognized by proteins that alter the compaction level of the chromatin. Other enzymes, called

histone deacetylases, can remove the acetyl groups and thereby restore a tighter interaction. As discussed further in Chapter 15, the competing actions of acetyltransferases and deacetylases play an important role in the transcriptional regulation of genes.

A second way to alter chromatin structure is through **ATP-dependent chromatin remodeling** (**Figure 12.15b**). In this process, the energy of ATP hydrolysis is used to drive a change in the structure of nucleosomes and thereby makes the DNA more or less accessible for transcription. Therefore, chromatin remodeling is important for both activation and repression of transcription. The remodeling process is carried out by a protein complex that recognizes nucleosomes and uses ATP to alter their configuration. The complexes that catalyze ATP-dependent chromatin remodeling are members of the **SWI/SNF family,** a large group of proteins found in all eukaryotic species. They were first identified in yeast as proteins that play a role in the activation of gene transcription. The abbreviations SWI and SNF refer to the effects that occur in yeast when these remodeling enzymes are defective. SWI mutants are defective in mating type switching, and SNF mutations create a sucrose nonfermenting phenotype.

As shown in Figure 12.15b, one result of ATP-dependent chromatin remodeling is a change in the locations of nucleosomes, involving either a shift in nucleosomes to a new location or an alteration in the spacing of nucleosomes over a long stretch of DNA. These effects on nucleosome location may substantially alter the ability of general transcription factors to recognize the promoter and also influence the ability of RNA polymerase to transcribe a given gene. More recently, research has shown that chromatin remodeling may also affect the configuration of the 30 nm fiber and/or the radial loop domains found in euchromatin.

12.4 RNA MODIFICATION

During the 1960s and 1970s, studies in bacteria established the physical structure of the gene. The analysis of bacterial genes showed that the sequence of DNA within the coding strand corresponds to the sequence of nucleotides in the mRNA, except that T is replaced with U. During translation, the sequence of codons in the mRNA is then read, providing the instructions for the correct amino acid sequence in a polypeptide. The one-to-one correspondence between the sequence of codons in the DNA coding strand and the amino acid sequence of the polypeptide has been termed the **colinearity** of gene expression.

The situation dramatically changed in the late 1970s, when the tools became available to study eukaryotic genes at the molecular level. The scientific community was astonished by the discovery that eukaryotic structural genes are not always colinear with their functional mRNAs. Instead, the coding sequences within many eukaryotic genes are separated by DNA sequences that are not translated into protein. The coding sequences are called **exons,** and the sequences that interrupt them are called **intervening sequences** or **introns.** During transcription, an RNA is made corresponding to the entire gene sequence. Subsequently, the sequences in the RNA that correspond to the introns are removed, while the RNA sequences derived from the exons are connected or spliced together. This process is called **RNA splicing.** Since the 1970s, research has revealed that splicing is a common genetic phenomenon in eukaryotic species. Splicing occurs occasionally in bacteria as well.

Aside from splicing, research has also shown that RNA transcripts can be modified in several other ways. **Table 12.3** describes the general types of RNA modifications. For example, rRNAs and tRNAs are synthesized as long transcripts that are processed into smaller functional pieces. In addition, most eukaryotic mRNAs have a cap attached to their 5' end and a tail attached at their 3' end. In this section, we will examine the molecular mechanisms that account for several types of RNA modifications and consider why they are functionally important.

Some Large RNA Transcripts Are Processed into Smaller Functional Transcripts by Enzymatic Cleavage

For many nonstructural genes, the RNA transcript initially made during gene transcription is processed or cleaved into smaller pieces. As an example, **Figure 12.16** shows the processing of mammalian ribosomal RNA. The ribosomal RNA gene is transcribed by RNA polymerase I to make a long primary transcript, known as 45S rRNA. The term 45S refers to the sedimentation characteristics of this transcript in Svedberg units. Following the synthesis of the 45S rRNA, cleavage occurs at several points to produce three fragments, termed 18S, 5.8S, and 28S rRNA. These are functional rRNA molecules that play a key role in creating the structure of the ribosome. In eukaryotes, the processing of 45S rRNA into smaller rRNAs and the assembly of ribosomal subunits occur in a structure within the cell nucleus known as the **nucleolus.**

The production of tRNA molecules requires processing via exonucleases and endonucleases. An **exonuclease** is a type of enzyme that cleaves a covalent bond between two nucleotides at one end of a strand. Starting at one end, an exonuclease can digest a strand, one nucleotide at a time. Some exonucleases can begin this digestion only from the 3' end, traveling in the 3' to 5' direction, while others can begin only at the 5' end and digest in the 5' to 3' direction. By comparison, an **endonuclease** can cleave the bond between two adjacent nucleotides within a strand.

Like ribosomal RNA, tRNAs are synthesized as large precursor tRNAs that must be cleaved to produce mature, functional tRNAs that bind to amino acids. This processing has been studied extensively in *E. coli*. **Figure 12.17** shows the processing of a precursor tRNA, which involves the action of two endonucleases and one exonuclease. The precursor tRNA is recognized by an enzyme known as RNaseP. This enzyme is an endonuclease that cuts the precursor tRNA. The action of RNaseP produces the correct 5' end of the mature tRNA. A different endonuclease cleaves the precursor tRNA to remove a 170-nucleotide segment from the 3' end. Next, an exonuclease, called RNaseD, binds to the 3' end and digests the RNA in the 3' to 5' direction. When it reaches an ACC sequence, the exonuclease stops digesting the precursor tRNA molecule. Therefore, all tRNAs in *E. coli* have

TABLE 12.3

Modifications That May Occur to RNAs

Modification	Description
Processing 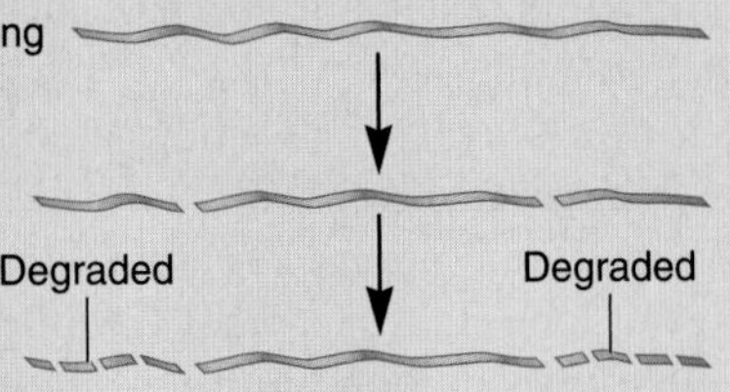 (Degraded / Degraded)	The cleavage of a large RNA transcript into smaller pieces. One or more of the smaller pieces becomes a functional RNA molecule. Processing occurs for rRNA and tRNA transcripts.
Splicing 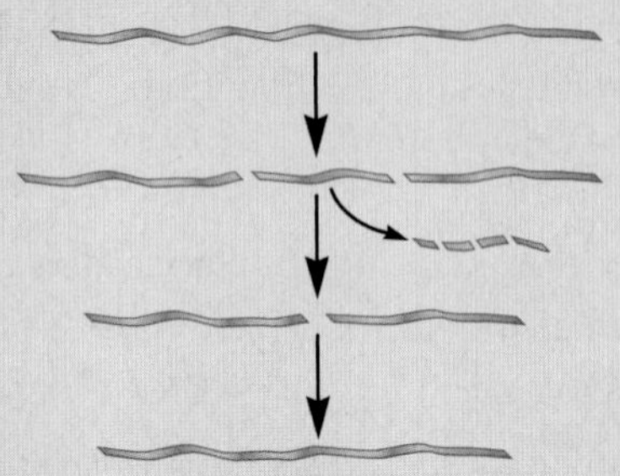	Splicing involves both cleavage and joining of RNA molecules. The RNA is cleaved at two sites, which allows an internal segment of RNA, known as an intron, to be removed. After the intron is removed, the two ends of the RNA molecules are joined together. Splicing is common among eukaryotic pre-mRNAs, and it also occurs occasionally in rRNAs, tRNAs, and a few bacterial RNAs.
5′ capping 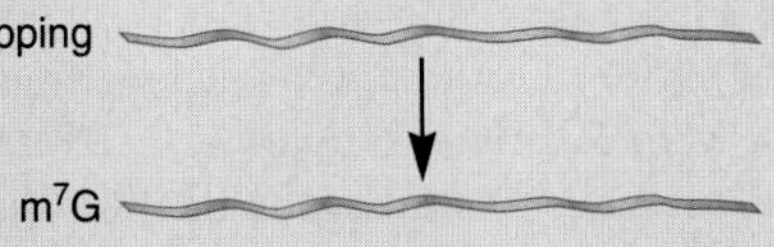 (m^7G)	The attachment of a 7-methylguanosine cap (m^7G) to the 5' end of mRNA. This occurs in eukaryotic mRNAs. The cap plays a role in the splicing of introns, the exit of mRNA from the nucleus, and the binding of mRNA to the ribosome.
3′ polyA tailing 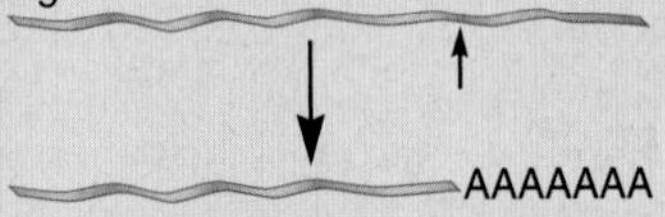 (AAAAAAA)	The attachment of a string of adenine-containing nucleotides to the 3' end of mRNA at a site where the mRNA is cleaved (see upward arrow). This occurs in eukaryotic mRNAs. It is important for RNA stability and translation.
RNA editing 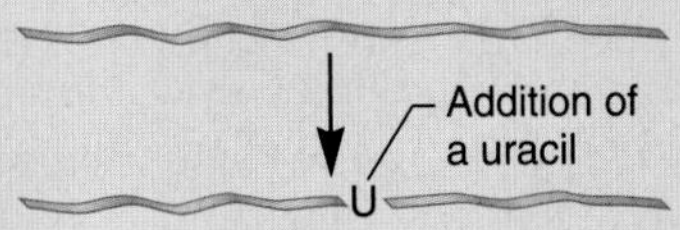 (Addition of a uracil; U)	The change of the base sequence of an RNA after it has been transcribed (described in Chapter 15).
Base modification 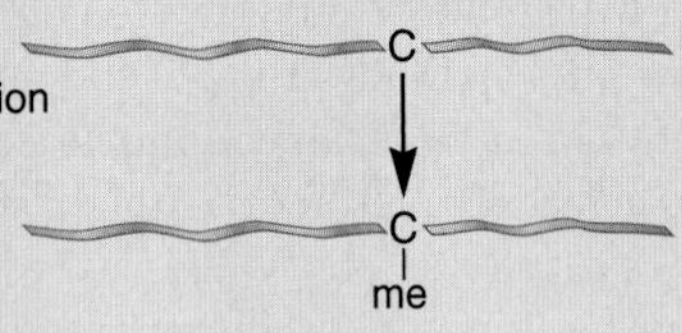 (C → C–me)	The covalent modification of a base within an RNA molecule. As described in Chapter 13, base modification commonly occurs in tRNA molecules.

an ACC sequence at their 3' ends. Finally, certain bases in tRNA molecules may be covalently modified to alter their structure. The functional importance of modified bases in tRNAs is discussed in Chapter 13.

As researchers studied tRNA processing, they discovered certain enzymatic features that were very unusual and exciting, changing the way biologists view the actions of enzymes. RNaseP has been found to be an enzyme that contains both RNA and protein subunits. In 1983, Sidney Altman and colleagues made the surprising discovery that the RNA portion of this enzyme, not the protein subunit, contains the catalytic ability to cleave the precursor tRNA. RNaseP is an example of a **ribozyme,** an RNA molecule with enzymatic activity. Prior to the study of RNaseP and the identification of self-splicing RNAs (discussed later), biochemists had staunchly believed that only protein molecules could function as enzymes.

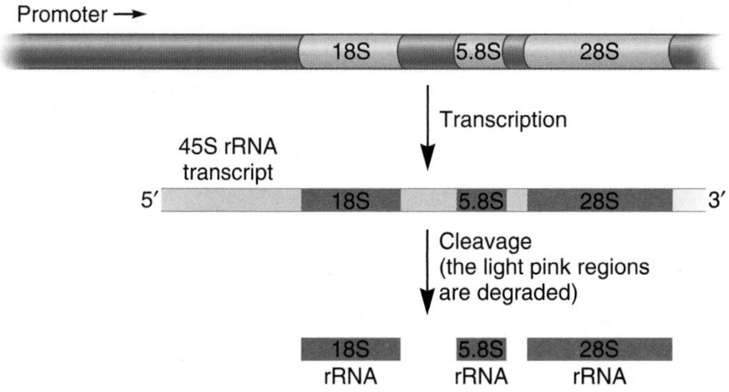

FIGURE 12.16 The processing of ribosomal RNA in eukaryotes. The large ribosomal RNA gene is transcribed into a long 45S rRNA primary transcript. This transcript is cleaved to produce 18S, 5.8S, and 28S rRNA molecules, which become associated with protein subunits in the ribosome. This processing occurs within the nucleolus of the cell.

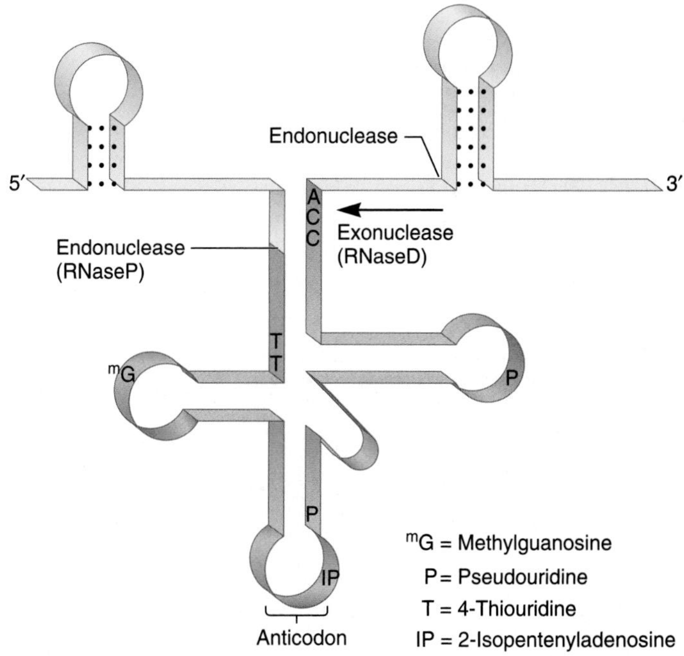

FIGURE 12.17 The processing of precursor tRNA molecules. RNaseP is an endonuclease that makes a cut that creates the 5' end of the mature tRNA. To produce the 3' end of mature tRNA, an endonuclease makes a cut, and then the exonuclease RNaseD removes nine nucleotides at the 3' end. In addition to these cleavage steps, several bases within the tRNA molecule are modified to other bases.

EXPERIMENT 12A

Introns Were Experimentally Identified Via Microscopy

Although the discovery of ribozymes was very surprising, the observation that tRNA and rRNA transcripts are processed to a smaller form did not seem unusual to geneticists and biochemists, because the enzymatic cleavage of RNA was similar to the cleavage that can occur for other macromolecules such as DNA and proteins. In sharp contrast, when splicing was detected in the 1970s, it was a novel concept. Splicing involves cleavage at two sites. An intron is removed, and—in a unique step—the remaining fragments are hooked back together again.

Eukaryotic introns were first detected by comparing the base sequence of viral genes and their mRNA transcripts during viral infection of mammalian cells by adenovirus. This research was carried out in 1977 by two groups headed by Philip Sharp and Richard Roberts. This pioneering observation led to the next question: Are introns a peculiar phenomenon that occurs only in viral genes, or are they found in eukaryotic genes as well?

In the late 1970s, several research groups, including those of Pierre Chambon, Bert O'Malley, and Phillip Leder, investigated the presence of introns in eukaryotic structural genes. The experiments of Leder used electron microscopy to identify introns in the β-globin gene. β globin is a polypeptide that is a subunit of hemoglobin, the protein that carries oxygen in red blood cells. To detect introns within the gene, Leder considered the possible effects of mRNA binding to a gene. Figure 12.18a considers the situation in which a gene does not contain an intron. In this experiment, a segment of double-stranded chromosomal DNA containing a gene was first denatured and mixed with mature mRNA encoded by that gene. Because the mRNA is complementary to the template strand of the DNA, the template strand and the mRNA will bind to each other to form a hybrid molecule. This event is called **hybridization.** Later, when the DNA is allowed to renature, the binding of the mRNA to the template strand of DNA prevents the two strands of DNA from forming a double helix. In the absence of any introns, the single-stranded DNA will form a loop. Because the RNA has displaced one of the DNA strands, this structure is known as an RNA displacement loop, or **R loop,** as shown in Figure 12.18a.

In contrast, Leder and colleagues realized that a different type of R loop structure would be formed if the gene contained an intron (Figure 12.18b). When mRNA is hybridized to a region of a gene containing one intron and then the other DNA strand is allowed to renature, two single-stranded R loops will form that are separated by a double-stranded DNA region. The intervening double-stranded region occurs because an intron has been spliced out of the mature mRNA, so the mRNA cannot hybridize to this segment of the gene.

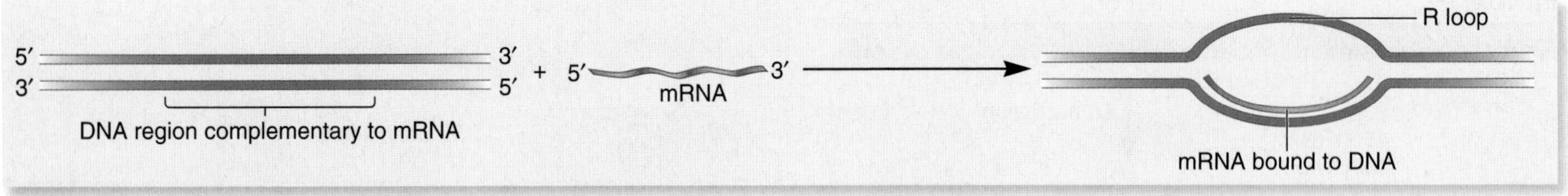

(a) No introns in the DNA

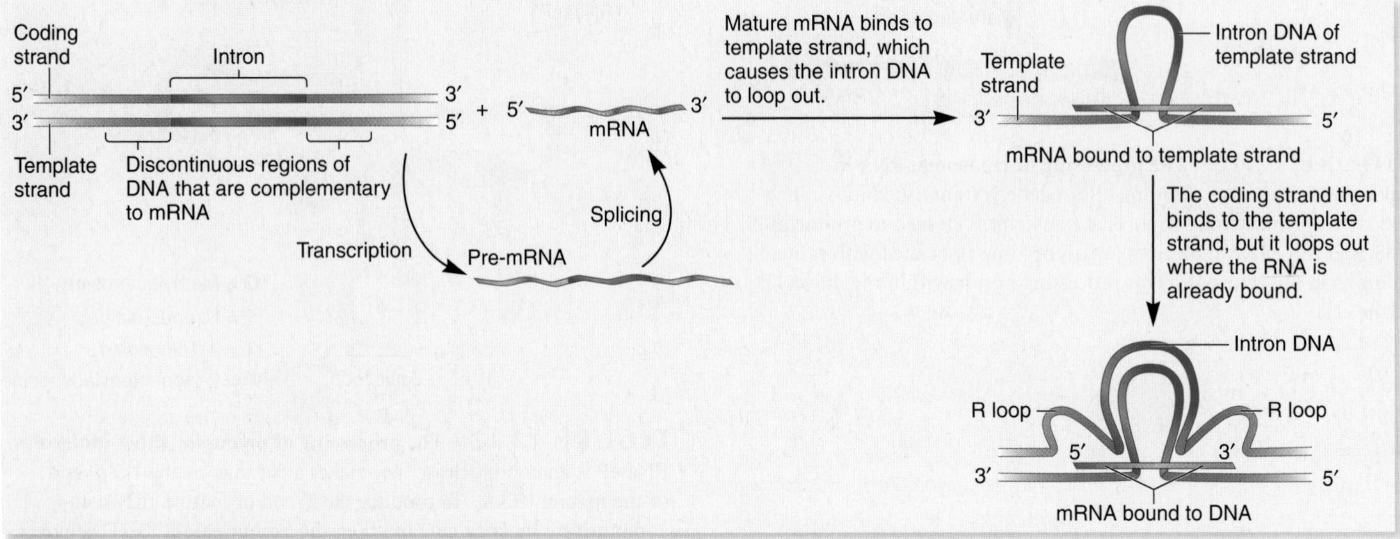

(b) One intron in the DNA. The intron in the pre-mRNA is spliced out.

FIGURE 12.18 **Hybridization of mRNA to double-stranded DNA.** In this experiment, the DNA is denatured and then allowed to renature under conditions that favor an RNA-DNA hybrid. (**a**) If the DNA does not contain an intron, the binding of the mRNA to the template strand of DNA prevents the two strands of DNA from forming a double helix. The single-stranded region of DNA will form an R loop. (**b**) When mRNA hybridizes to a gene containing one intron, two single-stranded R loops will form that are separated by a double-stranded DNA region. The intervening double-stranded region occurs because an intron has been spliced out of the mRNA and the mRNA cannot hybridize to this segment of the gene.

As shown in steps 1 through 4 of **Figure 12.19**, this hybridization approach was used to identify introns within the β-globin gene. Following hybridization, the samples were placed on a microscopy grid, shadowed with heavy metal, and then observed by electron microscopy.

THE HYPOTHESIS

The mouse β-globin gene contains one or more introns.

TESTING THE HYPOTHESIS — FIGURE 12.19 RNA hybridization to the β-globin gene reveals an intron.

Starting material: A cloned fragment of chromosomal DNA that contains the mouse β-globin gene.

1. Isolate mature mRNA for the mouse β-globin gene. Note: Globin mRNA is abundant in reticulocytes, which are immature red blood cells.

2. Mix together the β-globin mRNA and cloned DNA of the β-globin gene.

3. Separate the double-stranded DNA and allow the mRNA to hybridize. This is done using 70% formamide, at 52°C, for 16 hours.

4. Dilute the sample to decrease the formamide concentration. This allows the DNA to re-form a double-stranded structure. Note: The DNA cannot form a double-stranded structure in regions where the mRNA has already hybridized.

5. Spread the sample onto a microscopy grid.

6. Stain with uranyl acetate and shadow with heavy metal. Note: The technique of electron microscopy is described in the Appendix.

7. View the sample under the electron microscope.

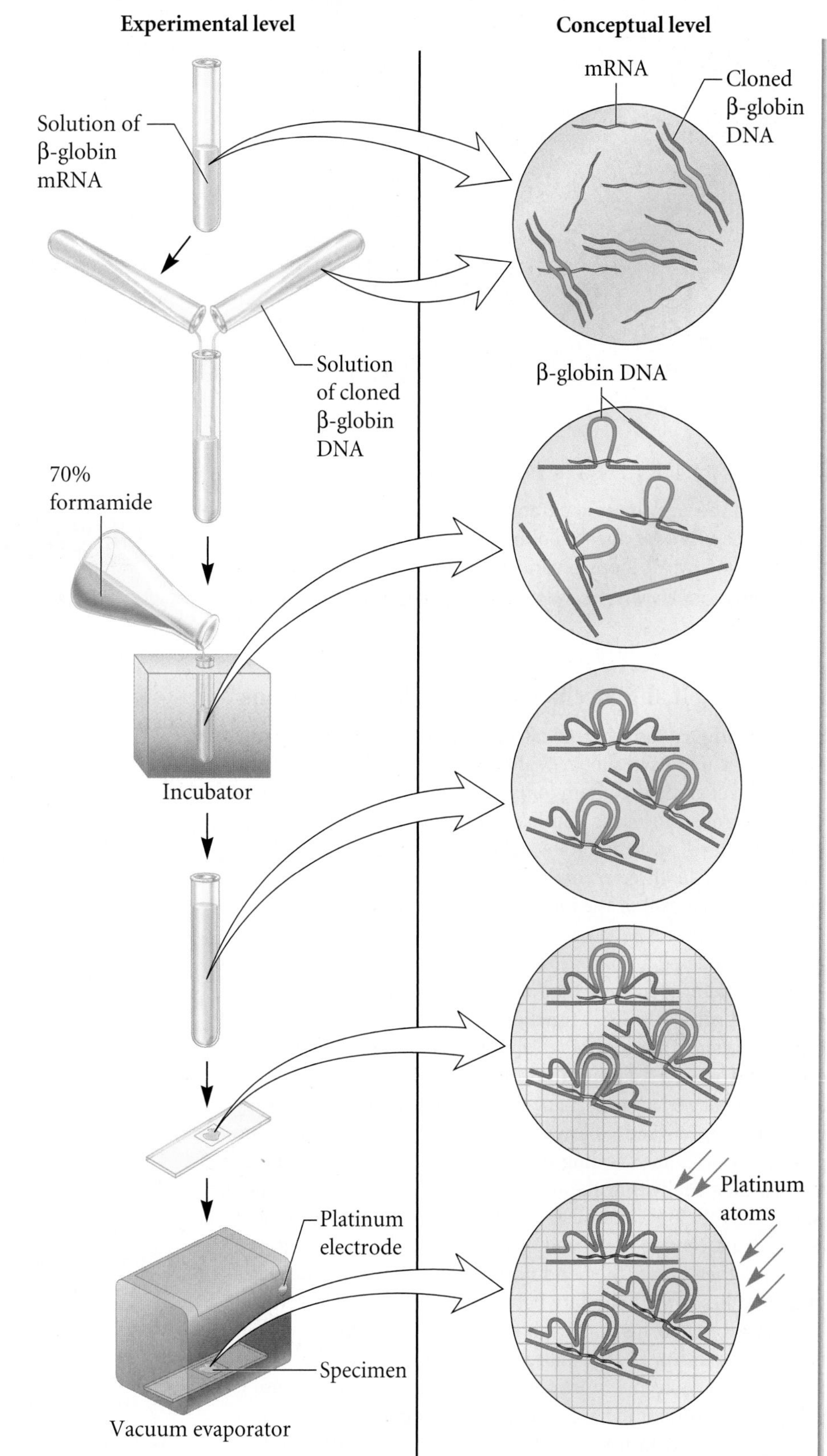

THE DATA

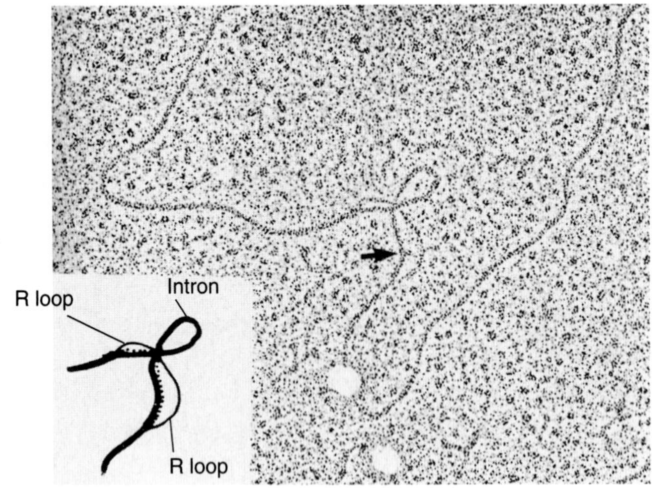

INTERPRETING THE DATA

As seen in the electron micrograph, the β-globin mRNA hybridized to the DNA of the β-globin gene, which resulted in the formation of two R loops separated by a double-stranded DNA region. These data were consistent with the idea that the DNA of the β-globin gene contains an intron. Similar results were obtained by Chambon and O'Malley for other structural genes. Since these initial discoveries, introns have been found in many eukaryotic genes. The prevalence and biological significance of introns are discussed later in this chapter.

Since the late 1970s, DNA sequencing methods have permitted an easier and more precise way of detecting introns. Researchers can clone a fragment of chromosomal DNA that contains a particular gene. This is called a **genomic clone.** In addition, mRNA can be used as a starting material to make a copy of DNA known as **complementary DNA (cDNA).** The cDNA will not contain introns, because the introns have been previously removed during RNA splicing. In contrast, if a gene contains introns, a genomic clone for a eukaryotic gene will also contain introns. Therefore, a comparison of the DNA sequences from genomic and cDNA clones can provide direct evidence that a particular gene contains introns. Compared to genomic DNA, the cDNA will be missing base sequences that were removed during splicing.

A self-help quiz involving this experiment can be found at the Online Learning Center.

Different Splicing Mechanisms Can Remove Introns

Since the original discovery of introns, the investigations of many research groups have shown that most structural genes among complex eukaryotes contain one or more introns. Less commonly, introns can occur within tRNA and rRNA genes. At the molecular level, different RNA splicing mechanisms have been identified. In the three examples shown in **Figure 12.20**, splicing leads to the removal of the intron RNA and the covalent connection of the exon RNA.

The splicing among **group I** and **group II introns** occurs via **self-splicing**—splicing that does not require the aid of other enzymes. Instead, the RNA functions as its own ribozyme. Group I and II differ in the ways that the introns are removed and the exons are connected. Group I introns that occur within the rRNA of *Tetrahymena* (a protozoan) have been studied extensively by Thomas Cech and colleagues. In this organism, the splicing process involves the binding of a single guanosine to a guanosine binding site within the intron (Figure 12.20a). This guanosine breaks the bond between the first exon and the intron and becomes attached to the 5' end of the intron. The 3'—OH group of exon 1 then breaks the bond next to a different guanine nucleotide that lies at the boundary between the end of the intron and exon 2; exon 1 forms a phosphoester bond with the 5' end of exon 2. The intron RNA is subsequently degraded. In this example, the RNA molecule functions as its own ribozyme, because it splices itself without the aid of a catalytic protein.

In group II introns, a similar splicing mechanism occurs, except the 2'–OH group on ribose found in an adenine nucleotide already within the intron strand begins the catalytic process (Figure 12.20b). Experimentally, group I and II self-splicing can occur in vitro without the addition of any proteins. However, in a living cell, proteins known as **maturases** often enhance the rate of splicing of group I and II introns.

In eukaryotes, the transcription of structural genes produces a long transcript known as **pre-mRNA,** which is located within the nucleus. These large RNA transcripts are also known as **heterogeneous nuclear RNA (hnRNA).** This pre-mRNA is usually altered by splicing and other modifications before it exits the nucleus. Unlike group I and II introns, which may undergo self-splicing, pre-mRNA splicing requires the aid of a multicomponent structure known as the **spliceosome.** As discussed shortly, this is needed to recognize the intron boundaries and to properly remove it.

Table 12.4 describes the occurrence of introns among the genes of different species. The biological significance of group I and II introns is not understood. By comparison, pre-mRNA splicing is a widespread phenomenon among complex eukaryotes. In mammals and flowering plants, most structural genes have at least one intron that can be located anywhere within the gene. In some cases, a single gene can have many introns. As an extreme example, the human dystrophin gene that, when mutant, causes Duchenne muscular dystrophy has 79 exons punctuated by 78 introns.

Pre-mRNA Splicing Occurs by the Action of a Spliceosome

As noted previously, the spliceosome is a large complex that splices pre-mRNA in eukaryotes. It is composed of several subunits known as **snRNPs** (pronounced "snurps"). Each snRNP

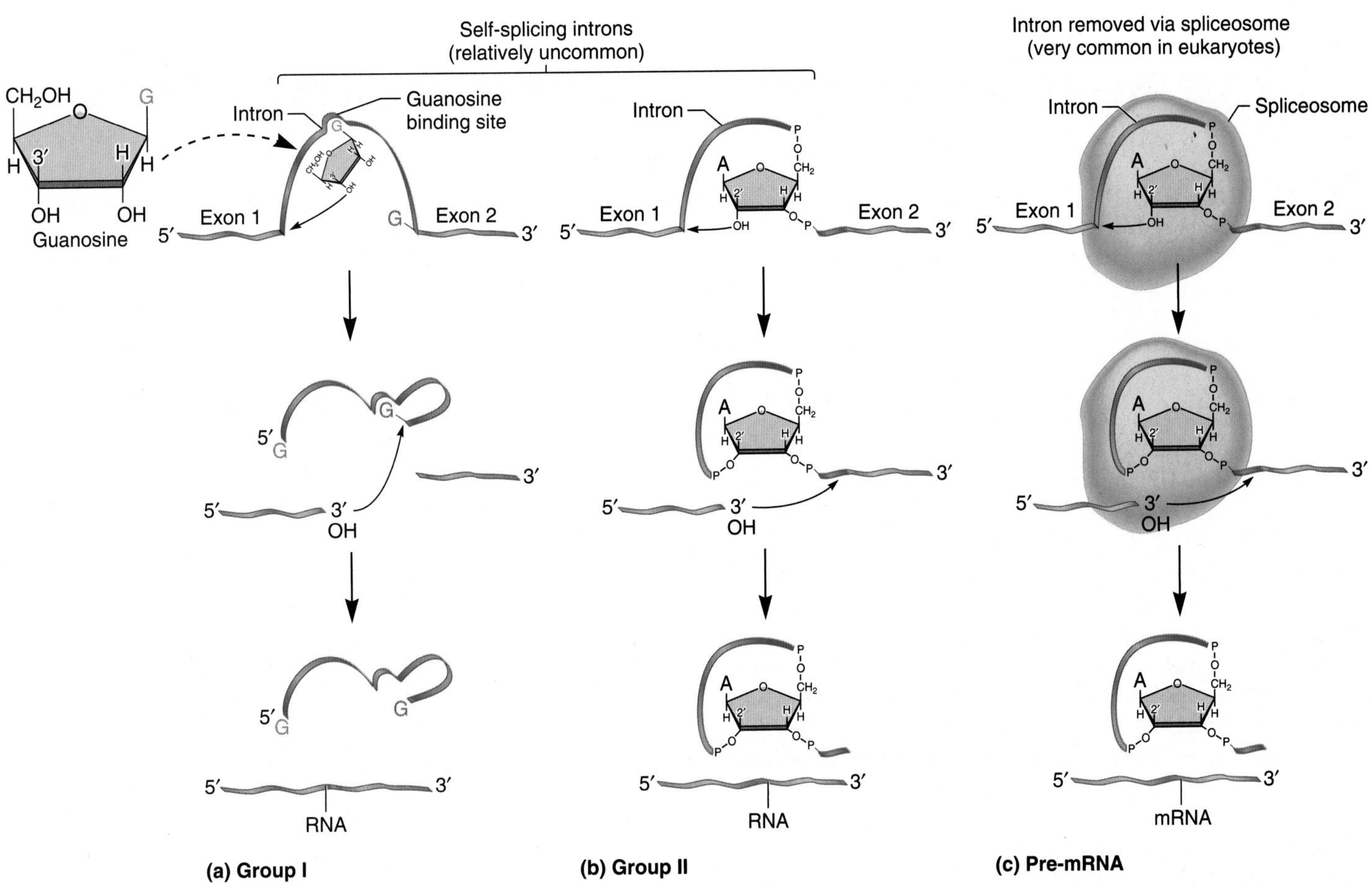

FIGURE 12.20 Mechanisms of RNA splicing. Group I and II introns are self-splicing. (**a**) The splicing of group I introns involves the binding of a free guanosine to a site within the intron, leading to the cleavage of RNA at the 3' end of exon 1. The bond between a different guanine nucleotide (in the intron strand) and the 5' end of exon 2 is cleaved. The 3' end of exon 1 then forms a covalent bond with the 5' end of exon 2. (**b**) In group II introns, a similar splicing mechanism occurs, except that the 2' –OH group on an adenine nucleotide (already within the intron) begins the catalytic process. (**c**) Pre-mRNA splicing requires the aid of a multicomponent structure known as the spliceosome.

TABLE 12.4

Occurrence of Introns

Type of Intron	Mechanism of Removal	Occurrence
Group I	Self-splicing	Found in rRNA genes within the nucleus of *Tetrahymena* and other simple eukaryotes. Found in a few structural, tRNA, and rRNA genes within the mitochondrial DNA (fungi and plants) and in chloroplast DNA. Found very rarely in tRNA genes within bacteria.
Group II	Self-splicing	Found in a few structural, tRNA, and rRNA genes within the mitochondrial DNA (fungi and plants) and in chloroplast DNA. Also found rarely in bacterial genes.
Pre-mRNA	Spliceosome	Very commonly found in structural genes within the nucleus of eukaryotes.

contains small nuclear RNA and a set of proteins. During splicing, the subunits of a spliceosome carry out several functions. First, spliceosome subunits bind to an intron sequence and precisely recognize the intron-exon boundaries. In addition, the spliceosome must hold the pre-mRNA in the correct configuration to ensure the splicing together of the exons. And finally, the spliceosome catalyzes the chemical reactions that cause the introns to be removed and the exons to be covalently linked.

Intron RNA is defined by particular sequences within the intron and at the intron-exon boundaries. The consensus sequences for the splicing of mammalian pre-mRNA are shown in **Figure 12.21**. These sequences serve as recognition sites for the binding of the spliceosome. The bases most commonly found at these sites—those that are highly conserved evolutionarily—are shown in bold. The 5' and 3' splice sites occur at the ends of the intron, while the branch site is somewhere in the middle. These sites are recognized by components of the spliceosome.

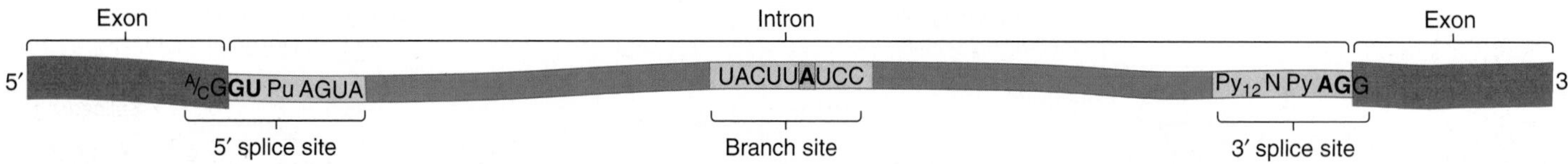

ONLINE ANIMATION **FIGURE 12.21** **Consensus sequences for pre-mRNA splicing in complex eukaryotes.** Consensus sequences exist at the intron-exon boundaries and at a branch site found within the intron itself. The adenine nucleotide shown in blue in this figure corresponds to the adenine nucleotide at the branch site in Figure 12.22. The nucleotides shown in bold are highly conserved. Designations: A/C = A or C, Pu = purine, Py = pyrimidine, N = any of the four bases. The 5' splice site is also called the splice donor site, and the 3' splice site is the splice acceptor site.

The molecular mechanism of pre-mRNA splicing is depicted in **Figure 12.22.** The snRNP designated U1 binds to the 5' splice site, while U2 binds to the branch site. This is followed by the binding of a trimer of three snRNPs: a U4/U6 dimer plus U5. The intron loops outward, and the two exons are brought closer together. The 5' splice site is then cut, and the 5' end of the intron becomes covalently attached to the 2'–OH group of a specific adenine nucleotide in the branch site. U1 and U4 are released. In the final step, the 3' splice site is cut, and then the exons are covalently attached to each other. The three snRNPs, U2, U5, and U6, remain attached to the intron, which is in a lariat configuration. Eventually, the intron will be degraded, and the snRNPs will be used again to splice other pre-mRNAs.

The chemical reactions that occur during pre-mRNA splicing are not completely understood. Though further research is needed, evidence is accumulating that certain snRNA molecules within the spliceosome may play an enzymatic role in the removal of introns and the connection of exons. In other words, snRNAs may function as ribozymes that cleave the RNA at the exon-intron boundaries and connect the remaining exons. Researchers have speculated that RNA molecules within U2 and U6 may have this catalytic function.

Before ending our discussion of pre-mRNA splicing, let's consider why it may be an advantage for a species to have genes that contain introns. One benefit is a phenomenon called **alternative splicing.** When a pre-mRNA has multiple introns, variation may occur in the pattern of splicing, so the resulting mRNAs contain alternative combinations of exons. The variation in splicing may happen in different cell types or during different stages of development. Such alternative splicing allows mature mRNAs to contain different patterns of exons.

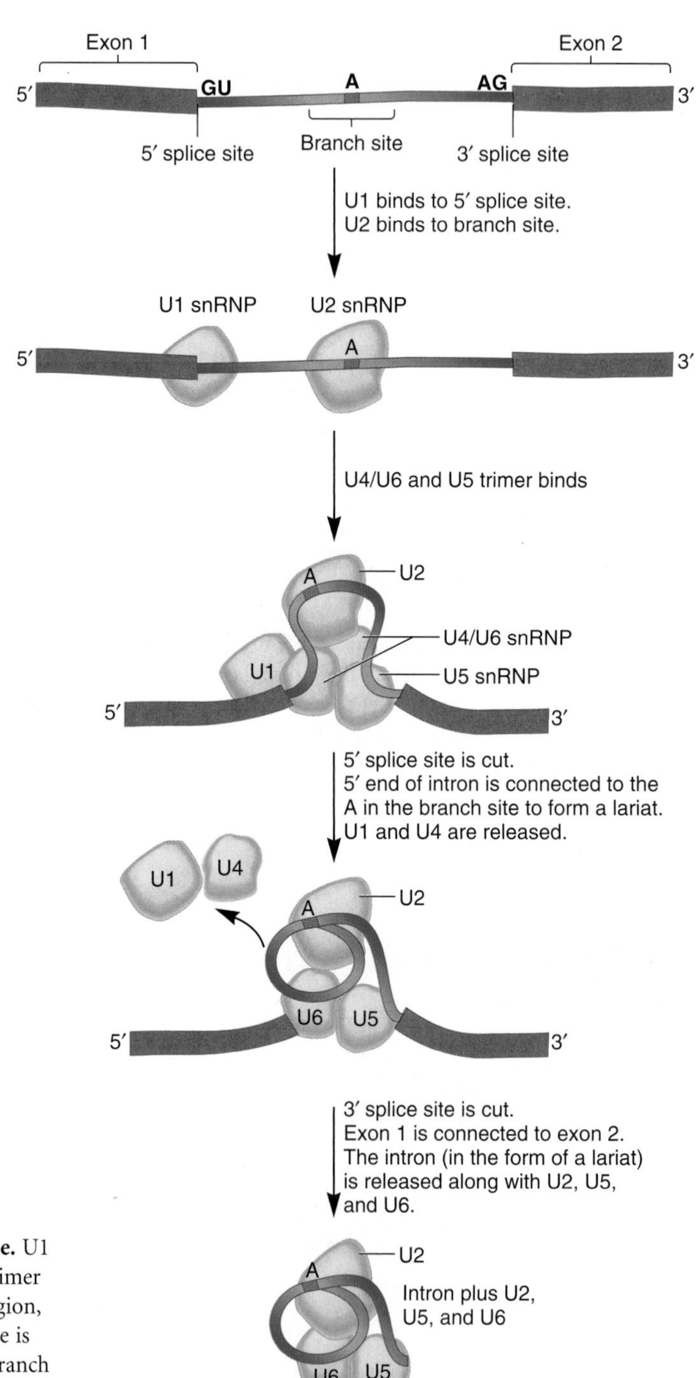

ONLINE ANIMATION **FIGURE 12.22** **Splicing of pre-mRNA via a spliceosome.** U1 binds to the 5' splice site, while U2 binds to the branch site. A trimer of three snRNPs, a U4/U6 dimer plus U5, binds to the intron region, causing it to form a loop. This brings the exons close together. The 5' splice site is cut, and the 5' end of the intron is attached to the adenine nucleotide in the branch site. U1 and U4 are released. In the final step, the 3' splice site is cut, and then the exons are connected together. The three snRNPs, U2, U5, and U6, remain bound to the intron, which is found in a lariat configuration. Later, the intron will be degraded, and the snRNPs will be used over again.

As an example, let's suppose a mammalian pre-mRNA contained seven exons. In one cell type (e.g., muscle cells), it may be spliced to produce a mature mRNA with the following pattern of exons: 1-2-4-5-6-7. In a different cell type (e.g., nerve cells), it might be spliced in an alternative pattern: 1-2-3-5-6-7. In this example, the mRNA from muscle cells contains exon 4, while the mRNA from nerve cells contains exon 3. When alternative splicing occurs, proteins with significant differences in their amino acid sequences are produced. In most cases, the alternative versions of a protein will have similar functions, because much of their amino acid sequences will be identical to each other. Nevertheless, alternative splicing produces differences in amino acid sequences that provide each protein with its own unique characteristics. The biological advantage of alternative splicing is that two or more different proteins can be derived from a single gene. This allows an organism to carry fewer genes in its genome. The molecular mechanism of alternative splicing is examined in greater detail in Chapter 15. It involves the actions of proteins (not shown in Figure 12.22) that influence whether or not U1 and U2 can begin the splicing process.

The Ends of Eukaryotic Pre-mRNAs Have a 5' Cap and a 3' Tail

In addition to splicing, pre-mRNAs in eukaryotes are also subjected to modifications at their 5' and 3' ends. At their 5' end, most mature mRNAs have a 7-methylguanosine covalently attached, an event known as **capping.** Capping occurs while the pre-mRNA is being made by RNA polymerase II, usually when the transcript is only 20 to 25 nucleotides in length. As shown in **Figure 12.23**, it is a three-step process. The nucleotide at the 5' end of the transcript has three phosphate groups. First, an enzyme called RNA 5'-triphosphatase removes one of the phosphates, and then a second enzyme, guanylyltransferase, uses GTP to attach a guanosine monophosphate (GMP) to the 5' end. Finally, a methyltransferase attaches a methyl group to the guanine base.

What are the functions of the 7-methylguanosine cap? The cap structure is recognized by cap-binding proteins, which perform various roles. For example, cap-binding proteins are required for the proper exit of most mRNAs from the nucleus. Also, the cap structure is recognized by initiation factors that are needed during the early stages of translation. Finally, the cap structure may be important in the efficient splicing of introns, particularly the first intron located nearest the 5' end.

Let's now turn our attention to the 3' end of the RNA molecule. Most mature mRNAs have a string of adenine nucleotides, referred to as a **polyA tail,** which is important for mRNA stability and in the synthesis of polypeptides. The polyA tail is not encoded in the gene sequence. Instead, it is added enzymatically after the pre-mRNA has been completely transcribed. This process is termed **polyadenylation.**

The steps required to synthesize a polyA tail are shown in **Figure 12.24.** To acquire a polyA tail, the pre-mRNA contains a polyadenylation sequence near its 3' end. In higher eukaryotes, the consensus sequence is AAUAAA. This sequence is downstream

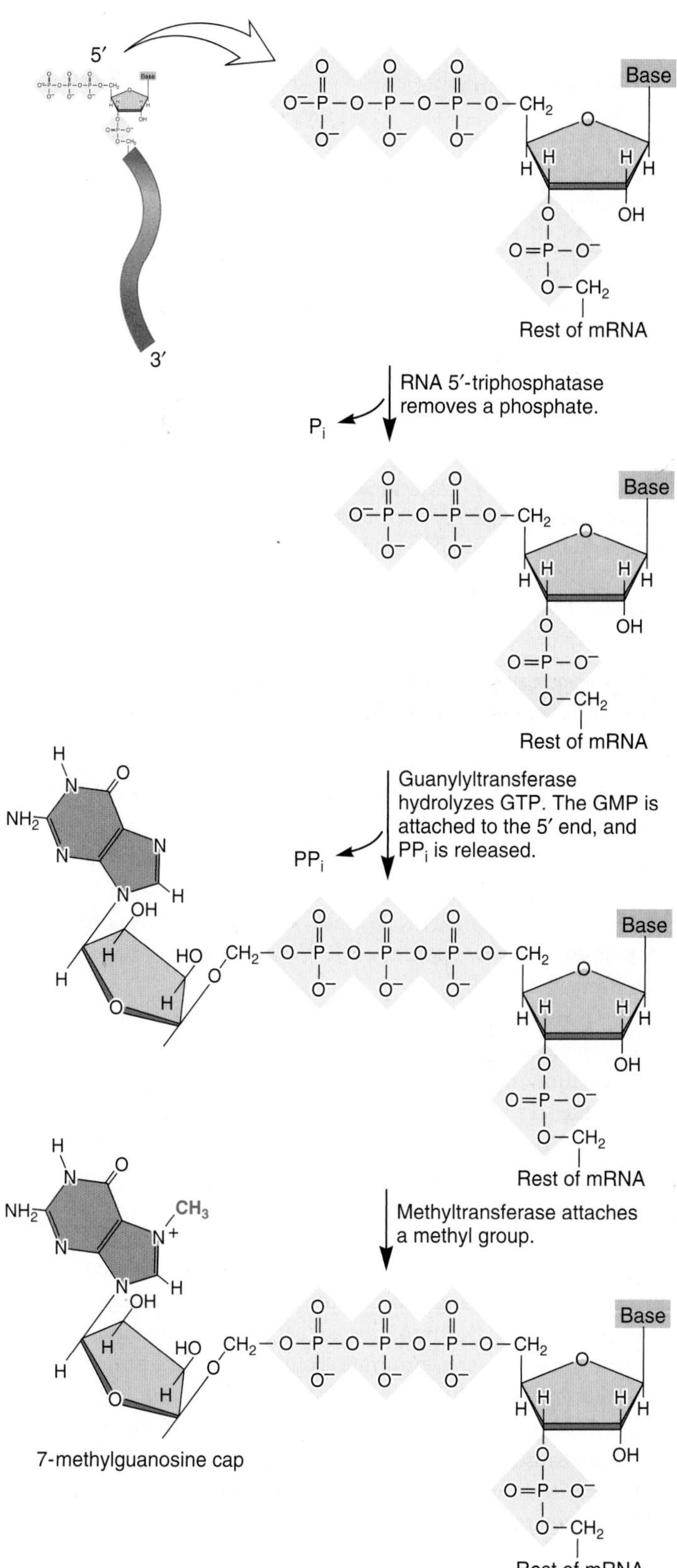

FIGURE 12.23 **Attachment of a 7-methylguanosine cap to the 5' end of mRNA.** When the transcript is about 20 to 25 nucleotides in length, RNA 5'-triphosphatase removes one of the three phosphates, and then a second enzyme, guanylyltransferase, attaches GMP to the 5' end. Finally, a methyltransferase attaches a methyl group to the guanine base.

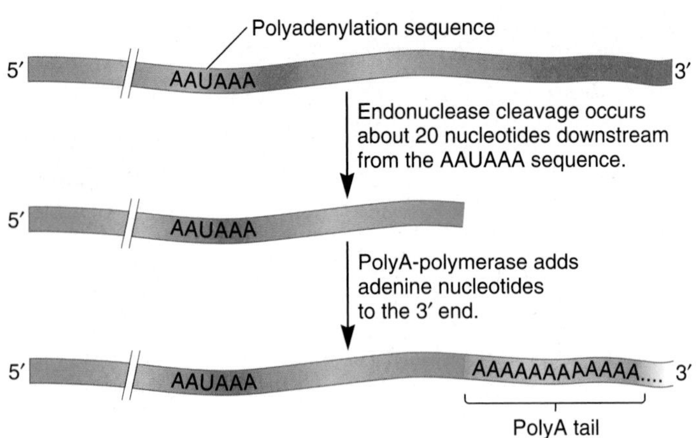

FIGURE 12.24 **Attachment of a polyA tail.** First, an endonuclease cuts the RNA at a location that is 11 to 30 nucleotides after the AAUAAA polyadenylation sequence, making the RNA shorter at its 3' end. Adenine-containing nucleotides are then attached, one at a time, to the 3' end by the enzyme polyA-polymerase.

(toward the 3' end) from the stop codon in the pre-mRNA. An endonuclease recognizes the polyadenylation sequence and cleaves the pre-mRNA at a location that is about 20 nucleotides beyond the 3' end of the AAUAAA sequence. The fragment beyond the 3' cut is degraded. Next, an enzyme known as polyA-polymerase attaches many adenine-containing nucleotides. The length of the polyA tail varies among different mRNAs, from a few dozen to several hundred adenine nucleotides. As discussed in Chapter 15, a long polyA tail increases the stability of mRNA and plays a role during translation.

CONCEPTUAL SUMMARY

Transcription is the process of RNA synthesis from a DNA template. Different types of RNA transcripts can be made. These include mRNA, which specifies the sequence of amino acids within a polypeptide, and tRNA and rRNA, which are necessary to translate mRNA. In addition, several other small RNA molecules, such as microRNA, 7S RNA, scRNA, the RNA component of RNaseP, snRNA, telomerase RNA, and snoRNA have various functions in the cell.

The synthesis of RNA during gene transcription involves three steps: initiation, elongation, and termination. During the initiation stage, proteins that bind to DNA recognize specific sequences within the promoter region. In the case of many bacterial promoters, the –35 and –10 sequences are recognized by σ factor, which is part of RNA polymerase holoenzyme. To begin the synthesis or elongation of a transcript, the DNA must be converted to an open complex by denaturing a small double-stranded region. At this point, RNA polymerase can slide down the DNA, synthesizing a complementary RNA molecule. At the end of the gene, a termination signal is found. In bacteria, termination signals can promote transcriptional termination in two different ways. In ρ-dependent termination, the ρ protein is responsible for dissociating the RNA transcript when a stem-loop structure causes RNA polymerase to pause. In ρ-independent termination, the formation of a stem-loop within the RNA causes RNA polymerase to stall and eventually fall off the DNA, but this occurs without the help of ρ protein.

Transcription in eukaryotes is similar but more complex compared to the process in bacteria. In eukaryotes, three RNA polymerases designated I, II, and III transcribe different types of genes. Transcriptional initiation of structural genes requires five general transcription factors that enable RNA polymerase II to recognize the TATA box in the –25 region. Transcription is initiated by an assembly process that leads to the formation of the open complex. A large protein complex termed mediator plays a role in the switch between initiation and elongation. For transcription to occur, the chromatin structure must be loosened so that RNA polymerase can transcribe the template DNA. This involves two mechanisms: covalent modifications of histone proteins and ATP-dependent chromatin remodeling.

Following transcription, a newly made RNA transcript may be modified in various ways. For example, tRNA transcripts are processed to smaller forms by endo- and exonucleases. Some genes contain introns that must be removed from the RNA via splicing. Group I and Group II introns are self-splicing, which means they catalyze their own removal. These types of introns are found in the rRNA genes of *Tetrahymena* and other simpler eukaryotes, in organelle DNA, and occasionally in bacterial genes. In contrast, introns are commonly found in eukaryotic pre-mRNA, where they are removed by a multicomponent structure known as a spliceosome. Many eukaryotic pre-mRNAs undergo alternative splicing, with the result that a single gene can produce more than one type of polypeptide. Besides splicing, most eukaryotic pre-mRNA is also capped with 7-methylguanosine at its 5' end, and a polyA tail is attached to its 3' end.

EXPERIMENTAL SUMMARY

Experimentally, many methods are used to study the transcription of genes at the molecular level. Most of these are described in Chapter 18. Several of these techniques are considered in the experimental questions at the end of this chapter. For example, the quantity of an RNA transcript can be determined by Northern blotting, in which RNA transcripts are identified through a DNA-RNA hybridization technique. This method can also be used to detect alternative splicing of mRNAs.

The binding and assembly of proteins during transcription can be examined in a variety of ways. In Chapter 11, we considered how Arthur Kornberg developed an in vitro system to study DNA replication. Similarly, researchers have succeeded in developing in vitro systems to study transcription. In this way, they have identified the basal transcriptional apparatus needed for transcription to occur.

In this chapter, we have also considered how hybridization and electron microscopy can be used to detect introns. In this approach, the hybridization of mRNA to a DNA sequence blocks the DNA's ability to re-form double-stranded regions. Using an electron microscope to examine the pattern of single-stranded DNA and double-stranded DNA regions, the hybridization of RNA to DNA allowed researchers to identify introns within the β-globin gene. Since the late 1970s, DNA sequencing methods that compare the DNA sequences of genomic clones and complementary DNA (cDNA) clones have provided an easier and more precise way of detecting whether a particular gene contains introns.

PROBLEM SETS & INSIGHTS

Solved Problems

S1. Describe the important events that occur during the three stages of gene transcription in bacteria. What proteins play critical roles in the three stages?

Answer: The three stages are initiation, elongation, and termination.

Initiation: RNA polymerase holoenzyme slides along the DNA until σ factor recognizes a promoter. σ factor binds tightly to this sequence, forming a closed complex. The DNA is then denatured to form a bubblelike structure known as the open complex.

Elongation: RNA polymerase core enzyme slides along the DNA, synthesizing RNA as it goes. The α subunits of RNA polymerase keep the enzyme bound to the DNA, while the β subunits are responsible for binding and for the catalytic synthesis of RNA. The ω (omega) subunit is also important for the proper assembly of the core enzyme. During elongation, RNA is made according to the AU/GC rule, with nucleotides being added in the 5' to 3' direction.

Termination: RNA polymerase eventually reaches a sequence at the end of the gene that signals the cessation of transcription. In ρ-independent termination, the properties of the termination sequences in the DNA are sufficient to cause termination. In ρ-dependent termination, the ρ (rho) protein recognizes a sequence within the RNA, binds there, and travels toward RNA polymerase. When the formation of a stem-loop structure causes RNA polymerase to pause, ρ catches up and separates the RNA-DNA hybrid region, releasing the RNA polymerase.

S2. What is the difference between a structural gene and a nonstructural gene?

Answer: Structural genes encode mRNA that is translated into a polypeptide sequence. Nonstructural genes encode RNAs that are never translated. Products of nonstructural genes include tRNA and rRNA, which function during translation; microRNA, which is involved in gene regulation; 7S RNA, which is part of a complex known as SRP; scRNA, small cytoplasmic RNA found in bacteria; the RNA of RNaseP; telomerase RNA, which is involved in telomere replication; snoRNA, which is involved in rRNA trimming; and snRNA, which is a component of spliceosome. In many cases, the RNA from nonstructural genes becomes part of a complex composed of RNA molecules and protein subunits.

S3. When RNA polymerase transcribes DNA, only one of the two DNA strands is used as a template. Take a look at Figure 12.4 and explain how RNA polymerase determines which DNA strand is the template strand.

Answer: The binding of σ factor and RNA polymerase depends on the sequence of the promoter. RNA polymerase binds to the promoter in such a way that the –35 sequence TTGACA and the –10 sequence TATAAT are within the coding strand, while the –35 sequence AACTGT and the –10 sequence ATATTA are within the template strand.

S4. The process of transcriptional termination is not as well understood in eukaryotes as it is in bacteria. Nevertheless, current evidence suggests several different mechanisms exist for termination. Like bacteria, the termination of certain genes appears to occur via intrinsic terminators (as in ρ-independent termination), while the termination of other genes may involve RNA-binding proteins (as in ρ-dependent termination). In eukaryotes, a third type of mechanism is found for the termination of rRNA genes by RNA polymerase I. In this case, a protein known as TTFI (transcription termination factor I) binds to the DNA downstream from the termination site. Discuss how the binding of a protein downstream from the termination site could promote transcriptional termination.

Answer: First, the binding of TTFI could act as a roadblock to the movement of RNA polymerase I. Second, TTFI could promote the dissociation of the RNA transcript and RNA polymerase I from the DNA; it may act like a helicase. Third, it could cause a change in the structure of the DNA that prevents RNA polymerase from moving past the termination site. Though multiple effects are possible, the third effect seems the most likely because TTFI is known to cause a bend in the DNA when it binds to the termination sequence.

Conceptual Questions

C1. Genes may be structural genes that encode polypeptides, or they may be nonstructural genes.

A. Describe three examples of genes that are not structural genes.

B. For structural genes, one DNA strand is called the template strand, and the complementary strand is called the coding strand. Are these two terms appropriate for nonstructural genes? Explain.

C. Do nonstructural genes have a promoter and terminator?

C2. In bacteria, what event marks the end of the initiation stage of transcription?

C3. What is the meaning of the term consensus sequence? Give an example. Describe the locations of consensus sequences within bacterial promoters. What are their functions?

C4. What is the consensus sequence of the following six DNA molecules?

GGCATTGACT

GCCATTGTCA

CGCATAGTCA

GGAAATGGGA

GGCTTTGTCA

GGCATAGTCA

C5. Mutations in bacterial promoters may increase or decrease the level of gene transcription. Promoter mutations that increase transcription are termed up promoter mutations, and those that decrease transcription are termed down promoter mutations. As shown in Figure 12.5, the sequence of the –10 region of the promoter for the *lac* operon is TATGTT. Would you expect the following mutations to be up promoter or down promoter mutations?

A. TATGTT to TATATT

B. TATGTT to TTTGTT

C. TATGTT to TATGAT

C6. According to the examples shown in Figure 12.5, which positions of the –35 sequence (i.e., first, second, third, fourth, fifth, and/or sixth) are more tolerant of changes? Do you think that these positions play a more or less important role in the binding of σ factor? Explain why.

C7. In Chapter 9, we considered the dimensions of the double helix (see Figure 9.17). In an α helix of a protein, there are 3.6 amino acids per complete turn. Each amino acid advances the α helix by 0.15 nm; a complete turn of an α helix is 0.54 nm in length. As shown in Figure 12.6, two α helices of a transcription factor occupy the major groove of the DNA. According to Figure 12.6, estimate the number of amino acids that bind to this region. How many complete turns of the α helices occupy the major groove of DNA?

C8. A mutation within a gene sequence changes the start codon to a stop codon. How will this mutation affect the transcription of this gene?

C9. What is the subunit composition of bacterial RNA polymerase holoenzyme? What are the functional roles of the different subunits?

C10. At the molecular level, describe how σ factor recognizes bacterial promoters. Be specific about the structure of σ factor and the type of chemical bonding.

C11. Let's suppose a DNA mutation changes the consensus sequence at the –35 location so that σ factor is no longer able to bind there. Explain how a mutation would prevent σ factor from binding to the DNA. Look at Figure 12.5 and describe two specific base substitutions you think would inhibit the binding of σ factor. Explain why you think your base substitutions would have this effect.

C12. What is the complementarity rule that governs the synthesis of an RNA molecule during transcription? An RNA transcript has the following sequence:

5'-GGCAUGCAUUACGGCAUCACACUAGGGAUC-3'

What is the sequence of the template and coding strands of the DNA that encodes this RNA? On which side (5' or 3') of the template strand is the promoter located?

C13. Describe the movement of the open complex along the DNA.

C14. Describe what happens to the chemical bonding interactions when transcriptional termination occurs. Be specific about the type of chemical bonding.

C15. Discuss the differences between ρ-dependent and ρ-independent termination.

C16. In Chapter 11, we discussed the function of DNA helicase, which is involved in DNA replication. The structure and function of DNA helicase and ρ protein are rather similar to each other. Explain how the function of these two proteins is similar and how it is different.

C17. Discuss the similarities and differences between RNA polymerase (described in this chapter) and DNA polymerase (described in Chapter 11).

C18. Mutations that occur at the end of a gene may alter the sequence of the gene and prevent transcriptional termination.

A. What types of mutations would prevent ρ-independent termination?

B. What types of mutations would prevent ρ-dependent termination?

C. If a mutation prevented transcriptional termination at the end of a gene, where would gene transcription end? Or would it end?

C19. If the following RNA polymerases were missing from a eukaryotic cell, what types of genes would not be transcribed?

A. RNA polymerase I

B. RNA polymerase II

C. RNA polymerase III

C20. What sequence elements are found within the core promoter of structural genes in eukaryotes? Describe their locations and specific functions.

C21. For each of the following transcription factors, how would eukaryotic transcriptional initiation be affected if it were missing?

A. TFIIB

B. TFIID

C. TFIIH

C22. Discuss how the binding of DNA to histones in a nucleosome structure may affect the process of transcription. Do you think that the nucleosome structure would affect the ability of transcription factors to recognize the promoter sequence in eukaryotic genes? Explain.

C23. Which eukaryotic transcription factor(s) shown in Figure 12.14 plays an equivalent role to σ factor found in bacterial cells?

C24. The initiation phase of eukaryotic transcription via RNA polymerase II is considered an assembly and disassembly process. Which types of biochemical interactions—hydrogen bonding, ionic bonding, covalent bonding, and/or hydrophobic interactions—would you expect to drive the assembly and disassembly process? How would temperature and salt concentration affect assembly and disassembly?

C25. A eukaryotic structural gene contains two introns and three exons: exon 1–intron 1–exon 2–intron 2–exon 3. The 5' splice site at the boundary between exon 2 and intron 2 has been eliminated by a small deletion in the gene. Describe how the pre-mRNA encoded by this mutant gene would be spliced. Indicate which introns and exons would be found in the mRNA after splicing occurs.

C26. Describe the processing events that occur during the production of tRNA in *E. coli.*

C27. Describe the structure and function of a spliceosome. Speculate why the spliceosome subunits contain snRNA. In other words, what do you think is/are the functional role(s) of snRNA during splicing?

C28. What is the unique feature of ribozyme function? Give two examples described in this chapter.

C29. What does it mean to say that gene expression is colinear?

C30. What is meant by the term self-splicing? What types of introns are self-splicing?

C31. In eukaryotes, what types of modification occur to pre-mRNA?

C32. What is alternative splicing? What is its biological significance?

C33. The processing of ribosomal RNA in eukaryotes is shown in Figure 12.16. Why is this called cleavage or processing but not splicing?

C34. In the splicing of group I introns shown in Figure 12.20, does the 5' end of the intron have a phosphate group? Explain.

C35. According to the mechanism shown in Figure 12.22, several snRNPs play different roles in the splicing of pre-mRNA. Identify the snRNP that recognizes the following sites:

A. 5' splice site

B. 3' splice site

C. Branch site

C36. After the intron (which is in a lariat configuration) is released during pre-mRNA splicing, a brief moment occurs before the two exons are connected to each other. Which snRNP(s) holds the exons in place so they can be covalently connected to each other?

C37. A lariat contains a closed loop and a linear end. An intron has the following sequence: 5'–GUPuAGUA–60 nucleotides–UACUUAUCC–100 nucleotides–$Py_{12}NPyAG$–3'. Which sequence would be found within the closed loop of the lariat, the 60-nucleotide sequence or the 100-nucleotide sequence?

Experimental Questions

E1. A research group has sequenced the cDNA and genomic DNA from a particular gene. The cDNA is derived from mRNA, so it does not contain introns. Here are the DNA sequences.

cDNA:

5'-ATTGCATCCAGCGTATACTATCTCGGGCCCAATTAAT-
GCCAGCGGCCAGACTATCACCCAACTCGGTTACCTACTAG-
TATATCCCATATACTAGCATATATTTTACCCATAATTTGTGTGT-
GGGTATACAGTATAATCATATA-3'

Genomic DNA (contains one intron):

5'-ATTGCATCCAGCGTATACTATCTCGGGCCCAAT-
TAATGCCAGCGGCCAGACTATCACCCAACTCG-
GCCCACCCCCCAGGTTTACACAGTCATACCATACA
TACAAAAATCGCAGTTACTTATCCCAAAAAAACCTAG-
ATACCCCACATACTATTAACTCTTTCTTTCTAGGTTACCTAC-
TAGTATATCCCATATACTAGCATATATTTTACCCATAATTTGT-
GTGTGGGTATACAGTATAATCATATA-3'

Indicate where the intron is located. Does the intron contain the normal consensus splice site sequences based on those described in Figure 12.21? Underline the splice site sequences, and indicate whether or not they fit the consensus sequence.

E2. What is an R loop? In an R loop experiment, to which strand of DNA does the mRNA bind, the coding strand or the template strand?

E3. If a gene contains three introns, draw what it would look like in an R loop experiment.

E4. Chapter 18 describes a technique known as Northern blotting that can be used to detect RNA transcribed from a particular gene. In this method, a specific RNA is detected using a short segment of cloned DNA as a probe. The DNA probe, which is radioactive, is complementary to the RNA that the researcher wishes to detect. After the radioactive probe DNA binds to the RNA, the RNA is visualized as a dark (radioactive) band on an X-ray film. As shown here, the method of Northern blotting can be used to determine the amount of a particular RNA transcribed in a given cell type. If one type of cell produces twice as much of a particular mRNA compared to another cell, the band will appear twice as intense. Also, the method can distinguish if alternative RNA splicing has occurred to produce an RNA that has a different molecular mass.

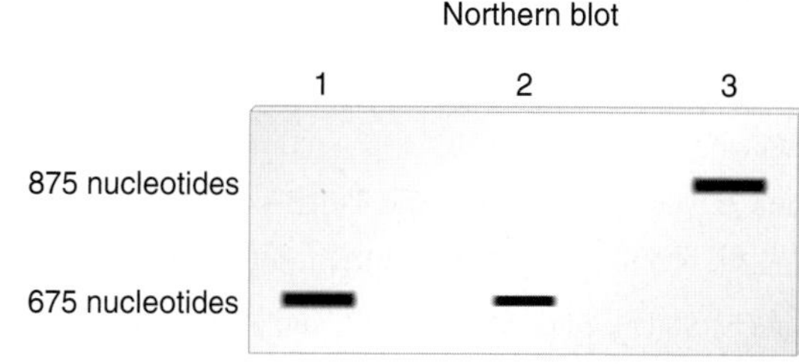

Lane 1 is a sample of RNA isolated from nerve cells.

Lane 2 is a sample of RNA isolated from kidney cells. Nerve cells produce twice as much of this RNA compared to kidney cells.

Lane 3 is a sample of RNA isolated from spleen cells. Spleen cells produce an alternatively spliced version of this RNA that is about 200 nucleotides longer than the RNA produced in nerve and kidney cells.

Let's suppose a researcher was interested in the effects of mutations on the expression of a particular structural gene in eukaryotes. The gene has one intron that is 450 nucleotides long. After this intron is removed from the pre-mRNA, the mRNA transcript is 1,100 nucleotides in length. Diploid somatic cells have two copies of this gene. Make a drawing that shows the expected results of a Northern blot using mRNA from the cytosol of somatic cells, which were obtained from the following individuals:

Lane 1: A normal individual

Lane 2: An individual homozygous for a deletion that removes the −50 to −100 region of the gene that encodes this mRNA

Lane 3: An individual heterozygous in which one gene is normal and the other gene had a deletion that removes the −50 to −100 region

Lane 4: An individual homozygous for a mutation that introduces an early stop codon into the middle of the coding sequence of the gene

Lane 5: An individual homozygous for a three-nucleotide deletion that removes the AG sequence at the 3' splice site

E5. A gel retardation assay can be used to study the binding of proteins to a segment of DNA. This method is described in Chapter 18. When a protein binds to a segment of DNA, it retards the movement of the DNA through a gel, so the DNA appears at a higher point in the gel (see the following).

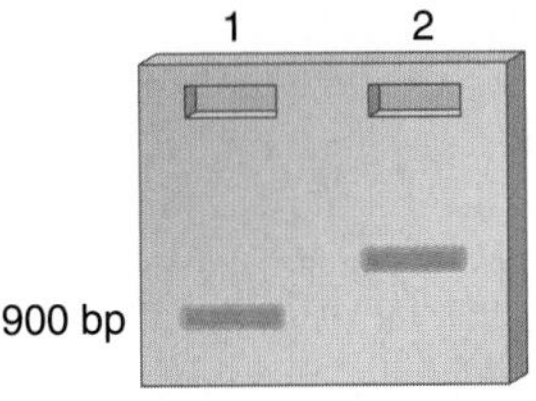

Lane 1: 900 bp fragment alone

Lane 2: 900 bp fragment plus a protein that binds to the 900 bp fragment

In this example, the segment of DNA is 900 bp in length, and the binding of a protein causes the DNA to appear at a higher point in the gel. If this 900 bp fragment of DNA contains a eukaryotic promoter for a structural gene, draw a gel that shows the relative locations of the 900 bp fragment under the following conditions:

Lane 1: 900 bp plus TFIID

Lane 2: 900 bp plus TFIIB

Lane 3: 900 bp plus TFIID and TFIIB

Lane 4: 900 bp plus TFIIB and RNA polymerase II

Lane 5: 900 bp plus TFIID, TFIIB, and RNA polymerase II/TFIIF

E6. As described in Chapter 18 and in experimental question E5, a gel retardation assay can be used to determine if a protein binds to DNA. This method can also determine if a protein binds to RNA. In the combinations described here, would you expect the migration of the RNA to be retarded due to the binding of a protein?

A. mRNA from a gene that is terminated in a ρ-independent manner plus ρ-protein

B. mRNA from a gene that is terminated in a ρ-dependent manner plus ρ-protein

C. pre-mRNA from a structural gene that contains two introns plus the snRNP called U1

D. Mature mRNA from a structural gene that contains two introns plus the snRNP called U1

E7. The technique of DNA footprinting is described in Chapter 18. If a protein binds over a region of DNA, it will protect chromatin in that region from digestion by DNase I. To carry out a DNA footprinting experiment, a researcher has a sample of a cloned DNA fragment. The fragments are exposed to DNase I in the presence and absence of a DNA-binding protein. Regions of the DNA fragment not covered by the DNA-binding protein will be digested by DNase I, and this will produce a series of bands on a gel. Regions of the DNA fragment not digested by DNase I (because a DNA-binding protein is preventing DNase I from gaining access to the DNA) will be revealed, because a region of the gel will not contain any bands.

In the DNA footprinting experiment shown here, a researcher began with a sample of cloned DNA 300 bp in length. This DNA contained a eukaryotic promoter for RNA polymerase II. For the sample loaded in lane 1, no proteins were added. For the sample loaded in lane 2, the 300 bp fragment was mixed with RNA polymerase II plus TFIID and TFIIB.

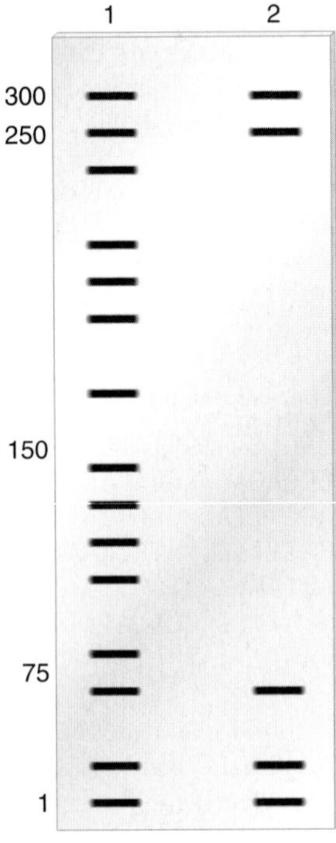

A. How long of a region of DNA is "covered up" by the binding of RNA polymerase II and the transcription factors?

B. Describe how this binding would occur if the DNA was within a nucleosome structure. (Note: The structure of nucleosomes is described in Chapter 10.) Do you think that the DNA is in a nucleosome structure when RNA polymerase and transcription factors are bound to the promoter? Explain why or why not.

E8. As described in Table 12.1, several different types of RNA are made, especially in eukaryotic cells. Researchers are sometimes interested in focusing their attention on the transcription of structural genes in eukaryotes. Such researchers want to study mRNA. One method that is used to isolate mRNA is column chromatography. (Note: See the Appendix for a general description of chromatography.) Researchers can covalently attach short pieces of DNA that contain stretches of thymine (i.e., TTTTTTTTTTTT) to the column matrix. This is called a poly-dT column. When a cell extract is poured over the column, mRNA binds to the column, while other types of RNA do not.

A. Explain how you would use a poly-dT column to obtain a purified preparation of mRNA from eukaryotic cells. In your description, explain why mRNA binds to this column and what you would do to release the mRNA from the column.

B. Can you think of ways to purify other types of RNA, such as tRNA or rRNA?

Questions for Student Discussion/Collaboration

1. Based on your knowledge of introns and pre-mRNA splicing, discuss whether or not you think alternative splicing fully explains the existence of introns. Can you think of other possible reasons to explain the existence of introns?

2. Discuss the types of RNA transcripts and the functional roles they play. Why do you think some RNAs form complexes with protein subunits?

Note: All answers appear at the website for this textbook; the answers to even-numbered questions are in the back of the textbook.

www.mhhe.com/brookergenetics3e

Visit the Online Learning Center for practice tests, answer keys, and other learning aids for this chapter. Enhance your understanding of genetics with our interactive exercises, quizzes, animations, and much more.

13 TRANSLATION OF mRNA

CHAPTER OUTLINE

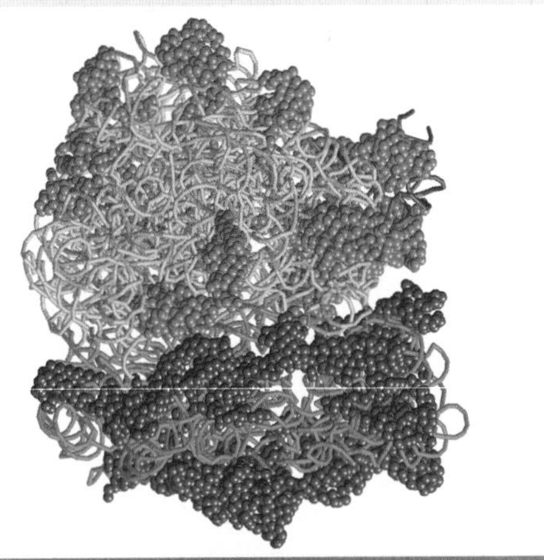

A molecular model for the structure of a ribosome. This is a model of ribosome structure based on X-ray crystallography. Ribosomes are needed to synthesize polypeptides, using mRNA as a template.

The synthesis of cellular proteins occurs via the translation of the sequence of codons within mRNA into a sequence of amino acids of a polypeptide. The general steps that occur in this process have already been outlined in Chapter 1. In this chapter, we will explore the current state of knowledge regarding translation, with an eye toward the specific molecular interactions responsible for this process. During the past few decades, the concerted efforts of geneticists, cell biologists, and biochemists have profoundly advanced our understanding of translation. Even so, many questions remain unanswered, and this topic continues to be an exciting area of research investigation.

We will begin by considering the classic experiments that revealed the purpose of some genes is to encode proteins that function as enzymes. Next, we will examine how the genetic code is used to decipher the information within mRNA to produce a polypeptide with a specific amino acid sequence. The rest of this chapter is devoted to a molecular understanding of translation as it occurs in living cells. This will involve an examination of the cellular components—including many different proteins, RNAs, and small molecules—needed for the translation process. We will consider the structure and function of tRNA molecules, which act as the translators of the genetic information within mRNA, and then examine the composition of ribosomes. Finally, we will explore the differences between translation in bacterial cells and eukaryotic cells.

13.1 THE GENETIC BASIS FOR PROTEIN SYNTHESIS

Proteins are critically important as active participants in cell structure and function. The primary role of DNA is to store the information needed for the synthesis of all the proteins that an organism makes. As we discussed in Chapter 12, genes that encode an amino acid sequence are known as **structural genes.** The RNA transcribed from structural genes is called **messenger RNA (mRNA).** The main function of the genetic material is to encode the production of cellular proteins in the correct cell, at the proper time, and in suitable amounts. This is an extremely complicated task because living cells make thousands of different proteins. Genetic analyses have shown that a typical bacterium can make a few thousand different proteins, and estimates for eukaryotes range from several thousand in simple eukaryote organisms, such as yeast, to tens of thousands in plants and animals.

In this section, we will begin by considering early experiments that showed the role of genes is to encode proteins. We will examine the general features of the genetic code—the sequence of bases in a codon that specifies an amino acid—and explore the experiments through which the code was deciphered or "cracked." Finally, we will look at the biochemistry of polypeptide synthesis to see how this determines the structure and function of proteins, which are ultimately responsible for the characteristics of living cells and an organism's traits.

Archibald Garrod Proposed That Some Genes Code for the Production of a Single Enzyme

The idea that a relationship exists between genes and the production of proteins was first suggested at the beginning of the twentieth century by Archibald Garrod, a British physician. Prior to Garrod's studies, biochemists had studied many metabolic pathways within living cells. These pathways consist of a series of metabolic conversions of one molecule to another, each step catalyzed by a specific enzyme. Each enzyme is a distinctly different protein that catalyzes a particular chemical reaction. Figure 13.1 illustrates part of the metabolic pathway for the degradation of phenylalanine, an amino acid commonly found in human diets. The enzyme phenylalanine hydroxylase catalyzes the conversion of phenylalanine to tyrosine, and a different enzyme, tyrosine aminotransferase, converts tyrosine into *p*-hydroxyphenylpyruvic acid, and so on. In all of the steps shown in Figure 13.1, a specific enzyme catalyzes a single type of chemical reaction.

Garrod studied patients who had defects in their ability to metabolize certain compounds. He was particularly interested in the inherited disease known as **alkaptonuria.** In this disorder, the patient's body accumulates abnormal levels of homogentisic acid (also called alkapton), which is excreted in the urine, causing it to appear black on exposure to air. In addition, the disease is characterized by bluish black discoloration of cartilage and skin (ochronosis). Garrod proposed that the accumulation of homogentisic acid in these patients is due to a missing enzyme, namely, homogentisic acid oxidase (see Figure 13.1).

How did Garrod realize that certain genes encode enzymes? He already knew that alkaptonuria is an inherited trait that follows a recessive pattern of inheritance. Therefore, an individual with alkaptonuria must have inherited the mutant gene that causes this disorder from both parents. From these observations, Garrod proposed that a relationship exists between the inheritance of the trait and the inheritance of a defective enzyme. Namely, if an individual inherited the mutant gene (which causes a loss of enzyme function), he or she would not produce any normal enzyme and would be unable to metabolize homogentisic acid. Garrod described alkaptonuria as an **inborn error of metabolism.** This hypothesis was the first suggestion that a connection exists between the function of genes and the production of enzymes. At the turn of the century, this idea was particularly insightful, because the structure and function of the genetic material were completely unknown.

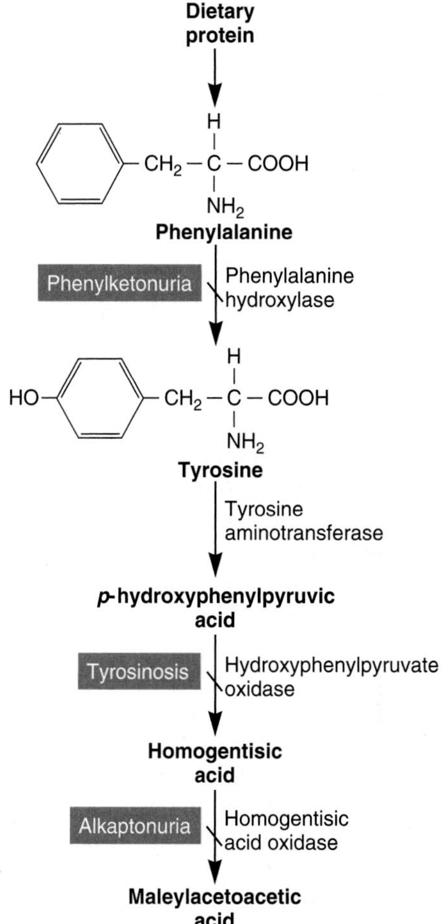

FIGURE 13.1 **The metabolic pathway of phenylalanine breakdown.** This diagram shows part of the pathway of phenylalanine metabolism, which consists of enzymes that successively convert one molecule to another. Certain human genetic diseases (shown in red boxes) are caused when enzymes in this pathway are missing or defective.

Genes→Traits When a person inherits two defective copies of the gene that encodes homogentisic acid oxidase, he or she cannot convert homogentisic acid into maleylacetoacetic acid. Such a person accumulates large amounts of homogentisic acid in the urine and has other symptoms of the disease known as alkaptonuria. Similarly, if a person has two mutant alleles of the gene encoding phenylalanine hydroxylase, he or she is unable to synthesize the enzyme phenylalanine hydroxylase and has the disease called phenylketonuria (PKU).

Beadle and Tatum's Experiments with *Neurospora* Led Them to Propose the One Gene–One Enzyme Hypothesis

In the early 1940s, George Beadle and Edward Tatum were also interested in the relationship among genes, enzymes, and traits. They developed an experimental system for investigating the relationship between genes and the production of particular enzymes. Consistent with the ideas of Garrod, the underlying assumption behind their approach was that a relationship exists between genes and the production of enzymes. However, the quantitative nature of this relationship was unclear. In particular, they asked the question, Does one gene control the production of one enzyme, or does one gene control the synthesis of many enzymes involved in a complex biochemical pathway?

At the time of their studies, many geneticists were trying to understand the nature of the gene by studying morphological traits. However, Beadle and Tatum realized that morphological traits are likely to be based on systems of biochemical reactions so complex as to make analysis exceedingly difficult. Therefore, they turned their genetic studies to the analysis of simple nutritional requirements in *Neurospora crassa*, a common bread mold. *Neurospora* can be easily grown in the laboratory and has few nutritional requirements: a carbon source (sugar), inorganic salts, and the vitamin biotin. Normal *Neurospora* cells produce many different enzymes that can synthesize the organic molecules, such as amino acids and other vitamins, which are essential for growth.

Beadle and Tatum wanted to understand how enzymes are controlled by genes. They reasoned that a mutation in a gene, causing a defect in an enzyme needed for the cellular synthesis of an essential molecule, would prevent that mutant strain from growing on minimal medium, which contains only a carbon source, inorganic salts, and biotin. For example, if a *Neurospora* mutant strain could not make the vitamin pantothenic acid, it would be unable to grow on minimal medium. However, it would grow if the growth medium was supplemented with pantothenic acid.

Beadle and Tatum analyzed more than 2,000 strains that had been irradiated to induce mutations. They found three strains that were unable to grow on minimal medium (**Table 13.1**). In each of these cases, the mutant strain required only a single vitamin to restore its growth on minimal medium. One of these strains required pyridoxine for growth, the second strain required thiamine, and the third required *p*-aminobenzoic acid. In the normal strain, these vitamins are synthesized by cellular enzymes. In the mutant strains, a genetic defect in one gene prevented the synthesis of a functional enzyme required for the synthesis of one particular vitamin. Therefore, Beadle and Tatum concluded that a single gene controlled the synthesis of a single enzyme. This was referred to as the **one gene–one enzyme hypothesis.**

TABLE **13.1**

Identification of Mutant Strains of *Neurospora crassa* That Were Unable to Synthesize Particular Vitamins

Single Vitamin Added to Minimal Media	Observed Growth: Strain 1	Strain 2	Strain 3
None	No	No	No
Riboflavin	No	No	No
Thiamine	No	**Yes**	No
Pantothenic acid	No	No	No
Pyridoxine	**Yes**	No	No
Niacin	No	No	No
Folic acid	No	No	No
p-aminobenzoic acid	No	No	**Yes**

In later decades, this idea had to be modified in two ways. First, enzymes are only one category of cellular proteins. All proteins are encoded by genes, and many of them do not function as enzymes. Second, some proteins are composed of two or more different polypeptides. Therefore, it is more accurate to say that a structural gene encodes a polypeptide. The term **polypeptide** refers to a structure; it is a linear sequence of amino acids. A structural gene encodes a polypeptide. By comparison, the term **protein** denotes function. Some proteins are composed of one polypeptide. In such cases, a single gene does encode a single protein. In other cases, however, a functional protein is composed of two or more different polypeptides. An example would be hemoglobin, which is composed of two α-globin and two β-globin polypeptides. In this case, the expression of two genes—the α-globin and β-globin genes—is needed to create one functional protein.

During Translation, the Genetic Code Within mRNA Is Used to Make a Polypeptide with a Specific Amino Acid Sequence

Let's now turn to a general description of translation. Why have researchers named this process translation? At the molecular level, **translation** involves an interpretation of one language—the language of mRNA, a nucleotide sequence—into the language of proteins—an amino acid sequence. The ability of mRNA to be translated into a specific sequence of amino acids relies on the **genetic code.** The sequence of bases within an mRNA molecule provides coded information that is read in groups of three nucleotides known as codons (**Figure 13.2**). The sequence of three bases in most codons specifies a particular amino acid. These codons are termed **sense codons.** For example, the codon AGC specifies the amino acid serine, while the codon GGG encodes the amino acid glycine. The codon AUG, which specifies methionine, is used as a **start codon;** it is usually the first codon that begins a polypeptide sequence. The AUG codon can also be used to specify additional methionines within the coding sequence. Finally, three codons are used to end the process of translation. These are UAA, UAG, and UGA, which are known as **stop codons.** They are also known as **termination** and **nonsense codons.**

The codons in mRNA are recognized by the anticodons in transfer RNA (tRNA) molecules (Figure 13.2). **Anticodons** are three-nucleotide sequences that are complementary to codons in mRNA. The tRNA molecules carry the amino acids that correspond to the codons in the mRNA. In this way, the order of codons in mRNA dictates the order of amino acids within a polypeptide.

The details of the genetic code are shown in **Table 13.2.** Because polypeptides are composed of 20 different kinds of amino acids, a minimum of 20 codons is needed in order to

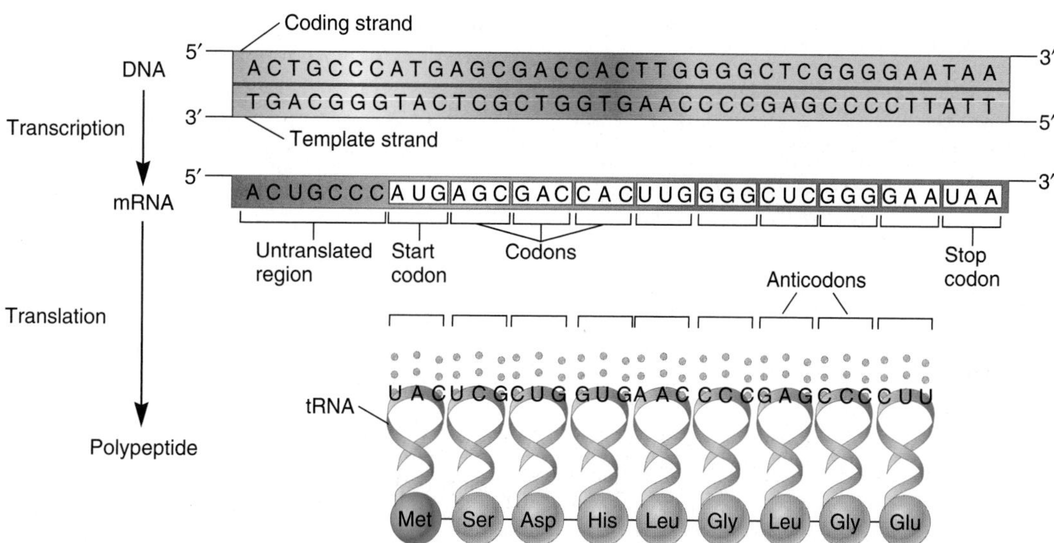

FIGURE 13.2 The relationships among the DNA coding sequence, mRNA codons, tRNA anticodons, and amino acids in a polypeptide. The sequence of nucleotides within DNA is transcribed to make a complementary sequence of nucleotides within mRNA. This sequence of nucleotides in mRNA is translated into a sequence of amino acids of a polypeptide. tRNA molecules act as intermediates in this translation process.

TABLE **13.2**

The Genetic Code

First base	Second base: U	Second base: C	Second base: A	Second base: G	Third base
U	UUU, UUC Phenylalanine (Phe)	UCU, UCC, UCA, UCG Serine (Ser)	UAU, UAC Tyrosine (Tyr)	UGU, UGC Cysteine (Cys)	U, C
U	UUA, UUG Leucine (Leu)		UAA Stop codon; UAG Stop codon	UGA Stop codon; UGG Tryptophan (Trp)	A, G
C	CUU, CUC, CUA, CUG Leucine (Leu)	CCU, CCC, CCA, CCG Proline (Pro)	CAU, CAC Histidine (His)	CGU, CGC, CGA, CGG Arginine (Arg)	U, C
C			CAA, CAG Glutamine (Gln)		A, G
A	AUU, AUC, AUA Isoleucine (Ile)	ACU, ACC, ACA, ACG Threonine (Thr)	AAU, AAC Asparagine (Asn)	AGU, AGC Serine (Ser)	U, C
A	AUG Methionine (Met); start codon		AAA, AAG Lysine (Lys)	AGA, AGG Arginine (Arg)	A, G
G	GUU, GUC, GUA, GUG Valine (Val)	GCU, GCC, GCA, GCG Alanine (Ala)	GAU, GAC Aspartic acid (Asp)	GGU, GGC, GGA, GGG Glycine (Gly)	U, C
G			GAA, GAG Glutamic acid (Glu)		A, G

specify each type. With four types of bases in mRNA (A, U, G, and C), a genetic code containing two bases in a codon would not be sufficient because it would only have 4^2, or 16, possible types. By comparison, a three-base codon system can specify 4^3, or 64, different codons. Because the number of possible codons exceeds 20—which is the number of different types of amino acids—the genetic code is termed **degenerate.** This means that more than one codon can specify the same amino acid. For example, the codons GGU, GGC, GGA, and GGG all specify the amino acid glycine. Such codons are termed **synonymous codons.** In most instances, the third base in the codon is the base that varies. The third base is sometimes referred to as the **wobble base.** This term is derived from the idea that the complementary base in the tRNA can "wobble" a bit during the recognition of the third base of the codon in mRNA. The significance of the wobble base will be discussed later in this chapter.

From the analysis of many different species, including bacteria, protists, fungi, plants, and animals, researchers have

TABLE 13.3

Examples of Exceptions to the Genetic Code*

Codon	Universal Meaning	Exception
AUA	Isoleucine	Methionine in yeast and mammalian mitochondria
UGA	Stop	Tryptophan in mammalian mitochondria
CUU, CUA, CUC, CUG	Leucine	Threonine in yeast mitochondria
AGA, AGG	Arginine	Stop codon in ciliated protozoa and in yeast and mammalian mitochondria
UAA, UAG	Stop	Glutamine in ciliated protozoa

*Several other exceptions, sporadically found among various species, are also known.

found that the genetic code is nearly universal. Relatively few exceptions to the genetic code have been noted (**Table 13.3**). The eukaryotic organelles known as mitochondria have their own DNA, which encodes a few structural genes. In mammals, the mitochondrial genetic code contains differences such as AUA = methionine and UGA = tryptophan. Also, in mitochondria and certain ciliated protists, AGA and AGG specify stop codons instead of arginine. Except for rare exceptions such as the ones listed in Table 13.3, the genetic code shown in Table 13.2 has been found to be universal in all species studied thus far.

The start codon (AUG) defines the **reading frame** of an mRNA—a sequence of codons determined by reading bases in groups of three, beginning with the start codon. This concept is best understood with a few examples. The mRNA sequence shown below encodes a short polypeptide with 7 amino acids:

5'-AUGCCCGGAGGCACCGUCCAAU-3'

Met-Pro-Gly-Gly-Thr-Val-Gln

If we remove one base (C) adjacent to the start codon, this changes the reading frame to produce a different polypeptide sequence:

5'-AUGCCGGAGGCACCGUCCAAU-3'

Met-Pro-Glu-Ala-Pro-Ser-Asn

Alternatively, if we remove three bases (CCC) next to the start codon, the resulting polypeptide has the same reading frame as the first polypeptide, though one amino acid (Pro, proline) has been deleted:

5'-AUGGGAGGCACCGUCCAAU-3'

Met-Gly-Gly-Thr-Val-Gln

How did researchers discover that the genetic code is read in triplets? The first evidence came from studies of Francis Crick and his colleagues in 1961. These experiments involved the isolation of mutants in a bacteriophage called T4. As described in Chapter 6, mutations in T4 genes that affect plaque morphology are easily identified (see Figure 6.15). In particular, loss-of-function mutations within certain T4 genes, designated *rII*, resulted in plaques that were larger and had a clear boundary. In comparison, wild-type phages, designated r^+, produced smaller plaques with a fuzzy boundary. Crick and colleagues exposed T4 phages to a chemical called proflavin, which causes single-nucleotide additions or deletions in gene sequences. The mutagenized phages were plated to identify large (*rII*) plaques. Though proflavin can cause either single-nucleotide additions or deletions, the first mutant strain that the researchers identified was arbitrarily called a (+) mutation. Many years later, when methods of DNA sequencing became available, it was determined that the (+) mutation is a single-nucleotide addition. **Table 13.4** shows a hypothetical wild-type sequence in a phage gene (first line) and considers how nucleotide additions and/or deletions could affect the resulting amino acid sequence. A single-nucleotide addition (+) would alter the reading frame beyond the point of insertion and thereby abolish the proper function of the encoded protein. This is called a frameshift mutation, because it has changed the reading frame. This mutation resulted in a loss-of-function for the protein encoded by this viral gene and thereby produced an *rII* plaque phenotype.

The (+) mutant strain was then subjected to a second round of mutagenesis via proflavin. Several plaques were identified that had reverted to a wild-type (r^+) phenotype. By analyzing these strains using methods described in Chapter 6, it was determined that each one contained a second mutation that was close to the original (+) mutation. These second mutations were designated (–) mutations. Three different (–) mutations, designated a, b, and c, were identified. Each of these (–) mutations was a single-nucleotide deletion that was close to the original (+) mutation. Therefore, it restored the reading frame and produced a protein with a nearly normal amino acid sequence.

The critical experiment that suggested the genetic code is read in triplets came by combining different (–) mutations together. Mutations in different phages can be brought together into the same phage via crossing over, as described in Chapter 6. Using such an approach, the researchers constructed strains containing one, two, or three (–) mutations. The results showed that a wild-type plaque morphology was obtained only when three (–) mutations were combined in the same phage (Table 13.4). The three (–) mutations in the same phage restored the normal reading frame. These results were consistent with the hypothesis that the genetic code is read in multiples of three nucleotides.

TABLE 13.4

Evidence That the Genetic Code Is Read in Triplets*

Strain	Plaque Phenotype	DNA Coding Sequence/Polypeptide Sequence§	Downstream Sequence‡
Wild type	r^+	ATG GGG CCC GTC CAT CCG TAC GCC GGA ATT ATA Met Gly Pro Val His Pro Tyr Ala Gly Ile Ile----------	In frame
		↓A	
(+)	rII	ATG GGG ACC CGT CCA TCC GTA CGC CGG AAT TAT A Met Gly Thr Arg Pro Ser Val Arg Arg Asn Tyr---------	Out of frame
		↓A ↑C	
$(+)(-)_a$	r^+	ATG GGG ACC GTC CAT CCG TAC GCC GGA ATT ATA Met Gly Thr Val His Pro Tyr Ala Gly Ile Ile----------	In frame
		↓A ↑T	
$(+)(-)_b$	r^+	ATG GGG ACC CGC CAT CCG TAC GCC GGA ATT ATA Met Gly Thr Arg His Pro Tyr Ala Gly Ile Ile----------	In frame
		↓A ↑G	
$(+)(-)_c$	r^+	ATG GGG ACC CTC CAT CCG TAC GCC GGA ATT ATA Met Gly Thr Leu His Pro Tyr Ala Gly Ile Ile----------	In frame
		↑C	
$(-)_a$	rII	ATG GGG CCG TCC ATC CGT ACG CCG GAA TTA TA Met Gly Pro Ser Ile Arg Thr Pro Glu Leu----------------	Out of frame
		↑C ↑T	
$(-)_a(-)_b$	rII	ATG GGG CCG CCA TCC GTA CGC CGG AAT TAT A Met Gly Pro Pro Ser Val Arg Arg Asn Tyr----------------	Out of frame
		↑↑↑GTC	
$(-)_a(-)_b(-)_c$	r^+	ATG GGG CCC CAT CCG TAC GCC GGA ATT ATA Met Gly Pro His Pro Tyr Ala Gly Ile Ile--------------	In frame [Only Val is missing]

*This table shows only a small portion of a hypothetical coding sequence.

§A down arrow (↓) indicates the location of a single nucleotide addition; an up arrow (↑) indicates the location of a single nucleotide deletion.

‡The term downstream sequence refers to the remaining part of the sequence that is not shown in this figure. It could include hundreds of codons. An "in-frame" sequence would be wild type, whereas an "out-of-frame" sequence (caused by the addition or deletion of one or two base pairs) would not.

EXPERIMENT 13A

Synthetic RNA Helped to Decipher the Genetic Code

Having determined that the genetic code is read in triplets, how did scientists determine the functions of the 64 codons of the genetic code? During the early 1960s, three research groups headed by Marshall Nirenberg, Severo Ochoa, and H. Gobind Khorana set out to decipher the genetic code. Though they used different methods, all of these groups used synthetic mRNA in their experimental approaches to "crack the code." We will first consider the work of Nirenberg and his colleagues. Prior to their studies, several laboratories had already determined that extracts from bacterial cells, containing a mixture of components including ribosomes, tRNAs, and other factors required for translation, are able to synthesize polypeptides if mRNA and amino acids are added. This mixture is termed an in vitro or **cell-free translation system.** If radiolabeled amino acids are added to a cell-free translation system, the synthesized polypeptides are radiolabeled and easy to detect.

To decipher the genetic code, Nirenberg and colleagues needed to gather information regarding the relationship between mRNA composition and polypeptide composition. To accomplish this goal, they made mRNA molecules of a known base composition, added them to a cell-free translation system, and then analyzed the amino acid composition of the resultant polypeptides. For example, if an mRNA molecule consisted of a string of adenine-containing nucleotides (e.g., 5'–AAAAAAAAAAAAAAAA–3'), researchers could add this polyA mRNA to a cell-free translation system and ask the question, Which amino acid is specified by a codon that contains only adenine nucleotides? (As Table 13.2 shows, it is lysine.)

Before discussing the details of this type of experiment, let's consider how the synthetic mRNA molecules were made. To synthesize mRNA, an enzyme known as polynucleotide phosphorylase was used. In the presence of excess ribonucleoside diphosphates, also called nucleoside diphosphates (NDPs), this enzyme catalyzes the covalent linkage of nucleotides to make a polymer of RNA. Because it does not use a template, the order of the nucleotides is random. For example, if only uracil-containing diphosphates (UDPs) are added, then a polyU mRNA (5'–UUUUUUUUUUUUUUUU–3') is made. If nucleotides containing two different bases, such as uracil and guanine, are added, then the phosphorylase makes a random polymer containing both nucleotides (5'–GGGUGUGUGGUGGGUG–3'). An experimenter can control the amounts of the nucleotides that are added. For example, if 70% G and 30% U are mixed together with polynucleotide phosphorylase, the predicted amounts of the codons within the random polymer are as follows:

Codon Possibilities	*Percentage in the Random Polymer*
GGG	0.7 × 0.7 × 0.7 = 0.34 = 34%
GGU	0.7 × 0.7 × 0.3 = 0.15 = 15%
GUU	0.7 × 0.3 × 0.3 = 0.06 = 6%
UUU	0.3 × 0.3 × 0.3 = 0.03 = 3%
UUG	0.3 × 0.3 × 0.7 = 0.06 = 6%
UGG	0.3 × 0.7 × 0.7 = 0.15 = 15%
UGU	0.3 × 0.7 × 0.3 = 0.06 = 6%
GUG	0.7 × 0.3 × 0.7 = 0.15 = 15%
	100%

By controlling the amounts of the NDPs in the phosphorylase reaction, the relative amounts of the possible codons can be predicted.

The first experiment that demonstrated the ability to synthesize polypeptides from synthetic mRNA was performed by Marshall Nirenberg and J. Heinrich Matthaei in 1961. As shown in **Figure 13.3**, a cell-free translation system was added to 20 different tubes. An mRNA template made via polynucleotide phosphorylase was then added to each tube. In this example, the mRNA was made using 70% G and 30% U. Next, the 20 amino acids were added to each tube, but each tube differed with regard to the type of radiolabeled amino acid. For example, radiolabeled glycine would be found in only 1 of the 20 tubes. The tubes were incubated a sufficient length of time to allow translation to occur. The newly made polypeptides were then precipitated onto a filter by treatment with trichloroacetic acid. This precipitates polypeptides but not amino acids. A washing step caused amino acids that had not been incorporated into polypeptides to pass through the filter. Finally, the amount of radioactivity trapped on the filter was determined by liquid scintillation counting.

THE GOAL

The researchers assumed that the sequence of bases in mRNA determines the incorporation of specific amino acids into a polypeptide. The purpose of this experiment was to provide information that would help to decipher the relationship between base composition and particular amino acids.

ACHIEVING THE GOAL — FIGURE 13.3 Elucidation of the genetic code.

Starting material: A cell-free translation system that can synthesize polypeptides if mRNA and amino acids are added.

Experimental level | **Conceptual level**

1. Add the cell-free translation system to each of 20 tubes.

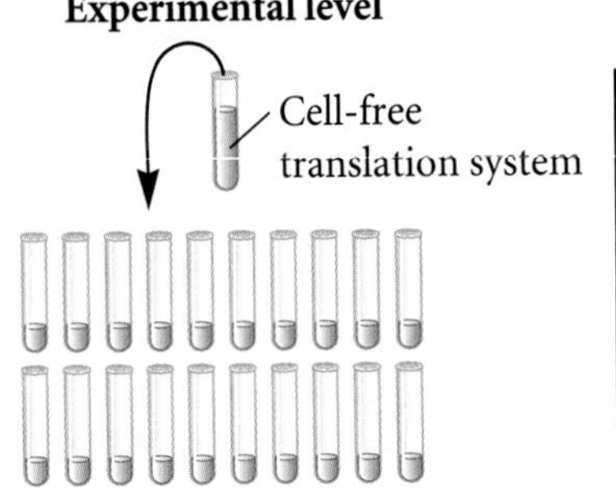

2. To each tube, add random mRNA polymers of G and U made via polynucleotide phosphorylase using 70% G and 30% U.

3. Add a different radiolabeled amino acid to each tube, and add the other 19 non-radiolabeled amino acids.

4. Incubate for 60 minutes to allow translation to occur.

5. Add 15% trichloroacetic acid (TCA), which precipitates polypeptides but not amino acids.

6. Place the precipitate onto a filter and wash to remove unused amino acids.

7. Count the radioactivity on the filter in a scintillation counter (see the Appendix for a description).

8. Calculate the amount of radiolabeled amino acids in the precipitated polypeptides.

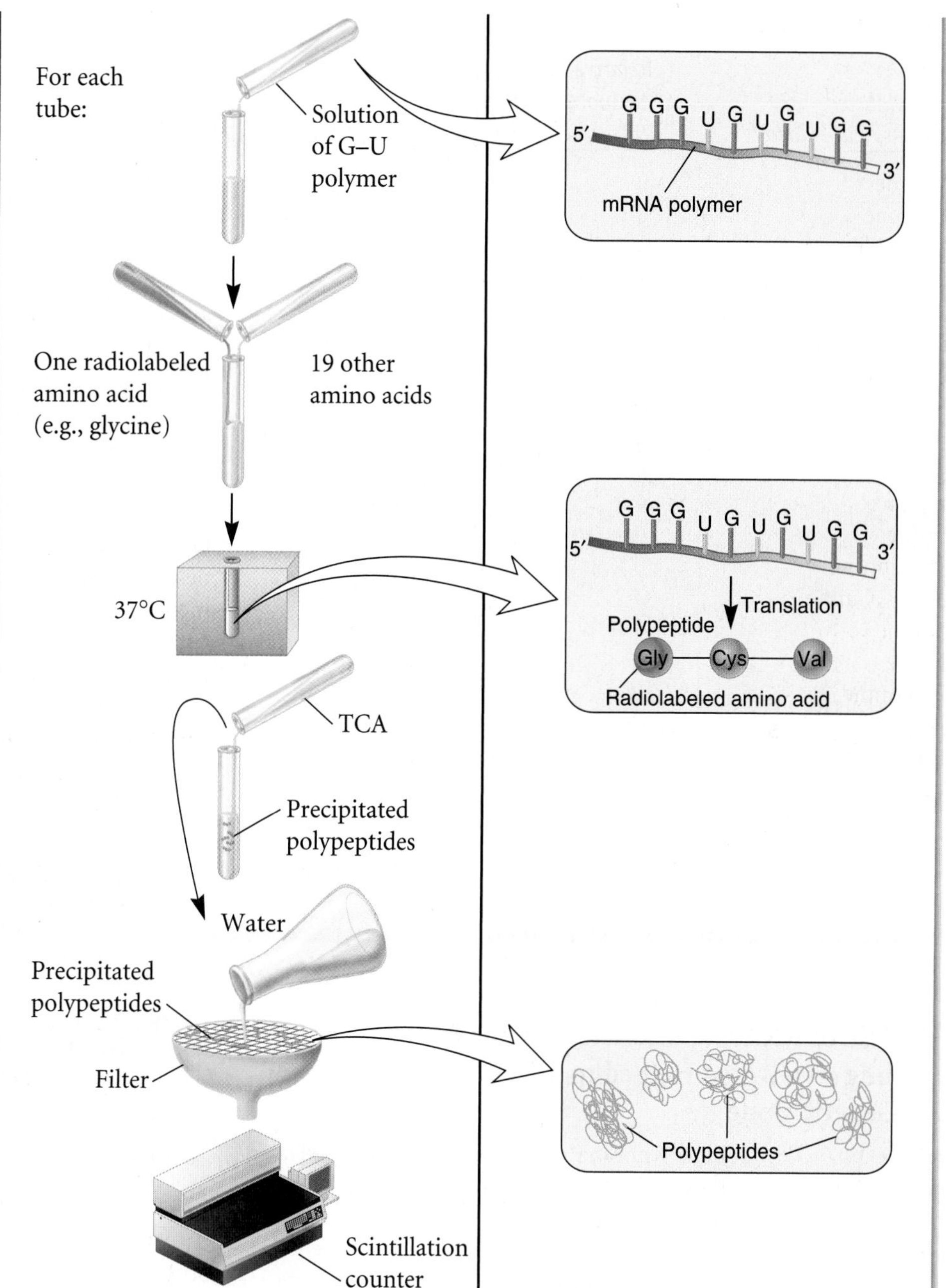

THE DATA

Radiolabeled Amino Acid Added	*Relative Amount of Radiolabeled Amino Acid Incorporated into Translated Polypeptides (% of total)*
Alanine	0
Arginine	0
Asparagine	0
Aspartic acid	0
Cysteine	6
Glutamic acid	0
Glutamine	0
Glycine	49
Histidine	0
Isoleucine	0
Leucine	6
Lysine	0
Methionine	0
Phenylalanine	3
Proline	0
Serine	0
Threonine	0
Tryptophan	15
Tyrosine	0
Valine	21

INTERPRETING THE DATA

According to the calculation previously described, codons should occur in the following percentages: 34% GGG, 15% GGU, 6% GUU, 3% UUU, 6% UUG, 15% UGG, 6% UGU, and 15% GUG. In the data shown in Figure 13.3, the value of 49% for glycine is due to two codons: GGG (34%) and GGU (15%). The 6% cysteine is due to UGU, and so on. It is important to realize that the genetic code was not deciphered in a single experiment such as the one described here. Furthermore, this kind of experiment yields information regarding only the nucleotide content of codons, not the specific order of bases within a single codon. For example, this experiment indicates that a cysteine codon contains two U's and one G. However, it does not tell us that a cysteine codon is UGU. Based on these data alone, a cysteine codon could be UUG, GUU, or UGU. However, by comparing many different RNA polymers, the laboratories of Nirenberg and Ochoa established patterns between the specific base sequences of codons and the amino acids they encode. In their first experiments, Nirenberg and Matthaei showed that a random polymer containing only uracil produced a polypeptide containing only phenylalanine. From this result, they inferred that UUU specifies phenylalanine. This idea is consistent with the results shown in the data table. In the random 70% G and 30% U polymer, 3% of the codons will be UUU. Likewise, 3% of the amino acids within the polypeptides were found to be phenylalanine.

A self-help quiz involving this experiment can be found at the Online Learning Center.

The Use of RNA Copolymers and the Triplet Binding Assay Also Helped to Crack the Genetic Code

In the 1960s, H. Gobind Khorana and colleagues developed a novel method to synthesize RNA. They first created short RNA molecules, two to four nucleotides in length, that had a defined sequence. For example, RNA molecules with the sequence 5'–AUC–3' were synthesized chemically. These short RNAs were then linked together enzymatically, in a 5' to 3' manner, to create long copolymers with the sequence

5'-AUCAUCAUCAUCAUCAUCAUCAUCAUCAUC-3'

This is called a copolymer, because it is made from the linkage of several smaller molecules. Depending on the reading frame, such a copolymer would contain three different codons: AUC, UCA, and CAU. Using a cell-free translation system like the one described in Figure 13.3, such a copolymer produced polypeptides containing isoleucine, serine, and histidine. **Table 13.5** summarizes some of the copolymers that were made using this approach and the amino acids that were incorporated into polypeptides.

Finally, another method that helped to decipher the genetic code also involved the chemical synthesis of short RNA molecules. In 1964, Marshall Nirenberg and Philip Leder discovered that RNA molecules containing three nucleotides—a triplet—could stimulate ribosomes to bind a tRNA. In other words, the RNA triplet acted like a codon. Ribosomes were able to bind RNA triplets, and then a tRNA with the appropriate anticodon could

TABLE **13.5**

Examples of Copolymers That Were Analyzed by Khorana and Colleagues

Synthetic RNA*	Codon Possibilities	Amino Acids Incorporated into Polypeptides
UC	UCU, CUC	Serine, leucine
AG	AGA, GAG	Arginine, glutamic acid
UG	UGU, GUG	Cysteine, valine
AC	ACA, CAC	Threonine, histidine
UUC	UUC, UCU, CUU	Phenylalanine, serine, leucine
AAG	AAG, AGA, GAA	Lysine, arginine, glutamic acid
UUG	UUG, UGU, GUU	Leucine, cysteine, valine
CAA	CAA, AAC, ACA	Glutamine, asparagine, threonine
UAUC	UAU, AUC, UCU, CUA	Tyrosine, isoleucine, serine, leucine
UUAC	UUA, UAC, ACU, CUU	Leucine, tyrosine, threonine

*The synthetic RNAs were linked together to make copolymers.

subsequently bind to the ribosome. To establish the relationship between triplet sequences and specific amino acids, samples containing ribosomes and a particular triplet were exposed to tRNAs with different radiolabeled amino acids.

As an example, in one experiment the researchers began with a sample of ribosomes that were mixed with 5'–CCC–3' triplets. Portions of this sample were then added to 20 different tubes that had tRNAs with different radiolabeled amino acids. For example, one tube contained radiolabeled histidine, a second tube had radiolabeled proline, a third tube contained radiolabeled glycine, and so on. Only one radiolabeled amino acid was added to each tube. After allowing sufficient time for tRNAs to bind to the ribosomes, the samples were filtered; only the large ribosomes and anything bound to them were trapped on the filter (**Figure 13.4**). Unbound tRNAs would pass through the filter. Next, the researchers determined the amount of radioactivity trapped on each filter. If the filter contained a large amount of radioactivity, the results indicated that the added triplet encoded the amino acid that was radiolabeled.

Using the triple binding assay, Nirenberg and Leder were able to establish relationships between particular triplet sequences and the binding of tRNAs carrying specific (radiolabeled) amino acids. In the case of the 5'–CCC–3' triplet, they determined that tRNAs carrying radiolabeled proline were bound to the ribosomes. Unfortunately, in some cases, a triplet could not promote sufficient tRNA binding to yield unambiguous results. Nevertheless, the triplet binding assay was an important tool in the identification of the majority of codons.

A Polypeptide Chain Has Directionality from Its Amino Terminal to Its Carboxyl Terminal End

Let's now turn our attention to polypeptide biochemistry. Polypeptide synthesis has a directionality that parallels the order of codons in the mRNA. As a polypeptide is made, a **peptide bond** is formed between the carboxyl group in the last amino acid of the polypeptide chain and the amino group in the amino acid being added. As shown in **Figure 13.5a**, this occurs via a condensation reaction that releases a water molecule. The newest amino acid added to a growing polypeptide always has a free carboxyl group. **Figure 13.5b** compares the sequence of a very short polypeptide with the mRNA that encodes it. The first amino acid is said to be at the **N-terminus,** or **amino terminal end,** of the polypeptide. An amino group (NH_3^+) is found at this site. The term N-terminal refers to the presence of a nitrogen atom (N) at this end. The first amino acid is specified by a codon that is near the 5' end of the mRNA. By comparison, the last amino acid in a completed polypeptide is located at the **C-terminus,** or **carboxyl terminal end.** A carboxyl group (COO–) is always found at this site in the polypeptide chain. This last amino acid is specified by a codon that is closer to the 3' end of the mRNA.

The Amino Acid Sequences of Polypeptides Determine the Structure and Function of Proteins

Now that we understand how mRNAs encode polypeptides, let's consider the structure and function of the gene product, namely,

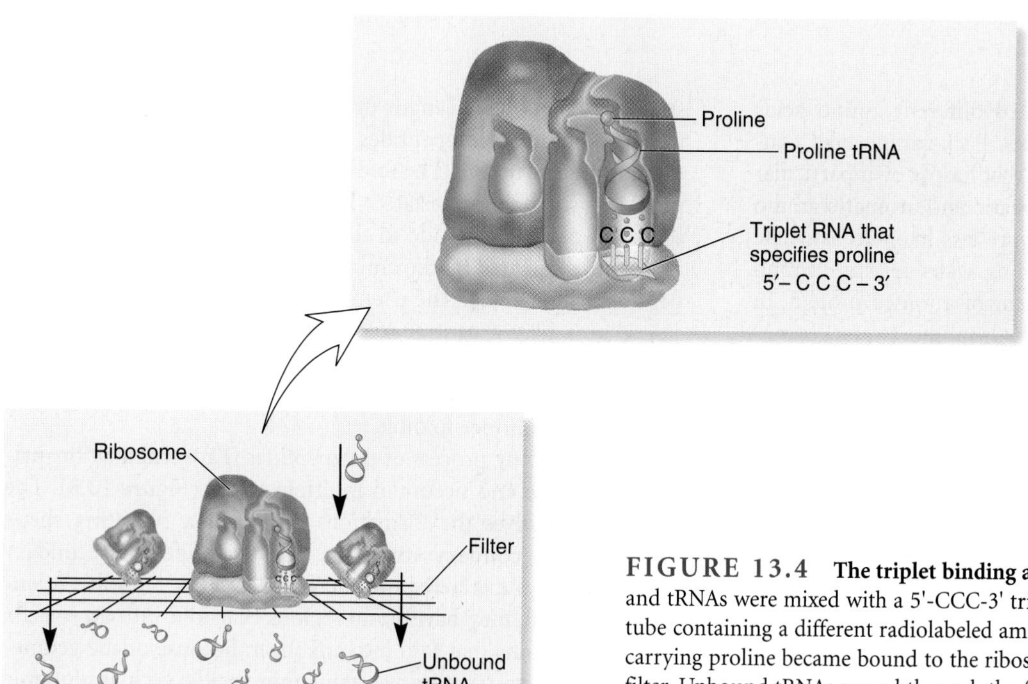

FIGURE 13.4 **The triplet binding assay.** In this experiment, ribosomes and tRNAs were mixed with a 5'-CCC-3' triplet in 20 separate tubes, with each tube containing a different radiolabeled amino acid (not shown). Only tRNAs carrying proline became bound to the ribosome, which became trapped on the filter. Unbound tRNAs passed through the filter. In this case, radioactivity would be trapped on the filter only from the tube in which radiolabeled proline was added.

Last peptide bond formed in the growing chain of amino acids

(a) Attachment of an amino acid to a peptide chain

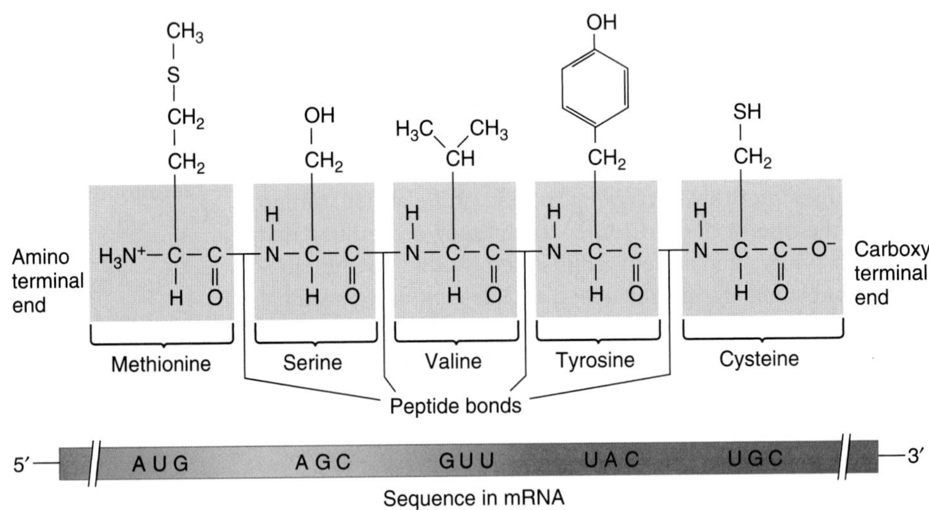

(b) Directionality in a polypeptide and mRNA

FIGURE 13.5 The directionality of polypeptide synthesis. (**a**) An amino acid is connected to a polypeptide chain via a condensation reaction that releases a water molecule. The letter R is a general designation for an amino acid side chain. (**b**) The first amino acid in a polypeptide chain (usually methionine) is located at the amino terminal end, and the last amino acid is at the carboxyl terminal end. Thus, the directionality of amino acids in a polypeptide chain is from the amino terminal end to the carboxyl terminal end, which corresponds to the 5′ to 3′ orientation of codons in mRNA.

polypeptides. Figure 13.6 shows the 20 different amino acids that may be found within polypeptides. Each amino acid contains a unique **side chain,** or **R group,** that has its own particular chemical properties. For example, aliphatic and aromatic amino acids are nonpolar, which means they are less likely to associate with water. These hydrophobic (meaning water-fearing) amino acids are often buried within the interior of a folded protein. In contrast, the polar amino acids are hydrophilic (water-loving) and are more likely to be on the surface of a protein, where they can favorably interact with the surrounding water. The chemical properties of the amino acids and their sequences in a polypeptide are critical factors that determine the unique structure of that polypeptide.

Following gene transcription and mRNA translation, the end result is a polypeptide with a defined amino acid sequence. This sequence is the **primary structure** of a polypeptide. Figure 13.7 shows the primary structure of an enzyme called lysozyme, a relatively small protein containing 129 amino acids. The primary structure of a typical polypeptide may be a few hundred or even a couple thousand amino acids in length. Within a living cell, a newly made polypeptide is not usually found in a long linear state for a significant length of time. Rather, to become a functional unit, most polypeptides quickly adopt a compact three-dimensional structure. The folding process begins while the polypeptide is still being translated. The progression from the primary structure of a polypeptide to the three-dimensional structure of a protein is dictated by the amino acid sequence within the polypeptide. In particular, the chemical properties of the amino acid side chains play a central role in determining the folding pattern of a protein. In addition, the folding of some polypeptides is aided by **chaperones**—proteins that bind to polypeptides and facilitate their proper folding.

This folding process of polypeptides is governed by the primary structure and occurs in multiple stages (Figure 13.8). The first stage involves the formation of a regular, repeating shape known as a **secondary structure.** The two types of secondary structures are the **α helix** and the **β sheet** (Figure 13.8b). A single polypeptide may have some regions that fold into an α helix and other regions that fold into a β sheet. Because of the geometry of secondary structures, certain amino acids, such as glutamic acid, alanine, and methionine, are good candidates to form an α helix. Other amino acids, such as valine, isoleucine, and tyrosine,

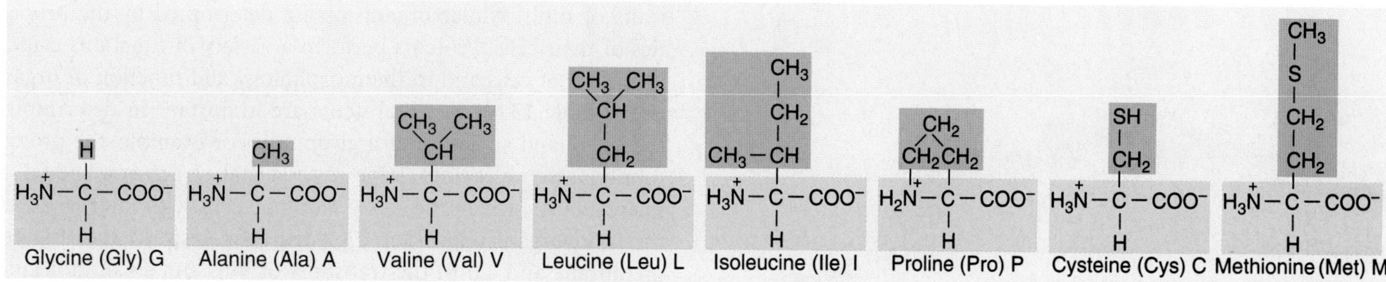

(a) Nonpolar, aliphatic amino acids

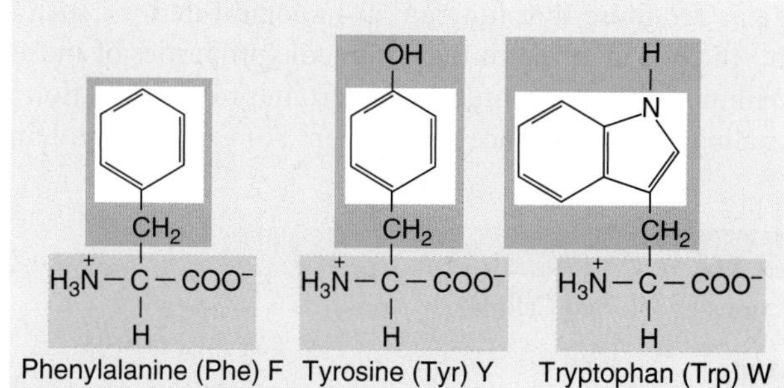

(b) Nonpolar, aromatic amino acids

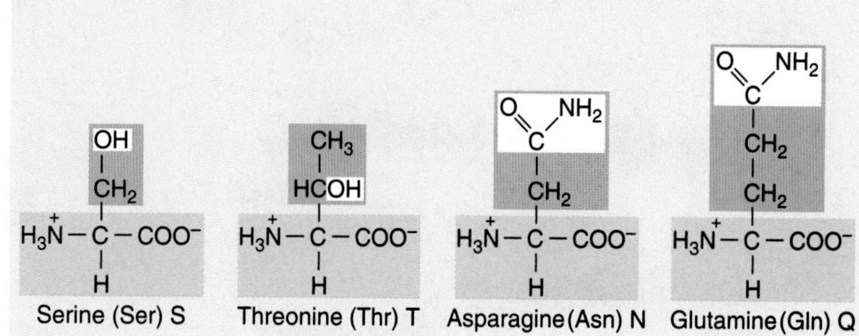

(c) Polar, neutral amino acids

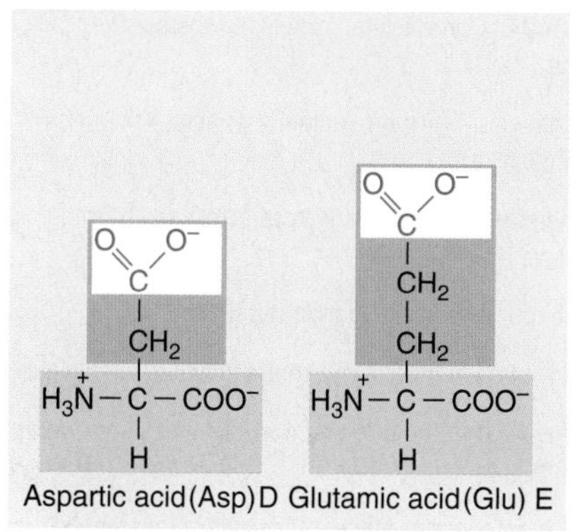

(d) Polar, acidic amino acids

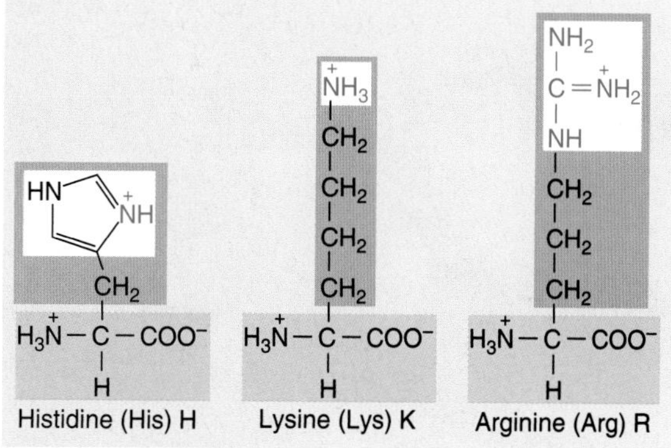

(e) Polar, basic amino acids

FIGURE 13.6 The 20 amino acids found within proteins.

are more likely to be found in a β-sheet conformation. Secondary structures within polypeptides are primarily stabilized by the formation of hydrogen bonds.

The short regions of secondary structure within a polypeptide are folded relative to each other to make the **tertiary structure** of a polypeptide. As shown in Figure 13.8c, α-helical regions and β-sheet regions are connected by irregularly shaped segments to determine the tertiary structure of the polypeptide. The folding of a polypeptide into its secondary and then tertiary conformation can usually occur spontaneously because it is a thermodynamically favorable process. The structure is determined by various interactions, including the tendency of hydrophobic amino acids to avoid water, ionic interactions among charged amino acids, hydrogen bonding among amino acids in the folded polypeptide, and weak bonding known as van der Waals interactions.

A protein is a functional unit that can be composed of one or more polypeptides. Some proteins are composed of a single polypeptide. Many proteins, however, are composed of two or more polypeptides that associate with each other to make a functional protein with a **quaternary structure** (Figure 13.8d). The individual polypeptides are called **subunits** of the protein, each of which has its own tertiary structure. The association of multiple subunits is the quaternary structure of a protein.

Cellular Proteins Are Primarily Responsible for the Characteristics of Living Cells and an Organism's Traits

Why is the genetic material largely devoted to storing the information to make proteins? To a great extent, the characteristics of a cell depend on the types of proteins that it makes. In turn, the

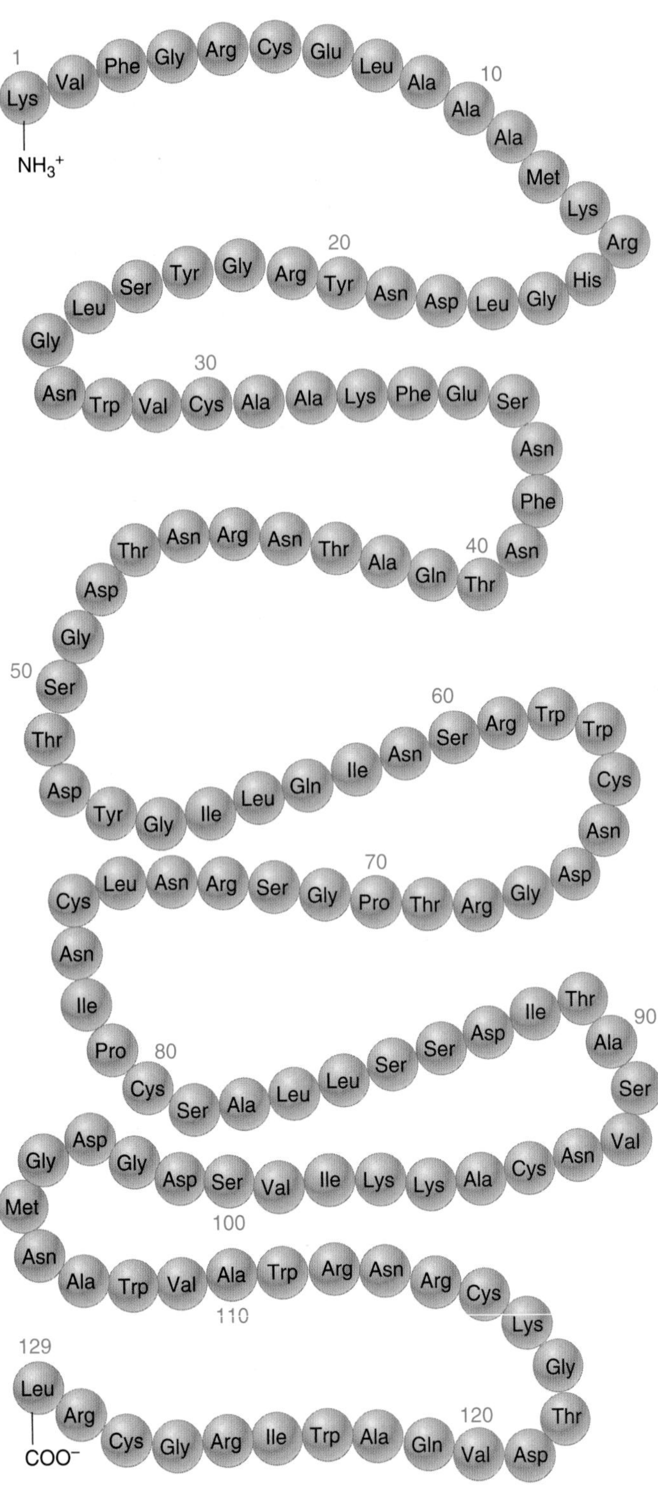

FIGURE 13.7 An example of a protein's primary structure. This is the amino acid sequence of the enzyme lysozyme, which contains 129 amino acids in its primary structure. As you may have noticed, the first amino acid is not methionine; instead, it is lysine. The first methionine residue in this polypeptide sequence is removed after or during translation. The removal of the first methionine occurs in many (but not all) proteins.

traits of multicellular organisms are determined by the properties of their cells. Proteins perform a variety of functions critical to the life of cells and to the morphology and function of organisms (**Table 13.6**). Some proteins are important in determining the shape and structure of a given cell. For example, the protein tubulin assembles into large cytoskeletal structures known as microtubules, which provide eukaryotic cells with internal structure and organization. Some proteins are inserted into the cell membrane and aid in the transport of ions and small molecules across the membrane. An example is a sodium channel that transports sodium ions into nerve cells. Another interesting category of proteins are those that function as biological motors, such as myosin, which is involved in the contractile properties of muscle cells. Within multicellular organisms, certain proteins function in cell signaling and cell surface recognition. For example, proteins,

TABLE **13.6**

Functions of Selected Cellular Proteins

Function	Examples
Cell shape and organization	Tubulin: Forms cytoskeletal structures known as microtubules
	Ankyrin: Anchors cytoskeletal proteins to the plasma membrane
Transport	Sodium channels: Transport sodium ions across the nerve cell membrane
	Lactose permease: Transports lactose across the bacterial cell membrane
	Hemoglobin: Transports oxygen in red blood cells
Movement	Myosin: Involved in muscle cell contraction
	Kinesin: Involved in the movement of chromosomes during cell division
Cell signaling	Insulin: A hormone that influences target cell metabolism and growth
	Epidermal growth factor: A growth factor that promotes cell division
	Insulin receptor: Recognizes insulin and initiates a cell response
Cell surface recognition	Integrins: Bind to large extracellular proteins
Enzymes	Hexokinase: Phosphorylates glucose during the first step in glycolysis
	β-galactosidase: Cleaves lactose into glucose and galactose
	Glycogen synthetase: Uses glucose molecules as building blocks to synthesize a large carbohydrate known as glycogen
	Acyl transferase: Links together fatty acids and glycerol phosphate during the synthesis of phospholipids
	RNA polymerase: Uses ribonucleotides as building blocks to synthesize RNA
	DNA polymerase: Uses deoxyribonucleotides as building blocks to synthesize DNA

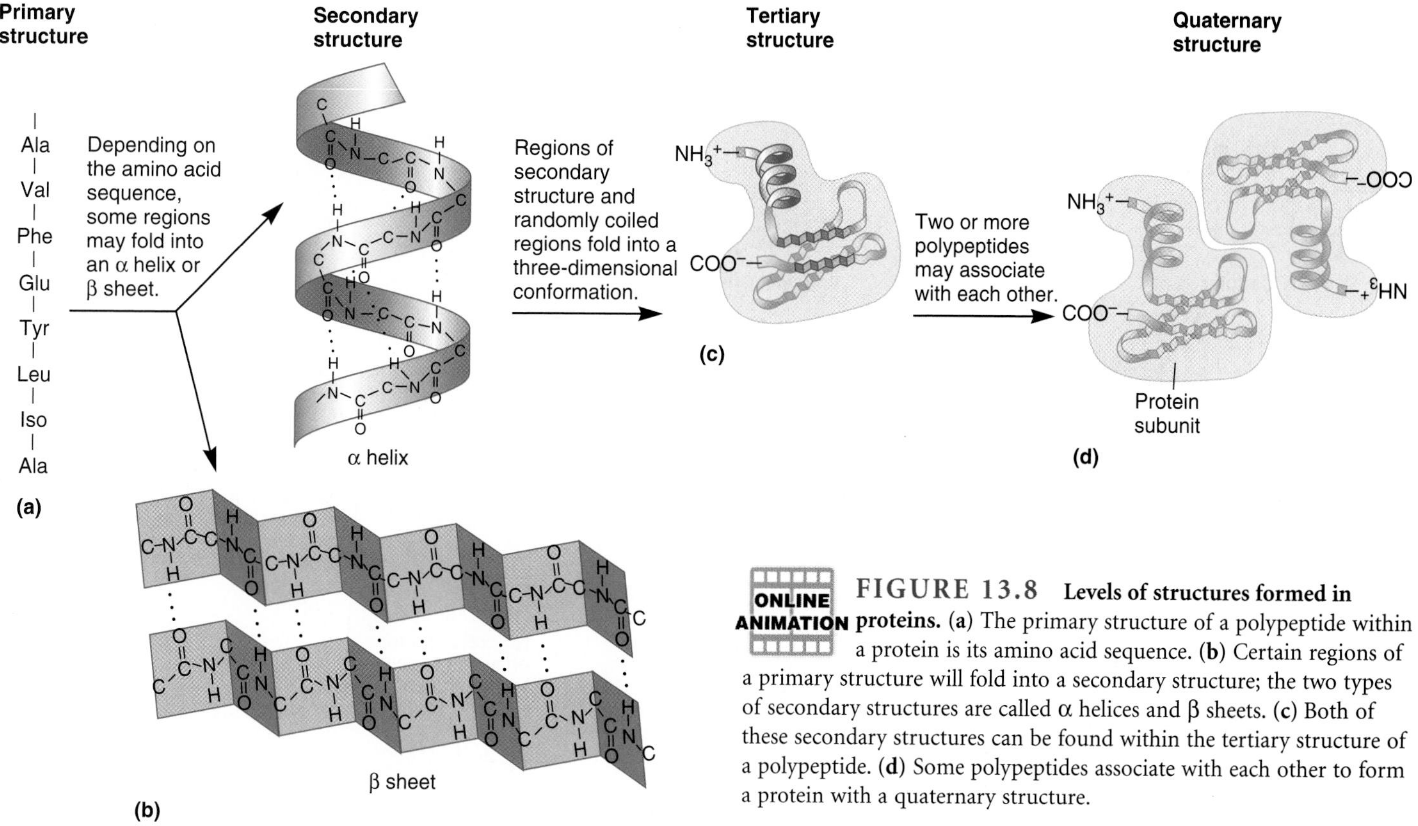
Primary structure
Ala
Val
Phe
Glu
Tyr
Leu
Iso
Ala
(a)
Depending on the amino acid sequence, some regions may fold into an α helix or β sheet.
Secondary structure
α helix
β sheet
(b)
Regions of secondary structure and randomly coiled regions fold into a three-dimensional conformation.
Tertiary structure
(c)
Two or more polypeptides may associate with each other.
Quaternary structure
Protein subunit
(d)

ONLINE ANIMATION **FIGURE 13.8 Levels of structures formed in proteins.** (**a**) The primary structure of a polypeptide within a protein is its amino acid sequence. (**b**) Certain regions of a primary structure will fold into a secondary structure; the two types of secondary structures are called α helices and β sheets. (**c**) Both of these secondary structures can be found within the tertiary structure of a polypeptide. (**d**) Some polypeptides associate with each other to form a protein with a quaternary structure.

such as the hormone insulin, are secreted by endocrine cells and bind to the insulin receptor proteins found within the plasma membrane of target cells.

A key category of proteins are **enzymes,** which function to accelerate chemical reactions within the cell. Some enzymes assist in the breakdown of molecules or macromolecules into smaller units. These are known as catabolic enzymes and are important in generating cellular energy. In contrast, anabolic enzymes function in the synthesis of molecules and macromolecules. Several anabolic enzymes are listed in Table 13.6, including DNA polymerase, which is required for the synthesis of DNA from nucleotide building blocks. Throughout the cell, the synthesis of molecules and macromolecules relies on enzymes and accessory proteins. Ultimately, then, the construction of a cell greatly depends on its anabolic enzymes because these are required to synthesize all cellular macromolecules.

A fundamental question in genetics is, How can we relate protein function to an organism's traits? Let's extend our understanding of protein function by considering an example. In *Drosophila*, the color of the eye provides an interesting case in point (**Figure 13.9**). The eye of the fruit fly is composed of several types of cells. Some of them are pigment cells that are responsible for eye color. In the pigment cells of the red-eyed fly, the X chromosomes contain the red-eye allele (w^+), which codes for a protein that is located in the cell membrane. In the 1990s, the cloning and DNA sequencing of this gene indicated that the encoded protein transports colorless pigment precursor molecules into the cell. In the pigment cells of the red eye, the w^+ genes are transcribed and translated to produce this transport protein. This allows the cells to take up the pigment precursor molecules. Once inside, additional enzymes (that are encoded by different genes) convert the colorless precursors into red pigment. Therefore, the eyes of this fly appear red. In contrast, the white allele (*w*) of this gene is an example of a defective or loss-of-function allele. In the pigment cells of the white-eyed fly, this gene is not expressed properly, so little or no functional transport protein is made. Without this transport protein, the pigment precursor molecules are not transported into the pigment cells of the eye, so the red pigment cannot be made, and the eyes of the fly remain white.

13.2 STRUCTURE AND FUNCTION OF tRNA

Thus far, we have considered the general features of translation and surveyed the structure and functional significance of cellular proteins. The rest of this chapter is devoted to a molecular understanding of translation as it occurs in living cells. Biochemical studies of protein synthesis and tRNA molecules began in the 1950s. As work progressed toward an understanding of translation, research revealed that different kinds of RNA molecules are involved in the incorporation of amino acids into growing polypeptides. Francis

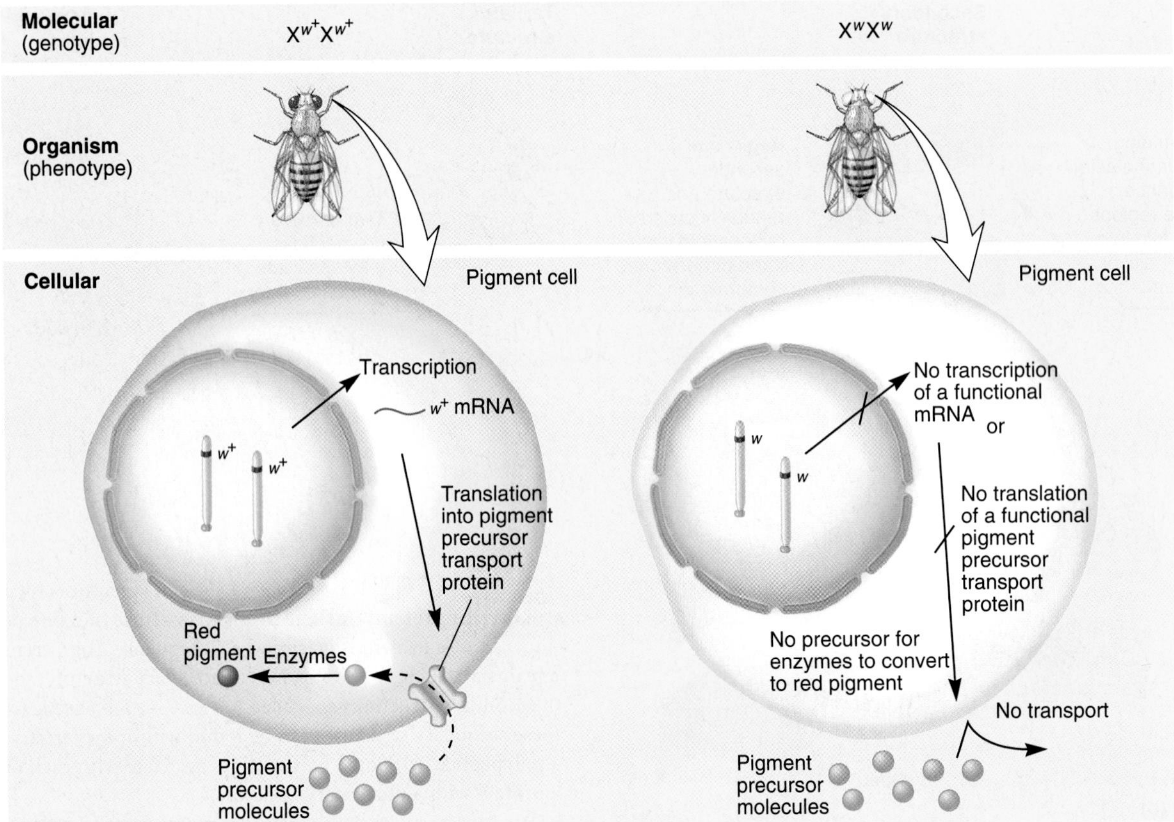

FIGURE 13.9 The relationship between genes and traits at the molecular, organism, and cellular levels.

Genes→Traits The wild-type (w^+) allele encodes a protein that transports a colorless pigment precursor into the pigment cells of the eye. This precursor is converted to a red pigment, so the eyes become red. The mutant *w* allele cannot produce a functional transporter, so little or none of the precursor is transported into the pigment cells. Therefore, the X^wX^w homozygous fly has colorless (white) eyes.

Crick proposed the **adaptor hypothesis.** According to this idea, the position of an amino acid within a polypeptide chain is determined by the binding between the mRNA and an adaptor molecule carrying a specific amino acid. Later, work by Paul Zamecnik and Mahlon Hoagland suggested that the adaptor molecule is tRNA. During translation, a tRNA has two functions. It recognizes a three-base codon sequence in mRNA, and it carries an amino acid specific for that codon. In this section, we will examine the general function of tRNA molecules. We will begin by considering an experiment that was critical in supporting the adaptor hypothesis and then explore some of the important structural features that underlie tRNA function.

The Function of a tRNA Depends on the Specificity Between the Amino Acid It Carries and Its Anticodon

The adaptor hypothesis proposes that tRNA molecules recognize the codons within mRNA and carry the correct amino acids to the site of polypeptide synthesis. During mRNA-tRNA recognition, the anticodon in a tRNA molecule binds to a codon in mRNA due to their complementary sequences (**Figure 13.10**). Importantly, the anticodon in the tRNA corresponds to the amino acid that it carries. For example, if the anticodon in the tRNA is 3'–AAG–5', it is complementary to a 5'–UUC–3' codon. According to the genetic code, described earlier in this chapter, the UUC codon specifies phenylalanine. Therefore, the tRNA with a 3'–AAG–5' anticodon must carry a phenylalanine. As another example, if the tRNA has a 3'–GGC–5' anticodon, it is complementary to a 5'–CCG–3' codon that specifies proline. This tRNA must carry proline.

Recall that the genetic code has 64 codons. Of these, 61 are sense codons that specify the 20 amino acids. Therefore, to synthesize proteins, a cell must produce many different tRNA molecules having specific anticodon sequences. To do so, the chromosomal DNA contains many distinct tRNA genes that encode tRNA molecules with different sequences. According to the adaptor hypothesis, the anticodon in a tRNA specifies the type of amino acid that it carries. Due to this specificity, tRNA molecules are named according to the type of amino acid that they carry. For example, a tRNA that attaches to phenylalanine is described as $tRNA^{phe}$, whereas a tRNA that carries proline is $tRNA^{pro}$.

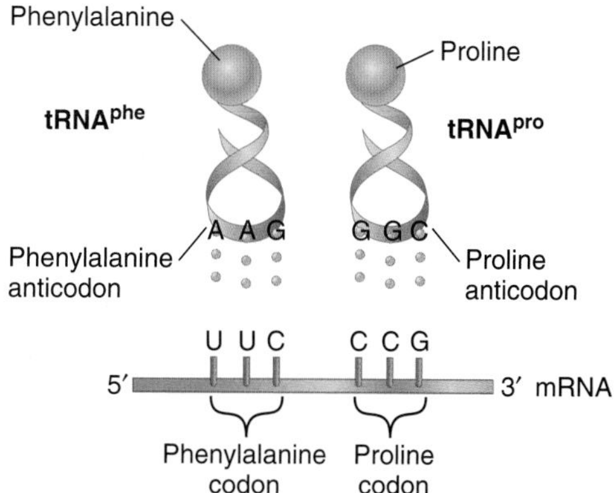

FIGURE 13.10 Recognition between tRNAs and mRNA. The anticodon in the tRNA binds to a complementary sequence in the mRNA. At its other end, the tRNA carries the amino acid that corresponds to the codon in the mRNA via the genetic code.

EXPERIMENT 13B

tRNA Functions as the Adaptor Molecule Involved in Codon Recognition

In 1962, François Chapeville and his colleagues conducted experiments aimed at testing the adaptor hypothesis. Their technical strategy was similar to that of the Nirenberg experiments that helped to decipher the genetic code (see Experiment 13A). In this approach, a cell-free translation system was made from cell extracts that contained the components necessary for translation. These components include ribosomes, tRNAs, and other translation factors. A cell-free translation system can synthesize polypeptides in vitro if mRNA and amino acids are added. Such a translation system can be used to investigate the role of specific factors by adding a particular mRNA template and varying individual components required for translation.

According to the adaptor hypothesis, the amino acid attached to a tRNA is not directly involved in codon recognition. Chapeville reasoned that if this were true, the alteration of an amino acid already attached to a tRNA should cause that altered amino acid to be incorporated into the polypeptide instead of the normal amino acid. For example, consider a tRNA^{cys} that carries the amino acid cysteine. If the attached cysteine were changed to an alanine, this tRNA^{cys} should insert an alanine into a polypeptide where it would normally put a cysteine. Fortunately, Chapeville could carry out this strategy because he had a reagent, known as Raney nickel, that can chemically convert cysteine to alanine.

An elegant aspect of the experimental design was the choice of the mRNA template. Chapeville and his colleagues synthesized an mRNA template that contained only U and G. Therefore, this template contained only the following codons (refer back to the genetic code in Table 13.2):

UUU = phenylalanine	GGU = glycine
UUG = leucine	GUU = valine
UGG = tryptophan	GUG = valine
GGG = glycine	UGU = cysteine

Among the eight possible codons, one cysteine codon occurs, but no alanine codons can be formed from a polyUG template.

As shown in the experiment of **Figure 13.11**, Chapeville began with a cell-free translation system that contained tRNA molecules. Amino acids, which would become attached to tRNAs, were added to this mixture. Of the 20 amino acids, only cysteine was radiolabeled. After allowing sufficient time for the amino acids to become attached to the correct tRNAs, the sample was divided into two tubes. One tube was treated with Raney nickel, while the control tube was not. As mentioned, Raney nickel converts cysteine into alanine by removing the –SH (sulfhydryl) group. However, it would not remove the radiolabel, which was a ^{14}C-label within the cysteine amino acid. Next, the polyUG mRNA was added as a template, and the samples were incubated to allow the translation of the mRNA into a polypeptide. In the control tube, we would expect the polypeptide to contain phenylalanine, leucine, tryptophan, glycine, valine, and cysteine, because these are the codons that contain only U and G. However, in the Raney nickel-treated sample, if the tRNA^{cys} was using its anticodon region to recognize the mRNA, we would expect to see alanine instead of cysteine.

Following translation, the polypeptides were isolated and hydrolyzed via a strong acid treatment, and then the individual amino acids were separated by column chromatography. The column separated cysteine from alanine; alanine eluted in a later fraction. The amount of radioactivity in each fraction was determined by liquid scintillation counting.

THE HYPOTHESIS

Codon recognition is dictated only by the tRNA anticodon; the chemical structure of the amino acid attached to the tRNA does not play a role.

TESTING THE HYPOTHESIS — FIGURE 13.11 Evidence that tRNA uses its anticodon sequence to recognize mRNA.

Starting material: A cell-free translation system that can synthesize polypeptides if mRNA and amino acids are added.

1. Place cell-free translation system into a tube. Note: This drawing emphasizes only $tRNA^{cys}$, even though the cell-free translation system contains all types of tRNAs, and other components, such as ribosomes. In the translation system, a substantial proportion of the tRNAs do not have an attached amino acid. The translation system also contains enzymes that attach amino acids to tRNAs. (These enzymes will be described later in the chapter.)

2. Add amino acids, including radiolabeled cysteine. An enzyme within the translation system will specifically attach the radiolabeled cysteine to $tRNA^{cys}$. The other tRNAs will have unlabeled amino acids attached to them. Incubate and divide into two tubes.

3. In one tube, treat the tRNAs with Raney nickel. This removes the –SH group from cysteine, converting it to alanine. In the control tube, do not add Raney nickel.

4. Add polyUG mRNA made via polynucleotide phosphorylase as a template. A polyUG mRNA contains cysteine codons but no alanine codons.

5. Allow translation to proceed.

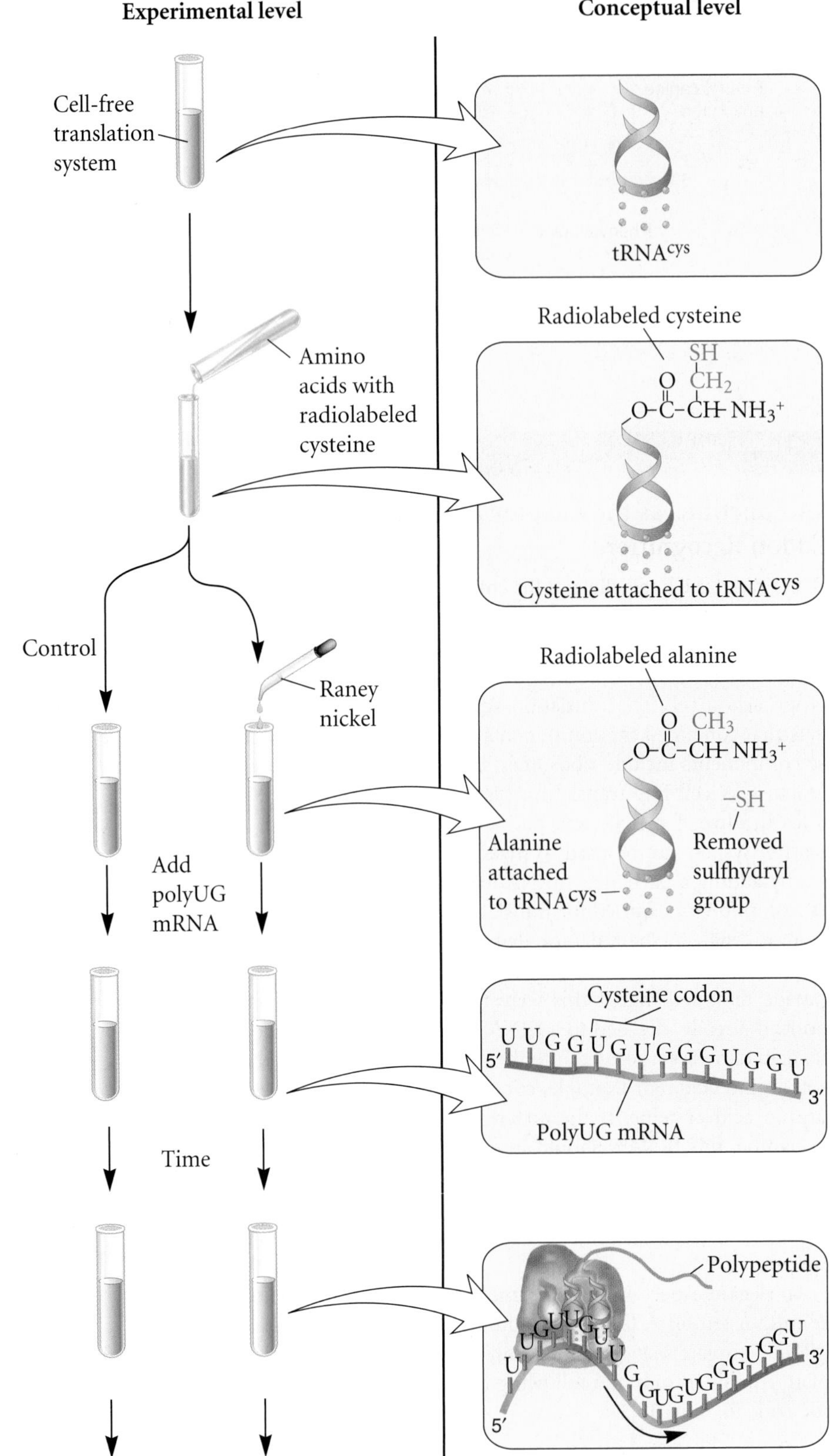

6. Precipitate the newly made polypeptides with trichloroacetic acid and then isolate the polypeptides on a filter.

7. Hydrolyze the polypeptides to their individual amino acids by treatment with a solution containing concentrated hydrochloric acid.

8. Run the sample over a column that separates cysteine and alanine. (See the Appendix for a description of column chromatography.) Separate into fractions. Note: Cysteine runs through the column more quickly and comes out in fraction 3. Alanine comes out later, in fraction 7.

9. Determine the amount of radioactivity in the fractions that contain alanine and cysteine.

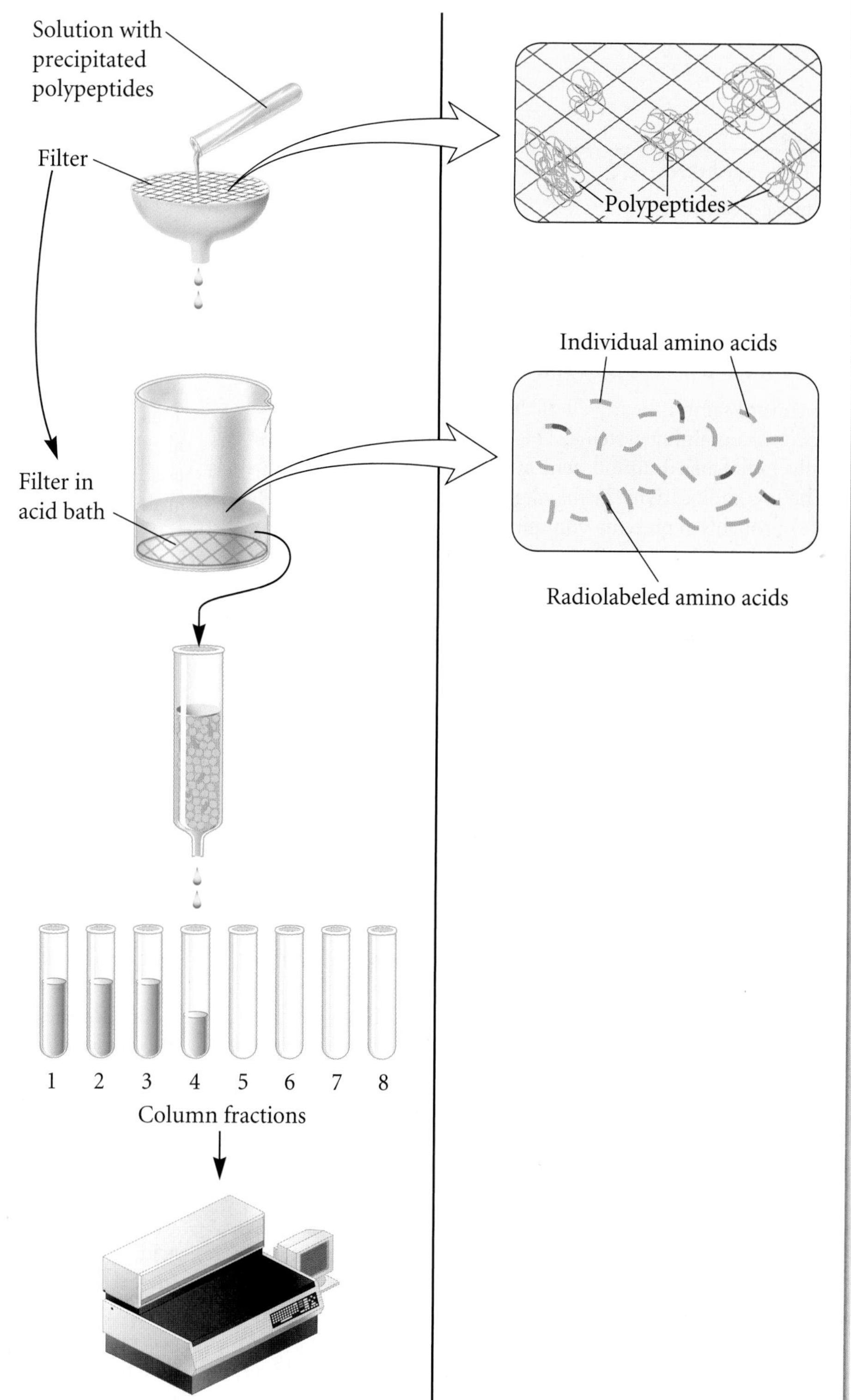

THE DATA

Conditions	*Amount of Radiolabeled Amino Acids Incorporated into Polypeptide (cpm)** Cysteine	Alanine	Total
Control, untreated tRNA	2,835	83	2,918
Raney nickel-treated tRNA	990	2,020	3,010

*Cpm is the counts per minute of radioactivity in the sample.
Adapted from Chapeville et al. (1962) *PNAS 48*, 1086–1092.

INTERPRETING THE DATA

In the control sample, nearly all of the radioactivity was found in the fraction containing cysteine. This result was expected because the only radiolabeled amino acid attached to tRNAs was cysteine. The low radioactivity in the alanine fraction (83 counts per minute) probably represents contamination of this fraction by a small amount of cysteine. By comparison, when the tRNAs were treated with Raney nickel, a substantial amount of radiolabeled alanine became incorporated into polypeptides. This occurred even though the mRNA template did not contain any alanine codons. How do we explain these results? They are consistent with the explanation that a tRNA^{cys}, which carried alanine instead of cysteine, incorporated alanine into the synthesized polypeptide. These observations indicate that the codons in mRNA are identified directly by the tRNA rather than the attached amino acid.

As seen in the data of Figure 13.11, the Raney nickel-treated sample still had 990 cpm of cysteine incorporated into polypeptides. This is about one-third of the total amount of radioactivity (namely, 990/3,010). In other experiments conducted in this study, the researchers showed that the Raney nickel did not react with about one-third of the tRNA^{cys}. Therefore, this proportion of the Raney nickel-treated tRNA^{cys} would still carry cysteine. This observation was consistent with the data shown here. Overall, the results of this experiment supported the adaptor hypothesis, indicating that tRNAs act as adaptors to carry the correct amino acid to the ribosome based on their anticodon sequence.

A self-help quiz involving this experiment can be found at the Online Learning Center.

Common Structural Features Are Shared by All tRNAs

To understand how tRNAs act as carriers of the correct amino acids during translation, researchers have examined the structural characteristics of these molecules in great detail. Though a cell makes many different tRNAs, all tRNAs share common structural features. As originally proposed by Robert Holley in 1965, the secondary structure of tRNAs exhibits a cloverleaf pattern. A tRNA has three stem-loop structures, a few variable sites, and an acceptor stem with a 3' single-stranded region (**Figure 13.12**). The acceptor stem is where an amino acid becomes attached to a tRNA (see inset). A conventional numbering system for the nucleotides within a tRNA molecule begins at the 5' end and proceeds toward the 3' end. Among different types of tRNA molecules, the variable sites (shown in blue) can differ in the number of nucleotides they contain. The anticodon is located in the second loop region.

The actual three-dimensional, or tertiary, structure of tRNA molecules involves additional folding of the secondary structure. In the tertiary structure of tRNA, the stem-loop regions are folded into a much more compact molecule. The ability of RNA molecules to form stem-loop structures and the tertiary folding of tRNA molecules are described in Chapter 9 (see Figure 9.24). Interestingly, in addition to the normal A, U, G, and C nucleotides, tRNA molecules commonly contain modified nucleotides within their primary structures. For example, Figure 13.12 illustrates a tRNA that contains several modified bases. Among many different species, researchers have found that more than 60 different nucleotide modifications can occur in tRNA molecules. We will explore the significance of modified bases in codon recognition later in this chapter.

Aminoacyl-tRNA Synthetases Charge tRNAs by Attaching the Appropriate Amino Acid

To function correctly, each type of tRNA must have the appropriate amino acid attached to its 3' end. How does an amino acid get attached to a tRNA with the correct anticodon? Enzymes in the cell known as **aminoacyl-tRNA synthetases** catalyze the attachment of amino acids to tRNA molecules. Cells produce 20 different aminoacyl-tRNA synthetase enzymes, one for each of the 20 distinct amino acids. Each aminoacyl-tRNA synthetase is named for the specific amino acid it attaches to tRNA. For example, alanyl-tRNA synthetase recognizes a tRNA with an alanine anticodon—tRNA^{ala}—and attaches an alanine to it.

Aminoacyl-tRNA synthetases catalyze a chemical reaction involving three different molecules: an amino acid, a tRNA molecule, and ATP. In the first step of the reaction, a synthetase recognizes a specific amino acid and also ATP (**Figure 13.13**). The ATP is hydrolyzed, and AMP becomes attached to the amino acid; pyrophosphate is released. During the second step, the correct tRNA binds to the synthetase. The amino acid becomes covalently attached to the 3' end of the tRNA molecule at the acceptor stem, and AMP is released. Finally, the tRNA with its attached amino acid is released from the enzyme. At this stage, the tRNA is called a **charged tRNA** or an **aminoacyl tRNA.** In a charged tRNA molecule, the amino acid is attached to the 3' end of the tRNA by a covalent bond (see Figure 13.12 inset).

The ability of the aminoacyl-tRNA synthetases to recognize tRNAs has sometimes been called the "second genetic code." This recognition process is necessary to maintain the fidelity of genetic information. The frequency of error for aminoacyl-tRNA synthetases is less than 10^{-5}. In other words, the wrong amino acid will be attached to a tRNA less than once in 100,000 times!

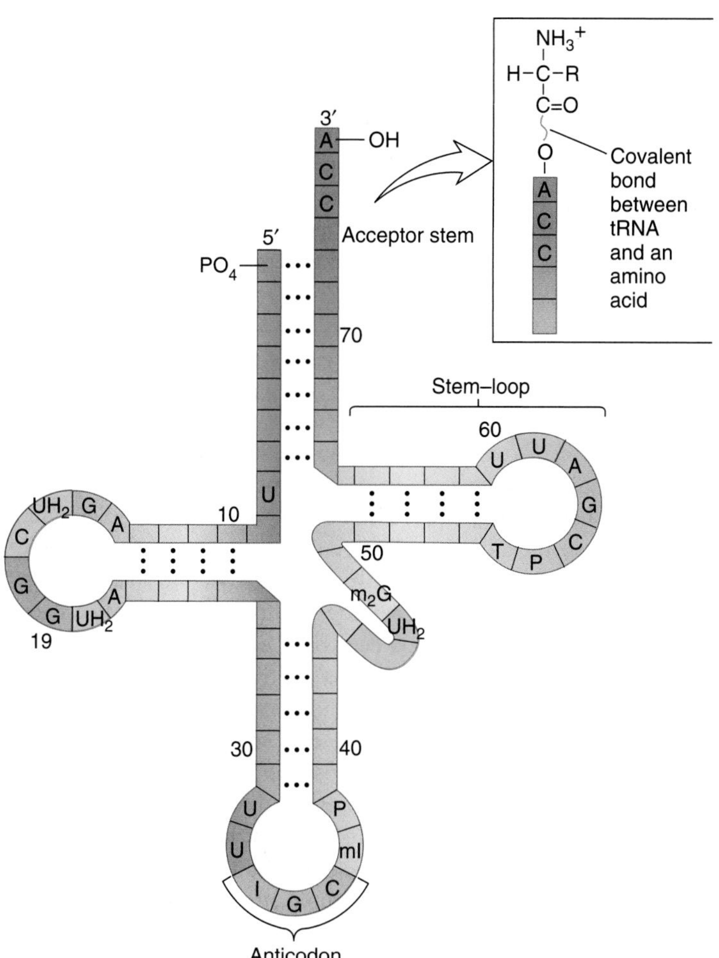

FIGURE 13.12 **Secondary structure of tRNA.** The conventional numbering of nucleotides begins at the 5' end and proceeds toward the 3' end. In all tRNAs, the nucleotides at the 3' end contain the sequence CCA. Certain locations can have additional nucleotides not found in all tRNA molecules; these variable sites are shown in blue. The figure also shows the locations of a few modified bases specifically found in a yeast tRNA that carries alanine. The modified bases are as follows: I = inosine, mI = methylinosine, T = ribothymidine, UH_2 = dihydrouridine, m_2G = dimethylguanosine, and P = pseudouridine. The inset shows an amino acid covalently attached to the 3' end of a tRNA.

As you might expect, the anticodon region of the tRNA is usually important for precise recognition by the correct aminoacyl-tRNA synthetase. In studies of *Escherichia coli* synthetases, 17 of the 20 types of aminoacyl-tRNA synthetases recognize the anticodon region of the tRNA. However, other regions of the tRNA are also important recognition sites. These include the acceptor stem and bases in the stem-loop regions.

As mentioned previously, tRNA molecules frequently contain bases within their structure that have been chemically modified. These modified bases can have important effects on tRNA function. For example, modified bases within tRNA molecules affect the rate of translation and the recognition of tRNAs by aminoacyl-tRNA synthetases. Positions 34 and 37 contain the largest variety of modified nucleotides; position 34 is the first base in the anticodon that matches the third base in the codon

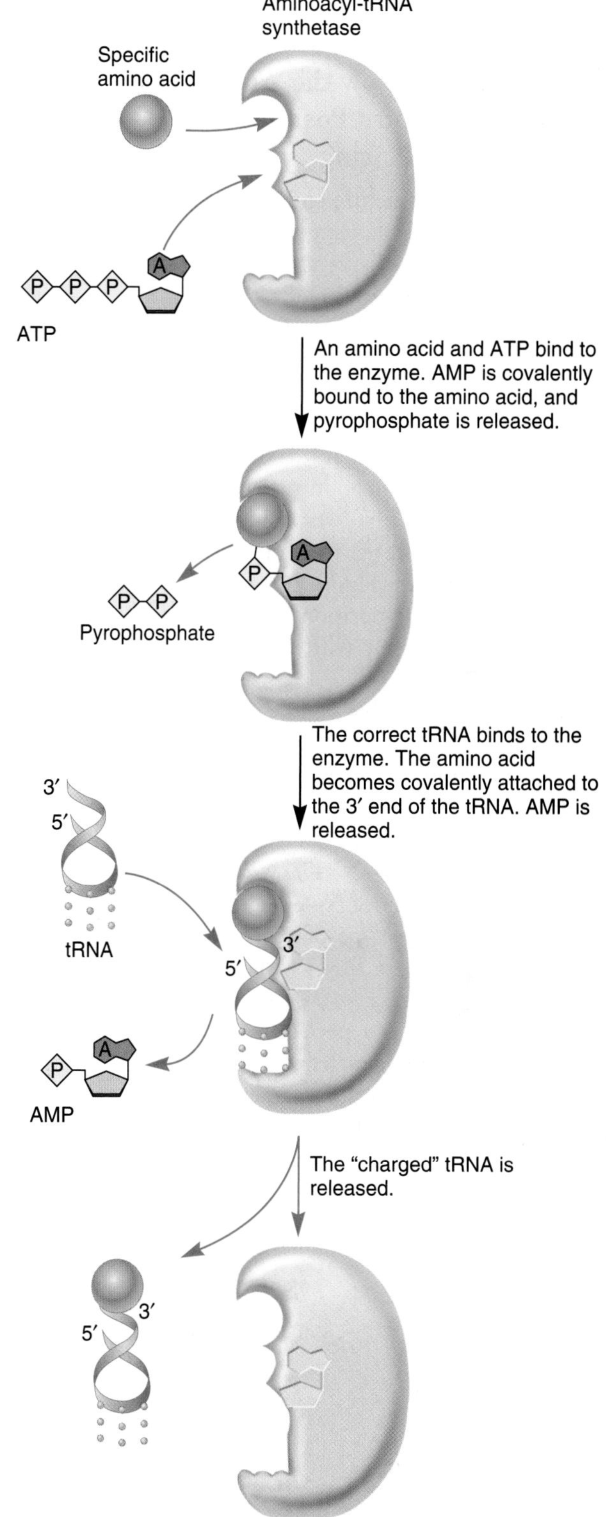

ONLINE ANIMATION

FIGURE 13.13 **Catalytic function of aminoacyl-tRNA synthetase.** Aminoacyl-tRNA synthetase has binding sites for a specific amino acid, ATP, and a particular tRNA. In the first step, the enzyme catalyzes the covalent attachment of AMP to an amino acid, yielding an activated amino acid. In the second step, the activated amino acid is attached to the appropriate tRNA.

of mRNA. As discussed next, a modified base at position 34 can have important effects on codon-anticodon recognition.

Mismatches That Follow the Wobble Rule Can Occur at the Third Position in Codon-Anticodon Pairing

After considering the structure and function of tRNA molecules, let's reexamine some subtle features of the genetic code. As discussed earlier, the genetic code is degenerate, which means that more than one codon can specify the same amino acid. Degeneracy usually occurs at the third position in the codon. For example, valine is specified by GUU, GUC, GUA, and GUG. In all four cases, the first two bases are G and U. The third base, however, can be U, C, A, or G. To explain this pattern of degeneracy, Francis Crick proposed in 1966 that it is due to "wobble" at the third position in the codon-anticodon recognition process. According to the **wobble rules,** the first two positions pair strictly according to the AU/GC rule. However, the third position can tolerate certain types of mismatches (**Figure 13.14**). This proposal suggested that the base at the third position in the codon does not have to hydrogen bond as precisely with the corresponding base in the anticodon.

Because of the wobble rules, some flexibility is observed in the recognition between a codon and anticodon during the process of translation. When two or more tRNAs that differ at the wobble base are able to recognize the same codon, these are termed **isoacceptor tRNAs.** As an example, tRNAs with an anticodon of 3'–CCA–5' or 3'–CCG–5' would be able to recognize a codon with the sequence of 5'–GGU–3'. In addition, the wobble rules enable a single type of tRNA to recognize more than one codon. For example, a tRNA with an anticodon sequence of 3'–AAG–5' can recognize a 5'–UUC–3' and a 5'–UUU–3' codon. The 5'–UUC–3' codon is a perfect match with this tRNA. The 5'–UUU–3' codon is mismatched according to the standard RNA-RNA hybridization rules (namely, G in the anticodon is mismatched to U in the codon), but the two can fit according to the wobble rules described in Figure 13.14. Likewise, the modification of the wobble base to an inosine allows a tRNA to recognize three different codons. At the cellular level, the ability of a single tRNA to recognize more than one codon makes it unnecessary for a cell to make 61 different tRNA molecules with anticodons that are complementary to the 61 possible sense codons. *E. coli* cells, for example, make a population of tRNA molecules that have just 40 different anticodon sequences.

13.3 RIBOSOME STRUCTURE AND ASSEMBLY

In Section 13.2, we examined how the structure and function of tRNA molecules are important in translation. According to the adaptor hypothesis, tRNAs bind to mRNA due to complementarity between the anticodons and codons. Concurrently, the tRNA molecules have the correct amino acid attached to their 3' ends.

To synthesize a polypeptide, additional events must occur. In particular, the bond between the 3' end of the tRNA and the amino acid must be broken, and a peptide bond must be formed between the adjacent amino acids. To facilitate these events, translation occurs on the surface of a macromolecular complex known as the **ribosome.** The ribosome can be thought of as the macromolecular arena where translation takes place.

In this section, we will begin by outlining the biochemical compositions of ribosomes in bacterial and eukaryotic cells. We will then examine the key functional sites on ribosomes for the translation process.

Bacterial and Eukaryotic Ribosomes Are Assembled from rRNA and Proteins

Bacterial cells have one type of ribosome that is found within the cytoplasm. Eukaryotic cells contain biochemically distinct ribosomes in different cellular locations. The most abundant type of ribosome functions in the cytosol, which is the region of

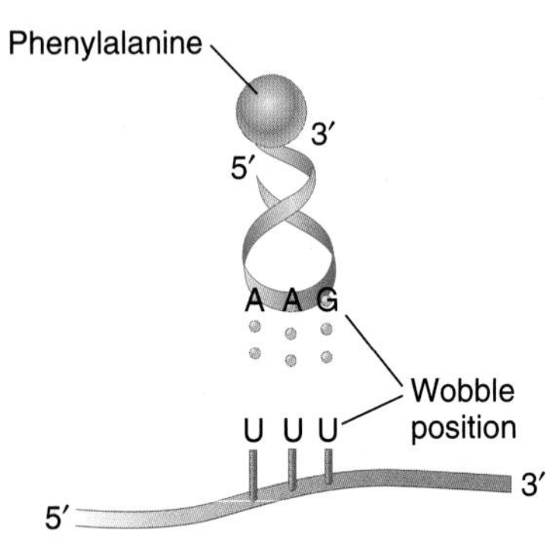

(a) Location of wobble position

Nucleotide of tRNA anticodon	Third nucleotide of mRNA codon
G	C, U
C	G
A	U, C, G, (A)
U	A, U, G, (C)
I	U, C, A
xm^5s^2U, xm^5Um, Um, xm^5U	A, (G)
xo^5U	U, A, G
k^2C	A

(b) Revised wobble rules

FIGURE 13.14 Wobble position and base pairing rules. **(a)** The wobble position occurs between the first base (meaning the first base in the 5' to 3' direction) in the anticodon and the third base in the mRNA codon. **(b)** The revised wobble rules are slightly different from those originally proposed by Crick. The standard bases found in RNA are G, C, A, and U. In addition, the structures of bases in tRNAs may be modified. Some modified bases that may occur in the wobble position in tRNA are I = inosine; xm^5s^2U = 5-methyl-2-thiouridine; xm5Um = 5-methyl-2'-O-methyluridine; Um = 2'-O-methyluridine; xm^5U = 5-methyluridine; xo^5U = 5-hydroxyuridine; k^2C = lysidine (a cytosine derivative). The mRNA bases in parentheses are recognized very poorly by the tRNA.

the eukaryotic cell that is inside the plasma membrane but outside the organelles. Besides the cytosolic ribosomes, all eukaryotic cells have ribosomes within the mitochondria. In addition, plant cells and algae have ribosomes in their chloroplasts. The compositions of mitochondrial and chloroplast ribosomes are quite different from that of the cytosolic ribosomes. Unless otherwise noted, the term eukaryotic ribosome refers to ribosomes in the cytosol, not to those found within organelles. Likewise, the description of eukaryotic translation refers to translation via cytosolic ribosomes.

Each ribosome is composed of structures called the large and small subunits. This term is perhaps misleading because each ribosomal subunit itself is formed from the assembly of many different proteins and RNA molecules called ribosomal RNA or rRNA. In bacterial ribosomes, the 30S subunit is formed from the assembly of 21 different ribosomal proteins and a 16S rRNA molecule; the 50S subunit contains 34 different proteins and 23S and 5S rRNA molecules (**Figure 13.15a**). The designations 30S and 50S refer to the rate that these subunits sediment when subjected to a centrifugal force. This rate is described as a sedimentation coefficient in Svedberg units (S), in honor of Theodor Svedberg, who invented the ultracentrifuge. Together, the 30S and 50S subunits form a 70S ribosome. (Note: Svedberg units do not add up linearly.) In bacteria, the ribosomal proteins and rRNA molecules are synthesized in the cytoplasm, and the ribosomal subunits are assembled there.

The synthesis of eukaryotic rRNA occurs within the nucleus, and the ribosomal proteins are made in the cytosol,

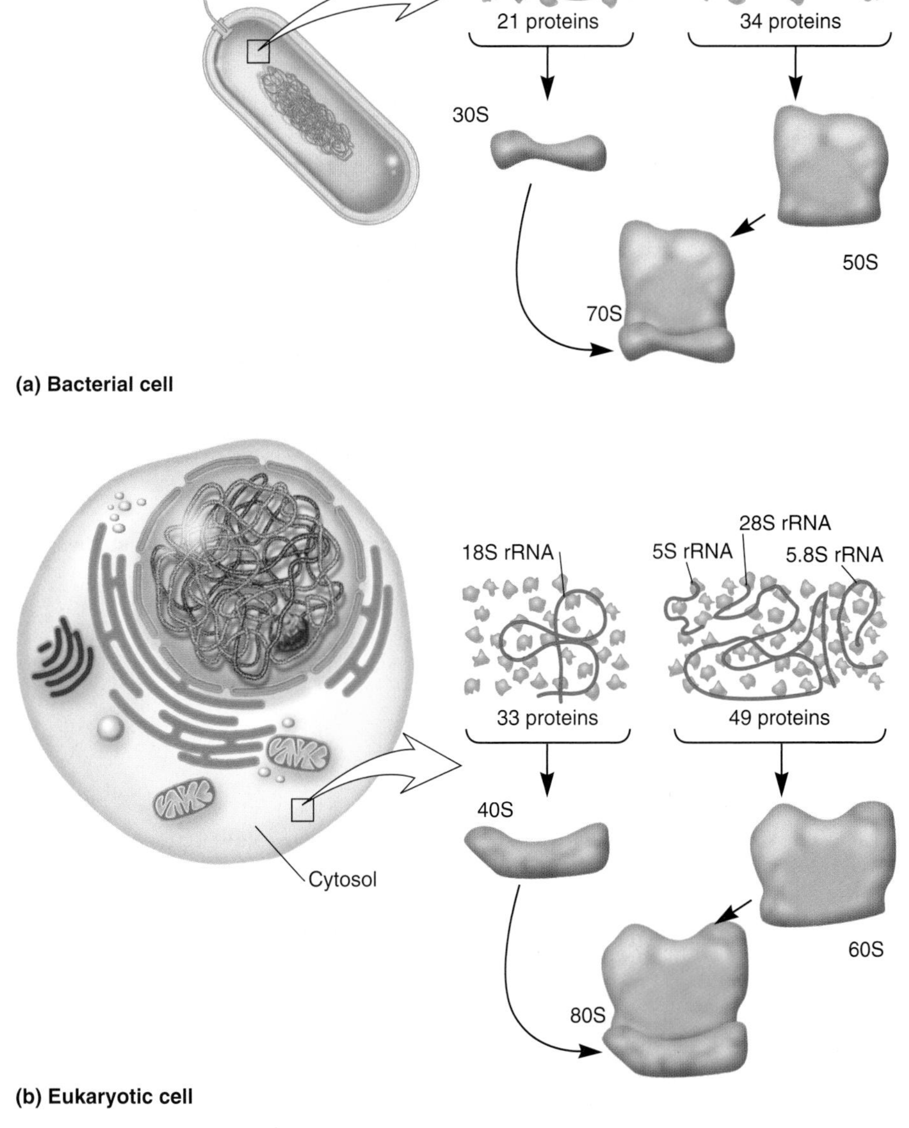

FIGURE 13.15 Composition of bacterial and eukaryotic ribosomes. (**a**) Bacterial ribosomes contain two subunits: 30S and 50S, which assemble to form a 70S ribosome. The 30S subunit contains 21 different proteins and 16S rRNA; the 50S subunit contains 34 proteins and 23S and 5S rRNAs. (**b**) Eukaryotic ribosomes found in the cytosol contain two subunits: 40S and 60S. The 40S subunit contains 33 proteins and 18S rRNA; the 60S subunit contains 49 proteins and 5S, 28S, and 5.8S rRNAs. The 40S and 60S subunits are formed in the nucleolus and then are exported into the cytosol, where they assemble to form an 80S ribosome.

where translation takes place. The 40S subunit is composed of 33 proteins and an 18S rRNA; the 60S subunit is made of 49 proteins and 5S, 5.8S, and 28S rRNAs (**Figure 13.15b**). The assembly of the rRNAs and ribosomal proteins to make the 40S and 60S subunits occurs within the **nucleolus,** a region of the nucleus specialized for this purpose. The 40S and 60S subunits are then exported into the cytosol, where they associate to form an 80S ribosome during translation.

Components of Ribosomal Subunits Form Functional Sites for Translation

To understand the structure and function of the ribosome at the molecular level, researchers must determine the locations and functional roles of the individual ribosomal proteins and rRNAs. In recent years, many advances have been made toward a molecular understanding of ribosomes. Microscopic and biophysical methods have been used to study ribosome structure. An electron micrograph of bacterial ribosomes is shown in **Figure 13.16a.** More recently, a few research groups have succeeded in crystallizing ribosomal subunits, and even intact ribosomes. This is an amazing technical feat, because it is difficult to find the right conditions under which large macromolecules will form highly ordered crystals. **Figure 13.16b** shows the crystal structure of bacterial ribosomal subunits. The overall shape of each subunit is largely determined by the structure of the rRNAs, which constitute most of the mass of the ribosome. The interface between the 30S and 50S subunits is primarily composed of rRNA. Ribosomal proteins cluster on the outer surface of the ribosome and on the periphery of the interface.

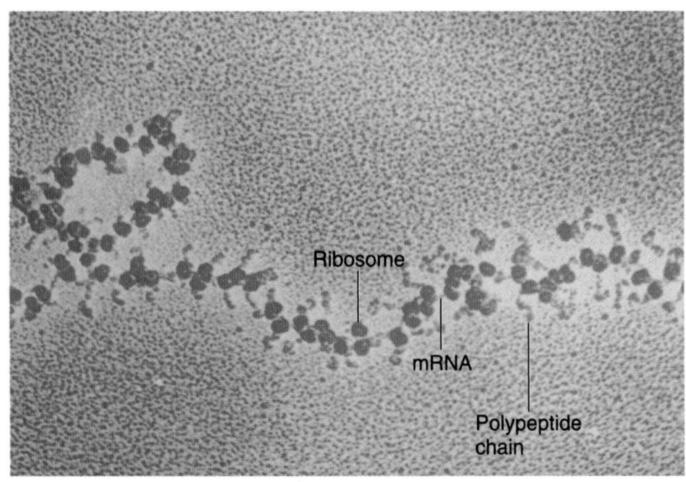

(a) Ribosomes as seen with electron microscope

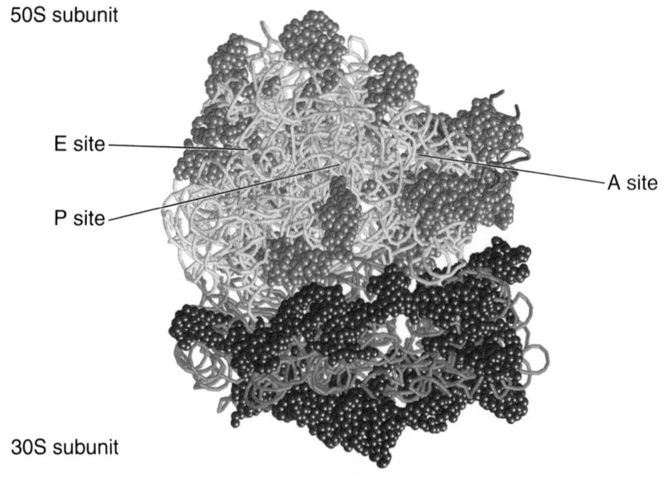

(b) Bacterial ribosome model based on X–ray diffraction studies

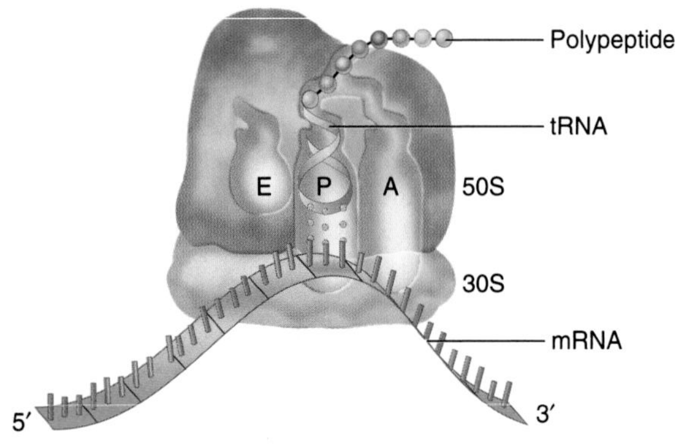

(c) Model for ribosome structure

FIGURE 13.16 **Ribosomal structure.** (**a**) Electron micrograph of ribosomes attached to an mRNA molecule. (**b**) Crystal structure of the 50S and 30S subunits in bacteria. This model shows the interface between the two subunits. The rRNA is shown in gray (50S subunit) and turquoise (30S subunit), and proteins are shown in violet (50S subunit) and dark blue (30S subunit). (**c**) A model depicting the sites where tRNA and mRNA bind to an intact ribosome. The mRNA lies on the surface of the 30S subunit. The E, P, and A sites are formed at the interface between the large and small subunits. The growing polypeptide chain exits through a hole in the 50S subunit.

During bacterial translation, the mRNA lies on the surface of the 30S subunit within a space between the 30S and 50S subunits. As the polypeptide is being synthesized, it exits through a channel within the 50S subunit (**Figure 13.16c**). Ribosomes contain discrete sites where tRNAs bind and the polypeptide is synthesized. In 1964, James Watson was the first to propose a two-site model for tRNA binding to the ribosome. These sites are known as the **peptidyl site (P site)** and **aminoacyl site (A site).** In 1981, Knud Nierhaus, Hans Sternbach, and Hans-Jorg Rheinberger proposed a three-site model. This model incorporated the observation that uncharged tRNA molecules can bind to a site on the ribosome that is distinct from the P and A sites. This third site is now known as the **exit site (E site).** The locations of the E, P, and A sites are shown in Figure 13.16c. Next, we will examine the roles of these sites during the three stages of translation.

13.4 STAGES OF TRANSLATION

Like transcription, the process of translation can be viewed as occurring in three stages: initiation, elongation, and termination. **Figure 13.17** presents an overview of these stages. During **initiation,** the ribosomal subunits, mRNA, and the first tRNA assemble to form a complex. After the initiation complex is formed, the ribosome slides along the mRNA in the 5' to 3' direction, moving over the codons. This is the **elongation** stage of translation. As the ribosome moves, tRNA molecules sequentially bind to the mRNA at the A site in the ribosome, bringing with them the appropriate amino acids. Therefore, amino acids are linked in the order dictated by the codon sequence in the mRNA. Finally, a stop codon is reached, signaling the **termination** of translation. At this point, disassembly occurs, and the newly made polypeptide is released.

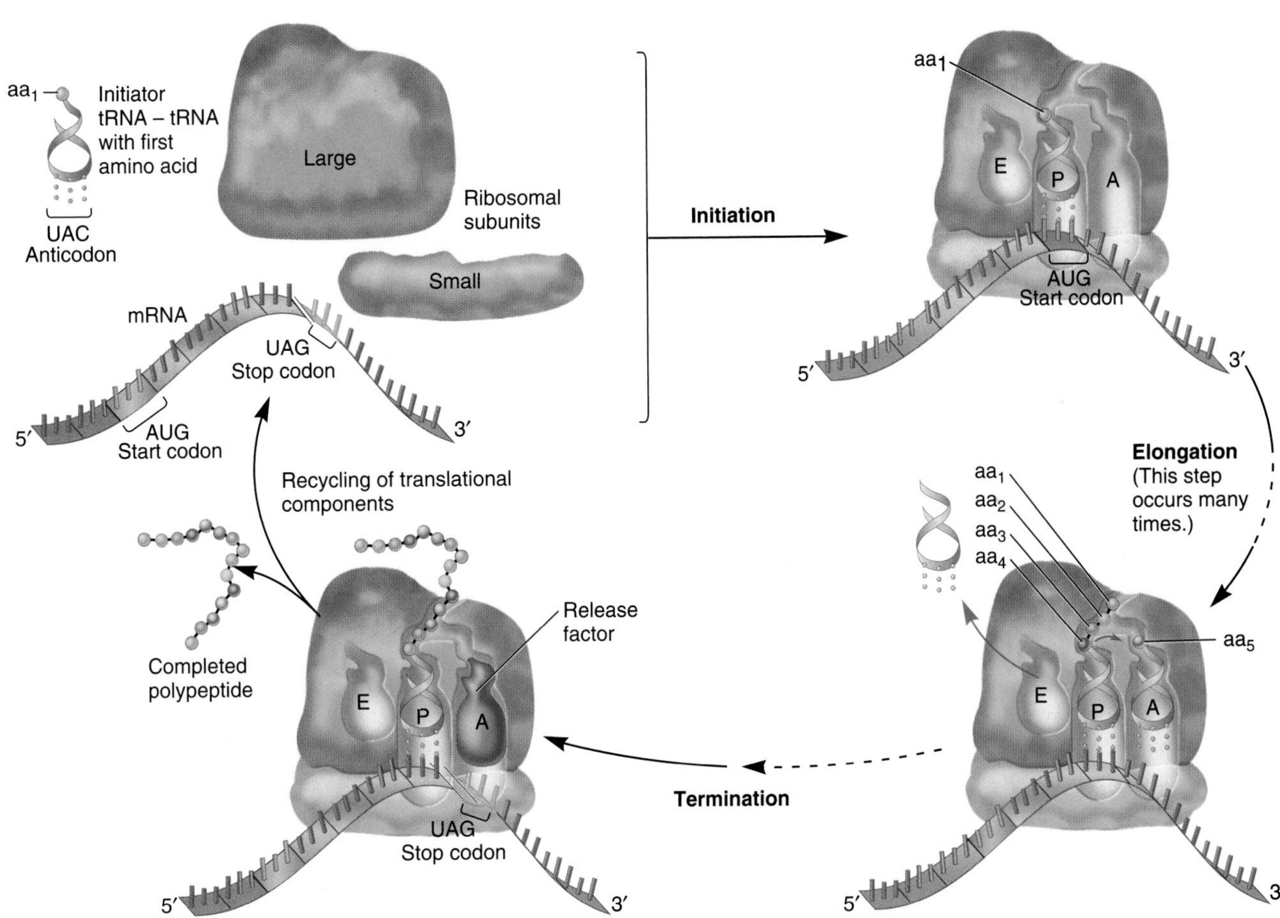

FIGURE 13.17 **Overview of the stages of translation.** The initiation stage involves the assembly of the ribosomal subunits, mRNA, and the initiator tRNA carrying the first amino acid. During elongation, the ribosome slides along the mRNA and synthesizes a polypeptide chain. Translation ends when a stop codon is reached and the polypeptide is released from the ribosome. (Note: In this and succeeding figures in this chapter, the ribosomes are drawn schematically to emphasize different aspects of the translation process. The structures of ribosomes are described in Figures 13.15 and 13.16.)

Genes→Traits The ability of genes to produce an organism's traits relies on the molecular process of gene expression. During translation, the codon sequence within mRNA (which is derived from a gene sequence during transcription) is translated into a polypeptide sequence. After polypeptides are made within a living cell, they function as proteins to govern an organism's traits. For example, once the β-globin polypeptide is made, it functions within the hemoglobin protein and provides red blood cells with the ability to carry oxygen, a vital trait for survival. Translation allows functional proteins to be made within living cells.

In this section, we will examine the components required for the translation process and consider their functional roles during the three stages of translation.

The Initiation Stage Involves the Binding of mRNA and the Initiator tRNA to the Ribosomal Subunits

During initiation, an mRNA and the first tRNA bind to the ribosomal subunits. A specific tRNA functions as the **initiator tRNA,** which recognizes the start codon in the mRNA. In bacteria, the initiator tRNA, which is also designated $tRNA^{fmet}$, carries a methionine that has been covalently modified to N-formylmethionine. In this modification, a formyl group (—CHO) is attached to the nitrogen atom in methionine after the methionine has been attached to the tRNA.

Figure 13.18 describes the initiation stage of translation in bacteria during which the mRNA, $tRNA^{fmet}$, and ribosomal subunits associate with each other to form an initiation complex. The formation of this complex requires the participation of three initiation factors: IF1, IF2, and IF3. First, the mRNA binds to the 30S ribosomal subunit. This binding is facilitated by IF3, which requires GTP for its function. In addition, complementarity between a short region of the mRNA and the 16S rRNA within the 30S subunit plays a key role. A nine-nucleotide sequence within bacterial mRNAs, known as the ribosomal-binding site, or **Shine-Dalgarno sequence,** is involved in the binding of the mRNA to the 30S subunit. The location of this sequence is shown in Figure 13.18 and in more detail in **Figure 13.19.** How does the Shine-Dalgarno sequence facilitate the binding of mRNA to the ribosome? The Shine-Dalgarno sequence is complementary to a short sequence within the 16S rRNA, which promotes the hydrogen bonding of the mRNA to the 30S subunit.

Next, $tRNA^{fmet}$ binds to the mRNA that is already attached to the 30S subunit (Figure 13.18). This step requires the function of IF2, which also uses GTP. The $tRNA^{fmet}$ binds to the start codon, which is typically a few nucleotides downstream from the Shine-Dalgarno sequence. The start codon is usually AUG, but in some cases it can be GUG or UUG. Even when the start codon is GUG (which normally encodes valine) or UUG (which normally encodes leucine), the first amino acid in the polypeptide is still a formylmethionine because only a $tRNA^{fmet}$ can initiate translation. During or after translation of the entire polypeptide, the formyl group or the entire formylmethionine may be removed. Therefore, some polypeptides may not have formylmethionine or methionine as their first amino acid.

The initiation stage of bacterial translation is completed when the 50S ribosomal subunit associates with the 30S subunit. For this to occur, the initiation factors must be released from the 30S subunit. Much later, after translation is completed, IF1 is necessary to dissociate the 50S and 30S ribosomal subunits so that the 30S subunit can reinitiate with another mRNA molecule. For this reason, it is viewed as an initiation factor. As noted in Figure 13.18, the $tRNA^{fmet}$ binds to the P site on the ribosome. During the elongation stage, which is discussed later, all of the other tRNAs initially bind to the A site.

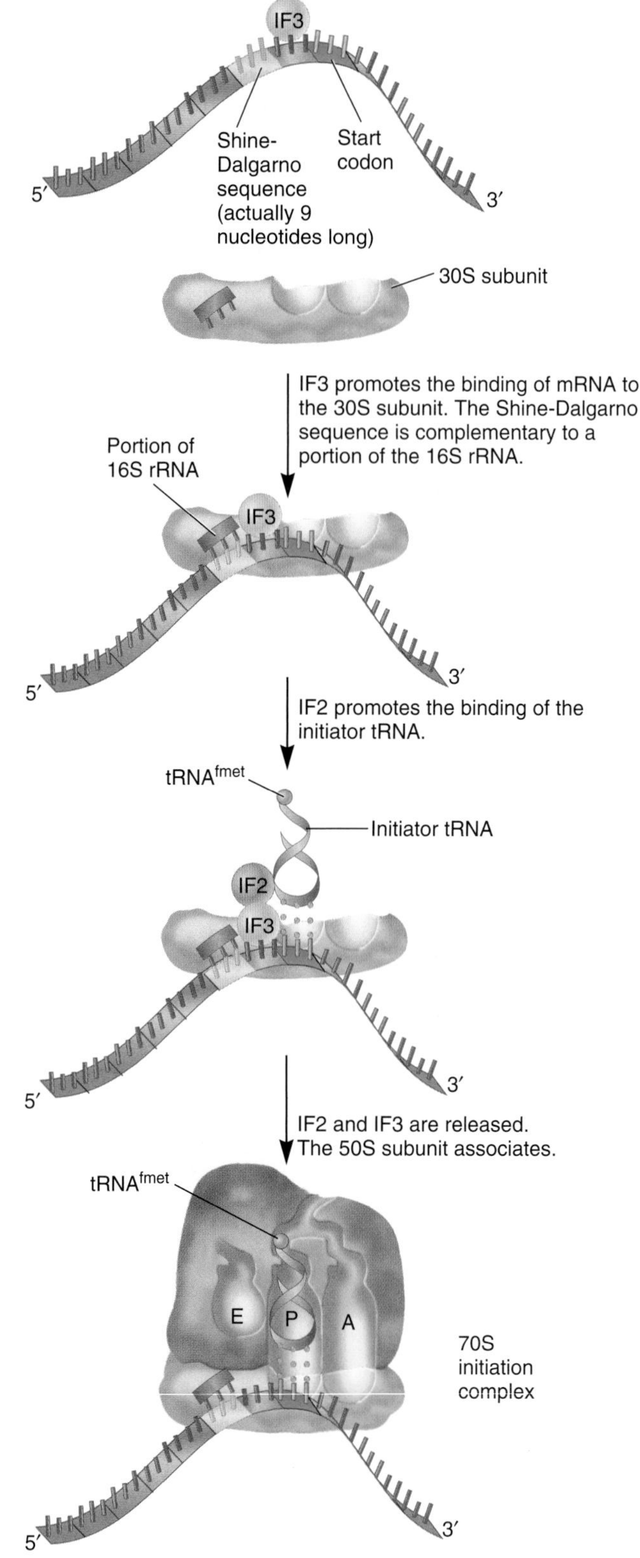

ONLINE ANIMATION **FIGURE 13.18 The initiation stage of translation in bacteria.** IF3 promotes the binding of mRNA to the 30S subunit. In addition, the Shine-Dalgarno sequence hydrogen bonds with a portion of the 16S rRNA. IF2 then binds to the initiator tRNA, which binds to the start codon in the P site. IF2 and IF3 are released, and the 50S subunit associates with the 30S subunit to form a 70S initiation complex. This marks the end of the initiation stage.

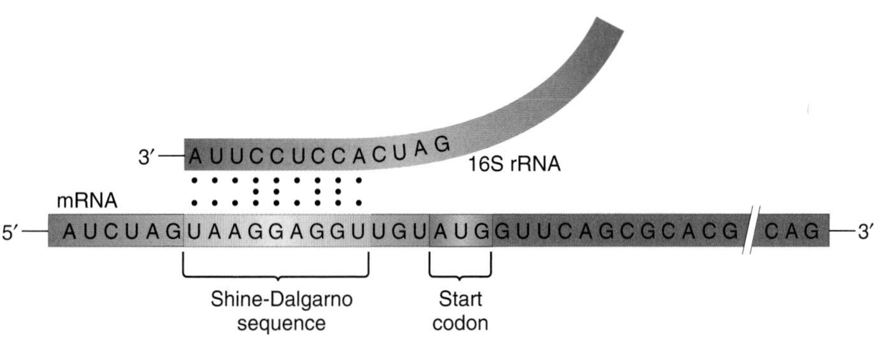

FIGURE 13.19 The locations of the Shine-Dalgarno sequence and the start codon in bacterial mRNA. The Shine-Dalgarno sequence is complementary to a sequence in the 16S rRNA. It hydrogen bonds with the 16S rRNA to promote initiation. The start codon is typically a few nucleotides downstream from the Shine-Dalgarno sequence.

In eukaryotes, the assembly of the initiation complex bears similarities to that in bacteria. However, as described in **Table 13.7**, additional factors are required for the initiation process. Note that the initiation factors are designated eIF (eukaryotic Initiation Factor) to distinguish them from bacterial initiation factors. The initiator tRNA in eukaryotes carries methionine rather than formylmethionine, as in bacteria. A eukaryotic initiation factor, eIF2, binds directly to $tRNA^{met}$ to recruit it to the 40S subunit. Eukaryotic mRNAs do not have a Shine-Dalgarno sequence. How then are eukaryotic mRNAs recognized by the ribosome? Several initiation factors (CBPI, eIF4A, eIF4B, eIF4F, and others) bind to the mRNA. Cap-binding protein I (CBPI) recognizes the 7-methylguanosine cap structure and aids in the recruitment of the other initiation factors. Collectively, these initiation factors unwind any secondary structure in the mRNA and promote its binding to the ribosome.

The identification of the correct AUG start codon in eukaryotes differs greatly from that in bacteria. After the initial binding of mRNA to the ribosome, the next step is to locate an AUG start codon that is somewhere downstream from the 5' cap structure. In 1986, Marilyn Kozak proposed that the ribosome begins at the 5' end and then scans along the mRNA in the 3' direction in search of an AUG start codon. In many, but not all, cases, the ribosome uses the first AUG codon that it encounters as a start codon. When a start codon is identified, the 60S subunit assembles with the aid of eIF5.

By analyzing the sequences of many eukaryotic mRNAs, researchers have found that not all AUG codons near the 5' end of mRNA can function as start codons. In some cases, the scanning ribosome passes over the first AUG codon and chooses an AUG farther down the mRNA. The sequence of bases around the AUG codon plays an important role in determining whether or not it will be selected as the start codon by a scanning ribosome. The consensus sequence for optimal start codon recognition is shown here.

						Start Codon			
G	C	C	(A/G)	C	C	A	U	G	G
–6	–5	–4	–3	–2	–1	+1	+2	+3	+4

Aside from an AUG codon itself, a guanine at the +4 position and a purine, preferably an adenine, at the –3 position are the most important sites for start codon selection. These rules for optimal translation initiation are called **Kozak's rules.**

TABLE 13.7

A Comparison of Translational Protein Factors in Bacteria and Eukaryotes

Bacterial Factors	Eukaryotic Factors	Function
Initiation Factors		
IF1	eIF2	Involved in forming the initiation complex
IF2	eIF3	Involved in forming the initiation complex
IF3	eIF4C	Involved in forming the initiation complex
	CBPI	Binds to the 7-methylguanosine cap
	eIF4A, eIF4B, eIF4F	Involved in the search for a start codon
	eIF5	Involved in forming the initiation complex; helps dissociate eIF2, eIF3, and eIF4C
	eIF6	Helps dissociate 60S subunit from inactive ribosomes
Elongation Factors		
EF-Tu	eEF1α	Binding of tRNAs to the A site
EF-Ts	eEF1$\beta\gamma$	Recycling factor for EF-Tu and eEF1α, respectively.
EF-G	eEF2	Required for translocation
Release Factors		
RF1	eRF	Recognizes stop codon and promotes termination
RF2		
RF3		

Polypeptide Synthesis Occurs During the Elongation Stage

During the elongation stage of translation, amino acids are added, one at a time, to the polypeptide chain (**Figure 13.20**). Even though this process is rather complex, it occurs at a remarkable rate. Under normal cellular conditions, a polypeptide chain

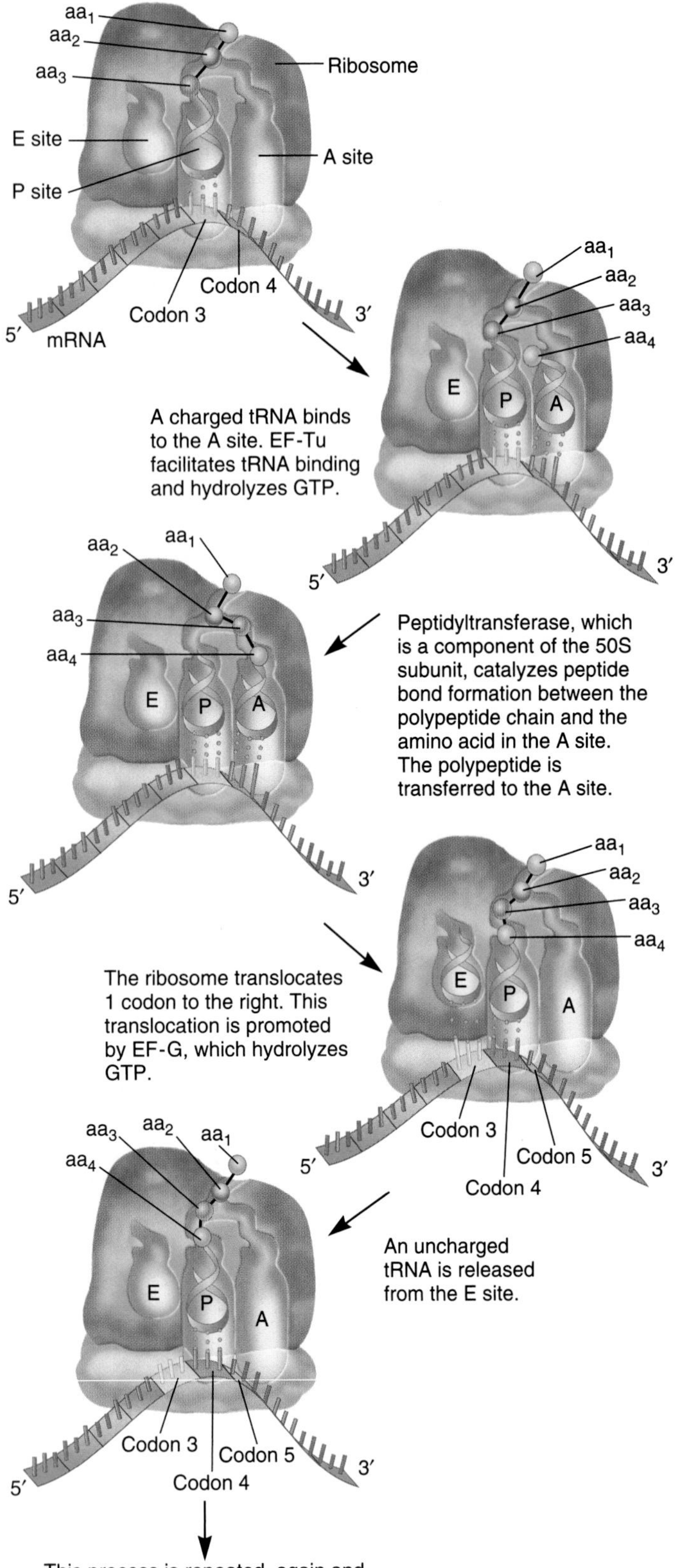

aa1
aa2
aa3
Ribosome
E site
A site
P site
Codon 4
Codon 3
5′
mRNA
3′
A charged tRNA binds to the A site. EF-Tu facilitates tRNA binding and hydrolyzes GTP.
aa1
aa2
aa3
aa4
E
P
A
5′
3′
Peptidyltransferase, which is a component of the 50S subunit, catalyzes peptide bond formation between the polypeptide chain and the amino acid in the A site. The polypeptide is transferred to the A site.
The ribosome translocates 1 codon to the right. This translocation is promoted by EF-G, which hydrolyzes GTP.
Codon 3
Codon 5
Codon 4
An uncharged tRNA is released from the E site.
This process is repeated, again and again, until a stop codon is reached.

ONLINE
ANIMATION

FIGURE 13.20 **The elongation stage of translation in bacteria.** This process begins with the binding of an incoming tRNA. The hydrolysis of GTP by EF-Tu provides the energy for the binding of the tRNA to the A site. A peptide bond is then formed between the incoming amino acid and the last amino acid in the growing polypeptide chain. This moves the polypeptide chain to the A site. The ribosome then translocates in the 3' direction so that the two tRNAs are moved to the E and P sites. The tRNA carrying the polypeptide is now back in the P site. This translocation requires the hydrolysis of GTP via EF-G. The uncharged tRNA in the E site is released from the ribosome. Now the process is ready to begin again. Each cycle of elongation causes the polypeptide chain to grow by one amino acid.

can elongate at a rate of 15 to 18 amino acids per second in bacteria and 6 amino acids per second in eukaryotes!

To begin elongation, a charged tRNA brings a new amino acid to the ribosome so that it can be attached to the end of the growing polypeptide chain. At the top of Figure 13.20, a short polypeptide is attached to the tRNA located at the P site of the ribosome. A charged tRNA carrying a single amino acid binds to the A site. This binding occurs because the anticodon in the tRNA is complementary to the codon in the mRNA. The hydrolysis of GTP by the elongation factor, EF-Tu, provides energy for the binding of a tRNA to the A site. In addition, the 16S rRNA, which is a component of the small 30S ribosomal subunit, plays a key role that ensures the proper recognition between the mRNA and correct tRNA. The 16S rRNA can detect when an incorrect tRNA is bound at the A site and will prevent elongation until the mispaired tRNA is released from the A site. This phenomenon, termed the **decoding function** of the ribosome, is important in maintaining high fidelity of mRNA translation. An incorrect amino acid is incorporated into a growing polypeptide at a rate of approximately one mistake per 10,000 amino acids or 10^{-4}.

The next step of elongation is the **peptidyl transfer** reaction—the polypeptide is removed from the tRNA in the P site and transferred to the amino acid at the A site. This transfer is accompanied by the formation of a peptide bond between the amino acid at the A site and the polypeptide chain, lengthening the chain by one amino acid. The peptidyl transfer reaction is catalyzed by a component of the 50S subunit known as **peptidyltransferase,** which is composed of several proteins and rRNA. Interestingly, based on the crystal structure of the 50S subunit, Thomas Steitz, Peter Moore, and their colleagues concluded that the 23S rRNA—not the ribosomal protein—catalyzes bond formation between adjacent amino acids. In other words, the ribosome is a ribozyme!

After the peptidyl transfer reaction is complete, the ribosome moves, or translocates, to the next codon in the mRNA. This moves the tRNAs at the P and A sites to the E and P sites, respectively. Finally, the uncharged tRNA exits the E site. You should notice that the next codon in the mRNA is now exposed in the unoccupied A site. At this point, a charged tRNA can enter the empty A site, and the same series of steps can add the next amino acid to the polypeptide chain. As you may have realized, the A, P, and E sites are named for the role of the tRNA that is usually found there. The A site binds an aminoacyl-tRNA (also called a charged tRNA), the P site usually contains the peptidyl-tRNA (a tRNA with an attached peptide), and the E site is where the uncharged tRNA exits.

Termination Occurs When a Stop Codon Is Reached in the mRNA

The final stage of translation, known as termination, occurs when a stop codon is reached in the mRNA. In most species, the three stop codons are UAA, UAG, and UGA. These codons are sometimes referred to as ochre (UAA), amber (UAG), and opal (UGA). (Note: The term amber, or brownstone, is the English translation of the name Bernstein, a graduate student who was involved in the discovery of the UAG codon. In keeping with this tradition, the lighthearted names ocher and opal were given to the other two stop codons.) The stop codons are not recognized by a tRNA with a complementary sequence. Instead, they are recognized by proteins known as **release factors** (see Table 13.7). Interestingly, the three-dimensional structure of release factor proteins mimics the structure of tRNAs. In bacteria, RF1 recognizes UAA and UAG, and RF2 recognizes UGA and UAA. A third release factor, RF3, is also required. In eukaryotes, a single release factor, eRF, recognizes all three stop codons.

Figure 13.21 illustrates the termination stage of translation in bacteria. At the top of this figure, the completed polypeptide chain is attached to a tRNA in the P site. A stop codon is located at the A site. In the first step, RF1 or RF2 binds to the stop codon at the A site and RF3 (not shown) binds at a different location on the ribosome. After RF1 (or RF2) and RF3 have bound, the bond between the polypeptide and the tRNA is hydrolyzed. The polypeptide and tRNA are then released from the ribosome. The final step in translational termination is the disassembly of ribosomal subunits, mRNA, and the release factors.

Bacterial Translation Can Begin Before Transcription Is Completed

While most of our knowledge concerning transcription and translation has come from genetic and biochemical studies, electron microscopy has also been an important tool in elucidating the mechanisms of transcription and translation. As described earlier in this chapter, electron microscopy (EM) has been a critical technique in facilitating our understanding of ribosome structure. In addition, EM has been employed to visualize genetic processes such as translation.

The first success in the EM observation of gene expression was achieved by Oscar Miller, Jr., and his colleagues in 1967. **Figure 13.22** shows an electron micrograph of a bacterial gene in the act of gene expression. Prior to this experiment, biochemical and genetic studies had suggested that the translation of a bacterial structural gene begins before the mRNA transcript is completed. In other words, as soon as an mRNA strand is long enough, a ribosome will attach to the 5' end and begin translation, even before RNA polymerase has reached the transcriptional termination site within the gene. This phenomenon is termed the coupling between transcription and translation in bacterial cells. Note that

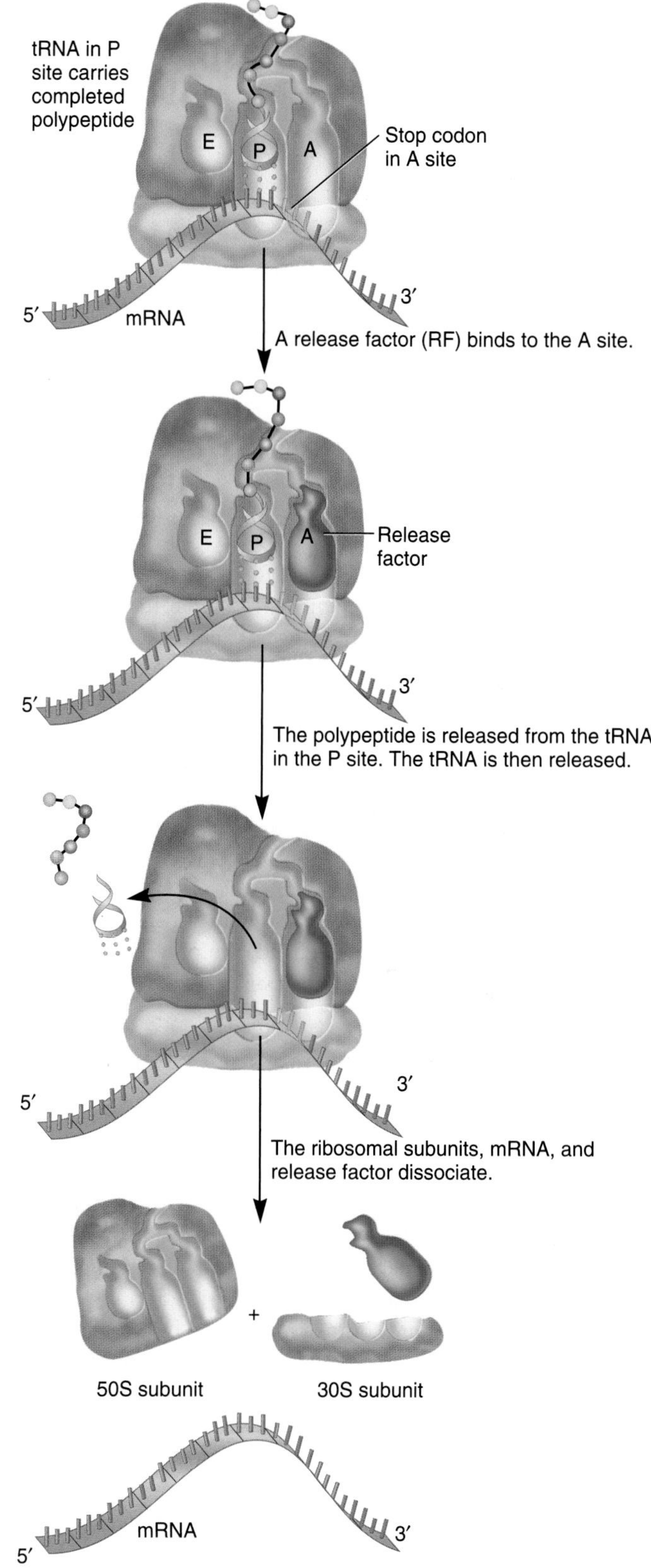

ONLINE ANIMATION

FIGURE 13.21 The termination stage of translation in bacteria. When a stop codon is reached, RF1 or RF2 binds to the A site. (RF3 binds elsewhere and uses GTP to facilitate the termination process.) The polypeptide is cleaved from the tRNA in the P site and released. The tRNA is released, and the rest of the components disassemble.

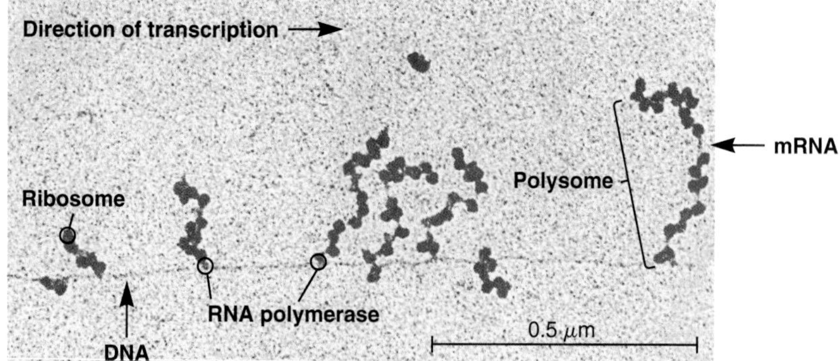

FIGURE 13.22 Coupling between transcription and translation in bacteria. An electron micrograph showing the simultaneous transcription and translation processes. The DNA is transcribed by many RNA polymerases that move along the DNA from left to right. Note that the RNA transcripts are getting longer as you go from left to right. Ribosomes attach to the mRNA, even before transcription is completed. The complex of many ribosomes bound to the same mRNA is called a polyribosome or a polysome. Several polyribosomes are seen here.

coupling of these processes does not usually occur in eukaryotes, because transcription takes place in the nucleus while translation occurs in the cytosol.

As shown in Figure 13.22, several RNA polymerase enzymes have recognized a gene and begun to transcribe it. Because the transcripts on the right side are longer than those on the left, Miller concluded that transcription was proceeding from left to right in the micrograph. This EM image also shows the process of translation. Relatively small mRNA transcripts, near the left side of the figure, have a few ribosomes attached to them. As the transcripts become longer, additional ribosomes are attached to them. The term **polyribosome,** or **polysome,** is used to describe an mRNA transcript that has many bound ribosomes in the act of translation. In this electron micrograph, the nascent polypeptide chains were too small for researchers to observe. In later studies, as EM techniques became more refined, the polypeptide chains emerging from the ribosome were also visible (see Figure 13.16a).

The Amino Acid Sequences of Some Proteins Contain Sorting Signals

The amino acid sequence of a protein may contain a sorting signal that will direct the protein to its correct location. In bacteria, for example, certain proteins may contain sorting signals that direct them across the plasma membrane. Also, some bacterial proteins have sorting signals that direct their attachment to the cell wall. In eukaryotes, many more sorting signals are necessary because eukaryotic cells are compartmentalized into many different membrane-bounded organelles, such as mitochondria, lysosomes, the nucleus, and so on. In general, any particular protein is meant to function in only a single location. For example, the ATP-synthase functions in the mitochondrion to make ATP and is not found in other locations in the cell. Cytoskeletal proteins such as tubulin and actin are found in the cytosol but not within the lumen of cellular organelles. To fulfill its function, each type of protein must be sorted into the correct cellular compartment. In eukaryotes, this sorting can occur either after translation is completed—**posttranslational sorting**—or during translation—**cotranslational sorting.**

Most eukaryotic proteins begin their synthesis on ribosomes in the cytosol (**Figure 13.23**). Translation is completed in the cytosol for those proteins destined for the cytosol, chloroplast, mitochondrion, nucleus, or peroxisome. The uptake of proteins into the chloroplast, mitochondrion, nucleus, and peroxisome then occurs posttranslationally. By comparison, the synthesis of other eukaryotic proteins begins in the cytosol and then is temporarily halted by a protein/RNA complex called the signal recognition particle (SRP). The SRP recognizes an amino acid sequence near the amino-terminal end of a polypeptide as it is being made. Translation resumes when the ribosome has become bound to the membrane of the endoplasmic reticulum (ER) and the polypeptide is synthesized into the ER lumen or ER membrane. This event is termed **cotranslational import** into the ER. Proteins destined for the ER, Golgi complex, lysosome, secretory vesicles, and the plasma membrane are first directed to the ER via this type of cotranslational sorting mechanism.

How is sorting achieved? A vital aspect of gene sequences is that the encoded proteins may contain short amino acid sequences that play a role in protein sorting. These short amino acid sequences are called **sorting signals.** **Table 13.8** provides a general description of the sorting signals contained within eukaryotic proteins. Each signal is recognized by specific cellular components that facilitate the sorting of the protein to its correct compartment, either posttranslationally or cotranslationally.

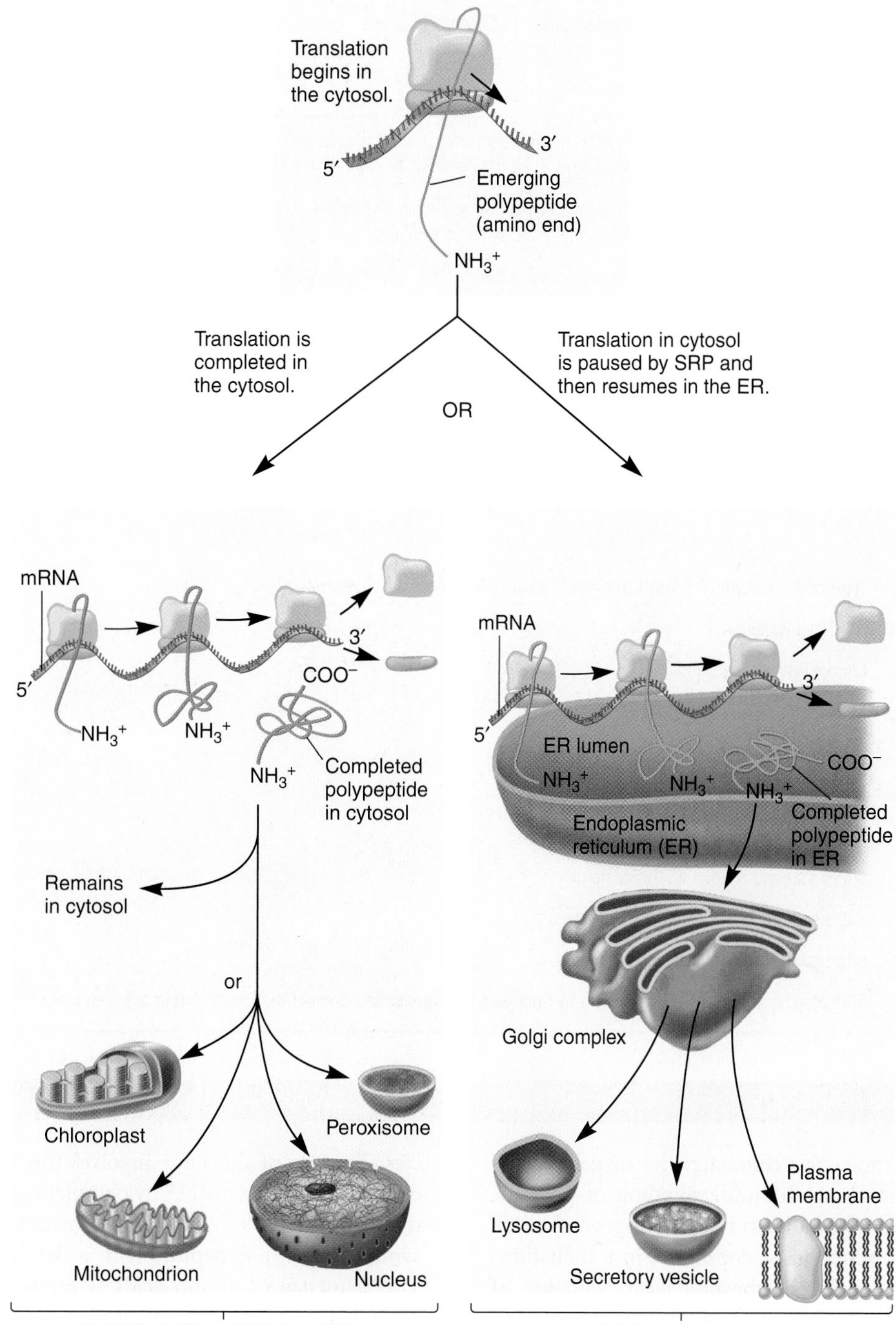

FIGURE 13.23 **Posttranslational and cotranslational protein sorting in eukaryotic cells.** The mRNA encoded by genes in the cell nucleus is exported from the nucleus and begins its translation in the cytosol. Proteins destined for the cytosol, chloroplasts, mitochondria, nucleus, and peroxisomes are completely synthesized in the cytosol and then are sorted posttranslationally. Other polypeptides contain an amino acid sequence at their amino terminal end that is recognized by the signal recognition particle (SRP). SRP directs these proteins to the ER, where they are sorted to the endoplasmic reticulum (ER), Golgi complex, lysosome, secretory vesicles, or plasma membrane.

TABLE 13.8

Sorting Signals in Eukaryotic Proteins

Type of Signal	Description
Mitochondrial-sorting signal	Usually a short sequence at the amino terminal end of a protein that contains several positively charged amino acids (i.e., lysine and arginine). This signal folds into an α helix in which the positive charges are on one face of the helix.
Nuclear-localization signal	Can be located almost anywhere in the polypeptide sequence. The signal is four to eight amino acids in length and contains several positively charged amino acids and usually one or more prolines.
Peroxisomal-sorting signal	Either a specific sequence of three amino acids (serine, lysine, leucine) that is located near the carboxyl terminal end of the protein or a 26-amino-acid sequence at the amino terminal end.
SRP signal	A sequence of ~20 amino acids near the amino terminal end that is composed of mostly nonpolar amino acids.
ER retention signal	A sequence of four amino acids (lysine, aspartic acid, glutamic acid, leucine) that is located at the carboxyl terminal end of the protein.
Golgi retention signal	A sequence of 20 hydrophobic amino acids that forms a transmembrane domain that is flanked by positively charged amino acids.
Lysosomal-sorting signal	A patch of amino acids within the polypeptide sequence. Positively charged amino acids within this patch are thought to play an important role. This patch causes lysosomal proteins to be covalently modified to contain a mannose-6-phosphate that directs the protein to the lysosome.

Destination of a Cellular Protein	Type of Signal the Protein Contains Within Its Amino Acid Sequence
Cytosol	No signal required.
Chloroplast	Chloroplast-sorting signal
Mitochondrion	Mitochondrial-sorting signal
Nucleus	Nuclear-localization signal
Peroxisome	Peroxisomal-sorting signal
ER	SRP signal and an ER retention signal
Golgi	SRP signal and a Golgi retention signal
Lysosome	SRP signal and a lysosomal-sorting signal
Secretory vesicles	SRP signal. No additional signal required.
Plasma membrane	SRP signal and the protein contains a hydrophobic transmembrane domain that anchors it in the membrane.

CONCEPTUAL SUMMARY

An appreciation for the molecular characteristics of gene structure and function is central to our understanding of genetics. Early ideas of Garrod and of Beadle and Tatum suggested a link between genes and the production of enzymes. Since their time, the molecular nature of that link has become clear. Segments of DNA within the chromosomes are organized into units called genes. Most genes are structural genes, which means that they provide the code for the amino acid sequence of a polypeptide. The function of most genes, therefore, is to store the information for the synthesis of a particular polypeptide—ultimately, for the synthesis of a functional protein. The action of proteins within living cells and organisms directly determines the outcome of an organism's traits.

During translation, the language of mRNA, which is a nucleotide sequence, is translated into the language of proteins, which is an amino acid sequence. The translation process employs a genetic code, which is nearly universal in all species. The use of this code involves the recognition of three-base codons within the mRNA by complementary anticodons in tRNA molecules. The tRNAs carry the correct amino acid, allowing the synthesis of a polypeptide with a defined amino acid sequence. The sequence of amino acids is known as a polypeptide's primary structure. The chemical properties of the amino acid side chains within a polypeptide cause it to fold into secondary and tertiary structures. Some proteins are composed of two of more polypeptide subunits that associate with each other and create a quaternary structure. The final three-dimensional structure of a protein determines its functional role.

The translation of mRNA into a polypeptide sequence requires many cellular components, including tRNAs, ribosomes, and protein factors. According to the adaptor hypothesis, the anticodons in tRNA molecules act as translators of the genetic code within mRNA. Prior to translation, enzymes known as aminoacyl-tRNA synthetases recognize tRNAs and attach the correct amino acid to their 3' end. At this stage, the tRNA is termed a charged tRNA. During translation, the anticodon in a tRNA binds to a codon in the mRNA due to their complementary sequences. Concurrently, the tRNA carries the correct amino acid, which is then added to the growing polypeptide chain. The third base in the codon of the mRNA is referred to as the wobble position because the complementary base in the tRNA anticodon does not require a precise fit when binding to the mRNA. The wobble rules enable a single type of tRNA to recognize more than one codon. In addition, modified bases in the tRNA can often be recognized by different bases in the mRNA.

The ribosomes are large macromolecular structures that provide a site for translation to occur. Each ribosome is composed of one small and one large subunit; each subunit contains several proteins and rRNAs that assemble together. In bacteria, both ribosomal assembly and translation take place in the cytoplasm. In eukaryotes, this assembly occurs within a region of the nucleus called the nucleolus; the ribosomal subunits are then exported into the cytosol, where translation occurs.

The translation process occurs in three stages, known as initiation, elongation, and termination. During initiation, mRNA and initiator tRNA bind to the ribosomal subunits. In bacteria, the ribosomal-binding site, or Shine-Dalgarno sequence, is located a few nucleotides upstream from the start codon. In eukaryotes, the ribosome binds at the 5' end of the mRNA and then scans along the mRNA until it finds a start codon. After the start codon has been identified, translation can proceed to the elongation phase, in which the polypeptide is made. During elongation, three sites on the ribosome are used for tRNA binding. An appropriate tRNA first binds to the A site, where its anticodon recognizes the codon in the mRNA. A peptide bond then forms between the amino acid attached to the tRNA at the A site and the growing polypeptide chain attached to a tRNA in the P site. The tRNA at the P site is then released from the ribosome at the E site. Finally, during the termination stage, the stop codon in the mRNA is reached. At this point, release factors bind to the ribosome and cause the disassembly of the completed polypeptide chain, tRNA, ribosomal subunits, and mRNA.

Some interesting differences are found between bacterial and eukaryotic translation due to their differing degrees of cell compartmentalization. Because bacteria contain a single intracellular compartment, translation of mRNA can begin before transcription is completed. This is referred to as coupling between transcription and translation. Coupling of these processes does not occur in eukaryotes because transcription takes place in the nucleus, while translation occurs in the cytosol. Moreover, compartmentalization in eukaryotic cells, due to the presence of different types of organelles, requires a more complicated system of sorting the proteins into their correct locations. This occurs via the presence of sorting signals within the translated polypeptides. The sorting of cytosol, chloroplast, mitochondrion, nucleus, and peroxisome proteins occurs after translation is completed and is called posttranslational sorting. The synthesis of proteins destined for the ER, Golgi, lysosomes, secretory vesicles, or plasma membrane occurs during translation and is called cotranslational sorting. In this case, translation begins in the cytosol and is temporarily halted; it resumes when the ribosome is bound to the membrane of the ER and the polypeptide is synthesized into the ER lumen or membrane.

EXPERIMENTAL SUMMARY

The first biochemical insights into the function of genes came from the experimental observations of Garrod. By making the connection between inherited defects in metabolism and the absence of a specific functional enzyme, he proposed that the function of genes is related to the production of enzymes. Later, Beadle and Tatum analyzed mutant genes in *Neurospora* and concluded that a single gene encodes a single enzyme, which is known as the one gene–one enzyme hypothesis. The experimental strategy used by Nirenberg and Ochoa established the relationship between the mRNA sequence and the polypeptide sequence. By synthesizing RNAs in vitro with a known nucleotide composition, they were able to correlate the predicted percentage of codons with the observed percentage of amino acids within a polypeptide chain. This helped to decipher the genetic code by determining the relationship between a codon's composition and the amino acid it specifies. Similarly, the use of RNA copolymers and the triplet binding assay were instrumental in cracking the genetic code.

Our molecular understanding of translation has come from several lines of experimentation, including biochemistry, genetics, and microscopy. The structure and composition of tRNAs, mRNAs, and ribosomes have been biochemically investigated for many decades. X-ray crystallography has aided our understanding of ribosome structure, and electron microscopy has provided a method to demonstrate that transcription and translation are coupled in bacteria.

The steps of translation have also been investigated at the molecular level. By modifying the amino acid attached to tRNA, Chapeville showed that the tRNA anticodon sequence—not the amino acid attached to the tRNA—is the determining factor during translation. This experimental observation supported the adaptor hypothesis initially proposed by Crick; the function of tRNA is to recognize a codon in mRNA and carry an amino acid specific for that codon. We have also learned a great deal about the translation process by examining genetic sequences. In addition, geneticists have identified the functional roles of sequences, such as the Shine-Dalgarno sequence in bacteria and Kozak's rules in eukaryotes, by analyzing mutations that interfere with translation.

PROBLEM SETS & INSIGHTS

Solved Problems

S1. The first amino acid in a certain bacterial polypeptide chain is methionine. The start codon in the mRNA is GUG, which codes for valine. Why isn't the first amino acid formylmethionine or valine?

Answer: The first amino acid in a polypeptide chain is carried by the initiator tRNA, which always carries formylmethionine. This occurs even when the start codon is GUG (valine) or UUG (leucine). The formyl group can be later removed to yield methionine as the first amino acid.

S2. A tRNA has the anticodon sequence 3'–CAG–5'. What amino acid does it carry?

Answer: Because the anticodon is 3'–CAG–5', it would be complementary to a codon with the sequence 5'–GUC–3'. According to the genetic code, this codon specifies the amino acid valine. Therefore, this tRNA must carry valine at its acceptor stem.

S3. In eukaryotic cells, the assembly of ribosomal subunits occurs in the nucleolus. As discussed in Chapter 12 (see Figure 12.16), a single 45S rRNA transcript is cleaved to produce the three rRNA fragments—18S, 5.8S, and 28S rRNA—that play a key role in creating the structure of the ribosome. The genes that encode the 45S precursor are found in multiple copies (i.e., they are moderately repetitive). The segments of chromosomes that contain the 45S rRNA genes align themselves at the center of the nucleolus. This site is called the nucleolar-organizing center. In this region, active transcription of the 45S gene takes place. Briefly explain how the assembly of the ribosomal subunits occurs.

Answer: In the nucleolar-organizing center, the 45S RNA is cleaved to the 18S, 5.8S, and 28s rRNAs. The other components of the ribosomal subunits, 5S rRNA and ribosomal proteins, must also be imported into the nucleolar region. Because proteins are made in the cytosol, they must enter the nucleus through the nuclear pores. When all the components are present, they assemble into 40S and 60S ribosomal subunits. Following assembly, the ribosomal subunits exit the nucleus through the nuclear pores and enter the cytosol.

S4. Throughout this chapter, we have seen that the general mechanism for bacterial and eukaryotic translation is very similar at the molecular level. In both cases, a polypeptide with a defined amino acid sequence is made on the surface of a ribosome due to the sequential interactions between the anticodons in tRNA molecules and the codons in an mRNA. In addition, we have also seen some interesting distinctions between these processes in bacteria and eukaryotes. Summarize the differences between bacterial and eukaryotic translation.

Answer: See table below.

A Comparison of Bacterial and Eukaryotic Translation

	Bacterial	Eukaryotic
Ribosome composition:	70S ribosomes: 30S subunit— 21 proteins + 1 rRNA 50S subunit— 34 proteins + 2 rRNAs	80S ribosomes: 40S subunit— 33 proteins + 1 rRNA 60S subunit— 49 proteins + 3 rRNAs
Initiator tRNA:	$tRNA^{fmet}$	$tRNA^{met}$
Formation of the initiation complex:	Requires IF1, IF2, and IF3	Requires more initiation factors compared to bacterial initiation
Initial binding of mRNA to the ribosome:	Requires a Shine-Dalgarno sequence	Requires a 7-methylguanosine cap
Selection of a start codon:	AUG, GUG, or UUG located just downstream from the Shine-Dalgarno sequence	According to Kozak's rules
Termination:	Requires RF1, RF2, and RF3	Requires eRF
Coupled to transcription:	Yes	Not usually
Coordinated with cell compartmentalization:	Less complex, due to a lack of internal organelles	Yes, via many different sorting signals, such as the SRP signal

S5. An antibiotic is a drug that kills or inhibits the growth of microorganisms. The use of antibiotics has been of great importance in the battle against many infectious diseases caused by microorganisms. For many antibiotics, their mode of action is to inhibit the translation process within bacterial cells. Certain antibiotics selectively bind to bacterial (70S) ribosomes but do not inhibit eukaryotic (80S) ribosomes. Their ability to inhibit translation can occur at different steps in the translation process. For example, tetracycline prevents the attachment of tRNA to the ribosome, while erythromycin inhibits the translocation of the ribosome along the mRNA. Why would an antibiotic bind to a bacterial ribosome but not to a eukaryotic ribosome? How does inhibition of translation by antibiotics such as tetracycline prevent bacterial growth?

Answer: Because bacterial ribosomes have a different protein and rRNA composition than eukaryotic ribosomes, certain drugs can recognize these different components, bind specifically to bacterial ribosomes, and thereby interfere with the process of translation. In other words, the surface of a bacterial ribosome must be somewhat different compared to the surface of a eukaryotic ribosome so that the drugs will bind to the surface of only bacterial ribosomes. If a bacterial cell is exposed to tetracycline or other antibiotics, it cannot synthesize new polypeptides. Because polypeptides form functional proteins needed for processes such as cell division, the bacterium will be unable to grow and proliferate.

Conceptual Questions

C1. An mRNA has the following sequence:

5'–GGCGAUGGGCAAUAAACCGGGCCAGUAAGC–3'

Identify the start codon and determine the complete amino acid sequence that would be translated from this mRNA.

C2. What does it mean when we say that the genetic code is degenerate? Discuss the universality of the genetic code.

C3. According to the adaptor hypothesis, are the following statements true or false?

A. The sequence of anticodons in tRNA directly recognizes codon sequences in mRNA, with some room for wobble.

B. The amino acid attached to the tRNA directly recognizes codon sequences in mRNA.

C. The amino acid attached to the tRNA affects the binding of the tRNA to codon sequences in mRNA.

C4. In bacteria, researchers have isolated strains that carry mutations within tRNA genes. These mutations can change the sequence of the anticodon. For example, a normal $tRNA^{trp}$ gene would encode a tRNA with the anticodon 3'–ACC–5'. A mutation could change this sequence to 3'–CCC–5'. When this mutation occurs, the tRNA still carries a tryptophan at its 3' acceptor stem, even though the anticodon sequence has been altered.

A. How would this mutation affect the translation of polypeptides within the bacterium?

B. What does this mutation tell you about the recognition between tryptophanyl-tRNA synthetase and $tRNA^{trp}$? Does the enzyme primarily recognize the anticodon or not?

C5. The covalent attachment of an amino acid to a tRNA is an endergonic reaction. In other words, it requires an input of energy for the reaction to proceed. Where does the energy come from to attach amino acids to tRNA molecules?

C6. The wobble rules for tRNA-mRNA pairing are shown in Figure 13.14. If we assume that the tRNAs do not contain modified bases, what is the minimum number of tRNAs needed to efficiently recognize the codons for the following types of amino acids?

A. Leucine

B. Methionine

C. Serine

C7. How many different sequences of mRNA could encode a peptide with the sequence proline-glycine-methionine-serine?

C8. If a tRNA molecule carries a glutamic acid, what are the two possible anticodon sequences that it could contain? Be specific about the 5' and 3' ends.

C9. A tRNA has an anticodon sequence 3'–GGU–5'. What amino acid does it carry?

C10. If a tRNA has an anticodon sequence 3'–CCI–5', what codon(s) can it recognize?

C11. Describe the anticodon of a single tRNA that could recognize the codons 5'–AAC–3' and 5'–AAU–3'. How would this tRNA need to be modified for it to also recognize 5'–AAA–3'?

C12. Describe the structural features that all tRNA molecules have in common.

C13. In the tertiary structure of tRNA, where is the anticodon region relative to the attachment site for the amino acid? Are they adjacent to each other?

C14. What is the role of aminoacyl-tRNA synthetase? The ability of the aminoacyl-tRNA synthetases to recognize tRNAs has sometimes been called the "second genetic code." Why has the function of this type of enzyme been described this way?

C15. What is an activated amino acid?

C16. Discuss the significance of modified bases within tRNA molecules.

C17. How and when does formylmethionine become attached to the initiator tRNA in bacteria?

C18. Is it necessary for a cell to make 61 different tRNA molecules, corresponding to the 61 codons for amino acids? Explain your answer.

C19. List the components required for translation. Describe the relative sizes of these different components. In other words, which components are small molecules, macromolecules, or assemblies of macromolecules?

C20. Describe the components of eukaryotic ribosomal subunits and where the assembly of the subunits occurs within living cells.

C21. The term subunit can be used in a variety of ways. Compare the use of the term subunit in proteins versus ribosomal subunit.

C22. Do the following events during bacterial translation occur primarily within the 30S subunit, within the 50S subunit, or at the interface between these two ribosomal subunits?

A. mRNA-tRNA recognition

B. Peptidyl transfer reaction

C. Exit of the polypeptide chain from the ribosome

D. Binding of initiation factors IF1, IF2, and IF3

C23. What are the three stages of translation? Discuss the main events that occur during these three stages.

C24. Describe the sequence in bacterial mRNA that promotes recognition by the 30S subunit.

C25. For each of the following initiation factors, how would eukaryotic initiation of translation be affected if it were missing?

A. eIF2

B. eIF4A

C. eIF5

C26. How does a eukaryotic ribosome select its start codon? Describe the sequences in eukaryotic mRNA that provide an optimal context for a start codon.

C27. For each of the following sequences, rank them in order (from best to worst) as sequences that could be used to initiate translation according to Kozak's rules.

GACGCCAUGG

GCCUCCAUGC

GCCAUCAAGG

GCCACCAUGG

C28. Explain the functional roles of the A, P, and E sites during translation.

C29. An mRNA has the following sequence: 5'–AUG UAC UAU GGG GCG UAA–3'. Describe the amino acid sequence of the polypeptide that would be encoded by this mRNA. Be specific about the amino and carboxyl terminal ends.

C30. For eukaryotic proteins, what is the function of a sorting signal? Describe an example.

C31. What is the difference between posttranslational and cotranslational protein sorting? What cellular components are required for cotranslational sorting?

C32. Which steps during the translation of bacterial mRNA involve an interaction between complementary strands of RNA?

C33. What is the function of the nucleolus?

C34. In which of the ribosomal sites, the A site, P site, and/or E site, could the following be found?

A. A tRNA without an amino acid attached

B. A tRNA with a polypeptide attached

C. A tRNA with a single amino acid attached

C35. What is a polysome?

C36. According to Figure 13.20, explain why the ribosome translocates along the mRNA in a 5' to 3' direction rather than a 3' to 5' direction.

C37. The lactose permease of *E. coli* is a protein composed of a single polypeptide that is 417 amino acids in length. By convention, the amino acids within a polypeptide are numbered from the amino terminal end to the carboxyl terminal end. Are the following questions about the lactose permease true or false?

A. Because the 64th amino acid is glycine and the 68th amino acid is aspartic acid, the codon for glycine-64 is closer to the 3' end of the mRNA compared to the codon for aspartic acid-68.

B. The mRNA that encodes the lactose permease must be greater than 1,241 nucleotides in length.

C38. An mRNA encodes a polypeptide that is 312 amino acids in length. The 53rd codon in this polypeptide is a tryptophan codon. A mutation in the gene that encodes this polypeptide changes this tryptophan codon into a stop codon. How many amino acids would be in the resulting polypeptide: 52, 53, 259, or 260?

C39. Explain what is meant by the coupling of transcription and translation in bacteria. Does coupling occur in bacterial and/or eukaryotic cells? Explain.

Experimental Questions

E1. In the experiment of Figure 13.3, what would be the predicted amounts of amino acids incorporated into polypeptides if the RNA was a random polymer containing 50% C and 50% G?

E2. With regard to the experiment described in Figure 13.11, answer the following questions:

A. Why was a polyUG mRNA template used?

B. Would you radiolabel the cysteine with the isotope ^{14}C or ^{35}S? Explain your choice.

C. What would be the expected results if the experiment was followed in the same way except that a polyGC template was used? Note: A polyGC template could contain two different alanine codons (GCC and GCG), but it could not contain any cysteine codons.

E3. An experimenter has a chemical reagent that modifies threonine to another amino acid. Following the protocol described in Figure 13.11, an mRNA is made composed of 50% C and 50% A. The amino acid composition of the resultant polypeptides is 12.5% lysine, 12.5% asparagine, 25% serine, 12.5% glutamine, 12.5% histidine, and 25% proline. One of the amino acids present in this polypeptide is due to the modification of threonine. Which amino acid is it? Based on the structure of the amino acid side chains, explain how the structure of threonine has been modified.

E4. Polypeptides can be translated in vitro. Would a bacterial mRNA be translated in vitro by eukaryotic ribosomes? Would a eukaryotic mRNA be translated in vitro by bacterial ribosomes? Why or why not?

E5. Discuss how the elucidation of the structure of the ribosome can help us to understand its function.

E6. Figure 13.22 shows an electron micrograph of a bacterial gene as it is being transcribed and translated. In this figure, label the 5' and 3' ends of the DNA and RNA strands. Place an arrow where you think the start codons are found in the mRNA transcripts.

E7. Chapter 18 describes a blotting method known as Western blotting that can be used to detect the production of a polypeptide that is translated from a particular mRNA. In this method, a protein is detected with an antibody that specifically recognizes and binds to its amino acid sequence. The antibody acts as a probe to detect the presence of the protein. In a Western blotting experiment, a mixture of cellular proteins is separated using gel electrophoresis according to their molecular masses. After the antibody has bound to the protein of interest within a blot of a gel, the protein is visualized as a dark band. For example, an antibody that recognizes the β-globin polypeptide could be used to specifically detect the β-globin polypeptide in a blot. As shown here, the method of Western blotting can be used to determine the amount and relative size of a particular protein that is produced in a given cell type.

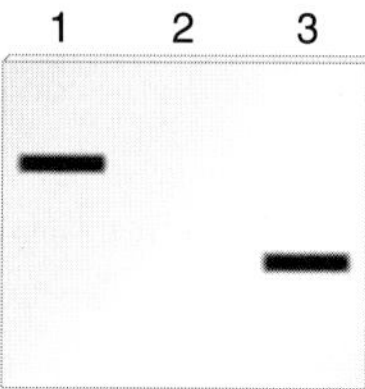

Lane 1 is a sample of proteins isolated from normal red blood cells.

Lane 2 is a sample of proteins isolated from kidney cells. Kidney cells do not produce β globin.

Lane 3 is a sample of proteins isolated from red blood cells from a patient with β-thalassemia. This patient is homozygous for a mutation that results in the shortening of the β-globin polypeptide.

Now here is the question. A protein called troponin contains 334 amino acids. Because each amino acid weighs 120 Daltons (on average), the molecular mass of this protein is about 40,000 Daltons, or 40 kDa. Troponin functions in muscle cells, and it is not expressed in nerve cells. Draw the expected results of a Western blot for the following samples:

Lane 1: Proteins isolated from muscle cells

Lane 2: Proteins isolated from nerve cells

Lane 3: Proteins isolated from the muscle cells of an individual who is homozygous for a mutation that introduces a stop codon at codon 177

E8. The technique of Western blotting can be used to detect proteins that are translated from a particular mRNA. This method is described in Chapter 18 and also in experimental question E7. Let's suppose a researcher was interested in the effects of mutations on the expression of a structural gene that encodes a protein we will call protein X. This protein is expressed in skin cells and contains 572 amino acids. Its molecular mass is approximately 68,600 Daltons, or 68.6 kDa. Make a drawing that shows the expected results of a Western blot using proteins isolated from the skin cells obtained from the following individuals:

Lane 1: A normal individual

Lane 2: An individual who is homozygous for a deletion, which removes the promoter for this gene

Lane 3: An individual who is heterozygous in which one gene is normal and the other gene has a mutation that introduces an early stop codon at codon 421

Lane 4: An individual who is homozygous for a mutation that introduces an early stop codon at codon 421

Lane 5: An individual who is homozygous for a mutation that changes codon 198 from a valine codon into a leucine codon

E9. The protein known as tyrosinase is needed to make certain types of pigments. Tyrosinase is composed of a single polypeptide with 511 amino acids. Because each amino acid weighs 120 Daltons (on average), the molecular mass of this protein is approximately 61,300 Daltons, or 61.3 kDa. People who carry two defective copies of the tyrosinase gene have the condition known as albinism. They are unable to make pigment in the skin, eyes, and hair. A blotting method known as Western blotting can be used to detect proteins that are translated from a particular mRNA. This method is described in Chapter 18 and also in experimental question E7. Skin samples were collected from a pigmented individual (lane 1) and from three unrelated albino individuals (lanes 2, 3, and 4) and subjected to a Western blot analysis using an antibody that recognizes tyrosinase. Explain the possible cause of albinism in the three albino individuals.

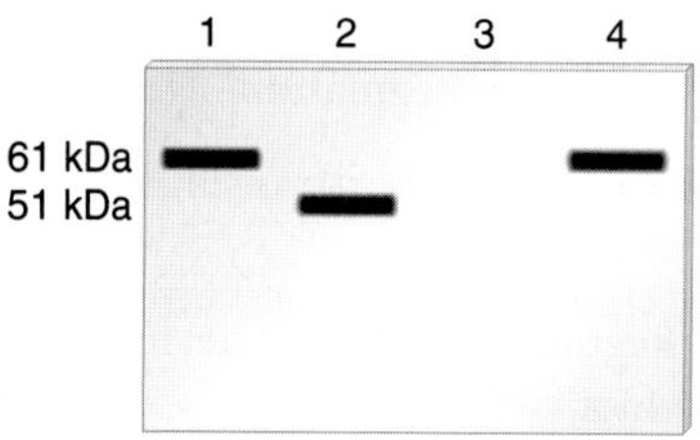

E10. Although 61 codons specify the 20 amino acids, most species display a codon bias. This means that certain codons are used much more frequently than other codons. For example, UUA, UUG, CUU, CUC, CUA, and CUG all specify leucine. In yeast, however, the UUG codon is used to specify leucine approximately 80% of the time.

A. The experiment of Figure 13.3 shows the use of an in vitro or cell-free translation system. In this experiment, the RNA, which was used for translation, was chemically synthesized. Instead of using a chemically synthesized RNA, researchers can isolate mRNA from living cells and then add the mRNA to the cell-free translation system. If a researcher isolated mRNA from kangaroo cells and then added it to a cell-free translation system that came from yeast cells, how might the phenomenon of codon bias affect the production of proteins?

B. Discuss potential advantages and disadvantages of codon bias for translation.

Questions for Student Discussion/Collaboration

1. Discuss why you think the ribosome needs to contain so many proteins and rRNA molecules. Does it seem like a waste of cellular energy to make so large a structure so that translation can occur?
2. Discuss and make a list of the similarities and differences in the events that occur during the initiation, elongation, and termination stages of transcription (see Chapter 12) and translation (Chapter 13).
3. Which events during translation involve molecular recognition of a nucleotide base sequence within RNA? Which events involve recognition between different protein molecules?

Note: All answers appear at the website for this textbook; the answers to even-numbered questions are in the back of the textbook.

www.mhhe.com/brookergenetics3e

Visit the Online Learning Center for practice tests, answer keys, and other learning aids for this chapter. Enhance your understanding of genetics with our interactive exercises, quizzes, animations, and much more.

14 GENE REGULATION IN BACTERIA AND BACTERIOPHAGES

CHAPTER OUTLINE

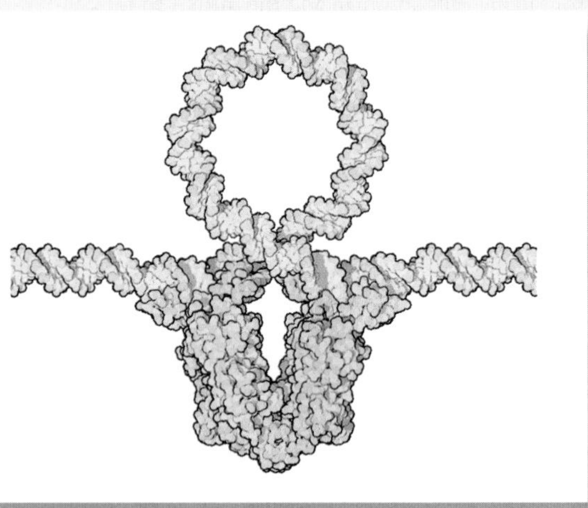

A model showing the binding of a genetic regulatory protein to DNA, which results in a DNA loop. The model shown here involves the lac repressor protein found in *E. coli* binding to the operator site in the *lac* operon.

Chromosomes of bacteria, such as *Escherichia coli*, contain a few thousand different genes. **Gene regulation** is a phenomenon in which the level of gene expression can vary under different conditions. In comparison, unregulated genes have essentially constant levels of expression in all conditions over time. Unregulated genes are also called **constitutive genes.** Frequently, constitutive genes encode proteins that are continuously needed for the survival of the bacterium. In contrast, the majority of genes are regulated so that the proteins they encode can be produced at the proper times and in the proper amounts.

The benefit of gene regulation is that the encoded proteins will be produced only when they are required. Therefore, the cell avoids wasting valuable energy making proteins it does not need. From the viewpoint of natural selection, this enables an organism such as a bacterium to compete as efficiently as possible for a limited amount of available resources. Gene regulation is particularly important because bacteria find themselves in an environment that is frequently changing with regard to temperature, nutrients, and many other factors. The following are a few common processes that are regulated at the genetic level:

1. **Metabolism:** Some proteins function in the metabolism of small molecules. For example, certain enzymes are needed for a bacterium to metabolize particular sugars. These enzymes are required only when the bacterium is exposed to such sugars in its environment.
2. **Response to environmental stress:** Certain proteins help a bacterium to survive environmental stress such as osmotic shock or heat shock. These proteins are required only when the bacterium is confronted with the stress.
3. **Cell division:** Some proteins are needed for cell division. These are necessary only when the bacterial cell is getting ready to divide.

The expression of structural genes, which encode polypeptides, ultimately leads to the production of functional cellular proteins. As we have seen in Chapters 12 and 13, gene expression is a multistep process that proceeds from transcription to translation, and it may involve posttranslational effects on protein structure and function. As shown in **Figure 14.1**, gene regulation can occur at any of these steps in the pathway of gene expression. In this chapter, we will examine the molecular mechanisms that account for these types of gene regulation. As you will learn, regulation allows bacteria to conserve their resources by expressing proteins only when they are needed.

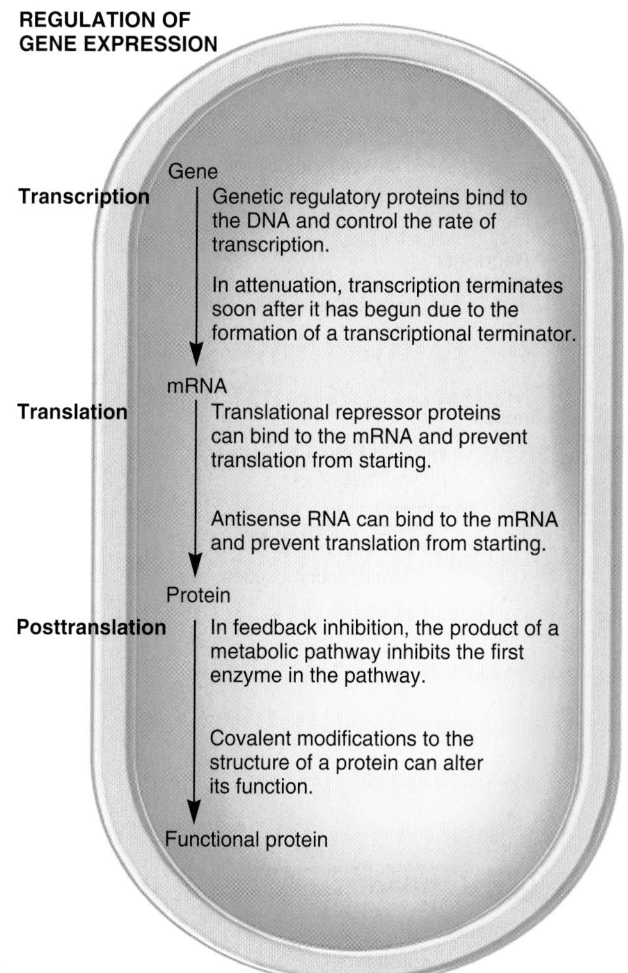

FIGURE 14.1 **Common points where regulation of gene expression in bacteria occurs.**

14.1 TRANSCRIPTIONAL REGULATION

In bacteria, the most common way to regulate gene expression is by influencing the rate at which transcription is initiated. Although we frequently refer to genes as being "turned on or off," it is more accurate to say that the level of gene expression is increased or decreased. At the level of transcription, this means that the rate of RNA synthesis can be increased or decreased.

In most cases, transcriptional regulation involves the actions of regulatory proteins that can bind to the DNA and affect the rate of transcription of one or more nearby genes. Two types of regulatory proteins are common. A **repressor** is a regulatory protein that binds to the DNA and inhibits transcription, whereas an **activator** is a regulatory protein that increases the rate of transcription. Transcriptional regulation by a repressor protein is termed **negative control,** and regulation by an activator protein is considered to be **positive control.**

In conjunction with regulatory proteins, small effector molecules often play a critical role in transcriptional regulation. However, small effector molecules do not bind directly to the DNA to alter transcription. Rather, an effector molecule exerts its effects by binding to an activator or repressor. The binding of the effector molecule causes a conformational change in the regulatory protein and thereby influences whether or not the protein can bind to the DNA. Genetic regulatory proteins that respond to small effector molecules have two functional domains. One domain is a site where the protein binds to the DNA; the other domain is the binding site for the effector molecule.

Regulatory proteins are given names describing how they affect transcription when they are bound to the DNA (repressor or activator). In contrast, small effector molecules are given names that describe how they affect transcription when they are present in the cell at a sufficient concentration to exert their effect (Figure 14.2). An **inducer** is a small effector molecule that causes transcription to increase. An inducer may accomplish this in two ways: It could bind to a repressor protein and prevent it from binding to the DNA, or it could bind to an activator protein and cause it to bind to the DNA. In either case, the transcription rate is increased. Genes that are regulated in this manner are called **inducible genes.**

Alternatively, the presence of a small effector molecule may inhibit transcription. This can also occur in two ways. A **corepressor** is a small molecule that binds to a repressor protein, thereby causing the protein to bind to the DNA. An **inhibitor** binds to an activator protein and prevents it from binding to the DNA. Both corepressors and inhibitors act to reduce the rate of transcription. Therefore, the genes they regulate are termed **repressible genes.** Unfortunately, this terminology can be confusing because a repressible system could involve an activator protein, or an inducible system could involve a repressor protein.

In this section, we will examine several examples where genes are regulated by the actions of genetic regulatory proteins that influence the rate of transcription. We will see how gene regulation provides a way for bacteria to synthesize proteins in an efficient manner.

The Phenomenon of Enzyme Adaptation Is Due to the Synthesis of Cellular Proteins

To a significant extent, our understanding of gene regulation can be traced back to the creative minds of François Jacob and Jacques Monod at the Pasteur Institute in Paris, France. Their research into genes and gene regulation stemmed from an interest in the phenomenon known as **enzyme adaptation,** which had been identified at the turn of the twentieth century. Enzyme adaptation refers to the observation that a particular enzyme appears within a living cell only after the cell has been exposed to the substrate for that enzyme. When a bacterium is not exposed to a particular substance, it does not make the enzymes needed to metabolize that substance.

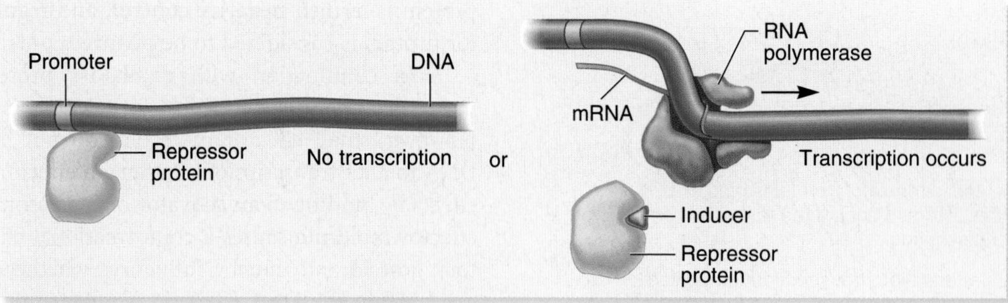

(a) Repressor protein, inducer molecule, inducible gene

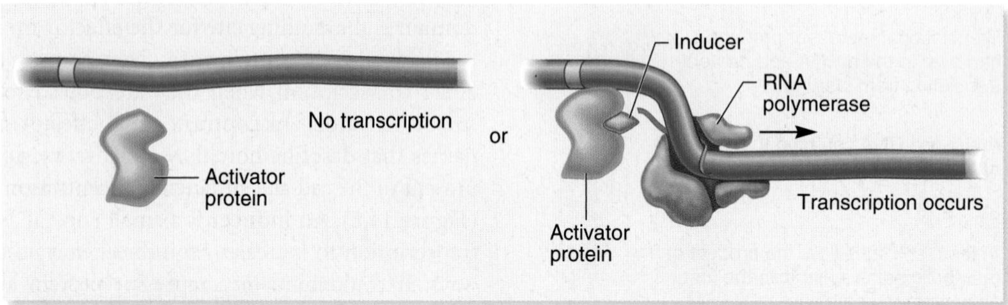

(b) Activator protein, inducer molecule, inducible gene

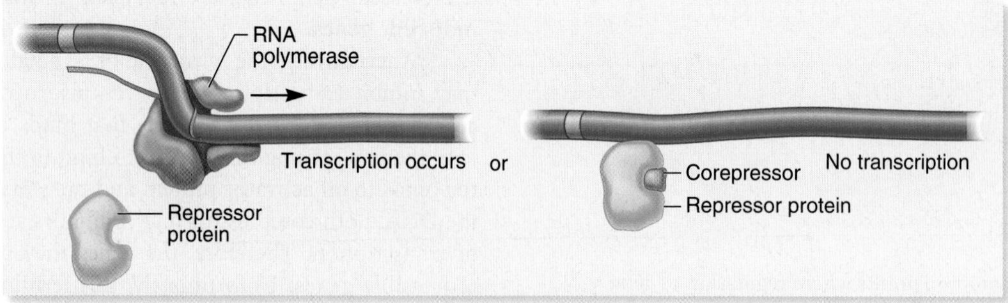

(c) Repressor protein, corepressor molecule, repressible gene

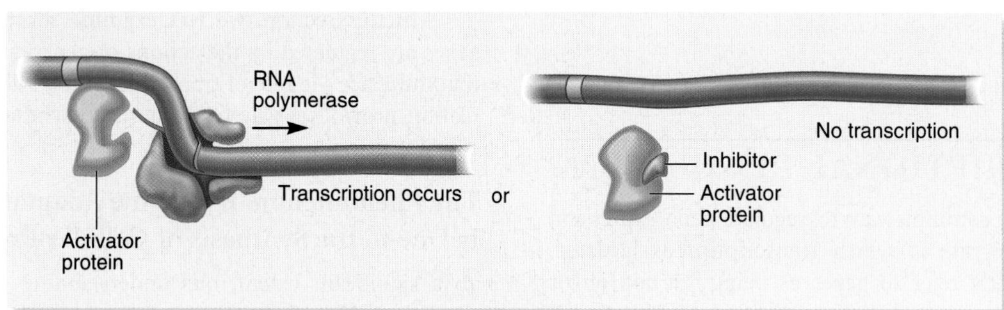

(d) Activator protein, inhibitor molecule, repressible gene

FIGURE 14.2 **Binding sites on a genetic regulatory protein.** In these examples, a regulatory protein has two binding sites: one for a small effector molecule and one for DNA. The binding of the small effector molecule will change the conformation of the regulatory protein, which alters the DNA-binding-site structure and thereby influences whether the protein can bind to the DNA. (**a**) In the absence of the inducer, a repressor protein shown here blocks transcription. The presence of an inducer causes a conformational change that inhibits the ability of a repressor protein to bind to the DNA. Transcription proceeds. (**b**) This activator protein cannot bind to the DNA unless an inducer is present. When the inducer is bound to the activator protein, this enables the activator protein to bind to the DNA and activate transcription. (**c**) In the absence of a corepressor, this repressor protein will not bind to the DNA. Therefore, transcription can occur. When the corepressor is bound to the repressor protein, this causes a conformational change that allows the protein to bind to the DNA and inhibit transcription. (**d**) Some activator proteins will bind to the DNA without the aid of an effector molecule. The presence of an inhibitor causes a conformational change that releases the activator protein from the DNA. This inhibits transcription.

To investigate this phenomenon, Jacob and Monod focused their attention on lactose metabolism in *E. coli.* Key experimental observations that led to an understanding of this genetic system are listed here.

1. The exposure of bacterial cells to lactose increased the levels of lactose-utilizing enzymes by 1,000- to 10,000-fold.
2. Antibody and labeling techniques revealed that the increase in the activity of these enzymes was due to the increased synthesis of the enzymes.
3. The removal of lactose from the environment caused an abrupt termination in the synthesis of the enzymes.
4. Mutations that prevented the synthesis of particular enzymes involved with lactose utilization showed that a separate gene encoded each enzyme.

These critical observations indicated to Jacob and Monod that enzyme adaptation is due to the synthesis of specific cellular proteins in response to lactose in the environment. As described next, we will examine how Jacob and Monod discovered that this phenomenon is due to the interactions between genetic regulatory proteins and small effector molecules. In other words, we will see that enzyme adaptation is due to the transcriptional regulation of genes.

The *lac* Operon Encodes Proteins Involved in Lactose Metabolism

In bacteria, it is common for a few structural genes to be arranged together in an **operon**—a group of two or more genes under the transcriptional control of a single promoter. An operon encodes a **polycistronic mRNA,** an mRNA that contains the coding sequence for two or more structural genes. Why do operons occur in bacteria? One biological advantage of an operon organization is that it allows a bacterium to coordinately regulate a group of genes that encode proteins with a common functional goal; the expression of the structural genes occurs as a single unit. For transcription to take place, an operon is flanked by a **promoter** that signals the beginning of transcription and a **terminator** that specifies the end of transcription. Two or more structural genes are found between these two sequences.

Figure 14.3a shows the organization of the genes involved with lactose utilization and their transcriptional regulation. Two distinct transcriptional units are present. The first unit, known

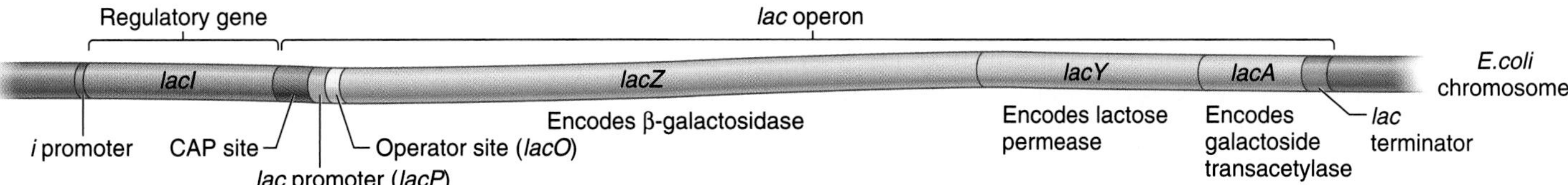

(a) Organization of DNA sequences in the *lac* region of the *E.coli* chromosome

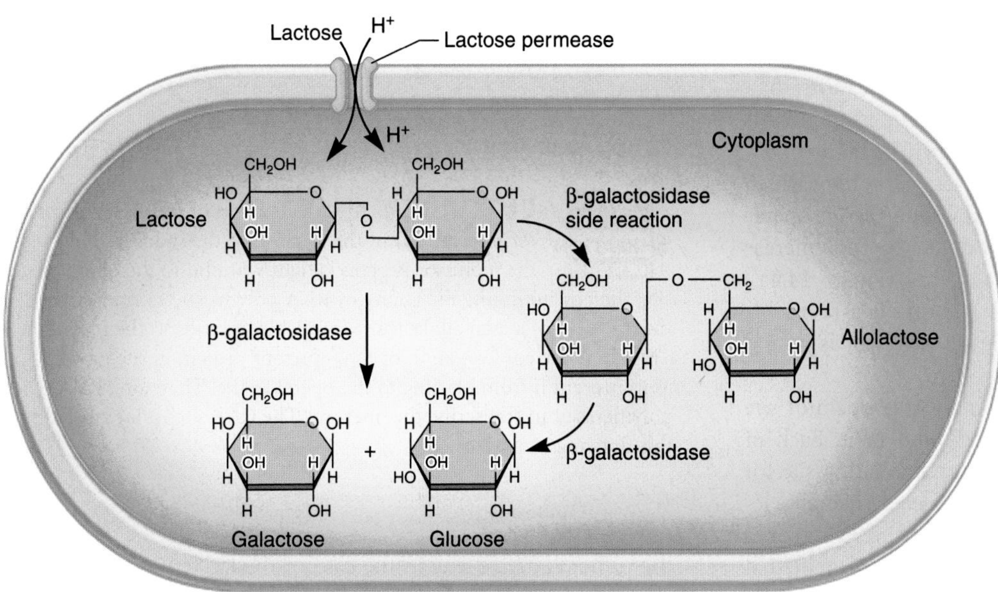

(b) Functions of lactose permease and β-galactosidase

ONLINE ANIMATION

FIGURE 14.3 Organization of the *lac* operon and other genes involved with lactose metabolism in *E. coli.* **(a)** The CAP site is the binding site for CAP; the operator site provides a binding site for the lac repressor. The promoter (*lacP*) is responsible for the transcription of the *lacZ, lacY,* and *lacA* genes as a single unit, which ends at the *lac* terminator. The *i* promoter is responsible for the transcription of the *lacI* gene. **(b)** Lactose permease is a membrane protein that allows the uptake of lactose into the bacterial cytoplasm. It cotransports lactose with H^+. Because bacteria maintain an H^+ gradient across their cytoplasmic membrane, this cotransport permits the active accumulation of lactose against a gradient. β-galactosidase is a cytoplasmic enzyme that cleaves lactose and related compounds into galactose and glucose. As a minor side reaction, β-galactosidase also converts lactose into allolactose. As shown here, allolactose can also be broken down into galactose and glucose.

as the *lac* operon, contains a CAP site; promoter (*lacP*); operator site (*lacO*); three structural genes, *lacZ*, *lacY*, and *lacA*; and a terminator. *LacZ* encodes the enzyme β-galactosidase, an enzyme that cleaves lactose into galactose and glucose. As a side reaction, β-galactosidase also converts a small percentage of lactose into allolactose, a structurally similar sugar (**Figure 14.3b**). As we will see later, allolactose acts as a small effector molecule to regulate the *lac* operon. The *lacY* gene encodes lactose permease, a membrane protein required for the active transport of lactose into the cytoplasm of the bacterium. The *lacA* gene encodes galactoside transacetylase, an enzyme that can covalently modify lactose and lactose analogues. Although the functional necessity of the transacetylase remains unclear, the acetylation of nonmetabolizable lactose analogues may prevent their toxic buildup within the bacterial cytoplasm.

The CAP site and the operator site are short DNA segments that function in gene regulation. The **CAP site** is a DNA sequence recognized by an activator protein called the **catabolite activator protein (CAP).** The **operator site** (also known simply as the **operator**) is a sequence of bases that provides a binding site for a repressor protein.

A second transcriptional unit involved in genetic regulation is the *lacI* gene (Figure 14.3a). The *lacI* gene is not part of the *lac* operon. The *lacI* gene encodes the **lac repressor,** a protein that is important for the regulation of the *lac* operon. The lac repressor functions as a homotetramer, a protein composed of four identical subunits. The *lacI* gene, which is constitutively expressed at fairly low levels, has its own promoter, the *i* promoter. The amount of lac repressor made is approximately 10 homotetramer proteins per cell. Only a small amount of the lac repressor protein is needed to repress the *lac* operon.

The *lac* Operon Is Regulated by a Repressor Protein

The *lac* operon can be transcriptionally regulated in more than one way. The first mechanism that we will examine is one that is inducible and under negative control. As shown in **Figure 14.4**, this form of regulation involves the lac repressor protein, which binds to the sequence of nucleotides found within the *lac* operator site. Once bound, the lac repressor prevents RNA polymerase from transcribing the *lacZ*, *lacY*, and *lacA* genes (Figure 14.4a). The binding of the repressor to the operator site is a reversible process. In the absence of allolactose, the lac repressor is bound to the operator site most of the time.

The ability of the lac repressor to bind to the operator site depends on whether or not allolactose is bound to it. Each of the repressor protein's four subunits has a single binding site for allolactose, the inducer. How does a small molecule like allolactose exert its effects? When allolactose binds to the repressor, a conformational change occurs in the lac repressor protein that prevents it from binding to the operator site. Under these conditions, RNA polymerase is now free to transcribe the operon (Figure 14.4b). In genetic terms, we would say that the operon has been **induced.** The action of a small effector molecule, such as allolactose, is called **allosteric regulation.** The effector molecule binds to the protein's **allosteric site,** which is a site other than

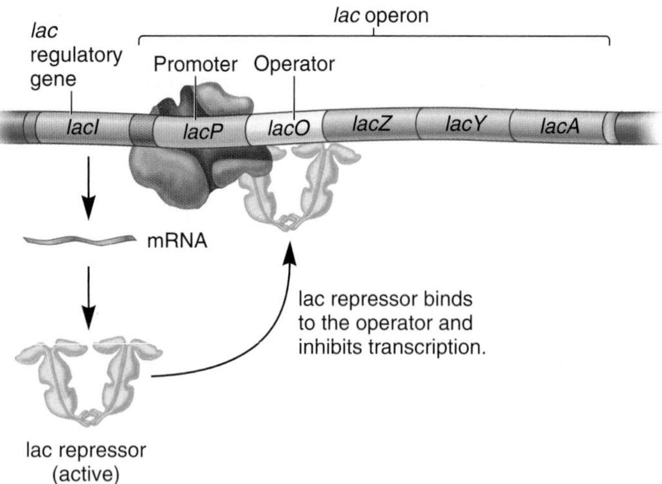

(a) No lactose in the environment

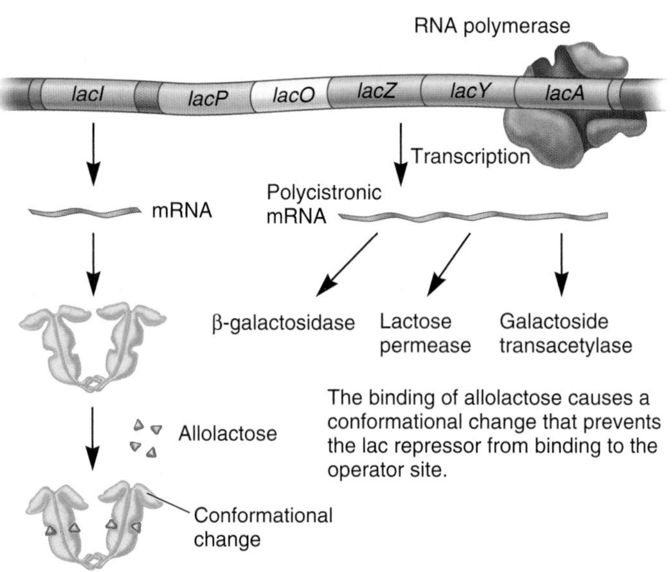

(b) Lactose present

ONLINE ANIMATION

FIGURE 14.4 Mechanism of induction of the *lac* operon. (a) In the absence of the inducer allolactose, the repressor protein is tightly bound to the operator site, thereby inhibiting the ability of RNA polymerase to transcribe the operon. **(b)** When allolactose is available, it binds to the repressor. This alters the conformation of the repressor protein in such a way as to prevent it from binding to the operator site. Therefore, RNA polymerase can transcribe the operon. (The CAP site is not labeled in this drawing.)

the protein's active site. In the case of the lac repressor, the active site is the part of the protein that binds to the DNA.

To better appreciate this form of regulation at the cellular level, let's consider the process as it occurs over time. **Figure 14.5** illustrates the effects of external lactose on the regulation of the *lac* operon. In the absence of lactose, no inducer is available to bind to the lac repressor. Therefore, the lac repressor binds to the

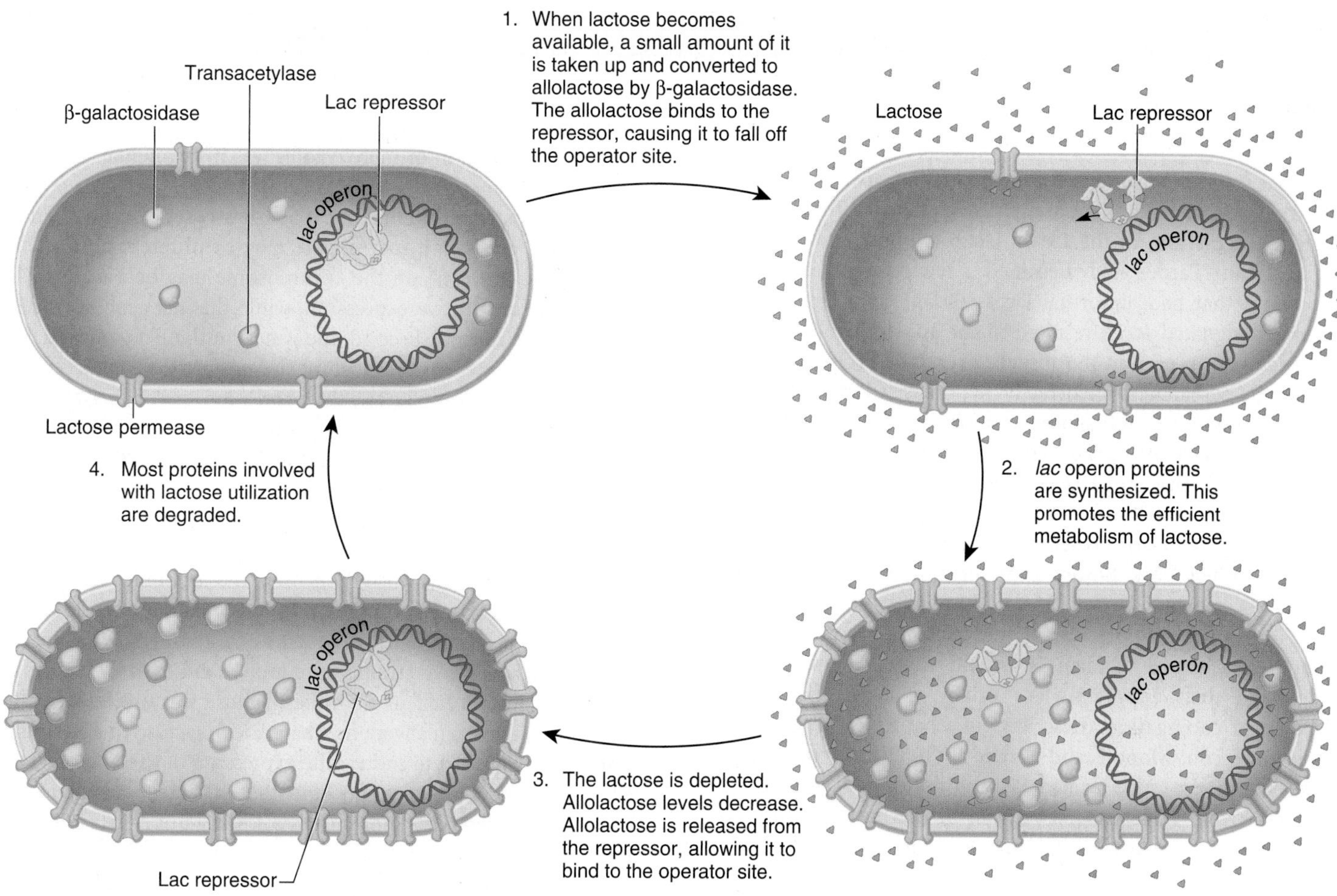

FIGURE 14.5 **The cycle of *lac* operon induction and repression.**

Genes→Traits The genes and the genetic regulation of the *lac* operon provide the bacterium with the trait of being able to metabolize lactose in the environment. When lactose is present, the genes of the *lac* operon are induced, and the bacterial cell can efficiently metabolize this sugar. When lactose is absent, these genes are repressed, so that the bacterium does not waste its energy expressing these genes.

operator site and inhibits transcription. In reality, the repressor does not completely inhibit transcription, so very small amounts of β-galactosidase, lactose permease, and transacetylase are made. However, the levels are far too low for the bacterium to readily use lactose. When the bacterium is exposed to lactose, a small amount can be transported into the cytoplasm via lactose permease, and β-galactosidase converts some of it to allolactose. As this occurs, the cytoplasmic level of allolactose gradually rises; eventually, allolactose binds to the lac repressor. The binding of allolactose promotes a conformational change that prevents the repressor from binding to the *lac* operator site and thereby allows transcription of the *lacZ*, *lacY*, and *lacA* genes to occur. Translation of the encoded polypeptides will produce the proteins needed for lactose uptake and metabolism.

To understand how the induction process is shut off in a lactose-depleted environment, let's consider the interaction between allolactose and the lac repressor. The lac repressor has a measurable affinity for allolactose. The binding of allolactose to the lac repressor is reversible. The likelihood that allolactose will bind to the repressor depends on the allolactose concentration. During induction of the operon, the concentration of allolactose rises and approaches the affinity for the repressor protein. This makes it likely that allolactose will bind to the lac repressor, thereby causing it to be released from the operator site. Later on, however, the bacterial cell metabolizes the sugars and thereby lowers the concentration of allolactose below its affinity for the repressor. At this point, the lac repressor is unlikely to be bound to allolactose. When allolactose is released, the lac repressor returns to the conformation that binds to the operator site. In this way, the binding of the repressor shuts down the *lac* operon when lactose is depleted from the environment. After repression occurs, the mRNA and proteins encoded by the *lac* operon will be degraded because they have a relatively short half-life (Figure 14.5).

EXPERIMENT 14A

The *lacI* Gene Encodes a Diffusible Repressor Protein

Now that we have an understanding of the *lac* operon, let's consider one of the experimental approaches that was used to elucidate its regulation. In the 1950s, Jacob, Monod, and their colleague Arthur Pardee had identified a few rare mutant strains of bacteria that had abnormal lactose adaptation. One type of mutant, designated *lacI*$^-$, resulted in the constitutive expression of the *lac* operon even in the absence of lactose. From this observation, the researchers incorrectly hypothesized that the *lacI*$^-$ mutation resulted in the synthesis of an internal inducer, making it unnecessary for cells to be exposed to lactose for induction (**Figure 14.6a**). By comparison, **Figure 14.6b** shows the correct explanation. A loss-of-function mutation in the *lacI* gene prevented the lac repressor protein from inhibiting transcription. At the time of their work, however, the function of the lac repressor was not yet known.

To further explore the nature of this mutation, Jacob, Monod, and Pardee applied a genetic approach. In order to understand their approach, let's briefly consider the process of bacterial conjugation (described in Chapter 6). The earliest studies of Jacob, Monod, and Pardee in 1959 involved matings between recipient cells, termed F$^-$, and donor cells, which were *Hfr* strains that transferred a portion of the bacterial chromosome. Later experiments in 1961 involved the transfer of circular segments of DNA known as F factors. We will consider the latter type of experiment here. Sometimes an F factor also carries genes that were originally found within the bacterial chromosome. These types of F factors are called F' factors (F prime factors). In their studies, Jacob, Monod, and Pardee identified F' factors that carried the *lacI* gene and portions of the *lac* operon. These F' factors can be transferred from one cell to another by bacterial conjugation. A strain of bacteria containing F' factor genes is called a **merozygote,** or partial diploid.

The production of merozygotes was instrumental in allowing Jacob, Monod, and Pardee to elucidate the function of the *lacI* gene. This experimental approach has two key points. First, the two *lacI* genes in a merozygote may be different alleles. For example, the *lacI* gene on the chromosome may be a *lacI*$^-$ allele that causes constitutive expression, while the *lacI* gene on the F' factor may be normal. Second, the genes on the F' factor and the genes on the bacterial chromosome are not physically adjacent to each other. As we now know, the expression of the *lacI* gene on an F' factor should produce repressor proteins that could diffuse within the cell and eventually bind to the operator site of the *lac* operon located on the chromosome.

Figure 14.7 shows one experiment of Jacob, Monod, and Pardee in which they analyzed a *lacI*$^-$ mutant strain that was already known to constitutively express the *lac* operon and compared it to the corresponding merozygote. The merozygote had a *lacI*$^-$ mutant gene on the chromosome and a normal *lacI* gene on an F' factor. These two strains were grown and then divided into two tubes each. In half of the tubes, lactose was omitted. In the other tubes, the strains were incubated with lactose to determine if lactose was needed to induce the expression of the operon. The cells were lysed by sonication, and then a lactose analogue, β-ONPG, was added. This molecule is colorless, but β-galactosidase cleaves it into a product that has a yellow color. Therefore, the amount of yellow color produced in a given amount of time is a measure of the amount of β-galactosidase that is being expressed from the *lac* operon.

THE HYPOTHESIS

The *lacI*$^-$ mutation results in the synthesis of an internal inducer.

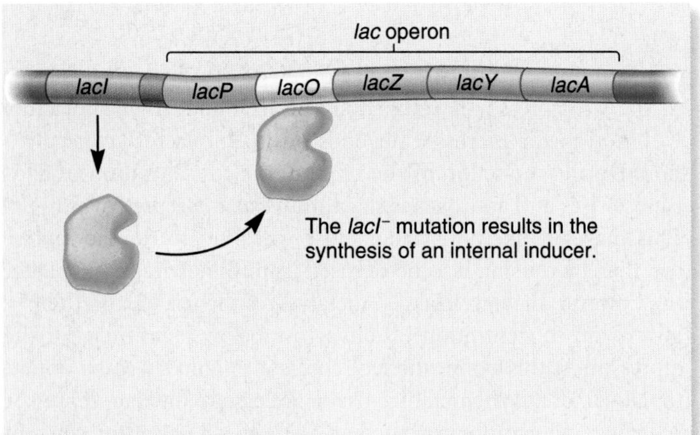

(a) Internal inducer hypothesis

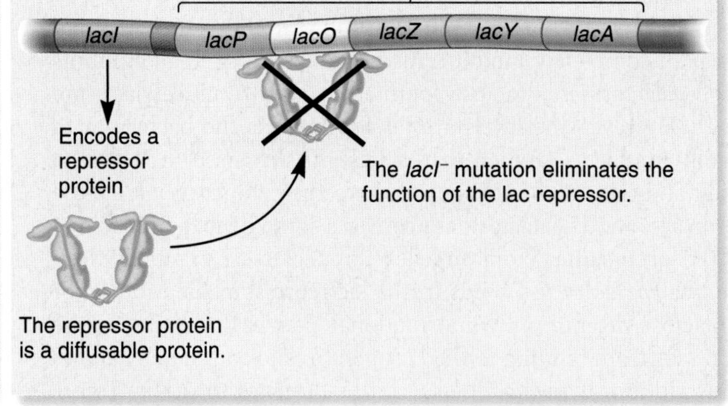

(b) Correct explanation

FIGURE 14.6 **Alternative ideas to explain how a *lacI*$^-$ mutation could cause the constitutive expression of the *lac* operon.** (a) The hypothesis of Jacob, Monod, and Pardee. In this case, the *lacI*$^-$ mutation would result in the synthesis of an internal inducer that turns on the *lac* operon. (b) The correct explanation in which the *lacI*$^-$ mutation eliminates the function of the lac repressor protein, which prevents it from repressing the *lac* operon.

TESTING THE HYPOTHESIS — FIGURE 14.7 Evidence that the *lacI* gene encodes a diffusible repressor protein.

Starting material: The genotype of the mutant strain was *lacI*$^-$ *lacZ*$^+$ *lacY*$^+$ *lacA*$^+$. The merozygote strain had an F' factor that was *lacI*$^+$ *lacZ*$^+$ *lacY*$^+$ *lacA*$^+$, which had been introduced into the mutant strain via conjugation.

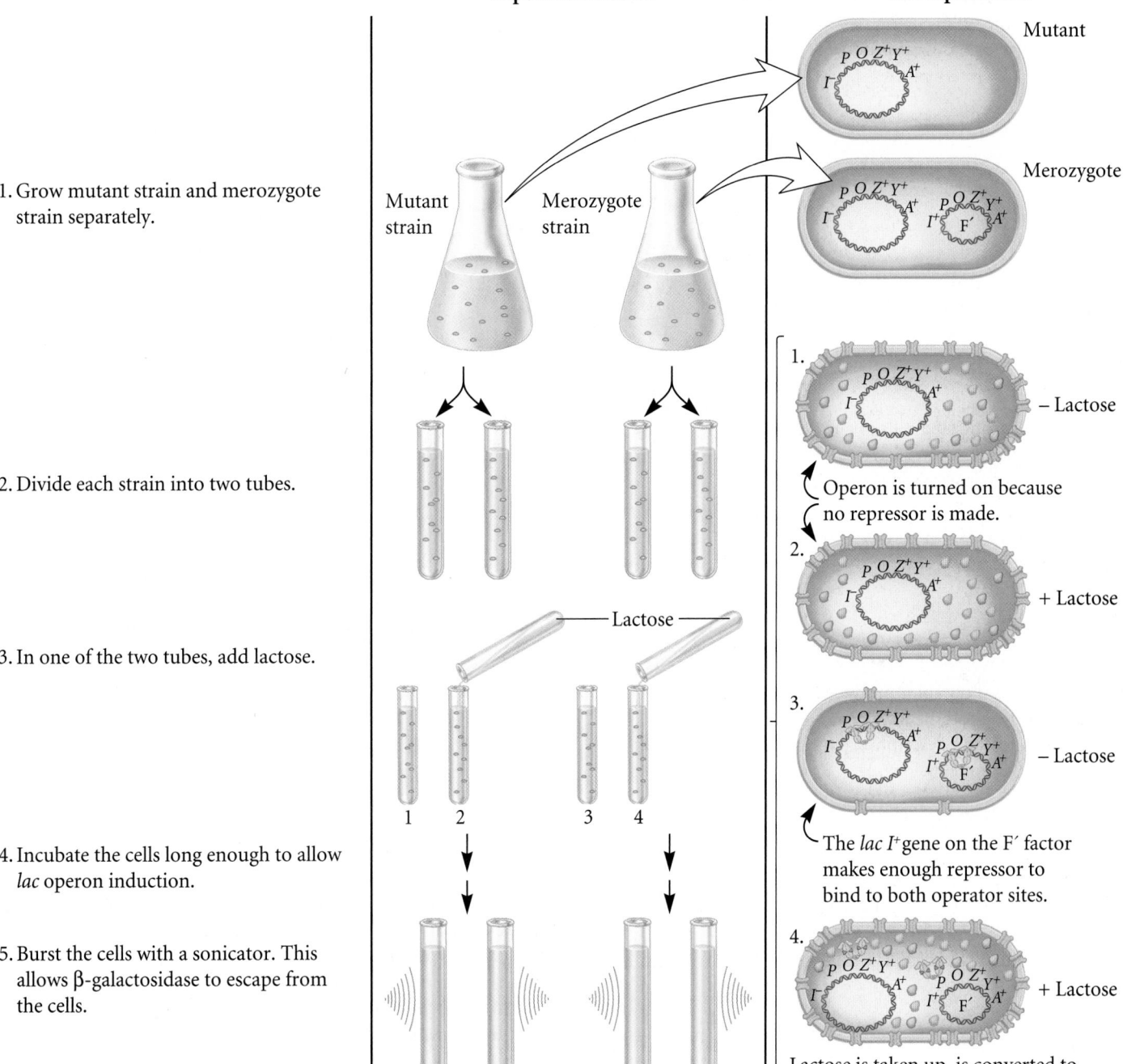

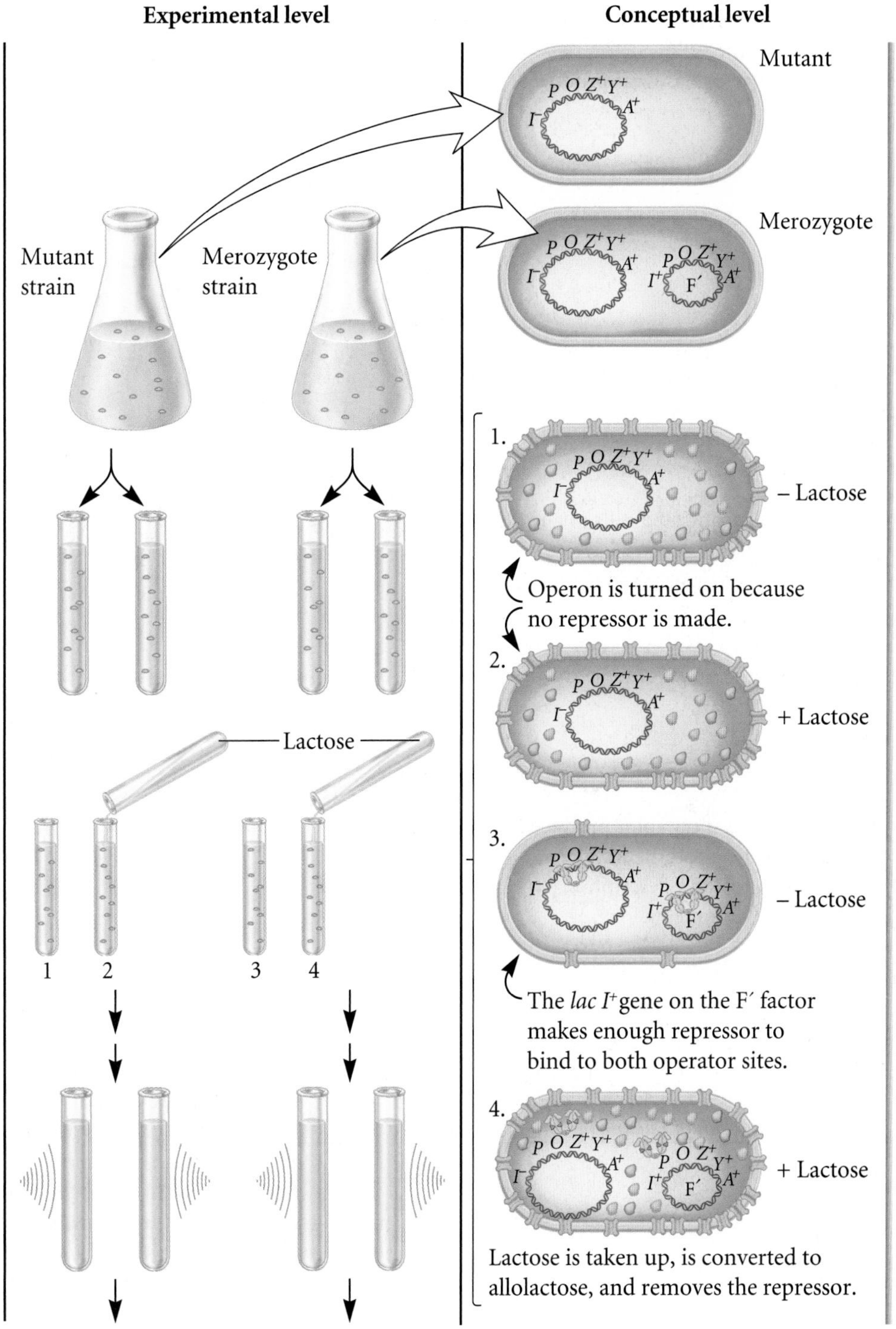

1. Grow mutant strain and merozygote strain separately.

2. Divide each strain into two tubes.

3. In one of the two tubes, add lactose.

4. Incubate the cells long enough to allow *lac* operon induction.

5. Burst the cells with a sonicator. This allows β-galactosidase to escape from the cells.

(continued)

6. Add β-*o*-nitrophenylgalactoside (β-ONPG). This is a colorless compound. β-galactosidase will cleave the compound to produce galactose and *o*-nitrophenol (O-NP). *O*-nitrophenol has a yellow color. The deeper the yellow color, the more β-galactosidase was produced.

7. Incubate the sonicated cells to allow β-galactosidase time to cleave β-*o*-nitrophenylgalactoside.

8. Measure the yellow color produced with a spectrophotometer. (See the Appendix for a description of spectrophotometry.)

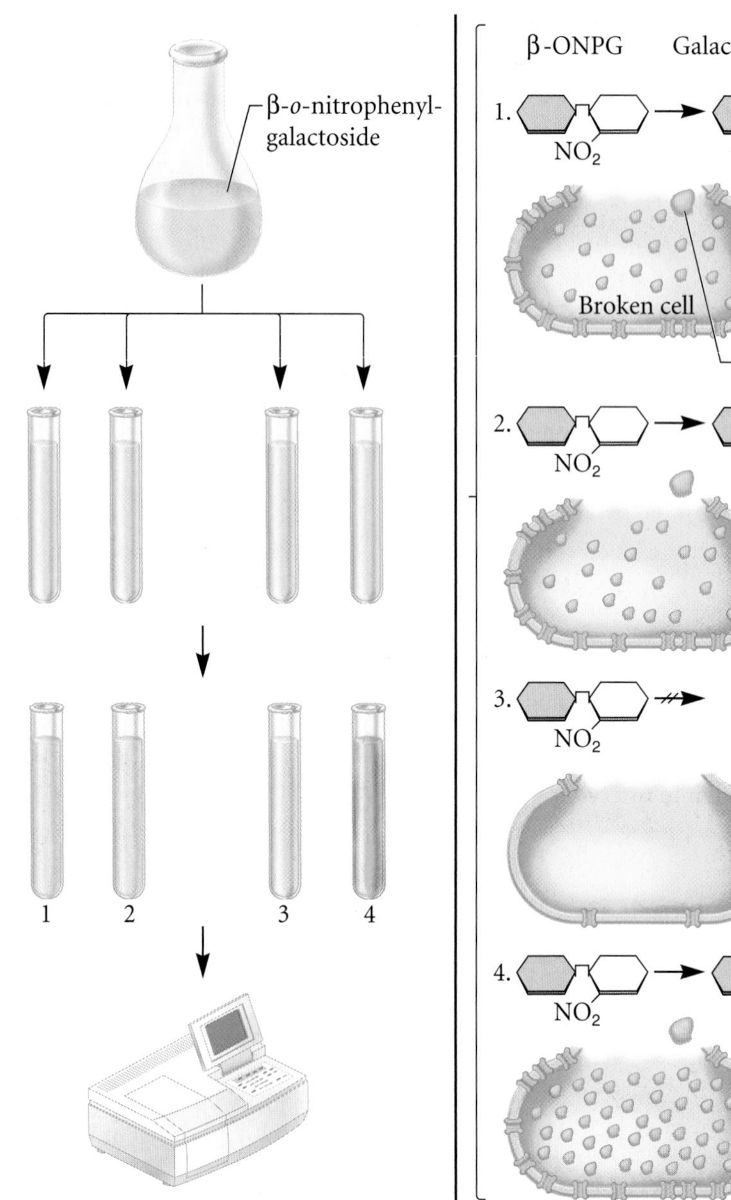

THE DATA

Strain	*Addition of Lactose*	*Amount of β-Galactosidase (percentage of parent strain)*
Mutant	No	100%
Mutant	Yes	100%
Merozygote	No	<1%
Merozygote	Yes	220%

INTERPRETING THE DATA

As seen in the data, the yellow production in the original mutant strain was the same in the presence or absence of lactose. This result is expected because the expression of β-galactosidase in the *lacI*$^-$ mutant strain was already known to be constitutive. In other words, the presence of lactose was not needed to induce the operon due to a defective *lacI* gene. In the merozygote strain, however, a different result was obtained. In the absence of lactose, the *lac* operons were repressed—even the operon on the bacterial chromosome. How do we explain these results? Because the normal *lacI* gene on the F' factor was not physically located next to the chromosomal *lac* operon, this result is consistent with the idea that the *lacI* gene codes for a repressor protein that can diffuse throughout the cell and bind to any *lac* operon. The hypothesis that the *lacI*$^-$ mutation resulted in the synthesis of an internal inducer was rejected. If that hypothesis had been correct, the merozygote strain would have still made an internal inducer, and the *lac* operons in the merozygote would have been expressed in the absence of lactose. This result was not obtained.

The interactions between regulatory proteins and DNA sequences illustrated in this experiment have led to the definition of two genetic terms. A ***trans*-effect** is a form of genetic regulation that can occur even though two DNA segments are not physically adjacent. The action of the lac repressor on the *lac* operon is a *trans*-effect. A regulatory protein, such as the lac repressor, is called a ***trans*-acting factor.** In contrast, a ***cis*-acting element** is a DNA segment that must be adjacent to the gene(s) that it regulates, and it is said to have a ***cis*-effect** on gene expression. The *lac* operator site is an example of a *cis*-acting element. A *trans*-effect is mediated by genes that encode regulatory proteins, whereas a *cis*-effect is mediated by DNA sequences that are bound by regulatory proteins.

Jacob and Monod also isolated constitutive mutants that affected the operator site, *lacO*. **Table 14.1** summarizes the effects of mutations based on their locations in the *lacI* regulatory gene versus *lacO* and their analysis in merozygotes. As seen here, a loss-of-function mutation in a gene encoding a repressor protein has the same effect as a mutation in an operator site that cannot bind a repressor protein. In both cases, the genes of the *lac* operon are constitutively expressed. In a merozygote, however, the results are quite different. When a normal *lacI* gene and a normal *lac* operon are introduced into a cell harboring a defective *lacI* gene, the normal *lacI* gene can regulate both operons. In contrast, when a *lac* operon with a normal operator site is introduced into a cell with a defective operator site, the operon with the defective operator site continues to be expressed without lactose present. Overall, a mutation in a *trans*-acting factor can be complemented by the introduction of a second gene with normal function. However, a mutation in a *cis*-acting element is not affected by the introduction of another *cis*-acting element with normal function into the cell.

TABLE 14.1

A Comparison of Loss-of-Function Mutations in the *lacI* Gene versus the Operator Site

		Expression of the *lac* Operon	
Chromosome	**F′ factor**	**With Lactose**	**Without Lactose**
Wild type	None	100%	<1%
lacI⁻	None	100%	100%
lacO⁻	None	100%	100%
lacI⁻	*lacI*⁺ and a normal *lac* operon	200%	<1%
lacO⁻	*lacI*⁺ and a normal *lac* operon	200%	100%

A self-help quiz involving this experiment can be found at the Online Learning Center.

The *lac* Operon Is Also Regulated by an Activator Protein

The *lac* operon can be transcriptionally regulated in a second way, known as **catabolite repression** (as we shall see, a somewhat imprecise term). This form of transcriptional regulation is influenced by the presence of glucose, which is a catabolite—a substance that is broken down inside the cell. The presence of glucose ultimately leads to repression of the *lac* operon. When exposed to both glucose and lactose, *E. coli* cells first use glucose, and catabolite repression prevents the use of lactose. Why is this an advantage? The explanation is efficiency. The bacterium does not have to express all the genes necessary for both glucose and lactose metabolism. If the glucose is used up, catabolite repression is alleviated, and the bacterium then expresses the *lac* operon. The sequential use of two sugars by a bacterium, known as **diauxic growth,** is a common phenomenon among many bacterial species. Typically, glucose, a more commonly encountered sugar, is metabolized preferentially, and then a second sugar is metabolized only after glucose has been depleted from the environment.

Glucose, however, is not itself the small effector molecule that binds directly to a genetic regulatory protein. Instead, this form of regulation involves a small effector molecule, **cyclic AMP (cAMP),** which is produced from ATP via an enzyme known as adenylyl cyclase. When a bacterium is exposed to glucose, the transport of glucose into the cell stimulates a signaling pathway that causes the intracellular concentration of cAMP to decrease because the pathway inhibits adenylyl cyclase, an enzyme needed for cAMP synthesis. The effect of cAMP on the *lac* operon is mediated by an activator protein called the catabolite activator protein (CAP). CAP is composed of two subunits, each of which binds one molecule of cAMP.

Figure 14.8 considers how the interplay between the lac repressor and CAP determines whether the *lac* operon is expressed in the presence or absence of lactose and/or glucose. When only lactose is present, cAMP levels are high (Figure 14.8a). The cAMP binds to CAP, and then CAP binds to the CAP site and stimulates the ability of RNA polymerase to begin transcription. In the presence of lactose, the lac repressor is not bound to the operator site, so transcription can proceed. In the absence of both lactose and glucose, cAMP levels are also high (Figure 14.8b). Under these conditions, however, the binding of the lac repressor inhibits transcription even though CAP is bound to the DNA.

Figure 14.8c considers the situation in which both sugars are present. The presence of lactose causes the lac repressor to be inactive, which prevents it from binding to the operator site. Even so, the presence of glucose decreases cAMP levels so that cAMP is released from CAP, which prevents CAP from binding to the CAP site. Because CAP is not bound to the CAP site, the *lac* operon is not expressed in the presence of both sugars. Finally, Figure 14.8d illustrates what happens when only glucose is present. The *lac* operon is shut off because the lac repressor is bound to the operator site and CAP is not bound to the CAP site due to low cAMP levels.

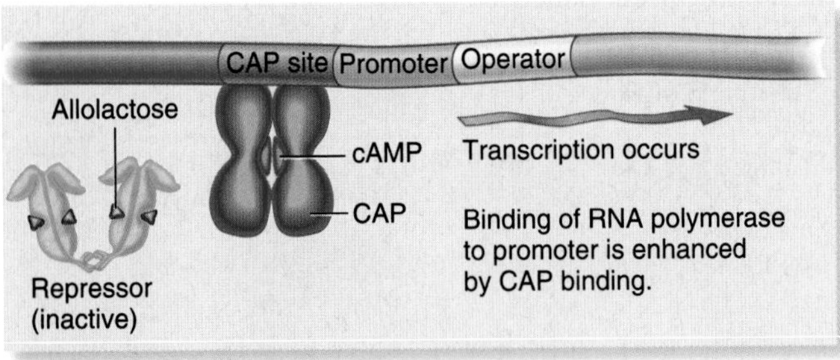

(a) Lactose, no glucose (high CAMP)

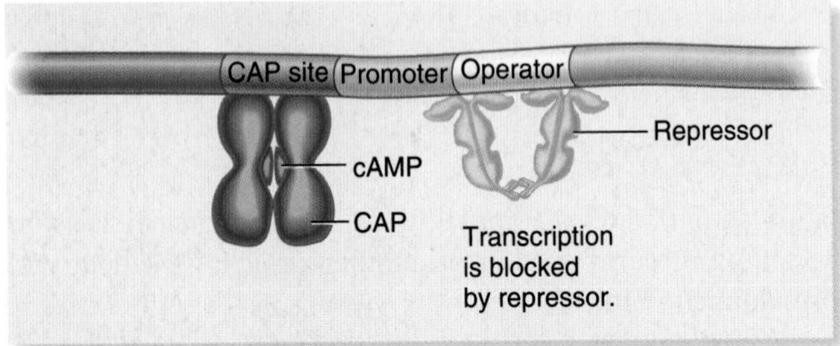

(b) No lactose or glucose (high cAMP)

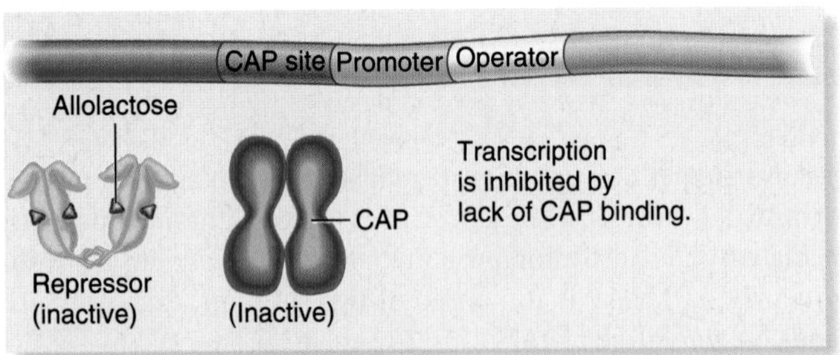

(c) Lactose and glucose (low cAMP)

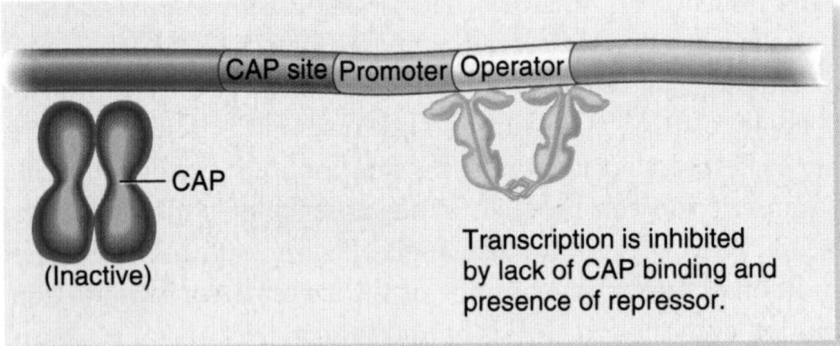

(d) Glucose, no lactose (low cAMP)

FIGURE 14.8 **The roles of the lac repressor and catabolite activator protein (CAP) in the regulation of the *lac* operon.** This figure illustrates how the *lac* operon will be regulated depending on its exposure to lactose or glucose.

Genes→Traits The mechanism of catabolite repression provides the bacterium with the trait of being able to choose between two sugars. When exposed to both glucose and lactose, the bacterium chooses glucose first. After the glucose is used up, it then expresses the genes necessary for lactose metabolism. This trait allows the bacterium to more efficiently use sugars from its environment.

The effect of glucose, called catabolite repression, may seem like a puzzling way to describe this process because this regulation involves the action of an inducer (cAMP) and an activator protein (CAP), not a repressor. The term was coined before the action of the cAMP-CAP complex was understood. At that time, the primary observation was that glucose (a catabolite) inhibited (repressed) lactose metabolism.

Many bacterial promoters that transcribe genes involved in the breakdown of other sugars, such as maltose, arabinose, and melibiose, also have binding sites for CAP. Therefore, when glucose levels are high, these operons are inhibited, thereby promoting diauxic growth.

Further Studies Have Revealed That the *lac* Operon Has Three Operator Sites for the lac Repressor

Our traditional view of the regulation of the *lac* operon has been modified as we have gained a greater molecular understanding of the process. In particular, detailed genetic and crystallography studies have shown that the binding of the lac repressor is more complex than originally realized. The site in the *lac* operon that is commonly called the operator site was first identified by mutations that prevented lac repressor binding. These mutations, called *lacO*$^-$ or *lacO*^{C} mutants, resulted in the constitutive expression of the *lac* operon even in strains that make a normal lac repressor protein. *LacO*^{C} mutations were localized in the *lac* operator site, which is now known as O_1. This led to the view that a single operator site was bound by the lac repressor to inhibit transcription, as in Figure 14.4.

In the late 1970s and 1980s, two additional operator sites were identified. As shown at the top of **Figure 14.9**, these sites are called O_2 and O_3. O_1 is the operator site slightly downstream from the promoter. O_2 is located farther downstream in the *lacZ* coding sequence, and O_3 is located slightly upstream from the promoter. The O_2 and O_3 operator sites were initially called pseudo-operators, because substantial repression occurred in the absence of either one of them. However, studies by Benno Müller-Hill and his colleagues revealed a surprising result. As shown in Figure 14.9 (fourth example down), if both O_2 and O_3 are missing, repression is dramatically reduced even when O_1 is present. When O_1 is missing, even in the presence of one of the other operator sites, repression is nearly abolished.

How were these results interpreted? The data of Figure 14.9 supported a hypothesis that the lac repressor must bind to O_1 and either O_2 or O_3 to cause full repression. According to this view, the lac repressor can readily bind to O_1 and O_2, or to O_1 and O_3, but not to O_2 and O_3. If either O_2 or O_3 were missing, maximal repression is not achieved because it is less likely for the repressor to bind when only two operator sites are present. If you take a look at Figure 14.9, you will notice that the operator sites are a fair distance away from each other. For this reason, it was proposed from these studies that the binding of the lac repressor to two operator sites requires the DNA to form a loop. A loop in the DNA would bring the operator sites closer together and thereby facilitate the binding of the repressor protein (**Figure 14.10a**).

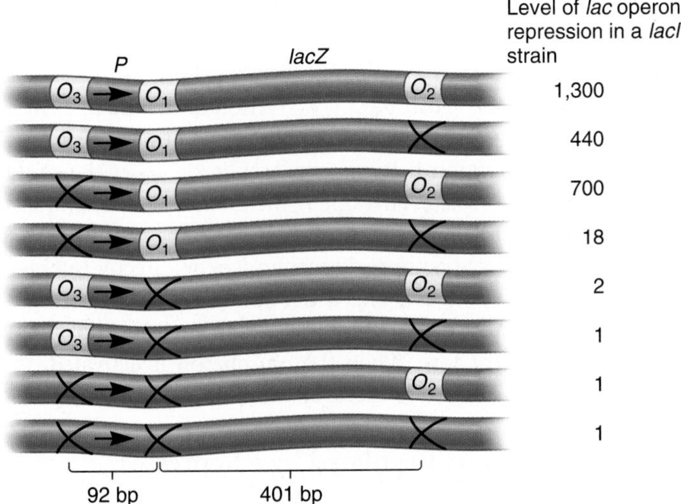

FIGURE 14.9 The identification of three *lac* operator sites. The top of this figure shows the locations of three *lac* operator sites, designated O_1, O_2, and O_3. O_1 is analogous to the *lac* operator site shown in previous figures. The arrows depict the starting site for transcription. Defective operator sites are indicated with an X. When all three operator sites are present, the repression of the *lac* operon is 1,300-fold; this means there is 1/1,300 the level of expression than when lactose is present. This figure also shows the amount of repression when one or more operator sites are removed. A repression value of 1.0 indicates that no repression is occurring. In other words, a value of 1.0 indicates constitutive expression.

In 1996, the proposal that the lac repressor binds to two operator sites was confirmed by studies in which the lac repressor was crystallized by Mitchell Lewis and his colleagues. The crystal structure of the lac repressor has provided exciting insights into its mechanism of action. As mentioned earlier in this chapter, the lac repressor is a tetramer of four identical subunits. The crystal structure revealed that each dimer within the tetramer recognizes one operator site. **Figure 14.10b** is a molecular model illustrating the binding of the lac repressor to the O_1 and O_3 sites. The amino acid side chains in the protein interact directly with bases in the major groove of the DNA helix. This is how genetic regulatory proteins recognize specific DNA sequences. Because each dimer within the tetramer recognizes a single operator site, the association of two dimers to form a tetramer requires that the two operator sites be close to each other. For this to occur, a loop must form in the DNA. The formation of this loop would dramatically inhibit the ability of RNA polymerase to slide past the O_1 site and transcribe the operon.

Figure 14.10b also shows the binding of the cAMP-CAP complex to the CAP site (see the blue protein within the loop). A particularly striking observation is that the binding of the cAMP-CAP complex to the DNA causes a 90° bend in the DNA structure. When the repressor is active—not bound to allolactose—the cAMP-CAP complex facilitates the binding of the lac repressor to the O_1 and O_3 sites. When the repressor is inactive, this bending also appears to be important in the ability of RNA polymerase to initiate transcription slightly downstream from the bend.

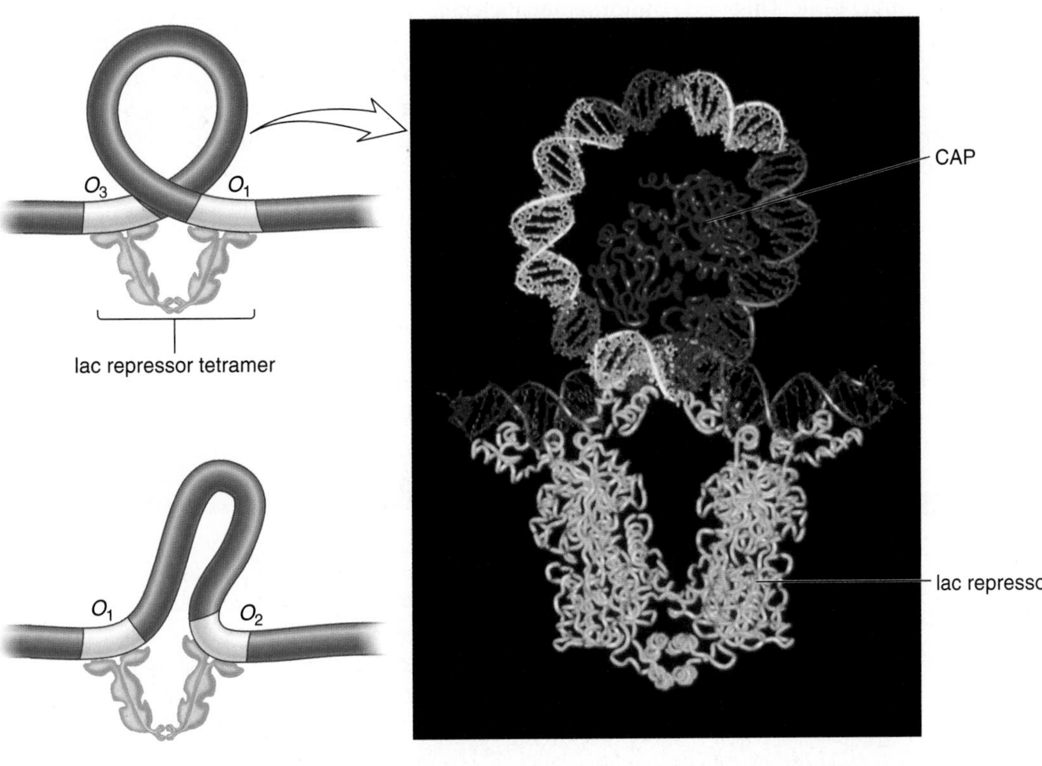

(a) DNA loops caused by the binding of the lac repressor

(b) Proposed model of the lac repressor binding to O_1 and O_3 based on crystallography studies

FIGURE 14.10 The binding of the lac repressor to two operator sites. (a) The binding of the lac repressor protein to the O_1 and O_3 or to the O_1 and O_2 operator sites. Because the two sites are far apart, a loop must form in the DNA. **(b)** A molecular model for the binding of the lac repressor to O_1 and O_3. Each repressor dimer binds to one operator site, so the repressor tetramer brings the two operator sites together. This causes the formation of a DNA loop in the intervening region. Note that the DNA loop contains the –35 and –10 regions (shown in green), which are recognized by σ factor of RNA polymerase. This loop also contains the binding site for the cAMP-CAP complex, which is the protein within the loop.

The *ara* Operon Can Be Regulated Positively or Negatively by the Same Regulatory Protein

Now that we have considered the regulation of the *lac* operon, let's compare its regulation to that of other genes in the bacterial chromosome. Another operon in *E. coli* involved in sugar metabolism is the *ara* (arabinose) operon. The sugar arabinose is a constituent of the cell walls of a few types of plants. As shown in **Figure 14.11**, the *ara* operon contains three structural genes, *araB*, *araA*, and *araD*, encoding a polycistronic mRNA for the three enzymes involved in arabinose metabolism. The actions of the three enzymes metabolize arabinose into D-xylulose-5-phosphate.

Like the *lac* operon, the *ara* operon contains a single promoter, designated P_{BAD}. The operon also contains a CAP site for the binding of the catabolite activator protein. The *araC* gene, which has its own promoter (P_C), is adjacent to the *ara* operon. *AraC* encodes a regulatory protein, called the AraC protein, that can bind to operator sites designated *araI*, $araO_1$, and $araO_2$.

As we have seen with the *lac* operon, some regulatory proteins such as the lac repressor inhibit transcription, whereas others such as CAP turn on transcription. AraC is a rather interesting and unusual protein, because it can act as either a negative or positive regulator of transcription, depending on whether or not arabinose is present. How can a protein function in both ways? In the absence of arabinose, the AraC protein binds to the *araI*, $araO_1$, and $araO_2$ operator sites (**Figure 14.12a**). An AraC protein dimer is bound to $araO_1$, while monomers are bound at $araO_2$ and *araI*. The binding of the AraC proteins to the $araO_1$ site inhibits the transcription of the *araC* gene. In other words, the AraC protein is a negative regulator of the *araC* gene. This keeps AraC protein levels fairly low. The AraC proteins bound at $araO_2$ and *araI* repress the *ara* operon, but only in the absence of arabinose. As shown in Figure 14.12a, the AraC proteins at $araO_2$ and *araI* can bind to each other by causing a loop in the DNA, as originally proposed by Robert Schleif and his colleagues in 1990. This DNA loop prevents RNA polymerase from binding to the DNA and transcribing the *ara* operon via P_{BAD}. Therefore, in the absence of arabinose, the *ara* operon is turned off.

Figure 14.12b illustrates the activation of the *ara* operon in the presence of arabinose. When arabinose is bound to the AraC protein, the interaction between the AraC proteins at the $araO_2$ and *araI* sites is broken. This opens the DNA loop. In addition, a second AraC protein binds at the *araI* site. This AraC dimer at the *araI* operator site activates transcription by directly interacting with RNA polymerase. This activation can occur in conjunction with the activation of the *ara* operon by CAP and cAMP if glucose levels are low. When the *ara* operon is activated, the bacterial cell can efficiently metabolize arabinose.

The *trp* Operon Is Regulated by a Repressor Protein and Also by Attenuation

The *trp* operon (pronounced "trip") encodes enzymes that are needed for the biosynthesis of the amino acid tryptophan. The *trpE*, *trpD*, *trpC*, *trpB*, and *trpA* genes encode enzymes involved in tryptophan biosynthesis. The *trpR* and *trpL* genes are involved in regulating the *trp* operon in two different ways. The *trpR* gene encodes the **trp repressor** protein. When tryptophan levels within the cell are very low, the trp repressor cannot bind to the operator site. Under these conditions, RNA polymerase transcribes the *trp* operon (**Figure 14.13a**). In this way, the cell expresses the genes required for the synthesis of tryptophan. When the tryptophan levels within the cell become high, tryptophan acts as a corepressor that binds to the trp repressor protein. This causes a conformational change in the trp repressor that allows it to bind to the *trp* operator site (**Figure 14.13b**). This inhibits the ability of RNA polymerase to transcribe the operon. Therefore, when a high level of tryptophan is present within the cell—when the cell does not need to make more tryptophan—the *trp* operon is turned off.

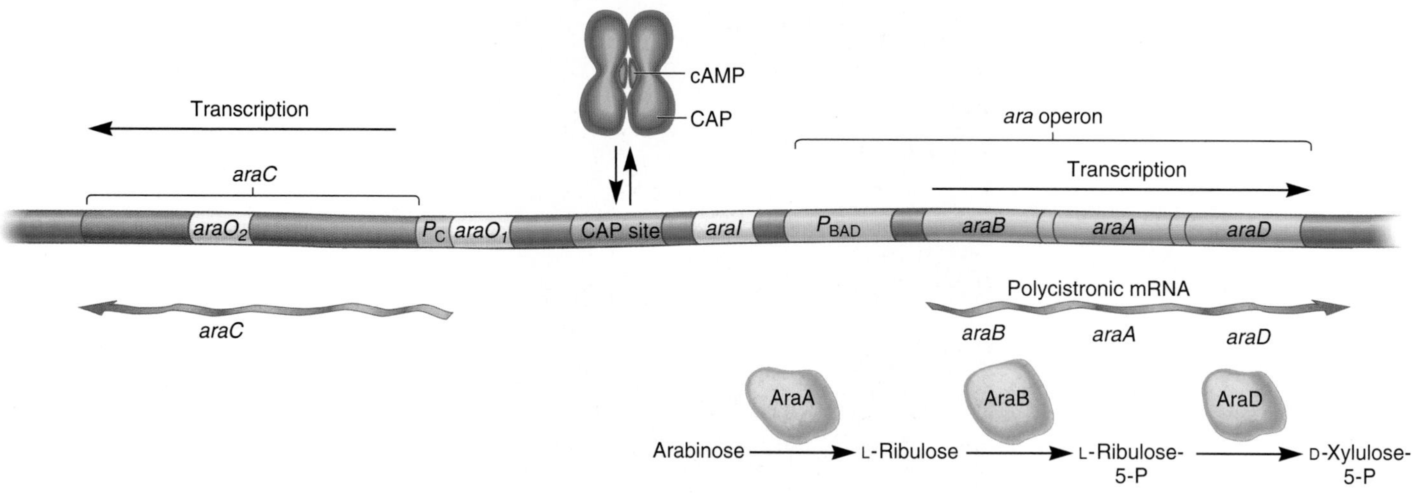

FIGURE 14.11 **Organization of the *ara* operon and other genes involved in arabinose metabolism.** *AraC* encodes a genetic regulatory protein called the AraC protein. This protein can bind to three different regulatory sites, called *araI*, $araO_1$, and $araO_2$. P_C is the promoter for the *araC* gene; P_{BAD} is the promoter for the *ara* operon, which contains three genes (*araB*, *araA*, and *araD*) encoding proteins involved in arabinose metabolism.

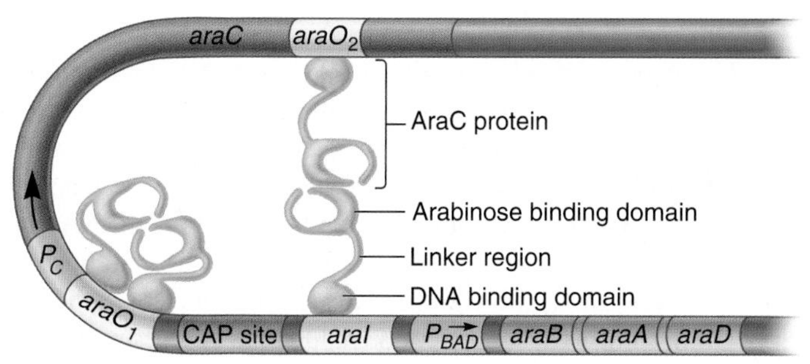

(a) Operon inhibited in the absence of arabinose

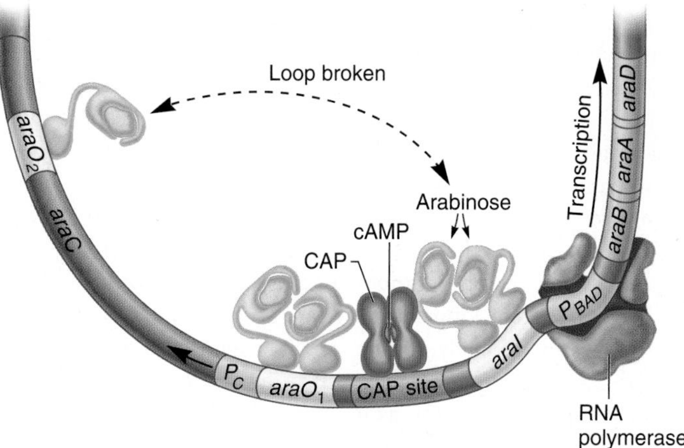

(b) Operon activated in the presence of arabinose

FIGURE 14.12 DNA looping and unlooping via the AraC protein. (a) In the absence of arabinose, an AraC protein binds to the *araO*$_2$ operator site and another to the *araI* site. These two AraC proteins interact to promote a loop in the DNA. This loop prevents RNA polymerase from transcribing the *ara* operon from the P_{BAD} promoter. (b) In the presence of arabinose, arabinose binds to the AraC proteins and causes a conformational change that breaks the interaction between the AraC proteins bound at *araO*$_2$ and *araI*. This causes the DNA loop to be broken. Under these conditions, an additional AraC protein binds to *araI*, and RNA polymerase can transcribe the *ara* operon.

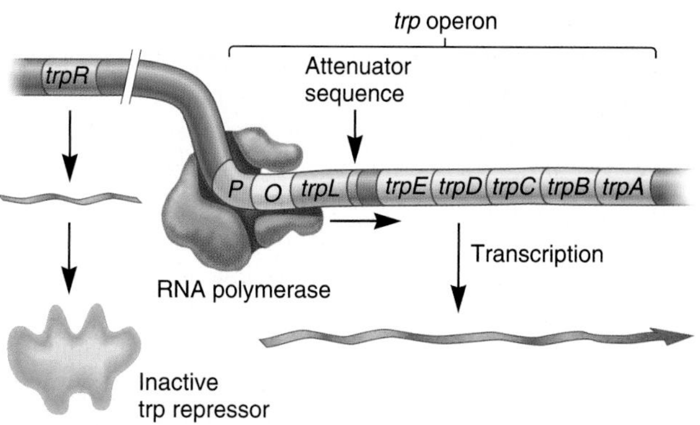

(a) Low tryptophan levels, transcription of the entire *trp* operon occurs

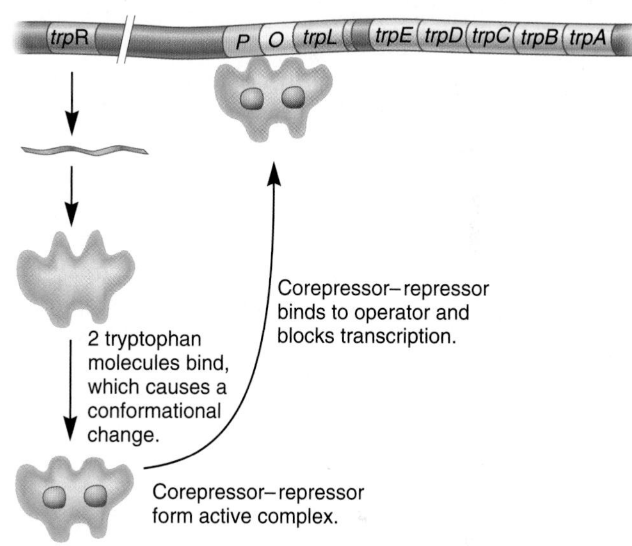

(b) High tryptophan levels, repression occurs

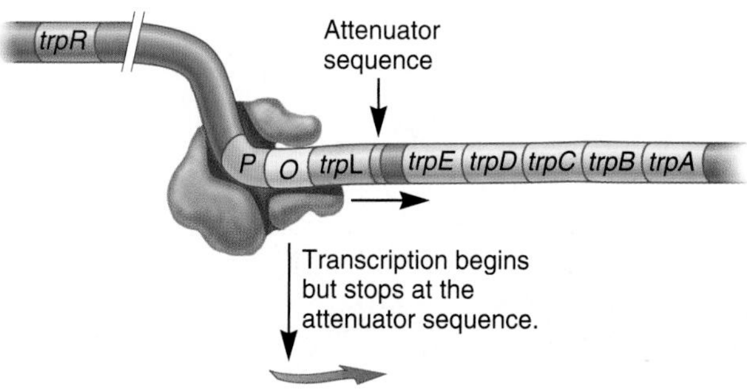

(c) High tryptophan levels, attenuation occurs

FIGURE 14.13 Organization of the *trp* operon and regulation via the trp repressor protein. (a) When tryptophan levels are low, tryptophan does not bind to the repressor protein, which prevents the repressor protein from binding to the operator site. Under these conditions, RNA polymerase can transcribe the operon, which leads to the expression of the *trpE, trpD, trpC, trpB*, and *trpA* genes. These genes encode enzymes involved in tryptophan biosynthesis. (b) When tryptophan levels are high, tryptophan acts as a corepressor that binds to the trp repressor protein. The tryptophan-trp repressor complex then binds to the operator site to inhibit transcription. (c) Another mechanism of regulation is attenuation. When attenuation occurs, the RNA is transcribed only to the attenuator sequence, and then transcription is terminated.

In the 1970s, after the action of the trp repressor was elucidated, Charles Yanofsky and coworkers made a few unexpected observations. Mutant strains were found that lacked the trp repressor protein. Surprisingly, these mutant strains still could inhibit the expression of the *trp* operon in the presence of tryptophan. In addition, *trp* operon mutations were identified in which a region including the *trpL* gene was missing from the operon. These mutations resulted in higher levels of expression of the other genes in the *trp* operon. As is often the case, unusual observations can lead scientists into productive avenues of study. By pursuing this research further, Yanofsky discovered a second regulatory mechanism in the *trp* operon, called **attenuation,** that is mediated by the region that includes the *trpL* gene (**Figure 14.13c**).

Attenuation can occur in bacteria because the processes of transcription and translation are coupled. As described in Chapter 13, bacterial ribosomes quickly attach to the 5' end of mRNA soon after its synthesis begins via RNA polymerase. During attenuation, transcription actually begins, but it is terminated before the entire mRNA is made. A segment of DNA, termed the **attenuator sequence,** is important in facilitating this termination. When attenuation occurs, the mRNA from the *trp* operon is made as a short piece that terminates shortly past the *trpL* gene (Figure 14.13c). Because this short mRNA has been terminated before RNA polymerase has transcribed the *trpE, trpD, trpC, trpB,* and *trpA* genes, it will not encode the proteins required for tryptophan biosynthesis. In this way, attenuation inhibits the further production of tryptophan in the cell.

The segment of the *trp* operon immediately downstream from the operator site plays a critical role during attenuation. The first gene in the *trp* operon is the *trpL* gene, which encodes a peptide containing 14 amino acids called the leader peptide. As shown in **Figure 14.14**, two features are key in the attenuation mechanism. First, two tryptophan (Trp) codons are found within the mRNA that encodes the *trp* leader peptide. What is the role of these codons? As we will see later, these two codons provide a way to sense whether or not the bacterium has sufficient tryptophan to synthesize its proteins. Second, the mRNA can form stem-loop structures. The type of stem-loop structure that forms underlies attenuation.

Different combinations of stem-loop structures are possible due to interactions among four sequences within the RNA transcript (see the yellow regions in Figure 14.14). Region 2 is complementary to region 1 and also to region 3. Region 3 is complementary to region 2 as well as to region 4. Therefore, three stem-loop structures are possible: 1–2, 2–3, and 3–4. Even so, keep in mind that a particular segment of RNA can participate in the formation of only one stem-loop structure. For example, if region 2 forms a stem-loop with region 1, it cannot (at the same time) form a stem-loop with region 3. Alternatively, if region 2 forms a stem-loop with region 3, then region 3 cannot form a stem-loop with region 4. Though three stem-loop structures are possible, the 3–4 stem-loop structure is functionally unique. The 3–4 stem-loop together with the U-rich attenuator sequence acts as an intrinsic terminator—a ρ-independent terminator, as described in Chapter 12. Therefore, the formation of the 3–4 stem-loop causes RNA polymerase to

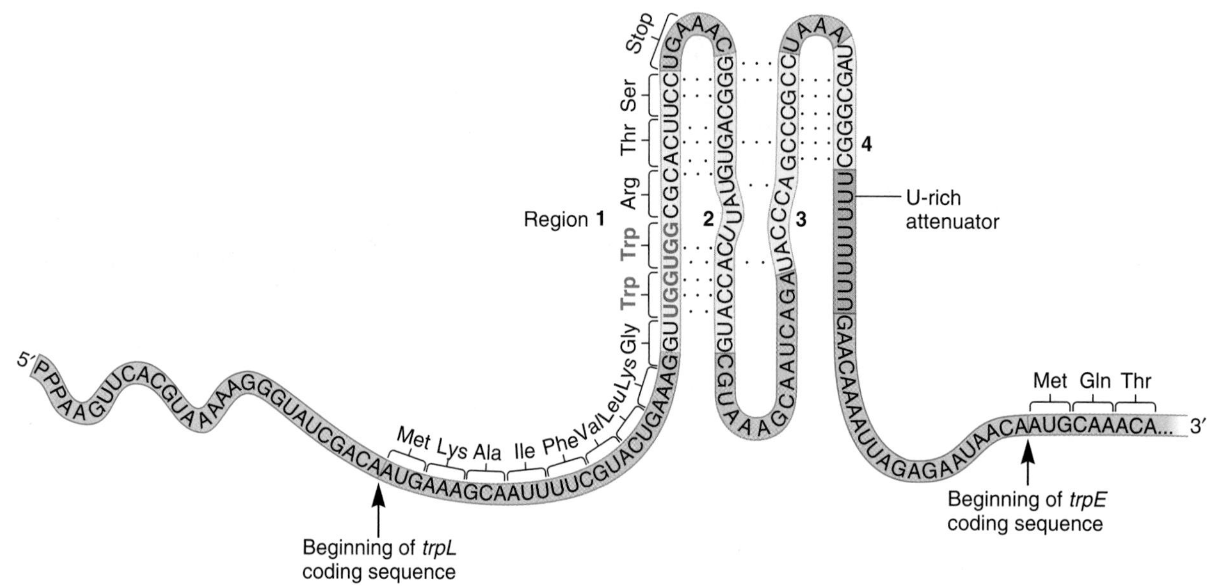

ONLINE ANIMATION

FIGURE 14.14 Sequence of the *trpL* mRNA produced during attenuation. A second method of regulation of the *trp* operon is attenuation. At high tryptophan levels, it may occasionally happen that the trp repressor protein does not bind to the operator site to inhibit transcription. If so, a short mRNA is made that includes the *trpL* region. As shown here, this mRNA has several regions that are complementary to each other. The hydrogen bonding between yellow regions 1 and 2, 2 and 3, and 3 and 4 is also shown. The last U in the purple attenuator sequence is the last nucleotide that would be transcribed during attenuation. At low tryptophan concentrations, however, transcription would occur beyond the end of *trpL* and proceed through the *trpE* gene and the rest of the *trp* operon.

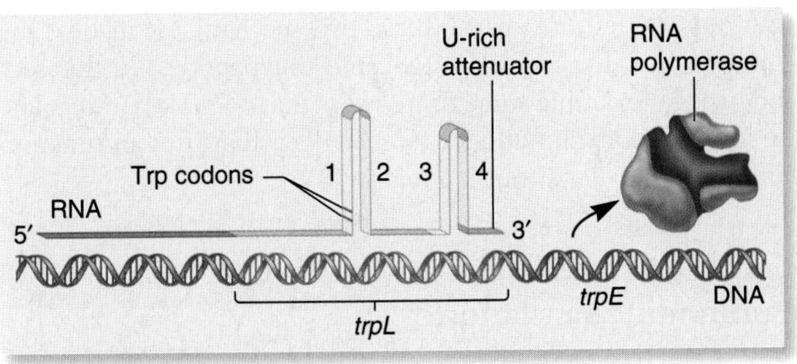

(a) No translation

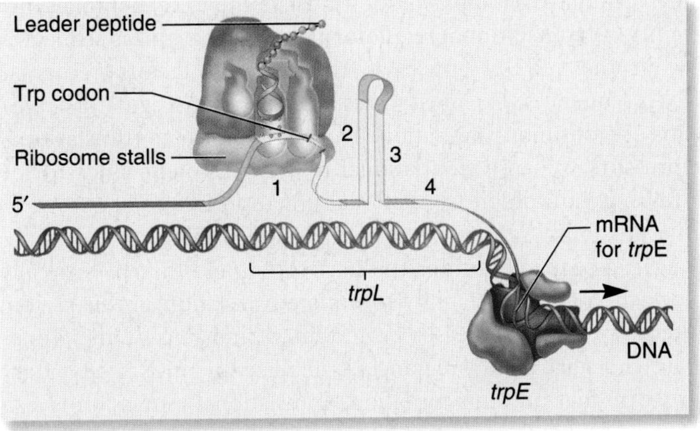

(b) Low tryptophan levels, 2–3 stem-loop forms

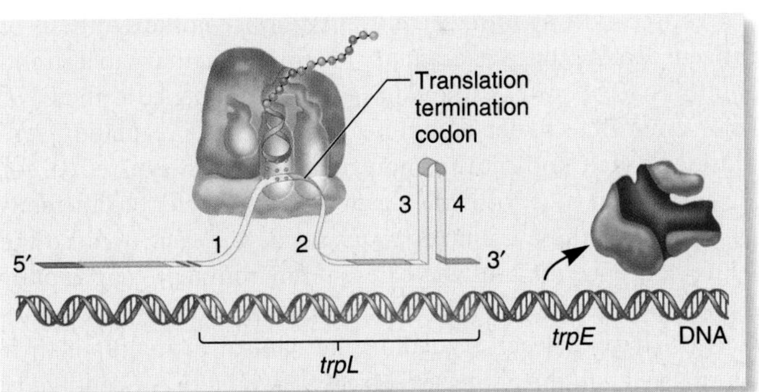

(c) High tryptophan levels, 3–4 stem-loop forms

ONLINE ANIMATION

FIGURE 14.15 **Possible stem-loop structures formed from *trpL mRNA* under different conditions of translation.** (a) When translation is not coupled with transcription, the most stable form of the mRNA occurs when region 1 hydrogen bonds to region 2 and region 3 hydrogen bonds to region 4. A terminator stem-loop forms, and transcription will be terminated just past the *trpL* gene. (b) Coupled transcription and translation have occurred under conditions in which the tryptophan concentration is low. The ribosome pauses at the Trp codons in the *trpL* gene because insufficient amounts of charged tRNA^{trp} are present. This pause blocks region 1 of the mRNA, so region 2 can hydrogen bond only with region 3. When this happens, the 3–4 stem-loop structure cannot form. Transcriptional termination does not occur, and RNA polymerase transcribes the rest of the operon. (c) Coupled transcription and translation occurs under conditions in which a sufficient amount of tryptophan is present in the cell. Translation of the *trpL* gene progresses to its stop codon, where the ribosome pauses. This blocks region 2 from hydrogen bonding with any region and thereby enables region 3 to hydrogen bond with region 4. This terminates transcription at the U-rich attenuator.

pause, and the U-rich sequence dissociates from the DNA. This terminates transcription at the U-rich attenuator. In comparison, if region 3 forms a stem-loop with region 2, transcription will not be terminated because a 3–4 stem-loop cannot form.

Conditions that favor the formation of the 3–4 stem-loop ultimately rely on the translation of the *trpL* gene. As shown in Figure 14.15, three scenarios are possible. In Figure 14.15a, translation is not coupled with transcription. Region 1 rapidly hydrogen bonds to region 2, and region 3 is left to hydrogen bond to region 4. Therefore, the terminator stem-loop forms, and transcription will be terminated just past the *trpL* gene at the U-rich attenuator.

In Figure 14.15b, coupled transcription and translation occur under conditions in which the tryptophan concentration is low. When tryptophan levels are low, the cell cannot make a sufficient amount of charged tRNA^{trp}. As we see in Figure 14.15b, the ribosome pauses at the Trp codons in the *trpL* mRNA because it is waiting for charged tRNA^{trp}. This pause occurs in such a way that the ribosome shields region 1 of the mRNA. This sterically prevents region 1 from hydrogen bonding to region 2. As an alternative, region 2 hydrogen bonds to region 3. Therefore, because region 3 is already hydrogen bonded to region 2, the 3–4 stem-loop structure cannot form. Under these conditions, transcriptional termination does not occur, and RNA polymerase transcribes the rest of the operon. This ultimately enables the bacterium to make more tryptophan.

Finally, in Figure 14.15c, coupled transcription and translation occur under conditions in which a sufficient amount of tryptophan is present in the cell. In this case, translation of the *trpL* mRNA progresses to its stop codon, where the ribosome pauses. The pausing at the stop codon prevents region 2 from hydrogen bonding with any region and thereby enables region 3 to hydrogen bond with region 4. As in Figure 14.15a, this terminates transcription. Of course, keep in mind that the *trpL* gene contains two tryptophan codons. For the ribosome to smoothly progress to the *trpL* stop codon, enough charged tRNA^{trp} must be available to translate this mRNA. It follows that the bacterium must have a sufficient amount of tryptophan. Under these conditions, the rest of the transcription of the operon is terminated.

Attenuation is a mechanism to regulate transcription that is found in several other operons involved with amino acid biosynthesis. In all cases, the mRNAs that encode the leader peptides are rich in codons for the particular amino acid that is synthesized by the enzymes encoded by the particular operon. For example, the mRNA that encodes the leader peptide of the histidine operon has seven histidine codons in its sequence, and the mRNA for the leader peptide of the leucine operon has four leucine codons. Like the *trp* operon, these other operons have alternative stem-loop structures, one of which is a transcriptional terminator.

Inducible Operons Encode Catabolic Enzymes, and Repressible Operons Usually Encode Anabolic Enzymes

Thus far, we have seen that bacterial genes can be transcriptionally regulated in a positive or negative way—and sometimes both. The *lac* operon and *ara* operon are inducible systems regulated by sugar molecules that activate transcription of these operons. By comparison, the *trp* operon is a repressible operon regulated by tryptophan, a corepressor that binds to the repressor and turns the operon off. In addition, an abundance of charged $tRNA^{trp}$ in the cytoplasm can turn the *trp* operon off via attenuation.

By studying the genetic regulation of many operons, geneticists have discovered a general trend concerning inducible versus repressible regulation. When the genes in an operon encode proteins that function in the catabolism or breakdown of a substance, they are usually regulated in an inducible manner. The substance to be broken down or a related compound often acts as the inducer. For example, allolactose and arabinose act as inducers of the *lac* and *ara* operons, respectively. An inducible form of regulation allows the bacterium to phenotypically express the appropriate genes only when they are needed to catabolize these sugars.

In contrast, other enzymes are important for the anabolism or synthesis of small molecules. The genes that encode these anabolic enzymes tend to be regulated by a repressible mechanism. The corepressor or inhibitor is commonly the small molecule that is the product of the enzymes' biosynthetic activities. For example, tryptophan is produced by the sequential action of several enzymes that are encoded by the *trp* operon. Tryptophan itself acts as a corepressor that can bind to the trp repressor protein when the intracellular levels of tryptophan become relatively high. This mechanism turns off the genes required for tryptophan biosynthesis when enough of this amino acid has been made. Therefore, genetic regulation via repression provides the bacterium with a way to prevent the overproduction of the product of a biosynthetic pathway.

14.2 TRANSLATIONAL AND POSTTRANSLATIONAL REGULATION

Though genetic regulation in bacteria is exercised predominantly at transcription, many examples are known in which regulation is exerted at a later stage in gene expression. In some cases, specialized mechanisms have evolved to regulate the translation of certain mRNAs. Recall that the translation of mRNA occurs in three stages: initiation, elongation, and termination. Genetic regulation of translation is usually aimed at preventing the initiation step.

The net result of translation is the synthesis of a protein. The activities of proteins within living cells and organisms ultimately determine an individual's traits. Therefore, to fully understand how proteins influence an organism's traits, researchers have investigated how the functions of proteins are regulated. The term **posttranslational** regulation refers to the functional control of proteins that are already present in the cell rather than regulation of transcription or translation. Posttranslational control can either activate or inhibit the function of a protein. Compared to transcriptional or translational regulation, posttranslational control can be relatively fast, occurring in a matter of seconds, which is an important advantage. In contrast, transcriptional and translational regulation typically require several minutes or even hours to take effect because these two mechanisms involve the synthesis and turnover of mRNA and polypeptides. In this section, we will examine some of the ways that bacteria can regulate the initiation of translation, as well as ways that protein function can be regulated posttranslationally.

Repressor Proteins and Antisense RNA Can Inhibit Translation

For some bacterial genes, the translation of mRNA is regulated by the binding of proteins or other RNA molecules that influence the ability of ribosomes to translate the mRNA into a polypeptide. A **translational regulatory protein** recognizes sequences within the mRNA, much as transcription factors recognize DNA sequences. In most cases, translational regulatory proteins act to inhibit translation. These are known as **translational repressors.** When a translational repressor protein binds to the mRNA, it can inhibit translational initiation in one of two ways. One possibility is that it can bind in the vicinity of the Shine-Dalgarno sequence and/or the start codon and thereby sterically block the ribosome's ability to initiate translation in this region. Alternatively, the repressor protein may bind outside the Shine-Dalgarno/start codon region but stabilize an mRNA secondary structure that prevents initiation. Translational repression is also a form of genetic regulation found in eukaryotic species, and we will consider specific examples in Chapter 15.

A second way to regulate translation is via the synthesis of **antisense RNA,** an RNA strand that is complementary to a strand of mRNA. (The mRNA strand has the same sequence as the DNA sense strand.) To understand this form of genetic regulation, let's consider a trait known as osmoregulation, which is essential for the survival of most bacteria. Osmoregulation refers to the ability to control the amount of water inside the cell. Because the solute concentrations in the external environment may rapidly change between hypotonic and hypertonic conditions, bacteria must have an osmoregulation mechanism to maintain their internal cell volume. Otherwise, bacteria would be susceptible to the harmful effects of lysis or shrinking.

In *E. coli,* an outer membrane protein encoded by the *ompF* gene is important in osmoregulation. At low osmolarity, the ompF protein is preferentially produced, whereas at high osmolarity, its synthesis is decreased. The expression of another gene, known as *micF,* is responsible for inhibiting the expression of the *ompF* gene at high osmolarity. As shown in **Figure 14.16**, the inhibition occurs because the *micF* RNA is complementary to the *ompF* mRNA; it is an antisense strand of RNA. When the *micF* gene is transcribed, its RNA product binds to the *ompF* mRNA via hydrogen bonding between their complementary regions. The binding of the *micF* RNA to the *ompF* mRNA prevents the *ompF* mRNA from being translated. The RNA transcribed from the *micF* gene is called antisense RNA because it is complementary to the *ompF* mRNA, which is a sense strand of mRNA that encodes a protein. The *micF* RNA does not encode a protein.

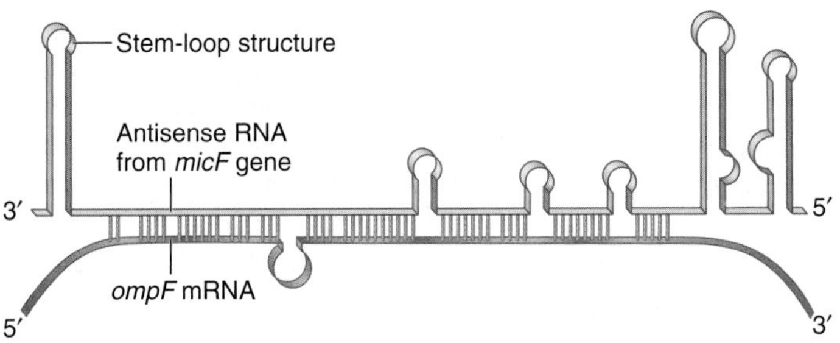

FIGURE 14.16 **The double-stranded RNA structure formed between the *micF* antisense RNA and the *ompF* mRNA.** Because they have regions that are complementary to each other, the *micF* antisense RNA binds to the *ompF* mRNA to form a double-stranded structure in which the *ompF* mRNA cannot be translated.

Posttranslational Regulation Can Occur Via Feedback Inhibition and Covalent Modifications

Let's now turn our attention to ways that protein function is regulated posttranslationally. A common mechanism to regulate the activity of metabolic enzymes is **feedback inhibition.** The synthesis of many cellular molecules such as amino acids, vitamins, and nucleotides occurs via the action of a series of enzymes that convert precursor molecules to a particular product. The final product in a metabolic pathway then inhibits an enzyme that acts early in the pathway.

Figure 14.17 depicts feedback inhibition in a metabolic pathway. Enzyme 1 is an example of an **allosteric enzyme,** an enzyme that contains two different binding sites. (The lac repressor is an allosteric protein, but not an enzyme.) The catalytic site is responsible for the binding of the substrate and its conversion to intermediate 1. The second site is a regulatory or allosteric site. This site binds the final product of the metabolic pathway. When bound to the regulatory site, the final product inhibits the catalytic ability of enzyme 1.

To appreciate feedback inhibition at the cellular level, we can consider the relationship between the product concentration and the regulatory site on enzyme 1. As the final product is made within the cell, its concentration will gradually increase. Once the final product concentration has reached a level that is similar to its affinity for enzyme 1, the final product is likely to bind to the regulatory site on enzyme 1 and inhibit its function. In this way, the net result is that the final product of a metabolic pathway inhibits the further synthesis of more product. Under these conditions, the concentration of the final product has reached a level sufficient for the purpose of the bacterium.

A second strategy to control the function of proteins is by the covalent modification of their structure, a process called **posttranslational covalent modification.** Certain types of modifications are involved primarily in the assembly and construction of a functional protein. These alterations include proteolytic processing; disulfide bond formation; and the attachment of prosthetic groups, sugars, or lipids. These are typically irreversible

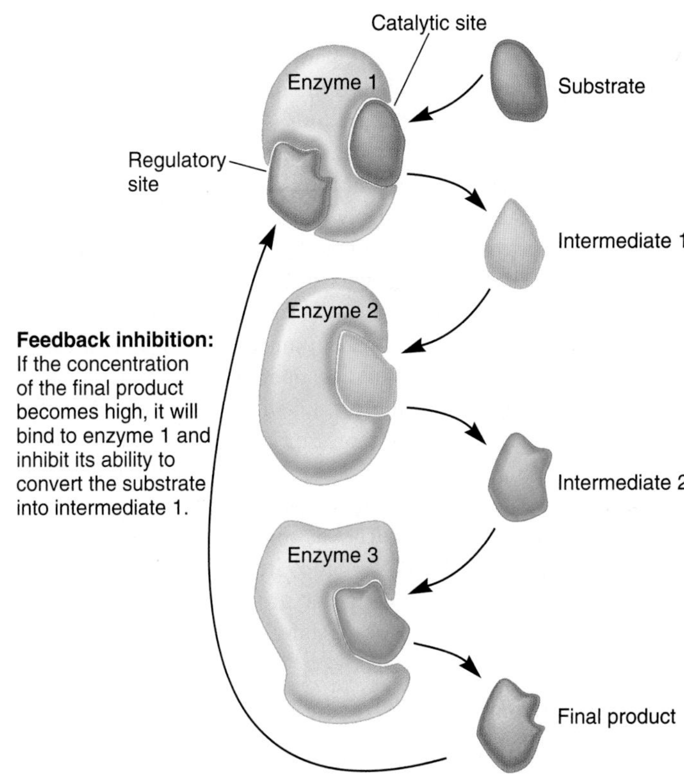

FIGURE 14.17 **Feedback inhibition in a metabolic pathway.** The substrate is converted to a product by the sequential action of three different enzymes. Enzyme 1 has a catalytic site that recognizes the substrate, and it also has a regulatory site that recognizes the final product. When the final product binds to the regulatory site, it inhibits enzyme 1.

changes required to produce a functional protein. In contrast, other types of modifications, such as phosphorylation ($—PO_4$), acetylation ($—COCH_3$), and methylation ($—CH_3$), are often reversible modifications that transiently affect the function of a protein.

14.3 GENE REGULATION IN THE BACTERIOPHAGE REPRODUCTIVE CYCLE

Viruses are small particles that contain genetic material surrounded by a protein coat. They can infect a living cell and then propagate themselves by using the energy and metabolic machinery of the host cell. For this to occur, the genetic material of viruses orchestrates an intricate series of steps, involving the expression of many viral genes. During the past several decades, the reproduction of viruses has presented an interesting and challenging problem for geneticists to investigate. The study of bacteriophages, which are viruses that infect bacteria, has greatly advanced our basic knowledge of how genetic regulatory proteins work. In addition, the study of viruses has been instrumental in our ability to devise medical strategies aimed at combating viral diseases. For example, our knowledge of the reproductive cycles of human viruses has led to the development of drugs that inhibit viral growth. Azidothymidine (AZT), which is used to combat human immunodeficiency virus (HIV), suppresses the production of viral DNA by inhibiting a viral gene product called reverse transcriptase that is involved in viral DNA synthesis.

In this section, we will focus on the function of bacteriophage genes that encode genetic regulatory proteins. The structural genes of bacteriophages are often arranged in operons. This enables all of the genes within an operon to be controlled by regulatory proteins that bind to operator sites and influence the function of nearby promoters. Like bacterial operons, phage operons can be controlled by repressor proteins or activator proteins. To understand how this works, we will carefully examine the two reproductive cycles—lytic and lysogenic—of a virus called phage λ (lambda), which was discovered by André Lwoff and his colleagues in the 1940s. Since its discovery, phage λ has been investigated extensively and has provided geneticists with a model on which to base our understanding of viral proliferation.

Phage λ Can Follow a Lytic or Lysogenic Cycle

Phage λ can bind to the surface of a bacterium and inject its genetic material into the bacterial cytoplasm. After this occurs, the phage proceeds along only one of two alternative cycles, known as the **lytic cycle** and the **lysogenic cycle.** This topic is also discussed in Chapter 6 (see Figure 6.9). During the lytic cycle, the genetic instructions of the bacteriophage direct the synthesis of many copies of the phage genetic material and coat proteins that are then assembled to make new phages. When synthesis and assembly are completed, the bacterial host cell is lysed, and the newly made phages are released into the environment.

Alternatively, in the lysogenic cycle, the phage can act as a **temperate phage,** which will usually not produce new phages and will not kill the bacterial cell that acts as its host. During the lysogenic cycle, phage λ integrates its genetic material into the chromosome of the bacterium. This integrated phage DNA is known as a **prophage.** A prophage can exist in a dormant state for a long time, during which no new bacteriophages are made. When a bacterium containing a lysogenic prophage divides to produce two daughter cells, it will copy the prophage's genetic material along with its own chromosome. Therefore, both daughter cells inherit the prophage. At some later time, in a process called induction, a prophage may become activated to excise itself from the bacterial chromosome and enter the lytic cycle. During induction, the phage promotes the synthesis of new phages and eventually lyses the host cell.

Inside the virus's head, phage λ DNA is linear. After injection into the bacterium, the two ends of the DNA become covalently attached to each other to form a circular piece of DNA. **Figure 14.18** shows the genome of phage λ. The organization of

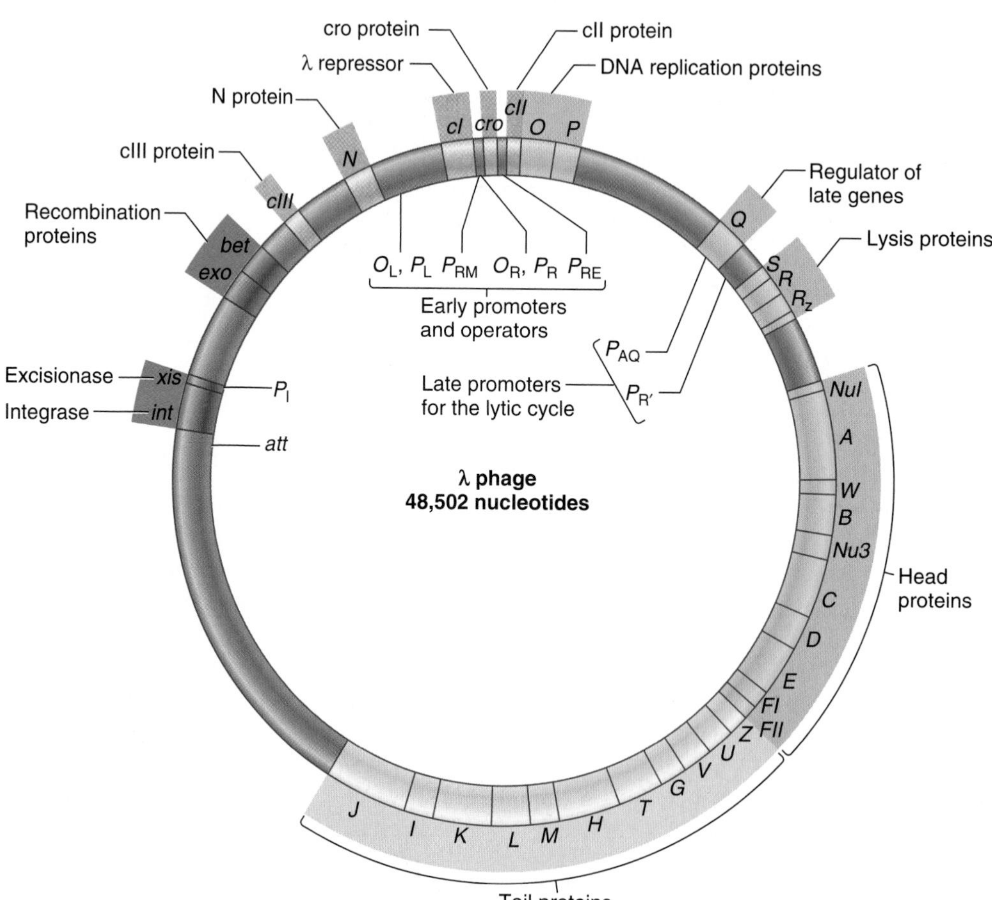

FIGURE 14.18 The genome of phage λ. The genes shaded in orange encode genetic regulatory proteins that determine whether the lysogenic or lytic cycle prevails. Genes shaded in green encode proteins necessary for the lysogenic cycle. The genes shaded in dark and light tan encode proteins required for the lytic cycle.

the genes within this circular structure reflects the two alternative cycles of this virus. The genes in the top center of the figure are transcribed very soon after infection. This occurs at the beginning of either reproductive cycle. As we will see, the expression pattern of these early genes determines whether the lytic or lysogenic cycle prevails.

The genes on the upper left side of Figure 14.18 encode proteins needed for the lysogenic cycle, while the genes on the right side and bottom promote the lytic cycle. If the lysogenic cycle prevails, the integrase (*int*) gene is subsequently turned on. The integrase gene encodes an enzyme that integrates the λ DNA into the bacterial chromosome. (The mechanism of phage λ integration is described in Chapter 17.) If the lysogenic cycle is not chosen, the genes on the right side and bottom of the figure will be transcribed. These genes are necessary for λ assembly and release. They encode replication proteins, coat proteins, proteins involved in coat assembly, proteins involved in packaging the DNA into the phage head, and enzymes that cause the bacterium to lyse.

During the Lysogenic Cycle, the cII Protein Activates Expression of the λ Repressor

Now that we have an understanding of the phage λ reproductive cycles and genome organization, let's examine how the decision is made between the lytic and lysogenic cycles. This choice depends on the actions of several genetic regulatory proteins. Our molecular understanding of the phage λ cycles represents an extraordinary accomplishment in the field of genetic regulation. This process is quite detailed because it involves a series of intricate steps in which regulatory proteins bind to several different sites in the λ genome. To simplify these events, we will begin by considering the steps that occur when the lysogenic cycle prevails.

Soon after the λ DNA enters the bacterial cell, two promoters, designated P_L and P_R, are used for transcription. This initiates a competition between the lytic and lysogenic cycles (**Figure 14.19**). Initially, transcription from P_L and P_R results in the synthesis of two short RNA transcripts that encode two proteins called the N protein and the cro protein, both of which are genetic regulatory proteins. The N protein is a genetic regulatory protein with an interesting function that we have not yet considered. Its function, known as **antitermination,** is to prevent transcriptional termination. The N protein inhibits termination at three sites, designated t_L, t_{R1}, and t_{R2}. The N protein actually binds to RNA polymerase and prevents transcriptional termination when these sites are being transcribed. When the N protein prevents termination at t_{R1} and t_{R2}, the transcript from P_R is extended to include the *cII*, *O*, *P*, and *Q* genes. The *cII* gene encodes an activator protein, the *O* and *P* genes encode enzymes needed for the initiation of λ DNA synthesis, and the *Q* gene encodes another antiterminator that is required for the lytic cycle. When the N protein prevents termination at t_L, the transcript from P_L is extended to include the *int*, *xis*, and *cIII* genes. The *int* gene encodes integrase, which is involved with integrating λ DNA into the *E. coli* chromosome, and the *xis* gene encodes excisionase, which can excise the λ DNA if a switch is made from the lysogenic to the lytic cycle. The *cIII* gene encodes the cIII protein, which forms a complex with cII that helps to stabilize the cII activator protein and make it less vulnerable to protease digestion.

As shown on the left side of Figure 14.19, if the cII-cIII complex accumulates to sufficient levels, the lysogenic cycle is favored. Once it is made, the cII protein activates two different promoters in the λ genome. When the cII protein binds to the promoter P_{RE}, it turns on the transcription of *cI*, a gene that encodes the λ repressor. The cII protein also activates the *int* gene by binding to the promoter P_I. The λ repressor and integrase proteins play central roles in promoting the lysogenic cycle. When the λ repressor is made in sufficient quantities, it binds to operator sites (O_L and O_R) that are adjacent to P_L and P_R. When the λ repressor is bound to O_R, it inhibits the expression of the genes required for the lytic cycle.

As you are looking at the left side of Figure 14.19, you may have noticed that the binding of the λ repressor to O_R will inhibit the expression of *cII*. This may seem counterintuitive, because the cII protein was initially required to activate the *cI* gene that encodes the λ repressor. You may be thinking that the inhibition of the *cII* gene will eventually prevent the expression of the *cI* gene and ultimately stop the synthesis of the λ repressor protein. What prevents the inhibition of *cI* gene expression? The explanation is that the *cI* gene actually has two promoters: P_{RE}, which is activated by the cII protein, and a second promoter called P_{RM}. Transcription from P_{RE} occurs early in the lysogenic cycle. P_{RE} gets its name because the use of this promoter results in the expression of the λ <u>R</u>epressor during the <u>E</u>stablishment of the lysogenic cycle. The transcript made from the use of P_{RE} is very stable and quickly leads to a buildup of the λ repressor protein. This causes an abrupt inhibition of the lytic cycle because the binding of the λ repressor protein to O_R blocks the P_R promoter. Later in the lysogenic cycle, it is no longer necessary to make a large amount of the λ repressor. At this point, the use of the P_{RM} promoter is sufficient to make enough <u>R</u>epressor protein to <u>M</u>aintain the lysogenic cycle. Interestingly, the P_{RM} promoter is activated by the λ repressor protein. The λ repressor was named when it was understood that it repressed the lytic cycle. Later studies revealed that it also activates its own transcription from P_{RM}.

The Lytic Cycle Depends on the Action of the Cro Protein

As we have just seen, the λ repressor protein binds to O_R and prevents the expression of the operons needed for the lytic cycle. For the lytic cycle to occur, the λ repressor must be prevented from inhibiting P_R. This is the role of the cro protein. If the activity of the cro protein exceeds the activity of the cII protein, the lytic cycle prevails (see the right side of Figure 14.19). As was mentioned, an early step in the expression of λ genes is the transcription from P_R to produce the cro protein. If the concentration of the cro protein builds to sufficient levels, it will bind to two operator regions, O_L and O_R. The binding of cro to O_L inhibits transcription from P_L; the binding of cro to O_R has several effects. When the cro protein binds to O_R, it inhibits transcription from P_{RM} in the leftward direction. This inhibition prevents the expression of the *cI* gene, which encodes the λ repressor; the λ repressor is

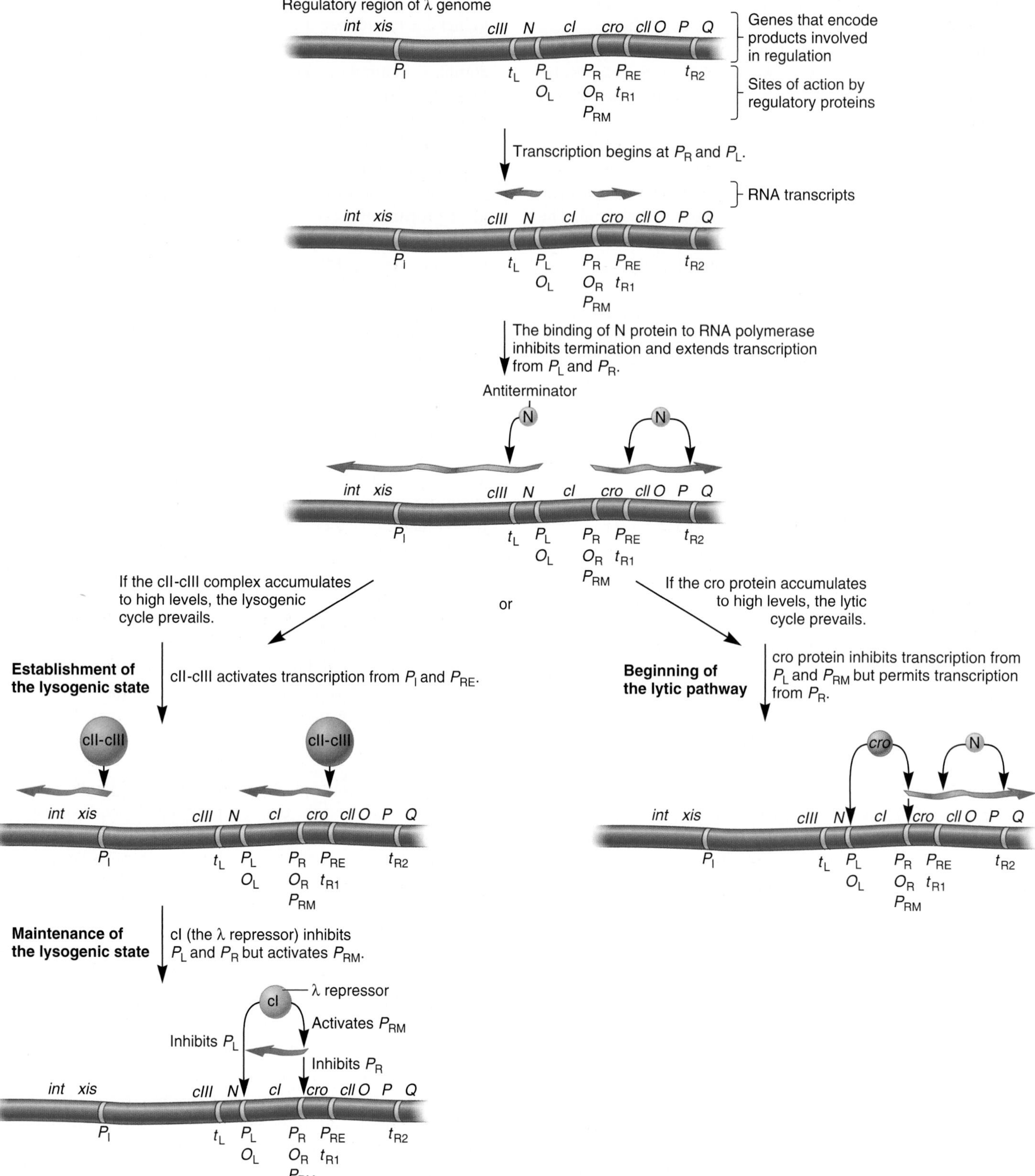
Regulatory region of λ genome
int xis cIII N cI cro cII O P Q
Genes that encode products involved in regulation
P_I t_L P_L O_L P_R O_R P_{RM} P_{RE} t_{R1} t_{R2}
Sites of action by regulatory proteins
Transcription begins at P_R and P_L.
RNA transcripts
The binding of N protein to RNA polymerase inhibits termination and extends transcription from P_L and P_R.
Antiterminator
N
If the cII-cIII complex accumulates to high levels, the lysogenic cycle prevails.
or
If the cro protein accumulates to high levels, the lytic cycle prevails.
Establishment of the lysogenic state
cII-cIII activates transcription from P_I and P_{RE}.
cII-cIII
Beginning of the lytic pathway
cro protein inhibits transcription from P_L and P_{RM} but permits transcription from P_R.
cro
Maintenance of the lysogenic state
cI (the λ repressor) inhibits P_L and P_R but activates P_{RM}.
cI
λ repressor
Activates P_{RM}
Inhibits P_L
Inhibits P_R

FIGURE 14.19 The sequence of events that occur during the lysogenic and lytic cycles of phage λ. The top part of this figure shows the region of the phage λ genome that regulates the choice between the lytic and lysogenic cycles. DNA is shown in blue. The names of genes that encode proteins are shown above the DNA. Promoters and operator sites are shown below the DNA. RNA transcripts are shown in red or green. The key regulatory proteins are indicated as spheres.

Immediately after infection, P_L and P_R are used to make two short mRNAs. These mRNAs encode two early proteins, designated N and cro. The N protein prevents transcriptional termination at three sites in the RNA (t_L, t_{R1}, and t_{R2}). This allows the transcription of several genes, which include *cIII*, *cII*, O, *P*, and *Q*. The expression of the *O*, *P*, and *Q* genes is necessary only for the lytic cycle. If the lysogenic cycle is chosen, transcription of these genes is abruptly inhibited.

The left side shows the steps that lead to the lysogenic cycle. The cII-cIII protein complex binds to P_{RE}. This activates the transcription of the *cI* gene, which encodes the λ repressor. The cII-cIII complex also binds to P_I to activate the transcription of the *int* gene. The λ repressor binds to O_L and O_R to inhibit transcription from P_L and P_R. This prevents the lytic cycle. The integrase protein catalyzes the integration of the λ DNA into the *E. coli* chromosome. At a later stage in the lysogenic cycle, the *cI* gene is transcribed from the P_{RM} promoter. This results in a low but steady synthesis of the λ repressor, which is necessary to further maintain the lysogenic state.

The right side shows the steps that lead to the lytic cycle. The cro protein binds to O_L and blocks transcription from P_L. Cro also binds to O_R and blocks transcription from P_{RM}. This prevents the synthesis of the λ repressor. Transcription from P_R is still allowed to occur at a low level, which leads to a buildup of the O, P, and Q proteins. The O and P proteins catalyze the replication of additional λ DNA. The Q protein allows transcription through $P_{R'}$ (not shown in this figure, but see Figure 14.18). This leads to the synthesis of many proteins that are necessary to make new λ phages and to lyse the cell.

Genes→Traits The ability to choose between two alternative reproductive cycles can be viewed as a trait of this bacteriophage. As described here, the choice between the two cycles depends on the pattern of gene regulation.

needed to maintain the lysogenic state. Therefore, the λ repressor cannot successfully shut down transcription from P_R.

The binding of the cro protein to O_R also allows a low level of transcription from P_R in the rightward direction. This enables the transcription of the *O*, *P*, and *Q* genes. The O and P proteins are necessary for the replication of the λ DNA. The Q protein is an antiterminator protein that permits transcription through another promoter, designated $P_{R'}$ (not to be confused with P_R). The $P_{R'}$ promoter controls a very large operon that encodes the proteins necessary for the phage coat, the assembly of the coat proteins, the packaging of the λ DNA, and the lysis of the bacterial cell (refer back to Figure 14.18). These proteins are made toward the end of the lytic cycle. The expression of these late genes leads to the synthesis and assembly of many new λ phages that are released from the bacterial cell when it lyses.

Cellular Proteases Influence the Choice Between the Lytic and Lysogenic Cycle

What factors determine whether the lytic or lysogenic cycle prevails? As you may have noticed from Figure 14.19, the first two steps of both cycles are identical. Whether the lytic or lysogenic cycle prevails depends on the steps that occur after the early genes are transcribed. In particular, the activity of the cII protein plays a key role in directing λ to one of the two cycles. A critical physiological issue is that the cII protein is easily degraded by cellular proteases that are produced by *E. coli*. Whether or not these proteases are made depends on the environmental conditions. If the growth conditions are very favorable, such as a rich growth medium, the intracellular protease levels are relatively high, and the cII protein tends to be degraded. When cII protein is degraded, P_{RE} cannot be activated, and therefore, the λ repressor is not made. Instead, the cro protein slowly accumulates to sufficient levels, as in the right side of Figure 14.19. The binding of the cro protein to O_R prevents transcription of the λ repressor gene from P_{RM} and, at the same time, allows the lytic cycle to proceed. In this way, environmental conditions that are favorable for growth promote the lytic cycle. This makes sense, because a sufficient supply of nutrients is necessary to synthesize new bacteriophages.

Alternatively, starvation conditions favor the lysogenic cycle. When nutrients are limiting, cellular proteases are relatively inactive. Under these conditions, the cII protein will build up much more quickly than the cro protein. Therefore, the cII protein will turn on P_{RE} and lead to the transcription of the λ repressor (as in the left side of Figure 14.19). This event favors the lysogenic cycle. From the perspective of the bacteriophage, lysogeny may be a better choice under starvation conditions because nutrients may be insufficient for the production of new λ phages.

After lysogeny is established, certain environmental conditions can also favor induction to the lytic cycle at some later time. For example, exposure to UV light promotes induction. This also is caused by the activation of cellular proteases. In this case, a cellular protein known as recA (a protein ordinarily involved in facilitating recombination between DNA molecules) detects the DNA damage from UV light and is activated to become a mediator of protein cleavage. RecA protein mediates cleavage of the λ repressor and thereby inactivates it. This allows transcription from P_R and eventually leads to the accumulation of the cro protein and thereby favors the lytic cycle. Under these conditions, it may be advantageous for λ to make new phages and lyse the cell, because the exposure to UV light may have already damaged the bacterium to the point where further bacterial growth and division are prevented.

The O_R Region Provides a Genetic Switch Between the Lytic and Lysogenic Cycles

Before we end this section on the λ reproductive cycles, let's consider how the O_R region acts as a genetic switch between the two cycles. Depending on the binding of genetic regulatory proteins to this region, the switch can be turned to favor the lytic or lysogenic cycle. How does this switch work? To understand how, we need to take a closer look at the O_R region (**Figure 14.20**).

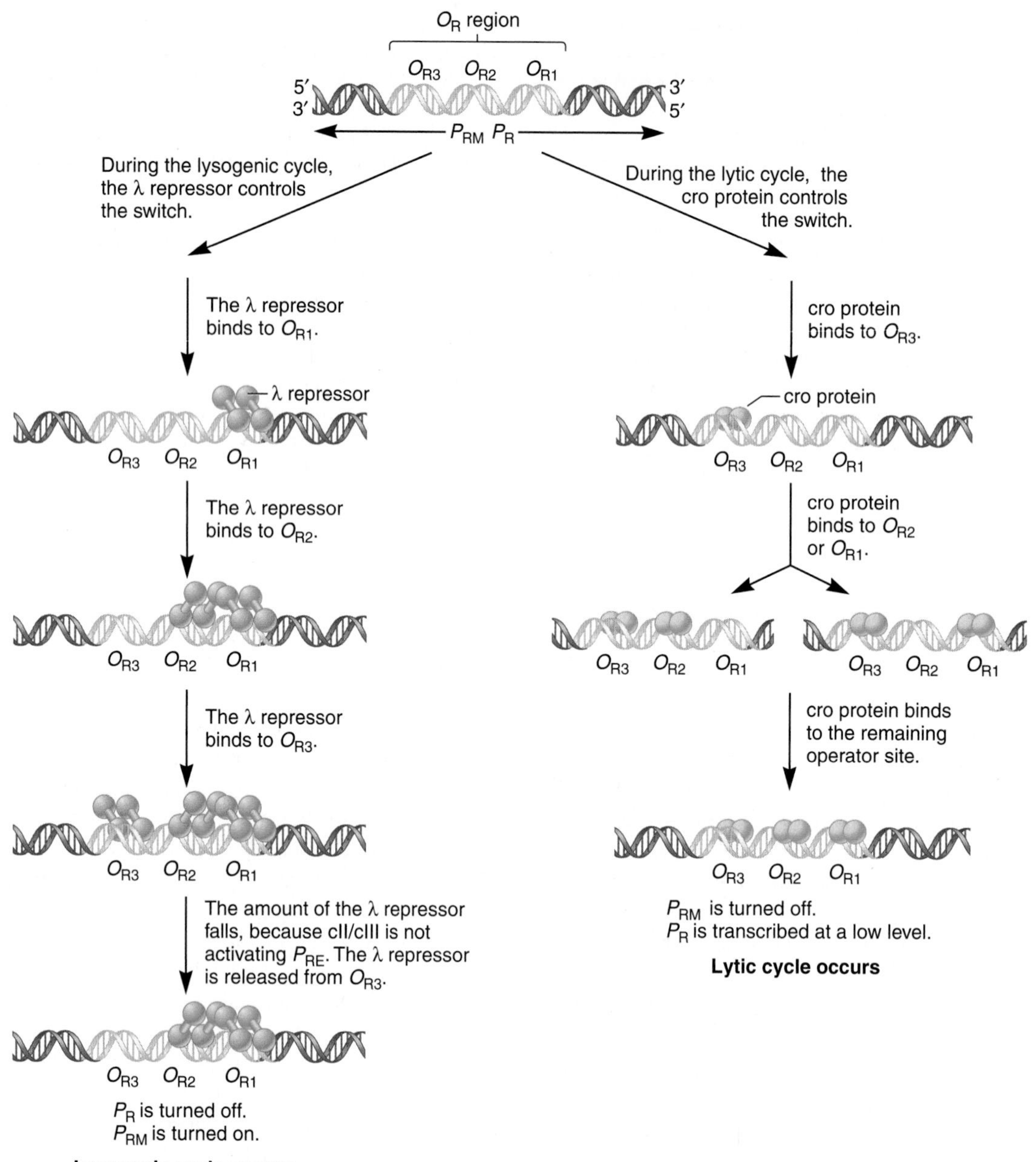

FIGURE 14.20 **The O_R region, the genetic switch between the lysogenic and lytic cycles.** The O_R region contains three operator sites and two promoters, P_{RM} and P_R, which transcribe in opposite directions. The left side of the figure depicts the events that promote the lysogenic cycle. The λ repressor protein is a dimer; each subunit contains two globular domains connected by a short link. The λ repressor protein dimer first binds to O_{R1} because it has the highest affinity for this site. A second λ repressor dimer binds to O_{R2}. This occurs very rapidly, because the binding of the first dimer to O_{R1} favors the binding of a second dimer to O_{R2}. The binding of the λ repressor to O_{R1} and O_{R2} inhibits transcription from P_R and thereby switches off the lytic cycle. Early in the lysogenic cycle, the λ repressor protein concentration may become so high that it occupies O_{R3}. Later, the λ repressor concentration begins to drop, because the inhibition of P_R decreases the synthesis of cII protein, which activates the λ repressor gene (from P_{RE}). As the λ repressor concentration gradually falls, it is first removed from O_{R3}. This allows transcription from P_{RM} and maintains the lysogenic cycle.

The right side depicts events at the O_R region that promote the lytic cycle. The cro repressor protein is a small globular protein that also binds to each operator site as a dimer. The cro protein has its highest affinity for O_{R3}, and so it binds there first. This blocks transcription from P_{RM} and thereby switches off the lysogenic cycle. The cro protein has a similar affinity for O_{R2} and O_{R1}, and so it may occupy either of these sites next. Later in the lytic cycle, the cro protein concentration will continue to rise, so eventually it binds to both O_{R2} and O_{R1}. This turns down the expression from P_R, which is not needed in the later stages of the lytic cycle.

The O_R region contains three operator sites, designated O_{R1}, O_{R2}, and O_{R3}. These operator sites control two promoters called P_{RM} and P_R that transcribe in opposite directions. The λ repressor protein or the cro protein can bind to any or all of the three operator sites. The binding of these two proteins at these sites governs the switch between the lysogenic and lytic cycles. Two critical issues influence this binding event. The first is the relative affinities that the regulatory proteins have for these operator sites. The second is the concentrations of the λ repressor protein and the cro protein within the cell.

Let's consider how an increasing concentration of the λ repressor protein can switch on the lysogenic cycle and switch off the lytic cycle (Figure 14.20, left side). This protein was first isolated by Mark Ptashne and his colleagues in 1967. Their studies showed that λ repressor binds with highest affinity to O_{R1}, followed by O_{R2} and O_{R3}. As the concentration of the λ repressor builds within the cell, a dimer of the λ repressor protein first binds to O_{R1} because it has the highest affinity for this site. Next, a second λ repressor dimer binds to O_{R2}. This occurs very rapidly, because the binding of the first dimer to O_{R1} favors the binding of a second dimer to O_{R2}. This is called a cooperative interaction. The binding of the λ repressor to O_{R1} and O_{R2} inhibits transcription from P_R and thereby switches off the lytic cycle.

Early in the lysogenic cycle, the λ repressor protein concentration may become so high that it occupies O_{R3}. Eventually, however, the λ repressor concentration begins to drop, because the inhibition of P_R decreases the synthesis of cII, which activates the λ repressor gene from P_{RE}. As the λ repressor concentration gradually falls, it is first removed from O_{R3}. This allows transcription from P_{RM}. As mentioned earlier, the term λ repressor is somewhat misleading because the binding of the λ repressor at only O_{R1} and O_{R2} acts as an activator of P_{RM}. The ability of the λ repressor to activate its own transcription allows the switch to the lysogenic cycle to be maintained.

In the lytic cycle (Figure 14.20, right side), the binding of the cro protein controls the switch. The cro protein has its highest affinity for O_{R3} and has a similar affinity for O_{R2} and O_{R1}. Under conditions that favor the lytic cycle, the cro protein accumulates, and a cro dimer first binds to O_{R3}. This blocks transcription from P_{RM} and thereby switches off the lysogenic cycle. Later in the lytic cycle, the cro protein concentration continues to rise, so eventually it binds to O_{R2} and O_{R1}. This turns down expression from P_R, which is not needed in the later stages of the lytic cycle.

Genetic switches, like the one just described for phage λ, represent an important form of genetic regulation. As we have just seen, a genetic switch can be used to control two alternative reproductive cycles of a bacteriophage. In addition, genetic switches are also important in the developmental pathways for bacteria and eukaryotes. For example, certain species of bacteria can grow in a vegetative state when nutrients are abundant but will produce spores when conditions are unfavorable for growth. The choice between vegetative growth and sporulation involves genetic switches. Likewise, genetic switches operate in the developmental pathways in eukaryotes. As we will discover in Chapter 23, they are key events in the initiation of cell differentiation during development. Studies of the phage λ life cycle have provided fundamental information with which to understand how these other switches can operate at the molecular level.

CONCEPTUAL SUMMARY

Genes can be regulated at any step in the pathway of gene expression. In bacterial cells, it is common for structural genes to be organized into a multigene unit called an operon. An operon typically consists of a promoter, an operator site, two or more structural genes, and a terminator. The organization of an operon allows two or more genes to be regulated as a single unit. In most cases, transcriptional regulation involves the actions of regulatory proteins that bind to the operon and can greatly influence the rate of transcription. A repressor is a protein that inhibits transcription, while an activator is a protein that increases the rate of transcription. Furthermore, the binding of small effector molecules to regulatory proteins can influence whether or not the regulatory proteins can bind to the DNA. Some regulatory proteins that exert negative control, such as the lac repressor and trp repressor, inhibit transcription. Others, such as the catabolite activator protein (CAP), exert positive control by enhancing transcription. The AraC protein is interesting because it can act as both a repressor and an activator, depending on the presence or absence of arabinose. Experiments involving the use of F' factors showed that regulatory proteins are synthesized and can diffuse within the cell to ultimately bind to a distant operator site and cause repression; this form of genetic regulation is called a *trans*-effect. In contrast, a *cis*-effect on gene expression is due to DNA sequences adjacent to the gene.

Our understanding of transcriptional regulation has been greatly aided by studying the reproductive cycles of viruses. In this chapter, we considered the ability of phage λ to proceed along one of two alternative cycles: the lytic and lysogenic cycles. This choice is determined by the actions of several genetic regulatory proteins that can bind to the λ DNA and influence the transcription of nearby genes. The O_R region provides a genetic switch for the lytic or lysogenic cycle. If the λ repressor controls the switch, the lysogenic cycle is favored, whereas binding of the cro protein to this switch favors the lytic cycle. Cellular proteases, which are influenced by environmental conditions, play a key role in the choice between the lytic and lysogenic cycles.

In addition to direct transcriptional regulation, gene expression in bacteria can be affected at later stages in the expression process. During attenuation of the *trp* operon, for example, transcription actually begins, but it is terminated before the entire mRNA is made; the length of the transcript is determined by the translation of the *trpL* gene. This form of regulation is unique to bacteria because they couple the processes of transcription and translation. The translation of a complete mRNA transcript can also be regulated. One form of translational regulation involves the binding of translational repressor proteins that prevent translational initiation. A second form involves the binding of antisense RNA to an mRNA to block translation. Finally, the function of proteins that have already been made can be influenced posttranslationally. Feedback inhibition involves the noncovalent binding of molecules to an allosteric site on a protein. This inhibition blocks the function of an enzyme that is required during an early step in a metabolic pathway. In addition, the function of proteins can be controlled by irreversible and reversible covalent modifications.

EXPERIMENTAL SUMMARY

Early insights into the molecular mechanisms of gene expression came from experiments involving lactose metabolism in *E. coli.* Jacob and Monod examined the phenomenon of enzyme adaptation and concluded that the proteins involved in lactose metabolism were made after the bacterium was exposed to lactose. This observation led them to investigate the genes involved in this regulation process. By constructing merozygotes, strains of bacteria containing F' factor genes, they discovered that the *lacI* gene encodes a repressor protein that in the absence of lactose is bound to the operator site. More recently, studies by Müller-Hill revealed the existence of three operator sites in the *lac* operon. The crystal structure of the lac repressor tetramer bound to two operator sites has been determined.

A different form of genetic regulation, known as attenuation, was discovered by Yanofsky and colleagues. They identified mutant strains of bacteria that were missing the trp repressor yet still could repress the *trp* operon when tryptophan levels were high. Additional mutations in the *trpL* region destroyed the ability to attenuate transcription. An analysis of the DNA sequence in this region revealed the ability of the mRNA to form alternative stem-loop structures. Of the three possible stem-loop structures, only the 3–4 stem-loop structure acts as a transcription terminator and thereby prevents the further transcription of the *trp* operon.

An analysis of genetic sequences also revealed a translational method of gene regulation involving the synthesis of antisense RNA. In *E. coli,* a gene known as *micF* inhibits expression of the *ompF* gene. Researchers discovered that the *micF* RNA is complementary to the *ompF* mRNA. The binding of the *micF* RNA (the antisense RNA) to *ompF* mRNA (the sense strand of mRNA that encodes a protein) inhibits the translation of the *ompF* mRNA.

We concluded the chapter with a description of gene regulation in bacteriophage λ. Many of the same kinds of experimental approaches were used to elucidate the molecular mechanisms that underlie the choice between the lysogenic and lytic cycles. In particular, many researchers have studied how mutations in particular genes in the λ genome favor the lysogenic or lytic cycles. In addition, biochemical and biophysical studies have analyzed the detailed interactions between λ regulatory proteins and the operator sites that they recognize.

PROBLEM SETS & INSIGHTS

Solved Problems

S1. Researchers have identified mutations in the promoter region of the *lacI* gene that make it more difficult for the *lac* operon to be induced. These are called *lacI*^{Q} mutants, because a greater Quantity of lac repressor protein is made. Explain why an increased transcription of the *lacI* gene makes it more difficult to induce the *lac* operon.

Answer: An increase in the amount of lac repressor protein makes it easier to repress the *lac* operon. When the cells become exposed to lactose, allolactose levels slowly rise. Some of the allolactose binds to the lac repressor protein and causes it to be released from the operator site. If many more lac repressor proteins are found within the cell, more allolactose is needed to ensure that no unoccupied repressor proteins can repress the operon.

S2. Explain how the pausing of the ribosome in the presence or absence of tryptophan affects the formation of a terminator stem-loop.

Answer: The key issue is the location where the ribosome stalls. In the absence of tryptophan, it stalls over the Trp codons in the *trpL* mRNA. Stalling at this site shields region 1 in the attenuator region. Because region 1 is unavailable to hydrogen bond with region 2, region 2 hydrogen bonds with region 3. Therefore, region 3 cannot form a terminator stem-loop with region 4. Alternatively, if tryptophan levels in the cell are sufficient, the ribosome pauses over the stop codon in the *trpL* RNA. In this case, the ribosome shields region 2. Therefore, regions 3 and 4 hydrogen bond with each other to form a terminator stem-loop, which abruptly halts the continued transcription of the *trp* operon.

S3. With regard to the key proteins that affect the choice between the lytic and lysogenic cycles, which one(s) may be degraded by cellular proteases?

Answer: After infection, the key protein affected by cellular proteases is cII. If protease levels are high, as under good growth conditions, cII is degraded. This promotes the lytic cycle. Under starvation conditions, the protease levels are low. This prevents the degradation of cII and thereby promotes the lysogenic cycle. After lysogeny has been established, the key protein affected by cellular proteases is the λ repressor. Agents such as UV light activate cellular proteases that digest the λ repressor. This permits induction of the lytic cycle.

S4. In bacteria, it is common for two or more structural genes to be arranged together in an operon. Discuss the arrangement of genetic sequences within an operon. What is the biological advantage of an operon organization?

Answer: An operon contains several different DNA sequences that play specific roles. For transcription to take place, an operon is flanked by a promoter to signal the beginning of transcription and a terminator to signal the end of transcription. Two or more structural genes that encode different proteins are found between these two sequences. A key feature of an operon is that the expression of the structural genes occurs as a unit. When transcription takes place, a polycistronic mRNA is made that encodes all of the structural genes.

In order to control the ability of RNA polymerase to transcribe an operon, an additional DNA sequence, known as the operator site, is usually present. The base sequence within the operator site can serve as a binding site for genetic regulatory proteins called activators or repressors. The advantage of an operon organization is that it allows a bacterium to coordinately regulate a group of genes whose encoded proteins have a common function. For example, an operon may contain a group of genes involved in lactose breakdown or a group of genes involved in tryptophan synthesis. The genes within an operon usually encode proteins within a common metabolic pathway or cellular function.

S5. The sequential use of two sugars by a bacterium is known as diauxic growth. It is a common phenomenon among many bacterial species. When glucose is one of the two sugars available, it is typical that the bacterium metabolizes glucose first, and then a second sugar after the glucose has been used up. Among *E. coli* and related species, diauxic growth is regulated by intracellular cAMP levels and the catabolite activator protein (CAP). Summarize the effects of glucose and lactose on the ability of the lac repressor and the cAMP-CAP complex to regulate the *lac* operon.

Answer: In the absence of lactose, the lac repressor has the dominating effect of shutting off the *lac* operon. Even when glucose is also absent and the cAMP-CAP complex is formed, the presence of the bound lac repressor prevents the expression of the *lac* operon. The effects of the cAMP-CAP complex are exerted only in the presence of lactose. When lactose is present and glucose is absent, the cAMP-CAP complex acts to enhance the rate of transcription. However, when both lactose and glucose are present, the inability of CAP to bind to the *lac* operon decreases the rate of transcription. The table shown here summarizes these effects.

***lac* Operon Regulation**

Sugar Present	Transcription of the *lac* Operon
None	The operon is turned off due to the dominating effect of the lac repressor protein.
Lactose	The operon is maximally turned on. The repressor protein is removed from the operator site, and the cAMP-CAP complex is bound to the CAP site.
Glucose	The operon is turned off due to the dominating effect of the lac repressor protein.
Lactose and glucose	The expression of the *lac* operon is greatly decreased. The lac repressor is removed from the operator site, and the catabolite activator protein is not bound to the CAP site. The absence of CAP at the CAP site makes it difficult for RNA polymerase to begin transcription. However, a little more transcription occurs under these conditions than in the absence of lactose, when the repressor is bound.

S6. The ability of DNA-binding proteins to promote a loop in DNA structure is an interesting phenomenon that is important in the structure and function of DNA. Besides the regulation of genes and operons, DNA looping is required in the compaction of DNA within the nucleoid of a bacterium and the nucleus of a eukaryotic cell (see Chapter 10). In addition, DNA looping is frequently involved in the expression of eukaryotic genes (see Chapter 15). In this solved problem, we will examine an experimental approach that made it possible for Robert Schleif and his colleagues to determine that the AraC protein causes a loop to form in the DNA. This work relied on the mobility of DNA in an acrylamide gel. A segment of DNA that contains a loop is more compact than the same DNA segment without a loop. Therefore, when these two alternative structures (looped versus unlooped) are run through a gel, the looped structure migrates more quickly to the bottom of the gel because it can more easily penetrate the gel matrix. As a starting material, Schleif had a sample of DNA that contained a portion of the *ara* operon including both the *araI* and *araO*$_2$ sites. In the gel show here, this segment of DNA was exposed to the following conditions before it was run on the gel:

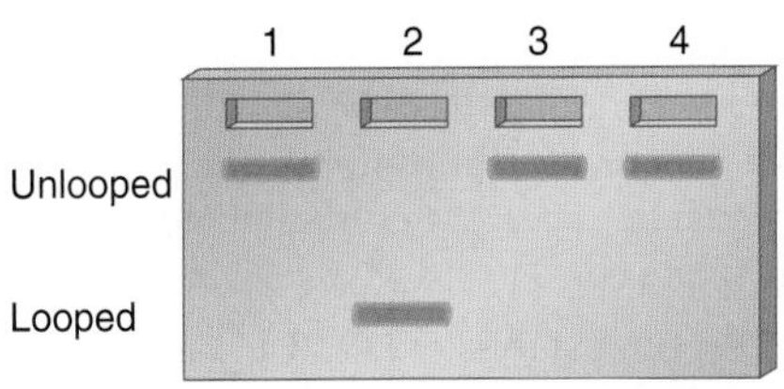

Explain these results.

Answer: In lane 2, AraC protein was added and arabinose was not. As expected from the DNA looping hypothesis, a DNA loop was formed as evidenced from the faster mobility on the acrylamide gel. In lane 1, no AraC protein was added, so no DNA loop was able to form. These results confirm the idea that the AraC protein causes a loop to form in the sample loaded into lane 2. In lane 3, only the sugar arabinose is present but not the AraC protein. Because the DNA remains unlooped, the sugar by itself has no effect. In the sample loaded into lane 4, AraC protein was added, and arabinose was added as well. In this case, no DNA loop is observed. These results are consistent with the idea that arabinose binds to the AraC protein and breaks the DNA loop that is promoted by the AraC protein.

Conceptual Questions

C1. What is the difference between a constitutive gene and a regulated gene?

C2. In general, why is it important to regulate genes? Discuss examples of situations in which it would be advantageous for a bacterial cell to regulate genes.

C3. If a gene is repressible and under positive control, describe what kind of effector molecule and regulatory protein are involved. Explain how the binding of the effector molecule affects the regulatory protein.

C4. Transcriptional regulation often involves a regulatory protein that binds to a segment of DNA and a small effector molecule that binds to the regulatory protein. Do the following terms apply to a regulatory protein, a segment of DNA, or a small effector molecule?

A. Repressor
B. Inducer
C. Operator site
D. Corepressor
E. Activator
F. Attenuator
G. Inhibitor

C5. An operon is repressible—a small effector molecule turns off transcription. Which combinations of small effector molecules and regulatory proteins could be involved?

A. An inducer plus a repressor

B. A corepressor plus a repressor

C. An inhibitor plus an activator

D. An inducer plus an activator

C6. Some mutations have a *cis*-effect, whereas others have a *trans*-effect. Explain the molecular differences between *cis*- and *trans*-mutations. Which type of mutation (*cis* or *trans*) can be complemented in a merozygote experiment?

C7. What is enzyme adaptation? From a genetic point of view, how does it occur?

C8. In the *lac* operon, how would gene expression be affected if one of the following segments were missing?

A. *lac* operon promoter

B. Operator site

C. *lacA* gene

C9. If an abnormal repressor protein could still bind allolactose but the binding of allolactose did not alter the conformation of the repressor protein, how would this affect the expression of the *lac* operon?

C10. What is diauxic growth? Explain the roles of cAMP and the catabolite activator protein in this process.

C11. Mutations may have an effect on the expression of the *lac* operon, the *ara* operon, and the *trp* operon. Would the following mutations have a *cis*- or *trans*-effect on the expression of the structural genes in the operon?

A. A mutation in the operator site that prevents the lac repressor from binding to it

B. A mutation in the *lacI* gene that prevents the lac repressor from binding to DNA

C. A mutation in the *araC* gene that prevents two AraC proteins from binding to each other and forming a loop

D. A mutation in *trpL* that prevents attenuation

C12. Would a mutation that inactivated the lac repressor and prevented it from binding to the *lac* operator site result in the constitutive expression of the *lac* operon under all conditions? Explain. What is the disadvantage to the bacterium of having a constitutive *lac* operon?

C13. Describe the function of the AraC protein. How does it positively and negatively regulate the *ara* operon?

C14. Explain how a mutation would affect the regulation of the *ara* operon if the mutation prevented AraC protein from binding to the following sites:

A. $araO_2$

B. $araO_1$

C. *araI*

D. $araO_2$ and *araI*

C15. What is meant by the term attenuation? Is it an example of gene regulation at the level of transcription or translation? Explain your answer.

C16. As described in Figure 14.14, four regions within the *trpL* gene can form stem-loop structures. Let's suppose that mutations have been previously identified that prevent the ability of a particular region to form a stem-loop structure with a complementary region. For example, a region 1 mutant cannot form a 1–2 stem-loop structure, but it can still form a 2–3 or 3–4 structure. Likewise, a region 4 mutant can form a 1–2 or 2–3 stem-loop but not a 3–4 stem-loop. Under the following conditions, would attenuation occur?

A. Region 1 is mutant, tryptophan is high, and translation is not occurring.

B. Region 2 is mutant, tryptophan is low, and translation is occurring.

C. Region 3 is mutant, tryptophan is high, and translation is not occurring.

D. Region 4 is mutant, tryptophan is low, and translation is not occurring.

C17. As described in Chapter 13, enzymes known as aminoacyl-tRNA synthetases are responsible for attaching amino acids to tRNAs. Let's suppose that tryptophanyl-tRNA synthetase was partially defective at attaching tryptophan to tRNA; its activity was only 10% of that found in a normal bacterium. How would that affect attenuation of the *trp* operon? Would it be more or less likely to be attenuated? Explain your answer.

C18. The 3–4 stem-loop and U-rich attenuator found in the *trp* operon (see Figure 14.14) is an example of ρ-independent termination. The function of ρ-independent terminators is described in Chapter 12. Would you expect attenuation to occur if the tryptophan levels were high and mutations at the end of the *trpL* gene changed the UUUUUUUU sequence to UGGUUGUC? Explain why or why not.

C19. Mutations in tRNA genes can create tRNAs that recognize stop codons. Because stop codons are sometimes called nonsense codons, these types of mutations that affect tRNAs are called nonsense suppressors. For example, a normal $tRNA^{gly}$ has an anticodon sequence CCU that recognizes a glycine codon in mRNA (GGA) and puts in a glycine during translation. However, a mutation in the gene that encodes $tRNA^{gly}$ could change the anticodon to ACU. This mutant $tRNA^{gly}$ would still carry glycine, but it would recognize the stop codon UGA. Would this mutation affect attenuation of the *trp* operon? Explain why or why not. Note: To answer this question, you need to look carefully at Figure 14.14 and see if you can identify any stop codons that may exist beyond the UGA stop codon that is found after region 1.

C20. Translational control is usually aimed at preventing the initiation of translation. With regard to cellular efficiency, why do you think this is the case?

C21. What is antisense RNA? How does it affect the translation of a complementary mRNA?

C22. A species of bacteria can synthesize the amino acid histidine so that it does not require histidine in its growth medium. A key enzyme, which we will call histidine synthetase, is necessary for histidine biosynthesis. When these bacteria are given histidine in their growth media, they stop synthesizing histidine intracellularly. Based on this observation alone, propose three different regulatory mechanisms to explain why histidine biosynthesis ceases when histidine is in the growth medium. To explore this phenomenon further, you measure the amount of intracellular histidine synthetase protein when cells are grown in the presence and absence of histidine. In both conditions, the amount of this protein is identical. Which mechanism of regulation would be consistent with this observation?

C23. Using three examples, describe how allosteric sites are important in the function of genetic regulatory proteins.

C24. In what ways are the actions of the lac repressor and trp repressor similar and different? In other words, discuss the similarities and differences with regard to their binding to operator sites, their effects on transcription, and the influences of small effector molecules.

C25. Transcriptional repressor proteins (e.g., lac repressor), antisense RNA, and feedback inhibition are three different mechanisms that can turn off the expression of genes and gene products. Which of these three mechanisms would be most effective in each of the following situations?

A. Shutting down the synthesis of a polypeptide

B. Shutting down the synthesis of mRNA

C. Shutting off the function of a protein

For your answers in parts A–C that have more than one mechanism, which mechanism would be the fastest or the most efficient?

C26. What are key features that distinguish the lytic and lysogenic cycles?

C27. With regard to promoting the lytic or lysogenic cycle, what would happen if the following genes were missing from the λ genome?

A. *cro*

B. *cI*

C. *cII*

D. *int*

E. *cII* and *cro*

C28. How do the λ repressor and the cro protein affect the transcription from P_R and P_{RM}? Explain where these proteins are binding to cause their effects.

C29. In your own words, explain why it is necessary for the *cI* gene to have two promoters. What would happen if it had only P_{RE}?

C30. A mutation in P_R causes its transcription rate to be increased 10-fold. Do you think that this mutation would favor the lytic or lysogenic cycle? Explain your answer.

C31. When an *E. coli* bacterium already has a λ prophage integrated into its chromosome, another λ phage cannot usually infect the cell and establish the lysogenic or lytic cycle. Based on your understanding of the genetic regulation of the λ life cycles, why do you think the other phage would be unsuccessful?

C32. If a bacterium were exposed to a drug that inhibited the N protein, what would you expect to happen if the bacterium was later infected by phage λ? Would phage λ follow the lytic cycle, the lysogenic cycle, or neither? Explain your answer.

C33. Figure 14.20 shows a genetic switch that controls the choice between the lytic and lysogenic cycles of phage λ. What is a genetic switch? Compare the roles of a genetic switch and a simple operator site (like the one found in the *lac* operon) in gene regulation.

C34. This question combines your knowledge of conjugation (described in Chapter 6) and the genetic regulation that directs the phage λ reproductive cycles. When donor *Hfr* strains conjugate with recipient F^- bacteria that are lysogenic for phage λ, the conjugated cells survive normally. However, if donor *Hfr* strains that are lysogenic for phage λ conjugate with recipient F^- bacteria that do not contain any phage λ, the conjugated cells often lyse, due to the induction of λ into the lytic cycle. Based on your knowledge of the regulation of the two λ cycles, explain this observation.

Experimental Questions

E1. Answer the following questions that pertain to the experiment of Figure 14.7.

A. Why was β-ONPG used? Why was no yellow color observed in one of the four tubes? Can you propose alternative methods to measure the level of expression of the *lac* operon?

B. The optical density values were twice as high for the mated strain as for the parent strain. Why was this result obtained?

E2. Chapter 18 describes a blotting method known as Northern blotting, which can be used to detect RNA transcribed from a particular gene or a particular operon. In this method, a specific RNA is detected by using a short segment of cloned DNA as a probe. The DNA probe, which is radioactive, is complementary to the RNA that the researcher wishes to detect. After the radioactive probe DNA binds to the RNA within a blot of a gel, the RNA is visualized as a dark (radioactive) band on an X-ray film. For example, a DNA probe complementary to the mRNA of the *lac* operon could be used to specifically detect the *lac* operon mRNA on a gel blot. As shown here, the method of Northern blotting can be used to determine the amount of a particular RNA transcribed under different types of growth conditions. In this Northern blot, bacteria containing a normal *lac* operon were grown under different types of conditions, and then the mRNA was isolated from the cells and subjected to a Northern blot, using a probe that is complementary to the mRNA of the *lac* operon.

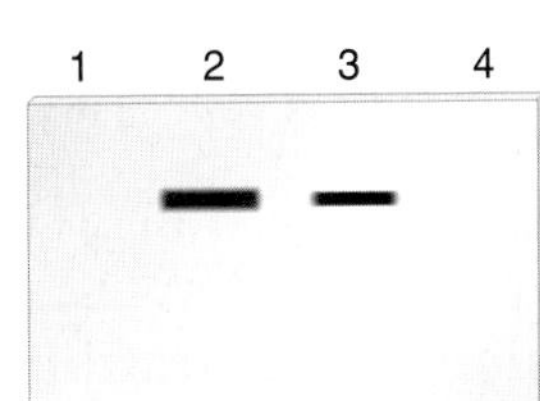

Lane 1. Growth in media containing glucose
Lane 2. Growth in media containing lactose
Lane 3. Growth in media containing glucose and lactose
Lane 4. Growth in media that doesn't contain glucose or lactose

Based on your understanding of the regulation of the *lac* operon, explain these results. Which is more effective at shutting down the *lac* operon, the binding of the *lac* repressor or the removal of CAP? Explain your answer based on the results shown in the Northern blot.

E3. As described in experimental question E2 and also in Chapter 18, the technique of Northern blotting can be used to detect the transcription of RNA. Draw the results you would expect from a Northern blot if bacteria were grown in media containing lactose (and no glucose) but had the following mutations:

Lane 1. Normal strain

Lane 2. Strain with a mutation that inactivates the lac repressor

Lane 3. Strain with a mutation that prevents allolactose from binding to the lac repressor

Lane 4. Strain with a mutation that inactivates CAP

How would your results differ if these bacterial strains were grown in media that did not contain lactose or glucose?

E4. An absentminded researcher follows the protocol described in Figure 14.7 and (at the end of the experiment) does not observe any yellow color in any of the tubes. Yikes! Which of the following mistakes could account for this observation?

A. Forgot to sonicate the cells

B. Forgot to add lactose to two of the tubes

C. Forgot to add β-ONPG to the four tubes

E5. Explain how the data shown in Figure 14.9 indicate that two operator sites are necessary for repression of the *lac* operon. What would the results have been if all three operator sites were required for the binding of the lac repressor?

E6. A mutant strain has a defective *lac* operator site that results in the constitutive expression of the *lac* operon. Outline an experiment you would carry out to demonstrate that the operator site must be physically adjacent to the genes that it influences. Based on your knowledge of the *lac* operon, describe the results you would expect.

E7. Let's suppose you have isolated a mutant strain of *E. coli* in which the *lac* operon is constitutively expressed. To understand the nature of this defect, you create a merozygote in which the mutant strain contains an F' factor with a normal lac operon and a normal *lacI* gene. You then compare the mutant strain and the merozygote with regard to their β-galactosidase activities in the presence and absence of lactose. You obtain the following results:

	Addition of Lactose	Amount of β-Galactosidase (percentage of mutant strain in the presence of lactose)
Mutant	No	100
Mutant	Yes	100
Merozygote	No	100
Merozygote	Yes	200

Explain the nature of the defect in the mutant strain.

E8. In the experiment of Figure 14.7, a *lacI*$^-$ mutant was conjugated to a strain that had a functional *lacI* gene on an F' factor. The results of this experiment were important in determining the action of the lac repressor protein. What results would you expect if you used the same approach to investigate the regulation of the *ara* operon via AraC? In other words, what would be the level of expression of the *ara* operon in an araC$^-$ strain in the presence and absence of arabinose, and how would the level of expression change when a functional *araC* gene was introduced into the strain via conjugation?

E9. A segment of DNA that contains a loop is more compact than the same DNA segment without a loop. Therefore, when these two alternative structures (looped versus unlooped) are electrophoresed through a gel, the looped structure migrates more quickly to the bottom of the gel because it can more easily penetrate the gel matrix (see solved problem S6). Let's suppose that a mutant *E. coli* strain has been identified in which the AraC protein represses the *ara* operon, even in the presence of arabinose. In the experiment shown here, mutant or normal AraC protein was mixed with a segment of DNA containing the *ara* operon in the absence or presence of arabinose and then run on a gel.

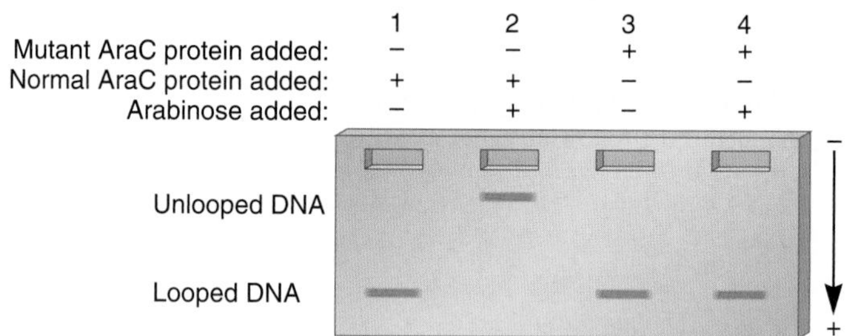

Describe the defect in this mutant AraC protein.

Questions for Student Discussion/Collaboration

1. Discuss the advantages and disadvantages of genetic regulation at the different levels described in Figure 14.1.

2. As you look at Figure 14.10, discuss possible "molecular ways" that the cAMP-CAP complex and lac repressor may influence RNA polymerase function. In other words, try to explain how the bending and looping in DNA may affect the ability of RNA polymerase to initiate transcription.

3. Certain environmental conditions such as UV light are known to activate lysogenic λ prophages and cause them to progress into the lytic cycle. UV light initially causes the repressor protein to be proteolytically degraded. Make a flow diagram that describes the subsequent events that would lead to the lytic cycle. Note: The *xis* gene codes for an enzyme that is necessary to excise the λ prophage from the *E. coli* chromosome. The integrase enzyme is also necessary to excise the λ prophage.

Note: All answers appear at the website for this textbook; the answers to even-numbered questions are in the back of the textbook.

www.mhhe.com/brookergenetics3e

Visit the Online Learning Center for practice tests, answer keys, and other learning aids for this chapter. Enhance your understanding of genetics with our interactive exercises, quizzes, animations, and much more.

GENE REGULATION IN EUKARYOTES

15

CHAPTER OUTLINE

Gene regulation provides many benefits to eukaryotic organisms, a category that includes protists, fungi, plants, and animals. Like their prokaryotic counterparts, eukaryotic cells need to adapt to changes in their environment. For example, eukaryotic cells can respond to changes in nutrient availability by enzyme adaptation, much as prokaryotic cells do. Eukaryotic cells also respond to environmental stresses such as ultraviolet (UV) radiation by inducing genes that provide protection against harmful environmental agents. An example is the ability of humans to develop a tan. The tanning response helps to protect a person's cells against the damaging effects of UV rays.

Among plants and animals, multicellularity and a more complex cell structure also demand a much greater level of gene regulation. The life cycle of complex eukaryotic organisms involves the progression through several developmental stages to achieve a mature organism. Some genes are expressed only during early stages of development, such as the embryonic stage, whereas others are expressed in the adult. In addition, complex eukaryotic species are composed of many different tissues that contain a variety of cell types. Gene regulation is necessary to ensure the differences in structure and function among distinct cell types. It is amazing that the various cells within a multicellular organism usually contain the same genetic material, yet phenotypically may look quite different. For example, the appearance of a human nerve cell seems about as similar to a muscle cell as an amoeba is to a paramecium. In spite of these phenotypic differences, a human nerve cell and muscle cell actually contain the same complement of human chromosomes. Nerve and muscle cells are strikingly different because of gene regulation rather than differences in DNA content. Many genes are expressed in the nerve cell and not the muscle cell, and vice versa.

The molecular mechanisms that underlie gene regulation in eukaryotes bear many similarities to the ways that bacteria regulate their genes. As in prokaryotes, regulation in eukaryotes can occur at any step in the pathway of gene expression (**Figure 15.1**). We will begin this chapter with an exploration of gene regulation at the level of transcription, an important form of control. In addition, research in the past few decades has revealed that eukaryotic organisms frequently regulate gene expression at points other than transcription. We will discuss some well-studied examples in which genes are regulated at many of these other control points.

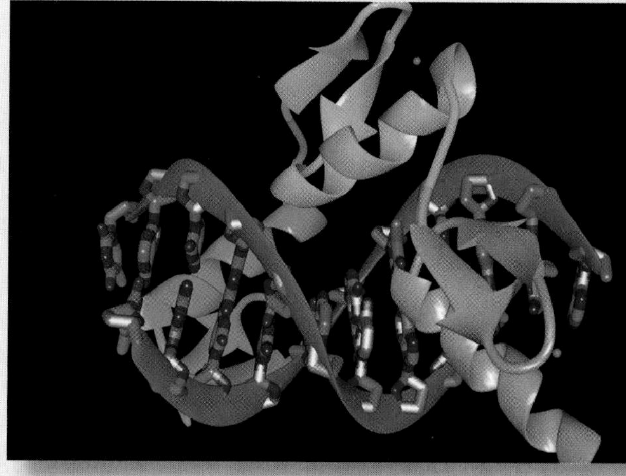

Certain proteins, known as regulatory transcription factors, have the ability to bind into the major groove of DNA and regulate gene transcription.

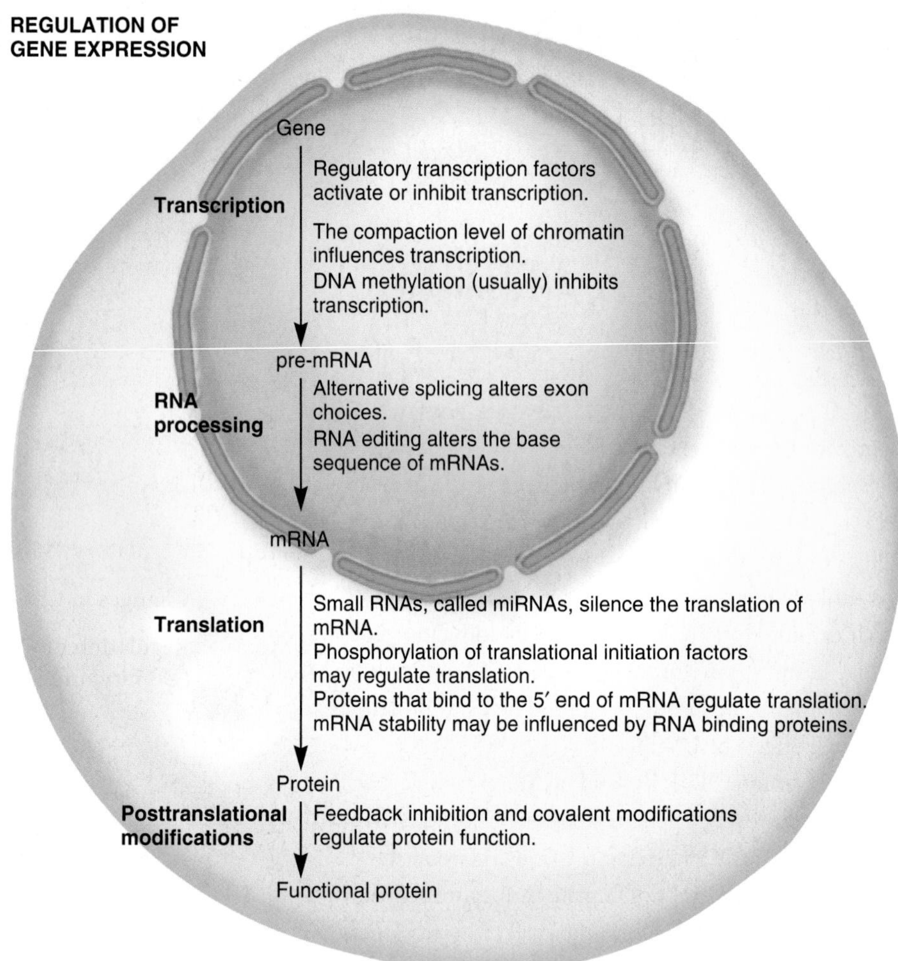

FIGURE 15.1 **Levels of gene expression commonly subject to regulation.**

15.1 REGULATORY TRANSCRIPTION FACTORS

The term **transcription factor** is broadly used to describe proteins that influence the ability of RNA polymerase to transcribe a given gene. We will focus our attention on transcription factors that affect the ability of RNA polymerase to begin the transcription process. Such transcription factors can regulate the binding of the transcriptional apparatus to the core promoter and/or control the switch from the initiation to the elongation stage of transcription. Two categories of transcription factors play a key role in these processes. In Chapter 12, we considered **general transcription factors,** which are required for the binding of RNA polymerase to the core promoter and its progression to the elongation stage. General transcription factors are necessary for a basal level of transcription. In addition, eukaryotic cells possess a diverse array of **regulatory transcription factors** that serve to regulate the rate of transcription of target genes. The importance of transcription factors is underscored by the number of genes that encode this category of proteins. A sizeable portion of the genomes of complex organisms such as plants and animals is devoted to the process of gene regulation. For example, in *Arabidopsis thaliana*, a plant that is used as a model organism by plant geneticists, approximately 5% of its genome encodes transcription factors. This species has more than 1,500 different genes that encode proteins that regulate the transcription of other genes. Similarly, 2 to 3% of the genes in the human genome encode transcription factors.

As discussed in this section, regulatory transcription factors exert their effects by influencing the ability of RNA polymerase to begin transcription of a particular gene. They typically recognize *cis*-acting elements that are located in the vicinity of the core promoter. These DNA sequences are analogous to the operator sites found near bacterial promoters. In eukaryotes, these DNA sequences are generally known as **control elements** or **regulatory elements.** When a regulatory transcription factor binds to a regulatory element, it affects the transcription of an associated gene. For example, the binding of regulatory transcription factors may enhance the rate of transcription (**Figure 15.2a**). Such a transcription factor is termed an **activator,** and the sequence it binds to is called an **enhancer.** Alternatively, regulatory transcription factors may act as **repressors** by binding to elements called **silencers** and preventing transcription from occurring (**Figure 15.2b**).

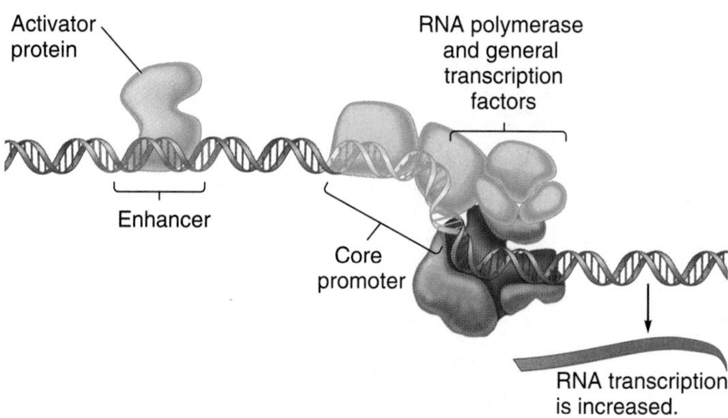

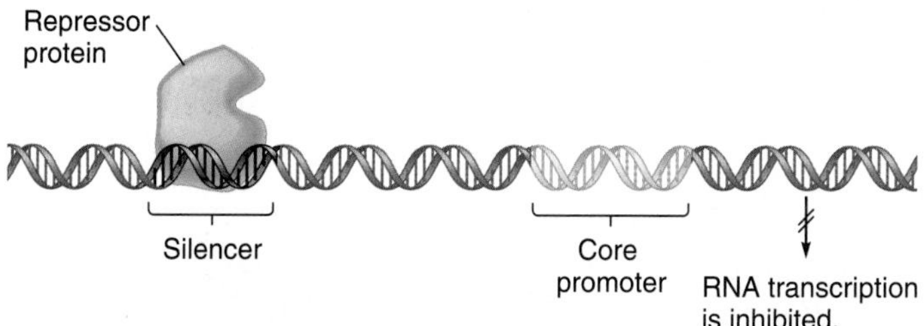

FIGURE 15.2 **Overview of transcriptional regulation by regulatory transcription factors.** These proteins can act as either (**a**) an activator to increase the rate of transcription or (**b**) a repressor to decrease the rate of transcription.

By studying transcriptional regulation, researchers have discovered that most eukaryotic genes, particularly those found in multicellular species, are regulated by many factors. This phenomenon is called **combinatorial control** because the combination of many factors determines the expression of any given gene. At the level of transcription, the following are common factors that contribute to combinatorial control:

1. One or more activator proteins may stimulate the ability of RNA polymerase to initiate transcription.
2. One or more repressor proteins may inhibit the ability of RNA polymerase to initiate transcription.
3. The function of activators and repressors may be modulated in a variety of ways. These include the binding of small effector molecules, protein–protein interactions, and covalent modifications.
4. Activator proteins may promote the loosening of chromatin compaction in the chromosome where a gene is located, thereby making it easier for the gene to be recognized and transcribed by RNA polymerase.
5. DNA methylation may inhibit transcription, either by preventing the binding of an activator protein or by recruiting proteins that cause the chromatin to become more compact.

All five of these factors can contribute to the regulation of a single gene, or possibly only three or four will play a role. In most cases, transcriptional regulation is aimed at controlling the initiation of transcription at the promoter. In this section, we will survey the first three factors that affect transcription. In the following section, we will consider the fourth and fifth factors.

Structural Features of Regulatory Transcription Factors Allow Them to Bind to DNA

Genes that encode general and regulatory transcription factor proteins have been identified and sequenced from a wide variety of eukaryotic species, including yeast, plants, and animals. Several different families of evolutionarily related transcription factors have been discovered. In recent years, the molecular structures of transcription factor proteins have become an area of intense research. Transcription factor proteins contain regions, called **domains,** that have specific functions. For example, one domain of a transcription factor may have a DNA-binding function, while another may provide a binding site for a small effector molecule. When a domain or portion of a domain has a very similar structure in many different proteins, such a structure is called a **motif.**

Figure 15.3 depicts several different domain structures found in transcription factor proteins. The protein secondary structure known as an α helix is frequently found in transcription factors. Why is the α helix common in such proteins? The explanation is that the α helix is the proper width to bind into the major groove of the DNA double helix. In helix-turn-helix and helix-loop-helix motifs, an α helix called the recognition helix makes contact with and recognizes a base sequence along the major groove of the DNA (Figure 15.3a and b). Recall that the major groove is a region of the DNA double helix where the bases contact the surrounding water in the cell. Hydrogen bonding between an α helix and nucleotide bases is one way that a transcription factor can bind to a specific DNA sequence. In addition, the recognition helix often contains many positively charged amino acids (e.g., arginine and lysine) that favorably interact with the DNA backbone, which is negatively charged. Such basic domains are a common feature of many DNA-binding proteins.

A zinc finger motif is composed of one α helix and two β-sheet structures that are held together by a zinc (Zn^{2+}) metal ion (Figure 15.3c). The zinc finger also can recognize DNA sequences within the major groove.

A second interesting feature of certain motifs is that they promote protein dimerization. The leucine zipper (Figure 15.3d) and helix-loop-helix motif (Figure 15.3b) mediate protein dimerization. For example, Figure 15.3d depicts the dimerization and DNA binding of two proteins that have several leucine amino acids (a zipper). Alternating leucines in both proteins interact ("zip up"), resulting in protein dimerization. Two identical transcription factor proteins will come together to form a **homodimer,** or two different transcription factors can form a **heterodimer.** As discussed later, the dimerization of transcription factors can be an important way to modulate their function.

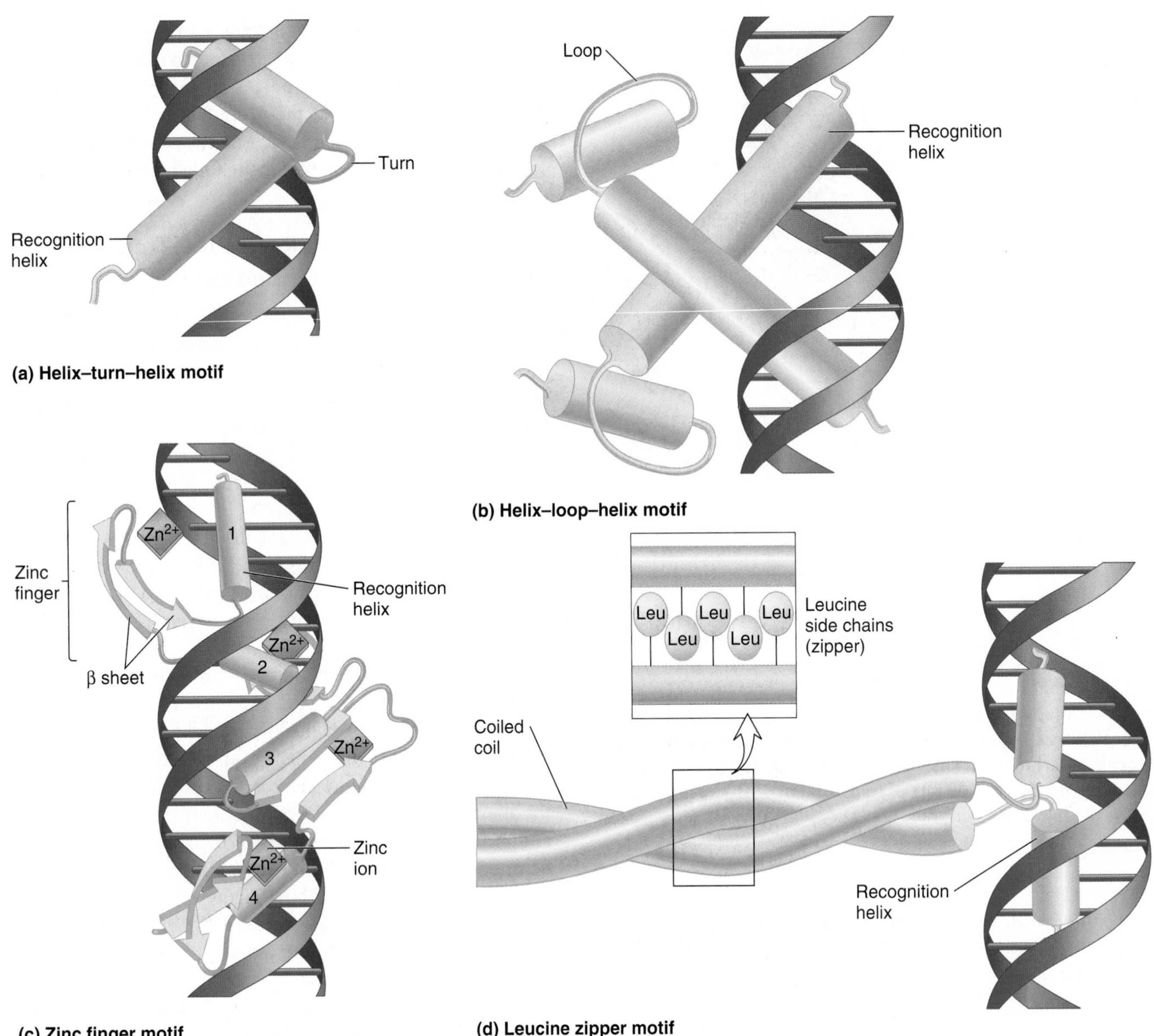

FIGURE 15.3 Structural motifs found in transcription factor proteins. Certain types of protein secondary structure are found in many different transcription factors. In this figure, α helices are shown as cylinders and β sheets as flattened arrows. (**a**) Helix-turn-helix motif: Two α helices are connected by a turn. The α helices lie in the major groove of the DNA. (**b**) Helix-loop-helix motif: A short α helix is connected to a longer α helix by a loop. In this illustration, a dimer is formed from the interactions of two helix-loop-helix motifs, and the longer helices are binding to the DNA. (**c**) Zinc finger motif: Each zinc finger is composed of one α helix and two antiparallel β sheets. A zinc ion (Zn^{2+}), shown in red, holds the zinc finger together. This illustration shows four zinc fingers in a row. (**d**) Leucine zipper motif: The leucine zipper promotes the dimerization of two transcription factor proteins. Two α helices (termed a coiled coil) are intertwined via the leucines (see inset).

Regulatory Transcription Factors Recognize Regulatory Elements That Function as Enhancers or Silencers

As mentioned previously, when the binding of a regulatory transcription factor to a regulatory element increases transcription, the regulatory element is known as an enhancer. Such elements can stimulate transcription 10- to 1,000-fold; this is termed **up regulation.** Alternatively, regulatory elements that serve to inhibit transcription are called silencers, and their action is called **down regulation.**

Many regulatory elements are **orientation independent** or **bidirectional.** This means that the regulatory element can function in the forward or reverse direction. For example, if the forward orientation of an enhancer is

5'-GATA-3'
3'-CTAT-5'

this enhancer will also be bound by a regulatory transcription factor and enhance transcription even when it is oriented in the reverse direction:

5'-TATC-3'
3'-ATAG-5'

Striking variation is also found in the location of regulatory elements relative to a gene's promoter. Regulatory elements are often located in a region within a few hundred base pairs upstream from the promoter site. However, they can be quite distant from the promoter, even 100,000 base pairs away, yet exert strong effects on the ability of RNA polymerase to initiate transcription at the core promoter! Regulatory elements were first discovered by Susumu Tonegawa and coworkers in the 1980s. While studying genes that play a role in immunity, they identified a region far away from the core promoter, but needed for high levels of transcription to take place. In some cases, regulatory elements are located downstream from the promoter site and may even be found within introns, the noncoding parts of genes. As you may imagine, the variation in regulatory element orientation and location profoundly complicates the efforts of geneticists to identify the regulatory elements that affect the expression of any given gene.

Regulatory Transcription Factors May Exert Their Effects Through TFIID and Mediator

Different mechanisms have been discovered that explain how a regulatory transcription factor can bind to a regulatory element and thereby affect gene transcription. Indeed, more than one mechanism is typically involved. The net effect of a regulatory transcription factor is to influence the ability of RNA polymerase to transcribe a given gene. However, most regulatory transcription factors do not bind directly to RNA polymerase. How then do most regulatory transcription factors exert their effects? In many cases, their mechanism of action is to influence the function of RNA polymerase by interacting with other proteins that directly bind to RNA polymerase. Two protein complexes that communicate the effects of regulatory transcription factors are TFIID and mediator.

Figure 15.4 depicts how regulatory transcription factors may control transcription. In some cases, regulatory transcription factors bind to a regulatory element and then influence the function of TFIID (Figure 15.4a). As discussed in Chapter 12, **TFIID** is a general transcription factor that binds to the TATA box and is needed to recruit RNA polymerase to the core promoter. Activator proteins are expected to enhance the ability of TFIID to initiate transcription. One possibility is that activator proteins could help recruit TFIID to the TATA box or they could enhance the function of TFIID in a way that facilitates its ability to bind RNA polymerase. In contrast, repressors inhibit the function of TFIID. They could exert their effects by preventing the binding of TFIID to the TATA box or by inhibiting the ability of TFIID to recruit RNA polymerase to the core promoter.

A second way that regulatory transcription factors control RNA polymerase is via mediator, a protein complex discovered by Roger Kornberg and colleagues in 1990 (Figure 15.4b). The term **mediator** refers to the observation that it mediates the interaction between RNA polymerase and regulatory transcription factors. As discussed in Chapter 12, mediator controls the ability of RNA polymerase to progress to the elongation stage of transcription. Transcriptional activators stimulate the ability of mediator to facilitate the switch between the initiation and elongation stages, whereas repressors have the opposite effect. When a repressor protein interacts with mediator, RNA polymerase cannot progress to the elongation stage of transcription.

A third way that regulatory transcription factors can influence transcription is by recruiting proteins to the promoter region that affect chromatin compaction. For example, certain transcriptional activators can recruit proteins to the promoter region and thereby promote the conversion of chromatin from a closed to an open conformation. We will return to this topic later in this chapter.

The Function of Regulatory Transcription Factor Proteins Can Be Modulated in Three Ways

Thus far, we have considered the structures of regulatory transcription factors and the molecular mechanisms that account for their abilities to control transcription. The functions of the regulatory transcription factors themselves must also be modulated. Why is this necessary? The answer is that the genes they control must be turned on at the proper time, in the correct cell type, and under the appropriate environmental conditions. Therefore, eukaryotes have evolved different ways to modulate the functions of these proteins.

The functions of regulatory transcription factor proteins are controlled in three common ways, through (1) the binding of a small effector molecule, (2) protein–protein interactions, and (3) covalent modifications. **Figure 15.5** depicts these three mechanisms of modulating regulatory transcription factor function. Usually, one or more of these modulating effects are important in determining whether a transcription factor can bind to the DNA and/or influence transcription by RNA polymerase. For example, a small effector molecule may bind to a regulatory transcription factor and promote its binding to DNA (Figure 15.5a). We will see that steroid hormones function in this manner. Another important way is via protein–protein interactions (Figure 15.5b). The formation of homodimers and heterodimers is a fairly common means of controlling transcription. Finally, the function of a regulatory transcription factor can be affected by covalent modifications such as the attachment of a phosphate group (Figure 15.5c). As discussed later, the phosphorylation of activators can control their ability to stimulate transcription.

Steroid Hormones Exert Their Effects by Binding to a Regulatory Transcription Factor

Now that we have a general understanding regarding the structure and function of transcription factors, let's turn our attention to specific examples that illustrate how regulatory transcription factors carry out their roles within living cells. Our first example

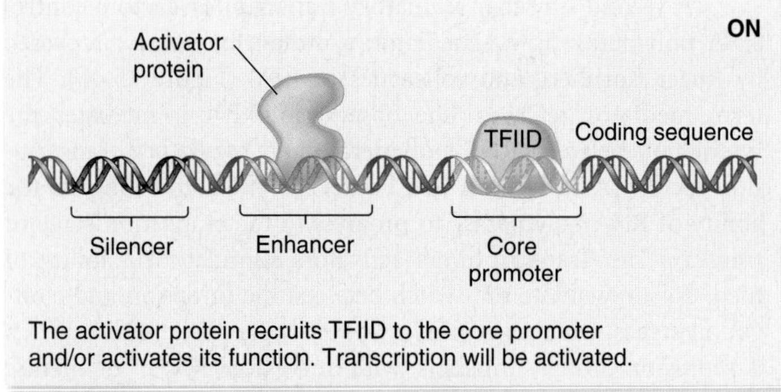
ON
Activator protein
TFIID
Coding sequence
Silencer
Enhancer
Core promoter
The activator protein recruits TFIID to the core promoter and/or activates its function. Transcription will be activated.

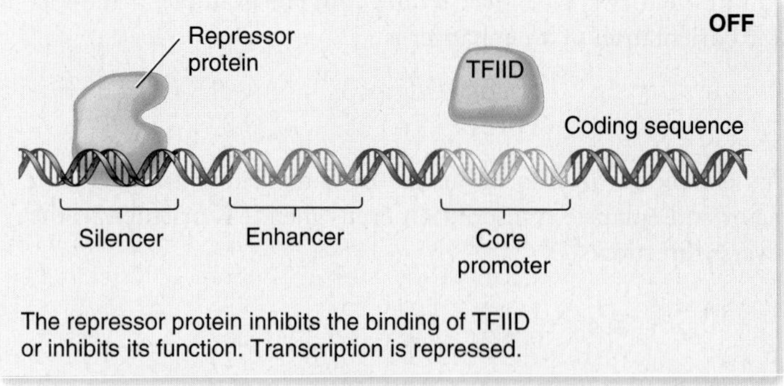
OFF
Repressor protein
TFIID
Coding sequence
Silencer
Enhancer
Core promoter
The repressor protein inhibits the binding of TFIID or inhibits its function. Transcription is repressed.

(a) Regulatory transcription factors and TFIID

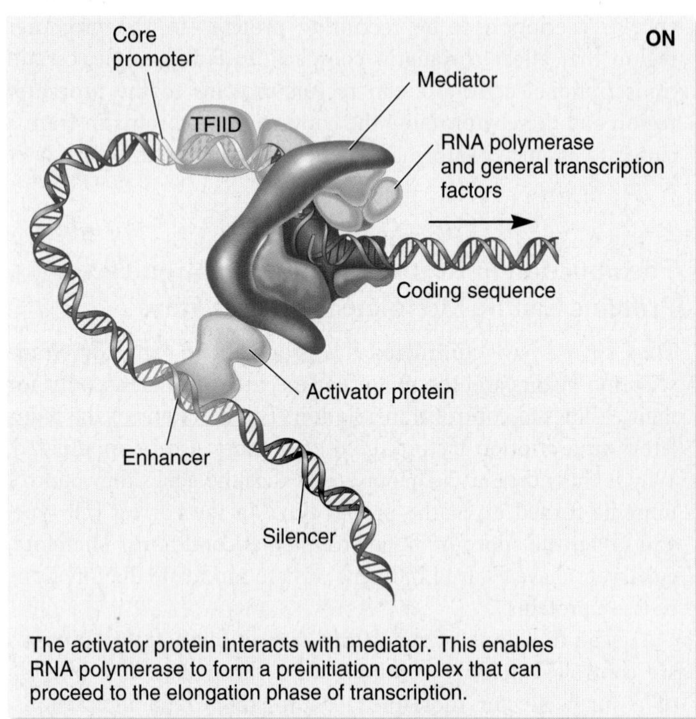
ON
Core promoter
TFIID
Mediator
RNA polymerase and general transcription factors
Coding sequence
Activator protein
Enhancer
Silencer
The activator protein interacts with mediator. This enables RNA polymerase to form a preinitiation complex that can proceed to the elongation phase of transcription.

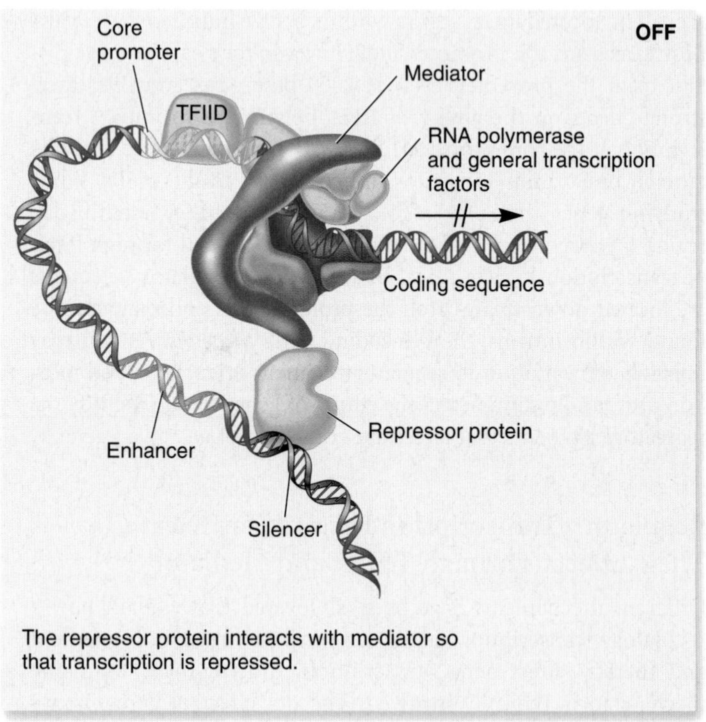
OFF
Core promoter
TFIID
Mediator
RNA polymerase and general transcription factors
Coding sequence
Repressor protein
Enhancer
Silencer
The repressor protein interacts with mediator so that transcription is repressed.

(b) Regulatory transcription factors and mediator

FIGURE 15.4 **Ability of regulatory transcription factors to affect transcription.** (**a**) Some regulatory transcription factors exert their effects through TFIID. Transcriptional activators may recruit TFIID to the core promoter and/or activate its function (left), while repressors inhibit TFIID binding or its activity (right). (**b**) Regulatory transcription factors may also exert their effects via mediator. By interacting with mediator in different ways, activators stimulate transcription (left), while repressors inhibit transcription (right). In this example, the enhancer and silencer are relatively far away from the core promoter. Therefore, a loop must form in the DNA so that the regulatory transcription factor can interact with mediator.

is a category that responds to steroid hormones. This type of regulatory transcription factor is known as a **steroid receptor,** because the steroid hormone binds directly to the protein.

The ultimate action of a steroid hormone is to affect gene transcription. Steroid hormones act as signaling molecules that are synthesized by endocrine glands of animals and secreted into the bloodstream. The hormones are then taken up by cells that can respond to the hormones in different ways. For example, glucocorticoid hormones influence nutrient metabolism in most body cells. Other steroid hormones, such as estrogen and testosterone, are called gonadocorticoids because they influence the growth and function of the gonads.

Figure 15.6 shows the stepwise action of glucocorticoid hormones, which are produced in mammals. In this example, the hormone enters the cytosol of a cell by diffusing through the plasma membrane. Once inside, the hormone specifically binds to **glucocorticoid receptors.** Prior to hormone binding, the glucocorticoid receptor is complexed with proteins known as heat shock proteins (HSP), one example being HSP90. After the hormone binds to the glucocorticoid receptor, HSP90 is released. This exposes a nuclear localization signal (NLS)—a signal that directs a protein into the nucleus. Two glucocorticoid receptors form a homodimer and then travel through a nuclear pore into the nucleus.

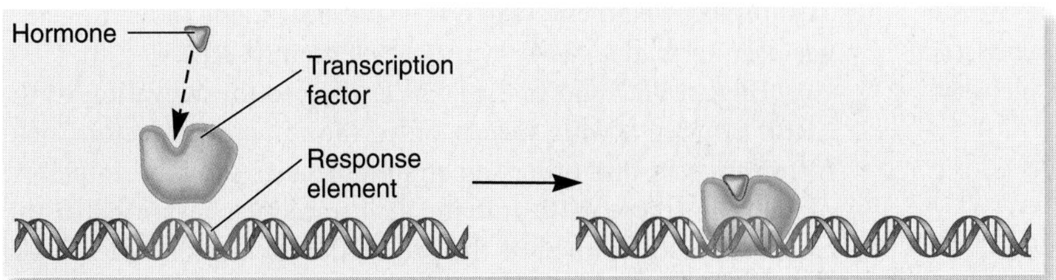

(a) Binding of a small effector molecule such as a hormone

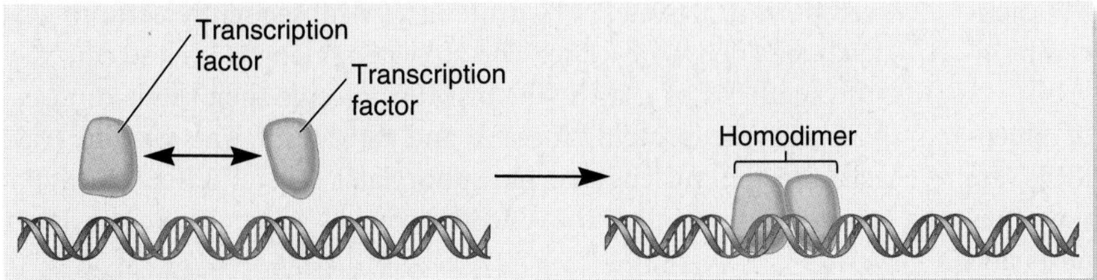

(b) Protein–protein interaction

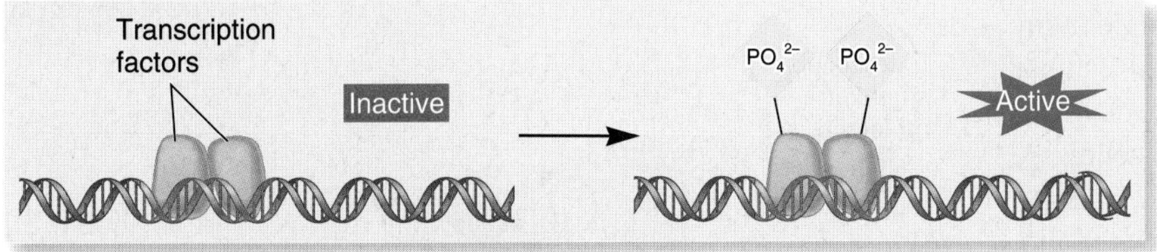

(c) Covalent modification such as phosphorylation

FIGURE 15.5 Common ways to modulate the function of regulatory transcription factors. (**a**) The binding of an effector molecule such as a hormone may influence the ability of a transcription factor to bind to the DNA. (**b**) Protein–protein interactions among transcription factor proteins may influence their functions. (**c**) Covalent modifications such as phosphorylation may alter transcription factor function.

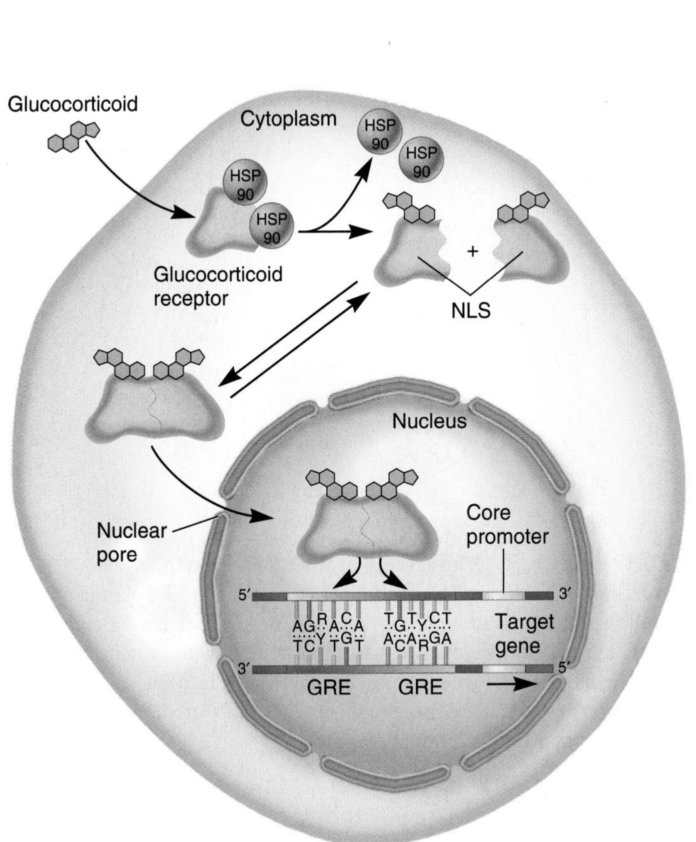

ONLINE ANIMATION

FIGURE 15.6 The action of glucocorticoid hormones. Once inside the cell, the glucocorticoid hormone binds to the glucocorticoid receptor, releasing it from a protein known as HSP90. This exposes a nuclear localization signal (NLS). Two glucocorticoid receptors then form a dimer and travel into the nucleus, where the dimer binds to glucocorticoid response elements (GREs) that are next to particular genes. The binding of the glucocorticoid receptors to the GREs activates the transcription of the adjacent target gene.

Genes→Traits Glucocorticoid hormones are produced by the endocrine glands in response to fasting and activity. They enable the body to regulate its metabolism properly. When glucocorticoids are produced, they are taken into cells and bind to the glucocorticoid receptors. This eventually leads to the activation of genes that encode proteins involved in the synthesis of glucose, the breakdown of proteins, and the mobilization of fats.

How does the glucocorticoid receptor regulate the expression of particular genes? In the nucleus, the glucocorticoid receptor homodimer binds to glucocorticoid response elements (GREs) with the following consensus sequence:

5'-AGRACA-3'
3'-TCYTGT-5'

where R is a purine and Y is pyrimidine. This sequence is found next to many genes and functions as an enhancer. The binding of the glucocorticoid receptor homodimer to adjacent GREs activates the transcription of the nearby gene, eventually leading to the synthesis of the encoded protein.

Mammalian cells usually have a large number of glucocorticoid receptors within the cytoplasm. Because GREs are located near dozens of different genes, the uptake of many hormone molecules can activate many glucocorticoid receptors and thereby stimulate the transcription of many different genes. For this reason, a cell can respond to the presence of the hormone in a very complex way. Glucocorticoid hormones stimulate many genes that encode proteins involved in several different cellular processes, including the synthesis of glucose, the breakdown of proteins, and the mobilization of fats. Although the genes are not physically connected to each other, the regulation of multiple genes via glucocorticoid hormones is much like the ability of bacterial operons to simultaneously control the expression of several genes.

The CREB Protein Is an Example of a Regulatory Transcription Factor Modulated by Covalent Modification

As we have just seen, steroid hormones function as signaling molecules that bind directly to regulatory transcription factors to alter their function. This enables a cell to respond to a hormone by up regulating a particular set of genes. Most extracellular signaling molecules, however, do not enter the cell or bind directly to transcription factors. Instead, most signaling molecules must bind to receptors in the plasma membrane. This binding activates the receptor and may lead to the synthesis of an intracellular signal that causes a cellular response. One type of cellular response is to affect the transcription of particular genes within the cell.

As our second example of regulatory transcription factor function within living cells, we will examine the **cAMP response element–binding (CREB) protein.** The CREB protein is a regulatory transcription factor that becomes activated in response to cell-signaling molecules that cause an increase in the cytoplasmic concentration of the molecule cyclic adenosine monophosphate (cAMP). This transcription factor recognizes a response element with the following consensus sequence.

5'-TGACGTCA-3'
3'-ACTGCAGT-5'

This response element, which is found near many different genes, has been termed a **cAMP response element (CRE).**

Figure 15.7 shows the steps leading to the activation of the CREB protein. A wide variety of hormones, growth factors, neurotransmitters, and other signaling molecules can bind to plasma membrane receptors to initiate an intracellular response. In this case, the response involves the production of a second messenger, cAMP. The extracellular signaling molecule itself is considered the primary messenger. When the signaling molecule binds to the receptor, it activates a G protein that subsequently activates the enzyme adenylyl cyclase. The activated adenylyl cyclase catalyzes the synthesis of cAMP. The cAMP molecule then binds to a second enzyme, protein kinase A, and activates it. This enzyme travels into the nucleus and phosphorylates several different cellular proteins, including the CREB protein. When phosphorylated,

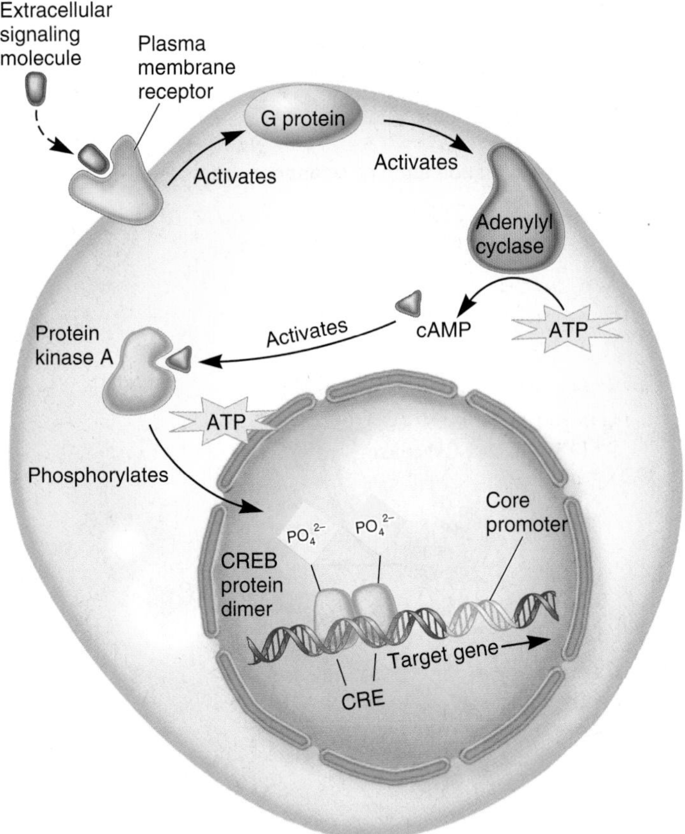

FIGURE 15.7 The activity of the CREB protein. When an extracellular signaling molecule binds to a receptor in the plasma membrane, this activates a G protein, which activates adenylyl cyclase, leading to the synthesis of cAMP. Next, cAMP binds to protein kinase A, which activates it. Protein kinase A then travels into the nucleus and phosphorylates the CREB protein. Once phosphorylated, the CREB protein acts as a transcriptional activator.

CREB proteins stimulate transcription. In contrast, unphosphorylated CREB proteins can still bind to CREs but do not activate RNA polymerase.

15.2 CHANGES IN CHROMATIN STRUCTURE

Changes in chromatin structure can involve changes in the level of chromosomal compaction and/or direct alterations in the structure of DNA. Because genes are segments of DNA, it is not surprising that such alterations can affect gene expression in a variety of ways (**Table 15.1**).

In eukaryotes, changes in chromatin compaction are required for gene transcription to take place. For this reason, the regulation of gene transcription is not simply dependent on the activity of regulatory transcription factors that influence RNA polymerase via TFIID and mediator to turn genes on or off. Transcription must also involve changes in chromatin structure that affect the ability of transcription factors to gain access to and bind their target sequences in the promoter region. If the chromatin is very tightly packed or in a **closed conformation,** transcription may be difficult or impossible. By comparison, chromatin that is in an **open conformation** is more easily accessible to transcription factors and RNA polymerase so that transcription can take place.

Gene regulation may also involve changes in DNA structure. **Gene amplification,** an increase in the number of copies of a gene, is a possible but not common method of gene regulation. Another change in DNA that affects gene expression is the rearrangement of the DNA structure. In Chapter 17, we will see that **gene rearrangement** occurs in the DNA of immune system cells to generate a diverse array of antibody proteins. A more common way that alterations of DNA structure affect gene regulation is **DNA methylation.**

In this section, we will begin by considering how the conversion between the closed and open conformations of chromatin is associated with the expression of genes. We will then examine molecular mechanisms that explain how chromosome compaction in the vicinity of specific genes is altered. Finally, we will examine how DNA methylation occurs and how it can regulate genes in a tissue-specific manner.

TABLE **15.1**

Gene Regulation That Occurs via Changes in Chromatin or DNA Structure

Modification	Description
Chromatin compaction	This is a common mechanism of eukaryotic gene regulation. For a gene to be transcribed, it must be converted from a closed (highly compacted) conformation to an open conformation through the actions of enzymes such as histone acetyltransferases and ATP-dependent remodeling enzymes. Transcriptional activators recruit these enzymes to promoter regions to activate gene transcription.
Gene amplification	Though not a common mechanism of gene regulation, gene amplification involves an increase in gene number. For example, in the oocytes of *Xenopus laevis* (a South African frog), the number of rRNA genes is increased through gene amplification. The mechanism involves an unusual DNA replication event in which the rRNA genes are replicated to form many copies of circular DNA molecules called minichromosomes.
Gene rearrangement	This is not a common mechanism of gene regulation. As discussed in Chapter 17, gene rearrangement occurs within immunoglobulin genes to generate a diverse array of genes that encode antibody proteins.
DNA methylation	This is a common DNA modification in vertebrates and plants that involves the attachment of methyl groups to cytosine bases. When DNA methylation occurs in the promoter region, it usually inhibits transcription. This inhibition may be due to the inability of transcription factors to recognize the promoter region and/or a conversion of the chromatin to the closed conformation.

Gene Accessibility Can Be Controlled by Changes in Chromatin Compaction

Chromatin is composed of DNA and proteins that are organized into a compact structure that fits inside the nucleus of the cell. As discussed in Chapter 10, the DNA is wound around histone proteins to form nucleosomes that are 11 nm in diameter. This 11 nm fiber is condensed further to a 30 nm fiber, which is the predominant form of euchromatin found within the nucleus during interphase, when gene expression primarily occurs. Nevertheless, chromatin is a very dynamic structure that can alternate between highly condensed and highly extended conformations. The dynamic nature of chromatin is important in regulating gene transcription.

Variations in the degree of chromatin packing occur along the length of eukaryotic chromosomes during interphase. Tightly packed chromatin in a closed conformation cannot be transcribed. During gene activation, such chromatin must be converted to an open conformation that is less tightly packed than a 30 nm fiber. In certain cells, researchers can microscopically observe the decondensation of the 30 nm fiber when transcription is occurring. **Figure 15.8** shows photomicrographs of a chromosome from an amphibian oocyte, the genes of which are being actively transcribed. This chromosome does not form a uniform, compact 30 nm fiber. Instead, many decondensed loops radiate from the central axis of the chromosome. These loops are regions of DNA in which the genes are being actively transcribed. These chromosomes have been named lampbrush chromosomes because their feathery appearance resembles the brushes that were once used to clean kerosene lamps.

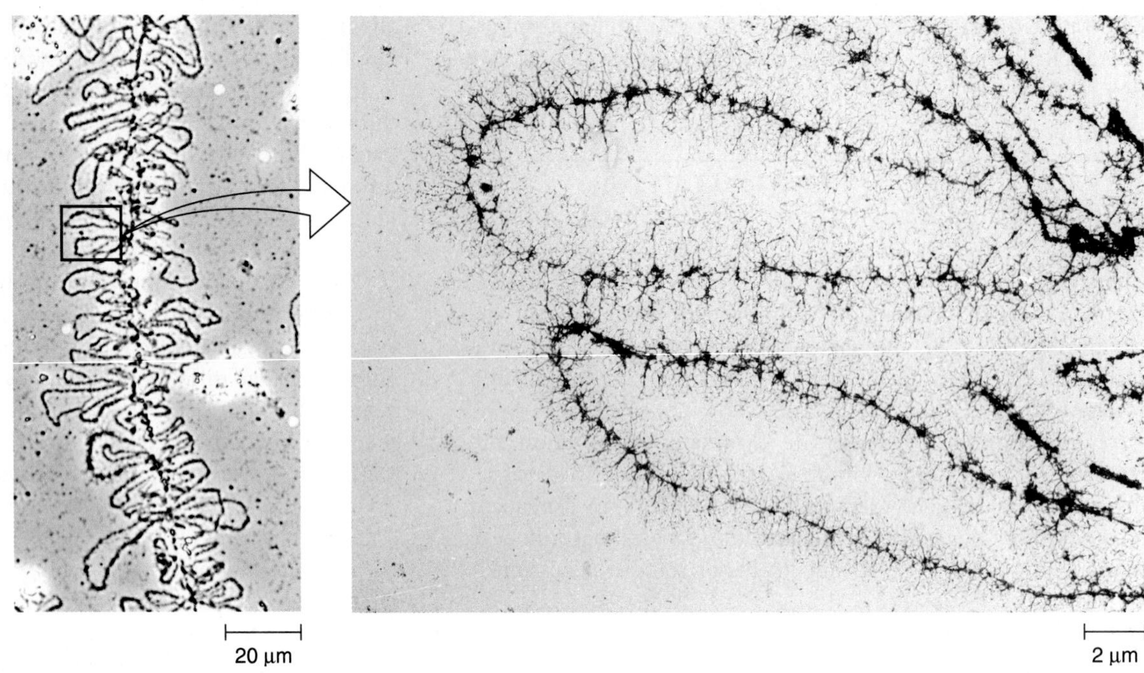

FIGURE 15.8 A lampbrush chromosome from an amphibian oocyte. The loops that radiate from the chromosome are regions of DNA containing genes that are being actively transcribed.

EXPERIMENT 15A

DNase I Sensitivity Can Be Used to Study Changes in Chromatin Compaction

To understand the interconversions between closed and open chromatin conformations, researchers need methods to evaluate the degree of chromatin packing as it occurs in living cells. An important technique is the use of DNase I to monitor DNA conformation. Recall from Chapter 10 (see Figure 10.15) that DNase I is an endonuclease that cleaves DNA. DNase I is much more likely to cleave DNA in an open conformation than a closed conformation because a looser conformation allows greater accessibility of DNase I to the DNA. When DNA is in an open conformation and able to be digested by DNase I, such DNA is said to exhibit **DNase I sensitivity.**

In 1976, Harold Weintraub and Mark Groudine used DNase I sensitivity as a tool to evaluate differences in chromatin structure that occur when a gene is being actively transcribed. In particular, they focused their attention on the β-globin gene. Humans possess several different genes that encode globin polypeptides. These polypeptides are the subunits of the oxygen-carrying protein hemoglobin. Prior to this work, it had been well established that globin genes are specifically expressed in reticulocytes (immature red blood cells) but not in other cell types such as brain cells and fibroblasts. Weintraub and Groudine asked the question, Is there a difference in the chromatin packing of globin genes in cells that can actively transcribe the globin genes compared with cells in which the globin genes are turned off? To answer this question, they used DNase I sensitivity to compare the degree of globin gene packing in reticulocytes versus brain cells and fibroblasts.

Before discussing the steps in their experimental protocol, let's consider the rationale behind their experimental approach (**Figure 15.9**). Because the globin genes are only a small part of the total chromosomal DNA, having a way to specifically monitor the DNase I digestion of the β-globin gene was a vital aspect of their protocol. They accomplished this by using a probe—a segment of DNA complementary to the β-globin gene. In Weintraub and Groudine's experiments, the DNA probe was radiolabeled to allow its detection. Following the exposure of the chromatin to DNase I, the chromosomal DNA was denatured and the probe was added. If DNase I had digested the chromosomal β-globin gene, the probe DNA would not be able to hybridize with the β-globin gene (Figure 15.9a). Alternatively, if the chromosomal β-globin gene had not been digested by DNase I, the probe would hybridize to it (Figure 15.9b).

After hybridization, the samples were then exposed to another enzyme, known as S1 nuclease. This enzyme digests DNA, but only when it is single stranded, not when it is double stranded. As shown in Figure 15.9a, S1 nuclease digests the radiolabeled DNA probe because it had not hybridized with the complementary chromosomal strand. In contrast, S1 nuclease does not digest the probe if the chromosomal DNA remains intact and hybridizes with the probe (Figure 15.9b). In this way, an intact strand of chromosomal DNA protects the probe from S1 nuclease digestion.

After S1 incubation, Weintraub and Groudine reasoned that their sample would contain an intact radiolabeled DNA probe if

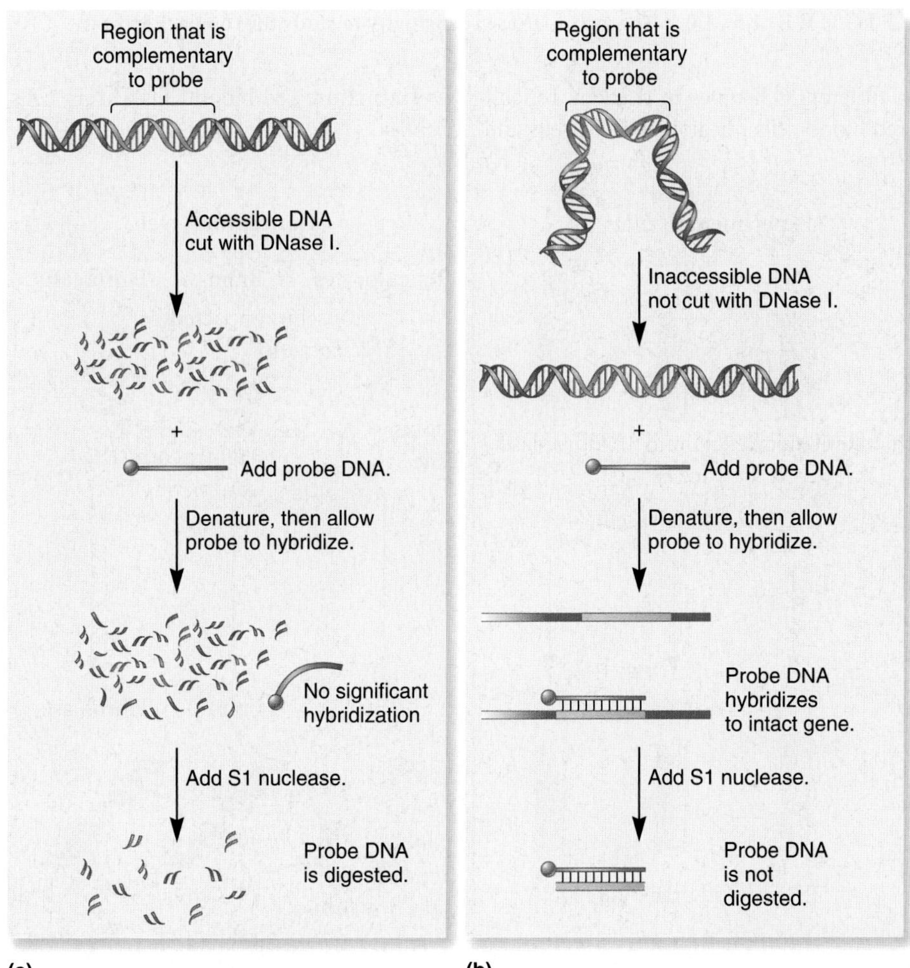

FIGURE 15.9 The use of DNase I and S1 nuclease to probe the chromatin structure of the β-globin gene. (**a**) Accessible DNA has been digested into small pieces via DNase I. These small pieces cannot hybridize with the probe DNA. When S1 nuclease is added, the single-stranded probe DNA is digested. (**b**) Due to compaction, the DNA is not cut with DNase I. The probe DNA can hybridize with a strand of chromosomal DNA after it has been denatured. When S1 nuclease is added, it will digest only the single-stranded regions of DNA, not the double-stranded region where the probe DNA and chromosomal DNA are bound to each other.

the chromosomal DNA had been in a closed conformation. This is because DNase I would not digest the chromosomal gene, and therefore, the β-globin gene would be available to hybridize to the radiolabeled DNA probe. However, if the chromosomal globin gene was in an open conformation, DNase I would digest the chromosomal DNA, preventing it from hybridizing with the radiolabeled DNA probe. In this case, the radiolabeled DNA strand would be digested by S1 nuclease. Therefore, the susceptibility of the radiolabeled DNA to S1 nuclease digestion allowed them to evaluate whether the chromosomal globin gene was in an open or closed conformation.

With these concepts in mind, let's examine the steps in their experimental procedure (**Figure 15.10**). They began with three cell types: reticulocytes, brain cells, and fibroblasts. Only the reticulocytes express the β-globin gene. They hypothesized that if the globin gene was expressed, the DNA would be in an open conformation and accessible to digestion. The nuclei were extracted from these cells and incubated with DNase I. The DNA was then isolated from the three types of nuclei. Following the isolation procedure, the chromosomal DNA was broken into fragments by sonication, with the average fragment size larger than the probe DNA. The DNA fragments and probe were mixed together, denatured, and then allowed to hybridize. The cooled samples were then divided into two tubes each. Into one of the two tubes, S1 nuclease was added. A key point is that the probe would not be digested by S1 nuclease if it had hybridized to a complementary strand from the globin gene and thus was double stranded. The DNA fragments were then precipitated using trichloroacetic acid. The samples were subjected to centrifugation, and the amounts of radioactivity in the pellets were determined by scintillation counting.

THE HYPOTHESIS

A loosening of chromatin structure occurs when the β-globin gene is transcriptionally active.

TESTING THE HYPOTHESIS — FIGURE 15.10 The use of DNase I sensitivity to study the compaction of the β-globin gene.

Starting material: Nuclei were isolated from three different cell types in chicken: reticulocytes (immature red blood cells), brain cells, and fibroblasts. The globin genes are expressed in red blood cells but not in brain cells and fibroblasts.

1. Treat nuclei from three types of cells with the same amount of DNase I.

2. Extract the DNA from the nuclei. This involves lysing the nuclei with detergent, removing the protein by treatment with a phenol–chloroform mixture, and then precipitating the DNA by adding ethanol. The precipitated DNA forms a pellet at the bottom of a test tube following centrifugation. The DNA at the bottom of the tube can then be resuspended in an appropriate solution for the next step.

3. Subject the DNA to sound waves (i.e., sonication) to break the DNA into fragments of an average length of 500 bp.

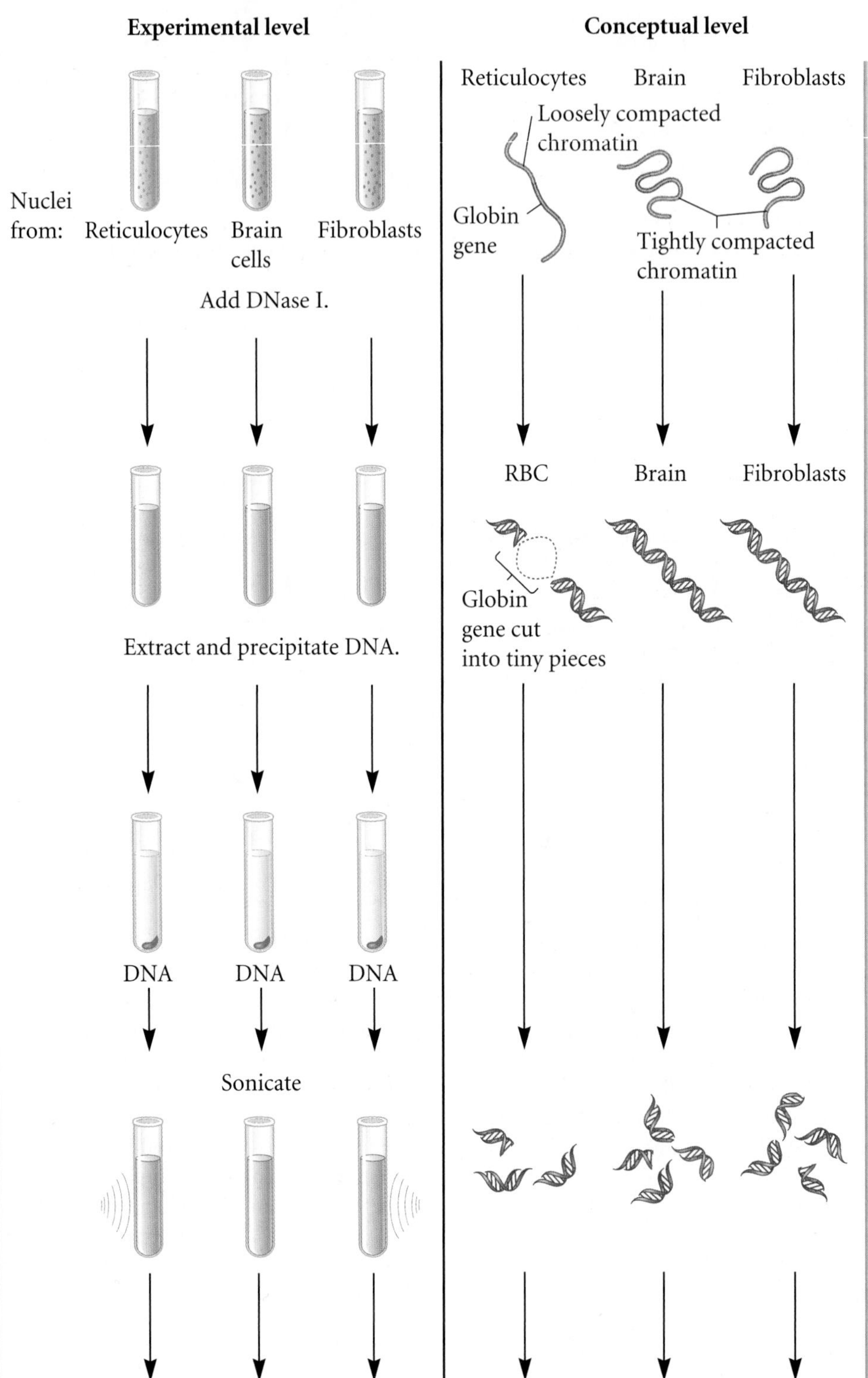

4. Add a radiolabeled DNA probe that is complementary to the β-globin gene.

5. Denature the DNA into single strands by treatment with high temperature. Then cool it to 65°C to allow the complementary DNA strands to hybridize with each other.

6. Reduce the temperature further and divide each sample into two tubes. Add S1 nuclease to one tube of each sample. Omit S1 nuclease from the second tube.

7. Precipitate the DNA with trichloroacetic acid. Subject to centrifugation.

8. Count the amount of radioactivity in the pellet. (The technique of scintillation counting is described in the Appendix.) The amount of radioactivity in the presence of S1 nuclease divided by the amount of radioactivity in the absence of S1 nuclease provides a measure of the percentage of radiolabeled DNA that has hybridized to the chromosomal DNA.

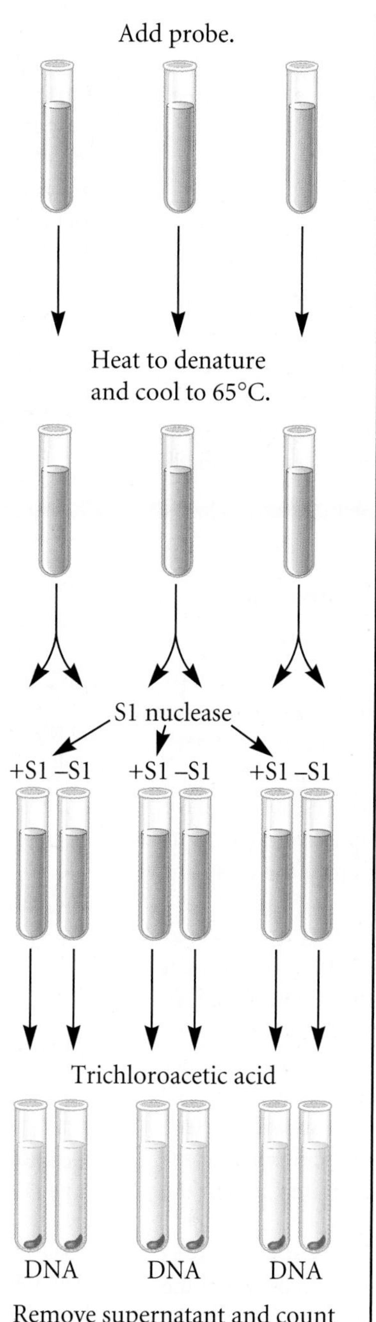

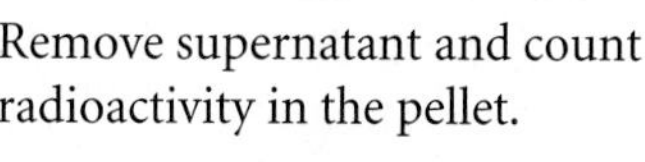

Remove supernatant and count radioactivity in the pellet.

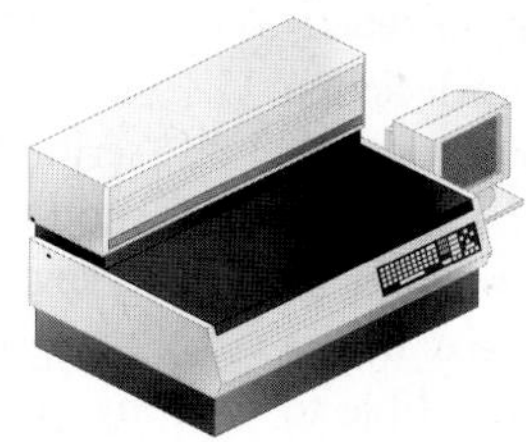

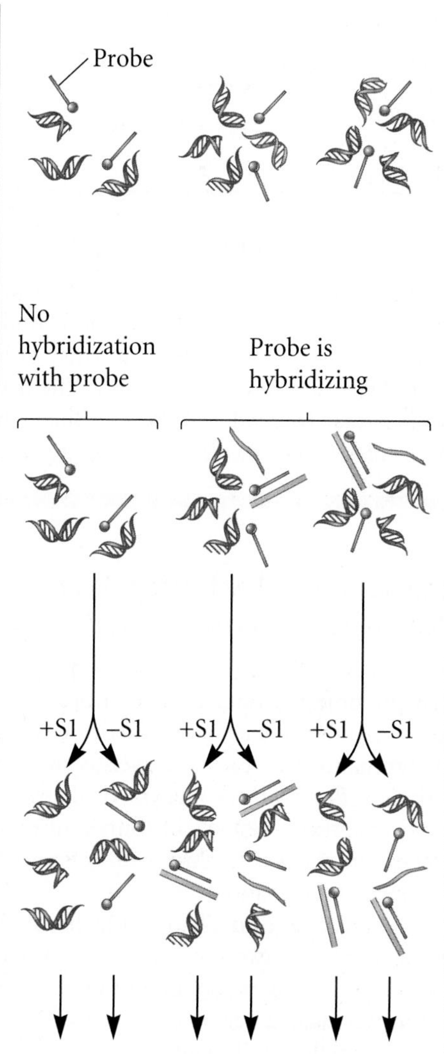

Single- and double-stranded DNA is in the pellet in the absence of S1; only double-stranded DNA is in the pellet in the presence of S1, because S1 digests single-stranded DNA.

THE DATA

Source of Nuclei	*% Hybridization of DNA Probe*
Reticulocytes	25%
Brain cells	>94%
Fibroblasts	>94%

INTERPRETING THE DATA

As shown in the data of Figure 15.10, a much smaller percentage of the radiolabeled DNA probe hybridized to the chromosomal DNA in reticulocytes compared to brain cells and fibroblasts. What do these results mean? The data are consistent with the idea that DNase I had digested the globin gene in the chromatin of reticulocytes into small fragments that were too small to hybridize to the radiolabeled DNA probe. In other words, the globin genes in reticulocytes are more sensitive to DNase I digestion. By comparison, the globin genes in brain cells and fibroblasts are relatively resistant. Because the globin genes are expressed in reticulocytes but not in brain cells and fibroblasts, these results are consistent with the hypothesis that the globin gene is less tightly packed when it is being expressed. In cells where the globin genes should not be expressed, such as brain cells and fibroblasts, the chromatin containing the β-globin gene is tightly compacted. Changes in chromatin compaction provide one way to regulate globin gene expression among different cell types.

A self-help quiz involving this experiment can be found at the Online Learning Center.

Transcriptional Activators Recruit Chromatin-Remodeling Enzymes to the Promoter Region

In recent years, geneticists have been trying to identify the steps that promote the interconversion between the closed and open conformations of chromatin. As we saw for globin genes, changes in chromatin structure are associated with the level of gene transcription. Based on the analysis of many genes, researchers have discovered that a key role of some transcriptional activators is to orchestrate changes in chromatin compaction from the closed to the open conformation.

As discussed in Chapter 12, chromatin structure is altered in eukaryotes in two common ways: ATP-dependent chromatin remodeling and covalent modification of histones. The SWI/SNF family is a major category of proteins found in all eukaryotic species that catalyze ATP-dependent chromatin remodeling. The net result of ATP-dependent chromatin remodeling is a change in the locations of nucleosomes. This may involve a shift in nucleosomes to a new location or a change in the spacing of nucleosomes over a long stretch of DNA, both of which may loosen the level of chromatin compaction.

Nucleosomes have been shown to change position in cells that normally express a particular gene but not in cells where the gene is inactive. For example, in fibroblasts that do not express the β-globin gene, nucleosomes are positioned at regular intervals from nucleotides –3,000 to +1,500 (**Figure 15.11a**). However, in reticulocytes that express the β-globin gene, a disruption in nucleosome positioning occurs in the region from nucleotide –500 to +200 (**Figure 15.11b**). This disruption is a key step in gene activation. The position of a nucleosome may greatly influence whether or not a gene can be transcribed. For example, if the TATA box is tightly bound to a core histone protein, it may be inaccessible to general transcription factors and RNA polymerase.

A second mechanism that leads to an alteration in chromatin structure involves the covalent modification of histones. A histone code appears to play a key role in controlling the level of chromatin compaction. The amino terminal ends of histone proteins are covalently modified in several ways, including the acetylation of lysines, methylation of lysines, and phosphorylation of serines (see Chapter 10, Figure 10.20). Because these covalent modifications influence the compaction of chromatin, they play a key role in transcriptional regulation. For example, positively charged lysines within the core histone proteins can be acetylated by enzymes called **histone acetyltransferases.** The attachment of the acetyl group ($-COCH_3$) may have two effects. First, it eliminates the positive charge on the lysine side chain and thereby disrupts the electrostatic attraction between the histone protein and the negatively charged DNA backbone. Secondly, covalently modified histones are recognized by DNA-binding proteins that promote changes in the compaction level of the chromatin. In the case of acetylation, the net effect is to diminish the interactions

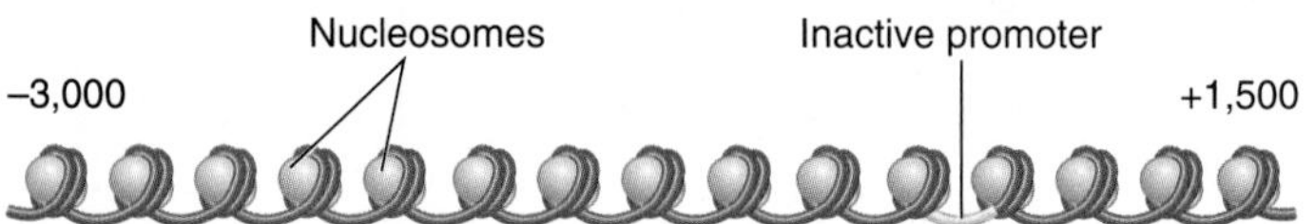

(a) Nucleosomes in fibroblasts, which do not express the β-globin gene

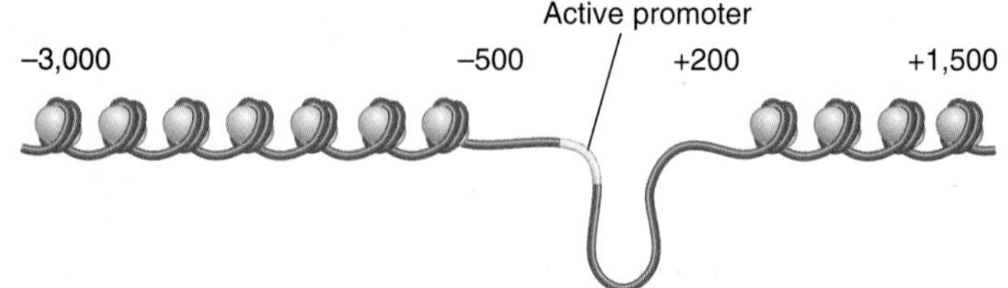

(b) Nucleosomes in reticulocytes, which express the β-globin gene

FIGURE 15.11 **Changes in nucleosome position during the activation of the β-globin gene.** **(a)** In fibroblasts, which do not express β globin, nucleosome positioning is uninterrupted. **(b)** In reticulocytes, which express this gene, disruption occurs in the positions of nucleosomes from the –500 to +200 region. (Position +1 is the beginning of transcription.) It is not yet clear whether the histones are removed, the histones are partially displaced, and/or other proteins are bound to this region.

between DNA and nucleosomes and thereby loosen the compaction of the 30 nm fiber. This would facilitate the ability of RNA polymerase to transcribe a gene. Some studies suggest that histones are completely displaced, whereas others suggest they are loosened but remain attached to the DNA.

An important role of certain transcriptional activators is to recruit ATP-dependent remodeling enzymes and histone-modifying enzymes to the promoter region. Though the order of recruitment may differ among specific transcriptional activators, this appears to be a critical step that is necessary to initiate transcription. A well-studied example involves a gene in yeast that is involved in mating. Yeast can exist in two mating types, termed *a* and α. The yeast *HO* gene encodes an enzyme that is required for the switch from one mating type to the other.

Prior to transcription, much of the region surrounding the core promoter of the *HO* gene is in a closed conformation. How is this region loosened up so that transcription can occur? A simplified mechanism for the transcription of this gene is shown in **Figure 15.12**. A regulatory transcription factor (SWI5P) binds to an accessible enhancer in the vicinity of the core promoter. This transcription factor then recruits an ATP-dependent remodeling enzyme (SWI/SNF) to the region, which promotes the conversion of the chromatin from the closed to the open conformation. Next, a histone acetyltransferase (SAGA) is recruited to this region, which serves to further open the chromatin conformation. Overall, the actions of the ATP-dependent remodeling enzyme and the histone acetyltransferase remodel a region around the promoter that is approximately 1,000 bp in length. A second regulatory transcription factor (SBP) is then able to bind to an enhancer near the core promoter and thereby initiate the binding of general transcription factors and RNA polymerase.

DNA Methylation Usually Inhibits Gene Transcription

We now turn our attention to a change in chromatin structure that silences gene expression. As discussed in previous chapters, DNA structure can be modified by the covalent attachment of methyl groups. DNA methylation is common in some but not all eukaryotic species. For example, yeast and *Drosophila* have little or no detectable methylation of their DNA, whereas DNA methylation in vertebrates and plants is relatively abundant. In mammals, approximately 2 to 7% of the DNA is methylated. As shown in **Figure 15.13**, eukaryotic DNA methylation occurs via an enzyme called **DNA methyltransferase,** which attaches

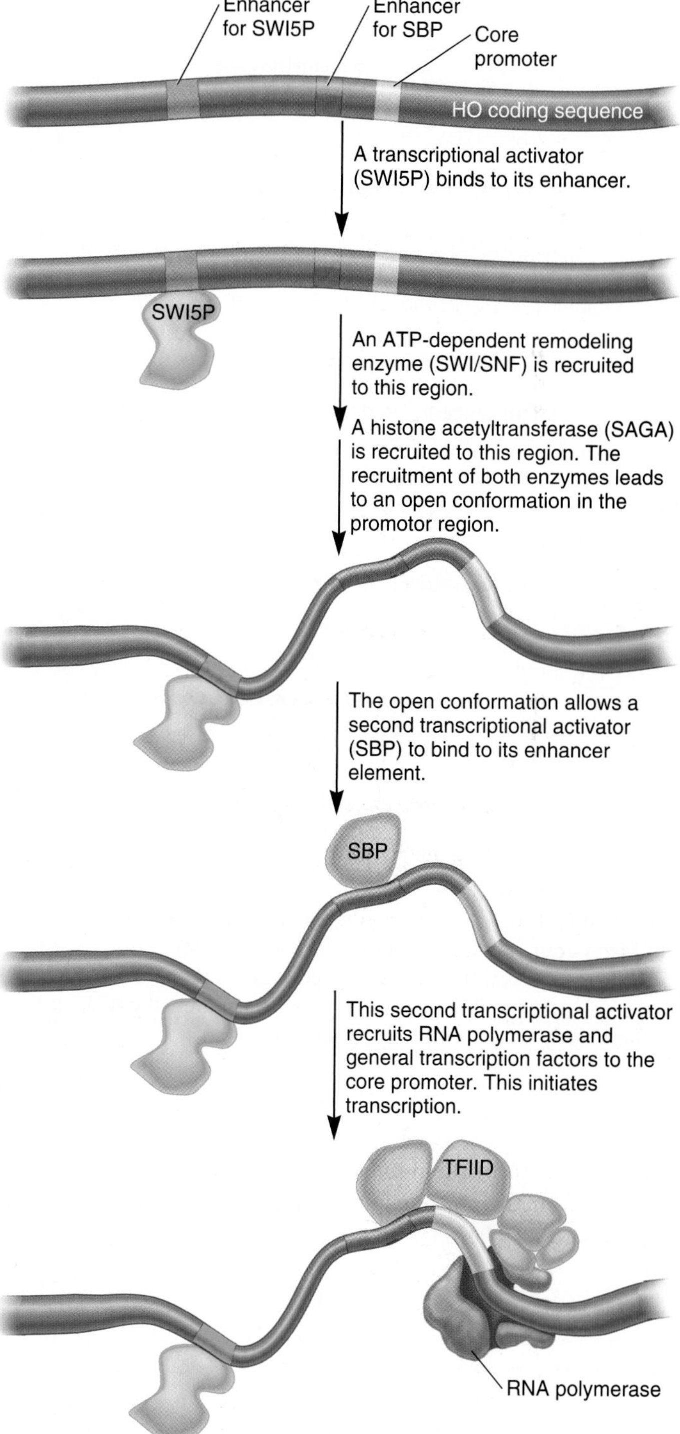
Enhancer for SWI5P
Enhancer for SBP
Core promoter
HO coding sequence
A transcriptional activator (SWI5P) binds to its enhancer.
SWI5P
An ATP-dependent remodeling enzyme (SWI/SNF) is recruited to this region.
A histone acetyltransferase (SAGA) is recruited to this region. The recruitment of both enzymes leads to an open conformation in the promotor region.
The open conformation allows a second transcriptional activator (SBP) to bind to its enhancer element.
SBP
This second transcriptional activator recruits RNA polymerase and general transcription factors to the core promoter. This initiates transcription.
TFIID
RNA polymerase

ONLINE ANIMATION

FIGURE 15.12 **An example of chromatin remodeling of a promoter region that leads to transcriptional activation.** The *HO* gene is involved in yeast mating type switching. For this gene to be transcribed, the promoter region must be converted from a closed to an open conformation. A transcription factor termed SWI5P (SWI refers to mating type switching) binds to an enhancer in the promoter region and recruits an ATP-dependent chromatin remodeling enzyme (of the SWI/SNF family) to the region. This begins the opening process. A histone acetyltransferase termed SAGA is then recruited to the region and further decompacts the chromatin. SAGA is an acronym for Spt/Ada/GCN5/Acetyltransferase. *Spt, Ada,* and *GCN5* are the names of three genes that had been shown to be transcriptionally regulated by this histone acetyltransferase. A second transcription factor termed SBP (an acronym for a mating type switching cell cycle box protein) binds to an enhancer in the promoter region and initiates transcription by recruiting general transcription factors and RNA polymerase.

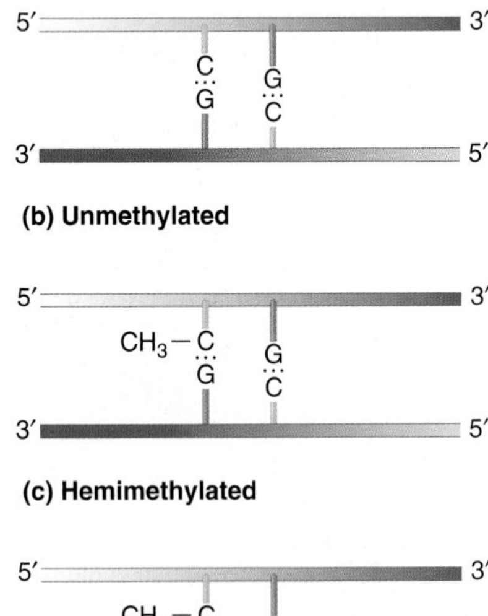

FIGURE 15.13 DNA methylation on cytosine bases. (**a**) Methylation occurs via an enzyme known as DNA methyltransferase, which attaches a methyl group to the number 5 carbon on cytosine. The CG sequence can be (**b**) unmethylated, (**c**) hemimethylated, or (**d**) fully methylated.

a methyl group to the number 5 position of the cytosine base, forming 5-methylcytosine. The sequence that is methylated is shown here.

$$\begin{array}{c} CH_3 \\ | \\ 5'\text{–CG–}3' \\ 3'\text{–GC–}5' \\ \quad | \\ \quad CH_3 \end{array}$$

Note that this sequence contains cytosines in both strands. Methylation of the cytosine in both strands is termed full methylation, whereas methylation of only one strand is called hemimethylation.

DNA methylation usually inhibits the transcription of eukaryotic genes, particularly when it occurs in the vicinity of the promoter. In vertebrates and plants, **CpG islands** occur near many promoters of genes. (Note: CpG refers to a dinucleotide of C and G in DNA that is connected by a phosphodiester linkage.) These CpG islands are commonly 1,000 to 2,000 base pairs in length and contain a high number of CpG sites. In the case of **housekeeping genes**—genes that encode proteins required in most cells of a multicellular organism—the cytosine bases in the CpG islands are unmethylated. Therefore, housekeeping genes tend to be expressed in most cell types. By comparison, other genes are highly regulated and may be expressed only in a particular cell type. These are **tissue-specific genes.** In some cases, it has been found that the expression of such genes may be silenced by the methylation of CpG islands. Overall, evidence is accumulating that unmethylated CpG islands are correlated with active genes, whereas suppressed genes contain methylated CpG islands. In this way, DNA methylation is thought to play an important role in the silencing of tissue-specific genes so they are not expressed in the wrong tissue.

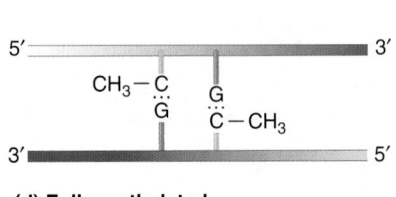

Methylation can inhibit transcription in two general ways. First, methylation of CpG islands may prevent the binding of transcription factors to the promoter region. For example, methylated CG sequences could prevent the binding of an activator protein to an enhancer element, presumably by the methyl group protruding into the major groove of the DNA (Figure 15.14a). The inability of an activator protein to bind to the DNA would inhibit the initiation of transcription. However, CG methylation does not slow down the movement of RNA polymerase along a gene. In vertebrates and plants, coding regions downstream from the core promoter usually contain methylated CG sequences, but these do not hinder the elongation phase of transcription. This suggests that methylation must occur in the vicinity of the promoter to have an inhibitory effect on transcription.

A second way that methylation inhibits transcription is by converting chromatin from an open to a closed conformation (Figure 15.14b). Proteins known as **methyl-CpG-binding proteins** bind methylated sequences. These proteins contain a domain called the methyl-binding domain that specifically recognizes a methylated CG sequence. Once bound to the DNA, the methyl-CpG-binding protein recruits other proteins to the region that cause the chromatin to become very compact. For example, methyl-CpG-binding proteins may recruit histone deacetylase to a methylated CpG island near a promoter. Histone deacetylation removes acetyl groups from the histone proteins and thereby favors a chromatin conformation that is very compact.

DNA Methylation Is Heritable

Methylated DNA sequences are inherited during cell division. Experimentally, if fully methylated DNA is introduced into a plant or vertebrate cell, the DNA will remain fully methylated even in subsequently produced daughter cells. However, if the same

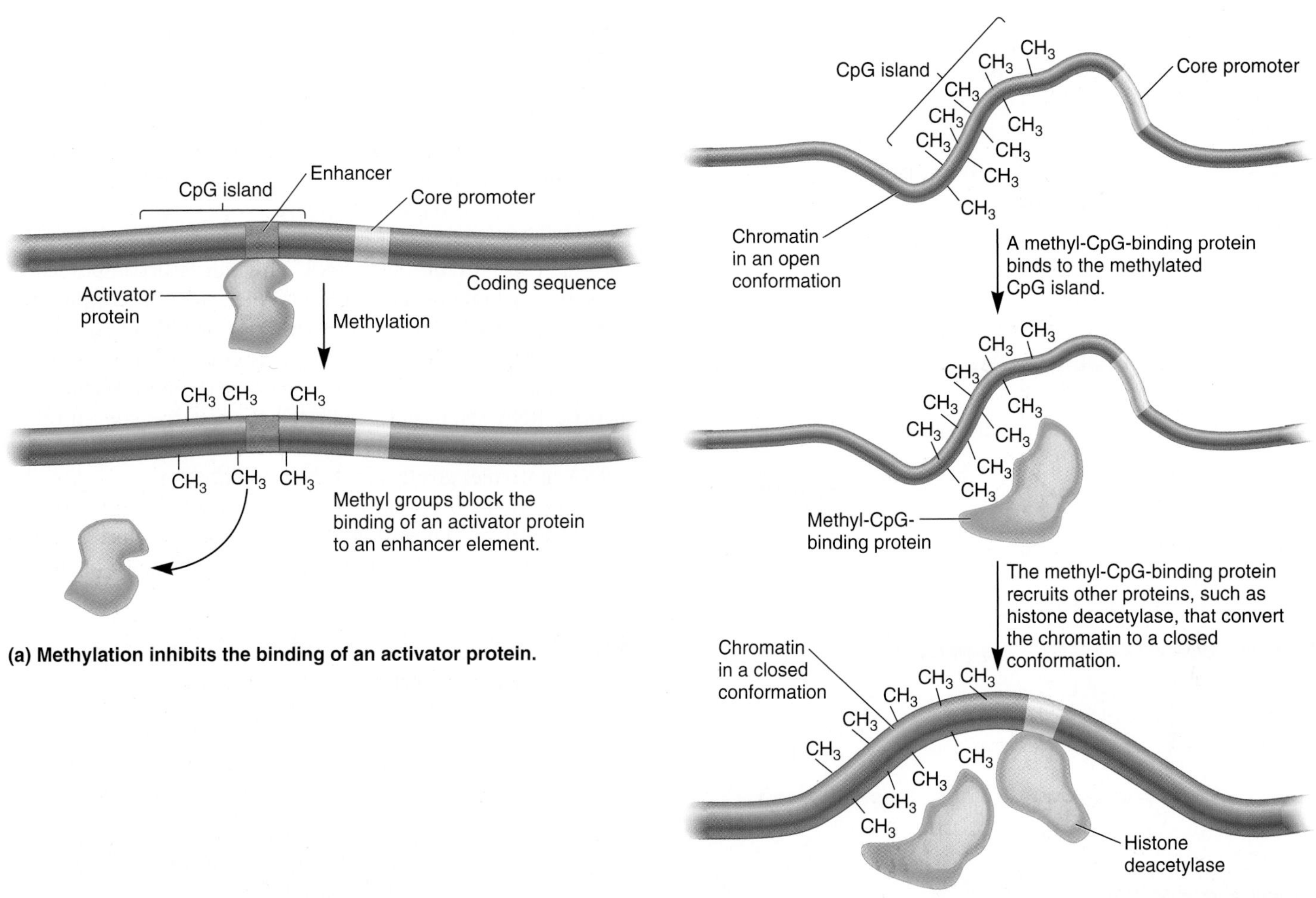
CpG island
Enhancer
Core promoter
Activator protein
Coding sequence
Methylation
CH3 CH3 CH3
CH3 CH3 CH3
Methyl groups block the binding of an activator protein to an enhancer element.
(a) Methylation inhibits the binding of an activator protein.
CpG island
Core promoter
Chromatin in an open conformation
A methyl-CpG-binding protein binds to the methylated CpG island.
Methyl-CpG-binding protein
The methyl-CpG-binding protein recruits other proteins, such as histone deacetylase, that convert the chromatin to a closed conformation.
Chromatin in a closed conformation
Histone deacetylase
(b) Methyl-CpG-binding protein recruits other proteins that cause the region to become more compact.

FIGURE 15.14 **Transcriptional silencing via methylation.** (**a**) The methylation of a CpG island may inhibit the binding of transcriptional activators to the promoter region. (**b**) The binding of a methyl-CpG-binding protein to a CpG island may lead to the recruitment of proteins, such as histone deacetylase, that convert chromatin to a closed conformation and thus suppress transcription.

sequence of nonmethylated DNA is introduced into a cell, it will remain nonmethylated in the daughter cells. These observations indicate that the pattern of methylation is retained following DNA replication and, therefore, is inherited in future daughter cells.

How can methylation be inherited from cell to cell? Figure 15.15 illustrates a molecular model that explains this process, which was originally proposed by Arthur Riggs, Robin Holliday, and J. E. Pugh. The DNA in a particular cell may become methylated by ***de novo* methylation**—the methylation of DNA that was previously unmethylated. When a fully methylated segment of DNA replicates in preparation for cell division, the newly made daughter strands contain unmethylated cytosines. Because only one strand is methylated, such DNA is called hemimethylated. This hemimethylated DNA is efficiently recognized by DNA methyltransferase, which makes it fully methylated. This process is called **maintenance methylation,** because it preserves the methylated condition in future cells. However, maintenance methylation does not act on unmethylated DNA. Overall, maintenance methylation appears to be an efficient process that routinely occurs within vertebrate and plant cells. By comparison, *de novo* methylation and demethylation are infrequent and highly regulated events. According to this view, the initial methylation or demethylation of a given gene can be regulated so that it occurs in a specific cell type or stage of development. Once methylation has occurred, it can then be transmitted from mother to daughter cells via maintenance methylation.

The methylation mechanism shown in Figure 15.15 can explain the phenomenon of genomic imprinting, which is described in Chapter 7. In this case, specific genes are methylated during oogenesis or spermatogenesis, but not both. Following fertilization, the pattern of methylation is maintained in the offspring. For example, if a gene is methylated only during spermatogenesis, the allele that is inherited from the father will be methylated in the somatic cells of the offspring, while the maternal allele

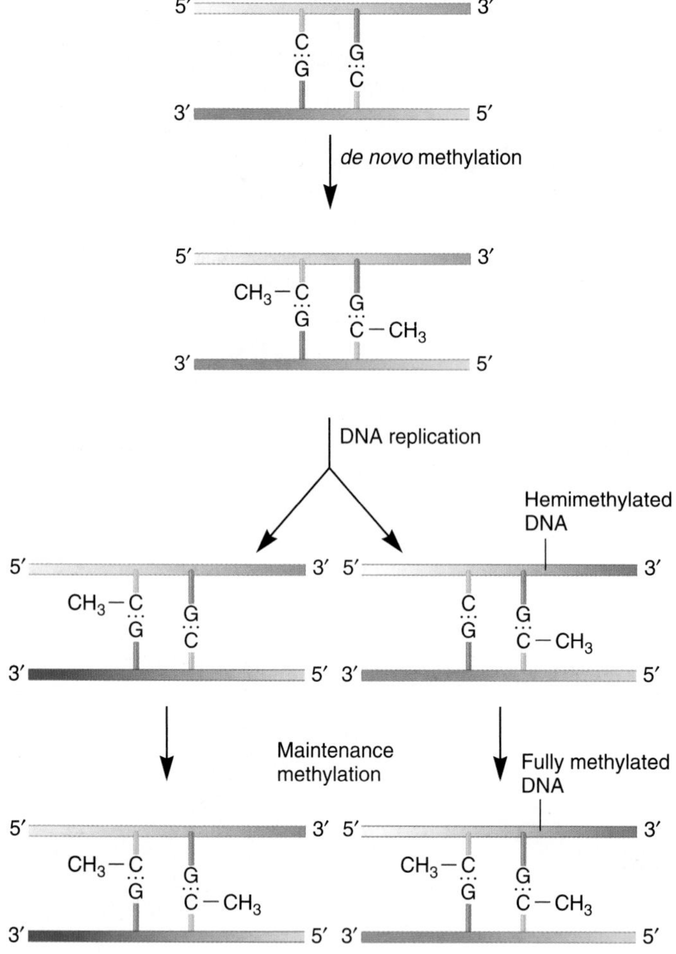

FIGURE 15.15 A molecular model for the inheritance of DNA methylation. The DNA initially undergoes *de novo* methylation, which is a rare, highly regulated event. Once this occurs, DNA replication produces hemimethylated DNA molecules, which are then fully methylated by DNA methyltransferase. This process, called maintenance methylation, is a routine event that is expected to occur for all hemimethylated DNA.

will remain unmethylated. Along these lines, geneticists are also eager to determine how variations in DNA methylation patterns may be important for cell differentiation. It may be a key way to silence genes in different cell types. However, additional research will be necessary to understand how specific genes may be targeted for *de novo* methylation or demethylation during different developmental stages or in specific cell types.

15.3 REGULATION OF RNA PROCESSING, RNA STABILITY, AND TRANSLATION

Thus far, we have considered a variety of mechanisms that regulate the level of gene transcription. These mechanisms control the amount of RNA transcribed from a given gene. In addition, eukaryotic gene expression is commonly regulated after the RNA is made (**Table 15.2**). One mechanism is pre-mRNA processing. Following transcription, a pre-mRNA transcript is processed before it becomes a functional mRNA. These processing events, which include splicing, capping, polyA tailing, and RNA editing, were described in Chapter 12. We will begin this section by examining how the processes of alternative splicing and RNA editing are regulated at the RNA level.

Another strategy for regulating gene expression is to influence the concentration of mRNA. This can be accomplished by regulating the rate of transcription. When the transcription of a gene is increased, a higher concentration of the corresponding RNA results. In addition, RNA concentration is greatly affected by the stability or half-life of a particular RNA. Factors that increase RNA stability are expected to raise the concentration of that RNA molecule. We will explore how sequences within mRNA molecules greatly affect their stability.

We will also consider a newly discovered mechanism of mRNA silencing, known as RNA interference, which involves double-stranded RNAs that may direct the breakdown of specific

TABLE 15.2

Gene Regulation via RNA Processing and Translation

Effect	Description
Alternative splicing	Certain pre-mRNAs can be spliced in more than one way, leading to polypeptides that have different amino acid sequences. Alternative splicing is often cell specific so that a protein can be fine-tuned to function in a particular cell type. It is an important form of gene regulation in multicellular eukaryotic species.
RNA editing	The sequence of RNAs can be altered after the RNA is made. This usually involves the addition or deletion of one or a few bases, or a change of C to U, or A to I. This does not appear to be a widespread phenomenon, though the study of RNA editing is still relatively recent.
RNA stability	The amount of RNA is greatly influenced by the half-life of RNA transcripts. A long polyA tail on mRNAs promotes their stability due to the binding of polyA-binding protein. Some RNAs with a relatively short half-life contain sequences that target them for rapid destruction. Some RNAs are stabilized by specific RNA-binding proteins that usually bind near the 3′ end.
RNA interference	Double-stranded RNA can mediate the degradation of homologous mRNAs in the cell. This is a mechanism of gene regulation. Also, it probably provides eukaryotic cells with protection from invasion by certain types of viruses and may prevent the movement of transposable elements.
General regulation of translation	The function of translational initiation factors may be regulated to permit or inhibit translation. This regulation affects the translation of all cellular mRNAs. Inhibition of translation is desirable if a cell has been exposed to a virus or to toxic materials.
Translational regulation of specific mRNAs	Some mRNAs are regulated via binding proteins that inhibit the ability of the ribosomes to initiate translation. These proteins usually bind at the 5′ end of the mRNA and thereby prevent the ribosome from binding.

RNAs or inhibit their translation. In addition to RNA interference, the process of mRNA translation may be regulated at the level of the translational machinery such as ribosomes. In eukaryotes, this can occur by directly controlling the function of translational initiation factors. Also, RNA-binding proteins can prevent ribosomes from initiating the translation process for specific mRNAs. In this section, we will examine these interesting mechanisms for regulating mRNA translation.

Alternative Splicing Regulates Which Exons Occur in an RNA Transcript, Allowing Different Polypeptides to Be Made from the Same Structural Gene

When it was first discovered, the phenomenon of splicing seemed like a rather wasteful process. During transcription, energy is used to synthesize intron sequences. Likewise, energy is also used to remove introns via a large spliceosome complex. This observation intrigued many geneticists, because natural selection tends to eliminate wasteful processes. Therefore, instead of simply viewing splicing as a wasteful process, many geneticists expected to find that pre-mRNA splicing has one or more important biological roles. In recent years, one very important biological advantage has become apparent. This is **alternative splicing,** which refers to the phenomenon that a pre-mRNA can be spliced in more than one way.

What is the advantage of alternative splicing? To understand the biological effects of alternative splicing, remember that the sequence of amino acids within a polypeptide determines the structure and function of a protein. Alternative splicing produces two or more polypeptides with differences in their amino acid sequences, leading to possible changes in their functions. In most cases, the alternative versions of the protein will have similar functions, because much of their amino acid sequences are identical to each other. Nevertheless, alternative splicing produces differences in amino acid sequences that will provide each polypeptide with its own unique characteristics. Because alternative splicing allows two or more different polypeptide sequences to be derived from a single gene, some geneticists have speculated that an important advantage of this process is that it allows an organism to carry fewer genes in its genome.

The degree of splicing and alternative splicing varies greatly among different species. Baker's yeast, for example, contains about 6,300 genes and approximately 300 (i.e., approximately 5%) encode pre-mRNAs that are spliced. Of these, only a few have been shown to be alternatively spliced. Therefore, in this unicellular eukaryote, alternative splicing is not a major mechanism to generate protein diversity. In comparison, complex multicellular organisms seem to rely on alternative splicing to a great extent. Humans contain approximately 20,000 to 25,000 different genes, and most of these contain one or more introns. Recent estimates suggest that about 70% of all human pre-mRNAs are alternatively spliced. Furthermore, certain pre-mRNAs are alternatively spliced to an extraordinary extent. In fact, some pre-mRNAs can be alternatively spliced to produce dozens of different mRNAs. This provides a much greater potential for human cells to create protein diversity.

Figure 15.16 considers an example of alternative splicing for a gene that encodes a protein known as α-tropomyosin. This protein functions in the regulation of cell contraction. It is located along the thin filaments found in smooth muscle cells, such as those in the uterus and small intestine, and in striated muscle cells that are found in cardiac and skeletal muscle. Alpha-tropomyosin is also synthesized in many types of nonmuscle cells but in lower amounts. Within a multicellular organism, different types of cells must regulate their contractibility in subtly different ways. One way that this may be accomplished is by the production of different forms of α-tropomyosin.

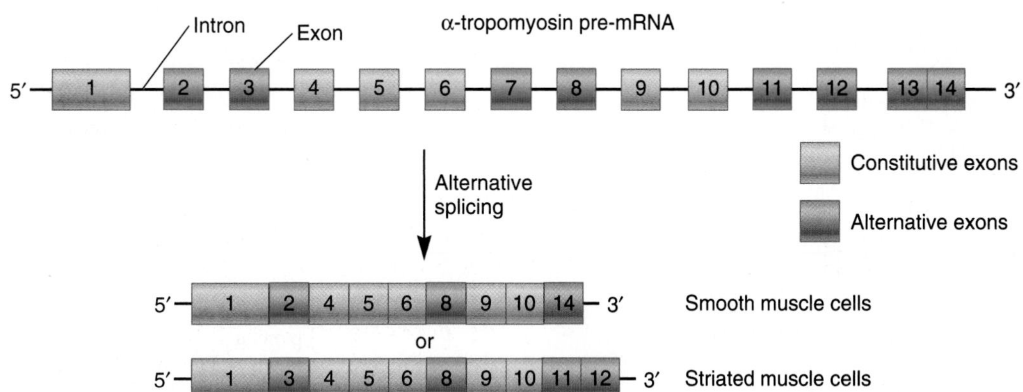

FIGURE 15.16 Alternative ways that the rat α-tropomyosin pre-mRNA can be spliced. The top part of this figure depicts the structure of the rat α-tropomyosin pre-mRNA. Exons are shown as colored boxes, and introns are illustrated as connecting black lines. The lower part of the figure describes the final mRNA products in smooth and striated muscle cells. Note: Exon 8 is found in the final mRNA of smooth and striated muscle cells, but not in the mRNA of certain other cell types. The junction between exons 13 and 14 contains a 3' splice so that exon 13 can be separated from exon 14.

Genes→Traits α-tropomyosin functions in the regulation of cell contraction in muscle and nonmuscle cells. Alternative splicing of the pre-mRNA provides a way to vary contractibility in different types of cells by modifying the function of α-tropomyosin. As shown here, the alternatively spliced versions of the pre-mRNA produce α-tropomyosin proteins that differ slightly from each other in their structure (i.e., amino acid sequence). These alternatively spliced versions vary in function to meet the needs of the cell type in which they are found. For example, the sequence of exons 1–2–4–5–6–8–9–10–14 produces an α-tropomyosin protein that functions suitably in smooth muscle cells. Overall, alternative splicing affects the traits of an organism by allowing a single gene to encode several versions of a protein, each optimally suited to the cell type in which it is made.

The intron-exon structure of the rat α-tropomyosin pre-mRNA and two alternative ways that the pre-mRNA can be spliced are described in Figure 15.16. The pre-mRNA contains 14 exons, six of which are **constitutive exons** (shown in red), which are always found in the mature mRNA from all cell types. Presumably, constitutive exons encode polypeptide segments of the α-tropomyosin protein that are necessary for its general structure and function. By comparison, **alternative exons** (shown in green) are not always found in the mRNA after splicing has occurred. The polypeptide sequences encoded by alternative exons may subtly change the function of α-tropomyosin to meet the needs of the cell type in which it is found. For example, Figure 15.16 shows the predominant splicing products found in smooth muscle cells and striated muscle cells. Exon 2 encodes a segment of the α-tropomyosin protein that alters its function to make it suitable for smooth muscle cells. By comparison, the α-tropomyosin mRNA found in striated muscle cells does not include exon 2. Instead, this mRNA contains exon 3, which is more suitable for that cell type.

Alternative splicing is not a random event. Rather, the specific pattern of splicing is regulated in any given cell. The molecular mechanism for the regulation of alternative splicing involves proteins known as **splicing factors.** Such splicing factors play a key role in the choice of particular splice sites. **SR proteins** are an example of a type of splicing factor. SR proteins contain a domain at their carboxyl terminal end that is rich in serines (S) and arginines (R) and is involved in protein-protein recognition. They also contain an RNA-binding domain at their amino terminal end.

As discussed in Chapter 12, components of the spliceosome recognize the 5' and 3' splice sites and then remove the intervening intron. The key effect of splicing factors is to modulate the ability of the spliceosome to choose 5' and 3' splice sites. This can occur in two ways. Some splicing factors act as repressors that inhibit the ability of the spliceosome to recognize a splice site. In **Figure 15.17a**, a splicing repressor binds to a 3' splice site and prevents the spliceosome from recognizing the site. Instead, the spliceosome binds to the next available 3' splice site. The splicing

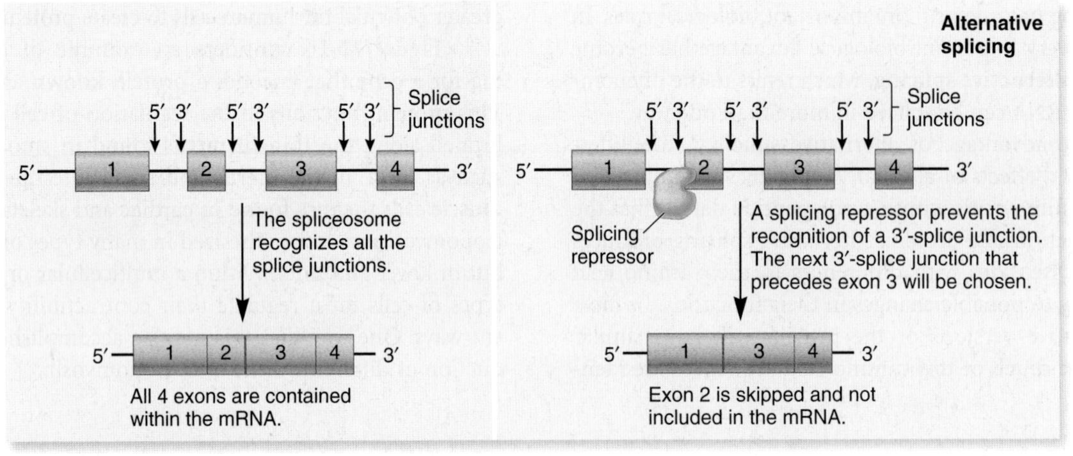

(a) Splicing repressors

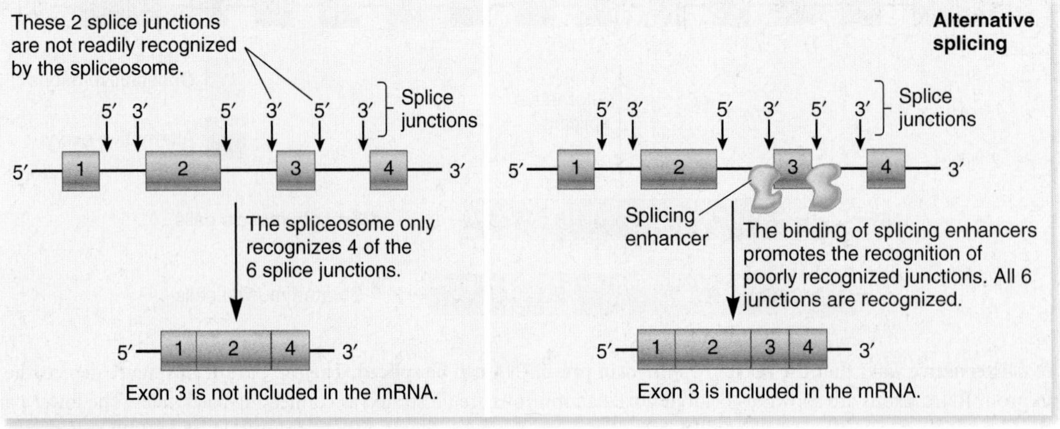

(b) Splicing enhancers

FIGURE 15.17 **The roles of splicing factors during alternative splicing.** (**a**) Splicing factors can act as repressors to prevent the recognition of splice sites. In this example, the presence of the splicing repressor causes exon 2 to be skipped and thus not included in the mRNA. (**b**) Other splicing factors can enhance the recognition of splice sites. In this example, the splicing enhancers promote the recognition of sites that flank exon 3, thereby causing its inclusion in the mRNA.

repressor causes exon 2 to be spliced out of the mature mRNA, an event called **exon skipping.** Alternatively, other splicing factors enhance the ability of the spliceosome to recognize particular splice sites. In **Figure 15.17b**, splicing enhancers bind to the 3' and 5' splice sites that flank exon 3, which results in the inclusion of exon 3 in the mature mRNA.

Alternative splicing in different tissues is thought to occur because each cell type has its own characteristic concentration of many kinds of splicing factors. Furthermore, much like transcription factors, splicing factors may be regulated by the binding of small effector molecules, protein–protein interactions, and covalent modifications. Overall, the differences in the composition of splicing factors, and the regulation of their activities, form the basis for alternative splicing decisions.

The Nucleotide Sequence of RNA Can Be Modified by RNA Editing

The term **RNA editing** refers to a change in the nucleotide sequence of an RNA molecule that involves additions or deletions of particular bases, or a conversion of one type of base to another, such as a cytosine to a uracil. In the case of mRNAs, editing can have various effects, such as generating start codons, generating stop codons, and changing the coding sequence for a polypeptide.

The phenomenon of RNA editing was first identified in trypanosomes, the protists that cause sleeping sickness. As with the discovery of RNA splicing, the initial finding of RNA editing was met with great skepticism. Since that time, however, RNA editing has been shown to occur in various organisms and in a variety of ways, although its functional significance is slowly emerging (**Table 15.3**).

The molecular mechanisms of RNA editing have been the subject of many investigations. In the specific case of trypanosomes, which are single-celled protists, the editing process involves the participation of a **guide RNA.** This guide RNA directs the addition of one or more uracil nucleotides into an RNA or the deletion of one or more uracils. As shown in **Figure 15.18**, the guide RNA has two important characteristics. First, its 5' anchor is complementary to the RNA that is to be edited. Second, it has uracil nucleotides at its 3' end. During the editing process, the 5' anchor binds to the target RNA. The RNA is cleaved at a defined location by an endonuclease, and the 3' end of the guide RNA becomes displaced from the target RNA. For additions, an enzyme called terminal U-transferase then attaches uracil-containing nucleotides. For deletions, an enzyme called 3'-U-exonuclease removes one or more uracils. Finally, RNA ligase rejoins the two pieces of RNA.

A more widespread mechanism for RNA editing involves changes of one type of base to another. In this form of editing, a base in the RNA is deaminated—an amino group is removed from the base. When cytosine is deaminated, uracil is formed, and when adenine is deaminated, inosine is formed (**Figure 15.19**). Inosine is recognized as guanine during translation.

An example of RNA editing occurs in mammals involving an mRNA that encodes a protein called apolipoprotein B. In the liver, the RNA editing process produces apolipoprotein B-100, a protein essential for the transport of cholesterol in the blood. In intestinal cells, the mRNA may be edited so that a single C is changed to a U. What is the significance of this base substitution? This change converts a glutamine codon (CAA) to a stop codon (UAA) and thereby results in a shorter apolipoprotein. In this case, RNA editing produces an apolipoprotein B with an altered structure. Therefore, RNA editing can produce two proteins from the same gene, much like the phenomenon of alternative splicing described earlier in this chapter.

TABLE 15.3

Examples of RNA Editing

Organism	Type of Editing	Found in
Trypanosomes (protozoa)	Primarily additions but occasionally deletions of uracil nucleotides	Many mitochondrial mRNAs
Land plants	C-to-U conversion	Many mitochondrial and chloroplast mRNAs, tRNAs, and rRNAs
Slime mold	C additions	Many mitochondrial mRNAs
Mammals	C-to-U conversion	Apolipoprotein B mRNA, and NFI mRNA, which encodes a tumor-suppressor protein
	A-to-I conversion	Glutamate receptor mRNA, many tRNAs
Drosophila	A-to-I conversion	mRNA for calcium and sodium channels

The Stability of mRNA Influences mRNA Concentration

In eukaryotes, the stability of mRNAs can vary considerably. Certain mRNAs have very short half-lives, such as several minutes, whereas others can persist for many days or even months. In some cases, the stability of an mRNA can be regulated so that its half-life is shortened or lengthened. A change in the stability of mRNA can greatly influence the cellular concentration of that mRNA molecule. In this way, mechanisms that control mRNA stability can dramatically affect gene expression.

Various factors can play a role in mRNA stability. One important structural feature is the length of the polyA tail. As you may recall from Chapter 12, most newly made mRNAs contain a polyA tail that averages 200 nucleotides in length. The polyA tail is recognized by the **polyA-binding protein.** The binding of this protein enhances RNA stability. However, as an mRNA ages, its polyA tail tends to be shortened by the action of cellular exonucleases. Once it becomes less than 10 to 30 adenosines in length, the polyA-binding protein can no longer bind, and the mRNA is rapidly degraded by exo- and endonucleases.

Certain mRNAs, particularly those with short half-lives, contain sequences that act as destabilizing elements. While these destabilizing elements can be located anywhere within the mRNA,

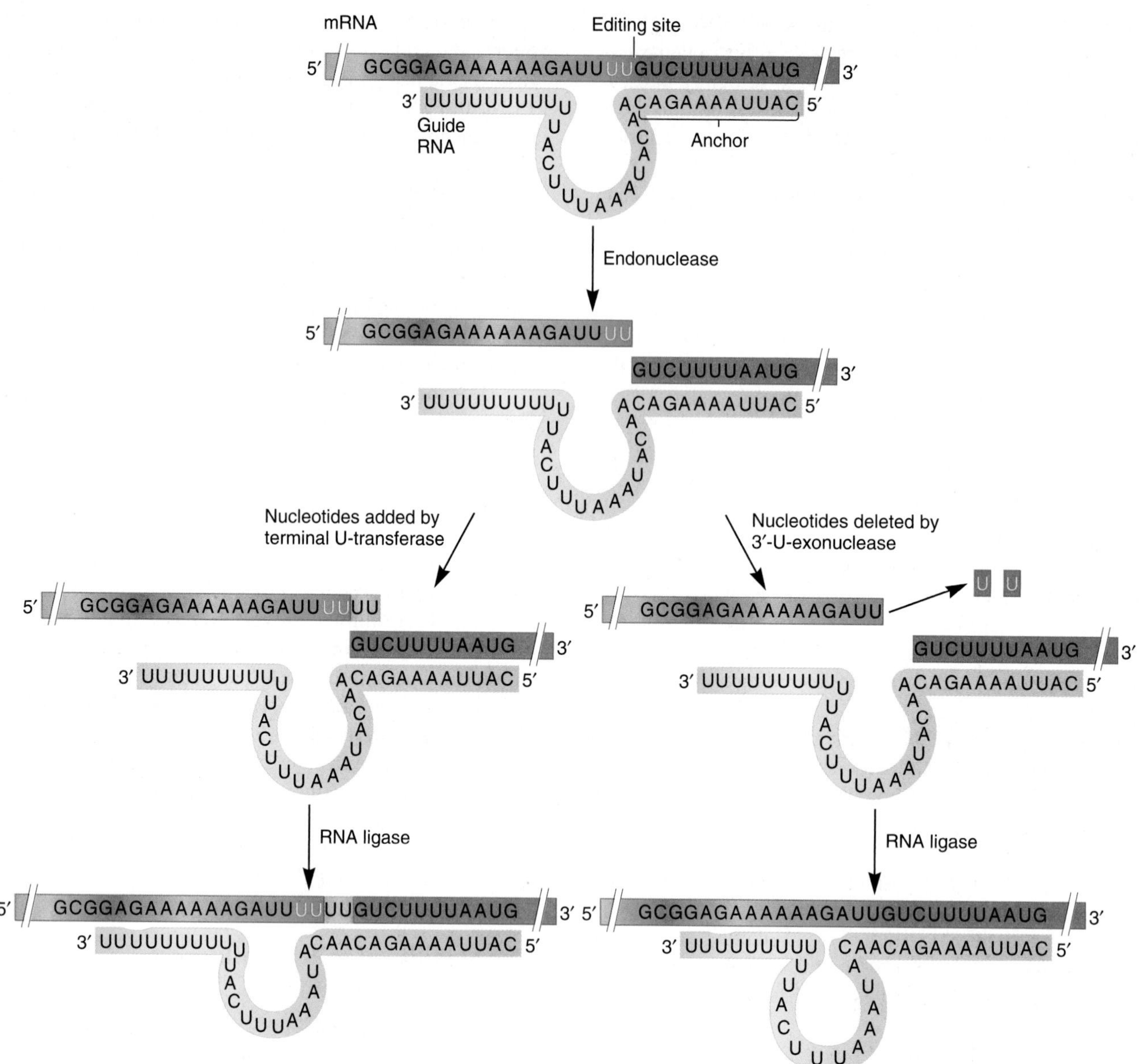

FIGURE 15.18 **RNA editing in trypanosomes.** The 5' end of the guide RNA (the anchor) binds to the target mRNA. The mRNA is then cleaved at the editing site by an endonuclease. On the left side, uracil residues are added by terminal U-transferase. On the right side, uracils are removed by 3'-U-exonuclease. In both cases, the ends of the target mRNA are rejoined by RNA ligase.

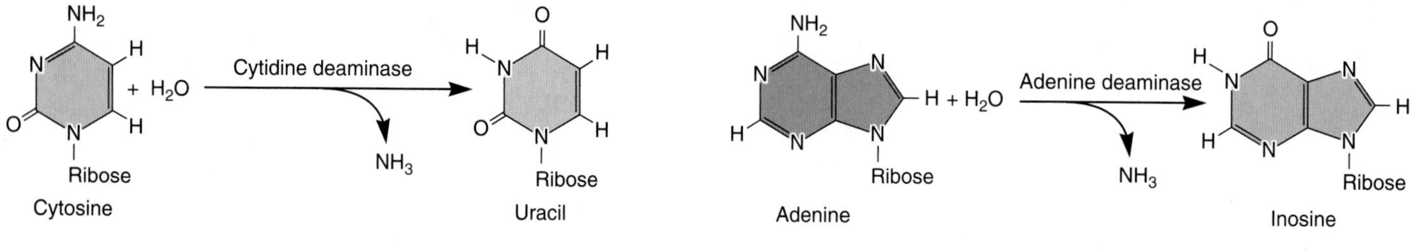

Cytosine + H_2O —Cytidine deaminase→ Uracil + NH_3

Adenine + H_2O —Adenine deaminase→ Inosine + NH_3

FIGURE 15.19 **RNA editing by deamination.** A cytidine deaminase can remove an amino group from cytosine, thereby creating uracil. An adenine deaminase can remove an amino group from adenine to make inosine.

they are most commonly located near the 3' end between the stop codon and the polyA tail. This region of the mRNA is known as the **3'-untranslated region (3'-UTR).** An example of a destabilizing element is the **AU-rich element (ARE)** that is found in many short-lived mRNAs (**Figure 15.20**). This element, which contains the consensus sequence AUUUA, is recognized by cellular proteins that bind to the ARE and thereby influence whether or not the mRNA is rapidly degraded.

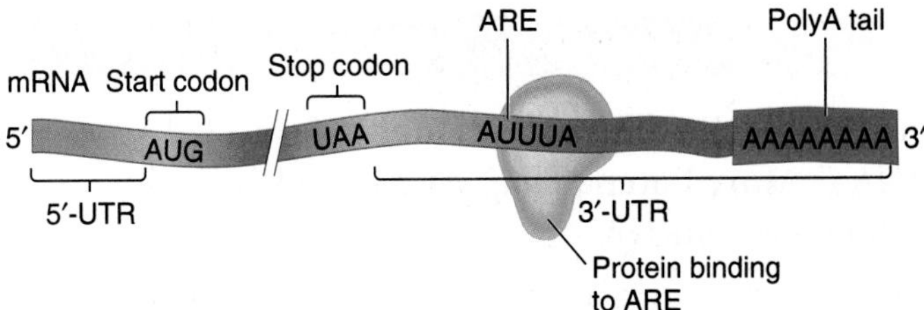

FIGURE 15.20 The location of AU-rich elements (AREs) within mRNAs. One or more AREs are commonly found within the 3'-UTRs (untranslated regions) of mRNAs with short half-lives. The 5'-UTR is the untranslated region of the mRNA that precedes the start codon. AREs are recognized by cellular proteins that bind and thereby influence whether or not the mRNA is rapidly degraded.

Double-Stranded RNA Can Silence the Expression of mRNA

Another way that specific RNAs can be targeted for degradation or translational inhibition is via a recently discovered mechanism involving double-stranded RNA. Research in plants and the nematode *Caenorhabditis elegans* led to the discovery that double-stranded RNA can silence the expression of particular genes. Studies of plant viruses were one avenue of research that identified this mechanism of gene regulation. Certain plant viruses produce double-stranded RNA as part of their life cycle. In addition, these plant viruses may carry genes very similar to genes that already exist in the genome of the plant cell. When such viruses infect plant cells, they silence the expression of the plant gene that is similar to the viral gene.

Plant research that involved the production of transgenic plants also was consistent with the idea that double-stranded RNA can cause mRNA to be degraded or inhibited. Using molecular techniques described in Chapter 19, cloned genes can be introduced into the genome of plants. Surprisingly, researchers observed that when cloned genes were introduced in multiple copies, the expression of the gene was often silenced. How were these results explained? We now know that this may be due to the formation of double-stranded RNA. When a cloned gene randomly inserts into a genome, it may, as a matter of random chance, happen to insert itself next to a promoter for a plant gene that is already present in the genome (**Figure 15.21**). The cloned gene itself has a promoter to make a copy of the sense strand. If it randomly inserts next to a plant gene promoter that is oriented in the opposite direction, both strands of the cloned gene would be transcribed, thereby generating double-stranded RNA. As more copies of a cloned gene are introduced into a genome, the likelihood is greater that the scenario described in Figure 15.21 may occur. This event silences the expression of the cloned gene even though it is present in multiple copies. Furthermore, if the cloned gene is homologous to a plant gene that is already present in the plant cell, this phenomenon will also silence the endogenous plant gene. Therefore, the curious observation made by researchers was that increasing the number of copies of a cloned gene frequently led to gene silencing.

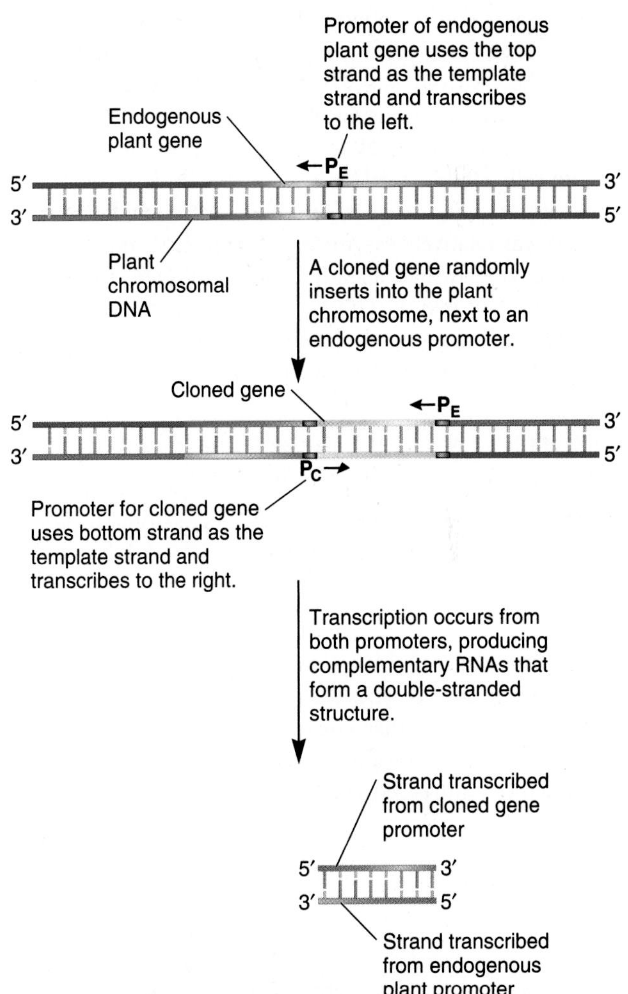

FIGURE 15.21 Gene insertion leading to the production of double-stranded RNA. When a cloned gene with its own promoter is introduced into a plant cell, it will randomly insert into the chromosomal DNA. In some cases, the cloned gene may insert next to a promoter for an endogenous plant gene, so the promoter for the endogenous gene will transcribe the antisense strand of the cloned gene. Therefore, the two promoters will transcribe the sense strand and the antisense strand to produce complementary RNAs that will form double-stranded RNA.

EXPERIMENT 15B

Fire and Mello Show That Double-Stranded RNA Is More Potent Than Antisense RNA at Silencing mRNA

The discovery that double-stranded RNA inhibits mRNA function came about as a result of studies involving gene expression. As a research tool to investigate the function of particular genes, researchers had often introduced antisense RNA (RNA that is complementary to mRNA) into cells as a way to inhibit mRNA translation. Because antisense RNA is complementary to mRNA, the antisense RNA would bind to the mRNA and thereby prevent translation. Oddly, in some experiments, researchers introduced sense RNA (RNA with the same sequence as mRNA) into cells, and this also inhibited mRNA translation. Another curious observation was that the effects of antisense RNA often persisted for a very long time, much longer than would have been predicted by the relatively short half-lives of most RNA molecules in the cell. These two unusual observations caused Andrew Fire, Craig Mello, and colleagues to investigate how the injection of RNA into cells inhibits mRNA.

They used *C. elegans* as their experimental organism because it was relatively easy to inject with RNA and the expression of many genes had already been established. In 1998, Fire and Mello investigated the effects of several injected RNAs known to be complementary to cellular mRNAs. In the investigation of **Figure 15.22**, we will focus on one of their experiments involving a gene called *mex-3*, which had already been shown to be highly expressed in early embryos. They started with the cloned gene for *mex-3*. This gene was genetically engineered using techniques described in Chapter 18 so that one version had a promoter that would result in the synthesis of the normal mRNA, which is the sense RNA. They made another version of *mex-3* in which a promoter directed the synthesis of the opposite strand, which produced antisense RNA. To make RNA in vitro, they added RNA polymerase and nucleotides to these cloned genes to make sense or antisense RNA.

Next, they injected RNA into *C. elegans* eggs and observed the effects in developing embryos. They either injected antisense RNA or they mixed sense and antisense RNA and injected double-stranded RNA. They also used uninjected eggs as controls. To determine the expression of *mex-3*, they incubated the resulting embryos with a probe complementary to the *mex-3* mRNA. The probe was labeled so it could be observed under the microscope. After this incubation step, any probe that was not bound to this mRNA was washed away.

THE GOAL

The goal was to further understand how the experimental injection of RNA was responsible for the silencing of particular mRNAs.

ACHIEVING THE GOAL — FIGURE 15.22 Injection of antisense and double-stranded RNA into *C. elegans* to compare their effects on mRNA silencing.

Starting material: The researchers used eggs from *C. elegans*. They also had the cloned *mex-3* gene, which had been previously shown to be highly expressed in the embryo.

1. Make sense and antisense *mex-3* RNA in vitro using cloned genes for *mex-3* with promoters on either side of the gene. RNA polymerase and nucleotides are added to synthesize the RNAs.

2. Inject *C. elegans* eggs (or early embryos) with either *mex-3* antisense RNA or a mixture of *mex-3* sense and antisense RNA. As a control, do not inject any RNA.

Experimental level

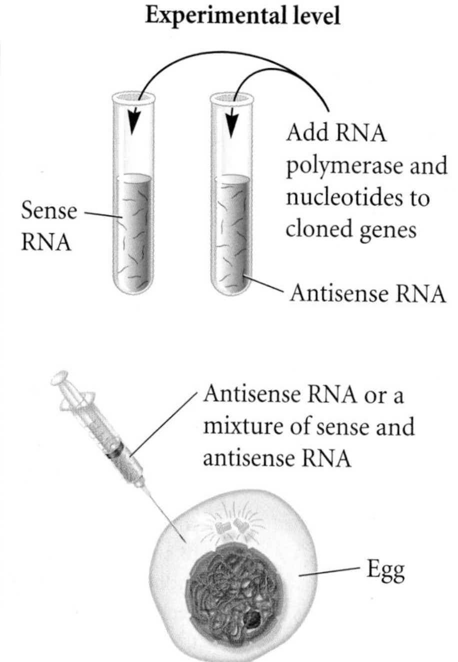

Conceptual level

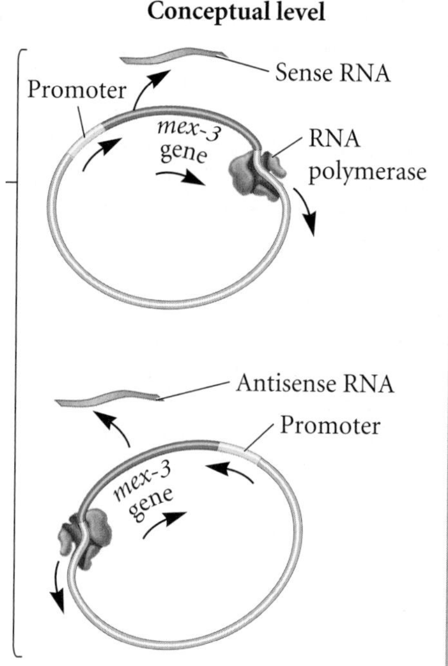

3. Incubate and then subject early embryos to *in situ* hybridization. In this method, a labeled probe is added that is complementary to *mex-3* mRNA. If cells express *mex-3*, the mRNA in the cells will bind to the probe and become labeled. After incubation with a labeled probe, the cells are washed to remove unbound probe.

4. Observe embryos under the microscope.

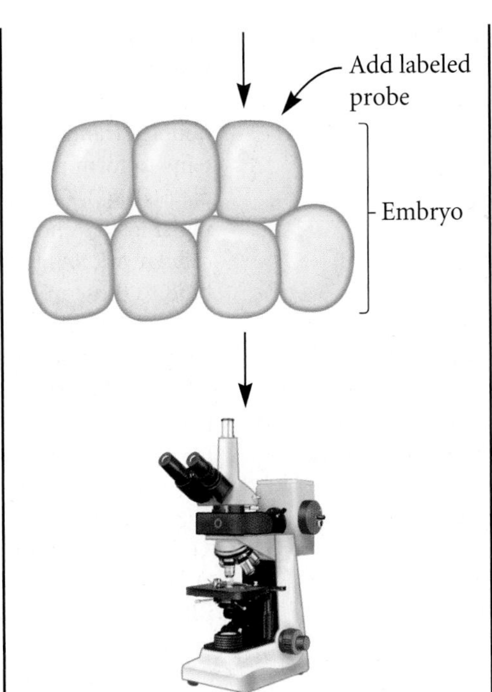

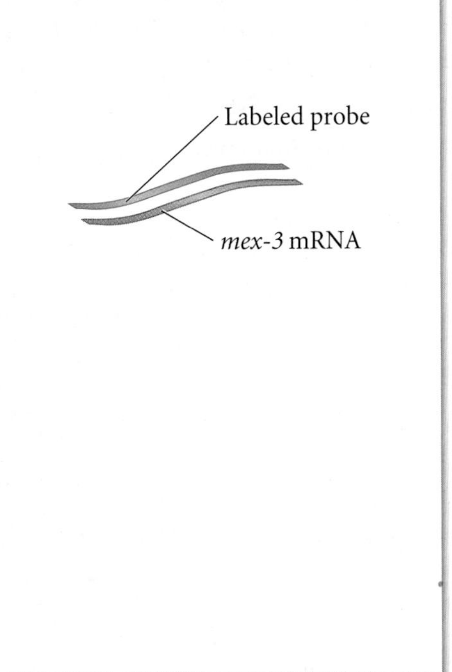

THE DATA

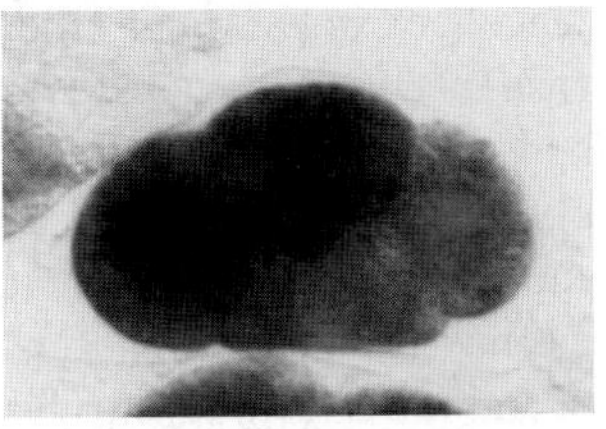

Control

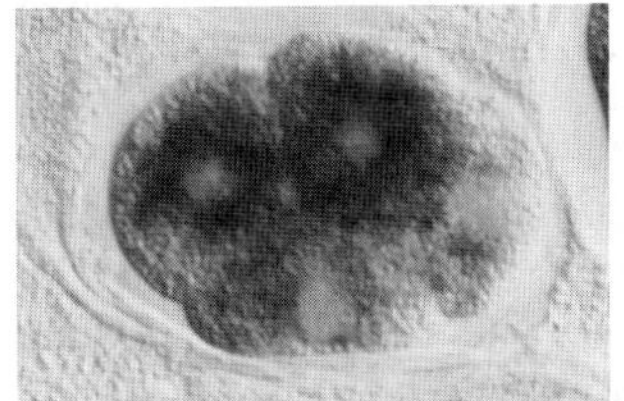

Injected with *mex-3* antisense RNA

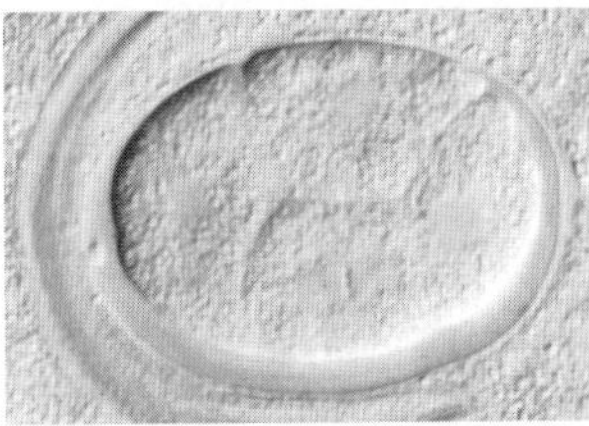

Injected with both *mex-3* sense and antisense RNA

Photos reprinted by permission from Macmillan Publishers Ltd. *Nature.* Potent and specific genetic interference by double-stranded RNA in *Caenohabditis elegans.* Andrew Fire et al. 391:6669, 806–811, 1998.

INTERPRETING THE DATA

As seen in the data of Figure 15.22, the control embryos were very darkly staining. These results indicated that the *mex-3* gene is highly expressed, which was known from previous research. In the embryos injected with antisense RNA, *mex-3* mRNA levels were decreased, but detectable. Remarkably, in samples injected with double-stranded RNA, no *mex-3* mRNA was detected! These results indicated that double-stranded RNA is more potent at silencing mRNA than is antisense RNA. In this case, the double-stranded RNA caused the *mex-3* mRNA to be degraded. They used the term **RNA interference (RNAi)** to describe the phenomenon in which double-stranded RNA causes the silencing of mRNA. This surprising observation led researchers to investigate the underlying molecular mechanism that accounts for this phenomenon, as described next.

A self-help quiz involving this experiment can be found at the Online Learning Center.

RNA Interference Is Mediated by MicroRNAs and the RNA-Induced Silencing Complex

RNA interference is widely found in eukaryotic species. Double-stranded RNA can come from a variety of sources. As mentioned, viruses sometimes produce double-stranded RNA as part of their reproductive cycle, and the insertion of multiple copies of a gene into the genome can also result in double-stranded RNA. In addition, the genomes of eukaryotic organisms have been found to contain genes that encode RNA molecules that cause RNA interference. These RNA molecules, called **microRNAs** (abbreviated **miRNAs**), are small RNA molecules, typically 21 to 23 nucleotides in length, that silence the expression of specific mRNAs. In 1993, Victor Ambros and his colleagues, who were interested in the developmental stages that occur in the worm *C. elegans*, determined that the transcription of a particular gene produced a small RNA, now called a microRNA, that does not encode a protein. Instead, this miRNA was found to be complementary to an mRNA, and it inhibited the translation of the mRNA. Since this study, researchers have discovered that genes encoding miRNAs are widely found in animals and plants. In humans, for example, approximately 200 different genes encode miRNAs.

MiRNAs represent an important mechanism of mRNA silencing. How do miRNAs cause the silencing of specific mRNAs? **Figure 15.23** shows how an miRNA, which is encoded by a gene, leads to RNA interference. In this example, the miRNA is first synthesized as a pre-miRNA, which forms a hairpin structure that is cut by an endonuclease called **dicer.** This releases a double-stranded RNA molecule that associates with cellular proteins to form a complex called the **RNA-induced silencing complex (RISC).** One of the RNA strands is degraded. The remaining single-stranded miRNA, which is complementary to specific cellular mRNAs, will allow the RISC to specifically recognize and bind to those mRNA molecules.

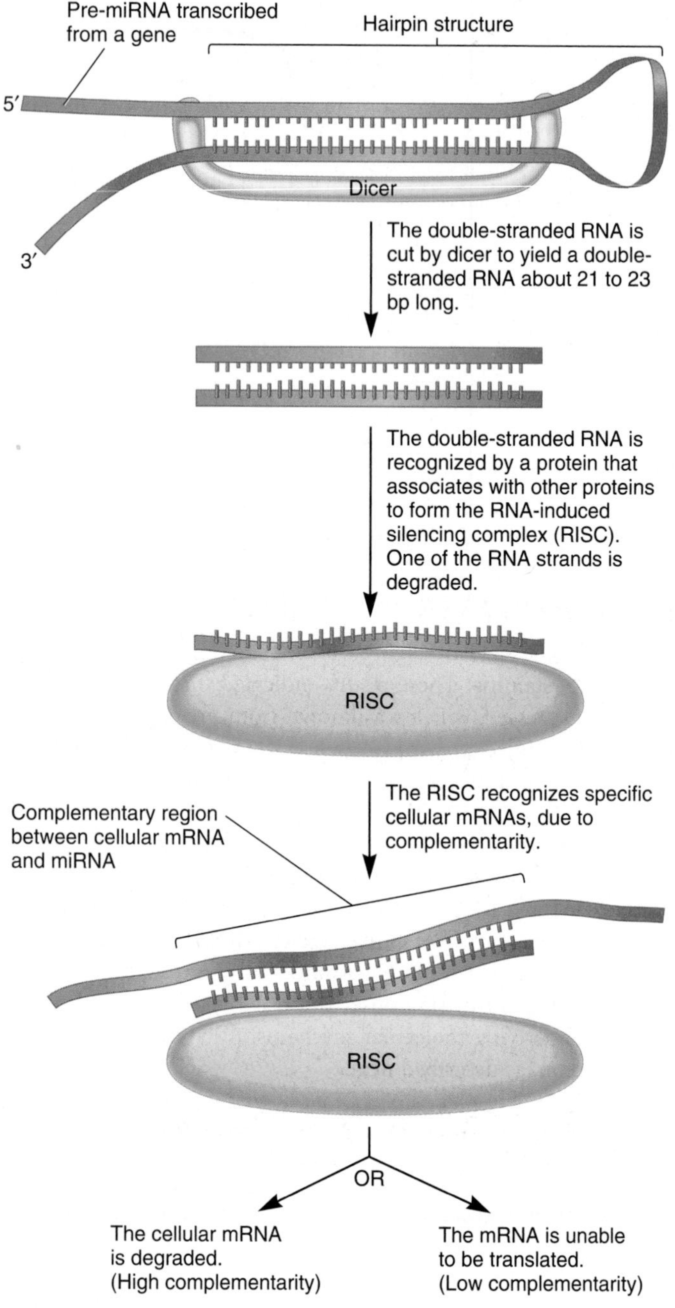
Pre-miRNA transcribed from a gene
Hairpin structure
5′
3′
Dicer
The double-stranded RNA is cut by dicer to yield a double-stranded RNA about 21 to 23 bp long.
The double-stranded RNA is recognized by a protein that associates with other proteins to form the RNA-induced silencing complex (RISC). One of the RNA strands is degraded.
RISC
The RISC recognizes specific cellular mRNAs, due to complementarity.
Complementary region between cellular mRNA and miRNA
RISC
OR
The cellular mRNA is degraded. (High complementarity)
The mRNA is unable to be translated. (Low complementarity)
ONLINE ANIMATION

FIGURE 15.23 **Mechanism of RNA interference.** A double-stranded region of RNA within a pre-miRNA is recognized by an endonuclease called dicer that digests the RNA to produce a small piece, 21 to 23 bp in length. This small RNA binds with proteins to form an RNA-induced silencing complex (RISC). One strand of the RNA will be degraded. The remaining RNA strand in RISC will bind to complementary mRNAs. After RISC binding, endonucleases within the complex may cut the mRNA and thereby inactivate it. It should be emphasized that the miRNAs are mRNA specific. For example, if the miRNA within RISC was complementary to the β-globin mRNA, only the β-globin mRNA would be silenced.

Upon binding with the mRNA, two different things may happen. In some cases, the RISC may direct the degradation of the mRNA. This tends to occur when the miRNA and mRNA are a perfect match or are highly complementary. Alternatively, the RISC may inhibit translation. This is more common when the miRNA and mRNA are not a perfect match or are only partially complementary. In either case, the expression of the mRNA is silenced. The effect is termed RNA interference because the miRNA interferes with the proper expression of an mRNA. Likewise, dicer and RISC may act upon double-stranded RNA that comes from viruses or from the insertion of multiple copies of a gene into the genome (as shown at the bottom of Figure 15.21). In 2006, Fire and Mello received the Nobel Prize for their discovery of this phenomenon.

RNA interference is believed to have at least three benefits. First, this phenomenon represents a newly identified form of gene regulation. When genes encoding pre-miRNAs are turned on, the production of miRNAs will silence the expression of other genes. Second, RNAi may offer a defense mechanism against certain viruses. In particular, this mechanism may inhibit the proliferation of viruses that have a double-stranded RNA genome and viruses that produce double-stranded RNAs as part of their reproductive cycles. After entering the host cell, the viral RNA would be degraded, as in Figure 15.23, and the cell would survive the infection. Third, researchers have speculated that RNAi may also play a role in the silencing of certain transposable elements. As discussed in Chapter 17, transposable elements are DNA segments that have the capacity to move throughout the genome, an event termed transposition. Transposable elements carry genes, such as transposase, that are needed in the transposition process. The random insertion of many transposable elements in a cell may ultimately lead to gene silencing due to the scenario described in Figure 15.21. The silencing of genes within transposable elements via RNAi would protect the organism against the potentially harmful effects of transposition.

Phosphorylation of Ribosomal Initiation Factors Can Alter the Rate of Translation

Let's now turn our attention to another regulatory mechanism that affects translation. Modulation of translational initiation factors is widely used to control fundamental cellular processes. Under certain conditions, it is advantageous for a cell to stop synthesizing proteins. For example, if a virus infects a cell, the inhibition of protein synthesis can prevent viral proliferation by inhibiting the production of new viral proteins. Likewise, if critical nutrients are in short supply, it is beneficial for a cell to conserve its resources by inhibiting unnecessary protein synthesis.

Translational initiation factors are required to begin protein synthesis. The phosphorylation of many different initiation factors has been found to affect translation in eukaryotic cells. Two factors, eIF2 and eIF4F, appear to play a central role in controlling the initiation of translation. The functions of these two translational initiation factors are modulated by phosphorylation in opposite ways. When the α subunit of eIF2 (known as eIF2α) is phosphorylated, translation is inhibited, whereas the phosphorylation of eIF4F increases the rate of translation.

Figure 15.24 shows the events leading to translational inhibition by eIF2α. A variety of conditions can lead to a shutdown of protein synthesis, including viral infection, nutrient deprivation, heat shock, and the presence of toxic heavy metals. These conditions promote the activation of protein kinases known as eIF2α protein kinases. Several eIF2α protein kinases have been identified. Once activated, an eIF2α protein kinase can phosphorylate eIF2α. The phosphorylation of eIF2α causes it to bind tightly to another initiation factor subunit called eIF2B. Functional eIF2B is necessary so that eIF2 can promote the binding of the initiator $tRNA^{met}$ to the 40S ribosomal subunit. However, when the phosphorylated eIF2α binds to eIF2B, it prevents eIF2B from functioning. Therefore, the initiator $tRNA^{met}$ does not bind to the 40S subunit, and translation is inhibited.

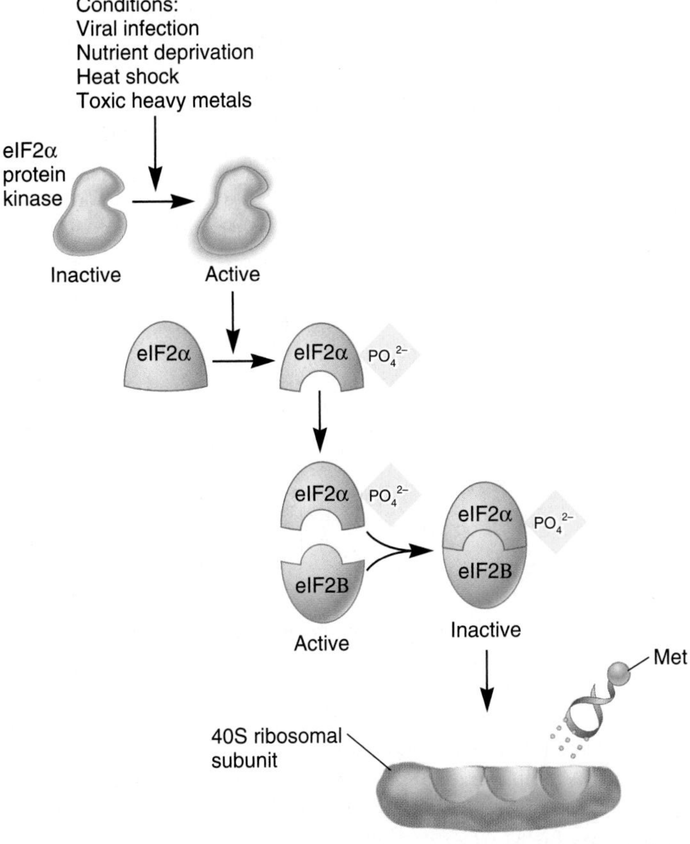

FIGURE 15.24 **The pathway that leads to the phosphorylation of eIF2α (eukaryotic initiation factor α) and the inhibition of translation.**

A second important way to control translation is via the eIF4F translation factor that modulates the binding of mRNA to the ribosomal initiation complex. The function of eIF4F is stimulated by phosphorylation. A variety of conditions have been shown to cause eIF4F to become phosphorylated. These include the presence of growth factors, insulin, and other signaling molecules that promote cell proliferation. Conversely, conditions such as heat shock and viral infection decrease the level of eIF4F phosphorylation and thereby inhibit translation.

The Regulation of Iron Assimilation Is an Example of the Regulatory Effect of RNA-Binding Proteins on Translation

As we have just seen, the phosphorylation of translational initiation factors can modulate the translation of mRNA. Because these initiation factors are necessary to translate all of a cell's mRNA, this form of regulation affects the expression of many mRNAs. By comparison, specific mRNAs are sometimes regulated by RNA-binding proteins that directly affect translational initiation or RNA stability. The regulation of iron assimilation provides a well-studied example in which both of these phenomena occur. Before discussing this form of translational control, let's consider the biology of iron metabolism.

Iron is an essential element for the survival of living organisms because it is required for the function of many different enzymes. The pathway by which mammalian cells take up iron is depicted in **Figure 15.25**. Iron ingested by an animal is absorbed

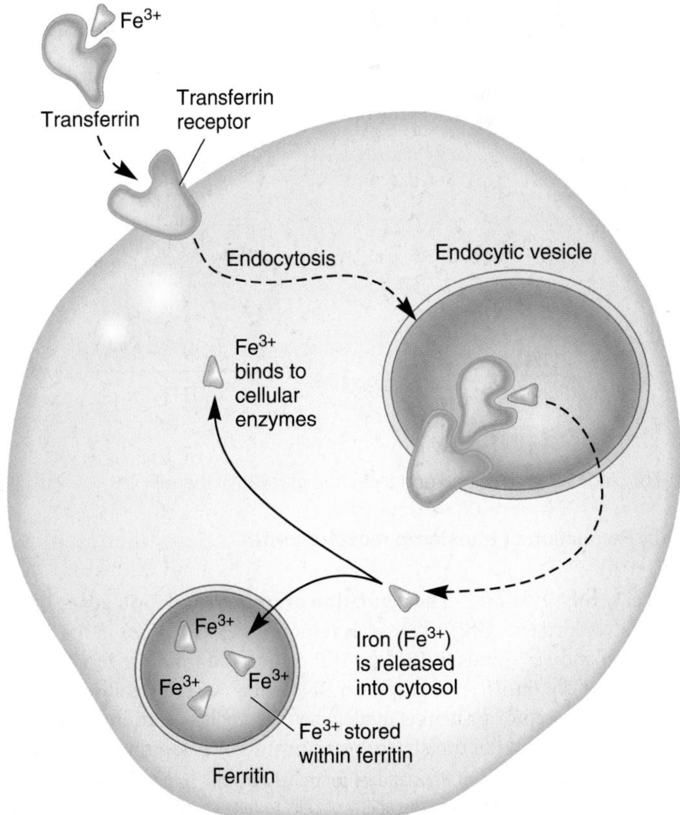

FIGURE 15.25 **The uptake of iron (Fe^{3+}) into mammalian cells.**

into the bloodstream and becomes bound to transferrin, a protein that carries iron through the bloodstream. The transferrin-Fe^{3+} complex is recognized by a transferrin receptor on the surface of cells; the complex binds to the receptor and then is transported into the cytosol by endocytosis. Once inside, the iron is then released from transferrin. At this stage, Fe^{3+} may bind to cellular enzymes that require iron for their activity. Alternatively, if too much iron is present, the excess iron is stored within a hollow, spherical protein known as ferritin. The storage of excess iron within ferritin helps to prevent the toxic buildup of too much iron within the cell.

Because iron is a vital yet potentially toxic substance, mammalian cells have evolved an interesting way to regulate iron assimilation. The two mRNAs that encode ferritin and the transferrin receptor are both influenced by an RNA-binding protein known as the **iron regulatory protein (IRP)**. How does IRP exert its effects? This protein binds to a regulatory element within these two mRNAs known as the **iron response element (IRE).** The ferritin mRNA has an IRE in its 5'-untranslated region (5'-UTR). When IRP binds to this IRE, it inhibits the translation of the ferritin mRNA (**Figure 15.26a**, left). However, when iron is abundant in the cytosol, the iron binds directly to IRP and prevents it from binding to the IRE. Under these conditions, the ferritin mRNA is translated to make more ferritin protein (Figure 15.26a, right), which prevents the toxic buildup of iron within the cytosol.

The transferrin receptor mRNA also contains an iron response element, but it is located in the 3'-UTR. When IRP binds to this IRE, it does not inhibit translation. Instead, the binding of IRP increases the stability of the mRNA by blocking the action

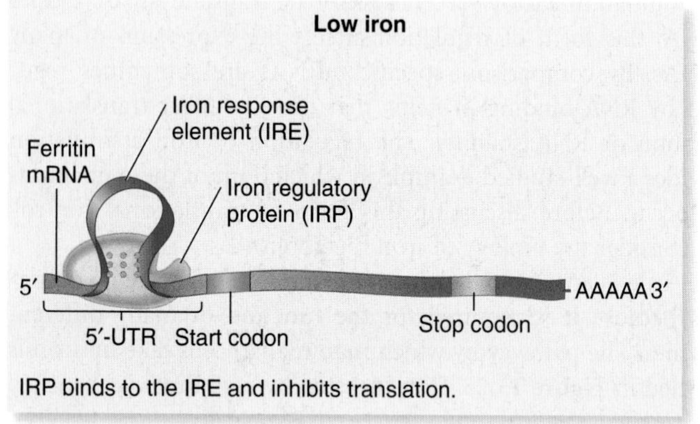

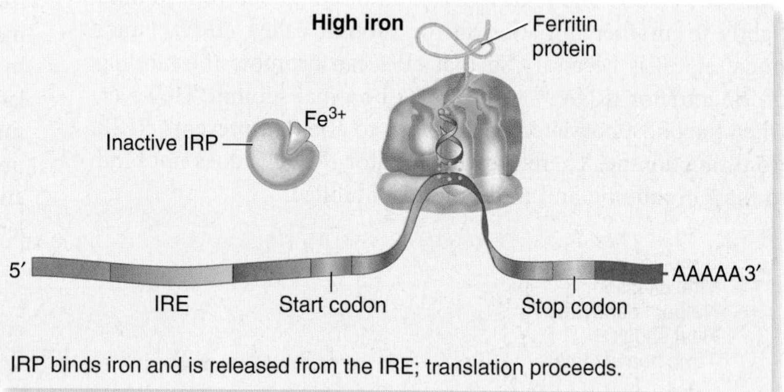

(a) Regulation of ferritin mRNA

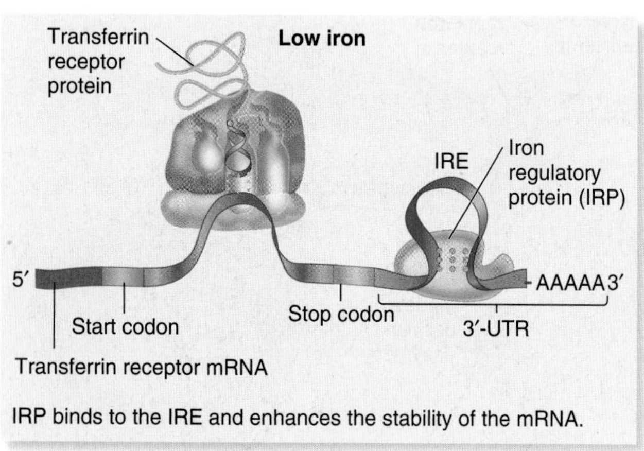

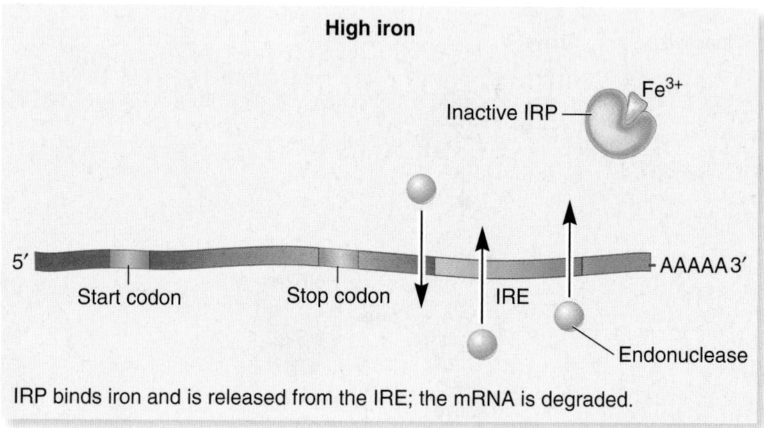

(b) Regulation of transferrin receptor mRNA

FIGURE 15.26 The regulation of iron assimilation genes by IRP and IRE. (**a**) When Fe^{3+} concentrations are low, the binding of the iron regulatory protein (IRP) to the iron response element (IRE) in the 5'-UTR of ferritin mRNA inhibits translation (left). When Fe^{3+} concentrations are high and Fe^{3+} binds to IRP, the IRP is removed from the ferritin mRNA, so translation can proceed (right). (**b**) The binding of IRP to IRE in the 3'-UTR of the transferrin receptor mRNA enhances the stability of the mRNA and leads to a higher concentration of this mRNA. Therefore, more transferrin receptor protein is made when Fe^{3+} concentration is low (left). When Fe^{3+} levels are high and this metal binds to IRP, the IRP dissociates from the IRE, and the transferrin receptor mRNA is rapidly degraded (right).

Genes→Traits This form of translational control allows cells to use iron appropriately. When the cellular concentration of iron is low, the translation of the transferrin receptor is increased, thereby enhancing the ability of cells to take up more iron. Also, the translation of ferritin mRNA is inhibited, because ferritin is not needed to store excess iron. By comparison, when the cellular concentration of iron is high, the translation of ferritin mRNA is enhanced. This leads to the synthesis of ferritin and prevents the toxic buildup of iron. When iron is high, the transferrin receptor mRNA is degraded, which decreases further uptake of iron.

of RNA-degrading enzymes. This leads to increased amounts of transferrin receptor mRNA within the cell when the cytosolic levels of iron are very low (**Figure 15.26b**, left). Under these conditions, more transferrin receptor is made. This promotes the uptake of iron when it is in short supply. In contrast, when iron is abundant within the cytosol, IRP is removed from the transferrin receptor mRNA, and the mRNA becomes rapidly degraded (Figure 15.26b, right). This leads to a decrease in the amount of transferrin receptor and thereby helps to prevent the uptake of too much iron into the cell.

CONCEPTUAL SUMMARY

In this chapter, we have surveyed a wide variety of molecular mechanisms through which eukaryotic organisms regulate gene expression. Eukaryotic cells have an array of regulatory transcription factors that influence the ability of RNA polymerase to begin transcription of a particular gene. Such transcription factors recognize and bind to short DNA sequences called regulatory elements that control a particular gene. The binding of regulatory transcription factors may enhance (up regulate) the rate of transcription, in which case the transcription factor is termed an activator and the sequence it binds to is called an enhancer. Conversely, the binding may prevent transcription from occurring (down regulate), in which case the transcription factor is termed a repressor and the sequence, a silencer. Regulatory elements can be bidirectional and function at a fairly large distance away from the core promoter site.

Regulatory transcription factors may influence the function of RNA polymerase by interacting with protein complexes called TFIID and mediator. We considered common ways that the functions of regulatory transcription factors themselves are modulated. The glucocorticoid hormone is an example of an effector molecule that binds to a regulatory transcription factor (the glucocorticoid receptor) and alters its function. Once the hormone is bound, it activates several different genes by binding to glucocorticoid response elements (GREs) that are next to the genes. By comparison, the CREB protein is a regulatory transcription factor that is activated in response to cell-signaling molecules that cause increased intracellular levels of cAMP. The CREB protein binds to a response element known as CRE. When phosphorylated, CREB proteins stimulate transcription.

Gene regulation may also involve changes in the level of chromatin compaction or changes in DNA structure. For transcription to occur, the chromatin cannot be in a closed conformation. Instead, it must be in a loosely packed or open conformation. Transcriptional activators play a role in the recruitment of enzymes to the promoter region that convert the chromatin from a closed to an open conformation. These include ATP-dependent chromatin remodeling enzymes and histone acetyltransferase. The DNA itself can be altered by gene amplification, gene rearrangement, and DNA methylation. The first two mechanisms are uncommon and not examined in this chapter. In contrast, DNA methylation in vertebrates and higher plants appears to be an important mechanism to turn genes off in a tissue-specific or developmentally specific manner. CpG islands, regions that are commonly 1,000 to 2,000 base pairs in length and contain a high concentration of CpG sites, may be found in the vicinity of promoters. The methylation of CpG islands is one way to inhibit transcription. Methylation may prevent the binding of transcription factors and/or promote compaction. Methylated DNA sequences are passed from parent to daughter cell during cell division, a process called maintenance methylation. This preserves the methylated condition in future cells and appears to be a process that routinely occurs within vertebrate and plant cells.

Gene regulation can also occur at the level of RNA processing and translation. For example, a pre-mRNA transcript can be regulated by alternative splicing, an RNA processing event that allows different proteins to be made from a single type of pre-mRNA. RNA editing involves an addition or deletion of bases or a conversion of one type of base to another. The stability of RNA can also be affected by particular sequences within the mRNA. One important feature influencing stability is the length of the polyA tail; one example of a destabilizing element is the AU-rich element found in mRNAs. RNA can also be a target for degradation by double-stranded RNA, a mechanism termed RNA interference (RNAi).

We explored two ways in which the rate of mRNA translation can be controlled. First, the phosphorylation of two initiation factors, eIF2α and eIF4F, can affect translation in eukaryotic cells. The functions of these factors are modulated in opposite ways. When eIF2α is phosphorylated, translation is inhibited, whereas the phosphorylation of eIF4F increases the rate of translation. Second, specific RNA-binding proteins can directly affect the ability of certain mRNAs to be translated. The regulation of iron assimilation genes by iron regulatory proteins (IRPs) and iron response elements (IREs) is a form of translational control that allows cells to respond to changing levels of iron in the cell.

EXPERIMENTAL SUMMARY

The regulation of gene expression in eukaryotes has been studied via many different techniques. In Chapter 15, we have considered a few approaches to study this phenomenon. In one example, we saw how Weintraub and Groudine used DNase I sensitivity to evaluate the degree of globin gene packing in reticulocytes versus brain cells and fibroblasts. This approach exploits the fact that tightly packed chromatin is less susceptible to DNase I digestion than is DNA in an open conformation.

Other studies have focused on the identification of regulatory transcription factors that influence the expression of particular genes. By comparing the structures of many different transcription factors, researchers have found that they tend to have common domains that either interact with small effector molecules or bind to DNA. The sequences of many regulatory elements have also been determined. In a few cases, the combined efforts of many research groups have elucidated the detailed molecular mechanisms for the regulation of particular genes. For example, we examined how the glucocorticoid receptor and CREB protein regulate genes at the level of transcription.

Insights into how specific RNAs can be targeted for degradation or translational inhibition came from a study by Fire and Mello that showed how double-stranded RNA is more potent than antisense RNA at silencing mRNA. Their research led to the discovery that this phenomenon, termed RNA interference, is mediated by microRNAs and an RNA-induced silencing complex (RISC). The solved problems and experimental questions at the end of Chapter 15 also serve to illustrate several techniques that have been used to understand the molecular mechanisms of eukaryotic gene regulation.

PROBLEM SETS & INSIGHTS

Solved Problems

S1. Describe how the tight packing of chromatin in a closed conformation may prevent the initiation of gene transcription.

Answer: The tight packing of chromatin may physically inhibit transcription in a few ways. First, it may prevent transcription factors and/or RNA polymerase from binding to the major groove of the DNA. Second, tight packing may prevent RNA polymerase from forming an open complex, which is necessary to begin transcription.

S2. What are the two alternative ways that IRP can affect gene expression at the RNA level?

Answer: The ferritin mRNA has an IRE in its 5'-UTR. When IRP binds to this IRE, it inhibits the translation of the ferritin mRNA. This decreases the amount of ferritin protein, which is not needed when iron levels are low. However, when iron is abundant in the cytosol, the iron binds directly to IRP and prevents it from binding to the IRE. This allows the ferritin mRNA to be translated, producing more ferritin protein. An IRE is also located in the transferrin receptor mRNA in the 3'-UTR. When IRP binds to this IRE, it increases the stability of the mRNA, which leads to an increase in the amount of transferrin receptor mRNA within the cell when the cytosolic levels of iron are very low. Under these conditions, more transferrin receptor is made to promote the uptake of iron, which is in short supply. When the iron is found in abundance within the cytosol, IRP is removed from the mRNA, and the mRNA becomes rapidly degraded. This leads to a decrease in the amount of the transferrin receptor.

S3. Eukaryotic regulatory elements are often orientation independent and can function in a variety of locations. Explain the meaning of this statement.

Answer: Orientation independence means that the regulatory element can function in the forward or reverse direction. In addition, regulatory elements can function at a variety of locations that may be upstream or downstream from the core promoter. A loop in the DNA must form to bring the regulatory transcription factors bound at the regulatory elements and the core promoter in close proximity with one another.

S4. To gain a molecular understanding of how the glucocorticoid receptor works, geneticists have attempted to "dissect" the protein to identify smaller domains that play specific functional roles. Using recombinant DNA techniques described in Chapter 18, particular segments in the coding region of the glucocorticoid receptor gene can be removed. The altered gene can then be expressed in a living cell to see if functional aspects of the receptor have been changed or lost. For example, the removal of the portion of the gene encoding the carboxyl terminal half of the protein causes a loss of glucocorticoid binding. These results indicate that the carboxyl terminal portion contains a domain that functions as a glucocorticoid-binding site. The figure shown here illustrates the locations of several functional domains within the glucocorticoid receptor relative to the entire amino acid sequence.

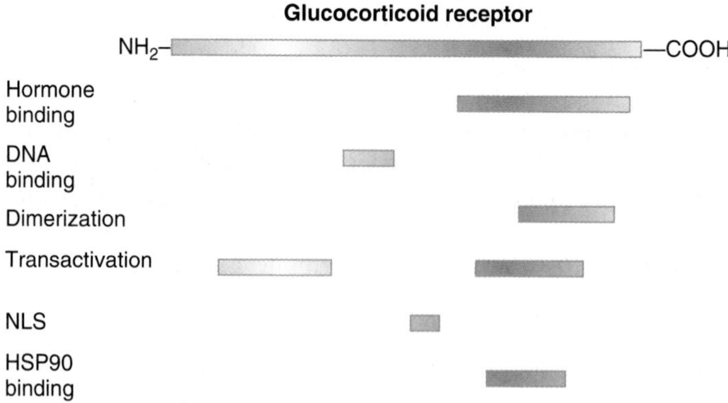

Based on your understanding of the mechanism of glucocorticoid receptor function described in Figure 15.6, explain the functional roles of the different domains in the glucocorticoid receptor.

Answer: The hormone-binding domain is located in the carboxyl terminal half of the protein. This part of the protein also contains the region necessary for HSP90 binding and receptor dimerization. A nuclear localization sequence (NLS) is located near the center of the protein. After hormone binding, the NLS is exposed on the surface of the protein and allows it to be targeted to the nucleus. The DNA-binding domain, which contains zinc fingers, is also centrally located in the primary amino acid sequence. Zinc fingers promote DNA binding to the major groove. Finally, two separate regions of the protein, one in the amino terminal half and one in the carboxyl terminal half, are necessary for the transactivation of RNA polymerase. If these domains are removed from the receptor, it can still bind to the DNA, but it cannot activate transcription.

S5. A common approach to identify genetic sequences that play a role in the transcriptional regulation of a gene is the strategy that is sometimes called **promoter bashing.** This approach requires gene cloning methods, which are described in Chapter 18. A clone is obtained that has the coding region for a structural gene as well as the region that is upstream from the core promoter. This upstream region is likely to contain genetic regulatory elements such as enhancers and silencers. The diagram shown here depicts a cloned DNA region that contains the upstream region, the core promoter, and the coding sequence for a protein that is expressed in human liver cells. The upstream region may be several thousand base pairs in length.

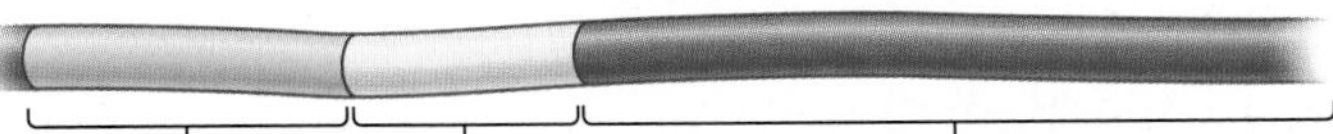

To determine if promoter bashing has an effect on transcription, it is helpful to have an easy way to measure the level of gene expression. One way to accomplish this is to swap the coding sequence of the gene of interest with the coding sequence of another gene. For example, the coding sequence of the *lacZ* gene, which encodes β-galactosidase, is frequently swapped because it is easy to measure the activity of β-galactosidase using an assay for its enzymatic activity. The *lacZ* gene is called a "reporter gene" because it is easy to measure its activity. As shown here, the coding sequence of the *lacZ* gene has been swapped with the coding sequence of the liver-specific gene. In this new genetic construct, the expression and transcriptional regulation of the *lacZ* gene is under the control of the core promoter and upstream region of the liver-specific gene.

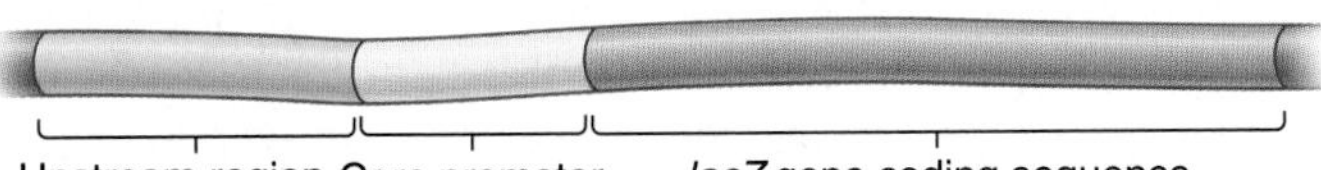

Now comes the "bashing" part of the experiment. Different segments of the upstream region are deleted, and then the DNA is transformed into living cells. In this case, the researcher would probably transform the DNA into liver cells, because those are the cells where the gene is normally expressed. The last step is to measure the β-galactosidase activity in the transformed liver cells

In the diagram shown here, the upstream region and the core promoter have been divided into five regions, labeled A–E.

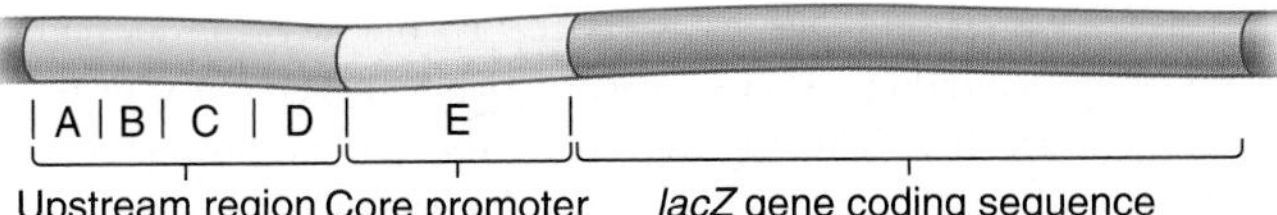

One of these regions was deleted (i.e., bashed out), and the rest of the DNA segment was transformed into liver cells. The data shown here are from this experiment.

Region Deleted	Percentage of β-Galactosidase Activity*
None	100
A	100
B	330
C	100
D	5
E	<1

*The amount of β-galactosidase activity in the cells carrying an undeleted upstream and promoter region was assigned a value of 100%. The amounts of activity in the cells carrying a deletion were expressed relative to this 100% value.

Explain what these results mean.

Answer: The amount of β-galactosidase activity found in liver cells that do not carry a deletion reflects the amount of expression under normal circumstances. If the core promoter (region E) is deleted, very little expression is observed. This is expected because a core promoter is needed for transcription. If enhancers are deleted, the activity should be less than 100%. It appears that one or more enhancers are found in region D. If a silencer is deleted, the activity should be above 100%. From the data shown it appears that one or more silencers are found in region B. Finally, if a deletion has no effect, there may not be any regulatory elements there. This was observed for regions A and C.

Note: The deletion of an enhancer will have an effect on β-galactosidase activity only if the cell is expressing the regulatory transcription factor that binds to the enhancer and activates transcription. Likewise, the deletion of a silencer will have an effect only if the cell is expressing the repressor protein that binds to the silencer and inhibits transcription. In the problem described, the liver cells must be expressing the activators and repressors that recognize the regulatory elements found in regions B and D, respectively.

Conceptual Questions

C1. Discuss the common points of control in eukaryotic gene regulation.

C2. Discuss the structure and function of regulatory elements. Where are they located relative to the core promoter?

C3. What is meant by the term transcription factor modulation? List three general ways that this can occur.

C4. What are the functions of transcriptional activator proteins and repressor proteins? Explain how they work at the molecular level.

C5. Are the following statements true or false?

A. An enhancer is a type of regulatory element.

B. A core promoter is a type of regulatory element.

C. Regulatory transcription factors bind to regulatory elements.

D. An enhancer may cause the down regulation of transcription.

C6. Transcription factors usually contain one or more motifs that play key roles in their function. What is the function of the following motifs?

A. Helix-turn-helix

B. Zinc finger

C. Leucine zipper

C7. The binding of an effector molecule, protein-protein interactions, and covalent modifications are three common ways to modulate the activities of transcription factors. Which of these three mechanisms are used by steroid receptors and the CREB protein?

C8. Describe the steps that occur for the glucocorticoid receptor to bind to a GRE.

C9. Let's suppose a mutation in the glucocorticoid receptor does not prevent the binding of the glucocorticoid hormone to the protein but prevents the ability of the receptor to activate transcription. Make a list of all the possible defects that may explain why transcription cannot be activated.

C10. Explain how phosphorylation affects the function of the CREB protein.

C11. A particular drug inhibits the protein kinase that is responsible for phosphorylating the CREB protein. How would this drug affect the following events?

A. The ability of the CREB protein to bind to CREs

B. The ability of extracellular hormone to enhance cAMP levels

C. The ability of the CREB protein to stimulate transcription

D. The ability of the CREB protein to dimerize

C12. The glucocorticoid receptor and the CREB protein are two examples of transcriptional activators. These proteins bind to response elements and activate transcription. (Note: The answer to this question is not directly described in Chapter 15. You have to rely on your understanding of the functioning of other proteins that are modulated by the binding of effector molecules, such as the lac repressor.)

A. How would the function of the glucocorticoid receptor be shut off?

B. What type of enzyme would be needed to shut off the activation of transcription by the CREB protein?

C13. Transcription factors such as the glucocorticoid receptor and the CREB protein form homodimers and activate transcription. Other transcription factors form heterodimers. For example, a transcription factor known as myogenic bHLH forms a heterodimer with a protein called the E protein. This heterodimer activates the transcription of genes that promote muscle cell differentiation. However, when myogenic bHLH forms a heterodimer with a protein called the Id protein, transcriptional activation does not occur. (Note: Id stands for Inhibitor of differentiation.) Which of the following possibilities would best explain this observation? Only one possibility is correct.

	Myogenic bHLH	E Protein	Id Protein
Possibility 1			
DNA-binding domain:	Yes	No	No
Leucine zipper:	Yes	No	Yes
Possibility 2			
DNA-binding domain:	Yes	Yes	No
Leucine zipper:	Yes	Yes	Yes
Possibility 3			
DNA-binding domain:	Yes	No	Yes
Leucine zipper:	Yes	No	No

C14. An enhancer, located upstream from a gene, has the following sequence:

5'–GTAG–3'
3'–CATC–5'

This enhancer is orientation independent. Which of the following sequences would also work as an enhancer?

A. 5'–CTAC–3'
3'–GATG–5'

B. 5'–GATG–3'
3'–CTAC–5'

C. 5'–CATC–3'
3'–GTAG–5'

C15. The DNA-binding domain of each CREB protein recognizes the sequence 5'–TGACGTCA–3'. As a matter of random chance, how often would you expect this sequence to occur in the human genome, which contains approximately 3 billion base pairs? Actually, only a few dozen genes are activated by the CREB protein. Does the value of a few dozen agree with the number of random sites found in the human genome? If the number of random sites in the human genome is much higher than a few dozen, provide at least one explanation why the CREB protein is not activating more than a few dozen genes.

C16. Solved problem S4 shows the locations of domains in the glucocorticoid receptor relative to the amino and carboxyl terminal ends of the protein. Make a drawing that illustrates the binding of a glucocorticoid receptor dimer to the DNA. In your drawing, label the amino and carboxyl ends, the hormone-binding domains, the DNA-binding domains, the dimerization domains, and the transactivation domains.

C17. The gene that encodes the enzyme called tyrosine hydroxylase is known to be up regulated by the CREB protein. Tyrosine hydroxylase is expressed in nerve cells and is involved in the synthesis of catecholamine, a neurotransmitter. The exposure of cells to adrenaline normally up regulates the transcription of the tyrosine hydroxylase gene. A mutant cell line was identified in which the tyrosine hydroxylase gene was not up regulated when exposed to adrenaline. List all the possible mutations that could explain this defect. How would you explain the defect if only the tyrosine hydroxylase gene was not up regulated by the CREB protein and other genes having CREs were properly up regulated in response to adrenaline in this cell line?

C18. What is the predominant form of chromatin in the eukaryotic nucleus? What must happen to its structure for transcription to take place? Discuss what is believed to cause this change in chromatin structure.

C19. Explain how the acetylation of core histones may loosen chromatin packing.

C20. Figure 15.11 shows the nucleosome structure of the β-globin gene in an active and inactive state. Suggest two or more possible reasons why the arrangement shown in Figure 15.11a would be transcriptionally inactive.

C21. What is DNA methylation? When we say that DNA methylation is heritable, what do we mean? How is it passed from a mother to a daughter cell?

C22. Let's suppose that a vertebrate organism carries a mutation that causes some cells that would normally differentiate into nerve cells to differentiate into muscle cells. A molecular analysis of this mutation revealed that it was in a gene that encodes a methyltransferase. Explain how an alteration in a methyltransferase could produce this phenotype.

C23. What is a CpG island? Where would you expect one to be located? How does the methylation of CpG islands affect gene expression?

C24. What is the function of a splicing factor? Explain how splicing factors can regulate the tissue-specific splicing of mRNAs.

C25. Figure 15.16 shows the products of alternative splicing for the α-tropomyosin pre-mRNA. Let's suppose that smooth muscle cells produce splicing factors that are not produced in other cell types. Explain where you think such splicing factors bind and how they influence the splicing of the α-tropomyosin pre-mRNA.

C26. Let's suppose a person is homozygous for a mutation in the IRP gene that changed the structure of the iron regulatory protein in such a way that it could not bind iron, but it could still bind to IREs. How would this mutation affect the regulation of ferritin and transferrin receptor mRNAs? Do you think such a person would need more iron in his/her diet compared to normal individuals? Do you think that excess iron in his/her diet would be more toxic compared to normal individuals? Explain your answer.

C27. In response to potentially toxic substances (e.g., high levels of iron), eukaryotic cells often use translational or posttranslational regulatory mechanisms to prevent cell death, rather than using transcriptional regulatory mechanisms. Explain why.

C28. What are the advantages and disadvantages of mRNAs with a short half-life compared to mRNAs with a long half-life?

C29. What is the phenomenon of RNA interference (RNAi)? During RNAi, explain how the double-stranded RNA is processed and how it leads to the silencing of a complementary mRNA.

C30. With regard to RNAi, what are three possible sources for double-stranded RNA?

C31. What conditions lead to the phosphorylation of eIF2α? Discuss why a cell would want to shut down translation when these conditions occur.

C32. What is the relationship between mRNA stability and mRNA concentration? What factors affect mRNA stability?

C33. Describe how the binding of the iron regulatory protein affects the mRNAs for ferritin and the transferrin receptor. How does iron influence this process?

Experimental Questions

E1. Why is a DNA site able to be cleaved by DNase I? With regard to DNase I cleavage, what do you expect will happen when a gene is converted from an inactive to a transcriptionally active state?

E2. In the experiment of Figure 15.10, explain how the use of S1 nuclease makes it possible to determine whether or not the β-globin gene is DNase I sensitive. In this experiment, why is it necessary to precipitate the DNA in step 7?

E3. Researchers can isolate a sample of cells, such as skin fibroblasts, and grow them in the laboratory. This procedure is called a cell culture. A cell culture can be exposed to a sample of DNA. If the cells are treated with agents that make their membranes permeable to DNA, the cells may take up the DNA and incorporate the DNA into their chromosomes. This process is called transformation or transfection. Experimenters have transformed human skin fibroblasts with methylated DNA and then allowed the fibroblasts to divide for several cellular generations. The DNA in the daughter cells was then isolated, and the segment that corresponded to the transformed DNA was examined. This DNA segment in the daughter cells was also found to be methylated. However, if the original skin fibroblasts were transformed with unmethylated DNA, the DNA found in the daughter cells was also unmethylated. Do fibroblasts undergo *de novo* methylation, maintenance methylation, or both? Explain your answer.

E4. Restriction enzymes, described in Chapter 18, are enzymes that recognize a particular DNA sequence and cleave the DNA (along the DNA backbone) at that site. The restriction enzyme known as *Not*I recognizes the sequence

5'–GCGGCCGC–3'
3'–CGCCGGCG–5'

However, if the cytosines in this sequence have been methylated, *Not*I will not cleave the DNA at this site. For this reason, *Not*I is commonly used to investigate the methylation state of CpG islands.

A researcher has studied a gene, which we will call gene *T*, that is found in corn. This gene encodes a transporter involved in the uptake of phosphate from the soil. A CpG island is located near the core promoter of gene *T*. The CpG island has a single *Not*I site. The arrangement of gene *T* is shown here.

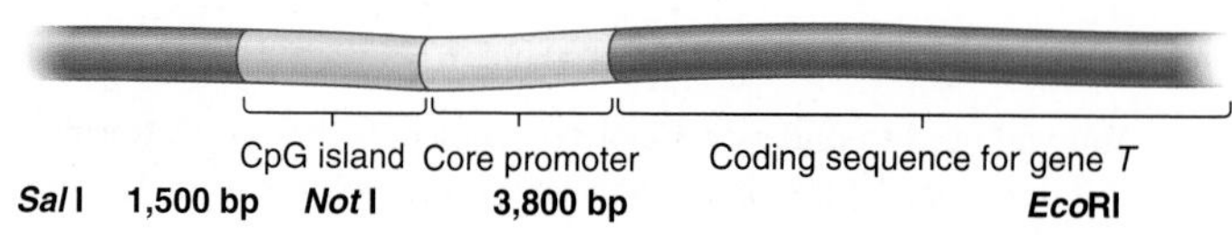

A *Sal*I restriction site is located upstream from the CpG island, and an *Eco*RI restriction site is located near the end of the coding sequence for gene *T*. The distance between the *Sal*I and *Not*I sites is 1,500 bp, and the distance between the *Not*I and *Eco*RI sites is 3,800 bp. No other sites for *Sal*I, *Not*I, or *Eco*RI are found in this region.

Here is the question. Let's suppose a researcher has isolated DNA samples from four different tissues in a corn plant. These include the leaf, the tassel, a section of stem, and a section of root. The DNA was then digested with all three restriction enzymes, separated by gel electrophoresis, and then probed with a DNA fragment complementary to the gene *T* coding sequence. The results are shown here.

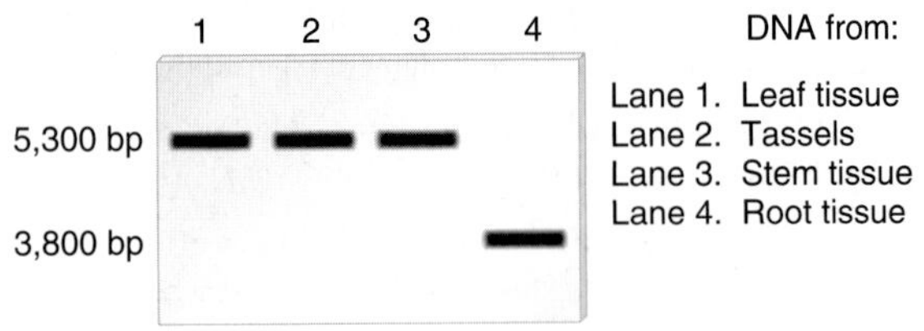

In which type of tissue is the CpG island methylated? Does this make sense based on the function of the protein encoded by gene *T*?

E5. You will need to understand solved problem S5 before answering this question. A muscle-specific gene was cloned and then subjected to promoter bashing as described in solved problem S5. As shown here, six regions, labeled A–F, were deleted, and then the DNA was transformed into muscle cells.

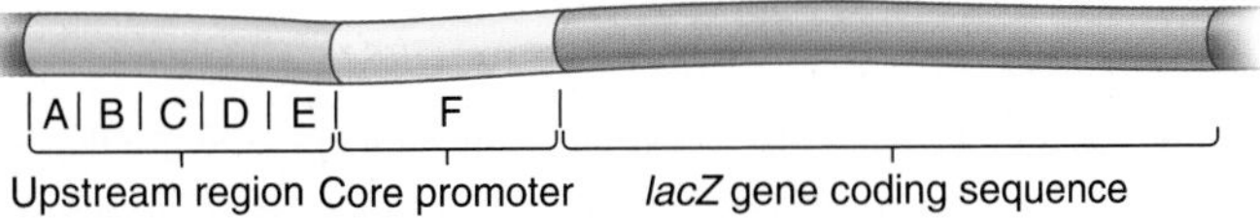

The data shown here are from this experiment.

Region Deleted	Percentage of β-Galactosidase Activity
None	100
A	20
B	330
C	100
D	5
E	15
F	<1

Explain these results.

E6. You will need to understand solved problem S5 before answering this question. A gene that is normally expressed in pancreatic cells was cloned and then subjected to promoter bashing as described in solved problem S5. As shown here, four regions, labeled A–D, were individually deleted, and then the DNA was transformed into pancreatic cells or into kidney cells.

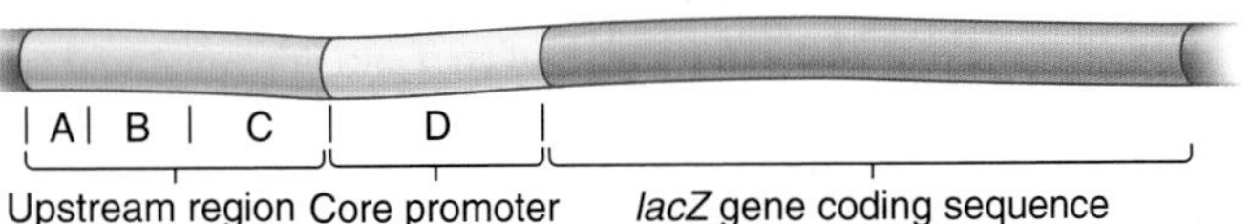

The data shown here are from this experiment.

Region Deleted	Cell Type Transformed	Percentage of β-Galactosidase Activity
None	Pancreatic	100
A	Pancreatic	5
B	Pancreatic	100
C	Pancreatic	100
D	Pancreatic	<1
None	Kidney	<1
A	Kidney	<1
B	Kidney	100
C	Kidney	<1
D	Kidney	<1

If we assume that the upstream region has one silencer and one enhancer, answer the following questions:

A. Where are the silencer and enhancer located?

B. Why don't we detect the presence of the silencer in the pancreatic cells?

C. Why isn't this gene normally expressed in kidney cells?

E7. A gel retardation assay can be used to determine if a protein binds to a segment of DNA. When a segment of DNA is bound by a protein, its mobility will be retarded, and the DNA band will appear higher in the gel. In the gel retardation assay shown here, a cloned gene fragment that is 750 bp in length contains a regulatory element that is recognized by a transcription factor called protein X. Previous experiments have shown that the presence of hormone X results in transcriptional activation by protein X. The results of a gel retardation assay are shown here.

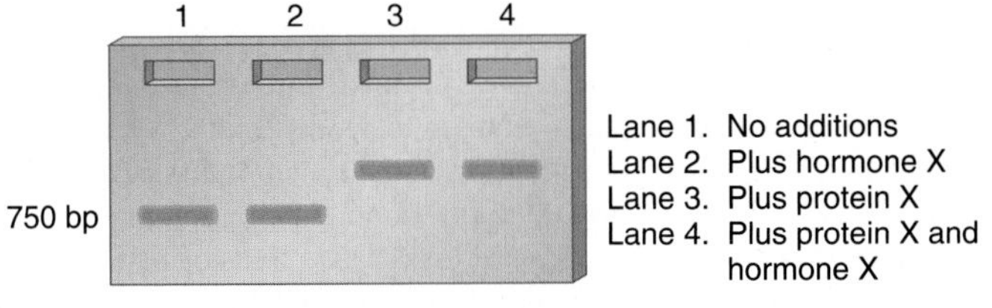

Explain the action of hormone X.

E8. Explain how the data of Fire and Mellow suggested that double-stranded RNA is responsible for the silencing of the *mex-3* gene.

E9. Chapter 18 describes a blotting method known as Northern blotting, in which a short segment of cloned DNA is used as a probe to detect RNA that is transcribed from a particular gene. The DNA probe, which is radioactive, is complementary to the RNA that the researcher wishes to detect. After the radioactive probe DNA binds to the RNA within a blot of a gel, the RNA is visualized as a dark (radioactive) band on an X-ray film. The method of Northern blotting can be used to determine the amount of a particular RNA transcribed in a given cell type. If one type of cell produces twice as much of a particular mRNA compared to another cell, the band appears twice as intense.

For this question, a researcher has a DNA probe complementary to the ferritin mRNA. This probe can be used to specifically detect the amount of ferritin mRNA on a gel. A researcher began with two flasks of human skin cells. One flask contained a very low concentration of iron, and the other flask had a high concentration of iron. The mRNA was isolated from these cells and then subjected to Northern blotting, using a probe complementary to the ferritin mRNA. The sample loaded in lane 1 was from the cells grown in a low concentration of iron, and the sample in lane 2 was from the cells grown in a high concentration of iron. Three Northern blots are shown here, but only one of them is correct. Based on your understanding of ferritin mRNA regulation, which blot (a, b, or c) would be your expected result? Explain. Which blot (a, b, or c) would be your expected result if the gel had been probed with a DNA segment complementary to the transferrin receptor mRNA?

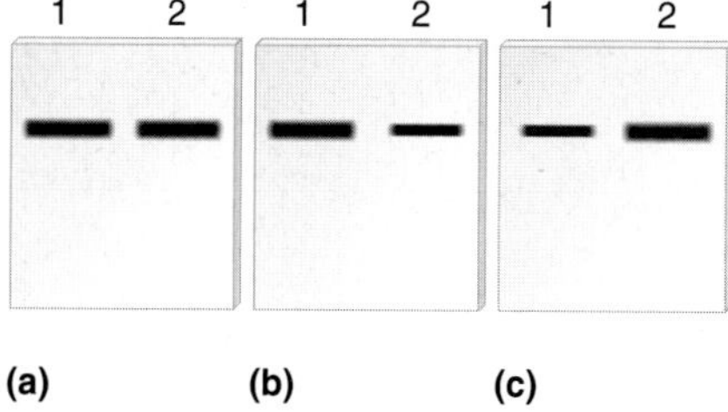

Questions for Student Discussion/Collaboration

1. Explain how DNA methylation could be used to regulate gene expression in a tissue-specific way. When and where would *de novo* methylation occur, and when would demethylation occur? What would occur in the germ line?
2. Enhancers can be almost anywhere and affect the transcription of a gene. Let's suppose you have a gene cloned on a piece of DNA, and the DNA fragment is 50,000 bp in length. Using cloning methods described in Chapter 18, you can cut out short segments from this 50,000 bp fragment and then reintroduce the smaller fragments into a cell that can express the gene. You would like to know if any enhancers are within the 50,000 bp region that may affect the expression of the gene. Discuss the most efficient strategy you can think of to trim your 50,000 bp fragment and thereby locate enhancers. You can assume that the coding sequence of the gene is in the center of the 50,000 bp fragment and that you can trim the 50,000 bp fragment into any size piece you want using molecular techniques described in Chapter 18.
3. How are regulatory transcription factors and regulatory splicing factors similar in their mechanism of action? In your discussion, consider the domain structures of both types of proteins. How are they different?

Note: All answers appear at the website for this textbook; the answers to even-numbered questions are in the back of the textbook.

www.mhhe.com/brookergenetics3e

Visit the Online Learning Center for practice tests, answer keys, and other learning aids for this chapter. Enhance your understanding of genetics with our interactive exercises, quizzes, animations, and much more.

16

GENE MUTATION AND DNA REPAIR

CHAPTER OUTLINE

The effects of a mutation. A mutation during embryonic development has caused this sheep to have a black spot on its side.

The primary function of DNA is to store information for the synthesis of cellular proteins. A key aspect of the gene expression process is that the DNA itself does not normally change. This allows DNA to function as a permanent storage unit. However, on relatively rare occasions, a mutation can occur. The term **mutation** refers to a heritable change in the genetic material. This means that the structure of DNA has been changed permanently, and this alteration can be passed from mother to daughter cells during cell division. If a mutation occurs in reproductive cells, it may also be passed from parent to offspring.

The topic of mutation is centrally important in all fields of genetics, including molecular genetics, Mendelian inheritance, and population genetics. Mutations provide the allelic variation that we have discussed throughout this textbook. For example, phenotypic differences, such as tall versus dwarf pea plants, are due to mutations that alter the expression of particular genes. With regard to their phenotypic effects, mutations can be beneficial, neutral, or detrimental. On the positive side, mutations are essential to the continuity of life. They provide the variation that enables species to change and adapt to their environment. Mutations are the foundation for evolutionary change. On the negative side, however, new mutations are much more likely to be harmful rather than beneficial to the individual. The genes within each species have evolved to work properly. They have functional promoters, coding sequences, terminators, and so on, that allow the genes to be expressed. Random mutations are more likely to disrupt these sequences rather than improve their function. For example, many inherited human diseases result from mutated genes. In addition, diseases such as skin and lung cancer can be caused by environmental agents that are known to cause DNA mutations. For these and many other reasons, understanding the molecular nature of mutations is a deeply compelling area of research. In this chapter, we will consider the nature of mutations and their consequences on gene expression at the molecular level.

Because mutations can be quite harmful, organisms have developed several ways to repair damaged DNA. DNA repair systems reverse DNA damage before it results in a mutation that could potentially have negative consequences. DNA repair systems have been studied extensively in many organisms, particularly *Escherichia coli*, yeast, mammals, and plants. A variety of systems repair different types of DNA lesions. In this chapter, we will examine the ways that several of these DNA repair systems operate.

16.1 CONSEQUENCES OF MUTATION

How do mutations affect phenotype? To answer this question, we must appreciate how changes in DNA structure can ultimately affect DNA function. Much of our understanding of mutation has come from the study of experimental organisms, such as bacteria, yeast, and *Drosophila*. Researchers can expose these organisms to environmental agents that cause mutation and then study the consequences of the induced mutations. In addition, because these organisms have a short generation time, researchers can investigate the effects of mutation when they are passed from cell to cell and from parent to offspring.

As discussed in Chapter 8, changes in chromosome structure and number are important occurrences within natural populations of eukaryotic organisms. These types of changes are considered to be mutations because the genetic material has been altered in a way that can be inherited. In comparison, a gene mutation is a relatively small change in DNA structure that affects a single gene. In this section, we will be primarily concerned with the ways that mutations may affect the molecular and phenotypic expression of single genes. We will also consider how the timing of mutations during an organism's development has important consequences.

Gene Mutations Are Molecular Changes in the DNA Sequence of a Gene

A gene mutation occurs when the sequence of the DNA within a gene is altered in a permanent way. A gene mutation can involve a base substitution in the sequence within a gene or a removal or addition of one or more base pairs.

A **point mutation** is a change in a single base pair within the DNA. For example, the DNA sequence shown here has been altered by a **base substitution,** in which one base is substituted for another base.

```
                      ↓
5'-AACGCTAGATC-3'    5'-AACGCGAGATC-3'
                  →
3'-TTGCGATCTAG-5'    3'-TTGCGCTCTAG-5'
```

A change of a pyrimidine to another pyrimidine, such as C to T, or a purine to another purine, such as A to G, is called a **transition.** This type of mutation is more common than a **transversion,** in which purines and pyrimidine are interchanged. The example just shown is a transversion (a T to G change), not a transition.

Besides base substitutions, a short sequence of DNA may be deleted from or added to the chromosomal DNA:

```
                       ↓
5'-AACGCTAGATC-3'    5'-AACGCTC-3'
                  →
3'-TTGCGATCTAG-5'    3'-TTGCGAG-5'
```

(Deletion of 4 bp)

```
                        _↓_
5'-AACGCTAGATC-3'    5'-AACAGTCGCTAGATC-3'
                  →
3'-TTGCGATCTAG-5'    3'-TTGTCAGCGATCTAG-5'
```

(Addition of 4 bp)

As we will see next, small deletions or additions to the sequence of a gene can significantly affect its function.

Gene Mutations Can Alter the Coding Sequence Within a Gene

How might a mutation within the coding sequence of a structural gene affect the amino acid sequence of the polypeptide that is encoded by the gene? **Table 16.1** describes the possible effects of point mutations. **Silent mutations** are those that do not alter the amino acid sequence of the polypeptide even though the nucleotide sequence has changed. Because the genetic code is degenerate, silent mutations can occur in certain bases within a codon, such as the third base, so that the specific amino acid is not changed. In contrast, **missense mutations** are base substitutions in which an amino acid change does occur. An example of a missense mutation occurs in the human disease known as sickle-cell anemia. This disease involves a mutation in the β-globin gene, which alters the polypeptide sequence so that the sixth amino acid is changed from a glutamic acid to valine. This single amino acid substitution alters the structure and function of the hemoglobin protein. One consequence of this alteration is that the red blood cells sickle under conditions of low oxygen (**Figure 16.1**). In this case, a single amino acid substitution has a profound effect on the phenotype of cells and even causes a serious disease.

Nonsense mutations involve a change from a normal codon to a stop codon. This terminates the translation of the polypeptide earlier than expected, producing a truncated polypeptide (see Table 16.1). When a nonsense mutation occurs in a bacterial operon, it may also inhibit the expression of downstream genes. This phenomenon, termed **polarity,** is described in solved problem S4 at the end of this chapter. Finally, **frameshift mutations** involve the addition or deletion of a number of nucleotides that is not divisible by three. Because the codons are read in multiples of three, this shifts the reading frame, Therefore, translation of the mRNA will result in a completely different amino acid sequence downstream from the mutation.

Except for silent mutations, new mutations are more likely to produce polypeptides that have reduced rather than enhanced function. For example, nonsense mutations will produce polypeptides that are substantially shorter and, therefore, unlikely to function properly. Likewise, frameshift mutations dramatically alter the amino acid sequence of polypeptides and are thereby likely to disrupt function. Missense mutations are less likely to alter function because they involve a change of a single amino acid within polypeptides that typically contain hundreds of amino acids. When a missense mutation has no detectable effect on protein function, it is referred to as a **neutral mutation.** A missense mutation that substitutes an amino acid with a similar chemistry as the original amino acid is likely to be neutral or nearly neutral. For example, a missense mutation that substitutes a glutamic acid for an aspartic acid is likely to be neutral because both amino acids are negatively charged and have similar side chain structures. Silent mutations are also considered a type of neutral mutation.

TABLE **16.1**

Consequences of Point Mutations Within the Coding Sequence

Type of Change	Mutation in the DNA	Example*	Amino Acids Altered	Likely Effect on Protein Function
None	None	5′–A–T–G–A–C–C–G–A–C–C–C–G–A–A–A–G–G–G–A–C–C–3′ Met – Thr – Asp – Pro – Lys – Gly – Thr –	None	None
Silent	Base substitution	↓ 5′–A–T–G–A–C–C–G–A–C–C–C–C–A–A–A–G–G–G–A–C–C–3′ Met – Thr – Asp – Pro – Lys – Gly – Thr –	None	None
Missense	Base substitution	↓ 5′–A–T–G–C–C–C–G–A–C–C–C–G–A–A–A–G–G–G–A–C–C–3′ Met – Pro – Asp – Pro – Lys – Gly – Thr –	One	Neutral or inhibitory
Nonsense	Base substitution	↓ 5′–A–T–G–A–C–C–G–A–C–C–C–G–T–A–A–G–G–G–A–C–C–3′ Met – Thr – Asp – Pro – STOP!	Many	Inhibitory
Frameshift	Addition/deletion	↓ 5′–A–T–G–A–C–C–G–A–C–G–C–C–G–A–A–A–G–G–G–A–C–C–3′ Met – Thr – Asp – Ala – Glu – Arg – Asp –	Many	Inhibitory

*DNA sequence in the coding strand. Note that this sequence is the same as the mRNA sequence except that the RNA contains uracil (U) instead of thymine (T). The 3-base codons are shown in alternating black and red colors. Mutations are shown in green.

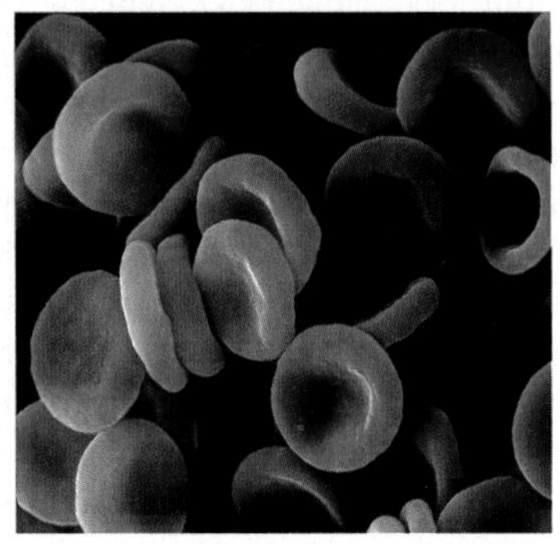

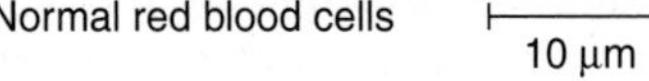

(a) Micrographs of red blood cells

NORMAL: NH_2– VALINE – HISTIDINE – LEUCINE – THREONINE – PROLINE – GLUTAMIC ACID – GLUTAMIC ACID...

SICKLE CELL: NH_2– VALINE – HISTIDINE – LEUCINE – THREONINE – PROLINE – VALINE – GLUTAMIC ACID...

(b) A comparison of the amino acid sequence between normal β-globin and sickle-cell β-globin

FIGURE 16.1 **Missense mutation in sickle-cell anemia.** (**a**) Normal red blood cells (left) and sickled red blood cells (right). (**b**) A comparison of the amino acid sequence of the normal β-globin polypeptide and the polypeptide encoded by the sickle-cell allele. This figure shows only a portion of the polypeptide sequence, which is 146 amino acids long. As seen here, a missense mutation changes the sixth amino acid from a glutamic acid to a valine.

Genes→Traits A missense mutation alters the structure of β globin, which is a subunit of hemoglobin, the oxygen-carrying protein in the red blood cells. When an individual is homozygous for this allele, this missense mutation causes the red blood cells to sickle under conditions of low oxygen concentration. The sickling phenomenon is a description of the trait at the cellular level. At the organism level, the sickled cells can clog the capillaries, thereby causing painful episodes, called crises, that can result in organ damage. The shortened life span of the red blood cells leads to symptoms of anemia.

Mutations can occasionally produce a polypeptide that has an enhanced ability to function. While these favorable mutations are relatively rare, they may result in an organism with a greater likelihood to survive and reproduce. If this is the case, natural selection may cause such a favorable mutation to increase in frequency within a population. This topic will be discussed later in this chapter and also in Chapter 24.

Gene Mutations Are Also Given Names That Describe How They Affect the Wild-Type Genotype and Phenotype

Thus far, several genetic terms have been introduced that describe the molecular effects of mutations. Genetic terms are also used to describe the effects of mutations relative to a wild-type genotype or phenotype. In a natural population, the **wild type** is a relatively prevalent genotype. For some genes with multiple alleles, a population may have two or more wild-type alleles. A mutation may change the wild-type genotype by altering the DNA sequence of a gene. When such a mutation is rare in a population, it is generally referred to as a **mutant allele.** A reverse mutation, more commonly called a **reversion,** changes a mutant allele back to a wild-type allele.

Another way to describe a mutation is based on its influence on the wild-type phenotype. Mutants are often characterized by their differential ability to survive. As mentioned, a neutral mutation does not alter protein function, so it does not affect survival or reproductive success. A **deleterious mutation** will decrease the chances of survival and reproduction. The extreme example of a deleterious mutation is a **lethal mutation,** which results in death to the cell or organism. On the other hand, a **beneficial mutation** will enhance the survival or reproductive success of an organism. In some cases, an allele may be either deleterious or beneficial depending on the genotype and/or the environmental conditions. An example is the sickle-cell allele. In the homozygous state, the sickle-cell allele lessens the chances of survival. However, when an individual is heterozygous for the sickle-cell allele and wild-type allele, this increases the chances of survival due to malarial resistance. Finally, some mutations are called **conditional mutants** because they affect the phenotype only under a defined set of conditions. Geneticists often study conditional mutants in microorganisms; a common example is a temperature-sensitive (*ts*) mutant. A bacterium that has a *ts* mutation grows normally in one temperature range—the permissive temperature range—but exhibits defective growth at a different temperature range—the nonpermissive temperature range. For example, an *E. coli* strain carrying a *ts* mutation may be able to grow from 33 to 38°C but not between 40 to 42°C, whereas the wild-type strain can grow at either temperature range.

A second mutation will sometimes affect the phenotypic expression of a first mutation. As an example, let's consider a mutation that causes an organism to grow very slowly. A second mutation at another site in the organism's DNA may restore the normal growth rate, converting the mutant back to the wild-type phenotype. Geneticists call these second-site mutations **suppressors** or **suppressor mutations.** This name is meant to indicate that a suppressor mutation acts to suppress the phenotypic effects of another mutation. A suppressor mutation differs from a reversion, because it occurs at a DNA site that is distinct from the first mutation.

Suppressor mutations are classified according to their relative locations with regard to the mutation they suppress (**Table 16.2**). When the second mutant site is within the same gene as the first mutation, the mutation is termed an **intragenic suppressor.** This type of suppressor often involves a change in protein structure that compensates for an abnormality in protein structure caused by the first mutation. Researchers often isolate suppressor mutations to obtain information about protein structure and function. For example, Robert Brooker and colleagues have isolated many intragenic suppressors in the *lacY* gene of *E. coli*, which encodes the lactose permease described in Chapter 15. This protein must undergo conformational changes to transport lactose across the cell membrane. They began with single mutations that altered amino acids on transmembrane regions, which inhibited this conformational change and thereby prevented growth on lactose. Suppressor mutations were then isolated that allowed growth on lactose by restoring transport function. By analyzing the locations of these suppressor mutations, the researchers were able to determine that certain transmembrane regions in the protein are critical for conformational changes required for lactose transport.

Alternatively, a suppressor mutation can be in a different gene from the first mutation—an **intergenic suppressor.** Researchers often study intergenic suppressors as a way to gain information about proteins that have similar or redundant functional roles, proteins that participate in a common pathway, multimeric proteins with two or more subunits, and the regulation of protein expression by transcription factors. How do intergenic suppressors work? These suppressor mutations usually involve a change in the expression of one gene that compensates for a loss-of-function mutation affecting another gene (see Table 16.2). For example, a first mutation may cause one protein to be partially or completely defective. An intergenic suppressor mutation in a different structural gene might overcome this defect by altering the structure of a second protein so that it could take over the functional role the first protein cannot perform. Alternatively, intergenic suppressors may involve proteins that participate in a common cellular pathway. When a first mutation decreases the activity of a protein, a suppressor mutation could enhance the function of a second protein involved in this pathway and thereby overcome the defect in the first protein.

In some cases, intergenic suppressors involve multimeric proteins, with each subunit encoded by a different gene. A mutation in one subunit that inhibits function may be compensated for by a mutation in another subunit. Another type of intergenic suppressor is one that involves mutations in genetic regulatory proteins such as transcription factors. When a first mutation causes a protein to be defective, a suppressor mutation may occur in a gene that encodes a transcription factor. The mutant transcription factor transcriptionally activates another gene that can compensate for the loss-of-function mutation in the first gene.

TABLE **16.2**

Examples of Suppressor Mutations

Type	No Mutation	First Mutation	Second Mutation	Description
Intragenic	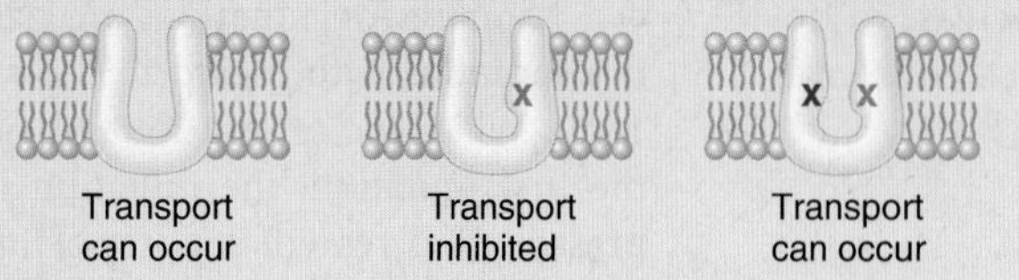 Transport can occur	Transport inhibited	Transport can occur	A first mutation disrupts normal protein function and a suppressor mutation affecting the same protein restores function. In this example, the first mutation inhibits lactose transport function, and the second mutation restores lactose transport.
Intergenic Redundant function	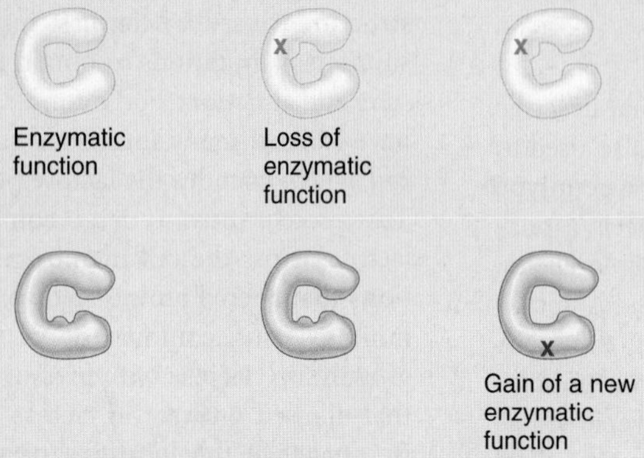 Enzymatic function	Loss of enzymatic function	Gain of a new enzymatic function	A first mutation inhibits the function of a protein, and a second mutation alters a different protein to carry out that function. In this example, the proteins function as enzymes.
Common pathway	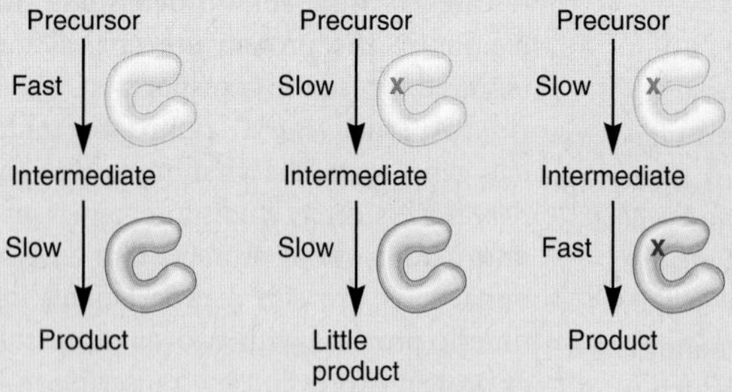 Precursor Fast Intermediate Slow Product	Precursor Slow Intermediate Slow Little product	Precursor Slow Intermediate Fast Product	Two or more different proteins may be involved in a common pathway. A mutation that causes a defect in one protein may be compensated for by a mutation that alters the function of a different protein in the same pathway.
Multimeric protein	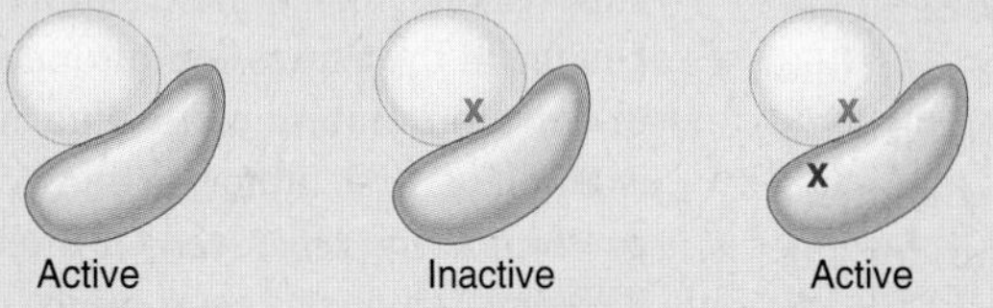 Active	Inactive	Active	A mutation in a gene encoding one protein subunit that inhibits function may be suppressed by a mutation in a gene that encodes a different subunit. The double mutant has restored function.
Transcription factor	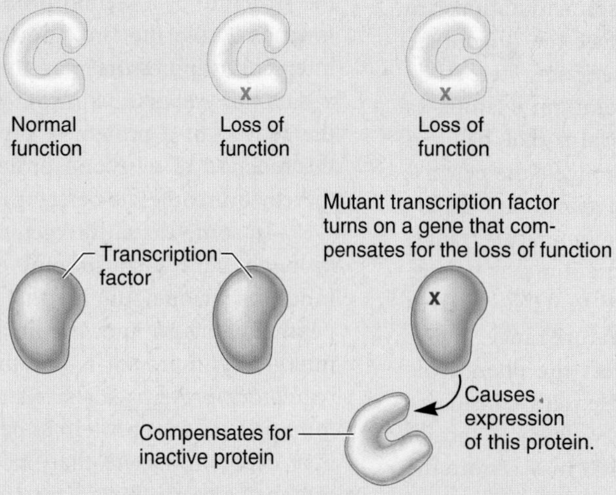 Normal function Transcription factor	Loss of function Compensates for inactive protein	Loss of function Mutant transcription factor turns on a gene that compensates for the loss of function Causes expression of this protein.	A first mutation causes loss of function of a particular protein. A second mutation may alter a transcription factor and cause it to activate the expression of another gene. This other gene encodes a protein that can compensate for the loss of function caused by the first mutation.

Less commonly, intergenic suppression involves mutations in nonstructural genes that enable the cell to defy the genetic code. Examples are suppressor tRNA mutants, which have been identified in microorganisms. Suppressor tRNA mutants have a change in the anticodon region that causes the tRNA to behave contrary to the genetic code. Nonsense suppressors are mutant tRNAs that recognize a stop codon, and instead of stopping translation, put an amino acid into the growing polypeptide chain. This type of mutant tRNA can suppress a nonsense mutation in a structural gene. However, such bacterial strains grow poorly because they may also suppress stop codons in normal genes.

Gene Mutations Can Occur Outside of the Coding Sequence and Still Influence Gene Expression

Thus far, we have focused our attention primarily on mutations in the coding regions of genes and their effects on polypeptide structure and protein function. In previous chapters, we learned how various sequences outside the coding sequence play important roles during the process of gene expression. A mutation can occur within noncoding sequences and thereby affect gene expression (**Table 16.3**). For example, a mutation may alter the sequence within the core promoter of a gene. Promoter mutations that increase transcription are termed **up promoter mutations.** Mutations that make a sequence more like the consensus sequence are likely to be up promoter mutations. In contrast, in a **down promoter mutation,** a mutation causes the promoter to become less like the consensus sequence, decreasing its affinity for regulatory factors and decreasing the transcription rate.

In Chapter 14, we considered how mutations could affect regulatory elements. For example, mutations in the *lac* operator site, called *lacO*^{C} mutations, prevent the binding of the lac repressor protein. This causes the *lac* operon to be constitutively expressed even in the absence of lactose. Bacteria strains with *lacO*^{C} mutations are at a selective disadvantage compared to wild-type *E. coli* strains because they waste their energy expressing the *lac* operon even when these proteins are not needed. As noted in Table 16.3, mutations can also occur in other noncoding regions of a gene and alter gene expression in a way that may affect phenotype. For example, mutations that affect the untranslated regions of mRNA—the 5'-UTR and 3'-UTR—may affect gene expression if they alter its ability to be translated or its stability. In addition, mutations in eukaryotic genes can alter splice junctions and affect the order and/or number of exons that are contained within mRNA.

TABLE **16.3**

Possible Consequences of Gene Mutations Outside of the Coding Sequence

Sequence	Effect of Mutation
Promoter	May increase or decrease the rate of transcription
Regulatory element/operator site	May disrupt the ability of the gene to be properly regulated
5'-UTR/3'-UTR	May alter the ability of mRNA to be translated; may alter mRNA stability
Splice recognition sequence	May alter the ability of pre-mRNA to be properly spliced

DNA Sequences Known as Trinucleotide Repeats Are Hotspots for Mutation

Researchers have discovered several human genetic diseases caused by an unusual form of mutation known as **trinucleotide repeat expansion (TNRE).** The term refers to the phenomenon in which a repeated sequence of three nucleotides can readily increase in number from one generation to the next. In normal individuals, certain genes and chromosomal locations contain regions where trinucleotide sequences are repeated in tandem. These sequences are transmitted normally from parent to offspring without mutation. However, in persons with TNRE disorders, the length of a trinucleotide repeat has increased above a certain critical size and becomes prone to frequent expansion. This phenomenon is depicted here, where the trinucleotide repeat of CAG has expanded from 11 tandem copies to 18 copies.

CAGCAGCAGCAGCAGCAGCAGCAGCAGCAGCAG n = 11

to

CAGCAGCAGCAGCAGCAGCAGCAGCAGCAGCAGCAGCAGCAG-
CAGCAGCAGCAG n = 18

Several human diseases have been discovered that involve these types of expansions, including spinal and bulbar muscular atrophy (SBMA), Huntington disease (HD), spinocerebellar ataxia (SCA1), fragile X syndromes (FRAXA and FRAXE), and myotonic muscular dystrophy (DM) (**Table 16.4**). In some cases, the expansion is within the coding sequence of the gene. Typically, such an expansion is a CAG repeat. Because CAG encodes a glutamine codon, these repeats cause the encoded proteins to contain long tracks of glutamine. Although the reasons for the disease symptoms in TNRE disorders are not well understood, the presence of glutamine tracts causes the proteins to aggregate. This aggregation of proteins or protein fragments carrying glutamine repeats is correlated with the progression of the disease. In other TNRE disorders, the expansions are located in the noncoding regions of genes. In the case of the two fragile X syndromes, the repeat produces CpG islands that become methylated. As discussed in Chapter 15, methylation can lead to chromatin compaction and thereby silence gene transcription. For myotonic muscular dystrophy, it has been hypothesized that these expansions cause abnormal changes in RNA structure and thereby produce disease symptoms.

TNRE disorders have the unusual feature that their severity tends to worsen in future generations—a phenomenon called **anticipation.** However, anticipation does not occur with all TNRE disorders and usually depends on whether the disease is inherited from the mother or father. In the case of Huntington disease, anticipation is likely to occur if the mutant gene is inherited from the father. In contrast, myotonic muscular dystrophy is

TABLE 16.4

TNRE Disorders

Disease	SBMA	HD	SCA1	FRAXA	FRAXE	DM
Repeat Sequence	CAG	CAG	CAG	CGG	GCC	CTG
Location of Repeat	Coding sequence	Coding sequence	Coding sequence	5'-UTR	5'-UTR	3'-UTR
Number of Repeats in Unaffected Individuals	11–33	6–37	6–44	6–53	6–35	5–37
Number of Repeats in Affected Individuals	36–62	27–121	43–81	>200	>200	>200
Pattern of Inheritance	X linked	Autosomal dominant	Autosomal dominant	X linked	X linked	Autosomal dominant
Disease Symptoms	Neuro-degenerative	Neuro-degenerative	Neuro-degenerative	Mental impairment	Mental impairment	Muscle disease
Anticipation *	None	Male	Male	Female	None	Female

*Indicates the parent in which anticipation occurs most prevalently.

more likely to get worse if the gene is inherited from the mother. These results suggest that TNRE can happen more frequently during oogenesis or spermatogenesis, depending on the particular gene involved. Overall, TNRE is a newly discovered form of mutation that is receiving a lot of attention by the research community. It poses many challenging questions in molecular genetics. The phenomenon of TNRE also makes it particularly difficult for genetic counselors to advise couples as to the severity of these diseases if they are passed to their children.

At the DNA level, the cause of TNRE is not well understood. Researchers have speculated that tandem repeats, such as trinucleotide repeats, produce alterations in DNA structure, including stem-loop formation and DNA strand slippage, and this may lead to errors in DNA replication. However, future research will be necessary to understand the underlying mechanism that causes TNRE. Nevertheless, it is well established that TNRE within certain genes alters the expression of the gene and thereby produces the disease symptoms.

Changes in Chromosome Structure Can Affect the Expression of a Gene

Thus far, we have considered small changes in the DNA sequence of particular genes. A change in chromosome structure can also be associated with an alteration in the expression of specific genes. Quite commonly, an inversion or translocation has no obvious phenotypic consequence. However, in 1925, Alfred Sturtevant was the first to recognize that chromosomal rearrangements in *Drosophila melanogaster* can influence phenotypic expression (namely, eye morphology). In some cases, a chromosomal rearrangement may affect a gene because a chromosomal **breakpoint**—the region where two chromosome pieces break and rejoin with other chromosome pieces—occurs within a gene. A breakpoint within the middle of a gene is very likely to inhibit gene function because it will separate the gene into two pieces.

In other cases, a gene may be left intact, but its expression may be altered when it is moved to a new location. When this occurs, the change in gene location is said to have a **position effect.** How do position effects alter gene expression? Researchers have discovered two common explanations. **Figure 16.2** depicts a schematic example in which a piece of one chromosome has been inverted or translocated to a different chromosome. One possibility is that a gene may be moved next to regulatory sequences, such as silencers or enhancers, that influence the expression of the relocated gene (Figure 16.2a). Alternatively, a chromosomal rearrangement may reposition a gene from a less condensed or euchromatic chromosome to a very highly condensed or heterochromatic chromosome. When the gene is moved to a heterochromatic region, its expression may be turned off (Figure 16.2b). This second type of position effect may produce a variegated phenotype in which the expression of the gene is variable. For genes that affect pigmentation, this produces a mottled appearance rather than an even color. **Figure 16.3** shows a position effect that alters eye color in *Drosophila.* Figure 16.3a depicts a normal red-eyed fruit fly, and Figure 16.3b shows a mutant fly that has inherited a chromosomal rearrangement in which a gene affecting eye color has been relocated to a heterochromatic chromosome. The variegated appearance of the eye occurs because the degree of heterochromatin formation varies across different regions of the eye. In cells where heterochromatin formation has turned off the eye color gene, a white phenotype occurs, while other cells allow this same region to remain euchromatic and produce a red phenotype.

Mutations Can Occur in Germ-Line or Somatic Cells

In this section, we have considered many different ways that mutations affect gene expression. For multicellular organisms, the timing of mutations also plays an important role. A mutation can occur very early in life, such as in a gamete or a fertilized egg, or it may occur later in life, such as in the embryonic or adult stages. The exact time when mutations occur can be important with regard to the severity of the genetic effect and whether they are passed from parent to offspring.

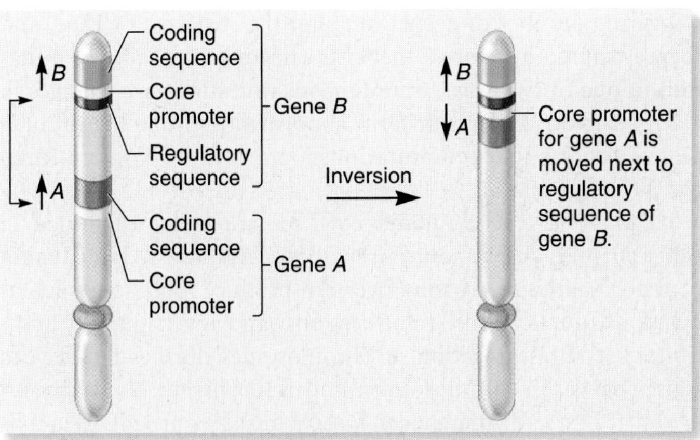

(a) Position effect due to regulatory sequences

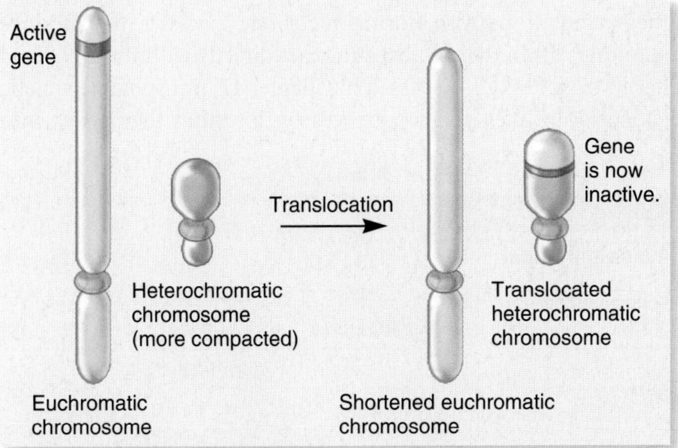

(b) Position effect due to translocation to a heterochromatic chromosome

FIGURE 16.2 **Causes of position effects.** **(a)** A chromosomal inversion has repositioned the core promoter of gene *A* next to the regulatory sequences for gene *B*. Because regulatory sequences are often bidirectional, the regulatory sequences for gene *B* may regulate the transcription of gene *A*. **(b)** A translocation has moved a gene from a euchromatic to a heterochromatic chromosome. This type of position effect prevents the expression of the relocated gene.

Geneticists classify the cells of animals into two types—the germ line and the somatic cells. The term **germ line** refers to cells that give rise to the gametes such as eggs and sperm. A **germ-line mutation** can occur directly in a sperm or egg cell, or it can occur in a precursor cell that produces the gametes. If a mutant gamete participates in fertilization, all cells of the resulting offspring will contain the mutation (**Figure 16.4a**). Likewise, when an individual with a germ-line mutation produces gametes, the mutation may be passed along to future generations of offspring.

The **somatic cells** comprise all cells of the body excluding the germ-line cells. Examples include muscle cells, nerve cells, and skin cells. Mutations can also happen within somatic cells at early or late stages of development. **Figure 16.4b** illustrates the consequences of a mutation that took place during the embryonic stage. In this example, a **somatic mutation** has occurred

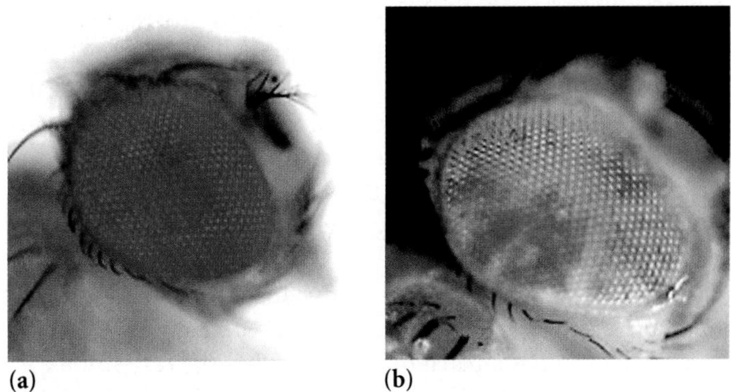

FIGURE 16.3 **A position effect that alters eye color in *Drosophila*.** **(a)** A normal red eye. **(b)** An eye in which an eye color gene has been relocated to a heterochromatic chromosome. This can inactivate the gene in some cells and produces a variegated phenotype.

Genes→Traits Variegated eye color occurs because the degree of heterochromatin formation varies throughout different regions of the eye. In some cells, heterochromatin formation occurs and turns off the eye color gene, thereby leading to the white phenotype. In other patches of cells, the region containing the eye color allele remains euchromatic, yielding a red phenotype.

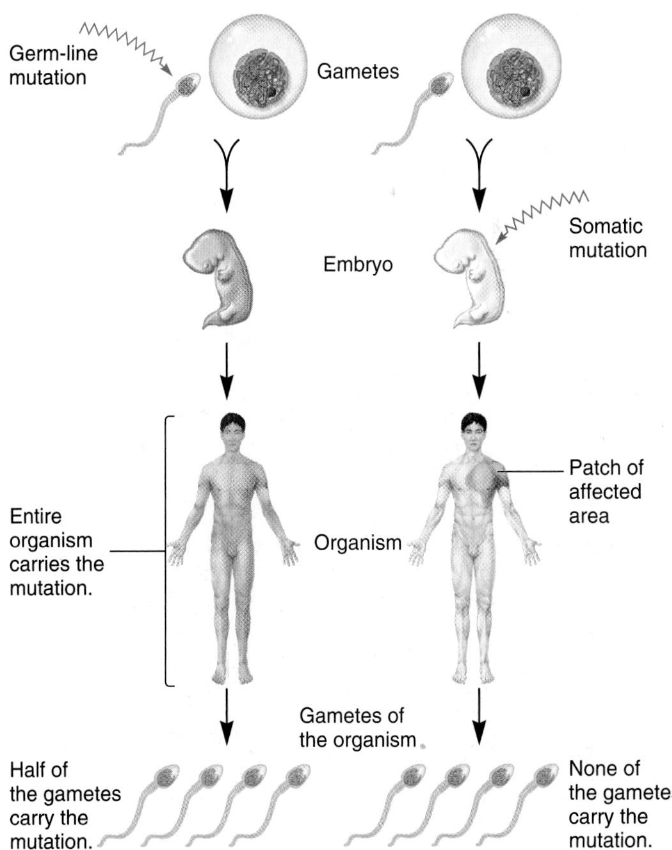

FIGURE 16.4 **The effects of germ-line versus somatic mutations.**

within a single embryonic cell. As the embryo grows, this single cell will be the precursor for many cells of the adult organism. Therefore, in the adult, a portion of the body will contain the mutation. The size of the affected region will depend on the timing of the mutation. In general, the earlier the mutation occurs during development, the larger the region. An individual that has somatic regions that are genotypically different from each other is called a **genetic mosaic.**

Figure 16.5 illustrates an individual who had a somatic mutation occur during an early stage of development. In this case, the person has a patch of white hair, but the rest of the hair is pigmented. Presumably, this individual initially had a single mutation occur in an embryonic cell that ultimately gave rise to a patch of scalp that produced the white hair. Although a patch of white hair is not a harmful phenotypic effect, mutations during early stages of life can be quite detrimental, especially if they disrupt essential developmental processes. Therefore, even though it is smart to avoid environmental agents that cause mutations during all stages of life, the possibility of somatic mutations is a rather compelling reason to avoid them during the very early stages of life such as fetal development, infancy, and early childhood. For example, the possibility of somatic mutations in an embryo is a reason why women are advised to avoid getting X-rays during pregnancy.

16.2 OCCURRENCE AND CAUSES OF MUTATION

As we have seen, mutations can have a wide variety of effects on the phenotypic expression of genes. For this reason, geneticists have expended a great deal of effort identifying the causes of mutations. This task has been truly challenging, because a staggering number of agents can alter the structure of DNA and thereby cause mutation. Geneticists categorize the cause of mutation in one of two ways. **Spontaneous mutations** are changes in DNA structure that result from abnormalities in biological processes, whereas **induced mutations** are caused by environmental agents (**Table 16.5**).

Many causes of spontaneous mutations are examined in other chapters throughout this textbook. As discussed in Chapter 8, abnormalities in crossing over can produce mutations such as deletions, duplications, translocations, and inversions. In addition, aberrant segregation of chromosomes during meiosis can cause changes in chromosome number. In Chapter 11, we discovered that DNA polymerase can make a mistake during DNA replication by putting the wrong base in a newly synthesized daughter strand. Errors in DNA replication are usually infrequent except in certain viruses, such as HIV, that have relatively high rates of spontaneous mutations. Also, normal metabolic processes may produce chemicals within the cell that can react directly with the DNA and alter its structure. As discussed in Chapter 17, transposable genetic elements can alter gene sequences by inserting themselves into

FIGURE 16.5 Example of a somatic mutation.

Genes→Traits This person has a patch of white hair because a somatic mutation occurred in a single cell during embryonic development that prevented pigmentation of the hair. This cell continued to divide to produce a patch of white hair.

TABLE 16.5

Causes of Mutations

Common Causes of Mutations	Description
Spontaneous	
Aberrant recombination	Abnormal crossing over may cause deletions, duplications, translocations, and inversions (see Chapter 8).
Aberrant segregation	Abnormal chromosomal segregation may cause aneuploidy or polyploidy (see Chapter 8).
Errors in DNA replication	A mistake by DNA polymerase may cause a point mutation (see Chapter 11).
Toxic metabolic products	The products of normal metabolic processes may be chemically reactive agents that can alter the structure of DNA.
Transposable elements	Transposable elements can insert themselves into the sequence of a gene (see Chapter 17).
Depurination	On rare occasions, the linkage between purines (i.e., adenine and guanine) and deoxyribose can spontaneously break. If not repaired, it can lead to mutation.
Deamination	Cytosine and 5-methylcytosine can spontaneously deaminate to create uracil or thymine.
Tautomeric shifts	Spontaneous changes in base structure can cause mutations if they occur immediately prior to DNA replication.
Induced	
Chemical agents	Chemical substances may cause changes in the structure of DNA.
Physical agents	Physical phenomena such as UV light and X-rays can damage the DNA.

genes. In this section, we will examine how spontaneous changes in nucleotide structure, such as depurination, deamination, and tautomeric shifts, can cause mutation. Overall, a distinguishing feature of spontaneous mutations is that their underlying cause originates within the cell. By comparison, the cause of induced mutations originates outside the cell. Induced mutations are produced by environmental agents, either chemical or physical, that enter the cell and lead to changes in DNA structure. Agents known to alter the structure of DNA are called **mutagens.**

We will begin by examining the random nature of spontaneous mutations and general features of the mutation rate. Then we will explore several mechanisms by which mutagens can alter the structure of DNA. Finally, laboratory tests that can identify potential mutagens will be described.

Spontaneous Mutations Are Random Events

For a couple of centuries, biologists had questioned whether heritable changes occur purposefully as a result of behavior or exposure to particular environmental conditions or whether they are spontaneous events that may happen randomly. In the nineteenth century, the naturalist Jean Baptiste Lamarck proposed that physiological events—such as the use or disuse of muscles—determine whether traits are passed along to offspring. For example, his hypothesis suggested that an individual who practiced and became adept at a physical activity and developed muscular legs would pass that characteristic on to the next generation. The alternative point of view is that genetic variation exists in a population as a matter of random chance and natural selection results in the differential reproductive success of organisms that are better adapted to their environments. Those individuals who happen to have beneficial mutations will be more likely to survive and pass these genes to their offspring. These opposing ideas of the nineteenth century—one termed physiological adaptation and the other termed random mutation—were tested in bacterial studies in the 1940s and 1950s, two of which are described here.

Salvadore Luria and Max Delbrück were interested in the ability of bacteria to become resistant to infection by a bacteriophage called T1. When a population of *E. coli* cells is exposed to T1, a small percentage of bacteria are found to be resistant to T1 infection and pass this trait to their progeny. Luria and Delbrück were interested in whether such resistance, called *ton*^{r} (T one resistance), is due to the occurrence of random mutations or whether it is a physiological adaptation that occurs at a low rate within the bacterial population.

According to the physiological adaptation hypothesis, the rate of mutation should be a relatively constant value and depend on the exposure to the bacteriophage. Therefore, when comparing different populations of bacteria, the number of *ton*^{r} bacteria should be an essentially constant proportion of the total population. In contrast, the random mutation hypothesis depends on the timing of mutation. If a *ton*^{r} mutation occurs early within the proliferation of a bacterial population, many *ton*^{r} bacteria will be found within that population because mutant bacteria will have time to grow and multiply. However, if it occurs much later in population growth, fewer *ton*^{r} bacteria will be observed. In general, a random mutation hypothesis predicts a much greater fluctuation in the number of *ton*^{r} bacteria among different populations.

To distinguish between the physiological adaptation and random mutation hypotheses, Luria and Delbrück inoculated one large flask and 20 individual tubes with *E. coli* cells and grew them in the absence of T1 phage. The flask was grown to produce a very large population of cells, while each individual culture was grown to a smaller population of approximately 20 million cells. They then plated the individual cultures onto media containing T1 phage. Likewise, 10 subsamples, each consisting of 20 million bacteria, were removed from the large flask and plated onto media with T1 phage.

The results of the Luria and Delbrück experiment are shown in **Figure 16.6**. Within the smaller individual cultures, a great fluctuation was observed in the number of *ton*^{r} mutants. This test, therefore, has become known as the **fluctuation test.** Which hypothesis do these results support? They are consistent with a random mutation hypothesis in which the timing of a mutation during the growth of a culture greatly affects the number of mutant cells. For example, in tube 14, many *ton*^{r} bacteria were observed. Luria and Delbrück reasoned that a mutation occurred randomly in one bacterium at an early stage of the population growth, before the bacteria were exposed to T1 on plates. This mutant bacterium then divided to produce many daughter cells that inherited the *ton*^{r} trait. In other tubes, such as 1 and 3, this spontaneous mutation did not occur, so none of the bacteria had a *ton*^{r} phenotype. By comparison, the cells plated from the large flask tended toward a relatively constant and intermediate number of *ton*^{r} bacteria. Because the large flask had so many cells, several independent *ton*^{r} mutations were likely to have occurred during different stages of its growth. In a single flask, however, these independent events would be mixed together to give an average value of *ton*^{r} cells.

Several years later, Joshua and Esther Lederberg were also interested in the relationship between mutation and the environmental conditions that select for mutation. To distinguish between the physiological adaptation and random mutation hypotheses, they developed a technique known as **replica plating** in the 1950s. As shown in **Figure 16.7**, they plated a large number of bacteria onto a master plate that did not contain any selective agent (namely, the T1 phage). A sterile piece of velvet cloth was lightly touched to this plate in order to pick up a few bacterial cells from each colony. This replica was then transferred to two secondary plates that contained an agent that selected for the growth of bacterial cells with a particular genotype.

In the example shown in Figure 16.7, the secondary plates contained T1 bacteriophages. On these plates, only those mutant cells that are *ton*^{r} could grow. On the secondary plates, a few colonies were observed. Strikingly, they occupied the same location on each plate. These results indicated that the mutations conferring *ton*^{r} occurred randomly while the cells were growing on the nonselective master plate. The presence of the T1 phage in the secondary plates simply selected for the growth of previously occurring *ton*^{r} mutants. These results supported the random mutation hypothesis. In contrast, the physiological adaptation hypothesis

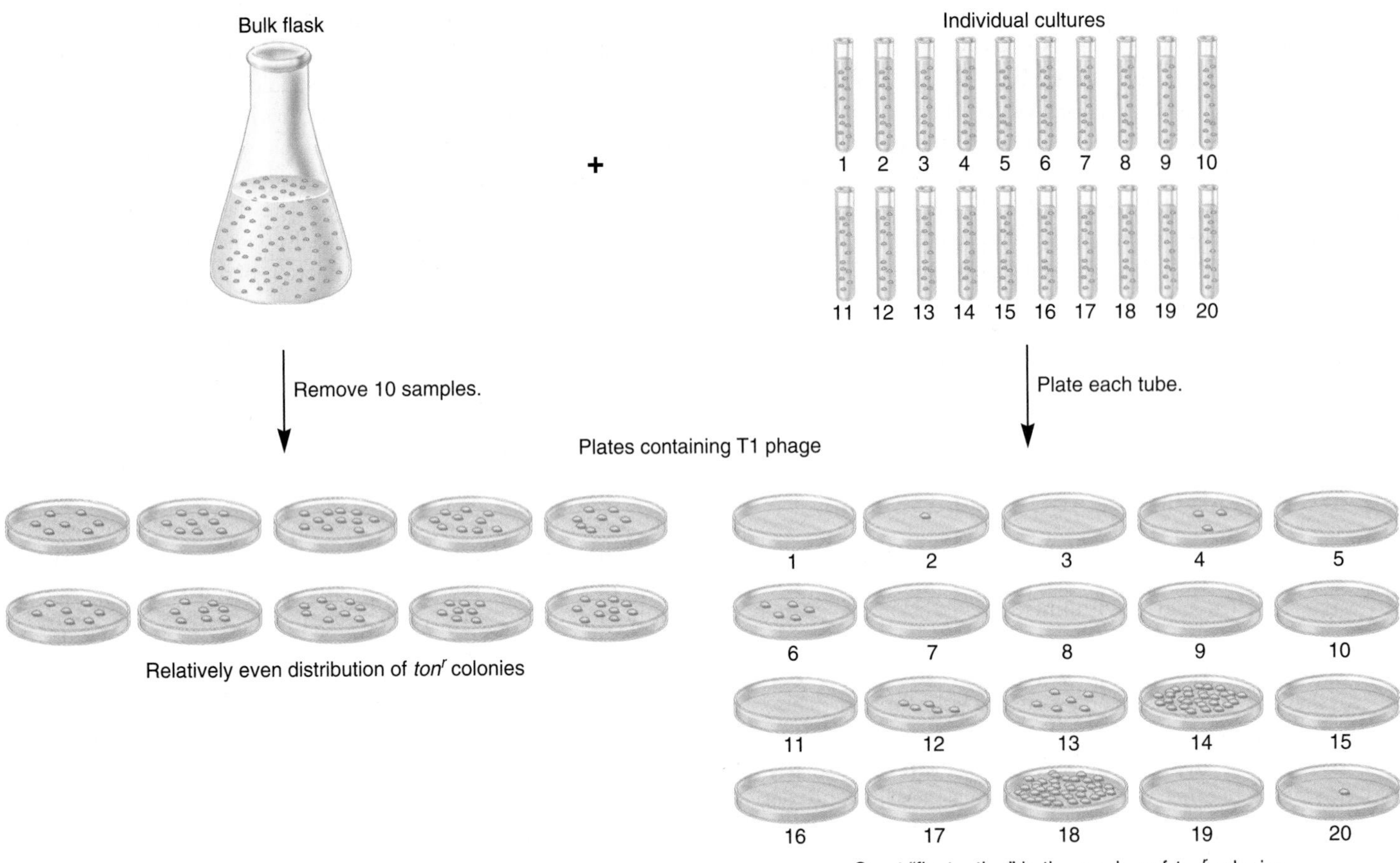

FIGURE 16.6 The Luria-Delbrück fluctuation test.

would have predicted that ton^r bacterial mutants would occur after exposure to the selective agent. If that had been the case, the colonies would not be expected to arise in identical locations on different secondary plates but rather in random patterns.

Taken together, the results of Luria and Delbrück and those of the Lederbergs supported the random mutation hypothesis, now known as the **random mutation theory.** According to this theory, mutations are a random process—they can occur in any gene and do not involve exposure of an organism to a particular condition that selects for specific types of mutations. In some cases, a random mutation may provide a mutant organism with an advantage, such as resistance to T1 phage. Although such mutations occur as a matter of random chance, growth conditions may select for organisms that happen to carry them.

As researchers have learned more about mutation at the molecular level, the view that mutations are a totally random process has required some modification. Within the same individual, some genes mutate at a much higher rate than other genes. Why does this happen? Some genes are larger than others, which provides a greater chance for mutation. Also, the relative locations of genes within a chromosome may cause some genes to be more susceptible to mutation compared to others. Even within a single gene, **hot spots** are usually found—certain regions of a gene that are more likely to mutate compared to other regions (refer back to Chapter 6, Figure 6.20).

Mutation Rates and Frequencies Are Ways to Quantitatively Assess Mutation in a Population

Because mutations occur spontaneously among populations of living organisms, geneticists are greatly interested in learning how prevalent they are. The term **mutation rate** is the likelihood that a gene will be altered by a new mutation. This rate is commonly expressed as the number of new mutations in a given gene per cell generation. In general, the spontaneous mutation rate for a particular gene is in the range of 1 in 100,000 to 1 in 1 billion, or 10^{-5} to 10^{-9} per cell generation. These numbers tell us that it is very unlikely that a particular gene will mutate due to natural causes. However, the mutation rate is not a constant number. The presence of certain environment agents, such as X-rays, can increase the rate of induced mutations to a much higher value than the spontaneous mutation rate. In addition, mutation rates vary substantially from species to species and even within different strains of the same species. One explanation for this variation is that there are many different causes of mutations (refer back to Table 16.5).

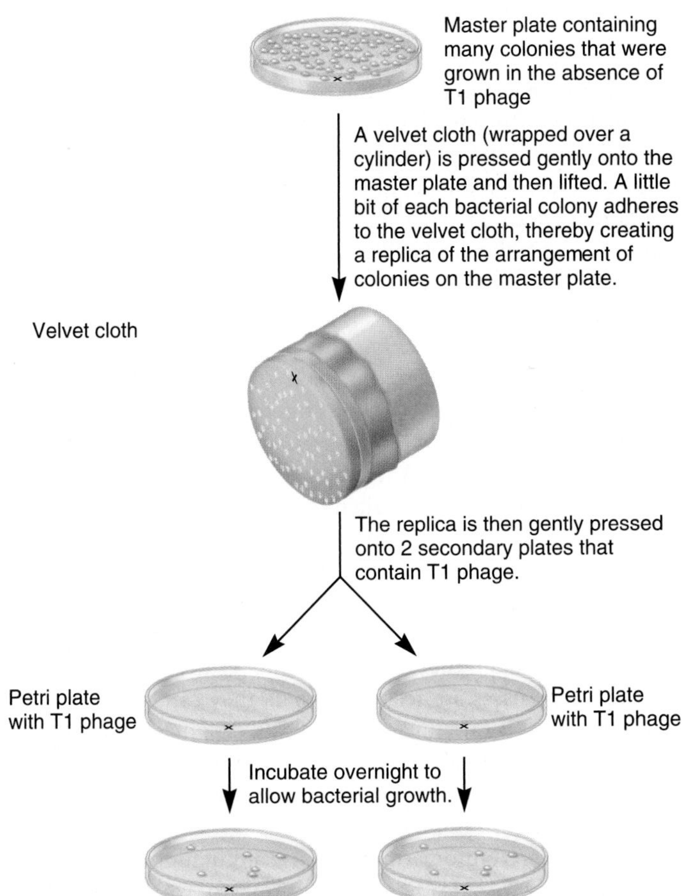
Master plate containing many colonies that were grown in the absence of T1 phage
A velvet cloth (wrapped over a cylinder) is pressed gently onto the master plate and then lifted. A little bit of each bacterial colony adheres to the velvet cloth, thereby creating a replica of the arrangement of colonies on the master plate.
Velvet cloth
The replica is then gently pressed onto 2 secondary plates that contain T1 phage.
Petri plate with T1 phage
Petri plate with T1 phage
Incubate overnight to allow bacterial growth.

FIGURE 16.7 **Replica plating.** Bacteria were first plated on a master plate under nonselective conditions. A sterile velvet cloth was used to make a replica of the master plate. This replica was gently pressed onto two secondary plates that contained a selective agent. In this case, the two secondary plates contained T1 bacteriophage. Only those mutant cells that are *ton*^{r} (resistant to T1) could grow to form visible colonies. Note: The black x indicates the alignment of the velvet and the plates.

Before we end our discussion of mutation rate, let's distinguish the rate of new mutation from the concept of **mutation frequency.** The mutation frequency for a gene is the number of mutant genes divided by the total number of genes within a population. If 1 million bacteria were plated and 10 were found to be mutant, the mutation frequency would be 1 in 100,000, or 10^{-5}. As we have seen, Luria and Delbrück showed that among the bacteria in the 20 tubes, the timing of mutations influenced the mutation frequency within any particular tube. Some tubes had a high frequency of mutation, while others did not. Therefore, the mutation frequency depends not only on the mutation rate but also on the timing of mutation, and on the likelihood that a mutation will be passed to future generations. The mutation frequency is an important genetic concept, particularly in the field of population genetics. As we will see in Chapter 24, mutation frequencies may rise above the mutation rate due to evolutionary forces such as natural selection and genetic drift.

Spontaneous Mutations Can Arise by Depurination, Deamination, and Tautomeric Shifts

Thus far, we have considered the random nature of mutation and the quantitative difference between mutation rate and mutation frequency. We now turn our attention to the molecular changes in DNA structure that can cause mutation. Our first examples concern changes that can occur spontaneously, albeit at a low rate. The most common type of chemical change that occurs naturally is **depurination,** which involves the removal of a purine (adenine or guanine) from the DNA. The covalent bond between deoxyribose and a purine base is somewhat unstable and occasionally undergoes a spontaneous reaction with water that releases the base from the sugar, thereby creating an **apurinic site** (**Figure 16.8a**). In a typical mammalian cell, approximately 10,000 purines are lost from the DNA in a 20-hour period at 37°C. The rate of loss is higher if the DNA is exposed to agents that cause certain types of base modifications such as the attachment of alkyl groups (methyl or ethyl groups). Fortunately, as discussed later in this chapter, apurinic sites are recognized by DNA base excision repair enzymes that repair the site. If the repair system fails, however, a mutation may result during subsequent rounds of DNA replication. What happens at an apurinic site during DNA replication? Because a complementary base is not present to specify the incoming base for the new strand, any of the four bases will be added to the

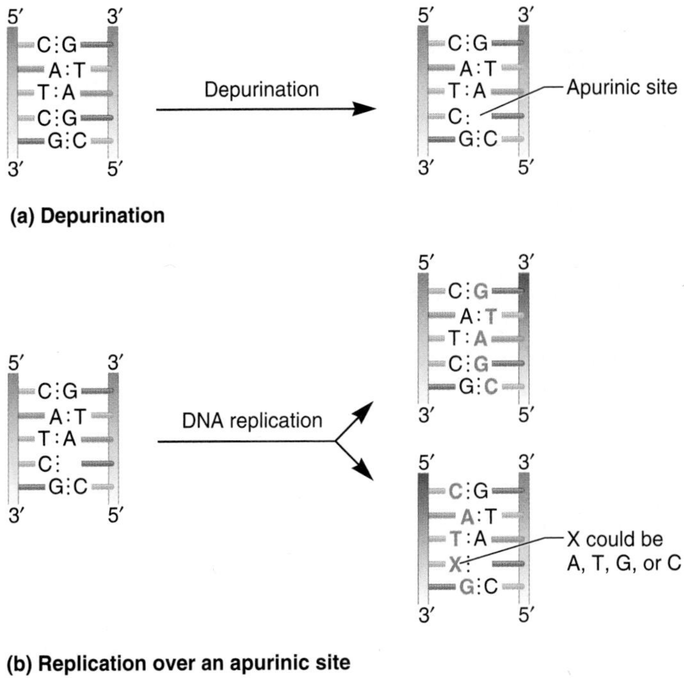

5′ 3′ C:G A:T T:A C:G G:C 3′ 5′ — Depurination → 5′ 3′ C:G A:T T:A C: G:C 3′ 5′ Apurinic site

(a) Depurination

5′ 3′ C:G A:T T:A C: G:C 3′ 5′ — DNA replication → 5′ 3′ C:G A:T T:A C:G G:C 3′ 5′ ; 5′ 3′ C:G A:T T:A X: G:C 3′ 5′ X could be A, T, G, or C

(b) Replication over an apurinic site

FIGURE 16.8 **Spontaneous depurination.** (**a**) The bond between guanine and deoxyribose is broken, thereby releasing the base. This leaves an apurinic site in the DNA. (**b**) If an apurinic site remains in the DNA as it is being replicated, any of the four nucleotides can be added to the newly made strand. Because three out of four (A, T, and G) are the incorrect base, the chance of causing a mutation is 75%.

new strand in the region that is opposite the apurinic site (**Figure 16.8b**). This may lead to a new mutation.

A second spontaneous lesion that may occur in DNA is the **deamination** of cytosines. The other bases are not readily deaminated. As shown in **Figure 16.9a**, deamination involves the removal of an amino group from the cytosine base. This produces uracil. As discussed later, DNA repair enzymes can recognize uracil as an inappropriate base within DNA and subsequently remove it. However, if such repair does not take place, a mutation may result because uracil hydrogen bonds with adenine during DNA replication. Therefore, if a DNA template strand has uracil instead of cytosine, a newly made strand will incorporate adenine into the daughter strand instead of guanine.

Figure 16.9b shows the deamination of 5-methylcytosine. As discussed in Chapter 15, the methylation of cytosine occurs in many eukaryotic species. It also occurs in prokaryotes. If 5-methylcytosine is deaminated, the resulting base is thymine, which is a normal constituent of DNA. Therefore, this poses a problem for DNA repair enzymes because they cannot distinguish which is the incorrect base—the thymine that was produced by deamination or the guanine in the opposite strand that originally base-paired with the methylated cytosine. For this reason, methylated cytosine bases tend to create hot spots for mutation. As an example, researchers analyzed 55 spontaneous mutations that occurred within the *lacI* gene of *E. coli* and determined that 44 of them involved changes at sites that were originally occupied by a methylated cytosine base.

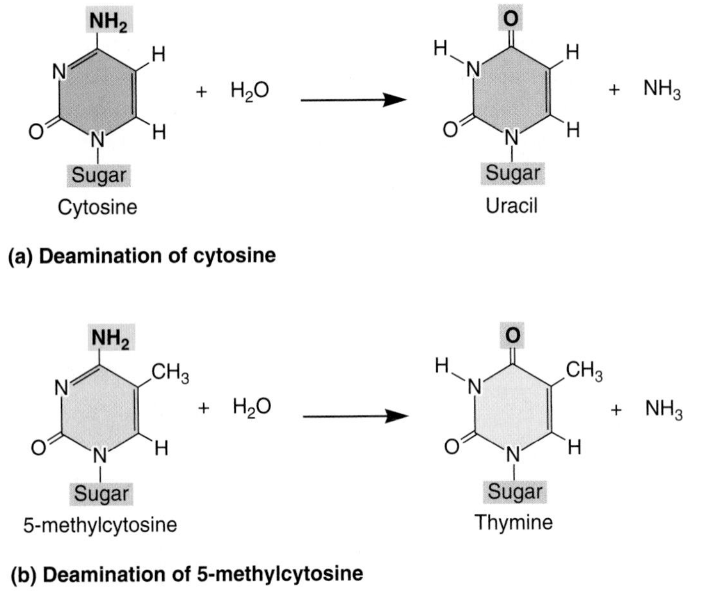

A third way that mutations may arise spontaneously involves a temporary change in base structure called a **tautomeric shift.** In this case, the **tautomers** are bases, which exist in keto and enol or amino and imino forms. These forms can interconvert by a chemical reaction that involves the migration of a hydrogen atom and a switch of a single bond and an adjacent double bond. The common, stable form of guanine and thymine is the keto form; the common form of adenine and cytosine is the amino form (**Figure 16.10a**). At a low rate, G and T can interconvert to an enol form, and A and C can change to an imino form. Though the relative amounts of the enol and imino forms of these bases are relatively small, they can cause a mutation because these rare forms of the bases do not conform to the AT/GC rule of base pairing. Instead, if one of the bases is in the enol or imino form, hydrogen bonding will promote TG and CA base pairs, as shown in **Figure 16.10b**.

How does a tautomeric shift cause a mutation? The answer is that it must occur immediately prior to DNA replication. When DNA is in a double-stranded condition, the base pairing usually holds the bases in their more stable forms. After the strands unwind, however, a tautomeric shift may occur. In the example shown in **Figure 16.10c**, a thymine base in the template strand has undergone a tautomeric shift just prior to the replication of the complementary daughter strand. During replication, the daughter strand will incorporate a guanine opposite this thymine, creating a base mismatch. This mismatch could be repaired via the proofreading function of DNA polymerase or via a mismatch repair system (discussed later in this chapter). However, if these repair mechanisms fail, the next round of DNA replication will create a double helix with a CG base pair, while the correct base pair should be TA. As shown in the right side of Figure 16.10c, one of four daughter cells will inherit this CG mutation.

NH2 + H_2O → + NH_3

Cytosine → Uracil

(a) Deamination of cytosine

5-methylcytosine + H_2O → Thymine + NH_3

(b) Deamination of 5-methylcytosine

FIGURE 16.9 **Spontaneous deamination of cytosine and 5-methylcytosine.** (a) The deamination of cytosine produces uracil. (b) The deamination of 5-methylcytosine produces thymine.

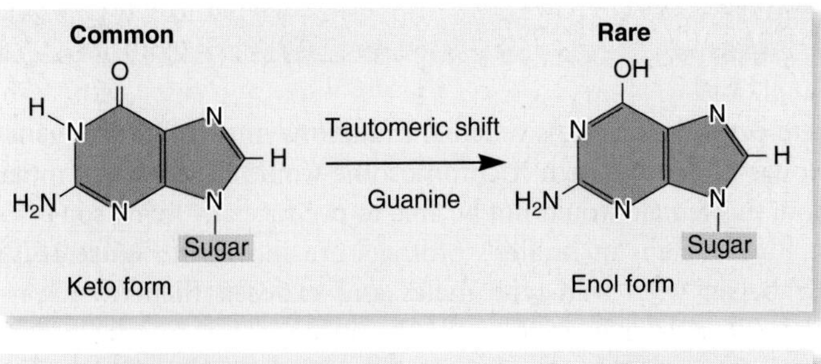

Common
Rare
Tautomeric shift
Thymine
Sugar
Keto form
Enol form

Tautomeric shift
Adenine
Sugar
Amino form
Imino form

Tautomeric shift
Cytosine
Sugar
Amino form
Imino form

(a) Tautomeric shifts that occur in the 4 bases found in DNA

Sugar
Thymine (enol)
Guanine (keto)
Cytosine (imino)
Adenine (amino)

(b) Mis–base pairing due to tautomeric shifts

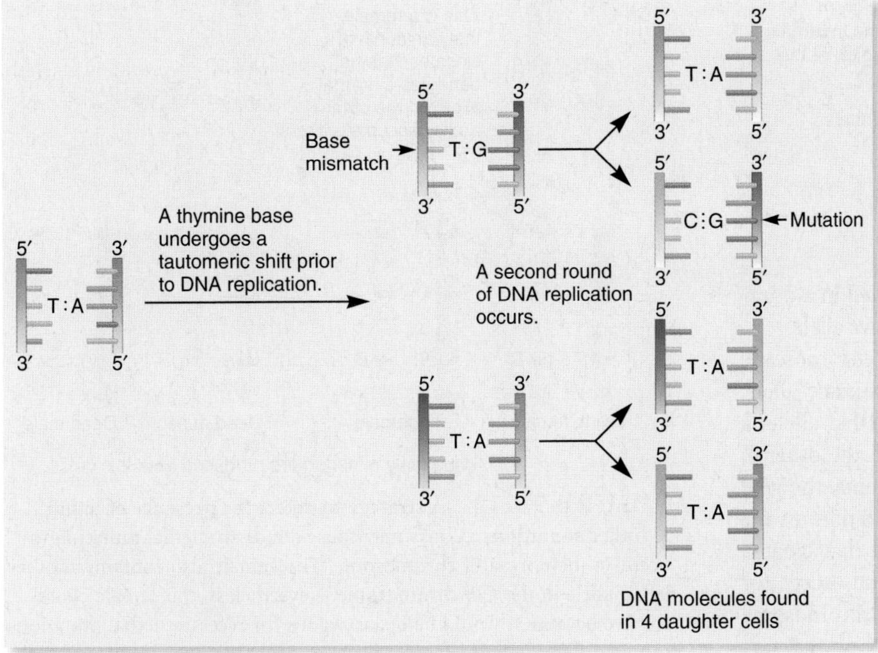

(c) Tautomeric shifts and DNA replication can cause mutation

FIGURE 16.10 **Tautomeric shifts and their ability to cause mutation.** **(a)** The common forms of the bases are shown on the left, and the rare forms produced by a tautomeric shift are shown on the right. **(b)** On the left, the rare enol form of thymine pairs with the common keto form of guanine (instead of adenine); on the right, the rare imino form of cytosine pairs with the common amino form of adenine (instead of guanine). **(c)** A tautomeric shift occurred in a thymine base just prior to replication, causing the formation of a TG base pair. If not repaired, a second round of replication will lead to the formation of a permanent CG mutation. Note: A tautomeric shift is a very temporary situation. During the second round of replication, the thymine base that shifted prior to the first round of DNA replication is likely to have shifted back to its normal form. Therefore, during the second round of replication, an adenine base will be found opposite this thymine.

EXPERIMENT 16A

X-Rays Were the First Environmental Agent Shown to Cause Induced Mutations

As shown in Table 16.5, changes in DNA structure can also be caused by environmental agents, either chemical or physical agents. These agents are called mutagens, and the mutations they cause are referred to as induced mutations. In 1927, Hermann Müller devised an approach to show that X-rays can cause induced mutations in *Drosophila melanogaster.* Müller reasoned that a mutagenic agent might cause some genes to become defective. His experimental approach focused on the ability of a mutagen to cause defects in X-linked genes that result in a recessive lethal phenotype.

To determine if X-rays increase the rate of recessive, X-linked lethal mutations, Müller sought an easy way to detect the occurrence of such mutations. He cleverly realized that he had a laboratory strain of fruit flies that could make this possible. In particular, he conducted his crosses in such a way that a female fly that inherited a new mutation causing a recessive X-linked lethal allele would not be able to produce any male offspring. This made it very easy for him to detect lethal mutations; he had to count only the number of female flies that could not produce sons. To understand Müller's crosses, we need to take a closer look at a peculiar version of one of the X chromosomes in a strain of flies that he used in his crosses. This X chromosome, designated *ClB*, had three important genetic alterations.

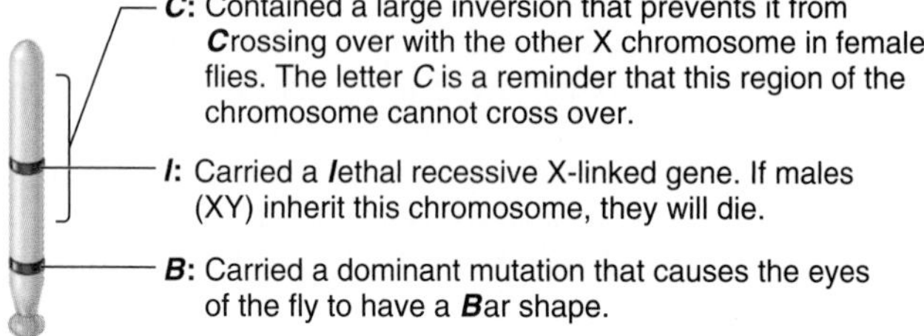
C: Contained a large inversion that prevents it from *C*rossing over with the other X chromosome in female flies. The letter *C* is a reminder that this region of the chromosome cannot cross over.
l: Carried a *l*ethal recessive X-linked gene. If males (XY) inherit this chromosome, they will die.
B: Carried a dominant mutation that causes the eyes of the fly to have a *B*ar shape.

(Note: *C* and *B* are uppercase because they are inherited in a dominant manner, and *l* is lowercase because it is a recessive allele.)

A female fly that has one copy of this X chromosome would have bar-shaped eyes, because bar is a dominant allele. Even though this X chromosome has a recessive lethal allele, a female fly can survive if the corresponding gene on the other X chromosome is a normal allele. In Müller's experiments, the goal was to determine if exposure to X-rays caused a mutation on the normal X chromosome (not the *ClB* chromosome) that created a recessive lethal allele in any essential X-linked gene except for the gene that already had a lethal allele on the *ClB* chromosome (**Figure 16.11**). If a recessive lethal mutation occurred on the normal X chromosome, this female could survive because it would be heterozygous for recessive lethal mutations in two different genes. However, because each X chromosome would have a lethal mutation, this female would not be able to produce any living sons.

The steps in Müller's protocol are shown in **Figure 16.12**. He began with wild-type males and exposed them to X-rays. These X-rays may mutate the X chromosome in sperm cells, resulting in a recessive lethal allele. These males, and a control group of males that were not exposed to X-rays, were then mated to females carrying the *ClB* chromosome. Daughters with bar eyes were saved from this cross and mated to nonirradiated males. You should look carefully at this cross and realize that if these daughters also contained a lethal allele on the X chromosome they inherited from their father (e.g., an irradiated male in step 1), they would not be able to produce living sons.

THE HYPOTHESIS

The exposure of flies to X-rays will increase the rate of mutation.

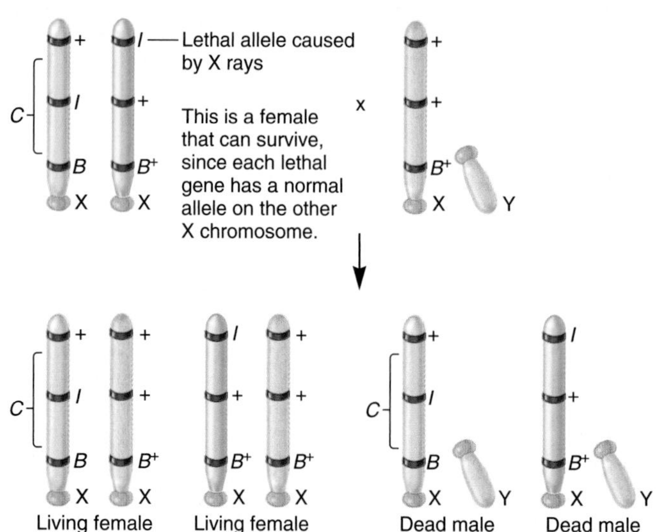
Lethal allele caused by X rays
This is a female that can survive, since each lethal gene has a normal allele on the other X chromosome.
C
+
l
B
B^+
X
Y
Living female
Living female
Dead male
Dead male
No living male offspring are produced from this cross.

FIGURE 16.11 A strategy to detect the presence of lethal X-linked mutations. X-rays may cause a recessive lethal mutation to occur in the normal X chromosome. This female also contains another lethal allele in the *ClB* chromosome. Nevertheless, this female could survive because it would be heterozygous for recessive lethal mutations in two different genes. Because each X chromosome would have a lethal mutation, this female would not be able to produce any living sons.

TESTING THE HYPOTHESIS — FIGURE 16.12 Evidence that X-rays cause mutation.

Starting material: The female flies used in this study had one normal X chromosome and a *ClB* X chromosome. The male flies had a normal X chromosome.

1. Expose male flies to X-rays. Also, have a control group that is not exposed to X-rays.

2. Mate the male flies to female flies carrying one normal X chromosome and one *ClB* X chromosome.

3. Save about 1,000 daughters with bar eyes. Note: These females contain a *ClB* X chromosome from their mothers and an X chromosome from their fathers that may (or may not) have a recessive lethal mutation.

4. Mate each bar-eyed daughter with normal (nonirradiated) males. Note: This is done in (1,000) individual tubes. (Only six are shown.)

5. Count the number of crosses that do not contain any male offspring. These crosses indicate that the bar-eyed female parent contained an X-linked lethal recessive mutation on the non-*ClB* X chromosome.

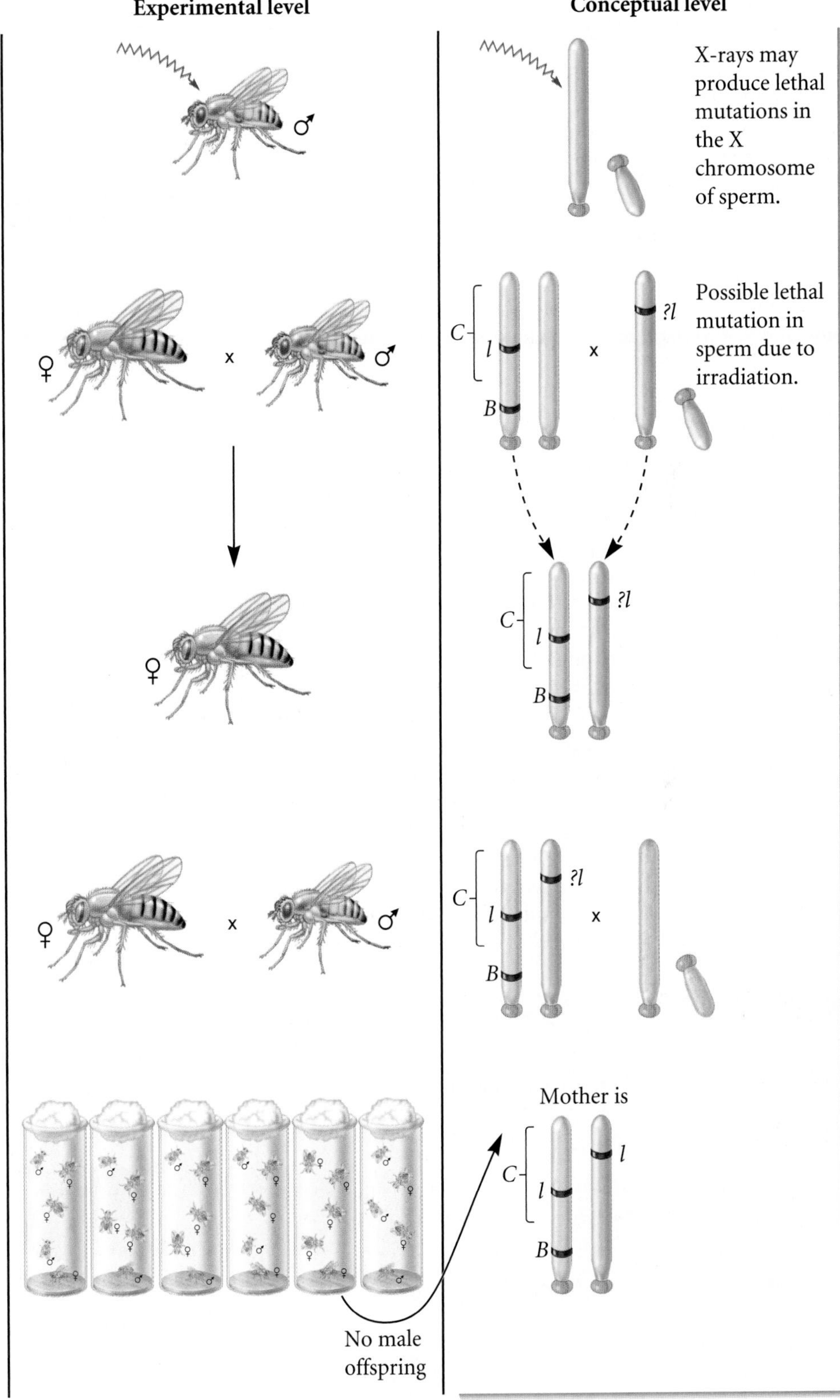

THE DATA

Treatment of Fathers of the *ClB* Daughters	Number of *ClB* Daughters Crossed to Normal Males*	Number of Tubes Containing Any Offspring†	Number of Tubes with Female Offspring but Lacking Male Offspring
Control	1,011	947	1
X-ray treated	1,015	783	91

*See step 4.
†The reason why these values are less than the numbers of mated *ClB* daughters is because some crosses did not produce living offspring.

INTERPRETING THE DATA

As shown in the data of Figure 16.12, in the absence of X-ray treatment, only 1 cross in approximately 1,000 was unable to produce male offspring. This means that the spontaneous rate for any X-linked lethal mutation was relatively low. By comparison, X-ray treatment of the fathers that gave rise to these *ClB* females resulted in 91 crosses without male offspring. Because these females inherited their non-*ClB* chromosome from irradiated fathers, these results indicate that X-rays greatly increase the rate of X-linked, recessive lethal mutations. This conclusion has been confirmed in many subsequent studies, which have shown that the increase in mutation rate is correlated with the amount of exposure to X-rays.

A self-help quiz involving this experiment can be found at the Online Learning Center.

Mutagens Alter DNA Structure in Different Ways

Since this pioneering study of Müller, researchers have found that an enormous array of agents can act as mutagens to permanently alter the structure of DNA. We often hear in the news media that we should avoid these agents in our foods and living environment. For example, we use products such as sunscreens to help us avoid the mutagenic effects of ultraviolet (UV) rays. The public is concerned about mutagens for two important reasons. First, mutagenic agents are often involved in the development of human cancers. In addition, because new mutations may be deleterious, people want to avoid mutagens to prevent gene mutations that may have harmful effects in their future offspring.

Mutagenic agents are usually classified as chemical or physical mutagens. Examples of both types of agents are listed in **Table 16.6**. In some cases, chemicals that are not mutagenic can be altered to a mutagenically active form after they have been ingested into the body. Cellular enzymes such as oxidases have been shown to activate some mutagens. Certain foods contain chemicals that act as antioxidants. Many scientists are investigating whether particular antioxidants found in food can counteract the effects of oxidases and thereby lower the cancer rate. Colorful fruits and vegetables, including prunes, grapes, blueberries, cranberries, citrus fruits, spinach, broccoli, beets, beans, red peppers, carrots, and strawberries are usually high in antioxidants.

Mutagens can alter the structure of DNA in various ways. Some mutagens act by covalently modifying the structure of bases. For example, **nitrous acid** (HNO_2) replaces amino groups with keto groups ($=NH_2$ to $=O$), a process called deamination. This can change cytosine to uracil and adenine to hypoxanthine. When this altered DNA replicates, the modified bases do not pair with the appropriate bases in the newly made strand. Instead, uracil pairs with adenine, and hypoxanthine pairs with cytosine (**Figure 16.13**).

TABLE **16.6**

Examples of Mutagens

Mutagen	Effect(s) on DNA Structure
Chemical	
Nitrous acid	Deaminates bases
Nitrogen mustard	Alkylating agent
Ethyl methanesulfonate	Alkylating agent
Proflavin	Intercalates within DNA helix
5-bromouracil	Base analogue
2-aminopurine	Base analogue
Physical	
X-rays	Cause base deletions, single-stranded nicks in DNA, cross-linking, and chromosomal breaks
UV light	Promotes pyrimidine dimer formation, such as thymine dimers

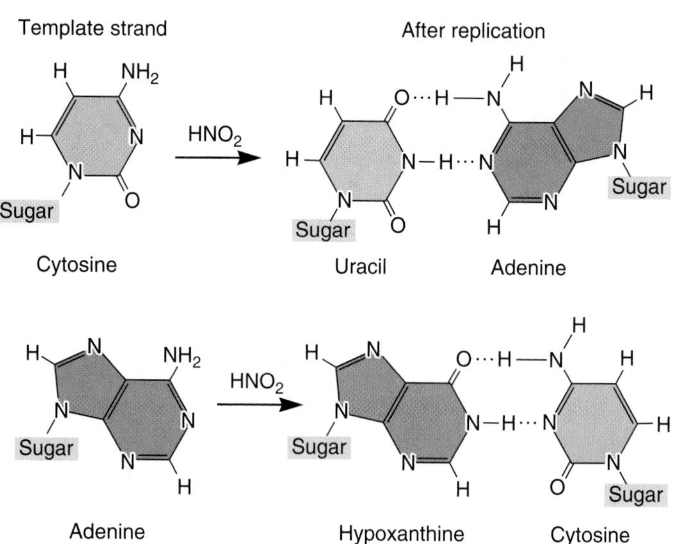

FIGURE 16.13 **Mispairing of modified bases that have been deaminated by nitrous acid.** Nitrous acid converts cytosine to uracil, and adenine to hypoxanthine. During DNA replication, uracil pairs with adenine, and hypoxanthine pairs with cytosine. This will create mutations in the newly replicated strand during DNA replication.

Other chemical mutagens can also disrupt the appropriate pairing between nucleotides by alkylating bases within the DNA. During alkylation, methyl or ethyl groups are covalently attached to the bases. Examples of alkylating agents include **nitrogen mustard** (a type of mustard gas) and **ethyl methanesulfonate (EMS).** Mustard gas was used as a chemical weapon during World War I. Such agents damage the skin, eyes, mucous membranes, lungs, and blood-forming organs.

Some mutagens exert their effects by directly interfering with the DNA replication process. For example, **acridine dyes** such as **proflavin** contain flat structures that intercalate, or insert themselves, between adjacent base pairs, thereby distorting the helical structure. When DNA containing these mutagens is replicated, single-nucleotide additions and/or deletions can be incorporated into the newly made daughter strands, creating a frameshift mutation.

Compounds such as **5-bromouracil (5BU)** and **2-aminopurine** are base analogues that become incorporated into daughter strands during DNA replication. 5BU is a thymine analogue that can be incorporated into DNA instead of thymine. Like thymine, 5BU base-pairs with adenine. However, at a relatively high rate, 5BU will undergo a tautomeric shift and base-pair with guanine (**Figure 16.14a**). When this occurs during DNA replication, 5BU causes a mutation in which a TA base pair is changed to a 5BU-G base pair (**Figure 16.14b**). This is a transition, because the adenine has been changed to a guanine, both of which are purines. During the next round of DNA replication, the template strand containing the guanine base will create a GC base pair. In this way, 5-bromouracil can promote a change of an AT base pair into a GC base pair.

Compounds like 5BU are sometimes used in chemotherapy for cancer. The rationale is that these compounds will be incorporated only into the DNA of actively replicating cells such as cancer cells. When incorporated, these compounds tend to cause many mutations in the cells, hopefully leading to the death of cancer cells. Unfortunately, other actively dividing cells, such as those in the skin and the lining of the digestive tract, will also incorporate 5BU. This leads to unwanted side effects of chemotherapy such as hair loss and a diminished appetite.

DNA molecules are also sensitive to physical agents such as radiation. In particular, radiation of short wavelength and high energy, known as ionizing radiation, can alter DNA structure. This type of radiation includes X-rays and gamma rays. Ionizing radiation can penetrate deeply into biological materials, where it creates chemically reactive molecules known as free radicals. These molecules can alter the structure of DNA in a variety of ways. Exposure to high doses of ionizing radiation can cause base deletions, single nicks in DNA strands, cross-linking, and even chromosomal breaks.

Nonionizing radiation, such as UV light, contains less energy, and so it penetrates only the surface of an organism, such as the skin. Nevertheless, UV light is known to cause DNA mutations. As shown in **Figure 16.15**, UV light causes the formation of cross-linked **thymine dimers.** Thymine dimers interfere with transcription and DNA replication. This can result in a mutation when that DNA strand is replicated. Plants, in particular, must have effective ways to prevent UV damage because they are exposed to sunlight throughout the day. Also, sun tanning greatly increases a person's exposure to UV light, raising the potential for thymine dimers and mutation. This explains the higher incidence of skin cancer among people who have been exposed to large amounts of sun during their lifetime. Because of the link between skin cancer and sun exposure, people now apply sunscreen to their skin to prevent the harmful effects of UV light. Most sunscreens contain organic compounds that absorb UV light, such as oxybenzone, and/or opaque ingredients that reflect UV light, such as zinc oxide.

Testing Methods Can Determine If an Agent Is a Mutagen

To determine if an agent is mutagenic, researchers use testing methods that can monitor whether or not an agent increases the rate of mutation. Many different kinds of tests have been used to evaluate mutagenicity. One commonly used test is the **Ames**

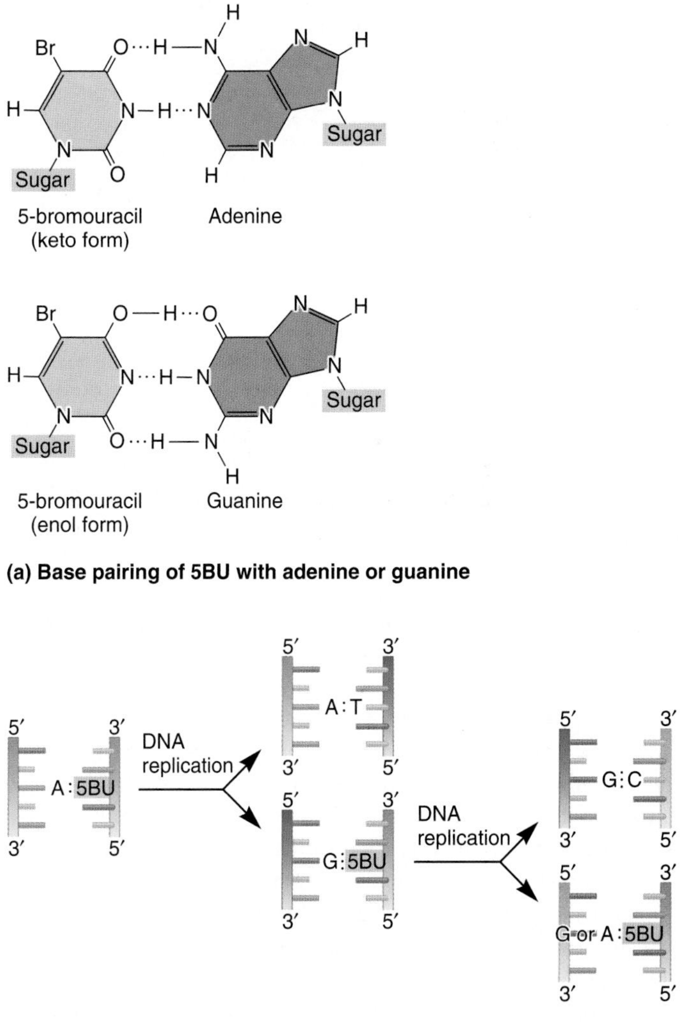

(a) Base pairing of 5BU with adenine or guanine

(b) How 5BU causes a mutation in a base pair during DNA replication

FIGURE 16.14 **Structure and effects of the mispairing of 5-bromouracil with guanine.** (**a**) In its keto form, 5BU bonds with adenine; in its enol form, it bonds with guanine. (**b**) During DNA replication, guanine may be incorporated into a newly made strand by pairing with 5BU. After a second round of replication, the DNA will contain a GC base pair instead of the original AT base pair.

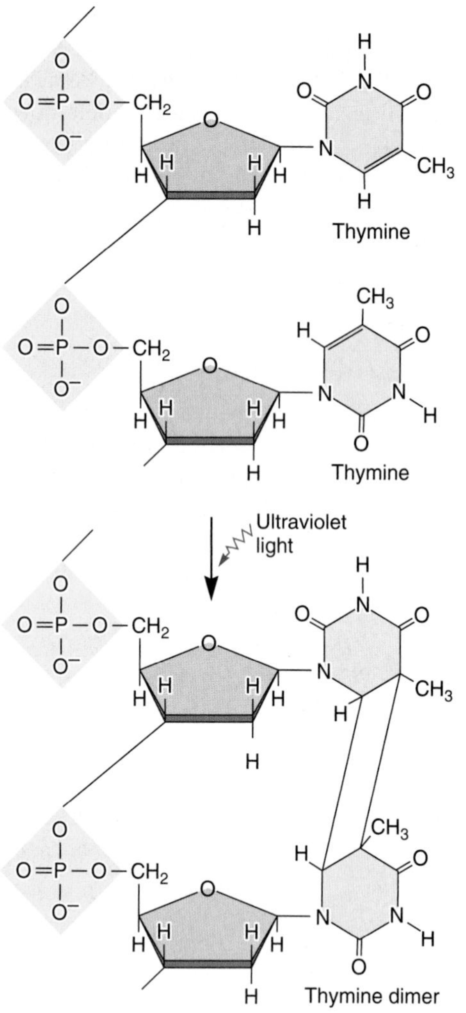

FIGURE 16.15 **Formation and structure of a thymine dimer.**

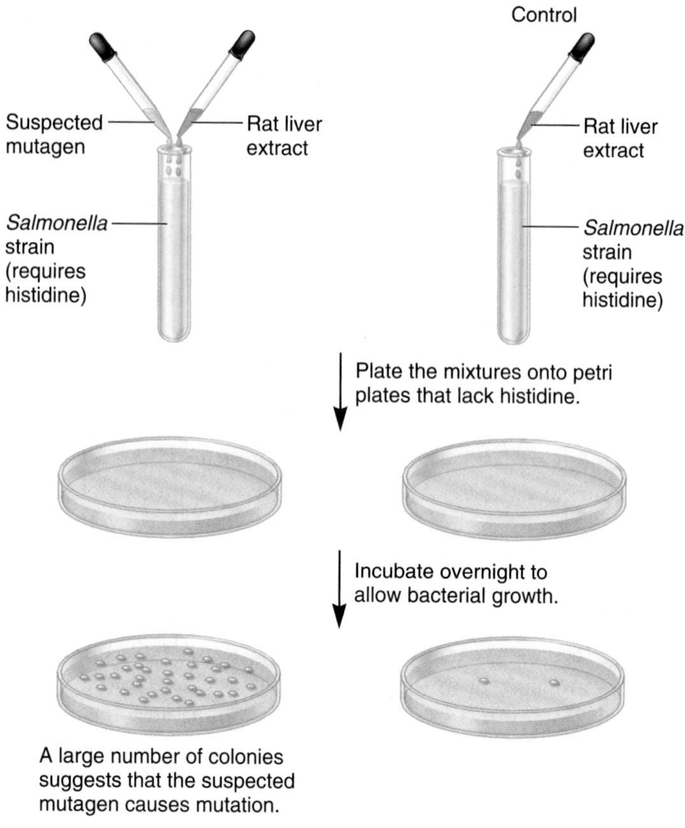

FIGURE 16.16 **The Ames test for mutagenicity.**

test, which was developed by Bruce Ames. This test uses strains of a bacterium, *Salmonella typhimurium*, that cannot synthesize the amino acid histidine. These strains contain a point mutation within a gene that encodes an enzyme required for histidine biosynthesis. The mutation renders the enzyme inactive. Therefore, the bacteria cannot grow on petri plates unless histidine has been added to the growth medium. However, a second mutation—a reversion—may occur that restores the ability to synthesize histidine. In other words, a second mutation can cause a reversion back to the wild-type condition. The Ames test monitors the rate at which this second mutation occurs and thereby indicates whether an agent increases the mutation rate above the spontaneous rate.

Figure 16.16 outlines the steps in the Ames test. The suspected mutagen is mixed with a rat liver extract and a strain of *Salmonella* that cannot synthesize histidine. A mutagen may require activation by cellular enzymes, which are provided by the rat liver extract. This step improves the ability of the test to identify agents that may cause mutation in mammals. After the incubation period, a large number of bacteria are then plated on a growth medium that does not contain histidine. The *Salmonella* strain is not expected to grow on these plates. However, if a mutation has occurred that allows the strain to synthesize histidine, it can grow on these plates to form a visible bacterial colony. To estimate the mutation rate, the colonies that grow on the media are counted and compared to the total number of bacterial cells that were originally streaked on the plate. For example, if 10,000,000 bacteria were plated and 10 growing colonies were observed, the rate of mutation is 10 out of 10,000,000; this equals 1 in 10^6, or simply 10^{-6}. As a control, bacteria that have not been exposed to the mutagen are also tested, because a low level of spontaneous mutations is expected to occur.

How do we judge if an agent is a mutagen? Researchers compare the mutation rate in the presence and absence of the suspected mutagen. The experimental approach shown in Figure 16.16 is conducted several times. If statistics reveal that the mutation rate in the experimental and control samples are significantly different, researchers may tentatively conclude that the agent may be a mutagen. Interestingly, many studies have been conducted in which researchers used the Ames test to compare the urine from cigarette smokers to that from nonsmokers. This research has shown that the urine from smokers contains much higher levels of mutagens. Furthermore, heavy smokers have more mutagens in their urine compared to light smokers.

As discussed in solved problem S3 at the end of this chapter, several *his*$^-$ strains of Salmonella are available that carry different kinds of mutations within the coding sequence. This makes it possible to determine if a mutagen causes transitions, transversions, or frameshift mutations.

16.3 DNA REPAIR

Because most mutations are deleterious, DNA repair systems are vital to the survival of all organisms. If DNA repair systems did not exist, spontaneous and environmentally induced mutations would be so prevalent that few species, if any, would survive. The necessity of DNA repair systems becomes evident when they are missing. Bacteria contain several different DNA repair systems. Yet, when even a single system is absent, the bacteria have a much higher rate of mutation. In fact, the rate of mutation is so high that these bacterial strains are sometimes called mutator strains. Likewise, in humans, an individual who is defective in only a single DNA repair system may manifest various disease symptoms, including a higher risk of skin cancer. This increased risk is due to the inability to repair UV-induced mutations.

Living cells contain several DNA repair systems that can fix different types of DNA alterations (**Table 16.7**). Each repair system is composed of one or more proteins that play specific roles in the repair mechanism. In most cases, DNA repair is a multistep process. First, one or more proteins in the DNA repair system detect an irregularity in DNA structure. Next, the abnormality is removed by the action of DNA repair enzymes. Finally, normal DNA is synthesized via DNA replication enzymes. In this section, we will examine several different repair systems that have been characterized in bacteria, yeast, mammals, and plants. Their diverse ways of repairing DNA underscore the extreme necessity for the structure of DNA to be maintained properly.

TABLE 16.7

Common Types of DNA Repair Systems

System	Description
Direct repair	An enzyme recognizes an incorrect alteration in DNA structure and directly converts it back to a correct structure.
Base excision repair and nucleotide excision repair	An abnormal base or nucleotide is first recognized and removed from the DNA. A segment of DNA in this region is excised, and then the complementary DNA strand is used as a template to synthesize a normal DNA strand.
Mismatch repair	Similar to excision repair except that the DNA defect is a base pair mismatch in the DNA, not an abnormal nucleotide. The mismatch is recognized, and a segment of DNA in this region is removed. The parental strand is used to synthesize a normal daughter strand of DNA.
Homologous recombination repair	Occurs at double-strand breaks or when DNA damage causes a gap in synthesis during DNA replication. The strands of a normal sister chromatid are used to repair a damaged sister chromatid.
Nonhomologous end joining	Occurs at double-strand breaks. The broken ends are recognized by proteins that keep the ends together; the broken ends are eventually rejoined.

Damaged Bases Can Be Directly Repaired

In a few cases, the covalent modification of nucleotides by mutagens can be reversed by specific cellular enzymes. As discussed earlier in this chapter, UV light causes the formation of thymine dimers. Yeast, such as *Saccharomyces cerevisiae*, and most plants produce an enzyme called **photolyase** that can repair thymine dimers by splitting the dimers, which returns the DNA to its original condition (**Figure 16.17a**). This process directly restores the structure of DNA. Because plants are exposed to sunlight throughout the day, photolyase is a critical DNA repair enzyme for many plant species.

A protein known as **alkyltransferase** can remove methyl or ethyl groups from guanine bases that have been mutagenized by alkylating agents such as nitrogen mustard and ethyl methanesulfonate. This protein is called alkyltransferase because it transfers the methyl or ethyl group from the base to a cysteine side chain within the alkyltransferase protein (**Figure 16.17b**). Surprisingly, this permanently inactivates alkyltransferase, which means it can be used only once!

Base Excision Repair Removes a Damaged Base

A second type of repair system, called **base excision repair (BER),** involves the function of a category of enzymes known as **DNA *N*-glycosylases.** This type of enzyme can recognize an abnormal base and cleave the bond between it and the sugar in the DNA backbone, creating an apurinic or apyrimidinic site (**Figure 16.18**). Depending on the species, this repair system can eliminate abnormal bases such as uracil, 3-methyladenine, 7-methylguanine, and pyrimidine dimers.

Figure 16.18 illustrates the general steps involved in DNA repair via *N*-glycosylase. In this example, the DNA contains a uracil in its sequence. This could have happened spontaneously or by the action of a chemical mutagen. *N*-glycosylase recognizes a uracil within the DNA and cleaves (nicks) the bond between the sugar and base. This releases the uracil base and leaves behind an apyrimidinic nucleotide. This abnormal nucleotide is recognized by a second enzyme, **AP endonuclease,** which makes a cut on the 5' side. DNA polymerase, which has a 5' to 3' exonuclease activity, removes the abnormal region and, at the same time, replaces it with normal nucleotides. This process is called **nick translation** (although DNA replication, not mRNA translation, actually occurs). Finally, DNA ligase closes the nick.

Nucleotide Excision Repair Systems Remove Segments of Damaged DNA

An important general process for DNA repair is the **nucleotide excision repair** (**NER**) system. This type of system can repair many different types of DNA damage, including thymine dimers, chemically modified bases, missing bases, and certain types of cross-links. In NER, several nucleotides in the damaged strand are removed from the DNA, and the intact strand is used as a

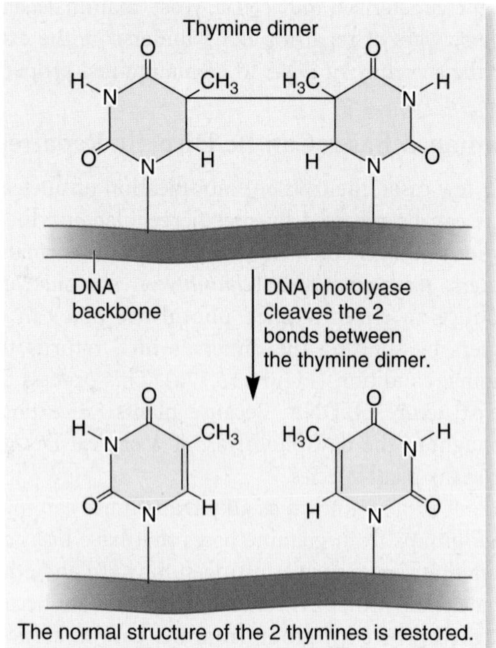

(a) Direct repair of a thymine dimer

(b) Direct repair of a methylated base

FIGURE 16.17 **Direct repair of damaged bases in DNA.** (**a**) The repair of thymine dimers by photolyase. (**b**) The repair of methylguanine by the transfer of the methyl group to alkyltransferase.

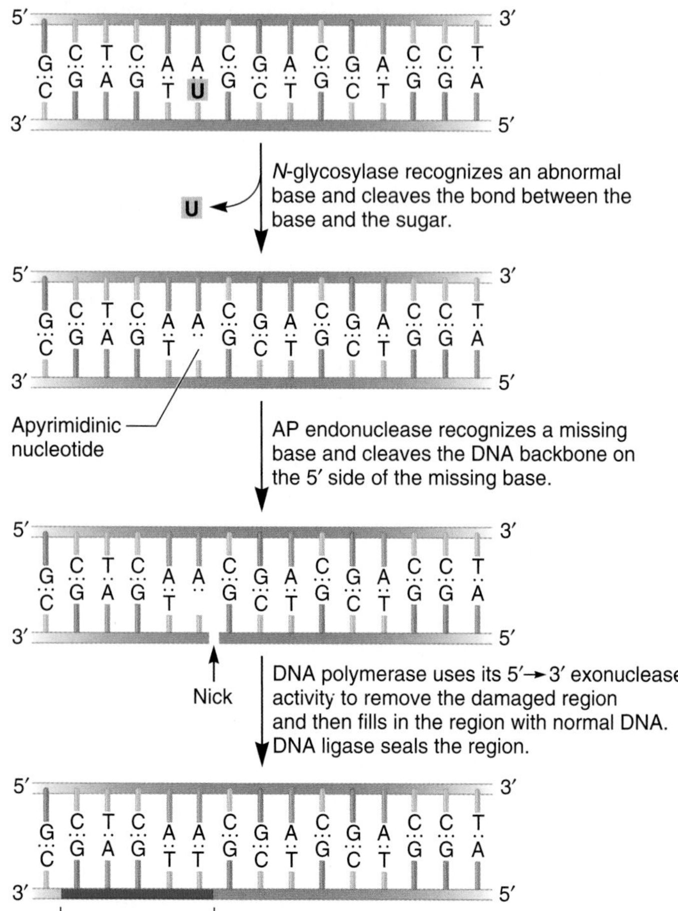

template for resynthesis of a normal complementary strand. NER is found in all eukaryotes and prokaryotes, although its molecular mechanism is better understood in prokaryotic species.

In *E. coli*, the NER system requires four key proteins, designated UvrA, UvrB, UvrC, and UvrD, plus the help of DNA polymerase and DNA ligase. UvrA, B, C, and D recognize and remove a short segment of a damaged DNA strand. They are named Uvr because they are involved in Ultraviolet light repair of pyrimidine dimers, although the UvrA–D proteins are also important in repairing chemically damaged DNA.

Figure 16.19 outlines the steps involved in the *E. coli* NER system. A protein complex consisting of two UvrA molecules and one UvrB molecule tracks along the DNA in search of damaged DNA. Such DNA will have a distorted double helix, which is sensed by the UvrA/UvrB complex. When a damaged segment is identified, the two UvrA proteins are released, and UvrC binds to the site. The UvrC protein makes cuts in the damaged strand on both sides of the damaged site. Typically, the damaged strand is cut four to five nucleotides away from the 3' side of the damage and eight nucleotides from the 5' end. After this process, UvrD, which is a helicase, recognizes the region and separates the two

FIGURE 16.18 **Base excision repair.** An abnormal base is initially recognized by an enzyme known as *N*-glycosylase, which cleaves the bond between the base and the sugar. Depending on whether a purine or pyrimidine is removed, this creates an apurinic or apyrimidinic site, respectively. In either case, the region is recognized by AP endonuclease, which makes a cut at the 5' side of the abnormal nucleotide. DNA polymerase then removes the abnormal region and fills it in with normal DNA. Finally, DNA ligase seals the repaired strand.

strands of DNA. This releases a short DNA segment that contains the damaged region, and UvrB and UvrC are also released. Following the excision of the damaged DNA, DNA polymerase fills in the gap, using the undamaged strand as a template. Finally, DNA ligase makes the final covalent connection between the newly made DNA and the original DNA strand.

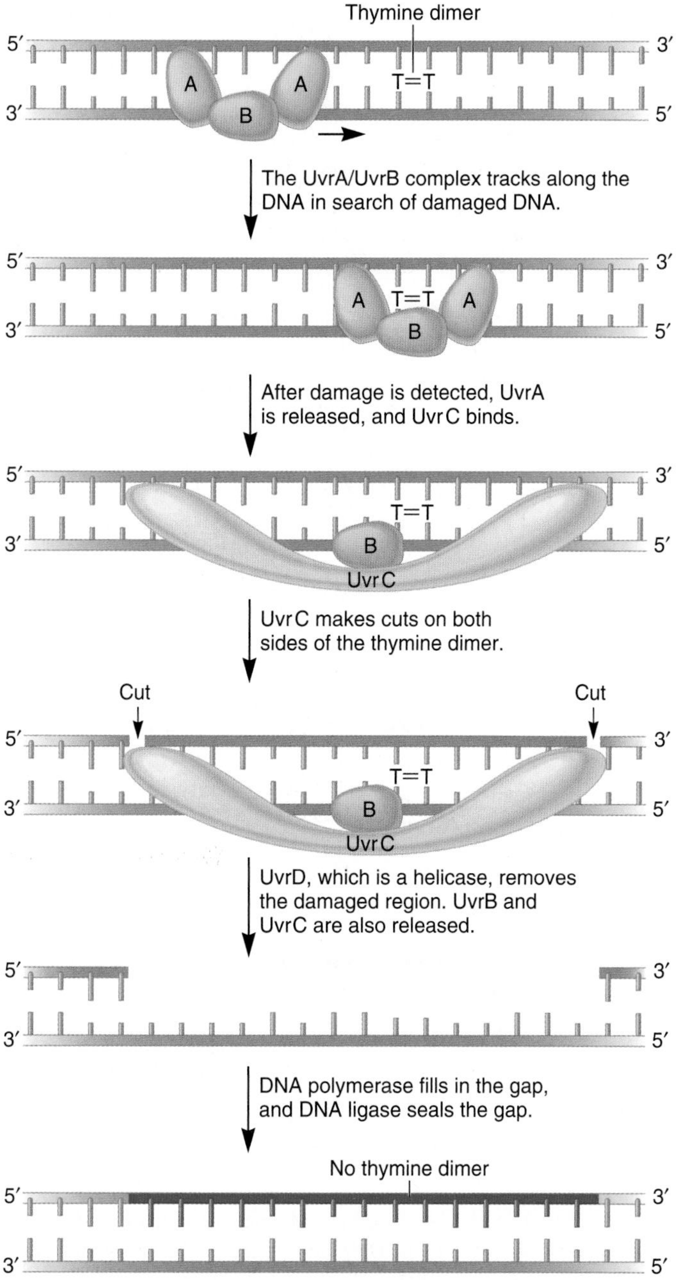

Thymine dimer

5′ 3′

A A T=T

3′ B 5′

The UvrA/UvrB complex tracks along the DNA in search of damaged DNA.

5′ 3′

A T=T A

3′ B 5′

After damage is detected, UvrA is released, and UvrC binds.

5′ 3′

T=T

3′ B 5′

UvrC

UvrC makes cuts on both sides of the thymine dimer.

Cut Cut

5′ 3′

T=T

3′ B 5′

UvrC

UvrD, which is a helicase, removes the damaged region. UvrB and UvrC are also released.

5′ 3′

3′ 5′

DNA polymerase fills in the gap, and DNA ligase seals the gap.

No thymine dimer

5′ 3′

3′ 5′

ONLINE ANIMATION

FIGURE 16.19 Nucleotide excision repair in *E. coli.* UvrA/UvrB complex scans along the DNA. When the complex encounters a thymine dimer in the DNA, the UvrA dimer is released. UvrC binds to UvrB. UvrC make two cuts in the damaged DNA strand. One cut is made eight nucleotides 5' to the site, the other approximately five nucleotides 3' from the site. UvrD (not shown in this figure) binds to the site. UvrD is a helicase that causes the damaged strand to be removed. During this process, UvrB and UvrC are released. DNA polymerase resynthesizes a complementary strand. DNA ligase makes the final connection.

Several human diseases are due to inherited defects in genes involved in nucleotide excision repair. These include xeroderma pigmentosum (XP), Cockayne syndrome (CS), and PIBIDS. (PIBIDS is an acronym for a syndrome with symptoms that include photosensitivity; ichthyosis, a skin abnormality; brittle hair; impaired intelligence; decreased fertility; and short stature.) A common characteristic in all three syndromes is an increased sensitivity to sunlight because of an inability to repair UV-induced lesions. **Figure 16.20** shows a photograph of an individual with xeroderma pigmentosum. Such individuals have pigmentation abnormalities, many premalignant lesions, and a high predisposition to skin cancer. They may also develop early degeneration of the nervous system.

Genetic analyses of patients with XP, CS, and PIBIDS have revealed that these syndromes result from defects in a variety of different genes that encode NER proteins. For example, xeroderma pigmentosum can be caused by defects in any of seven different NER genes. In all cases, individuals have a defective

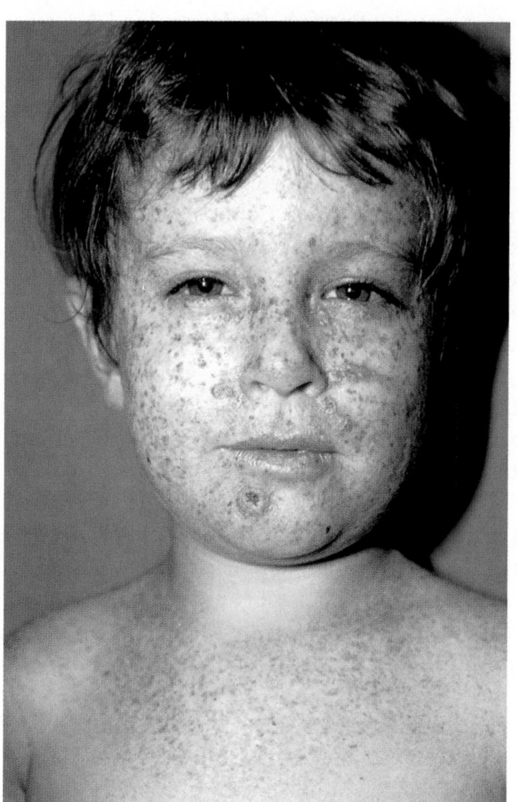

FIGURE 16.20 An individual affected with xeroderma pigmentosum.

Genes→Traits This disease involves a defect in genes that are involved with nucleotide excision repair. Xeroderma pigmentosum can be caused by defects in seven different NER genes. Affected individuals have an increased sensitivity to sunlight because of an inability to repair UV-induced DNA lesions. In addition, they may also have pigmentation abnormalities, many premalignant lesions, and a high predisposition to skin cancer.

NER pathway. In recent years, several human NER genes have been successfully cloned and sequenced. Although more research is needed to completely understand the mechanisms of DNA repair, the identification of NER genes has helped unravel the complexities of NER pathways in human cells.

Mismatch Repair Systems Recognize and Correct a Base Pair Mismatch

Thus far, we have considered several DNA repair systems that recognize abnormal nucleotide structures within DNA, including thymine dimers, alkylated bases, and the presence of uracil in the DNA. Another type of abnormality that should not occur in DNA is a **base mismatch.** The structure of the DNA double helix obeys the AT/GC rule of base pairing. During the normal course of DNA replication, however, an incorrect nucleotide may be added to the growing strand by mistake. This creates a mismatch between a nucleotide in the parental and newly made strand. Various DNA repair mechanisms can recognize and remove this mismatch. For example, as described in Chapter 11, DNA polymerase has a 3' to 5' proofreading ability that can detect mismatches and remove them. However, if this proofreading ability fails, cells contain additional DNA repair systems that can detect base mismatches and fix them. An interesting DNA repair system that exists in all species is the **mismatch repair system.**

In the case of a base mismatch, how does a DNA repair system determine which base to remove? If the mismatch is due to an error in DNA replication, the newly made daughter strand contains the incorrect base, while the parental strand is normal. Therefore, an important aspect of mismatch repair is that it specifically repairs the newly made strand rather than the parental template strand. Prior to DNA replication, the parental DNA has already been methylated. Immediately after DNA replication, some time must pass before a newly made strand is methylated. Therefore, newly replicated DNA is hemimethylated—only the parental DNA strand is methylated. Hemimethylation provides a way for a DNA repair system to distinguish between the parental DNA strand and the daughter strand.

The molecular mechanism of mismatch repair has been studied extensively in *E. coli.* As shown in **Figure 16.21**, three proteins, designated MutS, MutL, and MutH, detect the mismatch and direct the removal of the mismatched base from the newly made strand. These proteins are named Mut because their absence leads to a much higher mutation rate compared to normal strains of *E. coli.* The role of MutS is to locate mismatches. Once a mismatch is detected, MutS forms a complex with MutL. MutL acts as a linker that binds to MutH by a looping mechanism. This stimulates MutH, which is bound to a hemimethylated site, to make a cut in the newly made, nonmethylated DNA strand. After the strand is cut, MutU, which functions as helicase, separates the strands, and an exonuclease then digests the nonmethylated DNA strand in the direction of the mismatch and proceeds beyond the mismatch site. This leaves a gap in the daughter strand that is repaired by DNA polymerase and DNA ligase. The net result is that the mismatch has been corrected by removing the incorrect region in the daughter strand and then

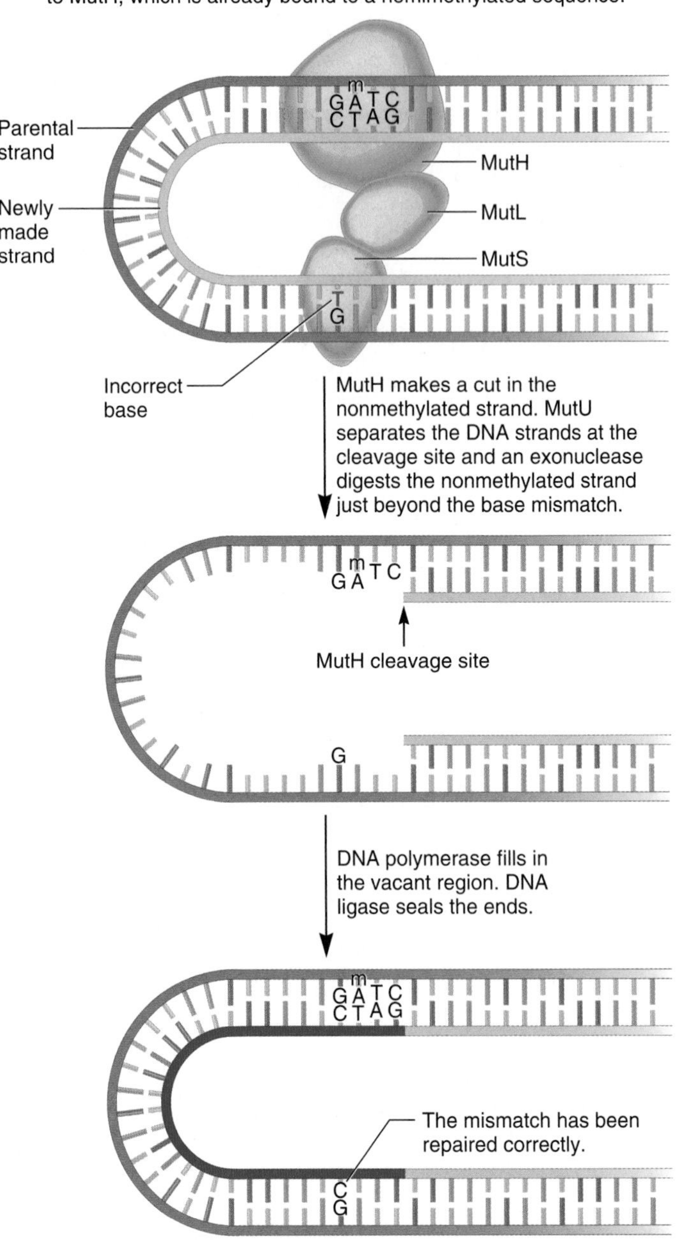
The MutS protein finds a mismatch. The MutS/MutL complex binds to MutH, which is already bound to a hemimethylated sequence.
Parental strand
Newly made strand
m
GATC
CTAG
MutH
MutL
MutS
T
G
Incorrect base
MutH makes a cut in the nonmethylated strand. MutU separates the DNA strands at the cleavage site and an exonuclease digests the nonmethylated strand just beyond the base mismatch.
m
GATC
MutH cleavage site
G
DNA polymerase fills in the vacant region. DNA ligase seals the ends.
m
GATC
CTAG
The mismatch has been repaired correctly.
C
G

ONLINE ANIMATION

FIGURE 16.21 Mismatch repair in *E. coli.* MutS slides along the DNA and recognizes base mismatches in the double helix. MutL binds to MutS and acts as a linker between MutS and MutH. The DNA must loop for this interaction to occur. The role of MutH is to identify the methylated strand of DNA, which is the nonmutated parental strand. MutH can be bound to a methylated site on either side of the base mismatch. In this example, the methylated adenine is designated with an m. MutH makes a cut in the nonmethylated strand. MutU (not shown in the figure) begins to separate the strands at the cleavage site, and then an exonuclease digests the nonmethylated strand until it passes the MutS/MutL region. This leaves a gap, which is filled in by DNA polymerase and sealed by DNA ligase.

resynthesizing the correct sequence using the parental DNA as a template.

Eukaryotic species also have homologs to MutS and MutL, along with many other proteins that are needed for mismatch repair. However, a MutH homolog has not been identified in eukaryotes, and the mechanism by which the eukaryotic mismatch repair system distinguishes between the parental and daughter strands is not well understood. As with defects in nucleotide excision repair systems, mutations in the human mismatch repair system are associated with particular types of cancer. For example, mutations in two human mismatch repair genes, *hMSH2* and *hMLH1*, play a role in the development of a type of colon cancer known as hereditary nonpolyposis colorectal cancer.

Double-Strand Breaks Can Be Repaired by Homologous Recombination Repair and by Nonhomologous End Joining

Of the many types of DNA damage that can occur within living cells, the breakage of chromosomes—called a DNA double-strand break (DSB)—is perhaps the most dangerous. DSBs can be caused by ionizing radiation (X-rays or gamma rays), chemical mutagens, and certain drugs used for chemotherapy. In addition, free radicals that are the byproducts of cellular metabolism can cause double-strand breaks. Surprisingly, researchers estimate that naturally occurring double-strand breaks in a typical human cell occur at a rate of between 10 and 100 breaks per cell per day! Such breaks can be harmful in a variety of ways. First, DSBs can result in chromosomal rearrangements such as inversions and translocations (see Figure 8.2). In addition, DSBs can lead to terminal or interstitial deficiencies (see Figure 8.3). Such genetic changes have the potential to result in detrimental phenotypic effects.

How are DSBs repaired? The two main mechanisms are **homologous recombination repair (HRR)** and **nonhomologous end joining (NHEJ).** Homologous recombination repair, also called homology-directed repair, occurs when homologous DNA strands, usually from a sister chromatid, are used to repair a DSB in the other sister chromatid (**Figure 16.22**). First, the DSB is processed by the short digestion of DNA strands at the break site. This processing event is followed by the exchange of DNA strands between the broken and unbroken sister chromatids. The unbroken strands are then used as templates to synthesize DNA in the region where the break occurred. Finally, the criss-crossed strands are resolved, which means that they are broken and then rejoined in a way that produces separate chromatids. Because sister chromatids are genetically identical, an advantage is that homologous recombination repair can be an error-free mechanism to repair a DSB. A disadvantage is that sister chromatids are available only during the S and G_2 phases of the cell cycle in eukaryotes or following DNA replication in bacteria. Although sister chromatids are strongly preferred, HRR may also occur between homologous regions that are not identical. Therefore, HRR may occasionally happen when sister chromatids are unavailable. The proteins involved in homologous recombination repair are described in Chapter 17.

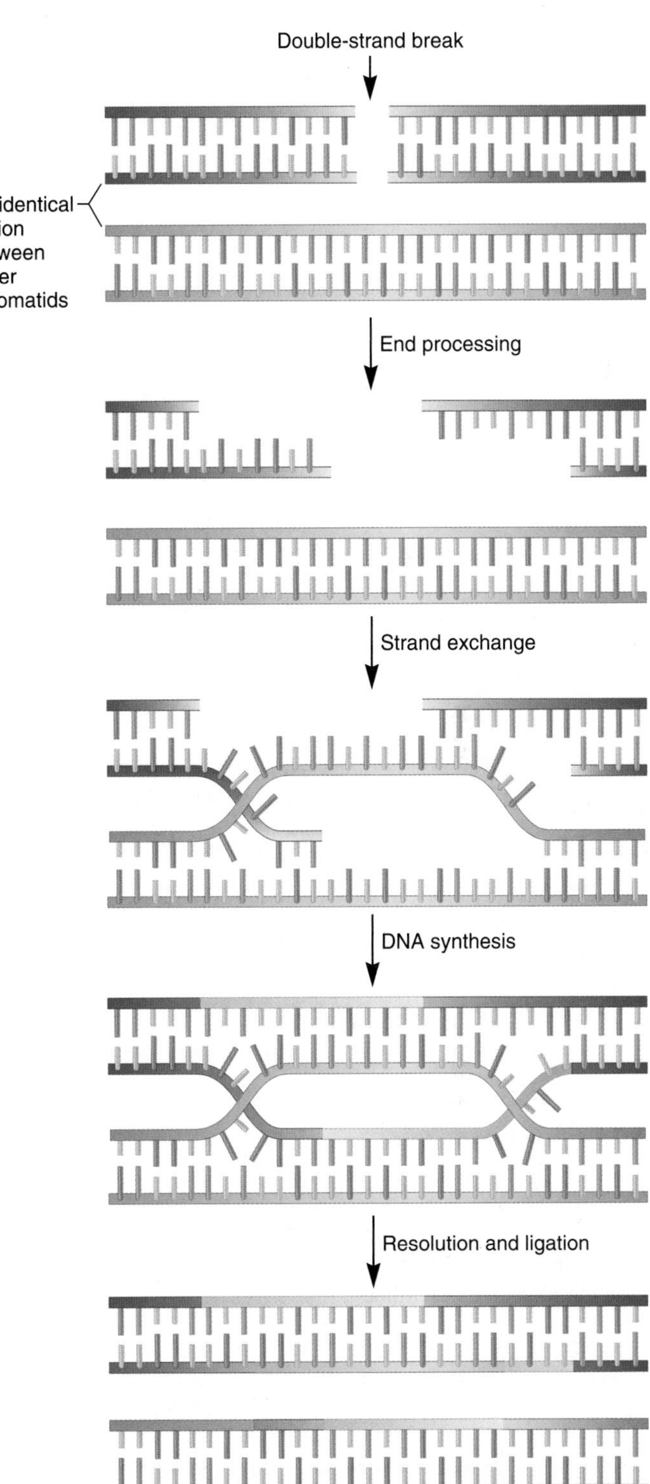

FIGURE 16.22 **DNA repair of double-strand breaks (DSBs) via homologous recombination repair.**

During nonhomologous end joining, the two broken ends of DNA are simply pieced back together (**Figure 16.23**). This mechanism requires the participation of several proteins that play key roles in the process. First, the DSB is recognized by end-binding proteins. These proteins then recognize additional

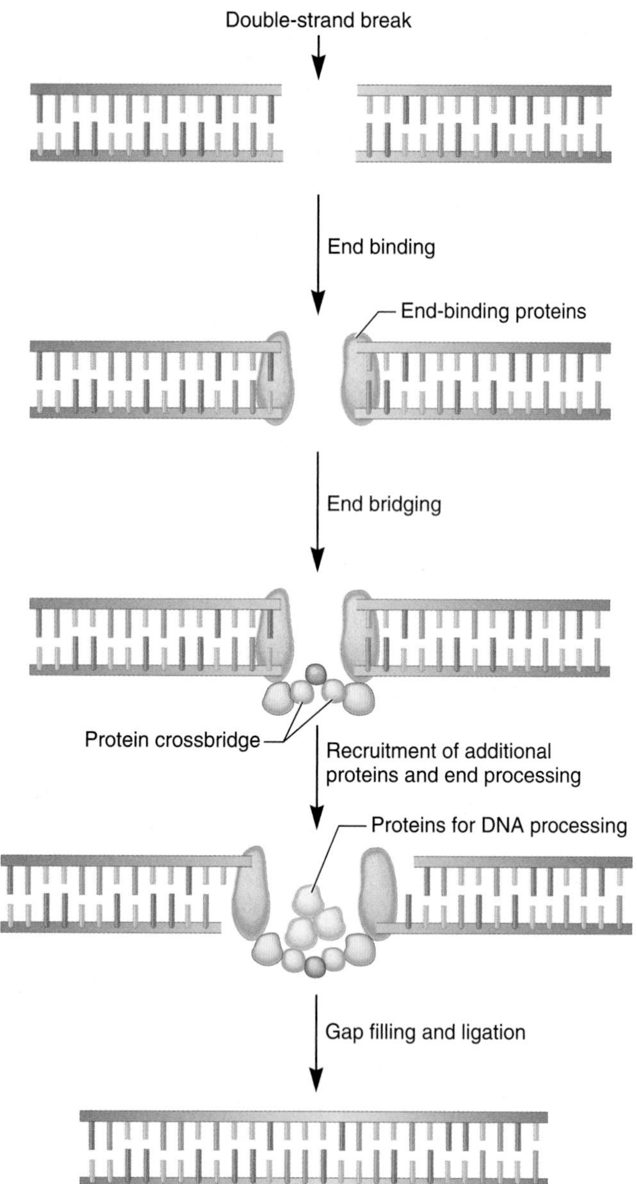

FIGURE 16.23 **DNA repair of double-strand breaks via nonhomologous end joining.**

proteins that form a crossbridge that prevents the two ends from drifting apart. Next, additional proteins are recruited to the region that may process the ends of the broken chromosome by digesting particular DNA strands. This processing may result in the deletion of a small amount of genetic material from the region. Finally, any gaps are filled in via DNA polymerase, and the DNA ends are ligated together. One advantage of NHEJ is that it doesn't involve the participation of a sister chromatid, so it can occur at any stage of the cell cycle. However, a disadvantage is that NHEJ can result in small deletions in the region that has been repaired.

Actively Transcribed DNA Is Repaired More Efficiently Than Nontranscribed DNA

Not all DNA is repaired at the same rate. In the 1980s, Philip Hanawalt and colleagues conducted experiments showing that actively transcribed genes in eukaryotes and prokaryotes are more efficiently repaired following radiation damage than is nontranscribed DNA. The targeting of DNA repair enzymes to actively transcribing genes may have several biological advantages. First, active genes are loosely packed and may be more vulnerable to DNA damage. Likewise, the process of transcription itself may make the DNA more susceptible to agents that cause damage. In addition, DNA regions that contain actively transcribed genes are more likely to be important for survival than nontranscribed regions. This may be particularly true in differentiated cells, such as nerve cells, which no longer divide but actively transcribe genes for many years. In nondividing cells, gene transcription rather than DNA replication is of utmost importance.

How are actively transcribed regions targeted by DNA repair systems? In *E. coli*, a protein known as **transcription-repair coupling factor** (**TRCF**) is responsible for targeting the nucleotide excision repair system to actively transcribed genes having damaged DNA. **Figure 16.24** illustrates a case where RNA polymerase had been transcribing a gene until it encountered a thymine dimer and became stalled. TRCF is a helicase that displaces the stalled RNA polymerase from the damaged template strand. TRCF also has a binding site for UvrA. Therefore, after RNA polymerase has been removed, TRCF recruits the nucleotide excision repair system to the damaged site by recognizing the UvrA/UvrB complex. After the UvrA/UvrB complex has bound to this site, TRCF is released from the damaged site. At this stage, the UvrA/UvrB complex can begin nucleotide excision repair, as described earlier in Figure 16.19.

In eukaryotes, the mechanism that couples DNA repair and transcription is not completely understood. Several different proteins have been shown to act as transcription-repair coupling factors. Some of these have been identified in people with high rates of mutation. As mentioned previously, Cockayne syndrome (CS) involves defects in genes that play a role in nucleotide excision repair. Two genes that are defective in CS, termed *CS-A* and *CS-B*, encode proteins that function as transcription-repair coupling factors. In addition, certain general transcription factors, such as TFIIH, play a role in both transcription and nucleotide excision repair.

Damaged DNA May Be Replicated by Translesion DNA Polymerases

Despite the efficient action of numerous repair systems that remove lesions in DNA in an error-free manner, it is inevitable that some lesions may escape these repair mechanisms. Such lesions may be present when DNA is being replicated. If so, replicative DNA polymerases, such as polIII in *E. coli*, are highly sensitive to geometric distortions in DNA and are unable to replicate through DNA lesions. During the past decade, researchers have discovered that cells are equipped with specialized DNA polymerases that

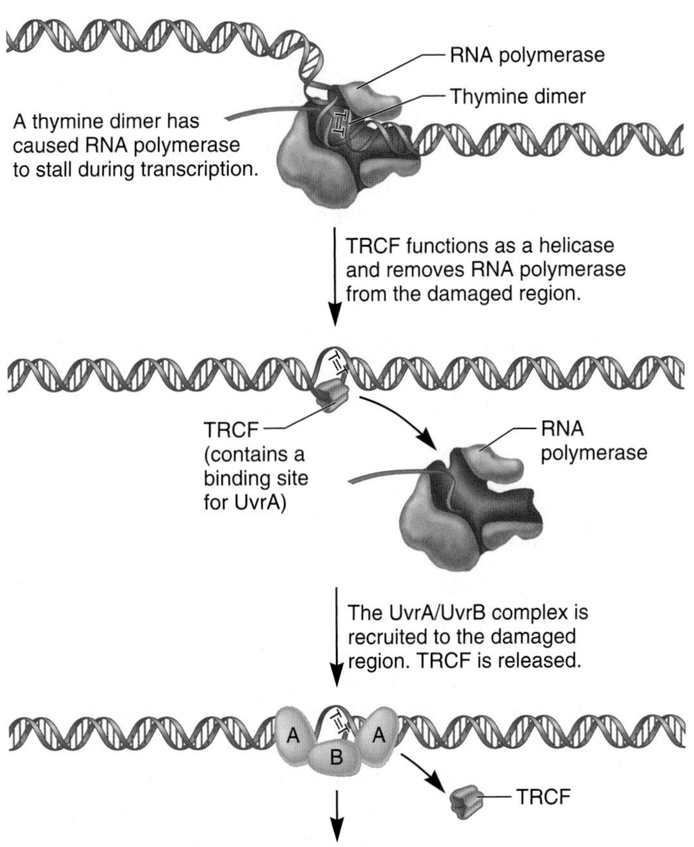
RNA polymerase
Thymine dimer
A thymine dimer has caused RNA polymerase to stall during transcription.
TRCF functions as a helicase and removes RNA polymerase from the damaged region.
TRCF (contains a binding site for UvrA)
RNA polymerase
The UvrA/UvrB complex is recruited to the damaged region. TRCF is released.
A
B
A
TRCF
The region is repaired as described in Figure 16.19.

FIGURE 16.24 **Targeting of the nucleotide excision repair system to an actively transcribing gene containing a DNA alteration.** When RNA polymerase encounters damaged DNA, it will stall over the damaged site. TRCF, which is a helicase, recognizes a stalled RNA polymerase and removes it from the damaged site. TRCF has a binding site for UvrA and thereby recruits the UvrA/UvrB complex to the region. This initiates the process of nucleotide excision repair.

will assist the replicative DNA polymerase during the process of **translesion synthesis (TLS)**—the synthesis of DNA over a template strand that harbors some type of DNA damage. These lesion-replicating polymerases, which are also described in Chapter 11 (see Table 11.4), contain an active site with a loose, flexible pocket that can accommodate aberrant structures in the template strand. When a replicative DNA polymerase encounters a damaged region, it is swapped with a lesion-replicating polymerase.

A negative consequence of translesion synthesis is low fidelity. Due to their flexible active site, lesion-replicating polymerases are much more likely to incorporate the wrong nucleotide in a newly made daughter strand. The mutation rate is typically in the range of 10^{-2} to 10^{-3}. By comparison, replicative DNA polymerases are highly intolerant of the geometric distortions imposed on DNA by the incorporation of incorrect nucleotides, and consequently, they incorporate wrong nucleotides with a very low frequency. In other words, they copy DNA with a high degree of fidelity.

In *E. coli*, translesion synthesis has been shown to occur under extreme conditions that promote DNA damage, an event termed the **SOS response.** Causes of the SOS response include high doses of UV light and other types of mutagens. Such environmental factors result in the up regulation of several genes that function to repair the DNA lesions, restore replication, and prevent premature cell division. When the SOS response occurs, the damaged DNA that has not been repaired is replicated by DNA polymerases II, IV, and V, which can replicate over damaged regions. As previously mentioned, this translesion DNA synthesis results in a high rate of mutation. Even though this result may be harmful, it nevertheless allows the bacteria to survive under conditions of extreme environmental stress. Furthermore, the high rate of mutation may provide genetic variability within the bacterial population, so certain cells may become resistant to the harsh conditions.

CONCEPTUAL SUMMARY

Mutations are heritable changes in the genetic material. They can result from small changes in the genetic material (single-gene mutations), alterations of chromosome structure, or changes in chromosome number. Mutations can be beneficial, neutral, or detrimental. While they provide the variation that enables species to change and adapt to their environment, mutations are more likely to disrupt sequences within genes rather than improve their function.

Mutations can have various consequences on gene expression. If a point mutation—a change in a single base pair—occurs within the coding sequence, it may alter the amino acid sequence of the encoded polypeptide. Missense mutations are base substitutions that cause a change in a single amino acid; nonsense mutations are base substitutions that change a normal codon to a stop codon, resulting in a truncated polypeptide; and frameshift mutations involve the addition or deletion of nucleotides and thus produce a shift in the mRNA codon reading frame.

Mutations within promoters, regulatory elements, splicing sequences, and noncoding regions of genes can have major consequences on the level of gene expression. If a mutation occurs within the germ line—those cells that give rise to the eggs and sperm—it will occur in all the cells of an organism and can be passed from parent to offspring. By comparison, mutations in somatic cells—all cells of the body excluding the germ-line cells—affect only a portion of the organism.

Geneticists categorize mutations as spontaneous mutations—changes in DNA structure that result from abnormalities in biological processes—and induced mutations—changes in DNA structure that are caused by environmental agents. Spontaneous mutations occur during normal cellular conditions. These can arise from depurination, the removal of a purine (guanine or adenine) from the DNA; deamination, the removal of an amino group from the cytosine base; and tautomeric shifts, a temporary change of a nitrogen base structure to an alternative form. In addition, a variety of environmental agents, known as chemical or physical mutagens, are known to induce mutations. Mutagens can alter DNA structure by modifying or removing bases, causing errors in DNA replication, forming thymine dimers, and producing breaks in DNA strands. Testing methods such as the Ames test can determine whether an agent is mutagenic.

Due to the prevalence of mutagens in the environment, organisms have evolved DNA repair systems that can detect and repair different types of DNA lesions. Some DNA repair systems can recognize altered bases and repair them directly. For example, photolyase is a critical DNA repair enzyme for many plant species that removes thymine dimers, and alkyltransferase is a protein that can remove methyl or ethyl groups from guanine bases that have been mutagenized by alkylating agents. Base excision repair and nucleotide excision repair systems recognize alterations in DNA structure and excise the abnormal base or region, respectively. Certain proteins, such as the transcription-repair coupling factor (TRCF) of *E. coli*, can target the nucleotide excision repair system to actively transcribing genes that have damaged DNA. Mismatch repair systems detect and repair base mismatches between nucleotides in parental and newly made strands. The breakage of chromosomes—called a DNA double-strand break (DSB)—is a dangerous type of DNA damage. The two main mechanisms for repair of DSBs are homologous recombination repair (HRR) and nonhomologous end joining (NHEJ). Finally, in translesion synthesis (TLS), specialized DNA polymerases are able to replicate over damaged DNA.

EXPERIMENTAL SUMMARY

Experimentally, the occurrence, causes, and consequences of mutation have been studied in many ways. Luria and Delbrück conducted a fluctuation test to explore whether mutations occur randomly in a population or whether they are a physiological adaptation. Similarly, the Lederbergs used replica plating to distinguish between the random mutation hypothesis and the physiological adaptation hypothesis. Results from both studies supported what is now known as the random mutation theory. According to this theory, mutations are a random process that can occur in any gene and do not involve exposure of an organism to particular conditions that select for specific types of mutations. However, certain regions of genes, called hot spots, occur where mutations are more frequent.

Müller was the first scientist to establish that environmental agents such as X-rays can cause induced mutations. His work showed that X-rays dramatically increased the likelihood of X-linked, recessive lethal mutations. Since these studies, biochemists, microbiologists, and geneticists have discovered an enormous array of agents that can act as mutagens that permanently alter the structure of DNA. Many kinds of testing methods, such as the commonly used Ames test, can ascertain whether or not an agent is a mutagen.

PROBLEM SETS & INSIGHTS

Solved Problems

S1. Mutant tRNAs may act as nonsense and missense suppressors. At the molecular level, explain how you think these suppressors work.

Answer: A suppressor is a second-site mutation that suppresses the phenotypic effects of a first mutation. Intergenic suppressor mutations in tRNA genes can act as nonsense or missense suppressors. For example, let's suppose a first mutation puts a stop codon into a structural gene. A second mutation in a tRNA gene can alter the anticodon region of a tRNA so that the anticodon recognizes a stop codon but inserts an amino acid at this site. A missense suppressor is a mutation in a tRNA gene that changes the anticodon so that it puts in the wrong amino acid at a normal codon that is not a stop codon. These mutant tRNAs are termed missense tRNAs. For example, a tRNA that normally recognizes glutamic acid may incur a mutation that changes its anticodon sequence so that it recognizes a glycine codon instead. Like nonsense suppressors, missense suppressors can be produced by mutations in the anticodon region of tRNAs so that the tRNA recognizes an incorrect codon. Alternatively, missense suppressors can also be produced by mutations in aminoacyl-tRNA synthetases that cause them to attach the incorrect amino acid to a tRNA.

S2. If the rate of mutation is 10^{-5} per gene, how many new mutations per gene would you expect in a population of 1 million bacteria?

Answer: If we multiply the mutation rate times the number of bacteria ($10^{-5} \times 10^{6}$), we obtain a value of 10 new mutations per gene in this population. This answer is correct, but it is an oversimplification of mutation rate. For any given gene, the mutation rate is based on a probability that an event will occur. Therefore, when we consider a particular population of bacteria, we should be aware that the actual rate of new mutation would vary. Even though the rate may be 10^{-5}, we would not be surprised if a population of 1 million bacteria had 9 or 11 new mutations per gene instead of the expected number of 10. We would be surprised if it had 5,000 new mutations per gene, because this value would deviate much too far from our expected number.

S3. In the Ames test, several *Salmonella* strains are used that contain different types of mutations within the gene that encodes an enzyme necessary for histidine biosynthesis. These mutations include transversions, transitions, and frameshift mutations. Why do you think it would be informative to test a mutagen with these different strains?

Answer: Different types of mutagens have different effects on DNA structure. For example, if a mutagen caused transversions, an experimenter would want to use a *Salmonella* strain in which a transversion would convert a *his*$^-$ strain into a *his*$^+$ strain. This type of strain would make it possible to detect the effects of the mutagen.

S4. In Chapter 14, we discussed how bacterial genes can be arranged in an operon structure in which a polycistronic mRNA contains the coding sequences for two or more genes. For genes in an operon, a relatively short distance occurs between the stop codon in the first gene and the start codon in the next gene. In addition to the ribosomal-binding sequence at the beginning of the first gene, a ribosomal-binding sequence is also found at the beginning of the second gene and at the beginning of all subsequent genes. After the ribosome has moved past the stop codon in the first gene, it quickly encounters a ribosomal-binding site in the next gene. This prevents the complete disassembly of the ribosome as would normally occur after a stop codon and leads to the efficient translation of the second gene.

Jacob and Monod, who studied mutations in the *lac* operon, discovered a mutation in the *lacZ* gene that had an unusual effect. This mutation resulted in a shorter version of the β-galactosidase protein, and it also prevented the expression of the *lacY* gene, even though the *lacY* gene was perfectly normal. Explain this mutation.

Answer: This mutation introduced a stop codon much earlier in the *lacZ* gene sequence. This type of mutation, which affects the translation of a gene downstream in an operon, is referred to as a **polar mutation,** and the phenomenon is called polarity. The explanation is that the early nonsense codon causes the ribosome to disassemble before it has a chance to reach the next gene sequence in the operon. The ribosomal-binding sequence that precedes the second gene is efficiently recognized only by a ribosome that has traveled to the normal stop codon in the first gene.

S5. A reverse mutation or a reversion is a mutation that returns a mutant codon back to a codon that gives a wild-type phenotype. At the DNA level, this can be an exact reversion or an equivalent reversion.

GAG (glutamic acid)	First → GTG mutation (valine)	Exact → GAG reversion (glutamic acid)
GAG (glutamic acid)	First → GTG mutation (valine)	Equivalent → GAA reversion (glutamic acid)
GAG (glutamic acid)	First → GTG mutation (valine)	Equivalent → GAT reversion (aspartic acid)

An equivalent reversion produces a protein equivalent to the wild type with regard to structure and function. This can occur in two ways. In some cases, the reversion produces the wild-type amino acid (in this case, glutamic acid), but it uses a different codon than the wild-type gene. Alternatively, an equivalent reversion may substitute an amino acid structurally similar to the wild-type amino acid. In our example, an equivalent reversion has changed valine to an aspartic acid. Because aspartic and glutamic acids are structurally similar—they are acidic amino acids—this type of reversion can restore the wild-type structure and function. Now here is the question. The template strand within the coding sequence of a gene has the following sequence:

3'-TACCCCTTCGACCCCGGA-5'

This template produces an mRNA, 5'–AUGGGGAAGCUGGGG CCA–3', that encodes a polypeptide with the sequence methionine–glycine–lysine–leucine–glycine–proline.

A mutation changes the template strand to 3'–TACCCCTA CGACCCCGGA–5'.

After the first mutation, another mutation occurs to change this sequence again. Would the following second mutations be an exact reversion, an equivalent reversion, or neither?

A. 3'-TACCCCTCCGACCCCGGA-5'

B. 3'-TACCCCTTCGACCCCGGA-5'

C. 3'-TACCCCGACGACCCCGGA-5'

Answer:

A. This is probably an equivalent reversion. The third codon, which encodes a lysine in the normal gene, is now an arginine codon. Arginine and lysine are both basic amino acids, so the polypeptide would probably function normally.

B. This is an exact reversion.

C. The third codon, which is a lysine in the normal gene, has been changed to a leucine codon. It is difficult to say if this would be an equivalent reversion or not. Lysine is a basic amino acid and leucine is a nonpolar, aliphatic amino acid. The protein may still function normally with a leucine at the third codon, or it may function abnormally. You would need to test the function of the protein to determine if this was an equivalent reversion or not.

Conceptual Questions

C1. For each of the following mutations, is it a transition, transversion, addition, or deletion? The original DNA strand is 5'–GGACTAGATAC–3'. (Note: Only the coding DNA strand is shown.)

A. 5'-GAACTAGATAC-3'

B. 5'-GGACTAGAGAC-3'

C. 5'-GGACTAGTAC-3'

D. 5'-GGAGTAGATAC-3'

C2. A gene mutation changes an AT base pair to a GC pair. This causes a gene to encode a truncated protein that is nonfunctional. An organism that carries this mutation cannot survive at high temperatures. Make a list of all the genetic terms that could be used to describe this type of mutation.

C3. What does a suppressor mutation suppress? What is the difference between an intragenic and intergenic suppressor?

C4. How would each of the following types of mutations affect the amount of functional protein that is expressed from a gene?

A. Nonsense

B. Missense

C. Up promoter mutation

D. Mutation that affects splicing

C5. X-rays strike a chromosome in a living cell and ultimately cause the cell to die. Did the X-rays produce a mutation? Explain why or why not.

C6. The lactose permease is encoded by the *lacY* gene of the lac operon. A mutation occurred at codon 64 that changed the normal glycine codon into a valine codon. The mutant lactose permease is unable to function. However, a second mutation, which changes codon 50 from an alanine codon to a threonine codon, is able to restore function. Are the following terms appropriate or inappropriate to describe this second mutation?

A. Reversion

B. Intragenic suppressor

C. Intergenic suppressor

D. Missense mutation

C7. Nonsense suppressors tend to be very inefficient at their job of allowing readthrough of a stop codon. How would it affect the cell if they were efficient at their job?

C8. Are each of the following mutations silent, missense, nonsense, or frameshift mutations? The original DNA strand is 5'–ATGGGACTAGATACC–3'. (Note: Only the coding strand is shown; the first codon is methionine.)

A. 5'-ATGGGTCTAGATACC-3'

B. 5'-ATGCGACTAGATACC-3'

C. 5'-ATGGGACTAGTTACC-3'

D. 5'-ATGGGACTAAGATACC-3'

C9. In Chapters 12 through 15, we discussed many sequences that are outside the coding sequence and are important for gene expression. Look up two of these sequences and write them out. Explain how a mutation could change these sequences and thereby alter gene expression.

C10. Explain two ways that a chromosomal rearrangement can cause a position effect.

C11. Is a random mutation more likely to be beneficial or harmful? Explain your answer.

C12. Which of the following mutations could be appropriately described as a position effect?

A. A point mutation at the –10 position in the promoter region prevents transcription.

B. A translocation places the coding sequence for a muscle-specific gene next to an enhancer that is turned on in nerve cells.

C. An inversion flips a gene from the long arm of chromosome 17 (which is euchromatic) to the short arm (which is heterochromatic).

C13. As discussed in Chapter 22, most forms of cancer are caused by environmental agents that produce mutations in somatic cells. Is an individual with cancer considered a genetic mosaic? Explain why or why not.

C14. Discuss the consequences of a germ-line versus a somatic mutation.

C15. Draw and explain how alkylating agents alter the structure of DNA.

C16. Explain how a mutagen can interfere with DNA replication to cause a mutation. Give two examples.

C17. What type of mutation (transition, transversion, and/or frameshift) would you expect each of the following mutagens to cause?

A. Nitrous acid

B. 5-bromouracil

C. Proflavin

C18. Explain what happens to the sequence of DNA during trinucleotide repeat expansion. If someone was mildly affected with a trinucleotide repeat expansion (TNRE) disorder, what issues would be important when considering whether to have offspring?

C19. Distinguish between spontaneous and induced mutations. Which are more harmful? Which are avoidable?

C20. Are mutations random events? Explain your answer.

C21. Give an example of a mutagen that can change cytosine to uracil. Which DNA repair system(s) would be able to repair this defect?

C22. If a mutagen causes bases to be removed from nucleotides within DNA, what repair system would fix this damage?

C23. Trinucleotide repeat expansions (TNREs) are associated with several different human inherited diseases. Certain types of TNREs produce a long stretch of glutamines (an amino acid) within the encoded protein. This long stretch of glutamines somehow inhibits the function of the protein and thereby causes a disorder. In cases where a TNRE exerts its detrimental effect by producing a glutamine stretch, are the following statements true or false?

A. The TNRE is within the coding sequence of the gene.

B. The TNRE prevents RNA polymerase from transcribing the gene properly.

C. The trinucleotide sequence is CAG.

D. The trinucleotide sequence is CCG.

C24. With regard to TNRE, what is meant by the term anticipation?

C25. What is the difference between the mutation rate and the mutation frequency?

C26. Achondroplasia is a rare form of dwarfism. It is caused by an autosomal dominant mutation within a single gene. Among 1,422,000 live births, the number of babies born with achondroplasia was 31. Among those 31 babies, 18 of them had one parent with achondroplasia. The remaining babies had two unaffected parents. What is the mutation frequency for this disorder among these 1,422,000 babies? What is the mutation rate for achondroplasia?

C27. A segment of DNA has the following sequence:

TTGGATGCTGG
AACCTACGACC

A. What would be the sequence immediately after reaction with nitrous acid? Let the letters H represent hypoxanthine and U represent uracil.

B. Let's suppose this DNA was reacted with nitrous acid. The nitrous acid was then removed, and the DNA was replicated for two generations. What would be the sequences of the DNA products after the DNA had replicated two times? Your answer should contain the sequences of four double helices.

C28. In the treatment of cancer, the basis for many types of chemotherapy and radiation therapy is that mutagens are more effective at killing dividing cells compared to nondividing cells. Explain why. What are possible harmful side effects of chemotherapy and radiation therapy?

C29. An individual contains a somatic mutation that changes a lysine codon into a glutamic acid codon. Prior to acquiring this mutation, the individual had been exposed to UV light, proflavin, and 5-bromouracil. Which of these three agents would be the most likely to have caused this somatic mutation? Explain your answer.

C30. Which of the following examples is likely to be caused by a somatic mutation?

A. A purple flower has a small patch of white tissue.

B. One child, in a family of seven, is an albino.

C. One apple tree, in a very large orchard, produces its apples two weeks earlier than any of the other trees.

D. A 60-year-old smoker develops lung cancer.

C31. How would nucleotide excision repair be affected if one of the following proteins were missing? Describe the condition of the DNA that had been repaired in the absence of the protein.

A. UvrA

B. UvrC

C. UvrD

D. DNA polymerase

C32. During mismatch repair, why is it necessary to distinguish between the template strand and the newly made daughter strand? How is this accomplished?

C33. What are the two main mechanisms by which cells repair double-strand breaks? Briefly describe each one.

C34. With regard to the repair of double-strand breaks, what are the advantages and disadvantages of homologous recombination repair versus nonhomologous end joining?

C35. When DNA *N*-glycosylase recognizes thymine dimers, it detects only the thymine located on the 5' side of the thymine dimer as being abnormal. Draw and explain the steps whereby a thymine dimer would be removed by the consecutive actions of DNA *N*-glycosylase, AP endonuclease, and DNA polymerase.

C36. Discuss the relationship between transcription and DNA repair. Why is it beneficial to repair actively transcribed DNA more efficiently than nontranscribed DNA?

C37. Three common ways to repair changes in DNA structure are nucleotide excision repair, mismatch repair, and homologous recombination repair. Which of these three mechanisms would be used to fix the following types of DNA changes?

A. A change in the structure of a base caused by a mutagen in a nondividing eukaryotic cell

B. A change in DNA sequence caused by a mistake made by DNA polymerase

C. A thymine dimer in the DNA of an actively dividing bacterial cell

C38. What is the underlying genetic defect that causes xeroderma pigmentosum? How can the symptoms of this disease be explained by the genetic defect?

C39. In *E. coli*, a methyltransferase enzyme encoded by the *dam* gene recognizes the sequence 5'–GATC–3' and attaches a methyl group to the N^6 position of adenine. *E. coli* strains that have the *dam* gene deleted are known to have a higher spontaneous mutation rate compared to normal strains. Explain why.

C40. Discuss the similarities and differences between the nucleotide excision repair and mismatch repair systems.

Experimental Questions

E1. The Luria-Delbrück fluctuation test is consistent with the random mutation theory. How would the results have been different if the physiological adaptation hypothesis had been correct?

E2. Explain how the technique of replica plating supports a random mutation theory but conflicts with the physiological adaptation hypothesis. Outline how you would use this technique to show that antibiotic resistance is due to random mutations.

E3. In the experiment of Figure 16.12, the *ClB* chromosome carries a large paracentric inversion, in which the centromere is found outside of the inverted region. This inversion prevents crossing over with the other X chromosome (which may carry an X-ray-induced lethal mutation). Does a paracentric inversion really prevent crossing over? Explain. You may wish to review Chapter 8 to answer this question.

E4. As you may have noticed in the data table of Figure 16.12, the daughters whose fathers were exposed to X-rays were less likely to produce any offspring compared to the control daughters (whose fathers were not exposed to X-rays). Among 1,015 daughters whose fathers had been exposed to X-rays, only 783 produced offspring. By comparison, among 1,011 control daughters, 947 were able to produce offspring. Suggest reasons why the daughters of irradiated fathers appear less likely to produce any offspring compared to the control daughters.

E5. In Müller's experiment with *ClB* chromosomes, is the experiment measuring the mutation rate within a single gene? Explain. If we divide 91 by 783, we obtain a mutation rate of 11.6%. In your own words, explain what this value means.

E6. Researchers were interested in whether two different chemicals were mutagens. Let's call them chemical A and chemical B. They followed the protocol described in the experiment of Figure 16.12 and obtained the following results:

Treatment	Number of *ClB* Daughters Crossed to Normal Males	Number of Tubes Containing Offspring	Number of Tubes Lacking Male Offspring
Control	2,108	2,077	3
Chemical A	1,402	1,378	2
Chemical B	4,203	3,100	77

Would you conclude that chemical A and/or chemical B is a mutagen? Explain.

E7. From an experimental point of view, is it better to use haploid or diploid organisms for mutagen testing? Consider both the Ames test and the experiment of Figure 16.12 when preparing your answer.

E8. How would you modify the Ames test to discover physical mutagens? Would it be necessary to add the rat liver extract? Explain why or why not.

E9. During an Ames test, bacteria were exposed to a potential mutagen. Also, as a control, another sample of bacteria was not exposed to the mutagen. In both cases, 10 million bacteria were plated and the following results were obtained:

No mutagen: 17 colonies

With mutagen: 2,017 colonies

Calculate the mutation rate in the presence and absence of the mutagen. How much does the mutagen increase the rate of mutation?

E10. Richard Boyce and Paul Howard-Flanders conducted an experiment that provided biochemical evidence that thymine dimers are removed from the DNA by a DNA repair system. In their studies, bacterial DNA was radiolabeled so that the amount of radioactivity reflected the amount of thymine dimers. The DNA was then subjected to UV light, causing the formation of thymine dimers. When radioactivity was found in the soluble fraction, thymine dimers had been excised from the DNA by a DNA repair system. But when the radioactivity was in the insoluble fraction, the thymine dimers had been retained within the DNA. The following table illustrates some of their results involving a normal strain of *E. coli* and a second strain that was very sensitive to killing by UV light:

Strain	Treatment	Radioactivity in the Insoluble Fraction (cpm)	Radioactivity in the Soluble Fraction (cpm)
Normal	No UV	<100	<40
Normal	UV-treated, incubated 2 hours at 37°C	357	940
Mutant	No UV	<100	<40
Mutant	UV-treated, incubated 2 hours at 37°C	890	<40

(Adapted from: Boyce, R. P., and Howard-Flanders, P. Release of ultraviolet light-induced thymine dimers from DNA in *E. coli* K-12. *Proc. Natl. Acad. Sci. USA* 51, 293–300.)

Explain the results found in this table. Why is the mutant strain sensitive to UV light?

Questions for Student Discussion/Collaboration

1. In *E. coli*, a variety of mutator strains have been identified in which the spontaneous rate of mutation is much higher than normal strains. Make a list of the types of abnormalities that could cause a strain of bacteria to become a mutator strain. Which abnormalities do you think would give the highest rate of spontaneous mutation?

2. Discuss the times in a person's life when it would be most important to avoid mutagens. Which parts of a person's body should be most protected from mutagens?

3. A large amount of research is aimed at studying mutation. However, there is not an infinite amount of research dollars. Where would you put your money for mutation research?

   A. Testing of potential mutagens

   B. Investigating molecular effects of mutagens

   C. Investigating DNA repair mechanisms

   D. Or some other place?

Note: All answers appear at the website for this textbook; the answers to even-numbered questions are in the back of the textbook.

www.mhhe.com/brookergenetics3e

Visit the Online Learning Center for practice tests, answer keys, and other learning aids for this chapter. Enhance your understanding of genetics with our interactive exercises, quizzes, animations, and much more.

17

RECOMBINATION AND TRANSPOSITION AT THE MOLECULAR LEVEL

In this chapter, we will examine a variety of molecular processes in which segments of DNA, such as those found in chromosomes, become rearranged. **Homologous recombination** is the process whereby DNA segments, which are similar or identical to each other, break and rejoin to form a new combination. As described in Chapter 3, homologous recombination occurs when chromosomes cross over during meiosis. Not only does homologous recombination enhance genetic diversity, it also helps to repair DNA and ensures the proper segregation of chromosomes.

We will also examine ways that nonhomologous segments of DNA may recombine with each other. During **site-specific recombination,** nonhomologous DNA segments are recombined at specific sites. This type of recombination occurs when certain viruses integrate their genomes into host cell DNA and also happens within genes that encode antibody polypeptides. In addition, we will consider a widespread form of recombination known as **transposition.** As you will learn, small segments of DNA called transposable elements can move themselves to multiple locations within the chromosomal DNA.

From a molecular viewpoint, homologous recombination, site-specific recombination, and transposition are important mechanisms for DNA rearrangement. These processes involve a series of steps that direct the breakage and rejoining of DNA fragments. Various cellular proteins are necessary for these steps to occur properly. The past few decades have seen many exciting advances in our understanding of recombination at the molecular level. In this chapter, we will consider the general concepts of recombination and examine molecular models that explain how recombination occurs.

CHAPTER OUTLINE

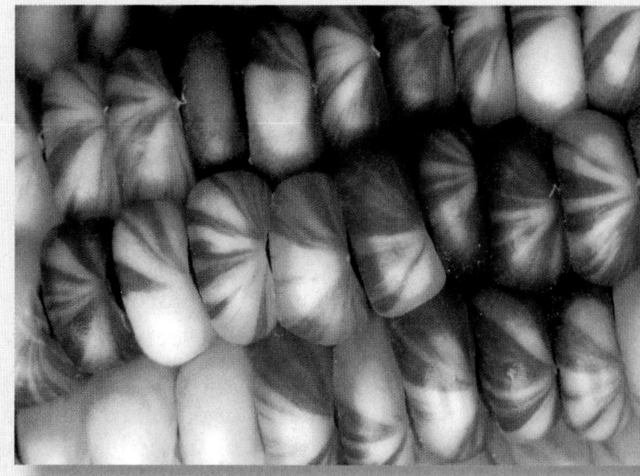

Speckled corn kernels. This speckling phenotype is due to the movement of DNA segments called transposable elements.

17.1 HOMOLOGOUS RECOMBINATION

Homologous recombination involves an exchange between DNA segments that are similar or identical. Eukaryotic chromosomes that have similar or identical sequences frequently participate in crossing over during meiosis I and occasionally during mitosis. Crossing over involves the alignment of a pair of homologous chromosomes, followed by the breakage of two chromatids at analogous locations, and the subsequent exchange of the corresponding segments (refer to Figure 3.10).

Figure 17.1 shows two types of crossing over that may occur between replicated chromosomes in a diploid species. When crossing over takes place between sister chromatids, the process is called **sister chromatid exchange (SCE)**. Because sister chromatids are genetically identical to each other, SCE

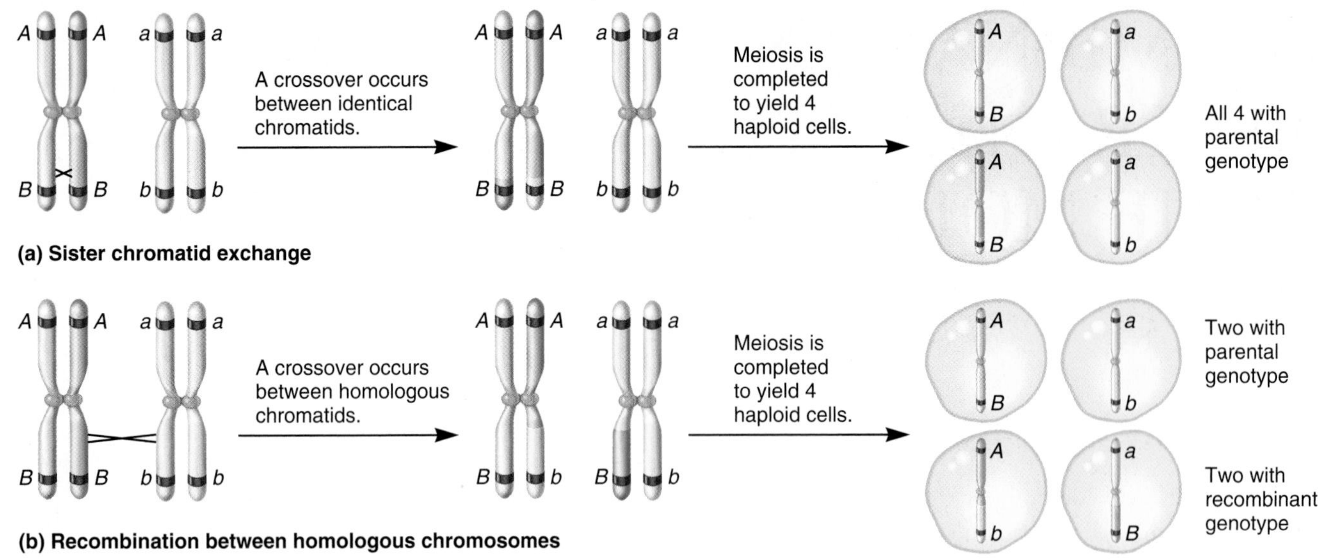

FIGURE 17.1 **Crossing over between eukaryotic chromosomes.** (**a**) Sister chromatid exchange occurs when genetically identical chromatids cross over. (**b**) Homologous recombination can also occur when homologous chromosomes cross over. This form of homologous recombination may lead to a new combination of alleles, which is called a recombinant (or nonparental) genotype.

Genes→Traits Homologous recombination is particularly important when we consider the relationships between multiple genes and multiple traits. For example, if the X chromosome in a female fruit fly carried alleles for red eyes and gray body and its homologue carried alleles for white eyes and yellow body, homologous recombination could produce recombinant chromosomes that carry alleles for red eyes and yellow body, or alleles for white eyes and gray body. Therefore, new combinations of two or more alleles can arise when homologous recombination takes place.

does not produce a new combination of alleles (Figure 17.1a). By comparison, it is common for crossing over to occur between homologous chromosomes. As shown in Figure 17.1b, this form of homologous recombination may produce a new combination of alleles in the resulting chromosomes. In this second case, homologous recombination has resulted in **genetic recombination,** which refers to the shuffling of genetic material to create a new genetic combination that differs from the original. Homologous recombination is an important mechanism to foster genetic recombination.

Although bacteria are usually haploid—having only one type of chromosome—they also can undergo homologous recombination. How can the exchange of DNA segments occur in a haploid organism? First, bacteria may have more than one copy of a chromosome per cell, though the copies are genetically identical. These copies can exchange genetic material via homologous recombination. Second, during DNA replication, the replicated regions may also undergo homologous recombination. In bacteria, homologous recombination is particularly important in the repair of DNA segments that have been damaged.

In this section, we will begin with an experimental approach to detect SCE, overcoming the obstacle that the exchanged chromosomes are genetically identical to each other. We will then focus our attention on the molecular mechanisms that underlie homologous recombination.

EXPERIMENT 17A

The Staining of Harlequin Chromosomes Can Reveal Recombination Between Sister Chromatids

Our understanding of crossing over and homologous recombination has come from a variety of experimental approaches, including genetic, biochemical, and cytological analyses. Chromosomal staining methods have allowed researchers to visualize the genetic exchange between eukaryotic chromosomes. In the 1970s, the Russian cytogeneticist A. F. Zakharov and colleagues developed methods that improved our ability to identify chromosomes. They made the interesting observation that chromosomes labeled with the nucleotide analog **5-bromodeoxyuridine (BrdU)** bind certain types of stain to a different degree compared to normal chromosomes. In 1974, Paul Perry and Sheldon Wolff extended this approach to differentially stain sister chromatids and microscopically identify SCEs.

Before we consider the experiment of Perry and Wolff, let's examine how their staining procedure allowed them to accurately distinguish the two sister chromatids. In their approach, eukaryotic cells were grown in a laboratory and exposed to BrdU for two rounds of DNA replication. After the second round of DNA replication, one of the sister chromatids in each pair contained one unlabeled strand and one BrdU-labeled strand. The other sister chromatid had two BrdU-labeled strands (**Figure 17.2**). When treated with two dyes, Hoechst 33258 and Giemsa, the sister chromatid containing two strands with BrdU stains very weakly and appears light, whereas the sister chromatid with only one strand containing BrdU stains much more strongly and

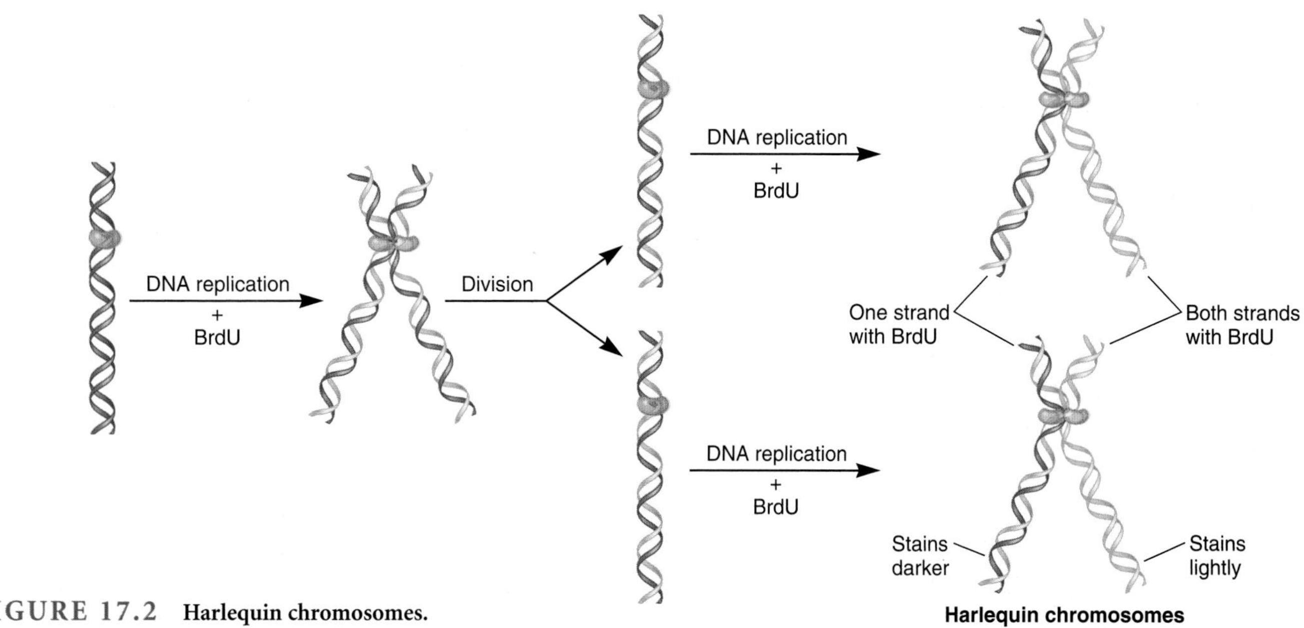

FIGURE 17.2 **Harlequin chromosomes.**

appears very dark. In this way, the two sister chromatids can be distinguished microscopically. Chromosomes stained in this way are referred to as harlequin chromosomes, because they are reminiscent of a harlequin character's costume with its variegated pattern of light and dark patches. In these chromosomes, SCEs can be clearly identified as exchanges between light and dark chromatids.

The steps in Perry and Wolff's protocol are shown in **Figure 17.3**. They began with Chinese hamster ovary cells, a commonly used mammalian cell line, and exposed the cells to BrdU for two rounds of DNA replication. Near the end of the second round, colcemid was added to prevent the completion of mitosis. The cells were treated with KCl to spread out the chromosomes, which were subsequently fixed and then stained with Hoechst 33258 and Giemsa.

THE HYPOTHESIS

Crossing over may occur between sister chromatids.

TESTING THE HYPOTHESIS — FIGURE 17.3 The staining of harlequin chromosomes reveals sister chromatid exchange.

Starting material: A laboratory cell line of Chinese hamster ovary (CHO) cells.

Experimental level | **Conceptual level**

1. Expose CHO cells to BrdU for two cell generations (approximately 24 hours). Note: CHO cells are Chinese hamster ovary cells, which is a commonly used mammalian cell line.

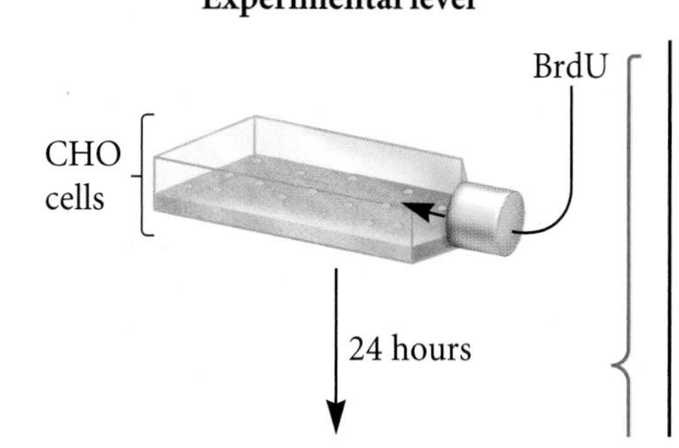

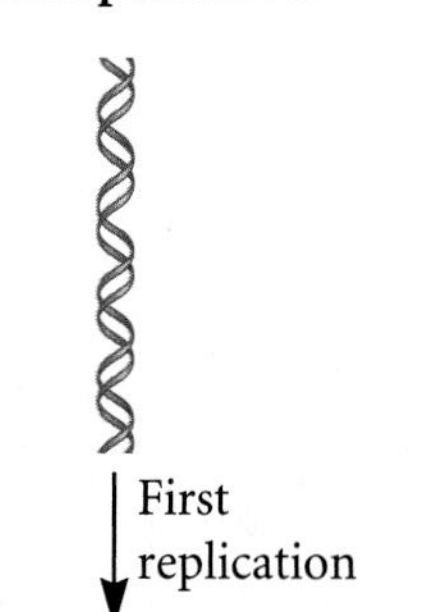

2. Near the end of the growth, expose the cells to colcemid. This prevents the cells from completing mitosis following the second round of DNA replication.

3. Add 0.075 M KCl to spread the chromosomes and then methanol/acetic acid to fix the cells.

4. Stain with Hoechst 33258, rinse, and later stain with Giemsa. Note: This refinement in the staining procedure greatly improved the ability to discern the sister chromatids.

5. View under a microscope.

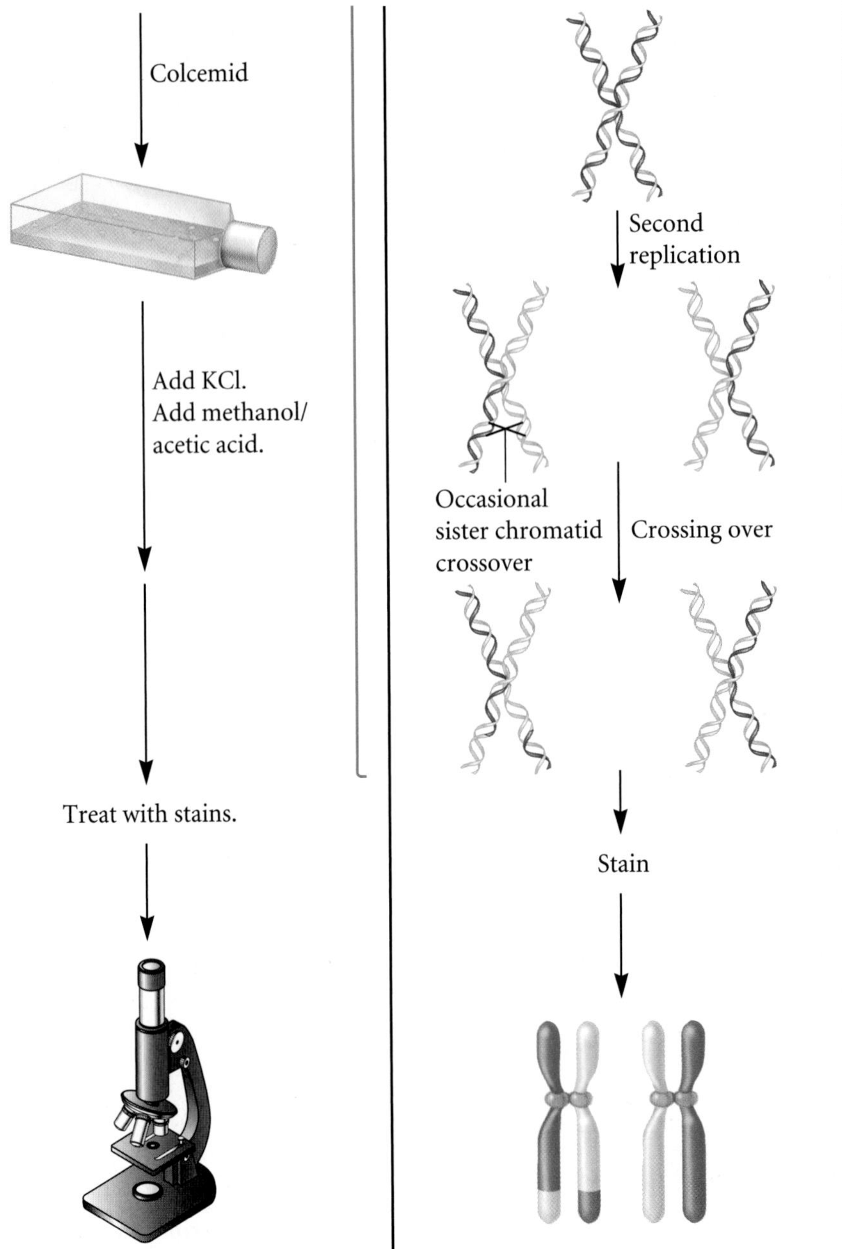

THE DATA

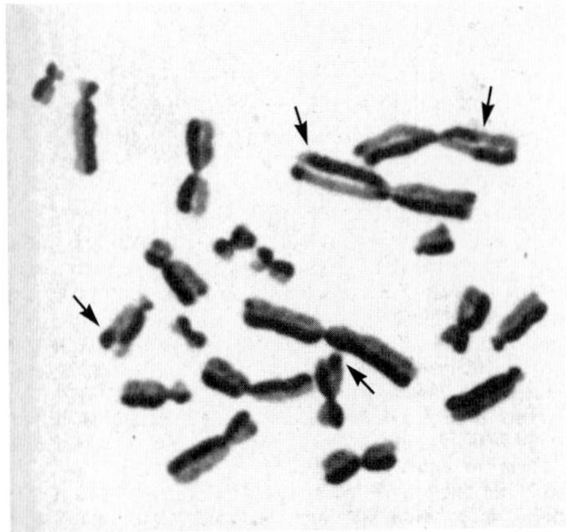

INTERPRETING THE DATA

A micrograph of their results is shown in the data of Figure 17.3. As seen here, the chromosomes show the classic harlequin appearance due to the differential staining of the sister chromatids. Furthermore, examples of SCE are clearly visible. The arrows depict regions where crossing over has taken place. In this study, Perry and Wolff found that SCEs occurred at a frequency of approximately 0.67 per chromosome. This method has provided an accurate (and dramatic) way to visualize genetic exchange between sister chromatids.

Many subsequent studies have used the harlequin staining method to study the effects of agents that may influence the frequency of genetic exchanges. Researchers have found that DNA damage caused by radiation and chemical mutagens tends to increase the level of genetic exchange. When cells are exposed to these types of mutagens, the technique of harlequin staining has revealed a substantial increase in the frequency of SCEs. In addition, certain genetic disorders that result in higher levels of chromosome breakage also show elevated SCEs. For example, Bloom syndrome is a rare autosomal recessive disorder characterized by short stature, skin abnormalities, and a predisposition for developing certain forms of cancer. The defect is associated with a gene that is involved with DNA replication. In the cells of Bloom syndrome patients, chromosome breaks are more frequent during DNA replication. Likewise, SCE is typically 10- to 15-fold more frequent in Bloom syndrome patients compared to unaffected individuals.

A self-help quiz involving this experiment can be found at the Online Learning Center.

The Holliday Model Describes a Molecular Mechanism for the Recombination Process

We now turn our attention to genetic exchange that occurs between homologous chromosomes. Perhaps it is surprising that the first molecular model of homologous recombination did not come from a biochemical analysis of DNA or from electron microscopy studies. Instead, it was deduced from the outcome of genetic crosses in fungi.

As discussed in Chapter 5, geneticists have learned a great deal from the analysis of fungal asci, because an ascus is a sac that contains the products of a single meiosis. When two haploid fungi that differ at a single gene are crossed to each other, the ascus is expected to contain an equal proportion of each genotype. For example, if a pigmented strain of *Neurospora* producing orange spores is crossed to an albino strain producing white spores, the resulting group of eight cells, or octad, should contain four orange spores and four white spores (refer to Figure 5.14). As early as 1934, H. Zickler noticed that unequal proportions of the spores sometimes occurred within asci. He occasionally observed octads with six orange spores and two white spores, or six white spores and two orange spores.

Zickler used the term **gene conversion** to describe the phenomenon in which one allele is converted to the allele on the homologous chromosome. Subsequent studies by several researchers confirmed this phenomenon in yeast and *Neurospora*. Gene conversion occurred at too high a rate to be explained by new mutations. In addition, research showed that gene conversion often occurs in a chromosomal region where a crossover has taken place.

Based on studies involving gene conversion, Robin Holliday proposed a model in 1964 to explain the molecular steps that occur during homologous recombination. We will first consider the steps in the Holliday model and then consider more recent models. Later, we will examine how the Holliday model can explain the phenomenon of gene conversion.

The **Holliday model** is shown in **Figure 17.4a**. At the beginning of the process, two homologous chromatids are aligned with each other. According to the model, a break or nick occurs at identical sites in one strand of each of the two homologous chromatids. The strands then invade the opposite helices and base pair with the complementary strands. This event is followed by a covalent linkage to create a **Holliday junction.** The cross in the Holliday junction can migrate in a lateral direction. As it does so, a DNA strand in one helix is swapped for a DNA strand in the other helix. This process is called **branch migration** because the branch connecting the two double helices migrates laterally. Because the DNA sequences in the homologous chromosomes are similar but not identical, the swapping of the DNA strands during branch migration may produce a **heteroduplex,** a region in the double-stranded DNA that contains base mismatches. In other words, because the DNA strands in this region are from homologous chromosomes, their sequences are not perfectly complementary, yielding mismatches.

The final two steps in the recombination process are collectively called **resolution** because they involve the breakage and rejoining of two DNA strands to create two separate chromosomes. In other words, the entangled DNA strands become resolved into two separate structures. The bottom left side of Figure 17.4a shows the Holliday junction viewed from two different planes. If breakage occurs in the same two DNA strands that were originally nicked at the beginning of this process, the subsequent joining of strands produces nonrecombinant chromosomes with a heteroduplex region. Alternatively, if breakage occurs in the strands that were not originally nicked, the rejoining process results in recombinant chromosomes, also with a heteroduplex region.

The Holliday model can account for the general properties of recombinant chromosomes formed during eukaryotic meiosis. As mentioned, the original model was based on the results of crosses in fungi where the products of meiosis are contained within a single ascus. Nevertheless, molecular research in many other organisms has supported the central tenets of the model. Particularly convincing evidence came from electron microscopy studies in which recombination structures could be visualized. **Figure 17.4b** shows an electron micrograph of two DNA fragments in the process of recombination. This structure has been called a chi form because its shape is similar to the Greek letter χ (chi).

More Recent Models Have Refined the Molecular Steps of Homologous Recombination

As more detailed studies of homologous recombination have become available, certain steps in the Holliday model have been reconsidered. In particular, more recent models have modified the initiation phase of recombination. Researchers now suggest that homologous recombination is not likely to involve nicks at identical sites in one strand of each homologous chromatid. Instead, it is more likely for a DNA helix to incur a break in both strands of one chromatid or a single nick. Both of these types of changes have been shown to initiate homologous recombination. Therefore, newer models have tried to incorporate these experimental

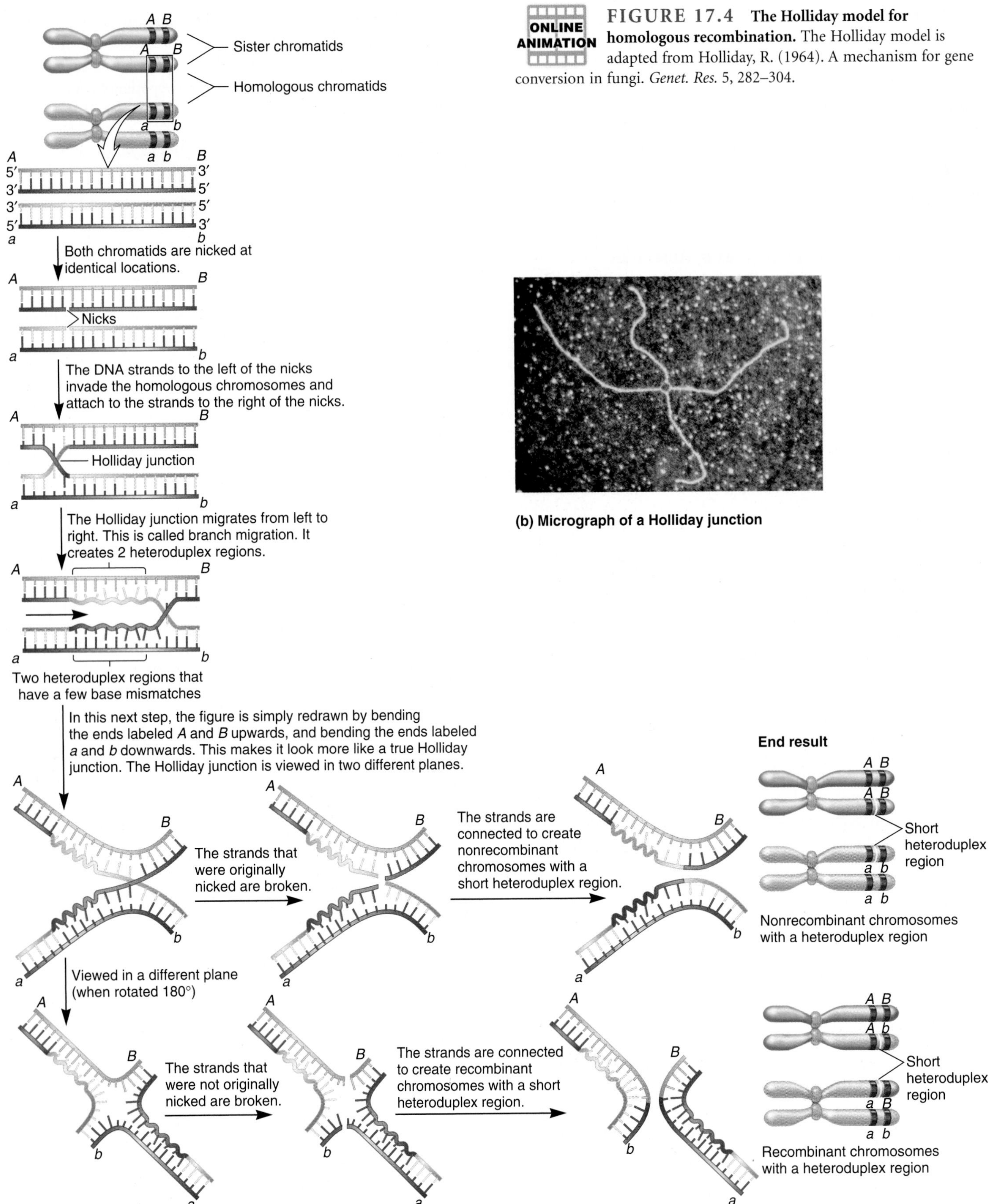
A B
Sister chromatids
Homologous chromatids
a b
5′ 3′
3′ 5′
3′ 5′
5′ 3′
Both chromatids are nicked at identical locations.
Nicks
The DNA strands to the left of the nicks invade the homologous chromosomes and attach to the strands to the right of the nicks.
Holliday junction
The Holliday junction migrates from left to right. This is called branch migration. It creates 2 heteroduplex regions.
Two heteroduplex regions that have a few base mismatches
In this next step, the figure is simply redrawn by bending the ends labeled A and B upwards, and bending the ends labeled a and b downwards. This makes it look more like a true Holliday junction. The Holliday junction is viewed in two different planes.
The strands that were originally nicked are broken.
The strands are connected to create nonrecombinant chromosomes with a short heteroduplex region.
End result
Short heteroduplex region
Nonrecombinant chromosomes with a heteroduplex region
Viewed in a different plane (when rotated 180°)
The strands that were not originally nicked are broken.
The strands are connected to create recombinant chromosomes with a short heteroduplex region.
Recombinant chromosomes with a heteroduplex region

(a) The Holliday model for homologous recombination

(b) Micrograph of a Holliday junction

ONLINE ANIMATION

FIGURE 17.4 The Holliday model for homologous recombination. The Holliday model is adapted from Holliday, R. (1964). A mechanism for gene conversion in fungi. *Genet. Res.* 5, 282–304.

observations. In 1975, a model proposed by Matthew Meselson and Charles Radding hypothesized that a single nick in one DNA strand initiates recombination. A second model, proposed by Jack Szostak, Terry Orr-Weaver, Rodney Rothstein, and Franklin Stahl, suggests that a double-strand break initiates the recombination process. This is called the **double-strand break model.** Though recombination may occur via more than one mechanism, recent evidence suggests that double-strand breaks commonly promote homologous recombination during meiosis and during DNA repair.

Figure 17.5 shows the general steps in the double-strand break model. As seen here, the top chromosome has experienced a double-strand break. A small region near the break is degraded, which generates a single-stranded DNA segment that can invade the intact double helix. The strand displaced by the invading segment forms a structure called a displacement loop (D-loop). After the D-loop is formed, two regions have a gap in the DNA. How is the problem fixed? DNA synthesis occurs in the relatively short gaps where a DNA strand is missing. This DNA synthesis is called **DNA gap repair synthesis.** Once this is completed, two Holliday junctions are produced. Depending on the way these are resolved, the end result is nonrecombinant or recombinant chromosomes containing a short heteroduplex. In eukaryotes such as yeast, recent evidence suggests that certain proteins bound to Holliday junctions may regulate the resolution step in a way that favors the formation of recombinant chromosomes rather than nonrecombinant chromosomes.

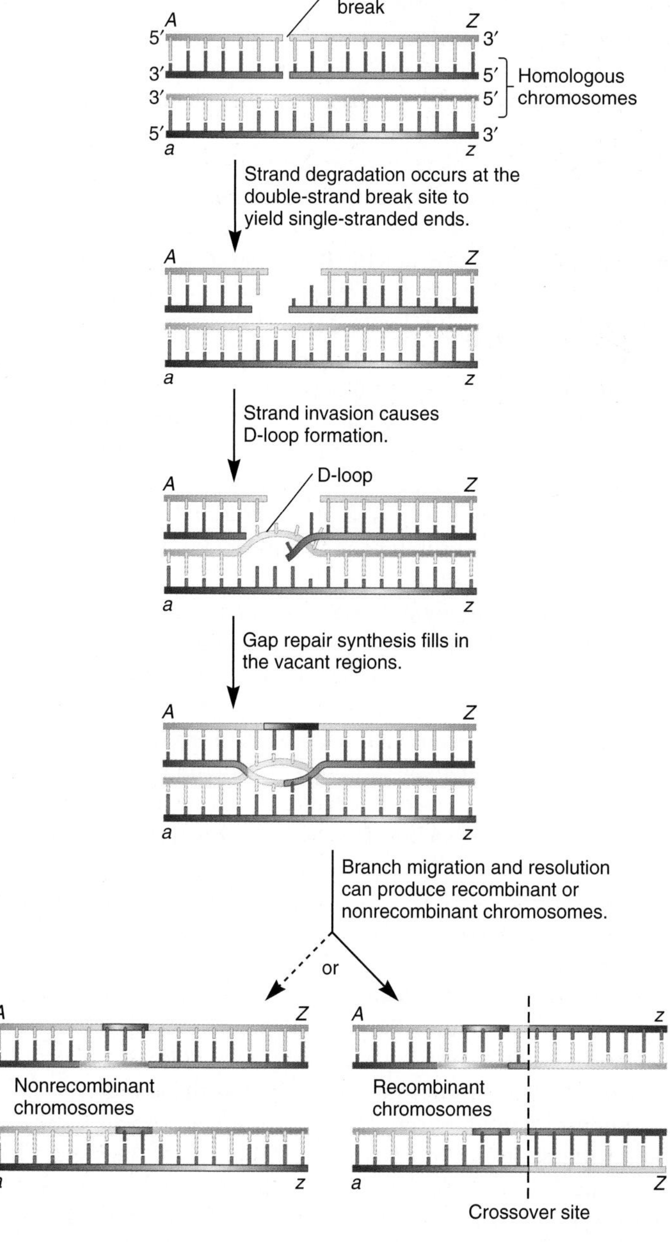

FIGURE 17.5 A simplified version of the double-strand break model. For simplicity, this illustration does not include the formation of heteroduplexes. The dashed arrow indicates that the pathway to the left may be less favored.

Various Proteins Are Necessary to Facilitate Homologous Recombination

The homologous recombination process requires the participation of many proteins that catalyze different steps in the recombination pathway. Homologous recombination is found in all species, and the types of proteins that participate in the steps outlined in Figure 17.5 are very similar. The cells of any given species may have more than one molecular mechanism to carry out homologous recombination. This process is best understood in *Escherichia coli*. Table 17.1 summarizes some of the *E. coli* proteins that play critical roles in one recombination pathway found in this species. Though it is beyond the scope of this textbook, *E. coli* has other pathways to carry out homologous recombination.

TABLE 17.1

***E. coli* Proteins That Play a Role in Homologous Recombination**

Protein	Description
RecBCD	A complex of three proteins that tracks along the DNA and recognizes double-strand breaks. The complex partially degrades the double-stranded regions to generate single-stranded regions that can participate in strand invasion. RecBCD is also involved in loading RecA onto single-stranded DNA. In addition, RecBCD can create single-strand breaks at chi sequences.
Single-stranded binding protein	Coats broken ends of chromosomes and prevents excessive strand degradation.
RecA	Binds to single-stranded DNA and promotes strand invasion, which enables homologous strands to find each other. It also promotes the displacement of the complementary strand to generate a D-loop.
RuvABC	This protein complex binds to Holliday junctions. RuvAB promotes branch migration. RuvC is an endonuclease that cuts the crossed or uncrossed strands to resolve Holliday junctions into separate chromosomes.
RecG	RecG protein can also promote branch migration of Holliday junctions.

RecBCD is a protein complex composed of the RecB, RecC, and RecD proteins. (The term Rec indicates that these proteins are involved with recombination.) The RecBCD complex plays an important role in the initiation of recombination involving double-strand breaks, as outlined in Figure 17.5. In this process, RecBCD recognizes a double-strand break within DNA and catalyzes DNA unwinding and strand degradation. The action of RecBCD produces single-stranded DNA ends that can participate in strand invasion and exchange. The single-stranded DNA ends are coated with single-stranded binding protein to prevent their further degradation. The RecBCD complex can also create breaks in the DNA at sites known as chi sequences. In *E. coli*, the chi sequence is 5'–GCTGGTGG–3'. As RecBCD tracks along the DNA, when it encounters a chi sequence, it cuts one of the DNA strands to the 3' side of this sequence. This generates a single nick that can initiate homologous recombination.

The function of the RecA protein is to promote strand invasion. To accomplish this task, it binds to the single-stranded ends of DNA molecules generated from the activity of RecBCD. A large number of RecA proteins bind to single-stranded DNA, forming a structure called a filament. During strand invasion, this filament makes contact with the unbroken chromosome. Initially, this contact is most likely to occur at nonhomologous regions. The contact point slides along the DNA until it reaches a homologous region. Once a homologous site is located, RecA catalyzes the displacement of one DNA strand, and the invading single-stranded DNA quickly forms a double helix with the other strand. This results in a D-loop, as shown in Figure 17.5. RecA proteins mediate the movement of the invading strand and the displacement of the complementary strand. This occurs in such a way that the displaced strand invades the vacant region of the broken chromosome.

Proteins that bind specifically to Holliday junctions have also been identified. These include a complex of proteins termed RuvABC and a protein called RecG. RuvA is a tetramer that forms a platform on which the Holliday junction is held in a square planar configuration. This platform also contains two hexameric rings of RuvB. Together, RuvA and RuvB act as a helicase that catalyzes an ATP-dependent migration of a four-way branched DNA junction in either the 5' $\rightarrow$ 3' or 3' $\rightarrow$ 5' direction. RuvC is an endonuclease that binds as a dimer to Holliday junctions and resolves these structures by making cuts in the DNA (see bottom of Figure 17.4). RecG is a helicase that also catalyzes branch migration of Holliday junctions.

Before ending this discussion regarding the molecular mechanism of homologous recombination, let's consider recombinational events during meiosis in eukaryotic cells. As described in Chapter 3, crossing over between homologous chromosomes is an important event during prophase of meiosis I. An intriguing question is, How are crossover sites chosen between two homologous chromosomes? While the answer is not entirely understood, molecular studies in two different yeast species, *Saccharomyces cerevisiae* and *Schizosaccharomyces pombe*, suggest that double-strand breaks initiate the homologous recombination that occurs during meiosis. In other words, double-strand breaks create sites where a crossover will occur. In *S. cerevisiae*, the formation of DNA double-strand breaks that initiate meiotic recombination requires at least 10 different proteins. One particular protein, termed Spo11 protein, is thought to be instrumental in cleaving the DNA and thereby creating a double-strand break. However, the roles of the other proteins, and the interactions among them, are not well understood. Once a double-strand break is made, homologous recombination can then occur according to the model described in Figure 17.5.

Gene Conversion May Result from DNA Mismatch Repair or DNA Gap Repair

As mentioned earlier, homologous recombination can lead to an event where one allele is converted to the allele on the homologous chromosome, a process known as gene conversion. The original Holliday model was based on the phenomenon of gene conversion.

How can homologous recombination account for gene conversion? Researchers have identified two possible ways that gene conversion can occur. One mechanism involves DNA mismatch repair, a topic that was described in Chapter 16. To understand how this works, let's take a closer look at the heteroduplexes formed during branch migration of a Holliday junction (see Figure 17.4a). A heteroduplex contains a DNA strand from each of the two original parental chromosomes. The two parental chromosomes may contain an allelic difference within this region. In other words, this short region may contain DNA sequence differences. If this is the case, the heteroduplex formed after branch migration will contain an area of base mismatch. Gene conversion occurs when recombinant chromosomes are repaired and result in two copies of the same allele.

As shown in **Figure 17.6**, mismatch repair of a heteroduplex may result in gene conversion. In this example, the two parental chromosomes had different alleles due to a single base-pair difference in their DNA sequences, as shown at the top of the figure. During recombination, branch migration has occurred across this region, thereby creating two heteroduplexes with base mismatches. As described in Chapter 16, DNA mismatches will be recognized by DNA repair systems and repaired to a double helix that obeys the AT/GC rule. These two mismatches can be repaired in four possible ways. As shown here, two possibilities produce no gene conversion, whereas the other two lead to gene conversion.

A second mechanism for gene conversion is via DNA gap repair synthesis. **Figure 17.7** illustrates how gap repair synthesis can lead to gene conversion according to the double-strand break model. The top chromosome, which carries the recessive *b* allele, has suffered a double-strand break in this gene. A gap is created by the digestion of the DNA in the double helix. This digestion eliminates the *b* allele. The two template strands used in gap repair synthesis are from one homologous chromatid. This helix carries the dominant *B* allele. Therefore, after gap repair synthesis takes place, the top chromosome will contain the *B* allele, as will the bottom chromosome. Gene conversion has changed the recessive *b* allele to a dominant *B* allele.

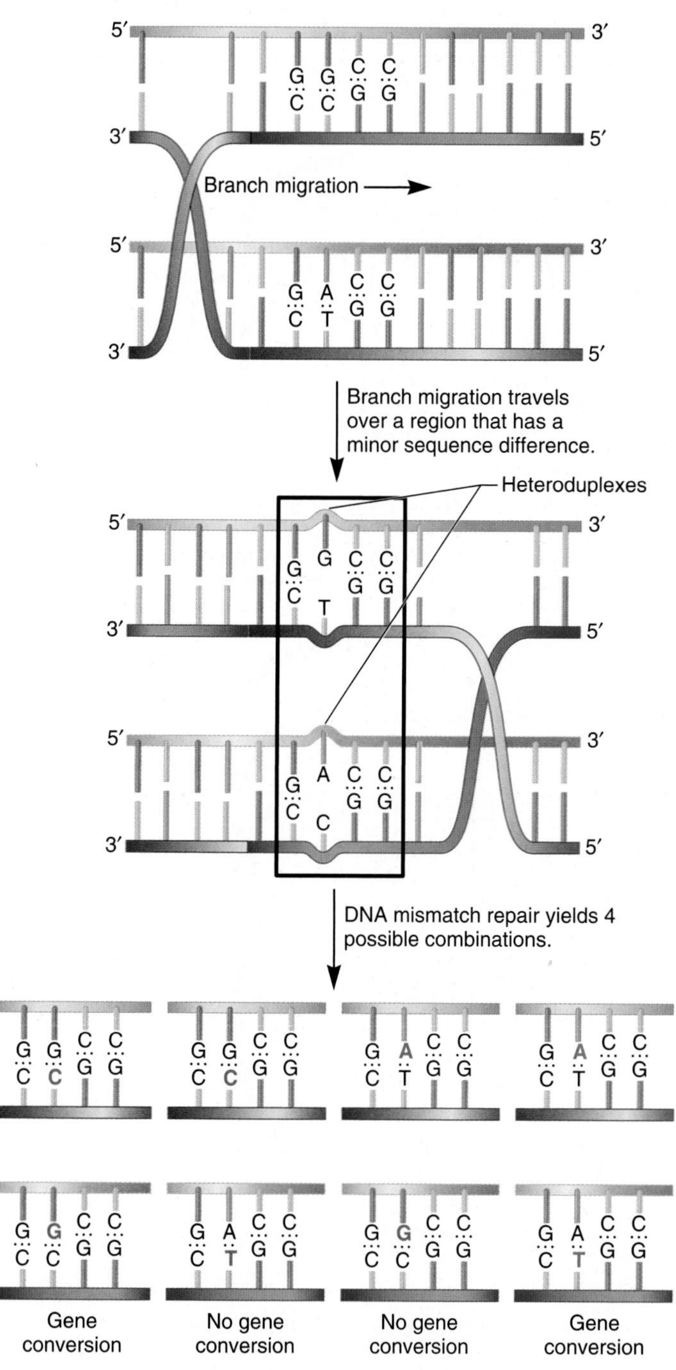

FIGURE 17.6 **Gene conversion by DNA mismatch repair.** A branch migrates past a homologous region that contains slightly different DNA sequences. This produces two heteroduplexes—DNA double helices with mismatches. The mismatches can be repaired in four possible ways by the mismatch repair system described in Chapter 16. Two of these ways result in gene conversion. The repaired base is shown in red.

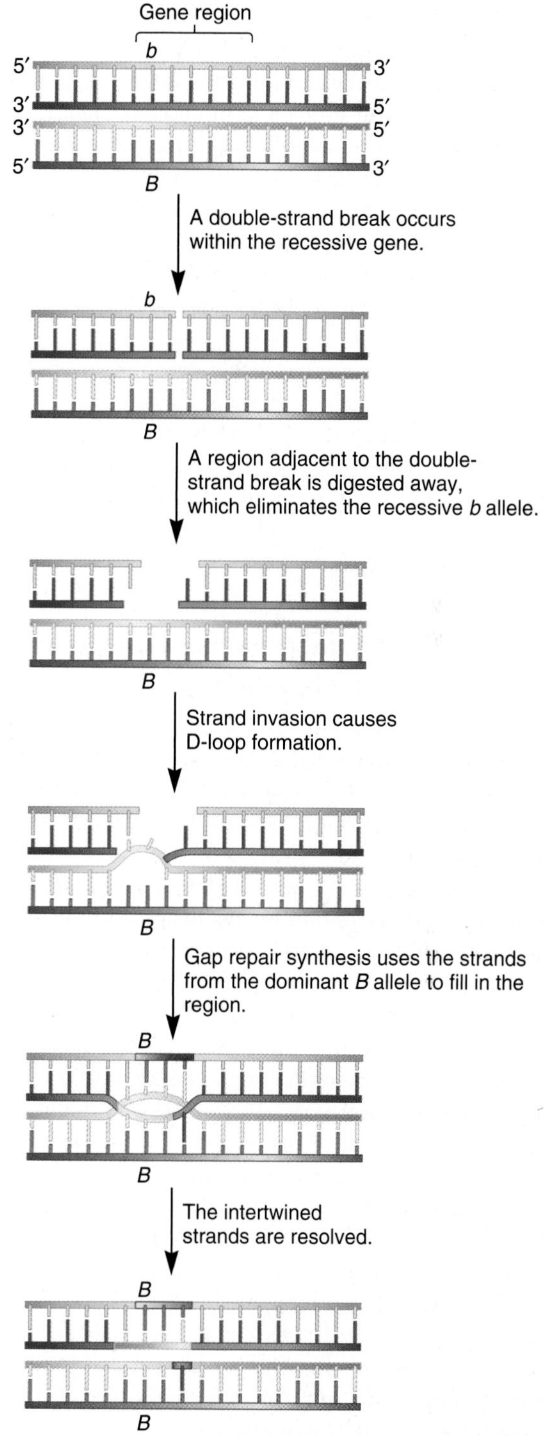

FIGURE 17.7 **Gene conversion by gap repair synthesis in the double-strand break model.** A gene is found in two alleles, designated *B* and *b*. A double-strand break occurs in the DNA encoding the *b* allele. Both of these DNA strands are digested away, thereby eliminating the *b* allele. A complementary DNA strand encoding the *B* allele migrates to this region and provides the template to synthesize a double-stranded region. Following resolution, both DNA double helices carry the *B* allele.

17.2 SITE-SPECIFIC RECOMBINATION

Thus far, we have examined recombination between segments of DNA similar or identical to each other. **Site-specific recombination** is another mechanism by which DNA fragments can recombine to make new genetic combinations. During this process, two DNA segments with little or no homology align themselves at specific sites. The sites are relatively short DNA sequences (a dozen or so nucleotides in length) that provide a specific location where recombination will occur. Chromosome breakage and reunion occur at these defined locations to create a recombinant chromosome. For this to occur, these sites are recognized by specialized enzymes that catalyze the breakage and rejoining of DNA fragments within the sites.

Certain viruses use site-specific recombination to insert their viral DNA into their host cell's chromosome. This process has been examined extensively in bacteriophage λ. In addition, mammalian genes that encode antibody polypeptides are rearranged by site-specific recombination that enables the generation of a diverse array of antibodies. In this section, we will consider both mechanisms.

The Integration of Viral Genomes Can Occur by Site-Specific Recombination

The reproductive cycle of some viruses involves the integration of viral DNA into host cell DNA. Certain bacteriophages, for example, can integrate their viral DNA into the bacterial chromosome, creating a prophage. The prophage can exist in a latent, or **lysogenic,** state for many generations. The two cycles of phage λ—lytic and lysogenic—are described in Chapter 14. Phage λ infects *E. coli.* The integration of this phage into the host chromosome is well understood and occurs by a mechanism involving site-specific recombination (**Figure 17.8**).

Integration of the λ DNA into the *E. coli* chromosome requires sequences known as **attachment sites.** As shown at the top of Figure 17.8, a common core sequence within the attachment site sequence is identical in the λ DNA and the *E. coli* chromosome. An enzyme known as integrase is encoded by a gene in the λ DNA. Several molecules of integrase recognize the core sequences and bring them close together. Integrase then makes staggered cuts in both the λ and *E. coli* attachment sites. The strands are then exchanged, and the ends are ligated together. In this way, the phage DNA is integrated into the host cell chromosome. As a prophage, the λ DNA may remain latent for many generations. Certain environmental conditions, such as exposure to UV light, may act to stimulate the excision of the prophage from the host DNA, thereby reactivating the virus. Excision also requires integrase, which catalyzes the reverse reaction, as well as a second protein known as excisionase.

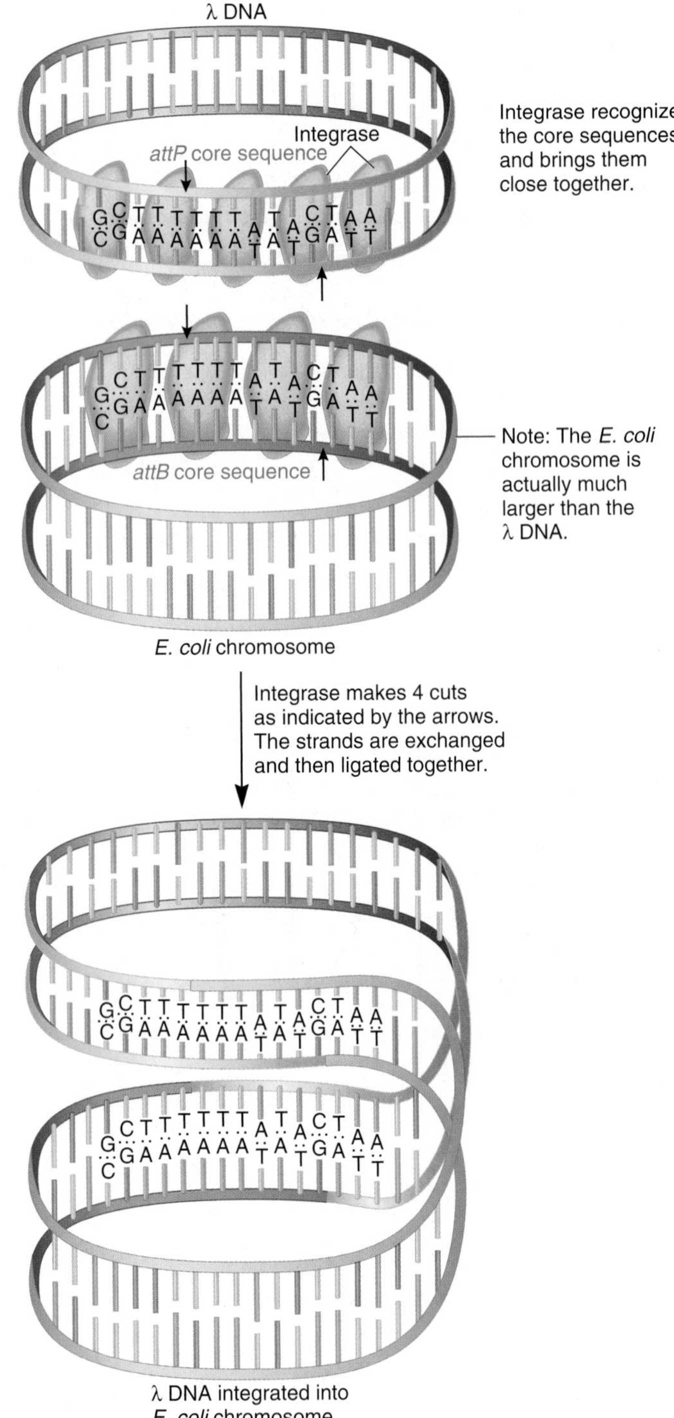

FIGURE 17.8 The integration of λ DNA into the *E. coli* chromosome. The core sequence within *attP* in the λ DNA attaches to the core sequence within *attB* in the *E. coli* chromosome. As noted here, the core sequences of *attP* and *attB* are identical to each other and thereby provide recognition sites for site-specific recombination.

Antibody Diversity in the Immune System Is Produced by Site-Specific Recombination

As we have just seen, viruses can integrate their DNA into the host cell chromosome by site-specific recombination. This process requires a viral enzyme, integrase, that recognizes the sites and catalyzes the recombination reaction. A similar process occurs in certain cells of the immune system. The DNA sequences within antibody genes are rearranged by enzymes that recognize specific sites within those genes and catalyze the breakage and reunion of DNA segments. Before we discuss the details of this mechanism, let's first consider the biology of antibodies.

Antibodies, or **immunoglobulins (Igs),** are proteins produced by the B cells of the immune system. Their function is to recognize foreign material, such as viruses and bacteria, and target them for destruction. A foreign material that elicits an immune response is called an **antigen.** Antibodies recognize sites within antigens. The recognition between an antibody and antigen is very specific. Within the immune system, each B cell produces a single type of antibody. However, our bodies have millions of B cells. Site-specific recombination allows each B cell to produce an antibody with a different amino acid sequence. These differences in the amino acid sequences of antibody proteins enable them to recognize different antigens. In this way, site-specific recombination plays an important role in the ability of the immune system to identify an impressive variety of foreign materials as being antigens and thereby target them for destruction.

From a genetic viewpoint, the production of millions of different antibodies poses an interesting problem. If a distinct gene was needed to produce each different antibody polypeptide, the genome would need to contain millions of different antibody genes. By comparison, consider that the entire human genome contains only about 20,000 to 25,000 different genes. How then is it possible for the genome to produce millions of different antibody proteins, which are encoded by genes? The answer to this question baffled geneticists for several decades until research revealed that antibodies with different polypeptide sequences are generated by an unusual mechanism in which the DNA is cut and reconnected by site-specific recombination. We might call this "DNA splicing," although the term splicing is usually reserved for the cutting and rejoining of RNA molecules. With this mechanism, only a few large antibody precursor genes are needed to produce millions of different antibodies. These precursor genes are spliced in many different ways to produce a vast array of polypeptides with differing amino acid sequences.

Figure 17.9 shows the site-specific recombination of an antibody precursor gene. Antibodies are tetrameric proteins composed of two heavy polypeptide chains and two light chains. One type of light chain is the κ (kappa) light chain, which is a component of a class of antibodies known as immunoglobulin G (IgG). The organization of the precursor gene for the κ light chain is shown at the top of Figure 17.9. At the left side, the gene has approximately 300 regions known as variable (V) sequences or domains. In addition, the gene contains four different joining (J) sequences and a single constant (C) sequence. Each variable domain or joining domain encodes a different amino acid sequence.

During the maturation of B cells, the κ light-chain precursor gene is cut and rejoined so that one variable domain becomes adjacent to a joining domain. At the end of every V domain and the beginning of every J domain is located a **recombination signal sequence** that functions as a recognition site for site-specific recombination between the V and J regions. The recombination event is initiated by two proteins called **RAG1** and **RAG2.** RAG is an acronym for recombination-activating gene. These proteins recognize recombination signal sequences and generate two

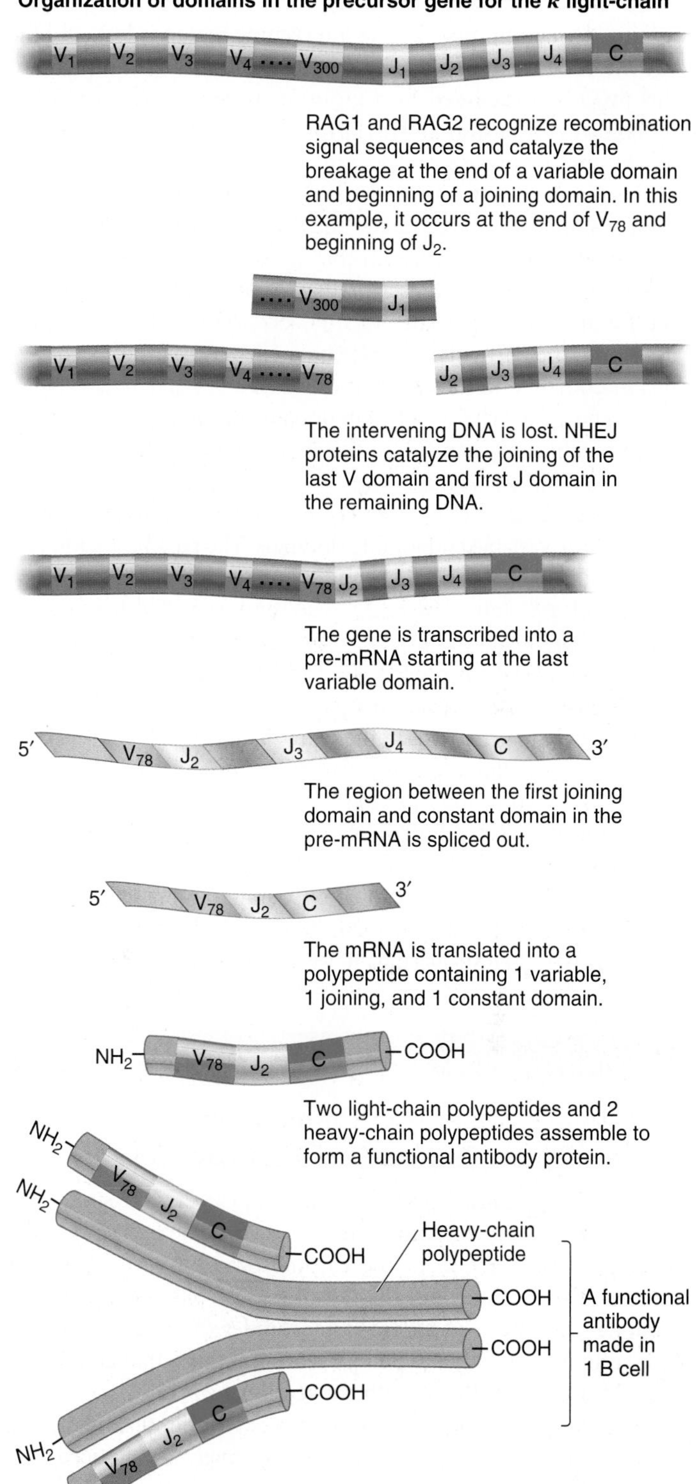
Organization of domains in the precursor gene for the κ light-chain
V1 V2 V3 V4 .... V300 J1 J2 J3 J4 C
RAG1 and RAG2 recognize recombination signal sequences and catalyze the breakage at the end of a variable domain and beginning of a joining domain. In this example, it occurs at the end of V78 and beginning of J2.
.... V300 J1
V1 V2 V3 V4 .... V78
J2 J3 J4 C
The intervening DNA is lost. NHEJ proteins catalyze the joining of the last V domain and first J domain in the remaining DNA.
V1 V2 V3 V4 .... V78 J2 J3 J4 C
The gene is transcribed into a pre-mRNA starting at the last variable domain.
5′ V78 J2 J3 J4 C 3′
The region between the first joining domain and constant domain in the pre-mRNA is spliced out.
5′ V78 J2 C 3′
The mRNA is translated into a polypeptide containing 1 variable, 1 joining, and 1 constant domain.
NH2 V78 J2 C COOH
Two light-chain polypeptides and 2 heavy-chain polypeptides assemble to form a functional antibody protein.
NH2
NH2
V78 J2 C COOH
Heavy-chain polypeptide
COOH
COOH
A functional antibody made in 1 B cell
COOH
C J2 V78
NH2
NH2

FIGURE 17.9 **Site-specific recombination within the precursor gene that encodes the κ light chain for immunoglobulin G (IgG) proteins.**

double-strand breaks, one at the end of a V domain and one at the beginning of a J domain. For example, in the recombination event shown in Figure 17.9, RAG1 and RAG2 have made cuts at the end of variable domain number 78 and the beginning of joining domain number 2. The intervening region is lost, and the two ends are then joined to each other. The connection phase of this process is catalyzed by a group of proteins termed **nonhomologous end-joining (NHEJ) proteins.** The fusion process may not be entirely precise, so a few nucleotides can be added or lost at the junction between the variable and joining domains. This imprecision further accentuates the diversity in antibody genes.

Following transcription, the fused VJ region is contained within a pre-mRNA transcript that is then spliced to connect the J and C domains. After this has occurred, a B cell will produce only the particular κ light chain encoded by the specific VJ fusion domain and the constant domain.

The heavy-chain polypeptides are produced by a similar recombination mechanism. In this case, the heavy-chain gene has about 500 variable domains and four joining domains. In addition, the gene encodes 12 diversity (D) domains, which are found between the variable and joining domains. The recombination first involves the connection of a D and J domain, followed by the connection of a V and DJ domain. The same proteins that catalyze VJ fusion are involved in the recombination with the heavy-chain gene. Collectively, this process is called **V(D)J recombination.** The D is in parentheses because this type of domain is found only in the heavy-chain genes, not in the light-chain genes.

The recombination process within immunoglobulin genes produces an enormous diversity in polypeptides. Even though it occurs at specific junctions within the antibody gene, the recombination is fairly random with regard to the particular V and J domains that can be joined. Overall, the possible number of functional antibodies that can be produced by V(D)J recombination is rather staggering. If we assume that any of the 300 different variable sequences can be spliced next to any of the four joining sequences, this results in 1,200 possible light-chain combinations. The number of heavy-chain possibilities is 500 × 12 × 4 = 24,000. Because any light-chain–heavy-chain combination is possible, this yields 1,200 × 24,000 = 28,800,000 possible antibody molecules from the random recombination within two precursor genes!

17.3 TRANSPOSITION

The last form of recombination we will consider is transposition. In some ways, transposition resembles the site-specific recombination we examined for phage λ. In that case, a segment of λ DNA was able to integrate itself into the *E. coli* chromosome. Transposition also involves the integration of small segments of DNA into the chromosome. Transposition, though, can occur at many different locations within the genome. The DNA segments that transpose themselves are known as **transposable elements (TEs).** TEs have sometimes been referred to as "jumping genes" because they are inherently mobile.

Transposable elements were first identified by Barbara McClintock in the early 1950s from her classic studies with corn plants. Since that time, geneticists have discovered many different types of TEs in organisms as diverse as bacteria, fungi, plants, and animals. The advent of molecular technology has allowed scientists to understand more about the characteristics of TEs that enable them to be mobile. In this section, we will examine the characteristics of TEs and explore the mechanisms that explain how they move. We will also discuss the biological significance of TEs and their uses as experimental tools.

EXPERIMENT 17B

McClintock Found That Chromosomes of Corn Plants Contain Loci That Can Move

Barbara McClintock began her scientific career as a student at Cornell University. Her interests quickly became focused on the structure and function of the chromosomes of corn plants, an interest that continued for the rest of her life. She spent countless hours examining corn chromosomes under the microscope. She was technically gifted and had a theoretical mind that could propose ideas that conflicted with conventional wisdom.

During her long career as a scientist, McClintock identified many unusual features of corn chromosomes. She noticed that one strain of corn had the strange characteristic that a particular chromosome, number 9, tended to break at a fairly high rate at the same site. McClintock termed this a **mutable site** or locus. This observation initiated a six-year study concerned with highly unstable chromosomal locations. In 1951, at the end of her study, McClintock proposed that these sites are actually locations where transposable elements have been inserted into the chromosomes. At the time of McClintock's studies, such an idea was entirely unorthodox.

McClintock focused her efforts on the relationship between a mutable locus and its phenotypic effects on corn kernels. Chromosome 9 with a mutable locus also carried several genes that affected the phenotype of corn kernels. Each gene existed in (at least) one dominant and one recessive allele. The mutable locus was termed *Ds* (for dissociation), because the locus was known to frequently cause chromosomal breaks. In the chromosome shown here, the *Ds* locus is located next to several genes affecting kernel traits.

In this case, there are three genes that exist as two or more alleles:

1. *C* is an allele for normal kernel color (dark red), *c* is a recessive allele of the same gene that causes a colorless kernel, while C^I is a third allele of this gene that is dominant to both *C* and *c* and causes a colorless kernel.
2. *Sh* is an allele that produces normal endosperm, while *sh* is a recessive allele that causes shrunken endosperm. (Note:

The endosperm is the storage material in the kernel that is used by the plant embryo to provide energy for growth.)

3. *Wx* is the allele that produces normal starch in the endosperm, while *wx* is a recessive allele that produces a waxy-appearing phenotype.

During her intensive work, in which she studied corn chromosomes under the microscope, McClintock identified strains of corn in which the *Ds* locus was found in different locations within the corn genome. She could determine the location of *Ds* because the movement of *Ds* occasionally causes chromosome breakage (**Figure 17.10**). Keep in mind that the endosperm of a kernel is triploid because it is derived from the fusion of two maternal haploid nuclei and one paternal haploid nucleus (refer back to Chapter 2, Figure 2.2c). The kernel shown in Figure 17.10 was produced by a cross in which the pollen carried the top chromosome—C^I *Sh Wx Ds*—while the two maternal chromosomes are *C sh wx*. This kernel is expected to be colorless, because the C^I allele is dominant and causes a colorless phenotype. However, as the kernel grows by cell division, the movement of the *Ds* locus out of its original location may occasionally cause a chromosome to break, and the distal part of this chromosome is lost. This chromosome breakage may happen in several cells, which continue to divide and grow as the kernel becomes larger. This process produces a sectoring phenotype—patches of cells occur in the kernel that are red, shrunken, and waxy.

By analyzing many kernels, McClintock was also able to identify cases in which *Ds* had moved to a new location. For example, if *Ds* had moved out of its original location and inserted between *Sh* and *Wx*, a break at *Ds* would produce the following combination:

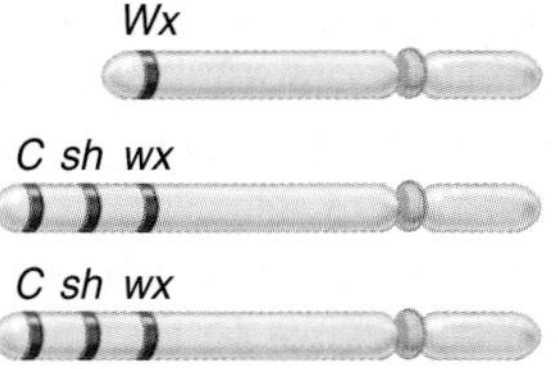

This genotype would produce patches on the kernel that are red and shrunken but not waxy. In this way, McClintock identified 20 independent cases in which the *Ds* locus had moved to a new location within this chromosome. Overall, the results from many crosses were consistent with the idea that *Ds* can transpose itself throughout the corn genome. McClintock also found that a second locus, termed *Ac* (for activator), was necessary for the *Ds* locus to move. Researchers later discovered that the *Ac* locus contains a gene that encodes an enzyme called transposase, which is necessary for *Ds* to move. We will discuss the function of transposase later in this chapter. Some strains of McClintock's corn contained the *Ac* locus, while others did not.

During her studies, McClintock noticed a particularly exciting and unusual event. By making the appropriate cross, she sought to produce kernels with the following genotype:

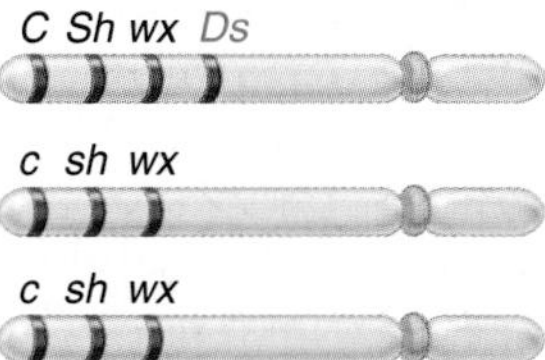

The kernels were expected to be red. Because the strain also contained the *Ac* locus, breakage would occur occasionally at the *Ds* locus to produce colorless patches. Among 4,000 kernels, she noticed one kernel with the opposite phenotype—a colorless kernel with red patches. This observation suggested that the inherited genotype in this case, which produced a colorless phenotype, was mutable to become *C*.

How did McClintock explain these results? She postulated that the colorless phenotype was due to a transposition of *Ds* into the *C* gene:

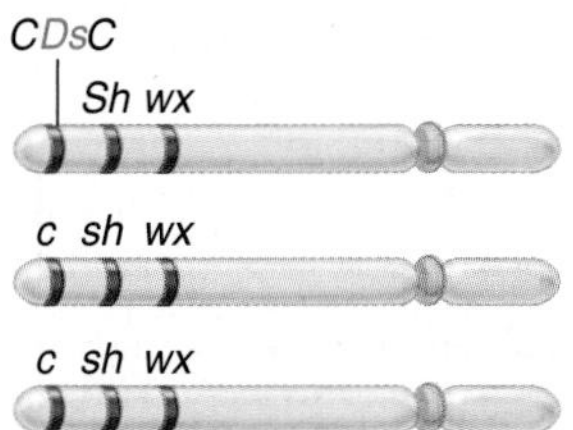

When *Ds* was located within the *C* gene, it inactivated the *C* gene, thereby resulting in the colorless phenotype. However, McClintock proposed that when *Ds* occasionally transposed out of the gene during kernel growth, the *C* allele would be restored and a red patch would result. In this case, the formation of patches, or sectoring, was due to the movement of *Ds* out of its

FIGURE 17.10 The sectoring trait in corn kernels.

Genes→Traits This kernel is expected to be colorless, because the C^I allele is dominant. On occasion, though, the movement of *Ds* may cause a chromosome break, thereby losing the C^I, *Sh*, and *Wx* alleles. As such a cell continues to divide, it will produce a patch of daughter cells that are red, shrunken, and waxy. Therefore, this sectoring trait arises from the loss of genes that occurs when the movement of *Ds* causes chromosome breakage. Note: The movement of *Ds* does not usually cause chromosome breakage. In most cases, the movement of *Ds* out of a site is followed by nonhomologous end joining (described in Chapter 16) in which the chromosome pieces are connected together.

location within the *C* gene and the rejoining of the two ends, not simply due to chromosome breakage. According to this hypothesis, the red phenotype should be associated with two observations. First, *Ds* should have moved to a new location. Second, the restored *C* allele should no longer be mutable because the mutable locus had been removed.

To further study this phenomenon, McClintock carried out the experiment shown in **Figure 17.11**. As seen here, a cross produced kernels that were not entirely colorless. In most cases, red sectoring occurred because the *Ds* element left the *C* gene in a few cells during the growth of the kernel. However, in the ear of corn shown in Figure 17.11, one kernel was completely red. This suggests that the *Ds* element transposed out of the *C* gene prior to kernel growth. In this case, the *Ds* element transposed out of the *C* gene during the formation of the haploid male gametophyte that produced the pollen nuclei that fertilized this particular red kernel. Therefore, all of the cells in this kernel would have a red phenotype. (Note: Every kernel in an ear of corn is equivalent to a distinct offspring, because each is produced from the union of haploid cells from the male and female gametophytes.)

THE HYPOTHESIS

The transposition of the *Ds* element into the normal *C* gene prevents kernel pigmentation. When the *Ds* element transposes back out of the *C* gene, the normal *C* allele is restored, which gives a red phenotype.

TESTING THE HYPOTHESIS — FIGURE 17.11 Evidence for transposable elements in corn.

Starting material: The male pollen of the corn plant was homozygous for a chromosome in which the *Ds* element had moved into the *C* gene: *CDsC Sh wx*. The male plant also contained the *Ac* locus. The plant contributing the female gamete was homozygous for a chromosome carrying *c sh Wx*.

1. Cross a plant that is homozygous for *c*, *sh*, and *Wx* to a plant that is homozygous for *CDsC*, *Sh*, *wx*.
2. In this cross, most kernels will have a colorless background with red sectoring. On rare occasions, *Ds* may have transposed during male gametophyte formation, producing pollen that gives rise to a completely red kernel.
3. Identify those occasional kernels that are completely red.
4. Germinate the solid red kernels.

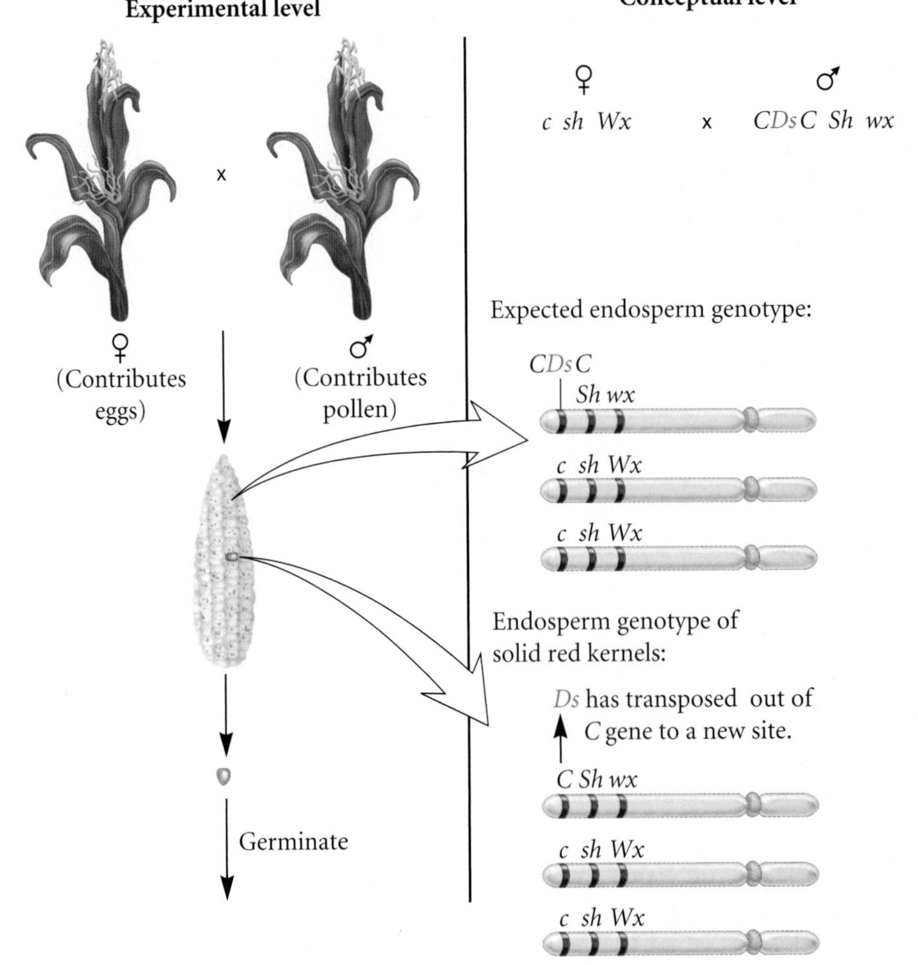

5. Conduct crosses to determine the location of *Ds* in the plants derived from the solid red kernels. Note: This can be done using the chromosomal markers and the strategy that was described at the beginning of this experiment. When observing the results of these crosses, determine whether the *C* gene is still mutable. (Is red sectoring occurring?)

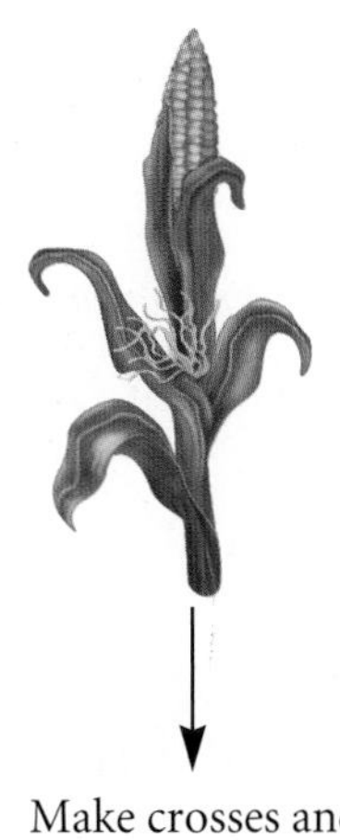

Make crosses and observe kernels.

THE DATA

Strain	Kernel Phenotype	Location of *Ds*	Mutability?
From parental cross (see step 2)	Colorless background with red sectoring	Within the *C* gene	Yes, red sectoring occurred in strains containing *Ac*.
From red kernels (see step 3)	Red kernels	*Ds* had moved out of the *C* gene to another location.	No, the *C* gene was stable; no sectoring was observed.

INTERPRETING THE DATA

By conducting the appropriate crosses, McClintock found that in the progeny of a solid red kernel, the *Ds* locus had moved out of the *C* gene to another location (see the data of Figure 17.11). In addition, the "restored" *C* gene behaved normally. In other words, the *C* gene was no longer highly mutable, and the kernels did not show a sectoring phenotype. Taken together, the results are consistent with the hypothesis that the *Ds* locus can move around the corn genome by transposition.

When McClintock published these results in 1951, they were met with great skepticism. Some geneticists of that time were unable to accept the idea that the genetic material was susceptible to frequent rearrangement. Instead, they believed that the genetic material was always very stable and permanent in its structure. Over the next several decades, the scientific community came to realize that transposable elements are a widespread phenomenon. Much like Gregor Mendel and Charles Darwin, Barbara McClintock was clearly ahead of her time. She was awarded the Nobel Prize in 1983, more than 30 years after her original discovery.

A self-help quiz involving this experiment can be found at the Online Learning Center.

Transposable Elements and Retroelements Move Via One of Three Transposition Pathways

Since the pioneering studies of Barbara McClintock, many different transposable elements (TEs) have been found in bacteria, fungi, plants, and animals. Three general types of transposition pathways have been identified (**Figure 17.12**). In **simple,** or **conservative transposition,** the TE is removed from its original site and transferred to a new target site. This mechanism is also called a cut-and-paste mechanism because the element is cut out of its original site and pasted into a new site. Transposable elements that move via simple transposition are widely found in bacterial and eukaryotic species. By comparison, **replicative transposition** involves the replication of the TE and insertion of the newly made copy into a second site. In this case, one of the TEs remains in its original location, and the other is inserted at another location. Replicative transposition is relatively uncommon and found only in bacterial species.

The first two categories of TEs we have considered thus far move as a DNA molecule from one site to another. Such TEs are called **transposons.** A third category of elements move via an RNA intermediate. This form of transposition, termed **retrotransposition,** is found in eukaryotic species, where it is very common. Transposable elements that move via retrotransposition are known as **retroelements, retrotransposons,** or **retroposons.** In retrotransposition, the element is transcribed into RNA. An enzyme called reverse transcriptase uses the RNA as a template to synthesize a DNA molecule that is integrated into a new region of the genome. Like replicative transposons, retroelements increase in number during retrotransposition.

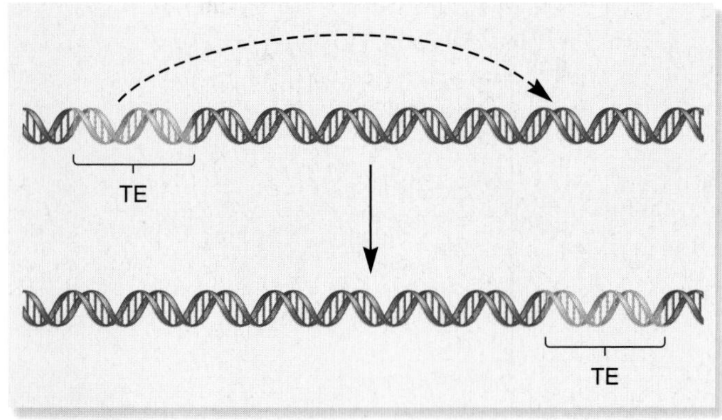

(a) Simple transposition

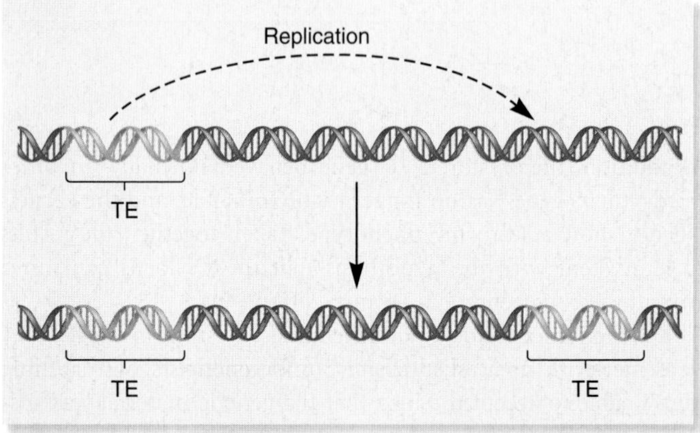

(b) Replicative transposition

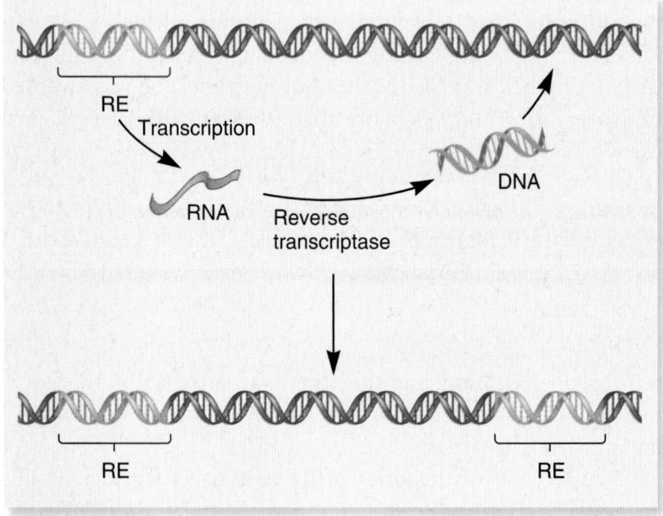

(c) Retrotransposition

FIGURE 17.12 **Three mechanisms of transposition.**

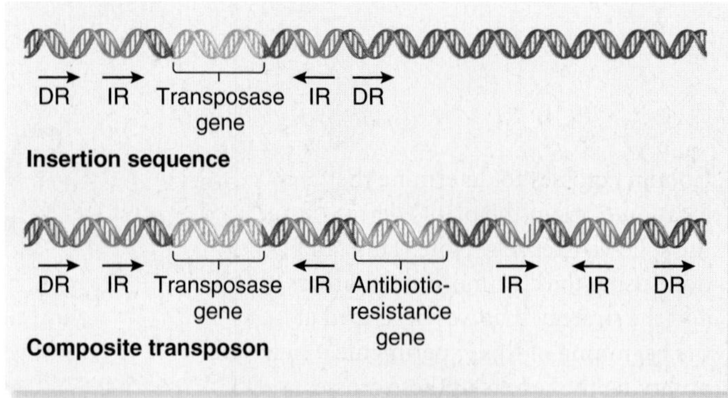

(a) Elements that move by simple transposition

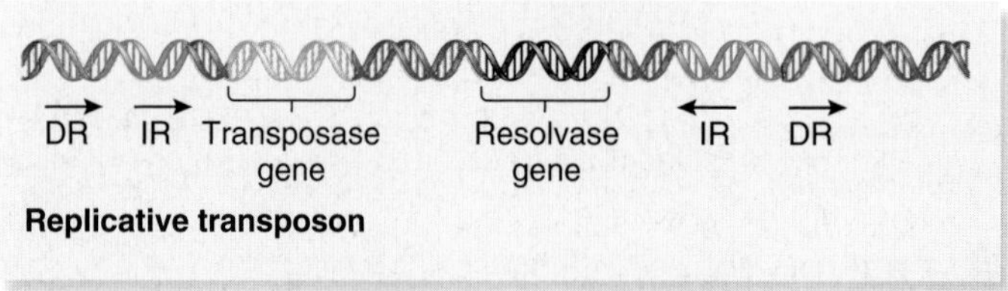

(b) An element that moves by replicative transposition

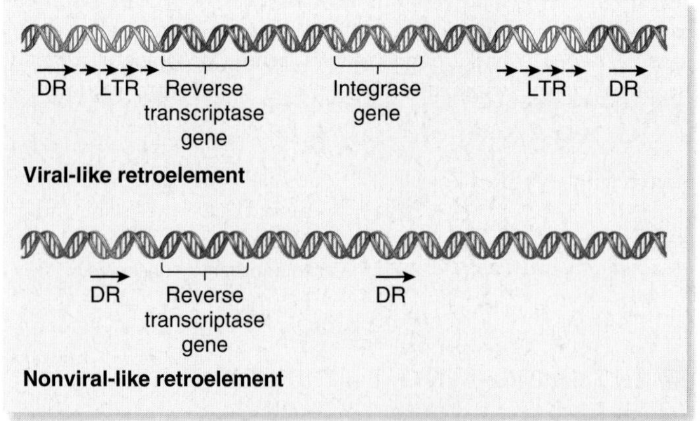

(c) Elements that move by retrotransposition (via an RNA intermediate)

FIGURE 17.13 **Common organizations of transposable elements.** Direct repeats (DRs) are found within the host DNA. Inverted repeats (IRs) are at the ends of most transposable elements. Long terminal repeats (LTRs) are regions containing a large number of tandem repeats.

Each Type of Transposable Element Has a Characteristic Pattern of DNA Sequences

As researchers have studied TEs from many species, they have found that DNA sequences within transposable elements are organized in several different ways. **Figure 17.13** describes a few ways that TEs are organized, although many variations are possible. All TEs are flanked by **direct repeats (DR),** which are identical nucleotide sequences that are in the same direction and repeated. Direct repeats are adjacent to both ends of the element. The simplest TE, which is commonly found in bacteria, is known as an **insertion sequence**. As shown in Figure 17.13a, an insertion sequence has two important characteristics. First, both ends of the insertion sequence contain **inverted repeats (IRs)**. Inverted repeats are DNA sequences that are identical (or very similar) but run in opposite directions, such as the following:

```
5'-CTGACTCTT-3'   and   5'-AAGAGTCAG-3'
3'-GACTGAGAA-5'         3'-TTCTCAGTC-5'
```

Depending on the particular element, the lengths of inverted repeats range from 9 to 40 bp in length. In addition, insertion sequences may contain a central region that encodes the enzyme **transposase,** which catalyzes the transposition event.

Composite transposons contain additional genes that are not necessary for transposition per se. They commonly contain genes that confer a selective advantage to the organism. Composite transposons are prevalent in bacteria, where they often contain genes that provide resistance to antibiotics or toxic heavy metals. For example, the composite transposon shown in Figure 17.13a contains two insertion sequences flanking a gene that confers antibiotic resistance. During transposition of a composite transposon, only the inverted repeats at the ends of the transposon are involved in the transpositional event. Whenever insertion sequences are found at both ends of a gene, they create a composite transposon. Both insertion sequences and composite transposons are elements that move via simple transposition, also called cut-and-paste transposition.

Replicative transposons, elements that move by replicative transposition, have a sequence organization that is similar to insertion sequences except that replicative transposons have a resolvase gene that is found between the inverted repeats (Figure 17.13b). As discussed later in this chapter, both transposase and resolvase are needed to catalyze the transposition of replicative transposons.

The organization of retroelements can be quite variable, and they are categorized based on their evolutionary relationship to retroviral sequences. Retroviruses are RNA viruses that make a DNA copy that integrates into the host's genome. The **viral-like retroelements** are evolutionarily related to known retroviruses. These transposable elements have retained the ability to move around the genome, though, in most cases, they do not produce mature viral particles. Viral-like retroelements contain **long terminal repeats (LTRs)** at both ends of the element (Figure 17.13c). The LTRs are typically a few hundred nucleotides in length. Like their viral counterparts, viral-like retroelements encode virally related proteins such as reverse transcriptase and integrase that are needed for the transposition process.

By comparison, **nonviral-like retroelements** appear less like retroviruses in their sequence, although some similarity, such as the occurrence of a reverse transcriptase gene, may be present (Figure 17.13c). Many nonviral-like retroelements, however, do not share any sequence similarity with known viruses. Instead, some nonviral-like retroelements are evolutionarily derived from normal eukaryotic genes. For example, the *Alu* family of repetitive sequences found in humans is derived from a single ancestral gene known as the *7SL RNA* gene (a component of the complex called signal recognition particle, which is described in Chapter 13). This gene sequence has been copied by retrotransposition to achieve the current number of approximately 1,000,000 copies.

Transposable elements are considered to be complete or **autonomous elements** when they contain all the information necessary for transposition or retrotransposition to take place. However, TEs are often incomplete or nonautonomous. A **nonautonomous element** typically lacks a gene such as transposase or reverse transcriptase that is necessary for transposition. The *Ds* locus described in the experiment of Figure 17.11 is a nonautonomous element, because it lacks a transposase gene. An element that is similar to *Ds* but contains a functional transposase gene is called the *Ac* locus or *Ac* element, which stands for Activator element. As mentioned earlier, an *Ac* locus provides a transposase gene that enables *Ds* to transpose. Therefore, nonautonomous TEs such as *Ds* can transpose only when the *Ac* locus is present at another region in the genome.

Transposase Catalyzes the Excision and Insertion of Transposable Elements

Now that we have an understanding of the typical organization of transposable elements, let's examine the steps of the transposition process. The enzyme transposase catalyzes the removal of a TE from its original site in the chromosome and its subsequent insertion at another location. A general scheme for simple transposition is shown in **Figure 17.14**. Transposase binds to the inverted repeat sequences at the ends of the TE and brings them close together. The DNA is cleaved at the ends of the TE, excising it from its original site within the chromosome. The transposase carries the TE to a new site and cleaves the target DNA sequence at staggered recognition sites. The TE is then inserted and ligated to the target DNA.

As noted in Figure 17.14, the ligation of the transposable element into its new site initially leaves short gaps in the target DNA. Notice that the DNA sequences in these gaps are complementary to each other (in this case, ATGCT and TACGA). Therefore, when they are filled in by DNA gap repair synthesis, the DNA base pair sequences that flank both ends of the TE are identical. These direct repeats are common features found adjacent to all TEs (see Figure 17.13).

Although the transposition process depicted in Figure 17.14 does not directly alter the number of transposable elements, simple transposition is known to increase the number of TEs in genomes, in some cases to fairly high levels. How can this happen? The answer is that transposition often occurs around the time of DNA replication (**Figure 17.15**). After a replication fork has passed a region containing a TE, two TEs will be found behind the fork—one in each of the replicated regions. One of these TEs could then transpose from its original location into a region ahead of the replication fork. After the replication fork has passed this second region and DNA replication is completed, two TEs will be found in one of the chromosomes and one TE in the other chromosome. In this way, simple transposition can lead to an increase in TEs. We will discuss the biological significance of transposon proliferation later in this chapter.

Replicative Transposition Requires Both Transposase and Resolvase

Replicative transposition has been studied in several bacterial transposons and in bacteriophage μ (mu), which behaves like a

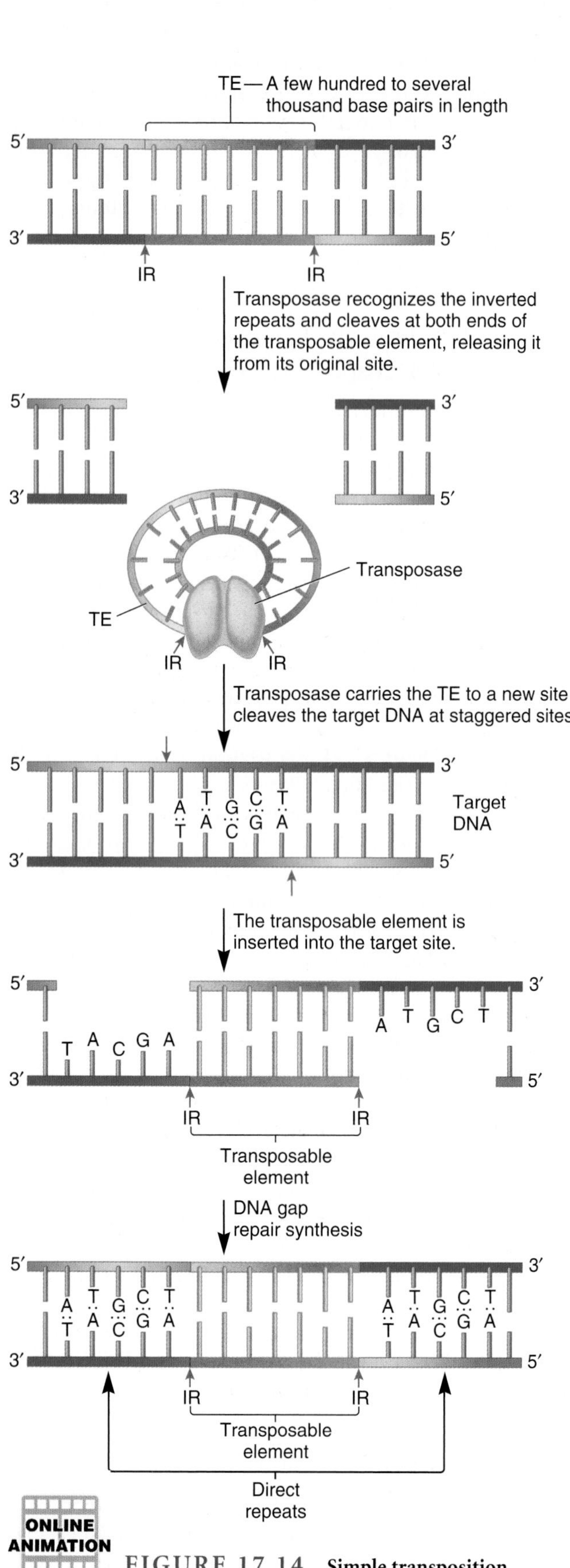

FIGURE 17.14 **Simple transposition.**

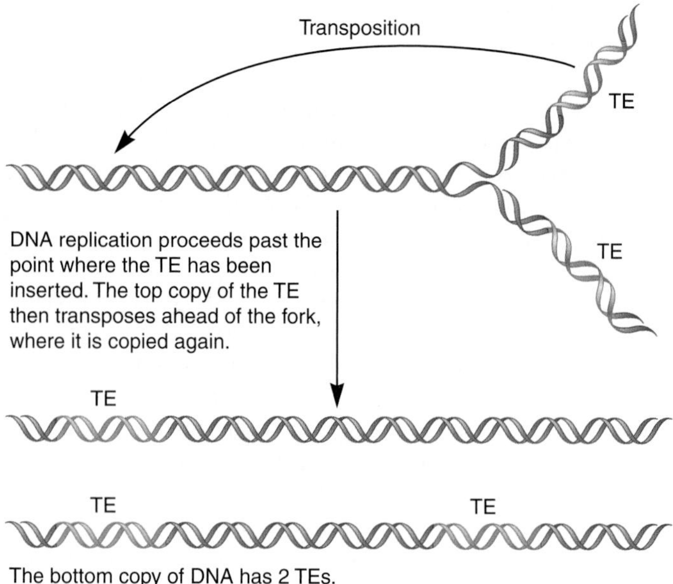

FIGURE 17.15 **Increase in TE copy number via simple transposition.** In this example, a TE that has already been replicated transposes to a new site that has not yet replicated. Following the completion of DNA replication, the TE has increased in number.

transposon. The net result of replicative transposition is that a transposable element occurs at a new site, and a TE also remains in its original location. **Figure 17.16** describes a model for replicative transposition between two circular DNA molecules. One DNA molecule already has a TE, whereas the other does not. In this mechanism, transposase initially makes one cut at each end of the TE and two cuts in the target DNA. Note that this differs from simple transposition, in which the transposase makes four cuts and completely removes the TE from its original site (refer back to Figure 17.14). In replicative transposition, the TE is left at its original location.

Following ligation, both the target DNA and the transposable element have a long gap. DNA gap repair synthesis copies the target DNA gap as well as the TE. This creates two copies of the TE within a large circular molecule known as a cointegrant. The enzyme resolvase catalyzes homologous recombination within the TEs so that the cointegrant can be resolved into two separate DNA molecules. One of these molecules contains the TE in its original location, and the other has a TE at a new location.

Retroelements Use Reverse Transcriptase and Integrase for Retrotransposition

Thus far, we have considered how DNA elements—transposons—can move throughout the genome. By comparison, retroelements use an RNA intermediate in their transposition mechanism. As shown in **Figure 17.17**, the movement of retroelements also requires two key enzymes: reverse transcriptase and integrase. In this example, the cell already contains a retroelement known as the *Alu* sequence within its genome. This retroelement, which

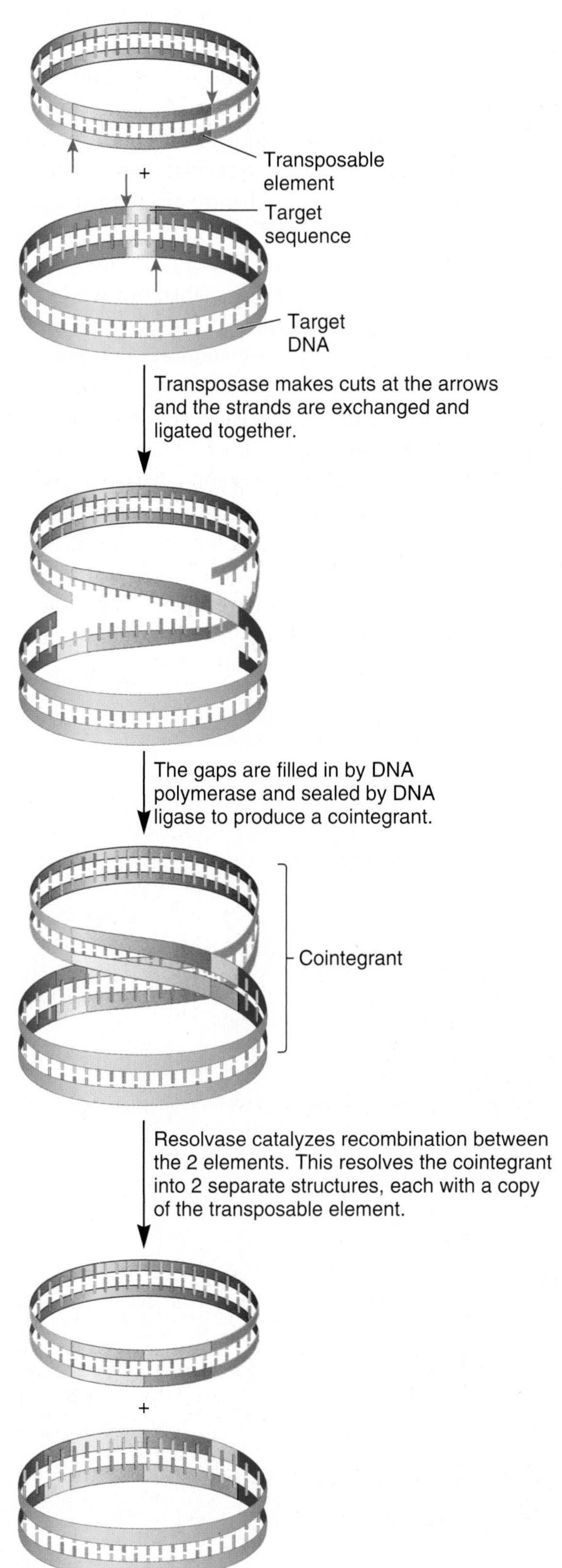

FIGURE 17.16 Replicative transposition. The end result of this process is that one TE remains in its original site and another TE is inserted at a new site.

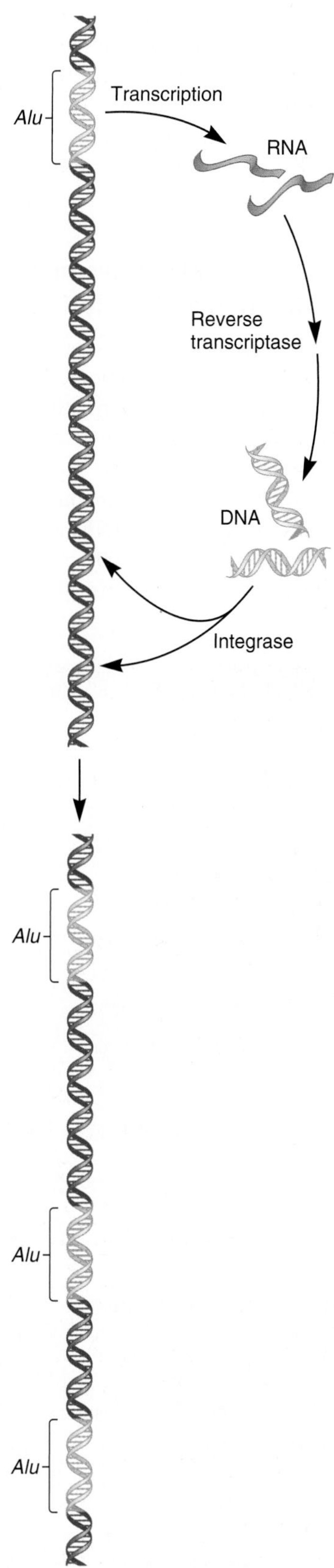

FIGURE 17.17 Retrotransposition.

behaves like a gene, is transcribed into RNA. In a series of steps, **reverse transcriptase** uses this RNA as a template to synthesize a double-stranded DNA molecule. The ends of the double-stranded DNA are then recognized by **integrase,** which catalyzes the insertion of the DNA into the target chromosomal DNA. The integration of retroelements can occur at many locations within the genome. Furthermore, because a single retroelement can be copied into many RNA transcripts, retroelements may accumulate rapidly within a genome.

Transposable Elements May Have Important Influences on Mutation and Evolution

Over the past few decades, researchers have found that transposable elements probably occur in the genomes of all species. **Table 17.2** describes a few TEs that have been studied in great detail. As discussed in Chapter 10, the genomes of eukaryotic species typically contain moderately and highly repetitive sequences. In some cases, these repetitive sequences are due to the proliferation of TEs. In mammals, for example, **LINEs** are long interspersed elements that are usually 1,000 to 5,000 bp in length and found in 20,000 to 100,000 copies per genome. **SINEs** are short interspersed elements that are less than 500 bp in length. A specific example of a SINE is the *Alu* sequence, present in about 1 million copies in the human genome. About 10% of the human genome is composed of this particular TE! The *Alu* sequence continues to proliferate in the human genome but at a fairly low rate. In about 1 live birth in 200, an *Alu* sequence has been inserted into a new site in the human genome. On rare occasions, a new insertion can disrupt a gene and cause phenotypic abnormalities.

The relative abundance of TEs varies widely among different species. As shown in **Table 17.3**, TEs can be quite prevalent in amphibians, mammals, and flowering plants, but tend to be less abundant is simpler organisms such as bacteria and yeast. The biological significance of TEs in the evolution of prokaryotic and eukaryotic species remains a matter of debate. According to the **selfish DNA hypothesis,** TEs exist because they contain characteristics that allow them to multiply within the chromosomal DNA of living cells. In other words, they resemble parasites in the sense that they inhabit a cell without offering any selective advantage to the organism. They can proliferate as long as they do not harm the organism to the extent that they significantly disrupt survival.

Alternatively, other geneticists have argued that most transpositional events are deleterious. Therefore, TEs would be eliminated

TABLE 17.2

Examples of Transposable Elements

Element	Type	Approximate Length (bp)	Description
Bacterial			
IS1	Insertion sequence	768	An insertion sequence that is commonly found in 5–8 copies in *E. coli*.
Mu	Replicative transposon	36,000	A true virus that can insert itself anywhere in the *E. coli* chromosome. Its name, *Mu*, is derived from its ability to insert into genes and mutate them.
Tn10	Composite transposon	9,300	One of many different bacterial transposons that carries antibiotic resistance.
Tn951	Composite transposon	16,600	A transposon that provides bacteria with genes that allow them to metabolize lactose.
Yeast			
Ty elements	Viral-like retroelement	6,200	A retroelement found in *S. cerevisiae* at about 35 copies per genome.
Drosophila			
P elements	Simple transposon	500–3,000	A transposon that may be found in 30–50 copies in P strains of *Drosophila*. It is absent from M strains.
Copia-like elements	Viral-like retroelement	5,000–8,000	A family of *copia*-like elements found in *Drosophila*, which vary slightly in their lengths and sequences. Typically, each family member is found at about 5–100 copies per genome.
Humans			
Alu sequence	Nonviral-like retroelement	300	A SINE that is abundantly interspersed throughout the human genome.
L1	Viral-like retroelement	6,500	A LINE found in 50,000–100,000 copies in the human genome.
Plants			
Ac/*Ds*	Simple transposon	4,500	*Ac* is an autonomous transposon found in corn and other plant species. It carries a transposase gene. *Ds* is a nonautonomous version that lacks a functional transposase gene.

TABLE 17.3

Abundance of TEs in the Genomes of Selected Species

Species	Percentage of the Total Genome Composed of Transposable Elements*
Frog (*Xenopus laevis*)	77
Corn (*Zea mays*)	60
Human (*Homo sapiens*)	45
Mouse (*Mus musculus*)	40
Fruit fly (*Drosophila melanogaster*)	20
Nematode (*Caenorhabditis elegans*)	12
Yeast (*Saccharomyces cerevisiae*)	4
Bacterium (*Escherichia coli*)	0.3

*In some cases, the abundance of TEs may vary somewhat among different strains of the same species. The values reported here are typical values.

TABLE 17.4

Possible Consequences of Transposition

Consequence	Cause
Chromosome Structure	
Chromosome breakage	Excision of a TE.
Chromosomal rearrangements	Homologous recombination between TEs located at different positions in the genome.
Gene Expression	
Mutation	Incorrect excision of TEs.
Gene inactivation	Insertion of a TE into a gene.
Alteration in gene regulation	Transposition of a gene next to regulatory sequences or the transposition of regulatory sequences next to a gene.
Alteration in the exon content of a gene	Insertion of exons into the coding sequence of a gene via TEs. This phenomenon is called exon shuffling.
Gene duplications	Creation of a composite transposon that transposes to another site in the genome.

from the genome by natural selection if they did not also offer a compensating advantage. Several potential advantages have been suggested. For example, TEs may cause greater genetic variability by promoting recombination. In addition, bacterial TEs often carry an antibiotic-resistance gene that provides the organism with a survival advantage (refer back to Figure 17.13a). Researchers have also suggested that transposition may cause the insertion of exons into the coding sequences of structural genes. This phenomenon, called **exon shuffling,** may lead to the evolution of genes with more diverse functions, as described in Chapter 24.

While this controversy remains unresolved, it is clear that transposable elements can rapidly enter the genome of an organism and proliferate quickly. In *Drosophila melanogaster*, for example, a TE known as the P element was probably introduced into this species in the 1950s. Laboratory stocks of *D. melanogaster* collected prior to this time do not contain P elements. Remarkably, in the last 50 years, the P element has expanded throughout *D. melanogaster* populations worldwide. The only strains without the P element are laboratory strains collected prior to the 1950s. This observation underscores the surprising ability of TEs to infiltrate a population of organisms.

Transposable elements have a variety of effects on chromosome structure and gene expression (**Table 17.4**). Because many of these outcomes are likely to be harmful, transposition is usually a highly regulated phenomenon that occurs only in a few individuals under certain conditions. Agents such as radiation, chemical mutagens, and hormones stimulate the movement of TEs. When it is not carefully regulated, transposition is likely to be potently detrimental. For example, in *D. melanogaster*, if females that lack P elements (M strain females) are crossed with males that contain numerous P elements (P strain males), the egg cells allow the P elements inherited via the sperm to transpose at a high rate. The resulting hybrid offspring exhibit a variety of abnormalities, which include a high rate of sterility, mutation, and chromosome breakage. This deleterious outcome, which is called **hybrid dysgenesis,** occurs because the P elements were able to insert into a variety of locations in the genome.

Transposons Have Become Important Tools in Molecular Biology

The unique and unusual features of transposons have made them an important experimental tool in molecular biology. For researchers, the introduction of a transposon into a cell is a convenient way to alter the expression of a particular gene. If a transposon "hops" into a gene, it is likely to inactivate the gene's function. This phenomenon can be used to clone a particular gene in an approach known as **transposon tagging.** In this strategy, researchers use transposons in an attempt to clone genes.

An early example of transposon tagging involved an X-linked gene in *Drosophila* that affects eye color. This X-linked gene can exist in the wild-type (red) allele and a loss-of-function allele that causes a white-eye phenotype. In 1981, Paul Bingham, in collaboration with Robert Levis and Gerald Rubin, used transposon tagging to clone this gene (**Figure 17.18**). Prior to their cloning work, a wild-type strain of *Drosophila* had been characterized that carried a transposable element called *copia*. From this red-eyed strain, a white-eyed strain was obtained in which the *copia* element had transposed into a region on the X chromosome that corresponded to where the eye color gene mapped.

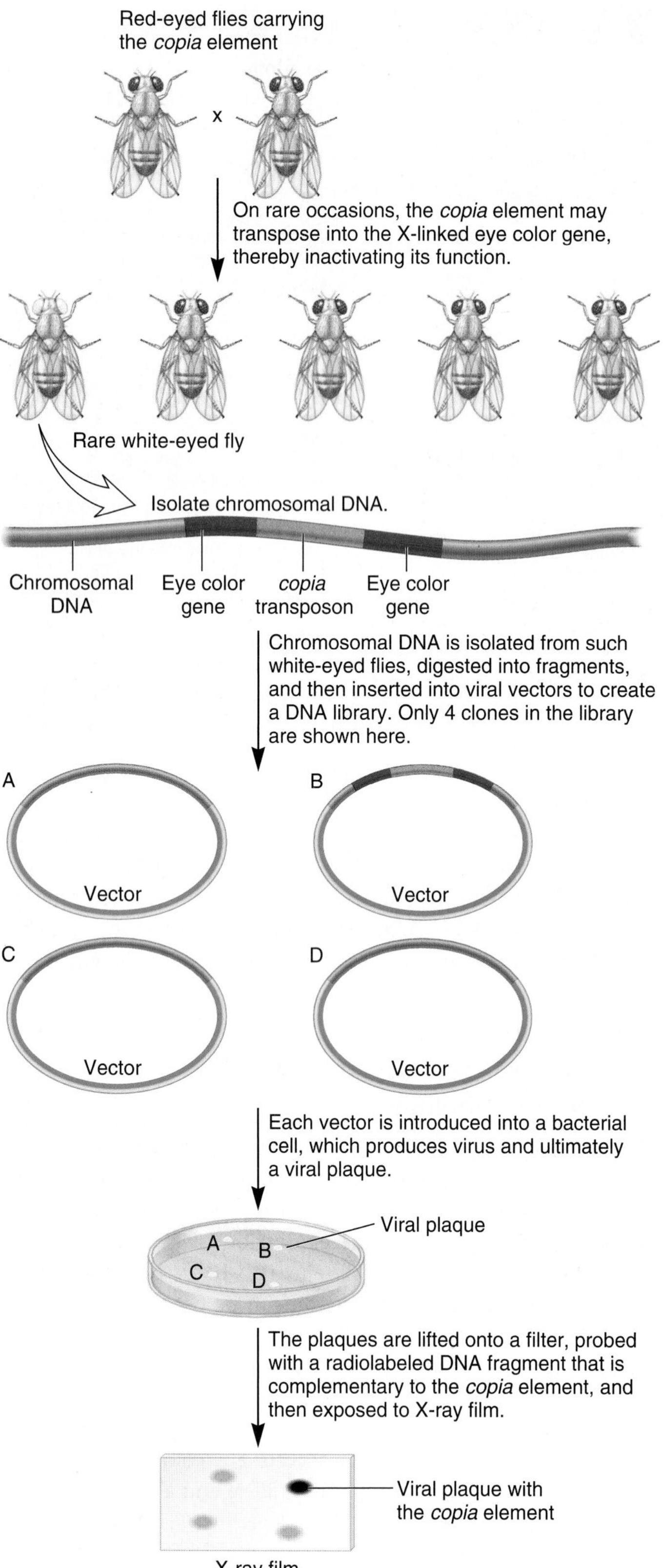
Red-eyed flies carrying the *copia* element
x
On rare occasions, the *copia* element may transpose into the X-linked eye color gene, thereby inactivating its function.
Rare white-eyed fly
Isolate chromosomal DNA.
Chromosomal DNA
Eye color gene
copia transposon
Eye color gene
Chromosomal DNA is isolated from such white-eyed flies, digested into fragments, and then inserted into viral vectors to create a DNA library. Only 4 clones in the library are shown here.
A
Vector
B
Vector
C
Vector
D
Vector
Each vector is introduced into a bacterial cell, which produces virus and ultimately a viral plaque.
A
B
C
D
Viral plaque
The plaques are lifted onto a filter, probed with a radiolabeled DNA fragment that is complementary to the *copia* element, and then exposed to X-ray film.
Viral plaque with the *copia* element
X-ray film

The researchers reasoned that the white-eye phenotype could be due to the insertion of the *copia* element into the wild-type gene, thereby inactivating it.

To clone the eye color gene, chromosomal DNA from this white-eyed strain was isolated, digested with restriction enzymes, and cloned into vectors that were derived from a naturally occurring virus (viral vectors). This procedure creates a DNA library, a collection of vectors that contain different pieces of chromosomal DNA (described in Chapter 18). If a transposon had "jumped" into the eye color gene, vectors that contain this gene will also contain the transposon sequence. In other words, the presence of the transposon tags the eye color gene. Each vector is introduced into a bacterial cell, which produces a virus and ultimately a viral plaque, a clear area where the bacteria have been lysed. A radiolabeled fragment of DNA that is complementary to the transposon sequence can be used as a probe to identify plaques that also contain the eye color gene. The method of using a probe to screen a DNA library is also described in Chapter 18. In the example of Figure 17.18, the method of transposon tagging was successful at cloning an eye color gene in *Drosophila*.

FIGURE 17.18 The procedure of transposon tagging.

Genes→Traits A white-eyed fly may occur due to the insertion of a transposable element into a gene that confers red eye color. As discussed in Chapter 13, the wild-type eye color gene encodes a protein that is necessary for red pigment production. When a TE inserts into this gene, it disrupts the coding sequence and thereby causes the gene to produce a nonfunctional protein. Therefore, no red pigment can be made, and a white-eye phenotype results. In many cases, transposons affect the phenotypes of organisms by inactivating individual genes.

CONCEPTUAL SUMMARY

Homologous recombination, an essential feature of all species, is a major category of genetic recombination. Sister chromatid exchange (SCE) is a type of homologous recombination that involves crossing over between sister chromatids and the exchange of identical genetic material. By comparison, homologous recombination can also occur when homologous chromosomes cross over. During this form of homologous recombination, homologous DNA regions align and exchange DNA segments. This enhances genetic variability by producing chromosomes with new combinations of alleles.

At the molecular level, several models have been proposed to explain the steps in homologous recombination. The Holliday model was the first example of a molecular explanation for the recombination process. This model was able to account for the phenomenon of gene conversion, in which one allele is converted to the allele on the homologous chromosome. More recently, other models, such as that proposed by Meselson and Radding and the double-strand break model, have more accurately described steps in the recombination pathway. In addition, much progress has been made toward identifying the many proteins that play important roles in recombination.

A second form of recombination is known as site-specific recombination, because the breakage and rejoining of nonhomologous DNA segments occur at particular DNA sequences. Site-specific recombination is responsible for the integration of certain bacteriophages such as λ into the host cell DNA. In addition, site-specific recombination within immunoglobulin genes is important in generating an astounding diversity in antibody polypeptides. During this process, regions in precursor immunoglobulin genes known as V, D, and J domains are randomly connected to each other via recombination. This mechanism enables the production of a very diverse array of antibody proteins.

In transposition, a third form of recombination, short segments of DNA known as transposable elements (TEs) integrate themselves into different locations within the genome. During simple transposition, a cut-and-paste mechanism occurs in which the TE is cut out of its original site and transferred to a new site. This process is catalyzed by transposase. In replicative transposition, a TE is duplicated, with one TE remaining in its original location and a new TE located at a new site. The process is catalyzed by the enzymes transposase and resolvase. In retrotransposition, a retroelement is transcribed into RNA. Reverse transcriptase uses this RNA as a template to synthesize DNA, which is then integrated into the chromosome by integrase. From an evolutionary viewpoint, all mechanisms of transposition can cause many different types of mutations.

EXPERIMENTAL SUMMARY

Homologous recombination, which may occur between sister chromatids or homologous chromosomes, can be detected by several techniques. Sister chromatid exchange (SCE) can be observed using staining methods that produce harlequin chromosomes, which distinguish the sister chromatids in which crossing over has occurred. At the molecular level, researchers have studied recombination that occurs between homologous chromosomes by identifying and characterizing the proteins and intermediates that facilitate the process. The Holliday model describes a molecular mechanism for the recombination process.

Likewise, researchers have elucidated the mechanisms of site-specific recombination and transposition using genetic and molecular techniques. McClintock conducted genetic crosses in corn that provided compelling evidence for the existence of transposable elements (TEs). The sequencing of TEs has revealed certain patterns of DNA sequences that are required for simple transposition, replicative transposition, and retrotransposition. Researchers have also identified the proteins necessary to promote the movement of TEs. In addition, molecular biologists have taken advantage of our knowledge of transposition and routinely use transposons to inactivate genes and to clone genes via transposon tagging.

PROBLEM SETS & INSIGHTS

Solved Problems

S1. Zickler was the first person to demonstrate gene conversion by observing unusual ratios in *Neurospora* octads. At first, it was difficult for geneticists to believe these results because they seemed to contradict the Mendelian concept that alleles do not physically interact with each other. However, work by Mary Mitchell provided convincing evidence that gene conversion actually takes place. She investigated three different genes in *Neurospora*. One *Neurospora* strain had three mutant alleles: *pdx-1* (pyridoxine-requiring), *pyr-1* (pyrimidine-requiring), and *col-4* (a mutation that affected growth morphology). The *pdx-l* gene had been previously shown to map between the *pyr-1* and *col-4* genes. As shown here, Mitchell crossed this strain to a wild-type *Neurospora* strain:

pyr-1 pdx-1 col-4 × *pyr-1*$^+$ *pdx-1*$^+$ *col-4*$^+$

She first analyzed many octads with regard to their requirement for pyridoxine. Out of 246 octads, two of them had an aberrant ratio in which two spores were *pdx-1*, and six were *pdx-1*$^+$. These same spores were then analyzed with regard to the other two genes. In both cases, the aberrant asci gave a normal 4:4 ratio of *pyr-1*:*pyr-1*$^+$ and *col-4*:*col-4*$^+$. Explain these results.

Answer: These results can be explained by gene conversion. The gene conversion took place in a limited region of the chromosome (within the *pdx-1* gene), but it did not affect the flanking genes (*pyr-1* and *col-4*) located on either side of the *pdx-1* gene. In the asci containing two *pdx-1* alleles and six *pdx-1*$^+$ alleles, a crossover occurred during meiosis I in the region of the *pdx-1* gene. Gene conversion changed the *pdx-1* allele into the *pdx-1*$^+$ allele. This gene conversion could have occurred by two mechanisms. If branch migration occurred across the *pdx-1* gene, a heteroduplex may have formed, and this could be repaired by mismatch DNA repair, as described in Figure 17.6. In the aberrant asci with two *pdx-1* and six *pdx-1*$^+$ alleles, the *pdx-1* allele was converted to *pdx-1*$^+$. Alternatively, gene conversion of *pdx-1* into *pdx-1*$^+$ could have taken place via gap repair synthesis, as described in Figure 17.7. In this case, the *pdx-1* allele would have been digested away, and the DNA encoding the *pdx-1*$^+$ allele would have migrated into the digested region and provided a template to make a copy of the *pdx-1*$^+$ allele. (Note: Since this pioneering work, additional studies have shed considerable light concerning the phenomenon of gene conversion. It occurs at a fairly high rate in fungi, approximately 0.1 to 1% of the time. It is not due to new mutations occurring during meiosis.)

S2. Recombination involves the pairing of identical or similar sequences, followed by crossing over and the resolution of the intertwined helices. On rare occasions, the direct repeats or the inverted repeats within a single transposable element can align and undergo homologous recombination. What are the consequences when the direct repeats recombine? What are the consequences when the inverted repeats recombine?

Answer:

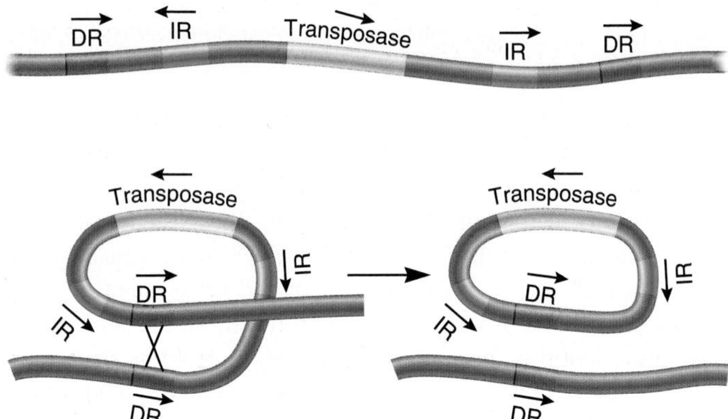

Most of the transposable element has been excised.

(a) Recombination between direct repeats

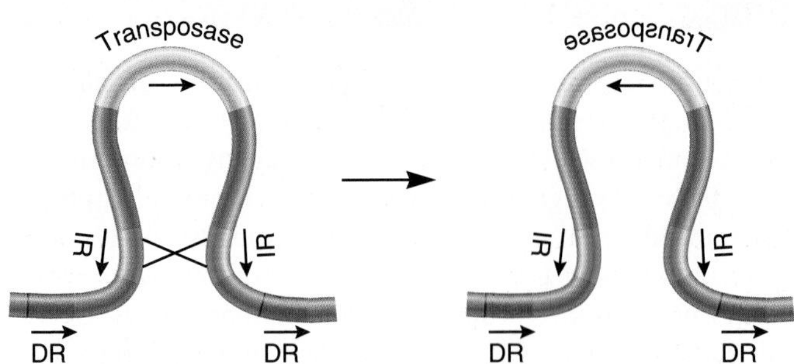

The sequence within the transposable element has been inverted. Note that the transposase gene has changed to the opposite direction.

(b) Recombination between inverted repeats

S3. A schematic drawing of an uncrossed Holliday junction is shown here. One chromatid is shown in red, and the homologous chromatid is shown in blue. The red chromatid carries a dominant allele labeled *A* and a recessive allele labeled *b*, whereas the blue chromatid carries a recessive allele labeled *a* and a dominant allele labeled *B*.

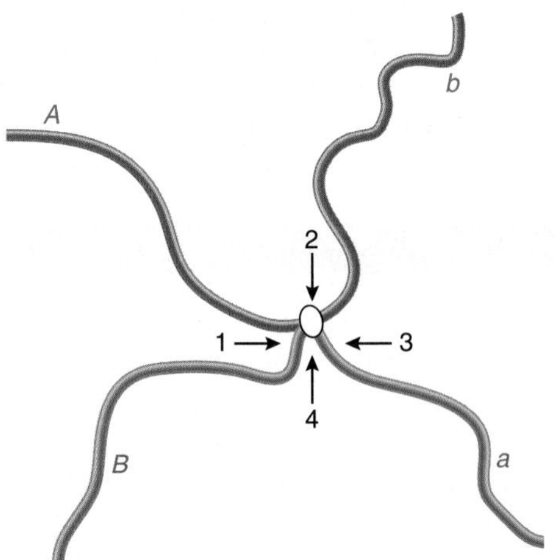

Where would the breakage of the crossed strands have to occur to get recombinant chromosomes? Would it have to occur at sites 1 and 3, or at sites 2 and 4? What would be the genotypes of the two recombinant chromosomes?

Answer:

Breakage would have to occur at the arrows labeled 2 and 4. This would connect the *A* allele with the *B* allele. The *a* allele in the other homologue would become connected with the *b* allele. In other words, one chromosome would be *AB*, and the homologue would be *ab*.

Conceptual Questions

C1. Describe the similarities and differences between homologous recombination involving sister chromatid exchange (SCE) versus that involving homologues. Would you expect the same types of proteins to be involved in both processes? Explain.

C2. The molecular mechanism of SCE is similar to homologous recombination between homologues except that the two segments of DNA are sister chromatids instead of homologous chromatids. If branch migration occurs during SCE, will a heteroduplex be formed? Explain why or why not. Can gene conversion occur during sister chromatid exchange?

C3. Which steps in the double-strand break model for recombination would be inhibited if the following proteins were missing? Explain the function of each protein required for the step that is inhibited.

A. RecBCD

B. RecA

C. RecG

D. RuvABC

C4. What are the two molecular mechanisms that can explain the phenomenon of gene conversion? Would both of these mechanisms occur in the double-strand break model?

C5. Is homologous recombination an example of mutation? Explain.

C6. What are recombinant chromosomes? How do they differ from the original parental chromosomes from which they are derived?

C7. In the Holliday model for homologous recombination (see Figure 17.4), the resolution steps can produce recombinant or nonrecombinant chromosomes. Explain how this can occur.

C8. What is gene conversion?

C9. Make a list of the differences between the Holliday model and the double-strand break model.

C10. In recombinant chromosomes, where is gene conversion likely to have taken place: near the breakpoint or far away from the breakpoint? Explain.

C11. What are the events that RecA protein facilitates?

C12. According to the double-strand break model, does gene conversion necessarily involve DNA mismatch repair? Explain.

C13. What type of DNA structure is recognized by RecG and RuvABC? Do you think these proteins recognize DNA sequences? Be specific about what type(s) of molecular recognition these proteins can perform.

C14. Briefly describe three ways that antibody diversity is produced.

C15. Describe the function of RAG1 and RAG2 proteins and NHEJ proteins.

C16. According to the scenario shown in Figure 17.9, how many segments of DNA (one, two, or three) are removed during site-specific recombination within the gene that encodes the κ (kappa) light chain for IgG proteins? How many segments are spliced out of the pre-mRNA?

C17. Describe the role that integrase plays during the insertion of λ DNA into the host chromosome.

C18. If you were examining a sequence of chromosomal DNA, what characteristics would cause you to believe that the DNA contained a transposable element?

C19. According to the model for replicative transposition shown in Figure 17.16, does the transposable element replicate before or after it transposes? Explain your answer.

C20. Why does transposition always produce direct repeats in the chromosomal DNA?

C21. Which types of transposable elements have the greatest potential for proliferation: insertion sequences, replicative TEs, or retroelements? Explain your choice.

C22. Do you consider transposable elements to be mutagens? Explain.

C23. Let's suppose that a species of mosquito has two different types of simple transposons that we will call X elements and Z elements. The X elements appear quite stable. When analyzing a population of 100 mosquitoes, every mosquito has six X elements, and they are always located in the same chromosomal locations among different individuals. In contrast, the Z elements seem to "move around" quite a bit. Within the same 100 mosquitoes, the number of Z elements ranges from 2 to 14, and the locations of the Z elements tend to vary considerably among different individuals. Explain how one simple transposon can be stable and another simple transposon can be mobile, within the same group of individuals.

C24. This chapter describes five different types of transposable elements including insertion sequences, composite transposons, replicative transposons, viral-like retroelements, and nonviral-like retroelements. Which of these five types of TEs would have the following features?

A. Require reverse transcriptase to transpose

B. Require transposase to transpose

C. Are flanked by direct repeats

D. Have inverted repeats

C25. What features distinguish a transposon from a retroelement? How are their sequences different and how are their mechanisms of transposition different?

C26. Solved problem S2 illustrates the consequences of crossing over between the direct and inverted repeats within a single transposable element. The drawing here shows the locations of two copies of the same transposable element within a single chromosome. The chromosome is depicted according to its G banding pattern. (Note: G bands are illustrated in Figure 8.1.) The direct and inverted repeats are labeled 1, 2, 3, and 4, from left to right.

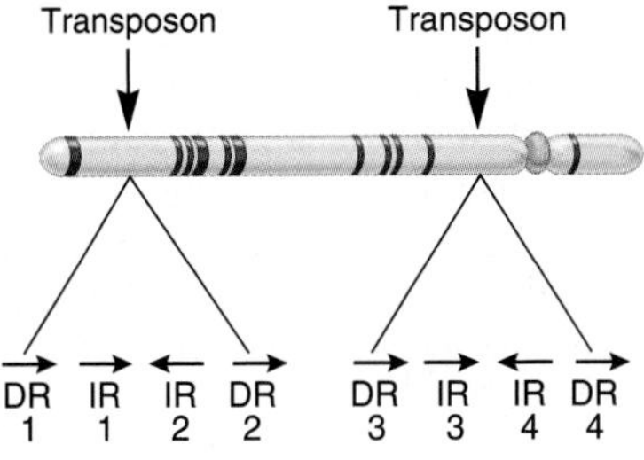

Draw the end result of recombination between the following sequences. Your drawing should include the banding pattern of the resulting chromosome.

A. DR-1 and DR-4

B. IR-1 and IR-4

C27. What is the difference between an autonomous and a nonautonomous transposable element? Is it possible for nonautonomous TEs to move? If yes, explain how.

C28. An operon in the bacterium *Salmonella typhimurium* has the following arrangement:

DR IR IR DR

↓ ↓ promoter→ ↓ ↓ *H2 rH1*

The promoter for this operon is contained within a transposable element. The *H2* gene encodes a protein that is part of the bacterial flagellum. The *rH1* gene encodes a repressor protein that represses the *H1* gene, which is found at another location in the bacterial chromosome. The *H1* gene also encodes a flagellar protein. When the promoter is found in the arrangement shown here, the *H2/rH1* operon is turned on. This results in flagella that contain the *H2* protein. The *H1* protein is not made because the *rH1* repressor prevents the transcription of the *H1* gene. At a frequency of approximately 1 in 10,000 (which is much higher than the spontaneous mutation rate), this strain of bacterium can "switch" its expression so that *H2* is turned off and *H1* is turned on. Bacteria that have *H1* turned on and *H2* turned off can also switch back to having *H2* turned on and *H1* turned off. This switch also occurs at a frequency of about 1 in 10,000. Based on your understanding of transposons and recombination, explain how switching occurs in *Salmonella typhimurium*. Hint: Take a look at solved problem S2.

C29. The occurrence of multiple transposons within the genome of organisms has been suggested as a possible cause of chromosomal rearrangements such as deletions, translocations, and inversions. How could the occurrence of transposons promote these kinds of structural rearrangements?

Experimental Questions

E1. With the harlequin staining technique, one sister chromatid appears to fluoresce more brightly than the other. Why?

E2. In the data shown here, harlequin staining was used to determine the frequency of sister chromatid exchanges (SCEs) in the presence of a suspected mutagen.

	Frequency of SCEs/Chromosome
No mutagen	0.67
With suspected mutagen	14.7

Would you conclude that this substance is a mutagen?

E3. In the experiment described in experimental question E2, at what point would you need to add the mutagen: before the first round of DNA replication, after the first round but before the second round, or after the second round?

E4. Let's suppose that a researcher followed the protocol described in the experiment of Figure 17.3 but exposed the cells to BrdU for three cell generations instead of two. Near the end of growth, the cells were exposed to colcemid to prevent them from completing mitosis following the third round of DNA replication. What would be the expected results if a parental cell contained a total of four chromosomes (two homologous pairs) and a single sister chromatid exchange occurred after the second replication? Your drawing should show four cells that contain four condensed chromosomes in each cell.

E5. Based on your understanding of the experiment of Figure 17.3, does BrdU enhance the binding of Giemsa to the chromatids or inhibit the binding of Giemsa? Explain.

E6. Briefly explain how McClintock determined that *Ds* was occasionally moving from one chromosomal location to another. Discuss the type of data she examined to arrive at this conclusion.

E7. In the data of Figure 17.3, is the solid red phenotype due to chromosome breakage or the excision of a transposable element? Explain how you have arrived at your conclusion.

E8. In your own words, explain the term transposon tagging.

E9. Tumor-suppressor genes are normal human genes that prevent uncontrollable cell growth. Starting with a normal laboratory human cell line, describe how you could use transposon tagging to identify tumor-suppressor genes. Note: When a transposable element hops into a tumor-suppressor gene, it may cause uncontrolled cell growth. This is detected as a large clump of cells among a normal monolayer of cells.

E10. As discussed in the experiment of Figure 17.11, the presence of a transposon can create a mutable site or locus that is subject to frequent chromosome breakage. Why do you think a transposon creates a mutable site? If chromosome breakage occurs, do you think the transposon has moved somewhere else? How would you experimentally determine if it has?

E11. Gerald Rubin and Allan Spradling devised a method of introducing a transposon into *Drosophila*. This approach has been important for the transposon tagging of many *Drosophila* genes. They began with a P element that had been cloned on a plasmid. (Note: Methods of cloning are described in Chapter 18.) Using cloning methods, they inserted the wild-type allele for the *rosy* gene into the P element in this plasmid. The recessive allele, *rosy*, results in a rosy eye color, while the wild-type allele, $rosy^+$, produces red eyes. This plasmid also has an intact transposase gene. The cloned DNA is shown here:

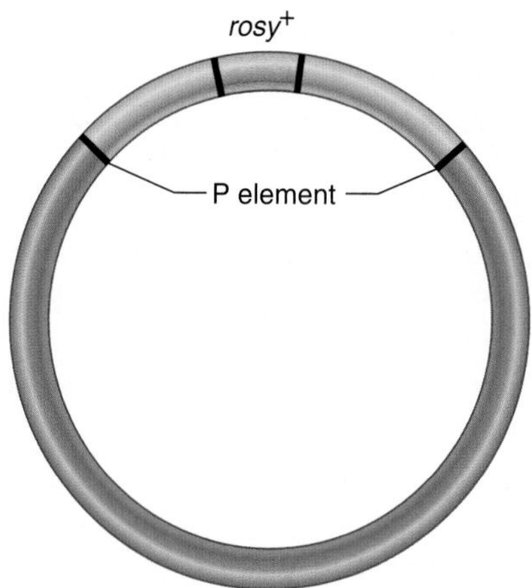

They used a micropipette to inject this DNA into regions of embryos that would later become reproductive cells. These embryos were originally homozygous for the recessive *rosy* allele. However, the P element carrying the *rosy*$^+$ allele could "hop" out of the plasmid and into a chromosome of the cells that were destined to become germ cells (i.e., sperm or egg cells). After they had matured to adults, these flies were then mated to flies that were homozygous for the recessive rosy allele. If offspring inherited a chromosome carrying the P element with the *rosy*$^+$ gene, such offspring would have red eyes. Therefore, the phenotype of red eyes provided a way to identify offspring that had a P element insertion.

Now here is the question. Let's suppose you were interested in identifying genes that play a role in wing development. Outline the experimental steps you would follow, using the plasmid with the P element containing the *rosy*$^+$ gene, as a way to transposon tag genes that play a role in wing development. Note: You should assume that the inactivation of a gene involved in wing development would cause an abnormality in wing shape. Also keep in mind that most P element insertions inactivate genes and may be inherited in a recessive manner.

Questions for Student Discussion/Collaboration

1. Make a list of the similarities and differences among homologous recombination, site-specific recombination, and transposition.
2. If no homologous recombination of any kind could occur, what would be the harmful and beneficial consequences?
3. Based on your current knowledge of genetics, discuss whether or not you think the selfish DNA hypothesis is correct.

Note: All answers appear at the website for this textbook; the answers to even-numbered questions are in the back of the textbook.

www.mhhe.com/brookergenetics3e

Visit the Online Learning Center for practice tests, answer keys, and other learning aids for this chapter. Enhance your understanding of genetics with our interactive exercises, quizzes, animations, and much more.

18 RECOMBINANT DNA TECHNOLOGY

CHAPTER OUTLINE

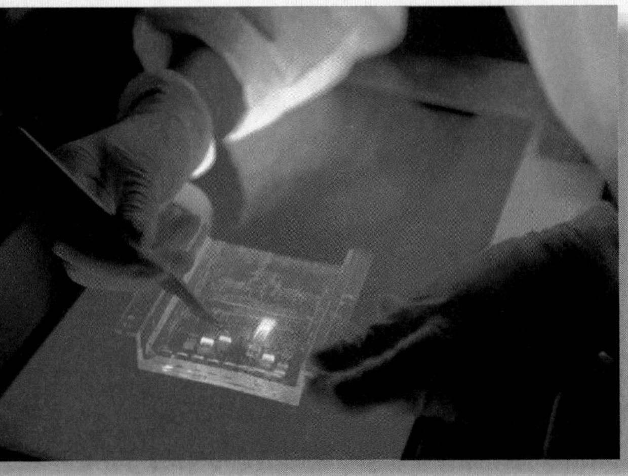

Visualization of DNA bands on a gel. DNA can be cut into fragments that can be separated via gel electrophoresis and visualized by staining. The cutting and pasting of DNA fragments allows researchers to clone genes.

PART V

Genetic Technologies

Recombinant DNA technology is the use of in vitro molecular techniques to isolate and manipulate fragments of DNA. In the early 1970s, the first successes in making recombinant DNA molecules were accomplished independently by two groups at Stanford University: David Jackson, Robert Symons, and Paul Berg; and Peter Lobban and A. Dale Kaiser. Both groups were able to isolate and purify pieces of DNA in a test tube and then covalently link DNA fragments from two different sources. In other words, they constructed molecules called **recombinant DNA molecules.** Shortly thereafter, researchers were able to introduce such recombinant DNA molecules into living cells. Once inside a host cell, the recombinant molecules can be replicated to produce many identical copies of a gene—a process called gene cloning.

Recombinant DNA technology and gene cloning have enabled geneticists to probe relationships between gene sequences and phenotypic consequences, and thereby have been fundamental to our understanding of gene structure and function. Most researchers in molecular genetics are familiar with recombinant DNA technology and apply it frequently in their work. Significant practical applications of recombinant DNA technology also have been developed. These include exciting advances such as gene therapy, screening for human diseases, recombinant vaccines, and the production of transgenic plants and animals in agriculture, in which a cloned gene from one species is transferred to some other species. Transgenic organisms have also been important in basic research.

In this chapter, we will focus primarily on the use of recombinant DNA technology as a way to further our understanding of gene structure and function. We will look at the materials and molecular techniques used in gene cloning, examine the first gene cloning experiment, and explore polymerase chain reaction (PCR), which can make many copies of DNA in a defined region. We then explore techniques to identify a specific gene or gene product, and methods to detect the binding of proteins to DNA sequences. Finally, we will examine how scientists analyze and alter DNA sequences through the techniques of DNA sequencing, a method that enables researchers to determine the base sequence of a DNA strand, and site-directed mutagenesis, a procedure that allows researchers to make mutations within a cloned segment of DNA.

In Chapter 19, we will consider many of the practical applications that have arisen as a result of these technologies. Chapters 20 and 21 are devoted to genomics, the molecular analysis of many genes and even the entire genome of a species.

18.1 GENE CLONING

Molecular biologists want to understand how the molecules within living cells contribute to cell structure and function. Because proteins are the workhorses of cells and because they are the products of genes, most molecular biologists focus their attention on the structure and function of proteins or the genes that encode them. Researchers may focus their efforts on the study of just one or perhaps a few different genes or proteins. At the molecular level, this poses a daunting task. In eukaryotic species, any given cell can express thousands of different proteins, making the study of any single gene or protein akin to a "needle in a haystack" exploration. To overcome this truly formidable obstacle, researchers frequently take the approach of cloning the genes that encode their proteins of interest. The term **gene cloning** refers to the phenomenon of isolating and making many copies of a gene in order to study it in detail. The laboratory methods necessary to clone a gene were devised during the early 1970s. Since then, many technical advances have enabled gene cloning to become a widely used procedure among scientists, including geneticists, cell biologists, biochemists, plant biologists, microbiologists, evolutionary biologists, clinicians, and biotechnologists.

Table 18.1 summarizes some of the common uses of gene cloning. In modern molecular biology, the myriad of uses for gene cloning is remarkable. For this reason, gene cloning has provided the foundation for critical technical advances in a variety of disciplines including molecular biology, genetics, cell biology, biochemistry, and medicine. In this section, we will examine the two general strategies used to make copies of a gene–the insertion of a gene into a vector that is then propagated in living cells, and cloning via polymerase chain reaction. Later sections in this chapter, as well as Chapters 19 through 21, will consider many of the uses of gene cloning that are described in Table 18.1.

TABLE **18.1**

Some Uses of Gene Cloning

Technique	Description
Gene sequencing	Cloned genes provide enough DNA to subject the gene to DNA sequencing (described later in this chapter). The sequence of the gene can reveal the gene's promoter, regulatory sequences, and coding sequence. Gene sequencing is also important in the identification of alleles that cause cancer and inherited human diseases.
Site-directed mutagenesis	A cloned gene can be manipulated to change its DNA sequence. Mutations within genes can help to identify gene sequences such as promoters and regulatory elements. The study of a mutant gene can also help to elucidate its normal function and how its expression may affect the roles of other genes. Mutations in the coding sequence can reveal which amino acids are important for a protein's structure and function.
Gene probes	Labeled DNA strands from a cloned gene can be used as probes to identify similar or identical genes or RNA. These methods, known as Southern and Northern blot analysis, are described later in this chapter. Probes can also be used to localize genes within intact chromosomes (see Chapter 20). DNA probes are used in techniques such as DNA fingerprinting to identify criminals (see Chapter 24).
Expression of cloned genes	Cloned genes can be introduced into a different cell type or different species. The expression of cloned genes has many uses: *Research* 1. The expression of a cloned gene can help to elucidate its cellular function. 2. The coding sequence of a gene can be placed next to an active promoter and then introduced into a culture of cells that will express a large amount of the protein. This greatly aids in the purification of large amounts of protein that may be needed for biochemical or biophysical studies. *Biotechnology* 1. Cloned genes can be introduced into bacteria to make pharmaceutical products such as insulin (see Chapter 19). 2. Cloned genes can be introduced into plants and animals to make transgenic species with desirable traits (see Chapter 19). *Clinical trials* 1. Cloned genes have been used in clinical trials involving gene therapy (see Chapter 19).

Cloning Experiments May Involve Two Kinds of DNA Molecules: Chromosomal DNA and Vector DNA

If a scientist wants to clone a particular gene, the source of the gene is the chromosomal DNA of the species that carries the gene. For example, if the goal is to clone the rat β-globin gene, this gene is found within the chromosomal DNA of rat cells. In this case, therefore, the rat's chromosomal DNA is one type of DNA needed in a cloning experiment. To prepare chromosomal DNA, an experimenter first obtains cellular tissue from the organism of interest. The preparation of chromosomal DNA then involves the breaking open of cells and the extraction and purification of the DNA using biochemical techniques such as chromatography and centrifugation (see the Appendix for a description of these techniques).

Let's begin our discussion of gene cloning by considering a recombinant DNA technology in which a gene is removed from its native site within a chromosome and inserted into a smaller segment of DNA known as a **vector**—a small DNA molecule that can replicate independently of host cell chromosomal DNA and produce many identical copies of an inserted gene. The purpose of vector DNA is to act as a carrier of the DNA segment that is to be cloned. (The term vector comes from a Latin term meaning carrier.) In cloning experiments, a vector may carry a small segment of chromosomal DNA, perhaps only a single gene. By comparison, a chromosome carries many more genes, perhaps a few thousand. Like a chromosome, a vector is replicated when it resides within a living cell; a cell that harbors a vector is called a **host cell.** When a vector is replicated within a host cell, the DNA that it carries is also replicated.

The vectors commonly used in gene cloning experiments were derived originally from two natural sources: plasmids or viruses. Some vectors are **plasmids,** which are small circular pieces of DNA. As discussed in Chapter 6, plasmids are found naturally in many strains of bacteria and occasionally in eukaryotic cells. Many naturally occurring plasmids carry genes that confer resistance to antibiotics or other toxic substances. These plasmids are called **R factors.** Some of the plasmids used in modern cloning experiments were derived from R factors.

Plasmids also contain a DNA sequence, known as an **origin of replication,** that is recognized by the replication enzymes of the host cell and allow it to be replicated. The sequence of the origin of replication determines whether or not the vector can replicate in a particular type of host cell. Some plasmids have origins of replications with a broad host range. Such a plasmid can replicate in the cells of many different species. Alternatively, many vectors used in cloning experiments have a limited host cell range. In cloning experiments, researchers must choose a vector that replicates in the appropriate cell types for their experiments. For example, if researchers want a cloned gene to be propagated in *Escherichia coli,* the vector they employ must have an origin of replication that is recognized by this species of bacterium. The origin of replication also determines the copy number of a plasmid. Some plasmids are said to have strong origins because they can achieve high copy number—perhaps 100 to 200 copies of the plasmid per cell. Others have weaker origins whereby only one or two plasmids are found per cell.

Commercially available plasmids have been genetically engineered for effective use in cloning experiments. They contain unique sites where geneticists can easily insert pieces of DNA. Another useful feature of cloning vectors is that they often contain resistance genes that provide host bacteria with the ability to grow in the presence of a toxic substance. Such a gene is called a **selectable marker,** because the expression of the gene selects for the growth of the bacterial cells. Many selectable markers are genes that confer antibiotic resistance to the host cell. For example, the gene amp^R encodes an enzyme known as β-lactamase. This enzyme degrades ampicillin, an antibiotic that normally kills bacteria. Bacteria containing the amp^R gene can grow on media containing ampicillin because they can degrade it. In a cloning experiment where the amp^R gene is found within the plasmid, the growth of cells in the presence of ampicillin identifies bacteria that contain the plasmid. These bacteria can grow and form visible colonies on solid growth media. In contrast, those cells that do not contain the plasmid are ampicillin sensitive and will not grow.

An alternative type of vector used in cloning experiments is a viral vector. As discussed in Chapter 6, viruses can infect living cells and propagate themselves by taking control of the host cell's metabolic machinery. When a chromosomal gene is inserted into a viral genome, the gene will be replicated whenever the viral DNA is replicated. Therefore, viruses can be used as vectors to carry other pieces of DNA. When a virus is used as a vector, the researcher may analyze viral plaques rather than bacterial colonies. The characteristics of viral plaques are described in Chapter 6 (see Figure 6.14).

Molecular biologists use hundreds of different vectors in cloning experiments. **Table 18.2** provides a general description of several different types of vectors that are commonly used to clone small segments of DNA. In addition, other types of vectors, such as cosmids, BACs, and YACs, are used to clone large pieces of DNA. These vectors are described in detail in Chapter 20. Vectors designed to introduce genes into plants and animals are discussed in Chapter 19.

Enzymes Are Used to Cut DNA into Pieces and Join the Pieces Together

A key step in a cloning experiment is the insertion of chromosomal DNA into a plasmid or viral vector. This requires the cutting and pasting of DNA fragments. To cut DNA, researchers use enzymes known as **restriction endonucleases,** or **restriction enzymes.** The restriction enzymes used in cloning experiments

TABLE 18.2

Some Vectors Used in Cloning Experiments

Example	Type	Description
pBluescript	Plasmid	A type of vector like the one shown in Figure 18.2. It is used to clone small segments of DNA and propagate them in *E. coli.*
YEp24	Plasmid	This plasmid is an example of a **shuttle vector,** which can replicate in two different host species, *E. coli* and *Saccharomyces cerevisiae.* It carries an origin of replication for both species.
λgt11	Viral	This vector is derived from the bacteriophage λ, which is described in Chapter 14. λgt11 also contains a promoter from the *lac* operon. When fragments of DNA are cloned next to this promoter, the DNA is expressed in *E. coli.* This is an example of an **expression vector.** An expression vector is designed to clone the coding sequence of genes so they will be transcribed and translated correctly.
SV40	Viral	This virus naturally infects mammalian cells. Genetically altered derivatives of the SV40 viral DNA are used as vectors for the cloning and expression of genes in mammalian cells that are grown in the laboratory.
Baculovirus	Viral	This virus naturally infects insect cells. In a laboratory, insect cells can be grown in liquid media. Unlike many other types of cells, insect cells often express large amounts of proteins that are encoded by cloned genes. When researchers want to make a large amount of a protein, they can clone the gene that encodes the protein into baculovirus and then purify the protein from insect cells.

bind to a specific base sequence and then cleave the DNA backbone at two defined locations, one in each strand. Proposed by Werner Arber in the 1960s and discovered by Hamilton Smith and Daniel Nathans in the 1970s, restriction enzymes are made naturally by many different species of bacteria and protect bacterial cells from invasion by foreign DNA, particularly that of bacteriophages. Researchers have isolated and purified restriction enzymes from many bacterial species and now use them in their cloning experiments.

Figure 18.1 shows the role of a restriction enzyme, called *Eco*RI, in producing a recombinant DNA molecule. Certain types of restriction enzymes are useful in cloning because they digest DNA into fragments with "sticky ends." As shown in Figure 18.1, the sticky ends are single-stranded regions of DNA that can hydrogen bond to a complementary sequence of DNA from a different source. The ends of two different DNA pieces will hydrogen bond to each other because of their complementary sticky ends.

The hydrogen bonding between the sticky ends of DNA fragments promotes a temporary interaction between the two fragments. However, this interaction is not stable because it involves only a few hydrogen bonds between complementary bases. How can this interaction be made more permanent? The answer is that the sugar–phosphate backbones within the DNA

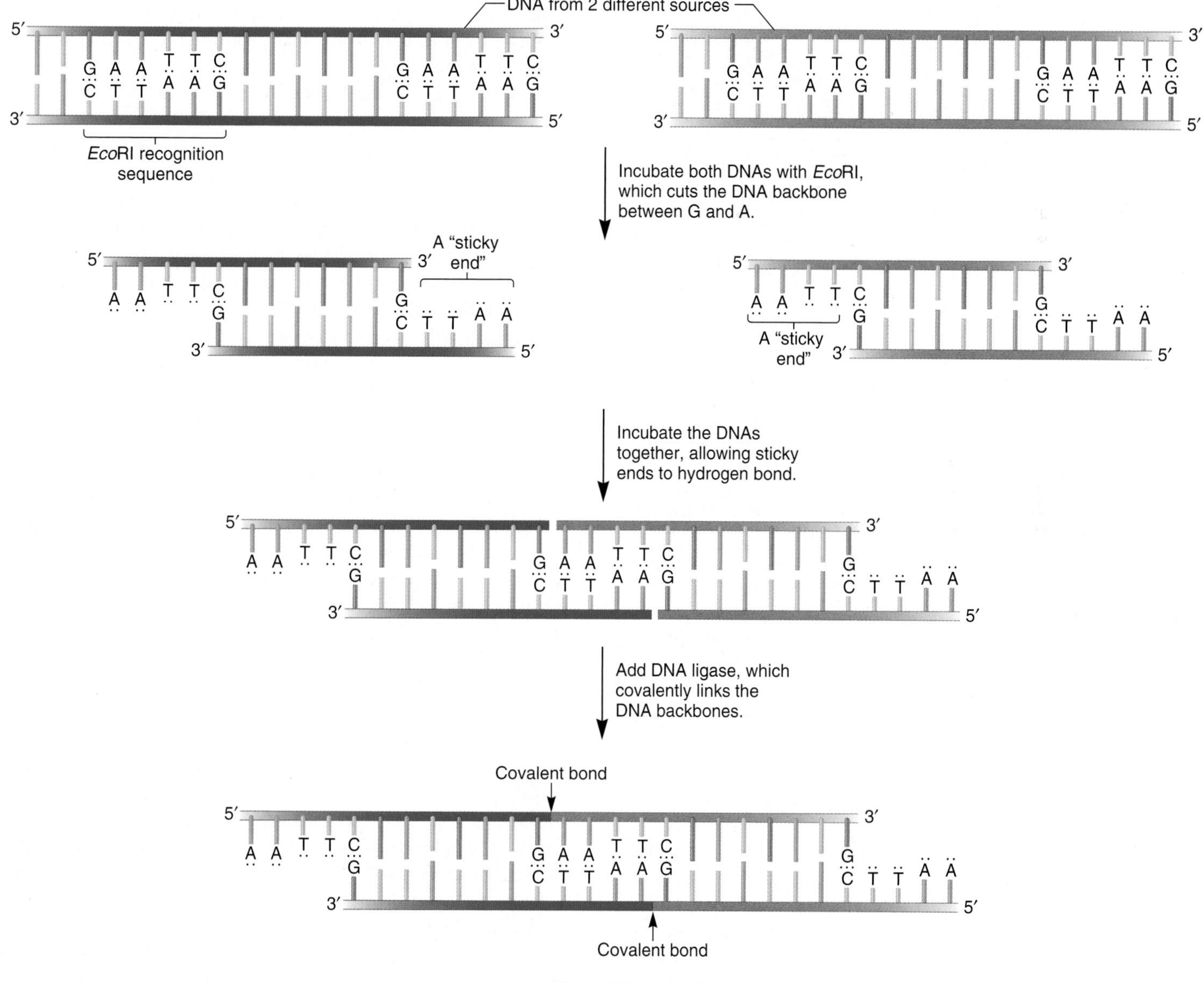

FIGURE 18.1 **The action of a restriction enzyme and the production of recombinant DNA.** The restriction enzyme *Eco*RI binds to a specific sequence, in this case 5'–GAATTC–3'. It then cleaves the DNA backbone between G and A, producing DNA fragments. The single-stranded ends of the DNA fragments can hydrogen bond with each other, because they have complementary sequences. The enzyme DNA ligase can then catalyze the formation of covalent bonds in the DNA backbones of the fragments.

TABLE 18.3

Some Restriction Enzymes Used in Gene Cloning

Restriction Enzyme*	Bacterial Source	Sequence Recognized†
*Bam*HI	*Bacillus amyloliquefaciens* H	↓ 5'–GGATCC–3' 3'–CCTAGG–5' ↑
*Cla*I	*Caryophanon latum*	↓ 5'–ATCGAT–3' 3'–TAGCTA–5' ↑
*Eco*RI	*E. coli* RY13	↓ 5'–GAATTC–3' 3'–CTTAAG–5' ↑
*Nae*I	*Nocardia aerocolonigenes*	↓ 5'–GCCGGC–3' 3'–CGGCCG–5' ↑
*Pst*I	*Providencia stuartii*	↓ 5'–CTGCAG–3' 3'–GACGTC–5' ↑
*Sac*I	*Streptomyces achromogenes*	↓ 5'–GAGCTC–3' 3'–CTCGAG–5' ↑

*Restriction enzymes are named according to the species in which they are found. The first three letters are italicized because they indicate the genus and species names. Because a species may produce more than one restriction enzyme, the enzymes are designated I, II, III, and so on, to indicate the order in which they were discovered in a given species. Some restriction enzymes, like *Eco*RI, produce a sticky end with a 5' overhang (see Figure 18.1), whereas others, such as *Pst*I, produce a 3' overhang. However, not all restriction enzymes cut DNA to produce sticky ends. For example, the enzyme *Nae*I cuts DNA to produce blunt ends.

†The arrows show the locations in the upper and lower DNA strands where the restriction enzymes cleave the DNA backbone. A complete list of restriction enzymes can be found at http://rebase.neb.com/rebase/rebase.html.

strands must be covalently linked together. This linkage is catalyzed by **DNA ligase.** Figure 18.1 illustrates the action of DNA ligase, which catalyzes covalent bond formation in the sugar–phosphate backbones of both DNA strands after the sticky ends have hydrogen bonded with each other.

Currently, several hundred different restriction enzymes from various bacterial species have been identified and are available commercially to molecular biologists. **Table 18.3** gives a few examples. Restriction enzymes usually recognize sequences that are **palindromic;** that is, the sequence in one strand is identical when read in the opposite direction in the complementary strand. For example, the sequence recognized by *Eco*RI is 5'–GAATTC–3' in the top strand. Read in the opposite direction in the bottom strand, this sequence is also 5'–GAATTC–3'.

Gene Cloning Involves the Insertion of DNA Fragments into Vectors, Which Are Then Propagated Within Host Cells

Now that we are familiar with the materials, let's outline the general strategy that is followed in a typical cloning experiment. In the procedure shown in **Figure 18.2**, the goal is to clone a chromosomal gene of interest into a plasmid vector that already carries the amp^R gene. To begin this experiment, the chromosomal DNA is isolated and digested with a restriction enzyme. This enzyme will cut the chromosomes into many small fragments. The plasmid DNA is also cut at one site with the same restriction enzyme, so the vector will have the same sticky ends as the chromosomal DNA fragments. The digested chromosomal DNA and plasmid DNA are mixed together and incubated under conditions that promote the binding of complementary sticky ends.

DNA ligase is then added to catalyze the covalent linkage between DNA fragments. In some cases, the two ends of the vector will simply ligate back together, restoring the vector to its original structure. This is called a recircularized vector. In other cases, a fragment of chromosomal DNA may become ligated to both ends of the vector. In this way, a segment of chromosomal DNA has been inserted into the vector. The vector containing a piece of chromosomal DNA is referred to as a **recombinant vector.**

Following ligation, the DNA is introduced into living cells treated with agents that render them permeable to DNA molecules. Cells that can take up DNA from the extracellular medium are called **competent cells.** This step in the procedure is commonly called **transformation** (see Chapter 6). In the experiment shown in Figure 18.2, a plasmid is introduced into bacterial cells that were originally sensitive to ampicillin. The bacteria are then streaked onto plates containing bacterial growth media and ampicillin. A bacterium that has taken up a plasmid carrying the amp^R gene will continue to divide and form a bacterial colony containing tens of millions of cells. Because each cell within a single colony is derived from the same original cell, all cells within a colony contain the same type of plasmid DNA.

In the experiment shown in Figure 18.2, how can the experimenter distinguish between bacterial colonies that contain a recircularized vector versus those with a recombinant vector carrying a piece of chromosomal DNA? As shown here, the chromosomal DNA has been inserted into a region of the vector that contains the *lacZ* gene, which encodes the enzyme β-galactosidase (see Chapter 14). The insertion of chromosomal DNA into the vector disrupts the *lacZ* gene so that it is no longer able to produce a functional enzyme. By comparison, a recircularized vector has a functional *lacZ* gene. The functionality of *lacZ* can be determined by providing the growth media with a colorless compound, X-Gal, which is cleaved by β-galactosidase into a blue dye. Bacteria grown in the presence of X-Gal and IPTG (an inducer of the *lacZ* gene) will form blue colonies if they have a functional *lacZ* gene and white colonies if they do not. In this experiment, therefore, bacterial colonies

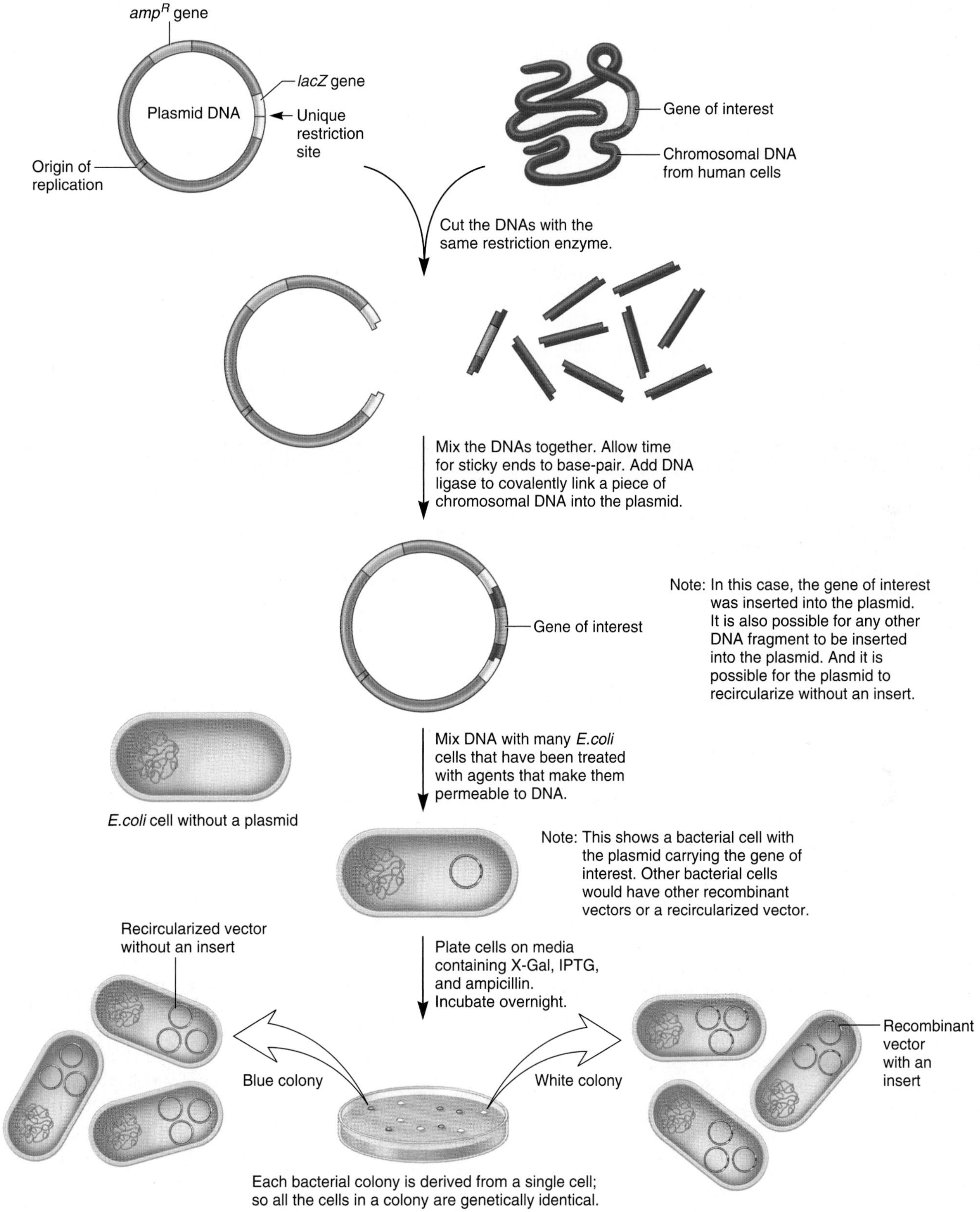

amp^R gene

Plasmid DNA

lacZ gene

Unique restriction site

Origin of replication

Gene of interest

Chromosomal DNA from human cells

Cut the DNAs with the same restriction enzyme.

Mix the DNAs together. Allow time for sticky ends to base-pair. Add DNA ligase to covalently link a piece of chromosomal DNA into the plasmid.

Gene of interest

Note: In this case, the gene of interest was inserted into the plasmid. It is also possible for any other DNA fragment to be inserted into the plasmid. And it is possible for the plasmid to recircularize without an insert.

Mix DNA with many *E.coli* cells that have been treated with agents that make them permeable to DNA.

E.coli cell without a plasmid

Note: This shows a bacterial cell with the plasmid carrying the gene of interest. Other bacterial cells would have other recombinant vectors or a recircularized vector.

Recircularized vector without an insert

Plate cells on media containing X-Gal, IPTG, and ampicillin. Incubate overnight.

Blue colony

White colony

Recombinant vector with an insert

Each bacterial colony is derived from a single cell; so all the cells in a colony are genetically identical.

FIGURE 18.2 The steps in gene cloning. Note: X-Gal refers to the colorless compound 5-bromo-4-chloro-3-indolyl-β-D-galactoside. IPTG is an acronym for isopropyl-β-D-thiogalactopyranoside, which is a nonmetabolizable lactose analogue that can induce the *lac* promoter.

containing recircularized vectors will form blue colonies, while colonies containing recombinant vectors will be white.

In the example of Figure 18.2, one of the white colonies contains cells with a recombinant vector that carries a human gene of interest; the segment containing the human gene is shown in red. The net result of gene cloning is to produce an enormous number of copies of a recombinant vector. During transformation, a single bacterial cell usually takes up a single copy of a recombinant vector. However, two subsequent events lead to the amplification of the cloned gene. First, because the vector has an origin of replication, the bacterial host cell replicates the recombinant vector to produce many copies per cell. Second, the bacterial cells divide approximately every 20 minutes. Following overnight growth, a population of many millions of bacteria will be obtained. Each of these bacterial cells will contain many copies of the cloned gene. For example, a bacterial colony may be composed of 10 million cells, with each cell containing 50 copies of the recombinant vector. Therefore, this bacterial colony would contain 500 million copies of the cloned gene!

The preceding description has acquainted you with the steps required to clone a gene. A misleading aspect of Figure 18.2 is that the digestion of the chromosomal DNA with restriction enzymes appears to yield only a few DNA fragments, one of which contains the gene of interest. In an actual cloning experiment, however, the digestion of the chromosomal DNA with a restriction enzyme produces tens of thousands of different pieces of chromosomal DNA, not just a single piece of chromosomal DNA that happens to be the gene of interest. Later, we will consider methods to identify bacterial colonies containing the specific gene that a researcher wants to clone.

Recombinant DNA technology can also be used to clone fragments of DNA that do not code for genes. For example, sequences such as telomeres, centromeres, and highly repetitive sequences have been cloned by this procedure.

EXPERIMENT 18A

In the First Gene Cloning Experiment, Cohen, Chang, Boyer, and Helling Inserted a *kanamycin*^{R} Gene into a Plasmid Vector

Now that we are familiar with the basic procedures followed in a cloning experiment, let's consider the first successful attempt at creating a recombinant DNA molecule and propagating that molecule in bacterial cells. This was accomplished in a collaboration between Stanley Cohen, Annie Chang, Herbert Boyer, and Robert Helling in 1973. Prior discoveries had led to their ability to clone a gene. In 1970, H. Gobind Khorana found that DNA ligase could covalently link DNA fragments together. In 1972, Janet Mertz, Ronald Davis, and Vittorio Sgaramella discovered that the digestion of DNA with the restriction enzyme *Eco*RI produced sticky ends that enabled the DNA fragments to hydrogen bond. With these two observations in mind, Cohen, Chang, Boyer, and Helling realized it might be possible to create recombinant DNA molecules by digesting DNA with *Eco*RI, allowing the fragments to hydrogen bond with each other, and then covalently linking these fragments together with DNA ligase.

One last condition was necessary for the researchers to succeed in cloning a gene. They needed to identify a vector that could independently replicate itself once inside a host cell. They chose a plasmid vector as their vehicle for gene cloning. In their collection of plasmids, they found one small plasmid, designated pSC101, that had a single *Eco*RI site and carried a gene for tetracycline resistance (*tet*^{R}).

*Eco*RI site
pSC101
tet^R

They reasoned that it should be possible to open up this plasmid with *Eco*RI and then insert another piece of DNA into this *Eco*RI site via the hydrogen bonding of the *Eco*RI sticky ends.

As a source of a gene to insert into pSC101, they obtained a second plasmid, which they called pSC102.

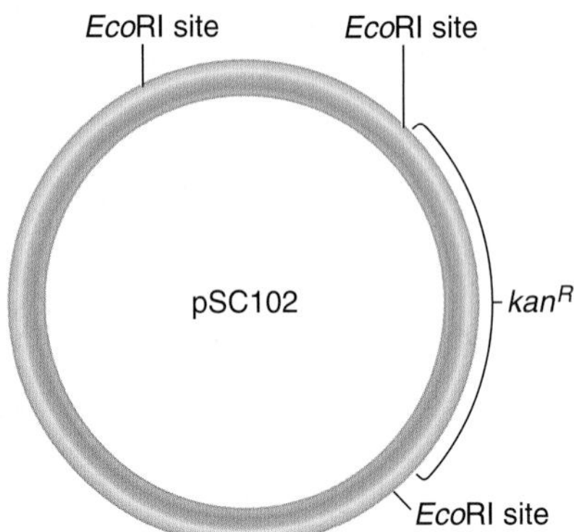

This plasmid carried a gene, *kan*^{R}, that provided resistance to the antibiotic kanamycin. However, this second plasmid was cut at three locations by *Eco*RI, yielding three DNA fragments. One of these DNA fragments would be expected to carry the kanamycin-resistance gene (unless, unluckily, *Eco*RI happened to cut in the middle of the *kan*^{R} gene). Their goal in this experiment was to clone the *kan*^{R} gene by inserting the DNA fragment carrying this gene into the *Eco*RI site of pSC101 and then introduce the recombinant plasmid into bacterial cells.

As shown in **Figure 18.3**, the researchers isolated plasmid DNA from bacterial strains carrying pSC101 or pSC102 (step 1). The purified plasmids were digested with *Eco*RI and then mixed together to allow the sticky ends to hydrogen bond with each other. Ligase was added to covalently seal the plasmids (step 4). Their goal was to obtain a recombinant vector in which pSC101 carried the kanamycin-resistance gene from pSC102. Such a plasmid would confer resistance to both tetracycline and kanamycin. To identify such a plasmid, the DNA sample was exposed to bacteria that were treated in such a way as to promote the uptake of DNA. Some colonies were obtained that were resistant to both antibiotics (step 7). Four of these colonies were picked, and DNA was isolated from them. How did the researchers confirm that they had constructed recombinant plasmids? The plasmid DNA from these colonies, termed pSC105, was then subjected to two different procedures. In step 9, some of the DNA was analyzed by cesium chloride density gradient centrifugation and scintillation counting. In step 10, some of the DNA was digested with *Eco*RI and subjected to gel electrophoresis. (See the Appendix for a description of the techniques of density gradient centrifugation, scintillation counting, and gel electrophoresis.) As a control, samples of pSC101 and pSC102 also underwent these last two steps.

THE HYPOTHESIS

A piece of DNA carrying a gene can be inserted into a plasmid vector using recombinant DNA techniques. If this recombinant plasmid is introduced into a bacterial host cell, it will be replicated and transmitted to daughter cells, producing many copies of the recombinant plasmid.

TESTING THE HYPOTHESIS — FIGURE 18.3 The first success at gene cloning.

Starting material: Three different strains of *E. coli:* one strain that did not carry any plasmid and two strains that carried pSC101 or pSC102.

1. Isolate and purify the two types of plasmid DNA:

   Grow the bacterial cells containing the plasmids.

   Break open the cells. One way to break open cells is to subject them to harsh sound waves, or sonication.

   Isolate each type of plasmid DNA by density gradient centrifugation (see the Appendix for a description of this procedure).

2. Digest the plasmid DNAs with *Eco*RI.

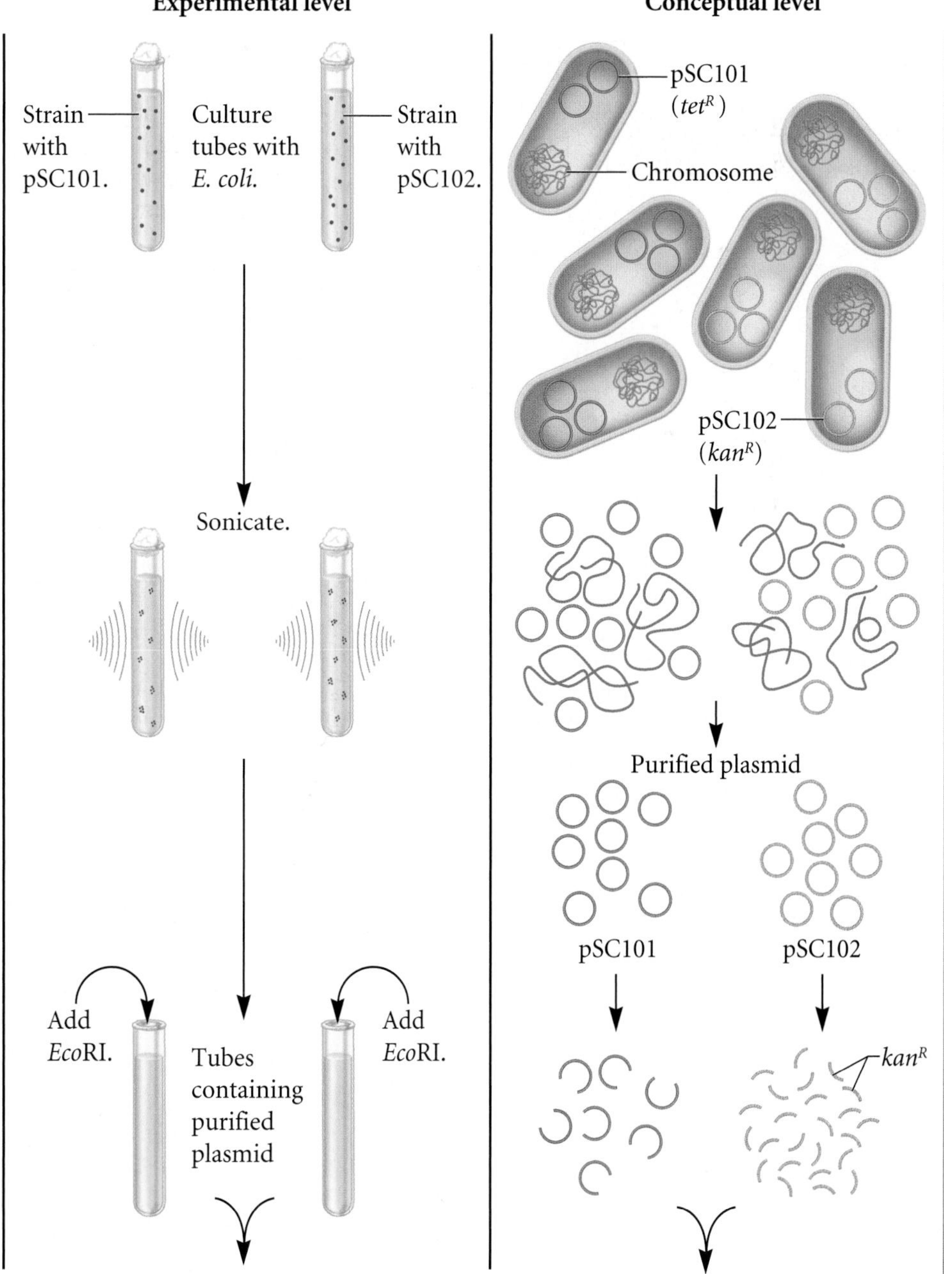

3. Mix together the two samples.

4. Add DNA ligase to covalently link DNA pieces.

5. Grow an *E. coli* strain that does not carry a plasmid. Treat the cells with $CaCl_2$ to make them permeable to DNA.

6. Add the ligated DNA samples from step 4 to the bacterial cells. Most cells do not take up any plasmid but an occasional cell will take up one plasmid.

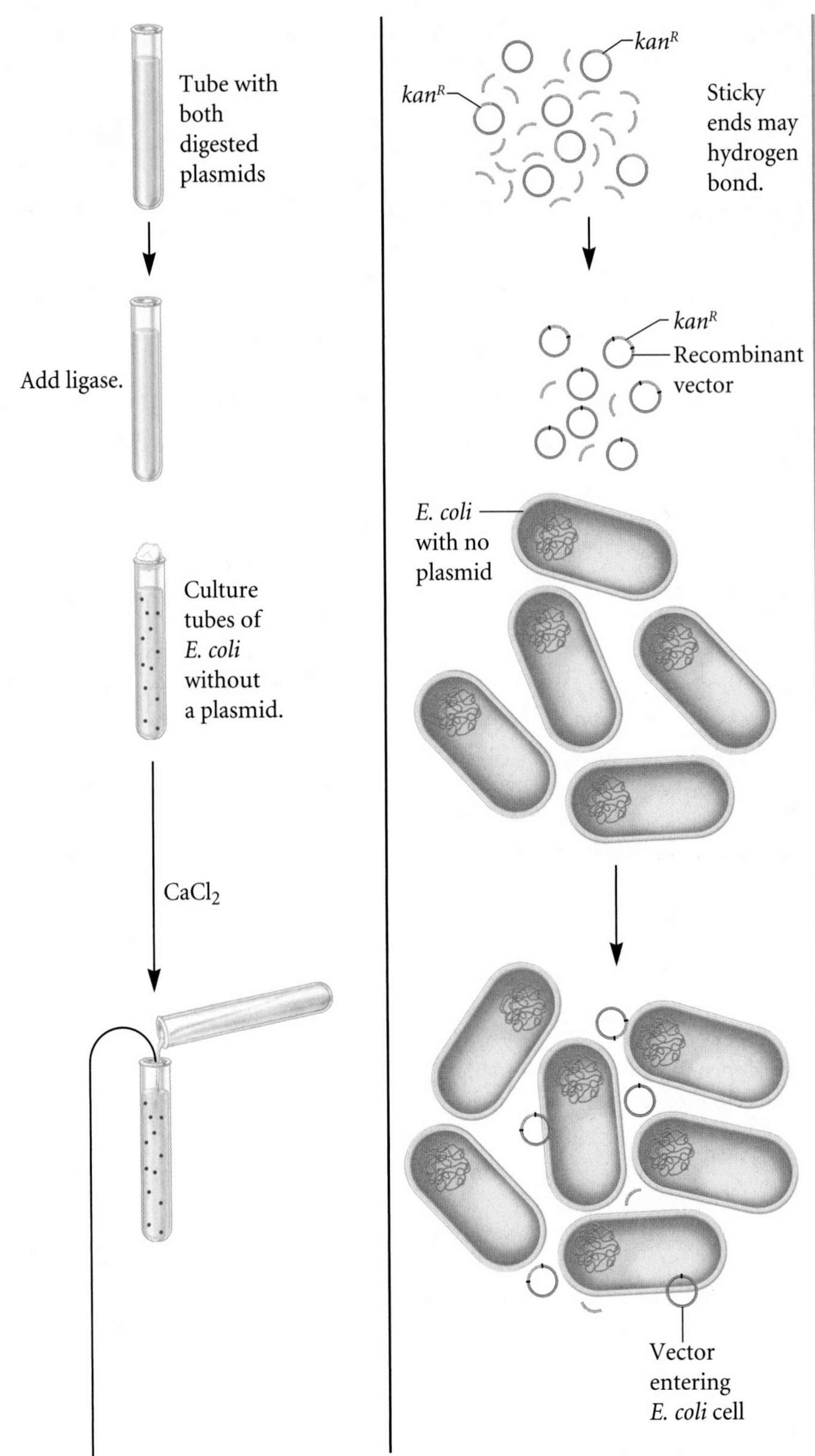

7. Plate the cells on growth media containing both tetracycline and kanamycin. Grow overnight to allow growth of visible bacterial colonies.

8. Pick four colonies from the plates. The plasmid found in these colonies is designated pSC105. Grow the colonies in liquid culture containing radiolabeled deoxyribonucleotides.

9. To isolate the radiolabeled plasmid DNA, break open the cells and subject the DNA to cesium chloride density gradient centrifugation. Collect fractions and count the amount of radioactivity in each fraction using a scintillation counter. (See data on the next page that is shown on the top left.) As a control, subject pSC101 and pSC102 to the same procedure. (See data on the bottom left.)

10. Digest the plasmid DNAs with *Eco*RI, and subject them to gel electrophoresis. (See Appendix for a description of gel electrophoresis.) (See data shown on the next page that is on the right.)

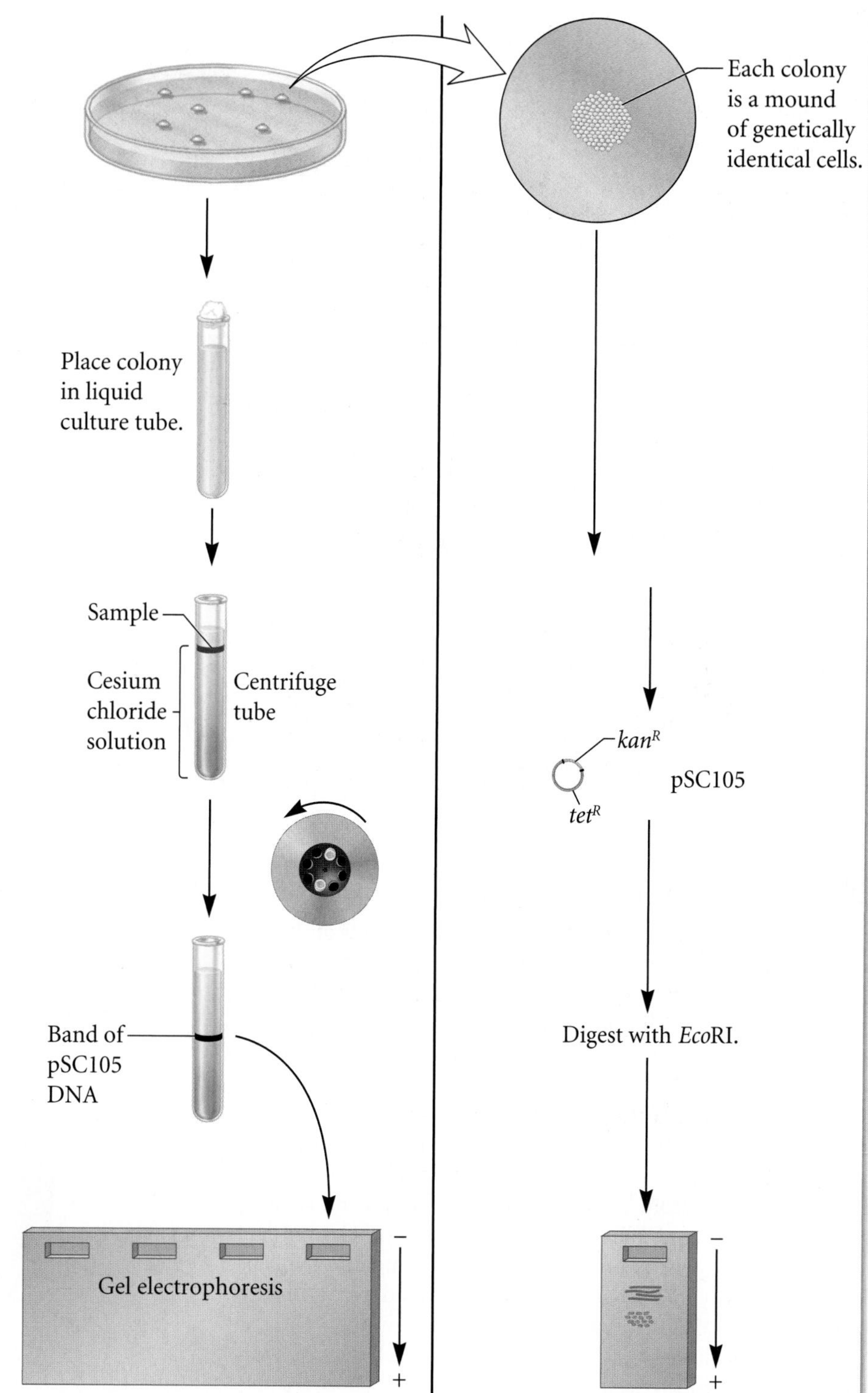

THE DATA

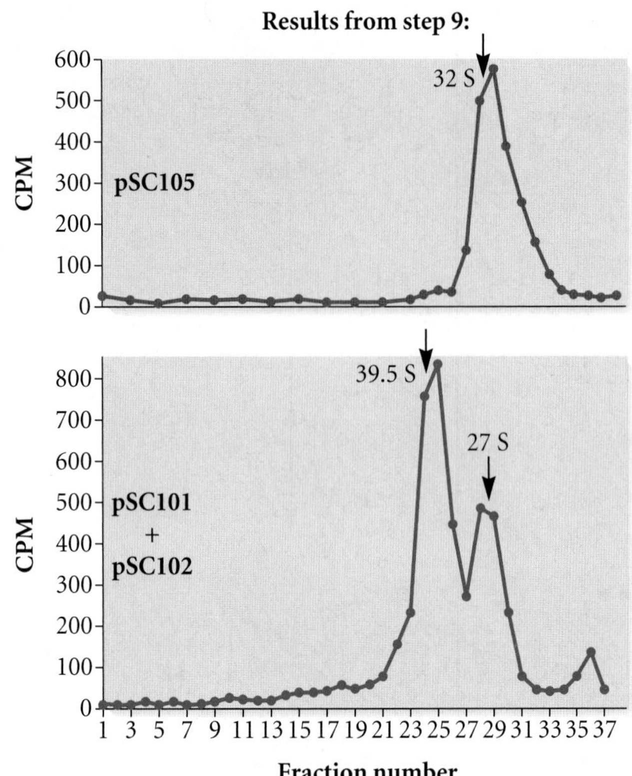

Results from step 10:

1 2 3 4

Lane 1: pSC101

2: pSC102

3: pSC101 + pSC102

4: pSC105

INTERPRETING THE DATA

With regard to their data, let's first consider the results of density gradient centrifugation. In the control experiment, the pSC101 and pSC102 plasmids were mixed together and then subjected to density gradient centrifugation. As shown in the data at the bottom left, this yielded two peaks at 27 S and 39.5 S. The pSC102 plasmid is larger than pSC101 and sediments at a density of 39.5 S, whereas the latter plasmid has a density of 27 S. (Note: The letter S refers to Svedberg units, a unit of centrifugation velocity; see the Appendix.) The top graph shows the results for pSC105 obtained from a bacterial colony that was resistant to both tetracycline and kanamycin. In this case, the plasmid had an intermediate density of 32 S. These results indicate that this bacterial colony contained a recombinant plasmid, not a mixture of pSC101 and pSC102 plasmids.

This conclusion was confirmed in the gel electrophoresis analysis. When digested with *Eco*RI, pSC101 yielded a single band, whereas pSC102 yielded three bands (see lanes 1 and 2). In lane 3, the experimenters had mixed together samples of pSC101 and pSC102 (isolated in step 1) and digested them with *Eco*RI. As expected, this produced four bands. Lane 4 shows the plasmid DNA, pSC105, from a bacterial colony that was tetracycline and kanamycin resistant. This plasmid showed the pSC101 band plus one other band that was found in pSC102. The results of lane 4 are consistent with the idea that the pSC105 plasmid is formed by the insertion of one fragment from pSC102 into the single *Eco*RI site of pSC101. This recombinant plasmid is shown here.

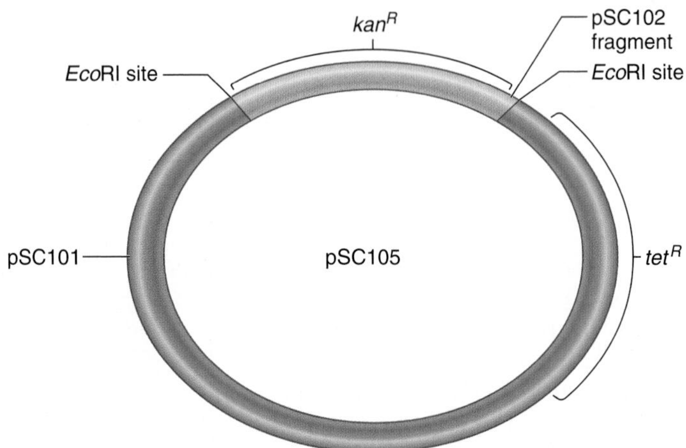

This study showed the scientific community that it is possible to create recombinant DNA molecules and then propagate them in bacterial cells. In other words, researchers were able to clone genes. This hallmark achievement ushered in the era of gene cloning.

A self-help quiz involving this experiment can be found at the Online Learning Center.

cDNA Can Be Made from mRNA Via Reverse Transcriptase

In the discussion of gene cloning described earlier in Figure 18.2, chromosomal DNA and plasmid DNA were used as the material to clone genes. Alternatively, a sample of RNA can provide a starting point to clone DNA. As mentioned in Chapter 17, the enzyme **reverse transcriptase** can use RNA as a template to make a complementary strand of DNA. This enzyme is encoded in the genome of retroviruses and provides a way for retroviruses to copy their RNA genome into DNA molecules that integrate into the host cell's chromosomes. Likewise, reverse transcriptase is encoded in viral-like retroelements and is needed in the retrotransposition of such elements.

Researchers can use purified reverse transcriptase in a strategy to clone genes, using mRNA as the starting material (**Figure 18.4**). To begin this experiment, RNA is purified from a sample of cells. The RNA is mixed with primers composed of a string of thymine-containing nucleotides. This short strand of DNA, or **oligonucleotide,** is called a poly-dT primer. Because eukaryotic mRNAs contain a polyA tail, poly-dT primers will be complementary to the 3' end of mRNAs. Reverse transcriptase and deoxyribonucleotides (dNTPs) are then added to make a DNA strand that is complementary to the mRNA. Next, RNaseH, DNA polymerase, and DNA ligase are added. One way to make the other DNA strand is to use RNaseH, which partially digests the RNA, generating short RNAs that are used as primers by DNA polymerase to make a second DNA strand that is complementary to the strand made by reverse transcriptase. Finally, DNA ligase will seal any nicks in this second DNA strand. When DNA is made from RNA as the starting material, the DNA is called **complementary DNA (cDNA).** The term originally referred to the single strand of DNA that is complementary to the RNA template. However, cDNA now refers to any DNA, whether it is single or double stranded, that is made using RNA as the starting material.

Why is cDNA cloning useful? From a research perspective, an important advantage of cDNA is that it lacks introns. Because introns can be quite large, it is much simpler to insert cDNAs into vectors if researchers want to focus their attention on the coding sequence of a gene. For example, if the primary goal was to determine the coding sequence of a structural gene, a researcher would insert cDNA into a vector and then determine the DNA sequence of the insert, as described later in this chapter. Similarly, if a scientist wanted to express an encoded protein of interest in a cell that would not splice out the introns properly (e.g., in a bacterial cell), it would be necessary to make cDNA clones of the respective gene.

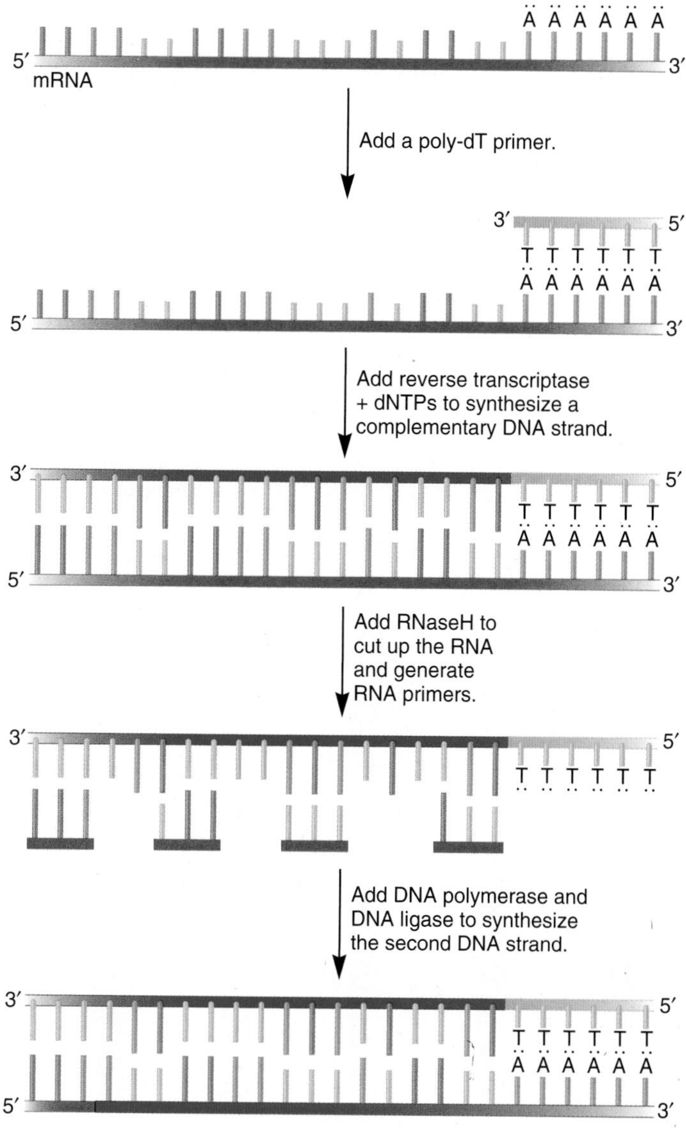

FIGURE 18.4 **Synthesis of cDNA.** A poly-dT primer anneals to the 3' end of mRNAs. Reverse transcriptase then catalyzes the synthesis of a complementary DNA strand (cDNA). RNaseH digests the mRNA into short pieces that are used as primers by DNA polymerase I to synthesize the second DNA strand. The 5' to 3' exonuclease function of DNA polymerase I removes all of the RNA primers except the one at the 5' end (because there is no primer upstream from this site). This RNA primer can be removed by the subsequent addition of an RNase. After the double-stranded cDNA is made, it can then be inserted into vectors as described in Figure 18.2. Prior to vector insertion, short pieces of DNA, called linkers, are covalently attached to both ends of the cDNA. The DNA sequence of the linkers contains unique restriction sites, making it easy to cut the cDNA at both ends and insert it into a vector.

Restriction Mapping Is Used to Locate the Restriction Sites Within a Vector

As we have seen, DNA or gene cloning involves the digestion of vector and chromosomal DNA with restriction enzymes and the subsequent ligation of DNA fragments into vectors. In this type of procedure, the locations of restriction enzyme sites, or simply restriction sites, are important for the design of experiments. In the vector, for example, it is desirable to have unique restriction sites for the insertion of chromosomal DNA.

A common approach to determine the locations of restriction sites is known as **restriction mapping.** **Figure 18.5** outlines the restriction mapping of a bacterial plasmid. To begin this experiment, the small circular plasmid DNA is isolated and purified

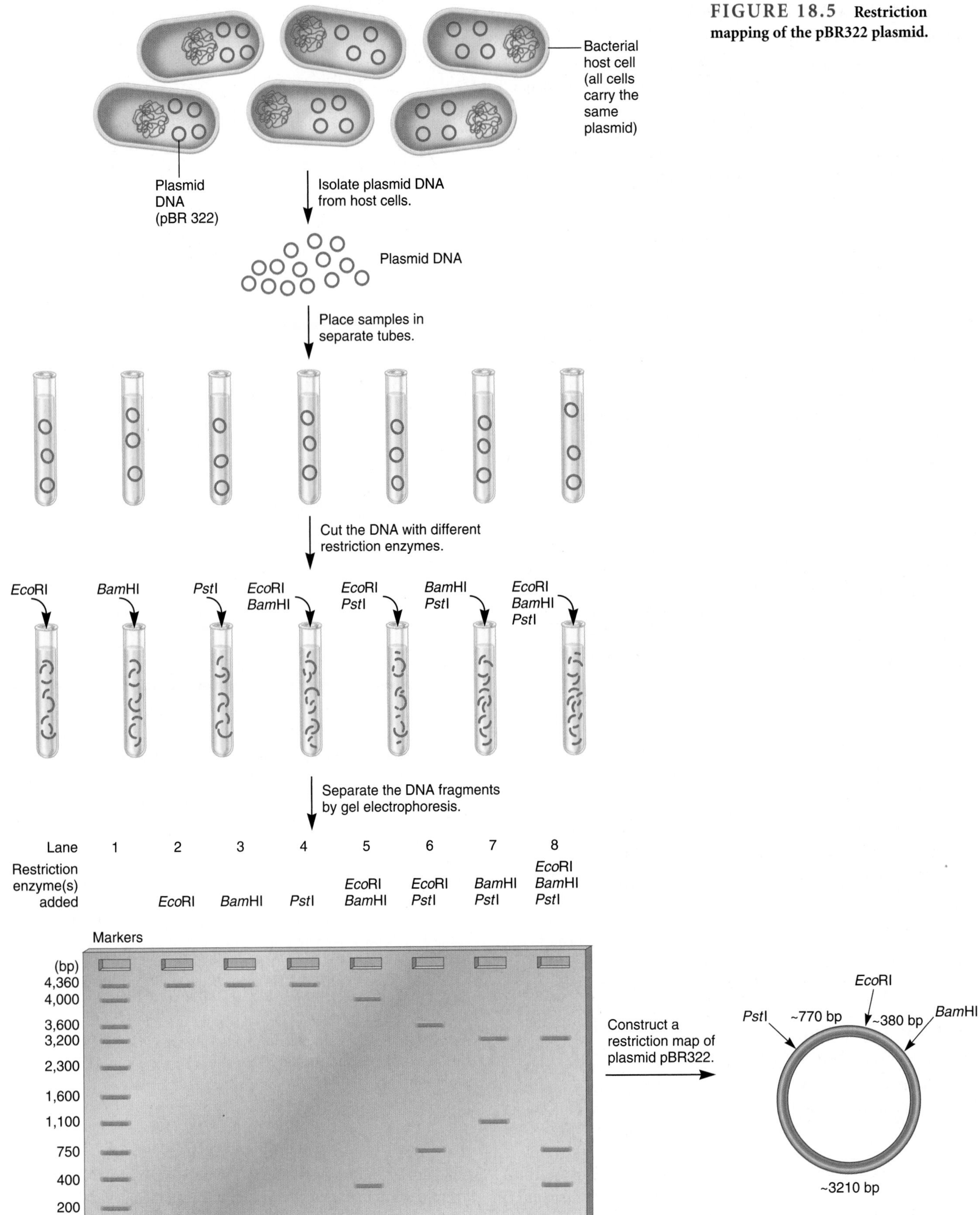

FIGURE 18.5 **Restriction mapping of the pBR322 plasmid.**

from host cells. Samples of the purified DNA are then placed in separate test tubes that contain a particular restriction enzyme or combination of enzymes. The plasmid DNA is incubated with the restriction enzymes long enough for digestion to occur. The DNA fragments are then separated by gel electrophoresis. In lane 1 of this experiment, a different DNA sample, known as a set of molecular markers, is also subjected to gel electrophoresis. This sample contains a mixture of DNA fragments with molecular masses that are known from previous experiments. (These markers are obtained from commercial sources or can be prepared in the laboratory.) To determine the sizes of the fragments obtained by digesting the plasmid, the fragments in lanes 2 to 8 are compared to the known markers in lane 1.

The restriction map shown in Figure 18.5 was deduced by comparing the sizes of fragments obtained from digestions with one, two, or all three of the restriction enzymes. The starting plasmid is a circular molecule 4,363 bp in length. A single digestion with any of the three enzymes yields a single linear fragment of size 4,363 bp. This means that *Eco*RI, *Bam*HI, and *Pst*I cut the plasmid at a single site. The double digestion with *Eco*RI and *Bam*HI yields two fragments of about 380 bp and 3,980 bp. This result indicates that the *Eco*RI and *Bam*HI sites are approximately 380 bp apart in one direction along the circle and 3,980 bp apart along the circle in the opposite direction. Likewise, the pairwise combinations of *Eco*RI/*Pst*I and *Bam*HI/*Pst*I indicate how far apart these sites are along the circular plasmid. Finally, the triple digestion confirms the locations of the single sites for *Eco*RI, *Bam*HI, and *Pst*I. Taken together, these results provide a map of the restriction sites within this plasmid. A similar approach can be used on a recombinant vector to determine the locations of restriction sites within a fragment of DNA that has been inserted into a plasmid.

Alternatively, another way to obtain a restriction map is via DNA sequencing, a technique described later in the chapter. If the DNA sequence of a recombinant vector has been determined, computer programs can scan the sequence and identify sites that are recognized by particular restriction enzymes. For example, *Eco*RI recognizes the sequence 5'–GAATTC–3'. If a recombinant vector contained this sequence, a computer program would identify this as an *Eco*RI site and place it on a map. Such a program would scan the DNA sequence for the recognition sequences of many different restriction enzymes and generate a detailed restriction map of a recombinant vector.

A second use of restriction enzymes is gene mapping. The technique known as restriction fragment length polymorphism (RFLP) enables researchers to map particular genes within a species' genome. This approach is discussed in Chapter 20.

Polymerase Chain Reaction (PCR) Can Also Be Used to Make Many Copies of DNA

In our previous discussions of gene cloning, the DNA of interest was inserted into a vector, which then was introduced into a host cell. The replication of the vector within the host cell, and the proliferation of the host cells, led to the production of many copies of the DNA. Another way to copy DNA, without the aid of vectors and host cells, is a technique called **polymerase chain reaction (PCR),** which was developed by Kary Mullis in 1985.

The PCR method is outlined in **Figure 18.6**. In this example, the starting material contains a sample of DNA that someone wants to clone. This DNA is called **template DNA.** The goal of PCR is to make many copies of the template DNA in a defined region. Several reagents are added to facilitate the synthesis of DNA. These include a high concentration of two oligonucleotide primers, which are complementary to sequences at the ends of the DNA fragment to be amplified; deoxyribonucleoside triphosphates (dNTPs); and a thermostable form of DNA polymerase such as ***Taq* polymerase,** isolated from the bacterium *Thermus aquaticus*. A thermostable form of DNA polymerase is necessary because PCR involves heating steps that would inactivate most other natural forms of DNA polymerase (which are thermolabile, or readily denatured by heat).

To make copies of the DNA, the template DNA is first denatured by heat treatment, causing the strands to separate (see Online Animation). As the temperature is lowered, oligonucleotide primers bind to the DNA in a process called **annealing.** Once the primers have annealed, *Taq* polymerase catalyzes the synthesis of complementary DNA strands, starting at the primers. This doubles the amount of the template DNA. The sequential process of denaturation—annealing—synthesis is then repeated to double the amount of template DNA many times in a row. This method is called a chain reaction because the products of each previous reaction (the newly made DNA strands) are used as reactants (the template strands) in subsequent reactions.

PCR is carried out in a thermal reactor, known as a **thermocycler,** that automates the timing of each cycle. The experimenter mixes the DNA sample, dNTPs, *Taq* polymerase, and an excess amount of primers together in a single tube. The tube is placed in a thermocycler, and the experimenter sets the machine to operate within a defined temperature range and number of cycles. During each cycle, the thermocycler increases the temperature to denature the DNA strands and then lowers the temperature to allow annealing and DNA synthesis to take place. Typically, each cycle lasts 2 to 3 minutes and is then repeated. A typical PCR run is likely to involve 20 to 30 cycles of replication and takes a couple of hours to complete. The PCR technique can amplify the amount of DNA by a staggering amount. After 30 cycles of amplification, the intervening region between the two primers will increase 2^{30}-fold, which is approximately a billion-fold!

In the PCR reaction shown in Figure 18.6, the outcome of PCR is the synthesis of many copies of a DNA sequence that is flanked by two primers. The sequences of the PCR primers are complementary to two specific sequences within the template DNA. How are the primers designed, and how are they made? To conduct this type of PCR experiment, a researcher must have prior knowledge about the sequence of the template DNA in order to design oligonucleotide primers complementary to two sites in the template sequence. This information might come from the sequencing of a species' genome. As discussed in Chapter 20, the DNA sequence of the entire genome of many organisms is

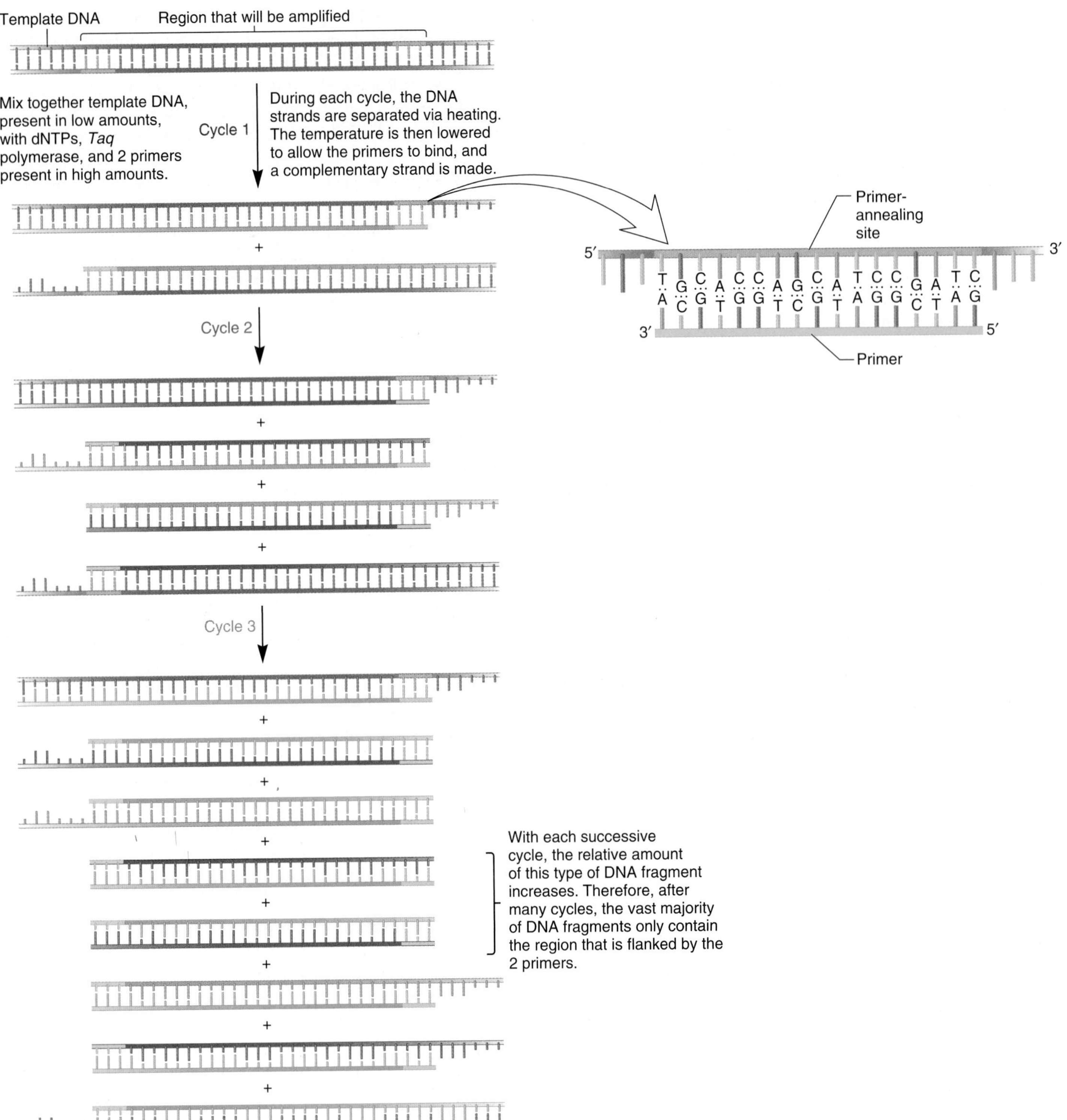

Template DNA
Region that will be amplified
Mix together template DNA, present in low amounts, with dNTPs, Taq polymerase, and 2 primers present in high amounts.
Cycle 1
During each cycle, the DNA strands are separated via heating. The temperature is then lowered to allow the primers to bind, and a complementary strand is made.
Primer-annealing site
5′
3′
3′
5′
Primer
Cycle 2
Cycle 3
With each successive cycle, the relative amount of this type of DNA fragment increases. Therefore, after many cycles, the vast majority of DNA fragments only contain the region that is flanked by the 2 primers.

ONLINE ANIMATION

FIGURE 18.6 The technique of polymerase chain reaction (PCR). During each cycle, oligonucleotides that are complementary to the ends of the targeted DNA sequence bind to the DNA and act as primers for the synthesis of this DNA region. The primers used in actual PCR experiments are usually about 20 nucleotides in length. The region between the two primers is typically hundreds of nucleotides in length, not just several nucleotides as shown here. The net result of PCR is the synthesis of many copies of DNA in the region flanked by the two primers. Note: The DNA template strands are purple, and those strands made during the first, second, and third cycles are blue, red, and gray, respectively.

known. Complete genome sequences have been obtained for many species of bacteria, yeast, and several plants and animals. Once a researcher decides which gene or DNA region to amplify, she/he must choose DNA sites that flank both sides of this region and specify short DNA primer sequences, typically about 20 nucleotides in length, that are complementary to those sites. After the sequences are chosen, researchers usually obtain such primers via biotechnology companies or academic facilities that synthesize them chemically.

When specific primers can be constructed, PCR can amplify a particular region of DNA from a very complex mixture of template DNA. For example, if a researcher uses two primers that anneal to the human β-globin gene, PCR can amplify just the β-globin gene from a DNA sample that contains all of the human chromosomes!

Alternatively, PCR can be used to amplify a sample of chromosomal DNA semispecifically or nonspecifically. As discussed in Chapter 24, this approach is used in DNA fingerprinting analysis. In a semispecific PCR experiment, the primers recognize a known repetitive DNA sequence found at several sites within the genome. When chromosomal DNA is used as a template, this will amplify many different DNA fragments. In a nonspecific approach, a mixture of short PCR primers with many different random sequences is used. These primers will anneal randomly throughout the genome and amplify most of the chromosomal DNA. Nonspecific DNA amplification is used to increase the total amount of DNA in very small samples, such as blood stains found at crime scenes.

PCR is also used to detect and quantitate the amount of specific RNAs in living cells. To accomplish this goal, RNA is isolated from a sample and mixed with reverse transcriptase and a primer that will bind near the 3' end of the RNA of interest. This generates a single-stranded cDNA, which then can be used as template DNA in a conventional PCR reaction. This method, called **reverse transcriptase PCR (RT-PCR),** is extraordinarily sensitive. RT-PCR can detect the expression of small amounts of RNA from a single cell!

18.2 DETECTION OF GENES, GENE PRODUCTS, AND PROTEIN-DNA INTERACTIONS

The advent of gene cloning has enabled scientists to investigate gene structure and function at the molecular level. However, unlike the first gene cloning experiment, in which a single gene was removed from one plasmid (pSC102) and cloned into another plasmid (pSC101), molecular geneticists usually want to study genes that are originally within the chromosomes of living species. This presents a problem, because chromosomal DNA contains thousands of different genes. For this reason, researchers need methods to specifically identify a gene within a mixture of many other genes or DNA fragments. Similarly, when studying gene expression at the molecular level, scientists also need techniques for the identification of gene products, such as the RNA that is transcribed from a particular gene or the protein that is encoded by an mRNA. In this section, we will consider the methodology and uses of several common techniques used to detect DNA, RNA, proteins, and protein-DNA interactions.

A DNA Library Is Constructed and Then Screened by Colony Hybridization to Identify a Cloned Gene

In a typical cloning experiment (refer back to Figure 18.2), the treatment of the chromosomal DNA with restriction enzymes yields tens of thousands of different DNA fragments. Therefore, after the DNA fragments are ligated individually to vectors, the researcher has a collection of recombinant vectors, with each vector containing a particular fragment of chromosomal DNA. A collection of recombinant vectors is known as a **DNA library.** When the starting material is chromosomal DNA, the library is called a **genomic library** (**Figure 18.7**). The library shown here uses a plasmid vector. Alternatively, a viral vector could be used, which would result in viral plaques rather than bacterial colonies.

It is also common for researchers to make a **cDNA library** that contains recombinant vectors with cDNA inserts. Recall that because cDNA is produced from mature mRNA via reverse transcriptase, it lacks any introns. A cDNA library could be made because a researcher wanted to express the encoded protein of interest in a cell that would not splice out the introns properly.

In many cloning experiments, the ultimate goal is to clone a specific gene. For example, let's suppose that a geneticist wishes to clone the rat β-globin gene. To begin a cloning experiment, chromosomal DNA would be isolated from rat cells. This chromosomal DNA would be digested with a restriction enzyme, yielding thousands of DNA fragments. The chromosomal fragments would then be ligated to vector DNA and transformed into bacterial cells. Unfortunately, only a small percentage of the recombinant vectors, perhaps one in a few thousand, would actually contain the rat β-globin gene. For this reason, researchers must have some way to identify those rare bacterial colonies that happen to contain the cloned gene of interest, in this case, a colony that contains the rat β-globin gene.

Figure 18.8 describes a method for identifying a bacterial colony that contains the gene of interest. This procedure is referred to as **colony hybridization.** The master plate shown at the top of Figure 18.8 has many bacterial colonies. Each bacterial colony is composed of bacterial cells containing a recombinant vector with a different piece of rat chromosomal DNA. The goal is to identify a colony that contains the gene of interest, in this case, the rat β-globin gene. To do so, a nylon membrane is laid gently onto the master plate containing many bacterial colonies. After the membrane is lifted, some cells from each colony are attached to it. In this way, the membrane contains a replica of the colonies on the master plate.

In this procedure, cells are treated with detergent, which dissolves the cell membrane and makes the DNA accessible. The DNA within the cells is then fixed (i.e., adhered) to the nylon membrane and denatured with NaOH. The membrane is then submerged in a solution containing a **probe,** which can be a oligonucleotide, a fragment of DNA from a cloned gene, or a specific RNA. In this case, the probe is single-stranded DNA with a base sequence that is complementary to one of the DNA strands of the rat β-globin gene. It is radiolabeled for ease of detection. The probe is given time to hybridize to the DNA on the membrane, which has already

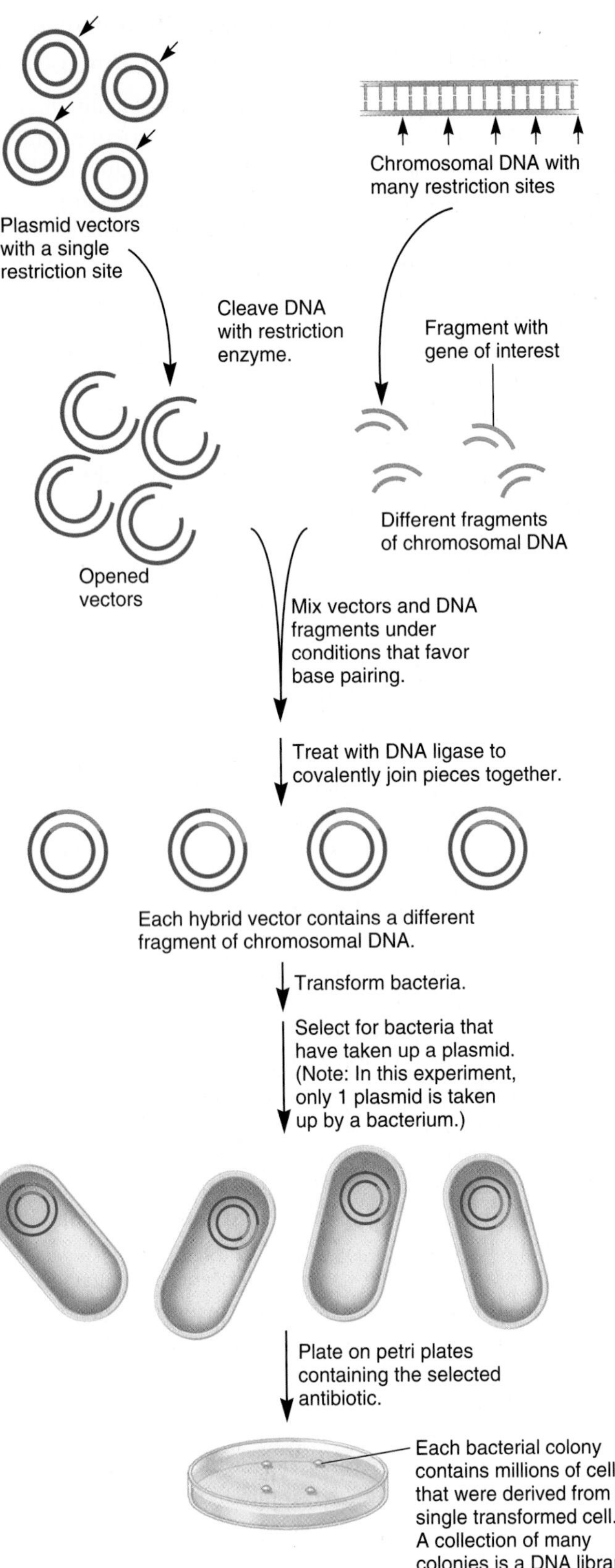

FIGURE 18.7 The construction of a DNA library. The digestion of chromosomal DNA produces many fragments. The fragment containing the gene of interest is highlighted in red. Following ligation, each vector contains a different piece of chromosomal DNA.

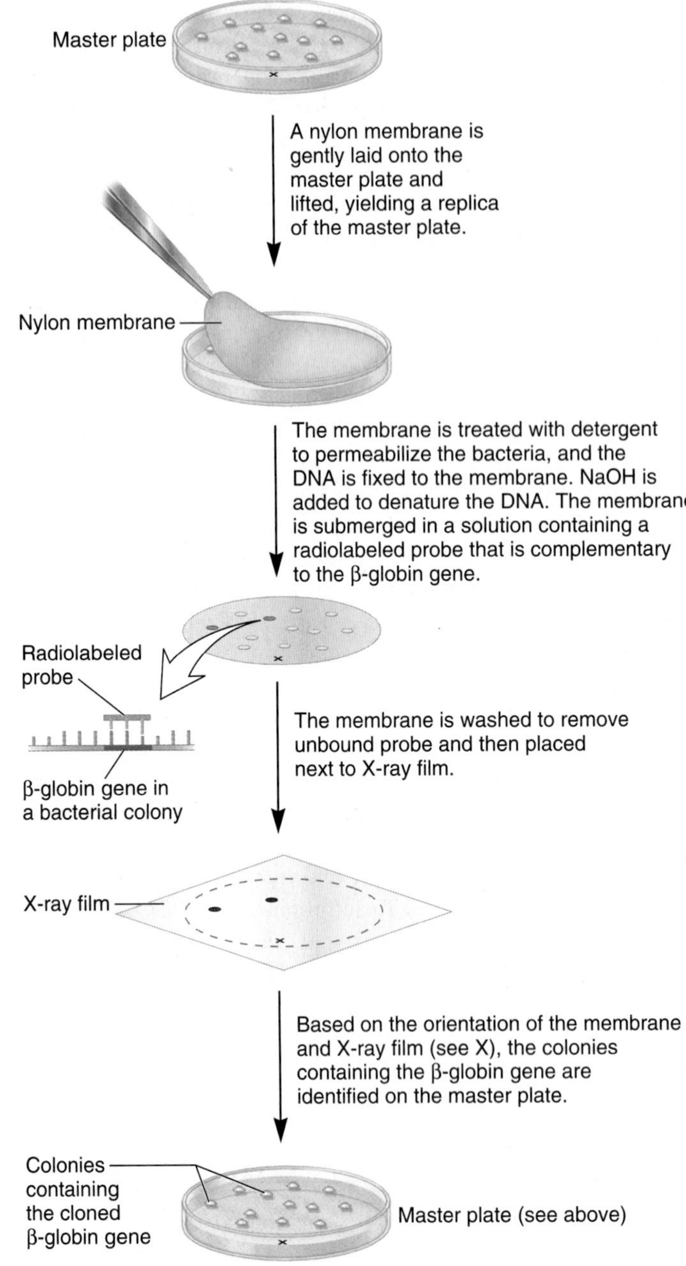

FIGURE 18.8 A colony hybridization experiment.
Note: If the vector used was a viral vector, the master plate would have viral plaques. The steps shown here could also be followed, in which case the experiment would be called plaque hybridization.

been denatured into single strands. If a bacterial colony contains the rat β-globin gene, the probe will hybridize to the DNA in this colony. Most bacterial colonies are not expected to contain this gene. The unbound probe is then washed away, and the membrane is placed next to X-ray film. If the DNA within a bacterial colony did hybridize to the probe, a dark spot will appear on the film in the corresponding location. The developed film is compared to the master plate to identify bacterial colonies carrying the rat β-globin gene. Following the identification of "hot" colonies, the experimenter can pick them from the plate, and grow bacteria containing the cloned gene.

A scientist might follow several different strategies to obtain a probe for a colony hybridization experiment. In some cases, the gene of interest may have already been cloned. A piece of the cloned gene then can be used as a radiolabeled probe. However, in many circumstances, a scientist may attempt to clone a gene that has never been cloned before. In such a case, one strategy is to use a probe that likely has a sequence similar to that of the gene of interest. For example, if the goal is to clone the β-globin gene from a South American rodent, it is expected that this gene is similar to the rat β-globin gene, which has already been cloned. Therefore, the cloned rat β-globin gene can be used as a DNA probe to "fish out" the β-globin gene from a similar species.

Can a researcher design a DNA probe for a novel type of gene that no one else has ever cloned before from any species? If the protein of interest has been previously isolated from living cells, fragments of the protein may be subjected to amino acid sequence analysis to obtain short amino acid sequences. Because the genetic code is known, a researcher can use this amino acid sequence information to design oligonucleotide probes that would correspond to the coding sequence of the gene that encodes the protein. Alternatively, the purified protein could be used to generate antibodies. In that case, the antibody would recognize clones that express the protein. Like DNA probes, labeled antibodies can be used as probes in a colony hybridization experiment. The use of antibodies as labeling tools in the technique known as Western blotting will be discussed later in this chapter.

Southern Blotting Is Used to Detect DNA Sequences

The technique of **Southern blotting** is used to detect the presence of a particular gene sequence within a mixture of many chromosomal DNA fragments. This method, developed by Edwin Southern in 1975, has several uses. It can determine the copy number of a gene within the genome of an organism. For example, Southern blotting has revealed that rRNA genes are found in multiple copies within a genome, whereas many structural genes are unique. In the study of human genetic diseases, Southern blotting can also detect small gene deletions that cannot be distinguished under the light microscope (see solved problem S4 at the end of this chapter). A common use of Southern blotting is to identify gene families. As discussed in Chapter 8, a gene family is a group of two or more genes derived from the same ancestral gene. The members of a gene family have similar but not identical DNA sequences; they are homologous genes. As we will see, Southern blotting can distinguish the homologous members of a gene family. Similarly, Southern blotting can identify homologous genes in different species. This is called a zoo blot.

Prior to a Southern blotting experiment, a gene or fragment of a gene of interest has been cloned. This cloned DNA is labeled (e.g., radiolabeled) in vitro, and then the labeled DNA is used as a probe to detect the presence of the gene or a homologous gene within a mixture of many DNA fragments. The basis for a Southern blotting experiment is that two DNA fragments will bind to each other only if they have complementary sequences. In such an experiment, the labeled strands from the cloned gene will pair specifically with complementary DNA strands, even if the complementary strands are found within a mixture of many other DNA pieces.

Figure 18.9a shows the Southern blotting procedure. The goal of this experiment is to determine if the chromosomal DNA contains a base sequence complementary to a specific probe. To begin the experiment, the chromosomal DNA is isolated and digested with a restriction enzyme. Because the restriction enzyme cuts the chromosomal DNA at many different sites within the chromosomes, this step produces thousands of DNA pieces of different sizes. Depending on the location of the restriction sites, each gene will be contained on one or more DNA pieces with specific masses. The chromosomal pieces are separated according to their masses by gel electrophoresis. The DNA pieces within the gel are denatured by soaking the gel in a NaOH solution, and then the DNA is transferred (i.e., blotted) to a nylon membrane.

As shown in **Figure 18.9b**, the traditional way to transfer the DNA to the membrane is to lay the gel on blotting paper in a transfer solution. A nylon membrane is placed on top of the gel, and then additional blotting paper and dry paper towels are layered above the nylon membrane. A glass plate and heavy weight are placed on the very top. The liquid from the transfer solution is drawn upward, carrying the DNA from the gel to the nylon membrane. When the DNA enters the nylon membrane, it remains bound to it. Alternatively, another way to transfer DNA is via a transfer apparatus in which the gel is placed on top of a nylon membrane and then an electric field causes the DNA fragments to electrophoretically move from the gel into the membrane (**Figure 18.9c**).

The procedures shown in Figure 18.9b and c result in a nylon membrane that contains many unlabeled DNA fragments that have been denatured and separated according to size. The next step is to determine if any of these unlabeled fragments from the chromosomal DNA contain sequences complementary to a labeled probe. One labeling method is to incorporate the radioisotope ^{32}P into the probe DNA, which labels the phosphate group in the DNA backbone. The nylon membrane, which has the unlabeled chromosomal DNA attached to it, is submerged in a solution containing the radiolabeled probe. If the radiolabeled probe and a fragment of chromosomal DNA are complementary, they will hydrogen bond to each other. Any unbound radiolabeled DNA is then washed away, and the membrane is exposed to X-ray film. Locations where radiolabeled probe has bound will appear as dark bands on the X-ray film.

Important variables in the Southern blotting procedure are the temperature and ionic strength of the hybridization and wash steps. If the procedure is done at very high temperatures and/or at low salt concentrations, then the probe DNA and chromosomal fragment must be very complementary—nearly a perfect match—in order to hybridize. This condition is called **high stringency.** Conditions of high stringency are used to detect close or exact complementarity between the probe and a chromosomal DNA fragment. However, if the temperature is lower and/or the salt concentration is higher, DNA sequences that are not perfectly complementary may hybridize to the probe. This is called **low stringency.** Conditions of low stringency are used to detect homologous genes with DNA sequences that are similar

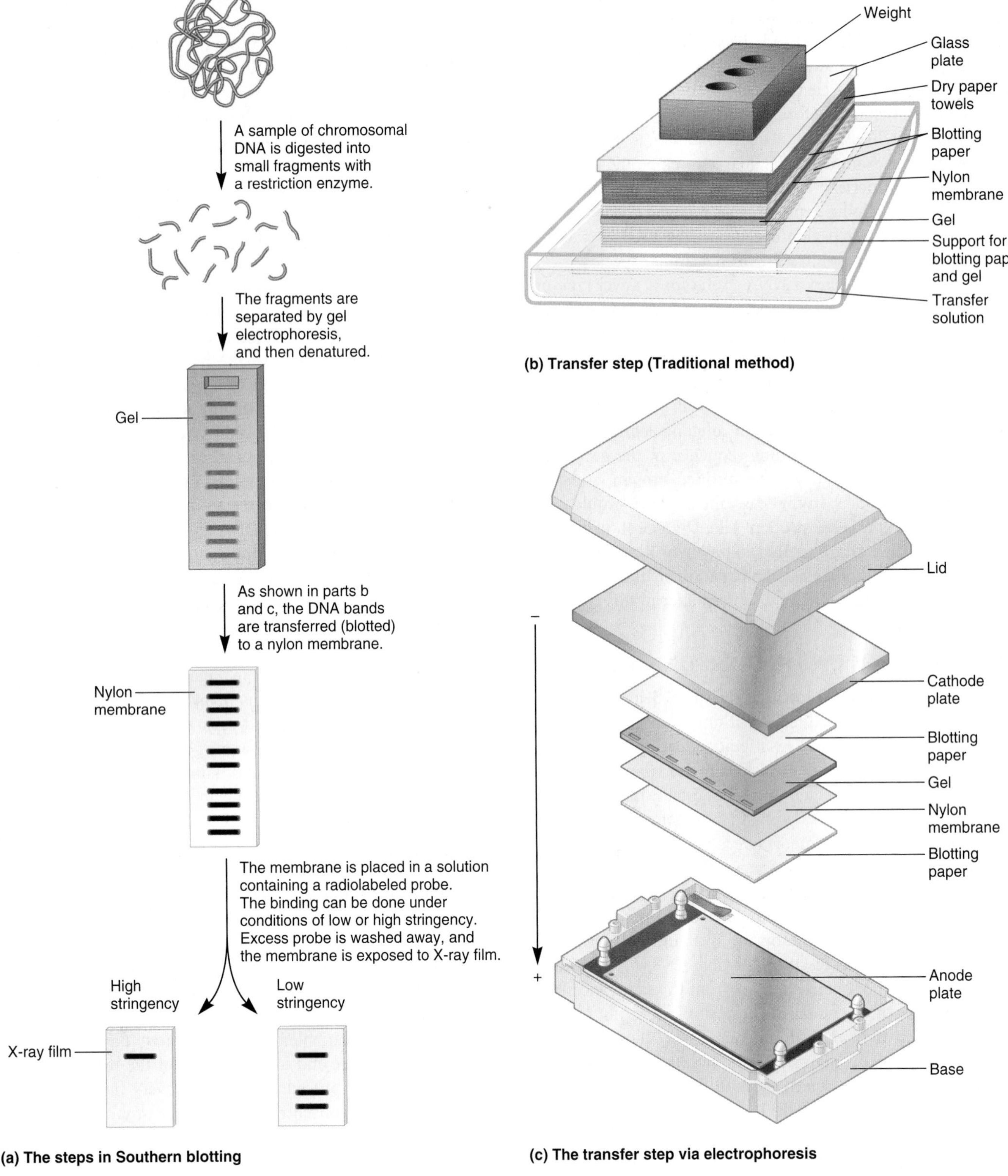
A sample of chromosomal DNA is digested into small fragments with a restriction enzyme.
The fragments are separated by gel electrophoresis, and then denatured.
Gel
As shown in parts b and c, the DNA bands are transferred (blotted) to a nylon membrane.
Nylon membrane
The membrane is placed in a solution containing a radiolabeled probe. The binding can be done under conditions of low or high stringency. Excess probe is washed away, and the membrane is exposed to X-ray film.
High stringency
Low stringency
X-ray film
(a) The steps in Southern blotting
Weight
Glass plate
Dry paper towels
Blotting paper
Nylon membrane
Gel
Support for blotting paper and gel
Transfer solution
(b) Transfer step (Traditional method)
Lid
Cathode plate
Blotting paper
Gel
Nylon membrane
Blotting paper
Anode plate
Base
–
+
(c) The transfer step via electrophoresis

FIGURE 18.9 **The technique of Southern blotting.** **(a)** Chromosomal DNA is isolated from a sample of cells and then digested with a restriction enzyme to yield many DNA fragments. The fragments are then separated by gel electrophoresis according to molecular mass and denatured with NaOH. The bands within the gel are transferred to a nylon membrane, a step known as blotting. The blot is then placed in a solution containing a radiolabeled DNA probe. When the DNA probe is hybridized to the mixture of chromosomal DNA fragments at high temperature and/or low ionic strength (i.e., high stringency), the probe recognizes only a single band. When hybridization is performed at lower temperature and/or higher ionic strength (i.e., low stringency), this band and two additional bands are produced. **(b)** The transfer step of DNA blotting (traditional method). **(c)** The transfer step of DNA blotting via electrophoresis.

but not identical to the cloned gene being used as a probe. In the results shown in Figure 18.9a, conditions of high stringency reveal that the gene of interest is unique in the genome. At low stringency, however, two other bands are detected. These results suggest that this gene is a member of a gene family composed of three distinct members.

Northern Blotting Is Used to Detect RNA

Let's now turn our attention to the technique known as Northern blotting, which is used to identify a specific RNA within a mixture of many RNA molecules. (Note: Even though Southern blotting was named after Edwin Southern, Northern blotting was not named after anyone called Northern! It was originally termed reverse-Southern blotting and later became known as Northern blotting.) Northern blotting is used to investigate the transcription of genes at the molecular level. This method can determine if a specific gene is transcribed in a particular cell type, such as nerve or muscle cells, or at a particular stage of development, such as fetal or adult cells. Also, Northern blotting can reveal if a pre-mRNA transcript is alternatively spliced into two or more mRNAs of different sizes.

From a technical viewpoint, Northern blotting is rather similar to Southern blotting with a few important differences. RNA is the starting material and is extracted and purified from living cells. This RNA can be isolated from a particular cell type under a given set of conditions or during a particular stage of development. Any given cell will produce thousands of different types of RNA molecules, because cells express many genes at any given time. After the RNA is extracted from cells and purified, it is loaded onto a gel that separates the RNA transcripts according to their size. The RNAs within the gel are then blotted onto a nylon membrane and probed with a radiolabeled fragment of DNA from a cloned gene. Using this method, RNAs that are complementary to the radiolabeled DNA fragment are detected as dark bands on an X-ray film. In **Figure 18.10**, the probe was complementary to an mRNA that encodes a protein called tropomyosin. In lane 1, the RNA was isolated from smooth muscle cells, in lane 2 from striated muscle cells, and in lane 3 from brain cells. As seen here, smooth and striated muscle cells contain a large amount of this mRNA. This result is expected because tropomyosin plays a role in the regulation of cell contraction. By comparison, brain cells have much less of this mRNA. In addition, we see that the molecular masses of the three mRNAs are slightly different among the three cell types. This observation indicates that the pre-mRNA is alternatively spliced to contain different combinations of exons.

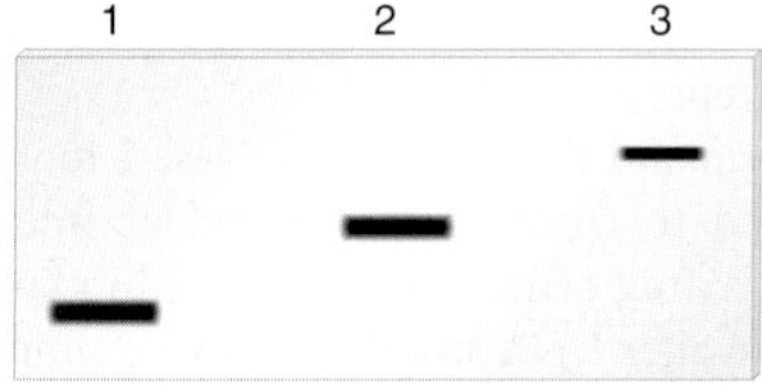

FIGURE 18.10 The results of Northern blotting. RNA is isolated from a sample of cells and then separated by gel electrophoresis. The separated RNA bands are blotted to a nylon membrane and then placed in a solution containing a radioactive probe. Following autoradiography, RNA molecules that are complementary to the probe appear as dark bands on the X-ray film.

As an alternative to Northern blotting, another way to detect a specific RNA is via RT-PCR, a method discussed previously in the chapter.

Western Blotting Is Used to Detect Proteins

For structural genes, the end result of gene expression is the synthesis of proteins. A particular protein within a mixture of many different protein molecules can be identified by **Western blotting,** a third detection procedure. This method can determine if a specific protein is made in a particular cell type or at a particular stage of development. Technically, this procedure is also similar to Southern and Northern blotting.

In a Western blotting experiment, proteins are extracted from living cells. As with RNA, any given cell will produce many different proteins at any time, because it is expressing many structural genes. After the proteins have been extracted from the cells, they are loaded onto a gel that separates them by molecular mass. To perform the separation step, the proteins are first dissolved in sodium dodecyl sulfate (SDS), a detergent that denatures proteins and coats them with negative charges. The negatively charged proteins are then separated in a gel made of polyacrylamide. This method of separating proteins is called SDS-PAGE (polyacrylamide gel electrophoresis).

Following SDS-PAGE, the proteins within the gel are blotted to a nylon membrane. The next step is to use a probe that will recognize a specific protein of interest. An important difference between Western blotting and either Southern or Northern blotting is the use of an **antibody** as a probe, rather than a labeled DNA strand. Antibodies bind to sites known as **epitopes.** An epitope has a three-dimensional structure that is recognized by an antibody. The term **antigen** refers to any molecule that is recognized by an antibody. An antigen contains one or more epitopes. In the case of proteins, an epitope is a short sequence of amino acids. Because the amino acid sequence is a unique feature of each protein, any given antibody will specifically recognize a particular protein. In a Western blot, this is called the primary antibody for that protein.

After the primary antibody has been given sufficient time to recognize the protein of interest, any unbound primary antibody is washed away, and a secondary antibody is added. A secondary antibody is an antibody that binds to the primary antibody. Secondary antibodies, which may be radiolabeled or conjugated to an enzyme, are used for convenience, as secondary antibodies are available commercially. In general, it is easier for researchers to obtain these antibodies from commercial sources rather than label their own primary antibodies. In a Western blotting experiment, the secondary antibody provides a way to detect

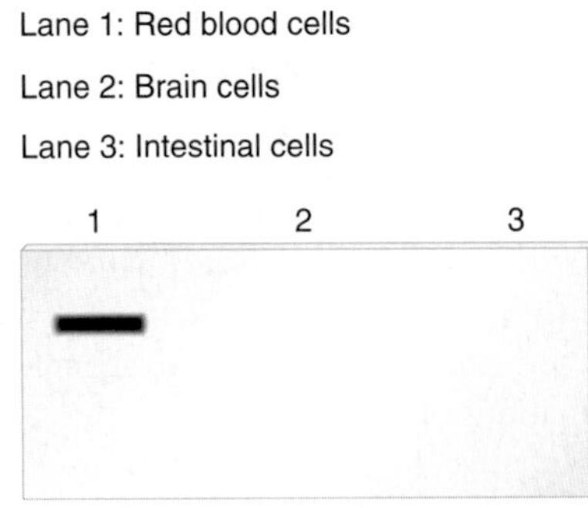

FIGURE 18.11 The results of Western blotting. The black band indicates where the primary antibody has recognized the protein of interest. In lane 1, proteins were isolated from mouse red blood cells. As seen here, the β-globin polypeptide is made in these cells. By comparison, lanes 2 and 3 were samples of protein from brain cells and intestinal cells, which do not synthesize β globin.

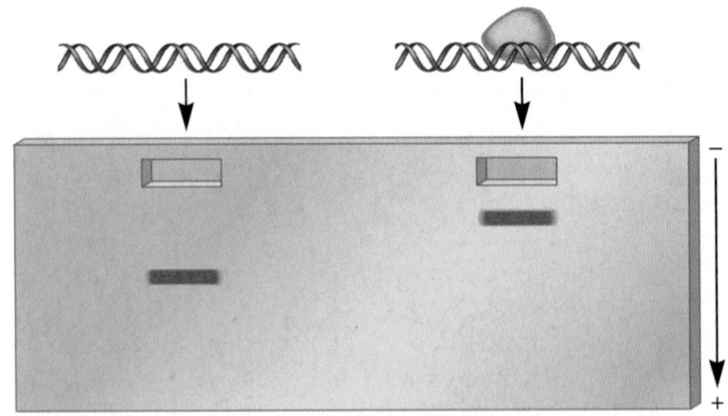

FIGURE 18.12 The results of a gel retardation assay. The binding of protein to a labeled fragment of DNA retards its rate of movement through a gel. For the results shown in the lane on the right, if the concentration of the DNA fragment was higher than the concentration of the protein, there would be two bands: one band with protein bound (at a higher molecular mass) and one band without protein bound (corresponding to the band found in the left lane).

the protein of interest in a gel blot. For example, it is common for the secondary antibody to be linked to the enzyme alkaline phosphatase. When the colorless dye, XP (5-bromo-4-chloro-3-indolyl-phosphate), is added to the blotting solution, alkaline phosphatase converts the dye to a black compound. Because the secondary antibody binds to the primary antibody, a protein band that is recognized by the primary antibody will become black. In the example shown in **Figure 18.11**, the primary antibody recognizes β globin. In lane 1, proteins were isolated from mouse red blood cells. The black band indicates that the β-globin polypeptide is made in these cells. By comparison, the proteins loaded into lanes 2 and 3 were from brain cells and intestinal cells, respectively. The absence of any bands indicates that these cell types do not synthesize β globin.

Techniques Can Be Used to Detect the Binding of Proteins to DNA Sequences

In addition to detecting the presence of genes and gene products using blotting techniques, researchers often want to study the binding of proteins to specific sites on a DNA molecule. For example, the molecular investigation of transcription factors requires methods that can identify interactions between transcription factor proteins and specific DNA sequences. A technically simple, widely used method for identifying this type of interaction is the **gel retardation assay,** also known as the **band shift assay.** This technique was used originally to study rRNA–protein interactions and quickly became popular after its success in studying protein–DNA interactions in the *lac* operon. Now it is commonly used as a technique to detect interactions between eukaryotic transcription factors and DNA regulatory elements.

The technical basis for a gel retardation assay is that the binding of a protein to a DNA fragment will retard the fragment's ability to move within a polyacrylamide or agarose gel. During electrophoresis, DNA fragments are pulled through the gel matrix toward the bottom of the gel by a voltage gradient. Smaller fragments of DNA migrate more quickly through a gel matrix than do larger fragments. As you might expect, therefore, the binding of a protein to a DNA fragment will retard the DNA's rate of movement through the gel matrix, because a protein–DNA complex has a higher mass. When comparing a DNA fragment and a protein–DNA complex after electrophoresis, the complex is shifted to a higher band than the DNA alone, because the complex migrates more slowly to the bottom of the gel (**Figure 18.12**).

A gel retardation assay must be carried out under nondenaturing conditions. This means that the buffers and gel cannot cause the unfolding of proteins or the separation of the DNA double helix. This is necessary so that the proteins and DNA retain their proper structure and thereby can bind to each other. The nondenaturing conditions of a gel retardation assay differ from the more common SDS–gel electrophoresis, in which the proteins are denatured by the detergent SDS.

A second method for studying protein–DNA interactions is **DNase I footprinting,** a technique described originally by David Galas and Albert Schmitz in 1978. A DNase I footprinting experiment attempts to identify one or more regions of DNA that interact with a DNA-binding protein. In their original study, Galas and Schmitz identified a site in the *lac* operon, known as the *lac* operator site, that is bound by a DNA-binding protein called the lac repressor.

To understand the basis of a DNase I footprinting experiment, we need to consider the molecular interactions among three things: a fragment of DNA, DNA-binding proteins, and agents that can alter DNA structure. As an example, let's examine the binding of RNA polymerase to a bacterial promoter, a topic discussed in Chapter 12. When RNA polymerase holoenzyme binds to the promoter to form a closed complex, it binds tightly to the –35 and –10 promoter regions, but the protein covers up an even larger region of the DNA. Therefore, holoenzyme bound at the promoter prevents other molecules from gaining access to this region of the DNA. The enzyme DNase I, which can cleave covalent bonds in the DNA backbone, is used as a reagent to determine if a DNA region has a protein bound to it. Galas and Schmitz reasoned that DNase I cannot cleave the DNA at locations where a protein is bound. In this example, it is expected that RNA polymerase holoenzyme will bind to a promoter and protect this DNA region from DNase I cleavage.

Figure 18.13 shows the results of a DNase I footprinting experiment. In this experiment, a sample of many identical DNA fragments, which are 150 bp in length, were radiolabeled at only one end. The sample of fragments was then divided into two tubes: tube A, which did not contain any holoenzyme, and tube B, which contained RNA polymerase holoenzyme. DNase I was then added to both tubes. The tubes were incubated long enough for DNase I to cleave the DNA at a single site in each DNA fragment. Each tube contained many 150 bp DNA fragments, and the cutting in any DNA strand by DNase I occurred randomly. Therefore, the DNase I treatment should produce a mixture of many smaller DNA fragments. A key point, however, is that DNase I cannot cleave the DNA in a region where RNA polymerase holoenzyme is bound. After DNase I treatment, the DNA fragments within the two tubes were separated by gel electrophoresis, and DNA fragments containing the labeled end were detected by autoradiography.

In the absence of RNA polymerase holoenzyme (tube A), DNase I should cleave the 150 bp fragments randomly at any single location. Therefore, a continuous range of sizes occurs (Figure 18.13). However, if we look at the gel lane from tube B, no bands are observed in the size range from 25 to 105 nucleotides. Why are these bands missing? The answer is that DNase I cannot cleave the DNA within the region where the holoenzyme is bound. The middle portion of the 150 bp fragment contains a promoter sequence that binds the RNA polymerase holoenzyme. Along the right side of the gel, the bases are numbered according to their position within the gene. (The site labeled +1 is where transcription begins.) As seen here, RNA polymerase covers up a fairly large region (its "footprint") of about 80 nucleotides, from the –50 region to the +30 region.

As illustrated in this experiment, DNase I footprinting can identify the DNA region that interacts with a DNA-binding protein. In addition to RNA polymerase–promoter binding, DNase I footprinting has been used to identify the binding sites for many other types of DNA-binding proteins, such as eukaryotic transcription factors and histones. This technique has greatly facilitated our understanding of protein-DNA interactions.

A newer method, called chromatin immunoprecipitation (ChIP), can determine if proteins bind to particular DNA sequences in a living cell. Because this method is sometimes done in conjunction with DNA microarrays, we will consider this approach in Chapter 21.

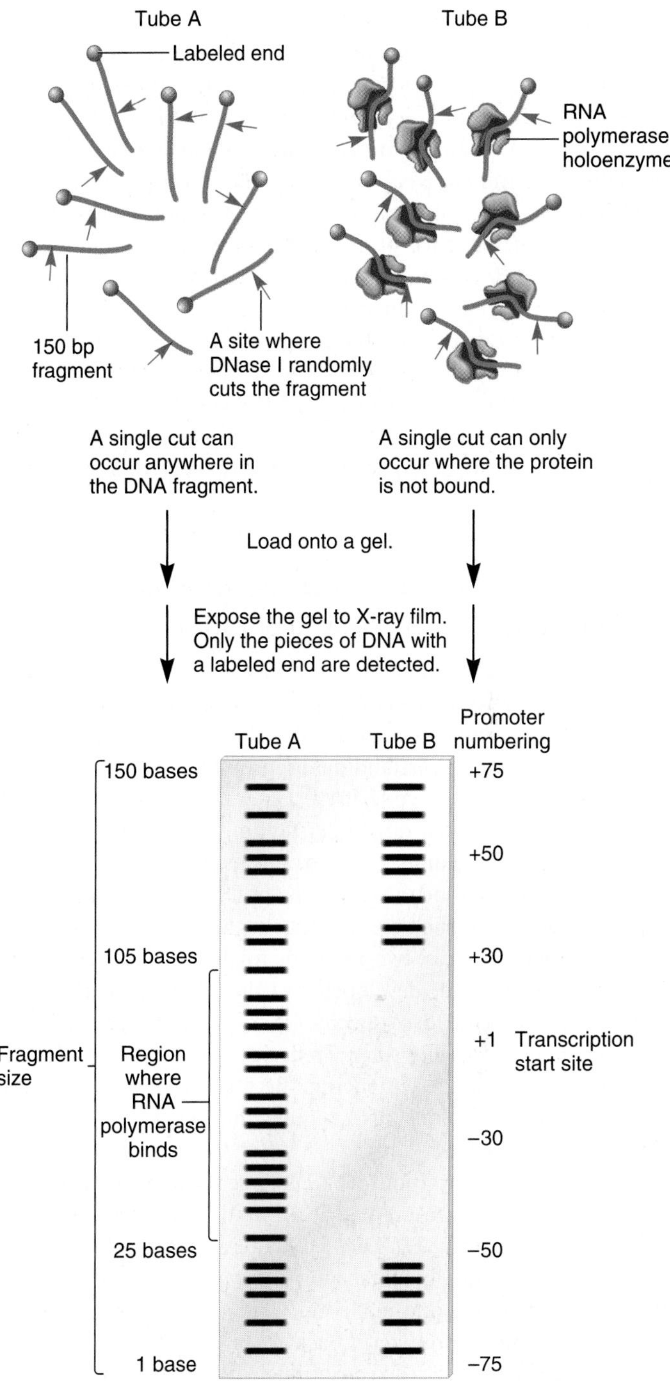

FIGURE 18.13 **A DNase I footprinting experiment.** Both tubes contained 150 bp fragments of DNA that were incubated with DNase I. Tube B also contained RNA polymerase holoenzyme. The binding of RNA polymerase holoenzyme protected a region of about 80 nucleotides (namely, the –50 region to the +30 region) from DNase I digestion. Note: The promoter numbering convention shown here is the same as that described previously in Figure 12.3.

18.3 ANALYSIS AND ALTERATION OF DNA SEQUENCES

As we have seen throughout this textbook, our knowledge of genetics can be largely attributed to an understanding of DNA structure and function. The feature that underlies all aspects of inherited traits is the DNA sequence. For this reason, analyzing and altering DNA sequences is a powerful approach in understanding genetics. In this last section, we will begin by examining a technique called **DNA sequencing.** This method enables researchers to determine the base sequence of DNA found in genes and other chromosomal regions. It is one of the most important tools for exploring genetics at the molecular level. DNA sequencing is practiced by scientists around the world, and the amount of scientific information contained within experimentally determined DNA sequences has become enormous. In Chapter 20, we will learn how researchers can determine the complete DNA sequence of entire genomes. In Chapter 21, we will consider how computers play an essential role in the storage and analysis of genetic sequences.

Not only can researchers determine DNA sequences, they can also use another technique, known as site-directed mutagenesis, to change the sequence of cloned DNA segments. At the end of this section, we will examine how site-directed mutagenesis is conducted and how it provides information regarding the function of genes.

The Dideoxy Method of DNA Sequencing Is Based on Our Knowledge of DNA Replication

Molecular geneticists often seek to determine base sequences as a first step toward understanding the function and expression of genes. For example, the investigation of genetic sequences has been vital to our understanding of promoters, regulatory elements, and the genetic code itself. Likewise, an examination of sequences has facilitated our understanding of origins of replication, centromeres, telomeres, and transposable elements. In this section, we will examine the technique of DNA sequencing, which is used to determine the base sequence within a DNA strand.

During the 1970s, two methods for DNA sequencing were devised. One method, developed by Allan Maxam and Walter Gilbert, involves the base-specific chemical cleavage of DNA. Another method, developed by Frederick Sanger and colleagues, is known as **dideoxy sequencing.** Because it has become the more popular method of DNA sequencing, we will consider the dideoxy method here.

The dideoxy procedure of DNA sequencing is based on our knowledge of DNA replication but uses a clever twist. As described in Chapter 11, DNA polymerase connects adjacent deoxyribonucleotides by catalyzing a covalent linkage between the 5' phosphate on one nucleotide and the 3' –OH group on the previous nucleotide (refer back to Figure 11.11). Chemists, though, can synthesize deoxyribonucleotides that are missing the –OH group at the 3' position (**Figure 18.14**). These synthetic nucleotides are called **dideoxyribonucleotides.** (Note: The prefix dideoxy- indicates that this sugar has two (di) removed (de) oxygens (oxy) compared to ribose; ribose has –OH groups at both the 2' and 3' positions.) Sanger reasoned that if a dideoxyribonucleotide is added to a growing DNA strand, the strand can no longer grow because the dideoxyribonucleotide is missing the 3' –OH group. The incorporation of a dideoxyribonucleotide into a growing strand is therefore referred to as **chain termination.**

Before describing the steps of this DNA-sequencing protocol, we need to become acquainted with the DNA segments that are used in a sequencing experiment. Prior to DNA sequencing, the segment of DNA to be sequenced must be obtained in large amounts. This is accomplished using gene cloning, which was described earlier in this chapter. In **Figure 18.15**, the segment of DNA to be sequenced, which we will call the target DNA, was cloned into a vector at a defined location. The target DNA was inserted next to a site in the vector where a primer will bind, which is called the primer-annealing site. The aim of the experiment is to determine the base sequence of the target DNA that has been inserted next to the primer-annealing site. In the experiment shown in Figure 18.15, the recombinant vector DNA has been previously denatured into single strands, usually via heat treatment. Only the strand needed for DNA sequencing is shown here.

With these ideas in mind, we can now consider the steps involved in DNA sequencing (Figure 18.15). First, a sample containing many copies of the single-stranded DNA of the recombinant vector is mixed with primers that will bind to the primer-annealing site. This annealing process is identical to hybridization, because the primer and primer-annealing site are complementary to each other. All four types of deoxyribonucleotides and DNA polymerase are then added to the annealed DNA fragments, and the mixture is divided into four separate tubes. Each of the four tubes has a low concentration of one dideoxyribonucleotide (ddGTP, ddATP, ddTTP, or ddCTP). The tubes are then incubated to allow DNA polymerase to make strands complementary to the target DNA sequence. In the third tube in this example, which contains ddTTP, chain termination can occasionally occur at the sixth or thirteenth position of the newly synthesized DNA strand if a ddT becomes incorporated at either of these sites. Note that the complementary A base is found at the sixth and thirteenth position in the target DNA. Therefore, in this tube, we would expect to make DNA strands that will terminate at the sixth or thirteenth positions. Likewise, in the second tube, which contains ddATP, chain termination will occur only at the second, seventh, eighth, or eleventh positions because a complementary T base is found at the corresponding positions in the target strand.

Within the four tubes, mixtures of DNA strands of different lengths will be made, depending on the number of nucleotides attached to the primer. These DNA strands can be separated according to their lengths by running them on an acrylamide gel. The shorter strands move to the bottom of the gel more quickly than the longer strands. To detect the newly made DNA strands, small amounts of radiolabeled deoxyribonucleotides are added to each reaction. This enables the strands to be visualized as bands when the gel is exposed to X-ray film. In Figure 18.15,

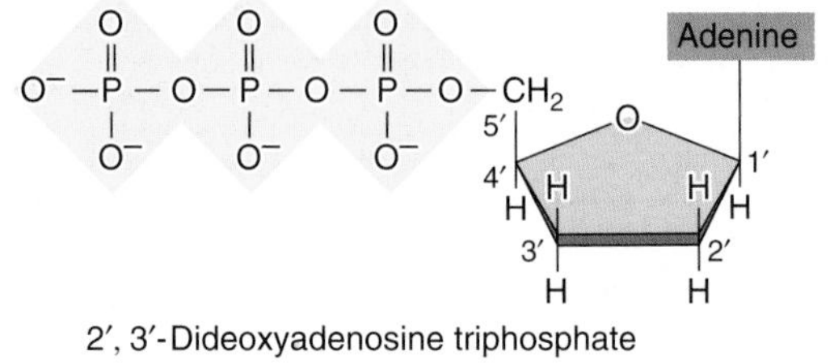

FIGURE 18.14 The structure of a dideoxyribonucleotide. Note that the 3' group is a hydrogen rather than an –OH group. For this reason, another nucleotide cannot be attached at the 3' position.

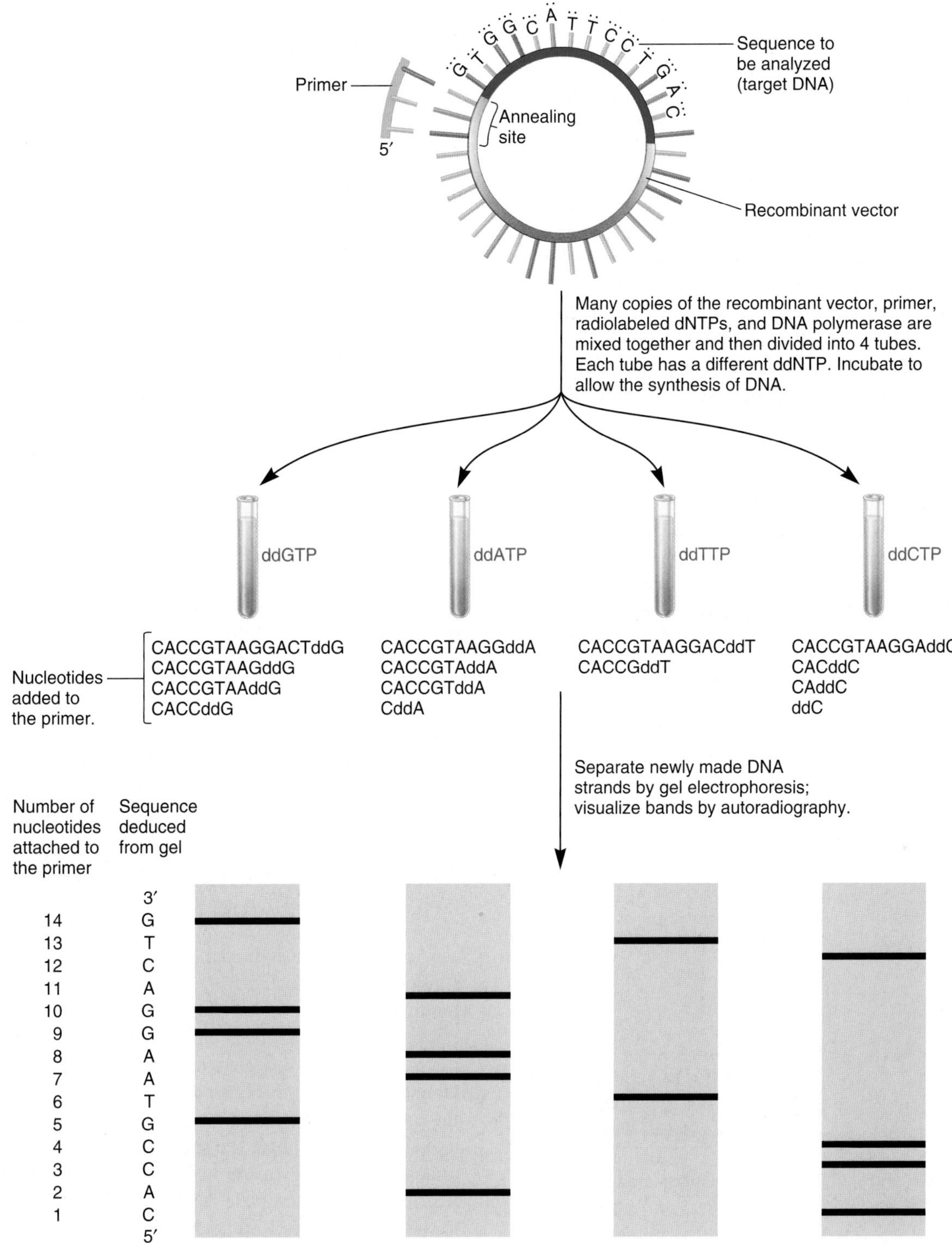

FIGURE 18.15 **The protocol for DNA sequencing by the dideoxy method.** The method shown here begins with single-stranded DNA in which the target DNA has been inserted into a viral vector. Alternatively, double-stranded DNA can be used as the template DNA. However, for the primer to bind, the template must be denatured into single-stranded DNA at the beginning of the experiment.

the DNA strands in the four tubes were run in separate lanes on an acrylamide gel. Because we know which dideoxyribonucleotide was added to each tube, we also know which base is at the very end of each DNA strand separated on this gel. Therefore, we can deduce the DNA sequence that is complementary to the target DNA by "reading" which base is at the end of every DNA strand and matching this sequence with the length of the strand. Reading the base sequence, from bottom to top, is much like climbing a ladder of bands. For this reason, the sequence obtained by this method is referred to as a **sequencing ladder.**

An important innovation in the method of dideoxy sequencing is **automated sequencing.** Instead of having four separate tubes, with a single type of dideoxyribonucleotide in each tube, automated sequencing uses one tube containing all four types of dideoxyribonucleotides. However, each type of dideoxyribonucleotide (ddGTP, ddATP, ddTTP, and ddCTP) has a different-colored fluorescent label attached. After incubating the target DNA with deoxyribonucleotides, the four types of fluorescent dideoxyribonucleotides, and DNA polymerase, the sample is then loaded into a single lane of a gel. A schematic example of a lane of automated sequencing is shown in **Figure 18.16a.**

In automated sequencing, a sample is loaded at the top of a gel, and then the fragments are separated by electrophoresis. Theoretically, it would be possible to read this sequence directly from the gel. From a practical perspective, however, it is more efficient to automate the procedure using a laser and fluorescence detector. Electrophoresis is continued until each band emerges from the bottom of the gel. As each band comes off, a laser excites the fluorescent dye, and a fluorescence detector records the amount of fluorescence emission. The detector reads the level of fluorescence at four wavelengths, corresponding to the four dyes. An example of the printout from the fluorescence detector is shown in **Figure 18.16b.** As seen here, the peaks of fluorescence correspond to the DNA sequence that is complementary to the target DNA.

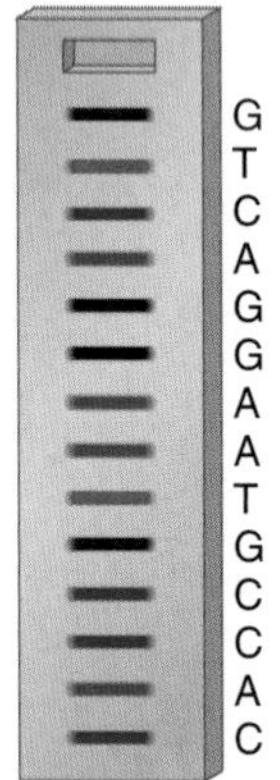

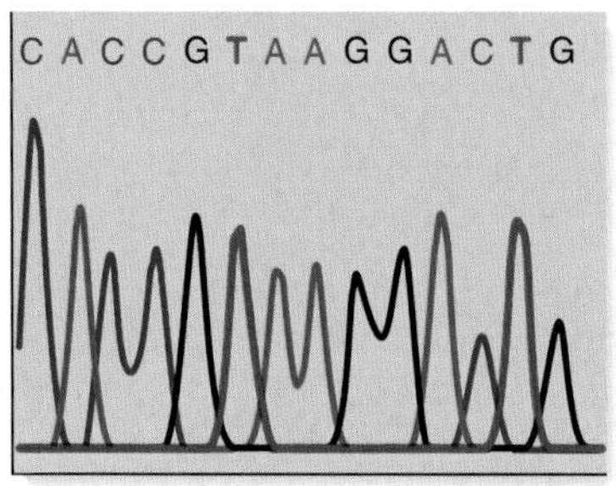

(a) Automated sequencing gel **(b) Output from automated sequencing**

FIGURE 18.16 **Automated DNA sequencing.** (**a**) This diagram schematically depicts a series of bands on a gel; the four colors of the bands occur because each type of dideoxynucleotide is labeled with a different colored fluorescent molecule. (**b**) The mixture of DNA fragments is electrophoresed off the end of the gel. As each band is released from the bottom of the gel, the fluorescent dye is excited by a laser, and the fluorescence emission is recorded by a fluorescence detector. The detector reads the level of fluorescence at four wavelengths, corresponding to the four dyes. As shown in the printout, the peaks of fluorescence correspond to the DNA sequence that is complementary to the target DNA described in Figure 18.15.

Site-Directed Mutagenesis Is a Technique to Alter DNA Sequences

As we have seen, dideoxy sequencing provides a way to determine the base sequence of DNA. To understand how the genetic material functions, researchers often analyze mutations that alter the normal DNA sequence and thereby affect the expression of genes and the outcome of traits. For example, geneticists have discovered that many inherited human diseases, such as sickle-cell disease and hemophilia, involve mutations within specific genes. These mutations provide insight into the function of the genes in unaffected individuals. Hemophilia, for example, involves deleterious mutations in genes that encode blood clotting factors.

Because the analysis of mutations can provide important information about normal genetic processes, researchers often wish to obtain mutant organisms. As we discussed in Chapter 16, mutations can arise spontaneously or can be induced by environmental agents. Mendel's pea plants are a classic example of allelic strains with different phenotypes that arose from spontaneous mutations. X-rays and UV light are physical agents that can cause induced mutations. In addition, experimental organisms can be treated with mutagens that increase the rate of mutations.

More recently, researchers have developed molecular techniques to make mutations within cloned genes or other DNA segments. One widely used method, known as **site-directed mutagenesis,** allows a researcher to produce a mutation at a specific site within a cloned DNA segment. For example, if a DNA sequence is 5'–AAATTTCTTTAAA–3', a researcher can use site-directed mutagenesis to change it to 5'–AAATTTGTTTAAA–3'. In this case, the researcher deliberately changed the seventh base from a C to a G. Why is this method useful? The site-directed mutant can then be introduced into a living organism to see how the mutation affects the expression of a gene, the function of a protein, and the phenotype of an organism.

The first successful attempts at site-directed mutagenesis involved changes in the sequences of viral genomes. These studies were conducted in the 1970s. Mark Zoller and Michael Smith also developed a protocol for the site-directed mutagenesis of DNA that has been cloned into a viral vector. Since these early studies, many approaches have been used to achieve site-directed mutagenesis. **Figure 18.17** describes the general steps in the procedure. Prior to this experiment, the DNA was denatured into single strands; only the single strand needed for site-directed mutagenesis is shown. As in PCR, this single-stranded DNA is referred to as the template DNA, because it is used as a template to synthesize a complementary strand.

As shown in Figure 18.17, an oligonucleotide primer is allowed to hybridize or anneal to the template DNA. The primer, typically 20 or so nucleotides in length, is synthesized chemically. (A shorter version of the primer is shown in Figure 18.17 for simplicity.) The scientist designs the base sequence of the primer. The primer has two important characteristics. First, most of the sequence of the primer is complementary to the site in the DNA where the mutation is to be made. However, a second feature is that the primer contains a region of mismatch where the primer and template DNA are not complementary. The mutation will

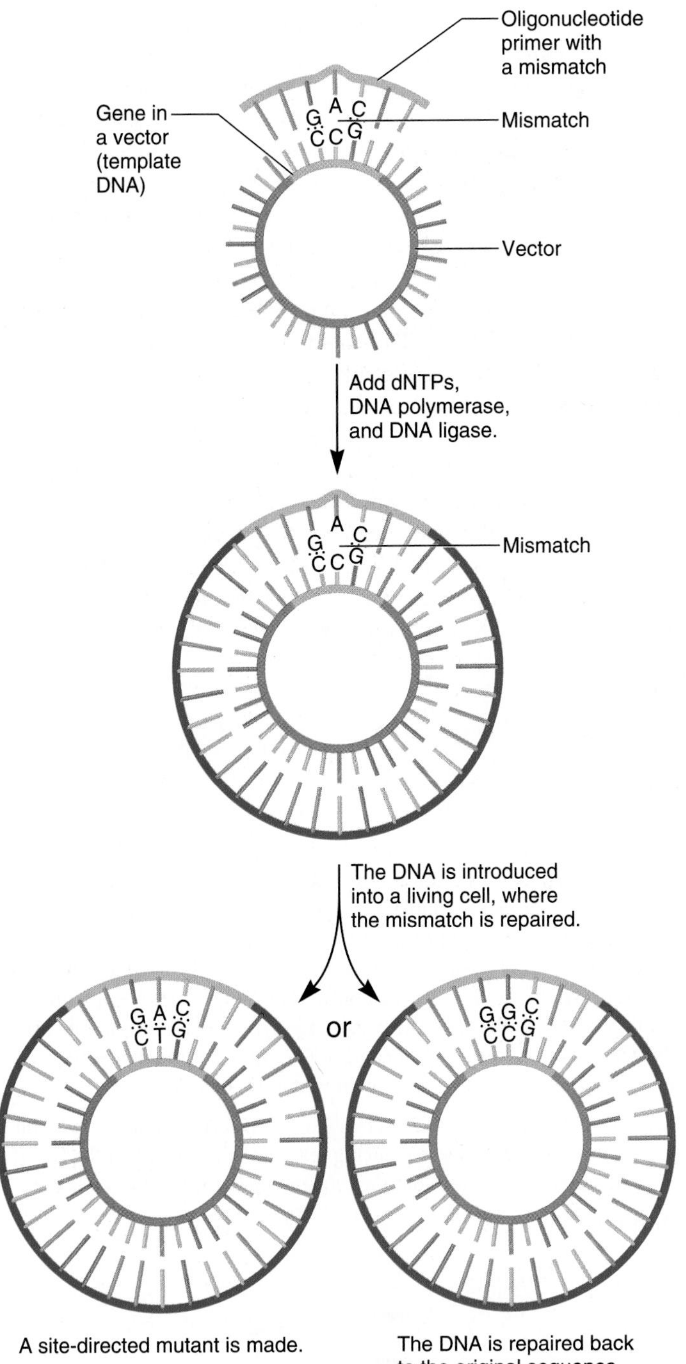

occur in this mismatched region. For this reason, site-directed mutagenesis is sometimes referred to as oligonucleotide-directed mutagenesis.

After the primer and template have annealed, the complementary strand is synthesized by adding deoxyribonucleoside triphosphates (dNTPs), DNA polymerase, and DNA ligase. This yields a double-stranded molecule that contains a mismatch only at the desired location. This double-stranded DNA is then introduced into a bacterial cell. Within the cell, the DNA mismatch will likely be repaired (see Chapter 16). Depending on which base is replaced, this may produce the mutant sequence or the original sequence. Clones containing the desired mutation can be identified by DNA sequencing and used for further studies.

After a site-directed mutation has been made within a cloned gene, its consequences are analyzed by introducing the mutant gene into a living cell or organism. As described earlier, recombinant vectors containing cloned genes can be introduced into bacterial cells. Following transformation into a bacterium, a researcher can study the differences in function between the mutant and wild-type genes and the proteins they encode. Similarly, mutant genes made via site-directed mutagenesis can be introduced into plants and animals, as discussed in Chapter 19.

FIGURE 18.17 The method of site-directed mutagenesis.

Genes→Traits To examine the relationship between genes and traits, researchers can alter gene sequences via site-directed mutagenesis. The altered gene can then be introduced into a living organism to examine how the mutation affects the organism's traits. For example, a researcher could introduce a nonsense mutation into the middle of the *lacY* gene in the *lac* operon. If this site-directed mutant was introduced into an *E. coli* bacterium that did not a have a normal copy of the *lacY* gene, the bacterium would be unable to use lactose. These results indicate that a functional *lacY* gene is necessary for bacteria to have the trait of lactose utilization.

CONCEPTUAL AND EXPERIMENTAL SUMMARY

Recombinant DNA technology, the development of which began in the early 1970s, has revolutionized our understanding of molecular genetics. Researchers can now use restriction enzymes to cut DNA fragments out of their native sites within a chromosome. The DNA fragments can then be ligated to a vector, either a plasmid or viral vector, which propagates within a living host cell. This technology, known as gene cloning, refers to the phenomenon of isolating and making many copies of a gene in order to study it in detail. Similarly, the goal of polymerase chain reaction (PCR) is to make many copies of the template DNA in a defined region. PCR can achieve gene cloning without the need for vectors and host cells and can amplify the amount of DNA by a staggering amount. Gene cloning has many practical applications, which will be described in Chapter 19.

In some cases, a researcher needs to detect a specific gene or gene product. For example, gene detection procedures, such as colony hybridization, are usually needed to identify particular cloned genes in a DNA library. In addition, detection of gene products can provide information regarding the expression pattern of a gene in particular cell types or during specific stages of development. Three common methods of detection are Southern blotting, Northern blotting, and Western blotting. In Southern blotting, a DNA probe is used to detect the presence of a gene or other DNA sequence within a mixture of many DNA fragments. Under conditions of high stringency, this technique can detect close or exact complementarity between the probe and a chromosomal DNA fragment. Under conditions of low stringency, it can detect homologous genes with DNA sequences that are similar but not identical to the DNA probe. The Northern blotting procedure is used to detect the amount of RNA from a specific gene. In this technique, a DNA probe is used to detect RNA. Both Southern and Northern blotting rely on sequence homology between the DNA probe and the DNA or RNA in the sample. They are called hybridization techniques because the labeled DNA probe forms a hybrid with a specific molecule in the sample. Finally, Western blotting is used to detect the protein product from a particular gene. In this technique, an antibody is used as a probe because antibodies bind to proteins very specifically. In addition to these detection methods, researchers can use a gel retardation assay and DNase I footprinting as a way to study protein-DNA interactions.

Recombinant DNA technology has also enabled researchers to determine the base sequence of DNA. In dideoxy sequencing, dideoxyribonucleotides are used that terminate the action of DNA polymerase at specific locations in the growing DNA strands. When the terminated DNA strands are separated on the basis of their size by gel electrophoresis, this creates a sequencing ladder of DNA fragments that are terminated at defined locations. The DNA sequence can then be determined by reading the bases along the ladder. The method of automated sequencing uses four different fluorescently labeled dideoxyribonucleotides and a fluorescence detector that reads the level of fluorescence at four wavelengths, corresponding to the four dyes. In addition, researchers can answer questions concerning the functional importance of DNA sequences by site-directed mutagenesis, a procedure that allows a researcher to produce a mutation at a particular site in a cloned fragment of DNA. The site-directed mutant can be introduced into a living organism to see how the mutation affects the expression of a gene, the function of a protein, and the phenotype of an organism.

PROBLEM SETS & INSIGHTS

Solved Problems

S1. RNA was isolated from four different cell types and probed with radiolabeled DNA strands from a cloned gene that is called gene *X*. The results are shown here.

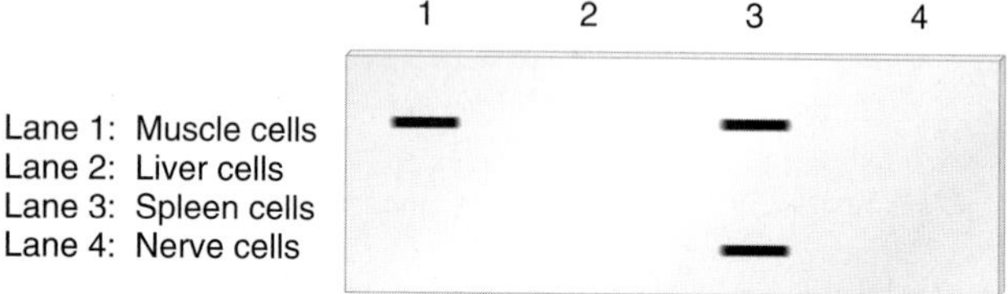

Explain the results of this experiment.

Answer: In this Northern blot, a dark band appears in those lanes where RNA was isolated from muscle and spleen cells but not from liver and nerve cells. These results indicate that the muscle and spleen cells contain a significant amount of RNA from gene *X*, but the liver and nerve cells do not. The muscle cells show a single band, whereas the spleen cells show this band plus a second band of lower molecular mass. An interpretation of these results is that the spleen cells can alternatively splice the RNA to produce a second RNA containing fewer exons.

S2. In the Western blotting experiment shown here, proteins were extracted from red blood cells obtained from tissue samples at different stages of human development. An equal amount of total cellular proteins was added to each lane. The primary antibody recognizes the β-globin polypeptide that is found in the hemoglobin protein.

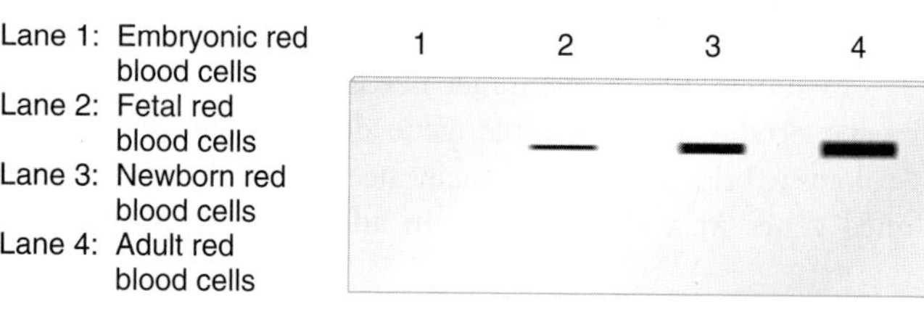

Explain these results.

Answer: As shown here, the amount of β globin increases during development. Little detectable β globin is produced during embryonic development. The amount increases significantly during fetal development and becomes maximal in the adult. These results indicate that the β-globin gene is "turned on" in later stages of development, leading to the synthesis of the β-globin polypeptide. This experiment illustrates how a Western blot can provide information concerning the relative amount of a specific protein within living cells.

S3. A DNA strand is 3'–ATACGACTAGTCGGGACCATATC–5'. If the primer in a dideoxy sequencing reaction anneals just to the left of this sequence, draw what the sequencing ladder would look like.

Answer:

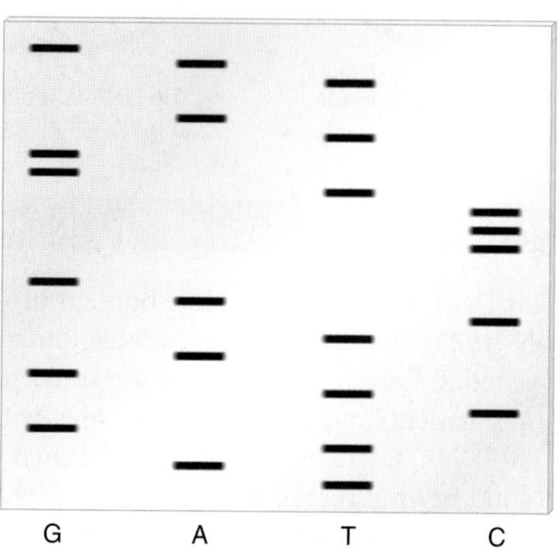

S4. The human genetic disease phenylketonuria (PKU) involves a defect in a gene that encodes the enzyme phenylalanine hydroxylase. It is inherited as a recessive autosomal disorder. Using a strand of DNA from the nonmutant phenylalanine hydroxylase gene as a probe, a Southern blot was carried out on a PKU patient, one of her parents, and an unaffected, unrelated person. In this example, the DNA fragments were subjected to acrylamide gel electrophoresis, rather than agarose gel electrophoresis, because acrylamide gel electrophoresis is better able to detect small deletions within genes. The following results were obtained:

Suggest an explanation for these results.

Answer: In the affected person, the PKU defect is caused by a small deletion within the PKU gene. The parents are heterozygous for the nonmutant gene and the deletion. The PKU-affected person carries only the deletion, which runs at a lower molecular mass than the nonmutant gene. The unrelated person carries two copies of the nonmutant (nondeleted) gene.

Conceptual Questions

C1. Discuss three important advances that have resulted from gene cloning.

C2. What is a restriction enzyme? What structure does it recognize? What type of chemical bond does it cleave? Be as specific as possible.

C3. Write a double-stranded sequence that is 20 bases long and is palindromic.

C4. What is cDNA? In eukaryotes, how would cDNA differ from genomic DNA?

C5. Explain and draw the structural feature of a dideoxyribonucleotide that causes chain termination.

Experimental Questions

E1. What is the functional significance of sticky ends in a cloning experiment? What type of bonding makes the ends sticky?

E2. Table 18.3 describes the cleavage sites of six different restriction enzymes. After these restriction enzymes have cleaved the DNA, five of them produce sticky ends that can hydrogen bond with complementary sticky ends, as shown in Figure 18.1. The efficiency of sticky ends binding together depends on the number of hydrogen bonds; more hydrogen bonds makes the ends "stickier" and more likely to stay attached. Rank these five restriction enzymes in Table 18.3 (from best to worst) with regard to the efficiency of their sticky ends binding to each other.

E3. Describe the important features of cloning vectors. Explain the purpose of selectable marker genes in cloning experiments.

E4. How does gene cloning produce many copies of a gene?

E5. In your own words, describe the series of steps necessary to clone a gene. Your answer should include the use of a probe to identify a bacterial colony that contains the cloned gene of interest.

E6. What is a recombinant vector? How is a recombinant vector constructed? Explain how X-Gal can be used in a method to identify recombinant vectors that contain segments of chromosomal DNA.

E7. In the experiment of Figure 18.3, would the researchers have been successful if pSC102 had an *Eco*RI site in the middle of the kan^R gene? Explain why or why not.

E8. In the experiment of Figure 18.3, would the results have been entirely convincing if they had done only gel electrophoresis (step 10) but not the density gradient centrifugation experiment (step 9)? Can you think of an alternative explanation for the results shown in step 10 that would not require the insertion of the kan^R gene into pSC101? In other words, can you think of an explanation for the results shown in step 10 that would not require the construction of a recombinant plasmid? Do the results shown in step 9 rule out your alternative explanation?

E9. A circular plasmid was digested with one or more restriction enzymes, run on a gel, and the following results were obtained:

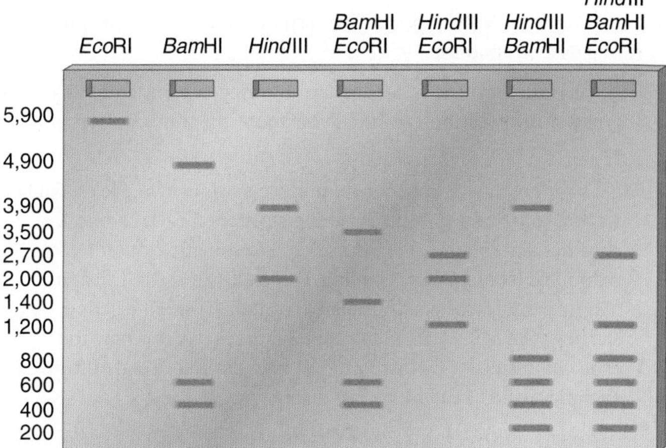

Construct a restriction map for this plasmid.

E10. If a researcher began with a sample that contained three copies of double-stranded DNA, how many copies would be present after 27 cycles of PCR?

E11. Why is a thermostable form of DNA polymerase (e.g., *Taq* polymerase) used in PCR? Is it necessary to use a thermostable form of DNA polymerase in the techniques of dideoxy DNA sequencing or site-directed mutagenesis?

E12. Reverse transcriptase is an enzyme that uses RNA as template to make a complementary strand of DNA. Experimentally, it is used to make cDNA. Reverse transcriptase can also be used in conjunction with PCR to amplify RNAs. This method is called reverse transcriptase PCR, or RT-PCR. In other words, it is possible to make many copies of double-stranded DNA using RNA as a template. Starting with a sample of RNA that contained the mRNA for the β-globin gene, explain how you could create many copies of the β-globin cDNA using RT-PCR.

E13. Let's suppose you have recently cloned a gene, which we will call gene *X*, from corn. You use a labeled DNA strand from this cloned gene to probe genomic DNA from corn in a Southern blot experiment under conditions of low and high stringency. The following results were obtained:

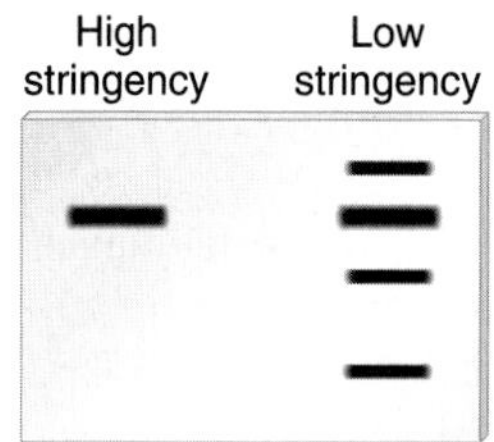

What do these results mean?

E14. What is a DNA library? Do you think this is an appropriate name?

E15. Some vectors used in cloning experiments contain bacterial promoters that are adjacent to unique cloning sites. This makes it possible to insert a gene sequence next to the bacterial promoter and express the gene in bacterial cells. These are called expression vectors. If you wanted to express a eukaryotic protein in bacterial cells, would you clone genomic DNA or cDNA into the expression vector? Explain your choice.

E16. Southern and Northern blotting depend on the phenomenon of hybridization. In these two techniques, explain why hybridization occurs. Which member of the hybrid is labeled?

E17. In Southern, Northern, and Western blotting, what is the purpose of gel electrophoresis?

E18. What is the purpose of a Northern blotting experiment? What types of information can it tell you about the transcription of a gene?

E19. Let's suppose an X-linked gene in mice exists as two alleles, which we will call *B* and *b*. X inactivation, a process in which one X chromosome is turned off, occurs in the somatic cells of female mammals (see Chapter 7). Allele *B* encodes an mRNA that is 900 nucleotides long, while allele *b* contains a small deletion that shortens the mRNA to a length of 825 nucleotides. Draw the expected results of a Northern blot using mRNA isolated from somatic tissue of the following mice:

Lane 1. mRNA from an X^bY male mouse

Lane 2. mRNA from an X^bX^b female mouse

Lane 3. mRNA from an X^BX^b female mouse. Note: The sample taken from the female mouse is not from a clone of cells. It is from a tissue sample, like the one shown at the beginning of the experiment of Figure 7.6.

E20. The method of Northern blotting can be used to determine the amount and size of a particular RNA transcribed in a given cell type. Alternative splicing (discussed in Chapter 15) can produce mRNAs from the same gene that have different lengths. A Northern blot is shown here using a DNA probe that is complementary to the mRNA encoded by a particular gene. The mRNA in lanes 1–4 was isolated from different cell types, and equal amounts of total cellular mRNA were added to each lane.

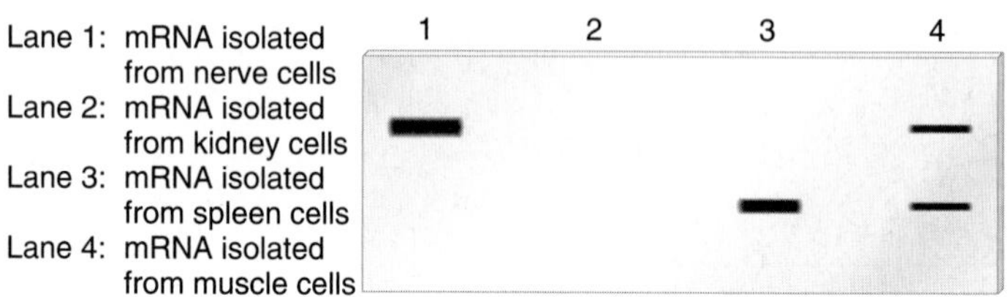

Explain these results.

E21. Southern blotting can be used to detect the presence of repetitive sequences, such as transposable elements, that are present in multiple copies within the chromosomal DNA of an organism. (Note: Transposable elements are described in Chapter 17.) In the Southern blot shown here, chromosomal DNA was isolated from three different strains of baker's yeast, digested with a restriction enzyme, run on a gel, blotted, and then probed with a radioactive DNA probe that is complementary to a transposable element called the *Ty* element.

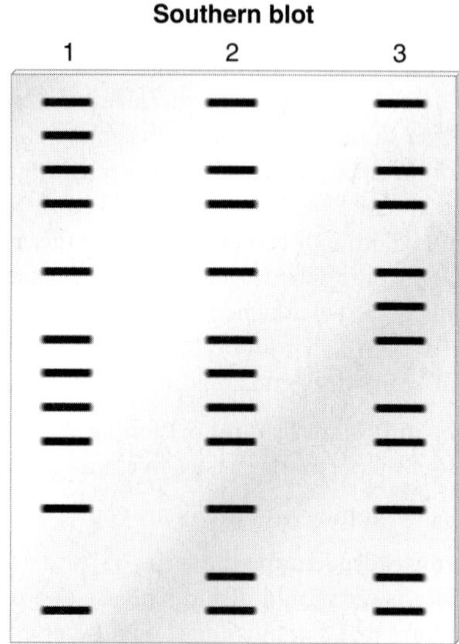

Explain, in a general way, why the banding patterns are not the same in lanes 1, 2, and 3.

E22. In Chapter 8, Figure 8.10 describes the evolution of the globin gene family. All of the genes in this family are homologous to each other, though the degree of sequence similarity varies depending on the time of divergence. Genes that have diverged more recently have sequences that are more similar. For example, the α_1 and α_2 genes have DNA sequences that are more similar to each other compared to the α_1 and ξ genes. In a Southern blotting experiment, the degree of sequence similarity can be discerned by varying the stringency of hybridization. At high temperature (i.e., high stringency), the probe will recognize genes that are only a perfect or very close match. At a lower temperature, however, homologous genes with lower degrees of similarities can be detected because slight mismatches are tolerated. If a Southern blot was conducted on a sample of human chromosomal DNA and a probe was used that was a perfect match to the β-globin gene, rank the following genes (from those that would be detected at high stringency down to those that would only be detected at low stringency) as they would appear in a Southern blot experiment: *Mb*, α_1, β, γ_A, δ, and ε.

E23. In the Western blot shown here, polypeptides were isolated from red blood cells and muscle cells from two different individuals. One individual was unaffected, and the other individual suffered from a disease known as thalassemia, which involves a defect in hemoglobin. In the Western blot, the gel blot was exposed to an antibody that recognizes β globin, which is one of the polypeptides that constitute hemoglobin. Equal amounts of total cellular proteins were added to each lane.

Lane 1: Proteins isolated from normal red blood cells
Lane 2: Proteins isolated from the red blood cells of a thalassemia patient
Lane 3: Proteins isolated from normal muscle cells
Lane 4: Proteins isolated from the muscle cells of a thalassemia patient

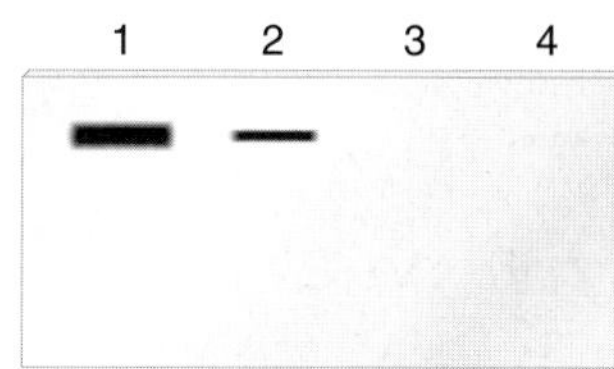

Explain these results.

E24. Let's suppose a researcher was interested in the effects of mutations on the expression of a structural gene that encodes a polypeptide 472 amino acids in length. This polypeptide is expressed in leaf cells of *Arabidopsis thaliana.* Because the average molecular mass of an amino acid is 120 Daltons, this protein has a molecular mass of approximately 56,640 Daltons. Make a drawing that shows the expected results of a Western blot using polypeptides isolated from the leaf cells that were obtained from the following individuals:

Lane 1. A plant homozygous for a nonmutant gene

Lane 2. A plant homozygous for a deletion that removes the promoter for this gene

Lane 3. A heterozygous plant in which one gene is nonmutant and the other gene has a mutation that introduces an early stop codon at codon 112

Lane 4. A plant homozygous for a mutation that introduces an early stop codon at codon 112

Lane 5. A plant homozygous for a mutation that changes codon 108 from a phenylalanine codon into a leucine codon

E25. If you wanted to know if a protein was made during a particular stage of development, what technique would you choose?

E26. Explain the basis for using an antibody as a probe in a Western blotting experiment.

E27. Starting with pig cells and a probe that is a labeled DNA strand from the human β-globin gene, describe how you would clone the β-globin gene from pigs. You may assume that you have available all the materials needed in a cloning experiment. How would you confirm that a putative clone really contained a β-globin gene?

E28. A cloned gene fragment contains a response element that is recognized by a regulatory transcription factor. Previous experiments have shown that the presence of a hormone results in transcriptional activation by this transcription factor. To study this effect, you conduct a gel retardation assay and obtain the following results:

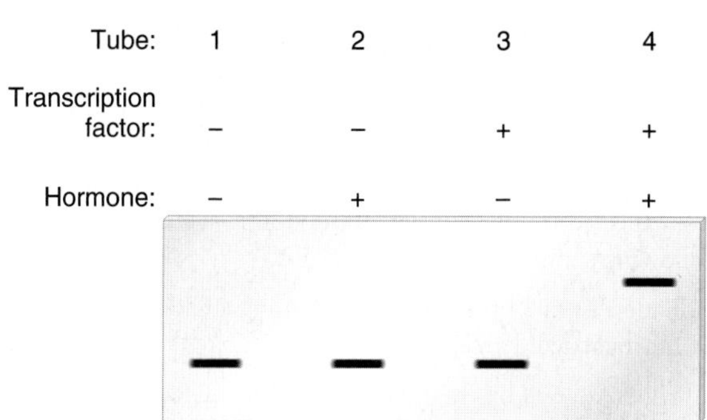

Explain the action of the hormone.

E29. Describe the rationale behind a gel retardation assay.

E30. Certain hormones, such as epinephrine, can increase the levels of cAMP within cells. Let's suppose you can pretreat cells with or without epinephrine and then prepare a cell extract that contains the CREB protein (see Chapter 15 for a description of the CREB protein). You then use a gel retardation assay to analyze the ability of the CREB protein to bind to a DNA fragment containing a cAMP response element (CRE). Describe what the expected results would be.

E31. A gel retardation assay can be used to study the binding of proteins to a segment of DNA. In the experiment shown here, a gel retardation assay was used to examine the requirements for the binding of RNA polymerase II (from eukaryotic cells) to the promoter of a structural gene. The assembly of RNA polymerase II at the core promoter is described in Chapter 12 (Figure 12.13). In this experiment, the segment of DNA containing a promoter sequence was 1,100 bp in length. The fragment was mixed with various combinations of proteins and then subjected to a gel retardation assay.

Lane 1: No proteins added
Lane 2: TFIID
Lane 3: TFIIB
Lane 4: RNA polymerase II
Lane 5: TFIID + TFIIB
Lane 6: TFIID + RNA polymerase II
Lane 7: TFIID + TFIIB + RNA polymerase II

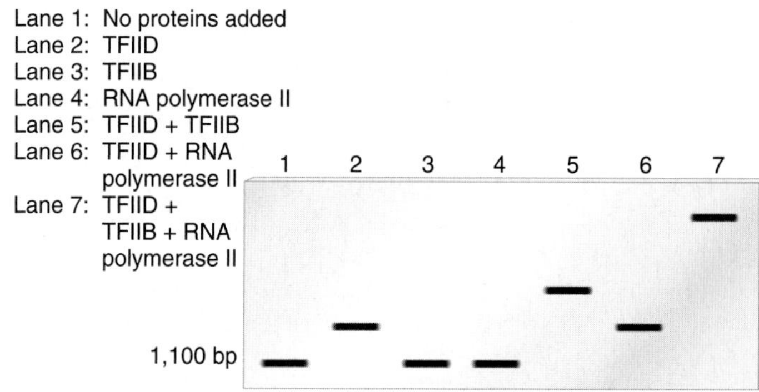

Explain which proteins (TFIID, TFIIB, and/or RNA polymerase II) are able to bind to this DNA fragment by themselves. Which transcription factors (i.e., TFIID and/or TFIIB) are needed for the binding of RNA polymerase II?

E32. As described in Chapter 15 (Figure 15.6), certain regulatory transcription factors bind to the DNA and activate RNA polymerase II. When glucocorticoid binds to the glucocorticoid receptor (a regulatory transcription factor), this changes the conformation of the receptor and allows it ultimately to bind to the DNA. The glucocorticoid receptor binds to a DNA sequence called a glucocorticoid response element (GRE). In contrast, other regulatory transcription factors, such as the CREB protein, do not require hormone binding in order to bind to DNA. The CREB protein can bind to the DNA in the absence of any hormone, but it will not activate RNA polymerase II unless the CREB protein is phosphorylated. (Phosphorylation is stimulated by certain hormones.) The CREB protein binds to a DNA sequence called a cAMP response element (CRE). With these ideas in mind, draw the expected results of a gel retardation assay conducted on the following samples:

Lane 1. A 600 bp fragment containing a GRE, plus the glucocorticoid receptor

Lane 2. A 600 bp fragment containing a GRE, plus the glucocorticoid receptor, plus glucocorticoid hormone

Lane 3. A 600 bp fragment containing a GRE, plus the CREB protein

Lane 4. A 700 bp fragment containing a CRE, plus the CREB protein

Lane 5. A 700 bp fragment containing a CRE, plus the CREB protein, plus a hormone (such as epinephrine) that causes the phosphorylation of the CREB protein

Lane 6. A 700 bp fragment containing a CRE, plus the glucocorticoid receptor, plus glucocorticoid hormone.

E33. In the technique of DNase I footprinting, the binding of a protein to a region of DNA will protect that region from digestion by DNase I by blocking the ability of DNase I to gain access to the phosphodiester linkages in the DNA. In the DNase I footprinting experiment shown here, a researcher began with a sample of cloned DNA 400 bp in length. This DNA contained a eukaryotic promoter for RNA polymerase II. The assembly of RNA polymerase II at the core promoter is described in Chapter 12. For the sample loaded in lane 1, no proteins were added. For the sample loaded in lane 2, the 400 bp fragment was mixed with RNA polymerase II plus TFIID and TFIIB.

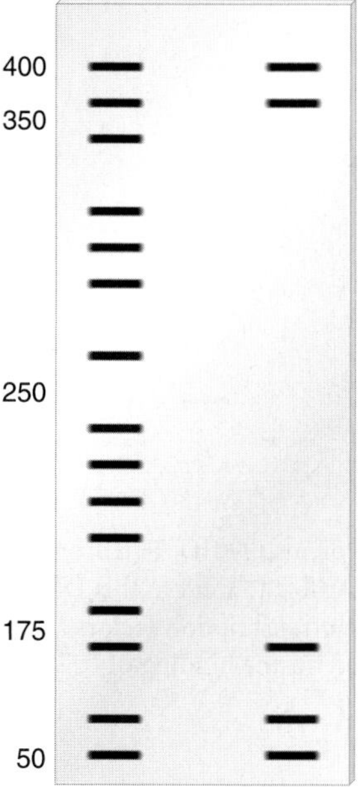

Which region of this 400 bp fragment of DNA is bound by RNA polymerase II and TFIID and TFIIB?

E34. Explain the rationale behind a DNase I footprinting experiment.

E35. DNA sequencing can help us to identify mutations within genes. The following data are derived from an experiment in which a normal gene and a mutant gene have been sequenced:

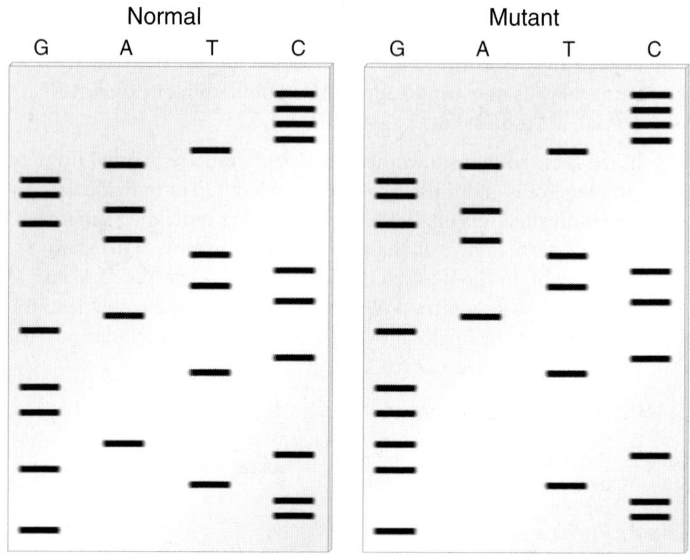

Locate and describe the mutation.

E36. A sample of DNA was subjected to automated DNA sequencing as shown here.

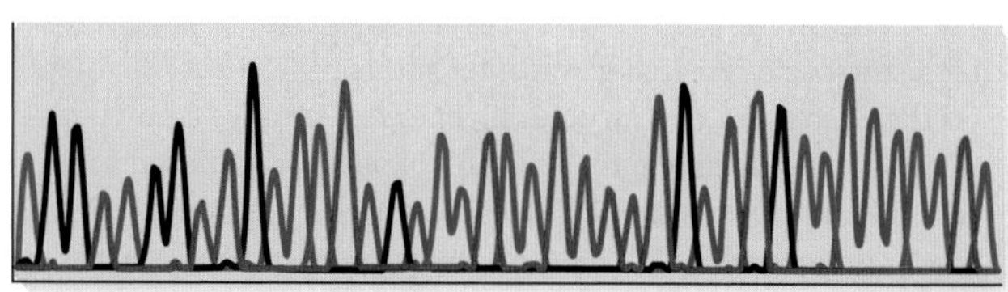

A. What is the sequence of this DNA segment?

B. Discuss the advantages of automated sequencing over conventional sequencing that uses radiolabeled nucleotides.

E37. A portion of the coding sequence of a cloned gene is shown here:

5'-GCCCCCGATCTACATCATTACGGCGAT-3'
3'-CGGGGGCTAGATGTAGTAATGCCGCTA-5'

This portion of the gene encodes a polypeptide with the amino acid sequence alanine–proline–aspartic acid–leucine–histidine–histidine–tyrosine–glycine–aspartic acid. Using the method of site-directed mutagenesis, a researcher wants to change the leucine codon into an arginine codon, with an oligonucleotide that is 19 nucleotides long. What is the sequence of the oligonucleotide that should be made? You should designate the 5' and 3' ends of the oligonucleotide in your answer. Note: The mismatch should be in the middle of the oligonucleotide, and a one-base mismatch is preferable over a two- or three-base mismatch. Use the bottom strand as the template strand for this site-directed mutagenesis experiment.

E38. Let's suppose you want to use site-directed mutagenesis to investigate a DNA sequence that functions as a response element for hormone binding. From previous work, you have narrowed down the response element to a sequence of DNA that is 20 bp in length with the following sequence:

```
5'-GGACTGACTTATCCATCGGT-3'
3'-CCTGACTGAATAGGTAGCCA-5'
```

As a strategy to pinpoint the actual response element sequence, you decide to make 10 different site-directed mutants and then analyze their effects by a gel retardation assay. What mutations would you make? What results would you expect to obtain?

E39. Site-directed mutagenesis can also be used to explore the structure and function of proteins. For example, changes can be made to the coding sequence of a gene to determine how alterations in the amino acid sequence affect the function of a protein. Let's suppose that you are interested in the functional importance of one glutamic acid residue within a protein you are studying. By site-directed mutagenesis, you make mutant proteins in which this glutamic acid codon has been changed to other codons. You then test the encoded mutant proteins for functionality. The results are as follows:

	Functionality
Normal protein	100%
Mutant proteins containing	
Tyrosine	5%
Phenylalanine	3%
Aspartic acid	94%
Glycine	4%

From these results, what would you conclude about the functional significance of the glutamic acid residue within the protein?

Questions for Student Discussion/Collaboration

1. Discuss and make a list of some of the reasons why it would be informative for a geneticist to determine the amount of a gene product. Use specific examples of known genes (e.g., β-globin and other genes) when making your list.

2. Make a list of all the possible genetic questions that could be answered using site-directed mutagenesis.

Note: All answers appear at the website for this textbook; the answers to even-numbered questions are in the back of the textbook.

www.mhhe.com/brookergenetics3e

Visit the Online Learning Center for practice tests, answer keys, and other learning aids for this chapter. Enhance your understanding of genetics with our interactive exercises, quizzes, animations, and much more.

19 BIOTECHNOLOGY

CHAPTER OUTLINE

The sheep named Dolly, which was cloned using genetic material from a somatic cell.

Biotechnology is broadly defined as the application of technologies that involve the use of living organisms, or products from living organisms, for the development of products that benefit humans. Biotechnology is not a new topic. It began about 12,000 years ago when humans began to domesticate animals and plants for the production of food. Since that time, many species of microorganisms, animals, and plants have become routinely used by people. More recently, the term biotechnology has become associated with molecular genetics. Since the 1970s, molecular genetic tools have provided novel ways to make use of living organisms for products and services. As discussed in Chapter 18, recombinant DNA techniques can be used to genetically engineer microorganisms. In addition, recombinant methods enable the introduction of genetic material into animals and plants. **Genetically modified organisms (GMOs)** have received genetic material via recombinant DNA technology. If an organism has received genetic material from a different species, it is called a **transgenic organism.** A gene from one species that is introduced into another species is called a **transgene.**

In the 1980s, court rulings made it possible to patent recombinant organisms such as transgenic animals and plants. This was one factor that contributed to the growth of many biotechnology industries. In this chapter, we will examine how molecular techniques have expanded our knowledge of the genetic characteristics of commercially important species. We will also discuss examples in which recombinant microorganisms and transgenic animals and plants have been given characteristics that are useful in the treatment of disease or in agricultural production. These include recombinant bacteria that make human insulin, transgenic livestock that produce human proteins in their milk, and transgenic tomatoes with a longer shelf life. In addition, the topics of mammalian cloning and stem cell research will be examined from a technical point of view. Likewise, the current and potential use of human gene therapy—the introduction of cloned genes into living cells in the treatment of a disease—will be addressed. In the process, we will also touch upon some of the ethical issues associated with these technologies.

19.1 THE USES OF MICROORGANISMS IN BIOTECHNOLOGY

Microorganisms are used to benefit humans in various ways (**Table 19.1**). In this section, we will examine how molecular genetic tools have become increasingly important for improving our use of microorganisms. Such tools can produce recombinant microorganisms with genes that have been manipulated in vitro. Why are recombinant organisms useful? Recombinant techniques can improve strains of microorganisms and have even yielded strains that make products not normally produced by microorganisms. For example, human genes have been introduced into bacteria to produce medically important products such as insulin and human growth hormone. As discussed in this section, several recombinant strains are in widespread use. However, in some areas of biotechnology and in some parts of the world, the commercialization of recombinant strains has proceeded very slowly. This is particularly true for applications in which recombinant microorganisms may be used to produce food products or where they are released into the environment. In such cases, safety and environmental concerns and negative public perceptions have slowed or even halted the commercial use of recombinant microorganisms. Nevertheless, molecular genetic research continues, and many biotechnologists expect an expanding use of recombinant microbes in the future.

TABLE **19.1**

Common Uses of Microorganisms

Application	Examples
Production of medicines	Antibiotics
	Synthesis of human insulin in recombinant *E. coli*
Food fermentation	Cheese, yogurt, vinegar, wine, and beer
Biological control	Control of plant diseases, insect pests, and weeds
	Symbiotic nitrogen fixation
	Prevention of frost formation
Bioremediation	Cleanup of environmental pollutants such as petroleum hydrocarbons and synthetics that are difficult to degrade

EXPERIMENT 19A

Somatostatin Was the First Human Peptide Hormone Produced by Recombinant Bacteria

During the 1970s, geneticists became aware of the great potential of recombinant DNA technology to produce therapeutic agents to treat certain human diseases. Healthy individuals possess many different genes that encode short peptide and longer polypeptide hormones. Diseases can result when an individual is unable to produce these hormones.

In 1976, Robert Swanson and Herbert Boyer formed Genentech Inc. The aspiration of this company was to engineer bacteria to synthesize useful products, particularly peptide and polypeptide hormones. Their first contract was with researchers Keiichi Itakura and Arthur Riggs. Their intent was to engineer a bacterial strain that would produce somatostatin, a human hormone that inhibits the secretion of a number of other hormones, including growth hormone, insulin, and glucagon. Somatostatin was not chosen for its commercial potential. Instead, it was chosen because the researchers thought it would be technically less difficult than other hormones. Somatostatin is very small (only 14 amino acids long), which requires a short coding sequence, and it can be detected easily.

Before discussing the details of this experiment, let's consider the researchers' approach to constructing the somatostatin gene. To express somatostatin in bacteria, the coding sequence for somatostatin must be inserted next to a bacterial promoter that is contained within a plasmid. Rather than obtaining the gene from the human genome that encodes this 14-amino acid hormone, the researchers took a different approach. As shown below, they chemically synthesized short (single-stranded) oligonucleotides that would hydrogen bond with each other to form the coding sequence for this gene:

Eight separate oligonucleotides (labeled A through H) were synthesized chemically. Due to base complementarity within their sequences, the oligonucleotides hydrogen bonded to each other forming a longer double-stranded DNA fragment with two important characteristics. First, its single-stranded ends (i.e., overhangs) allowed it to be inserted into *Eco*RI and *Bam*HI restriction sites within plasmid DNA. Second, the middle of this DNA fragment encodes the amino acid sequence of the somatostatin peptide hormone. (Today, oligonucleotide synthesis methods have greatly improved, making it unnecessary to synthesize several, short oligonucleotides. Instead, researchers now could make an oligonucleotide that would span the entire length of the somatostatin coding sequence.)

The coding sequence was constructed so that an extra methionine would be located at the amino terminal end of somatostatin. This methionine provided a link between somatostatin and a bacterial protein, β-galactosidase. As discussed in Chapter 14, this enzyme is encoded by the *lacZ* gene. Why was this link necessary? During the course of their experiments, the researchers learned that somatostatin made in bacteria is rapidly degraded by cellular proteases. To prevent this from happening, they linked the somatostatin sequence to the *lacZ* gene encoding β-galactosidase. When this linked gene is expressed in bacteria, a fusion protein is made between somatostatin and β-galactosidase. The fusion protein is not rapidly degraded. The researchers could then separate somatostatin from β-galactosidase by treatment with cyanogen bromide (CNBr), which cleaves polypeptides at the carboxyl terminal side of methionine. Because no methionines are found within the somatostatin sequence itself, this treatment does not degrade somatostatin.

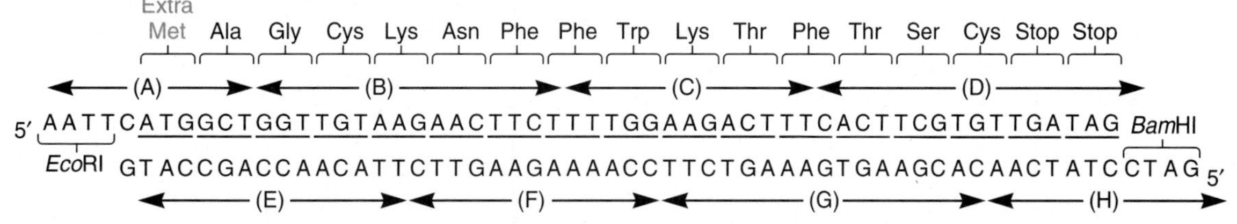

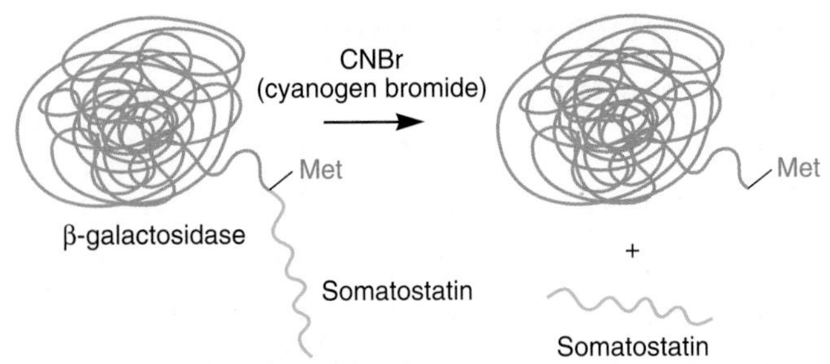

The steps in their protocol are shown in **Figure 19.1**. As described here, the researchers made a synthetic somatostatin gene that was flanked by unique restriction sites and had a methionine codon at the beginning of the somatostatin-coding sequence. This gene was then inserted into a plasmid at the end of the *lacZ* gene. As a control, they also inserted the somatostatin gene in the wrong orientation (shown on the right of step 3). The plasmid with the wrong orientation should not make any somatostatin. The plasmids were then introduced into *E. coli* cells. The plasmid contained the *lac* promoter (from the *lac* operon), which was induced with IPTG (a nonmetabolizable lactose analogue). In the cells harboring the somatostatin gene in the correct orientation, this would induce the synthesis of a fusion protein containing β-galactosidase and somatostatin. The bacterial cells were then collected by centrifugation and exposed to formic acid and cyanogen bromide (CNBr). As mentioned, the CNBr cleaves polypeptides next to methionine residues. Therefore, this treatment would break the link between β-galactosidase and somatostatin. The amount of somatostatin was then determined by a radioimmunoassay (RIA). (See the Appendix for a description of radioimmunoassay.)

THE GOAL

The researchers wanted to produce human somatostatin in a recombinant bacterium.

ACHIEVING THE GOAL — FIGURE 19.1 The production of human somatostatin in *E. coli*.

Starting material: A normal *E. coli* strain that was unable to synthesize somatostatin, and bacterial plasmids that carry the amp^R gene along with bacterial promoters.

1. Chemically synthesize eight oligonucleotides. When they are mixed together, the complementary oligonucleotides will hybridize to each other.

2. Treat with DNA ligase to covalently link the oligonucleotides, which will form the molecule shown at the right.

3. Using recombinant techniques described in Chapter 18 (Figure 18.2), insert this fragment into a plasmid by digesting the plasmid at unique sites for *Eco*RI and *Bam*HI.

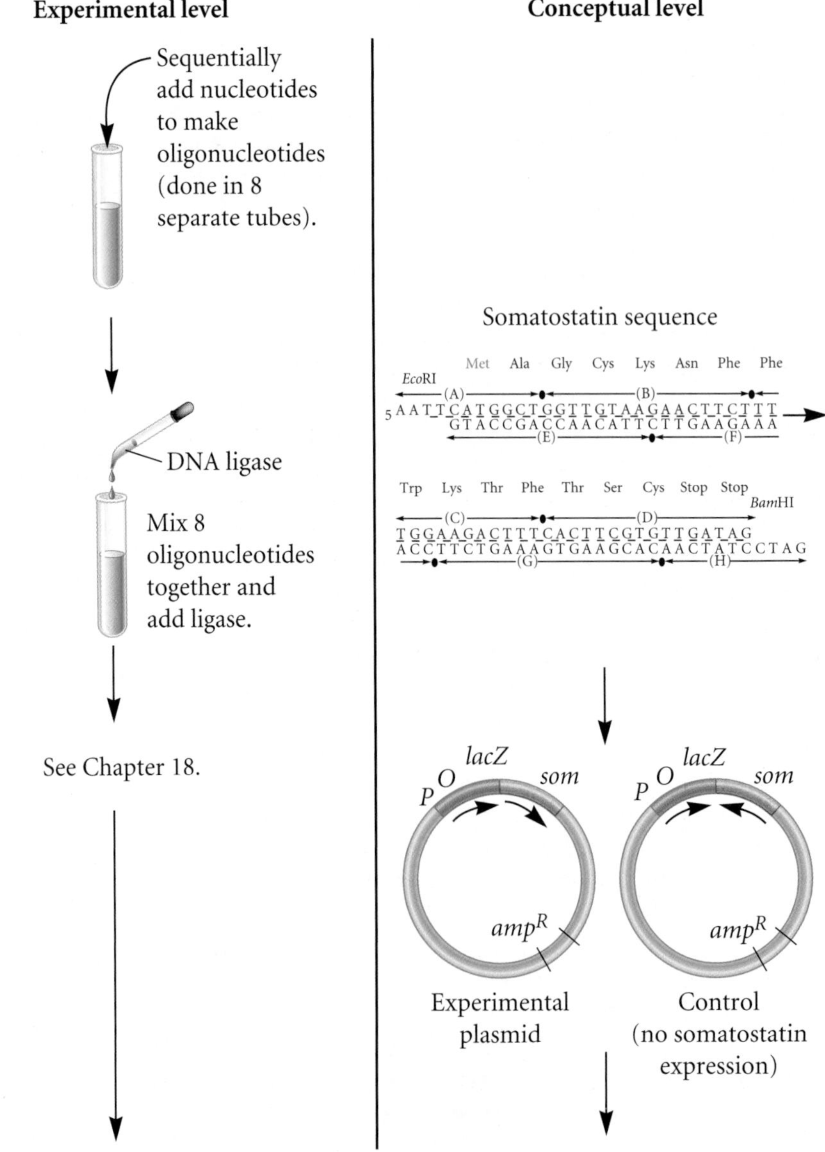

4. Transform the plasmids into *E. coli* by treatment with $CaCl_2$. The transformed cells are spread on plates containing ampicillin. Grow overnight to obtain bacterial colonies that are ampicillin resistant because they contain the plasmid.

5. Pick cells from a bacterial colony and grow these recombinant bacteria in liquid media; activate the *lac* promoter with isopropyl thiodigalactoside (IPTG).

   Note: The *lac* promoter in this plasmid is controlled by the lac repressor (see Chapter 15). IPTG is an inducer that activates transcription of the *lacZ* gene by removing the lac repressor.

6. Place in a tube and centrifuge to obtain a bacterial cell pellet.

7. Resuspend the pellet in 70% formic acid and cyanogen bromide (5 mg/ml).

   Note: This breaks open the cells and cleaves polypeptides at methionine residues.

8. Determine the amount of somatostatin using a radioimmunoassay. (See the Appendix for a description of this procedure.)

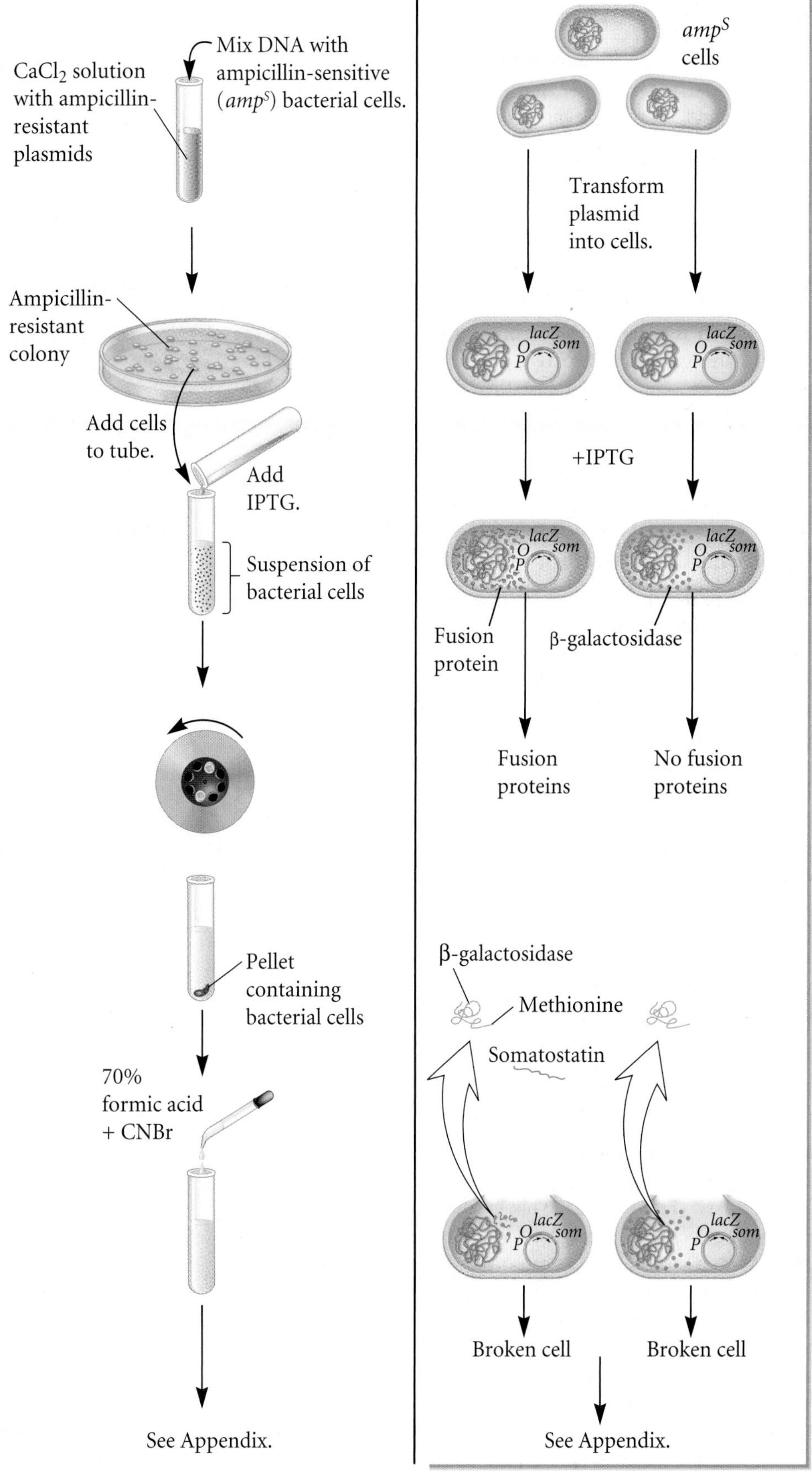

THE DATA

Plasmid Strain	Amount of Somatostatin (detected by RIA) (picograms of somatostatin/ milligram of bacterial proteins)
Correct orientation	8–320*
Incorrect orientation	<0.4

*The amount of somatostatin was determined in several independent experiments.

INTERPRETING THE DATA

As shown in the data table, recombinant bacteria carrying the somatostatin gene in the correct orientation produced this hormone. The amount of somatostatin varied from 8 to 320 picograms per milligram of bacterial proteins. This variability could be attributed to several factors, including protein degradation, incomplete cyanogen bromide cleavage, and unknown genetic changes in the plasmids during bacterial cell growth. In spite of this variability, the exciting result was the production of a human hormone in recombinant bacteria. By comparison, the plasmid with the incorrect orientation did not produce a significant amount of the hormone. This study was the first demonstration that recombinant bacteria could make products encoded by human genes. At the time, this was a major breakthrough that catalyzed the growth of the biotechnology industry!

A self-help quiz involving this experiment can be found at the Online Learning Center.

Many Important Medicines Are Produced by Recombinant Microorganisms

Since the pioneering study described in the experiment of Figure 19.1, recombinant DNA technology has been used to develop bacterial strains that synthesize several other medical agents, a few of which are described in **Table 19.2**.

In 1982, the U.S. Food and Drug Administration approved the sale of the first genetically engineered drug, human insulin, which was produced by Genentech and marketed by Eli Lilly. In nondiabetic individuals, insulin is produced by the β cells of the pancreas. Insulin functions to regulate several physiological processes, particularly the uptake of glucose into fat and muscle cells. Persons with insulin-dependent diabetes cannot synthesize an adequate amount of insulin due to a loss of their β cells. Prior to 1982, insulin was isolated from the pancreases removed from cattle and pigs. Unfortunately, in some cases, diabetic individuals became allergic to such insulin and had to use expensive combinations of insulin from other animals and human cadavers. Today, people with diabetes can use genetically engineered human insulin to treat their disease.

Insulin is a hormone composed of two polypeptide chains, called the A and B chains. To make this hormone using bacteria, the coding sequences of the A and B chains are placed next to the coding sequence of a native *E. coli* protein, β-galactosidase (**Figure 19.2**). This creates a fusion protein comprising β-galactosidase and the A or B chain. As with somatostatin, this step is necessary because the A and B chains are rapidly degraded when expressed in bacterial cells by themselves. The fusion proteins, however, are not. How are the two fusion proteins used to make human insulin? After the fusion proteins are expressed in bacteria, they can be purified and then treated with cyanogen bromide (CNBr) to separate β-galactosidase from the A or B chain. The A and B chains are then purified and mixed together under conditions in which they will refold and form disulfide bonds with each other to make an active insulin hormone.

TABLE 19.2

Examples of Medical Agents Produced by Recombinant Microorganisms

Drug	Action	Treatment
Insulin	A hormone that promotes glucose uptake	For diabetic patients
Tissue plasminogen activator (TPA)	Dissolves blood clots	For heart attack victims and other arterial occlusions
Superoxide dismutase	Antioxidant	For heart attack victims to minimize tissue damage
Factor VIII	Blood clotting factor	For certain types of hemophilia patients
Renin inhibitor	Lowers blood pressure	For hypertension
Erythropoietin	Stimulates the production of red blood cells	For anemia

Bacterial Species Can Be Used as Biological Control Agents

The term **biological control** refers to the use of living organisms or their products to alleviate plant diseases or damage from environmental conditions. During the past 20 years, interest in the biological control of plant diseases and insect pests as an alternative to chemical pesticides has increased. Biological control agents can prevent disease in several ways. In some cases, nonpathogenic microorganisms are used to compete effectively against pathogenic strains for nutrients or space. In other cases, microorganisms may produce a toxin that inhibits other pathogenic microorganisms or insects without harming the plant.

Biological control can also involve the use of microorganisms living in the field. A successful example is the use of *Agrobacterium radiobacter* to prevent crown gall disease caused by

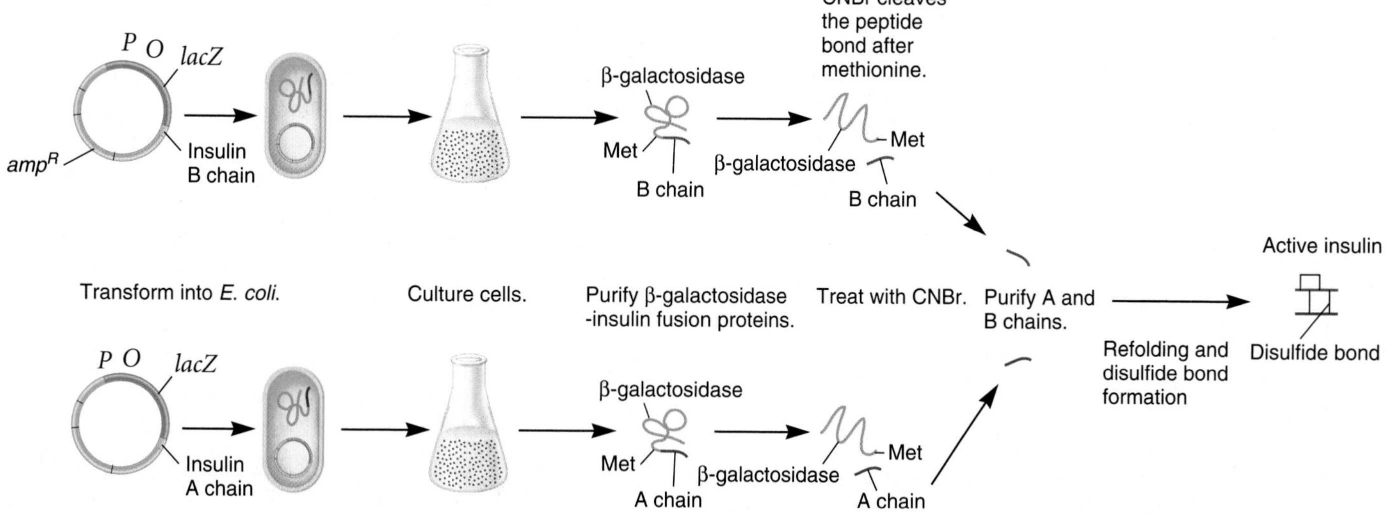

FIGURE 19.2 The use of bacteria to make human insulin.

Genes → Traits The synthesis of human insulin is not a trait that bacteria normally possess. However, genetic engineers can introduce the genetic sequences that encode the A and B chains of human insulin via recombinant DNA technology, yielding bacteria that make these polypeptides as fusion proteins with β-galactosidase. CNBr treatment releases the A and B polypeptides, which are then purified and mixed together under conditions in which they refold and form functional human insulin.

Agrobacterium tumefaciens. The disease gets its name from the large swellings (galls) produced by the plant in response to the bacteria. *A. radiobacter* produces agrocin 84, an antibiotic that kills *A. tumefaciens.* Molecular geneticists have determined that *A. radiobacter* contains a plasmid with genes responsible for agrocin 84 synthesis and resistance. Unfortunately, this plasmid is occasionally transferred from *A. radiobacter* to *A. tumefaciens* during interspecies conjugation. When this occurs, *A. tumefaciens* can gain resistance to agrocin 84. Researchers have identified *A. radiobacter* strains in which this plasmid has been altered genetically to prevent its transfer during conjugation. This conjugation-deficient strain is now used commercially worldwide to prevent crown gall disease.

Another biological control agent is *Bacillus thuringiensis*, usually referred to as Bt (pronounced "bee-tee"). This naturally occurring bacterium produces toxins that are lethal to many caterpillars and beetles that feed on trees, shrubs, flowers, and fruits. Bt is generally harmless to plants and other animals, such as humans, and does not usually harm beneficial insects that act as pollinators. Therefore, it is viewed as an environmentally friendly pesticide. Commercially, Bt is sold in a powder form that is used as a dust or mixed with water as a foliar spray. Bt is then dusted or sprayed on plants that are under attack by caterpillars or beetles so that the pests will ingest the bacteria as they eat the leaves, flowers, or fruits. The toxins produced by Bt bring about paralysis of the insect's digestive tract, causing it to stop feeding within hours and die within a few days. Geneticists have cloned the genes that encode Bt toxins, which are proteins. As discussed later, such genes have been introduced into crops, such as corn, to produce transgenic plants resistant to insect attack.

The Release of Recombinant Microorganisms into the Environment Is Sometimes Controversial

As we have seen, genetically altered strains can have commercial applications in the field. Whether or not a microorganism is recombinant has become an important issue in the use of biological control agents that are released into the environment. Alterations in the genetic characteristics of a microorganism, such as the acquisition of naturally occurring plasmids or mutagenesis by chemical agents and radiation, do not produce bacteria that are classified as recombinant strains. The *A. radiobacter* strain and Bt bacterium fall into this category. This distinction is important from the perspective of both governmental regulation and public perception, even though the organisms produced by nonrecombinant and recombinant approaches may be genetically identical or nearly identical.

Knowledge from molecular genetic research is used to develop both nonrecombinant and recombinant strains with desirable characteristics. Each year, many new strains of nonrecombinant microorganisms are analyzed in field tests for the biological control of plant diseases and insect pests. By comparison, the use of recombinant microorganisms in field tests has proceeded much more slowly. This slow progress is related to increased levels of governmental regulation and, in some cases, to negative public perception of recombinant microorganisms.

As an example of the controversial nature of this topic, let's consider the first field test of a recombinant bacterium, which involved the use of a genetically engineered strain of *Pseudomonas syringae* to control frost damage. Experiments by Steven Lindow and colleagues showed that the formation of ice on the surface of plants is enhanced by the presence of certain bacterial species. These *Ice*$^+$ species synthesize cellular proteins that promote ice nucleation—the initiation of ice crystals. Using recombinant DNA technology, Lindow constructed an *Ice*$^-$ strain of *P. syringae* in the early 1980s, which lacks the gene responsible for the production of ice-nucleation proteins. When applied to the surface of plants, an *Ice*$^-$ strain can compete with and thereby reduce the proliferation of *Ice*$^+$ bacteria, thus inhibiting the formation of frost.

Lindow sought approval for field tests of an *Ice*$^-$ recombinant strain in Tulelake, California. For several years, these tests were delayed because of a lawsuit from the Foundation on Economic Trends. During that time, Lindow made great efforts to ensure the safety of this project by studying the local environment. He also consulted with local townspeople where the field test was to take place. Initially, the idea was well received by the local residents. However, another company tested similar bacteria on the roof of an Oakland facility without Environmental Protection Agency (EPA) approval. The media reported this incident, and it caused many Tulelake townspeople to become apprehensive about the release of recombinant bacteria. Nevertheless, in 1987, approval was finally granted for the field testing of the recombinant *P. syringae* (**Figure 19.3**).

During the first test on several thousand strawberry plants, the plants were ripped out by vandals. In a second field test, the ability of *Ice*$^-$ bacteria to protect potato plants was tested. While some of the plants were destroyed by vandals, the results of this field experiment showed that the *Ice*$^-$ bacteria did protect potato plants from frost damage. In addition, soil sampling showed that the recombinant bacteria were contained at the field site and did not proliferate into surrounding areas. Even so, the release of recombinant microorganisms into the environment remains controversial. Since this first test, relatively few recombinant strains have been released. In the case of *Ice*$^-$ bacteria, further research was discouraged by a variety of factors, including governmental regulation and the expense of doing additional experiments. The recombinant *Ice*$^-$ bacteria were never commercialized.

FIGURE 19.3 **The release of recombinant *P. syringae* in a field test.**

Microorganisms Can Reduce Environmental Pollutants

The term **bioremediation** refers to the use of living organisms or their products to decrease pollutants in the environment. As its name suggests, this is a biological remedy for pollution. During bioremediation via microorganisms, enzymes produced by a microorganism modify a toxic pollutant by altering or transforming its structure. This event is called **biotransformation.** In many cases, biotransformation results in **biodegradation,** in which the toxic pollutant is degraded, yielding less complex, nontoxic metabolites. Alternatively, biotransformation without biodegradation can also occur. For example, toxic heavy metals can often be rendered less toxic by oxidation or reduction reactions carried out by microorganisms. Another way to alter the toxicity of organic pollutants is by promoting polymerization. In many cases, polymerized toxic compounds are less likely to leach from the soil and, therefore, are less environmentally toxic than their parent compounds.

Since the early 1900s, microorganisms have been used in the treatment and degradation of sewage. More recently, the field of bioremediation has expanded into the treatment of hazardous and refractory chemical wastes—chemicals that are difficult to degrade and usually associated with industrial activity. These pollutants include petroleum hydrocarbons, halogenated organic compounds, pesticides, herbicides, and organic solvents. Many new applications that use microorganisms to degrade these pollutants are being tested. The field of bioremediation has been fostered, to a large extent, by better knowledge of how pollutants are degraded by microorganisms, the identification of new and useful strains of microbes, and the ability to enhance bioremediation through genetic engineering.

Molecular genetic technology is key in identifying genes that encode enzymes involved in bioremediation. The characterization of the relevant genes greatly enhances our understanding of how microbes can modify toxic pollutants. In addition, recombinant strains created in the laboratory can be more efficient at degrading certain types of pollutants.

In 1980, in a landmark case (*Diamond v. Chakrabarty*), the U.S. Supreme Court ruled that a live, recombinant microorganism is patentable as a "manufacture or composition of matter." The first recombinant microorganism to be patented was an "oil-eating" bacterium that contained a laboratory-constructed plasmid. This strain can oxidize the hydrocarbons commonly found in petroleum. It grew faster on crude oil than did any of the natural isolates tested. However, it has not been a commercial success because this recombinant strain metabolizes only a limited number of toxic compounds; the number of compounds actually present in crude oil is over 3,000. Unfortunately, the recombinant strain did not degrade many higher-molecular-weight compounds, which tend to persist in the environment.

Currently, bioremediation should be considered a developing industry. This field will need well-trained molecular geneticists to conduct research aimed at elucidating the mechanisms whereby microorganisms degrade toxic pollutants. In the future, recombinant microorganisms may provide an effective way to decrease the levels of toxic chemicals within our environment. However, this approach will require careful studies to demonstrate that recombinant organisms are effective at reducing pollutants and are safe when released into the environment.

19.2 GENETICALLY MODIFIED ANIMALS

As mentioned at the beginning of this chapter, transgenic organisms contain recombinant DNA from another species that has been integrated into their genome. A dramatic example of this is shown in **Figure 19.4**. In this case, the gene that encodes the human growth hormone was introduced into the genome of a mouse. The larger mouse on the right is a transgenic mouse that expresses the human growth hormone gene to a very high level.

The production of transgenic animals is a relatively new area of biotechnology. In recent years, a few transgenic species have reached the stage of commercialization. Many researchers believe that this technology holds great promise for innovations in biotechnology. However, the degree to which this potential may be realized will depend, in part, on the public's concern about the creation and consumption of transgenic species.

In this section, we will begin by examining the mechanisms through which cloned DNA becomes integrated into the chromosomal DNA of animal cells. We will then explore the current techniques used to create transgenic animals and their potential uses. In addition, we will consider mammalian cloning and stem cell research. These topics have received enormous public attention due to the complex ethical issues that they raise.

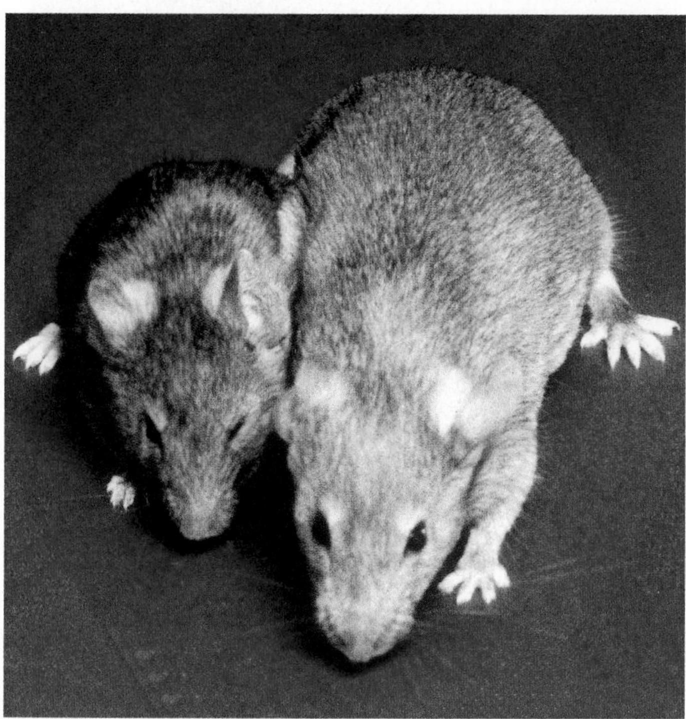

FIGURE 19.4 A comparison between a normal mouse and a transgenic mouse that carries a growth hormone gene.

Genes → Traits The transgenic mouse (on the right) carries a gene encoding human growth hormone. The introduction of the human gene into the mouse's genome has caused it to grow larger.

The Integration of a Cloned Gene into a Chromosome Can Result in Gene Replacement or Gene Addition

In Chapter 18, we considered methods to clone genes. A common approach is to insert a chromosomal gene into a vector and then propagate the vector in living microorganisms such as bacteria or yeast cells. Cloned genes can also be introduced into animal and plant cells. However, to be inherited stably from generation to generation, the cloned gene must become integrated into one (or more) of the chromosomes that reside in the cell nucleus. The integration of cloned DNA into a chromosome occurs by recombination, which is described in Chapter 17.

Figure 19.5 illustrates how a cloned gene can integrate into a chromosome by recombination. If the genome of the host cell carries the same type of gene and if the cloned gene is swapped with the normal chromosomal gene by homologous recombination, then the cloned gene will replace the normal gene within the chromosome (Figure 19.5a). This process is termed **gene replacement.** If the cloned gene has been rendered inactive by mutation and replaces the normal gene, researchers can study how the loss of normal gene function affects the organism. This is called a **gene knockout.** More commonly, a cloned gene may be introduced into a cell and become integrated into the genome by nonhomologous recombination (Figure 19.5b) in a process known as **gene addition.** As shown here, both the cloned and normal genes will be present following gene addition.

Researchers may also introduce a gene that is not already present in a particular species' genome. An interesting example of this type of gene addition involves the production of aquarium fish that "glow," which are aptly named **GloFish®.** The GloFish is a brand of transgenic zebrafish (*Danio rerio*) that glows with bright green, red, or yellow fluorescent color (**Figure 19.6**). How were these fish produced? In 1999, Dr. Zhiyuan Gong and his colleagues started with a gene from jellyfish that encodes a green fluorescent protein and inserted it into the zebrafish genome by gene addition, causing the zebrafish to glow green. By placing the gene next to a gene promoter that would turn the gene on in the presence of certain environmental toxins, their goal was to eventually develop a fish that could be used to detect water pollution. The researchers subsequently collaborated with a company to market the fish for aquarium use. They developed a red fluorescent zebrafish by adding a gene from a sea coral, and a yellow fluorescent zebrafish by adding a variant of the jellyfish gene. In 2003, the GloFish became the first genetically modified organism to be sold as a pet. GloFish have been successfully marketed in several countries including the United States, although the sale of GloFish is banned in California.

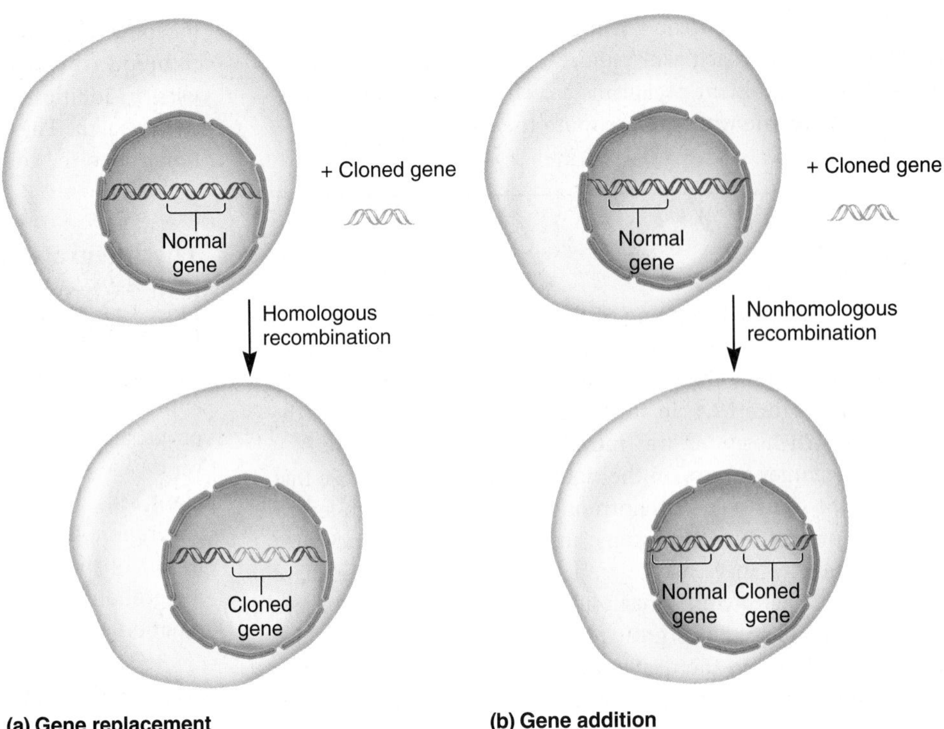

FIGURE 19.5 The introduction of a cloned gene into a cell can lead to gene replacement or gene addition. In these two examples, the cloned gene is similar to a normal gene already present in the genome of a cell. **(a)** If the cloned gene undergoes homologous recombination and replaces the normal gene, this event is called gene replacement. **(b)** Alternatively, the cloned gene may recombine nonhomologously at some other chromosomal location, leading to gene addition.

FIGURE 19.6 The use of gene addition to produce fish that glow. The aquarium fish shown here, which are named GloFish®, are transgenic organisms that have received a gene from jellyfish or sea corals that encodes a fluorescent protein, causing them to glow green, red, or yellow.

Molecular Biologists Can Produce Mice That Contain Gene Replacements

In bacteria and yeast, which have relatively small genomes, homologous recombination between cloned genes and the host cell chromosome occurs at a relatively high rate, so gene replacement is commonly achieved. Gene replacement is useful when a researcher or biotechnologist wants to make a gene mutation in the laboratory and then introduce the mutant gene into a living organism to compare the effects of the normal and mutant genes on the phenotype. However, in more complex eukaryotes with very large genomes, the introduction of cloned genes into cells is much more likely to result in gene addition rather than gene replacement. For example, when a cloned gene is introduced into a mouse cell, it will undergo homologous recombination less than 0.1% of the time. Most of the time (over 99.9%), gene addition occurs.

To produce mice with gene replacements, molecular biologists have devised laboratory procedures to preferentially select cells in which homologous recombination has occurred. One approach is shown in **Figure 19.7**. The gene of interest is found in a mouse chromosome and is cloned so that it can be manipulated in vitro. The cloned gene is shown at the top of Figure 19.7. The cloning procedure involves two selectable marker genes that influence whether or not mouse embryonic cells can grow in the presence of certain drugs. First, the gene of interest is inactivated by inserting a neomycin-resistance gene (called Neo^R) into the center of its coding sequence. Neo^R provides cells with resistance to neomycin. Next, a thymidine kinase gene, designated *TK*, is inserted adjacent to the gene of interest but not within the gene itself. The *TK* gene renders cells sensitive to killing by a drug called gancyclovir.

After the cloned gene has been modified in vitro, it is introduced into mouse embryonic stem cells. How do researchers identify the cells in which homologous recombination has occurred? When the cells are grown in the presence of neomycin and gancyclovir, most nonhomologous recombinants will be killed because they will also carry the *TK* gene. In contrast, homologous recombinants in which the normal gene has been partially replaced with the cloned gene contain only the Neo^R gene, and so they will be resistant to both drugs. The surviving embryonic cells can then be injected into blastocysts, early embryos that are obtained from a pregnant mouse. In the example shown in Figure 19.7, the embryonic cells are from a mouse with dark fur, and the blastocysts are from a mouse with white fur. The embryonic cells can mix with the blastocyst cells to create a **chimera,** an organism that contains cells from two different individuals. To identify chimeras, the injected blastocysts are reimplanted into the uterus of a female mouse and allowed to develop. When this mouse gives birth, chimeras are identified easily because they contain patches of white and dark fur (Figure 19.7).

Chimeric animals that contain a single-gene replacement can then be mated to other mice to produce offspring that carry the mutant gene. Because mice are diploid, researchers must make two or more subsequent crosses to create a strain of mice that contains both copies of the mutant target gene. However, homozygous strains for a gene knockout cannot be produced if the mutant gene is lethal in the homozygous state.

An alternative method for producing genetically modified mice is to inject the desired gene into a fertilized egg. To conduct this type of experiment, researchers obtain mouse eggs and fertilize them in vitro. Immediately following fertilization, the cloned gene is injected into the sperm pronucleus—the haploid nucleus that has not yet fused with the egg nucleus. The cloned DNA then integrates into the genome, and the two pronuclei fuse to form the diploid nucleus of a zygote. The zygote begins to divide and is introduced into the uterus of a female mouse, where it becomes implanted and grows. As discussed next, genetically modified mice are used in basic research and to study human diseases.

Gene Knockouts and Knockins Are Produced in Mice to Understand Gene Function and Human Disease

As we have seen, researchers can replace a normal mouse gene with one that has been inactivated by the insertion of an antibiotic-resistance gene. As mentioned, when a mouse is homozygous for an inactivated gene, this is called a gene knockout. The inactive mutant gene has replaced both copies of the normal gene. In other words, the function of the normal gene has been "knocked out." By creating gene knockouts, researchers can study how the loss of normal gene function affects the organism. Gene knockouts frequently have specific effects on the phenotype of a mouse, which helps researchers to determine that the function of a gene is critical within a particular tissue or during a specific stage of development. In many cases, however, a gene knockout produces no obvious phenotypic effect at all. One explanation is that a single gene may make such a small contribution to an organism's phenotype that its loss may be difficult to detect. Alternatively, another possible explanation for a lack of observable phenotypic change in a knockout mouse may involve **gene redundancy.** This means that when one type of gene is inactivated, another gene with a similar function may be able to compensate for the inactive gene.

A particularly exciting avenue of gene knockout research is its application in the study and treatment of human disease. How is this useful? Knocking out the function of a gene may provide clues about what that gene normally does. Because humans share many genes with mice, observing the characteristics of knockout mice gives researchers information that can be used to better understand how a similar gene may cause or contribute to a disease in humans. Examples of research areas in which knockout mice have been useful include cancer, obesity, heart disease, diabetes, and many inherited disorders. In 2006, the National Institutes of Health (NIH) launched the Knockout Mouse Project. The goal of this program is to build a comprehensive and publicly available resource of knockout mutations in mice. The NIH Knockout Mouse Project will collaborate with other large-scale efforts to produce mouse knockouts that are under way in Canada, called the North American Conditional Mouse Mutagenesis Project (NorCOMM), and in Europe, called the European Conditional Mouse Mutagenesis Program (EUCOMM). The collective goal of these programs is to create at least one loss-of-function mutation in each of the approximately 20,000 genes in the mouse genome.

In contrast to knockouts, researchers may introduce genes into the mouse genome to study the effects of gene overexpression or to examine the effects of particular alleles, such as those that may cause disease in humans. To accomplish this, researchers can produce a **gene knockin.** A gene knockin is a gene addition in which a gene of interest has been added to a particular site in the mouse genome (**Figure 19.8**). In this example, the cloned gene is inserted into the middle of a segment of DNA from a noncritical site in the mouse genome. The noncritical segment is very long, which allows the transgene to be targeted to that specific, noncritical integration site by homologous recombination after it is

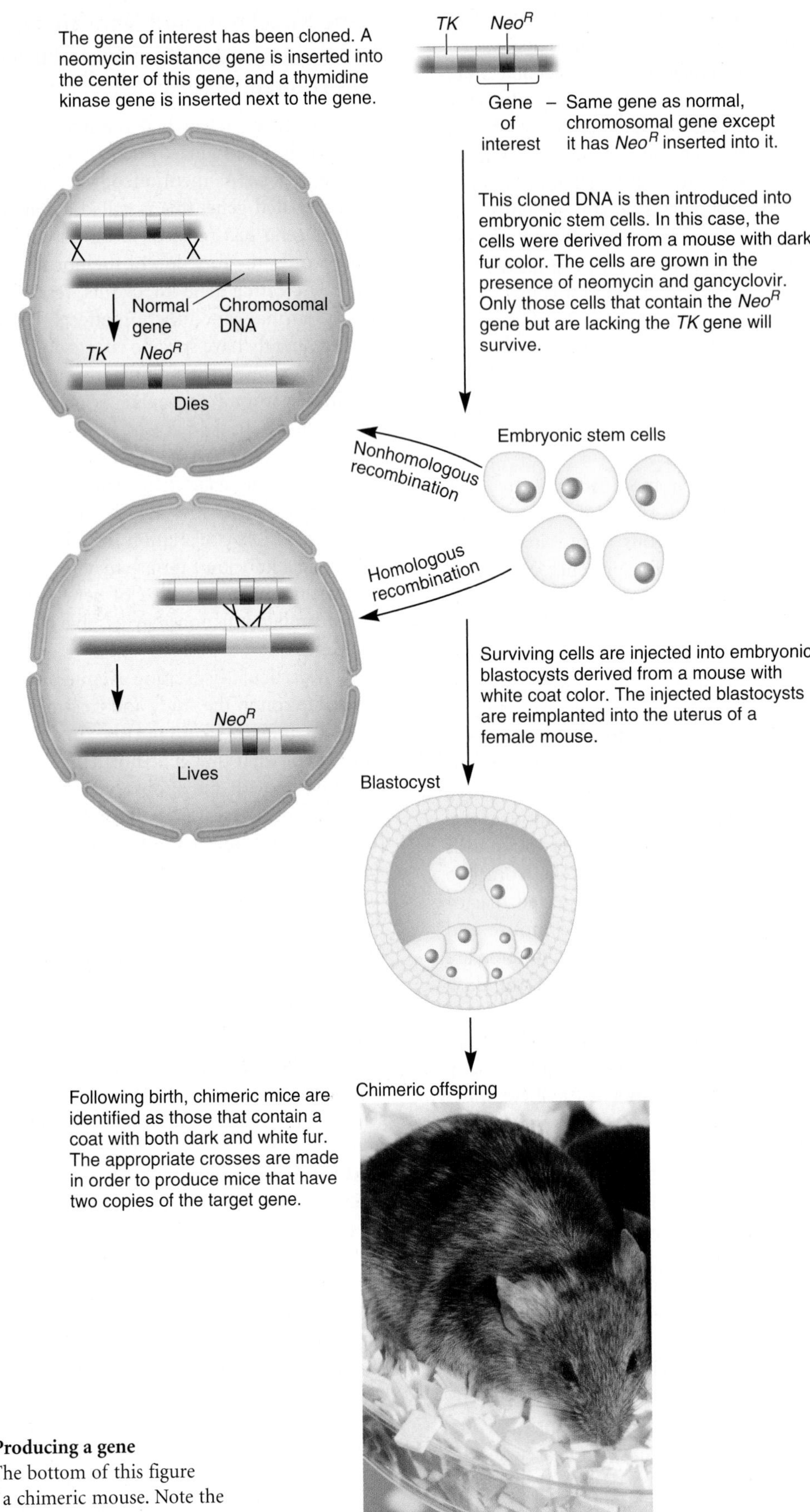
The gene of interest has been cloned. A neomycin resistance gene is inserted into the center of this gene, and a thymidine kinase gene is inserted next to the gene.
TK
NeoR
Gene of interest
– Same gene as normal, chromosomal gene except it has NeoR inserted into it.
This cloned DNA is then introduced into embryonic stem cells. In this case, the cells were derived from a mouse with dark fur color. The cells are grown in the presence of neomycin and gancyclovir. Only those cells that contain the NeoR gene but are lacking the TK gene will survive.
Normal gene
Chromosomal DNA
TK
NeoR
Dies
Embryonic stem cells
Nonhomologous recombination
Homologous recombination
NeoR
Lives
Surviving cells are injected into embryonic blastocysts derived from a mouse with white coat color. The injected blastocysts are reimplanted into the uterus of a female mouse.
Blastocyst
Following birth, chimeric mice are identified as those that contain a coat with both dark and white fur. The appropriate crosses are made in order to produce mice that have two copies of the target gene.
Chimeric offspring

FIGURE 19.7 Producing a gene replacement in mice. The bottom of this figure shows a photograph of a chimeric mouse. Note the patches of black and white fur.

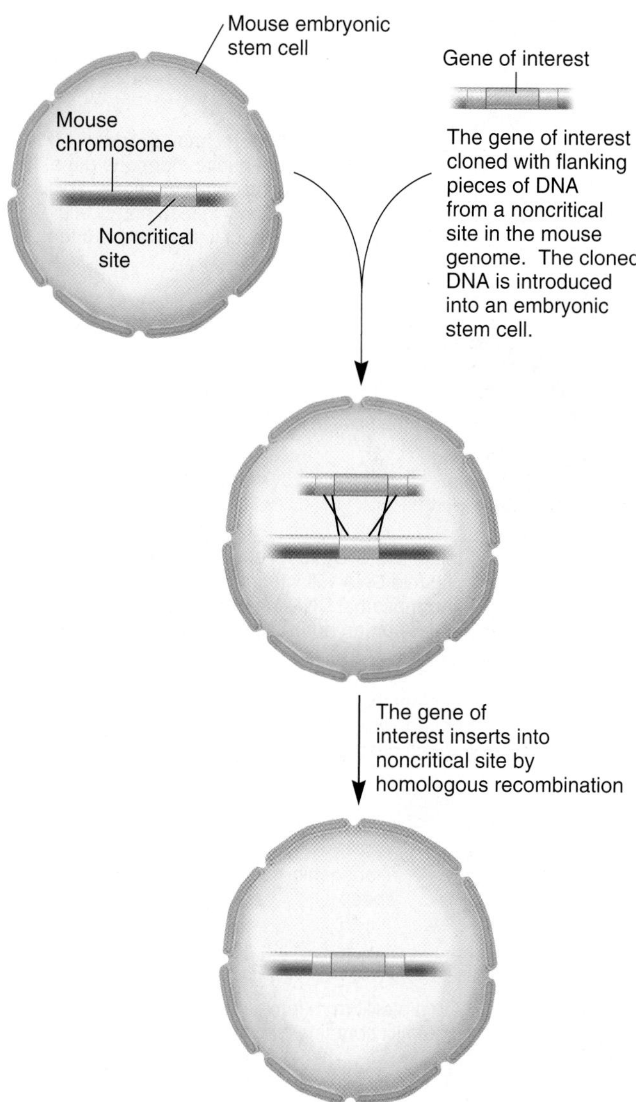

FIGURE 19.8 Producing a gene knockin in mice. This simplified diagram illustrates the strategy of producing a gene knockin. The gene of interest is cloned, and the cloned gene is flanked by DNA sequences that are homologous to a noncritical site in the mouse genome. When introduced into a mouse embryonic stem cell, the gene will insert into the genome by homologous recombination, which is an example of a targeted gene addition.

introduced into mouse cells. Such gene knockins tend to result in a more consistent level of expression of the transgene compared to gene additions that may occur randomly in another place in the genome. Also, because a targeted transgene is not interfering with a critical locus, the researcher can be more certain that any resulting phenotypic effect is due to the expression of the transgene.

To study human diseases, researchers have produced strains of transgenic mice that harbor both gene knockouts and gene knockins. A strain of mice engineered to carry a mutation that is analogous to a disease-causing mutation in a human gene is termed a mouse model. As an example, let's consider sickle-cell disease, which is due to a mutation in the human β-globin gene (refer back to Figure 4.7). This gene encodes a polypeptide called β globin; adult hemoglobin is composed of both α-globin and β-globin polypeptides. When researchers produced a gene knockin by introducing the mutant human β-globin gene into mice, the resulting mice showed only mild symptoms of the disease. However, Chris Paszty and Edward Rubin produced a mouse model with multiple gene knockins and gene knockouts. In particular, the mice had gene knockins for the normal human α-globin gene and the mutant β-globin gene from patients with sickle-cell anemia. The strain also had gene knockouts of the mouse α-globin gene and β-globin gene. Therefore, these mice made adult hemoglobin just like people with sickle-cell disease, but they did not produce any normal mouse hemoglobin. These transgenic mice exhibit the major features of sickle-cell disease—sickled red blood cells, anemia, and multiorgan pathology. They have been useful as a model to study the disease and to test potential therapies for treatment.

Biotechnology Holds Promise in Producing Transgenic Livestock

The technology for creating transgenic mice has been extended to other animals, and much research is under way to develop transgenic species of livestock, including fish, sheep, pigs, goats, and cattle. For some farmers, the ability to modify the characteristics of livestock via the introduction of cloned genes is an exciting prospect. In addition, work is currently under way to produce genetically modified pigs that are expected to be resistant to rejection mechanisms that occur following organ transplantation to humans. These strains may in time become a source of organs or cells for patients.

A novel avenue of research involves the production of medically important proteins in the mammary glands of livestock. This approach is sometimes called **molecular pharming.** (The term is also used to describe the manufacture of medical products by agricultural plants.) As shown in **Table 19.3**, several human proteins have been successfully produced in the milk of domestic livestock. Compared to the production of proteins in bacteria, one advantage is that certain proteins are more likely to function properly when expressed in mammals. This may be due to covalent modifications, such as the attachment of carbohydrate groups, which occur in eukaryotes but not in bacteria. In addition, certain proteins may be degraded rapidly or folded improperly when expressed in bacteria. Furthermore, the yield of recombinant proteins in milk can be quite large. Dairy cows, for example, produce about 10,000 liters of milk per year per cow. In most cases, a transgenic cow can produce approximately 1 g of the transgenic protein per liter of milk.

To introduce a human gene into an animal so that the encoded protein will be secreted into its milk, the strategy is to insert the gene next to a milk-specific promoter. Eukaryotic genes often are expressed in a tissue-specific fashion. In mammals, certain genes are expressed specifically within the mammary gland so that their protein product will be secreted into the milk. Examples of milk-specific genes include genes that encode milk proteins such as β-lactoglobulin, casein, and whey acidic protein.

TABLE 19.3

Proteins That Can Be Produced in the Milk of Domestic Animals

Protein	Host	Use
Lactoferrin	Cattle	Used as an iron supplement in infant formula
Tissue plasminogen activator (TPA)	Goat	Dissolves blood clots
Antibodies	Cattle	Used to combat specific infectious diseases
α-1-antitrypsin	Sheep	Treatment of emphysema
Factor IX	Sheep	Treatment of certain inherited forms of hemophilia
Insulin-like growth factor	Cattle	Treatment of diabetes

To express a human gene that encodes a protein hormone into a domestic animal's milk, the promoter for a milk-specific gene is linked to the coding sequence for the human gene (**Figure 19.9**). The DNA is then injected into an oocyte, where it will be integrated into the genome. The fertilized oocyte is then implanted into the uterus of a female animal, which later gives birth to a transgenic offspring. If the offspring is a female, the protein hormone encoded by the human gene will be expressed within the mammary gland and secreted into the milk. The milk can then be obtained from the animal, and the human hormone isolated.

Researchers Have Succeeded in Cloning Mammals from Somatic Cells

We now turn our attention to cloning as a way to genetically manipulate mammals. The term cloning has many different meanings. In Chapter 18, we discussed gene cloning, which involves methods that produce many copies of a gene. The cloning of an entire organism is a different matter. **Reproductive cloning** refers to methods that produce two or more genetically identical individuals. This happens occasionally in nature; identical twins are genetic clones that began from the same fertilized egg. Similarly, researchers can take mammalian embryos at an early stage of development (e.g., the two-cell to eight-cell stage), separate the cells, implant them into the uterus, and obtain multiple births of genetically identical individuals.

In the case of plants, cloning is an easier undertaking, as we will explore later in the chapter. Plants can be cloned from somatic cells. In most cases, it is easy to take a cutting from a plant, expose it to growth hormones, and obtain a separate plant that is genetically identical to the original. However, this approach has not been possible with mammals. For several decades, scientists believed that chromosomes within the somatic cells of mammals had incurred irreversible genetic changes that render them unsuitable for cloning. However, this hypothesis has proven to be incorrect. In 1997, Ian Wilmut and his colleagues at the Roslin Institute in Scotland announced that a sheep, named Dolly, had been cloned using the genetic material from somatic cells.

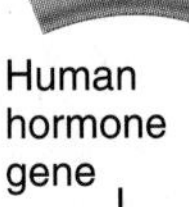
Human
hormone
gene
Using recombinant DNA technology (described in Chapter 18), clone a human hormone gene next to a sheep β-lactoglobulin promoter. This promoter is functional only in mammary cells so that the protein product is secreted into the milk.
β-lactoglobulin
promoter

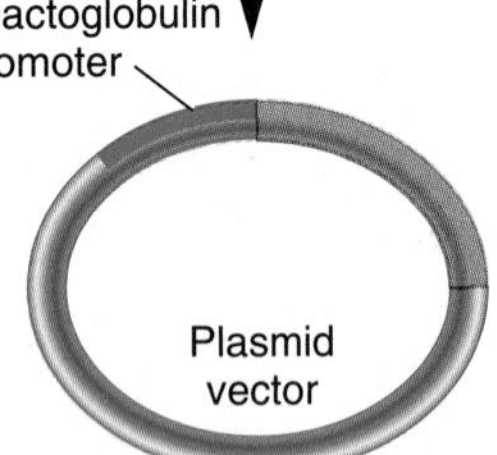
Plasmid
vector
Inject this DNA into a sheep oocyte. The plasmid DNA will integrate into the chromosomal DNA, resulting in the addition of the human hormone gene into the sheep's genome.
Implant the fertilized oocyte into a female sheep, which then gives birth to a transgenic sheep offspring.

Transgenic
sheep
Obtain milk from female transgenic sheep. The milk contains a human hormone.

Purify the hormone from the milk.

FIGURE 19.9 **Strategy for expressing human genes in a domestic animal's milk.** The β-lactoglobulin gene is normally expressed in mammary cells, whereas the human hormone gene is not. To express the human hormone gene in milk, the promoter from the milk-specific gene in sheep is linked to the coding sequence of the human hormone gene. In addition to the promoter, a short signal sequence may also be necessary so that the protein will be secreted from the mammary cells and into the milk.

Genes → Traits By using genetic engineering, researchers can give sheep the trait of producing a human hormone in their milk. This hormone can be purified from the milk and used to treat humans.

How was Dolly created? As shown in **Figure 19.10**, the researchers removed mammary cells from an adult female sheep and grew them in the laboratory. The researchers then extracted the nucleus from an egg cell of a different sheep and used electrical pulses to fuse the diploid mammary cell with the enucleated egg cell. After fusion, the zygote began embryonic development and the resulting embryo was implanted into the uterus of a surrogate mother sheep. One hundred and forty-eight days later, Dolly was born.

While Dolly was clearly a clone of the initial adult female sheep, tests conducted when she was three years old suggested that she was "genetically older" than her actual age indicated. As mammals age, chromosomes in somatic cells tend to shorten from the telomeres—the ends of eukaryotic chromosomes. Therefore, older individuals have shorter chromosomes in their somatic cells compared to younger ones. This shortening does not seem to occur in the cells of the germ line, however. When researchers analyzed the chromosomes in Dolly's somatic cells when she was about three years old, the lengths of her chromosomes were consistent with a sheep that was significantly older, say, nine or 10 years old. The sheep that donated the somatic cell that produced Dolly was six years old, and her mammary cells had been grown in culture for several cell doublings before a mammary cell was fused with an oocyte. This led researchers to suspect that Dolly's shorter telomeres were a result of chromosome shortening in the somatic cells of the sheep that donated the nucleus. In 2003, the Roslin Institute announced the decision to euthanize six-year-old Dolly after an examination showed progressive lung disease. Her death has raised concerns among experts that the techniques used to produce Dolly could have caused premature aging.

With regard to telomere length, research in mice and cattle has shown different results; the telomeres of these cloned animals appear to be the correct length. For example, cloning was conducted on mice via the method described in Figure 19.10 for six consecutive generations. The cloned mice of the sixth generation had normal telomeres. Further research will be necessary to determine if cloning via somatic cells has an effect on the length of telomeres in subsequent generations. However, other studies in mice point to various types of genetic flaws in cloned animals. For example, Rudolf Jaenisch and his colleagues used DNA microarray technology (described in Chapter 21) and analyzed the transcription patterns of over 10,000 genes in cloned mice. Up to 4% of those genes were not expressed normally. Furthermore, research has shown that cloned mice die at a younger age than their naturally bred counterparts.

Mammalian cloning is still at an early stage of development. Nevertheless, the breakthrough of creating Dolly has shown that it is technically possible. In recent years, cloning from somatic cells has been achieved in several mammalian species, including sheep, cattle, mice, goats, and pigs. In 2002, the first pet was cloned, which was named Carbon Copy, also called Copy Cat (**Figure 19.11**). Mammalian cloning has the potential to provide many practical applications. With regard to livestock, cloning would enable farmers to use the somatic cells from their best individuals to create genetically homogeneous herds. This could

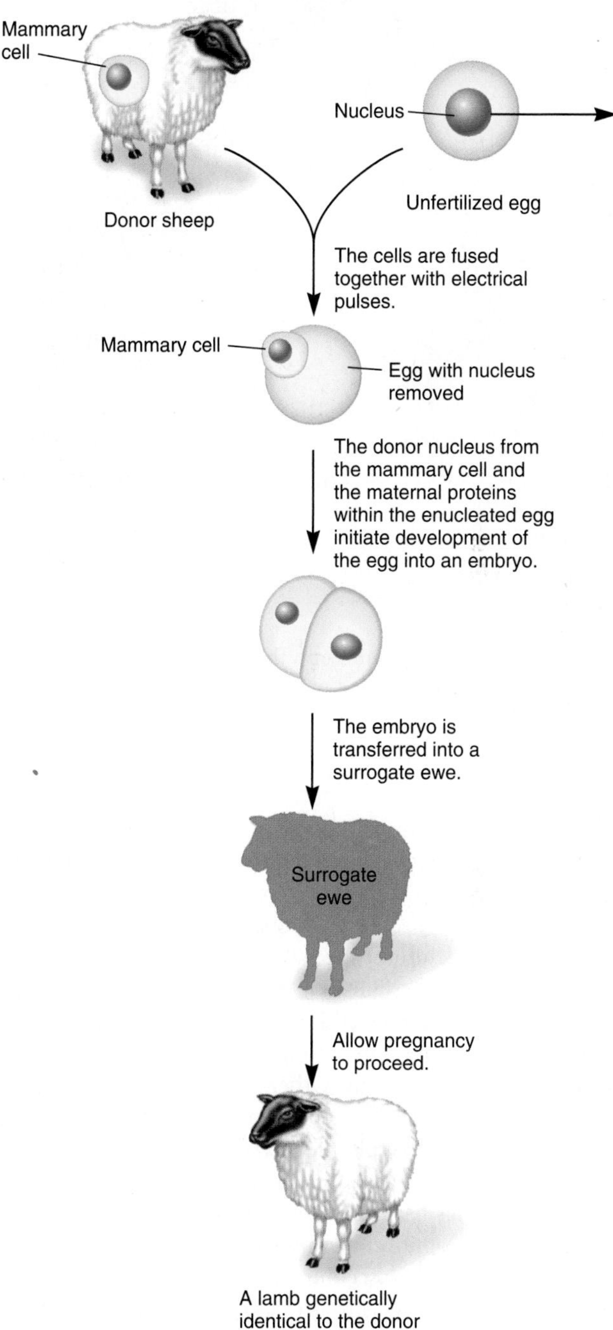

FIGURE 19.10 Protocol for the successful cloning of sheep.
Genes → Traits Dolly was (almost) genetically identical to the sheep that donated a mammary cell to create her. Dolly and the donor sheep were (almost) genetically identical in the same way that identical twins are; they carried the same set of genes and looked remarkably similar. However, they may have had minor genetic differences due to possible variation in their mitochondrial DNA and may have exhibited some phenotypic differences due to maternal effect or imprinted genes.

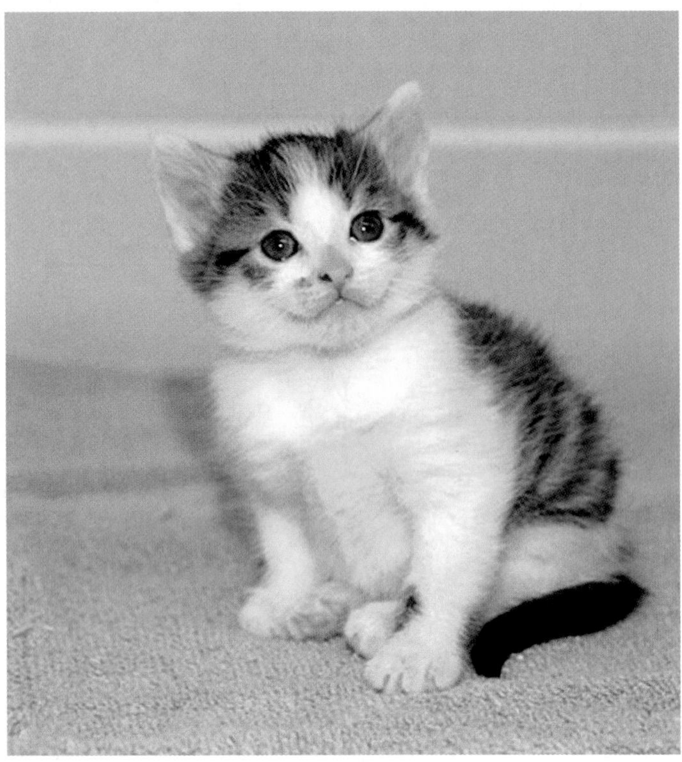

FIGURE 19.11 **Carbon Copy, the first cloned pet.** The animal shown here was produced using a procedure similar to the one shown in Figure 19.10.

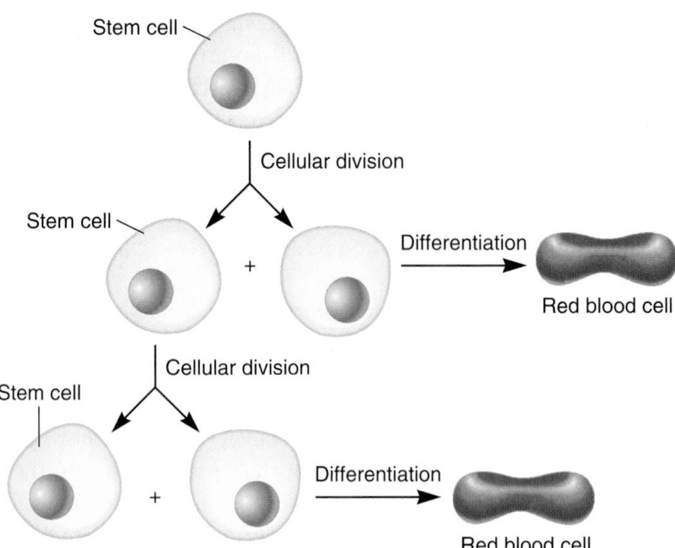

FIGURE 19.12 **Growth pattern of stem cells.** The two main traits that stem cells exhibit are an ability to divide and an ability to differentiate. When a stem cell divides, one of the two cells remains a stem cell, and the other daughter cell differentiates into a specialized cell type.

be advantageous with regard to agricultural yield, although such a genetically homogeneous herd may be more susceptible to rare diseases.

Though some people are concerned about the practical uses of cloning agricultural species, a majority have become very concerned with the possibility of human cloning. This prospect has raised a host of serious ethical questions. For example, some people feel that it is morally wrong and threatens the basic fabric of parenthood and family. Others feel that it is a technology that could offer a new avenue for reproduction, one that could be offered to infertile couples, for example. In the public sector, the sentiment toward human cloning has been generally negative. Many countries have issued an all-out ban on human cloning, while others permit limited research in this area. Because the technology for cloning exists, our society will continue to wrestle with the legal and ethical aspects of cloning as it applies not only to animals but also to people.

Stem Cells Have the Ability to Divide and Differentiate into Different Cell Types

Stem cells supply the cells that construct our bodies from a fertilized egg. In adults, stem cells also replenish worn-out or damaged cells. To accomplish this task, stem cells have two common characteristics. First, they have the capacity to divide, and second, they can differentiate into one or more specialized cell types. As shown in **Figure 19.12**, the two daughter cells produced from the division of a stem cell can have different fates. One of the cells may remain an undifferentiated stem cell, while the other daughter cell can differentiate into a specialized cell type. With this type of asymmetric division/differentiation pattern, the population of stem cells remains constant, yet the stem cells provide a population of specialized cells. In the adult, this type of mechanism is needed to replenish cells that have a finite life span, such as skin epithelial cells and red blood cells.

In mammals, stem cells are commonly categorized according to their developmental stage and their ability to differentiate (**Figure 19.13**). The ultimate stem cell is the fertilized egg, which, via multiple cellular divisions, can give rise to an entire organism. A fertilized egg is considered **totipotent,** because it can give rise to all the cell types in the adult organism. The early mammalian embryo contains **embryonic stem cells (ES cells),** which are found in the inner cell mass of the blastocyst. The blastocyst is the stage of embryonic development prior to uterine implantation—the preimplantation embryo. Embryonic stem cells are **pluripotent,** which means they can differentiate into almost every cell type of the body. However, a single embryonic stem cell has lost the ability to produce an entire, intact individual.

During the early fetal stage of development, the germ-line cells found in the gonads also are pluripotent. These cells are called **embryonic germ cells (EG cells)**. Interestingly, certain types of human cancers called teratocarcinomas arise from cells that are pluripotent. These bizarre tumors contain a variety of tissues including cartilage, neuroectoderm, muscle, bone, skin, ganglionic structures, and primitive glands. Due to the seemingly embryonic origin of teratocarcinoma cells, these cells are termed **embryonic carcinoma cells (EC cells).**

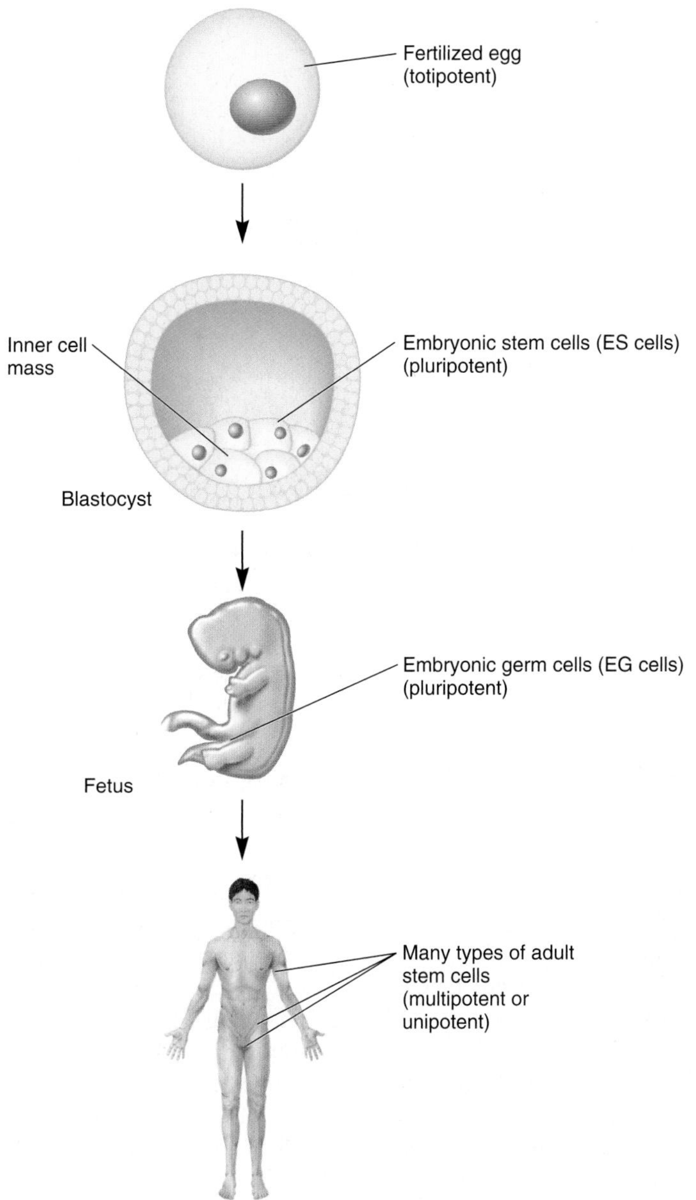

FIGURE 19.13 **Occurrence of stem cells at different stages of development.**

As mentioned, adults also contain stem cells, but these are thought to be multipotent or unipotent. A **multipotent** stem cell can differentiate into several cell types but far fewer than an embryonic stem cell. For example, hematopoietic stem cells (HSCs) found in the bone marrow can supply cells that populate two different tissues, namely, the blood and lymphoid tissues (**Figure 19.14**). Furthermore, each of these tissues contains several cell types. Multipotent hematopoietic stems cells can follow a pathway in which cell division produces a myeloid progenitor cell, which can then differentiate into a red blood cell, megakaryocyte, basophil, monocyte, eosinophil, neutrophil, or dendritic cell. Alternatively, an HSC can follow a path in which it becomes a lymphoid progenitor cell, which then differentiates into a T cell, B cell, natural killer cell, or dendritic cell. Other stem cells found in the adult seem to be **unipotent.** For example, primordial germ cells in the testis differentiate only into a single cell type, the sperm.

TABLE 19.4

Potential Uses of Stem Cells to Treat Diseases

Cell/Tissue Type	Disease Treatment
Neural	Implantation of cells into the brain to treat Parkinson disease
	Treatment of injuries such as spinal cord injuries
Skin	Treatment of burn victims and other types of skin disorders
Cardiac	Repair of heart damage associated with heart attacks
Cartilage	Repair of joints damaged by injury or arthritis
Bone	Repair of damaged bone or replacement with new bone
Liver	Repair or replacement of liver tissue that has been damaged by injury or disease
Skeletal muscle	Repair or replacement of damaged muscle

What are the potential uses of stem cells? Interest in stem cells centers around two main areas. Because stem cells have the capacity to differentiate into multiple cell types, the study of stem cells may help us to understand basic genetic mechanisms that underlie the process of development, the details of which are described in Chapter 23. A second compelling reason why people have become interested in stem cells is the potential to treat human diseases or injuries that cause cell and tissue damage. This application has already become a reality in certain cases. For example, bone marrow transplants are used to treat patients with certain forms of cancers. Such patients may be given radiation treatments that destroy their immune systems. When these patients are injected with bone marrow from a healthy person, the stem cells within the transplanted marrow have the ability to proliferate and differentiate within their bodies and provide them with a functioning immune system.

Renewed interest in the use of stem cells in the potential treatment of many other diseases was fostered in 1998 by studies of two separate teams, headed by James Thomson and John Gearhart, showing that embryonic cells, either ES or EG cells, can be successfully propagated in the laboratory. As mentioned, ES and EG cells are pluripotent and therefore have the capacity to produce many different kinds of tissue. As shown in **Table 19.4**, embryonic cells could potentially be used to treat a wide variety of diseases associated with cell and tissue damage. By comparison, it would be difficult, based on our modern knowledge, to treat these diseases with adult stem cells because of the inability to locate most types of adult stem cells within the body and successfully grow them in the laboratory. Even hematopoietic stem cells are elusive. In the bone marrow, about one cell in 10,000 is a stem cell, yet that is enough to populate all of the blood and

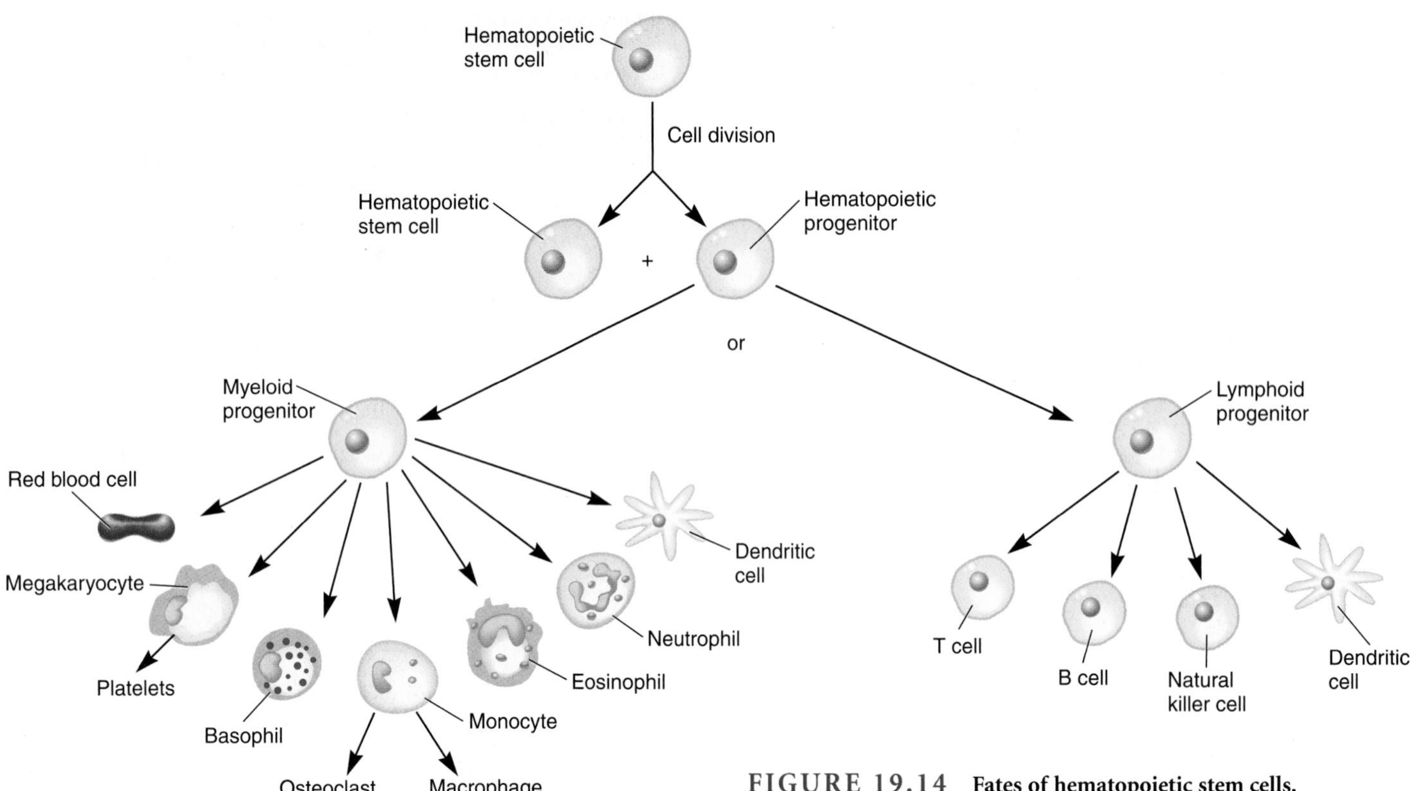

FIGURE 19.14 **Fates of hematopoietic stem cells.**

lymphoid cells of the body. The stem cells of most other adult tissues are equally difficult to locate, if not more so. In addition, with the exception of stem cells in the blood, other types of stem cells in the adult body are difficult to remove in sufficient numbers for transplantation. By comparison, ES and EG cells are easy to identify and have the great advantage of rapid growth in the laboratory. For these reasons, ES and EG cells offer a greater potential for transplantation, based on our current knowledge of stem cell biology.

For ES or EG cells to be used in transplantation, researchers will need to derive methods that cause them to differentiate into the appropriate type of tissue. For example, if the goal was to repair a spinal cord injury, ES or EG cells would need the appropriate cues that would cause them to differentiate into neural tissue. At present, much research is needed to understand and potentially control the fate of ES or EG cells. Currently, researchers speculate that a complex variety of factors determines the developmental fates of stem cells. These include internal factors within the stem cells themselves, as well as external factors such as the properties of neighboring cells and the presence of hormones and growth factors in the environment.

From an ethical perspective, the primary issue that raises debate is the source of the stem cells for research and potential treatments. Most ES cells have been derived from human embryos that were produced from in vitro fertilization and were subsequently not used. Most EG cells are obtained from aborted fetuses. Some feel that it is morally wrong to use such tissue in research and/or the treatment of disease or fear that this technology could lead to intentional abortions for the sole purpose of obtaining fetal tissues for transplantation. Alternatively, others feel that the embryos and fetuses that provide the ES and EG cells are not going to become living individuals, and therefore, it is beneficial to study these cells and use them in a positive way to treat human diseases and injury. It is not clear whether these two opposing viewpoints can reach a common ground.

As a compromise, many governments have enacted laws that limit or prohibit the use of embryos or fetuses to obtain stem cells, yet permit the use of stem cell lines that are already available in research laboratories. In the United States, for example, recent interpretations of federal laws prohibit the use of government funding for research projects that involve the destruction of embryos to obtain stem cells. As of 2007, government-sponsored research can be done on the 60 or so stem cell lines that were created prior to this legislation, and nongovernment-sponsored research is not subject to this limitation.

If stem cells could be obtained from adult cells and propagated in the laboratory, an ethical dilemma may be avoided because most people do not have serious moral objections to current procedures such as bone marrow transplantation. In 2006, work by Shinya Yamanaka and colleagues showed that adult mouse fibroblasts (a type of connective tissue cell) could become pluripotent via the injection of four different genes that encode transcription factors. In 2007, Yamanaka's laboratory and two other research groups were able to show that such induced pluripotent stem cells (iPS) can differentiate into all cell types when injected into mouse blastocysts and grown into baby mice. Though further research is still needed, these recent results indicate that adult cells can be reprogrammed to become embryonic stem cells.

19.3 GENETICALLY MODIFIED PLANTS

As we have seen, researchers have succeeded in making genetically modified animals for a variety of reasons. In this section, we will examine the methods that scientists follow to make transgenic plants.

A relatively new area of research is the use of transgenic species in agriculture. For centuries, agriculture has relied on selective breeding programs to produce plants and animals with desirable characteristics. For agriculturally important species, this often means the production of strains that are larger, have disease resistance, and yield high-quality food. Agricultural scientists can now complement traditional breeding strategies with modern molecular genetic approaches. In the mid-1990s, genetically modified crops first became commercialized. Since that time, their use has progressively increased. In 2006, roughly 20% of all agricultural crops were transgenic. Worldwide, more than 100 million hectares (247 million acres) of transgenic crops were planted. In this section, we will discuss some current and potential uses of transgenic plants in agriculture.

Agrobacterium tumefaciens and Other Methods Can Be Used to Make Transgenic Plants

As we have seen, the introduction of cloned genes into embryonic cells can produce transgenic animals. The production of transgenic plants is somewhat easier, because certain plant cells are totipotent, which means they are capable of developing into an entire organism. Therefore, a transgenic plant can be made by the introduction of cloned genes into somatic tissue, such as the tissue of a leaf. After the cells of a leaf have become transgenic, an entire plant can be regenerated by the treatment of the leaf with plant growth hormones, which cause it to form roots and shoots.

Molecular biologists can use the bacterium *Agrobacterium tumefaciens*, which naturally infects plant cells, to produce transgenic plants. A plasmid from the bacterium, known as the **Ti plasmid** (Tumor-inducing plasmid), naturally induces tumor formation after a plant has been infected (**Figure 19.15a**). A segment of the plasmid DNA, known as **T DNA** (for transferred DNA), is transferred from the bacterium to the infected plant cells. The T DNA from the Ti plasmid becomes integrated into the chromosomal DNA of the plant cell by recombination. After this occurs, genes within the T DNA that encode plant growth hormones cause uncontrolled plant cell growth. This produces a cancerous plant growth known as a crown gall tumor (**Figure 19.15b**).

Because *A. tumefaciens* inserts its T DNA into the chromosomal DNA of plant cells, it can be used as a vector to introduce cloned genes into plants. Molecular geneticists have been able to modify the Ti plasmid to make this an efficient process. The T DNA genes that cause the development of a gall have been identified. Fortunately for genetic engineers, when these genes are deleted, the T DNA is still taken up into plant cells and integrated within the plant chromosomal DNA. However, a gall does not form. In addition, geneticists have inserted selectable marker genes into the T DNA to allow selection of plant cells that have taken up the T DNA. A gene that provides resistance to the antibiotic kanamycin is a commonly used selectable marker. The Ti plasmids used in cloning experiments are also modified to contain unique restriction sites for the convenient insertion of any gene.

Figure 19.16 shows the general strategy for producing transgenic plants via T DNA-mediated gene transfer. A gene of interest is inserted into a genetically engineered Ti plasmid and then transformed into *A. tumefaciens*. Plant cells are exposed to the transformed *A. tumefaciens*. After allowing time for infection, the plant cells are exposed to the antibiotics kanamycin and carbenicillin. Carbenicillin kills *A. tumefaciens*, and kanamycin kills any plant cells that have not taken up the T DNA with the antibiotic-resistance gene. Therefore, the only surviving cells are those plant cells that have integrated the T DNA into their genome. Because the T DNA also contains the cloned gene of interest, the selected plant cells are expected to have received this cloned gene as well. The cells are then transferred to a medium that contains the plant growth hormones necessary for the regeneration of entire plants. These plants can then be analyzed to verify that they are transgenic plants containing the cloned gene.

A. tumefaciens infects a wide range of plant species, including most dicotyledonous plants, most gymnosperms, and some monocotyledonous plants. However, not all plant species are infected by this bacterium. Fortunately, other methods are available for introducing genes into plant cells. Another common way to produce transgenic plants is an approach known as **biolistic gene transfer.** In this method, plant cells are bombarded with high-velocity microprojectiles coated with DNA. When fired upon by this "gene gun," the microprojectiles penetrate the cell wall and membrane and thereby enter the plant cell. The cells that take up the DNA are identified with a selectable marker and regenerated into new plants.

Other methods are also available to introduce DNA into plant cells (and also animal cells). For example, DNA can enter plant cells by **microinjection**—the use of microscopic-sized needles—or by **electroporation**—the use of electrical current to create temporary pores in the plasma membrane. Because the rigid plant cell wall is a difficult barrier for DNA entry, other approaches involve the use of protoplasts, which are plant cells that have had their cell walls removed. DNA can be introduced into protoplasts using a variety of methods, including treatment with polyethylene glycol and calcium phosphate.

The production of transgenic plants has been achieved for many agriculturally important plant species. These include alfalfa, corn, cotton, soybean, tobacco, and tomato. Some of the applications of transgenic plants are described next.

Transgenic Plants Can Be Given Characteristics That Are Agriculturally Useful

Various traits can be modified in transgenic plants (**Table 19.5**). Frequently, transgenic research has sought to produce plant strains resistant to insects, disease, and herbicides. For example, transgenic plants highly tolerant of particular herbicides have

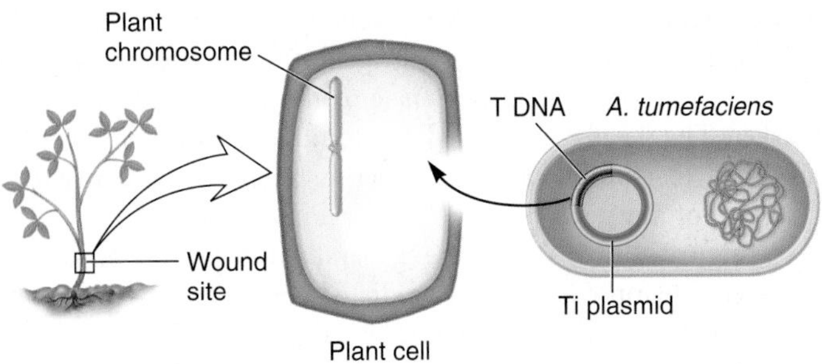

Agrobacterium tumefaciens is found within the soil. A wound on the plant enables the bacterium to infect the plant cells.

During infection, the T DNA within the Ti plasmid is transferred to the plant cell. The T DNA becomes integrated into the plant cell's DNA. Genes within the T DNA promote uncontrolled plant cell growth.

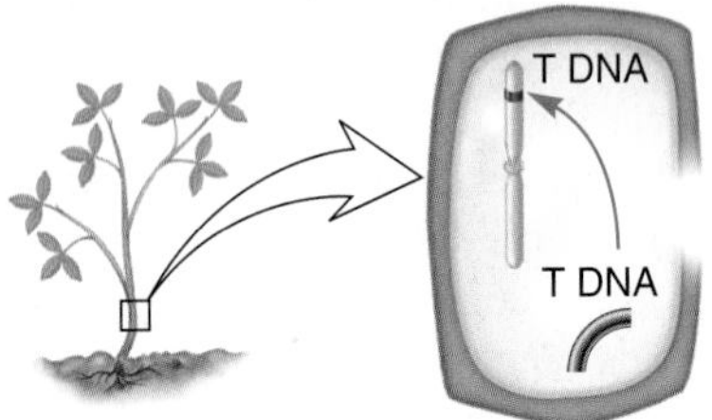

The growth of the recombinant plant cells produces a crown gall tumor.

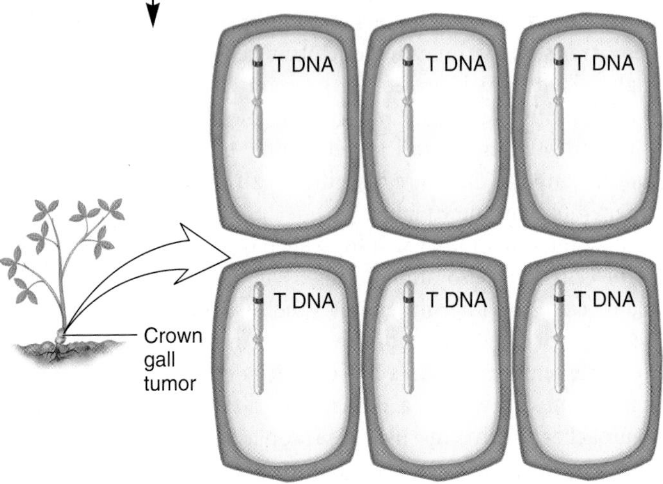

(a) The production of a crown gall tumor by *A. tumefaciens* infection

(b) A crown gall tumor on a pecan tree

FIGURE 19.15 ***Agrobacterium tumefaciens*** **infecting a plant and causing a crown gall tumor.**

been made. The Monsanto Company has produced transgenic plant strains tolerant of glyphosate, the active agent in the herbicide Roundup™. The herbicide remains effective against weeds, but the herbicide-resistant crop is spared (**Figure 19.17**).

Another important approach is to make plant strains that are disease resistant. In many cases, virus-resistant plants have been developed by introducing a gene that encodes a viral coat protein. When the plant cells express the viral coat protein, they become resistant to infection by that pathogenic virus.

Many transgenic plants have been approved for human consumption. The first example was the Flavr Savr tomato (**Figure 19.18**), which was developed by Calgene Inc., which is now part of Monsanto. In this technique, a tomato plant was given a gene that encodes an antisense RNA complementary to the mRNA that encodes the enzyme polygalacturonase. This enzyme, which is expressed during ripening, digests sugar linkages within the pectin found in plant cell walls and thus softens the tomato. The antisense RNA binds to the polygalacturonase mRNA, preventing it from being translated. In

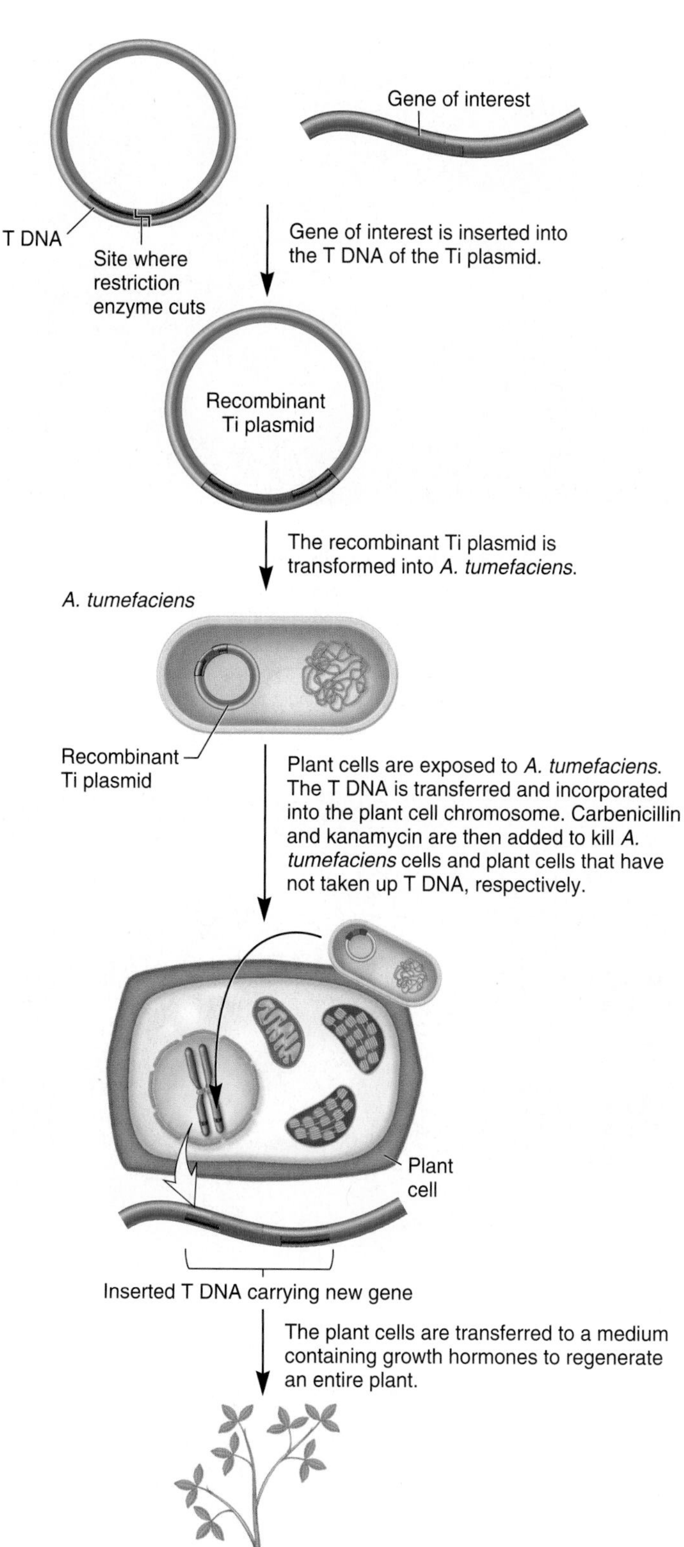

FIGURE 19.16 The transfer of genes into plants using the Ti plasmid from *A. tumefaciens* as a vector.

TABLE 19.5

Traits That Have Been Modified in Transgenic Plants

Trait	Examples
Plant Protection	
Resistance to viral, bacterial, and fungal pathogens	Transgenic plants that express the pokeweed antiviral protein are resistant to a variety of viral pathogens.
Resistance to insects	Transgenic plants that express the CryIA protein from *Bacillus thuringiensis* are resistant to a variety of insects (see Figure 19.19).
Resistance to herbicides	Transgenic plants can express proteins that render them resistant to particular herbicides (see Figure 19.17).
Plant Quality	
Improvement in storage	Transgenic plants can express antisense RNA that silences a gene involved in fruit softening (see Figure 19.18).
Change in plant composition	Transgenic strains of canola have been altered with regard to oil composition; the seeds of the Brazil nut have been rendered methionine-rich via transgenic technology.
New Products	
Biodegradable plastics	Transgenic plants have been made that can synthesize polyhydroxyalkanoates, which are used as biodegradable plastics.
Vaccines	Transgenic plants have been modified to produce vaccines in their leaves against many human and animal diseases including hepatitis B, cholera, and malaria.
Pharmaceuticals	Transgenic plants have been made that produce a variety of medicines including human interferon-α (to fight viral diseases and cancer), human epidermal growth factor (for wound repair), and human aprotinin (for transplantation surgery).
Antibodies	Human antibodies have been made in transgenic plants to battle various diseases such as non-Hodgkin's lymphoma.

addition, the double-stranded RNA is targeted for degradation (RNA interference), as discussed in Chapter 15 (Figure 15.23). Because the expression of polygalacturonase is silenced, the practical advantage is that the tomatoes do not soften as quickly as unmodified tomatoes. Therefore, these tomatoes can be allowed to ripen on the vine for a longer period of time, enhancing their flavor and extending their shelf life. Improved taste is an important consideration in the \$5-billion annual U.S. tomato market. By comparison, other commercial tomatoes commonly are picked while they are green and ripen later in order to maintain their firmness longer.

The Flavr Savr tomato was not a commercial success, and its sales were eventually discontinued. The failure of the Flavr Savr has been attributed to a variety of issues. In particular, the variety of tomato that was genetically engineered may have not been the best choice, and the antirotting trait was not as helpful

FIGURE 19.17 Transgenic plants that are resistant to glyphosate.

Genes → Traits This field of soybean plants has been treated with glyphosate. The plants on the left have been genetically engineered to contain a herbicide-resistance gene. They are resistant to killing by glyphosate. By comparison, the dead or stunted plants in the row with the orange stick do not contain this gene.

FIGURE 19.18 The genetically engineered Flavr Savr tomato.

Genes → Traits This transgenic tomato plant has been genetically engineered to contain an artificial gene that encodes an antisense RNA that will bind to the mRNA that encodes polygalacturonase. This prevents the mRNA from being translated, which inhibits the synthesis of polygalacturonase. Without this enzyme, the Flavr Savr tomato can be left on the vine to ripen longer than traditional tomatoes, thereby improving its flavor.

for the tomato business as Calgene had anticipated. Compared to unmodified, green tomatoes, another factor was that the ripe Flavr Savr tomatoes were more delicate and required the purchase of expensive handling equipment.

A much more successful example of the use of transgenic plants has involved the introduction of genes from *Bacillus thuringiensis* (Bt). As discussed earlier in this chapter, this bacterium produces toxins that kill certain types of caterpillars and beetles and has been widely used as a biological control agent for several decades. These toxins are proteins encoded in the genome of *B. thuringiensis.* Researchers have succeeded in cloning toxin genes from *B. thuringiensis* and transferring those genes into plants. Such Bt varieties of plants produce the toxins themselves and therefore are resistant to many types of caterpillars and beetles. Examples of commercialized crops include Bt corn (**Figure 19.19a**) and Bt cotton. Since their introduction in 1996, the commercial use of these two Bt crops has steadily increased (**Figure 19.19b**).

The introduction of transgenic plants into agriculture has been strongly opposed by some people. What are the perceived risks? One potential risk is that transgenes in commercial crops could endanger native species. For example, Bt crops may kill pollinators of native species. Another worry is that the planting of transgenic crops could potentially lead to the proliferation of resistant insects. To prevent this from happening, researchers

(a) A field of Bt corn

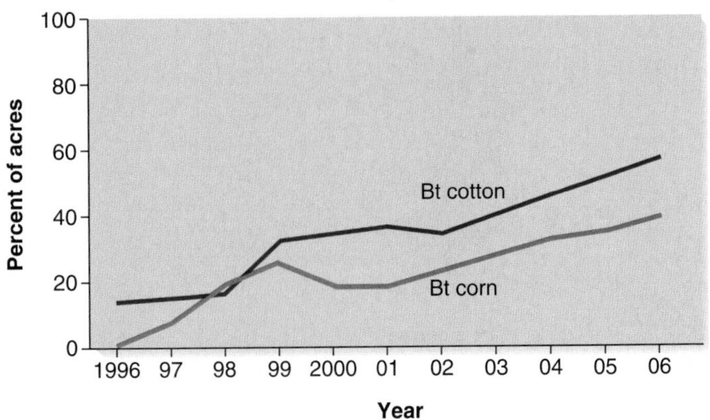

(b) Bt corn and Bt cotton usage since 1996

FIGURE 19.19 The production of Bt crops. (a) A field of Bt corn. These corn plants carry an endotoxin gene from *Bacillus thuringiensis* that provides them with resistance to insects such as corn borers, which are a major pest of corn plants. **(b)** A graph showing the increase in usage of Bt corn and Bt cotton since their commercial introduction in 1996.

are producing transgenic strains that carry more than one toxin gene, which makes it more difficult for insect resistance to arise. Despite these and other concerns, many farmers are embracing transgenic crops, and their use continues to rise.

19.4 HUMAN GENE THERAPY

Throughout this textbook, we have considered examples in which mutant genes cause human diseases. As discussed in Chapter 22, these include rare inherited disorders and more common diseases such as cancer. Because mutant genes cause disease, geneticists are actively pursuing the goal of using normal, cloned genes to compensate for defects in mutant genes. Gene therapy is the introduction of cloned genes into living cells in the treatment of disease. It is a potential method of treating a wide variety of illnesses.

Many current research efforts in gene therapy are aimed at alleviating inherited human diseases. Over 4,000 human genetic diseases are known to involve a single gene abnormality. Familiar examples include cystic fibrosis, sickle-cell disease, and hemophilia. In addition, gene therapies have also been aimed at treating diseases such as cancer and cardiovascular disease, which may occur later in life. Some scientists are even pursuing research that will use gene therapy to combat infectious diseases such as AIDS. Even though gene therapy is still at an early stage of development, a large amount of research has already been conducted. Unfortunately, success has been limited, and relatively few patients have been treated with gene therapy. Nevertheless, a few results have been somewhat promising. **Table 19.6** describes several types of diseases that are being investigated as potential targets for gene therapy. In this section, we will examine the approaches to gene therapy and how it may be used to treat human disease.

TABLE **19.6**

Future Prospects in Gene Therapy

Type of Disease	Treatment of
Blood	Sickle-cell disease, hemophilia, severe combined immunodeficiency disease (SCID)
Metabolic	Glycogen storage diseases, lysosomal storage diseases, and phenylketonuria
Muscular	Duchenne muscular dystrophy, myotonic muscular dystrophy
Lung	Cystic fibrosis
Cancer	Brain tumors, breast cancer, colorectal cancer, malignant melanoma, ovarian cancer, and several other types of malignancies
Cardiovascular	Atherosclerosis, essential hypertension
Infectious	AIDS, possibly other viral diseases that involve latent infections

Gene Therapy Involves the Introduction of Cloned Genes into Human Cells

A key step in gene therapy is the introduction of a cloned gene into the cells of people. The techniques to transfer a cloned gene into human cells can be categorized as nonviral and viral gene transfer methods. The most common nonviral technique involves the use of **liposomes,** which are lipid vesicles (**Figure 19.20a**). The DNA containing the gene of interest is complexed with liposomes that carry a positive charge (i.e., cationic liposomes). The DNA-liposome complexes are taken into cells via endocytosis, in which a portion of the plasma membrane invaginates and creates an intracellular vesicle known as an endosome; the liposome is degraded within the endosome. The DNA is then released into the cytosol, imported into the nucleus, and then integrated into a chromosome of the target cell. An advantage of gene transfer via liposomes is that the liposomes do not elicit an immune response. A disadvantage is that the efficiency of gene transfer may be very low.

A second way to transfer genes into human cells is via viruses. Commonly used viruses for gene therapy include retroviruses, adenoviruses, and parvoviruses. The genetic modification of these viral genomes has led to the development of gene therapy vectors with a capacity to infect cells or tissues, much like the ability of wild-type viruses to infect cells. However, in contrast to wild-type viruses, gene therapy viral vectors have been genetically engineered so they have lost their capacity for replication in target cells. Nevertheless, the genetically engineered viruses are naturally taken up by cells via endocytosis (**Figure 19.20b**). The viral coat disassembles, and the viral genome is released into the cytosol. In the case of retroviruses, the genome is RNA, which is reverse transcribed into DNA. The viral DNA, which carries a gene of interest, travels into the nucleus and is then integrated into a chromosome of the target cell.

A key advantage of viral vectors is their ability to efficiently transfer cloned genes to a variety of human cell types. However, a major disadvantage of viral-mediated gene therapy is the potential to evoke an undesirable immune response when injected into a patient. The inflammatory responses induced by adenovirus particles, for example, can be very strong and even fatal. Therefore, much effort has been aimed at preventing the inflammatory responses mediated by virus particles. These include the application of immunosuppressive drugs and the generation of less immunogenic viral vectors by further genetic modification within the viral genome.

Adenosine Deaminase Deficiency Was the First Inherited Disease Treated with Gene Therapy

Adenosine deaminase (ADA) is an enzyme involved in purine metabolism. If both copies of the *ADA* gene are defective, deoxyadenosine will accumulate within the cells of the individual. At high concentrations, deoxyadenosine is particularly toxic to lymphocytes in the immune system, namely, T cells and B cells. In affected individuals, the destruction of T and B cells leads to a form of severe combined immunodeficiency disease (SCID). If

FIGURE 19.20 Methods of gene transfer used in gene therapy. **(a)** In this example, the DNA containing the gene of interest is complexed with cationic liposomes. These complexes are taken into cells by endocytosis, in which a portion of the plasma membrane invaginates and creates an intracellular vesicle known as an endosome. After it is released from the endosome, the DNA may then integrate into the chromosomal DNA via recombination. **(b)** In this example, the gene of interest is cloned into a retrovirus. When the retrovirus infects a cell, the RNA genome is reverse transcribed into double-stranded DNA, which then integrates into the chromosome. Viruses used in gene therapy have been genetically altered so they cannot proliferate after entry into the target cell.

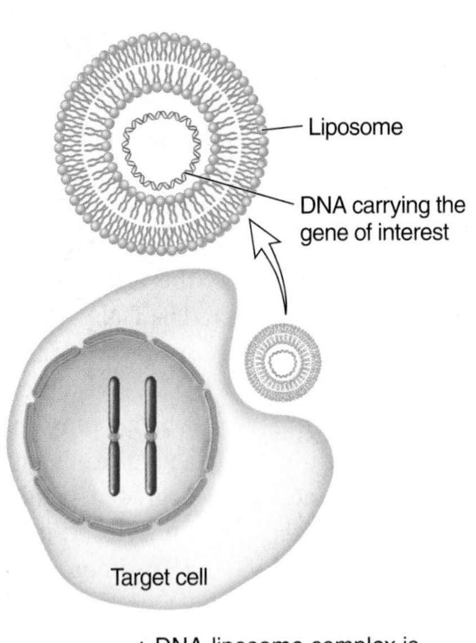
Liposome
DNA carrying the gene of interest
Target cell
DNA-liposome complex is taken into the target cell by endocytosis.
The liposome is degraded within the endosome and the DNA is released into the cytosol.
The DNA is imported into the cell nucleus.

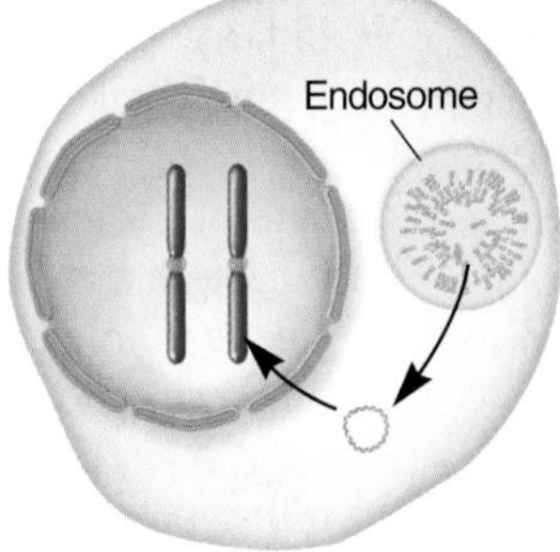
Endosome
By recombination, the DNA carrying the gene of interest is integrated into a chromosome of the target cell.

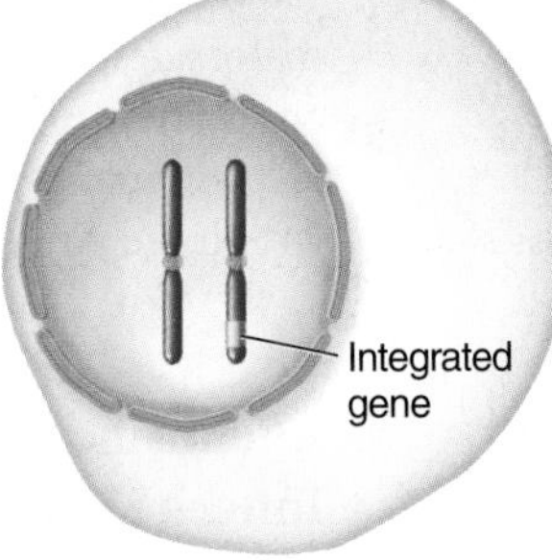
Integrated gene

(a) Nonviral approach

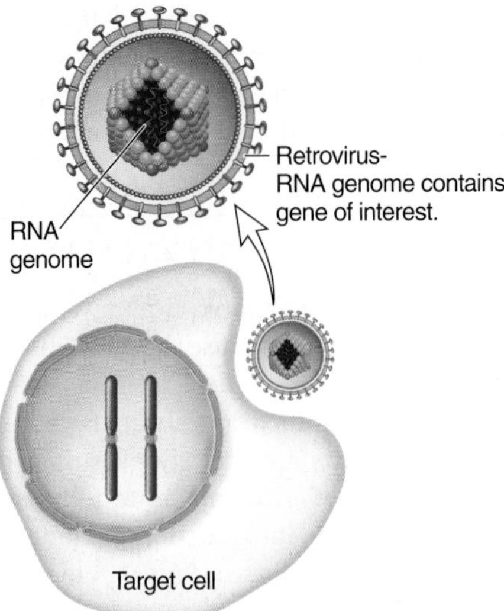
Retrovirus-RNA genome contains gene of interest.
RNA genome
Target cell
Retrovirus is taken into the target cell via endocytosis.
The viral coat is disassembled in the endosome, and two copies of the RNA genome are released into the cytosol.
The RNA is reverse transcribed into DNA, which travels into the nucleus.

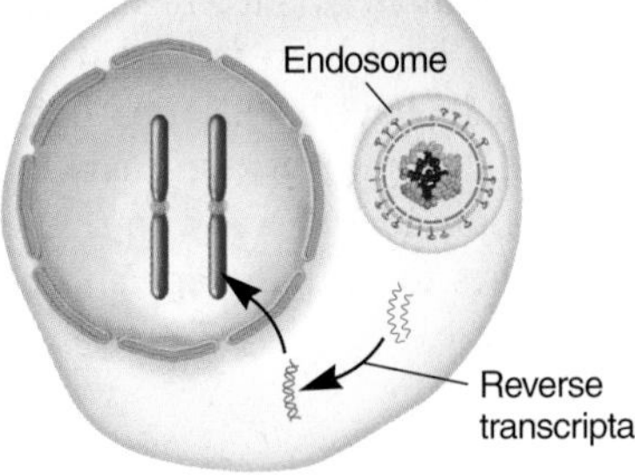
Endosome
Reverse transcriptase
By recombination, the viral DNA, carrying the gene of interest, is integrated into a chromosome of the target cell.

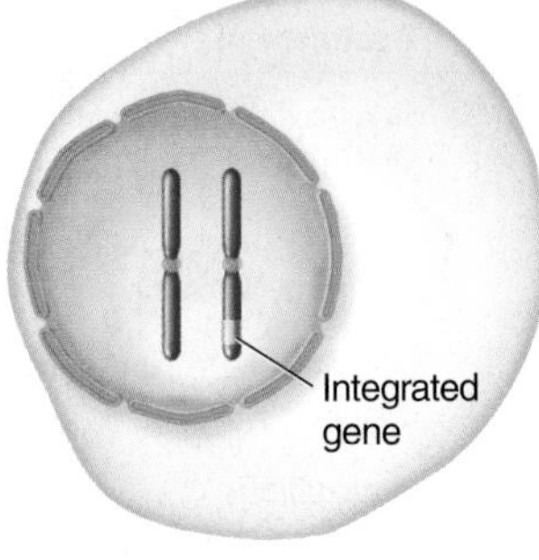
Integrated gene

(b) Viral approach

left untreated, SCID is typically fatal at an early age (generally, one to two years old), because the immune system of these individuals is severely compromised and cannot fight infections.

Three approaches can be used to treat ADA deficiency. In some cases, a patient may receive a bone marrow transplant from a compatible donor. A second method is to treat SCID patients with purified ADA that is coupled to polyethylene glycol (PEG). This PEG-ADA is taken up by lymphocytes and can correct the ADA deficiency. Unfortunately, these two approaches are not always available and/or successful. A third, more recent approach is to treat ADA patients with gene therapy.

On September 14, 1990, the first human gene therapy was approved for a young girl suffering from ADA deficiency. This work was carried out by a large team of researchers composed of R. Michael Blaese, Kenneth Culver, W. French Anderson, and colleagues. Prior to this clinical trial, the normal gene for ADA had been cloned into a retroviral vector that can infect lymphocytes. The general aim of this therapy was to remove lymphocytes from the blood of the young girl with SCID, introduce the normal *ADA* gene into her cells, and then return them to her bloodstream.

Figure 19.21 outlines the protocol for the experimental treatment. Lymphocytes (i.e., T cells) were removed and cultured in a laboratory. The lymphocytes were then transfected with a nonpathogenic retrovirus that had been genetically engineered to contain the normal *ADA* gene. During the life cycle of a retrovirus, the retroviral genetic material is inserted into the host cell's DNA. Therefore, because this retrovirus contained the normal ADA gene, this gene also was inserted into the chromosomal DNA of the girl's lymphocytes. After this had occurred in the laboratory, the cells were reintroduced back into the patient. This approach is called an **ex vivo approach** because the genetic manipulations occur outside the body, yet the products are reintroduced into the body.

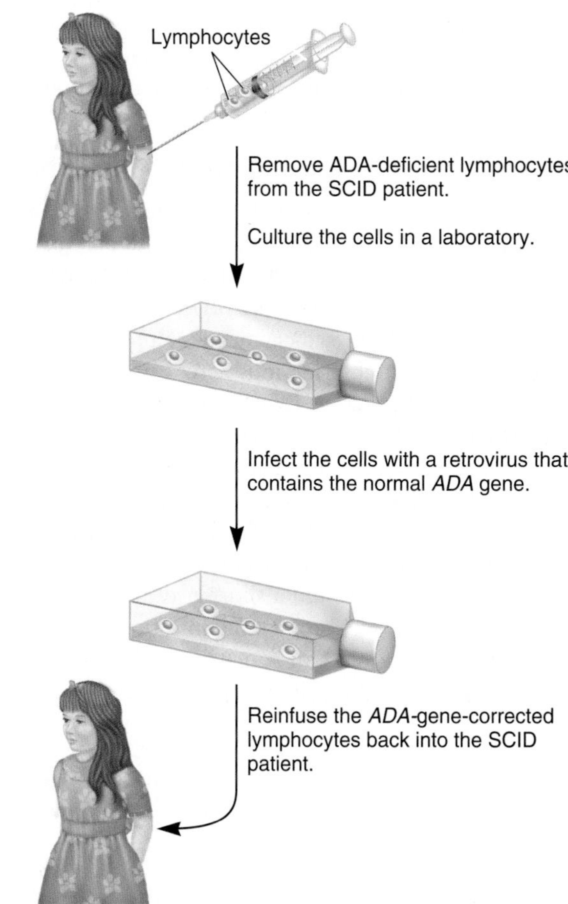
Lymphocytes
Remove ADA-deficient lymphocytes from the SCID patient.
Culture the cells in a laboratory.
Infect the cells with a retrovirus that contains the normal *ADA* gene.
Reinfuse the *ADA*-gene-corrected lymphocytes back into the SCID patient.

FIGURE 19.21 The first human gene therapy for adenosine deaminase (ADA) deficiency.

In this clinical trial, two patients were enrolled, and a third patient was later treated in Japan. Was the treatment a success? The results of this trial showed that the transfer of DNA into a large number of human cells is feasible. In at least one patient, T cells carrying the cloned gene were still detectable 8 to 10 years after they had been transferred. However, most of the circulating T cells were not found to contain the cloned gene. Because the individuals also received a low dose of PEG-ADA treatment, researchers could not determine whether or not gene transfer into T cells in itself was of significant clinical benefit.

Another form of SCID, termed SCID-X1, is inherited as an X-linked trait. SCID-X1 is characterized by a block in T cell growth and differentiation. This block is caused by mutations in the gene encoding the γ_c cytokine receptor, which plays a key role in the recognition of signals that are needed to promote the growth, survival, and differentiation of T cells. A gene therapy trial for SCID-X1 similar to the trial shown in Figure 19.21 was initiated in 2000 in which a normal γ_c cytokine receptor gene was cloned into a retroviral vector and then introduced into SCID-X1 patients' lymphocytes. The lymphocytes were then reintroduced back into their bodies. At a 10-month follow-up, T cells expressing the normal γ_c cytokine receptor were detected in two patients. Most importantly, the T cell counts in these two patients had risen to levels that were comparable to normal individuals. This clinical trial was the first clear demonstration that gene therapy can offer clinical benefit, providing in these cases what seemed to be a complete correction of the disease phenotype. However, in a French study involving 10 SCID-X1 patients, an unexpected and serious side effect occurred. Within three years of gene therapy treatment, three out of the 10 treated children developed leukemia—a form of cancer involving the proliferation of white blood cells. In these cases, the disease was caused by the integration of the retroviral vector next to a particular gene in the patients' genomes. The development of leukemia in these patients has halted many clinical trials involving gene therapy.

Aerosol Sprays May Be Used to Treat Cystic Fibrosis

Cystic fibrosis (CF) is a rare recessive disorder with debilitating consequences. About 1 in 3,000 babies whose parents are of northern European descent are affected with this disorder. CF is caused by a defect in a gene termed the cystic fibrosis transmembrane regulator (*CFTR*), which encodes a protein that functions in the transport of chloride ions across the plasma membrane of epithelial cells, such as cells lining the respiratory tract and the intestinal tract. A defect in membrane transport leads to an abnormality in salt and water balance, which causes a variety of symptoms, particularly an overaccumulation of mucus in the lungs. Even though great strides have been made in the treatment of the symptoms of CF, this disease remains associated with repeated lung infections and a shortened life span. In most cases, mortality results from chronic lung infections.

CF has been the subject of much gene therapy research. Clinical trials have tested the ability of gene therapy to improve the condition of patients suffering from CF. To implement CF gene therapy, it is necessary to deliver the normal *CFTR* gene to the lung cells. Unlike *ADA* gene therapy, in which the lymphocytes can be treated ex vivo, lung epithelial cells cannot be removed and then put back into the individual. Instead, researchers must design innovative approaches that can target the *CFTR* gene directly to the lung cells.

To achieve this goal, CF gene therapy methods have involved the use of an inhaled aerosol spray. In one protocol, the normal *CFTR* gene is cloned into an adenovirus, a virus that normally infects lung epithelial cells and causes a lung infection. This adenovirus, however, has been engineered so that it will gain entry into the epithelial cells but not cause a lung infection. In addition, the adenovirus has been engineered to contain the normal *CFTR* gene. In a second approach, the normal *CFTR* gene is complexed with liposomes. When inhaled by the patient via an aerosol spray, the lung epithelial cells take up this liposome complex.

Like *ADA* gene therapy, CF gene therapy is at an early stage of development. Researchers hope that gene therapy eventually will become an effective method to alleviate the symptoms associated with this disease.

CONCEPTUAL AND EXPERIMENTAL SUMMARY

Biotechnology, the application of technologies involving the use of living organisms in the development of products and services that benefit humans, has increasingly become associated with molecular genetics. For example, molecular genetic tools can engineer recombinant microorganisms used in the production of medically important products. The production of somatostatin by recombinant bacteria paved the way for the creation of genetically engineered medicines, including insulin and human growth hormone. Recombinant microorganisms have also been used as agents of biological control of plant diseases and for bioremediation to modify toxic pollutants.

Molecular tools are also being used to create transgenic organisms, which contain recombinant DNA from another species that has been integrated into their genome. We explored the current techniques employed to create transgenic animals and their potential uses. The introduction of cloned DNA into a cell can lead to gene replacement or gene addition. In this way, researchers can produce gene knockouts and knockins, which are powerful tools to explore gene function and human disease. Research is developing transgenic species of livestock that produce medically important proteins. In recent years, cloning from somatic cells has been achieved in several mammalian species, beginning with the first cloned sheep, Dolly. Research studies of stem cells can help us to understand basic genetic mechanisms underlying the process of development. Furthermore, stem cells have the potential to treat human diseases or injuries that cause cell and tissue damage.

A relatively new area of research is the use of transgenic species in agriculture. Transgenic plants can be produced through the use of the bacterium *Agrobacterium tumefaciens* as a vector or through the technique of biolistic gene transfer. An important goal is to make plant strains that are disease or herbicide resistant. While many people have concerns over the use of genetically modified organisms in agriculture, their use continues to rise. In the future, it is expected that the production and commercial success of transgenic organisms will become more prevalent.

Much scientific effort is going into gene therapy, the introduction of cloned genes into living cells in order to treat a wide variety of human diseases. In this approach, a cloned gene is introduced via liposomes or viruses into somatic cells as a way to compensate for a defective gene. While success has been limited and sobering risks have been identified, a few results have been somewhat promising. In the future, this may become an important way to alleviate many human disorders.

PROBLEM SETS & INSIGHTS

Solved Problems

S1. The protocol that was followed to produce somatostatin in recombinant *E. coli* cells would have been difficult to carry out if somatostatin was a long polypeptide. Explain why.

Answer: The researchers constructed the coding sequence for somatostatin by making several short oligonucleotides that were complementary to each other and formed the somatostatin coding sequence. This would have been difficult to achieve if the polypeptide sequence was much longer, say, 100 amino acids in length, because it would have required the synthesis of many short oligonucleotides.

S2. Which of the following would appropriately be described as a transgenic organism?

A. The sheep "Dolly," which was produced by cloning

B. A sheep that produces human α-1-antitrypsin in its milk

C. The Flavr Savr strain of tomato

D. A hybrid strain of corn produced from crossing two inbred strains of corn (The inbred strains were not transgenic.)

Answer:

A. No, Dolly was not produced using recombinant techniques. Pieces of DNA were not cut and combined in a new way.

B. Yes

C. Yes

D. No, the hybrids simply contain chromosomal genes from two different parental strains.

S3. Describe the strategy for producing human proteins in the milk of livestock.

Answer: Milk proteins are encoded by genes with promoters and regulatory sequences that direct the expression of these genes within the cells of the mammary gland. To get other proteins expressed in the mammary gland, the strategy is to link the promoter and regulatory sequences from a milk-specific gene to the coding sequence of the gene that encodes the human protein of interest. In some cases, it is also necessary to add a signal sequence to the amino terminal end of the target protein. A signal sequence is a short polypeptide that directs the secretion of a protein from a cell. If the target protein does not already have a signal sequence, it is possible to use a signal sequence from a milk-specific gene to promote the secretion of the target protein from the mammary cells and into the milk. During this process, the signal sequence is cleaved from the secreted protein.

S4. With regard to genetically modified organisms, describe two that have not been successful and two that have.

Answer: Two unsuccessful examples are oil-eating bacteria and the Flavr Savr tomato. Two very successful examples include bacteria that make human insulin and Bt varieties of agricultural crops.

Conceptual Questions

C1. What is a recombinant microorganism? Discuss examples of recombinant microbes.

C2. A conjugation-deficient strain of *A. radiobacter* is used to combat crown gall disease. Explain how this bacterium prevents the disease and the advantage of a conjugation-deficient strain.

C3. What is bioremediation? What is the difference between biotransformation and biodegradation?

C4. What is a biological control agent? Briefly describe three examples.

C5. As described in Table 19.2, several medical agents are now commercially produced by genetically engineered microorganisms. Discuss the advantages and disadvantages of making these agents this way.

C6. What is a mouse model for human disease?

C7. What is a transgenic organism? Describe three examples.

C8. What part of the *A. tumefaciens* DNA gets transferred to the genome of a plant cell during infection?

C9. Explain the difference between gene addition and gene replacement. Would the following descriptions be examples of gene addition or gene replacement?

A. A mouse model to study cystic fibrosis

B. Introduction of a pesticide-resistance gene into corn using the Ti plasmid of *A. tumefaciens*

C10. As described in Chapter 7, not all inherited traits are determined by nuclear genes (i.e., genes located in the cell nucleus) that are expressed during the life of an individual. In particular, maternal effect genes and mitochondrial genes are notable exceptions. With these ideas in mind, let's consider the cloning of sheep (e.g., Dolly).

A. With regard to maternal effect genes, would the phenotype of such a cloned animal be determined by the animal that donated the enucleated egg or by the animal that donated the somatic cell nucleus? Explain.

B. Would the cloned animal inherit extranuclear traits from the animal that donated the egg or by the animal that donated the somatic cell? Explain.

C. In what ways would you expect this cloned animal to be similar to or different from the animal that donated the somatic cell? Is it fair to call such an animal a "clone" of the animal that donated the nucleus?

C11. Discuss some of the worthwhile traits that can be modified in transgenic plants.

C12. Discuss the concerns that some people have with regard to the uses of genetically engineered organisms.

Experimental Questions

E1. Recombinant bacteria can produce hormones that are normally produced in humans. Briefly describe how this is accomplished.

E2. In the experiment of Figure 19.1, why did the plasmid with somatostatin in the incorrect orientation fail to produce somatostatin?

E3. *Bacillus thuringiensis* can make toxins that kill insects. This toxin must be applied several times during the growth season to prevent insect damage. As an alternative to repeated applications, one strategy is to apply bacteria directly to leaves. However, *B. thuringiensis* does not survive very long in the field. Other bacteria, such as *Pseudomonas syringae*, do. Propose a way to alter *P. syringae* so that it could be used as an insecticide. Discuss advantages and disadvantages of this approach compared to the repeated applications of the insecticide from *B. thuringiensis.*

E4. In the experiment of Figure 19.1, explain how the coding sequence for somatostatin was constructed. Why was it necessary to link this coding sequence to the sequence for β-galactosidase? How were these two polypeptides separated after the fusion protein was synthesized in *E. coli*?

E5. Explain how it is possible to select for homologous recombination in mice. What phenotypic marker is used to readily identify chimeric mice?

E6. To produce transgenic plants, plant tissue is exposed to *A. tumefaciens* and then grown in media containing kanamycin, carbenicillin, and plant growth hormones. Explain the purpose behind each of these three agents. What would happen if you left out the kanamycin?

E7. List and briefly describe five methods for the introduction of cloned genes into plants.

E8. What is a gene knockout? Is an animal or plant with a gene knockout a heterozygote or homozygote? What might you conclude if a gene knockout does not have a phenotypic effect?

E9. Nowadays, it is common for researchers to identify genes using cloning methods described in Chapters 18 and 19. A gene can be identified according to its molecular features. For example, a segment of DNA can be identified as a gene because it contains the right combination of sequences: a promoter, exons, introns, and a terminator. Or a gene can be identified because it is transcribed into mRNA. In the study of plants and animals, it is relatively common for researchers to identify genes using molecular techniques without knowing the function of the gene. In the case of mice, the function of the gene can be investigated by making a gene knockout. If the knockout causes a phenotypic change in the mouse, this may provide an important clue regarding the function of a gene. For example, if a gene knockout produced an albino mouse, this would indicate the gene knocked out probably plays a role in pigment formation. The experimental strategy of first identifying a gene based on its molecular properties and then investigating its function by making a knockout is called reverse genetics. Explain how this approach is opposite (or "in reverse") to the conventional way that geneticists study the function of genes.

E10. According to the methods described in Figure 19.7, can homologous recombination that results in gene replacement cause the integration of both the *TK* and Neo^R genes? Explain why or why not. Describe how the *TK* gene and Neo^R gene are used in a selection scheme that favors gene replacement.

E11. What is a chimera? How are chimeras made?

E12. Evidence (see Shiels et al., "Analysis of telomere lengths in cloned sheep," *Nature* 399, 316–17) suggested that Dolly may have been "genetically older" than her actual age would have suggested. As mammals age, the chromosomes in somatic cells tend to shorten from the telomeres. Therefore, older individuals have shorter chromosomes in their somatic cells compared to younger ones. When researchers analyzed the chromosomes in the somatic cells of Dolly when she was about three years old, the lengths of her chromosomes were consistent with a sheep that was significantly older, say, nine or ten years old. (Note: As described in the chapter, the sheep that donated the somatic cell that produced Dolly was six years old, and her mammary cells had been grown in culture for several cell doublings before a mammary cell was fused with an oocyte.)

A. Suggest an explanation why Dolly's chromosomes seemed so old.

B. Let's suppose that Dolly at age 11 gave birth to a lamb named Molly; Molly was produced naturally (by mating Dolly with a normal male). When Molly was eight years old, a sample of somatic cells was analyzed. How old would you expect Molly's chromosomes to appear, based on the phenomenon of telomere shortening? Explain your answer.

C. Discuss how the observation of chromosome shortening, which was observed in Dolly, might affect the popularity of reproductive cloning.

E13. When transgenic organisms are made, the transgene may integrate into multiple sites within the genome. Furthermore, the integration site may influence the expression of the gene. For example, if a transgene integrates into a heterochromatic (i.e., highly condensed) region of a chromosome, the transgene may not be expressed. For these reasons, it is important for geneticists to analyze transgenic organisms with regard to the number of transgene insertions and the expression levels of the transgenes. Chapter 18 describes three methods (Southern blotting, Northern blotting, and Western blotting) that can be used to detect genes and gene products. Which of these techniques would you use to determine the number of transgenes in a transgenic plant or animal? Why is it important to know the number of transgenes? (Hint: You may want to use a transgenic animal or plant as breeding stock to produce many more transgenic animals or plants.) Which technique would you use to determine the expression levels of transgenes?

E14. What is molecular pharming? Compared to the production of proteins by bacteria, why might it be advantageous?

E15. What is reproductive cloning? Are identical twins in humans considered to be clones? With regard to agricultural species, what are some potential advantages to reproductive cloning?

E16. Researchers have identified a gene in humans that (when mutant) causes severe dwarfism and mental retardation. This disorder is inherited in an autosomal recessive manner, and the mutant allele is known to be a loss-of-function mutation. The same gene has been found in mice, although a mutant version of the gene has not been discovered in mice. To develop drugs and an effective therapy to treat this disorder in humans, it would be experimentally useful to have a mouse model. In other words, it would be desirable to develop a strain of mice that carry the mutant allele in the homozygous condition. Experimentally, how would you develop such a strain?

E17. Treatment of adenosine deaminase (ADA) deficiency is an example of ex vivo gene therapy. Why is this therapy called ex vivo? Can ex vivo gene therapy be used to treat all inherited diseases? Explain.

E18. Describe the targeting methods used in cystic fibrosis gene therapy. Provided the *CFTR* gene gets to the patient's lung cells, would you expect this to be a permanent cure for the patient, or would it be necessary to perform this gene therapy on a regular basis (say, monthly)?

E19. Several research studies are under way that involve the use of gene therapies to inhibit the growth of cancer cells. As discussed in Chapter 22, oncogenes are mutant genes that are overexpressed and cause cancer. New gene therapies are aimed at silencing an oncogene by producing antisense RNA that recognizes the mRNA transcribed from an oncogene. Based on your understanding of antisense RNA (described in Chapters 14 and 15), explain how this strategy would prevent the growth of cancer cells.

Questions for Student Discussion/Collaboration

1. Discuss the advantages and disadvantages of gene therapy. Because a limited amount of funding is available for gene therapy research, make a priority list of the three top diseases that you would fund. Discuss your choices.
2. A commercially available strain of *P. syringae* marketed as Frostban B is used to combat frost damage. This is a naturally occurring *Ice*$^-$ strain. Discuss the advantages and disadvantages of using this strain compared with a recombinant version.
3. Make a list of the types of traits you would like to see altered in transgenic plants and animals. Suggest ways (i.e., what genes would you use?) to accomplish these alterations.

Note: All answers appear at the website for this textbook; the answers to even-numbered questions are in the back of the textbook.

www.mhhe.com/brookergenetics3e

Visit the Online Learning Center for practice tests, answer keys, and other learning aids for this chapter. Enhance your understanding of genetics with our interactive exercises, quizzes, animations, and much more.

20 GENOMICS I: ANALYSIS OF DNA

CHAPTER OUTLINE

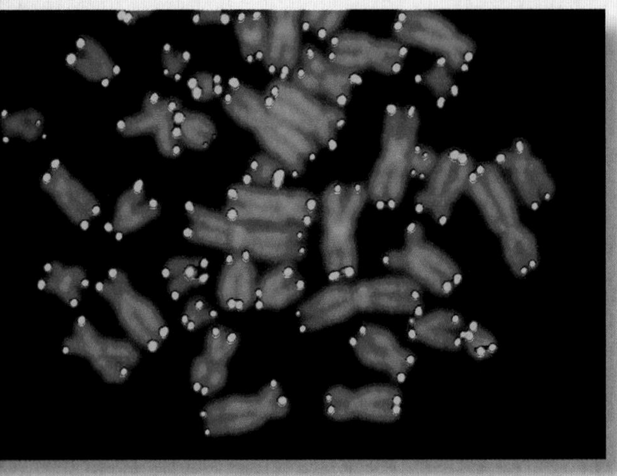

Labeling the ends of chromosomes. In this micrograph, the telomeric sequences at ends of chromosomes are labeled with an orange fluorescent probe, while the rest of the chromosomes are labeled in blue. This method, called fluorescent *in situ* hybridization, allows geneticists to identify particular sequences within intact chromosomes.

The term **genome** refers to the total genetic composition of an organism or species. For example, the nuclear genome of humans is composed of 22 different autosomes and an X and Y chromosome. In addition, humans have a mitochondrial genome composed of a single circular chromosome.

As genetic technology has progressed over the past few decades, researchers have gained an increasing ability to analyze the composition of genomes as a whole unit. The term **genomics** refers to the molecular analysis of the entire genome of a species. Genome analysis is a molecular dissection process applied to a complete set of chromosomes. Segments of chromosomes are analyzed in progressively smaller pieces, the locations of which are known on the intact chromosomes. This is the mapping phase of genome analysis. The mapping of the genome ultimately progresses to the determination of the complete DNA sequence, which provides the most detailed description of an organism's genome at the molecular level.

In 1995, a team of researchers headed by Craig Venter and Hamilton Smith obtained the first complete DNA sequence of the bacterial genome from *Haemophilus influenzae.* Its genome is composed of a single circular chromosome 1.83 million base pairs (bp) in length and contains approximately 1,743 genes (**Figure 20.1**). In 1996, the first entire DNA sequence of a eukaryote, *Saccharomyces cerevisiae* (baker's yeast), was completed. This work was carried out by a European-led consortium of more than 100 laboratories, including some in the United States, Japan, and Canada. The overall coordinator was Andre Goffeau in Belgium. The yeast genome contains 16 linear chromosomes; these 16 chromosomes have a combined length of about 12.1 million bp and contain approximately 6,300 genes. Since that time, genome sequences from many prokaryotes and eukaryotes have been completed.

In this chapter, we will focus on methods aimed at elucidating the organization of the sequences within a species' genome. This process begins with the mapping of regions along an organism's chromosomes and ends with the determination of the complete DNA sequence. We will consider three mapping strategies—cytogenetic, linkage, and physical mapping—and the approaches used to carry them out. We will then turn to exploring genome sequencing projects, research endeavors that have the ultimate goal of determining the sequence of DNA bases of the entire genome of a given species. We will consider the methods, goals, and results of these large undertakings, which include the Human Genome Project.

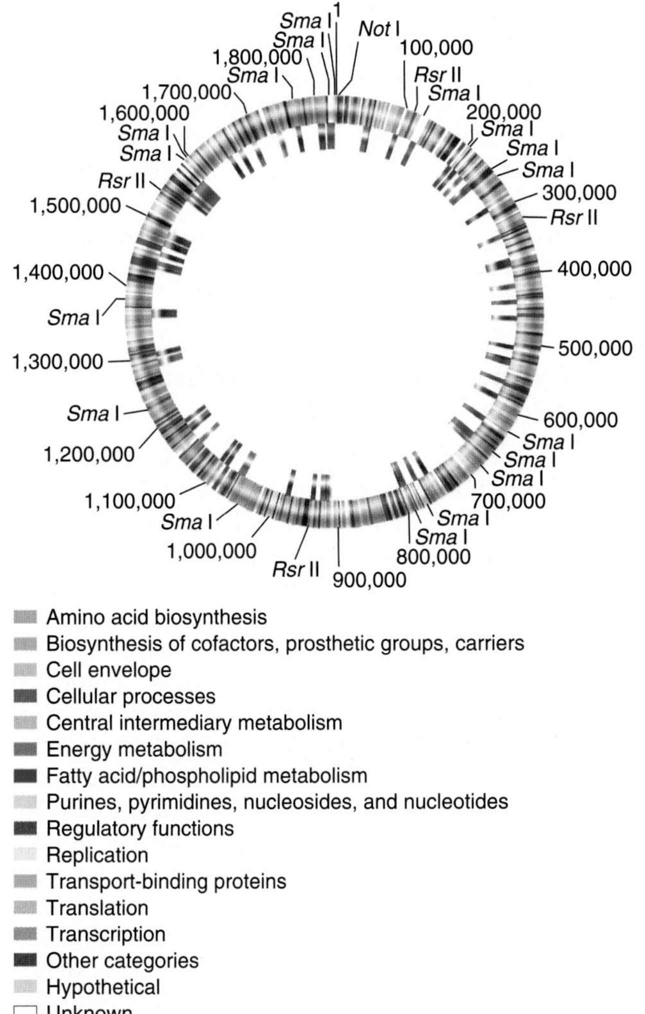

FIGURE 20.1 A complete map of the genome of the bacterium *Haemophilus influenzae*. On the outside of the circular chromosome, the numbering of nucleotides begins at the top of the circle and proceeds clockwise in 100,000-nucleotide increments. The map also shows the locations of restriction enzyme sites for *Not*I, *Sma*I, and *Rsr*II. As shown in the key, the general categories of genes are color coded.

Genes→Traits By mapping and DNA sequencing the entire genome of a species, researchers can identify most of the genes it possesses. It is the expression of these many genes that ultimately determines an organism's inherited traits.

Once a genome sequence is known, researchers can examine, at the level of many genes, how the components of a genome interact to produce the traits of an organism. This approach is called **functional genomics.** Ultimately, a long-term goal of researchers is to determine the roles of all cellular proteins, as well as the interactions that these proteins experience, to produce the characteristics of particular cell types and the traits of complete organisms. This is the research area known as **proteomics.** In this chapter, we will explore genomics at the level of DNA segments and sequences. In Chapter 21, we will consider the exciting prospects that functional genomics and proteomics have to offer regarding our future understanding of genetics.

20.1 CYTOGENETIC AND LINKAGE MAPPING

In genetics, the term **mapping** refers to the experimental process of determining the relative locations of genes or other segments of DNA along individual chromosomes. Researchers may follow three general approaches to map a chromosome: cytogenetic, linkage, and physical mapping strategies. Before we discuss these approaches in detail, let's compare them to each other. **Cytogenetic mapping** (also called cytological mapping) relies on the localization of gene sequences within chromosomes that are viewed microscopically. When stained, each chromosome of a given species has a characteristic banding pattern, and genes are mapped cytogenetically relative to a band location. By comparison, in Chapter 5, we considered how genetic crosses are conducted to map the relative locations of genes within a chromosome. Such genetic studies, which are called **linkage mapping** or genetic mapping, use the frequency of genetic recombination between different genes to determine their relative spacing and order along a chromosome. In eukaryotes, linkage mapping involves crosses among organisms that are heterozygous for two or more genes. The number of recombinant offspring provides a relative measure of the distance between genes, which is computed in map units (or centiMorgans). Finally, a third approach is **physical mapping.** DNA cloning techniques are used to determine the location of and distance between genes and other DNA regions. In a physical map, the distances are computed as the number of nucleotide base pairs between genes.

A **genetic map** is a chart that describes the relative locations of genes or other DNA segments along a chromosome. The term **locus** (plural, **loci**) refers to the site within a genetic map where a specific gene or other DNA segment is found. **Figure 20.2** compares genetic maps that show the loci for two X-linked genes, *sc* (scute, a gene affecting bristle morphology) and *w* (a gene affecting eye color), in *Drosophila melanogaster.* In the cytogenetic map (top), the *sc* gene is located at band 1A8, and the *w* gene is located at band 3B6. In the linkage map, genetic crosses indicate that the two genes are approximately 1.5 map units (mu) apart. The physical map shows that the two genes are approximately 2.4×10^6 bp apart from each other along the X chromosome. Correlations between cytogenetic, linkage, and physical maps often vary from species to species and from one region of the chromosome to another. For example, a distance of 1 mu may correspond to 1 to 2 million base pairs in one region of the chromosome, but other regions may recombine at a much lower rate, so a distance of 1 mu may be a much longer physical segment of DNA.

In this section, we will explore several techniques aimed at producing cytogenetic and linkage maps and consider physical mapping in the section that follows.

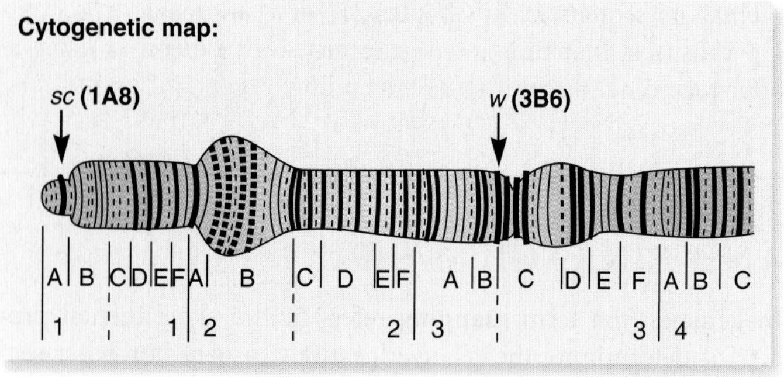

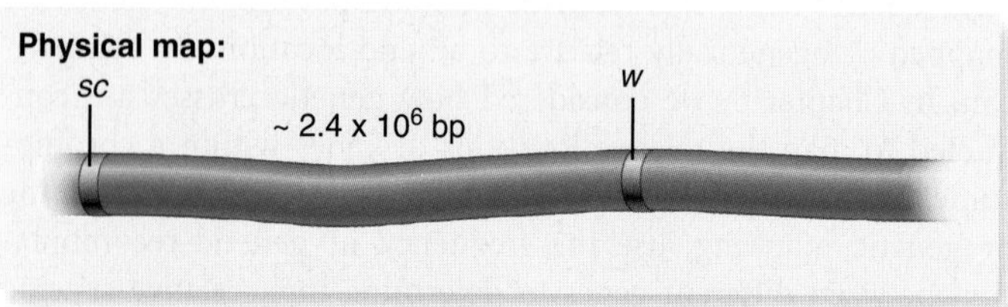

FIGURE 20.2 **A comparison of cytogenetic, linkage, and physical maps.** Each of these maps shows the distance between the *sc* and *w* genes along the X chromosome in *Drosophila melanogaster.* The cytogenetic map is that of the polytene chromosome.

A Goal of Cytogenetic Mapping Is to Determine the Location of a Gene Along an Intact Chromosome

Cytogenetic mapping is commonly used in eukaryotes, which have very large chromosomes compared to those of bacteria. Microscopically, eukaryotic chromosomes can be distinguished from one another by their size, centromeric location, and banding patterns (refer to Chapter 8, Figure 8.1). By treating chromosomal preparations with particular dyes, a discrete banding pattern is obtained for each chromosome. Cytogeneticists use this banding pattern as a way to describe specific regions along a chromosome.

Cytogenetic mapping attempts to determine the locations of particular genes relative to a banding pattern of a chromosome. For example, the human gene that encodes the cystic fibrosis transmembrane regulator, the protein that is defective in people with cystic fibrosis, is located on chromosome 7, at a specific site in the q3 region.

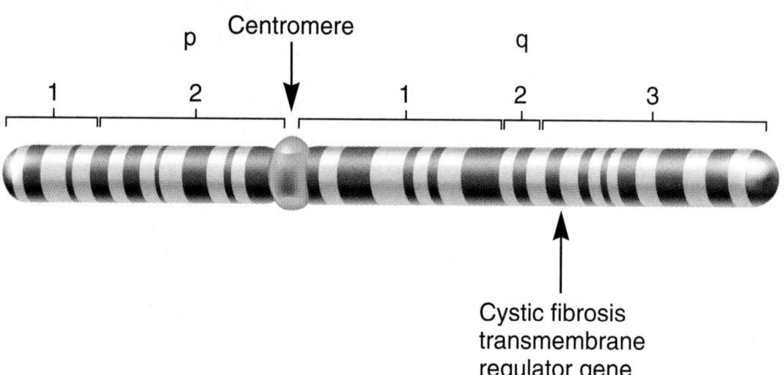

Cytogenetic mapping may be used as a first step in the localization of genes in plants and animals. However, because it relies on light microscopy, cytogenetic analysis has a fairly crude limit of resolution. In most species, cytogenetic mapping is accurate only within limits of approximately 5 million bp along a chromosome. In species that have large polytene chromosomes, such as *Drosophila*, the resolution is much better. A common strategy used by geneticists is to roughly locate a gene by cytogenetic analysis and then determine its location more precisely by the physical mapping methods described later.

In situ Hybridization Can Localize Genes Along Particular Chromosomes

The technique of ***in situ* hybridization** is widely used to cytogenetically map the locations of genes or other DNA sequences within large eukaryotic chromosomes. The term *in situ* (from the Latin for "in place") indicates that the procedure is conducted on chromosomes that are being held in place—adhered to a surface.

To map a gene via *in situ* hybridization, researchers use a probe to detect the location of the gene within a set of chromosomes. If the gene of interest has been cloned previously, as described in Chapter 18, the DNA of the cloned gene can be used as a probe. Because a DNA strand from a cloned gene, which is a very small piece of DNA relative to a chromosome, will hybridize only to its complementary sequence on a particular chromosome, this technique provides the ability to localize the gene of interest. For example, let's consider the gene that causes the white-eye phenotype in *Drosophila* when it carries a loss-of-function mutation. This gene has already been cloned. If a single-stranded piece of this cloned DNA is mixed with *Drosophila* chromosomes in which the DNA has been denatured, it will bind only to the X chromosome at the location corresponding to the site of the eye color gene.

The most common method of *in situ* hybridization uses fluorescently labeled DNA probes and is referred to as **fluorescence *in situ* hybridization (FISH)**. Figure 20.3 describes the steps of the FISH procedure. The cells are prepared using a technique that keeps the chromosomes intact. The cells are treated with agents that cause them to swell, and their contents are fixed to the slide. The chromosomal DNA is then denatured, and a DNA probe is added. For example, the added DNA probe might be single-stranded DNA that is complementary to a specific gene. In this case, the goal of a FISH experiment is to determine the location of the gene within a set of chromosomes. The probe will bind to a site in the chromosomes where the gene is located because the probe and chromosomal gene will line up and hydrogen bond with each other. To detect where the probe has bound to a chromosome, the probe is subsequently tagged with a fluorescent molecule. This is usually accomplished by first incorporating biotin-labeled nucleotides into the probe. Biotin, a small, nonfluorescent molecule, has a very high affinity for a protein called avidin. Fluorescently labeled avidin is added, which binds tightly to the biotin and thereby labels the probe as well.

How is the fluorescently labeled probe detected? A fluorescent molecule is one that absorbs light at a particular wavelength and then emits light at a longer wavelength. To detect the light emitted by a fluorescently labeled probe, a fluorescence microscope is

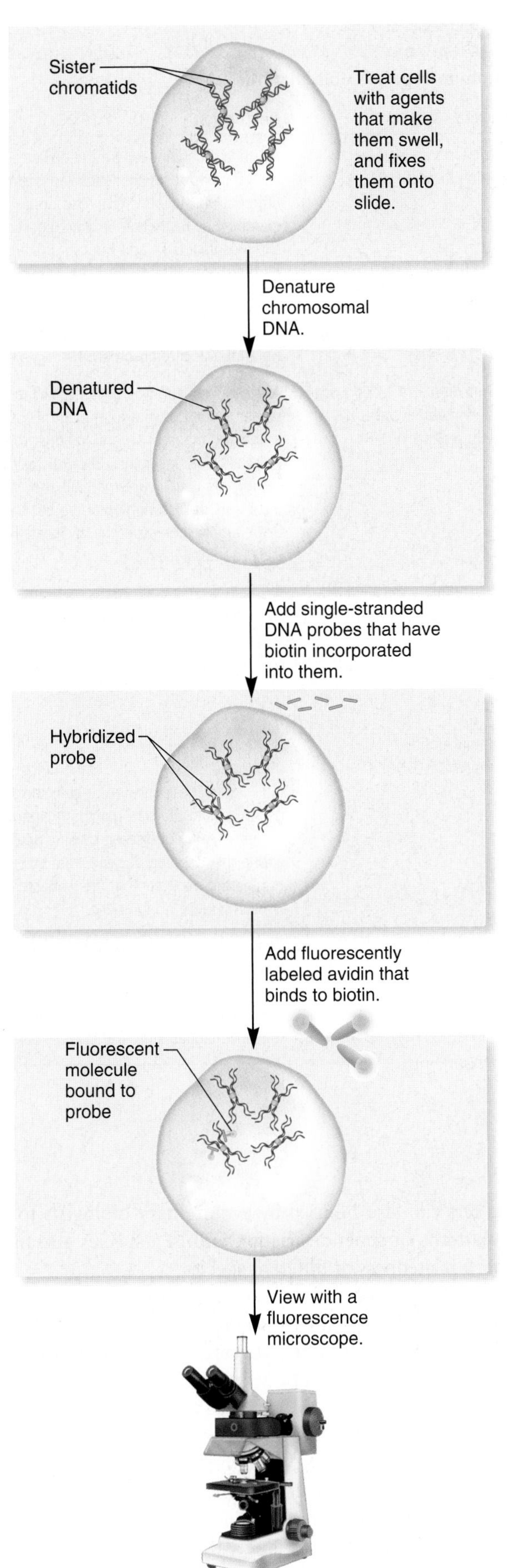
Sister chromatids
Treat cells with agents that make them swell, and fixes them onto slide.
Denature chromosomal DNA.
Denatured DNA
Add single-stranded DNA probes that have biotin incorporated into them.
Hybridized probe
Add fluorescently labeled avidin that binds to biotin.
Fluorescent molecule bound to probe
View with a fluorescence microscope.

ONLINE ANIMATION

FIGURE 20.3 The technique of fluorescence *in situ* hybridization (FISH). The probe hybridizes to the denatured chromosomal DNA only at specific, complementary sites in the genome. Note that the chromosomes are highly condensed metaphase chromosomes that have already replicated. Therefore, each X-shaped chromosome actually contains two copies of a particular gene. These are sister chromatids. Because the sister chromatids are identical, a probe that recognizes one sister chromatid will also bind to the other.

used. Such a microscope contains filters that allow the passage of light only within a defined wavelength range. The sample is illuminated at the wavelength of light that is absorbed by the fluorescent molecule. The fluorescent molecule then emits light at a longer wavelength. The fluorescence microscope has a filter that allows the transmission of the emitted light. Because only the emitted light is viewed, the background of the sample is dark, and the fluorescence is seen as a brightly glowing color on a dark background. For most FISH experiments, chromosomes are generally counterstained by a fluorescent dye that is specific for DNA. The most common dye is DAPI (4', 6-diamidino-2-phenyl-indol) that is excited by UV light. This provides all of the DNA with a blue background. The results of a FISH experiment are then compared to a sample of chromosomes that have been stained with Giemsa to produce banding, so the location of a probe can be mapped relative to the banding pattern.

Figure 20.4 illustrates the results of an experiment involving six different DNA probes. The six probes were strands of DNA corresponding to six different DNA segments of human chromosome 5. In this experiment, each probe was labeled with a different fluorescent molecule. This enabled researchers to distinguish the probes when they became bound to their corresponding locations on chromosome 5. In this experiment, computer-imaging methods were used to assign each fluorescently labeled

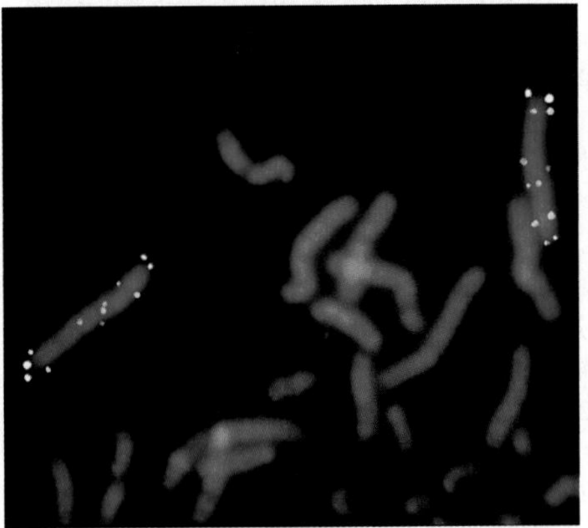

FIGURE 20.4 The results of a fluorescence *in situ* hybridization experiment. In this experiment, six different probes were used to locate six different sites along chromosome 5. The colors are due to computer imaging of fluorescence emission; they are not the actual colors of the fluorescent labels. Two spots are usually seen at each site because the probe binds to both sister chromatids.

probe a different color. In this way, fluorescence *in situ* hybridization discerns the sites along chromosome 5 corresponding to the six different probes. In a visual, colorful way, this technique was used here to determine the order and relative distances between several specific sites along a single chromosome. FISH is commonly used in genetics and cell biology research, and its use has become more widespread in clinical applications. For example, clinicians may use FISH to detect small changes in chromosome structure such as deletions and duplications, which may occur in patients with genetic disorders.

Linkage Mapping Can Use Molecular Markers

Let's now turn to linkage mapping, which relies on the frequency of recombinant offspring to determine the distance between sites located along the same chromosome. The linkage mapping methods described in Chapters 5 and 6 used allelic differences between genes to map the relative locations of those genes along a chromosome. As an alternative to gene mapping, geneticists have realized that regions of DNA that do not encode genes can be used as markers along a chromosome. A **molecular marker** is a segment of DNA found at a specific site along a chromosome and has properties that enable it to be uniquely recognized using molecular tools such as polymerase chain reaction (PCR) and gel electrophoresis. As with alleles, the molecular markers may be **polymorphic;** that is, within a population, they may vary from individual to individual. Therefore, the distances between linked molecular markers can be determined from the outcomes of crosses. Using molecular techniques, researchers have found it easier to identify many molecular markers within a given species' genome rather than identify many allelic differences among individuals. For this reason, geneticists have increasingly turned to molecular markers as points of reference along genetic maps. As described in **Table 20.1**, many different kinds of molecular markers are used by geneticists.

TABLE 20.1

Common Types of Molecular Markers

Marker	Description
Restriction fragment length polymorphism (RFLP)	A site in a genome where the distance between two restriction sites varies among different individuals. These sites are identified by restriction enzyme digestion of chromosomal DNA and the use of Southern blotting.
Amplified restriction fragment length polymorphism (AFLP)	The same as an RFLP except that the fragment is amplified via PCR instead of isolating the chromosomal DNA
Minisatellite, also called a site with a variable number of tandem repeats (VNTR)	A site in the genome that contains many repeat sequences. The total length is usually in the size range of several hundred to a few thousand base pairs. Because their total lengths are usually polymorphic, they were once used in DNA fingerprinting but have been largely superseded by microsatellites (see Chapter 24).
Microsatellite, also called a short tandem repeat (STR)	A site in the genome that contains many short tandem repeat sequences. The total length is usually in the size range of 100 to 500 bp, and their lengths may be polymorphic within a population. They are isolated via PCR.
Single-nucleotide polymorphism (SNP)	A site in the genome where a single nucleotide is polymorphic among different individuals. These sites occur commonly in all genomes, and they are gaining greater use in the mapping of disease-causing alleles and in the mapping of genes that contribute to quantitative traits that are valuable in agriculture (see Chapter 24).
Sequence-tagged site (STS)	This is a general term to describe any molecular marker that is found at a unique site in the genome and is amplified by PCR. AFLPs, microsatellites, and SNPs can provide sequence-tagged sites within a genome.

Researchers have constructed detailed genetic maps in which a series of many molecular markers have been identified along each chromosome of certain species. These species include humans, model organisms, agricultural species, and many others. Why are molecular markers useful? One key reason is that molecular markers can be used to determine the approximate location of an unknown gene that causes a human disease. As we will see in Experiment 20A, a particular molecular marker (an RFLP) was linked to the Hb^S allele that causes sickle-cell disease. Similarly, for inherited diseases in which the mutant genes have yet to be identified, human geneticists can sometimes follow the transmission patterns of polymorphic molecular markers in family pedigrees. The discovery of a particular marker in those who have the disease can indicate that the marker is close to the disease-causing allele. This may help researchers identify the gene by cloning methods, such as chromosome walking, which we will examine later in this chapter.

In addition, molecular markers may help researchers identify the locations of genes involved in quantitative traits, such as fruit yield and meat weight, that are valuable in agriculture. The use of molecular markers to identify such genes is described in Chapter 25 (see Figure 25.7). Genetic maps with a large number of markers can also be used by evolutionary biologists to determine patterns of genetic variation within a species and the evolutionary relatedness of different species.

All of the types of markers described in Table 20.1 have been used in linkage mapping studies. To exemplify their use, we will first consider **restriction fragment length polymorphisms (RFLPs)**, but similar mapping strategies can be followed for other types of markers. As discussed in Chapter 18, restriction enzymes recognize specific DNA sequences and cleave the DNA at those sequences. Along a very long chromosome, a particular restriction enzyme will recognize many sites. For example, a commonly used restriction enzyme, *Eco*RI, recognizes 5'-GAATTC-3'. Simply by chance, this six-nucleotide sequence is expected to occur (on average) every 4^{-6} bases, or once in every 4,096 bases. A chromosome composed of millions of nucleotides contains many *Eco*RI sites, which will be randomly distributed along the chromosome. Therefore, *Eco*RI will digest a chromosome into many smaller pieces of different lengths.

When comparing two different individuals, the digestion of chromosomal DNA by a given restriction enzyme may produce certain fragments that differ in their length, even though such fragments are found at the same chromosomal locations in the two individuals (**Figure 20.5**). In this case, geneticists would say there is polymorphism in the population with regard to the length of a particular DNA fragment. This variation can arise for several reasons. Among different individuals, genetic changes such as small deletions and duplications may subtract or add segments of DNA to a particular region of the chromosome. This tends to occur more frequently if a region contains a repetitive sequence. Alternatively, a mutation may change the DNA sequence in a way that alters a recognition site for a restriction enzyme. For example, in Figure 20.5, the elimination of a restriction site has occurred in the chromosome of individual 2. This has altered the sizes of certain DNA fragments that result from the digestion with *Eco*RI.

Figure 20.6 considers the application of RFLP analysis to a short chromosomal region among three diploid individuals. Due to slight differences in their DNA sequences, these individuals vary with regard to the existence of an *Eco*RI site shown in red. In certain individuals, a mutation may change the sequence at this site so that it is no longer recognized by *Eco*RI. In a diploid species, each individual has two copies of this region. In individual 1, both of these *Eco*RI sites are present, while in individual 2, both are absent. By comparison, individual 3 is a heterozygote. In one of the chromosomes, the site is present, but in the other, it is absent. In this experiment, the DNA was isolated and then cut by *Eco*RI. The DNA from individual 1 is cut at six sites by *Eco*RI, whereas individual 2 has only five locations where the DNA is cut. In individual 2, an *Eco*RI site is missing because the sequence of DNA in this region has been changed so that it is no longer GAATTC and is not recognized by *Eco*RI.

When subjected to gel electrophoresis, the change in the DNA sequence at a single *Eco*RI restriction site produces a variation in the sizes of homologous DNA fragments. The term polymorphism, meaning many forms, refers to the idea that the individuals within a population differ with regard to these particular DNA fragments. As noted in Figure 20.6, not all restriction sites are polymorphic. The three individuals share many DNA fragments that are identical in size. When a DNA segment is identical among all members of a population, it is said to be **monomorphic** (meaning one form). As a practical rule of thumb, a DNA segment is considered monomorphic when over 99% of the individuals in the population have identical sequences at that segment.

The preceding discussion was meant as an overview of the phenomenon of RFLPs. However, the experiment of Figure 20.6 is a technical oversimplification, because it considers only a very short region of chromosomal DNA. In an actual RFLP analysis,

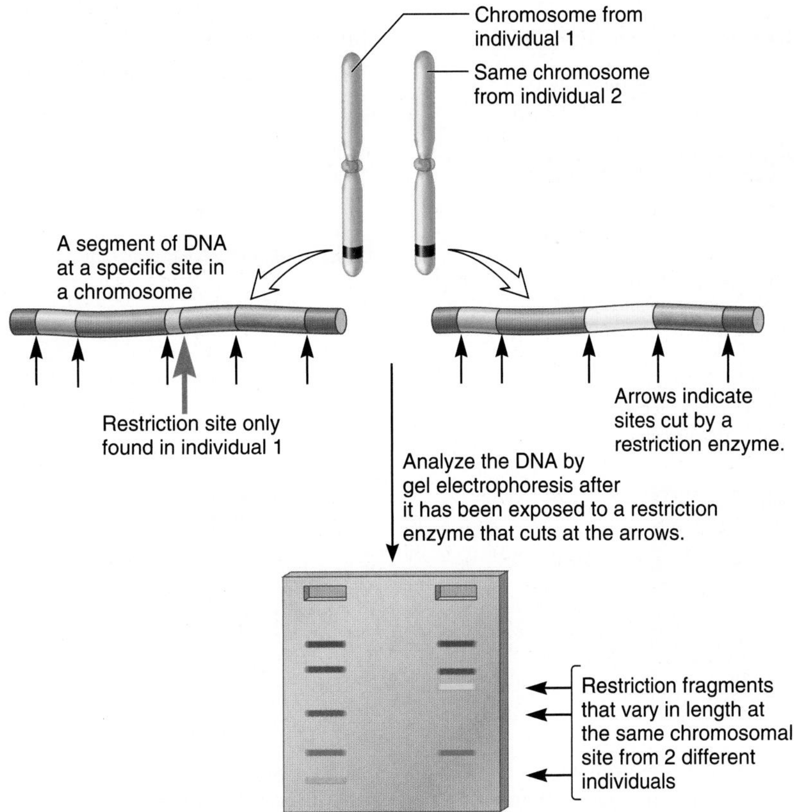

FIGURE 20.5 **Restriction fragment length polymorphisms (RFLPs).** For simplicity, only one chromosome is shown for each individual. In a diploid species, a cell would contain two copies of each chromosome, which may or may not have the same RFLP.

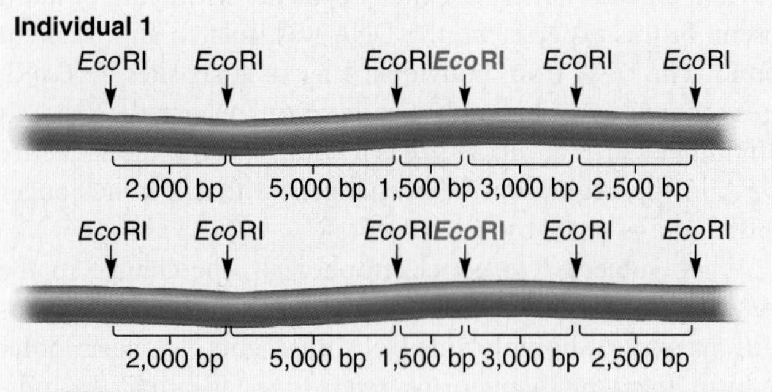

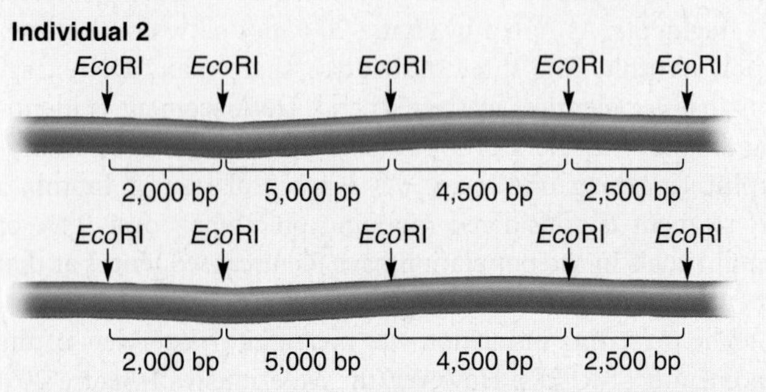

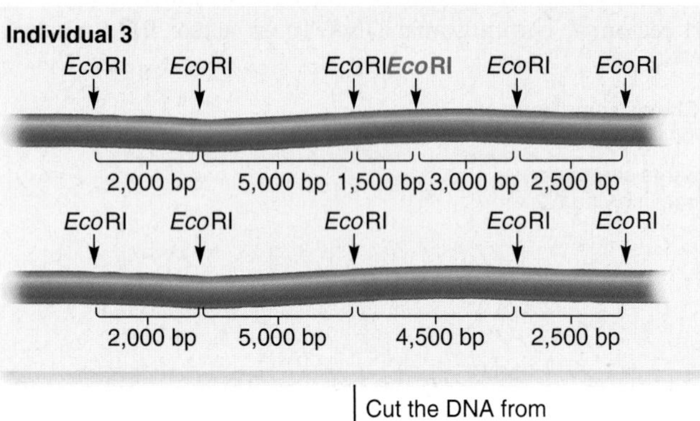

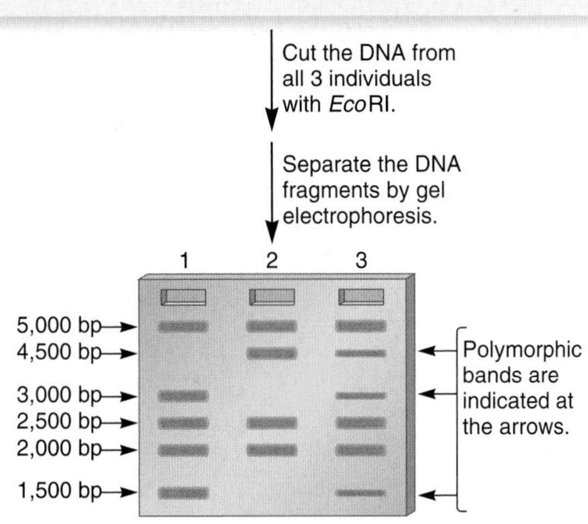

FIGURE 20.6 An RFLP analysis of chromosomal DNA from three different individuals. The *Eco*RI site shown in red is sometimes missing because a mutation has changed the DNA sequence in this region.

DNA samples containing all of the chromosomal DNA would be isolated from these three individuals. The digestion of the chromosomal DNA with *Eco*RI would then yield so many fragments that the results would be very difficult to analyze. To circumvent this problem, Southern blotting is used to identify specific RFLPs.

Figure 20.7 illustrates such an experiment that reconsiders the chromosomes from the three individuals described in Figure 20.6. A DNA strand from a cloned piece of DNA that corresponds to the region near the fourth *Eco*RI site is used as a radiolabeled probe in a Southern blot of the chromosomal DNA. When the blot is exposed to X-ray film, we will see only the DNA bands that can hybridize to the radiolabeled probe. As shown in Figure 20.7, individual 1 shows one band of 3,000 bp, and individual 2 has a band 4,500 bp long. As discussed in regard to Figure 20.6, this difference is due to the absence of an *Eco*RI site in individual 2. Individual 3 has both bands because of heterozygosity for this *Eco*RI site. In other words, the heterozygote will have two thinner bands of different lengths, whereas homozygotes will display only one band, because the RFLP procedure detects the molecular products of each chromosome. In an actual RFLP analysis, researchers would analyze hundreds or thousands of RFLPs by using a collection of probes that recognize RFLPs throughout the genome.

The Distance Between Two Linked RFLPs Can Be Determined

Now that we understand what an RFLP is, let's consider how RFLPs are mapped and used as chromosomal markers. In the linkage mapping experiments described in Chapter 5, we learned how the results of genetic crosses are used to map the distance

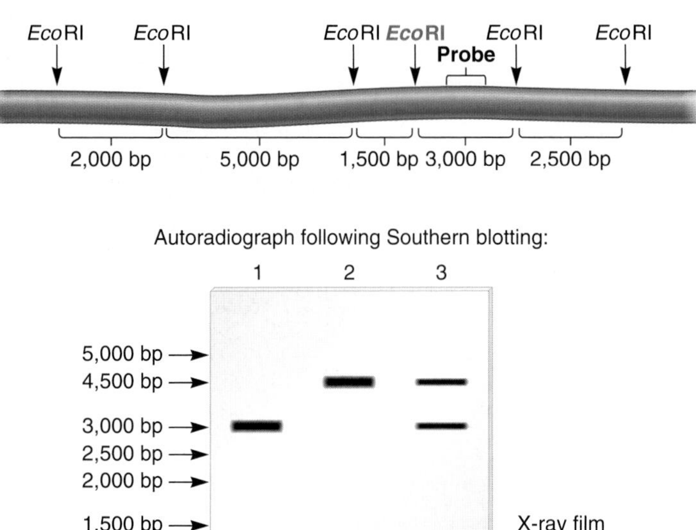

FIGURE 20.7 Southern blotting of a specific RFLP. In this experiment, a Southern blot was conducted on the chromosomal DNA from the same three individuals described in Figure 20.6. A labeled strand of DNA, just to the right of the fourth *Eco*RI site, was used as a probe.

between two genes. In that type of analysis, the proportions of offspring with recombinant phenotypes were used to calculate the map distance between genes. Likewise, we can map the distance between two RFLPs by making crosses and analyzing the offspring. In an RFLP analysis, however, we do not look at the phenotypic characteristics of the offspring (e.g., white eyes or miniature wings). Instead, we isolate DNA from a tissue sample and look at the DNA bands on a gel.

Figure 20.8 shows an example of how RFLP mapping is done. Let's suppose two RFLPs are located at different sites in the genome. One RFLP is detected with probe 1 and yields either a 4,500 bp band or a 6,500 bp band. A second RFLP is detected with probe 2 and yields either a 2,000 bp band or a 1,500 bp band. In the experiment of Figure 20.8, the goal is to determine whether or not the 4,500/6,500 and 2,000/1,500 sites are linked along the same chromosome and, if so, to find the map distance between them. The researcher begins with two strains that are homozygous: 4,500 and 2,000 versus 6,500 and 1,500. These strains are crossed to each other to produce a heterozygote. The heterozygote can then be crossed to a homozygote that carries only the 4,500 and 2,000 bands. This is analogous to a dihybrid test cross described in Chapter 5, in which an F_1 heterozygote is crossed to a homozygote (see Figure 5.9). The results of this cross depend on whether the RFLPs are closely linked or not. If an offspring inherits pairs of fragments like the parents (4,500 bp and 2,000 bp; 6,500 bp and 1,500 bp), such offspring will show a parental phenotype consisting of four bands of 4,500 bp, 2,000 bp, 6,500 bp, and 1,500 bp or two bands of 4,500 and 2,000 bp. The recombinant phenotypes are 4,500, 2,000 bp, and 6,500 bp or 4,500, 2,000, and 1,500 bp. If the RFLPs are not linked, we would expect a 1:1:1:1 ratio of the four types among the offspring. If the RFLPs are linked, we expect a higher percentage of offspring with a parental combination of fragments. Figure 20.8 describes the results of 100 offspring. Many more parental offspring were observed, which indicates that the RFLPs are closely linked.

In an actual analysis of RFLP data, researchers simultaneously analyze many RFLPs and decide on the likelihood of linkage between two RFLPs by using a statistical test called the **lod** (logarithm of the odds) **score method.** This method was devised by Newton Morton in 1955. Although the theoretical basis of this approach is beyond the scope of this textbook, computer programs analyze pooled data from a large number of pedigrees or crosses involving many RFLPs. The programs determine the probability, or odds, that any two markers exhibit a certain degree of linkage (that is, are within a particular number of map units apart) and also the probability that the data would have been obtained if the two markers were unlinked. The lod score is then calculated as the ratio of these values:

$$\text{Lod score} = \log_{10}\frac{\text{Probability of a certain degree of linkage}}{\text{Probability of independent assortment}}$$

For example, a lod score of 3 reflects a $\log_{10}$ of 1,000. This means the probability is 1,000-fold greater that the two markers are linked rather than assort independently. Traditionally, geneticists accept that two markers are linked if the lod score is 3 or greater.

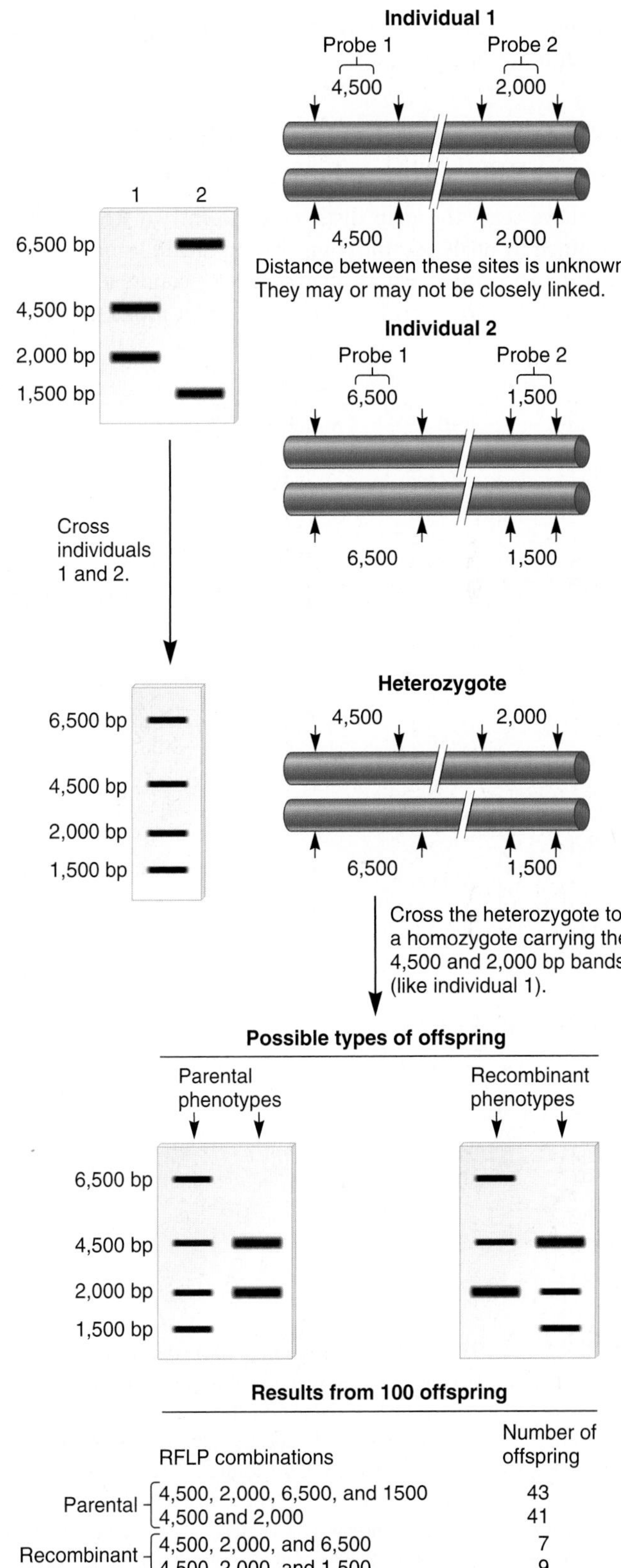

	RFLP combinations	Number of offspring
Parental	4,500, 2,000, 6,500, and 1500	43
	4,500 and 2,000	41
Recombinant	4,500, 2,000, and 6,500	7
	4,500, 2,000, and 1,500	9

FIGURE 20.8 **Linkage analysis of RFLP markers.**

Figure 20.8 illustrates the general type of approach that researchers can follow to map RFLP markers. If a lod score suggested that these two markers were linked, they would calculate the map distance by dividing the recombinant offspring (7 + 9 = 16) by the total number (100) and multiplying by 100, to obtain a map distance of 16 mu.

An RFLP Map Describes the Locations of Many Different RFLPs Throughout a Genome

As we have seen, the map distance between two RFLPs can be determined by analyzing the transmission of RFLP markers from parents to offspring. Such an analysis can be conducted on many different RFLPs to determine their relative locations throughout a genome. RFLPs are quite common in virtually all species, so geneticists can easily map the locations of many RFLPs within a genome. A linkage map composed of many RFLP markers is called an **RFLP map.** Figure 20.9 shows a simplified RFLP map of the plant *Arabidopsis thaliana* (a small plant in the mustard family), which is one of the favorite model organisms of plant molecular geneticists. Many RFLPs have been mapped to different locations along the five *Arabidopsis* chromosomes.

An RFLP map, like the one shown in Figure 20.9, can be used to locate functional genes within the genome. As an example, solved problem S2 illustrates how RFLP analysis can be used to map a gene that confers herbicide resistance. As described next, RFLP analysis can also be used to determine the likelihood that a person may carry a disease-causing allele.

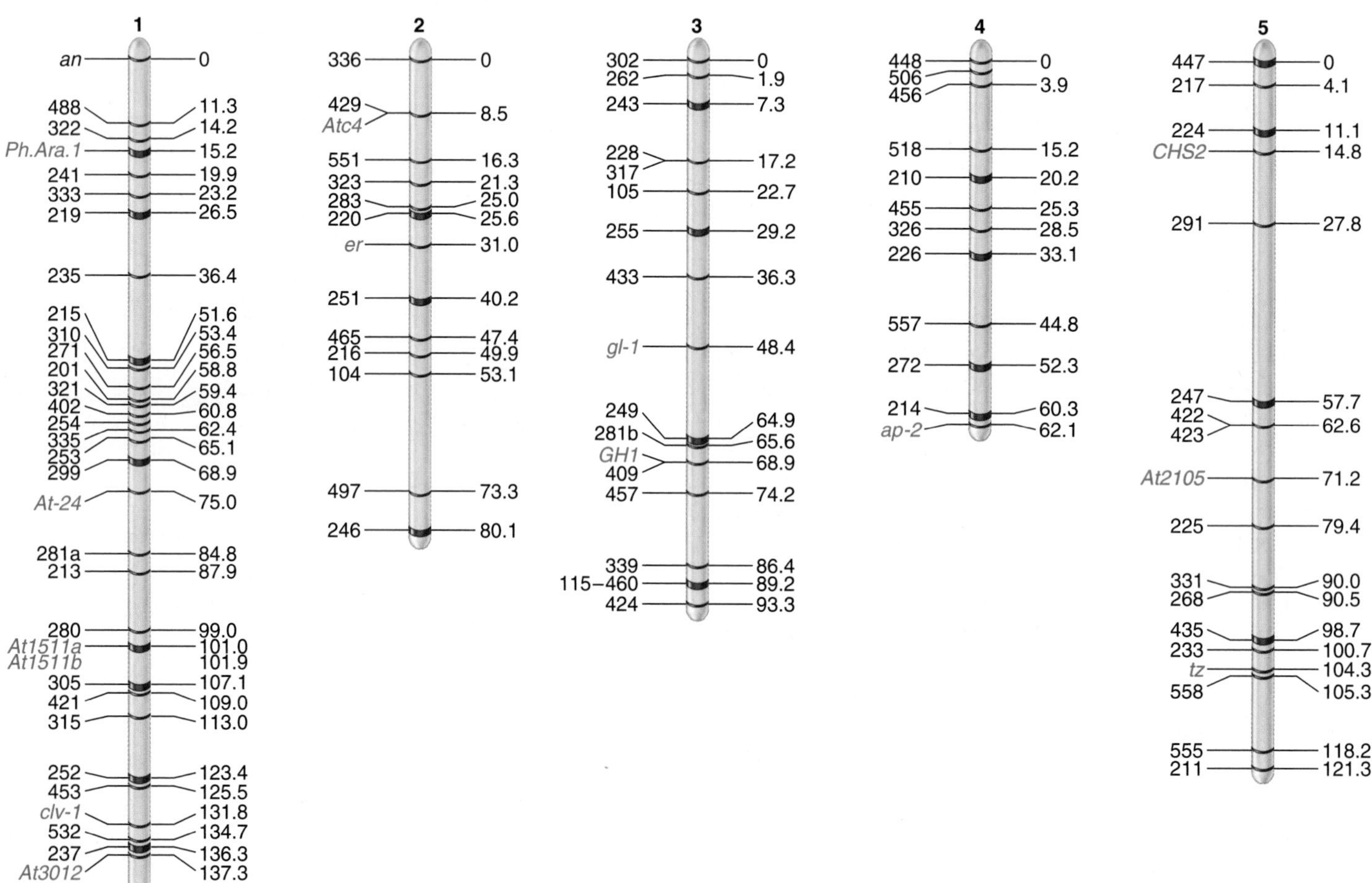

FIGURE 20.9 **An RFLP linkage map of *Arabidopsis thaliana*.** This plant has five different chromosomes. The left side of each chromosome describes the locations of RFLP markers. The numbers along the right side of each chromosome are the map distances in map units. For example, 219 and 235 are RFLPs located 9.9 mu apart on chromosome 1. The top marker at the end of each chromosome was arbitrarily assigned as the starting point (zero) for each chromosome. In addition, the map shows the locations of a few known genes (shown in red): *Ph.Ara.1* = phytochrome, *At-24* = nitrate reductase, *At1511a/b* = small RNA found in seeds, *clv-1* = clavata-1, *At3012* = alcohol dehydrogenase, *Atc4* = actin, *er* = erecta, *gl-1* = glabra-1, *GH1* = acetolactate synthase, *ap-2* = apelata-2, *CHS2* = chalcone synthase, *At2105* = 12S seed storage protein, and *tz* = thiazole-requiring allele.

Genes→Traits As a first step in mapping the locations of an organism's genes, researchers may initially determine the sites of RFLPs along the chromosomes. In this case, researchers determined the locations of many of these sites along the *Arabidopsis* chromosomes. By mapping these RFLP sites, it becomes easier to locate genes within the *Arabidopsis* genome. The identification of genes helps researchers to elucidate the relationship between genes and traits.

EXPERIMENT 20A

RFLP Analysis Can Be Used to Follow the Inheritance of Disease-Causing Alleles

As mentioned earlier, RFLP analysis can locate particular genes relative to an RFLP marker. This method may also be used to inform prospective parents of the likelihood that they are heterozygous for a recessive allele that may cause a genetic disease. This information would allow them to predict the likelihood of having children affected with the disorder. This genetic testing method is particularly useful for diseases in which the phenotype of the heterozygote may not differ from that of the unaffected homozygote.

The assumption behind this approach is that a disease-causing allele had its origin in a single individual known as a **founder,** who lived many generations ago. Since that time, the allele has spread throughout portions of the human population. In the case of recessive disorders, descendants of the founder who are affected with the disease have inherited two copies of the mutant allele. A second assumption is that the founder is likely to have had a polymorphic molecular marker that lies somewhere near the mutant allele. This is a reasonable assumption, because all people carry many polymorphic markers throughout their genomes. If a polymorphic marker lies very close to the disease-causing allele, it is unlikely that a crossover will occur in the intervening region. Therefore, such a polymorphic marker may be linked to the disease-causing allele for many generations. For this reason, an association between a particular polymorphic marker and a disease-causing allele may help to predict whether a person is a heterozygote for a recessive disease-causing allele.

As an example, let's consider the allele that causes sickle-cell disease. A couple may want to know if they are heterozygous carriers of the recessive allele that causes sickle-cell disease (see Chapter 4 for a description of this disease). While other techniques are now used to distinguish sickle-cell homozygotes and heterozygotes, the experiment described in **Figure 20.10** is a classic study conducted in 1978 that confirmed that RFLP markers can be used to predict the likelihood an individual carries a disease-causing allele. In this study, researchers Yuet Kan and Andree Dozy conducted an RFLP analysis in the region of the β-globin gene. This gene encodes the β-globin polypeptide, which is a subunit of hemoglobin. The common allele (Hb^A) results in the production of hemoglobin A, while a mutant form of the β-globin gene (Hb^S) encodes an abnormal polypeptide that results in the production of an altered hemoglobin form, hemoglobin S. Individuals who are homozygous for this mutant allele develop sickle-cell disease. Kan and Dozy set out to determine if RFLP differences could be detected between chromosomes containing these forms of the β-globin gene. In other words, they wanted to discover if the restriction digestion of chromosomal DNA in the vicinity of the β-globin gene was different in the Hb^A and Hb^S forms of the gene.

In the study described in Figure 20.10, the genotypes of 73 individuals were already known from a biochemical analysis of their hemoglobin proteins. The goal of this research was to see if an RFLP analysis could also distinguish individuals carrying the Hb^S allele from those carrying the Hb^A allele. As shown in Figure 20.10, they began with samples of white blood cells from these individuals and isolated the chromosomal DNA. Next, the chromosomal DNA was digested with the restriction enzyme *Hpa*I. The DNA fragments were separated by gel electrophoresis during the procedure of Southern blotting; strands that are complementary to the β-globin gene were used as probes. An RFLP was identified that produced three different sized fragments among a population of 73 individuals. The sizes of the fragments were 7,000 bp, 7,600 bp, and 13,000 bp, or 7.0, 7.6, and 13 kbp (kilobase pairs).

THE HYPOTHESIS

A mutant allele may be closely linked to a particular RFLP. People who carry the mutant allele can be identified by which sized fragment(s) they carry.

TESTING THE HYPOTHESIS — FIGURE 20.10 RFLP linkage to a disease-causing allele.

Starting materials: The β-globin gene had been cloned previously, and radiolabeled strands from this gene were used as probes. The researchers obtained blood samples from 73 individuals, some of whom were known to be affected with sickle-cell disease.

	Experimental level	Conceptual level
1. Take blood samples from 73 people. (Note: The *Hb* genotype was already known from a biochemical analysis of the subjects' hemoglobin.)	Take blood sample that contains white blood cells.	DNA in the nuclei

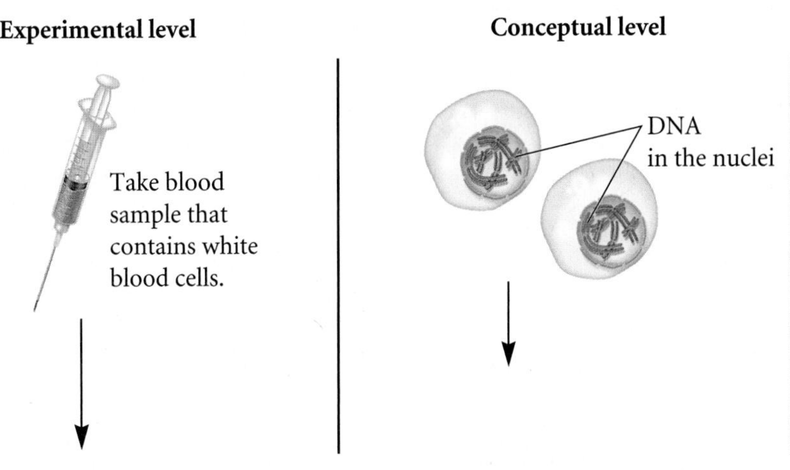

2. Isolate chromosomal DNA from the cells. This is similar to plasmid isolation. It involves breaking the cells and then using chromatography or centrifugation to purify the DNA. See Figure 19.1.

3. Digest the chromosomal DNA with the restriction enzyme *Hpa*I.

4. Use radiolabeled DNA strands from the β-globin gene as probes to conduct a Southern blot.

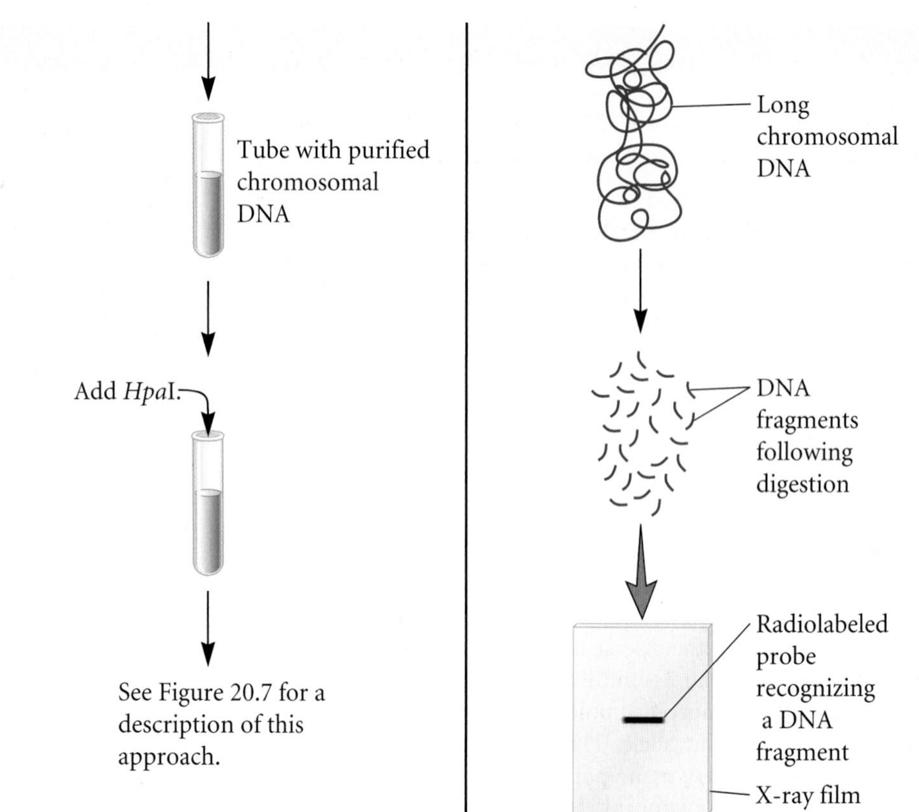

THE DATA

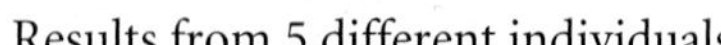

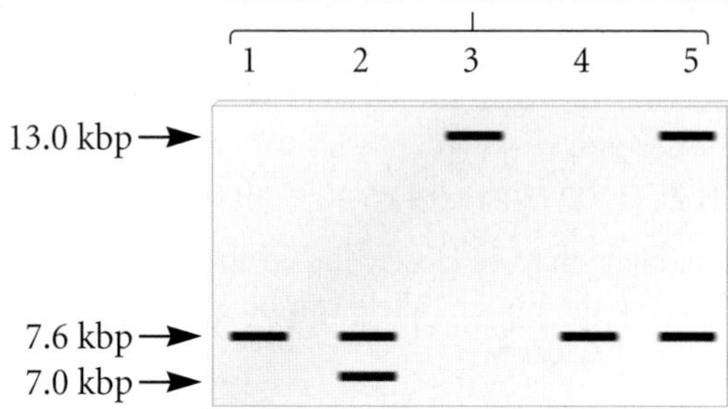

Summary of data from 73 people

Hemoglobin* Composition	*Hpa*I β-globin RFLP 7.6/7.6	7.6/7.0	7.0/13	7.6/13	13/13	Total	Frequency of the 13 kbp Fragment
African**							
AA	8	6	0	1	0	15	3%
AS	5	1	1	9	0	16	31%
SS	0	0	0	4	11	15	87%
Caucasian**							
AA	12	0	0	0	0	12	0%
Asian**							
AA	15	0	0	0	0	15	0%
						73	

*A is β-globin from Hb^A, and S is β-globin from Hb^S.

**Racial origin of the individuals used in this study.

Data from: Kan, Y.W., and Dozy, A.M. (1978) Polymorphism of DNA sequence adjacent to human β-globin structural gene: Relationship to sickle mutation. *Proc. Natl. Acad. Sci. USA* 75, 5631–5635.

INTERPRETING THE DATA

With regard to the Hb^A and Hb^S alleles, the occurrence of an RFLP existing in three different fragments was not random. Instead, the 13 kbp fragment was usually found in persons who were known to have at least one copy of the Hb^S allele. For example, in approximately 87% of the homozygotes (Hb^S Hb^S), the Hb^S allele was associated with the 13 kbp fragment. This observation is consistent with the diagram shown here.

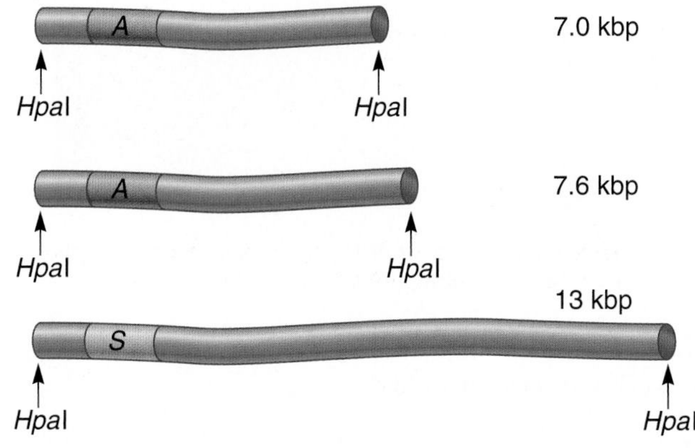

In most people with the Hb^A allele, the site where the normal β-globin gene is located has a distance of 7.0 or 7.6 kbp between the *Hpa*I sites. For people carrying the Hb^S allele, the mutant β-globin gene is found in a chromosomal region where the *Hpa*I sites are usually 13 kbp apart.

In most, but not all, cases, the Hb^S allele was found to be associated with a 13 kbp fragment. This type of information can be used as a predictive tool in human genetic analysis. For example, if an individual was found to have both 7.6 kbp and 13 kbp fragments, it would be fairly likely that the individual is a heterozygous ($Hb^A Hb^S$) carrier. By comparison, if an RFLP analysis revealed that an individual had only a 7.6 kbp fragment, that person would more likely be homozygous for the Hb^A allele.

A self-help quiz involving this experiment can be found at the Online Learning Center.

Linkage Mapping Also Uses Molecular Markers Called Microsatellites

To make a highly refined map of a genome, many different polymorphic sites must be identified and their transmission followed from parent to offspring over many generations. RFLPs were among the first molecular markers studied by geneticists. More recently, the use of other molecular markers has become more prevalent because they are easier to generate via PCR. As an example, let's consider **microsatellites,** short repetitive sequences that are abundantly interspersed throughout a species' genome and are quite variable in length among different individuals. For example, the most common microsatellite encountered in humans is a sequence $(CA)_n$, where n may range from 5 to more than 50. In other words, this dinucleotide sequence can be tandemly repeated 5 to 50 or more times. The $(CA)_n$ microsatellite is found, on average, about every 10,000 bases in the human genome. Researchers have identified thousands of different DNA segments that contain $(CA)_n$ microsatellites, located at many distinct sites within the human genome. Using primers complementary to the unique DNA sequences that flank a specific $(CA)_n$ region, a particular microsatellite can be amplified by PCR. In other words, the PCR primers will amplify only a particular microsatellite, but not the thousands of others that are interspersed throughout the genome (**Figure 20.11**).

When a pair of PCR primers amplifies a single site within a set of chromosomes, the amplified region is called a **sequence-tagged site (STS)**. In a diploid species, an individual will have two copies of a given STS. When an STS contains a microsatellite, the two PCR products will be identical only if the region is the same length in both copies (i.e., if the individual is homozygous for the microsatellite). However, if an individual has two copies that differ in the number of repeats in the microsatellite sequence (i.e., if the individual is heterozygous for the microsatellite), the two PCR products obtained will be slightly different in length (as in Figure 20.11).

Like RFLPs, microsatellites that have length polymorphisms allow researchers to follow their transmission from parent to offspring. PCR amplification of particular microsatellites provides an important strategy in the genetic analysis of human pedigrees, as shown in **Figure 20.12**. Prior to this analysis, a unique segment of DNA containing a microsatellite had been identified. Using PCR primers complementary to this microsatellite's unique flanking segments, two parents and their three offspring were tested for the inheritance of this microsatellite. A small sample of cells was obtained from each individual and subjected to PCR amplification, as described earlier in Figure 20.11. The amplified PCR products were then analyzed by high-resolution gel electrophoresis, which can detect small differences in the lengths of DNA fragments. The mother's PCR products were 154 and 150 bp in length; the father's were 146 and 140 bp. Their first offspring inherited the 154 bp product from the mother and the 146 bp from the father, the second inherited the 150 bp from the mother and the 146 bp from the father, and the third inherited the 150 bp from the mother and the 140 bp from the father. As shown in the figure, the transmission of polymorphic microsatellites is relatively easy to follow from generation to generation.

The simple pedigree analysis shown in Figure 20.12 illustrates the general method used to follow the transmission of a single microsatellite that is polymorphic in length. In linkage studies, the goal is to follow the transmission of many different microsatellites to determine those that are linked along the same chromosome versus those that are not. Those that are not

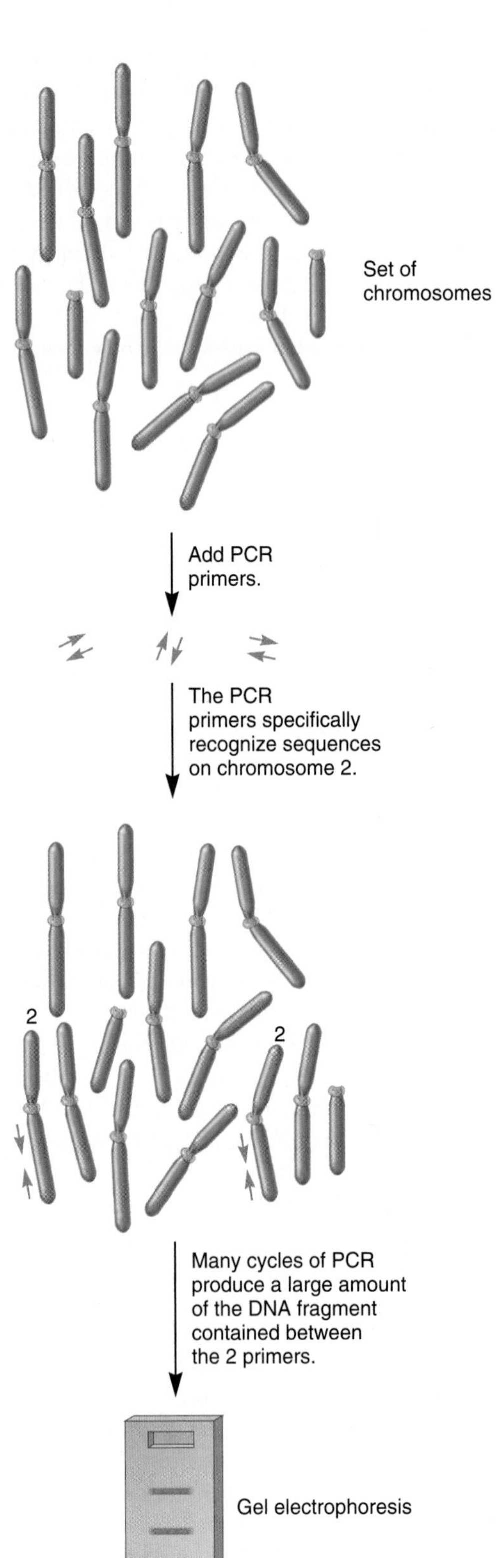
Set of
chromosomes
Add PCR
primers.
The PCR
primers specifically
recognize sequences
on chromosome 2.
2
2
Many cycles of PCR
produce a large amount
of the DNA fragment
contained between
the 2 primers.
Gel electrophoresis

FIGURE 20.11 **Identifying a microsatellite using PCR primers.**

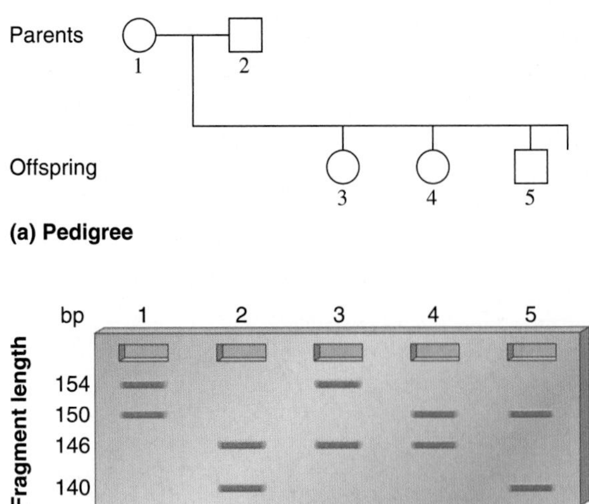
Parents
1
2
Offspring
3
4
5
(a) Pedigree
bp
1
2
3
4
5
Fragment length
154
150
146
140
(b) Electrophoretic gel of PCR products for a polymorphic microsatellite found in the family in (a).

FIGURE 20.12 **Inheritance pattern of a polymorphic microsatellite in a human pedigree.**

linked will independently assort from generation to generation. Those that are linked tend to be transmitted together to the same offspring. In a large pedigree, it is possible to identify cases where linked microsatellites have segregated due to crossing over. The frequency of crossing over provides a measure of the map distance, in this case between different microsatellites. This approach can help researchers to obtain a finely detailed genetic linkage map of the human chromosomes.

How can pedigree analysis involving STSs, such as polymorphic microsatellites, help researchers to identify the location of disease-causing alleles? By following the transmission of many polymorphic markers within large family pedigrees, it may be possible to determine that particular markers are found in people who carry specific disease-causing alleles. (An example is shown in Chapter 22, Figure 22.5.) As mentioned earlier, this is based on the assumption that the disease-causing allele had its origin in a founder and is likely to have a polymorphic marker nearby. After the identification of a closely linked marker, a disease-causing allele can be identified using a technique called chromosome walking, which is described later in this chapter.

20.2 PHYSICAL MAPPING

We now turn our attention to methods aimed at establishing a physical map of a species' genome. Physical mapping requires the cloning of many pieces of chromosomal DNA. The cloned DNA fragments are then characterized by size (that is, their length in base pairs), as well as the genes they contain and their relative locations along a chromosome.

As mentioned in Chapter 10, eukaryotic genomes are very large; the *Drosophila* genome is roughly 175 million bp long, and the human genome is approximately 3 billion bp in length. When making a physical map of a genome, researchers must characterize many DNA clones that contain much smaller pieces of the genome. For species with very large genomes, the physical mapping of the entire genome of a species is a massive undertaking carried out as a collaborative project by many genome research laboratories. In this section, we will examine the general strategies used in creating a physical map of a species' genome. We will also consider how physical mapping information can be used to clone genes.

A Physical Map of a Chromosome Is Constructed by Creating a Contiguous Series of Clones That Span a Chromosome

As discussed in Chapter 18, a DNA library contains a collection of hybrid vectors, with each vector containing a particular fragment of chromosomal DNA. In physical mapping studies, the goal is to determine the relative locations of the cloned chromosomal fragments from a DNA library, as they would occur in an intact chromosome. In other words, the members of the library must be organized according to their actual locations along a chromosome. To obtain a complete physical map of a chromosome, researchers need a series of clones that contain contiguous, overlapping pieces of chromosomal DNA. Such a collection of clones is known as a **contig** (**Figure 20.13**). A contig represents a physical map of a chromosome. As discussed later, cloning vectors known as BACs and cosmids are commonly used in the construction of a contig.

Different experimental strategies can be used to align the members of a contig. The general approach is to identify clones that contain overlapping regions. Historically, Southern blotting was used to determine if two different clones contain an overlapping region. In the example shown in Figure 20.13, the DNA from clone 1 could be radiolabeled in vitro and then used as a probe in a Southern blot of the other clones shown in this figure. Clone 1 would hybridize to clone 2, because they share identical DNA sequences in the overlapping region. Similarly, clone 2 could be used as a probe to show that it will hybridize to clone 1 and clone 3. By conducting Southern blots between many combinations of clones, researchers can determine which clones have common overlapping regions and thereby order them as they would occur along the chromosome.

Alternatively, other methods to order the members of a contig involve the use of molecular markers. For example, if many STSs have already been identified along a chromosome via linkage and/or cytogenetic analysis, the STSs can be used as probes to order the members of a contig. Another approach is to subject the members of a library to a restriction enzyme to produce a restriction digest pattern for each member. The patterns of fragments are then analyzed by computer programs, which can identify regions that are potentially overlapping.

An ultimate goal of physical mapping procedures is to obtain a complete contig for each type of chromosome within a full set. For example, in the case of humans, a complete physical map requires a contig for each of the 22 autosomes and for the X and Y chromosomes. Geneticists can also correlate cloned DNA fragments in a contig with locations along a chromosome obtained from linkage or cytogenetic mapping. For example, a member of a contig may contain a gene, RFLP, or STS that previously has been mapped by linkage analysis. **Figure 20.14** considers a situation in which two members of a contig carry genes previously mapped by linkage analysis to be approximately 1.5 mu apart on chromosome 11. In this example, clone 2 has an insert that carries gene *A*, while clone 7 has an insert that carries

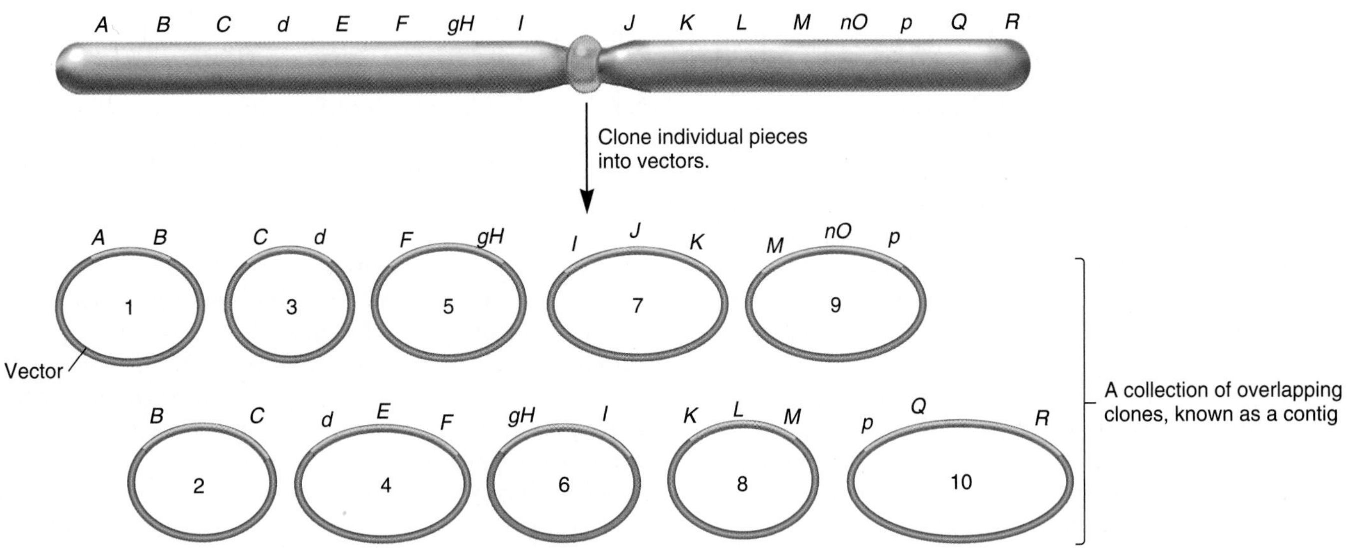

FIGURE 20.13 The construction of a contig. Large pieces of chromosomal DNA are cloned into vectors, and their order is determined by the identification of overlapping regions. For example, clone 3 ends with gene *d*, and clone 4 begins with gene *d*; clone 4 ends with gene *F*, and clone 5 begins with gene *F*.

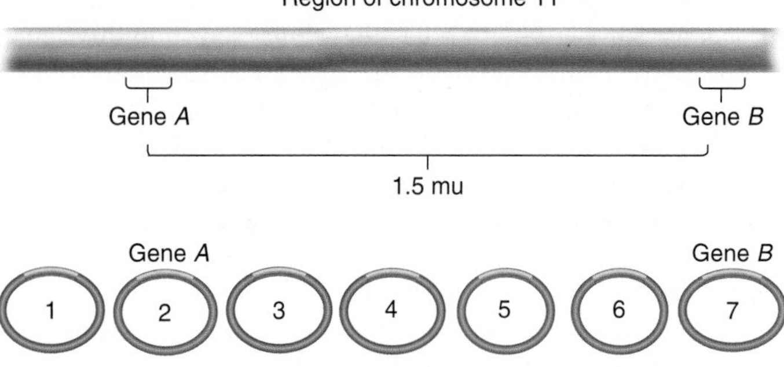

FIGURE 20.14 The use of genetic markers to align a contig. In this example, gene *A* and gene *B* had been mapped previously to specific regions of chromosome 11. Gene *A* was found within the insert of clone 2, gene *B* within the insert of clone 7. This made it possible to align the contig using gene *A* and gene *B* as genetic markers (i.e., reference points) along chromosome 11.

gene *B*. Because a contig is composed of overlapping members, a researcher can align the contig along chromosome 11, starting with gene *A* and gene *B* as reference points. In this example, genes *A* and *B* serve as markers that identify the location of specific clones within the contig.

YAC, BAC, and PAC Cloning Vectors Are Used to Make Contigs of Eukaryotic Chromosomes

For large eukaryotic genomes, researchers often begin with cloning vectors that can accept chromosomal DNA inserts of very large size. By having large insert sizes, a contig is more easily constructed and aligned because fewer recombinant vectors are needed. In general, most plasmid and viral vectors can accommodate inserts only a few thousand to perhaps tens of thousands of nucleotides in length. If a plasmid or viral vector has a DNA insert that is too large, it will have difficulty with DNA replication and is likely to suffer deletions in the insert.

By comparison, other cloning vectors, known as **artificial chromosomes,** can accommodate much larger sizes of DNA inserts. As their name suggests, they behave like chromosomes when inside of living cells. The first type to be made was the **yeast artificial chromosome (YAC),** which was developed by David Burke, Georges Carle, and Maynard Olson in 1987. An insert within a YAC can be several hundred thousand to perhaps 2 million base pairs in length. For an average human chromosome, a few hundred YACs are sufficient to create a contig with fragments that span the entire length of the chromosome. By comparison, it would take thousands or even tens of thousands of recombinant plasmid vectors to create such a contig.

At the molecular level, YACs have structural similarities to normal eukaryotic chromosomes, yet have characteristics that make them suitable for cloning. The general structure of a YAC vector is shown at the top left of **Figure 20.15**. The YAC vector contains two telomeres (*TEL*), a centromere (*CEN*), a bacterial origin of replication (*ORI*), a yeast origin of replication (known as an *ARS*, for autonomous replication sequence), selectable markers, and unique cloning sites that are each recognized by a single restriction enzyme. Without an insert, the circular form of this vector can replicate in *E. coli*. After a large fragment has been inserted, the linear form of the vector can replicate in yeast.

In the experiment shown in Figure 20.15, chromosomal DNA is digested with the restriction enzyme *Eco*RI at a low concentration so that only some of the restriction sites are cut. This partial digestion will result in only occasional cleavage of the chromosomal DNA to yield very large DNA fragments.

The circular YAC vector is also digested with *Eco*RI and a second restriction enzyme, *Bam*HI, to yield two arms of the YAC. The YAC arms are then ligated to the large fragments of chromosomal DNA and transformed into yeast cells. When ligation occurs in the desired way, a large piece of chromosomal DNA becomes ligated to both arms of the YAC. Because each arm contains a different selectable marker, it is possible to select for the growth of yeast cells that carry a YAC construct having both arms.

Newer types of cloning vectors called **bacterial artificial chromosomes (BACs)** and **P1 artificial chromosomes (PACs)** have been constructed. BACs were developed from bacterial F factors, which are described in Chapter 6, and PACs were developed from P1 bacteriophage chromosomes. BACs and PACs typically can contain inserts up to 300,000 bp and sometimes larger. These vectors are somewhat easier to use than YACs because the DNA is inserted into a circular molecule and transformed into *E. coli*. BACs and PACs have largely replaced YACs for the cloning of large DNA fragments.

Figure 20.16 shows a simplified drawing of a BAC cloning vector. The vector contains several genes that function in vector replication and segregation. The origin of replication is designated *oriS*, and the *repE* gene encodes a protein essential for replication at *oriS*. The *parA*, *parB*, and *parC* genes encode proteins required for the proper segregation of the vector into daughter cells. A chloramphenicol resistance gene, cm^R, provides a way to select for cells that have taken up the vector based on their ability to grow in the presence of the antibiotic chloramphenicol. The vector also contains unique restriction enzyme sites, such as *Hin*dIII, *Bam*HI, and *Sph*I, for the insertion of large fragments of DNA. These sites are located within the *lacZ* gene, which encodes the enzyme β-galactosidase. Vectors with DNA inserts can be determined by plating cells on media containing the compound X-Gal (as described for plasmid vectors in Chapter 18, Figure 18.2).

YAC, BAC, and PAC cloning vectors have been very useful in the construction of contigs that span long segments of chromosomes. They have been used as the first step in creating a rough physical map of a genome. While this is an important step in physical mapping, the large insert sizes make them difficult to use in gene cloning and sequencing experiments. Therefore, libraries containing hybrid vectors with smaller insert sizes are needed. Most commonly, a type of cloning vector called a **cosmid** is used. A cosmid is a hybrid between a plasmid vector and phage λ; its DNA can replicate in a cell like a plasmid or be packaged into a protein coat like a phage. Cosmid vectors typically can accept DNA fragments that are tens of thousands of bp in length.

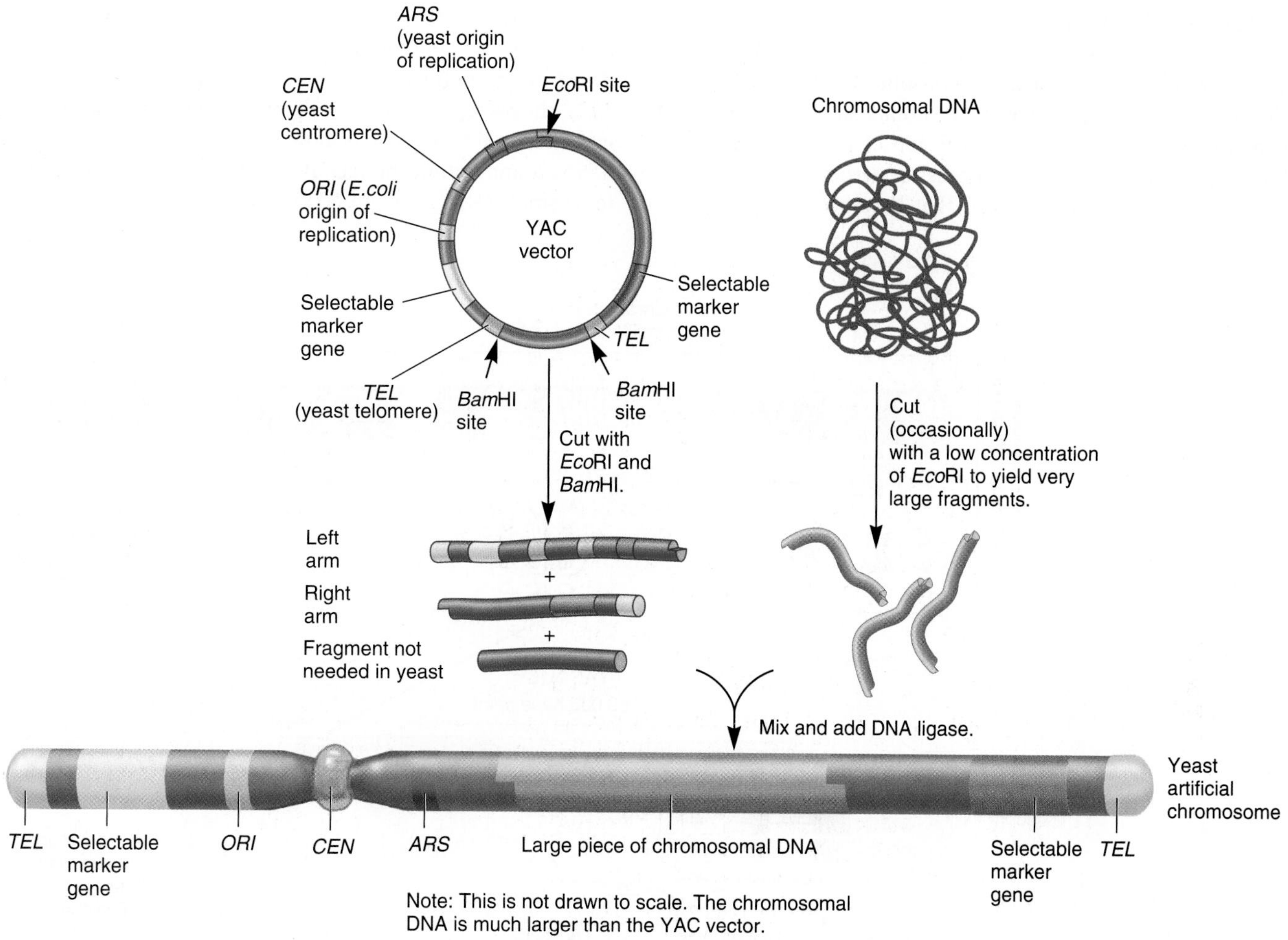

FIGURE 20.15 The use of a YAC vector in DNA cloning. Note: *Eco*RI and *Bam*HI both cut DNA to yield sticky ends. In this diagram, only the *Eco*RI ends are shown with a notch to emphasize that these are the sticky ends that bind to the chromosomal DNA.

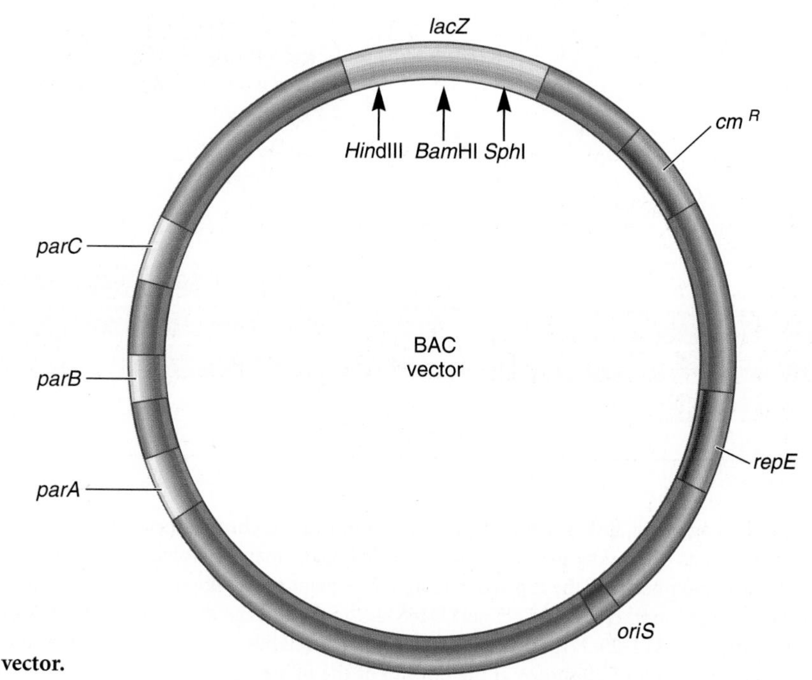

FIGURE 20.16 A BAC vector.

Figure 20.17 illustrates a comparison between cytogenetic, linkage, and physical maps of human chromosome 16. Actually, this is a very simplified map of chromosome 16. A much more detailed map is available, although it would take well over 10 pages of this textbook to print it! The top of this figure shows the banding pattern of this chromosome. Underneath the banded chromosome are molecular markers that have been mapped by linkage analysis. These same markers have been mapped cytogenetically. A complete contig of this chromosome has also been produced by generating a series of overlapping YACs. However, Figure 20.17 shows the location of only one YAC within the contig. Cosmids within this region are shown below the YAC. In addition, an STS is found within the region and provides a molecular marker for cosmids N16Y1-19 and N16Y1-10, as well as YAC N16Y1.

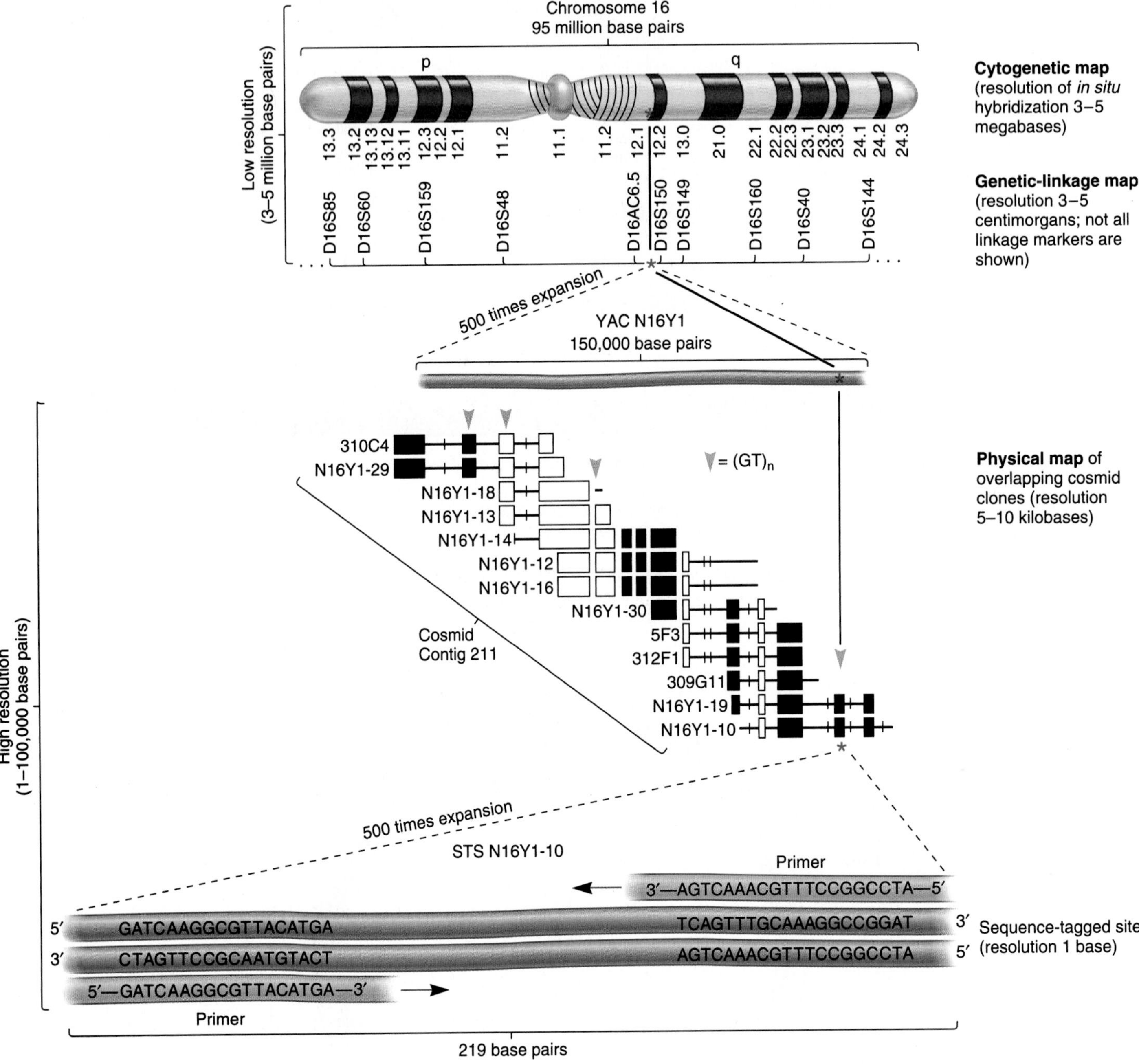

FIGURE 20.17 A correlation of cytogenetic, linkage, and physical maps of human chromosome 16. The top part shows the cytogenetic map of human chromosome 16 according to its G banding pattern. A very simple linkage map of molecular markers (D16S85, D16S60, etc.) is aligned below the cytogenetic map. A correlation between the linkage map and a segment of the physical map is shown below the linkage map. A YAC clone designated YAC N16Y1 is located between markers D16AC6.5 and D16S150 on the linkage map. Pieces of DNA from this YAC were subcloned into cosmid vectors, and the cosmids (310C4, N16Y1-29, N16Y1-18, etc.) were aligned relative to each other. One of the cosmids (N16Y1-10) was sequenced, and this sequence was used to generate an STS shown at the bottom of the figure.

Positional Cloning Can Be Achieved Using Chromosome Walking

The creation of a contig bears many similarities to a gene cloning strategy known as **positional cloning,** a strategy in which a gene is cloned based on its mapped position along a chromosome. This approach has been successful in the cloning of many human genes, particularly those that cause genetic diseases when mutated. These include genes involved in cystic fibrosis, Huntington disease, and Duchenne muscular dystrophy.

A common method used in positional cloning is known as **chromosome walking.** To initiate this type of experiment, a gene's position relative to a marker must be known from mapping studies. For example, a gene may be known to be fairly close to a previously mapped gene or RFLP marker. This provides a starting point to molecularly "walk" toward the gene of interest.

Figure 20.18 considers a chromosome walk in which the goal is to locate a gene that we will call gene *A*. In this example, linkage mapping studies have revealed that gene *A* is relatively close to another gene, called gene *B*, that has been previously cloned. Gene *A* and gene *B* have been deduced from genetic crosses to be approximately 1 mu apart. To begin this chromosome walk, a cloned DNA fragment that contains gene *B* and flanking sequences can be used as a starting point.

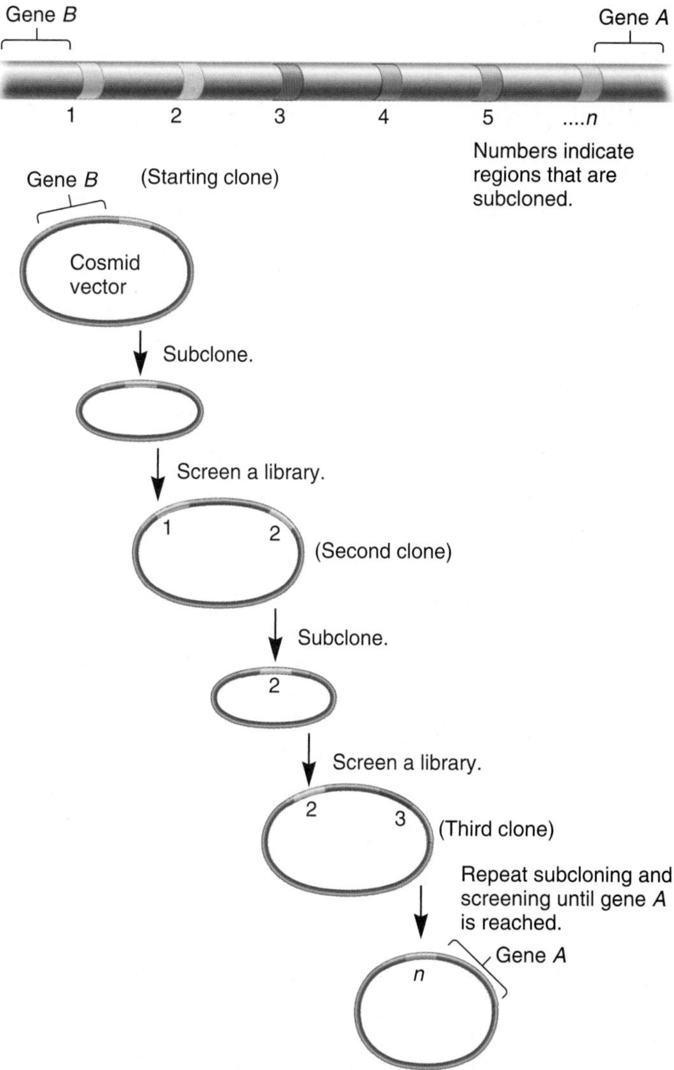

FIGURE 20.18 **The technique of chromosome walking.**

To walk from gene *B* to gene *A*, a series of library screening methods are followed. In this example, the starting materials are a cosmid library and a clone containing gene *B*. A small piece of DNA from the first cosmid vector containing gene *B* is inserted into another vector. This is called **subcloning.** The subclone is radiolabeled and used as a probe to screen a cosmid library. This will enable the researchers to identify a second clone that extends into the region that is closer to gene *A*. A subclone from this second clone is then used to screen the library a second time. This allows the researchers to identify a clone that is even closer to gene *A*. This repeated pattern of subcloning and library screening is used to reach gene *A*. The term chromosome walking is an appropriate description of this technique, because each clone takes you a step closer to the gene of interest. When starting at gene *B* in Figure 20.18, researchers would also want to have markers to the left of gene *B* to ensure they were not walking in the wrong direction.

The number of steps required to reach the gene of interest depends on the distance between the starting and ending points and on the sizes of the DNA inserts in the library. If the two points are 1 mu apart, they are expected to be approximately 1 million bp apart, although the correlation between map units and physical distances can be quite variable. In a typical walking experiment, each clone might have an average insert size of 50,000 bp. Therefore, it will take about 20 walking steps to reach the gene of interest. Researchers want to locate starting points in a chromosome walking experiment that are as close as possible to the gene they wish to identify.

How do researchers know when they have reached a gene of interest? In the case of a gene that causes a disease when mutant, researchers would conduct their walking steps on DNA from both an unaffected individual and an affected individual. Each set of clones would be subjected to DNA sequencing, and those DNA sequences would be compared to each other. When the researchers reach a spot where the DNA sequences differ between an unaffected individual and an affected individual, such a site may be within the gene of interest. However, this has to be confirmed by sequencing the region from several unaffected and affected individuals to be certain the change in DNA sequence is correlated with the disease.

Genomes Can Be Compared by Pulsed-Field Gel Electrophoresis

Researchers may also compare genomes using techniques that separate large DNA fragments and/or small chromosomes. One technique that can accomplish this goal is **pulsed-field gel electrophoresis (PFGE),** developed in 1984 by David Schwartz and Charles Cantor. Using this method, large pieces of chromosomes or entire small chromosomes can be separated and identified. Pulsed-field gel electrophoresis differs from conventional electrophoresis in that the DNA is driven through the gel using alternating pulses of electrical current at different angles. This can be achieved using electrophoresis devices that have two sets of electrodes. The alternating pulses can separate larger DNA fragments that vary in length from 50,000 bp up to 10 million bp.

For PFGE, large fragments of chromosomes or individual chromosomes must be isolated very carefully. To avoid mechanical breakage of the chromosomes, cells are embedded in agarose blocks (also called plugs). The agarose protects the chromosomal DNA from mechanical breakage. Once inside the plug, the cells are lysed, and if desired, the chromosomal DNA can be subjected to restriction digestion to obtain large DNA fragments. This can be achieved using a restriction enzyme that cuts very infrequently. To separate the large pieces of DNA, the plug is inserted into a well in an agarose gel and then subjected to PFGE.

Figure 20.19a illustrates a separation of yeast chromosomes that have not been subjected to restriction digestion. **Figure 20.19b** shows fragments of DNA that have been obtained by digesting a bacterial chromosome with three different restriction enzymes that cut the chromosome infrequently.

What are the applications of PFGE? In the past, PFGE was often used to isolate large pieces of chromosomes to construct contigs. However, in recent years, PFGE has been most commonly used to identify species of microorganisms or strains of the same species. For example, some strains of *E. coli* are pathogenic, while others are not. When subjected to PFGE, such strains may differ in their banding patterns. Therefore, this method has gained widespread use to distinguish different strains of microorganisms.

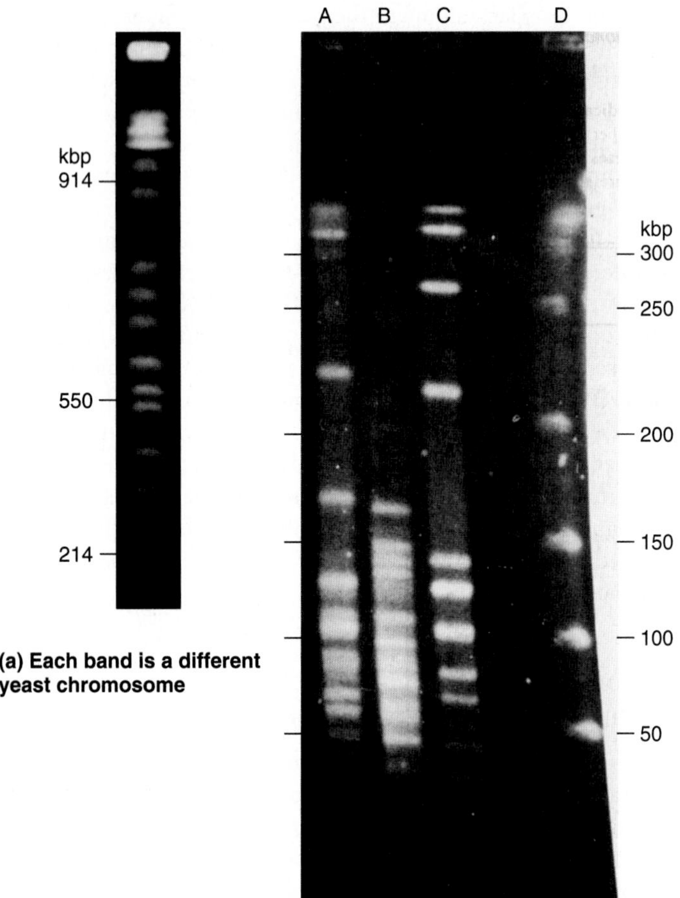

FIGURE 20.19 Pulsed-field gel electrophoresis (PFGE). (**a**) Yeast chromosomes that have not been digested with a restriction enzyme. The numbers at the left indicate the size of the chromosomes in kilobase pairs (kbp). (**b**) The *Haemophilus influenzae* chromosome digested with *Eag*I (lane A), *Nae*I (lane B), and *Sma*I (lane C). Lane D shows molecular weight markers.

20.3 GENOME SEQUENCING PROJECTS

Genome sequencing projects are research endeavors that have the ultimate goal of determining the sequence of DNA bases of the entire genome of a given species. Such projects involve many participants, including scientists who isolate DNA and perform DNA sequencing reactions, as well as theoreticians who gather the DNA sequence information and assemble it into a long DNA sequence for each chromosome. For bacteria and archaea, which usually have just one chromosome, the genome sequence is that of a single chromosome. For eukaryotes, each chromosome must be sequenced. Thus, for humans, the genome sequence includes sequences of 22 autosomes, 2 sex chromosomes, and the mitochondrial genome.

In just a couple of decades, our ability to map and sequence genomes has improved dramatically. By 2007, the complete genome sequences have been obtained from hundreds of different species, including over 400 bacteria, 30 archaea, 20 protists, 20 fungi, 6 plants, and 20 animals. Considering that the first genome sequence was generated in 1995, the progress of genome sequencing projects since then has been truly remarkable! In this section, we will examine the approaches that researchers follow when tackling such a large project. We will also survey some of the general goals of the Human Genome Project, the largest of its kind, and compare the results from the genome sequencing of various species.

Shotgun Sequencing Is Used to Determine the DNA Sequence of Entire Genomes

When sequencing an entire genome, researchers must consider factors such as genome size, the efficiency of the methods used to sequence DNA, and the costs of the project. Since genome sequencing projects began in the 1990s, researchers have learned that the most efficient and inexpensive way to sequence genomes is via an approach called **shotgun sequencing,** in which DNA fragments to be sequenced are randomly generated from larger DNA fragments. In this method, genomic DNA is isolated and sheared into smaller DNA fragments, typically 1,500 bp or longer in length. The researchers then use the technique of dideoxy sequencing, described in Chapter 18, to randomly sequence fragments from the genome. As a matter of chance, some of the fragments will be overlapping, as schematically shown at the top of the next page.

Overlapping Region

```
TTACGGTACCAGTTACAAATTCCAGACCTAGTACC
AATGCCATGGTCAATGTTTAAGGTCTGGATCATGG
                                GACCTAGTACCGGACTTATTCGATCCCCAATTTTGCAT
                                CTGGATCATGGCCTGAATAAGCTAGGGGTTAAAACGTA
```

The DNA sequence in two different fragments will be identical in the overlapping region. This allows researchers to order them as they are found in the intact chromosome.

An advantage of shotgun DNA sequencing is that it does not require extensive mapping, which can be very time consuming and expensive. A disadvantage is that researchers will waste some time sequencing the same region of DNA more times than needed.

To obtain a complete sequence of a genome with the shotgun approach, how do researchers decide how many fragments to sequence? We can calculate the probability that a base will not be sequenced using this approach with the following equation:

$$P = e^{-m}$$

where

P is the probability that a base will be left unsequenced

e is the base of the natural logarithm; $e = 2.72$

m is the number of bases sequenced divided by the total genome size

For example, in the case of a bacterial species with a genome size of 1.8 Mb (i.e., 1.8 million bp), if researchers sequenced 9 Mb, $m = 5$ (i.e., 9.0 Mb divided by 1.8 Mb):

$$P = e^{-m} = e^{-5} = 0.0067, \text{ or } 0.67\%$$

This means that if we randomly sequence 9.0 Mb, which is five times the length of a single genome, we are likely to miss only 0.67% of the genome. With a genome size of 1.8 Mb, we would miss about 12,000 nucleotides out of approximately 1,800,000. Such missed sequences are typically on small DNA fragments that—as a matter of random chance—did not happen to be sequenced. The missing links in the genome can be sequenced later using mapping methods such as chromosome walking.

For a genome sequencing project, researchers have typically followed two types of shotgun strategies—hierarchical shotgun sequencing or whole genome shotgun sequencing. In the late 1990s, when researchers first began to sequence the larger genomes from eukaryotic species, they first took the approach of **hierarchical shotgun sequencing** (**Figure 20.20a**). To begin this process, researchers clone large (e.g., 150,000 bp long) DNA fragments into BACs (or YACs) to generate a contig for each chromosome. For DNA sequencing, researchers choose a set of clones from each chromosome contig that have relatively short overlapping regions. Each BAC clone in a chosen set is then subjected to shotgun sequencing. A clone is sheared into smaller pieces (e.g., 2,000 bp), and those small pieces are cloned into vectors. The recombinant vectors are then randomly sequenced using a primer that anneals at a site next to the inserted DNA (see Figure 18.15). The method of dideoxy sequencing yields a DNA sequence, usually 500 to 1,000 bp in length, at one end of each insert. The resulting DNA sequences are analyzed by a computer program that identifies overlapping sequences between two DNA fragments and assembles the DNA sequence into one long sequence along a single chromosome.

In the hierarchical method, the nucleotide sequence is determined clone by clone. For this reason, the method is also called the clone-by-clone or BAC-to-BAC approach. This strategy has the advantage of making it easier to align the overlapping sequences because the DNA sequences are grouped according to the BAC clone from which the sequence was obtained. Another issue is that it minimizes the amount of DNA sequencing by reducing the amount of sequencing in overlapping regions. This was an advantage in the 1990s when DNA sequencing was an expensive and labor-intensive procedure. However, a disadvantage of this approach is that BAC mapping can be time-consuming and expensive.

In more recent years, advances in DNA sequencing technology have greatly diminished the time and expense of dideoxy sequencing. Modern genome sequencing projects more commonly follow a strategy called **whole genome shotgun sequencing.** In this method, the physical mapping step used in the hierarchical approach is bypassed. The DNA from the entire genome is isolated and sheared into small (e.g., 2,000 or 10,000 bp) and large (50,000 or 150,000 bp) fragments and cloned into vectors (**Figure 20.20b**). Each insert is usually sequenced at both ends, an approach called **double-barrel shotgun sequencing.** As in the hierarchical approach, the DNA sequences, which are 500 bp to 1,000 bp in length, are then analyzed by a computer program that identifies overlapping regions between two DNA fragments and assembles the data into one long sequence. The double-barrel strategy improves the ability of researchers to align their DNA sequences. Most of the DNA sequencing in a whole genome shotgun approach is carried out on small DNA fragments. However, due to the prevalence of repetitive sequences, particularly within the genomes of eukaryotic species, the alignment of small pieces may be difficult. The sequencing of both ends of longer fragments makes alignment easier, because the relative spacing of particular DNA sequences is known.

The Human Genome Project Was the Largest Genome Sequence Project in History

Due to its large size, the sequencing of the human genome was an enormous undertaking. Scientists had been discussing how to undertake this project since the mid-1980s. In 1988, the National Institutes of Health established an Office of Human Genome Research, with James Watson as its first director. The **Human Genome Project,** which officially began on October 1, 1990, was a 13-year effort coordinated by the U.S. Department of Energy

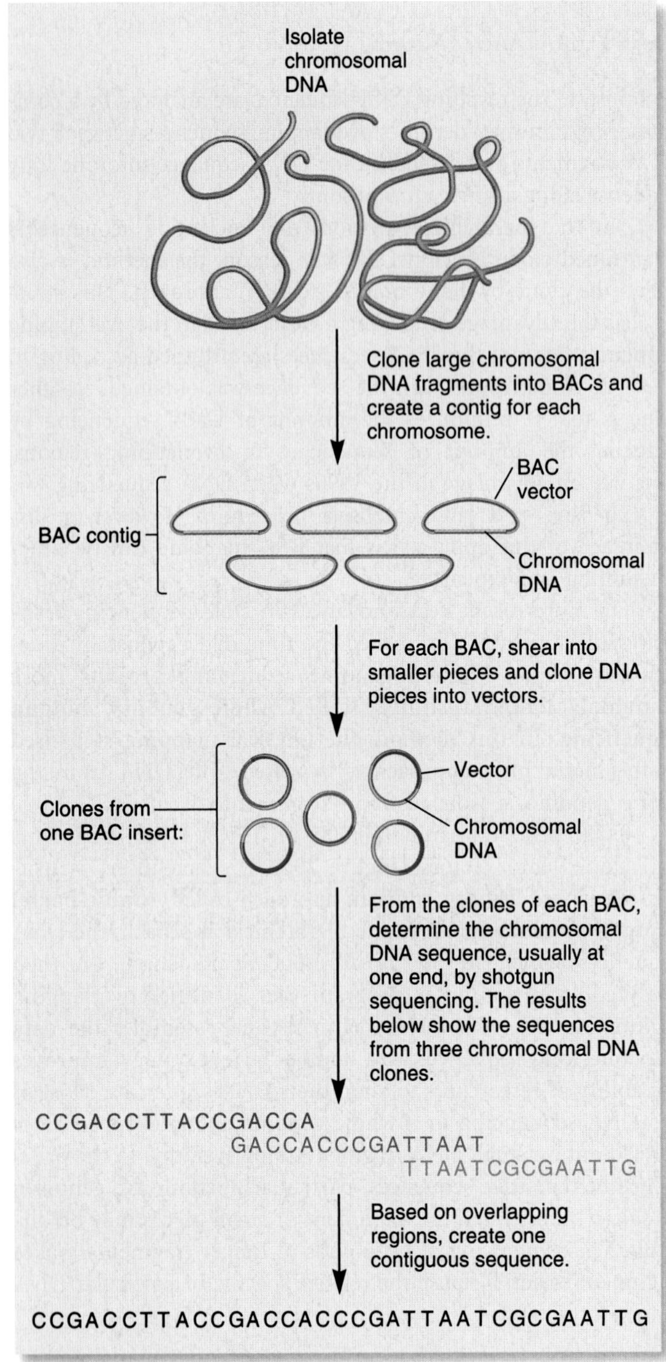

(a) Hierarchical genome shotgun sequencing

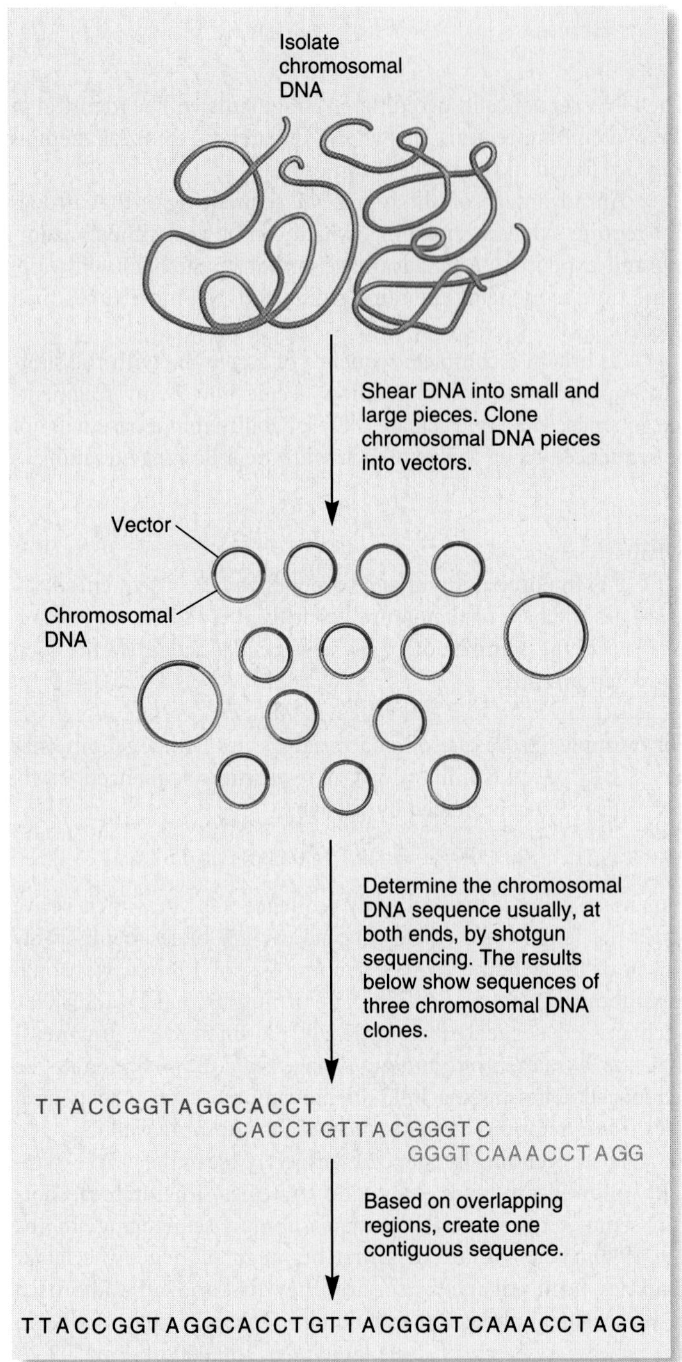

(b) Whole genome shotgun sequencing

FIGURE 20.20 Two approaches of shotgun sequencing of genomes. Note: the BAC vector and chromosomal DNA are not drawn to scale. The chromosomal DNA would be much longer than the BAC vector DNA.

and the National Institutes of Health. From its outset, the Human Genome Project had the following goals:

1. *To obtain a genetic linkage map of the human genome.* This has involved the identification of thousands of genetic markers and their localization along the autosomes and sex chromosomes.
2. *To obtain a physical map of the human genome.* This has required the cloning of many segments of chromosomal DNA into BACs, YACs, and cosmids.
3. *To obtain the DNA sequence of the entire human genome.* The first (nearly complete) sequence was published in February 2001. This was considered a first draft. A second draft was published in 2003, and the completed maps and sequences for all of the human chromosomes were published by 2006. The entire genome is approximately 3 billion nucleotide base pairs in length. If the entire human genome were typed in a textbook like this, it would be nearly 1 million pages long!
4. *To develop technology for the management of human genome information.* The amount of information obtained from this project is staggering, to say the least. The Human Genome Project developed user-friendly tools to provide scientists easy access to up-to-date information obtained from the project. The Human Genome Project also developed analytical tools to interpret genome information.
5. *To analyze the genomes of other model organisms.* These include bacterial species (e.g., *Escherichia coli* and *Bacillus subtilis*), *Drosophila melanogaster* (fruit fly), *Caenorhabditis elegans* (a nematode), *Arabidopsis thaliana* (a simple plant), and *Mus musculus* (mouse).
6. *To develop programs focused on understanding and addressing the ethical, legal, and social implications of the results obtained from the Human Genome Project.* The Human Genome Project has sought to identify the major genetic issues that will affect members of our society and to develop policies to address these issues. For example, what is an individual's right to privacy regarding genetic information? Some people are worried that their medical insurance company may discriminate against them if it is found that they carry a disease-causing or otherwise deleterious gene.
7. *To develop technological advances in genetic methodologies.* Some of the efforts of the Human Genome Project have involved improvements in molecular genetic technology such as gene cloning, contig construction, DNA sequencing, and so forth. The project has also been aimed at developing computer technology for data processing, storage, and the analysis of sequence information (see Chapter 21).

A great benefit expected from the characterization of the human genome is the ability to identify and study the sequences of our genes. Mutations in many different genes are known to be correlated with human diseases, which include cancer, heart disease, and many other abnormalities. The identification of mutant genes that cause inherited diseases was a strong motivation for the Human Genome Project. A detailed genetic and physical map has made it profoundly easier for researchers to locate such genes. Furthermore, a complete DNA sequence of the human genome provides researchers with insight into the types of proteins encoded by these genes. The cloning and sequencing of disease-causing alleles is expected to play an increasingly important role in the diagnosis and treatment of disease.

Many Genome Sequences Have Been Determined

How do researchers decide which genomes to sequence? Motivation behind genome sequencing projects comes from a variety of sources. For example, basic research scientists can greatly benefit from a genome sequence. It allows a scientist to know which genes a given species has, and it aids in the cloning and characterization of such genes. This has been the impetus for genome projects involving model organisms such as *Escherichia coli, Saccharomyces cerevisiae, Drosophila melanogaster, Caenorhabditis elegans, Arabidopsis thaliana,* and the mouse. A second impetus for genome sequencing has involved human disease. As noted previously, researchers expect that the sequencing of the human genome will aid in the identification of genes that, when mutant, play a role in disease. Likewise, the decision to sequence many bacterial, fungal, and protist genomes has been related to the role of these species in infectious diseases. Hundreds of microbial genomes have been sequenced, many of them from species that are pathogenic in humans. The sequencing of such genomes may help to elucidate the genes that play a role in the infectious process.

In addition, the genomes of agriculturally important species have been the subject of genome sequencing projects. An understanding of a species' genome may aid in the development of new strains of livestock and plant species that have improved traits from an agricultural perspective. Finally, genome sequencing projects help us to better understand the evolutionary relationships among living species. This approach, called **comparative genomics,** uses information from genome projects to understand the genetic variation among different populations. We will explore this topic in Chapter 26.

Table 20.2 describes the results of several genome-sequencing projects that have been completed. Newly completed genome sequences are emerging rapidly, particularly those of microbial species. As we obtain more genome sequences, it becomes progressively more interesting to compare them to each other as a way to understand the process of evolution. As we will discover in Chapter 21, the field of functional genomics enables researchers to probe the roles of many genes as they interact to generate the phenotypic traits of the species that contain them.

TABLE 20.2

Examples of Genomes That Have Been Sequenced

Species	Genome Size (bp)*	Number of Genes**	Description
Prokaryotic genomes			
Bacteria			
Mycoplasma genitalium	580,000	521	An inhabitant of the human genital tract with a very small genome.
Helicobacter pylori	1,668,000	1,590	A common inhabitant of the stomach that may cause gastritis, peptic ulcer, and gastric cancer.
Mycobacterium tuberculosis	4,412,000	4,294	The bacterial species that causes tuberculosis.
Escherichia coli	4,639,000	4,289	A widespread bacterial inhabitant of the gut of animals; also a model research organism.
Archaea			
Thermoplasma volcanium	1,580,000	1,494	An archaea with an optimal growth temperature of 60°C and pH optimum of <2.0.
Pyrococcus abyssi	1,760,000	1,765	Originally isolated from samples taken close to a hot spring situated 3,500 meters deep in the southeast Pacific. Optimal growth conditions are 103°C and 200 atmospheres pressure.
Sulfolobus solfataricus	2,990,000	2,977	Found in terrestrial volcanic hot springs with optimum growth occurring at a temperature of 75–80°C and pH 2–3.
Eukaryotic genomes			
Protists			
Plasmodium falciparum	22,900,000	5,268	A parasitic protist that causes malaria in humans.
Entamoeba histolytica	23,800,000	9,938	An amoeba that causes dysentery in humans.
Fungi			
Saccharomyces cerevisiae (baker's yeast)	12,100,000	6,294	A structurally simple eukaryotic species that has been extensively studied by researchers to understand eukaryotic genetics and cell biology.
Neurospora crassa	40,000,000	10,082	A common bread mold and also a structurally simple eukaryotic species that has been extensively studied by researchers.
Plants			
Arabidopsis thaliana	142,000,000	26,000	A model organism studied by plant biologists.
Oryza sativa (rice)	440,000,000	40,000	A cereal grain with a relatively small genome. It is very important worldwide as a food crop.
Populus trichocarpa (Balsam poplar)	550,000,000	45,555	A tree with a relatively small genome.
Animals			
Caenorhabditis elegans	97,000,000	19,000	A nematode worm that has been used as a model organism to study animal development.
Drosophila melanogaster (fruit fly)	175,000,000	14,000	Used as a model organism to study many genetic phenomena, including development.
Anopheles gambiae	278,000,000	13,683	A mosquito that carries the malaria parasite, *Plasmodium falciparum.*
Canis familiaris (dog)	2,400,000,000	19,300	A common house pet.
Mus musculus (mouse)	2,500,000,000	24,174	A rodent and a model organism studied by researchers.
Pan troglodytes (chimpanzee)	3,100,000,000	25,000	A primate and the closest living relative to humans.
Homo sapiens (human)	3,200,000,000	25,000	A primate.

*In some cases, the values indicate the estimated amount of DNA. For eukaryotic genomes, DNA sequencing is considered completed in the euchromatic regions. The DNA in certain heterochromatic regions is not able to be sequenced, and the total amount is difficult to estimate.

**The numbers of genes were predicted using computer methods described in Chapter 21. These numbers should be considered as estimates of the total gene number.

CONCEPTUAL AND EXPERIMENTAL SUMMARY

The molecular analysis of entire genomes relies on a variety of methods. Cytogenetic mapping uses light microscopy to locate genes or other DNA markers along intact chromosomes. A widely used method to cytogenetically map the locations of genes within large eukaryotic chromosomes is fluorescence *in situ* hybridization (FISH). In this technique, a small fluorescently labeled DNA probe is hybridized to intact chromosomes to detect where a specific DNA sequence is located in the genome. Linkage mapping uses the frequency of genetic recombination between different loci to determine their relative spacing and order along a chromosome. In Chapter 5, we studied linkage mapping that involved crosses between individuals heterozygous for two or more genes. In this chapter, we have considered how a molecular marker, a segment of DNA found at a specific site along a chromosome, can be used as a tool in mapping. It is now common for geneticists to construct maps of entire genomes using molecular markers such as restriction fragment length polymorphisms (RFLPs) and microsatellites. These maps determine the locations of genes by correlating the inheritance of the gene with the inheritance of a particular RFLP or microsatellite. The distance between two linked RFLPs can also be determined, and the likelihood of linkage between them can be evaluated by the lod score method.

Physical mapping relies on the molecular analysis of an organism's genome via DNA cloning methods. Yeast artificial chromosomes (YACs), bacterial artificial chromosomes (BACs), and P1 artificial chromosomes (PACs) have been used as cloning vectors that can accept large DNA inserts, which can be aligned to create a contig, a series of clones that contains contiguous, overlapping pieces of chromosomal DNA. A more refined contig is made by cloning smaller fragments of DNA into cosmid vectors. The methods of constructing a contig are similar in theory to those of positional cloning. A common method used in positional cloning is chromosome walking, in which one gene is first mapped to a region of the chromosome and a clone from that region is used as a starting point in a chromosome walk toward the gene of interest. This is accomplished by sequential hybridization methods. To initiate this type of experiment, a gene's position relative to a marker must be known from mapping studies.

The goal of genome sequencing projects is to determine the DNA sequence of the entire genome of a given species. Shotgun sequencing is the most efficient and inexpensive way to determine the DNA sequence of the entire genome of a given species. Chromosomal DNA is sheared into small fragments that are randomly sequenced, and then computer programs identify overlapping sequences and create one long sequence for each chromosome. In a hierarchical shotgun sequencing approach, large segments of DNA are cloned into BAC contigs, and a physical map is made. In a whole genome shotgun sequencing approach, the entire genome is fragmented into pieces, the ends of which are sequenced. The whole genome shotgun sequencing approach is now the more common method. To date, hundreds of genomes have been sequenced, and the list is increasing steadily. These include the genomes of model organisms, pathogenic microorganisms, and agriculturally important species. By 2006, the Human Genome Project has published completed maps and sequences for all of the human chromosomes. The many benefits of the project, the largest genome project in history, are just beginning to emerge.

PROBLEM SETS & INSIGHTS

Solved Problems

S1. An RFLP marker is located 1 million bp away from a gene of interest. Your goal is to start at this RFLP marker and walk to this gene. The average insert size in the library is 55,000 bp, and the average overlap at each end is 5,000 bp. Approximately how many steps will it take to get there?

Answer: Each step is only 50,000 bp (i.e., 55,000 minus 5,000), because you have to subtract the overlap between adjacent fragments, which is 5,000 bp, from the average inset size. Therefore, it will take about 20 steps to go 1 million bp.

S2. When many RFLPs have been mapped within the genome of a plant, an RFLP analysis can be used to map a herbicide-resistance gene. For example, let's suppose that an agricultural geneticist has two strains, one that is herbicide resistant and one that is herbicide sensitive. The two strains differ with regard to many RFLPs. The sensitive and resistant strains are crossed, and the F_1 offspring are allowed to self-fertilize. The F_2 offspring are then analyzed with regard to their herbicide sensitivity and RFLP markers. The following results are obtained:

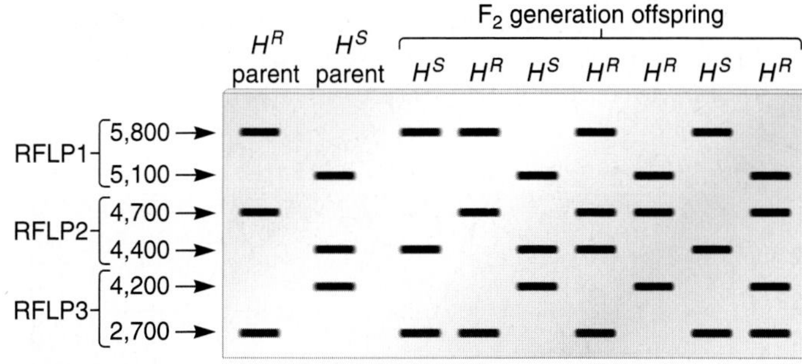

To which RFLP might the herbicide-resistance gene be linked?

Answer: The aim of this experiment is to correlate the presence of a particular RFLP with the herbicide-resistant phenotype. As shown in the data, the herbicide-resistant parent and all the herbicide-resistant offspring have an RFLP that is 4,700 bp in length. In an actual experiment, a thorough lod analysis would be conducted to determine if linkage is considered likely. If so, the 4,700 bp RFLP may either contain the gene that confers herbicide resistance, or, as is more likely, the two may be linked. If the 4,700 bp RFLP has already been mapped to a particular site in the plant's genome, the herbicide-resistance gene also maps to the same site or very close to it. This information may then be used by a plant breeder when making future crosses to produce herbicide-resistant plant strains.

S3. Does a molecular marker have to be polymorphic to be useful in mapping studies? Does a molecular marker have to be polymorphic to be useful in linkage mapping (i.e., involving family pedigree studies or genetic crosses)? Explain why or why not.

Answer: A molecular marker does not have to be polymorphic to be useful in mapping studies. Many sequence-tagged sites (STSs) that are used in physical or cytogenetic studies are monomorphic. Monomorphic markers can provide landmarks in mapping studies.

In linkage mapping studies, a marker must be polymorphic to be useful. Polymorphic molecular markers can be RFLPs, microsatellites, or SNPs. To compute map distances in linkage analysis, individuals must be heterozygous for two or more markers (or genes). For experimental organisms, heterozygotes are testcrossed to homozygotes, and then the number of recombinant offspring and nonrecombinant offspring are determined. For markers that do not assort independently (that is, linked markers), the map distance is computed as the number of recombinant offspring divided by the total number of offspring times 100.

S4. The distance between two molecular markers that are linked along the same chromosome can be determined by analyzing the outcome of crosses. This can be done in humans by analyzing the members of a pedigree. However, the accuracy of linkage mapping in human pedigrees is fairly limited because the number of people in most families is relatively small. As an alternative, researchers can analyze a population of sperm, produced from a single male, and compute linkage distance in this manner. As an example, let's suppose a male is heterozygous for two polymorphic sequence-tagged sites (STSs). STS-1 exists in two sizes: 234 bp and 198 bp. STS-2 also exists in two sizes: 423 bp and 322 bp. A sample of sperm was collected from this man, and individual sperm were placed into 40 separate tubes. In other words, there was one sperm in each tube. Believe it or not, PCR is sensitive enough to allow analysis of DNA in a single sperm! Into each of the 40 tubes were added the primers that amplify STS-1 and STS-2, and then the samples were subjected to PCR. The following results were obtained.

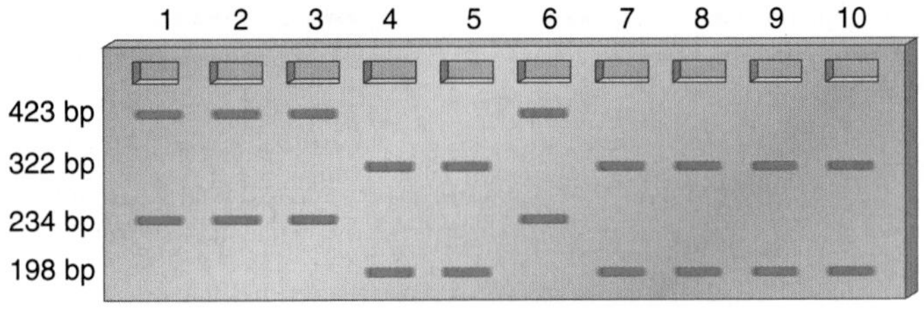

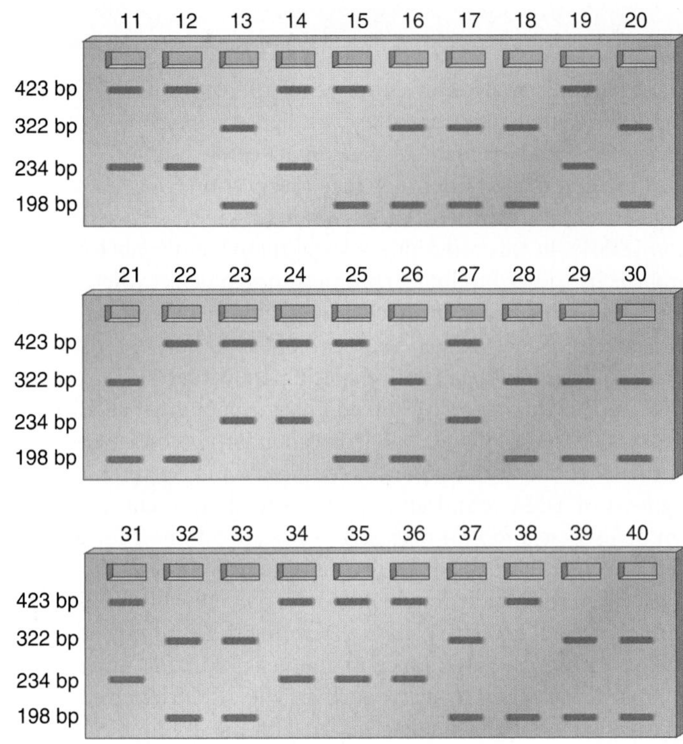

A. What is the arrangement of these two STSs in this individual?

B. What is the map distance between STS-1 and STS-2?

Answer: Keep in mind that mature sperm are haploid, so they will have only one copy of STS-1 and one copy of STS-2.

A. If we look at the 40 lanes, most of the lanes (i.e., 36 of them) have either the 234 bp and 423 bp STSs or the 198 bp and 322 bp. This is the arrangement of STSs in this male. One chromosome has STS-1 that is 234 bp and STS-2 that is 423 bp, and the homologous chromosome has STS-1 that is 198 bp and STS-2 that is 322 bp.

B. There are four recombinant sperm, shown in lanes 15, 22, 25, and 38.

$$\text{Map distance} = \frac{4}{40} \times 100$$
$$= 10.0 \text{ mu}$$

Note: This is a relatively easy experiment compared to a pedigree analysis, which would involve contacting lots of relatives and collecting samples from each of them.

S5. The technique of chromosome walking involves the stepwise cloning of adjacent DNA fragments until the gene of interest is reached. This method was used to clone the gene that (when mutant) causes cystic fibrosis (CF), an autosomal recessive disorder in humans. Linkage mapping studies first indicated that the *CF* gene was located between two markers designated MET and D7S8. Later mapping studies put the *CF* gene between IRP and D7S8. Walking was initiated from the D7S122 marker.

MET D7S122 IRP *CF* gene D7S8

(From Kommens, J. M., et al. 1989. Identification of the cystic fibrosis gene: Chromosome walking and jumping. *Science* 245: 1059–65.)

In this study, the researchers restriction mapped this region using two enzymes, *Not*I and *Xho*I. The locations of sites are shown here:

MET D7S122 IRP *CF* gene D7S8

↑A↑ B ↑C↑ D ↑E↑F↑

N X X X X X N

N = *Not*I, X = *Xho*I

How would the researchers know they were walking toward the *CF* gene and not away from it?

Answer: The general answer is that they had to determine if they were walking toward the IRP marker or toward the MET marker.

As depicted in the previous diagram, *Not*I would produce a large DNA fragment (including regions A–F) that contained both the D7S122 and IRP markers. Because the D7S122 and IRP markers were available as probes, they could confirm that this region contained both markers by Southern blotting of this *Not*I fragment to both markers. *Xho*I cuts this region into smaller pieces. The piece labeled B carried the D7S122 marker, and the piece labeled D carried the IRP marker. Again, this could be confirmed by hybridization. This map enabled the researchers to determine which direction to walk. They began at fragment B (which contained the D7S122 marker) and walked toward the adjacent fragment labeled C. They then walked to the fragment labeled D, which contains the IRP marker. If they had walked toward the fragment labeled A, they would then have walked to the MET marker. This is the incorrect direction, because they knew from their linkage mapping that IRP is closer to the *CF* gene compared to MET.

Conceptual Questions

C1. A person with a rare genetic disease has a sample of her chromosomes subjected to *in situ* hybridization using a probe that is known to recognize band p11 on chromosome 7. Even though her chromosomes look cytologically normal, the probe does not bind to this person's chromosomes. How would you explain these results? How would you use this information to positionally clone the gene that is related to this disease?

C2. For each of the following, decide if it could be appropriately described as a genome:

A. The *E. coli* chromosome

B. Human chromosome 11

C. A complete set of 10 chromosomes in corn

D. A copy of the single-stranded RNA packaged into human immunodeficiency virus (HIV)

C3. Which of the following statements are true about molecular markers?

A. All molecular markers are segments of DNA that carry specific genes.

B. A molecular marker is a segment of DNA that is found at a specific location in a genome.

C. We can follow the transmission of a molecular marker by analyzing the phenotype (i.e., the individual's bodily characteristics) of offspring.

D. We can follow the transmission of molecular markers using molecular techniques such as gel electrophoresis.

E. An STS is a molecular marker.

Experimental Questions

E1. Would the following methods be described as linkage, cytogenetic, or physical mapping?

A. Fluorescence *in situ* hybridization (FISH)

B. Conducting dihybrid crosses to compute map distances

C. Chromosome walking

D. Examination of polytene chromosomes in *Drosophila*

E. Use of RFLPs in crosses

F. Using BACs and cosmids to construct a contig

E2. In an *in situ* hybridization experiment, what is the relationship between the sequence of the probe DNA and the site on the chromosomal DNA where the probe binds?

E3. Describe the technique of *in situ* hybridization. Explain how it can be used to map genes.

E4. The cells from a malignant tumor were subjected to *in situ* hybridization using a probe that recognizes a unique sequence on chromosome 14. The probe was detected only once in each of these cells. Explain these results and speculate on their significance with regard to the malignant characteristics of these cells.

E5. Figure 20.3 describes the technique of FISH. Why is it necessary to "fix" the cells (and the chromosomes inside of them) to the slides? What does it mean to fix them? Why is it necessary to denature the chromosomal DNA?

E6. Explain how the use of DNA probes with different fluorescence emission wavelengths can be used in a single FISH experiment to map the locations of two or more genes. This method is called chromosome painting. Explain why this is an appropriate term.

E7. A researcher is interested in a gene found on human chromosome 21. Describe the expected results of a FISH experiment using a probe that is complementary to this gene. How many spots would you see if the probe was used on a sample from an individual with 46 chromosomes versus an individual with Down syndrome?

E8. What is a contig? Explain how you would determine that two clones in a contig are overlapping.

E9. Contigs are often made using BAC or cosmid vectors. What are the advantages and disadvantages of these two types of vectors? Which type of contig would you make first, a BAC or cosmid contig? Explain.

E10. Describe the molecular features of a BAC cloning vector. What is the primary advantage of a BAC compared to plasmid or viral vectors?

E11. In general terms, what is a polymorphism? Explain the molecular basis for a restriction fragment length polymorphism (RFLP). How is an RFLP detected experimentally? Why are RFLPs useful in physical mapping studies? How can they be used to clone a particular gene?

E12. Five RFLPs designated 1A and 1B, 2A and 2B, 3A and 3B, 4A and 4B, and 5A and 5B, are known to map along chromosome 4 of corn. A plant breeder has obtained a strain of corn that carries a pesticide-resistance gene that (from previous experiments) is known to map somewhere along chromosome 4. The plant breeder crosses this pesticide-resistance strain that is homozygous for RFLPs 1A, 2B, 3A, 4B, and 5A to a pesticide-sensitive strain that is homozygous for 1B, 2A, 3B, 4A, and 5B. The F_1 generation plants were allowed to self-hybridize to produce the following F_2 plants:

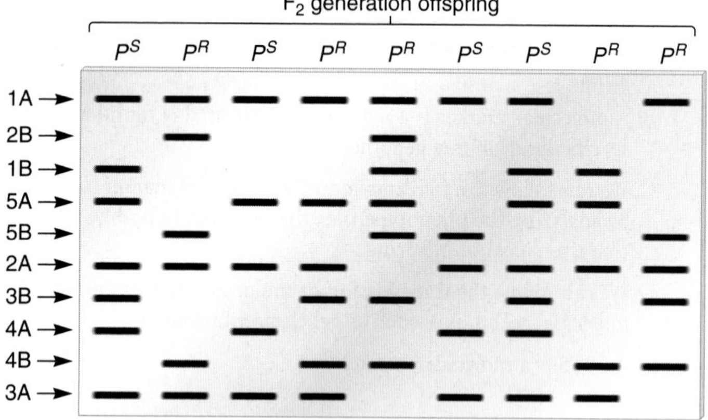

Based on these results, which RFLP does the pesticide-resistance gene map closest to?

E13. Let's suppose there are two different RFLPs in a species. RFLP 1 can be found in 4,500 bp and 5,200 bp lengths; RFLP 2 can be 2,100 bp and 3,200 bp. A homozygote, 4,500 bp and 2,100 bp, was crossed to a homozygote 5,200 bp and 3,200 bp. The F_1 offspring were then crossed to a homozygote containing the 4,500 bp and 2,100 bp fragments. The following results were obtained:

5,200, 4,500, and 2,100	40 offspring
5,200, 4,500, 3,200, 2,100	98 offspring
4,500 and 2,100	97 offspring
4,500, 2,100, and 3,200	37 offspring

Conduct a chi-square analysis to determine if these two RFLPs are linked. If so, calculate the map distance between them.

E14. Explain how a detailed RFLP map of an organism's genome can be helpful in mapping the location of genes. Describe an experimental strategy you would follow to map a gene near a particular RFLP.

E15. A woman has been married to two different men and produced five children. This group is analyzed with regard to three different STSs: STS-1 is 146 and 122 bp; STS-2 is 102 and 88 bp, and STS-3 is 188 and 204 bp. The mother is homozygous for all three STSs: STS-1 = 122, STS-2 = 88, and STS-3 = 188. Father 1 is homozygous for STS-1 = 122 and STS-2 = 102, and heterozygous for STS-3 = 188/204. Father 2 is heterozygous for STS-1 = 122/146, STS-2 = 88/102, and homozygous for STS-3 = 204. The five children have the following results:

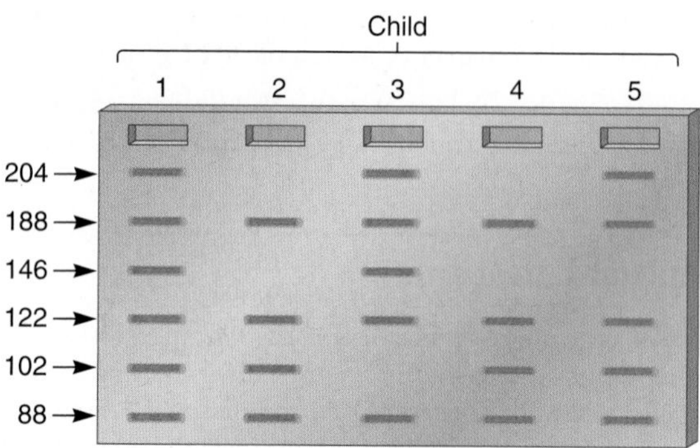

Which children can you definitely assign to father 1 and father 2?

E16. An experimenter used primers to nine different STSs to test their presence along five different BAC clones. The results are shown here.

Alignment of STSs and BACs

	STSs								
	1	2	3	4	5	6	7	8	9
BACs									
1	−	−	−	−	−	+	+	−	+
2	+	−	−	−	+	−	−	+	−
3	−	−	−	+	−	+	−	−	−
4	−	+	+	−	−	−	−	+	−
5	−	−	+	−	−	−	−	−	+

Make a contig map that describes the alignment of the five BACs.

E17. In the Human Genome Project, researchers have collected linkage data from many crosses in which the male was heterozygous for markers and many crosses where the female was heterozygous for markers. The distance between the same two markers, computed in map units or centiMorgans, is different between males and females. In other words, the linkage maps for human males and females are not the same. Propose an explanation for this discrepancy. Do you think the sizes of chromosomes (excluding the Y chromosome) in human males and females are different? How could physical mapping resolve this discrepancy?

E18. Take a look at solved problem S4. Let's suppose a male is heterozygous for two polymorphic sequence-tagged sites. STS-1 exists in two sizes: 211 bp and 289 bp. STS-2 also exists in two sizes: 115 bp and 422 bp. A sample of sperm was collected from this man, and individual sperm were placed into 30 separate tubes.

Into each of the 30 tubes were added the primers that amplify STS-1 and STS-2, and then the samples were subjected to PCR. The following results were obtained:

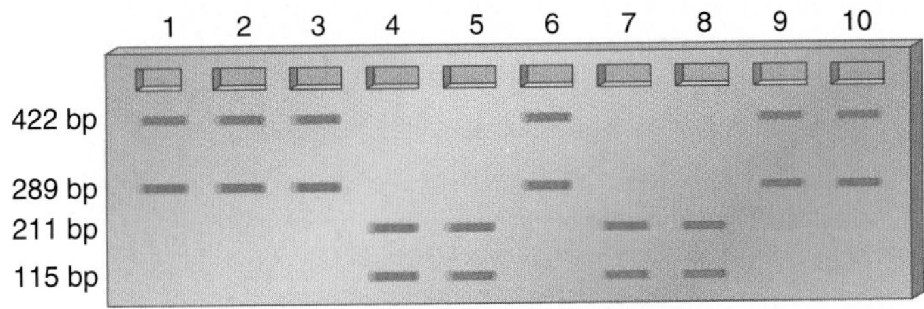

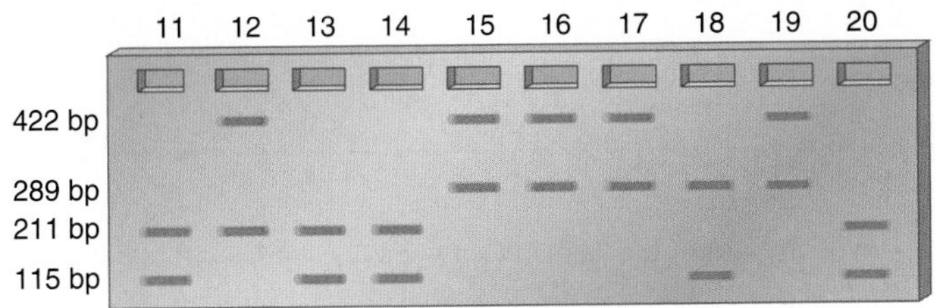

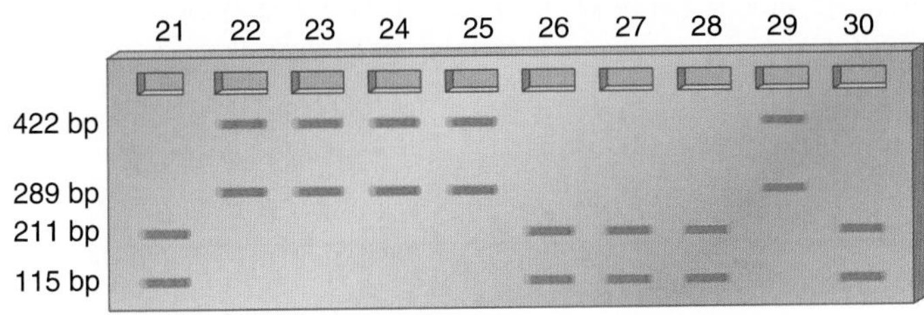

A. What is the arrangement of these two sequence-tagged sites in this individual?

B. What is the linkage distance between STS-1 and STS-2?

C. Could this approach of analyzing a population of sperm be applied to RFLPs?

E19. A gene affecting flower color in petunias is closely linked to an RFLP. A red allele for this gene is associated with a 4,000 bp RFLP, and a purple allele is linked to a 3,400 bp version of this same RFLP. A second gene in petunias affects flower size. An allele causing big flowers is linked to a 7,200 bp RFLP, and a small-flower allele is linked to the same RFLP that is 1,600 bp. A true-breeding strain with small, red flowers was crossed to a true-breeding strain with big, purple flowers. All the F_1 offspring had big, purple flowers. These F_1 offspring were then crossed to true-breeding petunias with small, red flowers. The following results were obtained:

Red, small	725
Red, big	111
Purple, small	109
Purple, big	729

Are these two genes linked to each other? If so, compute the map distance. What would be the expected outcome regarding the inheritance of the RFLPs among the offspring?

E20. An agricultural geneticist has studied a gene in alfalfa that affects pesticide resistance. It exists in three alleles that confer low, medium, and high levels of resistance. This gene has a significant impact on the yield of alfalfa, depending on seasonal variation in pest problems. This geneticist has followed the basic protocol in Figure 20.6, using *Eco*RI as the enzyme to digest the chromosomal DNA. Unfortunately, after tireless efforts and the analysis of thousands of offspring, it has not been possible to identify an RFLP that is associated with the three alleles of this pesticide-resistance gene. What should the geneticist do next? In other words, discuss ways to vary the RFLP method or propose alternative approaches to identify molecular markers that may be linked to the pesticide-resistance gene.

E21. Figure 20.8 describes the transmission of two RFLPs that were linked and 16 mu apart. If these two RFLPs had not been linked and 100 offspring had been analyzed, what would have been the expected results?

E22. Explain why it is necessary to use the technique of Southern blotting for RFLP mapping.

E23. Compared to a conventional plasmid, what additional sequences are required in a YAC vector so that it can behave like an artificial chromosome? Describe the importance of each required sequence.

E24. In the data of Figure 20.10, individuals who were homozygous for the sickle-cell allele had an 87% likelihood of carrying the 13.0 kbp RFLP. Based on this observation, is it likely that the Hb^S allele originated in an individual with a 13.0 kbp RFLP? How would you explain the observation that the Hb^S allele is found with the 7.6 kbp RFLP 13% of the time?

E25. In the experiment of Figure 20.10, explain the conclusion that the 13.0 kbp fragment is more closely linked to Hb^S compared to the 7.0 and 7.6 fragments. Is it always linked? Why or why not? How is this type of information useful?

E26. Explain the technique of pulsed-field gel electrophoresis (PFGE). What special precautions are needed to prevent the mechanical breakage of the chromosomes? What are the uses of PFGE?

E27. When conducting physical mapping studies, place the following methods in their most logical order:

A. Clone large fragments of DNA to make a BAC library.

B. Determine the DNA sequence of subclones from a cosmid library.

C. Subclone BAC fragments to make a cosmid library.

D. Subclone cosmid fragments for DNA sequencing.

E28. Four cosmid clones, which we will call cosmid A, B, C, and D, were subjected to a Southern blot in pairwise combinations. The insert size of each cosmid was also analyzed. The following results were obtained:

Cosmid	Insert Size (bp)	Hybridized to?
A	6,000	C
B	2,200	C, D
C	11,500	A, B, D
D	7,000	B, C

Draw a map that shows the order of the inserts within these four cosmids.

E29. What is an STS (sequence-tagged site)? How are STSs generated experimentally? What are the uses of STSs? Explain how a microsatellite can produce a polymorphic STS.

E30. A human gene, which we will call gene *X*, is located on chromosome 11 and is found as a normal allele and a recessive disease-causing allele. The location of gene *X* has been approximated on the map shown here that contains four STSs, labeled STS-1, STS-2, STS-3, and STS-4.

STS-1 **STS-2** **STS-3** **Gene *X*** **STS-4**

A. Explain the general strategy of positional cloning.

B. If you applied the approach of positional cloning to clone gene *X*, where would you begin? As you progressed in your cloning efforts, how would you know if you were walking toward gene *X* or away from gene *X*?

C. How would you know you had reached gene *X*? (Keep in mind that gene *X* exists as a normal allele and a disease-causing allele.)

E31. Describe how you would clone a gene by positional cloning. Explain how a (previously made) contig would make this task much easier.

E32. A bacterium has a genome size of 4.4 Mb. If a researcher carries out shotgun DNA sequencing and sequences a total of 19 Mb, what is the probability that a base will be left unsequenced? What percentage of the total genome will be left unsequenced?

E33. Discuss the general differences between hierarchical shotgun sequencing versus whole genome shotgun sequencing.

Questions for Student Discussion/Collaboration

1. How is it possible to obtain an RFLP linkage map? What kind of experiments would you conduct to correlate the RFLP linkage map with the positions of known genes that had already been cloned? Discuss the uses of RFLPs in genetic analyses.
2. What is a molecular marker? Give two examples. Discuss why it is easier to locate and map many molecular markers rather than functional genes.
3. Which goals of the Human Genome Project do you think are the most important? Why? Discuss the types of ethical problems that might arise as a result of identifying all of our genes.

Note: All answers appear at the website for this textbook; the answers to even-numbered questions are in the back of the textbook.

www.mhhe.com/brookergenetics3e

Visit the Online Learning Center for practice tests, answer keys, and other learning aids for this chapter. Enhance your understanding of genetics with our interactive exercises, quizzes, animations, and much more.

GENOMICS II: FUNCTIONAL GENOMICS, PROTEOMICS, AND BIOINFORMATICS

21

CHAPTER OUTLINE

In Chapter 20, we learned that genomics involves the mapping of an entire genome and, eventually, the determination of a species' complete DNA sequence. The amount of information found within a species' genome is enormous. The goal of **functional genomics** is to elucidate the roles of genetic sequences—DNA, RNA, and amino acid sequences—in a given species. In most cases, functional genomics is aimed at an understanding of gene function. At the genomic level, researchers can study genes as groups. For example, the information gained from a genome-sequencing project can help researchers study entire metabolic pathways. This provides a description of the ways in which gene products interact to carry out cellular processes. In addition, a study of genetic sequences can help to identify regions that play particular functional roles. For example, an analysis of certain species of bacteria helped to identify DNA sequences that promote the uptake of DNA during bacterial transformation.

Because most genes encode proteins, a goal of many molecular biologists is to understand the functional roles of all the proteins a species produces. The entire collection of proteins a given cell or organism can make is called its **proteome,** and the study of the function and interactions of these proteins is termed **proteomics.** An objective of researchers in the field of proteomics is to understand the interplay among many proteins as they function to create cells and, ultimately, the traits of a given species.

From a research perspective, functional genomics and proteomics can be broadly categorized in two ways: experimental and computational. The experimental approach involves the study of groups of genes or proteins using molecular techniques in the laboratory. In the first two sections in this chapter, we will focus on these techniques. By comparison, the computational strategy attempts to analyze genetic sequences using a mathematical approach. This area, which is called **bioinformatics,** has become an important branch of science. The tools of bioinformatics are computers, computer programs, and genetic sequences. In the last section of this chapter, we will consider the field of bioinformatics and discover how this area has provided great insights into the subjects of functional genomics and proteomics.

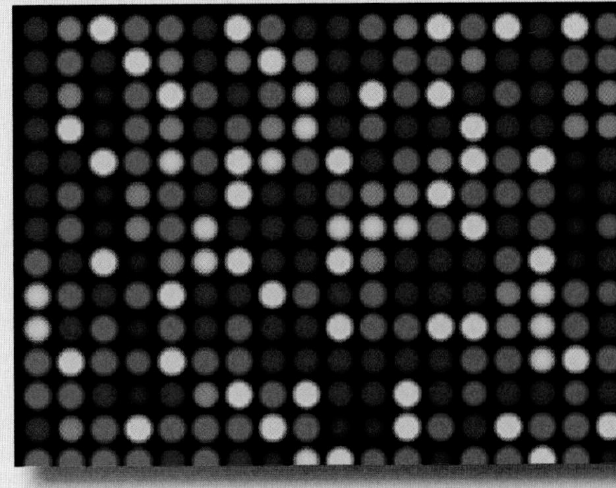

A DNA microarray. Each spot on the array corresponds to a specific gene. The color of the spots, which occurs via computer imaging techniques, indicates the amount of RNA transcribed from that gene. A DNA microarray allows researchers to simultaneously analyze the expression of many genes.

21.1 FUNCTIONAL GENOMICS

Though the rapid sequencing of genomes, particularly the human genome, has generated great excitement in the field of genetics, many would argue that an understanding of genomic function is fundamentally more interesting. In the past, our ability to study genes involved many of the techniques described in Chapter 18, such as gene cloning, Northern blotting, gel retardation assays, and site-directed mutagenesis. These approaches continue to provide a solid foundation for our understanding of gene function. More recently, genome-sequencing projects have enabled researchers to consider gene function at a more complex level. We now can analyze groups of many genes simultaneously to determine how they work as integrated units that produce the characteristics of cells and the traits of complete organisms. In this section, we will explore how a cDNA library is created and used to identify regions of DNA that contain genes. We also will examine the use of DNA microarrays, which enable researchers to monitor the expression of thousands of genes simultaneously.

Expressed Genes Can Be Identified in a cDNA Library

Chromosomal DNA contains regions where genes are found, as well as intergenic regions that do not contain genes. Therefore, a basic goal of genomic research is to definitively identify the regions of DNA that are actually genes. To do so, one approach is to show that a given region is transcribed into RNA. This can be accomplished by the generation of a **cDNA library.** As described in Chapter 18, cDNA—complementary DNA—is made using RNA as a template, which is reversely transcribed into DNA with reverse transcriptase. A cDNA library used in physical mapping studies is also called an **expressed sequence tag (EST) library,** because the sequences, which are expressed as mRNA, can also be used as molecular markers or tags. The members of an EST library are subjected to DNA sequencing, and EST sequences are then compared to a complete genome sequence. A particular EST sequence will match a genomic sequence, indicating that the genomic region encodes a gene. From the perspective of gene function, a cDNA or EST library provides a reliable identification of a region that is truly transcribed. The mere fact that a region is transcribed indicates that the corresponding genomic region is a gene.

The method of making a cDNA library can be extended to study gene regulation at the genomic level. One strategy is to isolate mRNA under different conditions and then identify particular mRNAs that are expressed only under a specific set of conditions. For example, in the experiment shown in **Figure 21.1**, samples of cells were incubated with and without a particular hormone. The goal is to identify those genes that are turned on by the presence of the hormone. The mRNAs from these cells were isolated and used to make two different groups of cDNAs. The cDNAs derived from cells not exposed to hormone were denatured and tightly bound to a column. (See the Appendix for a description of column chromatography.) The cDNAs derived from the cells exposed to hormone were also denatured and then run over the column. Any of these latter cDNAs will remain bound to the column if they match a cDNA that is already tightly bound to the column. In contrast, cDNAs derived only from the hormone-induced cells will not bind to the column because they do not have a complementary match that is tightly bound to the column. These cDNAs are eluted and cloned into vectors to

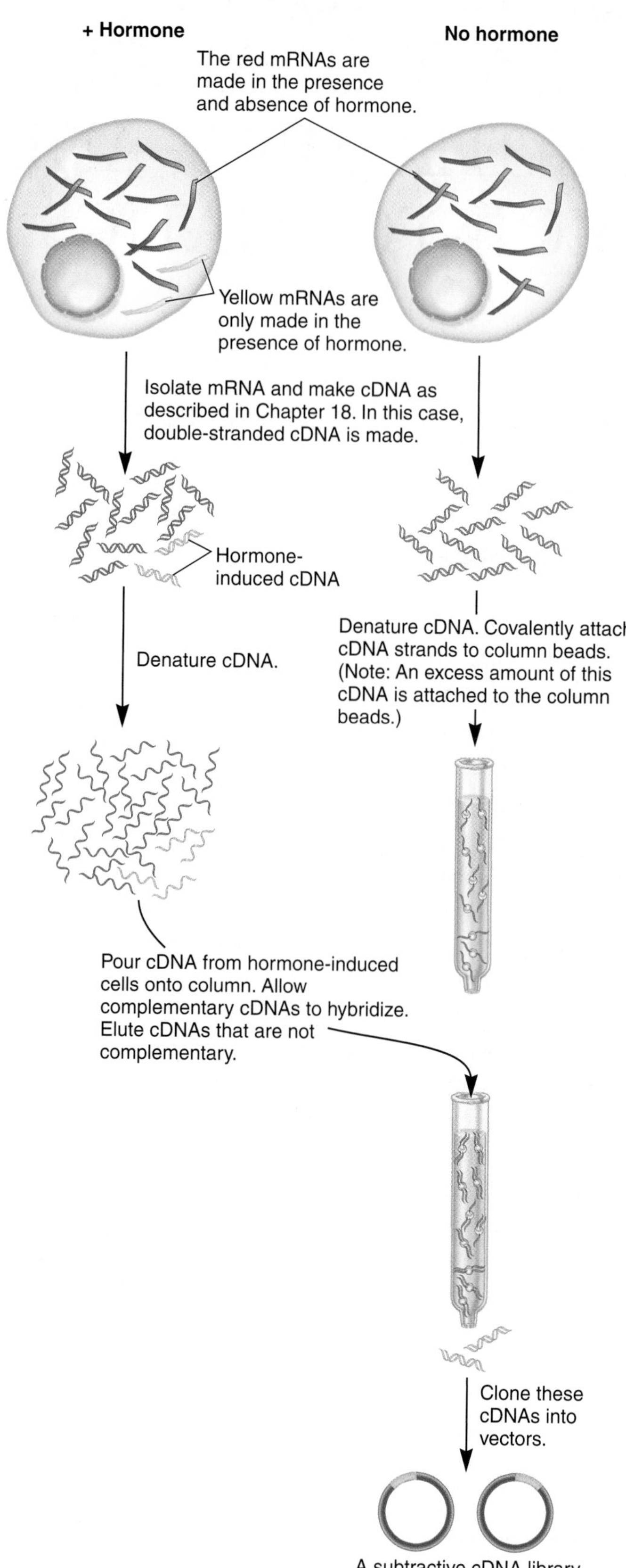
+ Hormone
No hormone
The red mRNAs are made in the presence and absence of hormone.
Yellow mRNAs are only made in the presence of hormone.
Isolate mRNA and make cDNA as described in Chapter 18. In this case, double-stranded cDNA is made.
Hormone-induced cDNA
Denature cDNA.
Denature cDNA. Covalently attach cDNA strands to column beads. (Note: An excess amount of this cDNA is attached to the column beads.)
Pour cDNA from hormone-induced cells onto column. Allow complementary cDNAs to hybridize. Elute cDNAs that are not complementary.
Clone these cDNAs into vectors.
A subtractive cDNA library

FIGURE 21.1 **Creation of a subtractive cDNA library.**

create a cDNA library termed a **subtractive cDNA library.** This approach is also called **subtractive hybridization.** The members of the subtractive cDNA library can be subjected to DNA sequencing to identify the genes that are turned on by the presence of the hormone.

A Microarray Can Identify Genes That Are Transcribed

Researchers have developed an exciting new technology called a **DNA microarray** (also called a **gene chip**), that makes it possible to monitor the expression of thousands of genes simultaneously. A DNA microarray is a small silica, glass, or plastic slide that is dotted with many different sequences of DNA, each corresponding to a short sequence within a known gene. For example, one spot in a microarray may correspond to a sequence within the β-globin gene, while another spot could correspond to a gene that encodes an iron transporter. A single slide may contain tens of thousands of different spots in an area the size of a postage stamp. The relative location of each spot is known.

How are microarrays made? Some are produced by spotting different samples of DNA onto a slide, much like the way an ink-jet printer works. Different DNA fragments, which are made synthetically (for example, by PCR), are individually spotted onto the slide. The DNA fragments are typically 500 to 5,000 bp in length, and a few thousand to tens of thousands are spotted to make a single array. Alternatively, other microarrays contain shorter DNA segments—oligonucleotides—that are directly synthesized on the surface of the slide. In this case, the DNA sequence at a given spot is produced by selectively controlling the growth of the oligonucleotide using narrow beams of light. Such oligonucleotides are typically 25 to 30 nucleotides in length. Hundreds of thousands of different spots can be found on a single array. Overall, the technology of making DNA microarrays is quite amazing.

Once a DNA microarray has been made, it is used as a hybridization tool, as shown in **Figure 21.2**. In this experiment, RNA was isolated from a sample of cells and then used to make fluorescently labeled cDNA. The labeled cDNAs were then layered onto a DNA microarray. Those cDNAs that are complementary to the DNAs in the microarray will hybridize and thereby remain bound to the microarray. The array is then washed with a buffer to remove any unbound cDNAs and placed in a type of a microscopic device called a laser scanner, which produces higher resolution images than a conventional optical microscope. The device scans each pixel—the smallest element in a visual image—and after correction for local background, the final fluorescence intensity for each spot is obtained by averaging across the pixels in each spot. This results in a group of fluorescent spots at defined locations in the microarray. If the fluorescence intensity in a spot is high, this result means that a large amount of cDNA in the sample hybridized to the DNA at this location. Because the DNA sequence of each spot is already known, a fluorescent spot identifies cDNAs that are complementary to those DNA sequences. Furthermore, because the cDNA was generated from mRNA, this technique identifies RNAs that have been made in a particular cell type under a given set of conditions.

The technology of DNA microarrays has found many important uses (**Table 21.1**). Thus far, its most common use is to study gene expression patterns. This helps us to understand how genes are regulated in a cell-specific manner and how environmental conditions can induce or repress the transcription of

TABLE 21.1

Applications of DNA Microarrays

Application	Description
Cell-specific gene expression	A comparison of microarray data using cDNAs derived from RNA of different cell types can identify genes that are expressed in a cell-specific manner.
Gene regulation	Environmental conditions play an important role in gene regulation. A comparison of microarray data may reveal genes that are induced under one set of conditions and repressed under another set of conditions.
Elucidation of metabolic pathways	Genes that encode proteins that participate in a common metabolic pathway are oftentimes expressed in a parallel manner. This can be revealed from a microarray analysis (as described in Figure 21.3). This application overlaps with the study of gene regulation via microarrays.
Tumor profiling	Different types of cancer cells exhibit striking differences in their profiles of gene expression. This can be revealed by a DNA microarray analysis. This approach is gaining widespread use to subclassify tumors that are sometimes morphologically indistinguishable. Tumor profiling may provide information that can improve a patient's clinical treatment.
Genetic variation	A mutant allele may not hybridize to a spot on a microarray as well as a wild-type allele. Therefore, microarrays are gaining widespread use as a tool to detect genetic variation. They have been used to identify disease-causing alleles in humans and mutations that contribute to quantitative traits in plants and other species. In addition, microarrays are used to detect chromosomal deletions and duplications. As discussed in Chapter 8, a method called comparative genomic hybridization (CGH) is used to determine if cancer cells carry deletions or duplications. In the traditional method, equal amounts of two fluorescently labeled DNA samples are denatured, mixed together, and applied to intact metaphase chromosomes in which the DNA has also been denatured. In a newer approach, called array comparative genomic hybridization (aCGH), the fluorescently labeled DNAs are applied to a microarray that has short segments of DNA from all of the chromosomes. This method is more sensitive in the detection of deletions and duplications.
Microbial strain identification	Microarrays can distinguish between closely related bacterial species and subspecies.
DNA-protein binding	Chromatin immunoprecipitation, which is described in Figure 21.4, can be used with DNA microarrays to determine where in the genome a particular protein binds to the DNA.

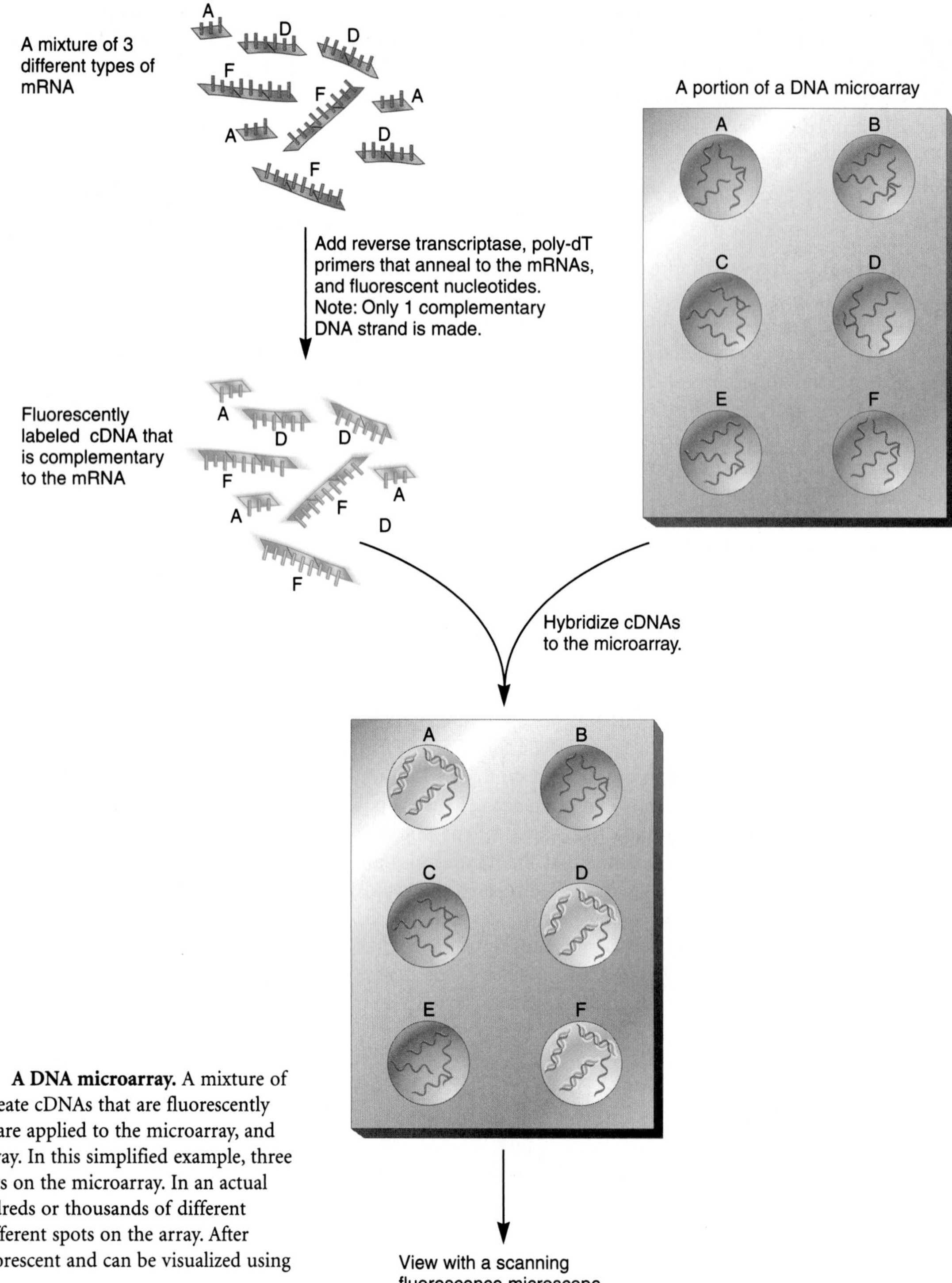
A mixture of 3 different types of mRNA
A portion of a DNA microarray
Add reverse transcriptase, poly-dT primers that anneal to the mRNAs, and fluorescent nucleotides.
Note: Only 1 complementary DNA strand is made.
Fluorescently labeled cDNA that is complementary to the mRNA
Hybridize cDNAs to the microarray.
View with a scanning fluorescence microscope.

ONLINE ANIMATION

FIGURE 21.2 A DNA microarray. A mixture of mRNAs is used to create cDNAs that are fluorescently labeled. The cDNAs are applied to the microarray, and any unbound cDNAs are washed away. In this simplified example, three cDNAs specifically hybridize to spots on the microarray. In an actual experiment, there are typically hundreds or thousands of different cDNAs and tens of thousands of different spots on the array. After hybridization, the spots become fluorescent and can be visualized using a laser scanner.

genes. In some cases, microarrays can even help to elucidate the genes encoding proteins that participate in a complicated metabolic pathway. Microarrays can also be used as identification tools. For example, gene expression patterns can aid in the categorization of tumor types. Such identification can be important in the treatment of the disease. Instead of using labeled cDNA, researchers can also hybridize labeled genomic DNA to a microarray. This can be used to identify mutant alleles in a population of individuals and to detect deletions and duplications. In addition, this technology is proving useful in the correct identification of closely related bacterial species and subspecies. Finally, microarrays can be used to study DNA–protein interactions as described later in this section.

EXPERIMENT 21A

The Coordinate Regulation of Many Genes Is Revealed by a DNA Microarray Analysis

One way that cells respond to environmental changes is via the coordinate regulation of genes. Under one set of environmental conditions, a particular set of genes may be induced, while under another set of conditions, those same genes may be repressed. In the past, researchers have been able to study this type of gene regulation using tools that can analyze the expression of a few genes at a time. The advent of microarrays, however, has made it possible to study the expression of the whole genome under different sets of environmental conditions.

One of the first studies using this approach involved the analysis of the yeast genome. The genome of baker's yeast, *Saccharomyces cerevisiae*, was the first eukaryotic genome to be sequenced. It encodes approximately 6,300 different genes. An important process in the growth of yeast cells, as well as cells of other species, is the ability to metabolize carbon sources using different metabolic pathways. When yeast cells have glucose available, they metabolize the glucose to smaller products in a process called glycolysis. If oxygen is present, these products can be broken down via the tricarboxylic acid cycle (TCA or citric acid cycle) that occurs in the mitochondria. Therefore, when yeast are first given glucose and then allowed to metabolize it in the presence of oxygen, they first metabolize the carbohydrate via glycolysis, and then, when the glucose is used up, they metabolize the products of glycolysis via the TCA cycle. The process of switching from glycolysis to the TCA cycle, called a diauxic shift, involves major changes in the expression of genes involved with carbohydrate metabolism. The goal of the experiment described in **Figure 21.3** was to identify genes that are induced and repressed as yeast cells shift from glycolysis to the TCA cycle. It was carried out by Joseph DeRisi, Vishwanath Iyer, and Patrick Brown in 1997.

As shown in Figure 21.3, yeast cells were initially given glucose as their carbon source for growth and then allowed to grow for several hours. Over time, the glucose was used up, and the cells shifted from glycolysis to the TCA cycle. At various time points, samples of cells were removed, and the RNA was isolated. The RNA was then exposed to reverse transcriptase, poly-dT primers, and deoxyribonucleotides, one of which was fluorescently labeled. This created fluorescently labeled cDNAs.

To determine the relative changes in RNA synthesis, two different fluorescent dyes were used. The RNA collected at the first time point—when glucose was at a high level—was used to make cDNA that contained a green fluorescent dye. The RNA collected at later time points was used to make cDNAs that contained a red fluorescent dye. A sample of green cDNA (from the first time point) was then mixed with a sample of red cDNA (from later time points), and the mixture was hybridized to a microarray containing about 6,200 yeast genes. A laser scanner was used to measure the amount of red fluorescence and green fluorescence at each spot in the microarray. The fluorescence ratio (red fluorescent units divided by green fluorescent units) provided a way to quantitatively determine how the expression of the genes was changing. For example, if the ratio was high, this means that a gene was being induced as glucose levels fall, because the amount of red cDNA would be higher than the amount of green cDNA. Alternatively, if the red to green ratio was low, this means that a gene was being repressed as glucose is used up.

THE GOAL

A diauxic switch from glycolysis to the TCA cycle will involve the induction of certain genes and the repression of other genes. The goal was to identify such groups of genes.

ACHIEVING THE GOAL — FIGURE 21.3 **The use of a DNA microarray to study carbohydrate metabolism in yeast.**

Starting material: A commonly used strain of *Saccharomyces cerevisiae* (baker's yeast) was used in this study. The researchers had made a DNA microarray containing (nearly) all of the known yeast genes.

1. Inoculate yeast into a media containing 2% glucose. Grow for up to 21 hours.

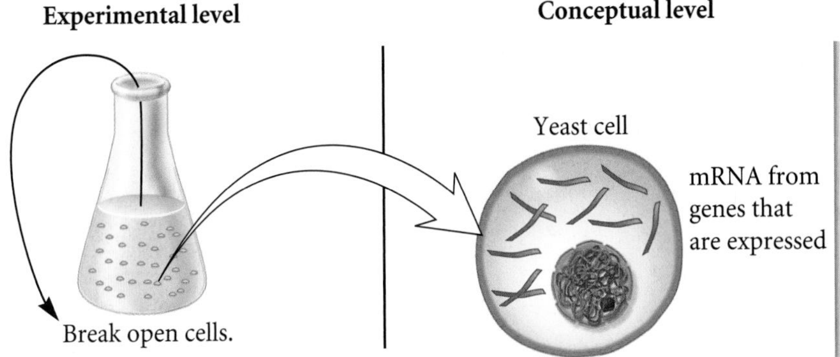

2. Beginning at 9 hours after inoculation, take out samples of cells and isolate mRNA. This involves breaking open cells and running the cell contents over a poly-dT column under high-salt conditions. Since mRNA has a polyA tail, it will bind to this column while other cell components flow through. The purified mRNA can be then eluted by adding a buffer to the top of the column that contains a low concentration of salt. This breaks the interaction between the poly-dT and polyA tails. In this experiment, mRNA samples were isolated at approximately 2-hour intervals, beginning at 9 hours and ending at 21 hours.

3. Add reverse transcriptase, poly-dT primers, and fluorescently labeled nucleotides to make complementary strands of fluorescently labeled cDNA. Note: The sample at 9 hours was used to make green cDNA, while mRNA samples collected at later time points were used to make red cDNA.

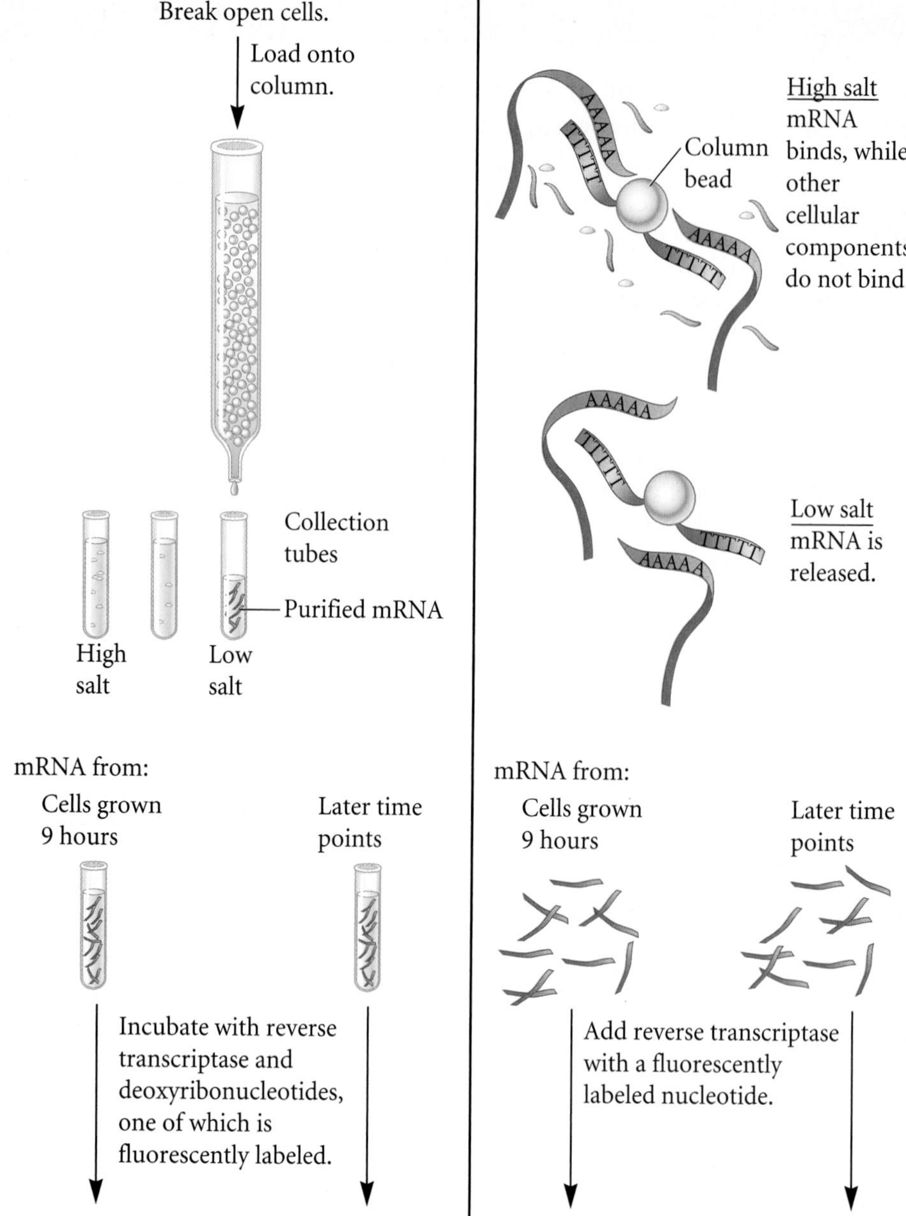
Break open cells.
Load onto column.
Collection tubes
Purified mRNA
High salt
Low salt
AAAAA
TTTTT
Column bead
High salt
mRNA binds, while other cellular components do not bind.
Low salt
mRNA is released.
mRNA from:
Cells grown 9 hours
Later time points
Incubate with reverse transcriptase and deoxyribonucleotides, one of which is fluorescently labeled.
mRNA from:
Cells grown 9 hours
Later time points
Add reverse transcriptase with a fluorescently labeled nucleotide.

(continued)

4. At each time point, mix together the cDNA derived from that time (i.e., red cDNA) with cDNA derived from the 9-hour time point (i.e., green cDNA).

5. Hybridize the mixture to the yeast DNA microarray as described in Figure 21.2.

6. Examine the DNA microarray with a laser scanner. The data are then analyzed by a computer, which can correlate expression levels among different genes. (This is a cluster analysis.)

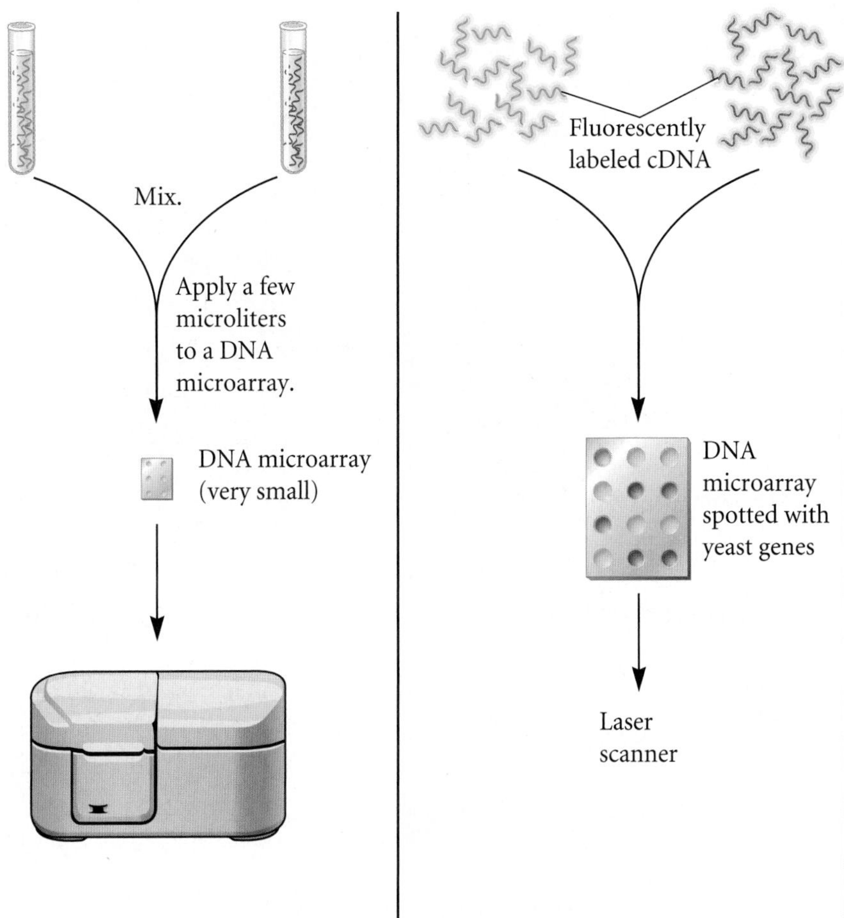

THE DATA*

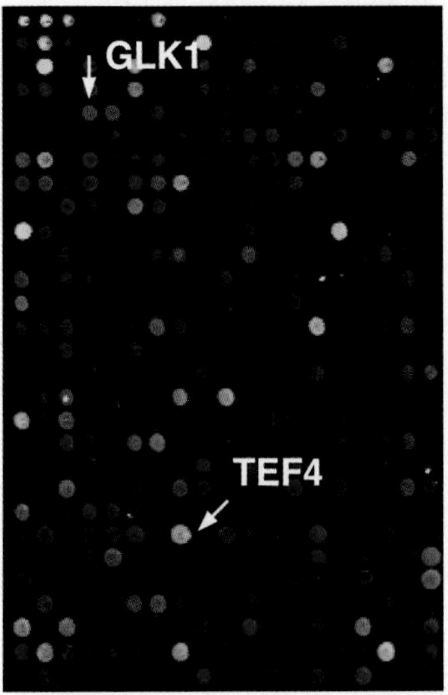

* Green spots are genes expressed early in growth, while red spots are expressed later. Yellow spots are expressed more evenly. Spots that are barely visible indicate genes that are not substantially expressed under these growth conditions.

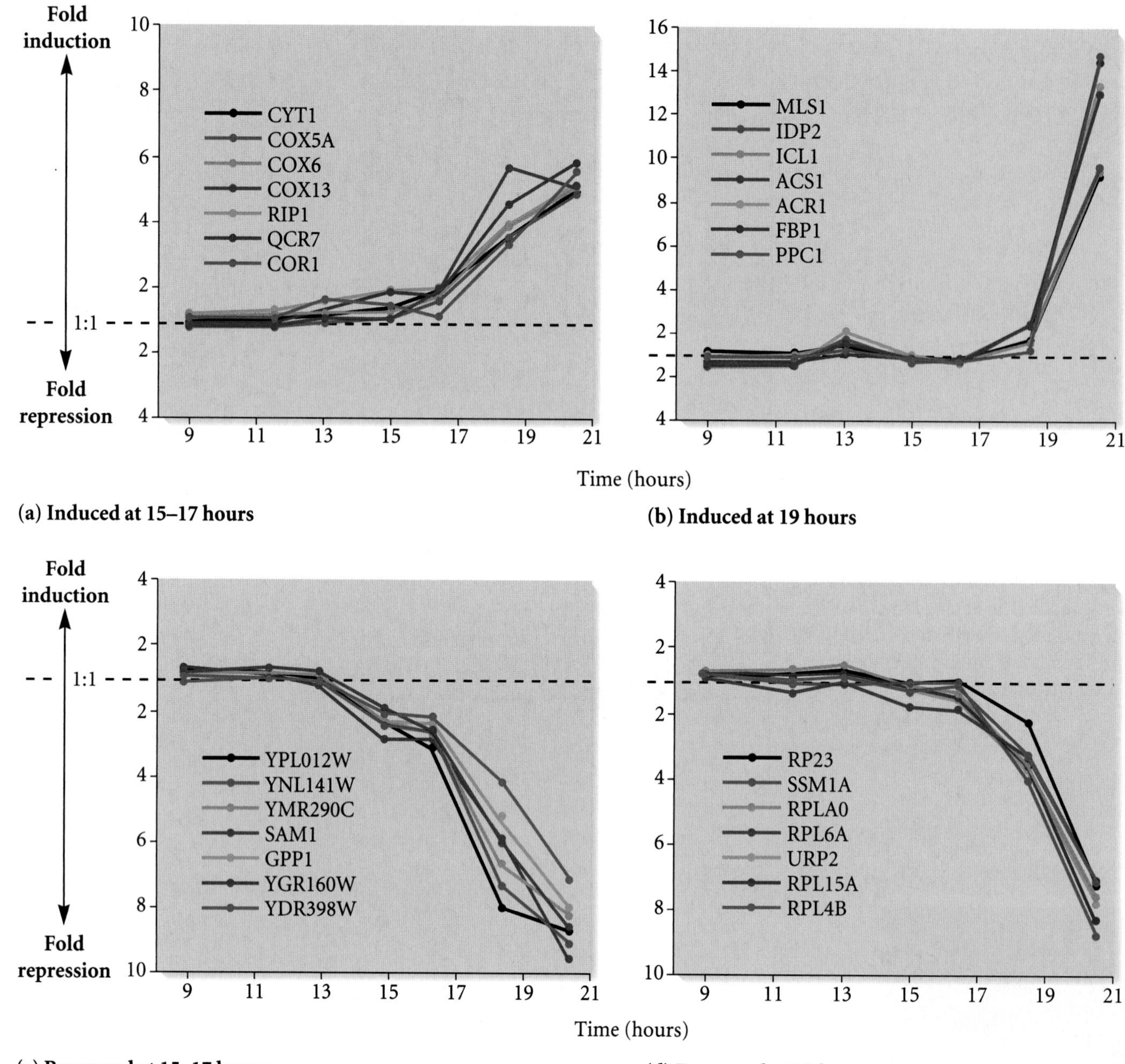

(a) Induced at 15–17 hours

(b) Induced at 19 hours

(c) Repressed at 15–17 hours

(d) Repressed at 19 hours

INTERPRETING THE DATA

A portion of one microarray is shown at the beginning of the data of Figure 21.3. As seen here, the array shows many spots, some of which are relatively green and some of which are relatively red. The green spots indicate genes that are expressed at higher levels during early stages of growth when glucose levels are high. An example is a gene designated *TEF4*, which is involved with protein synthesis. Red spots identify genes that are expressed when the glucose is depleted. The gene *GLK1* encodes an enzyme that phosphorylates glucose but is expressed only when glucose levels are low. In addition, many spots are not fluorescent, indicating there was not much cDNA in the sample to hybridize to the DNA strands in those locations. These spots would correspond to genes that are not greatly expressed under either condition. Judging from the numbers of red and green spots, the shift from glycolysis to the TCA cycle involved a great amount of gene induction and repression. By determining the red:green ratio at each spot, it was found that 710 genes were induced by at least a factor of two, while 1,030 genes were repressed by a factor of two or more. Therefore, it was found that a diauxic shift involves a staggering amount of gene regulation. Of the 6,200 yeast genes, 1,740, or roughly 28%, appeared to be regulated as a result of a diauxic shift.

Because the gene sequence in each spot was known, the next step was to relate the levels of gene expression based on the microarray data to specific genes. A common goal is to identify genes whose pattern of expression seems to strongly correlate with each other. This statistical technique is called a **cluster analysis.** The graphs shown in the data of Figure 21.3 illustrate how microarray data can be used to make this type of comparison. Though the sequences of these genes were known from the yeast genome-sequencing project, the functions of some of these genes were not known. Each time point involved the measurement of the red:green ratio at a particular spot. Figure 21.3 shows the analysis of 28 genes (seven genes in each panel) over the course of 9 to 21 hours following the addition of glucose. The diauxic shift occurred at approximately 15 hours.

Let's examine the data shown in Figure 21.3. In part (a), the genes were induced at 15 to 17 hours of growth. Due to their coordinate regulation, these genes may be controlled by the same transcription factor(s) and may participate in a common metabolic response to the induction of the TCA cycle. In fact, several of the genes in part (a) have already been studied, and they are known to play a role in the ability of the mitochondria to run the TCA cycle. This makes sense because the diauxic shift occurred around 15 hours of growth, which correlates with the time when the genes were induced. By comparison, the genes in part (b) were induced later, after the TCA cycle had operated for a few hours and when the carbon sources were becoming depleted. Therefore, it would seem that these genes were induced as a response to the operation of the TCA cycle or they were induced because the carbon sources in the media were low. Why is this information useful? Though further work needs to be done to elucidate the functions of some of the genes shown in parts (a) and (b), the data suggest a common regulation and metabolic function for particular groups of genes. It would seem more likely that the transcription factor(s) that regulate the genes in part (a) would be different from those that regulate the genes in part (b). Likewise, the proteins encoded by these genes may work in different cellular pathways.

The genes shown in parts (c) and (d) illustrate a similar phenomenon, except that the switch to the TCA cycle and the operation of the TCA cycle represses those genes. The genes shown in part (c) were active during glycolysis and then were repressed at the time of the diauxic shift, while those shown in part (d) were repressed after the TCA cycle had operated for a few hours. Most of the genes shown in part (d) were already known to function as ribosomal proteins. Therefore, as the carbon sources in the media were depleted, these results suggest that one of the cellular responses is to diminish the synthesis of ribosomes, which, in turn, would slow down the rate of protein synthesis. It would seem that the yeast cells were trying to conserve energy at this late stage of growth.

Overall, the data shown in the experiment of Figure 21.3 illustrate how a microarray analysis can shed light on gene function at the genomic level. It provides great insight regarding gene regulation and may help to identify groups of proteins (i.e., clusters) that share a common cellular function.

A self-help quiz involving this experiment can be found at the Online Learning Center.

DNA Microarrays Can Be Used to Identify DNA-Protein Binding at the Genome Level

As discussed throughout this textbook, the binding of proteins to specific DNA sites plays a critical role in a variety of molecular processes, including gene transcription and DNA replication. To study these processes, researchers have devised a variety of techniques to identify whether or not specific proteins bind to particular sites in the DNA. For example, the techniques of gel retardation and DNaseI footprinting, which are described in Chapter 18, are used for this purpose (see Figures 18.12 and 18.13).

More recently, a newer approach called **chromatin immunoprecipitation (ChIP)** has gained widespread use in the analysis of DNA–protein interactions. This method can determine whether proteins can bind to a particular region of DNA. A distinguishing feature of this method is that it analyzes DNA-protein interactions as they occur in the chromatin of living cells. In contrast, gel retardation and DNaseI footprinting are in vitro techniques, which typically use cloned DNA and purified proteins.

Figure 21.4 describes the steps of the chromatin immunoprecipitation protocol. Proteins in living cells, which are noncovalently bound to DNA, can be more tightly attached to the chromatin by the addition of formaldehyde or some other chemical that covalently cross-links the protein to the DNA. Following cross-linking, the cells are lysed, and the DNA is broken into pieces of approximately 200 to 1,000 bp by sonication. Next, an antibody is added that is specific for the protein of interest. To conduct a ChIP assay, a researcher must already suspect that a particular protein binds to the DNA and has previously had an antibody made against that protein. As discussed in Chapter 17, antibodies are molecules that specifically bind to antigens. In this case, the antigen is the protein of interest that is thought to be a DNA-binding protein. In Figure 21.4, the binding of antibodies to the DNA–protein complexes causes the complexes to aggregate and precipitate out of solution. Because an antibody is made by the immune system of an animal, this step is called immunoprecipitation. The sample is then subjected to centrifugation to collect the DNA–protein complexes in the pellet. (See the Appendix for a description of centrifugation.)

The next step is to identify the DNA to which the protein is covalently cross-linked. To do so, the protein is removed by treatment with chemicals that break the covalent cross-links. Because the DNA is usually present in very low amounts, researchers must amplify the DNA to analyze it. This is done using PCR, which is described in Chapter 18. If researchers already suspect that a protein binds to a known DNA region, they can use PCR primers that specifically flank the DNA region (see bottom left side of Figure 21.4).

If they get a PCR product, this means that the protein of interest must have been bound (either directly or through other proteins) to this DNA site in living cells.

Alternatively, a researcher may want to determine where the protein of interest binds across the whole genome. In this case, a DNA microarray can be used (see bottom right side of Figure 21.4). Because a DNA microarray is found on a chip, this is called a **ChIP-on-chip assay.** The ends of the precipitated DNA are first ligated to short DNA pieces called linkers. PCR primers are then added that are complementary to the linkers and, therefore, amplify the DNA regions between the linkers. During PCR, the DNA is fluorescently labeled. The labeled DNA is then hybridized to a DNA microarray. Because the DNA was isolated using an antibody against the protein of interest, the fluorescent spots on the microarray identify sites in the genome where the protein binds. In this way, researchers may be able to determine where a protein binds to locations in the genome, even if those site(s) had not been previously determined by other methods.

21.2 PROTEOMICS

Thus far, we have considered ways to characterize the genome of a given species and study its function. Because most genes encode proteins, a logical next step is to examine the functional roles of the proteins that a species can make. As mentioned, this field is called proteomics, and the entire collection of a species' proteins is its proteome.

Genomics represents only the first step in our comprehensive understanding of protein structure and function. Researchers often use genomic information to initiate proteomic studies, but such information must be followed up with research that involves the direct analysis of proteins. For example, as discussed in the preceding section, a DNA microarray may provide insights into the transcription of particular genes under a given set of conditions. However, mRNA levels may not provide an accurate measure of the abundance of a protein that is encoded by a given gene. Protein levels are greatly affected, not only by the level of mRNA, but also by the rate of mRNA translation and by the turnover rate of a given protein. Therefore, DNA microarray data must be corroborated using other methods, such as Western blotting (discussed in Chapter 18), which directly determine the abundance of a protein in a given cell type.

A second way that genomic data can elucidate the workings of the proteome is via homology. As discussed later in this chapter, homology between the genes of different species can be used to predict protein structure and/or function, when the structure or function of the protein in one of the species is already known. However, homology may not provide direct information regarding the regulation of protein structure and function. Also, it may not reveal potential types of protein–protein interactions in which a given protein may participate. Therefore, even though

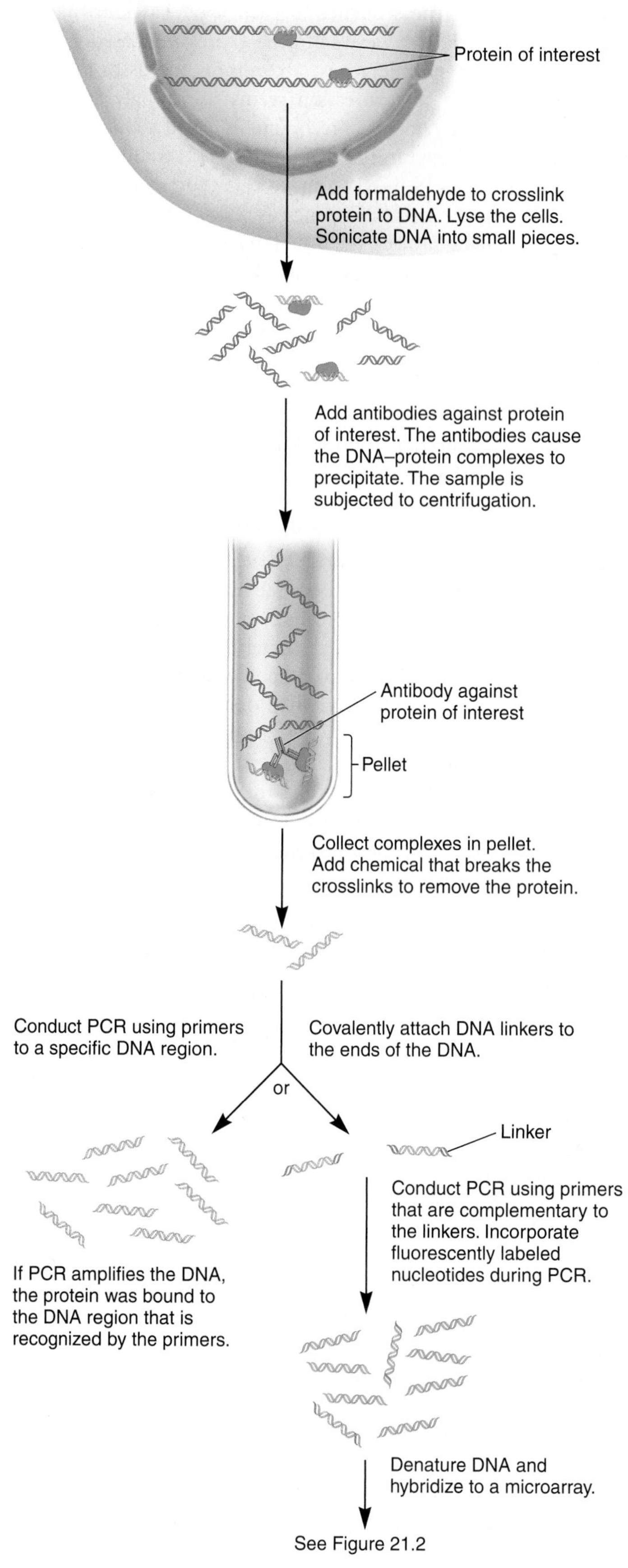
Protein of interest
Add formaldehyde to crosslink protein to DNA. Lyse the cells. Sonicate DNA into small pieces.
Add antibodies against protein of interest. The antibodies cause the DNA–protein complexes to precipitate. The sample is subjected to centrifugation.
Antibody against protein of interest
Pellet
Collect complexes in pellet. Add chemical that breaks the crosslinks to remove the protein.
Conduct PCR using primers to a specific DNA region.
or
Covalently attach DNA linkers to the ends of the DNA.
Linker
If PCR amplifies the DNA, the protein was bound to the DNA region that is recognized by the primers.
Conduct PCR using primers that are complementary to the linkers. Incorporate fluorescently labeled nucleotides during PCR.
Denature DNA and hybridize to a microarray.
See Figure 21.2

FIGURE 21.4 Chromatin immunoprecipitation (ChIP). This method can determine whether proteins bind to a particular region of DNA found within the chromatin of living cells.

homology is profoundly useful as a guiding tool, a full understanding of protein function must involve the direct analysis of proteins as they are found in living cells.

As we move into the twenty-first century, a key challenge facing molecular biologists is the study of proteomes. Much like genomic research, this will require the collective contributions of many research scientists, as well as improvements in technologies that are aimed at unraveling the complexities of the proteome. In this section, we will discuss the phenomena that increase protein diversity beyond genetic diversity. In addition, we will see how the techniques of two-dimensional gel electrophoresis and mass spectrometry can isolate and identify cellular proteins, and explore the use of protein microarrays to study protein expression and function.

The Proteome Is Much Larger Than the Genome

From the sequencing and analysis of an entire genome, researchers can identify all or nearly all of the genes from a given species. The size of the proteome, however, is larger than the genome, and its actual size is somewhat more difficult to determine. What phenomena help to account for the larger size of the proteome? First, changes in pre-mRNAs may ultimately affect the resulting amino acid sequence of a protein. The most important alteration that occurs commonly in eukaryotic species is **alternative splicing.** For many genes, a single pre-mRNA can be spliced into more than one version. The splicing is often cell specific, or it may be related to environmental conditions. As discussed in Chapter 15, alternative splicing is widespread, particularly among complex multicellular organisms. It can lead to the production of several or perhaps dozens of different polypeptide sequences from the same pre-mRNA, which greatly increases the number of potential proteins in the proteome. Similarly, the phenomenon of **RNA editing,** a change in the base sequence of RNA after it has been transcribed (see Chapter 15), can lead to changes in the coding sequence of an mRNA. However, RNA editing is much less common than alternative splicing.

Another process that greatly diversifies the composition of a proteome is the phenomenon of **posttranslational covalent modification.** Certain types of modifications are involved with the assembly and construction of a functional protein. These alterations include proteolytic processing, disulfide bond formation, and the attachment of prosthetic groups, sugars, or lipids. These are typically irreversible changes that are necessary to produce a functional protein. Other types of modifications, such as phosphorylation, acetylation, and methylation, are often reversible modifications that transiently affect the function of a protein. In addition, ubiquitination can promote protein degradation. Because a protein may be subject to several different types of covalent modification, this can greatly increase the forms of a particular protein that are found in a cell at any given time.

Two-Dimensional Gel Electrophoresis Is Used to Separate a Mixture of Cellular Proteins

As we have just learned, the proteome is usually much larger than a species' genome. Even so, any given cell within a complex multicellular organism will produce only a subset of the proteins found in the proteome of a species. For example, the human genome has approximately 20,000 to 25,000 different genes, yet a muscle cell makes only a subset, perhaps 15,000 types of proteins. The proteins a cell makes depend primarily on the cell type, the stage of development, and the environmental conditions. An objective of researchers in the field of proteomics is the identification and functional characterization of all the proteins a cell type will make. Because cells produce thousands of different proteins, this is a daunting task. Nevertheless, along with genomic research, the past decade has seen important advances in our ability to isolate and identify cellular proteins.

A common technique in the field of proteomics is **two-dimensional gel electrophoresis.** It is a technique that can separate hundreds or even thousands of different proteins within a cell extract. The steps in this procedure are shown in **Figure 21.5**. As its name suggests, the technique involves two different gel electrophoresis procedures. A sample of cells is lysed, and the proteins are loaded onto the top of a tube gel that separates proteins according to their net charge at a given pH. A protein migrates to the point in the gel where its net charge is zero. This is termed **isoelectric focusing.** After the tube gel has run, it is laid horizontally on top of a polyacrylamide slab gel that contains sodium dodecyl sulfate (SDS). The SDS coats each protein with negative charges and denatures them; proteins in the slab gel are separated according to their molecular mass. Smaller proteins move toward the bottom of the gel more quickly than larger ones. After the slab gel has run, the proteins within the gel can be stained with a dye. As seen in Figure 21.5, the end result is a collection of spots, each spot corresponding to a unique cellular protein, with proteins of a larger mass remaining higher in the gel. The resolving power of two-dimensional gel electrophoresis is extraordinary. Proteins that differ by a single charged amino acid can be resolved as two distinct spots using this method.

Various approaches can be followed to identify spots on a gel that may be of interest to researchers. One possibility is that a given cell type may show a few very large spots that are not found when proteins are analyzed from other cell types. The relative abundance of such spots may indicate that a particular protein is important to that cell's structure or function. Secondly, certain spots on a two-dimensional gel may be seen only under a given set of conditions. For example, a researcher may be interested in the effects of a hormone on the function of a particular cell type. Two-dimensional gel electrophoresis could be conducted on a sample in which the cells had not been exposed to the hormone versus a sample in which they had. This may reveal particular spots that are present only when the cells are exposed to a given hormone. Finally, abnormal cells often express proteins that are not found in normal cells. This is particularly true for cancer cells. A researcher may compare normal and cancer cells via two-dimensional gel electrophoresis to identify proteins expressed only in cancer cells.

Mass Spectrometry Is Used to Identify Proteins

Two-dimensional gel electrophoresis may be used as the first step in the separation of cellular proteins. The next goal is to correlate a

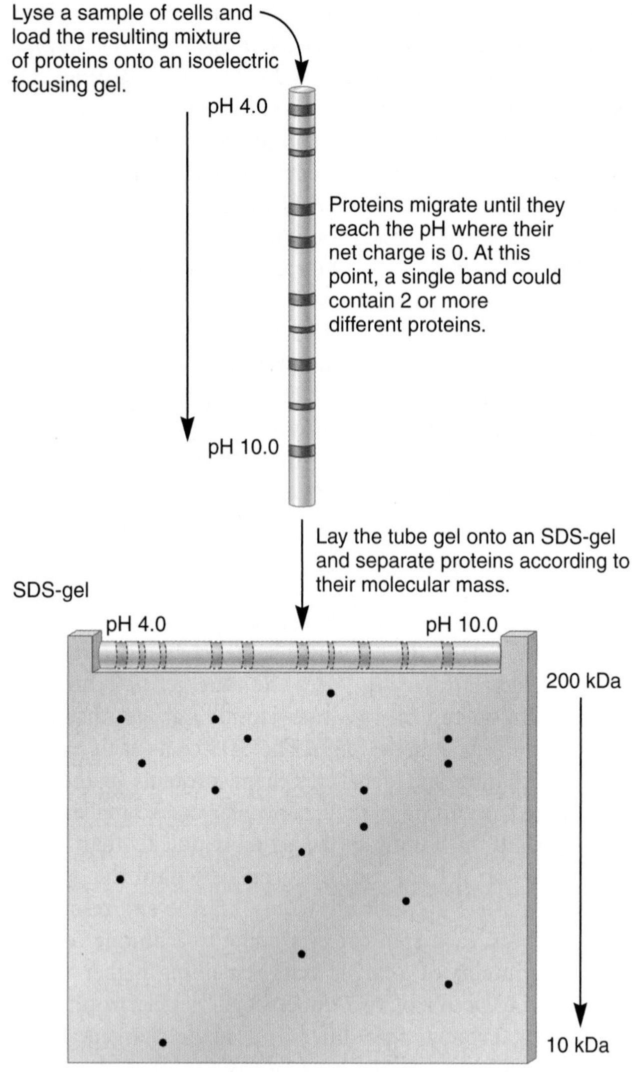
Lyse a sample of cells and load the resulting mixture of proteins onto an isoelectric focusing gel.
pH 4.0
Proteins migrate until they reach the pH where their net charge is 0. At this point, a single band could contain 2 or more different proteins.
pH 10.0
Lay the tube gel onto an SDS-gel and separate proteins according to their molecular mass.
SDS-gel
pH 4.0
pH 10.0
200 kDa
10 kDa

(a) The technique of two-dimensional gel electrophoresis

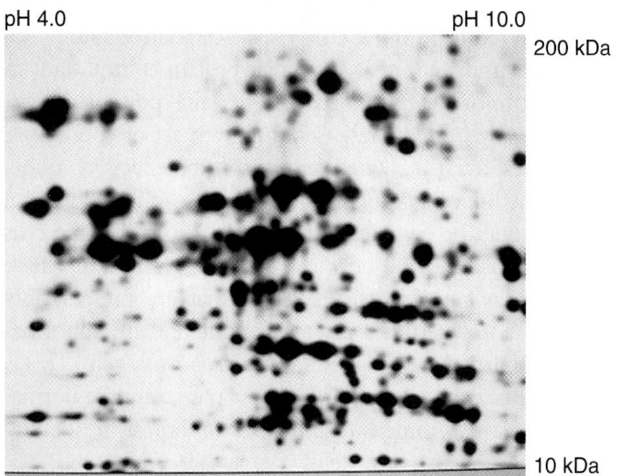
pH 4.0
pH 10.0
200 kDa
10 kDa

(b) A two-dimensional gel that has been stained to visualize proteins

ONLINE
ANIMATION

FIGURE 21.5 **Two-dimensional gel electrophoresis.** (**a**) The technique involves two electrophoresis steps. First, a mixture of proteins is separated on an isoelectric focusing gel that has the shape of a tube. Proteins migrate to the pH in the gel where their net charge is zero. This tube gel is placed into a long well on top of an SDS-polyacrylamide gel. This second gel separates the proteins according to their mass. In this diagram, only a few spots are seen, but an actual experiment would involve a mixture of hundreds or thousands of different proteins. (**b**) A photograph of an SDS-polyacrylamide gel that has been stained for proteins. Each spot represents a unique cellular protein.

given spot on a two-dimensional gel with a particular protein. To accomplish this goal, a spot on a two-dimensional gel can be cut out of the gel to obtain a tiny amount of the protein within the spot. In essence, the two-dimensional gel electrophoresis procedure purifies a small amount of the cellular protein of interest. The next step is to identify that protein. This can be accomplished via **mass spectrometry,** a technique that measures the mass of a molecule, such as a peptide fragment.

Figure 21.6 describes how mass spectrometry can determine the amino acid sequence of a protein. As shown here, the technique actually determines the mass of peptide fragments that are produced by digesting a purified protein with a protease that cuts the protein into small peptide fragments. The peptides are mixed with an organic acid and dried onto a metal slide. The sample is then struck with a laser. This causes the peptides to become ejected from the slide in the form of an ionized gas in which the peptide contains one or more positive charges. The charged peptides are then accelerated via an electric field and fly toward a detector. The time of flight is determined by their mass and net charge. A measurement of the flight time provides an extremely accurate way to determine the mass of a peptide.

An ultimate goal of mass spectrometry is to determine the amino acid sequence of a given peptide. This is accomplished using two mass spectrometers, a method called **tandem mass spectrometry.** The first mass spectrometer measures the mass of a given peptide. This same peptide is then analyzed by a second spectrometer after the peptide has been digested into smaller fragments. The differences in the masses of the peaks in the spectrum from the second spectrometer reveal the amino acid sequence of the peptide, because the masses of all 20 amino acids are known. For example, as shown in Figure 21.6, let's suppose a peptide had a mass of 1,652 daltons. If one amino acid at the end was removed and the smaller peptide had a mass that was 87 daltons less (i.e., 1,565 daltons), this would indicate that a serine is at one end of the peptide, because the mass of serine within a polypeptide chain is 87 daltons. If two amino acids were removed at one end and the mass was 224 daltons less, this would correspond to the removal of one serine (87 daltons) and one histidine (137 daltons). If three amino acids were removed and the mass was decreased by 337 daltons, this would correspond to the removal of one serine (87 daltons), one histidine (137 daltons), and one leucine (113 daltons). Therefore, from these measurements, we would conclude that the amino acid sequence from one end of the peptide was serine–histidine–leucine.

How does this information lead to the identification of a specific protein? After a researcher has obtained a few short peptide sequences from a given protein, genomic information can readily predict the entire amino acid sequence of the protein. For example, if a peptide had the sequence serine–histidine–leucine–asparagine–serine–asparagine, one could determine the possible codon sequences that could encode such a peptide. More than one sequence is possible due to the degeneracy of the genetic code. Using computer software described later, the codon sequences would be used as query sequences to search an entire genomic sequence. This program would locate a match between the predicted codon sequence and a specific gene within the genome. In this way, mass spectrometry makes it possible to identify the gene that encodes the entire protein. The gene sequence, in turn, can be used to predict the remaining amino acid sequence of the entire protein.

Mass spectrometry can also identify protein covalent modifications. For example, if an amino acid within a peptide was phosphorylated, the mass of the peptide would be increased by the mass of a phosphate group. This increase in mass can be determined via mass spectrometry.

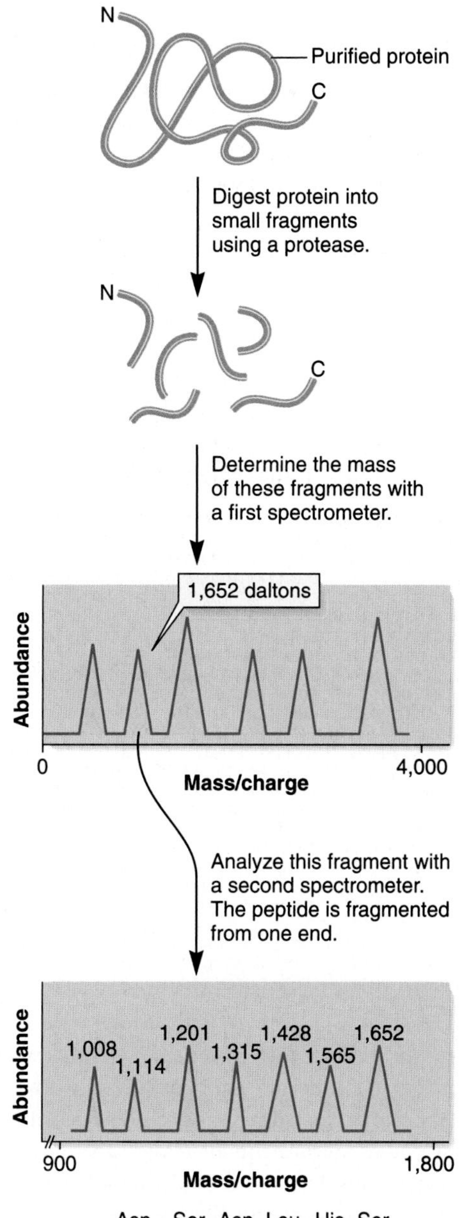

FIGURE 21.6 **The use of tandem mass spectrometry to determine the amino acid sequence of a peptide.**

Protein Microarrays Can Be Used to Study Protein Expression and Function

Earlier in this chapter, we learned about DNA microarrays, which have gained widespread use to study gene expression at the RNA level. The technology to make DNA microarrays is also being applied to make **protein microarrays.** In this type of technology, proteins, rather than DNA molecules, are spotted onto a glass or silica slide. The development of protein microarrays is more challenging because proteins are much more easily damaged by the manipulations that occur during microarray formation. For example, the three-dimensional structure of a protein may be severely damaged by drying, which usually occurs during the formation of a microarray. This has created additional challenges for researchers who are developing the technology of protein microarrays. In addition, the synthesis and purification of proteins tend to be more time-consuming compared to the production of DNA, which can be amplified by PCR or directly synthesized on the microarray itself. In spite of these technical difficulties, the last few years have seen progress in the production and uses of protein microarrays (**Table 21.2**).

TABLE **21.2**

Some Applications of Protein Microarrays

Application	Description
Protein expression	An antibody microarray can measure protein expression, because each antibody in a given spot recognizes a specific amino acid sequence. This can be used to study the expression of proteins in a cell-specific manner. It can also be used to determine how environmental conditions affect the levels of particular proteins.
Protein function	The substrate specificity and enzymatic activities of groups of proteins can be analyzed by exposing a functional protein microarray to a variety of substrates.
Protein–protein interactions	The ability of two proteins to interact with each other can be determined by exposing a functional protein microarray to fluorescently labeled proteins.
Pharmacology	The ability of drugs to bind to cellular proteins can be determined by exposing a functional protein microarray to different kinds of labeled drugs. This can help to identify the proteins within a cell to which a given drug may bind.

The two common types of protein microarray analyses are antibody microarrays and functional protein microarrays. The purpose of an **antibody microarray** is to study protein expression. As discussed in Chapter 17, antibodies are proteins that recognize antigens. One type of antigen that an antibody can recognize is a short peptide sequence found within another protein. Therefore, an antibody can specifically recognize a cellular protein. Researchers can produce thousands of different antibodies, each one recognizing a different peptide sequence. These can be spotted onto a microarray. Cellular proteins can then be isolated, fluorescently labeled, and exposed to the antibody microarray. When a given protein is recognized by an antibody on the microarray, it will be captured by the antibody and remain bound to that spot. The level of fluorescence at a given spot indicates the amount of a cellular protein that is recognized by a particular antibody.

The other type of array is a **functional protein microarray.** To make this type of array, researchers must purify cellular proteins and then spot them onto a microarray. The microarray can then be analyzed with regard to specific kinds of protein function. In 2000, for example, Heng Zhu, Michael Snyder, and colleagues purified 119 proteins from yeast that were known to function as protein kinases. These kinds of proteins attach phosphate groups onto other cellular proteins. A microarray was made consisting of different possible proteins that may or may not be phosphorylated by these 119 kinases, and then the array was exposed to each of the kinases in the presence of radiolabeled ATP. By following the incorporation of phosphate into the array, they determined the protein specificity of each kinase. On a much larger scale, the same group of researchers purified 5,800 different yeast proteins and spotted them onto a microarray. The array was then exposed to fluorescently labeled calmodulin, which is a regulatory protein that binds calcium ions. Several proteins in the microarray were found to bind calmodulin. While some of these were already known to be regulated by calmodulin, other proteins in the array that had not been previously known to bind calmodulin were identified.

21.3 BIOINFORMATICS

Geneticists use computers to collect, store, manipulate, and analyze data. Molecular genetic data is particularly amenable to computer analysis because it comes in the form of a sequence, such as a DNA, RNA, or amino acid sequence. The ability of computers to analyze data at a rate of millions or even billions of operations per second has made it possible to solve problems concerning genetic information that were thought intractable a few decades ago.

In recent years, the marriage between genetics and computational tools and approaches has yielded an important branch of science known as bioinformatics. Several scientific journals are largely devoted to this topic. Computer analysis of genetic sequences relies on three basic components: a computer, a computer program, and some type of data. In genetic research, the data consist of a particular genetic sequence or several sequences that a researcher or clinician wants to study. For example, this could be a DNA sequence derived from a cloned DNA fragment. A sequence of interest may be relatively short or thousands to millions of nucleotides in length. Experimentally, DNA sequences and related data are obtained using the techniques described in Chapters 18 and 20.

In this section, we will first consider the fundamental concepts that underlie the analysis of genetic sequences. We will then explore how these methods are used to provide insights regarding functional genomics and proteomics. Chapter 26 will describe applications of bioinformatics in the area of evolutionary biology. In addition, you may wish to actually run computer programs, which are widely available at university and government websites (for example, see www.ncbi.nlm.nih.gov/). This type of hands-on learning will help you to see how the computer has become a valuable tool to analyze genetic data.

Sequence Files Are Analyzed by Computer Programs

Most people are familiar with **computer programs,** which consist of a defined series of operations that can manipulate and analyze data in a desired way. For example, a computer program might be designed to take a DNA sequence and translate it into an amino acid sequence. A first step in the computer analysis of genetic data is the creation of a **computer data file** to store the data. This file is simply a collection of information in a form suitable for storage and manipulation on a computer. In genetic studies, a computer data file might contain an experimentally obtained DNA, RNA, or amino acid sequence. For example, a file could contain the DNA sequence of one strand of the *lacY* gene from *Escherichia coli*, as shown here. The numbers to the left represent the base number in the sequence file.

```
    1 ATGTACTATT TAAAAAACAC AAACTTTTGG ATGTTCGGTT TATTCTTTTT
   51 CTTTTACTTT TTTATCATGG GAGCCTACTT CCCGTTTTTC CCGATTTGGC
  101 TACATGACAT CAACCATATC AGCAAAGTG ATACGGGTAT TATTTTGCC
  151 GCTATTTCTC TGTTCTCGCT ATTATTCCAA CCGCTGTTTG GTCTGCTTC
  201 TGACAAACTC GGGCTGCGCA AATACCTGCT GTGGATTATT ACGGCATGT
  251 TAGTCATGTT TGCGCCGTTC TTTATTTTTA TCTTCGGGCC ACTGTTACAA
  301 TACAACATTT TAGTAGGATC GATTGTTGGT GGTATTTATC TAGGCTTTTG
  351 TTTTAACGCC GGTGCGCCAG CAGTAGAGGC ATTTATTGAG AAAGTCAGCC
  401 GTCGCAGTAA TTTCGAATTT GGTCGCGCGC GGATGTTTGG CTGTGTTGGC
  451 TGGGCGCTGT GTGCCTGAT TGTCGGCATC ATGTTCACCA TCAATAATCA
  501 GTTTGTTTTC TGGCTGGGCT CTGGCTGTGC ACTCATCCTC GCCGTTTTAC
  551 TCTTTTTCGC CAAAACGGAT GCGCCCTCTT CTGCCACGGT TGCCAATGCG
  601 GTAGGTGCCA ACCATTCGGC ATTTAGCCTT AAGTGGCAC TGGAACTGTT
  651 CAGACAGCCA AAACTGTGGT TTTGTCTACT GTATGTTATT GGCGTTTCCT
  701 GCACCTACGA TGTTTTTGAC CAACAGTTTG CTAATTTCTT TACTTCGTTC
  751 TTTGCTACCG GTGAACAGGG TACGCGGGTA TTTGGCTACG TAACGACAAT
  801 GGGCGCAATTA CTTAACGCCT CGATTATCTT CTTGCGCCA CTGATCATTA
  851 ATCGCATCGG TGGGAAAAAC GCCCTGCTGC TGGCTGGCAC TATTATCTCT
  901 CTACGTATTA TTGGCTCATC GTTCGCCACC TCAGCGCTGG AAGTGGTTAT
  951 TCTGAAAACG CTGCATATGT TTGAAGTACC GTTCCTGCTG GTGGGCTGCT
1,001 TTAAATATAT TACCAGCCAG TTTGAAGTGC GTTTTTCAGC GACGATTTAT
1,051 CTGGTCTGTT TCTGCTTCTT TAAGCAACTG GCATGATTT TTATGTCTGT
1,101 ACTGGCGGGC AATATGTATG AAAGCATCGG TTTCCAGGGC GCTTATCTGG
1,151 TGCTGGGTCT GGTGGCGCTG GGCTTCACCT TAATTTCCGT GTTCACGCTT
1,201 AGCGGCCCCG GCCCGCTTTC CCTGCTGCGT CGTCAGGTGA ATGAAGTCGC
1,251 TTAA
```

To store data in a computer data file, a scientist creates the file and enters the data, either by hand or, what is now more common, by laboratory instruments such as densitometers and fluorometers. These instruments have the capability to read data, such as a sequencing ladder, and enter the DNA sequence information directly into a computer file.

The purpose of making a computer file that contains a genetic sequence is to take advantage of the swift speed with which computers can analyze this information. Genetic sequence data in a computer file can be investigated in many different ways, corresponding to the many questions a researcher might ask about the sequence and its functional significance. These include the following:

1. Does a sequence contain a gene?
2. Where are functional sequences such as promoters, regulatory sites, and splice sites located within a particular gene?
3. Does a sequence encode a polypeptide? If so, what is the amino acid sequence of the polypeptide?
4. Does a sequence predict certain structural features for DNA, RNA, or proteins? For example, is a DNA sequence likely to be in a Z-DNA conformation? What is the secondary structure of an RNA sequence or polypeptide sequence?
5. Is a sequence homologous to any other known sequences?
6. What is the evolutionary relationship between two or more genetic sequences?

To answer these and many other questions, different computer programs have been written to analyze genetic sequences in particular ways. These programs have been devised by theoreticians who understand basic genetic principles and can design computational strategies for analyzing genetic sequences. When constructing a computer program, a theoretician has a goal that the program is meant to fulfill. For example, a theoretician may wish to write a program to translate a DNA sequence into an amino acid sequence. Based on knowledge of the genetic code, she/he can devise computational procedures, or algorithms, that relate a DNA sequence to an amino acid sequence. In a computer program, these procedures are executed as a stepwise "plan of operations" that manipulates the data in a sequence file. For a computer to perform the plan of operations, the program instructions must be written in a programming language the computer can decipher. After this has been accomplished, the program is tested to see if it works. Usually, errors (bugs) are found in the program that prevent it from working. After the program is debugged, it is ready for use. An ideal computer program is one that accurately manipulates data and is user-friendly.

As an example, let's consider a computer program aimed at translating a DNA sequence into an amino acid sequence and see how it might work in practice. The operation of the program as it would appear on a computer screen is shown in **Figure 21.7**. The geneticist—the user—has a DNA sequence file that she or he may want to have translated into an amino acid sequence. The user is sitting at a computer that is connected to a program that can translate a DNA sequence into an amino acid sequence. In this example, the program is named TRANSLATION. After the program is launched, it asks a series of questions. These are depicted on the screen in black. The typed answers of the user are shown in red. The first question asks which sequence file the user wants translated. In this case, the user wants the *lacY* gene sequence translated into an amino acid sequence. The name of this file is lacY.SEQ. The program then asks which codon translation table to use. As you may recall from Chapter 13, some species have slight variations in their usage of the genetic code. In this case, the user selects the standard genetic code. Next, the program requires the parameters of the sequence file that is to be translated. The user decides to begin the translation at the first nucleotide in the sequence file and end the translation at nucleotide number 1,254. The user wants the program to translate the sequence in all three forward reading frames and to show the longest reading frame—the longest amino acid sequence that is uninterrupted by a stop codon. This translated sequence is saved in a file that the user wants to be named lacY.PEP. (Note: The symbol PEP provides a reminder that it is a poly<u>pep</u>tide sequence.) The user clicks on the RUN button, and the translation program proceeds to translate the *lacY* sequence and stores the amino acid sequence in a file called lacY.PEP. The contents of the lacY.PEP file are shown here.

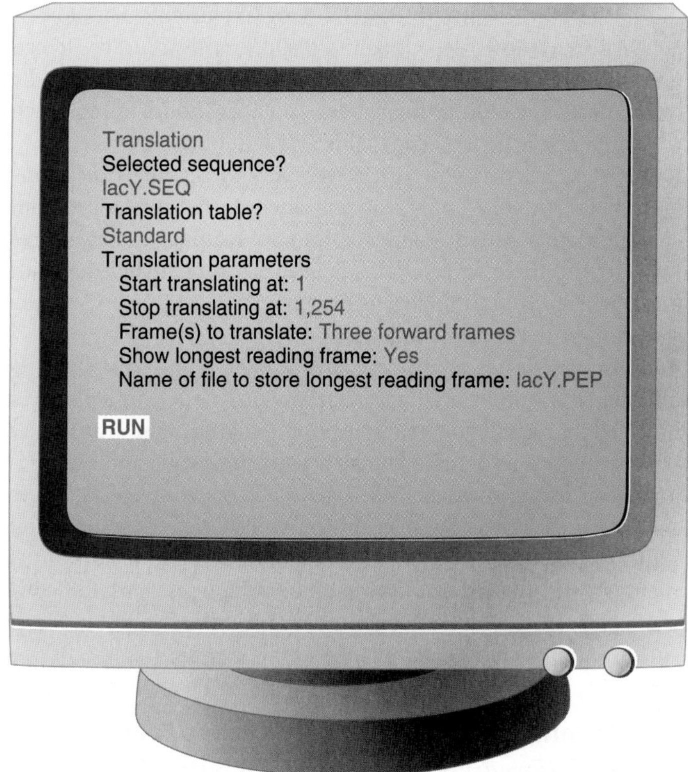

FIGURE 21.7 **The operation of a computer program as it would appear on a computer screen.**

```
  1 MYYLKNTNFW MFGLFFFFYF FIMGAYFPFF PIWLHDINHI SKSDTGIIFA
 51 AISLFSLLFQ PLFGLLSDKL GLRKYLLWII TGMLVMFAPF FIFIFGPLLQ
101 YNILVGSIVG GIYLGFCFNA GAPAVEAFIE KVSRRSNFEF GRARMFGCVG
151 WALCASIVGI MFTINNQFVF WLGSGCALIL AVLLFFAKTD APSSATVANA
201 VGANHSAFSL KLALELFRQP KLWFLSLYVI GVSCTYDVFD QQFANFFTSF
251 FATGEQGTRV FGYVTTMGEL LNASIMFFAP LIINRIGGKN ALLLAGTIMS
301 VRIIGSSFAT SALEVVIKLT LHMFEVPFLL VGCFKYITSQ FEVRFSATIY
351 LVCFCFFKQL AMIFMSVLAG NMYESIGFQG AYLVLGLVAL GFTLISVFTL
401 SGPGPLSLLR RQVNEVA
```

In this file, which was created by the computer program TRANSLATION, each of the 20 amino acids is given a single-letter abbreviation (see Figure 13.6).

Why is the TRANSLATION program useful? The advantages of running this program are speed and accuracy. It can translate a relatively long genetic sequence within seconds. By comparison, it would probably take you a few hours to look each codon up in the genetic code table and write the sequence out in the correct order. If you visit a website and actually run a program like TRANSLATION, you will discover that such a program can translate a genetic sequence into six reading frames—three forward and three reverse. This is useful if a researcher does not know where the start codon is located and/or does not know the direction of the coding sequence.

In genetic research, large software packages typically contain many computer programs that can analyze genetic sequences in different ways. For example, one program can translate a DNA sequence into an amino acid sequence, while another program can locate introns within genes. These software packages are found at universities, government facilities, hospitals, and industries. At such locations, a central computer with substantial memory and high-speed computational abilities runs the software, and individuals can connect to this central computer. Many such programs are freely available on the Internet.

The Scientific Community Has Collected Sequence Files and Stored Them in Large Computer Databases

In Chapter 18, we considered how researchers can clone and sequence genes. Likewise, in Chapter 20, we learned how scientists are investigating the genetic sequences of entire genomes from several species, including humans. The amount of genetic information generated by researchers has become enormous. The Human Genome Project, for example, has produced more data than any other undertaking in the history of biology. With these advances, scientists realize that another critical use of computers is to simply store the staggering amount of data produced from genetic research.

A large number of computer data files collected and stored in a single location is called a **database.** In addition to genetic sequences, the files within databases are **annotated,** which means they contain additional information such as a concise description of a sequence, the name of the organism from which this sequence was obtained, and the function of the encoded protein, if it is known. The file may also describe other features of significance and cite a published journal reference that describes the sequence.

The scientific community has collected the genetic information from thousands of research labs and created several large databases. **Table 21.3** describes some of the major genetic databases in use worldwide. These databases enable researchers to access and compare genetic sequences that are obtained by many laboratories. Later, we will learn how researchers can use databases to analyze genetic sequences.

TABLE 21.3

Examples of Major Computer Databases

Type	Description
Nucleotide sequence	DNA sequence data are collected into three internationally collaborating databases: GenBank (a USA database), EMBL (European Molecular Biology Laboratory Nucleotide Sequence Database), and DDBJ (DNA Databank of Japan). These databases receive sequence and sequence annotation data from genome projects, sequencing centers, individual scientists, and patent offices. These databases are accessed via the Internet and available on CD-ROM.
Amino acid sequence	Amino acid sequence data are collected into a few international databases including Swissprot (Swiss protein database), PIR (Protein Information Resource), Genpept (translated peptide sequences from the GenBank database), and TrEMBL (Translated sequences from the EMBL database).
Three-dimensional structure	PDB (Protein Data Bank) collects the three-dimensional structures of biological macromolecules with an emphasis on protein structure. These are primarily structures that have been determined by X-ray crystallography and nuclear magnetic resonance (NMR), but some models are included in the database. These structures are stored in files that can be viewed on a computer with the appropriate software.
Protein motifs	Prosite is a database containing a collection of amino acid sequence motifs that are characteristic of a protein family, domain structure, or certain posttranslational modifications. Pfam is a database of protein families with multiple amino acid sequence alignments.

The databases described in Table 21.3 collect genetic information from many different species. Scientists have also created more specialized databases, called **genome databases,** that focus on the genetic sequences and characteristics of a single species. Genome databases have been created for species of bacteria (e.g., *E. coli*), yeast (e.g., *Saccharomyces cerevisiae*), worms (e.g., *Caenorhabditis elegans*), fruit flies (e.g., *Drosophila melanogaster*), plants (e.g., *Arabidopsis thaliana*), and mammals (e.g., mice and humans). The primary aim of genome databases is to organize the information from sequencing and mapping projects for a single species. Genome databases identify an organism's known genes and describe their map location within the genome. In addition, a genome database may provide information concerning gene alleles, bibliographic information, a directory of researchers who study the species, and other pertinent information.

Different Computational Strategies Can Identify Functional Genetic Sequences

At the molecular level, the function of the genetic material is based largely on specific genetic sequences that play distinct roles. For example, codons are three-base sequences that specify particular amino acids, and promoters are sequences that provide a binding site for RNA polymerase to initiate transcription.

Computer programs can be designed to scan very long sequences, such as those obtained from genome sequencing projects, and locate meaningful features within them. To illustrate this concept, let's first consider the following sequence file, which contains an alphabetic sequence of 54 letters:

Sequence file:

GJTRLLAMAQLHEOGYLTOBWENTMNMTORXXXTGOODNTHEQ
ALLRTLSTORE

We will now compare how three different computer programs might analyze this sequence to identify meaningful features. The goal of our first program is to locate all the English words within this sequence. If we ran this program, we would obtain the following result:

GJTRLLAMAQLHEOGYLTOBWENTMNMTORXXXTGOODNTHEQ
ALLRTLSTORE

In this case, a computer program has identified locations where the sequence of letters forms a word. Several words (which are underlined) have been located within this sequence.

A second computer program could be aimed at locating a series of words that are organized in the correct order to form a grammatically logical English sentence. If we used our sequence file and ran this program, we would obtain the following result:

GJTRLLAMAQLHEOGYLTOBWENTMNMTORXXXTGOODNTHEQ
ALLRTLSTORE

The second program has identified five words that form a logical sentence.

Finally, a computer program might be used to identify patterns of letters, rather than words or sentences. For example, a computer program could locate a pattern of five letters that occurs in both the forward and reverse directions. If we applied this program to our sequence file, we would obtain the following:

GJTRLLAMAQLHEOGYLTOBWENTMNMTORXXXTGOODNTHEQ
ALLRTLSTORE

In this case, the program has identified a pattern where five letters are found in both the forward and reverse directions.

In the three previous examples, we can distinguish between **sequence recognition** (as in our first example) and **pattern recognition** (as in our third example). In sequence recognition, the program has the information that a specific sequence of symbols has a specialized meaning. This information must be supplied to the computer program. For example, the first program would have access to the information from a dictionary with all known English words. With this information, the first program can identify sequences of letters that make words. By comparison, the third program does not rely on specialized sequence information. Rather, it is looking for a pattern of symbols that can occur within any group of symbol arrangements.

Overall, the simple programs we have considered illustrate three general types of identification strategies:

1. *Locate specialized sequences within a very long sequence.* A specialized sequence with a particular meaning or function is called a **sequence element** or **motif.** The computer program has a list of predefined sequence elements and can identify such elements within a sequence of interest.
2. *Locate an organization of sequences.* As shown in the second program, this could be an organization of sequence elements. Alternatively, it could be an organization of a pattern of sequences.
3. *Locate a pattern of sequences.* The third program is an example of locating a pattern of sequences.

The great power of computer analysis is that these types of operations can be performed with great speed and accuracy on sequences that may be enormously long.

Now that we understand the general ways that computer programs identify sequences, let's consider specific examples. As we have discussed throughout this textbook, many short nucleotide sequences play specialized roles in the structure or function of genetic material. A geneticist may want to locate a short sequence element within a longer nucleotide sequence in a data file. For example, a sequence of chromosomal DNA might be tens of thousands of nucleotides in length, and a geneticist may want to know whether a sequence element, such as a TATA box, is found at one or more sites within the chromosomal DNA. To do so, a researcher could visually examine the long chromosomal DNA sequence in search of a TATA box. Of course, this process would be tedious and prone to error. By comparison, the appropriate computer program can locate a sequence element within seconds. Therefore, computers are very useful for this type of application. **Table 21.4** lists some examples of sequence elements that can be identified by computer analysis.

TABLE 21.4

Short Sequence Elements That Can Be Identified by Computer Analysis

Type of Sequence	Examples*
Promoter	Many *E. coli* promoters contain TTGACA (−35 site) and TATAAT (−10 site). Eukaryotic core promoters may contain CAAT boxes, GC boxes, TATA boxes, etc.
Response elements	Glucocorticoid response element (AGRACA), cAMP response element (GTGACGTRA)
Start codon	ATG
Stop codons	TAA, TAG, TGA
Splice site	GTRAGT————YNYTRAC(Y)$_n$AG
Polyadenylation signal	AATAAA
Highly repetitive sequences	Relatively short sequences that are repeated many times throughout a genome
Transposable elements	Usually characterized by a pattern in which direct repeats flank inverted repeats

*The sequences shown in this table would be found in the DNA. For gene sequences, only the coding strand is shown. R = purine (A or G); Y = pyrimidine (T or C); N = A, T, G, or C; U in RNA = T in DNA.

By comparing the amino acid sequences and known functions of proteins in thousands of cases, researchers have also found amino acid sequence motifs that carry out specialized functions within proteins. For example, researchers have determined that the amino acid sequence motif asparagine–X–serine (where X is any amino acid except proline) within eukaryotic proteins is a glycosylation site (i.e., it may have a carbohydrate attached to it). The Prosite database (refer back to Table 21.3) contains a collection of all amino acid sequence motifs known to be functionally important. Researchers can use computer programs to determine whether an amino acid sequence contains any of the motifs found in the Prosite database. This may help them to understand the role of a newly found protein of unknown function.

Several Computer-Based Approaches Can Identify Structural Genes Within a Nucleotide Sequence

A structural gene is composed of nucleotide sequences organized in a particular way. A typical gene contains a promoter, followed by a start codon, a coding sequence, a stop codon, and a transcriptional termination site. In addition, most genes contain regulatory sequences (e.g., eukaryotic response elements or prokaryotic operator sites), and eukaryotic genes are likely to contain introns. After researchers have sequenced a long segment of chromosomal DNA, they frequently want to know if the sequence contains any genes. In an attempt to answer this question, geneticists can use computer programs that are aimed at identifying genes in long genomic DNA sequences.

How do computer programs identify a gene within a long genetic sequence? These programs employ different strategies. A **search by signal** approach relies on known sequences such as promoters, start and stop codons, and splice sites to help predict whether or not a DNA sequence contains a structural gene. The program tries to locate an organization of known sequence elements that normally are found within a gene. It would try to locate a region that contains a promoter sequence, followed by a start codon, a coding sequence, a stop codon, and a transcriptional terminator.

A second strategy is a **search by content** approach. The goal here is to identify sequences with a nucleotide content that differs significantly from a random distribution. Within structural genes, this occurs primarily due to codon usage. Although there are 64 codons, most organisms display a **codon bias** within structural genes. This means that certain codons are used much more frequently than others. For example, UUA, UUG, CUU, CUC, CUA, and CUG all specify leucine. In yeast, however, 80% of the leucine codons are UUG. Codon bias allows organisms to more efficiently rely on a smaller population of tRNA molecules. A search by content strategy, therefore, attempts to locate coding regions by identifying regions where the nucleotide content displays a known codon bias.

A third way to locate coding regions within a DNA sequence is to examine translational reading frames. Recall that the reading frame is a sequence of codons determined by reading bases in groups of three. In a new DNA sequence, researchers must consider that the reading of codons (in groups of three nucleotides) could begin with the first nucleotide (reading frame 1), the second nucleotide (reading frame 2), or the third nucleotide (reading frame 3). An **open reading frame (ORF)** is a region of a nucleotide sequence that does not contain any stop codons. Because most proteins are several hundred amino acids in length, a relatively long reading frame is required to encode them. In prokaryotic species, long ORFs are contained within the chromosomal gene sequences. In eukaryotic genes, however, the coding sequence may be interrupted by introns. As described earlier, one way to determine eukaryotic ORFs is to clone and sequence cDNA, which is complementary to mRNA. Alternatively, a computer program can translate a genomic DNA sequence in all three reading frames, seeking to identify a long ORF. In **Figure 21.8**, a DNA sequence has been translated in all three reading frames. Only one of the three reading frames (3) contains a very long open reading frame without any stop codons, suggesting that this DNA sequence encodes a protein. In Figure 21.8, it was assumed that the reading frame was from left to right. In an uncharacterized genetic sequence, however, a reading frame could proceed from right to left. Therefore, in a newly discovered genetic sequence, six reading frames are possible—three in the forward direction and three in the reverse direction.

Even though computer programs are a valuable tool, they are not always accurate in their prediction of gene sequences. In particular, it is often difficult for programs to predict the correct start codon and the precise intron/exon boundaries. In some cases, computer programs may even suggest that a region encodes a gene when it does not. Therefore, while a bioinformatic approach is a relatively easy tool to identify potential genes, it should not be viewed as a definitive method. The confirmation that a DNA region encodes an actual gene requires laboratory experimentation to show that it is truly transcribed into RNA.

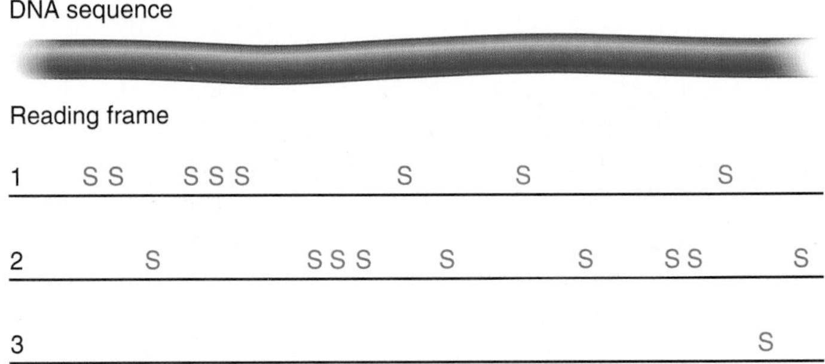

FIGURE 21.8 Translation of a DNA sequence in all three reading frames. The three lines represent the translation of a gene sequence in each of three forward reading frames; the reading frames proceed from left to right. The letter S indicates the location of a stop codon. Reading frame 3 has a very long open reading frame, suggesting that it may be the reading frame for a structural gene. Reading frames 1 and 2 are not likely to be the reading frames for a structural gene, because they contain many stop codons. During the cloning of DNA, the orientation of a gene may become flipped so that the coding sequence is inverted. Therefore, when analyzing many cloned DNA fragments, six reading frames (i.e., three forward and three reverse) are evaluated. Only the three forward frames are shown here.

Computer Programs Can Identify Homologous Sequences

Let's now turn our attention to the uses of computer technology to identify genes that are evolutionarily related. The ability to sequence DNA allows geneticists to examine evolutionary relationships at the molecular level. This has become an extremely powerful tool in the field of genomics. When comparing genetic sequences, researchers frequently find two or more sequences that are similar. For example, the sequence of the *lacY* gene that encodes the lactose permease in *E. coli* is similar to that of the *lacY* gene that encodes the lactose permease in a closely related bacterium, *Klebsiella pneumoniae.* As shown here, when segments of the two *lacY* genes are lined up, approximately 78% of their bases are a perfect match.

```
              151                                                    200
E. coli       TCTTTTTCTT TTACTTTTTT ATCATGGGAG CCTACTTCCC GTTTTTCCCG
K. pneumoniae TCTTTTTCTT TTACTATTTC ATTATGTCAG CCTACTTTCC TTTTTTTCCG

              201                                                    250
              ATTTGGCTAC ATGACATCAA CCATATCAGC AAAAGTGATA CGGGTATTAT
              GTGTGGCTGG CGGAAGTTAA CCATTTAACC AAAACCGAGA CGGGTATTAT
```

In this case, the two sequences are similar because the genes are **homologous** to each other. This means they have been derived from the same ancestral gene. This idea is shown schematically in **Figure 21.9**. An ancestral *lacY* gene was located in a bacterium that preceded the evolutionary divergence between *E. coli* and *K. pneumoniae.* After these two bacteria had diverged from each other, their *lacY* genes accumulated distinct mutations that produced somewhat different base sequences for this gene. Therefore, in these two species of bacteria, the *lacY* genes are similar but not identical. When two homologous genes are found in different species, these genes are termed **orthologs.**

Two or more homologous genes can also be found within a single organism. These are termed paralogous genes, or **paralogs.** As discussed in Chapter 8, abnormal gene duplication may happen several times during evolution, which results in multiple copies of a gene and ultimately leads to the formation of a gene family. A **gene family** consists of two or more paralogs within the genome of a single organism. When a gene family occurs, the concept of orthologs becomes more complex. For example, let's consider the globin gene family found in mammals (see Chapter 8, Figure 8.7). Researchers would say that the β-globin gene in humans is an ortholog to the β-globin gene found in mice. Likewise, α-globin genes found in both species would be considered

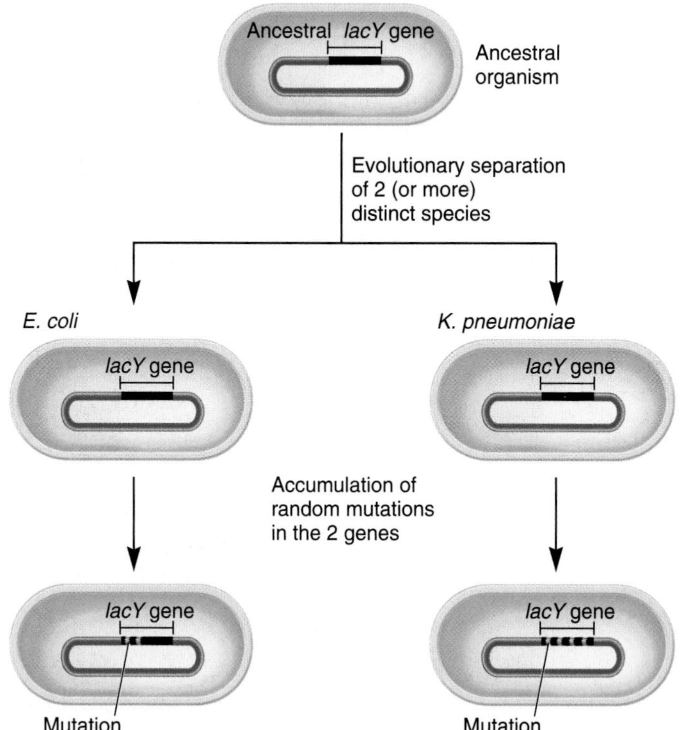

FIGURE 21.9 The origin of homologous *lacY* genes in *Escherichia coli* and *Klebsiella pneumoniae.* This figure emphasizes a single gene within an ancestral organism. During evolution, the ancestral organism diverged into two different species, *E. coli* and *K. pneumoniae.* After this divergence, the *lacY* gene in the two separate species accumulated mutations, yielding *lacY* genes with somewhat different sequences.

Genes→Traits After two species diverge evolutionarily, their genes will accumulate different random mutations. This example concerns the *lacY* gene, which encodes lactose permease. In both species, the function of lactose permease is to transport lactose into the cell. The *lacY* gene in these two species has accumulated different mutations that slightly alter the amino acid sequence of the protein. Researchers have determined that these two species transport lactose at significantly different rates. Therefore, the changes in gene sequences have affected the ability of these two species to transport lactose.

orthologs. However, the α-globin gene in humans would not be called an ortholog of the β-globin gene in mice, though they could be called homologous. The most closely related genes in two different species are considered orthologs.

Homologous genes, whether they are orthologs or paralogs, have similar sequences. What is the distinction between homology and sequence similarity? **Homology** implies a common ancestry. **Similarity** means that two sequences are similar to each other. In many cases, such as the *lacY* example, similarity is due to homology. However, this is not always the case. Short genetic sequences may be similar to each other even though two genes are not related evolutionarily. For example, many nonhomologous bacterial genes contain similar promoter sequences at the −35 and −10 regions.

A Simple Dot Matrix Can Compare the Degree of Similarity Between Two Sequences

To evaluate the similarity between two sequences, a matrix can be constructed. In a general way, Figure 21.10 illustrates the use of a simple dot matrix. In Figure 21.10a, the sequence GENETICSISCOOL is compared to itself. Each point in the grid corresponds to one position of each sequence. The matrix allows all such pairs to be compared simultaneously. Dots are placed where the same letter occurs in both sequences. The key observation is that regions of similarity are distinguished by the occurrence of many dots along a diagonal line within the dot matrix. In contrast, Figure 21.10b compares two unrelated sequences: GENETICSISCOOL and THECOURSEISFUN. In this comparison, no diagonal lines are seen. In some cases, two sequences may be related to each other but differ in length. Figure 21.10c compares the sequences GENETICSISCOOL and GENETICSISVERYCOOL. In this example, the second sequence is four letters longer than the first. In the dot matrix, two diagonal lines occur. To align these two lines, a gap must be created in the first sequence. Figure 21.10d shows the insertion of a gap that aligns the two sequences so that the two diagonal lines fall along the same line.

Overall, Figure 21.10 illustrates two important features of dot matrix methods. First, regions of homology are recognized by a series of dots that lie along a diagonal line. Second, gaps

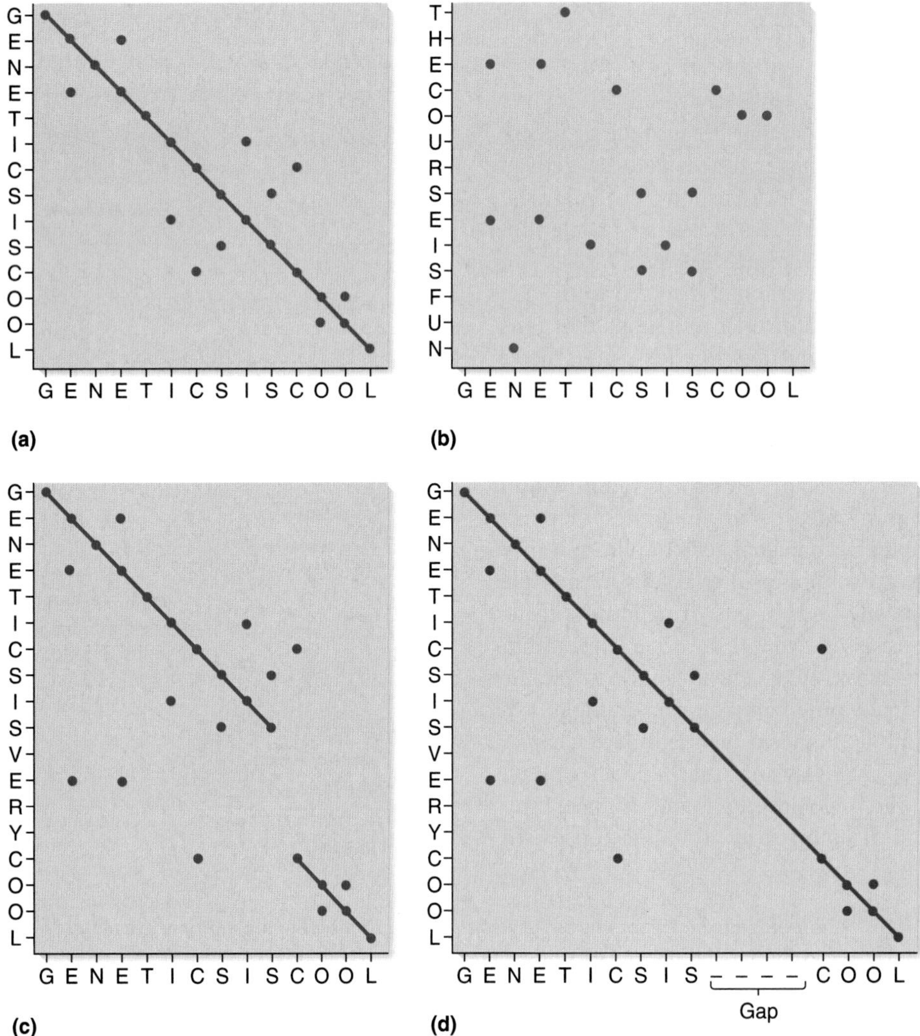

FIGURE 21.10 The use of dot matrices to evaluate similarity between two sequences. On a dot matrix, many points that fall on a diagonal line indicate regions of similarity.

can be inserted to align sequences of unequal length but that still have some similarity. These same concepts hold true when genetic sequences are compared to each other. Unfortunately, for anything but the shortest genetic sequences, a simple dot matrix approach is not adequate. Instead, dynamic programming methods are used to identify similarities between genetic sequences. These methods, originally proposed by Saul Needleman and Christian Wunsch, are theoretically similar to a dot matrix, but they involve complex mathematical operations beyond the scope of this textbook.

In their original work of 1970, Needleman and Wunsch demonstrated that whale myoglobin and human β hemoglobin have similar sequences. Since then, this approach has been extended to compare more than two genetic sequences. Newer computer programs can align several genetic sequences and sensibly put in gaps. This produces a **multiple sequence alignment.**

To illustrate the usefulness of a multiple sequence alignment, let's use the general methods of Needleman and Wunsch and apply them to the globin gene family. Hemoglobin, a protein found in red blood cells, is responsible for carrying oxygen through the bloodstream. In humans, nine paralogous globin genes are functionally expressed. (There are also several pseudogenes that are not expressed and one myoglobin gene.) The nine globin genes fall into two categories: the α chains and the β chains. The α-chain genes are α_1, α_2, θ, and ξ; the β-chain genes are β, δ, γ_A, γ_G, and ε. Each hemoglobin protein is composed of two α chains and two β chains.

Because the globin genes are expressed at different stages of human development, the composition of hemoglobin changes during the course of growth. For example, the ξ and ε genes are expressed during early embryonic development, whereas the α and β genes are expressed in the adult.

Insights into the structure and function of the hemoglobin polypeptide chains can be gained by comparing their sequences. In **Figure 21.11**, the sequences of the human globin polypeptides are compared in a multiple sequence alignment using dynamic programming methods. An inspection of a multiple sequence alignment may reveal important features concerning the similarities and differences within a gene family. In this alignment, dots are shown where it is necessary to create gaps to keep the amino acid sequences aligned. As we can see, the sequence similarity is very high between α_1, α_2, θ, and ξ. In fact, the amino acid sequences encoded by the α_1 and α_2 genes are identical. This suggests that the four types of α chains likely carry out very similar functions. Likewise, the β chains encoded by the β, δ, γ_A, γ_G, and ε genes are very similar to each other. In the globin gene family, the α chains are much more similar to each other than they are to the β chains, and vice versa.

In general, amino acids that are highly conserved within a gene family are more likely to be important functionally. The arrows in the multiple sequence alignment point to histidine amino acids that are conserved in all nine members of the hemoglobin gene family. These histidines are involved in the necessary function of binding the heme molecule to the globin polypeptides.

	1				50
β beta	VHLTPEEEKSA	VTALWGKV..	NVDEVGGEAL	GRLLVVYPWT	QRFFESFGDL
δ delta	VHLTPEEEKSA	VNALWGKV..	NVDAVGGEAL	GRLLVVYPWT	QRFFESFGDL
γ_A gamma-A	GHFTEEDKAT	ITSLWGKV..	NVEDAGGEAL	GRLLVVYPWT	QRFFESFGDL
γ_G gamma-G	GHFTEEDKAT	ITSLWGKV..	NVEDAGGEAL	GRLLVVYPWT	QRFFESFGDL
ε epsilon	VHFTAEEKAA	VTSLWSKM..	NVEEAGGEAL	GRLLVVYPWT	QRFFESFGDL
α_1 alpha-1	VLSPADKTN	VKAAWGKVGA	HAGEGAEAL	ERMFLSFPTT	KTYFPHF.DL
α_2 alpha-2	VLSPADKTN	VKAAWGKVGA	HAGEGAEAL	ERMFLSFPTT	KTYFPHF.DL
θ theta	ALSAEDRAL	VRALWKKLGS	NVGVYTTEAL	ERTFLAFPAT	KTYFSHL.DL
ξ zeta	SLTKTERTI	IVSMWAKIST	QADTIGTETL	ERLFLSHPQT	KTYFPHF.DL
	51				100
β beta	STPDAVMGNP	KVKAHGKKVL	GAFSDGLAHL	DNLKGTFATL	SELHCDKLHV
δ delta	SSPDAVMGNP	KVKAHGKKVL	GAFSDGLAHL	DNLKGTFSQL	SELHCDKLHV
γ_A gamma-A	SSASAIMGNP	KVKAHGKKVL	TSLGDAIKHL	DDLKGTFAQL	SELHCDKLHV
γ_G gamma-G	SSASAIMGNP	KVKAHGKKVL	TSLGDAIKHL	DDLKGTFAQL	SELHCDKLHV
ε epsilon	SSPSAILGNP	KVKAHGKKVL	TSFGDAIKNM	DNLKPAFAKL	SELHCDKLHV
α_1 alpha-1	SHGSA.....	QVKGHGKKVA	DALTNAVAHV	DDMPNALSAL	SDLHAHKLRV
α_2 alpha-2	SHGSA.....	QVKGHGKKVA	DALTNAVAHV	DDMPNALSAL	SDLHAHKLRV
θ theta	SPGSS.....	QVKAHGQKVA	DALSLAVERL	DDLPHALSAL	SHLHACQLRV
ξ zeta	HPGSA.....	QLRAHGSKVV	AAVGDAVKSI	DDIGGALSKL	SELHAYQLRV
		↑			↑
	101				148
β beta	DPENFRLLGN	VLVCVLAHHF	GKEFTPPVQA	AYQKVVAGVA	NALAHKYH
δ delta	DPENFRLLGN	VLVCVLARNF	GKEFTPQMQA	AYQKVVAGVA	NALAHKYH
γ_A gamma-A	DPENFRLLGN	VLVCVLAIHF	GKEFTPEVQA	SWQKMVTAVA	SALSSRYH
γ_G gamma-G	DPENFRLLGN	VLVCVLAIHF	GKEFTPEVQA	SWQKMVTAVA	SALSSRYH
ε epsilon	DPENFRLLGN	VMVIILATHF	GKEFTPEVQA	AWQKLVSAVA	IALAHKYH
α_1 alpha-1	DPVNFKLLSH	CLLVTLAAHL	PAEFTPAVHA	SLDKFLASVS	TVLTSKYR
α_2 alpha-2	DPVNFKLLSH	CLLVTLAAHL	PAEFTPAVHA	SLDKFLASVS	TVLTSKYR
θ theta	DPASFQLLGH	CLLVTLARHL	PGDFSPALQA	SLDKFLSHSVI	SALVSEYR
ξ zeta	DPVNFKLLSH	CLLVTLAARF	PADFTAEAHA	AWDKFLSVVS	SVLTEKYR

FIGURE 21.11 **A multiple sequence alignment among selected members of the globin gene family in humans.**

Overall, the alignment shown in Figure 21.11 illustrates the type of information that can be derived from a multiple sequence alignment. In this case, multiple sequence alignment has shown that a group of nine genes falls into two closely related subgroups. The alignment has also identified particular amino acids within the proteins' sequences that are highly conserved. This conservation is consistent with an important role in protein function.

A Database Can Be Searched to Identify Homologous Sequences

Homologous genes usually encode proteins that carry out similar or identical functions. As we have just considered, the members of the globin gene family are all involved with carrying and transporting oxygen. Likewise, the *lacY* genes in *E. coli* and *K. pneumoniae* both encode lactose permease, a protein that transports lactose across the bacterial cell membrane.

A strong correlation is typically found between homology and function. How is this relationship useful with regard to bioinformatics? In many cases, the first indication of the function of a newly determined sequence is through homology to known sequences in a database. An example is the gene that is altered in cystic fibrosis (CF) patients. After the *CF* gene was identified in humans, a database search revealed it is homologous to several genes found in other species. Moreover, a few of the homologous genes were already known to encode proteins that function in the transport of ions and small molecules across the plasma membrane. This observation provided an important clue that cystic fibrosis involves a defect in ion transport.

The ability of computer programs to identify homology between genetic sequences provides a powerful tool for predicting the function of genetic sequences. In 1990, Stephen Altschul, David Lipman, and colleagues developed a program called **BLAST** (for **basic local alignment search tool**). The BLAST program has been described by many geneticists as the single most important bioinformatic tool. This type of computer program can start with a particular genetic sequence and then locate homologous sequences within a large database. Because there are 20 amino acids but only four bases, homology among protein sequences is easier to identify than DNA sequence homology. Among proteins, sequences that diverged more than 2.5 billion years ago can still be correlated. By comparison, it becomes difficult to identify homologous DNA sequences that diverged more than 100 million years ago.

To see how the BLAST program works, let's consider the human enzyme phenylalanine hydroxylase, which functions in the metabolism of phenylalanine, an amino acid. Recessive mutations in the gene that encodes this enzyme are responsible for the disease called phenylketonuria (PKU). The computational experiment shown in **Table 21.5** started with the amino acid sequence of this protein and used the BLAST program to search the Swissprot database, which contains hundreds of thousands of different protein sequences. The BLAST program can determine which sequences in this database are the closest matches to the amino acid sequence of human phenylalanine hydroxylase. Table 21.5 shows the results—the 10 best matches to human phenylalanine hydroxylase that were identified by the program. Because this enzyme is found in nearly all eukaryotic species, the program identified phenylalanine hydroxylase from many different species. The column to the right of the match number shows the percentage of amino acids that are identical between the species indicated and the human sequence. Because the human phenylalanine hydroxylase sequence is already in the Swissprot database, the closest match of human phenylalanine hydroxylase is to

TABLE 21.5

Results from a BLAST Program Comparing Human Phenylalanine Hydroxylase with Database Sequences

Match	% of Identical Amino Acids*	Species	Function of Sequence†
1	100	Human (*Homo sapiens*)	Phenylalanine hydroxylase
2	99	Orangutan (*Pongo pygmaeus*)	Phenylalanine hydroxylase
3	95	Mouse (*Mus musculus*)	Phenylalanine hydroxylase
4	95	Rat (*Rattus norvegicus*)	Phenylalanine hydroxylase
5	89	Chicken (*Gallus gallus*)	Phenylalanine hydroxylase
6	82	Pipid frog (*Xenopus tropicalis*)	Phenylalanine hydroxylase
7	82	Green pufferfish (*Tetraodon nigroviridis*)	Phenylalanine hydroxylase
8	82	Zebrafish (*Danio rerio*)	Phenylalanine hydroxylase
9	80	Japanese pufferfish (*Takifugu rubripes*)	Phenylalanine hydroxylase
10	75	Fruit fly (*Drosophila melanogaster*)	Phenylalanine hydroxylase

*The number indicates the percentage of amino acids that are identical to the amino acid sequence of human phenylalanine hydroxylase.

†In some cases, the function of the sequence was determined by biochemical assay. In other cases, the function was inferred due to the high degree of sequence similarity with other species.

itself (100%). The next nine sequences are in order of similarity. The next most similar sequence is from the orangutan (99%), a close relative of humans. This is followed by two mammals, the mouse and rat, and then five vertebrates that are not mammals. The tenth best match is from *Drosophila*, an invertebrate.

You can see two trends in Table 21.5. First, the order of the matches follows the evolutionary relatedness of the various species to humans. The similarity between any two sequences is related to the time that has passed since they diverged from a common ancestor. Among the species listed in this table, the human sequence is most similar to the orangutan, a closely related primate. The next most similar sequences are found in other mammals, followed by other vertebrates, and finally invertebrates. A second trend you may have noticed is that several of the matches involve species that are important from a research, medical, or agricultural perspective. Currently, our genetic databases are biased toward organisms that are of interest to humans, particularly model organisms such as mice and *Drosophila.* Over the next several decades, the sequencing of genomes from many different species will tend to lessen this bias.

The results shown in Table 21.5 illustrate the remarkable computational abilities of current computer technology. In minutes, the human phenylalanine hydroxylase sequence can be compared to hundreds of thousands of different sequences.

Genetic Sequences Can Be Used to Predict the Structure of RNA and Proteins

Another topic in which bioinformatics has impacted functional genomics and proteomics is the area of structure prediction. The function of macromolecules such as DNA, RNA, and proteins relies on their three-dimensional structure, which, in turn, depends on the linear sequences of their building blocks. In the case of DNA and RNA, this means a linear sequence of nucleotides; proteins are composed of a linear sequence of amino acids. Currently, the three-dimensional structure of macromolecules is determined primarily through the use of biophysical techniques such as X-ray crystallography and nuclear magnetic resonance (NMR). These methods are technically difficult and very time-consuming. DNA sequencing, by comparison, requires much less effort. Therefore, because the three-dimensional structure of macromolecules depends ultimately on the linear sequence of their building blocks, it would be far easier if we could predict the structure (and function) of DNAs, RNAs, and proteins from their sequence of building blocks.

RNA molecules typically are folded into a secondary structure, which commonly contains double-stranded regions. This secondary structure is further folded and twisted to adopt a tertiary conformation. Such structural features of RNA molecules are functionally important. For example, the folding of RNA into secondary structures, such as stem-loops, affects transcriptional termination and other regulatory events. Therefore, geneticists are interested in the secondary and tertiary structures that RNA molecules can adopt.

Many approaches are available for investigating RNA structure. In addition to biophysical and biochemical techniques, computer modeling of RNA structure has become an important tool. Modeling programs can consider different types of information. For example, the known characteristics of RNA secondary structure, such as the ability to form double-stranded regions, can provide parameters for use in a modeling program.

A comparative approach can also be used in RNA structure prediction. This method assumes that RNAs of similar function and sequence have a similar structure. For example, the genes that encode certain types of RNAs, such as the 16S rRNA that makes up most of the small ribosomal subunit, have been sequenced from many different species. Among different species, the 16S rRNAs have similar but not identical sequences. Computer programs can compare many different 16S rRNA sequences to aid in the prediction of secondary structure. **Figure 21.12** illustrates a secondary structural model for 16S rRNA from *E. coli* based on a comparative sequence analysis of many bacterial 16S rRNA sequences. This large RNA contains 45 stem-loop regions. As you can imagine, it would be rather difficult to deduce such a model without the aid of a computer! In addition, RNA secondary structure prediction may be aimed at predicting the lowest energy state of a folded molecule. This approach, called free energy minimization, is also related to the base sequence of an RNA molecule.

Structure prediction is also used in the area of proteomics. As described in Chapter 12, proteins contain repeating secondary structural patterns known as α helices and β sheets. Several

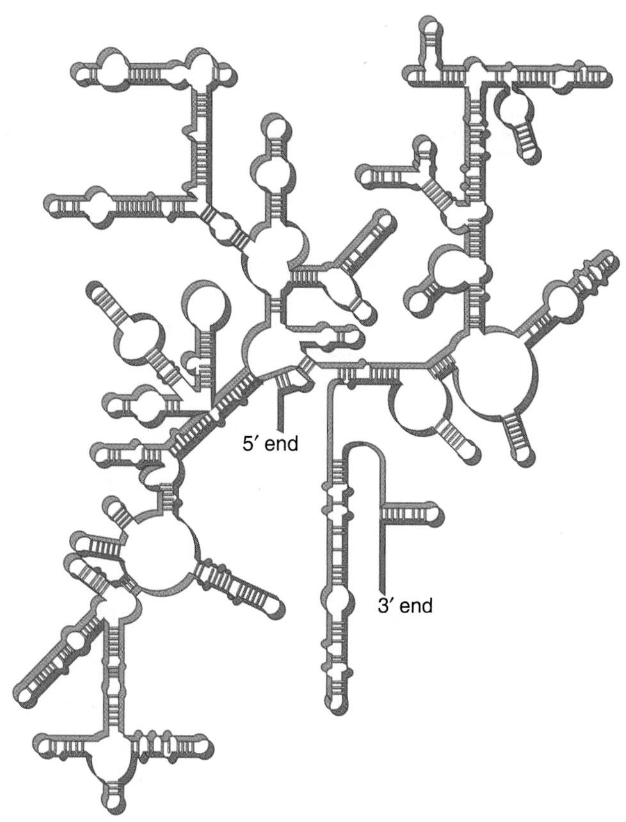

FIGURE 21.12 A secondary structural model for *E. coli* 16S rRNA.

computer-based approaches attempt to predict secondary structure from the primary amino acid sequence. These programs base their predictions on different types of parameters. Some programs rely on the physical and energetic properties of the amino acids and the polypeptide backbone. More commonly, however, secondary structure predictions are based on the statistical frequency of amino acids within secondary structures that have already been crystallized.

For example, Peter Chou and Gerald Fasman have compiled X-ray crystallographic data to calculate the likelihood that an amino acid will be found in an α helix or a β sheet. Certain amino acids, such as glutamic acid and alanine, are likely to be found in an α helix; others, such as valine and isoleucine, are more likely to be found in a β-sheet structure. Such information can be used to predict whether a sequence of amino acids within a protein is likely to be folded in an α-helix or β-sheet conformation. Secondary structure prediction is correct for approximately 60 to 70% of all sequences. While this degree of accuracy is promising, it is generally not sufficient to predict protein secondary structure reliably. Therefore, one must be cautious in interpreting the results of a secondary structure prediction program.

In recent years, an exciting computer methodology known as a neural network has been applied to protein secondary structure prediction. A computer neural network is a large number of calculation units organized into interconnected layers; this structure is reminiscent of the organization of neurons in the brain. The input layer receives data and may (or may not) transmit that information to the next layer. A neural network can adjust the parameters that define the interconnections among its units in response to data; the network can thus be trained to identify complex patterns coming from the input data. For example, the amino acid sequences of proteins with known crystal structures can be used to train a network (i.e., adjust its parameters) to predict secondary structures for new amino acid sequences. Thus far, neural networks have yielded small improvements in the prediction of protein secondary structure. In the future, a combination of innovative predictive approaches and increased information concerning the biophysical properties of amino acids may make secondary structure prediction a reliable strategy.

While the three-dimensional structure of a protein is extremely difficult to predict solely from its amino acid sequence, researchers have had some success in predicting tertiary structure using a comparative approach. This strategy requires that the protein of interest be homologous to another protein, the tertiary structure of which already has been solved by X-ray crystallography. In this situation, the crystal structure of the known protein can be used as a starting point to model the three-dimensional structure of the protein of interest. This approach is known as homology-based modeling, knowledge-based modeling, or comparative homology. As an example, two research groups independently predicted the three-dimensional structure of a protein, now known as HIV protease, that is encoded by a gene in HIV. The protein is homologous to other proteases with structures that had been solved by X-ray crystallography. Two similar models of the HIV protease were predicted before its actual crystal structure was determined. Both models turned out to be fairly accurate representations of the actual structure, which was later solved by X-ray crystallography. Knowledge of the structure of HIV protease helped scientists synthesize compounds known as HIV protease inhibitors, which are an important part of drug therapy to treat HIV/AIDS.

CONCEPTUAL AND EXPERIMENTAL SUMMARY

The field of functional genomics attempts to elucidate the roles of genetic sequences in a given species. This involves the analysis of groups of genes and even the entire genome. A cDNA or EST library can aid in the identification of genes, because the cDNAs are derived from cellular mRNA (a product of gene transcription). A subtractive cDNA library can identify genes that are expressed only under particular environmental conditions. A DNA microarray is a slide that contains numerous DNA fragments, each corresponding to a short sequence within a known gene. Some arrays contain DNA from every known gene within a species' genome. The arrays are used as hybridization tools to measure the expression levels of thousands of genes simultaneously. Fluorescently labeled cDNA is typically used, although researchers can also hybridize labeled genomic DNA to a microarray. By comparing the results of DNA microarray analyses at different time points or under different sets of environmental conditions, researchers can identify genes whose patterns of expression correlate with each other. This is termed a cluster analysis. These genes often encode proteins that participate in common cellular functions. One of the first studies using a DNA microarray involved the analysis of carbohydrate metabolism in yeast.

The entire collection of proteins a species can produce is called its proteome. The study of the function and interaction of these proteins is termed proteomics. Due to alternative splicing and the posttranslational covalent modification of proteins, the proteome may be much larger than the genome. A common technique to detect proteins is via two-dimensional gel electrophoresis, the end result of which is a collection of spots, each representing a unique cellular protein. Mass spectrometry is used to correlate a given spot with a particular protein. Spots from a two-dimensional gel can be removed, and the amino acid sequences of peptide fragments can be determined via tandem mass spectrometry. Recently, microarray analysis has been applied to proteins. In this type of technology, proteins, rather than DNA molecules, are spotted onto a slide. The two common types of protein microarray analysis are antibody microarrays and functional protein microarrays. An antibody microarray consists of a collection of antibodies spotted onto a microarray. Cellular proteins are isolated and exposed to the antibody array, and the level of protein expression from a given cell type can be assessed. A functional protein microarray consists of many different cellular proteins spotted onto an array. This array can probe the function of proteins, such as their substrate specificity, interactions with drugs, or interactions with other cellular proteins.

The computer has become an invaluable tool in genetic studies and a central instrument in the field of bioinformatics, for both the analysis and the storage of the data produced by genetic research. Genetic sequences contained within computer data files can be analyzed by many different programs. For example, a computer program can scan data and identify meaningful features, such as short sequence elements within very long genetic sequences. Likewise, computer programs can be used to find structural genes within nucleotide sequences and compare several homologous genetic sequences in a multiple sequence alignment. In addition, geneticists may take newly identified sequences and search a database to identify homologous sequences using the BLAST program. The identification of homologous genes within a database can provide important clues concerning the function of a newly determined sequence.

Computer programs have been developed to predict the structure of RNA and proteins. These programs rely on known biochemical and biophysical techniques as well as computer modeling and may also use a comparative approach that requires the RNA or protein of interest to be homologous to another RNA or protein of known structure. Some predictive programs are more reliable than others. Nevertheless, future research is expected to improve the predictive capabilities of such computer programs.

PROBLEM SETS & INSIGHTS

Solved Problems

S1. When a cell experiences a change in its environment, it may activate one or more genes as a way to respond to the changes. For example, when confronted with a toxic substance, a cell may turn on genes that encode proteins that can degrade or export the toxin. Let's suppose a researcher is interested in the ability of mouse liver cells to protect themselves against toxic heavy metals, such as mercury. To understand the cellular response, the researcher grows mouse liver cells in the laboratory and divides them into two samples. One of the samples is exposed to mercury, while the other is not. Samples of cDNA were made from these cells, and then a subtractive cDNA library was made as described in Figure 21.1. In this case, the cDNAs derived from the cells that were not exposed to mercury were the cDNAs bound to the column. The results showed that the subtractive library contained seven different types of cDNAs.

A. What do these results mean?

B. What would you do next, assuming that the entire mouse genome is contained within a database?

Answer:

A. The results mean that seven different genes are turned on when liver cells are exposed to mercury.

B. Take the seven different cDNAs and subject them to DNA sequencing. Next, the cDNA sequences are used (via computer programs) to find their genomic matches. This will make it possible to determine the entire gene sequences. Then take the entire coding sequence, for each of the seven genes, and look for homology between each mouse gene sequence and sequences within a database. In other words, you would do a BLAST database search, using each of the seven sequences as the query sequence. If you discovered that any of the seven genes were homologous to other genes whose function is already known, this would give you direct information regarding the probable function of particular genes. For example, if one of the genes turned out to be homologous to a gene in bacteria that encodes a protein already known to function in the export of metal ions, this would suggest that one way that mouse liver cells try to avoid the toxic effects of mercury is to try to pump it out.

S2. To answer this question, you will need to look back at the evolution of the globin gene family, which is shown in Chapter 8, Figure 8.7. Throughout the evolution of this gene family, mutations have occurred that have resulted in globin polypeptides with similar but significantly different amino acid sequences. If we look at the sequence alignment in Figure 21.11, we can make logical guesses regarding the timing of mutations, based on a comparison of the amino acid sequences of family members. What is/are the most probable time(s) that mutations occurred to produce the following amino acid differences? Note: You will have to examine the alignment in Figure 21.11 and the evolutionary timescale in Figure 8.10 to answer this question.

A. Val-111 and Cys-111

B. Met-112 and Leu-112

C. Ser-141, Asn-141, Ile-141, and Thr-141

Answer:

A. We do not know if the original globin gene encoded a cysteine or valine at codon 111. The mutation could have changed cysteine to valine or valine to cysteine. The mutation probably occurred after the duplication that produced the α-globin family and β-globin family (about 300 million years ago) but before the gene duplications that occurred in the last 200 millions years to produce the multiple copies of the globin genes on chromosome 11 and chromosome 16. Therefore, all of the globin genes on chromosome 11 have a valine at codon 111, while all of the globin genes on chromosome 16 have a cysteine.

B. Met-112 occurs only in the ε-globin polypeptide; all of the other globin polypeptides contain a leucine at position 112. Therefore, the primordial globin gene probably contained a leucine codon at position 112. After the gene duplication that produced the ε-globin gene, a mutation occurred that changed this leucine codon into a methionine codon. This would have occurred since the evolution of primates (i.e., within the last 10 or 20 million years).

C. When we look at the possible codons at position 141 (i.e., Ser-141, Asn-141, Ile-141, and Thr-141), we notice that a serine codon is found in θ globin, ξ globin, and γ globin. Because the θ- and ξ-globin genes are found on chromosome 16 and the γ-globin genes are found on chromosome 11, it is probable that serine is the primordial codon and that the other codons (asparagine, isoleucine, and threonine) arose later by mutation of the serine codon. If this is correct, the Thr-141 codon arose after the gene duplications that produced the θ- and ξ-globin genes. The Asn-141 and Ile-141 mutations arose after the gene duplications that produced the γ-globin genes. Therefore, the Thr-141, Asn-141, and Ile-141 arose since the evolution of primates (i.e., within the last 10 or 20 million years).

S3. Using a comparative sequence analysis, the secondary structures of rRNAs have been predicted. Among many homologous rRNAs, one stem-loop usually has the following structure:

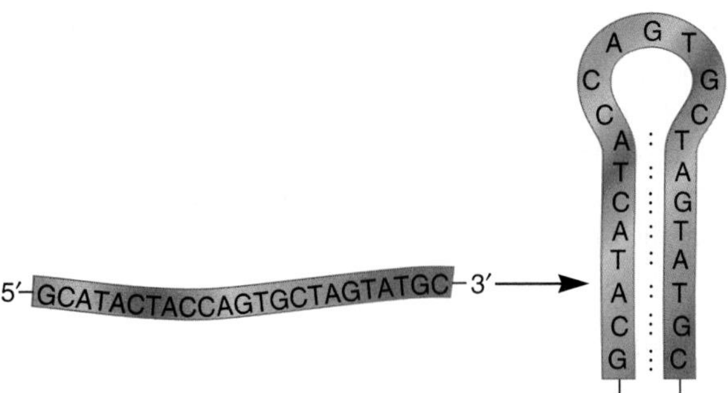

You have sequenced a homologous rRNA from a new species and have obtained most of its sequence, but you cannot read the last five bases on your sequencing gel.

5'-GCATTCTACCAGTGCTAG?????-3'

Of course, you will eventually repeat this experiment to determine the last five bases. However, before you get around to doing this, what do you expect will be the sequence of the last five bases?

Answer: AATGC–3'

This will also form a similar stem-loop structure.

S4. How can codon bias be used to search for structural genes within uncharacterized genetic sequences?

Answer: Most species exhibit a bias in the codons they use within the coding sequence of structural genes. This causes the base content within coding sequences to differ significantly from that of noncoding DNA regions. By knowing the codon bias for a particular species, researchers can use a computer to locate regions that display this bias and thereby identify what are likely to be the coding regions of structural genes.

Conceptual Questions

C1. Discuss the meaning of the following terms: structural genomics, functional genomics, and proteomics.

C2. Discuss the reasons why the proteome is larger than the genome of a given species.

C3. What is a database? What types of information are stored within a database? Where does the information come from? Discuss the objectives of a genome database.

C4. Besides the examples listed in Table 21.4, list five types of short sequences that a geneticist might want to locate within a DNA sequence.

C5. Discuss the distinction between sequence recognition and pattern recognition.

C6. A multiple sequence alignment of five homologous proteins is shown here:

```
  1                                                  50
1 MLAFLNQVRK PTLDLPLEVR RKMWFKPFM. QSYLVVFIGY LTMYLIRKNF
2 MLAFLNQVRK PTLDLALDVR RKMWFKPFM. QSYLVVFIGY LTMYLIRKNF
3 MLPFLKAPAD APL.MTDKYE IDARYRYWRR HILLTIWLGY ALFYFTRKSF
4 MLSFLKAPAN APL.ITDKHE VDARYRYWRR HILITIWLGY ALFYFTRKSF
5 MLSIFKPAPH KAR.LPAA.E IDPTYRRLRW QIFLGIFFGY AAYYLVRKNF

  51                                                100
1 NIAQNDMIST YGLSMTQLGM IGLGFSITYG VGKTLVSYYA DGKNTKQFLP
2 NIAQNDMIST YGLSMTELGM IGLGFSITYG VGKTLVSYYA DGKNTKQFLP
3 NAAVPEILAN GVLSRSDIGL LATLFYITYG VSKFVSGIVS DRSNARYFMG
4 NAAAPEILAS GILTRSDIGL LATLFYITYG VSKFVSGIVS DRSNARYFMG
5 ALAMPYLVEQ .GFSRGDLGF ALSGISIAYG FSKFIMGSVS DRSNPRVFLP
```

Discuss some of the interesting features that this alignment reveals.

C7. What is the difference between similarity and homology?

C8. Which of the following statements uses the term homologous correctly?

A. The two X chromosomes in female mammalian cells are homologous to each other.

B. The α-tubulin gene in *Saccharomyces cerevisiae* is homologous to the α-tubulin gene in *Arabidopsis thaliana.*

C. The promoter of the *lac* operon is homologous to the promoter of the *trp* operon.

D. The *lacY* gene of *E. coli* and *Klebsiella pneumoniae* are approximately 60% homologous to each other.

C9. When comparing (i.e., aligning) two or more genetic sequences, it is sometimes necessary to put in gaps. Explain why. Discuss two changes (i.e., two types of mutations) that could happen during the evolution of homologous genes that would explain the occurrence of gaps in multiple sequence alignments.

Experimental Questions

E1. Explain how a subtractive cDNA library can provide information regarding gene regulation.

E2. Take a look at solved problem S1 regarding a subtractive DNA library. Would you want to load a relatively small amount of cDNA from the cells that had been exposed to mercury, compared to the amount of cDNA that is attached to the column? Explain. What would happen if you ran too much cDNA over the column from the cells that had been exposed to mercury?

E3. With regard to DNA microarrays, answer the following questions:

A. What is attached to the slide? Be specific about the number of spots, the lengths of DNA fragments, and the origin of the DNA fragments.

B. What is hybridized to the microarray?

C. How is hybridization detected?

E4. In the experiment of Figure 21.3, explain how the ratio of red: green fluorescence provides information regarding gene regulation.

E5. What is meant by the term cluster analysis? How is this approach useful?

E6. For two-dimensional gel electrophoresis, what physical properties of proteins promote their separation in the first dimension and the second dimension?

E7. Can two-dimensional gel electrophoresis be used as a purification technique? Explain.

E8. Explain how tandem mass spectroscopy can be used to determine the sequence of a peptide. Once a peptide sequence is known, how is this information used to determine the sequence of the entire protein?

E9. Describe the two general types of protein microarrays. What are their possible applications?

E10. Discuss the strategies that can be used to identify a structural gene using bioinformatics.

E11. What is a motif? Why is it useful for computer programs to identify functional motifs within amino acid sequences?

E12. Discuss why it is useful to search a database to identify sequences that are homologous to a newly determined sequence.

E13. The secondary structure of 16S rRNA has been predicted using a computer-based sequence analysis. In general terms, discuss what type of information is used in a comparative sequence analysis, and explain what assumptions are made concerning the structure of homologous RNAs.

E14. Discuss the basis for secondary structure prediction in proteins. How reliable is it?

E15. To reliably predict the tertiary structure of a protein based on its amino acid sequence, what type of information must be available?

E16. In Figure 21.7, we considered a computer program that can translate a DNA sequence into a polypeptide sequence. A researcher has a sequence file that contains the amino acid sequence of a polypeptide and runs a program that is opposite to the TRANSLATION program. This other program is called BACKTRANSLATE. It can take an amino acid sequence file and determine the sequence of DNA that would encode such a polypeptide. How does this program work? In other words, what are the genetic principles that underlie this program? What type of sequence file would this program generate: a nucleotide sequence or an amino acid sequence? Would the BACKTRANSLATE program produce only a single sequence file? Explain why or why not.

E17. In experimental question E16, we considered a computer program that can translate a DNA sequence into a polypeptide sequence. Instead of running this program, a researcher could simply look the codons up in a genetic code table and determine the sequence by hand. What are the advantages of running this program compared to the old-fashioned way of doing it by hand?

E18. To identify the following types of genetic occurrences, would a program use sequence recognition, pattern recognition, or both?

A. Whether a segment of *Drosophila* DNA contains a P element (which is a specific type of transposable element)

B. Whether a segment of DNA contains a stop codon

C. In a comparison of two DNA segments, whether there is an inversion in one segment compared to the other segment

D. Whether a long segment of bacterial DNA contains one or more genes

E19. The goal of many computer programs is to identify "sequence elements" within a long segment of DNA. What is a sequence element? Give two examples. How is the specific sequence of a sequence element determined? In other words, is it determined by the computer program or by genetic studies? Explain.

E20. Take a look at the multiple sequence alignment in Figure 21.11 of the globin polypeptides from amino acids 101–148.

A. Which of these amino acids are likely to be most important for globin structure and function? Explain why.

B. Which are likely to be least important?

E21. See solved problem S2 before answering this question. Based on the sequence alignment in Figure 21.11, what is/are the most probable time(s) that mutations occurred in the human globin gene family to produce the following amino acid differences?

A. His-119 and Arg-119

B. Gly-121 and Pro-121

C. Glu-103, Val-103, and Ala-103

E22. Below is short nucleotide sequence within a gene. Via the Internet (e.g., see www.ncbi.nlm.nih.gov/), determine what gene this sequence in found within. Also, determine the species in which this gene sequence is found.

```
5'-GGGCGCAATTACTTAACGCCTCGATT
ATCTTCTTGCGCCACTGATCATTA-3'
```

E23. Take a look at solved problem S2 and the codon table found in Chapter 13 (Table 13.2). Assuming that a mutation involving a single-base change is more likely than a double-base change, propose how the Asn-141, Ile-141, and Thr-141 codons arose. In your answer, describe which of the six possible serine codons is/are likely to be the primordial serine codon of the globin gene family and how that codon changed to produce the Asn-141, Ile-141, and Thr-141 codons.

E24. Membrane proteins often have transmembrane regions that span the membrane in an α-helical conformation. These transmembrane segments are about 20 amino acids long and usually contain amino acids with nonpolar (i.e., hydrophobic) amino acid side chains. Researchers can predict whether a polypeptide sequence has transmembrane segments based on the occurrence of segments that contain 20 nonpolar amino acids. To do so, each amino acid is assigned a hydropathy value, based on the chemistry of its amino acid side chain. Amino acids with very nonpolar side chains are given a high (positive) value, whereas amino acids that are charged and/or polar are given low (negative) values. The hydropathy values usually range from about +4 to −4.

Computer programs have been devised that scan the amino acid sequence of a polypeptide and calculate values based on the hydropathy values of the amino acid side chains. The program usually scans a window of seven amino acids and assigns an average hydropathy value. For example, the program would scan amino acids 1–7 and give an average value, then it would scan 2–8 and give a value, then it would scan 3–9 and give a value, and so on, until it reached the end of the polypeptide sequence (i.e., until it reached the carboxyl terminus).

The program then produces a figure, known as a hydropathy plot, which describes the average hydropathy values throughout the entire polypeptide sequence. An example of a hydropathy plot is shown here.

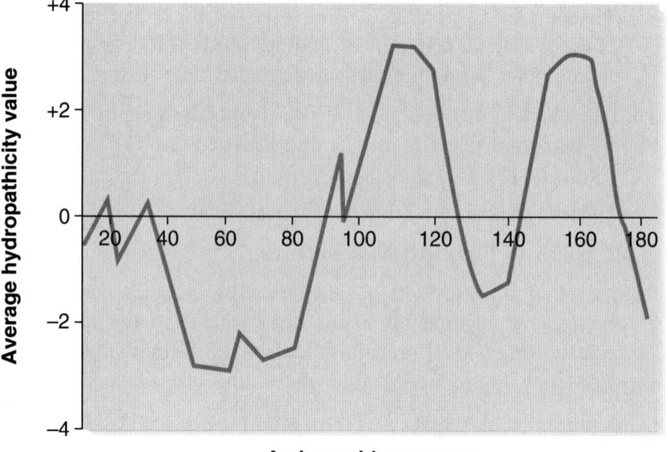

A. How many transmembrane segments are likely in this polypeptide?

B. Draw the structure of this polypeptide if it were embedded in the plasma membrane. Assume that the amino terminus is found in the cytoplasm of the cell.

E25. Explain how a computer program can predict RNA secondary structure. What is the underlying genetic concept used by the program to predict secondary structure?

E26. Are the following statements about protein structure prediction true or false?

A. The prediction of secondary structure relies on information regarding the known occurrence of amino acid residues in α helices or β sheets from X-ray crystallographic data.

B. The prediction of secondary structure is highly accurate, nearly 100% correct.

C. To predict the tertiary structure of a protein based on its amino acid sequence, it is necessary that the protein of interest is homologous to another protein whose tertiary structure is already known.

Questions for Student Discussion/Collaboration

1. Let's suppose you are in charge of organizing and publicizing a genomic database for the mouse genome. Make a list of innovative strategies you would initiate to make the mouse genome database useful and effective.

2. Let's suppose a 5-year-old told you that she was interested in pursuing a career studying the three-dimensional structure of proteins. (Okay, so she's a bit precocious.) Would you advise her to become a geneticist, a mathematical theoretician, or a biophysicist?

3. If you have access to the necessary computer software, make a sequence file and analyze it in the following ways: What is the translated sequence in all three reading frames? What is the longest open reading frame? Is the sequence homologous to any known sequences? If so, does this provide any clues about the function of the sequence?

Note: All answers appear at the website for this textbook; the answers to even-numbered questions are in the back of the textbook.

www.mhhe.com/brookergenetics3e

Visit the Online Learning Center for practice tests, answer keys, and other learning aids for this chapter. Enhance your understanding of genetics with our interactive exercises, quizzes, animations, and much more.

MEDICAL GENETICS AND CANCER

22

CHAPTER OUTLINE

Genetic information is highly personal and unique. Our genes underlie every aspect of human health, both in function and dysfunction. Obtaining a detailed understanding of how genes work together and interact with environmental factors ultimately will allow us to appreciate the differences between the events in normal cellular processes and those that occur in disease pathogenesis. Such knowledge will have a profound impact on the way many diseases are defined, diagnosed, treated, and prevented. Genetic insight is expected to bring about revolutionary changes in medical practices. In fact, changes are already beginning. Currently, several hundred genetic tests are in clinical use, with many more under development. Most of these tests detect mutations associated with rare genetic disorders that follow Mendelian inheritance patterns. These include Duchenne muscular dystrophy, cystic fibrosis, sickle-cell disease, and Huntington disease. In addition, genetic tests are available to detect the predisposition to develop certain forms of cancer.

Approximately 5,000 genetic diseases are known to afflict people, but this is almost certainly an underestimate. Given enough time and effort, scientists can learn how to prevent or treat a great many of them. Most of the genetic disorders discussed in the first part of this chapter are the direct result of a mutation in one gene. However, many diseases have a complex pattern of inheritance involving several genes. These include common medical disorders such as diabetes, asthma, and mental illness. In these cases, a single mutant gene does not determine whether a person has a disease. Instead, a number of genes may each make a subtle contribution to a person's susceptibility to a disease. Unraveling these complexities will be a challenge for some time to come. The availability of the human genome sequence, discussed in Chapter 20, will be of great help.

In this chapter, we will focus our attention on ways that mutant genes contribute to human disease. In the first part of the chapter, we will explore the molecular basis of several genetic disorders and their patterns of inheritance. We will also examine how genetic testing can determine if an individual carries a defective allele. The second part of the chapter will concern cancer, a disease that involves the uncontrolled growth of somatic cells. We will examine the underlying genetic basis for cancer and discuss the roles that many different genes may play in the development of this disease.

Cigarette smoking and lung cancer. Cigarette smoke contains chemicals that are known to mutate genes in the cells of a person's lungs, thereby leading to lung cancer. Lung cancer remains the top cause of cancer death in the United States, with 87% of those deaths linked to smoking.

22.1 GENETIC ANALYSIS OF HUMAN DISEASES

Human genetics is a topic hard to resist. Almost everyone who looks at a newborn is tempted to speculate whether the baby resembles the mother, the father, or perhaps a distant relative. In this section, we will focus primarily on the inheritance of human genetic diseases rather than common traits found in the general population. Even so, the study of human genetic diseases provides insights regarding our traits. The disease hemophilia illustrates this point. Hemophilia (also spelled haemophilia) is a condition in which an individual's blood will not clot properly. By analyzing people with this disorder, researchers have identified genes that participate in the process of blood clotting. The study of hemophilia has helped to elucidate a clotting pathway involving several different proteins. Therefore, as with the study of mutants in model organisms such as *Drosophila,* mice, and yeast, when we study the inheritance of genetic diseases, we often learn a great deal about the genetic basis for normal physiological processes as well.

Because thousands of human diseases have an underlying genetic basis, human genetic analysis is of great medical importance. In this section, we will examine the causes and inheritance patterns of human genetic diseases that result from defects in single genes. As you will learn, the mutant genes that cause these diseases often follow simple Mendelian inheritance patterns.

A Genetic Basis for a Human Disease May Be Suggested from a Variety of Observations

When we view the characteristics of people, we usually think that some traits are inherited, whereas others are caused by environmental factors. For example, when the facial features of two related individuals look strikingly similar, we think that this similarity has a genetic basis. The profound resemblance between identical twins is an obvious example. By comparison, other traits are governed by the environment. If we see a person with purple hair, we likely suspect that he or she has used hair dye as opposed to showing an unusual genetic trait.

For human diseases, geneticists would like to know the relative contributions from genetics and the environment. Is a disease caused by a pathogenic microorganism, a toxic agent in the environment, or a faulty gene? Unlike the case with experimental organisms, we cannot conduct human crosses to elucidate the genetic basis for diseases. Instead, we must rely on analyzing the occurrence of a disease in families that already exist. As described in the following list, several observations are consistent with the idea that a disease is caused, at least in part, by the inheritance of mutant genes. When the occurrence of a disease correlates with several of these observations, a geneticist will become increasingly confident that it has a genetic basis.

1. *When an individual exhibits a disease, this disorder is more likely to occur in genetic relatives than in the general population.* For example, if someone has cystic fibrosis, his or her relatives are more likely to have this disease than will randomly chosen members of the general population.
2. *Identical twins share the disease more often than nonidentical twins.* Identical twins, also called **monozygotic (MZ) twins,** are genetically identical to each other, because they were formed from the same sperm and egg. By comparison, nonidentical twins, also called fraternal, or **dizygotic (DZ) twins,** are formed from separate pairs of sperm and egg cells. Fraternal twins share, on average, 50% of their genetic material, the same as any two siblings. When a disorder has a genetic component, a pair of identical twins is more likely to exhibit the disorder than are two fraternal twins.

   Geneticists evaluate a disorder's **concordance,** the degree to which it is inherited, by calculating the percentage of twin pairs in which both twins exhibit the disorder relative to pairs where only one twin shows the disorder. Theoretically, for diseases caused by a single gene, concordance among identical twins should be 100%. For fraternal twins, concordance for dominant disorders is expected to be 50%, assuming only one parent is a heterozygous individual. For recessive diseases, concordance among fraternal twins would be 25% if we assume both parents are heterozygous carriers. However, the actual concordance values observed for most single-gene disorders are usually less than such theoretical values for a variety of reasons. Some disorders are not completely penetrant, meaning that the symptoms associated with the disorder are not always produced. Also, one twin may have a disorder due to a new mutation that occurred after fertilization; it would be very unlikely for the other twin to have the same mutation.
3. *The disease does not spread to individuals sharing similar environmental situations.* Inherited disorders cannot spread from person to person. The only way genetic diseases can be transmitted is from parent to offspring during sexual reproduction.
4. *Different populations tend to have different frequencies of the disease.* Due to evolutionary forces, the frequencies of traits usually vary among different populations of humans. For example, the frequency of the disease sickle-cell anemia is highest among certain African and Asian populations and relatively low in other parts of the world (see Chapter 24, Figure 24.11).
5. *The disease tends to develop at a characteristic age.* Many genetic disorders exhibit a characteristic **age of onset** at which the disease appears. Some mutant genes exert their effects during embryonic and fetal development, so their effects are apparent at birth. Other genetic disorders tend to develop much later in life.
6. *The human disorder may resemble a disorder that is already known to have a genetic basis in an animal.* In animals, where we can conduct experiments, various traits are known to be governed by genes. For example, the albino phenotype is found in humans as well as in many animals (**Figure 22.1**).

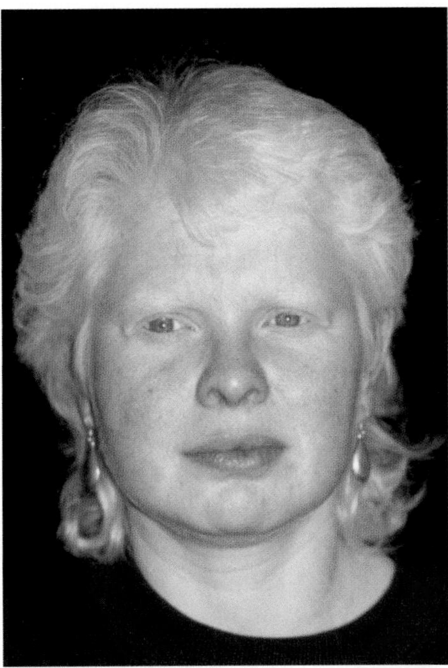

FIGURE 22.1 The albino phenotype in humans and other mammals.

Genes→Traits Certain enzymes (encoded by genes) are necessary for the production of pigment. A homozygote with two defective alleles in these pigmentation genes exhibits an albino phenotype. This phenotype can occur in humans and other animals.

7. *A correlation is observed between a disease and a mutant human gene or a chromosomal alteration.* A particularly convincing piece of evidence that a disease has a genetic basis is the identification of altered genes or chromosomes that occur only in people exhibiting the disorder. When comparing two individuals, one with a disease and one without, we expect them to have differences in their genetic material if the disorder has a genetic component. Alterations in gene sequences can be determined by gene cloning and DNA sequencing techniques (see Chapter 18). Also, changes in chromosome structure and number can be detected by the microscopic examination of chromosomes (see Chapter 8).

Inheritance Patterns of Human Diseases May Be Determined via Pedigree Analysis

When a human disorder is caused by a mutation in a single gene, the pattern of inheritance can be deduced by analyzing human pedigrees. How is this accomplished? A geneticist must obtain data from many large pedigrees containing several individuals who exhibit the disorder and then follow its pattern of inheritance from generation to generation. To appreciate the basic features of pedigree analysis, we will examine a few large pedigrees that involve diseases inherited in different ways. You may wish to review Figure 2.10 for the organization and symbols of pedigrees.

The pedigree shown in **Figure 22.2** concerns a genetic disorder called Tay-Sachs disease (TSD), first described by Warren Tay, a British ophthalmologist, and Bernard Sachs, an American neurologist, in the 1880s. Affected individuals appear healthy at birth but then develop neurodegenerative symptoms at 4 to 6 months of age. The primary characteristics are cerebral degeneration, blindness, and loss of motor function. Individuals with TSD typically die in the third or fourth year of life. This disease is particularly prevalent in Ashkenazi (eastern European) Jewish populations, in which it has a frequency of about 1 in 3,600 births, which is over 100 times more frequent than in most other human populations.

At the molecular level, the mutation that causes TSD is in a gene that encodes the enzyme hexosaminidase A (hexA). HexA is responsible for the breakdown of a category of lipids called G_{M2}-gangliosides, which are prevalent in the cells of the central nervous system. A defect in the ability to break down this lipid leads to its excessive accumulation in nerve cells and eventually causes the neurodegenerative symptoms characteristic of TSD. This defect in lipid breakdown was recognized long before the *hexA* gene was identified as being defective in these patients.

As illustrated in Figure 22.2, Tay-Sachs disease is inherited in an autosomal recessive manner. The four common features of autosomal recessive inheritance are as follows:

1. *Frequently, an affected offspring will have two unaffected parents.* For rare recessive traits, the parents are usually unaffected, meaning they do not exhibit the disease. For deleterious alleles that cause early death or infertility, the two parents must be unaffected. This is always the case in TSD.
2. *When two unaffected heterozygotes have children, the percentage of affected children is (on average) 25%.*
3. *Two affected individuals will have 100% affected children.* This observation can be made only when a recessive trait produces fertile, viable individuals. In the case of TSD, the

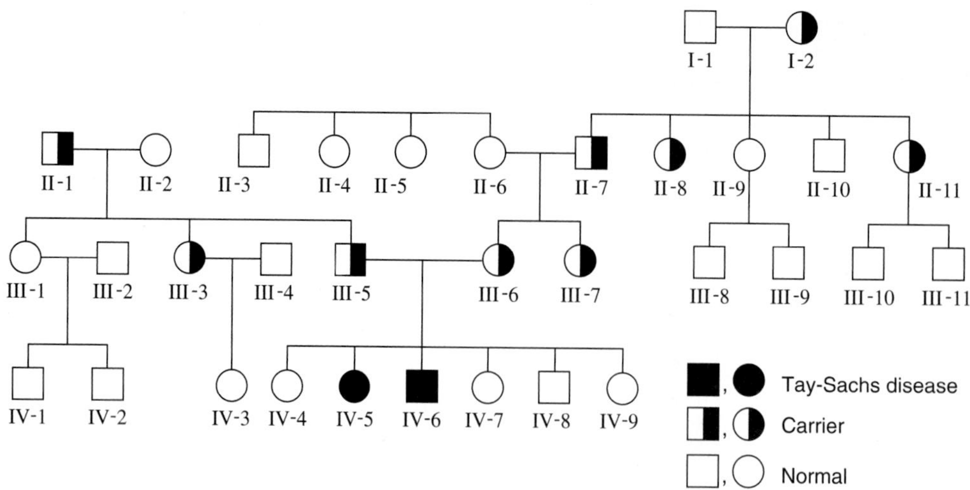

INTERACTIVE EXERCISE **FIGURE 22.2** **A family pedigree of Tay-Sachs disease, indicating recessive inheritance.** Heterozygotes were determined by genetic testing.

affected individual dies in early childhood, and so it is not possible to observe crosses between two affected people.

4. *The trait occurs with the same frequency in both sexes.*

Autosomal recessive inheritance is a common mode of transmission for genetic disorders, particularly those that involve defective enzymes. Human recessive alleles are often caused by mutations that result in a loss of function in the encoded enzyme. In the case of Tay-Sachs disease, a heterozygous carrier has approximately 50% of the functional enzyme, which is sufficient for a normal phenotype. However, for other genetic diseases, the level of functional protein may vary due to effects of gene regulation. Hundreds of human genetic diseases are inherited in a recessive manner, and in many cases, the mutant genes have been identified. Several of these diseases are described in **Table 22.1.**

Now let's examine a human pedigree involving an autosomal dominant disease (**Figure 22.3**). In this example, the affected individuals have a disorder called Huntington disease (also called Huntington chorea). The major symptoms of this disease, which usually occurs during middle age, are due to the degeneration of certain types of neurons in the brain, leading to personality changes, dementia, and early death. In 1993, the gene involved in Huntington disease was identified and sequenced. It encodes a protein called huntingtin that is expressed in neurons but is also found in some cells not affected in Huntington disease. In persons with this disorder, a mutation has added a polyglutamine tract—many glutamines in a row—to the amino acid sequence of the huntingtin protein. This causes an aggregation of the protein in neurons. However, additional research will be needed to elucidate the molecular relationship between the abnormality in the huntingtin protein and the disease symptoms. Five common features of autosomal dominant inheritance are as follows:

1. *An affected offspring usually has one or both affected parents.* However, this is not always the case. Some dominant traits show incomplete penetrance (see Chapter 4), so a heterozygote may not exhibit the trait even though it may be passed to offspring who do exhibit the trait. Also, a

TABLE **22.1**

Examples of Human Disorders Inherited in an Autosomal Recessive Manner

Disorder	Chromosomal Location of Gene	Gene Product	Effective of Disease-Causing Allele
Adenosine deaminase deficiency	20q	Adenosine deaminase	Defective immune system and skeletal and neurological abnormalities
Albinism (type I)	11q	Tyrosinase	Inability to synthesize melanin, resulting in white skin, hair, etc.
Cystic fibrosis (CF)	7q	CF transmembrane conductance regulator	Water imbalance in tissues of the pancreas, intestine, sweat glands, and lungs due to impaired ion transport; leads to lung damage
Phenylketonuria (PKU)	12q	Phenylalanine hydroxylase	Foul-smelling urine, neurological abnormalities, mental impairment; may be remedied by diet modification at birth
Sickle-cell anemia	11p	β globin	Anemia, blockages in blood circulation
Tay-Sachs disease (TSD)	15q	Hexosaminidase A	Progressive neurodegeneration

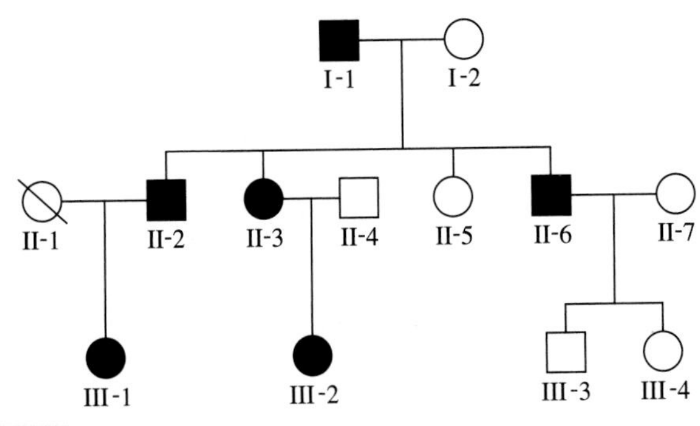

INTERACTIVE EXERCISE **FIGURE 22.3 A family pedigree of Huntington disease, indicating dominant inheritance.**

dominant mutation may occur during gametogenesis, so two unaffected parents may produce an affected offspring.

2. *An affected individual with only one affected parent is expected to produce 50% affected offspring (on average).*
3. *Two affected, heterozygous individuals will have (on average) 25% unaffected offspring.*
4. *The trait occurs with the same frequency in both sexes.*
5. *For most dominant, disease-causing alleles, the homozygote is more severely affected with the disorder. In some cases, a dominant allele may be lethal in the homozygous condition.*

Numerous autosomal dominant diseases have been identified in humans (**Table 22.2**). In these disorders, 50% of the normal protein is not sufficient to produce a normal phenotype. The three common explanations for dominant disorders are haploinsufficiency, a gain-of-function mutation, or a dominant-negative mutation. Let's consider examples of all three types.

The term **haploinsufficiency** refers to the phenomenon in which a person has only a single functional copy of a gene, and that single functional copy does not produce a normal phenotype. Haploinsufficiency shows a dominant pattern of inheritance because a heterozygote (with one normal allele and one inactive allele) will have the disease. An example is aniridia, which is a rare disorder that results in an absence of the iris of the eye. Aniridia leads to visual impairment and blindness in severe cases.

A second category of dominant disorders involves **gain-of-function mutations.** Such mutations change the gene product so that it "gains" a new or abnormal function. An example is achondroplasia, which is characterized by abnormal bone growth that results in short stature with relatively short arms and legs. This disorder is caused by a point mutation that occurs in the fibroblast growth factor receptor-3 gene. In achondroplasia, the mutant form of the receptor is overactive. This overactivity disrupts the normal signaling pathway and leads to severely shortened bones.

A third category of dominant disorders is characterized by **dominant-negative mutations** in which the altered gene product acts antagonistically to the normal gene product. In humans, Marfan syndrome, which is due to a mutation in the *fibrillin-1* gene, is an example. The *fibrillin-1* gene encodes a glycoprotein that is a structural component of the extracellular matrix that provides structure and elasticity to tissues. The mutant gene encodes a fibrillin-1 protein that antagonizes the effects of the normal protein, thereby weakening the elasticity of certain body parts. For example, the walls of the major arteries such as the aorta, the large artery that leaves the heart, are often affected.

Let's now turn to another inheritance pattern common in humans that is called X-linked recessive inheritance (**Table 22.3**). With regard to recessive genetic diseases, X-linked inheritance poses a special problem for males. Why are males more likely to be affected? Most X-linked genes lack a counterpart on the Y

TABLE 22.2

Examples of Human Disorders Inherited in an Autosomal Dominant Manner

Disorder	Chromosomal Location of Gene	Gene Product	Effects of Disease-Causing Allele
Aniridia	11p	Pax6 transcription factor	An absence of the iris of the eye, leading to visual impairment and sometimes blindness
Achondroplasia	4p	Fibroblast growth factor receptor-3	A common form of dwarfism associated with a defect in the growth of long bones
Marfan syndrome	15q	Fibrillin-1	Tall and thin individuals with abnormalities in the skeletal, ocular, and cardiovascular systems due to a weakening in the elasticity of certain body parts
Osteoporosis	7q	Collagen (type 1α2)	Brittle, weakened bones
Familial hypercholesterolemia	19p	LDL receptor	Very high serum levels of low-density lipoprotein (LDL), a predisposing factor in heart disease
Huntington disease	4p	Huntingtin	Neurodegeneration that occurs relatively late in life, usually in middle age
Neurofibromatosis I	17q	Neurofibromin	Individuals may exhibit spots of abnormal pigmentation (café-au-lait spots) and growth of noncancerous tumors in the nervous system

TABLE 22.3

Examples of Human Disorders Inherited in an X-linked Recessive Manner

Disorder	Gene Product	Effects of Disease-Causing Allele
Duchenne muscular dystrophy	Dystrophin	Progressive degeneration of muscles that begins in early childhood
Hemophilia A	Clotting factor VIII	Defect in blood clotting
Hemophilia B	Clotting factor IX	Defect in blood clotting
Androgen insensitivity syndrome	Androgen receptor	Missing male steroid hormone receptor; XY individuals have external features that are feminine but internally have undescended testes and no uterus

chromosome. Males are hemizygous—have a single copy—for these genes. Therefore, a female heterozygous for an X-linked recessive gene will pass this trait to 50% of her sons, as shown in the following Punnett square for hemophilia. In this example, $X^{h\text{-}A}$ is the chromosome that carries a mutant allele causing hemophilia.

♀ \ ♂	X^H	Y
X^H	X^HX^H Normal female	X^HY Normal male
$X^{h\text{-}A}$	$X^HX^{h\text{-}A}$ Carrier female	$X^{h\text{-}A}Y$ Male with hemophilia

As mentioned previously, hemophilia is a disorder in which the blood cannot clot properly when a wound occurs. For individuals with this trait, a minor cut may bleed for a very long time, and small injuries can lead to large bruises, because internal broken capillaries may leak blood profusely before they are repaired. For hemophiliacs, common injuries pose a threat of severe internal or external bleeding. Hemophilia A, also called classical hemophilia, is caused by a defect in an X-linked gene that encodes the protein clotting Factor VIII. This disease has also been called the "Royal disease," because it affected many members of European royal families. The pedigree shown in **Figure 22.4** illustrates the prevalence of hemophilia A among the descendants of Queen Victoria of England. The pattern of X-linked recessive inheritance is revealed by the following observations:

1. *Males are much more likely to exhibit the trait.*
2. *The mothers of affected males often have brothers or fathers who are affected with the same trait.*
3. *The daughters of affected males will produce, on average, 50% affected sons.*

Many Genetic Disorders Exhibit Locus Heterogeneity

Hemophilia can be used to illustrate another concept in genetics called **locus heterogeneity.** This term refers to the phenomenon in which a particular type of disease may be caused by mutations in two or more different genes. For example, blood clotting involves the participation of several different proteins that take part in a cellular cascade that leads to the formation of a clot. Hemophilia is usually caused by a defect in one of three different clotting factors. In hemophilia A, also called classic hemophilia, a protein called Factor VIII is missing. Hemophilia B is a deficiency in a different clotting factor, called Factor IX. Both Factor VIII and IX are encoded by different genes on the X chromosome. These two types of hemophilia show an X-linked recessive pattern of inheritance. By comparison, hemophilia C is due to a Factor XI deficiency. The gene encoding Factor XI is found on chromosome 4, and this form of hemophilia follows an autosomal recessive pattern of inheritance.

In hemophilia, locus heterogeneity arises from the participation of several proteins in a common cellular process. Another mechanism that may lead to locus heterogeneity occurs when proteins are composed of two or more different subunits, with each subunit being encoded by a different gene. The disease thalassemia is an example of locus heterogeneity caused by a mutation in a protein composed of multiple subunits. This potentially life-threatening disease involves defects in the ability of the red blood cells to transport oxygen. The underlying cause is an alteration in hemoglobin. In adults, hemoglobin is a tetrameric protein composed of two α-globin and two β-globin subunits; α globin and β globin are encoded by separate genes (namely, the α-globin and β-globin genes). Two main types of thalassemia have been discovered in human populations: α thalassemia, in which the α-globin subunit of hemoglobin is defective; and β thalassemia, in which the β-globin subunit is defective.

Unfortunately, locus heterogeneity may greatly confound pedigree analysis. For example, a human pedigree might contain individuals with X-linked hemophilia and other individuals with hemophilia C. A geneticist who assumed all affected individuals had defects in the same gene would be unable to explain the resulting pattern of inheritance. For disorders such as hemophilia and thalassemia, pedigree analysis is not a major problem because the biochemical basis for these diseases is well understood. However, for rare diseases that are poorly understood at the molecular level, locus heterogeneity may profoundly obscure the pattern of inheritance.

Disease-Causing Mutant Genes Are Identified by Mapping and DNA Sequencing

How do geneticists identify genes that cause disease when they are mutant? While a variety of approaches may be followed, the hunt often begins with family pedigrees. As described in Chapter 20, the human genome has been extensively mapped with many molecular markers such as microsatellites (refer back to Figure 20.12). By comparing the transmission patterns of many molecular markers with the occurrence of an inherited disease, researchers can pinpoint particular markers that are closely linked to the disease-causing mutant gene.

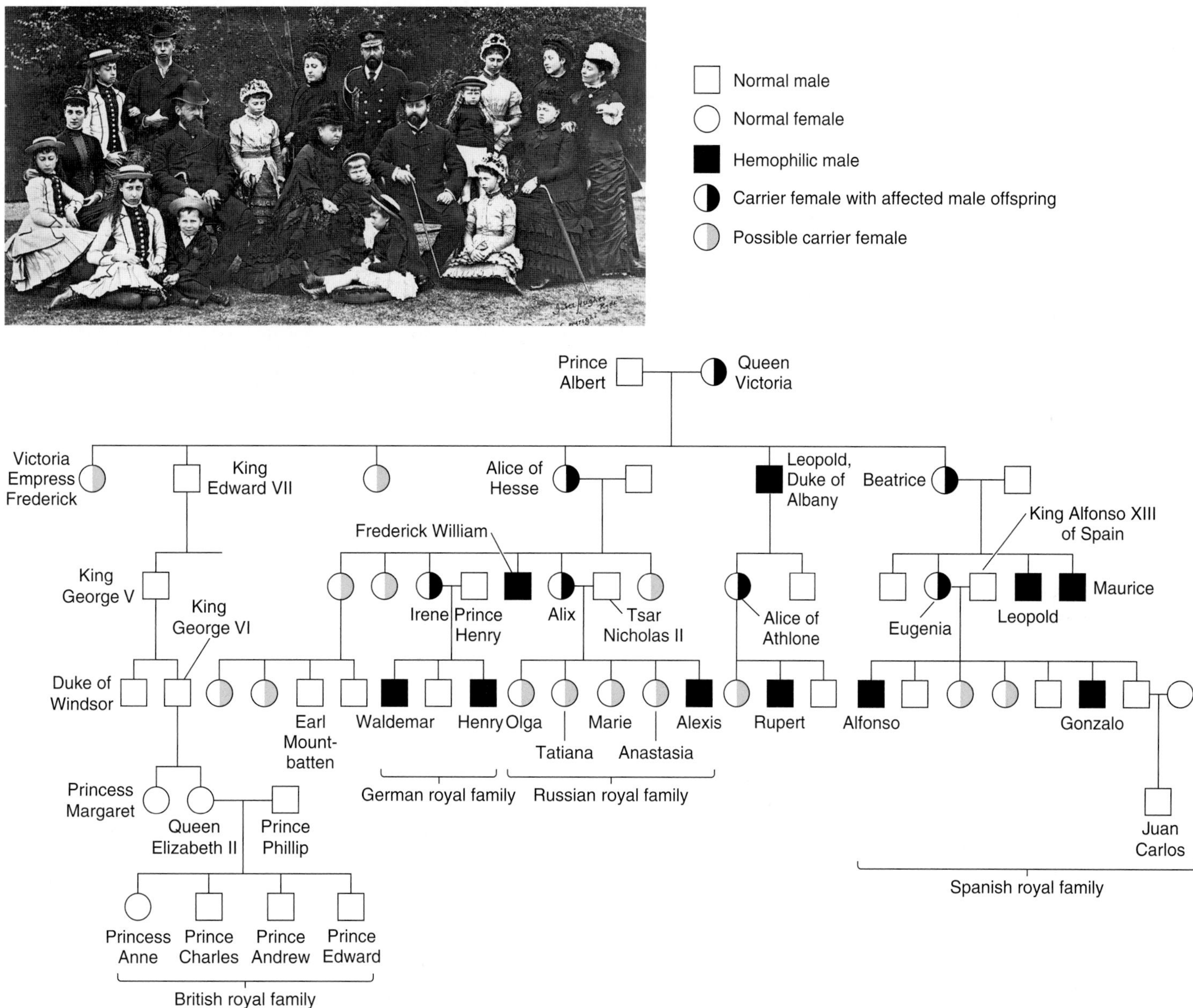

INTERACTIVE EXERCISE

FIGURE 22.4 A family tree of hemophilia A in the royal families of Europe, indicating an X-linked recessive inheritance. Pictured are Queen Victoria and Prince Albert of Great Britain with some of their descendants.

As a classic example, Nancy Wexler, whose own mother died of Huntington disease, has studied this disorder among a population of related individuals in Venezuela. To date, over 15,000 individuals have been analyzed. **Figure 22.5** shows a simplified version of a large family pedigree from this Venezuelan population in which many individuals were affected with Huntington disease. A specific polymorphic marker, called G8, is found in four different versions, named A, B, C, or D. In this Venezuelan population, pedigree analysis revealed that the G8-C marker, which is located near the tip of the short arm of chromosome 4, is almost always associated with the mutant gene causing Huntington disease. In other words, the G8-C marker is closely linked to the Huntington allele.

Once a disease-causing mutant gene has been mapped to a chromosome site, the next step is to determine which gene in the region is responsible for the disease. Modern mapping methods can localize a gene to a chromosome region that is typically about 1 Mb (1,000,000 bp) in length. One way to identify a disease-causing allele is chromosome walking (refer back to Figure 20.18). This is how the *Huntingtin* gene was identified. However, the technique of chromosome walking is no longer widely used because the entire human genome has been sequenced and most genes have been identified. Therefore, researchers can analyze the 1 Mb region where a gene has been mapped to determine if a mutant gene in this region is responsible for a disease. In the human genome, a 1 Mb region will usually contain about 5 to 10

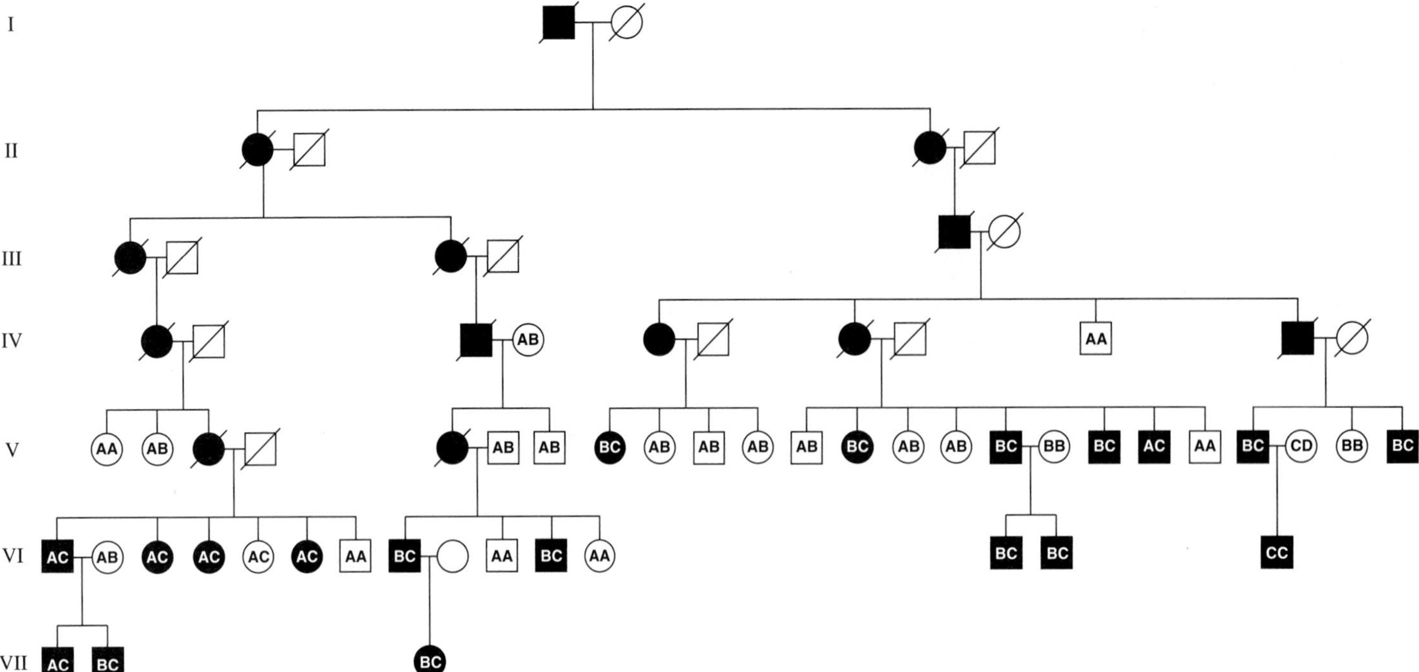

FIGURE 22.5 **The transmission pattern of a molecular marker for Huntington disease.** Affected members are shown with black symbols. The letters under each person indicate the forms of the G8 marker (A, B, C, or D) the individual carries. Affected individuals always carry the C version of this marker. Symbols with slashes indicate deceased individuals. *Data from Gusella et al. (1983) "A polymorphic DNA marker genetically linked to Huntington's disease."* Nature *306, 234–239.*

different genes, though the number can vary greatly. Therefore, mapping does not definitively tell researchers which gene may play a role in human disease, but it usually narrows down the list to a few candidate genes.

How does a researcher determine which of the candidate genes is the correct one? To further narrow down the list, researchers may also consider biological function. As the scientific community explores the functions of genes experimentally, the data are published in the research literature and placed into databases. In some cases, this information may help to narrow down the list of candidate genes. For example, if the disease of interest is neurological, researchers may discover that only certain genes in the mapped region are expressed in nerve cells. Also, researchers may compare data from other organisms. If a mutant gene in a mouse causes neurological problems and a human homolog of the mouse gene is found in the mapped region, this human gene would be a good candidate for being responsible for the human disease symptoms.

After researchers have narrowed down the number of candidate genes to as short a list as possible, the next phase is to sequence the candidate gene(s) from many affected and unaffected individuals, using the dideoxy sequencing method described in Chapter 18. The goal is to identify a gene in which affected individuals always carry a mutation. Once a gene has been identified that has a genetic change found in affected individuals, this is strong evidence that the candidate gene is causing the disease symptoms.

Why is it useful to identify disease-causing mutant genes? In many cases, the identification of these genes will help us to understand how genes contribute to pathogenesis and may even aid in developing strategies aimed at the treatment of the disease. As described next, the identification of such genes may also result in the development of genetic tests that can enable people to determine if they are carrying disease-causing alleles.

Genetic Testing Can Identify Many Inherited Human Diseases

Because genetic abnormalities occur in the human population at a significant level, people have sought ways to determine whether individuals possess disease-causing alleles. The term **genetic testing** refers to the use of testing methods to determine if an individual carries a genetic abnormality. **Table 22.4** describes several different testing strategies. The term **genetic screening** refers to population-wide genetic testing.

In many cases, single-gene mutations that affect the function of cellular proteins can be examined at the protein level. If a gene encodes an enzyme, biochemical assays to measure that enzyme's activity may be available. As mentioned earlier, Tay-Sachs disease involves a defect in the enzyme hexosaminidase A (hexA). Enzymatic assays for this enzyme involve the use of an artificial substrate in which 4-methylumbelliferone (MU) is covalently linked to *N*-acetylglucosamine (GlcNAc). HexA cleaves this covalent bond and releases MU, which is fluorescent.

$$\underset{\text{(nonfluorescent)}}{\text{MU–GlcNAc}} \xrightarrow{\text{hexA}} \underset{\text{(fluorescent)}}{\text{MU}} + \text{GlcNAc}$$

TABLE 22.4

Testing Methods for Genetic Abnormalities

Method	Description
Protein Level	
Biochemical	As described in this chapter for Tay-Sachs disease, the enzymatic activity of a protein can be assayed in vitro.
Immunological	The presence of a protein can be detected using antibodies that specifically recognize that protein. Western blotting is an example of this type of technique (see Chapter 18).
DNA or Chromosomal Level	
RFLP analysis	As described in the experiment of Figure 20.10, RFLP analysis can also be used to determine the likelihood that a person may carry a disease-causing allele.
Altered restriction site	A gene mutation may alter the presence of a particular restriction enzyme site. A sample of DNA from an individual can be subjected to restriction enzyme digestion and gel electrophoresis (see Chapter 18) to see whether or not the restriction enzyme site is present.
DNA sequencing	If the normal gene has already been identified and sequenced, it is possible to design PCR primers that can amplify the gene from a sample of cells. The amplified DNA segment can then be subjected to DNA sequencing, as described in Chapter 18.
In situ hybridization	A DNA probe that hybridizes to a particular gene or gene segment can be used to determine if the gene is present, absent, or altered in an individual. The technique of fluorescence *in situ* hybridization (FISH) is described in Chapter 20 (see Figures 20.3, 20.4).
Karyotyping	The chromosomes from a sample of cells can be stained and then analyzed microscopically for abnormalities in chromosome structure and number (see Figure 3.2).
DNA microarrays	This method, which is described in Chapter 21, can be used to determine the expression levels of genes, such as those that are mutant in certain forms of cancer. In addition, microarrays are used to detect polymorphisms found in the human population that are associated with diseases.

To perform this assay, a small sample of cells is collected and incubated with MU–GlcNAc, and the fluorescence is measured with a device called a fluorometer. Individuals affected with Tay-Sachs, who do not produce the hexA enzyme, will produce little or no fluorescence, whereas individuals who are homozygous for the normal *hexA* allele produce a high level of fluorescence. Heterozygotes, who have 50% hexA activity, produce intermediate levels of fluorescence.

An alternative and more common approach is to detect single-gene mutations at the DNA level. To apply this testing strategy, researchers must have previously identified the mutant gene using molecular techniques. The identification of many human genes, such as those involved in Duchenne muscular dystrophy, cystic fibrosis, and Huntington disease, has made it possible to test for affected individuals or those who may be carriers of these diseases. Table 22.4 describes several ways to test for gene mutations. These laboratory techniques are described in Chapters 18 and 20.

Many human genetic abnormalities involve changes in chromosome number and/or structure. In fact, changes in chromosome number are a common class of human genetic abnormality. Most of these result in spontaneous abortions. However, approximately 1 in 200 live births are aneuploid—have an abnormal number of chromosomes (see Chapter 8, Table 8.1). About 5% of infant and childhood deaths are related to such genetic abnormalities. With regard to testing, changes in chromosome number and many changes in chromosome structure can be detected by karyotyping the chromosomes with a light microscope.

In the United States, genetic screening for certain disorders has become common medical practice. For example, pregnant women over 35 years old often have tests conducted to see if their fetuses are carrying chromosomal abnormalities. As discussed in Chapter 8, these tests are indicated because the rate of such defects increases with the age of the mother. Another example is the widespread screening for phenylketonuria (PKU). An inexpensive test can determine if newborns have this disease. Those who test positive can then be given a low-phenylalanine diet to avoid PKU's devastating effects.

Genetic screening also has been conducted on specific populations in which a genetic disease is prevalent. For example, in 1971, community-based screening for heterozygous carriers of Tay-Sachs disease was begun among specific Ashkenazi Jewish populations. With the use of this screening, over the course of one generation, the incidence of TSD births was reduced by 90%. For most rare genetic abnormalities, however, genetic screening is not routine practice. Rather, genetic testing is performed only when a family history reveals a strong likelihood that a couple may produce an affected child. This typically involves a couple that already has an affected child or has other relatives with a genetic disease.

Genetic testing can be performed prior to birth, which may affect a woman's decision to terminate a pregnancy. The two common ways of obtaining cellular material from a fetus for the purpose of genetic testing are **amniocentesis** and **chorionic villus sampling.** In amniocentesis, a doctor removes amniotic fluid containing fetal cells, using a needle that is passed through the abdominal wall (**Figure 22.6**). The fetal cells are cultured for several weeks and then karyotyped to determine the number of chromosomes per cell and whether changes in chromosome structure have occurred. In chorionic villus sampling, a small piece of the chorion (the fetal part of the placenta) is removed, and a karyotype is prepared directly from the collected cells. Chorionic villus sampling can be performed earlier

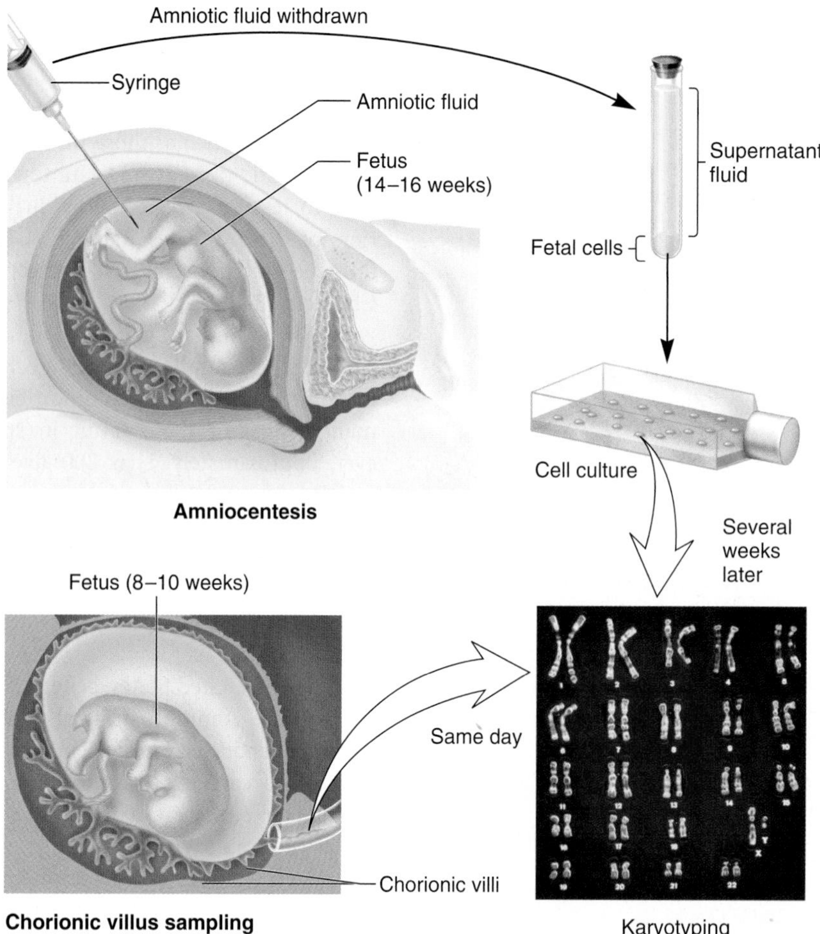

FIGURE 22.6 Techniques to determine genetic abnormalities before birth. In amniocentesis, amniotic fluid is withdrawn, and fetal cells are collected by centrifugation. The cells are then allowed to grow in laboratory culture media for several weeks prior to karyotyping. In chorionic villus sampling, a small piece of the chorion is removed. These cells can be prepared directly for karyotyping.

during pregnancy than amniocentesis, usually around the eighth to tenth week, compared with the fourteenth to sixteenth week for amniocentesis, and results are available sooner. Weighed against these advantages, however, is that chorionic villus sampling may pose a slightly greater risk of causing a miscarriage.

Genetic testing and screening are medical practices with many social and ethical dimensions. For example, people must decide whether or not they want to make use of available tests, particularly when the disease in question has no cure. For example, Huntington disease typically does not affect people until their 50s and can last 20 years. People who learn they are carriers of genetic diseases such as Huntington disease can be devastated by the news. Some argue that people have a right to know about their genetic makeup; others assert that it does more harm than good. Another issue is privacy. Who should have access to personal genetic information, and how could it be used? Could routine genetic testing lead to discrimination by employers or medical insurance companies? In the coming years, we will gain an ever-increasing awareness of our genetic makeup and the underlying causes of genetic diseases. As a society, establishing guidelines for the uses of genetic testing will be a necessary, yet very difficult, task.

Prions Are Infectious Particles That Alter Protein Function Posttranslationally

Before we end this section, let's consider an unusual mechanism in which agents known as prions cause disease. As shown in Table 22.5, prions cause several types of neurodegenerative diseases affecting humans and livestock, including mad cow disease. Recent evidence has shown that prions also exist in yeast. In the 1960s, British researchers Tikvah Alper and J. S. Griffith discovered that preparations from animals with certain neurodegenerative diseases remained infectious even after exposure to radiation that would destroy any DNA or RNA. They suggested that the infectious agent was a protein. Furthermore, Alper and Griffith speculated that the protein usually preferred one folding pattern but could sometimes misfold and then catalyze other proteins to do the same. In the early 1970s, Stanley Prusiner, moved by the death of a patient from a neurodegenerative disease, began to search for the causative agent. In 1982, he isolated a disease-causing agent composed entirely of protein, which he called a **prion.** The term emphasizes the prion's unusual character as a proteinaceous infectious agent. Before the discovery of prions, all known infectious agents such as viruses and bacteria contained their own genetic material (either DNA or RNA).

TABLE 22.5

Neurodegenerative Diseases Caused by Prions*

Disease	Description
Infectious Diseases	
Kuru	A human disease that was once common in New Guinea. It begins with a loss of coordination, usually followed by dementia. Infection was spread by cannibalism, a practice that ended in 1958.
Scrapie	A disease of sheep and pigs characterized by intense itching in which the animals tend to scrape themselves against trees, followed by neurodegeneration
Mad cow disease	Begins with changes in posture and temperament, followed by loss of coordination and neurodegeneration
Human Inherited Diseases	
Creutzfeldt-Jakob disease	Characterized by loss of coordination and dementia
Gerstmann-Straussler-Scheinker disease	Characterized by loss of coordination and dementia
Familial fatal insomnia	Begins with sleeping and autonomic nervous system disturbances followed by insomnia and dementia

*All of these diseases are eventually fatal.

Prion-related diseases arise from the ability of the prion protein to exist in two conformational states: a normal form PrP^C, which does not cause disease, and an abnormal form, PrP^{Sc}, which does. (Note: The superscript C refers to the normal conformation, while the superscript Sc refers to the abnormal conformation, such as the one found in the disease called scrapie.) The gene encoding the prion protein (*PrP*) is found in humans and other mammals, and the protein is expressed at low levels in certain types of cells such as nerve cells. The abnormal conformation of the prion protein can come from two sources. An individual can be infected with the abnormal protein by taking the abnormal protein into their bodies. For example, someone may eat products from an animal that had the disease. Alternatively, some people carry alleles of the *PrP* gene that cause their prion protein to convert spontaneously to the abnormal conformation at a very low rate. These individuals have an inherited predisposition to develop a prion-related disease. An example of an inherited prion disease is familial fatal insomnia (Table 22.5).

What is the molecular mechanism through which prions cause disease? As noted, the prion protein can exist in two conformations, PrP^C and PrP^{Sc}. As shown in **Figure 22.7**, the abnormal conformation, PrP^{Sc}, acts as a catalyst to convert normal prion proteins within the cell to the misfolded conformation. As a prion disease progresses, the PrP^{Sc} protein forms dense aggregates in the cells of the brain and peripheral nervous tissues. This deposition is correlated with the disease symptoms affecting the nervous system. Some of the abnormal prion protein is also excreted from infected cells and can travel through the nervous system to infect other cells.

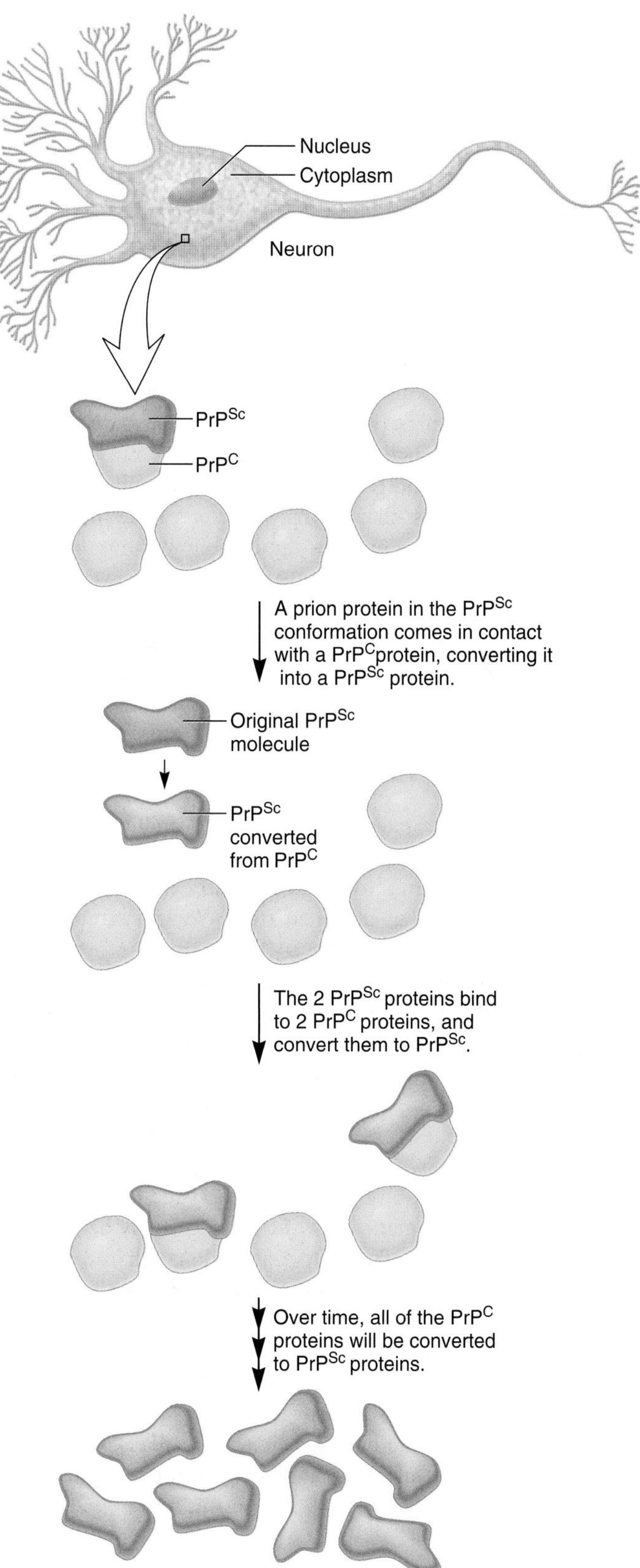

ONLINE ANIMATION

FIGURE 22.7 A proposed molecular mechanism of prion diseases. The PrP^{Sc} protein catalyzes the conversion of PrP^C to PrP^{Sc}. Over time, the PrP^{Sc} conformation will accumulate to high levels, leading to symptoms of the prion disease.

22.2 GENETIC BASIS OF CANCER

Cancer is a disease characterized by uncontrolled cell division. It is a genetic disease at the cellular level. More than 100 kinds of human cancers have been identified, and they are classified according to the type of cell that has become cancerous. Though cancer is a diverse collection of many diseases, some characteristics are common to all cancers.

1. Most cancers originate in a single cell. This single cell, and its line of daughter cells, undergoes a series of genetic changes that accumulate during cell division. In this regard, a cancerous growth can be considered to be **clonal** in origin. A hallmark of a cancer cell is that it divides to produce two daughter cancer cells.
2. At the cellular and genetic levels, cancer usually is a multistep process that begins with a precancerous genetic change—a **benign** growth—and is followed by additional genetic changes that lead to cancerous cell growth (**Figure 22.8**).
3. When cells have become cancerous, their growth is described as **malignant.** Cancer cells are **invasive**—they can invade healthy tissues—and **metastatic**—they can migrate to other parts of the body and cause secondary tumors.

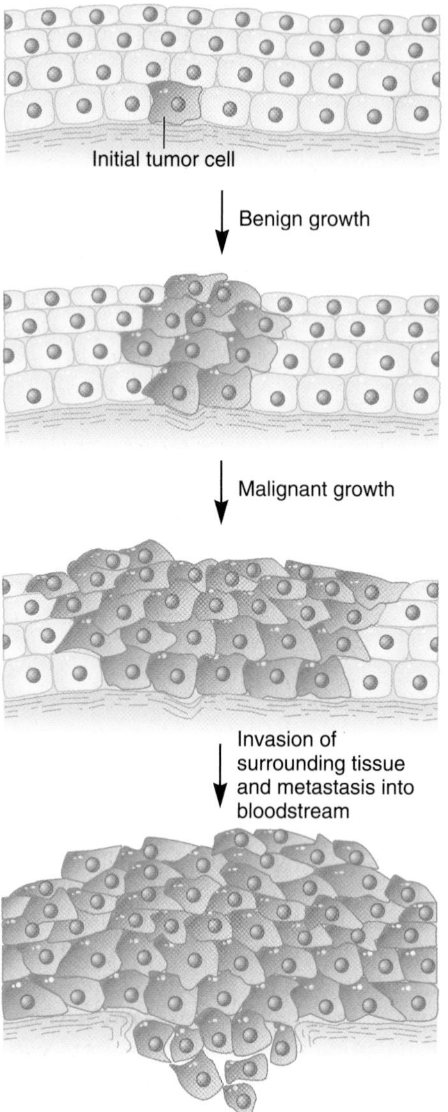

FIGURE 22.8 Progression of cellular growth leading to cancer.
Genes→Traits In a healthy individual, an initial gene mutation converts a normal cell into a tumor cell. This tumor cell divides to produce a benign tumor. Additional genetic changes in the tumor cells may occur, leading to malignant growth. At a later stage in malignancy, the tumor cells will invade surrounding tissues, and some malignant cells may metastasize by traveling through the bloodstream to other parts of the body where they can grow and cause secondary tumors. As a trait, cancer can be viewed as a series of genetic changes that eventually lead to uncontrolled cell growth.

In the United States, approximately 1 million people are diagnosed with cancer each year, and about half that number will die from the disease. In 5 to 10% of all cases, a predisposition to develop the cancer is an inherited trait. We will examine some inherited forms of cancer later in this chapter. Most cancers, though, perhaps 90 to 95%, are not passed from parent to offspring. Rather, cancer is usually an acquired condition that typically occurs later in life. While some cancers are caused by spontaneous mutations and viruses, at least 80% of all human cancers are related to exposure to agents that promote genetic changes in somatic cells. These environmental agents, such as UV light and certain chemicals, are mutagens that alter the DNA in a way that affects the function of normal genes. If the DNA is permanently modified in somatic cells, such changes may be transmitted during cell division. These DNA alterations can lead to effects on gene expression that ultimately affect cell division and thereby lead to cancer. An environmental agent that causes cancer in this manner is called a **carcinogen.**

In this section, we will begin by considering some early experimental observations that suggested genes play a role in cancer. We will then explore how genetic abnormalities, which affect the functions of particular cellular proteins, can lead to cancer.

Certain Viruses Can Cause Cancer by Carrying Viral Oncogenes into the Cell

As mentioned, most types of cancers are caused by mutagens that alter the structure and expression of genes. A few viruses, however, are known to cause cancer in plants, animals, and humans. We begin our discussion here, because early studies of cancer-causing viruses identified the first genes that play a role in cancer. Many of these viruses can also infect normal laboratory-grown cells and convert them into malignant cells. In cancer biology, the process of converting a normal cell into a malignant cell is called **transformation.**

Most cancer-causing viruses are not very potent at inducing cancer. An organism must be infected for a long period of time before a tumor actually develops. Furthermore, most viruses are inefficient at transforming or are unable to transform normal cells grown in the laboratory. By comparison, a few types of viruses rapidly induce tumors in animals and efficiently transform cells in culture. These are called **acutely transforming viruses (ACTs).** About 40 such viruses have been isolated from chickens, turkeys, mice, rats, cats, and monkeys. The first virus of this type, the Rous sarcoma virus (RSV), was isolated from chicken sarcomas by Peyton Rous in 1911.

During the 1970s, RSV research led to the identification of the first gene known to cause cancer. Researchers investigated RSV by infecting chicken fibroblast cells in the laboratory. This causes the chicken fibroblasts to grow like cancer cells. During the course of their studies, researchers identified mutant RSV strains that infected and proliferated within chicken cells without transforming them into malignant cells. These RSV strains were determined to contain a defective viral gene. In contrast, in other strains where this gene is functional, cancer occurs. This viral gene was designated *src* for sarcoma, the type of cancer it causes. The *src* gene is known as v-*src* because it is found within a viral genome. The v-*src* gene was the first example of an **oncogene,** a gene that promotes cancer.

Because the viral oncogene of RSV is not necessary for viral replication, researchers were curious as to why the virus has this v-*src* oncogene. Harold Varmus and Michael Bishop, in collaboration with Peter Vogt, soon discovered that normal host cells contain a copy of the *src* gene in their chromosomes. This normal copy of the *src* gene found in the chromosomal DNA of the host is termed c-*src*, for cellular *src*. The c-*src* gene does not cause cancer. However, once incorporated into a viral genome, this gene can become a viral oncogene that promotes cancer. Why does cancer occur when the gene is in a viral genome? Three explanations of this phenomenon are possible. First, the many copies of the virus made during viral replication may lead to overexpression of the *src* gene. Second, the incorporation of the *src* gene next to viral regulatory sequences may cause it to be overexpressed. A third possibility is that the v-*src* gene may accumulate additional mutations that convert it to an oncogene.

RSV has acquired the *src* gene by capturing it from a host cell's chromosome. This can occur during the RSV life cycle. RSV is a retrovirus with an RNA genome. During its life cycle, a retrovirus uses reverse transcriptase to make a DNA copy of its genome, which becomes integrated as a provirus into the host cell genome. This integration could occur next to the c-*src* gene. During transcription of the proviral DNA, the *src* gene could be included in the RNA transcript that becomes the retroviral genome.

Since these early studies of RSV, many other retroviruses carrying oncogenes have been investigated. The characterization of such oncogenes has led to the identification of several genes with cancer-causing potential. In addition to retroviruses, several viruses with DNA genomes cause tumors, and some of these are known to cause cancer in humans (**Table 22.6**). Researchers estimate that up to 15% of all human cancers are associated with viruses.

TABLE **22.6**

Examples of Viruses That May Cause Cancer

Virus	Description
Retroviruses	
Rous sarcoma virus	Causes sarcomas in chickens.
Simian sarcoma virus	Causes sarcomas in monkeys.
Abelson leukemia virus	Causes leukemia in mice.
Hardy-Zuckerman-4 feline sarcoma virus	Causes sarcomas in cats.
DNA tumor viruses	
Hepatitis B, SV40, polyomavirus	Causes liver cancer in several species including humans. In some cases, these viruses do not cause cancer in their natural hosts but can transform cells in culture.
Papillomavirus	Causes benign tumors and malignant carcinomas in several species, including humans. Causes cervical cancer in humans.
Adenovirus	Does not cause cancer in its natural host but can transform cells in culture.
Herpesvirus	Causes carcinoma in frogs and T-cell lymphoma in chickens. A human herpesvirus, Epstein-Barr virus, is a causative agent in Burkitt's lymphoma, which occurs primarily in immunosuppressed individuals such as AIDS patients.

EXPERIMENT 22A

DNA Isolated from Malignant Mouse Cells Can Transform Normal Mouse Cells into Malignant Cells

The study of retroviruses and other tumor-producing viruses led to the identification of a few dozen viral oncogenes, which were the first genes implicated in causing cancer. However, most cancers are not caused by viruses but instead are due to environmental mutagens, which cause mutations that alter the expression of normal cellular genes. Therefore, researchers also wanted to identify cellular genes that have been altered in a way that leads to malignancy. Methods developed in the study of viral oncogenes proved valuable in this search. Particularly useful were studies conducted in 1971 by Miroslav Hill and Jana Hillova. These experiments demonstrated that the purified DNA of RSV-infected cells could be taken up by chicken fibroblasts and would transform them into malignant cells.

In 1979, Robert Weinberg and colleagues wanted to determine if chromosomal DNA purified from cells that had become malignant due to exposure to mutagens could transform normal cells into malignant cells. If so, this would be the first step in the identification of chromosomally located genes that had been converted to oncogenes. Before we discuss this experiment, let's consider how a researcher can identify malignant cells in a laboratory. A widely used assay relies on the ability to recognize a clump of transformed cells as a distinct **focus,** or pile of raised cells, on a culture dish (**Figure 22.9**). Unlike normal cells, which grow as a single layer, or monolayer, on culture dishes, a malignant focus piles up over the monolayer to form a mass of cells. The focus of cells is a clone derived from a single cell that had become malignant and then proliferated. At the microscopic level, the malignant cells also have altered morphologies.

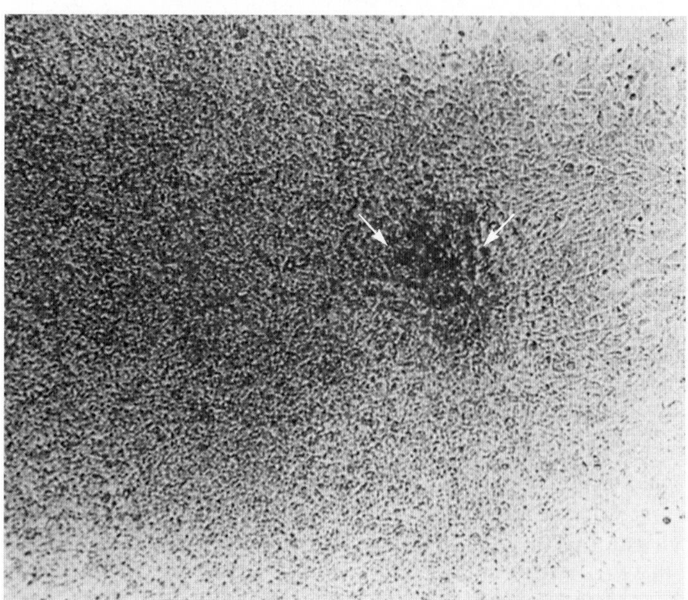

FIGURE 22.9 **The malignant growth of cells.** The growth of normal cells leads to the formation of a monolayer, while certain transformed cells grow as a focus (see arrows) or a raised pile of cells. (Reprinted by permission from Macmillan Publishers Ltd. *Nature.* Transforming activity of DNA of chemically transformed and normal cells. Geoffrey M. Cooper, S. Okenquist & L. Silverman. 284:5755, 418-421, 1980.)

Now that we understand how malignant cells are identified in a laboratory, let's examine the steps of the experiment of Weinberg and coworkers. They began with several malignant cell lines that had been previously characterized, along with normal cell lines. As shown in **Figure 22.10**, they isolated the DNA from the malignant and normal cells, and then separately mixed this DNA with normal mouse fibroblast cells. Calcium phosphate was added to promote the uptake of the DNA by the mouse fibroblast cells. Once inside the cells, the DNA can be incorporated into the chromosomes of the cells by recombination. The cells were incubated for two to three weeks, and then the researchers examined the plates for the growth of malignant foci of cells. In other words, they looked to see if any of the normal mouse cells had been transformed into malignant cells.

THE HYPOTHESIS

Cellular DNA isolated from malignant cells will be taken up by normal mouse fibroblast cells and will transform them into malignant cells.

TESTING THE HYPOTHESIS — FIGURE 22.10 Identification of chromosomal oncogenes.

Starting material: Several mouse cell lines. Some of the cell lines were normal, whereas others were malignant due to exposure to chemical or physical mutagens. It was known that none of the cell lines in this experiment were infected with oncogenic viruses.

1. Extract the chromosomal DNA from normal or malignant cell lines.

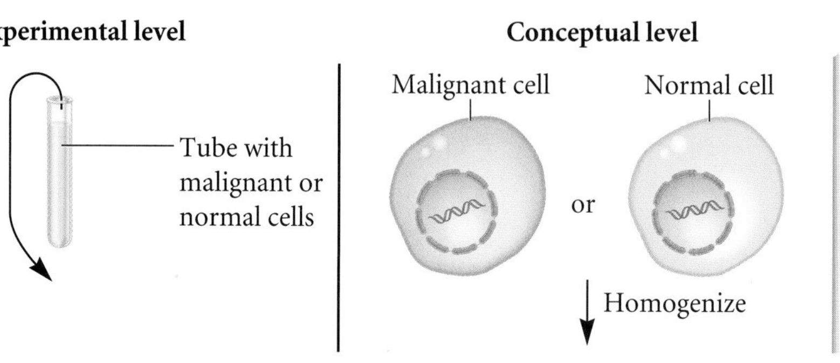

Homogenize the cells to break them open.

Isolate the nuclei by differential centrifugation. See the Appendix.

Dissolve the nuclear membrane with detergent, and isolate the chromosomal DNA.

2. Mix the DNA from the normal cells or from the malignant cells with normal mouse fibroblast cells that are growing on a tissue culture plate.

3. Add a buffer containing calcium ions and phosphate ions. This buffer makes the cells permeable to DNA.

4. Incubate for 14–20 days.

5. Examine the plates under a light microscope for cells growing as transformed foci.

Note: Normal cells in culture grow as a monolayer and very rarely form foci.

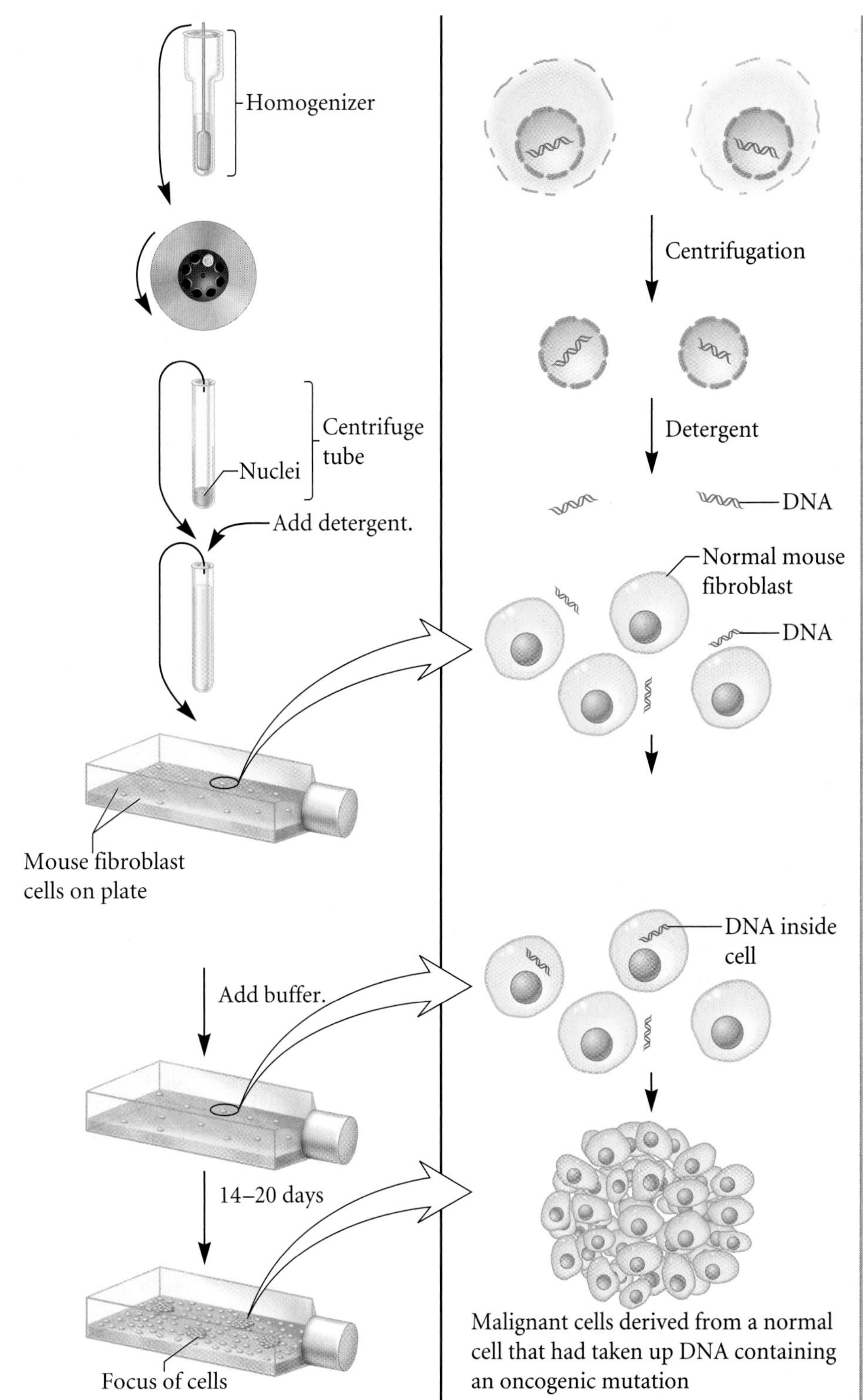

THE DATA

Source of DNA	*Recipient Cells*	*Number of Malignant Foci Found on 12 Plates*
Malignant Cell Lines		
MC5-5-0	NIH3T3 (normal fibroblasts)	48*
MCA16	"	5
MB66 MCA ad 36	"	8
MB66 MCA ACL 6	"	0
MB66 MCA ACL 13	"	0
Normal Cell Lines		
NIH3T3	"	1
C3H10T1/2	"	0

*In this experiment, two of the plates were contaminated, so this is 48 foci on 10 plates.

INTERPRETING THE DATA

As shown in the data, the DNA isolated from some (but not all) malignant cell lines could transform normal mouse cells, which then proliferated and produced malignant foci. These results are consistent with the hypothesis that oncogenes had been taken up and expressed in the normal mouse cells, converting them into malignant cells. By comparison, the DNA isolated from normal cells did not cause a significant amount of transformation. From these experiments, it is not clear why the DNA from two of the malignant cell lines, MB66 MCA ACL 6 and MB66 MCA ACL 13, could not transform the normal mouse cells. One possibility is that some malignancies are caused by dominant oncogenes, while others involve genes that act recessively. Recessive genes would be unable to transform normal mouse cells that already contain the normal (nonmalignant) dominant allele. Later, we will learn how another category of genes involved in cancer, called tumor-suppressor genes, act recessively.

Just two years later, in 1981, the laboratories of Weinberg and Geoffrey Cooper identified the first cellular oncogene in humans. They isolated chromosomal DNA from human bladder carcinoma cells and used it to transform mouse cells in vitro. This chromosomal DNA contained a human oncogene that was the result of a mutation. The identity of the gene was eventually determined by transposon tagging, as described in solved problem S2 at the end of this chapter. This observation paved the way for the isolation of many human cellular oncogenes.

A self-help quiz involving this experiment can be found at the Online Learning Center.

Many Oncogenes Have Abnormalities That Affect Proteins Involved in Cell Division Pathways

As described in Chapter 3, eukaryotic cells destined to divide progress through a series of stages known as the cell cycle (**Figure 22.11a**). The phases consist of G_1 (first gap), S (synthesis of DNA, the genetic material), G_2 (second gap), and M phase (mitosis and cytokinesis). The G_1 phase is a period in a cell's life when it may become committed to divide. Depending on the conditions, a cell in the G_1 phase may accumulate molecular changes that cause it to progress through the rest of the cell cycle. When this occurs, cell biologists say that a cell has reached a special control point called the restriction point. The commitment to divide is based on a variety of factors. For example, environmental conditions, such as the presence of sufficient nutrients, are important for cell division. In addition, multicellular organisms rely on signaling molecules to coordinate cell division throughout the body. These signaling molecules are often called **growth factors** because they promote cell division.

As researchers began to identify oncogenes, they wanted to understand how these mutant genes promote abnormal cell division. In parallel with cancer research, cell biologists have studied the roles that normal cellular proteins play in cell division. As mentioned, the cell cycle is regulated in part by growth factors that bind to cell surface receptors and initiate a cascade of cellular events that lead eventually to cell division. **Figure 22.11b** considers a protein called epidermal growth factor (EGF) that is secreted from endocrine cells and stimulates epidermal cells, such as skin cells, to divide. As seen here, EGF binds to its receptor, leading to the activation of an intracellular signaling pathway. This pathway, also known as a signal cascade, leads to a change in gene transcription. In other words, the transcription of specific genes is activated in response to the growth hormone. Once these genes are transcribed and translated, the gene products function to promote the progression through the cell cycle. Figure 22.11b is just one example of a pathway between a growth factor and gene activation. Eukaryotic species produce many different growth factors, and the signaling pathways are often more complex than the one shown here.

What is the relationship between normal genes and oncogenes? A normal, nonmutated gene that has the potential to become an oncogene is termed a **proto-oncogene.** To become an oncogene, a proto-oncogene must incur a mutation that causes its expression to be abnormally active. The mutation typically has one of three possible effects:

1. The amount of the encoded protein is greatly increased.
2. A change occurs in the structure of the encoded protein that causes it to be overly active.
3. The encoded protein is expressed in a cell type where it is not normally expressed.

The mutations that convert proto-oncogenes into oncogenes have been analyzed in many types of cancers. Oncogenes commonly encode proteins that function in cell growth signaling pathways (**Table 22.7**). These include growth factors, growth

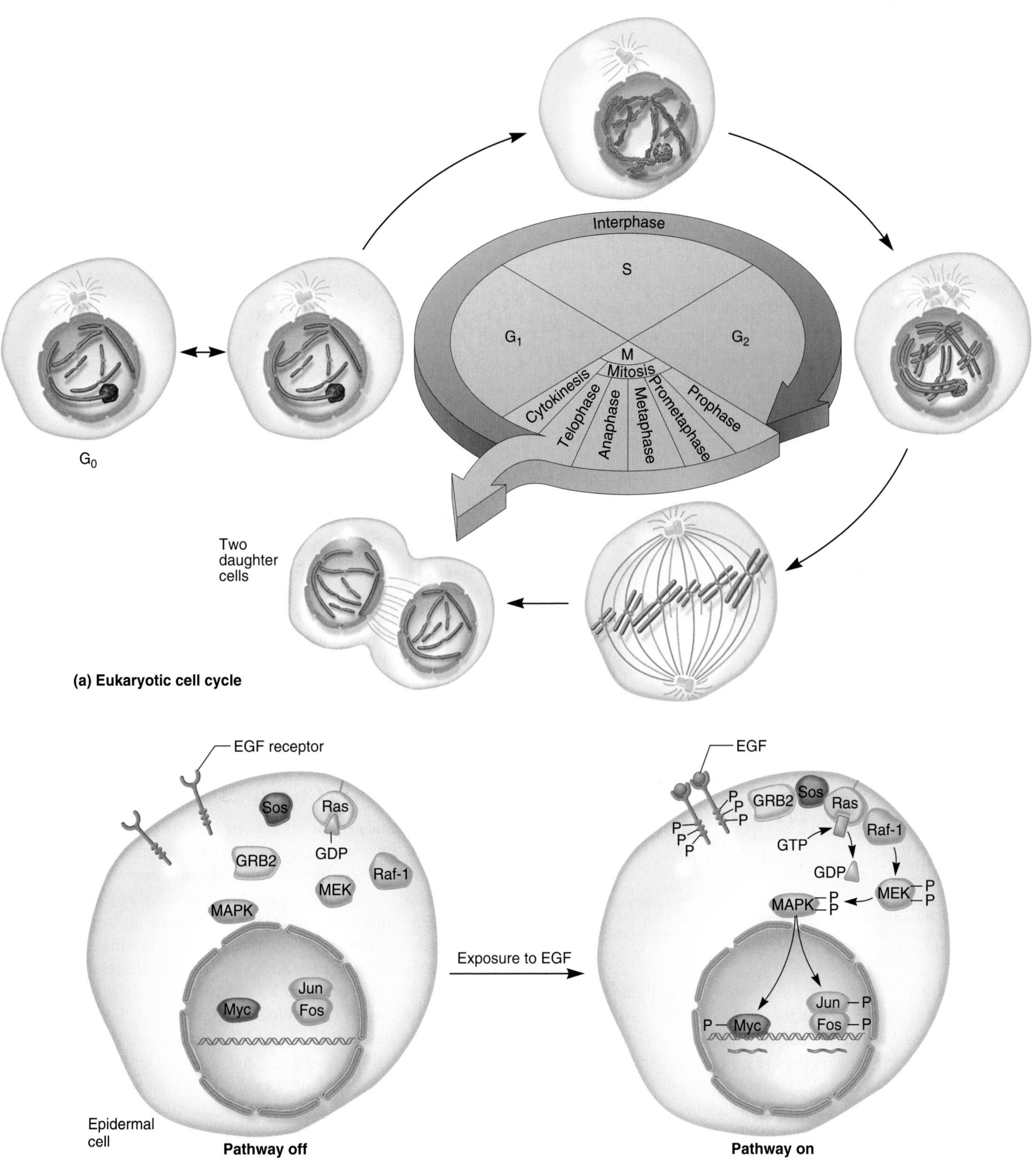

FIGURE 22.11 **The eukaryotic cell cycle and activation of a cell signaling pathway by a growth factor.** **(a)** The cell cycle involves a progression through the G_1, S, G_2, and M phases. Progression through the cell cycle is often stimulated by growth factors. (Note: Chromosomes are not condensed during G_0, G_1, S, and G_2 phases. They are shown that way in this figure so they can be counted.) **(b)** In this example, epidermal growth factor (EGF) binds to two EGF receptors, causing them to dimerize and phosphorylate each other. An intracellular protein called GRB2 is attracted to the phosphorylated EGF receptor, and it is subsequently bound by another protein called Sos. The binding of Sos to GRB2 enables Sos to activate a protein called Ras. This activation involves the release of GDP and the binding of GTP. The activated Ras/GTP complex then activates Raf-1, which is a protein kinase. Raf-1 phosphorylates MEK, and then MEK phosphorylates MAPK. More than one MAPK may be involved. Finally, the phosphorylated form of MAPK activates transcription factors. such as Myc, Jun, and Fos. This leads to the transcription of genes, which encode proteins that promote cell division.

TABLE 22.7

Examples of Proto-Oncogenes That Can Mutate into Oncogenes

Gene	Cellular Function
Growth Factors*	
sis	Platelet-derived growth factor
int-2	Fibroblast growth factor
Growth Factor Receptors	
erbB	Growth factor receptor for EGF (epidermal growth factor)
trk	Growth factor receptor for CSF-1 (cytostatic factor that inhibits cell division)
fms	Growth factor receptor for NGF (nerve growth factor)
K-sam	Growth factor receptor for FGF (fibroblast growth factor)
Intracellular Signaling Proteins	
ras	GTP/GDP-binding protein
raf	Serine/threonine kinase
src	Tyrosine kinase
abl	Tyrosine kinase
gsp	G-protein α subunit
Transcription Factors	
myc	Transcription factor
jun	Transcription factor
fos	Transcription factor
gli	Transcription factor
erbA	Steroid receptor (which functions as a transcription factor)

*The genes described in this table are found in humans as well as other eukaryotic species. Many of these genes were initially identified in retroviruses. Most of the genes have been given three-letter names that are abbreviations for the type of cancer the oncogene causes or the type of virus in which the gene was first identified.

factor receptors, proteins involved in the intracellular signaling pathways, and transcription factors. Let's consider two examples. Research groups headed by Robert Gallo and Mark Groudine showed that a *myc* gene, designated c-*myc*, was amplified about 10-fold in a human promyelocytic leukemia cell line. Since that time, research has shown that *myc* genes are overexpressed in many forms of cancer, including breast cancer, lung cancer, and colon carcinoma. The overexpression of this transcription factor leads to the transcriptional activation of genes that promote cell division.

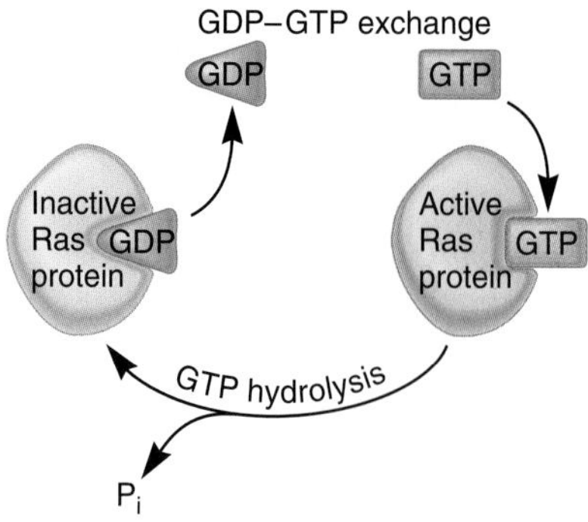

FIGURE 22.12 Functional cycle of the Ras protein. The binding of GTP to Ras activates the function of Ras and promotes cell division. The hydrolysis of GTP to GDP and P_i converts the active form of Ras to an inactive form.

As a second example, mutations that alter the amino acid sequence of the Ras protein have been shown to cause functional abnormalities. The Ras protein is a GTPase, which hydrolyzes GTP to GDP + P_i (**Figure 22.12**). Therefore, after it has been activated, the Ras protein returns to its inactive state by hydrolyzing GTP. Mutations that convert the normal *ras* gene into an oncogene either decrease the ability of the Ras protein to hydrolyze GTP or increase the rate of exchange of bound GDP for GTP. Both of these functional changes result in a greater amount of the active GTP-bound form of the Ras protein. In this way, these mutations keep the signaling pathway turned on and thereby stimulate the cell to divide.

Genetic Changes in Proto-Oncogenes Convert Them to Oncogenes

How do specific genetic alterations convert proto-oncogenes into oncogenes? By isolating and studying oncogenes at the molecular level, researchers have discovered four main ways this occurs (**Table 22.8**). These changes can be categorized as missense mutations, gene amplifications (i.e., an increase in copy number), chromosomal translocations, and viral integrations.

As mentioned previously, changes in the structure of the Ras protein can cause it to become permanently activated. These changes are caused by a missense mutation in the *ras* gene. The human genome contains four different but evolutionarily related *ras* genes: *ras*H, *ras*N, *ras*K-4a, and *ras*K-4b. All four homologous genes encode proteins with very similar amino acid sequences containing a total of 188 or 189 amino acids. Missense mutants in these normal *ras* genes are associated with particular forms of cancer. For example, a missense mutation in *ras*H that changes a glycine to a valine is responsible for the conversion of *ras*H into an oncogene:

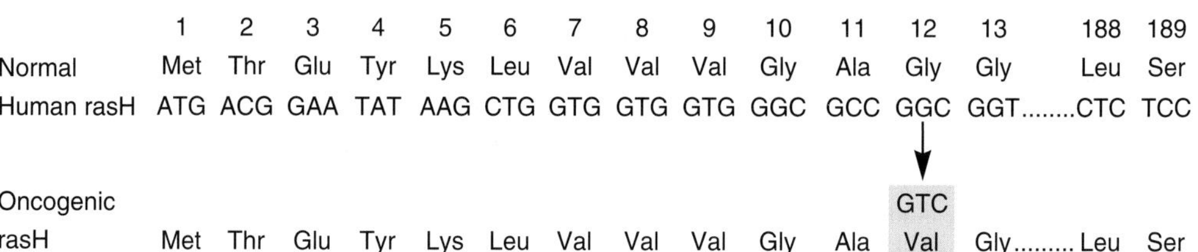

1 2 3 4 5 6 7 8 9 10 11 12 13 188 189
Normal Met Thr Glu Tyr Lys Leu Val Val Val Gly Ala Gly Gly Leu Ser
Human rasH ATG ACG GAA TAT AAG CTG GTG GTG GTG GGC GCC GGC GGT........CTC TCC
GTC
Oncogenic rasH Met Thr Glu Tyr Lys Leu Val Val Val Gly Ala Val Gly......... Leu Ser

TABLE 22.8

Genetic Changes That Convert Proto-Oncogenes into Oncogenes

Type of Change	Description and Examples
Missense mutation	A change in the amino acid sequence of a proto-oncogene protein may cause it to function in an abnormal way. Missense mutations can convert *ras* genes into oncogenes.
Gene amplification	The copy number of a proto-oncogene may be increased by gene duplication. *Myc* genes have been amplified in human leukemias; breast, stomach, lung, and colon carcinomas; and neuroblastomas and glioblastomas. *ErbB* genes have been amplified in glioblastomas, squamous cell carcinomas, and breast, salivary gland, and ovarian carcinomas.
Chromosomal translocations	A piece of chromosome may be translocated to another chromosome and affect the expression of genes at the breakpoint site. In Burkitt's lymphoma, a region of chromosome 8 is translocated to either chromosome 2, 14, or 22. The breakpoint in chromosome 8 causes the overexpression of the c-*myc* gene.
Viral integration	When a virus integrates into the chromosome, it may enhance the expression of nearby proto-oncogenes. In avian lymphomas, the integration of the avian leukosis virus can enhance the transcription of the c-*myc* gene.

Experimentally, chemical carcinogens have been shown to cause these missense mutations and thereby lead to cancer.

Another genetic event that may occur in cancer cells is gene amplification, or an abnormal increase in the copy number of a proto-oncogene. An increase in gene copy number is expected to increase the amount of the encoded protein and thereby contribute to malignancy. Gene amplification does not normally happen in mammalian cells, but it is a common occurrence in cancer cells. As mentioned previously, Gallo and Groudine discovered that c-*myc* was amplified in a human leukemia cell line. Many human cancers are associated with the amplification of particular oncogenes. In such cases, the extent of oncogene amplification may be correlated with the progression of tumors to increasing malignancy. These include the amplification of N-*myc* in neuroblastomas and *erbB-2* in breast carcinomas. In other types of malignancies, gene amplification is more random and may be a secondary event that increases the expression of oncogenes previously activated by other genetic changes.

A third type of genetic alteration that can lead to cancer is a chromosomal translocation. While structural abnormalities are common in cancer cells, very specific types of chromosomal translocations have been identified in certain types of tumors. In 1960, Peter Nowell and David Hungerford discovered that chronic myelogenous leukemia (CML) is correlated with the presence of a shortened version of chromosome 22, which they called the Philadelphia chromosome after the city where it was discovered. Rather than a deletion, this shortened chromosome is the result of a reciprocal translocation between chromosomes 9 and 22. Later studies revealed that this translocation activates a proto-oncogene, *abl*, in an unusual way (**Figure 22.13**). The reciprocal translocation involves breakpoints within the *abl* and *bcr* genes. Following the reciprocal translocation, the coding sequence of the *abl* gene fuses with the promoter and coding sequence of the *bcr* gene. This yields an oncogene that encodes an abnormal fusion protein, which contains the polypeptide sequences encoded from both genes. The *abl* gene encodes a tyrosine kinase enzyme, which uses ATP to attach phosphate groups onto target proteins. This phosphorylation activates certain proteins involved with cell division. Normally, the *abl* gene is highly

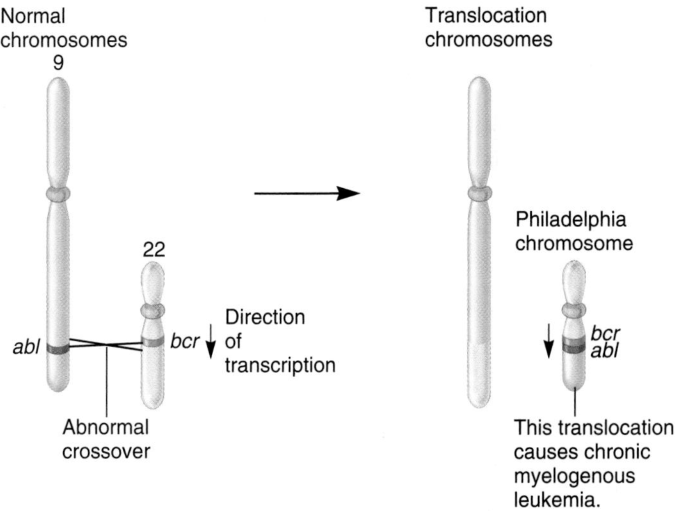

FIGURE 22.13 The reciprocal translocation commonly found in people with chronic myelogenous leukemia.

Genes→Traits In healthy individuals, the *abl* gene is located on chromosome 9, and the *bcr* gene is on chromosome 22. In certain forms of myelogenous leukemia, a reciprocal translocation causes the *abl* gene to fuse with the *bcr* gene. This combined gene, under the control of the *bcr* promoter, encodes an abnormal fusion protein that overexpresses the tyrosine kinase function of *abl* and leads to leukemia.

regulated. However, in the Philadelphia chromosome, the fusion gene is controlled by the *bcr* promoter, which is active in white blood cells. This leads to an overexpression of the tyrosine kinase function in such cells and thereby explains why this fusion causes a type of cancer called a leukemia, which involves a proliferation of white blood cells.

Interestingly, the study of the *abl* gene has led to an effective treatment for CML. Until recently, the only successful treatment was to destroy the patient's bone marrow and then restore blood-cell production by infusing stem cells from the bone marrow of a healthy donor. With knowledge about function of the ABL protein, researchers have developed the drug imatinib mesylate (Gleevec®) that appears to dramatically improve survival. This molecule fits into the active site of the ABL protein, preventing ATP from binding there. Without ATP, the ABL protein cannot phosphorylate its target proteins. This prevents the ABL protein from stimulating cell division. In a clinical trial, almost 90% of the CML patients treated with the drug showed no further progression of their disease!

Other forms of cancer also involve chromosomal translocations that cause an overexpression of an oncogene. In Burkitt's lymphoma, for example, a region of chromosome 8 is translocated to either chromosome 2, 14, or 22. The breakpoint in chromosome 8 is near the c-*myc* gene, and the sites on chromosomes 2, 14, and 22 correspond to locations of different immunoglobulin genes that are normally expressed in lymphocytes. The translocation of the c-*myc* gene near the immunoglobulin genes leads to the overexpression of the c-*myc* gene and thereby promotes malignancy in lymphocytes.

A fourth way that oncogenes may occur is via viral integration. As part of their reproductive cycle, certain viruses integrate their genomes into the chromosomal DNA of their host cell. If the integration occurs next to a proto-oncogene, a viral promoter or enhancer sequence may cause the proto-oncogene to be overexpressed. For example, in certain lymphomas that occur in birds, the genome of the avian leukosis virus has been found to be integrated next to the c-*myc* gene and enhances its level of transcription.

Tumor-Suppressor Genes Play a Role in Preventing the Proliferation of Cancer Cells

Thus far, we have considered how oncogenes promote cancer. An oncogene is an abnormally activated gene that leads to uncontrolled cell growth. We will now turn our attention to a second category of genes called **tumor-suppressor genes.** As the name suggests, the role of a tumor-suppressor gene is to prevent cancerous growth. Therefore, when a tumor-suppressor gene becomes inactivated by mutation, it becomes more likely that cancer will occur.

The first identification of a human tumor-suppressor gene involved studies of retinoblastoma, a tumor that occurs in the retina of the eye. Some people have inherited a predisposition to develop this disease within the first few years of life. By comparison, the noninherited form of retinoblastoma, which is caused by environmental agents, tends to occur later in life but only rarely.

Based on these differences, in 1971, Alfred Knudson proposed a "two-hit" model for retinoblastoma. According to this model, retinoblastoma requires two mutations to occur. People with the hereditary form already have received one mutant gene from one of their parents. They need only one additional mutation in the other copy of this tumor-suppressor gene to develop the disease. Because the retina has more than 1 million cells, it is relatively likely that a mutation may occur in one of these cells at an early age, leading to the disease. However, people with the noninherited form of the disease must have two mutations in the same retinal cell to cause the disease. Because two rare events are much less likely to occur than a single such event, the noninherited form of this disease is expected to occur much later in life and only rarely. Therefore, this hypothesis explains the different populations typically affected by the inherited and noninherited forms of retinoblastoma.

Since Knudson's original hypothesis, molecular studies have confirmed the two-hit hypothesis for retinoblastoma. In this case, the gene in which mutations occur is designated *rb* (for retinoblastoma). This tumor suppressor gene is found on the long arm of chromosome 13. Most people have two normal copies of the *rb* gene. Persons with hereditary retinoblastoma have inherited one normal and one defective copy. In nontumorous cells throughout the body, they have one functional copy and one defective copy of *rb*. However, in retinal tumor cells, the normal *rb* gene has also suffered the second hit (i.e., a mutation), which renders it defective. Without the tumor-suppressor ability, cells are allowed to grow and divide in an unregulated manner, which ultimately leads to cancer. (In contrast, as discussed later, most other forms of cancer involve mutations in several genes.)

More recent studies have revealed how the Rb protein suppresses the proliferation of cancer cells (**Figure 22.14**). The Rb protein regulates a transcription factor called E2F, which activates genes required for cell cycle progression. (The eukaryotic cell cycle is described earlier in Figure 22.11a.) The binding of the Rb protein to E2F inhibits its activity and prevents the cell from progressing through the cell cycle. As discussed later in this chapter, when a normal cell is supposed to divide, cellular proteins called cyclins bind to cyclin-dependent protein kinases. This activates

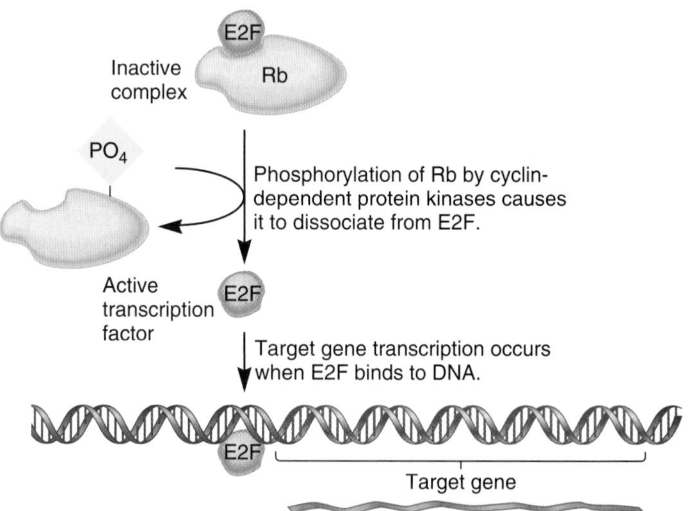

FIGURE 22.14 **Interactions between the Rb and E2F proteins.** The binding of the Rb protein to the transcription factor E2F inhibits the ability of E2F to function. This prevents cell division. For cell division to occur, cyclins bind to cyclin-dependent protein kinases, which then phosphorylate the Rb protein. The phosphorylated Rb protein is released from E2F. The free form of E2F can activate target genes needed to progress through the cell cycle.

the kinases, which then leads to the phosphorylation of the Rb protein. The phosphorylated form of the Rb protein is released from E2F, thereby allowing E2F to activate genes needed to progress through the cell cycle. What happens when both copies of the *rb* gene are rendered inactive by mutation? The answer is that the E2F protein is always active. This explains why uncontrolled cell division will occur.

The Vertebrate *p53* Gene Is a Master Tumor-Suppressor Gene That Senses DNA Damage

After the *rb* gene, the second tumor-suppressor gene discovered was the *p53* gene. The *p53* gene is the most commonly altered gene in human cancers. About 50% of all human cancers are associated with defects in *p53*. These include malignant tumors of the lung, breast, esophagus, liver, bladder, and brain as well as sarcomas, lymphomas, and leukemias. For this reason, an enormous amount of research has been aimed at elucidating the functions of the p53 protein.

A primary role of the p53 protein is to determine if a cell has incurred DNA damage. If damage is detected, p53 can promote three types of cellular pathways aimed at preventing the proliferation of cells with damaged DNA. First, when confronted with DNA damage, the cell can try to repair its DNA. This may prevent the accumulation of mutations that activate oncogenes or inactivate tumor-suppressor genes. Second, if a cell is in the process of dividing, it can arrest itself in the cell cycle. By stopping the cell cycle, a cell has more time to repair its DNA and avoid producing two mutant daughter cells. For this to happen, p53 stimulates the expression of another protein termed p21. The p21 protein inhibits cyclin/CDK protein complexes that are needed to progress from the G_1 phase of the cell cycle to the S phase.

The third, and most drastic event, is that a cell can initiate a series of events called **apoptosis,** or programmed cell death. In response to DNA-damaging agents, a cell may self-destruct. Apoptosis is an active process that involves cell shrinkage, chromatin condensation, and DNA degradation. This process is facilitated by proteases known as **caspases.** These types of proteases are sometimes called the "executioners" of the cell. Caspases digest selected cellular proteins such as microfilaments, which are components of the intracellular cytoskeleton. This causes the cell to break down into small vesicles that are eventually phagocytized by cells of the immune system. Apoptosis occurs in some cells as a normal process of embryonic development (see Chapter 23). In addition, it is an important way by which an adult organism can eliminate cells with cancer-causing potential.

Figure 22.15 summarizes how *p53* plays a central role in all three processes. As shown here, the expression of the *p53* gene is caused by the formation of damaged DNA. The inducing signal appears to be double-strand DNA breaks. The p53 protein functions as a transcription factor. It contains a DNA-binding domain and a transcriptional activation domain. Experimental studies have shown that p53 can activate the transcription of several specific target genes. As shown in Figure 22.15, it can activate genes that promote DNA repair, arrest the cell cycle, and promote apoptosis. In addition, p53 appears to act as a negative regulator by interacting with general transcription factors. This decreases the general expression of many other structural genes. This inhibition may also help prevent the cell from dividing.

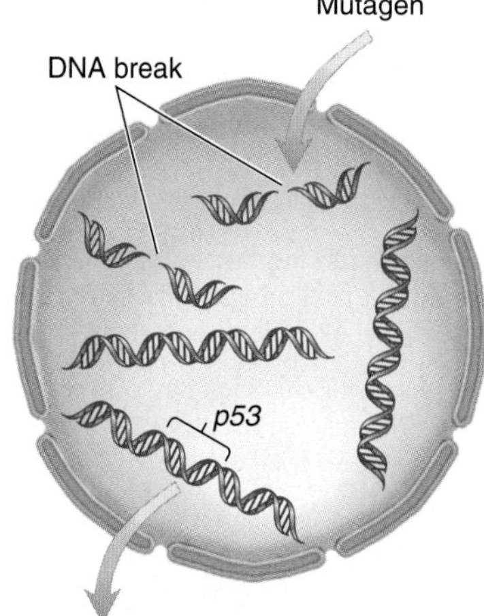
Environmental mutagen causes DNA damage, such as double-strand breaks.
Mutagen
DNA break
p53
Induction of the p53 gene by double-strand breaks leads to the synthesis of the p53 protein, which functions as a transcription factor. This transcription factor can:
1. Activate genes that promote DNA repair.
2. Activate genes that arrest cell division and may generally repress other genes that are required for cell division.
3. Activate genes that promote apoptosis.

FIGURE 22.15 Central role of *p53* in preventing the proliferation of cancer cells. Expression of the *p53* gene, which encodes a transcription factor, is induced by double-strand DNA breaks. The p53 transcription factor may activate genes that promote DNA repair, activate genes that arrest cell division, repress genes required for cell division, and activate genes that promote apoptosis.

Overall, p53 activates the expression of a few specific cellular genes and, at the same time, inhibits the expression of many other genes in the cell. Through the regulation of gene transcription, the p53 protein can prevent the proliferation of cells that have incurred DNA damage.

Tumor-Suppressor Genes Can No Longer Inhibit Cancer When Their Function Is Lost

During the past three decades, researchers have identified many tumor-suppressor genes that, when defective, contribute to the development and progression of cancer (**Table 22.9**). What are

TABLE 22.9

Functions of Selected Tumor-Suppressor Genes

Gene	Function
Genes that negatively regulate cell division	
rb	The Rb protein is a negative regulator of E2F (see Figure 22.14). The inhibition of E2F prevents the transcription of certain genes required for DNA replication and cell division.
p16	A protein kinase that negatively regulates cyclin-dependent kinases. This protein controls the transition from the G1 phase of the cell cycle to the S phase.
NF1	The NF1 protein stimulates Ras to hydrolyze its GTP to GDP. Loss of NF1 function causes the Ras protein to be overactive, which promotes cell division.
APC	A negative regulator of a cell-signaling pathway called the Wnt pathway. The Wnt pathway leads to the activation of genes that promote cell division.
VHL	VHL protein functions in the targeting of certain proteins for (ubiquitin-mediated) protein degradation. When these proteins are not degraded, the cell cycle may proceed in an unregulated manner.
MTS1	The MTS1 protein acts as an inhibitor of cyclin-dependent protein kinases. The activities of these protein kinases promote the progression through the cell cycle.
Genes that maintain genome integrity	
p53	p53 is a transcription factor that is a checkpoint protein that positively regulates a few specific target genes and negatively regulates others in a general manner. It acts as a sensor of DNA damage. It can prevent the progression through the cell cycle and also can promote apoptosis.
BRCA-1, BRCA-2	BRCA1 and BRCA2 proteins are both involved in the cellular defense against DNA damage. These proteins facilitate DNA repair and act as checkpoint proteins.

the general functions of these genes? They tend to fall into two broad categories—genes that negatively regulate cell division or genes that maintain genome integrity.

Some tumor-suppressor genes encode proteins that have direct effects on the regulation of cell division. An example is the *rb* gene. As mentioned earlier, the Rb protein negatively regulates E2F. If both copies of the *rb* gene are inactivated, the growth of cells will be accelerated. Therefore, loss of function of these kinds of negative regulators has a direct impact on the abnormal cell division rates seen in cancer cells. In other words, when the Rb protein is lost, a cell becomes more likely to divide.

Alternatively, other tumor-suppressor genes play a role in the proper maintenance of the integrity of the genome. The term **genome maintenance** refers to cellular mechanisms that either prevent mutations from occurring and/or prevent mutant cells from surviving or dividing. The proteins encoded by such genes help to ensure that gene mutations or changes in chromosome structure and number do not occur and are not transmitted to daughter cells. The proteins that participate in genome maintenance can be subdivided into two classes: checkpoint proteins and those involved directly with DNA repair.

Checkpoint proteins are vital for the detection of abnormalities such as DNA breaks and improperly segregated chromosomes. When such abnormalities are detected, the proteins participate in regulatory pathways that prevent cell division. These proteins are called checkpoint proteins because their role is to check the integrity of the genome and prevent cells from progressing past a certain point in the cell cycle if genetic abnormalities are detected.

Figure 22.16 shows a simplified diagram of cell cycle control. Proteins called cyclins and cyclin-dependent protein kinases (Cdks) are responsible for advancing a cell through the four phases of the cell cycle. For example, an activated G_1-cyclin/Cdk complex is necessary to advance from the G_1 phase to the S phase. Human cells produce several types of cyclins and Cdks. The formation of activated cyclin/Cdk complexes is regulated by a variety of factors. Earlier we considered how growth factor signaling pathways can stimulate cell division (see Figure 22.11b). These pathways operate, in part, by regulating cyclins and Cdks. In addition, checkpoint proteins monitor the state of the cell and stop the progression through the cell cycle if abnormalities are detected. Checkpoint proteins also may exert their effects by regulating the activities of cyclins and Cdks.

Several checkpoint proteins regulate the cell cycle of human cells. Figure 22.16 shows three of the major checkpoints where these proteins exert their effects. Both the G_1 and G_2 checkpoints involve the functions of proteins that can sense if the DNA has incurred damage. If so, these checkpoint proteins, such as p53, can prevent the formation of active cyclin/Cdk complexes. This would stop the progression of the cell cycle. A checkpoint also exists in metaphase. This checkpoint is monitored by proteins that can sense if a chromosome is not correctly attached to the spindle apparatus, making it likely that it will be improperly segregated.

Overall, checkpoint proteins prevent the division of cells that may have incurred DNA damage or harbor abnormalities in chromosome attachment. This provides a mechanism to stop the accumulation of genetic abnormalities that could produce cancer cells within the body. When checkpoint genes are lost, cell division may not be directly accelerated. However, the loss of checkpoint protein function makes it more likely that undesirable genetic changes will occur that could cause cancerous growth.

A second class of proteins involved with genome maintenance consists of DNA repair enzymes, which were discussed in Chapter 16 (refer back to Table 16.6). The genes encoding such enzymes are inactivated in certain forms of cancer. The loss of a DNA repair enzyme makes it more likely for a cell to accumulate mutations that could create an oncogene and/or eliminate the function of a tumor-suppressor gene. In Chapter 16, we considered how defects in the nucleotide excision repair process are responsible for the disease called xeroderma pigmentosum, which results in a predisposition to developing skin cancer (refer back to Figure 16.20). As discussed later in this chapter, defects in DNA mismatch repair enzymes can contribute to colorectal cancer. In these cases, the loss of a DNA repair enzyme contributes to a higher mutation rate, which makes it more likely for other genes to incur cancer-causing mutations.

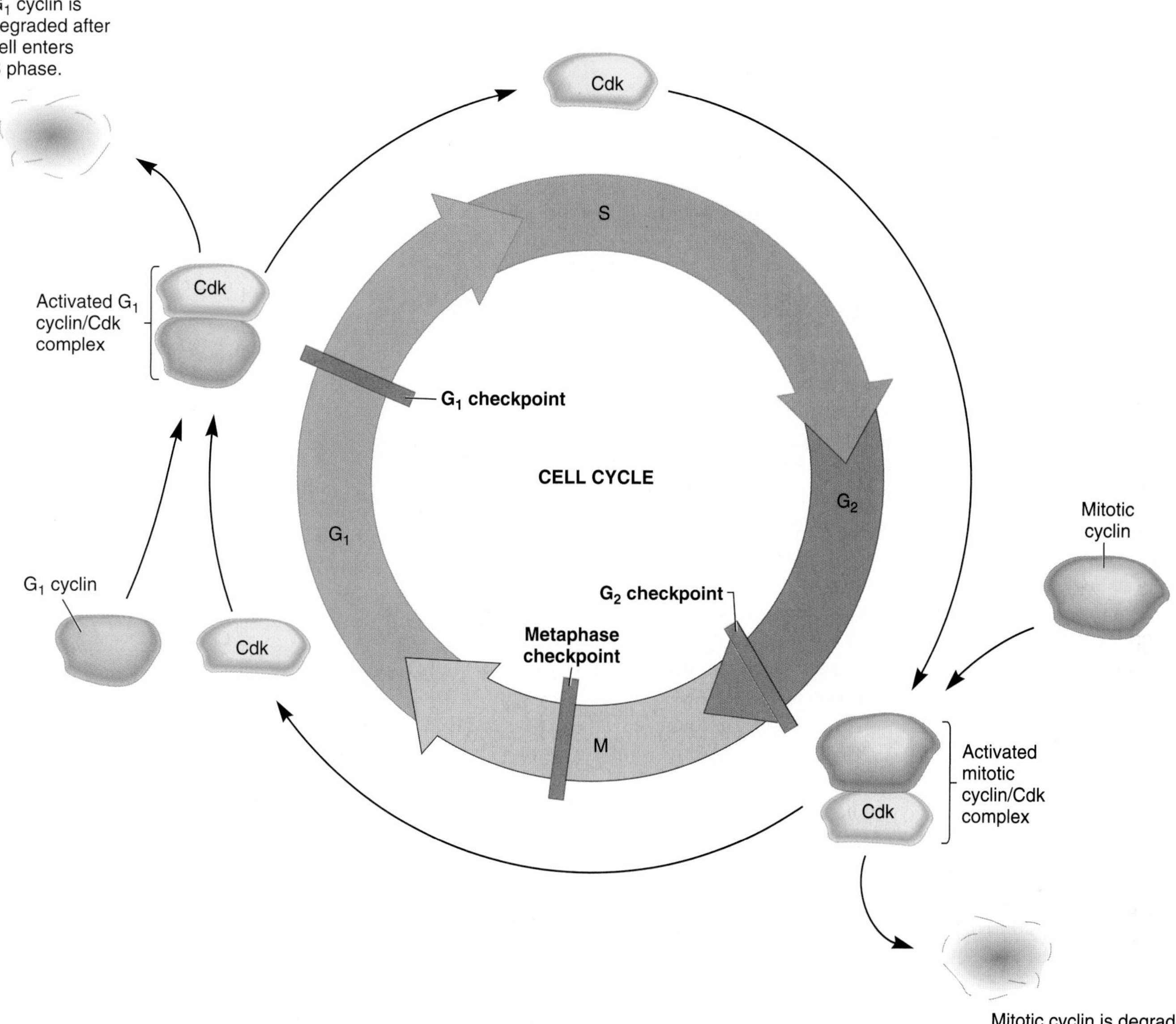

FIGURE 22.16 **Cell cycle control.** Progression through the cell cycle requires the formation of activated cyclin/Cdk complexes. A cell may produce several different types of cyclin proteins, which are typically degraded after the cell has progressed to the next phase. The formation of activated cyclin/Cdk complexes is regulated by checkpoint proteins. If these proteins detect DNA damage, they will prevent the formation of activated cyclin/Cdk complexes. In addition, other checkpoint proteins can detect chromosomes that are incorrectly attached to the spindle apparatus and prevent further progression through metaphase. Note that this is a simplified diagram of the cell cycle of humans.

Tumor-Suppressor Genes Can Be Silenced in a Variety of Ways

Thus far, we have considered the functions of proteins that are encoded by tumor-suppressor genes. Cancer biologists would also like to understand how tumor-suppressor genes are inactivated, because this knowledge may aid in the prevention of cancer. Researchers have identified three common ways that the function of tumor-suppressor genes can be lost. First, a mutation can occur specifically within a tumor-suppressor gene to inactivate its function. For example, a mutation could inactivate the promoter of a tumor-suppressor gene or introduce an early stop codon in the coding sequence. Either of these would prevent the expression of a functional protein. A second way that tumor-suppressor genes are inhibited is via DNA methylation. As discussed in Chapter 15, DNA methylation usually inhibits the transcription of eukaryotic genes, particularly when it occurs in the vicinity of the promoter. The methylation of CpG islands near the promoters of tumor-suppressor genes has been found in many types of tumors, suggesting that this form of gene inactivation plays an important role in the formation and/or progression of malignancy. However, further research will be needed to determine why tumor-suppressor genes are aberrantly methylated in cancer cells. Third, many types of cancer are associated with aneuploidy. As discussed in Chapter 8, aneuploidy involves the loss or addition of one or more chromosomes, so the total number of chromosomes is not an even multiple of a set. In some cases, chromosome loss may contribute to the progression of cancer because the lost chromosome carries one or more tumor-suppressor genes.

Most Forms of Cancer Involve Multiple Genetic Changes Leading to Malignancy

The discovery of oncogenes and tumor-suppressor genes, along with molecular techniques that can detect genetic alterations, has enabled researchers to study the progression of certain forms of cancer at the molecular level. Many cancers begin with a benign genetic alteration that, over time and with additional mutations, progresses eventually to malignancy. Furthermore, a malignancy can continue to accumulate genetic changes that make it even more difficult to treat. For example, some tumors may acquire mutations that cause them to be resistant to chemotherapeutic agents.

In 1990, Eric Fearon and Bert Vogelstein proposed a series of genetic changes that lead to colorectal cancer, the second most common cancer in the United States. As shown in **Figure 22.17**, colorectal cancer is derived from cells in the mucosa of the colon. The loss of function of *APC*, a tumor-suppressor gene on chromosome 5, leads to an increased proliferation of mucosal cells and the development of a benign polyp, a noncancerous growth. Additional genetic changes involving the loss of other tumor-suppressor genes and the activation of an oncogene (namely, *ras*) lead eventually to the development of a carcinoma. In Figure 22.17, the genetic changes that lead to colon cancer are portrayed as occurring in an orderly sequence. While the growth of a tumor often begins with mutations in *APC*, it is the total number of genetic changes, not their exact order, that is important. A key issue is that multiple mutations are needed to disable checkpoints.

Among different types of tumors, researchers have identified a large number of genes that are mutated in cancer cells. Though not all of these mutant genes have been directly shown to affect the growth rate of cells, such mutations are likely to be found in tumors because they provide some type of growth advantage for the cell population from which the cancer developed. For example, certain mutations may enable cells to metastasize to neighboring locations. These mutations may not affect growth rate, but they provide the growth advantage that cancer cells are not limited to growing in a particular location, but can migrate to new locations.

Researchers have estimated that about 300 different genes may play a role in the development of human cancer. With an approximate genome size of 20,000 to 25,000 genes, this observation indicates that over 1% of our genes have the potential to promote cancer if their function is altered by a mutation.

In addition to mutations within specific genes, another common genetic change associated with cancer are abnormalities in chromosome structure and number. **Figure 22.18** compares the

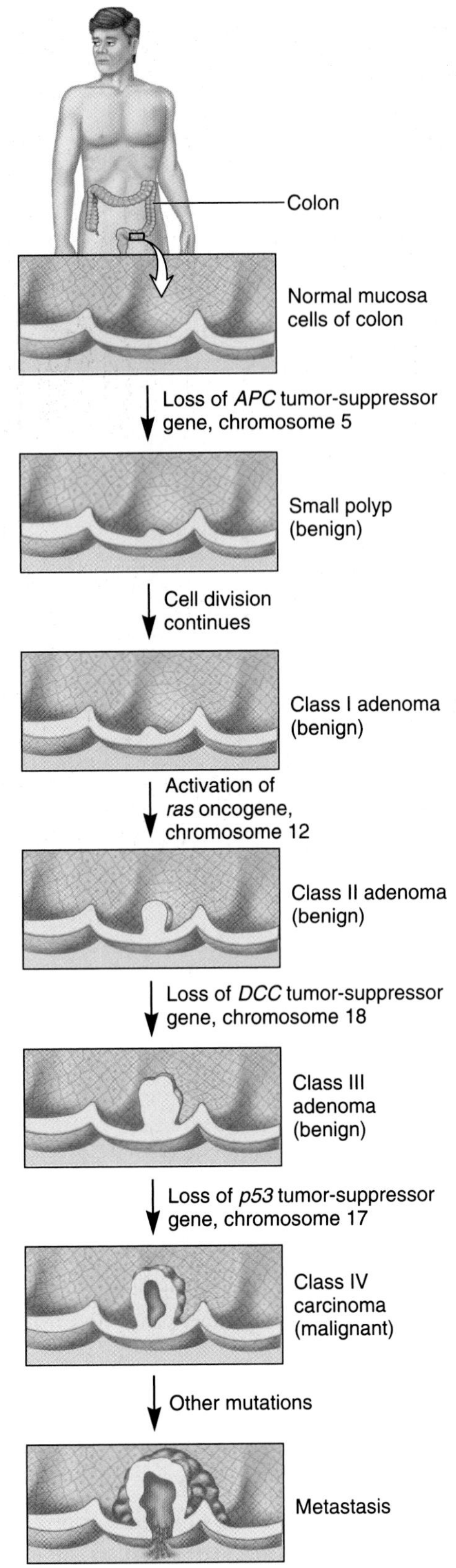

FIGURE 22.17 **Multiple genetic changes leading to colorectal cancer.**

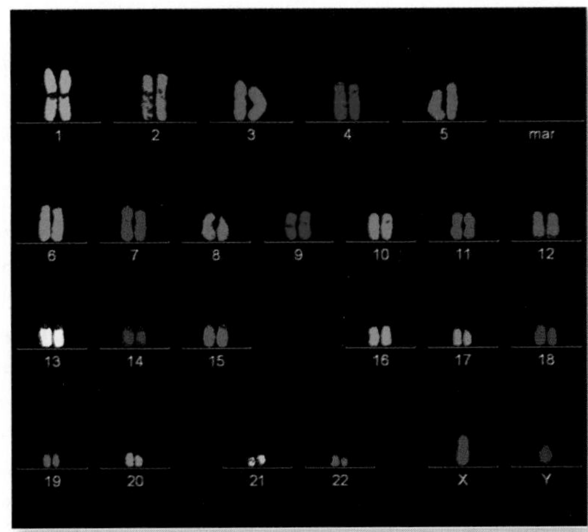

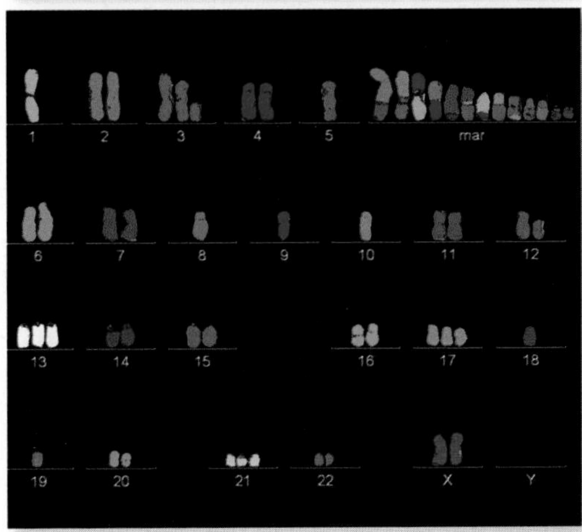

FIGURE 22.18 A comparison between chromosomes found in a normal human cell and a cancer cell from the same person. The bottom set found in a cancer cell is highly abnormal, with extra copies of some chromosomes and lost copies of others. Chromosomes made of fused pieces of chromosomes (designated mar in this figure) are also common in cancer cells.

chromosome composition of a normal male cell and a tumor cell taken from the same person. The normal composition for this person is 22 pairs of chromosomes plus two sex chromosomes (X and Y). By comparison, the chromosome composition of the tumor cell is quite bizarre, including the fact that the tumor cell has two X chromosomes, which is characteristic of females. There are many cases where chromosomes are missing. If tumor-suppressor genes were on these missing chromosomes, their function is lost as well. Figure 22.18 also shows a few cases of extra chromosomes. If these chromosomes contain proto-oncogenes, the expression of those genes may be overactive. Finally, tumor cells often contain chromosomes that have translocations. Such translocations may create fused genes (as in the case of the Philadelphia chromosome discussed earlier in this chapter), or they may place two genes close together so that the regulatory sequences of one gene affect the expression of the other gene.

DNA Microarrays Are Used to Classify Tumors

As we have just seen, cancer cells are usually the result of multiple genetic alterations that cause the activation of oncogenes and the loss of function of tumor suppressor genes. Therefore, each type of tumor is characterized by a particular set of gene and chromosome alterations. Traditionally, different types of tumors have not been identified on the basis of genetic changes but instead have been classified according to their appearance under a microscope. While this approach is useful, a major drawback is that two tumors may have a very similar microscopic appearance but yet have very different underlying genetic changes and clinical outcomes. For this reason, researchers and clinicians are turning to methods that enable them to understand the molecular changes that occur in diseases such as cancer. This general approach is called **molecular profiling.**

In cancer biology, molecular profiling involves the identification of the genes that play a role in the development of cancer. Why is this useful? First, molecular profiling can distinguish between tumors that look very similar under the microscope. Second, the classification of different tumors via molecular tools may help to predict the clinical outcome. Some tumors respond well to certain treatments, while others do not. Molecular profiling can be very effective in distinguishing such differences. Finally, researchers are optimistic that molecular profiling may lead to improved treatment options. As we gain a better understanding of the genetic changes associated with particular types of cancers, researchers may be able to develop drugs that specifically target the proteins that are encoded by cancer-causing gene mutations. As discussed earlier, the drug imatinib mesylate, which is used to treat chronic myelogenous leukemia, was developed in this way.

DNA microarrays, which are described in Chapter 21, are increasingly used as a tool in the molecular profiling of tumors. The goal is to identify those genes whose pattern of expression correlates with each other—an approach called cluster analysis (see Figure 21.3). In the study of cancer, researchers can compare cancer cells to normal cells and identify groups (clusters) of genes that are turned on in the cancer cells and off in the normal cells, and other groups of genes that are turned off in the cancer cells and on in the normal cells. Likewise, researchers can compare two different types of tumors and identify groups of genes that show different patterns of expression.

As an example, **Figure 22.19a** shows a computer-generated image that illustrates the results of a microarray analysis of 47 samples, most of which came from the tumors of patients with a type of cancer called diffuse large B-cell lymphoma (DLBCL). Each column represents the expression pattern of a set of genes from a particular sample. Genes that are expressed are shown in red; those that are not expressed are shown in green. During the course of these studies, the researchers identified two different patterns of gene expression. The tumor samples on the left side showed a set of genes (next to the orange bar) that tended to be turned on in the tumor and another set of genes (next to the blue bar) that tended to be turned off in the tumor. This pattern of gene expression was similar to the pattern found in a type of B cell called germinal center B cells. In contrast, the tumors

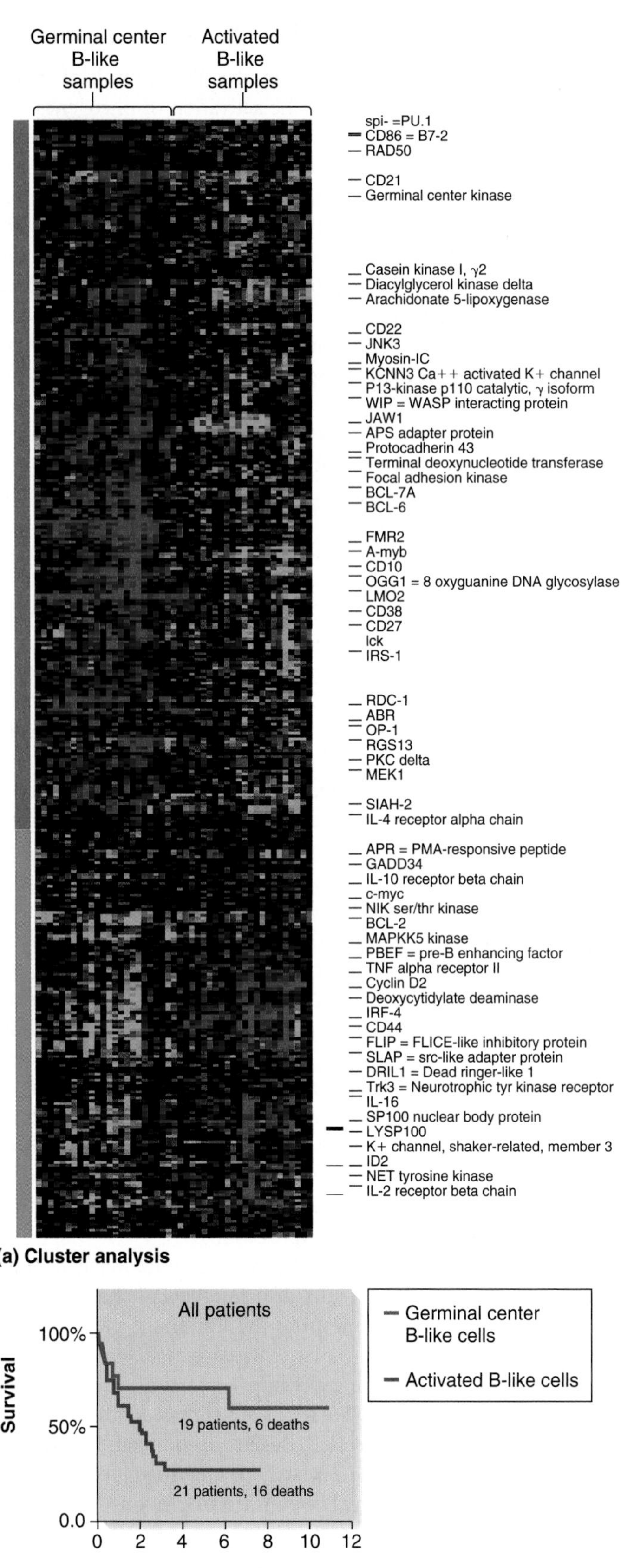

FIGURE 22.19 **The use of DNA microarrays to classify types of tumors.** **(a)** 47 samples, mostly from patients with DLBCL, were subjected to a DNA microarray analysis. The DNA microarray data were then subjected to a cluster analysis to identify genes that are coordinately expressed. The figure shown here is a graphical illustration of a cluster analysis. Each column represents one sample; each row represents the expression of a particular gene. The names of some of the genes are shown along the right side. (Note: The rows and columns are not easily resolved in this illustration). Genes highly expressed are shown in red, while those not expressed are shown in green. One group of samples had an expression pattern similar to that found in germinal center B cells; the other group had an expression pattern typical of activated B cells in the peripheral blood. **(b)** Survival of patients with DLBCL. (a. Reprinted by permission from Macmillan Publishers Ltd. *Nature.* Distinct types of diffuse large B-cell lymphoma identified by gene expression profiling. Ash Alizadeh et al. 403:6769, 503–511, 2000. Image courtesy of Ash Alizadeh.)

on the right side showed the opposite pattern. The upper genes tended to be turned off in these patients, while the lower genes were turned on. These samples showed a gene expression pattern found in normal activated peripheral blood B cells. These results suggest that the two groups of tumors may have originated in B cells at different stages of development—those on the left originated in germinal center B cells, while those on the right originated in activated B cells. Furthermore, the patients from whom these tumors were derived also appeared to have very different clinical outcomes (**Figure 22.19b**). The patients whose tumors had a pattern of gene expression similar to activated B cells had a significantly lower overall survival compared to the other patients.

Inherited Forms of Cancers May Be Caused by Defects in Tumor-Suppressor Genes and DNA Repair Genes

Before we end our discussion of the genetic basis of cancer, let's consider which genes are most likely to be affected in inherited forms of the disease. As mentioned previously, about 5 to 10% of all cases of cancer involve inherited (germ-line) mutations. These familial forms of cancer occur because people have inherited mutations from one or both parents that give them an increased susceptibility to develop cancer. This does not mean they will definitely get cancer, but they are more likely to develop the disease than are individuals in the general population. When individuals have family members who have developed certain forms of cancer, they may be tested to determine if they also carry a mutant gene. For example, von Hippel-Lindau disease and familial adenomatous polyposis are examples of syndromes for which genetic testing to identify at-risk family members is considered the standard of care.

What types of genes are mutant in familial cancers? Most inherited forms of cancer involve a defect in a tumor-suppressor gene (**Table 22.10**). In these cases, the individual is heterozygous, with one normal and one inactive allele. A mutation in the remaining normal allele may occur in a somatic cell, thereby producing a cell with two inactive copies of the

TABLE 22.10

Inherited Mutant Genes That Confer a Predisposition to Develop Cancer

Gene	Type of Cancer*
Tumor-Suppressor Genes**	
VHL	Causes von Hippel-Lindau disease, which is typically characterized by a clear cell renal carcinoma
APC	Familial adenomatous polyposis and familial colon cancer
rb	Retinoblastoma
p53	Li-Fraumeni syndrome, which is characterized by a wide spectrum of tumors including soft-tissue and bone sarcomas, brain tumors, adenocortical tumors, and premenopausal breast cancers
Pten	Causes Cowden syndrome, which is characterized by the presence of noncancerous, tumor-like growths called hamartomas in the skin, breast, thyroid, gastrointestinal tract, and central nervous system and an increased risk of breast and thyroid carcinomas
BRCA-1	Familial breast cancer
BRCA-2	Familial breast cancer
WT1	Wilm's tumor, which is a nephroblastoma
NF1	Neurofibromatosis
MTS1	Hereditary malignant melanoma
MSH2	Nonpolyposis colorectal cancer
MLH1	Nonpolyposis colorectal cancer
XP-A to *XP-G*	UV-sensitive forms of cancer such as basal cell carcinoma
Oncogenes	
RET	Multiple endocrine neoplasia type 2

*Many of the genes described in this table are mutated in more than one type of cancer. The cancers listed are those in which it has been firmly established that a predisposition to develop the disease is commonly due to germ-line mutations in the designated gene.

***MSH2, MLH1*, and *XP-A* to *XP-G* encode proteins that are involved in DNA repair. Some geneticists put these in a separate category from tumor-suppressor genes. However, in this textbook, they are considered tumor-suppressor genes because their loss of function promotes the development of cancer, as shown by the examples in this table.

tumor-suppressor gene. At the phenotypic level, a predisposition for developing cancer is inherited in a dominant fashion because a heterozygote exhibits this predisposition. However, the actual development of cancer is recessive because a second somatic mutation that results in two abnormal alleles (in the same cell) is necessary to promote malignancy. As noted in Table 22.10, not all hereditary forms of cancer are due to defective tumor-suppressor genes. For example, multiple endocrine neoplasia type 2 is due to the activation of an oncogene.

CONCEPTUAL SUMMARY

Throughout this chapter, we have examined many human diseases that have a genetic basis. To discover whether a disease has a genetic basis, researchers rely on pedigree analysis and the identification of altered genes or chromosomes in those who exhibit the disorder. Through pedigree analysis, geneticists have discovered that some diseases are inherited in an autosomal recessive manner, some are autosomal dominant, and some are X linked. Many genetic disorders, such as hemophilia, are characterized by locus heterogeneity, meaning the disease is caused by mutations in two or more different genes. Researchers use mapping and gene sequencing to identify mutant genes and their location on the chromosome. Once a gene is known, genetic testing can determine whether an individual possesses disease-causing alleles.

Cancer is also a human disease with a genetic basis that is characterized by uncontrolled cell division. Cancer can be caused by the inheritance of mutant genes, by spontaneous mutations, and by oncogenic viruses, such as RSV and other acutely transforming viruses (ACVs). However, most forms of cancer are due to environmentally induced mutations in somatic cells.

Cancer-causing mutations may involve two types of genes: oncogenes and tumor-suppressor genes. An oncogene, which is derived from a normal, unmutated proto-oncogene, is an abnormally activated gene that may stimulate cell growth. At the molecular level, a proto-oncogene may be converted to an oncogene by a missense mutation, gene amplification, chromosomal translocation, or viral integration. By comparison, a tumor-suppressor gene normally inhibits cell growth or maintains genome integrity. If tumor-suppressor genes are rendered inactive, cancerous growth may ensue. A master tumor-suppressor gene in vertebrates, called *p53*, plays a critical role in determining if a cell has incurred DNA damage and preventing cell division or eliminating cells that have been damaged. Tumor-suppressor genes can be silenced in many ways, including mutations within the gene, DNA methylation, or chromosomal loss.

EXPERIMENTAL SUMMARY

While researchers cannot experiment with humans as they can with model organisms such as *Drosophila* and mice, they can gain knowledge of human genetic disorders by conducting pedigree analysis and in vitro tests on human cell samples. Often, when a human disease is newly identified, researchers would like to know if it is caused, wholly or in part, by a mutation in a gene. Various observations point to a genetic cause. These include a higher frequency of affected individuals among family members and in identical versus fraternal twins, the inability of the disease to be spread by contact, different rates of the disease among different populations, a characteristic age of onset, a similar disease of known genetic origin in animals, and a correlation between a disease and a mutant gene.

Once genetics has become established as a cause of disease, researchers attempt to determine the pattern of transmission from parents to offspring. Many genetic diseases follow a simple Mendelian pattern of inheritance, and pedigree analysis may indicate whether a disease is transmitted in a recessive, dominant, or X-linked manner. Geneticists also use mapping and DNA sequencing to identify disease-causing genes. Genetic testing and screening involves the use of testing methods to discover if individuals or populations carry genetic abnormalities. Two forms—amniocentesis and chorionic villus sampling—can be done prior to birth.

Research involving the Rous sarcoma virus led to identification of the first oncogene, a gene that promotes cancerous cell growth. Experiments showed that DNA isolated from malignant cells can transform normal cells into cancerous cells. This approach paved the way for the identification of many human cellular oncogenes. The first identification of a human tumor-suppressor gene resulted from studies of retinoblastoma. For this form of cancer, Knudson proposed that cancer required two hits to occur. Research subsequently discovered that tumor-suppressor genes function to suppress the proliferation of cancer cells and can be silenced in a variety of ways. While tumors have traditionally been classified by their appearance under a microscope, newer approaches such as DNA microarrays are helping scientists profile the molecular changes that occur in cancer.

PROBLEM SETS & INSIGHTS

Solved Problems

S1. The pedigree shown here concerns a human disease known as familial hypercholesterolemia.

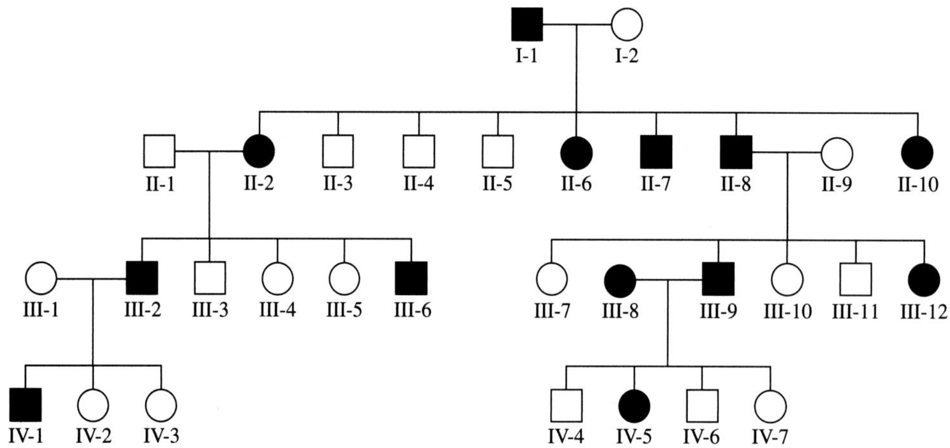

This disorder is characterized by an elevation of serum cholesterol in the blood. Though relatively rare, this genetic abnormality can be a contributing factor to heart attacks. At the molecular level, this disease is caused by a defective gene that encodes a protein called the low-density lipoprotein receptor (LDLR). In the bloodstream, serum cholesterol is bound to a carrier protein known as low-density lipoprotein (LDL). LDL binds to LDLR so that cells can absorb cholesterol. When LDLR is defective, it becomes more difficult for the cells to absorb cholesterol. This explains why the levels of LDL blood cholesterol remain high. Based on the pedigree, what is the most likely pattern of inheritance of this disorder?

Answer: The pedigree is consistent with a dominant pattern of inheritance. An affected individual always has an affected parent. Also, individuals III-8 and III-9, who are both affected, produced unaffected offspring. If this trait was recessive, two affected parents should always produce affected offspring. However, because the trait is dominant, two heterozygous parents can produce homozygous unaffected offspring. The ability of two affected parents to have unaffected offspring is a striking characteristic of dominant inheritance. On average, we would expect that two heterozygous parents should produce 25% unaffected offspring. In the family containing IV-4, IV-5, IV-6, and IV-7, three out of four offspring are actually unaffected. This higher-than-expected proportion of unaffected offspring is not too surprising because the family is a very small group and may deviate substantially from the expected value due to random sampling error.

S2. One way to identify a human cellular oncogene is to use human DNA from a malignant cell to transform a mouse cell. A mouse cell that has been transformed by human DNA will have a human oncogene incorporated into its genome; it also is likely to have *Alu* sequences that are closely linked to this human oncogene. (Note: As discussed in Chapter 10, the human genome contains *Alu* sequences interspersed every 5,000 to 6,000 base pairs. *Alu* sequences are not found in the mouse genome.) Discuss how the *Alu* sequence can provide a way to clone human oncogenes.

Answer: One approach is transposon tagging, described in Chapter 17. When a mouse cell is transformed with a human DNA fragment containing an oncogene, that fragment is likely to contain an *Alu* sequence as well. To clone the human oncogene, the chromosomal DNA can be isolated from the transformed mouse cells, digested with a restriction enzyme, and cloned into vectors to create a library of DNA fragments. The members of the library that carry the human oncogene can be identified using a probe complementary to the *Alu* sequence because this sequence is not found in the mouse genome. Using this strategy, researchers have identified several human cellular oncogenes. In human bladder carcinoma, for example, the human cellular oncogene called *ras* was identified this way.

S3. Oncogenes sometimes result from genetic rearrangements (e.g., translocations) that produce gene fusions. An example is the Philadelphia chromosome, in which a reciprocal translocation between chromosomes 9 and 22 leads to fusion of the first part of the *bcr* gene with the *abl* gene. Suggest two different reasons why a gene fusion could create an oncogene.

Answer: An oncogene is derived from a genetic change that abnormally activates the expression of a gene that plays a role in cell division. When a genetic change creates a gene fusion, this can abnormally activate the expression of the gene in two ways.

The first way is at the level of transcription. The promoter and part of the coding sequence of one gene may become fused with the coding sequence of another gene. For example, the promoter and part of the coding sequence of the *bcr* gene may fuse with the coding sequence of the *abl* gene. After this has occurred, the *abl* gene is now under the control of the *bcr* promoter, rather than its own normal promoter. Because the *bcr* promoter is turned on in different cells compared to the *abl* promoter, this leads to the overexpression of the abl protein in certain cell types compared to its normal level of expression.

A second way that a gene fusion can cause abnormal activation is at the level of protein structure. A fusion protein has parts of two different polypeptides. The first portion of a fusion protein may affect the structure of the second portion of the polypeptide in such a way that the second portion becomes abnormally active, or vice versa.

Conceptual Questions

C1. With regard to pedigree analysis, make a list of the patterns that distinguish recessive, dominant, and X-linked genetic diseases from each other.

C2. Explain, at the molecular level, why human genetic diseases often follow a simple Mendelian pattern of inheritance, whereas most normal traits, such as the shape of your nose or the size of your head, are governed by multiple gene interactions.

C3. Many genetic disorders exhibit locus heterogeneity. Define and give two examples of locus heterogeneity. How does locus heterogeneity confound a pedigree analysis?

C4. In general, why do changes in chromosome structure and/or number tend to affect an individual's phenotype? Explain why some changes in chromosome structure, such as reciprocal translocations, do not.

C5. We often speak of diseases such as phenylketonuria (PKU) and achondroplasia as having a "genetic basis." Explain whether the following statements are accurate with regard to the genetic basis of any human disease (not just PKU and achondroplasia).

A. An individual must inherit two copies of a mutant allele to have disease symptoms.

B. A genetic predisposition means that an individual has inherited one or more alleles that make it more likely for them to develop disease symptoms compared to other individuals.

C. A genetic predisposition to develop a disease may be passed from parents to offspring.

D. The genetic basis for a disease is always more important than the environment.

C6. Figure 22.1 illustrates albinism in different species. Describe two other genetic disorders found in both humans and animals.

C7. Discuss why a genetic disease might have a particular age of onset. Would an infectious disease have an age of onset? Explain why or why not.

C8. Gaucher disease (type I) is due to a defect in a gene that encodes a protein called acid β glucosidase. This enzyme plays a role in carbohydrate metabolism within the lysosome. The gene is located on the long arm of chromosome 1. Persons who inherit two defective copies of this gene exhibit Gaucher disease, the major symptoms of which include an enlarged spleen, bone lesions, and changes in skin pigmentation. Let's suppose a phenotypically unaffected woman, whose father had Gaucher disease, has a child with a phenotypically unaffected man, whose mother had Gaucher disease.

A. What is the probability that this child will have the disease?

B. What is the probability that this child will have two normal copies of this gene?

C. If this couple has five children, what is the probability that one of them will have Gaucher disease and four will be phenotypically unaffected?

C9. Ehler-Danlos syndrome is a relatively rare disorder due to a mutation in a gene that encodes a protein called collagen (type 3 A1). Collagen is a protein found in the extracellular matrix that plays an important role in the formation of skin, joints, and other connective tissues. Persons with this syndrome have extraordinarily flexible skin and very loose joints. The pedigree shown here contains several members affected with Ehler-Danlos syndrome, shown with black symbols. Based on this pedigree, does this syndrome appear to be an autosomal recessive, autosomal dominant, X-linked recessive, or X-linked dominant trait? Explain your reasoning.

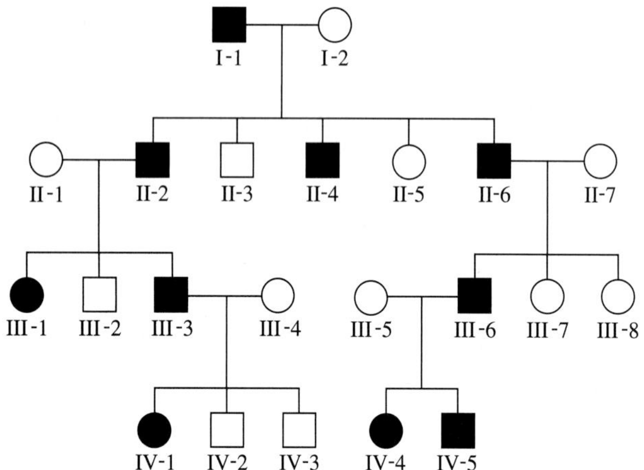

C10. Hurler syndrome is due to a mutation in a gene that encodes a protein called α-L-iduronidase. This protein functions within the lysosome as an enzyme that breaks down mucopolysaccharides (a type of polysaccharide that has many acidic groups attached). When this enzyme is defective, excessive amounts of the mucopolysaccharides dermatan sulfate and heparin sulfate accumulate within the lysosomes, especially in liver cells and connective tissue cells. This leads to symptoms such as an enlarged liver and spleen, bone abnormalities, corneal clouding, heart problems, and severe neurological problems. The pedigree shown here contains three members affected with Hurler syndrome, indicated with black symbols. Based on this pedigree, does this syndrome appear to be an autosomal recessive, autosomal dominant, X-linked recessive, or X-linked dominant trait? Explain your reasoning.

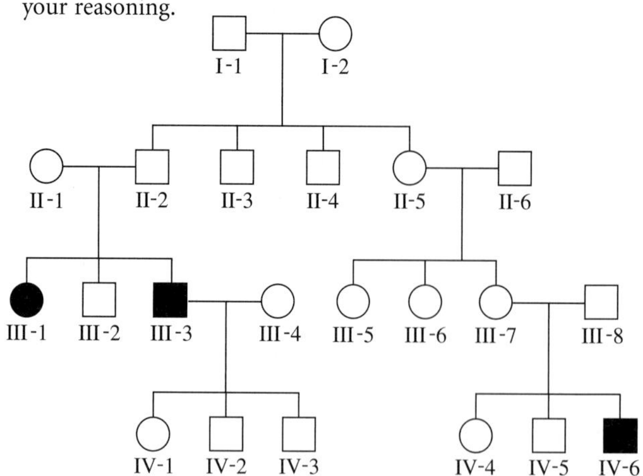

C11. Like Hurler syndrome, Fabry disease involves an abnormal accumulation of substances within lysosomes. However, the lysosomes of individuals with Fabry disease show an abnormal accumulation of lipids. The defective enzyme is α-galactosidase A, which is a lysosomal enzyme that functions in lipid metabolism. This defect causes cell damage, especially to the kidneys, heart, and eyes. The gene that encodes α-galactosidase A is found on the X chromosome. Let's suppose a phenotypically unaffected couple produces two sons with Fabry disease and one phenotypically unaffected daughter. What is the probability that the daughter will have an affected son?

C12. Achondroplasia is a rare form of dwarfism caused by an autosomal dominant mutation that affects the gene that encodes a fibroblast growth factor receptor. Among 1,422,000 live births, the number of babies born with achondroplasia was 31. Among those 31 babies, 18 of them had one parent with achondroplasia. The remaining babies had two unaffected parents. How do you explain these 13 babies, assuming that the mutant allele has 100% penetrance? What are the odds that these 13 individuals will pass this mutant gene to their offspring?

C13. Lesch-Nyhan syndrome is due to a mutation in a gene that encodes a protein called hypoxanthine-guanine phosphoribosyltransferase (HPRT). HPRT is an enzyme that functions in purine metabolism. Persons afflicted with this syndrome have severe neurodegeneration and loss of motor control. The pedigree shown here contains several members with Lesch-Nyhan syndrome. Affected members are shown with black symbols. Based on this pedigree, does this syndrome appear to be an autosomal recessive, autosomal dominant, X-linked recessive, or X-linked dominant trait? Explain your reasoning.

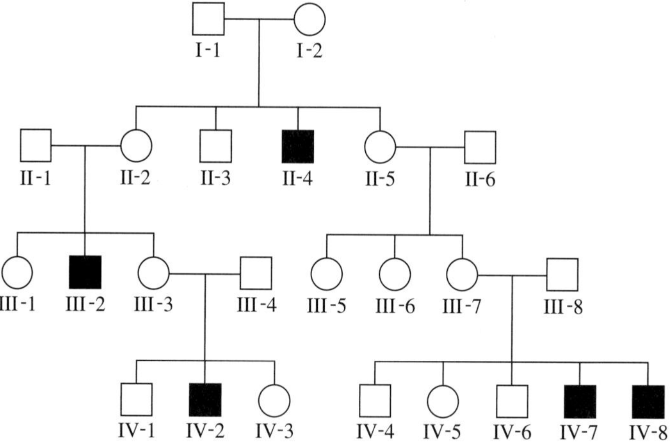

C14. Marfan syndrome is due to a mutation in a gene that encodes a protein called fibrillin-1. It is inherited as a dominant trait. The fibrillin-1 protein is the main constituent of extracellular microfibrils. These microfibrils can exist as individual fibers or associate with a protein called elastin to form elastic fibers. The gene that encodes fibrillin-1 is located on the long arm of chromosome 15. It has been suggested that Abraham Lincoln may have been afflicted with this disorder. Let's suppose a phenotypically unaffected woman has a child with a man who has Marfan syndrome.

A. What is the probability this child will have the disease?

B. If this couple has three children, what is the probability none of them will have Marfan syndrome?

C15. Sandhoff disease is due to a mutation in a gene that encodes a protein called hexosaminidase B. This disease has symptoms that are similar to Tay-Sachs disease. Weakness begins in the first six months of life. Individuals exhibit early blindness and progressive mental and motor deterioration. The pedigree shown here contains several members with Sandhoff disease. Affected members are shown with black symbols.

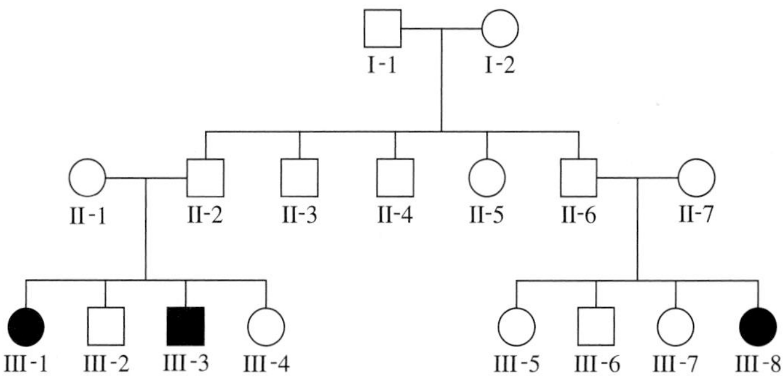

A. Based on this pedigree, does this syndrome appear to be an autosomal recessive, autosomal dominant, X-linked recessive, or X-linked dominant trait? Explain your reasoning.

B. What is the likelihood that II-1, II-2, II-3, II-4, II-5, II-6, and II-7 carry a mutant allele for the hexosaminidase B gene?

C16. What is a prion? Explain how a prion relies on normal cellular proteins to cause a disease such as mad cow disease.

C17. Some people have a genetic predisposition for developing prion diseases. Examples are described in Table 22.5. In the case of Gerstmann-Straussler-Scheinker disease, the age of onset is typically at 30 to 50 years, and the duration of the disease (which leads to death) is about five years. Suggest a possible explanation why someone can live for a relatively long time without symptoms and then succumb to the disease in a relatively short time.

C18. Familial fatal insomnia (described in Table 22.5) is a prion disease inherited as an autosomal dominant trait. Researchers have identified the *PrP* gene, located on the short arm of chromosome 20, and tried to understand the relationship between the *PrP* gene sequence and the molecular basis of familial fatal insomnia.

A. It has been found that a rare mutation at codon 178, changing an aspartic acid to an asparagine, can cause this disease. In addition, codon 129 seems to play a role. The human population is polymorphic at codon 129, which may encode valine or methionine. If codon 178 is the normal aspartic acid (Asp) codon, it does not seem to matter if valine or methionine is found at position 129. In other words, if codon 178 is aspartic acid, Met-129 and Val-129 codons are not associated with disease symptoms. However, if codon 178 specifies asparagine (Asn), then it does matter. Familial fatal insomnia seems to require an asparagine at codon 178 and a methionine at codon 129 in the *PrP* gene sequence. Although the actual biochemical mechanism is not known, suggest a possible reason why this is the case.

B. Also, researchers have compared the sequences of the *PrP* gene in many people with familial fatal insomnia. Keep in mind that this is a dominant autosomal trait, so people with this disorder have one mutant copy of the *PrP* gene and one normal copy. People with familial fatal insomnia must have one abnormal copy of the gene that contains Asn-178 and Met-129. The second copy of the gene has Asp-178, and it may have Met-129 or Val-129. Some results suggest that people having a Met-129 codon in this second copy of the *PrP* gene causes the disease to develop more rapidly than people who have Val-129 in the second copy. Propose an explanation why disease symptoms may occur more rapidly when Met-129 is found in the second copy of the *PrP* gene.

C19. What is the difference between an oncogene and a tumor-suppressor gene? Give two examples.

C20. What is a proto-oncogene? What are the typical functions of proteins encoded by proto-oncogenes? At the level of protein function, what are the general ways that proto-oncogenes can be converted into oncogenes?

C21. What is a retroviral oncogene? Is it necessary for viral infection and proliferation? How have retroviruses acquired oncogenes?

C22. A genetic predisposition to develop cancer is usually inherited as a dominant trait. At the level of cellular function, are the alleles involved actually dominant? Explain why some individuals who have inherited these dominant alleles do not develop cancer during their lifetimes.

C23. Describe the types of genetic changes that commonly convert a proto-oncogene into an oncogene. Give three examples. Explain how the genetic changes are expected to alter the activity of the gene product.

C24. Relatively few inherited forms of cancer involve the inheritance of mutant oncogenes. Instead, most inherited forms of cancer are defects in tumor-suppressor genes or DNA repair genes. Give two or more reasons why we seldom see inherited forms of cancer involving activated oncogenes.

C25. The *rb* gene encodes a protein that inhibits E2F, a transcription factor that activates several genes involved in cell division. Mutations in *rb* are associated with certain forms of cancer, such as retinoblastoma. Under each of the following conditions, would you expect cancer to occur?

A. One copy of *rb* is defective; both copies of *E2F* are functional.

B. Both copies of *rb* are defective; both copies of *E2F* are functional.

C. Both copies of *rb* are defective; one copy of *E2F* is defective.

D. Both copies of *rb* and E2F are defective.

C26. A *p53* knockout mouse in which both copies of *p53* are defective has been produced by researchers. This type of mouse appears normal at birth. However, it is highly sensitive to UV light. Based on your knowledge of *p53*, explain the normal appearance at birth and the high sensitivity to UV light.

C27. With regard to cancer cells, which of the following statements are true?

A. Cancer cells are clonal, which means they are derived from a single mutant cell.

B. To become cancerous, cells usually accumulate multiple genetic changes that eventually result in uncontrolled growth.

C. Most cancers are caused by oncogenic viruses.

D. Cancer cells have lost the ability to properly regulate cell division.

C28. When the DNA of a human cell becomes damaged, this leads to the activation of the *p53* gene. What is the general function of the p53 protein? Is it an enzyme, transcription factor, cell cycle protein, or something else? Describe three ways that the synthesis of the p53 protein affects cellular function. Why is it beneficial for these three things to happen when a cell's DNA has been damaged?

Experimental Questions

E1. Which of the following experimental observations would suggest that a disease has a genetic basis?

A. The frequency of the disease is less likely in relatives that live apart compared to relatives that live together.

B. The frequency of the disease is unusually high in a small group of genetically related individuals who live in southern Spain.

C. The disease symptoms usually begin around the age of 40.

D. It is more likely that both monozygotic twins will be affected by the disease compared to dizygotic twins.

E2. At the beginning of Chapter 22, we discussed the types of experimental observations that suggest a disease is inherited. Which of these observations do you find the least convincing? Which do you find the most convincing? Explain your answer.

E3. What is meant by the term genetic testing? What is different between testing at the protein level versus testing at the DNA level? Describe five different techniques used in genetic testing.

E4. A particular disease is found in a group of South American Indians. During the 1920s, many of these people migrated to Central America. In the Central American group, the disease is never found. Discuss whether or not you think the disease has a genetic component. What types of further observations would you make?

E5. Chapter 18 describes a blotting method known as Western blotting that can be used to detect a polypeptide that is translated from a particular mRNA. In this method, a particular polypeptide or protein is detected by an antibody that specifically recognizes a segment of its amino acid sequence. After the antibody binds to the polypeptide within a gel, a secondary antibody (which is labeled) is used to visualize the polypeptide as a dark band. For example, an antibody that recognizes α-galactosidase A could be used to specifically detect the amount of α-galactosidase A protein on a gel. The enzyme α-galactosidase A is defective in individuals with Fabry disease, which shows an X-linked recessive pattern of inheritance. Amy, Nan, and Pete are siblings, and Pete has Fabry disease. Aileen, Jason, and Jerry are brothers and sister, and Jerry has Fabry disease. Amy, Nan, and Pete are not related to Aileen, Jason, and Jerry. Amy, Nan, and Aileen are concerned that they could be carriers of a defective α-galactosidase A gene. A sample of cells was obtained from each of these six individuals and subjected to Western blotting, using an antibody against α-galactosidase A. Samples were also obtained from two unrelated normal females (lanes 7 and 8). The results are shown here.

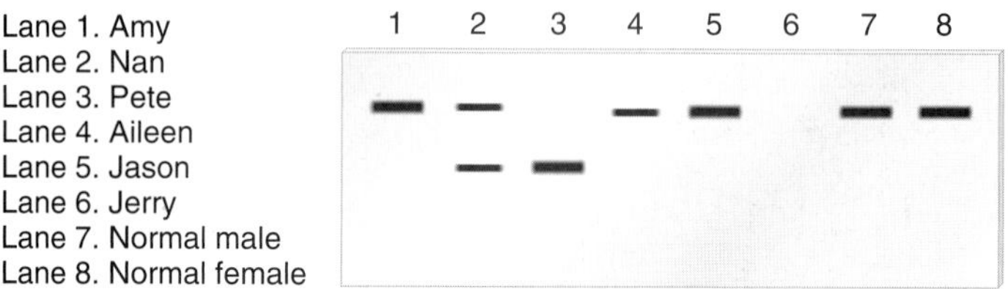

Note: Due to X inactivation in females, the amount of expression of genes on the single X chromosome in males is equal to the amount of expression from genes on both X chromosomes in females.

A. Explain the type of mutation (i.e., missense, nonsense, promoter, etc.) that causes Fabry disease in Pete and Jerry.

B. What would you tell Amy, Nan, and Aileen regarding the likelihood that they are carriers of the mutant allele and the probability of having affected offspring?

E6. An experimental assay for the blood clotting protein called Factor IX is available. A blood sample was obtained from each member of the pedigree shown here. The amount of Factor IX protein is shown within the symbol of each member and is expressed as a percent of the amount observed in normal individuals who do not carry a mutant copy of the gene.

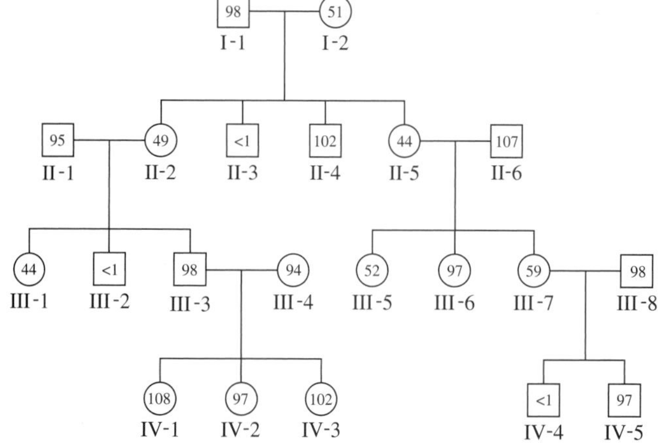

What are the likely genotypes of each member of this pedigree?

E7. Let's suppose a cell line has become malignant because it has accumulated mutations that inactivate two different tumor-suppressor genes. A researcher followed the protocol described in the experiment of Figure 22.10 and isolated DNA from this mutant cell line. The DNA was used to transform normal fibroblast (NIH3T3) cells. What results would you expect? Would you expect a high number of malignant foci or not? Explain your answer.

E8. What is a transformed cell? Describe three different methods to transform cells in a laboratory.

E9. In the experiment of Figure 22.10, what would be the results if the DNA sample had been treated with either DNase, RNase, or protease prior to the treatment with calcium and phosphate ions?

E10. Explain how the experimental study of cancer-causing viruses has increased our understanding of cancer.

E11. Discuss ways to distinguish whether a particular form of cancer involves an inherited predisposition or is due strictly to (postzygotic) somatic mutations. In your answer, consider that only one mutation may be inherited, but the cancer might develop only after several somatic mutations.

E12. The codon change (Gly-12 to Val-12) in human *ras*H that converts it to oncogenic *ras*H has been associated with many types of cancers. For this reason, researchers would like to develop drugs to inhibit oncogenic *ras*H. Based on your molecular understanding of the Ras protein, what types of drugs might you develop? In other words, what would be the structure of the drugs, and how would they inhibit Ras protein? How would you test the efficacy of the drugs? What might be some side effects?

E13. Describe how normal mammalian cells grow on solid media in a laboratory and how cancer cells grow. Is a malignant focus derived from a single cancer cell or from many independent cancer cells that happen to be in the same region of a tissue culture plate?

Questions for Student Discussion/Collaboration

1. Make a list of the benefits that may arise from genetic testing as well as possible negative consequences. Discuss the items on your list.
2. Our government has finite funds to devote to cancer research. Discuss which aspects of cancer biology you would spend the most money pursuing.

   A. Identifying and characterizing oncogenes and tumor-suppressor genes

   B. Identifying agents in our environment that cause cancer

   C. Identifying viruses that cause cancer

   D. Devising methods aimed at killing cancer cells in the body

   E. Informing the public of the risks involved in exposure to carcinogens

   In the long run, which of these areas would you expect to be the most effective in decreasing mortality due to human cancer?

Note: All answers appear at the website for this textbook; the answers to even-numbered questions are in the back of the textbook.

www.mhhe.com/brookergenetics3e

Visit the Online Learning Center for practice tests, answer keys, and other learning aids for this chapter. Enhance your understanding of genetics with our interactive exercises, quizzes, animations, and much more.

23

DEVELOPMENTAL GENETICS

CHAPTER OUTLINE

The fruit fly (*Drosophila melanogaster*), a model organism in the study of developmental genetics.

Multicellular organisms, such as animals and plants, begin their lives with a fairly simple organization, namely, a fertilized egg, and then proceed step by step to a much more complex arrangement. As this occurs, cells divide, migrate, and change their characteristics as they become highly specialized units within a multicellular individual. Each cell in an organism has its own particular role. In animals, for example, muscle cells allow an organism to move, while intestinal cells facilitate the absorption of nutrients. This division of labor among the various cells and organs of the body promotes the survival of the individual.

Developmental genetics is concerned with the roles that genes play in orchestrating the changes that occur during development. In this chapter, we will examine how the sequential actions of genes provide a program for the development of an organism from a fertilized egg to an adult. The last couple of decades have seen staggering advances in our understanding of developmental genetics at the molecular level. Scientists have chosen a few experimental organisms, such as the fruit fly (*Drosophila melanogaster*), nematode (*Caenorhabditis elegans*), mouse (*Mus musculus*), and a flowering plant (*Arabidopsis thaliana*), and worked toward the identification and characterization of the genes required for running their developmental programs. In certain organisms, notably *Drosophila*, most of the genes that play a critical role in the early stages of development have been identified. Researchers are now studying how the proteins encoded by these genes control the course of development. In this chapter, we will explore the body plans of invertebrates, vertebrates, and plants, and consider several examples in which geneticists understand how the actions of genes govern the developmental process.

23.1 INVERTEBRATE DEVELOPMENT

We will begin our discussion of multicellular development by considering two model organisms, *Drosophila melanogaster* and *Caenorhabditis elegans*, that have been pivotal in our understanding of developmental genetics. As we first saw in Chapter 3, *Drosophila* has been a favorite subject of geneticists since 1910, when Thomas Hunt Morgan isolated his first white-eyed mutant. The fruit fly has been used to determine many of the fundamental principles of genetics, including the chromosome theory of inheritance, the random mutation theory, and linkage mapping, to mention only a few.

In developmental genetics, *Drosophila* is also useful for a variety of reasons. First, the techniques for generating and analyzing mutants in this organism are highly advanced, and researchers have identified many mutant strains with altered developmental pathways. Second, at the embryonic and larval stages, *Drosophila* is large enough to conduct transplantation experiments, yet small enough to determine where particular genes are expressed at critical stages of development.

By comparison, *C. elegans* is used by developmental geneticists because of its simplicity. The adult organism is a small transparent worm composed of only about 1,000 somatic cells. Starting with the fertilized egg, the pattern of cell division and the fate of each cell are completely known.

In this section, we will begin by examining the general features of *Drosophila* development. We will then focus our attention on embryonic development (embryogenesis), because it is during this stage that the overall body plan is determined. We will see how the expression of particular genes and the localization of gene products within the embryo influence the developmental process. We will then briefly consider development in *C. elegans*. In this organism, we will examine how the timing of gene expression plays a key role in determining the developmental fate of particular cells.

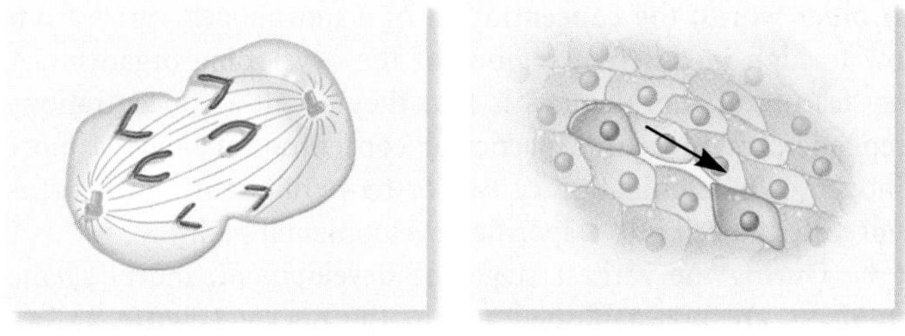

(a) Cell division (b) Cell migration

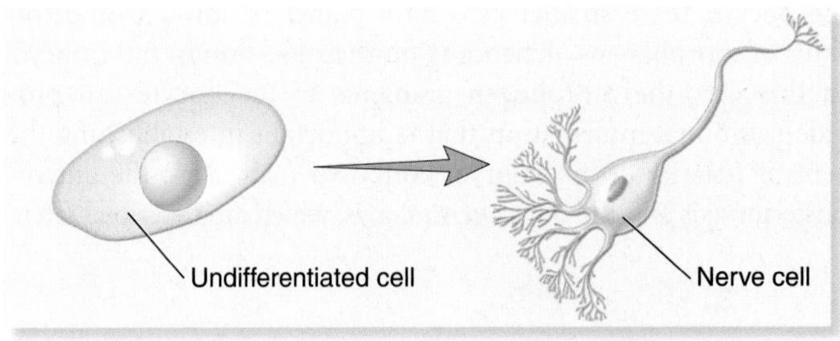

(c) Cell differentiation

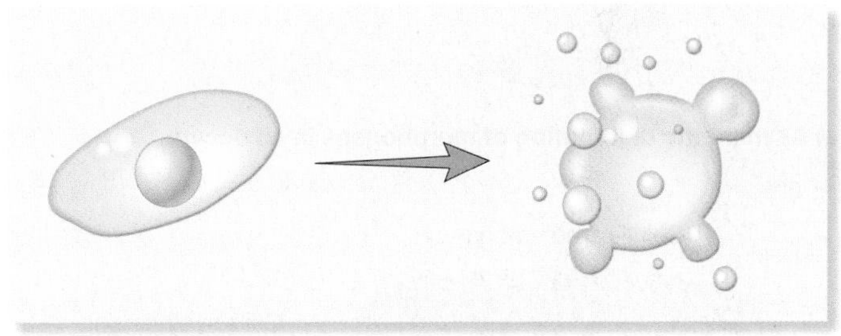

(d) Cell death (apoptosis)

FIGURE 23.1 Four events that may happen to cells during the course of development.

The Generation of a Body Pattern Depends on the Positional Information That Each Cell Receives During Development

Multicellular development in animals and plants follows a body plan or pattern. The term **pattern** refers to the spatial arrangement of different regions of the body. At the cellular level, the body pattern is due to the arrangement of cells and their specialization.

The progressive growth of a fertilized egg into an adult organism involves four types of cellular events: cell division, cell migration, cell differentiation, and cell death (**Figure 23.1**). The coordination of these four events leads to the formation of a body with a particular pattern. As we will see, the temporal expression of genes and the localization of gene products at precise regions in the fertilized egg and early embryo are the critical phenomena that underlie this coordination.

Before we consider how genes affect development, let's consider a central concept in developmental biology known as **positional information.** For an organism to develop a body pattern with unique morphological and cellular features, each cell of the body must receive signals—positional information—that cause it to become the appropriate cell type based on its position relative to the other cells.

How may a cell respond to positional information? Various responses are possible. For example, positional information may cause a cell or group of cells to divide into two daughter cells or migrate in a particular direction from one region of the embryo to another. In addition, positional information may stimulate a cell to differentiate into a particular cell type. Finally, positional information or other stimuli may cause a cell to die. This process, known as **apoptosis,** is a necessary event during normal development to sculpt tissues and organs. For example, in the later stages of development of the retina in *Drosophila*, excess cells are eliminated by apoptosis. During mammalian development, individual fingers and toes are created when embryonic cells that lie between each future finger and toe undergo this type of cell death.

A **morphogen** is a molecule that conveys positional information and promotes developmental changes. It is generally a diffusible molecule that acts in a concentration-dependent manner to influence the developmental fate of a cell. Morphogens provide the positional information that stimulate a cell to divide, migrate, differentiate, or die. For example, when regions of a *Drosophila* embryo are exposed to a high concentration of the morphogen known as Bicoid, they differentiate into structures characteristic of the anterior region of the body. Many such morphogens function as transcription factors that regulate the expression of many genes. This topic will be described in greater detail later. How do morphogens control pattern development? Within an oocyte and during embryonic development, morphogens typically are distributed along a concentration gradient.

In other words, the concentration of a morphogen varies from low to high in different regions of the developing organism. A key feature of morphogens is that they act in a concentration-dependent manner. In a particular concentration range, a morphogen or a combination of two or more different morphogens will restrict a cell into a specific developmental pathway.

During the earliest stages of development, morphogenic gradients are preestablished within the oocyte (**Figure 23.2a**). Following fertilization, the zygote subdivides into many smaller cells. Due to the preestablished gradient of morphogens within the oocyte, these smaller cells have higher or lower concentrations of morphogens, depending on their location in the embryo. In this way, the morphogen gradients in the oocyte can provide positional information that is important in establishing the general polarity of an embryo along two main axes: the antero-posterior axis and the dorso-ventral axis, which are described later.

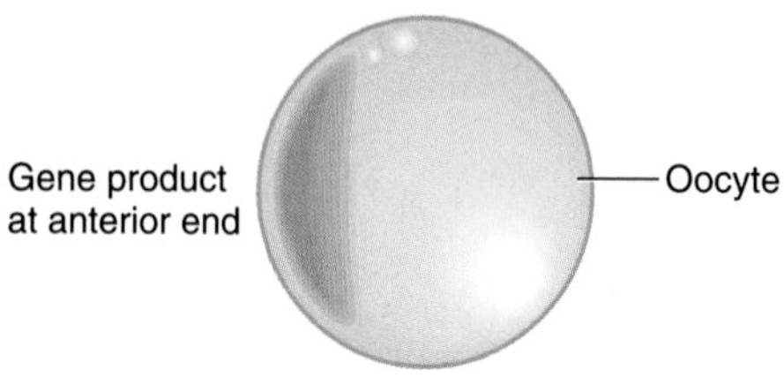

(a) Asymmetric distribution of morphogens in an oocyte

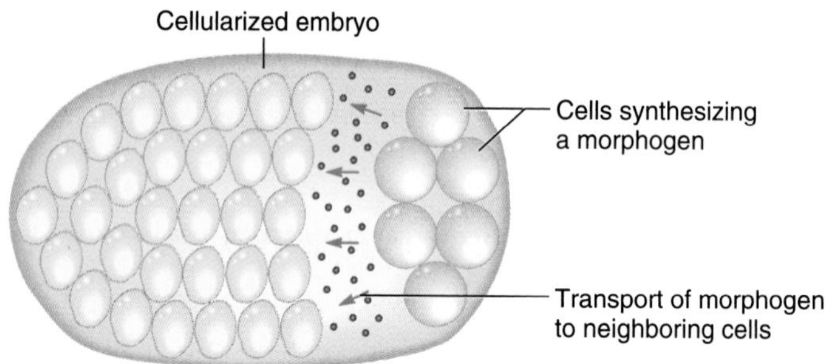

(b) Asymmetric synthesis and extracellular distribution of a morphogen in an embryo

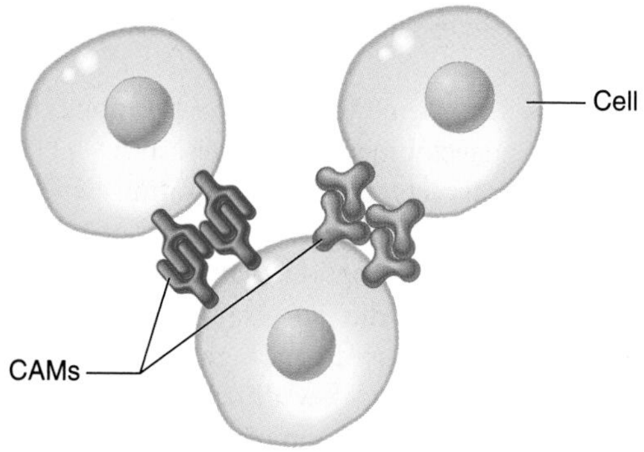

(c) Cell-to-cell contact conveys positional information

FIGURE 23.2 **Three molecular mechanisms of positional information.**

A morphogen gradient can also be established in the embryo by cell secretion and transport. A certain cell or group of cells may synthesize and secrete a morphogen at a specific stage of development. After secretion, the morphogen may be transported to neighboring cells. The concentration of the morphogen is usually highest near the cells that secrete it (**Figure 23.2b**). The morphogen may then influence the developmental fate of the cells exposed to it. The process by which a cell or group of cells governs the developmental fate of neighboring cells is known as **induction.** An example of a secreted morphogen in *Drosophila* is a protein known as Hedgehog. This protein is secreted from cells on the posterior side of body segments and forms a gradient from the posterior to anterior region of each segment. Hedgehog plays a role in creating the morphological differences between the anterior and posterior parts of individual body segments and also influences the formation of certain adult structures. Hedgehog mutants result in larvae that are short and stubby and have a higher density of denticles, which are cuticle spikes projecting from the body surface. The appearance of the stubby and spikey larvae inspired the name hedgehog.

In addition to morphogens, positional information is conveyed by **cell adhesion** (**Figure 23.2c**). Each cell makes its own collection of surface receptors that enable it to adhere to other cells and/or to the extracellular matrix (ECM), which consists primarily of carbohydrates and fibrous proteins. Such receptors are known as **cell adhesion molecules (CAMs).** A cell may gain positional information via the combination of contacts it makes with other cells or with the ECM.

The phenomenon of cell adhesion, and its role in multicellular development, was first recognized by Henry Wilson in 1907. He took multicellular sponges and passed them through a sieve, which disaggregated them into individual cells. Remarkably, the cells actively migrated until they adhered to one another to form a new sponge, complete with the chambers and canals that characterize a sponge's internal structure! When sponge cells from different species were mixed, they sorted themselves properly, adhering only to cells of the same species. Overall, these results indicate that cell adhesion plays an important role in governing the position that a cell will adopt during development.

The Study of Mutants with Disrupted Developmental Patterns Has Identified Genes That Control Development

Mutations that alter the course of development in experimental organisms, such as *Drosophila*, have greatly contributed to our understanding of the normal process of development. For example, **Figure 23.3** shows a photograph of a fly that carries multiple mutations in a complex of genes called *bithorax*. This mutant fly has four wings instead of two; the halteres (balancing organs that resemble miniature wings), which are found on the third thoracic segment, are changed into wings, normally found on the second thoracic segment. (The term bithorax refers to the observation that the characteristics of the second thoracic segment are duplicated.)

FIGURE 23.3 The bithorax mutation in *Drosophila*.

Genes→Traits A normal fly contains two wings on the second thoracic segment and two halteres on the third thoracic segment. However, this mutant fly contains multiple mutations in a complex of genes called the *bithorax* complex. In this fly, the third thoracic segment has the same characteristics as the second thoracic segment, thereby producing a fly with four wings instead of the normal number of two.

TABLE 23.1

Examples of *Drosophila* Genes That Play a Role in Pattern Development

Description	Examples
Some genes play a role in determining the axes of development. Also certain genes govern the formation of the extreme terminal (anterior and posterior) regions.	Anterior: *bicoid, exuperantia, swallow, staufen* Posterior: *nanos, cappuccino, oskar pumilio, spire, staufen, tudor, vasa* Terminal: *torso, torsolike, Trunk, NTF-1* Dorso-ventral: *Toll, cactus, dorsal, easter, gurken, nudel, pelle, pipe, snake, spatzle*
Some genes play a role in promoting the subdivision of the embryo into segments. These are called segmentation genes. The three types are gap genes, pair-rule genes, and segment-polarity genes.	Gap genes: *empty spiracles, giant, huckebein, hunchback, knirps, Krüppel, tailless, orthodenticle* Pair-rule genes: *even-skipped, hairy, runt, fushi tarazu, paired* Segment-polarity genes: *frizzled, frizzled-2, engrailed, patched, smoothened, hedgehog, wingless, gooseberry*
Some genes play a role in determining the fate of particular segments. These are known as homeotic genes. *Drosophila* has two clusters of homeotic genes known as the *Antennapedia* complex and the *bithorax* complex	*Antennapedia* complex: *labial, proboscipedia, Deformed, Sex combs reduced, Antennapedia P* *Bithorax* complex: *Ultrabithorax, abdominal A, Abdominal B*

Edward Lewis, a pioneer in the genetic study of development, became interested in the bithorax phenotype and began investigating it in 1946. Researchers later discovered that the mutant chromosomal region actually contained a complex of three genes involved in specifying developmental pathways in the fly. A gene that plays a central role in specifying the final identity of a body region is called a **homeotic gene.** We will discuss particular examples of homeotic genes later in this chapter.

During the 1960s and 1970s, interest in the relationship between genetics and embryology blossomed as biologists began to appreciate the role of genetics at the molecular and cellular levels. It soon became clear that the genomes of multicellular organisms contain groups of genes that initiate a program of development involving networks of gene regulation. By identifying mutant alleles that disrupt development, geneticists have begun to unravel the pattern of gene expression that underlies the normal pattern of multicellular development.

In *Drosophila*, the establishment of the body axes and division of the body into segments involves the participation of a few dozen genes. **Table 23.1** lists many of the important genes governing pattern formation during embryonic development. These genes were identified by characterizing mutants that had altered development patterns. They are often given interesting names based on the observed phenotype when they are mutant. It is beyond the scope of this textbook to describe how all of these genes exert their effects during embryonic development. Instead, we will consider a few examples that illustrate how the expression of a particular gene and the localization of its gene product have a defined effect on the pattern of development.

The Early Stages of Embryonic Development Determine the Pattern of Structures in the Adult Organism

Figure 23.4 illustrates the general sequence of events in *Drosophila* development. The oocyte is the most critical cell in determining the pattern of development in the adult organism. It is an elongated cell with preestablished axes (Figure 23.4a). After fertilization takes place, the zygote goes through a series of nuclear divisions that are not accompanied by cytoplasmic division. Initially, the resulting nuclei are scattered throughout the yolk, but eventually they migrate to the periphery of the cytoplasm. This is the syncytial blastoderm stage (Figure 23.4b).

After the nuclei have lined up along the cell membrane, individual cells are formed as portions of the cell membrane surround each nucleus, creating a cellular blastoderm (Figure 23.4c). This structure is composed of a sheet of cells on the outside with yolk in the center. In this arrangement, the cells are distributed asymmetrically. At the posterior end are a group of cells called the pole cells, the primordial germ cells that eventually give rise to gametes in the adult organism. After blastoderm formation is complete, dramatic changes occur during gastrulation (Figure 23.4d). This stage produces three cell layers known as the ectoderm, mesoderm, and endoderm.

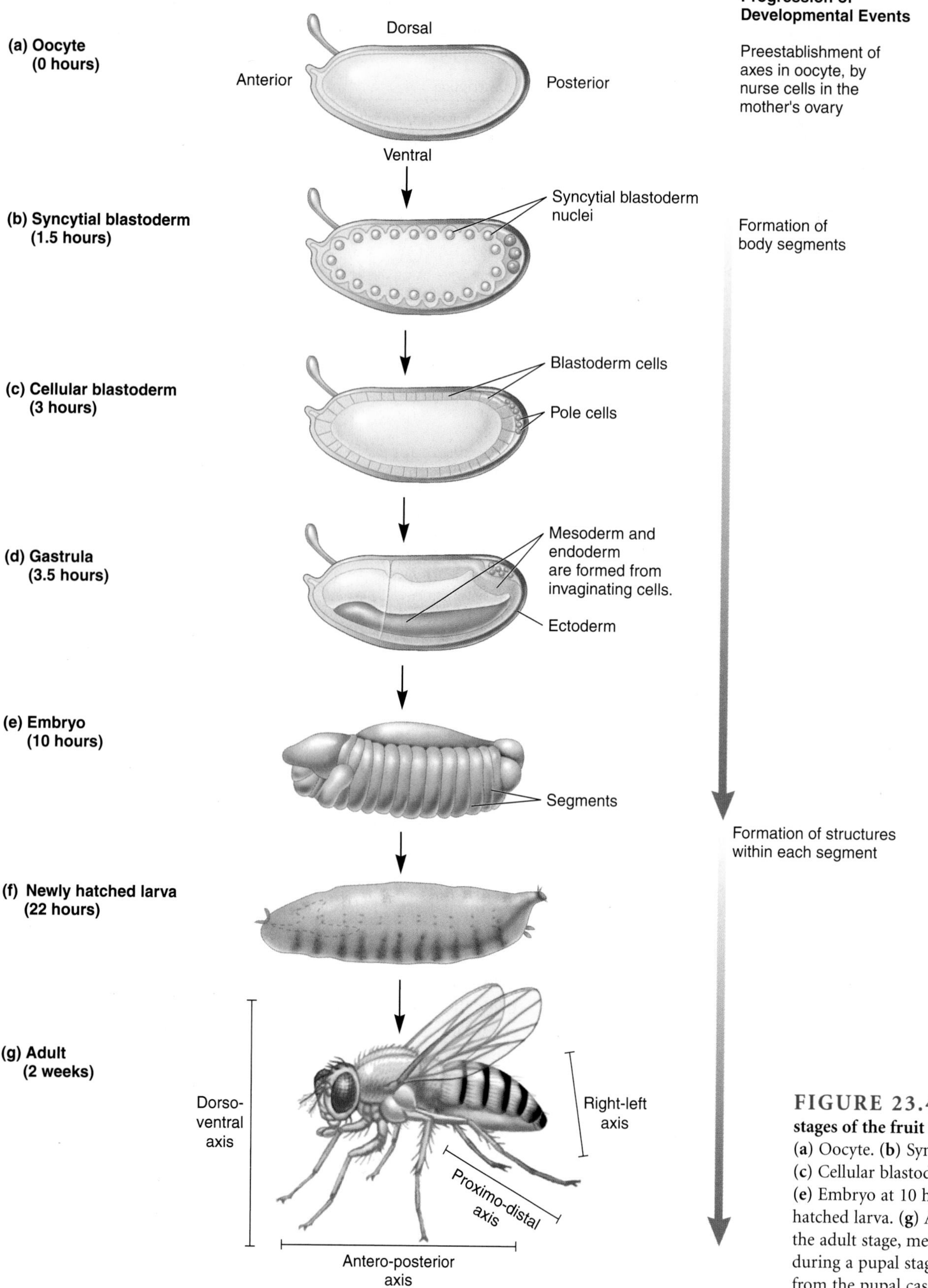

FIGURE 23.4 Developmental stages of the fruit fly *Drosophila*. (**a**) Oocyte. (**b**) Syncytial blastoderm. (**c**) Cellular blastoderm. (**d**) Gastrula. (**e**) Embryo at 10 hours. (**f**) Newly hatched larva. (**g**) Adult. Note: Prior to the adult stage, metamorphosis occurs during a pupal stage. The adult emerges from the pupal case.

As this process occurs, the embryo begins to be subdivided into detectable units. Initially, shallow grooves divide the embryo into 14 **parasegments,** which are only transient subdivisions. A short time later, these grooves disappear, and new boundaries are formed that divide the embryo into morphologically discrete **segments.** Figure 23.4e shows the segmented pattern of a *Drosophila* embryo about 10 hours after fertilization. Later in this section, we will explore how the coordination of gene expression underlies the formation of these parasegments and segments.

At the end of **embryogenesis,** which lasts about 18 to 22 hours, a larva hatches from the egg (Figure 23.4f) and begins feeding on its own. *Drosophila* has three larval stages, separated by molts. During molting, the larva sheds its cuticle, a hardened extracellular shell that is secreted by the epidermis.

After the third larval stage, *Drosophila* forms a pupa that undergoes metamorphosis. During metamorphosis, groups of cells called imaginal disks that were produced earlier in development differentiate into the structures found in the adult fly, such as the head, wings, legs, and abdomen. The adult fly then emerges from the pupal case. The adult is produced about two weeks after fertilization. In *Drosophila,* as in all bilateral animals, the final result of development is an adult body organized along three axes: the **antero-posterior axis,** the **dorso-ventral axis,** and the **right-left axis** (Figure 23.4g). An additional axis, used mostly for designating limb parts, is the **proximo-distal axis.**

Although many interesting developmental events occur during the three larval stages and the pupal stage, we will focus most of our attention on the events that occur during embryonic development. Even before hatching, the embryo develops the basic body plan that will be found in the adult organism. In other words, during the early stages of development, the embryo is divided into segments that correspond to the segments of the larva and adult. Therefore, an understanding of how these segments form in the embryo is critical to our understanding of pattern formation.

The Gene Products of Maternal Effect Genes Are Deposited Asymmetrically into the Oocyte and Establish the Antero-posterior and Dorso-ventral Axes at a Very Early Stage of Development

The first stage in *Drosophila* embryonic pattern development is the establishment of the body axes. This occurs before the embryo becomes segmented. In fact, the morphogens necessary to establish these axes are distributed prior to fertilization. During oogenesis, gene products such as mRNA, which are important in early developmental stages, are deposited asymmetrically within the egg. Later, after the egg has been fertilized and development begins, these gene products will establish independent developmental programs that govern the formation of the body axes of the embryo. Proper development is dependent on the products of several different genes.

As shown in **Figure 23.5**, a few gene products act as key morphogens, or receptors for morphogens, that initiate changes in embryonic development. As shown here, these gene products are deposited asymmetrically in the egg. For example, the product of the *bicoid* gene is necessary to initiate development of the anterior structures of the organism. During oogenesis, the mRNA for *bicoid* accumulates in the anterior region of the oocyte. In contrast, the mRNA from the *nanos* gene accumulates in the posterior end. Later in development, the *nanos* mRNA will be translated into protein, which functions to influence posterior development. *Nanos* is required for the formation of the abdomen.

In addition to morphogens such as Bicoid and Nanos, the development of the structures at the extreme anterior and posterior ends of the embryo are regulated in part by a receptor protein called Torso. This receptor, which is activated by ligand binding only at the anterior and posterior ends of the egg, is necessary

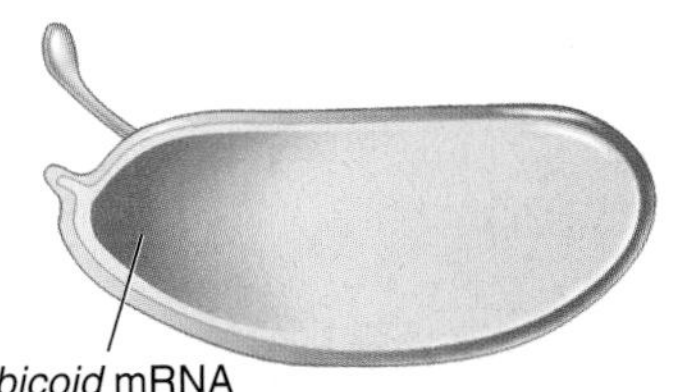

(a) Anterior distribution of *bicoid* mRNA

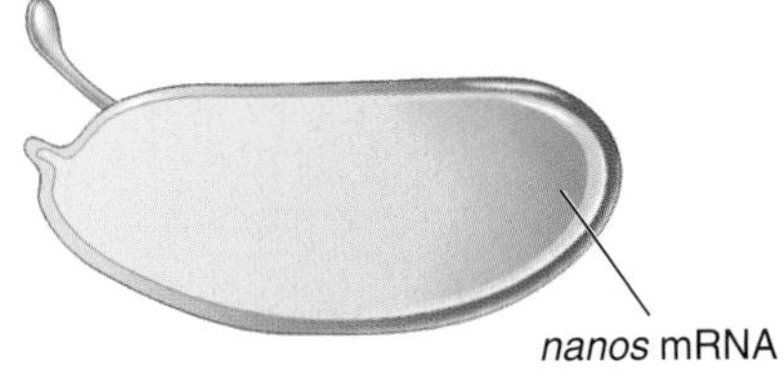

(b) Posterior distribution of *nanos* mRNA

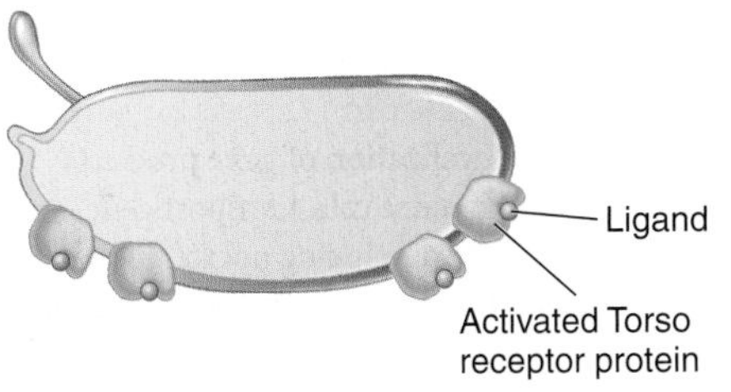

(c) Terminal distribution of Torso receptor protein

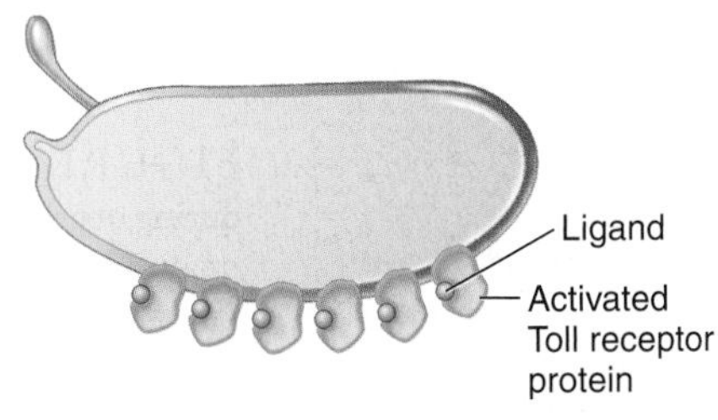

(d) Ventral distribution of Toll receptor protein

FIGURE 23.5 **The establishment of the axes of polarity in the *Drosophila* embryo.** This figure shows some of the maternal gene products deposited in the oocyte that are critical in the establishment of the antero-posterior, terminal, and dorso-ventral axes. **(a)** *Bicoid* mRNA is distributed in the anterior end of the oocyte and promotes the formation of anterior structures. **(b)** *Nanos* mRNA is localized to the posterior end and promotes the formation of posterior structures. **(c)** The Torso receptor protein is found in the membrane and is activated by ligand binding at either end of the oocyte. It causes the formation of structures that are found only at the ends of the organism. **(d)** The Toll receptor protein is activated by ligand binding at the ventral side of the embryo and establishes the dorso-ventral axis. Note: The Torso and Toll receptor proteins are distributed throughout the plasma membranes of the oocyte and embryo, but they are activated by ligand binding only in the regions shown in this figure.

for the formation of the terminal ends of the embryo. The dorsoventral axis is governed by the actions of several proteins, including a receptor protein known as Toll. This receptor is found in plasma membranes throughout the embryo but is activated by ligand binding only along the ventral midline of the embryo.

Let's now take a closer look at the molecular mechanism of one morphogen, Bicoid. The *bicoid* gene got its name because a larva defective in this gene develops with two posterior ends (**Figure 23.6**). This allele exhibits a maternal effect pattern of inheritance, in which the genotype of the mother determines the phenotypic traits of the offspring (see Chapter 7). A female fly that is phenotypically normal (because its mother was heterozygous for the normal *bicoid* allele), but genotypically homozygous for an inactive *bicoid* allele (because it inherited an inactive allele from both its mother and father), will produce 100% abnormal offspring even when mated to a male that is homozygous for the normal *bicoid* allele. In other words, the genotype of the mother determines the phenotype of the offspring. This occurs because the *bicoid* gene product is provided to the oocyte via the nurse cells.

In the ovaries of female flies, the nurse cells are localized asymmetrically toward the anterior end of the oocyte. During oogenesis, gene products are transferred from nurse cells into the oocyte via cell-to-cell connections called cytoplasmic bridges. Maternally encoded gene products enter one side of the oocyte, which will eventually become the anterior end of the embryo (**Figure 23.7a**). The *bicoid* gene is actively transcribed in the nurse cells, and *bicoid* mRNA is transported into the anterior end of the oocyte. The 3' end of *bicoid* mRNA contains a signal that is recognized by RNA-binding proteins thought necessary for the transport of this mRNA into the oocyte. After it enters the oocyte, the *bicoid* mRNA is trapped at the anterior end.

How can researchers determine the location of *bicoid* mRNA in the oocyte and resulting zygote? **Figure 23.7b** shows an *in situ* hybridization experiment in which a *Drosophila* egg was examined via a probe complementary to the *bicoid* mRNA. (The technique of *in situ* hybridization is described in Chapter 20.) As seen here, the *bicoid* mRNA is highly concentrated near the anterior end of the egg cell. When the *bicoid* mRNA subsequently is translated, a gradient of Bicoid protein is established as shown in **Figure 23.7c**.

After fertilization occurs, the Bicoid protein functions as a transcription factor. A remarkable feature of this protein is that its ability to influence gene expression is tuned exquisitely to its concentration. Depending on the distribution of the Bicoid protein, this transcription factor will activate genes only in certain regions of the embryo. For example, Bicoid stimulates the transcription of a gene called *hunchback* in the anterior half of the embryo, but its concentration is too low in the posterior half to activate the *hunchback* gene there.

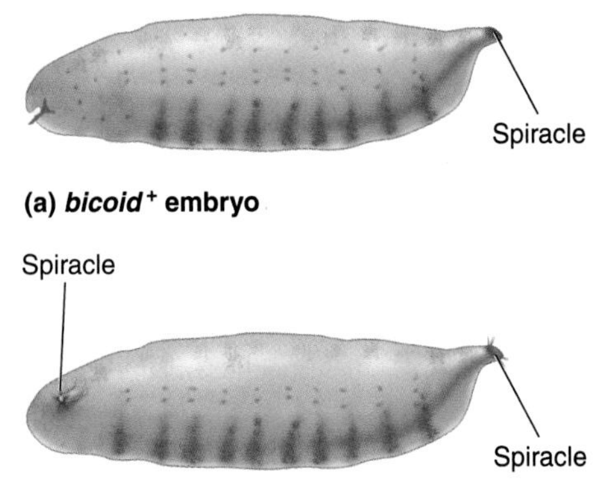

FIGURE 23.6 The bicoid mutation in *Drosophila*. **(a)** A normal *bicoid*$^+$ embryo. **(b)** A *bicoid*$^-$ embryo in which both ends of the larva develop posterior structures. For example, both ends develop a spiracle, which normally is found only at the posterior end. This is a lethal condition.

Gap, Pair-Rule, and Segment-Polarity Genes Act Sequentially to Divide the *Drosophila* Embryo into Segments

After the antero-posterior and dorso-ventral regions of the embryo have been established by maternal effect genes, the next develop-

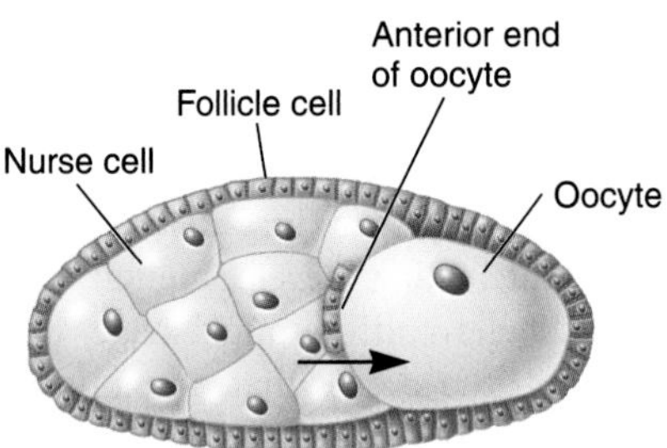

(a) Transport of maternal effect gene products into the oocyte

(b) *In situ* hybridization of *bicoid* mRNA

(c) Immunostaining of Bicoid protein

FIGURE 23.7 Asymmetrical localization of gene products during oogenesis in *Drosophila*. **(a)** The nurse cells transport gene products into the anterior (left) end of the developing oocyte. **(b)** An *in situ* hybridization experiment showing that the *bicoid* mRNA is trapped near the anterior end. **(c)** The *bicoid* mRNA is translated into protein soon after fertilization. The location of the Bicoid protein is revealed by immunostaining using an antibody that specifically recognizes this protein.

mental process organizes the embryo transiently into parasegments and then permanently into segments. The segmentation pattern of the embryo is shown in **Figure 23.8**. This pattern of positional information will be maintained, or "remembered," throughout the rest of development. In other words, each segment of the embryo gives rise to unique morphological features in the adult. For example, T2 becomes a thoracic segment with a pair of legs and a pair of wings, and A8 becomes a segment of the abdomen.

Figure 23.8 shows the overlapping relationship between parasegments and segments. As seen here, the boundaries of the segments are out of register with the boundaries of the parasegments. An appreciation of this feature is critical to our understanding of segmentation. From the viewpoint of genes, we will see that the parasegments are the locations where gene expression is controlled spatially. The anterior compartment of each segment coincides with the posterior region of a parasegment; the posterior compartment of a segment coincides with the anterior region of the next parasegment. The pattern of gene expression that occurs in the posterior region of one parasegment and the anterior region of an adjacent parasegment results in the formation of a segment.

Now that we have a general understanding of the way the *Drosophila* embryo is subdivided, we can examine how particular genes cause it to become segmented into this pattern. Three classes of genes, collectively called **segmentation genes,** play a role in the formation of body segments: gap genes, pair-rule genes, and segment-polarity genes. The expression and activation patterns of these genes in specific regions of the embryo cause it to become segmented.

How were the three classes of segmentation genes discovered? In the 1970s, segmentation genes were identified by Christiane Nüsslein-Volhard and Eric Wieschaus, who undertook a systematic search for mutations affecting embryonic development. Their pioneering effort identified most of the genes required for the embryo to develop a segmented pattern.

Figure 23.9 represents a few of the phenotypic effects that occur when a particular segmentation gene is defective. The gray boxes indicate the regions missing in the resulting larvae. In flies

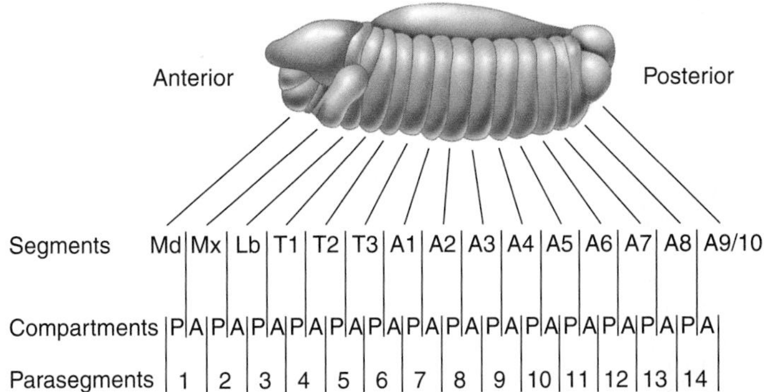

FIGURE 23.8 A comparison of segments and parasegments in the *Drosophila* embryo. Note that the parasegments and segments are out of register. The posterior (P) and anterior (A) regions are shown for each segment.

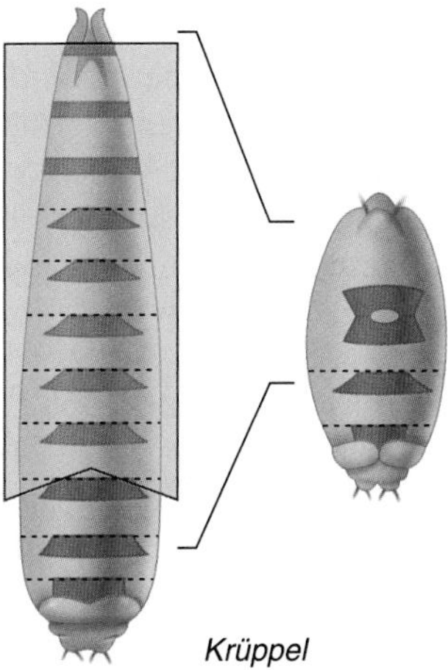

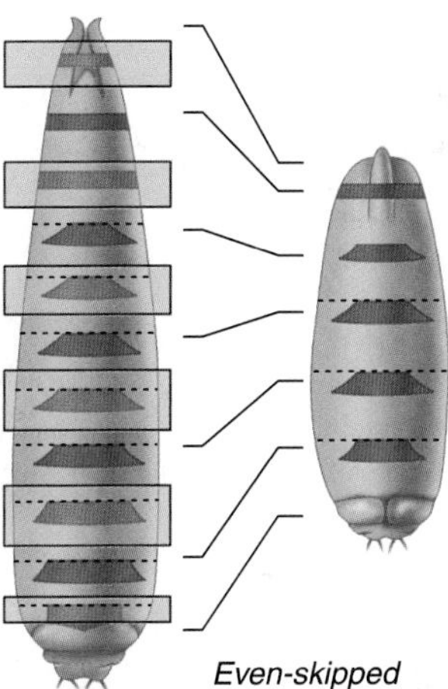

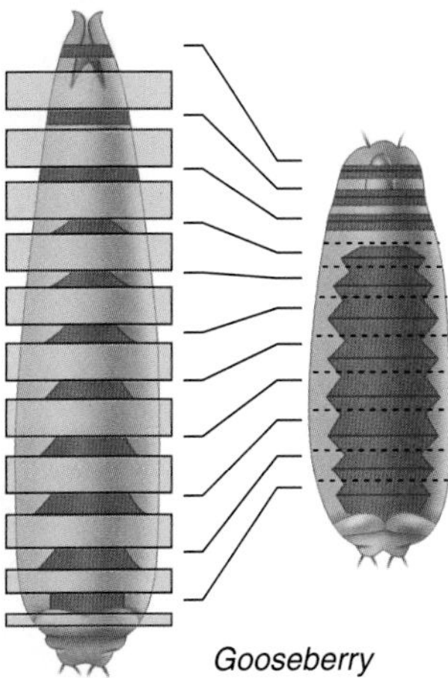

FIGURE 23.9 Phenotypic effects in *Drosophila* larvae with mutations in segmentation genes. The effects shown here are caused by mutations in **(a)** a gap gene, **(b)** a pair-rule gene, and **(c)** a segment-polarity gene.

with a mutation in a gap gene known as *Krüppel*, a contiguous section of the larva is missing (Figure 23.9a). In other words, a gap of several segments has occurred. By comparison, a defect in a pair-rule gene causes alternating regions to be deleted (Figure 23.9b). For example, when the *even-skipped* gene is defective, portions of alternating segments in the resulting larva are missing. Finally, segment-polarity mutations cause individual segments to be missing either an anterior or a posterior region. Figure 23.9c shows a mutation in a segmentation gene known as *gooseberry*. When this gene is defective, the anterior portion of each segment is missing from the larva. In the case of segment-polarity mutants, the segments adjacent to the deleted regions exhibit a mirror-image duplication. Overall, the phenotypic effects of mutant segmentation genes provided geneticists with important clues regarding the roles of these genes in the developmental process of segmentation.

Figure 23.10 presents a partial, simplified scheme of the genetic hierarchy that leads to a segmented pattern in the *Dro-*

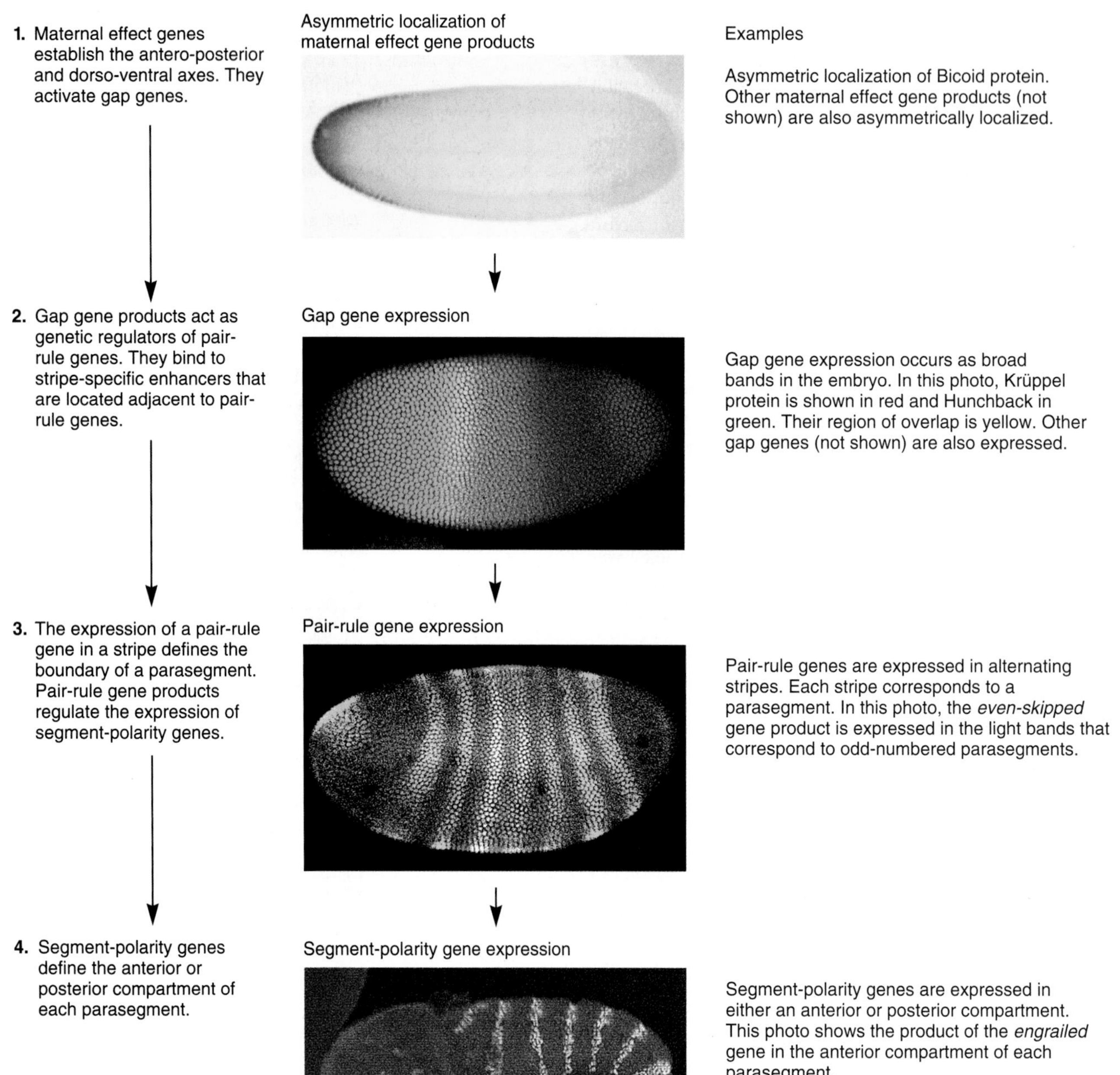

1. Maternal effect genes establish the antero-posterior and dorso-ventral axes. They activate gap genes.
Asymmetric localization of maternal effect gene products
Examples
Asymmetric localization of Bicoid protein. Other maternal effect gene products (not shown) are also asymmetrically localized.
2. Gap gene products act as genetic regulators of pair-rule genes. They bind to stripe-specific enhancers that are located adjacent to pair-rule genes.
Gap gene expression
Gap gene expression occurs as broad bands in the embryo. In this photo, Krüppel protein is shown in red and Hunchback in green. Their region of overlap is yellow. Other gap genes (not shown) are also expressed.
3. The expression of a pair-rule gene in a stripe defines the boundary of a parasegment. Pair-rule gene products regulate the expression of segment-polarity genes.
Pair-rule gene expression
Pair-rule genes are expressed in alternating stripes. Each stripe corresponds to a parasegment. In this photo, the *even-skipped* gene product is expressed in the light bands that correspond to odd-numbered parasegments.
4. Segment-polarity genes define the anterior or posterior compartment of each parasegment.
Segment-polarity gene expression
Segment-polarity genes are expressed in either an anterior or posterior compartment. This photo shows the product of the *engrailed* gene in the anterior compartment of each parasegment.

FIGURE 23.10 **Overview of the genetic hierarchy leading to segmentation in *Drosophila*.** Note: When comparing steps 3 and 4, the embryo has undergone a 180° turn, folding back on itself.

sophila embryo. Keep in mind that while this figure presents the general sequence of events that occurs during the early stages of embryonic development, many more genes are actually involved in this process (refer back to Table 23.1). As described in this figure, the following steps occur:

1. Maternal effect gene products, such as *bicoid* mRNA, are deposited asymmetrically into the oocyte. These gene products form a gradient that will later influence the formation of axes, such as the antero-posterior axis.
2. In contrast to maternal effect genes, which are expressed during oogenesis, **zygotic genes** are expressed after fertilization. The first zygotic genes to be activated are the gap genes. Maternal effect genes are responsible for activating the gap genes. As shown in step 2 of Figure 23.10, gap genes divide the embryo into a series of broad bands or regions. These bands do not correspond to parasegments or segments within the embryo.
3. The gap genes and maternal effect genes then activate the pair-rule genes. The photograph shown in step 3 of Figure 23.10 illustrates the alternating pattern of expression of the *even-skipped* gene. The expression of pair-rule genes in stripes defines the boundaries of parasegments. *Even-skipped* is expressed in the odd-numbered parasegments. Solved problem S4 at the end of the chapter examines how the *even-skipped* gene can be expressed in this pattern of alternating stripes. The pair-rule genes divide the broad regions established by gap genes into seven bands or stripes.
4. Once the pair-rule genes are activated in an alternating banding arrangement, their gene products then regulate the expression of segment-polarity genes. The segment-polarity genes divide the embryo into 14 stripes, one within each parasegment. As shown in step 4 of Figure 23.10, the segment-polarity gene *engrailed* is expressed in the anterior region of each parasegment. Another segment-polarity gene, *wingless*, is expressed in the posterior region. Later in development, the anterior end of one parasegment and the posterior end of another parasegment will develop into a segment with particular morphological characteristics.

The Expression of Homeotic Genes Controls the Phenotypic Characteristics of Segments

Thus far, we have considered how the *Drosophila* embryo becomes organized along axes and then into a segmented body pattern. Now let's examine how each segment develops its unique morphological features. Geneticists often use the term **cell fate** to describe the ultimate morphological features that a cell or group of cells will adopt. For example, the fate of the cells in segment T2 in the *Drosophila* embryo is to develop into a thoracic segment containing two legs and two wings. In *Drosophila*, the fate of the cells in each segment of the body is determined at a very early stage of embryonic development, long before the morphological features become apparent.

Our understanding of developmental fate has been greatly aided by the identification of mutant genes that alter cell fates. In animals, the first mutant of this type was described by the German entomologist Ernst Gustav Kraatz in 1876. He observed a sawfly (*Cimbex axillaris*) in which part of an antenna was replaced with a leg. During the late nineteenth century, the English zoologist William Bateson collected many of these types of observations and published them in 1894 in a book entitled *Materials for the Study of Variation Treated with Especial Regard to Discontinuity in the Origin of Species*. In this book, Bateson coined the term **homeotic** to describe mutants in which one body part is replaced by another, such as the transformation of the antennae of insects into legs.

As mentioned earlier, Edward Lewis began to study strains of *Drosophila* having homeotic mutations. This work, which began in 1946, was the first systematic study of homeotic genes. Each homeotic gene controls the fate of a particular region of the body. *Drosophila* contains two clusters of homeotic genes called the *Antennapedia* complex and the *bithorax* complex. **Figure 23.11** shows the organization of genes within these complexes.

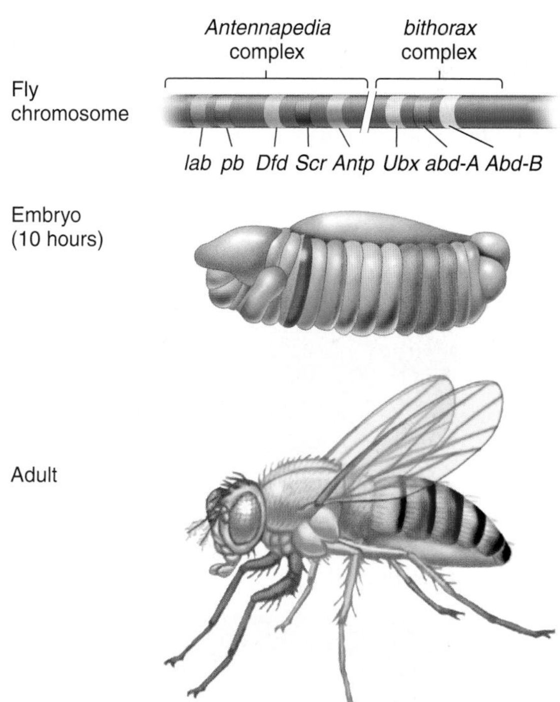

FIGURE 23.11 Expression pattern of homeotic genes in *Drosophila*. The order of homeotic genes, *labial* (*lab*), *proboscipedia* (*pb*), *Deformed* (*Dfd*), *Sex combs reduced* (*Scr*), *Antennapedia* (*Antp*), *Ultrabithorax* (*Ubx*), *abdominal A* (*abd-A*), and *Abdominal B* (*Abd-B*), correlates with the order of expression in the embryo. The expression pattern of seven of these genes is shown. *Lab* (purple) is expressed in the region that will eventually give rise to mouth structures. *Dfd* (light green) is expressed in the region that will form much of the head. *Scr* (forest green) and *Antp* (blue) are expressed in embryonic segments that give rise to thoracic segments. *Ubx* (gray), *abd-A* (red), and *Abd*-B (yellow) are expressed in posterior segments that will form the abdomen. The order of gene expression, from anterior to posterior, parallels the order of genes along the chromosome. The expression pattern of the *pb* gene is not well understood.

The *Antennapedia* complex contains five genes, designated *lab*, *pb*, *Dfd*, *Scr*, and *Antp*. The *bithorax* complex has three genes, *Ubx*, *abd-A*, and *Abd-B*. Both of these complexes are located on chromosome 3, but a large segment of DNA separates them.

As noted in Figure 23.11, the order of these genes along chromosome 3 correlates with their pattern of gene expression along the antero-posterior axis of the body. For example, *lab* is expressed in an anterior segment and governs the formation of mouth structures. The *Antp* gene is expressed strongly in the thoracic region during embryonic development and controls the formation of thoracic structures. Transcription of the *Abd-B* gene occurs in the posterior region of the embryo. This gene controls the formation of the posterior-most abdominal segments.

The role of homeotic genes in determining the identity of particular segments has been revealed by mutations that alter their function. For example, the *Antp* gene is normally expressed in the thoracic region. The *Antennapedia* mutation causes the *Antp* gene to also be expressed in the region where the antennae are made. These mutant flies have legs in the place of antennae (Figure 23.12)! This is an example of a **gain-of-function mutation.** In this case, the *Antp* gene is also expressed abnormally in the anterior segment that normally gives rise to the antennae.

Investigators have also studied many loss-of-function mutations in homeotic genes. When a particular homeotic gene is defective, the region it normally governs will usually be controlled by the homeotic gene that acts in the adjacent anterior region. For example, the *Ubx* gene functions within parasegments 5 and 6. If this gene is missing, this section of the fly becomes converted to the structures found in parasegment 4.

The homeotic genes are part of the genetic hierarchy that produces the morphological characteristics of the fly. How are they regulated? The homeotic genes are controlled by gap genes and pair-rule genes, and they are also regulated by interactions among themselves. Other genes maintain the patterns of homeotic gene expression after the initial patterns are established by segmentation genes. A group of genes known as *Polycomb* genes represses the expression of homeotic genes in regions of the embryo where they should not act. This is accomplished by the remodeling of chromatin structure to convert it to a closed conformation in which transcription is difficult or impossible. Alternatively, in regions where the homeotic genes are active, the products of *Trithorax* genes maintain an open conformation in which transcription can take place. Overall, the concerted actions of many gene products cause the homeotic genes to be expressed only in the appropriate region of the embryo, as shown in Figure 23.11.

Because they are part of a genetic hierarchy in which genes activate other genes, perhaps it is not surprising that homeotic genes encode transcription factors. The coding sequence of homeotic genes contains a 180 base pair consensus sequence known as a **homeobox** (Figure 23.13a). This sequence was first discovered in the *Antp* and *Ubx* genes, and it has since

(a) Normal fly

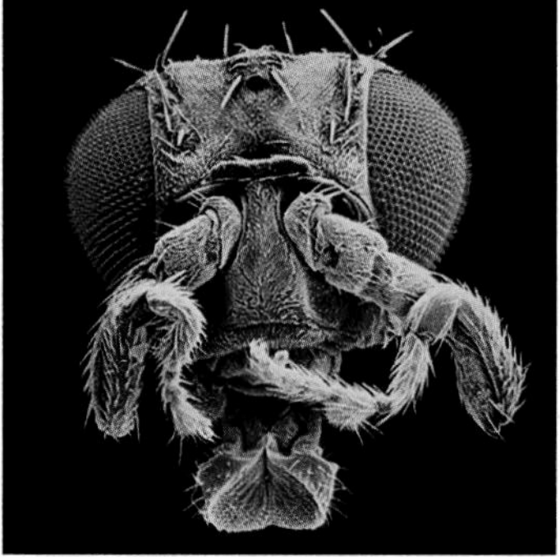

(b) *Antennapedia* mutant

FIGURE 23.12 **The *Antennapedia* mutation in *Drosophila*.** Genes→Traits **(a)** A normal fly with antennae. **(b)** This mutant fly has a gain-of-function mutation in which the *Antp* gene is expressed in the embryonic segment that normally gives rise to antennae. The expression of *Antp* causes this region to have legs rather than antennae.

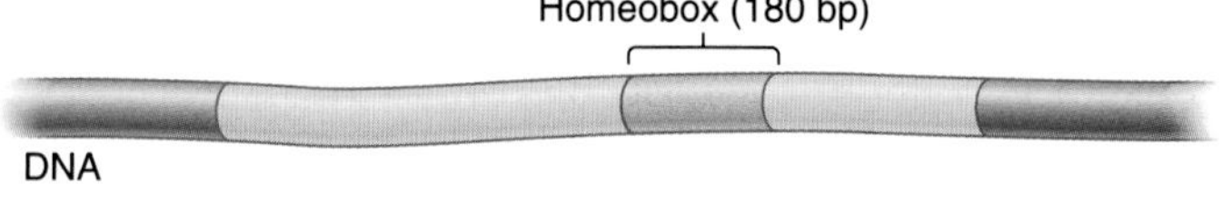

(a) Homeotic gene

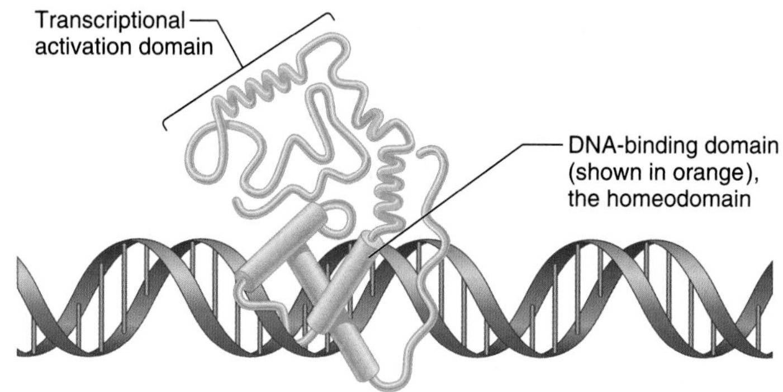

(b) Homeotic protein bound to DNA

FIGURE 23.13 **Molecular features of homeotic proteins.** **(a)** A homeotic gene (shown in tan and orange) contains a 180-bp sequence called the homeobox (shown in orange). **(b)** When a homeotic gene is expressed, it produces a protein that functions as a transcription factor. The homeobox encodes a region of the protein called a homeodomain, which binds to the major groove of DNA. These DNA-binding sites are found within genetic regulatory elements (i.e., enhancers). The enhancers are found in the vicinity of promoters that are turned on by homeotic proteins. For this to occur, the homeotic protein also contains a transcriptional activation domain, which will activate the transcription of a gene after the homeodomain has bound to the DNA.

been found in all *Drosophila* homeotic genes and in some other genes affecting pattern development, such as *bicoid.* The protein domain encoded by the homeobox is called a **homeodomain.** The arrangement of α helices within the homeodomain promotes the binding of the protein to the major groove of DNA (**Figure 23.13b**). In this way, homeotic proteins can bind to DNA in a sequence-specific manner. In addition to DNA-binding ability, homeotic proteins also contain a transcriptional activation domain that functions to activate the genes to which the homeodomain can bind.

The transcription factors encoded by homeotic genes activate genes encoding proteins that produce the morphological characteristics of each segment. Much current research attempts to identify these genes and determine how their expression in particular regions of the embryo leads to morphological changes in the embryo, larva, and adult. In some cases, these genes also encode transcription factors that control the expression of other sets of genes that will alter the morphological characteristics of cells. In other cases, genes that are activated by homeotic proteins play a role in cell-to-cell signaling pathways. The activation of these signaling pathways enables cells and groups of cells to adopt their correct morphologies. It is expected that research during the next few decades will shed considerable light on the pathways by which genes control morphological changes in the fruit fly.

The Developmental Fate of Each Cell in the Nematode *Caenorhabditis elegans* Is Known

We now turn our attention to another invertebrate, *C. elegans*, a nematode that has been the subject of numerous studies in developmental genetics. The embryo develops within the eggshell and hatches when it reaches a size of 550 cells. After hatching, it continues to grow and mature as it passes through four successive molts. It takes about three days for a fertilized egg to develop into an adult worm about 1 mm in length. With regard to sex, *C. elegans* can be a male (and only produce sperm) or a hermaphrodite (capable of producing sperm and egg cells). An adult male is composed of 1,031 somatic cells and produces about 1,000 sperm. A hermaphrodite consists of 959 somatic cells and produces about 2,000 gametes (both sperm and eggs).

A remarkable feature of this organism is that the pattern of cellular development is extremely constant from worm to worm. In the early 1960s, Sydney Brenner pioneered the effort to study the pattern of cell division in *C. elegans* and establish it as a model organism. Because *C. elegans* is transparent and composed of relatively few cells, researchers can follow cell division step by step under the microscope, beginning with a fertilized egg and ending with an adult worm. Researchers can identify a particular cell at an embryonic stage, follow that cell as it divides, and observe where its descendant cells will be located in the adult. An illustration that depicts how cell division proceeds is called a **lineage diagram.** It describes the cell division patterns and fates of any cell's descendants.

Figure 23.14 shows a partial lineage diagram for a *C. elegans* hermaphrodite. At the first cell division, the egg divides to produce two cells, called AB and P_1. AB then divides into two cells, ABa and ABp; and P_1 divides into two cells, EMS and P_2. The EMS cell then divides into two cells, called MS and E. The cellular descendants of the E cell give rise to the worm's intestine.

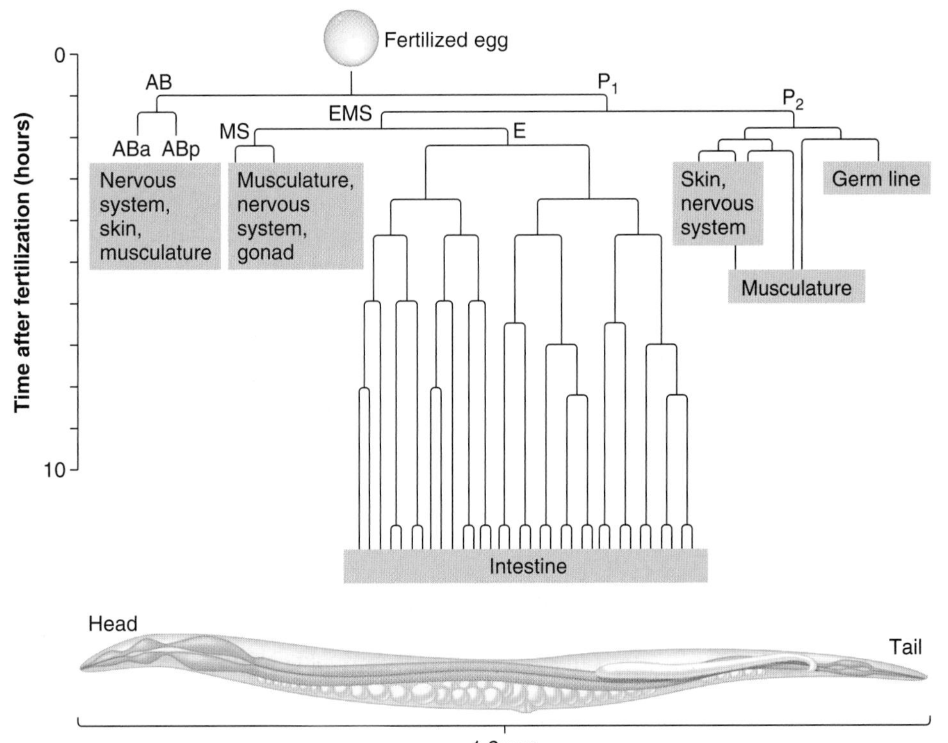

FIGURE 23.14 A cell lineage diagram of the nematode *Caenorhabditis elegans.* This partial lineage diagram illustrates how the cells divide to produce different regions of the adult. The fate of the intestinal cell lineage is shown in greater detail than that of other cell lineages. A complete lineage diagram is known for this organism, although its level of detail is beyond the scope of this textbook. Note: The lowercase letters "a" and "p" refer to anterior and posterior, respectively, and thereby indicate the relative positions of the daughter cells.

In other words, the fate of the E cell's descendants is to develop into intestinal cells. This diagram also illustrates the concept of a **cell lineage**—a series of cells that are derived from each other by cell division. For example, the EMS cell, E cell, and the intestinal cells are all part of the same cell lineage.

Why is a lineage diagram for an organism an important experimental advantage? It allows researchers to investigate how gene expression in any cell, at any stage of development, may affect the outcome of a cell's fate. In the experiment described next, we will see how the timing of gene expression is an important factor in the fate of a cell's descendants.

EXPERIMENT 23A

Heterochronic Mutations Disrupt the Timing of Developmental Changes in *C. elegans*

Our discussion of *Drosophila* development focused on how the spatial expression and localization of gene products can lead to a particular pattern of embryonic development. Another important issue in development is timing. The cells of a multicellular organism must know when to divide and when to differentiate into a particular cell type. If the timing of these processes is not coordinated, certain tissues will develop too early or too late, disrupting the developmental process.

In *C. elegans*, the timing of developmental events can be examined carefully at the cellular level. As mentioned, the fate of each cell in *C. elegans* has been determined. Using a microscope, a researcher can focus on a particular cell within this transparent worm and watch it divide into two cells, then four cells, and so forth. Therefore, a scientist can judge whether a cell is behaving as it should during the developmental process.

To identify genes that play a role in the timing of cell fates, researchers have searched for mutant alleles that disrupt the normal timing process. In a collaboration in the late 1970s, H. Robert Horvitz and John Sulston set out to identify mutant alleles in *C. elegans* that disrupt cell fates or the timing of cell fates. Using a microscope, they screened thousands of worms for altered morphologies that might indicate an abnormality in development. During this screening process, one of the phenotypic abnormalities they found was a defective egg-laying phenotype. They reasoned that because the egg-laying system depends on a large number of cell types (vulval cells, muscle cells, and nerve cells), an abnormality in any of the cell lineages leading to these cell types might cause an inability to lay eggs.

In *C. elegans*, a defective egg-laying phenotype is easy to identify, because the hermaphrodite will be able to fertilize its own eggs but will be unable to lay them. When this occurs, the eggs actually hatch within the hermaphrodite's body. This leads to the death of the hermaphrodite as it becomes filled with hatching worms. This defective egg-laying phenotype, in which the hermaphrodite becomes filled with its own offspring, is called a "bag of worms." Eventually, the newly hatched larvae eat their way out and can be saved for further study.

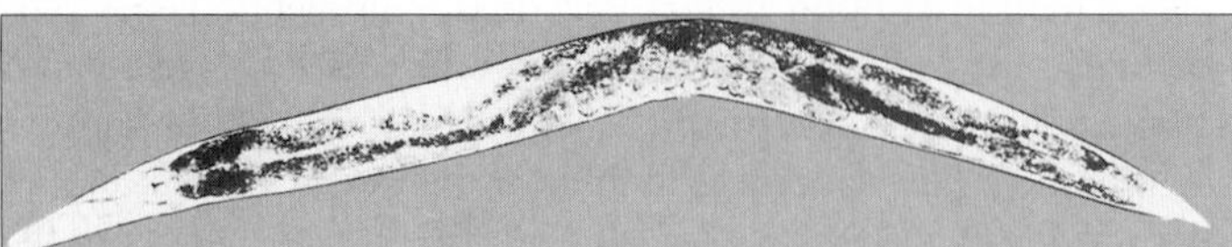

Normal *C. elegans*

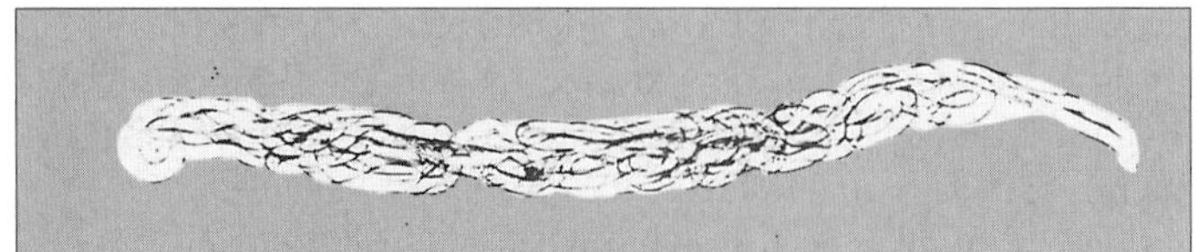

C. elegans with the defective egg-laying phenotype, the "bag of worms"

In their initial study, published in 1980, the defective egg-laying phenotype yielded several mutant strains that were defective in particular cell lineages. A few years later, in the experiment described in **Figure 23.15**, Victor Ambros and Horvitz took this same approach and were able to identify genes that play a key role in the timing of cell fate. They began with wild-type *C. elegans* and three mutant lines designated *n536*, *n355*, and *n540*. All three of the mutant lines showed an egg-laying defect. Right after hatching, they observed the fates of particular cells via microscopy. This involved spending hours looking at the nuclei of specific cells (which are relatively easy to see in this transparent worm) and timing the patterns of cell division. The patterns in the mutant and wild-type strains were then used to create lineage diagrams for particular cells.

THE HYPOTHESIS

Mutations that cause a defective egg-laying phenotype may affect the timing of cell lineages.

TESTING THE HYPOTHESIS — FIGURE 23.15 Identification of heterochronic mutations in *C. elegans.*

Starting material: Prior to this work, many laboratories had screened thousand of *C. elegans* worms and identified many different mutant strains that were defective egg-laying. (Note: When mutated, many different genes may cause a defective egg-laying phenotype. Only some of them are expected to be genes that alter the timing of cell fate within a particular cell lineage.)

1. Obtain a large number of *C. elegans* strains that have a defective egg-laying phenotype. The wild-type strain was also studied as a control.

2. Right after hatching, observe the fate of particular cells via microscopy. In this example, a researcher began watching a cell called the T cell and monitored its division pattern, and the pattern of subsequent daughter cells, during the larval stages. These patterns were examined in both wild-type and defective egg-laying worms.

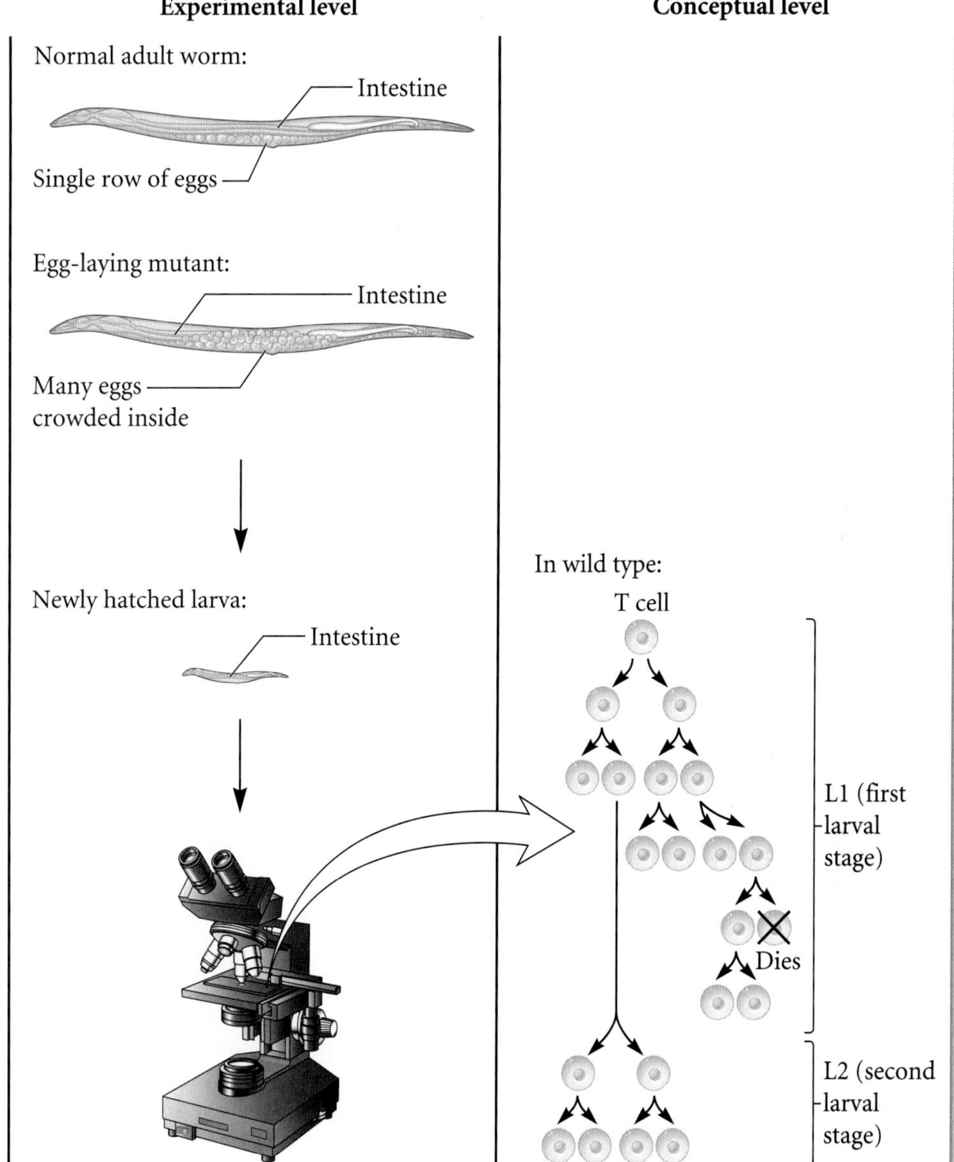

THE DATA

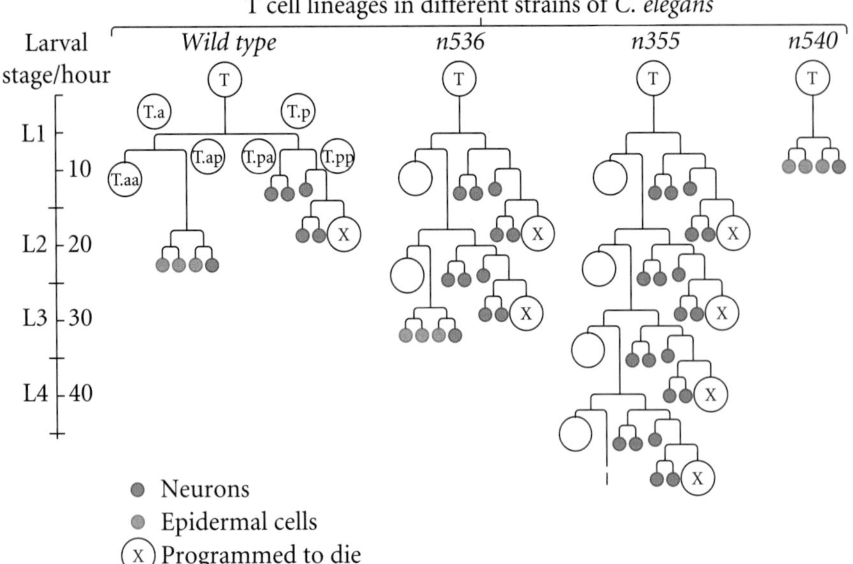

INTERPRETING THE DATA

Lineage diagrams involving cells derived from a cell called a T cell are shown in the data of Figure 23.15. As seen here, the wild-type strain follows a particular pattern of cell division for the T cell lineage. Each division event occurs at specific times during the L1 and L2 larval stages. In the normal strain, the T cell divides during the L1 larval stage to produce a T.a and T.p cell. The T.a cell also divides during L1 to produce a T.aa and T.ap cell. The T.p cell divides during L1 to produce T.pa and T.pp. These cells also divide during L1, eventually producing five neurons (labeled in blue) and one cell that is programmed to die (designated with an X). During the L2 larval stage, the T.ap cell resumes division to produce four cells: three epidermal cells (labeled in red) and one neuron.

The other T cell lineages in the data are from worms that carry mutations causing an egg-laying defect. Later research revealed that these three mutations are located in a gene called *lin-14*. The allele designated *n536* has caused the reiteration of the normal events of L1 during the L2 larval stage. In L2, the only cell of this lineage that is supposed to divide is T.ap. In worms carrying the *n536* allele, however, this cell behaves as if it were a T cell, rather than a T.ap cell, and produces a group of cells identical to what a T cell produces during the L1 stage. In the L3 stage, the cells in the *n536* strain behave as if they were in L2. Besides the egg-laying defect, the phenotypic outcome of this irregularity in the timing of cell fates is a worm that has several additional cells derived from the T cell lineage.

An allele that causes multiple reiterations is the *n355* allele. This strain continues to reiterate the normal events of L1 during the L2, L3, and L4 stages. In contrast, the *n540* allele has an opposite effect on the T cell lineage. During the L1 larval stage, the T cell behaves as if it were a T.ap cell in the L2 stage. In this case, it skips the divisions and cell fates of the L1 and proceeds directly to cell fates that occur during the L2 stage.

The types of mutations described here are called **heterochronic mutations.** The term heterochrony refers to a change in the relative timing of developmental events. In heterochronic mutations, the timing of the fate of particular cell lineages is not synchronized with the development of the rest of the organism. More recent molecular data have shown that this is due to an irregular pattern of gene expression. In wild-type worms, the LIN-14 protein accumulates during the L1 stage and promotes the T cell division pattern shown for the wild type. During L2, the LIN-14 protein diminishes to negligible levels. The *n536* and *n355* alleles are examples of gain-of-function mutations. In strains with these alleles, the LIN-14 protein persists during later larval stages. For the *n536* allele, it is made for one additional cell division, while the *n355* allele continues to express *lin-14* for several cell divisions. By comparison, the *n540* allele is a loss-of-function mutation. This allele causes LIN-14 to be inactive during L1, so it cannot promote the normal L1 pattern of cell division and cell fate.

Overall, the results described in this experiment are consistent with the idea that the precise timing of *lin-14* expression during development is necessary to correctly control the fate of particular cells in *C. elegans.* Mutations that alter the expression of *lin-14* lead to phenotypic abnormality, including the inability to lay eggs. This detrimental phenotypic consequence illustrates the importance of the correct timing for cell division and differentiation during development.

A self-help quiz involving this experiment can be found at the Online Learning Center.

23.2 VERTEBRATE DEVELOPMENT

Biologists have studied the morphological features of development in many vertebrate species. Historically, amphibians and birds have been studied extensively, because their eggs are rather large and easy to manipulate. For example, certain developmental stages of the frog and chicken have been described in great detail. In more recent times, the successes obtained in *Drosophila* have shown the great power of genetic analyses in elucidating the underlying molecular mechanisms that govern biological development. With this knowledge, many researchers are attempting to understand the genetic pathways that govern the development of the more complex body structure found in vertebrate organisms.

Several vertebrate species have been the subject of developmental studies. These include the mouse (*Mus musculus*), the frog (*Xenopus laevis*), and the small aquarium zebrafish (*Danio rerio*). In this section, we will primarily discuss the genes that are important in mammalian development, particularly those that have been characterized in the mouse, one of the best-studied mammals. As we will see, several genes affecting its developmental pathways have been cloned and characterized. In this section, we will examine how these genes affect the course of vertebrate development.

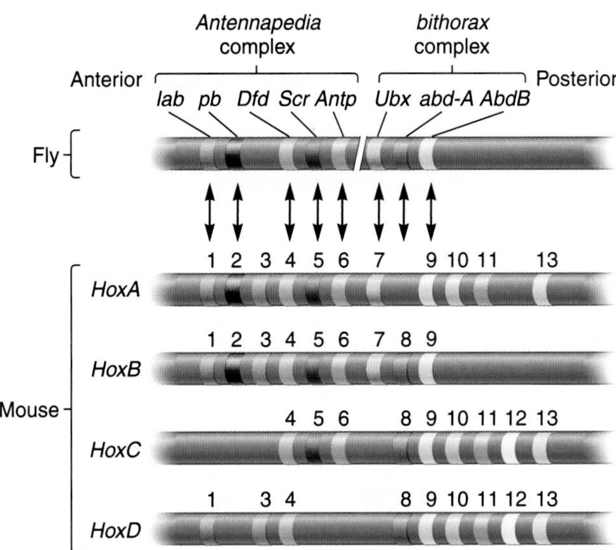

FIGURE 23.16 A comparison of homeotic genes in *Drosophila* and the mouse. The mouse contains four gene clusters, *HoxA–D*, that correspond to certain homeotic genes found in *Drosophila*. Thirteen different types of homeotic genes are found in the mouse, although each *Hox* gene cluster does not contain all 13 genes. In this drawing, orthologous genes are aligned in columns. For example, *lab* is the ortholog to *HoxA-1*, *HoxB-1*, and *HoxD-1*; *Ubx* is orthologous to *HoxA-7* and *HoxB-7*.

Researchers Have Identified Homeotic Genes in Vertebrates

Vertebrates typically have long generation times and produce relatively few offspring. Therefore, it is usually not practical to screen large numbers of embryos or offspring in search of mutant phenotypes with developmental defects. As an alternative, a successful way of identifying genes that affect vertebrate development has been the use of molecular techniques to identify vertebrate genes similar to those that control development in simpler organisms such as *Drosophila*.

As discussed in Chapters 21 and 26, species that are evolutionarily related to each other often contain genes with similar DNA sequences. When two or more genes have similar sequences because they are derived from the same ancestral gene, they are called **homologous genes.** Homologous genes found in different species are termed **orthologs.** Experimentally, a DNA strand from one gene will hybridize to a complementary strand of a homologue, because they have similar sequences.

Researchers initially followed a strategy of using cloned *Drosophila* genes as probes to identify homologous vertebrate genes, which has been quite successful. With this method, researchers have found complexes of homeotic genes in many vertebrate species that are homologous to those in the fruit fly. These groups of adjacent homeotic genes are called ***Hox* complexes.** As shown in Figure 23.16, the mouse has four *Hox* complexes, designated *HoxA* (on chromosome 6), *HoxB* (on chromosome 11), *HoxC* (on chromosome 15), and *HoxD* (on chromosome 2). A total of 38 genes are found in the four complexes. Thirteen different types of homeotic genes occur within the four *Hox* complexes, although none of the four complexes contains representatives of all 13 types of genes. The addition of *Hox* genes into the genomes of animals has allowed animals to develop more complex body plans. We will consider the evolutionary origin of the *Hox* genes and their importance in the evolution of animal species in Chapter 26.

Remarkably, several of the homeotic genes in fruit flies and mammals are strikingly similar. Among the first six types of genes, five of them are homologous to genes found in the *Antennapedia* complex of *Drosophila*. Among the last seven, three are homologous to the genes of the *bithorax* complex. These results indicate there are fundamental similarities in the ways that animals as different as fruit flies and mammals undergo embryonic development. This suggests that a "universal body plan" underlies animal development.

Like the *Antennapedia* and *bithorax* complexes in *Drosophila*, the arrangement of *Hox* genes along the mouse chromosomes reflects their pattern of expression from the anterior to the posterior end (Figure 23.17a). This phenomenon is seen in more detail in Figure 23.17b, which shows the expression pattern for a group of *HoxB* genes in a mouse embryo. Overall, these results are consistent with the idea that the *Hox* genes play a role in determining the fates of segments along the anteroposterior axis.

Currently, researchers are trying to understand the functional roles of the genes within the *Hox* complexes in vertebrate development. In *Drosophila*, great advances in developmental genetics have been made by studying mutant alleles in genes that control development. In mice, however, few natural mutations have been identified that affect development. This has made it

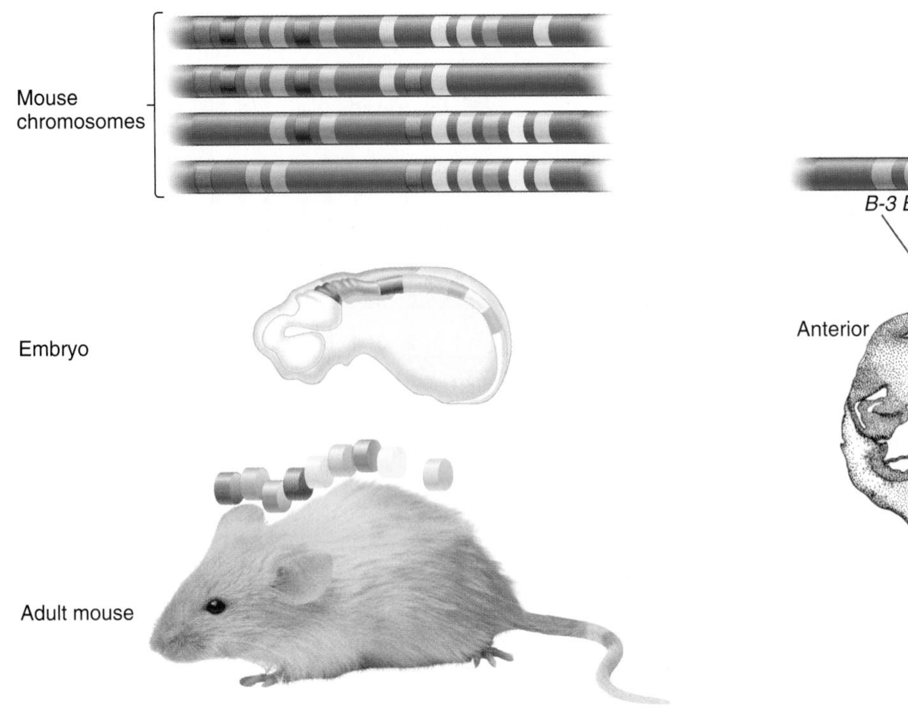

(a) Correlation between *Hox* gene arrangement and expression

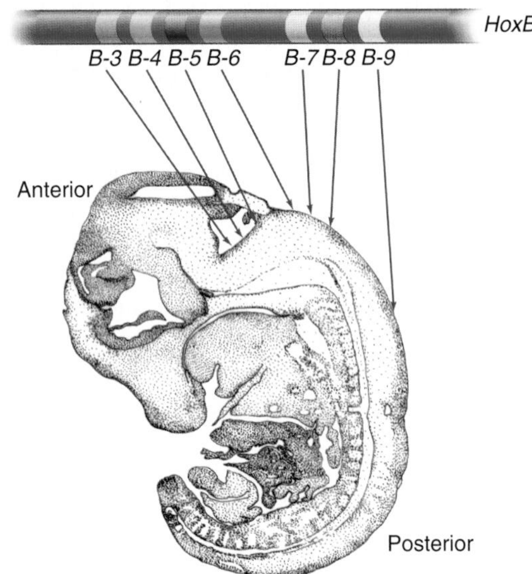

(b) Anterior expression boundaries for a series of *HoxB* genes

FIGURE 23.17 Expression pattern of *Hox* genes in the mouse. (a) A schematic illustration of the *Hox* gene expression in the embryo and the corresponding regions in the adult. (b) A more-detailed description of *HoxB* expression in a mouse embryo. The arrows indicate the anterior-most boundaries for the expression of *HoxB-3* to *HoxB-9*. The order of *Hox* gene expression, from the anterior end to posterior of the embryo, is in the same order as the genes are found along the chromosome.

difficult to understand the role that genetics plays in the development of the mouse and other vertebrate organisms.

How are researchers attempting to overcome this problem? Geneticists are taking an approach known as **reverse genetics.** In this strategy, researchers first identify the wild-type gene using cloning methods. In this case, the *Hox* genes in vertebrates have been cloned using *Drosophila* genes as probes. The next step is to create a mutant version of a *Hox* gene in vitro. This mutant allele is then reintroduced into a mouse using gene replacement techniques described in Chapter 19. When the function of the wild-type gene is eliminated, it is called a **gene knockout.** In this way, researchers can determine how the mutant allele affects the phenotype of the mouse.

The term reverse genetics reflects the idea that the experimental steps occur in an order opposite to that in the conventional approach used in *Drosophila* and other organisms. In the fly, mutant alleles were identified by their phenotype first, and then they were cloned. In the mouse, the genes were cloned first, the mutations were made in vitro, and then the mutated genes were introduced into the mouse to observe their phenotypic effects.

In recent years, several laboratories have used a reverse genetic approach to understand how many different genes, including *Hox* genes, affect vertebrate development. In *Drosophila*, loss-of-function alleles for homeotic genes usually show an anterior transformation. This means that the segment where the defective homeotic gene is expressed now exhibits characteristics that resemble the adjacent anterior segment. Similarly, certain gene knockouts (e.g., *HoxA-2*, *B-4*, and *C-8*) also show anterior transformations within particular regions of the mouse. However, knockouts of other *Hox* genes (e.g., *A-11*) have posterior transformations, and knockouts of *A-3* and *A-1* exhibit abnormalities in morphology but no clear homeotic transformations. Interestingly, a *HoxA-5* knockout in mice shows evidence of both anterior and posterior transformations, which are also seen in *Drosophila* when its ortholog, *Scr*, is knocked out.

Overall, the current picture indicates that the *Hox* genes in vertebrates play a key role in patterning the antero-posterior axis. During the evolution of animals, increases in the number of *Hox* genes and changes in their patterns of expression have had an important impact on their morphologies. As an example, let's consider the expression of the *HoxC-6* gene. Changes in its pattern of expression among vertebrate species are associated with changes in the boundary between cervical (neck) vertebrae and thoracic (chest) vertebrae. The *HoxC-6* gene is expressed during embryonic development prior to vertebrae formation. Differences in the relative position of its expression correlate with the number of neck vertebrae produced (**Figure 23.18**). In the mouse, which has a relatively short neck, *HoxC-6* expression begins in the region of the early embryo that will later develop into vertebrae 7 and 8.

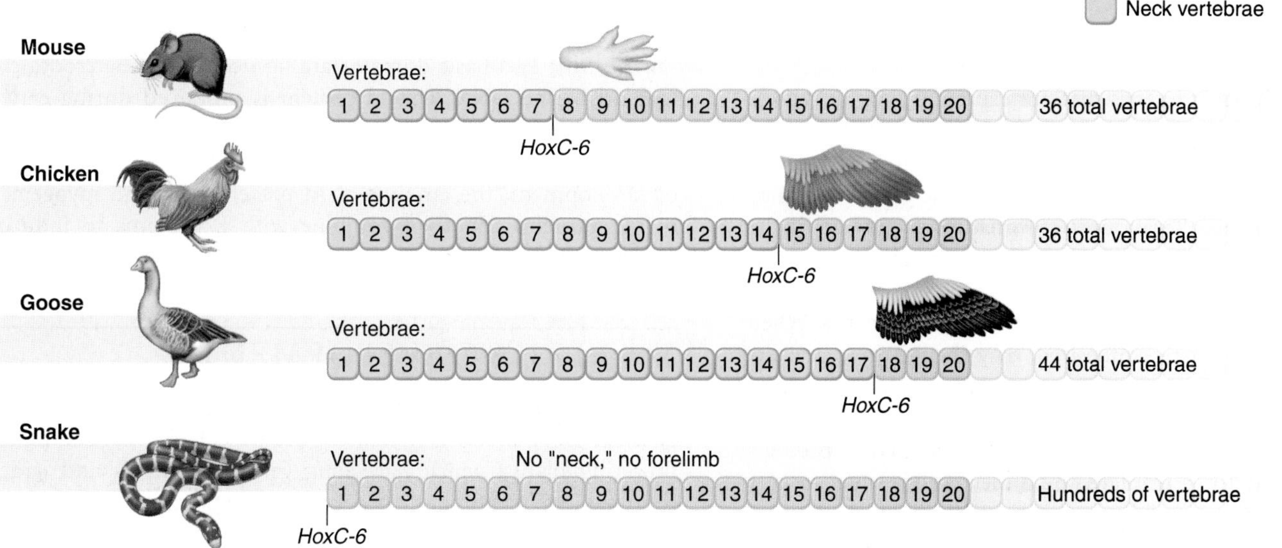

FIGURE 23.18 **Expression of the *HoxC-6* gene in different species of vertebrates.** This figure indicates the anterior boundary where the *HoxC-6* gene is expressed in the mouse, chicken, goose, and snake.

In contrast, *HoxC-6* expression in the chicken and goose begins much farther back, between vertebrae 14 and 15, or 17 and 18, respectively. The forelimbs also arise at this boundary in all vertebrates. However, snakes, which have no neck or forelimbs, do not have such a boundary because *HoxC-6* expression begins toward their heads.

Genes That Encode Transcription Factors Also Play a Key Role in Cell Differentiation

Thus far, we have focused our attention on patterns of gene expression that occur during the very early stages of development. These genes control the basic body plan of the organism. As this process occurs, cells become **determined.** This term refers to the phenomenon that a cell will be destined to become a particular cell type. In other words, its fate has been predetermined to eventually become a particular type of cell such as a nerve cell. This occurs long before a cell becomes **differentiated.** This later term means that a cell's morphology and function have changed, usually permanently, into a highly specialized cell type. For example, an undifferentiated mesodermal cell may differentiate into a specialized muscle cell, or an ectodermal cell may differentiate into a nerve cell.

At the molecular level, the profound morphological differences between muscle cells and nerve cells arise from gene regulation. Though muscle and nerve cells contain the same set of genes, they regulate the expression of those genes in very different ways. Certain genes that are transcriptionally active in muscle cells are completely inactive in nerve cells, and vice versa. Therefore, muscle and nerve cells express different proteins, which affect the morphological and physiological characteristics of the respective cells in distinct ways. In this manner, differential gene regulation underlies cell differentiation.

We learned earlier that a hierarchy of gene regulation is responsible for establishing the body pattern in *Drosophila.* Maternal effect genes control the expression of gap genes, which control the expression of pair-rule genes, and so forth. A similar type of hierarchy is thought to underlie cell differentiation in vertebrates. Researchers have identified specific genes that cause cells to differentiate into particular cell types. These genes trigger undifferentiated cells to differentiate and follow their proper cell fates.

In 1987, Harold Weintraub and colleagues identified a gene, which they called *MyoD.* This gene plays a key role in skeletal muscle cell differentiation. Experimentally, when the cloned *MyoD* gene was introduced into fibroblast cells in a laboratory, the fibroblasts differentiated into skeletal muscle cells. This result was particularly remarkable because fibroblasts normally differentiate into osteoblasts (bone cells), chondrocytes (cartilage cells), adipocytes (fat cells), and smooth muscle cells, but in vivo, they never differentiate into skeletal muscle cells.

Since this initial discovery, researchers have found that *MyoD* belongs to a small group of genes that initiate muscle development. Besides *MyoD,* these include *Myogenin, Myf5,* and *Mrf4.* All four of these genes encode transcription factors that contain a **basic domain** and a **helix-loop-helix domain (bHLH).** The basic domain is responsible for DNA binding and the activation of skeletal muscle-cell-specific genes. The helix-loop-helix domain is necessary for dimer formation between transcription factor proteins. Because of their common structural features and their role in muscle differentiation, MyoD, Myogenin, Myf5, and Mrf4 constitute a family of proteins called **myogenic bHLH proteins.** They are found in all vertebrates and have been identified in several invertebrates, such as *Drosophila* and *C. elegans.* In all cases, the myogenic bHLH genes are activated during skeletal muscle cell development.

At the molecular level, certain key features enable myogenic bHLH proteins to promote muscle cell differentiation. The basic domain binds to a muscle-cell-specific enhancer sequence; this sequence is adjacent to genes that are expressed only in muscle cells (**Figure 23.19**). Therefore, when myogenic bHLH proteins are activated, they can bind to these enhancers and activate the expression of many different muscle-cell-specific genes. They may exert their effects via alterations in chromatin structure or via the activation of RNA polymerase to a transcriptionally active state. In this way, myogenic bHLH proteins function as master switches that activate the expression of many muscle-specific genes. When the encoded proteins are synthesized, they change the characteristics of an undifferentiated cell into those of a highly specialized skeletal muscle cell.

Another important aspect of myogenic bHLH proteins is that their activity is regulated by dimerization, which occurs via the helix-loop-helix domain. As shown in Figure 23.19, heterodimers—dimers formed from two different proteins—may be activating or inhibitory. When a heterodimer forms between a myogenic bHLH protein and an E protein, which also contains a basic domain, the heterodimer binds to the DNA and activates gene expression (Figure 23.19a). However, when a heterodimer forms between a myogenic bHLH protein and a protein called Id (for inhibitor of differentiation), the heterodimer cannot bind to DNA because the Id protein lacks a basic domain (Figure 23.19b). Two basic domains are needed for the heterodimer to bind to the DNA. The Id protein is produced during early stages of development and prevents myogenic bHLH proteins from promoting muscle differentiation too soon. At later stages of development, the amount of Id protein falls, and myogenic bHLH proteins can then combine with E proteins to induce muscle-cell differentiation.

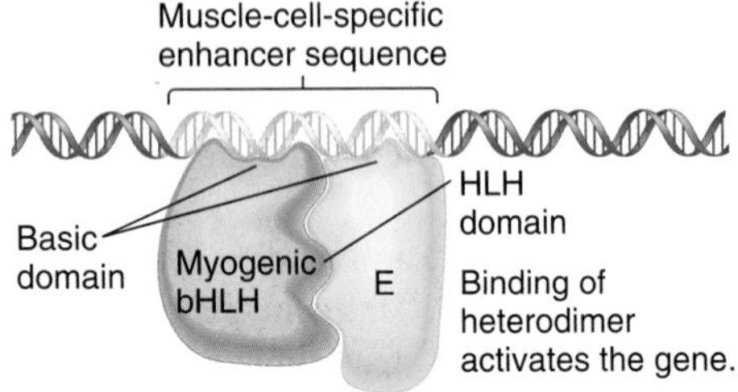

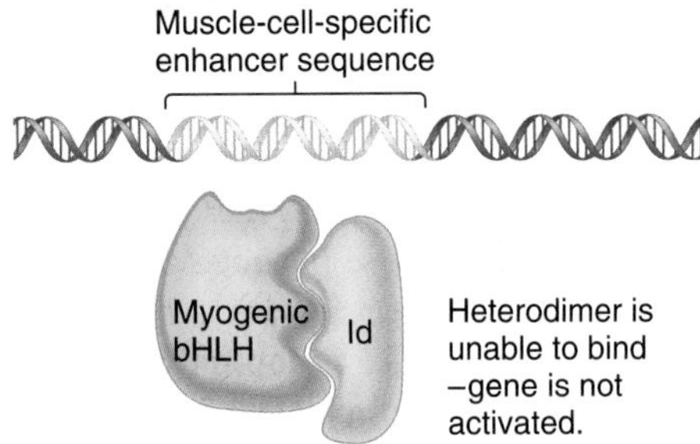

FIGURE 23.19 **Regulation of muscle-cell-specific genes by myogenic bHLH proteins.** (**a**) A heterodimer formed from a myogenic bHLH protein and an E protein can bind to a muscle-cell-specific enhancer sequence and activate gene expression. (**b**) When a myogenic bHLH protein forms a heterodimer with an Id protein, it cannot bind to the DNA and therefore does not activate gene transcription.

23.3 PLANT DEVELOPMENT

In developmental plant biology, the model organism for genetic analysis is *Arabidopsis thaliana* (**Figure 23.20**). Unlike most flowering plants, which have long generation times and large genomes, *Arabidopsis* has a generation time of about two months

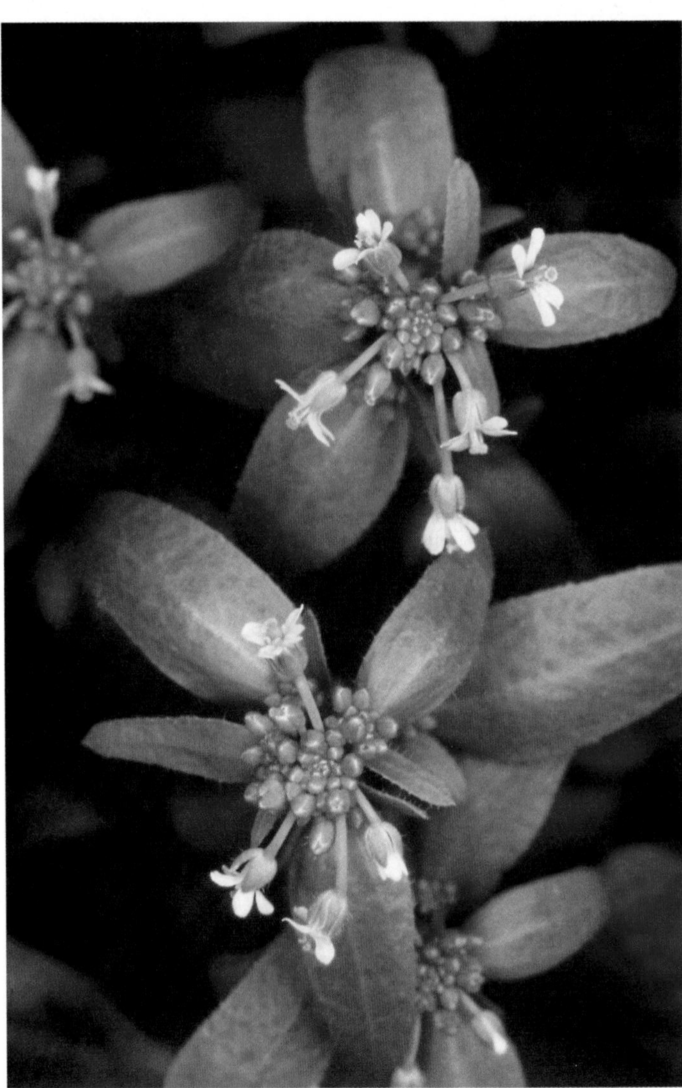

FIGURE 23.20 **The model organism *Arabidopsis*.** The plant is relatively small, making it easy to grow many of them in the laboratory.

and a genome size of 14×10^7 bp, which is similar to *Drosophila* and *C. elegans*. A flowering *Arabidopsis* plant produces a large number of seeds and is small enough to be grown in the laboratory. Like *Drosophila*, *Arabidopsis* can be subjected to mutagens to generate mutations that alter developmental processes. The small genome size of this organism makes it relatively easy to map these mutant alleles and eventually clone the relevant genes (as described in Chapters 18 and 20).

The morphological patterns of growth are markedly different between animals and plants. As described previously, animal embryos become organized along antero-posterior, dorso-ventral, and right-left axes, and then they subdivide into segments. By comparison, the form of plants has two key features. The first is the root-shoot axis. Most plant growth occurs via cell division near the tips of the shoots and the bottoms of the roots.

Second, this growth occurs in a well-defined radial pattern. For example, early in *Arabidopsis* growth, a rosette of leaves or flowers is produced from buds that emanate in a spiral pattern directly from the main shoot (see Figure 23.20). Later, the shoot generates branches that will also produce leaf buds as they grow. Overall, the radial pattern in which a plant shoot gives off the buds that produce leaves, flowers, and branches is an important mechanism that determines much of the general morphology of the plant.

At the cellular level, too, plant development differs markedly from animal development. For example, cell migration does not occur during plant development. In addition, the development of a plant does not rely on morphogens that are deposited asymmetrically in the oocyte. In plants, an entirely new individual can be regenerated from many types of somatic cells. In other words, many plant cells are **totipotent,** meaning that they have the ability to differentiate into every cell type and to produce an entire individual. By comparison, animal development typically relies on the organization within an oocyte as a starting point for development.

In spite of these apparent differences, the underlying molecular mechanisms of pattern development in plants still share some similarities with those in animals. In this section, we will consider a few examples in which the genes encoding transcription factors play a key role in plant development.

Plant Growth Occurs from Meristems Formed During Embryonic Development

Figure 23.21 illustrates a common sequence of events that takes place in the development of seed plants such as *Arabidopsis*. After fertilization, the first cellular division is asymmetrical and produces a smaller cell, called the apical cell, and a larger basal cell (Figure 23.21a). The apical cell will give rise to most of the embryo, and it will later develop into the shoot of the plant. The basal cell will give rise to the root, along with the suspensor that produces extraembryonic tissue required for seed formation. At the heart stage, which is composed of only about 100 cells, the basic organization of the plant has been established (Figure 23.21d). The **shoot meristem** will arise from a group of cells located between the cotyledons. These cells are the precursors that will produce the shoot of the plant, along with lateral structures such as leaves and flowers. The **root meristem** is located at the opposite side and will create the root.

A meristem contains an organized group of actively dividing stem cells. As discussed in Chapter 19, stem cells retain the ability to divide and differentiate into multiple cell types. As they grow, meristems produce offshoots of proliferating cells. On a shoot meristem, for example, these offshoots or buds give rise to structures such as leaves and flowers. The organization of a shoot meristem is shown in Figure 23.22. It is organized into three areas called the **organizing center,** the **central zone,** and the **peripheral zone.** The role of the organizing center is to ensure the proper organization of the meristem and preserve the correct number

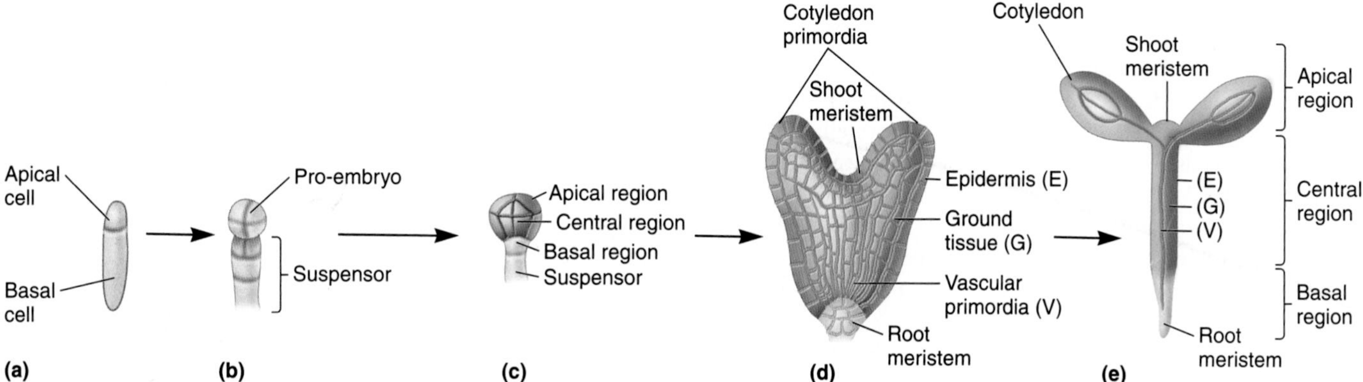

FIGURE 23.21 **Developmental steps in the formation of a plant embryo.** (a) The two-cell stage consists of the apical cell and basal cell. (b) The eight-cell stage consists of a pro-embryo and a suspensor. The suspensor gives rise to extraembryonic tissue, which is needed for seed formation. (c) At this stage of embryonic development, the three main regions of the embryo (i.e., apical, central, and basal) have been determined. (d) At the heart stage, all of the plant tissues have begun to form. Note that the shoot meristem is located between the future cotyledons, and the root meristem is on the opposite side. (e) A seedling.

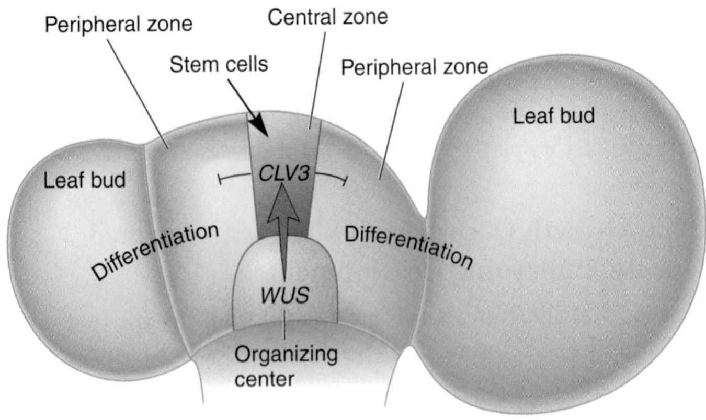

FIGURE 23.22 Organization of a shoot meristem. The organization of a shoot meristem is controlled by the *WUS* and *CLV3* genes, which are acronyms for *Wuschel* and *clavata*, respectively. The *WUS* gene is expressed in the organizing center and induces the cells in the central zone to become undifferentiated stem cells. The red arrow indicates that the WUS protein induces these stem cells to turn on the *CLV3* gene, which encodes a secreted protein that binds to receptors in the cells of the peripheral zone. The black lines with a vertical slash indicate that the CLV3 protein prevents the cells in the peripheral zone from expressing the *WUS* gene. This limits the area of *WUS* gene expression to the underlying organizing center and thereby maintains a small population of stem cells at the growing tip. The cells in the peripheral zone are allowed to grow and differentiate into lateral structures such as leaves.

of actively dividing stem cells. The central zone is an area where undifferentiated stem cells are always maintained. The peripheral zone contains dividing cells that will eventually differentiate into plant structures. For example, the peripheral zone may form a bud, which will produce a leaf or flower.

In *Arabidopsis*, the organization of a shoot meristem is controlled by two critical genes termed *WUS* and *CLV3*. The *WUS* gene encodes a transcription factor that is expressed in the organizing center (Figure 23.22). The expression of the *WUS* gene induces the adjacent cells in the central zone to become undifferentiated stem cells. These stem cells then turn on the *CLV3* gene, which encodes a secreted protein. The CLV3 protein binds to receptors in the cells of the peripheral zone, preventing them from expressing the *WUS* gene. This limits the area of *WUS* gene expression to the underlying organizing center and thereby maintains a small population of stem cells at the growing tip. A shoot meristem in *Arabidopsis* contains only about 100 cells. The inhibition of *WUS* expression in the peripheral cells also allows them to embark on a path of cell differentiation so they can produce structures such as leaves and flowers.

In the seedling shown earlier in Figure 23.21e, three main regions are observed. The **apical region** produces the leaves and flowers of the plant. The **central region** (not to be confused with the central zone) creates the stem. The radial pattern of cells in the central region causes the radial growth observed in plants. Finally, the **basal region** produces the roots. Each of these three regions develops differently as indicated by their unique cell division patterns and distinct morphologies. In addition, by analyzing mutants that disrupt the developmental process, researchers have discovered that these three regions express different sets of genes. Plant biologists have identified a category of genes, known as **apical-basal-patterning genes,** that are important in the early stages of development. As described in **Table 23.2**, defects in apical-basal-patterning genes cause dramatic effects in each of the three main regions. For example, the *gurke* gene is necessary for apical development. When it is defective, the embryo lacks apical structures. Currently, a great amount of effort is directed toward the identification of genes that govern pattern formation in the three regions of *Arabidopsis* and other plants.

TABLE 23.2

Examples of *Arabidopsis* Apical-Basal-Patterning Genes That Affect the Development of the Apical, Central, or Basal Region

Gene	Description
Apical	
Gurke	Minor loss-of-function alleles in the *Gurke* gene produce seedlings with highly reduced or no cotyledons. Complete loss-of-function eliminates the entire shoot.
Pin1	Encodes a putative auxin exporter that is expressed in the peripheral zone. It plays a role in the location of bud sites, which give rise to leaves and flowers.
Aintegumenta	Encodes a transcription factor that is also expressed in the peripheral zone. Its expression maintains the proliferative cell state during the growth of lateral buds.
Central	
Fackel	Encodes an enzyme involved in plant sterol synthesis. Loss-of-function alleles show a severe defect in stem growth.
Scarecrow	Encodes a transcription factor protein that plays a role in the asymmetric division that produces the radial pattern of growth in the stem. Note: The scarecrow protein also affects cell division patterns in roots and plays a role in sensing gravity.
Basal	
Monopterous	Encodes a transcription factor. When this gene is defective, the plant embryo cannot initiate the formation of root structures, but root structures can be formed postembryonically under the correct growth conditions. This gene seems to be required for organizing root formation in the embryo but is not required for root formation per se.
Hobbit	Encodes a subunit of a protein that functions as a cell cycle checkpoint during anaphase. *Hobbit* function may be required to couple cell division to cell differentiation in the root meristem or to restrict the response to growth hormones such as auxin. Loss-of-function alleles result in plants that are incapable of forming a root.
Overall Organization	
Gnom	Plays a role in the stable fixation of the apical-basal axis of the *Arabidopsis* embryo.

Plant Homeotic Genes Control Flower Development

Although the term homeotic was coined by William Bateson to describe homeotic mutations in animals, the first known mutations in homeotic genes were observed in plants. In ancient Greece and Rome, for example, double flowers in which stamens were replaced by petals were noted. In current research, geneticists have been studying these types of mutations to better understand developmental pathways in plants. Many homeotic mutations affecting flower development have been identified in *Arabidopsis* and also in the snapdragon (*Antirrhinum majus*).

Examples of homeotic mutants in *Arabidopsis* are shown in **Figure 23.23**. A normal *Arabidopsis* flower is composed of four concentric whorls of structures (Figure 23.23a). The outer whorl contains four sepals, which protect the flower bud before it opens. The second whorl is composed of four petals, and the third whorl contains six stamens, the structures that make the male gametophyte, pollen. Finally, the innermost whorl contains two carpels, which are fused together. The carpel produces the female gametophyte. The homeotic mutants shown in Figure 23.23 have undergone transformations of particular whorls. For example, in Figure 23.23b, the sepals have been transformed into carpels, and the petals into stamens.

By analyzing the effects of many different homeotic mutations in *Arabidopsis*, Elliot Meyerowitz and colleagues proposed the **ABC model** for flower development. In this model, three classes of genes, called *A*, *B*, and *C*, govern the formation of sepals, petals, stamens, and carpels. More recently, a fourth category of genes called the *Sepallata* genes (*SEP* genes) have been found to be required for this process. **Figure 23.24** illustrates how these genes affect normal flower development in *Arabidopsis*. Gene *A* products are made in the outermost whorl (whorl 1), promoting sepal formation. In whorl 2, gene *A*, gene *B*, and *SEP* gene products are made, which promotes petal formation. The expression of gene *B*, gene *C*, and *SEP* genes in whorl 3 causes stamens to be made. Finally, in whorl 4, gene *C* and *SEP* genes promote carpel formation.

What are the phenotypic effects of homeotic mutants affecting the *A*, *B*, *C*, or *SEP* genes? In the original ABC model, it was proposed that genes *A* and *C* repress each other's expression, and gene *B* functions independently. In a mutant defective in gene *A* expression, gene *C* will also be expressed in whorls 1 and 2. This produces a carpel-stamen-stamen-carpel arrangement. When gene *B* is defective, a flower cannot make petals or stamens. Therefore, a gene *B* defect yields a flower with a sepal-sepal-carpel-carpel arrangement. When gene *C* is defective, gene *A* is expressed in all four whorls. This results in a sepal-petal-petal-sepal pattern. If the expression of *SEP* genes is defective, the flower consists entirely of sepals, which is the origin of the gene's name.

Overall, the types of genes described in Figure 23.24 promote cell differentiation that leads to sepal, petal, stamen, or carpel structures. But what happens if genes *A*, *B*, and *C* are all defective? As shown in Figure 23.23c, this produces a "flower" that is composed entirely of leaves! These results indicate that the leaf structure is the default pathway and that the *A*, *B*, *C*, and *SEP* genes cause development to deviate from a leaf structure in order to make something else. In this regard, the sepals, petals,

(a) Normal flower

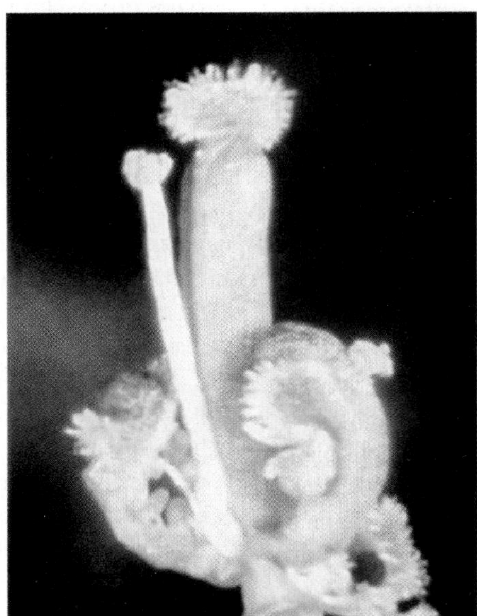

(b) Single homeotic mutant

(c) Triple mutant

FIGURE 23.23 Examples of homeotic mutations in *Arabidopsis*.

Genes→Traits **(a)** A normal flower. It is composed of four concentric whorls of structures: sepals, petals, stamens, and carpel. **(b)** A homeotic mutant in which the sepals have been transformed into carpels and the petals have been transformed into stamens. **(c)** A triple mutant in which all of the whorls have been changed into leaves.

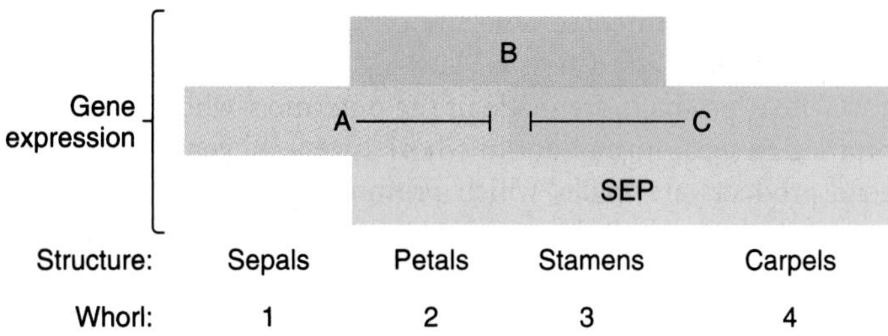

FIGURE 23.24 **The ABC model of homeotic gene action in *Arabidopsis*.** Note: This is a revised model based on the recent identification of *SEP* genes. The black lines with a vertical slash indicate that the *A* gene product represses the *C* gene product, and vice versa.

stamens, and carpels can be viewed as modified leaves. Interestingly, the German philosopher and poet Johann Goethe originally proposed this idea—that flower formation comes from modifications of the leaf—over 200 years ago.

Arabidopsis has two types of gene *A* (*apetala1* and *apetala2*), two types of gene *B* (*apetala3* and *pistillata*), one type of gene *C* (*agamous*), and three *SEP* genes (*SEP1*, *SEP2*, and *SEP3*). All of these plant homeotic genes encode transcription factor proteins that contain a DNA-binding domain and a dimerization domain. However, the *Arabidopsis* homeotic genes do not contain a sequence similar to the homeobox found in animal homeotic genes.

Like the *Drosophila* homeotic genes, plant homeotic genes are part of a hierarchy of gene regulation. Genes that are expressed within a flower bud produce proteins that activate the expression of these homeotic genes. Once they are transcriptionally activated, the homeotic genes then regulate the expression of other genes, the products of which promote the formation of sepals, petals, stamens, or carpels.

23.4 SEX DETERMINATION IN ANIMALS AND PLANTS

To end our discussion of development, let's consider how genetics plays a role in the development of male and female individuals. In the animal kingdom, the existence of two sexes is nearly universal. In plants, most species produce male and female haploid reproductive cells on the same individual. However, some plant species are sexually dimorphic. Such species produce two distinct types of individuals that have different abilities to produce male and female gametophytes. We will consider such types of plants in this section.

The underlying factors that distinguish female versus male development vary widely. In animals, sex determination is often caused by differences in chromosomal composition (**Table 23.3**). In flies and worms, the ratio of the number of X chromosomes to the number of sets of autosomes determines sex. By comparison, in mammals, the presence of the *Sry* gene on the Y chromosome causes maleness. Similarly, plants that are sexually dimorphic typically have sex chromosomes in which the plant that produces the male gametophyte has two types of sex chromosomes, similar to the XY system in mammals. In birds, the female has heteromorphic sex chromosomes: ZW birds are female, and ZZ birds are male. Finally, sex determination is not always caused by two types of sex chromosomes. In bees, for example, males are haploid, and females are diploid.

TABLE 23.3

Mechanisms of Sex Determination in Selected Species

Species	Mechanism
Drosophila, C. elegans	Ratio of the number of X chromosomes to the number of sets of autosomes determines sex. An X:A ratio of 1.0 results in females (or hermaphrodites in *C. elegans*), and an X:A ratio of 0.5 produces males.
Mammals	The presence of the *Sry* gene on the Y chromosome causes maleness.
Birds	Females are ZW, and males are ZZ.
Bees	Males are haploid, and females (workers and queen bees) are diploid.
Certain species of reptiles, fish, and turtles	Environmental conditions, usually temperature, influence the ratio of male and female offspring.
Plants	The male plant produces pollen. In some sexually dimorphic species, the males are XY. In other species, there are not distinct sex chromosomes, but the male plant is heterozygous for a dominant allele that suppresses the female pathway.

In many animal species, variation in chromosome composition is not the underlying factor that distinguishes female from male development. Sex determination in many species of reptiles and fish is controlled by environmental factors such as temperature. For example, in the American alligator (*Alligator mississippiensis*), temperature controls sex determination. Eggs incubated at 33°C typically result in ~100% male individuals, whereas eggs incubated at lower or higher temperatures such as 30°C or 34.5°C result in ~100% or 95% females, respectively.

The adoption of one of two sexual fates is an event that has been studied in great detail in several species. Researchers have discovered that sex determination is a process controlled genetically by a hierarchy of genes that exert their effects in early embryonic development. In this section, we will consider features of these hierarchies in *Drosophila*, *C. elegans*, mammals, and plants.

In *Drosophila*, Sex Determination Involves a Regulatory Cascade That Includes Alternative Splicing

In diploid fruit flies, XX flies develop into females, and X0 flies become males. The ratio of the number of X chromosomes to

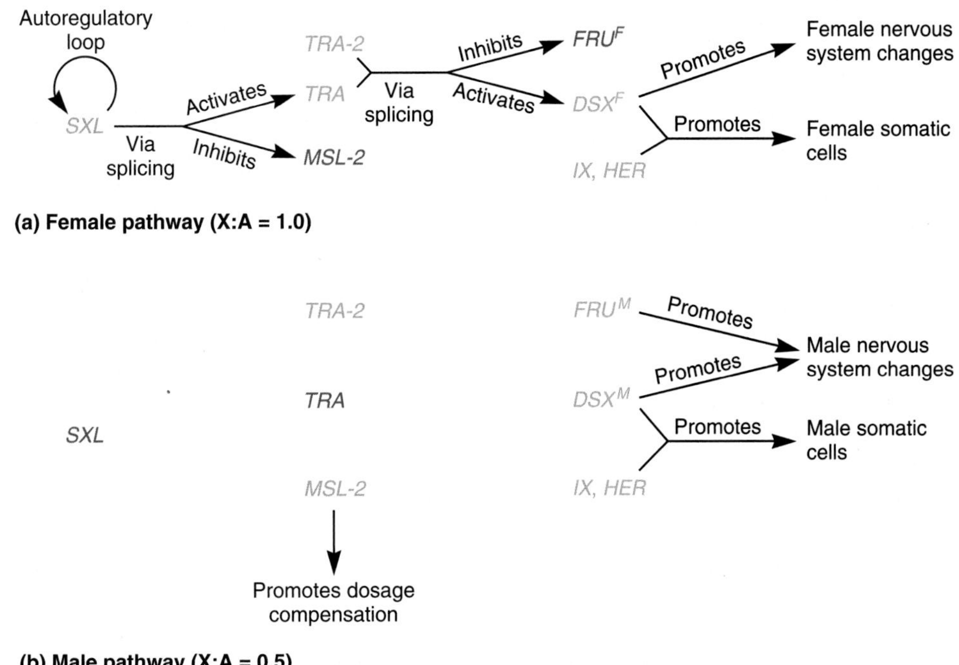

FIGURE 23.25 **Sex determination pathway for *Drosophila melanogaster*.** Genes or gene products that are functionally expressed are shown in light orange, while those that are not expressed are shown in blue. The gene names are acronyms for the phenotypes that result from mutations that cause loss-of-function or aberrant expression. These are as follows: *SXL* (sex lethal), *MSL* (male sex lethal), *TRA* (transformer), *DSX* (double sex), *FRU* (fruitless), *IX* (intersex), and *HER* (hermaphrodite). Note: This is a simplified pathway. More gene products are involved than shown here.

the number of sets of autosomes is the determining factor. In a diploid fly that carries two sets of chromosomes, this ratio is 1.0 in females versus 0.5 in males. Although male fruit flies usually carry a Y chromosome, it is not necessary for male development. The mechanism of sex determination begins in early embryonic development and involves a regulatory cascade composed of several genes. Females and males follow one of two alternative pathways. Simplified versions of these pathways are depicted in **Figure 23.25**.

Let's begin with the pathway that produces female flies. In females, the higher ratio of X chromosomes results in the embryonic expression of a gene designated *SXL* (Figure 23.25a). The *SXL* gene product is a protein that functions in the splicing of pre-mRNA. In female embryos, the Sxl protein enhances its own expression by splicing its own pre-mRNA, an event termed an **autoregulatory loop.** In addition, it splices the pre-mRNA from two other genes called *MSL-2* and *TRA*. The Sxl protein promotes the splicing of the *MSL-2* pre-mRNA in a way that introduces an early stop codon in the coding sequence and thereby produces a shortened version of the Msl-2 protein that is functionally inactive. By comparison, the Sxl protein promotes the splicing of *TRA* pre-mRNA to produce an mRNA that is translated into a functional protein. Therefore, *SXL* activates *TRA*.

The *TRA* gene product and a constitutively expressed product from a gene called *TRA-2* are also splicing factors. In the female, they cause the alternative splicing of the pre-mRNAs that are expressed from the *FRU* and *DSX* genes. The tra and tra-2 proteins cause these pre-mRNAs to be spliced into mRNAs designated FRU^F and DSX^F, respectively. The female-specific FRU^F mRNA is not translated into a sex-specific gene product. However, the DSX^F mRNA, together with two other gene products from the *IX* and *HER* genes, promotes female sexual development and controls some aspects of female-specific behavior via the central nervous system. The dsx^F protein is known to be a transcription factor that regulates certain genes that promote these changes.

How are males produced? In X0 or XY flies, the *SXL* gene is transcriptionally activated, but it is spliced in a way that places an early stop codon in the coding sequence. Therefore, a functional Sxl protein is not made (Figure 23.25b). This permits the expression of *MSL-2*, which promotes dosage compensation. In fruit flies, dosage compensation is accomplished by turning up the expression of X-linked genes in the male to a level that is twofold higher. Therefore, even though the male has only one X chromosome, the expression of X-linked genes in the male and female are approximately equal. The absence of *SXL* expression in male embryos also promotes the development of maleness. Without *SXL*, the *TRA* mRNA is not properly spliced, so *TRA* is not expressed. Without the tra protein, the *FRU* and *DSX* mRNAs are spliced in a different way to produce mRNAs designated FRU^M and DSX^M. Along with *IX* and *HER* gene products, the dsx^M protein promotes male development as well as male-specific behavior. Like the dsx^F protein, the dsx^M protein is a transcription factor that regulates certain genes. Because DSX^M is spliced differently from DSX^F, the dsx^M protein's structure is different from the dsx^F protein, and this difference alters the regulation pattern of dsx^M. In addition, the FRU^M gene product is necessary for the regulation of genes involved in male-specific behaviors.

In *C. elegans* the Ratio of X Chromosomes to Sets of Autosomes Initiates a Regulatory Cascade That Determines Sex

C. elegans has two sexes: hermaphrodites and males. During larval development, the hermaphrodites produce sperm that are stored in a structure called the spermathecae. In adulthood, hermaphrodites are anatomically female. They produce oocytes that are fertilized when they are forced through the spermathecae by muscular contractions. Fertilization can take place in either of two ways. One possibility is that a sperm can fertilize an oocyte from the same worm. In other words, the hermaphrodite is capable of self-fertilization. Alternatively, a hermaphrodite can mate with a male worm, which produces only sperm.

The sexual identity of *C. elegans*, hermaphrodite versus male, is a trait determined very early in embryonic development. As in *Drosophila*, sexual identity is controlled by the activities of many genes that interact in a regulatory cascade that determines the sex of the worm and controls dosage compensation. In *C. elegans*, the expression of genes on the X chromosome in the hermaphrodite, which carries two X chromosomes, is decreased to 50% compared to males, which carry one X chromosome.

As in *Drosophila*, the ratio of the number of X chromosomes to the number of sets of autosomes (the X:A ratio) is the factor that causes the regulatory cascade to follow one of two alternative pathways (**Figure 23.26**). Let's first consider the events that occur during early embryonic development in the hermaphrodite. Because hermaphrodites have two X chromosomes, this results in an X:A ratio of 1.0, compared to the ratio of 0.5 for males. This higher X:A ratio results in an enhanced expression of *fox-1* and *sex-1*, which are located on the X chromosome. In the XX hermaphrodite, the expression of *fox-1* and *sex-1* inhibits the expression of *xol-1*. The protein product of the *xol-1* gene inhibits the expression of three genes called *sdc-1*, *sdc-2*, and *sdc-3*. Because *xol-1* activity is inhibited in hermaphrodites, this permits the expression of the *sdc* genes. The expression of *sdc-1*, *sdc-2*, and *sdc-3* has two effects. First, these three genes are necessary for dosage compensation, and second, they are necessary to permit the downstream expression of *tra-1*, which is needed to promote hermaphrodite development. This second effect occurs via a chain of events that are also shown in Figure 23.26a. The *sdc* gene products inhibit the expression of *her-1*. When *her-1* is inhibited, this permits the expression of *tra-2*. When *tra-2* is active, this inhibits the activities of *fem-1*, *fem-2*, and *fem-3*. When the *fem* gene products are inhibited, this permits the expression of *tra-1*. It is the *tra-1* gene product that promotes hermaphrodite development and inhibits male development.

Figure 23.26b describes the pathway for male development. Because males contain a single X chromosome, the expression of *fox-1* and *sex-1* is insufficient to inhibit the expression of *xol-1*. When *xol-1* is expressed, its protein product inhibits the expression of the *sdc* genes. This prevents dosage compensation and permits the expression of *her-1*. The *her-1* gene product

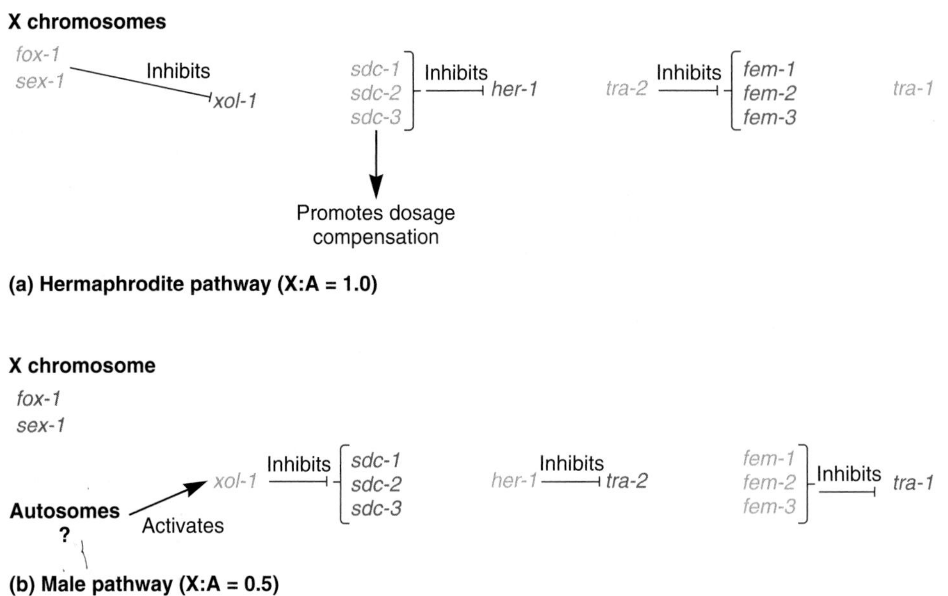

FIGURE 23.26 **Sex determination pathway for *C. elegans*.** Genes or gene products that are expressed at relatively high levels are shown in light orange, while those that are not expressed or are expressed at low levels are shown in blue. The gene names are acronyms that usually refer to the effects of mutations. These are as follows: *fox* (feminizing on X), *sex* (sex determination), *xol* (XO lethal), *sdc* (sex determination and dosage compensation), *her* (hermaphrodite), *tra* (transformer), and *fem* (feminization). Note: This is a simplified pathway.

then inhibits *tra-2*. This, in turn, permits the expression of the *fem* genes, which inhibit the expression of *tra-1*. Without *tra-1* expression, the worm develops into a male instead of a hermaphrodite. Experimentally, it is has been shown that XX worms lacking a functional *tra-1* gene develop into males.

In Mammals, the *Sry* Gene on the Y Chromosome Determines Maleness

In most mammals, such as humans, mice, and marsupials, the presence of the *Sry* gene on the Y chromosome determines maleness. In cases of abnormal sex chromosome composition such as XXY, an individual develops into a male. The *Sry* gene, which is located on the Y chromosome, causes the sex determination pathway to follow a male developmental scheme. The *Sry* gene encodes a protein that contains a DNA-binding domain called an HMG box, which is found in a broad category of DNA-binding proteins known as the high mobility group. In several of these proteins, the HMG box is known to cause DNA bends. Thus far, the ability of the Sry protein to promote male sex determination is not well understood, but it may act as a transcription factor and/or promote changes in DNA architecture, much like the DNA remodeling proteins discussed in Chapters 10 and 15.

Sex determination in mammals, like that of fruit flies and worms, involves a cascade that is initiated in early embryonic development. However, the details of the pathway are not completely elucidated. **Figure 23.27** illustrates a simplified pathway.

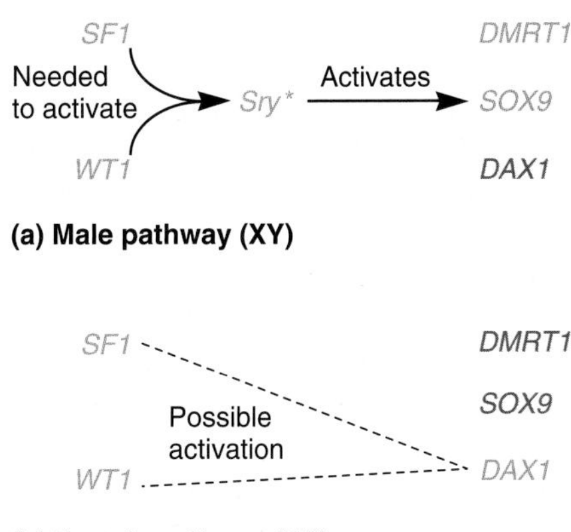

FIGURE 23.27 Sex determination pathway for mammals. Genes or gene products that are expressed at relatively high levels are shown in light orange, while those that are not expressed or are inactive are shown in blue. The regulation of the *DAX1* gene is not well understood, but it may involve activation via *SF1* and/or *WT1* gene products. The gene names are acronyms that usually refer to the effects of mutations and/or their relatedness to other genes or chromosomal regions. These are as follows: *SF1* (steroidogenic factor-1), *WT1* (Wilm's tumor gene), *Sry* (sex determining region of the Y chromosome), *DMRT1* (doublesex- and mab-3-related transcription factor), *SOX9* (Sry-like HMG box), and *DAX1* (dosage-sensitive sex-reversal, adrenal hypoplasia congenital, X chromosome). Note: This is a simplified pathway.

Several genes in mammals have been identified that are expressed very early in embryonic development and may be directly or indirectly involved in turning on the *Sry* gene. For example, two genes, designated *SFI* and *WTI*, are expressed in the early embryo prior to sexual differentiation and may regulate *Sry* expression. Once activated, researchers postulate that *Sry* expression causes the expression of other genes that promote male development. However, the target genes that are turned on by the Sry protein have not been definitively identified. One likely candidate is a gene called *SOX9*. Like the *Sry* gene, *SOX9* encodes a protein with an HMG domain. In addition, the SOX9 protein is larger than the Sry protein because it also contains two transcriptional activation domains, indicating that it is a transcriptional activator. *SOX9* expression is necessary for male development, because a loss-of-function allele of *SOX9* results in female development in XY animals. Furthermore, when researchers produced XX mice with three copies of the *SOX9* gene, such individuals developed into males.

Two other genes, designated *DAX1* and *DMRT1*, also play a role in sex determination. The *DAX1* gene is X linked, and its gene product is thought to prevent male development. In XX animals, its expression remains high, in contrast to XY males. XY animals with two copies of the *DAX1* gene develop into females when *Sry* gene expression is low due to a weak allele in the *Sry* gene. This suggests that *DAX1* can inhibit the effects of the *Sry* gene. However, the *DAX1* gene is not needed for female development because XX mice lacking the *DAX1* gene develop as normal females. *DAX1* encodes a hormone-receptor protein. Further research is needed to understand the pathways that are activated via this receptor. Finally, the *DMRT1* gene in mammals is evolutionarily related to the *mab-3* gene in *C. elegans* and the *DSX* gene in *Drosophila*. These genes encode transcription factors involved in the differentiation of the gonads. In mammals, *DMRT1* expression is gonad specific, and it is expressed at higher levels in the testes. XY mice that are lacking the *DMRT1* gene develop as males, but they are infertile due to severe defects in testis structure and the inability to produce sperm cells.

The mechanism of dosage compensation in mammals involves the process of X inactivation. In normal females, one of the two X chromosomes in females is compacted into a Barr body. This topic is described in Chapter 7.

In Sexually Dimorphic Plants, the Male Plant Is Usually Heteromorphic

As mentioned earlier, most species of plants, about 95%, are sexually monomorphic. Only a single type of individual is produced that can make both male and female gametophytes. By comparison, sexually dimorphic species have two separate types of individuals. The most common types are **dioecious,** in which one kind of individual produces male gametophytes, while the other produces female gametophytes. A few plants species are gynodioecious, which produce hermaphrodite plants and female plants. Alternatively, a few species are androdioecious and produce hermaphrodite plants and male plants.

The genetics of sex determination in dioecious plant species is beginning to emerge. Because *Arabidopsis* is sexually monomorphic, researchers have had to turn to other species to study sex determination in plants. For example, the white campion, *Silene latifolia*, has been the subject of numerous investigations. In this species, sex chromosomes, designated X and Y, are responsible for sex determination. The male plant has heteromorphic sex chromosomes, XY; the female plant is XX. In other dioecious species, cytological examination of the chromosomes does not always reveal distinct types of sex chromosomes. Nevertheless, the male plants appear to be the heterozygote. This has been determined because male plants are often "inconstant," which means that they produce an occasional fruit. When a male plant undergoes such self-fertilization, the seeds from the fruit produce a 3:1 ratio of male to female plants. This is the expected result if the male parent was heterozygous for a dominant gene that promotes maleness.

Researchers are beginning to identify genes that are important for sex determination in plants, though further studies are needed to unravel the genetic hierarchy that promotes sexual development. A chromosomal site designated Su^F plays a role in this process. The Su^F locus is believed to contain a gene that acts as a dominant suppressor of the female pathway and prevents carpel development. It is responsible for producing the 3:1 ratio when male plants undergo self-fertilization. In addition, other loci that appear to control early and late anther development have been discovered. Along with Su^F, these loci are linked on the Y chromosome in *S. latifolia*. Eventually, the discovery of more genes that control sexual development will allow researchers to propose alternative pathways that promote male and female development in dioecious plant species.

CONCEPTUAL SUMMARY

In this chapter, we have examined the role that genetics plays in the development of multicellular organisms. Each organism has its own developmental program, driven by a genetic hierarchy that encodes transcription factors and other types of regulatory proteins. In *Drosophila*, the developmental program begins in the oocyte. It is here that a few key gene products act as morphogens, molecules that convey positional information and promote developmental changes. Morphogens provide positional information that initiates body plan development. Maternal effect gene products are deposited asymmetrically in the oocyte and serve to organize the antero-posterior and dorso-ventral axes. Once these regions have been established, the next step of the developmental process organizes the embryo into segments.

Three categories of genes, known as gap genes, pair-rule genes, and segment-polarity genes, act sequentially in the formation of body segments. Gap genes divide the embryo into broad regions; the pair-rule genes control the identity of each parasegment; and the segment-polarity genes divide each parasegment into anterior and posterior regions. A segment, which is a morphological feature, is formed from the posterior region of one parasegment and the anterior region of an adjacent parasegment. The expression of homeotic genes dictates the phenotypic characteristics of each segment. These genes encode transcription factors that activate other genes responsible for determining the morphological characteristics of cells. A protein domain called the homeodomain promotes the binding of homeotic proteins to DNA.

In *C. elegans*, the availability of a complete lineage diagram has allowed geneticists to identify heterochronic mutations that alter the timing of cell fate and lead to abnormalities in morphology. This phenomenon illustrates the importance of coordinated cell division in determining cell fate during development.

In vertebrates, much less is known about the programs that promote development. Nevertheless, they appear to bear many similarities to the programs in simpler invertebrates. For example, *Hox* complexes in mice contain groups of genes that are homologous to homeotic genes in *Drosophila*. Furthermore, experiments with gene knockouts of particular *Hox* genes indicate that they play a key role in patterning the antero-posterior axis. Researchers have identified genes involved in the differentiation of cells into highly specialized cell types. As an example, the myogenic bHLH proteins play a critical role in the differentiation of skeletal muscle cells.

Morphologically, plant development differs markedly from animal development. Plants are organized along a root-shoot axis and growth occurs in a well-defined radial pattern. Plant geneticists have identified some of the genes that control the development of the meristems, organized groups of actively dividing cells. Geneticists have also discovered many homeotic genes that dictate the morphology of particular structures later in development. In the ABC model, four classes of homeotic genes, *A*, *B*, *C*, and *SEP*, are involved in the formation of the four whorls that make up a flower. These results show that plant development is dictated by a genetic program that bears some similarities to the programs found in animals.

Sex determination in animals and plants relies on a regulatory cascade that begins in early embryonic development. Different factors provide the basis for choosing between the female- and male-specific pathways. The hierarchy in any given species may involve a variety of control mechanisms including regulation of transcription factors, alternative splicing, and dosage compensation. The outcome of such regulation is to promote dosage compensation and the development of female or male structures and behaviors.

EXPERIMENTAL SUMMARY

We have seen that a genetic approach, namely, the identification of mutant alleles, has been essential to our understanding of development. Researchers, particularly those studying invertebrates and *Arabidopsis*, have identified many different mutant alleles that alter key steps in the developmental process. In *Drosophila*, for example, these include mutations in maternal effect genes, segmentation genes, and homeotic genes. The molecular characterization of these genes has shown that many of them encode transcription factors that influence the expression of other genes. The study of developmental mutations has enabled scientists to piece together a genetic hierarchy that underlies many developmental programs.

The characterization of developmental mutants has also been correlated with the spatial localization of gene products using techniques such as *in situ* hybridization. This enables researchers to determine when and where a gene product is made during oogenesis or embryogenesis. This approach has shown that some mutations that alter development are gain-of-function alleles, which cause a gene to be expressed in the wrong place or at the wrong time. Other mutations are loss-of-function alleles, which cause a defect in gene expression. In *C. elegans*, the availability of a lineage diagram has also allowed researchers to examine how heterochronic mutations, such as those that occur in the *lin-14* gene, can affect the timing of developmental steps.

The key experimental advantage of studying invertebrates is the ability to identify mutations that alter developmental steps. This approach is not as easy in vertebrates, but a reverse genetics approach is proving to be successful. Using invertebrate genes as probes, researchers have identified vertebrate genes, such as the *Hox* genes in mice, that are homologous to the homeotic genes of invertebrates. To understand their role, the wild-type *Hox* genes have been mutated in vitro and reintroduced into mice, replacing the normal gene. This is called a gene knockout. The results of this method are consistent with the idea that the *Hox* genes function as the mouse homologues of the *Drosophila* homeotic genes. However, further research will be needed to understand the individual roles of these 38 genes.

In plants, a genetic approach is also proving effective in elucidating the development process. Early in development, a category of genes known as the apical-basal-patterning genes appear necessary for the formation of the apical and basal regions of the embryo. These genes were identified by investigating the effects of mutations on the developing plant embryo. Similarly, mutations in plant homeotic genes have been discovered by their ability to abnormally transform one plant structure into another (e.g., sepals into petals). A comparison of single, double, and triple homeotic mutations provided the framework for the ABC model of flower development.

PROBLEM SETS & INSIGHTS

Solved Problems

S1. Discuss and distinguish the functional roles of the maternal effect genes, gap genes, pair-rule genes, and segment-polarity genes in *Drosophila*.

Answer: These genes are involved in pattern formation of the *Drosophila* embryo. The asymmetric distribution of maternal effect gene products in the oocyte establishes the antero-posterior and dorso-ventral axes. These gene products also control the expression of the gap genes, which are expressed as broad bands in certain regions of the embryo. The overlapping expression of maternal effect genes and gap genes controls the pair-rule genes, which are expressed in alternating stripes. A stripe corresponds to a parasegment. Within each parasegment, the expression of segment-polarity genes defines an anterior and posterior compartment. With regard to morphology, an anterior compartment of one parasegment and the posterior compartment of an adjacent parasegment will form a segment of the fly.

S2. With regard to genes affecting development, what are the phenotypic effects of gain-of-function mutations versus loss-of-function mutations?

Answer: Gain-of-function mutations cause a gene to be expressed in the wrong place, at the wrong time, or at an abnormal level. When the gene is expressed in the wrong place, that region may develop into an inappropriate structure. For example, when *Antp* is abnormally expressed in an anterior segment, this segment develops legs in place of antennae. When gain-of-function mutations cause a gene to be expressed at the wrong time, this can also disrupt the development process. Gain-of-function heterochronic mutations cause cell lineages to be reiterated and thereby alter the course of development. By comparison, loss-of-function mutations result in a defect in the expression of a gene with the result that protein function is reduced or abolished. This usually will disrupt the developmental process, because the cells in the region where the gene is normally expressed will not be directed to develop along the correct pathway.

S3. Mutations in genes that control the early stages of development are often lethal (e.g., see Figure 23.6b). To circumvent this problem, developmental geneticists may try to isolate temperature-sensitive developmental mutants or *ts* alleles. If an embryo carries a *ts* allele, it will develop correctly at the permissive temperature (e.g., 25°C) but will fail to develop if incubated at the nonpermissive temperature (e.g., 30°C). In most cases, *ts* alleles have missense mutations that slightly alter the amino acid sequence of a protein, causing a change in its structure that prevents it from working properly at the nonpermissive temperature. *Ts* alleles are particularly useful because they can provide insight regarding the stage of development when the protein is necessary. Researchers can take groups of embryos that carry a *ts* allele and expose them to the permissive and nonpermissive temperature at different stages of development. In the experiment described next, embryos were divided into five groups and exposed to the permissive or nonpermissive temperature at different times after fertilization.

Time After Fertilization (hours) **Group:**	**1**	**2**	**3**	**4**	**5**
0–1	25°C	25°C	25°C	25°C	25°C
1–2	25°C	30°C	25°C	25°C	25°C
2–3	25°C	25°C	30°C	25°C	25°C
3–4	25°C	25°C	25°C	30°C	25°C
4–5	25°C	25°C	25°C	25°C	30°C
5–6	25°C	25°C	25°C	25°C	25°C
SURVIVAL:	Yes	Yes	Yes	No	Yes

Explain these results.

Answer: By varying the temperature during different stages of development, researchers can pinpoint the stage when the function of the protein encoded by this *ts* allele is critical. As shown, embryos fail to survive if they are subjected to the nonpermissive temperature between

3 and 4 hours after fertilization, but they do survive if subjected to the nonpermissive temperature at other times of development. These results indicate that this protein plays a crucial role at the 3–4-hour stage of development.

S4. An intriguing question in developmental genetics is, How can a particular gene, such as *even-skipped*, be expressed in a multiple banding pattern as seen in Figure 23.10? Another way of asking this question is, How is the positional information within the broad bands of the gap genes able to be deciphered in a way that causes the pair-rule genes to be expressed in this alternating banding pattern? The answer lies in a complex mechanism of genetic regulation. Certain pair-rule genes have several stripe-specific enhancers that are controlled by multiple transcription factors. A stripe-specific enhancer is typically a short segment of DNA, 300 to 500 bp in length, that contains binding sequences recognized by several different transcription factors. This term is a bit misleading because a stripe-specific enhancer is a regulatory region that contains both enhancer and silencer elements.

In 1992, Michael Levine and his colleagues investigated stripe-specific enhancers that are located near the promoter of the *even-skipped* gene. A segment of DNA, termed the stripe 2 enhancer, controls the expression of the *even-skipped* gene; this enhancer is responsible for the expression of the *even-skipped* gene in stripe 2, which corresponds to parasegment 3 of the embryo. The stripe 2 enhancer is a segment of DNA that contains binding sites for four transcription factors that are the products of the *Krüppel, bicoid, hunchback,* and *giant* genes. The Hunchback and Bicoid transcription factors bind to this enhancer and activate the transcription of the *even-skipped* gene. In contrast, the transcription factors encoded by the *Krüppel* and *giant* genes bind to the stripe 2 enhancer and repress transcription. The figure shown below describes the concentrations of these four transcription factor proteins in the region of parasegment 3 (i.e., stripe 2) in the *Drosophila* embryo.

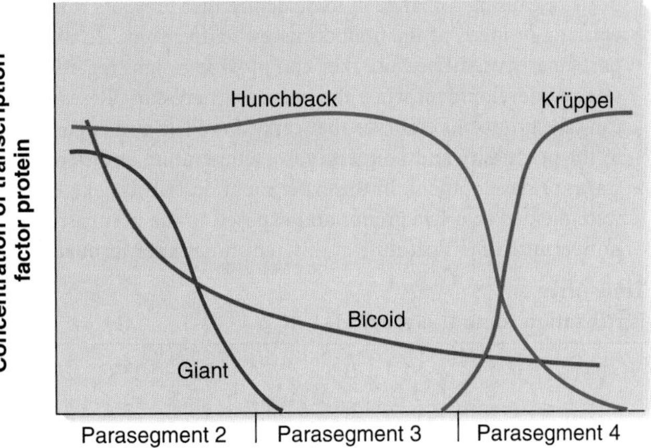

To study stripe-specific enhancers, researchers have constructed artificial genes in which the enhancer is linked to a reporter gene, the expression of which is easy to detect. The next figure shows the results of an experiment in which an artificial gene was made by putting the stripe 2 enhancer next to the β-galactosidase gene. This artificial gene was introduced into *Drosophila*, and then embryos containing this gene were analyzed for β-galactosidase activity. If a region of the embryo is expressing β galactosidase, the region will stain darkly because β galactosidase converts a colorless compound into a dark blue compound.

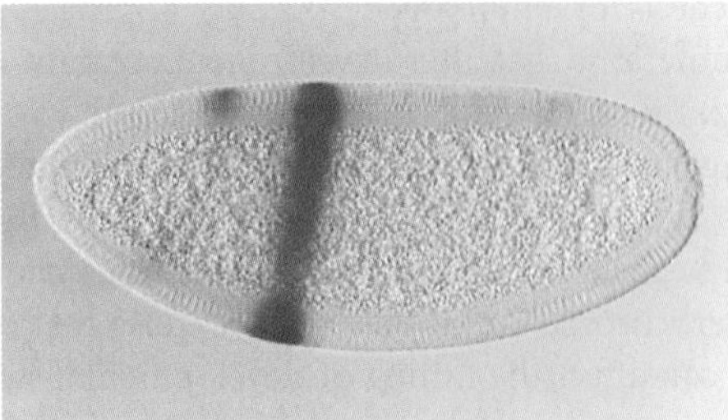

Explain these results.

Answer: As shown in the first figure to this problem, the concentrations of the Hunchback and Bicoid transcription factors are relatively high in the region of the embryo corresponding to stripe 2 (which is parasegment 3). The levels of Krüppel and Giant are very low in this region. Therefore, the high levels of activators and low levels of repressors cause the *even-skipped* gene to be transcribed. In this experiment, β galactosidase was made only in stripe 2 (i.e., parasegment 3). These results show that the stripe-2-specific enhancer controls gene expression only in parasegment 3. Because we know that the *even-skipped* gene is expressed as several alternating stripes (as seen in Figure 23.10), the *even-skipped* gene must contain other stripe-specific enhancers that allow it to be expressed in these other alternating parasegments.

Conceptual Questions

C1. In the case of multicellular animals, what are the four types of cellular processes that occur so a fertilized egg can develop into an adult organism? Briefly discuss the role of each process.

C2. The arrangement of body axes of the fruit fly are shown in Figure 23.4g. Are the following statements true or false with regard to body axes in the mouse?

A. Along the antero-posterior axis, the head is posterior to the tail.

B. Along the dorso-ventral axis, the vertebrae of the back are dorsal to the stomach.

C. Along the dorso-ventral axis, the feet are dorsal to the hips.

D. Along the proximo-distal axis, the toes on the hind legs are distal to the hips of the hind legs.

C3. If you observed fruit flies with the following developmental abnormalities, would you guess that a mutation has occurred in a segmentation gene or a homeotic gene? Explain your guess.

A. Three abdominal segments were missing.

B. One abdominal segment had legs.

C. A fly with the correct number of segments had two additional thoracic segments and two fewer abdominal segments.

C4. Which of the following statements are true with regard to positional information in *Drosophila*?

A. Morphogens are a type of molecule that conveys positional information.

B. Morphogenetic gradients are established only in the oocyte, prior to fertilization.

C. Cell adhesion molecules also provide a way for a cell to obtain positional information.

C5. Discuss the morphological differences between the parasegments and segments of *Drosophila*. Discuss the evidence, providing specific examples, that suggests the parasegments of the embryo are the subdivisions for the organization of gene expression.

C6. Here are schematic diagrams of mutant larvae.

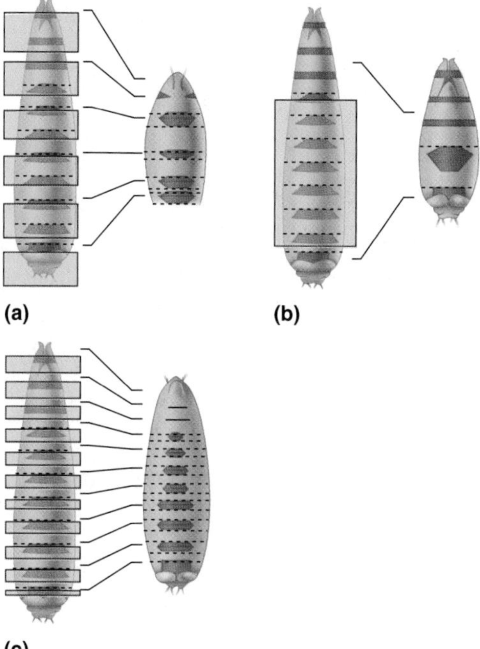

The left side of each pair shows a wild-type larva, with gray boxes showing the sections that are missing in the mutant larva. Which type of gene is defective in each larva: a gap gene, a pair-rule gene, or a segment-polarity gene?

C7. Describe what a morphogen is and how it exerts its effects. What do you expect will happen when a morphogen is expressed in the wrong place in an embryo? List five examples of morphogens that function in *Drosophila*.

C8. What is the meaning of positional information? Discuss three different ways that cells can obtain positional information. Which of these three ways do you think is the most important for the formation of a segmented body pattern in *Drosophila*?

C9. Gradients of morphogens can be preestablished in the oocyte. Also, later in development, morphogens can be secreted from cells. How are these two processes similar and different?

C10. Discuss how the anterior portion of the antero-posterior axis is established. What aspects of oogenesis are critical in establishing this axis? What do you think would happen if the *bicoid* mRNA was not trapped at the anterior end but instead diffused freely throughout the oocyte?

C11. Describe the function of the Bicoid protein. Explain how its ability to exert its effects in a concentration-dependent manner is a critical feature of its function.

C12. With regard to development, what are the roles of the maternal effect genes versus the zygotic genes? Which types of genes are needed earlier in the development process?

C13. Discuss the role of homeotic genes in development. Explain what happens to the phenotype of a fly when a gain-of-function homeotic gene mutation occurs in an abnormal region of the embryo. What are the consequences of a loss-of-function mutation in such a gene?

C14. Describe the molecular features of the homeobox and homeodomain. Explain how these features are important in the function of homeotic genes.

C15. What would you predict to be the phenotype of a *Drosophila* larva whose mother was homozygous for a loss-of-function allele in the *nanos* gene?

C16. Based on the photographs in Figure 23.12, in which segments would the *Antp* gene normally be expressed?

C17. If a mutation in a homeotic gene produced the following phenotypes, would you expect the mutation to be a loss-of-function or a gain-of-function allele? Explain your answer.

A. An abdominal segment has antennae attached to it.

B. The most anterior abdominal segment resembles the most posterior thoracic segment.

C. The most anterior thoracic segment resembles the most posterior abdominal segment.

C18. Explain how loss-of-function mutations in the following categories of genes would affect the morphologies of *Drosophila* larvae:

A. Gap genes

B. Pair-rule genes

C. Segment-polarity genes

C19. What is the difference between a maternal effect gene and a zygotic gene? Of the following genes that play a role in *Drosophila* development, which would be maternal effect genes and which would be zygotic? Explain your answer.

A. *nanos*

B. *Antp*

C. *bicoid*

D. *lab*

C20. Cloning of mammals (such as Dolly) is described in Chapter 19. Based on your understanding of animal development, explain why an enucleated egg is needed to clone mammals. In other words, what features of the oocyte are essential for animal development?

C21. A hypothetical cell lineage is shown here.

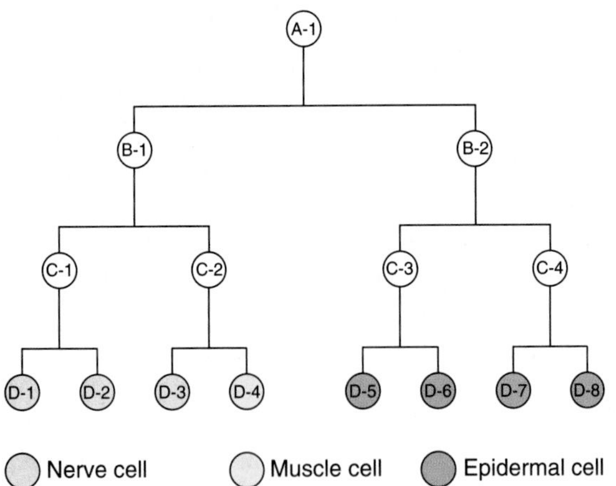

A gene, which we will call gene *X*, is activated in the B-1 cell, so the B-1 cell will progress through the proper developmental stages to produce three nerve cells (D-1, D-2, and D-3) and one muscle cell (D-4). Gene *X* is normally inactivated in A-1, C-1, and C-2 cells, as well as the four D cells. Draw the expected cell lineages if a heterochronic mutation had the following effects:

A. Gene *X* is turned on one cell division too early.

B. Gene *X* is turned on one cell division too late.

C22. What is a heterochronic mutation? How does it affect the phenotypic outcome of an organism? What phenotypic effects would you expect if a heterochronic mutation affected the cell lineage that determines the fates of intestinal cells?

C23. Discuss the similarities and differences between the *bithorax* and *Antennapedia* complexes in *Drosophila* and the *Hox* gene complexes in mice.

C24. What is cell differentiation? Discuss the role of bHLH proteins in the differentiation of muscle cells. Explain how they work at the molecular level. In your answer, indicate how protein dimerization is a key feature in gene regulation.

C25. The *MyoD* gene in mammals plays a role in muscle cell differentiation, whereas the *Hox* genes are homeotic genes that play a role in the differentiation of particular regions of the body. Explain how the functions of these genes are similar and different.

C26. What is a totipotent cell? In the following examples, which cells would be totipotent? In the case of multicellular organisms, such as humans and corn, explain which cell types would be totipotent.

A. Humans

B. Corn

C. Yeast

D. Bacteria

C27. What is a meristem? Explain the role of meristems in plant development.

C28. Discuss the morphological differences between animals and plants. How are they different at the cellular level? How are they similar at the genetic level?

C29. Predict the phenotypic consequences of each of the following mutations:

A. *apetala1* defective

B. *pistillata* defective

C. *apetala1* and *pistillata* defective

Experimental Questions

E1. Researchers have used the cloning methods described in Chapter 18 to clone the *bicoid* gene and express large amounts of the Bicoid protein. The Bicoid protein was then injected into the posterior end of a zygote immediately after fertilization. What phenotypic results would you expect to occur? What do you think would happen if the Bicoid protein was injected into a segment of a larva?

E2. Compare and contrast the experimental advantages of *Drosophila* and *C. elegans* in the study of developmental genetics.

E3. What is meant by the term cell fate? What is a lineage diagram? Discuss the experimental advantage of having a lineage diagram. What is a cell lineage?

E4. Explain why a lineage diagram is necessary to determine if a mutation is heterochronic.

E5. Explain the rationale behind the use of the "bag of worms" phenotype as a way to identify heterochronic mutations.

E6. Here are the results of cell lineage analyses of hypodermal cells in wild-type and mutant strains of *C. elegans*.

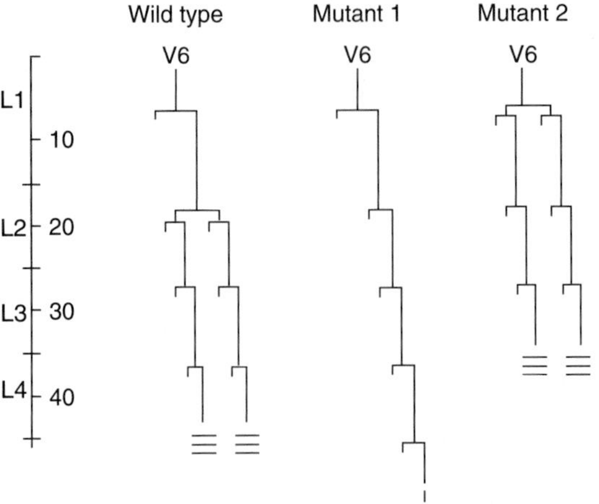

Explain the nature of the mutations in the altered strains.

E7. Take a look at solved problem S3 before answering this question. *Drosophila* embryos carrying a *ts* mutation were exposed to the permissive (25°C) or nonpermissive (30°C) temperature at different stages of development. Explain these results.

Time After Fertilization (hours) Group:	1	2	3	4	5
0–1	25°C	25°C	25°C	25°C	25°C
1–2	25°C	30°C	25°C	25°C	25°C
2–3	25°C	25°C	30°C	25°C	25°C
3–4	25°C	25°C	25°C	30°C	25°C
4–5	25°C	25°C	25°C	25°C	30°C
5–6	25°C	25°C	25°C	25°C	25°C
SURVIVAL:	Yes	No	No	Yes	Yes

E8. All of the homeotic genes in *Drosophila* have been cloned. As discussed in Chapter 18, cloned genes can be manipulated in vitro. They can be subjected to cutting and pasting, site-directed mutagenesis, etc. After *Drosophila* genes have been altered in vitro, they can be inserted into a *Drosophila* transposon vector (i.e., a P element vector), and then the genetic construct containing the altered gene within a P element can be injected into *Drosophila* embryos. The P element will then transpose into the chromosomes and thereby introduce one or more copies of the altered gene into the *Drosophila* genome. This method is termed P element transformation.

With these ideas in mind, how would you make a mutant gene with a "gain-of-function" in which the *Antp* gene would be expressed where the *abd-A* gene is normally expressed? What phenotype would you expect for flies that carried this altered gene?

E9. You will need to understand solved problem S4 before answering this question. If the artificial gene containing the stripe 2 enhancer and the β-galactosidase gene was found within an embryo that also contained the following loss-of-function mutations, what results would you expect? In other words, would there be a stripe or not? Explain why.

A. *Krüppel*

B. *bicoid*

C. *hunchback*

D. *giant*

E10. Two techniques commonly used to study the expression patterns of genes that play a role in development are Northern blotting and *in situ* hybridization. As described in Chapter 18, Northern blotting can be used to detect RNA that is transcribed from a particular gene. In this method, a specific RNA is detected by using a short segment of cloned DNA as a probe. The DNA probe, which is radioactive, is complementary to the RNA that the researcher wishes to detect. After the radioactive probe DNA binds to the RNA within a blot of a gel, the RNA is visualized as a dark (radioactive) band on an X-ray film. For example, a DNA probe that is complementary to the *bicoid* mRNA could be used to specifically detect the amount and size of the mRNA in a blot.

A second technique, termed fluorescence *in situ* hybridization (FISH), can be used to identify the locations of genes on chromosomes. This technique can also be used to locate gene products within oocytes, embryos, and larvae. For this reason, it has been commonly used by developmental geneticists to understand the expression patterns of genes during development. The photograph in Figure 23.7b is derived from the application of the FISH technique. In this case, the probe was complementary to bicoid mRNA.

Now here is the question. Suppose a researcher has three different *Drosophila* strains that have loss-of-function mutations in the *bicoid* gene. We will call them *bicoid-A*, *bicoid-B*, and *bicoid-C*; the wild type is designated *bicoid*$^+$. To study these mutations, phenotypically normal female flies that are homozygous for the bicoid mutation were obtained, and their oocytes were analyzed using these two techniques. A wild-type strain was also analyzed as a control. In other words, RNA was isolated from some of the oocytes and analyzed by Northern blotting, and some oocytes were subjected to *in situ* hybridization. In both cases, the probe was complementary to the bicoid mRNA. The results are shown here.

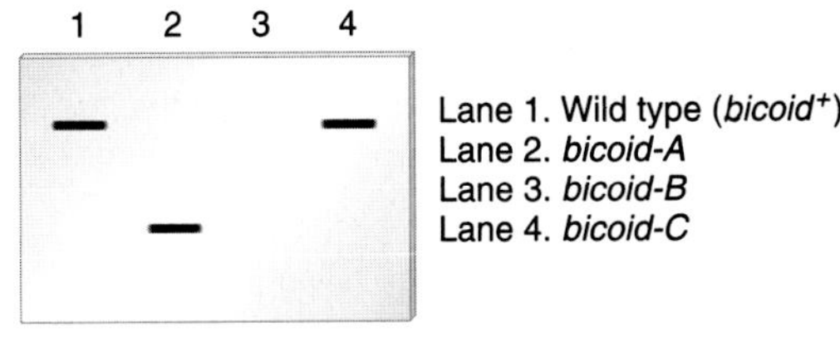

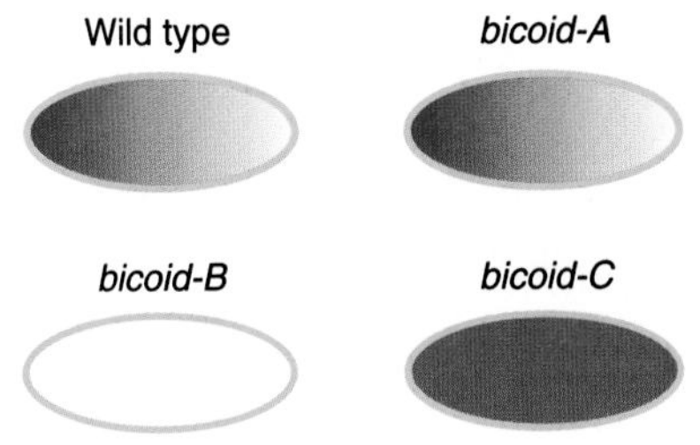

A. How can phenotypically normal female flies be homozygous for a loss-of-function allele in the *bicoid* gene?

B. Explain the type of mutation (e.g., deletion, point mutation, etc.) in each of the three strains. Explain how the mutation may cause a loss of normal function for the *bicoid* gene product.

C. Discuss how the use of both techniques provides more definitive information compared to the application of just one of the two techniques.

E11. Explain one experimental strategy you could follow to determine the functional role of the mouse *HoxD-3* gene.

E12. In the experiment of Figure 23.15, suggest reasons why the *n536*, *n355*, and *n540* strains have an egg-laying defect.

E13. Another way to study the role of proteins (e.g., transcription factors) that function in development is to microinject the mRNA that encodes a protein, or the purified protein itself, into an oocyte or embryo, and then determine how this affects the subsequent development of the embryo, larva, and adult. For example, if Bicoid protein is injected into the posterior region of an oocyte, the resulting embryo will develop into a larva that has anterior structures at both ends. Based on your understanding of the function of these developmental genes, what would be the predicted phenotype if the following proteins or mRNAs were injected into normal oocytes?

A. *Nanos* mRNA injected into the anterior end of an oocyte

B. Antp protein injected into the posterior end of an embryo

C. *Toll* mRNA injected the dorsal side of an early embryo

E14. Why have geneticists been forced to use reverse genetics to study the genes involved in vertebrate development? Explain how this strategy differs from traditional genetic analyses like those done by Mendel.

Questions for Student Discussion/Collaboration

1. Compare and contrast the experimental advantages and disadvantages of *Drosophila*, *C. elegans*, mammals, and *Arabidopsis*.
2. It seems that developmental genetics boils down to a complex network of gene regulation. Try to draw how this network is structured for *Drosophila*. How many genes do you think are necessary to describe a complete developmental network for the fruit fly? How many genes do you think are needed for a network to specify one segment? Do you think it is more difficult to identify genes that are involved in the beginning, middle, or end of this network? Knowing what you know about *Drosophila* development, suppose you were trying to identify all the genes needed for development in a chicken. Would you first try to identify genes necessary for early development, or would you begin by identifying genes involved in cell differentiation?
3. At the molecular level, how do you think a gain-of-function mutation in a developmental gene might cause it to be expressed in the wrong place or at the wrong time? Explain what type of DNA sequence would be altered.

Note: All answers appear at the website for this textbook; the answers to even-numbered questions are in the back of the textbook.

www.mhhe.com/brookergenetics3e

Visit the Online Learning Center for practice tests, answer keys, and other learning aids for this chapter. Enhance your understanding of genetics with our interactive exercises, quizzes, animations, and much more.

POPULATION GENETICS

24

CHAPTER OUTLINE

The African cheetah. This species has a relatively low level of genetic variation because the population was reduced to a small size approximately 10,000 to 12,000 years ago.

Until now, we have primarily focused our attention on genes within individuals and their related family members. In this chapter and Chapters 25 and 26, we will turn to the study of genes in a population or species. The field of **population genetics** is concerned with changes in genetic variation within a group of individuals over time. Population geneticists want to know the extent of genetic variation within populations, why it exists, and how it changes over the course of many generations. The field of population genetics emerged as a branch of genetics in the 1920s and 1930s. Its mathematical foundations were developed by theoreticians who extended the principles of Gregor Mendel and Charles Darwin by deriving formulas to explain the occurrence of genotypes within populations. These foundations can be largely attributed to three scientists: Sir Ronald Fisher, Sewall Wright, and J. B. S. Haldane. As we will see, support for their mathematical theories was provided by several researchers who analyzed the genetic composition of natural and experimental populations. More recently, population geneticists have used techniques to probe genetic variation at the molecular level. In addition, staggering advances in computer technology have aided population geneticists in the analysis of their genetic theories and data. In this chapter, we will explore the genetic variation that occurs in populations and consider the reasons why the genetic composition of populations may change over the course of several generations.

24.1 GENES IN POPULATIONS, AND THE HARDY-WEINBERG EQUATION

Population genetics may seem like a significant departure from other topics in this textbook, but it is a direct extension of our understanding of Mendel's laws of inheritance, molecular genetics, and the ideas of Darwin. In the field of population genetics, the focus shifts away from the individual and toward the population of which the individual is a member. Conceptually, all of the alleles of every gene in a population make up the **gene pool.** In this regard, each member of the population is viewed as receiving its genes from its parents, which, in turn, are members of the gene pool. Furthermore, individuals that reproduce contribute to the gene pool of the next generation. Population geneticists study the genetic variation within the gene pool and how such variation changes from one generation to the next. The emphasis is often on allelic variation. In this introductory section, we will examine some of the general features of populations and gene pools.

A Population Is a Group of Interbreeding Individuals That Share a Gene Pool

In genetics, the term population has a very specific meaning. With regard to sexually reproducing species, a **population** is a group of individuals of the same species that occupy the same region and can interbreed with one another. Many species occupy a wide geographic range and are divided into discrete populations. For example, distinct populations of a given species may be located on different continents, or populations on the same continent could be divided by a geographical feature such as a large mountain range.

A large population usually is composed of smaller groups called **local populations** or **demes.** The members of a local population are far more likely to breed among themselves than with other members of the general population. Local populations are often separated from each other by moderate geographic barriers. As shown in **Figure 24.1**, two groups of Douglas fir (*Pseudotsuga menziesii*) are separated by a wide river, where the fir trees cannot grow. Trees on opposite sides of this river constitute local populations. Breeding is much more apt to occur among members of each local population than between members of neighboring populations. On relatively rare occasions, however, pollen can be blown across the river, which allows breeding between these two local populations.

Populations typically are dynamic units that change from one generation to the next. A population may change its size, geographic location, and genetic composition. With regard to size, natural populations commonly go through cycles of "feast or famine," during which the population swells or shrinks. In addition, natural predators or disease may periodically decrease the size of a population to significantly lower levels; the population later may rebound to its original size. Populations or individuals within populations may migrate to a new site and establish a distinct population in this location. The environment at this new geographic location may differ from the original site. What are the consequences of such changes? As population sizes and locations change, their genetic composition generally changes as well. As described later, population geneticists have developed mathematical theories that predict how the gene pool will change in response to fluctuations in size, migration, and new environments.

FIGURE 24.1 **Two local populations of Douglas fir.** A wide river separates the two local populations. A much higher frequency of breeding occurs among members of the same local population than between the two different local populations.

At the Population Level, Some Genes May Be Monomorphic, But Most Are Polymorphic

In population genetics, the term **polymorphism** (meaning many forms) refers to the observation that many traits display variation within a population. Historically, polymorphism first referred to the variation in traits that are observable with the naked eye. Polymorphisms in color and pattern have long attracted the attention of population geneticists. These include studies involving yellow and red varieties of the elder-flowered orchid, and brown, pink, and yellow land snails, which are discussed later in this chapter. **Figure 24.2** illustrates a striking example of polymorphism in the Hawaiian happy-face spider (*Theridion grallator*). The three individuals shown in this figure are from the same species, but they differ in alleles that affect color and pattern.

What is the underlying cause of polymorphism? At the DNA level, polymorphism may be due to two or more alleles that influence the phenotype of the individual that inherits them. In other words, it is due to genetic variation. Geneticists also use the term **polymorphic** to describe a gene that commonly exists as two or more alleles in a population. By comparison, a **monomorphic** gene exists predominantly as a single allele in a population. By convention, when a single allele is found in at least 99% of all cases, the gene is considered monomorphic. (Some geneticists view an allele frequency of 95% or greater to be monomorphic.)

At the level of a particular gene, a polymorphism may involve various types of changes such as a deletion of a significant region of the gene, a duplication of a region, or a change in a single nucleotide. This last phenomenon is called a **single-nucleotide polymorphism (SNP).** SNPs are the smallest type of genetic change that can occur within a given gene and are also the most common. In humans, for example, SNPs represent 90% of all the variation in DNA sequences that occurs among different people. SNPs are found very frequently in genes. In the human population, a gene that is 2,000 to 3,000 bp in length will, on average, contain 10 different sites that are polymorphic. The high frequency of SNPs indicates that polymorphism is the norm for most human genes. Likewise, relatively large, healthy populations of nearly all species exhibit a high level of genetic variation as evidenced by the occurrence of SNPs within most genes.

Within a population, the alleles of a given gene may arise by different types of genetic changes. **Figure 24.3** considers a gene that exists in multiple forms in humans. This example is a short segment of DNA found within the human β-globin gene. The top sequence is an allele designated Hb^A, whereas the middle sequence is called Hb^S. These alleles differ from each other

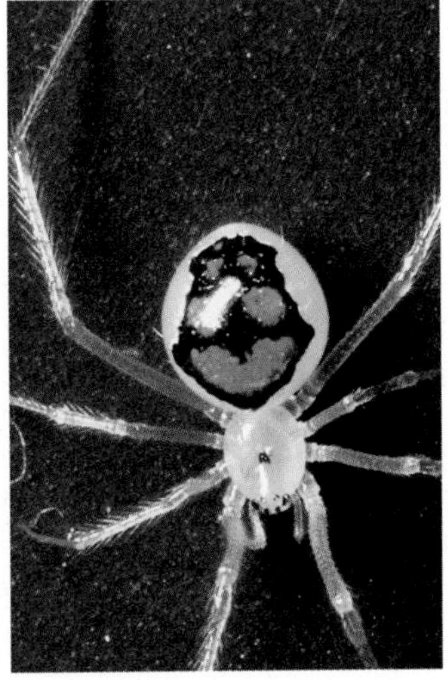

FIGURE 24.2 Polymorphism in the Hawaiian happy-face spider.

Genes→Traits These three spiders are members of the same species and carry the same genes. However, several genes that affect pigmentation patterns are polymorphic, meaning there is more than one allele for each gene within the population. This polymorphism within the Hawaiian happy-face spider population produces members that look quite different from each other.

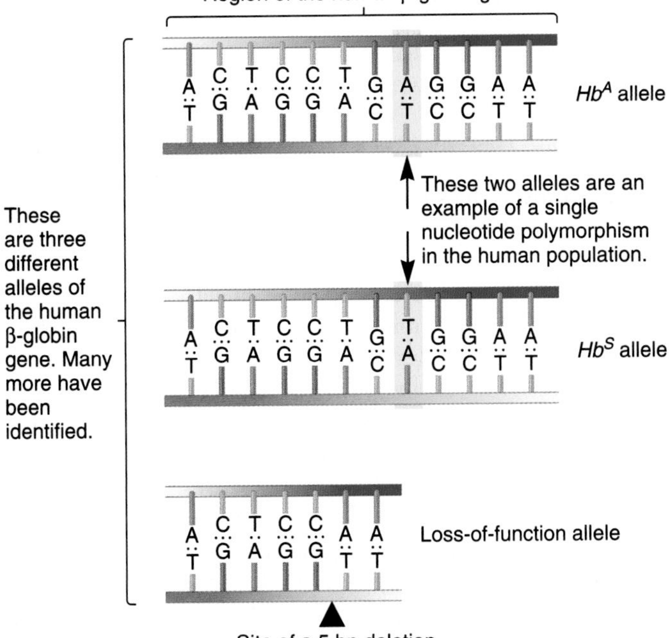

FIGURE 24.3 The relationship between alleles and various types of mutations. The DNA sequence shown here is a small portion of the β-globin gene in humans. Mutations have altered the gene to create the three different alleles in this figure. The top two alleles differ by a single base pair and are referred to as a single nucleotide polymorphism (SNP). The bottom allele has a 5 bp deletion that results in a nonfunctional polypeptide. It is a loss-of-function allele.

by a single nucleotide, so they are an example of a single nucleotide polymorphism. As discussed in Chapter 4, the Hb^S allele causes sickle-cell disease in a homozygote. The bottom sequence contains a short, 5 bp deletion compared to the other two alleles. This deletion results in a nonfunctional β-globin polypeptide. Therefore, the bottom sequence is an example of a loss-of-function allele.

Population Genetics Is Concerned with Allele and Genotype Frequencies

As we have seen, population geneticists want to understand the prevalence of polymorphic genes within populations. Much of their work evaluates the frequency of alleles in a quantitative way. Two fundamental calculations are central to population genetics: **allele frequencies** and **genotype frequencies.** The allele and genotype frequencies are defined as

$$\text{Allele frequency} = \frac{\text{Number of copies of an allele in a population}}{\text{Total number of all alleles for that gene in a population}}$$

$$\text{Genotype frequency} = \frac{\text{Number of individuals with a particular genotype in a population}}{\text{Total number of individuals in a population}}$$

Though these two frequencies are related, a clear distinction between them must be kept in mind. As an example, let's consider a population of 100 pea plants with the following genotypes:

64 tall plants with the genotype *TT*

32 tall plants with the genotype *Tt*

4 dwarf plants with the genotype *tt*

When calculating an allele frequency, homozygous individuals have two copies of an allele, whereas heterozygotes have only one. For example, in tallying the *t* allele, each of the 32 heterozygotes has one copy of the *t* allele, and each dwarf plant has two copies. The allele frequency for *t* equals

$$t = \frac{32 + 2(4)}{2(64) + 2(32) + 2(4)}$$

$$= \frac{40}{200} = 0.2, \text{ or } 20\%$$

This result tells us that the allele frequency of *t* is 20%. In other words, 20% of the alleles for this gene in the population are the *t* allele.

Let's now calculate the genotype frequency of *tt* (dwarf) plants:

$$tt = \frac{4}{64 + 32 + 4}$$

$$= \frac{4}{100} = 0.04, \text{ or } 4\%$$

We see that 4% of the individuals in this population are dwarf plants.

Allele and genotype frequencies are always less than or equal to 1 (i.e., less than or equal to 100%). If a gene is monomorphic, the allele frequency for the single allele will equal or be close to a value of 1.0. For polymorphic genes, if we add up the frequencies for all the alleles in the population, we should obtain a value of 1.0. In our pea plant example, the allele frequency of *t* equals 0.2. The frequency of the other allele, *T*, equals 0.8. If we add the two together, we obtain a value of $0.2 + 0.8 = 1.0$.

The Hardy-Weinberg Equation Can Be Used to Calculate Genotype Frequencies Based on Allele Frequencies

Now that we have a general understanding of genes in populations, we can begin to relate these concepts to mathematical expressions as a way to examine whether allele and genotype frequencies will change over the course of many generations. In 1908, a British mathematician, Godfrey Harold Hardy, and a German physician, Wilhelm Weinberg, independently derived a simple mathematical expression that predicted stability of allele and genotype frequencies from one generation to the next. It is called the **Hardy-Weinberg equilibrium,** because (under a given set of conditions, described later) the allele and genotype frequencies do not change over the course of many generations. This relationship establishes a framework on which to understand changes in allele frequencies within a population when such an equilibrium is violated.

Let's begin by considering a situation in which a gene is polymorphic and exists as two different alleles, *A* and *a*. If the allele frequency of *A* is denoted by the variable p, and the allele frequency of *a* by q, then

$$p + q = 1$$

For example, if $p = 0.8$, then q must be 0.2. In other words, if the allele frequency of *A* equals 80%, the remaining 20% of alleles must be *a*, because together they equal 100%.

The Hardy-Weinberg equation is used to relate allele frequencies and genotype frequencies. For a gene that exists in two alleles:

$$p^2 + 2pq + q^2 = 1 \text{ } \textbf{(Hardy-Weinberg equation)}$$

If this equation is applied to a gene that exists in alleles designated *A* and *a*, then

p^2 equals the genotype frequency of *AA*

$2pq$ equals the genotype frequency of *Aa*

q^2 equals the genotype frequency of *aa*

If $p = 0.8$ and $q = 0.2$ and if the population is in Hardy-Weinberg equilibrium, then

$$AA = p^2 = (0.8)^2 = 0.64$$

$$Aa = 2pq = 2(0.8)(0.2) = 0.32$$

$$aa = q^2 = (0.2)^2 = 0.04$$

In other words, if the allele frequency of *A* is 80% and the allele frequency of *a* is 20%, the genotype frequency of *AA* is 64%, *Aa* is 32%, and *aa* is 4%.

To illustrate the relationship between allele frequencies and genotypes, **Figure 24.4** compares the Hardy-Weinberg equation with the way that gametes combine randomly with each other to produce offspring. In a population, the frequency of a gamete carrying a particular allele is equal to the allele frequency in that population. In this example, the frequency of a gamete carrying the *A* allele equals 0.8.

We can use the product rule to determine the frequency of genotypes. For example, the frequency of producing an *AA* homozygote is $0.8 \times 0.8 = 0.64$, or 64%. Likewise, the probability of inheriting both *a* alleles is $0.2 \times 0.2 = 0.04$, or 4%. As seen in Figure 24.4, heterozygotes can be produced in two different ways. An offspring could inherit the *A* allele from its father and *a* from its mother, or *A* from its mother and *a* from its father. Therefore, the frequency of heterozygotes is $pq + pq$, which equals $2pq$; in our example, this is $2(0.8)(0.2) = 0.32$, or 32%.

One useful feature of the Hardy-Weinberg equation is that it allows us to determine the frequency of heterozygotes for recessive genetic diseases. As an example, let's consider cystic fibrosis, which involves a gene that encodes a chloride transporter. Persons afflicted with this disorder have an irregularity in salt and water

♀ \ ♂	*A* 0.8	*a* 0.2
A 0.8	*AA* (0.8)(0.8) = 0.64	*Aa* (0.8)(0.2) = 0.16
a 0.2	*Aa* (0.8)(0.2) = 0.16	*aa* (0.2)(0.2) = 0.04

AA genotype = 0.64 = 64%
Aa genotype = 0.16 + 0.16 = 0.32 = 32%
aa genotype = 0.04 = 4%

FIGURE 24.4 A comparison between allele frequencies and the union of alleles in a Punnett square. In a population in Hardy-Weinberg equilibrium, the frequency of gametes carrying a particular allele is equal to the allele frequency in the population.

balance. One of the symptoms is thick mucus in the lungs that can contribute to repeated lung infections. In Caucasian populations, the frequency of affected individuals is approximately 1 in 2,500. Because this is a recessive disorder, affected individuals are homozygotes. If q represents the allele frequency of the disease-causing allele, then

$$q^2 = 1/2{,}500$$
$$q^2 = 0.0004$$

We take the square root to determine q:

$$q = \sqrt{0.0004}$$
$$q = 0.02$$

If p represents the normal allele,

$$p = 1 - q$$
$$p = 1 - 0.02 = 0.98$$

The frequency of heterozygous carriers is

$$2pq = 2(0.98)(0.02) = 0.0392, \text{ or } 3.92\%$$

The Hardy-Weinberg equation predicts an equilibrium—unchanging allele and genotype frequencies from generation to generation—if certain conditions are met in a population. With regard to the gene of interest, these conditions are as follows:

1. No new mutations: The gene of interest does not incur any new mutations.
2. No genetic drift: The population is so large that allele frequencies do not change due to random sampling effects.
3. No migration: Individuals do not travel between different populations.
4. No natural selection: All of the genotypes have the same reproductive success.
5. Random mating: With respect to the gene of interest, the members of the population mate with each other without regard to their phenotypes and genotypes.

The Hardy-Weinberg equation provides a quantitative relationship between allele and genotype frequencies in a population. **Figure 24.5** describes this relationship for different allele frequencies of *a* and *A*. As expected, when the allele frequency of *a* is very low, the *AA* genotype predominates; when the *a* allele frequency is high, the *aa* homozygote is most prevalent in the population. When the allele frequencies of *a* and *A* are intermediate in value, the heterozygote predominates.

In reality, no population satisfies the Hardy-Weinberg equilibrium completely. Nevertheless, in large natural populations with little migration and negligible natural selection, the Hardy-Weinberg equilibrium may be nearly approximated for certain genes. In addition, the Hardy-Weinberg equilibrium can be extended to situations in which a single gene exists in three or more alleles, as described in solved problem S4 at the end of the chapter.

As discussed in Chapter 2, the chi square test can determine if observed and expected data are in agreement. Therefore, we can use a chi square test to determine whether a population exhibits Hardy-Weinberg equilibrium for a particular gene. To do so, it is necessary to distinguish between the homozygotes and heterozygotes, either phenotypically or at the molecular level. This is necessary so that we can determine both the allele and genotype frequencies. As an example, let's consider a human blood type called the MN type. In this case, the blood type is determined by two codominant alleles, *M* and *N*. In an Inuit population in East

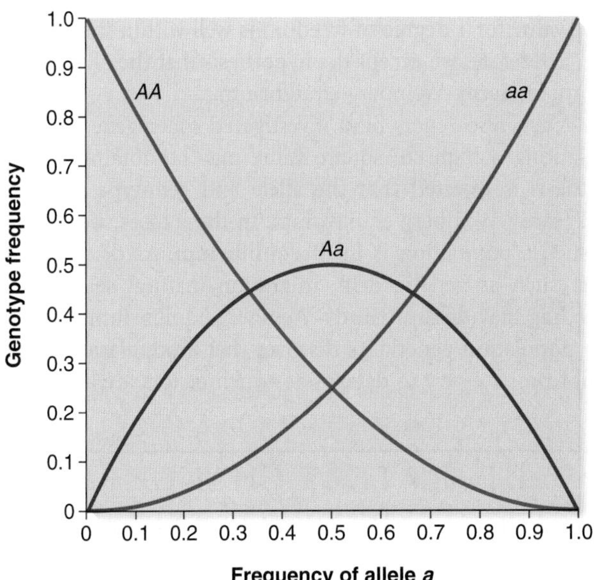

FIGURE 24.5 The relationship between allele frequencies and genotype frequencies according to the Hardy-Weinberg equilibrium. This graph assumes that *a* and *A* are the only two alleles for this gene.

Greenland, it was found that among 200 people, 168 were *MM*, 30 were *MN*, and 2 were *NN*. We can use these observed data to calculate the expected number of each genotype based on the Hardy-Weinberg equation.

$$\text{Allele frequency of } M = \frac{2(168) + 30}{400} = 0.915$$

$$\text{Allele frequency of } N = \frac{2(2) + 30}{400} = 0.085$$

Expected frequency of $MM = p^2 = (0.915)^2 = 0.837$

Expected number of MM individuals $= 0.837 \times 200 = 167.4$ (or 167 if rounded to the nearest individual)

Expected frequency of $NN = q^2 = (0.085)^2 = 0.007$

Expected number of NN individuals $= 0.007 \times 200 = 1.4$ (or 1 if rounded to the nearest individual)

Expected frequency of $MN = 2pq = 2(0.915)(0.085) = 0.156$

Expected number of MN individuals $= 0.156 \times 200 = 31.2$ (or 31 if rounded to the nearest individual)

$$\chi^2 = \frac{(O_1 - E_1)^2}{E_1} + \frac{(O_2 - E_2)^2}{E_2} + \frac{(O_3 - E_3)^2}{E_3}$$

$$\chi^2 = \frac{(168 - 167)^2}{167} + \frac{(30 - 31)^2}{31} + \frac{(2 - 1)^2}{1}$$

$$= 1.04$$

If we refer back to Table 2.1 in Chapter 2, the calculated chi square value for 1 degree of freedom is well within the acceptable range. Therefore, we accept the hypothesis that the alleles for this gene are in Hardy-Weinberg equilibrium.

When researchers have investigated other genes in various populations, a high chi square value may be obtained, and the hypothesis is rejected that the allele and genotype frequencies are in Hardy-Weinberg equilibrium. In these cases, we would say that such a population is in **disequilibrium.** As discussed next, factors such as genetic drift, migration, natural selection, and inbreeding may disrupt Hardy-Weinberg equilibrium. Therefore, when population geneticists discover that a population is not in equilibrium, they try to determine which factors are at work.

24.2 FACTORS THAT CHANGE ALLELE AND GENOTYPE FREQUENCIES IN POPULATIONS

The genetic variation in natural populations typically changes over the course of many generations. The term **microevolution** describes changes in a population's gene pool from generation to generation. Such change is rooted in two related phenomena (**Table 24.1**). First, the introduction of new genetic variation into a population is one essential aspect of microevolution. As discussed later in this chapter, gene variation can originate by a variety of molecular mechanisms. For example, new alleles of preexisting genes can arise by random mutations. Such events provide a continuous source of new variation to populations. However, due to their low rate of occurrence, mutations do not act as a major force in promoting widespread changes in a population. If mutations were the only type of change occurring in a population, that population would not evolve at a significant rate because mutations are so rare.

Microevolution also involves the action of evolutionary mechanisms that alter the prevalence of a given allele or genotype in a population. These mechanisms are random genetic drift, migration, natural selection, and nonrandom mating (see Table 24.1). The collective contributions of these evolutionary mechanisms over the course of many generations have the potential to promote widespread genetic changes in a population. In this section, we will examine how these mechanisms may affect the

TABLE 24.1

Factors that Govern Microevolution

Source of New Allelic Variation*	
Mutation	In this section, we will consider allelic variation. Random mutations within pre-existing genes introduce new alleles into populations, but at a very low rate. New mutations may be beneficial, neutral, or deleterious. For new alleles to rise to a significant percentage in a population, evolutionary mechanisms (i.e., random genetic drift, migration, natural selection) must operate on them.
Mechanisms That Alter Existing Genetic Variation	
Random genetic drift	This is a change in genetic variation from generation to generation due to random sampling error. Allele frequencies may change as a matter of chance from one generation to the next. This tends to have a greater impact in a small population.
Migration	Migration can occur between two different populations that have different allele frequencies. The introduction of migrants into a recipient population may change the allele frequencies of that population.
Natural selection	This is the phenomenon in which the environment selects for individuals that possess certain traits. Natural selection can favor the survival of members with beneficial traits or disfavor the survival of individuals with unfavorable traits.
Nonrandom mating	This is the phenomenon in which individuals select mates based on their phenotypes or genetic lineage. This can alter the relative proportion of homozygotes and heterozygotes predicted by the Hardy-Weinberg equation but will not change allele frequencies.

*Allelic variation is just one source of new genetic variation. Section 24.3 considers a variety of mechanisms through which new genetic variation can occur.

type of genetic variation that occurs when a gene exists in two or more alleles in a population. As you will learn, these mechanisms may cause a particular allele to be favored, or they may create a balance where two or more alleles are maintained in a population.

Mutations Provide the Source of Genetic Variation

As discussed in Chapters 8 and 16, mutations involve changes in gene sequences, chromosome structure, and/or chromosome number. Mutations are random events that occur spontaneously at a low rate or are caused by mutagens at a higher rate. In 1926, the Russian geneticist Sergei Chetverikov was the first to suggest that mutational variability provides the raw material for evolution but does not constitute evolution itself. In other words, mutation can provide new alleles to a population but does not substantially alter allele frequencies. Chetverikov proposed that populations in nature absorb mutations like a sponge and retain them in a heterozygous condition, thereby providing a source of variability for future change.

Population geneticists often consider how new mutations affect the survival and reproductive potential of the individual that inherits them. A new mutation may be deleterious, neutral, or beneficial. For genes that encode proteins, the effects of new mutations will depend on their impact on protein function. Deleterious and neutral mutations are far likelier to occur than beneficial mutations. For example, alleles can be altered in many different ways that render an encoded protein defective. As discussed in Chapter 16, deletions and point mutations such as frameshift mutations, missense mutations, and nonsense mutations all may cause a gene to express a protein that is nonfunctional or less functional than the wild-type protein. Also, mutations in noncoding regions can alter gene expression (refer back to Table 16.3). Neutral mutations can also occur in several different ways. For example, a neutral mutation can change the wobble base without affecting the amino acid sequence of the encoded protein, or it can be a missense mutation that has no effect on protein function. Such point mutations occur at specific sites within the coding sequence. Neutral mutations can also occur within introns, the noncoding sequences of genes. By comparison, beneficial mutations are relatively uncommon. To be advantageous, a new mutation could alter the amino acid sequence of a protein to yield a better-functioning product. While such mutations do occur, they are expected to be very rare for a population in a stable environment.

The **mutation rate** is defined as the probability that a gene will be altered by a new mutation. The rate is typically expressed as the number of new mutations in a given gene per generation. A common value for the mutation rate is in the range of 1 in 100,000 to 1 in 1,000,000, or 10^{-5} to 10^{-6} per generation. However, mutation rates vary depending on species, cell types, chromosomal location, and gene size. Furthermore, in experimental studies, the mutation rate is usually measured by following the change of a normal (functional) gene to a deleterious (nonfunctional) allele. The mutation rate producing beneficial alleles is expected to be substantially less.

It is clear that new mutations provide genetic variability, but population geneticists also want to know how much the mutation rate affects the allele frequencies in a population. Can random mutations have a large impact on allele frequencies over time? To answer this question, let's take the simple case where a gene exists in a functional allele, *A*; the allele frequency of *A* is denoted by the variable *p*. A deleterious mutation can convert the *A* allele into a nonfunctional allele, *a*. The allele frequency of *a* is designated by *q*. The conversion of the *A* allele into the *a* allele by mutation will occur at a rate that is designated μ. If we assume that the rate of the reverse mutation (*a* to *A*) is negligible, the increase in the frequency of the *a* allele after one generation will be

$$\Delta q = \mu p$$

For example, let's consider the following conditions:

$p = 0.8$ (i.e., frequency of *A* is 80%)

$q = 0.2$ (i.e., frequency of *a* is 20%)

$\mu = 10^{-5}$ (i.e., the mutation rate of converting *A* to *a*)

$\Delta q = (10^{-5})(0.8) = (0.00001)(0.8) = 0.000008$

Therefore, in the next generation (designated $n + 1$),

$q_{n+1} = 0.2 + 0.000008 = 0.200008$

$p_{n+1} = 0.8 - 0.000008 = 0.799992$

As we can see from this calculation, new mutations do not significantly alter the allele frequencies in a single generation.

We can use the following equation to calculate the change in allele frequency after any number of generations:

$$(1 - \mu)^t = \frac{p_t}{p_0}$$

where

μ is the mutation rate of the conversion of *A* to *a*

t is the number of generations

p_0 is the allele frequency of *A* in the starting generation

p_t is the allele frequency of *A* after *t* generations

As an example, let's suppose that the allele frequency of *A* is 0.8, $\mu = 10^{-5}$, and we want to know what the allele frequency will be after 1,000 generations ($t = 1{,}000$). Plugging these values into the preceding equation and solving for p_t,

$$(1 - 0.00001)^{1{,}000} = \frac{p_t}{0.8}$$

$$p_t = 0.792$$

Therefore, after 1,000 generations, the frequency of *A* has dropped only from 0.8 to 0.792. Again, these results point to how slowly the occurrence of new mutations will change allele frequencies. In natural populations, the rate of new mutation is rarely a significant catalyst in shaping allele frequencies. Instead, other processes such as genetic drift, migration, and natural selection have far greater effects on allele frequencies. Next, we will examine how these other factors work.

In Small Populations, Allele Frequencies Can Be Altered by Random Genetic Drift

In the 1930s, geneticist Sewall Wright played a key role in developing the concept of **random genetic drift,** which refers to changes in allele frequencies in a population due to chance fluctuations. As a matter of random chance, the frequencies of alleles found in gametes that unite to form zygotes will vary from generation to generation. Over the long run, genetic drift favors either the loss of an allele or its fixation at 100% in the population. The rate at which this occurs depends on the population size and on the initial allele frequencies. Figure 24.6 illustrates the potential consequences of genetic drift in one large ($N = 1{,}000$) and five small ($N = 20$) populations. At the beginning of this hypothetical simulation, all of these populations have identical allele frequencies: $A = 0.5$ and $a = 0.5$. In the five small populations, this allele frequency fluctuates substantially from generation to generation. Eventually, one of the alleles is eliminated and the other is fixed at 100%. At this point, the allele has become monomorphic and cannot fluctuate any further. By comparison, the allele frequencies in the large population fluctuate much less, because random sampling error is expected to have a smaller impact. Nevertheless, genetic drift will lead to homozygosity even in large populations, but this will take many more generations to occur.

Now let's ask two questions:

1. How many new mutations do we expect in a natural population?
2. How likely is it that any new mutation will be either fixed in, or eliminated from, a population due to random genetic drift?

With regard to the first question, the average number of new mutations depends on the mutation rate (μ) and the number of individuals in a population (N). If each individual has two copies of the gene of interest, the expected number of new mutations in this gene is

$$\text{Expected number of new mutations} = 2N\mu$$

From this, we see that a new mutation is more likely to occur in a large population than in a small one. This makes sense, because the larger population has more copies of the gene to be mutated. With regard to the second question, the probability of fixation of a newly arising allele due to genetic drift is

$$\text{Probability of fixation} = \frac{1}{2N} \quad \text{(assuming equal numbers of males and females contribute to the next generation)}$$

In other words, the probability of fixation is the same as the initial allele frequency in the population. For example, if $N = 20$, the probability of fixation of a new allele equals $1/(2 \times 20)$, or 2.5%. Conversely, a new allele may be lost from the population.

$$\text{Probability of elimination} = 1 - \text{probability of fixation}$$
$$= 1 - \frac{1}{2N}$$

If $N = 20$, the probability of elimination equals $1 - 1/(2 \times 20)$, or 97.5%. As you may have noticed, the value of N has opposing effects with regard to new mutations and their eventual fixation in a population. When N is very large, new mutations are much more likely to occur. Each new mutation, however, has a greater chance of being eliminated from the population due to random genetic drift. On the other hand, when N is small, the probability of new mutations is also small, but if they occur, the likelihood of fixation is relatively large.

Now that we have an appreciation for the phenomenon of genetic drift, we can ask a third question:

3. If fixation of a new allele does occur, how many generations is it likely to take?

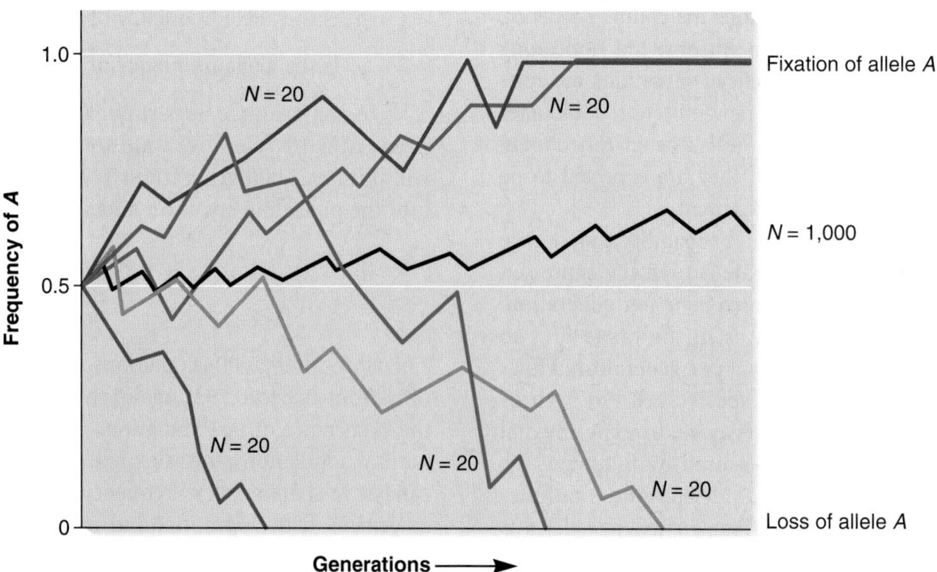

FIGURE 24.6 **A hypothetical simulation of random genetic drift.** In all cases, the starting allele frequencies are $A = 0.5$ and $a = 0.5$. The colored lines illustrate five populations in which $N = 20$; the black line shows a population in which $N = 1{,}000$.

The formula for calculating this also depends on the number of individuals in the population:

$$\bar{t} = 4N$$

where

$\bar{t}$ equals the average number of generations to achieve fixation

N equals the number of individuals in the population, assuming that males and females contribute equally to each succeeding generation

As you may have expected, allele fixation will take much longer in large populations. If a population has 1 million breeding members, it will take, on average, 4 million generations, perhaps an insurmountable period of time, to reach fixation. In a small group of 100 individuals, however, fixation will take only 400 generations, on average. As discussed in Chapter 26, the drifting of neutral alleles among different populations and species provides a way to measure the rate of evolution and can be used to determine evolutionary relationships.

The preceding discussion of random genetic drift has emphasized two important points. First, genetic drift ultimately operates in a random manner with regard to allele frequency and, over the long run, leads to either allele fixation or elimination. The process is random with regard to particular alleles. Genetic drift can lead to the fixation of deleterious, neutral, or beneficial alleles. A second important feature of genetic drift is that its impact is greatly affected by population size. Genetic drift may lead more quickly to allele loss or fixation in a small population.

In nature, there are different ways that geography and population size influence how genetic drift affects the genetic composition of a species. Some species occupy wide ranges in which small, local populations become geographically isolated from the rest of the species. The allele frequencies within these small populations are more susceptible to genetic drift. Because this is a random process, small isolated populations tend to be more genetically disparate in relation to other populations.

Changes in population size may influence genetic drift via the **bottleneck effect.** In nature, a population can be reduced dramatically in size by events such as earthquakes, floods, drought, or human destruction of habitat. Such events may randomly eliminate most of the members of the population without regard to genetic composition. The time period of the bottleneck, when the population size is very small, may be greatly influenced by genetic drift because the surviving members may have allele frequencies that differ from those of the original population. In addition, allele frequencies are expected to drift substantially during the generations when the population size is small. In extreme cases, alleles may even be eliminated. Eventually, the bottlenecked population may regain its original size (**Figure 24.7**). However, the new population will have less genetic variation than the original large population. As an example, the African cheetah population lost a substantial amount of its genetic variation due to a bottleneck effect. DNA analysis by population geneticists has suggested that a severe bottleneck occurred approximately 10,000 to 12,000 years ago when the population size was dramatically reduced. The population eventually rebounded, but the bottleneck significantly decreased the genetic variation.

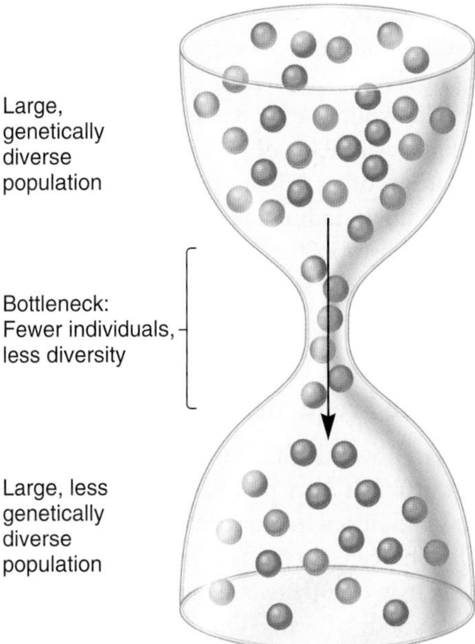

(a) Bottleneck effect

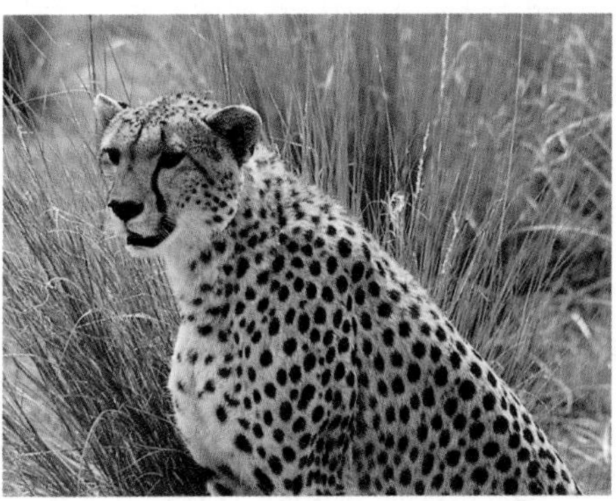

(b) An African cheetah

FIGURE 24.7 **The bottleneck effect, an example of genetic drift.** **(a)** A representation of the bottleneck effect. Note that the genetic variation denoted by the green balls has been lost. **(b)** The African cheetah. The modern species has low genetic variation due to a genetic bottleneck that is thought to have occurred about 10,000 to 12,000 years ago.

Geography and population size may also influence genetic drift via the **founder effect.** Compared to the bottleneck effect, the key difference is that the founder effect involves migration; a small group of individuals separates from a larger population and establishes a colony in a new location. For example, a few individuals may migrate from a large continental population and become the founders of an island population. The founder effect has two important consequences. First, the founding population is expected to have less genetic variation than the original population from which it was derived. Second, as a matter of chance, the allele frequencies in the founding population may differ markedly from those of the original population.

Population geneticists have studied many examples where isolated populations have been started from a few members of another population. In the 1960s, Victor McKusick studied allele frequencies in the Old Order Amish of Lancaster County, Pennsylvania. At that time, this was a group of about 8,000 people, descended from just three couples that immigrated to the United States in 1770. Among this population of 8,000, a genetic disease known as the Ellis-van Creveld syndrome (a recessive form of dwarfism) was found at a frequency of 0.07, or 7%. By comparison, this disorder is extremely rare in other human populations, even the population from which the founding members had originated. The high frequency of dwarfism in the Lancaster County population is a chance occurrence due to the founder effect.

Migrations Between Two Populations Can Alter Allele Frequencies

We have just seen how migration to a new location by a relatively small group can result in a population with an altered genetic composition. In addition, migration between two different established populations can alter allele frequencies. For example, a species of birds may occupy two geographic regions that are separated by a large body of water. On rare occasions, the prevailing winds may allow birds from the western population to fly over this body of water and become members of the eastern population. If the two populations have different allele frequencies and if migration occurs in sufficient numbers, this may alter the allele frequencies in the eastern population.

After migration has occurred, the new (eastern) population is called a conglomerate. To calculate the allele frequencies in the conglomerate, we need two kinds of information. First, we must know the original allele frequencies in the donor and recipient populations. Second, we must know the proportion of the conglomerate population that is due to migrants. With these data, we can calculate the change in allele frequency in the conglomerate population using the following equation:

$$\Delta p_C = m(p_D - p_R)$$

where

Δp_C is the change in allele frequency in the conglomerate population

p_D is the allele frequency in the donor population

p_R is the allele frequency in the original recipient population

m is the proportion of migrants that make up the conglomerate population

$$m = \frac{\text{number of migrants in the conglomerate population}}{\text{total number of individuals in the conglomerate population}}$$

As an example, let's suppose the allele frequency of *A* is 0.7 in the donor population and 0.3 in the recipient population. A group of 20 individuals migrates and joins the recipient population, which originally had 80 members. Thus,

$$m = \frac{20}{20 + 80} = 0.2$$

$$\Delta p_C = m(p_D - p_R) = 0.2(0.7 - 0.3) = 0.08$$

We can now calculate the allele frequency in the conglomerate:

$$p_C = p_R + \Delta p_C = 0.3 + 0.08 = 0.38$$

Therefore, in the conglomerate population, the allele frequency of *A* has changed from 0.3 (its value before migration) to 0.38. This increase in allele frequency arises from the higher allele frequency of *A* in the donor population. **Gene flow** occurs whenever individuals migrate between populations having different allele frequencies, and the migrants are able to breed successfully with the members of the recipient population. Gene flow depends not only on migration, but also on the ability of the migrants' alleles to be passed to subsequent generations.

In our previous example, we considered the consequences of a unidirectional migration from a donor to a recipient population. In nature, it is common for individuals to migrate in both directions. What are the main consequences of bidirectional migration? Depending on its rate, migration tends to reduce differences in allele frequencies between neighboring populations. In fact, population geneticists can analyze allele frequencies in two different populations to evaluate the rate of migration between them. Populations that frequently mix their gene pools via migration tend to have similar allele frequencies, whereas isolated populations are expected to be more disparate. In addition, migration can enhance genetic diversity within a population. As discussed earlier, new mutations are relatively rare events. Therefore, a particular mutation may arise only in one population. Migration may then introduce this new allele into neighboring populations.

Natural Selection Is Based on the Relative Reproductive Success of Genotypes

In the 1850s, Charles Darwin and Alfred Russel Wallace independently proposed the theory of **natural selection.** We will discuss the phenotypic consequences of natural selection in greater detail in Chapter 26. According to this theory, the conditions found in nature result in the selective survival and reproduction of individuals whose characteristics make them well adapted to their environment. These surviving individuals are more likely to reproduce and contribute offspring to the next generation. Natural selection can be related not only to differential survival but also to mating efficiency and fertility.

A modern restatement of the principles of natural selection can relate our knowledge of molecular genetics to the phenotypes of individuals.

1. Within a population, allelic variation arises in various ways, such as through random mutations that cause differences in DNA sequences. A mutation that creates a new allele may alter the amino acid sequence of the

encoded protein, which, in turn, may alter the function of the protein.

2. Some alleles may encode proteins that enhance an individual's survival or reproductive capability compared to that of other members of the population. For example, an allele may produce a protein that is more efficient at a higher temperature, conferring on the individual a greater probability of survival in a hot climate.
3. Individuals with beneficial alleles are more likely to survive and contribute to the gene pool of the next generation.
4. Over the course of many generations, allele frequencies of many different genes may change through this process, thereby significantly altering the characteristics of a species. The net result of natural selection is a population that is better adapted to its environment and more successful at reproduction.

As mentioned at the beginning of the chapter, Fisher, Wright, and Haldane developed mathematical relationships to explain the theory of natural selection. As our knowledge of the process of natural selection has increased, it has become apparent that it operates in many different ways. In this chapter, we will consider a few examples of natural selection involving a single trait or a single gene that exists in two alleles. In reality, however, natural selection acts on populations of individuals in which many genes are polymorphic and each individual contains thousands or tens of thousands of different genes.

To begin our quantitative discussion of natural selection, we must examine the concept of **Darwinian fitness**—the relative likelihood that a genotype will contribute to the gene pool of the next generation as compared to other genotypes. Natural selection acts on phenotypes that are derived from individuals' genotypes. Although Darwinian fitness often correlates with physical fitness, the two ideas should not be confused. Darwinian fitness is a measure of reproductive superiority. An extremely fertile genotype may have a higher Darwinian fitness than a less fertile genotype that appears more physically fit.

To consider Darwinian fitness, let's use our example of a gene existing in the *A* and *a* alleles. If the three genotypes have the same level of mating success and fertility, we can assign fitness values to each of the three genotype classes based on their likelihood of surviving to reproductive age. For example, let's suppose that the relative survival to adulthood of each of the three genotype classes is as follows: For every five *AA* individuals that survive, four *Aa* individuals survive, and one *aa* individual survives. By convention, the genotype with the highest reproductive ability is given a fitness value of 1.0. Relative fitness values are denoted by the variable *W*. The fitness values of the other genotypes are assigned values relative to this 1.0 value:

$$\text{Fitness of } AA\text{: } W_{AA} = 1.0$$

$$\text{Fitness of } Aa\text{: } W_{Aa} = 4/5 = 0.8$$

$$\text{Fitness of } aa\text{: } W_{aa} = 1/5 = 0.2$$

Keep in mind that differences in reproductive achievement among genotypes may stem from various reasons. In this case, the fittest genotype is more likely to survive to reproductive age. In other situations, the most fit genotype is more likely to mate. For example, a bird with brightly colored feathers may have an easier time attracting a mate than a bird with duller plumage. Finally, a third possibility is that the fittest genotype may be more fertile. It may produce a higher number of gametes or gametes that are more successful at fertilization.

By studying species in their native environments, population geneticists have discovered that natural selection can occur in several ways. The patterns of natural selection depend on the relative fitness values of the different genotypes and on the variation of environmental effects. The four patterns of natural selection that we will consider are called directional, stabilizing, disruptive, and balancing selection. In most of the examples described next, natural selection leads to adaptation so that a species is better able to survive to reproductive age.

Directional Selection Favors the Extreme Phenotype

Directional selection favors individuals at one extreme of a phenotypic distribution that are more likely to survive and reproduce in a particular environment. Different phenomena may initiate the process of directional selection. One way that directional selection may arise is that a new allele may be introduced into a population by mutation, and the new allele may promote a higher fitness in individuals that carry it (**Figure 24.8**). If the homozygote carrying the favored allele has the highest fitness value, directional selection may cause this favored allele to eventually become the predominant allele in the population, perhaps even becoming a monomorphic allele.

Another possibility is that a population may be exposed to a prolonged change in its living environment. Under the new environmental conditions, the relative fitness values may change to favor one genotype, and this will promote the elimination of other genotypes. As an example, let's suppose a population of finches on the mainland already has genetic variation in beak size. A small number of birds migrate to an island where the seeds are generally larger than they are on the mainland. In this new environment, birds with larger beaks would have a higher fitness because they would be better able to crack open the larger seeds and thereby survive to reproductive age. Over the course of many generations, directional selection would produce a population of birds carrying alleles that promote larger beak size.

In the case of directional selection, allele frequencies may change in a step-by-step, generation-per-generation way. To appreciate how this occurs, let's take a look at how fitness affects the Hardy-Weinberg equilibrium and allele frequencies. Again, let's suppose a gene exists in two alleles, *A* and *a*. The three fitness values, which are based on relative survival levels, are

$$W_{AA} = 1.0$$

$$W_{Aa} = 0.8$$

$$W_{aa} = 0.2$$

In the next generation, we expect that the Hardy-Weinberg equilibrium will be modified in the following way due to directional selection:

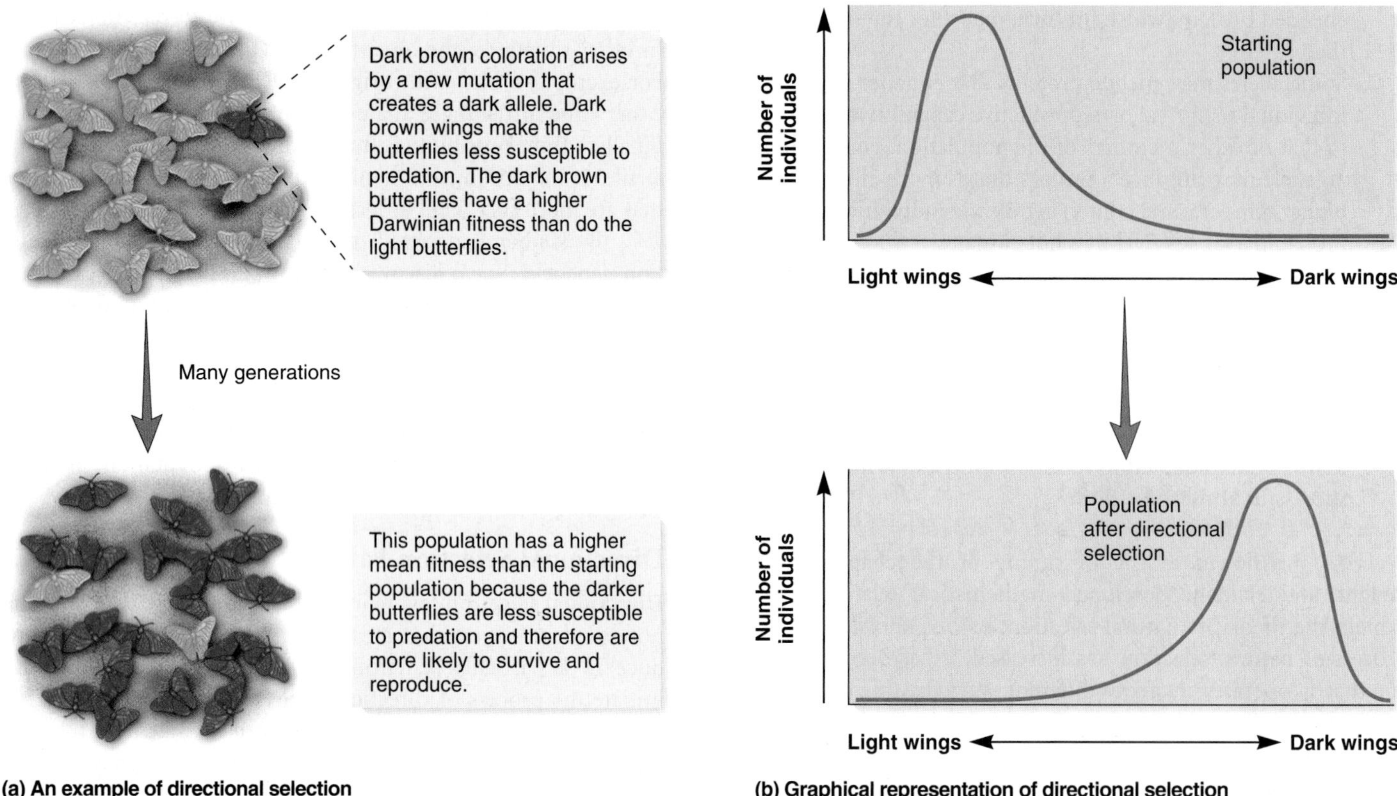
Dark brown coloration arises by a new mutation that creates a dark allele. Dark brown wings make the butterflies less susceptible to predation. The dark brown butterflies have a higher Darwinian fitness than do the light butterflies.
Many generations
This population has a higher mean fitness than the starting population because the darker butterflies are less susceptible to predation and therefore are more likely to survive and reproduce.
(a) An example of directional selection
Number of individuals
Starting population
Light wings
Dark wings
Number of individuals
Population after directional selection
Light wings
Dark wings
(b) Graphical representation of directional selection

FIGURE 24.8 **Directional selection.** (**a**) A new mutation arises in a population that confers higher Darwinian fitness. In this example, butterflies with dark wings are more likely to survive and reproduce. Over many generations, directional selection will favor the prevalence of darker individuals. (**b**) A graphical representation of directional selection.

Frequency of *AA*: p^2W_{AA}

Frequency of *Aa*: $2pqW_{Aa}$

Frequency of *aa*: q^2W_{aa}

In a population that is changing due to natural selection, these three terms may not add up to 1.0, as they would in the Hardy-Weinberg equilibrium. Instead, the three terms sum to a value known as the **mean fitness of the population** ($\overline{W}$):

$$p^2W_{AA} + 2pqW_{Aa} + q^2W_{aa} = \overline{W}$$

Dividing both sides of the equation by the mean fitness of the population,

$$\frac{p^2W_{AA}}{\overline{W}} + \frac{2pqW_{Aa}}{\overline{W}} + \frac{q^2W_{aa}}{\overline{W}} = 1$$

Using this equation, we can calculate the expected genotype and allele frequencies after one generation of directional selection:

Frequency of *AA* genotype: $\frac{p^2W_{AA}}{\overline{W}}$

Frequency of *Aa* genotype: $\frac{2pqW_{Aa}}{\overline{W}}$

Frequency of *aa* genotype: $\frac{q^2W_{aa}}{\overline{W}}$

Allele frequency of *A*: $p_A = \frac{p^2W_{AA}}{\overline{W}} + \frac{pqW_{Aa}}{\overline{W}}$

Allele frequency of *a*: $q_a = \frac{q^2W_{aa}}{\overline{W}} + \frac{pqW_{Aa}}{\overline{W}}$

As an example, let's suppose that the starting allele frequencies are $A = 0.5$ and $a = 0.5$, and use fitness values of 1.0, 0.8, and 0.2 for the three genotypes, *AA*, *Aa*, and *aa*, respectively. We begin by calculating the mean fitness of the population:

$$p^2W_{AA} + 2pqW_{Aa} + q^2W_{aa} = \overline{W}$$

$$\overline{W} = (0.5)^2(1) + 2(0.5)(0.5)(0.8) + (0.5)^2(0.2)$$

$$\overline{W} = 0.25 + 0.4 + 0.05 = 0.7$$

After one generation of directional selection,

Frequency of *AA* genotype: $\frac{p^2W_{AA}}{\overline{W}} = \frac{(0.5)^2(1)}{0.7} = 0.36$

Frequency of *Aa* genotype: $\frac{2pqW_{Aa}}{\overline{W}} = \frac{2(0.5)(0.5)(0.8)}{0.7} = 0.57$

Frequency of *aa* genotype: $\frac{q^2W_{aa}}{\overline{W}} = \frac{(0.5)^2(0.2)}{0.7} = 0.07$

Allele frequency of *A*: $p_A = \frac{p^2W_{AA}}{\overline{W}} + \frac{pqW_{Aa}}{\overline{W}}$

$$= \frac{(0.5)^2(1)}{0.7} + \frac{(0.5)(0.5)(0.8)}{0.7} = 0.64$$

Allele frequency of *a*: $q_a = \frac{q^2W_{aa}}{\overline{W}} + \frac{pqW_{Aa}}{\overline{W}}$

$$= \frac{(0.5)^2(0.2)}{0.7} + \frac{(0.5)(0.5)(0.8)}{0.7} = 0.36$$

After one generation, the allele frequency of *A* has increased from 0.5 to 0.64, while the frequency of *a* has decreased from 0.5 to 0.36. This has occurred because the *AA* genotype has the highest fitness, while the *Aa* and *aa* genotypes have lower fitness values. Another interesting feature of natural selection is that it raises the mean fitness of the population. If we assume the individual fitness values are constant, the mean fitness of this next generation is

$$\begin{aligned}\overline{W} &= p^2W_{AA} + 2pqW_{Aa} + q^2W_{aa}\\ &= (0.64)^2(1) + 2(0.64)(0.36)(0.8) + (0.36)^2(0.2)\\ &= 0.80\end{aligned}$$

The mean fitness of the population has increased from 0.7 to 0.8.

What are the consequences of natural selection at the population level? This population is better adapted to its environment than the previous one. Another way of viewing this calculation is that the subsequent population has a greater reproductive potential than the previous one. We could perform the same types of calculations to find the allele frequencies and mean fitness value in the next generation. If we assume the individual fitness values remain constant, the frequencies of *A* and *a* in the next generation are 0.85 and 0.15, respectively, and the mean fitness increases to 0.931. As we can see, the general trend is to increase *A*, decrease *a*, and increase the mean fitness of the population.

In the previous example, we considered the effects of natural selection by beginning with allele frequencies at intermediate levels (namely, $A = 0.5$ and $a = 0.5$). **Figure 24.9** illustrates what would happen if a new mutation introduced the *A* allele into a population that was originally monomorphic for the *a* allele. As before, the *AA* homozygote has a fitness of 1.0, the *Aa* heterozygote 0.8, and the recessive *aa* homozygote 0.2. Initially, the *A* allele is at a very low frequency in the population. If it is not lost initially due to genetic drift, its frequency slowly begins to rise and then, at intermediate values, rises much more rapidly.

Eventually, this type of natural selection may lead to fixation of a beneficial allele. However, a new beneficial allele is in a

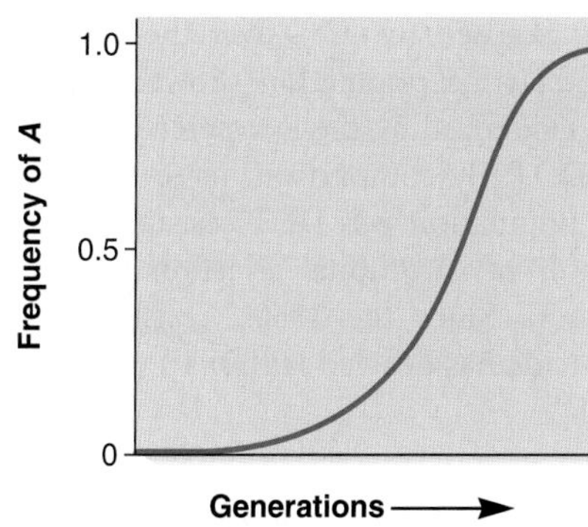

FIGURE 24.9 **The fate of a beneficial allele that is introduced as a new mutation into a population.** A new allele (*A*) is beneficial in the homozygous condition: $W_{AA} = 1.0$. The heterozygote, *Aa* ($W_{Aa} = 0.8$), and homozygote, *aa* ($W_{aa} = 0.2$), have lower fitness values.

precarious situation when its frequency is very low. As we have seen, random genetic drift is likely to eliminate new mutations, even beneficial ones, due to chance fluctuations.

Researchers have identified many examples of directional selection in nature. As mentioned in Chapter 6, resistance to antibiotics is a growing concern among members of the medical profession. The selection of bacterial strains resistant to one or more antibiotics typically occurs in a directional manner. Similarly, the resistance of insects to pesticides, such as DDT (dichlorodiphenyltrichloroethane), occurs in a directional manner. DDT usage began in the 1940s as a way to decrease the populations of mosquitoes and other insects. However, certain insect species can become resistant to DDT by a dominant mutation in a single enzyme-encoding gene. The mutant enzyme detoxifies DDT, making it harmless to the insect. **Figure 24.10** shows the results of an experiment in which mosquito larva (*Aedes*

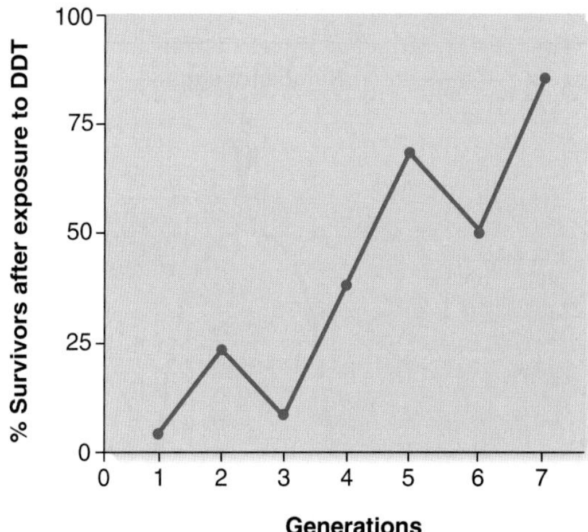

FIGURE 24.10 **Directional selection for DDT-resistance in a mosquito population.** In this experiment, mosquito larvae (*A. aegypti*) were exposed to 10 mg/liter of DDT. The percentage of survivors was recorded, and then the survivors of each generation were used as parents for the next generation.

aegypti) were exposed to DDT over the course of seven generations. The starting population showed a low level of DDT resistance, as evidenced by the low percentage of survivors after exposure to DDT. By comparison, in seven generations, nearly 100% of the population was DDT resistant. These results illustrate the power of directional selection in promoting change in a population. Since the 1950s, resistance to nearly every known insecticide has evolved within 10 years of its commercial introduction!

Stabilizing Selection Favors Individuals with Intermediate Phenotypes

In **stabilizing selection,** the extreme phenotypes for a trait are selected against, and those individuals with the intermediate phenotypes have the highest fitness values. Stabilizing selection tends to decrease genetic diversity for a particular gene because it eliminates alleles that cause a greater variation in phenotypes. An example of stabilizing selection involves clutch size in birds, which was first proposed by British biologist David Lack in 1947. Under stabilizing selection, birds that lay too many or too few eggs will have lower fitness values compared to those that lay an intermediate value (**Figure 24.11**). Laying too many eggs means that many offspring will die due to inadequate parental care and food. In addition, the strain on the parents themselves may decrease their likelihood of survival and therefore their ability to produce more offspring. Having too few offspring, on the other hand, does not contribute many individuals to the next generation. Therefore, the most successful parents are those that produce an intermediate clutch size. In the 1980s, Swedish evolutionary biologist Lars Gustafsson and colleagues examined the phenomenon of stabilizing selection in the collared flycatcher, *Ficedula albicollis*, on the island of Gotland, which is southeast of the mainland of Sweden. They discovered that Lack's hypothesis that clutch size is subject to the action of stabilizing selection also appears to be true for this species.

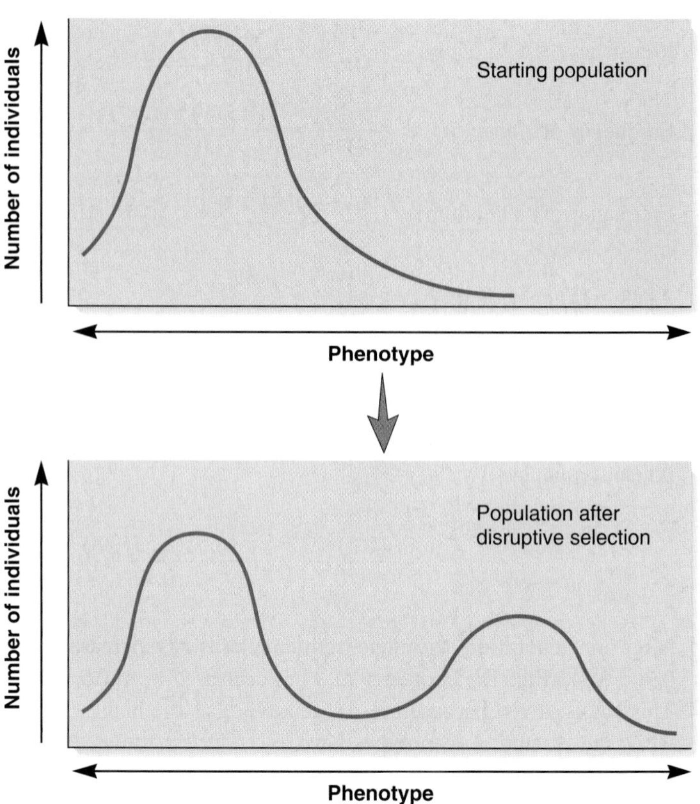

FIGURE 24.12 **Disruptive selection.** Over time, this form of selection favors two or more phenotypes due to heterogeneous environments.

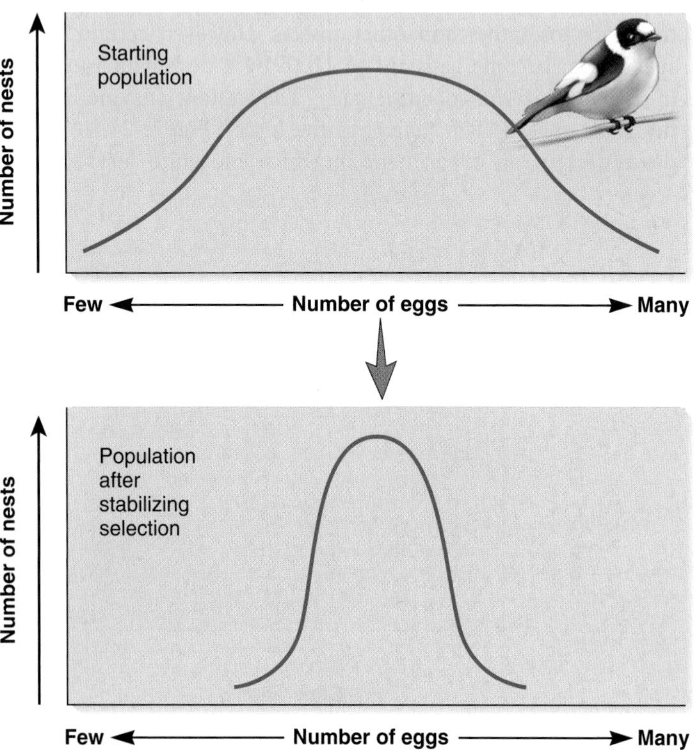

FIGURE 24.11 **Stabilizing selection.** In this pattern of natural selection, the extremes of a phenotypic distribution are selected against. Those individuals with intermediate traits have the highest fitness. This results in a population with less diversity and more uniform traits.

Disruptive Selection Favors Multiple Phenotypes

Disruptive selection, also known as diversifying selection, favors the survival of two or more different genotypes that produce different phenotypes (**Figure 24.12**). In disruptive selection, the fitness values of a particular genotype are higher in one environment and lower in a different environment. Disruptive selection is likely to occur in populations that occupy diverse environments so that some members of the species will survive in each type of environmental condition.

As an example, **Figure 24.13a** shows a photograph of land snails, *Cepaea nemoralis*, that live in woods and open fields. This snail is polymorphic in color and banding patterns. In 1954, Arthur Cain and Phillip Sheppard found that snail color was correlated with the environment. As shown in **Figure 24.13b**, the highest frequency of brown shell color was found in snails in the

(a) Land snails

Habitat	*Brown*	*Pink*	*Yellow*
Beechwoods	0.23	0.61	0.16
Deciduous woods	0.05	0.68	0.27
Hedgerows	0.05	0.31	0.64
Rough herbage	0.004	0.22	0.78

(b) Frequency of snail color

FIGURE 24.13 **Polymorphism in the land snail, *Cepaea nemoralis.*** (**a**) This species of snail can exist in several different colors and banding patterns. (**b**) Coloration of the snails is correlated with the specific environments where they are located.

Genes→Traits Snail coloration is an example of genetic polymorphism due to heterogeneous environments; the genes governing shell coloration are polymorphic. The predation of snails is correlated with their ability to be camouflaged in their natural environment. Snails with brown shells are most prevalent in beechwoods, where the soil is dark. Pink snails are most abundant in the leaf litter of beechwoods and deciduous woods. Yellow snails are most prevalent in more sunny locations, such as hedgerows and rough herbage.

beechwoods, where there are wide expanses of dark soil. Their frequency was substantially less in other environments. By comparison, pink snails are most common in the leaf litter of forest floors, and the yellow snails are most abundant in the sunny, grassy areas of hedgerows and rough herbage. Researchers have suggested that this disruptive selection can be explained by different levels of predation by thrushes. Depending on the environment, certain snail phenotypes may be more easily seen by their predators compared to others. Migration can occasionally occur between the snail populations, which keeps the polymorphism in balance among these different environments.

Balanced Polymorphisms May Occur Due to Heterozygote Advantage or Negative Frequency-Dependent Selection

As we have just seen, polymorphisms may occur when a species occupies a diverse environment. Researchers have discovered other patterns of natural selection that favor the maintenance of two or more alleles in a more homogeneous environment. This pattern, called **balancing selection,** results in a genetic polymorphism in a population.

For genetic variation involving a single gene, balancing selection may arise when the heterozygote has a higher fitness than either corresponding homozygote, a situation called **heterozygote advantage.** In this case, an equilibrium is reached in which both alleles are maintained in the population. If the fitness values are known for each of the genotypes, the allele frequencies at equilibrium can be calculated. To do so, we must consider the **selection coefficient (*s*),** which measures the degree to which a genotype is selected against.

$$s = 1 - W$$

By convention, the genotype with the highest fitness has an s value of zero. Genotypes at a selective disadvantage have s values that are greater than zero but less than or equal to 1.0. An extreme case is a recessive lethal allele. It would have an s value of 1.0 in the homozygote, while the s value in the heterozygote could be zero.

Let's consider the following case of relative fitness, where

$$W_{AA} = 0.7$$

$$W_{Aa} = 1.0$$

$$W_{aa} = 0.4$$

The selection coefficients are

$$s_{AA} = 1 - 0.7 = 0.3$$

$$s_{Aa} = 1 - 1.0 = 0$$

$$s_{aa} = 1 - 0.4 = 0.6$$

The population will reach an equilibrium when

$$s_{AA}p = s_{aa}q$$

If we take this equation, let $q = 1 - p$, and then solve for p:

$$p = \text{Allele frequency of } A = \frac{s_{aa}}{s_{AA} + s_{aa}}$$

$$= \frac{0.6}{0.3 + 0.6} = 0.67$$

If we let $p = 1 - q$ and then solve for q:

$$q = \text{Allele frequency of } a = \frac{s_{AA}}{s_{AA} + s_{aa}}$$

$$= \frac{0.3}{0.3 + 0.6} = 0.33$$

In this example, balancing selection maintains the two alleles in the population at frequencies in which A equals 0.67 and a equals 0.33.

Heterozygote advantage can sometimes explain the high frequency of alleles that are deleterious in a homozygous condition. A classic example is the Hb^S allele of the human β-globin gene. A homozygous Hb^SHb^S individual displays sickle-cell disease, a disorder that leads to the sickling of the red blood cells. The Hb^SHb^S homozygote has a lower fitness than a homozygote with two normal copies of the β-globin gene, Hb^AHb^A. However, the heterozygote, Hb^AHb^S, has a higher level of fitness than either

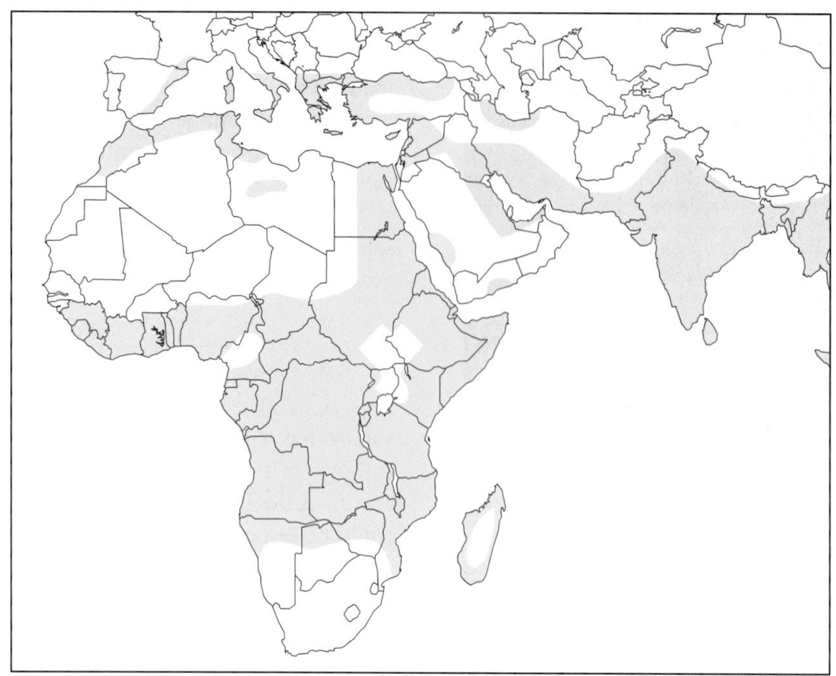

(a) Malaria prevalence

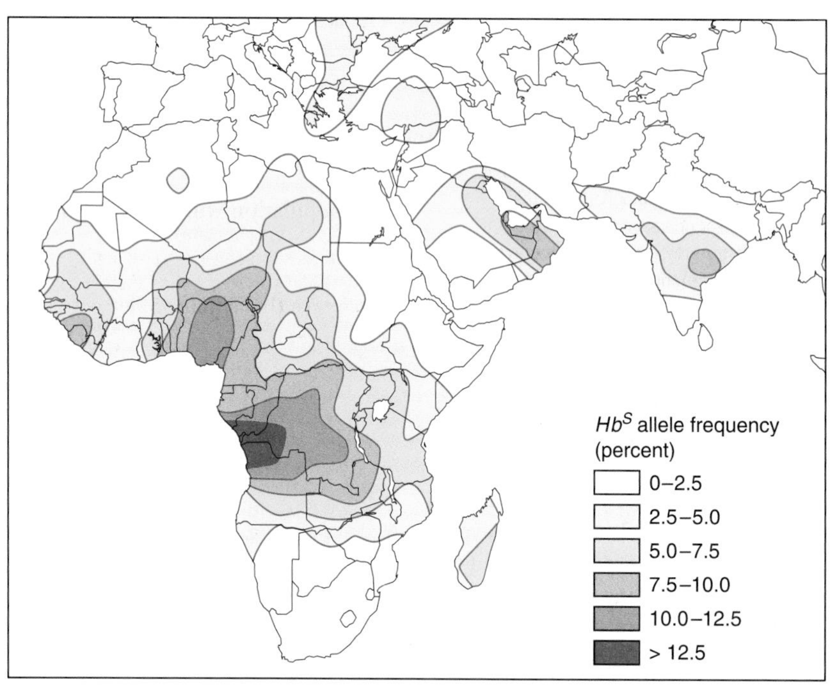

(b) Hb^S allele frequency

FIGURE 24.14 The geographic relationship between malaria and the frequency of the sickle-cell allele in human populations. (**a**) The geographic prevalence of malaria in Africa and surrounding areas. (**b**) The frequency of the Hb^S allele in the same areas.

Genes→Traits The sickle-cell allele of the β-globin gene is maintained in human populations as a balanced polymorphism. In areas where malaria is prevalent, the heterozygote carrying one copy of the Hb^S allele has a greater fitness than either of the corresponding homozygotes (Hb^AHb^A and Hb^SHb^S). Therefore, even though the Hb^SHb^S homozygotes suffer the detrimental consequences of sickle-cell disease, this negative aspect is balanced by the beneficial effects of malarial resistance in the heterozygotes.

homozygote in areas where malaria is endemic (**Figure 24.14**). Compared to Hb^AHb^A homozygotes, heterozygotes have a 10 to 15% better chance of survival if infected by the malarial parasite, *Plasmodium falciparum*. Therefore, the Hb^S allele is maintained in populations where malaria is prevalent, even though the allele is detrimental in the homozygous state.

In addition to sickle-cell disease, other gene mutations that cause human disease in the homozygous state are thought to be prevalent because of heterozygote advantage. For example, the high prevalence of the allele causing cystic fibrosis may be related to this phenomenon, but the advantage that a heterozygote may possess is not understood.

Negative frequency-dependent selection is a second mechanism of balancing selection. In this pattern of natural selection, the fitness of a genotype decreases when its frequency becomes higher. In other words, rare individuals have a higher fitness than more common individuals. Therefore, rare individuals are more likely to reproduce, while common individuals are less likely, thereby producing a balanced polymorphism in which no genotype becomes too rare or too common.

An interesting example of negative frequency-dependent selection involves the elder-flowered orchid, *Dactylorhiza sambucina* (**Figure 24.15**). Throughout its range, both yellow- and red-flowered individuals are prevalent. The explanation for this polymorphism is related to its pollinators, which are mainly bumblebees such as *Bombus lapidarius* and *B. terrestris*. The pollinators increase their visits to the flower color of *D. sambucina* as it becomes less common in a given area. One reason why this may occur is because *D. sambucina* is a rewardless flower; that is, it does not provide its pollinators with any reward for visiting, such as sweet nectar. Pollinators learn that the more common color of *D. sambucina* in a given area does not offer a reward, and they increase their visits to the less-common flower. Thus, the relative fitness of the less-common flower increases.

FIGURE 24.15 **The two color variations found in the elder-flowered orchid, *D. sambucina*.** The two colors are maintained in the population due to negative frequency-dependent selection.

EXPERIMENT 24A

The Grants Have Observed Natural Selection in Galápagos Finches

Let's now turn to a study that demonstrates natural selection in action. Since 1973, Peter Grant, Rosemary Grant, and their colleagues have studied the process of natural selection in finches found on the Galápagos Islands. For over 30 years, the Grants have focused much of their research on one of the Galápagos Islands known as Daphne Major (**Figure 24.16**). This small island (0.34 square kilometer) has a moderate degree of isolation (8 km from the nearest island), an undisturbed habitat, and a resident population of finches, including the medium ground finch, *Geospiza fortis*.

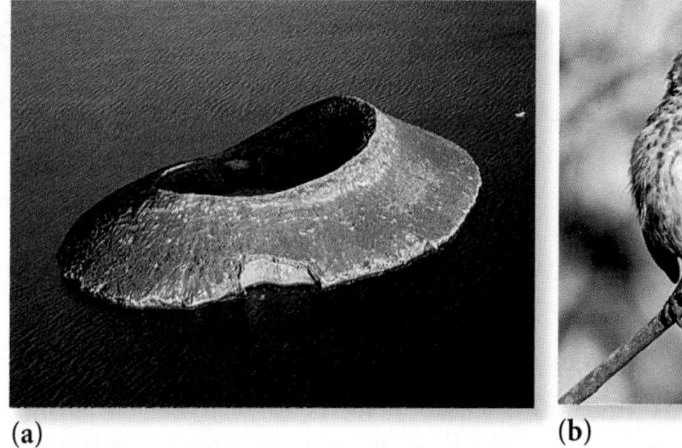

FIGURE 24.16 **Location and species used in the Grants' investigation of natural selection in finches.** **(a)** Daphne Major, one of the Galápagos Islands. **(b)** A medium ground finch (*Geospiza fortis*) that populates this island.

To study natural selection, the Grants observed various traits in the medium ground finch, including beak size, over the course of many years. The medium ground finch has a relatively small crushing beak, suitable for breaking open small, tender seeds. The Grants quantified beak size among the medium ground finches of Daphne Major by carefully measuring beak depth (a measurement of the beak from top to bottom, at its base) on individual birds. During the course of their studies, they compared the beak sizes of parents and offspring by examining many broods over several years. The depth of the beak was transmitted from parents to offspring, regardless of environmental conditions, indicating that differences in beak sizes are due to genetic differences in the population. In other words, they found that beak depth is a heritable trait.

By measuring many birds every year, the Grants were able to assemble a detailed portrait of natural selection from generation to generation. In the study shown in **Figure 24.17**, they measured beak depth in 1976 and 1978.

THE HYPOTHESIS

Beak size will be influenced by natural selection. Environments that produce larger seeds will select for birds with large beaks.

TESTING THE HYPOTHESIS — FIGURE 24.17 Natural selection in medium ground finches of Daphne Major.

	Experimental level	Conceptual level
1. In 1976, measure beak depth in parents and offspring of the species *G. fortis*.	Capture birds and measure beak depth. 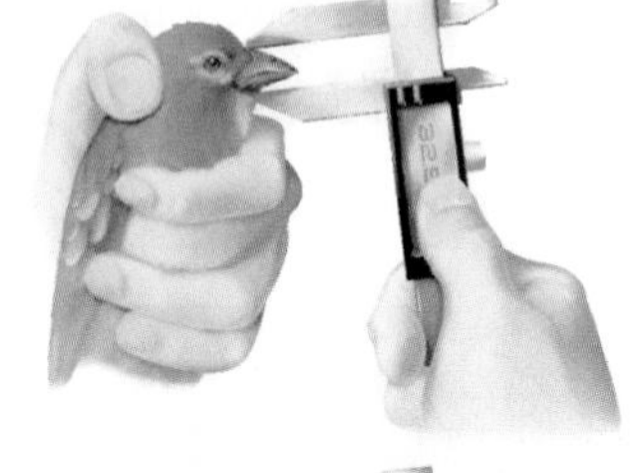	This is a way to measure a trait that may be subject to natural selection.
2. Repeat the procedure on offspring that were born in 1978 and had reached mature size. A drought had occurred in 1977 that caused plants on the island to produce mostly larger seeds and relatively few small seeds.	Capture birds and measure beak depth. 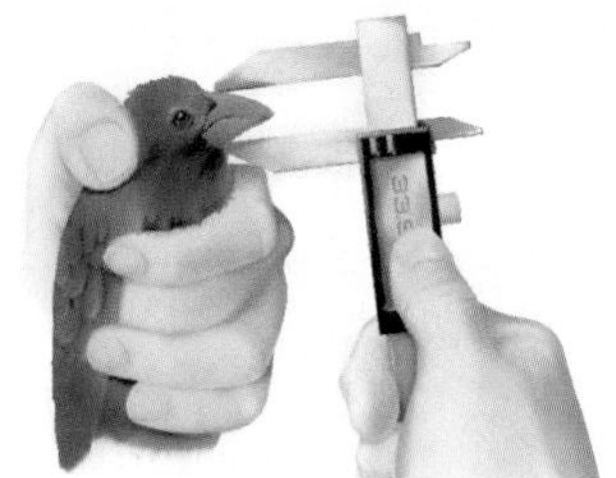	This is a way to measure a trait that may be subject to natural selection.

THE DATA

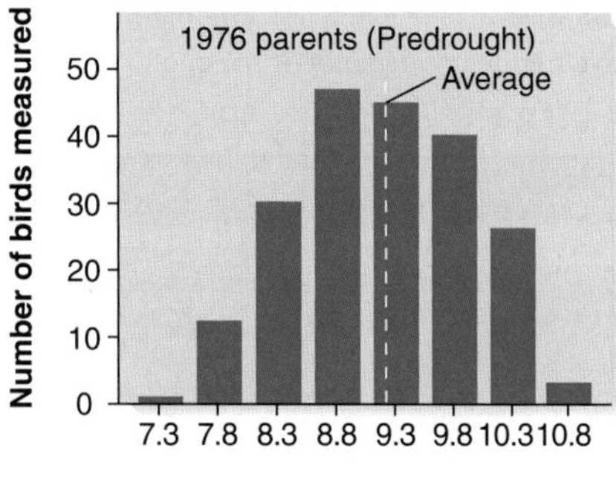

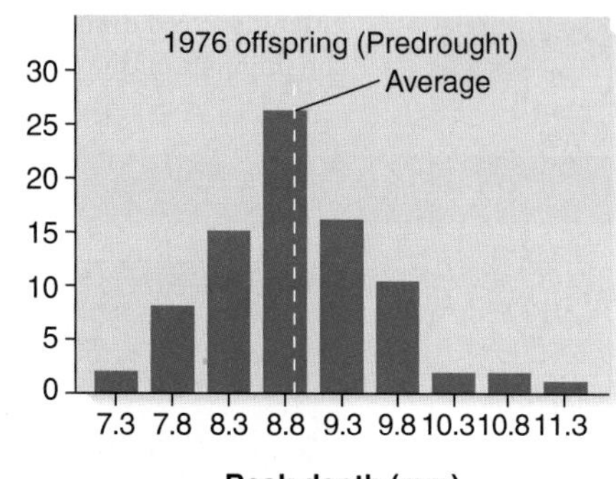

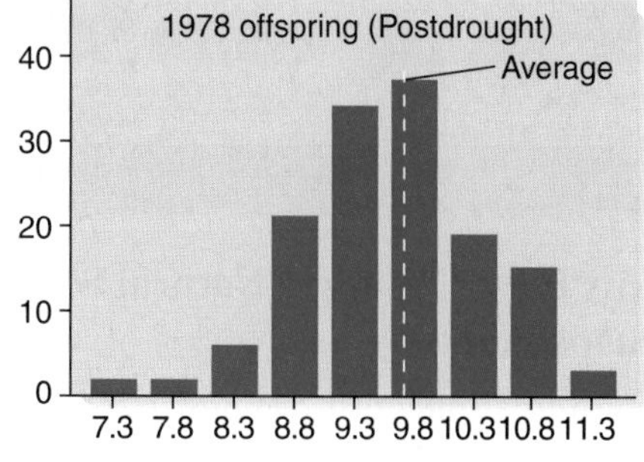

INTERPRETING THE DATA

In the wet year of 1976, the plants of Daphne Major produced the small seeds that these finches were able to eat in abundance. However, a drought occurred in 1977. During this year, the plants on Daphne Major tended to produce few of the smaller seeds, which the finches rapidly consumed. To survive, the finches resorted to eating larger, drier seeds, which are harder to crush. As a result, the birds that survived tended to have larger beaks, because they were better able to break open these large seeds. In the year after the drought, the average beak depth of offspring in the population increased to approximately 9.8 mm because the surviving birds with larger beaks passed this trait on to their offspring. This is likely to be due to directional selection (see Figure 24.8), although genetic drift could also contribute to these data. Overall, these results illustrate the power of natural selection to alter the nature of a trait, in this case, beak depth, in a given population.

A self-help quiz involving this experiment can be found at the Online Learning Center.

Nonrandom Mating May Occur in Populations

As mentioned earlier, one of the conditions required to establish the Hardy-Weinberg equilibrium is random mating. This means that individuals choose their mates irrespective of their genotypes and phenotypes. In many cases, particularly in human populations, this condition is violated frequently.

When mating is nonrandom in a population, the process is called **assortative mating.** Positive assortative mating occurs when individuals with similar phenotypes choose each other as mates. The opposite situation, where dissimilar phenotypes mate preferentially, is called negative assortative mating. In addition, individuals may choose a mate that is part of the same genetic lineage. The mating of two genetically related individuals, such as cousins, is called **inbreeding.** This is also termed consanguinity. Inbreeding sometimes occurs in human societies and is more likely to take place in nature when population size becomes very limited. In Chapter 25, we will examine how inbreeding is a useful strategy for developing agricultural breeds or strains with desirable characteristics. Conversely, **outbreeding,** which involves mating between unrelated individuals, can create hybrids that are heterozygous for many genes.

In the absence of other evolutionary processes, inbreeding and outbreeding do not affect allele frequencies in a population. However, these patterns of mating do disrupt the balance of genotypes that is predicted by the Hardy-Weinberg equation. Let's first consider inbreeding in a family pedigree. Figure 24.18 illustrates a human pedigree involving a mating between cousins. Individuals III-2 and III-3 are cousins and have produced the daughter labeled IV-1. She is said to be inbred, because her parents are genetically related to each other.

During inbreeding, the gene pool is smaller, because the parents are related genetically. In the 1940s, Gustave Malécot developed methods to quantify the degree of inbreeding. The **inbreeding coefficient** is the probability that two alleles in a particular individual will be identical for a given gene because both copies are due to descent from a common ancestor. An inbreeding coefficient (F) can be computed by analyzing the degree of relatedness within a pedigree.

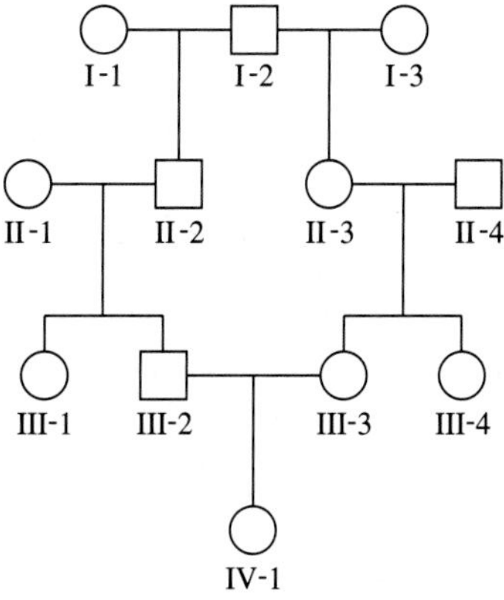

FIGURE 24.18 A human pedigree containing inbreeding. Individual IV-1 is the result of inbreeding because her parents are related.

As an example, let's determine the inbreeding coefficient for individual IV-1. To begin this problem, we must first identify all of this individual's common ancestors. A common ancestor is anyone who is an ancestor to both of an individual's parents. In Figure 24.18, IV-1 has one common ancestor, I-2, her great-grandfather. I-2 is the grandfather of III-2 and III-3.

Our next step is to determine the inbreeding paths. An inbreeding path for an individual is the shortest path through the pedigree that includes both parents and the common ancestor. In a pedigree, there is an inbreeding path for each common ancestor. The length of each inbreeding path is calculated by adding together all of the individuals in the path except the individual of interest. In this case, there is only one path because IV-1 has only one common ancestor. To add the members of the path, we begin with individual IV-1, but we do not count her. We then move to her father (III-2); to her grandfather (II-2); to I-2, her great-grandfather (the common ancestor); back down to her other grandmother (II-3); and finally to her mother (III-3). This path has five members. Finally, to calculate the inbreeding coefficient, we use the following formula:

$$F = \Sigma(1/2)^n(1 + F_A)$$

where

F is the inbreeding coefficient of the individual of interest

n is the number of individuals in the inbreeding path, excluding the inbred offspring

F_A is the inbreeding coefficient of the common ancestor

Σ indicates that we add together $(1/2)^n(1 + F_A)$ for each inbreeding path

In this case, there is only one common ancestor and, therefore, only one inbreeding path. Also, we do not know anything about the heritage of the common ancestor, so we assume that F_A is zero. Thus, in our example of Figure 24.18,

$$F = \Sigma(1/2)^n(1 + 0)$$

$$= (1/2)^5 = 1/32 = 3.125\%$$

What does this value mean? Our inbreeding coefficient, 3.125%, tells us the probability that a gene in the inbred individual (IV-1) is homozygous due to its inheritance from a common ancestor (I-2). In this case, therefore, each gene in individual IV-1 has a 3.125% chance of being homozygous because she has inherited the same allele twice from her great-grandfather (I-2), once through each parent.

As an example, let's suppose that the common ancestor (I-2) is heterozygous for the gene involved with cystic fibrosis. His genotype would be *Cc*, where *c* is the recessive allele that causes cystic fibrosis. There is a 3.125% probability that the inbred individual (IV-1) is homozygous (*CC* or *cc*) for this gene because she has inherited both copies from her great-grandfather. She has a 1.56% probability of inheriting both normal alleles (*CC*) and a 1.56% probability of inheriting both mutant alleles (*cc*). The inbreeding coefficient is denoted by the letter F (for Fixation) because it is the probability that an allele will be fixed in the homozygous condition. The term fixation signifies that the homozygous individual can pass only one type of allele to their offspring.

In other pedigrees, an individual may have two or more common ancestors. In this case, the inbreeding coefficient F is calculated as the sum of the inbreeding paths. Such an example is described in solved problem S2 at the end of the chapter.

The effects of inbreeding and outbreeding can also be considered within a population as well as within individuals. For example, let's consider the situation in which the frequency of $A = p$ and the frequency of $a = q$. In a given population, the genotype frequencies are determined in the following way:

$p^2 + fpq$ equals the frequency of *AA*

$2pq(1 - f)$ equals the frequency of *Aa*

$q^2 + fpq$ equals the frequency of *aa*

where f is a measure of how much the genotype frequencies deviate from Hardy-Weinberg equilibrium due to nonrandom mating. The value of f ranges from -1 to $+1$. When inbreeding occurs, the value is greater than zero. When outbreeding occurs, the value is less than zero.

As an example, let's suppose that $p = 0.8$, $q = 0.2$, and $f = 0.25$. We can calculate the frequencies of the *AA*, *Aa*, and *aa* genotypes under these conditions as follows:

$$AA = p^2 + fpq = (0.8)^2 + (0.25)(0.8)(0.2) = 0.68$$

$$Aa = 2pq(1 - f) = 2(0.8)(0.2)(1 - 0.25) = 0.24$$

$$aa = q^2 + fpq = (0.2)^2 + (0.25)(0.8)(0.2) = 0.08$$

There will be 68% *AA* homozygotes, 24% heterozygotes, and 8% *aa* homozygotes. If mating had been random (i.e., $f = 0$), the genotype frequencies of *AA* would be p^2, which equals 64%, and *aa* would be q^2, which equals 4%. The frequency of heterozygotes would be $2pq$, which equals 32%. When comparing these numbers, we see that inbreeding raises the proportions of homozygotes and decreases the proportion of heterozygotes. In natural populations, the value of f tends to become larger as a population becomes smaller, because each individual has a more limited choice in mate selection.

What are the consequences of inbreeding in a population? From an agricultural viewpoint, it results in a higher proportion of homozygotes, which may exhibit a desirable trait. For example, an animal breeder may use inbreeding to produce animals that are larger because they have become homozygous for alleles promoting larger size. On the negative side, many genetic diseases are inherited in a recessive manner (see Chapter 22). For these disorders, inbreeding increases the likelihood that an individual will be homozygous and therefore afflicted with the disease. Also, in natural populations, inbreeding will lower the mean fitness of the population if homozygous offspring have a lower fitness value. This can be a serious problem as natural populations become smaller due to human habitat destruction. As the population shrinks, inbreeding becomes more likely because individuals have fewer potential mates from which to choose. The inbreeding, in turn, produces homozygotes that are less fit, thereby decreasing the reproductive success of the population. This phenomenon is called **inbreeding depression.** Conservation biologists sometimes try to circumvent this problem by introducing individuals from one population into another. For example, the endangered Florida panther (*Felis concolor coryi*) suffers from inbreeding-related defects, which include poor sperm quality and quantity, and morphological abnormalities. To help alleviate these effects, panthers of the same species from Texas have been introduced into the Florida population.

24.3 SOURCES OF NEW GENETIC VARIATION

In the previous section, we primarily focused on genetic variation in which a single gene exists in two or more alleles. This simplified scenario allows us to appreciate the general principles behind evolutionary mechanisms. As researchers have analyzed genetic variation at the molecular, cellular, and population level, they have come to understand that new genetic variation occurs in many ways (**Table 24.2**). Among eukaryotic species, sexual reproduction is an important way that new genetic variation occurs among offspring. In Chapters 3 and 5, we considered how independent assortment and crossing over during sexual reproduction may produce new combinations of alleles among different genes and thereby produce new genetic variation in the resulting offspring. Similarly, in Chapter 26, we will consider how breeding between members of different species may produce hybrid offspring that harbor new combinations of genetic material. Such hybridization events have been important in the evolution of new species, particularly those in the plant kingdom. Though prokaryotic species reproduce asexually, they also possess mechanisms for gene transfer, such as conjugation, transduction, and transformation (see Chapter 6). These mechanisms are important for fostering genetic variation among bacterial and archaeal populations.

Rare mutations in DNA may also give rise to new types of variation (see Table 24.2). As discussed earlier in this chapter (see Figure 24.3) and in Chapter 16, mutations may occur within a particular gene to create new alleles of that gene. Such allelic variation is common in natural populations. Also, as described in Chapter 8, gene duplications may create a gene family; each family member will acquire independent mutations and often times will evolve more specialized functions. An example is the globin gene family (see Figure 8.7). Changes in chromosome structure and number are also important in the evolution of new species (see Chapter 26). In this section, we will examine some additional mechanisms through which an organism can acquire new genetic variation. These include exon shuffling, horizontal gene transfer, and changes in repetitive sequences. The myriad of mechanisms that foster genetic variation underscores its profound importance

TABLE 24.2

Sources of New Genetic Variation That Occur in Populations

Type	Description
Independent assortment	The independent segregation of different homologous chromosomes may give rise to new combinations of alleles in offspring (see Chapter 3).
Crossing over	Recombination (crossing over) between homologous chromosomes can also produce new combinations of alleles that are located on the same chromosome (see Chapter 5).
Interspecies crosses	On occasion, members of different species may breed with each other to produce hybrid offspring. This topic is discussed in Chapter 26.
Prokaryotic gene transfer	Prokaryotic species possess mechanisms of genetic transfer such as conjugation, transduction, and transformation (see Chapter 6).
New alleles	Point mutations can occur within a gene to create single nucleotide polymorphisms (SNPs). In addition, genes can be altered by small deletions and additions. Gene mutations are also discussed in Chapter 16.
Gene duplications	Events, such as misaligned crossovers, can add additional copies of a gene into a genome and lead to the formation of gene families. This topic is discussed in Chapter 8.
Chromosome structure and number	Chromosome structure may be changed by deletions, duplications, inversions, and translocations. Changes in chromosome number result in aneuploid, polyploid, and alloploid offspring. These mechanisms are discussed in Chapters 8 and 26.
Exon shuffling	New genes can be created when exons of preexisting genes are rearranged to make a gene that encodes a protein with a new combination of protein domains.
Horizontal gene transfer	Genes from one species can be introduced into another species and become incorporated into that species' genome.
Changes in repetitive sequences	Short repetitive sequences are common in genomes due to the occurrence of transposable elements and due to tandem arrays. The number and lengths of repetitive sequences tend to show considerable variation in natural populations.

in the evolution of species that are well adapted to their native environments and successful at reproduction.

New Genes Are Created in Eukaryotes Via Exon Shuffling

Sources of new genetic variation are revealed when the parts of genes that encode protein domains are compared within a single species. Many proteins, particularly those found in eukaryotic species, have a modular structure composed of two or more domains with different functions. For example, certain transcription factors have discrete domains involved with hormone binding, dimerization, and DNA binding. As described in Chapter 15, the glucocorticoid receptor has a domain that binds the hormone, a second domain that facilitates protein dimerization, and a third domain that allows the glucocorticoid receptor to bind to glucocorticoid response elements (GREs) next to genes (see Figure 15.6). By comparing the modular structure of eukaryotic proteins with the genes that encode them, geneticists have discovered that each domain tends to be encoded by one coding sequence, or exon, or by a series of two or more adjacent exons.

During the evolution of eukaryotic species, many new genes have been created by a process known as **exon shuffling,** in which an exon and its flanking introns are inserted into a gene, thereby producing a new gene that encodes a protein with an additional domain (**Figure 24.19**). This process may also involve the duplication and rearrangement of exons. Exon shuffling results in novel genes that express proteins with diverse functional modules. Such proteins can then alter traits in the organism and may be acted upon by evolutionary forces such as genetic drift and natural selection.

Exon shuffling may occur by more than one mechanism. As described in Chapter 17, transposable elements may promote the insertion of exons into the coding sequences of other genes. Alternatively, a double crossover event could promote the insertion of an exon into another gene (this is the case in Figure 24.19). This is called nonhomologous recombination because the two regions involved in the crossover are not homologous to each other.

New Genes Are Acquired Via Horizontal Gene Transfer

Species also accumulate genetic changes by a process called **horizontal gene transfer,** which involves the exchange of genetic material among different species. **Figure 24.20** illustrates one possible mechanism for horizontal gene transfer. In this example, a eukaryotic cell has engulfed a bacterium by endocytosis. During the degradation of the bacterium, a bacterial gene escapes to the nucleus of the cell, where it is inserted into one of the chromosomes. In this way, a gene has been transferred from a bacterial species to a eukaryotic species. By analyzing gene sequences among many different species, researchers have discovered that horizontal gene transfer is a common phenomenon. This process can occur from prokaryotes to eukaryotes, from eukaryotes to prokaryotes, between different species of prokaryotes, and between different species of eukaryotes.

Gene transfer among bacterial species is relatively widespread. As discussed in Chapter 6, bacterial species may carry out three natural mechanisms of gene transfer known as conjugation, transduction, and transformation. By analyzing the genomes of bacterial species, scientists have determined that many genes within a given bacterial genome are derived from horizontal gene transfer. Genome studies have suggested that as much as 20 to 30% of the variation in the genetic composition of modern prokaryotic species can be attributed to this process. For example, in

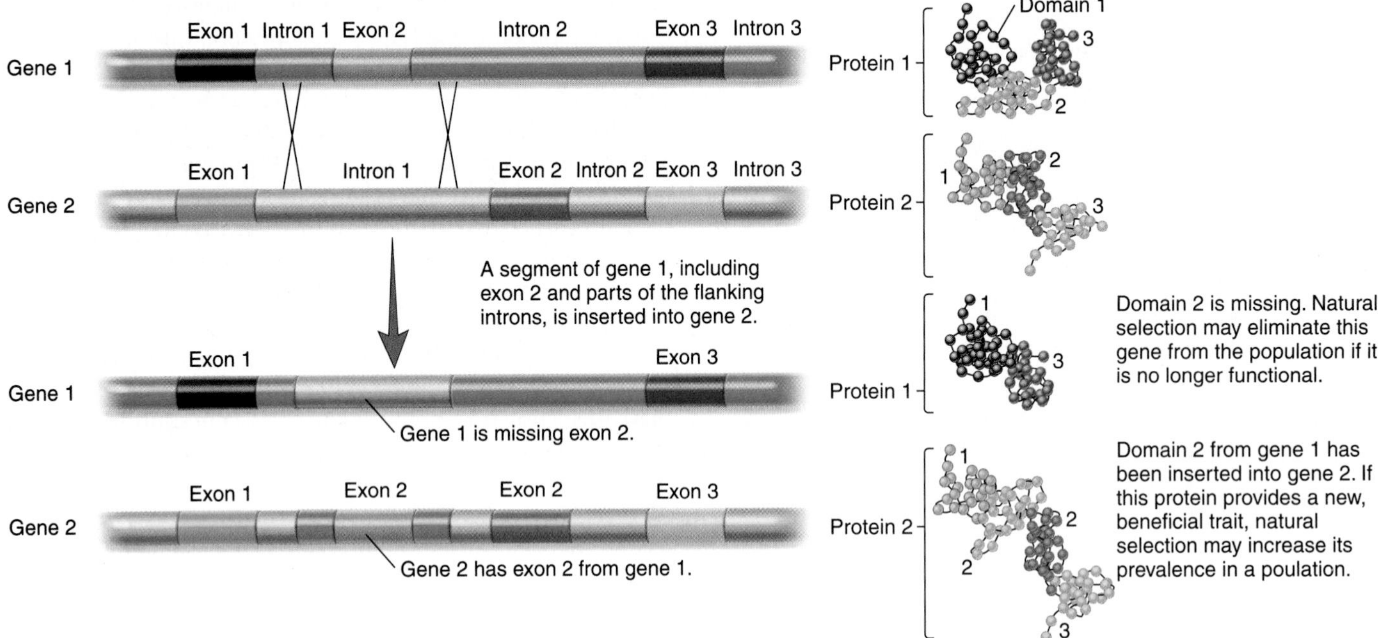

FIGURE 24.19 The process of exon shuffling. In this example, a segment of one gene containing an exon and its flanking introns has been inserted into another gene. A rare, abnormal crossing over event called nonhomologous recombination may cause this to happen. This results in proteins that have new combinations of domains and possibly new functions.

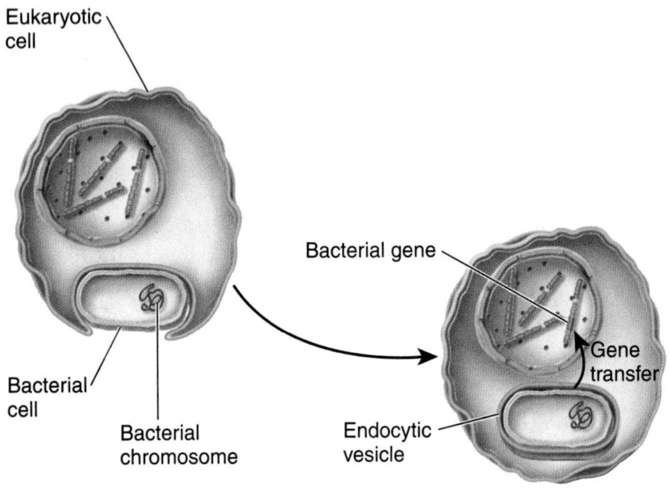

ONLINE ANIMATION **FIGURE 24.20 Horizontal gene transfer from a bacterium to a eukaryote.** In this example, a bacterium is engulfed by a eukaryotic cell, and a bacterial gene is transferred to one of the eukaryotic chromosomes.

E. coli and *Salmonella typhimurium*, roughly 17% of their genes have been acquired via horizontal gene transfer during the past 100 million years. The roles of these acquired genes are quite varied, though they commonly involve functions that are readily acted upon by natural selection. These include genes that confer antibiotic resistance, the ability to degrade toxic compounds, and pathogenicity (the ability to cause disease).

Genetic Variation Is Produced Via Changes in Repetitive Sequences

Another source of genetic variation comes from changes in **repetitive sequences**—short sequences typically a few base pairs to a few thousand base pairs that are repeated many times within a species' genome. Repetitive sequences usually come from two types of sources. First, transposable elements are genetic sequences that can move from place to place in a species' genome (see Chapter 17). The prevalence and movement of transposable elements provides a great deal of genetic variation between species and within a single species. In certain eukaryotic species, transposable elements have become fairly abundant (see Table 17.3).

A second type of repetitive sequence is nonmobile and involves short sequences that are tandemly repeated. In a **microsatellite** (also called short tandem repeats, STRs), the repeat unit is usually 1 to 6 bp, and the whole tandem repeat is less than a couple hundred bp in length. For example, the most common microsatellite encountered in humans is a sequence $(CA)_N$, where N may range from 5 to more than 50. In other words, this dinucleotide sequence can be tandemly repeated 5 to 50 or more times. The $(CA)_N$ microsatellite is found, on average, about every 10,000 bases in the human genome. In a **minisatellite,** the repeat unit is typically 6 to 80 bp in length, and the size of the minisatellite ranges from 1 kbp to 20 kbp. An example of a minisatellite in humans is telomeric DNA. In a human sperm cell, for example, the repeat unit is 6 bp and the size of a telomere is about 15 kbp. (Note: Tandem repeat sequences are called satellites because they sediment away from the rest of the chromosomal DNA during equilibrium gradient centrifugation.)

Tandem repetitive sequences such as microsatellites and minisatellites tend to undergo mutation in which the number of tandem repeats changes. For example, a microsatellite with a 4 bp repeat unit and a length of 64 bp may undergo a mutation that adds three more repeat units and become 76 bp long. The mechanism whereby micro- and minisatellites change in length is not well understood, but it may involve errors in DNA replication and homologous recombination.

Because repetitive sequences tend to vary within a population, they have become a common tool that geneticists use in a variety of ways. For example, as described in Chapters 20 and 22, microsatellites can be used as molecular markers to map the locations of genes (see Figure 22.5). Likewise, population geneticists analyze microsatellites or minisatellites to study variation at the population level and to determine the relationships among individuals and neighboring populations. The sizes of microsatellites and minisatellites found in closely related individuals tend to be more similar compared to unrelated individuals. As described next, this phenomenon is the basis for DNA fingerprinting.

DNA Fingerprinting Is Used for Identification and Relationship Testing

The technique of **DNA fingerprinting,** also known as **DNA profiling,** analyzes individuals based on the occurrence of repetitive sequences in their genome. When subjected to traditional DNA fingerprinting, the chromosomal DNA gives rise to a series of bands on a gel (**Figure 24.21**). The sizes and order of bands is an individual's DNA fingerprint. Like the human fingerprint, the DNA of each individual has a distinctive pattern. It is the unique pattern of these bands that makes it possible to distinguish individuals.

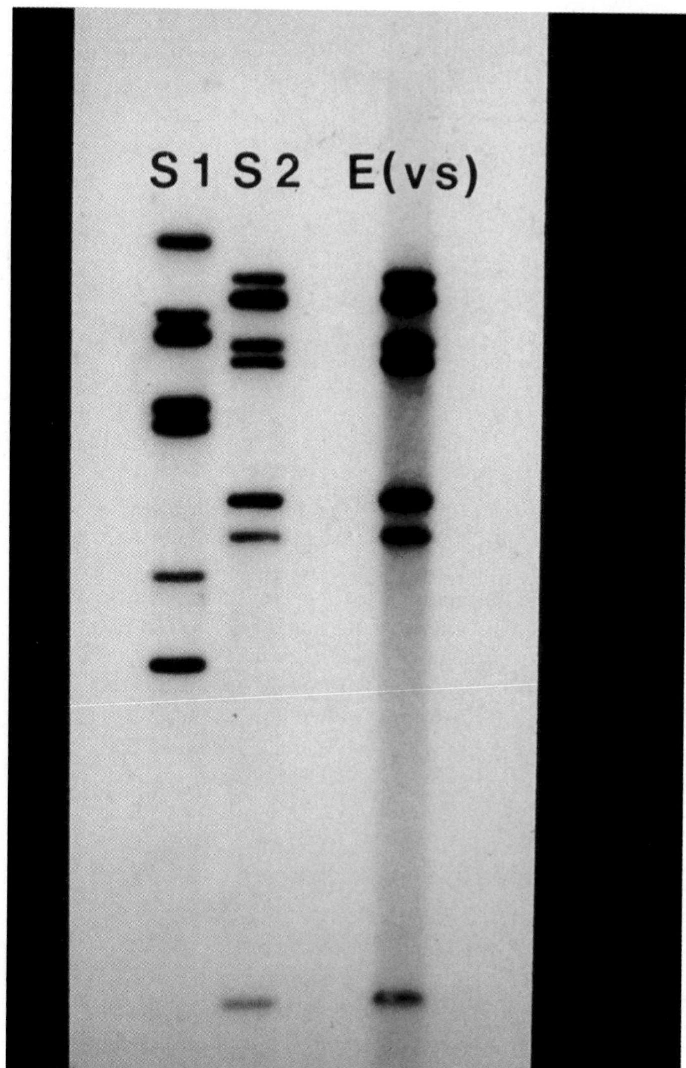

FIGURE 24.21 **A comparison of two DNA fingerprints.** The chromosomal DNA from two different individuals (Suspect 1—S1, and Suspect 2—S2) was subjected to DNA fingerprinting. The DNA evidence at a crime scene, E(vs), was also subjected to DNA fingerprinting. Following the hybridization of a radiolabeled probe, the DNA appears as a series of bands on a gel. The dissimilarity in the pattern of these bands distinguishes different individuals, much as the differences in physical fingerprint patterns can be used for identification. As seen here, S2 matches the DNA found at the crime scene.

A comparison of the DNA fingerprints among different individuals has found two applications. First, DNA fingerprinting can be used as a method of identification. In forensics, DNA fingerprinting can identify a crime suspect. In medicine, the technique can identify the type of bacterium that is causing an infection in a particular patient. A second use of DNA fingerprinting is relationship testing. Closely related individuals have more similar fingerprints compared to distantly related ones (see solved problem S6). In humans, this can be used in paternity testing. In population genetics, DNA fingerprinting can provide evidence regarding the degree of relatedness among members of a population. Such information may help geneticists determine if a population is likely to be suffering from inbreeding depression.

The development of DNA fingerprinting has relied on the identification of DNA fragments that are quite variable among members of a population. This naturally occurring variation causes each individual to have a unique DNA fingerprint. In the 1980s, Alec Jeffries and his colleagues found that certain locations within human chromosomes are particularly variable in their lengths. These minisatellites tend to vary within the human population due to changes in the number of tandem repeats at each minisatellite. For this reason, minisatellites are also referred to as loci with a variable number of tandem repeats (VNTRs).

The occurrence of minisatellites is shown schematically in **Figure 24.22**, where the chromosomal DNA from two individuals is compared. The diagram emphasizes two features of the chromosomal DNA. First, it shows the sites recognized by a particular restriction enzyme (designated with an arrow). As a matter of chance, these sites are interspersed throughout the genome of both individuals. Second, minisatellites (designated by their orange color with a series of repeat units labeled as R's) are also found. For the minisatellite shown on the left, the two individuals have the same number of repeats. For the other two minisatellites, the number of repeats differs substantially. The variation in the sizes of minisatellites affects the sizes of fragments produced when the chromosomal DNAs are digested with the restriction

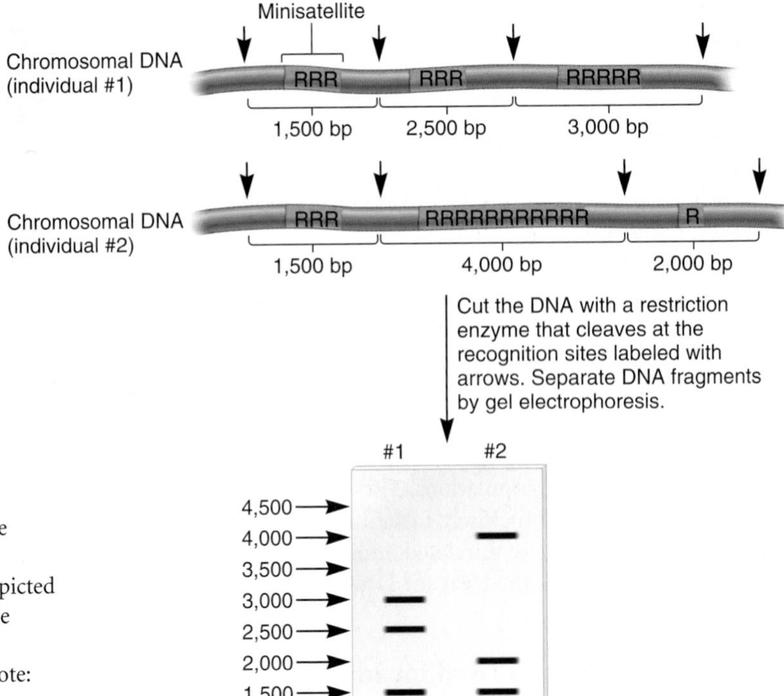

FIGURE 24.22 A comparison of minisatellites between two individuals. The restriction sites found in both individuals are represented by arrows. Each individual carries three minisatellites, shown in orange. The repeat units within each minisatellite are depicted with the letter R. The variation in the number of repeats affects the sizes of the DNA fragments produced when the DNA is digested with the restriction enzyme that cleaves at the designated sites. (Note: Minisatellites actually have more repeat units than shown here.)

enzyme; these differences yield a distinct pattern of DNA fragments when analyzed via gel electrophoresis.

Figure 24.22 depicts only a short segment of DNA. When the chromosomal DNA from a sample of actual cells is digested with a restriction enzyme, this would yield too many DNA fragments to analyze. In traditional DNA fingerprinting, DNA probes were used that hybridized specifically to the repeat sequence located within selected minisatellites. DNA fingerprinting probes were made that recognized a selected minisatellite sequence found at a relatively small number of sites (say, 5 to 30 sites) in the human genome. Using such probes, a DNA fingerprinting analysis examined the length of the DNA fragments at the sites of the chosen minisatellite sequence. This involved the analysis of 5 to 30 bands that correspond to minisatellite sites, as in Figure 24.21.

Figure 24.23a outlines the steps in a traditional DNA fingerprinting experiment. This procedure is a Southern blot using a radiolabeled probe complementary to a selected minisatellite sequence. The chromosomal DNA is isolated from a sample and digested with a restriction enzyme. The resulting DNA fragments are then separated by gel electrophoresis. The fragments in the gel are blotted onto a nylon membrane, the DNA is denatured, and the membrane is exposed to the radiolabeled probe. Because the probe is complementary to a selected minisatellite sequence, it hybridizes to approximately 5 to 30 fragments of DNA that contain this sequence and thereby labels 5 to 30 bands. In Figure 24.23a, the results of a DNA fingerprinting experiment on two samples are compared. Because the pattern of band sizes does not match between the two samples, it is concluded that the samples came from different individuals.

In the past decade, the technique of DNA fingerprinting has become automated, much like the automation that changed the procedure of DNA sequencing described in Chapter 18. DNA fingerprinting is now done using the technique of polymerase chain reaction (PCR), which amplifies microsatellites. Like minisatellites, microsatellites are found in multiple sites in the genome of humans and other species and are variable among different individuals. In this procedure, the microsatellites from a sample of DNA are amplified by PCR using primers that flank the repetitive region and then separated by gel electrophoresis according to their molecular masses. As in automated DNA sequencing, the amplified microsatellite fragments are fluorescently labeled. A laser excites the fluorescent molecule within a microsatellite, and a detector records the amount of fluorescence emission for each microsatellite. As shown in **Figure 24.23b**, this type of DNA fingerprint yields a series of peaks, each peak having a characteristic molecular mass. In this automated approach, the pattern of peaks rather than bands constitutes an individual's DNA fingerprint.

Isolate DNA from cells
(blood, skin, hair roots,
or semen, etc.)

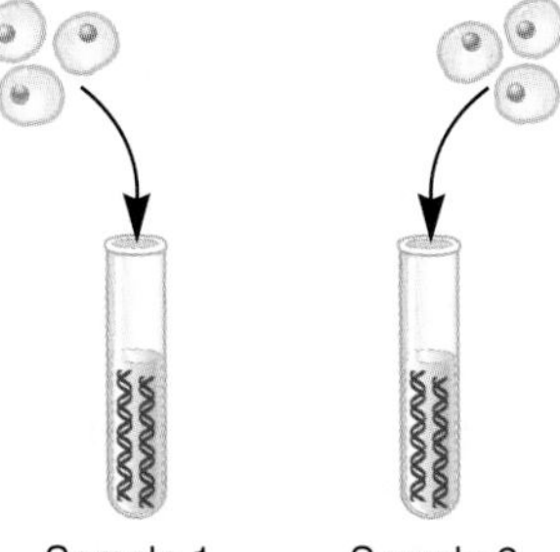

Sample 1
Sample 2
Cut DNA with a restriction enzyme and separate the fragments by gel electrophoresis.

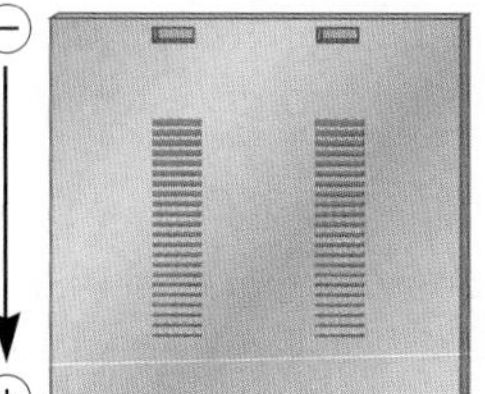

Digestion of chromosomal DNA yields thousands of fragments with varying sizes.
−
+
Blot the gel to a nylon membrane. Denature the DNA, and add a radiolabeled probe that is complementary to a selected minisatellite sequence. Allow the probe to hybridize, and then wash away the excess probe.

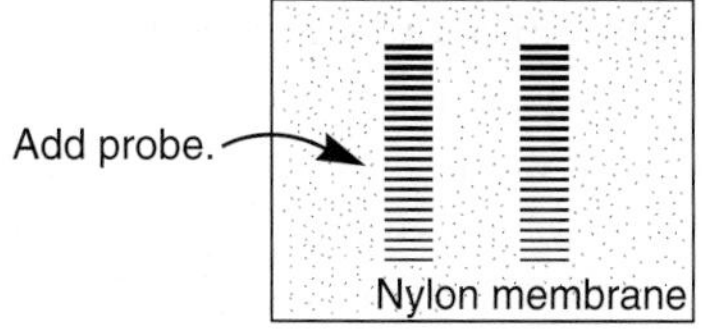

Add probe.
Nylon membrane
Expose the filter to X-ray film, and compare the results. The outcome of this experiment indicates that the 2 samples came from different individuals.

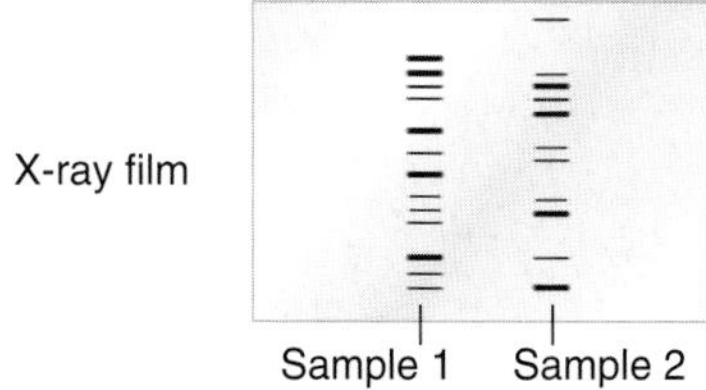

X-ray film
Sample 1
Sample 2

(a) Traditional DNA fingerprinting

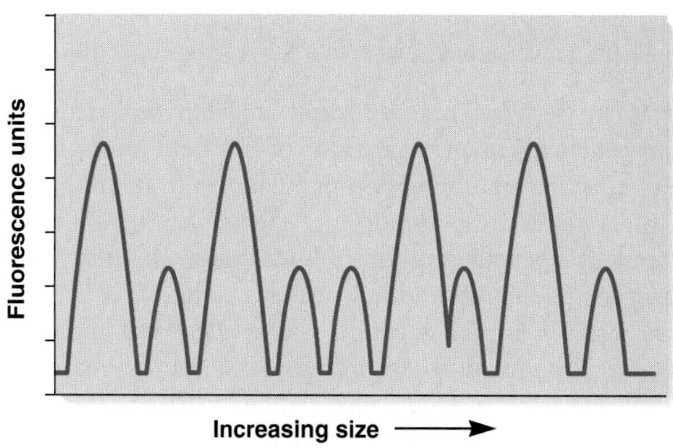

Fluorescence units
Increasing size

(b) Automated DNA fingerprinting

ONLINE ANIMATION

FIGURE 24.23 Protocol for DNA fingerprinting. (**a**) In traditional DNA fingerprinting, chromosomal DNA is isolated from a sample of cells (often a blood sample, hair roots, or semen) and cut with a restriction enzyme, and then the DNA fragments are separated by gel electrophoresis. The DNA fragments within the gel are then transferred to a nylon membrane. The DNA is denatured, and a radiolabeled probe is added that is complementary to a selected minisatellite sequence. The probe is given sufficient time to hybridize to any complementary DNA fragments, and any unbound probe is washed away. Finally, the membrane is exposed to X-ray film. The outcome of this experiment indicates that the two samples came from different individuals. (**b**) In automated DNA fingerprinting, a sample of DNA is amplified, using primers that recognize the ends of microsatellites. The microsatellite fragments are fluorescently labeled and then separated by gel electrophoresis. The fluorescent molecules within each microsatellite are excited with a laser, and the amount of fluorescence is measured via a fluorescence detector. A printout from the detector is shown here.

CONCEPTUAL SUMMARY

In this chapter, we have surveyed some of the critical issues in population genetics. The primary focus of this field is an understanding of genetic variation within populations. In natural populations, most genes are polymorphic, existing in two or more alleles. Population geneticists want to understand why this variation in the gene pool exists and how it may change.

The Hardy-Weinberg equation relates allele and genotype frequencies in a population. According to the Hardy-Weinberg equilibrium, allele frequencies remain stable as long as several conditions are met. These include no new mutations, no genetic drift, no migration, no natural selection, and random mating. While these conditions are never truly met, they can be approximated in certain large populations.

However, we know that many processes in nature promote genetic change. The term microevolution describes changes in a population's gene pool over generations. The initial source of genetic variation is mutation. Evolutionary forces may act to increase the frequency of some new alleles. Genetic drift can alter allele frequencies due to random sampling error; a new mutation may be eliminated from a population or it may become fixed in it. Genetic drift can occur via a bottleneck effect, when a population is reduced in size due to negative environmental events, or via a founder effect, when a small population moves to a new site. Allele frequencies can also be altered by migration between two populations that differ in their allele frequencies, a process known as gene flow.

Natural selection favors individuals with the greatest reproductive potential (i.e., the greatest Darwinian fitness). These may be individuals who are more likely to survive, to mate, or to be more fertile. Depending on the relative fitness values of genotypes and the complexities of the environmental effects, natural selection may follow different patterns, including directional, stabilizing, disruptive, and balancing,. Finally, the process of nonrandom mating, including inbreeding, can alter genotype frequencies in a population.

EXPERIMENTAL SUMMARY

A central approach in population genetics is to measure variation experimentally and then examine the mathematical relationships between genetic variation and the occurrence of genotypes and phenotypes within populations. In this chapter, we have focused on how population geneticists explain phenotypic variation via mathematical equations. Two fundamental calculations are allele frequencies and genotype frequencies. Experimentally, phenotypic variation can be measured by tallying the organisms within a population that display particular phenotypes. For example, we can count the number of dark snails and light snails that live in a particular location. This provides a measure of the phenotypic variation within this population. In Chapter 26, we will see that many other molecular approaches, particularly DNA sequencing, allow population geneticists to measure the genetic variation within populations.

After variation is measured, population geneticists attempt to develop and apply mathematical relationships that provide insight into the causes and dynamics of the variation. Although geneticists understand that mutation is the source of new variation, their mathematical formulas indicate that the rate of new mutations is too low to explain the experimentally observed variation in natural populations. Instead, other processes, such as genetic drift, migration, and natural selection, must alter the frequency of new mutations after they have occurred. For example, the mathematical theories of population geneticists allow us to predict how natural selection may be directional, leading to the eventual fixation or elimination of an allele, or promote a balanced polymorphism within a population. As we have seen, Peter and Rosemary Grant observed natural selection in action by tracking changes in beak size in a population of medium ground finches over the course of many years. Their results indicated that beak size is subject to directional selection.

DNA fingerprinting relies on variation in the length of microsatellites and minisatellites that are found at multiple sites in a given species' genome. This method is used to identify individuals based on their pattern of bands and also to study genetic relationships between individuals (e.g., paternity testing) and among members of a population.

PROBLEM SETS & INSIGHTS

Solved Problems

S1. The phenotypic frequency of people who cannot taste phenylthiocarbamide (PTC) is approximately 0.3. The inability to taste this bitter substance is due to a recessive allele. If we assume there are only two alleles in the population (namely, tasters, *T*, and nontasters, *t*) and that the population is in Hardy-Weinberg equilibrium, calculate the frequencies of these two alleles.

Answer: Let p = allele frequency of the taster allele and q = the allele frequency of the nontaster allele. The frequency of nontasters is 0.3. This is the frequency of the genotype *tt*, which in this case is equal to q^2:

$$q^2 = 0.3$$

To determine the frequency q of the nontaster allele, we take the square root of both sides of this equation:

$$q = 0.55$$

With this value, we can calculate the frequency p of the taster allele:

$$p = 1 - q$$
$$= 1 - 0.55 = 0.45$$

S2. In the pedigree shown here, answer the following questions with regard to individual VII-1:

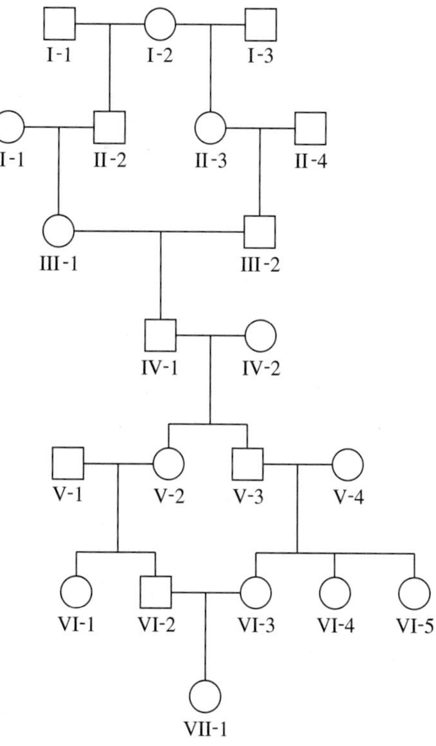

A. Who are the common ancestors of her parents?

B. What is the inbreeding coefficient?

Answer:

A. The common ancestors are IV-1 and IV-2. They are the grandparents of VI-2 and VI-3, who are the parents of VII-1.

B. The inbreeding coefficient is calculated using the formula

$$F = \Sigma(1/2)^n(1 + F_A)$$

In this case, there are two common ancestors, IV-1 and IV-2. Also, IV-1 is inbred, because I-2 is a common ancestor to both of IV-1's parents. The first step is to calculate F_A, the inbreeding coefficient for this common ancestor. The inbreeding path for IV-1 contains five people: III-1, II-2, I-2, II-3, and III-2. Therefore,

$$n = 5$$

$$F_A = (1/2)^5 = 0.03$$

Now we can calculate the inbreeding coefficient for VII-1. Each inbreeding path contains five people: VI-2, V-2, IV-1, V-3, and VI-3; and VI-2, V-2, IV-2, V-3, and VI-3. Thus,

$$F = (1/2)^5\,(1 + 0.03) + (1/2)^5\,(1 + 0)$$

$$= 0.032 + 0.031 = 0.063$$

S3. The Hardy-Weinberg equation provides a way to predict genotype frequency based on allele frequency. In the case of mammals, males are hemizygous for X-linked genes, whereas females have two copies. Among males, the frequency of any X-linked trait will equal the frequency of males with the trait. For example, if an allele frequency for an X-linked disease-causing allele was 5%, then 5% of all males would be affected with the disorder. Female genotype frequencies are computed using the Hardy-Weinberg equation.

As a specific example, let's consider the human X-linked trait known as hemophilia A (see Chapter 22 for a description of this disorder). In human populations, the allele frequency of the hemophilia A allele is approximately 1 in 10,000, or 0.0001. The other allele for this gene is the normal allele. Males can be affected or unaffected, whereas females can be affected, unaffected carriers, or unaffected noncarriers.

A. What are the allele frequencies for the mutant and normal allele in the human population?

B. Among males, what is the frequency of affected individuals?

C. Among females, what is the frequency of affected individuals and heterozygous carriers?

D. Within a population of 100,000 people, what is the expected number of affected males? In this same population, what is the expected number of carrier females?

Answer: Let p represent the normal allele and q represent the allele that causes hemophilia.

A. X^H normal allele, frequency $= 0.9999 = p$

X^h hemophilia allele, frequency $= 0.0001 = q$

B. X^hY genotype frequency of affected males $= q = 0.0001$

C. X^hX^h genotype frequency of affected females $= q^2 = (0.0001)^2 = 0.00000001$

X^HX^h genotype frequency of carrier females $= 2pq = 2(0.9999)(0.0001) = 0.0002$

D. We will assume this population is composed of 50% males and 50% females.

Number of affected males $= 50{,}000 \times 0.0001 = 5$

Number of carrier females $= 50{,}000 \times 0.0002 = 10$

S4. The Hardy-Weinberg equation can be modified to include situations of three or more alleles. In its standard (two-allele) form, the Hardy-Weinberg equation reflects the Mendelian notion that each individual inherits two copies of each allele, one from both parents. For a two-allele situation, it can also be written as

$(p + q)^2 = 1$ (Note: The number 2 in this equation reflects the idea that the genotype is due to the inheritance of two alleles, one from each parent.)

This equation can be expanded to include three or more alleles. For example, let's consider a situation in which a gene exists as three alleles: *A1*, *A2*, and *A3*. The allele frequency of *A1* is designated by the letter p, *A2* by the letter q, and *A3* by the letter r. Under these circumstances, the Hardy-Weinberg equation becomes

$$(p + q + r)^2 = 1$$

$$p^2 + q^2 + r^2 + 2pq + 2pr + 2qr = 1$$

where

p^2 is the genotype frequency of *A1A1*

q^2 is the genotype frequency of *A2A2*

r^2 is the genotype frequency of *A3A3*

$2pq$ is the genotype frequency of *A1A2*

$2pr$ is the genotype frequency of *A1A3*

$2qr$ is the genotype frequency of *A2A3*

Now here is the question. As discussed in Chapter 4, the gene that affects human blood type can exist in three alleles. In a Japanese population, the allele frequencies are

$I^A = 0.28$

$I^B = 0.17$

$i = 0.55$

Based on these allele frequencies, calculate the different possible genotype frequencies and blood type frequencies.

Answer: If we let p represent I^A, q represent I^B, and r represent i, then

p^2 is the genotype frequency of I^AI^A, which is type A blood $= (0.28)^2 = 0.08$

q^2 is the genotype frequency of I^BI^B, which is type B blood $= (0.17)^2 = 0.03$

r^2 is the genotype frequency of ii, which is type O blood $= (0.55)^2 = 0.30$

$2pq$ is the genotype frequency of I^AI^B, which is type AB blood $= 2(0.28)(0.17) = 0.09$

$2pr$ is the genotype frequency of I^Ai, which is type A blood $= 2(0.28)(0.55) = 0.31$

$2qr$ is the genotype frequency of I^Bi, which is type B blood $= 2(0.17)(0.55) = 0.19$

Type A $= 0.08 + 0.31 = 0.39$, or 39%

Type B $= 0.03 + 0.19 = 0.22$, or 22%

Type O $= 0.30$, or 30%

Type AB $= 0.09$, or 9%

S5. Let's suppose that pigmentation in a species of insect is controlled by a single gene existing in two alleles, *D* for dark and *d* for light. The heterozygote *Dd* is intermediate in color. In a heterogeneous environment, the allele frequencies are $D = 0.7$ and $d = 0.3$. This polymorphism is maintained because the environment contains some dimly lit forested areas and some sunny fields. During a hurricane, a group of 1,000 insects is blown to a completely sunny area. In this environment, the fitness values are $DD = 0.3$, $Dd = 0.7$, and $dd = 1.0$. Calculate the allele frequencies in the next generation.

Answer: The first step is to calculate the mean fitness of the population:

$$p^2W_{DD} + 2pqW_{Dd} + q^2W_{dd} = \overline{W}$$

$$\overline{W} = (0.7)^2(0.3) + 2(0.7)(0.3)(0.7) + (0.3)^2(1.0)$$

$$= 0.15 + 0.29 + 0.09 = 0.53$$

After one generation of selection, we get

$$\text{Allele frequency of } D\text{: } p_D = \frac{p^2W_{DD}}{\overline{W}} + \frac{pqW_{Dd}}{\overline{W}}$$

$$= \frac{(0.7)^2(0.3)}{0.53} + \frac{(0.7)(0.3)(0.7)}{0.53}$$

$$= 0.55$$

$$\text{Allele frequency of } d\text{: } q_d = \frac{q^2W_{dd}}{\overline{W}} + \frac{pqW_{Dd}}{\overline{W}}$$

$$= \frac{(0.3)^2(1.0)}{0.53} + \frac{(0.7)(0.3)(0.7)}{0.53}$$

$$= 0.45$$

After one generation, the allele frequency of *D* has decreased from 0.7 to 0.55, while the frequency of *d* has increased from 0.3 to 0.45.

S6. An important application of DNA fingerprinting is relationship testing. Persons who are related genetically will have some bands or peaks in common. The number they share depends on the closeness of their genetic relationship. For example, an offspring is expected to receive half of his or her minisatellites from one parent and the rest from the other. The diagram below schematically shows a traditional DNA fingerprint of an offspring, mother, and two potential fathers.

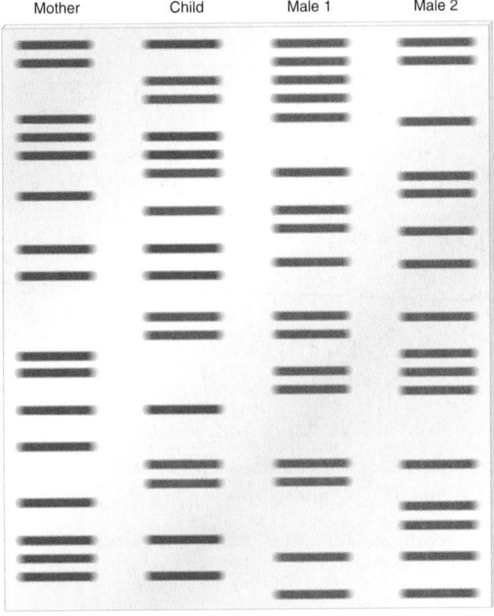

In paternity testing, the offspring's DNA fingerprint is first compared to that of the mother. The bands that the offspring has in common with the mother are depicted in purple. The bands that are not similar between the offspring and the mother must have been inherited from the father. These bands are depicted in red. Which male could be the father?

Answer: Male 2 does not have many of the paternal bands. Therefore, he can be excluded as being the father of this child. However, male 1 has all of the paternal bands. He is very likely to be the father.

Geneticists can calculate the likelihood that the matching bands between the offspring and a prospective father could occur just as a matter of random chance. To do so, they must analyze the frequency of each band in a reference population (e.g., Caucasians living in the United States). For example, let's suppose that DNA fingerprinting analyzed 40 bands. Of these, 20 bands matched with the mother and 20 bands matched with a prospective father. If the probability of each of these bands in a reference population was 1/4, the likelihood of such a match occurring by random chance would be $(1/4)^{20}$, or roughly 1 in 1 trillion. Therefore, a match between two samples is rarely a matter of random chance.

Conceptual Questions

C1. What is the gene pool? How is a gene pool described in a quantitative way?

C2. In genetics, what does the term population mean? Pick any species you like and describe how its population might change over the course of many generations.

C3. What is a genetic polymorphism? What is the source of genetic variation?

C4. For each of the following, state whether it is an example of an allele, genotype, and/or phenotype frequency:

A. Approximately 1 in 2,500 Caucasians is born with cystic fibrosis.

B. The percentage of carriers of the sickle-cell allele in West Africa is approximately 13%.

C. The number of new mutations for achondroplasia, a genetic disorder, is approximately 5×10^{-5}.

C5. The term polymorphism can refer to both genes and traits. Explain the meaning of a polymorphic gene and a polymorphic trait. If a gene is polymorphic, does the trait that the gene affects also have to be polymorphic? Explain why or why not.

C6. Cystic fibrosis is a recessive autosomal trait. In certain Caucasian populations, the number of people born with this disorder is about 1 in 2,500. Assuming Hardy-Weinberg equilibrium for this trait,

A. What are the frequencies for the normal and CF alleles?

B. What are the genotype frequencies of homozygous normal, heterozygous, and homozygous affected individuals?

C. Assuming random mating, what is the probability that two phenotypically unaffected heterozygous carriers will choose each other as mates?

C7. Does inbreeding affect allele frequencies? Why or why not? How does it affect genotype frequencies? With regard to rare recessive diseases, what are the consequences of inbreeding in human populations?

C8. For a gene existing in two alleles, what are the allele frequencies when the heterozygote frequency is at its maximum value? What if there are three alleles?

C9. In a population, the frequencies of two alleles are $B = 0.67$ and $b = 0.33$. The genotype frequencies are $BB = 0.50$, $Bb = 0.37$, and $bb = 0.13$. Do these numbers suggest inbreeding? Explain why or why not.

C10. The ability to roll your tongue is inherited as a recessive trait. The frequency of the rolling allele is approximately 0.6, and the dominant (nonrolling) allele is 0.4. What is the frequency of individuals who can roll their tongues?

C11. In the pedigree shown here, answer the following questions for individual VI-1:

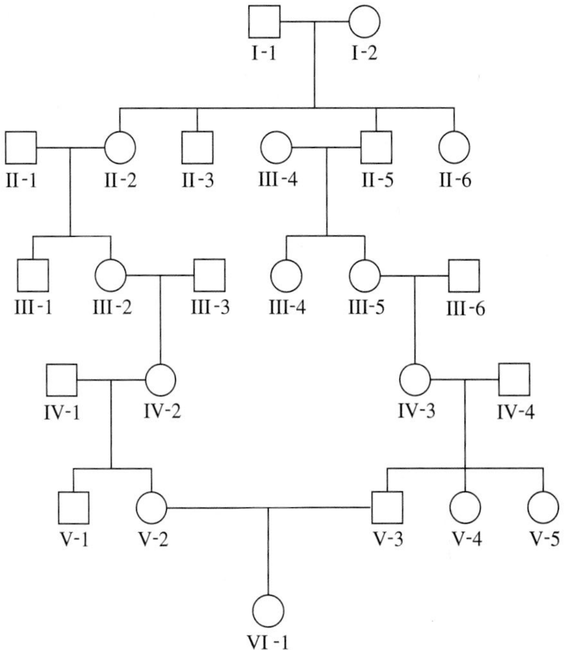

A. Is this individual inbred?

B. If so, who are her common ancestor(s)?

C. Calculate the inbreeding coefficient for VI-1.

D. Are the parents of VI-1 inbred?

C12. A family pedigree is shown here.

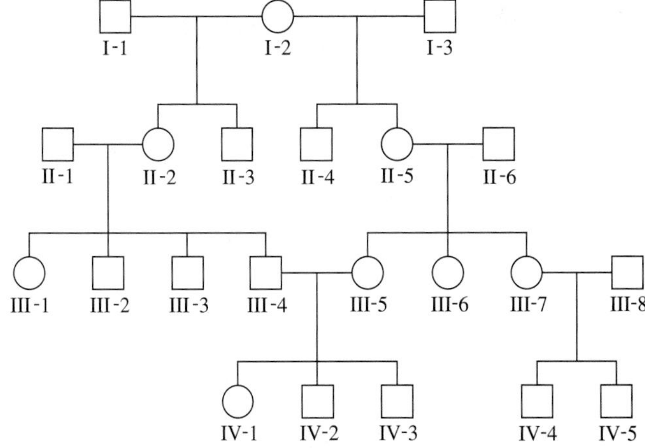

A. What is the inbreeding coefficient for individual IV-3?

B. Based on the data shown in this pedigree, is individual IV-4 inbred?

C13. A family pedigree is shown here.

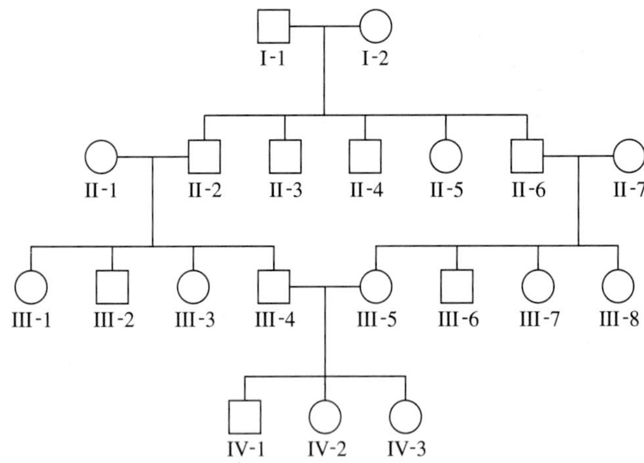

A. What is the inbreeding coefficient for individual IV-2? Who is/are her common ancestors?

B. Based on the data shown in this pedigree, is individual III-4 inbred?

C14. What evolutionary forces can cause allele frequencies to change and possibly lead to a genetic polymorphism? Discuss the relative importance of each type of process.

C15. In the term genetic drift, what is drifting? Why is this an appropriate term to describe this phenomenon?

C16. What is the difference between a random and an adaptive evolutionary process? Describe two or more examples of each. At the molecular level, explain how mutations can be random or adaptive.

C17. Let's suppose the mutation rate for converting a *B* allele into a *b* allele is 10^{-4}. The current allele frequencies are $B = 0.6$ and $b = 0.4$. How long will it take for the allele frequencies to equal each other, assuming that no genetic drift is taking place?

C18. Why is genetic drift more significant in small populations? Why does it take longer for genetic drift to cause allele fixation in large populations than in small ones?

C19. A group of four birds flies to a new location and initiates the formation of a new colony. Three of the birds are homozygous *AA*, and one bird is heterozygous *Aa*.

A. What is the probability that the *a* allele will become fixed in the population?

B. If fixation occurs, how long will it take?

C. How will the growth of the population, from generation to generation, affect the answers to parts A and B? Explain.

C20. Describe what happens to allele frequencies during the bottleneck effect. Discuss the relevance of this effect with regard to species that are approaching extinction.

C21. When two populations frequently intermix due to migration, what are the long-term consequences with regard to allele frequencies and genetic variation?

C22. Discuss the similarities and differences among directional, disruptive, balancing, and stabilizing selection.

C23. What is Darwinian fitness? What types of characteristics can promote high fitness values? Give several examples.

C24. What is the intuitive meaning of the mean fitness of a population? How does its value change in response to natural selection?

C25. Antibiotics are commonly used to combat bacterial and fungal infections. During the past several decades, however, antibiotic-resistant strains of microorganisms have become alarmingly prevalent. This has undermined the ability of physicians to treat many types of infectious disease. Discuss how the following processes that alter allele frequencies may have contributed to the emergence of antibiotic-resistant strains:

A. Random mutation

B. Genetic drift

C. Natural selection

C26. With regard to genetic drift, are the following statements true or false? If a statement is false, explain why.

A. Over the long run, genetic drift will lead to allele fixation or loss.

B. When a new mutation occurs within a population, genetic drift is more likely to cause the loss of the new allele rather than the fixation of the new allele.

C. Genetic drift promotes genetic diversity in large populations.

D. Genetic drift is more significant in small populations.

C27. Two populations of antelope are separated by a mountain range. The antelope are known to occasionally migrate from one population to the other. Migration can occur in either direction. Explain how migration will affect the following phenomena:

A. Genetic diversity in the two populations

B. Allele frequencies in the two populations

C. Genetic drift in the two populations

C28. Do the following examples describe directional, disruptive, balancing, or stabilizing selection?

A. Polymorphisms in snail color and banding pattern as described in Figure 24.13

B. Thick fur among mammals exposed to cold climates

C. Birth weight in humans

D. Sturdy stems and leaves among plants exposed to windy climates

Experimental Questions

E1. You will need to be familiar with the techniques described in Chapter 18 to answer this question. Gene polymorphisms can be detected using a variety of cellular and molecular techniques. Which techniques would you use to detect gene polymorphisms at the following levels?

A. DNA level

B. RNA level

C. Polypeptide level

E2. You will need to understand solved problem S4 to answer this question. As described in Chapter 4, the gene for coat color in rabbits can exist in four alleles termed C (full coat color), c^{ch} (chinchilla), c^h (Himalayan), and c (albino). In a population of rabbits in Hardy-Weinberg equilibrium, the allele frequencies are

$C = 0.34$

$c^{ch} = 0.17$

$c^h = 0.44$

$c = 0.05$

Assume that C is dominant to the other three alleles. c^{ch} is dominant to c^h and c, and c^h is dominant to c.

A. What is the frequency of albino rabbits?

B. Among 1,000 rabbits, how many would you expect to have a Himalayan coat color?

C. Among 1,000 rabbits, how many would be heterozygotes with a chinchilla coat color?

E3. In a large herd of 5,468 sheep, 76 animals have yellow fat, compared to the rest of the members of the herd, which have white fat. Yellow fat is inherited as a recessive trait. This herd is assumed to be in Hardy-Weinberg equilibrium.

A. What are the frequencies of the white and yellow fat alleles in this population?

B. Approximately how many sheep with white fat are heterozygous carriers of the yellow allele?

E4. The human MN blood group is determined by two codominant alleles, M and N. The following data were obtained from various human populations:

		Percentages		
Population	**Place**	***MM***	***MN***	***NN***
Inuit	East Greenland	83.5	15.6	0.9
Navajo Indians	New Mexico	84.5	14.4	1.1
Finns	Karajala	45.7	43.1	11.2
Russians	Moscow	39.9	44.0	16.1
Aborigines	Queensland	2.4	30.4	67.2

(Data from Speiss, E. B. (1990). *Genes in Populations*, 2d ed. NY: Wiley-Liss.)

A. Calculate the allele frequencies in these five populations.

B. Which populations appear to be in Hardy-Weinberg equilibrium?

C. Which populations do you think have had significant intermixing due to migration?

E5. You will need to understand solved problem S4 before answering this question. In an island population, the following data were obtained regarding the numbers of people with each of the four blood types:

Type O	721
Type A	932
Type B	235
Type AB	112

Is this population in Hardy-Weinberg equilibrium? Explain your answer.

E6. In a donor population, the allele frequencies for the normal (Hb^A) and sickle-cell alleles (Hb^S) are 0.9 and 0.1, respectively. A group of 550 individuals migrates to a new population containing 10,000 individuals; in the recipient population, the allele frequencies are $Hb^A = 0.99$ and $Hb^S = 0.01$.

A. Calculate the allele frequencies in the conglomerate population.

B. Assuming that the donor and recipient populations are each in Hardy-Weinberg equilibrium, calculate the genotype frequencies in the conglomerate population prior to further mating between the donor and recipient populations.

C. What will be the genotype frequencies of the conglomerate population in the next generation, assuming that it achieves Hardy-Weinberg equilibrium in one generation?

E7. A recessive lethal allele has achieved a frequency of 0.22 due to genetic drift in a very small population. Based on natural selection, how would you expect the allele frequencies to change in the next three generations? Note: Your calculation can assume that genetic drift is not altering allele frequencies in either direction.

E8. Among a large population of 2 million gray mosquitoes, one mosquito is heterozygous for a body color gene; this mosquito has one gray allele and one blue allele. There is no selective advantage or disadvantage between gray and blue body color. All the other mosquitoes carry the gray allele.

A. What is the probability of fixation of the blue allele?

B. If fixation happens to occur, how many generations is it likely to take?

C. Qualitatively, how would the answers to parts A and B be affected if the blue allele conferred a slight survival advantage?

E9. Resistance to the poison warfarin is a genetically determined trait in rats. Homozygotes carrying the resistance allele (*WW*) have a lower fitness because they suffer from vitamin K deficiency, but heterozygotes (*Ww*) do not. However, the heterozygotes are still resistant to warfarin. In an area where warfarin is applied, the heterozygote has a survival advantage. Due to warfarin resistance, the heterozygote is also more fit than the normal homozygote (*ww*). If the relative fitness values for *Ww*, *WW*, and *ww* individuals are 1.0, 0.37, and 0.19, respectively, in areas where warfarin is applied, calculate the allele frequencies at equilibrium. How would this equilibrium be affected if the rats were no longer exposed to warfarin?

E10. Describe, in as much experimental detail as possible, how you would test the hypothesis that snail color distribution is due to predation.

E11. In the Grants' study of the medium ground finch, do you think the pattern of natural selection was directional, stabilizing, disruptive, or balancing? Explain your answer. If the environment remained dry indefinitely (for many years), what do you think would be the long-term outcome?

E12. Here are traditional DNA fingerprints of five people: a child, mother, and three potential fathers:

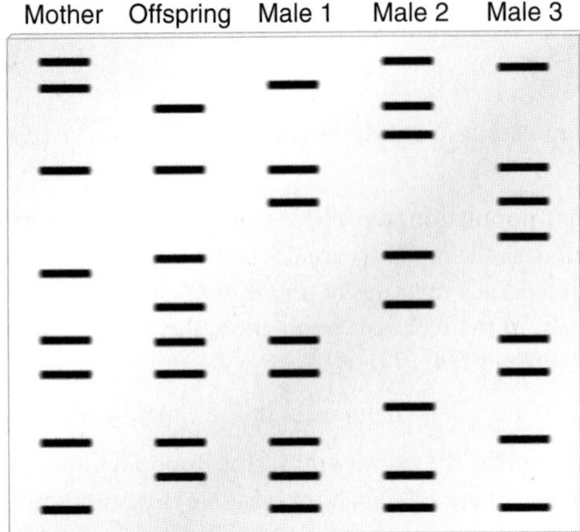

Which males can be ruled out as being the father? Explain your answer. If one of the males could be the father, explain the general strategy for calculating the likelihood that he could match the offspring's DNA fingerprint by chance alone. (See solved problem S6 before answering this question.)

E13. What is DNA fingerprinting? How can it be used in human identification?

E14. What roles do PCR and Southern blotting play in the analysis of DNA fingerprints?

E15. What is a minisatellite (also called a VNTR)? Discuss the relationship between minisatellites and DNA fingerprinting.

E16. When analyzing the DNA fingerprints of a father and his biological daughter, a technician examined 50 bands and found that 30 of them were a perfect match. In other words, 30 out of 50 bands, or 60%, were a perfect match. Is this percentage too high, or would you expect a value of only 50%? Explain why or why not.

E17. What would you expect to be the minimum percentage of matching bands in a DNA fingerprint for the following pairs of individuals?

A. Mother and son

B. Sister and brother

C. Uncle and niece

D. Grandfather and grandson

Questions for Student Discussion/Collaboration

1. Discuss examples of positive and negative assortative mating in natural populations, human populations, and agriculturally important species.

2. Discuss the role of mutation in the origin of genetic polymorphisms. Suppose that a genetic polymorphism has two alleles at frequencies of 0.45 and 0.55. Describe three different scenarios to explain these observed allele frequencies. You can propose that the alleles are neutral, beneficial, or deleterious.

3. Most new mutations are detrimental, yet rare beneficial mutations can be adaptive. With regard to the fate of new mutations, discuss whether you think it is more important for natural selection to select against detrimental alleles or to select in favor of beneficial alleles. Which do you think is more significant in human populations?

Note: All answers appear at the website for this textbook; the answers to even-numbered questions are in the back of the textbook.

www.mhhe.com/brookergenetics3e

Visit the Online Learning Center for practice tests, answer keys, and other learning aids for this chapter. Enhance your understanding of genetics with our interactive exercises, quizzes, animations, and much more.

QUANTITATIVE GENETICS

25

In this chapter, we will examine **complex traits**—characteristics that are determined by several genes and are significantly influenced by environmental factors. Many complex traits are viewed as **quantitative traits** because they can be described numerically. In humans, quantitative traits include height, the shape of our noses, and the rate at which we metabolize food, to name a few examples. The field of genetics that studies the mode of inheritance of complex or quantitative traits is called **quantitative genetics.** Quantitative genetics is an important branch of genetics for several reasons. In agriculture, most of the key characteristics of interest to plant and animal breeders are quantitative traits. These include traits such as weight, fruit size, resistance to disease, and the ability to withstand harsh environmental conditions. As we will see later in this chapter, genetic techniques have improved our ability to develop strains of agriculturally important species with desirable quantitative traits. In addition, many human diseases are viewed as complex traits that are influenced by several genes.

Quantitative genetics is also important in the study of evolution. Many of the traits that allow a species to adapt to its environment are quantitative. Examples include the long neck of a giraffe, the swift speed of the cheetah, and the sturdy branches of trees in windy climates. The importance of quantitative traits in the evolution of species will be discussed in Chapter 26. In this chapter, we will examine how genes and the environment contribute to the phenotypic expression of complex or quantitative traits. We will begin with an examination of quantitative traits and how to analyze them using statistical techniques. We will then look at the inheritance of polygenic traits and at quantitative trait loci, locations on chromosomes containing genes that affect the outcome of quantitative traits. Advances in genetic mapping strategies have enabled researchers to identify these genes. Last, we look at heritability and consider various ways of calculating and modifying the genetic variation that affects phenotype.

CHAPTER OUTLINE

Domesticated wheat. The color of wheat ranges from a dark red to white, which is an example of a complex or quantitative trait.

25.1 QUANTITATIVE TRAITS

When we compare characteristics among members of the same species, the differences are often quantitative rather than qualitative. Humans, for example, all have the same basic anatomical features (two eyes, two ears, and so on), but they differ in quantitative ways. People vary with regard to height, weight, the shape of facial features, pigmentation, and many other characteristics. As shown in Table 25.1, quantitative traits can be categorized as anatomical, physiological, and behavioral. In addition, many human diseases exhibit characteristics and inheritance patterns analogous to those of quantitative traits. Three of the leading causes of death worldwide—heart disease, cancer, and diabetes—are considered complex traits.

TABLE 25.1

Types of Quantitative Traits

Trait	Examples
Anatomical traits	Height, weight, number of bristles in *Drosophila*, ear length in corn, and the degree of pigmentation in flowers and skin
Physiological traits	Metabolic traits, speed of running and flight, ability to withstand harsh temperatures, and milk production in mammals
Behavioral traits	Mating calls, courtship rituals, ability to learn a maze, and the ability to grow or move toward light
Diseases	Atherosclerosis, hypertension, cancer, diabetes, arthritis, and obesity

In many cases, quantitative traits are easily measured and described numerically. Height and weight can be measured in centimeters (or inches) and kilograms (or pounds). Speed can be measured in kilometers per hour, and metabolic rate can be assessed as the grams of glucose burned per minute. Behavioral traits can also be quantified. A mating call can be evaluated with regard to its duration, sound level, and pattern. The ability to learn a maze can be described as the time and/or repetitions it takes to learn the skill. Finally, complex diseases such as diabetes can also be studied and described via numerical parameters. For example, the severity of the disease can be assessed by the age of onset or by the amount of insulin needed to prevent adverse symptoms.

From a scientific viewpoint, the measurement of quantitative traits is essential when comparing individuals or evaluating groups of individuals. It is not very informative to say that two people are tall. Instead, we are better informed if we know that one person is 6 feet 4 inches and the other is 6 feet 7 inches. In this branch of genetics, the measurement of a quantitative trait is how we describe the phenotype.

In the early 1900s, Francis Galton in England and his student Karl Pearson showed that many traits in humans and domesticated animals are quantitative in nature. To understand the underlying genetic basis of these traits, they founded what became known as the **biometric field** of genetics, which involved the statistical study of biological traits. During this period, Galton and Pearson developed various statistical tools for studying the variation of quantitative traits within groups of individuals; many of these tools are still in use today. In this section, we will examine how quantitative traits are measured and how statistical tools are used to analyze their variation within groups.

Quantitative Traits Exhibit a Continuum of Phenotypic Variation That May Follow a Normal Distribution

In Part II of this textbook, we discussed many traits that fall into discrete categories. For example, fruit flies might have white eyes or red eyes, and pea plants might have wrinkled seeds or smooth seeds. The alleles that govern these traits affect the phenotype in a qualitative way. In analyzing crosses involving these types of traits, each offspring can be put into a particular phenotypic category. Such attributes are called **discontinuous traits.**

In contrast, quantitative traits show a continuum of phenotypic variation within a group of individuals. For such traits, it is often impossible to place organisms into a discrete phenotypic class. For example, **Figure 25.1a** is a classic photograph from 1914 showing the range of heights of 175 students at the Connecticut Agricultural College. Though height is found at minimum and maximum values, the range of heights between these values is fairly continuous.

How do geneticists describe traits that show a continuum of phenotypes? Because quantitative traits do not naturally fall into a small number of discrete categories, an alternative way to describe them is a **frequency distribution.** To construct a frequency distribution, the trait is divided arbitrarily into a number of convenient, discrete phenotypic categories. For example, in Figure 25.1, the range of heights is partitioned into 1-inch intervals. Then a graph is made that shows the numbers of individuals found in each of the categories.

Figure 25.1b shows a frequency distribution for the heights of students pictured in Figure 25.1a. The measurement of height is plotted along the x-axis, and the number of individuals who exhibit that phenotype is plotted on the y-axis. The values along the x-axis are divided into the discrete 1-inch intervals that define the phenotypic categories, even though height is essentially continuous within a group of individuals. For example, in Figure 25.1a, 22 students were between 64.5 and 65.5 inches in height, which is plotted as the point (65 inches, 22 students) on the graph in Figure 25.1b. This type of analysis can be conducted on any group of individuals who vary with regard to a quantitative trait.

The line in the frequency distribution depicts a **normal distribution,** a distribution for a large sample in which the trait of interest varies in a symmetric way around an average value. The distribution of measurements of many biological characteristics is approximated by a symmetrical bell curve like that in Figure 25.1b. Normal distributions are common when the phenotype is determined by the cumulative effect of many small independent factors. We will consider the significance of this type of distribution next.

Statistical Methods Are Used to Evaluate a Frequency Distribution Quantitatively

Statistical tools are used to analyze a normal distribution in a number of ways. One measure that you are probably familiar with is a parameter called the **mean,** which is the sum of all the values in the group divided by the number of individuals in the group. The mean is computed using the following formula:

$$\overline{X} = \frac{\Sigma X}{N}$$

where

$\overline{X}$	is the mean
ΣX	is the sum of all the values in the group
N	is the number of individuals in the group

Number of students	1	0	0	1	5	7	7	22	25	26	27	17	11	17	4	4	1
Height (inches)	58	59	60	61	62	63	64	65	66	67	68	69	70	71	72	73	74

(a)

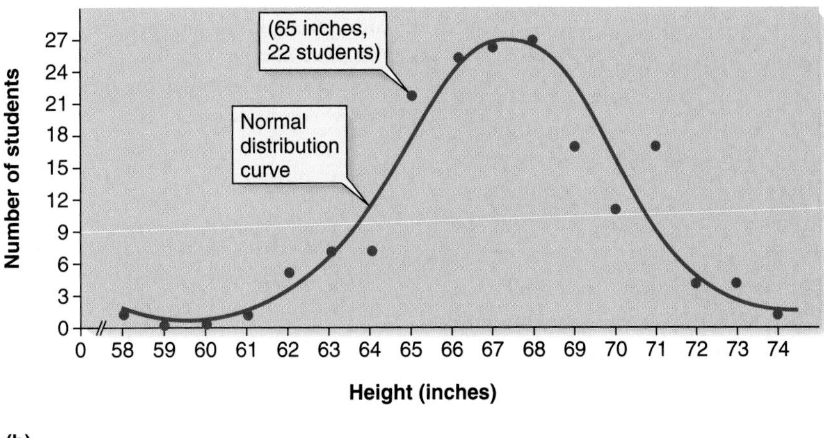

(b)

FIGURE 25.1 Normal distribution of a quantitative trait. (a) The distribution of heights in 175 students at the Connecticut Agricultural College in 1914. **(b)** A frequency distribution for the heights of students shown in (a).

A more generalized form of this equation can be used:

$$\overline{X} = \frac{\Sigma f_i X_i}{N}$$

where

$\overline{X}$ is the mean

$\Sigma f_i X_i$ is the sum of all the values in the group; each value in the group is multiplied by its frequency (f_i) in the group

N is the number of individuals in the group

For example, suppose a bushel of corn had ears with the following lengths (rounded to the nearest centimeter): 15, 14, 13, 14, 15, 16, 16, 17, 15, and 15. Then

$$\overline{X} = \frac{4(15) + 2(14) + 13 + 2(16) + 17}{10}$$

$$\overline{X} = 15 \text{ cm}$$

In genetics, we are often interested in the amount of phenotypic variation that exists in a group. As we will see later in this chapter and in Chapter 26, variation lies at the heart of breeding experiments and of evolution. Without variation, selective breeding is not possible, and natural selection cannot favor one phenotype over another. A common way to evaluate variation within a population is with a statistic called the **variance,** which is a measure of the variation around the mean. The variance is the sum of the squared deviations from the mean divided by the degrees of freedom (*df* equals $N - 1$; see Chapter 2 for a review of degrees of freedom).

$$V_X = \frac{\Sigma f_i(X_i - \overline{X})^2}{N - 1}$$

where

V_X is the variance

$X_i - \overline{X}$ is the difference between each value and the mean

N equals the number of observations

For example, if we use the values given previously for the lengths of ears of corn, the variance in this group is calculated as follows:

$$\Sigma f_i (X_i - \overline{X})^2 = 4(15 - 15)^2 + 2(14 - 15)^2 + (13 - 15)^2 + 2(16 - 15)^2 + (17 - 15)^2$$

$$\Sigma f_i(X_i - \overline{X})^2 = 0 + 2 + 4 + 2 + 4$$

$$\Sigma f_i(X_i - \overline{X})^2 = 12 \text{ cm}^2$$

$$V_X = \frac{\Sigma f_i(X_i - \overline{X})^2}{N - 1}$$

$$V_X = \frac{12 \text{ cm}^2}{9}$$

$$V_X = 1.33 \text{ cm}^2$$

Because the variance is computed from squared deviations, it is a statistic that may be difficult to understand intuitively. For example, weight can be measured in grams; the corresponding variance would be measured in square grams. Even so, variances are centrally important in the analysis of quantitative traits because they are additive under certain conditions. This means that the variances for different factors that contribute to a quantitative trait, such as genetic and environmental factors, can be added together to predict the total variance for that trait. Later, we will examine how this property is useful in predicting the outcome of genetic crosses.

To gain a more intuitive grasp of variation, we can take the square root of the variance. This statistic is called the **standard deviation (SD).** Again, using the same values for length, the standard deviation is

$$SD = \sqrt{V_X} = \sqrt{1.33}$$

$$SD = 1.15 \text{ cm}$$

If the values in a population follow a normal distribution, it is easier to appreciate the amount of variation by considering the standard deviation. **Figure 25.2** illustrates the relationship between the standard deviation and the percentages of individuals that deviate from the mean. Approximately 68% of all individuals have values within one standard deviation from the mean, either in the positive or negative direction. About 95% are within two standard deviations, and 99.7% are within three standard deviations. When a quantitative characteristic follows a normal distribution, less than 0.3% of the individuals will have values that are more or less than three standard deviations from the mean of the population. In our corn example, three standard deviations equal 3.45 cm. Therefore, we would expect that approximately 0.3% of the ears of corn would be less than 11.55 cm or greater than 18.45 cm, assuming that length follows a normal distribution.

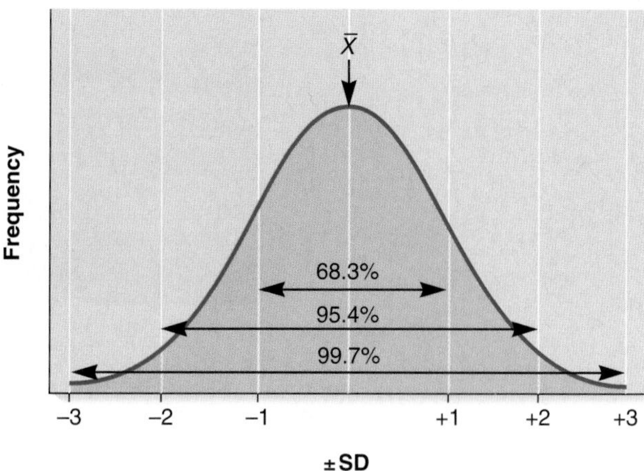

FIGURE 25.2 The relationship between the standard deviation and the proportions of individuals in a normal distribution. For example, approximately 68% of the individuals in a population are between the mean and one standard deviation above or below the mean.

Some Statistical Methods Compare Two Variables to Each Other

In many biological problems, it is useful to compare two different variables. For example, we may wish to compare the occurrence of two different phenotypic traits. Do obese animals have larger hearts? Are brown eyes more likely to occur in people with dark skin pigmentation? A second type of comparison is between traits and environmental factors. Does insecticide resistance occur more frequently in areas that have been exposed to insecticides? Is heavy body weight more prevalent in colder climates? Finally, a third type of comparison is between traits and genetic relationships. Do tall parents tend to produce tall offspring? Do women with diabetes tend to have brothers with diabetes?

To gain insight into such questions, a statistic known as the correlation coefficient is often applied. To calculate this statistic, we first need to determine the **covariance,** which describes the relationship between two variables within a group. The covariance is similar to the variance, except that we multiply together the deviations of two different variables rather than squaring the deviations from a single factor.

$$CoV_{(X,Y)} = \frac{\Sigma[(X - \overline{X})(Y - \overline{Y})]}{N - 1}$$

where

$CoV_{(X,Y)}$ is the covariance between X and Y values

X represents the values for one variable, and $\overline{X}$ is the mean value in the group

Y represents the values for another variable, and $\overline{Y}$ is the mean value in that group

N is the total number of pairs of observations

As an example, let's compare the weight of cows and that of their adult female offspring. A farmer might be interested in this relationship to determine if genetic variation plays a role in the weight of cattle. The data here describe the weights at 5 years of age for 10 different cows and their female offspring.

Mother's Weight (kg)	Offspring's Weight (kg)	$X-\overline{X}$	$Y-\overline{Y}$	$(X-\overline{X})(Y-\overline{Y})$
570	568	−26	−30	780
572	560	−24	−38	912
599	642	3	44	132
602	580	6	−18	−108
631	586	35	−12	−420
603	642	7	44	308
599	632	3	34	102
625	580	29	−18	−522
584	605	−12	7	−84
575	585	−21	−13	273
$\overline{X} = 596$	$\overline{Y} = 598$			$\Sigma = 1{,}373$
$SD_X = 21.1$	$SD_Y = 30.5$			

$$CoV_{(X,Y)} = \frac{\Sigma[(X - \overline{X})(Y - \overline{Y})]}{N - 1}$$

$$CoV_{(X,Y)} = \frac{1{,}373}{10 - 1}$$

$$CoV_{(X,Y)} = 152.6$$

After we have calculated the covariance, we can evaluate the strength of the association between the two variables by calculating a **correlation coefficient** (***r***). This value, which ranges between −1 and +1, indicates how two factors vary in relation to each other. The correlation coefficient is calculated as

$$r_{(X,Y)} = \frac{CoV_{(X,Y)}}{SD_X SD_Y}$$

A positive r value means that two factors tend to vary in the same way relative to each other; as one factor increases, the other will increase with it. A value of zero indicates that the two factors do not vary in a consistent way relative to each other; the values of the two factors are not related. Finally, an inverse correlation, in which the correlation coefficient is negative, indicates that the two factors tend to vary in opposite ways to each other; as one factor increases, the other will decrease.

Let's use the data of 5-year weights for mother and offspring to calculate a correlation coefficient.

$$r_{(X,Y)} = \frac{152.6}{(21.1)(30.5)}$$

$$r_{(X,Y)} = 0.237$$

The result is a positive correlation between the 5-year weights of mother and offspring. In other words, the positive correlation value suggests that heavy mothers tend to have heavy offspring and that lighter mothers have lighter offspring.

How do we evaluate the value of r? After a correlation coefficient has been calculated, one must consider whether the r value represents a true association between the two variables or whether it could be simply due to chance. To accomplish this, we can test the hypothesis that there is no real correlation (i.e., the null hypothesis, $r = 0$). The null hypothesis is that the r value differs from zero due only to random sampling error. We followed a similar approach in the chi square analysis described in Chapter 2. Like the chi square value, the significance of the correlation coefficient is directly related to sample size and the degrees of freedom (*df*). In testing the significance of correlation coefficients, *df* equals $N - 2$, which is one less than the degrees of freedom of variance (i.e., *df* for variance equals $N - 1$). **Table 25.2** shows the relationship between the r values and degrees of freedom at the 5% and 1% confidence intervals.

The use of Table 25.2 is valid only if several assumptions are met. First, the values of X and Y in the study must have been obtained by an unbiased sampling of the entire population. In addition, this approach assumes that the scores of X and Y follow a normal distribution, like that of Figure 25.1, and that the relationship between X and Y is linear.

TABLE **25.2**

Values of *r* at the 5% and 1% Confidence Intervals

Degrees of Freedom (*df*)	5%	1%	Degrees of Freedom (*df*)	5%	1%
1	.997	1.000	24	.388	.496
2	.950	.990	25	.381	.487
3	.878	.959	26	.374	.478
4	.811	.917	27	.367	.470
5	.754	.874	28	.361	.463
6	.707	.834	29	.355	.456
7	.666	.798	30	.349	.449
8	.632	.765	35	.325	.418
9	.602	.735	40	.304	.393
10	.576	.708	45	.288	.372
11	.553	.684	50	.273	.354
12	.532	.661	60	.250	.325
13	.514	.641	70	.232	.302
14	.497	.623	80	.217	.283
15	.482	.606	90	.205	.267
16	.468	.590	100	.195	.254
17	.456	.575	125	.174	.228
18	.444	.561	150	.159	.208
19	.433	.549	200	.138	.181
20	.423	.537	300	.113	.148
21	.413	.526	400	.098	.128
22	.404	.515	500	.088	.115
23	.396	.505	1,000	.062	.081

Note: *df* equals $N - 2$.
From Spence, J. T. et al. (1976). *Elementary Statistics*. Prentice-Hall, Englewood Cliffs, New Jersey.

To illustrate the use of Table 25.2, let's consider the correlation we have just calculated for 5-year weights of cows and their female offspring. In this case, we obtained a value of 0.237 for r, and the value of N was 10. Under these conditions, df equals 8. To be valid at a 5% confidence interval, the value of r would have to be 0.632 or higher. Because the value that we obtained is much less than this, it is fairly likely that this value could have occurred as a matter of random sampling error. In this case, we cannot reject the null hypothesis, and, therefore, we cannot conclude the positive correlation is due to a true association between the weights of mothers and offspring.

In an actual experiment, however, a researcher would examine many more pairs of cows and offspring, perhaps 500 to 1,000. If a correlation of 0.237 was observed for $N = 1{,}000$, the value would be significant at the 1% level. We would therefore reject the null hypothesis that weights are not associated with each other. Instead, we would conclude that a real association occurs between the weights of mothers and their offspring. In fact, these kinds of experiments have been done for cattle weights, and the correlations between mothers and offspring have often been found to be significant.

If a statistically significant correlation is obtained, how do we interpret its meaning? An r value that is statistically significant need not imply a cause-and-effect relationship. When parents and offspring display a significant correlation for a trait, we should not jump to the conclusion that genetics is the underlying cause of the positive association. In many cases, parents and offspring share similar environments, so the positive association might be rooted in environmental factors. In general, correlations are quite useful in identifying positive or negative associations between two variables. We should use caution, however, because this statistic, by itself, cannot prove that the association is due to cause and effect. When it has been established that the variables are related due to cause and effect—that one variable (the independent variable) affects the outcome of another (the dependent variable)—researchers may use a **regression analysis** to predict how much the dependent variable will change in response to the independent variable. This approach is described in solved problem S4 at the end of the chapter.

25.2 POLYGENIC INHERITANCE

In Section 25.1, we saw that quantitative traits tend to show a continuum of variation and can be analyzed with various statistical tools. At the beginning of the 1900s, a great debate focused on the inheritance of quantitative traits. The biometric school, founded by Francis Galton and Karl Pearson, argued that these types of traits are not controlled by discrete genes that affect phenotypes in a predictable way. To some extent, the biometric school favored the idea of blending inheritance, which had been proposed many years earlier (see Chapter 2).

Alternatively, the followers of Mendel, led by William Bateson in England and William Castle in the United States, held firmly to the idea that traits are governed by genes, which are inherited as discrete units. As we know now, Bateson and Castle were correct. However, as we will see in this section, the difficulty of studying quantitative traits lies in the fact that these traits are controlled by multiple genes and substantially influenced by environmental factors.

Most quantitative traits are polygenic and exhibit a continuum of phenotypic variation. The term **polygenic inheritance** refers to the transmission of a trait governed by two or more different genes. The location on a chromosome that harbors one or more genes that affect the outcome of a quantitative trait is called a **quantitative trait locus (QTL)**. As discussed later, QTLs are chromosomal regions that are identified by genetic mapping. Because such mapping usually locates the QTL to a relatively large chromosomal region, a QTL may contain a single gene or two or more closely linked genes that affect a quantitative trait.

Just a few years ago, it was extremely difficult for geneticists to determine the inheritance patterns for genes underlying polygenic traits, particularly those determined by three or more genes having multiple alleles for each gene. Recently, however, molecular genetic tools (described in Chapters 19 and 20) have greatly enhanced our ability to find regions in the genome where QTLs are likely to reside. This has been a particularly exciting advance in the field of quantitative genetics. In some cases, the identification of QTLs may allow the improvement of quantitative traits in agriculturally important species.

Polygenic Inheritance and Environmental Factors Create Overlaps Between Genotypes and Phenotypes

The first experiment demonstrating that continuous variation is related to polygenic inheritance was conducted by the Swedish geneticist Herman Nilsson-Ehle in 1909. He studied the inheritance of red pigment in the hull of bread wheat, *Triticum aestivum* (**Figure 25.3a**). When true-breeding plants with white hulls were crossed to a variety with red hulls, the F_1 generation had an intermediate color. When the F_1 generation was allowed to self-fertilize, great variation in redness was observed in the F_2 generation, ranging from white, light red, intermediate red, medium red, and dark red. An unsuspecting observer might conclude that this F_2 generation displayed a continuous variation in hull color. However, as shown in **Figure 25.3b**, Nilsson-Ehle carefully categorized the colors of the hulls and discovered that they followed a 1:4:6:4:1 ratio. He concluded that this species is diploid for two different genes that control hull color, each gene existing in a red or white allelic form. He hypothesized that these two loci must contribute additively to the color of the hull; the contribution of each red allele to the color of the hull is additive.

Later, researchers discovered a third gene that also affects hull color. The strains that Nilsson-Ehle had used in his original experiments must have been homozygous for the white allele of this third gene. It makes sense that wheat would be diploid for three genes that affect hull color because we now know that *T. aestivum* is a hexaploid derived from three closely related diploid species, as discussed in Chapter 8. Therefore, *T. aestivum* has six copies of many genes.

As we have just seen, Nilsson-Ehle categorized wheat hull colors into several discrete genotypic categories. However, for many polygenic traits, this is difficult or impossible. In general, as the number of genes controlling a trait increases and the influ-

(a) Red and white hulls of wheat

♀ \ ♂	*R1R2*	*R1r2*	*r1R2*	*r1r2*
R1R2	*R1R1R2R2* Dark red	*R1R1R2r2* Medium red	*R1r1R2R2* Medium red	*R1r1R2r2* Intermediate red
R1r2	*R1R1R2r2* Medium red	*R1R1r2r2* Intermediate red	*R1r1R2r2* Intermediate red	*R1r1r2r2* Light red
r1R2	*R1r1R2R2* Medium red	*R1r1R2r2* Intermediate red	*r1r1R2R2* Intermediate red	*r1r1R2r2* Light red
r1r2	*R1r1R2r2* Intermediate red	*R1r1r2r2* Light red	*r1r1R2r2* Light red	*r1r1r2r2* White

(b) *R1r1R2r2* x *R1r1R2r2*

FIGURE 25.3 **The Nilsson-Ehle experiment studying how continuous variation is related to polygenic inheritance in wheat.** (**a**) Red (top) and white (bottom) varieties of wheat, *Triticum aestivum.* (**b**) Nilsson-Ehle carefully categorized the colors of the hulls in the F_2 generation and discovered that they followed a 1:4:6:4:1 ratio. This occurs because the contributions of the red alleles are additive.

Genes→Traits In this example, two genes, with two alleles each (red and white), govern hull color. Offspring can display a range of colors, depending on how many copies of the red allele they inherit. If an offspring is homozygous for the red allele of both genes, it will have very dark red hulls. By comparison, if it carries three red alleles and one white allele, it will be medium red (which is not quite as deep in color). In this way, this polygenic trait can exhibit a range of phenotypes from dark red to white.

ence of the environment increases, the categorization of phenotypes into discrete genotypic classes becomes increasingly difficult, if not impossible. Therefore, a Punnett square cannot be used to analyze most quantitative traits. Instead, statistical methods, which are described later, must be employed.

Figure 25.4 illustrates how genotypes and phenotypes may overlap for polygenic traits. In this example, the environment (sunlight, soil conditions, and so forth) may affect the phenotypic outcome of a trait in plants (namely, seed weight). Figure 25.4a considers a situation in which seed weight is controlled by one gene with light (*w*) and heavy (*W*) alleles. A heterozygous plant (*Ww*) is allowed to self-fertilize. When the weight is only slightly influenced by variation in the environment, as seen on the left, the light, intermediate, and heavy seeds fall into separate, well-defined categories. When the environmental variation has a greater impact on seed weight, as shown on the right, more phenotypic variation is found in seed weight within each genotypic class. The variance in the frequency distribution on the right is much higher. Even so, most individuals can be classified into the three main categories.

By comparison, Figure 25.4b illustrates a situation in which seed weight is governed by three genes instead of one, each existing in light and heavy alleles. When the environmental variation is low and/or plays a minor role in the outcome of this trait, the expected 1:6:15:20:15:6:1 ratio is observed. As shown in the upper illustration in Figure 25.4b, nearly all individuals fall within a phenotypic category that corresponds to their genotype. When the environment has a more variable effect on phenotype, as shown in the lower illustration, the situation becomes more ambiguous. For example, individuals with one *w* allele and five *W* alleles have a phenotype that overlaps with that of individuals having six *W* alleles or two *w* alleles and four *W* alleles. Therefore, it becomes difficult to categorize each phenotype into a unique genotypic class. Instead, the trait displays a continuum ranging from light to heavy seed weight.

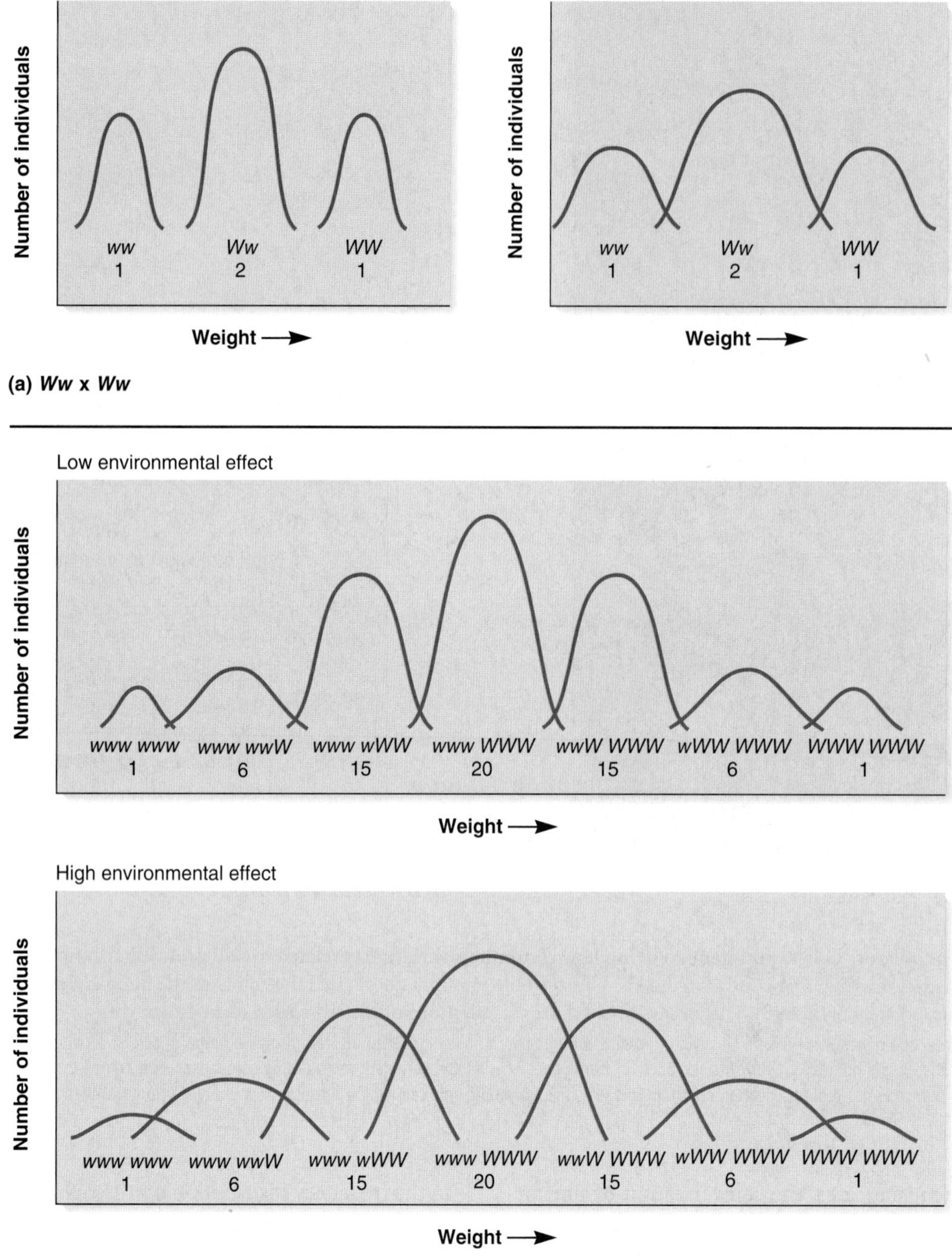

FIGURE 25.4 How genotypes and phenotypes may overlap for polygenic traits. (**a**) Situations in which seed weight is controlled by one gene, existing in light (*w*) and heavy (*W*) alleles. (**b**) Situations in which seed weight is governed by three genes instead of one, each existing in light and heavy alleles. Note: The 1:2:1 and 1:6:15:20:15:6:1 ratios were derived by analyzing these crosses using a Punnett square.

Genes→Traits The ability of geneticists to correlate genotype and phenotype depends on how many genes are involved and how much the environment causes the phenotype to vary. In **(a)**, a single gene influences weight. In the graph on the left side, the environment does not cause much variation in weight. This makes it easy to distinguish the three genotypes. There is no overlap in the weights of *ww*, *Ww*, and *WW* individuals. In the graph on the right side, the environment causes more variation in weight. In this case, a few individuals with *ww* genotypes will have the same weight as a few individuals with *Ww* genotypes; and a few *Ww* genotypes will have the same weight as *WW* genotypes. As shown in **(b)**, it becomes even more difficult to distinguish genotype based on phenotype when three genes are involved. The overlaps are minor when the environment does not cause much weight variation. However, when the environment causes substantial phenotypic variation, the overlaps between genotypes and phenotypes are very pronounced and greatly confound genetic analysis.

EXPERIMENT 25A

Polygenic Inheritance Explains DDT Resistance in *Drosophila*

As we have just learned, the phenotypic overlap for a quantitative trait may be so great that it may not be possible to establish discrete phenotypic classes. This is particularly true if many genes contribute to the trait. One way to identify the genes affecting polygenic inheritance is to look for linkage between genes affecting quantitative traits and genes affecting discontinuous traits. This approach was first studied in *Drosophila melanogaster* because many alleles had been identified and mapped to particular chromosomes.

In 1957, James Crow conducted one of the earliest studies to show linkage between genes affecting quantitative traits and genes affecting discontinuous traits. Crow, who was interested in evolution, spent time studying insecticide resistance in *Drosophila.* He noted, "Insecticide resistance is an example of evolutionary change, the insecticide acting as a powerful selective sieve for concentrating resistant mutants that were present in low frequencies in the population." His aim was to determine the genetic basis for insecticide resistance in *Drosophila melanogaster.* Many alleles were already known in this species, and these could serve as **genetic markers** for each of the four different chromosomes. Dominant alleles are particularly useful because they allow the experimenter to determine which chromosomes are inherited from either parent. The general strategy in identifying QTLs is to cross two strains that are homozygous for different genetic markers and also differ with regard to the quantitative trait of interest. This produces an F_1 generation that is heterozygous for the markers and usually exhibits an intermediate phenotype for the quantitative trait. The next step is to backcross the F_1 offspring to the parental strains. This backcross produces a population of F_2 offspring that differ with regard to their combinations of parental chromosomes. A few offspring may contain all of their chromosomes from one parental strain or the other, but most offspring will contain a few chromosomes from one parental strain and the rest from the other strain. The genetic markers on the chromosomes provide a way to determine whether particular chromosomes were inherited from one parental strain or the other.

To illustrate how genetic markers work, **Figure 25.5** considers a situation in which two strains differ with regard to a quantitative trait, resistance to DDT, and also differ with regard to dominant alleles on chromosome 3. The dominant alleles serve as markers for this chromosome. One strain is resistant to DDT and carries a dominant allele that causes minute bristles (M), while another strain is sensitive to DDT and carries a dominant allele that causes a rough eye (R). The wild-type alleles, which are recessive, produce long bristles (m) and smooth eyes (r). At the start of this experiment, it is not known if alleles affecting DDT resistance are located on this chromosome. If offspring from a backcross inherit both copies of chromosome 3 from the DDT-resistant strain, they will have smooth eyes and minute bristles. If they inherit both copies from the DDT-sensitive strain, they will have rough eyes and long bristles. By comparison, a fly with

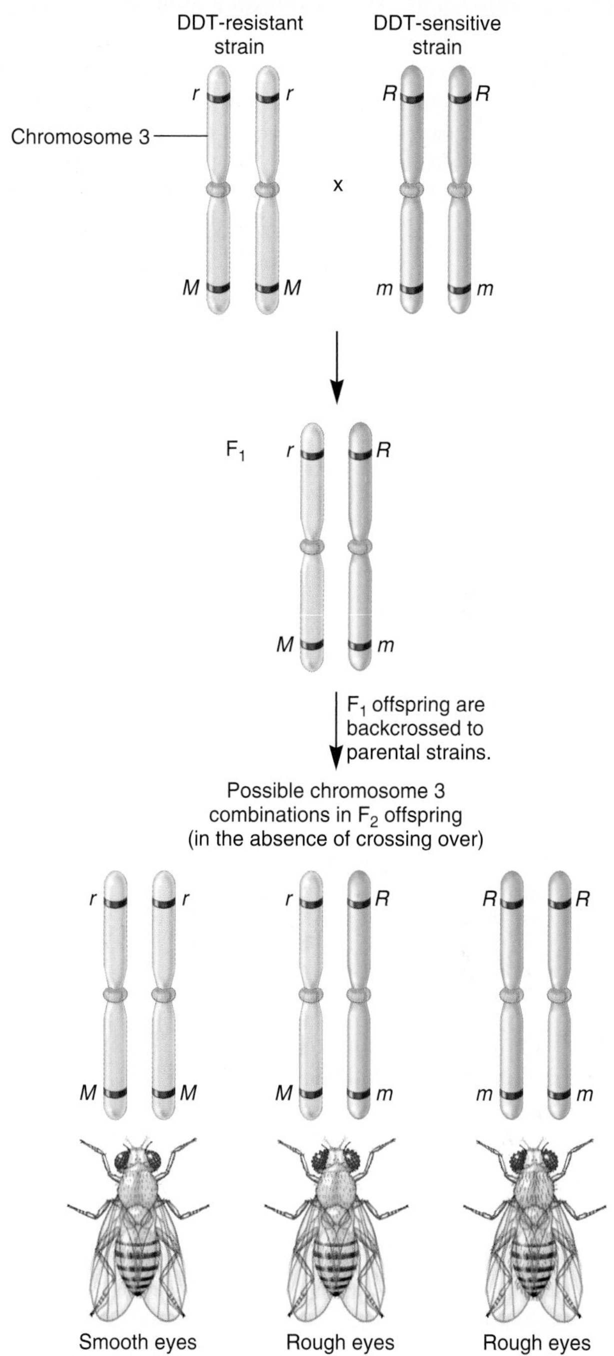

FIGURE 25.5 The use of genetic markers to map a QTL affecting DDT resistance. One strain is DDT resistant. On chromosome 3, it also carries a dominant allele that causes minute bristles (M). The other strain is DDT sensitive and carries a dominant allele that causes a roughness to the eye (R). The wild-type alleles, which are recessive, produce long bristles (m) and smooth eyes (r). F_2 offspring can have either both copies of chromosome 3 from the DDT-resistant strain, both from the sensitive strain, or one of each. This can be discerned by the phenotypes of the F_2 offspring.

rough eyes (*R*) and minute bristles (*M*) inherited one copy of chromosome 3 from the DDT-resistant strain and one copy from the DDT-sensitive strain. The transmission of the other *Drosophila* chromosomes can also be followed in a similar way. Therefore, the phenotypes of the offspring from the backcross provide a way to discern whether particular chromosomes were inherited from the DDT-resistant or DDT-sensitive strain.

Figure 25.6 shows the protocol followed by James Crow. He began with a DDT-resistant strain that had been produced by exposing flies to DDT for many generations. This DDT-resistant strain was crossed to a sensitive strain. As described previously in Figure 25.5, the two strains had allelic markers that made it possible to determine the origins of the different *Drosophila* chromosomes. Recall that *Drosophila* has four chromosomes. In this study, only chromosomes X, 2, and 3 were marked with alleles. Chromosome 4 was neglected due to its very small size. The F_1 flies were backcrossed to both parental strains, and then the F_2 female progeny were examined in two ways. First, their phenotypes were examined to determine whether particular chromosomes were inherited from the DDT-resistant or DDT-sensitive strain. Next, the female flies were exposed to filter paper impregnated with DDT. It was then determined if the flies survived this exposure for 18 to 24 hours.

THE HYPOTHESIS

DDT resistance is a polygenic trait.

TESTING THE HYPOTHESIS — FIGURE 25.6 Polygenic inheritance of DDT-resistance alleles in *Drosophila melanogaster.*

Starting material: DDT-resistant and DDT-sensitive strains of fruit flies.

1. Cross the DDT-resistant strain to the sensitive strain. In each strain, chromosomes X, 2, and 3 were marked with alleles that provided easily discernible phenotypes.

2. Take the F_1 flies and backcross to both parental strains.

3. Identify the origin of the chromosomes in the F_2 flies according to their phenotypes.

4. Expose the F_2 female flies to DDT on a filter paper for 18–24 hours.

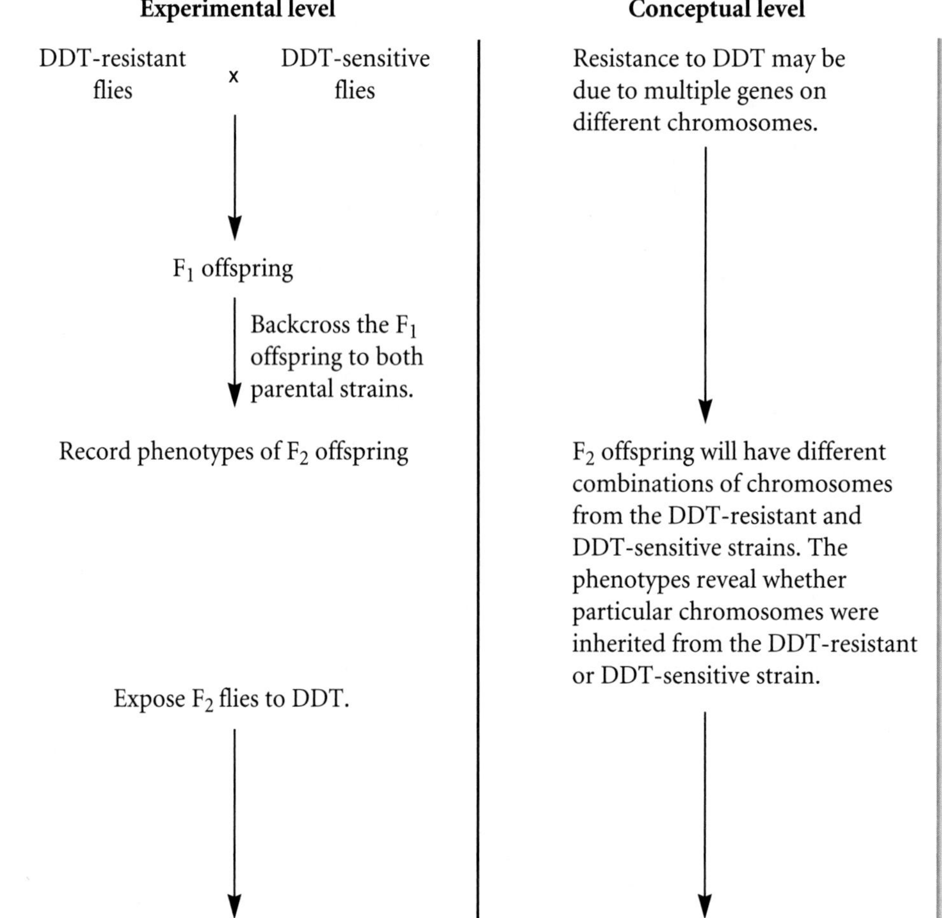
Experimental level
Conceptual level
DDT-resistant flies
x
DDT-sensitive flies
Resistance to DDT may be due to multiple genes on different chromosomes.
F_1 offspring
Backcross the F_1 offspring to both parental strains.
Record phenotypes of F_2 offspring
F_2 offspring will have different combinations of chromosomes from the DDT-resistant and DDT-sensitive strains. The phenotypes reveal whether particular chromosomes were inherited from the DDT-resistant or DDT-sensitive strain.
Expose F_2 flies to DDT.

5. Record the number of survivors.

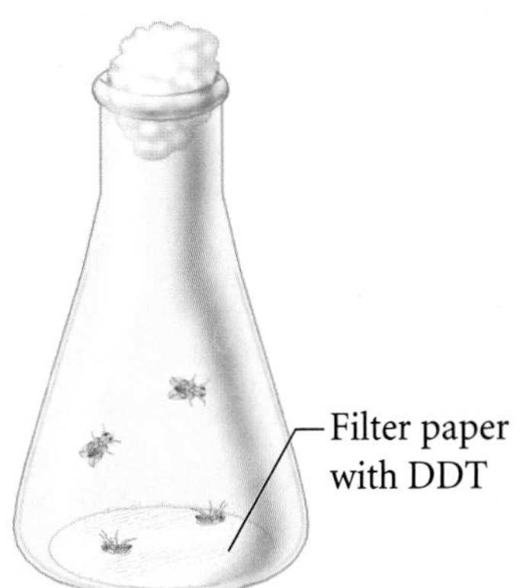

The goal is to determine if survival is correlated with the inheritance of specific chromosomes from the DDT-resistant strain.

THE DATA

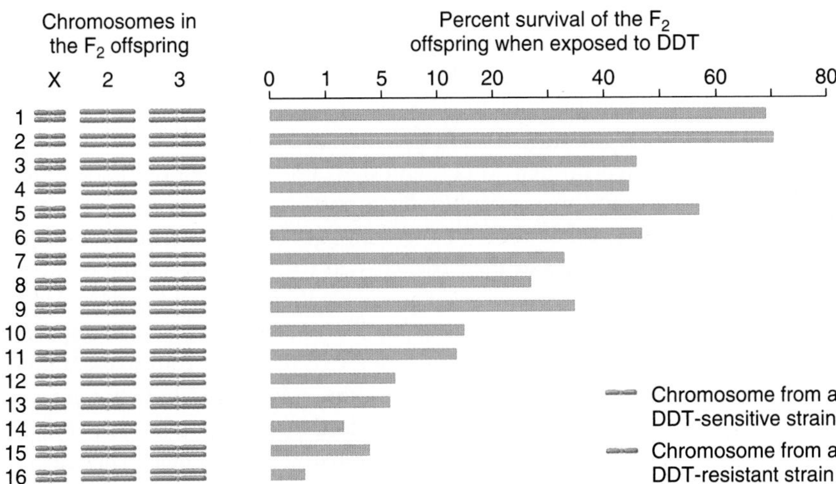

INTERPRETING THE DATA

The results of this analysis are shown in the data of Figure 25.6. As a matter of chance, some offspring contained all of their chromosomes from one parental strain or the other, but most offspring contained a few chromosomes from one parental strain and the rest from the other. The data in Figure 25.6 suggest that each copy of the X chromosome and chromosomes 2 and 3 from the DDT-resistant strain confer a significant amount of insecticide resistance. A general trend was observed in which flies inheriting more chromosomes from the DDT-resistant strain had greater levels of resistance. (However, exceptions to the trend did occur; compare examples 1 and 2.) Overall, the results are consistent with the hypothesis that insecticide resistance is a polygenic trait involving multiple genes that reside on the X chromosome and on chromosomes 2 and 3.

A self-help quiz involving this experiment can be found at the Online Learning Center.

Quantitative Trait Loci (QTLs) Are Now Mapped by Linkage to Molecular Markers

In the previous experiment, we have seen how the minimum number of genes affecting a quantitative trait, such as DDT resistance, was determined by the linkage of such unknown genes to known genes on the *Drosophila* chromosomes. In the past few years, newer research techniques have identified molecular markers, such as RFLPs and microsatellites, that serve as reference points along chromosomes. This topic is discussed in Chapter 20. These markers have been used to construct detailed genetic maps of several species' genomes. Once a genome map is obtained, it becomes much easier to determine the number of genes that affect a quantitative trait. In addition to model organisms such as *Drosophila*, *Arabidopsis*, *Caenorhabditis elegans*, and mice, detailed molecular maps have been obtained for many species of agricultural importance. These include crops such as corn, rice, and tomatoes, as well as livestock such as cattle, pigs, sheep, and chickens.

To map the genes in a eukaryotic species, researchers now determine their locations by identifying molecular markers that are close to such genes. In 1989, Eric Lander and David Botstein extended this technique to identify QTLs that govern a quantitative trait. The basis of **QTL mapping** is the association between genetically determined phenotypes for quantitative traits and molecular markers such as RFLPs, microsatellites, and single nucleotide polymorphisms (SNPs). The general strategy for QTL mapping is shown in Figure 25.7. This figure depicts two different strains of a diploid species with four chromosomes per set. The strains are highly inbred, which means they are homozygous for most molecular markers and genes. They differ in two important ways. First, the two strains differ with regard to many molecular markers. These markers are designated 1A and 1B, 2A and 2B, and so forth. The markers 1A and 1B would mark the same chromosomal location in this species, namely, the upper tip of chromosome 1. However, the two markers would be distinguishable in the two strains at the molecular level. For example, 1A might be a microsatellite that is 148 bp, while 1B might be 212 bp. Second, the two strains differ with regard to a quantitative trait of interest. In this example, the strain on the left produces large fruit, while the strain on the right produces small fruit. The unknown genes affecting this trait are designated with the letter X. A black X indicates a QTL that harbors alleles that promote large fruit, and a blue X is the same site that carries alleles that promote small fruit. Prior to conducting their crosses, researchers would not know the chromosomal locations of the QTLs shown in this figure. The purpose of the experiment is to determine their locations.

With these ideas in mind, the protocol shown in Figure 25.7 begins by mating the two inbred strains to each other and then backcrossing the F_1 offspring to both parental strains. This would produce an F_2 generation with a great degree of variation. The F_2 offspring would then be characterized in two ways. First, they would be examined for their fruit size, and second, a sample of cells would be analyzed to determine which molecular markers were found in their chromosomes. The goal is to find an association between particular molecular markers and fruit size. For example, 2A would be strongly associated with large size, while 2B would be strongly associated with small size. By comparison, 9A and 9B would not be associated with large or small size, because a QTL affecting this trait is not found on this chromosome. Also, markers such as 14A and 14B, which are fairly far away from a QTL, would not be strongly associated with a particular QTL. Markers that are on the same chromosome but far away from a QTL would often be separated from the QTL in the F_1 heterozygote due to crossing over. Only closely linked markers will be strongly associated with a particular QTL.

Overall, QTL mapping involves the analysis of a large number of markers and offspring. The data are analyzed by computer programs that can statistically associate the phenotype (e.g., fruit size) with particular markers. Markers found throughout the genome of a species provide a way to identify the locations of several different genes that possess allelic differences that may affect the outcome of a quantitative trait.

As an actual example of QTL mapping, in 1988, Andrew Paterson and his colleagues examined quantitative trait inheritance in the tomato. They studied a domestic strain of tomato and a South American green-fruited variety. These two strains differed in their RFLPs, and they also exhibited dramatic differences in three agriculturally important characteristics: fruit mass, soluble solids content, and fruit pH. The researchers crossed the two strains and then backcrossed the offspring to the domestic tomato. A total of 237 plants were then examined with regard to 70 known RFLP markers. In addition, between 5 and 20 tomatoes from each plant were analyzed with regard to fruit mass, soluble solids content, and fruit pH. Using this approach, they were able to map genes contributing much of the variation in these traits to particular intervals along the tomato chromosomes. They identified six loci causing variation in fruit mass, four affecting soluble solids content, and five with effects on fruit pH.

25.3 HERITABILITY

As we have just seen, recent approaches in molecular mapping have enabled researchers to identify the genes that contribute to a quantitative trait. The other key factor that affects the phenotypic outcomes of quantitative traits is the environment. All traits of biological organisms are influenced by genetics and the environment, and this is particularly pertinent in the study of quantitative traits. Researchers want to understand how variation, both genetic and environmental, will affect the phenotypic results.

The term **heritability** refers to the amount of phenotypic variation within a group of individuals that is due to genetic variation. Genes play a role in the development of essentially all of an organism's traits. Even so, variation of a trait in a population may be due entirely to environmental variation, entirely to genetic variation, or to a combination of the two. If all the phenotypic variation in a group is due to genetic variation, the heritability would have a value of 1. If all the variation is due to environmental effects, the heritability would equal 0. For most groups of organisms, the heritability for a given trait lies between

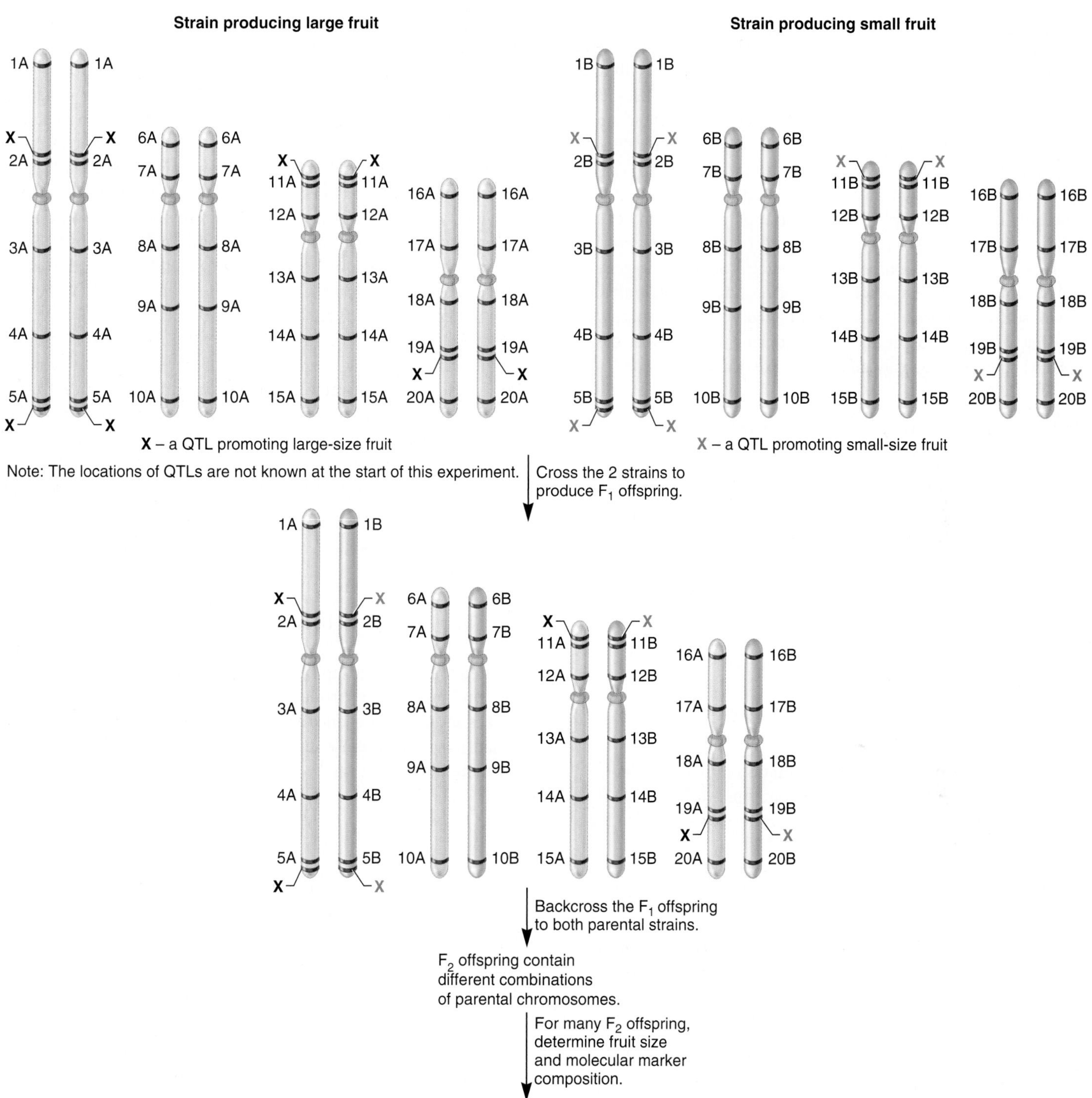
Strain producing large fruit
Strain producing small fruit
X – a QTL promoting large-size fruit
X – a QTL promoting small-size fruit
Note: The locations of QTLs are not known at the start of this experiment.
Cross the 2 strains to produce F1 offspring.
Backcross the F1 offspring to both parental strains.
F2 offspring contain different combinations of parental chromosomes.
For many F2 offspring, determine fruit size and molecular marker composition.
2A, 5A, 11A, and 19A are strongly associated with large fruit size. This suggests that 4 QTLs lie close to these markers.

FIGURE 25.7 The general strategy for QTL mapping via molecular markers. Two different inbred strains have four chromosomes per set. The strain on the left produces large fruit, and the strain on the right produces small fruit. The goal of this mapping strategy is to locate the unknown genes affecting this trait, which are designated with the letter X. A black X indicates a site promoting large fruit, and a blue X is a site promoting small fruit. The two strains differ with regard to many molecular markers designated 1A and 1B, 2A and 2B, and so forth. The two strains are mated, and then the F_1 offspring are backcrossed to the parental strains. Many F_2 offspring are then examined for their fruit size and to determine which molecular markers are found in their chromosomes. The data are analyzed by computer programs that can statistically associate the phenotype (e.g., fruit size) with particular markers. Markers found throughout the genome of this species provide a way to locate many different genes that may affect the outcome of a single quantitative trait. In this case, the analysis would predict four QTLs promoting heavier fruit weight that would be linked to regions of the chromosomes containing the following markers: 2A, 5A, 11A, and 19A.

these two extremes. For example, both genes and diet affect the size that an individual will attain. Some individuals will inherit alleles that tend to make them large, and a proper diet will also promote larger size. Other individuals will inherit alleles that make them small, and an inadequate diet may contribute to small size. Taken together, both genetics and the environment influence the phenotypic results.

In the study of quantitative traits, a primary goal is to determine how much of the phenotypic variation arises from genetic variation and how much comes from environmental variation. In this section, we will examine how geneticists analyze the genetic and environmental components that affect quantitative traits. As we will see, this approach has been applied with great success in breeding strategies to produce domesticated species with desirable and commercially valuable characteristics.

Genetic Variance and Environmental Variance Both May Contribute to Phenotypic Variance

Earlier, we examined the amount of phenotypic variation within a group by calculating the variance. In studying quantitative trait variation, geneticists partition this variation into components that are attributable to different causes. These include

Genetic variation (V_G)
Environmental variation (V_E)
Variation due to interactions between genetic and environmental factors ($V_{G \times E}$)
Variation due to associations between genetic and environmental factors ($V_{G \leftrightarrow E}$)

Let's begin by considering a simple situation in which V_G and V_E are the only factors that determine phenotypic variance and they are independent of each other. If so, then the total variance for a trait in a group of individuals is

$$V_T = V_G + V_E$$

where

V_T is the total variance. It reflects the amount of variation that is measured at the phenotypic level.

V_G is the relative amount of variance due to genetic variation.

V_E is the relative amount of variance due to environmental variation.

Why is this equation useful? The partitioning of variance into genetic and environmental components allows us to estimate their relative importance in influencing the variation within a group. If V_G is very high and V_E is very low, genetics plays a greater role in promoting variation within a group. Alternatively, if V_G is low and V_E is high, environmental factors underlie much of the phenotypic variation. As described later in this chapter, a livestock breeder might want to apply selective breeding if V_G for an important (quantitative) trait is high. In this way, the characteristics of the herd may be improved. Alternatively, if V_G is negligible, it would make more sense to investigate (and manipulate) the environmental causes of phenotypic variation.

With experimental and domesticated species, one possible way to determine V_G and V_E is by comparing the variation in traits between genetically identical and genetically disparate groups. For example, researchers have followed the practice of **inbreeding** to develop genetically homogeneous strains of mice. Inbreeding in mice involves many generations of brother-sister matings, which eventually produces strains that are **monomorphic** for all of their genes. The term monomorphic means that all the members of a population are homozygous for the same allele of a given gene. Within such an inbred strain of mice, V_G equals zero. Therefore, all phenotypic variation is due to V_E. When studying quantitative traits such as weight, an experimenter might want to know the genetic and environmental variance for a different, genetically heterogeneous group of mice. To do so, the genetically homogeneous and heterogeneous mice could be raised under the same environmental conditions and their weights measured. The phenotypic variance for weight could then be calculated as described earlier. Let's suppose we obtained the following results:

$V_T = 0.30$ sq oz for the group of genetically homogeneous mice

$V_T = 0.52$ sq oz for the group of genetically heterogeneous mice

In the case of the homogeneous mice, $V_T = V_E$, because V_G equals zero. Therefore, V_E equals 0.30 sq oz. To estimate V_G for the heterogeneous group of mice, we could assume that V_E (i.e., the environmentally produced variance) is the same for them as it is for the homogeneous mice, because the two groups were raised in identical environments. This assumption allows us to calculate the genetic variance for the heterogeneous mice.

$$V_T = V_G + V_E$$
$$0.52 = V_G + 0.30$$
$$V_G = 0.22 \text{ sq oz}$$

This result tells us that some of the phenotypic variance in the genetically heterogeneous group is due to the environment (namely, 0.30 sq oz) and some (0.22 sq oz) is due to genetic variation in alleles that affect weight.

Phenotypic Variation May Also Be Influenced by Interactions and Associations Between Genotype and the Environment

Thus far, we have considered the simple situation in which genetic variation and environmental variation are independent of each other and affect the phenotypic variation in an additive way. As another example, let's suppose that three genotypes, *TT*, *Tt*, and *tt*, affect height, producing tall, medium, and dwarf plants, respectively. Greater sunlight makes the plants grow taller regardless of their genotypes. In this case, our assumption that $V_T = V_G + V_E$ would be reasonably valid.

However, let's consider a different environmental factor such as minerals in the soil. Perhaps the *Tt* and *tt* plants are shorter because they cannot take up a sufficient supply of certain minerals to support maximal growth, while the *TT* plants may not be limited by mineral uptake. According to this hypothetical scenario, adding minerals to the soil would enhance the growth rate of *tt* plants by a large amount and the *Tt* plants by a smaller amount (**Figure 25.8**). The height of *TT* plants would not be affected by mineral supplementation. When the environmental effects on phenotype differ according to genotype, this phenomenon is called a **genotype-environment interaction.** Variation due to interactions between genetic and environmental factors is termed $V_{G \times E}$ as noted earlier.

Interactions of genetic and environmental factors are common. As an actual example, **Table 25.3** shows results from a study conducted in 2000 by Cristina Vieira, Trudy Mackay, and colleagues in which they investigated genotype-environment interaction for quantitative trait loci affecting life span in *Drosophila melanogaster*. The data seen here compare the life span in days of male and female flies from two different strains of *D. melanogaster* raised at different temperatures. Because males and females differ in their sex chromosomes and gene expression patterns, they can be viewed as having different genotypes. The effects of environmental changes depended greatly on the strain and the sex of the flies. Under standard culture conditions, the females of strain A had the longest life span, while females of strain B had the shortest. In strain A, high temperature increased the longevity of males and decreased the longevity of females. In contrast, under hotter conditions, the longevity of males of strain B was dramatically reduced, while females of this same strain were not significantly affected. Lower growth temperature also had different effects in these two strains. While low temperature increased the longevity of both strains, the effects were most dramatic in the males of strain A and the females of strain B. Taken together, these results illustrate the potential complexity of genotype-environmental interaction when measuring a quantitative trait such as life span.

Another issue confronting geneticists is that genotypes may not be randomly distributed in all possible environments. When certain genotypes are preferentially found in particular environments, this phenomenon is called a **genotype-environment association** ($V_{G \leftrightarrow E}$). When such an association occurs, the effects of genotype and environment are not independent of each other, and the association needs to be considered when determining the effects of genetic and environmental variation on the total phenotypic variation. Genotype-environment associations are very common in the study of human genetics, in which large families tend to have more similar environments compared to the population as a whole. One way to evaluate this effect is to compare different genetic relationships, such as identical versus fraternal twins. We will examine this approach later in the chapter. Another strategy that geneticists might follow is to analyze siblings that have been adopted by different parents at birth. Their environment conditions tend to be more disparate, and this may help to minimize the effects of genotype-environment association.

TABLE 25.3

Longevity of Two Strains of *Drosophila melanogaster**

	Strain A		Strain B	
Temperature	Male	Female	Male	Female
Standard	33.6	39.5	37.5	28.9
High	36.3	33.9	23.2	28.6
Low	77.5	48.3	45.8	77.0

*Longevity was measured in the mean number of days of survival. Strains A and B were inbred strains of *D. melanogaster* called Oregon and 2b, respectively. The standard, high, and low temperature conditions were 25°C, 29°C, and 14°C, respectively.

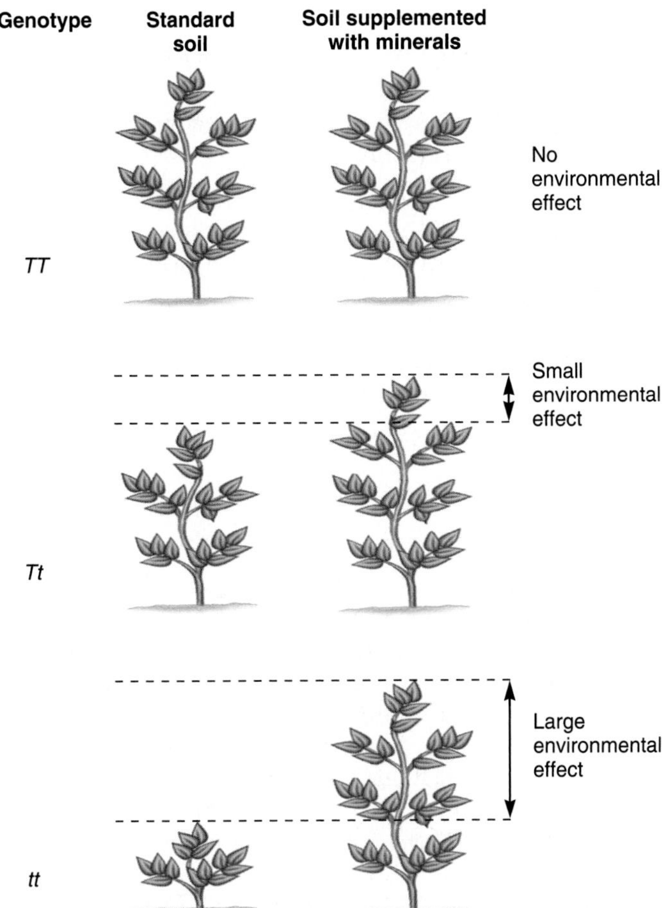

FIGURE 25.8 **A schematic example of genotype-environment interaction.** When grown in standard soil, the three genotypes *TT*, *Tt*, and *tt* show large, medium, and small heights, respectively. When the soil is supplemented with minerals, a great effect is seen on the *tt* genotype and a smaller effect on the *Tt* genotype. The *TT* genotype is unaffected by the environmental change.

Heritability Is the Relative Amount of Phenotypic Variation That Is Due to Genetic Variation

Another way to view variance is to focus our attention on the genetic contribution to phenotypic variation. Heritability is the proportion of the phenotypic variance that is attributable to genetic variation. If we assume again that environment and genetics are independent and the only two factors affecting phenotype, then

$$h_B^2 = V_G / V_T$$

where

h_B^2 is the heritability in the broad sense

V_G is the variance due to genetics

V_T is the total phenotypic variance, which equals $V_G + V_E$

The heritability defined here, h_B^2, called the **broad-sense heritability,** takes into account different types of genetic variation that may affect the phenotype. As we have seen throughout this textbook, genes can affect phenotypes in various ways. As described earlier, the Nilsson-Ehle experiment showed that the alleles determining hull color in wheat affect the phenotype in an additive way. Alternatively, alleles affecting other traits may show a dominant/recessive relationship. In this case, the alleles are not strictly additive, because the heterozygote has a phenotype closer to, or perhaps the same as, the homozygote containing two copies of the dominant allele. For example, both *TT* and *Tt* pea plants show a tall phenotype. In addition, another complicating factor is epistasis (described in Chapter 4), in which the alleles for one gene can mask the phenotypic expression of the alleles of another gene. To account for these differences, geneticists usually subdivide V_G into these three different genetic categories:

$$V_G = V_A + V_D + V_I$$

where

V_A is the variance due to additive alleles. A heterozygote shows a phenotype that is intermediate between the respective homozygotes.

V_D is the variance due to alleles that follow a dominant/recessive pattern of inheritance.

V_I is the variance due to genes that interact in an epistatic manner.

In analyzing quantitative traits, geneticists may focus on V_A and neglect the contributions of V_D and V_I. They do this for scientific as well as practical reasons. For some quantitative traits, the additive effects of alleles may play a primary role in the phenotypic outcome. In addition, when the alleles behave additively, we can predict the outcomes of crosses based on the quantitative characteristics of the parents. The heritability of a trait due to the additive effects of alleles is called the **narrow-sense heritability:**

$$h_N^2 = V_A / V_T$$

The narrow-sense heritability (h_N^2) may be an inaccurate measure of the actual heritability if V_D and V_I are not small values. However, for certain quantitative traits, the value of V_A may be relatively large compared to V_D and V_I. In such cases, the broad-sense and the narrow-sense heritabilities are similar to each other.

How can the narrow-sense heritability be determined? In this chapter, we will consider two common ways. As discussed later, one way to estimate the narrow-sense heritability involves selective breeding practices, which are done with agricultural species. A second common strategy to determine h_N^2 involves the measurement of a quantitative trait among groups of genetically related individuals. For example, agriculturally important traits, such as egg weight in poultry, can be analyzed in this way. To calculate the heritability, a researcher would determine the observed egg weights between individuals whose genetic relationships are known, such as a mother and her female offspring. These data could then be used to compute a correlation between the parent and offspring, using the methods described earlier. The narrow-sense heritability is then calculated as

$$h_N^2 = r_{obs} / r_{exp}$$

where

r_{obs} is the observed phenotypic correlation between related individuals

r_{exp} is the expected correlation based on the known genetic relationship

In our example, r_{obs} is the observed phenotypic correlation between parent and offspring. In particular research studies, the observed phenotypic correlation for egg weights between mothers and daughters has been found to be about 0.25 (although this varies among strains). The expected correlation, r_{exp}, is based on the known genetic relationship. A parent and child share 50% of their genetic material, so r_{exp} equals 0.50. So,

$$\begin{aligned} h_N^2 &= r_{obs} / r_{exp} \\ &= 0.25/0.50 \\ &= 0.50 \end{aligned}$$

(Note: For siblings, $r_{exp} = 0.50$; for identical twins, $r_{exp} = 1.0$; and for an uncle-niece relationship, $r_{exp} = 0.25$.)

According to this calculation, about 50% of the phenotypic variation in egg weight is due to additive genetic variation; the other half is due to the environment.

When calculating heritabilities from correlation coefficients, keep in mind that this computation assumes that genetics and the environment are independent variables. However, this is not always the case. The environments of parents and offspring are often more similar to each other than they are to those of unrelated individuals. As mentioned earlier, there are several ways to minimize this confounding factor. First, in human studies, one may analyze the heritabilities from correlations between adopted children and their biological parents. Alternatively, one can examine a variety of relationships (uncle-niece, identical twins versus fraternal twins, and so on) and see if the heritability values are roughly the same in all cases. This approach was applied in the study that is described next.

EXPERIMENT 25B

Heritability of Dermal Ridge Count in Human Fingerprints Is Very High

Fingerprints are inherited as a quantitative trait. It has long been known that identical twins have fingerprints that are very similar, whereas fraternal twins show considerably less agreement. Galton was the first researcher to study fingerprint patterns, but this trait became more amenable to genetic studies in the 1920s, when Kristine Bonnevie, a Norwegian geneticist, developed a method for counting the number of ridges within a human fingerprint.

As shown in **Figure 25.9**, human fingerprints can be categorized as having an arch, loop, or whorl, or a combination of these patterns. The primary difference among these patterns is the number of triple junctions, each known as a triradius (Figure 25.9b and c). At a triradius, a ridge emanates in three different directions. An arch has zero triradii, a loop has one, and a whorl has two. In Bonnevie's method of counting, a line is drawn from a triradius to the center of the fingerprint. The ridges that touch this line are then counted. (Note: The triradius ridge itself is not counted, and the last ridge is not counted if it forms the center of the fingerprint.) With this method, one can obtain a ridge count for all 10 fingers. Bonnevie conducted a study on a small population and found that ridge count correlations were relatively high in genetically related individuals.

Sarah Holt, who was also interested in the inheritance of this quantitative trait, carried out a more exhaustive study of ridge counts in a British population. As shown in the two graphs below, in groups of 825 males and 825 females, the ridge count on all 10 fingers varied from 0 to 300, with mean values of approximately 145 for males ($SD = 51.1$) and 127 for females ($SD = 52.5$).

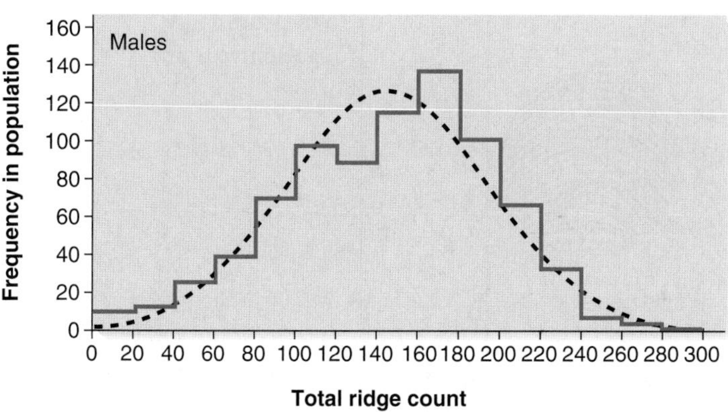

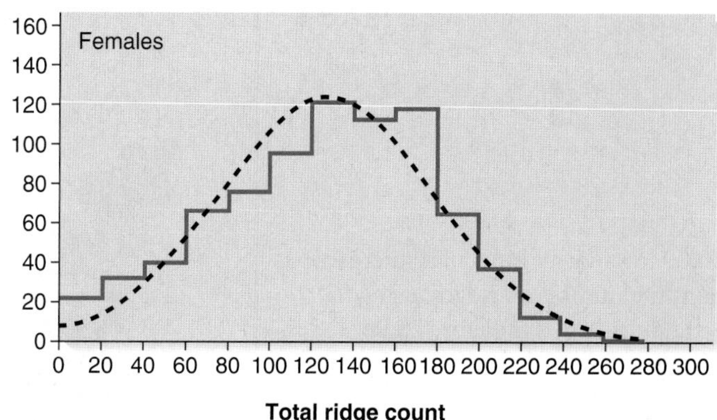

Based on these results, Holt decided to conduct a more detailed analysis of ridge counts by examining the fingerprint patterns of a large group of people and their close relatives. In the experiment of **Figure 25.10**, the ridge counts for pairs of related individuals were determined by the method described in Figure 25.9. The correlation coefficients for ridge counts were then calculated among the pairs of related or unrelated individuals. To estimate the narrow-sense heritability, the observed correlations were then divided by the expected correlations based on the known genetic relationships.

THE HYPOTHESIS

Dermal ridge count has a genetic component. The goal of this experiment is to determine the contribution of genetics in the variation of dermal ridge counts.

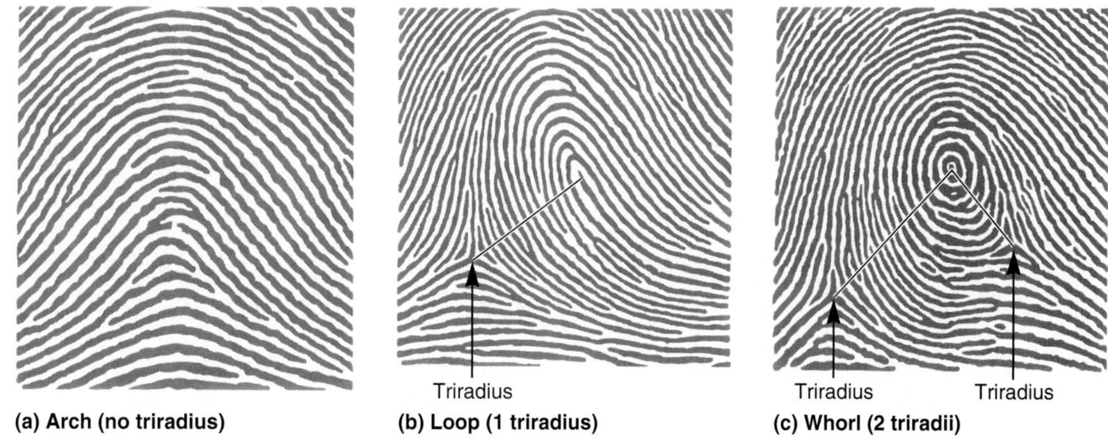

FIGURE 25.9 Human fingerprints and the ridge count method of Bonnevie. (**a**) This print has an arch rather than a triradius. The ridge count is zero. (**b**) This print has one triradius. A straight line is drawn from the triradius to the center of the print. The number of ridges dissecting this straight line is 13. (**c**) This print has two triradii. Straight lines are drawn from both triradii to the center. There are 16 ridges touching the left line and 7 touching the right line, giving a total ridge count of 23.

TESTING THE HYPOTHESIS — FIGURE 25.10 Heritability of human fingerprint patterns.

Starting material: A group of human subjects from Great Britain.

	Experimental level	**Conceptual level**
1. Take a person's finger and blot it onto an ink pad.		
2. Roll the person's finger onto a recording surface to obtain a print.		This is a method to measure a quantitative trait.
3. With a low-power binocular microscope, count the number of ridges, using the method described in Figure 25.9.		
4. Calculate the correlation coefficients between different pairs of individuals as described earlier in this chapter.	See the data.	The correlation coefficient provides a way to determine the heritability for the quantitative trait.

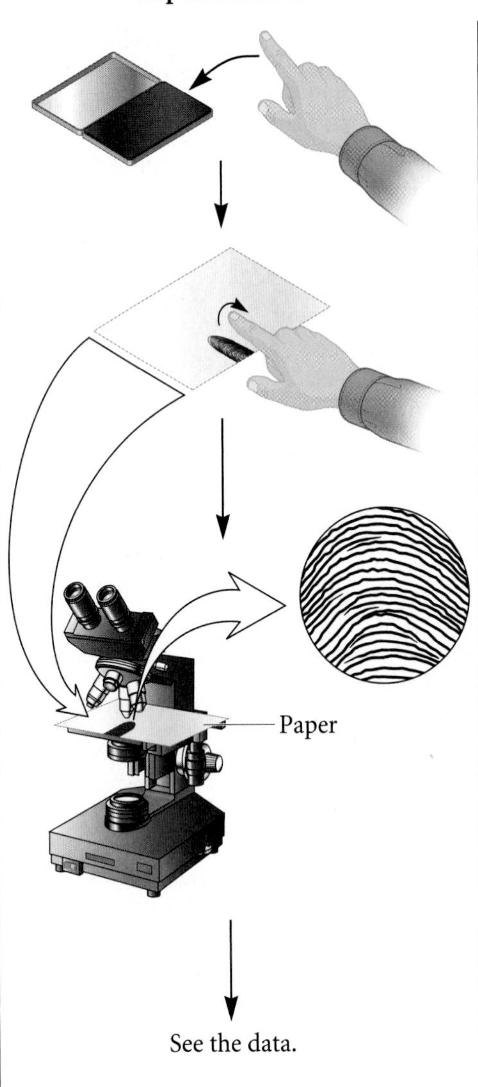

THE DATA

Type of Relationship	*Number of Pairs Examined*	*Correlation Coefficient* (r_{obs})	*Heritability* r_{obs}/r_{exp}
Parent-child	810	0.48 ± 0.04*	0.96
Parent-parent	200	0.05 ± 0.07	—†
Sibling-sibling	642	0.50 ± 0.04	1.00
Identical twins	80	0.95 ± 0.01	0.95
Fraternal twins	92	0.49 ± 0.08	0.98
			0.97 average heritability

*± = Standard error of the mean.

†Note: We cannot calculate a heritability value because the value for r_{exp} is not known. Nevertheless, the value for r_{obs} is very low, suggesting that there is a negligible correlation between unrelated individuals.

Adapted from Holt, S. B. (1961). Quantitative Genetics of Finger-Print Patterns. *British Med. Bull.* 17, 247–250.

INTERPRETING THE DATA

As seen in the data table, the results indicate that genetics plays the major role in explaining the variation in this trait. Genetically unrelated individuals (namely, parent-parent relationships) have a negligible correlation for this trait. By comparison, individuals who are genetically related have a substantially higher correlation. When the observed correlation coefficient is divided by the expected correlation coefficient based on the known genetic relationships, the average heritability value is 0.97, which is very close to 1.0.

What do these high heritability values mean? They indicate that nearly all of the variation in fingerprint pattern is due to genetic variation. Significantly, fraternal and identical twins have substantially different observed correlation coefficients, even though we expect that they have been raised in very similar environments. These results support the idea that genetics is playing the major role in promoting variation and that the results are not biased heavily by environmental similarities that may be associated with genetically related individuals. From an experimental viewpoint, the results show us how the determination of correlation coefficients between related and unrelated individuals can provide insight regarding the relative contributions of genetics and environment to the variation of a quantitative trait.

A self-help quiz involving this experiment can be found at the Online Learning Center.

Heritability Values Are Relevant Only to Particular Groups Raised in a Particular Environment

Table 25.4 describes heritability values that have been calculated for traits in particular populations. Unfortunately, heritability is a widely misunderstood concept. Heritability describes the amount of phenotypic variation due to genetic variation for a particular population raised in a particular environment. The words *variation, particular population,* and *particular environment* cannot be overemphasized. For example, in one population of cattle, the heritability for milk production may be 0.35, while in another group (with less genetic variation), the heritability may be 0.1. Second, if a group displays a heritability of 1.0 for a particular trait, this does not mean that the environment is unimportant in affecting the outcome of the trait. A heritability value of 1.0 only means that the amount of variation within this group is due to genetics. Perhaps the group has been raised in a relatively homogeneous environment, so the environment has not caused a significant amount of variation. Nevertheless, the environment may be quite important. It just is not causing much variation within this particular group.

TABLE 25.4

Examples of Heritabilities for Quantitative Traits

Trait	Heritability Value*
Humans	
Stature	0.65
IQ testing ability	0.60
Cattle	
Body weight	0.65
Butterfat, %	0.40
Milk yield	0.35
Mice	
Tail length	0.40
Body weight	0.35
Litter size	0.20
Poultry	
Body weight	0.55
Egg weight	0.50
Egg production	0.10

*As emphasized in this chapter, these values apply to particular populations raised in particular environments. The value for IQ testing ability is an average value from many independent studies. The other values were taken from Falconer, D. S. (1989). *Introduction to Quantitative Genetics,* 3d ed. Longman, Essex, England.

As a hypothetical example, let's suppose that we take a species of rodent and raise a group on a poor diet; we find their weights range from 1.5 to 2.5 pounds, with a mean weight of 2 pounds. We allow them to mate and then raise their offspring on a healthy diet of rodent chow. The weights of the offspring range from 2.5 to 3.5 pounds, with a mean weight of 3 pounds. In this hypothetical experiment, we might find a positive correlation in which the small parents tended to produce small offspring, and the large parents produce large offspring. The correlation of weights between parent and offspring might be, say, 0.5. In this case, the heritability for weight would be calculated as r_{obs}/r_{exp}, which equals 0.5/0.5, or 1.0. The value of 1.0 means that all the variation within the groups is due to genetic factors. The offspring vary from 2.5 to 3.5 pounds because of genetic variation, and also the parents range from 1.5 to 2.5 pounds because of genetics. However, as we see here, environment has played an important role. Presumably, the mean weight of the offspring is higher because of their better diet. This example is meant to emphasize the point that heritability tells us only the relative contributions of genetic variation and environment in influencing phenotypic *variation* in a *particular population* in a *particular environment.* Heritability does not describe the relative importance of these two factors in determining the outcomes of traits. When a heritability value is high, it does not mean that a change in the environment cannot have a major impact on the outcome of the trait.

With regard to the roles of genetics and environment (sometimes referred to as nature versus nurture), the topic of human intelligence has been hotly debated. As a trait, intelligence is difficult to define or to measure. Nevertheless, performance on an IQ test has been taken by some people as a reflection of intelligence ever since 1916 when Alfred Binet's test was used in the United States. Even though such tests may have inherent bias and consider only a limited subset of human cognitive abilities, IQ tests still remain a method of assessing intelligence. By comparing IQ scores among related and unrelated individuals, various studies have attempted to estimate heritability values in selected human populations. These values have ranged from 0.3 to 0.8. A heritability value of around 0.6 is fairly common among many studies. Such a value indicates that over half of the heritability for IQ testing ability is due to genetic factors.

Let's consider what a value of 0.6 means, and what it does not mean. It means that 60% of the variation in IQ testing ability is due to genetic variation in a selected population raised in a particular environment. It does not mean that 60% of an individual's IQ testing ability is due to genetics and 40% is due to the environment. Heritability is meaningless at the level of a single individual. Furthermore, even at the population level, a heritability value of 0.6 does not mean that 60% of the IQ testing ability is due to genetics and 40% is due to the environment. Rather, it means that in the selected population that was examined, 60% of the variation in IQ testing ability is due to genetics, while 40% of the variation is due to the environment. Heritability is strictly a population value that pertains to variation.

Selective Breeding of Species Can Alter Quantitative Traits Dramatically

The term **selective breeding** refers to programs and procedures designed to modify phenotypes in species of economically important plants and animals. This phenomenon, also called **artificial selection,** is related to natural selection, discussed in Chapter 24. In forming his theory of natural selection, Charles Darwin was influenced by his observations of selective breeding by pigeon fanciers and other breeders. The primary difference between artificial and natural selection is how the parents are chosen. Natural selection is due to natural variation in reproductive success. In artificial selection, the breeder chooses individuals that possess traits that are desirable from a human perspective.

For centuries, humans have been practicing selective breeding to obtain domestic species with interesting or agriculturally useful characteristics. The common breeds of dogs and cats have been obtained by selective breeding strategies (**Figure 25.11**). As shown here, it is very striking how selective breeding can modify the quantitative traits in a species. When comparing a greyhound with a bulldog, the magnitude of the differences is fairly amazing. They hardly look like members of the same species.

Likewise, most of the food we eat is obtained from species that have been modified profoundly by selective breeding strategies. This includes products such as grains, fruits, vegetables, meat, milk, and juices. **Figure 25.12** illustrates how certain characteristics in the wild mustard plant (*Brassica oleracea*) have been modified by selective breeding to create several varieties of important domesticated crops. This plant is native to Europe and Asia, and plant breeders began to modify its traits approximately

Greyhound

German shepherd

Bulldog

Cocker spaniel

FIGURE 25.11 Some common breeds of dogs that have been obtained by selective breeding.

Genes→Traits By selecting parents carrying the alleles that influence certain quantitative traits in a desired way, dog breeders have produced breeds with distinctive sets of traits. For example, the bulldog has alleles that give it short legs and a flat face. By comparison, the corresponding genes in a German shepherd are found in alleles that produce longer legs and a more pointy snout. All the dogs shown in this figure carry the same kinds of genes (e.g., many genes that affect their sizes, shapes, and fur color). However, the alleles for many of these genes are different among these dogs, thereby producing breeds with strikingly different phenotypes.

4,000 years ago. As seen here, certain quantitative traits in the domestic strains, such as stems and lateral buds, differ considerably from those of the original wild species.

The phenomenon that underlies selective breeding is variation. Within a group of individuals, allelic variation may affect the outcome of quantitative traits. The fundamental strategy of the selective breeder is to choose parents that will pass on alleles to their offspring that produce desirable phenotypic characteristics. For example, if a breeder wants large cattle, the largest members of the herd will be chosen as parents for the next generation. These large cattle will transmit an array of alleles to their offspring that confer large size. The breeder will often choose genetically related individuals (e.g., brothers and sisters) as the parental stock. As mentioned previously, the practice of mating between genetically related individuals is known as inbreeding. Some of the consequences of inbreeding are also described in Chapter 24.

What is the outcome when selective breeding is conducted for a quantitative trait? **Figure 25.13a** shows the results of a program begun at the Illinois Agricultural Experiment Station in 1896, even before the rediscovery of Mendel's laws. This experiment began with 163 ears of corn with an oil content ranging from 4 to 6%. In each of 80 succeeding generations, corn plants

Wild mustard plant

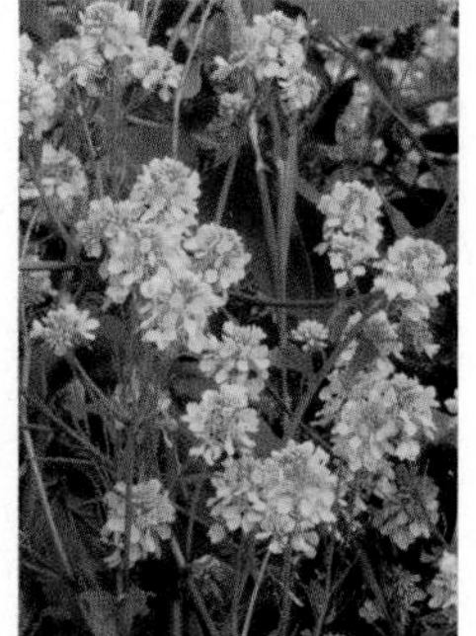

Strain	Modified trait
Kohlrabi	Stem
Kale	Leaves
Broccoli	Flower buds and stem
Brussels sprouts	Lateral leaf buds
Cabbage	Terminal leaf bud
Cauliflower	Flower buds

FIGURE 25.12 Crop plants developed by selective breeding of the wild mustard plant (*Brassica oleracea*).

Genes→Traits The wild mustard plant carries a large amount of genetic (i.e., allelic) variation, which was used by plant breeders to produce modern strains that are agriculturally desirable and economically important. For example, by selecting for alleles that promote the formation of large lateral leaf buds, the strain of Brussels sprouts was created. By selecting for alleles that alter the leaf morphology, kale was developed. Although these six agricultural plants look quite different from each other, they carry many of the same alleles as the wild mustard. However, they differ in alleles affecting the formation of stems, leaves, flower buds, and leaf buds.

were divided into two separate groups. In one group, members with the highest oil content were chosen as parents of the next generation. In the other group, members with the lowest oil content were chosen. After 80 generations, the oil content in the first group rose to over 18%; in the other group, it dropped to less than 1%. These results show that selective breeding can modify quantitative traits in a very directed manner.

Similar results have been obtained for many other quantitative traits. **Figure 25.13b** shows an experiment by Kenneth Mather conducted in the 1940s, in which flies were selected on the basis of their bristle number. The starting group had an average of 40 bristles for females and 35 bristles for males. After eight generations, the group selected for high bristle number had an average of 46 bristles for females and 40 for males, while the group selected for low bristle number had an average of 36 bristles for females and 30 for males.

When comparing the curves in Figure 25.13, keep in mind that quantitative traits are often at an intermediate value in unselected populations. Therefore, artificial selection can increase or decrease the magnitude of the trait. Oil content can go up or down, and bristle number can increase or decrease. Artificial selection tends to be the most rapid and effective in changing the frequency of alleles that are at intermediate range in a starting population, such as 0.2 to 0.8.

Figure 25.13 also shows the phenomenon known as a **selection limit**—after several generations a plateau is reached where artificial selection is no longer effective. A selection limit may occur for two reasons. Presumably, the starting population possesses a large amount of genetic variation, which contributes to the diversity in phenotypes. By carefully choosing the parents, each succeeding generation has a higher proportion of the desirable alleles. However, after many generations, the population may be nearly monomorphic for all or most of the desirable alleles that affect the trait of interest. At this point, additional selective breeding will have no effect. When this occurs, the heritability for the trait is near zero, because nearly all genetic variation for the trait of interest has been eliminated from the population. Without the introduction of new mutations into the population, further selection is not possible. A second reason for a selection limit is related to fitness. Some alleles that accumulate in a population due to artificial selection have a negative impact on the population's overall fitness. A selection limit is reached in which the desired effects of artificial selection are balanced by the negative effects on fitness.

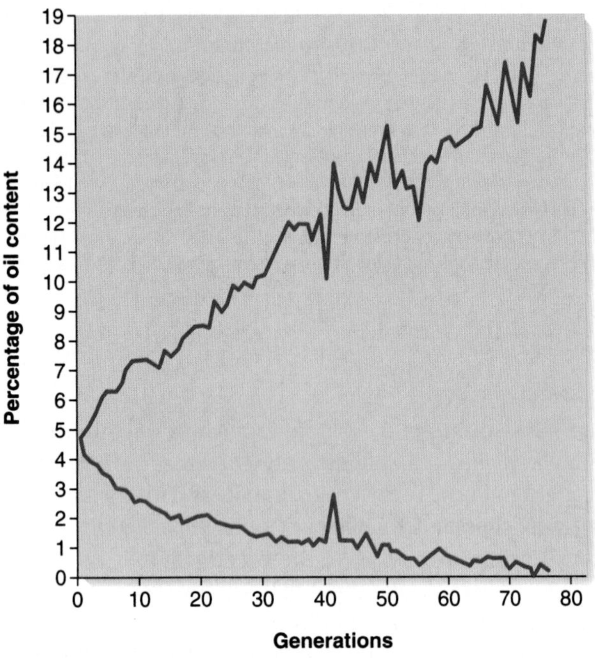

(a) Results of selective breeding for high and low oil content in corn

(b) Results of selective breeding for high and low bristle number in flies

FIGURE 25.13 **Common results of selective breeding for a quantitative trait.** (**a**) Selection for high and low oil content in corn. (**b**) Selection for high and low bristle number in flies.

Using artificial selection experiments, the response to selection is a common way to estimate the narrow-sense heritability in a starting population. The narrow-sense heritability measured in this way is also called the **realized heritability.** It is calculated as

$$h_N^2 = \frac{R}{S}$$

where

R is the response in the offspring to selection, or the difference between the mean of the offspring and the mean of the population of the parents' generation.

S is the selection differential in the parents, or the difference between the mean of the parents and the mean of the population.

Here,

$$R = \overline{X}_O - \overline{X}$$

$$S = \overline{X}_P - \overline{X}$$

where

$\overline{X}$ is the mean of the starting population
$\overline{X}_O$ is the mean of the offspring
$\overline{X}_P$ is the mean of the parents

So,

$$h_N^2 = \frac{\overline{X}_O - \overline{X}}{\overline{X}_P - \overline{X}}$$

The narrow-sense heritability is the proportion of the variance in phenotype that can be used to predict changes in the population mean when selection is practiced.

As an example, let's suppose we began with a population of fruit flies in which the average bristle number for both sexes was 37.5. The parents chosen from this population had an average bristle number of 40. The offspring of the next generation had an average bristle number of 38.7. With these values, the realized heritability is

$$h_N^2 = \frac{38.7 - 37.5}{40 - 37.5}$$

$$h_N^2 = \frac{1.2}{2.5}$$

$$h_N^2 = 0.48$$

This result tells us that about 48% of the phenotypic variation is due to the additive effects of alleles.

An important aspect of narrow-sense heritabilities is their ability to predict the outcome of selective breeding. Solved problem S1 at the end of the chapter illustrates this idea.

Heterosis May Be Explained by Dominance or Overdominance

As we have just seen, selective breeding can alter the phenotypes of domesticated species in a highly directed way. An unfortunate consequence of inbreeding, however, is that it may inadvertently promote homozygosity for deleterious alleles. This phenomenon is called **inbreeding depression.** In addition, genetic drift, described in Chapter 24, may contribute to the loss of beneficial alleles. In agriculture, it is widely observed that when two different inbred strains are crossed to each other, the resulting offspring are more vigorous (e.g., larger or longer-lived) than either of the inbred parental strains. This phenomenon is called **heterosis,** or **hybrid vigor.**

In modern agricultural breeding practices, many strains of plants and animals consist of hybrids produced by crossing two different inbred lines. In fact, much of the success of agricultural breeding programs is founded in heterosis. In rice, for example, hybrid strains have a 15 to 20% yield advantage over the best conventional inbred varieties under similar cultivation conditions.

As shown in **Figure 25.14**, two different phenomena may contribute to heterosis. In 1908, Charles Davenport developed the dominance hypothesis, in which the effects of dominant alleles explain the favorable outcome in a heterozygote. He suggested that highly inbred strains have become homozygous for one or more recessive genes that are somewhat deleterious (but not lethal). Because the homozygosity occurs by chance, two different inbred strains are likely to be homozygously recessive for different genes. Therefore, when they are crossed to each other, the resulting hybrids are heterozygous and do not suffer the consequences of homozygosity for deleterious recessive alleles. In other words, the benefit of the dominant alleles explains the observed heterosis. Steven Tanksley, working with colleagues in China, found that heterosis in rice seems to be due to the phenomenon of dominance. This is a common explanation for heterosis.

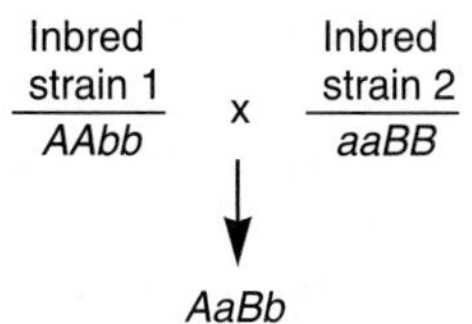
Inbred strain 1
AAbb
x
Inbred strain 2
aaBB
AaBb
The recessive alleles (a and b) are slightly harmful in the homozygous condition.
The hybrid offspring is more vigorous, because the harmful effects of the recessive alleles are masked by the dominant alleles.
The Dominance Hypothesis

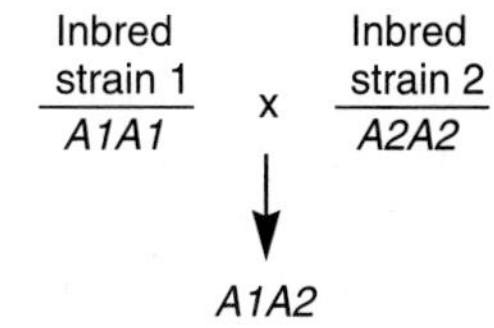
Inbred strain 1
A1A1
x
Inbred strain 2
A2A2
A1A2
Neither the A1 nor A2 allele is recessive.
The hybrid offspring is more vigorous, because the heterozygous combination of alleles exhibits overdominance. This means that the A1A2 heterozygote is more vigorous than either the A1A1 or A2A2 homozygote.
The Overdominance Hypothesis

FIGURE 25.14 **Mechanisms to explain heterosis.** The two common explanations are the dominance hypothesis and the overdominance hypothesis.

In 1908, George Shull and Edward East proposed a second hypothesis, known as the overdominance hypothesis (Figure 25.14). As described in Chapter 4, overdominance occurs when the heterozygote is more vigorous than either corresponding homozygote. According to this idea, heterosis can occur because the resulting hybrids are heterozygous for one or more genes that display overdominance. The heterozygote is more vigorous than either homozygote. In corn, Charles Stuber and his colleagues have found that several QTLs for grain yield support the overdominance hypothesis.

Finally, it should be pointed out that overdominance is very difficult to distinguish from **pseudo-overdominance,** a phenomenon initially suggested by James Crow. Pseudo-overdominance is really the same as dominance, except that the chromosomal region contains two or more genes that are very closely linked. For example, a QTL may be identified in a mapping experiment to be close to a particular molecular marker. However, a QTL could contain two genes, both affecting the same quantitative trait. For example, at a single QTL, two genes, *a* and *B*, may be closely linked in one strain, while *A* and *b* would be closely linked in another strain. The hybrid is really heterozygous (*AaBb*) for two different genes, but this may be difficult to discern in mapping experiments because the genes are so close together. If a researcher assumed there was only one gene at a QTL, the overdominance hypothesis would be favored, whereas if two genes were actually present, the dominance hypothesis would be correct. Therefore, without very fine mapping, which is rarely done for QTLs, it is hard to distinguish between overdominance and pseudo-overdominance.

CONCEPTUAL SUMMARY

Quantitative genetics is the study of traits that vary in a continuous way in populations. Such complex or quantitative traits can be anatomical, physiological, or behavioral, and some diseases exhibit characteristics and inheritance patterns analogous to those of quantitative traits. Typically, such traits are determined by two or more genes and are said to be polygenic. Within populations, quantitative traits commonly exhibit a continuum of phenotypic variation that follows a normal distribution. Statistical methods can be used to analyze such a distribution. The mean ($\overline{X}$) describes the average value among the population; the variance (V_X) measures the spread of values from the mean. The standard deviation (SD) is the square root of the variance. About 68% of all values will be within one SD above or below the mean, and about 95% are within two SDs of the mean. The correlation coefficient (r) evaluates the strength of association between two variables. In genetics, it is often used to see if there are phenotypic correlations between genetically related individuals.

The chromosomal location of one or more genes that influence quantitative traits is called a quantitative trait locus (QTL). The reasons most quantitative traits show a continuum are that each trait is governed by several genes existing in two or more alleles, and the environment contributes a substantial amount of variation to the phenotypic outcome.

In the simplest situation, phenotypic variance is due to the additive effects of genetic variance and environmental variance. More commonly, however, interactions and associations may occur between genetic and environmental factors. Heritability is the fraction of the phenotypic variance that is attributable to genetic variation; it describes the amount of phenotypic variation that is due to genetic variation for a particular population raised in a particular environment. Broad-sense heritability takes into account all of the genetic categories that could affect the phenotype. By comparison, the heritability of a trait due to the additive effects of alleles is called the narrow-sense heritability. For many quantitative traits, the broad-sense and narrow-sense heritabilities are similar.

The term selective breeding refers to programs and procedures designed to modify phenotypes in species of economically important plants and animals in a highly directed way. The fundamental strategy is to choose parents that will pass on alleles to their offspring that produce desirable phenotypic characteristics. Selective breeding typically involves the repeated breeding of related individuals to produce strains that are highly inbred. Much of the success of agricultural breeding programs is founded in heterosis, or hybrid vigor, the production of offspring that are more vigorous than either of the inbred parental strains. Heterosis may be due to the dominant effects of masking harmful recessive alleles or to overdominance of alleles in the heterozygous state.

EXPERIMENTAL SUMMARY

Geneticists measure the quantitative traits of an individual and describe them numerically. In a population, experimentally obtained numerical values from each individual are used to calculate statistics such as the mean, standard deviation, variance, and correlation coefficient.

Several methods can be used to determine the number of QTLs that influence a quantitative trait. Crow used genetic markers to explore the linkage between genes affecting quantitative traits and genes affecting discontinuous traits. His results showed that insecticide resistance is a polygenic trait involving multiple genes on different chromosomes. Newer research techniques have used molecular markers, such as RFLPs, microsatellites, and single nucleotide polymorphisms (SNPs), to determine the regions in the genome where QTLs for a given quantitative trait are likely to reside.

A central issue for quantitative geneticists is the relative contributions of genetics and the environment that underlie the phenotypic variation seen in quantitative traits. In this chapter, we have considered two experimental methods to measure the heritability of quantitative traits. One method is to compare the correlation coefficients among related and unrelated individuals. Holt's experiments with fingerprint patterns showed how the determination of correlation coefficients between related and unrelated individuals can provide insight regarding the relative contributions of genetics and environment to the variation of a quantitative trait. A second method is selective breeding, which is frequently used to improve quantitative traits in agriculturally important species. Quantitative traits are often at an intermediate value in unselected populations. Therefore, artificial selection can increase or decrease the magnitude of a trait. The calculation of realized heritability can be used to predict changes in the population mean when selection is practiced.

PROBLEM SETS & INSIGHTS

Solved Problems

S1. The narrow-sense heritability for potato weight in a starting population of potatoes is 0.42, and the mean weight is 1.4 lb. If a breeder crosses two strains with average potato weights of 1.9 and 2.1 lb, respectively, what is the predicted average weight of potatoes in the offspring?

Answer: The mean weight of the parental strains is 2.0 lb. To solve for the mean weight of the offspring:

$$h_N^2 = \frac{\overline{X}_O - \overline{X}}{\overline{X}_P - \overline{X}}$$

$$0.42 = \frac{\overline{X}_O - 1.4}{2.0 - 1.4}$$

$$\overline{X}_O = 1.65 \text{ lb}$$

S2. A farmer wants to increase the average body weight in a herd of cattle. She begins with a herd having a mean weight of 595 kg and chooses individuals to breed that have a mean weight of 625 kg. Twenty offspring were obtained, having the following weights in kilograms: 612, 587, 604, 589, 615, 641, 575, 611, 610, 598, 589, 620, 617, 577, 609, 633, 588, 599, 601, and 611. Calculate the realized heritability in this herd with regard to body weight.

Answer:

$$h_N^2 = \frac{R}{S}$$

$$= \frac{\overline{X}_O - \overline{X}}{\overline{X}_P - \overline{X}}$$

We already know the mean weight of the starting herd (595 kg) and the mean weight of the parents (625 kg). The only calculation missing is the mean weight of the offspring, $\overline{X}_O$.

$$\overline{X}_O = \frac{\text{Sum of the offspring's weights}}{\text{Number of offspring}}$$

$$\overline{X}_O = 604 \text{ kg}$$

$$h_N^2 = \frac{604 - 595}{625 - 595}$$

$$= 0.3$$

S3. The following data describe the 6-week weights of mice and their offspring of the same sex:

Parent (g)	Offspring (g)
24	26
21	24
24	22
27	25
23	21
25	26
22	24
25	24
22	24
27	24

Calculate the correlation coefficient.

Answer: To calculate the correlation coefficient, we first need to calculate the means and standard deviations for each group:

$$\overline{X}_{parents} = \frac{24 + 21 + 24 + 27 + 23 + 25 + 22 + 25 + 22 + 27}{10} = 24$$

$$\overline{X}_{offspring} = \frac{26 + 24 + 22 + 25 + 21 + 26 + 24 + 24 + 24 + 24}{10} = 24$$

$$SD_{parents} = \sqrt{\frac{0 + 9 + 0 + 9 + 1 + 1 + 4 + 1 + 4 + 9}{9}} = 2.1$$

$$SD_{offspring} = \sqrt{\frac{4 + 0 + 4 + 1 + 9 + 4 + 0 + 0 + 0 + 0}{9}} = 1.6$$

Next, we need to calculate the covariance.

$$CoV_{(parents,\, offspring)} = \frac{\Sigma[(X_P - \overline{X}_P)(X_O - \overline{X}_O)]}{N - 1}$$

$$= \frac{0 + 0 + 0 + 3 + 3 + 2 + 0 + 0 + 0 + 0}{9}$$

$$= 0.9$$

Finally, we calculate the correlation coefficient:

$$r_{(parent,\, offspring)} = \frac{CoV_{(P,O)}}{SD_P SD_O}$$

$$r_{(parent,\, offspring)} = \frac{0.9}{(2.1)(1.6)}$$

$$r_{(parent,\, offspring)} = 0.27$$

S4. As described in this chapter, the correlation coefficient provides a way to determine the strength of association between two variables. When the variables are related due to cause and effect (i.e., one variable affects the outcome of another variable), researchers may use a regression analysis to predict how much one variable will change in response to the other variable. This is easier to understand if we plot the data for two variables. The graph shown here compares mothers' and offspring's body weights in cattle. The line running through the data points is called a regression line. It is the straight line that is closest to all the data points. It is the minimal distance away from the squared vertical distance of all the points.

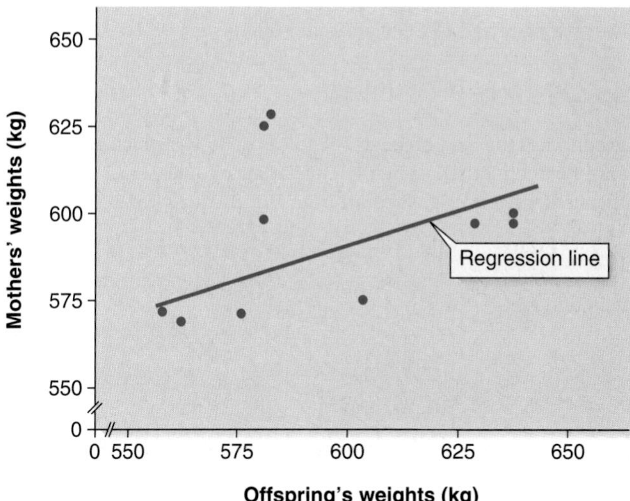

For many types of data, particularly those involving quantitative traits, the regression line can be represented by the equation

$$Y = bX + a$$

where

b is the regression coefficient

a is a constant

In this example, X is the value of an offspring's weight, and Y is the value of its mother's weight. The value of b, known as the **regression coefficient,** represents the slope of the regression line. The value of a is the y-intercept (i.e., the value of Y when X equals zero). The equation shown is very useful because it allows us to predict the value of X at any given value of Y, and vice versa. To do so, we first need to determine the values of b and a. This can be accomplished in the following manner:

$$b = \frac{CoV_{(X,Y)}}{V_X}$$

$$a = \overline{Y} - b\overline{X}$$

Once the values of a and b have been computed, we can use them to predict the values of X or Y by using the equation

$$Y = bX + a$$

For example, if $b = 0.5$, $a = 2$, and $Y = 31$, this equation can be used to compute a value of X that equals 58. It is important to keep in mind that this equation predicts the average value of X. As we see in the preceding figure, the data points tend to be scattered around the regression line. Deviations may occur between the data points and the line. The equation predicts the values that are the most likely to occur. In an actual experiment, however, some deviation will occur between the predicted values and the experimental values due to random sampling error.

Now here is the question. Using the data found in this chapter regarding weight in cattle, what is the predicted weight of an offspring if its mother weighed 660 lb?

Answer: We first need to calculate a and b.

$$b = \frac{CoV_{(X,Y)}}{V_X}$$

We need to use the data on page 701 to calculate V_X, which is the variance for the mothers' weights. The variance equals 445.1. The covariance is already calculated on page 701; it equals 152.6.

$$b = \frac{152.6}{445.1}$$

$$b = 0.34$$

$$a = \overline{Y} - b\overline{X}$$

$$a = 598 - (0.34)(596) = 395.4$$

Now we are ready to calculate the predicted weight of the offspring using the equation

$$Y = bX + a$$

In this problem, $X = 660$ pounds

$$Y = 0.34(660) + 395.4$$

$$Y = 619.8 \text{ lb}$$

The average weight of the offspring is predicted to be 619.8 lb.

S5. Genetic variance can be used to estimate the number of genes affecting a quantitative trait by using the following equation:

$$n = \frac{D^2}{8V_G}$$

where

n is the number of genes affecting the trait

D is the difference between the mean values of the trait in two strains that have allelic differences at every gene that influences the trait

V_G is the genetic variance for the trait; it is calculated using data from both strains

For this method to be valid, several assumptions must be met. In particular, the alleles of each gene must be additive, each gene must contribute equally to the trait, all the genes must assort independently, and the two strains must be homozygous for alternative alleles of each gene. For example, if three genes affecting a quantitative trait exist in two alleles each, one strain could be *AA bb CC* and the other would be *aa BB cc*. In addition, the strains must be raised under the same environmental conditions. Unfortunately, these assumptions are not typically met with regard to most quantitative traits. Even so, when one or more assumptions are invalid, the calculated value of n is smaller than the actual number. Therefore, this calculation can be used to estimate the minimum number of genes that affect a quantitative trait.

Now here is the question. The average bristle number in two strains of flies was 35 and 42. The genetic variance for bristle number calculated for both strains was 0.8. What is the minimum number of genes that affect bristle number?

Answer: We apply the equation described previously.

$$n = \frac{D^2}{8V_G}$$

$$n = \frac{(35 - 42)^2}{8(0.8)}$$

$$n = 7.7 \text{ genes}$$

Because genes must come in whole numbers and because this calculation is a minimum estimate, one would conclude there must be at least eight genes that affect bristle number.

S6. Are the following statements regarding heritability true or false?

A. Heritability applies to a specific population raised in a particular environment.

B. Heritability in the narrow sense takes into account all types of genetic variance.

C. Heritability is a measure of the amount that genetics contributes to the outcome of a trait.

Answer:

A. True

B. False. Narrow-sense heritability considers only the effects of additive alleles.

C. False. Heritability is a measure of the amount of phenotypic variation that is due to genetic variation; it applies to the variation of a specific population raised in a particular environment.

Conceptual Questions

C1. Give several examples of quantitative traits. How are these quantitative traits described within groups of individuals?

C2. At the molecular level, explain why quantitative traits often exhibit a continuum of phenotypes within a population. How does the environment help produce this continuum?

C3. What is a normal distribution? Discuss this curve with regard to quantitative traits within a population. What is the relationship between the standard deviation and the normal distribution?

C4. Explain the difference between a continuous trait and a discontinuous trait. Give two examples of each. Are quantitative traits likely to be continuous or discontinuous? Explain why.

C5. What is a frequency distribution? Explain how the graph is made for a quantitative trait that is continuous.

C6. The variance for weight in a particular herd of cattle is 484 lb^2. The mean weight is 562 lb. How heavy would an animal have to be if it was in the top 2.5% of the herd? The bottom 0.13%?

C7. Two different varieties of potatoes both have the same mean weight of 1.5 lb. One group has a very low variance, and the other has a much higher variance.

A. Discuss the possible reasons for the differences in variance.

B. If you were a potato farmer, would you rather raise a variety with a low or high variance? Explain your answer from a practical point of view.

C. If you were a potato breeder and you wanted to develop potatoes with a heavier weight, would you choose the variety with a low or high variance? Explain your answer.

C8. If an r value equals 0.5 and $N = 4$, would you conclude a positive correlation is found between the two variables? Explain your answer. What if $N = 500$?

C9. What does it mean when a correlation coefficient is negative? Can you think of examples?

C10. When a correlation coefficient is statistically significant, what do you conclude about the two variables? What do the results mean with regard to cause and effect?

C11. What is polygenic inheritance? Discuss the issues that make polygenic inheritance difficult to study.

C12. What is a quantitative trait locus (QTL)? Does a QTL contain one gene or multiple genes? What technique is commonly used to identify QTLs?

C13. Let's suppose that weight in a species of mammals is polygenic, and each gene exists as a heavy and light allele. If the allele frequencies in the population were equal for both types of allele (i.e., 50% heavy alleles and 50% light alleles), what percentage of individuals would be homozygous for the light alleles at all of the genes affecting this trait, if the trait was determined by the following number of genes?

A. Two

B. Three

C. Four

C14. The broad-sense heritability for a trait equals 1.0. In your own words, explain what this value means. Would you conclude that the environment is unimportant in the outcome of this trait? Explain your answer.

C15. Compare and contrast the dominance and overdominance hypotheses. Based on your knowledge of mutations and genetics, which do you think tends to be the more common explanation for heterosis?

C16. What is hybrid vigor (also known as heterosis)? Give examples that you might find in a vegetable garden.

C17. From an agricultural point of view, discuss the advantages and disadvantages of selective breeding. It is common for plant breeders to take two different, highly inbred strains, which are the product of many generations of selective breeding, and cross them to make hybrids. How does this approach overcome some of the disadvantages of selective breeding?

C18. Many beautiful varieties of roses have been produced, particularly in the last few decades. These newer varieties often have very striking and showy flowers, making them desirable as horticultural specimens. However, breeders and novices alike have noticed that some of these newer varieties do not have very fragrant flowers compared to the older, more traditional varieties. From a genetic point of view, suggest an explanation why some of these newer varieties with superb flowers are not as fragrant.

C19. In your own words, explain the meaning of the term heritability. Why is a heritability value valid only for a particular population of individuals raised in a particular environment?

C20. What is the difference between broad-sense heritability and narrow-sense heritability? Why is narrow-sense heritability such a useful concept in the field of agricultural genetics?

C21. The heritability for egg weight in a group of chickens on a farm in Maine is 0.95. Are the following statements regarding heritability true or false? If a statement is false, explain why.

A. The environment in Maine has very little impact on the outcome of this trait.

B. Nearly all of the phenotypic variation for this trait in this group of chickens is due to genetic variation.

C. The trait is polygenic and likely to involve a large number of genes.

D. Based on the observation of the heritability in the Maine chickens, it is reasonable to conclude that the heritability for egg weight in a group of chickens on a farm in Montana is also very high.

C22. In a fairly large population of people living in a commune in the southern United States, everyone cares about good nutrition. All the members of this population eat very nutritional foods, and their diets are very similar to each other. With regard to height, how do you think this commune population would compare to the general population in the following categories?

A. Mean height

B. Heritability for height

C. Genetic variation for alleles that affect height

C23. When artificial selection is practiced over many generations, it is common for the trait to reach a plateau in which further selection has little effect on the outcome of the trait. This phenomenon is illustrated in Figure 25.13. Explain why.

C24. Discuss whether a natural population of wolves or a domesticated population of German shepherds is more likely to have a higher heritability for the trait of size.

C25. With regard to heterosis, would the following statements be consistent with the dominance hypothesis, the overdominance hypothesis, or both?

A. Strains that have been highly inbred have become monomorphic for one or more recessive alleles that are somewhat detrimental to the organism.

B. Hybrid vigor occurs because highly inbred strains are monomorphic for many genes, while hybrids are more likely to be heterozygous for those same genes.

C. If a gene exists in two alleles, hybrids are more vigorous because heterozygosity for the gene is more beneficial than homozygosity of either allele.

Experimental Questions

E1. Here are data for height and weight among 10 male college students.

Height (cm)	Weight (kg)
159	48
162	50
161	52
175	60
174	64
198	81
172	58
180	74
161	50
173	54

A. Calculate the correlation coefficients for this group.

B. Is the correlation coefficient statistically significant? Explain.

E2. The abdomen length (in millimeters) was measured in 15 male *Drosophila*, and the following data were obtained: 1.9, 2.4, 2.1, 2.0, 2.2, 2.4, 1.7, 1.8, 2.0, 2.0, 2.3, 2.1, 1.6, 2.3, and 2.2. Calculate the mean, standard deviation, and variance for this population of male fruit flies.

E3. You will need to understand solved problem S5 before answering this question. The average weights for two varieties of cattle were 514 kg and 621 kg. The genetic variance for weight calculated for both strains was 382 kg^2. What is the minimum number of genes that affect weight variation in these two varieties of cattle?

E4. Using the same strategy as the experiment of Figure 25.6, the following data are the survival of F_2 offspring obtained from backcrosses to insecticide-resistant and control strains:

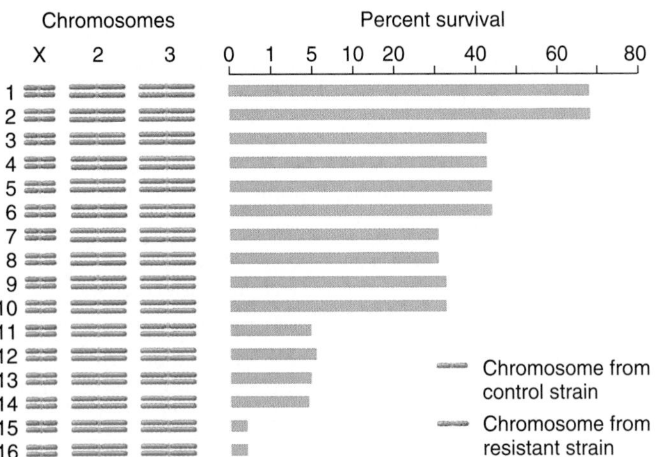

Interpret these results with regard to the locations of QTLs.

E5. In one strain of cabbage, you conduct an RFLP analysis with regard to head weight; you determine that seven QTLs affect this trait. In another strain of cabbage, you find that only four QTLs affect this trait. Note that both strains of cabbage are from the same species, although they may have been subjected to different degrees of inbreeding. Explain how one strain can have seven QTLs and another strain four QTLs for exactly the same trait. Is the second strain missing three genes?

E6. From an experimental viewpoint, what does it mean to say that an RFLP is associated with a trait? Let's suppose that two strains of pea plants differ in two RFLPs that are linked to two genes governing pea size. RFLP-1 is found in 2,000 bp and 2,700 bp bands, and RFLP-2 is found in 3,000 bp and 4,000 bp bands. The plants producing large peas have RFLP-1 (2,000 bp) and RFLP-2 (3,000 bp); those producing small peas have RFLP-1 (2,700 bp) and RFLP-2 (4,000 bp). A cross is made between these two strains, and the F_1 offspring are allowed to self-fertilize. Five phenotypic classes are observed: small peas, small-medium peas, medium peas, medium-large peas, and large peas. We assume that each of the two genes makes an equal contribution to pea size and that the genetic variance is additive. Draw a gel and explain what RFLP banding patterns you would expect to observe for these five phenotypic categories. Note: Certain phenotypic categories may have more than one possible banding pattern.

E7. Let's suppose that two strains of pigs differ in 500 RFLPs. One strain is much larger than the other. The pigs are crossed to each other, and the members of the F_1 generation are also crossed among themselves to produce an F_2 generation. Three distinct RFLPs are associated with F_2 pigs that are larger. How would you interpret these results?

E8. Outline the steps you would follow to determine the number of genes that influence the yield of rice. Describe the results you might get if rice yield is governed by six different genes.

E9. A researcher has two highly inbred strains of mice. One strain is susceptible to infection by a mouse leukemia virus, while the other strain is resistant. Susceptibility/resistance is a polygenic trait. The two strains were crossed together, and all the F_1 mice were resistant. The F_1 mice were then allowed to interbreed, and 120 F_2 mice were obtained. Among these 120 mice, 118 were resistant to the viral pathogen, and 2 were sensitive. Discuss how many different genes may be involved in this trait. How would your answer differ if none of the F_2 mice had been susceptible to the leukemia virus? Hint: You should assume that the inheritance of one viral-resistance allele is sufficient to confer resistance.

E10. In a wild strain of tomato plants, the phenotypic variance for tomato weight is 3.2 g^2. In another strain of highly inbred tomatoes raised under the same environmental conditions, the phenotypic variance is 2.2 g^2. With regard to the wild strain,

A. Estimate V_G.

B. What is $h_B{}^2$?

C. Assuming all the genetic variance is additive, what is $h_N{}^2$?

E11. The average thorax length in a *Drosophila* population is 1.01 mm. You want to practice selective breeding to make larger *Drosophila*. To do so, you choose 10 parents (five males and five females) of the following sizes: 0.97, 0.99, 1.05, 1.06, 1.03, 1.21, 1.22, 1.17, 1.19, and 1.20. You mate them and then analyze the thorax sizes of 30 offspring (half male and half female):

0.99, 1.15, 1.20, 1.33, 1.07, 1.11, 1.21, 0.94, 1.07, 1.11, 1.20, 1.01, 1.02, 1.05, 1.21, 1.22, 1.03, 0.99, 1.20, 1.10, 0.91, 0.94, 1.13, 1.14, 1.20, 0.89, 1.10, 1.04, 1.01, 1.26

Calculate the realized heritability in this group of flies.

E12. In a strain of mice, the average 6-week body weight is 25 g and the narrow-sense heritability for this trait is 0.21.

A. What would be the average weight of the offspring if parents with a mean weight of 27 g were chosen?

B. What weight of parents would you have to choose to obtain offspring with an average weight of 26.5 g?

E13. Two tomato strains, A and B, both produce fruit that weighs, on average, 1 lb each. All of the variance is due to V_G. When these two strains are crossed to each other, the F_1 offspring display heterosis with regard to fruit weight, with an average weight of 2 lb. You take these F_1 offspring and backcross them to strain A. You then grow several plants from this cross and measure the weights of their fruit. What would be the expected results for each of the following scenarios?

A. Heterosis is due to a single overdominant gene.

B. Heterosis is due to two dominant genes, one in each strain.

C. Heterosis is due to two overdominant genes.

D. Heterosis is due to dominance of several genes each from strains A and B.

E14. You will need to understand solved problem S4 before answering this question. With regard to height, the variance for fathers (in square inches) was 112, the variance for sons was 122, and the covariance was 144. The mean height for fathers was 68 in., and the mean height for sons was 69 in. If a father had a height of 70 in., what is the most probable height of his son?

E15. A danger in computing heritability values from studies involving genetically related individuals is the possibility that these individuals share more similar environments than do unrelated individuals. In the experiment of Figure 25.10, which data are the most compelling evidence that ridge count is not caused by genetically related individuals sharing common environments? Explain.

E16. A large, genetically heterogeneous group of tomato plants was used as the original breeding stock by two different breeders, named Mary and Hector. Each breeder was given 50 seeds and began an artificial selection strategy, much like the one described in Figure 25.13. The seeds were planted, and the breeders selected the 10 plants with the highest mean tomato weights as the breeding stock for the next generation. This process was repeated over the course of 12 growing seasons, and the following data were obtained:

	Mean Weight of Tomatoes (lb)	
Year	Mary's Tomatoes	Hector's Tomatoes
1	0.7	0.8
2	0.9	0.9
3	1.1	1.2
4	1.2	1.3
5	1.3	1.3
6	1.4	1.4
7	1.4	1.5
8	1.5	1.5
9	1.5	1.5
10	1.5	1.5
11	1.5	1.5
12	1.5	1.5

(continued)

A. Explain these results.

B. Another tomato breeder, named Martin, got some seeds from Mary's and Hector's tomato strains (after 12 generations), grew the plants, and then crossed them to each other. The mean weight of the tomatoes in these hybrids was about 1.7 lb. For a period of five years, Martin subjected these hybrids to the same experimental strategy that Mary and Hector had followed, and he obtained the following results:

	Mean Weight of Tomatoes (lb)
Year	Martin's Tomatoes
1	1.7
2	1.8
3	1.9
4	2.0
5	2.0

Explain Martin's data. Is heterosis occurring? Why was Martin able to obtain tomatoes heavier than 1.5 lb, while Mary's and Hector's strains appeared to plateau at this weight?

E17. The correlations for height were determined for 15 pairs of individuals with the following genetic relationships:

Mother/daughter: 0.36

Mother/granddaughter: 0.17

Sister/sister: 0.39

Sister/sister (fraternal twins): 0.40

Sister/sister (identical twins): 0.77

What is the average heritability for height in this group of females?

E18. An animal breeder had a herd of sheep with a mean weight of 254 lb at 3 years of age. He chose animals with mean weights of 281 lb as parents for the next generation. When these offspring reached 3 years of age, their mean weights were 269 lb.

A. Calculate the narrow-sense heritability for weight in this herd.

B. Using the heritability value that you calculated in part A, what weight of animals would you have to choose to get offspring that weigh 275 lb (at 3 years of age)?

E19. The trait of blood pressure in humans has a frequency distribution that is similar to a normal distribution. The following graph shows the ranges of blood pressures for a selected population of people. The red line depicts the frequency distribution of the systolic pressures for the entire population. Several individuals with high blood pressure were identified, and the blood pressures of their relatives were determined. This frequency distribution is depicted with a blue line. (Note: The blue line does not include the people who were identified with high blood pressure; it includes only their relatives.)

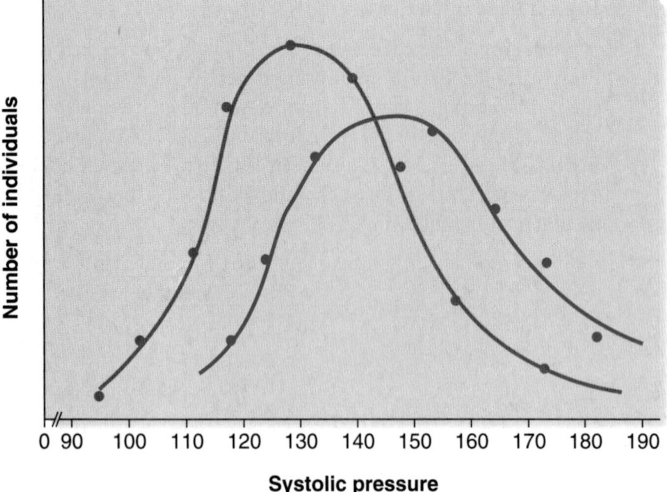

What do these data suggest with regard to a genetic basis for high blood pressure? What statistical approach could you use to determine the heritability for this trait?

Questions for Student Discussion/Collaboration

1. Discuss why heritability is an important phenomenon in agriculture. Discuss how it is misunderstood.
2. From a biological viewpoint, speculate as to why many traits seem to fit a normal distribution. Students with a strong background in math and statistics may want to explain how a normal distribution is generated, and what it means. Can you think of biological examples that do not fit a normal distribution?
3. What is heterosis? Discuss whether it is caused by a single gene or several genes. Discuss the two major hypotheses proposed to explain heterosis. Which do you think is more likely to be correct?

Note: All answers appear at the website for this textbook; the answers to even-numbered questions are in the back of the textbook.

www.mhhe.com/brookergenetics3e

Visit the Online Learning Center for practice tests, answer keys, and other learning aids for this chapter. Enhance your understanding of genetics with our interactive exercises, quizzes, animations, and much more.

EVOLUTIONARY GENETICS

26

CHAPTER OUTLINE

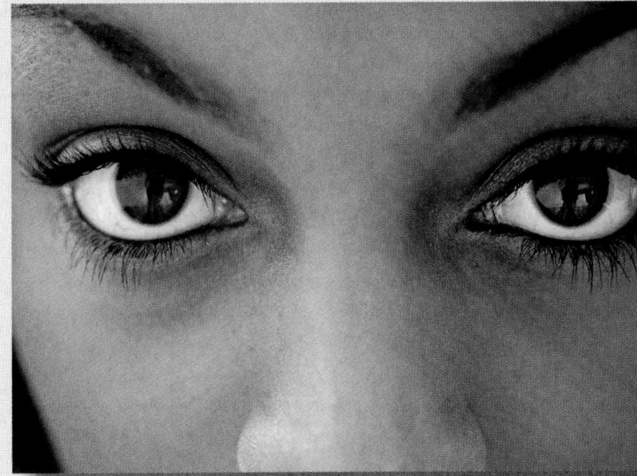

The evolution of eyes. Developmental biologists have recently discovered that the eyes of many diverse species, including fruit flies, frogs, mice, and people, are under the control of the homologous gene called *Pax6*, suggesting that animal eyes may have evolved from the same ancestral species.

Biological evolution is a heritable change in one or more characteristics of a population or species across many generations. Evolution can be viewed on a small scale as it relates to a single gene or it can be viewed on a larger scale as it relates to the formation of new species. In Chapter 24, we examined several factors that cause allele frequencies to change in populations. This process, also known as **microevolution,** concerns the changing composition of gene pools with regard to particular alleles over measurable periods of time. As we have seen, several evolutionary mechanisms, such as mutation, genetic drift, migration, natural selection, and inbreeding, affect the allele and genotypic frequencies within natural populations. On a microevolutionary scale, evolution can be viewed as a change in allele frequency over time.

The goal of this chapter is to relate phenotypic changes that occur during evolution to the underlying genetic changes that cause them to happen. In the first part of the chapter, we will be concerned with evolution on a large scale, which leads to the origin of new species. The question of how species form has been central to the development of evolutionary theory. The term **macroevolution** refers to large-scale evolutionary changes that create new species and higher taxa. It concerns the establishment of the diversity of organisms over long periods of time through the accumulated evolution and extinction of many species.

In the second part of this chapter, we will link molecular genetics to the evolution of species. The development of techniques for analyzing chromosomes and DNA sequences has greatly enhanced our understanding of evolutionary processes at the molecular level. The term **molecular evolution** refers to the molecular changes in genetic material that underlie the phenotypic changes associated with evolution. In this section, we will see how molecular data can provide information about the phylogenetic relationships among different organisms. Finally, as discussed in the third part of the chapter, we will see how genes that affect embryonic development can have a dramatic impact on the phenotypes of organisms. The field of evolutionary developmental biology (evo-devo) focuses on the role of developmental genes in the formation of traits that are important in the evolution of new species.

The topics of molecular evolution and evo-devo are a fitting way to end our discussion of genetics because they integrate the ongoing theme of this textbook—the relationship between molecular genetics and traits—in the broadest and most profound ways. Theodosius Dobzhansky, an influential evolutionary scientist, once said, "Nothing in biology makes sense except in the light of evolution." The extraordinarily diverse and seemingly bizarre array of species on our planet can be explained naturally within the context of evolution. An examination of molecular evolution allows us to make sense of the existence of these species at both the population and the molecular levels.

26.1 ORIGIN OF SPECIES

Charles Darwin, a British naturalist born in 1809, proposed the theory of evolution and provided evidence that existing species have evolved from preexisting species. Like many great scientists, Darwin had a broad background in science, which enabled him to see connections among different disciplines. His thinking was influenced by theories of geology indicating that the Earth is very old and that slow geological processes can lead eventually to substantial changes in the Earth's characteristics.

Darwin's own experimental observations also had a great impact on his thinking. His famous voyage on the *HMS Beagle*, which lasted from 1832 to 1836, involved a careful examination of many different species. He observed the similarities among many discrete species, yet noted the differences that enabled them to be adapted to their environmental conditions. He was particularly struck by the distinctive adaptations of island species. For example, the finches found on the Galápagos Islands had unique phenotypic characteristics compared to those of similar finches found on the mainland.

A third important influence on Darwin was a paper published in 1798, "Essay on the Principle of Population," by Thomas Malthus, an English economist. Malthus asserted that the population size of humans can, at best, increase arithmetically due to increased land usage and improvements in agriculture, whereas the reproductive potential of humans can increase geometrically. He argued that famine, war, and disease work to limit population growth, especially among the poor.

With these three ideas in mind, Darwin had largely formulated his theory of evolution by the mid-1840s. He then spent several years studying barnacles without having published his ideas. The geologist Charles Lyell, who had greatly influenced Darwin's thinking, strongly encouraged Darwin to publish his theory of evolution. In 1856, Darwin began to write a long book to explain his ideas. In 1858, however, Alfred Wallace, a naturalist working in the East Indies, sent Darwin an unpublished manuscript to read prior to its publication. In it, Wallace proposed the same ideas concerning evolution. Darwin therefore quickly excerpted some of his own writings on this subject, and two papers, one by Darwin and one by Wallace, were published in the *Proceedings of the Linnaean Society of London*. These papers were not widely recognized. A short time later, however, Darwin finished his book, *On the Origin of Species by Means of Natural Selection*, which expounded his ideas in greater detail and with experimental support. This book, which received high praise from many scientists and scorn from others, started a great debate concerning evolution. Although some of his ideas were incomplete because the genetic basis of traits was not understood at that time, Darwin's work represents one of the most important contributions to our understanding of biology.

Darwin called evolution "the theory of descent with modification through variation and natural selection." As its name suggests, evolution is based on two fundamental principles: genetic variation and natural selection. A modern interpretation of evolution can view these two principles at the species level (macroevolution) and at the level of genes in populations (microevolution).

1. Genetic variation at the species level: As we have seen in Chapter 24, genetic variation is a consistent feature of natural populations. Darwin observed that many species exhibit a great amount of phenotypic variation. Although the theory of evolution preceded Mendel's pioneering work in genetics, Darwin (as well as many other people before him) observed that offspring resemble their parents more than they do unrelated individuals. Therefore, he assumed that traits are passed from parent to offspring. However, the genetic basis for the inheritance of traits was not understood at that time.

   At the gene level: Genetic variation can involve allelic differences in genes, changes in chromosome structure, and alterations in chromosome number. These differences are caused by random mutations. Alternative alleles may affect the functions of the proteins they encode and thereby affect the phenotype of the organism. Likewise, changes in chromosome structure and number may affect gene expression and thereby influence the phenotype of the individual.
2. Natural selection at the species level: Darwin agreed with Malthus that most species produce many more offspring than will survive and reproduce, creating an ever-present struggle for existence. Over the course of many generations, those individuals who happen to possess the most favorable traits will dominate the composition of the population. The result of natural selection is to make a species better adapted to its environment and more successful at reproduction.

   At the gene level: Some alleles encode proteins that provide the individual with a selective advantage. Over time, natural selection may change the allele frequencies of genes and thereby lead to the fixation of beneficial alleles and the elimination of detrimental alleles.

In this section, we will examine the features of evolution as it occurs in natural populations over time.

Species Concepts Are Used in the Identification of Species

Before we begin to consider how biologists study the evolution of new species, we need to consider how species are defined and identified. A **species** refers to a group of organisms that maintains a distinctive set of attributes in nature. The difficulty of identifying whether certain groups constitute unique species is often rooted in the phenomenon that a single species may exist in two or more distinct populations that are in the slow process of evolving into different species. The amount of time that populations are separated from each other will have an important impact. If the time is short, the populations are likely to be very similar, so they would be considered the same species. If the time is long, significant changes may have occurred, so the different populations would show unequivocal differences that allow them to maintain their distinctive set of features in nature. When studying natural populations, evolutionary biologists are often confronted with situations where some differences between two populations are apparent, but it is difficult to decide whether they truly represent separate species. When two or more groups within the same species display one or more traits that are somewhat different but not enough to warrant their placement into different species, biologists sometimes classify such groups as subspecies.

Biologists adopt methods of species identification that are based on their own experience with the organisms they wish to study. The characteristics that a biologist uses to identify a species depend, in large part, on the species in question. For example, the traits used to distinguish insect species would be quite different from those used to identify bacterial species. The most commonly used characteristics are physical or morphological traits, the ability to interbreed, evolutionary lineages, and ecological factors. These various ways of considering species have led to the use of different approaches for distinguishing species called **species concepts** (**Table 26.1**). A comparison of these concepts will help you to appreciate the various factors involved in categorizing the species on our planet.

Phylogenetic Species Concept According to the **phylogenetic species concept,** as advocated by Quentin Wheeler and Norman Platnick, the members of a single species are identified by having a unique combination of character states.[1] Historically, the first way to categorize species was based on their physical characteristics. Organisms are classified as the same species if their anatomical traits appear to be very similar. Likewise, microorganisms can be classified according to morphological traits at the cellular level. In addition, molecular features, such as DNA sequences, can now be used to compare organisms. An advantage of the phylogenetic species concept is that it can be applied to all types of organisms.

Biological Species Concept In the late 1920s, Theodosius Dobzhansky proposed that each species is reproductively isolated from other species. Such **reproductive isolation** prevents one species from successfully interbreeding with other species. In 1942, Ernst Mayr expanded on the ideas of Dobzhansky to provide a biological definition of a species. According to Mayr's **biological species concept,** a species is a group of individuals whose members have the potential to interbreed with one another in nature to produce viable, fertile offspring, but cannot successfully interbreed with members of other species. How does reproductive isolation occur? **Table 26.2** describes several ways. These are classified as **prezygotic mechanisms,** which prevent the formation

TABLE 26.1

Species Concepts

Concept	Description
Phylogenetic species concept	Various physical characteristics can be analyzed to distinguish between species. These often include morphological (anatomical) traits. In the case of unicellular organisms, characteristics such as cell wall structure and other cellular traits may be examined. Molecular characteristics can also be compared.
Biological species concept	Two species are often judged to be separate species if they are unable to interbreed in nature to produce viable, fertile offspring.
Evolutionary species concept	An analysis of ancestry may help biologists determine if two groups are members of the same species or represent evolutionarily distinct species.
Ecological species concept	The ability of organisms to successfully occupy their own ecological niche or habitat, including their use of resources and impact on the environment, may be used to distinguish species.

[1] The phylogenetic species concept is related to other species concepts called the morphological species concept and typological species concept.

TABLE 26.2

Types of Reproductive Isolation Among Different Species

Prezygotic Mechanisms	
Habitat isolation	Species may occupy different habitats so that they never come in contact with each other.
Temporal isolation	Species have different mating or flowering seasons, mate at different times of day, or become sexually active at different times of the year.
Sexual isolation	Sexual attraction between males and females of different animal species is limited due to differences in behavior, physiology, or morphology.
Mechanical isolation	The anatomical structures of genitalia prevent mating between different species.
Gametic isolation	Gametic transfer takes place, but the gametes fail to unite with each other. This can occur because the male and female gametes fail to attract, because they are unable to fuse, or because the male gametes are inviable in the female reproductive tract of another species.
Postzygotic Mechanisms	
Hybrid inviability	The egg of one species is fertilized by the sperm from another species, but the fertilized egg fails to develop past early embryonic stages.
Hybrid sterility	The interspecies hybrid survives, but it is sterile. For example, the mule, which is sterile, is a cross between a female horse (*Equus caballus*) and a male donkey (*Equus asinus*).
Hybrid breakdown	The F_1 interspecies hybrid is viable and fertile, but succeeding generations (i.e., F_2, etc.) become increasingly inviable. This is usually due to the formation of less fit genotypes by genetic recombination.

of a zygote, and **postzygotic mechanisms,** which prevent the development of a viable and fertile individual after fertilization has taken place. Reproductive isolation in nature may be circumvented when species are kept in captivity. For example, different species of the genus *Drosophila* rarely mate with each other in nature. In the laboratory, however, it is fairly easy to produce interspecies hybrids.

Evolutionary Species Concept In 1961, American paleontologist George Gaylord Simpson proposed a species concept based on ancestry. According to the **evolutionary species concept,** a species is derived from a single lineage that is distinct from other lineages and has its own evolutionary tendencies and historical fate. A lineage is the genetic relationship between an individual or group of individuals and their ancestors. The evolutionary species concept is a theoretical viewpoint that is focused on the pathway that has led to the formation of each distinct species. In this regard, it can be applied to the formation of all species. One drawback of the evolutionary species concept is that it does not provide an easy way to identify a unique species. In most cases, lineages are difficult to examine and evaluate quantitatively. The interpretation of lineages involving fossils may be particularly difficult to evaluate due to incomplete fossil remains.

Ecological Species Concept The **ecological species concept,** described by American evolutionary biologist Leigh Van Valen in 1976, is a viewpoint that considers a species within its native environment. Each species occupies an ecological niche, which is the unique set of habitat resources that a species requires and the influence a species has on the environment and other species. Within their own niche, members of a given species compete with each other for survival. If two organisms are very similar, their needs will overlap, which results in competition. Such competing individuals are likely to be of the same species. According to this concept, species are formed because evolutionary mechanisms control how each type of species uses resources. This species concept is particularly useful in distinguishing bacterial species that do not reproduce sexually. Bacterial cells of the same species are likely to use the same types of resources (such as sugars and vitamins) and grow under the same types of conditions (temperature, pH, and so on).

Speciation Often Occurs Via a Branching Process Called Cladogenesis

Speciation is the process by which new species are formed via evolution. By examining the fossil record, evolutionary biologists have found two different patterns of speciation. During **anagenesis** (from the Greek *ana*, up, and *genesis*, origin), a single species evolves into a different species over the course of many generations (**Figure 26.1a**). During this process, the characteristics of the species change due to both random genetic drift and natural selection. As a result of natural selection, the new species may be better adapted to survive in its original environment, or the environment may have changed so that the new species is better adapted to the new surroundings.

By comparison, **cladogenesis** (from the Greek *clados*, branch) involves the division of a species into two or more species (**Figure 26.1b**). This form of speciation increases species diversity. Although cladogenesis is usually thought of as a splitting process, it commonly occurs as a budding process, which results in the original species plus one or more new species with different characteristics. If we view evolution as a tree, the new species bud from the original species and develop characteristics that prevent them from breeding with the original one.

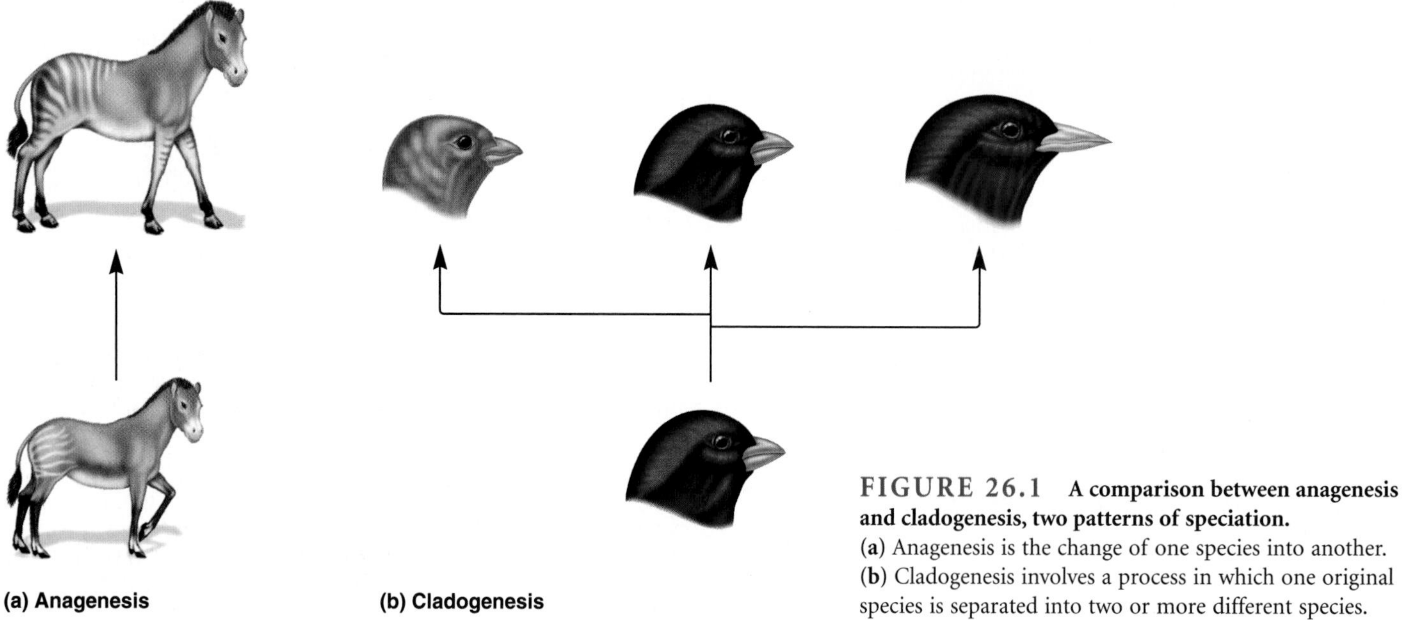

FIGURE 26.1 **A comparison between anagenesis and cladogenesis, two patterns of speciation.**
(**a**) Anagenesis is the change of one species into another. (**b**) Cladogenesis involves a process in which one original species is separated into two or more different species.

Cladogenesis Can Be Allopatric, Parapatric, or Sympatric

Depending on the geographic locations of the evolving population(s), cladogenesis is categorized as allopatric, parapatric, or sympatric (**Table 26.3**). **Allopatric speciation** (from the Greek *allos*, other, and Latin *patria*, homeland) is thought to be the most prevalent way for a species to diverge. It occurs when members of a species become geographically separated from the other members. This form of speciation can occur by the geographic subdivision of large populations via geological processes. For example, a mountain range may emerge and split a species that occupies the lowland regions, or a creeping glacier may divide a population. **Figure 26.2** shows an interesting example in which geological separation promoted speciation. Two species of antelope squirrels occupy opposite rims of the Grand Canyon. On the south rim is Harris's antelope squirrel (*Ammospermophilus harrisi*), while a closely related white-tailed antelope squirrel (*Ammospermophilus leucurus*) is found on the north rim. Presumably, these two species evolved from a common species that existed before the canyon was formed. Over time, the accumulation of genetic changes in the two separated populations led to the formation of two morphologically distinct species. Interestingly, birds that can easily fly across the canyon have not diverged into different species on the opposite rims.

Allopatric speciation can also occur via a second mechanism, known as the **founder effect,** which is thought to be more rapid and frequent than allopatric speciation caused by geological events. The founder effect, which was discussed in Chapter 24, occurs when a small group migrates to a new location that is geographically separated from the main population. For example, a storm may force a small group of birds from the mainland to a distant island. In this case, the migration of individuals between the island and the mainland is a very infrequent event. In a relatively short period of time, the founding population on the island may evolve into a new species. Two evolutionary mechanisms may contribute to this rapid evolution. First, genetic drift may quickly lead to the random fixation of certain alleles and the elimination of other alleles from the population. Another factor is natural selection. The environment on an island may differ significantly from the mainland environment. For this reason, natural selection on the island may favor different types of alleles.

A. harrisi

A. leucurus

FIGURE 26.2 An example of allopatric speciation: two closely related species of antelope squirrels that occupy opposite rims of the Grand Canyon.

Genes→Traits Harris's antelope squirrel (*Ammospermophilus harrisi*) is found on the south rim of the Grand Canyon, while the white-tailed antelope squirrel (*Ammospermophilus leucurus*) is found on the north rim. These two species evolved from a common species that existed before the canyon was formed. After the canyon was formed, the two separated populations accumulated genetic changes due to mutation, genetic drift, and natural selection that eventually led to the formation of two distinct species.

Parapatric speciation (from the Greek *para*, beside) occurs when members of a species are separated partially or when a species is very sedentary. In these cases, the geographic separation is not complete. For example, a mountain range may divide a species into two populations but have breaks in the range where the two groups are connected physically. In these zones of contact, the members of two populations can interbreed, although this tends to occur infrequently. Likewise, parapatric speciation may occur among very sedentary species even though no geographic isolation exists. Certain organisms are so sedentary that 100 to 1,000 meters may be sufficient to limit the interbreeding between neighboring groups. Plants, terrestrial snails, rodents, grasshoppers, lizards, and many flightless insects may speciate in a parapatric manner.

TABLE **26.3**

Common Genetic Mechanisms That Underlie Allopatric, Parapatric, and Sympatric Speciation

Type of Speciation	Common Genetic Mechanisms Responsible for Speciation
Allopatric—two large populations are separated by geographic barriers	Many small genetic differences may accumulate over a long period of time, leading to reproductive isolation. Some of these genetic differences may be adaptive, while others will be neutral.
Allopatric—a small founding population separates from the main population	Genetic drift may lead to the rapid formation of a new species. If the group has moved to an environment that is different from its previous environment, natural selection is expected to favor beneficial alleles and eliminate harmful alleles.
Parapatric—two populations occupy overlapping ranges, so a limited amount of interbreeding occurs	A new combination of alleles or chromosomal rearrangement may rapidly limit the amount of gene flow between neighboring populations because hybrid offspring have a very low fitness.
Sympatric—within a population occupying a single habitat in a continuous range, a small group evolves into a reproductively isolated species	An abrupt genetic change leads to reproductive isolation. For example, a mutation may affect gamete recognition. In plants, the formation of a tetraploid often leads to the formation of a new species because the interspecies hybrid (e.g., diploid × tetraploid) is triploid and sterile.

During parapatric speciation, **hybrid zones** may exist where two populations can interbreed. For speciation to occur, the amount of gene flow within the hybrid zones must become very limited. In other words, there must be selection against the offspring produced in the hybrid zone. One way that this can happen is that each of the two parapatric populations may accumulate different chromosomal rearrangements, such as inversions and balanced translocations. How do chromosomal rearrangements, such as inversions, prevent interbreeding? As discussed in Chapter 8, if a hybrid individual has one chromosome with a large inversion and one that does not carry the inversion, crossing over during meiosis can lead to the production of grossly abnormal chromosomes. Therefore, such an individual is substantially less fertile. By comparison, an individual homozygous for two normal chromosomes or for two chromosomes carrying the same inversion will be fertile, because crossing over can proceed normally.

Sympatric speciation (from the Greek *sym*, together) occurs when a new species arises in the same geographic area as the species from which it was derived. In plants, a common way for sympatric speciation to occur is the formation of polyploids. As discussed in Chapter 8, complete nondisjunction of chromosomes during gamete formation can increase the number of chromosome sets within a single species (autopolyploidy) or between different species (allopolyploidy). Polyploidy is so frequent in plants that it is a major form of speciation. In ferns and flowering plants, at least 30% of the species are polyploid. By comparison, polyploidy is much less common in animals, but it can occur. For example, roughly 30 species of reptiles and amphibians have been identified that are polyploids derived from diploid relatives.

The formation of a polyploid can abruptly lead to reproductive isolation. As an example, let's consider the probable events that led to the formation of a natural species of common hemp nettle known as *Galeopsis tetrahit*. This species is thought to be an allotetraploid derived from two diploid species, *Galeopsis pubescens* and *Galeopsis speciosa*. As shown in **Figure 26.3a**, *G. tetrahit* has 32 chromosomes, whereas the two diploid species contain 16 chromosomes each ($2n = 16$). **Figure 26.3b** illustrates what would happen in crosses between the allotetraploid

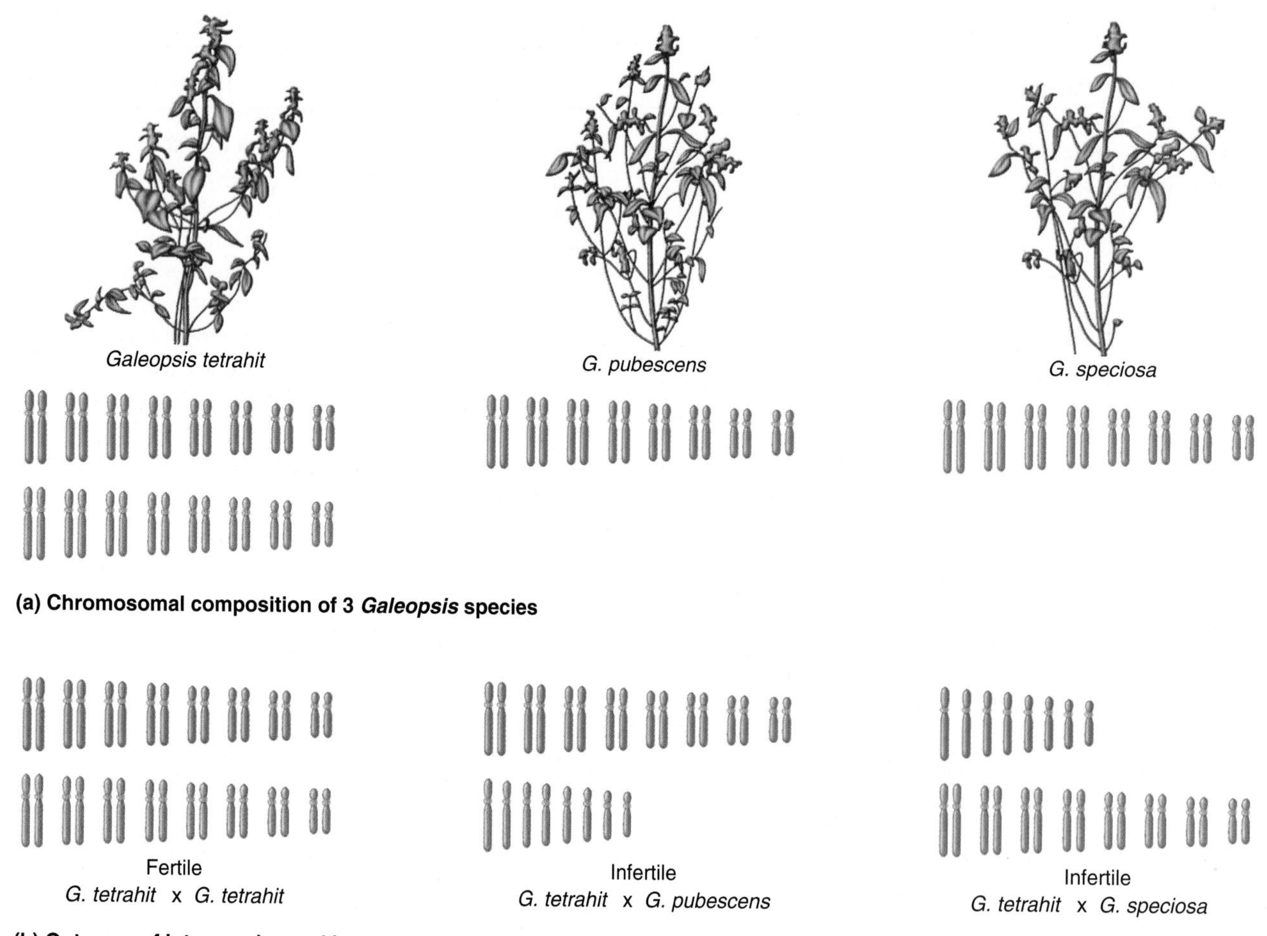

FIGURE 26.3 **A comparison of crosses between three natural species of hemp nettle with different ploidy levels.** *Galeopsis tetrahit* is an allotetraploid that is thought to be derived from *Galeopsis pubescens* and *Galeopsis speciosa*. If *G. tetrahit* is mated with the other two species, the F_1 hybrid offspring will be monoploid for one chromosome set and diploid for the other set. The F_1 offspring are likely to be sterile, because they will produce highly aneuploid gametes.

and the diploid species. The allotetraploid crossed to another allotetraploid produces an allotetraploid. The allotetraploid is fertile, because all of its chromosomes occur in homologous pairs that can segregate evenly during meiosis. However, a cross between an allotetraploid and a diploid produces an offspring that is monoploid for one chromosome set and diploid for the other set. These offspring are expected to be sterile, because they will produce highly aneuploid gametes that have incomplete sets of chromosomes. This hybrid sterility renders the allotetraploid reproductively isolated from the diploid species.

In the 1930s, Arne Müntzing first proposed that *Galeopsis tetrahit* is an allotetraploid that arose from an interspecies cross between *G. pubescens* and *G. speciosa.* To test this hypothesis, he performed a series of crosses between *G. pubescens* and *G. speciosa* and succeeded in producing an allotetraploid that had two sets of chromosomes from both species. This artificial *G. tetrahit* had traits similar to the natural *G. tetrahit* species but different from *G. pubescens* and *G. speciosa.* Furthermore, the artificial and natural *G. tetrahit* strains could be mated to each other to produce fertile offspring. Overall, Müntzing came to the conclusion that "not only the artificial tetraploid but probably also natural *G. tetrahit* represents a synthesis of *pubescens*- and *speciosa*-genomes."

Speciation Can Be Gradual or Punctuated by Periods of Rapid Change

As we have seen, many different genetic mechanisms give rise to new species. For this reason, the rates of evolutionary change are not constant, although the degree of inconstancy has been debated since the time of Darwin. Even Darwin himself suggested that evolution can occur at fast and slow paces. **Figure 26.4** illustrates contrasting views concerning the rates of evolutionary change. These views are not mutually exclusive but represent two different ways to consider the tempo of evolution. The concept of **gradualism** suggests that each new species evolves gradually over long spans of time (Figure 26.4a). The principal idea is that large phenotypic differences that produce the emergence of new species are due to the slow accumulation of many small genetic changes. By comparison, the concept of **punctuated equilibrium,** proposed by Niles Eldredge and Stephen Jay Gould, suggests that the tempo of evolution is more sporadic (Figure 26.4b). According to this model, species exist relatively unchanged for many generations. During this period, the species is in equilibrium with its environment. These long periods of equilibrium are punctuated by relatively short periods (on a geological timescale) during which evolution occurs at a far more rapid rate.

In reality, neither of the views presented in Figure 26.4 fully account for evolutionary change. The phenomenon of punctuated equilibrium is often supported by the fossil record. Paleontologists

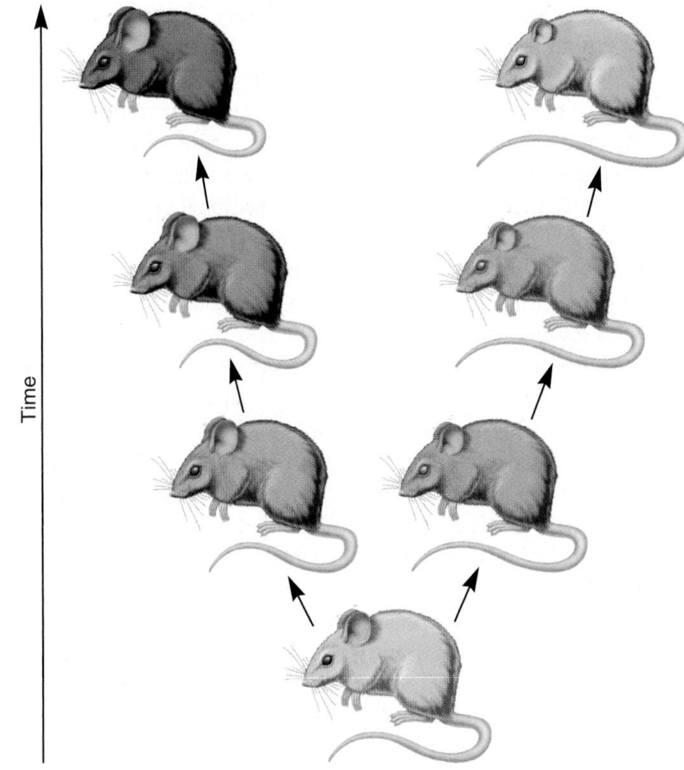

(a) Gradualism

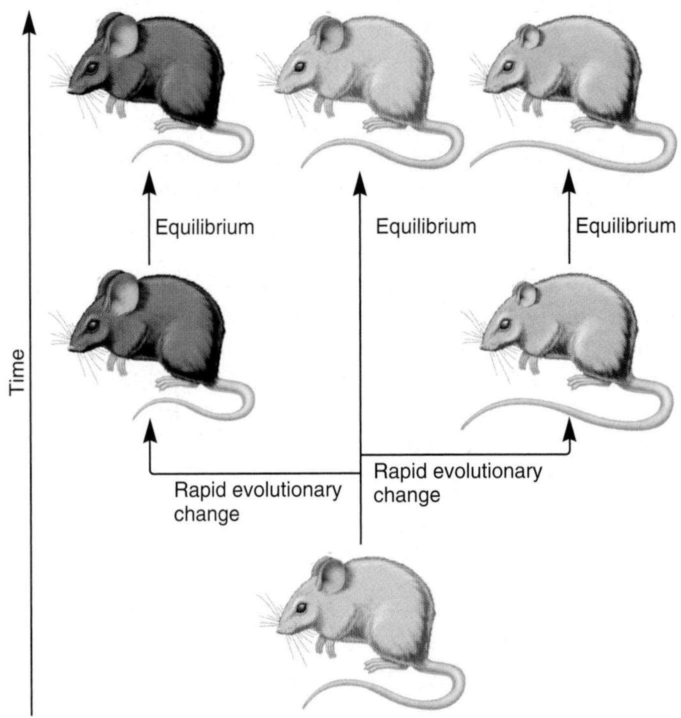

(b) Punctuated equilibrium

FIGURE 26.4 A comparison of gradualism and punctuated equilibrium. (**a**) During gradualism, a species gradually changes due to the accumulation of small genetic changes. (**b**) During punctuated equilibrium, a species exists essentially unchanged for long periods of time, during which it is in equilibrium with its environment. These equilibrium periods are punctuated by short periods of evolutionary change, during which its features may change rapidly.

rarely find a gradual transition of fossil forms. Instead, it is much more common to observe species appearing as new forms rather suddenly in a layer of rocks, persisting relatively unchanged for a very long period of time, and then suddenly becoming extinct. It is presumed that the transition period during which a previous species evolved into a new, morphologically distinct species was so short that few, if any, of the transitional members were preserved as fossils. Even so, these rapid periods of change were likely followed by long periods of equilibrium that involved the additional accumulation of many small genetic changes, consistent with gradualism.

As discussed earlier, rapid evolutionary change can be explained by genetic phenomena. As we have seen throughout this textbook, single-gene mutations can have dramatic effects on phenotypic characteristics. Therefore, only a small number of new mutations may be required to alter phenotypic characteristics, eventually producing a group of individuals that make up a new species. Likewise, genetic events such as changes in chromosome structure (e.g., inversions and translocations) or chromosome number may abruptly create individuals with new phenotypic traits. On an evolutionary timescale, these types of events can be rather rapid because one or only a few genetic changes can have a major impact on the phenotype of the organism.

In conjunction with genetic changes, species may also be subjected to sudden environmental shifts that quickly drive the gene pool in a particular direction via natural selection. For example, a small group may migrate to a new environment in which certain alleles provide better adaptation to the new surroundings. Alternatively, a species may be subjected to a relatively sudden environmental event that has a major impact on survival. There may be a change in climate, or a new predator may infiltrate the geographic range of the species. Natural selection may lead to rapid evolution of the gene pool by favoring those genetic changes that allow members of the population to survive the climatic change or more easily avoid the predator.

Overall, the fossil record and known genetic phenomena tend to support the idea that the tempo of evolution can be quite variable. In some cases, rapid evolutionary change has taken place and led to the formation of new species. During other periods, smaller phenotypic changes may occur over a longer timescale. In conjunction with phenotypic changes, the gradual accumulation of variations in gene sequences has been revealed by molecular analyses of DNA. Next, we will explore the use of molecular data to analyze evolution.

26.2 PHYLOGENETIC TREES AND MOLECULAR EVOLUTION

Thus far, we have considered the various factors that play a role in the formation of new species. In this section, we will examine **phylogeny**—the sequence of events involved in the evolutionary development of a species or group of species. A systematic approach is followed to produce a **phylogenetic tree,** which is a diagram that describes a phylogeny. Such a tree is a hypothesis of the evolutionary relationships among various species, based on the information available to and gathered by biologists termed **systematists.**

Historically, morphological differences have been used to construct evolutionary trees. In this approach, species that are more similar in appearance tend to be placed closer together on the tree. In addition, species have been categorized based on differences in physiology, biochemistry, and even behavior. While these approaches continue to be used, systematists are increasingly using molecular data to infer evolutionary relationships. With this approach, species are identified and placed into related groups based on the properties of their genetic material (i.e., their genotypes). Those species that are closely related evolutionarily are expected to have greater similarities in their genetic material than are distantly related species.

The advent of molecular approaches for analyzing DNA and gene products has revolutionized the field of evolution. Differences in nucleotide sequences are quantitative and can be analyzed using mathematical principles in conjunction with computer programs. Evolutionary changes at the DNA level can be objectively compared among different species to establish evolutionary relationships. Furthermore, this approach can be used to compare any two existing organisms, no matter how greatly they differ in their morphological traits. For example, we can compare DNA sequences between humans and bacteria, or between plants and fruit flies. Such comparisons would be very difficult at a morphological level. In this section, we will examine how phylogenetic trees are constructed, with an emphasis on the use of molecular data.

A Phylogenetic Tree Depicts the Evolutionary Relationships Among Different Species

Let's first take a look at what information is found within a phylogenetic tree and the form in which it is presented. **Figure 26.5** shows a hypothetical phylogenetic tree of the relationships between various insect species in which the species (butterflies) are labeled A through K. The vertical axis represents time, with the oldest species at the bottom.

As discussed previously, new species can form through anagenesis, in which a single species evolves into a different species, or through cladogenesis, in which a species diverges into two or more species. In cladogenesis, the original species may remain in existence when a second new species is formed. The nodes or branch points in a phylogenetic tree illustrate times when cladogenesis has occurred. For example, approximately 12 million years ago, species A diverged into species A and species B by cladogenesis. Figure 26.5 also shows anagenesis. After species B split into species B and D, species D then evolved into species G by anagenesis. The tips of branches represent either species that became extinct in the past, such as species C, B, and E, or modern species, such as F, I, G, J, H, and K, that are at the top of the tree. Species A and D are also extinct, but gave rise to modern species.

By studying the branch points of a phylogenetic tree, researchers can group species according to common ancestry. A **monophyletic group,** also known as a **clade,** is a group of species

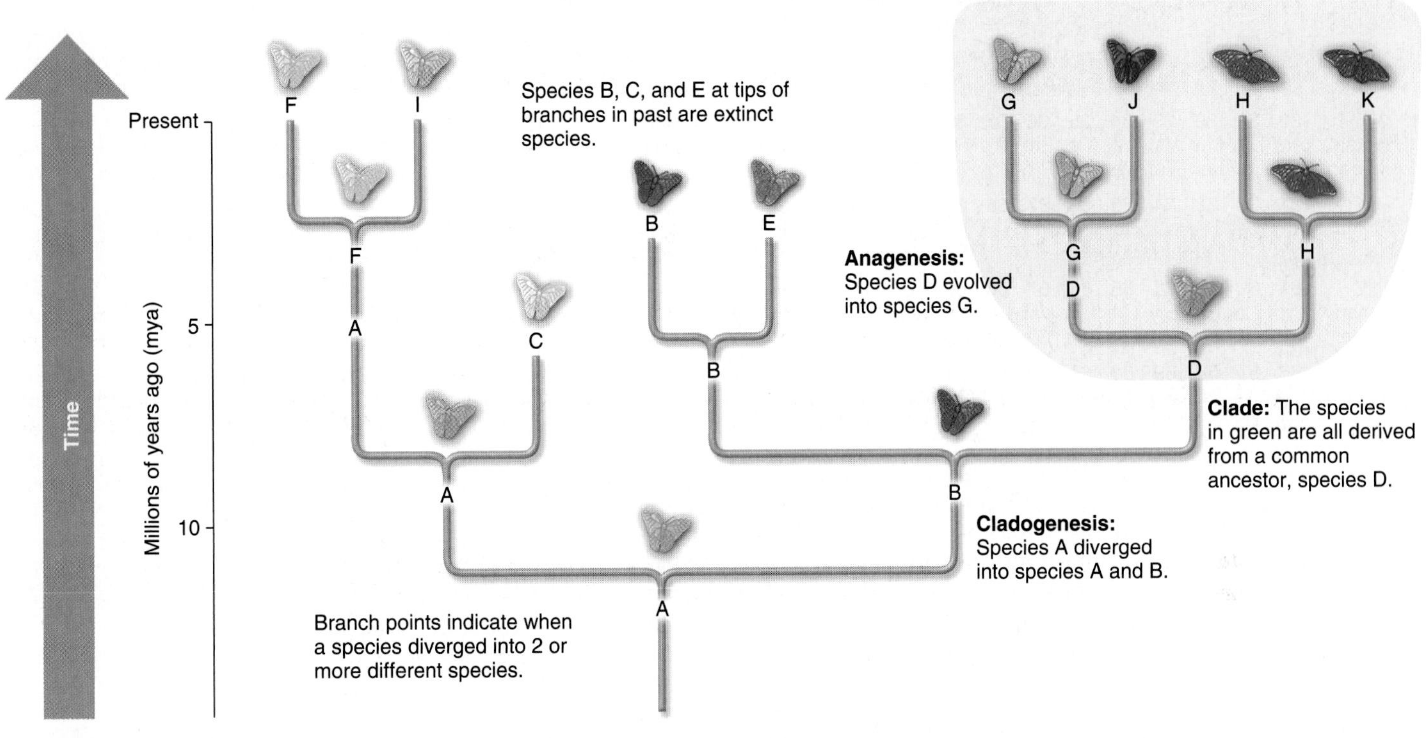
Species F – K, at the tips of branches, are modern species.
Time
Present
Millions of years ago (mya)
5
10
F
I
F
A
C
A
A
Species B, C, and E at tips of branches in past are extinct species.
B
E
B
B
G
J
H
K
G
D
H
D
Anagenesis:
Species D evolved into species G.
Clade: The species in green are all derived from a common ancestor, species D.
Cladogenesis:
Species A diverged into species A and B.
Branch points indicate when a species diverged into 2 or more different species.

FIGURE 26.5 **How to read a phylogenetic tree.** This hypothetical tree shows the proposed relationships among various butterfly species.

consisting of all descendents of the group's most common ancestor. For example, the group highlighted in light green is a clade derived from the common ancestor labeled D. The present-day descendents of a common ancestor can also be called a clade. In this case, species G, J, H, and K would form a modern clade. Likewise, the entire tree shown in Figure 26.5 forms a clade, with species A as a common ancestor. As we see in this figure, smaller and more recent clades are subsets of larger clades.

Homologous Genes Are Derived from a Common Ancestral Gene

A phylogenetic tree is based on **homology,** which refers to similarities among various species that occur because the species are derived from a common ancestor. Attributes that are the result of homology are said to be homologous. For example, the wing of a bat, the arm of a human, and the front leg of a cat are homologous structures. When constructing phylogenetic trees, researchers identify homologous features that are shared by some species but not by others. This allows them to group species based on their similarities.

Researchers typically study homology at the level of morphological traits or at the level of genes. We will focus on genetic homology. Two genes are said to be **homologous** if they are derived from the same ancestral gene. During evolution, a single species may become divided into two or more different species. When two homologous genes are found in different species, these genes are termed **orthologs.** In Chapter 23, we considered orthologs of *Hox* genes. In that case, several homologous genes were identified in the fruit fly and the mouse. In addition, two or more homologous genes can be found within a single species. These are termed paralogous genes, or **paralogs.** As discussed in Chapter 8, this can occur because abnormal gene duplication events can produce multiple copies of a gene and ultimately lead to the formation of a **gene family.** A gene family consists of two or more paralogs within the genome of a particular species.

Figure 26.6 shows examples of both orthologs and paralogs in the globin gene family. Hemoglobin is an oxygen-carrying protein found in all vertebrate species. It is composed of two different subunits, encoded by the α-globin and β-globin genes. Figure 26.6 shows the deduced amino acid sequences encoded by these genes. The sequences are homologous between humans and horses because of their evolutionary relationship. We would say that the human and horse α-globin genes are orthologs of each other, as are the human and horse β-globin genes. The α-globin and β-globin genes in humans are paralogs of each other.

If you scan the sequences shown in Figure 26.6, you may notice that the sequences of the orthologs are more similar to each other than the paralogs. For example, the sequences of human and horse β globins are much more similar than the sequences of the human α globin and human β globin. What do these results mean? They indicate that the gene duplications that

Hemoglobin

β	1	2	3	4	5	6	7	8	9	10	11	12	13	14	15	16	17	18	—	—	19	20	21	22	23	24	25	26	27	28	29
Human	Val	His	Leu	Thr	Pro	Glu	Glu	Lys	Ser	Ala	Val	Thr	Ala	Leu	Trp	Gly	Lys	Val	—	—	Asn	Val	Asp	Glu	Val	Gly	Gly	Glu	Ala	Leu	Gly
Horse	Val	Gln	Leu	Ser	Gly	Glu	Glu	Lys	Ala	Ala	Val	Leu	Ala	Leu	Trp	Asp	Lys	Val	—	—	Asn	Glu	Glu	Glu	Val	Gly	Gly	Glu	Ala	Leu	Gly
α	1	—	2	3	4	5	6	7	8	9	10	11	12	13	14	15	16	17	18	19	20	21	22	23	24	25	26	27	28	29	30
Human	Val	—	Leu	Ser	Pro	Ala	Asp	Lys	Thr	Asn	Val	Lys	Ala	Ala	Trp	Gly	Lys	Val	Gly	Ala	His	Ala	Gly	Glu	Tyr	Gly	Ala	Glu	Ala	Leu	Glu
Horse	Val	—	Leu	Ser	Ala	Ala	Asp	Lys	Thr	Asn	Val	Lys	Ala	Ala	Trp	Ser	Lys	Val	Gly	Gly	His	Ala	Gly	Glu	Val	Gly	Ala	Glu	Ala	Leu	Glu

β	30	31	32	33	34	35	36	37	38	39	40	41	42	43	44	45	46	47	48	49	50	51	52	53	54	55	56	57	58	59	60
Human	Arg	Leu	Leu	Val	Val	Tyr	Pro	Trp	Thr	Gln	Arg	Phe	Phe	Glu	Ser	Phe	Gly	Asp	Leu	Ser	Thr	Pro	Asp	Ala	Val	Met	Gly	Asn	Pro	Lys	Val
Horse	Arg	Leu	Leu	Val	Val	Tyr	Pro	Trp	Thr	Gln	Arg	Phe	Phe	Asp	Ser	Phe	Gly	Asp	Leu	Ser	Asn	Pro	Gly	Ala	Val	Met	Gly	Asn	Pro	Lys	Val
α	31	32	33	34	35	36	37	38	39	40	41	42	43	44	45	46	—	47	48	49	50	—	—	—	—	—	51	52	53	54	55
Human	Arg	Met	Phe	Leu	Ser	Phe	Pro	Thr	Thr	Lys	Thr	Tyr	Phe	Pro	His	Phe	—	Asp	Leu	Ser	His	—	—	—	—	—	Gly	Ser	Ala	Gln	Val
Horse	Arg	Met	Phe	Leu	Gly	Phe	Pro	Thr	Thr	Lys	Thr	Tyr	Phe	Pro	His	Phe	—	Asp	Leu	Ser	His	—	—	—	—	—	Gly	Ser	Ala	Gln	Val

β	61	62	63	64	65	66	67	68	69	70	71	72	73	74	75	76	77	78	79	80	81	82	83	84	85	86	87	88	89	90	91
Human	Lys	Ala	His	Gly	Lys	Lys	Val	Leu	Gly	Ala	Phe	Ser	Asp	Gly	Leu	Ala	His	Leu	Asp	Asn	Leu	Lys	Gly	Thr	Phe	Ala	Thr	Leu	Ser	Glu	Leu
Horse	Lys	Ala	His	Gly	Lys	Lys	Val	Leu	His	Ser	Phe	Gly	Glu	Gly	Val	His	His	Leu	Asp	Asn	Leu	Lys	Gly	Thr	Phe	Ala	Ala	Leu	Ser	Glu	Leu
α	56	57	58	59	60	61	62	63	64	65	66	67	68	69	70	71	72	73	74	75	76	77	78	79	80	81	82	83	84	85	86
Human	Lys	Gly	His	Gly	Lys	Lys	Val	Ala	Asp	Ala	Leu	Thr	Asn	Ala	Val	Ala	His	Val	Asp	Asp	Met	Pro	Asn	Ala	Leu	Ser	Ala	Leu	Ser	Asp	Leu
Horse	Lys	Ala	His	Gly	Lys	Lys	Val	Gly	Asp	Ala	Leu	Thr	Leu	Ala	Val	Gly	His	Leu	Asp	Asp	Leu	Pro	Gly	Ala	Leu	Ser	Asp	Leu	Ser	Asn	Leu

β	92	93	94	95	96	97	98	99	100	101	102	103	104	105	106	107	108	109	110	111	112	113	114	115	116	117	118	119	120	121	122
Human	His	Cys	Asp	Lys	Leu	His	Val	Asp	Pro	Glu	Asn	Phe	Arg	Leu	Leu	Gly	Asn	Val	Leu	Val	Cys	Val	Leu	Ala	His	His	Phe	Gly	Lys	Glu	Phe
Horse	His	Cys	Asp	Lys	Leu	His	Val	Asp	Pro	Glu	Asn	Phe	Arg	Leu	Leu	Gly	Asn	Val	Leu	Ala	Val	Val	Leu	Ala	Arg	His	Phe	Gly	Lys	Asp	Phe
α	87	88	89	90	91	92	93	94	95	96	97	98	99	100	101	102	103	104	105	106	107	108	109	110	111	112	113	114	115	116	117
Human	His	Ala	His	Lys	Leu	Arg	Val	Asp	Pro	Val	Asn	Phe	Lys	Leu	Leu	Ser	His	Cys	Leu	Leu	Val	Thr	Leu	Ala	Ala	His	Leu	Pro	Ala	Glu	Phe
Horse	His	Ala	His	Lys	Leu	Arg	Val	Asp	Pro	Val	Asn	Phe	Lys	Leu	Leu	Ser	His	Cys	Leu	Leu	Ser	Thr	Leu	Ala	Val	His	Leu	Pro	Asn	Asp	Phe

β	123	124	125	126	127	128	129	130	131	132	133	134	135	136	137	138	139	140	141	142	143	144	145	146
Human	Thr	Pro	Pro	Val	Gln	Ala	Ala	Tyr	Gln	Lys	Val	Val	Ala	Gly	Val	Ala	Asn	Ala	Leu	Ala	His	Lys	Tyr	His
Horse	Thr	Pro	Glu	Leu	Gln	Ala	Ser	Tyr	Gln	Lys	Val	Val	Ala	Gly	Val	Ala	Asn	Ala	Leu	Ala	His	Lys	Tyr	His
α	118	119	120	121	122	123	124	125	126	127	128	129	130	131	132	133	134	135	136	137	138	139	140	141
Human	Thr	Pro	Ala	Val	His	Ala	Ser	Leu	Asp	Lys	Phe	Leu	Ala	Ser	Val	Ser	Thr	Val	Leu	Thr	Ser	Lys	Tyr	Arg
Horse	Thr	Pro	Ala	Val	His	Ala	Ser	Leu	Asp	Lys	Phe	Leu	Ser	Ser	Val	Ser	Thr	Val	Leu	Thr	Ser	Lys	Tyr	Arg

FIGURE 26.6 A comparison of the α- and β-globin polypeptides from humans and horses. This figure shows the deduced amino acid sequences obtained by sequencing the exon portions of the corresponding genes. The gaps indicate where additional amino acids are found in the sequence of myoglobin, another member of this gene family.

created the α-globin and β-globin genes occurred long before the evolutionary divergence that produced different species of mammals. For this reason, there was a greater amount of time for the α- and β-globin genes to accumulate changes compared to the amount of time that has elapsed since the evolutionary divergence of mammalian species. This idea is schematically shown in **Figure 26.7.**

Based on the analysis of genetic sequences, evolutionary biologists have estimated that the gene duplication that created the α-globin and β-globin gene lineages occurred approximately 400 million years ago, while the speciation events that created different species of mammals occurred less than 200 million years ago. Therefore, the α-globin and β-globin genes have had much more time to accumulate changes relative to each other.

Genetic Variation at the Molecular Level Is Associated with Neutral Changes in Gene Sequences

As we have seen, the globin genes exhibit variation in their sequences. Researchers have asked the question, Is such variation due primarily to mutations that are favored by natural selection or due to random genetic drift?

A **nonneutral mutation** is one that affects the phenotype of the organism and can be acted on by natural selection. Such a mutation may only subtly alter the phenotype of an organism, or it may have a major impact. According to Darwin, natural selection is the agent that leads to evolutionary change in populations. It selects for individuals with the highest Darwinian fitness and often promotes the establishment of beneficial alleles and the elimination of deleterious ones. Therefore, many geneticists have assumed that natural selection is the dominant factor in changing the genetic composition of natural populations, thereby leading to variation.

In opposition to this viewpoint, in 1968, Motoo Kimura proposed the **neutral theory of evolution.** According to this theory, most genetic variation observed in natural populations is due to the accumulation of neutral mutations that do not affect the phenotype of the organism and are not acted on by natural selection. For example, a mutation within a structural gene that changes a glycine codon from GGG to GGC would not affect the amino acid sequence of the encoded protein. Because neutral mutations do not affect phenotype, they spread throughout a population according to their frequency of appearance and to random genetic drift. This theory has been called the "survival of the luckiest" and also **non-Darwinian evolution** to contrast

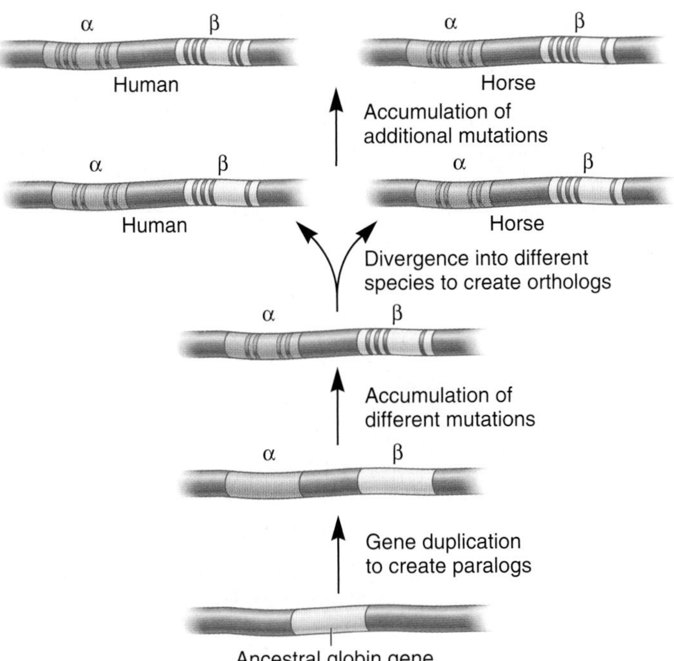

FIGURE 26.7 Evolution of paralogs and orthologs. In this schematic example, the ancestral globin gene duplicated to create the α- and β-globin genes, which are paralogous genes. (Note: Actually, gene duplications occurred several times to create a large globin gene family, but this is a simplified example that shows only one gene duplication event. Also, chromosomal rearrangements have placed the α- and β-globin genes on different chromosomes.) Over time, these two paralogs accumulated different mutations, designated with red and blue lines. At a later point in evolution, a species divergence occurred to create different mammalian species including humans and horses. Over time, the orthologs also accumulated different mutations, designated with green lines. As noted here, the orthologs have fewer differences compared to the paralogs because the gene duplication occurred prior to the species divergence.

it with Darwin's theory, which focuses on fitness. Kimura agreed with Darwin that natural selection is responsible for adaptive changes in a species during evolution. His main argument is that most modern variation in gene sequences is neutral with respect to natural selection.

In support of the theory, Kimura and his colleague Tomoko Ohta outlined five principles that govern the evolution of genes at the molecular level:

1. For each protein, the rate of evolution, in terms of amino acid substitutions, is approximately constant with regard to neutral substitutions that do not affect protein structure or function.

   Evidence: As an example, the amount of genetic variation between the coding sequence of the human α-globin and β-globin genes is approximately the same as the difference between the α-globin and β-globin genes in the horse (shown previously in Figure 26.6). This type of comparison holds true in many different genes compared among many different species.
2. Proteins that are functionally less important for the survival of an organism, or parts of a protein that are less important for its function, tend to evolve faster than more important proteins or regions of a protein. In other words, during evolution, less important proteins accumulate amino acid substitutions more rapidly than important proteins.

   Evidence: Certain proteins are critical for survival, and their structure is precisely suited to their function. Examples are the histone proteins necessary for nucleosome formation in eukaryotes. Histone genes tolerate very few mutations and have evolved extremely slowly. By comparison, fibrinopeptides, which bind to fibrinogen to form a blood clot, evolve very rapidly. Presumably, the sequence of amino acids in this polypeptide is not very important for allowing it to aggregate and form a clot. Another example concerns the amino acid sequences of enzymes. Amino acid substitutions are very rare within the active site, which is critical for function, but are more frequent in other parts of the protein.
3. Amino acid substitutions that do not significantly alter the existing structure and function of a protein are found more commonly than disruptive amino acid changes.

   Evidence: When comparing the coding sequences within homologous genes of modern species, nucleotide differences are more likely to be observed in the wobble base than in the first or second base within a codon. Mutations in the wobble base are often silent because they do not change the amino acid sequence of the protein. In addition, conservative substitutions (i.e., a substitution with a similar amino acid, such as a nonpolar amino acid for another nonpolar amino acid) are fairly common. By comparison, nonconservative substitutions—those that significantly alter the structure and function of a protein—are less frequent. Nonsense and frameshift mutations are very rare within the coding sequences of genes. Also, intron sequences evolve more rapidly than exon sequences.
4. Gene duplication often precedes the emergence of a gene having a new function.

   Evidence: When a single copy of a gene exists in a species, it usually plays a functional role similar to that of the homologous gene found in another species. Gene duplications have created gene families in which each family member can evolve somewhat different functional roles. Examples include the globin family described in Chapter 8 and the *Hox* genes described in Chapter 23.
5. Selective elimination of definitely deleterious mutations and the random fixation of selectively neutral or very

slightly deleterious alleles occur far more frequently in evolution than selection of advantageous mutants.

Evidence: As mentioned in principle 3, silent and conservative mutations are much more common than nonconservative substitutions. Presumably these nonconservative mutations usually have a negative effect on the phenotype of the organism, so they are effectively eliminated from the population by natural selection. On rare occasions, however, an amino acid substitution due to a mutation may have a beneficial effect on the phenotype. For example, a nonconservative mutation in the β-globin gene produces the Hb^S allele, which gives an individual resistance to malaria in the heterozygous condition.

In general, the DNA sequencing of hundreds of thousands of different genes from thousands of species has provided compelling support for these five principles of gene evolution at the molecular level. When it was first proposed, the neutral theory sparked a great debate. Some geneticists, called selectionists, strongly opposed the neutralist theory of evolution. However, the debate largely cooled after Ohta incorporated the concept of nearly neutral mutations into the theory. Nearly neutral mutations have a minimal impact on phenotype—they may be slightly beneficial or slightly detrimental. Ohta suggested that the prevalence of such alleles can depend mostly on natural selection or mostly on genetic drift, depending on the population size.

Why do evolutionary biologists care about neutral or nearly neutral mutations? One reason is that their prevalence is used as a tool to construct phylogenetic trees. This topic is discussed next.

Molecular Clocks Can Be Used to Date the Divergence of Species

In 1963, Linus Pauling and Emile Zuckerkandl were the first to suggest the use of molecular data to establish evolutionary relationships. When comparing homologous genes in different species, the DNA sequences from closely related species are more similar to each other than are the sequences from distantly related species. According to the neutral theory of evolution, most of the observed variation is due to neutral mutations. In a sense, the relatively constant rate of neutral mutations acts as a **molecular clock** on which to measure evolutionary time. According to this idea, neutral mutations will become fixed in a population at a rate that is proportional to the rate of mutation per generation. On this basis, the genetic divergence between species that is due to neutral mutations reflects the time elapsed since their last common ancestor.

Figure 26.8 illustrates the concept of a molecular clock. The y-axis is a measure of the number of nucleotide sequence differences between pairs of species. The x-axis plots the amount of time that has elapsed since a pair of species shared a common ancestor. As seen in this diagram, the number of sequence differences is higher when two species shared a common ancestor in the very distant past compared to pairs that shared a more recent common ancestor. The explanation for this phenomenon is that species accumulate independent mutations after they have diverged from each other; a longer period of time since their divergence allows for a greater accumulation of mutations that makes their sequences different.

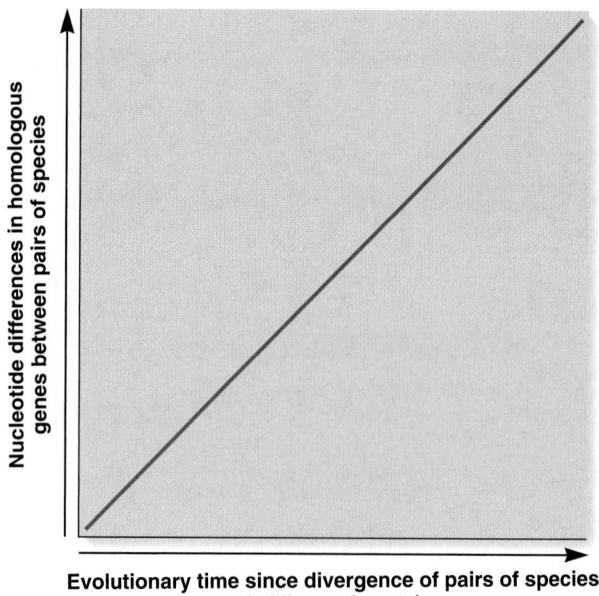

FIGURE 26.8 **A molecular clock.** According to the concept of a molecular clock, neutral mutations accumulate over evolutionary time. When comparing homologous genes between species, those species that diverged more recently tend to have fewer differences compared to those whose common ancestor occurred in the very distant past.

Figure 26.8 suggests a linear relationship between the number of sequence changes and the time of divergence. Such a relationship indicates that the observed rate of neutral mutations remains constant over millions of years. For example, a linear relationship predicts that a pair of species showing 20 nucleotide differences in a given gene sequence would have a (most recent) common ancestor that existed twice as long ago compared to another pair showing 10 nucleotide differences. While actual data sometimes show a relatively linear relationship over a defined time period, evolutionary biologists do not believe that molecular clocks are perfectly linear over very long periods of time. Several factors can contribute to nonlinearity of molecular clocks. These include differences in the generation times of the species being analyzed, the presence of mutations that are acted upon by natural selection, and variation in mutation rates between different species.

To obtain reliable data, researchers must calibrate their molecular clocks. How much time does it take to accumulate a certain percentage of nucleotide changes? To perform such a calibration, researchers must have information regarding the date when two species shared a common ancestor. Such information could come from the fossil record, for instance. The genetic differences between those species are then divided by the time elapsed since their last common ancestor to calculate a rate of change. For example, research suggests that humans and chimpanzees diverged from a common ancestor approximately 6 million years ago. The percentage of nucleotide differences in mitochondrial DNA between humans and chimpanzees is 12%. From these data, the molecular clock for changes in mitochondrial DNA sequences of primates is calibrated at roughly 2% nucleotide changes per million years.

To understand the concept of a molecular clock, let's consider the evolution of some species of primates. **Figure 26.9** illustrates a phylogenetic tree for several species that was derived by comparing DNA sequences in a mitochondrial gene that encodes a protein called cytochrome oxidase subunit II, which is involved with cellular respiration. The vertical scale represents time, and the branch points that are labeled with letters represent common ancestors. Let's take a look at three branch points (labeled A, D, and E) and relate them to the concept of a molecular clock. The common ancestor labeled A diverged into two species that ultimately gave rise to siamangs and the other five species. Since this divergence, a long time has elapsed (approximately 23 million years), allowing the siamang genome to accumulate random neutral changes that would be different from the random changes that have occurred in the genomes of the other five species (see yellow bar in Figure 26.9). Therefore, the cytochrome oxidase gene in the siamangs is more different compared with the genes in the other five species. Now let's compare humans and chimpanzees. The common ancestor that gave rise to these species is labeled D. This species diverged into two species that eventually gave rise to humans and chimpanzees. This divergence occurred a moderate time ago, approximately 6 million years ago, as illustrated by the red bar. Compared to humans and chimpanzees, humans and siamangs have more differences in their gene sequences because there has been more time for them to accumulate random neutral mutations. Finally, let's consider the two species of chimpanzees, whose common ancestor is labeled E. The time for them to accumulate random mutations has been relatively short, as depicted by the green bar. Therefore, the two modern species of chimpanzees have very similar gene sequences.

A Phylogenetic Tree Can Be Constructed Using a Cladistic Approach

Now that we appreciate the concept of a molecular clock, let's now turn our attention to how evolutionary biologists actually compare gene sequences to construct phylogenetic trees. A **cladistic approach** reconstructs a phylogenetic tree, also called a **cladogram,** by considering the various possible pathways of evolution and then choosing the most plausible tree. Cladistics is now the most commonly used method for the construction of phylogenetic trees.

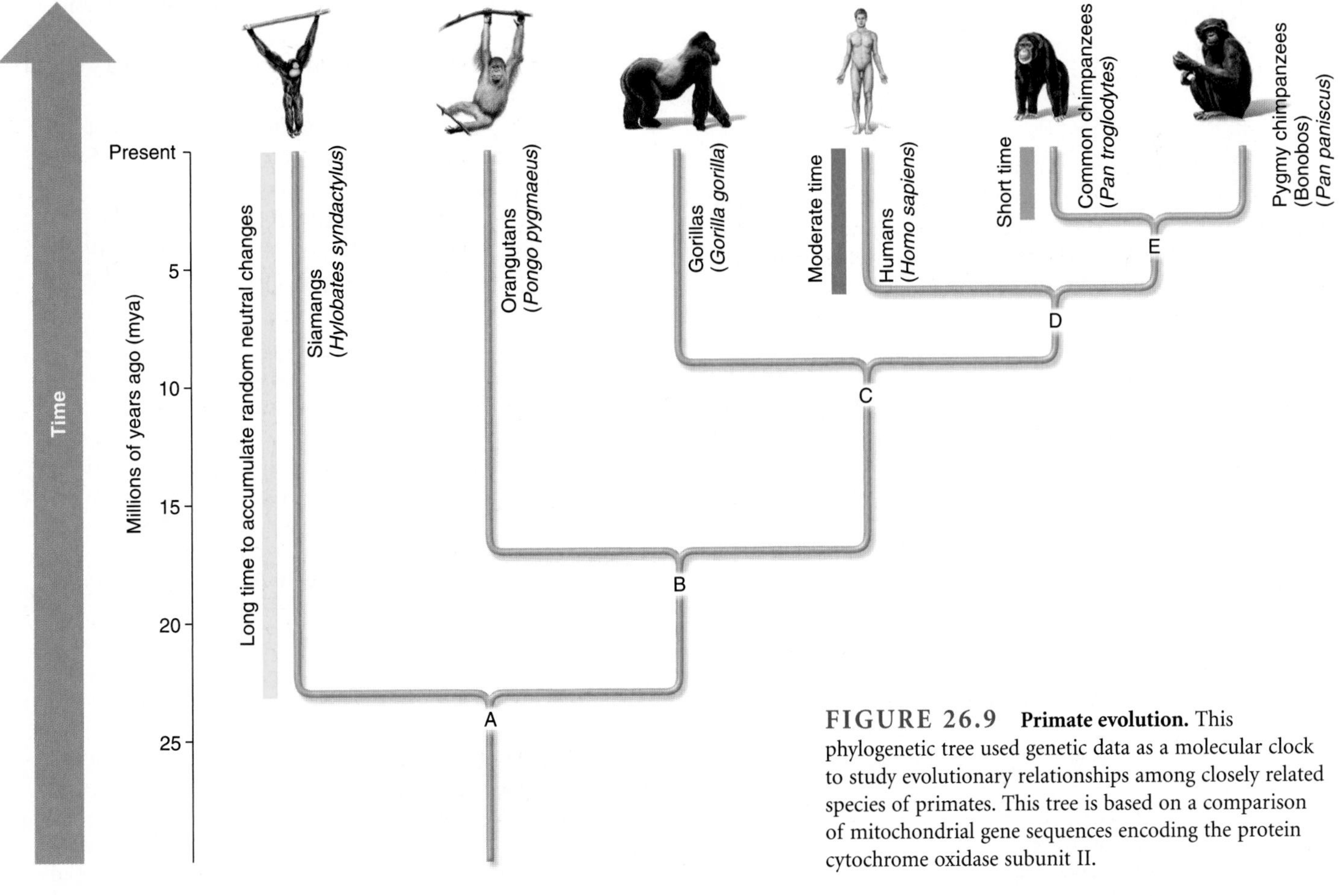

FIGURE 26.9 **Primate evolution.** This phylogenetic tree used genetic data as a molecular clock to study evolutionary relationships among closely related species of primates. This tree is based on a comparison of mitochondrial gene sequences encoding the protein cytochrome oxidase subunit II.

A cladistic approach compares traits, also called characters, that are either shared or not shared by different species. These can be morphological traits, such as the shapes of birds' beaks, or they can be molecular traits, such as sequences of homologous genes. Such characters may come in different versions called character states. Those that are shared with a distant ancestor are called **ancestral,** or **primitive characters.** Such characters are viewed as being older—ones that arose earlier in evolution. In contrast, a **shared derived character,** or **synapomorphy,** is a trait that is shared by a group of organisms but not by a distant common ancestor. Compared to primitive characters, shared derived characters are more recent traits on an evolutionary timescale. For example, among mammals, only some species have flippers, such as whales and dolphins. In this case, flippers were derived from the two front limbs of an ancestral species. The word "derived" refers to the observation that evolution involves the modification of traits in preexisting species. In other words, the features of newer populations of organisms are derived from changes in preexisting populations. The basis of the cladistic approach is to analyze many shared derived characters among groups of species to deduce the pathway that gave rise to the species.

To understand the concept of primitive and shared derived characters, let's consider how a cladogram can be constructed based on molecular data such as a sequence of a gene. Our example uses molecular data obtained from seven different hypothetical species called A through G. In these species, a homologous gene was sequenced; a portion of the gene sequence is shown in the following.

A: GATAGTACCC	E: GGTATAACCC
B: GATAGTTCCC	F: GGTAGTACCA
C: GATAGTTCCG	G: GGTAGTACCC
D: GGTATTACCC	

In a cladogram, an **ingroup** is a monophyletic group in which we are interested. By comparison, an **outgroup** is a species or group of species that is most closely related to an ingroup. The root of a cladogram is placed between the outgroup and the ingroup. In the cladogram of **Figure 26.10**, the outgroup is species E. This may have been inferred because the other species may share traits that are not found in species E. The other species (A, B, C, D, F, and G) form the ingroup. For these data, a mutation that changes the DNA sequence is analogous to a modification of a characteristic. Species that share such genetic changes possess shared derived characters because the new genetic sequence was derived from a more primitive sequence.

Now that we have an understanding of some of the general principles of cladistics, let's consider the steps a researcher would follow to construct a cladogram using this approach.

1. **Choose the species in whose evolutionary relationships you are interested.** In a simple cladogram, individual species are compared to each other. In more complex cladograms, species may be grouped into larger taxa (e.g., families) and compared with each other. If such grouping is done, the groups must be clades for the results to be reliable.
2. **Choose characters for comparing different species.** As mentioned, a character is a general feature of an organism. Characters may come in different versions called character states. For example, a base at a particular location in a gene can be considered a character, and this character could exist in different character states such as A, T, G, or C, due to mutations.
3. **Determine the polarity of character states.** In other words, determine if a character state is primitive or derived. In the case of morphological traits, this information may be available by examining the fossil record. If possible, identify an outgroup.
4. **Group species (or higher taxa) based on shared derived characters.**
5. **Build a cladogram based on the following principles:**
   - All species (or higher taxa) are placed on tips in the phylogenetic tree, not at branch points.
   - Each cladogram branch point should have a list of one or more shared derived characters that are common to all species above the branch point unless the character is later modified.

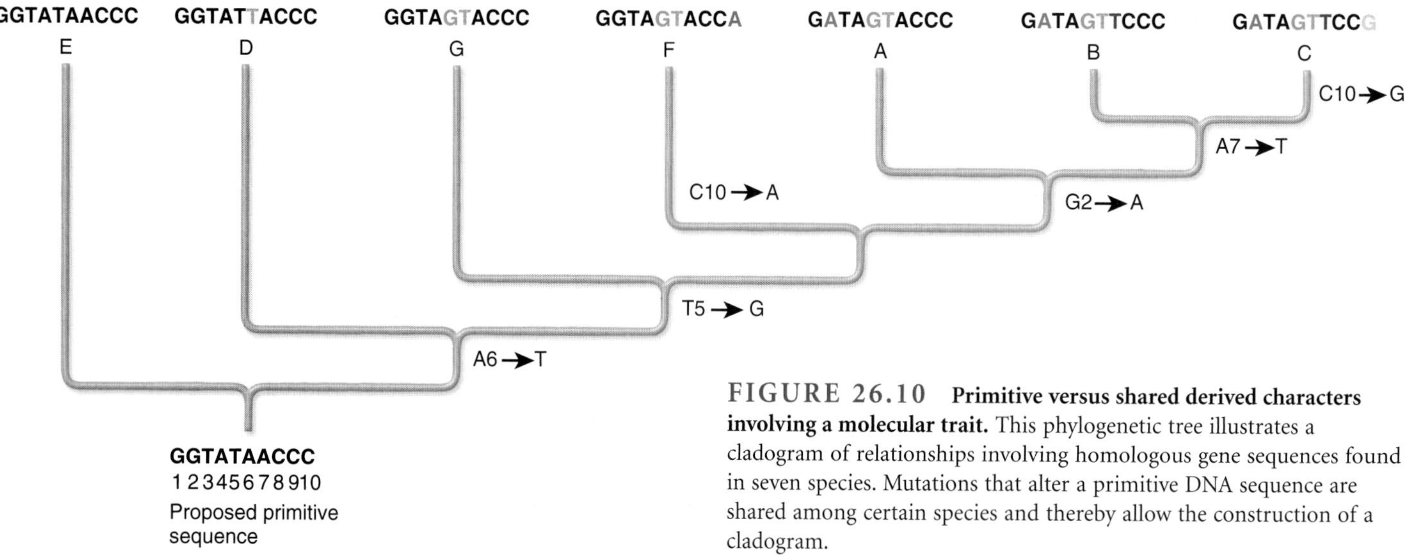

FIGURE 26.10 Primitive versus shared derived characters involving a molecular trait. This phylogenetic tree illustrates a cladogram of relationships involving homologous gene sequences found in seven species. Mutations that alter a primitive DNA sequence are shared among certain species and thereby allow the construction of a cladogram.

- All shared derived characters appear together only once in a cladogram unless they independently arose during evolution more than once in the ancestors of different clades.

6. **Choose the best cladogram among possible options.** When grouping species (or higher taxa), more than one cladogram may be possible. Therefore, analyzing the data and producing the best possible cladogram is a key aspect of this process. As described next, different theoretical approaches can be followed to achieve this goal.

The greatest challenge in a cladistic approach is to determine the correct order of events. It may not always be obvious which traits are primitive and came earlier and which are derived and came later in evolution. Different approaches can be used to deduce the correct order. We will discuss three of them here. First, for morphological traits, a common way to deduce the order of events is to analyze fossils and determine the relative dates that certain traits arose. A second strategy that can be used to deduce the correct order is to assume that the best hypothesis is the one that requires the fewest number of evolutionary changes. This concept, called the **principle of parsimony,** states that the preferred hypothesis is the one that is the simplest. For example, if two species possess a tail, we would initially assume that a tail arose once during evolution and that both species have descended from a common ancestor with a tail. Such a hypothesis is simpler than assuming that tails arose twice during evolution and that the tails in the two species are not due to descent from a common ancestor. A third strategy is to incorporate known features of genetic changes into the analysis. This method, known as **maximum likelihood,** allows researchers to choose the best trees based on inferences about the most likely ways that DNA incurs mutations. For example, a researcher might assume that the rate of neutral genetic changes is constant over a given time period or that neutral mutations are more likely to occur in the third base of a codon because of the redundancy of the genetic code. This approach can analyze different phylogenetic trees and identify which tree has the maximum likelihood of occurring, based on assumptions about how DNA changes over the course of evolution. When constructing their trees, researchers often have preferences about the kinds of strategies they employ. More than one strategy may be used to construct a cladogram.

Let's consider a simple example to illustrate the process. Our example uses molecular data obtained from four different hypothetical species called A through D. In these species, a homologous region of DNA was sequenced as shown in the following.

1 2 3 4 5
A: GTACA
B: GACAG
C: GTCAA
D: GACCG

Given this information, three different trees are shown in **Figure 26.11,** although more are possible. In these examples, tree 1 requires seven mutations, tree 2 requires six, and tree 3 requires only five. Because tree 3 requires the fewest number of mutations, it is considered the most parsimonious. Based on the principle of parsimony, tree 3 would be the more likely choice.

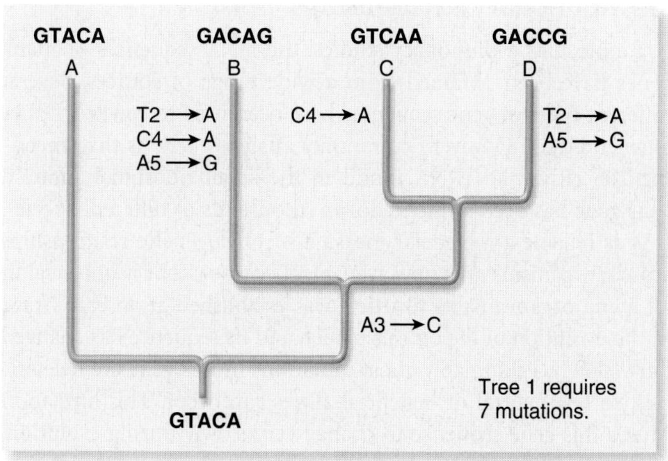

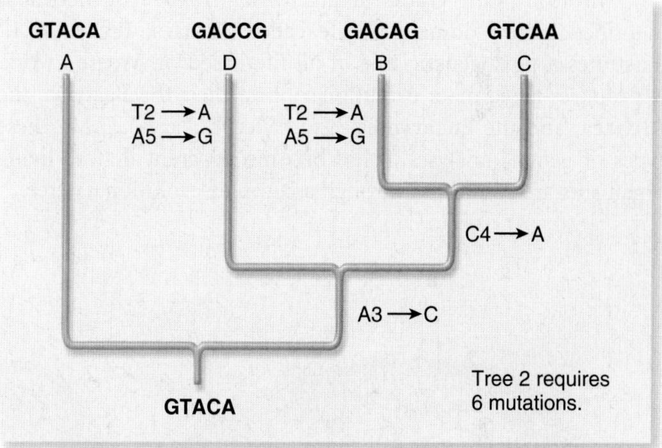

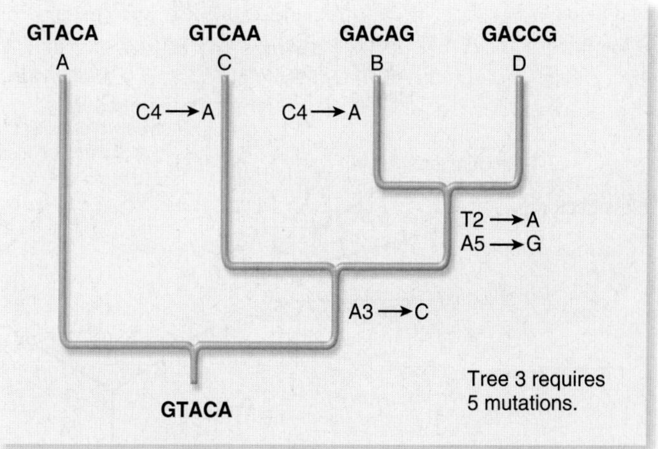

FIGURE 26.11 **The cladistic approach: choosing a cladogram from molecular genetic data.** This figure shows three different phylogenetic trees for the evolution of a short DNA sequence, but many more are possible. According to the principle of parsimony, the cladogram shown in tree 3 is the more plausible choice because it requires only five mutations. When constructing cladograms based on long genetic sequences, researchers use computers to generate trees with the fewest possible genetic changes.

Phylogenetic Trees Refine Our Understanding of Evolutionary Relationships

For molecular evolutionary studies, the DNA sequences of many genes have been obtained from a wide range of sources. Several different types of gene sequences have been used to construct phylogenetic trees. One very commonly analyzed gene is that encoding 16S rRNA, an rRNA found in the small ribosomal subunit. This gene has been sequenced from thousands of different species. It is as reliable a molecular measure of phylogenetic relationships among organisms as is now available. Because rRNA is universal in all living organisms, its function was established at an early stage in the evolution of life on this planet, and its sequence has changed fairly slowly. Presumably, most mutations in this gene are deleterious, so few neutral or beneficial alleles can occur. This limitation causes this gene sequence to change very slowly during evolution. Furthermore, 16S rRNA is a rather large molecule, and therefore it contains a large amount of sequence information.

In 1977, Carl Woese analyzed 16S rRNA sequences and identified a new domain of life called Archaea. Figure 26.12 illustrates a phylogenetic tree of all life based on Woese's work. It proposes three main evolutionary branches: the **Bacteria,** the **Archaea,** and the **Eukaryotes** (also called Eukarya). From these types of genetic analyses, it has become apparent that all living organisms are connected through a complex evolutionary tree.

Although the work of Woese was a breakthrough in our appreciation of evolution, more recent molecular genetic data have shed new light regarding the classification of species. With regard to eukaryotic species, biologists once categorized them into four kingdoms: protists (Protista), fungi (Fungi), animals (Animalia), and plants (Plantae). However, recent models propose several major groups, called **supergroups,** as a way to organize eukaryotes into monophyletic groups. Figure 26.13 shows a diagram that hypothesizes several supergroups. Of the four traditional kingdoms, both Fungi and Animalia are within the Opisthokonta supergroup, while Plantae is found within Archaeplastida. The remaining branches and supergroups used to be classified within the single kingdom Protista. As seen in this figure, molecular data and newer ways of building trees reveal that protists played a key role in the evolution of many diverse groups of eukaryotic species, producing several large monophyletic supergroups.

Horizontal Gene Transfer Also Contributes to the Evolution of Species

The types of phylogenetic trees considered thus far are examples of **vertical evolution,** in which species evolve from preexisting species by the accumulation of gene mutations and by changes in

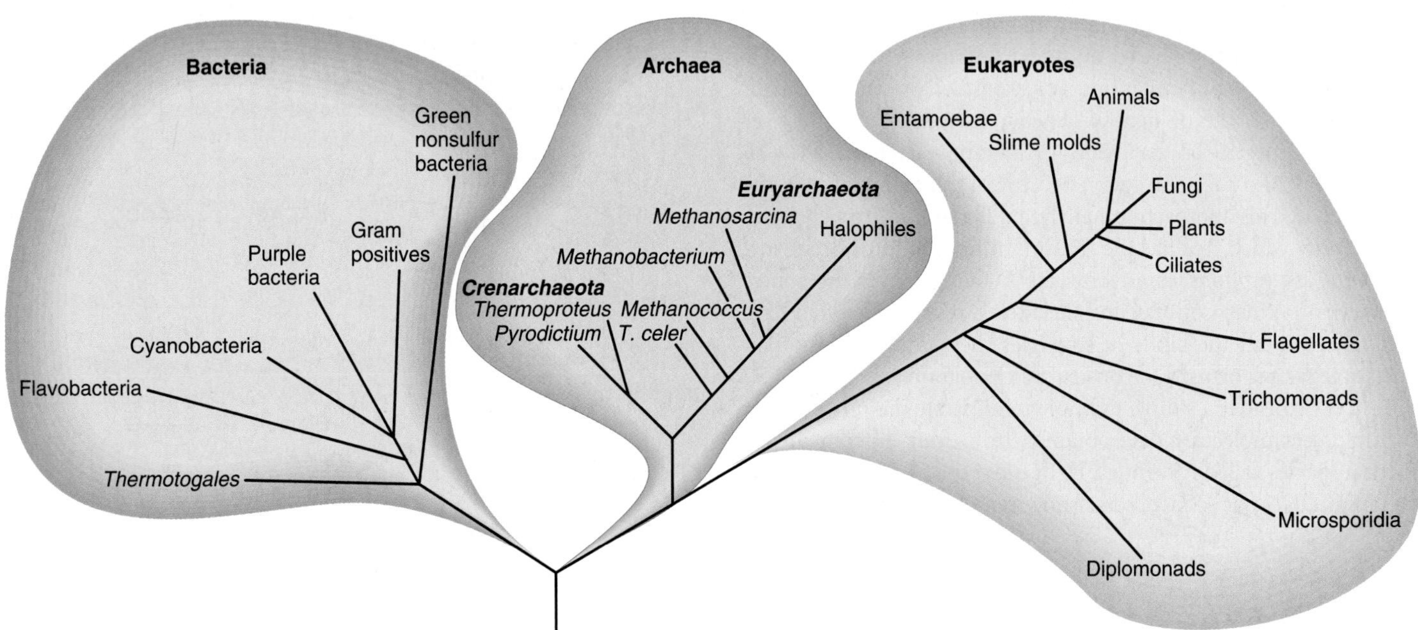

FIGURE 26.12 **A phylogenetic tree of all life on Earth based on 16S rRNA sequences.** The relationships in this figure among eukaryotic species have been substantially revised based on newer molecular data (see Figure 26.13).

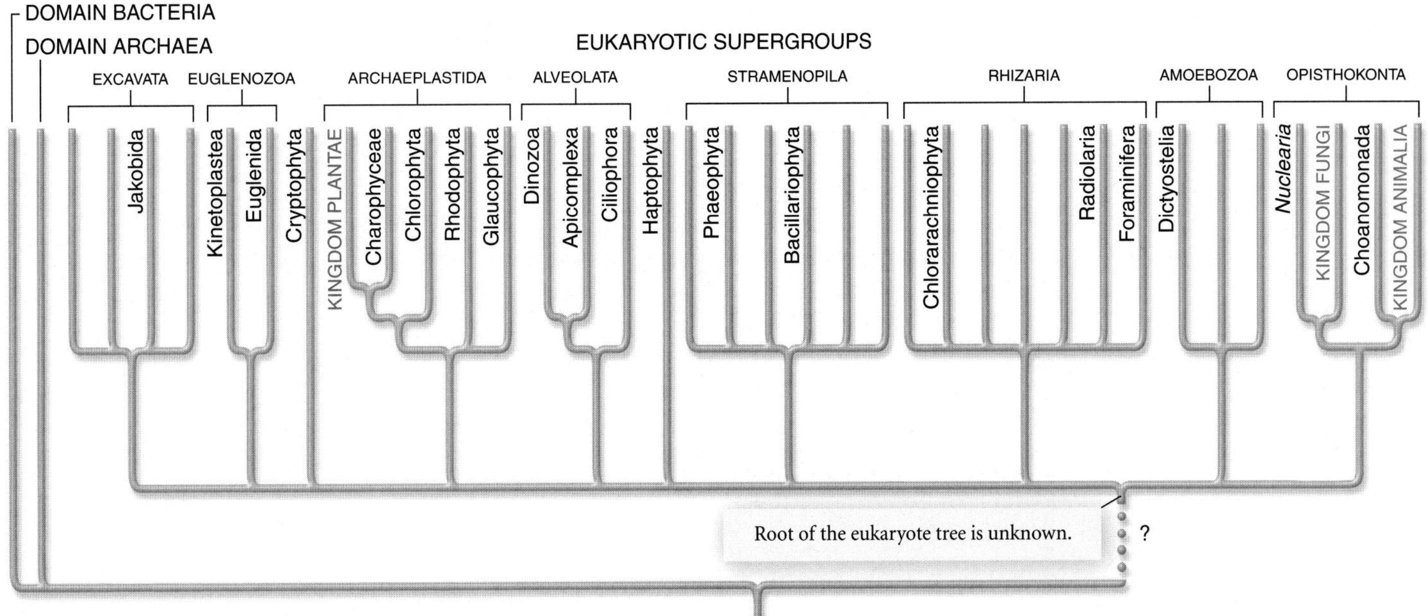

FIGURE 26.13 **A modern cladogram for eukaryotes.** Each of the supergroups shown here are hypothesized to be monophyletic. This drawing should be considered a working hypothesis. The arrangement of these supergroups relative to each other is not entirely certain.

chromosome structure and number. Vertical evolution involves genetic changes in a series of ancestors that form a lineage. In addition to vertical evolution, however, species accumulate genetic changes by another process called **horizontal gene transfer.** As discussed in Chapter 24, this involves the exchange of genetic material among different species.

An analysis of many genomes suggests that horizontal gene transfer was prevalent during the early stages of evolution, when all organisms were unicellular, but continued even after the divergence of the three major domains of life. With regard to modern organisms, horizontal gene transfer remains prevalent among prokaryotic species. By comparison, this process is less common among eukaryotes, though it does occur. Researchers have speculated that multicellularity and sexual reproduction have presented barriers to horizontal gene transfer in many eukaryotes. For a gene to be transmitted to eukaryotic offspring, it would have to be transferred into a eukaryotic cell that is a gamete or a cell that gives rise to gametes.

How has horizontal gene transfer impacted our understanding of evolution? In the past few decades, scientists have debated the role of horizontal gene transfer in the earliest stages of evolution, prior to the emergence of the two prokaryotic domains. The traditional viewpoint was that the three domains of life arose from a single type of prokaryotic cell (or pre-prokaryotic cell) called the universal ancestor. However, genomic research has suggested that horizontal gene transfer may have been particularly common during the early stages of evolution on Earth, when all species were unicellular. Rather than proposing that all life arose from a single type of prokaryotic cell, horizontal gene transfer may have been so prevalent that the universal ancestor may have actually been an ancestral community of cell lineages that evolved as a whole. If that were the case, the tree of life cannot be traced back to a single universal ancestor.

Figure 26.14 illustrates a schematic scenario for the evolution of life on Earth that includes the roles of both vertical evolution and horizontal gene transfer. This has been described as a "web of life" rather than a "tree of life." Instead of a universal ancestor, a web of life began with a community of primitive cells that transferred genetic material in a horizontal fashion. Horizontal gene transfer was also prevalent during the early evolution of bacteria and archaea and when eukaryotes first emerged as unicellular species. While horizontal gene transfer remains a prominent way to foster evolutionary change in modern bacteria and archaea, the region of the diagram that contains eukaryotic species has a more treelike structure, because horizontal gene transfer has become much less common in these species.

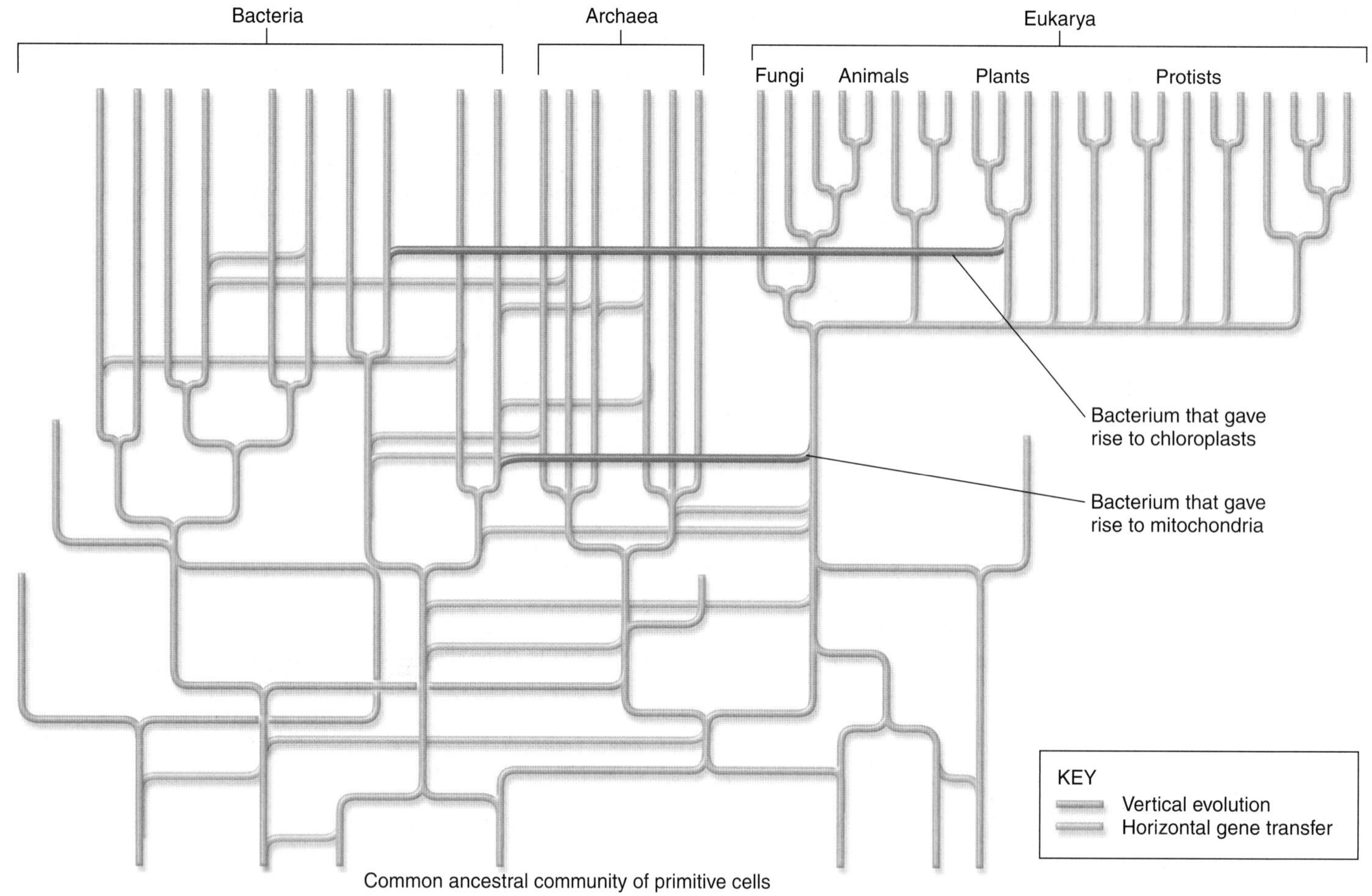

FIGURE 26.14 A revised view regarding the evolution of life, incorporating the concept of horizontal gene transfer. This phylogenetic tree shows a classification of life on Earth that includes the contribution of horizontal gene transfer in the evolution of species on our planet. This phenomenon was prevalent during the early stages of evolution when all organisms were unicellular. Horizontal gene transfer continues to be a prominent factor in the speciation of Bacteria and Archaea. Note: This tree is meant to be schematic. Figure 26.13 is a more realistic representation of the evolutionary relationships among modern species.

EXPERIMENT 26A

Scientists Can Analyze Ancient DNA to Examine the Relationships Between Living and Extinct Flightless Birds

The majority of phylogenetic trees have been constructed from molecular data using DNA samples collected from living species. With this approach, we can infer the prehistoric changes that gave rise to present-day DNA sequences. As an alternative, scientists have discovered that it is occasionally possible to obtain DNA sequence information from species that have lived in the past. In 1984, the first successful attempt at determining DNA sequences from an extinct species was accomplished by groups at the University of California at Berkeley and the San Diego Zoo, including Russell Higuchi, Barbara Bowman, Mary Freiberger, Oliver Ryder, and Allan Wilson. They obtained a sample of dried muscle from a museum specimen of the quagga (*Equus quagga*), a zebralike species that became extinct in 1883. This piece of muscle tissue was obtained from an animal that had died 140 years ago. A sample of its skin and muscle had been preserved in salt in the Museum of Natural History at Mainz, Germany. The researchers extracted DNA from the sample, cloned pieces of it into vectors, and then sequenced hybrid vectors containing the quagga DNA. This pioneering study opened the field of **ancient DNA analysis,** also known as **molecular paleontology.**

Since the mid-1980s, many researchers have become excited about the information that might be derived from sequencing DNA obtained from older specimens. Currently there is debate concerning how long DNA can remain significantly intact after an organism has died. Over time, the structure of DNA is degraded by hydrolysis and the loss of purines. Nevertheless, under certain conditions (e.g., cold temperature, low oxygen), DNA samples may remain stable for as long as 50,000 to 100,000 years and perhaps longer.

In most studies involving prehistoric specimens (in particular, those that are much older than the salt-preserved quagga sample), the ancient DNA is extracted from bone, dried muscle, or preserved skin. These samples are often obtained from museum specimens that have been gathered by archaeologists. However, it is unlikely that enough DNA can be extracted to enable a researcher to directly clone the DNA into a vector. Since 1985, however, the advent of PCR technology, described in Chapter 18, has made it possible to amplify very small amounts of DNA using PCR primers that flank a region within the 12S rRNA gene, a slowly changing gene. In recent years, this approach has been used to elucidate the phylogenetic relationships between modern and extinct species.

In the experiment described in **Figure 26.15**, Alan Cooper, Cécile Mourer-Chauviré, Geoffrey Chambers, Arndt von Haeseler, Allan Wilson, and Svante Pääbo investigated the evolutionary relationships among some extinct and modern species of flightless birds. Two groups of flightless birds, the moas and the kiwis, existed in New Zealand during the Pleistocene era. The moas are now extinct, although 11 species were formerly present. In this study, the researchers investigated the phylogenetic relationships among four extinct species of moas that were available as museum samples, three kiwis of New Zealand, and several other (nonextinct) species of flightless birds. These included the emu and the cassowary (found in Australia and New Guinea), the ostrich (found in Africa and formerly Asia), and two rheas (found in South America).

The samples from the various species were subjected to PCR to amplify the 12S rRNA gene. This provided enough DNA to subject the gene to DNA sequencing. The sequences of the genes were aligned using computer programs described in Chapter 21.

THE GOAL

Because DNA is a relatively stable molecule, it can be amplified by PCR from a preserved sample of a deceased organism and subjected to DNA sequencing. A comparison of these DNA sequences with modern species may help elucidate the phylogenetic relationships between extinct and modern species.

ACHIEVING THE GOAL — FIGURE 26.15 **DNA analysis reveals phylogenetic relationships among extinct and modern flightless birds.**

Starting material: Tissue samples from four extinct species of moas were obtained from museum specimens. Tissue samples were also obtained from three species of kiwis, one emu, one cassowary, one ostrich, and two species of rhea.

1. For soft tissue samples, treat with proteinase K (which digests protein) and a detergent that dissolves cell membranes. This releases the DNA from the cells.

2. Individually, mix the DNA samples with a pair of PCR primers that are complementary to the 12S rRNA gene.

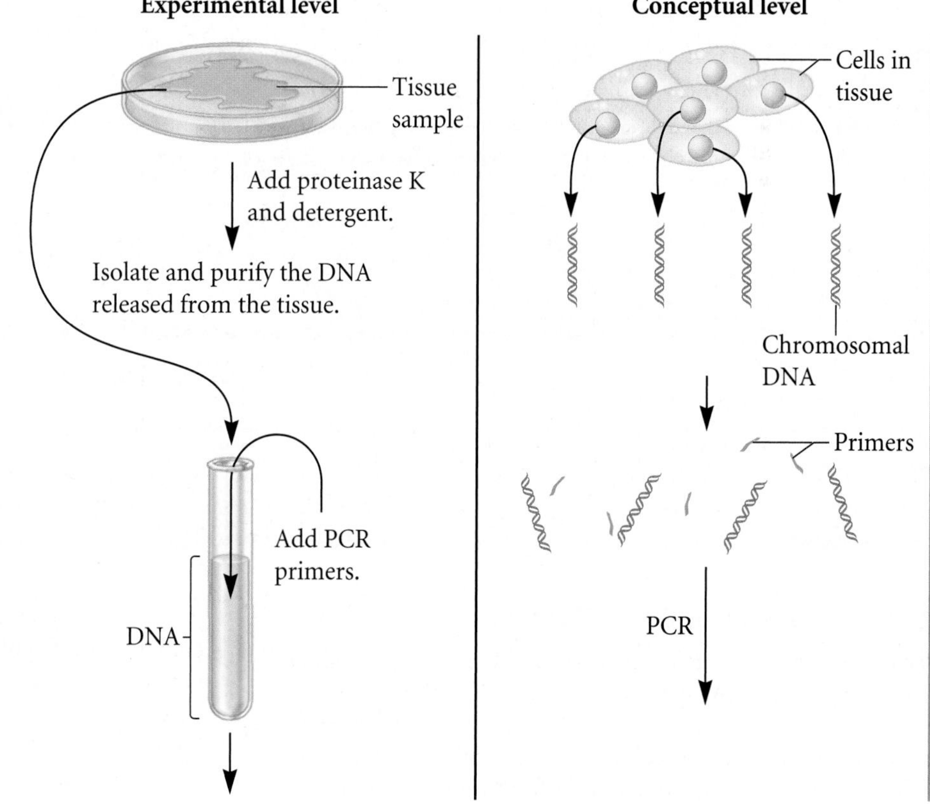

3. Subject the samples to PCR. See chapter 18 (Figure 18.6) for a description of PCR.

4. Subject the amplified DNA fragments to DNA sequencing. See chapter 18 for a description of DNA sequencing.

5. Align the DNA sequences to each other. Methods of DNA sequence alignment are described in chapter 21.

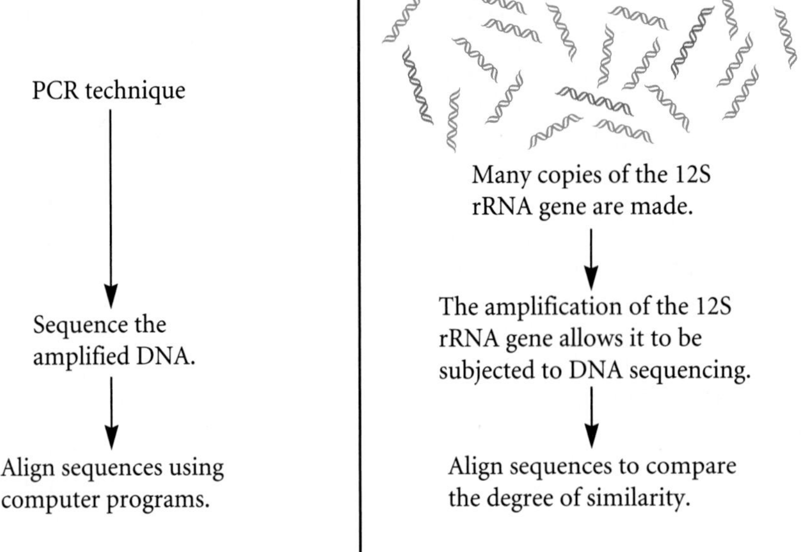
PCR technique
Sequence the amplified DNA.
Align sequences using computer programs.
Many copies of the 12S rRNA gene are made.
The amplification of the 12S rRNA gene allows it to be subjected to DNA sequencing.
Align sequences to compare the degree of similarity.

THE DATA

```
MOA 1      GCTTAGCCCTAAATCCAGATACTTACCCTACACAAGTATCCGCCCGAGAACTACGAGCACAAACGCTTAAAACTCTAAGGACTTGGCGGTGCCCCAAACCCACCTAGAGGAGCCTGTTCTATAATCGATAATCCACGATA
MOA 2      ············································································································································································
MOA 3      ····G··········T····························T···············································································································
MOA 4      ····························C···············································································································································
KIWI 1     ················T·G·······GT···CT·····C·······················································T····································C········
KIWI 2     ················T·G·G·····AT···CT·····C·······················································T····································C········
KIWI 3     ················T·G·G···G·AT···C······C·······················································T····································C········
EMU        ···············TT······C···T···CAG··C·······T·················································T····································C········
CASSOWARY  ···············TT······CG·TA···CTG············································································T····································C········
OSTRICH    ········T······AT···········C··CT·············································································T············································T
RHEA 1     ················T···········C··CT·············································································T····································C········
RHEA 2     ················C···········C··C··············································································T····································C········

MOA 1      CACCCGACCATCCCTCGCCCGT–GCAGCCTACATACCGCCGTCCCCAGCCCGCCT––AATGAAAG–AACAATAGCGAGCACAACAGCCCTCCCCCGCTAACAAGACAGGTCAAGGTATAGCATATGAGATGGAAGAAATG
MOA 2      ····················A·–································––········–······················TCA–················································
MOA 3      ············T··T····A·–································––········–·····················TA·––·T··············································
MOA 4      ············T··T····A·–································––········–·················T····AC––················································
KIWI 1     ·····A······T··T···AAC–A·······T···········G·····T·····AA·····G··–·····C·····A·········TA··–··A··························C··················
KIWI 2     ·····A······T··T···AAC–A·······T···········G·····T·····AA·····G··–·····C·····A·········TA··–··A··························C··················
KIWI 3     ·····A······T··T···AAC–A···················G···········AA········–····GC···············TACA–··A··························CC·C·······G·······
EMU        ·····AG·····T··T···AA·–A···················G···········––········–·················T····AC––TT···························G··················
CASSOWARY  ·····A······T··T···AA·TA···················G···········––·G···G··–··············T·······AC––T····························G··················
OSTRICH    ·····A····C····T···A––T····················G···········C––····G··–·················T·····A––··································GAG···········
RHEA 1     ············T··T····A·–································TA·G······–·····C···AG···T··T···TA–––··································G·············
RHEA 2     ············T··T····A·–································TA·····G··–·····C···A····T··T···TA–––·······G········································

MOA 1      GGCTACATTTTCTAACATAGAACACCC–––––––––––––ACGAAAGAGAAGGTGAAACCCTCCTCAAAAGGCGGATTTAGCAGTAAAATAGAACAAGAATGCCTATTTTAAGCCCGGCCCTGGGGC
MOA 2      ···························–––––––––––––············A······T·······G···············································T···········
MOA 3      ··························T–––––––––––––···························G························G·······C·····C········T···········
MOA 4      ···························–––––––––––––············A··············G························G·······C··C···········T···········
KIWI 1     ···············A······T··T––––––––––––––······A··GGT······T·–C···T·G·····················C····T····GA·T··········–·T······A····
KIWI 2     ···············A······T··T––––––––––––––······A··GGT······T··C···T·G·····················C····T····GA·T··········–·T······A····
KIWI 3     ···············A······T··T––––––––––––––······A··GGTA·····T··C···T·G···A·················C····T·····A·T··········–········A····
EMU        ······················T·T·––––––––––––––········AG·T······T·AC·T···G·····················C····T····GA·T·········A–·T···T··A····
CASSOWARY  ······················T···––––––––––––––······A··G·T······T·A····T·G·····················C·········GA·T·········A–········A····
OSTRICH    ······················T··A––––––––––––––·········G·TA·····T·A······G··························T····GA·T··········–T···T···A····
RHEA 1     ···············TC·······A·––––––––––––––···G·····GGCA······–AC····CG····················G··G·TC···A··C·C·········–········A····
RHEA 2     ··············GTC·······G·––––––––––––––·········GGCA······–AC····CG····················G·G·G·TC···A··C·C········–········A····
```

INTERPRETING THE DATA

The data of Figure 26.15 illustrate a multiple sequence alignment of the amplified DNA sequences. The first line shows the DNA sequence of one extinct moa species, and underneath it are the sequences of the other species. When the other sequences are identical to the first sequence, a dot is placed in the corresponding position. When the sequences are different, the nucleotide base (A, T, G, or C) is placed there. In a few regions, the genes are different lengths. In these cases, a dash is placed at the corresponding position.

As you can see from the large number of dots, the sequences among these flightless birds are very similar. To establish evolutionary relationships, researchers focus on the few differences that occur. Some surprising results were obtained. The sequences from the kiwis (a New Zealand species) are actually more similar to the sequence from the ostrich (an African species) than they are to those of the moas, which were once found in New Zealand. Likewise, the kiwis are more similar to the emu and cassowary (found in Australia and New Guinea) than they are to the moas. Contrary to their original expectations, the authors concluded that the kiwis

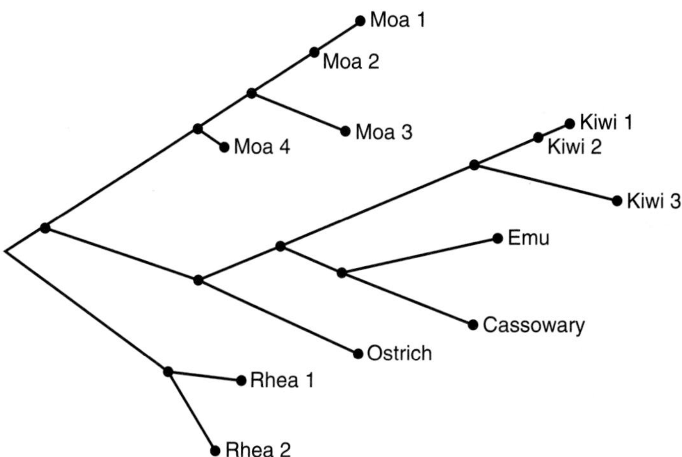

FIGURE 26.16 A revised phylogenetic tree of moas, kiwis, emus, cassowaries, ostriches, and rheas.

are more closely related to Australian and African flightless birds than they are to the moas. They proposed that New Zealand was colonized twice by ancestors of flightless birds. As shown in Figure 26.16, the researchers constructed a new evolutionary tree to illustrate the relationships among these modern and extinct species.

Since these early studies, sequences of ancient DNA have been derived from a variety of species. Figure 26.17 shows some extinct organisms from which DNA sequences have been determined. Many of these samples were tens of thousands of years old. For example, the sample of a Neanderthal man was approximately 30,000 years old. The oldest samples are likely to be in the range of 50,000 to 100,000 years old.

FIGURE 26.17 Extinct organisms from which DNA sequences have been obtained. Right panel, from bottom right to top right: quagga, marsupial wolf, sabre-toothed cat, moa, mammoth, cave bear, blue antelope, giant ground sloth, and Aurochs. Left panel, from bottom left to top right: mastodon, New Zealand coot, South Island piopio, Steller's sea cow, Neanderthal man, South Island adzebill (*Aptornis defossor*), Shasta ground sloth, pig-footed bandicoot, moa-nalo, and Balearic Islands cave goat (*Myotragus balearicus*). (Adapted from Hofreiter et al. [2001], "Ancient DNA," *Nature Reviews Genetics*, vol. 2, p. 357.)

A self-help quiz involving this experiment can be found at the Online Learning Center.

Speciation Is Associated with Changes in Chromosome Structure and Number

In this section, we have focused on mutations that alter the DNA sequences within genes. In addition to gene mutations, other types of changes, such as gene duplications, transpositions, inversions, translocations, and changes in chromosome number, are important features of evolution.

As discussed earlier, changes in chromosome structure and/or number may not always be adaptive, but they can lead to reproductive isolation and the origin of new species. As an example of variation in chromosome structure among closely related species, Figure 26.18 compares the banding pattern of the three largest chromosomes in humans and the corresponding chromosomes in chimpanzees, gorillas, and orangutans. The banding patterns are strikingly similar because these species are closely related evolutionarily. However, some interesting differences are observed. Humans have one large chromosome 2, but this chromosome is divided into two separate chromosomes in the other three species. This explains why humans have 23 types of chromosomes and the other species have 24. This may have occurred by a fusion of the two smaller chromosomes during the development of the human lineage. Another interesting change in chromosome structure is seen in chromosome 3. The banding patterns among humans, chimpanzees, and gorillas are very similar, but the orangutan has a large inversion that flips the arrangement of bands in the centromeric region.

With the advent of molecular techniques, researchers can analyze the chromosomes of two or more different species and identify regions that contain the same groups of linked genes, which are called **synteny groups.** Within a particular synteny group, the same types of genes are found in the same order. In 1995, Graham Moore and colleagues analyzed the locations of molecular markers along the chromosomes of several cereal grasses including rice (*Oryza sativa*), wheat (*Triticum aestivum*), maize (*Zea mays*), foxtail millet (*Setaria italica*), sugarcane (*Saccharum officinarum*), and sorghum (*Sorghum vulgare*). From this analysis, they were able to identify several large synteny groups that are common to most of these species (Figure 26.19). As an example, let's compare rice (12 chromosomes per set) to wheat

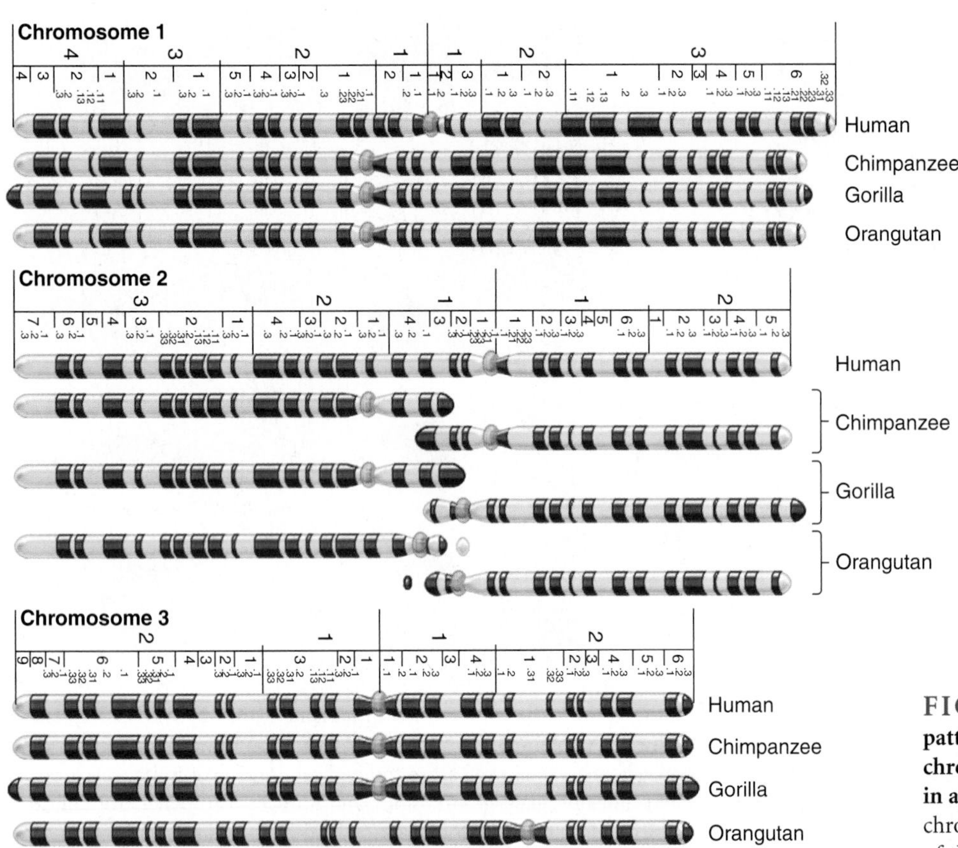

FIGURE 26.18 A comparison of banding patterns among the three largest human chromosomes and the corresponding chromosomes in apes. This is a schematic drawing of late prophase chromosomes. The conventional numbering system of the banding patterns is shown next to the human chromosomes.

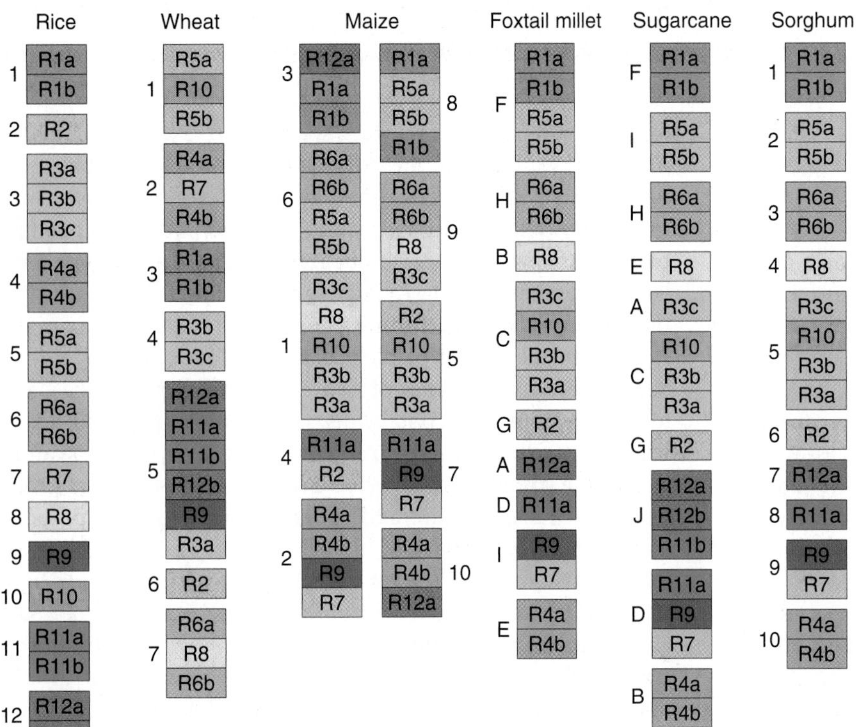

FIGURE 26.19 Synteny groups in cereal grasses. The synteny groups are named according to their relative arrangement in rice, which is shown on the left. In some cases, a synteny group or portion of a group may have incurred an inversion, but these are not shown in this figure. Also, the genome of maize contains two copies of most synteny groups, suggesting that it is derived from an ancestral polyploid species.

(7 chromosomes per set). In rice, chromosome 6 contains two synteny groups designated R6a and R6b, while chromosome 8 contains a single synteny group, R8. In wheat, chromosome 7 consists of R6a and R6b at either end, while R8 is sandwiched in the middle. One possible explanation for this difference could be a chromosomal rearrangement during the evolution of wheat in which R8 became inserted into the middle of a chromosome containing R6a and R6b. Overall, the evolution of cereal grass species has maintained most of the same types of genes, but many chromosomal rearrangements have occurred. These types of rearrangements would promote reproductive isolation.

26.3 EVO-DEVO: EVOLUTIONARY DEVELOPMENTAL BIOLOGY

The genetic changes that result in morphological and physiological differences distinguish one species from another. In recent years, researchers have begun to investigate how genetic variation produces species and groups of species with novel shapes and forms. The underlying basis for these changes is often rooted in the developmental pathways that control an organism's morphology. **Evolutionary developmental biology** (informally referred to as **evo-devo**) is a relatively new field of biology that compares the development of species in an attempt to understand ancestral relationships between different species and the developmental mechanisms that bring about evolutionary change. During the past few decades, developmental geneticists have gained a better understanding of biological development at the molecular level. Much of this work has involved the discovery of genes that control development in experimental organisms. As more and more organisms have been analyzed, researchers have become interested in the similarities and differences that occur between closely related and distantly related species. Evolutionary developmental biology has arisen in response to this trend.

How do new morphological forms come into being? For example, how does a nonwebbed foot evolve into a webbed foot? Or how does a new organ, such as an eye, come into existence? As we will learn, such novelty arises through genetic changes, also called genetic innovations. Certain types of genetic innovation have been so advantageous that they have resulted in groups of many new species. For example, the innovation of feathered wings resulted in the evolution of many different species of birds. In this section, we will learn that proteins that control developmental changes, such as cell-signaling proteins and transcription factors, often play a key role in promoting the morphological changes that occur during evolution.

Variation in the Expression Patterns of Developmental Genes Can Dramatically Affect Morphology

As discussed in Chapter 23, genes that play a role in development may influence cell division, cell differentiation, cell migration, and cell death. These four processes create an organism with a specific body form and function. Developmental genes affect traits such as the size of a flower or the shape of your nose. In recent years, researchers have determined that developmental genes are key players in the evolution of many types of traits. Changes in such genes affect traits that can be acted on by natural selection. Variation in the expression of development genes may be commonly involved in the acquisition of new traits that promote the formation of new species.

To appreciate this concept, let's compare the formation of a chicken's foot with that of a duck. Developmental biologists have discovered that the morphological differences between a webbed and a nonwebbed foot are due to the differential expression of two cell-signaling proteins called bone morphogenetic protein 4 (BMP4) and gremlin. As shown in **Figure 26.20a**, the *BMP4* gene is expressed throughout the developing limb of both chickens

Chicken Duck

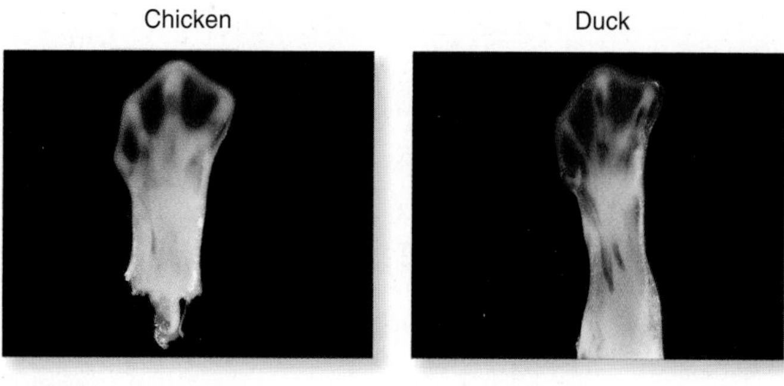

(a) BMP4 protein levels

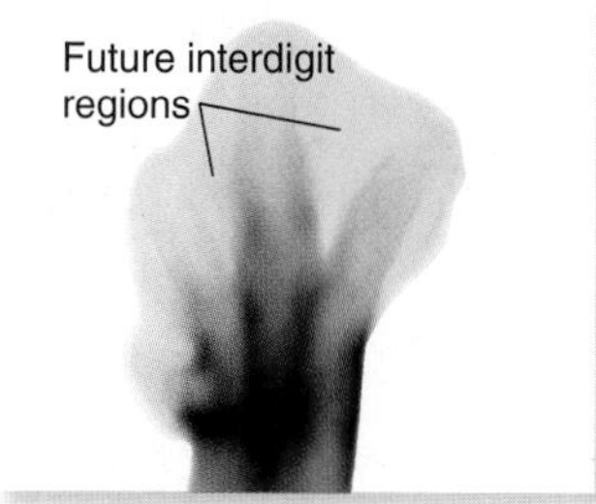

(b) Gremlin protein levels

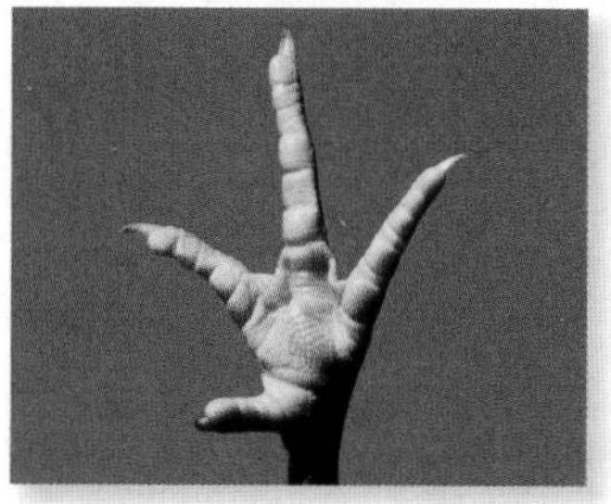

(c) Comparison of a chicken and a duck foot

FIGURE 26.20 The role of cell-signaling proteins in the morphology of birds' feet. (**a**) Expression of the *BMP4* gene in the developing limbs. BMP4 protein is stained blue. (**b**) Expression of the *gremlin* gene in the developing limbs. Gremlin protein is stained brown and is expressed in the interdigit region only in the duck. BMP4 causes programmed cell death; Gremlin inhibits BMP4. (**c**) Because BMP4 is not inhibited in the interdigit regions in the chicken, the cells in this region die, and the foot is not webbed. By comparison, inhibition of BMP4 in the interdigit regions in the duck results in a webbed foot.

and ducks. The BMP4 protein, which is stained in blue, causes cells to undergo apoptosis and die. The gremlin protein, which is stained brown in **Figure 26.20b**, inhibits the function of BMP4 and thereby allows cells to survive. In the developing chicken limb, the gremlin gene is expressed throughout the limb, except in the regions between each digit. Therefore, these cells die, and a chicken develops a nonwebbed foot (**Figure 26.20c**). In contrast, *gremlin* expression in the duck occurs throughout the entire limb, including the interdigit regions, resulting in a webbed foot. Interestingly, researchers have been able to introduce gremlin protein into the interdigit regions of developing chicken limbs. This produces a chicken with webbed feet!

How do these observations relate to bird evolution? As we have seen, variation in the expression of these genes determines whether or not a bird's feet are webbed. At some point in the evolution of birds, mutations occurred that provided variation in the expression of the *BMP4* and *gremlin* genes, which resulted in nonwebbed or webbed feet. In terrestrial settings, having nonwebbed feet is an advantage because these are more effective at holding on to perches, running along the ground, and snatching prey. Therefore, natural selection would maintain nonwebbed feet in terrestrial environments. This process explains the occurrence of nonwebbed feet in chickens, hawks, crows, and many other terrestrial birds. In aquatic environments, webbed feet are an advantage because they act as paddles for swimming, so genetic variation that produced webbed feet would have been promoted by natural selection. Over the course of many generations, this gave rise to webbed feet that are now found in ducks, geese, penguins, and other aquatic birds.

How does having webbed or nonwebbed feet influence speciation? This trait may not directly affect the ability of two individuals to mate. However, due to natural selection, birds with webbed feet would become more prevalent in aquatic environments, while birds with nonwebbed feet would be found in terrestrial locations. Therefore, reproductive isolation would occur because the populations would occupy different environments.

The Evolution of Animal Body Plans Is Related to Changes in *Hox* Gene Number and Expression

Hox genes, which are discussed in Chapter 23, are found in all animals. Developmental biologists have speculated that genetic variation in the *Hox* genes may have spawned the formation of many new body types, yielding many different animal species. As shown in **Figure 26.21**, the number and arrangement of *Hox* genes varies considerably among different types of animals.

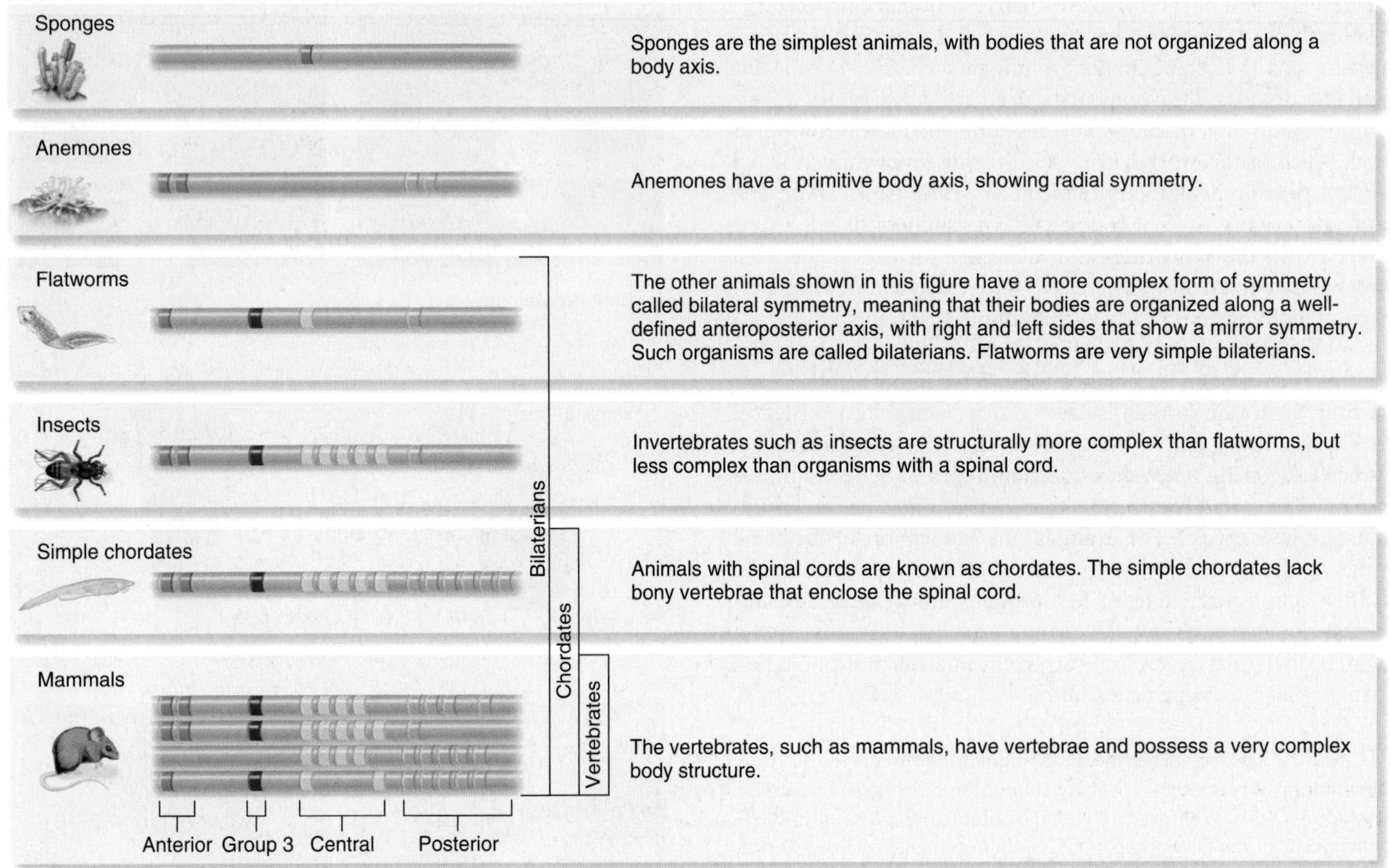

Sponges
Sponges are the simplest animals, with bodies that are not organized along a body axis.
Anemones
Anemones have a primitive body axis, showing radial symmetry.
Flatworms
The other animals shown in this figure have a more complex form of symmetry called bilateral symmetry, meaning that their bodies are organized along a well-defined anteroposterior axis, with right and left sides that show a mirror symmetry. Such organisms are called bilaterians. Flatworms are very simple bilaterians.
Insects
Invertebrates such as insects are structurally more complex than flatworms, but less complex than organisms with a spinal cord.
Simple chordates
Animals with spinal cords are known as chordates. The simple chordates lack bony vertebrae that enclose the spinal cord.
Mammals
The vertebrates, such as mammals, have vertebrae and possess a very complex body structure.
Bilaterians
Chordates
Vertebrates
Anterior
Group 3
Central
Posterior

FIGURE 26.21 ***Hox* gene composition in different types of animals.** Researchers speculate that the duplication of *Hox* genes and *Hox* gene clusters played a key role in the evolution of more complex body plans in animals. The *Hox* genes are divided into four groups, called anterior, group 3, central, and posterior, based on their relative similarities. Each group is represented by a different color in this figure.

Sponges, the simplest of animals, have at least one *Hox* gene, whereas insects typically have nine or more. In most cases, multiple *Hox* genes occur in a cluster in which the genes are close to each other along a chromosome. In mammals, such *Hox* gene clusters have been duplicated twice during the course of evolution to form four clusters with a total of 38 genes.

How would an increase in *Hox* genes enable more complex body forms to evolve? Part of the answer lies in the spatial expression of the *Hox* genes. In fruit flies, for example, different *Hox* genes are expressed in different segments of the body along the anteroposterior axis (see Figure 23.16). Therefore, an increase in the number of *Hox* genes allows each of these master control genes to become more specialized in the region that it controls. One segment in the middle of the fruit fly body can be controlled by a particular *Hox* gene and form wings and legs, while a segment in the head region can be controlled by a different *Hox* gene and develop antennae. Therefore, research suggests that one way that new, more complex body forms evolved was through an increase in the number of *Hox* genes, thereby making it possible to form many specialized parts of the body that are organized along a body axis.

Three lines of evidence support the idea that *Hox* gene complexity has been instrumental in the evolution and speciation of animals with different body patterns. First, *Hox* genes are known to control body development. Second, as described in Figure 26.21, a general trend is observed in which animals with a more complex body structure tend to have more *Hox* genes and more *Hox* clusters in their genomes than do simpler animals. Finally, a comparison of *Hox* gene evolution and animal evolution bears striking parallels. Researchers can analyze *Hox* gene sequences among modern species and make estimates regarding the timing of past events via molecular clock data. Geneticists can estimate when the first *Hox* gene arose by gene innovation; the date is difficult to pinpoint but is well over 600 million years ago. The single *Hox* gene found in the sponge has descended from this primordial *Hox* gene. In addition, gene duplications of this primordial gene produced clusters of *Hox* genes in other species. Clusters, such as those found in modern insects, were likely to have arisen approximately 600 million years ago. A duplication of that cluster is estimated to have occurred around 520 million years ago. Remarkably, these estimates of *Hox* gene origins correlate with major speciation events in the history of animals. The Cambrian explosion, which occurred from 533 to 525 million years ago, saw a phenomenal diversification in the body plan of invertebrate species. This diversification occurred after the *Hox* cluster was formed and was possibly undergoing its first duplication to create two *Hox* clusters. Also, approximately 420 million years ago, a second duplication produced species with four *Hox* clusters. This event precedes the proliferation of tetrapods—vertebrates with four limbs—that occurred approximately 417 to 354 million years ago. Modern tetrapods, such as mammals, have four *Hox* clusters. This second duplication may have been a critical event that led to the evolution of complex terrestrial vertebrates with four limbs, such as lizards, bears, and humans.

The Study of the *Pax6* Gene Indicates That Different Types of Eyes May Have Evolved from a Simpler Form

Explaining how a complex organ comes into existence is another major challenge for evolutionary biologists. While it is relatively easy to understand how a limb could undergo evolutionary modifications to become a wing, flipper, or arm, it is more difficult to understand how a body structure comes into being in the first place. In his book *The Origin of Species*, Charles Darwin addressed this question and pointed out that the existence of complex organs, such as the eye, was difficult to understand. As Darwin noted, the eye of vertebrate species is exceedingly complex, being able to adjust focus, let in different amounts of light, and detect a spectrum of colors. He suggested that such a complex eye must have evolved from a simpler structure through the process of descent with modification. With amazing insight, he speculated that a very simple eye could be composed of just two cells, a photoreceptor cell and an adjacent pigment cell. The photoreceptor cell, which is a type of nerve cell, is able to absorb light and respond to it. The function of the pigment cell is to stop the light from reaching one side of the photoreceptor cell. This primitive, two-cell arrangement would allow an organism to sense both light and the direction from which the light comes.

How would natural selection play a role in the evolution of eyes? A primitive eye would provide an additional way for a mobile organism to sense its environment, possibly allowing it to avoid predators or locate food. Vision is nearly universal among animals, which indicates that there must be a strong selective advantage to better eyesight. Over time, eyes could become more complex by enhancing the ability to absorb different amounts and wavelengths of light and also by refinements in structures such as the addition of lenses that detect the direction of light.

Since the time of Darwin, many evolutionary biologists have wrestled with the question of eye evolution. From an anatomical point of view, researchers have discovered many different types of eyes. For example, the eyes of fruit flies, squid, and humans are quite different from each other. Furthermore, species that are closely related evolutionarily sometimes have different types of eyes. This observation led evolutionary biologists Luitfried von Salvini-Plawen and Ernst Mayr to propose that eyes may have independently arisen many different times during evolution. Based solely on morphology, such a hypothesis seemed reasonable, and for many years was accepted by the scientific community.

The situation took a dramatic turn when geneticists began to study eye development. Researchers identified a gene, *Pax6*,[2] that influences eye development in both rodents and humans. In mice and rats, a mutation in this gene results in small eyes. In humans, a mutation in the *Pax6* gene results in an eye disorder called aniridia. In heterozygotes carrying one defective copy of the gene, the iris does not form. Researchers subsequently discovered a gene in *Drosophila*, named *eyeless*, that also causes a defect in eye development when mutant. DNA sequencing revealed the *eyeless* and *Pax6* genes are homologous; they are derived from the same ancestral gene.

[2] *Pax* is an acronym for paired box. The protein encoded by this gene contains a domain called a paired box.

In 1995, Walter Gehring and his colleagues were able to show experimentally that the abnormal expression of the *eyeless* gene in other parts of the fruit fly body could promote the formation of additional eyes. For example, using genetic engineering techniques, they were able to express the *eyeless* gene in the region where antennae should form. As seen in **Figure 26.22a**, this resulted in the formation of fruit fly eyes where antennae are normally found! Remarkably, the expression of the mouse *Pax6* gene in *Drosophila* can also cause the formation of eyes in unusual places. **Figure 26.22b** shows the formation of an eye on the leg of a fruit fly. This eye was caused by the expression of the mouse *Pax6* gene in this region.

The mouse *Pax6* gene switches on eye formation in *Drosophila*, but the eye produced is a *Drosophila* eye, not a mouse eye. This happens because the genes activated by the *Pax6* gene are all from the *Drosophila* genome. In *Drosophila*, the *eyeless* gene switches on a cascade involving 2,500 different genes required for eye morphogenesis. The *Pax6* gene and its *Drosophila* homolog, *eyeless*, are master control genes that promote the formation of an eye. In addition, Gehring has suggested that the eyes of *Drosophila* and mammals are evolutionarily derived from the modification of an eye that arose once during evolution. If *Drosophila* and mammalian eyes had arisen independently, the *Pax6* gene from mice would not be expected to induce the formation of eyes in *Drosophila*.

Since the initial discovery of the *Pax6* and *eyeless* genes, homologs of this gene have been discovered in many different species. In all cases where it has been tested, this gene directs eye development. The *Pax6* gene and its homologs encode a transcription factor protein that controls the expression of many different genes. Gehring and colleagues have hypothesized that the eyes from many different species all evolved from a common ancestral form consisting of, as proposed by Darwin, one photoreceptor cell and one pigment cell (**Figure 26.23**). As mentioned previously, such a very simple eye can accomplish some rudimentary form of vision by detecting light and its direction. Eyes such as these are still found in modern species such as the larvae of certain types of mollusks. Over the course of evolution, simple eyes were transformed into more complex types of eyes by modifications that resulted in the addition of more types of cells such as lens and muscle cells.

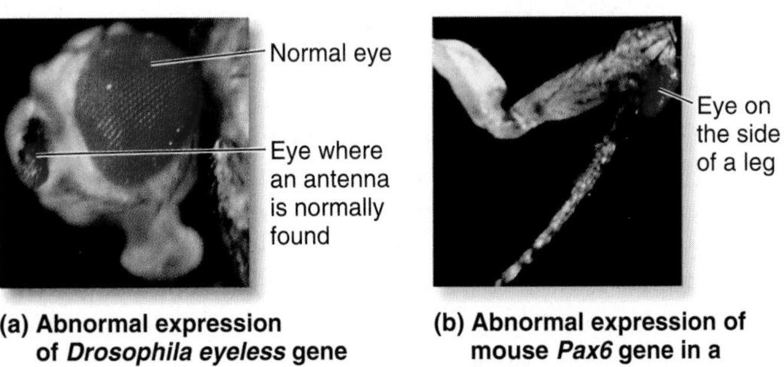

FIGURE 26.22 Formation of additional eyes in *Drosophila* due to the abnormal expression of a master control gene for eye morphogenesis. (**a**) When the *Drosophila eyeless* gene is expressed in the antenna region, eyes are formed where antennae should be located. (**b**) When the mouse *Pax6* gene is expressed in the leg region of *Drosophila*, a small eye is formed there.

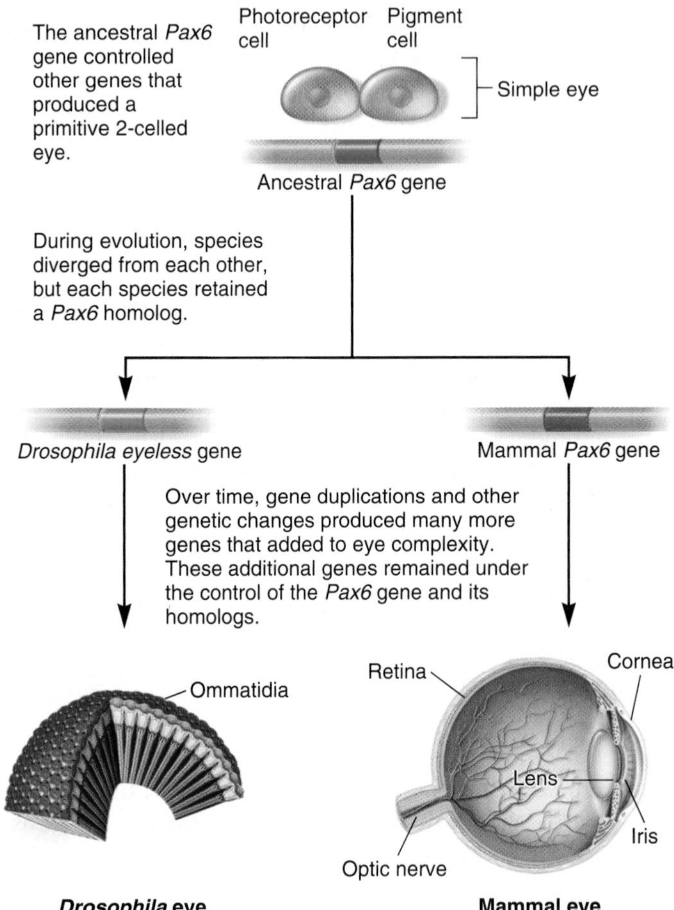

FIGURE 26.23 Eye evolution. In this diagram, genetic changes, under control of the ancestral *Pax6* gene, led to the evolution of different types of eyes. Ommatidia are units of the insect compound eye.

CONCEPTUAL SUMMARY

Biological evolution involves heritable changes in one or more characteristics in a population or species over the course of many generations. Charles Darwin's theory of evolution is based on two fundamental principles: genetic variation and natural selection. In the first part of this chapter, we were concerned with how these features of evolution result in the formation of new species. Biologists use four different species concepts—the phylogenetic, biological, evolutionary, and ecological species concepts—to define and identify species. Speciation is often a branching process (cladogenesis), although anagenesis (transformation of a single species) occasionally occurs. Speciation may be allopatric, parapatric, or sympatric, depending on whether or not geographic barriers are present. Allopatric speciation, which involves geographic barriers, is thought to be the most widespread form of speciation. It can occur slowly, due to the gradual formation of a geographic barrier, or rapidly, due to the founder effect. Parapatric speciation is similar to allopatric speciation, except that the geographic barriers are less complete, with ever-decreasing levels of interbreeding taking place in hybrid zones. Sympatric speciation does not require geographic barriers. Instead, an abrupt genetic event may lead to reproductive isolation. In plants, the formation of polyploids is a common form of sympatric speciation. The fossil record suggests that speciation occurs through punctuated equilibrium, a relatively rapid process that punctuates long periods during which a species is in equilibrium with its environment. In contrast, gradualism proposes a slower but steady rate of evolution due to the accumulation of many small genetic changes.

Molecular evolution is the study of the molecular changes in the genetic material during evolution. Molecular data are increasingly used to construct phylogenetic trees, which are diagrams that describe the evolutionary relationships among different species. Homologous genes are derived from a common ancestral gene. The analysis of DNA sequences and deduced amino acid sequences of organisms are the most commonly used methods to make phylogenetic trees. Those species with greater similarity in genetic material generally are assumed to be closely related evolutionarily. The cladistic approach constructs a phylogenetic tree, called a cladogram, by considering the various possible pathways of evolution and then choosing the most plausible route. According to the neutral theory of evolution, most variation within and between species is due to the accumulation of neutral or nearly neutral mutations. Kimura and Ohta proposed five principles of molecular evolution that are consistent with our present knowledge of gene sequences. The accumulation of neutral mutations can be used as a molecular clock on which to measure evolutionary time.

In addition to random changes in gene sequences, other mechanisms contribute to the evolution of new species. For example, horizontal gene transfer is particularly important in the evolution of prokaryotic species and was probably very prevalent during the earliest stages of evolution, when all organisms were unicellular. Also, changes in chromosome structure and number may promote reproductive isolation.

A relatively new area of evolutionary studies involves the study of how developmental genes play a role in evolutionary processes, a field called evolutionary developmental biology (evo-devo). Researchers have identified several genes that control developmental pathways. For example, *BMP4* and *gremlin* are key genes that control webbed versus nonwebbed feet in birds. In addition, an increase in the number of *Hox* genes appears to be a critical factor that led to the development of animals with more complex body plans. Finally, we considered how the study of the *Pax6* gene, a master control gene, has revealed that all types of eyes most likely evolved from a single type of eye, perhaps a simple two-celled eye composed of a photoreceptor cell and a pigment cell.

EXPERIMENTAL SUMMARY

In a broad sense, evolutionary biologists would like to know how genetic changes have led to the phenotypic characteristics of present-day species. At the heart of this question is speciation: How do we explain the existence of so many different species? To answer this question, several species concepts can be applied. Biologists also try to understand how two different, but closely related, species have become reproductively isolated. By observing species in nature, they have identified several prezygotic and postzygotic mechanisms that prevent the production of viable, interspecies offspring. Researchers have also correlated reproductive isolation with experimentally observable genetic changes. For example, parapatric speciation may be associated with the accumulation of chromosomal inversions that prevent the production of fertile offspring within hybrid zones of contact. Sympatric speciation can occur via the formation of allotetraploids. This speciation mechanism can be confirmed by producing "artificial" polyploids.

Evolutionary biologists are also interested in the rates of evolutionary change. Experimentally, the fossil record suggests that it is more common for evolution to follow a pattern of punctuated equilibrium. According to this idea, a species is well adapted to its environment for a long period of time (the equilibrium). This equilibrium is punctuated by short periods in which the species undergoes rapid evolutionary change. These short periods involve abrupt genetic changes (e.g., allopolyploidism or new alleles that have a dramatic effect on phenotype), or there may be a short period of strong selective pressure (e.g., the founder effect, a new predator in the region, or a significant environmental change).

At the molecular level, evolution is due to alterations in DNA sequences and changes in chromosome structure and number. Geneticists have found a great amount of variation within most species. The analysis of DNA sequences is now the most common way to study molecular evolution. To construct a phylogenetic tree, researchers usually follow a cladistic approach in which species are compared with regard to shared derived characters and primitive characters. Often times, the analysis involves homologous gene sequences among different species. It has even been possible to analyze DNA sequences from extinct species.

PROBLEM SETS & INSIGHTS

Solved Problems

S1. A codon for leucine is UUA. A mutation causing a single-base substitution in a gene can change this codon in the transcribed mRNA into GUA (valine), AUA (isoleucine), CUA (leucine), UGA (stop), UAA (stop), UCA (serine), UUG (leucine), UUC (phenylalanine), or UUU (phenylalanine). According to the neutral theory, which of these mutations would you expect to see within the genetic variation of a natural population? Explain.

Answer: The neutral theory proposes that neutral mutations will accumulate to the greatest extent in a population. Leucine is a nonpolar amino acid. For a UUA codon, single-base changes of CUA and UUG are silent, so they would be the most likely to occur in a natural population. Likewise, conservative substitutions to other nonpolar amino acids such as isoleucine (AUA), valine (GUA), and phenylalanine (UUC and UUU) may not affect protein structure and function, so they may also occur and not be eliminated rapidly by natural selection. The polar amino acid serine (UCA) is a nonconservative substitution; one would predict that it is more likely to disrupt protein function. Therefore, it may be less likely to be found. Finally, the stop codons, UGA and UAA, would be expected to diminish or eliminate protein function, particularly if they occur early in the coding sequence. These types of mutations are selected against and, therefore, are not usually found in natural populations.

S2. Explain why homologous genes have sequences that are similar but not identical.

Answer: Homologous genes are derived from the same ancestral gene. Therefore, as a starting point, they had identical sequences. Over time, however, each gene accumulates random mutations that the other homologous genes did not acquire. These random mutations change the gene from its original sequence. Therefore, much of the sequence between homologous genes remains identical, but some of the sequence will be altered due to the accumulation of independent random mutations.

S3. Explain why plants are more likely to evolve by sympatric speciation compared to animals.

Answer: A common way for sympatric speciation to occur is by the formation of polyploids. For example, if one species is diploid ($2n$), nondisjunction could produce an individual that is tetraploid ($4n$). If the tetraploid individual was a plant and if it was monoecious (i.e., produces both pollen and egg cells), the plant could multiply to produce many tetraploid offspring. These offspring would be reproductively isolated from the diploid plants that are found in the same geographic area. This isolation occurs because the offspring of a cross between a diploid and tetraploid plant would be infertile. For example, if the pollen from a diploid plant fertilized the egg from a tetraploid plant, this would produce triploid offspring. The triploid offspring might be viable, but it is very likely that triploids would be sterile because they would produce highly aneuploid gametes. The gametes would be aneuploid because there are an odd number of homologous chromosomes. In this case, there would be three copies of each homologous chromosome, and these could not be equally distributed into gametes. Therefore, when a tetraploid is produced, it can immediately create its own unique species that is reproductively isolated from other species.

In contrast, polyploidy rarely occurs in animals. Perhaps the main reason is because animals usually cannot tolerate polyploidy. Tetraploid animals typically die during early stages of development. In addition, most species of animals have male and female sexes. To develop a tetraploid species from a diploid species, nondisjunction would have to produce both a male and female offspring that could reproduce with each other. The chance of producing one tetraploid offspring is relatively rare. The chances of producing two tetraploid offspring that happen to be male and female and happen to mate with each other to produce many offspring would be extremely rare.

S4. As described in Figure 26.18, evolution is associated with changes in chromosome structure and number. As seen here, chromosome 2 in humans is divided into two distinct chromosomes in chimpanzees, gorillas, and orangutans. In addition, chromosome 3 in the orangutan has a large inversion not found in the other three primates. Discuss the potential role of these types of changes in the evolution of these primate species.

Answer: As discussed in Chapter 8, changes in chromosome structure, such as inversions and balanced translocations, may not have any phenotypic effects. Likewise, the division of a single chromosome into two distinct chromosomes may not have any phenotypic effect as long as the total amount of genetic material remains the same. Overall, the types of changes in chromosome structure and number shown in Figure 26.18 may not have caused any changes in the phenotypes of primates. However, the changes would be expected to promote reproductive isolation. For example, if a gorilla mated with an orangutan, the offspring would be an inversion heterozygote for chromosome 3. As shown in Figure 8.10, crossing over during gamete formation in an inversion heterozygote may produce chromosomes that have too much or too little genetic material. This is particularly likely if the inversion is fairly large (such as the one shown in Figure 26.18). The inheritance of too much or too little genetic material is likely to be detrimental or even lethal. For this reason, the hybrid offspring of a gorilla and orangutan would probably not be fertile. (Note: In reality, there are several other reasons why interspecies matings between gorillas and orangutans do not produce viable offspring.)

Overall, the primary effect of changes in chromosome structure and number, like the ones shown in Figure 26.18, is to promote reproductive isolation. Once two populations become reproductively isolated, they will accumulate different mutations, and over the course of many generations, this will lead to two different species that have distinct characteristics.

It should be noted that changes in chromosome number in plants are more likely to have effects that abruptly lead to the formation of new species. This idea is discussed in solved problem S3.

Conceptual Questions

C1. Discuss the two principles on which evolution is based.

C2. Evolution, which involves genetic change in a population of organisms over time, is often described as the unifying theme in biology. Discuss how evolution is unifying at the molecular and cellular levels.

C3. What is meant by the term reproductive isolation? Give several examples.

C4. Briefly describe four different species concepts.

C5. Would the following examples of reproductive isolation be considered a prezygotic or postzygotic mechanism?

A. Horses and donkeys can interbreed to produce mules, but the mules are infertile.

B. Three species of the orchid genus *Dendrobium* produce flowers 8 days, 9 days, and 11 days after a rainstorm. The flowers remain open for one day.

C. Two species of fish release sperm and eggs into seawater at the same time, but the sperm of one species do not fertilize the eggs of the other species.

D. Two tree frogs, *Hyla chrysoscelis* (diploid) and *H. versicolor* (tetraploid), can produce viable offspring, but the offspring are sterile.

C6. Distinguish between anagenesis and cladogenesis. Which type of speciation is more prevalent? Why?

C7. Describe three or more genetic mechanisms that may lead to the rapid evolution of a new species. Which of these genetic mechanisms are influenced by natural selection, and which are not?

C8. Explain the type of speciation (allopatric, parapatric, or sympatric) most likely to occur under each of the following conditions:

A. A pregnant female rat is transported by an ocean liner to a new continent.

B. A meadow containing several species of grasses is exposed to a pesticide that promotes nondisjunction.

C. In a very large lake containing several species of fish, the water level gradually falls over the course of several years. Eventually, the large lake becomes subdivided into smaller lakes, some of which are connected by narrow streams.

C9. Alloploids are created by crosses involving two different species. Explain why alloploids are reproductively isolated from the two original species from which they were derived. Explain why alloploids are usually sterile, whereas allotetraploids (containing a diploid set from each species) are commonly fertile.

C10. Discuss the evidence in favor of the punctuated equilibrium model of evolution. What mechanisms could account for this pattern of evolution? In contrast, what type of genetic changes are consistent with gradualism?

C11. Discuss whether the phenomenon of reproductive isolation applies to bacteria, which reproduce asexually. How would a geneticist divide bacteria into separate species?

C12. Discuss the major differences between allopatric, parapatric, and sympatric speciation.

C13. The following are two DNA sequences from homologous genes:

TTGCATAGGCATACCGTATGATATCGAAAACTAGAAAAATAGGGCGATAGCTA

GTATGTTATCGAAAAGTAGCAAAATAGGGCGATAGCTACCCAGACTACCGGAT

The two sequences, however, do not begin and end at the same location. Try to line them up according to their homologous regions.

C14. What is meant by the term molecular clock? How is this concept related to the neutral theory of evolution?

C15. Would the rate of deleterious or beneficial mutations be a good molecular clock? Why or why not?

C16. Which would you expect to exhibit a faster rate of evolutionary change, the nucleotide sequence of a gene or the amino acid sequence of the encoded polypeptide of the same gene? Explain your answer.

C17. When comparing the coding region of structural genes among closely related species, it is commonly found that certain regions of the gene have evolved more rapidly (i.e., have tolerated more changes in sequence) compared to other regions of the gene. Explain why different regions of a structural gene evolve at different rates.

C18. Plant seeds contain storage proteins that are encoded by plant genes. When the seed germinates, these proteins are rapidly hydrolyzed (i.e., the covalent bonds between amino acids within the polypeptides are broken), which releases amino acids for the developing seedling. Would you expect the genes that encode plant storage proteins to evolve slowly or rapidly compared to genes that encode enzymes? Explain your answer.

C19. Figure 26.12 shows a phylogenetic tree of all life on Earth based on 16S rRNA data. Based on your understanding of molecular genetics (in Chapter 26 and other chapters), describe three or more observations that suggest that all life-forms on Earth evolved from a common ancestor or group of ancestors.

C20. Take a look at the α-globin and β-globin sequences in Figure 26.6. Which sequences are more similar, the α-globin in humans and the α-globin in horses, or the α-globin in humans and the β-globin in humans? Based on your answer, would you conclude that the gene duplication that gave rise to the α-globin and β-globin genes occurred before or after the divergence of humans and horses? Explain your reasoning.

C21. Compare and contrast the neutral theory of evolution versus the Darwinian (i.e., selectionist) theory of evolution. Explain why the neutral theory of evolution is sometimes called non-Darwinian evolution.

C22. For each of the following examples, discuss whether it would be the result of neutral mutation or mutation that has been acted upon by natural selection, or both:

A. When comparing sequences of homologous genes, differences in the coding sequence are most common at the wobble base (i.e., the third base in each codon).

B. For a structural gene, the regions that encode portions of the polypeptide that are vital for structure and function are less likely to incur mutations compared to other regions of the gene.

C. When comparing the sequences of homologous genes, introns usually have more sequence differences compared to exons.

C23. As discussed in Chapter 24, genetic variation is prevalent in natural populations. This variation is revealed in the DNA sequencing of genes. According to the neutral theory of evolution, discuss the relative importance of natural selection against detrimental mutations, natural selection in favor of beneficial mutations, and neutral mutations in accounting for the genetic variation we see in natural populations.

C24. If you were comparing the karyotypes of species that are closely related evolutionarily, what types of similarities and differences would you expect to find?

C25. In the developing bud that gives rise to a hand in a human embryo, where would you expect the *gremlin* gene to be expressed?

C26. Discuss how *Hox* gene number is related to body complexity.

Experimental Questions

E1. Two populations of snakes are separated by a river. The snakes cross the river only on rare occasions. The snakes in the two populations look very similar to each other except that the members of the population on the eastern bank of the river have a yellow spot on the top of their head, while the members of the western population have an orange spot on the top of their head. Discuss two experimental methods that you might follow to determine whether the two populations are members of the same species or members of different species.

E2. Sympatric speciation by allotetraploidy has been proposed as a common mechanism for speciation. Let's suppose you were interested in the origin of certain grass species in Southern California. Experimentally, how would you go about determining if some of the grass species are the result of allotetraploidy?

E3. Two diploid species of closely related frogs, which we will call species A and species B, were analyzed with regard to genes that encode an enzyme called hexokinase. Species A has two distinct copies of this gene: *A1* and *A2*. In other words, this diploid species is *A1A1 A2A2*. The other species has three copies of the hexokinase gene, which we will call *B1*, *B2*, and *B3*. A diploid individual of species B would be *B1B1 B2B2 B3B3*. These hexokinase genes from the two species were subjected to DNA sequencing, and the percentage of sequence identity was compared among these genes. The results are shown here:

Percentage of DNA Sequence Identity

	A1	*A2*	*B1*	*B2*	*B3*
A1	100	62	54	94	53
A2	62	100	91	49	92
B1	54	91	100	67	90
B2	94	49	67	100	64
B3	53	92	90	64	100

If we assume that hexokinase genes were never lost in the evolution of these frog species, how many distinct hexokinase genes do you think there were in the most recent ancestor that preceded the divergence of these two species? Explain your answer. Also explain why species B has three distinct copies of this gene, whereas species A has only two.

E4. A researcher sequenced a portion of a bacterial gene and obtained the following sequence, beginning with the start codon, which is underlined:

<u>ATG</u> CCG GAT TAC CCG GTC CCA AAC AAA ATG
ATC GGC CGC CGA ATC TAT CCC

The bacterial strain that contained this gene has been maintained in the laboratory and grown serially for many generations. Recently, another person working in the laboratory isolated DNA from the bacterial strain and sequenced the same region. The following results were obtained.

<u>ATG</u> CCG GAT TAT CCG GTC CCA AAT AAA ATG
ATC GGC CGC CGA ATC TAC CCC

Explain why these sequencing differences may have occurred.

E5. F_1 hybrids between two species of cotton, *Gossypium barbadense* and *G. hirsutum*, are very vigorous plants. However, F_1 crosses produce many seeds that do not germinate and a low percentage of very weak F_2 offspring. Suggest two reasons for these observations.

E6. A species of antelope contains 20 chromosomes per set. The species is divided by a mountain range into two separate populations, which we will call the eastern and western population. When comparing the karyotypes between these two populations, it was discovered that the members of the eastern population are homozygous for a large inversion within chromosome 14. How would this inversion affect the interbreeding between the two populations? Could such an inversion play an important role in speciation?

E7. Explain why molecular techniques were needed as a way to provide evidence for the neutral theory of evolution.

E8. Prehistoric specimens often contain minute amounts of ancient DNA. What technique can be used to increase the amount of DNA in an older sample? Explain how this technique is performed and how it increases the amount of a specific region of DNA.

E9. In the experiment of Figure 26.15, explain how we know that the kiwis are more closely related to the emu and cassowary than to the moas. Cite particular regions in the sequences that support your answer.

E10. In Chapter 20, we learned about a technique called fluorescence *in situ* hybridization (FISH), during which a labeled piece of DNA is hybridized to a set of chromosomes. Let's suppose that we cloned a piece of DNA from *G. pubescens* (see Figure 26.3) and used it as a labeled probe for *in situ* hybridization. What would you expect to happen if we hybridized it to the *G. speciosa*, the natural *G. tetrahit*, or the artificial *G. tetrahit* strains? Describe your expected results.

E11. A team of researchers has obtained a dinosaur bone (*Tyrannosaurus rex*) and has attempted to extract ancient DNA from it. Using primers to the 12S rRNA gene, they have used PCR and obtained a DNA segment that yields a sequence homologous to crocodile DNA. Other scientists are skeptical that this sequence is really from the dinosaur. Instead, they believe that it may be due to contamination from more recent DNA, such as the remains of a reptile that lived much more recently. What criteria might you use to establish the credibility of the dinosaur sequence?

E12. Discuss how the principle of parsimony can be used in a cladistics approach of constructing a phylogenetic tree.

E13. As discussed in this chapter and Chapter 24, genes are sometimes transferred between different species via horizontal gene transfer. Discuss how horizontal gene transfer might give misleading results when constructing a phylogenetic tree. How could you overcome this problem?

E14. If a researcher used genetic engineering techniques to express the *Drosophila eyeless* gene in the embryo at the region that will become the tip of the mouse's tail, what results would you expect in the resulting offspring?

Questions for Student Discussion/Collaboration

1. The raw material for evolution is random mutation. Discuss whether or not you view evolution as a random process.
2. Compare the forms of speciation that are slow with those that occur more rapidly. Make a list of the slow and fast forms. With regard to mechanisms of genetic change, what features do slow and rapid speciation have in common? What features are different?
3. Do you think that Darwin would object to the neutral theory of evolution?

Note: All answers appear at the website for this textbook; the answers to even-numbered questions are in the back of the textbook.

www.mhhe.com/brookergenetics3e

Visit the Online Learning Center for practice tests, answer keys, and other learning aids for this chapter. Enhance your understanding of genetics with our interactive exercises, quizzes, animations, and much more.

APPENDIX
EXPERIMENTAL TECHNIQUES

OUTLINE

A.1 METHODS OF CELL GROWTH

Researchers often grow cells in a laboratory as a way to study their properties. This is known as a **cell culture.** Cell culturing offers several technical advantages. The primary advantage is that the growth medium is defined and can be controlled. Minimal growth medium contains the bare essentials for cell growth: salts, a carbon source, an energy source, essential vitamins, amino acids, and trace elements. In their experiments, geneticists often compare strains that can grow in minimal media and mutant strains that cannot grow unless the medium is supplemented with additional components. A rich growth medium contains many more components than are required for growth.

Researchers also add substances to the culture medium for other experimental reasons. For example, radioactive isotopes can be added to the culture medium to radiolabel cellular macromolecules. In addition, an experimenter could add a hormone to the growth medium and then monitor the cells' response to the hormone. In all of these cases, cell culturing is advantageous because the experimenter can control and vary the composition of the growth medium.

The first step in creating a cell culture is the isolation of a cell population that the researchers wish to study. For bacteria, such as *Escherichia coli,* and eukaryotic microorganisms, such as yeast and *Neurospora,* the researchers simply obtain a sample of cells from a colleague or a stock center. For animal or plant tissues, the procedure is a bit more complicated. When cells are contained within a complex tissue, they must first be dispersed by treating the tissue with agents that separate it into individual cells to create a cell suspension.

Once a desired population of cells has been obtained, researchers can grow them in a laboratory (i.e., in vitro) either suspended in a liquid growth medium or attached to a solid surface such as agar. Both methods have been used commonly in the experiments considered throughout this textbook. Liquid culture is often used when researchers want to obtain a large quantity of cells and isolate individual cellular components, such as nuclei or DNA. By comparison, **Figure A.1** shows animal cells and bacteria

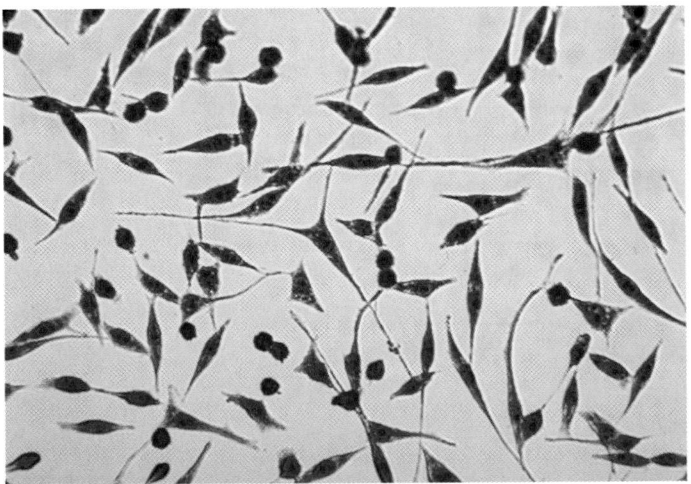

(a) Fibroblast (animal cell) culture

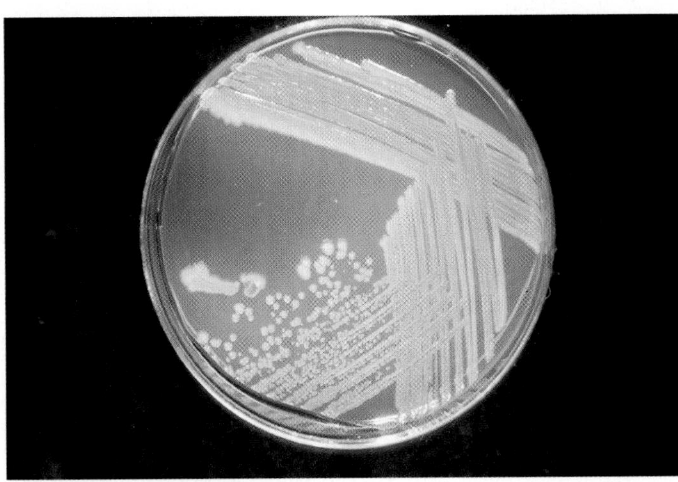
(b) Bacterial colonies

FIGURE A.1 **Growth of cells on solid growth media.** (**a**) This micrograph shows fibroblasts growing as a monolayer on a solid growth medium. (**b**) Bacterial cells form colonies that are a clonal population of cells derived from a single cell.

cells that are grown on solid growth media. As shown in Figure 22.9, solid media are used to study cancer cells, because such cells can be distinguished by the formation of foci in which malignant cells pile up on top of each other. In gene cloning experiments with bacteria and yeast, solid media are also used. Each colony of cells is a clone of cells that is derived from a single cell that divided to produce many cells (**Figure A.1b**). As discussed in Chapter 18, a solid medium is used in the isolation of individual clones that contain a desired gene.

A.2 MICROSCOPY

Microscopy is a technique to observe things that are not visible (or are hardly visible) with the naked eye. A key concept in microscopy is **resolution,** which is defined as the minimum distance between two objects that can be seen as separate from each other. The ability to resolve two points as being separate depends on several factors, including the wavelength of the illumination source (light or electron beam), the medium in which the sample is immersed, and the structural features of the microscope (which are beyond the scope of this textbook).

As shown in **Figure A.2**, two widely used kinds of microscopes are the optical (light) microscope and the transmission electron microscope (TEM). The light microscope is used to resolve cellular structures to a limit of approximately 0.3 μm. (For comparison, a typical bacterium is about 1 μm long.) At this resolution, the individual cell organelles in eukaryotic cells can be discerned easily, and chromosomes are also visible. Karyotyping is accomplished via light microscopy after the chromosomes have been treated with stains. A variation of light microscopy known as fluorescence microscopy is often used to highlight a particular feature of a chromosome or cellular structure. The technique of fluorescence *in situ* hybridization (FISH; see Chapter 20) makes use of this type of microscope. Also, optical modifications in certain light microscopes (e.g., phase contrast and differential interference) can be used to exaggerate the differences in densities between neighboring cells or cell structures. These kinds of light microscopes are useful in monitoring cell division in living (unstained) cells or in transparent worms (as in Figure 23.15).

The structural details of large macromolecules such as DNA and ribosomes are not observable by light microscopy. The coarse topology of these macromolecules can be determined by electron microscopy. Electron microscopes have a limit of resolution of about 2 nm, which is about 100 times finer than the best light microscopes. The primary advantage of electron microscopy over light microscopy is its better resolution. Disadvantages include a much higher expense and more extensive sample preparation. In transmission electron microscopy, the sample is bombarded with an electron beam. This requires that the sample be dried, fixed, and usually coated with a heavy metal that absorbs electrons.

A.3 SEPARATION METHODS

Biologists often wish to take complex systems and separate them into less complex components. For example, the cells within a complex tissue can be separated into individual cells, or the macromolecules within cells can be separated from the other cellular components. In this section, we will focus primarily on methods aimed at separating and purifying macromolecules.

Disruption of Cellular Components

In many experiments described in this textbook, researchers have obtained a sample of cells and then wish to isolate particular components from the cells. For example, a researcher may want to purify a protein that functions as a transcription factor. To do so,

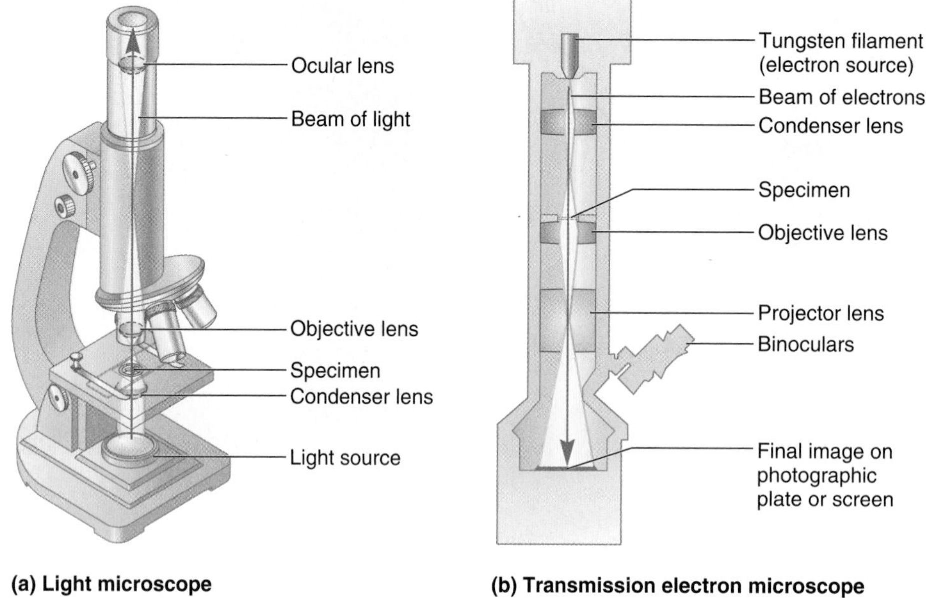

FIGURE A.2 **Design of (a) optical (light) and (b) transmission electron microscopes.**

the researcher would begin with a sample of cells that synthesize this protein and then break open the cells using one of the methods described in **Table A.1**. In eukaryotes, the breakage of cells releases the soluble proteins from the cell; it also dissociates the cell organelles that are bounded by membranes. This mixture of proteins and cell organelles can then be isolated and purified by centrifugation and chromatographic methods, which are described next.

Centrifugation

Centrifugation is a method commonly used to separate cell organelles and macromolecules. A **centrifuge** contains a motor, which causes a rotor holding centrifuge tubes to spin very rapidly. As the rotor spins, particles move toward the bottom of the centrifuge tube; the rate at which they move depends on several factors, including their densities, sizes, and shapes and the viscosity of the medium. The rate at which a macromolecule or cell organelle sediments to the bottom of a centrifuge tube is called its **sedimentation coefficient,** which is normally expressed in Svedberg units (S). A sedimentation coefficient has the units of seconds: $1\ S = 1 \times 10^{-13}$ seconds.

When a sample contains a mixture of macromolecules or cell organelles, it is likely that different components will sediment at different rates. This phenomenon, known as **differential centrifugation,** is shown in **Figure A.3**. As seen here, particles with large sedimentation coefficients reach the bottom of the tube more quickly than those with smaller coefficients. Researchers can follow two different strategies that use differential centrifugation as a separation technique. One way is to separate the **supernatant** from the **pellet** following centrifugation. The pellet is a collection of particles found at the bottom of the tube, and the supernatant is the liquid found above the pellet. In Figure A.3, when the experimenter had subjected the sample to a low-speed spin, most of the particles with large sedimentation coefficients would be found in the pellet while most of the particles with small and intermediate coefficients would be found in the supernatant. A high-speed spin of the supernatant would then separate the small and intermediate particles. Therefore, differential centrifugation provides a way to segregate these three types of particles.

A second way to separate particles using centrifugation is to collect fractions. A **fraction** is a portion of the liquid contained

TABLE **A.1**

Common Methods of Cell Disruption

Method	Description
Sonication	The exposure of cells to intense sound waves, which breaks the cell membranes.
French press	The passage of cells through a small aperture under high pressure, which breaks the cell membranes and cell wall.
Homogenization	Cells are placed in a tube that contains a pestle. When the pestle is spun, the cells are squeezed through the small space between the pestle and the glass wall of the tube, thereby breaking them.
Osmotic shock	The transfer of cells into a hypo-osmotic medium. The cells take up water and eventually burst.

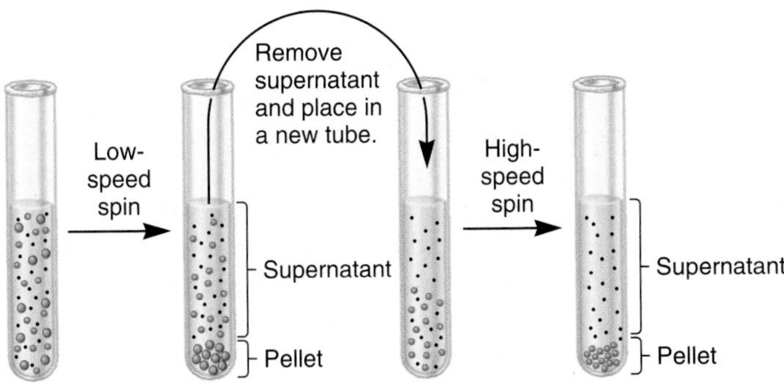

FIGURE A.3 The method of differential centrifugation. A sample containing a mixture of particles with different sedimentation coefficients is placed in a centrifuge tube. The tube is subjected to a low-speed spin that pellets the particles with large sedimentation coefficients. After a high-speed spin of the supernatant, the particles with an intermediate sedimentation coefficient are found in the pellet, while those with a small sedimentation coefficient are in the liquid supernatant.

within a centrifuge tube. The collection of fractions is done when the solution within the centrifuge tube contains a gradient. For example, as shown in **Figure A.4**, the solution at the top of the tube has a lower concentration of cesium chloride (CsCl) than that at the bottom. In this experiment, a sample is layered on the top of the gradient and then centrifuged. In this example, the DNA and RNA separate from each other, because they have different sedimentation coefficients. The experimenter then punctures the bottom of the tube and collects fractions. The DNA fragments, which are heavier, come out of the tube in the earlier fractions; the RNA molecules will be collected in later fractions.

A type of gradient centrifugation that may also be used to separate macromolecules and organelles is **equilibrium density centrifugation.** In this method, the particles will sediment through the gradient, reaching a position where the density of the particle matches the density of the solution. At this point, the particle is at equilibrium and does not move any farther toward the bottom of the tube.

Chromatography and Gel Electrophoresis

Chromatography is a method to separate different macromolecules and small molecules based on their chemical and physical properties. In this method, a sample is dissolved in a liquid solvent and exposed to some type of matrix, such as a column containing beads or a thin strip of paper. The degree to which the molecules interact with the matrix depends on their chemical and physical characteristics. For example, a positively charged molecule will bind tightly to a negatively charged matrix, whereas a neutral molecule will not.

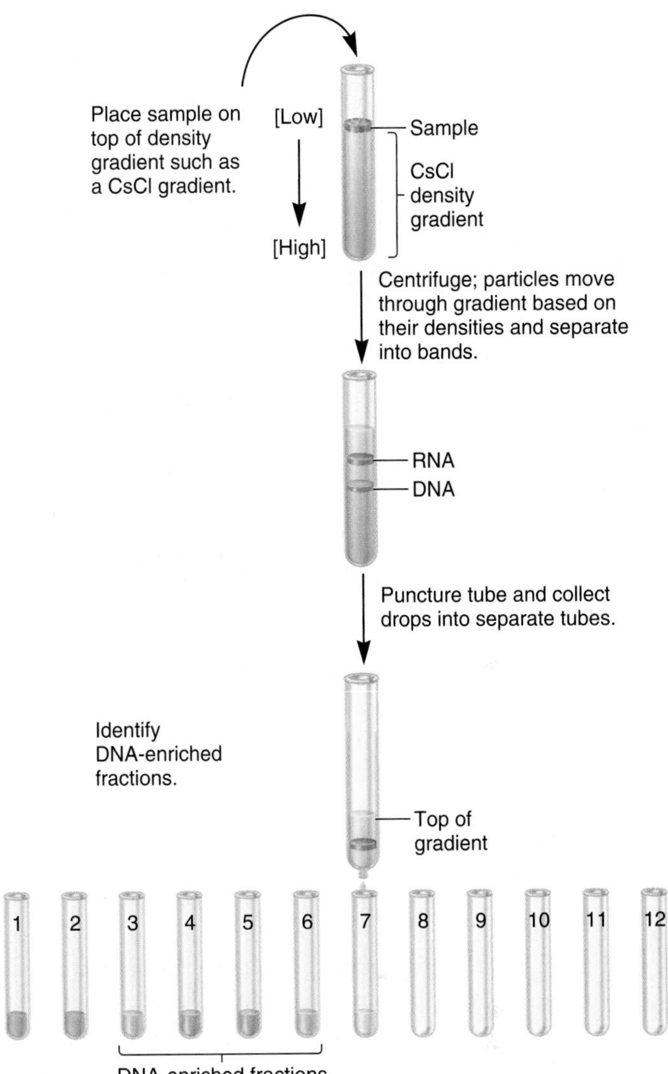

FIGURE A.4 Gradient centrifugation and the collection of fractions.

Figure A.5 illustrates how column chromatography can be used to separate molecules that differ with respect to charge. Prior to this experiment, a column is packed with beads that are positively charged. There is plenty of space between the beads for molecules to flow from the top of the column to the bottom. However, if the molecules are negatively charged, they will spend some of their time binding to the positive charges on the surface of the beads. In the example shown in Figure A.5, the red proteins are positively charged and, therefore, flow rapidly from the top of the column to the bottom. They emerge in the fractions that are collected early in this experiment. The blue proteins, however, are negatively charged and tend to bind to the beads. The binding of the blue proteins to the beads can be disrupted by changing the ionic strength or pH of the solution that is added to the column. Eventually, the blue proteins will be eluted (i.e., leave the column) in later fractions.

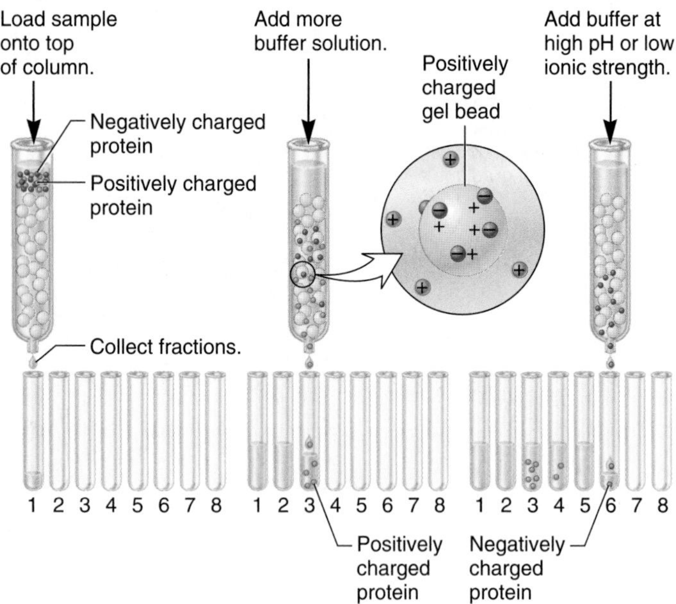

FIGURE A.5 **Ion-exchange chromatography.**

Researchers use many variations of chromatography to separate molecules and macromolecules. The type shown in Figure A.5 is called ion-exchange chromatography, because its basis for separation depends on the charge of the molecules. In another type of column chromatography, known as gel filtration chromatography, the beads are porous. Small molecules are temporarily trapped within the beads, while large molecules flow between the beads. In this way, gel filtration separates molecules on the basis of size. To separate different types of macromolecules, such as proteins, researchers may use another type of bead; this bead has a preattached molecule that binds specifically to the protein they want to purify. For example, if a transcription factor binds a particular DNA sequence as part of its function, the beads within a column may have this DNA sequence preattached to them. Therefore, the transcription factor will bind tightly to the DNA attached to these beads, while all other proteins will be eluted rapidly from the column. This form of chromatography is called affinity chromatography, because the beads have a special affinity for the macromolecule of interest.

Besides column chromatography, in which beads are packed into a column, a matrix can be made in other ways. In paper chromatography, molecules pass through a matrix composed of paper. The rate of movement of molecules through the paper depends on their degree of interaction with the solvent and paper. In thin-layer chromatography, a matrix is spread out as a very thin layer on a rigid support such as a glass plate. In general, paper and thin-layer chromatography are effective at separating small molecules, whereas column chromatography is used to separate macromolecules such as DNA fragments or proteins.

Gel electrophoresis combines chromatography and electrophoresis to separate molecules and macromolecules. As its name suggests, the matrix used in gel electrophoresis is composed of a gel. As shown in **Figure A.6**, samples are loaded into wells at one end of the gel, and an electric field is applied across the gel. This electric field causes charged molecules to migrate from one side of the gel to the other. The migration of molecules in response to an electric field is called **electrophoresis.** In the examples of gel electrophoresis found in this textbook, the macromolecules within the sample migrate toward the positive end of the gel. In most forms of gel electrophoresis, a mixture of macromolecules is separated according to their molecular masses. Small proteins or DNA fragments move to the bottom of the gel more quickly than larger ones. Because the samples are loaded in rectangular wells at the top of the gel, the molecules within the sample are separated into bands within the gel. These bands of separated macromolecules can be visualized with stains. For example, ethidium bromide is a stain that binds to DNA and RNA and can be seen under ultraviolet light.

The two most commonly used gels are polymers made from acrylamide or agarose. Proteins typically are separated on polyacrylamide gels, whereas DNA fragments are separated on agarose gels. Occasionally, researchers use polyacrylamide gels to separate DNA fragments that are relatively small (namely, less than 1,000 bp in length).

A.4 METHODS TO MEASURE CONCENTRATIONS OF MOLECULES AND TO DETECT RADIOISOTOPES AND ANTIGENS

To understand the structure and function of cells, researchers often need to detect the presence of molecules and macromolecules and to measure their concentrations. In this section, we will consider a variety of methods to detect and measure the concentrations of biological molecules and macromolecules.

Spectroscopy

Macromolecules found in living cells, such as proteins, DNA, and RNA, are fairly complex molecules that can absorb radiation (i.e., light). Likewise, small molecules such as amino acids and nucleotides can also absorb light. A device known as a **spectrophotometer** is used by researchers to determine how much radiation at various wavelengths a sample can absorb. The amount of absorption can be used to determine the concentration of particular molecules within a sample, because each type of molecule or macromolecule has its own characteristic wavelength(s) of absorption, called its absorption spectrum.

A spectrophotometer typically has two light sources, which can emit ultraviolet or visible light. As shown in **Figure A.7**, the light source is passed through a monochromator, which emits the light at a desired wavelength. This incident light then strikes a sample contained within a cuvette. Some of the incident light is absorbed, and some is not. The amount and wavelengths of light that are absorbed depend on the concentration and structures of the molecules and macromolecules in the cuvette. The

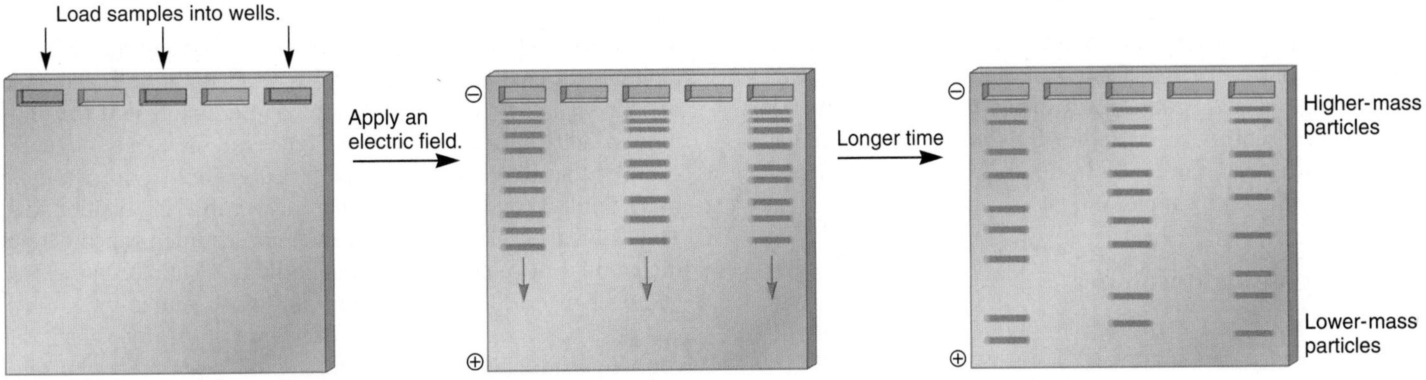

(a) Separation of a mixture of particles by gel electrophoresis

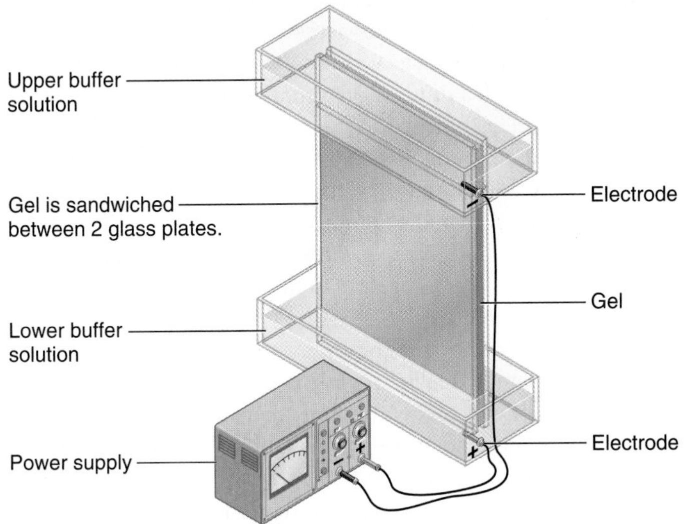

(b) Apparatus used in gel electrophoresis

FIGURE A.6 Acrylamide gel electrophoresis of DNA fragments.

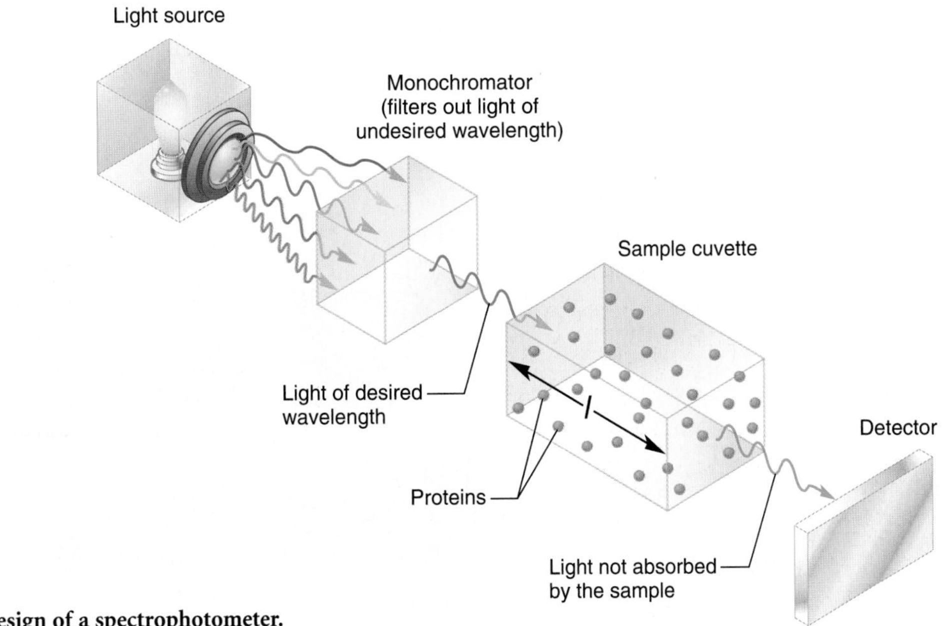

FIGURE A.7 Design of a spectrophotometer.

unabsorbed light passes through the sample and is detected by the spectrophotometer. The amount of light that strikes the detector is subtracted from the amount of incident light, yielding the measure of absorption. In this way, the spectrophotometer provides an absorption reading for the sample. This reading can be used to calculate the concentration of particular molecules or macromolecules in a sample.

Detection of Radioisotopes

A radioisotope is an unstable form of an atom that decays to a more stable form by emitting α, β, or γ rays, which are types of ionizing radiation. In research, radioisotopes that are β and/or γ emitters are commonly used. A β ray is an emitted electron, and a γ ray is an emitted photon. Some radioisotopes commonly used in biological experiments are shown in **Table A.2**.

Experimentally, radioisotopes are used because they are easy to detect. Therefore, if a particular compound is radiolabeled, its presence can be detected specifically throughout the course of the experiment. For example, if a nucleotide is radiolabeled with ^{32}P, a researcher can determine whether the isotope becomes incorporated into newly made DNA or whether it remains as the free nucleotide. Researchers commonly use two different methods to detect radioisotopes: scintillation counting and autoradiography.

The technique of **scintillation counting** permits a researcher to count the number of radioactive emissions from a sample containing a population of radioisotopes. In this approach, the sample is dissolved in a solution (called the scintillant) that contains organic solvents and one or more compounds known as fluors. When radioisotopes emit ionizing radiation, the energy is absorbed by the fluors in the solvent. This excites the fluor molecules, causing their electrons to be boosted to higher energy levels. The excited electrons return to lower, more stable energy levels by releasing photons of light. When a fluor is struck by ionizing radiation, it also absorbs the energy and then releases a photon of light within a particular wavelength range. The role of a device known as a scintillation counter is to count the photons of light emitted by the fluor. **Figure A.8** shows a scintillation counter. To use this device, a researcher dissolves her or his sample in a scintillant and then places the sample in a scintillation vial. The vial is then placed in the scintillation counter, which detects the amount of radioactivity. The scintillation counter has a digital meter that displays the amount of radioactivity in the sample, and it also provides a printout of the amount of radioactivity in counts per minute. A scintillation counter contains several rows for the loading and analysis of many scintillation vials. After they have been loaded, the scintillation counter will count the amount of radioactivity in each vial and provide the researcher with a printout of the amount of radioactivity in each vial.

A second way to detect radioisotopes is via **autoradiography.** This technique is not as quantitative as scintillation counting, because it does not provide the experimenter with a precise measure of the amount of radioactivity in counts per minute. However, autoradiography has the great advantage that it can detect the location of radioisotopes as they are found in macromolecules or cells. For example, autoradiography is used to detect a particular band on a gel or to map the location of a gene within an intact chromosome.

To conduct autoradiography, a sample containing a radioisotope is fixed and usually dried. If it is a cellular sample, it also may be thin sectioned. The sample is then pressed next to X-ray film (in the dark) and placed in a lightproof cassette. When a radioisotope decays, it will emit a β or γ ray, which may strike a thin layer of photoemulsion next to the film. The photoemulsion contains silver salts such as AgBr. When a radioactive particle is emitted and strikes the photoemulsion, a silver grain is deposited on the film. This produces a dark spot on the film, which correlates with the original location of the radioisotope in the sample. In this way, the dark image on the film reveals the location(s) of the radioisotopes in the sample. Figure 11.4b shows how autora-

TABLE **A.2**

Some Useful Isotopes in Genetics

Isotope	Stable or Radioactive	Emission	Half-life
^{2}H	Stable		
^{3}H	Radioactive	β	12.3 years
^{13}C	Stable		
^{14}C	Radioactive	β	5,730 years
^{15}N	Stable		
^{18}O	Stable		
^{24}Na	Radioactive	β (and γ)	15 hours
^{32}P	Radioactive	β	14.3 days
^{35}S	Radioactive	β	87.4 days
^{45}Ca	Radioactive	β	164 days
^{59}Fe	Radioactive	β (and γ)	45 days
^{131}I	Radioactive	β (and γ)	8.1 days

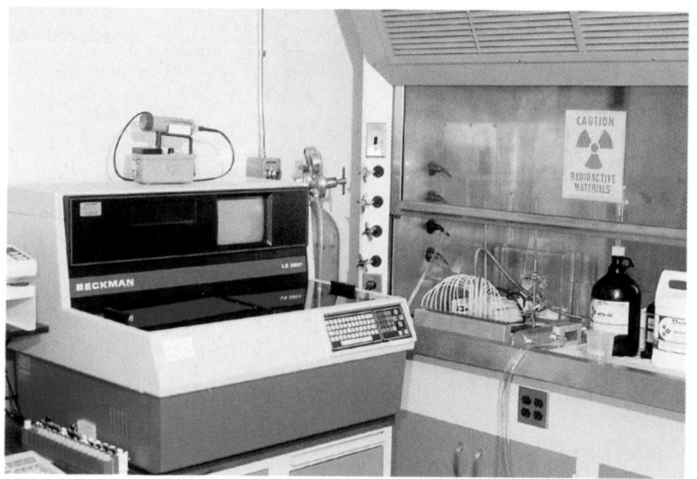

FIGURE A.8 **A scintillation counter.**

diography can be used to visualize the process of bacterial chromosome replication. In this case, radiolabeled nucleotides were incorporated into the DNA, making it possible to picture the topology of the chromosome as two replication forks pass around the circular chromosome.

Detection of Antigens by Radioimmunoassay

Antibodies, also known as **immunoglobulins,** are proteins that are used to ward off infection by foreign substances; they are produced by cells of the immune system. Antibodies bind to structures on the surface of foreign substances known as **epitopes;** the foreign substance is called an **antigen.** A particular antibody binds to a particular antigen with a very high degree of specificity. For this reason, antibodies have been used extensively by researchers to detect particular antigens. For example, a human protein such as hemoglobin can be injected into a rabbit. Human hemoglobin is a foreign substance in the rabbit's bloodstream. Therefore, the rabbit will make antibodies that specifically recognize human hemoglobin and are designed to destroy it. Researchers can isolate and purify these antibodies from a sample of the rabbit's blood and then use them to detect human hemoglobin in their experiments.

A **radioimmunoassay** is a method to measure the amount of an antigen in a biological sample. The steps in this method are shown in Figure A.9a. The researcher begins with two tubes that have a known amount of radiolabeled antigen (shown in blue). An unknown amount of the same antigen, which is not radiolabeled (shown in orange), is added to the tube on the right. The nonradiolabeled antigen comes from a biological sample; the goal of this experiment is to determine how much of this antigen is contained within the sample. Next, a known amount of antibody is added to each of the two tubes. The amount of the antibody is less than the amount of the antigen, so the nonlabeled and radiolabeled antigens compete with each other for binding to the antibody. After binding, a precipitating agent such as an anti-immunoglobulin antibody is added, and the precipitate is centrifuged to the bottom of the tube. The radioactivity in the precipitate is then determined by scintillation counting.

To calculate the amount of antigen in the sample being assayed, the researcher must determine the percentage of antibody that has bound to nonlabeled antigen. To do so, a second component of the experiment is to develop a standard curve in which a fixed amount of radiolabeled antigen is mixed with varying amounts of unlabeled antigen (Figure. A.9b). Using this standard curve, a researcher can determine how much antigen is found in the unknown sample. For example, as shown in the dashed line, if the unknown sample had about 45% of the antibody bound, then the concentration of antigen in the sample would be between 50 and 75 nanomolar.

Radioimmunoassays are used to determine the concentrations of many different kinds of antigens. This includes small molecules such as hormones (as in Figure 19.1) or macromolecules such as proteins.

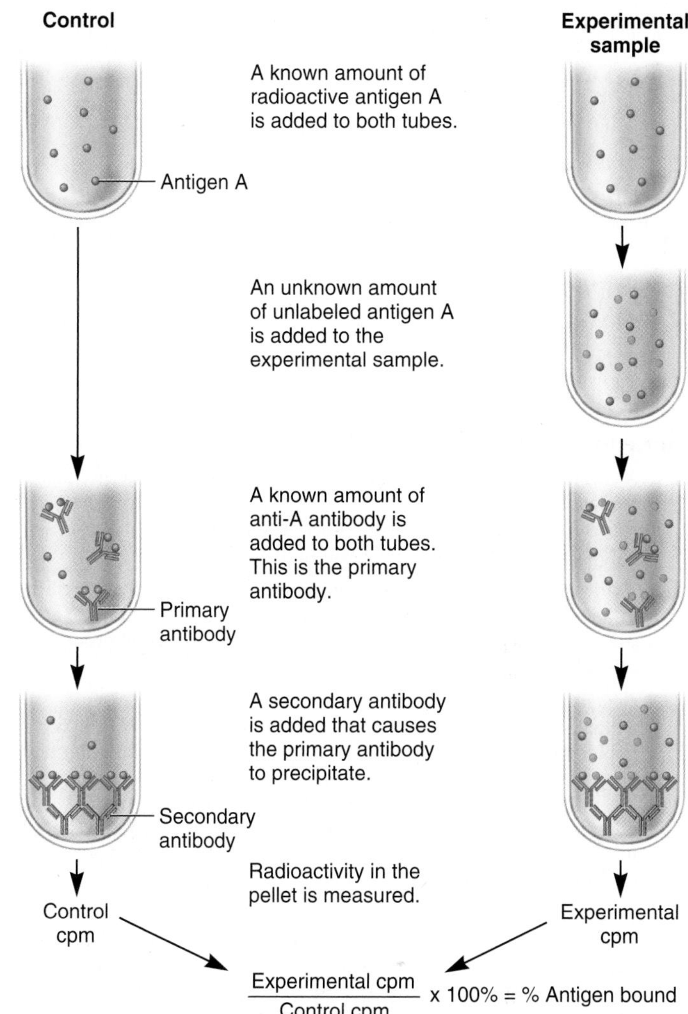

(a) A radioimmunoassay

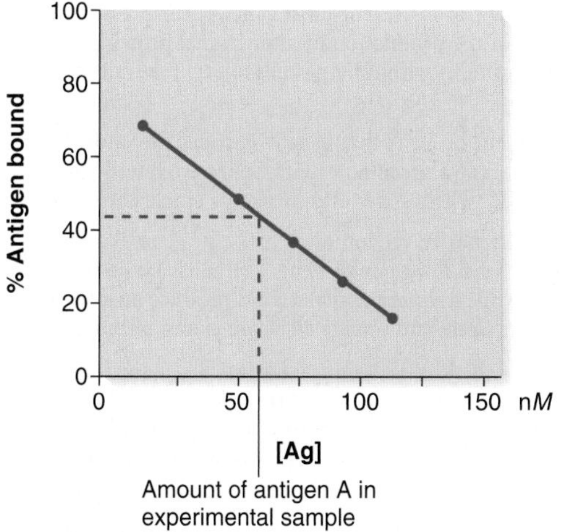

(b) Standard curve

FIGURE A.9 The method of radioimmunoassay (a) and the construction of a standard curve (b). In (**b**), the dashed line corresponds to the amount of antigen (Ag) bound by an unknown sample. This amounts to a concentration between 50 and 75 nanomolar of antigen.

SOLUTIONS TO EVEN-NUMBERED PROBLEMS

B

CHAPTER 1

Conceptual Questions

C2. A chromosome is a very long polymer of DNA. A gene is a specific sequence of DNA within that polymer; the sequence of bases creates a gene and distinguishes it from other genes. Genes are located in chromosomes, which are found within living cells.

C4. At the molecular level, a gene (a sequence of DNA) is first transcribed into RNA. The genetic code within the RNA is used to synthesize a protein with a particular amino acid sequence. This second process is called translation.

C6. Genetic variation involves the occurrence of genetic differences within members of the same species or different species. Within any population, variation may occur in the genetic material. Variation may occur in particular genes so that some individuals carry one allele and other individuals carry a different allele. An example would be differences in coat color among mammals. There also may be variation in chromosome structure and number. In plants, differences in chromosome number can affect disease resistance.

C8. You could pick almost any trait. For example, flower color in petunias would be an interesting choice. Some petunias are red and others are purple. There must be different alleles in a flower color gene that affect this trait in petunias. In addition, the amount of sunlight, fertilizer, and water also affects the intensity of flower color.

C10. A DNA sequence is a sequence of nucleotides. Each nucleotide may have one of four different bases (i.e., A, T, G, or C). When we speak of a DNA sequence, we focus on the sequence of bases.

C12. A. A gene is a segment of DNA. For most genes, the expression of the gene results in the production of a functional protein. The functioning of proteins within living cells affects the traits of an organism.

B. A gene is a segment of DNA that usually encodes the information for the production of a specific protein. Genes are found within chromosomes. Many genes are found within a single chromosome.

C. An allele is an alternative version of a particular gene. For example, suppose a plant has a flower color gene. One allele could produce a white flower, while a different allele could produce an orange flower. The white allele and orange allele are alleles of the flower color gene.

D. A DNA sequence is a sequence of nucleotides. The information within a DNA sequence (which is transcribed into an RNA sequence) specifies the amino acid sequence within a protein.

C14. A. How genes and traits are transmitted from parents to offspring

B. How the genetic material functions at the molecular and cellular levels

C. Why genetic variation exists in populations, and how it changes over the course of many generations

Experimental Questions

E2. This would be used primarily by molecular geneticists. The sequence of DNA is a molecular characteristic of DNA. In addition, as we will learn throughout this textbook, the sequence of DNA is interesting to transmission and population geneticists as well.

E4. A. Transmission geneticists. Dog breeders are interested in how genetic crosses affect the traits of dogs.

B. Molecular geneticists. This is a good model organism to study genetics at the molecular level.

C. Both transmission geneticists and molecular geneticists. Fruit flies are easy to cross and study the transmission of genes and traits from parents to offspring. Molecular geneticists have also studied many genes in fruit flies to see how they function at the molecular level.

D. Population geneticists. Most wild animals and plants would be the subject of population geneticists. In the wild, you cannot make controlled crosses. But you can study genetic variation within populations and try to understand its relationship to the environment.

E. Transmission geneticists. Agricultural breeders are interested in how genetic crosses affect the outcome of traits.

CHAPTER 2

Conceptual Questions

C2. In the case of plants, cross-fertilization occurs when the pollen and eggs come from different plants, whereas in self-fertilization, they come from the same plant.

C4. A homozygote that has two copies of the same allele

C6. Diploid organisms contain two copies of each type of gene. When they make gametes, only one copy of each gene is found in a gamete. Two alleles cannot stay together within the same gamete.

C8. Genotypes: 1:1 *Tt* and *tt*

Phenotypes: 1:1 Tall and dwarf

C10. *c* is the recessive allele for constricted pods; *Y* is the dominant allele for yellow color. The cross is $ccYy \times CcYy$. Follow the directions for setting up a Punnett square, as described in Chapter 2. The genotypic ratio is 2 *CcYY* : 4 *CcYy* : 2 *Ccyy* : 2 *ccYY* : 4 *ccYy* : 2 *ccyy*. This 2:4:2:2:4:2 ratio could be reduced to a 1:2:1:1:2:1 ratio.

The phenotypic ratio is 6 smooth pods, yellow seeds : 2 smooth pods, green seeds : 6 constricted pods, yellow seeds : 2 constricted pods, green seeds. This 6:2:6:2 ratio could be reduced to a 3:1:3:1 ratio.

C12. Offspring with a nonparental phenotype are consistent with the idea of independent assortment. If two different traits were always transmitted together as unit, it would not be possible to get nonparental phenotypic combinations. For example, if a true-breeding parent had two dominant traits and was crossed to a true-breeding parent having the two recessive traits, the F_2 generation could not have offspring with one recessive and one dominant phenotype. However, because independent assortment can occur, it is possible for F_2 offspring to have one dominant and one recessive trait.

C14. A. Barring a new mutation during gamete formation, the chance is 100% because they must be heterozygotes in order to produce a child with a recessive disorder.

B. Construct a Punnett square. There is a 50% chance of heterozygous children.

C. Use the product rule. The chance of being phenotypically normal is 0.75 (i.e., 75%), so the answer is $0.75 \times 0.75 \times 0.75 = 0.422$, which is 42.2%.

D. Use the binomial expansion equation, where $n = 3$, $x = 2$, $p = 0.75$, $q = 0.25$. The answer is 0.422, or 42.2%.

C16. First construct a Punnett square. The chances are 75% of producing a solid pup and 25% of producing a spotted pup.

A. Use the binomial expansion equation, where $n = 5$, $x = 4$, $p = 0.75$, $q = 0.25$. The answer is $0.396 = 39.6\%$ of the time.

B. You can use the binomial expansion equation for each litter. For the first litter, $n = 6$, $x = 4$, $p = 0.75$, $q = 0.25$; for the second litter, $n = 5$, $x = 5$, $p = 0.75$, $q = 0.25$. Because the litters are in a specified order, we use the product rule and multiply the probability of the first litter times the probability of the second litter. The answer is 0.070, or 7.0%.

C. To calculate the probability of the first litter, we use the product rule and multiply the probability of the first pup (0.75) times the probability of the remaining four. We use the binomial expansion equation to calculate the probability of the remaining four, where $n = 4$, $x = 3$, $p = 0.75$, $q = 0.25$. The probability of the first litter is 0.316. To calculate the probability of the second litter, we use the product rule and multiply the probability of the first pup (0.25) times the probability of the second pup (0.25) times the probability of the remaining five. To calculate the probability of the remaining five, we use the binomial expansion equation, where $n = 5$, $x = 4$, $p = 0.75$, $q = 0.25$. The probability of the second litter is 0.025. To get the probability of these two litters occurring in this order, we use the product rule and multiply the probability of the first litter (0.316) times the probability of the second litter (0.025). The answer is 0.008, or 0.8%.

D. Because this is a specified order, we use the product rule and multiply the probability of the firstborn (0.75) times the probability of the second born (0.25) times the probability of the remaining four. We use the binomial expansion equation to calculate the probability of the remaining four pups, where $n = 4$, $x = 2$, $p = 0.75$, $q = 0.25$. The answer is 0.040, or 4.0%.

C18. A. Use the product rule:

$(^1/_4)(^1/_4) = {^1/_{16}}$

B. Use the binomial expansion equation:

$n = 4$, $p = {^1/_4}$, $q = {^3/_4}$, $x = 2$

$P = 0.21 = 21\%$

C. Use the product rule:

$(^1/_4)(^3/_4)(^3/_4) = 0.14$, or 14%

C20. A. $^1/_4$

B. 1, or 100%

C. $(^3/_4)(^3/_4)(^3/_4) = {^{27}/_{64}} = 0.42$, or 42%

D. Use the binomial expansion equation, where

$n = 7$, $p = {^3/_4}$, $q = {^1/_4}$, $x = 3$

$P = 0.058$, or 5.8%

E. The probability that the first plant is tall is $^3/_4$. To calculate the probability that among the next four, any two will be tall, we use the binomial expansion equation, where $n = 4$, $p = {^3/_4}$, $q = {^1/_4}$, and $x = 2$.

The probability P equals 0.21.

To calculate the overall probability of these two events:

$(^3/_4)(0.21) = 0.16$, or 16%

C22. It violates the law of segregation because two copies of one gene are in the gamete. The two alleles for the *A* gene did not segregate from each other.

C24. Based on this pedigree, it is likely to be dominant inheritance because an affected child always has an affected parent. In fact, it is a dominant disorder.

C26. It is impossible for the F_1 individuals to be true-breeding because they are all heterozygotes.

C28. 2 *TY*, *tY*, 2 *Ty*, *ty*, *TTY*, *TTy*, 2 *TtY*, 2 *Tty*

It may be tricky to think about, but you get 2 *TY* and 2 *Ty* because either of the two *T* alleles could combine with *Y* or *y*. Also, you get 2 *TtY* and 2 *Tty* because either of the two *T* alleles could combine with *t* and then combine with *Y* or *y*.

C30. The genotype of the F_1 plants is *Tt Yy Rr*. According to the laws of segregation and independent assortment, the alleles of each gene will segregate from each other, and the alleles of different genes will randomly assort into gametes. A *Tt Yy Rr* individual could make eight types of gametes: *TYR*, *TyR*, *Tyr*, *TYr*, *tYR*, *tyR*, *tYr*, and *tyr*, in equal proportions (i.e., $^1/_8$ of each type of gamete). To determine genotypes and phenotypes, you could make a large Punnett square that would contain 64 boxes. You would need to line up the eight possible gametes across the top and along the side and then fill in the 64 boxes. Alternatively, you could use one of the two approaches described in solved problem S3. The genotypes and phenotypes would be

1 *TT YY RR*
2 *TT Yy RR*
2 *TT YY Rr*
2 *Tt YY RR*
4 *TT Yy Rr*
4 *Tt Yy RR*
4 *Tt YY Rr*
8 *Tt Yy Rr* = 27 tall, yellow, round
1 *TT yy RR*
2 *Tt yy RR*
2 *TT yy Rr*
4 *Tt yy Rr* = 9 tall, green, round
1 *TT YY rr*
2 *TT Yy rr*
2 *Tt YY rr*
4 *Tt Yy rr* = 9 tall, yellow, wrinkled
1 *tt YY RR*
2 *tt Yy RR*
2 *tt YY Rr*
4 *tt Yy Rr* = 9 dwarf, yellow, round
1 *TT yy rr*
2 *Tt yy rr* = 3 tall, green, wrinkled
1 *tt yy RR*
2 *tt yy Rr* = 3 dwarf, green, round
1 *tt YY rr*
2 *tt Yy rr* = 3 dwarf, yellow, wrinkled
1 *tt yy rr* = 1 dwarf, green, wrinkled

C32. The wooly-haired male is a heterozygote, because he has the trait and his mother did not. (He must have inherited the normal allele from his mother.) Therefore, he has a 50% chance of passing the wooly allele to his offspring; his offspring have a 50% of passing the allele to their offspring; and these grandchildren have a 50% chance of passing the allele to their offspring (the wooly-haired man's great-grandchildren). Because this is an ordered sequence of independent events, we use the product rule: $0.5 \times 0.5 \times 0.5 = 0.125$, or 12.5%. Because no other Scandinavians are on the island, the chance is 87.5% for the offspring being normal (because they could not inherit the wooly hair allele from anyone else). We use the binomial expansion equation to determine the likelihood that one out of eight great-grandchildren will have wooly hair, where $n = 8$, $x = 1$, $p = 0.125$, $q = 0.875$. The answer is 0.393, or 39.3%, of the time.

C34. Use the product rule. If the woman is heterozygous, there is a 50% chance of having an affected offspring: $(0.5)^7 = 0.0078$, or 0.78%, of the time. This is a pretty small probability. If the woman has an eighth child who is unaffected, however, she has to be a heterozygote, because it is a dominant trait. She would have to pass a normal allele to an unaffected offspring. The answer is 100%.

Experimental Questions

E2. The experimental difference depends on where the pollen comes from. In self-fertilization, the pollen and eggs come from the same plant. In cross-fertilization, they come from different plants.

E4. According to Mendel's law of segregation, the genotypic ratio should be 1 homozygote dominant : 2 heterozygotes : 1 homozygote recessive. This data table considers only the plants with a dominant phenotype. The genotypic ratio should be 1 homozygote dominant : 2 heterozygotes. The homozygote dominants would be true-breeding, while the heterozygotes would not be true-breeding. This 1:2 ratio is very close to what Mendel observed.

E6. All three offspring had black fur. The ovaries from the albino female could produce eggs with only the dominant black allele (because they were obtained from a true-breeding black female). The actual phenotype of the albino mother does not matter. Therefore, all offspring would be heterozygotes (*Bb*) and have black fur.

E8. If we construct a Punnett square according to Mendel's laws, we expect a 9:3:3:1 ratio. Because a total of 556 offspring were observed, the expected number of offspring are

$556 \times {}^{9}/_{16} = 313$ round, yellow

$556 \times {}^{3}/_{16} = 104$ wrinkled, yellow

$556 \times {}^{3}/_{16} = 104$ round, green

$556 \times {}^{1}/_{16} = 35$ wrinkled, green

If we plug the observed and expected values into the chi square equation, we get a value of 0.51. With four categories, our degrees of freedom equal $n - 1$, or 3. If we look up the value of 0.51 in the chi square table (see Table 2.1), we see that it falls between the *P* values of 0.80 and 0.95. This means that the probability is 80% to 95% that any deviation between observed results and expected results was caused by random sampling error. Therefore, we accept the hypothesis. In other words, the results are consistent with the law of independent assortment.

E10. A. If we let c^+ represent normal wings and c represent curved wings, and e^+ represents gray body and e represents ebony body,

Parental Cross: $cce^+e^+ \times c^+c^+ee$.

F_1 generation is heterozygous c^+ce^+e

An F_1 offspring crossed to flies with curved wings and ebony bodies is

$c^+ce^+e \times ccee$

The F_2 offspring would be a 1:1:1:1 ratio of flies:

$c^+ce^+e : c^+cee : cce^+e : ccee$

B. The phenotypic ratio of the F_2 flies would be a 1:1:1:1 ratio of flies:

normal wings, gray body : normal wings, ebony bodies : curved wings, gray bodies : curved wings, ebony bodies

C. From part B, we expect ${}^{1}/_{4}$ of each category. There are a total of 444 offspring. The expected number of each category is ${}^{1}/_{4} \times 444$, which equals 111.

$$\chi^2 = \frac{(114 - 111)^2}{111} + \frac{(105 - 111)^2}{111} + \frac{(111 - 111)^2}{111} + \frac{(114 - 111)^2}{111}$$

$\chi^2 = 0.49$

With 3 degrees of freedom, a value of 0.49 or greater is likely to occur between 80% and 95% of the time. Therefore, we accept our hypothesis.

E12. Follow through the same basic chi square strategy as before. We expect a 3:1 ratio, or ${}^{3}/_{4}$ of the dominant phenotype and ${}^{1}/_{4}$ of the recessive phenotype.

The observed and expected values are as follows (rounded to the nearest whole number):

Observed*	Expected	$\frac{(O-E)^2}{E}$
5,474	5,493	0.066
1,850	1,831	0.197
6,022	6,017	0.004
2,001	2,006	0.012
705	697	0.092
224	232	0.276
882	886	0.018
299	295	0.054
428	435	0.113
152	145	0.338
651	644	0.076
207	215	0.298
787	798	0.152
277	266	0.455
		$\chi^2 = 2.15$

*Due to rounding, the observed and expected values may not add up to precisely the same number.

Because $n = 14$, there are 13 degrees of freedom. If we look up this value in the chi square table, we have to look between 10 and 15 degrees of freedom. In either case, we would expect the value of 2.15 or greater to occur more than 99% of the time. Therefore, we accept the hypothesis.

E14. The dwarf parent with terminal flowers must be homozygous for both genes, because it is expressing these two recessive traits: *ttaa*, where *t* is the recessive dwarf allele, and *a* is the recessive allele for terminal flowers. The phenotype of the other parent is dominant for both traits. However, because this parent was able to produce dwarf offspring with axial flowers, it must have been heterozygous for both genes: *TtAa*.

E16. Our hypothesis is that blue flowers and purple seeds are dominant traits and they are governed by two genes that assort independently. According to this hypothesis, the F_2 generation should yield a ratio of 9 blue flowers, purple seeds : 3 blue flowers, green seeds : 3 white flowers, purple seeds : 1 white flower, green seeds. Because a total of 300 offspring are produced, the expected numbers would be

$^9/_{16} \times 300 = 169$ blue flowers, purple seeds

$^3/_{16} \times 300 = 56$ blue flowers, green seeds

$^3/_{16} \times 300 = 56$ white flowers, purple seeds

$^1/_{16} \times 300 = 19$ white flowers, green seeds

$$\chi^2 = \frac{(103 - 169)^2}{169} + \frac{(49 - 56)^2}{56} + \frac{(44 - 56)^2}{56} + \frac{(104 - 19)^2}{19}$$

$$\chi^2 = 409.5$$

If we look up this value in the chi square table under 3 degrees of freedom, the value is much higher than would be expected 1% of the time by chance alone. Therefore, we reject the hypothesis. The idea that the two genes are assorting independently seems to be incorrect. The F_1 generation supports the idea that blue flowers and purple seeds are dominant traits.

Questions for Student Discussion/Collaboration

2. If we construct a Punnett square, the following probabilities will be obtained:

   tall with axial flowers, $^3/_8$

   dwarf with terminal flowers, $^1/_8$

   To calculate the probability of being tall with axial flowers or dwarf with terminal flowers, we use the sum rule:

   $^3/_8 + {}^1/_8 = {}^4/_8 = {}^1/_2$

   We use the product rule to calculate the ordered events of the first three offspring being tall/axial or dwarf/terminal, and the fourth offspring being tall axial:

   $(^1/_2)(^1/_2)(^1/_2)(^3/_8) = {}^3/_{64} = 0.047 = 4.7\%$

CHAPTER 3

Conceptual Questions

C2. The term homologue refers to the members of a chromosome pair. Homologues are usually the same size and carry the same types and order of genes. They may differ in that the genes they carry may be different alleles.

C4. Metaphase is the organization phase, and anaphase is the separation phase.

C6. In metaphase I of meiosis, each pair of chromatids is attached to only one pole via the kinetochore microtubules. In metaphase of mitosis, there are two attachments (i.e., to both poles). If the attachment was lost, a chromosome would not migrate to a pole and may not become enclosed in a nuclear membrane after telophase. If left out in the cytoplasm, it would eventually be degraded.

C8. The reduction occurs because there is a single DNA replication event but two cell divisions. Because of the nature of separation during anaphase I, each cell receives one copy of each type of chromosome.

C10. It means that the maternally derived and paternally derived chromosomes are randomly aligned along the metaphase plate during metaphase I. Refer to Figure 3.17.

C12. There are three pairs of chromosomes. The number of different, random alignments equals 2^n, where n equals the number of chromosomes per set. So the possible number of arrangements equals 2^3, which is 8.

C14. The probability would be much lower because pieces of maternal chromosomes would be mixed with the paternal chromosomes. Therefore, it is unlikely to inherit a chromosome that was completely paternally derived.

C16. During interphase, the chromosomes are greatly extended. In this conformation, they might get tangled up with each other and not sort properly during meiosis and mitosis. The condensation process probably occurs so that the chromosomes easily align along the equatorial plate during metaphase without getting tangled up.

C18. During prophase II, your drawing should show four replicated chromosomes (i.e., four structures that look like Xs). Each chromosome is one homologue. During prophase of mitosis, there should be eight replicated chromosomes (i.e., eight Xs). During prophase of mitosis, there are pairs of homologues. The main difference is that prophase II has a single copy of each of the four chromosomes, whereas prophase of mitosis has four pairs of homologues. At the end of meiosis I, each daughter cell has received only one copy of a homologous pair, not both. This is due to the alignment of homologues during metaphase I and their separation during anaphase I.

C20. DNA replication does not take place during interphase II. The chromosomes at the end of telophase I have already replicated (i.e., they are found in pairs of sister chromatids). During meiosis II, the sister chromatids separate from each other, yielding individual chromosomes.

C22. A. 20

B. 10

C. 30

D. 20

C24. A. Dark males and light females; reciprocal: all dark offspring

B. All dark offspring; reciprocal: dark females and light males

C. All dark offspring; reciprocal: dark females and light males

D. All dark offspring; reciprocal: dark females and light males

C26. To produce sperm, a spermatogonial cell first goes through mitosis to produce two cells. One of these remains a spermatogonial cell and the other progresses through meiosis. In this way, the testes continue to maintain a population of spermatogonial cells.

C28. A. *A B C, A B c, A b C, A b c, a B C, a b C, a B c, a b c*

B. *A B C, A b C*

C. *A B C, A B c, a B C, a B c*

D. *A b c, a b c*

C30. A. The fly is a male because the ratio of X chromosomes to sets of autosomes is $^1/_2$, or 0.5.

B. The fly is female because the ratio of X chromosomes to sets of autosomes is 1.0.

C. The fly is male because the ratio of X chromosomes to sets of autosomes is 0.5.

D. The fly is female because the ratio of X chromosomes to sets of autosomes is 1.0.

Experimental Questions

E2. Perhaps the most convincing observation was that all of the white-eyed flies of the F_2 generation were males. This suggests a link between sex determination and the inheritance of this trait. Because sex determination in fruit flies is determined by the number of X chromosomes, this suggests a relationship between the inheritance of the X chromosome and the inheritance of this trait.

E4. The basic strategy is to set up a pair of reciprocal crosses. The phenotype of sons is usually the easiest way to discern the two patterns. If it is Y linked, the trait will be passed only from father to son. If it is X linked, the trait will be passed from mother to son.

E6. The 3:1 sex ratio occurs because the female produces 50% gametes that are XX (and must produce female offspring) and 50% that are X (and produce half male and half female offspring). The original female had one X chromosome carrying the red allele and two other

X chromosomes carrying the eosin allele. Set up a Punnett square assuming that this female produces the following six types of gametes: $X^{w+}X^{w-e}$, $X^{w+}X^{w-e}$, $X^{w-e}X^{w-e}$, X^{w+}, X^{w-e}, X^{w-e}. The male of this cross is $X^{w}Y$.

Male gametes

Female gametes ♀ \ ♂	X^{w}	Y
$X^{w^+}X^{w\text{-}e}$	$X^{w^+}X^{w\text{-}e}X^{w}$ Red, female	$X^{w^+}X^{w\text{-}e}Y$ Red, female
$X^{w^+}X^{w\text{-}e}$	$X^{w^+}X^{w\text{-}e}X^{w}$ Red, female	$X^{w^+}X^{w\text{-}e}Y$ Red, female
$X^{w\text{-}e}X^{w\text{-}e}$	$X^{w\text{-}e}X^{w\text{-}e}X^{w}$ Eosin female	$X^{w\text{-}e}X^{w\text{-}e}Y$ Eosin female
X^{w^+}	$X^{w^+}X^{w}$ Red, female	$X^{w^+}Y$ Red, male
$X^{w\text{-}e}$	$X^{w\text{-}e}X^{w}$ Light-eosin female	$X^{w\text{-}e}Y$ Light-eosin male
$X^{w\text{-}e}$	$X^{w\text{-}e}X^{w}$ Light-eosin female	$X^{w\text{-}e}Y$ Light-eosin male

E8. If we use the data from the F_1 mating (i.e., F_2 results), there were 3,470 red-eyed flies. We would expect a 3:1 ratio between red- and white-eyed flies. Therefore, assuming that all red-eyed offspring survived, there should have been about 1,157 (i.e., $^{3,470}/_{3}$) white-eyed flies. However, there were only 782. If we divide 782 by 1,157, we get a value of 0.676, or a 67.6% survival rate.

E10. You need to make crosses to understand the pattern of inheritance of traits (determined by genes) from parents to offspring. And you need to microscopically examine cells to understand the pattern of transmission of chromosomes. The correlation between the pattern of transmission of chromosomes during meiosis, and Mendel's laws of segregation and independent assortment is what led to the chromosome theory of inheritance.

E12. Originally, individuals who had abnormalities in their composition of sex chromosomes provided important information. In mammals, X0 individuals are females, whereas in flies, X0 individuals are males. In mammals, XXY individuals are males, while in flies, XXY individuals are females. These results indicate that the presence of the Y chromosome causes maleness in mammals, but it does not in flies. A further analysis of flies with abnormalities in the number of sets of autosomes revealed that it is the ratio between the number of X chromosomes and the number of sets of autosomes that determines sex in flies.

Questions for Student Discussion/Collaboration

2. It's not possible to give a direct answer, but the point is for students to be able to draw chromosomes in different configurations and understand the various phases. The chromosomes may or may not be

   1. In homologous pairs
   2. Connected as sister chromatids
   3. Associated in bivalents
   4. Lined up in metaphase
   5. Moving toward the poles.

   And so on.

CHAPTER 4

Conceptual Questions

C2. Sex-influenced traits are influenced by the sex of the individual even though the gene that governs the trait may be autosomally inherited. Pattern baldness in people is an example. Sex-limited traits are an extreme example of sex influence. The expression of a sex-limited trait is limited to one sex. For example, colorful plumage in certain species of birds is limited to the male sex. Sex-linked traits involve traits whose genes are found on the sex chromosomes. Examples in humans include hemophilia and color blindness.

C4. If the normal allele is dominant, it tells you that one copy of the gene produces a sufficient amount of the protein encoded by the gene. Having twice as much of this protein, as in the normal homozygote, does not alter the phenotype. If the allele is incompletely dominant, this means that one copy of the normal allele does not produce the same trait as the homozygote.

C6. The ratio would be 1 normal : 2 star-eyed individuals.

C8. If individual 1 is ii, individual 2 could be I^Ai, I^AI^A, I^Bi, I^BI^B, or I^AI^B.

If individual 1 is I^Ai or I^AI^A, individual 2 could be I^Bi, I^BI^B, or I^AI^B.

If individual 1 is I^Bi or I^BI^B, individual 2 could be I^Ai, I^AI^A, or I^AI^B.

Assuming individual 1 is the parent of individual 2:

If individual 1 is ii, individual 2 could be I^Ai or I^Bi.

If individual 1 is I^Ai, individual 2 could be I^Bi or I^AI^B.

If individual 1 is I^AI^A, individual 2 could be I^AI^B.

If individual 1 is I^Bi, individual 2 could be I^Ai or I^AI^B.

If individual 1 is I^BI^B, individual 2 could be I^AI^B.

C10. The father could not be I^AI^B, I^BI^B, or I^AI^A. He is contributing the O allele to his offspring. Genotypically, he could be I^Ai, I^Bi, or ii and have type A, B, or O blood, respectively.

C12. Perhaps it should be called codominant at the "hair level" because one or the other allele is dominant with regard to a single hair. However, this is not the same as codominance in blood types, in which every cell can express both alleles.

C14. A. X-linked recessive (unaffected mothers transmit the trait to sons)

B. Autosomal recessive (affected daughters and sons are produced from unaffected parents)

C16. First set up the following Punnett square:

Male gametes

Female gametes ♀ \ ♂	X^H	Y
X^H	X^HX^H	X^HY
X^h	X^HX^h	X^hY

There is a $^1/_4$ probability of each type of offspring.

A. $^1/_4$

B. $(^3/_4)(^3/_4)(^3/_4)(^3/_4) = {}^{81}/_{256}$

C. $^3/_4$

D. The probability of an affected offspring is $^1/_4$, and the probability of an unaffected offspring is $^3/_4$. For this problem, you use the binomial expansion equation where $x = 2$, $n = 5$, $p = {}^1/_4$, and $q = {}^3/_4$. The answer is 0.26, or 26%, of the time.

C18. 1 affected daughter : 1 unaffected daughter : 1 affected son : 1 unaffected son

C20. On a standard vegetarian diet, the results would be 50% white and 50% yellow.

On a xanthophyll-free diet, the results would be 100% white.

C22. Set up a Punnett square, but keep in mind that the eosin gene is X linked.

♀ \ ♂	$CX^{w\text{-}e}$	CY
CX^w	$CCX^{w\text{-}e}X^w$	CCX^wY
c^aX^w	$Cc^aX^{w\text{-}e}X^w$	Cc^aX^wY

Because *C* is dominant, the phenotypic ratios are two light-eosin females and two white males. Note: Females that have one eosin allele and one white allele have light-eosin eyes, whereas females with two eosin alleles or males with one eosin allele have eosin eyes, which are a bit darker than light-eosin eyes.

C24. A. Could be.

B. No, because an unaffected father has an affected daughter.

C. No, because two unaffected parents have affected children.

D. No, because an unaffected father has an affected daughter.

E. No, because both sexes exhibit the trait.

F. Could be.

C26. You would look at the pattern within families over the course of many generations. For a recessive trait, 25% of the offspring within a family are expected to be affected if both parents are unaffected carriers, and 50% of the offspring would be affected if one parent was affected. You could look at many families and see if these 25% and 50% values are approximately true. Incomplete penetrance would not necessarily predict such numbers. Also, for very rare alleles, incomplete penetrance would probably have a much higher frequency of affected parents producing affected offspring. For rare recessive disorders, it is most likely that both parents are heterozygous carriers. Finally, the most informative pedigrees would be situations in which two affected parents produce children. If they can produce an unaffected offspring, this would indicate incomplete penetrance. If all of their offspring were affected, this would be consistent with recessive inheritance.

C28. The probability of a heterozygote passing the allele to his/her offspring is 50%. The probability of an affected offspring expressing the trait is 80%. We use the product rule to determine the likelihood of these two independent events.

$(0.5)(0.8) = 0.4$, or 40% of the time

C30. This is an example of incomplete dominance. The heterozygous horses are palominos. For example, if *C* represents chestnut and *c* represents cremello, the chestnut horses are *CC*, the cremello horses are *cc*, and the palominos are *Cc*.

Experimental Questions

E2. Two redundant genes are involved in feathering. The unfeathered Buff Rocks are homozygous recessive for the two genes. The Black Langhans are homozygous dominant for both genes. In the F_2 generation (which is a double heterozygote crossed to another double heterozygote), 1 out of 16 offspring will be doubly homozygous for both recessive genes. All the others will have at least one dominant allele for one of the two (redundant) genes.

E4. The reason why all the puppies have black hair is because albino alleles are found in two different genes. If we let the letters *A* and *B* represent the two different pigmentation genes, then one of the dogs is *AAbb*, and the other is *aaBB*. Their offspring are *AaBb* and therefore are not albinos because they have one dominant copy of each gene.

E6. In general, you cannot distinguish between autosomal and pseudoautosomal inheritance from a pedigree analysis. Mothers and fathers have an equal probability of passing the alleles to sons and daughters. However, if an offspring had a chromosomal abnormality, you might be able to tell. For example, in a family tree involving the *Mic2* allele, an offspring that was X0 would have less of the gene product, and an offspring that was XXX or XYY or XXY would have extra amounts of the gene products. This may lead you to suspect that the gene is located on the sex chromosomes.

E8. One parent must be *RRPp*. The other parent could be *RRPp* or *RrPp*. All the offspring would inherit (at least) one dominant *R* allele. With regard to the other gene, $^3/_4$ would inherit at least one copy of the dominant *P* allele. These offspring would have a walnut comb. The other $^1/_4$ would be homozygous *pp* and have a rose comb (because they would also have a dominant *R* allele).

E10. Let's use the letters *A* and *B* for these two genes. Gene *A* exists in two alleles, which we will call *A* and *a*. Gene *B* exists in two alleles, *B* and *b*. The uppercase alleles are dominant to the lowercase alleles. The true-breeding long-shaped squash is *aabb*, and the true-breeding disk-shaped squash is *AABB*. The F_1 offspring are *AaBb*. You can construct a Punnett square, with 16 boxes, to determine the outcome of self-fertilization of the F_1 plants.

To get the disk-shaped phenotype, an offspring must inherit at least one dominant allele from both genes.

1 *AABB* + 2 *AaBB* + 2 *AABb* + 4 *AaBb* = 9 disk-shaped offspring

To get the round phenotype, an offspring must inherit at least one dominant allele for one of the two genes but must be homozygous recessive for only one of the two genes.

1 *aaBB* + 1 *AAbb* + 2 *aaBb* + 2 *Aabb* = 6 round-shaped offspring

To get the long phenotype, an offspring must inherit all recessive alleles:

1 *aabb* = 1 long-shaped offspring

E12. The results obtained when crossing two F_1 offspring appear to yield a 9:3:3:1 ratio, which would be expected if eye color is affected by two different genes that exist in dominant and recessive alleles. Neither gene is X linked. Let pr^+ represent the red allele of the first gene and *pr* the purple allele. Let sep^+ represent the red allele of the second gene and *sep* the sepia allele.

The first cross is $pr\,pr\,sep^+sep^+ \times pr^+pr^+sep\,sep$

All the F_1 offspring would be $pr^+pr\,sep^+sep$. They have red eyes because they have a dominant red allele for each gene. When the F_1 offspring are crossed to each other, the following results would be obtained:

♀ \ ♂	pr^+sep^+	pr^+sep	$pr\,sep^+$	$pr\,sep$
pr^+sep^+	pr^+pr^+ sep^+sep^+ Red	pr^+pr^+ sep^+sep Red	pr^+pr sep^+sep^+ Red	pr^+pr sep^+sep Red
pr^+sep	pr^+pr^+ sep^+sep Red	pr^+pr^+ $sep\,sep$ Sepia	pr^+pr sep^+sep Red	pr^+pr $sep\,sep$ Sepia
$pr\,sep^+$	pr^+pr sep^+sep^+ Red	pr^+pr sep^+sep Red	$pr\,pr$ sep^+sep^+ Purple	$pr\,pr$ sep^+sep Purple
$pr\,sep$	pr^+pr sep^+sep Red	pr^+pr $sep\,sep$ Sepia	$pr\,pr$ sep^+sep Purple	$pr\,pr$ $sep\,sep$ Pur/Sepia

In this case, one gene exists as the red (dominant) or purple (recessive) allele, and the second gene exists as the red (dominant) or sepia (recessive) allele. If an offspring is homozygous for the purple allele, it will have purple eyes. Similarly, if an offspring is homozygous for the sepia allele, it will have sepia eyes. An offspring homozygous for both recessive alleles has purplish sepia eyes. To have red eyes, it must have at least one copy of the dominant red allele for both genes. Based on an expected 9 red : 3 purple : 3 sepia : 1 purplish sepia, the observed and expected numbers of offspring are as follows:

Observed	*Expected*
146 purple eyes	148 purple eyes (791 × $^3/_{16}$)
151 sepia eyes	148 sepia eyes (791 × $^3/_{16}$)
50 purplish sepia eyes 4	49 purplish sepia eyes (791 × $^1/_{16}$)
<u>444 red eyes</u>	445 red eyes (791 × $^9/_{16}$)
791 total offspring	

If we plug the observed and expected values into our chi square formula, we obtain a chi square value of about 0.11. With 3 degrees of freedom, this is well within our expected range of values, so we cannot reject our hypothesis that purple and sepia alleles are in two different genes and that these recessive alleles are epistatic to each other.

E14. To see if the allele is X linked, the pink-eyed male could be crossed to a red-eyed female. All the offspring would have red eyes, assuming that the pink allele is recessive. When crossed to red-eyed males, the F_1 females will produce $^1/_2$ red-eyed daughters, $^1/_4$ red-eyed sons, and $^1/_4$ pink-eyed sons if the pink allele is X linked.

If the pink allele is X linked, then one could determine if it is in the same X-linked gene as the white and eosin alleles by crossing pink-eyed males to white-eyed females. (Note: We already know that white and eosin are alleles of the same gene.) If the pink and white alleles are in the same gene, the F_1 female offspring should have pink eyes (assuming that the pink allele is dominant over white). However, if the pink and white alleles are in different genes, the F_1 females will have red eyes (assuming that pink is recessive to red). This is because the F_1 females will be heterozygous for two genes, $X^{w+p} X^{wp+}$, in which the X^{w+} and X^{p+} alleles are the dominant wild-type alleles that produce red eyes, and the X^w and X^p alleles are recessive alleles for these two different genes, which produce white eyes and pink eyes, respectively.

Questions for Student Discussion/Collaboration

2. Perhaps the easiest way to solve this problem is to take one trait at a time. With regard to combs, all the F_1 generation would be *RrPp* or walnut comb. With regard to shanks, they would all be feathered, because they would inherit one dominant copy of a feathered allele. With regard to hen- or cock-feathering: 1 male cock-feathered : 1 male hen-feathered : 2 females hen-feathered. Overall then, we would have a 1 : 1 : 2 ratio of

walnut comb/feathered shanks/cock-feathered males

walnut comb/feathered shanks/hen-feathered males

walnut comb/feathered shanks/hen-feathered females

CHAPTER 5

C2. An independent assortment hypothesis is used because it enables us to calculate the expected values based on Mendel's ratios. Using the observed and expected values, we can calculate whether or not the deviations between the observed and expected values are too large to occur as a matter of chance. If the deviations are very large, we reject the hypothesis of independent assortment.

C4. If the chromosomes (on the right side) labeled 2 and 4 move into one daughter cell, that will lead to a patch that is albino and has long fur. The other cell will receive chromosomes 1 and 3, which will produce a patch that has dark, short fur.

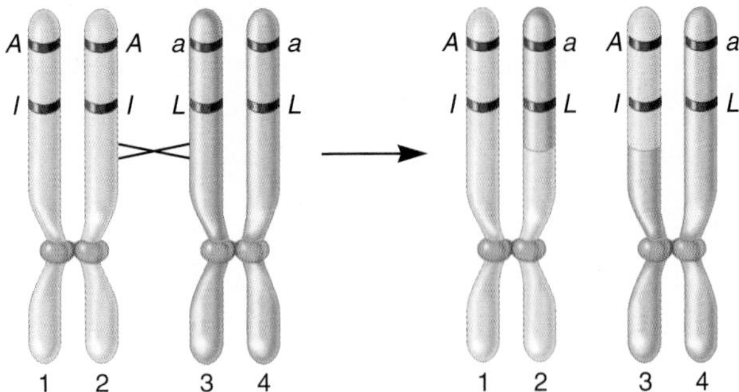

C6. A single crossover produces *A B C*, *A b c*, *a B C*, and *a b c*.

A. Between 2 and 3, between genes *B* and *C*

B. Between 1 and 4, between genes *A* and *B*

C. Between 1 and 4, between genes *B* and *C*

D. Between 2 and 3, between genes *A* and *B*

C8. The likelihood of scoring a basket would be greater if the basket was larger. Similarly, the chances of a crossover initiating in a region between two genes is proportional to the size of the region between the two genes. There are a finite number (usually a few) of crossovers that occur between homologous chromosomes during meiosis, and the likelihood that a crossover will occur in a region between two genes depends on how big that region is.

C10. The pedigree suggests a linkage between the dominant allele causing nail-patella syndrome and the I^B allele of the ABO blood type gene. In every case, the individual who inherits the I^B allele also inherits this disorder.

C12. <u>*Ass-1* 43 *Sdh-1* 5 *Hdc* 9 *Hao-1* 6 *Odc-2* 8 *Ada-1*</u>

C14. The inability to detect double crossovers causes the map distance to be underestimated. In other words, more crossovers occur in the region than we realize. When we have a double crossover, we do not get a recombinant offspring (in a dihybrid cross). Therefore, the second crossover cancels out the effects of the first crossover.

C16. The key feature is that all the products of a single meiosis are contained within a single sac. The spores in this sac can be dissected, and then their genetic traits can be analyzed individually.

C18. In an unordered ascus, the products of meiosis are free to move around. In an ordered octad (or tetrad), they are lined up according to their relationship to each other during meiosis and mitosis. An ordered octad can be used to map the distance between a single gene and its centromere.

C20. The percentage would be higher with respect to gene *A*. First-division segregation patterns occur when there is not a crossover between the centromere and the gene of interest. Because gene *A* is closer to the centromere compared to gene *B*, it would be less likely to have a crossover between gene *A* and the centromere. This would make it more likely to observe first-division segregation.

Experimental Questions (Includes Most Mapping Questions)

E2. They could have used a strain with two abnormal chromosomes. In this case, the recombinant chromosomes would either look normal or have abnormalities at both ends.

E4. A gene on the Y chromosome in mammals would be transmitted only from father to son. It would be difficult to genetically map Y-linked genes because a normal male has only one copy of the Y chromosome, so you do not get any crossing over between two Y chromosomes. Occasionally, abnormal males (XYY) are born with two Y chromosomes. If such males were heterozygous for alleles of Y-linked genes, one could examine the normal male offspring of XYY fathers and determine if crossing over has occurred.

E6. The answer is explained in solved problem S5. We cannot get more than 50% recombinant offspring because the pattern of multiple crossovers can yield an average maximum value of only 50%. When a testcross does yield a value of 50% recombinant offspring, it can mean two different things. Either the two genes are on different chromosomes or the two genes are on the same chromosome but at least 50 mu apart.

E8. If two genes are at least 50 mu apart, you would need to map genes between them to show that the two genes were actually in the same linkage group. For example, if gene *A* was 55 mu from gene *B*, there might be a third gene (e.g., gene *C*) that was 20 mu from *A* and 35 mu from *B*. These results would indicate that *A* and *B* are 55 mu apart, assuming dihybrid testcrosses between genes *A* and *B* yielded 50% recombinant offspring.

E10. Sturtevant used the data involving the following pairs: *y* and *w*, *w* and *v*, *v* and *r*, and *v* and *m*.

E12. A. Because they are 12 mu apart, we expect 12% (or 120) recombinant offspring. This would be approximately 60 *Aabb* and 60 *aaBb* plus 440 *AaBb* and 440 *aabb*.

B. We would expect 60 *AaBb*, 60 *aabb*, 440 *Aabb*, and 440 *aaBb*.

E14. Due to the large distance between the two genes, they will assort independently even though they are actually on the same chromosome. According to independent assortment, we expect 50% parental and 50% recombinant offspring. Therefore, this cross will produce 150 offspring in each of the four phenotypic categories.

E16. A. If we hypothesize two genes independently assorting, then the predicted ratio is 1:1:1:1. There are a total of 390 offspring. The expected number of offspring in each category is about 98. Plugging the figures into our chi square formula,

$$\chi^2 = \frac{(117-98)^2}{98} + \frac{(115-98)^2}{98} + \frac{(78-98)^2}{98} + \frac{(80-98)^2}{98}$$

$$\chi^2 = 3.68 + 2.95 + 4.08 + 3.31$$

$$\chi^2 = 14.02$$

Looking up this value in the chi square table under 1 degree of freedom, we reject our hypothesis, because the chi square value is above 7.815.

B. Map distance:

$$\text{Map Distance} = \frac{78 + 80}{117 + 115 + 78 + 80}$$
$$= 40.5 \text{ mu}$$

Because the value is relatively close to 50 mu, it is probably a significant underestimate of the true distance between these two genes.

E18. The percentage of recombinants for the green, yellow and wide, narrow is 7%, or 0.07; there will be 3.5% of the green, narrow and 3.5% of the yellow, wide. The remaining 93% parentals will be 46.5% green, wide and 46.5% yellow, narrow. The third gene assorts independently. There will be 50% long and 50% short with respect to each of the other two genes. To calculate the number of offspring out of a total of 800, we multiply 800 by the percentages in each category.

(0.465 green, wide)(0.5 long)(800) = 186 green, wide, long

(0.465 yellow, narrow)(0.5 long)(800) = 186 yellow, narrow, long

(0.465 green, wide)(0.5 short)(800) = 186 green, wide, short

(0.465 yellow, narrow)(0.5 short)(800) = 186 yellow, narrow, short

(0.035 green, narrow)(0.5 long)(800) = 14 green, narrow, long

(0.035 yellow, wide)(0.5 long)(800) = 14 yellow, wide, long

(0.035 green, narrow)(0.5 short)(800) = 14 green, narrow, short

(0.035 yellow, wide)(0.5 short)(800) = 14 yellow, wide, short

E20. Let's use the following symbols: *G* for green pods, *g* for yellow pods, *S* for green seedlings, *s* for bluish green seedlings, *C* for normal plants, *c* for creepers. The parental cross is *GG SS CC* crossed to *gg ss cc*.

The F_1 plants would all be *Gg Ss Cc*. If the genes are linked, the alleles *G*, *S*, and *C* would be linked on one chromosome, and the alleles *g*, *s*, and *c* would be linked on the homologous chromosome.

The testcross is F_1 plants, which are *Gg Ss Cc* crossed to *gg ss cc*.

To measure the distances between the genes, we can separate the data into gene pairs.

Pod color, seedling color

2,210 green pods, green seedlings—nonrecombinant

296 green pods, bluish green seedlings—recombinant

2,198 yellow pods, bluish green seedlings—nonrecombinant

293 yellow pods, green seedlings—recombinant

$$\text{Map Distance} = \frac{296 + 293}{2{,}210 + 296 + 2{,}198 + 293} \times 100 = 11.8 \text{ mu}$$

Pod color, plant stature

2,340 green pods, normal—nonrecombinant

166 green pods, creeper—recombinant

2,323 yellow pods, creeper—nonrecombinant

168 yellow pods, normal—recombinant

$$\text{Map Distance} = \frac{166 + 168}{2{,}340 + 166 + 2{,}323 + 168} \times 100 = 6.7 \text{ mu}$$

Seedling color, plant stature

2,070 green seedlings, normal—nonrecombinant

433 green seedlings, creeper—recombinant

2,056 bluish green seedlings, creeper—nonrecombinant

438 bluish green seedlings, normal—recombinant

$$\text{Map Distance} = \frac{433 + 438}{2,070 + 433 + 2,056 + 438} \times 100 = 17.4 \text{ mu}$$

The order of the genes is seedling color, pod color, and plant stature (or you could say the opposite order). Pod color is in the middle. If we use the two shortest distances to construct our map:

S —— 11.8 —— *G* —— 6.7 —— *C*

E22. To answer this question, we can consider genes in pairs. Let's consider the two gene pairs that are closest together. The distance between the wing length and eye color genes is 12.5 mu. From this cross, we expect 87.5% to have long wings and red eyes or short wings and purple eyes, and 12.5% to have long wings and purple eyes or short wings and red eyes. Therefore, we expect 43.75% to have long wings and red eyes, 43.75% to have short wings and purple eyes, 6.25% to have long wings and purple eyes, and 6.25% to have short wings and red eyes. If we have 1,000 flies, we expect 438 to have long wings and red eyes, 438 to have short wings and purple eyes, 62 to have long wings and purple eyes, and 62 to have short wings and red eyes (rounding to the nearest whole number).

The distance between the eye color and body color genes is 6 mu. From this cross, we expect 94% to have a parental combination (red eyes and gray body or purple eyes and black body) and 6% to have a nonparental combination (red eyes and black body or purple eyes and gray body). Therefore, of our 438 flies with long wings and red eyes, we expect 94% of them (or about 412) to have long wings, red eyes, and gray body and 6% of them (or about 26) to have long wings, red eyes, and black bodies. Of our 438 flies with short wings and purple eyes, we expect about 412 to have short wings, purple eyes, and black bodies and 26 to have short wings, purple eyes, and gray bodies.

Of the 62 flies with long wings and purple eyes, we expect 94% of them (or about 58) to have long wings, purple eyes, and black bodies and 6% of them (or about 4) to have long wings, purple eyes, and gray bodies. Of the 62 flies with short wings and red eyes, we expect 94% (or about 58) to have short wings, red eyes, and gray bodies and 6% (or about 4) to have short wings, red eyes, and black bodies.

In summary,

Long wings, red eyes, gray body	412
Long wings, purple eyes, gray body	4
Long wings, red eyes, black body	26
Long wings, purple eyes, black body	58
Short wings, red eyes, gray body	58
Short wings, purple eyes, gray body	26
Short wings, red eyes, black body	4
Short wings, purple eyes, black body	412

The flies with long wings, purple eyes, and gray bodies, or short wings, red eyes, and black bodies, are produced by a double-crossover event.

E24. Yes. Begin with females that have one X chromosome that is X^{Nl} and the other X chromosome that is X^{nL}. These females have to be mated to $X^{NL}Y$ males because a living male cannot carry the *n* or *l* allele. In the absence of crossing over, a mating between $X^{Nl}X^{nL}$ females to $X^{NL}Y$ males should not produce any surviving male offspring. However, during oogenesis in these heterozygous female mice, there could be a crossover in the region between the two genes, which would produce an X^{NL} chromosome and an X^{nl} chromosome. Male offspring inheriting these recombinant chromosomes will be either $X^{NL}Y$ or $X^{nl}Y$ (whereas nonrecombinant males will be $X^{nL}Y$ or $X^{Nl}Y$). Only the male mice that inherit $X^{NL}Y$ will live. The living males represent only half of the recombinant offspring. (The other half are $X^{nl}Y$, which are born dead.)

To compute map distance:

$$\text{Map Distance} = \frac{2(\text{Number of male living offspring})}{\text{Number of males born dead} + \text{Number of males born alive}} \times 100$$

E26.

$$\text{Map Distance} = \frac{(^1/_2)(\text{SDS})}{\text{Total}} \times 100$$

$$= \frac{(^1/_2)(22 + 21 + 21 + 23)}{22 + 21 + 21 + 21 + 451 + 23 + 455} \times 100$$

$$= 4.4 \text{ mu}$$

Questions for Student Discussion/Collaboration

2. The X and Y chromosomes are not completely distinct linkage groups. One might describe them as overlapping linkage groups having some genes in common, but most genes are not common to both.

CHAPTER 6

Conceptual Questions

C2. It is not a form of sexual reproduction, in which two distinct parents produce gametes that unite to form a new individual. However, conjugation is similar to sexual reproduction in the sense that the genetic material from two cells are somewhat mixed. In conjugation, there is not the mixing of two genomes, one from each gamete. Instead, there is a transfer of genetic material from one cell to another. This transfer can alter the combination of genetic traits in the recipient cell.

C4. An F^+ strain contains a separate, circular piece of DNA that has its own origin of transfer. An *Hfr* strain has its origin of transfer integrated into the bacterial chromosome. An F^+ strain can transfer only the DNA contained on the F factor. If given enough time, an *Hfr* strain can actually transfer the entire bacterial chromosome to the recipient cell.

C6. Sex pili promote the binding of donor and recipient cells and provide a passageway for the transfer of genetic material from the donor to the recipient cell.

C8. Though exceptions are common, interspecies genetic transfer via conjugation is not as likely because the cell surfaces do not interact correctly. Interspecies genetic transfer via transduction is also not very likely because each species of bacteria is sensitive to particular bacteriophages. The correct answer is transformation. A consequence of interspecies genetic transfer is that new genes can be introduced into a bacterial species from another species. For example, interspecies genetic transfer could provide the recipient bacterium with a new trait, such as resistance to an antibiotic. Evolutionary biologists call this horizontal gene transfer, while the passage of genes from parents to offspring is termed vertical gene transfer.

C10. Cotransduction is the transduction of two or more genes. The distance between the genes determines the frequency of cotransduction. When two genes are close together, the cotransduction frequency would be higher compared to two genes that are relatively farther apart.

C12. If a site that frequently incurred a breakpoint was between two genes, the cotransduction frequency of these two genes would be much less than expected. This is because the site where the breakage occurred would separate the two genes from each other.

C14. The transfer of conjugative plasmids such as F factor DNA

C16. A. If it occurred in a single step, transformation is the most likely mechanism because conjugation does not usually occur between different species, particularly distantly related species, and different species are not usually infected by the same bacteriophages.

B. It could occur in a single step, but it may be more likely to have involved multiple steps.

C. The use of antibiotics selects for the survival of bacteria that have resistance genes. If a population of bacteria is exposed to an antibiotic, those carrying resistance genes will survive, and their relative numbers will increase in subsequent generations.

C18. The term allele means alternative forms of the same gene. Therefore, mutations in the same gene among different phages are alleles of each other; the mutations may be at different positions within the same gene. When we map the distance between mutations in the same gene, we are mapping the distance between the mutations that create different alleles of the same gene. An intragenic map describes the locations of mutations within the same gene.

Experimental Questions

E2. Mix the two strains together and then put some of them on plates containing streptomycin and some of them on plates without streptomycin. If mated colonies are present on both types of plates, then the thr^+, leu^+, and thi^+ genes were transferred to the met^+ bio^+ thr^- leu^- thi^- strain. If colonies are found only on the plates that lack streptomycin, then the met^+ and bio^+ genes are being transferred to the met^- bio^- thr^+ leu^+ thi^+ strain. This answer assumes a one-way transfer of genes from a donor to a recipient strain.

E4. An interrupted mating experiment is a procedure in which two bacterial strains are allowed to mate, and then the mating is interrupted at various time points. The interruption occurs by agitation of the solution in which the bacteria are found. This type of study is used to map the locations of genes. It is necessary to interrupt mating so that you can vary the time and obtain information about the order of transfer; which gene transferred first, second, and so on.

E6. Mate unknown strains *A* and *B* to the F^- strain in your lab that is resistant to streptomycin and cannot use lactose. This is done in two separate tubes (i.e., strain *A* plus your F^- strain in one tube, and strain *B* plus your F^- strain in the other tube). Plate the mated cells on growth media containing lactose plus streptomycin. If you get growth of colonies, the unknown strain had to be strain A, the F^+ strain that had lactose utilization genes on its F factor.

E8. A. If we extrapolate these lines back to the x-axis, the *hisE* intersects at about 3 minutes, and the *pheA* intersects at about 24 minutes. These are the values for the times of entry. Therefore, the distance between these two genes is 21 minutes (i.e., 24 minus 3).

B.

↑ 4 ↑ 17 ↑

hisE *pabB* *pheA*

E10. One possibility is that you could treat the P1 lysate with DNase I, an enzyme that digests DNA. (Note: If DNA were digested with DNase I, the function of any genes within the DNA would be destroyed.) If the DNA were within a P1 phage, it would be protected from DNase I digestion. This would allow you to distinguish between transformation (which would be inhibited by DNase I) versus transduction (which would not be inhibited by DNase I). Another possibility is that you could try to fractionate the P1 lysate. Naked DNA would be smaller than a P1 phage carrying DNA. You could try to filter the lysate to remove naked DNA, or you could subject the lysate to centrifugation and remove the lighter fractions that contain naked DNA.

E12. Cotransduction frequency $= (1 - d/L)^3$

For the normal strain,

Cotransduction frequency $= (1 - {}^{0.7}/_{2})^3 = 0.275$, or 27.5%

For the new strain,

Cotransduction frequency $= (1 - {}^{0.7}/_{5})^3 = 0.64$, or 64%

The experimental advantage is that you could map genes that are farther than 2 minutes apart. You could map genes that are up to 2 minutes apart.

E14. Cotransduction frequency $= (1 - {}^{d}/_{L})^3$

$$0.53 = (1 - {}^{d}/_{2}\text{ minutes})^3$$

$$(1 - {}^{d}/_{2}\text{ minutes}) = \sqrt[3]{0.53}$$

$$(1 - {}^{d}/_{2}\text{ minutes}) = 0.81$$

$$d = 0.38\text{ minutes}$$

E16. A. We first need to calculate the cotransformation frequency, which equals 2/70, or 0.029.

Contransformation frequency $= (1 - {}^{d}/_{L})^3$

$$0.029 = (1 - {}^{d}/_{2}\text{ minutes})^3$$

$$d = 1.4\text{ minutes}$$

B. Contransformation frequency $= (1 - {}^{d}/_{L})^3$

$$= (1 - {}^{1.4}/_{4})^3$$

$$= 0.27$$

As you may have expected, the cotransformation frequency is much higher when the transformation involves larger pieces of DNA.

E18. Benzer could use this observation as a way to evaluate if intragenic recombination had occurred. If two *rII* mutations recombined to make a wild-type gene, the phage would produce plaques in this *E. coli* K12(λ) strain.

E20. *rIIA*: L47 and L92

rIIB: L33, L40, L51, L62, L65, and L91

E22. Benzer first determined the individual nature of each gene by showing that mutations within the same gene did not complement each other. He then could map the distance between two mutations within the same gene. The map distances defined each gene as a linear, divisible unit. In this regard, the gene is divisible due to crossing over.

Questions for Student Discussion/Collaboration

2. Flower color in sweet peas is another example. When two different genes affect flower color, the white allele of one gene may be epistatic to the purple allele of another gene. A heterozygote for both genes would have purple flowers. In other words, the two purple alleles are complementing the two white alleles. Other similar examples are discussed in Chapter 4 in the section on Gene Interactions.

CHAPTER 7

Conceptual Questions

C2. A maternal effect gene is one in which the genotype of the mother determines the phenotype of the offspring. At the cellular level, this happens because maternal effect genes are expressed in diploid nurse cells and then the gene products are transported into the egg. These gene products play key roles in the early steps of embryonic development.

C4. The genotype of the mother must be $bic^- bic^-$. That is why it produces abnormal offspring. Because the mother is alive and able to produce offspring, its mother (the maternal grandmother) must have been $bic^+ bic^-$ and passed the bic^- allele to its daughter (the mother in this problem). The maternal grandfather also must have passed the bic^- allele to its daughter. The maternal grandfather could be either $bic^+ bic^-$ or $bic^- bic^-$.

C6. The mother must be heterozygous. She is phenotypically abnormal because her mother must have been homozygous for the abnormal recessive allele. However, because she produces all normal offspring, she must have inherited the normal dominant allele from her father. She produces all normal offspring because this is a maternal effect gene, and the gene product of the normal dominant allele is transferred to the egg.

C8. Maternal effect genes exert their effects because the gene products are transferred from nurse cells to eggs. The gene products, mRNA and proteins, do not last a very long time before they are eventually degraded. Therefore, they can exert their effects only during early stages of embryonic development.

C10. Dosage compensation refers to the phenomenon that the level of expression of genes on the sex chromosomes is the same in males and females, even though they have different numbers of sex chromosomes. In many species it seems necessary so that the balance of gene expression between the autosomes and sex chromosomes is similar between the two sexes.

C12. In mammals, one of the X chromosomes is inactivated in females; in *Drosophila*, the level of transcription on the X chromosome in males is doubled; in *C. elegans*, the level of transcription of the X chromosome in hermaphrodites is decreased by 50% of that of males.

C14. X inactivation begins with the counting of Xics. If there are two X chromosomes, in the process of initiation, one is targeted for inactivation. During embryogenesis, this inactivation begins at the Xic locus and spreads to both ends of the X chromosome until it becomes a highly condensed Barr body. The *Tsix* gene may play a role in the choice of the X chromosome that remains active. The *Xist* gene, which is located in the Xic region, remains transcriptionally active on the inactivated X chromosome. It is thought to play an important role in X inactivation by coating the inactive X chromosome. After X inactivation is established, it is maintained in the same X chromosome in somatic cells during subsequent cell divisions. In germ cells, however, the X chromosomes are not inactivated, so an egg can transmit either copy of an active (noncondensed) X chromosome.

C16. A. One

B. Zero

C. Two

D. Zero

C18. The offspring inherited X^B from its mother and X^O and Y from its father. It is an XXY animal, which is male (but somewhat feminized).

C20. Erasure and reestablishment of the imprint occurs during gametogenesis. It is necessary to erase the imprint because each sex will transmit either inactive or active alleles of a gene. In somatic cells, the two alleles for a gene are imprinted according to the sex of the parent from which the allele was inherited.

C22. A person born with paternal uniparental disomy 15 would have Angelman syndrome, because this individual would not have an active copy of the *AS* gene; the paternally inherited copies of the *AS* gene are silenced. This individual would have normal offspring, because she does not have a deletion in either copy of chromosome 15.

C24. In some species, such as marsupials, X inactivation depends on the sex. This is similar to imprinting. Also, once X inactivation occurs during embryonic development, it is remembered throughout the rest of the life of the organism, which is also similar to imprinting. X inactivation in mammals is different from genomic imprinting, in that it is not sex dependent. The X chromosome that is inactivated could be inherited from the mother or the father. There was no marking process on the X chromosome that occurred during gametogenesis. In contrast, genomic imprinting always involves a marking process during gametogenesis.

C26. The term reciprocal cross refers to two parallel crosses that involve the same genotypes of the two parents, but their sexes are opposite in the two crosses. For example, the reciprocal cross of female *BB* × male *bb* is the cross female *bb* × male *BB*. Autosomal inheritance gives the same result because the autosomes are transmitted from parent to offspring in the same way for both sexes. However, for extranuclear inheritance, the mitochondria and plastids are not transmitted via the gametes in the same way for both sexes. For maternal inheritance, the reciprocal crosses would show that the gene is always inherited from the mother.

C28. The phenotype of a petite mutant is that it forms small colonies on growth media, as opposed to wild-type strains that formed larger colonies. These mutants are unable to grow when the cells only have an energy source that requires mitochondrial function. Because nuclear and mitochondrial genes are necessary for mitochondrial function, it is possible for a petite mutation to involve a gene in the nucleus or in the mitochondrial genome. The difference between neutral and suppressive petites is that neutral petites lack most of their mitochondrial DNA, whereas suppressive petites usually lack small segments of the mitochondrial genetic material.

C30. The mitochondrial and chloroplast genomes are composed of a circular chromosome found in one or more copies in a region of the organelle known as the nucleoid. The number of genes per chromosome varies from species to species. Chloroplasts genomes tend to be larger than mitochondria genomes. See Table 7.3 for examples of the variation among mitochondrial and chloroplast genomes.

C32. A. Yes

B. Yes

C. No, it is determined by a gene in the chloroplast genome.

D. No, it is determined by a gene in the mitochondria.

C34. Biparental extranuclear inheritance would resemble Mendelian inheritance in that offspring could inherit alleles of a given gene from both parents. It differs, however, when you think about it from the perspective of heterozygotes. For a Mendelian trait, the law of segregation tells us that a heterozygote passes one allele for a given gene to an offspring, but not both. In contrast, if a parent has a mixed population of mitochondria (e.g., some carrying a mutant gene and some carrying a normal gene), that parent could pass both types of genes (mutant and normal) to a single offspring, because more than one mitochondrion could be contained within a sperm or egg cell.

Experimental Questions

E2. The first type of observation was based on cytological studies. The presence of the Barr body in female cells was consistent with the idea that one of the X chromosomes was highly condensed. The second type of observation was based on genetic mutations. A variegated phenotype that is found only in females is consistent with the idea that certain patches express one allele and other patches express the other allele. This variegated phenotype would occur only if the inactivation of one X chromosome happened at an early stage of embryonic development and was inherited permanently thereafter.

E4. The pattern of inheritance is consistent with imprinting. In every cross, the allele that is inherited from the father is expressed in the offspring, but the allele inherited from the mother is not.

E6. We assume that the snails in the large colony on the second island are true-breeding, *DD*. Let the male snail from the deserted island mate with a female snail from the large colony. Then let the F_1 snails mate with each other to produce an F_2 generation. Then let the F_2 generation mate with each other to produce an F_3 generation. Here are the expected results:

Female *DD* × Male *DD*

All F_1 snails coil to the right.

All F_2 snails coil to the right.

All F_3 snails coil to the right.

Female *DD* × Male *Dd*

All F_1 snails coil to the right.

All F_2 snails coil to the right because all of the F_1 females are *DD* or *Dd*.

$^{15}/_{16}$ of F_3 snails coil to the right; $^{1}/_{16}$ of F_3 snails coil to the left (because $^{1}/_{16}$ of the F_2 females are *dd*).

Female *DD* × Male *dd*.

All F_1 snails coil to the right.

All F_2 snails coil to the right because all of the F_1 females are *Dd*.

$^{3}/_{4}$ of F_3 snails coil to the right; $^{1}/_{4}$ of F_3 snails coil to the left (because $^{1}/_{4}$ of the F_2 females are *dd*).

E8. Let's first consider the genotypes of male A and male B. Male A must have two normal copies of the *Igf2* gene. We know this because male A's mother was *Igf2 Igf2*; the father of male A must have been a heterozygote *Igf2* $Igf2^-$ because half of the litter that contained male A also contained dwarf offspring. But because male A was not dwarf, it must have inherited the normal allele from its father. Therefore, male A must be *Igf2 Igf2*. We cannot be completely sure of the genotype of male B. It must have inherited the normal *Igf2* allele from its father because male B is phenotypically normal. We do not know the genotype of male B's mother, but she could be either $Igf2^-$ $Igf2^-$ or *Igf2* $Igf2^-$. In either case, the mother of male B could pass the $Igf2^-$ allele to an offspring, but we do not know for sure if she did. So, male B could be either *Igf2* $Igf2^-$ or *Igf2 Igf2*.

For the *Igf2* gene, we know that the maternal allele is inactivated. Therefore, the genotypes and phenotypes of females A and B are irrelevant. The phenotype of the offspring is determined only by the allele that is inherited from the father. Because we know that male A has to be *Igf2 Igf2*, we know that it can produce only normal offspring. Because both females A and B both produced dwarf offspring, male A cannot be the father. In contrast, male B could be either *Igf2 Igf2* or *Igf2* $Igf2^-$. Because both females gave birth to dwarf babies (and because male A and male B were the only two male mice in the cage), we conclude that male B must be *Igf2* $Igf2^-$ and is the father of both litters.

E10. In fruit flies, the expression of a male's X-linked genes is turned up twofold. In mice, one of the two X chromosomes is inactivated; that is why females and males produce the same total amount of mRNA for most X-linked genes. In *C. elegans*, the expression of hermaphrodite X-linked genes is turned down twofold. Overall, the total amount of expression of X-linked genes is the same in males and females (or hermaphrodites) of these three species. In fruit flies and *C. elegans*, heterozygous females and hermaphrodites express 50% of each allele compared to a homozygous male, so that heterozygous females and hermaphrodites produce the same total amount of mRNA from X-linked genes compared to males. Note: In heterozygous females of fruit flies, mice, and *C. elegans*, there is 50% of each gene product (compared to hemizygous males and homozygous females).

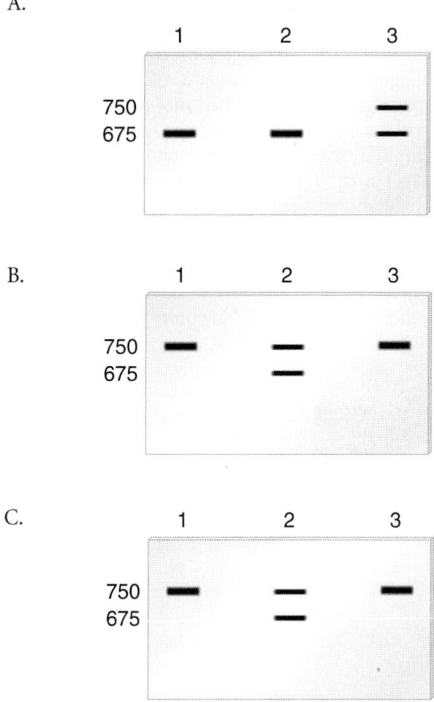

E12. In the absence of UV light, we would expect all sm^r offspring. With UV light, we would expect a greater percentage of sm^s offspring.

Questions for Student Discussion/Collaboration

2. An infective particle is something in the cytoplasm that contains its own genetic material and isn't an organelle. Some symbiotic infective particles, such as those found in killer paramecia, are similar to mitochondria and chloroplasts in that they contain their own genomes and are known to be bacterial in origin. The observation that these endosymbiotic relationships can initiate in modern species tells us that endosymbiosis can spontaneously happen. Therefore, it is reasonable that it happened a long time ago and led to the evolution of mitochondria and chloroplasts.

CHAPTER 8

Conceptual Questions

C2. Small deletions and duplications are less likely to affect phenotype simply because they usually involve fewer genes. If a small deletion did have a phenotypic effect, you would conclude that a gene or genes in this region is required to have a normal phenotype.

C4. A gene family is a group of genes that are derived from the process of gene duplications. They have similar sequences, but the sequences have some differences due to the accumulation of mutations over many generations. The members of a gene family usually encode proteins with similar but specialized functions. The specialization may occur in different cells or at different stages of development.

C6. It has a pericentric inversion.

C8. There are four products from meiosis. One would be a normal chromosome, and one would contain the inversion shown in the drawing to conceptual question C6. The other two chromosomes would be dicentric or acentric with the following order of genes:

centromere centromere

A ↓ *B C D E F G H I J D C B* ↓ *A* Dicentric

M L K J I H G F E K L M Acentric

C10. In the absence of crossing over, alternate segregation would yield half the cells with two normal chromosomes and half with a balanced translocation. For adjacent-1 segregation, all cells will be unbalanced. Two cells would be

A B C D E + *A I J K L M*

And the other two cells would be

H B C D E + *H I J K L M*

C12. One of the parents may carry a balanced translocation between chromosomes 5 and 7. The phenotypically abnormal offspring has inherited an unbalanced translocation due to the segregation of translocated chromosomes during meiosis.

C14. A deficiency and an unbalanced translocation are more likely to have phenotypic effects because they create genetic imbalances. For a deficiency, there are too few copies of several genes, and for an unbalanced translocation, there are too many.

C16. It is due to a crossover within the inverted region. You should draw the inversion loop (as is done in Figure 8.12a). The crossover occurred between *P* and *U*.

C18. This person has a total of 46 chromosomes. However, this person would be considered aneuploid rather than euploid, because one of the sets is missing a sex chromosome and one set has an extra copy of chromosome 21.

C20. It may be related to genetic balance. In aneuploidy, there is an imbalance in gene expression between the chromosomes found in their normal copy number versus those that are either too many or too few. In polyploidy, the balance in gene expression is still maintained.

C22. The male offspring is the result of nondisjunction during oogenesis. The female produced an egg without any sex chromosomes. The male parent transmitted a single X chromosome carrying the red allele. This produces an X0 male offspring with red eyes.

C24. Trisomies 13, 18, and 21 survive because the chromosomes are small and probably contain fewer genes compared to the larger chromosomes. Individuals with abnormal numbers of X chromosomes can survive because the extra copies are converted to transcriptionally inactive Barr bodies. The other trisomies are lethal because they cause a great amount of imbalance between the level of gene expression on the normal diploid chromosomes relative to the chromosomes that are trisomic.

C26. Endopolyploidy means that a particular somatic tissue is polyploid even though the rest of the organism is not. The biological significance is not entirely understood, although it has been speculated that an increase in chromosome number in certain cells may enhance their ability to produce specific gene products that are needed in great abundance.

C28. In certain types of cells, such as salivary cells, the homologous chromosomes pair with each other and then replicate approximately nine times to produce a polytene chromosome. The centromeres from each chromosome aggregate with each other at the chromocenter. This structure has six arms that arise from the single arm of two telocentric chromosomes (the X and 4) and two arms each from the metacentric chromosomes 2 and 3.

C30. The turtles are two distinct species that appear phenotypically identical. The turtles with 48 chromosomes are polyploid relatives (i.e., tetraploids) of the species with 24 chromosomes. In animals, it is somewhat hard to imagine how this could occur because animals cannot self-fertilize, so there had to be two animals (i.e., one male and one female) that became tetraploids. It is easy to imagine how one animal could become a tetraploid; complete nondisjunction could occur during the first cell division of a fertilized egg, thereby creating a tetraploid cell that continued to develop into a tetraploid animal. This would have to happen independently (i.e., in two individuals of opposite sex) to create a tetraploid species. If you mated a tetraploid turtle with a diploid turtle, the offspring would be triploid and probably phenotypically normal. However, the triploid offspring would be sterile because they would make highly aneuploid gametes.

C32. Polyploid, triploid, and euploid should not be used.

C34. The boy carries a translocation involving chromosome 21: probably a translocation in which nearly all of chromosome 21 is translocated to chromosome 14. He would have one normal copy of chromosome 14, one normal copy of chromosome 21, and the translocated chromosome that contains both chromosome 14 and chromosome 21. This boy is phenotypically normal because the total amount of genetic material is normal, although the total number of chromosomes is 45 (because chromosome 14 and chromosome 21 are fused into a single chromosome). His sister has familial Down syndrome because she has inherited the translocated chromosome, but she also must have one copy of chromosome 14 and two copies of chromosome 21. She has the equivalent of three copies of chromosome 21 (i.e., two normal copies and one copy fused with chromosome 14). This is why she has familial Down syndrome. One of the parents of these two children is probably normal with regard to karyotype (i.e., the parent has 46 normal chromosomes). The other parent would have a karyotype that would be like the phenotypically normal boy.

C36. Nondisjunction is a mechanism whereby the chromosomes do not segregate equally into the two daughter cells. This can occur during meiosis to produce cells with altered numbers of chromosomes, or it can occur during mitosis to produce a genetic mosaic individual. A third way to alter chromosome number is by interspecies crosses that produce an alloploid.

C38. A mutation occurred during early embryonic development to create the blue patch of tissue. One possibility is a mitotic nondisjunction in which the two chromosomes carrying the *b* allele went to one cell and the two chromosomes carrying the *B* allele went to the other daughter cell. A second possibility is that the chromosome carrying the *B* allele was lost. A third possibility is that the *B* allele was deleted. This would cause the recessive *b* allele to exhibit pseudodominance.

C40. Homeologous chromosomes are chromosomes from two species that are evolutionarily related to each other. For example, chromosome 1 in the sable antelope and the roan antelope are homeologous; they carry many of the same genes.

C42. In general, Turner syndrome could be due to nondisjunction during either oogenesis or spermatogenesis. However, the Turner individual with color blindness is due to nondisjunction during spermatogenesis. The sperm lacked a sex chromosome, due to nondisjunction, and the egg carried an X chromosome with the recessive color blindness allele. This X chromosome had to be inherited from the mother, because the father was not color-blind. The mother must be heterozygous for the recessive color blindness allele, and the father is hemizygous for the normal allele. Therefore, the mother must have transmitted a single X chromosome carrying the color blindness allele to her offspring, indicating that nondisjunction did not occur during oogenesis.

Experimental Questions

E2. Due to the mix up, the ratio of green/red fluorescence would be 0.5 in regions where the cancer cells had a normal amount of DNA. If a duplication occurred on both chromosomes of cancer cells, the ratio would be 1.0. If a deletion occurred on a single chromosome, the ratio would be 0.25.

E4. Colchicine interferes with the spindle apparatus and thereby causes nondisjunction. At high concentrations, it can cause complete nondisjunction and produce polyploid cells.

E6. The primary purpose is to generate diploid strains that are homozygous for all of their genes. Because the pollen is haploid, it has only one copy of each gene. The strain can later be made diploid (and homozygous) by treatment with colchicine.

E8. First, you would cross the two strains together. It is difficult to predict the phenotype of the offspring. Nevertheless, you would keep crossing offspring to each other and backcrossing them to the parental strains until you obtained a great-tasting tomato strain that was resistant to heat and the viral pathogen. You could then make this strain tetraploid by treatment with colchicine. If you crossed the tetraploid strain with your great-tasting diploid strain that was resistant to heat and the viral pathogen, you may get a triploid that had these characteristics. This triploid would probably be seedless.

E10. A polytene chromosome is formed when a chromosome replicates many times, and the chromatids lie side by side, as shown in Figure 8.21. The homologous chromosomes also lie side by side. Therefore, if there is a deletion, there will be a loop. The loop is the segment that is not deleted from one of the two homologues.

E12. Because the plant giving the pollen is heterozygous for many genes, some of the pollen grains may be haploid for recessive alleles, which are nonbeneficial or even lethal. However, some pollen grains may inherit only the dominant (beneficial) alleles and grow quite well.

Questions for Student Discussion/Collaboration

2. There are many possibilities. The students could look in agriculture and botany books to find many examples. In the insect world, there are interesting examples of euploidy affecting gender determination. Among amphibians and reptiles, there are also several examples of closely related species that have euploid variation.

4.
   1. Polyploids are often more robust and disease resistant.
   2. Allopolyploids may have useful combinations of traits.
   3. Hybrids are often more vigorous; they can be generated from monoploids.
   4. Strains with an odd number of chromosome sets (e.g., triploids) are usually seedless.

CHAPTER 9

Conceptual Questions

C2. The transformation process is described in Chapter 6.

1. A fragment of DNA binds to the cell surface.
2. It penetrates the cell wall/cell membrane.
3. It enters the cytoplasm.
4. It recombines with the chromosome.
5. The genes within the DNA are expressed (i.e., transcription and translation).
6. The gene products create a capsule. That is, they are enzymes that synthesize a capsule using cellular molecules as building blocks.

C4. The building blocks of a nucleotide are a sugar (ribose or deoxyribose), a nitrogenous base, and a phosphate group. In a nucleotide, the phosphate is already linked to the 5' position on the sugar. When two nucleotides are hooked together, a phosphate on one nucleotide forms a covalent bond with the 3' hyrdroxyl group on another nucleotide.

C6. The structure is a phosphate group connecting two sugars at the 3' and 5' positions, as shown in Figure 9.11.

C8. 3'–CCGTAATGTGATCCGGA–5'

C10. A drawing of a DNA helix with 10 bp per turn would look like Figure 9.17 in the textbook. To make 15 bp per turn, you would have to add 5 more base pairs, but the helix should still make only one complete turn.

C12. The nucleotide bases occupy the major and minor grooves. Phosphate and sugar are found in the backbone. If a DNA-binding protein does not recognize a nucleotide sequence, it probably is not binding in the grooves but instead is binding to the DNA backbone (i.e., sugar–phosphate sequence). DNA-binding proteins that recognize a base sequence must bind into a major or minor groove of the DNA, which is where the bases would be accessible to a DNA-binding protein. Most DNA-binding proteins that recognize a base sequence fit into the major groove. By comparison, other DNA-binding proteins, such as histones, which do not recognize a base sequence, bind to the DNA backbone.

C14. The structure is shown in Figure 9.8. You begin numbering at the carbon that is to the right of the ring oxygen and continue to number the carbon atoms in a clockwise direction. Antiparallel means that the backbones are running in the opposite direction. In one strand, the sugar carbons are oriented in a 3' to 5' direction, while in the other strand they are oriented in a 5' to 3' direction.

C16. Double-stranded RNA is more like A DNA than B DNA. See the text for a discussion of A-DNA structure.

C18. Its nucleotide base sequence

C20. G = 32%, C = 32%, A = 18%, T = 18%.

C22. One possibility is a sequential mechanism. First, the double helix could unwind and replicate itself as described in Chapter 11. This would produce two double helices. Next, the third strand (bound in the major groove) could replicate itself via a semiconservative mechanism. This new strand could be copied to make a copy that is identical to the strand that lies in the major groove. At this point, you would have two double helices and two strands that could lie in the major groove. These could assemble to make two triple helices.

C24. Lysines and arginines, and also polar amino acids

C26. This DNA molecule contains 280 bp. There are 10 base pairs per turn, so there are 28 complete turns.

C28. A hydroxyl group is at the 3' end, and a phosphate group is at the 5' end.

C30. Not necessarily. The AT/GC rule is required only of double-stranded DNA molecules.

C32. The first thing we need to do is to determine how many base pairs are in this DNA molecule. The linear length of 1 base pair is 0.34 nm, which equals 0.34×10^{-9} m. One centimeter equals 10^{-2} meters.

$$\frac{10^{-2}}{0.34 \times 10^{-9}} = 2.9 \times 10^{7}\text{ bp}$$

There are approximately 2.9×10^7 bp in this DNA molecule, which equals 5.8×10^7 nucleotides. If 15% are adenine, then 15% must also be thymine. This leaves 70% for cytosine and guanine. Because cytosine and guanine bind to each other, there must be 35% cytosine and 35% guanine. If we multiply 5.8×10^7 times 0.35, we get

$(5.8 \times 10^7)(0.35) = 2.0 \times 10^7$ cytosines, or about 20 million cytosines

C34. The methyl group is not attached to one of the atoms that hydrogen bonds with guanine, so methylation would not directly affect hydrogen bonding. It could indirectly affect hydrogen bonding if it perturbed the structure of DNA. Methylation may affect gene expression because it could alter the ability of proteins to recognize DNA sequences. For example, a protein might bind into the major groove by interacting with a sequence of bases that includes one or more cytosines. If the cytosines are methylated, this may prevent a protein from binding into the major groove properly. Alternatively, methylation could enhance protein binding. In Chapter 7, we considered DNA-binding proteins that were influenced by the methylation of DMRs (differentially methylated regions) that occur during genomic imprinting.

Experimental Questions

E2. A. There are different possible reasons why most of the cells were not transformed.

1. Most of the cells did not take up any of the type IIIS DNA.
2. The type IIIS DNA was usually degraded after it entered the type IIR bacteria.
3. The type IIIS DNA was usually not expressed in the type IIR bacteria.

B. The antibody/centrifugation steps were used to remove the bacteria that had not been transformed. It enabled the researchers to determine the phenotype of the bacteria that had been transformed. If this step was omitted, there would have been so many colonies on the plate it would have been difficult to identify any transformed bacterial colonies, because they would have represented a very small proportion of the total number of bacterial colonies.

C. They were trying to demonstrate that it was really the DNA in their DNA extract that was the genetic material. It was possible that the extract was not entirely pure and could contain contaminating RNA or protein. However, treatment with RNase and protease did not prevent transformation, indicating that RNA and protein were not the genetic material. In contrast, treatment with DNase blocked transformation, confirming that DNA is the genetic material.

E4. A. There are several possible explanations why about 35% of the DNA is in the supernatant. One possibility is that not all of the DNA was injected into the bacterial cells. Alternatively, some of the cells may have been broken during the shearing procedure, thereby releasing the DNA.

B. If the radioactivity in the pellet had been counted instead of the supernatant, the following figure would be produced:

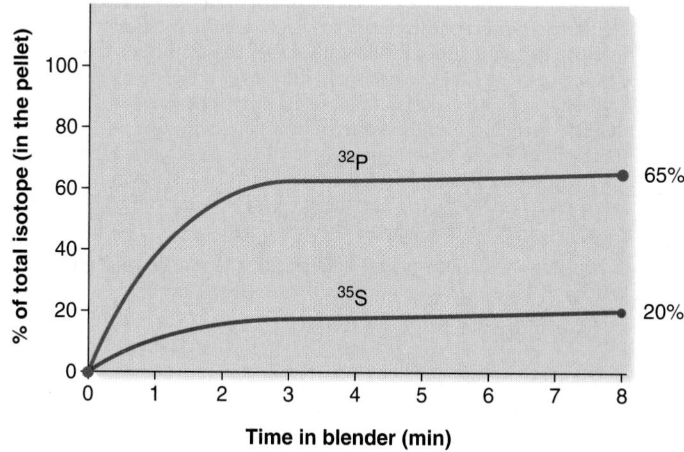

C. ^{32}P and ^{35}S were chosen as radioisotopes to label the phages because phosphorous is found in nucleic acids, while sulfur is found only in proteins.

D. There are multiple reasons why less than 100% of the phage protein is removed from the bacterial cells during the shearing process. For example, perhaps the shearing just is not strong enough to remove all of the phages, or perhaps the tail fibers remain embedded in the bacterium and only the head region is sheared off.

E6. This is really a matter of opinion. The Avery, MacLeod, and McCarty experiment seems to indicate directly that DNA is the genetic material, because DNase prevented transformation and RNase and protease did not. However, one could argue that the DNA is required for the rough bacteria to take up some other contaminant in the DNA preparation. It would seem that the other contaminant would not be RNA or protein. The Hershey and Chase experiments indicate that DNA is being injected into bacteria, although quantitatively the results are not entirely convincing. Some ^{35}S-labeled protein was not sheared off, so the results do not definitely rule out the possibility that protein could be the genetic material. But the results do indicate that DNA is the more likely candidate.

E8. A. The purpose of chromatography was to separate the different types of bases.

B. It was necessary to separate the bases and determine the total amount of each type of base. In a DNA strand, all the bases are found within a single molecule, so it is difficult to measure the total amount of each type of base. When the bases are removed from the strand, each type can be purified, and then the total amount of each type of base can be measured by spectroscopy.

C. Chargaff's results would probably not be very convincing if done on a single species. The strength of his data was that all species appeared to conform to the AT/GC rule, suggesting that this is a consistent feature of DNA structure. In a single species, the observation is that A = T and G = C could occur as a matter of chance.

Questions for Student Discussion/Collaboration

2. There are many possibilities. You could use a DNA-specific chemical and show that it causes heritable mutations. Perhaps you could inject an oocyte with a piece of DNA and produce a mouse with a new trait.

CHAPTER 10

Conceptual Questions

C2. Viruses also need sequences that enable them to be replicated. These sequences are equivalent to the origins of replication found in bacterial and eukaryotic chromosomes.

C4. A bacterium with two nucleoids is similar to a diploid eukaryotic cell because it would have two copies of each gene. The bacterium is different, however, with regard to alleles. A eukaryotic cell can have two different alleles for the same gene. For example, a cell from a pea plant could be heterozygous, *Tt*, for the gene that affects height. By comparison, a bacterium with two nucleoids has two identical chromosomes. Therefore, a bacterium with two nucleoids is homozygous for its chromosomal genes. Note: As discussed in Chapter 6, a bacterium can contain another piece of DNA, called an F' factor, that can carry a few genes. The alleles on an F' factor can be different from the alleles on the bacterial chromosome.

C6. A. One loop is 40,000 bp. One base pair is 0.34 nm, which equals 0.34×10^{-3} µm. If we multiply the two together:

$$(40{,}000)(0.34 \times 10^{-3}) = 13.6\ \mu m$$

B. Circumference = πD

$$13.6\ \mu m = \pi D$$

$$D = 4.3\ \mu m$$

C. No, it is too big to fit inside of *E. coli.* Supercoiling is needed to make the loops more compact.

C8. These drugs would diminish the amount of negative supercoiling in DNA. Negative supercoiling is needed to compact the chromosomal DNA, and it also aids in strand separation. Bacteria might not be able to survive and/or transmit their chromosomes to daughter cells if their DNA was not compacted properly. Also, because negative supercoiling aids in strand separation, these drugs would make it more difficult for the DNA strands to separate. Therefore, the bacteria would have a difficult time transcribing their genes and replicating their DNA, because both processes require strand separation. As discussed in Chapter 11, DNA replication is needed to make new copies of the genetic material to transmit from mother to daughter cells. If DNA replication was inhibited, the bacteria could not grow and divide into new daughter cells. As discussed in Chapters 12–14, gene transcription is necessary for bacterial cells to make proteins. If gene transcription was inhibited, the bacteria could not make many proteins that are necessary for survival.

C10.

Gyrase
Negative supercoil

C12. The centromere is the attachment site for the kinetochore, which attaches to the spindle. If a chromosome is not attached to the spindle, it is free to "float around" within the cell, and it may not be near a pole when the nuclear membrane re-forms during telophase. If a chromosome is left outside of the nucleus, it is degraded during interphase. That is why the chromosome without a centromere may not be found in daughter cells.

C14. Highly repetitive DNA, as its name suggests, is a DNA sequence that is repeated many times, from tens of thousands to millions of times throughout the genome. It can be interspersed in the genome or found clustered in a tandem array, in which a short nucleotide sequence is repeated many times in a row. In DNA renaturation studies, highly repetitive DNA renatures at a much faster rate because there are many copies of the complementary sequences.

C16. During interphase (i.e., G_1, S, and G_2), the euchromatin is found primarily as a 30 nm fiber in a radial loop configuration. Most interphase chromosomes also have some heterochromatic regions where the radial loops are more highly compacted. During M phase, each chromosome becomes entirely heterochromatic, which is needed for the proper sorting of the chromosomes during nuclear division.

C18.

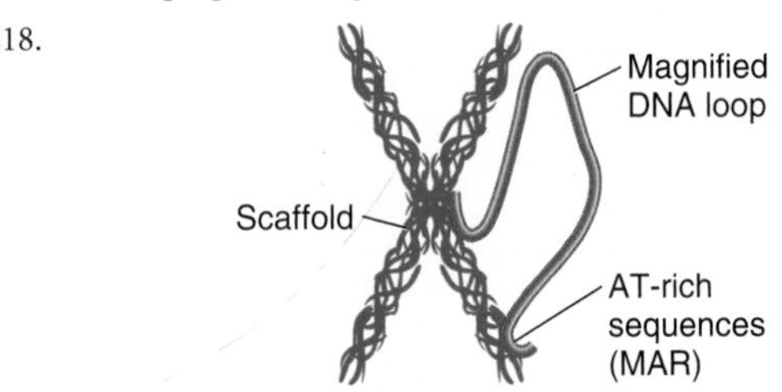

C20. During interphase, the chromosomes are found within the cell nucleus. They are less tightly packed and are transcriptionally active. Segments of chromosomes are anchored to the nuclear matrix. During M phase, the chromosomes become highly condensed, and the nuclear membrane is fragmented into vesicles. The chromosomal DNA remains anchored to a scaffold, formed from the nuclear matrix. The chromosomes eventually become attached to the spindle apparatus via microtubules that attach to the kinetochore, which is attached to the centromere.

C22. There are 146 bp around the core histones. If the linker region is 54 bp, we expect 200 bp of DNA (i.e., 146 + 54) for each nucleosome and linker region. If we divide 46,000 bp by 200 bp, we get 230. Because there are two molecules of H2A for each nucleosome, there would be 460 molecules of H2A in a 46,000-bp sample of DNA.

C24. The role of the core histones is to form the nucleosomes. In a nucleosome, the DNA is wrapped 1.65 times around the core histones. Histone H1 binds to the linker region. It may play a role in compacting the DNA into a 30 nm fiber.

C26. The answer is B and E. A Barr body is composed of a type of highly compacted chromatin called heterochromatin. Euchromatin is not so compacted. A Barr body is not composed of euchromatin. A Barr body is one chromosome, the X chromosome. The term genome refers to all the types of chromosomes that make up the genetic composition of an individual.

C28. First, they may directly influence interactions between nucleosomes. Second, histone modifications provide binding sites that are recognized by proteins. The pattern of covalent modifications of amino terminal tails acts much like a code in specifying alterations in chromatin structure. One pattern of histone modification may attract proteins that cause the chromatin to become even more compact, which would silence the transcription of genes in the region. Alternatively, a different combination of histone modifications may attract proteins that serve to loosen the chromatin and thereby promote gene transcription.

Experimental Questions

E2. This type of experiment gives the relative proportions of highly repetitive, moderately repetitive, and unique DNA sequences within the genome. The highly repetitive sequences renature at a fast rate, the moderately repetitive sequences renature at an intermediate rate, and the unique sequences renature at a slow rate.

E4. Supercoiled DNA would look curled up into a relatively compact structure. You could add different purified topoisomerases and see how they affect the structure via microscopy. For example, DNA gyrase relaxes positive supercoils, and topoisomerase I relaxes negative supercoils. If you added topoisomerase I to a DNA preparation and it became less compacted, then the DNA was negatively supercoiled.

E6.

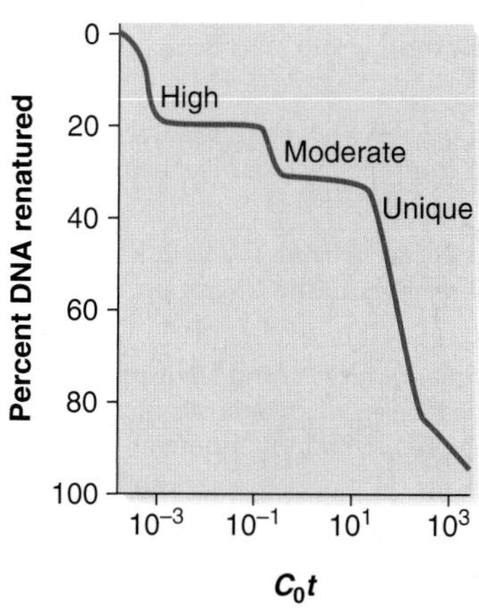

E8. You would get DNA fragments of about 446 to 496 bp (i.e., 146 bp plus 300 to 350 bp).

E10. Histones are positively charged, and DNA is negatively charged. They bind to each other by these ionic interactions. Salt is composed of positively charged ions and negatively charged ions. For example, when dissolved in water, NaCl becomes individual ions of Na^+ and Cl^-. When chromatin is exposed to a salt such as NaCl, the positively charged Na^+ could bind to the DNA, and the negatively charged Cl^- could bind to the histones. This would prevent the histones and DNA from binding to each other.

Questions for Student Discussion/Collaboration

2. This is a matter of opinion. It seems strange to have so much DNA that seems to have no obvious function. It's a waste of energy. Perhaps it has a function that we don't know about yet. On the other hand, evolution does allow bad things to accumulate within genomes, such as genes that cause diseases, etc. Perhaps this is just another example of the negative consequences of evolution.

CHAPTER 11

Conceptual Questions

C2. Bidirectional replication refers to DNA replication in both directions starting from one origin.

C4. A.

```
TTGGHTGUTGG
HHUUTHUGHUU
```

B.

```
TTGGHTGUTGG
HHUUTHUGHUU
      ↓
TTGGHTGUTGG  CCAAACACCAA
AACCCACAACC  HHUUTHUGHUU
      ↓
TTGGHTGUTGG  TTGGGTGTTGG  CCAAACACCAA  CCAAACACCAA
AACCCACAACC  AACCCACAACC  GGTTTGTGGTT  HHUUTHUGHUU
```

C6. Let's assume there are 4,600,000 bp of DNA and that DNA replication is bidirectional at a rate of 750 nucleotides per second.

If there were just a single replication fork,

4,600,000/750 = 6,133 seconds, or 102.2 minutes.

Because replication is bidirectional, $^{102.2}/_2$ = 51.1 minutes.

Actually, this is an average value based on a variety of growth conditions. Under optimal growth conditions, replication can occur substantially faster.

With regard to errors, if we assume an error rate of one mistake per 100,000,000 nucleotides,

4,600,000 × 1,000 bacteria = 4,600,000,000 nucleotides of replicated DNA.

4,600,000,000/100,000,000 = 46 mistakes.

When you think about it, this is pretty amazing. In this population, DNA polymerase would cause only 46 single mistakes in a total of 1,000 bacteria, each containing 4.6 million bp of DNA.

C8. DNA polymerase would slide from right to left. The new strand would be 3'–CTAGGGCTAGGCGTATGTAAATGGTCTAGTGGTGG–5'

C10. A. When looking at Figure 11.5, the first, second, and fourth DnaA boxes are running in the same direction, and the third and fifth are running in the opposite direction. Once you realize that, you can see the sequences are very similar to each other.

B. According to the direction of the first DnaA box, the consensus sequence is

```
TGTGGATAA
ACACCTATT
```

C. This sequence is nine nucleotides long. Because there are four kinds of nucleotides (i.e., A, T, G, and C), the chance of this sequence occurring by random chance is 4^{-9}, which equals once every 262,144 nucleotides. Because the *E. coli* chromosome is more than 10 times longer than this, it is fairly likely that this consensus sequence occurs elsewhere. The reason why there are not multiple origins, however, is because the origin has five copies of the consensus sequence very close together. The chance of having five copies of this consensus sequence occurring close together (as a matter of random chance) is very small.

C12. 1. According to the AT/GC rule, a pyrimidine always hydrogen bonds with a purine. A transition still involves a pyrimidine hydrogen bonding to a purine, but a transversion causes a purine to hydrogen bond with a purine or a pyrimidine to hydrogen bond with a pyrimidine. The structure of the double helix makes it much more difficult for this latter type of hydrogen bonding to occur.

2. The induced-fit phenomenon of the active site of DNA polymerase makes it unlikely for DNA polymerase to catalyze covalent bond formation if the wrong nucleotide is bound to the template strand. A transition mutation creates a somewhat bad interaction between the bases in opposite strands, but it is not as bad as the fit caused

by a transversion mutation. In a transversion, a purine is opposite another purine, or a pyrimidine is opposite a pyrimidine. This is a very bad fit.

3. The proofreading function of DNA polymerase is able to detect and remove an incorrect nucleotide that has been incorporated into the growing strand. A transversion is going to cause a larger distortion in the structure of the double helix and make it more likely to be detected by the proofreading function of DNA polymerase.

C14. Primase and DNA polymerase are able to knock the single-strand binding proteins off the template DNA.

C16. A. The right Okazaki fragment was made first. It is farthest away from the replication fork. The fork (not seen in this diagram) would be to the left of the three Okazaki fragments, and moving from right to left.

B. The RNA primer in the right Okazaki fragment would be removed first. DNA polymerase would begin by elongating the DNA strand of the middle Okazaki fragment and removing the right RNA primer with its 5' to 3' exonuclease activity. DNA polymerase I would use the 3' end of the DNA of the middle Okazaki fragment as a primer to synthesize DNA in the region where the right RNA primer is removed. If the middle fragment was not present, DNA polymerase could not fill in this DNA (because it needs a primer).

C. You need DNA ligase only at the right arrow. DNA polymerase I begins at the end of the left Okazaki fragment and synthesizes DNA to fill in the region as it removes the middle RNA primer. At the left arrow, DNA polymerase I is simply extending the length of the left Okazaki fragment. No ligase is needed here. When DNA polymerase I has extended the left Okazaki fragment through the entire region where the RNA primer has been removed, it hits the DNA of the middle Okazaki fragment. This occurs at the right arrow. At this point, the DNA of the middle Okazaki fragment has a 5' end that is a monophosphate. DNA ligase is needed to connect this monophosphate with the 3' end of the region where the middle RNA primer has been removed.

D. As mentioned in the answer to part C, the 5' end of the DNA in the middle Okazaki fragment is a monophosphate. It is a monophosphate because it was previously connected to the RNA primer by a phosphoester bond. At the location of the right arrow, there was only one phosphate connecting this deoxyribonucleotide to the last ribonucleotide in the RNA primer. For DNA polymerase to function, the energy to connect two nucleotides comes from the hydrolysis of the incoming triphosphate. In this location shown at the right arrow, however, the nucleotide is already present at the 5' end of the DNA, and it is a monophosphate. DNA ligase needs energy to connect this nucleotide with the left Okazaki fragment. It obtains energy from the hydrolysis of ATP or NAD^+.

C18. 1. It recognizes the origin of replication.

2. It initiates the formation of a replication bubble.

3. It recruits helicase to the region.

C20. The picture would depict a ring of helicase proteins traveling along a DNA strand and separating the two helices, as shown in Figure 11.6.

C22. The leading strand is primed once, at the origin, and then DNA polymerase III synthesizes DNA continuously in the direction of the replication fork. In the lagging strand, many short pieces of DNA (Okazaki fragments) are made. This requires many RNA primers. The primers are removed by DNA polymerase I, which then fills in the gaps with DNA. DNA ligase then covalently connects the Okazaki fragments. Having the enzymes within a complex such as a primosome or replisome provides coordination among the different steps in the replication process and thereby allows it to proceed faster and more efficiently.

C24. A processive enzyme is one that remains clamped to one of its substrates. In the case of DNA polymerase, it remains clamped to the template strand as it makes a new daughter strand. This is important to ensure a fast rate of DNA synthesis.

C26. The reason is the inability to synthesize DNA in the 3' to 5' direction and the need for a primer prevent replication at the 3' end of the DNA strands. Telomerase is different than DNA polymerase in that it uses a short RNA sequence, which is part of its structure, as a template for DNA synthesis. Because it uses this sequence many times in row, it produces a tandemly repeated sequence in the telomere at the 3' ends of linear chromosomes.

C28. Fifty, because two replication forks emanate from each origin of replication. DNA replication is bidirectional.

C30. A. Both reverse transcriptase and telomerase use an RNA template to make a complementary strand of DNA.

B. Because reverse transcriptase does not have a proofreading function, it is more likely for mistakes to occur. This creates many mutant strains of the virus. Some mutations might prevent the virus from proliferating. However, other mutations might prevent the immune system from battling the virus. These kinds of mutations would enhance the proliferation of the virus.

Experimental Questions

E2. A. You would probably still see a band of DNA, but you would only see a heavy band.

B. You would probably not see a band because the DNA would not be released from the bacteria. The bacteria would sediment to the bottom of the tube.

C. You would not see a band. UV light is needed to see the DNA, which absorbs light in the UV region.

E4. If you started with single-stranded DNA, you would need to add a primer (or primase), dNTPs, and DNA polymerase. If you started with double-stranded DNA, you would also need helicase. Adding single-strand binding protein and topoisomerase may also help.

E6. This is a critical step because you need to separate the radioactivity in the free nucleotides from the radioactivity in the newly made DNA strands. If you used an acid that precipitated free nucleotides and DNA strands, all of the radioactivity would be in the pellet. You would get the same amount of radioactivity in the pellet no matter how much DNA was synthesized into newly made strands. For this reason, the perchloric acid step is very critical. It separates radioactivity in the free nucleotides from radioactivity in the newly made strands.

E8. A. The left end is the 5' end. If you flip the sequence of the first primer around, you will notice it is complementary to the right end of the template DNA. The 5' end of the first primer binds to the 3' end of the template DNA.

B. The sequence would be 3'–CGGGGCCATG–5'. It could not be used because the 3' end of the primer is at the end of the template DNA. There wouldn't be any place for nucleotides to be added to the 3' end of the primer and bind to the template DNA strand.

E10. A. Heat is used to separate the DNA strands, so you do not need helicase.

B. Each primer must be a sequence that is complementary to one of the DNA strands. There are two types of primers, and each type binds to one of the two complementary strands.

C. A thermophilic DNA polymerase is used because DNA polymerases isolated from nonthermophilic species would be permanently inactivated during the heating phase of the PCR cycle. Remember that DNA polymerase is a protein, and most proteins are denatured by heating. However, proteins from thermophilic organisms have evolved to withstand heat, which is how thermophilic organisms survive at high temperatures.

D. With each cycle, the amount of DNA is doubled. Because there are initially 10 copies of the DNA, there will be 10×2^{27} copies after 27 cycles. $10 \times 2^{27} = 1.34 \times 10^9 = 1.34$ billion copies of DNA. As you can see, PCR can amplify the amount of DNA by a staggering amount!

Questions for Student Discussion/Collaboration

2. Basically, the idea is to add certain combinations of enzymes and see what happens. You could add helicase and primase to double-stranded DNA, along with radiolabeled ribonucleotides. Under these conditions, you would make short (radiolabeled) primers. If you also added DNA polymerase and deoxyribonucleotides, you would make DNA strands. Alternatively, if you added DNA polymerase plus an RNA primer, you wouldn't need to add primase. Or, if you used single-stranded DNA as a template rather than double-stranded DNA, you wouldn't need to add helicase.

CHAPTER 12

Conceptual Questions

C2. The release of sigma factor marks the transition to the elongation stage of transcription.

C4. GGCATTGTCA

C6. The most highly conserved positions are the first, second, and sixth. In general, when promoter sequences are conserved, they are more likely to be important for binding. That explains why changes are not found at these positions; if a mutation altered a conserved position, the promoter would probably not work very well. By comparison, changes are occasionally tolerated at the fourth position and frequently at the third and fifth positions. The positions that tolerate changes are less important for binding by sigma factor.

C8. This will not affect transcription. However, it will affect translation by preventing the initiation of polypeptide synthesis.

C10. Sigma factor can slide along the major groove of the DNA. In this way, it is able to recognize base sequences that are exposed in the groove. When it encounters a promoter sequence, hydrogen bonding between the bases and the sigma factor protein can promote a tight and specific interaction.

C12. DNA-G/RNA-C

DNA-C/RNA-G

DNA-A/RNA-U

DNA-T/RNA-A

The template strand is 3'–CCGTACGTAATGCCGTAGTGTGATCCCTAG–5' and the coding strand is 5'–GGCATGCATTACGGCATCACACTAGGGATC–3'. The promoter would be to the left (in the 3' direction) of the template strand.

C14. Transcriptional termination occurs when the hydrogen bonding is broken between the DNA and the part of the newly made RNA transcript that is located in the open complex.

C16. DNA helicase and ρ protein bind to a nucleic acid strand and travel in the 5' to 3' direction. When they encounter a double-stranded region, they break the hydrogen bonds between complementary strands. ρ protein is different from DNA helicase in that it moves along an RNA strand, while DNA helicase moves along a DNA strand. The purpose of DNA helicase function is to promote DNA replication; the purpose of ρ protein function is to promote transcriptional termination.

C18. A. Mutations that alter the uracil-rich region by introducing guanines and cytosines, and mutations that prevent the formation of the stem-loop structure.

B. Mutations that alter the termination sequence, and mutations that alter the ρ recognition site

C. Eventually, somewhere downstream from the gene, another transcriptional termination sequence would be found, and transcription would terminate there. This second termination sequence might be found randomly, or it might be at the end of an adjacent gene.

C20. Eukaryotic promoters are somewhat variable with regard to the pattern of sequence elements that may be found. In the case of structural genes that are transcribed by RNA polymerase II, it is common to have a TATA box, which is about 25 bp upstream from a transcriptional start site. The TATA box is important in the identification of the transcriptional start site and the assembly of RNA polymerase and various transcription factors. The transcriptional start site defines where transcription actually begins.

C22. It is primarily an accessibility problem. When the DNA is tightly wound around histones, it becomes difficult for large proteins, like RNA polymerase and transcription factors, to recognize the correct base sequence in the DNA and to catalyze the movement of the open complex. It is thought that a loosening or complete disruption of the nucleosome structure is necessary for transcription to occur.

C24. Hydrogen bonding is usually the predominant type of interaction when proteins and DNA follow an assembly and disassembly process. In addition, ionic bonding and hydrophobic interactions could occur. Covalent interactions would not occur. High temperature and high salt concentrations tend to break hydrogen bonds. Therefore, high temperature and high salt would inhibit assembly and stimulate disassembly.

C26. In bacteria, the 5' end of the tRNA is cleaved by RNase P. The 3' end is cleaved by a different endonuclease, and then a few nucleotides are digested away by an exonuclease that removes nucleotides until it reaches a CCA sequence.

C28. A ribozyme is an enzyme whose catalytic part is composed of RNA. Examples are RNase P and self-splicing group I and II introns. It is thought that the spliceosome may contain catalytic RNAs as well.

C30. Self-splicing means that an RNA molecule can splice itself without the aid of a protein. Group I and II introns can be self-splicing, although proteins can also enhance the rate of processing.

C32. In alternative splicing, variation occurs in the pattern of splicing, so the resulting mRNAs contain alternative combinations of exons. The biological significance is that two or more different proteins can be produced from a single gene. This is a more efficient use of the genetic material. In multicellular organisms, alternative splicing is often used in a cell-specific manner.

C34. As shown at the left side of Figure 12.20, the guanosine, which binds to the guanosine-binding site, does not have a phosphate group attached to it. This guanosine is the nucleoside that winds up at the 5' end of the intron. Therefore, the intron does not have a phosphate group at its 5' end.

C36. U5

Experimental Questions

E2. An R loop is a loop of DNA that occurs when RNA is hybridized to double-stranded DNA. While the RNA is hydrogen bonding to one of the DNA strands, the other strand does not have a partner to hydrogen bond with, so it bubbles out as a loop. RNA is complementary to the template strand, so that is the strand it binds to.

E4. The 1,100-nucleotide band would be observed from a normal individual (lane 1). A deletion that removed the −50 to −100 region would greatly diminish transcription, so the homozygote would produce hardly any of the transcript (just a faint amount, as shown in lane 2), and the heterozygote would produce roughly half as much of the 1,100-nucleotide transcript (lane 3) compared to a normal individual. A nonsense codon would not have an effect on transcription; it affects only translation. So the individual with this mutation would produce a normal amount of the 1,100-nucleotide transcript (lane 4). A mutation that removed the splice acceptor site would prevent splicing. Therefore, this individual would produce a 1,550-nucleotide transcript (actually, 1,547 to be precise, 1,550 minus 3). The Northern blot is shown here:

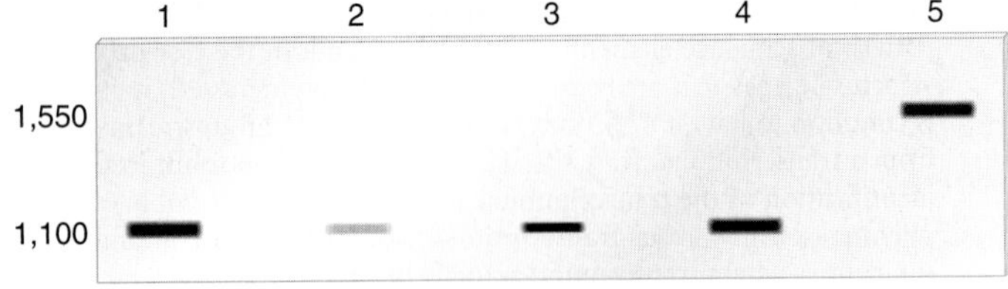

E6. A. It would not be retarded because ρ protein would not bind to the mRNA encoded by a gene that is terminated in a ρ-independent manner. The mRNA from such genes does not contain the sequence near the 3' end that acts as a recognition site for the binding of ρ protein.

B. It would be retarded because ρ protein would bind to the mRNA.

C. It would be retarded because U1 would bind to the pre-mRNA.

D. It would not be retarded because U1 would not bind to mRNA that has already had its introns removed. U1 binds only to pre-mRNA.

E8. A. mRNA molecules would bind to this column because they have a polyA tail. The string of adenine nucleotides in the polyA tail is complementary to stretch of thymine in the poly-dT column, so the two would hydrogen bond to each other. To purify mRNAs, one begins with a sample of cells; the cells need to be broken open by some technique such as homogenization or sonication. This would release the RNAs and other cellular macromolecules. The large cellular structures (organelles, membranes, etc.) could be removed from the cell extract by a centrifugation step. The large cellular structures would be found in the pellet, while soluble molecules such as RNA and proteins would stay in the supernatant. At this point, you would want the supernatant to contain a high salt concentration and neutral pH. The supernatant would then be poured over the poly-dT column. The mRNAs would bind to the poly-dT column and other molecules (i.e., other types of RNAs and proteins) would flow through the column. Because the mRNAs would bind to the poly-dT column via hydrogen bonds, to break the bonds, you could add a solution that contains a low salt concentration and/or a high pH. This would release the mRNAs, which would then be collected in a low salt/high pH solution as it dripped from the column.

B. The basic strategy is to attach a short stretch of DNA nucleotides to the column matrix that is complementary to the type of RNA that you want to purify. For example, if an rRNA contained a sequence 5'–AUUCCUCCA–3', a researcher could chemically synthesize an oligonucleotide with the sequence 3'–TAAGGAGGT–5' and attach it to the column matrix. To purify rRNA, one would use this 3'–TAAGGAGGT–5' column and follow the general strategy described in part A.

Questions for Student Discussion/Collaboration

2. RNA transcripts come in two basic types: those that function as RNA (e.g., tRNA, rRNA, etc.) versus those that are translated (i.e., mRNA). As described in this chapter, they play a myriad of functional roles. RNAs that form complexes with proteins carry out some interesting roles. In some cases, the role is to bind other types of RNA molecules. For example, rRNA in bacteria plays a role in binding mRNA. In other cases, the RNA plays a catalytic role. An example is RNaseP. The structure and function of RNA molecules may be enhanced by forming a complex with proteins.

CHAPTER 13

Conceptual Questions

C2. When we say the genetic code is degenerate, it means that more than one codon can specify the same amino acid. For example, GGG, GGC, GGA, and GGU all specify glycine. In general, the genetic code is nearly universal, because it is used in the same way by viruses, prokaryotes, fungi, plants, and animals. As discussed in Table 13.3, there are a few exceptions, which occur primarily in protists and yeast and mammalian mitochondria.

C4. A. This mutant tRNA would recognize glycine codons in the mRNA but would put in tryptophan amino acids where glycine amino acids are supposed to be in the polypeptide chain.

B. This mutation tells us that the aminoacyl-tRNA synthetase is primarily recognizing other regions of the tRNA molecule besides the anticodon region. In other words, tryptophanyl-tRNA synthetase (the aminoacyl-tRNA synthetase that attaches tryptophan) primarily recognizes other regions of the $tRNA^{trp}$ sequence (that is, other than the anticodon region), such as the T- and D-loops. If aminoacyl-tRNA synthetases recognized only the anticodon region, we would expect glycyl-tRNA synthetase to recognize this mutant tRNA and attach glycine. That is not what happens.

C6. A. The answer is three. There are six leucine codons: UUA, UUG, CUU, CUC, CUA, and CUG. The anticodon AAU would recognize UUA and UUG. You would need two other tRNAs to efficiently recognize the other four leucine codons. These could be GAG and GAU or GAA and GAU.

B. The answer is one. There is only one codon, AUG, so you need only one tRNA with the anticodon UAC.

C. The answer is three. There are six serine codons: AGU, AGC, UCU, UCC, UCA, and UCG. You would need only one tRNA to recognize AGU and AGC. This tRNA could have the anticodon UCG or UCA. You would need two tRNAs to efficiently recognize the other four tRNAs. These could be AGG and AGU or AGA and AGU.

C8. 3'–CUU–5' or 3'–CUC–5'

C10. It can recognize 5'–GGU–3', 5'–GGC–3', and 5'–GGA–3'. All of these specify glycine.

C12. All tRNA molecules have some basic features in common. They all have a cloverleaf structure with three stem-loop structures. The second stem-loop contains the anticodon sequence that recognizes the codon sequence in mRNA. At the 3' end, there is an acceptor stem, with the sequence CCA, that serves as an attachment site for an amino acid. Most tRNAs also have base modifications that occur within their nucleotide sequences.

C14. The role of aminoacyl-tRNA synthetase is to specifically recognize tRNA molecules and attach the correct amino acid to them. This ability is sometimes described as the second genetic code because the specificity of the attachment is a critical step in deciphering the genetic code. For example, if a tRNA has a 3'–GGG–5' anticodon, it will recognize a 5'–CCC–3' codon, which should specify proline. It is essential that the aminoacyl-tRNA synthetase known as prolyl-tRNA-synthetase recognizes this tRNA and attaches proline to the 3' end. The other aminoacyl-tRNA synthetases should not recognize this tRNA.

C16. Bases that have been chemically modified can occur at various locations throughout the tRNA molecule. The significance of all of these modifications is not entirely known. However, within the anticodon region, base modification alters base pairing to allow the anticodon to recognize two or more different bases within the codon.

C18. No, it is not. Due to the wobble rules, the 5' base in the anticodon of a tRNA can recognize two or more bases in the third (3') position of the mRNA. Therefore, any given cell type synthesizes far fewer than 61 types of tRNAs.

C20. The assembly process is very complex at the molecular level. In eukaryotes, 33 proteins and one rRNA assemble to form a 40S subunit, and 49 proteins and three rRNAs assemble to form a 60S subunit. This assembly occurs within the nucleolus.

C22. A. On the surface of the 30S subunit and at the interface between the two subunits

B. Within the 50S subunit

C. From the 50S subunit

D. To the 30S subunit

C24. Most bacterial mRNAs contain a Shine-Dalgarno sequence, which is necessary for the binding of the mRNA to the small ribosomal subunit. This sequence, UUAGGAGGU, is complementary to a sequence in the 16S rRNA. Due to this complementarity, these sequences will hydrogen bond to each other during the initiation stage of translation.

C26. The ribosome binds at the 5' end of the mRNA and then scans in the 3' direction in search of an AUG start codon. If it finds one that reasonably obeys Kozak's rules, it will begin translation at that site. Aside from an AUG start codon, two other important features are a guanosine at the +4 position and a purine at the −3 position.

C28. The A (aminoacyl) site is the location where a tRNA carrying a single amino acid initially binds. The only exception is the initiator tRNA, which binds to the P (peptidyl) site. The growing polypeptide chain is removed from the tRNA in the P site and transferred to the amino acid attached to the tRNA in the A site. The ribosome translocates in the 3' direction, with the result that the two tRNAs in the P and A sites are moved to the E (exit) and P sites, and the uncharged tRNA in the E site is released.

C30. Sorting signals direct a protein to its correct location (i.e., compartment) within the cell. Proteins destined for the ER, Golgi complex, lysosomes, plasma membrane, or secretory vesicles have an SRP sorting signal at their amino terminal end. Nuclear proteins have a nuclear-localization sequence (NLS), and so on. These sorting signals are recognized by specific cellular components that sort the proteins to their correct destination.

C32. The initiation phase involves the binding of the Shine-Dalgarno sequence to the rRNA in the 30S subunit. The elongation phase involves the binding of anticodons in tRNA to codons in mRNA.

C34. A. The E site and P sites. (Note: A tRNA without an amino acid attached is only briefly found in the P site, just before translocation occurs.)

B. P site and A site. (Note: A tRNA with a polypeptide chain attached is only briefly found in the A site, just before translocation occurs.)

C. Usually the A site, except the initiator tRNA, which can be found in the P site.

C36. The tRNAs bind to the mRNA because their anticodon and codon sequences are complementary. When the ribosome translocates in the 5' to 3' direction, the tRNAs remain bound to their complementary codons, and the two tRNAs shift from the A site and P site to the P site and E site. If the ribosome moved in the 3' direction, it would have to dislodge the tRNAs and drag them to a new position where they would not (necessarily) be complementary to the mRNA.

C38. 52

Experimental Questions

E2. A. There could have been other choices, but this template would be predicted to contain a cysteine codon, UGU, but would not contain any alanine codons.

B. You do not want to use ^{35}S because the radiolabel would be removed during treatment with Raney nickel.

C. There would not be a significant amount of radioactivity incorporated into newly made polypeptides with or without Raney nickel treatment. The only radiolabeled amino acid in this experiment was cysteine, which became attached to $tRNA^{cys}$. When exposed to Raney nickel, these cysteines were converted to alanine but only after they were already attached to $tRNA^{cys}$. If there were not any cysteine codons in the mRNA template, the $tRNA^{cys}$ would not recognize this mRNA. Therefore, we would not expect to see much radioactivity in the newly made polypeptides.

E4. The initiation phase of translation is very different in bacteria and in eukaryotes, so they would not be translated very efficiently. A bacterial mRNA would not be translated very efficiently in a eukaryotic translation system, because it lacks a cap structure attached to its 5' end. A eukaryotic mRNA would not have a Shine-Dalgarno sequence near its 5' end, so it would not be translated very efficiently in a bacterial translation system.

E6. Looking at the figure, the 5' end of the template DNA strand is toward the right side. The 5' ends of the mRNAs are farthest from the DNA, and the 3' ends of the mRNAs are closest to the DNA. The start codons are slightly downstream from the 5' ends of the RNAs.

E8.

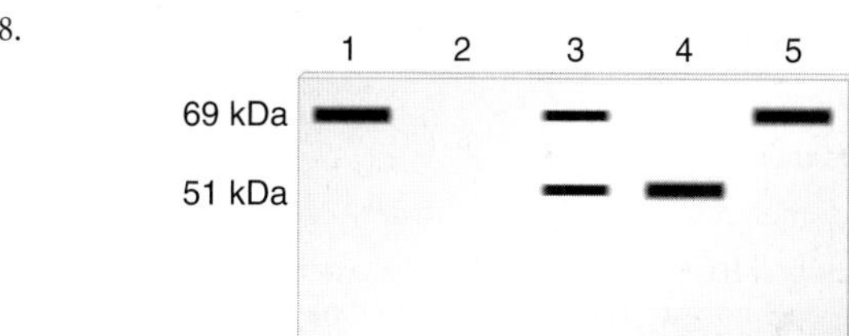

E10. A. If codon usage were significantly different between kangaroo and yeast cells, this would inhibit the translation process. For example, if the preferred leucine codon in kangaroos was CUU, translation would probably be slow in a yeast translation system. We would expect the cell-free translation system from yeast cells to primarily contain leucine tRNAs with an anticodon sequence that is AAC, because this $tRNA^{leu}$ would match the preferred yeast leucine codon, which is UUG. In a yeast translation system, there probably would not be a large amount of tRNA with an anticodon of GAA, which would match the preferred leucine codon, CUU, of kangaroos. For this reason, kangaroo mRNA would not be translated very well in a yeast translation system, but it probably would be translated to some degree.

B. The advantage of codon bias is that a cell can rely on a smaller population of tRNA molecules to efficiently translate its proteins. A disadvantage is that mutations, which do not change the amino acid sequence but do change a codon (e.g., UUG to UUA), may inhibit the production of a polypeptide if a preferred codon is changed to a nonpreferred codon.

Questions for Student Discussion/Collaboration

2. This could be a very long list. There are similarities along several lines:

1. There is a lot of molecular recognition going on, either between two nucleic acid molecules or between proteins and nucleic acid molecules. Students may see these as similarities or differences, depending on their point of view.

2. There is biosynthesis going on in both processes. Small building blocks are being connected together. This requires an input of energy.

3. There are genetic signals that determine the beginning and ending of these processes.

There are also many differences:

1. Transcription produces an RNA molecule with a similar structure to the DNA, whereas translation produces a polypeptide with a structure that is very different from RNA.

2. Depending on your point of view, it seems that translation is more biochemically complex, requiring more proteins and RNA molecules to accomplish the task.

CHAPTER 14

Conceptual Questions

C2. In bacteria, gene regulation greatly enhances the efficiency of cell growth. It takes a lot of energy to transcribe and translate genes. Therefore, a cell is much more efficient and better at competing in its environment if it expresses genes only when the gene product is needed. For example, a bacterium will express only the genes that are necessary for lactose metabolism when a bacterium is exposed to lactose. When the environment is missing lactose, these genes are turned off. Similarly, when tryptophan levels are high within the cytoplasm, the genes required for tryptophan biosynthesis are repressed.

C4. A. Regulatory protein
B. Effector molecule
C. DNA segment
D. Effector molecule
E. Regulatory protein
F. DNA segment
G. Effector molecule

C6. A mutation that has a *cis*-effect is within a genetic regulatory sequence, such as an operator site, that affects the binding of a genetic regulatory protein. A *cis*-effect mutation affects only the adjacent genes that the genetic regulatory sequence controls. A mutation having a *trans*-effect is usually in a gene that encodes a genetic regulatory protein. A *trans*-effect mutation can be complemented in a merozygote experiment by the introduction of a normal gene that encodes the regulatory protein.

C8. A. No transcription would take place. The *lac* operon could not be expressed.
B. No regulation would take place. The operon would be continuously turned on.
C. The rest of the operon would function normally, but none of the transacetylase would be made.

C10. Diauxic growth refers to the phenomenon in which a cell first uses up one type of sugar (such as glucose) before it begins to metabolize a second sugar (such as lactose). In this case, it is caused by gene regulation. When a bacterial cell is exposed to both sugars, the uptake of glucose causes the cAMP levels in the cell to fall. When this occurs, the catabolite activator protein (CAP) is removed from the *lac* operon, so it is not able to be activated by CAP.

C12. A mutation that prevented the lac repressor from binding to the operator would make the *lac* operon constitutive only in the absence of glucose. However, this mutation would not be entirely constitutive because transcription would be inhibited in the presence of glucose. The disadvantage of constitutive expression of the *lac* operon is that the bacterial cell would waste a lot of energy transcribing the genes and translating the mRNA when lactose was not present.

C14. A. Without *ara*O_2, the repression of the *ara* operon could not occur. The operon would be constitutively expressed at high levels because AraC protein could still activate transcription of the *ara* operon by binding to *araI*. The presence of arabinose would have no effect. Note: The binding of arabinose to AraC is not needed to form an AraC dimer at *araI*. The dimer is able to form because the loop has been broken. This point may be figured out if you notice that an AraC dimer is bound to *ara*O_1 in the presence and absence of arabinose (see Figure 14.12).
B. Without *ara*O_1, the AraC protein would be overexpressed. It would probably require more arabinose to alleviate repression. In addition, activation might be higher because there would be more AraC protein available.
C. Without *araI*, transcription of the *ara* operon cannot be activated. You might get a very low level of constitutive transcription.
D. Without *ara*O_2, the repression of the *ara* operon could not occur. However, without *araI*, transcription of the *ara* operon cannot be activated. You might get a very low level of constitutive transcription.

C16. A. Attenuation will not occur because loop 2–3 will form.
B. Attenuation will occur because 2–3 cannot form, so 3–4 will form.
C. Attenuation will not occur because 3–4 cannot form.
D. Attenuation will not occur because 3–4 cannot form.

C18. The addition of Gs and Cs into the U-rich sequence would prevent attenuation. The U-rich sequence promotes the dissociation of the mRNA from the DNA, when the terminator stem-loop forms. This causes RNA polymerase to dissociate from the DNA and thereby causes transcriptional termination. The UGGUUGUC sequence would probably not dissociate because of the Gs and Cs. Remember that GC base pairs have three hydrogen bonds and are more stable than AU base pairs, which have only two hydrogen bonds.

C20. It takes a lot of cellular energy to translate mRNA into a protein. A cell wastes less energy if it prevents the initiation of translation rather than a later stage such as elongation or termination.

C22. One mechanism is that histidine could act as corepressor that shuts down the transcription of the histidine synthetase gene. A second mechanism would be that histidine could act as an inhibitor via feedback inhibition. A third possibility is that histidine inhibits the ability of the mRNA encoding histidine synthetase to be translated. Perhaps it induces a gene that encodes an antisense RNA. If the amount of histidine synthetase protein was identical in the presence and absence of extracellular histidine, a feedback inhibition mechanism is favored, because this affects only the activity of the histidine synthetase enzyme, not the amount of the enzyme. The other two mechanisms would diminish the amount of this protein.

C24. The two proteins are similar in that both bind to a segment of DNA and repress transcription. They are different in three ways. (1) They recognize different effector molecules (i.e., the lac repressor recognizes allolactose, and the trp repressor recognizes tryptophan. (2) Allolactose causes the lac repressor to release from the operator, while tryptophan causes the trp repressor to bind to its operator. (3) The sequences of the operator sites that these two proteins recognize are different from each other. Otherwise, the lac repressor could bind to the *trp* operator, and the trp repressor could bind to the *lac* operator.

C26. In the lytic cycle, the virus directs the bacterial cell to make more virus particles until eventually the cell lyses and releases them. In the lysogenic cycle, the viral genome is incorporated into the host cell's genome as a prophage. It remains there in a dormant state until some stimulus causes it to excise itself from the bacterial chromosome and enter the lytic cycle.

C28. The O_R region contains three operator sites and two promoters. P_{RM} and P_R transcribe in opposite directions. The λ repressor will first bind to O_{R1} and then O_{R2}. The binding of the λ repressor to O_{R1} and O_{R2} inhibits transcription from P_R and thereby switches off the lytic cycle. Early in the lysogenic cycle, the λ repressor protein concentration may become so high that it will occupy O_{R3}. Later, when the λ repressor concentration begins to drop, it will first be removed from O_{R3}. This allows transcription from P_{RM} and maintains the lysogenic cycle. By comparison, the cro protein has its highest affinity for O_{R3}, and so it binds there first. This blocks transcription from P_{RM} and thereby switches off the lysogenic cycle. The cro protein has a similar affinity for O_{R2} and O_{R1}, and so it may occupy either of these sites next. It will bind to both O_{R2} and O_{R1}. This turns down the expression from P_R, which is not needed in the later stages of the lytic cycle.

C30. It would first increase the amount of cro protein, so the lytic cycle would be favored.

C32. Neither cycle could be followed. As shown in Figure 14.19, N protein is needed to make a longer transcript from P_L for the lysogenic cycle and also to make a longer transcript from P_R for the lytic cycle.

C34. If the F^- strain is lysogenic for phage λ, the λ repressor is already being made in that cell. If the F^- strain receives genetic material from an *Hfr* strain, you would not expect it to have an effect on the lysogenic cycle, which is already established in the F^- cell. However, if the *Hfr* strain is lysogenic for λ and the F^- strain is not, the *Hfr* strain could transfer the integrated λ DNA (i.e., the prophage) to the F^- strain. The cytoplasm of the F^- strain would not contain any λ repressor. Therefore, this λ DNA could choose between the lytic and lysogenic cycle. If it follows the lytic cycle, the F^- recipient bacterium will lyse.

Experimental Questions

E2. In samples loaded in lanes 1 and 4, we expect the repressor to bind to the operator because no lactose is present. In the sample loaded into lane 4, the CAP protein could still bind cAMP because there is no glucose. However, there really is no difference between lanes 1 and 4, so it does not look like the CAP can activate transcription when the lac repressor is bound. If we compare samples loaded into lanes 2 and 3, the lac repressor would not be bound in either case, and the CAP would not be bound in the sample loaded into lane 3. There is less transcription in lane 3 compared to lane 2, but because there is some transcription seen in lane 3, we can conclude that the removal of the CAP (because cAMP levels are low) is not entirely effective at preventing transcription. Overall, the results indicate that the binding of the lac repressor is more effective at preventing transcription of the *lac* operon compared to the removal of the CAP.

E4. A. Yes, if you do not sonicate, then β-galactosidase will not be released from the cell, and not much yellow color will be observed. (Note: You may observe a little yellow color because some β-ONPG may be taken into the cell.)

B. No, you should still get yellow color in the first two tubes even if you forgot to add lactose because the unmated strain does not have a functional lac repressor.

C. Yes, if you forgot to add β-ONPG, you could not get yellow color because the cleavage of β-ONPG by β-galactosidase is what produces the yellow color.

E6. You could mate a strain that has an F' factor carrying a normal *lac* operon and a normal *lacI* gene to this mutant strain. Because the mutation is in the operator site, you would still continue to get expression of β-galactosidase, even in the absence of lactose.

E8. In this case, things are more complex because AraC acts as a repressor and an activator protein. If AraC were missing due to mutation, there would not be repression or activation of the *ara* operon in the presence or absence of arabinose. It would be expressed constitutively at low levels. The introduction of a normal *araC* gene into the bacterium on an F' factor would restore normal regulation (i.e., a *trans*-effect).

Questions for Student Discussion/Collaboration

2. A DNA loop may inhibit transcription by preventing RNA polymerase from recognizing the promoter. Or, it may inhibit transcription by preventing the formation of the open complex. Alternatively, a bend may enhance transcription by exposing the base sequence that the sigma factor of RNA polymerase recognizes. The bend may expose the major groove in such a way that this base sequence is more accessible to binding by sigma factor and RNA polymerase.

CHAPTER 15

Conceptual Questions

C2. Regulatory elements are relatively short genetic sequences that are recognized by regulatory transcription factors. After the regulatory transcription factor has bound to the regulatory element, it will affect the rate of transcription, either activating it or repressing it, depending on the action of the regulatory protein. Regulatory elements are typically located in the upstream region near the promoter, but they can be located almost anywhere (i.e., upstream and downstream) and even quite far from the promoter.

C4. Transcriptional activation occurs when a regulatory transcription factor binds to a regulatory element and activates transcription. Such proteins, also called transactivators, may interact with TFIID and/or mediator to promote the assembly of RNA polymerase and general transcription factors at the promoter region. They also could alter the structure of chromatin so that RNA polymerase and transcription factors are able to gain access to the promoter. Transcriptional inhibition occurs when a regulatory transcription factor inhibits transcription. Such repressors also may interact with TFIID and/or mediator to inhibit RNA polymerase.

C6. A. DNA binding

B. DNA binding

C. Protein dimerization

C8. For the glucocorticoid receptor to bind to a GRE, a steroid hormone must first enter the cell. The hormone then binds to the glucocorticoid receptor, which releases HSP90. The release of HSP90 exposes a nuclear localization signal (NLS) within the receptor, which enables it to dimerize and then enter the nucleus. Once inside the nucleus, the dimer binds to a pair of GREs, which activates transcription of the adjacent genes.

C10. Phosphorylation of the CREB protein causes it to act as a transcriptional activator. The unphosphorylated CREB protein can still bind to CREs, but it does not stimulate transcription.

C12. A. Eventually, the glucocorticoid hormone will be degraded by the cell. The glucocorticoid receptor binds the hormone with a certain affinity. The binding is a reversible process. Once the concentration of the hormone falls below the affinity of the hormone for the receptor, the receptor will no longer have the glucocorticoid hormone bound to it. When the hormone is released, the glucocorticoid receptor will change its conformation, and it will no longer bind to the DNA.

B. An enzyme known as a phosphatase will eventually cleave the phosphate groups from the CREB protein. When the phosphates are removed, the CREB protein will stop activating transcription.

C14. The enhancer found in A would work, but the ones found in B and C would not. The sequence that is recognized by the transcriptional activator is 5'–GTAG–3' in one strand and 3'–CATC–5' in the opposite strand. This is the same arrangement found in A. In B and C, however, the arrangement is 5'–GATG–3' and 3'–CATC–5'. In the arrangement found in B and C, the two middle bases (i.e., A and T) are not in the correct order.

C16.

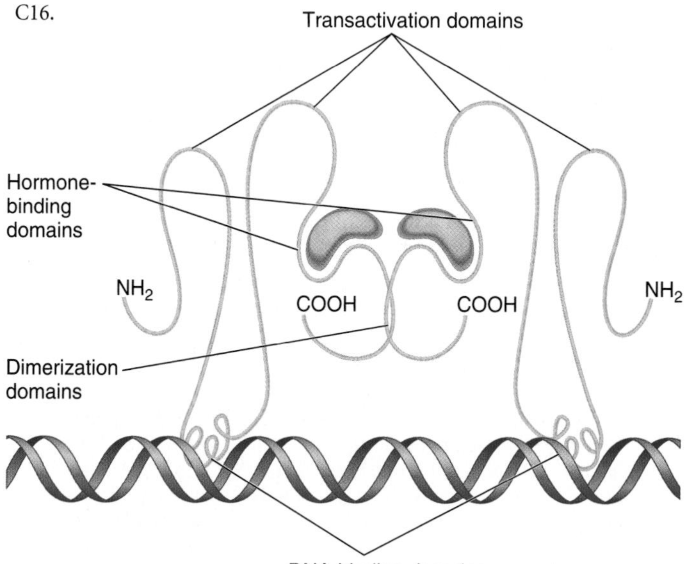

This is one hypothetical drawing of the glucocorticoid receptor. As discussed in your textbook, the glucocorticoid receptor forms a homodimer. The dimer shown here has rotational symmetry. If you flip the left side around, it has the same shape as the right side. The hormone-binding, DNA-binding, dimerization, and transactivation domains are labeled. The glucocorticoid hormone is shown in orange.

C18. The 30 nm fiber is the predominant form of chromatin during interphase. The chromatin must be converted from a closed conformation to an open conformation for transcription to take place. This "opening" process involves less packing of the chromatin and may involve changes in the locations of histone proteins. Transcriptional activators recruit histone acetyltransferase and ATP-dependent remodeling enzymes to the region, which leads to a conversion to the open conformation.

C20. The binding of the DNA to the core histones could

1. Prevent transcriptional activators from recognizing enhancer elements that are required for transcriptional activation.
2. Prevent RNA polymerase from binding to the core promoter.
3. Prevent RNA polymerase from forming an open complex in which the DNA strands separate.

C22. Perhaps the methyltransferase is responsible for methylating and inhibiting a gene that causes a cell to become a muscle cell. The methyltransferase is inactivated by the mutation.

C24. The function of splicing factors is to influence the selection of splice sites in RNA. In certain cell types, the concentration of particular splicing factors is higher than in other tissues. The high concentration of particular splicing factors, and the regulation of their activities, may promote the selection of particular splice sites and thereby lead to tissue-specific splicing.

C26. This person would be unable to make ferritin, because the IRP would always be bound to the IRE. The amount of transferrin receptor mRNA would be high, even in the presence of high amounts of iron, because the IRP would always remain bound to the IRE and stabilize the transferrin receptor mRNA. Such a person would not have any problem taking up iron into his/her cells. In fact, this person would take up a lot of iron via the transferrin receptor, even when the iron concentrations were high. Therefore, this person would not need more iron in the diet. However, excess iron in the diet would be very toxic for two reasons. First, the person cannot make ferritin, which prevents the toxic buildup of iron in the cytosol. Second, when iron levels are high, the person would continue to synthesize the transferrin receptor, which functions in the uptake of iron.

C28. A disadvantage of mRNAs with a short half-life is that the cells probably waste a lot of energy making them. If a cell needs the protein encoded by a short-lived mRNA, the cell has to keep transcribing the gene that encodes the mRNA because the mRNAs are quickly degraded. An advantage of short-lived mRNAs is that the cell can rapidly turn off protein synthesis. If a cell no longer needs the polypeptide encoded by a short-lived mRNA, it can stop transcribing the gene, and the mRNA will be quickly degraded. This will shut off the synthesis of more proteins rather quickly. With most long-lived mRNAs, it will take much longer to shut off protein synthesis after transcription has been terminated.

C30. First, the double-stranded miRNA could come from the transcription of a gene, as pre-miRNA. Second, it could come from a virus. Third, multiple insertions of the same gene in the genome may lead to both DNA strands being transcribed, thereby forming double-stranded RNA.

C32. If mRNA stability is low, this means it is degraded more rapidly. Therefore, low stability results in a low mRNA concentration. The length of the polyA tail is one factor that affects stability. A longer tail makes mRNA more stable. Certain mRNAs have sequences that affect their half-lives. For example, AU-rich elements (AREs) are found in many short-lived mRNAs. The AREs are recognized by cellular proteins that cause the mRNAs to be rapidly degraded.

Experimental Questions

E2. S1 nuclease cuts single-stranded DNA but not double-stranded DNA. If a gene is in an open conformation and cut by DNase I, it would not hybridize to the probe that is complementary to it. In this case, the single-stranded probe would be digested by the S1 nuclease. In contrast, if the gene is in a closed conformation and resistant to DNase I, the probe will hybridize to the gene and will not be digested by the S1 nuclease. Therefore, the ability of S1 nuclease to digest the probe tells you whether or not the gene is being cleaved by DNase I.

The precipitation step is needed to separate the DNA fragments from free nucleotides. If the radioactive probe is bound to a complementary DNA strand, it will be protected from degradation by S1 nuclease, and it will be found in the pellet. This occurs when the globin gene is in a closed conformation. In contrast, if the radioactive probe does not bind to a complementary DNA strand, it will be digested into free nucleotides, and the free nucleotides will remain in the supernatant. This occurs when the globin gene is in an open conformation.

E4. If the DNA band is 3,800 bp, it means the site is unmethylated, because it has been cut by *Not*I (*Not*I cannot cut methylated DNA). If the DNA fragment is 5,300 bp, the DNA is not cut by *Not*I, so we assume it is methylated. Now let's begin our interpretation of these data with lane 4. When the gene is isolated from root tissue, the DNA is not methylated because it runs at 3,800 bp. This suggests that gene *T* is expressed in root tissue. In the other samples, the DNA runs at 5,300 bp, indicating that the DNA is methylated. The pattern of methylation seen here is consistent with the known function of gene *T*. We would expect it to be expressed in root cells, because it functions in the uptake of phosphate from the soil. It would be silenced in the other parts of the plant via methylation.

E6. A. Based on the transformation of kidney cells, a silencer is in region B. Based on the transformation of pancreatic cells, an enhancer is in region A.

B. The pancreatic cells must not express the repressor protein that binds to the silencer in region B.

C. The kidney cells must express a repressor protein that binds to region B and represses transcription. Kidney cells express the downstream gene only if the silencer is removed. As mentioned in part B, this repressor is not expressed in pancreatic cells.

E8. When they injected just antisense *mex-3* RNA, they observed lower but detectable levels of *mex-3* mRNA. However, the injection of both sense and antisense RNA, which would form a double-stranded structure, resulted in a complete loss of *mex-3* mRNA in the cells. In this way, double-stranded RNA would silence the expression of the *mex-3* gene.

Questions for Student Discussion/Collaboration

2. Probably the most efficient method would be to systematically make deletions of progressively smaller sizes. For example, you could begin by deleting 20,000 bp on either side of the gene and see if that affects transcription. If you found that only the deletion on the 5' end of the gene had an effect, you could then start making deletions from the 5' end, perhaps in 10,000 bp or 5,000 bp increments until you localized response elements. You would then make smaller deletions in the putative region until it was down to a hundred or a few dozen nucleotides. At this point, you might conduct site-directed mutagenesis, as described in Chapter 18, as a way to specifically identify the regulatory element sequence.

CHAPTER 16

Conceptual Questions

C2. It is a gene mutation, a point mutation, a base substitution, a transition mutation, a deleterious mutation, a mutant allele, a nonsense mutation, a conditional mutation, and a temperature-sensitive lethal mutation.

C4. A. It would probably inhibit protein function, particularly if it was not near the end of the coding sequence.

B. It may or may not affect protein function, depending on the nature of the amino acid substitution and whether the substitution is in a critical region of the protein.

C. It would increase the amount of functional protein.

D. It may affect protein function if the alteration in splicing changes an exon in the mRNA that results in a protein with a perturbed structure.

C6. A. Not appropriate, because the second mutation is at a different codon

B. Appropriate

C. Not appropriate, because the second mutation is in the same gene as the first mutation

D. Appropriate

C8. A. Silent, because the same amino acid (glycine) is encoded by GGA and GGT

B. Missense, because a different amino acid is encoded by CGA compared to GGA

C. Missense, because a different amino acid is encoded by GTT compared to GAT

D. Frameshift, because an extra base is inserted into the sequence

C10. One possibility is that a translocation may move a gene next to a heterochromatic region of another chromosome and thereby diminish its expression, or it could be moved next to a euchromatic region and increase its expression. Another possibility is that the translocation breakpoint may move the gene next to a new promoter or regulatory sequences that may now influence the gene's expression.

C12. A. No; the position (i.e., chromosomal location) of a gene has not been altered.

B. Yes; the expression of a gene has been altered because it has been moved to a new chromosomal location.

C. Yes; the expression of a gene has been altered because it has been moved to a new chromosomal location.

C14. If a mutation within the germ line is passed to an offspring, all the cells of the offspring's body will carry the mutation. A somatic mutation affects only the somatic cell in which it originated and all of the daughter cells that the somatic cell produces. If a somatic mutation occurs early during embryonic development, it may affect a fairly large region of the organism. Because germ-line mutations affect the entire organism, they are potentially more harmful (or beneficial), but this is not always the case. Somatic mutations can cause quite harmful effects, such as cancer.

C16. A thymine dimer can interfere with DNA replication because DNA polymerase cannot slide past the dimer and add bases to the newly growing strand. Alkylating mutagens such as nitrous acid will cause DNA replication to make mistakes in the base pairing. For example, an alkylated cytosine will base-pair with adenine during DNA replication, thereby creating a mutation in the newly made strand. A third example is 5-bromouracil, which is a thymine analogue. It may base-pair with guanine instead of adenine during DNA replication.

C18. During TNRE, a trinucleotide repeat sequence gets longer. If someone was mildly affected with a TNRE disorder, he or she might be concerned that an expansion of the repeat might occur during gamete formation, yielding offspring more severely affected with the disorder, a phenomenon called anticipation. This phenomenon may depend on the sex of the parent with the TNRE.

C20. According to the random mutation theory, spontaneous mutations can occur in any gene and do not involve exposure of the organism to a particular environment that selects for specific types of mutation. However, the structure of chromatin may cause certain regions of the DNA to be more susceptible to random mutations. For example, DNA in an open conformation may be more accessible to mutagens and more likely to incur mutations. Simlarly, hot spots—certain regions of a gene that are more likely to mutate than other regions—can occur within a single gene. Also, another reason that some genes mutate at a higher rate is that some genes are larger than others, which provides a greater chance for mutation.

C22. Excision repair systems could fix this damage. Also, homologous recombination repair could fix the damage.

C24. Anticipation means that the TNRE expands even further in future generations. Anticipation may depend on the sex of the parent with the TNRE.

C26. The mutation frequency is the total number of mutant alleles divided by the total number of alleles in the population. If there are 1,422,000 babies, there are 2,844,000 copies of this gene (because each baby has two copies). The mutation frequency is $^{31}/_{2,844,000}$, which equals 1.09×10^{-5}. The mutation rate is the number of new mutations per generation. There are 13 babies who did not have a parent with achondroplasia; thus, 13 is the number of new mutations. If we calculate the mutation rate as the number of new mutations in a given gene per generation, then we should divide 13 by 2,844,000. In this case, the mutation rate would be 4.6×10^{-6}.

C28. The effects of mutations are cumulative. If one mutation occurs in a cell, this mutation will be passed to the daughter cells. If a mutation occurs in the daughter cell, now there will be two mutations. These two mutations will be passed to the next generation of daughter cells, and so forth. The accumulation of many mutations eventually kills the cells. That is why mutagens are more effective at killing dividing cells compared to nondividing cells. It is because the number of mutations accumulates to a lethal level.

There are two main side effects to this treatment. First, some normal (noncancerous) cells of the body, particularly skin cells and intestinal cells, are actively dividing. These cells are also killed by chemotherapy and radiation therapy. Secondly, it is possible that the therapy may produce mutations that will cause noncancerous cells to become cancerous. For these reasons, there is a maximal dose of chemotherapy or radiation therapy that is recommended.

C30. A. Yes

B. No; the albino trait affects the entire individual.

C. No; the early apple-producing trait affects the entire tree.

D. Yes

C32. Mismatch repair is aimed at eliminating mismatches that may have occurred during DNA replication. In this case, the wrong base is in the newly made strand. The binding of MutH, which occurs on a hemimethylated sequence, provides a sensing mechanism to distinguish between the unmethylated and methylated strands. In other words, MutH binds to the hemimethylated DNA in a way that allows the mismatch repair system to distinguish which strand is methylated and which is not.

C34. Because sister chromatids are genetically identical, an advantage of homologous recombination is that it can be an error-free mechanism to repair a DSB. A disadvantage, however, is that it occurs only during the S and G_2 phase of the cell cycle in eukaryotes or following DNA replication in bacteria. An advantage of NHEJ is that it doesn't involve the participation of a sister chromatid, so it can occur at any stage of the cell cycle. However, a disadvantage is that NHEJ can result in small deletions in the region that has been repaired. Overall, NHEJ is a quick but error-prone repair mechanism, while HR is a more accurate method of repair that is limited to certain stages of the cell cycle.

C36. In *E. coli*, the TRCF recognizes when RNA polymerase is stalled on the DNA. This stalling may be due to DNA damage such as a thymine dimer. The TRCF removes RNA polymerase and recruits the excision DNA repair system to the region, thereby promoting the repair of the template strand of DNA. It is beneficial to preferentially repair actively transcribed DNA because it is functionally important. It is a DNA region that encodes a gene.

C38. The underlying genetic defect that causes xeroderma pigmentosum is a defect in one of the genes that encode a polypeptide involved with nucleotide excision repair. These individuals are defective in repairing DNA abnormalities such as thymine dimers and abnormal bases. Therefore, they are very sensitive to environmental agents such as UV light. Because they are defective at repair, UV light is more likely to cause mutations in these people compared to unaffected individuals. For this reason, people with XP develop pigmentation abnormalities and premalignant lesions and have a high predisposition to skin cancer.

C40. Both types of repair systems recognize an abnormality in the DNA and excise the abnormal strand. The normal strand is then used as a template to synthesize a complementary strand of DNA. The systems differ in the types of abnormalities they detect. The mismatch repair system detects base pair mismatches, while the excision repair system recognizes thymine dimers, chemically modified bases, missing bases, and certain types of cross-links. The mismatch repair system operates immediately after DNA replication, allowing it to distinguish between the daughter strand (which contains the wrong base) and the parental strand. The excision repair system can operate at any time in the cell cycle.

Experimental Questions

E2. When cells from a master plate were replica plated onto two plates containing selective media with the T1 phage, T1-resistant colonies were observed at the same locations on both plates. These results indicate that the mutations occurred randomly while on the master plate (in the absence of T1) rather than occurring as a result of exposure to T1. In other words, mutations are random events, and selective conditions may promote the survival of mutant strains that occur randomly.

To show that antibiotic resistance is due to random mutation, one could follow the same basic strategy except the secondary plates would contain the antibiotic instead of T1 phage. If the antibiotic resistance arose as a result of random mutation on the master plate, one would expect the antibiotic-resistant colonies to appear at the same locations on two different secondary plates.

E4. Perhaps the X-rays also produce mutations that make the *ClB* daughters infertile. Many different types of mutations could occur in the irradiated males and be passed to the *ClB* daughters. Some of these mutations could prevent the *ClB* daughters from being fertile. These mutations could interfere with oogenesis, etc. Such *ClB* daughters would be unable to have any offspring.

E6. You would conclude that chemical A is not a mutagen. The percentage of *ClB* daughters (whose fathers had been exposed to chemical A) that did not produce sons was similar to the control (compare 3 out of 2,108 with 2 out of 1,402). In contrast, chemical B appears to be a mutagen. The percentage of *ClB* daughters (whose fathers had been exposed to chemical B) that did not produce sons was much higher than the control (compare 3 out of 2,108 with 77 out of 4,203).

E8. You would expose the bacteria to the physical agent. You could also expose the bacteria to the rat liver extract, but it is probably not necessary for two reasons. First, a physical mutagen is not something that a person would eat. Therefore, the actions of digestion via the liver are probably irrelevant if you are concerned that the agent might be a mutagen. Second, the rat liver extract would not be expected to alter the properties of a physical mutagen.

E10. The results suggest that the strain is defective in excision repair. If we compare the normal and mutant strains that have been incubated for 2 hours at 37°C, much of the radioactivity in the normal strain has been transferred to the soluble fraction because it has been excised. In the mutant strain, however, less of the radioactivity has been transferred to the soluble fraction, suggesting that it is not as efficient at removing thymine dimers.

Questions for Student Discussion/Collaboration

2. The worst time to be exposed to mutagens would be at very early stages of embryonic development. An early embryo is most sensitive to mutation because it will affect a large region of the body. Adults must also worry about mutagens for several reasons. Mutations in somatic cells can cause cancer, a topic discussed in Chapter 22. Also, adults should be careful to avoid mutagens that may affect the ovaries or testes because these mutations could be passed to offspring.

CHAPTER 17

Conceptual Questions

C2. Branch migration will not create a heteroduplex during SCE because the sister chromatids are genetically identical. There should not be any mismatches between the complementary strands. Gene conversion cannot take place because the sister chromatids carry alleles that are already identical to each other.

C4. The two molecular mechanisms that can explain the phenomenon of gene conversion are mismatch DNA repair and gap repair synthesis. Both mechanisms could occur in the double-strand break model.

C6. A recombinant chromosome is one that has been derived from a crossover and contains a combination of alleles that is different from the parental chromosomes. A recombinant chromosome is a hybrid of the parental chromosomes.

C8. Gene conversion occurs when a pair of different alleles is converted to a pair of identical alleles. For example, a pair of *Bb* alleles could be converted to *BB* or *bb*.

C10. Gene conversion is likely to take place near the breakpoint. According to the double-strand break model, a gap may be created by the digestion of one DNA strand in the double helix. Gap repair synthesis may result in gene conversion. A second way that gene conversion can occur is by mismatch repair. A heteroduplex may be created after DNA strand migration and may be repaired in such a way as to cause gene conversion.

C12. No, it does not necessarily involve DNA mismatch repair; it could also involve gap repair synthesis. The double-strand break model involves a migration of DNA strands and the digestion of a gap. Therefore, in this gap region, only one chromatid is providing the DNA strands. As seen in Figure 17.7, the top chromosome is using the top DNA strand from the bottom chromosome in the gap region. The bottom chromosome is using the bottom DNA strand from the bottom chromosome. After DNA synthesis, both chromosomes may have the same allele.

C14. First, gene rearrangement of V, D, and J domains occurs within the light- and heavy-chain genes. Second, within a given B cell, different combinations of light and heavy chains are possible. And third, imprecise fusion may occur between the V, D, and J domains.

C16. One segment, which includes some variable (V) domains and perhaps one or more joining (J) domain, of DNA is removed from the κ light-chain gene. One segment, which may include one or more J domains and the region between the J domain and C domain, is removed during pre-mRNA splicing.

C18. The ends of a short region would be flanked by direct repeats. This is a universal characteristic of all transposable elements. In addition, many elements contain IRs or LTRs that are involved in the transposition process. One might also look for the presence of a transposase or reverse transcriptase gene, although this is not an absolute requirement, because nonautonomous transposable elements typically lack transposase or reverse transcriptase.

C20. Direct repeats occur because transposase or integrase produces staggered cuts in the two strands of chromosomal DNA. The transposable element is then inserted into this site, which temporarily leaves two gaps. The gaps are filled in by DNA polymerase. Because this gap filling is due to complementarity of the base sequences, the two gaps end up with the exact same sequence.

C22. Transposable elements are mutagens, because they alter (disrupt) the sequences of chromosomes and genes within chromosomes. They do this by inserting themselves into genes.

C24. A. Viral-like retroelements and nonviral-like retroelements

B. Insertion sequences, composite transposons, and replicative transposons

C. All five types have direct repeats

D. Insertion sequences, composite transposons, and replicative transposons

C26. A. As shown in solved problem S2a, a crossover between direct repeats (DRs) will excise the region between the repeats. In this case, it will excise a large chromosomal region, which does not contain a centromere, and will be lost. The remaining portion of the chromosome is shown here. It has one direct repeat that is formed partly from DR-1 and partly from DR-4. It is designated DR-$^1/_4$.

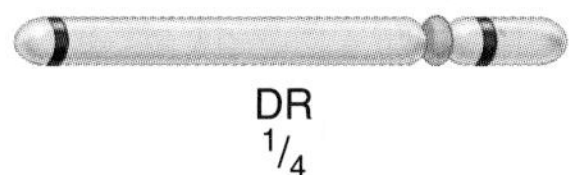

B. A crossover between IR-1 and IR-4 will cause an inversion between the intervening region. This is a similar effect to the crossover within a single transposable element, as described in solved problem S2b. The difference is that the crossover causes a large chromosomal region to be inverted.

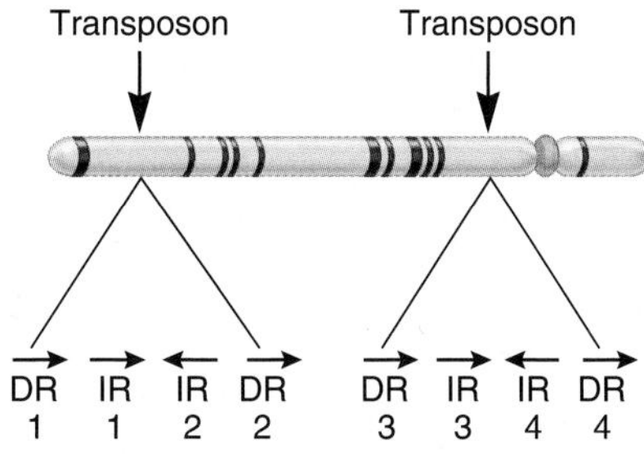

C28. At a frequency of about 1 in 10,000, recombination occurs between the inverted repeats within this transposon. As shown in solved problem S2b, recombination between inverted repeats causes the sequence within the transposon to be reversed. If this occurred in a strain that expressed the *H2/rH1* operon, the promoter would be flipped in the opposite direction, so the *H2* gene and the *rH1* gene would not be expressed. The *H2* flagellar protein would not be made, and the *H1* repressor protein would not be made. The *H1* flagellar protein would be made because the expression of the *H1* gene would not be repressed. A strain expressing the *H1* gene could also "switch" at a frequency of about 1 in 10,000 by the same mechanism. If a crossover occurred between the inverted repeats within the transposable element, the promoter would be flipped again, and the *H2* and *rH1* genes would be expressed again.

Experimental Questions

E2. You would conclude that the substance is a mutagen. Substances that damage DNA tend to increase the level of genetic exchange such as SCE.

E4. The drawing here shows the progression through three rounds of BrdU exposure. After one round, all of the chromosomes would be dark. After two rounds, all of the chromosomes would be harlequin. After three rounds, the number of light sister chromatids would be twice as much as the number of light sister chromatids found after two rounds of replication.

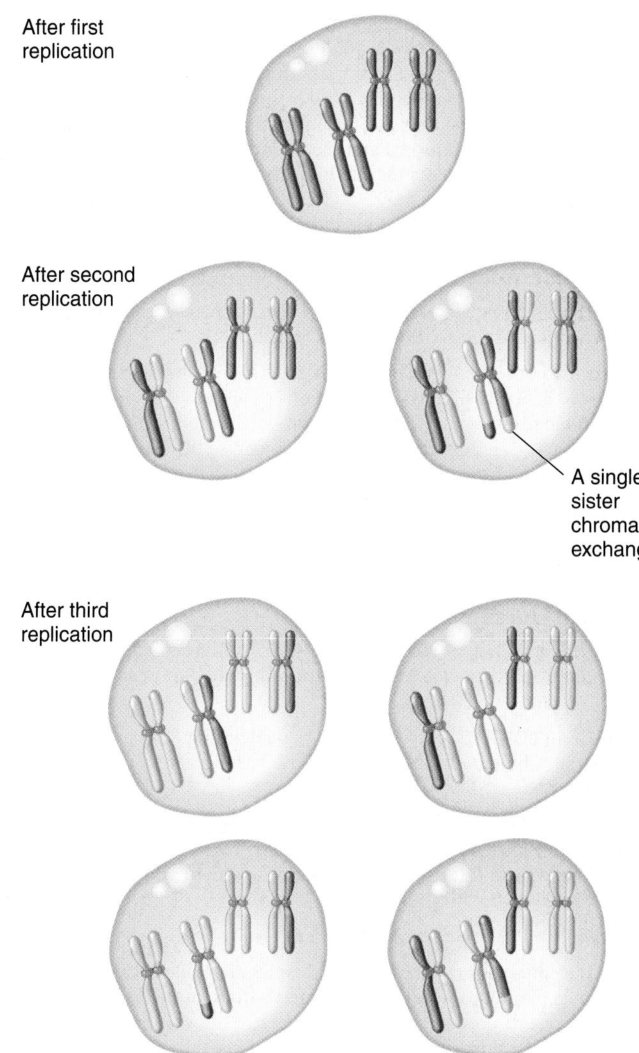

E6. When McClintock started with a colorless strain containing *Ds*, she identified 20 cases where *Ds* had moved to a new location to produce red and shrunken kernels. This identification was possible because the 20 strains had a higher frequency of chromosomal breaks at a specific site and because of the mutability of particular genes. She also had found a strain where *Ds* had inserted into the red-color-producing gene, resulting in the colorless phenotype, so its transposition out of the gene would produce a red patch. Overall, her analysis of the data showed that the sectoring (i.e., mutability) phenotype was consistent with the transposition of *Ds*.

E8. Transposon tagging is an experimental method aimed at cloning genes. In this approach, a transposon is introduced into a strain, and the researcher identifies individuals in which the function of a gene of interest has been inactivated. In many cases, the loss of function has occurred because the transposon has been inserted into the gene. If so, DNA can be isolated, digested into fragments, and cloned into viral vectors, creating a DNA library. The library is then probed using a radiolabeled fragment of DNA that is complementary to the transposon in order to identify clones in which the transposon has been inserted.

E10. A transposon creates a mutable site because the excision of a transposon causes chromosomal breakage if the two ends are not reconnected. After it has moved out of its original locus (causing chromosomal breakage), it may be inserted into a new locus somewhere else. You could experimentally determine this by examining a strain that has incurred a chromosomal breakage at the first locus. You could microscopically examine many cells that had such a broken chromosome and see if a new locus had been formed. This new locus would be the site into which the transposon had moved. On occasion, there would be chromosomal breakage at this new locus, which could be observed microscopically.

Questions for Student Discussion/Collaboration

2. Beneficial consequences: You wouldn't get (as many) translocations, inversions, and the accumulation of selfish DNA.

   Harmful consequences: The level of genetic diversity would be decreased, because linked combinations of alleles would not be able to recombine. You wouldn't be able to produce antibody diversity in the same way. Gene duplication could not occur, so the evolution of new genes would be greatly inhibited.

CHAPTER 18

Conceptual Questions

C2. A restriction enzyme recognizes and binds to a specific DNA sequence and then cleaves a (covalent) ester bond in each of two DNA strands.

C4. The term cDNA refers to DNA that is made using RNA as the starting material. Compared to genomic DNA, it lacks introns.

Experimental Questions

E2. Remember that AT base pairs form two hydrogen bonds, while GC base pairs form three hydrogen bonds. The order (from stickiest to least sticky) would be

*Bam*HI = *Pst*I = *Sac*I > *Eco*RI > *Cla*I

E4. In conventional gene cloning, many copies are made because the vector replicates to a high copy number within the cell, and the cells divide to produce many more cells. In PCR, the replication of the DNA to produce many copies is facilitated by primers, deoxyribonucleoside triphosphates (dNTPs), and *Taq* polymerase.

E6. A recombinant vector is a vector that has a piece of "foreign" DNA inserted into it. The foreign DNA came from somewhere else, such as the chromosomal DNA of some organism. To construct a recombinant vector, the vector and source of foreign DNA are digested with the same restriction enzyme. The complementary ends of the fragments are allowed to hydrogen bond to each other (i.e., sticky ends are allowed to bind), and then DNA ligase is added to create covalent bonds. In some cases, a piece of the foreign DNA will become ligated to the vector, thereby creating a recombinant vector. In other cases, the two ends of the vector ligate back together, restoring the vector to its original structure.

As described in Figure 18.2, the insertion of foreign DNA can be detected using X-Gal. As seen here, the insertion of the foreign DNA causes the inactivation of the *lacZ* gene. The *lacZ* gene encodes the enzyme β-galactosidase, which converts the colorless compound X-Gal to a blue compound. If the *lacZ* gene is inactivated by the insertion of foreign DNA, the enzyme will not be produced, and the bacterial colonies will be white. If the vector has simply recircularized and the *lacZ* gene remains intact, the enzyme will be produced, and the colonies will be blue.

E8. If the *Eco*RI fragment containing the kan^R gene also had an origin of replication, it is possible that this fragment could circularize and become a very small plasmid. The electrophoresis results would be consistent with the idea that the same bacterial cells could contain two different plasmids: pSC101 and a small plasmid corresponding to the segment of DNA (now circularized) that carried the kan^R gene. However, the results shown in step 9 rule out this possibility. The density gradient centrifugation showed a single peak, corresponding to a plasmid that had an intermediate size between pSC101 and pSC102. In contrast, if our alternative explanation had been correct (i.e., that bacterial cells contain two plasmids), there would be two peaks from the density gradient centrifugation. One peak would correspond to pSC101, and the other peak would indicate a very small plasmid (i.e., smaller than pSC101 and pSC102).

E10. 3×2^{27}, which equals 4.0×10^8, or about 400 million copies.

E12. Initially, the mRNA would be mixed with reverse transcriptase and nucleotides to create a complementary strand of DNA. Reverse transcriptase also needs a primer. This could be a primer that is known to be complementary to the β-globin mRNA. Alternatively, mature mRNAs have a polyA tail, so one could add a primer that consists of many Ts, called a poly-dT primer. After the complementary DNA strand has been made, the sample would then be mixed with primers, *Taq* polymerase, and nucleotides and subjected to the standard PCR protocol. Note: The PCR reaction would have two kinds of primers. One primer would be complementary to the 5' end of the mRNA and would be unique to the β-globin sequence. The other primer would be complementary to the 3' end. This second primer could be a poly-dT primer, or it could be a unique primer that would bind slightly upstream from the polyA-tail region.

E14. A DNA library is a collection of recombinant vectors that contain different pieces of DNA from a source of chromosomal DNA. Because it is a diverse collection of many different DNA pieces, the name library seems appropriate.

E16. Hybridization occurs due to the hydrogen bonding of complementary sequences. Due to the chemical properties of DNA and RNA strands, they form double-stranded regions when the base sequences are complementary. In a Southern and Northern experiment, the cloned DNA is labeled and used as a probe.

E18. The purpose of a Northern blotting experiment is to identify a specific RNA within a mixture of many RNA molecules, using a fragment of cloned DNA as a probe. It can tell you if a gene is transcribed in a particular cell or at a particular stage of development. It can also tell you if a pre-mRNA is alternatively spliced into two or more mRNAs of different sizes.

E20. It appears that this mRNA is alternatively spliced to create a high molecular mass and a lower molecular mass product. Nerve cells produce a very large amount of the larger mRNA, whereas spleen cells produce a moderate amount of the smaller mRNA. Both types are produced in small amounts by the muscle cells. It appears that kidney cells do not transcribe this gene.

E22.
1. β (detected at the highest stringency)
2. δ
3. γ_A and ϵ
4. α_1
5. *Mb* (detected only at the lowest stringency)

E24. The Western blot is shown here. The sample in lane 2 came from a plant that was homozygous for a mutation that prevented the expression of this polypeptide. Therefore, no protein was observed in this lane. The sample in lane 4 came from a plant that is homozygous for a mutation that introduces an early stop codon into the coding sequence; thus, the polypeptide is shorter than normal (13.3 kDa). The sample in lane 3 was from a heterozygote that expresses about 50% of each type of polypeptide. Finally, the sample in lane 5 came from a plant homozygous for a mutation that changed one amino acid to another amino acid. This type of mutation, termed a missense mutation, may not be detectable on a gel. However, a single amino acid substitution could affect polypeptide function.

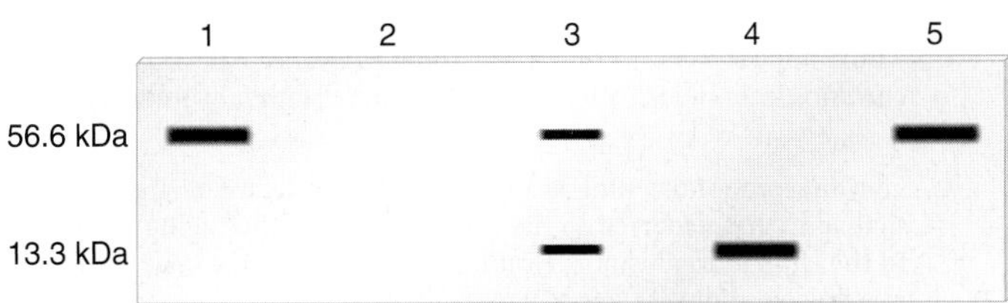

E26. The products of structural genes are proteins with a particular amino acid sequence. Antibodies can specifically recognize proteins due to their amino acid sequence. Therefore, an antibody can detect whether or not a cell is making a particular type of protein.

E28. In this case, the transcription factor binds to the response element when the hormone is present. Therefore, the hormone promotes the binding of the transcription factor to the DNA and thereby promotes transcriptional activation.

E30. The levels of cAMP affect the phosphorylation of CREB, and this affects whether or not it can activate transcription. However, CREB can bind to CREs whether or not it is phosphorylated. Therefore, in a gel retardation assay, we would expect CREB to bind to CREs and retard their mobility whether or not the cell extracts were pretreated with epinephrine.

E32. The glucocorticoid receptor will bind to GREs if glucocorticoid hormone is also present (lane 2). The glucocorticoid receptor does not bind without hormone (lane 1), and it does not bind to CREs (lane 6). The CREB protein will bind to CREs with or without hormone (lanes 4 and 5), but it will not bind to GREs (lane 3). The expected results are shown here. In this drawing, the binding of CREB protein to the 700 bp fragment results in a complex with a higher mass compared to the glucocorticoid receptor binding to the 600 bp fragment.

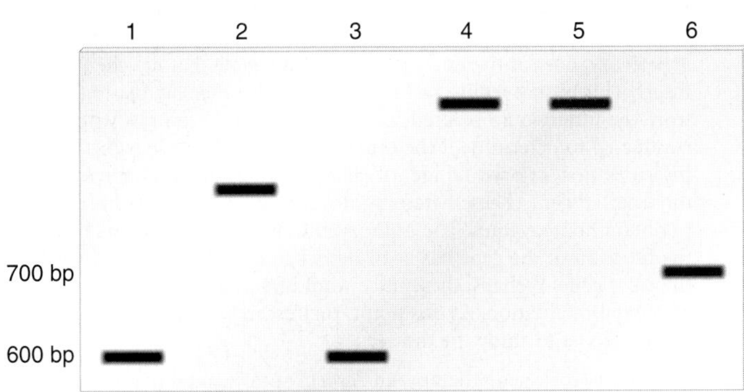

E34. The rationale behind a footprinting experiment has to do with accessibility. If a protein is bound to the DNA, it will cover up the part of the DNA where it is bound. This region of the DNA will be inaccessible to the actions of chemicals or enzymes that cleave the DNA, such as DNase I.

E36. A. AGGTCGGTTGCCATCGCAATAATTTCTGCCTGAACCCAATA

B. Automated sequencing has several advantages. First, the reactions are done in a single tube as opposed to four tubes. Second, the detector can "read" the sequence and provide the researcher with a printout of the sequence. This is much easier than looking at an X-ray film and writing the sequence out by hand. It also avoids human error.

E38. There are lots of different strategies one could follow. For example, you could mutate every other base and see what happens. It would be best to make very nonconservative mutations, such as a purine for a pyrimidine or a pyrimidine for a purine. If the mutation prevents protein binding in a gel retardation assay, then the mutation is probably within the response element. If the mutation has no effect on protein binding, it probably is outside the response element.

Questions for Student Discussion/Collaboration

2. A. Does a particular amino acid within a protein sequence play a critical role in the protein's structure or function?

B. Does a DNA sequence function as a promoter?

C. Does a DNA sequence function as a regulatory site?

D. Does a DNA sequence function as a splicing junction?

E. Is a sequence important for correct translation?

F. Is a sequence important for RNA stability?

And many others. . .

CHAPTER 19

Conceptual Questions

C2. *A. radiobacter* synthesizes an antibiotic that kills *A. tumefaciens.* The genes, which are necessary for antibiotic biosynthesis and resistance, are plasmid encoded and can be transferred during interspecies conjugation. If *A. tumefaciens* received this plasmid during conjugation, it would be resistant to killing. Therefore, the conjugation-deficient strain prevents the occurrence of *A. tumefaciens*–resistant strains.

C4. A biological control agent is an organism that prevents the harmful effects of some other agent in the environment. Examples include *Bacillus thuringiensis,* a bacterium that synthesizes compounds that act as toxins to kill insects, *Ice*$^-$ bacteria that inhibit the proliferation of *Ice*$^+$ bacteria, and the use of *Agrobacterium radiobacter* to prevent crown gall disease caused by *Agrobacterium tumefaciens.*

C6. A mouse model is a strain of mice that carries a mutation in a mouse gene that is analogous to a mutation in a human gene that causes disease. These mice can be used to study the disease and to test potential therapeutic agents.

C8. The T DNA gets transferred to the plant cell; it then is incorporated into the plant cell's genome.

C10. A. With regard to maternal effect genes, the phenotype would depend on the animal that donated the egg. It is the cytoplasm of the egg that accumulates the gene products of maternal effect genes.

B. The extranuclear traits depend on the mitochondrial genome. Mitochondria are found in the egg and in the somatic cell. So, theoretically, both cells could contribute extranuclear traits. In reality, however, researchers have found that the mitochondria in Dolly were from the animal that donated the egg. It is not clear why she had no mitochondria from the mammary cell.

C. The cloned animal would be genetically identical to the animal that donated the nucleus with regard to traits that are determined by nuclear genes, which are expressed during the lifetime of the organism. The cloned animal would/could differ from the animal that donated the nucleus with regard to traits that are determined by maternal effect genes and mitochondrial genes. Such an animal is not a true clone, but it is likely that it would greatly resemble the animal that donated the nucleus, because the vast majority of genes are found in the cell nucleus.

C12. Some people are concerned with the release of genetically engineered microorganisms into the environment. The fear is that such organisms may continue to proliferate and it may not be possible to "stop them." A second concern involves the use of genetically engineered organisms in the food we eat. Some people are worried that genetically engineered organisms may pose an unknown health risk. A third issue is ethics. Some people feel that it is morally wrong to tamper with the genetics of organisms. This opinion may also apply to genetic techniques such as cloning, stem cell research, and gene therapy.

Experimental Questions

E2. The plasmid with the incorrect orientation would not work because the coding sequence would be in the wrong direction relative to the promoter sequence. Therefore, the region containing the somatostatin sequence would not be translated into the somatostatin polypeptide.

E4. To construct the coding sequence for somatostatin, the researchers synthesized eight oligonucleotides, labeled A–H. When these oligonucleotides were mixed together, they would hydrogen bond to each other due to their complementary sequences. The addition of ligase would covalently link the DNA backbones. The left side of the coding sequence had an *Eco*RI site, and the right side had a *Bam*HI site; these sites made it possible to insert this sequence at the end of the β-galactosidase gene. You may also notice that an ATG codon (AUG in the mRNA) precedes the first alanine codon in somatostatin. This AUG codon specifies methionine, which allows the somatostatin to be cleaved from β-galactosidase using cyanogen bromide. This was necessary because somatostatin, by itself, is rapidly degraded in *E. coli*, whereas the fusion protein is not.

E6. A kanamycin-resistance gene is contained within the T DNA. Exposure to kanamycin selects for the growth of plant cells that have incorporated the T DNA into their genome. The carbenicillin kills the *A. tumefaciens*. The plant growth hormones promote the regeneration of an entire plant from somatic cells. If kanamycin were left out, it would not be possible to select for the growth of cells that had taken up the T DNA.

E8. The term gene knockout refers to an organism in which the function of a particular gene has been eliminated. For autosomal genes in animals and plants, a gene knockout is a homozygote for a defect in both copies of the gene. If a gene knockout has no phenotypic effect, the gene may be redundant. In other words, there may be multiple genes within the genome that can carry out the same function. Another reason why a gene knockout may not have a phenotypic effect is because of the environment. As an example, let's say a mouse gene is required for the synthesis of a vitamin. If the researchers were providing food that contained the vitamin, the knockout mouse that was lacking this gene would have a normal phenotype; it would survive just fine. Sometimes, researchers have trouble knowing the effects of a gene knockout until they modify the environmental conditions in which the animals are raised.

E10. Gene replacement occurs by homologous recombination. For homologous recombination to take place, two crossovers must occur, one at each end of the target gene. After homologous recombination, only the Neo^R gene, which is inserted into the target gene, can be incorporated into the chromosomal DNA of the embryonic stem cell. In contrast, nonhomologous recombination can involve two crossovers anywhere in the cloned DNA. Because the *TK* gene and target gene are adjacent to each other, nonhomologous recombination usually transfers both the *TK* gene and the Neo^R gene. If both genes are transferred to an embryonic stem cell, it will die because the cells are grown in the presence of gancyclovir. The product of the *TK* gene kills the cell under these conditions. In contrast, cells that have acquired the Neo^R gene due to homologous recombination but not the *TK* gene will survive. Stem cells, which have not taken up any of the cloned DNA, will die because they will be killed by the neomycin. In this way, the presence of gancyclovir and neomycin selects for the growth of stem cells that have acquired the target gene by homologous recombination.

E12. A. Dolly's chromosomes may seem so old because they were already old when they were in the nucleus that was incorporated into the enucleated egg. They had already become significantly shortened in the mammary cells. This shortening was not repaired by the egg.

B. Dolly's age does not matter. Remember that shortening does not occur in germ cells. However, Dolly's eggs are older than they seem by about six or seven years, because Dolly's germ-line cells received their chromosomes from a sheep that was 6 years old, and the cells were grown in culture for a few doublings before a mammary cell was fused with an enucleated egg. Therefore, the calculation would be 6 or 7 years (the age of the mammary cells that produced Dolly's germ-line cells) plus 8 years (the age of Molly), which equals 14 or 15 years. However, only half of Molly's chromosomes would appear to be 14 or 15 years old. The other half of her chromosomes, which she inherited from her father, would appear to be 8 years old.

C. Chromosome shortening is a bit disturbing, because it suggests that aging has occurred in the somatic cell, and this aging is passed to the cloned organism. If cloning was done over the course of many generations, this may eventually have a major impact on the life span of the cloned organism. It may die much earlier than a noncloned organism. However, chromosome shortening may not always occur. It did not seem to occur in mice that were cloned for six consecutive generations.

E14. The term molecular pharming refers to the practice of making transgenic animals that will synthesize human products in their milk. It also refers to the manufacture of medical products by agricultural plants. It can be advantageous when bacterial cells are unable to make a functional protein product from a human gene. For example, some proteins are posttranslationally modified by the attachment of carbohydrate molecules. This type of modification does not occur in bacteria, but it may occur correctly in transgenic animals and plants. Also, dairy cows produce large amounts of milk, which may improve the yield of the human product. Likewise, plants can produce a large amount of recombinant proteins.

E16. You would first need to clone the normal mouse gene. Cloning methods are described in Chapter 18. After the normal gene was cloned, you would then follow the protocol shown in Figure 19.7. The normal gene would be inactivated by the insertion of the Neo^R gene, and the *TK* gene would be cloned next to it. This DNA segment would be introduced into mouse embryonic stem cells and grown in the presence of neomycin and gancyclovir. This selects for homologous recombinants. The surviving embryonic stem cells would be injected into early embryos, which would then develop into chimeras. The chimeric mice would be identified by their patches of light and dark fur. At this point, if all has gone well, a portion of the mouse is heterozygous for the normal gene and the gene that has the Neo^R insert. This mouse would be bred, and then brothers and sisters from the litter would be bred to each other. Southern blots would be conducted to determine if the offspring carried the gene with the Neo^R insert. At first, one would identify heterozygotes that had one copy of the inserted gene. These heterozygotes would be crossed to each other to obtain homozygotes. The homozygotes are gene knockouts because the function of the gene has been "knocked out" by the insertion of the Neo^R gene. Perhaps, these mice would be dwarf and exhibit signs of mental retardation. At this point, the researcher would have a mouse model to study the disease.

E18. In cystic fibrosis gene therapy, an aerosol spray, containing the normal *CF* gene in a retrovirus or liposome, is used to get the normal *CF* gene into the lung cells. The epithelial cells on the surface of the lung will take up the gene. However, these surface epithelial cells have a finite life span, so it is necessary to have repeated applications of the aerosol spray. Ideally, scientists hope to devise methods whereby the normal *CF* gene will be able to penetrate more deeply into the lung tissue.

Questions for Student Discussion/Collaboration

2. From a genetic viewpoint, the recombinant and nonrecombinant strains are very similar. The main difference is their history. The recombinant strain has been subjected to molecular techniques to eliminate a particular gene. The nonrecombinant strain has had the same gene eliminated by a spontaneous mutation or via mutagens. The nonrecombinant strain has the advantage of a better public perception. People are less worried about releasing nonrecombinant strains into the environment.

CHAPTER 20

Conceptual Questions

C2. A. Yes

B. No; this is only one chromosome in the genome.

C. Yes

D. Yes

Experimental Questions

E2. They are complementary to each other.

E4. Because normal cells contain two copies of chromosome 14, one would expect that a probe would bind to complementary DNA sequences on both of these chromosomes. If a probe recognized only one of two chromosomes, this means that one of the copies of chromosome 14 has been lost, or it has suffered a deletion in the region where the probe binds. With regard to cancer, the loss of this genetic material may be related to the uncontrollable cell growth.

E6. After the cells and chromosomes have been fixed to the slide, it is possible to add two or more different probes that recognize different sequences (i.e., different sites) within the genome. Each probe has a different fluorescence emission wavelength. Usually, a researcher will use computer imagery that recognizes the wavelength of each probe and then assigns that probe a bright color. The color seen by the researcher is not the actual color emitted by the probe; it is a secondary color assigned by the computer. In a sense, the probes, with the aid of a computer, are "painting" the regions of the chromosomes that are recognized by a probe. An example of chromosome painting is shown in Figure 20.4. In this example, human chromosome 5 is painted with six different colors.

E8. A contig is a collection of clones that contain overlapping segments of DNA that span a particular region of a chromosome. To determine if two clones are overlapping, one could conduct a Southern blotting experiment. In this approach, one of the clones is used as a probe. If it is overlapping with the second clone, it will bind to it in a Southern blot. Therefore, the second clone is run on a gel, and the first clone is used as a probe. If the band corresponding to the second clone is labeled, this means that the two clones are overlapping.

E10. BAC cloning vectors have the replication properties of a bacterial chromosome and the cloning properties of a plasmid. To replicate like a chromosome, the BAC vector contains an origin of replication from an F factor. Therefore, in a bacterial cell, a BAC can behave as a chromosome. Like a plasmid, BACs also contain selectable markers and convenient cloning sites for the insertion of large segments of DNA. The primary advantage is the ability to clone very large pieces of DNA.

E12. The resistance gene appears to be linked to RFLP 4B.

E14. For most organisms, it is usually easy to locate many RFLPs throughout the genome. The RFLPs can be used as molecular markers to make a map of the genome. This is done using the strategy described in Figure 20.8. To map a functional gene, one would also follow the same general strategy described in Figure 20.8, except that the two strains would also have an allelic difference in the gene of interest. The experimenter would make crosses, such as dihybrid crosses, and determine the number of parental and recombinant offspring based on the alleles and RFLPs they had inherited. If an allele and RFLP are linked, there will be a much lower percentage (i.e., less than 50%) of the recombinant offspring.

E16. *Deduced Outcome*

STSs

4 6 7 9 3 2 8 5 1

YAC

3 ⟵⟶

1 ⟵⟶

5 ⟵⟶

4 ⟵⟶

2 ⟵⟶

E18. A. One homologue contains the STS-1 that is 289 bp and STS-2 that is 422 bp, while the other homologue contains STS-1 that is 211 bp and STS-2 that is 115 bp. This is based on the observation that 28 of the sperm have either the 289 bp and 422 bp bands or the 211 bp and 115 bp bands.

B. There are two recombinant sperm; see lanes 12 and 18. Because there are two recombinant sperm out of a total of thirty,

$$\text{Map Distance} = \frac{2}{30} \times 100$$

$$= 6.7 \text{ mu}$$

C. In theory, this method could be used. However, there is not enough DNA in one sperm to carry out an RFLP analysis unless the DNA is amplified by PCR.

E20. One possibility is that the geneticist could try a different restriction enzyme. Perhaps there is sequence variation in the vicinity of the pesticide-resistance gene that affects the digestion pattern of a restriction enzyme other than *Eco*RI. There are hundreds of different restriction enzymes that recognize a myriad of different sequences.

Alternatively, the geneticist could give up on the RFLP approach and try to identify one or more sequence-tagged sites that are in the vicinity of the pesticide-resistance gene. In this case, the geneticist would want to identify STSs that are also microsatellites. As shown in Figure 20.12, the transmission of microsatellites can be followed in genetic crosses. Therefore, if the geneticist could identify microsatellites in the vicinity of the pesticide-resistance gene, this would make it possible to predict the outcome of crosses. For example, let's suppose a microsatellite linked to the pesticide-resistance gene existed in three forms: 234 bp, 255 bp, and 311 bp. And let's also suppose that the 234 bp form was linked to the high-resistance allele, the 255 bp form was linked to the moderate-resistance allele, and the 311 bp form was linked to the low-resistance allele. According to this hypothetical example, the geneticist could predict the level of resistance in an alfalfa plant by analyzing the inheritance of these microsatellites.

E22. When chromosomal DNA is isolated and digested with a restriction enzyme, this produces thousands of DNA fragments of different sizes. This makes it impossible to see any particular band on a gel if you simply stained the gel for DNA. Southern blotting allows you to detect one or more RFLPs that are complementary to the radioactive probe that is used.

E24. Based on these results, it is likely that the sickle-cell allele originated in an individual with the 13.0 kbp RFLP. This would explain why the Hb^S allele is usually transmitted with the 13.0 kbp RFLP. On occasion, however, a crossover could occur in the region between the β-globin gene and the distal *Hpa*I site.

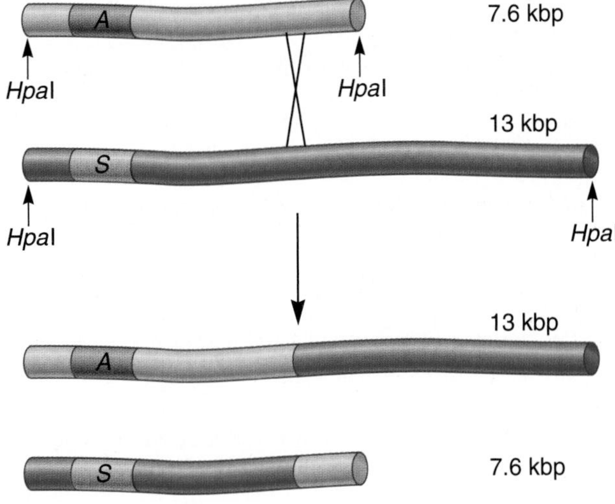

After crossing over, the Hb^S allele is now linked to a 7.6 kbp RFLP.

E26. PFGE is a method of electrophoresis that is used to separate small chromosomes and large DNA fragments. The electrophoresis devices used in PFGE have two sets of electrodes. The two sets of electrodes produce alternating pulses of current, and this facilitates the separation of large DNA fragments.

It is important to handle the sample gently to prevent the breakage of the DNA due to mechanical forces. Cells are first embedded in agarose blocks, and then the blocks are loaded into the wells of the gel. The agarose keeps the sample very stable and prevents shear forces that might mechanically break the DNA. After the blocks are in the gel, the cells within the blocks are lysed, and, if desired, restriction enzymes can be added to digest the DNA. For PFGE, a restriction enzyme that cuts very infrequently might be used.

PFGE can be used as a preparative technique to isolate and purify individual chromosomes or large DNA fragments. PFGE can also be used, in conjunction with Southern blotting, as a mapping technique.

E28. Note that the insert of cosmid B is contained completely within the insert of cosmid C.

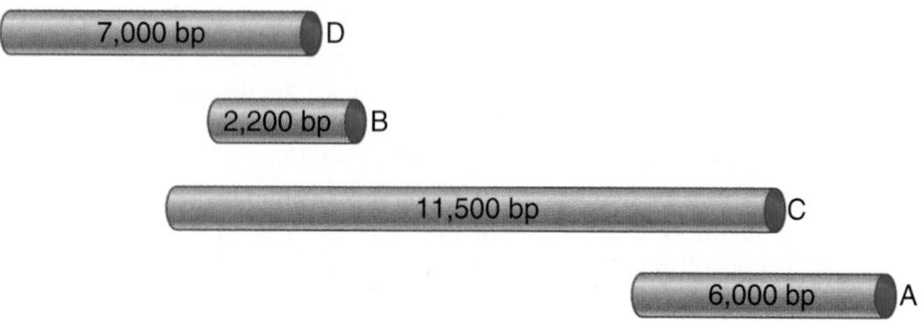

E30. A. The general strategy is shown in Figure 20.18. The researcher begins at a certain location and then walks toward the gene of interest. You begin with a clone that has a marker that is known to map relatively close to the gene of interest. A piece of DNA at the end of the insert is subcloned and then used in a Southern blot to identify an adjacent clone in a cosmid DNA library. This is the first "step." The end of this clone is subcloned to make the next step. And so on. Eventually, after many steps, you will arrive at your gene of interest.

B. In this example, you would begin at STS-3. If you walked a few steps and happened upon STS-2, you would know that you were walking in the wrong direction.

C. This is a difficult aspect of chromosome walking. Basically, you would walk toward gene *X* using DNA from a normal individual and DNA from an individual with a mutant gene *X*. When you have found a site where the sequences are different between the normal and mutant individual, you may have found gene *X*. You would eventually have to confirm this by analyzing the DNA sequence of this region and determining that it encodes a functional gene.

E32. We can calculate the probability that a base will not be sequenced using this approach with the following equation:

$$P = e^{-m}$$

where

P is the probability that a base will be left unsequenced

e is the base of the natural logarithm; $e = 2.72$

m is the number of bases sequenced divided by the total genome size

In this case, $m = 19$ divided by 4.4, which equals 4.3

$$P = e^{-m} = e^{-4.3} = 0.0136 = 1.36\%$$

This means that if we randomly sequence 19.0 Mb, we are likely to miss only 1.36% of the genome. With a genome size of 4.4 Mb, we would miss about 57,840 base pairs out of approximately 4,400,000.

Questions for Student Discussion/Collaboration

2. A molecular marker is a segment of DNA, not usually encoding a gene, that has a known location within a particular chromosome. It marks the location of a site along a chromosome. RFLPs and STSs are examples. It is easier to use these types of markers because they can be readily identified by molecular techniques such as restriction digestion analysis and PCR. The locations of functional genes is usually more difficult because this relies on conventional genetic mapping approaches whereby allelic differences in the gene are mapped by making crosses or following a pedigree. For monomorphic genes, this approach doesn't work.

CHAPTER 21

Conceptual Questions

C2. There are two main reasons why the proteome is larger than the genome. The first reason involves the processing of pre-mRNA, a phenomenon that occurs primarily in eukaryotic species. RNA splicing and editing can alter the codon sequence of mRNA and thereby produce alternative forms of proteins that have different amino acid sequences. The second reason for protein diversity is posttranslational modifications. There are many ways that a given protein's structure can be covalently modified by cellular enzymes. These include proteolytic processing, disulfide bond formation, glycosylation, attachment of lipids, phosphorylation, methylation, and acetylation, to name a few.

C4. Centromeric sequences, origins of replication, telomeric sequences, repetitive sequences, and enhancers. Other examples are possible.

C6. There are a few interesting trends. Sequences 1 and 2 are similar to each other, as are sequences 3 and 4. There are a few places where amino acid residues are conserved among all five sequences. These amino acids may be particularly important with regard to function.

C8. A. Correct

B. Correct

C. This is not correct. These are short genetic sequences that happen to be similar to each other. The *lac* operon and *trp* operon are not derived from the same primordial operon.

D. This is not correct. The two genes are homologous to each other. It is correct to say that their sequences are 60% identical.

Experimental Questions

E2. As described in solved problem S1, one reason for making a subtractive DNA library is to determine which RNAs are produced when environmental conditions change. You want to load a small amount of cDNA on the column that was derived from cells that had been exposed to mercury. Remember that the cDNA, which is derived from mRNA that is made in the absence of mercury, is already bound to the column. You want the cDNA that is made in the presence of mercury to bind to this cDNA if it is complementary. If too much of this cDNA is loaded, all of the cDNAs will have complementary cDNA bound to them, and some of them will not bind to the column, even though they may be complementary to the cDNAs. In other words, if you load too much cDNA (derived from the mercury-exposed cells), you will have saturated the binding sites for cDNAs that are made in the absence of mercury. You do not want this to happen, because you want only the cDNAs that are derived from mRNAs that are specifically expressed in the presence of mercury to flow through the column. These cDNAs are not complementary to any cDNAs attached to the column.

E4. The cDNA labeled with a green dye is derived from mRNA obtained from cells at an early time point, when glucose levels were high. The other samples of cDNA were derived from cells collected at later time points, when glucose levels were falling and when the diauxic shift was occurring. These were labeled with a red dye. The green fluorescence provides a baseline for gene expression when glucose is high. At later time points, if the red : green ratio is high (i.e., greater than one), this means that a gene is induced as glucose levels fall, because there is more red cDNA compared to green cDNA. If the ratio is low (i.e., less than one), this means that a gene is being repressed.

E6. In the first dimension (i.e., in the tube gel), proteins migrate to the point in the gel where their net charge is zero. In the second dimension (i.e., the slab gel), proteins are coated with SDS and separated according to their molecular mass.

E8. In tandem mass spectroscopy, the first spectrometer determines the mass of a peptide fragment from a protein of interest. The second spectrometer determines the masses of progressively smaller fragments derived from that peptide. Because the masses of each amino acid are known, the molecular masses of these smaller fragments reveal the amino acid sequence of the peptide. With peptide sequence information, it is possible to use the genetic code and produce DNA sequences that could encode such a peptide. More than one sequence is possible due to the degeneracy of the genetic code. These sequences are used as query sequences to search a genomic database. This program will (hopefully) locate a match. The genomic sequence can then be analyzed to determine the entire coding sequence for the protein of interest.

E10. One strategy is a search by signal approach, which relies on known sequences such as promoters, start and stop codons, and splice sites to help predict whether or not a DNA sequence contains a structural gene. It attempts to identify a region that contains a promoter sequence, then a start codon, a coding sequence, and a stop codon. A second strategy is a search-by-content approach. The goal is to identify sequences whose nucleotide content differs significantly from a random distribution, which is usually due to codon bias. A search-by-content approach attempts to locate coding regions by identifying regions where the nucleotide content displays a bias. A third method to locate structural genes is to search for long open reading frames within a DNA sequence. An open reading frame is a sequence that does not contain any stop codons.

E12. By searching a database, one can identify genetic sequences that are homologous to a newly determined sequence. In most cases, homologous sequences carry out identical or very similar functions. Therefore, if one identifies a homologous member of a database whose function is already understood, this provides an important clue regarding the function of the newly determined sequence.

E14. The basis for secondary structure prediction is that certain amino acids tend to be found more frequently in α helices or β sheets. This information is derived from the statistical frequency of amino acids within secondary structures that have already been crystallized. Predictive methods are perhaps 60 to 70% accurate, which is not very good.

E16. The BACKTRANSLATE program works by knowing the genetic code. Each amino acid has one or more codons (i.e., three-base sequences) that are specified by the genetic code. This program would produce a sequence file that was a nucleotide base sequence. The BACKTRANSLATE program would produce a degenerate base sequence because the genetic code is degenerate. For example, lysine can be specified by AAA or AAG. The program would probably store a single file that had degeneracy at particular positions. For example, if the amino acid sequence was lysine–methionine–glycine–glutamine, the program would produce the following sequence:

5'-AA(A/G)ATGGG(T/C/A/G)CA(A/G)

The bases found in parentheses are the possible bases due to the degeneracy of the genetic code.

E18. A. To identify a specific transposable element, a program would use sequence recognition. The sequence of P elements is already known. The program would be supplied with this information and scan a sequence file looking for a match.

B. To identify a stop codon, a program would use sequence recognition. There are three stop codons that are specific three-base sequences. The program would be supplied with these three sequences and scan a sequence file to identify a perfect match.

C. To identify an inversion of any kind, a program would use pattern recognition. In this case, the program would be looking for a pattern in which the same sequence was running in opposite directions in a comparison of the two sequence files.

D. A search by signal approach uses both sequence recognition and pattern recognition as a means to identify genes. It looks for an organization of sequence elements that would form a functional gene. A search-by-content approach identifies genes based on patterns, not on specific sequence elements. This approach looks for a pattern in which the nucleotide content is different from a random distribution. The third approach to identify a gene is to scan a genetic sequence for long open reading frames. This approach is a combination of sequence recognition and pattern recognition. The program is looking for specific sequence elements (i.e., stop codons), but it is also looking for a pattern in which the stop codons are far apart.

E20. A. The amino acids that are most conserved (i.e., the same in all of the family members) are most likely to be important for structure and/or function. This is because a mutation that changed the amino acid might disrupt structure and function, and these kinds of mutations would be selected against. Completely conserved amino acids are found at the following positions: 101, 102, 105, 107, 108, 116, 117, 124, 130, 134, 139, 143, and 147.

B. The amino acids that are least conserved are probably not very important because changes in the amino acid does not seem to inhibit function. (If it did inhibit function, natural selection would eliminate such a mutation.) At one location, position 118, there are five different amino acids.

E22. This sequence is within the *lacY* gene of the *lac* operon of *E. coli.* It is found on page 589, nucleotides 801–850.

E24. A. This sequence has two regions that are about 20 amino acids long and very hydrophobic. Therefore, it is probable that this polypeptide has two transmembrane segments.

B.

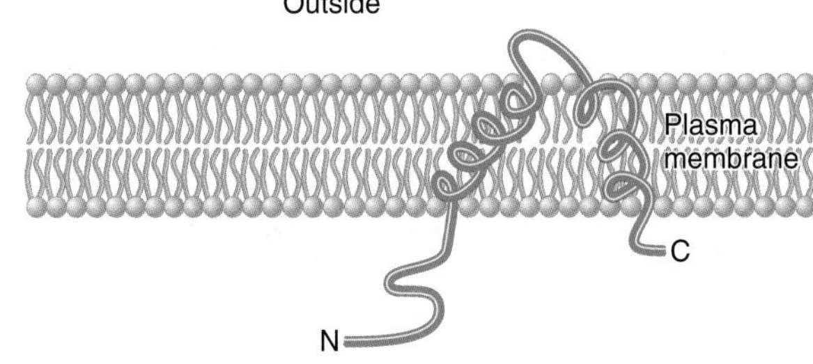

E26. A. True

B. False. The programs are only about 60 to 70% accurate.

C. True

Questions for Student Discussion/Collaboration

2. This is a very difficult question. In 20 years, we may have enough predictive information so that the structure of macromolecules can be predicted from their genetic sequences. If so, it would be better to be a mathematical theoretician with some genetics background. If not, it is probably better to be a biophysicist with some genetics background.

CHAPTER 22

Conceptual Questions

C2. When a disease-causing allele affects a trait, it is causing a deviation from normality, but the gene involved is not usually the only gene that governs the trait. For example, an allele causing hemophilia prevents the normal blood clotting pathway from operating correctly. It follows a simple Mendelian pattern because a single gene affects the phenotype. Even so, it is known that normal blood clotting is due to the actions of many genes.

C4. Changes in chromosome number and unbalanced changes in chromosome structure tend to affect phenotype because they create an imbalance of gene expression. For example, in Down syndrome, there are three copies of chromosome 21 and, therefore, three copies of all the genes on chromosome 21. This leads to a relative overexpression of genes that are located on chromosome 21 compared to the other chromosomes. Balanced translocations and inversions often are without phenotypic consequences because the total amount of genetic material is not altered, and the level of gene expression is not significantly changed.

C6. There are lots of possible answers; here are a few. Dwarfism occurs in people and dogs. Breeds like the dachshund and basset hound are types of dwarfism in dogs. There are diabetic people and mice. There are forms of inherited obesity in people and mice. Hip dysplasia is found in people and dogs.

C8. A. Because a person must inherit two defective copies of this gene and it is known to be on chromosome 1, the mode of transmission is autosomal recessive. Both members of this couple must be heterozygous, because they have one affected parent (who had to transmit the mutant allele to them) and their phenotypes are unaffected (so they must have received the normal allele from their other parent). Because both parents are heterozygotes, there is a $^1/_4$ chance of producing an affected child (a homozygote) with Gaucher disease. If we let *G* represent the nonmutant allele and *g* the mutant allele:

♀ \ ♂	*G*	*g*
G	*GG* Normal	*Gg* Normal
g	*Gg* Normal	*gg* Gaucher

B. From this Punnett square, we can also see that there is a $^1/_4$ chance of producing a homozygote with both normal copies of the gene.

C. We need to apply the binomial expansion equation to solve this problem (see Chapter 2 for a description of this equation). In this problem, $n = 5$, $x = 1$, $p = 0.25$, $q = 0.75$. The answer is 0.396, or 39.6%.

C10. The mode of transmission is autosomal recessive. All of the affected individuals do not have affected parents. Also, the disorder is found in both males and females. If it were X-linked recessive, individual III-1 would have to have an affected father, which she does not.

C12. The 13 babies have acquired a new mutation. In other words, during spermatogenesis or oogenesis, or after the egg was fertilized, a new mutation occurred in the fibroblast growth factor gene. These 13 individuals have the same chances of passing the mutant allele to their offspring as the 18 individuals who inherited the mutant allele from a parent. The chance is 50%.

C14. Because this is a dominant trait, the mother must have two normal copies of the gene, and the father (who is affected) is most likely to be a heterozygote. (Note: The father could be a homozygote, but this is extremely unlikely because the dominant allele is very rare.) If we let *M* represent the mutant Marfan allele and *m* the normal allele, the following Punnett square can be constructed:

♀ \ ♂	*M*	*m*
m	*Mm* Marfan	*mm* Normal
m	*Mm* Marfan	*mm* Normal

A. There is a 50% chance that this couple will have an affected child.

B. We use the product rule. The odds of having an unaffected child are 50%. So if we multiply $0.5 \times 0.5 \times 0.5$, this equals 0.125, or a 12.5% chance of having three unaffected offspring.

C16. A prion is a protein that behaves like an infectious agent. The infectious form of the prion protein has an abnormal conformation. This abnormal conformation is termed PrP^{Sc}, and the normal conformation of the protein is termed PrP^{C}. An individual can be "infected" with the abnormal conformation of the protein by eating products from another animal that had the disease, or the prion protein may convert spontaneously to the normal conformation. A prion protein in the PrP^{Sc} conformation can bind to a prion protein in the PrP^{C} conformation and convert it to the PrP^{Sc} form. An accumulation of prions in the PrP^{Sc} form is what causes the disease symptoms.

C18. A. Keep in mind that a conformational change is a stepwise process. It begins in one region of a protein and involves a series of small changes in protein structure. The entire conformational change from PrP^{C} to PrP^{Sc} probably involves many small changes in protein structure that occur in a stepwise manner. Perhaps the conformational change (from PrP^{C} to PrP^{Sc}) begins in the vicinity of position 178 and then proceeds throughout the rest of the protein. If there is a methionine at position 129, the complete conformational change (from PrP^{C} to PrP^{Sc}) can take place. However, perhaps the valine at position 129 somehow blocks one of the steps needed to complete the conformational change.

B. Once the PrP^{Sc} conformational change is completed, a PrP^{Sc} protein can bind to another prion protein in the PrP^{C} conformation. Perhaps it begins to convert it (to the PrP^{Sc} conformation) by initiating a small change in protein structure in the vicinity of position 178. The conversion would then proceed in a stepwise manner until the PrP^{Sc} conformation has been achieved. If a valine is at position 129, this could somehow inhibit one of the steps that are needed to complete the conformational change. If an individual had Val-129 in the polypeptide encoded by the second *PrP* gene, half of their prion proteins would be less sensitive to conversion by PrP^{Sc}, compared to individuals who had Met-129. This would explain why individuals with Val-129 in half of the prion proteins would have disease symptoms that would progress more slowly.

C20. A proto-oncogene is a normal cellular gene that typically plays a role in cell division. It can be altered by mutation to become an oncogene and thereby cause cancer. At the level of protein function, a proto-oncogene can become an oncogene by synthesizing too much of a protein or synthesizing the same amount of a protein that is abnormally active.

C22. The predisposition to develop cancer is inherited in a dominant fashion because the heterozygote has the higher predisposition. The mutant allele is actually recessive at the cellular level. But because we have so many cells in our bodies, it becomes relatively likely that a defective mutation will occur in the single normal gene and lead to a cancerous cell. Some heterozygous individuals may not develop the disease as a matter of chance. They may be lucky and not get a defective mutation in the normal gene, or perhaps their immune system is better at destroying cancerous cells once they arise.

C24. If an oncogene was inherited, it may cause uncontrollable cell growth at an early stage of development and thereby cause an embryo to develop improperly. This could lead to an early spontaneous abortion and thereby explain why we do not observe individuals with inherited oncogenes. Another possibility is that inherited oncogenes may adversely affect gamete survival, which would make it difficult for them to be passed from parent to offspring. A third possibility would be that oncogenes could affect the fertilized zygote in a way that would prevent the correct implantation in the uterus.

C26. The role of *p53* is to sense DNA damage and prevent damaged cells from proliferating. Perhaps, prior to birth, the fetus is in a protected environment, so DNA damage may be minimal. In other words, the fetus may not really need *p53*. After birth, agents such as UV light may cause DNA damage. At this point, *p53* is important. A *p53* knockout is more sensitive to UV light because it cannot repair its DNA properly in response to this DNA-damaging agent, and it cannot kill cells that have become irreversibly damaged.

C28. The *p*53 protein is a regulatory transcription factor; it binds to DNA and influences the transcription rate of nearby genes. This transcription factor (1) activates genes that promote DNA repair; (2) activates a protein that inhibits cyclin/CDK protein complexes that are required for cell division; and (3) activates genes that promote apoptosis. If a cell is exposed to DNA damage, it has a greater potential to become malignant. Therefore, an organism wants to avoid the proliferation of such a cell. When exposed to an agent that causes DNA damage, a cell will try to repair the damage. However, if the damage is too extensive, the *p*53 protein will stop the cell from dividing and program it to die. This helps to prevent the proliferation of cancer cells in the body.

Experimental Questions

E2. Perhaps the least convincing is the higher incidence of the disease in particular populations. Because populations living in specific geographic locations are exposed to their own unique environment, it is difficult to distinguish genetic versus environmental causes for a particular disease. The most convincing evidence might be the higher incidence of a disease in related individuals and/or the ability to correlate a disease with the presence of a mutant gene. Overall, however, the reliability that a disease has a genetic component should be based on as many observations as possible.

E4. You would probably conclude that it is less likely to have a genetic component. If it were rooted primarily in genetics, it would be likely to be found in the Central American population. Of course, there is a chance that very few or none of the people who migrated to Central America were carriers of the mutant gene, which is somewhat unlikely for a large migrating population. By comparison, one might suspect that an environmental agent present in South America but not present in Central America may underlie the disease. Researchers could try to search for this environmental agent (e.g., a pathogenic organism).

E6. Males I-1, II-1, II-4, II-6, III-3, III-8, and IV-5 have a normal copy of the gene. Males II-3, III-2, and IV-4 are hemizygous for an inactive mutant allele. Females III-4, III-6, IV-1, IV-2, and IV-3 have two normal copies of the gene, whereas females I-2, II-2, II-5, III-1, III-5, and III-7 are heterozygous carriers of a mutant allele.

E8. A transformed cell is one that has become malignant. In a laboratory, this can be done in three ways. First, the cells could be treated with a mutagen that would convert a proto-oncogene into an oncogene. Second, cells could be exposed to the DNA from a malignant cell line. Under the appropriate conditions, this DNA can be taken up by the cells and integrated into their genome so that they become malignant. A third way to transform cells is by exposure to an oncogenic virus.

E10. By comparing oncogenic viruses with strains that have lost their oncogenicity, researchers have been able to identify particular genes that cause cancer. This has led to the identification of many oncogenes. From this work, researchers have also learned that normal cells contain proto-oncogenes that usually play a role in cell division. This suggests that oncogenes exert their effects by upsetting the cell division process. In particular, it appears that oncogenes are abnormally active and keep the cell division cycle in a permanent "on" position.

E12. One possible category of drugs would be GDP analogues (i.e., compounds that resemble the structure of GDP). Perhaps one could find a GDP analogue that binds to the Ras protein and locks it in the inactive conformation.

One way to test the efficacy of such a drug would be to incubate the drug with a type of cancer cell that is known to have an overactive Ras protein and then plate the cells on solid media. If the drug locked the Ras protein in the inactive conformation, it should inhibit the formation of malignant growth or malignant foci.

There are possible side effects of such drugs. First, they might block the growth of normal cells, because Ras protein plays a role in normal cell proliferation. Second (if you have taken a cell biology course), there are many GTP/GDP-binding proteins in cells, and the drugs could somehow inhibit cell growth and function by interacting with these proteins.

Questions for Student Discussion/Collaboration

2. There isn't a clearly correct answer to this question, but it should stimulate a large amount of discussion.

CHAPTER 23

Conceptual Questions

C2. A. False; the head is anterior to the tail.
B. True
C. False; the feet are ventral to the hips.
D. True

C4. A. True
B. False; because gradients are also established after fertilization during embryonic development.
C. True

C6. A. This is a mutation in a pair-rule gene (*runt*).
B. This is a mutation in a gap gene (*knirps*).
C. This is a mutation in a segment-polarity gene (*patched*).

C8. Positional information refers to the phenomenon whereby the spatial locations of morphogens and cell adhesion molecules (CAMs) provide a cell with information regarding its position relative to other cells. In *Drosophila*, the formation of a segmented body pattern relies initially on the spatial location of maternal gene products. These gene products lead to the sequential activation of the segmentation genes. Positional information can come from morphogens that are found within the oocyte, from morphogens secreted from cells during development, and from cell-to-cell contact. Although all three are important, morphogens in the oocyte have the greatest impact on the overall body structure.

C10. The anterior portion of the antero-posterior axis is established by the action of Bicoid. During oogenesis, the mRNA for Bicoid enters the anterior end of the oocyte and is sequestered there to establish an anterior (high) to posterior (low) gradient. Later, when the mRNA is translated, the Bicoid protein in the anterior region establishes a genetic hierarchy that leads to the formation of anterior structures. If Bicoid was not trapped in the anterior end, it is likely that anterior structures would not form.

C12. Maternal effect gene products influence the formation of the main body axes, including the antero-posterior, dorso-ventral, and terminal regions. They are expressed during oogenesis and needed very early in development. Zygotic genes, particularly the three classes of the segmentation genes, are necessary after the axes have been established. The segmentation genes are expressed after fertilization.

C14. The coding sequence of homeotic genes contains a 180-bp consensus sequence known as a homeobox. The protein domain encoded by the homeobox is called a homeodomain. The homeodomain contains three conserved sequences that are folded into α-helical conformations. The arrangement of these α helices promotes the binding of the protein to the major groove of the DNA. Helix III is called the recognition helix because it recognizes a particular nucleotide sequence within the major groove. In this way, homeotic proteins are able to bind to DNA in a sequence-specific manner and thereby activate particular genes.

C16. It would normally be expressed in the three thoracic segments that have legs (T1, T2, and T3).

C18. A. When a mutation inactivates a gap gene, a contiguous section of the larva is missing.

B. When a mutation inactivates a pair-rule gene, some regions that are derived from alternating parasegments are missing.

C. When a mutation inactivates a segment-polarity gene, portions are missing at either the anterior or posterior end of the segments.

C20. Proper development in mammals is likely to require the products of maternal effect genes that play a key role in initiating embryonic development. The adult body plan is merely an expansion of the embryonic body plan, which is established in the oocyte. Because the starting point for the development of an embryo is the oocyte, this explains why an enucleated oocyte is needed to clone mammals.

C22. A heterochronic mutation is one that alters the timing when a gene involved in development is normally expressed. The gene may be expressed too early or too late, which causes certain cell lineages to be out of sync with the rest of the animal. If a heterochronic mutation affected the intestine, the animal may end up with too many intestinal cells if it is a gain-of-function mutation or too few if it is a loss-of-function mutation. In either case, the effects might be detrimental because the growth of the intestine must be coordinated with the growth of the rest of the animal.

C24. Cell differentiation is the specialization of a cell into a particular cell type. In the case of skeletal muscle cells, the bHLH proteins play a key role in the initiation of cell differentiation. When bHLH proteins are activated, they are able to bind to enhancers and activate the expression of many different muscle-specific genes. In this way, myogenic bHLH proteins turn on the expression of many muscle-specific proteins. When these proteins are synthesized, they change the characteristics of the cell into those of a muscle cell. Myogenic bHLH proteins are regulated by dimerization. When a heterodimer forms between a myogenic bHLH protein and an E protein, it activates gene expression. However, when a heterodimer forms between myogenic bHLH proteins and a protein called Id, the heterodimer is unable to bind to DNA. The Id protein is produced during early stages of development and prevents myogenic bHLH proteins from promoting muscle differentiation too soon. At later stages of development, the amount of Id protein falls, and myogenic bHLH proteins can combine with E proteins to induce muscle differentiation.

C26. A totipotent cell is a cell that has the potential to create a complete organism.

A. In humans, a fertilized egg is totipotent, and the cells during the first few embryonic divisions are totipotent. However, after several divisions, embryonic cells lose their totipotency and, instead, are determined to become particular tissues within the body.

B. In plants, many living cells are totipotent.

C. Because yeast are unicellular, one cell is a complete individual. Therefore, yeast cells are totipotent; they can produce new individuals by cell division.

D. Because bacteria are unicellular, one cell is a complete individual. Therefore, bacteria are totipotent; they can produce new individuals by cell division.

C28. Animals begin their development from an egg and then form antero-posterior and dorso-ventral axes. The formation of an adult organism is an expansion of the embryonic body plan. Plants grow primarily from two meristems: shoot and root meristems. At the cellular level, plant development is different in that it does not involve cell migration, and most plant cells are totipotent. Animals require the organization within an oocyte to begin development. At the genetic level, however, animal and plant development are similar in that a genetic hierarchy of transcription factors governs pattern formation and cell specialization.

Experimental Questions

E2. *Drosophila* has an advantage in that researchers have identified many more mutant alleles that alter development in specific ways. The hierarchy of gene regulation is particularly well understood in the fruit fly. *C. elegans* has the advantage of simplicity and a complete knowledge of cell fate. This enables researchers to explore how the timing of gene expression is critical to the developmental process.

E4. To determine that a mutation is affecting the timing of developmental decisions, a researcher needs to know the normal time or stage of development when cells are supposed to divide and what types of cells will be produced. A lineage diagram provides this. With this information, one can then determine if particular mutations alter the timing when cell division occurs.

E6. Mutant 1 is a gain-of-function allele; it keeps reiterating the L1 pattern of division. Mutant 2 is a loss-of-function allele; it skips the L1 pattern and immediately follows an L2 pattern.

E8. As discussed in Chapter 15, most eukaryotic genes have a core promoter that is adjacent to the coding sequence; regulatory elements that control the transcription rate at the promoter are typically upstream from the core promoter. Therefore, to get the *Antp* gene product expressed where the *abd-A* gene product is normally expressed, you would link the upstream genetic regulatory region of the *abd-A* gene to the coding sequence of the *Antp* gene. This construct would be inserted into the middle of a P element (see next). The construct shown here would then be introduced into an embryo by P element transformation.

P element	***abd-A* regulatory region**	/	***Antp* coding sequence**	P element

The *Antp* gene product is normally expressed in the thoracic region and produces segments with legs, as illustrated in Figure 23.11. Therefore, because the *abd-A* gene product is normally expressed in the anterior abdominal segments, one might predict that the genetic construct shown above would produce a fly with legs attached to the segments that are supposed to be the anterior abdominal segments. In other words, the anterior abdominal segments might resemble thoracic segments with legs.

E10. A. The female flies must have had mothers that were heterozygous for a (dominant) normal allele and the mutant allele. Their fathers were either homozygous for the mutant allele or heterozygous. The female flies inherited a mutant allele from both their father and mother. Nevertheless, because their mother was heterozygous for the normal (dominant) allele and mutant allele, and because this is a maternal effect gene, their phenotype is based on the genotype of their mother. The normal allele is dominant, so they have a normal phenotype.

B. *Bicoid-A* appears to have a deletion that removes part of the sequence of the gene and thereby results in a shorter mRNA. *Bicoid-B* could also have a deletion that removes all of the sequence of the *bicoid* gene, or it could have a promoter mutation that prevents the expression of the bicoid gene. *Bicoid-C* seems to have a point mutation that does not affect the amount of the bicoid mRNA.

With regard to function, all three mutations are known to be loss-of-function mutations. *Bicoid-A* probably eliminates function by truncating the Bicoid protein. The Bicoid protein is a transcription factor. The *bicoid-A* mutation probably shortens this protein and thereby inhibits its function. The *bicoid-B* mutation prevents expression of the bicoid mRNA. Therefore, none of the Bicoid protein would be made, and this would explain the loss of function. The *bicoid-C* mutation seems to prevent the proper localization of the *bicoid* mRNA in the oocyte. There must be proteins within the oocyte that recognize specific sequences in the *bicoid* mRNA and trap it in the anterior end of the oocyte. This mutation must change these sequences and prevent these proteins from recognizing the bicoid mRNA.

E12. An egg-laying defect is somehow related to an abnormal anatomy. The *n540* strain has fewer neurons compared to a normal worm. Perhaps the *n540* strain is unable to lay eggs because it is missing neurons that are needed for egg laying. The *n536* and *n355* strains have an abnormal abundance of neurons. Perhaps this overabundance also interferes with the proper neural signals needed for egg laying.

E14. Geneticists interested in mammalian development have used reverse genetics because it has been difficult for them to identify mutations in developmental genes based on phenotypic effects in the embryo. This is because it is difficult to screen a large number of mammalian embryos in search of abnormal ones that carry mutant genes. It is easy to have thousands of flies in a laboratory, but it is not easy to have thousands of mice. Instead, it is easier to clone the normal gene based on its homology to invertebrate genes and then make mutations in vitro. These mutations can be introduced into a mouse to create a gene knockout. This strategy is opposite to that of Mendel, who characterized genes by first identifying phenotypic variants (e.g., tall versus dwarf, green seeds versus yellow seeds, etc.).

Questions for Student Discussion/Collaboration

2. In this problem, the students should try to make a flow diagram that begins with maternal effect genes, then gap genes, pair-rule genes, and segment polarity genes. These genes then lead to homeotic genes and finally genes that promote cell differentiation. It's almost impossible to make an accurate flow diagram because there are so many gene interactions, but it is instructive to think about developmental genetics in this way. It is probably easier to identify mutant phenotypes that affect later stages of development because they are less likely to be lethal. However, modern methods can screen for conditional mutants as described in solved problem S3. To identify all of the genes necessary for chicken development, you may begin with early genes, but this assumes you have some way to identify them. If they had been identified, you would then try to identify the genes that they stimulate or repress. This could be done using molecular methods described in Chapters 14, 15, and 18.

CHAPTER 24

Conceptual Questions

C2. A population is a group of interbreeding individuals. Let's consider a squirrel population in a forested area. Over the course of many generations, several things could happen to this population. A forest fire, for example, could dramatically decrease the number of individuals and thereby cause a bottleneck. This would decrease the genetic diversity of the population. A new predator may enter the region and natural selection may select for the survival of squirrels that are best able to evade the predator. Another possibility is that a group of squirrels within the population may migrate to a new region and found a new squirrel population.

C4. A. Phenotype frequency and genotype frequency

B. Genotype frequency

C. Allele frequency

C6. A. The genotype frequency for the *CF* homozygote is $^1/_{2,500}$, or 0.004. This would equal q^2. The allele frequency is the square root of this value, which equals 0.02. The frequency of the corresponding normal allele equals $1 - 0.02 = 0.98$.

B. The frequency for the *CF* homozygote is 0.004; for the unaffected homozygote, $(098)^2 = 0.96$; and for the heterozygote, $2(0.98)(0.02)$, which equals 0.039.

C. If a person is known to be a heterozygous carrier, the chances that this particular person will happen to choose another as a mate is equal to the frequency of heterozygous carriers in the population, which equals 0.039, or 3.9%. The chances that two randomly chosen individuals will choose each other as mates equals $0.039 \times 0.039 = 0.0015$, or 0.15%.

C8. For two alleles, the heterozygote is at a maximum when they are 0.5 each. For three alleles, the two heterozygotes are at a maximum when each allele is 0.33.

C10. Because this is a recessive trait, only the homozygotes for the rolling allele will be able to roll their tongues. If p equals the rolling allele and q equals the nonrolling allele, the Hardy-Weinberg equation predicts that the frequency of homozygotes who can roll their tongues would be p^2. In this case, $p^2 = (0.6)^2 = 0.36$, or 36%.

C12. A. The inbreeding coefficient is calculated using the formula

$$F = \Sigma\, (^1/_2)^n\, (1 + F_A)$$

In this case, there is one common ancestor, I-2. Because we have no prior history on I-2, we assume she is not inbred, which makes $F_A = 0$. The inbreeding loop for IV-3 contains five people, III-4, II-2, I-2, II-5, and III-5. Therefore, $n = 5$.

$$F = (^1/_2)^5(1 + 0) = {}^1/_{32} = 0.031$$

B. Based on the data shown in this pedigree, individual IV-4 is not inbred.

C14. Migration, genetic drift, and natural selection are the driving forces that alter allele frequencies within a population. Natural selection acts to eliminate harmful alleles and promote beneficial alleles. Genetic drift involves random changes in allele frequencies that may eventually lead to elimination or fixation of alleles. It is thought to be important in the establishment of neutral alleles in a population. Migration is important because it introduces new alleles into neighboring populations. According to the neutral theory, genetic drift is largely responsible for the variation seen in natural populations.

C16. A random force is one that alters allele frequencies without any regard to whether the changes are beneficial or not. Genetic drift and migration are the two main ways this can occur. An example is the founder effect, when a group of individuals migrate to a new location such as an island. Adaptive forces increase the reproductive success of a species. Natural selection is the adaptive force that tends to eliminate harmful alleles from a population and increase the frequency of beneficial alleles. An example would be the long neck of the giraffe, which enables it to feed in tall trees. At the molecular level, beneficial mutations may alter the coding sequence of a gene and change the structure and function of the protein in a way that is beneficial. For example, the sickle-cell anemia allele alters the structure of hemoglobin, and in the heterozygous condition, this inhibits the sensitivity of red blood cells to the malaria pathogen.

C18. Genetic drift is due to sampling error, and the degree of sampling error depends on the population size. In small populations, the relative proportion of sampling error is much larger. If genetic drift is moving an allele toward fixation, it will take longer in a large population because the degree of sampling error is much smaller.

C20. During the bottleneck effect, allele frequencies are dramatically altered due to genetic drift. In extreme cases, some alleles are lost, while others may become fixed at 100%. The overall effect is to decrease genetic diversity within the population. This may make it more difficult for the species to respond in a positive way to changes in the environment. Species that are approaching extinction also face a bottleneck as their numbers decrease. The loss of genetic diversity may make it even more difficult for the species to rebound.

C22. In all cases, these forms of natural selection favor one or more phenotypes because such phenotypes have a reproductive advantage. However, the patterns differ with regard to whether a single phenotype or multiple phenotypes are favored and whether the favored phenotype is in the middle of the phenotypic range or at one or both extremes. Directional selection favors one phenotype at a phenotypic extreme. Over time, natural selection is expected to favor the fixation of alleles that cause these phenotypic characteristics. Disruptive selection favors two or more phenotypic categories. It will lead to a population with a balanced polymorphism for the trait. Examples of balancing selection are heterozygote advantage and negative-frequency dependent selection. These promote a stable polymorphism in a population. Stabilizing selection favors individuals with intermediate phenotypes. It tends to decrease genetic diversity because alleles that favor extreme phenotypes are eliminated.

C24. The intuitive meaning of the mean fitness of a population is the relative likelihood that members of a population will reproduce. If the mean fitness is high, it is likely that an average member will survive and produce offspring. Natural selection increases the mean fitness of a population.

C26. A. True

B. True

C. False; it causes allele loss or fixation, which results in less diversity.

D. True

C28. A. Disruptive. There are multiple environments that favor different phenotypes.

B. Directional. The thicker the fur, the more likely that survival will occur.

C. Stabilizing. Low birth weight is selected against because it results in low survival. Also, very high birth weight is selected against because it could cause problems in delivery, which also could decrease the survival rate.

D. Directional. Sturdy stems and leaves will promote survival in windy climates.

Experimental Questions

E2. Solved problem S4 shows how the Hardy-Weinberg equation can be modified to include situations of three or more alleles. In this case

$(p + q + r + s)^2 = 1$

$p^2 + q^2 + r^2 + s^2 + 2pq + 2qr + 2qs + 2rp + 2rs + 2sp = 1$

Let $p = C$, $q = c^{ch}$, $r = c^h$, and $s = c$.

A. The frequency of albino rabbits is s^2:

$s^2 = (0.05) = 0.0025 = 0.25\%$

B. Himalayan is dominant to albino but recessive to full and chinchilla. Therefore, Himalayan rabbits would be represented by r^2 and by $2rs$:

$r^2 + 2rs = (0.44)^2 + 2(0.44)(0.05) = 0.24 = 24\%$

Among 1,000 rabbits, about 240 would have a Himalayan coat color.

C. Chinchilla is dominant to Himalayan and albino but recessive to full coat color. Therefore, heterozygotes with chinchilla coat color would be represented by $2qr$ and by $2qs$.

$2qr + 2qs = 2(0.17)(0.44) + 2(0.17)(0.05) = 0.17$, or 17%

Among 1,000 rabbits, about 170 would be heterozygotes with chinchilla fur.

E4. A.

Inuit	$M = 0.913$	$N = 0.087$
Navajo	$M = 0.917$	$N = 0.083$
Finns	$M = 0.673$	$N = 0.327$
Russians	$M = 0.619$	$N = 0.381$
Aborigines	$M = 0.176$	$N = 0.824$

B. To determine if these populations are in equilibrium, we can use the Hardy-Weinberg equation and calculate the expected number of individuals with each genotype. For example

Inuit $MM = (0.913)^2 = 0.833 = 83.3\%$

$MN = 2(0.913)(0.087) = 0.159 = 15.9\%$

$NN = (0.087)^2 = 0.0076 = 0.76\%$

In general, the values agree pretty well with an equilibrium. The same is true for the other four populations.

C. Based on similar allele frequencies, the Inuit and Navajo Indians seem to have interbred, as well as the Finns and Russians.

E6. A. $\Delta p_C = m(p_D - p_R)$

With regard to the sickle-cell allele,

$\Delta p_C = (550/10{,}550)(0.1 - 0.01) = 0.0047$

$p_C = p_R + \Delta p_C = 0.01 + 0.0047 = 0.0147$

B. We need to calculate the genotypes separately:

For the 550 migrating individuals,

$Hb^AHb^A = (0.9)2 = 0.81$, or 81%
We expect (0.81)550 = 445.5 individuals to have this genotype.

$Hb^AHb^S = 2(0.9)(0.1) = 0.18$
We expect (0.18)550 = 99 heterozygotes.

$Hb^SHb^S = (0.1)^2 = 0.01$
We expect (0.01)550 = 5.5 Hb^SHb^S.

For the original recipient population,

$Hb^AHb^A = (0.99)^2 = 0.98$
We expect 9,801 individuals to have this genotype.

$Hb^AHb^S = 2(0.99)(0.01) = 0.0198$
We expect 198 with this genotype.

$Hb^SHb^S = (0.01)^2 = 0.0001$
We expect 1 with this genotype.

To calculate the overall population,

$(445.5 + 9801)/10{,}550 = 0.971$ Hb^AHb^A homozygotes

$(99 + 198)/10{,}550 = 0.028$ heterozygotes

$(5.5 + 1)/10{,}550 = 0.00062$ Hb^SHb^S homozygotes

C. After one round of mating, the allele frequencies in the conglomerate (calculated in part A), should yield the expected genotype frequencies according to the Hardy-Weinberg equilibrium.

Allele frequency of $Hb^S = 0.0147$, so $Hb^A = 0.985$.

$Hb^AHb^A = (0.985)^2 = 0.97$

$Hb^AHb^S = 2(0.985)(0.0147) = 0.029$

$Hb^SHb^S = (0.0147)^2 = 0.0002$

E8. A. Probability of fixation = $^1/_2N$ (Assuming equal numbers of males and females contributing to the next generation)

Probability of fixation = $^1/_2(2{,}000{,}000)$

= 1 in 4,000,000 chance

B. $\overline{t} = 4N$

Where $\overline{t}$ = the average number of generations to achieve fixation

N = the number of individuals in population, assuming that males and females contribute equally to each succeeding generation

$\overline{t} = 4$(2 million) = 8 million generations

C. If the blue allele had a selective advantage, the value calculated in part A would be slightly larger; there would be a higher chance of allele fixation. The value calculated in part B would be smaller; it would take a shorter period of time to reach fixation.

E10. You could mark snails with a dye and release equal numbers of dark and light snails into dimly lit forested regions and sunny fields. At a later time, recapture the snails and count them. It would be important to have a method of unbiased recapture because the experimenter would have an easier time locating the light snails in a forest and the dark snails in a field. Perhaps one could bait the region with something that the snails like to eat and only collect snails that are at the bait. In addition to this type of experiment, one could also observe predation as it occurs.

E12. Male 2 is the potential father, because he contains the bands found in the offspring but not found in the mother. To calculate the probability, one would have to know the probability of having each of the types of bands that match. In this case, for example, male 2 and the offspring have four bands in common. As a simple calculation, we could eliminate the four bands the offspring shares with the mother. If the probability of having each paternal band is $^1/_4$, the chances that this person is not the father are $(^1/_4)^4$.

E14. PCR is used to amplify DNA if there is only a small amount of it (e.g., a small sample at a crime scene). It is also used to amplify microsatellites. Southern blotting, using a probe complementary to a minisatellite, is needed to specifically identify a limited number of bands (20 or so) that are variable within human populations.

E16. This percentage is not too high. Based on their genetic relationship, we expect that a father and daughter must share at least 50% of the same bands in a DNA fingerprint. However, the value can be higher than that because the mother and father may have some bands in common, even though they are not genetically related. For example, at one site in the genome, the father may be heterozygous for a 4,100 bp and 5,200 bp microsatellite, and the mother may also be heterozygous in this same region and have 4,100 bp and 4,700 bp microsatellites. The father could pass the 5,200 bp band to his daughter, and the mother could pass the 4,100 bp band. Thus, the daughter would inherit the 4,100 bp and 5,200 bp bands. This would be a perfect match to both of the father's bands, even though the father transmitted only the 5,200 bp band to his daughter. The 4,100 bp band matches because the father and mother happened to have a microsatellite in common. Therefore, the 50% estimate of matching bands in a DNA fingerprint based on genetic relationships is a minimum estimate. The value can be higher than that.

Questions for Student Discussion/Collaboration

2. Mutation is responsible for creating new alleles, but the rate of new mutations is so low that it cannot explain allele frequencies in this range. Let's call the two alleles *B* and *b* and assume that *B* was the original allele and *b* is more recent allele that arose as a result of mutation. Three scenarios to explain the allele frequencies:

   1. The *b* allele is neutral and reached its present frequency by genetic drift. It hasn't reached elimination or fixation yet.

   2. The *b* allele is beneficial, and its frequency is increasing due to natural selection. However, there hasn't been enough time to reach fixation.

   3. The *Bb* heterozygote is at a selective advantage, leading to a balanced polymorphism.

CHAPTER 25

Conceptual Questions

C2. At the molecular level, quantitative traits often exhibit a continuum of phenotypic variation because they are usually influenced by multiple genes that exist as multiple alleles. A large amount of environmental variation will also increase the overlap between genotypes and phenotypes for polygenic traits.

C4. A discontinuous trait is one that falls into discrete categories. Examples include brown eyes versus blue eyes in humans and purple versus white flowers in pea plants. A continuous trait is one that does not fall into discrete categories. Examples include height in humans and fruit weight in tomatoes. Most quantitative traits are continuous; the trait falls within a range of values. The reason why quantitative traits are continuous is because they are usually polygenic and greatly influenced by the environment. As shown in Figure 25.4b, this tends to create ambiguities between genotypes and a continuum of phenotypes.

C6. To be in the top 2.5% is about two standard deviation units. If we take the square root of the variance, the standard deviation would be 22 lb. To be in the top 2.5%, an animal would have to weigh at least 44 lb heavier than the mean, which equals 606 lb. To be in the bottom 0.13%, an animal would have to be three standard deviations lighter, which would be at least 66 lb lighter than the mean, or 496 lb.

C8. There is a positive correlation, but it could have occurred as a matter of chance alone. According to Table 25.2, this value could have occurred through random sampling error. You would need to conduct more experimentation to determine if there is a significant correlation, such as examining a greater number of pairs of individuals. If $N =$ 500, the correlation would be statistically significant, and you would conclude that the correlation did not occur as a matter of random chance. However, you could not conclude cause and effect.

C10. When a correlation coefficient is statistically significant, it means that the association is likely to have occurred for reasons other than random sampling error. It may indicate cause and effect but not necessarily. For example, large parents may have large offspring due to genetics (cause and effect). However, the correlation may be related to the sharing of similar environments rather than cause and effect.

C12. Quantitative trait loci are sites within chromosomes that contain genes that affect a quantitative trait. It is possible for a QTL to contain one gene, or it may contain two or more closely linked genes. QTL mapping, which involves linkage to known molecular markers, is commonly used to determine the locations of QTLs.

C14. If the broad sense heritability equals 1.0, it means that all of the variation in the population is due to genetic variation rather than environmental variation. It does not mean that the environment is unimportant in the outcome of the trait. Under another set of environmental conditions, the trait may have turned out quite differently.

C16. Hybrid vigor is the phenomenon in which an offspring produced from two inbred strains is more vigorous than the corresponding parents. Tomatoes and corn are often the products of hybrids.

C18. When a species is subjected to selective breeding, the breeder is focusing his/her attention on improving one particular trait. In this case, the rose breeder is focused on the size and quality of the flowers. Because the breeder usually selects a small number of individuals (e.g., the ones with best flowers) as the breeding stock for the next generation, this may lead to a decrease in the allelic diversity at other genes. For example, several genes affect flower fragrance. In an unselected population, these genes may exist as "fragrant alleles" and "nonfragrant alleles." After many generations of breeding for large flowers, the fragrant alleles may be lost from the population, just as a matter of random chance. This is a common problem of selective breeding. As you select for an improvement in one trait, you may inadvertently diminish the quality of an unselected trait.

Others have suggested that the lack of fragrance may be related to flower structure and function. Perhaps the amount of energy that a flower uses to make beautiful petals somehow diminishes its capacity to make fragrance.

C20. Broad-sense heritability takes into account all genetic factors that affect the phenotypic variation in a trait. Narrow-sense heritability considers only alleles that behave in an additive fashion. In many cases, the alleles affecting quantitative traits appear to behave additively. More importantly, if a breeder assumes that the heritability of a trait is due to the additive effects of alleles, it is possible to predict the outcome of selective breeding. This is also termed the *realized heritability.*

C22. A. Because of their good nutrition, you may speculate that they would grow to be taller.

B. If the environment is rather homogeneous, then heritability values tend to be higher because the environment contributes less to the amount of variation in the trait. Therefore, in the commune, the heritability might be higher, because they uniformly practice good nutrition. On the other hand, because the commune is a smaller size than the general population, the amount of genetic variation might be less, so this would make the heritability lower. However, because the problem states that the commune population is large, we would probably assume that the amount of genetic variation is similar to that in the general population. Overall, the best guess would be that the heritability in the commune population is higher because of the uniform nutrition standards.

C. As stated in part B, the amount of variation would probably be similar, because the commune population is large. As a general answer, larger populations tend to have more genetic variation. Therefore, the general population probably has a bit more variation.

C24. A natural population of animals is more likely to have a higher genetic diversity compared to a domesticated population. This is because domesticated populations have been subjected to many generations of selective breeding, which decreases the genetic diversity. Therefore, V_G is likely to be higher for the natural population. The other issue is the environment. It is difficult to say which group would have a more homogeneous environment. In general, natural populations tend to have a more heterogeneous environment, but not always. If the environment is more heterogeneous, this tends to cause more phenotypic variation, which makes V_E higher.

Heritability $= V_G/V_T$

$= V_G/(V_G + V_E)$

When V_G is high, heritability increases. When V_E is high, heritability decreases. In the natural wolf population, we would expect that V_G would be high. In addition, we would guess that V_E might be high as well (but that is less certain). Nevertheless, if this were the case, the heritability of the wolf population might be similar to the domestic population. This is because the high V_G in the wolf population is balanced by its high V_E. On the other hand, if V_E is not that high in the wolf population, or if it is fairly high in the domestic population, then the wolf population would have a higher heritability for this trait.

Experimental Questions

E2. To calculate the mean, we add the values together and divide by the total number.

$$\text{Mean} = \frac{1.9 + 2(2.4) + 2(2.1) + 3(2.0) + 2(2.2) + 1.7 + 1.8 + 2(2.3) + 1.6}{15}$$

Mean = 2.1

The variance is the sum of the squared deviations from the mean divided by $N - 1$. The mean value of 2.1 must be subtracted from each value, and then the square is taken. These 15 values are added together and then divided by 14 (which is $N - 1$).

$$\text{Variance} = \frac{0.85}{14}$$

$= 0.061$

The standard deviation is the square root of the variance.

Standard deviation = 0.25

E4. The results are consistent with the idea that there are QTLs for this trait on chromosomes 2 and 3 but not on the X chromosome.

E6. When we say an RFLP is associated with a trait, we mean that a gene that influences a trait is closely linked to an RFLP. At the chromosomal level, the gene of interest is so closely linked to the RFLP that a crossover almost never occurs between them.

Note: Each plant inherits four RFLPs, but it may be homozygous for one or two of them.

Small: 2,700 and 4,000 (homozygous for both)

Small-medium: 2,700 (homozygous), 3,000, and 4,000; or 2,000, 2,700, and 4,000 (homozygous)

Medium: 2,000 and 4,000 (homozygous for both); or 2,700 and 3,000 (homozygous for both); or 2,000, 2,700, 3,000, and 4,000

Medium-large: 2,000 (homozygous), 3,000, and 4,000; or 2,000, 2,700, and 3,000 (homozygous)

Large: 2,000 and 3,000 (homozygous for both)

E8. Let's assume there is an extensive molecular marker map for the rice genome. We would begin with two strains of rice, one with a high yield and one with a low yield, that greatly differ with regard to the molecular markers they carry. We would make a cross between these two strains to get F_1 hybrids. We would then backcross the F_1 hybrids to either of the parental strains and then examine hundreds of offspring with regard to their rice yields and molecular markers. In this case, our expected results would be that six different markers in the high-producing strain would be correlated with offspring that produce higher yields. We might get fewer than six bands if some of these genes are closely linked and associate with the same marker. We also might get fewer than six if the two parental strains have the same marker that is associated with one or more of the genes that affect yield.

E10. A. If we assume that the highly inbred strain has no genetic variance,

V_G (for the wild strain) $= 3.2\ g^2 - 2.2\ g^2 = 1.0\ g^2$

B. $h_B^2 = 1.0\ g^2/3.2\ g^2 = 0.31$

C. It is the same as h_B^2, so it also equals 0.31.

E12. A. $h_N^2 = \dfrac{\overline{X}_O - \overline{X}}{\overline{X}_P - \overline{X}}$

$0.21 = (26.5\ g - 25\ g)/(27\ g - 25\ g)$

$\overline{X}_O - 25\ g = 2\ g\ (0.21)$

$\overline{X}_O = 25.42\ g$

B. $0.21 = (26.5\ g - 25\ g)/(\overline{X}_P - 25\ g)$

$(\overline{X}_P - 25\ g)(0.21) = 1.5\ g$

$\overline{X}_P = 32.14\ g$ parents

However, because this value is so far from the mean, there may not be 32.14 g parents in the population of mice that you have available.

E14. We first need to calculate a and b. In this calculation, X represents the height of fathers, and Y represents the height of sons.

$b = \dfrac{144}{112} = 1.29$

$a = 69 - (1.29)(68) = -18.7$

For a father who is 70 in. tall,

$Y = (1.29)(70) + (-18.7) = 71.6$

The most likely height of the son would be 71.6 in.

E16. A. After six or seven generations, the selective breeding seems to have reached a plateau. This suggests that the tomato plants have become monomorphic for the alleles that affect tomato weight.

B. There does seem to be heterosis, because the first generation has a weight of 1.7 lb, which is heavier than either Mary's or Hector's tomatoes. This partially explains why Martin has obtained tomatoes heavier than 1.5 lb. However, heterosis is not the whole story; it does not explain why Martin obtained tomatoes that weigh 2 lb. Even though Mary's and Hector's tomatoes were selected for heavier weight, they may not have all of the "heavy alleles" for each gene that controls weight. For example, let's suppose there are 20 genes that affect weight, with each gene existing in a light and heavy allele. During the early stages of selective breeding, when Mary and Hector picked their 10 plants as seed producers for the next generation, as a matter of random chance, some of these plants may have been homozygous for the light alleles at a few of the 20 genes that control weight. Therefore, just as a matter of chance, they probably "lost" a few of the heavy alleles that affect weight. So, after 12 generations of breeding, they have predominantly heavy alleles but also have light alleles for some of the genes. If we represent heavy alleles with a capital letter and light alleles with a lowercase letter, Mary's and Hector's strains could be the following:

Mary's strain: *AA BB cc DD EE FF gg hh II JJ KK LL mm NN OO PP QQ RR ss TT*

Hector's strain: *AA bb CC DD EE ff GG HH II jj kk LL MM NN oo PP QQ RR SS TT*

As we see here, Mary's strain is homozygous for the heavy allele at 15 of the genes but carries the light allele at the other 5. Similarly, Hector's strain is homozygous for the heavy allele at 15 genes and carries the light allele at the other 5. It is important to note, however, that the light alleles in Mary's and Hector's strains are not in the same genes. Therefore, when Martin crosses them together, he will initially get the following:

Martin's F_1 offspring: *AA Bb Cc DD EE Ff Gg Hh II Jj Kk LL Mm NN Oo PP QQ RR Ss* TT

If the alleles are additive and contribute equally to the trait, we would expect about the same weight (1.5 lb), because this hybrid has a total of 10 light alleles. However, if heterosis is occurring, genes (which were homozygous recessive in Mary's and Hector's strains) will become heterozygous in the F_1 offspring, and this may make the plants healthier and contribute to a higher weight. If Martin's F_1 strain is subjected to selective breeding, the 10 genes that are heterozygous in the F_1 offspring may eventually become homozygous for the heavy allele. This would explain why Martin's tomatoes achieved a weight of 2.0 pounds after five generations of selective breeding.

E18. A. $$h_N^2 = \frac{\overline{X}_O - \overline{X}}{\overline{X}_P - \overline{X}}$$

$$h_N^2 = \frac{269 - 254}{281 - 254} = 0.56$$

B. $$0.56 = \frac{275 - 254}{\overline{X}_P - 254}$$

$$\overline{X}_P = 291.5 \text{ lb}$$

Questions for Student Discussion/Collaboration

2. Most traits depend on the influence of many genes. Also, genetic variation is a common phenomenon in most populations. Therefore, most individuals have a variety of alleles that contribute to a given trait. For quantitative traits, some alleles may make the trait bigger, and other alleles may make the trait turn out smaller. If a population contains many different genes and alleles that govern a quantitative trait, most individuals will have an intermediate phenotype because they will have inherited some large and some small alleles. Fewer individuals will inherit a predominance of large alleles or a predominance of small alleles. An example of a quantitative trait that does not fit a normal distribution is snail pigmentation. The dark snails and light snails are favored rather than the intermediate colors because they are less susceptible to predation.

CHAPTER 26

Conceptual Questions

C2. Evolution is unifying because all living organisms on this planet evolved from the same primordial organism. At the molecular level, all organisms have a great deal in common. With the exception of a few viruses, they all use DNA as their genetic material. This DNA is found within chromosomes, and the sequence of the DNA is organized into units called genes. Most genes are structural genes that encode the amino acid sequence of polypeptides. Polypeptides fold to form functional units called proteins. At the cellular level, all living organisms also share many similarities. For example, living cells share many of the same basic features including a plasma membrane, ribosomes, enzymatic pathways, and so on. In addition, as discussed in Chapter 7, the mitochondria and chloroplasts of eukaryotic cells are evolutionarily derived from bacterial cells.

C4. 1. Phylogenetic—Each species has distinct morphological, cellular, and/or molecular traits.

2. Biological—Each species is considered distinct if it is reproductively isolated from other species.

3. Evolutionary—Each species is derived from a single lineage that is distinct from other lineages and has its own evolutionary tendencies and historical fate.

4. Ecological—Each species occupies its own ecological niche.

C6. Anagenesis is the evolution of one species into another, whereas cladogenesis is the divergence of one species into two or more species. Of the two, cladogenesis is more prevalent. There may be many reasons why. It is common for an abrupt genetic change such as alloploidy to produce a new species from a preexisting one. Also, migrations of a few members of species into a new region may lead to the formation of a new species in the region (i.e., allopatric speciation).

C8. A. Allopatric

B. Sympatric

C. At first, it may involve parapatric speciation with a low level of intermixing. Eventually, when smaller lakes are formed, allopatric speciation will occur.

C10. The main evidence in favor of punctuated equilibrium is the fossil record. Paleontologists rarely find a gradual transition of fossil forms. The transition period in which environment pressure and genetic changes cause a previous species to evolve into a new species is thought to be so short that few, if any, of the transitional members would be preserved as fossils. Therefore, the fossil record primarily contains representatives from the long equilibrium periods. Also, rapid evolutionary change is consistent with known genetic phenomena, including single-gene mutations that have dramatic effects on phenotypic characteristics, the founder effect, and genetic events such as changes in chromosome structure (e.g., inversions and translocations) or chromosome number, which may abruptly create individuals with new phenotypic traits. In some cases, however, gradual changes are observed in certain species over long periods of time. In addition, the gradual accumulation of mutations is known to occur from the molecular analyses of DNA.

C12. Allopatric speciation involves a physical separation of a species into two or more separate populations. Over time, each population accumulates mutations that alter the characteristics of each population. Because the populations are separated, each will evolve different characteristics and eventually become distinct species. In parapatric speciation, there is some physical separation of two or more populations, but the separation is not absolute. On occasion, members of different populations can interbreed. Even so, the (somewhat) separated populations will tend to accumulate different genetic changes (e.g., inversions) that will ultimately lead to reproductive isolation among the different populations. In sympatric speciation, members of a population are not physically separated, but something happens (e.g., polyploidy) that abruptly results in reproductive isolation between members of the population. For example, a species could be diploid and a member of the population could become tetraploid. The tetraploid member would be reproductively isolated from the diploid members because hybrid offspring would be triploid and sterile. Therefore, the tetraploid individual has become a separate species.

C14. The relatively constant rate of neutral mutations acts as a molecular clock on which to measure evolutionary time. Neutral mutations occur at a rate proportional to the rate of mutation per generation. Therefore, the genetic divergence between species that is due to neutral mutations reflects the time elapsed since their last common ancestor. The concept is related to the neutral theory of evolution because it assumes that most genetic variation is due to the accumulation of neutral or nearly neutral mutations.

C16. A gene sequence can evolve more rapidly. The purpose of structural genes is to encode a polypeptide with a defined amino acid sequence. Many nucleotide changes will have no effect on the amino acid sequence of the polypeptide. For example, mutations in intron sequences and mutations at the wobble base may not affect the amino acid sequence of the encoded polypeptide. These neutral mutations will happen rather rapidly on an evolutionary timescale because natural selection will not remove them from the population. In contrast, changes in the amino acid sequence may alter the structure and function of the polypeptide. Most random mutations that affect the polypeptide sequence are more likely to be detrimental than beneficial, and detrimental mutations will be eliminated by natural selection. This makes it more difficult for the amino acid sequence of the polypeptide to evolve. Only neutral changes and beneficial changes will happen rapidly, and these are less likely to occur in the amino acid sequence compared to the gene sequence.

C18. You would expect the sequences of plant storage proteins to evolve rapidly. The polypeptide sequence is not particularly important for the structure or function of the protein. The purpose of the protein is to provide nutrients to the developing embryo. Changing the sequence would likely be tolerated. However, major changes in the amino acid composition (not the sequence) may be selected against. For example, the storage protein would have to contain some cysteine in its amino acid sequence because the embryo would need some cysteine to grow. However, the location of cysteine codons within the amino acid sequence would not be important; it would only be important that the gene sequence have some cysteine codons.

C20. The α-globin sequences in humans and horses are more similar to each other, compared to the α-globin in humans and the β-globin in humans. This suggests that the gene duplication that produced the α-globin and β-globin genes occurred first. After this gene duplication occurred, each gene accumulated several different mutations that caused the sequences of the two genes to diverge. At a much later time, during the evolution of mammals, a split occurred that produced different branches in the evolutionary tree of mammals. One branch eventually led to the formation of horses and a different branch led to the formation of humans. During the formation of these mammalian branches (which has been more recent), some additional mutations could occur in the α- and β-globin genes. This would explain why the α-globin gene in humans and horses is not exactly the same. However, it is more similar than the α- and β-globin genes within humans because the divergence of humans and horses occurred much more recently than the gene duplication that produced the α- and β-globin genes. In other words, there has been much less time for the α-globin gene in humans to diverge from the α-globin gene in horses.

C22. A. This is an example of neutral mutation. Mutations in the wobble base are neutral when they do not affect the amino acid sequence.

B. This is an example of natural selection. Random mutations that occur in vital regions of a polypeptide sequence are likely to inhibit function. Therefore, these types of mutations are eliminated by natural selection. That is why they are relatively rare.

C. This is a combination of neutral mutation and natural selection. The prevalence of mutations in introns is due to the accumulation of neutral mutations. Most mutations within introns do not have any effect on the expression of the exons, which contain the polypeptide sequence. In contrast, mutations within the exons are more likely to be affected by natural selection. As mentioned in the answer to part B, mutations in vital regions are likely to inhibit function. Natural selection tends to eliminate these mutations. Therefore, mutations within exons are less likely than mutations within introns.

C24. Generally, one would expect a similar number of chromosomes with very similar banding patterns. However, there may be a few notable differences. An occasional translocation could change the size or chromosomal number between two different species. Also, an occasional inversion may alter the banding pattern between two species.

C26. Animals with very simple body plans, such as the sponge, have relatively few *Hox* genes. In contrast, animals with more complicated bodies, such as mammals, have many. Researchers speculate that each *Hox* gene can govern the morphological features of a particular region of the body. By having multiple *Hox* genes, different regions of the body can become more specialized, and therefore, more complex.

Experimental Questions

E2. Perhaps the easiest way to determine allotetraploidy is by the chromosomal examination of closely related species. A researcher could karyotype the chromosomes from many different species and look for homologous chromosomes that have similar banding patterns. This may enable them to identify allotetraploids that contain a diploid set of chromosomes from two different species.

E4. The mutations that have occurred in this sequence are neutral mutations. In all cases, the wobble base has changed, and this change would not affect the amino acid sequence of the encoded polypeptide. Therefore, a reasonable explanation is that the gene has accumulated random neutral mutations over the course of many generations. This observation would be consistent with the neutral theory of evolution. A second explanation would be that one of these two researchers made a few experimental mistakes when determining the sequence of this region.

E6. Inversions do not affect the total amount of genetic material. Usually, inversions do not affect the phenotype of the organism. Therefore, if members of the two populations were to interbreed, the offspring would probably be viable because they would have inherited a normal amount of genetic material from each parent. However, such offspring would be inversion heterozygotes. As described in Chapter 8 (see Figure 8.12), crossing over during meiosis may create chromosomes that have too much or too little genetic material. If these unbalanced chromosomes are passed to the next generation of offspring, the offspring may not survive. For this reason, inversion heterozygotes (that are phenotypically normal) may not be very fertile because many of their offspring will die. Because inversion heterozygotes are less fertile, this would tend to keep the eastern and western populations reproductively isolated. Over time, this would aid in the independent evolution of the two populations and would ultimately promote the evolution of the two populations into separate species.

E8. The technique of PCR is used to amplify the amount of DNA in a sample. To accomplish this, one must use oligonucleotide primers that are complementary to the region that is to be amplified. For example, as described in the experiment of Figure 26.15, PCR primers that were complementary to and flank the 12S rRNA gene can be used to amplify the 12S rRNA gene. The technique of PCR is described in Chapter 18.

E10. We would expect the probe to hybridize to the natural *G. tetrahit* and also the artificial *G. tetrahit*, because both of these strains contain two sets of chromosomes from *G. pubescens*. We would expect two bright spots in the *in situ* experiment. Depending on how closely related *G. pubescens* and *G. speciosa* are, the probe may also hybridize to two sites in the *G. speciosa* genome, but this is difficult to predict *a priori*. If so, the *G. tetrahit* species would show four spots.

E12. The principle of parsimony chooses a phylogenetic tree that requires the fewest number of evolutionary changes. When using molecular data, researchers can use computer programs that compare DNA sequences from homologous genes of different species and construct a tree that requires the fewest numbers of mutations. Such a tree is the most likely pathway for the evolution of such species.

E14. Possibly, the mouse would have an eye at the tip of its tail!

Questions for Student Discussion/Collaboration

2. The founder effect and allotetraploidy are examples of rapid forms of evolution. In addition, some single gene mutations may have a great impact on phenotype and lead to the rapid evolution of new species by cladogenesis. Geological processes may promote the slower accumulation of alleles and alter a species' characteristics more gradually. In this case, it is the accumulation of many phenotypically minor genetic changes that ultimately leads to reproductive isolation. Slow and fast mechanisms of evolution have the common theme that they result in reproductive isolation. This is a prerequisite for the evolution of new species. Fast mechanisms tend to involve small populations and a few number of genetic changes. Slower mechanisms may involve larger populations and involve the accumulation of a large number of genetic changes that each contributes in a small way.

GLOSSARY

A

A an abbreviation for adenine.
ABC model a model for flower development.
acentric describes a chromosome without a centromere.
acquired antibiotic resistance the acquisition of antibiotic resistance because a bacterium has taken up a gene or plasmid from another bacterial strain.
acridine dye a type of chemical mutagen that causes frameshift mutations.
acrocentric describes a chromosome with the centromere significantly off center, but not at the very end.
activator a transcriptional regulatory protein that increases the rate of transcription.
acutely transforming virus (ACT) a virus that readily transforms normal cells into malignant cells, when grown in a laboratory.
adaptor hypothesis a hypothesis that proposes a tRNA has two functions: recognizing a three-base codon sequence in mRNA and carrying an amino acid that is specific for that codon.
adenine a purine base found in DNA and RNA. It base-pairs with thymine in DNA.
A DNA a right-handed DNA double helix with 11 base pairs per turn. Does not occur in living cells.
age of onset for alleles that cause genetic diseases, the time of life at which disease symptoms appear.
alkaptonuria a human genetic disorder involving the accumulation of homogentisic acid due to a defect in homogentisic acid oxidase.
alkyltransferase an enzyme that can remove methyl or ethyl groups from guanine bases.
allele an alternative form of a specific gene.
allele frequency the number of copies of a particular allele in a population divided by the total number of all alleles for that gene in the population.
allelic variation genetic variation in a population that involves the occurrence of two or more different alleles for a particular gene.
allodiploid an organism that contains one set of chromosomes from two different species.
allopatric speciation (Greek, *allos*, "other"; Latin, *patria*, "homeland") an evolutionary phenomenon in which speciation occurs when members of a species become geographically separated from the other members.
alloploid an organism that contains chromosomes from two (or more) different species.
allopolyploid an organism that contains two (or more) sets of chromosomes from two (or more) species.
allosteric enzyme an enzyme that contains two binding sites—a catalytic site and a regulatory site.
allosteric regulation the phenomenon that an effector molecule binds to a noncatalytic site on a protein and causes a conformational change that regulates its function.
allosteric site the site on a protein where a small effector molecule binds to regulate the function of the protein.
allotetraploid an organism that contains two sets of chromosomes from two different species.
allozymes two or more enzymes (encoded by the same type of gene) with alterations in their amino acid sequences, which may affect their gel mobilities.
α helix a type of secondary structure found in proteins.
alternative exon an exon that is not always found in mRNA. It is only found in certain types of alternatively spliced mRNAs.
alternative splicing refers to the phenomenon in which a pre-mRNA can be spliced in more than one way.
amber a stop codon with the sequence UAG.
Ames test a test using strains of a bacterium, *Salmonella typhimurium*, to determine if a substance is a mutagen.
amino acid a building block of polypeptides and proteins. It contains an amino group, a carboxyl group, and a side chain.
aminoacyl site (A site) a site on the ribosome where a charged tRNA initially binds.
aminoacyl tRNA a tRNA molecule that has an amino acid covalently attached to its 3' end.
aminoacyl-tRNA synthetase an enzyme that catalyzes the attachment of a specific amino acid to the correct tRNA.
2-aminopurine a base analogue that acts as a chemical mutagen.
amino terminus the location of the first amino acid in a polypeptide chain. The amino acid at the amino terminus still retains a free amino group that is not covalently attached to the second amino acid.
amniocentesis a method of obtaining cellular material from a fetus for the purpose of genetic testing.
amplified restriction fragment length polymorphism (AFLP) a RFLP that is amplified via PCR.
anabolic enzyme an enzyme involved in connecting organic molecules to create larger molecules.
anagenesis (Greek, *ana*, "up," and *genesis*, "origin") the evolutionary phenomenon in which a single species is transformed into a different species over the course of many generations.
anaphase the fourth stage of M phase. As anaphase proceeds, half of the chromosomes move to one pole, and the other half move to the other pole.
ancestral character see *primitive character.*
ancient DNA analysis analysis of DNA that is extracted from the remains of extinct species.
aneuploid not euploid. Refers to a variation in chromosome number such that the total number of chromosomes is not an exact multiple of a set or *n* number.
annealing the process in which two complementary segments of DNA bind to each other.
annotated in files involving genetic sequences, annotation is a description of the known function and features of the sequence, as well as other pertinent information.
antero-posterior axis in animals, the axis that runs from the head (anterior) to the tail or base of the spine (posterior).
anther the structure in flowering plants that gives rise to pollen grains.
anther culture the generation of monoploid plants by cold-shock treatment of anthers.
antibodies proteins produced by the B cells of the immune system that recognize foreign substances (namely, viruses, bacteria, and so forth) and target them for destruction.
antibody microarray a small silica, glass, or plastic slide that is dotted with many different antibodies, which recognize particular amino acid sequences within proteins.
anticipation the phenomenon in which the severity of an inherited disease tends to get worse in future generations.
anticodon a three-nucleotide sequence in tRNA that is complementary to a codon in mRNA.
antigens foreign substances that are recognized by antibodies.
antiparallel an arrangement in a double helix in which one strand is running in the 5' to 3' direction, while the other strand is 3' to 5'.
antisense RNA an RNA strand that is complementary to a strand of mRNA.
antisense strand also called the template strand. It is the strand of DNA that is used as a template for RNA synthesis.
antitermination the function of certain proteins, such as N protein in bacteria, that prevents transcriptional termination.
AP endonuclease a DNA repair enzyme that recognizes a DNA region that is missing a base, and makes a cut in the DNA backbone near that site.
apical–basal-patterning gene one of several plant genes that play a role in embryonic development.
apical region in plants, the region that produces the leaves and flowers.
apoptosis programmed cell death.
apurinic site a site in DNA that is missing a purine base.
archaea also called archaebacteria; one of the three domains of life. Archaea are prokaryotic species. They tend to live in extreme environments and are less common than bacteria (also called eubacteria).
ARS elements DNA sequences found in yeast that function as origins of replication.
artificial chromosomes cloning vectors that can accommodate large DNA inserts and behave like chromosomes when inside of living cells.
artificial selection see *selective breeding.*
artificial transformation transformation of bacteria that occurs via experimental treatments.
ascus (pl. asci) a sac that contains haploid spores of fungi (i.e., yeast or molds).
asexual reproduction a form of reproduction that does not involve the union of gametes; at the cellular level, a preexisting cell divides to produce two new cells.
A site see *aminoacyl site.*
assortative mating breeding in which individuals preferentially mate with each other based on their phenotypes.
AT/GC rule in DNA, the phenomenon in which an adenine base in one strand always hydrogen bonds with a thymine base in the opposite strand, and a guanine base always hydrogen bonds with a cytosine.
ATP-dependent chromatin remodeling see *chromatin remodeling.*
attachment site a site in a host cell chromosome where a virus will integrate during site-specific recombination.
attenuation a mechanism of genetic regulation, seen in the *trp* operon, in which a short RNA is made but its synthesis is terminated before RNA polymerase can transcribe the rest of the operon.
attenuator sequence a sequence found in certain operons (e.g., *trp* operon) in bacteria that stops transcription soon after it has begun.
AU-rich element (ARE) a sequence found in many short-lived mRNAs that contains the consensus sequence AUUUA.
automated sequencing the use of fluorescently labeled dideoxyribonucleotides and a fluorescence detector to sequence DNA.
autonomous transposable element a transposable element that contains all of the information necessary for transposition or retroposition to take place.
autopolyploid a polyploid produced within a single species due to nondisjunction.
autoradiography a technique, which involves the use of X-ray film, to detect the location of radioisotopes as they are found in macromolecules or cells. It is used to detect a particular band on a gel or to map the location of a gene within an intact chromosome.

autoregulatory loop a form of gene regulation in which a protein, such as a splicing factor or a transcription factor, regulates its own expression.
autosomes chromosomes that are not sex chromosomes.
auxotroph a strain that cannot synthesize a particular nutrient and needs that nutrient supplemented in its growth medium or diet.

B

BAC see *bacterial artificial chromosome.*
backbone the portion of a DNA or RNA strand that is composed of the repeated covalent linkage of the phosphates and sugar molecules.
backcross in genetics, this usually refers to a cross of F_1 hybrids to individuals that have genotypes of the parental generation.
bacterial artificial chromosome (BAC) a cloning vector that propagates in bacteria and is used to clone large fragments of DNA.
bacteriophages (or **phages**) viruses that infect bacteria.
balanced polymorphism when natural selection favors the maintenance of two or more alleles in a population.
balanced translocation a translocation, such as a reciprocal translocation, in which the total amount of genetic material is normal or nearly normal.
balancing selection a pattern of natural selection that favors the maintenance of two or more alleles. Examples include heterozygote advantage and negative frequency-dependent selection.
band shift assay see *gel retardation assay.*
Barr body a structure in the interphase nuclei of somatic cells of female mammals that is a highly condensed X chromosome.
basal region in plants, the region that produces the roots.
basal transcription in eukaryotes, a low level of transcription via the core promoter. The binding of transcription factors to enhancer elements may increase transcription above the basal level.
basal transcription apparatus the minimum number of proteins needed to transcribe a gene.
base a nitrogen-containing molecule that is a portion of a nucleotide in DNA or RNA. Examples of bases are adenine, thymine, guanine, cytosine, and uracil.
base excision repair a type of DNA repair in which a modified base is removed from a DNA strand. Following base removal, a short region of the DNA strand is removed, which is then resynthesized using the complementary strand as a template.
base mismatch when two bases opposite each other in a double helix do not conform to the AT/GC rule. For example, if A is opposite C, that would be a base mismatch.
base pair the structure in which two nucleotides in opposite strands of DNA hydrogen bond with each other. For example, an AT base pair is a structure in which an adenine-containing nucleotide in one DNA strand hydrogen bonds with a thymine-containing nucleotide in the complementary strand.
base substitution a point mutation in which one base is substituted for another base.
basic domain a protein domain containing several basic amino acids, which is often involved in binding to DNA.
B DNA the predominant form of DNA in living cells. It is a right-handed DNA helix with 10 base pairs per turn.
behavioral trait a trait that involves behavior. An example would be the ability to learn a maze.
beneficial mutation a mutation that is beneficial with regard to its effect on phenotype.
benign refers to a noncancerous tumor that is not invasive and cannot metastasize.
β sheet a type of secondary structure found in proteins.
bHLH refers to a structure found in transcription factor proteins with a basic domain involved in DNA binding and a helix-loop-helix domain involved in dimerization.
bidirectionally means that two replication forks move in opposite directions outward from the origin.
bidirectional replication the phenomenon in which two DNA replication forks emanate in both directions from an origin of replication.
bilateral gynandromorph an animal in which one side is phenotypically male and the other side is female.
binary fission the physical process whereby a bacterial cell divides into two daughter cells. During this event, the two daughter cells become divided by the formation of a septum.
binomial expansion equation an equation to solve genetic problems involving two types of unordered events.
biodegradation the breakdown of a larger molecule into a smaller molecule via cellular enzymes.
bioinformatics literally, this term means the study of biological information. Recently, this term has been associated with the analysis of genetic sequences, using computers and computer programs.
biolistic gene transfer the use of microprojectiles to introduce DNA into plant cells.
biological control the use of microorganisms or products from microorganisms to alleviate plant diseases or damage from undesirable environmental conditions (e.g., frost damage).
biological evolution the accumulation of genetic changes in a species or population over the course of many generations.
biological species concept definition of a species as a group of individuals whose members have the potential to interbreed with one another in nature to produce viable, fertile offspring, but who cannot interbreed successfully with members of other species.
biometric field a field of genetics that involves the statistical study of biological traits.
bioremediation the use of microorganisms to decrease pollutants in the environment.
biotechnology technologies that involve the use of living organisms, or products from living organisms, as a way to benefit humans.
biotransformation the conversion of one molecule into another via cellular enzymes. This term is often used to describe the conversion of a toxic molecule into a nontoxic molecule.
bivalent a structure in which two pairs of homologous sister chromatids have synapsed (i.e., aligned) with each other.
BLAST (basic local alignment search tool) a computer program that can start with a particular genetic sequence and then locate homologous sequences within a large database.
blending hypothesis of inheritance an early, incorrect hypothesis of heredity. According to this view, the seeds that dictate hereditary traits are able to blend together from generation to generation. The blended traits would then be passed to the next generation.
bottleneck effect a type of genetic drift that occurs when most members of a population are eliminated without any regard to their genetic composition.
box in genetics, a term used to describe a sequence with a specialized function.
branch migration the lateral movement of a Holliday junction.
breakpoint the region where two chromosome pieces break and rejoin with other chromosome pieces.
broad sense heritability heritability that takes into account all genetic factors.
5-bromodeoxyuridine a base analogue that can be incorporated into chromosomes during DNA replication. The presence of this analogue can affect the ability of the chromosomes to absorb certain dyes. This is the basis for the staining of harlequin chromosomes.
5-bromouracil a base analogue that acts as a chemical mutagen.

C

C an abbreviation for cytosine.
cAMP see *cyclic AMP.*
cAMP response element (CRE) a short DNA sequence found next to certain eukaryotic genes that is recognized by the cAMP response element–binding (CREB) protein.
cancer a disease characterized by uncontrolled cell division.
cancer cell a cell that has lost its normal growth control. Cancer cells are invasive (i.e., they can invade normal tissues) and metastatic (i.e., they can migrate to other parts of the body).
CAP an abbreviation for the catabolite activator protein, a genetic regulatory protein found in bacteria.
capping the covalent attachment of a 7-methylguanosine nucleotide to the 5' end of mRNA in eukaryotes.
CAP site the sequence of DNA that is recognized by CAP.
carbohydrate organic molecules with the general formula $C(H_2O)$. An example of a simple carbohydrate would be the sugar glucose. Large carbohydrates are composed of multiple sugar units.
carboxyl terminus the location of the last amino acid in a polypeptide chain. The amino acid at the carboxyl terminus still retains a free carboxyl group that is not covalently attached to another amino acid.
carcinogen an agent that can cause cancer.
caspases proteolytic enzymes that play a role in apoptosis.
catabolic enzyme an enzyme that is involved in the breakdown of organic molecules into smaller units.
catabolite activator protein see *CAP.*
catabolite repression the phenomenon in which a catabolite (such as glucose) represses the expression of certain genes (such as the *lac* operon).
catenane interlocked circular molecules.
cDNA see *complementary DNA.*
cDNA library a DNA library made from a collection of cDNAs.
cell adhesion when the surfaces of cells bind to each other or to the extracellular matrix.
cell adhesion molecule (CAM) a molecule (e.g., surface protein or carbohydrate) that plays a role in cell adhesion.
cell culture refers to the growth of cells in a laboratory.
cell cycle in eukaryotic cells, a series of stages through which a cell progresses in order to divide. The phases are G for growth, S for synthesis (of the genetic material), and M for mitosis. There are two G phases, G_1 and G_2.
cell fate the final morphological features that a cell or group of cells will adopt.
cell-free translation system an experimental mixture that can synthesis polypeptides.
cell fusion describes the process in which individual cells are mixed together and made to fuse with each other.
cell lineage a series of cells that are descended from a cell or group of cells by cell division.
cell plate the structure that forms between two daughter plant cells that leads to the separation of the cells by the formation of an intervening cell wall.
cellular trait a trait that is observed at the cellular level. An example would be the shape of a cell.
centiMorgans (cM) (same as a map unit) a unit of map distance obtained from genetic crosses. Named in honor of Thomas Hunt Morgan.
central dogma of molecular biology the idea that the usual flow of genetic information is from DNA to RNA to polypeptide (protein). In addition, DNA replication serves to copy the information so that it can be transmitted from cell to cell and from parent to offspring.
central region in plants, it is the region that creates the stem. It is the radial pattern of cells in the central region that causes the radial growth observed in plants.
central zone in plants, an area in the meristem where undifferentiated stem cells are always maintained.
centrifugation a method to separate cell organelles and macromolecules in which samples are placed in tubes and spun very rapidly. The rate at which particles move toward the bottom of the tube depends on their densities, sizes, shapes, and the viscosity of the medium.
centrifuge a machine that contains a motor, which causes a rotor holding centrifuge tubes to spin very rapidly.
centromere a segment of eukaryotic chromosomal DNA that provides an attachment site for the kinetochore.
centrosome a cellular structure from which microtubules emanate.
chain termination refers to the stoppage of growth of a DNA strand, RNA strand, or polypeptide sequence.
chaperone a protein that aids in the folding of polypeptides.
character in genetics, this word has the same meaning as trait.

Chargaff's rule see *AT/GC rule.*
charged tRNA a tRNA that has an amino acid attached to its 3' end by an ester bond.
checkpoint protein a protein that monitors the conditions of DNA and chromosomes and may prevent a cell from progressing through the cell cycle if an abnormality is detected.
chiasma (pl. **chiasmata**) the site where crossing over occurs between two chromosomes. It resembles the Greek letter chi, χ.
chimera an organism composed of cells that are embryonically derived from two different individuals.
ChIP-on-chip assay a form of chromatin immunoprecipitation that utilizes a microarray to determine where in the genome a particular protein binds.
chi square (χ^2) test a commonly used statistical method to determine the goodness of fit. This method can be used to analyze population data in which the members of the population fall into different categories. It is particularly useful for evaluating the outcome of genetic crosses, because these usually produce a population of offspring that differ with regard to phenotypes.
chloroplast DNA (cpDNA) the genetic material found within a chloroplast.
chorionic villus sampling a method to obtain cellular material from a fetus for the purpose of genetic testing.
chromatid following chromosomal replication in eukaryotes, the two copies remain attached to each other in the form of sister chromatids.
chromatin the association between DNA and proteins that is found within chromosomes.
chromatin immunoprecipitation (ChIP) a method to determine whether proteins bind to a particular region of DNA. This method analyzes DNA–protein interactions as they occur in the chromatin of living cells.
chromatin remodeling a change in chromatin structure that alters the degree of compaction and/or the spacing of nucleosomes.
chromatography a method to separate different macromolecules and small molecules based on their chemical and physical properties. A sample is dissolved in a liquid solvent and exposed to some type of matrix, such as a gel, a column containing beads, or a thin strip of paper.
chromocenter the central point where polytene chromosomes aggregate.
chromomere a dark band within a polytene chromosome.
chromosome the structures within living cells that contain the genetic material. Genes are physically located within the structure of chromosomes. Biochemically, chromosomes contain a very long segment of DNA, which is the genetic material, and proteins, which are bound to the DNA and provide it with an organized structure.
chromosome territory in the cell nucleus, each chromosome occupies a nonoverlapping region called a chromosome territory.
chromosome theory of inheritance a theory of Sutton and Boveri, which indicated that the inheritance patterns of traits can be explained by the transmission patterns of chromosomes during gametogenesis and fertilization.
chromosome walking a common method used in positional cloning in which a mapped gene or RFLP marker provides a starting point to molecularly walk toward a gene of interest via overlapping clones.
***cis*-acting element** a sequence of DNA, such as a regulatory element, that exerts a *cis*-effect.
***cis*-effect** an effect on gene expression due to genetic sequences that are within the same chromosome and often are immediately adjacent to the gene of interest.
cistron refers to the smallest genetic unit that produces a positive result in a complementation experiment. A cistron is equivalent to a gene.
clade a group of species consisting of all descendents of the group's most common ancestor.
cladistic approach a way to construct a phylogenetic tree, also called a cladogram, by considering the various possible pathways of evolution and then choosing the most plausible tree.
cladogenesis (Greek, *clados*, "branch") during evolution, a form of speciation that involves the division of a species into two or more species.
cladogram a phylogenetic tree that has been constructed using a cladistic approach.
cleavage furrow a constriction that causes the division of two animal cells during cytokinesis.
clonal an adjective to describe a clone. For example, a clonal population of cells is a group of cells that are derived from the same cell.
clone the general meaning of this term is to make many copies of something. In genetics, this term has several meanings: (1) a single cell that has divided to produce a colony of genetically identical cells; (2) an individual that has been produced from a somatic cell of another individual, such as the sheep Dolly; (3) many copies of a DNA fragment that are propagated within a vector or produced by PCR.
closed complex the complex between transcription factors, RNA polymerase, and a promoter before the DNA has denatured to form an open complex.
closed conformation a tightly packed conformation of chromatin that cannot be transcribed.
cluster analysis the analysis of microarray data to determine if certain groups (i.e., clusters) of genes are expressed under the same conditions.
cM an abbreviation for centiMorgans; also see *map unit.*
coding strand the strand in DNA that is not used as a template for mRNA synthesis.
codominance a pattern of inheritance in which two alleles are both expressed in the heterozygous condition. For example, a person with the genotype I^AI^B will have the blood type AB and will express both surface antigens A and B.
codon a sequence of three nucleotides in mRNA that functions in translation. A start codon, which usually specifies methionine, initiates translation, and a stop codon terminates translation. The other codons specify the amino acids within a polypeptide sequence according to the genetic code.
codon bias the phenomenon that, in a given species, certain codons are used more frequently than other codons.
coefficient of inbreeding (F) see *inbreeding coefficient.*
cohesin a protein complex that facilitates the alignment of sister chromatids.
colinearity the correspondence between the sequence of codons in the DNA coding strand and the amino acid sequence of a polypeptide.
colony hybridization a technique in which a probe is used to identify bacterial colonies that contain a hybrid vector with a gene of interest.
combinatorial control refers to the phenomenon that is common in eukaryotes in which the combination of many factors determines the expression of any given gene.
common ancestor someone who is an ancestor to both of an individual's parents.
comparative genomic hybridization (CGH) a hybridization technique to determine if cells (e.g., cancer cells) have changes in chromosome structure, such as deletions or duplications.
comparative genomics uses information from genome projects to understand the genetic variation between different populations and evolutionary relationships among different species.
competence factors proteins that are needed for bacterial cells to become naturally transformed by extracellular DNA.
competence-stimulating peptide (CSP) a peptide secreted by certain species of bacteria that allow them to become competent for transformation.
competent cells cells that can be transformed by extracellular DNA.
complementary describes sequences in two DNA strands that match each other according to the AT/GC rule. For example, if one strand has the sequence of ATGGCGGATTT, then the complementary strand must be TACCGCCTAAA.
complementary DNA (cDNA) DNA that is made from an RNA template by the action of reverse transcriptase.
complementation a phenomenon in which the presence of two different mutant alleles in the same organism produces a wild-type phenotype. It usually happens because the two mutations are in different genes, so the organism carries one copy of each mutant allele and one copy of each wild-type allele.
complementation test an experimental procedure in which the goal is to determine if two different mutations that affect the same trait are in the same gene or in two different genes.
complete nondisjunction during meiosis or mitosis, when all of the chromosomes fail to disjoin and remain in one of the two daughter cells.
complete transposable element see *autonomous transposable element.*
complex traits characteristics that are determined by several genes and are significantly influenced by environmental factors.
composite transposon a transposon that contains additional genes, such as antibiotic resistance genes, that are not necessary for transposition *per se.*
computer data file a file (a collection of information) stored by a computer.
computer program a series of operations that can analyze data in a defined way.
concordance in genetics, the degree to which pairs of individuals (e.g., identical twins or fraternal twins) exhibit the same trait.
condensation a change in chromatin structure to become more compact.
condense refers to chromosomes forming a more compact structure
condensin a protein complex that plays a role in the condensation of interphase chromosomes to become metaphase chromosomes.
conditional alleles alleles in which the phenotypic expression depends on the environmental conditions. An example is temperature-sensitive alleles, which affect the phenotype only at a particular temperature.
conditional lethal allele an allele that is lethal, but only under certain environmental conditions.
conditional mutant a mutant whose phenotype depends on the environmental conditions, such as a temperature-sensitive mutant.
conglomerate a population composed of members of an original population plus new members that have migrated from another population.
conjugation a form of genetic transfer between bacteria that involves direct physical interaction between two bacterial cells. One bacterium acts as donor and transfers genetic material to a recipient cell.
conjugation bridge a connection between two bacterial cells that provides a passageway for DNA during conjugation.
conjugative plasmid a plasmid that can be transferred to a recipient cell during conjugation.
consensus sequence the most commonly occurring bases within a sequence element.
conservative model an incorrect model for DNA in which both strands of parental DNA remain together following DNA replication.
conservative transposition see *simple transposition.*
constitutive exon an exon that is always found in mRNA following splicing.
constitutive gene a gene that is not regulated and has essentially constant levels of expression over time.
constitutive heterochromatin regions of chromosomes that are always heterochromatic and are permanently transcriptionally inactive.
contig a series of clones that contain overlapping pieces of chromosomal DNA.
control element see *regulatory sequence or element.*
core enzyme the subunits of an enzyme that are needed for catalytic activity, as in the core enzyme of RNA polymerase.
corepressor a small effector molecule that binds to a repressor protein, thereby causing the repressor protein to bind to DNA and inhibit transcription.
core promoter a DNA sequence that is absolutely necessary for transcription to take place. It provides the binding site for general transcription factors and RNA polymerase.
correlation coefficient (*r*) a statistic with a value that ranges between -1 and 1. It describes how two factors vary with regard to each other.
cosmid a vector that is a hybrid between a plasmid vector and phage λ. Cosmid DNA can replicate in a cell like a plasmid or be packaged into a protein coat like a

phage. Cosmid vectors can accept fragments of DNA that are typically tens of thousands of base pairs in length.

C_0t curve a plot of C/C_0 versus C_0t.

cotransduction the phenomenon in which bacterial transduction transfers a piece of DNA carrying two closely linked genes.

cotransformation the phenomenon in which bacterial transformation transfers a piece of DNA carrying two closely linked genes.

cotranslational events that occur during translation.

cotranslational import during the synthesis of certain eukaryotic proteins, translation begins in the cytosol and then is temporarily halted by the signal recognition particle (SRP). Translation resumes when the ribosome has become bound to the ER membrane and the polypeptide is synthesized into the ER lumen or ER membrane.

cotranslational sorting refers to the sorting of proteins into the ER. The protein is actually translated into the ER lumen or ER membrane.

covariance a statistic that describes the degree of variation between two variables within a group.

cpDNA an abbreviation for chloroplast DNA.

CpG island a group of CG sequences that may be clustered near a promoter region of a gene. The methylation of the cytosine bases usually inhibits transcription.

CREB protein (cAMP response element-binding protein) a regulatory transcription factor that becomes activated in response to specific cell-signaling molecules that cause the synthesis of cAMP.

cross a mating between two distinct individuals. An analysis of their offspring may be conducted to understand how traits are passed from parent to offspring.

cross-fertilization same meaning as *cross*. It requires that the male and female gametes come from separate individuals.

crossing over a physical exchange of chromosome pieces that most commonly occurs during prophase of meiosis I.

C-terminus see *carboxyl terminus*.

cyclic AMP (cAMP) in bacteria, a small effector molecule that binds to CAP (catabolite activator protein). In eukaryotes, cAMP functions as a second messenger in a variety of intracellular signaling pathways; in some cases, it binds to transcription factors such as the CREB protein.

cyclin a type of protein that plays a role in the regulation of the eukaryotic cell cycle.

cyclin-dependent protein kinases (CDKs) enzymes that are regulated by cyclins and can phosphorylate other cellular proteins by covalently attaching a phosphate group.

cytogeneticist a scientist who studies chromosomes under the microscope.

cytogenetic mapping the mapping of genes or genetic sequences using microscopy.

cytogenetics the field of genetics that involves the microscopic examination of chromosomes.

cytokinesis the division of a single cell into two cells. The two nuclei produced in M phase are segregated into separate daughter cells during cytokinesis.

cytological mapping see *cytogenetic mapping*.

cytoplasmic inheritance (also known as *extranuclear inheritance*) refers to the inheritance of genetic material that is not found within the cell nucleus.

cytosine a pyrimidine base found in DNA and RNA. It base-pairs with guanine in DNA.

D

Dam see *DNA adenine methyltransferase*.

Darwinian fitness the relative likelihood that a phenotype will survive and contribute to the gene pool of the next generation as compared with other phenotypes.

database a computer storage facility that stores many data files such as those containing genetic sequences.

daughter strand in DNA replication, the newly made strand of DNA.

deamination the removal of an amino group from a molecule. For example, the removal of an amino group from cytosine produces uracil.

decoding function the ability of the 16S rRNA to detect when an incorrect tRNA is bound at the A site and prevent elongation until the mispaired tRNA is released from the A site.

decondensed refers to chromosomes forming a less compact structure.

deficiency condition in which a segment of chromosomal material is missing.

degeneracy in genetics, this term means that more that one codon specifies the same amino acid. For example, the codons GGU, GGC, GGA, and GGG all specify the amino acid glycine.

degrees of freedom in a statistical analysis, the number of categories that are independent of each other.

deleterious mutation a mutation that is detrimental with regard to its effect on phenotype.

deletion condition in which a segment of DNA is missing.

deletion mapping the use of strains carrying deletions within a defined region to map a mutation of unknown location.

deme see *local population*.

***de novo* methylation** the methylation of DNA that has not been previously methylated. This is usually a highly regulated event.

deoxyribonucleic acid (DNA) the genetic material. It is a double-stranded structure, with each strand composed of repeating units of deoxyribonucleotides.

deoxyribose the sugar found in DNA.

depurination the removal of a purine base from DNA.

determined cell a cell that is destined to differentiate into a specific cell type.

developmental genetics the area of genetics concerned with the roles of genes in orchestrating the changes that occur during development.

diakinesis the fifth stage of prophase of meiosis I.

diauxic growth the sequential use of two sugars by a bacterium.

dicentric describes a chromosome with two centromeres.

dicentric bridge the region between the two centromeres in a dicentric chromosome.

dicer an endonuclease that makes a cut in double-stranded RNA.

dideoxyribonucleotide a nucleotide used in DNA sequencing that is missing the 3'—OH group. If a dideoxyribonucleotide is incorporated into a DNA strand, it stops any further growth of the strand.

dideoxy sequencing a method of DNA sequencing that uses dideoxyribonucleotides to terminate the growth of DNA strands.

differential centrifugation a form of centrifugation involving a series of centrifugation steps in which the supernatant or pellet is used in each subsequent centrifugation step.

differentially methylated region (DMR) in the case of imprinting, a site that is methylated during spermatogenesis or oogenesis, but not both.

differentiated cell a cell that has become a specialized type of cell within a multicellular organism.

dihybrid cross a cross in which an experimenter follows the outcome of two different traits.

dihybrid testcross a cross in which an experimenter crosses an individual that is heterozygous for two genes to an individual that is homozygous recessive for the same two genes.

dimeric DNA polymerase a complex of two DNA polymerase proteins that move as a unit during DNA replication.

dioecious in plants, a species in which male and female gametophytes are produced on a single (sporophyte) individual.

diploid an organism or cell that contains two copies of each type of chromosome.

diplotene the fourth stage of prophase of meiosis I.

directionality in DNA and RNA, refers to the 5' to 3' arrangement of nucleotides in a strand; in proteins, refers to the linear arrangement of amino acids from the N-terminal to C-terminal ends.

directional selection natural selection that favors an extreme phenotype. This usually leads to the fixation of the favored allele.

direct repeat (DR) short DNA sequences that flank transposable elements in which the DNA sequence is repeated in the same direction.

discontinuous trait a trait in which each offspring can be put into a particular phenotypic category.

discovery-based science experimentation that does not require a preconceived hypothesis. In some cases, the goal is to collect data to be able to formulate a hypothesis.

disequilibrium in population genetics, refers to a population that is not in Hardy-Weinberg equilibrium.

dispersive model an incorrect model for DNA replication in which segments of parental DNA and newly made DNA are interspersed in both strands following the replication process.

disruptive selection natural selection that favors both extremes of a phenotypic category. This results in a balanced polymorphism.

dizygotic twins also known as fraternal twins; twins formed from separate pairs of sperm and egg cells.

DNA the abbreviation for deoxyribonucleic acid.

DNA adenine methyltransferase an enzyme in bacteria that attaches methyl groups to the adenine base in DNA that is found within the sequence GATC.

DnaA box sequence serves as a recognition site for the binding of the DnaA protein, which is involved in the initiation of bacterial DNA replication.

DnaA protein a protein that binds to the dnaA box sequence at the origin of replication in bacteria and initiates DNA replication.

DNA fingerprinting a technology to identify a particular individual based on the properties of their DNA.

DNA gap repair synthesis the synthesis of DNA in a region where a DNA strand has been previously removed, usually by a DNA repair enzyme or by an enzyme involved in homologous recombination.

DNA gyrase also known as topoisomerase II; an enzyme that introduces negative supercoils into DNA using energy from ATP. Gyrase can also relax positive supercoils when they occur.

DNA helicase an enzyme that separates the two strands of DNA.

DNA library a collection of many hybrid vectors, each vector carrying a particular fragment of DNA from a larger source. For example, each hybrid vector in a DNA library might carry a small segment of chromosomal DNA from a particular species.

DNA ligase an enzyme that catalyzes a covalent bond between two DNA fragments.

DNA methylation the phenomenon in which an enzyme covalently attaches a methyl group (—CH_3) to a base (usually adenine or cytosine) in DNA.

DNA methyltransferase the enzyme that attaches methyl groups to adenine or cytosine bases.

DNA microarray a small silica, glass, or plastic slide that is dotted with many different sequences of DNA, corresponding to short sequences within known genes.

DNA-N-glycosylase an enzyme that can recognize an abnormal base and cleave the bond between it and the sugar in the DNA backbone.

DNA polymerase an enzyme that catalyzes the covalent attachment of nucleotides together to form a strand of DNA.

DNA primase an enzyme that synthesizes a short RNA primer for DNA replication.

DNA probe in a hybridization experiment, a single-stranded DNA fragment with a base sequence that is complementary to a gene of interest. The DNA probe is labeled to detect a gene of interest.

DNA profiling see *DNA fingerprinting*.

DNA replication the process in which original DNA strands are used as templates for the synthesis of new DNA strands.

DNA replication licensing in eukaryotes, occurs when MCM helicase is bound at an origin, enabling the formation of two replication forks.

DNase an enzyme that cuts the sugar–phosphate backbone in DNA.

DNase I an endonuclease that cleaves DNA.

DNase I footprinting a method to study protein–DNA interactions in which the binding of a protein to DNA protects the DNA from digestion by DNase I.

DNase I sensitivity the phenomenon in which DNA is in an open conformation and able to be digested by DNase I.

DNA sequencing a method to determine the base sequence in a segment of DNA.
DNA supercoiling the formation of additional coils in DNA due to twisting forces.
DNA uptake signal sequences DNA sequences found in certain species of bacteria that are needed for a DNA fragment to be taken up during transformation.
domain a segment of a protein that has a specific function.
dominant describes an allele that determines the phenotype in the heterozygous condition. For example, if a plant is *Tt* and has a tall phenotype, the *T* (tall) allele is dominant over the *t* (dwarf) allele.
dominant negative mutation a mutation that produces an altered gene product that acts antagonistically to the normal gene product. Shows a dominant pattern of inheritance.
dorso-ventral axis in animals, the axis from the spine (dorsal) to the stomach (ventral).
dosage compensation refers to the phenomenon that in species with sex chromosomes, one of the sex chromosomes is altered so that males and females will have similar levels of gene expression, even though they do not contain the same complement of sex chromosomes.
double-barrel shotgun sequencing a type of shotgun sequencing in which DNA fragments are randomly sequenced from both ends.
double helix the arrangement in which two strands of DNA (and sometimes RNA) interact with each other to form a double-stranded helical structure.
double-strand break model a model for homologous recombination in which the event that initiates recombination is a double-strand break in one of the double helices.
down promoter mutation a mutation in a promoter that inhibits the rate of transcription.
down regulation genetic regulation that leads to a decrease in gene expression.
duplication the copying of a segment of DNA.

E

ecological species concept a species concept in which each species occupies an ecological niche, which is the unique set of habitat resources that a species requires, as well as the species' influence on the environment and other species.
editosome a complex that catalyzes RNA editing.
egg cell also known as an ovum; it is a female gamete that is usually very large and nonmotile.
electrophoresis the migration of ions or molecules in response to an electric field.
electroporation the use of electric current that creates transient pores in the plasma membrane of a cell to allow entry of DNA.
elongation (1) in transcription, the synthesis of RNA using DNA as a template; (2) in translation, the synthesis of a polypeptide using the information within mRNA.
embryogenesis an early stage of animal and plant development leading to the production of an embryo with organized tissue layers and a body plan organization.
embryonic carcinoma cell (EC cell) a type of pluripotent stem cell found in a specific type of human tumor.
embryonic germ cell (EG cell) a type of pluripotent stem cell found in the gonads of the fetus.
embryonic stem cell (ES cell) a type of pluripotent stem cell found in the early blastocyst.
embryo sac in flowering plants, the female gametophyte that contains an egg cell.
empirical approach a strategy in which experiments are designed to determine quantitative relationships as a way to derive laws that govern biological, chemical, or physical phenomena.
empirical laws laws that are discovered using an empirical (observational) approach.
endonuclease an enzyme that can cut in the middle of a DNA strand.
endopolyploidy in a diploid individual, the phenomenon in which certain cells of the body may be polyploid.
endosperm in flowering plants, the material in the seed, which is $3n$, that nourishes the developing embryo.
endosymbiosis a symbiotic relationship in which the symbiont actually lives inside (*endo*) the larger of the two species.
endosymbiosis theory the theory that the ancient origin of plastids and mitochondria was the result of certain species of bacteria taking up residence within a primordial eukaryotic cell.
enhancer a DNA sequence that functions as a regulatory element. The binding of a regulatory transcription factor to the enhancer increases the level of transcription.
environment the surroundings an organism experiences.
enzyme a protein that functions to accelerate chemical reactions within the cell.
enzyme adaptation the phenomenon in which a particular enzyme appears within a living cell after the cell has been exposed to the substrate for that enzyme.
epigenetic inheritance an inheritance pattern in which a modification to a nuclear gene or chromosome alters gene expression in an organism, but the expression is not changed permanently over the course of many generations.
episome a segment of bacterial DNA that can exist as an F factor and also integrate into the chromosome.
epistasis an inheritance pattern where one gene can mask the phenotypic effects of a different gene.
epitope the structure on the surface of an antigen that is recognized by an antibody.
equilibrium density centrifugation a form of centrifugation in which the particles will sediment through the gradient, reaching a position where the density of the particle matches the density of the solution.
E site see *exit site.*
essential gene a gene that is essential for survival.
EST library see *expressed sequence tagged library.*
ethyl methanesulfonate (EMS) a type of chemical mutagen that alkylates bases (i.e., attaches methyl or ethyl groups).
eubacteria one of the three domains of life. Eubacteria, more commonly known as bacteria, are prokaryotic species.
euchromatin DNA that is not highly compacted and may be transcriptionally active.
eukaryotes (Greek, "true nucleus") one of the three domains of life. A defining feature of these organisms is that their cells contain nuclei bounded by cell membranes. Some simple eukaryotic species are single-celled protists and yeast; more complex multicellular species include fungi, plants, and animals.
euploid describes an organism in which the chromosome number is an exact multiple of a chromosome set.
evo-devo see *evolutionary developmental biology.*
evolution see *biological evolution.*
evolutionary developmental biology (evo-devo) a field of biology that focuses on the role of developmental genes in the formation of traits that are important in the evolution of new species.
evolutionary species concept a species concept in which a species is derived from a single lineage that is distinct from other lineages and has its own evolutionary tendencies and historical fate.
excisionase an enzyme that excises a prophage from a host cell's chromosome.
exit site (E site) a site on the ribosome from which an uncharged tRNA exits.
exon a segment of RNA that is contained within the RNA after splicing has occurred. In mRNA, the coding sequence of a polypeptide is contained within the exons.
exon shuffling the phenomenon that exons have been transferred between different genes during evolution. Some researchers believe that transposable elements have played a role in this phenomenon.
exon skipping when an exon is spliced out of a pre-mRNA.
exonuclease an enzyme that digests an RNA or DNA strand from the end.
expressed sequence tagged (EST) library a DNA library containing many clones that have different cDNA inserts.
expression vector a cloning vector that contains a promoter so that the gene of interest will be transcribed into RNA when the vector is introduced into a host cell.
expressivity the degree to which a trait is expressed. For example, flowers with deep red color would have a high expressivity of the red allele.
extragenic suppressor a mutation in a gene that suppresses the phenotypic effect of a mutation in a different gene.
extranuclear inheritance (also known as *cytoplasmic inheritance*) refers to the inheritance of genetic material that is not found within the nucleus.
ex vivo approach in the case of gene therapy, refers to genetic manipulations that occur outside the body.

F

F see *inbreeding coefficient.*
facultative heterochromatin heterochromatin that is derived from the conversion of euchromatin to heterochromatin.
fate map a diagram that depicts how cell division proceeds in an organism.
feedback inhibition the phenomenon in which the final product of a metabolic pathway inhibits an enzyme that acts early in the pathway.
fertilization the union of gametes (e.g., sperm and egg) to begin the life of a new organism.
F factor a fertility factor found in certain strains of bacteria in addition to their circular chromosome. Strains of bacteria that contain an F factor are designated F^+; strains without F factors are F^-.
F' factors an F factor that also carries genes derived from the bacterial chromosome.
F_1 generation the offspring produced from a cross of the parental generation.
F_2 generation the offspring produced from a cross of the F_1 generation.
fidelity a term used to describe the accuracy of a process. If there are few mistakes, a process has a high fidelity.
fine structure mapping also known as intragenic mapping; the aim of fine structure mapping is to ascertain the distances between two (or more) different mutations within the same gene.
first-division segregation (FDS) in an ordered octad, a 4:4 arrangement of spores that occurs because the two alleles have segregated from each other after the first meiotic division.
fitness see *Darwinian fitness.*
fluctuation test the experimental procedure of Luria and Delbrück in which there was a greater fluctuation in the number of random mutations in small bacterial cultures compared to one large culture.
fluorescence *in situ* hybridization (FISH) a form of *in situ* hybridization in which the probe is fluorescently labeled.
focus with regard to cancer cells, it means a clump of raised cells that grow without regard to contact inhibition.
footprinting see *DNase I footprinting.*
fork see *replication fork.*
forked-line method a method to solve independent assortment problems in which lines are drawn to connect particular genotypes.
forward mutation a mutation that changes the wild-type genotype into some new variation.
founder with regard to genetic diseases, an individual who lived many generations ago and was the person in which a genetic disease originated.
founder effect changes in allele frequencies that occur when a small group of individuals separates from a larger population and establishes a colony in a new location.
fraction (1) following centrifugation, a portion of the liquid contained within a centrifuge tube; (2) following column chromatography, a portion of the liquid that has been eluted from a column.
frameshift mutation a mutation that involves the addition or deletion of nucleotides not in a multiple of three and thereby shifts the reading frame of the codon sequence downstream from the mutation.
frequency distribution a graph that describes the numbers of individuals that are found in each of several phenotypic categories.
functional genomics the study of gene function at the genome level. It involves the study of many genes simultaneously.

G

G an abbreviation for guanine.

gain-of-function mutation a mutation that causes a gene to be expressed in an additional place where it is not normally expressed or during a stage of development when it is not normally expressed.

gamete a reproductive cell (usually haploid) that can unite with another reproductive cell to create a zygote. Sperm and egg cells are types of gametes.

gametogenesis the production of gametes (e.g., sperm or egg cells).

gametophyte the haploid generation of plants.

gap gene one category of segmentation genes.

G bands the chromosomal banding pattern that is observed when the chromosomes have been treated with the chemical dye Giemsa.

gel electrophoresis a method that combines chromatography and electrophoresis to separate molecules and macromolecules. Samples are loaded into wells at one end of the gel, and an electric field is applied across the gel that causes charged molecules to migrate from one side of the gel to the other.

gel retardation assay a technique to study protein–DNA interactions in which the binding of protein to a DNA fragment retards it mobility during gel electrophoresis.

gene a unit of heredity that may influence the outcome of an organism's traits. At the molecular level, a gene contains the information to make a functional product, either RNA or protein.

gene addition the addition of a cloned gene into a site in a chromosome of a living cell.

gene amplification an increase in the copy number of a gene.

gene chip see *DNA microarray.*

gene cloning the production of many copies of a gene using molecular methods such as PCR or the introduction of a gene into a vector that replicates in a host cell.

gene conversion the phenomenon in which one allele is converted to another allele due to genetic recombination and DNA repair.

gene dosage effect when the number of copies of a gene affects the phenotypic expression of a trait.

gene duplication an increase in the copy number of a gene. Can lead to the evolution of gene families.

gene expression the process in which the information within a gene is accessed, first to synthesize RNA (and proteins), and eventually to affect the phenotype of the organism.

gene family two or more different genes within a single species that are homologous to each other because they were derived from the same ancestral gene.

gene flow changes in allele frequencies due to migration.

gene interaction when two or more different genes influence the outcome of a single trait.

gene knockin a type of gene addition in which a gene of interest has been added to a particular site in the mouse genome.

gene knockout when both copies of a normal gene have been replaced by an inactive mutant gene.

gene modifer effect when the allele of one gene modifies the phenotypic effect of the allele of a different gene.

gene mutation a relatively small mutation that affects only a single gene.

gene pool the totality of all genes within a particular population.

generalized transduction a form of transduction in which any piece of the bacterial chromosomal DNA can be incorporated into a phage.

general transcription factor one of several proteins that are necessary for basal transcription at the core promoter.

gene rearrangement a rearrangement in segments of a gene, as occurs in antibody precursor genes.

gene redundancy the phenomenon in which an inactive gene is compensated for by another gene with a similar function.

gene regulation the phenomenon in which the level of gene expression can vary under different conditions.

gene replacement the swapping of a cloned gene made experimentally with a normal chromosomal gene found in a living cell.

gene therapy the introduction of cloned genes into living cells in an attempt to cure or alleviate disease.

genetically modified organism (GMO) an organism that has received genetic material via recombinant DNA technology.

genetic approach in research, refers to the study of mutant genes that have abnormal function. By studying mutant genes, researchers may better understand normal genes and normal biological processes.

genetic code the correspondence between a codon (i.e., a sequence of three bases in an mRNA molecule) and the functional role that the codon plays during translation. Each codon specifies a particular amino acid or the end of translation.

genetic cross a mating between two individuals and the analysis of their offspring in an attempt to understand how traits are passed from parent to offspring.

genetic drift random changes in allele frequencies due to sampling error.

genetic map a chart that describes the relative locations of genes or other DNA segments along a chromosome.

genetic mapping any method used to determine the linear order of genes as they are linked to each other along the same chromosome. This term is also used to describe the use of genetic crosses to determine the linear order of genes.

genetic marker any genetic sequence that is used to mark a specific location on a chromosome.

genetic mosaic see *mosaicism.*

genetic polymorphism when two or more alleles occur in population; each allele is found at a frequency of 1% or higher.

genetic recombination (1) the process in which chromosomes are broken and then rejoined to form a novel genetic combination. (2) the process in which alleles are assorted and passed to offspring in combinations that are different from the parents.

genetics the study of heredity.

genetic screening the use of testing methods to determine if an individual is a heterozygous carrier for or has a genetic disease.

genetic testing the analysis of individuals with regard to their genes or gene products. In many cases, the goal is to determine if an individual carries a mutant gene.

genetic transfer describes the physical transfer of genetic material from one bacterial cell to another.

genetic variation genetic differences among members of the same species or among different species.

genome all of the chromosomes and DNA sequences that an organism or species can possess.

genome database a database that focuses on the genetic sequences and characteristics of a single species.

genome maintenance refers to cellular mechanisms that either prevent mutations from occurring and/or prevent mutant cells from surviving or dividing.

genome sequencing projects research endeavors that have the ultimate goal of determining the sequence of DNA bases of the entire genome of a given species.

genomic clone a clone made from the digestion and cloning of chromosomal DNA.

genomic imprinting a pattern of inheritance that involves a change in a single gene or chromosome during gamete formation. Depending on whether the modification occurs during spermatogenesis or oogenesis, imprinting governs whether an offspring will express a gene that has been inherited from its mother or father.

genomic library a DNA library made from chromosomal DNA fragments.

genomics the molecular analysis of the entire genome of a species.

genotype the genetic composition of an individual, especially in terms of the alleles for particular genes.

genotype-environment association when certain genotypes are preferentially found in particular environments.

genotype-environment interaction when the environmental effects on phenotype differ according to genotype.

genotype frequency the number of individuals with a particular genotype in a population divided by the total number of individuals in the population.

germ cells the gametes (i.e., sperm and egg cells).

germ line a lineage of cells that gives rise to gametes.

germ-line mutation a mutation in a cell of the germ line.

GloFish genetically modified aquarium fish that glow due to the introduction of genes that encode fluorescent proteins.

glucocorticoid receptor a type of steroid receptor that functions as a regulatory transcription factor.

goodness of fit the degree to which the observed data and expected data are similar to each other. If the observed and predicted data are very similar, the goodness of fit is high.

gradualism an evolutionary hypothesis suggesting that each new species evolves continuously over long spans of time. The principal idea is that large phenotypic differences that cause the divergence of species are due to the accumulation of many small genetic changes.

grande normal (large-sized) yeast colonies.

grooves in DNA, the indentations where the atoms of the bases are in contact with the surrounding water. In B DNA, there is a smaller minor groove and a larger major groove.

group I intron a type of intron found in self-splicing RNA that uses free guanosine in its splicing mechanism.

group II intron a type of intron found in self-splicing RNA that uses an adenine nucleotide within the intron itself in its splicing mechanism.

growth factors protein factors that influence cell division.

guanine a purine base found in DNA and RNA. It base-pairs with cytosine in DNA.

guide RNA in trypanosome RNA editing, an RNA molecule that directs the addition of uracil residues into the mRNA.

gyrase see *DNA gyrase.*

H

haplodiploid a species, such as certain bees, in which one sex is haploid (e.g., male) and the other sex is diploid (e.g., female).

haploid describes the phenomenon that gametes contain half the genetic material found in somatic cells. For a species that is diploid, a haploid gamete contains a single set of chromosomes.

haploinsufficiency the phenomenon in which a person has only a single functional copy of a gene, and that single functional copy does not produce a normal phenotype. Shows a dominant pattern of inheritance.

Hardy-Weinberg equation $p^2 + 2pq + q^2 = 1$.

Hardy-Weinberg equilibrium the phenomenon that, under certain conditions, allele frequencies will be maintained in a stable condition and genotypes can be predicted according to the Hardy-Weinberg equation.

helicase see *DNA helicase.*

helix–loop–helix domain a domain found in transcription factors that enables them to dimerize.

helix-turn-helix motif a structure found in transcription factor proteins that promotes binding to the major groove of DNA.

hemizygous describes the single copy of an X-linked gene in the male. A male mammal is said to be hemizygous for X-linked genes.

heritability the amount of phenotypic variation within a particular group of individuals that is due to genetic factors.

heterochromatin highly compacted DNA. It is usually transcriptionally inactive.

heterochronic mutation a mutation that alters the timing of expression of a gene and thereby alters the outcome of cell fates.

heterodimer when two polypeptides encoded by different genes bind to each other to form a dimer.

heteroduplex a double-stranded region of DNA that contains one or more base mismatches.

heterogametic sex in species with two types of sex chromosomes, the heterogametic sex produces two types of gametes. For example, in mammals, the male is the heterogametic sex, because a sperm can contain either an X or a Y chromosome.

heterogamous describes a species that produces two morphologically different types of gametes (i.e., sperm and eggs).
heterogeneity see *locus heterogeneity.*
heterogeneous nuclear RNA (hnRNA) same as pre-mRNA.
heterokaryon a cell produced from cell fusion that contains two separate nuclei.
heteroplasmy when a cell contains variation in a particular type of organelle. For example, a plant cell could contain some chloroplasts that make chlorophyll and other choroplasts that do not.
heterosis the phenomenon in which hybrids display traits superior to either corresponding parental strain. Heterosis is usually different from overdominance, because the hybrid may be heterozygous for many genes, not just a single gene, and because the superior phenotype may be due to the masking of deleterious recessive alleles.
heterozygote an individual who is heterozygous.
heterozygote advantage a pattern of inheritance in which a heterozygote is more vigorous than either of the corresponding homozygotes.
heterozygous describes a diploid individual who has different copies (i.e., two different alleles) of the same gene.
***Hfr* strain** (for **high frequency of recombination**) a bacterial strain in which an F factor has become integrated into the bacterial chromosome. During conjugation, an *Hfr* strain can transfer segments of the bacterial chromosome.
hierarchical shotgun sequencing a genome sequencing strategy in which small DNA fragments are mapped prior to DNA sequencing.
highly repetitive sequences sequences that are found tens of thousands or even millions of times throughout the genome.
high stringency refers to highly selective hybridization conditions that promote the binding of DNA or RNA fragments that are perfect or almost perfect matches.
histone acetyltransferase an enzyme that attaches acetyl groups to the amino terminal tails of histone proteins.
histone code hypothesis the hypothesis that the pattern of histone modification acts much like a language or code in specifying alterations in chromatin structure.
histone deacetylase an enzyme that removes acetyl groups from the amino terminal tails of histone proteins.
histones a group of proteins involved in forming the nucleosome structure of eukaryotic chromatin.
hnRNA an abbreviation for heterogeneous nuclear RNA.
holandric gene a gene on the Y chromosome.
Holliday junction a site where an unresolved crossover has occurred between two homologous chromosomes.
Holliday model a model to explain the molecular mechanism of homologous recombination.
holoenzyme an enzyme containing all of its subunits, as in the holoenzyme of RNA polymerase that has σ factor along with the core enzyme.
homeobox a 180 base pair consensus sequence found in homeotic genes.
homeodomain the protein domain encoded by the homeobox. The homeodomain promotes the binding of the protein to the DNA.
homeologous describes the homologous chromosomes from closely related species.
homeotic an adjective that was originally used to describe mutants in which one body part is replaced by another.
homeotic gene a gene that functions in governing the developmental fate of a particular region of the body.
homoallelic describes two or more alleles in different organisms that are due to mutations at exactly the same base within a gene.
homodimer when two polypeptides encoded by the same gene bind to each other to form a dimer.
homogametic sex in species with two types of sex chromosomes, the homogametic sex produces only one type of gamete. For example, in mammals, the female is the homogametic sex, because an egg can only contain an X chromosome.
homologous in the case of genes, this term describes two genes that are derived from the same ancestral gene. Homologous genes have similar DNA sequences. In the case of chromosomes, the two homologues of a chromosome pair are said to be homologous to each other.
homologous recombination the exchange of DNA segments between homologous chromosomes.
homologous recombination repair also called homology-directed repair, occurs when the DNA strands from a sister chromatid are used to repair a lesion in the other sister chromatid.
homologue one of the chromosomes in a pair of homologous chromosomes.
homology structures that are similar to each other because they evolved from a common ancestor.
homozygous describes a diploid individual who has two identical alleles of a particular gene.
horizontal gene transfer the transfer of genes between different species.
host cell a cell that is infected with a virus or bacterium.
host range the spectrum of host species that a virus or other pathogen can infect.
hot spots sites within a gene that are more likely to be mutated than other locations.
housekeeping gene a gene that encodes a protein required in most cells of a multicellular organism.
***Hox* complexes** a group of several *Hox* genes located in a particular chromosomal region.
***Hox* genes** mammalian genes that play a role in development. They are homologous to homeotic genes found in *Drosophila.*
Human Genome Project a worldwide collaborative project that provided a detailed map of the human genome and obtained a complete DNA sequence of the human genome.
hybrid (1) an offspring obtained from a hybridization experiment; (2) a cell produced from a cell fusion experiment in which the two separate nuclei have fused to make a single nucleus.
hybrid cell a cell produced from the fusion of two different cells, usually from two different species.
hybrid dysgenesis a syndrome involving defective *Drosophila* offspring, due to the phenomenon that P elements can transpose freely.
hybridization (1) the mating of two organisms of the same species with different characteristics; (2) the phenomenon in which two single-stranded molecules renature together to form a hybrid molecule.
hybrid vector a cloning vector that contains an insert of foreign DNA.
hybrid vigor see *heterosis.*
hybrid zones during parapatric speciation, places where two populations can interbreed.
hypothesis testing using statistical tests to determine if the data from experimentation are consistent with a hypothesis.

I

illegitimate recombination see *nonhomologous recombination.*
immunoglobulin (IgG) see *antibodies.*
imprinting see *genomic imprinting.*
inborn error of metabolism a genetic disease that involves a defect in a metabolic enzyme.
inbreeding the practice of mating between genetically related individuals.
inbreeding coefficient (F) the probability that two alleles in a particular individual will be identical for a given gene because both copies are due to descent from a common ancestor.
inbreeding depression the phenomenon in which inbreeding produces homozygotes that are less fit, thereby decreasing the reproductive success of a population.
incomplete dominance a pattern of inheritance in which a heterozygote that carries two different alleles exhibits a phenotype that is intermediate to the corresponding homozygous individuals. For example, an *Rr* heterozygote may be pink, while the *RR* and *rr* homozygotes are red and white, respectively.
incomplete penetrance a pattern of inheritance in which a dominant allele does not always control the phenotype of the individual.
incomplete transposable element see *nonautonomous transposable element.*
induced refers to a gene that has been transcriptionally activated by an inducer.
induced mutation a mutation caused by environmental agents.
inducer a small effector molecule that binds to a genetic regulatory protein and thereby increases the rate of transcription.
inducible gene a gene that is regulated by an inducer, which is a small effector molecule that causes transcription to increase.
induction (1) the effects of an inducer on increasing the transcription of a gene; (2) the process by which a cell or group of cells governs the developmental fate of neighboring cells.
infective particle genetic material found within the cytoplasm of eukaryotic cells that differs from the genetic material normally found in cell organelles.
ingroup in cladistics, a group of species in which a researcher is interested.
inhibitor a small effector molecule that binds to an activator protein, causing the protein to be released from the DNA and thereby inhibiting transcription.
initiation (1) in transcription, the stage that involves the initial binding of RNA polymerase to the promoter in order to begin RNA synthesis; (2) in translation, the formation of a complex between mRNA, the initiator tRNA, and the ribosomal subunits.
initiator tRNA during translation, the tRNA that binds to the start codon.
insertion sequences the simplest transposable elements. They are commonly found in bacteria.
***in situ* hybridization** a technique used to cytologically map the locations of genes or other DNA sequences within large eukaryotic chromosomes. In this method, a complementary probe is used to detect the location of a gene within a set of chromosomes.
integrase an enzyme that functions in the integration of viral DNA or retroelements into a chromosome.
interference see *positive interference.*
intergenic region in a chromosome, a region of DNA that lies between two different genes.
intergenic suppressor a suppressor mutation that is in a different gene from the gene that contains the first mutation.
internal nuclear matrix a network of irregular protein fibers and other proteins that is connected to the nuclear lamina and fills the interior of the nucleus.
interphase the series of phases G_1, S, and G_2, during which a cell spends most of its life.
interrupted mating a method used in conjugation experiments in which the length of time that the bacteria spend conjugating is stopped by a blender treatment or other type of harsh agitation.
interstitial deficiency see *interstitial deletion.*
interstitial deletion when an internal segment is lost from a linear chromosome.
intervening sequence also known as an *intron.* A segment of RNA that is removed during RNA splicing.
intragenic mapping see *fine structure mapping.*
intragenic suppressor a suppressor mutation that is within the same gene as the first mutation that it suppresses.
intrinsic termination transcriptional termination that does not require the function of the rho protein.
intron intervening sequences that are found between exons. Introns are spliced out of the RNA prior to translation.
invasive refers to a tumor that can invade surrounding tissue.
inversion a change in the orientation of genetic material along a chromosome such that a segment is flipped in the reverse direction.
inversion heterozygote a diploid individual that carries one normal chromosome and a homologous chromosome with an inversion.
inversion loop the loop structure that is formed when the homologous chromosomes of an inversion heterozygote attempt to align themselves (i.e., synapse) during meiosis.
inverted repeats DNA sequences found in transposable elements that are identical (or very similar) but run in the opposite directions.
iron regulatory protein a translational regulatory protein that recognizes iron response elements that are found in specific mRNAs. It may inhibit translation or stabilize the mRNA.

iron response element an RNA sequence that is recognized by the iron regulatory protein.

isoacceptor tRNAs two different tRNAs that can recognize the same codon.

isoelectric focusing a form of gel electrophoresis in which a protein migrates to the point in the gel where its net charge is zero.

isogamous describes a species that makes morphologically similar gametes.

isolating mechanism a mechanism that favors reproductive isolation. These can be prezygotic or postzygotic.

K

karyotype a photographic representation of all the chromosomes within a cell. It reveals how many chromosomes are found within an actively dividing somatic cell.

kinetochore a group of cellular proteins that attach to the centromere during meiosis and mitosis.

knockin see *gene knockin.*

knockout see *gene knockout.*

Kozac's rules a set of rules that describes the most favorable types of bases that flank a eukaryotic start codon.

L

lac repressor a protein that binds to the operator site of the *lac* operon and inhibits transcription.

lagging strand a strand during DNA replication that is synthesized as short Okazaki fragments in the direction away from the replication fork.

lariat an excised intron structure composed of a circle and a tail.

law of independent assortment see *Mendel's law of independent assortment.*

law of segregation see *Mendel's law of segregation.*

leading strand a strand during DNA replication that is synthesized continuously toward the replication fork.

leptotene the first stage of prophase of meiosis I.

lesion-replicating polymerase a type of DNA polymerase that can replicate over a DNA region that contains an abnormal structure (i.e., a lesion).

lethal allele an allele that may cause the death of an organism.

lethal mutation a mutation that produces a lethal allele that causes the death of an organism.

library see *DNA library.*

ligase see *DNA ligase.*

lineage diagram in developmental biology, a depiction of a cell and all of its descendants that are produced by cell division.

LINEs in mammals, long interspersed elements that are usually 1 to 5 kbp in length and found in 20,000 to 100,000 copies per genome.

linkage refers to the occurrence of two or more genes along the same chromosome.

linkage group a group of genes that are linked together because they are found on the same chromosome.

linkage mapping the mapping of genes or other genetic sequences along a chromosome by analyzing the outcome of crosses.

lipid a general name given to an organic molecule that is insoluble in water. Cell membranes contain a large amount of lipid.

liposome a vesicle that is surrounded by a phospholipid bilayer.

local population a segment of a population that is slightly isolated. Members of a local population are more likely to breed with each other than with members that are outside of the local population.

locus (pl. **loci**) the physical location of a gene within a chromosome.

locus control region (LCR) a segment of DNA that is involved in the regulation of chromatin opening and closing.

locus heterogeneity the phenomenon in which a particular type of disease or trait may be caused by mutations in two or more different genes.

lod score method a method that analyzes pooled data from a large number of pedigrees or crosses to determine the probability that two genetic markers exhibit a certain degree of linkage. A lod score value of >3 or higher is usually accepted as strong evidence that two markers are linked.

long terminal repeats (LTRs) sequences containing many short segments that are tandemly repeated. They are found in retroviruses and viral-like retroelements.

loop domain a segment of chromosomal DNA that is anchored by proteins, so it forms a loop.

loss-of-function allele an allele of a gene that encodes an RNA or protein that is nonfunctional or compromised in function.

loss-of-function mutation a change in a genetic sequence that creates a loss-of-function allele.

low stringency refers to hybridization conditions in which DNA or RNA fragments that have some mismatches are still able to recognize and bind to each other.

LTRs see *long terminal repeats.*

Lyon hypothesis a hypothesis to explain the pattern of X inactivation seen in mammals. Initially, both X chromosomes are active. However, at an early stage of embryonic development, one of the two X chromosomes is randomly inactivated in each somatic cell.

lysis cell breakage.

lysogenic cycle a type of growth cycle for a phage in which the phage integrates its genetic material into the chromosome of the bacterium. This integrated phage DNA can exist in a dormant state for a long time, during which no new bacteriophages are made.

lytic cycle a type of growth cycle for a phage in which the phage directs the synthesis of many copies of the phage genetic material and coat proteins. These components then assemble to make new phages. When synthesis and assembly is completed, the bacterial host cell is lysed, and the newly made phages are released into the environment.

M

macroevolution evolutionary changes at or above the species level involving relatively large changes in form and function that are sufficient to produce new species and higher taxa.

macromolecule a large organic molecule composed of smaller building blocks. Examples include DNA, RNA, proteins, and large carbohydrates.

MADS box a domain found in several transcription factors that play a role in plant development.

maintenance methylation the methylation of hemimethylated DNA following DNA replication.

major groove a wide indentation in the DNA double helix in which the bases have access to water.

malignant describes a tumor composed of cancerous cells.

mammalian cloning experimentally, this refers to the use of somatic cell nuclei and enucleated eggs to create a clone of a mammal.

map distance the relative distance between sites (e.g., genes) along a single chromosome.

mapping the experimental process of determining the relative locations of genes or other segments of DNA along individual chromosomes.

map unit (mu) a unit of map distance obtained from genetic crosses. One map unit is equivalent to 1% recombinant offspring in a testcross.

mass spectrometry a technique to accurately measure the mass of molecule, such as a peptide fragment.

maternal effect an inheritance pattern for certain nuclear genes in which the genotype of the mother directly determines the phenotypic traits of her offspring.

maternal inheritance inheritance of DNA that occurs through the cytoplasm of the egg.

matrix-attachment region (MAR) a site in the chromosomal DNA that is anchored to the nuclear matrix or scaffold.

maturase a protein that enhances the rate of splicing of Group I and II introns.

maximum likelihood a strategy for choosing the best phylogenetic tree based on assumptions about the most likely ways that DNA incurs mutations.

MCM helicase a group of eukaryotic proteins needed to complete a process called DNA replication licensing, which is necessary for the formation of two replication forks at an origin of replication.

mean the sum of all the values in a group divided by the number of individuals in the group.

mean fitness of the population the average fitness of a population that is calculated by considering the frequencies and fitness values for all genotypes.

mediator a protein complex that interacts with RNA polymerase II and various regulatory transcription factors. Depending on its interactions with regulatory transcription factors, mediator may stimulate or inhibit RNA polymerase II.

meiosis a form of nuclear division in which the sorting process results in the production of haploid cells from a diploid cell.

meiotic nondisjunction the event in which chromosomes do not segregate equally during meiosis.

Mendelian inheritance the common pattern of inheritance observed by Mendel, which involves the transmission of eukaryotic genes that are located on the chromosomes found within the cell nucleus.

Mendel's law of independent assortment two different genes will randomly assort their alleles during gamete formation (if they are not linked).

Mendel's law of segregation the two copies of a gene segregate from each other during transmission from parent to offspring.

meristem in plants, an organized group of actively dividing cells.

merozygote a partial diploid strain of bacteria containing F' factor genes.

messenger RNA (mRNA) a type of RNA that contains the information for the synthesis of a polypeptide.

metacentric describes a chromosome with the centromere in the middle.

metaphase the third stage of M phase. The chromosomes align along the center of the spindle apparatus, and the formation of the spindle apparatus is complete.

metaphase plate the plane at which chromosomes align during metaphase.

metastatic describes cancer cells that migrate to other parts of the body.

methylation see *DNA methylation.*

methyl-CpG-binding protein a protein that binds to a CpG island when it is methylated.

methyl-directed mismatch repair a DNA repair system that detects a mismatch and specifically removes the segment from the newly made daughter strand.

methyltransferase see *DNA methyltransferase.*

microarray see *DNA microarray, protein microarray,* or *antibody microarray.*

microevolution changes in the gene pool with regard to particular alleles that occur over the course of many generations.

microinjection the use of microscopic-sized needles or pipettes to inject a substance into cells, such as DNA.

microsatellite short simple sequences (typically a couple hundred base pairs in length) that are interspersed throughout a genome and are quite variable in length among different individuals. They can be amplified by PCR.

microscopy the use of a microscope to view cells or subcellular structures.

minichromosomes a structure formed from many copies of circular DNA molecules.

minisatellite a repetitive sequence that was formerly used in DNA fingerprinting. Its use has been largely superceded by smaller repetitive sequences called microsatellites.

minor groove a narrow indentation in the DNA double helix in which the bases have access to water.

minute a unit of measure in bacterial conjugation experiments. This unit refers to the relative time it takes for genes to first enter a recipient strain during conjugation.

mismatch repair system see *methyl-directed mismatch repair.*

missense mutation a base substitution that leads to a change in the amino acid sequence of the encoded polypeptide.

mitochondrial DNA the DNA found within mitochondria.

mitosis a type of nuclear division into two nuclei, such that each daughter cell will receive the same complement of chromosomes.
mitotic nondisjunction an event in which chromosomes do not segregate equally during mitosis.
mitotic recombination crossing over that occurs during mitosis.
mitotic spindle apparatus (also known as the *mitotic spindle*) the structure that organizes and separates the chromosomes during M phase of the eukaryotic cell cycle.
model organism an organism studied by many scientists so that researchers can more easily compare their results and begin to unravel the properties of a given species.
moderately repetitive sequences sequences that are found a few hundred to several thousand times in the genome.
molecular clock refers to the phenomenon that the rate of neutral mutations can be used as a tool to measure evolutionary time.
molecular evolution the molecular changes in the genetic material that underlie the phenotypic changes associated with evolution.
molecular genetics an examination of DNA structure and function at the molecular level.
molecular marker a segment of DNA that is found at a specific site in the genome and has properties that enable it to be uniquely recognized using molecular tools such as gel electrophoresis.
molecular paleontology the analysis of DNA sequences from extinct species.
molecular pharming a recombinant technology that involves the production of medically important proteins in the mammary glands of livestock.
molecular profiling methods that enable researchers to understand the molecular changes that occur in diseases such as cancer.
molecular trait a trait that is observed at the molecular level. An example would be the amount of a given protein in a cell.
monoallelic expression in the case of imprinting, refers to the phenomenon that only one of the two alleles of a given gene is transcriptionally expressed.
monohybrid an individual produced from a monohybrid cross.
monohybrid cross a cross in which an experimenter is following the outcome of only a single trait.
monomorphic a term used to describe a gene that is found as only one allele in a population.
monophyletic group see *clade.*
monoploid an organism with a single set of chromosomes within its somatic cells.
monosomic a diploid cell that is missing a chromosome (i.e., $2n - 1$).
monozygotic twins twins that are genetically identical to each other because they were formed from the same sperm and egg.
morph a form or phenotype in a population. For example, red eyes and white eyes are different eye color morphs.
morphogen a molecule that conveys positional information and promotes developmental changes.
morphological trait a trait that affects the morphology (physical form) of an organism. An example would be eye color.
mosaicism when the cells of part of an organism differ genetically from the rest of the organism.
motif the name given to a domain or amino acid sequence that functions in a similar manner in many different proteins.
M phase a general name given to nuclear division that can apply to mitosis or meiosis. It is divided into prophase, prometaphase, metaphase, anaphase, and telophase.
mRNA see *messenger RNA.*
mtDNA an abbreviation for mitochondrial DNA.
multinomial expansion equation an equation to solve genetic problems involving three or more types of unordered events.
multiple alleles when the same gene exists in two or more alleles within a population.
multiple sequence alignment an alignment of two or more genetic sequences based on their homology to each other.
multiplication method a method to solve independent assortment problems in which the probabilities of the outcome for each gene are multiplied together.
multipotent a type of stem cell that can differentiate into several different types of cells.
mutable site a site in a chromosome that tends to break at a fairly high rate due to the presence of a transposable element.
mutagen an agent that causes alterations in the structure of DNA.
mutant alleles alleles that have been created by altering a wild-type allele by mutation.
mutation a permanent change in the genetic material that can be passed from cell to cell or from parent to offspring.
mutation frequency the number of mutant genes divided by the total number of genes within the population.
mutation rate the likelihood that a gene will be altered by a new mutation.
myogenic bHLH protein a type of transcription factor involved in muscle cell differentiation.

N

n an abbreviation that designates the number of chromosomes in a set. In humans, $n = 23$, and a diploid cell has $2n = 46$ chromosomes.
narrow sense heritability heritability that takes into account only those genetic factors that are additive.
natural selection refers to the process whereby differential fitness acts on the gene pool. When a mutation creates a new allele that is beneficial, the allele may become prevalent within future generations because the individuals possessing this allele are more likely to reproduce and pass the beneficial allele to their offspring.
natural transformation a natural process of transformation that occurs in certain strains of bacteria.
negative control transcriptional regulation by repressor proteins.
negative frequency-dependent selection a pattern of natural selection in which the fitness of a genotype decreases when its frequency becomes higher.
neutral mutation a mutation that has no detectable effect on protein function and/or no detectable effect on the survival of the organism.
neutral theory of evolution the theory that most genetic variation observed in natural populations is due to the accumulation of neutral mutations.
nick translation the phenomenon in which DNA polymerase uses its 5' to 3' exonuclease activity to remove a region of DNA and, at the same time, replaces it with new DNA.
nitrogen mustard an alkylating agent that can cause mutations in DNA.
nitrous acid a type of chemical mutagen that deaminates bases, thereby changing amino groups to keto groups.
nonautonomous element a transposable element that lacks a gene such as transposase or reverse transcriptase that is necessary for transposition.
noncoding strand the strand of DNA within a structural gene that is complementary to the mRNA. The noncoding strand is used as a template to make mRNA.
noncomplementation the phenomenon in which two mutant alleles in the same organism do not produce a wild-type phenotype.
non-Darwinian evolution see *neutral theory of evolution.*
nondisjunction event in which chromosomes do not segregate properly during mitosis or meiosis.
nonessential genes genes that are not absolutely required for survival, although they are likely to be beneficial to the organism.
nonhomologous end-joining protein (NHEJ) a protein that joins the ends of DNA fragments that are not homologous. This occurs during site-specific recombination of immunoglobulin genes.
nonhomologous recombination the exchange of DNA between nonhomologous segments of chromosomes or plasmids.
nonneutral mutation a mutation that affects the phenotype of the organism and can be acted on by natural selection.
nonparental see *nonrecombinant.*
nonparental ditype (NPD) an ascus that contain cells that all have a nonparental combination of alleles.
nonrecombinant in a testcross, refers to a phenotype or arrangement of alleles on a chromosome that is not found in the parental generation.
nonsense codon a stop codon.
nonsense mutation a mutation that involves a change from a sense codon to a stop codon.
nonviral-like retroelement a type of retroelement in which the sequence does not resemble a modern virus.
normal distribution a distribution for a large sample in which the trait of interest varies in a symmetrical way around an average value.
norm of reaction the effects of environmental variation on an individual's traits.
Northern blotting a technique used to detect a specific RNA within a mixture of many RNA molecules.
N-terminus see *amino terminus.*
nuclear genes genes that are located on chromosomes found in the cell nucleus of eukaryotic cells.
nuclear lamina a collection of fibers that line the inner nuclear membrane.
nuclear matrix (or **nuclear scaffold**) a group of proteins that anchor the loops found in eukaryotic chromosomes.
nucleic acid RNA or DNA. A macromolecule that is composed of repeating nucleotide units.
nucleoid a darkly staining region that contains the genetic material of mitochondria, chloroplasts, or bacteria.
nucleolus a region within the nucleus of eukaryotic cells where the assembly of ribosomal subunits occurs.
nucleoprotein a complex of DNA (or RNA) and protein.
nucleoside structure in which a base is attached to a sugar, but no phosphate is attached to the sugar.
nucleosome the repeating structural unit within eukaryotic chromatin. It is composed of double-stranded DNA wrapped around an octamer of histone proteins.
nucleotide the repeating structural unit of nucleic acids, composed of a sugar, phosphate, and base.
nucleotide excision repair (NER) a DNA repair system in which several nucleotides in the damaged strand are removed from the DNA and the undamaged strand is used as a template to resynthesize a normal strand.
nucleus a membrane-bound organelle in eukaryotic cells where the linear sets of chromosomes are found.
null hypothesis a hypothesis that assumes there is no real difference between the observed and expected values.

O

ocher a stop codon with the sequence UGA.
octad a group of eight fungal spores contained within an ascus.
Okazaki fragments short segments of DNA that are synthesized in the lagging strand during DNA replication.
oligonucleotide a short strand of DNA, typically a few or a few dozen nucleotides in length.
oncogene a mutant gene that promotes cancer.
one gene–one enzyme hypothesis the idea, which later needed to be expanded, that one gene encodes one enzyme.
oogenesis the production of egg cells.
opal a stop codon with the sequence UAA.
open complex the region of separation of two DNA strands produced by RNA polymerase during transcription.
open conformation a loosely packed chromatin structure that is capable of transcription.
open reading frame a genetic sequence that does not contain stop codons.
operator (or **operator site**) a sequence of nucleotides in bacterial DNA that provides a binding site for a genetic regulatory protein.
operon an arrangement in DNA where two or more structural genes are found within a regulatory unit that is under the transcriptional control of a single promoter.
ordered octad an ascus composed of eight cells whose order depends on crossing over during meiosis.

ordered tetrad an ascus composed of four cells whose order depends on crossing over during meiosis.
ORF see *open reading frame.*
organelle a large specialized structure within a cell, which is often surrounded by a single or double membrane.
organismal cloning the cloning of whole organisms. An example is the cloning of the sheep named Dolly.
organism level when the level of observation or experimentation involves a whole organism.
organizing center in plants, a region of the meristem that ensures the proper organization of the meristem and preserves the correct number of actively dividing stem cells.
orientation independent refers to certain types of genetic regulatory elements that can function in the forward or reverse direction. Certain enhancers are orientation independent.
origin of replication a nucleotide sequence that functions as an initiation site for the assembly of several proteins required for DNA replication.
origin of transfer the location on an F factor or within the chromosome of an Hfr strain that is the initiation site for the transfer of DNA from one bacterium to another during conjugation.
origin recognition complex (ORC) a complex of six proteins found in eukaryotes that is necessary to initiate DNA replication.
ortholog homologous genes in different species that were derived from the same ancestral gene.
outbreeding mating between genetically unrelated individuals.
outgroup in cladistics, a species or group of species that is most closely related to the ingroup.
ovary (1) in plants, the structure in which the ovules develop; (2) in animals, the structure that produces egg cells and female hormones.
overdominance an inheritance pattern in which a heterozygote is more vigorous than either of the corresponding homozygotes.
ovule the structure in higher plants where the female gametophyte (i.e., embryo sac) is produced.
ovum a female gamete, also known as an egg cell.

P

p an abbreviation for the short arm of a chromosome.
pachytene the third stage of prophase of meiosis I.
pair-rule gene one category of segmentation genes.
palindromic when a sequence is the same in the forward and reverse direction.
pangenesis an incorrect hypothesis of heredity. It suggested that hereditary traits could be modified depending on the lifestyle of the individual. For example, it was believed that a person who practiced a particular skill would produce offspring that would be better at that skill.
paracentric inversion an inversion in which the centromere is found outside of the inverted region.
paralogs homologous genes within a single species that constitute a gene family.
parapatric speciation (Greek, *para*, "beside"; Latin, *patria*, "homeland") a form of speciation that occurs when members of a species are only partially separated or when a species is very sedentary.
parasegments transient subdivisions that occur in the *Drosophila* embryo prior to the formation of segments.
parental in a testcross, refers to a phenotype or arrangement of alleles on a chromosome that is the same as one or both members of the parental generation.
parental ditype (PD) an ascus that contains four spores with the parental combinations of alleles.
parental generation in a genetic cross, the first generation in the experiment. In Mendel's studies, the parental generation was true-breeding with regard to particular traits.
parental strand in DNA replication, the DNA strand that is used as a template.
parthenogenesis the formation of an individual from an unfertilized egg.
particulate theory of inheritance a theory proposed by Mendel. It states that traits are inherited as discrete units that remain unchanged as they are passed from parent to offspring.
P1 artificial chromosome (PAC) an artificial chromosome developed from P1 bacteriophage chromosomes.
paternal leakage the phenomenon that in species where maternal inheritance is generally observed, the male parent may, on rare occasions, provide mitochondria or chloroplasts to the zygote.
pattern the spatial arrangement of different regions of the body. At the cellular level, the body pattern is due to the arrangement of cells and their specialization.
pattern recognition in bioinformatics, this term refers to a program that recognizes a pattern of symbols.
PCR see *polymerase chain reaction.*
pedigree analysis a genetic analysis using information contained within family trees. In this approach, the aim is to determine the type of inheritance pattern that a gene follows.
pellet a collection of particles found at the bottom of a centrifuge tube.
peptide bond a covalent bond formed between the carboxyl group in one amino acid in a polypeptide chain and the amino group in the next amino acid.
peptidyl site (P site) a site on the ribosome that carries a tRNA along with a polypeptide chain.
peptidyl transfer the step during the elongation stage of translation in which the polypeptide is removed from the tRNA in the P site and transferred to the amino acid at the A site.
peptidyltransferase a complex that functions during translation to catalyze the formation of a peptide bond between the amino acid in the A site of the ribosome and the growing polypeptide chain.
pericentric inversion an inversion in which the centromere is located within the inverted region of the chromosome.
peripheral zone in plants, an area in the meristem that contains dividing cells that will eventually differentiate into plant structures.
petites mutant strains of yeast that form small colonies due to defects in mitochondrial function.
PFGE see *pulsed-field gel electrophoresis.*
P generation the parental generation in a genetic cross.
phage see *bacteriophage.*
pharming see *molecular pharming.*
phenotype the observable traits of an organism.
phenylketonuria (PKU) a human genetic disorder arising from a defect in phenylalanine hydroxylase.
phosphodiester linkage in a DNA or RNA strand, a linkage in which a phosphate group connects two sugar molecules together.
photolyase an enzyme found in yeast and plants that can repair thymine dimers by splitting the dimers, which returns the DNA to its original condition.
phylogenetic species concept a species concept in which the members of a single species are identified by having a unique combination of character states.
phylogenetic tree a diagram that describes the evolutionary relationships among different species.
phylogeny the sequence of events involved in the evolutionary development of a species or group of species.
physical mapping the mapping of genes or other genetic sequences using DNA cloning methods.
physiological trait a trait that affects a cellular or body function. An example would be the rate of glucose metabolism.
PKU see *phenylketonuria.*
plaque a clear zone within a bacterial lawn on a petri plate. It is due to repeated cycles of viral infection and bacterial lysis.
plasmid a general name used to describe circular pieces of DNA that exist independently of the chromosomal DNA. Some plasmids are used as vectors in cloning experiments.
pleiotrophy refers to the multiple effects of a single gene on the phenotype of an organism.
pluripotent a type of stem cell that can differentiate into all or nearly all the types of cells of the adult organism.
point mutation a change in a single base pair within DNA.
polarity in genetics, the phenomenon in which a nonsense mutation in one gene will affect the translation of a downstream gene in an operon.
pollen also, pollen grain; the male gametophyte of flowering plants.
polyA-binding protein a protein that binds to the 3' polyA tail of mRNAs and protects the mRNA from degradation.
polyadenylation the process of attachment of a string of adenine nucleotides to the 3' end of eukaryotic mRNAs.
polyA tail the string of adenine nucleotides at the 3' end of eukaryotic mRNAs.
polycistronic mRNA an mRNA transcribed from an operon that encodes two or more proteins.
polygenic inheritance refers to the transmission of traits that are governed by two or more different genes.
polymerase chain reaction (PCR) the method to amplify a DNA region involving the sequential use of oligonucleotide primers and *Taq* polymerase.
polymerase switch during DNA replication, when one type of DNA polymerase (such as α) is switched for another type (such as β).
polymorphic a term used to describe a trait or gene that is found in two or more forms in a population.
polymorphism (1) the prevalence of two or more phenotypic forms in a population; (2) the phenomenon in which a gene exists in two or more alleles within a population.
polypeptide a sequence of amino acids that is the product of mRNA translation. One or more polypeptides will fold and associate with each other to form a functional protein.
polyploid an organism or cell with three or more sets of chromosomes.
polyribosome an mRNA transcript that has many bound ribosomes in the act of translation.
polysome see *polyribosome.*
polytene chromosome chromosomes that are found in certain cells, such as *Drosophila* salivary cells, in which the chromosomes have replicated many times and the copies lie side by side.
population a group of individuals of the same species that are capable of interbreeding with one another.
population genetics the field of genetics that is primarily concerned with the prevalence of genetic variation within populations.
population level when the level of observation or experimentation involves a population of organisms.
positional cloning a cloning strategy in which a gene is cloned based on its mapped position along a chromosome.
positional information chemical substances and other environmental cues that enable a cell to deduce its position relative to other cells.
position effect a change in phenotype that occurs when the position of a gene is changed from one chromosomal site to a different location.
positive control genetic regulation by activator proteins.
positive interference the phenomenon in which a crossover that occurs in one region of a chromosome decreases the probability that another crossover will occur nearby.
posttranslational describes events that occur after translation is completed.
posttranslational covalent modification the covalent attachment of a molecule to a protein after it has been synthesized via ribosomes.
posttranslational sorting refers to protein sorting to an organelle that occurs after the protein has been completely synthesized in the cytosol.
postzygotic isolating mechanism a mechanism of reproductive isolation that prevents an offspring from being viable or fertile.
preinitiation complex the stage of transcription in which the assembly of RNA polymerase and general transcription factors occurs at the core promoter, but the DNA has not yet started to unwind.
pre-mRNA in eukaryotes, the transcription of structural genes produces a long transcript known as pre-mRNA, which is located within the nucleus. This pre-mRNA is usually altered by splicing and other modifications before it exits the nucleus.
prereplication complex (preRC) in eukaryotes, an assembly of at least 14 different proteins, including a group of six proteins called the origin recognition complex (ORC), that acts as the initiator of eukaryotic DNA replication.

prezygotic isolating mechanism a mechanism for reproductive isolation that prevents the formation of a zygote.
Pribnow box the TATAAT sequence that is often found at the −10 region of bacterial promoters.
primary structure with regard to proteins, the linear sequence of amino acids.
primase an enzyme that synthesizes a short RNA primer during DNA replication.
primitive character a trait that is shared with a distant ancestor.
primosome a multiprotein complex composed of DNA helicase, primase, and several accessory proteins.
principle of parsimony the principle that the preferred hypothesis is the one that is the simplest.
prion an infectious particle that causes several types of neurodegenerative diseases affecting humans and livestock. It is composed entirely of protein.
probability the chance that an event will occur in the future.
processive enzyme an enzyme, such as RNA and DNA polymerase, which glides along the DNA and does not dissociate from the template strand as it catalyzes the covalent attachment of nucleotides.
product rule the probability that two or more independent events will occur is equal to the products of their individual probabilities.
proflavin a type of chemical mutagen that causes frameshift mutations.
prokaryotes (Greek, "prenucleus") another name for bacteria and archaea. The term refers to the fact that their chromosomes are not contained within a separate nucleus of the cell.
prometaphase the second phase of M phase. During this phase, the nuclear membrane vesiculates, and the mitotic spindle is completely formed.
promoter a sequence within a gene that initiates (i.e., promotes) transcription.
promoter bashing the approach of making deletions in the vicinity of a promoter as a way to identify the core promoter and response elements.
proofreading function the ability of DNA polymerase to remove mismatched bases from a newly made strand.
prophage phage DNA that has been integrated into the bacterial chromosome.
prophase the first stage of M phase. The chromosomes have already replicated and begin to condense. The mitotic spindle starts to form.
protease an enzyme that digests the polypeptide backbone found in proteins.
protein a functional unit composed of one or more polypeptides.
protein microarray a small silica, glass, or plastic slide that is dotted with many different proteins.
proteome the collection of all proteins that a given cell or species can make.
proteomics the study of protein function at the genome level. It involves the study of many proteins simultaneously.
proto-oncogene a normal cellular gene that does not cause cancer but which may incur a gain-of-function mutation or become incorporated into a viral genome and thereby lead to cancer.
protoplast a plant cell without a cell wall.
prototroph a strain that does not need a particular nutrient supplemented in its growth medium or diet.
provirus viral DNA that has been incorporated into the chromosome of a host cell.
proximo-distal axis in animals, an axis involving limbs in which the part of the limb attached to the trunk is located proximal, while the end of the limb is located distal.
pseudoautosomal inheritance the inheritance pattern of genes that are found on both the X and Y chromosomes. Even though such genes are located physically on the sex chromosomes, their pattern of inheritance is identical to that of autosomal genes.
pseudodominance a pattern of inheritance that occurs when a single copy of a recessive allele is phenotypically expressed because the second copy of the gene has been deleted from the homologous chromosome.
pseudo-overdominance occurs when two closely linked genes exhibit heterosis. The heterosis is really due to the masking of recessive alleles.
P site see *peptidyl site.*
pulse/chase experiment an experimental strategy in which cells are given a radiolabeled compound for a brief period of time—a pulse—followed by an excess amount of unlabeled compound—a chase.
pulsed-field gel electrophoresis (PFGE) a method of gel electrophoresis used to separate small chromosomes or very large pieces of chromosomes.
punctuated equilibrium an evolutionary theory proposing that species exist relatively unchanged for many generations. These long periods of equilibrium are punctuated by relatively short periods during which evolution occurs at a relatively rapid rate.
Punnett square a diagrammatic method in which the gametes that two parents can produce are aligned next to a square grid as a way to predict the types of offspring the parents will produce and in what proportions.
purine a type of nitrogenous base that has a double-ring structure. Examples are adenine and guanine.
***P* value** in a chi square table, the probability that the deviations between observed and expected values are due to random chance.
pyrimidine a type of nitrogenous base that has a single-ring structure. Examples are cytosine, thymine, and uracil.

Q

q an abbreviation for the long arm of a chromosome.
QTL see *quantitative trait loci.*
QTL mapping the determination of the location of QTLs using mapping methods such as genetic crosses coupled with the analysis of molecular markers.
quantitative genetics the area of genetics concerned with traits that can be described in a quantitative way.
quantitative trait a trait, usually polygenic in nature, that can be described with numbers.
quantitative trait loci (QTLs) the locations on chromosomes where the genes that influence quantitative traits reside.
quaternary structure in proteins, refers to the binding of two or more polypeptides to each other to form a protein.

R

radial loop domains organization of chromatin during interphase into loops, often 25,000 to 200,000 bp in size, which are anchored to the nuclear matrix at matrix-attachment regions.
radioimmunoassay a method to measure the amount of an antigen in a biological sample.
RAG1/RAG2 enzymes that recognize recombination signal sequences and make double-stranded cuts. In the case of V/J recombination in immunoglobulin genes, a cut is made at the end of one V region and the beginning of one J region.
random genetic drift see *genetic drift.*
random mutation theory according to this theory, mutations are a random process—they can occur in any gene and do not involve exposure of an organism to a particular condition that selects for specific types of mutations.
random sampling error the deviation between the observed and expected outcomes due to chance.
reading frame a series of codons beginning with the start codon as a frame of reference.
realized heritability a form of narrow sense heritability that is observed when selective breeding is practiced.
recessive describes a trait or gene that is masked by the presence of a dominant trait or gene.
recessive epistasis a form of epistasis in which an individual must be homozygous for either recessive allele to mask a particular phenotype.
reciprocal crosses a pair of crosses in which the traits of the two parents differ with regard to sex. For example, one cross could be a red-eyed female fly and a white-eyed male fly, and the reciprocal cross would be a red-eyed male fly and a white-eyed female fly.
reciprocal translocation when two different chromosomes exchange pieces.
recombinant (1) refers to combinations of alleles or traits that are not found in the parental generation; (2) describes DNA molecules that are produced by molecular techniques in which segments of DNA are joined to each other in ways that differ from their original arrangement in their native chromosomal sites. The cloning of DNA into vectors is an example.
recombinant DNA molecules molecules that are produced in a test tube by covalently linking DNA fragments from two different sources.
recombinant DNA technology the use of in vitro molecular techniques to isolate and manipulate different pieces of DNA.
recombinant vector a vector that contains an inserted fragment of DNA, such as a gene from a chromosome.
recombination see *genetic recombination.*
recombinational repair DNA repair via homologous recombination.
recombination signal sequence a specific DNA sequence that is involved in site-specific recombination. Such sequences are found in immunoglobulin genes.
redundancy see *gene redundancy.*
regression analysis the approach of using correlated data to predict the outcome of one variable when given the value of a second variable.
regulatory sequence or element a sequence of DNA (or possibly RNA) that binds a regulatory protein and thereby influences gene expression. Bacterial operator sites and eukaryotic enhancers and silencers are examples.
regulatory transcription factor a protein or protein complex that binds to a regulatory element and influences the rate of transcription via RNA polymerase.
relaxosome a protein complex that recognizes the origin of transfer in F factors and other conjugative plasmids, cuts one DNA strand, and aids in the transfer of the T DNA.
release factor a protein that recognizes a stop codon and promotes translational termination and the release of the completed polypeptide.
repetitive sequences DNA sequences that are present in many copies in the genome.
replica plating a technique in which a replica of bacterial colonies is transferred to a new petri plate.
replication see *DNA replication.*
replication fork the region in which two DNA strands have separated and new strands are being synthesized.
replicative transposition a form of transposition in which the end result is that the transposable element remains at its original site and also is found at a new site.
replicative transposon a transposable element that increases in number during the transposition process.
replisome a complex that contains a primosome and dimeric DNA polymerase.
repressible gene a gene that is regulated by a corepressor or inhibitor, which are small effector molecules that cause transcription to decrease.
repressor a regulatory protein that binds to DNA and inhibits transcription.
reproductive isolation the inability of a species to successfully interbreed with other species.
resolution the last stage of homologous recombination, in which the entangled DNA strands become resolved into two separate structures.
resolvase an endonuclease that makes the final cuts in DNA during the resolution phase of recombination.
response element see *regulatory sequence or element.*
restriction endonuclease (or **restriction enzyme**) an endonuclease that cleaves DNA. The restriction enzymes used in cloning experiments bind to specific base sequences and then cleave the DNA backbone at two defined locations, one in each strand.
restriction fragment length polymorphism (RFLP) genetic variation within a population in the lengths of DNA fragments that are produced when chromosomes are digested with particular restriction enzymes.
restriction mapping a technique to determine the locations of restriction endonuclease sites within a segment of DNA.

restriction point a point in the G_1 phase of the cell cycle that causes a cell to progress to cell division.
retroelement a type of transposable element that moves via an RNA intermediate.
retroposon see *retroelement.*
retrospective testing a procedure in which a researcher formulates a hypothesis and then collects observations as a way to confirm or refute the hypothesis. This approach relies on the observation and testing of already existing materials, which themselves are the result of past events.
retrotransposition a form of transposition in which the element is transcribed into RNA. The RNA is then used as a template via reverse transcriptase to synthesize a DNA molecule that is integrated into a new region of the genome via integrase.
retrotransposon see *retroelement.*
reverse genetics an experimental strategy in which researchers first identify the wild-type gene using cloning methods. The next step is to make a mutant version of the wild-type gene, introduce it into an organism, and see how the mutant gene affects the phenotype of the organism.
reverse mutation see *reversion.*
reverse transcriptase an enzyme that uses an RNA template to make a complementary strand of DNA.
reverse transcriptase PCR (RT-PCR) a modification of PCR in which the first round of replication involves the use of RNA and reverse transcriptase to make a complementary strand of DNA.
reversion a mutation that returns a mutant allele back to the wild-type allele.
R factor a type of plasmid found commonly in bacteria that confers resistance to a toxic substance such as an antibiotic.
RFLP see *restriction fragment length polymorphism.*
RFLP map a genetic map composed of many RFLP markers.
RFLP mapping the mapping of a gene or other genetic sequence relative to the known locations of RFLPs within a genome.
R group the side chain of an amino acid.
rho (ρ) a protein that is involved in transcriptional termination for certain bacterial genes.
rho-dependent termination transcriptional termination that requires the function of the rho protein.
rho-independent termination transcription termination that does not require the rho protein. It is also known as intrinsic termination.
ribonucleic acid (RNA) a nucleic acid that is composed of ribonucleotides. In living cells, RNA is synthesized via the transcription of DNA.
ribose the sugar found in RNA.
ribosomal-binding site a sequence in bacterial mRNA that is needed to bind to the ribosome and initiate translation.
ribosome a large macromolecular structure that acts as the catalytic site for polypeptide synthesis. The ribosome allows the mRNA and tRNAs to be positioned correctly as the polypeptide is made.
ribozyme an RNA molecule with enzymatic activity.
right-left axis in bilateral animals, this axis determines the two sides of the body relative to the anteroposterior axis.
R loop experimentally, a DNA loop that is formed because RNA is displacing it from its complementary DNA strand.
RNA see *ribonucleic acid.*
RNA editing the process in which a change occurs in the nucleotide sequence of an RNA molecule that involves additions or deletions of particular bases, or a conversion of one type of base to a different type.
RNA-induced silencing complex (RISC) the complex that mediates RNA interference.
RNA interference the phenomenon that double-stranded RNA targets complementary RNAs within the cell for degradation.
RNA polymerase an enzyme that synthesizes a strand of RNA using a DNA strand as a template.
RNA primer a short strand of RNA, made by DNA primase, that is used to elongate a strand of DNA during DNA replication.
RNase an enzyme that cuts the sugar–phosphate backbone in RNA.
RNaseP a bacterial enzyme that is an endonuclease and cuts precursor tRNA molecules. RNaseP is a ribozyme, which means that its catalytic ability is due to the action of RNA.
RNA splicing the process in which pieces of RNA are removed and the remaining pieces are covalently attached to each other.
Robertsonian translocation the structure produced when two telocentric chromosomes fuse at their short arms.
root meristem an actively dividing group of cells that gives rise to root structures.
RT-PCR see *reverse transcriptase PCR.*

S

satellite DNA in a density centrifugation experiment, a peak of DNA that is separated from the majority of the chromosomal DNA. It is usually composed of highly repetitive sequences.
scaffold a collection of proteins that holds the DNA in place and gives chromosomes their characteristic shapes.
scaffold-attachment region (SAR) a site in the chromosomal DNA that is anchored to the nuclear matrix or scaffold.
science a way of knowing about our natural world. The science of genetics allows us to understand how the expression of genes produces the traits of an organism.
scientific method a basis for conducting science. It is a process that scientists typically follow so that they may reach verifiable conclusions about the world in which they live.
scintillation counting a technique that permits a researcher to count the number of radioactive emissions from a sample containing a population of radioisotopes.
search by content in bioinformatics, an approach to predict the location of a gene because the nucleotide content of a particular region differs significantly (due to codon bias) from a random distribution.
search by signal in bioinformatics, an approach that relies on known sequences such as promoters, start and stop codons, and splice sites to help predict whether or not a DNA sequence contains a gene.
secondary structure a regular repeating pattern of molecular structure, such as the DNA double helix or the α helix and β sheet found in proteins.
second-division segregation (SDS) in an ordered octad, a 2:2:2:2 or 2:4:2 arrangement of spores that occurs because two alleles do not segregate until the second meiotic division is completed.
sedimentation coefficient a measure of centrifugation that is normally expressed in Svedberg units (S). A sedimentation coefficient has the units of seconds: $1\ S = 1 \times 10^{-13}$ seconds.
segmentation gene in animals, a gene, whose encoded product is involved in the development of body segments.
segment polarity gene one category of segmentation genes.
segments anatomical subdivisions that occur during the development of species such as *Drosophila.*
segregate when two things are kept in separate locations. For example, homologous chromosomes segregate into different gametes.
selectable marker a gene that provides a selectable phenotype in a cloning experiment. Many selectable markers are genes that confer antibiotic resistance.
selection coefficient one minus the fitness value.
selectionists scientists who oppose the neutral theory of evolution.
selection limit the phenomenon in which several generations of artificial selection results in a plateau where artificial selection is no longer effective.
selective breeding refers to programs and procedures designed to modify the phenotypes in economically important species of plants and animals.
self-fertilization fertilization that involves the union of male and female gametes derived from the same parent.
selfish DNA hypothesis the idea that transposable elements exist because they possess characteristics that allow them to multiply within the host cell DNA and inhabit the host without offering any selective advantage.
self-splicing refers to RNA molecules that can remove their own introns without the aid of other proteins or RNA.
semiconservative model the correct model for DNA replication in which the newly made double-stranded DNA contains one parental strand and one daughter strand.
semiconservative replication refers to the net result of DNA replication in which the DNA contains one original strand and one newly made strand.
semilethal alleles lethal alleles that kill some individuals but not all.
semisterility when an individual has a lowered fertility.
sense codon a codon that encodes a specific amino acid.
sense strand the strand of DNA within a structural gene that has the same sequence as mRNA except that T is found in the DNA instead of U.
sequence complexity the number of times a particular base sequence appears throughout the genome of a given species.
sequence element in genetics, a sequence with a specialized function.
sequence recognition in bioinformatics, this term refers to a program that recognizes particular sequence elements.
sequence-tagged site (STS) a short segment of DNA, usually between 100 and 400 base pairs long, the base sequence of which is found to be unique within the entire genome. Sequence-tagged sites are identified by PCR.
sequencing see *DNA sequencing.*
sequencing ladder a series of bands on a gel that can be followed in order (e.g., from the bottom of the gel to the top of the gel) to determine the base sequence of DNA.
sex chromosomes a pair of chromosomes (e.g., X and Y in mammals) that determines sex in a species.
sex-influenced inheritance an inheritance pattern in which an allele is dominant in one sex but recessive in the opposite sex. In humans, pattern baldness is an example of a sex-influenced trait.
sex-limited inheritance an inheritance pattern in which a trait is found in only one of the two sexes. An example would be beard development in men.
sex-limited traits traits that occur in only one of the two sexes.
sex linkage the phenomenon that certain genes are found on one of the two types of sex chromosomes but not both.
sex-linked gene a gene that is located on one of the sex chromosomes.
sex pilus (pl. **pili**) a structure on the surface of bacterial cells that acts as an attachment site to promote the binding of bacteria to each other. The sex pilus provides a passageway for the movement of DNA during conjugation.
sexual reproduction the process whereby parents make gametes (e.g., sperm and egg) that fuse with each other in the process of fertilization to begin the life of a new organism.
sexual selection natural selection that acts to promote characteristics that give individuals a greater chance of reproducing.
shared derived character a trait shared by a group of organisms but not by a distant common ancestor.
Shine–Dalgarno sequence a sequence in bacterial mRNAs that functions as a ribosomal binding site.
shoot meristem an actively dividing group of cells that gives rise to shoot structures.
short tandem repeat sequences (STRs) short DNA sequences that are repeated many times in a row. Often found in centromeric and telomeric regions.
shotgun sequencing a genome sequencing strategy in which DNA fragments to be sequenced are randomly generated from larger DNA fragments.
shuttle vector a cloning vector that can propagate in two or more different species, such as *E. coli* and yeast.
side chain in an amino acid, the chemical structure that is attached to the carbon atom (i.e., the α carbon) that is located between the amino group and carboxyl group.

sigma factor a transcription factor that recognizes bacterial promoter sequences and facilitates the binding of RNA polymerase to the promoter.
silencer a DNA sequence that functions as a regulatory element. The binding of a regulatory transcription factor to the silencer decreases the level of transcription.
silent mutation a mutation that does not alter the amino acid sequence of the encoded polypeptide even though the nucleotide sequence has changed.
similarity with regard to DNA, refers to a comparison of DNA sequences that have regions where the bases match up. Similarity may be due to homology.
simple Mendelian inheritance an inheritance pattern involving a simple, dominant/recessive relationship that produces observed ratios in the offspring that readily obey Mendel's laws.
simple translocation when one piece of a chromosome becomes attached to a different chromosome.
simple transposition a cut-and-paste mechanism for transposition in which a transposable element is removed from one site and then inserted into another site.
SINEs in mammals, short interspersed elements that are less than 500 base pairs in length.
single-factor cross see *monohybrid cross.*
single-nucleotide polymorphism a genetic polymorphism within a population in which two alleles of the gene differ by a single nucleotide.
single-strand binding protein a protein that binds to both of the single strands of DNA during DNA replication and prevents them from re-forming a double helix.
sister chromatid exchange (SCE) the phenomenon in which crossing over occurs between sister chromatids, thereby exchanging identical genetic material.
sister chromatids pairs of replicated chromosomes that are attached to each other at the centromere. Sister chromatids are genetically identical to each other.
site-directed mutagenesis a technique that enables scientists to change the sequence of cloned DNA segments.
site-specific recombination when two different DNA segments break and rejoin with each other at a specific site. This occurs during the integration of certain viruses into the host chromosome and during the rearrangement of immunoglobulin genes.
SMC proteins proteins that use energy from ATP to catalyze changes in chromosome structure.
snRNP refers to a complex containing small nuclear RNA and a set of proteins, which are components of the spliceosome.
somatic cell refers to any cell of the body except for germ-line cells that give rise to gametes.
somatic mutation a mutation in a somatic cell.
sorting signal an amino acid sequence or posttranslational modification that directs a protein to the correct region of the cell.
SOS response a response to extreme environmental stress in which bacteria replicate their DNA using DNA polymerases that are likely to make mistakes.
Southern blotting a technique used to detect the presence of a particular genetic sequence within a mixture of many chromosomal DNA fragments.
speciation the process by which new species are formed via evolution.
species a group of organisms that maintains a distinctive set of attributes in nature.
species concepts different approaches for distinguishing species.
spectrophotometer a device used by researchers to determine how much radiation at various wavelengths a sample can absorb.
spermatids immature sperm cells produced from spermatogenesis.
spermatogenesis the production of sperm cells.
sperm cell a male gamete. Sperm are small and usually travel relatively far distances to reach the female gamete.
spindle see *mitotic spindle apparatus.*
spliceosome a multisubunit complex that functions in the splicing of eukaryotic pre-mRNA.
splicing see *RNA splicing.*
splicing factor a protein that regulates the process of RNA splicing.
spontaneous mutation a change in DNA structure that results from random abnormalities in biological processes.
spores haploid cells that are produced by certain species such as fungi (i.e., yeast and molds).
sporophyte the diploid generation of plants.
SR protein a type of splicing factor.
stabilizing selection natural selection that favors individuals with an intermediate phenotype.
stamen the structure found in the flower of higher plants that produces the male gametophyte (i.e., pollen).
standard deviation a statistic that is computed as the square root of the variance.
start codon a three-base sequence in mRNA that initiates translation. It is usually 5'-AUG-3' and encodes methionine.
stem cell a cell that has the capacity to divide and to differentiate into one or more specific cell types.
steroid receptor a category of transcription factors that respond to steroid hormones. An example is the glucocorticoid receptor.
stigma the structure in flowering plants on which the pollen land and the pollen tube starts to grows so that sperm cells can reach the egg cells.
stop codon a three-base sequence in mRNA that signals the end of translation of a polypeptide. The three stop codons are 5'–UAA–3', 5'–UAG–3', and 5'–UGA–3'.
strain a variety that continues to produce the same characteristic after several generations.
strand in DNA or RNA, nucleotides covalently linked together to form a long, linear polymer.
stripe-specific enhancer in *Drosophila*, a regulatory region that controls the expression of a gene so that it occurs only in a particular parasegment during early embryonic development.
STRs see *short tandem repeat sequences.*
structural gene a gene that encodes the amino acid sequence within a particular polypeptide or protein.
STS see *sequence-tagged site.*
subcloning the procedure of making smaller DNA clones from a larger one.
submetacentric describes a chromosome in which the centromere is slightly off center.
subtractive cDNA library a cDNA library that contains cDNA inserts derived from mRNA that is expressed only under certain conditions.
subtractive hybridization a method used to create a subtractive cDNA library.
subunit this term may have multiple meanings. In a protein, each subunit is a single polypeptide.
sum rule the probability that one of two or more mutually exclusive events will occur is equal to the sum of their individual probabilities.
supercoiling see *DNA supercoiling.*
supergroups a relatively recent way for evolutionary biologists to subdivide the eukaryotic domain.
supernatant following centrifugation, the fluid that is found above the pellet.
suppressor (or **suppressor mutation**) a mutation at a second site that suppresses the phenotypic effects of another mutation.
SWI/SNF family a group of related proteins that catalyze chromatin remodeling.
sympatric speciation (Greek, *sym*, "together"; Latin, *patria*, "homeland") a form of speciation that occurs when members of a species diverge while occupying the same habitat within the same range.
synapomorphy see *shared derived character.*
synapsis the event in which homologous chromosomes recognize each other and then align themselves along their entire lengths.
synaptonemal complex a complex of proteins that promote the interconnection between homologous chromosomes during meiosis.
synonymous codons two different codons that specify the same amino acid.
synteny group a group of genes that are found in the same order on the chromosomes of different species.
systematists biologists who study the evolutionary relationships among different species.

T

T an abbreviation for thymine.
tandem array (or **tandem repeat**) a short nucleotide sequence that is repeated many times in a row.
tandem mass spectrometry the sequential use of two mass spectrometers. It can be used to determine the sequence of amino acids in a polypeptide.
***Taq* polymerase** a thermostable form of DNA polymerase used in PCR experiments.
TATA box a sequence found within eukaryotic core promoters that determines the starting site for transcription. The TATA box is recognized by a TATA-binding protein, which is a component of TFIID.
tautomer refers to the forms of certain small molecules, such as bases, which can spontaneously interconvert between chemically similar forms.
tautomeric shift a change in chemical structure such as an alternation between the keto- and enol-forms of the bases that are found in DNA.
T DNA a segment of DNA found within a Ti plasmid that is transferred from a bacterium to infected plant cells. The T DNA from the Ti plasmid becomes integrated into the chromosomal DNA of the plant cell by recombination.
TE see *transposable element.*
telocentric describes a chromosome with its centromere at one end.
telomerase the enzyme that recognizes telomeric sequences at the ends of eukaryotic chromosomes and synthesizes additional numbers of telomeric repeat sequences.
telomeres specialized DNA sequences found at the ends of linear eukaryotic chromosomes.
telophase the fifth stage of M phase. The chromosomes have reached their respective poles and decondense.
temperate phage a bacteriophage that usually exists in the lysogenic cycle.
temperature-sensitive allele an allele in which the resulting phenotype depends on the environmental temperature.
temperature-sensitive (ts) lethal allele an allele that is lethal at a certain environmental temperature.
template DNA a strand of DNA that is used to synthesize a complementary strand of DNA or RNA.
template strand see *template DNA.*
terminal deficiency see *terminal deletion.*
terminal deletion when a segment is lost from the end of a linear chromosome.
termination (1) in transcription, the release of the newly made RNA transcript and RNA polymerase from the DNA; (2) in translation, the release of the polypeptide and the last tRNA and the disassembly of the ribosomal subunits and mRNA.
termination codon see *stop codon.*
termination sequences (ter sequences) in *E. coli*, a pair of sequences in the chromosome that bind a protein known as the termination utilization substance (Tus), which stops the movement of the replication forks.
terminator a sequence within a gene that signals the end of transcription.
tertiary structure the three-dimensional structure of a macromolecule, such as the tertiary structure of a polypeptide.
testcross an experimental cross between a recessive individual and an individual whose genotype the experimenter wishes to determine.
tetrad (1) the association among four sister chromatids during meiosis; (2) a group of four fungal spores contained within an ascus.
tetraploid having four sets of chromosomes (i.e., $4n$).
tetratype (T) an ascus that has two parental cells and two nonparental cells.
TFIID a type of general transcription factor in eukaryotes that is needed for RNA polymerase II function. It binds to the TATA box and recruits RNA polymerase II to the core promoter.
thermocycler a device that automates the timing of temperature changes in each cycle of a PCR experiment.
30 nm fiber the association of nucleosomes to form a more compact structure that is 30 nm in diameter.

thymine a pyrimidine base found in DNA. It base pairs with adenine in DNA.
thymine dimer a DNA lesion involving a covalent linkage between two adjacent thymine bases in a DNA strand.
Ti plasmid a tumor-inducing plasmid found in *Agrobacterium tumefaciens.* It is responsible for inducing tumor formation after a plant has been infected.
tissue-specific gene a gene that is highly regulated and is expressed in a particular cell type.
topoisomerase an enzyme that alters the degree of supercoiling in DNA.
topoisomers DNA conformations that differ with regard to supercoiling.
totipotent a cell that possesses the genetic potential to produce an entire individual. A somatic plant cell or a fertilized egg is totipotent.
traffic signal see *sorting signal.*
trait any characteristic that an organism displays. Morphological traits affect the appearance of an organism. Physiological traits affect the ability of an organism to function. A third category of traits are those that affect an organism's behavior (behavioral traits).
***trans*-acting factor** a regulatory protein that binds to a regulatory element in the DNA and exerts a *trans* effect.
transcription the process of synthesizing RNA from a DNA template.
transcriptional start site the site in a gene where transcription begins.
transcription factors a broad category of proteins that influence the ability of RNA polymerase to transcribe DNA into RNA.
transcription-repair coupling factor (TRCF) a protein that recognizes when RNA polymerase is stalled over a damaged region of DNA and recruits DNA repair enzymes to fix the damaged site.
transduction a form of genetic transfer between bacterial cells in which a bacteriophage transfers bacterial DNA from one bacterium to another.
***trans*-effect** an effect on gene expression that occurs even though two DNA segments are not physically adjacent to each other. *Trans*-effects are mediated through diffusible genetic regulatory proteins.
transfection (1) when a viral vector is introduced into a bacterial cell; (2) the introduction of any type of recombinant DNA into a eukaryotic cell.
transfer RNA (tRNA) a type of RNA used in translation that carries an amino acid. The anticodon in tRNA is complementary to a codon in the mRNA.
transformation (1) when a plasmid vector or segment of chromosomal DNA is introduced into a bacterial cell; (2) when a normal cell is converted into a malignant cell.
transgene a gene from one species that is introduced into another species.
transgenic organism an organism that has DNA from another organism incorporated into its genome via recombinant DNA techniques.
transition a point mutation involving a change of a pyrimidine to another pyrimidine (e.g., C to T) or a purine to another purine (e.g., A to G).
translation the synthesis of a polypeptide using the codon information within mRNA.
translational regulatory protein a protein that regulates translation.
translational repressor a protein that binds to mRNA and inhibits its ability to be translated.
translesion synthesis (TLS) the synthesis of DNA over a template strand that harbors some type of DNA damage. This occurs via lesion-replicating polymerases.
translocation (1) when one segment of a chromosome breaks off and becomes attached to a different chromosome; (2) when a ribosome moves from one codon in an mRNA to the next codon.
translocation cross the structure that is formed when the chromosomes of a reciprocal translocation attempt to synapse during meiosis. This structure contains two normal (nontranslocated chromosomes) and two translocated chromosomes. A total of eight chromatids are found within the cross.
transposable element (TE) a small genetic element that can move to multiple locations within the chromosomal DNA.
transposase the enzyme that catalyzes the movement of transposable elements.
transposition the phenomenon of transposon movement.
transposon see *transposable element.*
transposon tagging a technique for cloning genes in which a transposon inserts into a gene and inactivates it. The transposon-tagged gene is then cloned using a complementary transposon as a probe to identify the gene.
transversion a point mutation in which a purine is interchanged with a pyrimidine, or vice versa.
trihybrid cross a cross in which an experimenter follows the outcome of three different traits.
trinucleotide repeat expansion (TNRE) a type of mutation that involves an increase in the number of tandemly repeated trinucleotide sequences.
triplex DNA a double-stranded DNA that has a third strand wound around it to form a triple-stranded structure.
triploid an organism or cell that contains three sets of chromosomes.
trisomic a diploid cell with one extra chromosome (i.e., $2n + 1$).
tRNA see *transfer RNA.*
trp repressor a protein that binds to the operator site of the *trp* operon and inhibits transcription.
true-breeding line a strain of a particular species that continues to produce the same trait after several generations of self-fertilization (in plants) or inbreeding.
tumor-suppressor gene a gene that functions to inhibit cancerous growth.
two-dimensional gel electrophoresis a technique to separate proteins that involves isoelectric focusing in the first dimension and SDS-gel electrophoresis in the second dimension.
two-factor cross see *dihybrid cross.*

U

U an abbreviation for uracil.
unbalanced translocation a translocation in which a cell has too much genetic material compared to a normal cell.
unipotent a type of stem cell that can differentiate into only a single type of cell.
universal in genetics, this term refers to the phenomenon that nearly all organisms use the same genetic code with just a few exceptions.
unordered octad an ascus composed of eight unordered cells.
unordered tetrad an ascus composed of four unordered cells.
up promoter mutation a mutation in a promoter that increases the rate of transcription.
up regulation genetic regulation that leads to an increase in gene expression.
uracil a pyrimidine base found in RNA.
UTR an abbreviation for the untranslated region of mRNA.
U-tube a U-shaped tube that has a filter at bottom of the U. The pore size of the filter allows the passage of small molecules (e.g., DNA molecules) from one side of the tube to the other, but restricts the passage of bacterial cells.

V

variance the sum of the squared deviations from the mean divided by the degrees of freedom.
variants individuals of the same species that exhibit different traits. An example is tall and dwarf pea plants.
V(D)J recombination site-specific recombination that occurs within immunoglobulin genes.
vector a small segment of DNA that is used as a carrier of another segment of DNA. Vectors are used in DNA cloning experiments.
vertical evolution refers to the phenomenon that species evolve from preexisting species by the accumulation of gene mutations and by changes in chromosome structure and number. Vertical evolution involves genetic changes in a series of ancestors that form a lineage.
vertical gene transfer the transfer of genetic material from parents to offspring or from mother cell to daughter cell.
viral genome the genetic material of a virus.
viral-like retroelements retroelements that are evolutionarily related to known retroviruses. These transposable elements have retained the ability to move around the genome, though, in most cases, mature viral particles are not produced.
viral vector a vector used in gene cloning that is derived from a naturally occurring virus.
virulent phage a phage that follows the lytic cycle.
virus a small infectious particle that contains nucleic acid as its genetic material, surrounded by a capsid of proteins. Some viruses also have an envelope consisting of a membrane embedded with spike proteins.
VNTRs also called minisatellites; segments of DNA that are located in several places in a genome and have a variable number of tandem repeats. The pattern of VNTRs was originally used in DNA fingerprinting.

W

Western blotting a technique used to detect a specific protein among a mixture of proteins.
whole genome shotgun sequencing a genome sequencing strategy that bypasses the mapping step. The whole genome is subjected to shotgun sequencing.
wild-type allele an allele that is fairly prevalent in a natural population, generally greater than 1% of the population. For polymorphic genes, there may be more than one wild-type allele.
wobble base the first base (from the 5' end) in an anticodon. This term suggests that the first base in the anticodon can wobble a bit to recognize the third base in the mRNA.
wobble rules rules that govern the binding specificity between the third base in a codon and the first base in an anticodon.

X

X chromosomal controlling element (Xce) a region adjacent to Xic that influences the choice of the active X chromosome during the process of X inactivation.
X inactivation a process in which mammals equalize the expression of X-linked genes by randomly turning off one X chromosome in the somatic cells of females.
X-inactivation center (Xic) a site on the X chromosome that appears to play a critical role in X inactivation.
X-linked genes (alleles) genes (or alleles of genes) that are physically located within the X chromosome.
X-linked inheritance an inheritance pattern in certain species that involves genes that are located only on the X chromosome.
X-linked recessive an allele or trait in which the gene is found on the X chromosome and the allele is recessive relative to a corresponding dominant allele.

Y

YAC see *yeast artificial chromosome.*
yeast artificial chromosome (YAC) a cloning vector propagated in yeast that can reliably contain very large insert fragments of DNA.
Y-linked genes (alleles) genes (or alleles of genes) that are located only on the Y chromosome.

Z

Z DNA a left-handed DNA double helix that is found occasionally in the DNA of living cells.
zygote a cell formed from the union of a sperm and egg.
zygotene the second stage of prophase of meiosis I.
zygotic gene a gene that is expressed after fertilization.

CREDITS

Photographs

Chapter 1
Opener: © Photo courtesy of the College of Veterinary Medicine, Texas A&M University/CORBIS; **1.2(left):** © Roslin Institute; **1.3a(right):** Advanced Cell Technology, Inc., Worcester, Massachusetts; **1.3b:** Photo taken by Flaminia Catteruccia, Jason Benton, and Andrea Crisanti, and assembled by www.luciariccidesign.com; **1.4:** © Biophoto Associates/Photo Researchers; **1.5:** © CNRI/SCIENCE PHOTO LIBRARY/Photo Researchers; **1.8(top left):** © Edmund D. Brodie III; **1.9a(top right):** © Joseph Sohm/ChromoSohm Inc./CORBIS; **1.9b:** © Paul Edmondson/Photodisc Red/Getty Images; **1.10:** © March of Dimes Birth Defects Foundation; **1.13a:** © George Musil/Visuals Unlimited; **1.13b:** © SciMAT/Photo Researchers, Inc.; **1.13c:** © Steve Hopkin/ARDEA LONDON; **1.13d:** © Brad Mogen/Visuals Unlimited; **1.13e:** © Mark Smith/Photo Researchers, Inc.; **1.13f:** © J-M. Labat/Photo Researchers, Inc.; **1.13g:** © Wally Eberhart/Visuals Unlimited.

Chapter 2
Opener: © CHRIS MARTIN BAHR/SCIENCE PHOTO LIBRARY; **2.1:** © SPL/Photo Researchers, Inc.; **2.2:** © Nigel Cattlin/Photo Researchers, Inc.

Chapter 3
Opener: © Photomicrographs by Dr. Conly L. Rieder, Wadsworth Center, Albany, New York 12201-0509; **3.2b(right top):** © Burger/Photo Researchers; **3.2c(right bottom):** © Leonard Lessin/Peter Arnold; **3.6a(left):** © Leonard Lessin/Peter Arnold; **3.6a(right):** © Biophoto Associates/Photo Researchers, Inc.; **3.8a-f:** © Photomicrographs by Dr. Conly L. Rieder, Wadsworth Center, Albany, New York 12201-0509; **3.9a(1):** © Dr. David M. Phillips/Visuals Unlimited; **3.9a(2):** © Carolina Biological Supply Company/Phototake.com; **3.11:** © Diter von Wettstein.

Chapter 4
Opener: © Robert Calentine/Visuals Unlimited; **4.1:** © Blickwinkel/Alamy; **4.5b:** © Bob Shanley/The Palm Beach Post; **4.6a(1):** © PETER WEIMANN/Animals Animals-Earth Scenes; **4.6a(2):** © Tom Walker/Visuals Unlimited; **4.6b:** © Sally Haugen/Virginia Schuett, www.pkunews.org; **4.6c:** © PHOTOTAKE Inc./Alamy; **4.7a(left), 4.7b(right):** © Stan Flegler/Visuals Unlimited; **4.9b:** © John T. Fowler/Alamy; **4.9c:** © Wegner, P./Peter Arnold, Inc.; **4.9d:** © Gary Randall/Visuals Unlimited; **4.9a(top left):** © Zig Leszcynski/AnimalsAnimals-Earth Scenes; **4.10(top right):** © Alan & Sandy Carey/Photo Researchers; **4.13a:** © AP Images; **4.15a(top left), 4.15b(top right), 4.15c(bottom left):** National Parks Service, Adams National Historical Park; **4.15d(bottom right):** © Bettmann/CORBIS ; **4.17a(left), 4.17b(right):** © Robert Maier/AnimalsAnimals-Earth Scenes; **4.18a:** © Jane Burton/naturepl.com.

Chapter 5
Opener: © Carolina Biological/Visuals Unlimited.

Chapter 6
Opener: © David Scharf/Peter Arnold; **6.4b:** © Dr. L. Caro/Science Photo Library/Photo Researchers; **6.15a(left), 6.15b(right):** © Carolina Biological Supply/Phototake.

Chapter 7
Opener: © John Mendenhall, Institute for Cellular and Molecular Biology, University of Texas at Austin; **7.3a(1), 7.3a(2):** Courtesy of I. Solovei, University of Munich (LMU); **7.6:** © G.W. Willis, MD/Visuals Unlimited; **TA 7.1:** Ronald G. Davidson, Harold M. Nitowsky, and Barton Childs. "Demonstration of Two Populations of Cells in the Human Female Heterozygous for Glucose-6-Phosphate Dehydrogenase Variants." *PNAS* 50 (1963) f. 2, p. 484. Courtesy Harold M. Nitowsky; **7.9:** Courtesy of Dr. Argiris Efstratiadis; **7.13a(left):** Reproduced with permission from *The Journal of Cell Biology,* 1977, 72:687-694. Copyright 1977 The Rockefeller University Press; **7.13b(right):** Gibbs, SP., Mak, R., Ng, R., & Slankis, T. "The chloroplast nucleoid in Ochromonas danica, II. Evidence for an increase in plastid DNA during greening." *J Cell Sci.* 1974 Dec;16(3):579-91. Fig. 1. By permission of the Company of Biologists Limited.

Chapter 8
Opener: © BSIP/Phototake, Inc.; **8.1a(top left):** © Scott Camazine/Photo Researchers; **8.1a(center):** © Michael Abbey/Photo Researchers; **8.1a(top right):** © Carlos R Carvalho/Universidade Federal de Viçosa; **8.1c(bottom left):** © C.N.R.I./Phototake; **8.4a(left):** © Biophoto Associates/Science Source/Photo Researchers; **8.4b(right):** © Jeff Noneley; **8.12b(left):** © Paul Benke/University of Miami School of Medicine; **8.12c(right):** © Will Hart/PhotoEdit; **8.18a(top), 8.18b(bottom):** © A. B. Sheldon; **8.19b:** © David M. Phillips/Visuals Unlimited; **8.20a(left):** © James Steinberg/Photo Researchers; **8.20a(top right), 8.20b(bottom right):** © Biophoto Associates/Science Source/Photo Researchers; **8.26:** Robinson, T.J. & Harley, E.H. "Absence of geographic chromosomal variation in the roan and sable antelope and the cytogenetics of a naturally occurring hybrid." *Cytogenet Cell Genet.* 1995; 71(4):363-9. Permission granted by S. Karger AG, Basel.

Chapter 9
Opener: © Ken Eward/Photo Researchers; **9.4(left):** © Omikron/Photo Researchers; **9.12b(left):** © Pictorial Parade/Getty Images; **9.16a(left):** © Barrington Brown/Photo Researchers; **9.16b(right):** © Hulton Archive by Getty Images; **9.18b:** © Michael Freeman/Phototake; **9.24b:** © Alfred Pasieka/Photo Researchers, Inc.

Chapter 10
Opener: © Dr. Gopal Murti/Visuals Unlimited; **10.3:** ©American Society for Microbiology; **10.10b(top):** © Simpson's Nature Photography; **10.10c(bottom):** © William Leonard; **10.13c:** Image courtesy of Timothy J. Richmond. Figure 1 from" Crystal structure of the nucleosome core particle at 2.8 Å resolution." *Nature* 1997 Sept 18; 389(6648): 251-60. Reprinted by permission from Macmillan Publishers Ltd.; **TA 10.14:** Reprinted by permission from Macmillan Publishers Ltd. *Nature* Subunit structure of chromatin. Markus Noll. 251: 5472, 249-251. 1974.; **10.17b, 10.17c:** Nickerson et al. "The nuclear matrix revealed by eluting chromatin from a cross-linked nucleus." *PNAS* 94: 4446-4450. Figure 2a &b. © 1997 National Academy of Sciences, U.S.A.; **10.18a(left), 10.18b(right):** Reprinted by permission from Macmillan Publishers Ltd. *Nature Reviews/Genetics.* Chromosome territories, nuclear architecture and gene regulation in mammalian cells. Cremer, T. & Cremer, C., 2: 4, 292-301, 2001; **10.21a(2nm):** © Dr. Gopal Murit/Visuals Unlimited; **10.21a(11nm):** © Olins and Olins/Biological Photo Service; **10.21b(30nm):** Jerome Rattner/University of Calgary; **10.21c(300nm):** This article was published in *Cell.* Nov. 12(3). Paulson, JR. & Laemmli, UK. "The structure of histone-depleted metaphase chromosomes." 817-2 8. f. 5. Copyright Elsevier, 1977; **10.21d(700nm), 10.21d(1400nm), 10.22a:** © Peter Engelhardt/Department of Virology, Haartman Institute.; **10.22b:** © Dr. Donald Fawcett/Visuals Unlimited.

Chapter 11
Opener: © Clive Freeman, The Royal Institution/Photo Researchers; **TA 11.1:** Meselson M, Stahl, F. "The Replication of DNA in Escherichia Coli." *PNAS* Vol. 44, 1958. f. 4a, p. 673. Courtesy of M. Meselson; **11.4b:** From *Cold Spring Harbor Symposia of Quantitative Biology,* 28, p. 43 (1963). Copyright holder is Cold Spring Habour Laboratory Press; **11.8b:** Reprinted by permission from Macmillan Publishers Ltd. *The Embo Journal.* "Crystal structures of open and closed forms of binary and ternary complexes of the large fragment of Thermus aquaticus DNA polymerase I: structural basis for nucleotide incorporation." Ying Li et al. 17: 24, 7514-7525, 1998.; **11.20:** This article was published in *Journal of Molecular Biology.* Mar 14; 32(2). Huberman JA. Riggs AD. "Links On the mechanism of DNA replication in mammalian chromosomes":327-41. Copyright Elsevier, 1968; **11.21b:** Henry J. Kriegstein and David S. Hogness "Mechanism of DNA Replication in Drosophila Chromosomes: Structure of Replication Forks and Evidence for Bidirectionality." *PNAS.* January 1, 1974. vol. 71, no. 1. 135-139. Permission courtesy D. Hogness.

Chapter 12
Opener: From Patrick Cramer, David A. Bushnell, Roger D. Kornberg. "Structural Basis of Transcription: RNA Polymerase II at 2.8 Angstrom Resolution." *Science,*

Vol. 292:5523, 1863-1876, June 8, 2001. Reprinted with permission of AAAS; **TA 12.1:** Tilghman et al. "Intervening sequence of DNA identified in the structural protein of a mouse ß-glovin gene." *PNAS* 1978, Vol 75. f. 2, p. 727. This image is in the public domain; **12.12a(left):** From Seth Darst, Bacterial RNA polymerase. *Current Opinion* in Structural Biology; **12.12b(right):** Image and permission courtesy of David Bushnell.

Chapter 13
Opener: © Tom Pantages; **13.16a:** © E. Kiseleva and Donald Fawcett/Visuals Unlimited; **13.16b:** © Tom Pantages; **13.22:** Miller, O. L. *Scientific American.* Vol 228:3, 1973. p.35. Image courtesy O.L. Miller.

Chapter 14
14.10b: © Mitchell Lewis, University of Pennsylvania Medical Center/Science Photo Library/Photo Researchers.

Chapter 15
Opener: Zif268 zinc fingers bound to DNA (Pavletich and Pabo, 1991). Courtesy Song Tan, Penn State University; **15.8(left):** Courtesy of Dr. Joseph G. Gall; **15.8(right):** Courtesy Dr. Oscar L. Miller; **15.22(a), 15.22(b), 15.22(c):** Reprinted by permission from Macmillan Publishers Ltd. *Nature.* Andrew Fire et al. Potent and specific genetic interference by double-stranded RNA in Caenorhabditis elegans, 391: 6669. 806-811. 1998.

Chapter 16
Opener: © Robert Brooker; **16.1(left), 16.1(right):** © PHOTOTAKE Inc./Alamy; **16.3a, 16.3b:** Aulner et al. "The AT-Hook Protein D1 Is Essential for Drosophila melanogaster Development and Is Implicated in Position-Effect Variegation." Molecular and Cellular Biology. 22: 4. 1218-1232, February 2002. Figure. 7. © 2002 American Society for Microbiology and E. Kas; **16.5:** ©Scott Aitken, www.scottpix.com; **16.20:** © Dr. Kenneth Greer/Visuals Unlimited.

Chapter 17
Opener: © Matt Meadows/Peter Arnold Inc.; **17.3(the data):** Reprinted by permission from Macmillan Publishers Ltd. *Nature.* New Giemsa method for the differential staining of sister chromatids. Perry, P. & Wolff, S. 251:5471, 156-158, 1974; **17.4b:** © Dr. John D. Cunningham/Visuals Unlimited.

Chapter 18
Opener: © Argus Fotoarchive/Peter Arnold Inc.; **TA18.4:** Cohen, Stanely et al."Construction of Biologically Functional Bacterial Plasmids In Vitro." *PNAS* 1973. Vol. 40. Figure 4. p. 3242. © Stanley Cohen.

Chapter 19
Opener: © Najlah Feanny/SABA/CORBIS; **19.3:** © M. Greenlar/Image Works; **19.4:** R. L. Brinster and R. E. Hammer, School of Veterinary Medicine, University of Pennsylvania; **19.6:** Glofish © www.glofish.com; **19.7:** © Yoav Levy/Phototake; **19.11:** © Photo courtesy of the College of Veterinary Medicine, Texas A&M University/CORBIS; **19.15:** © Jack Bostrack/Visuals Unlimited; **19.17:** Courtesy Monsanto; **19.18:** © Richard T. Nowitz/Phototake; **19.19a:** © Bill Barksdale/AGStockUSA.

Chapter 20
Opener: © Dr. Peter Lansdorp/Visuals Unlimited; **20.4:** From Ried, T., Baldini, A., Rand, T.C., and Ward, D.C. "Simultaneous visualization of seven different DNA probes by in situ hybridization using combinatorial fluorescence and digital imaging microscopy. PNAS 89: 4.1388-92. 1992. Courtesy Thomas Ried; **20.19a:** This image was published in *Pulsed Field Gel* Electrophoresis, A Practical Guide. Bruce Birren and Eric Lai. Figure .1, p. 109. Copyright Elsevier 1993; **20.19b:** From Peter D. Butler and E. Richard Moxon. "A physical map of the genome of Haemophilus influenzae type b." *J Gen Microbiol.* 1990 Dec; 136 (Pt 12):2333-42. Reprinted by permission of the Society for General Microbiology.

Chapter 21
Opener: © Alfred Pasieka/Photo Researchers; **TA21.1:** Courtesy Joseph DeRisi; **21.5b:** © Medical School, University of Newcastle upon Tyne/Simon Fraser/Photo Researchers.

Chapter 22
Opener: © Royalty-Free/CORBIS; **22.1(top):** © Dr. P. Marazzi/Photo Researchers; **22.1(middle):** © Hiroya Minakuchi/Seapics.com; **22.1(bottom):** © Mitch Reardon/Photo Researchers; **22.4:** © Hulton-Deutsch Collection/Corbis; **22.6:** © CNRI/SCIENCE PHOTO LIBRARY/Photo Researchers; **22.9:** Reprinted by permission from Macmillan Publishers Ltd. *Nature.* Transforming activity of DNA of chemically transformed and normal cells. Authors: Geoffrey M. Cooper, S. Okenquist & L. Silverman. 284: 5755, 418-421, 1980; **22.18(top), 22.18(bottom):** Dr. Ruhong Li, Molecular and Cell Biology, University of California at Berkeley. Permission granted by The Duesberg Laboratory. Appeared in *Science,* Vol. 297, Number 5581, Issue 26, Jul 2002, p. 544. Author: Jean Marx, Title: "Debate Surges Over the Origins of Genomic Defects in Cancer"; **22.19a:** Reprinted by permission from Macmillan Publishers Ltd. *Nature.* Distinct types of Diffuse large B-cell lymphoma Identified by Gene Expression Profiling. Ash Alizadeh et al. 403:6769, 503-511, 2000. Image courtesy Ash Alizadeh.

Chapter 23
Opener: © Herman Eisenbeiss/Photo Researchers, Inc.; **TA 23.1, TA 23.2:** Horvitz, H.R. & Sulston, J. (1980) Isolation and genetic characterization of cell lineage mutants of the nematode caenorhabditis elegans. *Genetics* 96, 435-454. fig. 1a&b (1980). Courtesy Dr. Horvitz; **23.3:** Courtesy of E. B. Lewis, California Institute of Technology; **TA 23.3:** Courtesy Stephen Small/New York University; **23.7b, 23.7c, 23.10(1):** Christiane Nusslein-Volhard, *Development,* Supplement 1, 1991. © The Company of Biologists Limited; **23.10(2). 23.10(3), 23.10(4):** Jim Langeland, Steve Paddock and Sean Carroll/University of Wisconsin-Madison; **23.12a:** © Juergen Berger/Photo Researchers; **23.12b:** © F. R. Turner, Indiana University/Visuals Unlimited; **23.17:** © Photodisc/Object Series/Vol. 50; **23.20:** © Jeremy Burgess/Photo Researchers; **23.23a,c:** Elliott Meyerowitz and John Bowman. Development. Vol. 112:1-20, 1991. Courtesy of Elliott Meyerowitz; **23.23b:** Courtesy of Elliott Meyerowitz.

Chapter 24
Opener: © PhotoDisc website; **24.1:** © Color-Pic/Animals Animals-Earth Scenes; **24.2(1), 24.2(2), 24.2(3):** © Geoff Oxford; **24.7b:** © Corbis/RF website; **24.13a:** © OSF/Animals Animals-Earth Scenes; **24.15:** © Paul Harcourt Davies/SPL/Photo Researchers, Inc; **24.16(left):** © D. Parer & E .Parer-Cook/ardea.com; **24.21:** Leonard Lessin/Peter Arnold Inc.; **24.26(right):** © Gerald & Buff Corsi/Visuals Unlimited.

Chapter 25
Opener: © PhotoDisc/Vol. 74; **25.1a:** Image originally appeared in: Albert and Blakeslee, "Corn and Man," *Journal of Heredity.* 1914, Vol. 5, pg. 51. Oxford University Press. This image is in the public domain; **25.3a(1):** © GRANT HEILMAN /Grant Heilman Photography; **25.3a(2):** © Nigel Cattlin/Visuals Unlimited; **25.11(top left):** © Henry Ausloos/Animals Animals-Earth Scenes; **25.11(top right):** © Roger Tidman/Corbis; **25.11(bottom left):** © Philip Gould/Corbis; **25.11(bottom right):** © Juniors Bildarchiv/Alamy; **25.12(1):** © Inga Spence/Visuals Unlimited; **25.12(2):** © Nigel Cattlin/Photo Researchers Inc.; **25.12(3):** © Valerie Giles/ Photo Researchers Inc.; **25.12(4-7):** © Michael P. Gadomski/Photo Researchers Inc.

Chapter 26
Opener: © LWA/Stephen Welstead/Blend Images/Corbis Royalty-Free; **26.2(left):** © Paul & Joyce Berquist/Animals Animals-Earth Scenes; **26.2(right):** © Gerald and Buff Corsi/Visuals Unlimited; **26.20a(left), 26.20a(right):** Courtesy Ed Laufer; **26.20b(left), 26.20b(right), 26.20c(left), 26.20c(right):** Courtesy of Dr. J.M. Hurle (Originally published in *Development.* 1999 Dec; 126 (23):5515-22.); **26.22a, 26.22b:** © Prof. Walter J. Gehring, University of Basel.

Appendixes
1a: © Michael Gabridge/Visuals Unlimited; **1b:** © Fred Hossler/Visuals Unlimited; **8:** © Richard Wehr/Custom Medical Stock Photo.

Line Art

Chapter 1
1.1a, b: Courtesy of the U.S. Department of Energy Genome Programs. http://genomics.energy.gov.

Chapter 9
9.19: Illustration, Irving Geis. Rights owned by Howard Hughes Medical Institute. Not to be reproduced without permission.

Chapter 12
12.12: Reprinted from Structural Biology, vol. 11, Darst, "Current Opinion," page 157, Copyright © 2001, with permission from Elsevier.

Chapter 14
14.10: The binding of the lac repressor to two operator sites, from M. Lewis et al., *Science,* vol. 271, pp. 1247-1254, March 1, 1996. Reprinted by permission.

Chapter 20
20.1: A complete map of the bacterium Hemophilus influenzae, from R. D. Fleishman, *Science,* vol. 269, p.28, 1995. Reprinted by permission.

Chapter 22
22.19a,b: Reprinted by permission from Macmillan Publishers Ltd: Nature, Alizadeh et al., February 2000, vol. 403, pp. 503-511, Copyright 2000.

Chapter 26
26.12: FASEB Journal by Olsen and Woesl. Copyright 1993 by Fedn of Am Societies for Experimental Bio (FASEB). Reproduced with permission of Fedn of Am Societies for Experimental Bio (FASEB) in the format Textbook via Copyright Clearance Center.

INDEX

Page numbers in *italics* indicate figures and tables.

E

F

O

P

Q

R

S

T

FIFTH EDITION

ABNORMAL PSYCHOLOGY IN A CHANGING WORLD

JEFFREY S. NEVID
St. John's University

SPENCER A. RATHUS
New York University

BEVERLY GREENE
St. John's University

Upper Saddle River, New Jersey 07458

Library of Congress Cataloging-in-Publication Data

Nevid, Jeffrey S.
Abnormal psychology in a changing world / Jeffrey S. Nevid, Spencer A. Rathus, Beverly Greene. — 5th ed.
p. cm.
Includes bibliographical references and index.
ISBN 0-13-048176-9
1. Psychology, Pathological. 2. Psychiatry. I. Rathus, Spencer A. II. Greene, Beverly. III. Title.

RC454 .N468 2002
616.89—dc21

2002069884

Editor-in-Chief: Leah Jewell
Executive Editor: Stephanie Johnson
Editorial Assistant: Catherine Fox
Executive Marketing Manager: Sheryl Adams
Marketing Assistant: Ron Fox
AVP/Director of Production and Manufacturing: Barbara Kittle
Senior Production Editor: Shelly Kupperman
Manufacturing Manager: Nick Sklitsis
Prepress and Manufacturing Buyer: Tricia Kenny
Line Art Coordinator: Guy Ruggiero
Artist: Maria Piper
Creative Design Director: Leslie Osher
Interior and Cover Designer: Anne DeMarinis
Director, Image Resource Center: Melinda Reo
Image Specialist: Beth Boyd
Photo Researcher: Melinda Alexander
Cover Art: Naoki Okamoto, Untitled, Living—Asian. B. 1990

This book was set in 10/12 Minion by TSI Graphics Group and was printed and bound by Von Hoffmann. The cover was printed by Coral Graphics.

Acknowledgments begin on page 603, which constitutes a continuation of this copyright page.

Printed in the United States of America
10 9 8 7 6 5 4 3 2 1

ISBN 0-13-048176-9

Pearson Education LTD., London
Pearson Education Australia PTY, Limited, Sydney
Pearson Education Singapore, Pte. Ltd
Pearson Education North Asia Ltd, Hong Kong
Pearson Education Canada, Ltd., Toronto
Pearson Educación de Mexico, S.A. de C.V.
Pearson Education – Japan, Tokyo
Pearson Education Malaysia, Pte. Ltd
Pearson Education, Upper Saddle River, New Jersey

BRIEF CONTENTS

CONTENTS

CHAPTER SIX

Anxiety Disorders 158

CHAPTER SEVEN

Dissociative and Somatoform Disorders 194

CHAPTER EIGHT

Mood Disorders and Suicide 222

CHAPTER NINE

Personality Disorders 266

CHAPTER TEN

Substance Abuse and Dependence 298

13 CHAPTER THIRTEEN
Schizophrenia and Other Psychotic Disorders 400

14 CHAPTER FOURTEEN
Abnormal Behavior in Childhood and Adolescence 438

15 CHAPTER FIFTEEN
Cognitive Disorders and Disorders Related to Aging 478

16

CHAPTER SIXTEEN

Violence and Abuse 500

CHAPTER SEVENTEEN

Abnormal Psychology and Society 536

PREFACE

Integrated Approach

With the fifth edition of *Abnormal Psychology in a Changing World*, we launch the first fully integrated textbook in abnormal psychology. We bring to our writing a focus on integrating issues of diversity, theoretical perspectives, and multimedia content.

Integration of Diversity

We are gratified by the many comments we have received over the years about our incorporation of material on diversity and multicultural issues in the field of abnormal psychology. We believe that issues of diversity are inseparable from the discussion of disorders and their treatments, and as a result, we now integrate material relating to diversity directly within the body of the text rather than setting it off in boxed features.

Integration of Theoretical Perspectives

Students often feel as though one theoretical perspective must ultimately be right and all the others wrong. The "Tying It Together" features helps students integrate the theoretical discussions and examine possible causal pathways involving interactions of psychological, sociocultural, and biological factors in explaining many forms of abnormal behavior, including anxiety disorders, dissociative disorders, mood disorders, substance abuse and dependence, and schizophrenia. We hope to impress upon students the importance of taking a broader view of complex problems by considering the influences of multiple factors and their interactions.

Integration of Multimedia

With this edition, we have enhanced the quality of our multimedia supplements *and* made them easier to use. We've reorganized and strengthened the material on our Companion Website™ to make it a more useful study tool for students. In addition, every copy of the fifth edition comes with a free CD-ROM that includes video interviews with patients, clinicians, and researchers.

QUIZ

RESEARCH UPDATE

VIDEO

WEB LINK

Web sites and CD-ROMs won't help, though, if it isn't clear how and when you should use them. That's why we've integrated the content on the Companion Website and CD-ROM with the textbook through the use of marginal icons. Throughout the text, you'll notice four different types of green icons and a brief title in the margins; the icons signal Quizzes, Web Links, Research Updates, and Videos. So when you see a green icon, you can go to the Companion Website or CD-ROM to find material that relates to that section of the text. Each icon is numbered to make it easier to find the corresponding material on the site. For example, as you're reading through Chapter 7 and come to an icon for "Quiz 7.1," go to the Nevid fifth edition Companion Website, click on Chapter 7, and then click on the "Quiz" button. This will bring up all of the quizzes for Chapter 7, including the first one—"Quiz 7.1, Dissociative Disorders."

CD-ROM with Video Case Examples

New to this edition is a free CD-ROM featuring 20 video case vignettes. The video case examples are indicated in the margins of the text with a Video/CD-ROM icon, so students can easily see how the case examples relate to the text material. (See the front endpapers of this text for a full list of video cases found on the CD-ROM.) These video case examples provide students with an opportunity to see and hear individuals who are diagnosed with a range of psychological disorders, from depression to schizophrenia. Students can now read about the clinical features of specific disorders and with a few clicks of a computer mouse see a case example illustrating the concepts discussed in the text. The video case examples supplement the many case examples we include in the text itself. Putting a human face on the subject matter helps make complex material more accessible.

Companion Website™

Icons in the margins of the text also highlight related Web content that helps students reinforce and expand their learning. By visiting the Companion Website, students can take quizzes that help them review what they've learned and find related resources on the Web that expand their learning. We also provide a new Web-based feature, "Research Update," which provides abstracts of the latest research developments in the field that have appeared since the book was published. Students may use these updates to write term papers, and instructors may find them helpful in keeping abreast of the latest research findings.

Also New to the Fifth Edition

As with previous editions of this book, we believe that a textbook should do more than offer a portrait of a field of knowledge. It should primarily be a teaching device—a means of presenting information that arouses student interest and encourages understanding and critical thinking. This new edition of *Abnormal Psychology in a Changing World* helps make the material more accessible to a new generation of students in a rapidly changing world. We sought to make this edition an even better learning experience for students and teaching experience for their instructors by incorporating the following new chapter organization and additional pedagogical features.

New Chapter Organization

We have reorganized the early chapters to provide students with a firm foundation in current theoretical perspectives on abnormal behavior and methods of assessment and treatment before they begin examining specific clinical disorders. The first 15 chapters cover the essential bases of abnormal psychology and all of the major diagnostic groupings. Chapter 16 ("Violence and Abuse") expands the focus to consider the roots of aggressive behavior and how violent and abusive behavior represent forms of abnormal behavior. In Chapter 17 ("Abnormal Psychology and Society") we review the legal and ethical issues that relate to the interface of society and abnormal behavior, as in, for example, the insanity defense and psychiatric commitment.

NEW Feature: "Think About It"

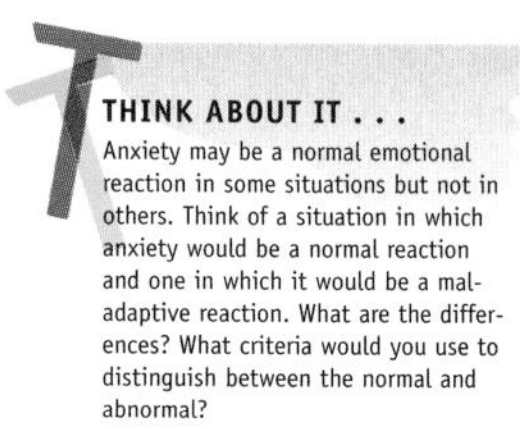
THINK ABOUT IT . . .
Anxiety may be a normal emotional reaction in some situations but not in others. Think of a situation in which anxiety would be a normal reaction and one in which it would be a maladaptive reaction. What are the differences? What criteria would you use to distinguish between the normal and abnormal?

With this edition we do a good deal more to promote critical thinking skills. In Chapter 1 we outline the skills of critical thinking and invite students to apply these skills in answering the new "Think About It" features that are interspersed throughout the text. Students are encouraged to think critically about the topics they encounter and to examine how these topics relate to their own personal experiences.

This new feature poses questions that encourage critical thinking, asks students to review key concepts, and helps them see the connections between the text material and their own experiences. Here are some examples of these "Think About It" features:

Critical Thinking Questions

- How would you recognize abnormal behavior? What criteria would you use to distinguish abnormal behavior from normal behavior? (Chapter 1)
- How do judgments about abnormal behavior reflect the cultural context in which they are made? In your answer, give at least one specific example. (Chapter 1)
- Anxiety may be a normal emotional reaction in some situations but not in others. Think of a situation in which anxiety would be a normal reaction and one in which it would be a maladaptive reaction. What are the differences? What criteria would you use to distinguish between the normal and abnormal? (Chapter 6)
- Why is the diagnosis of dissociative identity disorder controversial? Do you believe that people with dissociative identity disorder are merely playing a role they have learned? Why or why not? (Chapter 7)
- Are some personality disorders more likely to be diagnosed in men or in women because of societal expectations rather than because of real underlying pathology? Have you ever assumed that women are "just dependent or hysterical" or that men are "just narcissists or antisocial"? What kinds of problems do these underlying assumptions pose for clinicians and researchers? (Chapter 9)
- Where should we draw the line between normal and abnormal sexual behavior? What criteria should we use? (Chapter 12)

Review Questions

- What cultural issues do therapists need to consider when working with members of diverse cultural and racial groups? (Chapter 4)
- What role does childhood physical and sexual abuse play in dissociative disorders? (Chapter 7)
- What is Alzheimer's disease? What have we learned about the biological underpinnings of the disease? What don't we know? (Chapter 15)
- What is the difference between the concept of competency to stand trial and the legal defense of insanity? (Chapter 17)

Personal Reflection Questions

- Whom would you consider to be a self-actualizer? What about yourself? Are you a self-actualizer? Why or why not? (Chapter 2)
- What type of therapy would you prefer if you were seeking treatment for a psychological disorder? Why? (Chapter 4)
- Can you think of examples in your own life in which you have been hampered by performance anxiety of one kind or another? What did you do about it? (Chapter 12)
- Have you ever been subjected to sexual harassment? What was the outcome? Would you do anything differently if it were to happen again? (Chapter 16)

NEW Feature: Marginal Glossary

New to this edition is a running glossary in the text margins so students will not need to thumb to the back of the book for definitions of key terms. Key terms are in boldface type in the text and are defined in both the glossary in the back of the book and in the margins of the text on the two-page spread where the terms appear. The origins of key terms are often discussed. By learning to attend to commonly found Greek and Latin word origins, students can acquire skills that will help them decipher the meanings of new words. These decoding skills are a valuable objective for general education as well as a specific asset for the study of abnormal psychology.

NEW Feature: "Overview" Feature

Overview of Mood Disorders

TYPES OF MOOD DISORDERS	Description	Features
Major Depressive Disorder	Episodes of severe depression	• A range of features may be present, from downcast mood to appetite and sleep disturbance, to lack of interest and motivation • Seasonal affective disorder and postpartum depression are subtypes of major depression
Dysthymic Disorder	Long-standing mild depression	• Feeling "down in the dumps" most of the time, but not as severely depressed as people with major depressive disorder • Double depression is characterized by major depressive episodes occurring during the course of dysthymia
Bipolar Disorder	Mood swings between extreme elation and severe depression	• The two general subtypes are bipolar I disorder and bipolar II disorder • In rapid cycling, mania and major depression alternate without intervening periods of normal mood
Cyclothymia	Milder mood swings than bipolar disorder	• Chronic, cyclical pattern of shifting mood states from hypomanic episodes to states of mild depression • Frequent periods of depressed mood or loss of interest or pleasure in activities, but not at the level [illegible]

One of the most challenging aspects of this course is mastering and organizing the vast amounts of information about all of the disorders. To help students accomplish this more successfully, we have included a new "Overview" feature that helps students organize information by providing a visual summary of the major types of disorders, their causal factors, and the treatment approaches available. The "Overviews" are presented in a consistent format, so students can see "at a glance" how the information presented in the chapter fits together.

NEW Feature: Summing Up

In this new edition we developed a question-and-answer format for the chapter summary, in place of the traditional narrative summary. We believe that students will learn better when material is framed in terms of questions to answer rather than just narrative rewording of the text.

Maintaining Our Focus

Abnormal Psychology in a Changing World, 5th edition, is a complete learning and teaching package that represents our focus on four major objectives: (1) engaging student interest by incorporating student-oriented features; (2) integrating an interactionist or biopsychosocial model of abnormal behavior; (3) underscoring the importance of issues of diversity to the understanding and treatment of psychological disorders; and (4) maintaining currency with recent developments in the field.

Focus on Student-Oriented Features

We continue to provide students with opportunities to gain insight into their attitudes and behavior patterns. In addition to the new "Think About It" features described above, there are other student-focused features, as illustrated below.

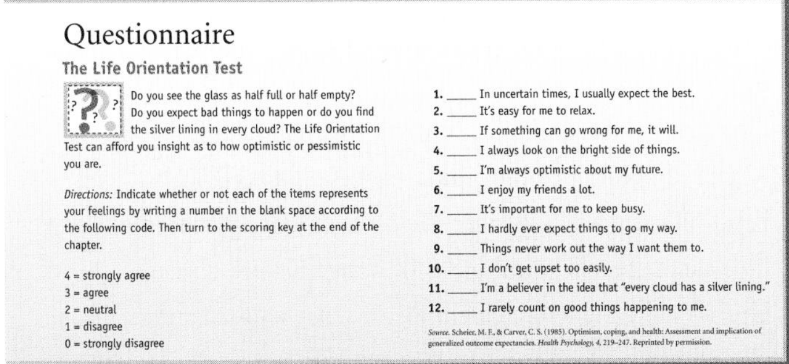

Questionnaire

The Life Orientation Test

Do you see the glass as half full or half empty? Do you expect bad things to happen or do you find the silver lining in every cloud? The Life Orientation Test can afford you insight as to how optimistic or pessimistic you are.

Directions: Indicate whether or not each of the items represents your feelings by writing a number in the blank space according to the following code. Then turn to the scoring key at the end of the chapter.

4 = strongly agree
3 = agree
2 = neutral
1 = disagree
0 = strongly disagree

1. _____ In uncertain times, I usually expect the best.
2. _____ It's easy for me to relax.
3. _____ If something can go wrong for me, it will.
4. _____ I always look on the bright side of things.
5. _____ I'm always optimistic about my future.
6. _____ I enjoy my friends a lot.
7. _____ It's important for me to keep busy.
8. _____ I hardly ever expect things to go my way.
9. _____ Things never work out the way I want them to.
10. _____ I don't get upset too easily.
11. _____ I'm a believer in the idea that "every cloud has a silver lining."
12. _____ I rarely count on good things happening to me.

Source. Scheier, M. F., & Carver, C. S. (1985). Optimism, coping, and health: Assessment and implication of generalized outcome expectancies. *Health Psychology, 4*, 219–247. Reprinted by permission.

Self-Scoring Questionnaires

Questionnaires involve students in the discussion at hand and permit them to evaluate their own attitudes and behavior patterns. In some cases, students may become more aware of troubling concerns, such as states of depression or problems with drug or alcohol use, which they may wish to bring to the attention of a professional. We have screened the questionnaires to ensure that they will provide students with useful information upon which to reflect as well as to serve as a springboard for class discussion. This allows students to have immediate scoring and feedback. Questionnaires include:

- The Life Orientation Test (Optimism Scale) (Chapter 5)
- Are You Type A? (Chapter 5)
- The Dissociative Experiences Scale (Chapter 7)
- Are You Depressed? (Chapter 8)
- The Fear of Fat Scale (Chapter 11)
- Examining Your Attitudes Toward Aging (Chapter 15)

This book also contains a number of features that are intended to keep it "on the beam" as a vehicle for instruction and learning, as with the following features.

"Truth or Fiction?" Chapter Openers

Each chapter begins with "Truth or Fiction?" statements that whet students' appetites for the subject matter within the chapter. Some items challenge preconceived ideas and common folklore and debunk myths and misconceptions, while others highlight new research in the field. Throughout the chapter, we give students feedback about the accuracy of their preconceptions by revisiting the "Truth or Fiction?" statements, at the points where the topics are discussed. Instructors and students have repeatedly reported that these items stimulate and challenge students. Examples include:

- Researchers find that people report more personal problems when interviewed by human interviewers than by computers.
- In some ways, many "mentally healthy" people see things less realistically than people who are depressed.
- People who threaten suicide are only seeking attention.
- Being able to "hold your liquor" better than most people helps prevent the development of problem drinking.
- The male sex hormone testosterone is produced by the body in women as in men.
- Dementia is a normal part of the aging process.
- Women are more likely to be raped by men they know than by strangers.

"A Closer Look" Feature

These features highlight new developments and controversies in the field and offer applications or resources that students can apply in their own lives. Examples include:

- Thinking Critically About Abnormal Psychology (Chapter 1)
- How Do I Find a Psychologist? (Chapter 4)
- Internet Counseling: Psychological Help May Be Only a Few Mouse Clicks Away (Chapter 4)
- Psychological Methods for Lowering Arousal (Chapter 5)
- EMDR: A Fad or a Find? (Chapter 6)
- Coping with a Panic Attack (Chapter 6)
- The Recovered Memory Controversy (Chapter 7)

- St. John's wort—A Natural "Prozac"? (Chapter 8)
- Suicide Prevention (Chapter 8)
- Did Samson Have Antisocial Personality Disorder? (Chapter 9)
- Gender Identity Disorder: A Disorder or Culture-Specific Creation? (Chapter 12)
- Anger Management (Chapter 16)
- Rape Prevention (Chapter 16)
- The Duty to Warn (Chapter 17)

Focus on Interactionist Approaches

We approach our writing with the belief that a better understanding of abnormal psychology is gained by adopting a biopsychosocial orientation that takes into account the roles of biological, psychological, and sociocultural factors and their interactions in the development of abnormal behavior patterns. We emphasize the value of taking an interactionist approach as a running theme throughout the text.

Focus on Issues of Diversity

We examine relationships between abnormal behavior patterns and issues of diversity relating to ethnicity, cultural factors, gender, sexual orientation, and socioeconomic status. We believe that students need to understand how issues of diversity affect both the conceptualization of abnormal behavior as well as the diagnosis and treatment of psychological disorders. Among the many issues relating to diversity that we explore in the text are the following:

- Cultural bases of abnormal behavior (Chapter 1)
- Cultural issues in diagnostic evaluation (Chapter 3)
- Surgeon General's 2001 report on culture, race, ethnicity, and mental health (Chapter 4)
- Multicultural issues in psychotherapy (Chapter 4)
- Relationship between acculturation and mental health (Chapter 5)
- Culture-bound syndromes (Chapters 3 and 7)
- Gender differences in prevalence of depression (Chapter 8)
- Sociocultural factors in substance use and abuse (Chapter 10)
- Sociocultural factors in eating disorders (Chapter 11)
- Cultural differences in expressed emotion (Chapter 13)
- Cultural differences in beliefs about abnormal behavior in childhood (Chapter 14)
- Sociocultural perspectives on aggression (Chapter 16)

Focus on Recent Developments in the Field

We help the reader keep abreast of the ever-changing subject matter, in the field. This new edition has also been thoroughly updated to incorporate information from the 2000 edition of the *DSM-IV-TR*. Here is a small sample of the extensive coverage of new research included in this edition:

- The 1999 U.S. Surgeon General's report on the state of the nation's mental health
- Computerized clinical interviews and computer-assisted therapy
- Controversies over online counseling services
- New findings on validity of projective tests and comparisons of validities of psychological and medical tests
- Use of fMRI brain imaging in studying brain function
- Effectiveness of psychotherapy in ordinary clinical practice
- New evidence on specific versus nonspecific factors in psychotherapy
- Update on effects of stress on the immune system functioning
- New evidence on effects of expressive writing on psychological and physical well-being
- Genetic links between OCD and tic disorders
- Update on vulnerability factors in PTSD
- Classical conditioning model of panic disorder
- New information on role of GABA in anxiety disorders
- Update on treatment approaches for anxiety disorders and mood disorders
- Update on EMDR
- Update on St. John's wort
- Update on childhood abuse and development of dissociative disorders
- Update on phototherapy for SAD
- Relationships between dysfunctional attitudes and proneness to depression and between stress and depression
- Update on genetic factors in mood disorders
- Update on effectiveness of psychological and pharmacological treatment for depression
- The 2001 U.S. Surgeon General's report on suicide
- Recent developments in the treatment of personality disorders
- Updated prevalence rates of drug use and abuse
- Abuse of prescription opioids, especially OxyContin
- Psychological and physical effects of the drug known as ecstasy
- Update on biochemical bases of drug use and abuse
- Update on genetic factors in alcoholism and substance abuse
- Update on sociocultural factors in drug use in adolescents
- Update on treating substance abuse and dependence
- Update on factors underlying eating disorders
- Biochemical causes of narcolepsy
- Need evidence of brain abnormalities in schizophrenia
- Latest estimates of the prevalence of autism
- MRI and PET scan studies of autistic children and adults

- New evidence on brain abnormalities in children with ADHD
- New evidence on combining drug treatment and behavior modification in treating ADHD
- Latest research on causal factors and treatment of Alzheimer's disease
- Research evidence on psychological and drug treatment of geriatric depression
- New evidence on the psychological profile of batterers
- Update on effects of child abuse and child sexual abuse on psychological functioning
- Update on the accuracy of clinician predictions of dangerousness

Ancillaries

No matter how comprehensive a textbook is, today's instructors and students require a complete teaching package to advance teaching and comprehension. *Abnormal Psychology in a Changing World, Fifth Edition*, is accompanied by the following ancillaries:

For Students

Videos in Abnormal Psychology CD-ROM

With every new copy of *Abnormal Psychology, Fifth Edition,* students will receive a free CD-ROM with video clips showing skilled clinicians interviewing real patients who have been diagnosed with various disorders. Disorders represented include panic disorder, schizophrenia, anorexia, bulimia, and others. Icons in the text margins indicate when students should go to the CD-ROM to find video case interviews.

Companion Website™ (www.prenhall.com/nevid)

Fred Whitford of Montana State University, along with the text authors, has carefully created and selected all of the resources on the text-matched Companion Website to reinforce students' understanding of the concepts in the fifth edition. The Web resources are tightly integrated with the text through icons found in the text margins. These icons indicate when material on the Web site corresponds to material in the text. Students can take online quizzes and get immediate scoring and feedback; link to related Web sites; and find research updates to keep up with recent developments in the field. Students can also link from the Companion Website to the video segments found on the *Abnormal Psychology* CD-ROM.

ContentSelect Research Database

Prentice Hall and EBSCO, the world leader in online journal subscription management, have developed a customized research database for students of psychology. The database provides unlimited access to the text of dozens of peer-reviewed psychology publications. Student access codes can be packaged *free* with *Abnormal Psychology, Fifth Edition.* To see for yourself how this site works, ask your local Prentice Hall representative for a free Instructor Access Code.

Study Guide (0-13-049508-5)

Gwen Parsons of Hillsborough Community College has created a study guide that includes numerous review and study questions, crossword puzzles, and other learning aids to help reinforce students' understanding of the concepts covered in the text.

Print Supplements for Instructors

Instructor's Resource Manual (0-13-049509-3)

Nancy Simpson of Trident Technical College has prepared an instructor's resource manual with many useful features, including "Lecture and Discussion" suggestions, "Student Activities," "Classroom Demonstration" ideas, and interesting suggestions on how to integrate the new *Abnormal Psychology* CD-ROM into your course.

Test Item File (0-13-049505-0)

Developed by Joanne Karpinen of Hope College, this comprehensive test bank has been updated to include new questions on revised text material. It contains over 4,000 multiple-choice, true/false, short answer, and essay questions.

Color Transparencies for Abnormal Psychology, Series II (0-13-080451-7)

This set of full-color transparencies includes illustrations, figures, and graphs from the text as well as images from a variety of other sources. It has been designed with lecture hall visibility and convenience in mind.

Media and Online Resources for Instructors

Test Manager (0-13-049507-7)

One of the best-selling test-generating software programs on the market, Test Manager is available in Windows and Macintosh formats (both of which are included on one CD-ROM). The Test Manager includes a Gradebook, On-Line Network Testing, and many tools to help you edit and create tests quickly and easily.

PsychologyCentral Web Site (www.prenhall.com/psychology)

This site is password-protected for instructors' use only and allows you online access to all Prentice Hall psychology supplements at any time. You'll find a multitude of resources (both text-specific and non-text-specific) for teaching abnormal psychology—and many other psychology courses too. From this site, you can download any of the key supplements for Nevid et al., fifth edition: Instructor's Resource Manual, Test Item File, and PowerPoint presentations. Contact your local sales representative for the User ID and Password to access this site.

PowerPoint Presentations (on the Companion Website and the *PsychologyCentral* Web site)

Fred Whitford of Montana State University has created two sets of PowerPoint presentations—one with graphics and one without—

to give you even greater flexibility in using PowerPoint in your lectures. Both presentations highlight all of the key points in the fifth edition of *Abnormal Psychology*.

Online Course Management

For instructors interested in using online course management, Prentice Hall offers fully customizable courses in BlackBoard and Course Compass to accompany this textbook. These online courses are preloaded with material for *Abnormal Psychology, Fifth Edition,* including the Test Item File. Contact your local Prentice Hall representative or visit www.prenhall.com/demo for more information.

Video Resources for Instructors

ABC News Videos for Abnormal Psychology, Series III

Qualified adopters can obtain this series consisting of segments from the *ABC Nightly News* with Peter Jennings, *Nightline, 20/20, Prime Time Live*, and *The Health Show*. The programs cover issues such as drugs and alcoholism, psychotherapy, autism, crime motivation, depression, and others. Contact your Prentice Hall representative for more details.

Patients as Educators: Video Cases in Abnormal Psychology

This video was created by James H. Scully, Jr., M.D., and Alan M. Dahms, Ph.D., Colorado State University, and is available to qualified adopters. It includes a series of ten patient interviews that illustrate a range of disorders. Each interview is preceded by a brief history of the patient and a synopsis of some major symptoms of the disorder and ends with a summary and brief analysis. Contact your local sales representative for more details.

Acknowledgments

The field of abnormal psychology is a moving target, as the literature base that informs our understanding is continually expanding. We are deeply indebted to a number of talented individuals who helped us hold our camera steady in taking a portrait of the field, focus in on the salient features of our subject matter, and develop our snapshots through prose.

First, our professional colleagues, who reviewed our manuscript through the first several versions and continue to help us refine and strengthen the material:

Reviewers of the Fifth Edition

Heinz Fischer, *Long Beach City College*
John H. Forthman, *Vermilion Community College*
Pam Gibson, *James Madison University*
John K. Hall, *University of Pittsburgh*
Shay McCordick, *San Diego State University*
Linda L. Morrison, *University of New England*
Ari Solomon, *Williams College*
Theresa Wadkins, *University of Nebraska at Kearney*

Reviewers of Previous Editions

Sally Bing, *University of Maryland Eastern Shore*
Christiane Brems, *University of Alaska Anchorage*
Bernard Gorman, *Nassau Community College*
Gary Greenberg, *Connecticut College*
Bob Hill, *Appalachian State University*
Robert Kapche, *California State University, Long Beach*
Stuart Keeley, *Bowling Green State University*
Joseph J. Palladino, *University of Southern Indiana*
Carol Pandey, *Los Angeles Pierce College*
J. Langhinrichsen-Rohling, *University of Nebraska-Lincoln*
Esther D. Rosenblum, *University of Vermont*
Harold Siegel, *Nassau Community College*
Larry Stout, *Nicholls State University*
Max Zwanziger, *Central Washington University*

Second, but by no means second-rate, are the publishing professionals at Prentice Hall who helped guide the development of this edition, especially Stephanie Johnson, Executive Editor; Rochelle Diogenes, Associate Editor-in-Chief for Development; Elaine Silverstein, Developmental Editor; Shelly Kupperman, Senior Production Editor; Sheryl Adams, Executive Marketing Manager; Karen Branson, Media Editor; Mindy DePalma, Media Editor; Katie Fox, Editorial Assistant; and Ronald Fox, Marketing Assistant. A special thanks to Leslie Osher, Creative Design Director, and Anne DeMarinis, freelance designer, for the beautiful new design of the fifth edition. We would also like to thank the entire Prentice Hall sales force for their enthusiastic and diligent efforts on behalf of our text.

Finally, we especially wish to thank two people without whose inspiration and support this effort would never have materialized or been completed, Judith Wolf-Nevid and Lois Fichner-Rathus.

J.S.N.

S.A.R.

B.A.G.

ABOUT THE AUTHORS

Jeffrey S. Nevid is a professor of psychology at St. John's University in New York, where he directs the Doctoral Program in Clinical Psychology; teaches graduate courses in research methods, psychological assessment, and behavior therapy; and supervises doctoral students in clinical practicum work. He received his doctorate in clinical psychology from the State University of New York at Albany and was awarded a National Institute of Mental Health (NIMH) Postdoctoral Fellowship in Mental Health Evaluation Research. He has published numerous articles in the areas of clinical and community psychology, health psychology, training models in clinical psychology, and methodological issues in clinical research. He holds a diplomate in clinical psychology from the American Board of Professional Psychology, is a fellow of the Academy of Clinical Psychology (FAClinP), has served on the editorial boards of several journals, and is presently an associate editor of *The Journal of Consulting and Clinical Psychology*. He is also an author of several leading textbooks in psychology and related fields and was keynote speaker in 2002 at the 16th Annual Conference on Undergraduate Teaching of Psychology.

Spencer A. Rathus received his doctorate from the State University of New York at Albany. He is on the faculty at the New York University School of Continuing and Professional Studies. His areas of interest include psychological assessment, cognitive behavior therapy, and deviant behavior. He is the originator of the Rathus Assertiveness Schedule, which has become a Citation Classic. He has authored several books, including *Psychology in the New Millennium, Essentials of Psychology*, and *The World of Children*. He has coauthored *Making the Most of College* with Lois Fichner-Rathus; *AIDS: What Every Student Needs to Know* with Susan Boughn; *Behavior Therapy, Psychology and the Challenges of Life* and *Health in the New Millennium* with Jeffrey S. Nevid; and *Human Sexuality in a World of Diversity* with Jeffrey S. Nevid and Lois Fichner-Rathus. His professional activities include service on the American Psychological Association Task Force on Diversity Issues at the Precollege and Undergraduate Levels of Education in Psychology and on the Advisory Panel, American Psychological Association, Board of Educational Affairs (BEA) Task Force on Undergraduate Psychology Major Competencies.

Beverly Greene is a professor of psychology at St. John's University and is a fellow of the American Psychological Association, the American Orthopsychiatric Association, and the Academy of Clinical Psychology. She holds a diplomate in clinical psychology from the American Board of Professional Psychology and serves on the editorial boards of numerous scholarly journals. She received her doctorate in clinical psychology from Adelphi University and is founding coeditor of *Psychological Perspectives on Lesbian, Gay and Bisexual Issues* and coeditor of *Education, Research and Practice in Lesbian, Gay, Bisexual and Transgendered Psychology: A Resource Manual* (Vol. 5, 2000). The author of nearly 70 professional publications, Dr. Greene was the recipient of the 1996 Outstanding Achievement Award from the American Psychological Association's Committee on Lesbian, Gay and Bisexual Concerns, the 2000 Heritage Award from the APA Division of the Psychology of Women, and the 1995, 1996, and 2000 (co-recipient) of the Psychotherapy with Women Research Award from The Society for the Psychology of Women. One of her papers was honored with the 2000 Women of Color Psychologies Publication Award, an award she previously received in 1991 and 1995. Her coedited book, *Psychotherapy with African American Women: Innovations in Psychodynamic Perspectives and Practice,* was also the recipient of the Association for Women in Psychology's 2001 Distinguished Publication Award.

ABNORMAL PSYCHOLOGY IN A CHANGING WORLD

CHAPTER ONE

Introduction and Methods of Research

Wassily Kandinsky
Sweet Pink

Truth OR Fiction?

- Psychological disorders actually affect only relatively few of us. (p. 4)
- Behavior deemed abnormal in one society may be perceived as perfectly normal in another. (p. 6)
- The modern medical model of abnormal behavior can be traced to the work of a Greek physician some 2,500 years ago. (p. 10)
- Innocent people were drowned in medieval times as a way of certifying they were not possessed by the devil. (p. 11)
- A night's entertainment in London a few hundred years ago might have included peering at the inmates at the local asylum. (p. 12)
- Finding that two variables are closely linked together means that one is a cause of the other. (p. 21)
- A survey of 1,500 Americans may provide a more accurate reflection of voting attitudes and preferences of the American public than one based on millions of participants. (p. 25)
- Case studies have been conducted on people who have been dead for hundreds of years. (p. 27)

Abnormal behavior might seem the concern of only a few. After all, only a minority of the population will ever be admitted to a psychiatric hospital. Most people never seek the help of a **psychologist** or **psychiatrist.** Only a few people plead not guilty to crimes on grounds of insanity. Many of us have what we call an "eccentric" relative, but few of us have relatives we would consider truly bizarre.

The truth of the matter is that abnormal behavior affects virtually everyone in one way or another. Abnormal behavior patterns that involve a disturbance of psychological functioning or behavior are classified by mental health professionals as **psychological disorders,** or *mental disorders.* The term *mental illness* refers collectively to all of the diagnosable mental disorders, including anxiety disorders, mood disorders, schizophrenia, sexual dysfunctions, and substance use disorders (USDHHS, 1999a). If we confine our definition of abnormal behavior to diagnosable mental disorders, about one in two of us have been directly affected (R. C. Kessler, 1994). About one in five people in the United States are affected by mental disorders in a given year (USDDHS, 1999a). Figure 1.1 shows the lifetime and past-year rates of several major classes of psychological (or mental) disorders based on a representative sample of the adult U.S. population (R. C. Kessler et al., 1993, 1994).[1] Psychological disorders were most common among people in the 25- to 32-year-old age range and declined with increasing age. Problems involving anxiety and depression were more common among women. Alcohol and substance abuse problems more commonly affected men. If we also include the mental health problems of our family members, friends, and coworkers, and take into account those who foot the bill for treatment in the form of taxes and health insurance premiums and lost productivity due to sick days, disability leaves, and impaired job performance inflating product costs, then perhaps none of us remains unaffected ("Mental Health Problems," 2000).

In December 1999, the U.S. Surgeon General issued a major report on the state of the nation's mental health. Here let us highlight some key conclusions in the report (USDHHS, 1999b; Satcher, 2000):

- Mental health and illness reflects a complex interaction of brain functioning and environmental influences.
- A range of effective treatments exists for most mental disorders, including psychological interventions such as psychotherapy and counseling, and psychopharmacologic or drug therapies. Treatment is often more effective when psychological and pharmacological treatments are combined.
- Progress in developing effective prevention programs in the mental health field has been slow, as we continue to lack clear knowledge about the underlying causes of mental disorders or ways of altering known casual influences, such as genetic predispositions. Nonetheless, some effective prevention programs have been developed.
- Though about 15% of adult Americans receive some form of mental health assistance in a given year, a critical gap continues to exist between the numbers of people needing help and those receiving it.
- Mental health problems are best understood when we take a broader view and consider the social and cultural contexts in which they occur.
- Mental health services need to be designed and delivered in a culturally sensitive manner that takes into account the viewpoints and needs of racial and ethnic minorities.

The Surgeon General's findings form a backdrop for our study of abnormal psychology. As we shall see throughout the text, the understanding of abnormal behavior is best viewed through a lens that takes into account complex interactions involving biological and environmental factors. We shall also see the importance of social and cultural (or *sociocultural*) factors in attempting to understand mental disorders and in developing men-

[1]To be precise, the survey was not quite national, as the sample was limited to people residing in the 48 contiguous U.S. states.

Chapter Outline

WWW **Web Link 1.1**
Mental Disorders in America

psychologist A person with advanced graduate training in psychology.

psychiatrist A physician who specializes in the diagnosis and treatment of emotional disorders.

psychological disorders Abnormal behavior patterns that involve a disturbance of psychological functioning or behavior.

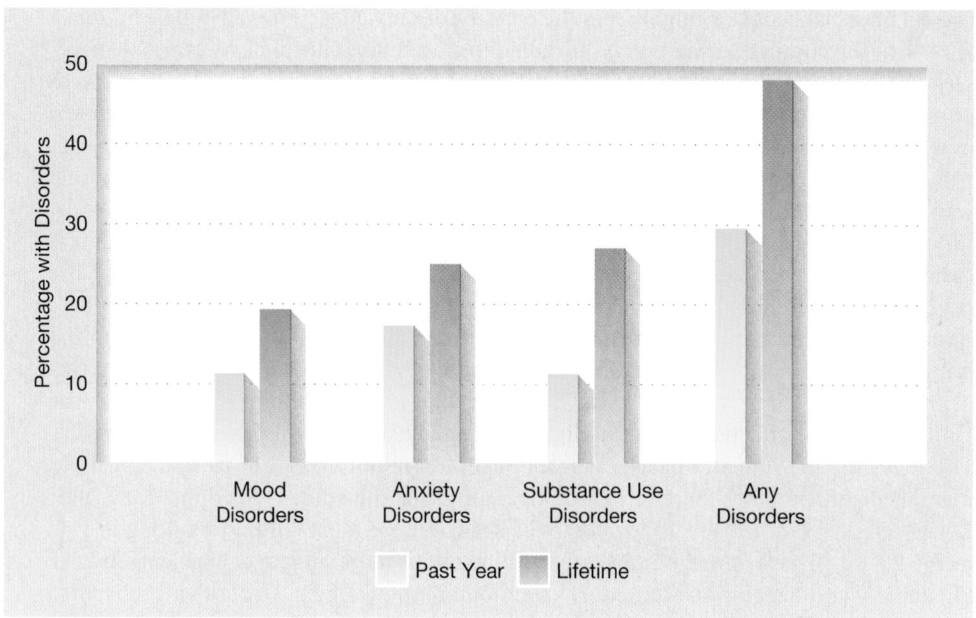

FIGURE 1.1 Lifetime and past-year prevalences of psychological disorders. This graph shows the percentages of adults in the United States in the 15- to 49-year-old age range who show evidence of having diagnosable psychological disorders in either the past year or lifetime. The data are shown for several major diagnostic categories. The mood disorders category includes major depressive episode, manic episode, and dysthymia (discussed in Chapter 8). Anxiety disorders includes panic disorder, agoraphobia without panic disorder, social phobia, specific phobia, and generalized anxiety disorder (discussed in Chapter 6). Substance use disorders include abuse or dependence disorders involving alcohol or other drugs (discussed in Chapter 10).

Source. National Comorbidity Survey, (Kessler et al., 1994).

Truth OR Fiction? REVISITED

Psychological disorders actually affect only relatively few of us.

FALSE. In one way or another, psychological disorders affect virtually all of us.

Web Link 1.2
Mental Health: A Report of the Surgeon General WWW

abnormal psychology The branch of psychology that deals with the description, causes, and treatment of abnormal behavior patterns.

medical model A biological perspective in which abnormal behavior is viewed as symptomatic of underlying illness.

tal health services for people from diverse cultural backgrounds. We shall also survey the range of effective treatment approaches that are now available to help people who are affected by mental disorders.

Abnormal psychology is the branch of psychology that seeks to understand abnormal behavior patterns and ways of helping people who are affected by them. Abnormal psychology encompasses a broader view of abnormal behavior than the study of mental (or psychological) disorders. For example, rape is certainly a form of abnormal behavior, even though it is not classified as a psychological disorder. However, our attention throughout the text is focused on understanding the nature and treatment of mental disorders.

Let us pause for a moment to consider our use of terms. Though the terms *mental disorder* and *psychological disorder* are often used interchangeably, we generally prefer using the term *psychological disorder.* There are several reasons we have adopted this approach. First, the term *psychological disorder* puts the study of abnormal behavior squarely within the purview of the field of psychology. Another reason is that the term *mental disorder* is generally associated with the **medical model** perspective that considers abnormal behavior patterns to be symptoms of underlying illnesses or disorders. Although the medical model remains a prominent perspective for understanding abnormal behavior patterns, we will show that other perspectives, including psychological and sociocultural perspectives, also inform our understanding of abnormal behavior. In addition, the term *mental disorder* reinforces the traditional distinction between mental and physical phenomena. As we'll see, there is increasing awareness of the interrelationships between the body and the mind that calls into question this distinction.

In this chapter we first address the task of defining abnormal behavior. We see that throughout history, and even in prehistory, abnormal behavior has been viewed from different perspectives, or models. We chronicle the development of concepts of abnormal behavior and its treatment. We see that, historically speaking, treatment usually referred to what was done *to*, rather than *for*, people with abnormal behavior. Finally, we describe the ways in which psychologists and other scholars study abnormal behavior today.

Cultural sensitivity in a diverse society. The recent Surgeon General's report on mental health emphasizes the importance of providing mental health services in a culturally sensitive manner.

How Do We Define Abnormal Behavior?

Most of us become anxious or depressed from time to time, but our behavior is not deemed abnormal. It is normal to become anxious in anticipation of an important job interview or a final examination. It is appropriate to feel depressed when you have lost someone close to you or when you have failed at a test or on the job. But when do we cross the line between normal and abnormal behavior?

One answer is that emotional states such as anxiety and depression may be considered abnormal when they are not appropriate to the situation. It is normal to feel down because of failure on a test, but not when one's grades are good or excellent. It is normal to feel anxious during a college admissions interview, but not whenever entering a department store or boarding a crowded elevator.

Abnormal behavior may also be suggested by the magnitude of the problem. Although some anxiety is normal enough before a job interview, feeling that your heart is hammering away so relentlessly that it might leap from your chest—and consequently canceling the interview—are not. Nor is it normal to feel so anxious in this situation that your clothing becomes soaked with perspiration.

Criteria for Determining Abnormality

Mental health professionals apply various criteria in making judgments about whether behavior is abnormal. The most commonly used criteria include the following:

1. *Behavior is unusual.* Behavior that is unusual is often considered abnormal. Only a few of us report seeing or hearing things that are not really there; "seeing things" and "hearing things" are almost always considered abnormal in our culture, except, perhaps, in cases of certain religious experiences in which "hearing voices" or "seeing visions" of religious figures are not unusual (USDHHS, 1999a). Moreover, "hearing voices" and others forms of hallucinations under some circumstances are not considered unusual in some preliterate societies.

Becoming overcome with feelings of panic when entering a department store or when standing in a crowded elevator is uncommon and considered abnormal in our culture. But uncommon behavior is not in itself abnormal. Only one person can hold the record for swimming or running the fastest 100 meters. The record-holding athlete differs from the

When is anxiety abnormal? Negative emotions such as anxiety are considered abnormal when they are judged to be excessive or inappropriate to the situation. Anxiety is generally regarded as normal when it is experienced during a job interview (left), so long as it is not so severe that it prevents the interviewee from performing adequately. Anxiety is deemed to be abnormal if it is experienced whenever one boards an elevator (right).

paranoid Referring to irrational suspicions.

rest of us but, again, is not considered abnormal. Thus rarity or statistical deviance is not a sufficient basis for labeling behavior abnormal; nevertheless, it is one yardstick often used to judge abnormality.

2. *Behavior is socially unacceptable or violates social norms.* All societies have norms (standards) that define the kinds of behaviors acceptable in given contexts. Behavior deemed normal in one culture may be viewed as abnormal in another. In our society, standing on a soapbox in a park and repeatedly shouting "Kill!" to passersby would be labeled abnormal; shouting "Kill!" in the grandstands at an important football game is usually within normal bounds, however tasteless it may seem. Although the use of norms remains one of the important standards for defining abnormal behavior, we should be aware of some limitations of this definition.

Truth OR Fiction? REVISITED

Behavior deemed abnormal in one society may be perceived as perfectly normal in another.

TRUE. Behavior that is deemed normal in one culture may be viewed as abnormal in another.

One implication of basing the definition of abnormal behavior on social norms is that norms reflect relative standards, not universal truths. What is normal in one culture may be abnormal in another. For example, Americans who assume strangers are devious and try to take advantage are usually regarded as distrustful, perhaps even **paranoid.** But such suspicions were justified among the Mundugumor, a tribe of cannibals studied by anthropologist Margaret Mead (1935). Within that culture, male strangers, even the male members of one's own family, *were* typically malevolent toward others.

Clinicians need to weigh cultural differences in determining what is normal and abnormal. In the case of the Mundugumor, this need is more or less obvious. Sometimes, however, differences are more subtle. For example, what is seen as normal, outspoken behavior by most American women might be interpreted as brazen behavior when viewed in the context of another, more traditional culture. Moreover, what strikes one generation as abnormal may be considered by others to fall within the normal spectrum. For example, until the mid-1970s homosexuality was classified as a mental disorder by the psychiatric profession (see Chapter 11). Today, however, the psychiatric profession no longer considers homosexuality a mental disorder, and many people argue that contemporary societal norms should include homosexuality as a normal variation in behavior.

Another implication of basing normality on compliance with social norms is the tendency to brand nonconformists as mentally disturbed. We may come to brand behavior we do not like or understand as "sick" rather than accept that that behavior may be normal, even if it offends or puzzles us.

3. *Perception or interpretation of reality is faulty.* Normally speaking, our sensory systems and cognitive processes permit us to form accurate mental representations of the

Is this abnormal? One of the criteria used to determine whether or not behavior is abnormal is whether it deviates from acceptable standards of conduct or social norms. The behavior and attire of these men might be considered abnormal in a classroom or workplace, but not perhaps at a football game.

environment. But seeing things and hearing voices that are not present are considered **hallucinations,** which in our culture are often taken as signs of an underlying disorder. Similarly, holding unfounded ideas or **delusions,** such as **ideas of persecution** that the CIA or the Mafia are out to get you, may be regarded as signs of mental disturbance—unless, of course, they *are.* (A former secretary of state is credited with having remarked that he might, indeed, be paranoid; his paranoia, however, did not mean he was without enemies.)

It is normal in the United States to say that one "talks" to God through prayer. If, however, a person claims to have literally seen God or heard the voice of God—as opposed to, say, being divinely inspired—we may come to regard her or him as mentally disturbed.

4. *The person is in significant personal distress.* States of personal distress caused by troublesome emotions, such as anxiety, fear, or depression, may be considered abnormal. As we noted earlier, however, anxiety and depression are sometimes appropriate responses to the situation. Real threats and losses occur from time to time, and *lack* of an emotional response to them would be regarded as abnormal. Appropriate feelings of distress are not considered abnormal unless they become prolonged or persist long after the source of anguish has been removed (after most people would have adjusted) or if they are so intense that they impair the individual's ability to function.

5. *Behavior is maladaptive or self-defeating.* Behavior that leads to unhappiness rather than self-fulfillment can be regarded as abnormal. Behavior that limits our ability to function in expected roles, or to adapt to our environments, may also be considered abnormal. According to these criteria, heavy alcohol consumption that impairs health or social and occupational functioning may be viewed as abnormal. **Agoraphobic** behavior, characterized by intense fear of venturing into public places, may be considered abnormal in that it is both uncommon and also maladaptive because it impairs the individual's ability to fulfill work and family responsibilities.

6. *Behavior is dangerous.* Behavior that is dangerous to oneself or other people may be considered abnormal. Here, too, the social context is crucial. In wartime, people who sacrifice themselves or charge the enemy with little apparent concern for their own safety may be characterized as courageous, heroic, and patriotic. But people who threaten or attempt suicide because of the pressures of civilian life are usually considered abnormal.

Football and hockey players, even adolescent boys who occasionally get into altercations, may be normal enough. Given the cultural demands of the sports, unaggressive football and hockey players would not last long in college or professional ranks. But individuals involved in frequent unsanctioned fights may be regarded as abnormal. Physically aggressive behavior is most often maladaptive in modern life. Moreover, outside the contexts of sports and warfare, physical aggression is discouraged as a way of resolving interpersonal conflicts—although it is by no means uncommon.

Abnormal behavior thus has multiple definitions. Depending on the case, some criteria may be weighted more heavily than others. But in most cases, a combination of these criteria is used to define abnormality.

It is one thing to recognize and label behavior as abnormal; it is another to understand and explain it. Philosophers, physicians, natural scientists, and psychologists have used various approaches, or *models,* in the effort to explain abnormal behavior. Some approaches have been based on superstition; others have invoked religious explanations. Some current views are predominantly biological; others, psychological. We consider various historical and contemporary approaches to understanding abnormal behavior. First, let us look further at the importance of cultural beliefs and expectations in determining which behavior patterns are deemed abnormal.

Cultural Bases of Abnormal Behavior

Behavior that is considered normal in one culture may be deemed abnormal in another. Hallucinations (hearing voices or seeing things that are not in fact present) are a common experience among Australian aborigines but are generally taken as a sign of abnormality in our culture. Aborigines also believe they can communicate with the spirits of their ancestors

hallucination A perception that occurs in the absence of an external stimulus and that is confused with reality.

delusion A firmly held but inaccurate belief that persists despite evidence that it has no basis in reality.

ideas of persecution A form of delusional thinking characterized by false beliefs that one is being persecuted or victimized by others.

agoraphobic Relating to excessive, irrational fear of open places.

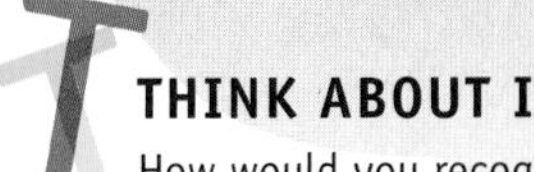

THINK ABOUT IT

How would you recognize abnormal behavior? What criteria would you use to distinguish abnormal behavior from normal behavior?

WWW **Web Link 1.3**
Online Guide to Mental Health

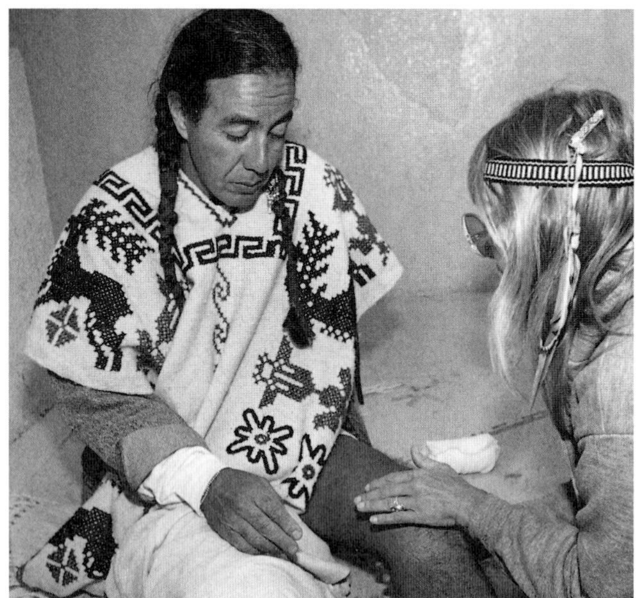

A traditional Native American healer. Many traditional Native Americans distinguish between illnesses that are believed to arise from influences external to their own culture (thus "White man's sicknesses") and those that emanate from a lack of harmony with traditional tribal life and thought ("Indian sicknesses"). Tribal healers (such as the one shown here) may be called on to treat "Indian sickness," whereas "White man's medicine" may be sought for people with problems that have causes seen as lying outside the traditional community, such as alcoholism and drug addiction.

and that dreams are shared among people, especially close relatives. Such beliefs may be regarded in Western culture as delusions (fixed false beliefs). Hallucinations and delusions are taken to be common features of schizophrenia in Western culture. Should we thus conclude that aborigines are seriously disturbed or have schizophrenia? What standards should be applied in judging abnormal behavior in other cultures? Even aborigines perceive "madness" in some members of their community, although the criteria they use to label someone as mentally disturbed may differ from those used by health professionals in Western society.

Kleinman (1987) offers an example of "hearing voices" among Native Americans to underscore the ways in which judgments about abnormal behavior are embedded within a cultural context:

> Ten psychiatrists trained in the same assessment technique and diagnostic criteria who are asked to examine 100 American Indians shortly after the latter have experienced the death of a spouse, a parent or a child may determine with close to 100% consistency that those individuals report hearing, in the first month of grieving, the voice of the dead person calling to them as the spirit ascends to the afterworld. [While such judgments may be consistent across observers] the determination of whether such reports are a sign of an abnormal mental state is an interpretation based on knowledge of this group's behavioural norms and range of normal experiences of bereavement. (p. 453)

To many Native Americans, bereaved people who report hearing the spirits of the deceased calling to them as they ascend to the afterlife are regarded as normal. Kleinman's example leads us to recognize that behavior should not be considered abnormal when it is normative within the cultural setting in which it occurs. Concepts of health and illness may also have different meanings in different cultures. Many traditional Native American cultures distinguish between illnesses that are believed to arise from influences outside the culture, called "White man's sicknesses," such as alcoholism and drug addiction, from those that emanate from a lack of harmony with traditional tribal life and thought, which are called "Indian sicknesses" (Trimble, 1991). Traditional healers, shamans, and medicine men and women are called on to treat and cure "Indian sickness." When the problem is thought to have its cause outside the community, help may be sought from "White man's medicine."

Web Link **1.4**
Mental Health: Culture, Race, and Ethnicity
WWW

The very words that Western cultures use to describe psychological disorders—words such as depression or even mental health—may have very different meanings in other cultures or no equivalent meaning at all. In many non-Western societies, depression may be closer in meaning to the concept of "soul loss" than to Western concepts involving a sense of loss of purpose and meaning in life (Shweder, 1985).

THINK ABOUT IT

How do judgments about abnormal behavior reflect the cultural context in which they are made? Give at least one specific example.

Abnormal behavior patterns may also take different forms in different cultures (USDHHS, 1999a). Westerners may experience anxiety, for example, in the form of excessive worrying about paying the mortgage, losing a job, and so on. Yet "[I]n a number of African cultures, anxiety is expressed as fears of failure in procreation, in dreams and complaints about witchcraft" (Kleinman, 1987). Some Australian aborigines develop intense fears of sorcery, which may be accompanied by the belief that one is in mortal danger from evil spirits (D. J. Spencer, 1983). Trancelike states in which young aboriginal women are mute, immobile, and unresponsive are also quite common. If women do not recover from the trance within hours or, at most, a few days, they may be brought to a sacred site for healing.

Depression may also be expressed differently in different cultures (Bebbington, 1993; Thakker & Ward, 1998). This doesn't mean that depression doesn't exist in other cultures. Rather, it suggests we need to consider how people in different cultures experience states of emotional distress, including depression and anxiety, rather than imposing our perspectives on their experiences. Among some Far Eastern peoples, such as the Chinese, depression is

demonology The idea that abnormal behavior is caused by supernatural forces.

worldview The prevailing view of the times (English translation of German term *Weltanschaung*).

often expressed through physical symptoms, such as headaches, fatigue, or weakness, rather than by feelings of guilt or sadness that are common in Western cultures (American Psychiatric Association, 2000; Parker, Gladstone, & Chee, 2001).

Cultural differences in how abnormal behavior patterns are expressed lead us to realize that we must determine that our concepts of abnormal behavior are recognizable and valid before they are applied to other cultures (Bebbington, 1993). The reverse is equally true. The concept of "soul loss" characterizes psychological distress in some non-Western societies, but it has little or no relevance to middle-class North Americans. Evidence from multinational studies conducted by the World Health Organization (WHO) in the 1960s and 1970s shows that the behavior pattern we characterize as schizophrenia exists in countries as far flung as Colombia, India, China, Denmark, Nigeria, and the former Soviet Union, among others (Jablensky et al., 1992). Rates of schizophrenia and its general features were actually quite similar among the countries studied. However, some differences have been observed in the specific features of schizophrenia across cultures (Thakker & Ward, 1998).

Societal views of abnormal behavior vary across cultures. In our culture, models based on medical disease and psychological factors have achieved prominence in explaining abnormal behavior. But in traditional cultures, concepts of abnormal behavior often invoke supernatural causes, such as possession by demons or the devil (Lefley, 1990). In Filipino folk society, for example, psychological problems are often attributed to the influence of "spirits" or the possession of a "weak soul" (Edman & Johnson, 1999). The notion of supernatural causation, or **demonology,** also held prominence in Western society until the Age of Enlightenment.

possession A superstitious belief in which abnormal behavior is taken as a sign that the person is possessed by demons or the devil.

trephination A harsh, prehistoric practice of cutting a hole in a person's skull, possibly in an attempt to release demons.

demonological model The model that explains abnormal behavior in terms of supernatural forces.

THINK ABOUT IT

What behaviors have you observed in members of other cultural groups that might be considered abnormal in your own? What behaviors in your own cultural group might members of other groups consider abnormal?

Quiz 1.1
How Do We Define Abnormal Behavior?

Historical Perspectives on Abnormal Behavior

Throughout the history of Western culture, concepts of abnormal behavior have been shaped, to some degree, by the prevailing **worldview** of the time. Throughout much of history, beliefs in supernatural forces, demons, and evil spirits held sway. Abnormal behavior was often taken as a sign of **possession.** In more modern times, the predominant—but by no means universal—worldview has shifted toward beliefs in science and reason. Abnormal behavior has come to be viewed in our culture as the product of physical and psychosocial factors, not demonic possession.

The Demonological Model

Let us begin our journey with an example from prehistory. Archaeologists have unearthed human skeletons from the Stone Age with egg-sized cavities in the skull. One interpretation of these holes is that our prehistoric ancestors believed abnormal behavior reflected the invasion of evil spirits. Perhaps they used the harsh method—called **trephination**—of creating a pathway through the skull to provide an outlet for those irascible spirits. Fresh bone growth indicates that some people managed to survive the ordeal.

Threat of *trephining* may have persuaded people to comply with group or tribal norms to the best of their abilities. Because no written records or accounts of the purposes of trephination exist, other explanations are possible. Perhaps trephination was used as a primitive form of surgery to remove shattered pieces of bone or blood clots that resulted from head injuries (Maher & Maher, 1985).

Explanation of abnormal behavior in terms of supernatural or divine causes is termed the **demonological model.** The ancients explained natural forces in terms of divine will and spirits. The

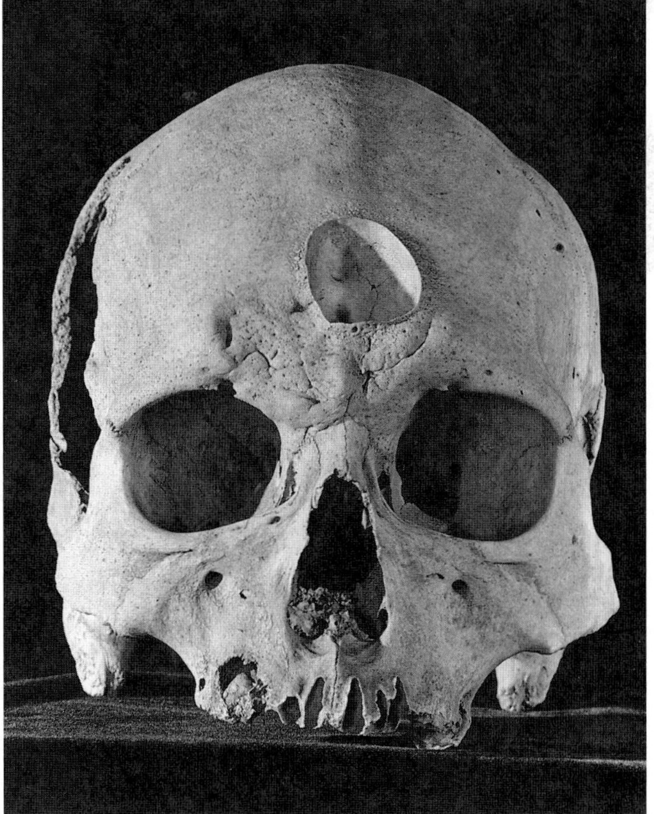

Trephination. Trephination refers to a practice of some prehistoric cultures by which a hole was chipped into a person's skull. Some investigators speculate that the practice represented an ancient form of surgery. Perhaps trephination was intended to release demons that were believed responsible for abnormal behavior.

humors According to the ancient Hippocratic belief system, the vital bodily fluids (phlegm, black bile, blood, yellow bile).

phlegmatic Slow and solid.

melancholia A state of severe depression.

sanguine Having a cheerful disposition.

choleric Having or showing bad temper.

exorcism A ritual intended to expel demons from a person believed to be possessed.

ancient Babylonians believed the movements of the stars and the planets were fashioned by the adventures and conflicts of the gods. The ancient Greeks believed their gods toyed with humans; when aroused to wrath, they could unleash forces of nature to wreak havoc on disrespectful or arrogant humans, even cloud their minds with madness.

In ancient Greece, people who behaved abnormally were often sent to temples dedicated to Aesculapius, the god of healing. Priests believed that Aesculapius would visit the afflicted persons while they slept in the temple and offer them restorative advice through dreams. Rest, a nutritious diet, and exercise were also believed to contribute to treatment. Incurables might be driven from the temple by stoning.

Origins of the Medical Model: In "Ill Humor"

Not all ancient Greeks believed in the demonological model. The seeds of naturalistic explanations of abnormal behavior were sown by Hippocrates and developed by other physicians in the ancient world, especially Galen.

Hippocrates (ca. 460–377 B.C.E.), the celebrated physician of the Golden Age of Greece, challenged the prevailing beliefs of his time by arguing that illnesses of the body and mind were the result of natural causes, not possession by supernatural spirits. He believed the health of the body and mind depended on the balance of **humors,** or vital fluids, in the body: phlegm, black bile, blood, and yellow bile. An imbalance of humors, he thought, accounted for abnormal behavior. A lethargic or sluggish person was believed to have an excess of phlegm, from which we derive the word **phlegmatic.** An overabundance of black bile was believed to cause depression, or **melancholia.** An excess of blood created a **sanguine** disposition: cheerful, confident, and optimistic. An excess of yellow bile made people "bilious" and **choleric**—quick-tempered, that is.

Truth OR Fiction? REVISITED

The modern medical model of abnormal behavior can be traced to the work of a Greek physician some 2,500 years ago.

TRUE. The foundations of what is known as the modern medical model can be traced to the Greek physician Hippocrates, who lived some 2,500 years ago.

Though we no longer subscribe to Hippocrates's theory of bodily humors, his theory is of historical importance because of its break from demonology. It also foreshadowed the development of the modern medical model, the view that abnormal behavior results from underlying biological processes. Hippocrates made many contributions to modern thought and, indeed, to modern medical practice. Hippocrates had even begun to classify abnormal behavior patterns, using three main categories that find some equivalents today: *melancholia* to characterize excessive depression, *mania* to refer to exceptional excitement, and *phrenitis* (from the Greek "inflammation of the brain") to characterize the bizarre kinds of behavior that might today typify schizophrenia. Medical schools continue to pay homage to Hippocrates by having new physicians swear the Hippocratic oath in his honor.

Galen (ca. 130–200 C.E.), a Greek physician who attended Roman emperor-philosopher Marcus Aurelius, adopted and expanded on the teachings of Hippocrates. Among Galen's contributions was the discovery that arteries carry blood, not air, as had been formerly believed.

Exorcism. This medieval woodcut illustrates the practice of exorcism, which was used to expel evil spirits who were believed to have possessed people.

Medieval Times

The Middle Ages, or medieval times, cover the millennium of European history from about 476 C.E. through 1450 C.E. After the passing of Galen, belief in supernatural causes, especially the doctrine of possession, increased in influence and eventually dominated medieval thought. The doctrine of possession held that abnormal behaviors were a sign of possession by evil spirits or the devil. This belief was embodied within the teachings of the Roman Catholic Church, which became the unifying force in western Europe following the decline of the Roman Empire. Although belief in possession antedated the Church and is found in ancient Egyptian and Greek writings, the Church revitalized it. The treatment of choice for abnormal behavior was **exorcism.** Exorcists were employed to persuade evil spirits that the bodies of their intended victims were basically uninhabitable. Methods included prayer, waving a cross at the victim, beating and flogging, even starving the victim. If the victim still displayed unseemly behavior, there were yet more powerful remedies, such as the rack, a device of torture. It seems clear that recipients of these "remedies" would be motivated to conform their behavior to social expectations as best they could.

The Renaissance—the great European revival in learning, art, and literature—began in Italy in the 1400s and spread gradually throughout Europe. The Renaissance is considered the transition from the medieval world to the modern. Therefore, it is ironic that fear of witches also reached its height during the Renaissance.

The water-float test. This so-called test was one way in which medieval authorities sought to detect possession and witchcraft. Managing to float above the water line was deemed a sign of impurity. In the lower right-hand corner, you can see the bound hands and feet of one poor unfortunate who failed to remain afloat, but whose drowning would have cleared away any suspicions of possession.

Witchcraft

The late 15th through the late 17th centuries were especially dangerous times to be unpopular with your neighbors. These were times of massive persecutions of people, particularly women, who were accused of witchcraft. Officials of the Roman Catholic Church believed witches made pacts with the devil, practiced satanic rituals, and committed heinous acts, such as eating babies and poisoning crops. In 1484, Pope Innocent VIII decreed that witches be executed. Two Dominican priests compiled a manual for witch-hunting, called the *Malleus Maleficarum* (The Witches' Hammer), to help inquisitors identify suspected witches. Over 100,000 accused witches were killed in the next two centuries.

There were also creative "diagnostic" tests for detecting possession and witchcraft. In the case of the water-float test, innocent people were drowned in medieval times as a way of certifying they were not possessed by the devil. The water-float test was based on the principle that pure metals settle to the bottom during smelting, whereas impurities bob up to the surface. Suspects who sank and were drowned were ruled pure. Suspects who were able to keep their heads above water were regarded as being in league with the devil. Then they were in real trouble. This trial is the source of the phrase, "Damned if you do and damned if you don't."

Modern scholars once believed the so-called witches of the Middle Ages and the Renaissance were actually people who were mentally disturbed. They were believed to be persecuted because their abnormal behavior was taken as evidence they were in league with the devil. It is true that many suspected witches confessed to impossible behaviors, such as flying or engaging in sexual intercourse with the devil. At face value such confessions might suggest disturbances in thinking and perception that are consistent with a modern diagnosis of a psychological disorder, such as schizophrenia. Most of these confessions can be discounted, however, because they were extracted under torture by inquisitors who were bent on finding evidence to support accusations of witchcraft (Spanos, 1978). In other cases, the threat of torture and other forms of intimidation were sufficient to extract false confessions. Although some of those who were persecuted as witches probably did show abnormal behavior patterns, most did not (Schoenman, 1984). Rather, accusations of witchcraft appeared to be a convenient means of disposing of social nuisances and political rivals, of seizing property, and of suppressing heresy (Spanos, 1978). In English villages, many of the accused were poor, unmarried elderly women who were forced to beg their neighbors for food. If misfortune befell people who declined to help, the beggar might be accused of causing the misery by having cast a curse on the uncharitable family. If the woman was generally unpopular, accusations of witchcraft were more likely to be followed up.

Although demons were believed to play roles both in abnormal behavior and witchcraft, there was a difference between the two. Victims of possession may have been perceived to be afflicted as retribution for wrongdoing, but some people who showed abnormal behavior were considered to be innocent victims of demonic possession. Witches, however, were believed to have voluntarily entered into a pact with the devil and renounced God. Witches were generally seen as more deserving of torture and execution (Spanos, 1978).

Historical trends do not follow straight lines. Although the demonological model held sway during the Middle Ages and much of the Renaissance, it did not universally

Truth OR Fiction? REVISITED

Innocent people were drowned in medieval times as a way of certifying they were not possessed by the devil.

TRUE. Drowning was thought to be evidence that a person was not possessed by the devil.

"Bedlam." The bizarre antics of the patients at St. Mary's of Bethelem Hospital in London in the 18th century were a source of entertainment for the well-heeled gentry of the town, such as the two well-dressed women in the middle of the painting.

supplant belief in naturalistic causes (Schoenman, 1984). In medieval England, for example, demonic possession was only rarely invoked as the cause of abnormal behavior in cases in which a person was held to be insane by legal authorities (Neugebauer, 1979). Most explanations for unusual behavior involved natural causes, such as physical illness or trauma to the brain. In England, in fact, some disturbed people were kept in hospitals until they were restored to sanity (Allderidge, 1979). The Renaissance Belgian physician Johann Weyer (1515–1588) also took up the cause of Hippocrates and Galen by arguing that abnormal behavior and thought patterns were caused by physical problems.

Asylums

By the late 15th and early 16th centuries, asylums, or madhouses, began to crop up throughout Europe. Many were former leprosariums, which were no longer needed because of a decline in leprosy that occurred during the late Middle Ages. Asylums often gave refuge to beggars as well as the disturbed, and conditions were generally appalling. Residents were often chained to their beds and left to lie in their own waste or wander about unassisted. Some asylums became public spectacles. In one asylum in London, St. Mary's of Bethlehem Hospital—from which the word *bedlam* is derived—the public could buy tickets to observe the bizarre antics of the inmates, much as we would view a sideshow in a circus or animals at a zoo.

The Reform Movement and Moral Therapy

Truth OR Fiction? REVISITED

A night's entertainment in London a few hundred years ago might have included peering at the inmates at the local asylum.

TRUE. A night on the town for the gentry of London might have included a visit to a local asylum, St. Mary's of Bethlehem Hospital, a name from which we get the word *bedlam*.

The modern era of treatment can be traced to the efforts of individuals such as the Frenchmen Jean-Baptiste Pussin and Philippe Pinel in the late 18th and early 19th centuries. They argued that people who behave abnormally suffer from diseases and should be treated humanely. This view was not at all popular at the time. Deranged people were generally regarded as threats to society, not as sick people in need of treatment.

From 1784 to 1802, Pussin, a layman, was placed in charge of a ward for people considered "incurably insane" at La Bicêtre, a large mental hospital in Paris. Although Pinel is often credited with freeing the inmates of La Bicêtre from their chains, Pussin was actually the first official to unchain a group of the "incurably insane." These unfortunates had been considered too dangerous and unpredictable to be left unchained. But Pussin believed that if they were treated with kindness, there would be no need for chains. As he predicted, most of the shut-ins became manageable and calm when their chains were removed. They could walk the hospital grounds and take in fresh air. Pussin also forbade the staff from treating the residents harshly, and he discharged any employees who disregarded his directives.

Pinel (1745–1826) became medical director for the incurables' ward at La Bicêtre in 1793 and continued the humane treatment Pussin had begun. He stopped harsh practices, such as bleeding and purging, and moved patients from darkened dungeons to well-ventilated, sunny rooms. Pinel also spent hours talking to inmates, in the belief that showing understanding and concern would help restore them to normal functioning.

The philosophy of treatment that emerged from these efforts was labeled **moral therapy.** It was based on the belief that providing humane treatment in a relaxed and decent environment could restore functioning. Similar reforms were instituted at about this time in England by William Tuke and later in the United States by Dorothea Dix. Another influential figure was the American physician Benjamin Rush (1745–1813)—also a signatory to the Declaration of Independence and an early leader of the antislavery movement (Farr, 1994). Rush, considered the father of American psychiatry, penned the first American textbook on psychiatry in 1812: *Medical Inquiries and Observations Upon the Diseases of the Mind.* He believed that madness is caused by engorgement of the blood vessels of the brain. To relieve pressure, he recommended bloodletting and other

moral therapy A 19th-century treatment approach that emphasized treating hospitalized patients with care and understanding.

harsh treatments such as purging and ice-cold baths. But he did advance humane treatment by encouraging the staff of his Philadelphia Hospital to treat patients with kindness, respect, and understanding. He also favored the therapeutic use of occupational therapy, music, and travel (Farr, 1994). His hospital became the first in the United States to admit patients for psychological disorders.

Dorothea Dix (1802–1887), a Boston schoolteacher, traveled about the country decrying the deplorable conditions in the jails and almshouses where deranged people were often placed. As a direct result of her efforts, 32 mental hospitals were established throughout the United States.

A Step Backward

In the latter half of the 19th century, however, the belief that abnormal behaviors could be successfully treated or cured by moral therapy fell into disfavor (USDHHS, 1999a). A period of apathy ensued in which patterns of abnormal behavior were deemed incurable (Grob, 1994). Mental institutions in the United States grew in size and came to provide little more than custodial care. Conditions deteriorated. Mental hospitals became frightening places. It was not uncommon to find residents "wallowing in their own excrements," in the words of a New York State official of the time (Grob, 1983). Straitjackets, handcuffs, cribs, straps, and other devices were used to restrain excitable or violent patients.

Deplorable hospital conditions remained commonplace through the middle of the 20th century. By the mid-1950s, the population in mental hospitals had risen to half a million patients. Although some good state hospitals provided decent and humane care, many were described as little more than human *snakepits*. Residents were crowded into wards that lacked even rudimentary sanitation. Mental patients in back wards were essentially *warehoused*. That is, they were left to live out their lives with little hope or expectation of recovery or return to the community. Many received little professional care and were abused by poorly trained and supervised staffs. By the mid-20th century, the appalling conditions that many mental patients were forced to endure led to increasing calls for reforms of the mental health system.

The unchaining of inmates at La Bicêtre by 18th-century French reformer Philippe Pinel. Continuing the work of Jean-Baptiste Pussin, Pinel stopped harsh practices, such as bleeding and purging, and moved inmates from darkened dungeons to sunny, airy rooms. Pinel also took the time to converse with inmates, in the belief that understanding and concern would help restore them to normal functioning.

deinstitutionalization The policy of discharging hospitalized mental patients into the community and of reducing the need for new admissions through alternative treatment approaches.

phenothiazines A group of antipsychotic drugs ("major tranquilizers") used to treat schizophrenia.

The Community Mental Health Movement: The Exodus from State Hospitals

In response to the growing call for reform of the mental health system, Congress in 1963 established a nationwide system of community mental health centers (CMHCs) that was intended to offer an alternative to long-term custodial care in bleak institutions. CMHCs were charged with providing continuing support and care to former hospital residents who were released from state mental hospitals under a policy of **deinstitutionalization.** Another factor that spurred the exodus from mental hospitals was the advent of a new class of drugs—the *phenothiazines*. The **phenothiazines,** a group of antipsychotic drugs that helped quell the most flagrant behavior patterns associated with schizophrenia, were introduced in the 1950s. They reduced the need for indefinite hospital stays and permitted many people with schizophrenia to be discharged to less restrictive living arrangements in the community, such as halfway houses, group homes, and independent living. The mental hospital population across the United States plummeted from more than 550,000 in 1955 to fewer than 130,000 by the late 1980s (D. Braddock, 1992; Kiesler & Sibulkin, 1987). Some mental hospitals were closed entirely (Salokangas & Saarinen, 1998).

The community mental health movement was predicated on the belief—the hope perhaps—that mental patients could return to their communities and assume more independent and fulfilling lives. Critics contend that the exodus from state hospitals abandoned tens of thousands of marginally functioning people to communities that lacked adequate housing and other forms of support. Many homeless people we see wandering city streets and sleeping in bus terminals and train stations are discharged mental patients. In Chapter 4, we take a closer look at the policy of deinstitutionalization and the problems faced by the psychiatric homeless population.

THINK ABOUT IT

How have beliefs about abnormal behavior changed over time? What changes have occurred in how society treats people considered mentally disturbed?

Contemporary Perspectives on Abnormal Behavior: From Demonology to Science

Beliefs in possession or demonology persisted until the rise of the natural sciences in the late 17th and 18th centuries. Society at large began to turn toward reason and science as ways of explaining natural phenomena and human behavior. The nascent sciences of biology, chemistry, physics, and astronomy offered promise that knowledge could be derived from scientific methods of observation and experimentation. The 18th and 19th centuries witnessed rapid developments in medical science. Scientific discoveries uncovered the microbial causes of some kinds of diseases and gave rise to preventive measures. Models of abnormal behavior also began to emerge, including models representing biological, psychological, sociocultural, and biopsychosocial perspectives. We will briefly discuss each of these models here, particularly in terms of their historical background, which will lead to a fuller discussion in Chapter 2.

The Biological Perspective Against the backdrop of advances in medical science, the German physician Wilhelm Griesinger (1817–1868) argued that abnormal behavior was rooted in diseases of the brain. Griesinger's views influenced another German physician, Emil Kraepelin (1856–1926), who wrote an influential textbook on psychiatry in 1883 in which he likened mental disorders to physical diseases. Griesinger and Kraepelin paved the way for the development of the modern medical model, which attempts to explain abnormal behavior on the basis of underlying biological defects or abnormalities, not evil spirits. According to the medical model, people behaving abnormally suffer from mental illnesses or disorders that can be classified, like physical illnesses, according to their distinctive causes and symptoms. Not all adopters of the medical model believe every pattern of abnormal behavior is a product of defective biology, but they maintain that patterns of abnormal behavior can be likened to physical illnesses in that their features can be conceptualized as symptoms of underlying disorders, whatever their cause.

Kraepelin specified two main groups of mental disorders or diseases: **dementia praecox** (from roots meaning "precocious [premature] insanity"), which we now call schizophrenia, and manic-depressive psychosis, which is now labeled *bipolar disorder*. Kraepelin believed that dementia praecox was caused by a biochemical imbalance and manic-depressive psychosis by an abnormality in body metabolism. But his major contribution was the development of a classification system that forms the cornerstone for current diagnostic systems.

Much of the terminology in current use reflects the influence of the medical model. Because of the medical model, many professionals and laypeople speak of people whose behavior is deemed abnormal as being mentally *ill*. It is because of the medical model that so many speak of the *symptoms* of abnormal behavior, rather than the features or characteristics of abnormal behavior. Other terms spawned by the medical model include *mental health, syndrome, diagnosis, patient, mental patient, mental hospital, prognosis, treatment, therapy, cure, relapse*, and *remission*.[2]

The medical model is a major advance over demonology. It inspired the idea that abnormal behavior should be treated by learned professionals rather than punished. Compassion supplanted hatred, fear, and persecution.

dementia praecox The term given by Kraepelin to the disorder now called schizophrenia.

hypnosis A trancelike state, induced by suggestion, in which the individual responds to the commands of the hypnotist.

psychodynamic model The theoretical model of Freud and his followers, in which abnormal behavior is viewed as the product of clashing forces within the personality.

The Psychological Perspective Although the medical model was gaining influence in the 19th century, there were those who believed organic factors alone could not explain the many forms of abnormal behavior. In Paris, a highly respected neurologist, Jean-Martin Charcot (1825–1893), experimented with the use of **hypnosis** in treating *hysteria*, a condition in which people present with physical symptoms like paralysis or numbness that could not be explained by any underlying physical cause. The thinking at the time was that they must have an affliction of the nervous system, which caused their symptoms. Yet Charcot and his associates demonstrated that these symptoms could be removed in hysterical patients or actually induced in normal patients by means of hypnotic suggestions.

Among those who attended Charcot's demonstrations was a young Austrian physician named Sigmund Freud (1856–1939). Freud reasoned that if hysterical symptoms could be made to disappear or appear through hypnosis—the mere "suggestion of ideas"—they must be psychological in origin (E. Jones, 1953). He concluded that whatever psychological factors give rise to hysteria, they must lie outside the range of conscious awareness. This was the kernel of the idea that underlies the first psychological perspective on abnormal behavior—the **psychodynamic model.** Freud believed that the causes of abnormal behavior lie in the interplay of forces within the unconscious mind. "I received the proudest impression," Freud wrote of his experience with Charcot, "of the possibility that there could be powerful mental processes which nevertheless remained hidden from the consciousness of men" (as cited in Sulloway, 1983, p. 32).

Freud was also influenced by the Viennese physician, Joseph Breuer (1842–1925), 14 years his senior. Breuer too had used hypnosis to treat a 21-year-old woman, Anna O., with hysterical complaints for which there was no apparent medical basis, such as paralysis in her limbs, numbness, and disturbances of vision and hearing (E. Jones, 1953). A

Charcot's teaching clinic. Parisian neurologist Jean-Martin Charcot presents a woman patient who exhibits the highly dramatic behavior associated with hysteria, such as falling faint at a moment's notice. Charcot was an important influence on the young Sigmund Freud.

[2]Because the medical model is not the only way of viewing abnormal behavior patterns, we adopt a more neutral language in this text in describing abnormal behavior patterns. For example, we often refer to "features" or "characteristics" of abnormal behavior patterns or psychological disorders rather than to "symptoms." But our adoption of nonmedical jargon is not an absolute rule. In some cases, there may be no handy substitutes for terms that derive from the medical model, such as the term *remission* or the reference to patients in mental hospitals as "mental patients." In other cases we may use terms such as *disorder, therapy*, and *treatment* because they are commonly used by psychologists who "treat" "mental disorders" with psychological "therapies."

catharsis The purging or free expression of feelings.

"paralyzed" muscle in her neck prevented her from turning her head. Immobilization of the fingers of her left hand made it all but impossible for her to feed herself. Breuer believed there was a strong psychological component to the symptoms. He treated her by encouraging her to talk about them, sometimes under hypnosis. Recalling and talking about events connected with the appearance of the symptoms—especially events that evoked feelings of fear, anxiety, or guilt—appeared to provide symptom relief, at least for a time. Anna referred to the treatment as the "talking cure" or, when joking, as "chimney sweeping."

The hysterical symptoms were taken to represent the transformation of these blocked-up emotions, forgotten but not lost, into physical complaints. In Anna's case, the symptoms seemed to disappear once the emotions were brought to the surface and "discharged." Breuer labeled the therapeutic effect **catharsis,** a Greek term meaning purgation or purification of feelings. Cases of hysteria, such as that of Anna O., seemed to have been a common occurrence in the later Victorian period, but are relatively rare today (Spitzer et al., 1989).

THINK ABOUT IT

What role did hypnosis play in the development of a psychological model of abnormal behavior?

Freud's theoretical model was the first major psychological model of abnormal behavior. As we'll see in Chapter 2, other psychological perspectives on abnormal behavior soon followed based on behavioral, humanistic, and cognitive models. We'll also see that each of these perspectives, well as the contemporary medical model, spawned particular forms of therapy to treat psychological disorders.

The Sociocultural Perspective Sociocultural theorists believe that we must consider the broader social contexts in which behavior occurs to understand the roots of abnormal behavior. They believe the causes of abnormal behavior may be found in the failures of society rather than in the person. Psychological problems may be rooted in the social ills of society, such as poverty, social decay, racial and gender discrimination, and lack of economic opportunity.

According to the more radical sociocultural theorists, such as the psychiatrist Thomas Szasz, mental illness is a myth—a label used to stigmatize and subjugate people whose behavior is socially deviant (T. S. Szasz, 1961, 2000). Szasz states that so-called mental illnesses are really "problems in living," not actual diseases like influenza, AIDS, and cancer. Szasz argues that people who offend others or engage in socially deviant behavior are perceived as threats by the establishment. Labeling them as sick allows others to deny the validity of their problems and to put them away in institutions.

Sigmund Freud and Bertha Pappenheim. Freud is shown here at around age 30, a few years after he treated Bertha Pappenheim. Pappenheim (1859–1936) is known more widely in the psychological literature as "Anna O." Freud believed that her hysterical symptoms represented the transformation of blocked-up emotions into physical complaints.

Sociocultural theorists maintain that once the label of "mental illness" is applied to a person, it is difficult to remove. The label also affects other people's responses to the "patient." Mental patients are stigmatized and socially degraded. Job opportunities may be denied, friendships may dissolve, and the "patient" may become increasingly alienated from society. Szasz argues that treating people as mentally ill strips them of their dignity because it denies them responsibility for their own behavior and choices. He claims that troubled people should be encouraged to take more responsibility for managing their lives and solving their problems.

Although not all sociocultural theorists subscribe to Szasz's radical views, all do emphasize the importance of taking sociocultural factors into account in understanding people whose behavior leads them to be perceived as mentally ill or abnormal. Sociocultural factors may include those related to gender, race, ethnicity, lifestyle, or social ills such as poverty and discrimination.

The Biopsychosocial Perspective Many leading scholars today believe that patterns of abnormal behavior are too complex to be understood from any one model or perspective. They endorse the view that abnormal behavior is best understood by taking into account the interaction of multiple causes representing the biological, psychological, and sociocultural domains. The biopsychosocial perspective, or *interactionist model*, informs the approach we take in this text toward understanding the origins of abnormal behavior. We believe we need to consider the interplay of biological, psychological, and sociocultural factors in the development of psychological disorders. Though we recognize that our understanding of these causal factors may be incomplete, we encourage the reader to consider possible causal pathways that take into account the influences of multiple factors and their interactions.

Perspectives on psychological disorders provide a framework not only for explanation but also for treatment (see Chapter 4). They also lead to formulations of predictions, or *hypotheses*, that guide research. The medical model, for example, fosters inquiry into genetic and biochemical treatment methods. In the next section, we consider the ways in which psychologists and other mental health professionals study abnormal behavior.

Research Methods in Abnormal Psychology

Abnormal psychology is a branch of the scientific discipline of psychology, which means research in the field is based on the application of the **scientific method.** Here we examine how researchers apply the scientific method in investigating abnormal behavior.

Let us begin by asking you to imagine you are a brand-new graduate student in psychology and are sitting in your research methods course on the first day of the term. The professor, a distinguished woman of about 50, enters the class. She is carrying a small wire-mesh cage that holds a white rat. She smiles and sets the cage on her desk.

The professor removes the rat from the cage and places it on the desk. She asks the class to observe its behavior. As a serious student, you attend closely. The animal moves to the edge of the desk, pauses, peers over the edge, and seems to jiggle its whiskers at the floor below. It maneuvers along the edge of the desk, tracking the perimeter. Now and then the rat pauses and vibrates its whiskers downward in the direction of the floor.

The professor picks up the rat and returns it to the cage. She asks the class to describe the animal's *behavior.*

A student responds, "The rat seems to be looking for a way to escape."

Another student says, "It is reconnoitering its environment, examining it." "Reconnoitering?" you think. That student has seen too many war movies.

The professor writes each response on the blackboard. Another student raises her hand. "The rat is making a visual search of the environment," she says. "Maybe it's looking for food."

The professor prompts other students for their descriptions.

"It's looking around," says one.

scientific method A method of conducting scientific research in which theories or assumptions are examined in the light of evidence.

"Trying to escape," says another.

Your turn arrives. Trying to be scientific, you say, "We can't say what its motivation might be. All we know is that it's scanning its environment."

"How so?" the professor asks.

"Visually," you reply, confidently.

The professor writes the response and then turns to the class, shaking her head. "Each of you observed the rat," she said, "but none of you described its *behavior*. Each of you made certain *inferences*, that the rat was 'looking for a way down' or 'scanning its environment' or 'looking for food,' and the like. These are not unreasonable inferences, but they are inferences, not descriptions. They also happen to be wrong. You see, the rat is blind. It's been blind since birth. It couldn't possibly be looking around, at least not in a visual sense."

Description, Explanation, Prediction, and Control: The Objectives of Science

Description is one of the primary objectives of science. To understand abnormal behavior, we must first learn to describe it. Description allows us to recognize abnormal behavior and provides the basis for explaining it.

Descriptions should be clear, unbiased, and based on careful observation. Our anecdote about the blind rat illustrates the point that our observations and our attempts to describe them can be influenced by our expectations, or biased. Our expectations reflect our models of behavior, and they may incline us to perceive events—such as the rat's movements and other people's behavior—in certain ways. Describing the rat in the classroom as "scanning" and "looking" for something is an **inference,** or conclusion, we draw from our observations based on our model of how animals explore their environments. In contrast, description would involve a precise accounting of the animal's movements around the desk, measuring how far in each direction it moves, how long it pauses, how it bobs its head from side to side, and so on.

Inference is also important in science, however. Inference allows us to jump from the particular to the general—to suggest laws and principles of behavior that can be woven into a model or **theory** of behavior. In Chapter 2 we consider the major theoretical perspectives or models of abnormal behavior. Here let us note that without a way of organizing our descriptions of phenomena in terms of models and theories, we would be left with a buzzing confusion of unconnected observations.

Theories help scientists explain puzzling behavior and predict future behavior. Prediction entails the discovery of factors that anticipate the occurrence of events. Geology, for example, seeks clues in the forces affecting the earth that can forecast natural events such as earthquakes and volcanic eruptions. Scientists who study abnormal behavior seek clues in overt behavior, biological processes, family interactions, and so forth, to predict the development of abnormal behaviors as well as factors that might predict response to various treatments. It is not sufficient for theoretical models to help us explain or make sense of events or behaviors that have already occurred. Useful theories must allow us to predict the occurrence of particular behaviors.

The idea of controlling human behavior—especially the behavior of people with serious problems—is controversial. The history of societal response to abnormal behaviors, including abuses such as exorcism and cruel forms of physical restraint, render the idea particularly distressing. Within science, however, the word *control* does not imply that people are coerced into doing the bidding of others, like puppets dangling on strings. Psychologists, for example, are committed to the dignity of the individual, and the concept of human dignity requires that people be free to make decisions and exercise choices. Within this context, *controlling behavior* means using scientific knowledge to help people shape their own goals and more efficiently use their resources to accomplish them. Today, in the United States, even when helping professionals restrain people who are violently disturbed, their goal is to assist them to overcome their agitation and regain the ability to exercise

description The representation of observations without making interpretations or drawing inferences.

inference A conclusion that is drawn from data.

theory A formulation of the relationships underlying observed events.

meaningful choices in their lives.[3] Ethical standards prohibit the use of injurious techniques in research or practice.

Psychologists and other scientists use the *scientific method* to advance the description, explanation, prediction, and control of abnormal behavior.

hypothesis An assumption that is tested through experimentation.

significant In statistics, a magnitude of difference that is taken as indicating meaningful differences between groups.

The Scientific Method

The scientific method involves systematic attempts to test our assumptions and theories about the world through gathering objective evidence. Various means are used in applying the scientific method, including observational and experimental methods. Here let us focus on the basic steps involved in using the scientific method in experimentation:

1. *Formulating a research question.* Scientists derive research questions from their observations and theories of events and behavior. For instance, based on their clinical observations and understandings of the underlying mechanisms in depression, they may formulate questions about whether certain experimental drugs or particular types of psychotherapy can help people overcome depression.
2. *Framing the research question in the form of a hypothesis.* A **hypothesis** is a precise prediction about behavior that is examined through research. For example, scientists might hypothesize that people who are clinically depressed will show greater improvement on measures of depression if they are given an experimental drug than if they receive an inert placebo ("sugar pill").
3. *Testing the hypothesis.* Scientists test hypotheses through carefully controlled observation and experimentation. They might test the hypothesis about the experimental drug by setting up an experiment in which one group of people with depression is given the experimental drug and another group is given the placebo. They would then administer tests to see if the people who received the active drug showed greater improvement over a period of time than those who received the placebo.
4. *Drawing conclusions about the hypothesis.* In the final step, scientists draw conclusions from their findings about the correctness of their hypotheses. Psychologists use statistical methods to determine the likelihood that differences between groups are **significant** as opposed to chance fluctuations. Psychologists are reasonably confident that group differences are significant—that is, not due to chance—when the probability that chance alone can explain the difference is less than 5%. When well-designed research findings fail to bear out hypotheses, scientists can modify the theories from which the hypotheses are derived. Research findings often lead to modifications in theory, new hypotheses, and, in turn, subsequent research.

Let us consider the major research methods used by psychologists and others in studying abnormal behavior: the naturalistic-observation, correlational, experimental, epidemiological, kinship, and case-study methods. Before we do so, however, let us consider some of the principles that guide ethical conduct in research.

Ethics in Research

Ethical principles are designed to promote the dignity of the individual, protect human welfare, and preserve scientific integrity (McGovern, 1991). Psychologists are prohibited by the ethical standards of their profession from using methods that cause psychological or physical harm to subjects or clients (APA, 1992). Psychologists also must follow ethical guidelines that protect animal subjects in research.

Institutions such as universities and hospitals have review committees, called *institutional review boards* (IRBs), that review proposed research studies in the light of ethical guidelines. Investigators must receive IRB approval before they are permitted to begin

[3]Here we are talking about violently confused and disordered behavior, not criminal behavior. Criminals and disturbed people may both be dangerous to others, but with criminals the intention of restraint is usually limited to protecting society.

their studies. Two of the major principles on which ethical guidelines are based are (1) *informed consent* and (2) *confidentiality*.

The principle of **informed consent** requires that people be free to choose whether they wish to participate in research studies. They must be given sufficient information in advance about the study's purposes and methods, and its risks and benefits, to allow them to make an informed decision about their participation. Subjects must also be free to withdraw from a study at any time without penalty. In some cases, researchers may withhold certain information until all the data are collected. For instance, subjects in placebo-control studies of experimental drugs are told that they may receive an inert placebo rather than the active drug. After the study is concluded, participants who received the placebo would be given the option of receiving the active treatment. In studies in which information was withheld or deception was used, subjects must be **debriefed** afterward. That is, they must receive an explanation of the true methods and purposes of the study and why it was necessary to keep them in the dark.

Web Link 1.5
Ethical Issues in Research Involving Human Participants WWW

Subjects also have a right to expect that their identities will not be revealed. Investigators are required to protect their **confidentiality** by keeping the records of their participation secure and by not disclosing their identities to others.

We turn now to discussion of the research methods used to investigate abnormal behavior.

The Naturalistic-Observation Method

The **naturalistic-observation method** is used to observe behavior in the field, where it happens. Anthropologists have lived in preliterate societies in order to study human diversity. Sociologists have followed the activities of adolescent gangs in inner cities. Psychologists have spent weeks observing the behavior of homeless people in train stations and bus terminals. They have even observed the eating habits of slender and overweight people in fast-food restaurants, searching for clues to obesity.

Scientists take every precaution to ensure their naturalistic observations are **unobtrusive,** so as to prevent any interference with the behavior they observe. Otherwise, the presence of the observer may distort the behavior that is observed.

Naturalistic observation provides a good deal of information on how subjects behave, but it does not necessarily reveal why they do so. Men who frequent bars and drink, for example, are more likely to get into fights than men who do not. But such observations do not show that alcohol *causes* aggression. As we see in the following pages, questions of cause and effect are best approached by means of controlled experiments.

The Correlational Method

A **correlation** is a statistical measure of the relationships between two factors, or **variables.** In the naturalistic-observation study that occurred in the fast-food restaurant, eating behaviors were related—or correlated—to patrons' weights. They were not directly manipulated. In other words, the investigators did not manipulate the weights or eating rates of their subjects, but merely measured the two variables in some fashion and examined whether they were statistically related to each other. When one variable (weight level) increases as the second variable (rate of eating) increases, there is a **positive correlation** between them. If one variable decreases as the other increases, there is a **negative correlation** between the variables.

The correlational method tests the statistical relationship between variables. However, it does not prove that correlated variables are causally related to each other. Sometimes there is no causal connection between variables that are merely correlated. For example, children's foot size is correlated with their vocabulary development, but changes in foot size certainly do not determine growth of vocabulary. Depression and negative thoughts are also correlated, as we shall see in Chapter 8. Though depression may be caused by negative thinking, it is possible that the direction of causality works

informed consent The principle that subjects should receive enough information about an experiment beforehand to decide freely whether to participate.

debriefed To be fully informed about an experiment after it takes place.

confidentiality Protection of the identity of participants by keeping records secure and not disclosing their identities.

naturalistic-observation method A research method in which subjects' behavior is observed and measured in their natural environments.

unobtrusive Not interfering or conspicuous.

correlation A relationship or association between variables.

variables Conditions that are measured (dependent variables) or manipulated (independent variables) in experiments.

positive correlation A statistical relationship between two variables such that increases in one are associated with increases in the other.

negative correlation A statistical relationship between two variables such that increases in one are associated with decreases in the other.

the other way—that depression gives rise to negative thinking. Or perhaps the direction of causality works both ways, with negative thinking contributing to depression and depression in turn influencing negative thinking. Then again, depression and negative thinking may both reflect a common causative factor, such as stress, and not be causally related to each other at all.

Although correlational research does not reveal cause and effect, it can be used to serve the scientific objective of prediction. When two variables are correlated, we can use one to predict the other. Knowledge of correlations among alcoholism, family history, and attitudes toward drinking helps us predict which adolescents are at great risk of developing problems with alcohol, although causal connections are complex and somewhat nebulous. But knowing which factors predict future problems may help us direct preventive efforts toward these high-risk groups to help prevent these problems from developing.

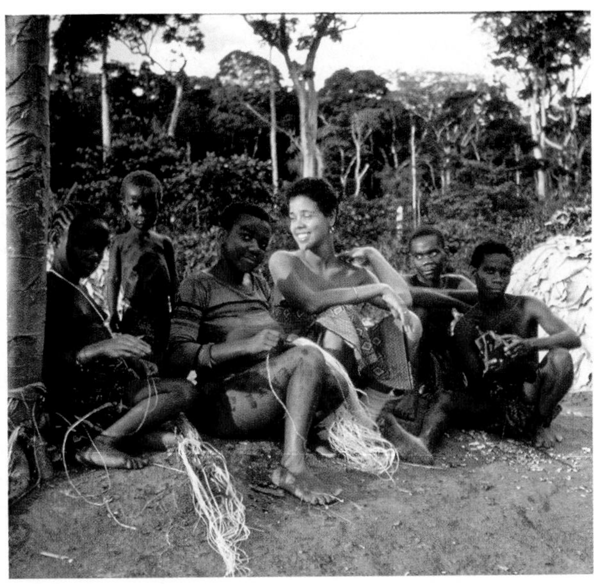

Naturalistic observation. Anthropologists learn about other cultures by observing how members of these other societies live from day to day, in some cases actually living for a time in the societies they study. Here an American anthropologist is shown sitting among members of an African pygmy tribe.

The Longitudinal Study One type of correlational study is the **longitudinal study,** in which subjects are studied at periodic intervals over lengthy periods, perhaps for decades. By studying people over time, researchers can investigate the events associated with the onset of abnormal behavior and, perhaps, learn to identify factors that predict the development of such behavior. However, this type of research is time consuming and costly. It requires a commitment that may literally outlive the original investigators. Therefore, long-term longitudinal studies are relatively uncommon. In Chapter 13 we examine one of the best known longitudinal studies, the Danish high-risk study that has tracked, since 1962, a group of children whose mothers had schizophrenia and who were therefore themselves at increased risk of developing the disorder (Mednick, Parnas, & Schulsinger, 1987; Parnas et al., 1993).

Truth OR Fiction? REVISITED

Finding that two variables are closely linked together means that one is a cause of the other.

FALSE. Two variables may be linked or correlated without being causally related to one another.

The Experimental Method

Prediction is based on the *correlation* between events or factors that are separated in time. As in other forms of correlational research, we must be careful not to infer *causation* from *correlation.* A *causal relationship* between two events involves a time-ordered relationship in which the second event is the direct result of the first. We need to meet two strict conditions to posit a causal relationship between two factors:

1. The effect must follow the cause in a time-ordered sequence of events.
2. Other plausible causes of the observed effects (rival hypotheses) must be eliminated.

The **experimental method** allows scientists to demonstrate causal relationships by first manipulating the causal factor and then measuring its effects under controlled conditions that minimize the risk of other factors explaining these effects.

The term *experiment* can cause some confusion. Broadly speaking, an experiment is a trial or test of a hypothesis. From this vantage point, any method that actually seeks to test a hypothesis could be considered experimental—including naturalistic observation and correlational studies. But investigators usually limit the use of the term *experimental method* to refer to studies in which researchers seek to uncover cause-and-effect relationships by manipulating possible causal factors directly.

The factors or variables hypothesized to play a causal role are manipulated or controlled by the investigator in experimental research. These are called the **independent variables.** The observed effects are labeled **dependent variables,** because changes in them are believed to depend on the independent or manipulated variable. Dependent variables are observed and measured, not manipulated, by the experimenter. Examples of independent and dependent variables of interest to investigators of abnormal behavior are shown in Table 1.1.

longitudinal study A research study in which subjects are followed over time.

experimental method A scientific method that aims to discover cause-and-effect relationships by manipulating independent variables and observing the effects on the dependent variables.

independent variables Factors that are manipulated in experiments.

dependent variables Outcomes of an experiment believed to be dependent on the effects of an independent variable.

TABLE 1.1 Examples of Independent and Dependent Variables in Experimental Research

Independent Variables	Dependent Variables
Type of treatment: for example, different types of drug treatments or psychological treatments	Behavioral variables: for example, measures of adjustment, activity levels, eating behavior, smoking behavior
Treatment factors: for example, brief vs. long-term treatment, inpatient vs. outpatient treatment	Physiological variables: for example, measures of physiological responses such as heart rate, blood pressure, and brain wave activity
Experimental manipulations: for example, types of beverage consumed (alcoholic vs. nonalcoholic)	Self-report variables: for example, measures of anxiety, mood, or marital or life satisfaction

THINK ABOUT IT

Why should we not assume that because two variables are correlated they are causally linked?

In an experiment, subjects are exposed to an *independent variable*, for example, the type of beverage (alcoholic vs. nonalcoholic) they consume in a laboratory setting. They are then observed or examined to determine whether the independent variable makes a difference in their behavior, or, more precisely, whether the independent variable affects the dependent variable—for example, whether they behave more aggressively if they consume alcohol.

Experimental and Control Subjects Well-controlled experiments assign subjects to experimental and control groups at random. **Experimental subjects** are given the experimental treatment; **control subjects,** are not. Care is taken to hold other conditions constant for each group. By using random assignment and holding other conditions constant, experimenters can be reasonably confident that the experimental treatment, and not uncontrolled factors, such as room temperature or differences between the types of subjects in the experimental and control groups, brought about the differences in outcome between the experimental and control groups (Ioannidis & Karassa, 2001).

Why should experimenters assign subjects to experimental and control groups at random? Consider a study intended to investigate the effects of alcohol on behavior. Let's suppose we allowed subjects themselves to decide whether or not they wanted to be in an experimental group that drank alcohol or a control group that drank a nonalcoholic beverage. If this were the case, then differences between the groups might be to due to an underlying selection factor rather than the experimental manipulation. For example, subjects who *chose* the alcoholic beverage might differ in their personalities from those who chose the control beverage. They might be more willing to explore or to take risks, for example. Therefore, we would not know whether the independent variable (type of beverage) or a **selection factor** (difference in the kinds of subjects making up the groups) was ultimately responsible for observed differences in behavior. Random assignment controls for selection factors by ensuring that subject characteristics are randomly distributed across groups. Thus it is reasonable to assume that differences between groups result from the treatments they receive rather than from differences between the subjects making up the groups. Still, it is possible that apparent treatment effects stem from subjects' expectancies about the treatments they receive rather than from the active components in the treatments themselves. In other words, knowing you are being given an alcoholic beverage to drink might affect your behavior, quite apart from the alcoholic content of the beverage itself.

Controlling for Subject Expectancies To control for subject expectancies, experimenters rely on procedures that render subjects **blind,** or uninformed about, as to what treatments they are receiving. For example, the taste of an alcoholic beverage such as vodka

experimental subjects In an experiment, subjects who receive the experimental treatment.

control subjects In an experiment, subjects who do not receive the experimental treatment.

selection factor A type of bias in which differences between experimental and control groups result from differences in the subjects placed in the groups, not from the independent variable.

blind A state of being unaware of whether one has received an experimental treatment.

may be masked by mixing it with tonic water in certain amounts, so as to keep subjects unaware of whether the drinks they receive contain alcohol or tonic water only. In this way, subjects who truly receive alcohol should have no different expectations than those receiving the nonalcoholic control beverage. Similarly, drug treatment studies are often designed to control for subjects' expectations by keeping subjects in the dark as to whether they are receiving the experimental drug or an *inert* placebo control.

placebo An inert medication or bogus treatment that is intended to control for expectancy effects.

The term **placebo** derives from the Latin meaning "I shall please," referring to the fact that belief in the effectiveness of a treatment (its pleasing qualities) may inspire hopeful expectations that help people mobilize themselves to overcome their problems—regardless of whether the substance they receive is chemically active or inert. In medical research on chemotherapy, a placebo—also referred to as a "sugar pill"—is an inert substance that physically resembles an active drug. By comparing the effects of the active drug with those of the placebo, the experimenter can determine whether the drug has specific effects beyond those accounted for by expectations.

In a *single-blind placebo-control study*, subjects are randomly assigned to treatment conditions in which they receive an active drug (experimental condition) or a placebo (placebo-control condition), but they are kept blind, or uninformed, about which drug they are receiving. It is also helpful to keep the dispensing researchers blind as to which substances the subjects are receiving, lest the researchers' expectations come to affect the results. So in the case of a *double-blind placebo design*, neither the researcher nor the subject is told whether an active drug or a placebo is being administered. Of course, this approach assumes the subjects and the experimenters cannot "see through" the blind. In some cases, however, telltale side effects or obvious drug effects may break the blind (Basoglu et al., 1997). Still, the double-blind placebo control is among the strongest and most popular experimental designs, especially in drug treatment research.

Though placebos are routinely used in clinical research, evidence suggests that the effects of placebos are generally weak (Hrobjartsson & Gotzsche, 2001; Bailar, 2001). Evidence of placebo effects is strongest in pain studies, presumably because pain is a subjective experience that may be influenced more by the power of suggestion than other medical conditions that rely on objective measures, such as blood pressure ("Doubt Cast," 2001).

Placebo-control groups have also been used in psychotherapy research to control for subject expectancies. Assume you were to study the effects of therapy method A on mood. It would be inadequate to assign the experimental group to therapy A randomly and the control group to a no-treatment waiting list. The experimental group might show improvement because of group participation, not because of therapy method A. Participation might raise expectations of success, and these expectations might be sufficient to engender improvement. Changes in control subjects placed on the "waiting list" would help to account for effects due to the passage of time, but they would not account for placebo effects, such as the benefits of therapy that result from instilling a sense of hope.

An *attention-placebo* control group design can be used to separate the effects of a particular form of psychotherapy from placebo effects. In an attention-placebo group, subjects are exposed to a believable or credible treatment that contains the nonspecific factors that all therapies share—such as the attention and emotional support of a therapist—but not the specific ingredients of therapy represented in the active treatment. Attention-placebo treatments commonly substitute general discussions of participants' problems for the specific ingredients of therapy contained in the experimental treatment. Unfortunately, although attention-placebo subjects may be kept blind as to whether or not they are receiving the experimental treatment, the therapists themselves are generally aware of which treatment is being administered. Therefore, the attention-placebo method may not control for therapists' expectations.

THINK ABOUT IT

Why is a double-blind experimental design preferable to a single-blind design? How can a therapist's knowledge or expectations affect the results of an attention-placebo study? How are these two problems related?

Experimental Validity Experimental studies are judged on whether they are valid, or sound. The concept of experimental validity has multiple meanings, and we consider three of them: *internal validity, external validity*, and *construct validity*. We will see in Chapter 3 that the term *validity* is also applied in the context of tests and measures to refer to the degree to which these instruments measure what they purport to measure.

internal validity The degree to which manipulation of the independent variables can be causally related to changes in the dependent variables.

external validity The degree to which experimental results can be generalized to other settings and conditions.

construct validity The degree to which treatment effects can be accounted for by the theoretical mechanisms (constructs) represented in the independent variables.

Experiments are said to have **internal validity** when the observed changes in the dependent variable(s) can be causally related to the independent or treatment variable. Assume a group of depressed subjects is treated with a new antidepressant medication (the independent variable), and changes in their mood and behavior (the dependent variables) are tracked over time. After several weeks of treatment, the researcher finds most subjects have improved and claims the new drug is an effective treatment for depression. "Not so fast," you think to yourself. "How does the experimenter know that the independent variable and not some other factor was causally responsible for the improvement? Perhaps the subjects improved naturally as time passed, or perhaps they were exposed to other events responsible for their improvement." Experiments lack internal validity to the extent they fail to control for other factors (called *confounds*, or threats to validity) that might pose rival hypotheses for the results.

Experimenters randomly assign subjects to treatment and control groups to help control for such rival hypotheses. Random assignment helps ensure that subjects' attributes—intelligence, motivation, age, race, and so on—and presumably the life events they experience are randomly distributed across the groups and are not likely to favor one group over the other. Through the random assignment to groups, researchers can be reasonably confident that significant differences between the treatment and control groups reflect the effects of independent (treatment) variables and not confounding selection factors.

External validity refers to the generalizability or applicability of the results of an experimental study to other subjects and settings and at other times. In most cases, researchers are interested in generalizing the results of a specific study (for example, effects of a new antidepressant medication on a sample of people who are depressed) to a larger population (people in general who are depressed). The external validity of a study is strengthened to the degree the *sample* is representative of the target population. In studying the problems of the urban homeless, it is essential to make the effort to recruit a representative sample of the homeless population, for example, rather than focusing on a few homeless people who happen to be available. One way of obtaining a representative sample is by means of random sampling. In a *random sample*, every member of the target population has an equal chance of being selected.

Researchers may seek to extend the results of a particular study by replication, which refers to the process of repeating the experiment in other settings, with samples drawn from other populations, or at other times. A treatment for hyperactivity may be helpful with economically deprived children in an inner city classroom but not with children in affluent suburbs or rural areas. The external validity of the treatment may be limited if its effects do not generalize to other samples or settings. That does not mean the treatment is less effective, but rather that its range of effectiveness may be limited to certain populations or situations.

Construct validity represents a conceptually higher level of validity—the degree to which treatment effects can be accounted for by the theoretical mechanisms or constructs represented in the independent variables. A drug, for example, may have predictable effects but not for the theoretical reasons claimed by the researchers.

Consider a hypothetical experimental study of a new antidepressant medication. The research may have internal validity in the form of solid controls and external validity in the form of generalizability across samples of seriously depressed people. However, it may lack construct validity if the drug does not work for the reasons proposed by the researchers. Perhaps the researchers assumed that the drug would work by raising the levels of certain chemicals in the nervous system, whereas the drug actually works by increasing the sensitivity of receptors for those chemicals. "So what?" you may think. After all, the drug still works. True enough—in terms of immediate clinical applications. However, a better understanding of why the drug works can advance theoretical knowledge of depression and give rise to the development of yet more effective treatments.

We can never be certain about the construct validity of research. Scientists recognize that their current theories about why their results occurred may eventually be toppled by other theories that better account for the findings.

The Epidemiological Method

The **epidemiological method** studies the rates of occurrence of abnormal behavior in various settings or population groups. One type of epidemiological study is the **survey method,** which relies on interviews or questionnaires. Surveys are used to ascertain the rates of occurrence of various disorders in the population as a whole and in various subgroups classified according to such factors as race, ethnicity, gender, or social class. Rates of occurrence of a given disorder are expressed in terms of **incidence,** the number of new cases occurring during a specific period of time, and **prevalence,** the overall number of cases of a disorder existing in the population during a given period of time. Prevalence rates, then, include both new and continuing cases.

Epidemiological studies may point to potential causal factors in illnesses and disorders, even though they lack the power of experiments. By finding that illnesses or disorders "cluster" in certain groups or locations, researchers may be able to identify distinguishing characteristics that place these groups or regions at higher risk. Yet such epidemiological studies cannot control for selection factors—that is, they cannot rule out the possibility that other unrecognized factors might play a causal role in putting a certain group at greater risk. Therefore they must be considered suggestive of possible causal influences that must be tested further in experimental studies.

Samples and Populations In the best of possible worlds, we would conduct surveys in which every member of the **population** of interest would participate. In that way, we could be sure the survey results accurately represent the population we wish to study. In reality, unless the population of interest is rather narrowly defined (say, for example, designating the population of interest as the students living on your dormitory floor), chances are it is extremely difficult, if not impossible, to survey every member of a given population. Even census takers can't count every head in the general population. Consequently, most surveys are based on a **sample,** or subset, of the population. Researchers take steps when constructing a sample to ensure that it *represents* the target population. A researcher who sets out to study smoking rates in a local community by interviewing people drinking coffee in late-night cafés will probably overestimate its true prevalence.

One method of obtaining a representative sample is random sampling. A **random sample** is drawn in such a way that each member of the population of interest has an equal probability of selection. Epidemiologists sometimes construct random samples by surveying at random a given number of households within a target community. By repeating this process in a random sample of U.S. communities, the overall sample can approximate the general U.S. population, based on even a tiny percentage of the overall population.

Random sampling is often confused with random assignment. Random sampling refers to the process of randomly choosing individuals within a target population to participate in a survey or research study. By contrast, random assignment refers to a process by which members of a research sample are assigned at random to different experimental conditions or treatments.

Truth OR Fiction? REVISITED

A survey of 1,500 Americans may provide a more accurate reflection of voting attitudes and preferences of the American public than one based on millions of participants.

TRUE. A representative nationwide sample of about 1,500 people may be more accurate than a haphazard sample of millions.

Kinship Studies

Kinship studies attempt to disentangle the roles of heredity and environment in determining behavior. Heredity plays a critical role in determining a wide range of traits. The structures we inherit make our behavior possible (humans can walk and run) and at the same time place limits on us (humans cannot fly without artificial equipment). Heredity plays a role in determining not only our physical characteristics (hair color, eye color, height, and the like) but also many of our psychological characteristics. The science of heredity is called **genetics.**

Genes are the basic building blocks of heredity. They regulate the development of traits. **Chromosomes,** rod-shaped structures that house our genes, are found in the nuclei of the body's cells. A normal human cell contains 46 chromosomes, organized into 23 pairs.

epidemiological method A method of research that involves tracking the rates of occurrence of a disorder among different groups.

survey method A research method in which large samples of people are questioned by means of a survey instrument.

incidence The number of new cases of a disorder that occurs within a specific period of time.

prevalence The overall number of cases of a disorder in a population within a specific period of time.

population A total group of people, other organisms, or events.

sample Part of a population.

random sample A sample that is drawn in such a way that every member of a population has an equal chance of being included.

genetics The science of heredity.

genes The units, found on chromosomes, that carry heredity.

chromosomes The structures found in the nuclei of cells that carry the units of heredity, or genes.

genotype The set of traits specified by an individual's genetic code.

phenotype An individual's actual or expressed traits.

proband The case first diagnosed of a given disorder.

monozygotic (MZ) twins Twins that develop from the same fertilized egg and therefore share identical genes.

dizygotic (DZ) twins Twins that develop from separate fertilized eggs.

concordance Agreement.

Chromosomes consist of large complex molecules of deoxyribonucleic acid (DNA). Genes occupy various segments along the length of chromosomes. There are an estimated 30,000 to 40,000 genes in the nucleus of a human body cell (N. Wade, 2001a, 2001b).

The set of traits specified by our genetic code is referred to as our **genotype.** Our appearance and behavior are not determined by our genotype alone. We are also influenced by environmental factors such as nutrition, learning, exercise, accident and illness, learning, and culture. The constellation of our actual or expressed traits is called our **phenotype.** Our phenotype represents the interaction of genetic and environmental influences. People who possess genotypes for particular psychological disorders are said to have a *genetic predisposition* that makes them more likely to develop the disorder in response to stress or other factors, such as physical or psychological trauma.

The more closely people are related, the more genes they have in common. Children receive half their genes from each parent. There is thus a 50% overlap in genetic heritage between each parent and his or her offspring. Siblings (brothers and sisters) similarly share half their genetic heritage. Aunts and uncles related by blood to their nephews and nieces have a 25% overlap; first cousins, a 12.5% overlap (see Figure 1.2).

To determine whether abnormal behavior runs in a family, as we would expect if genetics plays a role, researchers would locate a person with the disorder and then study how the disorder is distributed among the person's family members (Nestadt et al., 2000; Tillfors et al., 2001). The case first diagnosed is referred to as the index case, or **proband.** If the distribution of the disorder among family members of the proband approximates their degree of kinship, there may be a genetic involvement in the disorder. However, the closer their kinship, the more likely people also are to share environmental backgrounds. For this reason, twin and adoptee studies are of particular value.

Twin Studies Sometimes a fertilized egg cell (or *zygote*) divides into two cells that separate, so each develops into a separate person. In such cases, there is a 100% overlap in genetic makeup, and the offspring are known as identical twins, or **monozygotic (MZ) twins.** Sometimes a woman releases two egg cells, or ova, in the same month, and they are both fertilized. In such cases, the *zygotes* (fertilized egg cells) develop into fraternal twins, or **dizygotic (DZ) twins.** DZ twins overlap 50% in their genetic heritage, just as other siblings do.

Identical, or MZ, twins are important in the study of the relative influences of heredity and environment because differences between MZ twins are the result of environmental rather than genetic influences. MZ twins look more alike and are closer in height than DZ twins. In twin studies, researchers identify probands for a given disorder who are members of MZ or DZ twin pairs and then study the other twins in the pairs. A role for genetic factors is suggested when MZ twins are more likely than DZ twins to share a disorder. Differences in the rates of **concordance** (agreement for the given trait or disorder) for MZ versus DZ twins are found for some forms of abnormal behavior, such as schizophrenia and bipolar disorder. Even among MZ twins, though, environmental influences cannot be ruled out. Parents and teachers, for example, often encourage MZ twins to behave in similar ways. Put in another way: If one twin does X, everyone expects the other to do X also. Expectations have a way of influencing behavior and making for self-fulfilling prophecies. We should also note that twins might not be typical of the general population, so we need to be cautious when generalizing the results of twin studies to the larger population. Twins tend to have had shorter gestational periods, lower birth weights, and a greater frequency of congenital malformations than nontwins (Kendler, 1994). Perhaps differences in prenatal experiences influence their later development in ways that set them apart from nontwins.

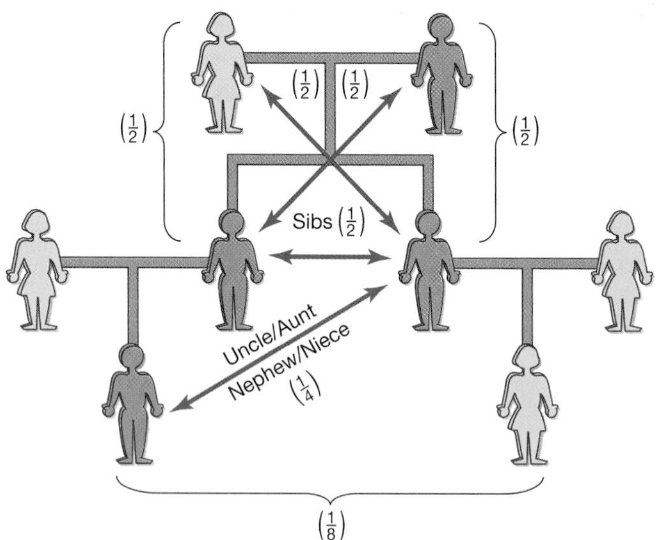

FIGURE 1.2 A family tree showing the proportion of shared inheritance among relatives.
The more closely people are related, the more genes they have in common. Kinship studies, including twin studies and adoptee studies, afford researchers insight into the heritability of various patterns of abnormal behavior.

Adoptee Studies **Adoptee studies** can provide powerful arguments for or against a role for genetic factors in the appearance of psychological traits and disorders. Assume that children are reared by adoptive parents from a very early age—perhaps from birth. The children share environmental backgrounds with their adoptive parents but not their genetic heritages. Then assume we compare the traits and behavior patterns of these children to those of their biological parents and their adoptive parents. If the children show a greater similarity to their biological parents than their adoptive parents on certain traits or disorders, we have strong evidence indeed for genetic factors in these traits and disorders.

The study of monozygotic twins reared apart might provide even more dramatic testimony to the relative roles of genetics and environment in shaping abnormal behavior. However, this situation is so uncommon that few examples exist in the literature. Although adoptee studies may represent the strongest source of evidence for genetic factors in explaining abnormal behavior patterns, we should recognize that adoptees, like twins, may not be typical of the general population. In later chapters we explore the role that adoptee and other kinship studies play in ferreting out genetic and environmental influences in many psychological disorders.

adoptee studies Studies that compare the traits and behavior patterns of adopted children to those of their biological parents and their adoptive parents.

case study A carefully drawn biography based on clinical interviews, observations, and psychological tests.

single-case experimental design A type of case study in which the subject is used as his or her own control.

reversal design An experimental design that consists of repeated measurement of a subject's behavior through a sequence of alternating baseline and treatment phases.

The Case-Study Method

Case studies have been important influences in the development of theories and treatment of abnormal behavior. Freud developed his theoretical model primarily on the basis of case studies, such as that of Anna O. Therapists representing other theoretical viewpoints have also reported cases studies.

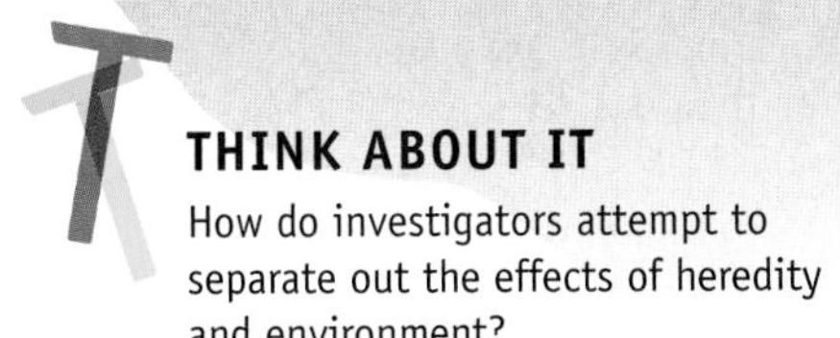

THINK ABOUT IT

How do investigators attempt to separate out the effects of heredity and environment?

Types of Case Studies **Case studies** involve intensive studies of individuals. Some case studies are based on historical material, involving subjects who have been dead for hundreds of years. Freud, for example, conducted a case study of the Renaissance artist and inventor Leonardo da Vinci. More commonly, case studies reflect an in-depth analysis of an individual's course of treatment. They typically include detailed histories of the subject's background and response to treatment. The therapist attempts to glean information from a particular client's experience in therapy that may be of help to other therapists treating similar clients.

Despite the richness of clinical material that case studies can provide, they are much less rigorous as research designs than experiments. Distortions or gaps in memory are bound to occur when people discuss historical events, especially those of their childhoods. Some people may intentionally color events to make a favorable impression on the interviewer; others aim to shock the interviewer with exaggerated or fabricated recollections. Interviewers themselves may unintentionally guide subjects into slanting the histories they report in ways that are compatible with their own theoretical perspectives.

Truth OR Fiction? REVISITED

Case studies have been conducted on people who have been dead for hundreds of years.

TRUE. Case studies have been conducted on people who have been dead for hundreds of years, such as Freud's study of Leonardo. Such studies rely on historical records rather than interviews.

Single-Case Experimental Designs The lack of control available in the traditional case-study method led researchers to develop more sophisticated methods, called **single-case experimental designs** (sometimes called *single-participant research designs*) in which subjects are used as their own controls (Morgan & Morgan, 2001). One of the most common forms of the single-case experimental design is the A-B-A-B, or **reversal design** (see Figure 1.3). The reversal design consists of the repeated measurement of clients' behavior across four successive phases:

1. A baseline phase (A). The baseline phase, which occurs prior to the inception of treatment and is characterized by repeated measurement of the target problem

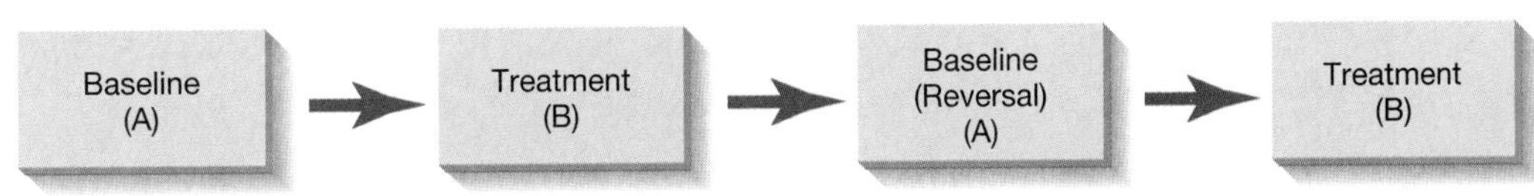

FIGURE 1.3 Diagram of an A-B-A-B reversal design.

baseline The rate at which a behavior occurs before treatment.

critical thinking Adoption of a questioning attitude and careful scrutiny of claims and arguments in the light of evidence.

behaviors at periodic intervals. This measurement allows the experimenter to establish a **baseline** rate for the behavior before treatment begins.

2. A treatment phase (B). Now the target behaviors are measured as the client undergoes treatment.
3. A second baseline phase (A, again). Treatment is now temporarily withdrawn or suspended. This is the reversal in the reversal design, and it is expected that the positive effects of treatment should now be reversed because the treatment has been withdrawn.
4. A second treatment phase (B, again). Treatment is reinstated and the target behaviors are assessed yet again.

Clients' target behaviors or response patterns are compared from one phase to the next to determine the effects of treatment. The experimenter looks for evidence of a correspondence between a subject's behavior and the particular phase of the design to determine whether the independent variable (that is, the treatment) has produced the intended effects. If the behavior improves whenever treatment is introduced (during the first and second treatment phases) but returns (or is reversed) to baseline levels during the reversal phase, the experimenter can be reasonably confident the treatment had the intended effect.

The method is illustrated by a case in which Azrin and Peterson (1989) used a controlled blinking treatment to eliminate a severe eye tic—a form of squinting in which her eyes shut tightly for a fraction of a second—in a 9-year-old girl. The tic occurred about 20 times a minute when the girl was at home. In the clinic, the rate of eye tics or squinting was measured for 5 minutes during a baseline period (A). Then the girl was prompted to blink her eyes softly every 5 seconds (B). The experimenters reasoned that voluntary "soft" blinking would activate motor (muscle) responses incompatible with those producing the tic, thereby suppressing the tic. As you can see in Figure 1.4, the tic was virtually eliminated in but a few minutes of practicing the incompatible, or competing, response ("soft" blinking) but returned to near baseline levels during the reversal phase (A) when the competing response was withdrawn. The positive effects were quickly reinstated during the second

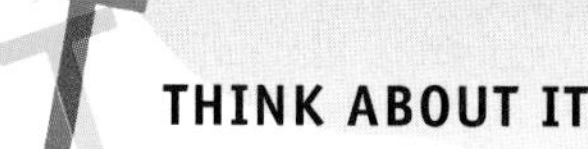

THINK ABOUT IT

What are the major methods of research used to study abnormal behavior? What are their strengths and weaknesses?

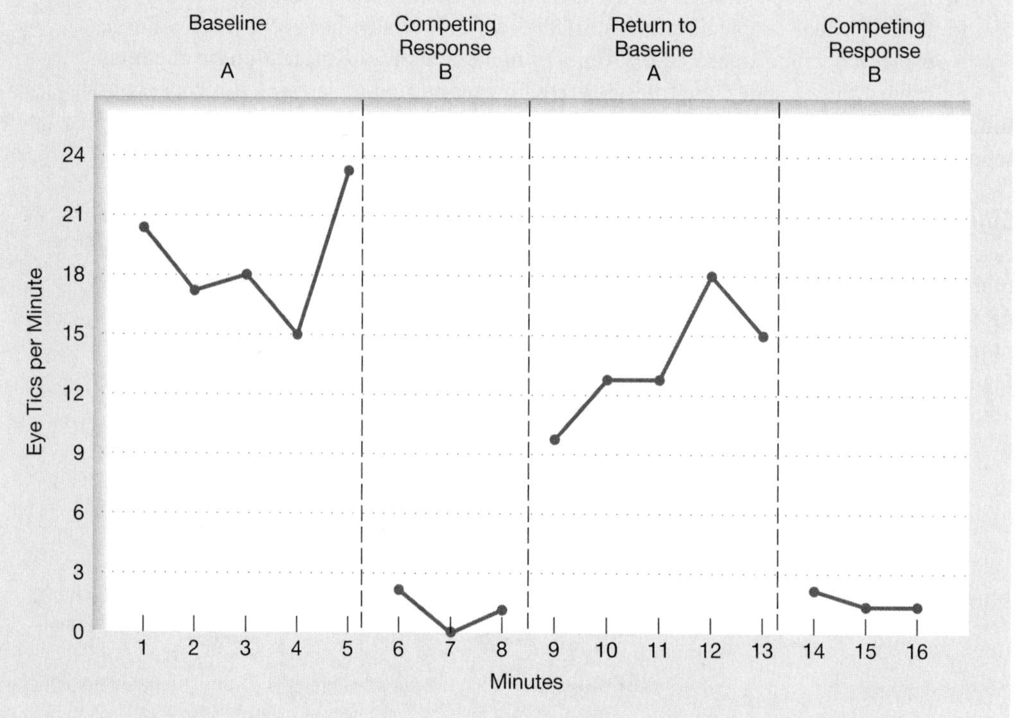

FIGURE 1.4 Treatment results from the Azrin and Peterson study (an A-B-A-B design). Notice how the target response, eye tics per minute, decreased when the competing response was introduced in the first "B" phase. The rate then increased to near baseline levels when the competing response was withdrawn during the second "A" phase. It decreased again when the competing response was reinstated in the second "B" phase.

treatment period (B). The child was also taught to practice the blinking response at home during scheduled 3-minute practice periods and whenever the tic occurred or she felt an urge to squint. The tic was completely eliminated during the first 6 weeks of the treatment program and remained absent at a follow-up evaluation 2 years later.

Although reversal designs offer better controls than traditional treatment case studies, it is not always possible or ethical to reverse certain behaviors or treatment effects. Participants in a stop-smoking program who reduce or quit smoking during treatment may not revert to their baseline smoking rates when treatment is temporarily withdrawn during a reversal phase.

The *multiple-baseline design* is a type of single-case experimental design that does not require a reversal phase. In a multiple-baseline design *across behaviors*, treatment is applied, in turn, to two or more behaviors following a baseline period. A treatment effect is inferred if changes in each of these behaviors corresponded to the time at which each was subjected to treatment. Because no reversal phase is required, many of the ethical and practical problems associated with reversal designs are avoided.

A multiple-baseline design was used to evaluate the effects of a social skills training program in the treatment of a shy, unassertive 7-year-old girl named Jane (Bornstein, Bellack, & Hersen, 1977). The program taught Jane to maintain eye contact, speak more loudly, and make requests of other people through modeling (therapist demonstration of the target behavior), rehearsal (practice), and therapist feedback regarding the effectiveness of practice. However, the behaviors were taught sequentially, not simultaneously. Measurement of each behavior and an overall rating of assertiveness were obtained during a baseline period from observations of Jane's role-playing of social situations with other children, such as playing social games at school and conversing in class. As shown in Figure 1.5, Jane's performance of each behavior that were improved following treatment. The rating of overall assertiveness showed more gradual improvement as the number of behaviors that were included in the program increased. Treatment gains were generally maintained at a follow-up evaluation.

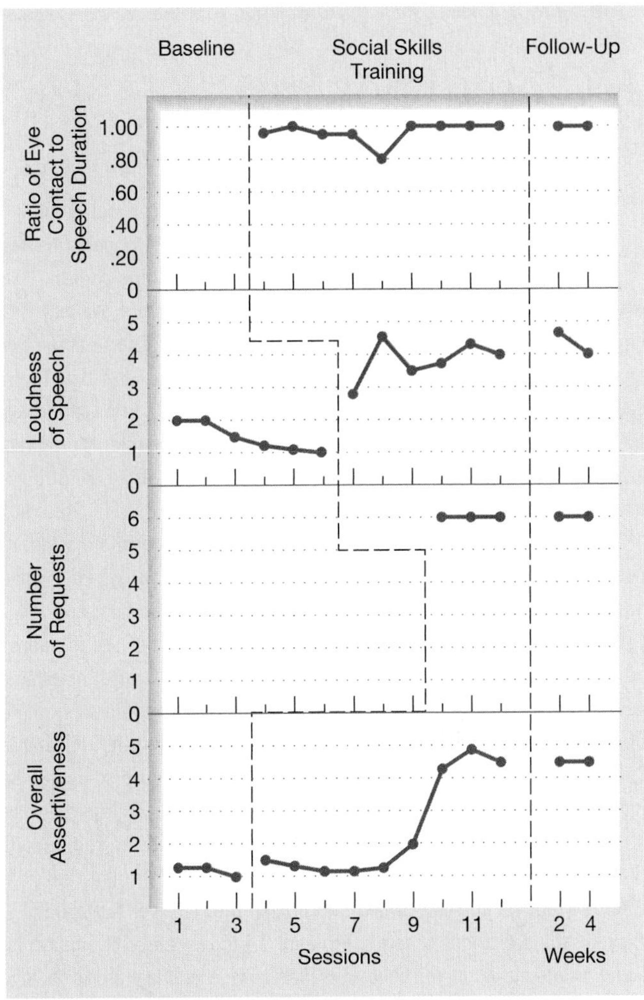

FIGURE 1.5 Treatment results from the study by Bornstein, Bellack, and Hersen.

The dotted line shows the point at which social skills training was applied to each of the targeted behaviors. Here we see that the targeted behaviors (eye contact, loudness of speech, and number of requests) improved only when they were subject to the treatment approach (social skills training). We thus have evidence that the treatment—and not another, unidentified factor—accounted for the results. The section on the bottom shows ratings of Jane's overall level of assertiveness during the baseline assessment period, the social skills training program, and the follow-up period.

Source. Bornstein, M. R., Bellack, A. S., & Hersen, M. (1977). Social-skills training for unassertive children: A multiple-baseline analysis. *Journal of Applied Behavior Analysis, 10,* pp. 183–195. Reprinted with permission.

To show a clear-cut treatment effect, changes in target behaviors should occur only when they are subjected to treatment. In some cases, however, changes in the treated behaviors may lead to changes in the yet untreated behaviors, apparently because of generalization of the effect. Fortunately though, generalization effects have tended to be the exception rather than the rule in experimental research (Kazdin, 1992).

No matter how tightly controlled the design, or how impressive the results, single-case designs suffer from weak external validity because they do not show whether a treatment effective for one person is effective for others. Replication with other individuals is essential to help strengthen external validity. If these results prove encouraging, they may lead to controlled experiments to provide even more convincing evidence of treatment effectiveness.

Scientists may use different methods to study phenomena of interest to them. But they share in common a skeptical, hard-nosed way of thinking called critical thinking. **Critical thinking** involves a willingness to challenge conventional wisdom and common knowledge that many of us take for granted. It also means finding *reasons* to support beliefs rather than relying on feelings or gut impressions. When people think critically, they maintain open minds. They suspend their beliefs until they have obtained and evaluated evidence that either supports or refutes them. In the "A Closer Look" feature on page 30, we examine the features of critical thinking and how they can be applied in our study of abnormal psychology.

Quiz 1.3
Research Methods in Abnormal Psychology

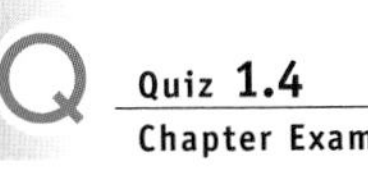

Quiz 1.4
Chapter Exam

A Closer Look

Thinking Critically About Abnormal Psychology

We are exposed to a flood of information about mental health streaming down to us through the popular media—television, radio, and print media, including books, magazines, and newspapers, and increasingly, the Internet. We may hear a news report touting a new drug as a "breakthrough" in the treatment of anxiety, depression, or obesity, only to learn some time later that the so-called breakthrough doesn't live up to expectations or carries serious side effects. Some reports in the media are accurate and reliable, while others are misleading, biased, or contain half-truths, exaggerated claims, or unsupported conclusions.

To sort through the welter of sometimes confusing information, we need to arm ourselves with the skills of critical thinking, which involves adopting a questioning attitude toward information you hear and read. Critical thinkers carefully weigh the available evidence to see if claims people make can stand up to scrutiny. Becoming a critical thinker means never taking claims at face value. It means looking carefully at both sides of the argument. Sad to say, most of us take certain "truths" for granted. Critical thinkers, however, never say, "This is true because so-and-so says it is true." They seek to evaluate assertions and claims for themselves.

We encourage you to apply critical thinking skills to the questions posed in the "Think About It" features in each chapter. Critical thinkers adopt a skeptical attitude toward information they receive. They carefully examine the definitions of terms, evaluate the logical bases of arguments, and evaluate claims in the light of available evidence. Here are some key features of critical thinking:

1. *Maintain a skeptical attitude.* Don't take anything at face value, not even claims made by respected scientists or textbook authors. Consider the evidence yourself and seek additional information to help you evaluate claims made by others.
2. *Consider the definitions of terms.* Statements may be true or false depending on how the terms that are used are defined. Consider the statement, "Stress is bad for you." If we define the concept *stress* in terms of hassles and work or family pressures that stretch to the max our ability to cope, then there is perhaps substance to the statement. However, if we define stress (see Chapter 5) to include any factors that impose a demand on us to adjust, including events such as a new marriage or the birth of a child, then perhaps certain types of stress can be positive, even if they are stressful. Perhaps, as we'll see, we all need some amount of stress to be energized and alert.
3. *Weigh the assumptions or premises on which arguments are based.* Consider a case in which we are comparing differences in the rates of psychological disorders across racial or ethnic groups in our society. Assuming we find differences, should we conclude that ethnicity or racial identity accounts for these differences? This conclusion might be valid if we can assume that all other factors that distinguish one racial or ethnic group from another are held constant. However, ethnic or racial minorities in the United States and Canada are disproportionately represented among the poor, and the poor are more apt to develop more severe psychological disorders. Differences among racial or ethnic groups may thus be a function of poverty, not race or ethnicity per se. These differences may also be due to negative stereotyping of racial minorities by clinicians in making diagnostic judgments rather than to differences in underlying rates of the disorder.
4. *Bear in mind that correlation is not causation.* Critical thinkers recognize that correlation is not causation. Consider the relationship between depression and stress. Evidence shows a positive correlation between these variables, which means depressed people tend to have higher levels of stress in their lives (Hammen & de Mayo, 1982; Pianta & Egeland, 1994). But does stress cause depression? Perhaps it does. Or perhaps depression leads to greater stress. After all, depressive symptoms may be stressful in themselves and may lead to additional stress as the person finds it increasingly difficult to meet life responsibilities, such as keeping up with work at school or on the job. It is also possible that the two variables are not causally linked at all but are linked through a third variable, perhaps an underlying genetic factor. It is conceivable that people inherit clusters of genes that make them more prone to encounter both depression and stress.
5. *Consider the kinds of evidence on which conclusions are based.* Some conclusions, even seemingly "scientific" conclusions, are based on anecdotes and personal endorsements. They are not founded on sound research. There is much controversy today about so-called recovered memories that may suddenly arise in adulthood, usually during psychotherapy or hypnosis, and usually involving incidents of sexual abuse committed during childhood by the person's parents or family members. But are such memories accurate? (See Chapter 7.)
6. *Do not oversimplify.* Consider the statement, "Alcoholism is inherited." In Chapter 10, we review evidence suggesting that genetic factors may create a predisposition to alcoholism, at least in males. But the origins of alcoholism, as well as of schizophrenia, depression, and physical health problems such as cancer and heart disease, are more complex, reflecting a complicated interplay of biological and environmental factors. In only a few cases are diseases the direct result of a single defective gene or genes. People may even inherit a predisposition to develop a particular psychological or physical disorder but can avoid developing it if they are raised in a supportive family environment and learn to manage stress effectively.
7. *Do not overgeneralize.* In Chapter 7, we consider evidence showing that a history of severe abuse in childhood figures prominently in the great majority of cases of people who later develop multiple personalities. Does this mean that all (or even most) abused children go on to develop multiple personalities? Actually, very few do.

Summing Up

How Do We Define Abnormal Behavior?

What criteria do mental health professionals use to determine that behavior is abnormal? Psychologists generally consider behavior abnormal when it meets some combination of the following criteria: (1) unusual or statistically infrequent; (2) socially unacceptable or in violation of social norms; (3) fraught with misperceptions or misinterpretations of reality; (4) associated with states of severe personal distress; (5) maladaptive or self-defeating; or (6) dangerous.

What are psychological disorders? Psychological disorders (also called *mental disorders*) involve abnormal behavior patterns associated with disturbances in mental health or psychological functioning.

How do views about abnormal behavior vary across cultures? Behaviors deemed normal in one culture may be considered abnormal in another. Concepts of health and illness may also have different meanings in different cultures. Abnormal behavior patterns may also take different forms in different cultures, and societal views or models explaining abnormal behavior also vary across cultures.

Historical Perspectives on Abnormal Behavior

How have views about abnormal behavior changed over time? Ancient societies attributed abnormal behavior to divine or supernatural forces. In medieval times, belief in possession held sway, and exorcists were used to rid people who behaved abnormally of the evil spirits that were believed to possess them. The 19th-century German physician Wilhelm Griesinger argued that abnormal behavior was caused by diseases of the brain. He, along with another German physician who followed him, Emil Kraepelin, were influential in the development of the modern medical model, which likens abnormal behavior patterns to physical illnesses.

How has the treatment of people with mental disorders changed over time? Asylums, or madhouses, began to crop up throughout Europe in the late 15th and early 16th centuries. Conditions in these asylums were dreadful. With the rise of moral therapy in the 19th century, conditions in mental hospitals improved. Proponents of moral therapy believed that mental patients could be restored to functioning if they were treated with dignity and understanding. The decline of moral therapy in the latter part of the 19th century led to a period of apathy and to the belief the "insane" could not be successfully treated. Conditions in mental hospitals deteriorated, and they offered little more than custodial care. Not until the middle of the 20th century did public outrage and concern about the plight of mental patients mobilize legislative efforts toward the development of community mental health centers as alternatives to long-term hospitalization.

What are the major contemporary models of abnormal behavior? The medical model conceptualizes abnormal behavior patterns, like physical diseases, in terms of clusters of symptoms, called syndromes, which have distinctive causes that are presumed to be biological in nature. Psychological models focus on the psychological roots of abnormal behavior and derive from psychoanalytic, behavioral, humanistic, and cognitive perspectives. The sociocultural model emphasizes a broader perspective that takes into account the social contexts in which abnormal behavior occurs. Today, many theorists subscribe to a biopsychosocial model that posits that multiple causes representing biological, psychological, and sociocultural domains interact in complex ways in the development of abnormal behavior patterns.

Research Methods in Abnormal Psychology

What are the basic objectives of the scientific method, and what steps are involved in applying it? The scientific approach focuses on four general objectives: description, explanation, prediction, and control. There are four steps to the scientific method: formulating a research question, framing the research question in the form of a hypothesis, testing the hypothesis, and drawing conclusions about the correctness of the hypothesis. Psychologists follow the ethical principles of the profession that govern research.

What are the methods psychologists use to study abnormal behavior? The naturalistic-observation method allows scientists to measure behavior under naturally occurring conditions. The correlational method explores the relationship between variables, which may help predict future behavior and suggest possible underlying causes of behavior. But correlational research does not directly test cause-and-effect relationships. Longitudinal research is a correlational method in which a sample of subjects is repeatedly studied at periodic intervals over long periods of time, sometimes spanning decades.

In the experimental method, the investigator directly controls (manipulates) the independent variable under controlled conditions to demonstrate cause-and-effect relationships. Experiments use random assignment as the basis for determining which subjects (called experimental subjects) receive an experimental treatment and which others (called control subjects) do not. Experiments are evaluated in terms of internal, external, and construct validity.

The epidemiological method examines the rates of occurrence of abnormal behavior in various population groups or settings. Kinship studies, such as twin studies and adoptee studies, attempt to disentangle the contributions of environment and heredity.

Case-study methods can provide a richness of clinical material, but they are limited by difficulties of obtaining accurate and unbiased client histories, by possible therapist biases, and by the lack of control groups. Single-case experimental designs are intended to help researchers overcome some of the limitations of the case-study method.

CHAPTER TWO

Contemporary Perspectives Abnormal Behavior

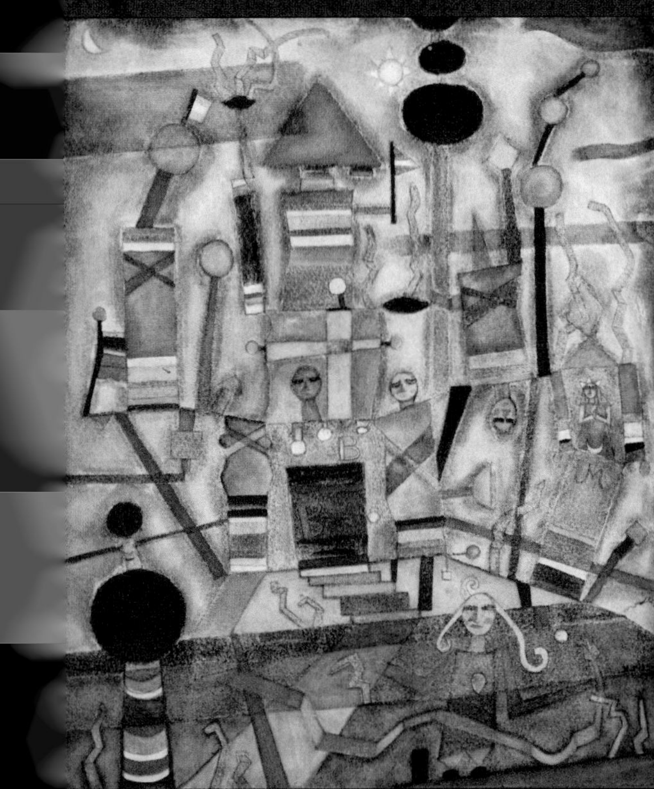

Alejandro Xul Solar
Patria B, 1925

Truth OR Fiction?

- Messages are transmitted through the nervous system by chemical messengers. (p. 34)
- Depressant drugs are drugs that cause depression. (p. 37)
- Freud likened the mind to a giant iceberg, with only the tip rising into conscious awareness. (p. 38)
- Freud believed that an ancient Greek legend about a king who slew his father and married his mother contained key insights into the nature of the development of the human psyche. (p. 43)
- Punishment does not eliminate undesirable behavior. (p. 51)
- Children may acquire a distorted self-concept that mirrors what others expect them to be but which does not reflect who they truly are. (p. 53)
- According to a leading cognitive theorist, states of emotional distress are caused by the beliefs people hold about their life experiences, not by the experiences themselves. (p. 55)
- Rates of mental disorders are generally higher among African Americans than Euro-Americans, even when we account for income differences between these groups. (p. 58)

Since earliest times humans have sought explanations for strange or deviant behavior. As we saw in Chapter 1, in ancient times and through the Middle Ages, beliefs about abnormal behavior centered on the role of demons and other supernatural forces. But even in ancient times, there were some scholars, such as Hippocrates and Galen, who sought natural explanations of abnormal behavior. In contemporary times, superstition and demonology have given way to theoretical models engendered by the natural and social sciences. These approaches pave the way not only for a scientifically based understanding of abnormal behavior but also for ways of treating people with psychological disorders.

In this chapter we examine major contemporary perspectives on abnormal behavior, including the biological, psychological, and sociocultural perspectives. Each of these major perspectives provides a window for examining abnormal behavior. Each contributes to our understanding of abnormal behavior, but none captures a complete view of our subject matter. Many scholars today believe that abnormal behavior patterns are complex phenomena that are best understood by adopting an interactionist or biopsychosocial perspective that takes into account the interaction of factors representing biological, psychological, and sociocultural domains.

Chapter Outline

The Biological Perspective

The medical model, inspired by physicians from Hippocrates through Kraepelin, remains a powerful force in contemporary understanding of abnormal behavior. The medical model represents a biological perspective on abnormal behavior. We prefer to use the term *biological perspective* rather than *medical model* to refer to approaches that emphasize the role of biological factors in explaining abnormal behavior and the use of biologically based treatments in treating psychological disorders. We can speak of biological perspectives without adopting the tenets of the medical model, which treats abnormal behavior patterns as *disorders* and their features as *symptoms.* For example, certain behavior patterns (shyness or a lack of musical ability) may have a strong genetic component but not be considered "symptoms" of underlying "disorders."

Knowledge of the biological underpinnings of abnormal behavior has grown with great speed in recent years. In Chapter 1 we focused on the methods of studying the role of heredity or genetics. Genetics plays a large role in many forms of abnormal behavior, as we shall see throughout the text.

We also know that other biological factors, especially the functioning of the nervous system, are involved in the development of abnormal behavior (Cravchik & Goldman, 2000). To better understand the role of the nervous system in abnormal behavior patterns, we first need to learn how the nervous system is organized and how nerve cells communicate with each other.

The Nervous System

Perhaps you could not be nervous if you did not have a nervous system—neither could you see, hear, perceive touch or pain, or move—but even calm people have nervous systems. The nervous system is made up of nerve cells called **neurons.** Neurons communicate with one another, or transmit "messages." These messages account for events as diverse as sensing an itch from a bug bite, coordinating a figure skater's vision and muscles, composing a symphony, solving an architectural equation, and, in the case of hallucinations, hearing or seeing things that are not really there.

Every neuron has a cell body, or **soma,** dendrites, and an axon (see Figure 2.1). The cell body contains the nucleus of the cell and metabolizes oxygen to carry out the work of the cell. Short fibers called **dendrites** project from the cell body to receive messages from adjoining neurons. Each neuron has a single **axon** projecting trunklike from the cell body. Axons can extend as long as several feet if they are conveying messages between the toes and the spinal cord. They may branch and project in various directions. Axons terminate

neurons Nerve cells.

soma A cell body.

dendrites The rootlike structures at the ends of neurons that receive nerve impulses from other neurons.

axon The long, thin part of a neuron along which nerve impulses travel.

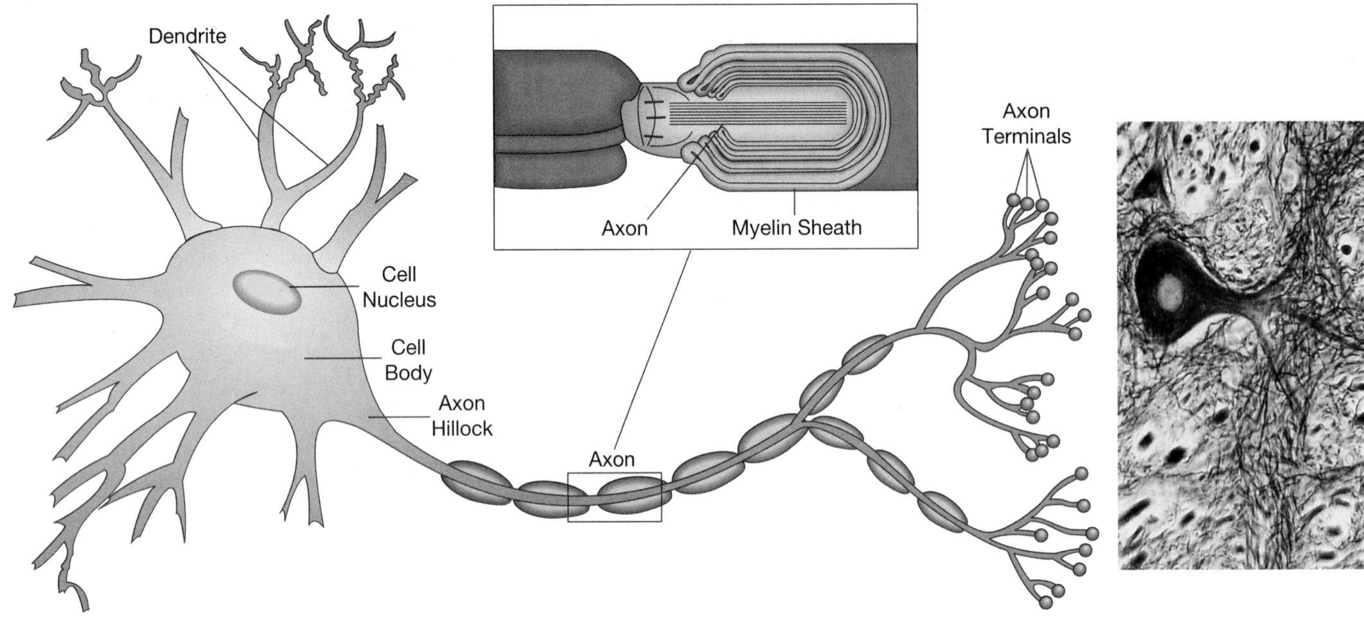

FIGURE 2.1 Anatomy of a neuron.
Neurons typically consist of cells bodies (or somas), dendrites, and one or more axons. The axon of this neuron is wrapped in a myelin sheath, which insulates it from the bodily fluids surrounding the neuron and facilitates transmission of neural impulses (messages that travel within the neuron).

in small branching structures that are aptly termed **terminals.** Swellings called **knobs** are found at the tips of axon terminals. Neurons convey messages in one direction, from the dendrites or cell body along the axon to the axon terminals. The messages are then conveyed from terminal knobs to other neurons, muscles, or glands.

Neurons transmit messages to other neurons by means of chemical substances called **neurotransmitters.** Neurotransmitters induce chemical changes in receiving neurons. These changes cause axons to conduct the messages in electrical form.

The junction between a transmitting neuron and a receiving neuron is termed a **synapse.** A transmitting neuron is termed *presynaptic.* A receiving neuron is said to be *postsynaptic.* A synapse consists of an axon terminal from a transmitting neuron, a dendrite of a receiving neuron, and a small fluid-filled gap between the two called the *synaptic cleft.* The message does not jump the synaptic cleft like a spark. Instead, axon terminals release neurotransmitters into the cleft like myriad ships casting off into the seas (Figure 2.2).

Each kind of neurotransmitter has a distinctive chemical structure. It will fit only into one kind of harbor, or **receptor site,** on the receiving neuron. Consider the analogy of a lock and key. Only the right key (neurotransmitter) operates the lock, causing the postsynaptic neuron to forward the message.

Once released, some molecules of a neurotransmitter reach port at receptor sites of other neurons. "Loose" neurotransmitters may be broken down in the synaptic clefts by enzymes or be reabsorbed by the axon terminal (a process termed *reuptake*), so as to prevent the receiving cell from continuing to fire.

Irregularities in the workings of neurotransmitter systems in the brain are closely related to abnormal behavior patterns (see Table 2.1). For example, depression is linked to dysfunctions involving the neurotransmitters *norepinephrine* and *serotonin* (see Chapter 8). Irregularities of serotonin functioning are also implicated in eating disorders (see Chapter 11). **Alzheimer's disease,** a brain disease in which there is a progressive loss of memory and cognitive functioning, is associated with reductions in the levels of the neurotransmitter *acetylcholine* in the brain. Irregularities involving the neurotransmitter *dopamine* appear to be involved in schizophrenia (see Chapter 13). People with schizophrenia may use more of

Truth OR Fiction? REVISITED

Messages are transmitted through the nervous system by chemical messengers.

TRUE. Chemicals called neurotransmitters carry messages from one neuron to another.

terminals The small branching structures at the tips of axons.

knobs The swollen endings of axon terminals.

neurotransmitters Chemical substances that transmit messages from one neuron to another.

synapse The junction between the terminal knob of one neuron and the dendrite or soma of another through which nerve impulses pass.

receptor site A part of a dendrite on a receiving neuron that is structured to receive a neurotransmitter.

Alzheimer's disease A progressive brain disease characterized by gradual loss of memory and intellectual functioning, personality changes, and eventual loss of ability to care for oneself.

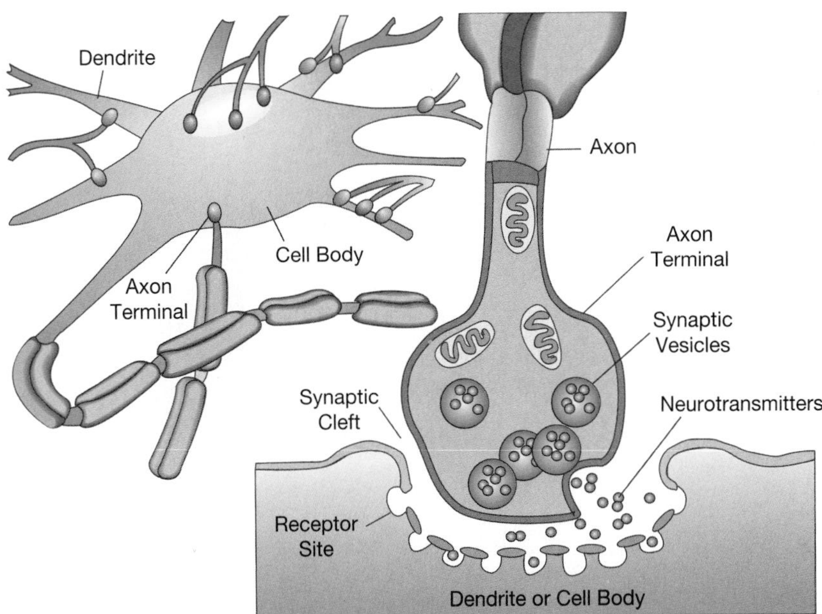

FIGURE 2.2 Transmission of neural impulses across the synapse.
The diagram here shows the structure of the neuron and the mode of transmission of neural impulses between neurons. Neurons transmit messages, or neural impulses, across synapses, which consist of the axon terminal of the transmitting neuron, the gap or synaptic cleft between the neurons, and the dendrite of the receiving neuron. The "message" consists of neurotransmitters that are released into the synaptic cleft and taken up by receptor sites on the receiving neuron. Somehow the patterns of firing of many thousands of neurons give rise to psychological events such as thoughts and mental images. Many patterns of abnormal behavior have been associated with irregularities in the transmission or reception of neural messages.

the dopamine that is available in their brains than do people without schizophrenia. The result may be hallucinations, incoherent speech, and delusional thinking. Antipsychotic drugs used to treat schizophrenia apparently work by blocking dopamine receptors in the brain. Serotonin is linked not only to depression but also to anxiety disorders, sleep disorders, and eating disorders (Lesch et al., 1996; Mann et al., 1996; McBride, Anderson, & Shapiro, 1996). Although neurotransmitter systems are implicated in many psychological disorders, the precise causal mechanisms remain to be determined.

Parts of the Nervous System The nervous system consists of two major parts, the **central nervous system** and the **peripheral nervous system.** The two parts are also divided. The central nervous system consists of the brain and spinal cord. The peripheral nervous system is made up of nerves that (1) receive and transmit sensory messages (messages from sense organs such as the eyes and ears) to the brain and spinal cord, and (2) transmit messages from the brain or spinal cord to the muscles, causing them to contract, and to glands, causing them to secrete hormones.

central nervous system The brain and spinal cord.

peripheral nervous system The somatic and autonomic nervous systems.

Central Nervous System We begin our overview of the parts of the central nervous system with the back of the head, where the spinal cord meets the brain, and work forward (see Figure 2.3). The lower part of the brain, or *hindbrain*, consists of the medulla, pons, and

TABLE 2.1 Neurotransmitter Functions and Relationships with Abnormal Behavior Patterns

Neurotransmitter	Functions	Associations with Abnormal Behavior
Acetylcholine (ACh)	Control of muscle contractions and formation of memories	Reduced levels found in patients with Alzheimer's disease
Dopamine	Regulation of muscle contractions and mental processes involving learning, memory, and emotions	Overutilization of dopamine in the brain may be involved in the development of schizophrenia
Norepinephrine	Mental processes involved in learning and memory	Imbalances linked with mood disorders such as depression
Serotonin	Regulation of mood states, satiety, and sleep	Irregularities may be involved in depression and eating disorders

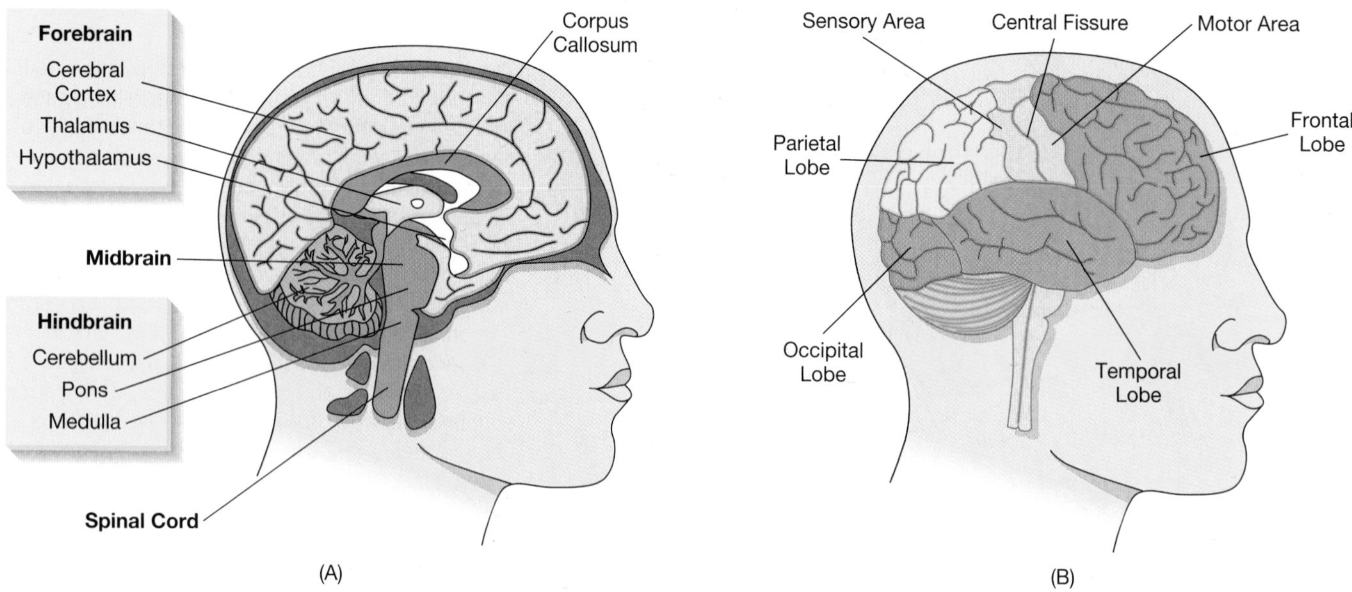

FIGURE 2.3 The geography of the brain.
Part A shows parts of the hindbrain, midbrain, and forebrain. Part B shows the four lobes of the cerebral cortex: frontal, parietal, temporal, and occipital. In B, the sensory (tactile) and motor areas lie across the central fissure from one another. Researchers are investigating the potential relationships between various patterns of abnormal behavior and irregularities in the formation or functioning of the structures of the brain.

Web Link 2.1 WWW
Neuroscience of Mental Health

medulla An area of the hindbrain involved in regulation of heartbeat and respiration.

pons A structure in the hindbrain involved in respiration.

cerebellum A structure in the hindbrain involved in coordination and balance.

reticular activating system Brain structure involved in processes of attention, sleep, and arousal.

comatose In a coma, a state of deep, prolonged unconsciousness.

thalamus A structure in the forebrain involved in relaying sensory information to the cortex and in processes related to sleep and attention.

hypothalamus A structure in the forebrain involved in regulating body temperature, emotion, and motivation.

limbic system A group of forebrain structures involved in learning, memory, and basic drives.

basal ganglia An assemblage of neurons located between the thalamus and cerebrum, involved in coordinating motor (movement) processes.

cerebellum. Many nerves that link the spinal cord to higher brain levels pass through the **medulla.** The medulla plays roles in such vital functions as heart rate, respiration, and blood pressure, and also in sleep, sneezing, and coughing. The **pons** transmits information about body movement and is involved in functions related to attention, sleep, and respiration.

Behind the pons is the **cerebellum** (Latin for "little brain"). The cerebellum is involved in balance and motor (muscle) behavior. Injury to the cerebellum may impair motor coordination and cause stumbling and loss of muscle tone.

The midbrain lies above the hindbrain and contains nerve pathways that link the hindbrain with the forebrain. The **reticular activating system** (RAS) starts in the hindbrain and rises through the midbrain into the lower forebrain. The RAS, which consists of a weblike network of neurons, plays vital roles in sleep, attention, and arousal. RAS injury may leave an animal **comatose.** RAS stimulation triggers messages that heighten alertness. *Depressant drugs*, such as alcohol, which dampen central nervous system activity, lower RAS activity.

Important areas in the frontal part of the brain, or *forebrain*, are the thalamus, hypothalamus, limbic system, basal ganglia, and cerebrum. The **thalamus** relays sensory information (such as touch and vision) to higher brain regions. The thalamus is also involved in sleep and attention, in coordination with other structures, such as the RAS.

The **hypothalamus** is a tiny structure located between the thalamus and the pituitary gland. The hypothalamus is vital in regulating body temperature, concentration of fluids, storage of nutrients, and motivation and emotion. By implanting electrodes in parts of the hypothalamus of animals and observing the effects when a current is switched on, researchers have found that the hypothalamus is involved in a range of motivational drives and behaviors, including hunger, thirst, sex, parenting behaviors, and aggression.

The hypothalamus, together with parts of the thalamus and other structures, make up the **limbic system.** The limbic system plays a role in memory and in regulating the more basic drives involving hunger, thirst, and aggression. The **basal ganglia** lie in front of the thalamus and help to regulate postural movements and coordination.

The **cerebrum** is your "crowning glory" and is responsible for the round shape of the human head. The surface of the cerebrum is convoluted with ridges and valleys. This surface is the **cerebral cortex,** the thinking, planning, and executive center of the brain. The two hemispheres, of the cerebral cortex are connected by the **corpus callosum,** a thick fiber bundle.

Peripheral Nervous System The peripheral nervous system connects the brain to the outer world. Without the peripheral nervous system, people could not perceive the world or act on it. The two main divisions of the peripheral nervous system are the *somatic nervous system* and the *autonomic nervous system.*

The **somatic nervous system** transmits messages about sights, sounds, smells, temperature, body position, and so on, to the brain. Messages from the brain and spinal cord to the somatic nervous system regulate intentional body movements, such as raising an arm, winking, or walking; breathing; and subtle movements that maintain posture and balance.

Psychologists are particularly interested in the **autonomic nervous system** (ANS) because its activities are linked to emotional response. *Autonomic* means "automatic." The ANS regulates the glands and **involuntary** activities such as heart rate, breathing, digestion, and dilation of the pupils of the eyes, even when we are sleep.

The ANS has two branches or subdivisions, the **sympathetic** and the **parasympathetic.** These branches have mostly opposing effects. Many organs and glands are served by both branches of the ANS. The sympathetic division is most involved in processes that mobilize the body's resources in times of stress, such as drawing energy from stored reserves to prepare the person to deal with imposing threats or dangers (see Chapter 5). When we face a threat or dangerous situation, the sympathetic branch of the ANS accelerates the heart rate and breathing rate, which helps prepare our bodies to either fight or flee. Sympathetic activation in the face of a threatening stimulus is associated with emotional responses such as fear or anxiety. When we relax, the parasympathetic branch decelerates the heart rate. The parasympathetic division is most active during processes that replenish energy reserves, such as digestion. Because the sympathetic branch dominates when we are fearful or anxious, fear or anxiety can lead to indigestion because activation of the sympathetic nervous system curbs digestive activity.

The Cerebral Cortex The human activities of thought and language involve the two hemispheres of the cerebrum. Each hemisphere is divided into four parts, or lobes, as shown in Figure 2.3. The *occipital lobe* is primarily involved in vision; the *temporal lobe* is involved in processing sounds or auditory stimuli. The *parietal lobe* is involved in determining sensations of touch, temperature, and pain. The *sensory area* of the parietal lobe receives messages from skin sensors all over the body. Neurons in the motor area (or *motor cortex*) of the *frontal lobe* are involved in controlling muscular responses, which enables us to move our limbs. The *prefrontal cortex* (the part of the frontal lobe that lies in front of the motor cortex) is involved in higher mental functions such as thinking, problem solving, and use of language.

Evaluating Biological Perspectives on Abnormal Behavior

There is no question that biological structures and processes are involved in many patterns of abnormal behavior, as we will see in later chapters. Factors such as disturbances in neurotransmitter functioning and underlying brain abnormalities or defects are implicated in many psychological disorders. For some disorders, such as Alzheimer's disease, biological processes play the direct causative role. Even then, however, the precise causes remain unknown. In other cases, such as schizophrenia, biological factors, especially genetics, appear to interact with stressful environmental factors in the development of the disorder.

Genetic influences are implicated in a wide range of psychological disorders, including schizophrenia, bipolar (manic-depressive) disorder, major depression, alcoholism, autism, dementia due to Alzheimer's disease, anxiety disorders, dyslexia, and antisocial personality disorder (DiLalla et al., 1996; Plomin et al., 1997). Where genetic

Truth OR Fiction? REVISITED

Depressant drugs are drugs that cause depression.

FALSE. Depressant drugs are not drugs that cause depression. Rather, they are drugs that dampen the level of activity of the central nervous system (brain and spinal cord).

cerebrum The large mass of the forebrain, consisting of the two cerebral hemispheres.

cerebral cortex The wrinkled surface area of the cerebrum, it is responsible for processing sensory stimuli and controlling higher mental functions, such as thinking and use of language.

corpus callosum A thick bundle of fibers that connects the two cerebral hemispheres.

somatic nervous system The division of the peripheral nervous system that relays information from the sense organs to the brain and transmits messages from the brain to the skeletal muscles.

autonomic nervous system The division of the peripheral nervous system that regulates the activities of the glands and involuntary functions.

involuntary Automatic or without conscious direction.

sympathetic Pertaining to the division of the autonomic nervous system whose activity leads to heightened states of arousal.

parasympathetic Pertaining to the division of the autonomic nervous system whose activity reduces states of arousal and regulates bodily processes that replenish energy reserves.

THINK ABOUT IT

Do you believe that abnormal behavior is more a function of nature (biology) or nurture (environment)? Explain.

Quiz 2.1
The Biological Perspective

Truth OR Fiction? REVISITED

Freud likened the mind to a giant iceberg, with only the tip rising into conscious awareness.

TRUE. Freud believed that the larger part of the mind remains below the surface of consciousness.

Web Link 2.2
The American Psychoanalytic Association

psychoanalytic theory The theoretical model of personality developed by Sigmund Freud; also called psychoanalysis.

conscious To Freud, the part of the mind that corresponds to our present awareness.

factors play a role, they involve a complex interaction of multiple genes. We have yet to find any specific mental health disorder that can be explained by defects or variations on a single gene (USDHHS, 1999a). Nor do genetic factors alone account for any particular mental health disorder (Carey & DiLalla, 1994). Environmental factors also play important roles.

While we continue to learn more about the biological foundations of abnormal behavior patterns, the interface between biology and behavior can be construed as a two-way street. Researchers have uncovered links between psychological factors and many physical disorders and conditions (see Chapter 5). Researchers are also investigating whether the combination of psychological and drug treatments for such problems as depression, anxiety disorders, and substance abuse disorders, among others, may increase the therapeutic benefits of either of the two approaches alone. Although American psychiatry has become increasingly medicalized in recent years, some within the psychiatric community have warned their colleagues not to overlook the role of psychological factors in explaining and treating mental health problems (e.g., van Praag, 1988).

The Psychological Perspective

At about the time that biological models of abnormal behavior were beginning to achieve prominence with the contributions of Kraepelin, Griesinger, and others, another approach to understanding the bases of abnormal behavior began to emerge. This approach emphasized the psychological roots of abnormal behavior and was most closely identified with the work of the Austrian physician Sigmund Freud. Over time other psychological models would emerge from the behaviorist, humanistic, and cognitivist traditions. Let us begin our study of psychological perspectives with Freud's contribution and the development of psychodynamic models.

Psychodynamic Models

Psychodynamic theory is based on the contributions of Sigmund Freud and his followers. The psychodynamic model espoused by Freud, called **psychoanalytic theory,** is based on the belief that psychological problems such as hysteria are derived from unconscious psychological conflicts that can be traced to childhood. Freud held that much of our behavior is driven by unconscious motives and conflicts of which we are unaware. These underlying conflicts revolve around primitive sexual and aggressive instincts or drives and the need to keep these primitive impulses out of direct awareness. Why? Because awareness of these primitive impulses, including murderous urges and incestuous im-pulses, would flood the conscious self with crippling anxiety. Within the Freudian view, abnormal behavior patterns such as hysteria represent "symptoms" of the dynamic struggles taking place within the mind. In the case of hysteria, the "symptom" represents the *conversion* of an unconscious psychological conflict into a physical problem.

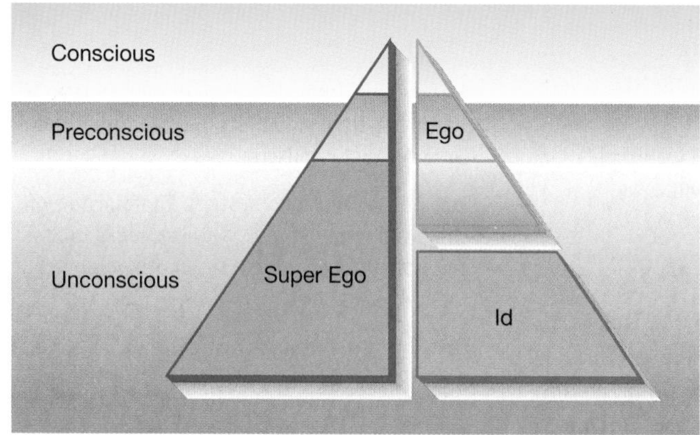

FIGURE 2.4 The parts of the mind, according to Freud. In psychodynamic theory, the mind is akin to an iceberg in that only a small part of it rises to conscious awareness at any moment in time. Although material in the preconscious mind may be brought into consciousness by focusing our attention on it, the impulses and ideas in the unconscious tend to remain veiled in mystery.

The Structure of the Mind Freud's clinical experiences led him to conclude that the mind is like an iceberg (Figure 2.4). Only the tip of an iceberg is visible above the surface of the water. The great mass of the iceberg lies below the surface. Freud came to believe that people, similarly, perceive but a few of the ideas, wishes, and impulses that dwell within them and determine their behavior. Freud held that the larger part of the mind, which includes our deepest wishes, fears, and instinctual urges, remains below the surface of consciousness. Freud labeled the region that corresponds to our present awareness the **conscious** part of the mind. The regions that lie beneath the surface of awareness were labeled the *preconscious* and the *unconscious.*

Ego and Id. Psychodynamic theorists believe the ego curbs the instinctual demands of id. The ego seeks socially acceptable ways of channeling these demands. When you share a meal with other people, you do not grab their food or snatch away the serving dish.

In the **preconscious** is the part of the mind are found memories of experiences that are not in awareness, but that can be brought into awareness by focusing on them. Your telephone number, for example, remains in the preconscious until you focus on it. The **unconscious** is the part of the mind, the largest part of the mind, remains shrouded in mystery. Its contents can only be brought to awareness with great difficulty, if at all. Freud believed the unconscious is the repository of biological drives, or instincts, such as sex and aggression.

The Structure of Personality According to Freud's **structural hypothesis,** the personality is divided into three mental entities, or **psychic** structures: the *id, ego,* and *superego*. Psychic structures cannot be seen or measured directly, but their presence is suggested by observable behavior and expressed in thoughts and emotions.

The **id** is the only psychic structure present at birth. It is the repository of our baser drives and instinctual impulses, including hunger, thirst, sex, and aggression. The id, which operates completely in the unconscious, was described by Freud as "a chaos, a cauldron of seething excitations" (1933/1964, p. 73). The id follows the **pleasure principle.** It demands instant gratification of instincts without consideration of social rules or customs or the needs of others. It operates by **primary process thinking,** which is a mode of relating to the world through imagination and fantasy. This enables the id to achieve gratification by conjuring up the mental image of the object of desire.

During the first year of life, the child discovers that every demand is not instantly gratified. He or she must learn to cope with the delay of gratification. The **ego** develops during this first year to organize reasonable ways of coping with frustration. Standing for "reason and good sense" (Freud, 1933/1964, p. 76), the ego seeks to curb the demands of the id and to direct behavior in keeping with social customs and expectations. Gratification can thus be achieved, but not at the expense of social disapproval. The id floods your consciousness with hunger pangs. Were it to have its way, the id might also prompt you to wolf down any food at hand or even to swipe someone else's plate. But the ego creates the ideas of walking to the refrigerator, making yourself a sandwich, and pouring a glass of milk.

The ego is governed by the **reality principle.** It considers what is practical and possible, as well as the urgings of the id. The ego engages in **secondary process thinking**—the remembering, planning, and weighing of circumstances that permit a compromise between the fantasies of the id and the realities of the world outside. The ego lays the groundwork for the development of the conscious sense of oneself as a distinct individual.

preconscious To Freud, the part of the mind whose contents lie outside of present awareness but can be brought into awareness by focusing attention.

unconscious To Freud, the part of the mind that lies outside the range of ordinary awareness and that contains instinctual urges.

structural hypothesis The belief that the clashing forces within the personality can be divided into three structures: id, ego, and superego.

psychic Relating to mental phenomena.

id The unconscious psychic structure, present at birth, that contains primitive instincts and is regulated by the pleasure principle.

pleasure principle The governing principle of the id, involving demands for immediate gratification of needs.

primary process thinking In infancy, the mental process by which the id seeks gratification by imagining that it possesses what it desires; thinking that is illogical or magical.

ego The psychic structure that corresponds to the concept of the self, governed by the reality principle and characterized by the ability to tolerate frustration.

reality principle The governing principle of the ego, which involves considerations of social acceptability and practicality.

secondary process thinking The reality-based thinking processes and problem-solving activities of the ego.

superego The psychic structure that incorporates the values of the parents and important others and that is governed by the moral principle; consists of two parts, the conscience and the ego ideal.

identification The process of incorporating the personality or behavior of others.

moral principle The principle that governs the superego to set and enforce moral standards.

ego ideal The set of higher social values and moral ideals embodied in the superego.

defense mechanisms The reality-distorting strategies used by the ego to shield the self from awareness of anxiety-provoking materials.

repression A defense mechanism involving the unconscious ejection of anxiety-provoking ideas, images, or impulses.

During middle childhood, the **superego** develops. The moral standards and values of parents and other key people become internalized through a process of **identification.** The superego operates according to the **moral principle;** it demands strict adherence to moral standards. The superego represents the moral values of an ideal self, called the **ego ideal.** It also serves as a conscience, or internal moral guardian, that monitors the ego and passes judgment on right and wrong. It metes out punishment in the form of guilt and shame when it finds that ego has failed to adhere to the superego's moral standards. Ego stands between the id and the superego. It endeavors to satisfy the cravings of the id without offending the moral standards of the superego.

Defense Mechanisms Although part of the ego rises to consciousness, some of its activity is carried out unconsciously. In the unconscious, the ego serves as a kind of watchdog, or censor, that screens impulses from the id. It uses **defense mechanisms** (psychological defenses) to prevent socially unacceptable impulses from rising into consciousness. If it were not for these defenses, or defense mechanisms, the darkest sins of our childhoods, the primitive demands of our ids, and the censures of our superegos might disable us psychologically. **Repression,** or motivated forgetting (the banishment of unacceptable ideas or motives to the unconscious), is considered the most basic of the defense mechanisms. A number of these defense mechanisms are described in Table 2.2.

A dynamic unconscious struggle thus takes place between the id and the ego. It pits biological drives that strive for expression (the id) against the ego, which seeks to restrain

TABLE 2.2 Some Defense Mechanisms of the Ego, According to Psychodynamic Theory

Defense Mechanism	Definition	Examples
Repression	The ejection of anxiety-evoking ideas from awareness.	A student forgets a difficult term paper is due.
		A patient in therapy forgets an appointment when anxiety-evoking material is to be discussed.
Regression	The return, under stress, to a form of behavior characteristic of an earlier stage of development.	An adolescent cries when forbidden to use the family car.
		An adult becomes highly dependent on his parents following the breakup of his marriage.
Rationalization	The use of self-deceiving justifications for unacceptable behavior.	A student blames her cheating on her teacher's leaving the room during a test.
		A man explains his cheating on his income tax by saying, "Everyone does it."
Displacement	The transfer of ideas and impulses from threatening or unsuitable objects onto less threatening objects.	A worker picks a fight with her spouse after being criticized sharply by her supervisor.
Projection	The thrusting of one's own unacceptable impulses onto others so that others are assumed to harbor them.	A hostile person perceives the world as being a dangerous place.
		A sexually frustrated person interprets innocent gestures of others as sexual advances.
Reaction formation	Assumption of behavior in opposition to one's genuine impulses in order to keep impulses repressed.	A person who is angry with a relative behaves in a "sickly sweet" manner toward that relative.
		A sadistic individual becomes a physician.
Denial	Refusal to accept the true nature of a threat.	Belief that one will not contract cancer or heart disease although one smokes heavily ("It can't happen to me").
Sublimation	The channeling of primitive impulses into positive, constructive efforts.	A person paints nudes for the sake of "beauty" and "art."
		A hostile person directs aggressive energies into competitive sports.

Source. From *Essentials of Psychology, 6th edition,* by S. A. Rathus, © 2001. Reprinted with permission of Brooks/Cole, an imprint of the Wadsworth Group, a division of Thomson Learning. FAX 800-730-2215.

them or channel them into socially acceptable outlets. When these conflicts are not resolved smoothly, they can lead to the development of symptoms or features associated with psychological disorders, such as hysterical symptoms, phobias, and behavioral problems. Because we cannot view the unconscious mind directly, Freud developed a method of mental detective work called *psychoanalysis*, which is described in Chapter 4.

The use of defense mechanisms to cope with feelings such as anxiety, guilt, and shame is considered normal. These mechanisms enable us to constrain impulses from the id as we go about our daily business. In his work *The Psychopathology of Everyday Life*, Freud noted that slips of the tongue and ordinary forgetfulness could represent hidden motives that are kept out of consciousness by repression. If a friend means to say, "I hear what you're saying," but it comes out, "I hate what you're saying," perhaps the friend is expressing a repressed emotion. If a lover storms out in anger but forgets his umbrella, perhaps he is unconsciously creating an excuse for returning. Defense mechanisms may also give rise to abnormal behavior, however. The person who regresses to an infantile state under pressures of enormous stress is clearly not acting adaptively to the situation.

Denial? Denial is a defense mechanism in which the ego fends off anxiety by preventing recognition of the true nature of a threat. Failing to take seriously warnings of health risks from cigarette smoking can be considered a form of denial.

Stages of Psychosexual Development Freud aroused heated controversy by arguing that sexual drives are the dominant factors in the development of personality, even in childhood. Freud believed that the child's basic relationship to the world in its first several years of life is organized around the pursuit of sensual or sexual pleasure. In Freud's view, all activities that are physically pleasurable, such as eating or moving one's bowels, are in essence "sexual." The word *sensuality* is probably closer in present-day meaning to what Freud meant by *sexuality*.

The drive for sexual pleasure represents, in Freud's view, the expression of a major life instinct, which he called **Eros**—the basic drive to preserve and perpetuate life. The energy contained in Eros that allows it to fulfill its function was termed **libido,** or sexual energy. Freud believed libidinal energy is expressed through sexual pleasure in different body parts, called **erogenous zones** as the child matures. In Freud's view, the stages of human development are **psychosexual** in nature, because they correspond to the transfer of libidinal energy from one erogenous zone to another. Freud proposed the existence of five psychosexual stages of development: oral, anal, phallic, latency, and genital.

The Oral Stage In the first year of life, the **oral stage,** infants achieve sexual pleasure by sucking their mothers' breasts and by mouthing anything that happens to be nearby. Oral stimulation, in the form of sucking and biting, is a source of both sexual gratification and nourishment.

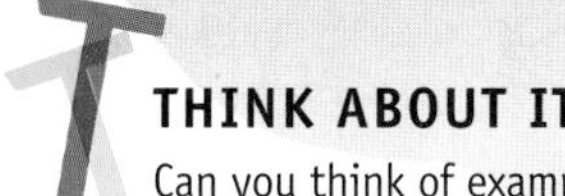

THINK ABOUT IT

Can you think of examples of behaviors in others (or yourself) in which defense mechanisms may have played a role?

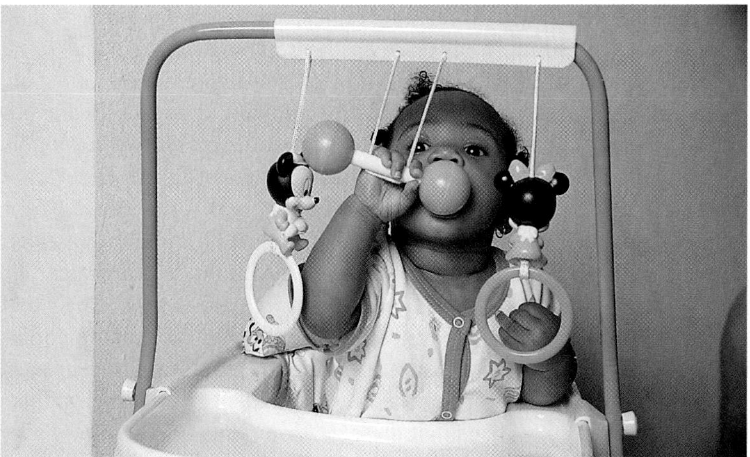

The oral stage of psychosexual development? According to Freud, the child's early encounters with the world are largely experienced through the mouth.

Eros Freud's concept of the basic life instinct that seeks to preserve and perpetuate life.

libido The energy of Eros; sexual drive or energy.

erogenous zone A part of the body that is sensitive to sexual stimulation.

psychosexual Pertaining to Freud's stages of development, in which libido becomes expressed through different erogenous zones during different stages.

oral stage The psychosexual stage during infancy in which pleasure is sought primarily through oral activities.

weaning The process of accustoming a child to eat solid food.

fixation Arrested development in the form of attachment to objects of an earlier developmental stage.

anal stage The psychosexual stage during toddlerhood in which pleasure is sought primarily through anal activities.

anal fixation Attachment to objects and behaviors characteristic of the anal stage.

anal retentive Excessive need for self-control and orderliness.

anal expulsive A loosening of restraint, as seen in extreme sloppiness of messiness.

phallic stage A psychosexual stage in early childhood during which pleasure is sought primarily through the phallic region and the child develops incestuous desires for the parent of the opposite sex.

Oedipus complex The conflict that occurs during the phallic stage of development, in which the young boy desires his mother and perceives his father as a rival.

One of Freud's central beliefs is that the child may encounter conflict during each of the psychosexual stages of development. Conflict during the oral stage centers around the issue of whether or not the infant receives adequate oral gratification. Too much gratification could lead the infant to expect that everything in life is given with little or no effort on his or her part. In contrast, early **weaning** might lead to frustration. Too little or too much gratification at any stage could lead to **fixation** in that stage, which leads to the development of personality traits characteristic of that stage. Oral fixations could include an exaggerated desire for "oral activities," which could become expressed in later life in smoking, alcohol abuse, overeating, and nail biting. Like the infant who depends on the mother's breast for survival and gratification of oral pleasure, orally fixated adults may also become clinging and dependent in their interpersonal relationships.

The Anal Stage During the **anal stage** of psychosexual development, the child experiences sexual gratification through contraction and relaxation of the sphincter muscles that control elimination of bodily waste. Although elimination had been controlled reflexively during much of the first year of life, the child now learns that she or he is able, although perhaps not reliably at first, to exercise voluntary muscular control over elimination.

Now the child begins to learn that she or he can delay gratification of the need to eliminate when the urge is felt. During toilet training, the issue of self-control may become a source of conflict between the parent and the child. **Anal fixations** that derive from this conflict are associated with two sets of traits. Harsh toilet training may lead to the development of **anal-retentive** traits, which involve excessive needs for self-control. These include perfectionism and extreme needs for orderliness, cleanliness, and neatness. By contrast, excessive gratification during the anal period might lead to **anal-expulsive** traits, which include carelessness and messiness.

Are young children interested in sex? According to Freud, even young children have sexual impulses. Freud's view of childhood sexuality shocked the scientific establishment of his day, and many of Freud's own followers believe that Freud placed too much emphasis on sexual motivation.

The Phallic Stage The next stage of psychosexual development, the **phallic stage,** generally begins during the third year of life. The major erogenous zone during the stage is the phallic region (the penis in boys, the clitoris in girls). Conflict between parent and child may occur over masturbation—the rubbing of the phallic areas for sexual pleasure—to which parents may react with threats and punishments. Perhaps the most controversial of Freud's beliefs was his suggestion that phallic-stage children develop unconscious incestuous wishes for the parent of the opposite gender and begin to view the parent of the same sex as a rival. Freud dubbed this conflict the **Oedipus complex,** after the legendary Greek king Oedipus who unwittingly slew his father and married his mother. The female version of the Oedipus complex has been named by some followers (although not by Freud himself) the **Electra complex,** after the character of Electra, who, according to Greek legend, avenged the death of her father, King Agamemnon, by slaying her father's murderers—her own mother and her mother's lover.

Freud believed the Oedipus conflict represents a central psychological conflict of early childhood, the resolution of which has far-reaching consequences in later development and in determining the acquisition of **gender roles.** He also believed that **castration anxiety** plays an important role in resolving the complex for boys. Adults sometimes threaten boys with castration to try to get them to stop touching themselves. Freud documented such castration threats from parents or nurses in several case studies. At some point boys discover that girls are different—they do not have penises. Freud conjectured that boys might imagine that girls had lost their penises as a form of punishment. Going further, Freud hypothesized that boys develop castration anxiety, based on the fantasy that their rivals for their mother's affections, namely their fathers, would seek to punish them for their incestuous wishes by removing the organ that has become connected with sexual pleasure. To prevent castration, Freud argued, boys repress their incestuous wishes for their mothers and identify with their fathers. Keep in mind that Freudian theory posits that these developments (incestuous wishes and castration anxiety) are largely unconscious and are part and parcel of normal development. Successful resolution of the Oedipus complex involves the boy repressing his incestuous wishes for his mother and identifying with his father. This identification leads to development of the aggressive, independent characteristics associated with the traditional masculine gender role.

The Oedipus complex in girls is somewhat of a mirror image of the one in boys. Freud believed little girls naturally become envious of boys' penises. This jealousy leads them to become resentful toward their mothers, whom they blame for bringing them into the world so "ill-equipped." Girls develop the desire to possess their fathers, in a way substituting their fathers' penises for their own missing ones. But the rivalry with their mothers for their fathers' affection places them in peril of losing their mothers' love and protection.

Successful resolution of the complex for the girl involves repression of the incestuous wishes for her father and identification with her mother, leading to the acquisition of the more passive, dependent characteristics traditionally associated with the feminine sex role. Eventually, the wish for a penis is transformed into the desire to marry a man and bear children, which represents the ultimate adjustment of surrendering the wish "to be a man" by accepting a baby as a form of penis substitute. Freud hypothesized that, in adulthood, women who retain the unconscious wish for a penis of their own ("to be a man") can become maladjusted and develop masculine-typed characteristics such as competitiveness and dominance and even a lesbian sexual orientation.

The Oedipus complex comes to a point of resolution, whether fully resolved or not, by about the age of 5 or 6. From the identification with the parent of the same gender comes the internalization of parental values in the form of the superego. Children then enter the **latency stage** of psychosexual development, a period of late childhood during which sexual impulses remain in a latent state. Interests become directed toward school and play activities.

The Genital Stage Sexual drives are once again aroused with the **genital stage,** beginning with puberty, which reaches fruition in mature sexuality, marriage, and the bearing of children. The sexual feelings toward the parent of the opposite gender that had remained repressed during the latency period emerge during adolescence but are displaced, or transferred, onto socially appropriate members of the opposite gender. Boys might still look for girls "just like the girl that married dear old dad," and, in parallel girls might still be attracted to boys who resemble their "dear old dads."

In Freud's view, successful adjustment during the genital stage involves attaining sexual gratification through sexual intercourse with someone of the opposite gender, presumably within the context of marriage. Other forms of sexual expression, such as oral or anal stimulation, masturbation, and homosexual activity, are considered **pregenital** fixations, or immature forms of sexual conduct.

Other Psychodynamic Theorists Freud left us a rich intellectual legacy that has stimulated the thinking of many theorists. Psychodynamic theory has been shaped over the years by the contributions of other psychodynamic theorists who shared certain central tenets in common with Freud, such as the belief that behavior reflects unconscious motivation, inner conflict, and the operation of defensive responses to anxiety. They tended to place lesser emphasis than Freud on the roles of basic instincts such as sex and aggression, and to place greater emphasis on roles for conscious choice, self-direction, and creativity. These theorists also differed from each other in various ways.

One of the most prominent of the early psychodynamic theorists was Carl Jung (1875–1961), a Swiss psychiatrist who was formerly a member of Freud's inner circle. His break with Freud came when he developed his own psychodynamic theory, which he called **analytical psychology.** Like Freud, Jung believed that unconscious processes are important in explaining behavior. Jung believed that an understanding of human behavior must incorporate the facts of self-awareness and self-direction as well as the impulses of the id and the mechanisms of defense. He believed that not only do we have a *personal* unconscious, a repository of repressed memories and impulses, but we also inherit a **collective unconscious.** To Jung, the collective unconscious represents the accumulated experience of humankind, which he believed is passed down genetically through the generations. The

Truth OR Fiction? REVISITED

Freud believed that an ancient Greek legend about a king who slew his father and married his mother contained key insights into the nature of the development of the human psyche.

TRUE. Freud dubbed this conflict the Oedipus complex.

THINK ABOUT IT

Underlying the psychodynamic approach is the belief that we are not usually aware of the deeper motives and impulses that drive our behavior. Do you agree? Why or why not?

Electra complex The conflict that occurs during the phallic stage of development, in which the young girl desires her father and perceives her mother as a rival.

gender roles The characteristic ways in which males and females are expected to behave in a given culture.

castration anxiety The young boy's unconscious fear that he will be castrated as punishment for his incestuous desire for his mother.

latency stage The psychosexual stage in middle childhood characterized by repression of sexual impulses.

genital stage The final stage of psychosexual development, characterized by expression of libido through sexual intercourse with an adult of the opposite sex.

pregenital Referring to characteristics typical of psychosexual stages that precede the genital stage.

analytical psychology Jung's theory, emphasizing the collective unconscious, archetypes, and the self as the unifying force of personality.

collective unconscious The storehouse of archetypes and racial memories.

archetypes Primitive images or concepts that reside in the collective unconscious.

inferiority complex The feelings of inferiority that Adler believed to be a central source of motivation.

powerful drive for superiority Complex of feelings that motivates us to achieve prominence and social dominance.

creative self To Adler, the self-aware part of the personality that strives to achieve its potential.

individual psychology Adler's psychodynamic theory.

neo-Freudians Theorists, such as Jung, Adler, Horney, and Sullivan, who, in comparison with Freud, placed greater emphasis on the importance of cultural and social influences and lesser emphasis on sexual impulses.

ego psychology Modern psychodynamic approach that focuses more on the conscious strivings of the ego than on the hypothesized unconscious functions of the id.

ego analysts Psychodynamically oriented therapists who are influenced by ego psychology.

Web Link 2.3 WWW
Jungian Resources

collective unconscious is believed to contain primitive images, or **archetypes,** which reflect upon the history of our species, including vague, mysterious mythical images like the all-powerful God, the fertile and nurturing mother, the young hero, the wise old man, and themes of rebirth or resurrection. Although archetypes remain unconscious, in Jung's view, they influence our thoughts, dreams, and emotions and render us responsive to cultural themes in stories and films.

Alfred Adler (1870–1937), like Jung, had held a place in Freud's inner circle, but broke away as he developed his own beliefs that people are basically driven by an **inferiority complex,** not by the sexual instinct as Freud had maintained. For some people, feelings of inferiority are based on physical deficits and the resulting need to compensate for them. But all of us, because of our small size during childhood, encounter feelings of inferiority to some degree. These feelings lead to a **powerful drive for superiority**, which motivates us to achieve prominence and social dominance. In the healthy personality, however, strivings for dominance are tempered by devotion to helping other people.

Adler, like Jung, believed self-awareness plays a major role in the formation of personality. Adler spoke of a **creative self,** a self-aware aspect of personality that strives to overcome obstacles and develop the individual's potential. With the hypothesis of the creative self, Adler shifted the emphasis of psychodynamic theory from the id to the ego. Because our potentials are uniquely individual, Adler's views have been termed **individual psychology.**

Many other psychodynamic models have arisen with the work of Freud's followers, who are sometimes referred to collectively as **neo-Freudians.** Some psychodynamic theorists, such as Karen Horney (1885–1952) and Harry Stack Sullivan (1892–1949), focused on the social context of psychological problems and stressed the importance of child-parent relationships in determining the nature of later interpersonal relationships. Sullivan, for example, maintained that children of rejecting parents tend to become self-doubting and anxious. These personality features persist and impede the development of close relationships in adult life.

More recent psychodynamic models also place a greater emphasis on the self or the ego and less emphasis on the sexual instinct than Freud. Today, most psychoanalysts see

The power of archetypes. One of the reasons we may find the *Star Wars* saga so compelling is that it features archetypes such as the struggle between good and evil characters.

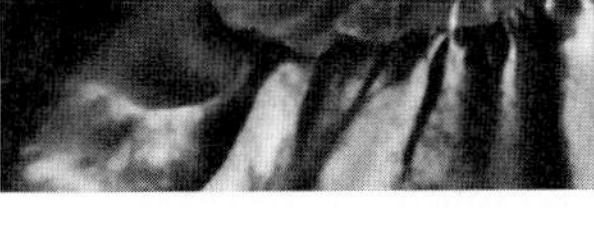

Karen Horney.

Erik Erikson.

Margaret Mahler.

people as motivated on two tiers: by the growth-oriented, conscious pursuits of the ego as well as by the more primitive, conflict-ridden drives of the id. Heinz Hartmann (1894–1970) was one of the originators of **ego psychology,** which posits that the ego has energy and motives of its own. Freud, remember, believed ego functions are fueled by the id, are largely defensive, and are perpetually threatened by the irrational. Hartmann and other **ego analysts** find Freud's views of the ego—and of people in general—too pessimistic and ignoble. Hartmann argued that the cognitive functions of the ego could be free of conflict. The choices to seek an education, dedicate oneself to art and poetry, and further humanity are not merely defensive forms of sublimation, as Freud had seen them.

Another ego analyst, Erik Erikson (1902–1994), attributed more importance to children's social relationships than to unconscious processes. Whereas Freud's developmental theory ends with the genital stage, beginning in early adolescence, Erikson focused on developmental processes that he believed continue throughout adulthood. The goal of adolescence, in Erikson's view, is not genital sexuality, but rather the attainment of **ego identity.** Adolescents who achieve ego identity develop a clearly defined and firm sense of who they are and what they believe in. Adolescents who drift without purpose or clarity of self remain in a state of **role diffusion** and are especially subject to negative peer influences.

One popular contemporary psychodynamic approach is termed **object-relations theory,** which focuses on how children come to develop symbolic representations of important others in their lives, especially their parents. One of the major contributors to object-relations theory was Margaret Mahler (1897–1985), who saw the process of separating from the mother during the first 3 years of life as crucial to personality development (discussed further in Chapter 9).

According to psychodynamic theory, we **introject,** or incorporate, into our own personalities, parts of parental figures in our lives. For example, you might introject your father's strong sense of responsibility or your mother's eagerness to please others. Introjection is more powerful when we fear losing others because of death or rejection. Thus we might be particularly apt to incorporate elements of people who *disapprove* of us or who see things differently.

In Mahler's view, these symbolic representations, which are formed from images and memories of others, come to influence our perceptions and behavior. We experience internal conflict as the attitudes of introjected people battle with our own. Some of our perceptions may be distorted or seem unreal to us. Some of our impulses and behavior may seem unlike us, as if they come out of the blue. With such conflict, we may not be able to tell where the influences of other people end and our "real selves" begin. The aim of Mahler's

ego identity The achievement of a firm sense of personal identity.

role diffusion A state of confusion, aimlessness, and heightened susceptibility to the suggestions of others, associated with failure to acquire a firm sense of identity during adolescence.

object-relations theory The psychodynamic viewpoint that focuses on the influences of internalized representations of the personalities of parents and other strong attachment figures (called "objects").

introject To unconsciously incorporate features of another person's personality into one's own ego structure.

neurosis A nonpsychotic form of disturbed behavior characterized by problems involving anxiety.

psychosis A severe form of disturbed behavior characterized by impaired ability to interpret reality and difficulty meeting the demands of daily life.

therapeutic approach was to help clients separate their own ideas and feelings from those of the introjected objects so they could develop as individuals—as their own persons.

Psychodynamic Views on Normality and Abnormality Freud believed that there is a thin line between the normal and the abnormal. Both normal and abnormal behavior are motivated or driven by irrational drives of the id. The difference may be largely a matter of degree. Mental health is a function of the dynamic balance among the psychic structures of id, ego, and superego (USDHHS, 1999a). In mentally healthy people, the ego is strong enough to control the instincts of the id and to withstand the condemnation of the superego. The presence of acceptable outlets for the expression of some primitive impulses, such as the expression of mature sexuality in marriage, decreases the pressures within the id and, at the same time, lessens the burdens of the ego in repressing the remaining impulses. Being reared by reasonably tolerant parents might prevent the superego from becoming overly harsh and condemnatory.

In people with psychological disorders, the balance among the psychic structures is lopsided. Some unconscious impulses may "leak," producing anxiety or leading to the development of **neuroses,** such as hysteria and phobias. The symptom expresses the conflict among the parts of the personality while it protects the self from recognizing the inner turmoil. A fear of knives, for example, shields the self from awareness of threatening unconscious impulses to use a knife to murder someone or attack the self. So long as the symptom is maintained (and the person avoids knives), the murderous or suicidal impulses are kept at bay. If the superego becomes overly powerful, it may create excessive feelings of guilt and lead to depression. An underdeveloped superego is believed to play a role in explaining the antisocial tendencies of people who intentionally hurt others without feelings of guilt.

Freud believed that the underlying conflicts in neuroses have childhood origins that are buried in the depths of the unconscious. Through psychoanalysis, he sought to help people uncover and learn to deal with these underlying conflicts to free themselves of the need to maintain the overt symptom.

Perpetual vigilance and defense take their toll. The ego can weaken and, in extreme cases, lose the ability to keep a lid on the id. **Psychosis** results when the urges of the id spill forth, untempered by an ego that is either weakened or underdeveloped. The fortress of the ego is overrun, and the person loses the ability to distinguish between fantasy and reality. Behavior becomes detached from reality. Primary process thinking and bizarre behavior rule the day. Psychoses are characterized, in general, by more severe disturbances of functioning than neuroses, by the appearance of bizarre behavior and thoughts, and by faulty perceptions of reality, such as hallucinations ("hearing voices" or seeing things that are not present). Speech may become incoherent and there may be bizarre posturing and gestures. The most prominent form of psychosis is schizophrenia, discussed in Chapter 13.

Freud equated psychological health with the *abilities to love and to work.* The normal person can care deeply for other people, find sexual gratification in an intimate relationship, and engage in productive work. To accomplish these ends, there must be an opportunity for sexual impulses to be expressed in a relationship with a partner of the opposite gender. Other impulses must be channeled (sublimated) into socially productive pursuits, such as work, enjoyment of art or music, or creative expression. When some impulses are expressed directly and others are sublimated, the ego has a relatively easy time repressing those that remain in the boiling cauldron.

Other psychodynamic theorists, such as Jung and Adler, emphasized the need to develop a differentiated self—the unifying force that provides direction to behavior and helps develop a person's potential. Adler also believed that psychological health involves efforts to compensate for feelings of inferiority by striving to excel in one or more of the arenas of human endeavor. For Mahler, similarly, abnormal behavior derives from failure to separate ourselves from those we have psychologically brought within us. The notion of a guiding self provides bridges between psychodynamic theories and other theories, such as humanistic theories (which also speak of a self and the fulfillment of inner potential) and social-cognitive theory (which speaks in terms of self-regulatory processes).

Evaluating Psychodynamic Models Psychodynamic theory has had a pervasive influence, not only on concepts of abnormal behavior but more broadly on art, literature, philosophy, and the general culture. It has focused attention on our inner lives—our dreams, fantasies, and hidden motives. People unschooled in Freud's writings nevertheless look for the symbolic meanings of each other's slips of the tongue and assume that abnormalities can be traced to early childhood. Terms like *ego* and *repression* have become commonplace, although their everyday meanings do not fully overlap with those intended by Freud.

One of the major contributions of the psychodynamic model was the increased awareness that people may be motivated by hidden drives and impulses of a sexual or aggressive nature. Freud's beliefs about childhood sexuality were both illuminating and controversial. Before Freud, children were perceived as *pure innocents*, free of sexual desire. Freud recognized, however, that young children, even infants, seek pleasure through stimulation of the oral and anal cavities and the phallic region. Yet his beliefs that primitive drives give rise to incestuous desires, intrafamily rivalries and conflicts, and castration anxiety and penis envy remain sources of controversy, even within psychodynamic circles. For one thing, these processes are deemed to occur largely if not entirely at an unconscious level and so are difficult if not impossible to study, let alone validate, by scientific means. For another, there is little evidence to support even the existence of the Oedipus complex, let alone its universality (Kupfersmid, 1995).

Freud's views of female psychosexual development have been roundly attacked by women and by modern-day psychoanalysts. One of the most prominent critics, the psychoanalyst Karen Horney, argued that Freud's views on women reflect underlying cultural prejudices in Western society. To Horney, cultural expectations play a greater role in shaping women's self-images than does penis envy. She argued that if women feel inferior, it is because they are relegated to second-class status in our society, not because they lack a penis. From birth, women are exposed to a male-dominated culture that treats women as an inferior sex. Horney's own views of child development take us in a different direction than Freud's. She emphasized the importance of parent-child relationships. She believed that when parents are harsh or uncaring, children may develop deep-seated feelings of anxiety and hostility that lead them to relate to others in unhealthy and rigid ways later in life.

In fairness to Freud, we should note that his theories should be viewed in the context of his day and time. In Freud's day, motherhood and family life were, by and large, the only socially proper avenues of fulfillment for women. Today, the choices available to women are more varied, and normality is not conceptualized in terms of rigidly defined gender roles.

Many critics, including some of Freud's followers, believe he placed too much emphasis on sexual and aggressive impulses and underemphasized social relationships. Critics have also argued that the psychic structures—the id, ego, and superego—may be little more than useful fictions, poetic ways to represent inner conflict. Many critics argue that Freud's hypothetical mental processes are not scientific concepts because they cannot be directly observed or tested. Therapists can speculate, for example, that a client "forgot" about an appointment because "unconsciously" she or he did not want to attend the session. Such unconscious motivation may not be subject to scientific verification, however. On the other hand, psychodynamically oriented researchers have developed scientific approaches that they believe make it possible to test many of Freud's concepts (Westen, 1998).

Learning Models

The psychodynamic models of Freud and his followers were the first major psychological theories of abnormal behavior, but other relevant psychologies were also taking shape early in the 20th century. Among the most important was the behavioral perspective, which is identified with contributions by the Russian physiologist Ivan Pavlov (1849–1936), the discoverer of the conditioned reflex, and the American psychologist John B. Watson (1878–1958), the father of **behaviorism.** The behavioral perspective focuses on the role of learning in explaining both normal and abnormal behavior. From a learning perspective, abnormal behavior represents the acquisition, or learning, of inappropriate, maladaptive behaviors.

behaviorism The school of psychology that defines psychology as the study of observable behavior.

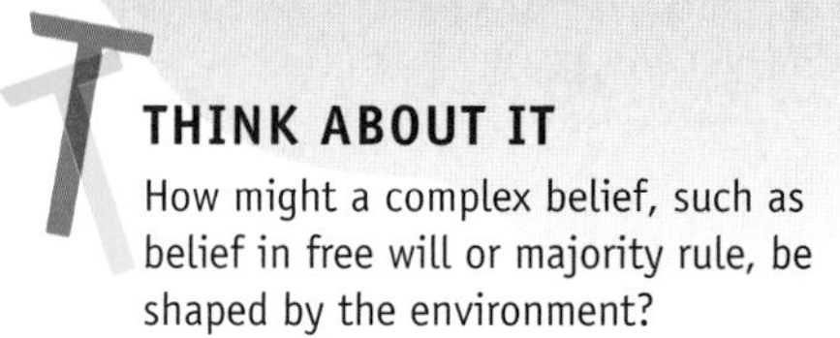

THINK ABOUT IT

How might a complex belief, such as belief in free will or majority rule, be shaped by the environment?

Ivan Pavlov. Russian physiologist Ivan Pavlov (center, with white beard) demonstrates his apparatus for classical conditioning to students. How might the principles of classical conditioning explain the acquisition of excessive irrational fears that we refer to as phobias?

From the medical and psychodynamic perspectives, abnormal behavior is *symptomatic*, respectively, of underlying biological or psychological problems. From the learning perspective, however, abnormal behavior need not be symptomatic of anything. The abnormal behavior itself is the problem. Abnormal behavior is regarded as learned in much the same way as normal behavior. Why, then, do some people behave abnormally? One reason is found in situational factors: Their learning histories, that is, might differ from most people's. For example, harsh punishment for early exploratory behavior, such as childhood sexual exploration in the form of masturbation, might give rise to adult anxieties over autonomy or sexuality. Poor child-rearing practices, such as the lack of praise or rewards for good behavior and harsh and capricious punishment for misconduct, might give rise to antisocial behavior. Then, too, children with abusive or neglectful parents might learn to pay more attention to inner fantasies than to the world outside, giving rise, at worst, to difficulty in distinguishing reality from fantasy.

B. F. Skinner.

Watson and other behaviorists, such as Harvard University psychologist B. F. Skinner (1904–1990), believed that human behavior is the product of genetic endowment and environmental or situational influences. Like Freud, Watson and Skinner discarded concepts of personal freedom, choice, and self-direction. But whereas Freud saw us as driven by irrational forces, behaviorists see us as products of environmental influences that shape and manipulate our behavior. To Watson and Skinner, even the belief that we have free will is determined by the environment. Behaviorists focus on the roles of two major forms of learning in shaping normal and abnormal behavior, classical conditioning and operant conditioning.

Role of Classical Conditioning The Russian physiologist Ivan Pavlov discovered the conditioned reflex (now called a *conditioned response*) quite by accident. In his laboratory, he harnessed dogs to an apparatus like that in Figure 2.5 to study their salivary response to food. Yet he observed that the animals would start salivating and secreting gastric juices even before they started eating. These responses appeared to be elicited by the sounds made by his laboratory assistants when they wheeled in the food cart. So Pavlov undertook a clever experimental program that showed that animals could learn to salivate in response to other stimuli, such as the sound of a bell, if these stimuli were *associated* with feeding.

FIGURE 2.5 The apparatus used in Ivan Pavlov's experiments on conditioning.
Pavlov used an apparatus such as this to demonstrate the process of conditioning. To the left is a two-way mirror, behind which a researcher rings a bell. After ringing the bell, meat is placed on the dog's tongue. Following several pairings of the bell and the meat, the dog learns to salivate in response to the bell. The animal's saliva passes through the tube to a vial, where its quantity may be taken as a measure of the strength of the conditioned response.

Because dogs don't normally salivate to the sound of bells, Pavlov reasoned that they had acquired this response, called a **conditioned response** (CR), or conditioned reflex, because it had been paired with a stimulus, called an **unconditioned stimulus** (US)—in this case, food—which naturally elicits salivation (see Figure 2.6). The salivation to food, an unlearned response, is called the **unconditioned response** (UR), and the bell, a previously neutral stimulus, is called the **conditioned stimulus** (CS).

Can you recognize classical conditioning in your everyday life? Do you flinch in the waiting room at the sound of the dentist's drill? The drill sounds may be conditioned stimuli for conditioned responses of fear and muscle tension.

Phobias or excessive fears may be acquired by classical conditioning. For instance, a person may develop a phobia for riding on elevators following a traumatic experience on an elevator. In this example, a previously neutral stimulus (elevator) becomes paired or associated with an aversive stimulus (trauma), which leads to the conditioned response (phobia).

conditioned response In classical conditioning, a learned response to a previously neutral stimulus.

unconditioned stimulus A stimulus that elicits an unlearned response.

unconditioned response An unlearned response.

conditioned stimulus A previously neutral stimulus that evokes a conditioned response after repeated pairings with an unconditioned stimulus that had previously evoked that response.

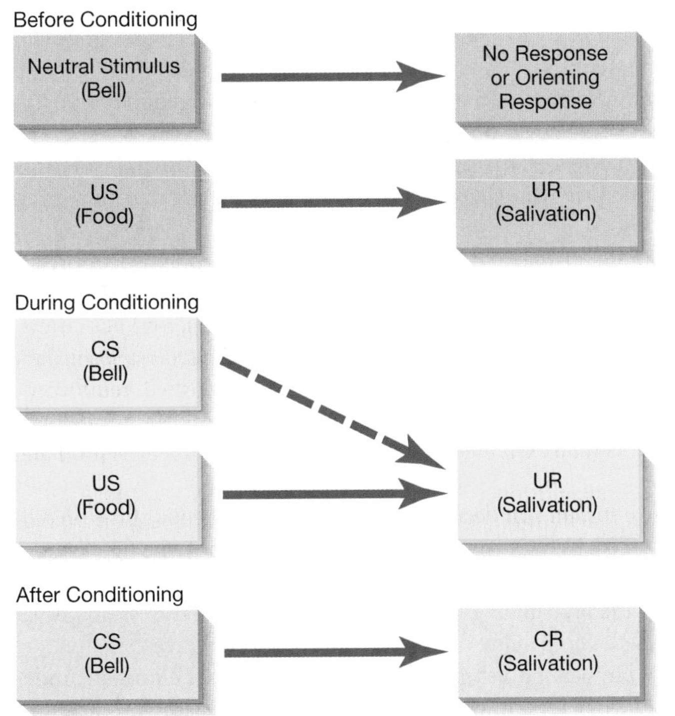

FIGURE 2.6 Schematic diagram of the process of classical conditioning.
Before conditioning, food (an unconditioned stimulus, or US) placed on a dog's tongue will naturally elicit salivation (an unconditioned response, or UR). The bell, however, is a neutral stimulus that may elicit an orienting response but not salivation. During conditioning, the bell (the conditioned stimulus, or CS) is rung while food (the US) is placed on the dog's tongue. After several conditioning trials have occurred, the bell (the CS) will elicit salivation (the conditioned response, or CR) when it is rung, even though it is not accompanied by food (the US). The dog is said to have been *conditioned,* or to have learned, to display the conditioned response (CR) in response to the conditioned stimulus (CS). Learning theorists have suggested that irrational excessive fears of harmless stimuli may be acquired through principles of classical conditioning.

reinforcement A stimulus or event that increases the frequency of the response that it follows.

reward A pleasant stimulus or event that increases the frequency of the response that it follows.

positive reinforcers Reinforcers that, when introduced, increase when the frequency of behavior.

negative reinforcers Reinforcers that on removal increase the frequency of behavior.

primary reinforcers Reinforcers that fulfill basic needs, such as water, food, warmth, and relief from pain.

secondary reinforcers Stimuli that gain reinforcement value through their association with established reinforcers, such as money and social approval.

punishments Unpleasant stimuli that reduce the frequency of the behaviors they follow.

From the learning perspective, normal behavior involves responding adaptively to stimuli, including conditioned stimuli. After all, if we do not learn to be afraid of drawing our hand too close to a hot stove after one or two experiences of being burned or nearly burned, we might suffer unnecessary burns. On the other hand, acquiring inappropriate and maladaptive fears on the basis of conditioning may cripple our efforts to function in the world. Chapter 6 explains how conditioning may help explain anxiety disorders such as phobias and posttraumatic stress disorder.

Role of Operant Conditioning Operant conditioning involves the acquisition of behaviors, called *operant behaviors,* that are emitted by the organism and that operate upon, or manipulate, the environment to produce certain effects. The psychologist B. F. Skinner (1938) showed that food-deprived pigeons would learn to peck buttons when food pellets drop into their cages as a result. It takes a while for the birds to happen on the first peck, but after a few repetitions of the association of button pecking and food, pecking behavior, an operant response, becomes fast and furious until the pigeons have had their fill.

In operant conditioning, organisms acquire responses or skills that lead to **reinforcement.** Reinforcers are changes in the environment (stimuli) that increase the frequency of the preceding behavior. A **reward** is a *pleasant* stimulus that increases the frequency of behavior, and so it is a type of reinforcer. Skinner found the concept of reinforcement to be preferable to that of reward because it is defined in terms of relationships between observed behaviors and environmental effects. In contrast to *reward,* the meaning of reinforcement does not depend on "mentalistic" conjectures about what is pleasant to another person or lower animal. Many psychologists use the words *reinforcement* and *reward* interchangeably, however.

Positive reinforcers boost the frequency of behavior when they are presented. Food, money, social approval, and the opportunity to mate are examples of positive reinforcers. **Negative reinforcers** increase the frequency of behavior when they are *removed.* Fear, pain, discomfort, and social disapproval are examples of negative reinforcers. We learn responses that lead to their removal (like learning to turn on the air conditioner to remove unpleasant heat and humidity from a room).

Adaptive, normal behavior involves learning responses or skills that permit us to obtain positive reinforcers and to avoid or remove negative reinforcers. Thus we learn adaptive behaviors that permit us to obtain such positive reinforcers as money, food, and social approval and to avoid or remove such negative reinforcers as fear, pain, and social condemnation. But if our early learning environments do not provide opportunities for learning new skills, we might be hampered in our efforts to develop the skills needed to obtain reinforcers. A lack of social skills, for example, may reduce opportunities for social reinforcement (approval or praise from others), especially when a person withdraws from social situations. This may lead to depression and social isolation. In Chapter 8, we examine links between changes in reinforcement levels and the development of depression.

We can also differentiate *primary* and *secondary,* or conditioned, reinforcers. **Primary reinforcers** influence behavior because they satisfy basic physical needs. We do not learn to respond to these basic reinforcers; we are born with that capacity. Food, water, sexual stimulation, and escape from pain are examples of primary reinforcers. **Secondary reinforcers** influence behavior through their association with established reinforcers. Thus we learn to respond to secondary reinforcers. People learn to seek money—a secondary reinforcer—because it can be exchanged for such primary reinforcers as food and heat (or air conditioning).

Punishments are aversive stimuli that decrease or suppress the frequency of the preceding behavior when they are applied. Negative reinforcers, by contrast, increase the frequency of the preceding behavior when they are removed. A loud noise, for example, can be either a punishment (if by its introduction the probability of the preceding behavior increases) or a negative reinforcer (if by its removal the probability of the preceding behavior decreases).

Web Link 2.4 WWW
B. F. Skinner Foundation

Punishment, especially physical punishment, may suppress but not eliminate undesirable behavior. The behavior may return when the punishment is withdrawn. One limitation of punishment is that it does not lead to the development of more desirable alternative behaviors. Another is that it may also encourage people to withdraw from such learning situations. Punished children may cut classes, drop out of school, or run away. Punishment may generate anger and hostility rather than constructive learning. Finally, because people also learn by observation, punishment may become imitated as a means for solving interpersonal problems.

Rewarding desirable behavior is thus generally preferable to punishing misbehavior. But rewarding good behavior requires paying attention to it, not just to misbehavior. Some children who develop conduct problems can gain the attention of other people only by misbehaving. They learn that when they act out, others will pay attention to them. To them punishment may actually serve as a positive reinforcer, increasing the rate of response of the behavior it follows. Learning theorists point out that it is not sufficient to expect good conduct from children. Instead, adults need to teach children proper behavior and regularly reinforce them for emitting it.

Let us now consider a contemporary model of learning, called *social-cognitive theory* (formerly called *social-learning theory*), which broadens the focus of traditional learning theory by considering the role of cognitive factors in learning and behavior.

Truth OR Fiction? REVISITED

Punishment does not eliminate undesirable behavior.

TRUE. Punishment does not eliminate an undesirable behavior, but only suppresses it. The behavior may return when it is no longer punished.

Social-Cognitive Theory **Social-cognitive theory** represents the contributions of theorists such as Albert Bandura, Julian B. Rotter, and Walter Mischel. Social-cognitive theorists emphasize the roles of thinking or cognition and of learning by observation, or **modeling**, in human behavior. For example, social-cognitive theorists suggest that phobias may be learned *vicariously*, by observing the fearful reactions of others in real life, on television, or in the movies.

Social-cognitive theorists view people as having an impact on their environments, just as their environment has an impact on them (Bandura, 2001). They see people as self-aware and purposeful learners who seek information about their environments, who do not just respond automatically to the stimuli that impinge on them. Social-cognitive theorists concur with traditional behaviorists that theories of human nature should be tied to observable behavior. They assert, however, that factors *within* the person should also be considered in explaining human behavior. Rotter (1990), for example, argues that behavior cannot be predicted from situational factors alone. Whether or not people behave in certain ways also depends on certain cognitive factors, such as the person's **expectancies** about the outcomes of behavior. For example, we see in Chapter 10 that people who hold more positive expectancies about the outcomes of using drugs are more likely to use them and to use them in larger quantities.

social-cognitive theory A learning-based theory that emphasizes observational learning and incorporates roles for both situational and cognitive variables in determining behavior.

modeling Learning by observing and imitating the behavior of others.

expectancies Beliefs about expected outcomes.

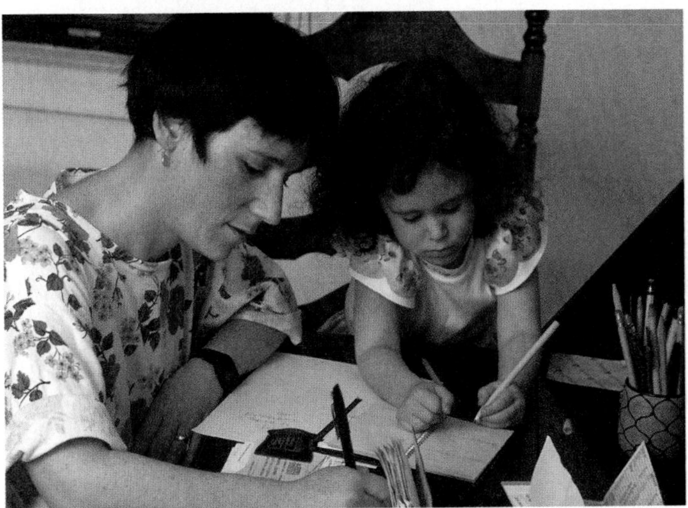

Observational learning. According to social-cognitive theory, much human behavior is acquired through modeling, or observational learning.

Carl Rogers and Abraham Maslow, two of the principal forces in humanistic psychology.

THINK ABOUT IT

How has your present behavior been influenced by your learning history? What learning principles (classical conditioning, operant conditioning, observational learning) can you use to account for your behavior, both normal and abnormal?

Evaluating Learning Models One of the principal values of learning models, in contrast to psychodynamic approaches, is their emphasis on observable behavior and environmental factors, such as rewards and punishments, that can be systematically manipulated to observe their effects on behavior. Learning perspectives have spawned a major model of therapy, called *behavior therapy* (also called *behavior modification*), which involves the systematic application of learning principles to help people make adaptive behavioral changes (see Chapter 4). Behavior therapy techniques have been applied to helping people overcome a wide range of psychological problems, including phobias and other anxiety disorders, sexual dysfunctions, and depression. Moreover, reinforcement-based programs are now widely used in helping parents learn better parenting skills and helping children learn in the classroom.

Web Link 2.5
Association for Humanistic Psychology

Critics contend that behaviorism cannot explain the richness of human behavior and that human experience cannot be reduced to observable responses. Many learning theorists, too—especially social-cognitive theorists—have been dissatisfied with the strict behavioristic view that environmental influences—rewards and punishments—mechanically control our behavior. Humans experience thoughts and dreams and formulate goals and aspirations; behaviorism does not seem to address much of what it means to be human. Social-cognitive theorists have broadened the scope of traditional behaviorism, but critics claim that social-cognitive theory places too little emphasis on genetic contributions to behavior and has failed to provide a meaningful account of self-awareness.

Self-Actualization. Humanistic theorists believe that there exists in each of us a drive toward self-actualization—to become all that we are capable of being. In the humanistic view, each of us, like artist Faith Ringgold—is unique. No two people follow quite the same pathway toward self-actualization.

Humanistic Models

A "third force" in modern psychology emerged during the mid-20th century—humanistic psychology. Humanistic theorists such as American psychologists Carl Rogers (1902–1987) and Abraham Maslow (1908–1970) believed that human behavior could not be explained as the product of either unconscious conflicts or simple conditioning. Rejecting the determinism implicit in these theories, these theorists saw people as *actors* in the drama of life, not *reactors* to instinctual or environmental pressures. They focused on the importance of subjective conscious experience and self-direction. Humanistic psychology is closely linked with the school of European philosophy called *existentialism*. The existentialists, notably the philosophers Martin Heidegger (1889–1976) and Jean-Paul Sartre (1905–1980), focused on the search for meaning and the importance of choice in human existence. Existentialists believe our humanness makes us responsible for the directions our lives will take.

The humanists maintain that people have an inborn tendency toward **self-actualization**—to strive to become all they are capable of being. Each of us possesses a singular cluster of traits and talents that gives rise to an individual set of feelings and needs and grants us a unique perspective on life. Despite the finality of death, we can each imbue our lives with meaning and purpose if we recognize and accept our genuine needs and feelings. By being true to ourselves, we live *authentically*. We may not decide to act out every wish and fancy, but self-awareness of authentic feelings and subjective experiences can help us make more meaningful choices.

To understand abnormal behavior, in the humanist's view, we need to understand the roadblocks that people encounter in striving for self-actualization and authenticity. To accomplish this, psychologists must learn to view the world from clients' own perspectives because their subjective views of their world lead them to interpret and evaluate their experiences in either self-enhancing or self-defeating ways. The humanistic viewpoint is sometimes called the *phenomenological* perspective because it involves the attempt to understand the subjective or phenomenological experience of others, the stream of conscious experiences people have of "being in the world."

self-actualization In humanistic psychology, the tendency to strive to become all that one is capable of being. The motive that drives one to reach one's full potential and express one's unique capabilities.

conditional positive regard Valuing other people on the basis of whether their behavior meets one's approval.

conditions of worth Standards by which one judges the worth or value of oneself or others.

Truth OR Fiction? REVISITED

Children may acquire a distorted self-concept that mirrors what others expect them to be but which does not reflect who they truly are.

TRUE. According to Rogers, children can develop a distorted self-concept that mirrors what others expect them to be but which is not true to themselves.

Humanistic Concepts of Abnormal Behavior Rogers developed the most influential humanistic account of abnormal behavior (Rogers, 1951). His central belief was that abnormal behavior results from the development of a distorted concept of the self. When parents show children **conditional positive regard**—accept them only when they behave in an approved manner—the children may learn to disown the thoughts, feelings, and behaviors their parents have rejected. With conditional positive regard, children may learn to develop **conditions of worth,** to think of themselves as worthwhile only if they behave in certain approved ways. For example, children who are valued by their parents only when they are compliant may deny to themselves ever having feelings of anger. Children in some families learn it is unacceptable to hold their own ideas, lest they depart from their parents' views. Parental disapproval causes them to see themselves as rebels and their feelings as wrong, selfish, or evil. If they wish to retain self-esteem, they may have to deny their genuine feelings or disown parts of themselves. They may thus develop a distorted *self-concept*, or view of themselves, and become strangers to their true selves.

Rogers believed that anxiety might arise when we begin to sense that our feelings and ideas are inconsistent with the distorted self-concept we have developed that mirrors what others expect us to be. Because anxiety is unpleasant, we may deny to ourselves that these feelings and ideas even exist. And so the actualization of our authentic self is bridled by the denial of important ideas and emotions. Psychological energy is channeled toward continued denial and self-defense, not toward growth. Under such conditions, we cannot hope to perceive our genuine values or personal talents, leading to frustration and setting the stage for abnormal behavior.

So we cannot fulfill all of the wishes of others and remain true to ourselves. This does not mean that self-actualization invariably leads to conflict. Rogers was more optimistic about human nature than Freud. Rogers believed that people hurt one another or become antisocial in their behavior only when they are frustrated in their endeavors to reach their unique potentials. But when parents and others treat children with love and tolerance for their differences, children, too, grow to be loving—even if some of their values and preferences differ from their parents' choices.

In Rogers's view, the pathway to self-actualization involves a process of self-discovery and self-acceptance, of getting in touch with our true feelings, accepting them as our own, and acting in ways that genuinely reflect them. These are the goals of Rogers's method of psychotherapy, called *client-centered therapy* or *person-centered therapy*.

The makings of unconditional positive regard. Rogers believed that parents can help their children develop self-esteem and set them on the road toward self-actualization by showing them unconditional positive regard—prizing them on the basis of their inner worth, regardless of their behavior of the moment.

Web Link 2.6 WWW
Online Self-Esteem Test

Evaluating Humanistic Models The strength of humanistic models in understanding abnormal behavior lies largely in their focus on conscious experience and their innovation of therapy methods that assist people along pathways of self-discovery and self-acceptance. The humanistic movement put concepts of free choice, inherent goodness, responsibility, and authenticity back on center stage and brought them into modern psychology. Ironically, the primary strength of the humanistic approach—its focus on conscious experience—may also be its primary weakness. Conscious experience is private and subjective. Therefore, the validity of formulating theories in terms of consciousness has been questioned. How can psychologists be certain they accurately perceive the world through the eyes of their clients?

Nor can the concept of self-actualization—which is so basic to Maslow and Rogers—be proved or disproved. Like a psychic structure, a self-actualizing force is not directly measurable or observable. It is inferred from its supposed effects. Self-actualization also yields circular explanations for behavior. When someone is observed engaging in striving, what do we learn by attributing striving to a self-actualizing tendency? The source of the tendency remains a mystery. Similarly, when someone is observed not to be striving, what do we gain by attributing the lack of endeavor to a blocked or frustrated self-actualizing tendency? We must still determine the source of frustration or blockage.

THINK ABOUT IT

Whom would you consider to be a self-actualizer? Are you a self-actualizer? Why or why not?

Cognitive Models

The word *cognitive* derives from the Latin *cognitio*, meaning "knowledge." Cognitive theorists study the cognitions—the thoughts, beliefs, expectations, and attitudes—that accompany and may underlie abnormal behavior. They focus on how reality is colored by our expectations, attitudes, and so forth, and how inaccurate or biased processing of information about the world—and our places within it—can give rise to abnormal behavior. Cognitive theorists believe that our interpretations of the events in our lives, and not the events themselves, determine our emotional states. Several of the more prominent cognitive models of abnormal behavior patterns are information-processing approaches and the models developed by psychologist Albert Ellis and psychiatrist Aaron Beck.

Information-Processing Models Many cognitive psychologists are influenced by concepts of computer science. Computers process information to solve problems. Information is fed into the computer (encoded so it can be accepted by the computer as input). Then it is placed in *memory* while it is manipulated. You can also place the information permanently in *storage*, on a floppy disk, a hard disk, or another device. Information-processing theorists apply these processes to human cognition. They think in terms such as *input* (based on perception), *manipulation* (interpreting or transforming information), *storage* (placing information in memory), *retrieval* (accessing information from memory), and *output* (acting on the information). They view psychological disorders as disturbances in these processes. Disturbances can be caused by the blocking or distortion of input or by faulty storage, retrieval, or manipulation of information. Any of these can lead to lack of output or to distorted output (e.g., bizarre behavior). People with schizophrenia, for example, frequently jump from topic to topic in a disorganized fashion, which may reflect problems in retrieving and manipulating information. They also seem to have difficulty focusing their attention and filtering out extraneous stimuli, such as distracting noises. This may represent problems relating to initial processing of input from their senses.

Manipulation of information may also be distorted by what cognitive therapists call *cognitive distortions*, or errors in thinking. For example, people who are depressed tend to develop an unduly negative view of their personal situation by exaggerating the importance of unfortunate events they experience (Meichenbaum, 1993). Cognitive theorists such as Albert Ellis and Aaron Beck have postulated that distorted or irrational thinking patterns can lead to emotional problems and maladaptive behavior.

Social-cognitive theorists, who share many basic ideas with the cognitive theorists, focus on the ways in which social information is encoded. For example, aggressive boys and adolescents are likely to incorrectly encode other people's behavior as threatening (see

Albert Ellis and Aaron Beck, two of the leading cognitive theorists.

Chapter 14). They assume other people intend them ill when they do not. Aggressive children and adults may behave in ways that elicit coercive or hostile behavior from others, which serves to confirm their aggressive expectations (Meichenbaum, 1993). Rapists, especially date rapists, may misread a woman's expressed wishes. They may wrongly assume, for example, that the woman who says "no" really means yes and is merely playing "hard to get."

Albert Ellis Psychologist Albert Ellis (1977b, 1993), a prominent cognitive theorist, believes that troubling events in themselves do not lead to anxiety, depression, or disturbed behavior. Rather, it is the irrational beliefs we hold about unfortunate experiences that foster negative emotions and maladaptive behavior. Consider someone who loses a job and becomes anxious and despondent about it. It may seem that being fired is the direct cause of the person's misery, but the misery actually stems from the person's beliefs about the loss, not directly from the loss itself.

Ellis uses an "ABC approach" to explain the causes of the misery. Being fired is an *activating event* (A). The ultimate outcome, or *consequence* (C), is emotional distress. But the activating event (A) and the consequences (C) are mediated by various *beliefs* (B). Some of these beliefs might include "That job was the major thing in my life," "What a useless washout I am," "My family will go hungry," "I'll never be able to find another job as good," "I can't do a thing about it." These exaggerated and irrational beliefs compound depression, nurture helplessness, and distract us from evaluating what to do. For instance, the beliefs "I can't do a thing about it" and "What a useless washout I am" promote helplessness.

The situation can be diagrammed like this:

Activating Event → Belief → Consequences

Ellis points out that apprehension about the future and feelings of disappointment are perfectly normal when people face losses. However, the adoption of irrational beliefs leads people to **catastrophize** their disappointments, leading to profound distress and states of depression. By intensifying emotional responses and nurturing feelings of helplessness, such beliefs impair coping ability. Examples of irrational beliefs include "I must have the love and approval of nearly everyone who is important to me or else I'm a worthless and unlovable person," and "I must be competent in virtually everything I do or else I'm an inadequate and incompetent person." In his later writings, Ellis emphasized the demanding nature of irrational or self-defeating beliefs—tendencies to impose "musts" and "shoulds" on ourselves (Ellis, 1993, 1997). Ellis notes that the desire for others' approval is understandable, but it is irrational to assume that one must have it to survive or to feel

Truth OR Fiction? REVISITED

According to a leading cognitive theorist, states of emotional distress are caused by the beliefs people hold about their life experiences, not by the experiences themselves.

TRUE. Ellis believed that emotional distress is determined by the beliefs we hold about events we experience, not by the events themselves.

WWW **Web Link 2.7**
Albert Ellis Institute

catastrophize To exaggerate the negative consequences of events.

THINK ABOUT IT

Can you think of examples from your personal life in which your thinking style reflected one or more of the cognitive distortions identified by Beck—selective abstraction, overgeneralization, magnification, or absolutist thinking? What effects did these thought patterns have on your moods? On your level of motivation? Do you think you can change how you think about your experiences? Why or why not?

Web Link 2.8 WWW
Beck Institute

Quiz 2.2 Q
The Psychological Perspective

worthwhile. It would be marvelous to excel in everything we do, but it's absurd to demand it of ourselves or believe that we couldn't stand it if we failed to measure up. Ellis developed a model of therapy, called *rational-emotive behavior therapy* (REBT), to help people dispute these irrational beliefs and substitute more rational ones. Ellis admits that childhood experiences are involved in the origins of irrational beliefs, but cognitive appraisal—the here and now—causes people misery. For most people who are anxious and depressed, the key to greater happiness does not lie in discovering and liberating deep-seated conflicts, but in recognizing and modifying irrational self-demands.

Aaron Beck Another prominent cognitive theorist, psychiatrist Aaron Beck, proposes that depression may result from "cognitive errors," such as judging oneself entirely on the basis of one's flaws or failures and interpreting events in a negative light (as though wearing blue-colored glasses) (A. T. Beck et al., 1979). Beck stresses the pervasive roles of four basic types of cognitive errors that contribute to emotional distress:

1. *Selective abstraction.* People may *selectively abstract* (focus exclusively on) the parts of their experiences that reflect on their flaws and ignore evidence of their competencies. For example, a student may focus entirely on the one mediocre grade he got on a math test and ignore all the higher grades.
2. *Overgeneralization.* People may *overgeneralize* from a few isolated experiences. For example, they may see their futures as hopeless because they were laid off or believe they will never marry because they were rejected by a dating partner.
3. *Magnification.* People may blow out of proportion, or *magnify*, the importance of unfortunate events. For example, a student may catastrophize a bad test grade by jumping to the conclusion that she will flunk out of college and her life will be ruined.
4. *Absolutist thinking.* Absolutist thinking is seeing the world in black and white terms, rather than in shades of gray. Absolutist thinkers may assume any grade less than a perfect "A," or a work evaluation less than a rave, is a total failure.

Like Ellis, Beck has developed a major model of therapy, called *cognitive therapy*, that focuses on helping individuals with psychological disorders identify and correct faulty ways of thinking.

Evaluating Cognitive Models As we'll see in later chapters, cognitive theorists have had an enormous impact on our understanding of abnormal behavior patterns and development of therapeutic approaches. The overlap between the learning-based and cognitive approaches is best represented by the emergence of *cognitive-behavioral therapy* (CBT), a form of therapy that focuses on modifying self-defeating beliefs in addition to overt behaviors.

A major issue concerning cognitive perspectives is their range of applicability. Cognitive therapists have largely focused on emotional disorders relating to anxiety and depression. They have had less impact on the development of treatment approaches, or conceptual models, of more severe forms of disturbed behavior, such as schizophrenia. Moreover, in the case of depression, it remains unclear, as we see in Chapter 8, whether distorted thinking patterns are causes of depression or are themselves effects of depression.

Roots of abnormal behavior? Sociocultural theorists believe that the roots of abnormal behavior are found not in the individual but in the social ills of society, such as poverty, social decay, discrimination based on race and gender, and lack of economic opportunity.

The Sociocultural Perspective

Does abnormal behavior arise from forces within the person, as the psychodynamic theorists propose, or from learning maladaptive behaviors, as the learning theorists suggest? Or, as the sociocultural perspective proposes, must a fuller accounting of abnormal behavior require that we consider the roles of social and cultural factors, including factors relating to ethnicity, gender, and social class? As we noted in Chapter 1, sociocultural theorists seek causes of abnormal behavior that may reside in the failures of society rather than in the person. Some of the more radical psychosocial theorists, like Thomas Szasz, even deny the existence of psychological disorders or mental illness. Szasz (1961, 2000) argues that "abnormal" is merely a label society attaches to people whose behavior deviates from accepted social norms. According to Szasz, this label is used to stigmatize and subjugate social deviants.

Throughout the text we examine relationships between abnormal behavior patterns and sociocultural factors such as gender, ethnicity, and socioeconomic status. Here let us examine recent research on relationships between ethnicity and mental health.

White 71.8%
Hispanic 11.4%
African American 12.2%
Asian American 3.9%
Native American 0.7%

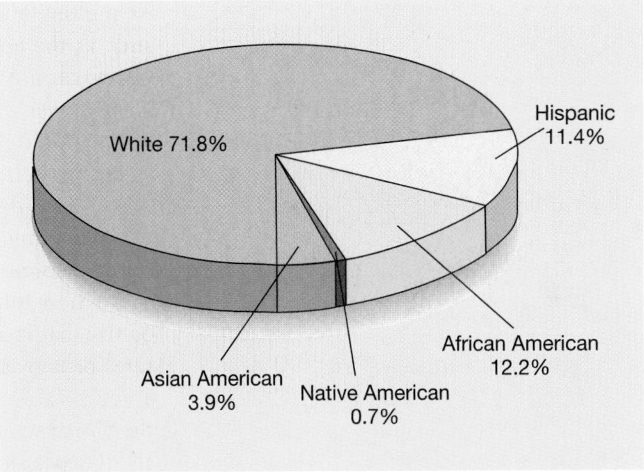

FIGURE 2.7 Ethnic/racial breakdown of the U.S. population in the year 2000.

Source. U.S. Bureau of the Census. (2000). *Resident population of the United States: Middle series projections, 1996–2000, by sex, race, and Hispanic origin, with median age.* Washington, DC: Author.

Ethnicity and Mental Health

When Europeans first arrived on America's shores, the land was populated solely by Native Americans. By the time the United States achieved nationhood, an ethnicity[1] profile would show that the numbers of people of European descent were approaching those of Native Americans. During the 19th century, the nation became predominantly populated by White people. Although Euro-Americans (also called European Americans or non-Hispanic White Americans) remain in the majority today, the nation is becoming increasingly ethnically diverse, as a result of both an excess of births over deaths among various U.S. ethnic groups and contemporary trends in immigration.

Figure 2.7 shows the ethnic composition of the U.S. population. African Americans, whose ancestors were forcibly brought to this country and enslaved, remain the largest non-White population group, constituting about 12% of the population. Hispanic Americans (Latinos) account for 11% of the population, while Asian Americans/Pacific Islanders account for 4%, and Native Americans for nearly 1%. During the early part of the 21st century, Hispanic Americans are expected to surpass African Americans as the country's largest ethnic minority group (Sachs, 2001; Rodriguez, 2001). The terms *Hispanic American* or *Latino(a)* refer to persons of Mexican, Puerto Rican, Cuban, or other Central and South American or Spanish origin (USDHHS, 1999a). The population of Asian Americans/Pacific Islanders—whose backgrounds and ancestries include peoples from areas as diverse as China, Japan, Korea, Indochina, Thailand, the Philippines, India, and Pakistan—represents the fastest growing ethnic minority group in the United States. ("America 2000," 2000; Clemetson, 2000). Their numbers in the U.S. population rose by more than 40% during the 1990s.

The term *minority* is quickly becoming something of a misnomer, as traditionally identified minority groups now comprise the majority in many U.S. cities and in the nation's most populous state, California (Purdum, 2001, Schmitt, 2001c). Euro-Americans will constitute but the barest majority of the population by the year 2050 (see Figure 2.8).

Ethnic designations, such as African American, Hispanic American or Latino, Asian American, Euro-American, or Native American, are general categories that encompass many different subgroups. For example, Latinos includes Spanish-speaking people who may trace their heritage to Mexico, Puerto Rico, or Colombia. Asian Americans may perceive themselves as Filipino Americans or Chinese Americans rather than Asian Americans per se. When considering racial or ethnic distinctions, we need to take into account differences among cultural and ethnic subgroups within our own culture. Yet traditional racial or ethnic distinctions are becoming blurry as increasing numbers of Americans and Canadians identify themselves as biracial, multiracial,

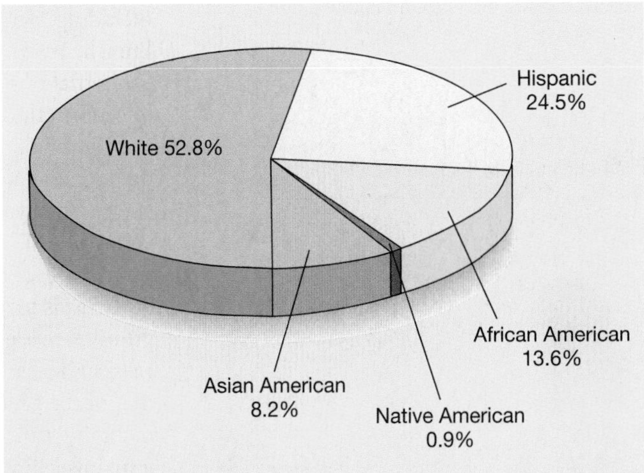

FIGURE 2.8 Projected ethnic/racial breakdown of the U.S. population in the year 2050.

Source. U.S. Bureau of the Census. (1996). *Resident population of the United States: Middle series projections, 2035–2050, by sex, race, and Hispanic origin, with median age.* Washington, DC: Author.

[1]The word *ethnicity* is derived from the Greek word *ethnikos*, meaning "people or nation."

or multiethnic (Meacham, 2000; Miller, 2000). Individuals of multiracial background, such as the golfer Tiger Woods, the baseball player Derek Jeter, and the singer Mariah Carey, cannot easily be classified according to traditional racial groupings (See, 1999). Nearly seven million U.S. residents described themselves as multiracial in the most recent population census (Schmitt, 2001a, 2001b). Young people today are twice as likely to consider themselves multiracial than were their parents (Takahashi, 2001).

Given the increasing ethnic diversity of the U.S. population, researchers have begun to study ethnic group differences in the prevalence of psychological disorders. Knowing that a disorder disproportionately affects one group or another can help planners direct prevention and treatment programs to the groups that are most in need. Researchers recognize that income level or socioeconomic status needs to be considered when comparing rates of a given diagnosis across ethnic groups. We also need to account for differences among ethnic subgroups, such as differences among the various subgroups that comprise the Hispanic American and Asian American populations. We find, for example, higher levels of depression among Hispanic immigrants to the United States from Central America than from Mexico, even when considering differences in educational backgrounds (Salgado de Snyder, Cervantes, & Padilla, 1990).

Some ethnic groups have been underrepresented in previous research. For example, no nationwide surveys of the prevalence of psychological disorders among Asian Americans have been reported (Sue et al., 1995). However, the available evidence indicates that psychological disorders are not significantly lower among Asian Americans than other ethnic groups, which stands in contrast to the popular perception of Asian Americans as a group generally free of mental health problems (Sue et al., 1995; Zane & Sue, 1991).

Truth OR Fiction? REVISITED

Rates of mental disorders are generally higher among African Americans than Euro-Americans, even when we account for income differences between these groups.

FALSE. Rates of mental disorders overall are no higher among African Americans than those of Euro-Americans when differences in socioeconomic levels are taken into account.

We should be cautious—and think critically—when interpreting ethnic group differences in rates of diagnoses of psychological disorders. Might these differences reflect ethnic or racial differences, or differences on other factors on which groups may vary, such as socioeconomic level, living conditions, or cultural backgrounds?

Prevalence rates of mental disorders are found to be higher among African Americans than among Euro-Americans (USDHHS, 1999a). However, these differences disappear once one controls for socioeconomic status (SES). In other words, the higher rates of mental disorder among African Americans are a function of their generally lower SES, not their race or ethnicity. Evidence also shows few differences in rates of mental disorders between Hispanic Americans and Euro-Americans (USDHHS, 1999a).

Native Americans, on the whole, are among the most impoverished ethnic groups in the United States and Canada. Like other socially and economically disadvantaged groups, Native Americans also suffer from a much greater prevalence of mental health problems, including alcoholism, depression, suicide, drug abuse, and delinquency (T. J. Young & French, 1996). Native Americans, for example, have alcohol-related disorders at a rate six times that of other Americans (Rabasca, 2000a). The death rate due to suicide among adolescents in the 10- to 14-year age range is about four times higher among Native Americans than among other ethnic groups. Male Native American adolescents and young adults have the highest suicide rates in the nation (USDHHS, 1999a).

When you envision stereotypes such as hula dancing, luaus, and wide tropical beaches, you may assume that Native Hawaiians are a carefree people. Reality paints a different picture, however. One reason for studying the relationship between ethnicity and abnormal behavior is to debunk erroneous stereotypes. Native Hawaiians, like other Native American groups, are economically disadvantaged and suffer a disproportionate share of physical diseases and mental health problems. The death rate for Native Hawaiians is 34% higher than that of the general U.S. population, largely because of an increased rate of serious diseases, including cancer and heart disease (Mokuau, 1990). Native Hawaiians also have a 5- to 10-year lower life expectancy than other groups in Hawaii. Compared to other Hawaiians, Native Hawaiians experience higher rates of mental health problems, including higher suicide rates among males, higher rates of alcoholism and drug abuse, and higher rates of antisocial behavior.

In addition to economic disadvantage, the mental health problems of Native Americans, including Native Hawaiians, may at least partly reflect alienation and disenfranchisement from the land and a way of life that resulted from colonization by European cultures (Rabasca, 2000a). Native peoples often attribute mental health problems, especially depression and alcoholism, to the collapse of their traditional culture brought about by colonization (Timpson et al., 1988). Researchers recount how a Native Canadian elder in northwestern Ontario explained depression in his people (Timpson et al., 1988, p. 6):

> Before the White Man came into our world we had our own way of worshipping the Creator. We had our own church and rituals. When hunting was good, people would gather together to give gratitude. This gave us close contact with the Creator. There were many different rituals depending on the tribe. People would dance in the hills and play drums to give recognition to the Great Spirit. It was like talking to the Creator and living daily with its spirit. Now people have lost this. They can't use these methods and have lost conscious contact with this high power. The more distant we are from the Creator the more complex things are because we have no sense of direction. We don't recognize where life is from.

The depression so common among indigenous or native peoples apparently reflects the loss of a relationship with the world that was based on maintaining harmony with nature (Timpson et al., 1988). The description of the loss of this special relationship reminds us of the Western concept of alienation.

Whatever the underlying differences in psychopathology among ethnic groups, members of ethnic minority groups tend to underutilize mental health services compared to European (non-Hispanic White) Americans (USDHHS, 1999a). Those who do seek services are more likely to drop out prematurely from treatment. In Chapter 4 we consider barriers that limit the utilization of mental health services by various ethnic minority groups in our society.

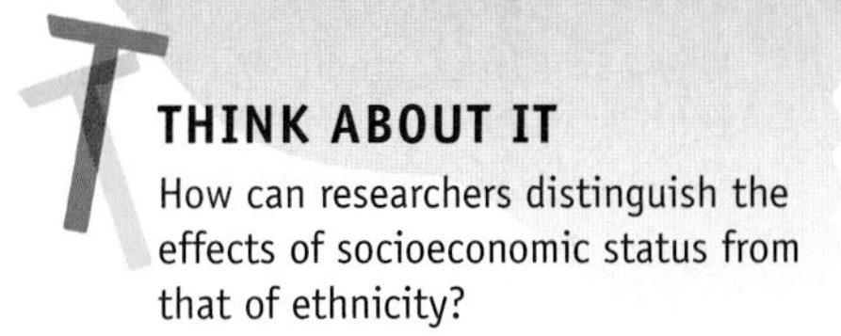

THINK ABOUT IT
How can researchers distinguish the effects of socioeconomic status from that of ethnicity?

Evaluating the Sociocultural Perspective

Lending support to the linkage between social class and psychological disturbance, classic research in New Haven, CT, showed that people from the lower socioeconomic groups were more likely to be institutionalized for psychiatric problems (Hollingshead & Redlich, 1958). One reason perhaps is that the poor have less access to private outpatient care.

An alternative view is that people from the lower socioeconomic groups may be at greater risk of severe behavior problems because living in poverty subjects them to a greater level of social stress than that faced by the more well-to-do people. Yet another view, the **downward drift hypothesis,** suggests that problem behaviors, such as alcoholism, may lead people to drift downward in social status, thereby explaining the linkage between low socioeconomic status and severe behavior problems.

Certainly it is desirable for social critics such as Szasz to focus our attention on the political implications of our responses to deviance. The views of Szasz and other critics of the mental health establishment have been influential in bringing about much needed changes to better protect the rights of mental patients in psychiatric institutions. Many professionals, however, believe that more radical sociocultural theorists like Szasz go too far in arguing that mental illness is merely a fabrication invented by society to stigmatize social deviants.

Sociocultural theorists have focused much needed attention on the social stressors that may lead to abnormal behavior. Throughout the text we consider how sociocultural factors relating to gender, race, ethnicity, and lifestyle inform our understanding of abnormal behavior and our response to people deemed mentally ill. In Chapter 4, we consider how issues relating to race, culture, and ethnicity impact the therapeutic process.

downward drift hypothesis The theory that explains the linkage between low socioeconomic status and behavior problems by suggesting that problem behaviors lead people to drift downward in social status.

The Biopsychosocial Perspective

THINK ABOUT IT
Do you believe the root causes of abnormal behavior lie in environment? In the person? In a combination of the two? Why?

Web Link 2.9 WWW
Biopsychosocial Model of Disease

We have seen that there are several models or perspectives for understanding and treating psychological disorders. The fact that there are different ways of looking at the same phenomenon doesn't mean that one model must be right and the others wrong. No one theoretical perspective can account for the many complex forms of abnormal behavior we encounter in this text. Each perspective contributes something to our understanding, but none offers a complete view. Many theorists today adopt a biopsychosocial perspective that considers how multiple factors representing biological, psychological, and sociocultural domains interact in the development of particular disorders. We are only beginning to ferret out the subtle and often complex interactions of multiple factors that give rise to abnormal behavior patterns.

The biopsychosocial perspective invites us to consider how biological, psychological, and social factors are linked in the development of abnormal behavior patterns (Kiesler, 1999). For some disorders, the causes may be primarily or even exclusively biological in nature. For instance, certain forms of mental retardation have clear-cut biological causes, such as chromosomal abnormalities (see Chapter 14) or maternal alcohol consumption during pregnancy (see Chapter 10). Other disorders may arise directly from learning experiences. For example, phobias may be acquired based on associations or pairings of particular objects or situations with traumatic or painful experiences (see Chapter 6). But in most psychological disorders, multiple causes, representing biological, psychological, and sociocultural domains, are involved. We must consider not only the contributions of multiple causes, but also the interactions among them. We are only beginning to unravel the complex web of factors that underlie many types of disorders. Here, let us consider a leading contemporary model that examines how the interaction of multiple causes may be involved in the development of psychological disorders—the diathesis-stress model.

THINK ABOUT IT
Why is it necessary to understand and consider multiple perspectives in explaining abnormal behavior?

The Diathesis-Stress Model

The **diathesis-stress model** holds that disorders arise from a combination or interaction of a **diathesis** (vulnerability or predisposition) with stress (see Figure 2.9). In most versions of the model, the diathesis is conceptualized as a biological vulnerability, generally genetic in nature, which increases the risks of developing a particular disorder. Yet whether the disorder actually develops depends on the type and severity of stressors the person experiences. These stressors may include prenatal or childhood trauma, birth complications, physical illness, childhood sexual or physical abuse and family conflict, prolonged unemployment, loss of loved ones, or other negative life circumstances or changes.

In some cases, people with a diathesis for a particular disorder may remain free of the disorder or develop a milder form of the disorder if the level of stress in their lives remains low or if they develop effective coping responses for handling the stress they encounter. However, the stronger the diathesis, the less stress is generally needed to produce the disorder. In some cases the diathesis may be so strong that the disorder develops even under the most benign life circumstances.

The diathesis-stress hypothesis was originally developed as an explanatory framework for understanding schizophrenia (see Chapter 13). It has since been applied to other psycho-

diathesis-stress model A model that posits that abnormal behavior problems involve the interaction of a vulnerability or predisposition and stressful life events or experiences.

diathesis A vulnerability or predisposition to a particular disorder.

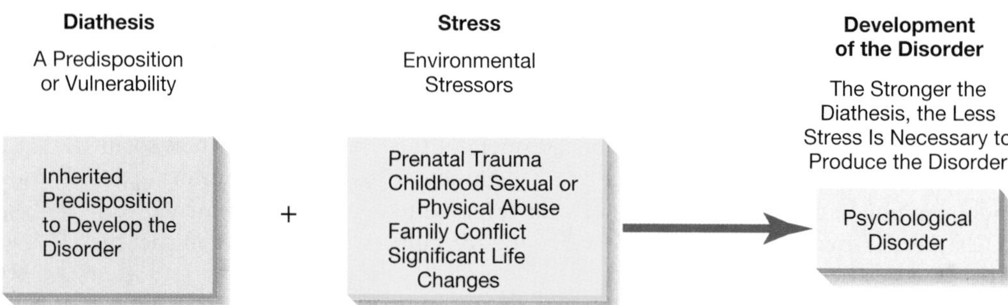

FIGURE 2.9 The diathesis-stress model.

logical disorders, including depression (Lewinsohn, Joiner, & Rohde, 2001; Ormel et al., 2001). The diathesis-stress model is not the only biopsychosocial account of how abnormal behavior patterns develop. We also consider other models that posit roles for biological and psychosocial influences, such as the cognitive model of panic disorder (see Chapter 6).

Not all forms of the diathesis-stress model are based on an interaction of a biological diathesis and life stress. Psychological diatheses may also be involved, such as dysfunctional thinking patterns or personality traits that increase the risk of developing a particular disorder in the face of life stress (Lewinsohn, Joiner, & Rohde, 2001). For example, the tendency to find fault with oneself for negative life events may put individuals at greater risk of developing depression following negative life events or stressors, such as divorce or job loss (see Chapter 8) (Just, Abramson, & Alloy, 2001).

THINK ABOUT IT

What does it mean to state that the diathesis-stress model as formulated here is an interactionist approach to explaining abnormal behavior? Give an example of how your own behavior reflects interactions of biological, psychological, and sociocultural influences.

Evaluating the Biopsychosocial Perspective

The strength of the biopsychosocial model—its very complexity—may also be its greatest weakness. The model endorses the view that with few exceptions, psychological disorders or other patterns of abnormal behavior are complex phenomena that arise from multiple causal factors. We cannot pinpoint any one cause that leads to the development of schizophrenia or panic disorder, for example. Adding to the complexity is the multiple causal pathways that are involved—that is, different people may develop the same disorder based on different sets of causal influences. Yet the complexity of understanding the interplay of underlying causes of abnormal behavior patterns should not deter us from the effort. The accumulation of a body of knowledge is a continuing process. We know a great deal more today than we did a few short years ago. We will surely know more in the years ahead.

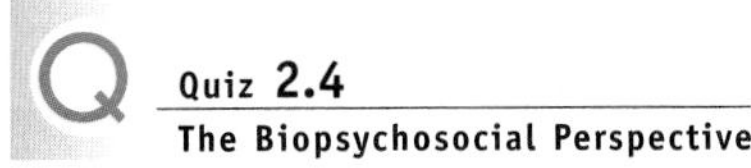

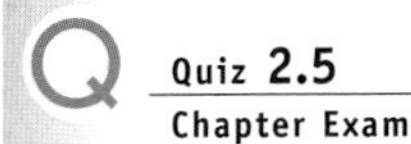

Summing Up

The Biological Perspective

How is the nervous system organized? The nervous system consists of two major parts, the central nervous system and the peripheral nervous system. The central nervous system consists of the brain and spinal cord. The peripheral nervous system consists of two major divisions, the somatic nervous system, which transmits messages between the central nervous system and the sense organs and muscles, and the autonomic nervous system, which controls involuntary bodily processes. The autonomic nervous system or ANS has two branches or subdivisions, the sympathetic and the parasympathetic. The nervous system is composed of nerve cells, or neurons, that communicate with one another through chemical messengers, called neurotransmitters, that transmit nerve impulses across the tiny gaps, or synapses, between neurons.

What are the biological underpinnings of abnormal behavior? Biological factors implicated in the development of abnormal behavior include disturbances in neurotransmitter functioning in the brain, heredity, and underlying brain abnormalities.

The Psychological Perspective

What are the major psychological models of abnormal behavior? Psychodynamic perspectives reflect the views of Freud and his followers, who believed that abnormal behavior stemmed from psychological causes based on underlying psychic forces within the personality. Learning theorists posit that the principles of learning can be used to explain both abnormal and normal behavior. Humanistic theorists believe it is important to understand the obstacles that people encounter as they strive toward self-actualization and authenticity. Cognitive theorists focus on the role of distorted and self-defeating thinking in explaining abnormal behavior.

The Sociocultural Perspective

What is the basic idea underlying the sociocultural perspective on abnormal behavior? Sociocultural theorists believe we need to broaden our outlook on abnormal behavior by taking into account the role of social ills in society, including poverty, racism, and lack of opportunity, in the development of abnormal behavior patterns.

The Biopsychosocial Perspective

What is the distinguishing feature of the biopsychosocial perspective? The biopsychosocial perspective seeks an understanding of abnormal behavior based on the interplay of biological, psychological, and sociocultural factors in abnormal behavior.

What is the diathesis-stress model? The diathesis-stress model holds that a person may have a predisposition, or diathesis, for a particular disorder, but whether the disorder actually develops depends on the interaction of the diathesis with stress-inducing life experiences.

CHAPTER THREE

Classification and Assessment of Abnormal Behavior

Jean Dubuffet
Ontogenese, 1975

Truth OR Fiction?

- Some men in India develop a psychological disorder involving excessive concerns or anxiety over losing semen. (p. 69)
- A psychological test must be valid in order to be reliable. (p. 73)
- Researchers find that people report more personal problems when interviewed by human interviewers than by computers. (p. 76)
- The most widely used personality inventory includes a number of questions that bear no obvious relationship to the traits the instrument purports to measure. (p. 80)
- Some clinicians examine how people interpret inkblots in order to uncover underlying aspects of their personalities. (p. 83)
- Weight-loss program participants who were more conscientious about monitoring what they ate lost more weight than those who were less reliable monitors. (p. 90)
- Despite advances in technology, physicians today still need to perform surgery to study the workings of the brain. (p. 92)

Systems of classification of abnormal behavior date to ancient times. Hippocrates classified abnormal behaviors on the basis of his theory of humors. Although his theory proved to be flawed, he arrived at some diagnostic categories that generally correspond to those in modern diagnostic systems. His description of melancholia, for example, is similar to present conceptions of depression. During the Middle Ages some so-called authorities classified abnormal behaviors according to those that represented demonic possession and those that represented natural causes. The 19th-century German psychiatrist Emil Kraepelin is generally considered the first modern theorist to develop a comprehensive model of classification based on the distinctive features, or "symptoms," associated with abnormal behavior patterns. The most commonly used classification system today is largely an outgrowth and extension of Kraepelin's work: the *Diagnostic and Statistical Manual of Mental Disorders (DSM)*, published by the American Psychiatric Association. The *DSM* classifies abnormal behavior patterns as mental disorders on the basis of specified diagnostic criteria.

Why is it important to classify abnormal behavior? For one thing, classification is the core of science. Without labeling and organizing patterns of abnormal behavior, researchers could not communicate their findings to one another, and progress toward understanding these disorders would come to a halt. Moreover, important decisions are made on the basis of classification. Certain psychological disorders respond better to one therapy than another or to one drug than another. Classification also helps clinicians predict behavior. Some patterns of abnormal behavior, such as schizophrenia, follow more or less predictable courses of development. Classification also helps researchers identify populations with similar patterns of abnormal behavior. By classifying groups of people as depressed, for example, researchers might be able to identify common factors that help explain the origins of depression.

This chapter reviews the classification and assessment of abnormal behavior, beginning with the major system clinicians use to classify abnormal behavior patterns: the *Diagnostic and Statistical Manual of Mental Disorders (DSM)*.

Chapter Outline

Classification of Abnormal Behavior Patterns

The *DSM* was first introduced in 1952. The latest version of the *DSM*, published in 2000 by the American Psychiatric Association, is the *DSM-IV-TR*, the Text Revision (TR) of the Fourth Edition (*DSM-IV*) (APA, 2000). We focus on the *DSM* as a method of classification because of its widespread adoption by mental health professionals. However, many psychologists and other professionals criticize the *DSM* on several grounds, such as relying too strongly on the medical model. Our focus on the *DSM* reflects recognition of its widespread use, not an endorsement.

In the *DSM*, abnormal behavior patterns are classified as "mental disorders." *Mental disorders* involve either emotional distress (typically depression or anxiety) and/or significant impairment in psychological functioning. *Impaired functioning* involves difficulties in meeting responsibilities at work, within the family, or within society at large. It also includes behavior that places people at risk for personal suffering, pain, or death.

Diagnosis of mental disorders within the *DSM* requires that the behavior pattern not represent an expected or culturally appropriate response to a stressful event, such as the loss of a loved one. People who show signs of bereavement or grief following the death of loved ones are not considered disordered, even if their behavior is significantly impaired. If their behavior remains significantly impaired over an extended period of time, however, a diagnosis of a mental disorder might become appropriate.

The *DSM* and Models of Abnormal Behavior

The *DSM* system adheres in some important respects to the medical model. It treats abnormal behaviors as signs or symptoms of underlying pathologies called mental disorders. Unlike the strictest form of the medical model, however, the manual does not assume that abnormal behaviors necessarily reflect biological causes or defects. It recognizes that the causes of most mental disorders remain uncertain: Some disorders may have purely biological causes. Some may have psychological causes. Still others are likely to reflect a multifactorial model, or the interaction of biological, psychological, social (socioeconomic, sociocultural, and ethnic), and physical environmental factors.

Nor does the *DSM* subscribe to a particular theory of abnormal behavior. With the introduction in 1980 of the third edition of the *DSM,* the *DSM-III,* terms linked to specific theories (such as *neurosis,* which was originally a psychodynamic term) were deemphasized in favor of descriptive terms such as *anxiety disorders* and *mood disorders.* Disorders are classified on the basis of their clinical features and behavior patterns, not on the basis of inferences about underlying theoretical mechanisms. Because the *DSM* does not endorse particular theoretical models unless evidence of causal factors is overwhelming, it can be used by practitioners of diverse theoretical persuasions. They can agree on the criteria for diagnosing various disorders, even if they disagree on their causes and proper treatments. Perhaps because of its emphasis on diagnostic criteria, critics contend that the *DSM* is something of a hodgepodge of disorders that are grouped together in various clusters without a consistent conceptual framework (Kutchins & Kirk, 1995).

Web Link 3.1 WWW
Diagnosis of Mental Disorders

The authors of the *DSM* recognize that their use of the term *mental disorder* is problematic because it perpetuates a long-standing but dubious distinction between mental and physical disorders (American Psychiatric Association, 1994, 2000). They point out that there is much that is "physical" in "mental" disorders and much that is "mental" in "physical" disorders. The diagnostic manual continues to use the term *mental disorder* because its developers have not been able to agree on an appropriate substitute. In this text we use the term *psychological disorder* in place of *mental disorder* because we feel it is more appropriate to place the study of abnormal behavior more squarely within a psychological context. Moreover, the term *psychological* has the advantage of encompassing behavioral patterns as well as strictly "mental" experiences such as emotions, thoughts, beliefs, and attitudes.

TABLE 3.1 Sample Diagnostic Criteria for Generalized Anxiety Disorder

1. Occurrence of excessive anxiety and worry on most days during a period of six months or longer.
2. Anxiety and worry are not limited to one or a few concerns or events.
3. Difficulty controlling feelings of worry.
4. The presence of a number of features associated with anxiety and worry, such as the following:
   a. experiencing restlessness or feelings of edginess
   b. becoming easily fatigued
   c. having difficulty concentrating or finding one's mind going blank
   d. feeling irritable
   e. having states of muscle tension
   f. having difficulty falling asleep or remaining asleep or having restless, unsatisfying sleep
5. Experiencing emotional distress or impairment in social, occupational, or other areas of functioning as the result of anxiety, worry, or related physical symptoms.
6. Worry or anxiety is not accounted for by the features of another disorder.
7. The disturbance does not result from the use of a drug of abuse or medication or a general medical condition and does not occur only in the context of another disorder.

Source. Adapted from *DSM-IV-TR* (APA, 2000).

Psychologist Jerome Wakefield (1992a, 1992b, 1997, 2001) proposed that the term *disorder* be conceptualized as "harmful dysfunction." A harmful dysfunction represents a failure of a mental or physical system to perform its natural function, resulting in negative consequences or harm to the individual. By this definition, dysfunction alone is not enough to constitute a disorder. For example, even though the body was naturally designed to have two kidneys, and it would be dysfunctional to have but one, a failure of one kidney to function properly (or even the loss of a kidney) may not be harmful to the individual's well-being. By contrast, a dysfunction involving a breakdown in the brain's ability to store or retrieve information would constitute a disorder if it leads to harmful consequences such as memory deficits that make it difficult for the person to function effectively. We find Wakefield's conceptualization of disorders as harmful dysfunctions to be instructive, although we recognize that not all psychologists share his viewpoint (see Lilienfeld & Marino, 1995). One problem is that we may lack agreement on what constitutes the "natural function" of mental systems (Bergner, 1997).

Finally, we should recognize that the *DSM* is used to classify disorders, not people. This is an important distinction that has a bearing on the terminology we use to describe people who display abnormal behavior patterns. Rather than classify someone as a *schizophrenic* or a *depressive,* we refer to *an individual with schizophrenia* or *a person with major depression.* This difference in terminology is not simply a matter of semantics. To label someone a schizophrenic carries an unfortunate and stigmatizing implication that a person's identity is defined in terms of a disorder he or she may have or exhibit.

Features of the *DSM* The *DSM* is descriptive, not explanatory. It describes the diagnostic features—or, in medical terms, symptoms—of abnormal behaviors rather than attempting to explain their origins. Let us consider a number of features of the *DSM* classification system.

1. *Specific diagnostic criteria are used.* The clinician arrives at a diagnosis by matching a client's behaviors with the criteria that define particular patterns of abnormal behavior ("mental disorders"). Diagnostic categories are described in terms of *essential features*

(criteria that must be present for the diagnosis to be made) and *associated features* (criteria often associated with the disorder but not essential to making a diagnosis). An example of diagnostic criteria for generalized anxiety disorder is shown in Table 3.1

2. *Abnormal behavior patterns that share clinical features are grouped together.* Abnormal behavior patterns are categorized according to their shared clinical features, not theoretical speculation about their causes. For example, abnormal behavior patterns chiefly characterized by anxiety are classified as anxiety disorders. Behaviors chiefly characterized by disruptions in mood are categorized as mood disorders.
3. *The system is multiaxial.* The *DSM* employs a *multiaxial*, or multidimensional, system of assessment that provides a broad range of information about the individual's functioning, not just a diagnosis (see Table 3.2). The system contains the following axes:
   a. *Axis I includes a classification of Clinical* ***Syndromes,*** which incorporates a wide range of diagnostic classes. These include anxiety disorders, mood disorders, schizophrenia and other psychotic disorders, adjustment disorders, and disorders usually first diagnosed during infancy, childhood, or adolescence (except for mental retardation, which is coded on Axis II). Axis I also includes a classification of *Other Conditions That May Be a Focus of Clinical Attention.* These are conditions or problems that may be the focus of diagnosis and treatment, such as relationship problems, academic or occupational problems, or bereavement, but that do not in themselves constitute definable psychological disorders. These conditions also include a category of psychological factors that affect medical conditions, such as

syndromes Clusters of symptoms that are characteristic of particular disorders.

TABLE 3.2 The Multiaxial Classification System of the *DSM-IV-TR*

Axis	Type of Information	Brief Description
Axis I	Clinical Disorders	The patterns of abnormal behavior ("mental disorders") that impair functioning and are stressful to the individual
	Other Conditions That May Be a Focus of Clinical Attention	Other problems that may be the focus of diagnosis or treatment but do not constitute mental disorders, such as academic, vocational, or social problems, and psychological factors that affect medical conditions (such as delayed recovery from surgery due to depressive symptoms)
Axis II	Personality Disorders Mental Retardation	Personality disorders involve excessively rigid, enduring, and maladaptive ways of relating to others and adjusting to external demands. Mental retardation involves a delay or impairment in the development of intellectual and adaptive abilities.
Axis III	General Medical Conditions	Chronic and acute illnesses and medical conditions that are important to the understanding or treatment of the psychological disorder or that play a direct role in causing the psychological disorder
Axis IV	Psychosocial and Environmental Problems	Problems in the social or physical environment that affect the diagnosis, treatment, and outcome of psychological disorders
Axis V	Global Assessment of Functioning	Overall judgment of current functioning with respect to psychological, social, and occupational functioning; the clinician may also rate the highest level of functioning occurring for at least a few months during the past year

Source. Adapted from the *DSM-IV-TR* (APA, 2000).

TABLE 3.3 **Psychosocial and Environmental Problems**

Problem Categories	Examples
Problems with Primary Support Group	Death of family members; health problems of family members; marital disruption in the form of separation, divorce, or estrangement; sexual or physical abuse within the family; child neglect; birth of a sibling
Problems Related to the Social Environment	Death or loss of a friend; social isolation or living alone; difficulties adjusting to a new culture (acculturation); discrimination; adjustment to transitions occurring during the life cycle, such as retirement
Educational Problems	Illiteracy; academic difficulties; problems with teachers or classmates; inadequate or impoverished school environment
Occupational Problems	Work-related problems including stressful workloads and problems with bosses or coworkers; changes in employment; job dissatisfaction; threat of loss of job; unemployment
Housing Problems	Inadequate housing or homelessness; living in an unsafe neighborhood; problems with neighbors or landlord
Economic Problems	Financial hardships or extreme poverty; inadequate welfare support
Problems with Access to Health Care Services	Inadequate health care services or availability of health insurance; difficulties with transportation to health care facilities
Problems Related to Interaction with the Legal System/Crime	Arrest or imprisonment; becoming involved in a lawsuit or trial; being a victim of crime
Other Psychosocial Problems	Natural or human-made disasters; war or other hostilities; problems with caregivers outside the family, such as counselors, social workers, and physicians; lack of availability of social service agencies

Source. Adapted from the *DSM-IV-TR* (APA, 2000).

hypothyroidism A physical condition, caused by deficiency of the hormone thyroxin, characterized by sluggishness and lowered metabolism.

anxiety that exacerbates an asthmatic condition, or depressive symptoms that delay recovery from surgery.

b. *Axis II, Personality Disorders*, includes the more enduring and rigid patterns of maladaptive behavior that generally impair interpersonal relationships and social adaptation, including antisocial, paranoid, narcissistic, and borderline personality disorders (see Chapter 9). Mental retardation is also coded on Axis II.

Separating the diagnostic categories into two axes provides greater flexibility in reaching diagnostic impressions. People may be given either Axis I or Axis II diagnoses, or a combination of the two when both apply. For example, a person may receive a diagnosis of an anxiety disorder (Axis I) and a second diagnosis of a personality disorder (Axis II) if the diagnostic criteria for both are met.

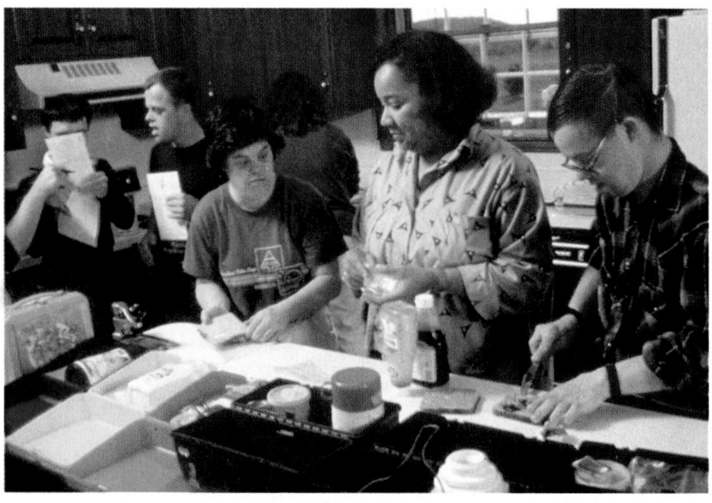

Assessment of level of functioning. The assessment of functioning takes into account the individual's ability to manage the responsibilities of daily living. Here we see a group home for people with mental retardation. The residents assume responsibility for household functions.

c. *Axis III, General Medical Conditions*, lists medical conditions and diseases that may be important to the understanding or treatment of the individual's mental disorder. For example, if **hypothyroidism** were a direct cause of an individual's mood disorder (such as major depression), it would be coded under Axis III. Medical conditions that affect the understanding or treatment of a mental disorder but are not direct causes of the disorder are also listed on Axis III. For instance, the presence of a heart condition may determine whether a particular course of drug therapy should be used with a depressed person.

d. *Axis IV, Psychosocial and Environmental Problems*, lists psychosocial and environmental problems believed to affect the diagnosis, treatment, or outcome of a mental disorder. Psychosocial and environmental problems include negative life events (such as a job termination or a marital separation or divorce), homelessness or inadequate housing, lack of social support, the death or loss of a friend, or exposure to war or disasters. Some positive life events may also be listed, such as a job promotion, but only when they create problems for the individual, such as difficulties adapting to a new job. A listing of these types of problems is found in Table 3.3.

e. *Axis V, Global Assessment of Functioning*, refers to the clinician's overall judgment of the client's psychological, social, and occupational functioning. Using a scale similar to that shown in Table 3.4, the clinician rates the client's current level of functioning and may also indicate the highest level of functioning achieved for at least a few months during the preceding year. The level of current functioning is taken to indicate the current need for treatment or intensity of care. The level of highest functioning is suggestive of the level of functioning that might be restored.

An example of a diagnosis in the *DSM* multiaxial system is shown in Table 3.5.

culture-bound syndrome A pattern of abnormal behavior that is found within only one or a few cultures.

taijin-kyofu-sho A psychiatric syndrome, found in Japan, involving excessive fear of offending or embarrassing others.

Culture-Bound Syndromes The *DSM* recognizes that some patterns of abnormal behavior, called **culture-bound syndromes,** occur in only one culture, or perhaps in only a few cultures. The fact that some forms of abnormal behavior flourish in some cultures but not in others suggests that culture and social environment have an important influence on the development of abnormal behavior (Osborne, 2001).

Culture-bound syndromes may reflect exaggerated forms of common folk superstitions and belief patterns within a particular culture. For example, the psychiatric disorder **taijin-kyofu-sho** (TKS) is common in Japan but rare elsewhere. The disorder is characterized by excessive fear that one may behave in ways that will embarrass or offend other

TABLE 3.4 Global Assessment of Functioning (GAF) Scale

Code	Severity of Symptoms	Examples
91–100	Superior functioning across a wide variety of activities of daily life	Lacks symptoms Handles life problems without them "getting out of hand"
81–90	Absent or minimal symptoms, no more than everyday problems or concerns	Mild anxiety before exams Occasional argument with family members
71–80	Transient and predictable reactions to stressful events, OR no more than slight impairment in functioning	Difficulty concentrating after argument with family Temporarily falls behind in schoolwork
61–70	Some mild symptoms, OR some difficulty in social, occupational, or school functioning, but functioning pretty well	Feels down, mild insomnia Occasional truancy or theft within household
51–60	Moderate symptoms, OR moderate difficulties in social, occupational, or school functioning	Occasional panic attacks Few friends, conflicts with coworkers
41–50	Serious symptoms, OR any serious impairment in social, occupational, or school functioning	Suicidal thoughts, frequent shoplifting Unable to hold job, has no friends
31–40	Some impairment in reality testing or communication, OR major impairment in several areas	Speech illogical Depressed man or woman unable to work, neglects family, and avoids friends
21–30	Strong influence on behavior of delusions or hallucinations, OR serious impairment in communication or judgment, OR inability to function in almost all areas	Grossly inappropriate behavior, speech sometimes incoherent Stays in bed all day, no job, home, or friends
11–20	Some danger of hurting self or others, OR occasionally fails to maintain personal hygiene, OR gross impairment in communication	Suicidal gestures, frequently violent Smears feces
1–10	Persistent danger of severely hurting self or others, OR persistent inability to maintain minimal personal hygiene, OR seriously suicidal act	Largely incoherent or mute Serious suicidal attempt, recurrent violence

Source. Adapted from the *DSM-IV-TR* (APA, 2000).

reliable In psychological assessment, the consistency of a measure or diagnostic instrument or system.

validity The degree to which a test or diagnostic system measures the traits or constructs it purports to measure.

predictive validity The degree to which a test score is predictive of some future behavior or outcome.

TABLE 3.5 Example of a Diagnosis in the Multiaxial *DSM*- System

Axis I	Generalized Anxiety Disorder
Axis II	Dependent Personality Disorder
Axis III	Hypertension
Axis IV	Problem with Primary Support Group (marital separation); Occupational Problem (unemployment)
Axis V	GAF = 62

people (McNally, Cassiday, & Calamari, 1990). People with TKS may dread blushing in front of others for fear of causing them embarrassment, not for fear of embarrassing themselves. In our culture, an excessive fear of social embarrassment is called a *social phobia* (see Chapter 6). Unlike people with TKS, however, people with social phobias have excessive concerns that they will be rejected by, or embarrassed in front of, others, not that they will embarrass other people. People with TKS may also fear mumbling thoughts aloud, lest they inadvertently offend others. The syndrome primarily affects young Japanese men and is believed to be related to an emphasis in Japanese culture on not embarrassing others as well as deep concerns over issues of shame (McNally et al., 1990; Spitzer et al., 1994). Chang (1984) reports that TKS is diagnosed in 7% to 36% of the people treated by psychiatrists in Japan. Table 3.6 lists some other culture-bound syndromes identified in the *DSM-IV-TR*.

Web Link 3.2 WWW
DSM-IV: Questions and Answers

We generally think of culture-bound syndromes as abnormal behavior patterns associated with folk cultures in non-Western societies. Yet some abnormal behavior patterns, such as anorexia nervosa (discussed in Chapter 11) and dissociative identity disorder (formerly called *multiple personality disorder*; discussed in Chapter 7), may be considered culture-bound syndromes occurring in some technological societies, including our own, but that are essentially unknown in less developed cultures.

Taijin-kyofu-sho. TKS is a culture-bound syndrome that is common in Japan. It is characterized by excessive fear that one may embarrass or offend other people. The syndrome primarily affects young Japanese men and is apparently connected with the Japanese cultural emphasis on being polite and avoiding embarrassing other people.

Evaluation of the *DSM* System Two basic criteria used in evaluating a diagnostic system such as the *DSM* are its reliability and validity. A diagnostic system may be considered **reliable,** or consistent, if various diagnosticians using the system are likely to arrive at the same diagnoses when they evaluate the same cases. The most appropriate test of the **validity** of a diagnostic system for psychological disorders is its correspondence with behavioral observations. Validity is often tested by examining whether people who receive a particular diagnosis display behaviors that correspond to those represented by the diagnostic category. Evidence supports the reliability and validity of most *DSM* anxiety disorder and mood disorder categories (T. A. Brown et al., 2001; Turner et al., 1986). However, the validity of some other diagnostic classes, such as personality disorders, remains a subject of debate in the scientific community, as does the validity of Axis V, Global Assessment of Functioning (Moos, McCoy, & Moos, 2000).

Another yardstick of validity, called **predictive validity,** is based on the ability of the diagnostic system to predict the course the disorder is likely to follow or its response to treatment. Evidence is accumulating that persons classified in certain categories respond better to certain types of medication. Persons with bipolar disorder, for example, respond reasonably well to the drug lithium (see Chapter 8). Specific forms of psychological treatment may also be more effective with certain diagnostic groupings. For example, persons who have specific phobias (such as fear of heights) are generally highly responsive to behavioral techniques for reducing fears (see Chapter 6).

Many observers (e.g., Eisenbruch, 1992; Fabrega, 1992) have argued that the *DSM* should become more sensitive to diversity in culture and ethnicity. The behaviors included as diagnostic criteria in the *DSM* are determined by consensus of mostly U.S.-trained psychiatrists, psychologists, and social workers. Had the American Psychiatric Association asked

Asian-trained or Latin American–trained professionals to develop their diagnostic manual, for example, there might have been some different or some revised diagnostic categories.

In fairness to the *DSM*, however, the latest edition does place greater emphasis than did earlier editions on weighing cultural factors when assessing abnormal behavior (DeAngelis, 1994b; Nathan, 1994). It recognizes that clinicians unfamiliar with an individual's cultural background may incorrectly classify the individual's behavior as abnormal when it in fact falls within the normal spectrum in the individual's culture. In Chapter 1 we noted the same behavior might be deemed normal in one culture but abnormal in another. The *DSM* specifies that for a diagnosis of a mental disorder to be made, the behavior in question must not merely represent a culturally expectable and sanctioned response to a particular event, even though it may seem odd in the light of the examiner's

Truth OR Fiction? REVISITED

Some men in India develop a psychological disorder involving excessive concerns or anxiety over losing semen.

TRUE. Some men in India develop dhat syndrome, a psychological disorder that is characterized by excessive fears over loss of semen.

TABLE 3.6 Examples of Culture-Bound Syndromes

Culture-Bound Syndrome	Description
amok	A disorder principally occurring in men in southeastern Asian and Pacific Island cultures, as well as in traditional Puerto Rican and Navajo cultures in the West, it describes a type of dissociative episode (a sudden change in consciousness or self-identity) in which an otherwise normal person suddenly goes berserk and strikes out at others, sometimes killing them. During these episodes, the person may have a sense of acting automatically or robotically. Violence may be directed at people or objects and is often accompanied by perceptions of persecution. A return to the person's usual state of functioning follows the episode. In the West, we use the expression "running amuck" to refer to an episode of losing oneself and running around in a violent frenzy. The word *amuck* is derived from the Malaysian word *amoq*, meaning "engaging furiously in battle." The word passed into the English language during colonial times when British colonial rulers in Malaysia observed this behavior among the native people.
ataque de nervios ("attack of nerves")	A way of describing states of emotional distress among Latin American and Latin Mediterranean groups, it most commonly involves features such as shouting uncontrollably, fits of crying, trembling, feelings of warmth or heat rising from the chest to the head, and aggressive verbal or physical behavior. These episodes are usually precipitated by a stressful event affecting the family (e.g., receiving news of the death of a family member) and are accompanied by feelings of being out of control. After the attack, the person returns quickly to his or her usual level of functioning, although there may be amnesia for events that occurred during the episode.
dhat syndrome	A disorder (described further in Chapter 7) affecting males found principally in India that involves intense fear or anxiety over the loss of semen through nocturnal emissions, ejaculations, or excretion with urine (despite the folk belief, semen doesn't actually mix with urine). In Indian culture, there is a popular belief that loss of semen depletes the man of his vital natural energy.
falling out or blacking out	Occurring principally among southern U.S. and Caribbean groups, the disorder involves an episode of sudden collapsing or fainting. The attack may occur without warning or be preceded by dizziness or feelings of "swimming" in the head. Although the eyes remain open, the individual reports an inability to see. The person can hear what others are saying and understand what is occurring but feels powerless to move.
ghost sickness	A disorder occurring among American Indian groups, it involves a preoccupation with death and with the "spirits" of the deceased. Symptoms associated with the condition include bad dreams, feelings of weakness, loss of appetite, fear, anxiety, and a sense of foreboding. Hallucinations, loss of consciousness, and states of confusion may also be present, among other symptoms.
koro	Found primarily in China and some other South and East Asian countries, the syndrome (also discussed further in Chapter 7) refers to an episode of acute anxiety involving the fear that one's genitals (the penis in men and the vulva and nipples in women) are shrinking and retracting into the body and that death may result.
zar	A term used in a number of countries in North Africa and the Middle East to describe the experience of spirit possession. Possession by spirits is often used in these cultures to explain dissociative episodes (sudden changes in consciousness or identity) that may be characterized by periods of shouting, banging of the head against the wall, laughing, singing, or crying. Affected people may seem apathetic or withdrawn or refuse to eat or carry out their usual responsibilities.

Source. Adapted from the *DSM-IV-TR* (APA, 2000); Osborne, 2001; and other sources.

own cultural standards. The *DSM-IV* also recognizes that abnormal behaviors may take different forms in different cultures and that some abnormal behavior patterns are culturally specific (see Table 3.6). All told, the *DSM-IV* is widely recognized as an improvement over previous editions, but questions still remain about the reliability and validity of certain diagnostic categories (Langenbucher et al., 2000; Thakker & Ward, 1998; Widiger & Clark, 2000).

Advantages and Disadvantages of the *DSM* System Many consider the major advantage of the *DSM* to be its designation of specific diagnostic criteria. The *DSM* permits the clinician to readily match a client's complaints and associated features with specific standards to see which diagnosis best fits the case. The multiaxial system paints a comprehensive picture of clients by integrating information concerning abnormal behaviors, medical conditions that affect abnormal behaviors, psychosocial and environmental problems that may be stressful to the individual, and level of functioning. The possibility of multiple diagnoses prompts clinicians to consider presenting current problems (Axis I) along with the relatively long-standing personality problems (Axis II) that may contribute to them.

Criticisms have also been leveled against the *DSM* system. As noted, questions remain about the system's reliability and validity. Some critics challenge specific diagnostic criteria, such as the requirement that major depression be present for 2 weeks before diagnosis (Kendler & Gardner, 1998). Others challenge the reliance on the medical model. In the *DSM* system, problem behaviors are viewed as symptoms of underlying mental disorders in much the same way that physical symptoms are signs of underlying physical disorders. The very use of the term *diagnosis* presumes the medical model is an appropriate basis for classifying abnormal behaviors. Some clinicians feel that behavior, abnormal or otherwise, is too complex and meaningful to be treated merely as symptomatic. They assert that the medical model focuses too much on what may happen within the individual and not enough on external influences on behavior, such as social factors (socioeconomic, sociocultural, and ethnic) and physical environmental factors.

THINK ABOUT IT

What are the advantages and disadvantages of using the *DSM* system to classify abnormal behavior patterns? Can you think of other ways we might classify abnormal behavior patterns?

Another concern is that the medical model focuses on categorizing psychological (or mental) disorders rather than describing people's behavioral strengths and weaknesses. Nor does the *DSM* attempt to place behavior within a contextual framework that examines the settings, situations, and cultural contexts in which behavior occurs (Follette & Houts, 1996; Wulfert, Greenway, & Dougher, 1996). To behaviorally oriented psychologists, the understanding of behavior, abnormal or otherwise, is best approached by examining the interaction between the person and the environment. The *DSM* aims to determine what "disorders" people "have"—not what they can "do" in particular situations. An alternative model of assessment, the behavioral model, focuses more on behaviors than on underlying processes—more on what people "do" than on what they "are" or "have." Behaviorists and behavior therapists also use the *DSM*, of course, in part because mental health centers and health insurance carriers require the use of a diagnostic code, in part because they want to communicate in a common language with practitioners of other theoretical persuasions. Many behavior therapists view the *DSM* diagnostic code as a convenient means of labeling patterns of abnormal behavior, a shorthand for a more extensive behavioral analysis of the problem.

Another concern about the *DSM* system is the potential for stigmatization of people labeled with psychiatric diagnoses. Our society is strongly biased against people who are labeled as mentally ill. They are often shunned by others, including even family members, and subjected to discrimination in housing and employment. The negative stereotyping of people identified as mentally ill is labeled **sanism** (Perlin, 1994), the counterpart to other forms of prejudice and discrimination that exist in our society, such as racism, sexism, and ageism.

The *DSM* system, despite its critics, has become part and parcel of the everyday practice of most U.S. mental health professionals. It may be the one reference manual found on the bookshelves of nearly all professionals and dog-eared from repeated use. Perhaps the *DSM* is best considered a work in progress, not a final product.

sanism The negative stereotyping of people who are identified as mentally ill.

Now let us consider various ways of assessing abnormal behavior. We begin by considering the basic requirements for methods of assessment—that they be reliable and valid.

Q Quiz 3.1
Classification of Abnormal Behavior Patterns

Issues of Reliability and Validity in Assessment

Important decisions are made on the basis of classification and assessment. For example, recommendations for specific treatment techniques vary according to our assessment of the problems that clients exhibit. Therefore, methods of assessment, like diagnostic categories, must be *reliable* and *valid.*

Reliability

The reliability of a method of assessment, like that of a diagnostic system, refers to its consistency. A gauge of height would be unreliable if people looked taller or shorter at every measurement. A reliable measure of abnormal behavior must also yield comparable results on different occasions. Also, different people should be able to check the yardstick and agree on the measured height of the subject. A yardstick that shrinks and expands markedly with the slightest change in temperature will be unreliable. So will one that is difficult to read.

There are three main approaches for demonstrating the reliability of assessment techniques.

Internal Consistency Correlational techniques are used to show whether the different parts or items of an assessment instrument, such as a personality scale or test, yield results that are consistent with one another and with the instrument as a whole. **Internal consistency** is crucial for tests intended to measure single traits or construct dimensions. When the individual items or parts of a test are highly correlated with each other, we can assume they are measuring a common trait or construct. For example, if responses to a set of items on a depression scale are not highly correlated with each other, there is no basis for assuming that the items measure a single common dimension or construct—in this case, depression.

Some tests are multidimensional in content. They contain subscales or factors that measure different construct dimensions. One such test is the Minnesota Multiphasic Personality Inventory (MMPI), which assesses various dimensions of abnormal behavior. In such cases, subscales within the test intended to measure individual traits, such as the hypochondriasis and depression subscales, are expected to show internal consistency. Subscales need not correlate with each other, however, unless the traits they are presumed to measure are interrelated.

Temporal Stability Reliable methods of assessment also have **temporal stability** (stability over time). They yield similar results on separate occasions. We would not trust a bathroom scale that yielded different results each time we weighed ourselves—unless we had stuffed or starved ourselves between weighings. The same principle applies to methods of psychological assessment. Temporal stability is measured by **test-retest reliability,** which represents the correlation between two administrations of the test separated by a period of time. The higher the correlation, the greater the temporal stability or test-retest reliability of the test.

Interrater Reliability **Interrater reliability**—also referred to as *interjudge reliability*—is usually of greatest importance for making diagnostic decisions and for measures requiring ratings of behavior. A diagnostic system is not reliable unless expert raters agree as to their diagnoses made on the basis of the system. For example, two teachers may be asked to use a behavioral rating scale to evaluate a child's aggressiveness, hyperactivity, and sociability. The level of agreement between the raters would be an index of the reliability of the scale.

internal consistency Cohesiveness or interrelationships of items on a test or scales.

temporal stability The consistency of test responses over time, as measured by test-retest reliability.

test-retest reliability A method of measuring the reliability of a test by means of comparing the scores of the same subjects on different occasions.

interrater reliability Consistency of or agreement between raters.

content validity The degree to which the content of a test covers a representative sample of the content it is designed to measure.

face validity The degree to which the content of a test bears an apparent relationship to the constructs it is designed to measure.

criterion validity The degree to which a test correlates with an independent, external criterion or standard.

concurrent validity A type of test validity based on the statistical relationship between the test and a criterion measure taken at the same time.

sensitivity The ability of a diagnostic instrument to correctly identify people who have the disorder the test is intended to detect.

specificity The ability of a diagnostic instrument to avoid classifying people as having a characteristic or disorder when they truly do not have the characteristic or disorder.

false negative An incorrect appraisal that a person is free of a disorder, when in fact he or she has the disorder.

false positive An incorrect appraisal that a person has a particular disorder.

construct validity The degree to which a test measures the hypothetical construct that it purports to measure.

Validity

The validity of assessment techniques or measures refers to the degree to which the instruments in question measure what they are intended to measure. There are various kinds of validity, such as *content, criterion,* and *construct validity.*

Content Validity The **content validity** of an assessment technique is the degree to which its content covers a representative sample of the behaviors associated with the construct dimension or trait in question. For example, depression includes features such as sadness and lack of participation in previously enjoyed activities. To have content validity, techniques assessing depression should thus have features or items that address these areas. One type of content validity, called **face validity,** is the degree to which questions or test items bear an apparent relationship to the constructs or traits they purport to measure. A face-valid item on a test of assertiveness could be, "I have little difficulty standing up for my rights." An item that lacks face validity as a measure of assertiveness could read, "I usually subscribe to magazines that contain features about world events."

The limitation of face validity is its reliance on subjective judgment in determining whether or not the test measures what it is supposed to measure. The apparent or face validity of an assessment technique is not sufficient to establish its scientific value. A scientific test may also be valid if its results relate to some standard or criterion, even though the items themselves do not have high face validity. This brings us to criterion validity.

Criterion Validity **Criterion validity** represents the degree to which the assessment technique correlates with an independent, external criterion (standard) of what the technique is intended to assess. There are two general types of criterion validity: concurrent validity and predictive validity.

Concurrent validity is the degree to which test responses predict scores on criterion measures taken at about the same time. Most psychologists presume intelligence is in part responsible for academic success. The concurrent validity of intelligence test scores is thus frequently studied by correlating test scores with criteria such as school grades and teacher ratings of academic abilities.

A test of depression might be validated in terms of its ability to identify people who meet diagnostic criteria for depression. Two related concepts are important here: **sensitivity** and **specificity.** Sensitivity refers to the degree to which a test correctly identifies people who have the disorder the test is intended to detect. Tests that lack sensitivity produce a high number of **false negatives**—individuals identified as not having the disorder who truly have the disorder. Specificity refers to the degree to which the test avoids classifying people as having a particular disorder who truly do not have the disorder. Tests that lack specificity produce a high number of **false positives**—people identified as having the disorder who truly do not have the disorder. By taking into account sensitivity and specificity of a given test, we can determine the ability of a test to classify individuals correctly.

We noted that predictive validity refers to the ability of a test to predict some future behavior outcome. A test of academic aptitude may be validated in terms of its ability to predict school performance in that particular area.

Construct Validity **Construct validity** is the degree to which a test corresponds to the theoretical model of the underlying construct or trait it purports to measure. Consider a test that purports to measure anxiety. Anxiety is not a concrete object or phenomenon. It can't be measured directly, counted, weighed, or touched. Anxiety is a theoretical construct that helps explain phenomena like a pounding heart or sudden inability to speak when you ask an attractive person out on a date. Anxiety may be indirectly measured by such means as self-report (rating one's own level of anxiety) and physiological techniques (measuring the level of sweat on the palms of one's hands).

The construct validity of a test of anxiety requires the results of the test to predict other behaviors that would be expected given your theoretical model of anxiety. Assume

THINK ABOUT IT

Suppose you wanted to develop a new psychological test or measure. How would you go about demonstrating that it was reliable and valid?

your theoretical model predicts that socially anxious college students would experience greater difficulties than calmer students in speaking coherently when asking someone for a date, but not when they are merely rehearsing the invitation in private. If the speech behavior of high and low scorers on a test purported to measure social anxiety fit these predicted patterns, we can say the evidence supports the construct validity of the test. Construct validity involves a continuing process of testing relationships among variables that are predicted from a theoretical framework. We can never claim to have proven the construct validity of a test because it is always possible to come up with an alternative theoretical account of these relationships.

A test may be reliable (give you consistent responses) but still not measure what it purports to measure (be invalid). A test of musical aptitude might have excellent reliability but be invalid as a measure of general intelligence. Nineteenth-century **phrenologists** believed they could gauge people's personalities by measuring the bumps on their heads. Their calipers provided reliable measures of their subjects' bumps and protrusions; the measurements, however, did not provide valid estimates of subjects' psychological traits. The phrenologists were bumping in the dark, so to speak.

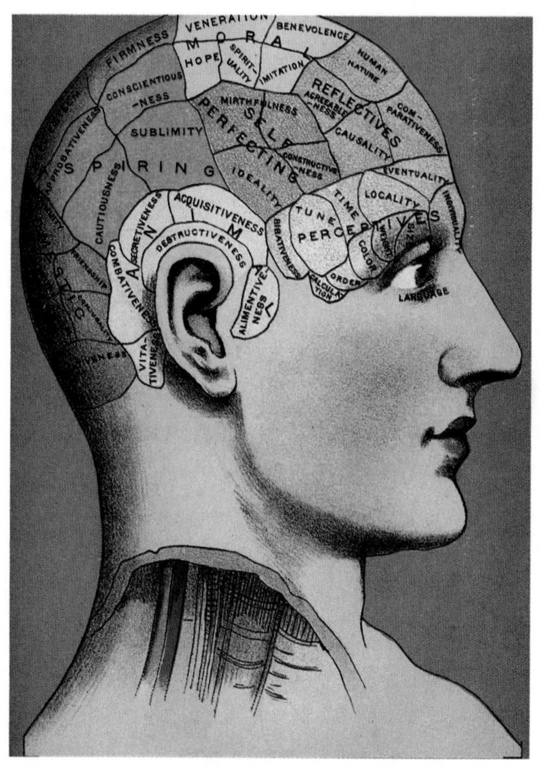

Phrenology. In the 19th century, some people believed that mental faculties and abilities were based in certain parts of the brain and that people's acumen in such faculties could be assessed by gauging the protrusions and indentations of the skull.

Sociocultural and Ethnic Factors in the Assessment of Abnormal Behavior

Researchers and clinicians also need to be aware of sociocultural and ethnic factors when they assess personality traits and psychological disorders. Assessment techniques may be reliable and valid within one culture but not within another, even when they are translated accurately (Bolton, 2001; Kleinman, 1987). In one study, Chan (1991) administered a Chinese-language version of the Beck Depression Inventory (BDI), a widely used inventory of depression in the United States, to a sample of Chinese students and psychiatric patients in Hong Kong. The Chinese BDI met tests of reliability, as judged by internal consistency, and of validity, as judged by its ability to distinguish people with depression from nondepressives among a small sample of Chinese psychiatric patients. Yet other investigators found that Chinese people in both Hong Kong and the People's Republic of China tended to achieve higher scores on a subscale of the Chinese MMPI, which is suggestive of deviant responses (Cheung, Song, & Butcher, 1991). These higher scores appeared to reflect cultural differences, however, rather than greater psychopathology (Cheung et al., 1991; Cheung & Ho, 1997).

A study in our own culture put the recently revised MMPI, called the MMPI-2, under a cultural microscope. Researchers found the test was as accurate in predicting the psychological adjustment of African Americans as of non-Hispanic White Americans (Timbrook & Graham, 1994). Moreover, researchers found small differences in the average test scores of African Americans and non-Hispanic White Americans when factors such as age, education, and income were taken into account. Another study found no evidence of cultural bias on the MMPI-2 between African American and non-Hispanic White clients at a mental health center (McNulty et al., 1997).

Other investigators found a greater prevalence of depression among Mexican Americans than among non-Hispanic White Americans in Los Angeles (Garcia & Marks, 1989). Here again the meaning of this difference was unclear. The difference may have reflected linguistic or sociocultural factors rather than differences in the prevalence of depression (Fabrega, 1992). Researchers thus need to disentangle psychopathology from sociocultural factors.

Most diagnostic schedules consider culture to some degree, but researchers believe they fail to provide adequate norms for different cultural and ethnic groups. Translations of instruments should not only translate words, but also provide instructions that encourage examiners to address the importance of cultural beliefs, norms, and values, so diagnosticians and interviewers will be prompted to consider the individual's background seriously when making assessments of abnormal behavior patterns.

Truth OR Fiction? REVISITED

A psychological test must be valid in order to be reliable.

FALSE. A test may be reliable (give you consistent responses) but still not measure what it purports to measure.

WWW Web Link 3.3
Phrenology Page

phrenologist Someone who studies the bumps on a person's head to determine the person's underlying traits.

Interviewers need also be sensitized to problems that can arise when interviews are conducted in a language other than the client's mother tongue. Hispanics, for example, often are judged more disturbed when interviewed in English (Fabrega, 1990). Problems can also arise when interviewers who use a second language fail to appreciate the idioms and subtleties of the language. The first author recalls a case in a U.S. mental hospital in which the interviewer, a foreign-born and -trained psychiatrist, reported that a patient exhibited the delusional belief that he was outside his body. This assessment was based on the patient's response to a question posed by the psychiatrist. The psychiatrist had asked the patient if he was feeling anxious and the patient replied, "Yes, Doc, I feel like I'm jumping out of my skin at times."

Quiz 3.2
Issues of Reliability and Validity in Assessment

THINK ABOUT IT
Why is it important for clinicians to take cultural factors into account when diagnosing psychological or mental disorders?

Methods of Assessment

Here we explore methods of assessment that clinicians use to arrive at diagnostic impressions, including interviews, psychological testing, self-report questionnaires, behavioral measures, and physiological measures. The role of assessment, however, goes further than classification. A careful assessment provides a wealth of information about clients' personalities and cognitive functioning. This information helps clinicians acquire a broader understanding of their clients' problems and recommend appropriate forms of treatment.

The Clinical Interview

The clinical interview is the most widely used means of assessment. It is employed by all helping professionals and paraprofessionals. The interview is usually the client's first face-to-face contact with a clinician. Clinicians often begin by asking clients to describe the presenting complaint in their own words. They may say something like, "Can you describe to me the problems you've been having lately?" (Therapists learn not to ask, "What brings you here?" to avoid receiving such answers as, "A car," "A bus," or "My social worker.") The clinician will then usually probe aspects of the presenting complaint, such as behavioral abnormalities and feelings of discomfort, the circumstances regarding the onset of the problem, history of past episodes, and how the problem affects the client's daily functioning. The clinician may explore possible precipitating events, such as changes in life circumstances, social relationships, employment, or schooling. The interviewer encourages the client to describe the problem in her or his own words in order to understand it from the client's viewpoint.

Although the format of the intake process may vary from clinician to clinician, most interviews cover topics such as these:

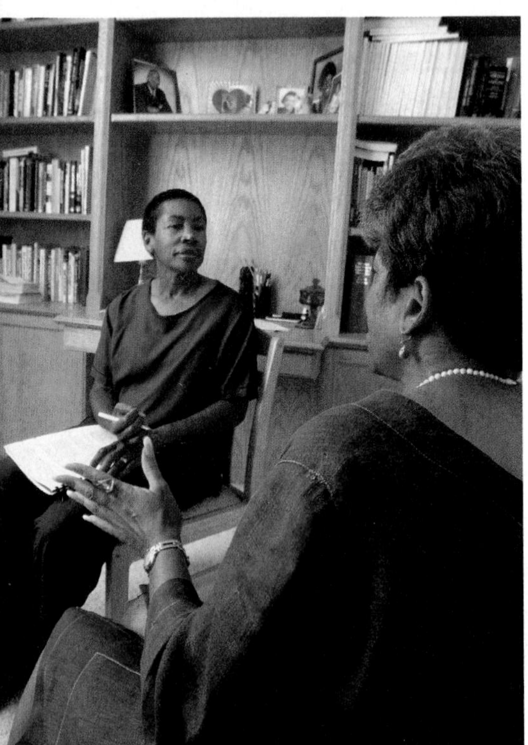

Building rapport. By developing rapport and feelings of trust with a client, the skillful interviewer helps put the client at ease and encourages candid communication.

1. *Identifying data.* Information regarding the client's sociodemographic characteristics: address and telephone number, marital status, age, gender, racial/ethnic characteristics, religion, employment, family composition, and so on.
2. *Description of the presenting problem(s).* How does the client perceive the problem? What troubling behaviors, thoughts, or feelings are reported? How do they affect the client's functioning? When did they begin?
3. *Psychosocial history.* Information describing the client's developmental history: educational, social, and occupational history; early family relationships.
4. *Medical/psychiatric history.* History of medical and psychiatric treatment and hospitalizations: Is the present problem a recurrent episode of a previous problem? How was the problem handled in the past? Was treatment successful? Why or why not?
5. *Medical problems/medication.* Description of present medical problems and present treatment, including medication. The clinician is alert to ways in which medical problems may affect the presenting psychological problem. For example, drugs for certain medical conditions can affect people's moods and general levels of arousal.

Interview Formats There are three general types of clinical interviews: unstructured interviews, semi-structured interviews, and structured interviews. In an **unstructured interview,** the clinician adopts his or her own style of questioning rather than following any standard format. In a **semi-structured interview,** the clinician follows a general outline of questions designed to gather essential information but is free to ask the questions in any particular order and to branch off into other directions in order to follow up clinically important information. In a **structured interview,** the interview follows a preset series of questions in a particular order.

The major advantage of the unstructured interview is its spontaneity and conversational style. There is an active give-and-take between the interviewer and the client, because the interviewer is not bound to follow any specific set of questions. The major disadvantage is the lack of standardization. Different interviewers may ask questions in different ways. For example, one interviewer might ask, "How have your moods been lately?" while another might pose the question, "Have you had any periods of crying or tearfulness during the past week or two?" The clients' responses may depend to a certain extent on how the questions are asked. Another drawback is that the conversational flow of the interview may fail to touch on important clinical information needed to form a diagnostic information. A semi-structured interview provides more structure and uniformity, but at the possible expense of spontaneity. Clinicians may seek to strike a balance by conducting a semi-structured interview in which they follow a general outline of questions but allow themselves the flexibility to depart from the interview protocol to pursue issues that seem important to them.

Structured interviews (also called *standardized interviews*) provide the highest level of reliability and consistency in reaching diagnostic judgments, which is why they are used frequently in research settings. Yet many clinicians prefer using a semi-structured approach because of its greater flexibility. A leading example of a structured interview protocol is the Structured Clinical Interview for the *DSM* (SCID). The SCID includes **closed-ended questions** to determine the presence of behavior patterns that suggest specific diagnostic categories and **open-ended questions** that allow clients to elaborate their problems and feelings. The SCID guides the clinician in testing diagnostic hypotheses as the interview progresses. Recent research supports the reliability of the SCID across various clinical settings (J. B. Williams et al., 1992).

In the course of the interview, the clinician may also conduct a more formal assessment of the client's cognitive functioning by administering a **mental status examination.** This involves a formal assessment of client appearance (appropriateness of attire and grooming), mood, attention, perceptual and thinking processes, memory, orientation (knowing who they are, where they are, and the present date), level of awareness or insight into their problems, and judgment in making life decisions.

The interviewer compiles all the information available from the interview and review of the client's background and presenting problems to arrive at a diagnostic impression.

unstructured interview Interview in which the clinician adopts his or her own style of questioning rather than following any standard format.

semi-structured interview Interview in which the clinician follows a general outline of questions designed to gather essential information but is free to ask them in any order and to branch off in other directions.

structured interview Interview that follows a preset series of questions in a particular order.

closed-ended questions Questionnaire or test items with a limited range of response options.

open-ended questions Questions that provide an unlimited range of response options.

mental status examination A structured clinical assessment to determine various aspects of the client's mental functioning.

Psychological Tests

Psychological tests are structured methods of assessment used to evaluate reasonably stable traits such as intelligence and personality. Tests are usually standardized on large numbers of subjects and provide norms that compare clients' scores with the average. By comparing test results from samples of people who are free of psychological disorders with those of people who have diagnosable psychological disorders, we may gain some insights into the types of response patterns that are indicative of abnormal behavior. A recent analysis showed that the validity of many psychological tests are comparable to those of many medical tests when judged by their ability to predict criterion variables, such as underlying conditions or future outcomes (Daw, 2001; Meyer et al., 2001) (see Figure 3.1). Here we examine two major types of psychological tests: intelligence tests and personality tests.

Intelligence Tests The assessment of abnormal behavior often includes an evaluation of intelligence. Formal tests of intelligence are used to help diagnose mental retardation.

A Closer Look

Would You Tell Your Problems to a Computer?

Picture yourself seated before a computer screen in the not-too-distant future. The message on the screen asks you to type in your name and press the return key. Not wanting to offend, you comply. This message then comes on the screen: "Hello, my name is Sigmund. I'm programmed to ask you a set of questions to learn more about you. May I begin?" You nod your head yes, momentarily forgetting that the computer can only "perceive" keystrokes. You type "yes" and the interview begins.

The future, as the saying goes, is now. Computerized clinical interviews have been used for more than 25 years. Computers offer some advantages over us traditional human interviewers (Farrell, Camplair, & McCullough, 1987):

1. Computers can be programmed to ask a specific set of questions in a predetermined order, whereas human beings may omit critical items or steer the interview toward less important topics.
2. The client may be less embarrassed about relating personal matters to the computer because computers do not show emotional responses to clients' responses.
3. Computerized interviews can free clinicians to spend more time providing direct clinical services.

Consider a computerized interview system named CASPER. Interview questions and response options, such as the following, are presented on the screen:

> *"About how many days in the past month did you have difficulty falling asleep, staying asleep, or waking too early (include sleep disturbed by bad dreams)?"*
>
> *"During the past month, how have you been getting along with your spouse/partner? (1) Very satisfactory; (2) Mostly satisfactory; (3) Sometimes satisfactory, sometimes unsatisfactory; (4) Mostly unsatisfactory; (5) Very unsatisfactory."*
>
> *Farrell et al., 1987, p. 692*

The subject presses a numeric key to respond to each item. CASPER is a branching program that follows up on problems suggested by the clients' responses. For example, if the client indicates difficulty in falling or remaining asleep, CASPER asks whether or not sleep has become a major problem—"something causing you great personal distress or interfering with your daily functioning" (p. 693). If the client indicates yes, the computer will return to the problem after other items have been presented and ask the client to rate the duration and intensity of the problem. Clients may also add or drop complaints—change their minds, that is.

Research with computer interviewing systems shows that people reveal as much if not more personal information to a computer than they do to a human interviewer (Kobak et al., 1997; Kalb, 2001). The computer interview may be especially helpful in identifying problems that clients are embarrassed or unwilling to report to humans. Perhaps people feel less self-conscious if someone isn't looking at them when they are interviewed (Kalb, 2001). Or perhaps the computer seems more willing to take the time to note all complaints.

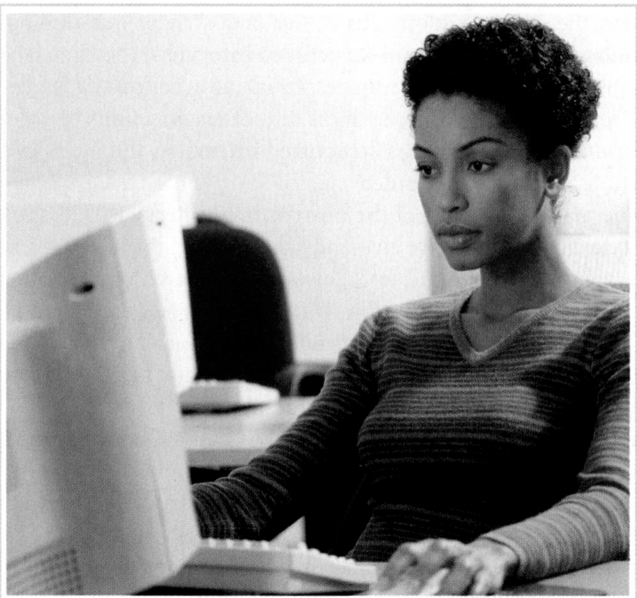

Interview by computer. Would you be more likely, or less likely, to tell your problems to a computer than to a person? Computerized clinical interviews have been used for more than 20 years, and some research suggests that the computer may be more sensitive than its human counterpart in teasing out problems.

Truth OR Fiction? REVISITED

Researchers find that people report more personal problems when interviewed by human interviewers than by computers.

FALSE. Evidence shows that people reveal as much if not more about themselves when interviewed by computers than by humans. Perhaps people are less concerned about being "judged" by computers.

Reviews of research suggest that computer programs are as capable as skilled clinicians at obtaining information from clients and reaching an accurate diagnosis, and are less expensive and more time efficient (Bloom, 1992; Kobak et al., 1996). It seems that most of the resistance to using computer interviews for this purpose comes from clinicians rather than clients.

Computer-assisted treatments for mental health problems have also become available, including systems designed to help people overcome phobias and test anxiety. Almost all of these are aids to therapy, not substitutes for a live therapist (Gilroy et al., 2000; I. Marks et al., 1998a). But with the rapid pace of technological developments today, there may well come a day when computerized programs with interactive voice technology will be used as stand-alone "therapists."

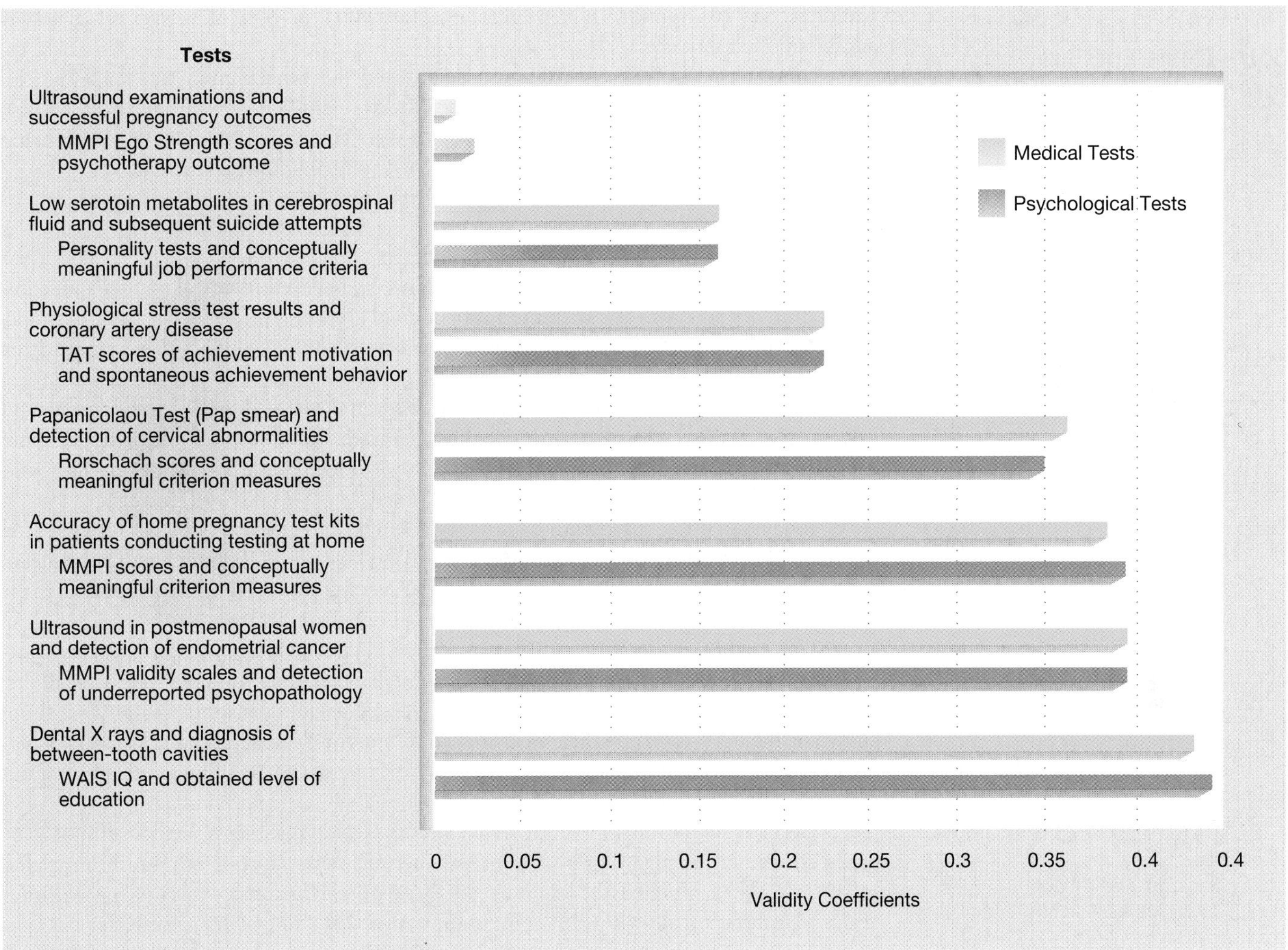

FIGURE 3.1 Comparisons of validity of psychological and medical tests.
Many psychological tests compare favorably with many medical tests in predicting many criterion variables.

Source. Adapted from Meyer et al., 2001.

They evaluate the intellectual impairment that may be caused by other disorders, such as organic mental disorders caused by damage to the brain. They also provide a profile of the client's intellectual strengths and weaknesses to help develop a treatment plan suited to the client's competencies.

Intelligence is a controversial concept in psychology, however. Even attempts at definition stir debate. David Wechsler (1975), the originator of a widely used series of intelligence tests, defined intelligence as "capacity . . . to understand the world . . . and . . . resourcefulness to cope with its challenges." From his perspective, intelligence has to do with the ways in which we (1) mentally represent the world, and (2) adapt to its demands. There are various intelligence tests, including group tests and those that are administered individually, such as the Stanford-Binet Intelligence Scale (SBIS) and the Wechsler scales. Individual tests allow examiners to observe the behavior of the respondent as well as record answers. Examiners can thus gain insight as to whether factors such as testing conditions, language problems, illness, or level of motivation contribute to a given test performance.

The SBIS was originated by the Frenchmen Alfred Binet and Theodore Simon in 1905 in response to the French public school system's need for a test that could identify children who might profit from special education. The initial Binet-Simon scale yielded a score called a **mental age** (MA) that represented the child's overall level of intellectual functioning. The child who received an MA of 8 was functioning like the typical 8-year-old.

intelligence (1) The capacity to understand the world and respond to its challenges; (2) the trait measured by intelligence tests.

mental age The age equivalent that corresponds to a person's level of intelligence as measured by the Stanford-Binet Intelligence Scale.

TABLE 3.7 Items Similar to Those on the Stanford-Binet

Age	Sample Item
2	"Point to your toes."
6	"Tell me what's next: A minute is short; an hour is _____."
10	"Try to repeat these numbers: 8-9-4-2-6-1."
11	"How are 'beginning' and 'end' alike?"
Adult	"What does this mean? 'The watched pot never boils'?"

Source. Fernald, D. (1997). *Psychology.* Upper Saddle River, NJ: Prentice Hall, p. 242.

Children received "months" of credit for correct answers, and their MAs were calculated by adding up their scores.

Louis Terman of Stanford University adapted the Binet-Simon test for American children in 1916, which is why it is now called the *Stanford-Binet Intelligence Scale* (SBIS). The SBIS yields an **intelligence quotient (IQ),** not an MA, which reflected the relationship between a child's MA and chronological age (CA), according to this formula:

$$\mathbf{IQ} = \frac{\mathbf{MA}}{\mathbf{CA}} \times \mathbf{100}$$

Examination of this formula shows that children who received identical mental age scores might differ markedly in IQ, with the younger child attaining the higher IQ. For example, an 8-year-old with a mental age of 10 would have an IQ of 125, while a 10-year-old with a mental age of 10 would have an IQ of 100.

Binet assumed that intelligence grew as children developed, so older children would obtain more answers that are correct. He thus age-graded his questions and arranged them according to difficulty level, a practice carried over into the Stanford-Binet, as shown in Table 3.7.

Today the SBIS is used with children and adults, and test takers' IQ scores are based on their deviation from the norms of their age group. A score of 100 is defined as the mean. People who answer more items correctly than the average obtain IQ scores above 100; those who answer fewer items correctly obtain scores of less than 100.

This method of deriving an IQ score, called the **deviation IQ,** was used by psychologist David Wechsler in developing various intelligence tests for children and adults, known as the Wechsler scales. The Wechsler scales group questions into subtests like those shown in Table 3.8, each of which measures a different intellectual task. The Wechsler scales are thus designed to offer insight into a person's relative strengths and weaknesses, and not simply yield an overall score.

Wechsler's scales include both *verbal* and *performance* subtests. Verbal subtests generally require knowledge of verbal concepts; performance subtests rely more on spatial relations skills. (Figure 3.2 shows items like those on performance scales of the Wechsler scales.) Wechsler's scales allow for computation of verbal and performance IQs.

TABLE 3.8 Examples of Subtests from the Wechsler Adult Intelligence Scale (WAIS)

Verbal Subtests

Information Who wrote *The Odyssey?*

Comprehension Why are people required to register their cars?

Arithmetic John wanted to buy a pair of pants that cost $47.25, but only had $19. How much more money would he need to buy the pants?

Similarities How are truth and honor alike?

Digit Span (Forward order) After listening to this series of numbers, repeat them in the same order: 5 2 4 9

(Backward order) After listening to this series of numbers, repeat them backward: 4 9 6 1

Vocabulary What does the word *augment* mean?

Performance Subtests

Digit Symbol Given a key showing a set of symbols that correspond to particular numbers, fill in the correct symbols for a series of numbers.

Picture Completion Identify the missing parts of a picture.

Block Design Use blocks like those in Figure 3.2 to match particular designs.

Picture Arrangement Place a set of storybook pictures in the correct order to tell a coherent story.

Object Assembly Arrange the pieces of a jigsaw puzzle so that they form a particular object.

Source. From "Subtests from the Wechsler Adult Intelligence Scale Revised (WAIS-R)" for S. A. Ratus, *Essentials of Pshychology* (6th ed.), © 2001. Reprinted with permission of Brooks/Cole, on imprint of the Wadsworth Group, a division of Thomson Learning. FAX 800-730-2215.

intelligence quotient (IQ) A measure of intelligence based on scores on an intelligence test; the ratio between a respondent's mental age and actual age.

deviation IQ An intelligence quotient obtained by determining the deviation between the person's score and the norm (mean).

Picture Arrangement
These pictures tell a story but they are in the wrong order. Put them in the right order so that they tell a story.

Picture Completion
What part is missing from this picture?

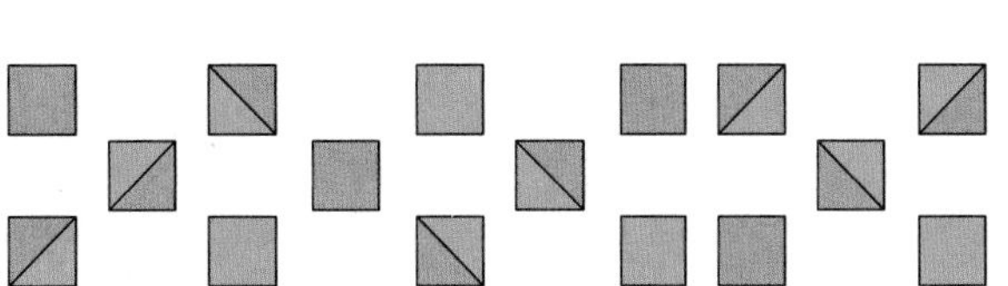
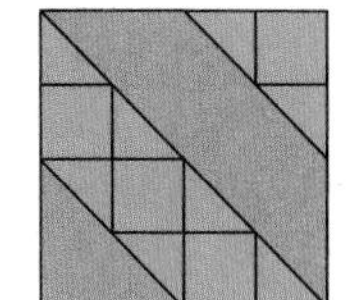

Block Design
Put the blocks together to make this picture.

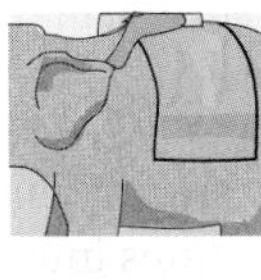

Object Assembly
Put the pieces together as quickly as you can.

FIGURE 3.2 Items similar to those found on the performance subtests of the Wechsler Intelligence Scale (WAIS).
The Wechsler scales yield verbal and performance IQs that are based on the extent to which an individual's test scores deviate from the norm for her or his age group.

Students from various backgrounds yield different profiles. College students, generally speaking, perform better on verbal subtests than on performance subtests. Australian Aboriginal children outperform White Australian children on performance-type tasks that involve visual-spatial skills (Kearins, 1981). Such skills are likely to foster survival in the harsh Australian outback. Intellectual attainments, like psychological adjustment, are connected with the demands of particular sociocultural and physical environmental settings.

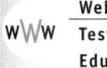

Web Link 3.4
Test Locator Services from Educational Testing Service (ETS)

Wechsler IQ scores are based on how respondents' answers deviate from those attained by their age-mates. The mean whole test score at any age is defined as 100. Wechsler distributed IQ scores so that 50% of the scores of the population would lie within a "broad average" range of 90 to 110.

Most IQ scores cluster around the mean (see Figure 3.3). Just 5% of them are above 130 or below 70. Wechsler labeled people who attained scores of 130 or above as "very superior" and those with scores below 70 as "intellectually deficient." IQ scores below 70 are one of the criteria used in diagnosing mental retardation.

Next we consider the tests psychologists use to assess personality. We consider two types of personality tests: *self-report personality tests* and *projective tests.*

self-report personality test A structured personality test in which individuals give information about themselves by responding to items that require a limited type of response, such as "yes-no" or "agree-disagree."

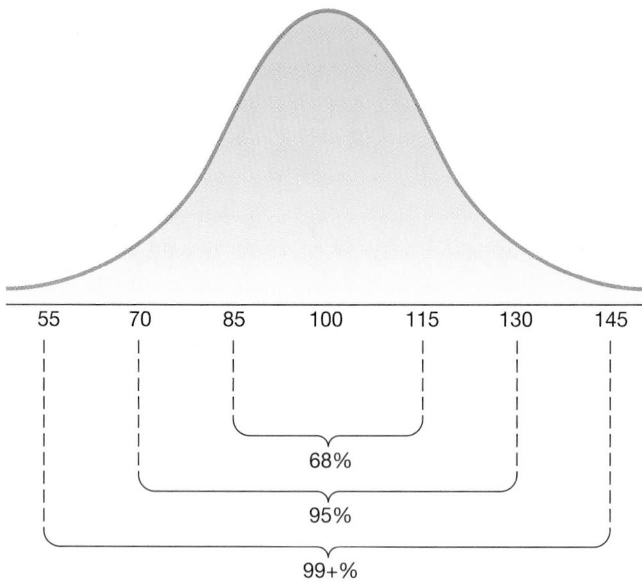

FIGURE 3.3 Normal distribution of IQ scores.
The distribution of IQ scores is based on a bell-shaped curve, which is referred to by psychologists as a *normal curve*. Wechsler defined the deviation IQ in such a way that 50% of the scores fall within the broad average range of 90 to 110.

Self-Report Personality Tests In **self-report personality tests,** individuals respond to a set of items about their feelings, thoughts, concerns, attitudes, interests, beliefs, and the like. Responses may be given in "yes-no," "true-false," or "agree-disagree" formats. Some self-report personality tests are intended to measure a particular trait or construct, such as anxiety or depression. The Beck Depression Inventory (BDI) (A. T. Beck et al., 1961), for instance, is a widely used measure of depression. In many clinic sites, such as the training clinic where the first author supervises clinical psychology students, the BDI is used routinely to screen new clients for depression. Some personality tests measure multiple dimensions of personality. Here we focus on two of the more widely used multidimensional personality tests in clinical settings, the Minnesota Multiphasic Personality Inventory (MMPI; now the MMPI-2) and the Millon Clinical Multiaxial Inventory (MCMI).

Do you like automobile magazines? Are you easily startled by noises in the night? Are you bothered by periods of anxiety or shakiness? Self-report inventories use structured items, similar to these, to measure personality traits such as anxiety, depression, emotionality, masculinity-femininity, and introversion. Comparison of clients' responses on scales measuring these traits to those of a normative sample reveals their relative standing.

Self-report personality inventories are also called **objective tests.** They are objective in that the range of possible responses to items is limited. Empirical objective standards—rather than psychological theory—are also used to derive test items. Tests might ask respondents to check adjectives that apply to them, to mark statements as true or false, to select preferred activities from lists, or to indicate whether items apply to them "always," "sometimes," or "never." Tests with **forced-choice formats** require respondents to mark which of a group of statements is truest for them, or to select their most preferred activity from a list. They cannot answer "none of the above."

With objective personality tests, items are selected according to some empirical standard. With the development of the original MMPI, the standard was whether items differentiated clinical diagnostic groups from normal comparison groups, not whether they bore an apparent relationship to the disorder itself.

Truth OR Fiction? REVISITED

The most widely used personality inventory includes a number of questions that bear no obvious relationship to the traits the instrument purports to measure.

TRUE. A leading personality inventory, the MMPI, contains items that bear no apparent relationship to the traits being measured. However, these items did differentiate between the response patterns of clinical diagnostic groups and normal reference groups.

Minnesota Multiphasic Personality Inventory (MMPI-2) The MMPI-2 contains more than 500 true-false statements that assess interest patterns, habits, family relationships, somatic complaints, attitudes, beliefs, and behaviors characteristic of psychological disorders. It is widely used as a test of personality as well as assisting in the diagnosis of abnormal behavior patterns. The MMPI-2 consists of a number of individual scales composed of items that tended to be answered differently by members of carefully selected diagnostic groups, such as patients diagnosed with schizophrenia or depression, than by members of normal comparison groups.

Consider a hypothetical item: "I often read detective novels." If groups of depressed people tended to answer the item in a direction different from normal groups, the item would be placed on the depression scale—regardless of whether or not the item had face validity. Many items that discriminate normal people from clinical groups are transparent in meaning, such as "I feel down much of the time." Some items are more subtle in meaning or bear no obvious relationship to the measured trait.

Derivation of scales on the basis of their ability to distinguish the response patterns of comparison groups such as clinical and normal groups is called the **contrasted groups approach.** The contrasted groups technique establishes concurrent validity; group membership is the criterion by which the validity of the test is measured.

Eight clinical scales were derived through the contrasted groups approach. Two additional clinical scales were developed by using nonclinical comparison groups: a scale

objective tests Tests that allow a limited range of response options and can therefore be scored objectively.

forced-choice formats Test questions that require respondents to select from among a limited number of choices.

contrasted groups approach A method of determining concurrent validity that uses a test's ability to differentiate between members of two or more comparison groups.

measuring masculine–feminine interest patterns and one measuring social introversion. The clinical scales are described in Table 3.9. The MMPI-2 also has **validity scales** that assess tendencies to distort test responses in a favorable ("faking good") or unfavorable ("faking bad") direction. The scale also contains additional validity scales and a set of scales, called *content scales*, which measure an individual's specific complaints and concerns, such as anxiety, anger, family problems, and problems of low self-esteem.

The respondent's raw score for each of the clinical scales on the MMPI scale is simply the number of items scored in a clinical direction. Raw scores are converted into **standard scores** with a mean of 50 and a standard deviation of 10. A standard score of 65 or higher on a particular scale places an individual at approximately the 92nd percentile or higher of the revised normative sample and is considered clinically significant.

The MMPI-2 is interpreted according to individual scale elevations and interrelationships among scales. For example, a "2–7 profile," commonly found among people

validity scales Groups of test items that are used to detect whether the results of a test are valid.

standard scores Scores that indicate the relative standing of raw scores in relation to the distribution of normative scores.

TABLE 3.9 Clinical Scales of the MMPI-2

Scale Number	Scale Label	Items Similar to Those Found on MMPI Scale	Sample Traits of High Scorers
1	Hypochondriasis	My stomach frequently bothers me. At times, my body seems to ache all over.	Many physical complaints, cynical defeatist attitudes, often perceived as whiny, demanding
2	Depression	Nothing seems to interest me anymore. My sleep is often disturbed by worrisome thoughts.	Depressed mood; pessimistic, worrisome, despondent, lethargic
3	Hysteria	I sometimes become flushed for no apparent reason. I tend to take people at their word when they're trying to be nice to me.	Naive, egocentric, little insight into problems, immature; develops physical complaints in response to stress
4	Psychopathic Deviate	My parents often disliked my friends. My behavior sometimes got me into trouble at school.	Difficulties incorporating values of society, rebellious, impulsive, antisocial tendencies; strained family relationships; poor work and school history
5	Masculinity-Femininity	I like reading about electronics. (M) I would like to work in the theater. (F)	Males endorsing feminine attributes: have cultural and artistic interests, effeminate, sensitive, passive Females endorsing male interests: Aggressive, masculine, self-confident, active, assertive, vigorous
6	Paranoia	I would have been more successful in life but people didn't give me a fair break. It's not safe to trust anyone these days.	Suspicious, guarded, blames others, resentful, aloof, may have paranoid delusions
7	Psychasthenia	I'm one of those people who have to have something to worry about. I seem to have more fears than most people I know.	Anxious, fearful, tense, worried, insecure, difficulties concentrating, obsessional, self-doubting
8	Schizophrenia	Things seem unreal to me at times. I sometimes hear things that other people can't hear.	Confused and illogical thinking, feels alienated and misunderstood, socially isolated or withdrawn, may have blatant psychotic symptoms such as hallucinations or delusional beliefs, or may lead detached, schizoid lifestyle
9	Hypomania	I sometimes take on more tasks than I can possibly get done. People have noticed that my speech is sometimes pressured or rushed.	Energetic, possibly manic, impulsive, optimistic, sociable, active, flighty, irritable, may have overly inflated or grandiose self-image or unrealistic plans
10	Social Introversion	I don't like loud parties. I was not very active in school activities.	Shy, inhibited, withdrawn, introverted, lacks self-confidence, reserved, anxious in social situations

seeking therapy, refers to a test pattern in which scores for scales 2 ("Depression") and 7 ("Psychasthenia") are clinically significant. Clinicians may refer to "atlases," or descriptions, of people who usually attain various profiles.

MMPI-2 scales are regarded as reflecting continua of personality traits associated with the diagnostic categories represented by the test. For example, a high score on *psychopathic deviation* suggests that the respondent holds a higher-than-average number of nonconformist beliefs and may be rebellious, which are characteristics often found in people with antisocial personality disorder. However, because it is not tied specifically to *DSM* criteria, this score cannot be used to establish a diagnosis of antisocial personality disorder or any other psychological disorder. Perhaps it is unfair to expect that the MMPI, which was originally developed in the 1930s and 1940s under a largely outmoded diagnostic system, should provide diagnostic judgments consistent with the current version of the *DSM* system. Even so, MMPI profiles may suggest possible diagnoses that can be considered in the light of other evidence. Moreover, many clinicians use the MMPI to gain general information about respondents' personality traits and attributes that may underlie their psychological problems, rather than a diagnosis per se.

The validity of the original MMPI and the MMPI-2 is supported by a large body of research findings (Kubiszyn et al., 2000). The test successfully discriminates between psychiatric patients and controls and between groups of people with different psychological disorders, such as anxiety versus depressive disorders (Ganellen, 1996; Graham, 2000). Moreover, the content scales of the MMPI-2 provide additional information to that provided by the clinical scales, which can help clinicians learn more about the client's specific problems (Graham, 2000; Strassberg, 1997).

The Millon Clinical Multiaxial Inventory (MCMI) The MCMI (Millon, 1982) was developed to help the clinician make diagnostic judgments within the multiaxial *DSM* system, especially in the personality disorders found on Axis II. The MCMI is the only objective personality test that focuses on personality style and disorders. The MMPI-2, in contrast, focuses on personality patterns associated with Axis I diagnoses, such as mood disorders, anxiety disorders, and schizophrenic disorders. Using the MCMI and MMPI-2 in combination may help the clinician make more subtle diagnostic distinctions than are possible with either test alone, because they assess different patterns of psychopathology (Antoni et al., 1986). However, relationships between the MCMI and the underlying personality disorders they are meant to assess remain under study. Though the MCMI may help clinicians discriminate among various Axis I and Axis II disorders (Ganellen, 1996; Kubiszyn et al., 2000), some researchers voice concern that it may overdiagnose personality disorders (Guthrie & Mobley, 1994; Wetzler & Marlowe, 1993).

Web Link 3.5 WWW
Online Links to Personality Tests

Evaluation of Self-Report Inventories Self-report tests have the benefits of relative ease and economy of administration. Once the examiner has read the instructions to clients and ascertained they can read and comprehend the items, clients can complete the tests unattended. Because the tests permit limited response options, such as marking items either true or false, they can be scored with high interrater reliability. Moreover, the accumulation of research findings on respondents provides a quantified basis for interpreting test responses. Such tests often reveal information that might not be revealed during a clinical interview or by observing the person's behavior.

A disadvantage of self-rating tests is that they rely on clients as the source of data. Test responses may therefore reflect underlying response biases, such as tendencies to answer items in a socially desirable direction, rather than accurate self-perceptions. For this reason, self-report inventories, like the MMPI, contain validity scales to help ferret out response biases. Yet even these validity scales may not detect all sources of bias (Bagby, Nicholson, & Buis, 1998; Nicholson et al., 1997). Examiners may also look for corroborating information, such as interviewing others who are familiar with the client's behavior.

Tests are also only as valid as the criteria that were used to validate them. The original MMPI was limited in its role as a diagnostic instrument by virtue of the obsolete diagnostic categories that were used to classify the original clinical groups. Moreover, if a test does nothing more than identify people who are likely to belong to a particular diagnostic category, its utility is usurped by more economical means of arriving at diagnoses, such as the structured clinical interview. We expect more from personality tests than diagnostic classification, and the MMPI has shown its value in showing personality characteristics associated with people with certain response patterns. Psychodynamically oriented critics suggest that self-report instruments tell us little about possible unconscious processes. The use of such tests may also be limited to relatively high functioning individuals who can read well, respond to verbal material, and focus on a potentially tedious task. Clients who are disorganized, unstable, or confused may not be able to complete the tests.

projective tests Psychological tests that present ambiguous stimuli onto which the examinee is thought to project his or her personality and unconscious motives.

Projective Tests **Projective tests,** unlike objective tests, offer no clear, specified response options. Clients are presented with ambiguous stimuli, such as vague drawings or inkblots, and asked to describe what the stimuli look like or to relate stories about them. The tests are called *projective* because they were derived from the psychodynamic projective hypothesis, the belief that people impose, or "project," their psychological needs, drives, and motives, much of which may lie in the unconscious, onto their interpretations of unstructured or ambiguous stimuli.

The psychodynamic model holds that potentially disturbing impulses and wishes, often of a sexual or aggressive nature, are often hidden from consciousness by defense mechanisms. Defense mechanisms may thwart direct probing of threatening material. Indirect methods of assessment, however, such as projective tests, may offer clues to unconscious processes. More behaviorally oriented critics contend, however, that the results of projective tests are based more on clinicians' subjective interpretations of test responses than on empirical evidence.

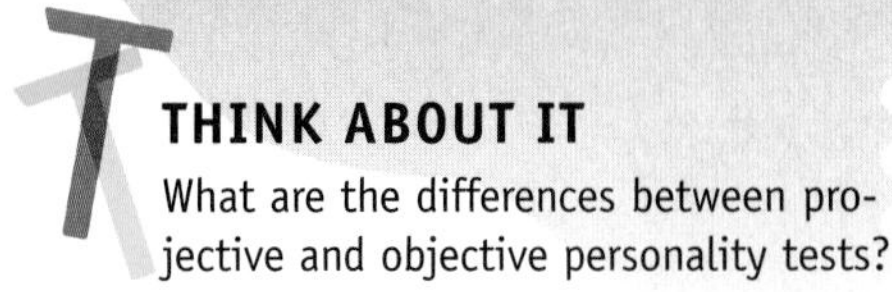

THINK ABOUT IT
What are the differences between projective and objective personality tests?

Many projective tests have been developed, including tests based on how people fill in missing words to complete sentence fragments or how they draw human figures and other objects. The two most prominent projective techniques are the Rorschach Inkblot Test and the Thematic Apperception Test (TAT).

Truth OR Fiction? REVISITED

Some clinicians examine how people interpret inkblots in order to uncover underlying aspects of their personalities.

TRUE. Some clinicians do use a psychological test—the Rorschach—in which a person's responses to inkblots are used to reveal aspects of his or her personality.

The Rorschach Test The Rorschach test, in which a person's responses to inkblots are used to reveal aspects of his or her personality, was developed by a Swiss psychiatrist, Hermann Rorschach (1884–1922). As a child, Rorschach was intrigued by the game of dripping ink on paper and folding the paper to make symmetrical figures. He noted that people saw different things in the same blot, and he believed their "percepts" reflected their personalities as well as the stimulus cues provided by the blot. Rorschach's fraternity nickname was *Klex*, which means "inkblot" in German ("Time Capsule," 2000). As a psychiatrist, Rorschach experimented with hundreds of blots to identify those that could help in the diagnosis of psychological problems. He finally found a group of 15 blots that seemed to do the job and could be administered in a single session. Ten blots are used today because Rorschach's publisher did not have the funds to reproduce all 15 blots in the first edition of the text on the subject. Rorschach never had the opportunity to learn how popular and influential his inkblot test would become. The year following its publication, at the age of 38, he died of complications from a ruptured appendix.

Five of the inkblots are black and white and the other five have color (see Figure 3.4). Each inkblot is printed on a separate card, which is handed to subjects in sequence. Subjects are asked to tell the examiner what the blot might be or what it reminds them of. A follow-up inquiry explores what features of the blot (its color, form, or texture) the person used in forming an impression of what it resembled.

FIGURE 3.4 An inkblot similar to those found on the Rorschach Inkblot Test. What does the blot look like to you? What could it be? Rorschach assumed that people project their personalities into their responses to ambiguous inkblots.

Clinicians who use the Rorschach tend to interpret responses in the following ways: Clients who use the entire blot in their responses show ability to perceive part-whole relationships and integrate events in meaningful ways. People whose responses are based solely on minor details of the blots may have obsessive-compulsive tendencies that, in psychodynamic theory,

reality testing The ability to perceive the world accurately and to distinguish between reality and fantasy.

protect them from having to cope with the larger issues in their lives. Clients who respond to the negative (white) spaces tend to see things in their own way, suggestive of negativism or stubbornness.

Relationships between form and color are suggestive of clients' capacity to control impulses. When clients use the color of the blot but are primarily guided by the form of the blots, they are believed capable of feeling deeply but also of holding their feelings in check. When color predominates—as in perceiving any reddened area as "blood"—clients may not be able to exercise control over impulses. A response consistent with the form or contours of the blot is suggestive of adequate **reality testing.** People who see movement in the blots may be revealing intelligence and creativity. Content analysis may shed light on underlying conflicts. For example, adult clients who see animals but no people may have problems relating to people. Clients who appear confused about whether or not percepts of people are male or female may, according to psychodynamic theory, be in conflict over their own gender identity.

VIDEO 3.1
Administration of Projective Tests:
Dr. Ruth Munroe

The Thematic Apperception Test (TAT) The Thematic Apperception Test (TAT) was developed by psychologist Henry Murray (1943) at Harvard University in the 1930s. *Apperception* is a French word that can be translated as "interpreting (new ideas or impressions) on the basis of existing ideas (cognitive structures) and past experience." The TAT consists of a series of cards, like that shown in Figure 3.5, each of which depicts an ambiguous scene. Respondents are asked to construct stories about the cards. It is assumed their tales reflect their experiences and outlooks on life—and, perhaps, also shed light on deep-seated needs and conflicts.

Respondents are asked to describe what is happening in each scene, what led up to it, what the characters are thinking and feeling, and what will happen next. Psychodynamically oriented clinicians assume that respondents identify with the protagonists in their stories and project their psychological needs and conflicts into the events they *apperceive*. On a more superficial level, the stories also suggest how respondents might interpret or behave in similar situations in their own lives. TAT results are also suggestive of clients' attitudes toward others, particularly family members and lovers.

The TAT has been used extensively in research on motivation as well as in clinical practice. For example, psychologist David McClelland (e.g., McClellan, Alexander, & Marks, 1982) pioneered use of the TAT in assessing social motives such as the needs for achievement and power. The rationales for this research are that we are likely to be somewhat preoccupied with our needs, and our needs are projected into our reactions to ambiguous stimuli and situations.

FIGURE 3.5 Thematic Apperception Test (TAT).
Psychologists ask test-takers to provide their impressions of what is happening in the scene depicted in the drawing. They ask test-takers what led up to the scene and how it will turn out. How might your responses reveal aspects of your own personality?

Evaluation of Projective Techniques The reliability and validity of projective techniques has been the subject of extensive research and debate. One problem is the lack of a standard scoring procedure. Interpretation of a person's responses depends to some degree on the subjective judgment of the examiner. For example, two examiners may interpret the same Rorschach or TAT response differently.

Recent attempts to develop a comprehensive scoring approach for the Rorschach, such as the Exner system (Exner, 1991, 1993), have advanced the effort to standardize scoring of responses. But the debate over the reliability of the Rorschach, including the Exner system, continues (see Acklin et al., 2000; G. J. Meyer, 1997; Wood, Nezworski, & Stejskal, 1996, 1997). Even if a Rorschach response can be scored reliably, the interpretation of the response—what it means—remains an open question.

Critics and even some proponents of the Rorschach technique, such as Hertz (1986), recognize that evidence is lacking to support the interpretation of particular responses. However, evidence has accumulated that supports the validity of some specific Rorschach responses (e.g., Blais et al., 2001; Kubiszyn et al., 2000; Leavitt & Labott, 1997; G. J. Meyer, 2001; Viglione, 1999). For example, investigators find that specific Rorschach indicators can distinguish between different types of psychological disorders (Kubiszyn et al., 2000), as well as predict psychotherapy outcomes (G. J. Meyer, 2000) and some types of behaviors, such as dependency behaviors (Bornstein, 1999). Though some reviewers (e.g., Meyer et al., 2001) find the validity of the Rorschach and TAT overall to be generally on par with that of other psychological tests, others claim that these tests have not yet met tests of scientific utility or validity (Hunsley & Bailey, 1999; Goode, 2001a; Lilienfeld, Wood, & Garb, 2000).

One criticism of the TAT is that the stimulus properties of some of the cards, such as cues depicting sadness or anger, may exert too strong a "stimulus pull" on the subject. The pictures themselves may pull for certain types of stories. If so, clients' responses may represent reactions to the stimulus cues rather than projections of their personalities (Murstein & Mathes, 1996). The validity of the TAT in eliciting deep-seated material or tapping underlying psychopathology also remains to be demonstrated. However, evidence does indicate that it can discriminate between different types of Axis I and Axis II disorders (Kubiszyn et al., 2000).

One general problem with projective instruments such as the TAT and Rorschach is that the more healthy test takers talk or see in response to projective instruments, the more likely they will be judged as having psychological problems (Murstein & Mathes, 1996). However, proponents of projective testing argue that in skilled hands, tests like the TAT and the Rorschach can yield meaningful material that might not be revealed in interviews or by self-rating inventories (Stricker & Gold, 1999). Moreover, allowing subjects freedom of expression through projective testing reduces the tendency of individuals to offer socially desirable responses. Despite the lack of direct evidence for the projective hypothesis, the appeal of projective tests among clinicians and internship training directors remains strong (Clemence & Handler, 2001; Lubin et al., 1985).

THINK ABOUT IT

Consider the debate over the use of projective tests. Do you believe that a person's response to inkblots or other unstructured stimuli might reveal aspects of his or her underlying personality? Why or why not?

Neuropsychological Assessment

Neuropsychological assessment is used to evaluate whether psychological problems reflect underlying neurological damage or brain defects. When neurological impairment is suspected, a neurological evaluation may be requested from a *neurologist*—a medical doctor who specializes in disorders of the nervous system. A clinical *neuropsychologist* may also be consulted to administer neuropsychological assessment techniques, such as behavioral observation and psychological testing, to reveal signs of possible brain damage. Neuropsychological testing may be used together with brain-imaging techniques such as the MRI and CT to shed light on relationships between brain function and underlying abnormalities (Fiez, 2001). The results of neuropsychological testing may not only suggest whether patients suffer from brain damage but also point to the parts of the brain that may be affected.

The Bender Visual Motor Gestalt Test One of the first neuropsychological tests to be developed was the Bender Visual Motor Gestalt Test (Bender, 1938). "The Bender" consists of geometric figures that illustrate various Gestalt principles of perception. The client is asked to copy nine geometric designs (see Figure 3.6). Signs of possible brain damage include rotation of the figures, distortions in shape, and incorrect sizing of the figures in relation to one another. The examiner then asks the client to reproduce the designs from memory, because neurological damage can impair memory functioning.

Although the Bender remains a convenient and economical means of uncovering possible organic impairment, it has been criticized for producing too many false negatives—that is, persons with neurological impairment who make satisfactory drawings (Bigler & Ehrhenfurth, 1981). In recent years, more sophisticated tests have been developed. Two of

neuropsychological Pertaining to the relationships between the brain and behavior.

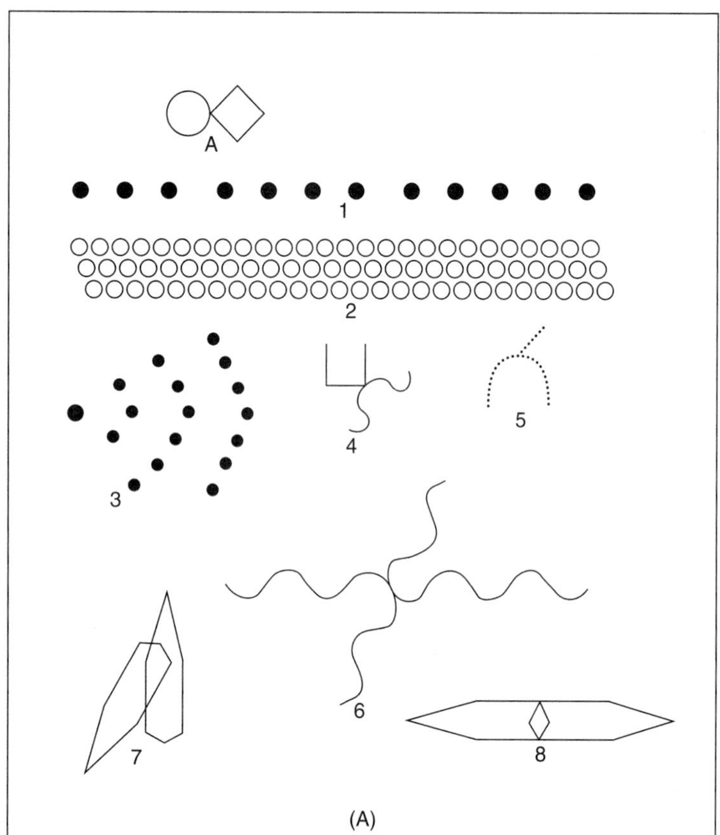

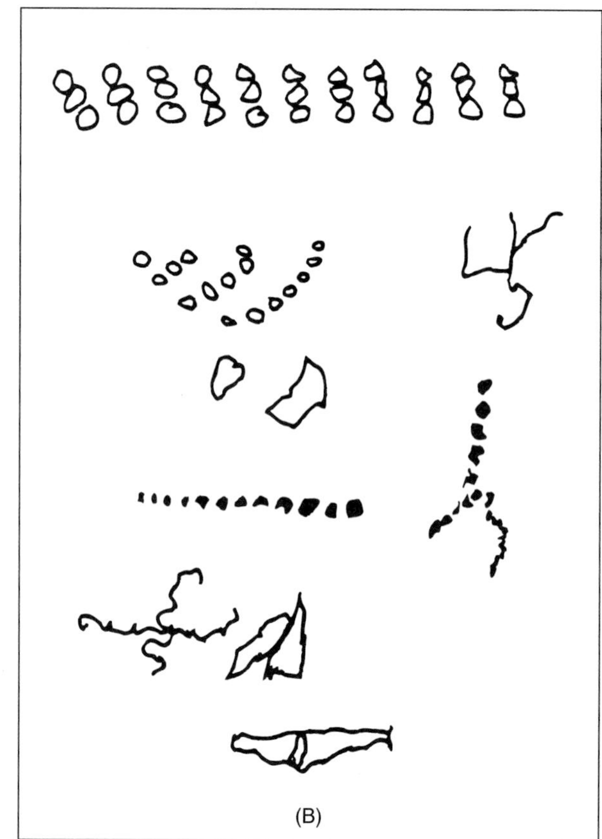

FIGURE 3.6 The Bender Visual Motor Gestalt Test.
The "Bender" is intended to assess organic impairment. Part A shows the series of figures respondents are asked to copy. Part B shows the drawings of a person who is known to be brain-damaged.

the more widely used neuropsychological inventories today are the Halstead-Reitan Neuropsychological Battery and the Luria Nebraska Test Battery.

The Halstead-Reitan Neuropsychological Battery Psychologist Ralph Reitan developed the battery by adapting tests used by his mentor, Ward Halstead, an experimental psychologist, to study brain-behavior relationships among organically impaired individuals. The battery contains tests that measure perceptual, intellectual, and motor skills and performance. A battery of tests permits the psychologist to observe patterns of results, and various patterns of performance deficits are suggestive of certain kinds of organic defects. The tests in the battery include the following:

1. *The Category Test.* This test measures abstract thinking ability, as indicated by the individual's proficiency at forming principles or categories that relate different stimuli to one another. A series of groups of stimuli that vary in shape, size, location, color, and other characteristics are flashed on a screen. The subject's task is to discern the principle that links them, such as shape or size, and to indicate which stimuli in each grouping represent the correct category by pressing a key. By analyzing the patterns of correct and incorrect choices, the subject normally learns to identify the principles that determine the correct choice. Performance on the test is believed to reflect functioning in the frontal lobe of the cerebral cortex.
2. *The Rhythm Test.* This is a test of concentration and attention. The subject listens to 30 pairs of tape-recorded rhythmic beats and indicates whether the beats in each pair are the same or different. Performance deficits are associated with damage to the right temporal lobe of the cerebral cortex.

uterized axial tomography) reveals abnormalities in shape and structure that may be suggestive of lesions, blood clots, or tumors. The computer enables scientists to integrate the measurements into a three-dimensional picture of the brain. Evidence of brain damage that was once detectable only by surgery may now be displayed on a monitor.

Another imaging method, **positron emission tomography (PET scan),** is used to study the functioning of various parts of the brain (Figure 3.9). In this method, a small amount of a radioactive compound or tracer is mixed with glucose and injected into the bloodstream. When it reaches the brain, patterns of neural activity are revealed by measurement of the positrons—positively charged particles—emitted by the tracer. The glucose metabolized by parts of the brain generates a computer image of neural activity. Areas of greater activity metabolize more glucose. The PET scan has been used to learn which parts of the brain are most active (metabolize more glucose) when we are listening to music, solving a math problem, or using language. It can also be used to reveal differences in brain activity in people with schizophrenia (see Chapter 13).

A third imaging technique is **magnetic resonance imaging (MRI).** In MRI, the person is placed in a donut-shaped tunnel that generates a strong magnetic field. Radio waves of certain frequencies are directed at the head. As a result, parts of the brain emit signals that can be measured from several angles. As with the CT scan, the signals are integrated into a computer-generated image of the brain, which can be used to investigate brain abnormalities associated with schizophrenia (see Chapter 13) and other disorders, such as obsessive-compulsive disorder. A new type of MRI, called **functional magnetic resonance imaging (fMRI),** is used to identify parts of the brain that become active when people engage in particular tasks, such as vision, memory, or use of speech (Carpenter, 2000; Ingram & Siegle, 2001; Stern & Silbersweig, 2001) (see Figure 3.10). A recent fMRI study showed that when cocaine-addicted subjects experienced cocaine cravings, they showed greater activity than healthy subjects did in parts of the brain that were engaged when healthy subjects watched depressing videotapes (Wexler et al., 2001). This suggests there may be a physiological link between depressive feelings and drug cravings.

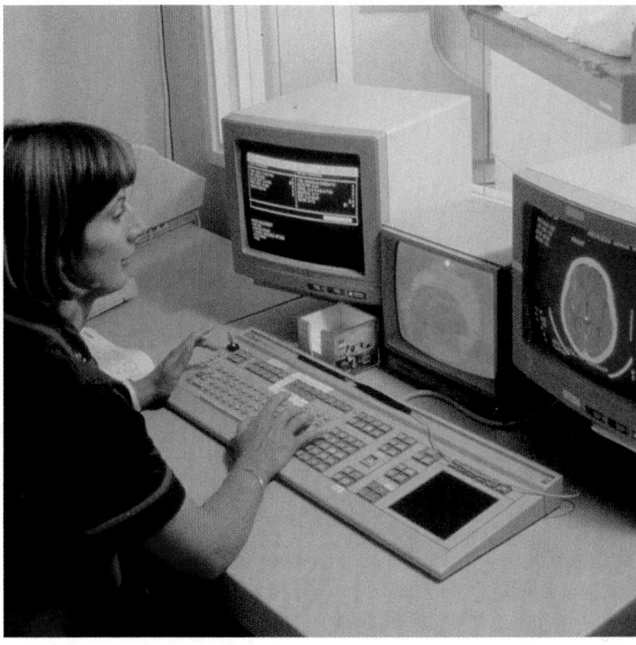

FIGURE 3.8 The Computed Tomography (CT) Scan. The CT scan aims a narrow X-ray beam at the head, and the resultant radiation is measured from multiple angles as it passes through. The computer enables researchers to consolidate the measurements into a three-dimensional image of the brain. The CT scan reveals structural abnormalities in the brain that may be implicated in various patterns of abnormal behavior.

positron emission tomography (PET scan) An imaging technique that forms a computer-generated image by tracing the amount of glucose used in various regions of the brain.

magnetic resonance imaging (MRI) A computer-generated image of the brain formed by measuring the signals emitted when the head is placed in a strong magnetic field.

functional magnetic resonance imaging (fMRI) Type of MRI used to identify parts of the brain that become active when people engage in particular tasks.

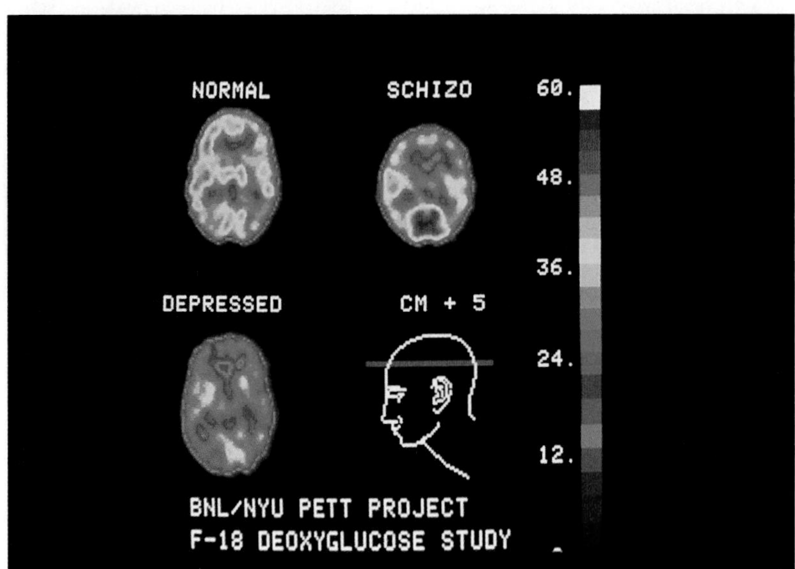

FIGURE 3.9 Positron Emission Tomography (PET) Scan. These PET scan images suggest differences in the metabolic processes of the brains of people with depression, schizophrenia, and controls who are free of psychological disorders.

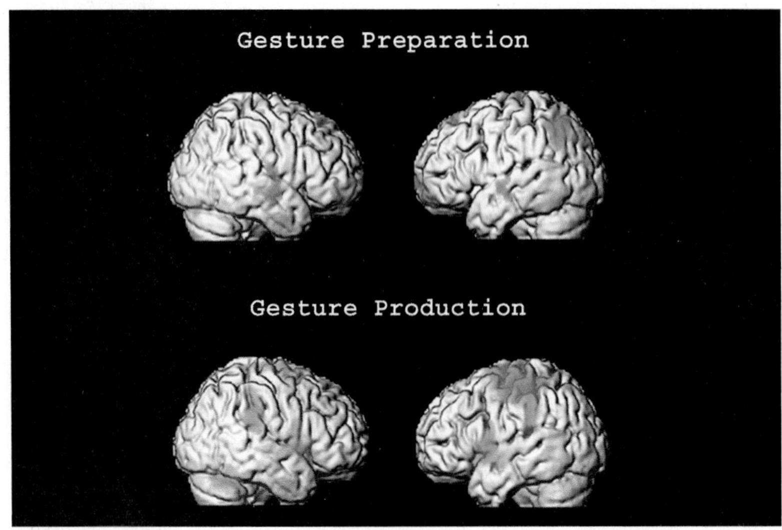

FIGURE 3.10 Functional Magnetic Resonance Imaging (fMRI). A fMRI is a specialized type of MRI that allows investigators to determine the parts of the brain that are activated during particular tasks. The areas depicted in red become activated when a person thinks about performing certain gestures (top), such as using a hammer or writing with a pen, and when the person actually performs these gestures (bottom). The right hemisphere is shown on the left side of the photographs and the left hemisphere is shown on the right side.

Brain electrical activity mapping (BEAM), a sophisticated type of EEG, uses the computer to analyze brain wave patterns and reveal areas of relative activity and inactivity from moment to moment (Figure 3.11) (F. H. Duffy, 1994; Silberstein et al., 1998). Twenty or more electrodes are attached to the scalp and simultaneously feed information about brain activity to a computer. The computer analyzes the signals and displays the pattern of brain activity on a color monitor, providing a vivid image of the electrical activity of the brain at work. BEAM and other similar techniques have been helpful in studying the brain activity of people with schizophrenia and children with attention-deficit hyperactivity disorder, among other physical and psychological disorders. In later chapters we see how modern imaging techniques are furthering our understanding of various patterns of abnormal behavior.

In conclusion, people's psychological problems, which are no less complex than people themselves, are thus assessed in many ways. Clients are generally asked to explain their problems as best they can, and sometimes a computer does the asking. Psychologists can also draw on batteries of tests that assess intelligence, personality, and neuropsychological integrity. Many psychologists prefer to observe people's behavior directly. Modern technology has provided several means of studying the structure and function of the brain. The methods of assessment selected by clinicians reflect the problems of their clients, their theoretical orientations, and their mastery of specialized technologies.

brain electrical activity mapping (BEAM) Imaging technique involving computer analysis of data from multiple electrodes to reveal areas of the brain with relatively high or low levels of activity.

THINK ABOUT IT

Jamie complains of feeling depressed since the death of her brother in a car accident last year. What methods of assessment might a psychologist use to evaluate her mental condition?

Web Link 3.6 — The Basics of Brain Imaging

Quiz 3.3 — Methods of Assessment

Quiz 3.4 — Chapter Exam

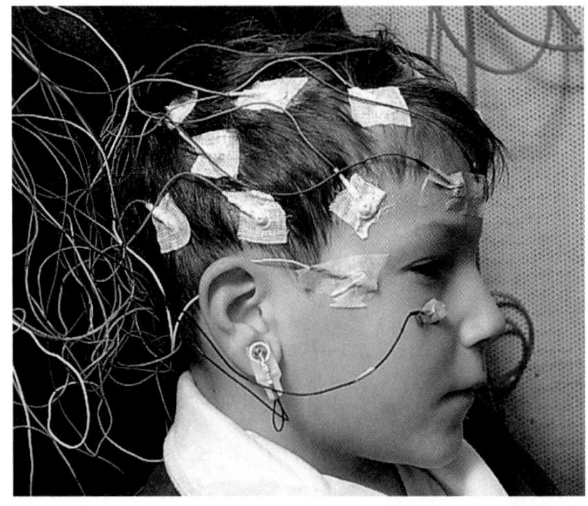

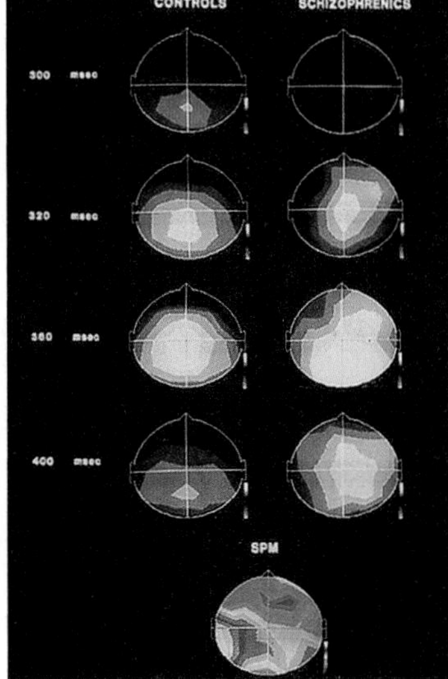

FIGURE 3.11 Brain Electrical Activity Mapping (BEAM). BEAM is a type of EEG in which electrodes are attached to the scalp (Part A) to measure electrical activity in various regions of the brain. The left column of Part B shows the average level of electrical activity in the brains of 10 normal people ("controls") at 4 time intervals. The column to the right shows the average level of activity of subjects with schizophrenia during the same intervals. Higher activity levels are represented in increasing order by yellows, reds, and whites. The computer-generated image in the bottom center summarizes differences in activity levels between the brains of normal subjects and those with schizophrenia. Areas of the brain depicted in blue show small differences between the groups. White areas represent larger differences.

Summing Up

Classification of Abnormal Behavior Patterns

What is the* DSM *and what are its major features? The *Diagnostic and Statistical Manual of Mental Disorders (DSM)* is the most widely accepted system for classifying mental disorders. The *DSM* uses specific diagnostic criteria to group patterns of abnormal behaviors that share common clinical features and a multiaxial system of evaluation.

Why is the* DSM *considered a multiaxial system? The *DSM* system consists of five axes of classification: Axis I (Clinical Syndromes), Axis II (Personality Disorders and Mental Retardation), Axis III (General Medical Conditions), Axis IV (Psychosocial and Environmental Problems), and Axis V (Global Assessment of Functioning).

What are the major strengths and weaknesses of the* DSM*? Strengths of the *DSM* include its use of specified diagnostic criteria and a multiaxial system to provide a comprehensive picture of the person's functioning. Weaknesses include questions about reliability and validity of certain diagnostic categories, and to some, the adoption of a medical model framework for classifying abnormal behavior patterns.

Issues of Reliability and Validity in Assessment

What are the standards by which methods of assessment are judged? Methods of assessment must be reliable and valid. Reliability of assessment techniques is shown in various ways, including internal consistency, temporal stability, and interrater reliability. Validity is measured by means of content validity, criterion validity, and construct validity.

Why is it important to take cultural or ethnic factors into account in psychological assessment? We need to ensure that tests that are validated in one culture are reliable and valid when used with members of another culture.

Methods of Assessment

What is a clinical interview? A clinical interview involves the use of a set of questions designed to elicit relevant information from people seeking treatment.

What are the three major types of clinical interviews? The three major types of clinical interviews are unstructured interviews (clinicians use their own style of questioning rather than follow a particular script), semi-structured interviews (clinicians follow a general outline in directing their questioning but are free to branch off in other directions), and structured interviews (clinicians strictly follow a preset order of questions).

What are psychological tests? Psychological tests are structured methods of assessment used to evaluate reasonably stable traits such as intelligence and personality.

What are the major types of psychological tests? Tests of intelligence, such as the Stanford-Binet Intelligence Scale and the Wechsler scales, are used for various purposes in clinical assessment, including determining evidence of mental retardation or cognitive impairment, and assessing strengths and weaknesses. Self-report personality inventories, such as the MMPI, use structured items to measure psychological characteristics or traits, such as anxiety, depression, and masculinity-femininity. These tests are considered objective in the sense that they make use of a limited range of possible responses to items and are based on an empirical, or objective, method of test construction. Projective personality tests, such as the Rorschach and TAT, require subjects to interpret ambiguous stimuli in the belief their answers may shed light on their unconscious processes.

What is neuropsychological assessment? Neuropsychological assessment involves the use of psychological tests to indicate possible neurological impairment or brain defects. The Halstead-Reitan Neuropsychological Battery and Luria Nebraska Test Battery measure perceptual skills, cognitive skills, and motor skills and performance that relate to specific areas of brain function.

What are some of the methods used in behavioral assessment? In behavioral assessment, test responses are taken as samples of behavior rather than as signs of underlying traits or dispositions. The behavioral examiner may conduct a functional assessment, which relates the problem behavior to its antecedents and consequents. Methods of behavioral assessment include behavioral interviewing, self-monitoring, use of analogue or contrived measures, direct observation, and behavioral rating scales.

What is cognitive assessment? Cognitive assessment focuses on the measurement of thoughts, beliefs, and attitudes in order to help identify distorted thinking patterns. Specific methods of assessment include the use of a thought record or diary and the use of rating scales such as the Automatic Thoughts Questionnaire (ATQ) and the Dysfunctional Attitudes Scale (DAS).

How do clinicians and researchers study physiological functioning? Measures of physiological functioning include heart rate, blood pressure, galvanic skin response (GSR), muscle tension, and brain wave activity. Brain-imaging and recording techniques such as EEG, CT scans, PET scans, MRI, and BEAM probe the inner workings and structures of the brain.

CHAPTER FOUR

Methods of Treatment

Paul Klee
Variation II

Truth OR Fiction?

- Some psychologists have been trained to prescribe drugs. (p. 98)
- In classical psychoanalysis, you are asked to express whatever thought happens to come to mind, no matter how seemingly trivial or silly. (p. 101)
- More psychotherapists identify with an eclectic approach than with any specific school of therapy. (p. 110)
- The average client who receives psychotherapy is no better off than control clients who go without it. (p. 113)
- Despite beliefs that it is a wonder drug, the antidepressant Prozac has not been shown to be more effective than the earlier generation of antidepressants. (p. 119)
- Severely depressed people who have failed to respond to other treatments may show rapid improvement from electroconvulsive therapy. (p. 121)

Carla, a 19-year-old college sophomore, had been crying more or less continuously for several days. She felt that her life was falling apart, that her college aspirations were in a shambles, and that she was a disappointment to her parents. The thought of suicide had crossed her mind. She could not seem to drag herself out of bed in the morning and had withdrawn from her friends. Her misery had seemed to descend on her from nowhere, although she could pinpoint some pressures in her life: a couple of poor grades, a recent breakup with a boyfriend, some adjustment problems with roommates.

The psychologist who examined her arrived at a diagnosis of major depressive disorder. Had she broken her leg, her treatment from a qualified professional would have followed a fairly standard course. Yet the treatment that Carla or someone else with a psychological disorder receives is likely to vary not only with the type of disorder involved but also with the therapeutic orientation and professional background of the helping professional. A psychiatrist might recommend a course of antidepressant medication, perhaps in combination with some form of psychotherapy. A cognitively oriented psychologist might suggest a program of cognitive therapy to help Carla identify dysfunctional thoughts that may underlie her depression, whereas a psychodynamic therapist might recommend she begin therapy to uncover inner conflicts originating in childhood that may lie at the root of her depression.

In this chapter we focus on ways of treating psychological disorders. About one out of every seven people in the United States receives some form of mental health treatment in a given year (USDHHS, 1999a). In later chapters we examine the kinds of treatment approaches applied to particular disorders, but here we focus on the treatments themselves. We will see that the biological and psychological perspectives have spawned corresponding approaches to treatment. First, however, we consider the major types of mental health professionals who treat psychological or mental disorders and the different roles they play.

Chapter Outline

Types of Mental Health Professionals

Many people are confused about the differences in qualifications and training of the various types of mental health providers (Farberman, 1997). It is little wonder people are confused, as there exist many different types of mental health professionals who represent a wide range of training backgrounds and areas of practice. The major professional groupings of mental health professionals include psychologists, psychiatrists, social workers, nurses, and counselors (see Table 4.1). Unfortunately, many states do not limit the use of the titles *therapist* or *psychotherapist* to trained professionals. In such states, anyone can set up shop as a psychotherapist and practice "therapy" without a license. Thus, people seeking help are advised to inquire about the training and licensure of helping professionals. If you or someone you know should seek the services of a psychologist, how would you find one? The nearby "A Closer Look" feature, "How Do I Find a Psychologist?" offers some suggestions.

Another reason for confusion is that all different types of mental health providers, such as psychologists, psychiatrists, and clinical social workers, practice psychotherapy, or "talk therapy"—a psychologically based method of treatment involving a series of verbal interchanges between clients and therapists taking place over a period of time, usually on a one-session-per-week basis. The particular approach used by psychotherapists reflects their theoretical orientation, such as psychodynamic, behavioral, humanistic, cognitive, and so on. Some therapists adopt an eclectic orientation, which means they draw on the theories and techniques espoused by two or more theoretical orientations.

We now consider the major types of psychotherapy and their relationships to the theoretical models from which they derive.

Psychotherapy

Psychotherapy is a systematic interaction between a client and a therapist that incorporates psychological principles to help bring about changes in the client's behaviors,

psychotherapy A structured form of treatment derived from a psychological framework which consists of one or more verbal interactions or treatment sessions between a client and a therapist.

TABLE 4.1 Major Types of Mental Health Professionals

Type	Description
Clinical Psychologists	Have earned a doctoral degree in psychology (either a Ph.D., or Doctor of Philosophy, a Psy.D., or Doctor of Psychology, or an Ed.D., Doctor of Education) from an accredited college or university. Training in clinical psychology typically involves 4 years of graduate coursework, followed by a year-long internship and completion of a doctoral dissertation. Clinical psychologists specialize in administering psychological tests, diagnosing psychological disorders, and practicing psychotherapy. Psychologists cannot prescribe psychiatric drugs, except for a handful of specially trained psychologists within a special government program (Sammons & Brown, 1997; Seppa, 1997). The granting of prescription privileges to psychologists remains a hotly contested issue between psychologists and psychiatrists and within the field of psychology itself (for example, see Foxhall, 2000a, 2000b, and Gutierrez & Silk, 1998).
Counseling Psychologists	Also hold doctoral degrees in psychology and have completed graduate training preparing them for careers in college counseling centers and mental health facilities. They typically provide counseling to people with psychological problems falling in a milder range of severity than those treated by clinical psychologists, such as difficulties adjusting to college or uncertainties regarding career choices.
Psychiatrists	Have earned a medical degree (M.D.) and completed a residency program in psychiatry. Psychiatrists are physicians who specialize in the diagnosis and treatment of psychological disorders. As licensed physicians, they can prescribe psychiatric drugs and may employ other medical interventions, such as electroconvulsive therapy (ECT). Many also practice psychotherapy based on training they receive during their residency programs or in specialized training institutes.
Clinical or Psychiatric Social Workers	Have earned a master's degree in social work (M.S.W.) and use their knowledge of community agencies and organizations to help people with severe mental disorders receive the services they need. For example, they may help people with schizophrenia make a more successful adjustment to the community once they leave the hospital. Many clinical social workers practice psychotherapy or specific forms of therapy, such as marital or family therapy.
Psychoanalysts	Typically are either psychiatrists or psychologists who have completed extensive additional training in psychoanalysis. They are required to undergo psychoanalysis themselves as part of their training.
Counselors	Have typically earned a master's degree by completing a graduate program in a counseling field. Counselors work in many settings, including public schools, college testing and counseling centers, and hospitals and health clinics. Many specialize in vocational evaluation, marital or family therapy, rehabilitation counseling, or substance abuse counseling. Counselors may focus on providing psychological assistance to people with milder forms of disturbed behavior or those struggling with a chronic or debilitating illness or recovering from a traumatic experience. Some are clergy members who are trained in pastoral counseling programs to help parishioners cope with personal problems.
Psychiatric Nurses	Typically are R.N.s who have completed a master's program in psychiatric nursing. They may work in a psychiatric facility or in a group medical practice where they treat people suffering from severe psychological disorders.

Source. Adapted from Nevid, J. S. (2003). *Psychology: Concepts & applications.* Boston: Houghton Mifflin.

Truth OR Fiction? REVISITED

Some psychologists have been trained to prescribe drugs.

TRUE. Some psychologists have been trained in an experimental program to prescribe psychotropic medications.

THINK ABOUT IT

What are the major types of mental health professionals? How do they differ in their training and the types of roles they perform?

thoughts, and feelings in order to help the client overcome abnormal behavior, solve problems in living, or develop as an individual. Let us take a closer look at these features of psychotherapy:

1. *Systematic interaction.* The process of psychotherapy involves systematic interactions between clients and therapists. "Systematic" means that therapists structure these interactions with plans and purposes that reflect their theoretical points of view.
2. *Psychological principles.* Psychotherapists draw on psychological principles, research, and theory in their practice.
3. *Behavior, thoughts, and feelings.* Psychotherapy may be directed at behavioral, cognitive, and emotional domains to help clients overcome psychological problems and lead more satisfying lives.
4. *Abnormal behavior, problem solving, and personal growth.* At least three groups of people are assisted by psychotherapy. First are people with abnormal behavior problems such as mood disorders, anxiety disorders, or schizophrenia. Second are people who seek help for personal problems that are not regarded as abnormal, such as social shyness or confusion about career choices. Third are people who seek personal

A Closer Look

How Do I Find a Psychologist?

To find a psychologist, ask your physician or another health professional. Call your local or state psychological association. Consult a local university or college department of psychology. Ask family and friends. Contact your area community mental health center. Inquire at your church or synagogue.

What to Consider When Making the Choice . . .

Psychologists and clients work together. The right match is important. Most psychologists agree that an important factor in determining whether or not to work with a particular psychologist, once that psychologist's credentials and competence are established, is your level of personal comfort with that psychologist. A good rapport with your psychologist is critical. Choose a psychologist with whom you feel comfortable and at ease.

Questions to Ask . . .

- Are you a licensed psychologist? How many years have you been practicing psychology?
- I have been feeling (anxious, tense, depressed, etc.), and I'm having problems (with my job, my marriage, eating, sleeping, etc.). What experience do you have helping people with these types of problems?
- What are your areas of expertise—for example, working with children and families?
- What kinds of treatments do you use, and have they been proven effective for dealing with my kind of problem or issue?
- What are your fees? (Fees are usually based on a 45- to 50-minute session.) Do you have a sliding-scale fee policy? How much therapy would you recommend?
- What types of insurance do you accept? Will you accept direct billing to/payment from my insurance company? Are you affiliated with any managed care organizations? Do you accept Medicare/Medicaid insurance?

Finances . . .

Many insurance companies provide coverage for mental health services. If you have private health insurance coverage (typically through an employer), check with your insurance company to see whether mental health services are covered and, if so, how you may obtain these benefits. This also applies to persons enrolled in HMOs and other types of managed care plans. Find out how much the insurance company will reimburse for mental health services and what limitations on the use of benefits may apply.

If you are not covered by a private health insurance plan or employee assistance program, you may decide to pay for psychological services out of pocket. Some psychologists operate on a sliding-scale fee policy, where the amount you pay depends on your income.

Another potential source of mental health services involves government-sponsored health care programs—including Medicare for individuals age 65 or older, as well as health insurance plans for government employees, military personnel, and their dependents. Community mental health centers throughout the country are another possible alternative for receiving mental health services. And some state Medicaid programs for economically disadvantaged individuals provide for limited mental health services from psychologists.

Credentials to Look For . . .

After graduation from college, psychologists spend an average of seven years in graduate education training and research before receiving a doctoral degree. As part of their professional training, they must complete a supervised clinical internship in a hospital or organized health setting and at least one year of postdoctoral supervised experience before they can practice independently in any health-care arena. It's this combination of doctoral-level training and clinical internship that distinguishes psychologists from many other mental health care providers.

Psychologists must be licensed by the state or jurisdiction in which they practice. Licensure laws are intended to protect the public by limiting licensure to those persons qualified to practice psychology as defined by state law. In most states, renewal of this license depends upon the demonstration of continued competence and requires continuing education. In addition, members of the American Psychological Association (APA) adhere to a strict code of professional ethics.

growth. For them, psychotherapy is a means of self-discovery that may help them reach their potentials as, for example, parents, creative artists, performers, or athletes.

Psychotherapies share other features as well. For one, psychotherapies involve verbal interactions. Psychotherapies are "talking therapies," forms of interchange between clients and therapists that involve talking or conversation. In some cases, there is much discussion between clients and therapists. In others, such as traditional psychoanalysis, clients do most of the talking. In each case, skillful therapists are attentive listeners. Attentive listening is an

WWW **Web Link 4.1** Overview of Psychotherapy

The therapeutic relationship. In the course of successful psychotherapy, a therapeutic relationship is forged between the therapist and patient. Therapists use attentive listening to understand as clearly as possible what the client is experiencing and attempting to convey. Skillful therapists are also sensitive to clients' nonverbal cues, such as gestures and posture, that may indicate underlying feelings or conflicts.

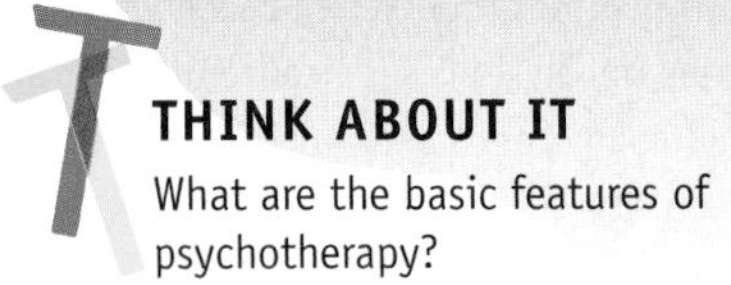

THINK ABOUT IT

What are the basic features of psychotherapy?

active, not a passive, activity. Therapists listen carefully to what clients are saying in order to understand as clearly as possible what they are experiencing and attempting to convey. Skillful therapists are also sensitive to clients' nonverbal cues, such as gestures that may indicate underlying feelings or conflicts. Therapists also seek to convey empathy through words as well as nonverbal gestures, such as establishing eye contact and leaning forward to indicate interest in what the client is saying. Therapist empathy is a consistent predictor of therapy outcome. Clients of therapists who are perceived as warmer and more empathic show greater improvement than clients of other therapists, whether the therapists are psychodynamic (Luborsky et al., 1988) or cognitive-behavioral (Burns & Nolen-Hoeksema, 1992) in their therapeutic approach.

Another common feature of psychotherapies is the instilling in clients of a sense of hope of improvement. Clients generally enter therapy with expectations of receiving help to overcome their problems. Responsible therapists do not promise results or guarantee cures. They do instill hope, however, that they can help clients deal with their problems. Positive expectancies can become a type of self-fulfilling prophecy by leading clients to mobilize their efforts toward overcoming their problems. Responses to positive expectancies are termed *placebo effects* or *expectancy effects.*

The common features of psychotherapy that are not specific to any one form of therapy, such as the encouragement of hope and the display of empathy and attentiveness on the part of the therapist, are often referred to as *nonspecific treatment factors.* Nonspecific factors may have therapeutic benefits in addition to the specific benefits of particular forms of therapy. We will discuss these factors in greater detail in the section on Evaluating Methods of Treatment.

Psychodynamic Therapy

Sigmund Freud was the first theorist to develop a psychological model—the *psychodynamic model*—of abnormal behavior (see Chapter 2). He was also the first to develop a model of psychotherapy, which he called **psychoanalysis,** to help people who suffered from psychological disorders. Psychoanalysis was the first **psychodynamic therapy.** Psychodynamic therapy helps individuals gain insight into, and resolve, the unconscious conflicts believed to lie at the root of abnormal behavior. Working through these conflicts, the ego would be freed of the need to maintain defensive behaviors—such as phobias, obsessive-compulsive behaviors, hysterical complaints, and the like—that shield it from recognition of inner turmoil.

Freud summed up the goal of psychoanalysis by saying, "Where id was, there shall ego be." This meant, in part, that psychoanalysis could help shed the light of awareness, represented by the conscious ego, on the inner workings of the id. But Freud did not expect, or intend, that clients should seek to become conscious of all repressed material—of all their impulses, wishes, fears, and memories. The aim, rather, was to replace defensive behavior with more adaptive behavior. By so doing, clients could find gratification without incurring social or self-condemnation.

Through this process a man with a phobia of knives might become aware he had been repressing impulses to vent a murderous rage against his father. His phobia keeps him from having contact with knives, thereby serving a hidden purpose of keeping his homicidal impulses in check. Another man might come to realize that unresolved anger toward his dominating or rejecting mother has sabotaged his intimate relationships with women during his adulthood. A woman with a loss of sensation in her hand that could not be explained medically might come to see that she harbored guilt over urges to masturbate. The loss of sensation might have prevented her from acting on these urges. Through confronting hidden impulses and the conflicts they produce, clients learn to sort out their feelings and find more constructive and socially acceptable ways of handling their impulses and wishes. The ego is then freed to focus on more constructive interests.

The major methods that Freud used to accomplish these goals were free association, dream analysis, and analysis of the transference relationship.

psychoanalysis The method of psychotherapy developed by Sigmund Freud.

psychodynamic therapy Therapy that helps individuals gain insight into, and resolve, unconscious conflicts.

Freud's consulting room in London. Here we see the consulting room used by Freud after his arrival in London. The patient lay on the couch, and Freud sat on the chair at the left, out of view.

free association The method of verbalizing thoughts as they occur without a conscious attempt to edit or censure them.

compulsion to utter The urge to verbally express repressed material.

resistance The blocking of thoughts or feelings that would evoke anxiety if they were consciously experienced.

insight The attainment of awareness and understanding of one's true motives and feelings.

manifest content The reported content or apparent meaning of dreams.

latent content The underlying or symbolic content of dreams.

Free Association You are asked to lie down on a couch and to say anything that enters your mind. The psychoanalyst (or *analyst* for short) sits in a chair behind you, out of direct view. For the next 45 or 50 minutes, you let your mind wander, saying whatever pops in, or saying nothing at all. The analyst remains silent most of the time, prompting you occasionally to utter whatever crosses your mind, no matter how seemingly trivial, no matter how personal. This process continues, typically for three or four sessions a week, for several years. At certain points in the process, the analyst offers an *interpretation*, drawing your attention to connections between your disclosures and unconscious conflicts.

Free association is the process of uttering uncensored thoughts as they come to mind. Free association is believed to gradually break down the defenses that block awareness of unconscious processes. Clients are told not to censor or screen out thoughts, but to let their minds wander "freely" from thought to thought. Psychoanalysts do not believe that the process of free association is truly free. Repressed impulses press for expression or release, leading to a **compulsion to utter.** Although free association may begin with small talk, the compulsion to utter eventually leads the client to disclose more meaningful material.

The ego, however, continues to try to avert the disclosure of threatening impulses and conflicts. Consequently, clients may show **resistance,** an unwillingness or inability to recall or discuss disturbing or threatening material. Clients might report that their minds suddenly go blank when they venture into sensitive areas. They might switch topics abruptly, or accuse the analyst of trying to pry into material that is too personal or embarrassing to talk about. Or they might conveniently "forget" the next appointment after a session in which sensitive material is touched upon. The analyst monitors the dynamic conflict between the "compulsion to utter" and resistance. Signs of resistance are often suggestive of meaningful material. Now and then, the analyst brings interpretations of this material to the attention of the client to help the client gain better **insight** into deep-seated feelings and conflicts.

Truth OR Fiction? REVISITED

In classical psychoanalysis, you are asked to express whatever thought happens to come to mind, no matter how seemingly trivial or silly.

TRUE. In classical psychoanalysis, clients are asked to report any thoughts that come to mind. The technique is called free association.

Dream Analysis To Freud, dreams represented the "royal road to the unconscious." During sleep, the ego's defenses are lowered and unacceptable impulses find expression in dreams. Because the defenses are not completely eliminated, the impulses take a disguised or symbolized form. In psychoanalytic theory, dreams have two levels of content:

1. **Manifest content:** the material of the dream the dreamer experiences and reports, and
2. **Latent content:** the unconscious material the dream symbolizes or represents.

Dream analysis. Freud believed that dreams represent the "royal road to the unconscious." Dream interpretation was one of the principal techniques that Freud used to uncover unconscious material.

displacing Transferring impulses toward threatening or unacceptable objects onto more acceptable or safer objects.

transference relationship The client's transfer onto the analyst of feelings or attitudes the client holds toward important figures in his or her life.

countertransference The transfer of feelings or attitudes that the analyst holds toward other persons onto the client.

A man might dream of flying in an airplane. Flying is the apparent or manifest content of the dream. Freud believed that flying may symbolize erection, so perhaps the latent content of the dream reflects unconscious issues related to fears of impotence. Such symbols may vary from person to person. Analysts therefore ask clients to free-associate to the manifest content of the dream to provide clues to the latent content. Though dreams may have a psychological meaning, as Freud believed, there remains no independent way of determining what dreams mean (Squier & Domhoff, 1998).

Transference Freud found that clients responded to him not only as an individual but also in ways that reflected their feelings and attitudes toward other important people in their lives. A young female client might respond to him as a father figure, **displacing,** or transferring, onto Freud her feelings toward her own father. A man might also view him as a father figure, responding to him as a rival in a manner that Freud believed might reflect the man's unresolved Oedipus complex.

The process of analyzing and working through the **transference relationship** is considered an essential component of psychoanalysis. Freud believed that the transference relationship provides a vehicle for the reenactment of childhood conflicts with parents. Clients may react to the analyst with the same feelings of anger, love, or jealousy they felt toward their own parents. Freud termed the enactment of these childhood conflicts the *transference neurosis*. This "neurosis" had to be successfully analyzed and worked through for clients to succeed in psychoanalysis.

Childhood conflicts usually involve unresolved feelings of anger, rejection, or need for love. For example, a client may interpret any slight criticism by the therapist as a devastating blow, transferring feelings of self-loathing that the client had repressed from childhood experiences of parental rejection. Transferences may also distort or color the client's relationships with others, such as a spouse or employer. Clients might relate to their spouses as they had to their parents, perhaps demanding too much from them or unjustly accusing them of being insensitive or uncaring. Or they might not give new friends or lovers the benefit of a fair chance, if they had been mistreated by others who played similar roles in their past. The analyst helps the client recognize transference relationships, especially the therapy transference, and work through the residues of childhood feelings and conflicts that lead to self-defeating behavior in the present.

According to Freud, transference is a two-way street. Freud felt he transferred his underlying feelings onto his clients, perhaps viewing a young man as a competitor or a woman as a rejecting love interest. Freud referred to the feelings that he projected onto clients as **countertransference.** Psychoanalysts in training are expected to undergo psychoanalysis themselves to help them uncover motives that might lead to countertransferences in their therapeutic relationships. In their training, psychoanalysts learn to monitor their own reactions in therapy, so as to become better aware of when and how countertransferences intrude on the therapy process.

Although the analysis of the therapy transference is a crucial element of psychoanalytic therapy, it generally takes months or years for a transference relationship to develop and be resolved. This is one reason why psychoanalysis is typically a lengthy process.

Modern psychodynamic psychotherapy. Modern psychodynamic therapists engage in more direct, face-to-face interactions with clients than do traditional Freudian psychoanalysts. Modern psychodynamic approaches are also generally briefer and focus more on the direct exploration of clients' defenses and transference relationships.

Modern Psychodynamic Approaches Although some psychoanalysts continue to practice traditional psychoanalysis in much the same manner as Freud, briefer and less intensive forms of psychodynamic treatment have emerged (Strupp, 1992). These newer approaches are often referred to as "psychoanalytic psychotherapy," "psychoanalytically oriented therapy", or "psychodynamic therapy." They are able to reach clients who are seeking briefer and less costly forms of treatment, perhaps once or twice a week.

Like Freudian psychoanalysis, the newer psychodynamic approaches aim to uncover unconscious motives and break down resistances and psychological defenses. Yet they focus more on the client's present relationships and encourage the client to make adaptive behavior changes. Because of the briefer format, therapy may entail a more open dialogue and direct exploration of the client's defenses and transference relationships than was traditionally the case (Messer, 2001b). Unlike the traditional approach, the client and therapist generally sit facing each other. Rather than offer an occasional interpretation, the therapist engages in more frequent verbal give-and-take with the client, as in the following vignette. Note how the therapist uses interpretation to help the client, Mr. Arianes, achieve insight into how his relationship with his wife involves a transference of his childhood relationship with his mother:

A Case Vignette of Psychodynamic Therapy

MR. ARIANES: I think you've got it there, Doc. We weren't communicating. I wouldn't tell her [his wife] what was wrong or what I wanted from her. Maybe I expected her to understand me without saying anything.

THERAPIST: Like the expectations a child has of its mother.

MR. ARIANES: Not my mother!

THERAPIST: Oh?

MR. ARIANES: No, I always thought she had too many troubles of her own to pay attention to mine. I remember once I got hurt on my bike and came to her all bloodied up. When she saw me she got mad and yelled at me for making more trouble for her when she already had her hands full with my father.

THERAPIST: Do you remember how you felt then?

MR. ARIANES: I can't remember, but I know that after that I never brought my troubles to her again.

THERAPIST: How old were you?

MR. ARIANES: Nine, I know that because I got that bike for my ninth birthday. It was a little too big for me still, that's why I got hurt on it.

THERAPIST: Perhaps you carried this attitude into your marriage.

MR. ARIANES: What attitude?

THERAPIST: The feeling that your wife, like your mother, would be unsympathetic to your difficulties. That there was no point in telling her about your experiences because she was too preoccupied or too busy to care.

MR. ARIANES: But she's so different from my mother. I come first with her.

THERAPIST: On one level you know that. On another, deeper level there may well be the fear that people—or maybe only women, or maybe only women you're close to—are all the same, and you can't take a chance at being rejected again in your need.

MR. ARIANES: Maybe you're right, Doc, but all that was so long ago, and I should be over that by now.

THERAPIST: That's not the way the mind works. If a shock or a disappointment is strong enough, it can permanently freeze our picture of ourselves and our expectations of the world. The rest of us grows up—that is, we let ourselves learn about life from experience and from what we see, hear, or read of the experiences of others, but that one area where we really got hurt stays unchanged. So what I mean when I say you might be carrying that attitude into your relationship with your wife is that when it comes to your hopes of being understood and catered to when you feel hurt or abused by life, you still feel very much like that nine-year-old boy who was rebuffed in his need and gave up hope that anyone would or could respond to him.

—From Doing psychotherapy *by M. F. Basch. Copyright © 1980 by Basic Books, pp. 29–30. Reprinted with permssion of Basic Books, a member of Perseus Books L. L. C.*

object relations The person's relationships to the internalized representations, or "objects," of others' personalities that have been introjected within the person's ego structure.

behavior therapy The therapeutic application of learning-based techniques.

systematic desensitization A behavior therapy technique for overcoming phobias by means of exposure to progressively more fearful stimuli while one remains deeply relaxed.

gradual exposure A behavior therapy technique for overcoming fears through direct exposure to increasingly fearful stimuli.

modeling A behavior therapy technique for helping an individual acquire a new behavior by means of having a therapist or another individual demonstrate a target behavior that is then imitated by the client.

Modeling. Modeling techniques are often used to help people overcome phobic behaviors. Here a woman models approaching and petting a dog to a phobic child. As the phobic child observes the woman harmlessly engage in the desired behavior, he is more likely to imitate the behavior.

Some modern psychodynamic therapies focus more on the role of the ego and less on the role of the id. Therapists adopting this view believe Freud placed too much emphasis on sexual and aggressive impulses and underplayed the importance of the ego. These therapists, such as Heinz Hartmann, are generally described as *ego analysts.* Other modern psychoanalysts, such as Melanie Klein and Margaret Mahler, are identified with object-relations approaches to psychodynamic therapy. **Object-relations** therapists focus on helping people separate their own ideas and feelings from the elements of others they have incorporated or introjected within themselves. They can then develop more as individuals—as their own persons, rather than trying to meet the expectations they believe others have of them.

Behavior Therapy

Behavior therapy is the systematic application of the principles of learning to the treatment of psychological disorders. Because the focus is on changing behavior—not on personality change or deep probing into the past—behavior therapy is relatively brief, lasting typically from a few weeks to a few months. Behavior therapists, like other therapists, seek to develop warm therapeutic relationships with clients, but they believe the special efficacy of behavior therapy derives from the learning-based techniques rather than from the nature of the therapeutic relationship.

Behavior therapy first gained widespread attention as a means of helping people overcome fears and phobias, problems that had proved resistant to insight-oriented therapies. Among these methods are systematic desensitization, gradual exposure, and modeling. **Systematic desensitization** involves a therapeutic program of exposure (in imagination or by means of pictures or slides) to progressively more fearful stimuli while one remains deeply relaxed. First the person uses a relaxation technique, such as progressive relaxation (discussed in Chapter 5), to become deeply relaxed. The client is then instructed to imagine (or perhaps view, as through a series of slides) progressively more anxiety-arousing scenes. If fear is evoked, the client focuses on restoring relaxation. The process is repeated until the scene can be tolerated without anxiety. The client then progresses to the next scene in the *fear-stimulus hierarchy.* The procedure is continued until the person can remain relaxed while imagining the most distressing scene in the hierarchy.

In **gradual exposure** (also called *in vivo,* meaning "in life," exposure), people troubled by phobias purposely expose themselves to the stimuli that evoke their fear. Like systematic desensitization, the person progresses at his or her own pace through a hierarchy of progressively more anxiety-evoking stimuli. The person with a fear of snakes, for example, might first look at a harmless, caged snake from across the room and then gradually approach and interact with the snake in a step-by-step process, progressing to each new step only when feeling completely calm at the prior step. Gradual exposure is often combined with cognitive techniques that focus on replacing anxiety-arousing irrational thoughts with calming rational thoughts.

In **modeling,** individuals learn desired behaviors by observing others perform them (Braswell & Kendall, 2001). For example, the client may observe and then imitate others who interact with fear-evoking situations or objects. After observing the model, the client may be assisted or guided by the therapist or the model in performing the target behavior. The client receives ample reinforcement from the therapist for each attempt. Modeling approaches were pioneered by Albert Bandura and his colleagues, who had remarkable success using modeling techniques with children to treat various phobias, especially fears of animals, such as snakes and dogs (Bandura, Jeffery, & Wright, 1974; Braswell & Kendall, 2001).

Behavior therapists also use techniques based on operant conditioning, or systematic use of rewards and punishments, to shape desired behavior. For example, parents and teachers may be trained to systematically reinforce children for appropriate behavior by showing appreciation and to extinguish inappropriate behavior by ignoring it. In institutional settings, **token economy** systems seek to increase adaptive behavior by allowing patients to earn tokens for performing appropriate behaviors, such as self-grooming and making their beds. The tokens can eventually be exchanged for desired rewards. Token systems have also been used to treat children with conduct disorder problems.

Other techniques of behavior therapy discussed in later chapters include aversive conditioning (used in the treatment of substance abuse problems like smoking and alcoholism), social skills training (used in the treatment of social anxieties and skills deficits associated with schizophrenia), and self-control techniques (used in helping people reduce excess weight and quit smoking).

token economy Behavioral treatment program in which a controlled environment is constructed such that people are reinforced for desired behaviors by receiving tokens that may be exchanged for desired rewards.

person-centered therapy The establishment of a warm, accepting therapeutic relationship that frees clients to engage in self-exploration and achieve self-acceptance.

client-centered therapy Another term for *person-centered therapy*.

unconditional positive regard The expression of unconditional acceptance of another person's basic worth as a person.

Humanistic Therapy

Psychodynamic therapists tend to focus on unconscious processes, such as internal conflicts. By contrast, humanistic therapists focus on clients' subjective, conscious experiences. Like behavior therapists, humanistic therapists also focus more on what clients are experiencing in the present—the here and now—than on the past. But there are also similarities between the psychodynamic and humanistic therapies. Both assume the past affects present behavior and feelings and both seek to expand clients' self-insight. The major form of humanistic therapy is **person-centered therapy** (also called **client-centered therapy**), which was developed by the psychologist Carl Rogers.

Person-Centered Therapy Rogers (1951) believed that people have natural motivational tendencies toward growth, fulfillment, and health. In Rogers's view, psychological disorders develop largely from the roadblocks that others place in the journey toward self-actualization. When others are selective in their approval of our childhood feelings and behavior, we may come to disown the criticized parts of ourselves. To earn social approval, we may don social masks or facades. We learn "to be seen and not heard" and may become deaf even to our own inner voices. Over time, we may develop distorted self-concepts that are consistent with others' views of us but are not of our own making and design. As a result, we may become poorly adjusted, unhappy, and confused as to who and what we are.

VIDEO 4.1
Client-Centered Therapy:
Dr. Carl Rogers

Well-adjusted people make choices and take actions consistent with their personal values and needs. *Person-centered therapy* creates conditions of warmth and acceptance in the therapeutic relationship that help clients become more aware and accepting of their true selves. Rogers was a major shaper of contemporary psychotherapy and was rated the single most influential psychotherapist in a survey of therapists (D. Smith, 1982). Rogers did not believe therapists should impose their own goals or values on their clients. His focus of therapy, as the name implies, is the person.

Person-centered therapy is *nondirective*. The client, not the therapist, takes the lead and directs the course of therapy. The therapist uses *reflection*—the restating or paraphrasing of the client's expressed feelings without interpreting them or passing judgment on them. This encourages the client to further explore his or her feelings and get in touch with deeper feelings and parts of the self that had become disowned because of social condemnation.

Rogers stressed the importance of creating a warm therapeutic relationship that would encourage the client to engage in self-exploration and self-expression. The effective therapist should possess four basic qualities or attributes: *unconditional positive regard, empathy, genuineness*, and *congruence*. First, the therapist must be able to express **unconditional positive regard** for clients. In contrast to the conditional approval the client may have received from parents and others in the past, the therapist must be unconditionally accepting of the client as a person, even if the therapist sometimes objects to the client's choices or behaviors. Unconditional positive regard provides clients with a sense of security that encourages them to explore their feelings without fear of disapproval. As clients feel accepted or prized for themselves, they are encouraged to accept themselves in turn. To Rogers, every human being has intrinsic worth and value. Traditional psychodynamic theory holds that people are basically motivated by primitive forces, such as sexual and aggressive impulses. Rogers believed, however, that people are basically good and are motivated to pursue *pro*social goals.

empathy The ability to understand someone's experiences and feelings from that person's point of view.

genuineness The ability to recognize and express one's true feelings.

congruence The fit between one's thoughts, behaviors, and feelings.

Therapists who display **empathy** are able to reflect or mirror accurately their clients' experiences and feelings. Therapists try to see the world through their clients' eyes or frames of reference. They listen carefully to clients and set aside their own judgments and interpretations of events. Showing empathy encourages clients to get in touch with feelings of which they may be only dimly aware.

Genuineness is the ability to be open about one's feelings. Rogers admitted he had negative feelings at times during therapy sessions, typically boredom, but he attempted to express these feelings openly rather than hide them (Bennett, 1985).

Congruence refers to the fit between one's thoughts, feelings, and behavior. The congruent person is one whose behavior, thoughts, and feelings are integrated and consistent. Congruent therapists serve as models of psychological integrity to their clients.

Here Rogers (C.R.) uses reflection to help a client focus more deeply on her inner feelings:

Reflection in Person-Centered Therapy

JILL: I'm having a lot of problems dealing with my daughter. She's 20 years old; she's in college; I'm having a lot of trouble letting her go. And I have a lot of guilt feelings about her; I have a real need to hang on to her.

C.R.: A need to hang on so you can kind of make up for the things you feel guilty about. Is that part of it?

JILL: There's a lot of that. Also, she's been a real friend to me, and filled my life. And it's very hard . . . a lot of empty places now that she's not with me.

C.R.: The old vacuum, sort of, when she's not there.

JILL: Yes. Yes. I also would like to be the kind of mother that could be strong and say, you know, "Go and have a good life," and this is really hard for me, to do that.

C.R.: It's very hard to give up something that's been so precious in your life, but also something that I guess has caused you pain when you mentioned guilt.

JILL: Yeah. And I'm aware that I have some anger toward her that I don't always get what I want. I have needs that are not met. And, uh, I don't feel I have a right to those needs. You know . . . she's a daughter; she's not my mother. Though sometimes I feel as if I'd like her to mother me . . . it's very difficult for me to ask for that and have a right to it.

C.R.: So, it may be unreasonable, but still, when she doesn't meet your needs, it makes you mad.

JILL: Yeah I get very angry, very angry with her.

C.R.: (*Pause*)You're also feeling a little tension at this point, I guess.

JILL: Yeah. Yeah. A lot of conflict . . . (C.R.: M-hm.). A lot of pain.

C.R.: A lot of pain. Can you say anything more about what that's about?

—From Farber, Brink, & Raskin, 1996, The psychotherapy of Carl Rogers: Cases and commentary, *pp. 74–75. Reprinted with permission of The Guilford Press.*

Cognitive Therapy

. . . there is nothing either good or bad, but thinking makes it so.

—Shakespeare, *Hamlet*

In these words, Shakespeare did not mean to imply that misfortunes or ailments are painless or easy to manage. His point, rather, is that the ways in which we evaluate upsetting events can heighten our discomfort and impair our ability to cope. Several hundred years later, cognitive therapists adopted this simple but elegant expression as a kind of motto for their approach to therapy.

Cognitive therapists focus on helping clients identify and correct maladaptive beliefs, automatic types of thinking, and self-defeating attitudes that create or compound emotional problems. They believe that negative emotions such as anxiety and depression are caused by the interpretations we place on troubling events, not on the events themselves. Here we focus on the contributions of two prominent types of cognitive therapy: Albert Ellis's rational-emotive behavior therapy, and Aaron Beck's cognitive therapy.

Rational-Emotive Behavior Therapy Albert Ellis (1977b, 1993, 2001; Dryden & Ellis, 2001) believes that the adoption of irrational, self-defeating beliefs gives rise to psychological problems and negative feelings. Consider the irrational belief that one must have the approval almost all of the time of the people who are important to you. Ellis finds it understandable to want other people's approval and love, but he argues that it is irrational to believe we cannot survive without it. Another irrational belief is that we must be thoroughly competent and achieving in virtually everything we seek to accomplish or else it would be awful and unbearable. We are doomed to eventually fall short of these irrational expectations. When we do fall short, we may experience negative emotional consequences, such as depression and lowered self-esteem. Emotional difficulties such as anxiety and depression are not directly caused by negative events, but rather by how we distort their meaning by viewing them through the dark-colored glasses of self-defeating beliefs. Imposing "musts" and "shoulds" on ourselves transforms challenging events, such as forthcoming examinations, into looming disasters (e.g., "It would be just so awful if I did poorly that I wouldn't be able to stand it."). In Ellis's **rational-emotive behavior therapy (REBT),** therapists actively *dispute* clients' irrational beliefs and the premises on which they are based and help clients to develop alternative, adaptive beliefs in their place.

Ellis and Dryden (1987) describe the case of a 27-year-old woman, Jane, who was socially inhibited and shy, particularly with attractive men. Through REBT, Jane identified some of her underlying irrational beliefs, such as "I must speak well to people I find attractive" and "When I don't speak well and impress people as I should, I'm a stupid, inadequate person!" (p. 68). REBT helped Jane discriminate between these irrational beliefs and rational alternatives, such as "If people do reject me for showing them how anxious I am, that will be most unfortunate, but I can stand it" (p. 68). REBT encouraged Jane to debate or dispute irrational beliefs by posing challenging questions to herself: (1) "*Why* must I speak well to people I find attractive?" and, (2) "When I don't speak well and impress people, how does that make me a *stupid and inadequate person*?" (p. 69). Jane learned to form rational responses to her self-questioning, for example, (1) "There is no reason I must speak well to people I find attractive, but it would be desirable if I do so, so I shall make an effort—but not kill myself—to do so," and, (2) "When I speak poorly and fail to impress people, that only makes me a *person who spoke unimpressively this time*—not a *totally stupid or inadequate person*" (p. 69).

Jane also rehearsed more rational ideas several times a day. Examples included, "I would like to speak well, but I never *have to*," and, "When people I favor reject me, it often reveals more about them and their tastes than about me" (pp. 69–70). After 9 months of REBT, Jane was able to talk comfortably to men she found attractive and was preparing to take a job as a teacher, a position she had previously avoided due to fear of facing a class.

Ellis recognizes that irrational beliefs may be formed on the basis of early childhood experiences. Changing them requires finding rational alternatives in the here and now, however. Rational-emotive behavior therapists also help clients substitute more effective interpersonal behavior for self-defeating or maladaptive behavior. Ellis often gives clients specific tasks or homework assignments, such as disagreeing with an overbearing relative or asking someone for a date. He assists them in practicing or rehearsing adaptive behaviors.

rational-emotive behavior therapy (REBT) A therapeutic approach that focuses on helping clients replace irrational, maladaptive beliefs with alternative, more adaptive beliefs.

Beck's Cognitive Therapy As formulated by psychiatrist Aaron Beck and his colleagues (Beck, 1976; Beck et al., 1979; DeRubeis, Tang, & Beck, 2001), cognitive therapy, like REBT, focuses on clients' maladaptive cognitions. Cognitive therapists encourage clients to recognize and change errors in their thinking, called *cognitive distortions*, which affect their moods and impair their behavior, such as tendencies to magnify negative events and minimize personal accomplishments.

Cognitive therapists have clients record the thoughts that are prompted by upsetting events they experience and note the connections between their thoughts and their emotional responses. They then help them to dispute distorted thoughts and replace them with rational alternatives. Therapists also use behavioral homework assignments, such as encouraging depressed clients to fill their free time with structured activities, such as gardening or completing work around the house. Carrying out such tasks serves to counteract the apathy and loss of motivation that tend to characterize depression and may also provide concrete evidence of competence, which helps combat self-perceptions of helplessness and inadequacy.

Another type of homework assignment involves reality testing. Clients are asked to test out their negative beliefs in the light of reality. For example, a depressed client who feels unwanted by everyone might be asked to call two or three friends on the phone to gather data about the friends' reactions to the calls. The therapist might then ask the client to report on the assignment: "Did they immediately hang up the phone? Or did they seem pleased you called? Did they express any interest at all in talking to you again or getting together sometime? Does the evidence support the conclusion that *no one* has any interest in you?" Such exercises help clients replace distorted beliefs with rational alternatives.

Consider this case in which a depressed man was encouraged to test his belief he was about to be fired from his job. The case also illustrates several cognitive distortions or errors in thinking, such as selectively perceiving only one's flaws (in this case, self-perceptions of laziness) and expecting the worst (expectations of being fired):

Kyle Tests His Irrational Beliefs

Kyle, a 35-year-old frozen foods distributor, had suffered from chronic depression since his divorce six years earlier. During the past year the depression had worsened and he found it increasingly difficult to call upon customers or go to the office. Each day that he avoided working made it more difficult for him to go to the office and face his boss. He was convinced that he was in imminent danger of being fired since he had not made any sales calls for more than a month. Since he had not earned any commissions in a while, he felt he was not adequately supporting his two daughters and was concerned that he wouldn't have the money to send them to college. He was convinced that his basic problem was laziness, not depression. His therapist pointed out the illogic in his thinking. First of all, there was no real evidence that his boss was about to fire him. His boss had actually encouraged him to get help and was paying for part of the treatment. His therapist also pointed out that judging himself as lazy was unfair because it overlooked the fact that he had been an industrious, successful salesman before he became depressed. While not fully persuaded, the client agreed to a homework assignment in which he was to call his boss and also make a sales call to one of his former customers. His boss expressed support and reassured him that his job was secure. The customer ribbed him about "being on vacation" during the preceding six weeks but placed a small order. The client discovered that the small unpleasantness he experienced in facing the customer and being teased paled in comparison to the intense depression he felt at home while he was avoiding work. Within the next several weeks he gradually worked himself back to a normal routine, calling upon customers and making future plans. This process of viewing himself and the world from a fresh perspective led to a general improvement in his mood and behavior.

—Adapted from Burns & Beck, 1978, pp. 124–126

■

REBT and Beck's cognitive therapy have much in common, especially the focus on helping clients replace self-defeating thoughts and beliefs with more rational ones. Perhaps the major difference between the two approaches is one of therapeutic style. REBT therapists tend to be more confrontational and forceful in their approach to disputing client's irrational beliefs (A. Ellis, Young, & Lockwood, 1989). Cognitive therapists tend to adopt a more gentle, collaborative approach in helping clients discover and correct the distortions in their thinking.

cognitive-behavioral therapy (CBT) A learning-based approach to therapy incorporating cognitive and behavioral techniques.

Cognitive-Behavioral Therapy

Today, most behavior therapists identify with a broader model of behavior therapy called **cognitive-behavioral therapy (CBT)** (also called *cognitive behavior therapy*). Cognitive-behavioral therapy attempts to integrate therapeutic techniques that focus on helping individuals make changes not only in their overt behavior but also in their underlying thoughts, beliefs, and attitudes. Cognitive-behavioral therapy draws on the assumption that thinking patterns and beliefs affect behavior and that changes in these cognitions can lead to desirable behavioral changes (Dobson & Dozois, 2001; McGinn & Sanderson, 2001).

Cognitive-behavioral therapists use an assortment of behavioral and cognitive techniques in therapy. The following case illustration shows how behavioral techniques (exposure to fearful situations) and cognitive techniques (changing maladaptive thoughts) were used in the treatment of *agoraphobia*, a type of anxiety disorder characterized by excessive fears of venturing out in public:

The Use of CBT to Treat Agoraphobia

Mrs. X was a 41-year-old woman with a 12-year history of agoraphobia. She feared venturing into public places alone and required her husband or children to accompany her from place to place. In-vivo (actual) exposure sessions were arranged in a series of progressively more fearful encounters—a fear-stimulus hierarchy. The first step in the hierarchy, for example, involved taking a shopping trip while accompanied by the therapist. After accomplishing this task, she gradually moved upwards in the hierarchy. By the third week of treatment, she was able to complete the last step in her hierarchy—shopping by herself in a crowded supermarket. Cognitive restructuring was conducted along with the exposure training. Mrs. X was asked to imagine herself in various fearful situations and to report the self-statements (self-talk) she experienced. The therapist helped her identify disruptive self-statements, such as "I am going to make a fool of myself." This particular self-statement was challenged by questioning whether it was realistic to believe that she would actually lose control, and, secondly, by disputing the belief that the consequences of losing control, were it to happen, would truly be disastrous. She progressed rapidly with treatment and became capable of functioning more independently. But she still harbored concerns about relapsing in the future. The therapist focused at this point on deeper cognitive structures involving her fears of abandonment by the people she loved if she were to relapse and be unable to attend to their needs. In challenging these beliefs, the therapist helped her realize that she was not as helpless as she perceived herself to be and that she was loved for other reasons than her ability to serve others. She also explored the question, "Who am I improving for?" She realized that she needed to find reasons to overcome her phobia that were related to meeting her own personal needs, not simply the needs of her loved ones. At a follow-up interview nine months after treatment, she was functioning independently, which allowed her to pursue her own interests, such as taking night courses and seeking a job.

—Adapted from Biran, 1988, pp. 173–176

■

eclectic therapy An approach to psychotherapy that incorporates principles or techniques from various systems or theories.

It could be argued that any behavioral method involving imagination or mental imagery, such as systematic desensitization, bridges behavioral and cognitive domains. Cognitive therapies such as Ellis's rational-emotive behavior therapy and Beck's cognitive therapy might also be regarded as forms of cognitive-behavioral therapy because they incorporate cognitive and behavioral treatment methods. The dividing lines between the psychotherapies may not be as clearly drawn as authors of textbooks—who are given the task of classifying them—might desire. Not only are traditional boundaries between therapies blurring, but many therapists today endorse an *eclectic* orientation in which they incorporate principles and techniques derived from different schools of therapy.

Eclectic Therapy

Each of the major psychological models of abnormal behavior—the psychodynamic, learning theory, humanistic and cognitive approaches—has spawned its own approaches to psychotherapy. Although many therapists identify with one or another of these schools of therapy, an increasing number of therapists practice **eclectic therapy,** in which they draw on techniques and teachings of multiple therapeutic approaches. Eclectic therapists look beyond the theoretical barriers that divide one school of psychotherapy from another in an effort to define what is common among the schools of therapy and what is useful in each of them. They seek to enhance their therapeutic effectiveness by incorporating principles and techniques from different therapeutic orientations. An eclectic therapist might use behavior therapy techniques to help a client change specific maladaptive behaviors, for example, along with psychodynamic techniques to help the client gain insight into the childhood roots of the problem.

Truth OR Fiction? REVISITED

More psychotherapists identify with an eclectic approach than with any specific school of therapy.

TRUE. Surveys of therapists show that more endorse an eclectic approach than any other therapeutic orientation.

A greater percentage of clinical and counseling psychologists identify with an eclectic or integrative orientation than any other therapeutic orientation (Bechtoldt et al., 2001; see Figure 4.1). Therapists who adopt an eclectic approach tend to be older and more experienced (Beitman, Goldfried, & Norcross, 1989). Perhaps they have learned through experience of the value of drawing on diverse contributions to the practice of therapy.

Eclecticism has different meanings for different therapists. Some therapists are *technical eclectics.* They draw on techniques from different schools of therapy without necessarily adopting the theoretical positions that spawned the techniques (Beutler, Harwood, & Caldwell, 2001; Lazarus, 1992). They assume a pragmatic approach in using techniques from different therapeutic approaches that they believe are most likely to work with a given client. The therapist attempts to match the therapeutic approach to the particular characteristics of the client, rather than apply the same therapeutic approach to all clients presenting with a given diagnosis or type of problem. Thus, the eclectic therapy offered to one client will be different than the eclectic therapy offered to another (Beutler, 1995).

Other eclectic therapists are *integrative eclectics.* They attempt to synthesize and integrate diverse theoretical approaches—to bring together different theoretical concepts and therapeutic approaches under the roof of one integrated model of therapy (Beutler, Harwood, & Caldwell, 2001; Stricker & Gold, 2001). Though various approaches to integrative

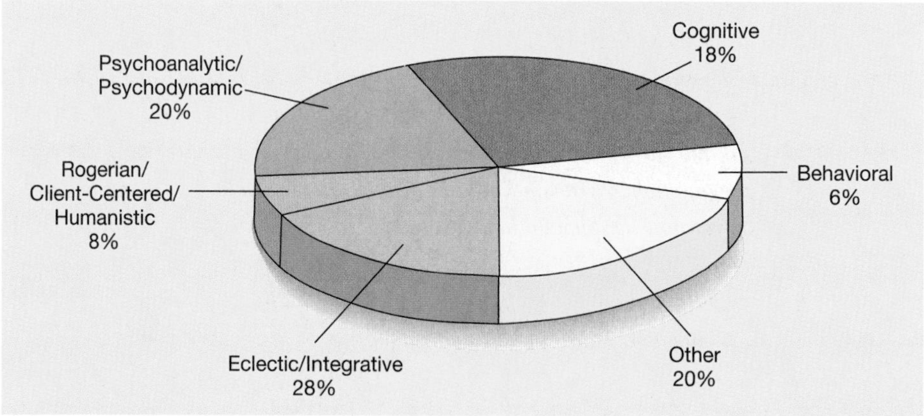

FIGURE 4.1 Therapeutic orientations of clinical and counseling psychologists. An eclectic/integrative orientation is the most widely endorsed therapeutic orientation among clinical and counseling psychologists today.

Source. Adapted from Bechtoldt, H., Norcross, J. C., Wyckoff, L. A., Pokrywa, M. L., & Campbell, L. F., 2001.

A Closer Look

Internet Counseling: Psychological Help May Be Only a Few Mouse Clicks Away

You can do most anything on the Internet these days, from ordering concert tickets to downloading music or whole books. You can also receive counseling or therapy services from an online therapist. By 2001, more than 250 Web sites were offering counseling services online (Kalb, 2001a). But as the numbers of online counseling services mushroom, many professionals voice concerns about the clinical, ethical, and legal issues regarding their use (Reed, McLaughlin, & Milholland, 2000; Rabasca, 2000b; Jacobs et al., 2001; Jerome et al., 2000).

One problem is that while psychologists are licensed in particular states, Internet communications easily cross state and international borders. It remains unclear whether psychologists or other mental health professionals can legally provide online services to residents of states in which they are not licensed. Then too are ethical problems and liability issues psychologists and other helping professionals may face in offering services to clients they never meet in person. Many therapists also express concerns that interacting with a client by computer would prevent them from evaluating nonverbal cues and gestures that might signal deeper levels of distress than are verbally reported or typed on a keyboard.

Yet another problem is that online therapists living at great distances from their clients may not be able to provide the more intensive services that clients need during times of emotional crisis. Professionals also express concern about the potential for unsuspecting clients to be victimized by unqualified practitioners or "quacks." We presently lack a system for ensuring that online therapists are licensed and otherwise qualified practitioners (Lauerman, 2000; Stamm & Perednia, 2000).

Despite these drawbacks, many professionals believe that online consultation and counseling services have potential value. For one thing, the distance of online consultation may encourage people to seek help who have hesitated out of shyness or embarrassment to consult a helping professional (Rabasca, 2000b). Online consultation may also make people feel more comfortable about receiving help and become a first step toward meeting a therapist in person. Online therapy may also provide people living in remote areas where finding a therapist is difficult, or those lacking mobility, with services they might not otherwise receive. Recently, investigators found that participants in an Internet-delivered behavioral program for weight loss achieved better results than a comparison group that was simply given a list of weight-loss education sites on the Web (Tate, Wing, & Winett, 2001).

The bottom line about Internet counseling, says Russ Newman, executive director for professional practice of the American Psychological Association, is the need for monitoring and evaluating this emerging technology (cited in Lauerman, 2000). Psychologists are not writing off so-called e-therapy, but they remain cautious in endorsing its widespread use.

psychotherapy have been proposed, there is as yet no clear agreement as to the principles and practices that constitute therapeutic integration (Garfield, 1994). Perhaps multiple approaches are needed (Safran & Messer, 1997).

Not all therapists subscribe to the view that therapeutic integration is a desirable or achievable goal. They believe that combining elements of different therapeutic approaches will lead to a hodgepodge of techniques that lack a cohesive conceptual framework. Still, interest in the professional community in therapeutic integration is growing, and we expect to see new models emerging that aim at tying together the contributions of different approaches.

Group, Family, and Marital Therapy

Some approaches to therapy expand the focus of treatment to include groups of people, families, and couples.

Group Therapy In **group therapy,** a group of clients meet together with a therapist or pair of therapists. Group therapy has several advantages over individual treatment. For one, group therapy is less costly to individual clients, because several clients are treated at the same time. Many clinicians also believe that group therapy is more effective in treating groups of clients who

group therapy Therapy method in which a group of clients meet together with a therapist.

Group therapy. What are some of the advantages of group therapy over individual therapy? What are some of its disadvantages?

family therapy Therapy in which the family, not the individual, is the unit of treatment.

couples therapy Therapy that focuses on resolving conflicts in distressed couples.

have similar problems, such as complaints relating to anxiety, depression, lack of social skills, or adjustment to divorce or other life stresses. The group format provides clients with the opportunity to learn how people with similar problems cope and provides the social support of the group as well as the therapist. Group therapy also provides members with opportunities to work through their problems in relating to others. For example, the therapist or other members may point out to a particular member when he or she acts in a bossy manner or tends to withdraw when criticized, patterns of behavior that may mirror the behavior the client shows in relationships with others outside the group. Group members may also rehearse social skills with one another in a supportive atmosphere.

Despite these advantages, clients may prefer individual therapy for various reasons. For one, clients might not wish to disclose their problems to others in a group. Some clients prefer the individual attention of the therapist. Others are too socially inhibited to feel comfortable in a group setting, even though they might be the ones who could most profit from a group experience. Because of such concerns, group therapists require that group disclosures be kept confidential, that group members relate to each other supportively and nondestructively, and that group members receive the attention they need.

Family Therapy In **family therapy,** the family, not the individual, is the unit of treatment. Family therapy aims to help troubled families resolve their conflicts and problems so the family functions better as a unit and individual family members are subjected to less stress from family conflicts.

Faulty patterns of communications within the family often contribute to problems. In family therapy, family members learn to communicate more effectively and to air their disagreements constructively. Family conflicts often emerge at transitional points in the life cycle, when family patterns are altered by changes in one or more members. Conflicts between parents and children, for example, often emerge when adolescent children seek greater independence or autonomy. Family members with low self-esteem may be unable to tolerate different attitudes or behaviors from other members of the family and may resist their efforts to change or become more independent. Family therapists work with families to resolve these conflicts and help them adjust to life changes.

THINK ABOUT IT

What type of therapy would you prefer if you were seeking treatment for a psychological disorder? Why?

Family therapists are sensitive to tendencies of families to scapegoat one family member as the source of the problem, or the "identified client." Disturbed families seem to adopt a sort of myth: Change the identified client, the "bad apple," and the "barrel," or family, will once again become functional. Family therapists encourage families to work together to resolve their disputes and conflicts, instead of resorting to scapegoating.

Family therapy. In family therapy, the family, not the individual, is the unit of treatment. Family therapists help family members communicate more effectively with one another, for example, to air their disagreements in ways that are not hurtful to individual members. Family therapists also try to prevent one member of the family from becoming the scapegoat for the family's problems.

Many family therapists adopt a *systems approach* to understanding the workings of the family and problems that may arise within the family. They see problem behaviors of individual family members as representing a breakdown in the system of communications and role relationships within the family. For example, a child may feel in competition with other siblings for a parent's attention and develop enuresis, or bed-wetting, as a means of securing attention. Operating from a systems perspective, the family therapist may focus efforts on helping family members understand the hidden messages in the child's behavior and then assist the family to make changes in their relationships to meet the child's needs more adequately.

Couples Therapy **Couples therapy** focuses on resolving conflicts in distressed couples, including married and unmarried couples. Like family therapy, couples therapy focuses on improving communication and analyzing role relationships in order to improve the relationship. For example, one partner may play a dominant role and resist any request to share power. The couples therapist helps bring these role relationships into the open, so that partners can explore alternative ways of relating to one another that would lead to a more satisfying relationship.

Evaluating Methods of Treatment

What, then, of the effectiveness of psychotherapy? Does psychotherapy work? Are some forms of therapy more effective than others? Are some forms of therapy more effective for some types of clients or for some types of problems than for others?

The effectiveness of psychotherapy receives strong support from the research literature. Reviews of the scientific literature often utilize a statistical technique called **meta-analysis,** which averages the results of a large number of studies in order to determine an overall level of effectiveness.

In the most frequently cited meta-analysis of psychotherapy research, M. L. Smith and Glass (1977) analyzed the results of some 375 controlled studies comparing various types of therapies (psychodynamic, behavioral, humanistic, etc.) against control groups. The results of their analyses showed that the average psychotherapy client in these studies was better off than 75% of the clients who remained untreated. In 1980, Smith and Glass and their colleague Miller reported the results of a larger analysis based on 475 controlled outcome studies, which showed the average person who received therapy was better off at the end of treatment than 80% of those who did not (M. L. Smith, Glass, & Miller, 1980).

Other meta-analyses also show positive outcomes for psychotherapy, including analyses of behavior therapy (Bowers & Clum, 1988; Lipsey & Wilson, 1993, 1995), brief psychodynamic therapy (E. M. Anderson & Lambert, 1995; Crits-Christoph, 1992), and group psychotherapy (McDermut, Miller, & Brown, 2001). Evidence indicates that psychotherapy is effective not only in the confines of clinical research centers, but also in settings that are more typical of ordinary clinical practice (Shadish et al., 2000). Although not all researchers endorse the use of meta-analysis as a methodological tool, the technique has achieved widespread acceptance within psychology and has provided some of the strongest evidence to date supporting the effectiveness of psychotherapy.

Evidence also shows that the greatest gains in psychotherapy are typically achieved in the first several months of treatment (Barkham et al., 1996; Howard et al., 1986). About 50% of patients show clinically significant change in about 3 or 4 months of treatment; by about 6 months, this figure rises to about 75% (E. M. Anderson & Lambert, 2001; Goode, 1998; Messer, 2001a).

Meta-analyses show only negligible differences, overall, in outcomes among the various therapies when such therapies are compared to control groups (Crits-Christoph, 1992; M. L. Smith, Glass, & Miller, 1980; Wampold et al., 1997a, 1997b). Such minor differences suggest that the effectiveness of psychotherapies may have more to do with the features they have in common than with the specific techniques that set them apart (M. J. Lambert & Bergin, 1994). These common features are called **nonspecific factors.** Nonspecific or common factors include expectations of improvement and features of the therapist–client relationship, including the following: (1) empathy, support, and attention shown by the therapist; (2) *therapeutic alliance*, or the attachment the client develops toward the therapist and the therapy process; and (3) the *working alliance*, or the development of an effective working relationship in which the therapist and client work together to identify and confront the important issues and problems the client faces (J. L. Binder & Strupp, 1997; Connors et al., 1997; M. J. Lambert & Okiishi, 1997; Perlman, 2001).

Should we conclude that different therapies are about equally effective? One possibility is that different therapies are about equal in their effects overall but may not be equal in their effects with every patient (Wampold et al., 1997a). That is, a given therapy may be more effective for a particular patient or for a particular type of problem. All in all, the question of whether differences exist in the relative effectiveness of different forms of therapy remains unresolved (Nathan, Stuart, & Dolan, 2000).

Another approach to the question of determining which therapies are effective for which types of problems was undertaken by a task force commissioned by the Clinical Psychology Division of the American Psychological Association. The task force concluded that enough evidence now exists from controlled trials to support the therapeutic efficacy of various psychological interventions (listed in Table 4.2) for specific psychological problems or disorders (Chambless & Ollendeck, 2001; Weisz et al., 2000).

meta-analysis A statistical technique for combining the results of different studies into an overall average.

nonspecific factors Factors not specific to any one form of psychotherapy, such as therapist attention and support, and the engendering of positive expectancies of change.

Truth OR Fiction? REVISITED

The average client who receives psychotherapy is no better off than control clients who go without it.

FALSE. The average psychotherapy client is better off than about 75% or 80% of control clients who do not receive psychotherapy.

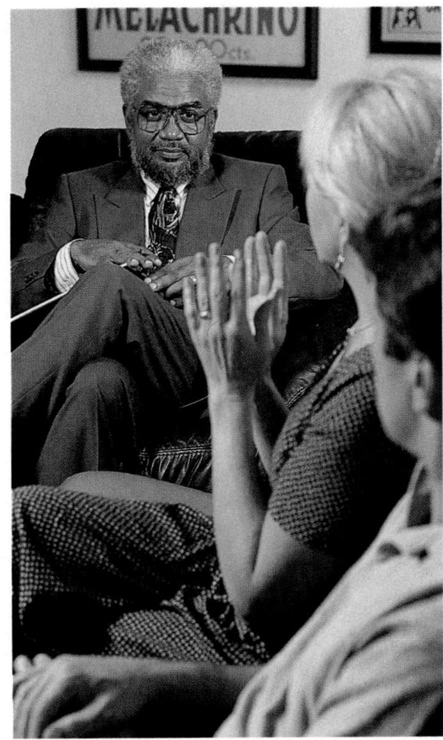

Nonspecific factors. Are the benefits of psychotherapy due to nonspecific factors that various psychotherapies share in common, such as the mobilization of hope, the attention and support provided by the therapist, and the development of a good working alliance between the client and therapist? It appears that both specific and nonspecific factors are involved in accounting for therapeutic change.

Other treatment interventions may be added to the list of these empirically supported treatments or ESTs as scientific evidence attesting to their effectiveness becomes available. We should caution you not to infer that the inclusion of a particular treatment guarantees that it is effective in every case.

The effort to develop a listing of ESTs comes at a time when health care professionals are facing increasing pressure to demonstrate the effectiveness of the treatments they use. As task force member William Sanderson ("Task Force," 1995) said, "More than ever before, psychologists—and all health care providers—are being called on to show the efficacy of their interventions. Society wants proof that a treatment works—whether it be medication, surgery or psychotherapy—before it is administered" (p. 5).

It is thus insufficient to ask which therapy works best. We must ask, Which therapy works best for which type of problem? Which clients are best suited for which type of therapy? What are the advantages and limitations of particular therapies? Behavior therapy, for example, has shown impressive results in treating various types of anxiety disorders, sleep disorders, and sexual dysfunctions and in improving the adaptive functioning of people with schizophrenia and mental retardation. Psychodynamic and humanistic approaches may be most effective in fostering self-insight and personality growth. Cognitive therapy has demonstrated impressive results in treating depression and anxiety disorders. By and large, however, the process of determining which treatment, practiced by whom, and under what conditions, is most effective for a given client remains a challenge.

All in all, psychotherapy is a complex process that incorporates common features along with specific techniques that foster adaptive change. A strong therapeutic alliance between client and therapist is associated with better therapy outcome (Barber et al., 2000; Martin, Garske, & Davis, 2000). But therapeutic gains are not accounted for entirely by nonspecific factors (Oei & Shuttlewood, 1996; Grissom, 1996). In fact, investigators believe that specific techniques may account for about twice the magnitude of therapeutic change as nonspecific factors that different therapies have in common (Stevens, Hynan, & Allen, 2000). In the final analysis, therapeutic change most likely

TABLE 4.2 Examples of Empirically Supported Treatments (ESTs)

Treatment	Conditions for Which Treatment Is Effective (Chapter in text where treatment is discussed is shown in parentheses.)
Cognitive therapy	Headache (Ch. 5) Depression (Ch. 8)
Behavior therapy or Behavior modification	Depression (Ch. 8) Persons with developmental disabilities (Ch. 14) Enuresis (Ch. 14)
Cognitive-behavior therapy	Panic disorder with and without agoraphobia (Ch. 6) Generalized anxiety disorder (Ch. 6) Smoking cessation (Ch. 10) Bulimia (Ch. 11)
Exposure treatment	Agoraphobia and specific phobia (Ch. 6)
Exposure and response prevention	Obsessive-compulsive disorder (Ch. 6)
Interpersonal psychotherapy	Depression (Ch. 8)
Parent training programs	Children with oppositional behavior (Ch. 14)

Source. Adapted from Chambless et al., 1998.

depends on the influence of specific and nonspecific factors, as well as their interactions (Ilardi & Craighead, 1994).

"IT'S YOUR INSURANCE COMPANY, THEY SAY YOU'RE CURED."

Managed Care or Managed Care Costs

Managed Care or Managed Costs? This is an appropriate juncture to note that the practice of psychotherapy has been influenced by changes in the general health care environment in recent years, especially the increasing role of **managed care systems,** such as health maintenance organizations (HMOs). Managed care systems typically impose limits on the number of treatment sessions they will approve for payment and the fees they will allow for reimbursement. Consequently, there is greater emphasis today on briefer, more direct forms of treatment, including cognitive-behavioral therapy and shorter-term psychodynamic therapies. Traditional long-term psychodynamic psychotherapy is likely to become a luxury that is available to only a very few (Strupp, 1992). Moreover, managed care has curbed costly inpatient mental health treatment, primarily through limiting the lengths of stay of patients in psychiatric hospitals (Wickizer, Lessler, & Travis, 1996; USDHHS, 1999a). However, many people express concerns about the risks of sacrificing quality of care in the interests of cutting costs.

Though health care providers understand the need to curtail the spiraling costs of care, they are understandably concerned that the cost-cutting emphasis of managed care plans may discourage needful people with identifiable psychological disorders from seeking help or receiving an adequate level of care (Landerman et al., 1994). According to the recent Surgeon General's report, "Excessively restrictive cost-containment strategies and financial incentives to providers and facilities to reduce specialty referrals, hospital admissions, or length or amount of treatment may ultimately contribute to lowered access and quality of care" (cited in USDDHS, 1999a, Chapter 6).

Overzealous cost-cutting policies may also be financially short-sighted because the failure to provide adequate mental health care when problems arise may lead to an increased need for more expensive care at some later point. Evidence shows that for people with severe psychological disorders, such as schizophrenia, bipolar disorder, and borderline personality disorder, psychotherapy actually reduces health care costs by reducing the need for hospitalization and reducing work impairment (Fraser, 1996; Gabbard et al., 1997).

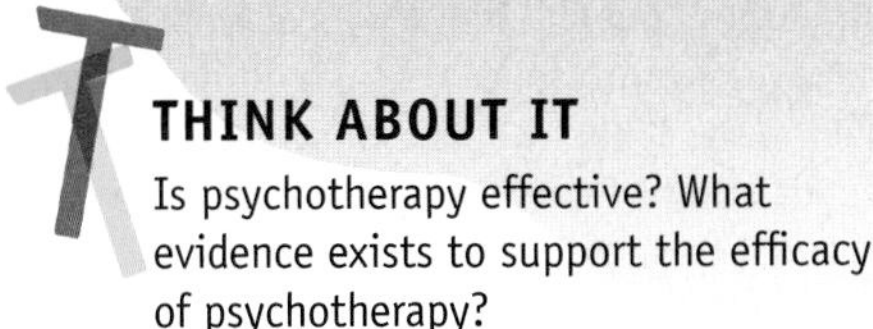
THINK ABOUT IT

Is psychotherapy effective? What evidence exists to support the efficacy of psychotherapy?

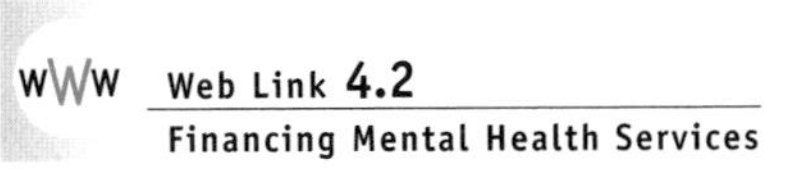
Web Link 4.2
Financing Mental Health Services

Multicultural Issues in Psychotherapy

We live in a multicultural society in which people bring to therapy not only their personal backgrounds and individual experiences but also their cultural learning and values. Therapists need to be sensitive to cultural differences and how they may affect the therapeutic process. They also need to avoid ethnic stereotyping and to demonstrate sensitivity to the values, languages, and cultural beliefs of members of racial or ethnic groups that are different than their own (Comas-Diaz & Griffith, 1988; Lee & Richardson, 1991). Let us touch on some of the issues involved in treating members of the major ethnic minority groups in our society: African Americans, Asian Americans, Hispanic Americans, and Native Americans.

African Americans The cultural history of African Americans must be understood in the context of a history of extreme racial discrimination (Boyd-Franklin, 1989; Greene, 1990). African Americans have needed to develop coping mechanisms for managing the pervasive racism they encounter in such areas as employment, housing, education, and access to health care (Greene, 1993a, 1993b). For example, the sensitivity of many African Americans to the potential for maltreatment and exploitation has been a survival tool that may take the form of a heightened level of suspiciousness or reserve (Greene, 1986). Therapists need to be aware of the tendency of African American clients to minimize their vulnerability by being less self-disclosing, especially in early stages of therapy (Ridley, 1984). Therapists should not confuse such suspiciousness with paranoia (Boyd-Franklin, 1989; Greene, 1986).

managed care systems Health care delivery systems that impose limits on the number of treatment sessions they will approve for payment and the fees they will allow for reimbursement.

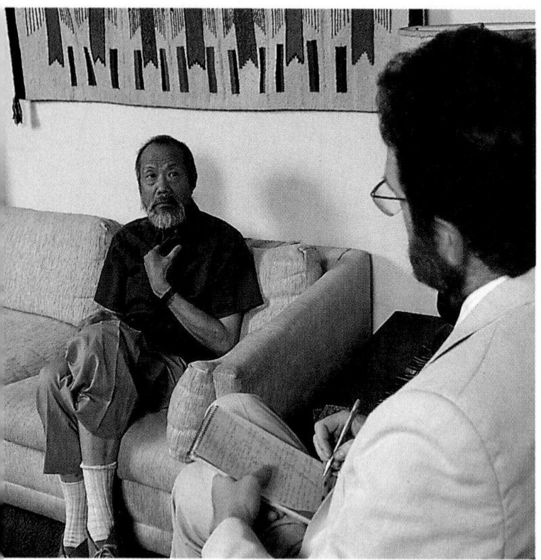

Cultural sensitivity. Therapists need to be sensitive to cultural differences and how they may affect the therapeutic process. They also need to avoid ethnic stereotyping and to demonstrate sensitivity to the values, languages, and cultural beliefs of members of racial or ethnic groups that are different than their own. Clients who are not fluent in English profit from having therapists who can conduct therapy in their own languages.

In addition to whatever psychological problems an African American client may present, the therapist often needs to help the client develop coping mechanisms to deal with societal racial barriers. Therapists also need to be attuned to tendencies of some African Americans to internalize within their self-concepts the negative stereotypes about Blacks that are perpetuated in the dominant culture (Greene, 1985, 1992b, 1992c; Pinderhughes, 1989; Nickerson, Helms, & Terrell, 1994).

To be culturally competent, therapists not only must develop a better awareness of the cultural traditions and languages of the groups with which they work, but also must come to an understanding of their own racial and ethnic attitudes and how their underlying attitudes affect their clinical practice (Greene, 1985, 1992b, 1992c; Nickerson, Helms, & Terrell, 1994; Pinderhughes, 1989). Therapists are exposed to the same negative stereotypes about African Americans as other people in society and must recognize how the incorporation of these stereotypes, if left unexamined, can become destructive to the therapeutic relationships they form with African American clients. In effect, therapists must be willing to confront their own racism and prejudices and work to replace these attitudes with more realistic appraisals of African Americans (Mays, 1985).

Therapists must also be aware of the cultural characteristics associated with African American families, such as strong kinship bonds between family members, often including people who are not biologically related (for example, a close friend of a parent may have some parenting role and may be addressed as "aunt"), strong religious and spiritual orientation, multigenerational households, adaptability and flexibility of gender roles (African American women have a long history of working outside the home), and distribution of child-care responsibilities among different family members (Boyd-Franklin, 1989; Collins, 1990; Ferguson-Peters, 1985; Greene, 1990; USDHHS, 1999a).

Asian Americans Culturally sensitive therapists not only understand the beliefs and values of other cultures but also integrate this knowledge within the therapy process. Generally speaking, Asian cultures, including Japanese culture, value restraint in talking about oneself and one's feelings. Therapists thus need to be patient and not expect instant self-disclosures from Asian clients (Henkin, 1985). Public expression of emotions is also discouraged in Asian cultures. Suppression of emotions, especially negative emotions, is valued, and failure to keep one's feelings to oneself is believed to reflect poorly on one's upbringing (Huang, 1994). Asian clients who appear emotionally restrained or constricted when judged by Western standards may be responding in ways that are culturally appropriate.

Clinicians also note that Asian clients often express psychological complaints in terms of physical symptoms. However, this tendency to *somaticize* emotional problems may be attributed in part to differences in communication styles (Zane & Sue, 1991). That is, Asians may use somatic terms to convey emotional distress.

In some cases, there may also be inherent role conflicts between the goals of therapy and the values of a particular culture. The individualism of American society, which becomes expressed in therapeutic interventions in Western society that focus on development of the self, contrasts sharply with the group- and family-centered values of Asian cultures (Huang, 1994). Therapeutic approaches that emphasize the importance of individuality and self-determination may be inappropriate when applied to Asian clients who adhere strongly to traditional Asian cultural values, which emphasize the importance of the group over the individual (Ching et al., 1995; Henkin, 1985).

Hispanic Americans Although Hispanic American subcultures differ in various respects, many of them share certain cultural values and beliefs, such as adherence to a strong patriarchal (male-dominated) family structure and strong kinship ties. De la Cancela and Guzman (1991) identify some other values shared by many Hispanic Americans:

> One's identity is in part determined by one's role in the family. The male, or *macho*, is the head of the family, the provider, the protector of the family honor, and the final decision maker. The woman's role (*marianismo*) is to care for the family and the children. Obviously, these roles are changing, with women entering the work force and achieving

> greater educational opportunities. Cultural values of *respeto* (respect), *confianza* (trust), *dignidad* (dignity), and *personalismo* (personalism) are highly esteemed and are important factors in working with many [Hispanic Americans]. (p. 60)

Therapists need to recognize that value conflicts may occur between the traditional Hispanic American value of interdependency on the family with the values of independence and self-reliance, which are stressed in the mainstream U.S. culture (De la Cancela & Guzman, 1991). Psychotherapeutic interventions should respect differences in values rather than attempt to impose values of majority cultures on people from ethnic minority groups. Therapists should also be trained to reach beyond the confines of their offices to work within the Hispanic American community itself, in settings that have an impact on the daily lives of Hispanic Americans, such as social clubs, *bodegas* (neighborhood groceries), and neighborhood beauty and barber shops. We can further break down barriers that may impede utilization of mental health services by Hispanic Americans by recruiting bicultural/bilingual staff and creating a welcoming therapeutic atmosphere that is accepting of Hispanic American cultural values (Guarnaccia & Rodriguez, 1996).

Native Americans Among all the ethnic minority groups in the United States, Native Americans may be most in need of effective mental health treatment. Lifetime prevalence of psychological disorders may exceed 50% of the population of some Native American tribes. Despite the need, Native Americans remain severely underserved by mental health professionals, in part because of the cultural gap that exists between providers and recipients of these services.

Kahn (1982) argues that if mental health professionals are to be successful in helping Native Americans, they must do so within a context that is relevant and sensitive to Native Americans' customs, culture, and values. For example, many Native Americans expect the therapist will do most of the talking and they will play a passive role in treatment. These expectations are in keeping with the traditional healer role but may conflict with the client-focused approach of many forms of conventional therapy. There may yet be other differences in gestures, eye contact, facial expression, and other modes of nonverbal expression that can impede effective communication between therapist and client (Renfrey, 1992).

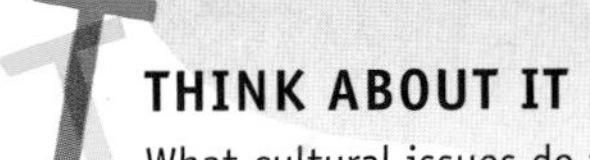

THINK ABOUT IT

What cultural issues do therapists need to consider when working with members of diverse cultural and racial groups?

Psychologists recognize the importance of bringing elements of tribal culture into mental health programs for American Indians (Rabasca, 2000a). For example, therapists can use indigenous ceremonies that are part of the client's cultural or religious traditions. To do so, mental health professionals need to become knowledgeable about traditional Native cultures as well as their own and attempt to integrate the two (Timpson et al., 1988). Lefley (1990) notes that purification and cleansing rites are therapeutic for many Native American peoples in the United States and elsewhere, as among the African Cuban *Santeria*, the Brazilian *umbanda*, and the Haitian *vodoun*. Cleansing rites are often sought by people who believe their problems are caused by failure to placate malevolent spirits or to perform mandatory rituals (Lefley, 1990).

Respect for cultural differences is a keynote feature of culturally sensitive therapies. Training in multicultural therapy is becoming more widely integrated into training programs for therapists (e.g., Neville et al., 1996). Culturally sensitive therapies adopt a respectful attitude that encourages people to tell their own personal story as well as the story of their culture (Coronado & Peake, 1992).

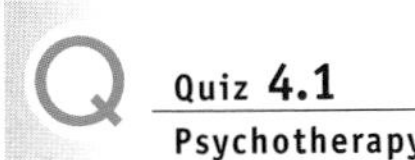

Quiz **4.1**
Psychotherapy

Biomedical Therapies

There is a growing emphasis in American psychiatry on biomedical therapies, especially the use of psychotherapeutic drugs (also called *psychotropic drugs*). Biomedical therapies are generally administered by medical doctors, many of whom have specialized training in psychiatry or **psychopharmacology.** Many family physicians or general practitioners also prescribe psychotherapeutic drugs for their patients, however.

psychopharmacology The field of study that examines the effects of therapeutic or psychiatric drugs.

Biomedical approaches have had dramatic success in treating some forms of abnormal behavior, though they also have their limitations. For one, drugs may have unwelcome

antianxiety drugs Drugs that combat anxiety and reduce states of muscle tension.

tolerance Physical habituation to use of a drug.

rebound anxiety The experiencing of strong anxiety following withdrawal from a tranquilizer.

antipsychotic drugs Drugs used to treat schizophrenia or other psychotic disorders.

Web Link 4.3 WWW
Psychiatric Medications

or dangerous side effects. There is also the potential for abuse. One of the most commonly prescribed minor tranquilizers, Valium, has become a major drug of abuse among people who become psychologically and physiologically dependent on it. Psychosurgery has been all but eliminated as a form of treatment because of serious harmful effects of earlier procedures.

Drug Therapy

Different classes of psychotropic drugs are used in the treatment of various types of mental health problems. These include antianxiety drugs, antipsychotic drugs, antidepressants, and lithium which is used to treat mood swings in people with bipolar disorder. The use of other psychotropic drugs, such as stimulants, will be discussed in later chapters.

Antianxiety Drugs **Antianxiety drugs** (also called *anxiolytics*, from the Greek *anxietas*, meaning "anxiety," and *lysis*, meaning "bringing to an end") are drugs that combat anxiety and reduce states of muscle tension. They include mild tranquilizers, such as those of the *benzodiazepines* class of drugs, including *diazepam* (Valium) and *alprazolam* (Xanax), as well as hypnotic-sedatives, such as *triazolam* (Halcion) and *flurazepam* (Dalmane).

Antianxiety drugs depress the level of activity in certain parts of the central nervous system (CNS). In turn, the CNS decreases the level of sympathetic nervous system activity, reducing the respiration rate and heart rate and lessening states of anxiety and tension. Mild tranquilizers such as Valium grew in popularity when physicians became concerned about the use of more potent depressants, such as barbiturates, which are highly addictive and extremely dangerous when taken in overdoses or mixed with alcohol. Unfortunately, it has become clear that these tranquilizers also can, and often do, lead to physiological dependence (addiction). People who are dependent on Valium may go into convulsions when they abruptly stop taking it. Deaths have been reported among people who mix mild tranquilizers with alcohol or who are unusually sensitive to them. There are other less severe side effects, such as fatigue, drowsiness, and impaired motor coordination, that might nonetheless impair the ability to function or to operate an automobile. Regular usage of benzodiazepines can also produce **tolerance,** a physiological sign of dependence, which refers to the need over time for increasing dosages of a drug to achieve the same effect. Quite commonly, patients become involved in tugs of war with their physicians as they demand increased dosages despite their physicians' concerns about the potential for abuse and dependence.

When used on a short-term basis, antianxiety drugs can be safe and effective in treating anxiety and insomnia. Yet drugs by themselves do not teach people more adaptive ways of solving their problems and may encourage them to rely on a chemical agent to cope with stress rather than develop active means of coping. Drug therapy is thus often combined with psychotherapy to help people with anxiety complaints deal with the psychological and situational bases of their problems. However, combining drug therapy and psychotherapy may present special problems and challenges. For one, drug-induced relief from anxiety may reduce clients' motivation to try to solve their problems. For another, medicated clients who develop skills for coping with stress in psychotherapy may fail to retain what they have learned once the tranquilizers are discontinued, or find themselves too tense to employ their newly acquired skills.

Rebound anxiety is another problem associated with regular use of tranquilizers. Many people who regularly use antianxiety drugs report that anxiety or insomnia returns in a more severe form once they discontinue them. For some, this may represent a fear of not having the drugs to depend on. For others, rebound anxiety might reflect changes in biochemical processes that are not well understood at present.

Antipsychotic Drugs **Antipsychotic drugs,** also called *neuroleptics*, are commonly used to treat the more flagrant features of schizophrenia or other psychotic disorders, such as hallucinations, delusions, and states of confusion. Many of these drugs, including *chlorpromazine* (Thorazine), *thioridazine* (Mellaril), and *fluphenazine* (Prolixin), belong to the *phenothiazine* class of chemicals. Phenothiazines appear to control psychotic features by blocking the action of the neurotransmitter dopamine at receptor sites in the brain. Although the underlying causes of schizophrenia remain unknown, researchers suspect an irregularity in the dopamine system in the brain may be involved (see Chapter 13). *Clozapine* (Clozaril), a neuroleptic of a different chemical class than the phenothiazines, has been shown to be effective in treating many people with schizophrenia whose symptoms were unresponsive to other neuroleptics (see Chapter 13). The use of clozapine must be carefully monitored, however, because of potentially dangerous side effects.

The use of neuroleptics has greatly reduced the need for more restrictive forms of treatment for severely disturbed patients, such as physical restraints and confinement in padded cells, and has lessened the need for long-term hospitalization. The introduction of the first generation of antipsychotic drugs in the mid-1950s was one of the major factors that led to a massive exodus of chronic mental patients from state institutions. Many formerly hospitalized patients have been able to resume family life and hold jobs while continuing to take their medications.

Neuroleptics are not without their problems, including potential side effects such as muscular rigidity and tremors. Although these side effects are generally controllable by use of other drugs, long-term use of antipsychotic drugs (possibly excepting clozapine) can produce a potentially irreversible and disabling motor disorder called *tardive dyskinesia* (see Chapter 13), which is characterized by uncontrollable eye blinking, facial grimaces, lip smacking, and other involuntary movements of the mouth, eyes, and limbs. Researchers are experimenting with lowered dosages, intermittent drug regimens, and use of new medications to reduce the risk of such complications.

Antidepressants Three major classes of **antidepressants** are used in treating depression: **tricyclics**, **monoamine oxidase (MAO) inhibitors**, and **selective serotonin-reuptake inhibitors (SSRIs).** The first two kinds of antidepressants, tricyclics and MAO inhibitors, increase the availability of the neurotransmitters norepinephrine and serotonin in the brain. Some of the more common tricyclics are *imipramine* (Tofranil), *amitriptyline* (Elavil), and *doxepin* (Sinequan). The MAO inhibitors include such drugs as *phenelzine* (Nardil) and *tranylcypromine* (Parnate). Tricyclic antidepressants (TCAs) are more commonly favored over MAO inhibitors because of potentially serious side effects associated with MAO inhibitors.

The third class of antidepressants, selective serotonin-reuptake inhibitors, or SSRIs, have more specific effects on serotonin function in the brain. Drugs in this class include *fluoxetine* (Prozac), now the most widely prescribed antidepressant on the market, and *sertraline* (Zoloft). They increase the availability of serotonin in the brain by interfering with its reuptake by the transmitting neuron.

By the latest estimates, it appears that slightly more than half of the people with clinically significant depression who are treated with antidepressants of the tricyclic class will respond favorably (Depression Guideline Panel, 1993b). A favorable response to treatment does not mean depression is relieved, however. Overall, the effects of tricyclic antidepressants (TCAs) appear to be modest (Greenberg et al., 1992). Nor does any particular antidepressant appear to be clearly more effective than any other (Depression Guideline Panel, 1993b). Even Prozac, which was hailed by some as a "wonder drug," produces about the same level of therapeutic benefit as the older generation of antidepressants, the TCAs (Greenberg et al., 1994). Prozac and other SSRIs may be preferred, however, because they are associated with fewer side effects, such as weight gain, and have a lower risk of lethal overdoses than the older tricyclics (Depression Guideline Panel, 1993b).

Antidepressants also have beneficial effects in treating a wide range of psychological disorders, including panic disorder (see Chapter 6), obsessive-compulsive disorder (also in Chapter 6), and eating disorders (see Chapter 11). As research into the underlying causes of these disorders continues, we may find that irregularities of neurotransmitter functioning in the brain plays a key role in their development.

antidepressants Drugs used to treat depression.

tricyclics A group of antidepressant drugs that increase the activity of norepinephrine and serotonin by interfering with the reuptake of these neurotransmitters.

monoamine oxidase (MAO) inhibitors A group of antidepressant drugs that increase the availability of neurotransmitters in the brain by inhibiting the actions of an enzyme that breaks down neurotransmitters.

selective serotonin-reuptake inhibitors (SSRIs) A group of antidepressant drugs that increase the availability of serotonin in the brain by interfering with its reuptake by the transmitting neuron.

Truth OR Fiction? REVISITED

Despite beliefs that it is a wonder drug, the antidepressant Prozac has not been shown to be any more effective than the earlier generation of antidepressants.

TRUE. Though Prozac may have fewer side effects than the older generation of antidepressants, it has not been shown to provide any greater therapeutic benefit.

THINK ABOUT IT
What problems do you see in taking pills to cope with anxiety or depression that may stem from academic or social difficulties?

Lithium Lithium carbonate, a salt of the metal lithium in tablet form, helps in many cases to stabilize the dramatic mood swings of patients with bipolar disorder (formerly manic depression) (discussed in Chapter 8). Like people with diabetes who must take insulin through their lifetimes to control their disease, people with bipolar disorder may have to continue using lithium indefinitely to control the disorder. Because of potential toxicity associated with lithium, the blood levels of patients maintained on the drug must be carefully monitored.

Table 4.3 lists psychotropic drugs according to their drug class and category.

TABLE 4.3 Major Psychotropic Drugs

Category	Drug Class	Generic Name	Trade Name
Antianxiety agents (also called anxiolytics)	Benzodiazepines	Diazepam	Valium
		Chlordiazepoxide	Librium
		Clorazepate	Tranxene
		Oxazepam	Serax
		Lorazepam	Ativan
		Alprazolam	Xanax
	Barbiturates	Meprobamate	Miltown Equanil
	Hypnotics	Flurazepam	Dalmane
		Triazolam	Halcion
		Zolpidem	Ambien
	Other anxiolytics	Busipirone	BuSpar
Antipsychotic drugs (also called neuroleptics or major tranquilizers)	Phenothiazines	Chlorpromazine	Thorazine
		Thioridazine	Mellaril
		Mesoridazine	Serentil
		Perphenazine	Trilafon
		Trifluoperazine	Stelazine
		Fluphenazine	Prolixin
	Thioxanthenes	Thiothixene	Navane
	Butyrophenones	Haloperidol	Haldol
	Dibenzoxazepines	Loxapine	Loxitane
	Dibenzodiazepines	Clozapine	Clozaril
Antidepressants	Tricyclic antidepressants (TCAs)	Imipramine	Tofranil
		Desipramine	Norpramin
		Amitriptyline	Elavil
		Doxepin	Sinequan
		Clomipramine	Anafranil
	MAO inhibitors (MAOIs)	Phenelzine	Nardil
		Tranylcypromine	Parnate
	Selective serotonin-reuptake inhibitors (SSRIs)	Fluoxetine	Prozac
		Sertraline	Zoloft
		Paroxetine	Paxil
		Fluvoxamine	Luvox
		Citalopram	Celexa
	Other antidepressants	Bupropion	Wellbutrin
		Nefazodone	Serzone
		Venlafaxine	Effexor
Antimanic agents		Lithium carbonate	Eskalith
		Carbamazepine	Tegretol
		Divalproex	Depakote
		Valproate	Depakene
Stimulants		Methylphenidate	Ritalin
		Pemoline	Cylert

Ethnic Differences in Response to Psychotropic Medication

Cultural or ethnic factors may contribute to differences in responsiveness to psychotropic medications (Lefley, 1990; Lesser, 1992). African Americans, for example, tend to show a better response to antidepressants and phenothiazines than other groups. Hispanic Americans tend to show lower effective dosage levels (Lawson, 1986). However, some research suggests that African Americans are at greater risk of experiencing potentially serious side effects from psychiatric drugs, especially phenothiazines (Jeste et al., 1996). Differences in response patterns and risks of side effects among ethnic groups brings into perspective the need to conduct psychopharmacological research on diverse groups. Unfortunately, people of color, including African Americans, have been underrepresented in drug trials (Lawson, 1996).

electroconvulsive therapy (ECT) A method of treating severe depression by administering electrical shock to the head.

prefrontal lobotomy A form of psychosurgery, no longer in use, in which certain neural pathways in the brain are severed in order to control disturbed behavior.

Electroconvulsive Therapy

In 1939, the Italian psychiatrist Ugo Cerletti introduced the technique of **electroconvulsive therapy (ECT)** in psychiatric treatment. Cerletti had observed the practice in some slaughterhouses of using electric shock to render animals unconscious. He observed that the shocks also produced convulsions. Cerletti incorrectly believed, as did other researchers in Europe at the time, that convulsions of the type found in epilepsy were incompatible with schizophrenia and that a treatment method that induced convulsions might be used to cure schizophrenia.

After the introduction of the phenothiazines in the 1950s, the use of ECT became generally limited to the treatment of severe depression. The introduction of antidepressants has limited the use of ECT even further today. However, evidence indicates that about 50% of people with major depression who fail to respond to antidepressants show significant improvement following ECT (Prudic et al., 1996).

ECT remains a source of controversy for several reasons. First, many people, including many professionals, are uncomfortable about the idea of passing an electric shock through a person's head, even if the level of shock is closely regulated and the convulsions are controlled by drugs. Second are the potential side effects. ECT often produces dramatic relief from severe depression, but concerns remain about its potential for inducing cognitive deficits, such as memory loss. Permanent loss of memory may occur for events that happen during the months that precede ECT and for several weeks afterwards (Glass, 2001). Third are questions of relative efficacy. The relative effectiveness of ECT as compared to antidepressant drugs, to sham (simulated) ECT, and to cognitive-behavioral therapy remains under study. Fourth, no one yet knows why ECT works, although it is suspected that it might help correct neurotransmitter imbalances in the brain. Fifth, evidence shows a high rate of relapse following ECT treatment (Sackeim et al., 2001).

Although controversies concerning the use of ECT persist, increasing evidence supports its effectiveness in helping people overcome severe depression that fails to respond to psychotherapy or antidepressant medication (Sackeim et al., 2001). However, ECT is usually considered a treatment of last resort, after less intrusive methods have been tried and failed.

Truth OR Fiction? REVISITED

Severely depressed people who have failed to respond to other treatments may show rapid improvement from electroconvulsive therapy.

TRUE. ECT is actually helpful in many cases of severe depression that do not respond to other forms of treatment.

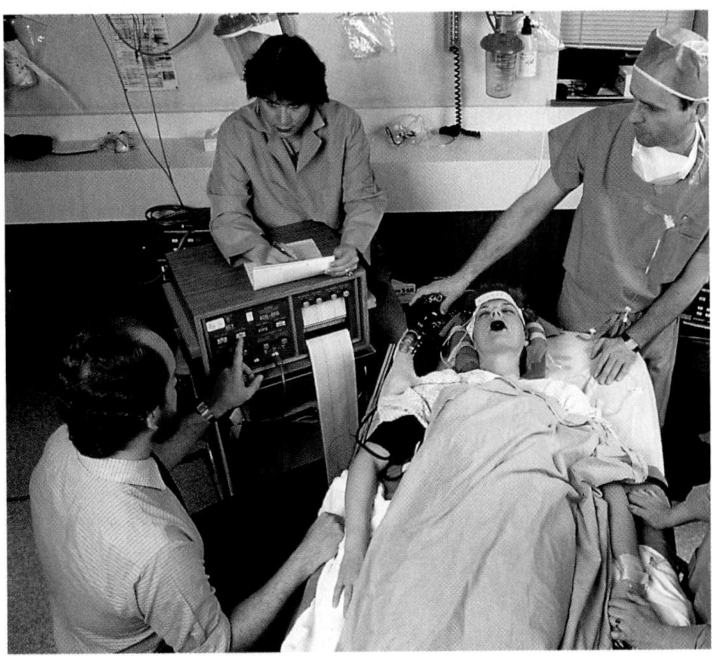

Electroconvulsive therapy (ECT). ECT is helpful in many cases of severe or prolonged depression that do not respond to other forms of treatment. Still, its use remains controversial.

Psychosurgery

Psychosurgery is yet more controversial than ECT and is rarely practiced today. Although no longer performed, the most widely used form of psychosurgery was the **prefrontal lobotomy,** in which the nerve pathways linking the thalamus to the prefrontal lobes of the brain are surgically severed. The operation was based on the theory that extremely disturbed patients suffer from overexcitation of emotional impulses that emanate from the lower brain centers, such as the thalamus and hypothalamus. It was believed that by severing the connections between the thalamus and the higher brain centers in the frontal lobe of the cerebral cortex, the patient's violent or aggressive tendencies could be controlled. The prefrontal lobotomy was developed by the Portuguese neurologist António Egas Moniz and was introduced to the United States in the 1930s. More than 1,000 mental patients received the operation by 1950. Although the operation did reduce violent and agitated behavior in many cases, it was not always successful. In a cruel ironic twist, a patient whom Moniz had treated later shot him, leaving him paralyzed from a bullet that lodged in his spine.

Many distressing side effects are associated with the prefrontal lobotomy, including hyperactivity, impaired learning ability and reduced creativity, distractibility, apathy, overeating, withdrawal, epileptic-type seizures, and even death. The occurrence of these side effects, combined with the introduction of the phenothiazines, led to the elimination of the operation.

More sophisticated psychosurgery techniques have been introduced in recent years. Generally speaking, they are limited to smaller parts of the brain and produce less damage than the prefrontal lobotomy. These operations have been performed to treat such problems as intractable aggression, depression, and psychotic behavior; chronic pain; some forms of epilepsy; and persistent obsessive-compulsive disorder (Baer et al., 1995; Irle et al., 1998; Sachdev & Hay, 1996). Follow-up studies of such procedures have shown marked improvement in about one-quarter to one-half of cases. But concerns about possible complications, including impaired intellectual functioning, have greatly reduced their use (Irle et al., 1998). Before leaving this topic, let us understand that psychosurgery should only be considered as a treatment of last resort.

Evaluation of Biological Approaches

There is little doubt that the use of psychotropic drugs has helped many people with severe psychological problems. Many thousands of people with schizophrenia who were formerly hospitalized are able to function more effectively in the community because of antipsychotic drugs. Antidepressant drugs have helped relieve depression in many cases and have shown therapeutic benefits in treating other disorders, such as panic disorder, obsessive-compulsive disorder, and eating disorders. ECT is helpful in relieving depression in many people who have been unresponsive to other treatments.

On the other hand, it may be that some forms of psychotherapy are as effective as drug therapy in treating anxiety disorders and depression (see Chapters 6 and 8). Moreover, problems persist with respect to the side effects of various psychotropic drugs. In addition, antianxiety agents, such as Valium, have often become drugs of abuse among people who become dependent on them for relieving the effects of stress rather than seeking more adaptive ways of solving their problems. Medical practitioners have often been too quick to use their prescription pads to help people with anxiety complaints, rather than to help them examine their lives or refer them for psychological treatment. Physicians often feel pressured, of course, by patients who seek a chemical solution to their life problems.

While we continue to learn more about the biological foundations of abnormal behavior patterns, the interface between biology and behavior can be construed as a two-way street. Researchers have uncovered links between psychological factors and many physical disorders and conditions (see Chapter 5). Researchers are also investigating whether the combination of psychological and drug treatments for such problems as depression, anxiety disorders, and substance abuse disorders, among others, may increase the therapeutic benefits of either of the two approaches alone.

Quiz 4.2
Biomedical Therapies

Hospitalization and Community-Based Care

People receive mental health services within various settings, including hospitals, outpatient clinics, community mental health centers, and private practices. In this section we explore the purposes of hospitalization and the movement toward community-based care. Due to

deinstitutionalization—the policy of shifting the burden of care from the state hospitals to community-based treatment settings—an exodus has taken place from state mental hospitals. We will see that deinstitutionalization has had a profound impact on the delivery of mental health services as well as on the larger community.

deinstitutionalization The policy of shifting care for patients with severe or chronic mental health problems from inpatient facilities to community-based facilities.

Roles for Hospitalization

Different types of hospitals provide different types of mental health treatment. State mental hospitals provide care to people with severe psychological problems. Municipal and community-based hospitals tend to focus on short-term care for people with serious psychological problems who need a structured hospital environment to help them through an acute crisis. In such cases, psychotropic drugs and other biological treatments, such as ECT for severe cases of depression, are often used in combination with short-term psychotherapy. Hospitalization may be followed by outpatient treatment. Many private care hospitals provide longer-term care or are specialized to help people withdraw safely from alcohol or drugs.

Most state hospitals today are better managed and provide more humane care than those of the 19th and early 20th centuries, but here and there deplorable conditions persist. Today's state hospital is generally more treatment oriented and focuses on preparing residents to return to community living. State hospitals today often function as part of an integrated, comprehensive approach to treatment. They provide the structured environment needed for people who are unable to function in a less restrictive community setting. When hospitalization restores patients to a higher level of functioning, the patients are reintegrated in the community and provided with follow-up care and transitional residences, if needed, to help them adjust to community living. Patients may be rehospitalized as needed in a state hospital if a community-based hospital is not available or if they require more extensive care than a community hospital can provide. For younger and less intensely disturbed people, the state hospital stay is typically briefer than it was in the past, lasting only until their condition allows them to reenter society. Older chronic patients may be unprepared to handle the most rudimentary tasks (shopping, cooking, cleaning, and so on) of independent life, however—in part because the state hospital may be the only home such patients have known as adults.

WWW **Web Link 4.4** Treatment Facilities

Modern psychiatric hospital. The modern psychiatric hospital is better managed and provides more humane care than those of earlier times, many of which housed patients under the most abysmal conditions. Here we see the bright and spacious dayroom in a modern psychiatric hospital.

halfway houses Supervised community residences that provide a bridge between institutional facilities and independent community living.

primary prevention Efforts designed to prevent problems from arising.

secondary prevention Efforts to ameliorate existing problems at an early stage.

The Community Mental Health Center

Community mental health centers (CMHCs) perform many functions in the effort to reduce the need for hospitalization of new patients and rehospitalization of formerly hospitalized patients. A primary function of the CMHC is to help discharged mental patients adjust to the community by providing continuing care and closely monitoring their progress. Unfortunately, not enough CMHCs have been established to serve the needs of the hundreds of thousands of ex-hospitalized patients and to try to prevent the need for hospitalization of new patients by providing intervention services and alternatives to full hospitalization, such as day hospital programs. Patients in day hospitals attend structured therapy and vocational rehabilitation programs in a hospital setting during the day but are returned to their families or homes at night. Many CMHCs also administer transitional treatment facilities in the community, such as **halfway houses,** which provide a sheltered living environment to help discharged mental patients gradually adjust to the community as well as to provide people in crisis with an alternative to hospitalization. CMHCs also serve in consultative roles to other professionals in the community, such as training police officers to handle disturbed people.

One of the major functions of the community mental health center is prevention. "An ounce of prevention is worth a pound of cure"—so goes the saying. Today we stockpile vital supplies such as grain and oil to lessen the effects of shortages. In medical science, the development of vaccines has helped protect people from contracting such diseases as smallpox and polio. In the mental health system, however, resources are generally directed toward treating mental health problems rather than attempting to prevent them from developing. A recent report by the prestigious Institute of Medicine (IOM) called for increased support for research on prevention and for development of programs to promote psychological well-being and reduce the risks of mental health disorders (Munoz, Mrazek, & Haggerty, 1996).

The Spectrum of Prevention Traditionally, the term *prevention* has been applied to interventions that run the gamut from programs designed to prevent the onset of mental disorders to those that attempt to reduce the impact of disorders once they develop (Kaplan, 2000). The IOM report, initiated by the U.S. Congress, limits the term *prevention* to interventions that occur before the onset of a diagnosable disorder (Munoz, Mrazek, & Haggerty, 1996). Interventions focusing on lessening the impact of already developed disorders are classified as *treatment interventions* rather than prevention.

The IOM report conceptualizes a mental health spectrum of interventions ranging from prevention efforts through treatment and maintenance interventions (see Figure 4.2). Three categories of prevention programs are identified: universal, selective, and indicated.

Universal preventive interventions are targeted toward the whole population or general public, such as programs designed to enhance prenatal health or childhood nutrition. *Selective preventive interventions* are targeted toward individuals or groups known to be at higher than average risk of developing mental disorders, such as children of schizophrenic parents. **Primary prevention** efforts—programs designed to prevent problems from arising in the first place, can be either universal or selective preventive interventions depending on whether they are focused on the general population or "at-risk" groups.

Indicated preventive interventions are directed toward individuals with early signs or symptoms that foreshadow the development of a mental disorder but don't yet meet diagnostic criteria for the particular disorder. This form of prevention, commonly called **secondary prevention,** attempts to nip in the bud developing problems. For example, secondary preventive programs aimed at changing the drinking habits of high-risk drinkers or early

An ounce of prevention. Obtaining good prenatal care can help prevent health problems in both the mother and the fetus. Mental health professionals face the challenge of developing programs to reduce the risks of psychological disorders.

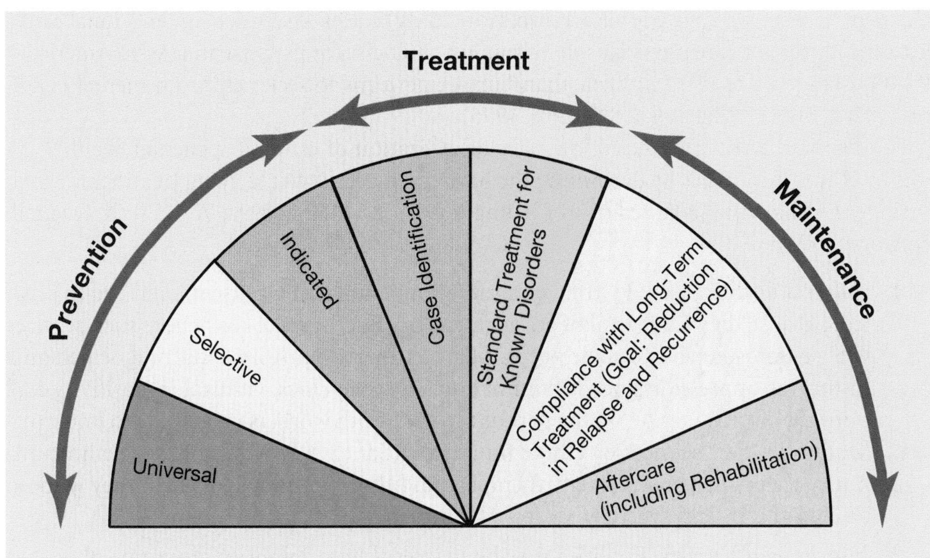

FIGURE 4.2 The mental health intervention spectrum for mental disorders.

Source. Mazrek, P. J., & Haggerty, R. J., *Reducing risks for mental disorders: Frontiers for preventive Intervention Research.* Copyright© 1994 by the National Academy of Sciences. Courtesy of the National Academy Press, Washington, D.C. Reprinted with permission.

problem drinkers may forestall the onset of more severe alcohol-related problems or alcohol dependence (Botelho & Richmond, 1996; Marlatt et al., 1998).

We have had some success in the health arena in developing effective prevention programs for reducing the risks of teenage pregnancies, sexually transmitted diseases, and some forms of drug abuse (e.g., Blackman, 1996; Stover et al., 1996). Psychologist Martin Seligman and his colleagues have shown that teaching cognitive skills involved in disputing catastrophic, negative thoughts reduced the risk of depression in both college students and school-age children (Jaycox et al., 1994; Seligman, 1998). Still, we have much to learn about developing effective prevention programs to prevent psychological disorders. The development of effective preventive programs rests in large part on expanding our knowledge base about the underlying causes of these disorders and mounting controlled investigations that examine ways of preventing them (Munoz, Mrazek, & Haggerty, 1996).

The challenge of preventing psychological disorders is before us. The question is whether the nation can muster the political will and financial resolve to meet the challenge.

THINK ABOUT IT

What are the roles of the community mental health center and contemporary mental hospital?

Ethnic Group Differences in Use of Mental Health Services

A 2001 report by the U.S. Surgeon General concluded that members of racial and ethnic minority groups typically have less access to mental health care and receive lower quality care than do other Americans (Goode, 2001f; USDHHS, 2001; see Table 4.4). A major reason for this disparity is that a disproportionate number of minority group members remain uninsured or underinsured, leaving many of them unable to afford mental health care. Consequently, minorities shoulder a greater burden of mental health problems that go undiagnosed and untreated (Stenson, 2001a).

Cultural factors are yet another reason for underutilization of mental health services by minority groups. Mental health clinics are not typically the first places where African Americans go for help for emotional problems. They are more likely to turn first to the church and second to the emergency room of the local general hospital (Lewis-Hall, 1992). The National Survey of African Americans found that slightly more than half (54%) of those who reported experiencing feelings of a "nervous breakdown" failed to consult any type of professional for help with their problems (Neighbors, 1992). Another study found that only about 1 in 10 African Americans in a community-based sample who experienced major depression consulted a mental health professional (D. R. Brown, Ahmed, Gary, & Milburn, 1995).

Hispanic Americans who encounter emotional problems are more likely to seek assistance from friends and relatives, or from spiritualists, than to reach out to mental health facilities, which they perceive as cold and impersonal institutions (De La Cancela &

Guzman, 1991). Hispanic people are also more likely to seek assistance for emotional problems from primary care physicians than from psychologists or psychiatrists. Asian American/Pacific Islanders are also less likely than Euro-Americans to seek help from mental health providers (Breaux, Matsuoka, & Ryujin, 1995).

We may better understand low rates of utilization of outpatient mental health services by ethnic minorities by examining the barriers that exist to receiving treatment, which include the following (adapted from Cheung, 1991; USDHHS, 1999a; Woodward, Dwinell, & Arons, 1992):

1. *Cultural mistrust.* People from minority groups often fail to use mental health services because they are fearful of government-operated services or believe such services will be unresponsive to their needs. Mistrust may stem from a cultural or personal history of oppression and discrimination. In some cases, cultural insensitivity and outright discrimination on the part of mental health workers contribute to underutilization of their services by ethnic minorities (Sanchez & Mohl, 1992). If ethnic minority clients perceive majority therapists and the institutions in which they work to be cold and insensitive, they are less likely to place their trust in them.
2. *Institutional barriers.* Facilities may be inaccessible to minority group members because they are located at a considerable distance from their homes or because of lack of public transportation. Most facilities only operate during daytime work hours, which means they are inaccessible to minority group members who are unable to take time off. Moreover, minority group members feel staff members often make them feel stupid for not being familiar with clinic procedures and their requests for assistance often become tangled in red tape.
3. *Cultural barriers.* Many recent immigrants, especially those from Southeast Asian countries, have had little, if any, previous contact with mental health professionals. They may hold different conceptions of mental health problems or view mental health problems as less severe than physical problems. In some ethnic minority subcultures, the family is expected to take care of members who have psychological problems and may resist seeking outside assistance because of guilt engendered by the belief that seeking outside help would represent rejection of the family member and would embarrass the family. Other cultural barriers include cultural differences between typically lower socioeconomic strata minority group members and mostly White, middle-class staff members and incongruence between the cultural practices of minority group members and techniques used by mental health professionals. For example, Asian immigrants may find little value in talking about their problems or may be uncomfortable expressing their feelings to strangers. In many ethnic minority groups, personal and interpersonal problems are brought to trusted elders in the family or religious leaders, not to outside professionals.
4. *Language barriers.* Differences in language make it difficult for minority group members to describe their problems or obtain needed services. Many mental health services do not have staff members who can communicate in the languages used by ethnic minority residents in their communities.
5. *Economic and accessibility barriers.* As mentioned above, financial barriers are often a major determination of underutilization of mental health services by ethnic minorities, many of whom live in economically distressed areas with limited resources. Moreover, many minority group members live in rural or isolated areas where mental health services may be lacking or inaccessible (USDHHS, 2001).

Cheung (1991) concludes that greater utilization of mental health services will depend to a great extent on the ability of the mental health system to develop programs that consider these cultural factors and build staffs that consist of culturally sensitive providers, including minority mental health professionals and paraprofessionals. Cultural mistrust of the mental health system among minority group members may be grounded in the perception that many mental health professionals are racially biased in how they evaluate and

TABLE 4.4 *Culture, Race, Ethnicity, and Mental Health:* Major Findings of the Surgeon General's Report

- The percentage of African Americans receiving needed care for mental health problems is only half that of non-Hispanic Whites. African Americans have less access to mental health care than Whites, partly because a greater percentage of African Americans lack health insurance.
- Of all American ethnic groups, Hispanic Americans are the least likely to have health insurance. Moreover, the limited availability of Spanish-speaking mental health professionals means that many Hispanic Americans who speak little or no English lack the opportunity to receive care from linguistically similar treatment providers.
- Largely because of lingering stigma and shame associated with mental illness, Asian Americans/Pacific Islanders often fail to seek care until their problems are more advanced than is the case with other groups. Moreover, accessibility is limited by scarcity of treatment providers with appropriate language skills.
- American Indians/Alaska Natives have a suicide rate that is 50% higher than the national average, but little is known about how many people within these groups receive needed care. In addition, the rural, isolated locations in which many Native Americans live places a severe constraint on the availability of mental health services.

Source. Adapted from Stenson, 2001a; USDHHS, 2001.

Web Link 4.5
Ending Discrimination in Health Insurance WWW

treat members of minority groups. Let's take a closer look at whether the evidence bears out this perception.

Racial Stereotyping and the Mental Health System

If you are African American, you are more likely to be admitted to a mental hospital, and more likely to be involuntarily committed, than if you are White (Lindsey & Paul, 1989). You are also more likely to be diagnosed with schizophrenia (Coleman & Baker, 1994; USDHHS, 1999a). The question is, why?

Relationships between ethnicity and diagnostic and admission practices are complex. They depend in part on differences in rates of mental disorders among different ethnic groups. If the rate of a given disorder is higher in a particular group, then it stands to reason that more members of the group will be diagnosed with the disorder. We know that African Americans as a group are no more likely to develop schizophrenia, a severe psychological disorder that often leads to hospitalization, than are Euro-Americans of the same socioeconomic level (USDHHS, 1999a). However, we also know that African Americans are overrepresented among lower socioeconomic groups in our society, and people in the lower strata on the socioeconomic ladder are more likely to have severe psychological disorders, such as schizophrenia. Thus, differences in socioeconomic backgrounds offer at least a partial explanation of ethnic/racial differences in diagnostic practices and rates of psychiatric hospitalization.

Ethnic stereotyping by mental health professionals may also contribute to an overdiagnosis of severe psychological problems requiring hospitalization. Evidence of clinician bias comes from research showing that whereas African Americans and Hispanic Americans are more likely than Euro-Americans to be diagnosed with schizophrenia, independent evidence fails to justify such differences (Garb, 1997; Lawson, 1994). African Americans are also more likely than Euro-Americans to receive psychiatric medication, including antipsychotic medication (Segal, Bola, & Watson, 1996). Investigators believe that clinician biases rather than clinical criteria may account for differences in prescription patterns (Frackiewicz et al., 1999).

How might clinician biases come to affect their clinical judgments? As the recent Surgeon General's report on mental health points out, diagnostic and treatment decisions in mental health settings are heavily weighted on behavioral signs and patient reporting of symptoms rather than more objective laboratory tests (USDHHS, 1999a). Consequently, clinician judgment plays an important role in determining whether or not someone receives a schizophrenia diagnosis and is deemed to be in need of hospitalization or antipsychotic medication.

Evaluation of Deinstitutionalization

Let us return to the issue of deinstitutionalization. Has this policy achieved its goal of successfully reintegrating mental patients into society, or does it remain a promise that is largely unfulfilled? Deinstitutionalization has often been criticized for failing to live up to its expectations. The criticism seems to be well founded. Among the most frequent criticisms is the charge that many hospital patients were merely dumped into the community and not provided with the community-based services they needed to adjust to demands of community living. A 1998 national study found that fewer than half of patients with schizophrenia were receiving adequate care (Winerip, 1999).

Though the community mental health movement has had some successes, a great many patients with severe and persistent mental health problems fail to receive the range of mental health and social services they need to adjust to life in the community (Jacobs, Newman, & Burns, 2001). One of the major challenges facing the community mental system is the problem of psychiatric homelessness.

Deinstitutionalization and the Psychiatric Homeless Population The federal government estimates that nearly one-third of the homeless people in the United States

Psychiatric homeless population. Many homeless people have severe psychological problems but fall through the cracks of the mental health and social service systems.

suffer from severe psychological disorders (Center for Mental Health Services, 1994). Many ex-hospitalized mental patients were essentially dumped into local communities following discharge and left with little if any support. Lacking adequate support, they often face more dehumanizing conditions on the street, under deinstitutionalization, than they did in the hospital. Many compound their problems by turning to illegal street drugs such as crack. Also, some of the younger psychiatric homeless population might have been hospitalized in earlier times but are now, in the wake of deinstitutionalization, directed toward community support programs, when they are available. The lack of available housing and transitional care facilities and effective case management play important roles in accounting for homelessness among people with psychiatric problems. Some homeless people with severe psychiatric problems are repeatedly hospitalized for brief stays in community-based hospitals during acute episodes. They move back and forth between the hospital and the community as though caught in a revolving door. Frequently, they are released from the hospital with inadequate arrangements for housing and community care. Some are left essentially to fend for themselves. While many state hospitals closed their doors and others slashed the number of beds, the states never funded the support services in the community that were supposed to replace the need for long-term hospitalization (Winerip, 1999).

Problems of homelessness are especially compounded for children in homeless families. Not surprisingly, homeless children tend to have more behavior problems than housed children (Schteingart et al., 1995). The problem of psychiatric homelessness is not limited to urban areas, although it is on our city streets that the problem is most visible. The pattern in rural areas tends to be one of inconsistent housing and unstable living arrangements, rather than outright homelessness (Drake et al., 1991).

The mental health system alone does not have the resources to resolve the multifaceted problems faced by the psychiatric homeless population. Helping the psychiatric homeless escape from homelessness requires an integrated effort involving mental health and alcohol and drug abuse programs; access to decent, affordable housing; and provision of other social services (Dixon et al., 1997). It also requires more effective means of evaluating the mental health needs of homeless people and matching services to their specific needs (Jacobs, Newman, & Burns, 2001; Tolomiczenko, Sota, & Goering, 2000).

THINK ABOUT IT

What do you believe should be done about the problem of psychiatric homelessness?

Another difficulty in helping meet the challenge of psychiatric homelessness is that homeless people with severe psychological problems typically do not seek out mental health services. More intensive outreach and intervention efforts that focus on helping homeless people connect with the types of services they need are likely to produce the best outcomes (Rosenheck, 2000). All in all, the problems of the psychiatric homeless population remain complex, vexing problems for the mental health system and society at large.

Deinstitutionalization: A Promise as Yet Unfulfilled Although the net results of deinstitutionalization may not have yet lived up to expectations, a number of successful community-oriented programs are available. However, they remain underfunded and unable to reach many people needing ongoing community support. Deinstitutionalization has worked best for those who experience acute episodes of disturbed behavior, who are hospitalized briefly and then returned to their homes, families, and jobs (Shadish et al., 1989). If deinstitutionalization is to eventually succeed, patients must be provided with continuing care and afforded opportunities for decent housing, gainful employment, and training in social and vocational skills.

New, promising services exist to improve community-based care for people with chronic psychological disorders—for example, psychosocial rehabilitation centers, family psychoeducational groups, supported housing and work programs, and social skills training. Unfortunately, too few of these services exist to meet the needs of many patients who might benefit from them. The community mental health movement continues to need expanded community support and adequate financial resources if it is to succeed in fulfilling its original promise.

Quiz 4.3
Hospitalization and Community-Based Care

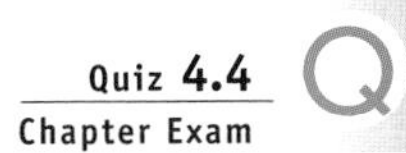

Quiz 4.4
Chapter Exam

Summing Up

Types of Mental Health Professionals

How do the three major groups of mental health professionals—clinical psychologists, psychiatrists, and psychiatric social workers—differ in their training backgrounds? Clinical psychologists complete graduate training in clinical psychology, typically at the doctoral level. Psychiatrists are medical doctors who specialize in psychiatry. Psychiatric social workers are trained in graduate schools of social work or social welfare, generally at the master's level.

Psychotherapy

What is psychotherapy? Psychotherapy involves a systematic interaction between a therapist and clients that incorporates psychological principles to help clients overcome abnormal behavior, solve problems in living, or develop as individuals.

What is psychodynamic therapy? Psychodynamic therapy originated with psychoanalysis, the approach to treatment developed by Freud. Psychoanalysts use techniques such as free association and dream analysis to help people gain insight into their unconscious conflicts and work through them in the light of their adult personalities. Contemporary psychodynamic therapy is typically briefer and more direct in its approach to exploring the patient's defenses and transference relationships.

What is behavior therapy? Behavior therapy applies the principles of learning to help people make adaptive behavioral changes. Behavior therapy techniques include systematic desensitization, gradual exposure, modeling, operant conditioning approaches, and social skills training. Cognitive-behavioral therapy integrates behavioral and cognitive approaches in treatment.

What is humanistic therapy? Humanistic therapy focuses on the client's subjective, conscious experience in the here and now. Rogers's person-centered therapy helps people increase their awareness and acceptance of inner feelings that had met with social condemnation and been disowned. The effective person-centered therapist possesses the qualities of unconditional positive regard, empathy, genuineness, and congruence.

What are two major approaches to cognitive therapy? Cognitive therapy focuses on modifying the maladaptive cognitions that are believed to underlie emotional problems and self-defeating behavior. Ellis's rational emotive behavior therapy focuses on disputing the irrational beliefs that occasion emotional distress and substituting adaptive beliefs and behavior. Beck's cognitive therapy focuses on helping clients identify, challenge, and replace distorted cognitions, such as tendencies to magnify negative events and minimize personal accomplishments.

What is cognitive-behavioral therapy? Cognitive-behavioral therapy is a broader form of behavior therapy that integrates cognitive and behavioral techniques in treatment.

What are the two major forms of eclectic therapy? These are technical eclecticism, a pragmatic approach that draws on techniques from different schools of therapy without necessarily subscribing to the theoretical positions represented by these schools, and integrative eclecticism, an approach that attempts to synthesize and integrate diverse theoretical approaches in an integrative model of therapy.

What are the general aims of group therapy, family therapy, and marital therapy? Group therapy provides opportunities for mutual support and shared learning experiences within a group setting to help individuals overcome psychological difficulties and develop more adaptive behaviors. Family therapists work with conflicted families to help them resolve their differences. Family therapists focus on clarifying family communications, resolving role conflicts, guarding against scapegoating individual members, and helping members develop greater autonomy. Marital therapists focus on helping couples improve their communications and resolve their differences.

Does psychotherapy work? Evidence from meta-analyses of psychotherapy outcome studies that compare psychotherapy with control groups supports the value of various approaches to psychotherapy. The question of whether there are differences in the effectiveness of different types of psychotherapy remains under study.

Biomedical Therapies

What are the major biomedical approaches to treating psychological disorders and how effective are they? The major biomedical therapies are drug therapy and electroconvulsive therapy (ECT). Antianxiety drugs, such as Valium, may relieve short-term anxiety but do not directly help people solve their problems or cope with stress. Antipsychotics help control flagrant psychotic symptoms, but regular use of these drugs are associated with the risk of serious side effects. Antidepressants can help relieve depression, and lithium is helpful in many cases in stabilizing mood swings in people with bipolar disorder. ECT often leads to dramatic relief from severe depression. Psychosurgery has all but disappeared as a form of treatment because of adverse consequences.

Hospitalization and Community-Based Care

What roles do mental hospitals and community mental health centers play in the mental health system? Mental hospitals provide structured treatment environments for people in acute crisis and for those who are unable to adapt to community living. Community mental health centers seek to prevent the need for psychiatric hospitalization by providing intervention services and alternatives to full hospitalization.

What factors account for underutilization of mental health services by racial or ethnic minorities in the United States? These include cultural factors regarding preferences for other forms of help, cultural mistrust of the mental health system, cultural barriers, linguistic barriers, and economic and accessibility barriers.

How successful is the policy of deinstitutionalization? Deinstitutionalization has greatly reduced the population of state mental hospitals, but it has not yet fulfilled its promise of providing the quality of care needed to restore discharged patients to a reasonable quality of life in the community. One example of the challenges yet to be met are the many homeless people with severe psychological problems who are not receiving adequate care in the community.

CHAPTER FIVE

Stress, chological Factors, and Health

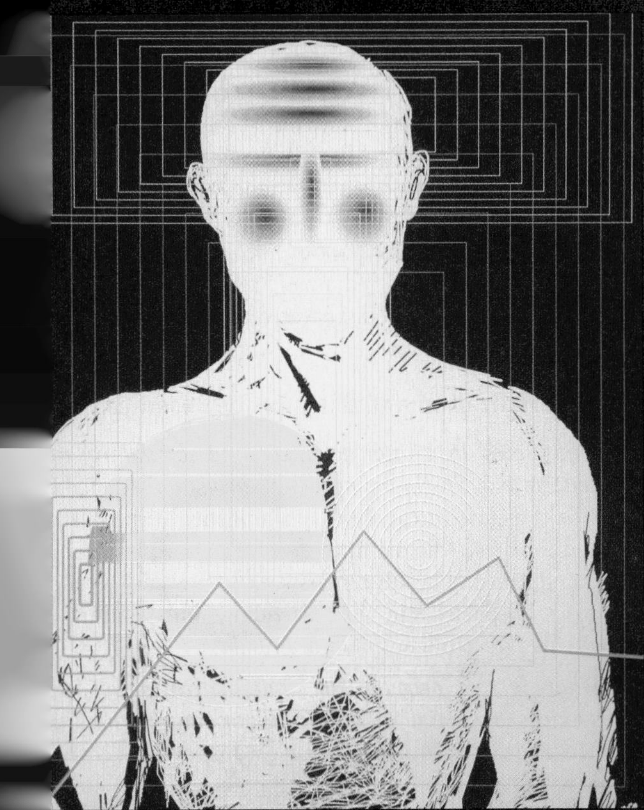

Diana Ong
Medicine Man B, 1940

Truth OR Fiction?

- If concentrating on your schoolwork has become difficult because of the breakup of a recent romance, you could be experiencing a psychological disorder. (p. 131)
- Stress actually makes you more resistant to developing the common cold. (p. 134)
- Writing about traumatic experiences may be good for one's health. (p. 135)
- Immigrants who become acculturated to their host culture are better off psychologically than those who maintain an identity with their original culture. (p. 139)
- Optimistic people recover more rapidly than pessimistic people from coronary artery bypass surgery. (p. 142)
- People can relieve the pain of migraine headaches by raising the temperature in a finger. (p. 145)
- Cancer patients who maintain a "fighting spirit" experience no better outcomes than those who resign themselves to their illness. (p. 153)

The relationship between the mind and the body is the subject of an age-old debate. Mental functioning certainly depends on the brain, but there is continuing temptation to regard them separately. The 17th-century French philosopher René Descartes (1596–1650) influenced modern thinking with his belief in *dualism*, or separateness, between the mind and body. Today, scientists and clinicians recognize that mind and body are more closely intertwined than would be suggested by a dualistic model—that psychological factors both influence and are influenced by physical functioning. In other words, mental health and physical health are inseparable (Kendler, 2001; USDHHS, 1999a). Psychologists who study the interrelationships between psychological factors and physical health are called **health psychologists** (Schneiderman et al., 2001).

We begin focusing on relationships between mind and body by examining the role of stress in both mental and physical functioning. The term *stress* refers to pressure or force placed on a body. In the physical world, tons of rocks that crash to the ground in a landslide cause stress on impact, forming indentations or craters when they land. In psychology, we use the term **stress** to refer to a pressure or demand that is placed on an organism to adapt or adjust. A **stressor** is a source of stress. Stressors (or stresses) include psychological factors, such as examinations in school and problems in social relationships, and life changes, such as the death of a loved one, divorce, or a job termination. They also include daily hassles, such as traffic jams, and physical environmental factors, such as exposure to extreme temperatures or noise levels. The term *stress* should be distinguished from **distress,** which refers to a state of physical or mental pain or suffering. Some degree of stress is probably healthy for us; it helps keep us active and alert. But stress that is prolonged or intense can overtax our coping ability and lead to emotional distress, such as states of anxiety or depression, and physical complaints, such as fatigue and headaches.

Stress is implicated in a wide range of physical and psychological problems. We begin our study of the effects of stress by discussing a category of psychological disorders called *adjustment disorders*, which involve maladaptive reactions to stress. We then consider the role of stress and other psychological and sociocultural factors in physical disorders.

Chapter Outline

Adjustment Disorders

Adjustment disorders are the first psychological disorders we discuss in this book, and they are among the mildest. An **adjustment disorder** is a maladaptive reaction to an identified stressor that develops within a few months of the onset of the stressor. The maladaptive reaction is characterized by significant impairment in social, occupational, or academic functioning, or by states of emotional distress that exceed those normally induced by the stressor. For the diagnosis to apply, the stress-related reaction must not be sufficient to meet the diagnostic criteria for other clinical syndromes, such as anxiety disorders or mood disorders. The maladaptive reaction may be resolved if the stressor is removed or the individual learns to cope with it. If the maladaptive reaction lasts for more than 6 months after the stressor (or its consequences) have been removed, the diagnosis may be changed.

If your relationship with someone comes to an end (an identified stressor) and your grades are falling off because you are unable to keep your mind on schoolwork, you may fit the bill for an adjustment disorder. If Uncle Harry has been feeling down and pessimistic since his divorce from Aunt Jane, he too may be diagnosed with an adjustment disorder. So too might Cousin Billy if he has been cutting classes and spraying obscene words on the school walls or showing other signs of disturbed conduct. There are several subtypes of adjustment disorders that vary in terms of the type of maladaptive reaction (see Table 5.1).

The concept of "adjustment disorder" as a *mental disorder* highlights some of the difficulties in attempting to define what is normal and what is not. When something important goes wrong in life, we should feel bad about it. If there is a crisis in business, if we are victimized by a crime, if there is a flood or a devastating hurricane, it is understandable that we might become anxious or depressed. There might, in fact, be something more seriously wrong with us if we did not react in a "maladaptive" way, at least temporarily. However, if our emotional reaction exceeds an expectable response, or our ability to function is impaired

Truth OR Fiction? REVISITED

If you have trouble concentrating on your schoolwork because of the breakup of a recent romance, you could be experiencing a psychological disorder.

TRUE. If you have trouble concentrating on your schoolwork following the breakup of a romantic relationship, you may have a mild type of psychological disorder called an adjustment disorder.

health psychologist A psychologist who studies the relationships between psychological factors and physical illness.

stress A demand made on an organism to adapt or adjust.

stressor A source of stress.

distress A state of physical or emotional pain or suffering.

adjustment disorder A maladaptive reaction to an identified stressor, which is characterized by impaired functioning or signs of emotional distress that exceed what would normally be expected.

TABLE 5.1 Subtypes of Adjustment Disorders

Disorder	Chief Features
Adjustment Disorder with Depressed Mood	Sadness, crying, and feelings of hopelessness.
Adjustment Disorder with Anxiety	Worrying, nervousness, and jitters (or in children, separation fears from primary attachment figures).
Adjustment Disorder with Mixed Anxiety and Depressed Mood	A combination of anxiety and depression.
Adjustment Disorder with Disturbance of Conduct	Violation of the rights of others or violation of social norms appropriate for one's age. Sample behaviors include vandalism, truancy, fighting, reckless driving, and defaulting on legal obligations (e.g., stopping alimony payments).
Adjustment Disorder with Mixed Disturbance of Emotions and Conduct	Both emotional disturbance, such as depression or anxiety, and conduct disturbance (as described above).
Adjustment Disorder Unspecified	A residual category that applies to cases not classifiable in one of the other subtypes.

Source. Adapted from the *DSM-IV-TR* (APA, 2000).

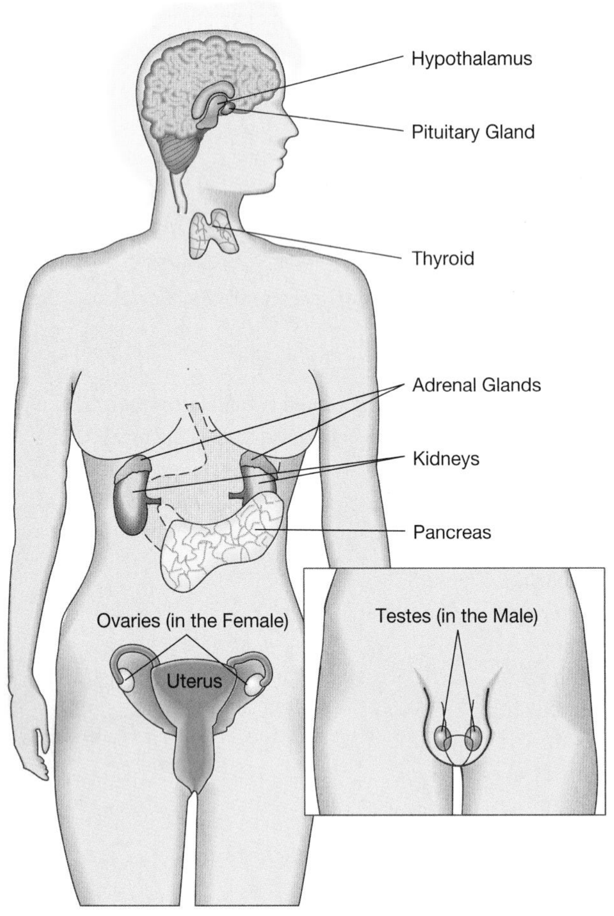

FIGURE 5.1 Major glands of the endocrine system. The glands of the endocrine pour their secretions—called hormones—directly into the bloodstream. Although hormones may travel throughout the body, they act only on specific receptor sites. Many hormones are implicated in stress reactions and various patterns of abnormal behavior.

(e.g., avoidance of social interactions, difficulty getting out of bed, or falling behind in our schoolwork), then a diagnosis of adjustment disorder may be indicated. Thus, if you are having trouble concentrating on your schoolwork following the breakup of a romantic relationship and your grades are slipping, you may have an adjustment disorder.

Stress and Illness

Psychological sources of stress not only diminish our capacity for adjustment, but also may adversely affect our health. Many visits to physicians, perhaps even most, can be traced to stress-related illness. Stress increases the risk of various types of physical illness, ranging from digestive disorders to heart disease (e.g., Cohen et al., 1993).

The field of **psychoneuroimmunology** studies relationships between psychological factors, especially stress, and the workings of the endocrine system, the immune system, and the nervous system (Kiecolt-Glaser & Glaser, 1992; Maier, Watkins, & Fleshner, 1994). Here we examine what we've learned about these relationships.

Stress and the Endocrine System

Stress has a domino effect on the **endocrine system,** the body's system of glands that release their secretions, called **hormones,** directly into the bloodstream. (Other glands, such as the salivary glands that produce saliva, release their secretions into a system of ducts.) The endocrine system consists of glands distributed throughout the body. Figure 5.1 shows the major endocrine glands in the body.

Several endocrine glands are involved in the body's response to stress. First, the hypothalamus, a small structure in the brain, releases a hormone that stimulates the nearby pituitary gland to secrete *adrenocorticotrophic hormone* (ACTH). ACTH, in turn, stimulates the adrenal glands, which are located above the kidneys. Under the influence of ACTH, the outer layer of the adrenal glands, called the *adrenal cortex*, releases a group of **steroids** (cortisol and cortisone are examples). These cortical steroids (also called *corticosteroids*) are hormones that have a number of different functions in the body. They boost resistance to stress; foster muscle development; and induce the liver to release sugar, which provides needed bursts of energy for responding to a threatening stressor (for example, a lurking predator or assailant) or an emergency situation. They also help the body defend against allergic reactions and inflammation.

The sympathetic branch of the autonomic nervous system, or ANS, stimulates the inner layer of the adrenal glands, called the *adrenal medulla*, to release a mixture of chemicals called **catecholamines**—epinephrine (adrenaline) and norepinephrine (noradrenaline). These chemicals function as hormones when released into the bloodstream. Norepinephrine is also produced in the nervous system and functions as a neurotransmitter. The mixture of epinephrine and norepinephrine mobilizes the body to deal with a threatening stressor by accelerating the heart rate and by also stimulating the liver to release stored glucose (sugar), making more energy available where it can be of use in protecting ourselves in a threatening situation.

The stress hormones produced by the adrenal glands help the body prepare to cope with an impending threat or stressor. Once the stressor has passed, the body returns to a normal state. During states of chronic stress, however, the body may continue to pump out stress hormones, which can have damaging effects throughout the body, including suppressing the ability of the immune system to protect us from various infections and disease ("Can Stress Make You Sick?" 1998).

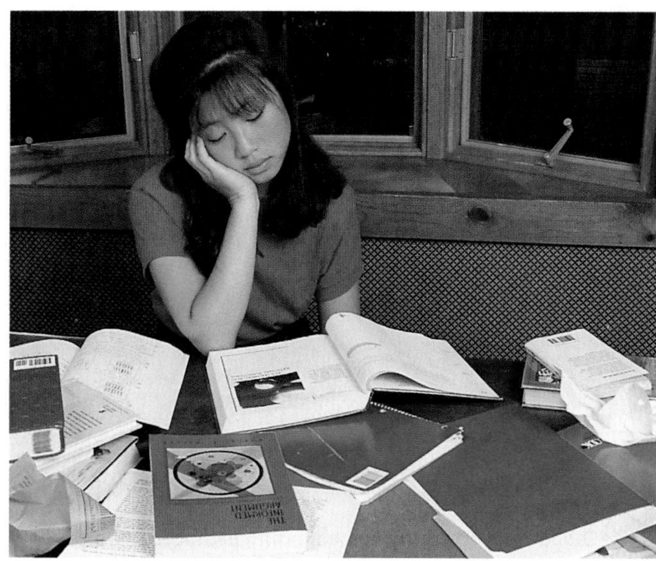

Difficulty in concentrating or adjustment disorder? An adjustment disorder is a maladaptive reaction to a stressor that may take the form of impaired functioning at school or at work, such as having difficulties keeping one's mind on one's studies.

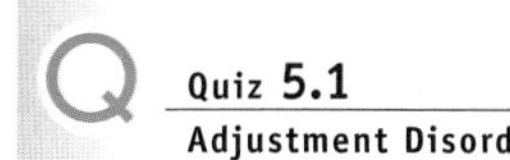

Quiz **5.1**
Adjustment Disorders

Stress and the Immune System

Given the intricacies of the human body and the rapid advance of scientific knowledge, we might consider ourselves dependent on highly trained medical specialists to contend with illness. Actually our bodies cope with most diseases on their own, through the functioning of the immune system.

The **immune system** is the body's system of defense against disease. It combats disease in a number of ways. Your body is constantly engaged in search-and-destroy missions against invading microbes, even as you're reading this page. Millions of white blood cells, or **leukocytes,** are the immune system's foot soldiers in this microscopic warfare. Leukocytes systematically envelop and kill **pathogens** like bacteria, viruses, and fungi; worn-out body cells; and cells that have become cancerous.

Leukocytes recognize invading pathogens by their surface fragments, called **antigens,** literally *anti*body *gen*erators. Some leukocytes produce **antibodies,** specialized proteins that attach to these foreign bodies, inactivate them, and mark them for destruction.

Special "memory lymphocytes" (lymphocytes are a type of leukocyte) are held in reserve rather than marking foreign bodies for destruction or going to war against them. They can remain in the bloodstream for years and form the basis for a quick immune response to an invader the second time around.

Evidence is accumulating that stress can make us more vulnerable to disease by weakening the immune system (Adler, 1999; Dougall & Baum, 2001; Sternberg, 2000). A weakened immune system can make us more vulnerable to common illnesses, such as colds and the flu, and may increase our risks of developing chronic diseases, including cancer.

Exposure to physical sources of stress such as cold or loud noise, especially when intense or prolonged, can dampen immunological functioning. So too can various psychological stressors ranging from sleep deprivation to final examinations (Maier, Watkins, & Fleshner, 1994). Medical students, for example, show poorer immune functioning during

psychoneuroimmunology The study of relationships between psychological factors and immunological functioning.

endocrine system The system of ductless glands that secrete hormones directly into the bloodstream.

hormones Substances secreted by endocrine glands that regulate body functions and promote growth and development.

steroids A group of hormones that includes testosterone, estrogen, progesterone, and corticosteroids.

catecholamines A group of substances that includes neurotransmitters (dopamine and norepinephrine) and hormones (epinephrine and norepinephrine).

immune system The body's system of defense against disease.

leukocytes White blood cells.

pathogens Disease-causing organisms.

antigens Substances that trigger an immune response.

antibodies Substances produced by white blood cells that identify and target antigens for destruction.

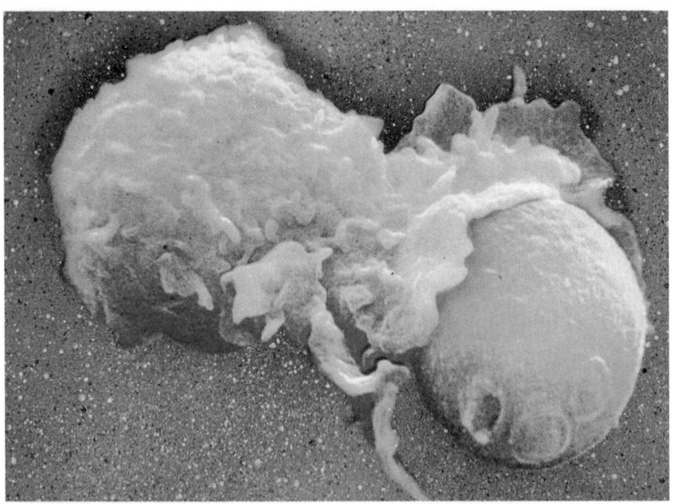

White blood cells attacking and engulfing pathogens. White blood cells, or *leukocytes,* form part of the body's immune system.

exam time than they do a month before exams, when their lives are less stressful (Glaser et al., 1987). Traumatic stress, such as exposure to earthquakes, hurricanes, or other natural or technological disasters, or to violence, can also dampen immunological functioning (Ironson et al., 1997; Solomon et al., 1997). Life stressors such as divorce and chronic unemployment can also take a toll on the immune system (O'Leary, 1990). Chronic stress may also make it take longer for wounds to heal (Kiecolt-Glaser et al., 1995).

Social support appears to moderate the harmful effects of stress on the immune system. For example, investigators find that medical and dental students with large numbers of friends show better immune functioning than students with fewer friends (Jemmott et al., 1983; Kiecolt-Glaser et al., 1984). Consider too that lonely students show a greater suppression of the immune response than do students with greater social support (Glaser et al., 1985). Newly separated and divorced people also show evidence of suppressed immune response, especially those who remain more attached to their ex-partners (Kiecolt-Glaser et al., 1987b, 1988).

Exposure to stress is linked to an increased risk of developing a common cold. In one study, people who reported higher levels of daily stress, such as pressures at work, showed lower levels in their blood streams of antibodies that fend off cold viruses (Stone et al., 1994). In another study, exposure to severe chronic stress lasting a month or longer of the type linked to underemployment, unemployment, or interpersonal problems with family members or friends was associated with a greater risk of developing a common cold after exposure to cold viruses (Cohen et al., 1998). Yet social support may boost resistance to the common cold. Researchers found that people who have more varied types of social relationships—with spouses, children, other relatives, friends, colleagues, members of organizations and religious groups, and so on—were less likely than others to come down with a cold after exposure to cold viruses (Cohen et al., 1997; Gilbert, 1997b). And when they did get sick, they tended to develop milder symptoms.

Truth OR Fiction? REVISITED

Stress actually makes you more resistant to developing the common cold.

FALSE. Stress increases the risk of developing a cold.

We should caution that much of the research in the field of psychoneuroimmunology is correlational in nature. Researchers examine immunological functioning in relation to different indices of stress, but do not (nor would they!) directly manipulate stress to observe its effect on subjects' immune systems or general health. Correlational research helps us better understand relationships between variables and may point to possible underlying causal factors, but does not in itself demonstrate causal connections.

Evidence indicates that writing about stressful events may enhance both psychological and physical well-being and perhaps even boost immune system responses (Carpenter, 2001b; Esterling et al., 1999; Smyth & Pennebaker, 2001). Writing about stressful or traumatic events has even reduced symptoms in asthma and arthritis patients (Smyth et al., 1999; Stone et al., 2000). Keeping thoughts and feelings about traumatic events tightly under wraps may place a stressful burden on the autonomic nervous system, which in turn may weaken the immune system, increasing susceptibility to certain stress-related disorders (Petrie, Booth, & Pennebaker, 1998). We should caution, however, that more research is needed before we can reach any definite conclusions about the effects of writing or other psychological interventions on the workings of the immune system (Miller & Cohen, 2001).

In the face of disaster. Exposure to traumatic stress, such as the World Trade Center disaster, can impair immunological functioning, increasing the risk of physical health problems.

The General Adaptation Syndrome

Stress researcher Hans Selye (1976) coined the term **general adaptation syndrome (GAS)** to describe a common biological response pattern to prolonged or excessive stress. Selye pointed out that our bodies respond similarly to many kinds of unpleasant stressors, whether the source of stress is an invasion of microscopic disease organisms, a divorce, or the aftermath of a flood. The GAS model suggests that our bodies, under stress, are like clocks with alarm systems that do not shut off until their energy is perilously depleted.

The GAS consists of three stages: the alarm reaction, the resistance stage, and the exhaustion stage. Perception of an immediate stressor (for example, a car that swerves in front of your own on the highway) triggers the **alarm reaction.** The alarm reaction mobilizes the body for defense. It is initiated by the brain and regulated by the endocrine system and the sympathetic branch of the autonomic nervous system (ANS). In 1929, Harvard University physiologist Walter Cannon termed this response pattern the **fight-or-flight reaction.** We noted earlier how the endocrine system responds to stress. During the alarm reaction, the adrenal glands, under control by the pituitary gland in the brain, pumps out cortical steroids and catecholamines that help mobilize the body's defenses (see Table 5.2).

The fight-or-flight reaction most probably helped our early ancestors cope with the many perils they faced. The reaction may have been provoked by the sight of a predator or by a rustling sound in the undergrowth. But our ancestors usually did not experience prolonged activation of the alarm reaction. Once a threat was eliminated, the body reinstates a lower level of arousal. Our ancestors fought off predators or they fled quickly; if not, they failed to contribute their genes to the genetic pools of their groups. In short, they died. Sensitive alarm reactions bestowed survival. Yet our ancestors did not invest years in the academic grind, struggle to balance the budget each month, or face any of the many daily stresses that repeatedly or persistently tax our body's ability to cope—everything from battling traffic in the morning to balancing school and work, or rushing from job to job. Consequently, much of the time our alarm system is turned on, which may eventually increase the likelihood of developing stress-related disorders.

When a stressor is persistent, we progress to the **resistance stage,** or adaptation stage, of the GAS. Endocrine and sympathetic system responses (release of stress hormones, for example) remain at high levels, but not quite as high as during the alarm reaction. During this stage the body tries to renew spent energy and repair damage. But when stressors continue to persist or new ones enter the picture, we may advance to the final or **exhaustion stage** of the GAS. Although there are individual differences in capacity to resist stress, all of us will eventually exhaust or deplete our bodily resources. The exhaustion stage is characterized by dominance of the parasympathetic branch of the ANS. Consequently, our heart and respiration rates decelerate. Do we benefit from the respite? Not necessarily. If the source of stress persists, we may

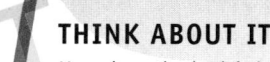

THINK ABOUT IT

How does the body's immune system help protect us from disease? What are the relationships between psychological factors, especially stress, and the functioning of the immune system?

Truth OR Fiction? REVISITED

Writing about traumatic experiences may be good for one's health.

TRUE. Talking or writing about your feelings may be good for the immune system and so bolster one's health.

TABLE 5.2 Stress-Related Changes in the Body Associated with the Alarm Reaction

Corticosteroids are released
Epinephrine and norepinephrine are released
Heart rate, respiration rate, and blood pressure increase
Muscles tense
Blood shifts from the internal organs to the skeletal muscles
Digestion is inhibited
Sugar is released by the liver
Blood-clotting ability is increased

Stress triggers the alarm reaction. The reaction is defined by secretion of corticosteroids, catecholamines, and activity of the sympathetic branch of the ANS. The alarm reaction mobilizes the body for combat or flight.

general adaptation syndrome (GAS) The body's three-stage response to states of prolonged or intense stress.

alarm reaction The first stage of the GAS, characterized by heightened sympathetic activity.

fight-or-flight reaction The inborn tendency to respond to a threat by either fighting or fleeing.

resistance stage The second stage of the GAS, involving the body's attempt to withstand prolonged stress and preserve resources.

exhaustion stage The third stage of the GAS, characterized by lowered resistance, increased parasympathetic activity, and possible physical deterioration.

develop what Selye termed "diseases of adaptation." These range from allergic reactions to heart disease—and, at times, even death. The lesson is clear: Chronic stress can damage our health, leaving us more vulnerable to a range of diseases and other physical health problems.

Cortical steroids are perhaps one reason that persistent stress may eventually lead to health problems. Although cortical steroids in some ways help the body cope with stress, persistent secretion of these steroids suppresses the activity of the immune system. Cortical steroids have negligible effects when they are only released periodically. Continuous secretion, however, weakens the immune system by disrupting the production of antibodies. As a result, we may become more vulnerable to various diseases, even the common cold (Cohen, Tyrrell, & Smith, 1991).

THINK ABOUT IT

What are the major changes in the body that occur during each of the phases of the general adaptation syndrome?

Although Selye's model speaks to the general response pattern of the body under stress, different biological processes may be involved in response to particular kinds of stressors. For example, persistent exposure to excessive noise may invoke different bodily processes than other sources of stress, such as overcrowding, or psychological sources of stress, such as divorce or separation.

Stress and Life Changes

Another way in which researchers have investigated the stress-illness connection is by quantifying life stress in terms of *life changes* (also called *life events*). Life changes become sources of stress when they impose demands on us to adjust. They include both positive events, such as getting married, and negative events, such as experiencing the death of a loved one. You can gain insight into the level of stressful life changes you may have experienced during the past year by completing the College Life Stress Inventory.

Investigators report links between exposure to life stressors, including life changes and daily hassles, and the risk of developing physical health problems, and even the risk of suffering sports injuries (Kanner et al., 1981; Smith, Smoll, & Ptacek, 1990; Stewart et al., 1994). Again, we need to be cautious in interpreting these findings. The reported links are correlational and not experimental. In other words, researchers did not (and would not!) assign subjects to conditions in which they were exposed to either a high or low level of life changes to see what effects these conditions might have on their health over time. Rather, existing data are based on observations of relationships, say, between life changes on the one hand and physical health problems on the other. Such relationships are open to other

Web Link 5.1 WWW
Stress: How and When to Get Help

For better or for worse. Life changes such as marriage and the death of loved ones are sources of stress that require adjustment. The death of a spouse may be one of the most stressful life changes that people ever face.

Questionnaire

Going Through Changes

How stressful has your life been lately? The College Life Stress Inventory contains a listing of stressful events that college students may face. Circle each of the events that you have experienced in the past year. Then compute your total, and look at the guide at the end of the chapter to interpreting your score.

Event	Stress Rating
Being raped	100
Finding out that you are HIV-positive	100
Being accused of rape	98
Death of a close friend	97
Death of a close family member	96
Contracting a sexually transmitted disease(other than AIDS)	94
Concerns about being pregnant	91
Finals week	90
Concerns about your partner being pregnant	90
Oversleeping for an exam	89
Flunking a class	89
Having a boyfriend or girlfriend cheat on you	85
Ending a steady dating relationship	85
Serious illness in a close friend or family member	85
Financial difficulties	84
Writing a major term paper	83
Being caught cheating on a test	83
Drunk driving	82
Sense of overload in school or work	82
Two exams in one day	80
Cheating on your boyfriend or girlfriend	77
Getting married	76
Negative consequences of drinking or drug use	75
Depression or crisis in your best friend	73

Event	Stress Rating
Difficulties with parents	73
Talking in front of a class	72
Lack of sleep	69
Change in housing situation (hassles, moves)	69
Competing or performing in public	69
Getting in a physical fight	66
Difficulties with a roommate	66
Job changes (applying, new job, work hassles)	65
Declaring a major or concerns about future plans	65
A class you hate	62
Drinking or use of drugs	61
Confrontations with professors	60
Starting a new semester	58
Going on a first date	57
Registration	55
Maintaining a steady dating relationship	55
Commuting to campus or work, or both	54
Peer pressures	53
Being away from home for the first time	53
Getting sick	52
Concerns about your appearance	52
Getting straight A's	51
A difficult class that you love	48
Making new friends; getting along with friends	47
Fraternity or sorority rush	47
Falling asleep in class	40
Attending an athletic event (e.g., football game)	20

Source. Renner, M. J., & Mackin, R. S. (1998). A life stress instrument for classroom use. *Teaching of Psychology, 25*, 46–48. Reprinted with permission.

interpretations. It could be that physical symptoms are sources of stress in themselves and lead to more life changes. Physical illness may cause disruptions of sleep or financial burdens, and so forth. Hence, in some cases at least, the causal direction may be reversed: Health problems may lead to life changes. Existing research does not allow us to tease out the possible cause-and-effect relationships (Suls, Wan, & Blanchard, 1994).

Although both positive and negative life changes can be stressful, positive life changes seem to be less disruptive than negative life changes (Thoits, 1983). In other words, marriage tends to be less stressful than divorce or separation. Or to put it another way, a change for the better may be a change, but it is less of a hassle. Let us also note that "eventlessness" (i.e.,

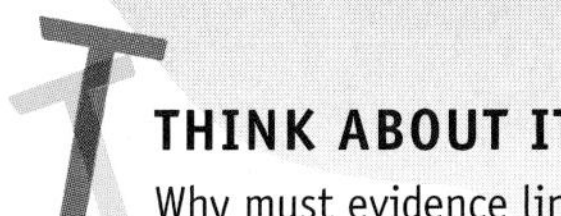

THINK ABOUT IT

Why must evidence linking life changes and stress be correlational rather than experimental?

Adapting to a new culture. The relationship between acculturation and mental health is complex and depends on such factors as financial status, economic opportunities, linguistic differences, and availability of strong family ties.

the absence of life changes) can also be stressful and may be as strongly linked to the risk of physical health problems as negative life events (Theorell, 1992).

Acculturative Stress: Making It in America

Should Hindu women who immigrate to the United States give up the sari in favor of California casuals? Should Soviet immigrants continue to teach their children Russian in the home? Should African American children be acquainted with the music and art of African peoples? Should women from traditional Islamic societies remove the veil and enter the competitive workplace? How do the stresses of acculturation affect the psychological well-being of immigrants and their families?

Sociocultural theorists have alerted us to the importance of accounting for social stressors in explaining abnormal behavior. One of the primary sources of stress imposed on immigrant groups, or on native groups living in the larger mainstream culture, is the need to adapt to a new culture. The term **acculturation** refers to the process of adaptation in which immigrants and native groups identify with the new culture through making behavioral and attitudinal changes (Rogler, Cortes, & Malgady, 1991).

Consider the challenges faced by Hispanic Americans. There are two general theories of the relationships between acculturation and adjustment (Griffith, 1983). One theory, dubbed the *melting pot theory*, holds that acculturation helps people adjust to living in the host culture. From this perspective, Hispanic Americans might adjust better by replacing Spanish with English and adopting the values and customs associated with mainstream American culture. A competing theory, the *bicultural theory*, holds that psychosocial adjustment is fostered by identification with both traditional and host cultures. That is, the ability to adapt to the ways of the new society, combined with a supportive cultural tradition and a sense of ethnic identity, may predict good adjustment. From a bicultural perspective, immigrants maintain their ethnic identity and traditional values while learning to adapt to the language and customs of the host culture.

We first must be able to measure acculturation if we are to investigate its relationship to mental health among immigrant and native groups. Measures of acculturation vary. In assessing acculturation among Hispanic Americans, for example, researchers assess variables such as the degree to which people favor English or Spanish in social situations, when reading, or while watching media such as TV; preferences for types of food and styles of clothing; and self-perceptions of ethnic identity. Using such measures, researchers find that the relationships between acculturation and adjustment are quite complex.

Let us summarize some of the principal findings concerning relationships between acculturation and psychological disorders in Hispanic Americans:

- Highly acculturated Hispanic American women in a large national survey were nine times more likely than relatively unacculturated women to be heavy drinkers (Caetano, 1987). In Latin American cultures, men tend to drink much more alcohol than women, largely because gender-based cultural prohibitions against drinking constrain alcohol use among women. These constraints appear to have loosened among Hispanic American women who adopt "mainstream" U.S. attitudes and values.
- Third-generation Mexican American male adolescents—who are more likely to be acculturated than first- or second-generation Mexican Americans—were at higher risk of delinquency (Buriel, Calzada, & Vasquez, 1982).

acculturation The process of adapting to a new culture.

- Acculturation is associated with an increased risk of smoking among Hispanic adolescents (Ribisl et al., 2000; Unger, Cruz, & Rohrbach, 2000).
- Studies show poorer mental health among U.S.-born Mexican Americans than among Mexican nationals and Mexican immigrants (Burnam et al., 1987; Escobar & Vega, 2000; Escobar, Hoyos Nervi, & Gara, 2000).
- More acculturated Hispanic Americans are more likely to experience a psychological disorder than their less acculturated counterparts (Ortega et al., 2000).
- Highly acculturated Hispanic American high school girls were more likely than their less acculturated counterparts to show test scores associated with anorexia (an eating disorder characterized by excessive weight loss and fears of becoming fat—see Chapter 11) on an eating attitudes questionnaire (Pumariega, 1986). Acculturation apparently made these girls more vulnerable to the demands of striving toward the contemporary American ideal of the (very!) slender woman.
- Despite associations of acculturation with mental health problems, researchers have found that Mexican Americans who are *less* proficient in English show *more* signs of depression and anxiety than those who are more proficient (Salgado de Snyder, 1987; Warheit et al., 1985).

Truth OR Fiction? REVISITED

Immigrants who become acculturated to their host culture are better off psychologically than those who maintain an identity with their original culture.

FALSE. Though the relationships between acculturation and mental health status are complex, evidence suggests that holding a bicultural identity is not associated with an increased risk of psychological problems.

In sum, evidence relating acculturation status to mental health outcomes is mixed. Inconsistencies in research results may partly reflect differences in the indices by which mental health is measured (problem drinking vs. depression, for example) and the complexities of acculturative processes. We need to recognize that relationships between mental health and psychological adjustment are complex. Among unacculturated groups, factors such as social stress resulting from financial hardship, limited opportunities, and linguistic differences may contribute to adjustment problems (e.g., Ryder, Alden, & Paulhus, 2000). On the other hand, the erosion of traditional family networks that may accompany acculturation might operate to increase the risk of psychological disorders in more acculturated groups (Ortega et al., 2000).

Consider a study of Mexican American elders (Zamanian et al., 1992). Those who were minimally acculturated showed higher levels of depression than did those who were either acculturated or bicultural. The bicultural and highly acculturated groups were similar in levels of depression. This evidence shows that low acculturation status was associated with a greater risk of depression. However, people who held a bicultural identity in which they maintained an identification with their original culture while adapting to the new culture experienced no greater vulnerability to depression.

Low acculturation status is often a marker for low socioeconomic status (SES). People who are minimally acculturated often face economic hardship. Financial difficulties add to the stress of adapting to the host culture, which can increase the risk of depression and other psychological problems. Yet SES isn't the only, or necessarily the most important, determinant of mental health in immigrant groups. In a northern California sample, researchers found better mental health profiles among Mexican immigrants than among people of Mexican descent born in the United States, despite the socioeconomic disadvantages faced by the immigrant group (Vega et al., 1998). Acculturation and "Americanization" may have damaging effects on the mental health of Mexican Americans, and the retention of cultural traditions may have a protective or "buffer" effect (Escobar, 1998).

Maintaining ethnic identity. Some studies point to psychological benefits in immigrant groups that adapt to the host culture while maintaining ethnic identity.

Other studies also point to the benefits of adapting to the host culture while maintaining ties to the

emotion-focused coping A coping style that attempts to minimize emotional responsiveness rather than deal with the stressor directly.

problem-focused coping A coping style that attempts to confront the stressor directly.

traditional culture. Among Asian Americans, establishing contacts with the majority culture while maintaining one's ethnic identity appears to generate less stress than withdrawal and separation from the host culture (Huang, 1994; Phinney, Lochner, & Murphy, 1990). Withdrawal fails to prepare the individual to make the adjustments necessary to function in a multicultural society, which often results in maladjustment. Maintaining one's ethnic identity also seems to hold a psychological benefit. Studies with Asian American adolescents show that those who have achieved an ethnic identity are better adjusted psychologically and have higher self-esteem (Huang, 1994; Phinney, 1989; Phinney & Alipuria, 1990).

Moreover, some outcomes need careful interpretation. For example, does the finding that highly acculturated Hispanic American women are more likely to drink heavily argue in favor of placing greater social constraints on women? The point would seem to be that a loosening of restraints is a double-edged sword, and that all people—male and female, Hispanic and non-Hispanic—may encounter adjustment problems when they gain new freedoms.

THINK ABOUT IT

Does the evidence presented in the text seem to argue for or against a melting-pot model of American culture? What evidence presented suggests that maintaining a strong ethnic identity may be beneficial?

Other research appears to bear out this point. There is a strong relationship in Hispanic immigrants between the stress of adapting to a new culture and environment and states of psychological distress. In one study, female immigrants showed higher levels of depression than male immigrants (Salgado de Snyder, Cervantes, & Padilla, 1990). Their depression may be linked to the greater level of stress women encountered in adjusting to changes in family and personal issues, such as the greater freedom of gender roles for men and women in U.S. society. Because they were reared in cultures in which men are expected to be breadwinners and women homemakers, immigrant women may encounter more family and internal conflict when they enter the workforce, regardless of whether their job entry results from economic necessity or personal choice. Given these factors, we shouldn't be surprised by recent findings that greater marital distress was reported by wives in more acculturated Mexican American couples (Negy & Snyder, 1997).

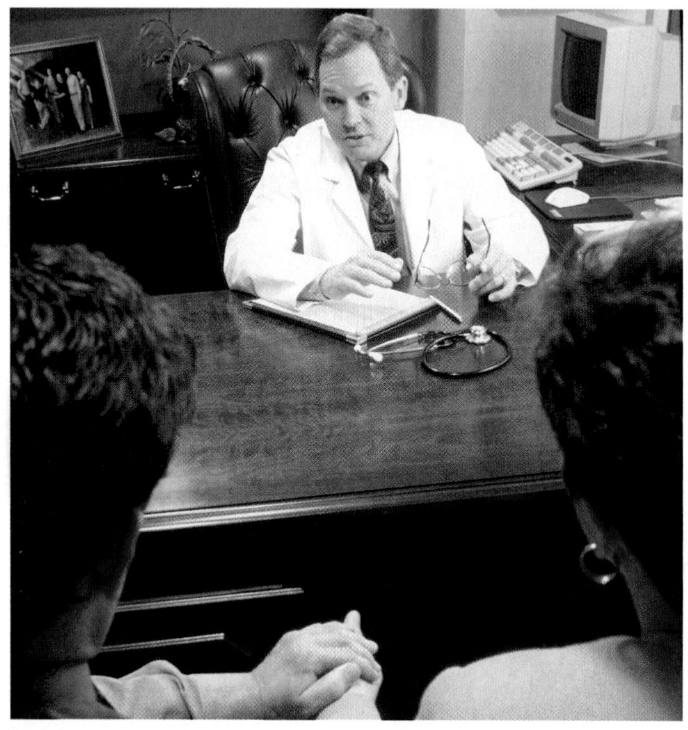

Problem-focused coping. Unlike emotion-focused coping, in which people attempt to distance themselves from sources of stress through denial or avoidance, problem-focused coping helps people meet their stressors head-on. When it comes to serious medical problems, problem-focused strategies such as seeking information and keeping a hopeful outlook may be adaptive and improve the chances of recovery.

Psychological Factors That Moderate Stress

Stress may be a fact of life, but the ways in which we handle stress help determine our ability to cope with it. Individuals react differently to stress depending on psychological factors such as the meaning they ascribe to stressful events. For example, whether a major life event, such as pregnancy, is a positive or negative stressor depends on a couple's desire for a child and their readiness to care for one. We can say the stress of pregnancy is moderated by the perceived value of children in a couple's eyes and their self-efficacy—their confidence in their ability to raise a child. As we see next, psychological factors such as coping styles, self-efficacy expectancies, psychological hardiness, optimism, social support, and ethnic identity may moderate or buffer the effects of stress.

Styles of Coping What do you do when faced with a serious problem? Do you pretend it does not exist? Like Scarlett O'Hara in *Gone With the Wind*, do you say to yourself "I'll think about it tomorrow" and then banish it from your mind? Or do you take charge and confront it squarely?

Pretending that problems do not exist is a form of denial. Denial is an example of **emotion-focused coping** (Lazarus & Folkman, 1984). In emotion-focused coping, people take measures that immediately reduce the impact of the stressor, such as denying its existence or withdrawing from the situation. Emotion-focused coping, however, does not eliminate the stressor (a serious illness, for example) or help the individual develop better ways of managing the stressor. In **problem-focused coping,** by contrast, people examine the stressors they face and do what they can to change them or modify their own reactions to render stressors less harmful. These basic styles of coping—emotion focused and problem focused—have been applied to ways in which people respond to illness.

Denial of illness can take various forms, including the following:

1. Failure to recognize the seriousness of the illness,
2. Minimization of the emotional distress the illness causes,
3. Misattribution of symptoms to other causes (for example, assuming the appearance of blood in the stool represents nothing more than a local abrasion), and
4. Ignoring threatening information about the illness.

Denial can be dangerous to your health, especially if it leads to avoidance of, or noncompliance with, needed medical treatment. Avoidance is another form of emotion-based coping. In one study, people who had a more avoidant style of coping with cancer (for example, by trying not to think or talk about it) showed greater disease progression when evaluated a year later than did people who more directly confronted the illness (Epping-Jordan, Compas, & Howell, 1994). Like denial, avoidance may deter people from complying with medical treatments, which can lead to a worsening of their medical conditions. It's also possible that avoidance may contribute to heightened emotional distress and arousal, which may impair immunological functioning.

Another form of emotion-focused coping, the use of wish-fulfillment fantasies, is also linked to poorer adjustment in coping with serious illness. Examples of wish-fulfillment fantasies include ruminating about what might have been had the illness not occurred and longing for better times. Wish-fulfillment fantasy offers the patient no means of coping with life's difficulties other than an imaginary escape.

Does this mean that people are invariably better off when they know all the facts concerning their illnesses? Not necessarily. Whether or not you will be better off knowing all the facts may depend on your preferred style of coping. A mismatch between the individual's style of coping and the amount of information provided may hamper recovery. In one study, cardiac patients with a repressive style of coping (relying on denial) who received information about their conditions showed a higher incidence of medical complications than repressors who were largely kept in the dark (Shaw et al., 1985). Sometimes ignorance helps people manage stress—at least temporarily.

Problem-focused coping involves strategies to deal directly with the source of stress, like seeking information about the illness through self-study and medical consultation. Information seeking may help the individual maintain a more optimistic frame of mind by creating an expectancy that the information will prove to be useful.

Self-Efficacy Expectancies Self-efficacy expectancies refer to our expectations regarding our abilities to cope with the challenges we face, to perform certain behaviors skillfully, and to produce positive changes in our lives (Bandura, 1982, 1986). We may be better able to manage stress, including the stress of coping with illness, if we feel confident (have higher self-efficacy expectancies) in our ability to cope effectively. A forthcoming exam may be more or less stressful depending on your confidence in your ability to achieve a good grade. Researchers find that spider-phobic women show high levels of the stress hormones epinephrine and norepinephrine when they interact with the phobic object, such as by allowing a spider to crawl on their laps (Bandura et al., 1985). As their confidence or self-efficacy expectancies for coping with these tasks increased, the levels of these stress hormones declined. Epinephrine and norepinephrine arouse the body by way of the sympathetic branch of the ANS. As a consequence, we are likely to feel shaky, to have "butterflies in the stomach" and general feelings of nervousness. Because high self-efficacy expectancies appear to be associated with lower secretion of catecholamines, people who believe they can cope with their problems may be less likely to feel nervous.

Psychological Hardiness **Psychological hardiness** refers to a cluster of traits that may help people manage stress. Research on the subject is largely indebted to Suzanne Kobasa (1979) and her colleagues who investigated business executives who resisted illness despite

psychological hardiness A cluster of stress-buffering traits characterized by commitment, challenge, and control.

internal locus of control Perception of one's ability to control reinforcements or affect outcomes.

heavy burdens of stress. Three key traits distinguished the psychologically hardy executives (Kobasa, Maddi, & Kahn, 1982, pp. 169–170):

1. The hardy executives were high in *commitment*. Rather than feeling alienated from their tasks and situations, they involved themselves fully. That is, they believed in what they were doing.
2. The hardy executives were high in *challenge*. They believed change was the normal state of things, not sterile sameness or stability for the sake of stability.
3. The hardy executives were also high in perceived *control* over their lives (Maddi & Kobasa, 1984). They believed and acted as though they were effectual rather than powerless in controlling the rewards and punishments of life. In terms suggested by social-cognitive theorist Julian Rotter (1966), psychologically hardy individuals have an **internal locus of control.**

Psychologically hardy people appear to cope more effectively with stress by using more active, problem-solving approaches (Williams, Wiebe, & Smith, 1992). They are also likely to report fewer physical symptoms and less depression in the face of stress than non-hardy people (Ouellette & DiPlacido, 2001; Pengilly & Thomas, 2000). Kobasa suggests that hardy people are better able to handle stress because they perceive themselves as *choosing* their stress-creating situations. They perceive the stressors they face as making life more interesting and challenging, not as simply burdening them with additional pressures. A sense of control is a key factor in psychological hardiness.

Optimism Research suggests that seeing the glass as half full is healthier than seeing it as half empty (Scheier & Carver, 1992). In one study on the relationships between optimism and health, Scheier and Carver (1985) administered a measure of optimism, the Life Orientation Test (LOT), to college students. The students also tracked their physical symptoms for 1 month. It turned out that those students who received higher optimism scores reported fewer symptoms such as fatigue, dizziness, muscle soreness, and blurry vision. (Subjects' symptoms at the beginning of the study were statistically taken into account, so it could not be argued that the study simply shows that healthier people are more optimistic.)

Other research also reveals links between optimism and better health outcomes. For example, pain patients who expressed more pessimistic thoughts during flare-ups of pain reported more severe pain and distress (Gil et al., 1990). The pessimistic thoughts included, "I can no longer do anything," "No one cares about my pain," and "It isn't fair I have to live this way." In a study of first-year law school students, optimism was associated with better mood and better immune system responses (Segerstrom et al., 1998). Among pregnant women, optimism is linked to a lower likelihood of postpartum depression (depression following childbirth) and higher infant birth weights (Carver & Gaines, 1987; Lobel et al., 2000). More optimistic women also suffered less depression and anxiety in the months following a diagnosis of breast cancer (Epping-Jordan et al., 1999). In separate studies, heart disease patients with more optimistic attitudes showed less depression when evaluated a year later (Shnek et al., 2001) and other patients undergoing coronary artery bypass procedure who had more optimistic attitudes about the procedure showed better outcomes (fewer complications requiring additional hospitalization or surgery) than did more pessimistic patients (Scheier et al., 1999).

Truth OR Fiction? REVISITED

Optimistic people recover more rapidly than pessimistic people from coronary artery bypass surgery.

TRUE. Investigators find that optimistic patients tend to recover more rapidly than pessimistic patients following coronary artery bypass surgery.

Research to date shows only correlational links between optimism and health. Perhaps we shall soon learn whether learning to alter attitudes—to learn to see the glass as half filled—plays a causal role in maintaining or restoring health. You can evaluate your own level of optimism by completing the nearby Life Orientation Test.

Social Support The role of social support as a buffer against stress is well documented (e.g., Wills & Filer Fegan, 2001). In one study, having a broader network of social contacts was associated with greater resistance to developing an infection following exposure to a common cold virus (Cohen et al., 1997). The investigators believe that having a wider range of social contacts may help protect the body's immune system by serving as a buffer against stress. Researchers in Sweden, as well as in the United States, find that people with a

Questionnaire

The Life Orientation Test

Do you see the glass as half full or half empty? Do you expect bad things to happen or do you find the silver lining in every cloud? The Life Orientation Test can afford you insight as to how optimistic or pessimistic you are.

Directions: Indicate whether or not each of the items represents your feelings by writing a number in the blank space according to the following code. Then turn to the scoring key at the end of the chapter.

4 = strongly agree
3 = agree
2 = neutral
1 = disagree
0 = strongly disagree

1. _____ In uncertain times, I usually expect the best.
2. _____ It's easy for me to relax.
3. _____ If something can go wrong for me, it will.
4. _____ I always look on the bright side of things.
5. _____ I'm always optimistic about my future.
6. _____ I enjoy my friends a lot.
7. _____ It's important for me to keep busy.
8. _____ I hardly ever expect things to go my way.
9. _____ Things never work out the way I want them to.
10. _____ I don't get upset too easily.
11. _____ I'm a believer in the idea that "every cloud has a silver lining."
12. _____ I rarely count on good things happening to me.

Source. Scheier, M. F., & Carver, C. S. (1985). Optimism, coping, and health: Assessment and implication of generalized outcome expectancies. *Health Psychology, 4,* 219–247. Reprinted by permission.

higher level of social support are likely to live longer (Goleman, 1993e). In the Swedish study, researchers followed middle-aged men who experienced a high level of emotional stress due to such factors as financial trouble or serious problems with a family member. Men who were highly stressed but lacked social support were three times more likely to die within a period of 7 years as were those whose lives were low in stress (Goleman, 1993e). Yet men with highly stressed lives who had ample amounts of emotional support in their lives showed no higher death rates. Having other people available may help people find alternative ways of coping with stressors or simply provide them with the emotional support they need during difficult times.

THINK ABOUT IT

Examine your own personality and behavior patterns. How might they be helping to promote your health? In what ways might these patterns damage your health or increase your risk of developing health-related problems? What changes can you make in your lifestyle to adopt healthier behaviors?

Ethnic Identity African Americans, on the average, stand a greater risk than Euro-Americans of developing chronic health problems, such as obesity, hypertension, heart disease, diabetes, and certain types of cancer (Angier, 2000b; Anderson, 1991). The particular stressors that African Americans often face, such as racism, poverty, violence, and overcrowded living conditions, may contribute to their heightened risks of serious health-related problems (Anderson, 1991). Yet African Americans often demonstrate a high degree of resilience in coping with stress (Cutrona et al., 2000). Among the factors that help buffer stress among African Americans are strong social networks of family and friends, beliefs in one's ability to handle stress (self-efficacy), coping skills, and ethnic identity. Ethnic identity appears to be more strongly related to psychological well-being among African Americans than it is among White Americans (Gray-Little & Hafdahl, 2000). Acquiring and maintaining pride

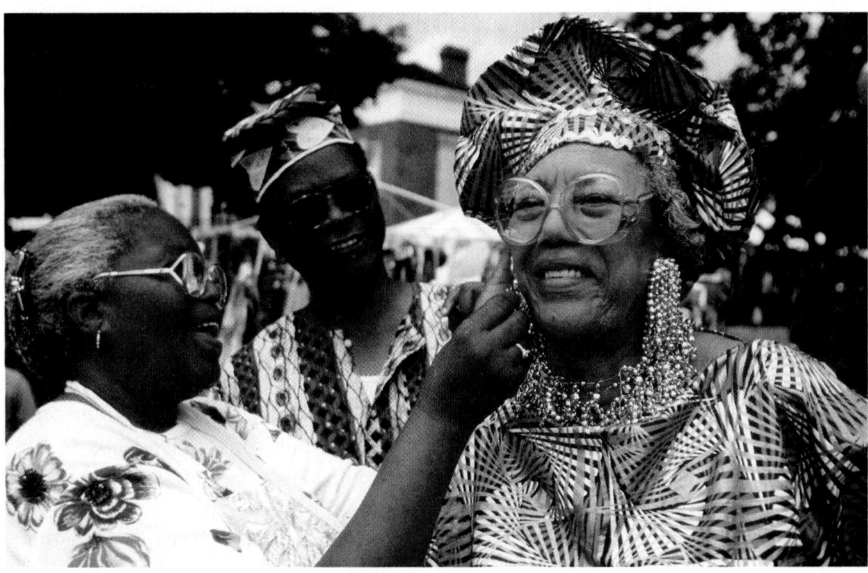

Ethnic pride as a moderator of the effects of stress. Pride in one's racial or ethnic identity may help the individual withstand the stress imposed by racism and intolerance.

in one's racial identity and cultural heritage may help African Americans and other ethnic minorities withstand stresses imposed by racism. Although more research is needed to elucidate the links among racial identity, self-esteem, and tolerance of stress, the available evidence suggests that African Americans who become alienated from their culture develop more negative self-images and stand a greater risk of developing not only physical and psychological disorders, but also academic underachievement and marital conflicts (Anderson, 1991).

Quiz 5.2
Stress and Illness

Psychological Factors and Physical Disorders

We noted at the start of the chapter that psychological factors can influence physical functioning; physical factors can also influence mental functioning. In these next sections we take a look at the role of psychological factors in various physical disorders. Physical disorders in which psychological factors are believed to play a causal or contributing role have traditionally been termed **psychosomatic** or *psychophysiological.* The term *psychosomatic* is derived from the Greek roots *psyche*, meaning "soul" or "intellect," and *soma*, which means "body." Disorders that involve psychological components range from asthma and headaches to heart disease.

Ulcers are another ailment traditionally identified as psychosomatic disorders. Ulcers affect about 1 in 10 people in the United States. However, their status as a psychosomatic disorder has been reevaluated in the light of recent landmark research that showed that a bacterium, *H. pylori*, not stress or diet, is the cause of the great majority of peptic ulcers (Boren et al., 1993; Mason, 1994). Researchers suspect that ulcers arise when the bacterium damages the protective lining of the stomach or intestines. Treatment with a regimen of antibiotics can help cure ulcers by attacking the bacterium directly (Altman, 1994a). We don't yet know why some people with the bacterium develop ulcers and others don't. The virulence of the particular strain of *H. pylori* may be involved in determining whether infected people develop peptic ulcers (Spechler, Fischbach, & Feldman, 2000). It is also conceivable that psychological stress is involved as well (Levenstein et al., 1999).

The field of psychosomatic medicine was developed to explore the possible health-related connections between the mind and the body. Today, evidence points to the importance of psychological factors in a much wider range of physical disorders than those traditionally identified as psychosomatic. In this section we discuss several of the traditionally identified psychosomatic disorders as well as two other diseases in which psychological factors may play a role in the course or treatment of the disease—cancer and AIDS.

Headaches

Headaches are symptomatic of many medical disorders. When they occur in the absence of other symptoms, however, they may be classified as stress related. By far the most frequent kind of headache is the tension headache (Mark, 1998). Stress can lead to persistent contractions of the muscles of the scalp, face, neck, and shoulders, giving rise to periodic or chronic tension headaches. Such headaches develop gradually and are generally characterized by dull, steady pain on both sides of the head and feelings of pressure or tightness. A survey in the Baltimore area showed that 38% of respondents complained of occasional tension headaches, with women reporting a 16% higher rate of these headaches than men (B. S. Schwartz et al., 1998).

Most other headaches, including the severe migraine headache, are believed to involve changes in the blood flow to the brain. Migraine headaches affect more than 28 million Americans ("Headache Coping," 2000; "New Research Could Open Doors," 2000). Typical migraines last for hours or days. They may occur as often as daily or as seldom as every other month. They are characterized by piercing or throbbing sensations on one side of the head only, or centered behind an eye. They can be so intense that they seem intolerable. Coping with the misery of brutal migraine attacks can take its toll, impairing the quality of life and leading to disturbances of sleep, mood, and thinking processes (Lipton et al., 2000b).

psychosomatic Pertaining to a physical disorder in which psychological factors play a causal or contributing role.

Migraine attacks typically last from 4 to 72 hours. There are two major types of migraines: migraine without aura (formerly called *common migraine*) and migraine with aura (formerly called *classic migraine*) (Olesen, 1994). An *aura* is a cluster of warning sensations that precedes the attack. Auras are typified by perceptual distortions, such as flashing lights, bizarre images, or blind spots. About 1 in 5 migraine sufferers experience auras. Other than the presence or absence of the aura, the two types of migraine are the same.

Theoretical Perspectives Why, under stress, do some people develop tension headaches? One possible answer is found in the principle of **individual response specificity,** which holds that people may respond to a stressor in idiosyncratic ways. In classic research, Malmo and Shagass (1949) induced pain in patients with muscular complaints (backaches) and in patients with hypertension (high blood pressure). The hypertensive patients responded to the stimulus with larger changes in the heart rate, whereas the backache group showed greater muscle contractions. Tension headache sufferers may thus be more likely to respond to stress by tensing the muscles of the forehead, shoulders, and neck.

The underlying causes of migraine headaches are poorly understood. Investigators suspect that imbalances of the brain chemical serotonin may be involved (Edelson, 1998). Falling levels of serotonin may cause blood vessels in the brain to contract (narrow) and then dilate (expand). This stretching stimulates nerve endings that give rise to the throbbing, piercing sensations associated with migraine.

Given a genetic predisposition to migraines, many factors may trigger an individual's migraine attacks. These include stress; stimuli such as bright lights; changes in barometric pressure; pollen; certain drugs; the chemical monosodium glutamate (MSG), which is often used to enhance the flavor of food; red wine; and even hunger (Martin & Seneviratne, 1997). Hormonal changes of the sort that affect women prior to and during menstruation can also trigger attacks, and the incidence of migraines among women is about twice that among men.

Treatment Commonly available pain relievers, such as aspirin, ibuprofen, and acetaminophen, may reduce or eliminate pain associated with tension headaches. A recent study reported that a combination of acetaminophen, aspirin, and caffeine (the ingredients in the over-the-counter pain reliever *Excedrin*) produced greater relief from the pain of migraine headaches than a placebo control (Lipton et al., 1998). Drugs that constrict dilated blood vessels in the brain or help regulate serotonin activity are used to treat the pain from migraine headache (Lipton et al., 2000a; Lohman, 2001; Silberstein et al., 2000).

Psychological treatment can also help relieve tension or migraine headache pain in many cases. These treatments include training in biofeedback, relaxation, coping skills training, and some forms of cognitive therapy (Blanchard & Diamond, 1996; Gatchel, 2001; Holroyd et al., 2001). **Biofeedback training (BFT)** helps people gain control over various bodily functions, such as muscle tension and brain waves, by giving them information (feedback) about these functions in the form of auditory signals (e.g., "bleeps") or visual displays. People learn to make the signal change in the desired direction. Training people to use relaxation skills combined with biofeedback has also been shown to be effective. *Electromyographic* (EMG) biofeedback is a form of BFT that involves relaying information about muscle tension in the forehead. EMG biofeedback thus heightens awareness of muscle tension in this region and provides cues that people can use to learn to reduce it.

Some people have relieved the pain of migraine headaches by raising the temperature in a finger. This biofeedback technique, called thermal BFT, modifies patterns of blood flow throughout the body, including blood flow to the brain, which helps to control migraine headaches (Blanchard et al., 1990; Gauthier, Ivers, & Carrier, 1996). One way of providing thermal feedback is by attaching a **thermistor** to a finger. A console "bleeps" more slowly[1] as the temperature rises. The temperature rises because more blood is flowing into the limb—away from the head. The client can imagine the finger growing warmer to bring about changes in the body's distribution of blood.

[1]Or more rapidly. The choice of direction is decided by the therapist or therapist and client.

Truth OR Fiction? REVISITED

People can relieve the pain of migraine headaches by raising the temperature in a finger.

TRUE. Some people have relieved migraine headaches by raising the temperature in a finger. This biofeedback technique modifies patterns of blood flow throughout the body.

individual response specificity The belief that people respond to the same stressor in different ways.

biofeedback training (BFT) A method of feeding back to the individual information about bodily functions so that the person can gain some degree of control over these functions.

thermistor A device for registering body temperature.

A Closer Look

Psychological Methods for Lowering Arousal

Stress induces bodily responses such as excessive levels of sympathetic nervous system arousal, which if persistent may impair our ability to function optimally and possibly increase the risk of stress-related illnesses. Psychological treatments have been shown to lower states of bodily arousal that may be prompted by stress. In this feature, we consider two widely used psychological methods of lowering arousal: meditation and progressive relaxation.

Meditation

Meditation comprises several ways of narrowing consciousness to moderate the stressors of the outer world. Yogis (adherents to Yoga philosophy) study the design on a vase or a mandala. The ancient Egyptians riveted their attention on an oil-burning lamp, which is the inspiration for the tale of Aladdin's lamp. In Turkey, Islamic mystics called whirling dervishes, fix on their motion and the cadences of their breathing.

There are many meditation methods, but they share the common thread of narrowing one's attention by focusing on repetitive stimuli. Through passive observation, the regular person–environment connection is transformed. Problem solving, worry, planning, and routine concerns are suspended, and consequently, levels of sympathetic arousal are reduced.

Many thousands of Americans regularly practice **transcendental meditation (TM),** a simplified kind of Indian meditation brought to the United States in 1959 by Maharishi Mahesh Yogi. Practitioners of TM repeat **mantras**—relaxing sounds like *ieng* and *om*.

Benson (1975) studied TM practitioners ages 17 to 41—students, businesspeople, artists. His subjects included relative novices and veterans of 9 years of practice. Benson found that TM yields a so-called relaxation response in many people. The relaxation response is typically characterized by a reduced heart rate and metabolic rate, and by reduced blood pressure in people with hypertension (Benson, Manzetta, & Rosner, 1973; Brody, 1996a; Gatchel, 2001). Meditators also produced more alpha waves, brain waves connected with relaxation. Critics of meditation do not hold that meditation is without value; they suggest, instead, that meditation may have no distinct effects when compared to a restful break from a stressful routine.

Meditation can also produce measurable health benefits. Evidence shows that it can lower blood pressure and actually reduce the amount of fatty deposits on artery walls, both of which are major risk factors for heart attacks and strokes (Ready, 2000).

Going with the flow. Meditation is a popular method of managing the stresses of the outside world by reducing states of bodily arousal. This young woman practices yoga, a form of meditation. She "goes with the flow," allowing the distractions of her environment to in a sense "pass through." Contrast her meditative state with the apparently stressful features of the young man sitting behind her.

Although there are differences among meditative techniques, the following suggestions illustrate some general guidelines:

1. Try meditation once or twice a day for 10 to 20 minutes at a time.
2. Keep in mind that when you're meditating, what you *don't* do is more important than what you do. So embrace a passive attitude: Tell yourself, "What happens, happens." In meditation, you take what you get. You don't *strive* for more. Striving of any kind hinders meditation.
3. Place yourself in a hushed, calming environment. For example, don't face a light directly.
4. Avoid eating for an hour before you meditate. Avoid caffeine (found in coffee, tea, many soft drinks, and chocolate) for at least 2 hours.
5. Get into a relaxed position. Modify it as needed. You can scratch or yawn if you feel the urge.
6. For a focusing device, you can concentrate on your breathing or sit in front of a serene object like a plant or incense. Benson suggests "perceiving" (not "mentally saying") the word *one* each time you breathe out. That is, think the word, but "less actively" than you

transcendental meditation (TM) A form of meditation that focuses on repeating a mantra to induce a meditative state.

mantra A word or phrase that is repeated to induce a state of relaxation and narrowing of consciousness.

Cardiovascular Disease

Cardiovascular disease (heart and artery disease) is the leading cause of death in the United States, claiming about 1 million lives annually and accounting for more than 4 in 10 deaths, most often as the result of heart attacks or strokes (NCHS, 1996b). *Coronary heart disease* (CHD) is the major form of cardiovascular disease, accounting for about 700,000

normally would. Other researchers suggest thinking the word *in* as you breathe in and *out*, or *ah-h-h*, as you breathe out. They also suggest mantras like *ah-nam, rah-mah*, and *shi-rim*.

7. When preparing for meditation, repeat your mantra aloud many times—if you're using a mantra. Enjoy it. Then say it progressively more softly. Close your eyes. Focus on the mantra. Allow thinking the mantra to become more and more "passive" so you "perceive" rather than think it. Again, embrace your "what happens, happens" attitude. Keep on focusing on the mantra. It may become softer or louder, or fade and then reappear.
8. If unsettling thoughts drift while you're meditating, allow them to "pass through." Don't worry about squelching them, or you may become tense.
9. Remember to take what comes. Meditation and relaxation cannot be forced. You cannot force the relaxing effects of meditation. Like sleep, you can only set the stage for it and then permit it to happen.
10. Let yourself drift. (You won't get lost.) What happens, happens.

Progressive Relaxation

Progressive relaxation was originated by University of Chicago physician Edmund Jacobson in 1938. Jacobson noticed that people tense their muscles under stress, intensifying their uneasiness. They tend to be unaware of these contractions, however. Jacobson reasoned that if muscle contractions contributed to tension, muscle relaxation might reduce tension. But clients who were asked to focus on relaxing muscles often had no idea what to do.

Jacobson's method of progressive relaxation teaches people how to monitor muscle tension and relaxation. With this method, people first tense, then relax, selected muscle groups in the arms; facial area; the chest, stomach, and lower back muscles; the hips, thighs, and calves; and so on. The sequence heightens awareness of muscle tension and helps people differentiate feelings of tension from relaxation. The method is progressive in that people progress from one group of muscles to another in practicing the technique. Since the 1930s, progressive relaxation has been used by a number of behavior therapists, including Joseph Wolpe and Arnold Lazarus (1966).

The following instructions from Wolpe and Lazarus (1966, pp. 177–178) illustrate how the technique is applied to relaxing the arms. Relaxation should be practiced in a favorable setting. Settle back on a recliner, a couch, or a bed with a pillow. Select a place and time when you're unlikely to be disturbed. Make the room warm and comfortable. Dim sources of light. Loosen tight clothing. Tighten muscles about two thirds as hard as you could if you were trying your hardest. If you sense that a muscle could have a spasm, you are tightening too much. After tensing, let go of tensions completely.

Relaxation of Arms (time: 4–5 minutes) *Settle back as comfortably as you can. Let yourself relax to the best of your ability . . . Now, as you relax like that, clench your right fist, just clench your fist tighter and tighter, and study the tension as you do so. Keep it clenched and feel the tension in your right fist, hand, forearm . . . and now relax. Let the fingers of your right hand become loose, and observe the contrast in your feelings . . . Now, let yourself go and try to become more relaxed all over . . . Once more, clench your right fist really tight . . . hold it, and notice the tension again . . . Now let go, relax; your fingers straighten out, and you notice the difference once more . . . Now repeat that with your left fist. Clench your left fist while the rest of your body relaxes; clench that fist tighter and feel the tension . . . and now relax. Again enjoy the contrast . . . Repeat that once more, clench the left fist, tight and tense . . . Now do the opposite of tension—relax and feel the difference. Continue relaxing like that for a while . . . Clench both fists tighter and together, both fists tense, forearms tense, study the sensations . . . and relax; straighten out your fingers and feel that relaxation. Continue relaxing your hands and forearms more and more . . . Now bend your elbows and tense your biceps, tense them harder and study the tension feelings . . . all right, straighten out your arms, let them relax and feel that difference again. Let the relaxation develop . . . Once more, tense your biceps; hold the tension and observe it carefully . . . Straighten the arms and relax; relax to the best of your ability . . . Each time, pay close attention to your feelings when you tense up and when you relax. Now straighten your arms, straighten them so that you feel most tension in the triceps muscles along the back of your arms; stretch your arms and feel that tension . . . And now relax. Get your arms back into a comfortable position. Let the relaxation proceed on its own. The arms should feel comfortably heavy as you allow them to relax . . . Straighten the arms once more so that you feel the tension in the triceps muscles; straighten them. Feel that tension . . . and relax. Now let's concentrate on pure relaxation in the arms without any tension. Get your arms comfortable and let them relax further and further. Continue relaxing your arms even further. Even when your arms seem fully relaxed, try to go that extra bit further; try to achieve deeper and deeper levels of relaxation.*

deaths annually, mostly from heart attacks. It may surprise you to learn that more women die from CHD than from breast cancer (Ansell, 2001).

About 10% of the population, some 22 million Americans, have CHD. In coronary heart disease, the flow of blood to the heart is insufficient to meet its needs. The underlying disease process in CHD is **arteriosclerosis,** or "hardening of the arteries," a condition in which artery walls become thicker, harder, and less elastic, which makes it

cardiovascular disease A disease or disorder of the cardiovascular system, such as coronary heart disease or hypertension.

arteriosclerosis A disease involving thickening and hardening of the arteries.

atherosclerosis The buildup of fatty deposits along artery walls that leads to the formation of artery-clogging plaque.

myocardial infarction A breakdown of heart tissue due to an obstruction in the blood vessels that supply blood to the heart.

stroke Blocking of a blood vessel that supplies the brain due to a blood clot.

Type A behavior pattern (TABP) A behavior pattern characterized by a sense of time urgency, competitiveness, and hostility.

Web Link 5.2 WWW
Test Your Heart Disease IQ

more difficult for blood to flow freely. The major underlying cause of arteriosclerosis is **atherosclerosis,** a process involving the build-up of fatty deposits along artery walls that leads to the formation of artery-clogging plaque. If a blood clot should form in an artery narrowed by plaque, it may nearly or completely block the flow of blood to the heart. The result is a heart attack (also called **myocardial infarction**), a life-threatening event in which heart tissue dies due to a lack of oxygen-rich blood. If a blood clot chokes off the supply of blood in an artery serving the brain, a **stroke** may occur, leading to death of brain tissue that can result in loss of function controlled by the part of the brain, coma, or even death.

Risk factors for CHD include some factors you can't control, such as age and family history. But a number of risk factors can be controlled through medical treatment or lifestyle changes—factors such as high cholesterol, hypertension, smoking, overeating, heavy drinking, consuming a high-fat diet, and leading a sedentary lifestyle (Fox, 2001; I-M. Lee et al., 2001; Pinel, Assanand, & Lehman, 2000; Noble, 2000; Stamler et al., 1999). Unfortunately, many of these factors remain uncontrolled. A recent study, for example, found that only about one in four adults with hypertension were taking medications to control their blood pressures (Chobanian, 2001; Hyman & Pavlik, 2001). Rates of uncontrolled hypertension were highest among older adults.

Psychological factors, such as negative emotional states like anger and anxiety, are also risk factors for cardiovascular disorders. Investigators have also identified a personality pattern, called the **Type A behavior pattern (TABP),** that poses yet another psychological risk factor in CHD.

Reducing Type A behavior. Slowing down the pace of your daily life and making time for loved ones are among the ways of reducing Type A behavior. Can you think of other ways that can help you decrease Type A behavior?

Type A Behavior Pattern The Type A behavior pattern, a style of behavior that characterizes people who are hard-driving, ambitious, impatient, and highly competitive, has been associated with a modestly higher risk of CHD (T. Q. Miller et al., 1991). Evidence indicates that psychological interventions focused on helping people reduce their Type A behavior can significantly reduce the risk of subsequent heart attacks in people who have already suffered one (Brody, 1996c; Friedman et al., 1986). Perhaps there is a lesson in this for us all.

Hostility—quickness to anger—is the element of the Type A behavior pattern most closely linked to cardiovascular risk (Donker, 2000; Matthews et al., 2000) (see "A Closer Look: Emotions and the Heart" feature, p. 150). People with TABP tend to have "short fuses" and are prone to get angry easily.

The questionnaire on page 152 "Are You Type A?" helps you assess whether or not you fit the Type A profile. If you would like to begin modifying Type A behavior, a good place to start is with lessening your sense of time urgency. Here are some suggestions (Friedman & Ulmer, 1984):

1. Increase social activity with family and friends.
2. Each day, spend a few minutes recalling distant events. Peruse photos of family and old friends.
3. Read books—biographies, literature, drama, politics, nature, science, science fiction. (Books on business and on climbing the corporate ladder are not recommended!)
4. Visit art galleries and museums. Consider works for their aesthetic value, not their prices.
5. Go to the movies, theater, concerts, ballet.
6. Write letters to family and old friends.
7. Take an art course; start violin or piano lessons.
8. Keep in mind that life is by nature unfinished. You needn't have all your projects finished by a certain date.
9. Ask family members what they did during the day. *Listen* to the answer.

Here are some additional suggestions for reducing anger and hostility, the elements believed to be the most toxic components of the Type A profile (Brody, 1996b; Friedman & Ulmer, 1984):

1. Don't get involved in discussions that you know lead to pointless arguments.
2. When others do things that disappoint you, consider situational factors that might explain their behavior. Don't jump to the conclusion that others intend to get you upset.
3. Focus on the beauty and pleasure in things.
4. Don't curse.
5. Express appreciation to people for their support and assistance.
6. Play for the fun of it, not to beat your opponent.
7. Check out your face in the mirror from time to time. Look for signs of anger and aggravation; ask yourself if you really need to look like that.
8. Don't sweat the small stuff. Let it go. Avoid grudges, and let bygones be bygones.

Social Environmental Stress Social environmental stress also appears to heighten the risk of CHD (Krantz et al., 1988). Such factors as overtime work, assembly-line labor, and exposure to conflicting demands are linked to increased risk of CHD (C. D. Jenkins, 1988). The stress-CHD connection is not straightforward, however. For example, the effects of demanding occupations may be moderated by factors such as psychological hardiness and whether or not people find their work meaningful (Krantz et al., 1988).

Other forms of stress are also linked to increased cardiovascular risk. Researchers in Sweden, for example, find that among women, marital stress triples the risk of recurrent cardiac events, including heart attacks and cardiac death (Foxhall, 2001b; Orth-Gomér et al., 2000).

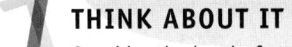

THINK ABOUT IT

Consider the level of stress in your own life. Might the stress you are encountering be affecting your psychological or physical health? In what ways? How might you reduce the level of stress in your life? How might you learn ways of coping more effectively with the stress you do encounter?

Ethnicity and CHD Coronary heart disease is not an equal opportunity destroyer. The disease burden and CHD-related death rates fall disproportionately on African Americans (see Figure 5.2).

Racial differences in deaths due to CHD reflect underlying differences in risk factors (Winkleby et al., 1998). Consider hypertension, a major risk factor for CHD. As shown in Figure 5.3 (p.151), non-Hispanic Black Americans are more likely than non-Hispanic White Americans to have hypertension (USDHHS, 1991b). In addition, African Americans have a higher prevalence of obesity and diabetes, two other major risk factors for CHD. Moreover, a dual standard of care limits access to quality health care of minority group members. Evidence shows that African Americans with CHD and who suffer heart attacks typically receive less aggressive and potentially life-saving treatments than do their White counterparts (Chen et al., 2001; Peterson et al., 1997; Stolberg, 2001). This dual standard of care may reflect discrimination as well as cultural factors limiting utilization of services, such as cultural mistrust of African Americans toward the medical establishment.

We finish this section with encouraging news. Americans have begun to take better care of their health. The incidence of CHD and deaths from heart disease have been declining steadily during the past 50 years, thanks largely to reductions in smoking, to improved treatment of heart patients, and perhaps also

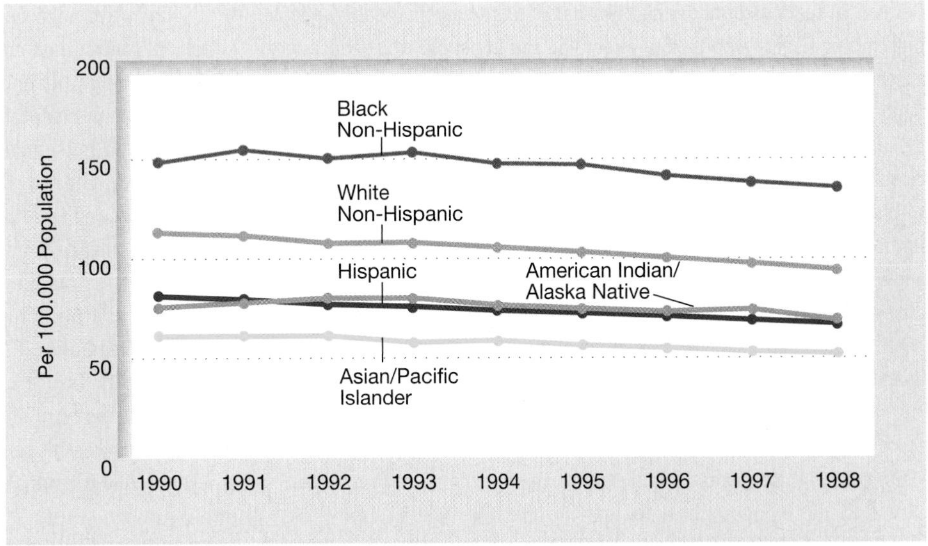

FIGURE 5.2 Coronary heart disease death rates in relation to race and ethnicity. Black (non-Hispanic) Americans are much more likely to die from CHD than other ethnic or racial groups in the United States. What factors might contribute to these differences?

Source. Centers for Disease Control/National Clearinghouse for Health Statistics, National Vital Statistics System.

A Closer Look

Emotions and the Heart

Might your emotions be putting you at risk of developing coronary heart disease? It appears so. Evidence shows that both anxiety and anger are hazardous to a person's cardiovascular health (Suinn, 2001).

The Anxious Heart

Investigators have linked phobic anxiety, the type of anxiety characterized by unfounded fears and panicky feelings, to a greater risk of death in men as the result of irregular heart rhythms. A study of some 34,000 men, none of whom were diagnosed at the outset of the study with coronary heart disease, showed those scoring at the high end of an index of phobic anxiety were six times more likely to suffer sudden coronary death over a 2-year period than were less anxious men (Hilchey, 1994; Kawachi et al., 1994). The researchers suspect that persistent, high levels of anxiety may produce "electrical storms" in the heart, resulting in irregular heart rhythms that may lead to sudden coronary death. Fortunately, the number of cardiac-related deaths during the 2-year study period was relatively small (only 16 among 34,000). Other investigators have also linked states of anxiety and tension with an increased risk of coronary symptoms and death in people with established CHD (Denollet et al., 1996; Gullette et al., 1997).

Researchers also find a connection between anxiety in middle-age men and the later risk of developing hypertension, a major risk factor for CHD (Markovitz et al., 1993). Highly anxious men were about twice as likely as their more relaxed counterparts to develop hypertension. We don't yet know whether this relationship also applies to women.

Anger and Hostility

Occasional feelings of anger may not damage the heart in healthy people, but chronic anger—the type of anger you see in people who seem angry all of the time—is linked to an increased risk of CHD and may even be as dangerous a risk factor as smoking, obesity, family history, or a high-fat diet (Brody, 1996c; Clay, 2001a; J. E. Williams et al., 2000). Anger is closely associated with hostility—an attitude characterized by tendencies to blame others and to perceive the world in negative terms (Eckhardt, Barbour, & Stuart, 1997). Hostile people are quick to anger and become angry more often and more intensely when they feel they have been mistreated than do nonhostile people. Young people with high levels of hostility stand an increased risk of developing early signs of coronary heart disease (Clay, 2001a; Matthews et al., 2000).

Although anger may not be a direct cause of heart disease, it is associated with an increased risk of death from cardiovascular disease (Suinn, 2001). Moreover, episodes of acute anger can actually trigger heart attacks and sudden cardiac death in people with established heart disease (Clay, 2001b).

Linking Emotions and the Heart

More research is needed to better understand the underlying mechanism linking negative emotions to heart disease, but investigators suspect that the stress hormones epinephrine and norepinephrine play significant roles (Januzzi & DeSanctis, 1999; Melani, 2001). Anxiety or anger triggers the adrenal glands to release these stress hormones, which then mobilize the body's resources to deal with threatening situations. They increase the heart rate, breathing rate, and blood pressure, which increases the flow of oxygen-rich blood to the muscles to prepare for defensive action—to fight or to flee—in the face of a threatening stressor. When people are persistently or repeatedly anxious or angry, the body may remain overaroused for long periods of time, continuing to pump out these stress hormones, which eventually may have damaging effects on the heart and blood vessels. Stress hormones also appear to increase the stickiness of the clotting factors in blood, which might increase the chances that potentially dangerous blood clots may form (Januzzi & DeSanctis, 1999).

Anxiety and anger may also compromise the cardiovascular system by increasing blood levels of cholesterol, the fatty substance that clogs arteries and increases the risk of heart attacks (Suinn, 2001). People with higher levels of hostility also tend to have higher blood pressures than their less hostile counterparts (Räikkönen et al., 1999). High blood pressure (hypertension) is a major risk factor for heart attacks and strokes.

Cognitive-behavioral therapists are helping chronically angry people learn to control their emotional responses in anxiety-provoking or angering situations (e.g., Deffenbacher et al., 2000). Helping angry people learn to remain calm in provocative situations may have beneficial effects on the heart as well as the mind (Gidron & Davidson, 1996). Along these lines, a recent study reported that men with CHD who received a hostility-reduction program showed less hostility and lower blood pressures after treatment than did controls (Gidron, Davidson, & Bata, 1999).

Investigators are finding additional links between coronary heart disease and other forms of emotional stress, including depression (Carney, Freedland, & Jaffe, 2001; Ferketich et al., 2000; Orth-Gomér et al., 2000). In one recent study, people without cardiac disease who were suffering from major depression were nearly four times more likely than nondepressed people to die from heart-related causes over a 4-year study period (Penninx et al., 2001).

to other changes in lifestyle habits, such as reduced overall intake of dietary fat (McGovern et al., 1996; Traven et al., 1995). Better educated people are also more likely to modify unhealthful behavior patterns and reap the benefits of change. Is there a message in there for you?

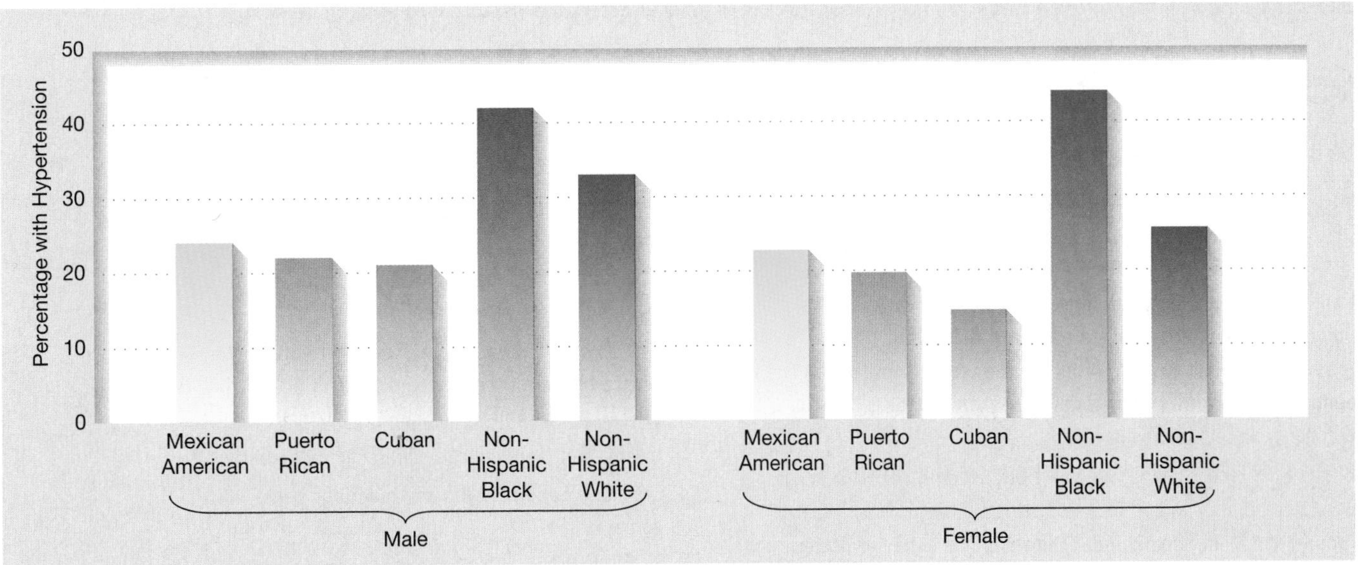

FIGURE 5.3 Hypertension among people ages 20 to 74, according to race/ethnicity. Non-Hispanic Black Americans are more likely than non-Hispanic White Americans to have high blood pressure, and non-Hispanic White Americans are more likely to have high blood pressure than Hispanic Americans. Except in the case of non-Hispanic Black Americans, men ages 20 to 74 are more likely than women in the same age group to have high blood pressure.

Source. USDHHS, Public Health Service (1991), *Health, United States 1990.* DHHS Pub. No. (PHS) 91-1232.

Asthma

Asthma is a respiratory disorder in which the main tubes of the windpipe—the bronchi—constrict and become inflamed, and large amounts of mucus are secreted. During asthma attacks, people wheeze, cough, and struggle to breathe in enough air. They may feel as though they are suffocating.

According to the Centers for Disease Control (CDC), nearly 15 million adults in the United States are affected by asthma ("Asthma Affects," 2001; CDC, 2001). About 5 million American children are also affected. Rates of asthma are on the rise, having doubled since 1980. Attacks can last from just a few minutes to several hours and vary notably in intensity. Series of attacks can harm the bronchial system, causing mucus to collect and muscles to lose their elasticity. Sometimes the bronchial system is weakened to the point where a subsequent attack is lethal.

Theoretical Perspectives Many causes are implicated in asthma, including allergic reactions; exposure to environmental pollutants, including cigarette smoke and smog; and genetic and immunological factors (Cookson & Moffatt, 1997; Giembycz & O'Connor, 2000). Asthmatic reactions in susceptible people can be triggered by exposure to allergens such as pollen, mold spores, and animal dander; by cold, dry air; and by emotional responses such as anger or even laughing too hard (Brody, 1988a). Psychological factors, such as emotional stress, loss of loved ones, and intense disappointment, appear to increase susceptibility to asthmatic attacks (Moran, 1991). Asthma, moreover, has psychological consequences. Some sufferers avoid strenuous activity, including exercise, for fear of increasing their demand for oxygen and tripping attacks.

Treatment Although asthma cannot be cured, it can be controlled by reducing exposure to allergens, by desensitization therapy ("allergy shots") to help the body acquire more resistance to allergens, by use of inhalers, and by drugs that open bronchial passages during asthma attacks (called *bronchodilators*) and others (called *anti-inflammatories)* that reduce future attacks by helping to keep bronchial tubes open. Psychological treatment may also

Questionnaire

Are You Type A?

People with the Type A behavior pattern are impatient, competitive, and aggressive. They feel rushed, under pressure; they keep one eye glued to the clock. They are prompt and often arrive early for appointments. They walk, talk, and eat rapidly. They grow restless when others work slowly.

Type A people don't just stroll out on the tennis court to bat the ball around. They scrutinize their form, polish their strokes, and demand consistent self-improvement.

Are you Type A? The following questionnaire may afford you insight.

Directions: Write a checkmark under the Yes if the behavior pattern described is typical of you. Place a checkmark under the No if it is not. Work rapidly and answer all items. Then check the scoring key at the end of the chapter.

DO YOU:	YES	NO
1. Strongly emphasize important words in your ordinary speech?	___	___
2. Walk briskly from place to place or meeting to meeting?	___	___
3. Think that life is by nature dog-eat-dog?	___	___
4. Get fidgety when you see someone complete a job slowly?	___	___
5. Urge others to complete what they're trying to express?	___	___
6. Find it exceptionally annoying to get stuck in line?	___	___
7. Envision all the things you have to do even when someone is talking to you?	___	___
8. Eat while you're getting dressed, or jot notes down while you're driving?	___	___
9. Catch up on work during vacations?	___	___

DO YOU:	YES	NO
10. Direct the conversation to things that interest you?	___	___
11. Feel as if things are going to pot because you're relaxing for a few minutes?	___	___
12. Get so wrapped up in your work that you fail to notice beautiful scenery passing by?	___	___
13. Get so wrapped up in money, promotions, and awards that you neglect expressing your creativity?	___	___
14. Schedule appointments and meetings back to back?	___	___
15. Arrive early for appointments and meetings?	___	___
16. Make fists or clench your jaws to drill home your views?	___	___
17. Think that you have achieved what you have because of your ability to work fast?	___	___
18. Have the feeling that uncompleted work must be done *now* and fast?	___	___
19. Try to find more efficient ways to get things done?	___	___
20. Struggle always to win games instead of having fun?	___	___
21. Interrupt people who are talking?	___	___
22. Lose patience with people who are late for appointments and meetings?	___	___
23. Get back to work right after lunch?	___	___
24. Find that there's never enough time?	___	___
25. Believe that you're getting too little done, even when other people tell you that you're doing fine?	___	___

Source. From Rathus, S. A. (1996).

play a role by helping asthma sufferers apply the skills of muscle relaxation to improve their breathing (Lehrer et al., 1994), and, for asthmatic children, family therapy that helps reduce family conflict (Lehrer et al., 1992).

Cancer

The word *cancer* is arguably the most feared word in the English language and rightly so: One of every four deaths in the United States is caused by cancer (Stolberg, 1998a). Cancer claims about a half a million lives in the United States annually, one every 90 seconds (Andersen, Golden-Kreutz, & DiLillo, 2001). Men have a one in two chance of developing cancer at some point in their lives; for women the odds are one in three. Yet there is good news to report: The number of new cancer cases and deaths from cancer are declining. Cancer cases are on the decline due largely to reductions in smoking, while the declining death rate is attributed largely to increased screening and better treatments ("Cancer Rates," 1999).

Cancer is characterized by development of aberrant, or mutant, cells that form growths (tumors) that spread to healthy tissue. Cancerous cells can take root anywhere—the blood, the bones, lungs, digestive tract, and genital organs. When they are not contained early, cancer may metastasize, or establish colonies throughout the body, leading to death.

There are many causes of cancer, including regular exposure to cancer-causing chemicals in the environment and genetic factors, such as defective or mutant genes. But many behavior patterns also contribute to the development of cancer, including dietary practices (high fat intake), heavy alcohol consumption, smoking, and sunbathing (ultraviolet light causes skin cancer). On the other hand, regular intake of a healthy daily supply of fruits and vegetables may lower the risk of some forms of cancer. Death rates from cancer are lower in Japan than in the United States, where people ingest more fat, especially animal fat. The difference is not genetic or racial, however, because Japanese Americans whose fat intake approximates that of other Americans show similar death rates from cancer.

Stress and Cancer A weakened or compromised immune system may increase susceptibility to cancer. We've seen that psychological factors, such as exposure to stress, may affect the immune system. Research with animals has shown that exposure to stress can hasten the onset of a virus-induced cancer (Riley, 1981). Might exposure to stress in humans increase the risk of cancer? Some studies show an increased incidence of stressful life events, such as the loss of loved ones, preceding the development of some forms of cancer (e.g., Levenson & Bemis, 1991). However, other studies show no linkage between exposure to stress and development of cancer (e.g., McKenna et al., 1999). Clearly, the links between stress and cancer require further study (Delahanty & Baum, 2001; Dougall & Baum, 2001).

Truth OR Fiction? REVISITED

Cancer patients who maintain a "fighting spirit" experience no better outcomes than those who resign themselves to their illness.

FALSE. In a sample of breast cancer patients, those who maintained a "fighting spirit" had higher survival rates than those who became resigned to their illness.

Psychological Factors in Treatment and Recovery Cancer is a physical disease treated medically by means of surgery, radiation, and chemotherapy. Yet psychologists and mental health professionals can play key roles in helping cancer patients deal with the emotional consequences of coping with the disease. Feelings of hopelessness and helplessness are common reactions to receiving a cancer diagnosis, but such feelings may hinder recovery (Andersen, 1992), perhaps by depressing the patient's immune system.

Evidence shows that breast cancer patients who maintain a "fighting spirit" experience better outcomes than those who resign themselves to their illness (Pettingale, 1985). This 10-year follow-up of breast cancer patients found that patients who met their diagnosis with anger and a fighting spirit rather than stoic acceptance showed significantly higher survival rates. The will to fight the illness may help to increase survival.

WWW **Web Link 5.3** Therapy and Cancer

Social support may also help. Women with metastatic breast cancer who participated in a group support program survived a year and half longer on the average than did women assigned to a no-treatment control group (Spiegel et al., 1989). However, how psychological approaches affect the course of cancer is unclear. One possible mode of action is enhancement of the immune system (Andersen, 1992).

Investigators have examined the value of training cancer patients to use coping skills, such as relaxation, stress management, and coping thoughts, to relieve the stress and pain of coping with cancer. These interventions may also help cancer patients cope with the anticipatory side effects of chemotherapy. Cues associated with chemotherapy, such as the hospital environment itself, may become conditioned stimuli that elicit nausea and vomiting even before the drugs are administered (Redd, 1995). By pairing relaxation, pleasant imagery, and attentional distraction with these cues, investigators find that nausea and vomiting can be lessened (Redd, 1995; Redd & Jacobsen, 2001). Playing video games as a form of distraction has also helped lessen the discomfort of chemotherapy in children with cancer (Kolko & Rickard-Figueroa, 1985).

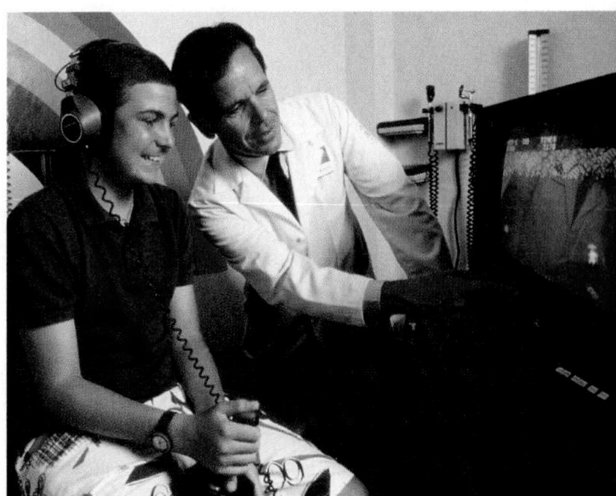

Coping with discomfort. The use of relaxation and distraction techniques may help cancer patients cope with the discomfort of chemotherapy.

Psychosocial interventions can also have positive effects on emotional and behavioral adjustment, and quality of life, of cancer patients (Andersen, Golden-Kreutz, & DiLillo, 2001; Compas et al., 1998; Meyer & Mark, 1995). It is too early to tell whether psychological interventions can prolong life expectancy of cancer patients, but preliminary evidence indicates that it may (Fawzy & Fawzy, 1994; Kogon et al., 1997).

acquired immunodeficiency syndrome (AIDS) An immunological disease caused by HIV.

human immunodeficiency virus (HIV) The virus that causes AIDS.

Learning to modify expectations is also important. Cancer patients who are able to maintain or restore their psychological well-being appear to be able to do so by readjusting their expectations of themselves in line with their present capabilities (Heidrich, Forsthoff, & Ward, 1994).

Acquired Immunodeficiency Syndrome (AIDS)

Acquired immunodeficiency syndrome (AIDS) is a disease caused by the **human immunodeficiency virus (HIV).** HIV attacks the person's immune system, leaving it helpless to fend off diseases it normally would hold in check. AIDS is one of history's worst epidemics, claiming nearly 22 million lives worldwide by 2001 and showing no signs of relenting (Altman, 2001; Begley, 2001a; Wren, 2001).

HIV is transmitted by sexual contact (vaginal and anal intercourse; oral-genital contact); direct infusion of contaminated blood, as from transfusions of contaminated blood, accidental pricks from needles used previously on an infected person, or needle sharing among injecting drug users; and from an infected mother to a child during pregnancy or childbirth or through breast-feeding. AIDS is not contracted by donating blood; by airborne germs; by insects; or by casual contact, such as using public toilets, holding or hugging infected people, sharing eating utensils with them, or living or going to school with them. Routine screening of the blood supplies for HIV have reduced the risk of infection from blood transfusions to virtually nil.

HIV infection and AIDS cut across all boundaries of race, ethnicity, income level, gender, sexual orientation, and drug use classification. You needn't be a sexually active gay male or an IV-drug user to become infected.

There is no cure or vaccine for HIV infection, but the introduction of highly active antiretroviral drugs has revolutionized treatment of the disease, raising hopes that it can become a chronic but manageable disease (Cowley, 2001c; Gallant, 2000; Sherbourne et al., 2000). However, hopes are tempered by the fact that many patients fail to derive or maintain any benefit from the newer antiviral drug combinations (Catz & Kelly, 2001).

The lack of a cure or effective vaccine means that prevention programs focusing on reducing or eliminating risky sexual and injection practices represent our best hope for controlling the epidemic (Begley, 2001a). Psychologists have become involved in the fight against AIDS because behavior is the major determinant of the risk of contracting the deadly virus and because AIDS, like cancer, has devastating psychological effects on persons affected by the disease, their families and friends, and society at large.

AIDS support group. AIDS support groups offer emotional support and assistance to people with HIV/AIDS, their families, and their friends.

Adjustment of People with HIV and AIDS Given the nature of the disease and the stigma suffered by people with HIV and AIDS, it is not surprising that many people with HIV, although certainly not all, develop psychological problems, most commonly anxiety and depression (Catz & Kelly, 2001; Ciesla & Roberts, 2001; Sherbourne et al., 2000). Recently, investigators reported that greater levels of depressive symptoms were associated with more rapid disease progression in women with HIV (Ickovics et al., 2001).

Psychological and Psychopharmacological Interventions Behavior change programs focus on reducing risky sexual and injection practices (Ickovics, Thayaparnan, & Ethier, 2001; Kelly et al., 1998). These training programs have been shown to be effective with groups of sexually active gay men (Kelly, Brasfield, & St. Lawrence, 1991) and with adolescents, including substance-dependent adolescents (St. Lawrence et al., 1995a, 1995b).

Psychological treatment, typically in the form of support groups, self-help groups, and organized therapy groups, have also been used to provide psychological assistance to people with HIV/AIDS and their families and friends. Treatment may incorporate training in active coping skills, such as stress management techniques like self-relaxation and positive mental imagery, and cognitive strategies to control intrusive negative thoughts and preoccupations. The importance of stress management skills is highlighted by recent findings that stressful life events and passive coping (use of denial) were associated with faster progression to AIDS in HIV-infected men (Leserman et al., 2000). Coping skills training and cognitive-behavioral therapy have been shown to help improve psychological functioning and ability to handle stress in people living with HIV or AIDS, and to reduce their feelings of depression and anxiety (Lutgendorf et al., 1997). Antidepressant medication has also been found to be helpful in treating depression in people with HIV (Elliott et al., 1998; Markowitz et al., 1998). Whether treatment of depression or coping skills training for handling stress can improve immunological functioning or prolong life in people with HIV and AIDS remains an open question.

The advent of AIDS presents the mental health community with unparalleled challenge to help prevent the spread of AIDS and to treat people who have been infected with HIV and who have developed AIDS. As frightening as AIDS may be, it is preventable, as noted in the next section.

THINK ABOUT IT

What are the psychological factors implicated in physical disorders and diseases such as headaches, cardiovascular disorders, asthma, cancer, and AIDS? What role do psychological techniques play in the treatment of physical health disorders and conditions?

Preventing AIDS For the first time, a generation of young people has come of age at a time when the threat of AIDS hangs over every sexual encounter. People may decrease the risk of being infected by HIV and other sexually transmitted diseases (STDs) by taking the following measures. Only the first two are sure paths to avoiding the sexual transmission of HIV. The others reduce the risk of infection, but cannot be certified as perfectly safe. If we are going to be sexually active without knowing (not guessing) whether we or our partners are infected with HIV or some other STD, we can speak only of safe(r) sex—not of perfectly safe sex.

1. *Maintaining lifelong celibacy.*
2. *Remaining in a lifelong monogamous relationship with an uninfected person who is doing the same thing.* Although these first two sexual career paths guarantee safety, they are not followed by the majority of students or other Americans.
3. *Being discerning in one's choice of sex partners.* Get to know another person before engaging in sexual activity. Still, getting to know a person is no guarantee the person is uninfected with HIV. Avoid contact with multiple partners, or with people who are likely to have multiple partners.
4. *Being assertive with sex partners.* It is important to communicate concerns about AIDS clearly and assertively with sex partners.
5. *Inspecting one's partner's sex organs.* There are no obvious signs of HIV infection, but people who are infected with HIV are often infected by other sexually transmitted diseases as well. It may be feasible to visually inspect your partner's sex organs for rashes, chancres, blisters, discharges, warts, and lice during foreplay. Consider any disagreeable odor a warning sign.
6. *Using latex condoms.* Condoms protect men from infected vaginal fluids and stop infected semen from entering women. All condoms (including so-called natural condoms made of animal intestines or "skins") act as barriers to sperm, but only latex condoms can prevent transmission of HIV.
7. *Using spermicides.* Spermicides containing the ingredient nonoyxnol-9 kill HIV as well as sperm. Spermicides should be used along with latex condoms, not as a substitute for condoms.
8. *Consulting a physician following suspected exposure to a sexually transmitted disease (STD).* Antibiotics following unprotected sex may guard against bacterial STDs, but they are of no use against viral STDs such as genital herpes and HIV/AIDS. Consult with a physician before using any medications, including medications you may have stored away in your medicine cabinet.

9. *Seeking regular medical checkups.* Checkups and appropriate laboratory tests enable you to learn about and treat disorders that might have gone unnoticed.
10. *Avoiding sexual activity if there are doubts about safety.* None of the safer sex practices listed guarantees protection. Why not avoid sexual activity when doubts of safety exist?

We end this section on a sobering note. Information about risk reduction alone is not sufficient to induce widespread changes in sexual behavior (Kelly et al., 1995). Despite awareness of the dangers, many people practice risky sexual and injection behaviors (Kalichman, 2000). People not only need to know about the dangers of unsafe sexual practices, they also need to know how to change their behavior (e.g., learning how to refuse invitations to engage in unsafe sex and how to communicate effectively with one's partner(s) about safer sex), and they must be motivated to change their risk behavior. Unfortunately, motivation has been waning and evidence points to increasingly risky behavior among the young (Begley, 2001a). Other factors not to be overlooked in prevention efforts are drug and alcohol use and peer group norms. The likelihood of people engaging in safer sex practices is linked to the avoidance of alcohol and drugs before sex and to the perception that safer sex practices are the social norm within one's peer group.

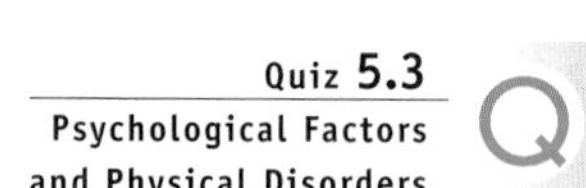

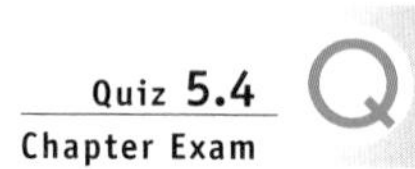

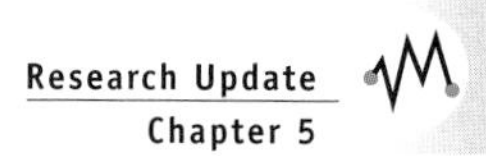

In this chapter we focused on relationships between stress and health, and on the psychological factors involved in health. Psychology has much to offer in the understanding and treatment of physical disorders. Psychological approaches may help in the treatment of such physical disorders as headaches and coronary heart disease. Psychologists also help people reduce the risks of contracting health problems such as cardiovascular disorders, cancer, and AIDS. Emerging fields like psychoneuroimmunology promise to further enhance our knowledge of the intricate relationships between mind and body.

Summing Up

Adjustment Disorders

What are adjustment disorders? Adjustment disorders are maladaptive reactions to identified stressors.

What are their features? Adjustment disorders are characterized by emotional reactions that are greater than normally expected given the circumstances or by evidence of significant impairment in functioning. Impairment usually takes the form of problems at work or school, or in social relationships or activities.

Stress and Illness

How is stress linked to physical illness? Evidence links exposure to stress to weakened immune system functioning, which in turn can increase vulnerability to physical illness. However, since this evidence is correlational, questions of cause and effect remain.

What is the general adaptation syndrome? This is the name given by Hans Selye to the generalized pattern of response of the body to persistent or enduring stress, as characterized by three stages: the alarm reaction, the resistance stage, and the exhaustion stage.

How are life changes related to physical health problems? Again, links are correlational, but evidence shows that people who experience more life stress in the form of life changes and daily hassles are at an increased risk of developing physical health problems.

What psychological factors buffer the effects of stress? These factors include coping styles, self-efficacy expectancies, psychological hardiness, optimism, and social support.

Psychological Factors and Physical Disorders

What roles do psychological factors play in the onset of headaches and their treatment? The most common headache is the muscle-tension headache, which is often stress related. Behavioral methods of relaxation training and biofeedback are of help in treating various types of headaches.

What behavioral or lifestyle factors increase the risk of coronary heart disease? Psychological factors that increase the risk of coronary heart disease include patterns of consumption, leading a sedentary lifestyle, Type A behavior pattern, and persistent negative emotions.

What role do psychological factors play in asthma? Psychological factors such as emotional stress, loss of loved ones, and sudden or intense disappointment may trigger asthma attacks in susceptible individuals.

What role do psychological factors play in the development of cancer and its treatment? Although relationships between stress and risk of cancer remain under study, behavioral risk factors for cancer include dietary practices (especially high fat intake), heavy alcohol use, smoking, and excessive sun exposure. Research shows that a fighting spirit may help people recover from cancer. Psychological interventions help cancer patients cope better with the symptoms of the disease and its treatment.

What roles do psychologists play in the prevention of HIV/AIDS and treatment of people with HIV? Our behavior patterns influence our risk for contracting AIDS. Psychologists have become involved in the prevention and treatment of AIDS because AIDS, like cancer, has devastating psychological effects on victims, their families and friends, and society at large, and because AIDS can be prevented through reducing risky behavior.

Scoring Key for "The College Life Stress Inventory" Though we have no national norms by which to compare your score, the test developers obtained an average (mean) score of 1,247 based on a sample of 257 introductory psychology students. About two of three students obtained scores in the range of 806 to 1,688.

Computing your total score helps you gauge how you compare to the students in the original study sample in terms of your overall stress level. Bear in mind, however, that the same level of stress may affect different people differently. Your ability to cope with stress depends on many factors, including your coping skills and the level of social support you have available. If you are experiencing a high level of stress, you may wish to examine the sources of stress in your life. Perhaps you can reduce the level of stress you experience or learn more effective ways of handling the sources of stress you can't avoid.

Scoring Key for "The Life Orientation Test" To arrive at your total score for the test, first *reverse* your score on items 3, 8, 9, and 12. That is,

4 is changed to 0
3 is changed to 1
2 remains the same
1 is changed to 3
0 is changed to 4

Now add the numbers of items 1, 3, 4, 5, 8, 9, 11, and 12. (Items 2, 6, 7, and 10 are "fillers"; that is, your responses are not scored as part of the test.) Your total score can vary from 0 to 32.

Scheier and Carver (1985) provide the following norms for the test, based on administration to 357 undergraduate men and 267 undergraduate women. The average (mean) score for men was 21.03 (standard deviation = 4.56), and the mean score for women was 21.41 (standard deviation = 5.22). All in all, approximately 2 out of 3 undergraduates obtained scores between 16 and 26. Scores above 26 may be considered quite optimistic, and scores below 16, quite pessimistic. Scores between 16 and 26 are within a broad average range, and higher scores within this range are relatively more optimistic.

Answer Key for "Are You Type A?" Yesses are suggestive of the Type A behavior pattern (TABP). In appraising whether or not you show the TABP, you need not be concerned with the precise number of yes answers. We have no normative data for you. As Friedman and Rosenman (1974, p. 85) note, however, you should have little trouble spotting yourself as "hard core" or "moderately afflicted"—that is, if you are honest with yourself.

CHAPTER SIX

Anxiety Disorders

Miriam Schapiro
Free Fall, 1985

Truth OR Fiction?

- Some people who experience panic attacks believe they are having a heart attack, even though there is nothing wrong with their heart. (p. 162)
- Some people are so fearful of leaving their homes that they are unable to venture outside even to mail a letter. (p. 168)
- It may take an hour or more for people with obsessive-compulsive disorder to leave the house. (p. 169)
- Men are more than twice as likely as women to develop posttraumatic stress disorder (PTSD). (p. 171)
- Some theorists believe we are genetically programmed to more readily acquire fears of some classes of stimuli, including snakes. (p. 175)
- Misinterpretations of bodily sensations may set into motion a spiraling cycle of anxiety that culminates in a full-fledged panic attack. (p. 179)
- The same drugs used to treat schizophrenia are also used to control panic attacks. (p. 183)
- Peering over a virtual ledge 20 stories up has helped some people overcome their fear of actual heights. (p. 188)

Anxiety is a generalized state of apprehension or foreboding. There is much to be anxious about—our health, social relationships, examinations, careers, international relations, and the condition of the environment are but a few sources of possible concern. It is normal, even adaptive, to be somewhat anxious about these aspects of life. Anxiety serves us when it prompts us to seek regular medical checkups or motivates us to study for tests. Anxiety is an appropriate response to threats, but anxiety can be abnormal when its level is out of proportion to a threat, or when it seems to come out of the blue—that is, when it is not in response to environmental changes. In extreme forms, anxiety can impair our daily functioning. Consider the following case:

Panic on the Long Island Railroad

Slowly the trains snake their way through the maze of tunnels that lie beneath the city, carrying the Dashing Dans and Danielles on their way to work each morning. Most commuters pass the time by reading the morning newspapers, sipping coffee, or catching a few last winks. For Dick, the morning commute was an exercise in terror on an ordinary day in July. At first, Dick noticed the perspiration clinging to his shirt. The air-conditioning seemed to be working fine, for a change. How then was he to account for the sweat? As the train entered the tunnel and darkness shrouded the windows, Dick was gripped by sheer terror. He sensed his heart beating faster, the muscles in his neck tightening. Queasiness soured his stomach. He felt as though he might pass out. Other commuters, engrossed in their morning papers or their private thoughts, paid no heed to Dick, nor did they seem concerned about the darkness that enveloped the train.

Dick had known these feelings all too well before. But now the terror was worse. Other days he could bear it. This time, it seemed to start earlier than usual, before the train entered the tunnel. "Just don't think about it," he told himself, hoping it would pass. "I must think of something to distract myself." He tried humming a song, but the panic grew worse. He tried telling himself that it would be all right, that at any moment the train would enter the station and the doors would open. Not this day, however. On this day, the train came to a screeching halt. The conductor announced a "signaling problem." Dick tried to calm himself: "It's only a short delay. We'll be moving soon." But the train did not start moving soon. More apologies from the conductor. A train had broken down further ahead in the tunnel. Dick realized it could be a long delay, hours perhaps. Suddenly, he felt the urgent desire to escape. But how? he wondered. There was barely room for a crawlspace outside the train. Then again, could he even break the window and crawl out of the train, if he had to?

He felt like he was losing control. Wild imaginings flooded his mind. He saw himself bolting down the aisles in a futile attempt to escape, bowling people over, trying vainly to pry open the doors. He was charged with a sense of doom. Something terrible was about to happen to him. "Is this the first sign of a heart attack?" he wondered anxiously. By now, the perspiration had soaked his clothes. His once neat tie hung awry. He felt his breathing become heavy and labored, drawing attention from other passengers. "What do they think of me?" he thought. "Will they help me if I need them?"

The train jerked into motion. He realized he would soon be free. "I'm going to be okay," he told himself, "the feelings will pass. I'm going to be myself again." The train pulled slowly into the station, twenty minutes late. The doors opened and the passengers hurried off. Stepping out himself, Dick adjusted his tie and readied himself to start the day. He felt as though he'd been in combat. Nothing that his boss could dish out could hold a candle to what he had experienced on the 7:30 train.

—From the Authors' Files

Chapter Outline

anxiety An emotional state characterized by physiological arousal, unpleasant feelings of tension, and a sense of apprehension or foreboding.

panic disorder A type of anxiety disorder characterized by repeated episodes of intense *anxiety* or panic.

anxiety disorder A type of mental disorder whose most prominent feature is anxiety.

THINK ABOUT IT

Anxiety may be a normal emotional reaction in some situations but not in others. Think of a situation in which anxiety would be a normal reaction and one in which it would be a maladaptive reaction. What are the differences? What criteria would you use to distinguish between the normal and abnormal?

Dick had suffered a panic attack, one of many he had experienced before seeking treatment. The attacks varied in frequency. Sometimes they occurred daily, sometimes once a week or so. He never knew whether an attack would occur on a particular day. He knew, however, that he couldn't go on living like this. He feared that one day he would suffer a heart attack on the train. He pictured some passengers trying vainly to revive him while others stared at him in the detached distant way that people stare at traffic accidents. He pictured emergency workers rushing to the train, bearing him on a stretcher through the darkened tunnels to an ambulance.

For a while he considered changing jobs, accepting a less remunerative job closer to home, one that would free him from the need to take the train. He also considered driving to work, but the roads were too thick with traffic. No choice, he figured; either commute by train or switch jobs. His wife, Jill, was unaware of his panic attacks. She wondered why his shirts were heavily stained with perspiration and why Dick was talking about changing jobs. She worried about making ends meet on a lower income. She had no idea it was the train ride, and not his job, that Dick was desperate to avoid.

Panic attacks, like that suffered by Dick, are a feature of **panic disorder,** a type of **anxiety disorder.** During a panic attack, one's level of anxiety can rise to the level of sheer terror. Panic attacks are an extreme form of anxiety. Anxiety encompasses a myriad of physical features, cognitions, and behaviors as shown in Table 6.1. Although anxious people do not often experience all of them, it is easy to see why anxiety is distressing.

TABLE 6.1 Some Features of Anxiety

Physical Features of Anxiety

Jumpiness, jitteriness
Trembling or shaking of the hands or limbs
Sensations of a tight band around the forehead
Tightness in the pit of the stomach or chest
Heavy perspiration
Sweaty palms
Light-headedness or faintness
Dryness in the mouth or throat
Difficulty talking
Difficulty catching one's breath
Shortness of breath or shallow breathing
Heart pounding or racing
Tremulousness in one's voice
Cold fingers or limbs
Dizziness
Weakness or numbness
Difficulty swallowing
A "lump in the throat"
Stiffness of the neck or back
Choking or smothering sensations
Cold, clammy hands
Upset stomach or nausea
Hot or cold spells
Frequent urination
Feeling flushed
Diarrhea
Feeling irritable or "on edge"

Behavioral Features of Anxiety

Avoidance behavior
Clinging, dependent behavior
Agitated behavior

Cognitive Features of Anxiety

Worrying about something
A nagging sense of dread or apprehension about the future
Belief that something dreadful is going to happen, with no clear cause
Preoccupation with bodily sensations
Keen awareness of bodily sensations
Feeling threatened by people or events that are normally of little or no concern
Fear of losing control
Fear of inability to cope with one's problems
Thinking the world is caving in
Thinking things are getting out of hand
Thinking things are swimming by too rapidly to take charge of them
Worrying about every little thing
Thinking the same disturbing thought over and over
Thinking that one must flee crowded places or else pass out
Finding one's thoughts jumbled or confused
Not being able to shake off nagging thoughts
Thinking that one is going to die, even when one's doctor finds nothing medically wrong
Worrying that one is going to be left alone
Difficulty concentrating or focusing one's thoughts

Types of Anxiety Disorders

The anxiety disorders, along with dissociative disorders and somatoform disorders (see Chapter 7), were classified as neuroses throughout most of the 19th century. The term *neurosis* derives from roots meaning "an abnormal or diseased condition of the nervous system." The Scottish physician William Cullen coined it in the 18th century. As the derivation implies, it was assumed neurosis had biological origins. It was seen as an affliction of the nervous system.

At the beginning of the 20th century, Cullen's organic assumptions were largely replaced by Sigmund Freud's psychodynamic views. Freud maintained that neurotic behavior stems from the threatened emergence of unacceptable anxiety-evoking ideas into conscious awareness. Various neurotic behavior patterns—anxiety disorders, somatoform disorders, and dissociative disorders—might look different enough on the surface. According to Freud, however, they all represent ways in which the ego attempts to defend itself against anxiety. Freud's **etiological** assumption, in other words, united the disorders as neuroses. Freud's concepts were so widely accepted in the early 1900s that they formed the basis for the classification systems found in the first two editions of the *Diagnostic and Statistical Manual of Mental Disorders (DSM).*

Since 1980, the *DSM* has not contained a category termed *neuroses.* The present *DSM* is based on similarities in observable behavior and distinctive features rather than on causal assumptions. Many clinicians continue to use the terms *neurosis* and *neurotic* in the manner in which Freud described them, however. Some clinicians use "neuroses" as a convenient means of grouping milder behavioral problems in which people maintain relatively good contact with reality. "Psychoses," such as schizophrenia, are typified by loss of touch with reality, and by the appearance of bizarre behavior, beliefs, and hallucinations. Anxiety is not limited to the diagnostic categories traditionally termed "neuroses," moreover. People with adjustment problems, depression, and psychotic disorders may also encounter problems with anxiety.

The present version of the *DSM* system, the *DSM-IV,* recognizes the following specific types of anxiety disorders: panic disorder; phobic disorders, such as specific phobia, social phobia, and agoraphobia; generalized anxiety disorder; obsessive-compulsive disorder; and acute and posttraumatic stress disorders. Table 6.2 lists the diagnostic features of anxiety disorders. The anxiety disorders are not mutually exclusive. People frequently meet diagnostic criteria for more than one of them.

WWW **Web Link 6.1**
NIMH Self-Screening for Anxiety Disorders

etiological Relating to cause or origin.

TABLE **6.2 Diagnostic Features of Anxiety Disorders**

Agoraphobia	Fear and avoidance of places or situations in which it would be difficult or embarrassing to escape, or in which help might be unavailable in the event of a panic attack or panic-type symptoms.
Panic Disorder Without Agoraphobia	Occurrence of recurrent, unexpected panic attacks in which there is persistent concern about them but without accompanying agoraphobia.
Panic Disorder with Agoraphobia	Occurrence of recurrent, unexpected panic attacks in which there is persistent concern about them and accompanying agoraphobia.
Generalized Anxiety Disorder	Persistent and excessive levels of anxiety and worry that is not tied to any particular object, situation, or activity.
Specific Phobia	Clinically significant anxiety relating to exposure to specific objects or situations, often accompanied by avoidance of these stimuli.
Social Phobia	Clinically significant anxiety relating to exposure to social situations or performance situations, often accompanied by avoidance of these situations.
Obsessive-Compulsive Disorder	Recurrent obsessions and/or compulsions.
Posttraumatic Stress Disorder	The reexperiencing of a highly traumatic event accompanied by heightened arousal and avoidance of stimuli associated with the event.
Acute Stress Disorder	Features similar to those of posttraumatic stress disorder but limited to the days and weeks following exposure to the trauma.

Note. All of these disorders are coded on Axis I in the *DSM-IV.*
Source. Adapted from *DSM-IV-TR* (APA, 2000).

Truth OR Fiction? REVISITED

Some people who experience panic attacks believe they are having a heart attack, even though there is nothing wrong with their heart.

TRUE People experiencing a panic attack may believe they are having a heart attack, even though their hearts are perfectly healthy.

VIDEO 6.1
Panic Disorder: *The Case of Jerry*

Web Link 6.2 WWW
Q&A: About Panic Disorder

Panic Disorder

Panic disorder involves the occurrence of repeated, unexpected panic attacks. Panic attacks involve intense anxiety reactions accompanied by physical symptoms such as a pounding heart; rapid respiration, shortness of breath, or difficulty breathing; heavy perspiration; and weakness or dizziness (Glass, 2000). There is a stronger bodily component to panic attacks than to other forms of anxiety. The attacks are accompanied by feelings of sheer terror and a sense of imminent danger or impending doom and by an urge to escape the situation. They are usually accompanied by thoughts of losing control, going crazy, or dying. People who experience panic attacks tend to be keenly aware of changes in their heart rates (Richards, Edgar, & Gibbon, 1996). They often believe they are having a heart attack even though there is nothing wrong with their hearts. But since the symptoms of panic attacks can mimic those of heart attacks or even severe allergic reactions, a thorough medical evaluation should be performed.

A panic attack occurs suddenly and builds to a peak of intensity within 10 to 15 minutes (USDHHS, 1999a). Attacks usually last for minutes, but can extend to hours, and are associated with a strong urge to escape the situation in which they occur. For a diagnosis of panic disorder to be made, there must be the presence of recurrent unexpected panic attacks—attacks that are not triggered by specific objects or situations. They seem to come out of the blue. Although the first attacks occur spontaneously or unexpectedly, over time they may become associated with certain situations or cues, such as entering a crowded department store, or, like Dick, riding on a train.

In many cases, people who experience panic attacks limit their activities to avoid places in which they fear attacks may occur or they are cut off from their usual supports. Panic disorder often leads to agoraphobia—fear of being in public places in which escape may be difficult or help unavailable (Glass, 2000). After a panic attack, the person may feel exhausted, as if he or she has survived a truly traumatic experience, as in the following case:

Panic. Panic attacks have stronger physical components—especially cardiovascular symptoms—than other types of anxiety reactions.

A Case of Panic Disorder

"I was inside a very busy shopping precinct and all of a sudden it happened; in a matter of seconds I was like a mad woman. It was like a nightmare, only I was awake; everything went black and sweat poured out of me—my body, my hands, and even my hair got wet through. All of the blood seemed to drain out of me; I went white as a ghost. I felt as if I was going to collapse; it was as if I had no control over my limbs; my back and legs were very weak and I felt as though it were impossible to move. It was as if I had been taken over by some stronger force. I saw all of the people looking at me—just faces, no bodies; all merged into one. My heart started pounding in my head and my ears; I thought that my heart was going to stop. I could see black and yellow lights. I could hear the voices of the people but from a long way off. I could not think of anything except the way that I was feeling and how I had to get out and run quickly or I would die. I must escape and get into fresh air. Outside it subsided a little but I felt limp and weak; my legs were like jelly as though I had run a race and lost; I had a lump in my throat like a golf ball. The incident seemed to me to have lasted hours. I was absolutely drained when I got home and I just broke down and cried; it took until the next day to feel normal again."

—Adapted from Hawkrigg, 1975, pp. 1280–1282

People often describe panic attacks as the worst experiences of their lives. Their coping abilities are overwhelmed. They may feel they must flee. If flight seems useless, they may freeze. There is a tendency to cling to others for help or support. Some people with panic attacks fear going out alone. Recurrent panic attacks may become so difficult to cope with that sufferers become suicidal. A study of community residents who suffered panic attacks found that 12% had attempted suicide (Weissman et al., 1989).

Table 6.3 lists the diagnostic features of panic attacks. Not all of these features need to be present. Not all panic attacks are signs of panic disorder; about 10% of otherwise healthy people in the population may experience an isolated attack in a given year (USDHHS, 1999a). A diagnosis of panic disorder is based on the following criteria: (1) encountering repeated (at least two) unexpected panic attacks; and (2) at least one of the attacks is followed by at least a month of persistent fear of subsequent attacks, or worry about the implications or consequences of the attack (e.g., fear of losing one's mind or "going crazy" or having a heart attack), or significant change in behavior (e.g., refusing to leave the house or venture into public for fear of having another attack) (APA, 2000). An estimated 1% to 4% of the population are affected by panic disorder at some point in their lives (APA, 2000; USDHHS, 1999a).

Panic disorder usually begins in late adolescence through the mid-30s (APA, 2000). Women are about twice as likely to develop panic disorder (USDHHS, 1999a) (see Figure 6.1). What little we know about the long-term course of panic disorder suggests it tends to follow a chronic course that waxes and wanes in severity over time (Ehlers, 1995).

TABLE 6.3 Diagnostic Features of Panic Attacks

A panic attack involves an episode of intense fear or discomfort in which at least four of the following features develop suddenly and reach a peak within 10 minutes:

1. Heart palpitations, pounding heart, tachycardia (rapid heart rate)
2. Sweating
3. Trembling or shaking
4. Shortness of breath or smothering sensations
5. Choking sensations
6. Chest pains or discomfort
7. Feelings of nausea or other signs of abdominal distress
8. Feelings of dizziness, unsteadiness, light-headedness, or faintness
9. Feelings of strangeness or unreality about one's surroundings (derealization) or detachment from oneself (depersonalization)
10. Fear of losing control or going crazy
11. Fear of dying
12. Numbness or tingling sensations
13. Chills or hot flushes

Source. Adapted from the *DSM-IV-TR* (APA, 2000).

Generalized Anxiety Disorder

Generalized anxiety disorder (GAD) is characterized by persistent feelings of anxiety that are not triggered by any specific object, situation, or activity, but rather seems to be what Freud labeled "free floating." The central feature of GAD is worry (Ruscio, Borkovec, & Ruscio, 2001). People with GAD are chronic worriers. They may worry excessively about their life circumstances, such as their finances, the well-being of their children, and their social relationships. Nine of 10 of them, according to one study, report excessive worrying about even minor things (Sanderson & Barlow, 1990). Children with generalized anxiety are more likely to be worried about academics, athletics, and other social aspects of school life. Other related features include restlessness; feeling tense, "keyed up," or "on edge"; becoming easily fatigued; having difficulty concentrating or finding one's mind going blank; irritability, muscle tension; and disturbances of sleep, such as difficulty falling asleep, staying asleep, or having restless and unsatisfying sleep (APA, 2000).

generalized anxiety disorder (GAD) A type of anxiety disorder characterized by general feelings of dread and foreboding and heightened states of bodily arousal.

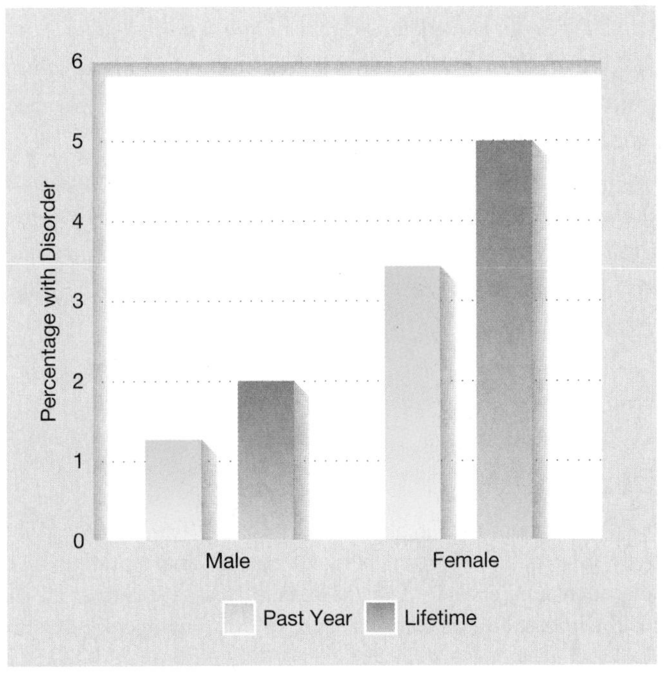

FIGURE 6.1 Prevalence of panic disorder by gender.
Panic disorder affects nearly two times as many women as men.
Source. National Comorbidity Survey (Kessler et al., 1994).

GAD tends to be a stable disorder that initially arises in the mid-teens to mid-20s and then follows a lifelong course (Rapee, 1991). The lifetime prevalence of GAD in the general U.S. population is estimated to be about 5% (APA, 2000). The disorder is believed to be about twice as common in women as in men (APA, 2000; USDHHS, 1999a).

Although GAD typically involves less intense physiological responses than panic disorder, the emotional distress associated with GAD is severe enough to interfere substantially with the person's daily life (Wittchen et al., 1994). GAD frequently occurs together (comorbidly) with other disorders, such as depression or other anxiety disorders like agoraphobia and obsessive-compulsive disorder.

In the following case, we find a number of features of generalized anxiety disorder:

A Case of Generalized Anxiety Disorder

Earl was a 52-year-old supervisor at the automobile plant. His hands trembled as he spoke. His cheeks were pale. His face was somewhat boyish, making his hair seem grayed with worry.

He was reasonably successful in his work, although he noted that he was not a "star." His marriage of nearly three decades was in "reasonably good shape," although sexual relations were "less than exciting—I shake so much that it isn't easy to get involved." The mortgage on the house was not a burden and would be paid off within 5 years, but "I don't know what it is; I think about money all the time." The three children were doing well. One was employed, one was in college, and one was in high school. But "With everything going on these days, how can you help worrying about them? I'm up for hours worrying about them."

"But it's the strangest thing," Earl shook his head. "I swear I'll find myself worrying when there's nothing in my head. I don't know how to describe it. It's like I'm worrying first and then there's something in my head to worry about. It's not like I start thinking about this or that and I see it's bad and then I worry. And then the shakes come, and then, of course, I'm worrying about worrying, if you know what I mean. I want to run away; I don't want anyone to see me. You can't direct workers when you're shaking."

Going to work had become a major chore. "I can't stand the noises of the assembly lines. I just feel jumpy all the time. It's like I expect something awful to happen. When it gets bad like that I'll be out of work for a day or two with shakes."

Earl had been worked up "for everything; my doctor took blood, saliva, urine, you name it. He listened to everything, he put things inside me. He had other people look at me. He told me to stay away from coffee and alcohol. Then from tea. Then from chocolate and Coca-Cola, because there's a little bit of caffeine [in them]. He gave me Valium [a minor tranquilizer] and I thought I was in heaven for a while. Then it stopped working, and he switched me to something else. Then that stopped working, and he switched me back. Then he said he was 'out of chemical miracles' and I better see a shrink or something. Maybe it was something from my childhood."

—From the Authors' Files

Phobic Disorders

The word *phobia* derives from the Greek *phobos,* meaning "fear." The concepts of fear and anxiety are closely related. *Fear* is the feeling of anxiety and agitation in response to a threat. Phobic disorders are persistent fears of objects or situations that are disproportionate to the threats they pose. To experience a sense of gripping fear when your car is about to go out of control is normal because there is an objective basis to the fear. In phobic disorders, however, the fear exceeds any reasonable appraisal of danger. People with a driving phobia, for example, might become fearful even when they are driving well below the speed limit on a sunny, uncrowded highway. Or they might be so afraid that they will not drive or even ride in a car. People with phobic disorders are not out of touch with reality; they generally recognize their fears are excessive or unreasonable.

A curious thing about phobias is that they usually involve fears of the ordinary events in life, not the extraordinary. People with phobias become fearful of ordinary experiences that most people take for granted, such as taking an elevator or driving on a highway. Phobias can become disabling when they interfere with such daily tasks as taking buses, planes, or trains; driving; shopping; or leaving the house.

Different types of phobias usually appear at different ages, as noted in Table 6.4. The ages of onset appear to reflect levels of cognitive development and life experiences. Fears of animals are frequent subjects of children's fantasies, for example. Agoraphobia, in contrast, often follows the development of panic attacks beginning in adulthood.

Here let us consider three types of phobic disorders classified within the *DSM* system: *specific phobia*, *social phobia*, and *agoraphobia*.

TABLE 6.4 Typical Age of Onset for Various Phobias

	No. of Cases	Mean Age of Onset
Animal phobia	50	7
Blood phobia	40	9
Injection phobia	59	8
Dental phobia	60	12
Social phobia	80	16
Claustrophobia	40	20
Agoraphobia	100	28

Source. Adapted from Öst (1987, 1992).

Specific Phobias **Specific phobias** are persistent, excessive fears of specific objects or situations, such as fear of heights (**acrophobia**), fear of enclosed spaces (**claustrophobia**), or fear of small animals such as mice or snakes and various other "creepy-crawlies." The person experiences high levels of fear and physiological arousal when encountering the phobic object, which prompts strong urges to avoid or escape the situation or avoid the feared stimulus, as in the following case:

A Case of Specific Phobia

Passing the bar exam was a significant milestone in Carla's life, but it left her feeling terrified at the thought of entering the county courthouse. She wasn't afraid of encountering a hostile judge or losing a case, but of climbing the stairs leading to a second floor promenade where the courtrooms were located. Carla, 27, suffered from acrophobia, or fear of heights. "It's funny, you know," Carla told her therapist. "I have no problem flying or looking out the window of a plane at 30,000 feet. But the escalator at the mall throws me into a tailspin. It's just any situation where I could possibly fall, like over the side of a balcony or banister." People with anxiety disorders look to avoid situations or objects they fear. Carla scouted out the courthouse before she was scheduled to appear. She was relieved to find a service elevator in the rear of the building she could use instead of climbing the stairs. She told her fellow attorneys with whom she was presenting the case that she suffered from a heart condition and couldn't climb stairs. Not suspecting the real reason she wanted to avoid the stairs, one of the attorneys turned to her and said, "This is great. I never knew this elevator existed. Thanks for finding it."

—From the Authors' Files

To rise to the level of a psychological disorder, the phobia must significantly impact the person's lifestyle or functioning, or cause significant distress. You may have a fear of snakes, but unless your fear interferes with your daily life or causes you significant emotional distress it would not warrant a diagnosis of phobic disorder.

Specific phobias often begin in childhood. Many children develop passing fears of specific objects or situations. Some, however, go on to develop chronic clinically significant phobias (Merckelbach et al., 1996). Claustrophobia seems to develop later than most other specific phobias, with a mean age of onset of 20 years (see Table 6.4).

specific phobia A persistent and excessive fear of a specific object or situation.

acrophobia Excessive, irrational fear of heights.

claustrophobia Excessive, irrational fear of small, enclosed spaces.

(A)

(B)

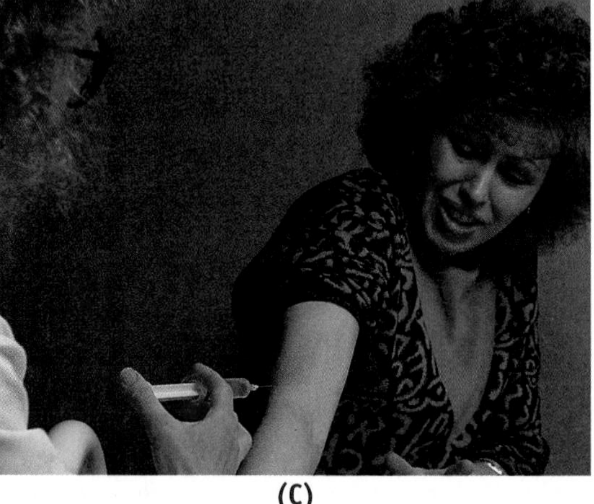

(C)

Three types of phobic disorder. The man in photo A has a specific phobia for dogs, a common phobia that may have an evolutionary origin. The young woman in photo B would like to join others but keeps to herself because of social phobia, an intense fear of social criticism and rejection. The woman in photo C has a specific phobia for injections. She does not fear the potential pain of the injection; rather, she cannot tolerate the idea of the needle sticking her.

Specific phobias are among the most common psychological disorders, affecting about 7% to 11% of the general population at some point in their lives (APA, 2000). Specific phobias tend to persist for years or decades unless they are treated successfully (USDHHS, 1999a). Women are about twice as likely to develop specific phobias (APA, 2000). This gender difference may to some degree reflect cultural factors that socialize women to be dependent on men for protection from threatening objects in the environment. Examiners also need to be aware of cultural factors when making diagnostic judgments. Fears of magic or spirits are common in some cultures and should not be considered a sign of a phobic disorder unless the fear is excessive in the light of the cultural context in which it occurs and leads to significant emotional distress or impaired functioning (APA, 2000).

Social Phobia It is not abnormal to experience some fear of social situations such as dating, attending parties or social gatherings, or giving a talk or presentation to a class or group. Yet people with **social phobia** (also called *social anxiety disorder*) have such an intense fear of social situations that they may avoid them altogether or endure them only with great distress. Underlying social phobia is an excessive fear of negative evaluations from others. People with social phobia fear doing or saying something humiliating or embarrassing. They may feel as if a thousand eyes are scrutinizing their every move. They tend to be severely critical of their social skills and become absorbed in evaluating their own performance when interacting with others. Some even experience full-fledged panic attacks in social situations.

Stage fright and speech anxiety are common types of social phobias. A random survey of some 500 residents of Winnipeg, Manitoba, found that about 1 in 3 had experienced excessive anxiety when speaking to a large audience that was significant enough to have had a detrimental impact on their lives (Stein, Walker, & Forde, 1996). People with social pho-

social phobia Excessive fear of social interactions or situations.

bias may find excuses for declining social invitations. They may lunch at their desks to avoid socializing with coworkers. Or they may find themselves in social situations and attempt a quick escape at the first sign of anxiety. Relief from anxiety negatively reinforces escape behavior, but escape prevents people with phobias from learning to cope with fear-evoking situations more adaptively. Leaving the scene before the anxiety dissipates only strengthens the association between the social situation and anxiety. Some people with social phobia are unable to order food in a restaurant for fear the server or their companions might make fun of the foods they order or how they pronounce them. Others fear meeting new people and dating.

Social phobias can severely impact daily functioning and the quality of life (Leibowitz et al., 2000; Olfson et al., 2000; Stein & Kean, 2000). They may prevent people from completing educational goals, advancing in their careers, or even holding a job in which they need to interact with others. The greater the number of feared situations, the greater the level of impairment tends to be (Stein, Torgrud, & Walker, 2000). People with social phobias often turn to tranquilizers or try to "medicate" themselves with alcohol when preparing for social interactions (see Figure 6.2). In extreme cases, they may become so fearful of interacting with others that they become essentially housebound.

Estimates of the lifetime prevalence of social phobia range from 3% to 13% (APA, 2000). The disorder appears to be more common among women than men, perhaps because of the greater social or cultural pressures placed on young women to please others and earn their approval.

Social phobia typically begins in childhood or adolescence and is often associated with a history of shyness (USDHHS, 1999a). People with social phobia typically report they were shy as children (Stemberger et al., 1995). Consistent with the *diathesis-stress model,* shyness may represent a diathesis or predisposition that makes one more vulnerable to develop social phobia in the face of stressful experiences, such as traumatic social encounters (e.g., being embarrassed in front of others). Once social phobia develops, it typically follows a chronic and persistent course throughout life.

Agoraphobia The word *agoraphobia* is derived from Greek words meaning "fear of the marketplace," which is suggestive of a fear of being out in open, busy areas. Agoraphobia involves fear of places and situations from which it might be difficult or embarrassing to escape in the event of panicky symptoms or a full-fledged panic attack, or of situations in which help may be unavailable if such problems should occur. People with agoraphobia may fear shopping in crowded stores; walking through crowded streets; crossing a bridge; traveling on a bus, train, or car; eating in restaurants; or even leaving the house. They may

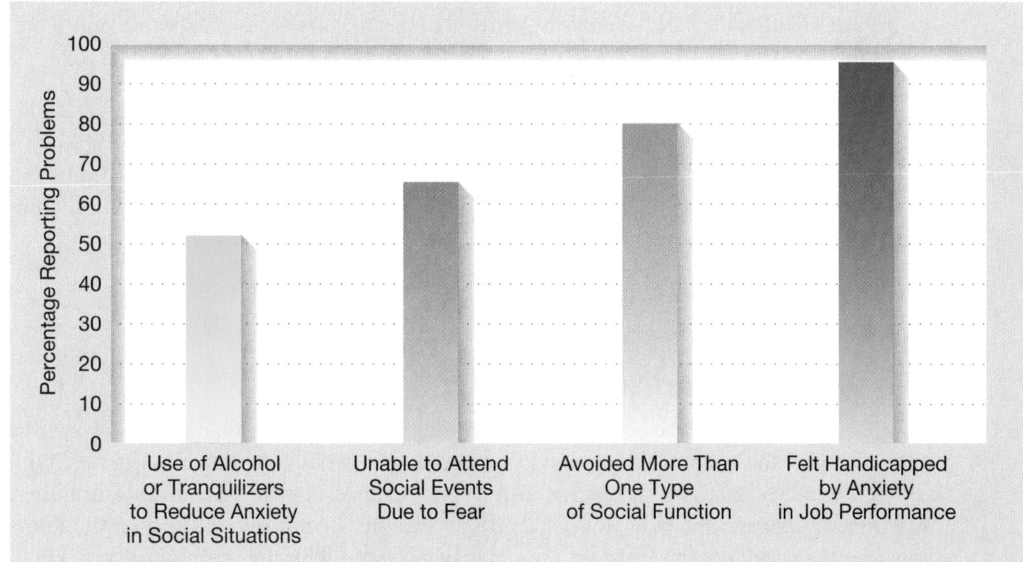

FIGURE 6.2 Percentages of people with social phobia reporting specific difficulties associated with their fears of social situations.
More than 90% of people with social phobia feel handicapped by anxiety in their jobs.
Source. Adapted from Turner & Beidel, 1989.

structure their lives around avoiding exposure to fearful situations and in some cases become housebound for months or even years, even to the extent of being unable to venture outside to mail a letter. Agoraphobia has the potential of becoming the most incapacitating type of phobia.

Agoraphobia is more common in women than men (USDHHS, 1999a). Frequently it begins in late adolescence or early adulthood. Approximately 6% of adult Americans have experienced agoraphobia at some point in their lives (Eaton, Dryman, & Weissman, 1991). Agoraphobia may occur with or without an accompanying panic disorder. In panic disorder with agoraphobia, the person may live in fear of recurrent attacks and avoid public places where attacks have occurred or might occur. Because panic attacks can descend from nowhere, some people restrict their activities for fear of making public spectacles of themselves or finding themselves without help. Others venture outside only with a companion. Still others forge ahead despite intense anxiety.

Truth OR Fiction? REVISITED

Some people are so fearful of leaving their homes that they are unable to venture outside even to mail a letter.

TRUE. Some people with agoraphobia become literally housebound and unable to venture outside even to mail a letter.

People with agoraphobia who have no history of panic disorder may experience mild panicky symptoms, such as dizziness, that lead them to avoid venturing away from places where they feel safe or secure. They too tend to become dependent on others for support. There is some evidence that people with agoraphobia without a history of panic disorder tend to function more poorly than do people who have both panic disorder and agoraphobia (Goisman et al., 1994). The following case of agoraphobia without a history of panic disorder illustrates the dependencies often associated with agoraphobia:

A Case of Agoraphobia

Helen, a 59-year-old widow, became increasingly agoraphobic after the death of her husband 3 years earlier. By the time she came for treatment, she was essentially housebound, refusing to leave her home except under the strongest urging of her daughter, Mary, age 32, and only if Mary accompanied her. Her daughter and 36-year-old son, Pete, did her shopping for her and took care of her other needs as best they could. However, the burden of caring for their mother, on top of their other responsibilities, was becoming too great for them to bear. They insisted that Helen begin treatment and Helen begrudgingly acceded to their demands.

Helen was accompanied to her evaluation session by Mary. She was a frail looking woman who entered the office clutching Mary's arm and insisted that Mary stay throughout the interview. Helen recounted that she had lost her husband and mother within 3 months of one another; her father had died 20 years earlier. Although she had never experienced a panic attack, she always considered herself an insecure, fearful person. Even so, she had been able to function in meeting the needs of her family until the deaths of her husband and mother left her feeling abandoned and alone. She had now become afraid of "just about everything" and was terrified of being out on her own, lest something bad would happen and she wouldn't be able to cope with it. Even at home, she was fearful that she might lose Mary and Pete. She needed constant reassurance from them that they too wouldn't abandon her.

—From the Authors' Files

THINK ABOUT IT

Do you have any specific phobias, such as fears of small animals, insects, heights, or enclosed spaces? What do you think contributed to the development of the phobia (or phobias)? How has the phobia impacted on your life? How have you coped with it?

Obsessive-Compulsive Disorder

An **obsession** is an intrusive and recurrent thought, idea, or urge that seems beyond the person's ability to control. Obsessions can be potent and persistent enough to interfere with daily life and can engender significant distress and anxiety. They include doubts, impulses, and mental images. One may wonder endlessly whether or not one has locked the doors and shut the windows, for example. One may be obsessed with the impulse to do harm to

obsession A recurrring thought or image that the individual cannot control.

one's spouse. One can harbor images, such as the recurrent fantasy of a young mother that her children had been run over by traffic on the way home from school.

A **compulsion** is a repetitive behavior (such as hand-washing or checking door locks) or mental acts (such as praying, repeating certain words, or counting) that the person feels compelled or driven to perform (APA, 2000). Compulsions often occur in response to obsessional thoughts and are frequent and forceful enough to interfere with daily life or cause significant distress. A compulsive hand-washer, Corinne, engaged in elaborate hand-washing rituals. She spent 3 to 4 hours daily at the sink and complained, "My hands look like lobster claws." Some people literally take hours checking and rechecking that all the appliances are off before they leave home, and then doubts still remain.

Most compulsions fall into two categories: checking rituals and cleaning rituals. Rituals can become the focal point of life. Checking rituals, such as repeatedly checking that the gas jets are turned off or the doors are securely locked before leaving the house, cause delays and annoy companions; cleaning can occupy several hours a day. Table 6.5 shows some relatively common obsessions and compulsions.

Compulsions often accompany obsessions and appear to at least partially relieve the anxiety created by obsessional thinking. By washing one's hands 40 or 50 times in a row each time a public doorknob is touched, the compulsive hand-washer may experience some relief from the anxiety engendered by the obsessive thought that germs or dirt still linger in the folds of skin. The person may believe the compulsive act will help prevent some dreaded event from occurring, even though there is no realistic basis to the belief or the behavior far exceeds what is reasonable under the circumstances. Compulsive rituals apparently also reduce the anxiety that would occur if they were prevented from being carried out (Foa, 1990).

An obsessive thought? One type of obsession involves recurrent, intrusive images of a calamity occurring as the result of one's own carelessness. For example, a person may not be able to shake the image of his or her house catching fire due to an electrical short in an appliance inadvertently left on.

Obsessive-compulsive disorder (OCD) affects between 2% and 3% of the general population at some point in their lives (APA, 2000; Taylor, 1995). A Swedish study found that while most OCD patients eventually showed some improvement, most also continued to have some symptoms of the disorder over the course of their lifetimes (Skoog & Skoog, 1999). The disorder occurs about equally often in men and women (APA, 2000; USDHHS, 1999a). The *DSM* diagnoses obsessive-compulsive disorder

VIDEO 6.2
Obsessive-Compulsive Disorder: *The Case of Ed*

TABLE 6.5 Examples of Obsessive Thoughts and Compulsive Behaviors

Obsessive Thought Patterns	Compulsive Behavior Patterns
Thinking that one's hands remain dirty despite repeated washing.	Rechecking one's work time and time again.
Difficulty shaking the thought that a loved one has been hurt or killed.	Rechecking the doors or gas jets before leaving home.
Repeatedly thinking that one has left the door to the house unlocked.	Constantly washing one's hands to keep them clean and germ free.
Worrying constantly that the gas jets in the house were not turned off.	
Repeatedly thinking that one has done terrible things to loved ones.	

Truth OR Fiction? REVISITED

It may take an hour or more for people with obsessive-compulsive disorder to leave the house.

TRUE. People with obsessive-compulsive disorder may be delayed in leaving the house for an hour or more as they carry out their checking rituals.

compulsion A repetitive or ritualistic behavior that the person feels compelled to perform.

Lady Macbeth. In Shakespeare's tragedy *Macbeth,* after spurring her husband on to murder the king and usurp the throne, Lady Macbeth obsessively washes her hands in an effort to cleanse herself of her crime.

when people are troubled by recurrent obsessions, compulsions, or both such that they cause marked distress, occupy more than an hour a day, or significantly interfere with normal routines or occupational or social functioning (APA, 2000). Many people with OCD, especially those who developed the disorder during childhood, also have a history of tic disorders. Investigators suspect there may be a genetic link between tic disorders and OCD, or at least child-onset OCD (Eichstedt & Arnold, 2001).

The line between obsessions and the firmly held but patently false beliefs that are labeled *delusions,* which are found in schizophrenia, is sometimes less than clear. Obsessions, such as the belief that one is contaminating other people, can, like delusions, become almost unshakable. Although adults with OCD may be uncertain at a given time about whether their obsessions or compulsions are unreasonable or excessive (Foa & Kozak, 1995), they will eventually concede that their concerns are groundless or excessive. True delusions fail to be shaken. Children with OCD may not come to recognize their concerns are groundless, however. The following case illustrates a checking compulsion:

A Case of Obsesssive-Compulsive Disorder

Jack, a successful chemical engineer, was urged by his wife Mary, a pharmacist, to seek help for "his little behavioral quirks," which she had found increasingly annoying. Jack was a compulsive checker. When they left the apartment, he would insist on returning to check that the lights or gas jets were off, or that the refrigerator doors were shut. Sometimes he would apologize at the elevator and return to the apartment to carry out his rituals. Sometimes the compulsion to check struck him in the garage. He would return to the apartment, leaving Mary fuming. Going on vacation was especially difficult for Jack. The rituals occupied the better part of the morning of their departure. Even then, he remained plagued by doubts.

Mary had also tried to adjust to Jack's nightly routine of bolting out of bed to recheck the doors and windows. Her patience was running thin. Jack realized that his behavior was impairing their relationship as well as causing himself distress. Yet he was reluctant to enter treatment. He gave lip service to wanting to be rid of his compulsive habits, but he also feared that surrendering his compulsions would leave him defenseless against the anxieties they helped ease.

—*From the Authors' Files*

Acute and Posttraumatic Stress Disorders

In adjustment disorders (discussed in Chapter 5), people have difficulty adjusting to life stressors, such as business or marital problems, chronic illness, or bereavement over a loss. Here we focus on stress-related disorders that arise from exposure to *traumatic* events. **Acute stress disorder (ASD)** is a maladaptive reaction that occurs during the initial month following the traumatic experience. **Posttraumatic stress disorder (PTSD)** is a prolonged maladaptive reaction to a traumatic experience. ASD is a major risk factor for PTSD, as many people with ASD later develop PTSD (Harvey & Bryant, 1999, 2000; Sharp & Harvey, 2001). In contrast to ASD, PTSD may persist for months, years, or even decades and may not develop until many months or years after exposure to the traumatic event (Zlotnick et al., 2001).

acute stress disorder (ASD) A traumatic stress reaction occurring in the days and weeks following exposure to a traumatic event.

posttraumatic stress disorder (PTSD) A prolonged maladaptive reaction to a traumatic event.

Both types of stress disorders have occurred among soldiers exposed to combat, rape survivors, victims of motor vehicle and other accidents, and people who have witnessed the destruction of their homes and communities by natural disasters such as floods, earthquakes, or tornadoes, or technological disasters such as railroad or airplane crashes.

In ASD and PTSD, the traumatic event involves either actual or threatened death or serious physical injury, or threat to one's own or another's physical safety. The person's response to the threat involves feelings of intense fear, helplessness, or a sense of horror. Children with PTSD may have experienced the threat differently, such as by showing confused or agitated behavior.

Exposure to trauma is quite common in the general population ("What Is PTSD?" 1996). A recent random sample of Americans showed that 72% reported some traumatic experience, such as exposure to natural disasters, death of a child, serious motor vehicle accidents, witnessing violence, or experiencing physical assault, rape, or physical or sexual abuse (Elliott, 1997).

Though most people who suffer trauma experience some degree of psychological distress (Sharp & Harvey, 2001), not all trauma survivors go on to develop ASD or PTSD. Yet many do. The prevalence of PTSD among trauma survivors remains an open question. In recent studies, investigators found that about 1 in 3 survivors of motor vehicle accidents developed PTSD within a year of the accident (Koren, Arnon, & Klein, 1999; Ursano et al., 1999). Yet some investigators find lower rates of PTSD among severely injured accident victims (e.g., Schnyder et al., 2000). A study of 255 adult survivors of the 1995 Oklahoma City bombing showed a prevalence rate for PTSD of 34% within 6 months of the disaster (North et al., 1999). Though we don't yet know how many people will develop PTSD in the aftermath of the 2001 terrorist attack on the World Trade Center in New York, we can expect the emotional toll to be staggering.

WWW **Web Link 6.3**
Coping with the Aftermath of a Disaster

Overall, investigators believe that about 8% of U.S. adults are affected by PTSD at some point in their lives (Kessler et al., 1995). About 2% of American adults currently show evidence of diagnosable PTSD. The prevalence of ASD in the general population is not known.

Vulnerability to PTSD may depend on such factors as resiliency and vulnerability to the effects of trauma, prior history of childhood sexual abuse, severity of the trauma and degree of exposure, availability of social support, use of active coping responses in dealing with the traumatic stressor, and feelings of shame (Andrews et al., 2000; Brewin, Andrews, & Valentine, 2000; Nishith, Mechanic, & Resick, 2000; Prigerson et al., 2001; Regehr, Hill, & Glancy, 2000; Sharkansky et al., 2000; Silva et al., 2000). Finding a sense of purpose or meaning in the traumatic experience, such as believing that the war one is fighting is just, may also bolster the person's ability to cope with the stressful circumstances and reduce the risk of traumatic stress reactions (Sutker et al., 1995).

Truth OR Fiction? REVISITED

Men are more than twice as likely as women to develop posttraumatic stress disorder (PTSD).

FALSE. Women are about twice as likely as men to develop PTSD.

Although men more often encounter traumatic experiences, women are more likely to develop PTSD in response to trauma (Ehlers, Mayou, & Bryant, 1998; Michaud, 2000). Overall, women are about twice as likely to develop the disorder during their lifetimes than are men. The risk of PTSD in women is also linked to a history of battering in marriage and childhood sexual abuse (Astin et al., 1995; Rodriguez et al., 1997). Researchers find that women who develop PTSD also tend to have an increased risk of suffering major depression and alcohol use disorders (Breslau et al., 1997b). PTSD may also occur among children exposed to traumatic experiences (Silva et al., 2000).

The *DSM-IV* loosened the criteria for PTSD to include reactions to a wider range of traumatic stressors, including receiving a diagnosis of a life-threatening illness. In breast cancer survivors, PTSD symptoms appear to be more common than would be expected in the general population (Cordova et al., 1995). A recent study of patients who had experienced severe traumatic brain injuries showed a 27% rate of PTSD (Harvey et al., 2000).

PTSD entered the popular vocabulary after the Vietnam conflict of the 1960s. Veterans with PTSD have a greater likelihood than other veterans to have problems at home; about 60% report a high level of marital problems (Jordan et al., 1992). They are also more likely to commit suicide (Bullman & Kang, 1994). Many veterans with PTSD also abuse alcohol and drugs and become violent or socially withdrawn (Chemtob et al., 1997). They also have high rates of other psychological disorders, including major depression, panic disorder, and social phobias (Orsillo et al., 1996).

Features of Traumatic Stress Reactions ASD and PTSD share many of the same features or symptoms (Bryant, 2001). Some common features are reexperiencing the traumatic event; avoidance of cues or stimuli associated with the event; a numbing of general or emotional responsiveness; heightened states of bodily arousal; and critical emotional distress or impairment of functioning. The major difference in the features of the two disorders is the emphasis in ASD on *dissociation*—feelings of detachment from oneself or one's environment (Bryant, 2001; USDHHS, 1999a). People with acute stress disorder may feel they are "in a daze" or that the world seems like a dreamlike or unreal place. In acute stress disorder, people may also be unable to perform necessary tasks, such as obtaining needed medical or legal assistance (APA, 2000).

Acute stress disorder frequently occurs in the context of combat or exposure to natural or technological disasters. A soldier may come through a horrific battle not remembering important features of the battle, and feeling numb and detached from the environment. People who are injured or who nearly lose their lives in a hurricane may walk around "in a fog" for days or weeks afterward; be bothered by intrusive images, flashbacks, and dreams of the disaster; or relive the experience as though it were happening again.

In acute and posttraumatic stress disorders, the traumatic event may be reexperienced in various ways. There can be intrusive memories, recurrent disturbing dreams, and the feeling the event is indeed recurring (as in "flashbacks" to the event). Exposure to events that resemble the traumatic experience can cause intense psychological distress. People with traumatic stress reactions tend to avoid stimuli that evoke recollections of the trauma. For example, they may not be able to handle a television account of it or a friend's wish to talk about it. They may have feelings of detachment or estrangement from other people. They may show less responsiveness to the external world after the traumatic event, losing the ability to enjoy previously preferred activities or to have loving feelings (Litz, 1992).

Have you ever been awakened by a nightmare and been reluctant to return to sleep for fear of reentering the orb of the dream? Nightmares in traumatic stress reactions often involve the reexperiencing of the traumatic event, which can lead to abrupt awakenings and difficulty going back to sleep—because of fear associated with the nightmare and elevated levels of arousal. Other features of heightened arousal include difficulty falling or staying asleep, irritability or angry outbursts, hypervigilance (being continuously on guard), difficulty concentrating, and an exaggerated startle response (jumping in response to sudden noises or other stimuli) (APA, 2000).

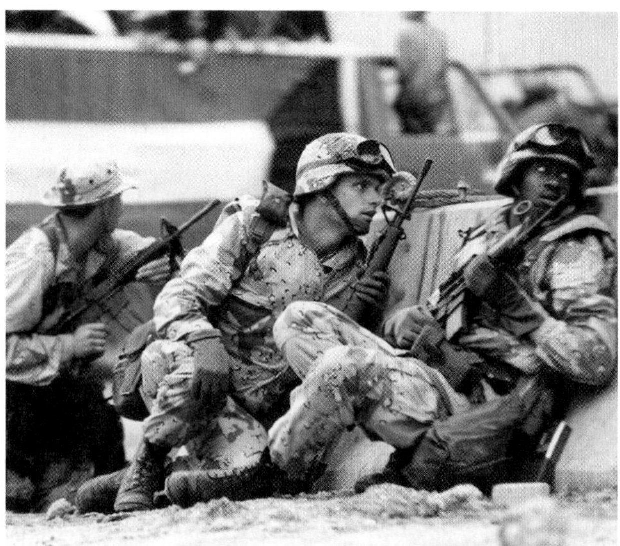

Trauma. Trauma may result from experiences in combat (left) or from terrorist violence (right), as in these survivors fleeing the World Trade Center disaster of September 11, 2001. In either case, stress-related problems may not develop until long after the experience, but may linger for years afterward in the form of posttraumatic stress disorder (PTSD).

Although PTSD often remits within a period of 6 months (USDHHS, 1999a), it can last for years, even decades (Bremmer et al., 1996; Kessler et al., 1995). Many World War II and Korean War veterans, for example, are found to meet diagnostic criteria for PTSD when evaluated four or five decades after their combat experience ended (Engdahl et al., 1997; Schnurr, Ford, & Friedman, 2000). Veterans with PTSD often present with a range of other problem behaviors, including substance abuse, marital problems, and poor work histories (Calhoun et al., 2000). Yet there is some good news to report: People who obtain treatment for PTSD typically recover sooner from the symptoms of PTSD than those who do not (Kessler et al., 1995). We can't say whether treatment shortens the duration of PTSD symptoms because it is possible that people who seek out treatment may differ in important ways from those who do not. Still, it suggests treatment has a positive influence.

Ethnic Differences in Anxiety Disorders

Although anxiety disorders have been the subject of extensive study, little attention has been directed toward examining ethnic differences in the prevalence of these disorders. Are anxiety disorders more common in certain racial/ethnic groups? We might think that stressors that African Americans in our society are more likely to encounter, such as racism and economic hardship, might contribute to a higher rates of anxiety disorders in this population group (Neal & Turner, 1991). On the other hand, it is possible that African Americans, by dint of having to cope with these hardships in early life, may have developed a resiliency in the face of stress that shields them from anxiety disorders.

The National Comorbidity Survey (NCS), which was based on a sample that closely represented the general U.S. adult population, found that anxiety disorders overall and specific anxiety disorders in particular were no more common among African Americans than among non-Hispanic White Americans (Eaton et al., 1994). Moreover, panic disorder was actually less common among African Americans in the 45- to 54-year age range than among non-Hispanic White Americans. Panic disorder was also less common among Hispanics than non-Hispanic Whites among people in the 35- to 44-year age range. Trivial differences were found among younger people across racial/ethnic lines. All in all, it appears that the rates of anxiety disorders are generally comparable across racial and ethnic groupings.

Anxiety disorders are not unique to our culture. Panic disorder, for example, is known to occur in many countries in the world, perhaps even universally (Amering & Katschnig, 1990). A multinational study of more than 40,000 people in 10 countries (Canada, Puerto Rico, France, United States, West Germany, Italy, Lebanon, Taiwan, Korea, and New Zealand) showed that rates of panic disorder were relatively consistent, ranging between 1% and 3% in all of these countries except Taiwan, where the rate was under 1%. However, the specific features of panic attacks, such as shortness of breath or fear of dying, may vary from culture to culture (Amering & Katschnig, 1990). Some culture-bound syndromes have features similar to panic attacks, such as *ataque de nervios* (see Chapter 3).

PTSD is also found in other cultures. High rates of PTSD were found in a southwestern U.S. Indian tribe, and among earthquake survivors in China, hurricane survivors in Nicaragua, Khmer refugees who survived the "killing fields" of the 1970s Pol Pot war in Cambodia, tortured Bhutanese refugees (van Ommeren et al., 2001), and survivors of the Balkan conflicts of the 1990s (Cardozo et al., 2000; Goenjian et al., 2001; Mitka, 2000; Mutler, 2000; Sack, Clarke, & Seeley, 1996; Wang et al., 2000; Weine et al., 2000). Cultural factors may play a role in determining how people manage and cope with trauma as well as their vulnerability to traumatic stress reactions and the specific form such a disorder might take (de Silva, 1993).

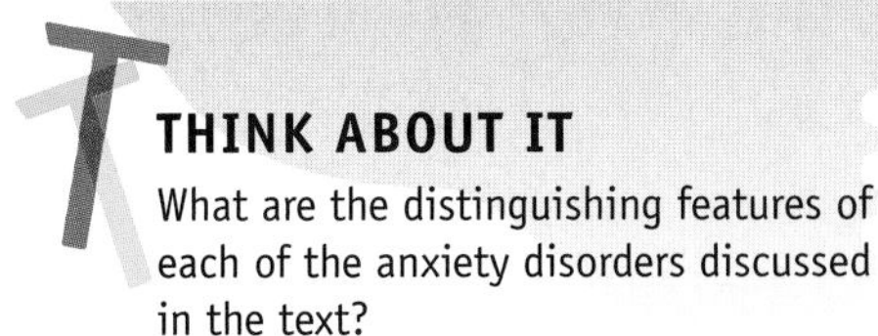

THINK ABOUT IT
What are the distinguishing features of each of the anxiety disorders discussed in the text?

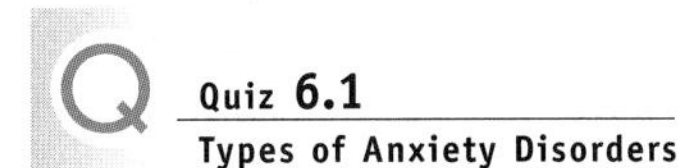

Theoretical Perspectives

The anxiety disorders offer something of a theoretical laboratory. Many theories of abnormal behavior were developed with these disorders in mind. Here we consider the contributions of these theoretical perspectives to our understanding of anxiety disorders.

Unconscious defense mechanisms? For psychodynamic theorists, phobias represent the operation of unconscious defense mechanisms such as projection and displacement. In their view, a fear of heights may represent the ego's attempt to defend itself against the emergence of threatening self-destructive impulses, such as an impulse to jump from a dangerous height. By avoiding heights, the person can maintain a safe distance from such threatening impulses. Because this process occurs unconsciously, the person may be aware of the phobia, not of the unconscious impulses that it symbolizes.

Psychodynamic Perspectives

From the psychodynamic perspective, anxiety is a danger signal that threatening impulses of a sexual or aggressive (murderous) nature are nearing the level of awareness. To fend off these threatening impulses, the ego tries to stem or divert the tide by mobilizing its defense mechanisms. For example, with phobias, the defense mechanisms of **projection** and **displacement** come into play. A phobic reaction is believed to involve the projection of the person's own threatening impulses onto the phobic object. For instance, a fear of knives or other sharp instruments may represent the projection of one's own destructive impulses onto the phobic object. The phobia serves a useful function. Avoiding contact with sharp instruments prevents these destructive wishes from becoming consciously realized or acted upon. The threatening impulses remain safely repressed. Similarly, people with acrophobia may harbor unconscious wishes to jump that are controlled by avoiding heights. The phobic object or situation symbolizes or represents these unconscious wishes or desires. The person is aware of the phobia, but not of the unconscious impulses that it symbolizes.

Freud's (1909/1959) historic case of "Little Hans," a 5-year-old boy who feared he would be bitten by a horse if he left his house, illustrates his principle of displacement. Freud hypothesized that Hans's fear of horses represented the displacement of an unconscious fear of his father. According to Freud's conception of the Oedipus complex, boys have unconscious incestuous desires to possess their mothers and fears of retribution from their fathers, whom they see as rivals in love. Hans's fear of being bitten by horses thus symbolized an underlying fear of castration.

Learning theorists view Hans's childhood fears as a case of classical conditioning (Wolpe & Rachman, 1960). They argue that Hans's fear had been learned from his being frightened by an accident involving a horse and a transport vehicle, which generalized to fears of horses. The story of Little Hans has sparked a spirited debate in the psychological annals.

Applying the psychodynamic model to other anxiety disorders, we might hypothesize that in generalized anxiety disorder, unconscious conflicts remain hidden, but anxiety leaks through to the level of awareness. The person is unable to account for the anxiety because its source remains shrouded in unconsciousness, however. In panic disorder, unacceptable sexual or aggressive impulses approach the boundaries of consciousness and the ego strives desperately to repress them, generating high levels of conflict that bring on a full-fledged panic attack. Panic dissipates when the impulse has been safely repressed.

Obsessions are believed to represent the leakage of unconscious impulses into consciousness, and compulsions are acts that help keep these impulses repressed. Obsessive thoughts about contamination by dirt or germs may represent the threatened emergence of unconscious infantile wishes to soil oneself and play with feces. The compulsion (in this case, cleanliness rituals) helps keep such wishes at bay or partly repressed.

The psychodynamic model remains largely speculative, in large part because of the difficulty (some would say impossibility) of arranging scientific tests to determine the existence of the unconscious impulses and conflicts believed to lie at the root of these disorders.

projection A defense mechanism in which one's own sexual or aggressive impulses are attributed to another person.

displacement A defense mechanism in which one transfers sexual or aggressive impulses toward less threatening or safer objects or persons.

two-factor model A theoretical model that accounts for the development of phobic reactions on the basis of classical and operant conditioning.

Learning Perspectives

From a learning perspective, anxiety disorders are acquired through the process of learning, specifically conditioning and observational learning. According to O. Hobart Mowrer's (1948) classic **two-factor model,** both classical and operant conditioning are involved in the development of phobias. The fear component of phobia is believed to be acquired by classical conditioning. It is assumed that previously neutral objects and situations gain the capacity to evoke fear by being paired with noxious or aversive stimuli. A child who is frightened by a barking dog may acquire a phobia for dogs. A child who receives a painful injection may develop a phobia for hypodermic syringes. Consistent with this model, evidence shows that many cases of acrophobia, claustrophobia, and blood and injection phobias involve earlier pairings of the phobic object with aversive experiences (e.g., Kendler et al., 1992c; Merckelbach et al., 1996).

As Mowrer pointed out, the avoidance component of phobias is acquired and maintained by operant conditioning. That is, relief from anxiety negatively reinforces avoiding fear-inducing stimuli. The person with an elevator phobia learns to avoid anxiety over riding the elevator by opting for the stairs instead. Avoiding the phobic stimulus thus lessens anxiety, which negatively reinforces the avoidance behavior. Yet there is a significant cost to avoiding the phobic stimulus. The person is not able to unlearn the fear via exposure to the phobic stimulus in the absence of any aversive consequences.

The development of panic disorder may represent a form of classical conditioning (Bouton, Mineka, & Barlow, 2001). In this view, both external cues (e.g., being in a crowd) and internal cues (e.g., heart palpitations or dizziness) may become conditioned stimuli (CSs) that elicit panicky feelings because they have been associated with the occurrence of panic attacks in the past.

Learning theorists have also noted the role of observational learning in acquiring fears. Modeling (observing parents or others react fearfully to a stimulus) and receiving negative information (hearing from others or reading that a particular stimulus—spiders, for example—are fearful or disgusting) may also lead to phobias (Merckelbach et al., 1996). In one study of 42 people with severe phobias for spiders, observational learning apparently played a more prominent role in fear acquisition than did conditioning (Merckelbach, Arnitz, & de Jong, 1991).

Some investigators suggest that people may be genetically prepared to more readily acquire phobic responses to certain classes of stimuli than others (McNally, 1987; Mineka, 1991; Seligman & Rosenhan, 1984). We're more likely to develop a fear of spiders than rabbits, for example. This model, called **prepared conditioning,** suggests that evolution would have favored the survival of human ancestors who were genetically predisposed to acquire fears of threatening objects, such as large animals, snakes, and other "creepy-crawlers"; heights; enclosed spaces; and even strangers. This model may explain why it is more likely for us to develop fears of spiders or heights than of objects that appeared much later on the evolutionary scene, such as guns or knives, even though these later-appearing objects pose more direct threats today to our survival.

prepared conditioning The belief that people are genetically prepared to acquire fear responses to certain stimuli, such as snakes or large animals.

Truth OR Fiction? REVISITED

Some theorists believe we are genetically programmed to more readily acquire fears of some classes of stimuli, including snakes.

TRUE. Some theorists believe that we are genetically predisposed to acquire certain fears, such as fears of large animals and snakes. The ability to readily acquire these fears may have had survival value to our ancestors.

Snakes and spiders. According to the concept of prepared conditioning, we are genetically predisposed to more readily acquire fears of the types of stimuli that would have threatened the survival of ancestral humans—stimuli such as large animals, snakes, and other creepy-crawlers.

PTSD may also be explained from a conditioning framework. From a classical conditioning perspective, traumatic experiences function as unconditioned stimuli that become paired with neutral (conditioned) stimuli such as the sights, sounds, and smells associated with the trauma scene—for example, the battlefield or the neighborhood in which a person has been raped or assaulted (Foy et al., 1987). Subsequent exposure to similar stimuli evokes the anxiety (a conditioned emotional response) associated with PTSD. The conditioned stimuli that reactivate the conditioned response include memories or dream images of the trauma or visits to the scene of the trauma. Consequently, the person avoids these stimuli. Avoidance is an operant response, which is reinforced by relief from anxiety. However, avoidance prolongs PTSD because sufferers do not have the opportunity to learn to manage their conditioned reactions. Extinction (gradual weakening or elimination) of conditioned anxiety may only occur when conditioned stimuli (e.g., cues associated with the trauma) are presented in a supportive therapeutic setting in the absence of the troubling unconditioned stimuli.

Web Link 6.4
National Anxiety Disorders Screening Day WWW

From a learning perspective, generalized anxiety is precisely that: a product of stimulus generalization. People concerned about broad life themes, such as finances, health, and family matters, are likely to experience their apprehensions in a variety of settings. Anxiety would thus become connected with almost any environment or situation. Similarly, agoraphobia would represent a kind of generalized anxiety. Anxiety would become triggered by cues associated with various social or vocational situations outside of the home in which the individual is expected to perform independently, as in traveling, going to work, even shopping. Some learning theorists similarly assume that panic attacks, which appear to descend out of nowhere, are triggered by cues that are subtle and not readily identified.

There are challenges to the learning theory account of phobias. Most specific phobias do not appear to develop from any specific traumatic incident, such as being bitten by a dog (USDHHS, 1999a). Learning theorists might counter that people may not recall traumatic or painful experiences from their early childhood. Yet phobias such as social phobias and agoraphobia develop at later ages and appear to involve cognitive processes usually related to an exaggerated appraisal of threat in social situations (excessive fears of embarrassment or criticism) or public places (perceptions of helplessness or fears of panic attacks) rather than the pairing of these situations with aversive experiences.

From the learning perspective, compulsive behaviors are operant responses that are negatively reinforced by relief of the anxiety engendered by obsessional thoughts. If a person obsesses that dirt or foreign bodies contaminate other people's hands, shaking hands or turning a doorknob may evoke powerful anxiety. Compulsive hand-washing following exposure to a possible contaminant provides some relief from anxiety. They thus become more likely to repeat the obsessive-compulsive cycle the next time they are exposed to anxiety-evoking cues, such as shaking hands or touching doorknobs.

The question remains why some people develop obsessive thoughts whereas others do not. Some theorists look to an interaction of learning and biological factors for answers. Perhaps people who develop obsessive-compulsive disorder are physiologically sensitized to overreact to minor cues of danger (Steketee & Foa, 1985).

Cognitive Factors in Anxiety Disorders

The focus of the cognitive perspective is on the role that distorted or dysfunctional ways of thinking may play in the development of anxiety disorders. Let us consider several styles of thinking that investigators have linked to anxiety disorders.

Overprediction of Fear People with anxiety disorders often overpredict how much fear or anxiety they will experience in anxiety-evoking situations (Rachman, 1994). The person with a snake phobia, for example, may expect to tremble on exposure to a snake. People with dental phobia tend to hold exaggerated expectations of the pain they will experience during dental visits (Marks & De Silva, 1994). Typically speaking, the actual fear or pain experienced during exposure to the phobic stimulus is a good deal less than what people had expected. Yet the tendency to expect the worst encourages avoidance of feared situations, which in turn prevents the individual from learning to manage and overcome anxiety. Overprediction of dental pain and fear may also lead people to postpone or cancel regular dental visits, which can contribute to more serious dental problems down the road. On the other hand, actual exposure to fearful situations tends to promote more accurate predictions of actual fear levels (Rachman & Bichard, 1988). A clinical implication is that with repeated exposure, people with anxiety disorders may come to anticipate their responses to fear-inducing stimuli more accurately, leading to reductions of fear expectancies. This in turn may reduce avoidance tendencies.

Self-Defeating or Irrational Beliefs Self-defeating thoughts can heighten and perpetuate anxiety and phobic disorders. When faced with fear-evoking stimuli, the person may think, "I've got to get out of here," or "My heart is going to leap out of my chest" (Meichenbaum & Deffenbacher, 1988). Thoughts like these intensify autonomic arousal, disrupt planning, magnify the aversiveness of stimuli, prompt avoidance behavior, and decrease self-efficacy expectancies concerning one's ability to control the situation.

People with phobias also tend to hold more of the sorts of irrational beliefs catalogued by Albert Ellis than nonfearful people do. Such beliefs often involve exaggerated needs to be approved of by everyone one meets and to avoid any situation in which negative appraisal from others might arise. Consider these beliefs: "What if I have an anxiety attack in front of other people? They might think I was crazy. I couldn't stand it if they looked at me that way." Results of one study may hit close to home: College men who believe it is awful (not just unfortunate) to be turned down when requesting a date show more social anxiety than those who are less likely to catastrophize rejection (Gormally et al., 1981).

Cognitive theorists relate obsessive-compulsive disorder to tendencies to exaggerate the risk of unfortunate events occurring (Bouchard, Rhϑaume, & Ladouceur, 1999). Because they expect terrible things to happen, people with OCD engage in rituals to prevent them. An accountant who imagines awful consequences for slight mistakes on a client's tax forms may feel compelled to repeatedly check her or his work. Another cognitive factor linked to the development of OCD is perfectionism, or belief that one must perform flawlessly (Shafran & Mansell, 2001). People who hold perfectionist beliefs exaggerate the consequences of turning in less than perfect work and may feel compelled to redo his or her efforts until every detail is flawless.

Oversensitivity to Threat An oversensitivity to threatening cues is a cardinal feature of anxiety disorders (Beck & Clark, 1997). People with phobias perceive danger in situations that most people consider safe, such as riding on elevators or driving over bridges. We all possess an internal alarm system that is sensitive to cues of threat. This system may have had evolutionary advantages to ancestral humans by increasing the chances of survival in a hostile environment (Beck & Clark, 1997). Ancestral humans who responded quickly to signs of threat, such as a rustling sound in the bush that may have indicated a lurking predator about to pounce, may have been better prepared to take defensive action (to fight or flee) than those with less sensitive alarm systems. The emotion of fear is a key element in this alarm system and may have motivated our ancestors to take defensive action, which in turn may have helped them survive. People today who have anxiety disorders may have inherited an acutely sensitive internal alarm that leads them to be overly responsive to cues of threat. Rather than helping them cope effectively with threats, it may lead to inappropriate anxiety reactions in response to a wide range of cues that actually pose no danger to them.

Anxiety Sensitivity **Anxiety sensitivity** is usually defined as a fear of anxiety and anxiety-related symptoms (Zinbarg et al., 2001). People with high levels of anxiety sensitivity have a fear of fear itself. They fear their emotions or associated bodily states of arousal will get out of control, leading to harmful consequences, such as having a heart

anxiety sensitivity A fear of anxiety and anxiety-related symptoms.

attack (Williams, Chambless, & Ahrens, 1997). They may be prone to panic whenever they experience bodily signs of anxiety, such as a racing heart or shortness of breath, because they take these symptoms to be signs of an impending catastrophe, such as a heart attack. Evidence suggests that anxiety sensitivity may have an inherited component (Stein, Jang, & Livesley, 1999). Anxiety sensitivity may also vary among cultural groups. Investigators found a higher level of anxiety sensitivity in a sample of American Indian and Alaska Native college students than in college students from the majority Caucasian culture (Zvolensky et al., 2001).

Anxiety sensitivity is an important risk factor for panic disorder (Lilienfeld, 1997). In one study, researchers used an anxiety sensitivity measure to predict which military recruits would be most likely to panic during a highly stressful period of basic training (Schmidt, Lerew, & Jackson, 1997). One in 5 recruits who scored in the top 10% on a measure of anxiety sensitivity experienced a panic attack, as compared to only 6% of the other recruits. As we see next, panic-prone individuals also tend to misattribute changes in their bodily sensations to dire consequences.

Misattributions of Bodily Cues Cognitive theorists point to the role that catastrophic misinterpretations of bodily sensations, such as heart palpitations, dizziness, or light-headedness, play in the escalation of panicky symptoms into full-fledged panic attacks (Clark, 1986; Zoellner, Craske, & Rapee, 1996). People with a proneness to panic disorder tend to attribute bodily cues like heart palpitations, dizziness, or light-headedness to an impending heart attack or other threatening event, such as loss of control or "going crazy." These bodily cues may occur as a consequence of unrecognized hyperventilation, temperature changes, or reactions to certain drugs or medications. Or they may be fleeting, normally occurring changes in bodily states that typically go unnoticed by most people. But in panic-prone individuals, these bodily cues may be misattributed to dire causes, setting in motion a vicious cycle that can bring on panic attacks.

A leading cognitive model of panic disorder focuses on the interaction of cognitive and physiological factors, as depicted in Figure 6.3. The model suggests that people with a proneness to panic disorder may perceive environmental cues or bodily cues as unduly threatening or dangerous, perhaps because they are overly sensitive to these cues or have associated these cues with earlier panic attacks (Clark, 1986; Zoellner, Craske, & Rapee, 1996). This sense of threat induces anxiety or feelings of apprehension, which is accompanied by sympathetic nervous system activation that leads to the release of epinephrine (adrenaline) by the adrenal glands (Wilkinson et al., 1998). Epinephrine intensifies physical sensations by inducing an accelerated heart rate, rapid breathing, and sweating. These changes in bodily sensations, in turn, become misinterpreted as signs of an impending panic attack or worse, as an impending catastrophe ("My God, I'm having a heart attack!"). Misattributions of bodily cues further reinforce perceptions of threat, which further heightens anxiety, leading to yet more anxiety-related bodily symptoms, and so on and so on in a vicious cycle that quickly spirals into a full-fledged panic attack.

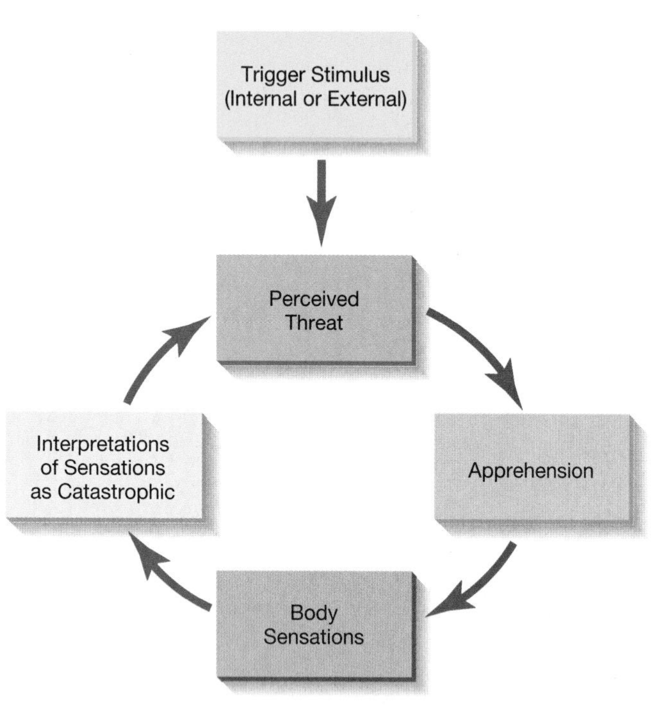

FIGURE 6.3 A cognitive model of panic disorder.
This model depicts the interaction of cognitive and physiological factors. In panic-prone people, perceptions of threat from internal or external cues lead to feelings of apprehension or anxiety, which lead to changes in body sensations (for example, cardiovascular symptoms). These changes lead, in turn, to catastrophic interpretations, thereby intensifying the perception of threat, further heightening anxiety, and so on in a vicious circle that may culminate in a full-blown panic attack.

Source. Adapted from Clark, 1986.

Low Self-Efficacy If you believe you lack the ability to handle the stressful challenges you face in life, the more anxious you are likely to feel in the face of these challenges (Bandura et al., 1985). On the other hand, if you feel capable of performing tasks you undertake, such as playing the piano, giving a speech in public, or riding on a train or driving over bridges without panicking, you are less likely to be troubled by anxiety or fear when you attempt them. People with low levels of self-efficacy (lack of belief in their ability to perform tasks successfully) tend to focus on their perceived inadequacies, as in the following case:

Anxiety and Low Self-Efficacy

Brenda, a 19-year-old sophomore, was plagued by anxiety almost from the moment she began her college studies. She had enrolled in a college several hundred miles away from home. While she had been away from home before—at sleep-away camp and on a teen tour through Europe—college life presented various challenges and stresses, which she felt a lack of ability to handle. She seemed to be most anxious when meeting new friends and when sitting in class, especially the small seminar classes in which she expected to be called on by the professor. She found herself becoming tongue tied and dripping with perspiration whenever she confronted these situations. What was more surprising and perplexing to her was that she had never had any trouble before either making new friends or talking in class.

In both situations, Brenda lost confidence in her ability to express herself. The ideas she wished to express were blocked by anxiety, which impaired her ability to think and speak clearly.The anxiety was maintained by an erroneous perception of herself as not capable of saying the right thing when called upon in class or when meeting new people. Brenda reported that she hadn't had any problems in high school either speaking up in class or making new friends. College, however, was a different experience. At college there were people she hadn't grow up with, and there were professors who had no tolerance, or so she believed, for any student who wasn't a budding genius. Her whole mental set had shifted into a defensive attitude in which self-doubts replaced self-confidence.

Brenda's history of social and academic success couldn't shield her from the nagging s elf-doubts she began to experience as she confronted the more demanding stresses of college life. She was not any less capable of coping with these challenges in college than she was in high school. She didn't suddenly lose her wits or her social skills when she entered college. What was different was that she began to perceive herself as unable to cope with the demands of a new environment that seemed both unsupportive and threatening. Appraising herself this way, it was little wonder that she experienced anxiety in class and social situations, which impaired her efforts to speak clearly. She then interpreted her speech difficulties as evidence of her inadequacies, feeding the vicious cycle of anxiety in which self-doubt leads to anxiety, which hampers performance, which occasions more self-doubts and anxiety, and so on.

—From the Authors' Files

We have increasing evidence of the importance of cognitive factors in anxiety disorders. For example, evidence shows that people with panic disorder do have a greater tendency to misinterpret bodily sensations as signs of impending catastrophe than do people without anxiety disorders or those with other types of anxiety disorders (Clark et al., 1997). Studies also show that panic-prone people have greater awareness of, and sensitivity to, their internal physiological cues, such as heart palpitations (Pauli et al., 1997; Richards, Edgar, & Gibbon, 1996). More research is needed, however, to determine the extent to which cognitive factors play a direct causal role in the development of panic disorder or other anxiety disorders.

Biological Factors in Anxiety Disorders

A growing body of evidence points to the importance of biological factors in anxiety disorders—factors such as heredity and biochemical imbalances in the brain.

Genetic Factors Genetic factors appear to play important roles in the development of many anxiety disorders, including panic disorder, generalized anxiety disorder, obsessive-compulsive disorder, and phobic disorders (APA, 2000; Gorman et al., 2000; Hettema,

Truth OR Fiction? REVISITED

Misinterpretations of bodily sensations may set into motion a spiraling cycle of anxiety that culminates in a full-fledged panic attack.

TRUE. According to cognitive theorists, catastrophic misinterpretations of bodily cues may set into motion a vicious cycle that brings on panic attacks.

How do self-doubts affect our performance? According to the self-efficacy model, we are likely to feel more anxious in situations in which we doubt our ability to perform competently. Anxiety may hamper our performance, making it more difficult for us to perform successfully. Even accomplished athletes may be seized with anxiety when they are under extreme pressure, as during slumps or when competing in championship games.

Neale, & Kendler, 2001; Kendler et al., 2001). Investigators have also linked a gene to **neuroticism**, a personality trait that may underlie proneness to developing anxiety disorders (Begley, 1998). The trait of neuroticism is characterized by anxiety, a sense of foreboding, and the tendency to avoid fear-inducing stimuli. Researchers estimate that about half of the variability among people in the general population on this underlying trait is due to genetic factors, with environmental factors accounting for the rest (Plomin, Owen, & McGuffin, 1994).

Neurotransmitters A number of neurotransmitters are implicated in anxiety reactions, including **gamma-aminobutyric acid (GABA).** GABA is an *inhibitory* neurotransmitter, which means that it tones down excess activity in the nervous system and helps quell stress responses (USDHHS, 1999a). When the action of GABA is inadequate, neurons can fire excessively, possibly bringing about seizures. In less dramatic cases, inadequate action of GABA may heighten states of anxiety. This view of the role of GABA is supported by findings that people with panic disorder show lower levels of GABA in some parts of the brain (Goddard et al., 2001). Also, we know that the group of antianxiety drugs called **benzodiazepines,** which include the well-known Valium and Librium, make GABA receptors more sensitive, thus enhancing GABA's calming (inhibitory) effects (Zorumski & Isenberg, 1991).

Irregularities or dysfunctions in serotonin and norepinephrine receptors in the brain are also implicated in anxiety disorders (Southwick et al., 1997). This may explain why antidepressant drugs that affect these neurotransmitter systems often have beneficial effects in treating some types of anxiety disorders, including panic disorder (Glass, 2000) and social phobia (Van Ameringen et al., 2001). Investigators also suspect that genes involved in regulation of serotonin may play a role in determining anxiety-related traits (Lesch et al., 1996).

Biochemical Aspects of Panic Disorder The strong physical components of panic disorder have led some theorists to speculate that panic attacks have biological underpinnings, perhaps involving a dysfunctional alarm system in the brain (Glass, 2000). Psychiatrist Donald Klein (1994) proposed that a defect in the brain's respiratory alarm system leads to a bodily overreaction in panic-prone individuals to cues of suffocation, perhaps involving mild changes in the levels of carbon dioxide in the blood. In Klein's model, cues of suffocation from hyperventilation or other causes trigger a respiratory alarm, which in turn produces the cascading sensations involved in the classic panic attack: shortness of breath, smothering sensations, dizziness, faintness, increased heart rate or palpitations, trembling, sensations of hot or cold flashes, and feelings of nausea. Klein's intriguing proposal has met with some support in the professional community (e.g., McNally et al., 1995; Taylor & Rachman, 1994), as well as some dissenting voices (e.g., Ley, 1997). Other researchers report that episodes of traumatic suffocation (near-drownings or near-chokings) may play a role in the development of panic disorder in some patients (Bouwer & Stein, 1997).

Support for a biological basis of panic disorder is found in studies showing that people with panic disorder are more likely than nonpatient controls to experience more anxious, panicky symptoms in response to biological challenges such as infusion of the chemical *sodium lactate* or manipulation of carbon dioxide (CO_2) levels in the blood either via intentional **hyperventilation** (which reduces levels of CO_2 in the blood) or inhalation of carbon dioxide (which increases CO_2 levels) (e.g., Gorman et al., 2001; Kent et al., 2001; Zvolensky & Eifert, 2001).

Cognitive theorists propose that cognitive factors may be involved in explaining these biological sensitivities. They point out that biological challenges produce intense physical sensations that may be catastrophically misinterpreted by panic-prone people as signs of an impending heart attack or loss of control (McNally & Eke, 1996; Schmidt, Trakowski, & Staab, 1997). Perhaps these misinterpretations—not underlying biological sensitivities—may in turn induce panic.

neuroticism A trait that involves characteristics such as anxious behavior, apprehension about the future, and avoidance behavior.

gamma-aminobutyric acid (GABA) An inhibitory neurotransmitter believed to play a role in anxiety.

benzodiazepines The class of antianxiety drugs that includes Valium and Xanax.

hyperventilation A pattern of overly rapid breathing associated with states of anxiety.

amygdala Limbic system structure involved in processing threatening stimuli.

Supportive evidence for the cognitivist perspective comes from a recent study showing that cognitive-behavioral therapy that focused on changing faulty interpretations of bodily sensations eliminated CO_2–induced panic in a majority of panic disorder patients (Schmidt, Trakowski, & Staab, 1997). Moreover, the results of another study showed that panic patients who underwent the CO_2 infusion with a safe person present did not experience more panicky symptoms than normal controls (Carter et al., 1995). Having a supportive person available may lead the person to appraise the situation cognitively as less threatening, which may avert the spiraling of anxiety that can lead to panic attacks. However, simply receiving reassurance about the safety of the CO_2 inhalation procedure does not seem to reduce the rate of panic (Welkowitz et al., 1999).

The fact that panic attacks often seem to come out of the blue also seems to support the belief that the attacks are biologically triggered. However, it is possible that the cues that set off many panic attacks may be internal, involving changes in bodily sensations, rather than external. Changes in physical cues, combined with catastrophic thinking, may lead to a spiraling of anxiety that culminates in a full-blown panic attack.

Biological Aspects of Obsessive-Compulsive Disorder Another biological model receiving attention of late suggests that obsessive-compulsive disorder may involve heightened arousal of a so-called *worry circuit*, a neural network in the brain involved in signaling danger. In OCD, the brain may be constantly sending messages that something is wrong and requires immediate attention, leading to obsessional worrisome thoughts and repetitive compulsive behaviors. This worry circuit incorporates parts of the *limbic system*, a set of structures located below the cerebral cortex that plays a key role in memory formation and processing of emotional responses. One structure in the limbic system, the almond-shaped **amygdala**, works as a kind of "emotional computer" in evaluating stimuli as to whether they represent a threat or danger (Davidson, 2000; Öhman & Mineka, 2001) (see Figure 6.4).

The compulsive aspects of OCD may involve disturbances in brain circuits that normally suppress repetitive behaviors. This disturbance may lead people to feel compelled to perform repetitive behaviors as though they were "stuck in gear" (Leocani et al., 2001). The

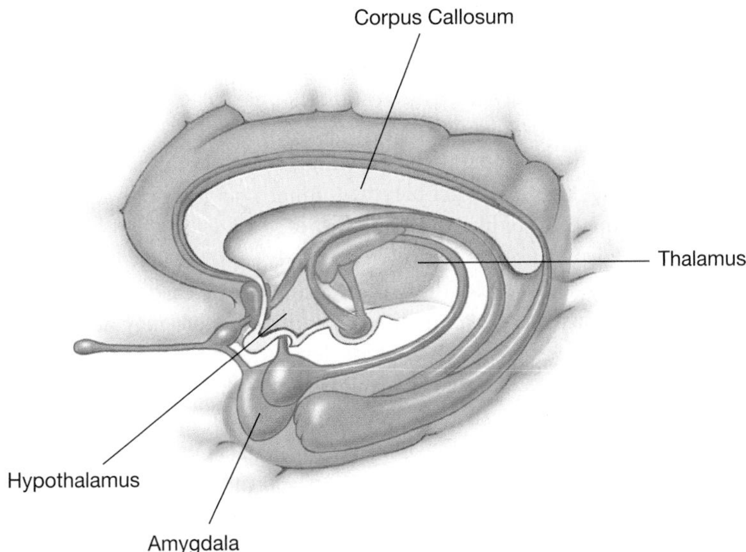

FIGURE 6.4 The amygdala and limbic system.
The amygdala is part of the limbic system, a set of interconnected structures in the brain involved in forming memories and processing emotional responses. The limbic system, which also consists of specific parts of the thalamus and hypothalamus and other nearby structures, is located in the forebrain below the cerebral cortex.

frontal lobes regulate brain centers in the lower brain that control bodily movement. A recent fMRI study showed abnormal patterns of activation in parts of the frontal lobes in OCD patients (Schwartz, 1998). Perhaps a disruption in these neural pathways explains the failure of people with compulsive behavior to inhibit these responses. Changes in frontal activation is also found among patients who respond favorably to cognitive-behavioral treatment, which suggests that CBT may directly affect parts of the brain implicated in OCD (Ingram & Siegle, 2001; Schwartz, 1998).

Tying It Together

Unraveling the complex interactions of environmental, physiological, and psychological factors in explaining how anxiety disorders develop remains a challenge. There may be different causal pathways at work. To illustrate, let us offer some possible causal pathways involved in phobic disorders and panic disorder.

Some people may develop phobias by way of classical conditioning—the pairing of a previously neutral stimulus with an unpleasant or traumatic experience. A person may develop a fear of small animals because of experiences in which they were bitten or nearly bitten. A fear of riding on elevators may arise from experiences of being trapped in elevators or other enclosed spaces.

Bear in mind that not all people who have traumatic experiences develop related phobias. Perhaps some people have a genetic predisposition that sensitizes them to more readily acquire conditioned responses to stimuli associated with aversive situations. Or perhaps people are more sensitized to these experiences because of an inherited predisposition to respond with greater negative arousal to aversive situations. Whatever factors may be involved in the acquisition of the phobia, people with persistent phobias may have learned to avoid any further contact with the phobic stimulus and so do not avail themselves of opportunities to unlearn the phobia through repeated uneventful contacts with the feared object or situation. Then there are people who acquire phobias without any prior aversive experiences with the phobic stimulus, or at least none they can recall. We can conjecture that cognitive factors, such as observing other people's aversive responses, may play a contributing role in these cases.

Possible causal pathways in panic disorder highlight roles for biological, cognitive, and environmental factors. Some people may inherit a genetic predisposition, or *diathesis,* that makes them more likely to panic in response to changes in bodily sensations. This genetic predisposition may involve an overly sensitive suffocation alarm system that is triggered by mild fluctuations in blood levels of carbon dioxide, perhaps resulting from unrecognized hyperventilation. Cognitive factors may also be involved. Cues associated with changing carbon dioxide levels, such as dizziness, tingling, or numbness, may be misconstrued as signs of an impending disaster—suffocation, a heart attack, or loss of control. This in turn may lead, like dominoes falling in line, to an anxiety reaction that quickly spirals into a full-fledged panic attack. Whether the anxiety reaction spirals into a state of panic may depend on another vulnerability factor, the individual's level of anxiety sensitivity. People with a high level of anxiety sensitivity (extreme fear of their own bodily sensations) may be more likely to panic in response to changes in their physical sensations. In some cases, anxiety sensitivity may be so high that panic ensues even in individuals without a genetic predisposition. Panic attacks may come to be triggered by exposure to internal or external cues (conditioned stimuli) that have been associated with panic attacks in the past, such as heart palpitations or boarding a train or elevator.

THINK ABOUT IT

How are anxiety disorders conceptualized within the psychodynamic, humanistic, biological, and learning-based perspectives?

Quiz **6.2**
Theoretical Perspectives

Treatment of Anxiety Disorders

Each of the major theoretical perspectives has spawned approaches for treating anxiety disorders. Psychological approaches may differ from one another in their techniques and expressed aims, but they seem to have one thing in common: In one way or another, they encourage clients to face rather than avoid the sources of their anxieties. The biological perspective, by contrast, has focused largely on the use of drugs that quell anxiety.

Psychodynamic Approaches

From the psychodynamic perspective, anxieties reflect the energies attached to unconscious conflicts and the ego's efforts to keep them repressed. Traditional psychoanalysis fosters awareness of how clients' anxiety disorders symbolize their inner conflicts, so the ego can be freed from expending its energy on repression. The ego can thus attend to more creative and enhancing tasks.

More modern psychodynamic therapies also foster clients' awareness of inner sources of conflict. They focus more so than traditional approaches on exploring sources of anxiety that arise from current rather than past relationships, however, and they encourage clients to develop more adaptive behaviors. Such therapies are briefer and more directive than traditional psychoanalysis. Though psychodynamic therapies may prove to be helpful in treating anxiety disorders, they lack extensive empirical support documenting their effectiveness (USDHHS, 1999a).

Humanistic Approaches

Humanistic theorists believe that many of our anxieties stem from social repression of our genuine selves. Anxiety occurs when the incongruity between one's true inner self and one's social facade draws closer to the level of awareness. The person senses something bad will happen, but is unable to say what it is because the disowned parts of oneself are not directly expressed in consciousness. Because of the disapproval of others, people may fail to develop their individual talents and recognize their authentic feelings. Humanistic therapies thus aim at helping people get in touch with and express their genuine talents and feelings. As a result, clients become free to discover and accept their true selves, rather than reacting with anxiety whenever their true feelings and needs begin to surface.

Biological Approaches

A variety of drugs are used to treat anxiety disorders. Among the most widely used drugs are mild tranquilizers such as the benzodiazepines Valium (generic name, *diazepam*) and Xanax (*alprazolam*). Though benzodiazepines have calming effects, they can lead to physical dependence (addiction) (USDHHS, 1999a). People who become dependent on them may experience a range of withdrawal symptoms if they stop using the drugs abruptly, including such symptoms as rebound anxiety, insomnia, and restlessness. These unpleasant symptoms may prompt people to resume using the drugs.

Antidepressant drugs have antianxiety and antipanic effects as well as antidepressant effects (Glass, 2000; Roy-Byrne & Cowley, 1998; USDHHS, 1999a). Antidepressants may help counter anxiety by normalizing the activity of neurotransmitters in the brain. Some antidepressants in common use for treating panic disorder include the tricyclics *imipramine* (brand name Tofranil) and *clomipramine* (brand name Anafranil) and the SSRIs *paroxetine* (brand name Paxil) and *sertraline* (brand name Zoloft). However, troublesome side effects may occur, such as heavy sweating and heart palpitations, which leads many patients prematurely to stop using the drugs. The high-potency tranquilizer *alprazolam* (Xanax), which is a type of benzodiazepine, is also helpful in treating panic disorder, social phobia, and generalized anxiety disorder (Barlow et al., 2000; Gould et al., 1997; van Balkom et al., 1997).

Truth OR Fiction? REVISITED

The same drugs used to treat schizophrenia are also used to control panic attacks.

FALSE. Drugs used to treat schizophrenia are not used to treat panic disorder. However, antidepressants have shown therapeutic benefits in helping to control panic attacks.

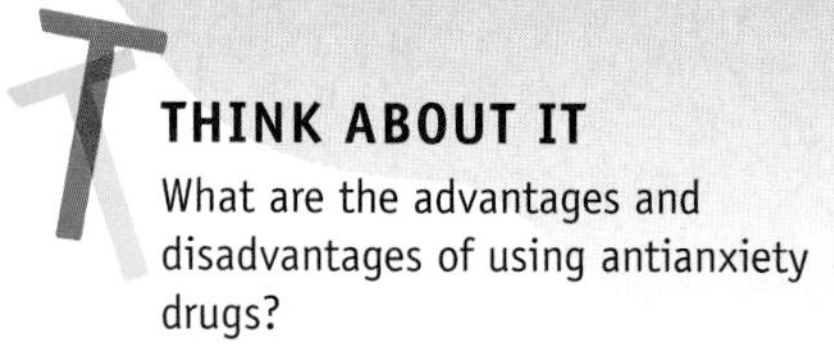

THINK ABOUT IT

What are the advantages and disadvantages of using antianxiety drugs?

Antidepressants may also be helpful in treating other anxiety disorders, including agoraphobia that accompanies panic disorder, social phobia, PTSD, obsessive-compulsive disorder, and generalized anxiety disorder (Brady et al., 2000; Davidson et al., 2001; van Ameringen et al., 2001).

Obsessive-compulsive disorder (OCD) appears to be especially responsive to SSRI-type antidepressants—drugs such as *fluoxetine* (Prozac), *clomipramine* (brand name Anafranil), and *fluvoxamine* (brand name Luvox). These drugs increase the availability in the brain of the neurotransmitter serotonin (Jenike et al., 1997; Riddle et al., 2001; USDHHS, 1999a). The effectiveness of these drugs leads researchers to suspect that problems with serotonin transmission play an important role in the development of OCD, at least in some cases (Hollander et al., 1992). Bear in mind, however, that some people with OCD fail to respond to these drugs, and among those who do respond, a complete remission of symptoms is uncommon (DeVeaugh-Geiss, 1994; Riddle et al., 2001).

A potential problem with drug therapy is that patients may attribute clinical improvement to the drugs and not their own resources. Nor do such drugs produce cures. Relapses are common after patients discontinue the medication (Spiegel & Bruce, 1997). Reemergence of panic is likely unless cognitive-behavioral treatment is provided to help panic patients modify their cognitive overreactions to their bodily sensations (Clark, 1986). Drug therapy is sometimes combined with cognitive-behavioral therapy. Evidence suggests that drug therapy does not interfere with the effectiveness of the cognitive-behavioral treatment (e.g., Bruce, 1996).

Learning-Based Approaches

A substantial body of research demonstrates the effectiveness of learning-based approaches in treating a range of anxiety disorders (USDHHS, 1999a). At the core of these approaches is the effort to help individuals learn to cope more effectively with objects or situations that elicit their fears and anxieties.

Systematic Desensitization

A Fear of Injections

Adam has a phobia for receiving injections. His behavior therapist treats him as he reclines in a comfortable padded chair. In a state of deep muscle relaxation, Adam observes slides projected on a screen. A slide of a nurse holding a needle has just been shown three times, 30 seconds at a time. Each time Adam has shown no anxiety. So now a slightly more discomforting slide is shown: one of the nurse aiming the needle toward someone's bare arm. After 15 seconds, our armchair adventurer notices twinges of discomfort and raises a finger as a signal (speaking might disturb his relaxation). The projector operator turns off the light, and Adam spends two minutes imagining his "safe scene"—lying on a beach beneath the tropical sun. Then the slide is shown again. This time Adam views it for 30 seconds before feeling anxiety.

—*From* Essentials of psychology *(6th ed.) by S. A. Rathus, p. 537.*

■

Adam is undergoing systematic desensitization, a fear-reduction procedure originated by psychiatrist Joseph Wolpe (1958) in the 1950s. Systematic desensitization is a gradual process. Clients learn to handle progressively more disturbing stimuli while they remain relaxed. About 10 to 20 stimuli are arranged in a sequence or hierarchy—called a **fear-stimulus hierarchy**—according to their capacity to evoke anxiety. By using their imagination or by viewing photos, clients are exposed to the items in the hierarchy, gradually imagining them-

fear-stimulus hierarchy An ordered series of increasingly fearful stimuli.

selves approaching the target behavior—be it ability to receive an injection or remain in an enclosed room or elevator—without undue anxiety.

Joseph Wolpe developed systematic desensitization on the assumption that phobias are learned or conditioned responses (Rachman, 2000). He assumed they can be unlearned by counterconditioning. In counterconditioning, a response incompatible with anxiety is made to appear under conditions that usually elicit anxiety. Muscle relaxation is generally used as the incompatible response, and followers of Wolpe usually use the method of progressive relaxation (described in Chapter 5) to help clients acquire relaxation skills. For this reason, Adam's therapist is teaching Adam to experience relaxation in the presence of (otherwise) anxiety-evoking slides of needles.

Behaviorally oriented therapists, like Wolpe, explain the benefits of systematic desensitization and similar therapies in terms of principles of counterconditioning. Cognitively oriented therapists note, however, that remaining in the presence of phobic imagery, rather than running from it, is also likely to enhance self-efficacy expectancies (i.e., self-perceptions of being able to manage the phobic stimuli without anxiety) (Galassi, 1988).

Gradual exposure. In gradual exposure, the client is exposed to a fear-stimulus hierarchy in real-life situations, often with a therapist or companion serving in a supportive role. To encourage the person to accomplish the exposure tasks increasingly on his or her own the therapist or companion gradually withdraws direct support. Gradual exposure is often combined with cognitive techniques that focus on helping the client replace anxiety-producing thoughts and beliefs with calming, rational alternatives.

Gradual Exposure This method helps people overcome phobias through a stepwise approach of actual exposure to the phobic stimuli. The effectiveness of exposure therapy is well established, making it the treatment of choice for specific phobias (Barlow, Esler, & Vitali, 1998; G. T. Wilson, 1997). Here exposure therapy was used in treating a patient's case of claustrophobia:

A Case of Claustrophobia

Claustrophobia (fear of enclosed spaces) is not very unusual, though Kevin's case was. Kevin's claustrophobia took the form of a fear of riding on elevators. What made his case so unusual was his occupation: He worked as an elevator mechanic. Kevin spent his work days repairing elevators. Unless it was absolutely necessary, however, Kevin managed to complete the repairs without riding in the elevator. He would climb the stairs to the floor where an elevator was stuck, make repairs, and hit the down button. He would then race downstairs to see that the elevator had operated correctly. When his work required an elevator ride, panic would seize him as the doors closed. Kevin tried to cope by praying for divine intervention to prevent him from passing out before the doors opened.

Kevin related the origin of his phobia to an accident three years earlier in which he had been pinned in his overturned car for nearly an hour. He remembered feelings of helplessness and suffocation. Kevin developed claustrophobia—a fear of situations from which he could not escape, such as flying on an airplane, driving in a tunnel, taking public transportation, and, of course, riding in an elevator. Kevin's fear had become so incapacitating that he was seriously considering switching careers, although the change would require considerable financial sacrifice. Each night he lay awake wondering whether he would be able to cope the next day if he were required to test-ride an elevator.

Kevin's therapy involved gradual exposure. Gradual exposure, like systematic desensitization, is a step-by-step procedure that involves a fear-stimulus hierarchy. In gradual exposure, however, the target behavior is approached in actuality rather than symbolically. Moreover, the individual is active rather than relaxed in a recliner.

A typical hierarchy for overcoming a fear of riding on an elevator might include the following steps:

1. *Standing outside the elevator.*
2. *Standing in the elevator with the door open.*
3. *Standing in the elevator with the door closed.*
4. *Taking the elevator down one floor.*
5. *Taking the elevator up one floor.*

6. *Taking the elevator down two floors.*
7. *Taking the elevator up two floors.*
8. *Taking the elevator down two floors and then up two floors.*
9. *Taking the elevator down to the basement.*
10. *Taking the elevator up to the highest floor.*
11. *Taking the elevator all the way down and then all the way up.*

Clients begin at step 1 and do not progress to step 2 until they are able to remain calm on the first. If they become bothered by anxiety, they remove themselves from the situation and regain calmness by practicing muscle relaxation or focusing on soothing mental imagery. The encounter is then repeated as often as necessary to reach and sustain feelings of calmness. They then proceed to the next step, repeating the process.

Kevin was also trained to practice self-relaxation and talk calmly and rationally to himself to help himself remain calm during his exposure trials. Whenever he began to feel even slightly anxious, he would tell himself to calm down and relax. He was able to counter the disruptive belief that he was going to fall apart if he was trapped in an elevator with rational self-statements such as, "Just relax. I may experience some anxiety, but it's nothing that I haven't been through before. In a few moments I'll feel relieved."

Kevin slowly overcame his phobia but still occasionally experienced some anxiety, which he interpreted as a reminder of his former phobia. He did not exaggerate the importance of these feelings. Now and then it dawned on him that an elevator he was servicing had once occasioned fear. One day following his treatment, Kevin was repairing an elevator, which serviced a bank vault 100 feet underground. The experience of moving deeper and deeper underground aroused fear, but Kevin did not panic. He repeated to himself, "It's only a couple of seconds and I'll be out." By the time he took his second trip down, he was much calmer.

—From the Authors' Files

■

Gradual exposure is also widely used in the treatment of agoraphobia (DeRubeis & Crits-Christoph, 1998; Mueser & Liberman, 1995). Treatment is stepwise and gradually exposes the agoraphobic individual to increasingly fearful stimulus situations, such as walking through congested streets or shopping in department stores. A trusted companion or perhaps the therapist may accompany the person during the exposure trials. The eventual goal is for the person to be able to handle each situation alone and without discomfort or an urge to escape. The benefits of gradual exposure are typically enduring. Overall, researchers find that about 6 in 10 people with agoraphobia show clinically meaningful improvement following exposure-based treatment (Jacobson, Wilson, & Tupper, 1988). Fewer than 1 in 3, however, are no longer agoraphobic by the end of treatment.

Flooding The method called **flooding** is a form of exposure therapy in which subjects are exposed to high levels of fear-inducing stimuli either in imagination or real-life situations. Why? The belief is that anxiety represents a conditioned response to a phobic stimulus and should extinguish if the individual remains in the phobic situation for a long enough period of time and no harmful consequences occur. Most individuals with phobias avoid confronting phobic stimuli or beat a hasty retreat at the first opportunity for escape if they cannot avoid them. Consequently, they lack the opportunity to unlearn the fear response through extinction. In one research example, 9 of 10 people with social phobia achieved at least moderate improvement through a flooding technique in which they directly faced fear-inducing situations, such as giving a talk before an expert audience (Turner, Beidel, & Jacob, 1994).

flooding A form of exposure therapy in which subjects are exposed to high levels of fear-inducing stimuli.

Cognitive Therapy Through his rational-emotive behavior therapy approach, Ellis might show people with social phobias how irrational needs for social approval and perfectionism produce unnecessary anxiety in social interactions. Eliminating exaggerated needs for social approval is apparently a key therapeutic factor (Butler, 1989). Beck's cognitive therapy seeks to identify and correct dysfunctional or distorted beliefs. For example, people with social phobias might think no one at a party will want to talk with them and that they will wind up lonely and isolated for the rest of their lives. Cognitive therapists help clients recognize the logical flaws in their thinking and assist them in viewing situations rationally. Clients may be asked to gather evidence to test out their beliefs, which may lead them to alter beliefs they find are not grounded in reality. Therapists may encourage clients with social phobias to test their beliefs that they are bound to be ignored, rejected, or ridiculed by others in social gatherings by attending a party, initiating conversations, and monitoring other people's reactions. Therapists may also help clients develop social skills to improve their interpersonal effectiveness and teach them how to handle social rejection, if it should occur, without catastrophizing.

cognitive restructuring A cognitive therapy method that involves replacing irrational thoughts with rational alternatives.

One example of cognitive techniques is **cognitive restructuring** (also called *rational restructuring*), a process in which therapists help clients pinpoint their self-defeating thoughts and generate rational alternatives so they learn to cope with anxiety-provoking situations. Kevin learned to replace self-defeating thoughts with rational alternatives and to practice speaking rationally and calmly to himself during his exposure trials. Consider the following case:

Getting Stuck on the Elevator

Phyllis, a 32-year-old writer and mother of two sons, had not been on an elevator in 16 years. Her life revolved around finding ways to avoid appointments and social events on high floors. She had suffered from fear of elevators since the age of 8, when she had been stuck between floors with her grandmother.

To help overcome her fear of elevators, Phyllis imagined herself getting stuck in an elevator and countering the self-defeating thoughts she might experience with rational self-statements. She closed her eyes and reported the thoughts that would come to mind. The psychologist encouraged her to create a rational counterpoint to each of them. She then repeated the exercise in imagination and practiced replacing the self-defeating thoughts with rational alternatives, as in the following examples:

Self-Defeating Thought	***Rational Alternative***
Oh, oh, I'm stuck. I'm going to lose control.	*Relax. Just think coolly, what do I have to do next?*
I can't take it. I'm going to pass out.	*Okay, practice your deep breathing. Help will be coming shortly.*
I'm having a panic attack. I can't stand it.	*You've experienced all these feelings before. Just let them pass through.*
If it takes hours, that would be horrible.	*That would be annoying, but it wouldn't necessarily be horrible. I've gotten stuck in traffic longer than that.*
I've got to get out of here.	*Stay calm. There's no real danger. I can just sit down and imagine I'm somewhere else until someone comes to help.*

—*From the Authors' Files*

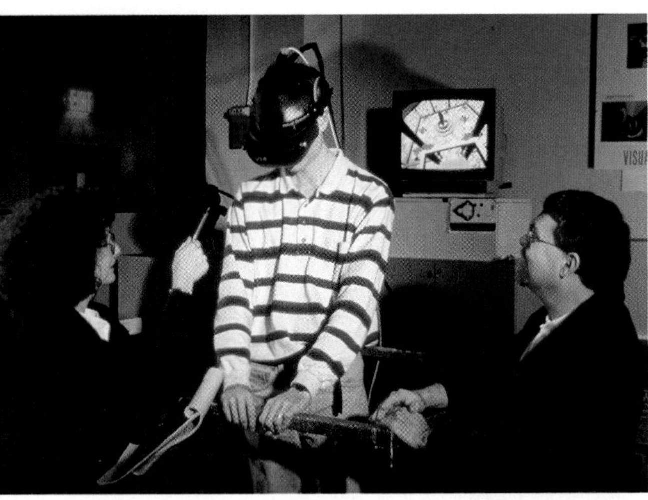

Overcoming fears with virtual reality. Virtual reality is now being used to help people overcome phobias. Using this technique, a person with a fear of heights, as pictured here, can learn to handle exposure to progressively more frightening stimuli in virtual situations. The hope is that this learning will transfer to real-life exposure to such stimuli.

Virtual Therapy for Phobias Virtual reality, the computer-generated simulated environment, has now become a therapeutic tool. By donning a specialized helmet and gloves that are connected to a computer, a person with a fear of heights, for example, can encounter frightening stimuli in this virtual world, such as riding a glass-enclosed elevator to the 49th floor, peering over a railing on a balcony on the 20th floor, or crossing a virtual Golden Gate Bridge (Goleman, 1995b; Steven, 1995). By a process of exposure to a series of increasingly more frightening virtual stimuli, while progressing only when fears at each preceding step diminish, people learn to overcome fears in much the same way they would had they followed a program of graduated exposure to phobic stimuli in real-life situations. The advantage of virtual reality is that it provides an opportunity to experience situations that might be difficult or impossible to arrange in reality (Yancey, 2000). Virtual therapy has been used successfully in helping people overcome phobias, including fears of heights and fear of flying (Rothbaum et al., 1995, 2000; Yancey, 2000). For virtual therapy to be effective, says psychologist Barbara Rothbaum who pioneered the use of the technique, the person must become immersed in the experience and believe at some level it is real and not like watching a videotape (as cited in Goleman, 1995b). "If the first person had put the helmet on and said, 'This isn't scary,' it wouldn't have worked, Dr. Rothbaum said. "But you get the same physiological changes—the racing heart, the sweat—that you would in the actual place" (Goleman, 1995b, p. C11).

We have only begun to explore the potential therapeutic uses of this new technology. Therapists are experimenting with virtual therapy to help people overcome other types of fears, such as fear of public speaking and agoraphobia. It has been used as a form of group therapy in which a group of people who are actually in different places can don virtual reality gear, log on to their computers at the same time, and meet electronically in a simulated therapy office. The virtual group members can see simulated faces of each other and communicate by typing messages directed at the group at large or to individual members (Goleman, 1995b). In other applications, virtual therapy may help clients work through unresolved conflicts with significant figures in their lives by allowing them to confront these "people" in a virtual environment. A family therapist envisions virtual family sessions in which participants can see things from the emotional and physical vantage points of each other member of the family (Steven, 1995). Other potential uses of virtual therapy include treating people with depression, social phobias, and obsessive-compulsive disorder; children with attention-deficit disorders; adults with fears of intimacy or sexual aversion; and people who have problems controlling their anger or aggressive behavior (Glantz et al., 1996; Steven, 1995). Self-help "therapy" modules, consisting of compact disks and virtual reality helmets and gloves, may even begin to appear on the shelves of your neighborhood computer software store in the not-too-distant future. With these self-help modules, people may be able in their own living rooms to confront objects or situations they fear, or learn to stop smoking or lose weight, all with the help and guidance of a "virtual therapist." Virtual therapy has recently been successfully extended to fear of spiders (Carlin, Hoffman, & Weghorst, 1997) and to fear of flying, in which the virtual environment simulates the experience of sitting in an airplane during takeoff and flight (Rothbaum, 1996).

Truth OR Fiction? REVISITED

Peering over a virtual ledge 20 stories up has helped some people overcome their fear of actual heights.

TRUE. Virtual reality therapy has been used successfully in helping people overcome phobias, including fear of heights.

Cognitive-Behavioral Therapy Cognitive-behavioral therapy (CBT) incorporates behavioral techniques, such as exposure, along with cognitive techniques, such as cognitive restructuring. Here we examine the use of CBT in treating several types of anxiety disorders: social phobia, posttraumatic stress disorder, generalized anxiety disorder, obsessive-compulsive disorder, and panic disorder.

THINK ABOUT IT

Select a particular anxiety disorder and, using specifics, describe how it would be treated from each of the major treatment approaches discussed in the text.

Social Phobia Exposure therapy is widely used successfully in treating social phobia (Barlow, Esler, & Vitali, 1998; DeRubeis & Crits-Christoph, 1998; Hoffman, 2000a, 2000b). Clients are instructed to enter increasingly stressful social situations and to remain in those situations until the urge to escape has lessened. The therapist may help guide them during exposure trials, gradually withdrawing direct support so that clients become capable of handling the situations on their own. The therapist may combine exposure treatment with cognitive techniques that assist clients in replacing maladaptive anxiety-inducing thoughts they may encounter in social situations with more adjustive thoughts. The gains achieved from cognitive-behavioral treatment of social phobia appear to be durable (Gould et al., 1997).

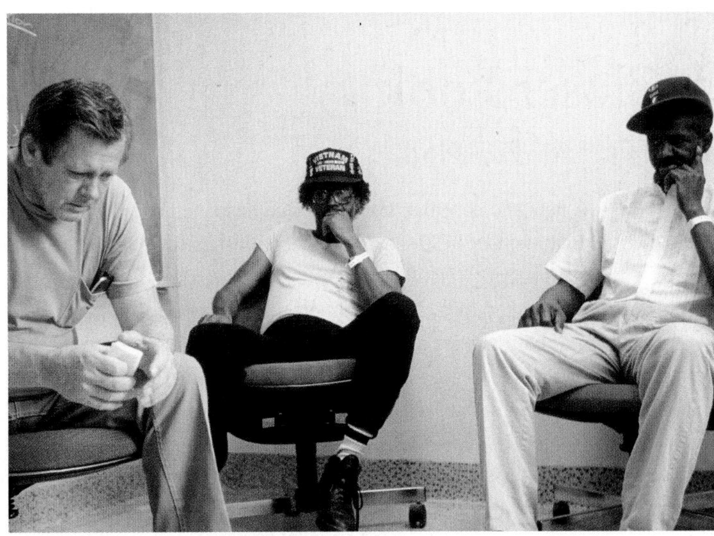

Counseling veterans with posttraumatic stress disorder. Storefront counseling centers have been established across the country to provide supportive services to combat veterans suffering from PTSD.

Posttraumatic Stress Disorder Evidence also supports the therapeutic use of cognitive-behavioral therapy in treating PTSD (DeRubeis & Crits-Christoph, 1998; Falsetti & Resnick, 2000; Taylor et al., 2001). A basic treatment component is exposure to cues associated with the trauma. The PTSD patient may be encouraged to talk about the trauma, reexperience parts of the trauma in imagination, view related slides or films, or visit the scene of the traumatic event (Foa et al., 1999; Keane, 1998; Tarrier et al., 2000). For combat-related PTSD, homework assignments may involve visiting war memorials or viewing war movies (Frueh et al., 1996). The person comes to gradually reexperience the traumatic event and accompanying anxiety in a safe setting that is free of negative consequences, which allows extinction to take its course. Exposure therapy may be supplemented with cognitive restructuring that focuses on replacing dysfunctional thoughts with rational alternatives (Marks et al., 1998b). Training in stress management skills, such as self-relaxation, may help enhance the client's ability to cope with the troubling features of PTSD, such as heightened arousal and the desire to run away from trauma-related stimuli. Training in anger management skills may also be helpful, especially with combat veterans with PTSD (Frueh et al., 1996).

Generalized Anxiety Disorder Cognitive-behavioral therapists also use a combination of techniques in treating generalized anxiety disorder (GAD). These techniques include training in self-relaxation skills; learning to substitute adaptive thoughts for intrusive, worrisome thoughts; and learning skills of decatastrophizing (avoiding tendencies to think the worst). Cognitive-behavioral approaches in treating GAD have produced greater benefits in controlled studies than either control conditions or alternative therapies in treating GAD (Barlow, Esler, & Vitali, 1998; DeRubeis & Crits-Christoph, 1998; Ladouceur et al., 2000).

Obsessive-Compulsive Disorder Therapists have achieved impressive results in treating obsessive-compulsive disorder with the technique of *exposure with response prevention* (Abramowitz & Foa, 2000; Franklin et al., 2000; McLean et al., 2001). The exposure component involves having clients intentionally place themselves in situations that evoke obsessive thoughts. For many people, such situations are hard to avoid. Leaving the house, for example, may trigger obsessive thoughts about whether or not the gas jets are turned off or the windows and doors are locked. Or clients may be instructed to purposely induce obsessive thoughts by leaving the house messy or rubbing their hands in dirt. Response prevention is the effort to prevent the compulsive behavior from occurring. Clients who rub their hands in dirt must avoid washing them for a designated period of time. The compulsive door-lock checker must avoid checking to see that the door was locked.

Through exposure with response prevention, people with OCD learn to tolerate the anxiety triggered by their obsessive thoughts while they are prevented from performing their compulsive rituals. With repeated trials, the anxiety eventually subsides and the person feels less compelled to perform the ritual. Extinction, or the weakening of the anxiety response following repeated presentation of the obsessional cues in the absence of any aversive consequences, is believed to underlie the treatment effect. Overall, about 4 of 5 people undergoing this therapy show significant improvement (Abramowitz, 1996; Foa, 1996).

Cognitive techniques are often combined with exposure therapy. The therapist focuses on helping the person correct cognitive distortions, such as tendencies to overestimate

A Closer Look

EMDR: A Fad or a Find?

A new and controversial technique has emerged in the treatment of PTSD—eye movement desensitization and reprocessing (EMDR) treatment (Shapiro, 1995). In EMDR, the client is asked to picture in mind an image associated with the trauma while the therapist rapidly moves a finger back and forth in front of the client's eyes for about 15 to 20 seconds. While holding the image in mind, the client is asked to move his or her eyes to follow the therapist's finger. The client then relates to the therapist the images, feelings, and thoughts that were experienced during the procedure. The procedure is then repeated until the client becomes desensitized to the emotional impact of this disturbing material. The technique remains controversial, in large part because we lack a compelling theoretical model to explain its effects (Keane, 1998). We do have evidence showing that EMDR can bring about therapeutic benefits in treating PTSD (e.g., DeBell & Jones, 1997a; Greenwald, 1996; Wilson, Becker, & Tinker, 1997). Yet the benefits produced by EMDR may reflect nonspecific factors common to most forms of therapy (expectancies of improvement, therapist attention) rather than the induction of rapid eye movements (Goldstein et al., 2000; Herbert et al., 2000). It is also conceivable that EMDR is merely a variant of exposure therapy, whereby the repeated presentation of the traumatic image in imagination is responsible for bringing about a reduction in fear (DeRubeis & Crits-Christoph, 1998). In support of this view, investigators who conducted a meta-analysis of research studies in the field reported that EMDR appeared to be no more effective than exposure therapy and that the eye movements that are believed to be an integral element of the technique are unnecessary (Davidson & Parker, 2001). In other words, EMDR may turn out to be but a novel way of conducting exposure-based therapy.

the likelihood and severity of feared consequences. Cognitive-behavioral techniques in treating OCD appear to be at least as effective as drug therapy (use of SSRI-type antidepressants) and may produce more lasting results (Rauch & Jenike, 1998; Stanley & Turner, 1995). Yet some patients may benefit from a combination of psychological and pharmacological treatment (USDHHS, 1999a).

THINK ABOUT IT

John has been experiencing sudden panic attacks on and off for the past few months. During the attacks, he has difficulty breathing and fears that his heart is racing out of control. His personal physician checked him out and told him the problem is with his nerves, not his heart. What treatment alternatives are available to John that might help him deal with this problem?

Panic Disorder Cognitive-behavioral therapists also use a variety of techniques in treating panic disorder, including coping skills for handling panic attacks without catastrophizing, breathing retraining and relaxation training to reduce states of heightened bodily arousal, and exposure to situations linked to panic attacks and bodily cues associated with panicky symptoms (Schmidt et al., 2000; Wilson, 1997). The therapist may assist clients with panic disorder to think differently about their bodily cues, such as sensations of dizziness or heart palpitations. By coming to recognize that these cues are fleeting sensations rather than signs of an impending heart attack or other catastrophe, clients learn to cope with them without panicking. Clients learn to replace catastrophizing thoughts and self-statements ("I'm having a heart attack") with calming, rational alternatives ("Calm down. These are panicky feelings that will soon pass."). Panic attack sufferers may also be reassured by having a medical examination to ensure that they are physically healthy and their physical symptoms are not signs of heart disease.

THINK ABOUT IT

Do you know anyone who has received treatment for an anxiety disorder? What was the outcome? What other treatment alternatives might be available? Which approach to treatment would you seek if you suffered from the same disorder?

Breathing retraining is a technique that aims at restoring a normal level of carbon dioxide in the blood by having clients breathe slowly and deeply from the abdomen, so as to avoid the shallow, rapid breathing (hyperventilation) that leads to breathing off too much carbon dioxide. In some treatment programs, people with panic disorder purposefully hyperventilate in the controled setting of the treatment clinic in order to discover for themselves the relationship between breathing off too much carbon dioxide and cardiovascular sensations. Through these firsthand experiences, they learn to calm themselves down and cope with these sensations rather than overreacting. Some commonly used elements in cognitive-behavioral therapy for panic disorder are shown in Table 6.6.

Quiz 6.3 Q
Treatment of Anxiety Disorders

Investigators find CBT to be an effective treatment for panic disorder (Barlow et al., 2000; DeRubeis & Crits-Christoph, 1998; Sanderson & Rego, 2000; Overholser, 2000). The results of a recent study show that nearly 90% of panic patients treated with CBT were free

A Closer Look

Coping with a Panic Attack

People who have panic attacks usually feel their hearts pounding such that they are overwhelmed and unable to cope. They typically feel an urge to flee the situation as quickly as possible. If escape is impossible, however, they may become immobilized and "freeze" until the attack dissipates. What can you do if you suffer a panic attack or an intense anxiety reaction? Let us suggest a few coping responses:

- Don't let your breathing get out of hand. Breathe slowly and deeply.
- Try breathing into a paper bag. The carbon dioxide in the bag may help you calm down by restoring a more optimal balance between oxygen and carbon dioxide.
- "Talk yourself down." Tell yourself to relax. Tell yourself you're not going to die. Tell yourself no matter how painful the attack is, it is likely to pass soon.
- Find someone to help you through the attack. Telephone someone you know and trust. Talk about anything at all until you regain control.
- Don't fall into the trap of making yourself housebound to avert future attacks.
- If you are uncertain about whether or not sensations such as pain or tightness in the chest have physical causes, seek immediate medical assistance. Even if you suspect your attack may "only" be one of anxiety, it is safer to have a medical evaluation than to diagnose yourself.

You need not suffer recurrent panic attacks and fears about loss of control. If your attacks are persistent or frightening, consult a professional. When in doubt, see a professional.

of panic attacks when evaluated at a follow-up assessment (Stuart et al., 2000). Despite the common belief that panic disorder is best treated with psychiatric drugs, CBT appears to produce about as good short-term results and even better long-term results after treatment termination than pharmacological approaches (Barlow et al., 2000; Otto, Pollack, & Maki, 2000). People who were treated with CBT may continue to use the skills they acquire even after treatment is completed, whereas those who receive psychiatric drugs may need the drugs to maintain the treatment effect (Glass, 2000). For some individuals, however, the effectiveness of CBT may be enhanced by the addition of antidepressant drugs (van Balkom et al., 1997).

In this chapter we have explored the diagnostic class of anxiety disorders. In the next chapter we examine dissociative and somatoform disorders, which historically have been linked to the anxiety disorders as neuroses.

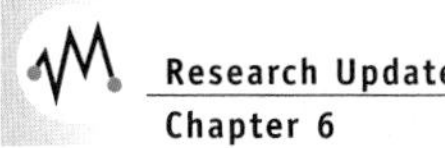

TABLE 6.6 Elements of Cognitive-Behavioral Programs for Treatment of Panic Disorder

Self-Monitoring	Keeping a log of panic attacks to help determine situational stimuli that might trigger them.
Exposure	A program of gradual exposure to situations in which panic attacks have occurred. During exposure trials, the person engages in self-relaxation and rational self-talk to prevent anxiety from spiraling out of control. In some programs, participants learn to tolerate changes in bodily sensations associated with panic attacks by experiencing these sensations within a controlled setting of the treatment clinic. The person may be spun around in a chair to induce feelings of dizziness, learning in the process that such sensations are not dangerous or signs of imminent harm.
Development of Coping Responses	Developing coping skills to interrupt the vicious cycle in which overreactions to anxiety cues or cardiovascular sensations culminate in panic attacks. Behavioral methods focus on deep, regular breathing and relaxation training. Cognitive methods focus on modifying catastrophic misinterpretations of bodily sensations. Breathing retraining may be used to help the individual avoid hyperventilation during panic attacks.

Sources. Adapted from Craske, Brown, & Barlow, 1991; Rapee, 1987; Turovsky & Barlow, 1995, and other sources.

Overview of Anxiety Disorders

TYPES OF ANXIETY DISORDERS

	Description	Features
Panic Disorder	Occurrence of repeated panic attacks, which are episodes of sheer terror accompanied by strong physiological symptoms, thoughts of imminent danger or impending doom, and an urge to escape	• Fears of recurring attacks may prompt avoidance of situations in which they occur or settings in which help might not be available • Panic attacks begin unexpectedly but may become associated with certain cues or specific situations
Generalized Anxiety Disorder	Persistent anxiety that is not limited to particular situations	• Excessive worrying is the keynote feature • Associated with heightened states of bodily arousal, tenseness, being "on edge"
Phobic Disorders	Excessive fears of particular objects or situations	• Carries a strong avoidance component in which the individual seeks to avoid contact with the phobic stimulus or situation • Subtypes include specific phobia (e.g., acrophobia, claustrophobia, fear of insects or snakes); social phobia (excessive fear of social interactions); and agoraphobia (fear of open, public places)
Obsessive-Compulsive Disorder	Recurrent obsessions (recurrent, intrusive thoughts) and/or compulsions (repetitive behaviors that the person feels compelled to perform)	• Two major types of compulsions: checking rituals and cleaning rituals • Obsessions generate anxiety that may be at least partially relieved by performance of the compulsive rituals
Traumatic Stress Disorders	Acute maladaptive reaction in the immediate aftermath of a traumatic event (acute stress disorder) or prolonged maladaptive reaction to a traumatic event (posttraumatic stress disorder)	• Reexperiencing the traumatic event, avoidance of cues or stimuli associated with the trauma, general or emotional numbing, hyperarousal, emotional distress, and impaired functioning • Vulnerability depends on such factors as severity of the trauma, degree of exposure, coping styles, and availability of social support

CAUSAL FACTORS **Anxiety disorders reflect an interplay of multiple causes**

Biological Factors	• Genetic predispositions • Irregularities in neurotransmitter functioning • Abnormalities in brain pathways signaling danger or inhibiting repetitive behaviors • Prepared conditioning
Social-Environmental Factors	• Exposure to threatening or traumatic events • Observing fear responses in others • Lack of social support
Behavioral Factors	• Pairing of aversive stimuli and previously neutral stimuli (classical conditioning) • Anxiety relief from performing compulsive rituals or avoiding phobic stimuli (operant conditioning) • Lack of extinction opportunities due to avoidance of feared objects or situations
Emotional and Cognitive Factors	• Unresolved psychological conflicts (Freudian or psychodynamic theory) • Cognitive factors, such as overprediction of fear, self-defeating or irrational beliefs, oversensitivity to threat, anxiety sensitivity, misattribution of bodily cues, and low self-efficacy

TREATMENT APPROACHES **Treatment may include one or more therapeutic approaches**

Drug Therapy	• To control anxiety symptoms
Cognitive-Behavioral Therapy	• To unlearn phobic reactions and develop more adaptive ways of thinking
Psychodynamic Therapy	• To gain insight into underlying conflicts that anxiety symptoms may symbolize
Humanistic Therapy	• To identify and come to accept one's genuine feelings and needs

Summing Up

Types of Anxiety Disorders

What are anxiety disorders? Anxiety, a generalized sense of apprehension or fear, is normal and desirable under some conditions, but it can become abnormal when it is excessive or inappropriate. Disturbed patterns of behavior in which anxiety is the most prominent feature are labeled anxiety disorders.

What is panic disorder? Panic disorder is characterized by often immobilizing, repeated panic attacks, which involve intense physical features, notably cardiovascular symptoms, that may be accompanied by sheer terror and fears of losing control, losing one's mind, or dying. Panic attack sufferers often limit their outside activity in fear of recurrent attacks. This can lead to agoraphobia, the fear of venturing into public places.

What is generalized anxiety disorder? Generalized anxiety disorder is a type of anxiety disorder involving persistent anxiety that seems to be "free floating" or not tied to specific situations.

What are phobic disorders? Phobias are excessive irrational fears of specific objects or situations. Phobias involve a behavioral component, avoidance of the phobic stimulus, in addition to physical and cognitive features. Specific phobias are excessive fears of particular objects or situations, such as mice, spiders, tight places, or heights. Social phobia involves an intense fear of being judged negatively by others. Agoraphobia involves fears of venturing into public places. Agoraphobia may occur with, or in the absence of, panic disorder.

What is obsessive-compulsive disorder? Obsessive-compulsive disorder, or OCD, involves recurrent patterns of obsessions, compulsions, or a combination of the two. Obsessions are nagging, persistent thoughts that create anxiety and seem beyond the person's ability to control. Compulsions are apparently irresistible repetitious urges to perform certain behaviors, such as repeated elaborate washing after using the bathroom.

What are the two types of traumatic stress disorders? There are two types of stress disorders—acute stress disorder and posttraumatic stress disorder. Both involve maladaptive reactions to traumatic stress. Acute stress disorder occurs in the days and weeks following exposure to a traumatic event. Posttraumatic stress disorder persists for months or even years or decades after the traumatic experience and may not begin until months or years after the event.

What relationships exist between ethnicity and the prevalence of anxiety disorders? Evidence from a nationally representative sample of U.S. adults showed that rates of anxiety disorders were generally comparable across racial and ethnic groupings.

Theoretical Perspectives

How are anxiety disorders conceptualized within the psychodynamic perspective? Psychodynamic theorists view anxiety disorders as attempts by the ego to control the conscious emergence of threatening impulses. Feelings of anxiety are warning signals that threatening impulses are nearing awareness. The ego mobilizes defense mechanisms to divert the impulses, thus leading to different anxiety disorders.

How do learning theorists view anxiety disorders? Learning theorists explain anxiety disorders through conditioning and observational learning. Mowrer's two-factor model incorporates classical and operant conditioning in the explanation of phobias. Phobias, however, appear to be moderated by cognitive factors, such as self-efficacy expectancies. The principles of reinforcement may help explain patterns of obsessive-compulsive behavior. People may be genetically predisposed to acquire certain types of phobias that may have had survival value for our prehistoric ancestors.

What cognitive factors are implicated in anxiety disorders? Cognitive factors may also play a role in the anxiety disorders, such as overpredictions of fear, self-defeating or irrational beliefs, oversensitivity to threatening cues and signs of anxiety, low self-efficacy expectations, and misattributions of bodily cues.

How do investigators explore the biological underpinnings of panic disorder? Investigators seek to uncover the biological underpinnings of anxiety disorders by studying the roles of genetic factors, neurotransmitters, and induction of panic by means of biological challenges.

Treatment of Anxiety Disorders

What are the major therapeutic approaches to treating anxiety disorders? Traditional psychoanalysis helps people work through unconscious conflicts that are believed to underlie anxiety disorders. Modern psychodynamic approaches focus more on current disturbed relationships in the client's life and encourage the client to develop more adaptive behavior patterns. Humanistic therapy focuses on helping clients identify and accept their true selves rather than reacting with anxiety whenever their genuine feelings and needs begin to surface. Drug therapy for anxiety disorders focuses on the use of benzodiazepines and antidepressants (which have more than just antidepressant effects). Learning-based approaches to treating anxiety disorders encompass a broad range of behavioral and cognitive-behavioral techniques, including exposure therapy, cognitive restructuring, exposure and response prevention, and relaxation skills training. Cognitive approaches, such as rational-emotive behavior therapy and cognitive therapy, help people identify and correct faulty thinking patterns that may underlie anxiety reactions. The cognitive-behavioral treatment of panic disorder incorporates self-monitoring, exposure, and development of coping responses to anxiety-inducing cues.

CHAPTER SEVEN

Dissociative and Somatoform Disorders

Naoki Okamoto
Untitled

Truth OR Fiction?

- In some reported cases, alternate personalities in people with multiple personalities had their own allergic reactions and eyeglass prescriptions. (p. 197)
- The term *split personality* refers to schizophrenia. (p. 200)
- Very few of us have episodes in which we feel strangely detached from our own bodies or thought processes. (p. 203)
- Most people with multiple personalities do not report any history of physical or sexual abuse during childhood. (p. 206)
- Some people show up repeatedly at hospital emergency rooms, feigning illness and seeking treatment for no apparent reason. (p. 211)
- Some people who have lost their ability to see or move their legs become strangely indifferent toward their physical condition. (p. 213)
- In China in the 1980s, more than 2,000 people fell prey to the belief that their genitals were shrinking and retracting into their bodies. (p. 216)

In the Middle Ages, the clergy used rites of exorcism to bring forth demons from people believed to be possessed. Using curious incantations, exorcists contended for the victims' souls against the demons believed to lurk within.

Curious phrasings were also heard in 20th-century Los Angeles. They were intended to evoke another sort of demon from Kenneth Bianchi, a suspect in a police inquest.

At one point, the question was put to Bianchi, "Part, are you the same thing as Ken or are you different?" The interviewer was not a member of the clergy, but a police psychiatrist. Bianchi had been dubbed the "Hillside strangler" by the press. He had terrorized the city, leaving prostitutes dead in the mountains that bank the metropolis.

Under hypnosis—not religious incantations—Bianchi claimed that a hidden personality, or "part," named "Steve," had committed the murders. He also claimed that "Ken" knew nothing of the murders and that he was suffering from multiple personality disorder (now called *dissociative identity disorder*), one of the intriguing but perplexing psychological disorders we explore in this chapter.

Dissociative identity disorder is classified as a *dissociative disorder,* a type of psychological disorder involving a change or disturbance in the functions of self—identity, memory, or consciousness—that make the personality whole. Normally speaking, we know who we are. We may not be certain of ourselves in an existential, philosophical sense, but we know our names, where we live, and what we do for a living. We also tend to remember the salient events of our lives. We may not recall every detail, and we may confuse what we had for dinner on Tuesday with what we had on Monday, but we generally know what we have been doing for the past days, weeks, and years. Normally speaking, there is a unity to consciousness that gives rise to a sense of self. We perceive ourselves as progressing through space and time. In the dissociative disorders, one or more of these aspects of daily living is disturbed—sometimes bizarrely so.

This chapter also focuses on *somatoform disorders,* a class of psychological disorders involving complaints of physical symptoms that are believed to reflect underlying psychological conflicts or issues. In some cases there is no apparent medical basis to the physical symptoms, such as in the form of hysterical blindness or numbness (now called *conversion disorder*). In other cases, people may hold an exaggerated view of the meaning of their physical symptoms, believing them to be signs of underlying serious illnesses despite the reassurances of their physicians to the contrary.

In early versions of the *DSM,* the dissociative and conversion disorders were grouped with the anxiety disorders under the general category of neurosis. The common grouping was based on the psychodynamic model, which holds that various disorders involve maladaptive ways of managing anxiety. In the anxiety disorders, the appearance of disturbing levels of anxiety was expressed directly in behavior, such as in a phobic reaction to an object or situation. But the role of anxiety in the dissociative and somatoform disorders was inferred rather than expressed in behavior. Persons with dissociative disorders may show no signs of overt anxiety. However, they manifest other psychological problems, such as loss of memory or changes in identity, that, according to the psychodynamic model, serve the purpose of keeping the underlying sources of anxiety out of awareness. Likewise, people with conversion disorder often show a strange indifference to physical problems (e.g., loss of vision) that would greatly concern most of us. Here, too, it was theorized that the "symptoms" mask unconscious sources of anxiety. Some theorists interpret indifference to symptoms to mean that those symptoms have an underlying benefit; that is, they help prevent anxiety from intruding into consciousness.

The *DSM* now separates the anxiety disorders from the other categories of neuroses—the dissociative and somatoform disorders—with which they were historically linked. Yet many practitioners continue to use the broad conceptualization of neuroses as a useful framework for classifying the anxiety, dissociative, and somatoform disorders.

Chapter Outline

dissociative disorder Any of a group of disorders characterized by a disruption, or dissociation, of the functions of identity, memory, or consciousness.

Kenneth Bianchi, the so-called Hillside Strangler. Bianchi claimed that a hidden personality had committed the murders of which he was accused.

Dissociative Disorders

The major **dissociative disorders** include *dissociative identity disorder, dissociative amnesia, dissociative fugue,* and *depersonalization disorder.* In each case there is a disruption or

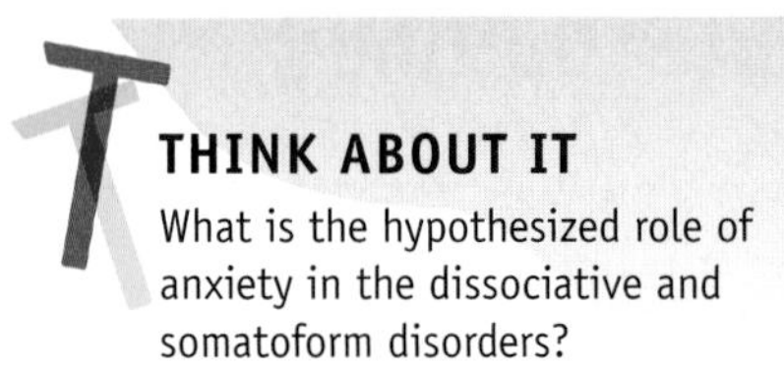

THINK ABOUT IT
What is the hypothesized role of anxiety in the dissociative and somatoform disorders?

dissociation ("splitting off") of the functions of identity, memory, or consciousness that normally make us whole.

Dissociative Identity Disorder

The Ohio State campus dwelled in terror as four college women were seized, coerced to cash checks or get money from automatic teller machines, then raped. A cryptic phone call led to the capture of Billy Milligan, a 23-year-old drifter who had been dishonorably discharged from the navy.

Not the Boy Next Door

Billy wasn't quite the boy next door. He tried twice to commit suicide while he was awaiting trial, so his lawyers requested a psychiatric evaluation. The psychologists and psychiatrists who examined Billy deduced that ten personalities dwelled inside of him. Eight were male and two were female. Billy's personality had been fractured by a brutal childhood. The personalities displayed diverse facial expressions, memories, and vocal patterns. They performed in dissimilar ways on personality and intelligence tests.

Arthur, a sensible but phlegmatic personality, conversed with a British accent. Danny, 14, was a painter of still lifes. Christopher, 13, was normal enough, but somewhat anxious. A 3-year-old English girl went by the name of Christine. Tommy, a 16-year-old, was an antisocial personality and escape artist. It was Tommy who had enlisted in the Navy. Allen was an 18-year-old con artist. Allen also smoked. Adelena was a 19-year-old introverted lesbian. It was she who had committed the rapes. It was probably David who had made the mysterious phone call. David was an anxious 9-year-old who wore the anguish of early childhood trauma on his sleeve. After his second suicide attempt, Billy had been placed in a straitjacket. When the guards checked his cell, however, he was sleeping with the straitjacket as a pillow. Tommy later explained that he was responsible for Billy's escape.

The defense argued that Billy was afflicted with multiple personality disorder. Several alternate personalities resided within him. The alternate personalities knew about Billy, but Billy was unaware of them. Billy, the core or dominant personality, had learned as a child that he could sleep as a way of avoiding the sexual and physical abuse of his father. A psychiatrist claimed that Billy had likewise been "asleep"—in a sort of "psychological coma"—when the crimes were committed. Therefore, Billy should be judged innocent by reason of insanity.

Billy was decreed not guilty by reason of insanity. He was committed to a mental institution. In the institution, 14 additional personalities emerged. Thirteen were rebellious and labeled "undesirables" by Arthur. The fourteenth was the "Teacher," who was competent and supposedly represented the integration of all the other personalities. Billy was released six years later.

—Adapted from Keyes, 1982

■

Billy was diagnosed with multiple personality disorder, which is now called **dissociative identity disorder.** In dissociative identity disorder, sometimes referred to as "split personality," two or more personalities—each with well-defined traits and memories—"occupy" one person. They may or may not be aware of one another. In some isolated cases, alternate personalities (also called *alter personalities*) may even show different EEG records, allergic reactions, responses to medication, and even different eyeglass prescriptions and pupil

dissociative identity disorder A dissociative disorder in which a person has two or more distinct, or alter, personalities.

sizes (Birnbaum, Martin, & Thomann, 1996; S. D. Miller et al., 1991). Or one personality may be color blind, whereas others are not (Braun, 1986). If such patterns stand up to further scientific scrutiny, they would offer a remarkable illustration of the diversity of perceptions and somatic patterns that are possible within the same person.

Celebrated cases of multiple personality have been depicted in the popular media. One became the subject of the 1950s film *The Three Faces of Eve*. In the film, Eve White is a timid housewife who harbors two other personalities: Eve Black, a sexually provocative, antisocial personality, and Jane, a balanced, developing personality who could balance her sexual needs with the demands of social acceptability. The three faces eventually merged into one—Jane, providing a "happy ending." The real-life Eve, whose name was Chris Sizemore, failed to maintain this integrated personality. Her personality reportedly split into 22 subsequent personalities. A second well-known case is that of Sybil. Sybil was played by Sally Field in the film of the same name and reportedly had 16 personalities.

Features In one of the largest studies on multiple personality to date, Ross, Norton, and Wozney (1989) collected 236 case reports of people with the disorder from 203 health professionals in Canada. Unlike reports of multiple personality in the 19th and early 20th centuries, in which most cases involved dual personalities, cases in the Canadian sample averaged 15 to 16 alter personalities each (Ross et al., 1989).

There are many variations. In some cases, the host (main) personality may be unaware of the existence of the other identities, while the other identities are aware of the existence of the host (Dorahy, 2001). In other cases, the different personalities are completely unaware of one another. Sometimes two personalities vie for control of the person. Sometimes there is one dominant or core personality and several subordinate personalities. Some of the more common alternate personalities (or "alter personalities") include children of various ages, adolescents of the opposite gender, prostitutes, and gay males and lesbians (Ross et al., 1989). Some of the personalities may show psychotic symptoms—a break with reality expressed in the form of hallucinations and delusional thinking.

All in all, the clusters of alter personalities serve as a microcosm of conflicting urges and cultural themes. Themes of sexual ambivalence (sexual openness vs. restrictiveness)

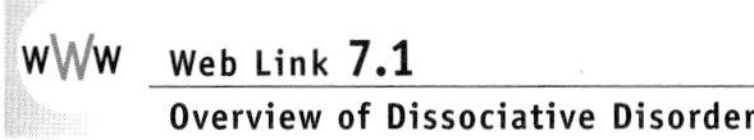

Web Link 7.1
Overview of Dissociative Disorders

Truth OR Fiction? REVISITED

In some reported cases, alternate personalities in people with multiple personalities had their own allergic reactions and eyeglass prescriptions.

TRUE. In a few reported cases, alternate personalities were reported to have had their own allergic reactions and eyeglass prescriptions that differed from those of other personalities within the same person.

VIDEO 7.1
Dissociative Identity Disorder:
The Three Faces of Eve

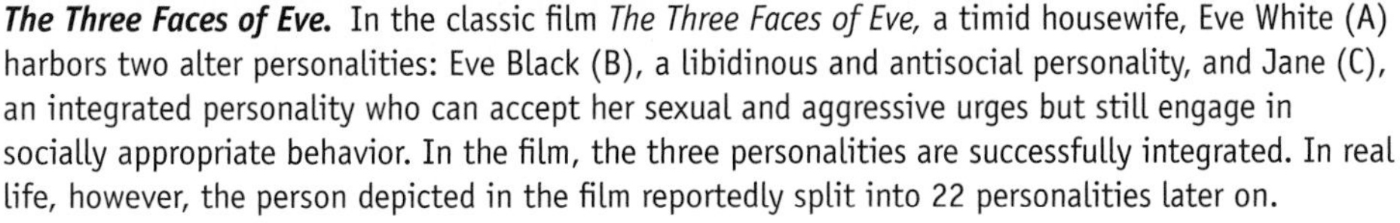

The Three Faces of Eve. In the classic film *The Three Faces of Eve,* a timid housewife, Eve White (A) harbors two alter personalities: Eve Black (B), a libidinous and antisocial personality, and Jane (C), an integrated personality who can accept her sexual and aggressive urges but still engage in socially appropriate behavior. In the film, the three personalities are successfully integrated. In real life, however, the person depicted in the film reportedly split into 22 personalities later on.

and shifting sexual orientations are particularly common. It is as if conflicting internal impulses cannot coexist or achieve dominance. As a result, each is expressed as the cardinal or steering trait of an alternate personality. The clinician can sometimes elicit alternate personalities by inviting them to make themselves known, as in asking, "Is there another part of you that wants to say something to me?" The following case illustrates the emergence of an alternate personality:

Harriet Emerges

[Margaret explained that] she often "heard a voice telling her to say things and do things." It was, she said, "a terrible voice" that sometimes threatened to "take over completely." When it was finally suggested to [Margaret] that she let the voice "take over," she closed her eyes, clenched her fists, and grimaced for a few moments during which she was out of contact with those around her. Suddenly she opened her eyes and one was in the presence of another person. Her name, she said, was "Harriet." Whereas Margaret had been paralyzed, and complained of fatigue, headache and backache, Harriet felt well, and she at once proceeded to walk unaided around the interviewing room. She spoke scornfully of Margaret's religiousness, her invalidism, and her puritanical life, professing that she herself liked to drink and "go partying" but that Margaret was always going to church and reading the Bible. "But," she said impishly and proudly, "I make her miserable—I make her say and do things she doesn't want to." At length, at the interviewer's suggestion, Harriet reluctantly agreed to "bring Margaret back," and after more grimacing and fist clenching, Margaret reappeared, paralyzed, complaining of her headache and backache, and completely amnesiac for the brief period of Harriet's release from prison.

—From Nemiah, 1978, pp. 179–180

As with Billy Milligan, Chris Sizemore, and Margaret, the dominant personality is often unaware of the existence of the alter personalities. It thus seems that the mechanism of dissociation is controlled by unconscious processes. Although the dominant personality lacks insight into the existence of the other personalities, she or he may vaguely sense that something is amiss. There may even be "interpersonality rivalry" in which one personality aspires to do away with another, usually in blissful ignorance of the fact that conferring the *coup de grace* on an alternate would result in the death of all.

Although women constitute the majority of cases of multiple personality (Ross et al., 1989; Schafer, 1986), the proportion of males diagnosed with the disorder has been on the rise (Goff & Summs, 1993). The numbers of reported alternates has also been increasing, rising to an average of 12 alternates during the 1980s from an average of 3 in earlier cases. Women with the disorder tend to have more alternate identities, averaging 15 or more, than do men, who average about 8 identities (APA, 2000). The reasons for this difference remain unknown.

The diagnostic features of dissociative identity disorder are listed in Table 7.1.

TABLE 7.1 Features of Dissociative Identity Disorder (Formerly Multiple Personality Disorder)

1. At least two distinct personalities exist within the person, with each having a relatively enduring and distinct pattern of perceiving, thinking about, and relating to the environment and the self.
2. Two or more of these personalities repeatedly take complete control of the individual's behavior.
3. There is a failure to recall important personal information too substantial to be accounted for by ordinary forgetfulness.
4. The disorder cannot be accounted for by the effects of a psychoactive substance or a general medical condition.

Source. Adapted from the *DSM-IV-TR* (APA, 2000).

Controversies Although multiple personality is generally considered rare, the very existence of the disorder continues to arouse debate. Only a handful of cases worldwide were reported from 1920 to 1970, but since then the number of reported cases has skyrocketed into the thousands (Spanos, 1994). This has led some practitioners to suggest that multiple personality may be more common than was earlier believed (Bliss & Jeppsen, 1985; Schafer, 1986). Others, however, are not so sure. Some professionals believe the disorder is overdiagnosed in highly suggestible people who might simply be following suggestions that they might have the disorder (APA, 2000). Increased public attention paid to the disorder in recent years may also account for the perception that its prevalence is greater than was commonly believed.

The disorder does appear to be culture bound and largely restricted to North America (Spanos, 1994). Relatively few cases have been reported elsewhere, even in such Western countries as Great Britain and France. A recent survey in Japan failed to find even one case, and in Switzerland, 90% of the psychiatrists polled had never seen a case of the disorder (Modestin, 1992; Spanos, 1994). Even in North America, few psychologists and psychiatrists have ever encountered a case of multiple personality. Most cases are reported by a relatively small number of investigators and clinicians who strongly believe in the existence of the disorder. Yet critics wonder, might they be helping to manufacture that which they are seeking?

Some leading authorities, such as the late psychologist Nicholas Spanos, believe so. Spanos and others have challenged the existence of dissociative identity (multiple personality) disorder (Reisner, 1994; Spanos, 1994). To Spanos, multiple personality is not a distinct disorder, but a form of role playing in which individuals first come to construe themselves as having multiple selves and then begin to act in ways that are consistent with their conception of the disorder. Eventually their role playing becomes so ingrained it becomes a reality to them. Perhaps their therapists or counselors unintentionally planted the idea in their minds that their confusing welter of emotions and behaviors may represent different personalities at work. Impressionable people may have learned how to enact the role of persons with the disorder by watching others enacting the role on television and in the movies. Films and TV shows like *The Three Faces of Eve* and *Sybil* have given detailed examples of the behaviors that characterize multiple personalities (Spanos, Weekes, & Bertrand, 1985). Or perhaps therapists provided cues about the features of multiple personality, enough to enact the role convincingly.

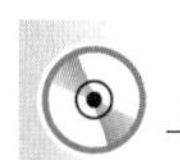

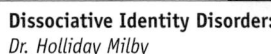

VIDEO 7.2
Dissociative Identity Disorder:
Dr. Holliday Milby

Many reinforcers may become contingent on enacting the role of a multiple personality. Receiving attention from others and evading accountability for unacceptable behavior are two possible sources of reinforcement (Spanos et al., 1985). This is not to suggest that people with multiple personalities are "faking," any more than you are faking when you perform different daily roles as student, spouse, or worker. You may enact the role of a student (e.g., sitting attentively in class, raising your hand when you wish to talk, etc.) because you have learned to organize your behavior according to the nature of the role and because you have been rewarded for doing so. People with multiple personalities may have come to identify so closely with the role that it becomes real for them.

In support of his belief that multiple personality represents a form of role playing, Spanos and his colleagues showed that with proper cues, college students in a laboratory simulation of the Bianchi-type interrogation could easily enact a multiple personality role, even attributing the blame to an alternate personality for a murder they were accused of committing (Spanos et al., 1985). Perhaps the manner in which the Bianchi interrogation was conducted had cued Bianchi to enact the multiple personality role in order to evade criminal responsibility. (It didn't work, as he was eventually convicted.)

Relatively few cases of multiple personality involve criminal behavior, in which enactment of a multiple personality role might relieve individuals of criminal responsibility for their behavior. But even in more typical cases, there may be more subtle incentives for enacting the role of a multiple personality, such as a therapist's expression of interest and excitement at discovering a multiple personality. People with multiple personalities were often highly imaginative during childhood. Accustomed to playing games of "make-believe," they may readily adopt alternate identities, especially if they learn how to enact the multiple personality role and there are external sources of validation, such as a clinician's interest and concern.

The social reinforcement model may help to explain why some clinicians seem to "discover" many more cases of multiple personality than others. These clinicians may be "multiple personality magnets." They may unknowingly cue clients to enact the multiple personality role and reinforce the performance with extra attention and concern. With the right set of cues, certain clients may adopt the role of a multiple personality to please their clinicians. The role-playing model has been challenged by some authorities (for example, Gleaves, 1996), and it remains to be seen how many cases of the disorder in clinical practice the model can explain.

dissociative amnesia A dissociative disorder in which a person experiences memory loss without any identifiable organic cause.

Whether multiple personality is a real phenomenon or a form of role playing, there is no question that people who display this behavior have serious emotional and behavioral difficulties. Moreover, the diagnosis may not be all that unusual among some subgroups in the population, such as psychiatric inpatients. In one study of 484 adult psychiatric inpatients, at least 5% showed evidence of multiple personality (Ross et al., 1991). We have noted a tendency for *claims* of multiple personality to spread on inpatient units. In one case, Susan, a prostitute admitted for depression and suicidal thoughts, claimed that she could only exchange sex for money when "another person" inside her emerged and took control. (Suicidal behavior is common among people with multiple personalities. Seventy-two percent of the cases in the Canadian study [Ross et al., 1989] had attempted suicide, and about 2% had succeeded.) Upon hearing this, another woman, Ginny—a child abuser who had been admitted for depression after her daughter had been removed from her home by social services—claimed that she only abused her daughter when another person inside of her assumed control of her personality. Susan's chart recommended that she be evaluated further for multiple personality disorder (the term used at the time to refer to the disorder), but Ginny was diagnosed with a depressive disorder and a personality disorder, not with multiple personality disorder.

Multiple personality, which often is called "split personality" by laypeople, should not be confused with schizophrenia. The term *split personality* refers to multiple personality, not schizophrenia. Schizophrenia (which comes from roots that mean "split brain") occurs much more commonly than multiple personality and involves the "splitting" of cognition, affect, and behavior. There may thus be little agreement between the thoughts and the emotions, or between the individual's perception of reality and what is truly happening. The person with schizophrenia may become giddy when told of disturbing events, or may experience hallucinations or delusions (see Chapter 13). In people with multiple personalities, the personality apparently divides into two or more personalities, but each of them usually shows more integrated functioning on cognitive, affective, and behavioral levels than is true of people with schizophrenia.

Dissociative Amnesia

Dissociative amnesia is believed to be the most common type of dissociative disorder (Maldonado, Butler, & Speigel, 1998). *Amnesia* derives from the Greek roots *a-*, meaning "not," and *mnasthai,* meaning "to remember." In **dissociative amnesia** (formerly called *psychogenic amnesia*), the person becomes unable to recall important personal information, usually involving traumatic or stressful experiences, in a way that cannot be accounted for by simple forgetfulness. Nor can the memory loss be attributed to a particular organic cause, such as a blow to the head or a particular medical condition, or to the direct effects of drugs or alcohol. Unlike some progressive forms of memory impairment (such as dementia associated with Alzheimer's disease; see Chapter 15), the memory loss in dissociative amnesia is reversible, although it may last for days, weeks, or even years. Recall of dissociated memories may happen gradually but often occurs suddenly and spontaneously, as when the soldier who has no recall of a battle for several days afterward suddenly recalls the experience after being transported to a hospital away from the battlefield.

Truth OR Fiction? REVISITED

The term *split personality* refers to schizophrenia.

FALSE. The term *split personality* refers to multiple personality, not schizophrenia.

Amnesia is not ordinary forgetfulness, such as forgetting someone's name or where you left your car keys. Memory loss in amnesia is more profound or wide-ranging. Most cases of dissociative amnesia take the form of *localized amnesia* in which events occurring during a specific time period are lost to memory. For example, the person cannot recall events for a number of hours or days after a stressful or traumatic incident, such as a battle or car accident. Other forms of dissociative amnesia include selective amnesia and generalized amnesia. In *selective amnesia,* people forget only the disturbing particulars that take place during a certain time period. A person may recall the period of life during which he conducted an extramarital affair, but not the guilt-arousing affair itself. A soldier may recall most of the battle, but not the death of his buddy. In *generalized amnesia,* people forget their entire lives—who they are, what they do, where they live, whom they live with. This form of amnesia is very rare, although you wouldn't think so if you watch day-

THINK ABOUT IT

Why is the diagnosis of dissociative identity disorder controversial? Do you believe that people with dissociative identity disorder are merely playing a role they have learned? Why or why not?

time soap operas. Persons with generalized amnesia cannot recall personal information, but they tend to retain their habits, tastes, and skills. If you had generalized amnesia, you would still know how to read, although you would not recall your elementary school teachers. You would still prefer French fries to baked potatoes—or vice versa.

People with dissociative amnesia usually forget events or periods of life that were traumatic—that generated strong negative emotions such as horror or guilt. Consider this case:

malingering Faking illness.

dissociative fugue A dissociative disorder in which a person suddenly flees from his or her life situation, travels to a new location, assumes a new identity, and has amnesia for personal material.

A Case of Dissociative Amnesia

He was brought to the emergency room of a hospital by a stranger. He was dazed and claimed not to know who he was or where he lived, and the stranger had found him wandering in the streets. Despite his confusion, it did not appear that he had been drinking or abusing drugs or that his amnesia could be attributed to physical trauma. After staying in the hospital for a few days, he awoke in distress. His memory had returned. His name was Rutger and he had urgent business to attend to. He wanted to know why he had been hospitalized and demanded to leave. At time of admission, Rutger appeared to be suffering from generalized amnesia: He could not recall his identity or the personal events of his life. But now that he was requesting discharge, Rutger showed localized amnesia for the period between entering the emergency room and the morning he regained his memory for prior events.

Rutger provided information about the events prior to his hospitalization that was confirmed by the police. On the day when his amnesia began, Rutger had killed a pedestrian with his automobile. There had been witnesses, and the police had voiced the opinion that Rutger—although emotionally devastated—was blameless in the incident. Rutger was instructed, however, to fill out an accident report and to appear at the inquest. Still nonplussed, Rutger filled out the form at a friend's home. He accidentally left his wallet and his identification there. After placing the form in a mailbox, Rutger became dazed and lost his memory.

Although Rutger was not responsible for the accident, he felt awful about the pedestrian's death. His amnesia was probably connected with feelings of guilt, the stress of the accident, and concerns about the inquest.

—*Adapted from Cameron, 1963, pp. 355–356*

People sometimes claim they cannot recall certain events of their lives, such as criminal acts, promises made to others, and so forth. Falsely claiming amnesia as a way of escaping responsibility is called **malingering,** which involves the attempt to fake symptoms or make false claims for personal gain. Our research methods cannot guarantee that we can distinguish people with dissociative amnesia from malingerers. But experienced clinicians can make reasonably well-educated guesses.

Dissociative Fugue

Fugue derives from the Latin *fugere,* meaning "flight." The word *fugitive* has the same origin. Fugue is like amnesia "on the run." In **dissociative fugue** (formerly called *psychogenic fugue*), the person travels suddenly and unexpectedly from his or her home or place of work, is unable to recall past personal information, and either becomes confused about his or her identity or assumes a new identity (either partially or completely) (APA, 2000). Despite these odd behaviors, the person may appear "normal" and show no other signs of mental disturbance (Maldonado et al., 1998). The person may not think about the past, or may report a past filled with false memories without recognizing them as false.

Whereas people with amnesia appear to wander aimlessly, people in a fugue state act more purposefully. Some stick close to home. They spend the afternoon in the park

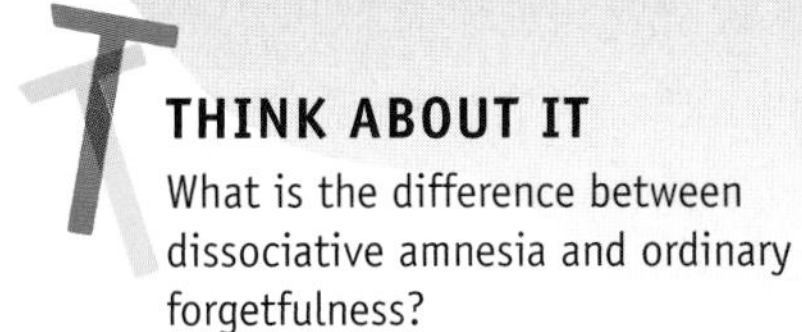

THINK ABOUT IT
What is the difference between dissociative amnesia and ordinary forgetfulness?

or in a theater, or they spend the night at a hotel under another name, usually having little if any contact with others during the fugue state. But the new identity is incomplete and fleeting and the individual's former sense of self returns in a matter of hours or a few days. Less common is a pattern in which the fugue state lasts for months or years and involves travel to distant places and assumption of a new identity. These individuals may assume an identity that is more spontaneous and sociable than their former selves, which were typically "quiet" and "ordinary." They may establish new families and successful businesses. Although these events may sound rather bizarre, the fugue state is not considered psychotic because people with the disorder can think and behave quite normally—in their new lives, that is. Then one day, quite suddenly, their awareness of their past identity returns to them, and they are flooded with old memories. Now they typically do not recall the events that occurred during the fugue state. The new identity, the new life—including all its involvements and responsibilities—vanish from memory.

Fugue, like amnesia, is relatively rare and is believed to affect only about 2 people in 1,000 within the general population (APA, 2000). It is most likely to occur in wartime (Loewenstein, 1991) or in the wake of another kind of disaster or extremely stressful event. The underlying notion is that dissociation in the fugue state protects one from traumatic memories or other sources of emotionally painful experiences or conflict (Maldonado et al., 1998).

Fugue can also be difficult to distinguish from malingering. That is, a number of persons who were dissatisfied with their former lives could claim to be amnesic when they are uncovered in their new locations and new identities.

Consider the following case, in which the evidence supports a diagnosis of dissociative fugue (Spitzer et al., 1989):

A Case of Dissociative Fugue?

The man told the police that his name was Burt Tate. "Burt," a 42-year-old white male, had gotten into a fight at the diner where he worked. When the police arrived, they found that he carried no identification. He told them he had drifted into town a few weeks earlier, but could not recall where he had lived or worked before arriving in town. While no charges were pressed against him, the police prevailed upon him to come to the emergency room for evaluation. "Burt" knew the town he was in and the current date, and recognized that it was somewhat unusual that he couldn't remember his past, but didn't seem to be concerned about it. There was no evidence of any physical injuries or head trauma, or of drug or alcohol abuse. The police made some inquiries and discovered that "Burt" fit the profile of a missing person, Gene Saunders, who had disappeared a month earlier from a city some 2,000 miles away. Mrs. Saunders was called in and confirmed that "Burt" was indeed her husband. She reported that her husband, who had worked in middle-level management in a manufacturing company, had been having difficulty at work before his disappearance. He was passed over for promotion and his supervisor was highly critical of his work. The job stress apparently affected his behavior at home. Once easygoing and sociable, he withdrew into himself and began to criticize his wife and children. Then, just before his disappearance, he had a violent argument with his 18-year-old son. His son called him a "failure" and stormed out the door. Two days later, the man disappeared. When he came face to face with his wife again, he claimed he didn't recognize her, but appeared visibly nervous.

—Adapted from Spitzer et al., 1994, pp. 254–255

■

Although the presenting evidence supported a diagnosis of dissociative fugue, clinicians can find it difficult to distinguish true amnesia from amnesia that is faked to allow a person to get a new start in life.

Depersonalization Disorder

Depersonalization involves a temporary loss or change in the usual sense of our own reality. In a state of depersonalization, people feel detached from themselves and their surroundings. They may feel as though they were dreaming or acting like a robot (Guralnik, Schmeidler, & Simeon, 2000; Maldonado, Butler, & Speigel, 1998).

Derealization—a sense of unreality about the external world involving strange changes in perception of surroundings, or in the sense of the passage of time—may also be present. People and objects may seem to change in size or shape; they may sound different. All these feelings can be associated with feelings of anxiety, including dizziness and fears of going insane, or with depression.

Although these sensations are strange, people with depersonalization maintain contact with reality. They can distinguish reality from unreality, even during the depersonalization episode. In contrast to generalized amnesia and fugue, they know who they are. Their memories are intact and they know where they are—even if they do not like their present state. Feelings of depersonalization usually come on suddenly and fade gradually.

Note that we have thus far described only normal feelings of depersonalization. According to the *DSM,* single brief episodes of depersonalization are experienced by about half of all adults, usually during times of extreme stress. Estimates are that 80% to 90% of the general population experiences dissociative experiences at one time or another (Gershuny & Thayer, 1999). Consider Richie's experience:

depersonalization Feelings of unreality or detachment from one's self or one's body.

derealization Loss of the sense of reality of one's surroundings, experienced in terms of strange changes in the environment or in the passage of time.

Depersonalization at Disneyworld

"We went to Orlando with the children after school let out. I had also been driving myself hard, and it was time to let go. We spent three days 'doing' Disneyworld, and it got to the point where we were all wearing shirts with mice and ducks on them and singing Disney songs. On the third day I began to feel unreal and ill at ease while we were watching these middle-American Ivory-soap teenagers singing and dancing in front of Cinderella's Castle. The day was finally cooling down, but I broke into a sweat. I became shaky and dizzy and sat down on the cement next to the 4-year-old's stroller without giving [my wife] an explanation. There were strollers and kids and [adults'] legs all around me, and for some strange reason I became fixated on the pieces of popcorn strewn on the ground. All of a sudden it was like the people around me were all silly mechanical creatures, like the dolls in the 'It's a Small World' [exhibit] or the animals on the 'Jungle Cruise.' Things sort of seemed to slow down, the way they do when you've smoked marijuana, and there was this invisible wall of cotton between me and everyone else.

"Then the concert was over and my wife was like 'What's the matter?' and did I want to stay for the Electrical Parade and the fireworks or was I sick? Now I was beginning to wonder if I was going crazy and I said I was sick, that my wife would have to take me by the hand and drive us back to the Sonesta Village [motel]. Somehow we got back to the monorail and turned in the strollers. I waited in the herd [of people] at the station like a dead person, my eyes glazed over, looking out over kids with Mickey Mouse ears and Mickey Mouse balloons. The mechanical voice on the monorail almost did me in and I got really shaky.

"I refused to go back to the Magic Kingdom. I went with the family to Sea World, and on another day I dropped [my wife] and the kids off at the Magic Kingdom and picked them up that night. My wife thought I was goldbricking or something, and we had a helluva fight about it, but we had a life to get back to and my sanity had to come first."

—*From the Authors' Files*

Truth OR Fiction? REVISITED

Very few of us have episodes in which we feel strangely detached from our own bodies or thought processes.

FALSE. About half of all adults, according to the *DSM,* at some time experience an episode of depersonalization in which they feel detached from their own bodies or mental processes.

TABLE 7.2 Diagnostic Features of Depersonalization Disorder

1. Recurrent or persistent experiences of depersonalization, which are characterized by feelings of detachment from one's mental processes or body, as if one were an outside observer of oneself. The experience may have a dreamlike quality.
2. The individual is able to maintain reality testing (i.e., distinguish reality from unreality) during the depersonalization state.
3. The depersonalization experiences cause significant personal distress or impairment in one or more important areas of functioning, such as social or occupational functioning.
4. Depersonalization experiences cannot be attributed to other disorders or to the direct effects of drugs, alcohol, or medical conditions.

Source. Adapted from the *DSM-IV-TR* (APA, 2000).

Richie's depersonalization experience was limited to the one episode and would not qualify for a diagnosis of **depersonalization disorder.** Depersonalization disorder is diagnosed only when such experiences are persistent or recurrent and cause marked distress (Steinberg, 1991). The *DSM* diagnoses depersonalization disorder according to the criteria shown in Table 7.2. Note the following case example:

A Case of Depersonalization Disorder

A 20-year-old college student feared that he was going insane. For two years, he had increasingly frequent experiences of feeling "outside" himself. During these episodes, he experienced a sense of "deadness" in his body, and felt wobbly, frequently bumping into furniture. He was more apt to lose his balance during episodes which occurred when he was out in public, especially when he was feeling anxious. During these episodes, his thoughts seemed "foggy," reminding him of his state of mind when he was given shots of a pain-killing drug for an appendectomy five years earlier. He tried to fight off these episodes when they occurred, by saying "stop" to himself and by shaking his head. This would temporarily clear his head, but the feeling of being outside himself and the sense of deadness would shortly return. The disturbing feelings would gradually fade away over a period of hours. By the time he sought treatment, he was experiencing these episodes about twice a week, each one lasting from three to four hours. His grades remained unimpaired, and had even improved in the past several months, since he was spending more time studying. However, his girlfriend, in whom he had confided his problem, felt that he had become totally absorbed in himself and threatened to break off their relationship if he didn't change. She had also begun to date other men.

—Adapted from Spitzer et al., 1994, pp. 270–271

depersonalization disorder A disorder characterized by persistent or recurrent episodes of depersonalization.

In terms of observable behavior and associated features, depersonalization may be more closely related to disorders such as phobias and panic than to dissociative disorders. Unlike other forms of dissociative disorders that seem to protect the self from anxiety, depersonalization can lead to anxiety and in turn to avoidance behavior, as we saw in the case of Richie.

Depersonalization. Episodes of depersonalization are characterized by feelings of detachment from oneself. During an episode, it may feel as if one were walking through a dream or observing the environment or oneself from outside one's body.

Culture-Bound Dissociative Syndromes

Similarities exist between the Western concept of dissociative disorder and certain culture-bound syndromes found in other parts of the world. For example, *amok* is a culture-bound syndrome occurring primarily in southeast Asian and Pacific island cultures that describes a trancelike state in which a person suddenly becomes highly excited and violently attacks other people or destroys objects (see Chapter 3). People who "run amuck" may later claim to have no memory of the episode or recall feeling as if they were acting like a robot. Another example is *zar,* a term used in countries in North Africa and the Middle East to describe spirit possession in people who experience dissociative states. During these states, individuals engage in unusual behavior, ranging from shouting to banging their heads against the wall. The behavior itself is not deemed abnormal, since it is believed to be controlled by spirits.

Questionnaire

The Dissociative Experiences Scale

Brief dissociative experiences, such as momentary feelings of depersonalization, are quite common. The great majority of us experience them at least some of the time (Gershuny & Thayer, 1999). Dissociative disorders, by contrast, involve more persistent and severe dissociative experiences. Researchers have developed a measure, the Dissociative Experiences Scale (DES), to offer clinicians a way of measuring dissociative experiences that occur in both the general population and among people with dissociative disorders (Bernstein & Putnam, 1986; Putnam & Carlson, 1994; Sanders & Green, 1995; Sar et al., 1996). Fleeting dissociative experiences are quite common, but those reported by people with dissociative disorders are more frequent and problematic than those in the general population (Waller & Ross, 1997).

The following is a listing of some of the types of dissociative experiences drawn from the Dissociative Experiences Scale that many people encounter from time to time. Bear in mind that transient experiences like these are reported by both normal and abnormal groups in varying frequencies. Let us also suggest that if these experiences become persistent or commonplace, or cause you concern or distress, then it might be worthwhile to discuss them with a professional.

Have You Ever Experienced the Following?

1. Suddenly realizing, when you are driving the car, that you don't remember what has happened during all or part of the trip.
2. Suddenly realizing, when you are listening to someone talk, that you did not hear part or all of what the person said.
3. Finding yourself in a place and having no idea how you got there.
4. Finding yourself dressed in clothes that you don't remember putting on.
5. Experiencing a feeling that seemed as if you were standing next to yourself or watching yourself do something and actually seeing yourself as if you were looking at another person.
6. Looking in a mirror and not recognizing yourself.
7. Feeling sometimes that other people, objects, and the world around you are not real.
8. Remembering a past event so vividly that it seems like you are reliving it in the present.
9. Having the experience of being in a familiar place but finding it strange and unfamiliar.
10. Becoming so absorbed in watching television or a movie that you are unaware of other events happening around you.
11. Becoming so absorbed in a fantasy or daydream that it feels as though it were really happening to you.
12. Talking out loud to yourself when you are alone.
13. Finding that you act so differently in a particular situation compared with another that it feels almost as if you were two different people.
14. Finding that you cannot remember whether or not you have just done something or perhaps had just thought about doing it (for example, not knowing whether you have just mailed a letter or have just thought about mailing it).
15. Feeling sometimes as if you were looking at the world through a fog such that people and objects appear faraway or unclear.

Source. Bernstein, E. M., & Putnam, F. W. (1986). Development, reliability, and validity of a dissociation scale. *Journal of Nervous and Mental Disease, 174,* 727–735. Copyright © Williams & Wilkins, 1986.

Theoretical Perspectives

The dissociative disorders are fascinating and perplexing phenomena. How can one's sense of personal identity become so distorted that one develops multiple personalities, blots out large chunks of personal memory, or develops a new self-identity? Although these disorders remain in many ways mysterious, clues have emerged that provide insights into their origins.

Psychodynamic Views Dissociative amnesia may serve an adaptive function of disconnecting or dissociating one's conscious from awareness of traumatic experiences or other sources of psychological pain or conflict (Dorahy, 2001). To psychodynamic theorists, dissociative disorders involve the massive use of repression, resulting in the "splitting off" from consciousness of unacceptable impulses and painful memories. In dissociative amnesia and fugue, the ego protects itself from becoming flooded with anxiety by blotting out disturbing memories or by dissociating threatening impulses of a sexual or aggressive nature. In multiple personality, people may express these unacceptable impulses through the development of alternate personalities. In depersonalization, people stand outside themselves—safely distanced from the emotional turmoil within.

Cognitive and Learning Views Learning and cognitive theorists view dissociation as a learned response that involves *not thinking* about disturbing acts or thoughts in order to avoid feelings of guilt and shame evoked by such experiences. The habit of not thinking about these matters is negatively reinforced by relief from anxiety, or by removal of feelings of guilt or shame. Some social cognitive theorists, such as the late Nicholas Spanos, believe that dissociative identity disorder is a form of role playing acquired through observational learning and reinforcement. This is not quite the same as pretending or malingering; people can honestly come to organize their behavior patterns according to particular roles they have observed. They might also become so absorbed in role playing that they "forget" they are enacting a role.

Brain Dysfunction Might dissociative behavior be connected with underlying brain dysfunction? Research along these lines is still in its infancy, but recent evidence showed differences in brain metabolic activity between people with depersonalization disorder and healthy subjects (Simeon et al., 2000). These findings, which point to a possible dysfunction in parts of the brain involved in body perception, may help account for the feeling of being disconnected from one's body that is associated with depersonalization.

Tying It Together

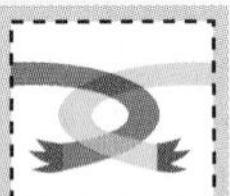

Although we have different conceptualizations of dissociative phenomena, psychologists recognize that a history of abuse in childhood often plays a pivotal role. The most widely held view of dissociative identity disorder is that it represents a means of coping with and surviving severe, repetitive childhood abuse, generally beginning before the age of 5 (Burton & Lane, 2001). The severely abused child may retreat into alter personalities as a psychological defense against unbearable abuse. The construction of these alter personalities allows such children to psychologically escape or distance themselves from their suffering (Burton & Lane, 2001). Dissociation offers a means of escape when no other means is available (Gershuny & Thayer, 1999). In the face of continued abuse, these alter personalities may become stabilized, making it difficult for the person to maintain a unified personality. In adulthood, people with multiple personalities may use their alter personalities to block out traumatic childhood memories and their emotional reactions to them, thus wiping the slate clean and beginning life anew in the guise of alter personalities (Schafer, 1986). The alter identities or personalities may also help the person cope with stressful situations or express deep-seated resentments that the individual is unable to integrate within his or her primary personality (Spanos, 1994).

Compelling evidence indicates that exposure to childhood trauma, usually by a relative or caretaker, is involved in the development of dissociative disorders, especially dissociative identity disorder. The great majority of people with multiple personalities report being physically or sexually abused as children (Lewis et al., 1997; Weaver & Clum, 1995). In one sample, 83% of people with dissociative identity disorder reported a history of childhood sexual abuse and 2 out of 3 reported both physical and sexual abuse (Putnam et al., 1986). In other samples, rates of childhood physical or sexual abuse have ranged from 76% to 95% of cases (Ross et al., 1990; Scroppo et al., 1998). Evidence of cross-cultural similarity comes from a study in Turkey, which showed that more than 3 out of 4 of 35 dissociative identity disorder patients reported sexual or physical abuse in childhood (Sar et al., 1996). Childhood trauma or abuse is also reported more often in cases of dissociative amnesia and depersonalization disorder than in control groups (Coons, Bowman, & Pellow, 1989; Simeon et al., 1997, 2001).

Truth OR Fiction? REVISITED

Most people with multiple personalities do not report any history of physical or sexual abuse during childhood.

FALSE. The great majority of people with multiple personalities do in fact report being physically or sexually abused as children.

Childhood abuse is not the only source of trauma linked to dissociative disorders. Exposure to the trauma of warfare among both civilians and soldiers plays a part in some cases of dissociative fugue and dissociative amnesia. In fugue, the stress of combat and the secondary gain of leaving the battlefield seem to be important contributors (Loewenstein, 1991). The stress of coping with severe financial problems and the wish to avoid punishment for socially unacceptable behavior are other possible antecedents to episodes of fugue (Riether & Stoudemire, 1988). Exposure to high levels of stress may also be linked to depersonalization disorder (Kluft, 1988).

Imaginary friends? Like the child in the photo, it is normal for children to have imaginary playmates. In the case of many multiple personalities, however, games of "make believe" and the invention of imaginary playmates may be used as psychological defenses against abuse. Research suggests that most people who develop multiple personalities were abused as children.

Diathesis-Stress Model Despite widespread evidence of childhood trauma in cases of dissociative identity disorder, very few abused children develop multiple personalities, even among those who suffer severe abuse. Consistent with the diathesis-stress model, certain personality traits, such as proneness to fantasize, high ability to be hypnotized, and openness to altered states of consciousness, may predispose individuals to develop dissociative experiences in the face of extreme stress, such as traumatic abuse. These personality traits themselves do not lead to dissociative disorders (Rauschenberger & Lynn, 1995). They are actually quite common in the population. However, they may increase the risk that people who experience severe trauma will develop dissociative phenomena as a survival mechanism (Butler et al., 1996). People who are low in fantasy proneness or hypnotizability may experience the kinds of anxious, intrusive thoughts characteristic of posttraumatic stress disorder (PTSD) in the aftermath of traumatic stress, rather than dissociative experiences (Kirmayer, Robbins, & Paris, 1994).

Perhaps most of us can divide our consciousness so that we become unaware of—at least temporarily—those events we normally focus on. Perhaps most of us can thrust the unpleasant from our minds and enact various roles—parent, child, lover, businessperson, soldier—that help us meet the requirements of our situations. Perhaps the marvel is *not* that attention can be splintered, but that human consciousness is normally integrated into a meaningful whole.

THINK ABOUT IT

What role does childhood physical and sexual abuse play in dissociative disorders?

THINK ABOUT IT

What psychological formulations have been proposed to explain the development of dissociative disorders? What do they have in common?

Treatment of Dissociative Disorders

Dissociative amnesia and fugue are usually fleeting experiences that end abruptly. Episodes of depersonalization can be recurrent and persistent, and they are most likely to occur when people are undergoing periods of mild anxiety or depression. In such cases, clinicians usually focus on managing the anxiety or the depression. Much of the attention in the research literature has focused on dissociative identity disorder and specifically on bringing together an integration of the alter personalities into a cohesive personality structure (Burton & Lane, 2001).

Psychoanalysts seek to help people with dissociative identity disorder uncover and learn to cope with early childhood traumas. They often recommend establishing direct contact with alter personalities (Burton & Lane, 2001). For instance, Wilbur (1986) points out that the analyst can work with whatever personality dominates the therapy session. Any and all personalities can be asked to talk about their memories and dreams as best they can. Any and all personalities can be assured that the therapist will help them make sense of their anxieties and to safely "relive" traumatic experiences and make them conscious. Wilbur enjoins therapists to keep in mind that anxiety experienced during a therapy session may lead to a switch in personalities, because alter personalities were presumably developed as a means to cope with intense anxiety. But if therapy is successful, the self will be

A Closer Look

The Recovered Memory Controversy

A high-level business executive's comfortable life fell apart one day when his 19-year-old daughter accused him of having repeatedly molested her throughout her childhood. The executive lost his marriage as well as his $400,000-a-year job. But he fought back against the allegations, which he insisted were untrue. He sued his daughter's therapists, who had assisted her in recovering these memories. A jury sided with the businessman, awarding him $500,000 in damages from the two therapists.

This case is but one of many involving adults who claim to have only recently become aware of memories of childhood sexual abuse. Hundreds of people throughout the country have been brought to trial on the basis of recovered memories of childhood abuse, with many of these cases resulting in convictions and long jail sentences, even in the absence of corroborating evidence. Such recovered memories often occur following suggestive probing by a therapist or hypnotist (Loftus, 1993). The issue of recovered memories continues to be hotly debated in psychology and the broader community. At the heart of the debate is the question, "Are recovered memories believable?" No one doubts that childhood sexual abuse is a major problem confronting our society. But should recovered memories be taken at face value?

Several lines of evidence lead us to question the validity of recovered memories. Experimental evidence shows that false memories can be created, especially under the influence of leading or suggestive questioning (Begley, 2001b; Loftus, 1997; Schacter, 1999; Zoellner et al., 2000). Memory for events that never happened may be induced in people's memories and may seem just as real as memories of events that really did occur (Zola, 1999). Moreover, although people who have experienced actual abuse in childhood may be somewhat sketchy on the details, total amnesia concerning the trauma is rare (Wakefield & Underwager, 1996). A leading memory expert, psychologist Elizabeth Loftus (1996, p. 356), writes of the dangers of taking recovered memories at face value:

> *After developing false memories, innumerable "patients" have torn their families apart, and more than a few innocent people have been sent to prison. This is not to say that people cannot forget horrible things that have happened to them; most certainly they can. But there is virtually no support for the idea that clients presenting for therapy routinely have extensive histories of abuse of which they are completely unaware, and that they can be helped only if the alleged abuse is resurrected from their unconscious.*

Should we conclude, then, that recovered memories are bogus? Not necessarily. It is possible for people in adulthood to recover memories of childhood (Melchert, 1996), including memories of abuse (Chu et al., 1999). Some recovered memories may be true; others may not be (Brown, 1997; Reisner, 1996; Rubin, 1996). Unfortunately we don't have the tools to distinguish the true memory from the false one (Loftus, 1993).

We shouldn't think of the brain as a kind of mental camera that stores snapshots of events as they actually happened in the form of memories. Memory is more of a reconstructive process in which bits of information are pieced together in ways that can sometimes lead to a distorted recollection of events, even though the person may be convinced the memory is accurate.

Web Link 7.2 WWW
Q&A: Memories of Childhood Abuse

able to work through the traumatic memories and will no longer need to escape into alternate "selves" to avoid the anxiety associated with the trauma. Thus, reintegration of the personality becomes possible.

Wilbur describes the formation of another treatment goal in the case of a woman with dissociative identity disorder:

A Case of Dissociative Identity Disorder

A 45-year-old woman had suffered from dissociative identity disorder throughout her life. Her dominant personality was timid and self-conscious, rather reticent about herself. But soon after she entered treatment, a group of "little ones" emerged, who cried profusely. The therapist asked to speak with someone in the personality system who could clarify the personalities that were present. It turned out that they included several children, all of whom were under 9 years of age and had suffered severe, painful sexual abuse at the hands of an uncle, a great-aunt, and a grandmother. The great-aunt was a lesbian with several voyeuristic lesbian friends. They would watch the sexual abuse, generating fear, pain, rage, humiliation, and shame.

It was essential in therapy for the "children" to come to understand that they should not feel ashamed because they had been helpless to resist the abuse.

—Adapted from Wilbur, 1986, pp. 138–139

■

Does therapy work? Coons (1986) followed 20 "multiples" aged from 14 to 47 at time of intake for an average of 3¼ years. Only 5 of the subjects showed a complete reintegration of their personalities. Other therapists report significant improvement in measures of dissociative symptoms and depressive symptoms in treated patients, even in those who failed to achieve integration. However, greater symptom improvement was reported for those who achieved integration (Ellason & Ross, 1997).

Reports of the effectiveness of psychoanalytic or of other forms of therapy, such as behavior therapy, rely on uncontrolled case studies. Controlled studies of treatments of dissociative identity disorder or other forms of dissociative disorder are yet to be reported (Maldonad et al., 1998). The relative infrequency of the disorder has hampered efforts to conduct controlled experiments that compare different forms of treatment with each other and with control groups. Nor do we have evidence showing psychiatric drugs or other biological approaches to be effective in bringing about an integration of various alternate personalities.

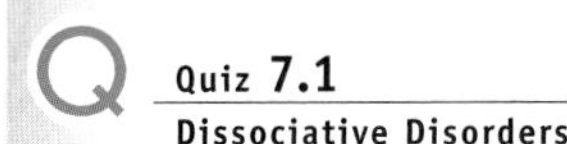

A Closer Look

The Truth Is Out There

The *X-Files,* one of television's most popular shows in recent years, featured FBI agents who were charged with investigating mysterious phenomena. One of the running themes in the show was the belief in alien abductions. The show picked up on the claims of hundred or thousands of real people who said they had been abducted by space aliens. Though details of alien abductions vary, they usually involve reports of being spirited away to an alien spaceship, whereupon various medical procedures or experiments are performed on them before they are returned to Earth. What are we to make of these reports? Some reports may be fabrications, concocted for publicity or in hopes of achieving fame or securing lucrative Hollywood contract for their stories. Yet many of these claims are difficult to ascribe to either lying or insanity (Newman & Baumeister, 1996).

These reports have not been subjected to formal scientific study, so our beliefs about them rest largely on theoretical speculation. Some psychologists view them as false memories derived from sleep-related hallucinations or nightmares that are pieced together under hypnosis and reinforced by a popular culture that gives credence to alien sightings (Clark & Loftus, 1996; Newman & Baumeister, 1996). Memory is a reconstructive process, not a photographic rendering of events. A number of experiments have shown that under the right circumstances people can be led to believe that they experienced events which did not actually take place (Clark & Loftus, 1996). Hypnosis is believed to play a part as memories of alien abductions often develop after hypnosis (Orne et al., 1996). However, we cannot simply ascribe memories of alien abductions to effects of hypnotic suggestion on reconstructed memories. Hypnosis doesn't play a part in many cases; in still others hypnosis was used to fill in some details after the person made a report of an alien abduction (Hall, 1996).

Others suggest that memories of alien abductions are individual delusions—false but strongly held beliefs that anyone can develop in the attempt to explain unusual events that happen to them (Banaji & Kihlstrom, 1996). Other avenues of speculation treat alien abduction phenomena as forms of mass delusion supported by a social network of people holding such deviant beliefs (Hall, 1996), or types of dissociative experiences or splitting of consciousness in response to extreme stress (Fisman & Takhar, 1996; Shopper, 1996), or attempts to escape from the self (Newman & Baumeister, 1996). Though speculation abounds, we lack a solid foundation of research evidence on which to judge these theoretical accounts (Arndt & Greenberg, 1996). Perhaps we will learn more as research on this intriguing phenomenon continues. Or perhaps reports of alien abductions will shortly disappear into the trash heap of discarded cultural trends. There is yet another, however unlikely explanation. Perhaps these people were truly abducted by space aliens. On this account, we'd best leave the investigation to the likes of the fictional X-file agents Fox Muldur and Dana Scully.

Overview of Dissociative Disorders

TYPES OF DISSOCIATIVE DISORDERS

	Description	Features
Dissociative Identity Disorder	Emergence of two or more distinct personalities	• Alternates may vie for control • Some cases reported of distinct physiological characteristics of alternates
Dissociative Amnesia	Inability to recall important personal material that cannot be accounted for by medical causes	• Information lost to memory is usually of traumatic or stressful experiences • Subtypes include localized amnesia, selective amnesia, and generalized amnesia
Dissociative Fugue	Amnesia "on the run"; the person travels to a new location and is unable to remember personal information or reports a past filled with false information that is not recognized as false	• Person may be confused about his or her personal identity or assumes a new identity • Person may start a new family or business
Depersonalization Disorder	Episodes of feeling detached from one's self or one's body or having a sense of unreality about one's surroundings (derealization)	• Person may feel as if he or she were living in a dream or acting like a robot • Episodes of depersonalization are persistent or recurrent and cause significant distress

CAUSAL FACTORS **A history of childhood trauma or abuse is implicated in many cases**

Biological Factors	• Not known
Social-Environmental Factors	• Childhood sexual or physical abuse (in dissociative identity disorder) • Other traumatic experiences, such as combat trauma (in dissociative amnesia and dissociative fugue)
Behavioral Factors	• Possible reinforcement (attention) for enacting the social role of a multiple personality
Emotional and Cognitive Factors	• Relief from anxiety by psychologically distancing oneself (dissociating) from troubling emotions or memories

TREATMENT APPROACHES **Dissociative identity disorder remains a challenge to treat; dissociative amnesia and dissociative fugue tend to resolve on their own. The relative infrequency of these disorders has limited efforts to mount controlled studies of other therapies**

Biomedical Treatment	• Drug therapy (SSRI-type antidepressants) may be helpful in treating depersonalization disorder
Psychodynamic Therapy	• For dissociative identity disorder, psychoanalytic therapy may be used to seek a reintegration of the personality

Somatoform Disorders

The word *somatoform* derives from the Greek *soma,* meaning "body." In the **somatoform disorders,** people have physical symptoms suggestive of physical disorders, but no organic abnormalities can be found to account for them. Moreover, there is evidence, or some reason to believe, that the symptoms reflect psychological factors or conflict. Some people complain of problems in breathing or swallowing, or of a "lump in the throat." Problems such as these can reflect overactivity of the sympathetic branch of the autonomic nervous system, which can be related to anxiety. Sometimes the symptoms take more unusual forms, as in a "paralysis" of a hand or leg that is inconsistent with the workings of the nervous system. In yet other cases, people are preoccupied with the belief that they have a serious disease, yet no evidence of a physical abnormality can be found. We consider several forms of somatoform disorders, including *conversion disorder, hypochondriasis,* and *somatization disorder.*

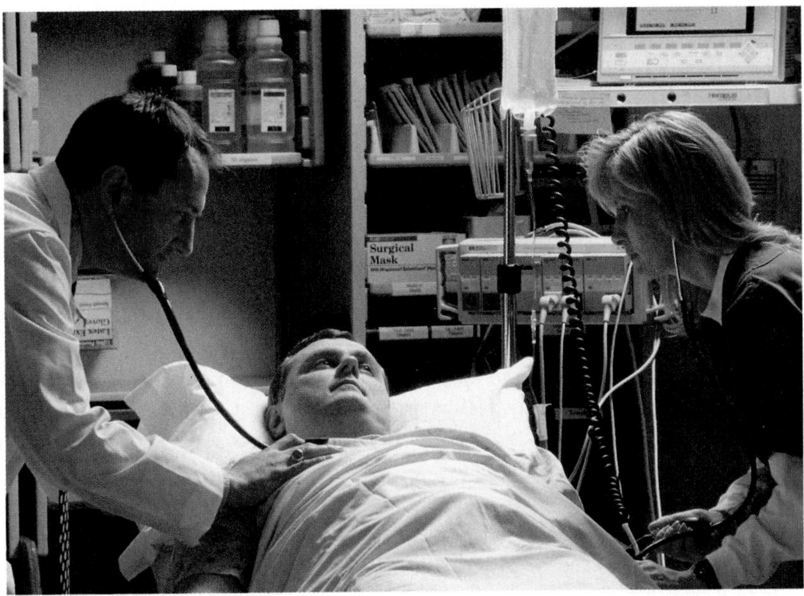

Is this patient really sick? Munchausen syndrome is characterized by the fabrication of medical complaints for no other apparent purpose than to gain admission to hospitals. Some Munchausen patients may produce life-threatening symptoms in their attempts to deceive doctors.

Somatoform disorders are distinguished from malingering, or purposeful fabrication of symptoms for obvious gain (such as avoiding work). They are also distinguished from a **factitious disorder,** the most common form of which is **Munchausen syndrome.** Munchausen is a form of feigned illness in which the person either fakes being ill or makes him- or herself ill (by ingesting toxic substances, for example). Some Munchausen patients go through unnecessary surgeries, even though they know there is nothing wrong with them. Yet unlike malingering, there is no apparent purpose to the fakery save for the attention the person receives from medical professionals. Because malingering is motivated by external incentives, it is not considered a mental disorder within the *DSM* framework. In factitious disorders, however, the symptoms are not connected with obvious gains. The absence of external incentives in these disorders suggests that they serve a psychological need; hence, they are considered mental disorders.

Why do patients with Munchausen syndrome feign illness or sometimes put themselves at grave risk by causing themselves to be sick or injured? Perhaps enacting the sick role in the protected hospital environment provides a sense of security that was lacking in childhood. Perhaps the hospital becomes a stage on which they can act out resentments against doctors and parents that have been brewing since childhood. Perhaps they are trying to identify with a parent who was often sick. Or perhaps they learned to enact a sick role in childhood to escape from repeated sexual abuse or other traumatic experiences and continue to enact the role to escape stressors in their adult lives (Trask & Sigmon, 1997). No one is really sure, and the disorder remains one of the more puzzling forms of abnormal behavior.

Here let us consider several of the major types of somatoform disorder: conversion disorder, hypochondriasis, body dysmorphic disorder, and somatization disorder.

Truth OR Fiction? REVISITED

Some people show up repeatedly at hospital emergency rooms, feigning illness and seeking treatment for no apparent reason.

TRUE. People with Munchausen syndrome may show up repeatedly at emergency rooms, feigning illness and demanding treatment. Their motives remain a mystery.

WWW **Web Link 7.3**
About Conversion Disorder

somatoform disorders A group of disorders characterized by complaints of physical problems or symptoms that cannot be explained by physical causes.

factitious disorder A disorder characterized by intentional fabrication of psychological or physical symptoms for no apparent gain.

Munchausen syndrome A type of factitious disorder characterized by the feigning of medical symptoms.

conversion disorder A type of somatoform disorder characterized by loss or impairment of physical function in the absence of any apparent organic cause.

Conversion Disorder

Conversion disorder is characterized by a major change in or loss of physical functioning, although no medical findings are found to account for the physical symptoms or deficits (see Table 7.3). The symptoms are not intentionally produced. The person is not malingering. The physical symptoms usually come on suddenly in stressful situations. A soldier's hand may become "paralyzed" during intense combat, for example. The fact that conversion symptoms first appear in the context of, or are aggravated by, conflicts or stressors the individual encounters gives credence to the view that they relate to psychological factors (APA, 2000). Reported rates of the disorder in the general population range from as few as 1.1 in 10,000 people to perhaps as many as 1 in 200 people (APA, 2000).

TABLE 7.3 Diagnostic Features of Conversion Disorder

1. At least one symptom or deficit involving voluntary motor or sensory functions that suggests the presence of a physical disorder.
2. Psychological factors are judged to be associated with the disorder because the onset or exacerbation of the physical symptom is linked to the occurrence of psychosocial stressors or conflict situations.
3. The person does not purposefully produce or fake the physical symptom.
4. The symptom cannot be explained as a cultural ritual or response pattern, nor can it be explained by any known physical disorder on the basis of appropriate testing.
5. The symptom causes significant emotional distress, impairment in one or more important areas of functioning, such as social or occupational functioning, or is sufficient to warrant medical attention.
6. The symptom is not restricted to complaints of pain or problems in sexual functioning, nor can it be accounted for by another mental disorder.

Source. Adapted from the *DSM-IV-TR* (APA, 2000).

la belle indifférence A French expression describing the lack of concern over one's symptoms displayed by some people with conversion disorder.

Conversion disorder is so named because of the psychodynamic belief that it represents the channeling, or *conversion,* of repressed sexual or aggressive energies into physical symptoms. Conversion disorder was formerly called *hysteria* or *hysterical neurosis,* and it played an important role in Freud's development of psychoanalysis (see Chapter 1). Hysterical or conversion disorders seem to have been more common in Freud's day than they are today, when they are relatively rare.

According to the *DSM,* conversion symptoms mimic neurological or general medical conditions involving problems with voluntary motor (movement) or sensory functions. Some of the "classic" symptom patterns involve paralysis, epilepsy, problems in coordination, blindness and tunnel vision, loss of the sense of hearing or of smell, or loss of feeling in a limb (anesthesia). The bodily symptoms found in conversion disorders often do not match the medical conditions they suggest. For example, conversion epileptics, unlike true epileptic patients, may maintain control over their bladders during an attack. People whose vision is supposedly impaired may wend their ways through the physician's office without bumping into the furniture. People who become "incapable" of standing or walking may nevertheless perform other leg movements normally. Nonetheless, hysteria may be incorrectly diagnosed in people who turn out to have underlying medical conditions. Perhaps as many as 80% of individuals given the diagnosis of conversion disorder have real neurological problems that go undiagnosed (Gould et al., 1986).

If you suddenly lost your vision, or if you could no longer move your legs, you would probably show understandable concern. But some people with conversion disorders, like those with dissociative amnesia, show a remarkable indifference to their symptoms, a phenomenon termed **la belle indifférence** ("beautiful indifference"). The *DSM* advises against relying on indifference to symptoms as a factor in making the diagnosis, however, because many people cope with real physical disorders by denying their pain or concern, which provides the semblance of indifference and relieves anxieties—at least temporarily.

Hypochondriasis

The core feature of hypochondriasis is a preoccupation or fear that one's physical symptoms are due to an underlying serious illness, such as cancer or a heart problem. The fear persists despite medical reassurances that it is groundless (see Table 7.4).

People with hypochondriasis do not consciously fake their physical symptoms. They generally experience physical discomfort, often involving the digestive system or an assortment of aches and pains. Unlike conversion disorder, hypochondriasis does not involve the loss or distortion of physical function. Unlike the attitude of indifference toward one's symptoms that is sometimes found in conversion disorders, people who develop hypochondriasis are very concerned, indeed unduly concerned, about their symptoms and

TABLE 7.4 Diagnostic Features of Hypochondriasis

1. The person is preoccupied with a fear of having a serious illness, or with the belief that one has a serious illness. The person interprets bodily sensations or physical signs as evidence of physical illness.
2. Fears of physical illness, or beliefs of having a physical illness, persist despite medical reassurances.
3. The preoccupations are not of a delusional intensity (the person recognizes the possibility that these fears and beliefs may be exaggerated or unfounded) and are not restricted to concerns about appearance.
4. The preoccupations cause significant emotional distress or interfere with one or more important areas of functioning, such as social or occupational functioning.
5. The disturbance has persisted for 6 months or longer.
6. The preoccupations do not occur exclusively within the context of another mental disorder.

Source. Adapted from the *DSM-IV-TR* (APA, 2000).

what they fear they may represent. Although the underlying rates of hypochondriasis remain unknown, the disorder appears to be about equally common in men and women. It most often begins between the ages of 20 and 30, although it can begin at any age.

People with hypochondriasis may be overly sensitive to benign changes in physical sensations, such as slight changes in heartbeat and minor aches and pains (Barsky et al., 2001). Anxiety about physical symptoms can produce its own physical sensations, however—for example, heavy sweating and dizziness, even fainting. Thus, a vicious cycle may ensue. People with hypochondriasis may become resentful when their doctors tell them how their own fears may be causing their physical symptoms. They frequently go "doctor shopping" in the hope that a competent and sympathetic physician will heed them before it is too late. Physicians, too, can develop hypochondriasis, as we see in the following case example:

A Case of Hypochondriasis

Robert, a 38-year-old radiologist, has just returned from a 10-day stay at a famous diagnostic center where he has undergone extensive testing of his entire gastrointestinal tract. The evaluation proved negative for any significant physical illness, but rather than feel relieved, the radiologist appeared resentful and disappointed with the findings. The radiologist has been bothered for several months with various physical symptoms, which he describes as symptoms of mild abdominal pain, feelings of "fullness," "bowel rumblings," and a feeling of a "firm abdominal mass." He has become convinced that his symptoms are due to colon cancer and has become accustomed to testing his stool for blood on a weekly basis and carefully palpating his abdomen for "masses" while lying in bed every several days. He has also secretly performed X-ray studies on himself after regular hours. There is a history of a heart murmur that was detected when he was 13 and his younger brother died of congenital heart disease in early childhood. When the evaluation of his murmur proved to be benign, he nonetheless began to worry that something might have been overlooked. He developed a fear that something was actually wrong with his heart, and while the fear eventually subsided, it has never entirely left him. In medical school he worried about the diseases that he learned about in pathology. Since graduating, he has repeatedly experienced concerns about his health that follow a typical pattern: noticing certain symptoms, becoming preoccupied with what the symptoms might mean, and undergoing physical evaluations that proved negative. His decision to seek a psychiatric consultation was prompted by an incident with his 9-year-old son. His son accidentally walked in on him while he was palpating his abdomen and asked, "What do you think it is this time, Dad?" He becomes tearful as he relates this incident, describing his feelings of shame and anger—mostly at himself.

—Adapted from Spitzer et al., 1994, pp. 88–90

People who develop hypochondriasis have more health worries, more psychiatric symptoms, and perceive their health to be worse than do other people (Noyes et al., 1993). They are also more likely than other psychiatric patients to report being sick as children, having missed school because of health reasons, and having experienced childhood trauma, such as sexual abuse or physical violence (Barsky et al., 1994). According to recent studies, most people who meet diagnostic criteria for hypochondriasis continue to show evidence of the disorder when reinterviewed 5 years later (Barsky et al., 1998). Most also have other psychological disorders, especially major depression and anxiety disorders (Barsky, Wyshak, & Klerman, 1992; Noyes et al., 1993).

Hypochondriasis is generally considered to be most common among elderly people. As noted by Paul Costa and Robert McCrae (1985) of the National Institute on Aging,

Truth OR Fiction? REVISITED

Some people who have lost their ability to see or move their legs become strangely indifferent toward their physical condition.

TRUE. Some people with conversion disorder show a remarkable lack of concern about their symptoms (la belle indifférence).

THINK ABOUT IT

Why is conversion disorder considered a treasure trove in the annals of abnormal psychology? What role did the disorder play in the development of psychological models of abnormal behavior?

WWW **Web Link 7.4**
Facts About Hypochondriasis

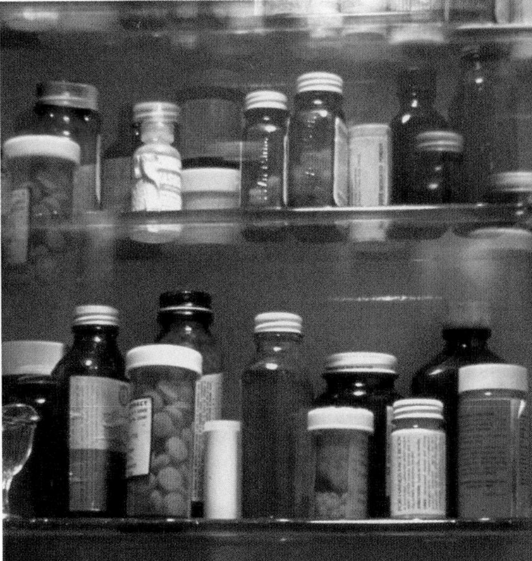

What to take? Hypochondriasis involves persistent concerns or fears that one is seriously ill, although no organic basis can be found to account for one's physical complaints. People with this disorder frequently medicate themselves with over-the-counter medications and find little if any reassurance in doctors' assertions that their health is not in jeopardy.

Can't you see it? A person with body dysmorphic disorder may spend hours in front of a mirror obsessing about an imagined or exaggerated physical defect in appearance.

however, authentic age-related health changes do occur, and most "hypochondriacal" complaints probably reflect these changes.

Body Dysmorphic Disorder

People with body dysmorphic disorder (BDD) are preoccupied with an imagined or exaggerated physical defect in their appearance (APA, 2000). They may spend hours examining themselves in the mirror and go to extreme measures to try to correct the perceived defect, even undergoing unnecessary plastic surgery. Others may remove any mirrors from their homes so as not to be reminded of the glaring flaw in their appearance. People with BDD may believe that others view them as ugly or deformed and that their unattractive physical appearance leads others to think negatively of their character or worth as a person (Rosen, 1996). The rates of BDD are not well established, since many people with this disorder fail to seek help or try to keep their symptoms a secret (Cororve & Gleaves, 2001). People with BDD often show a pattern of compulsive grooming or washing, or styling their hair, in an attempt to correct the perceived defect, as in the following case example:

THINK ABOUT IT

Do you know anyone you would consider to be a "hypochondriac"? What is the basis of your opinion? Did reading the text change your view?

A Case of Body Dysmorphic Disorder

For Claudia, a 24-year-old legal secretary, virtually every day was a bad hair day. She explained to her therapist, "When my hair isn't right, which is like every day, I'm not right." "Can't you see it," she went on to explain, "It's so uneven. This piece should be shorter and this one just lies there. People think I'm crazy but I can't stand looking like this. It makes me look like I'm deformed. It doesn't matter if people can't see what I'm talking about. I see it. That's what counts." Several months earlier Claudia had a haircut she described as a disaster. Shortly thereafter, she had thoughts of killing herself: "I wanted to stab myself in the heart. I just couldn't stand looking at myself."

Claudia checked her hair in the mirror innumerable times during the day. She would spend two hours every morning doing her hair and still wouldn't be satisfied. Her constant pruning and checking had become a compulsive ritual. As she told her therapist, "I want to stop pulling and checking it, but I just can't help myself."

—From the Authors' Files

Having a "bad hair day" for Claudia meant that she would not go out with her friends and would spend every second examining herself in the mirror and fixing her hair. Occasionally she would cut pieces of her hair herself in an attempt to correct the mistakes of her last haircut. But cutting it herself inevitably made it even worse, in her view. Claudia was forever searching for the perfect haircut that would correct defects only she could perceive. Several years earlier she had what she described as a perfect haircut. "It was just right. I was on top of the world. But it began to look crooked when it grew in." Forever in search of the perfect haircut, Claudia had obtained a hard-to-get appointment with a world-renowned hair stylist in Manhattan whose clientele included many celebrities. "People wouldn't understand paying this guy $375 for a haircut, especially on my salary, but they don't realize how important it is to me. I'd pay any amount I could." Unfortunately even this celebrated hair stylist disappointed her: "My $25 haircut from my old stylist on Long Island was better than this."

Claudia reported other fixations about her appearance earlier in life: "In high school, I felt my face was like a plate. It was just too flat. I didn't want any pictures taken of me. I couldn't help thinking what people thought of me. They won't tell you, you know. Even if they say there's nothing wrong, it doesn't mean anything. They were just lying to be polite." Claudia related that she was taught to equate physical beauty with happiness: "I was told that to be successful you had to be beautiful. How can I be happy if I look this way?"

WWW **Web Link 7.5**
Features of Body Dysmorphic Disorder

Somatization Disorder

Somatization disorder, formerly known as Briquet's syndrome, is characterized by multiple and recurrent somatic complaints that begin prior to the age of 30 (but usually during the teen years), persist for at least several years, and result either in the seeking of medical attention or in significant impairment in fulfilling social or occupational roles. Complaints usually involve different organ systems (Spitzer et al., 1989). Seldom a year passes without some physical complaint that prompts a trip to the doctor. People with somatization disorder are heavy users of medical services (G. R. Smith, 1994). Community surveys show that virtually all (95%) of the people with somatization disorder had visited a doctor during the past year and nearly half (45%) had been hospitalized (Swartz et al., 1991). The complaints cannot be explained by physical causes or exceed what would be expected from a known physical problem. Complaints seem vague or exaggerated, and the person frequently receives medical care from a number of physicians, sometimes at the same time.

Somatization disorder usually begins in adolescence or young adulthood and appears to be a chronic or even lifelong disorder (Kirmayer, Robbins, & Paris, 1994; Smith, 1994). It usually occurs in the context of other psychological disorders, especially anxiety disorders and depressive disorders (Swartz et al., 1991). Although not much is known about the childhood backgrounds of people with somatization disorder, one study reported that women with the disorder were significantly more likely to report sexual molestation in childhood than a matched comparison group of women with mood disorders (Morrison, 1989).

The essential feature of hypochondriasis is fear of disease, of what bodily symptoms may portend. Persons with somatization disorder, by contrast, are pestered by the symptoms themselves. Both diagnoses may be given to the same individual if the diagnostic criteria for both disorders are met.

Estimates are that 1 person in 1,000 in the United States is affected by somatization disorder, with 10 times as many cases found among women than men. The disorder is also 4 times more likely to occur among African Americans than other ethnic or racial groups (Swartz et al., 1991). Yet the disorder is controversial. Many patients, especially female patients, are misdiagnosed with psychological disorders, including somatization disorder, because of the failure of modern medicine to identify the underlying medical basis of their physical complaints (Klonoff & Landrine, 1997).

Koro and Dhat Syndromes: Far Eastern Somatoform Disorders?

In the United States, it is common for people who develop hypochondriasis to be troubled by the idea that they have serious illnesses, such as cancer. The koro and dhat syndromes of the Far East share some clinical features with hypochondriasis. Although these syndromes may seem foreign to most American readers, they are each connected with folklore within their Far Eastern cultures.

Koro Syndrome **Koro syndrome** is a culture-bound syndrome found primarily in China and some other Far Eastern countries (Sheung-Tak, 1996). People with koro syndrome fear that their genitals are shrinking and retracting into the body, which they believe will result in death (Fabian, 1991; Goetz & Price, 1994; Tseng et al., 1992). Koro is considered a

somatization disorder A type of somatoform disorder involving recurrent multiple complaints that cannot be explained by any physical cause.

koro syndrome A culture-bound somatoform disorder, found primarily in China, in which people fear that their genitals are shrinking.

dhat syndrome A culture-bound somatoform disorder, found primarily among Asian Indian males, characterized by excessive fears over the loss of seminal fluid.

culture-bound syndrome, although some cases have been reported outside China and the Far East (e.g., Chowdhury, 1996). The syndrome has been identified mainly in young men, although some cases have also been reported in women (Tseng et al., 1992). Koro syndrome tends to be short-lived and to involve episodes of acute anxiety that one's genitals are retracting. Physiological signs of anxiety that approach panic proportions are common, including profuse sweating, breathlessness, and heart palpitations. Men who suffer from koro have been known to use mechanical devices, such as chopsticks, to try to prevent the penis from retracting into the body (Devan, 1987).

Koro syndrome has been traced within Chinese culture as far back as 3000 B.C.E. (Devan, 1987). Epidemics involving hundreds or thousands of people have been reported in China, Singapore, Thailand, and India (Tseng et al., 1992). In Guangdong Province in China, an epidemic of koro involving more than 2,000 persons occurred during the 1980s (Tseng et al., 1992). Guangdong residents who did not fall victim to koro tended to be less superstitious, higher in intelligence, and less accepting of koro-related folk beliefs (such as the belief that shrinkage of the penis will be lethal) than those who fell victim to the epidemic (Tseng et al., 1992). Medical reassurance that such fears are unfounded often quell koro episodes (Devan, 1987). Medical reassurance generally fails to dent the concerns of Westerners who develop hypochondriasis, however. Koro episodes among those who do not receive corrective information tend to pass with time but may recur.

A number of investigators would like to see the koro syndrome incorporated into the *DSM* as a somatoform disorder (Bernstein & Gaw, 1990; Fishbain, 1991).

Truth OR Fiction? REVISITED

In China in the 1980s, more than 2,000 people fell prey to the belief that their genitals were shrinking and retracting into their bodies.

TRUE. An epidemic was reported in China in which some 2,000 people fell prey to the belief that their genitals were shrinking and retracting into their bodies. This condition, called koro, is classified by the *DSM* system as a culture-bound syndrome.

Dhat Syndrome **Dhat syndrome** is found among young Asian Indian males and involves excessive fears over the loss of seminal fluid during nocturnal emissions (Akhtar, 1988). Some men with this syndrome also believe (incorrectly) that semen mixes with urine and is excreted through urination. Men with dhat syndrome may roam from physician to physician seeking help to prevent nocturnal emissions or the (imagined) loss of semen mixed with excreted urine. There is a widespread belief within Indian culture (and other Near and Far Eastern cultures) that the loss of semen is harmful because it depletes the body of physical and mental energy (Chadda & Ahuja, 1990). Like other culture-bound syndromes, dhat must be understood within its cultural context:

> In India, attitudes toward semen and its loss constitute an organized, deep-seated belief system that can be traced back to the scriptures of the land ... [even as far back as the classic Indian sex manual, the Kama Sutra, which was believed to be written by the sage Vatsayana between the third and fifth centuries A.D.] ... Semen is considered to be the elixir of life, in both a physical and mystical sense. Its preservation is supposed to guarantee health and longevity.
>
> —*From Akhtar, 1988, p. 71*

THINK ABOUT IT

Does koro or dhat syndrome seem strange to you? How might your feelings depend on the culture in which you were raised? How might behaviors found in your culture be viewed as strange by members of other cultures?

It is a commonly held Hindu belief that it takes "forty meals to form one drop of blood; forty drops of blood to fuse and form one drop of bone marrow, and forty drops of this to produce one drop of semen" (Akhtar, 1988, p. 71). Based on the cultural belief in the life-preserving nature of semen, it is not surprising that some Indian males experience extreme anxiety over the involuntary loss of the fluid through nocturnal emissions (Akhtar, 1988). Dhat syndrome has also been associated with difficulty in achieving or maintaining erection, apparently due to excessive concern about loss of seminal fluid through ejaculation (Singh, 1985).

Dhat syndrome. Found principally in India, dhat syndrome describes men with an intense fear or anxiety over the loss of semen.

Theoretical Perspectives

Conversion disorder, or "hysteria," was known to Hippocrates, who attributed the strange bodily symptoms to a wandering uterus, which created internal chaos. The term *hysterical* derives from the Greek *hystera,* meaning "uterus." Hippocrates noticed that these complaints were less common among married than unmarried women. He prescribed marriage as a "cure" on the basis of these observations, and also on the theoretical assumption that pregnancy would satisfy uterine needs and fix the organ in place. Pregnancy fosters hormonal and structural changes that are of benefit to some women with menstrual complaints, but Hippocrates's belief in the "wandering uterus" has contributed throughout the centuries to degrading interpretations of complaints by women of physical problems. Despite Hippocrates's belief that hysteria is exclusively a female concern, it also occurs in men.

Modern theoretical accounts of the somatoform disorders, like those of the dissociative disorders, have most often sprung from psychodynamic and learning theories. Although not much is known about biological underpinnings of somatoform disorders, evidence indicates that somatization disorder tends to run in families, primarily among female members (Guze, 1993). This is suggestive of a genetic linkage, although we cannot rule out the possibility that family influences play a part in explaining this familial association.

primary gains Relief from underlying anxiety gained through the development of neurotic symptoms.

secondary gains Side benefits associated with neurotic or other disorders, such as expressions of sympathy, increased attention, and release from responsibilities.

Psychodynamic Theory Hysterical disorders provided an arena for some of the debate between the psychological and biological theories of the 19th century. The alleviation—albeit often temporarily—of hysterical symptoms through hypnosis by Charcot, Breuer, and Freud contributed to the belief that hysteria was rooted in psychological rather than physical causes and led Freud to the development of a theory of the unconscious mind. Freud held that the ego manages to control unacceptable or threatening sexual and aggressive impulses arising from the id through defense mechanisms such as repression. Such control prevents the outbreak of anxiety that would occur if the person were to become aware of these impulses. In some cases, the leftover emotion or energy that is "strangulated," or cut off, from the threatening impulses becomes *converted* into a physical symptom, such as hysterical paralysis or blindness. Although the early psychodynamic formulation of hysteria is still widely held, empirical evidence has been lacking. One problem with the Freudian view is that it does not explain how energies left over from unconscious conflicts become transformed into physical symptoms (E. Miller, 1987).

According to psychodynamic theory, hysterical symptoms are functional: They allow the person to achieve primary gains and secondary gains. The **primary gains** consist of allowing the individual to keep internal conflicts repressed. The person is aware of the physical symptom but not of the conflict it represents. In such cases, the "symptom" is symbolic of, and provides the person with a "partial solution" for, the underlying conflict. For example, the hysterical paralysis of an arm might symbolize and also prevent the individual from acting out on repressed unacceptable sexual (e.g., masturbatory) or aggressive (e.g., murderous) impulses. Repression occurs automatically, so the individual remains unaware of the underlying conflicts. *La belle indifférence,* first noted by Charcot, is believed to occur because the physical symptoms help relieve rather than cause anxiety. From the psychodynamic perspective, conversion disorders, like dissociative disorders, serve a purpose.

B-29 bombers on a bombing mission over Japan during World War II. Some World War II pilots were reported to have suffered from hysterical night blindness, which prevented them from carrying out dangerous nighttime missions. Their night blindness may have served the psychological purpose of shielding them from guilt over dropping bombs on civilian areas—a type of primary gain. It may also have served the secondary purpose of helping them avoid dangerous combat missions.

Secondary gains may allow the individual to avoid burdensome responsibilities and to gain the support—rather than condemnation—of those around them. For example, soldiers sometimes experience sudden "paralysis" of their hands, which prevents them from firing their guns in battle. They may then be sent to recuperate at a hospital rather than face enemy fire. The symptoms in such cases are not considered contrived, as would be the case in malingering. A number of bomber pilots during World War II suffered hysterical "night blindness" that prevented them from carrying out dangerous nighttime missions. In the psychodynamic view, their "blindness" may have achieved a primary gain of shielding them from guilt associated with dropping bombs on civilian areas. It may also have achieved a secondary purpose of helping them avoid dangerous missions.

Learning Theory Psychodynamic theory and learning theory concur that the symptoms in conversion disorders relieve anxiety. Psychodynamic theorists, however, seek the causes of anxiety in unconscious conflicts. Learning theorists focus on the more direct reinforcing properties of the symptom and its secondary role in helping the individual avoid or escape uncomfortable or anxiety-evoking situations.

From the learning perspective, the symptoms in conversion and other somatoform disorders may also carry the benefits, or reinforcing properties of, the "sick role." Persons with conversion disorders may be relieved of chores and responsibilities such as going to work or performing household tasks (Miller, 1987). Being sick also usually earns sympathy and support. People who received such reinforcers during past illnesses are likely to learn to adopt a sick role even when they are not ill (Kendell, 1983).

Differences in learning experiences may explain why conversion disorders were historically more often reported among women than men. It may be that women in Western culture are more likely than men to have been socialized to cope with stress by enacting a sick role (Miller, 1987). We are not suggesting that people with conversion disorders are fakers. We are merely pointing out that people may learn to adopt roles that lead to reinforcing consequences, regardless of whether they deliberately seek to enact these roles.

Some learning theorists link hypochondriasis and body dysmorphic disorder to obsessive-compulsive disorder (OCD; see Chapter 6) (e.g., Barsky et al., 1992; Cororve & Gleaves, 2001). In hypochondriasis, people are bothered by obsessive, anxiety-inducing thoughts about their health. Running from doctor to doctor may be a form of compulsive behavior that is reinforced by the temporary relief from anxiety they experience when they are reassured by their doctors that their fears are unwarranted. Yet the troublesome thoughts eventually return, prompting repeated consultations. The cycle then repeats. Similarly, with body dysmorphic disorder, the constant grooming and pruning in the attempt to "fix" the perceived physical defect may offer partial relief from anxiety, but the "fix" is never quite good enough to completely erase the underlying concerns. One possibility is that hypochondriasis and body dysmorphic disorder fall within a spectrum of OCD-type disorders.

Cognitive Theory Cognitive theorists have speculated that some cases of hypochondriasis may represent a type of self-handicapping strategy, a way of blaming poor performance on failing health (Smith, Snyder, & Perkins, 1983). In other cases, diverting attention to physical complaints can serve as a means of avoiding thinking about other life problems.

Another cognitive explanation focuses on the role of distorted thinking. People who develop hypochondriasis have a tendency to "make mountains out of molehills" by exaggerating the significance of minor physical complaints (Barsky et al., 2001). They misinterpret benign symptoms as signs of a serious illness, which creates anxiety that leads them to chase down one doctor after another in an attempt to uncover the dreaded disease they fear they have. The anxiety itself may lead to unpleasant physical symptoms, which are likewise exaggerated in importance, leading to more worrisome cognitions.

Cognitive theorists speculate that hypochondriasis and panic disorder, which often occur concurrently, may share a common cause: a distorted way of thinking that leads the person to misinterpret minor changes in bodily sensations as signs of pending catastrophe (Salkovskis & Clark, 1993). Differences between the two disorders may hinge on whether the misinterpretation of bodily cues carries a perception of imminent threat leading to a rapid spiraling of anxiety (panic disorder) or of a longer range threat in the form of an underlying disease process (hypochondriasis). Research into cognitive processes involved in hypochondriasis deserves further study. Given the linkages that may exist between hypochondriasis and anxiety disorders such as panic disorder and OCD, it remains unclear whether hypochondriasis should be classified as a somatoform disorder or an anxiety disorder (Barsky et al., 1992).

Treatment of Somatoform Disorders

The treatment approach that Freud pioneered, psychoanalysis, began with the treatment of hysteria, which is now termed conversion disorder. Psychoanalysis seeks to uncover and bring unconscious conflicts that originated in childhood into conscious awareness. Once the conflict is aired and worked through, the symptom is no longer needed as a "partial solution" and should disappear. The psychoanalytic method is supported by case studies, some reported by Freud and others by his followers. However, the infrequency of conversion disorders in contemporary times has made it difficult to mount controlled studies of the psychoanalytic technique.

The behavioral approach to treating conversion disorders and other somatoform disorders focuses on removing sources of secondary reinforcement (or secondary gain) that may become connected with physical complaints. Family members and others, for example, often perceive individuals with somatization disorder as sickly and infirm and as incapable of carrying normal responsibilities. Other people may be unaware of how they reinforce dependent and complaining behaviors when they relieve the sick person of responsibilities. The behavior therapist may teach family members to reward attempts to assume responsibility and ignore nagging and complaining. The behavior therapist may also work more directly with the person with a somatoform disorder, helping the person learn more adaptive ways of handling stress or anxiety (through relaxation and cognitive restructuring, for example). Exposure with response prevention (ERP, discussed in Chapter 6) and cognitive therapy have also achieved good success in treating hypochondriasis in recent trials (Clark et al., 1998; Visser & Bouman, 2001). Cognitive techniques, such as cognitive restructuring, are used to modify the patient's exaggerated illness-related beliefs, while exposure with response prevention is used to break the cycle of running to doctors for reassurance whenever minor physical complaints occur.

Cognitive-behavioral techniques, most often exposure with response prevention and cognitive restructuring, have also achieved encouraging results in treating body dysmorphic disorder (BDD) (Cororve & Gleaves, 2001). Exposure may take the form of purposefully revealing the perceived defect in public, rather than concealing it through use of make-up or clothing. Response prevention focuses on breaking compulsive rituals, such as mirror checking (for example, by covering mirrors at home) and excessive grooming. In cognitive restructuring, the therapist challenges clients' distorted beliefs about their physical appearance by encouraging them to evaluate their beliefs in the light of evidence.

Attention has recently turned to the use of antidepressants, especially fluoxetine (Prozac), in treating some types of somatoform disorder. Although we lack specific drug therapies for conversion disorder (Simon, 1998), a study of 16 patients with hypochondriasis showed significant reductions in hypochondriacal complaints over the course of a 12-week trial with Prozac (Fallon et al., 1993). Here, too, the lack of controlled drug-placebo studies prevents firm conclusions regarding the efficacy of drug therapy. We also lack any systematic studies of approaches to treating factitious disorder (Münchausen syndrome) and are limited to a few isolated case examples (Simon, 1998).

The dissociative and somatoform disorders remain among the most intriguing and least understood patterns of abnormal behavior.

Overview of Somatoform Disorders

TYPES OF SOMATOFORM DISORDERS

	Description	Features
Conversion Disorder	Change or loss of a physical function without medical cause	• Emerges in context of conflicts or stressful experiences, which lends credence to its psychological origins • May be associated with la belle indifférence (indifference to symptoms)
Hypochondriasis	Preoccupation with the belief that one is seriously ill	• Fear persists despite medical reassurance • Tendency to interpret physical sensations or minor aches and pains as signs of serious illness
Somatization Disorder	Recurrent, multiple complaint about physical symptoms that have no clear organic basis	• Symptoms prompt frequent medical visits or cause significant impairment of functioning
Body Dysmorphic Disorder	Preoccupation with an imagined or exaggerated physical defect	• Person may believe that others think less of them as a person because of the perceived defect • Person may engage in compulsive behaviors, such as excessive grooming, that aim to correct the perceived defect

CAUSAL FACTORS Multiple causes are involved

Biological Factors	• Possible genetic influences (somatization disorder)
Social-Environmental Factors	• Socialization of women into more dependent roles, such as the "sick role," that may be expressed in the form of somatoform disorders
Behavioral Factors	• Relief from ordinary responsibilities or escape or avoidance of uncomfortable or anxiety-laden situations (secondary gain) • Reinforcing properties of enacting a "sick role" • Compulsive behaviors associated with hypochondriasis or body dysmorphic disorder may partially relieve anxiety associated with preoccupation with health concerns or perceived physical defects
Emotional and Cognitive Factors	• Misinterpretations of bodily changes or physical symptoms as signs of serious illness (hypochondriasis) • In traditional Freudian theory, psychic energy that becomes cut off from unacceptable impulses is converted into physical symptoms (conversion disorder) • Blaming poor performance on failing health may be a self-handicapping strategy (hypochondriasis)

TREATMENT APPROACHES Treatment typically involves psychodynamic or cognitive-behavioral therapy

Biomedical Treatment	• Limited use of antidepressants in treating hypochondriasis
Cognitive-Behavioral Therapy	• May focus on removing sources of secondary reinforcement (secondary gain), promoting development of coping skills for handling stress, and correcting exaggerated or distorted beliefs about one's health or appearance
Psychodynamic Therapy	• Psychodynamic or insight-oriented therapy may be aimed at identifying and working through underlying unconscious conflicts

Summing Up

Dissociative Disorders

What are dissociative disorders? Dissociative disorders involve changes or disturbances in identity, memory, or consciousness that affect the ability to maintain an integrated sense of self. Thus, the symptoms are theorized to reflect psychological rather than organic factors.

What is dissociative identity disorder? In dissociative identity disorder, two or more distinct personalities, each possessing well-defined traits and memories, exist within the person and repeatedly take control of the person's behavior.

What is dissociative amnesia? In dissociative amnesia, the person experiences a loss of memory for personal information that cannot be accounted for by organic causes.

What is dissociative fugue? In dissociative fugue, the person travels suddenly away from home or place of work, shows a loss of memory for his or her personal past, and experiences identity confusion or takes on a new identity.

What is depersonalization disorder? In depersonalization disorder, the person experiences persistent or recurrent episodes of depersonalization of sufficient severity to cause significant distress or impairment in functioning.

How do theorists explain the development of dissociative disorders? Psychodynamic theorists view dissociative disorders as involving a form of psychological defense by which the ego defends itself against troubling memories and unacceptable impulses by blotting them out of consciousness. There is increasing documentation of a link between dissociative disorders and early childhood trauma, which lends support to the view that dissociation may serve to protect the self from troubling memories. To learning and cognitive theorists, dissociative experiences involve ways of learning not to think about certain troubling behaviors or thoughts that might lead to feelings of guilt or shame. Relief from anxiety negatively reinforces this pattern of dissociation. Some social-cognitive theorists suggest that multiple personality may represent a form of role-playing behavior.

What are the major treatment approaches for dissociative identity disorder? Psychotherapy seeks a reintegration of the personality by focusing on helping persons with dissociative identity disorder uncover and integrate dissociated painful experiences from childhood. Drug therapy may help treat the anxiety and depression often associated with the disorder, but cannot bring about reintegration of the personality.

Somatoform Disorders

What are somatoform disorders? In somatoform disorders, there are physical complaints that cannot be accounted for by organic causes. Thus, the symptoms are theorized to reflect psychological rather than organic factors. Three major types of somatoform disorders are conversion disorder, hypochondriasis, and somatization disorder.

What is conversion disorder? In conversion disorder, symptoms or deficits in voluntary motor or sensory functions occur that suggest an underlying physical disorder, but no apparent medical basis for the condition can be found to account for the condition.

What is hypochondriasis? Hypochondriasis is a preoccupation with the fear of having, or the belief that one has, a serious medical illness, although no medical basis for the complaints can be found and fears of illness persist despite medical reassurances.

What is body dysmorphic disorder? In body dysmorphic disorder, people are preoccupied with an imagined or exaggerated defect in their physical appearance.

What is somatization disorder? Somatization disorder involves multiple and recurrent complaints of physical symptoms that have persisted for many years and that cannot be accounted for by organic causes.

How are somatoform disorders conceptualized within the major theoretical perspectives? The psychodynamic view holds that conversion disorders represent the conversion into physical symptoms of the leftover emotion or energy cut off from unacceptable or threatening impulses that the ego has prevented from reaching awareness. The symptom is functional, allowing the person to achieve both primary gains and secondary gains. Learning theorists focus on reinforcements that are associated with conversion disorders, such as the reinforcing effects of adopting a "sick role." One learning theory model likens hypochondriasis to obsessive-compulsive behavior. Cognitive factors in hypochondriasis include possible self-handicapping strategies and cognitive distortions.

What are the major approaches to treating somatoform disorders? Psychodynamic therapists attempt to uncover and bring to the level of awareness the unconscious conflicts, originating in childhood, believed to be at the root of the problem. Once the conflict is uncovered and worked through, the symptoms should disappear because they are no longer needed as a partial solution to the underlying conflict. Behavioral approaches focus on removing underlying sources of reinforcement that may be maintaining the abnormal behavior pattern. More generally, behavior therapists assist people with somatoform disorders to learn to handle stressful or anxiety-arousing situations more effectively. In addition, a combination of cognitive-behavioral techniques, such as exposure with response prevention and cognitive restructuring, may be used in treating hypochondriasis and body dysmorphic disorder.

CHAPTER EIGHT

Mood Disorders and Suicide

Alexej von Jawlensky
Evening, 1929

Truth OR Fiction?

- Feeling sad or depressed is abnormal. (p. 223)
- Most people who experience a major depressive episode never have another one. (p. 226)
- The bleak light of winter casts some people into a diagnosable state of depression. (p. 228)
- For no apparent cause, some people experience dramatic mood swings from the depths of depression to the heights of elation. (p. 231)
- In some ways, many "mentally healthy" people see things *less* realistically than people who are depressed. (p. 243)
- The most widely used remedy for depression in Germany is not a drug, but an herb. (p. 254)
- The ancient Greeks and Romans used a chemical to curb turbulent mood swings that is still used today. (p. 254)
- People who threaten suicide are basically attention-seekers. (p. 259)

Life has its ups and downs. Most of us feel elated when we have earned high grades, a promotion, or the affections of Ms. or Mr. Right. Most of us feel down or depressed when we are rejected by a date, flunk a test, or suffer financial reverses. It is normal and appropriate to be happy about uplifting events. It is just as normal, just as appropriate, to feel depressed by dismal events. It might very well be "abnormal" if we were *not* depressed by life's miseries.

Moods are enduring states of feeling that color our psychological lives. Feeling down or depressed is not abnormal in the context of depressing events or circumstances. But people with **mood disorders** experience disturbances in mood that are unusually severe or prolonged and impair their ability to function in meeting their normal responsibilities. In any given year, about 7% of Americans suffer from mood disorders (USDHHS, 1999a). Some people become severely depressed even when things appear to be going well, or when they encounter mildly upsetting events that others take in stride. Still others experience extreme mood swings. They ride an emotional roller coaster with dizzying heights and abysmal depths when the world around them remains largely on an even keel.

Chapter Outline

Types of Mood Disorders

In this chapter we focus on several kinds of mood disorders, including two kinds of depressive disorders, major depressive disorder and dysthymic disorder, and two kinds of mood swing disorders, bipolar disorder and cyclothymic disorder (see Table 8.1). Table 8.2 lists some of the common features of depression. The depressive disorders are considered

TABLE 8.1 Types of Mood Disorders

Depressive Disorders (Unipolar Disorders)	
Major Depressive Disorder	Occurrence of one or more periods or episodes of depression (called major depressive episodes) without a history of naturally occurring manic or hypomanic episodes. People may have one major depressive episode, followed by a return to their usual state of functioning. The majority of people with a major depressive episode have recurrences that are separated by periods of normal or perhaps somewhat impaired functioning.
Dysthymic Disorder	A pattern of mild depression (but perhaps an irritable mood in children or adolescents) that occurs for an extended period of time—in adults, typically for many years.
Mood Swing Disorders (Bipolar Disorders)	
Bipolar Disorder	Disorders with one or more manic or hypomanic episodes (episodes of inflated mood and hyperactivity in which judgment and behavior are often impaired). Manic or hypomanic episodes often alternate with major depressive episodes with intervening periods of normal mood.
Cyclothymic Disorder	A chronic mood disturbance involving numerous hypomanic episodes (episodes with manic features of a lesser degree of severity than manic episodes) and numerous periods of depressed mood or loss of interest or pleasure in activities, but not of the severity to meet the criteria for a major depressive episode.

Source. Adapted from the *DSM-IV-TR* (APA, 2000).

Truth OR Fiction? REVISITED

Feeling sad or depressed is abnormal.

FALSE. Feeling depressed is not abnormal in the context of depressing events or circumstances.

mood The pervasive quality of an individual's emotional experience.

mood disorder A type of disorder characterized by disturbances of mood.

TABLE 8.2 **Common Features of Depression**

Changes in Emotional States	Changes in mood (persistent periods of feeling down, depressed, sad or blue) Tearfulness or crying Increased irritability, jumpiness, or loss of temper
Changes in Motivation	Feeling unmotivated, or having difficulty getting going in the morning or even getting out of bed Reduced level of social participation or interest in social activities Loss of enjoyment or interest in pleasurable activities Reduced interest in sex Failure to respond to praise or rewards
Changes in Functioning and Motor Behavior	Moving about or talking more slowly than usual Changes in sleep habits (sleeping too much or too little, awakening earlier than usual and having trouble getting back to sleep in early morning hours—so-called early morning awakening) Changes in appetite (eating too much or too little) Changes in weight (gaining or losing weight) Functioning less effectively than usual at work or school
Cognitive Changes	Difficulty concentrating or thinking clearly Thinking negatively about oneself and one's future Feeling guilty or remorseful about past misdeeds Lack of self-esteem or feelings of inadequacy Thinking of death or suicide

THINK ABOUT IT

How do clinicians distinguish between normal variations in mood and mood disorders?

unipolar because the disturbance lies in only one emotional direction or pole—down. Disorders that involve mood swings are **bipolar.** They involve excesses of both depression and elation, usually in an alternating pattern.

Many of us, probably most of us, have periods of sadness from time to time. We may feel down in the dumps, cry, lose interest in things, find it hard to concentrate, expect the worst to happen, or even consider suicide. A survey of a sample of college students at the University of North Iowa showed that about 30% of the students reported feeling at least mildly depressed (Wong & Whitaker, 1993). Downcast mood was greater among freshman than among seniors or graduate students, which may reflect the difficulties that many freshmen have adjusting to college life

For most of us, mood changes pass quickly or are not severe enough to interfere with our lifestyle or ability to function. Among people with mood disorders, including depressive disorders and bipolar disorders, mood changes are more severe or prolonged and affect daily functioning.

Major Depressive Disorder

The diagnosis of **major depressive disorder** (also called *major depression*) is based on the occurrence of one or more *major depressive episodes* in the absence of a history of **manic** or **hypomanic** episodes. In a major depressive episode, the person experiences either a depressed mood (feeling sad, hopeless, or "down in the dumps") or loss of interest or pleasure in all or virtually all activities for a period of at least 2 weeks (APA, 2000). The diagnostic features of a major depressive episode are listed in Table 8.3.

People with major depressive disorder may also have poor appetite, lose or gain substantial amounts of weight, have trouble sleeping or sleep too much, and become physically agitated or—at the other extreme—show a marked slowing down in their motor activity. Major depression impairs people's ability to meet the ordinary responsibility of everyday life (Judd et al., 2000a). People with major depression may lose interest in most of their usual activities and pursuits, have difficulty concentrating and making decisions, have pressing thoughts of death, and attempt suicide. Although depression is a diagnos-

unipolar Pertaining to a single pole, or direction.

bipolar Characterized by opposite ends of a dimension or continuum, as in bipolar disorder.

major depressive disorder A severe mood disorder characterized by major depressive episodes.

manic Relating to mania, as in the manic phase of bipolar disorder.

hypomanic Referring to a mild state of mania, or elation.

able psychological disorder, more than 40% of Americans polled in recent surveys perceive it to be a sign of personal weakness (Brody, 1992c). Many people don't seem to understand that people who are clinically depressed can't simply "shake it off" or "snap out of it." This attitude may explain why, despite the availability of safe and effective treatments, most people who are clinically depressed remain undiagnosed and untreated or fail to receive appropriate treatment (Gilbert, 1997a; Hirschfeld et al., 1997). Many people with untreated depression believe they can handle the problem themselves (Blumenthal & Endicott, 1997). Even for those who receive treatment, most receive inadequate or inappropriate care (Hirschfeld et al., 1997; Young et al., 2001).

Major depressive disorder is the most common type of diagnosable mood disorder, with estimates of lifetime prevalence ranging from 10% to 25% for women and from 5% to 12% for men (APA, 2000). An estimated 120 million people worldwide suffer from depression (E. Olson, 2001). About 1 in 20 people in the United States can be diagnosed with major depression at any given time (Blazer et al., 1994). Depression is so common that it has been dubbed the "common cold" of psychological problems (Seligman, 1973). The costs of depression incurred by employers, especially lost workdays, are as great if not greater than the costs of major medical illnesses such as heart disease and diabetes (Druss, Rosenheck, & Sledge, 2000). On the other hand, effective treatment for depression leads not only to psychological improvement but also to more stable employment and increased income, as people are able to return to a more productive level of functioning (Wells et al., 2000).

Major depression, particularly in more severe episodes, may be accompanied by psychotic features, such as delusions that one's body is rotting from illness (Coryell et al., 1996). People with severe depression may also experience hallucinations, such as "hearing" the voices of others, or of demons, condemning them for perceived misdeeds.

When are changes in mood considered abnormal? Although changes in mood in response to the ups and downs of everyday life may be quite normal, persistent or severe changes in mood, or cycles of extreme elation and depression, may suggest the presence of a mood disorder.

TABLE 8.3 Diagnostic Features of a Major Depressive Episode

A major depressive episode is denoted by the occurrence of five or more of the following features or symptoms during a 2-week period, which represents a change from previous functioning. At least one of the features must involve either (1) depressed mood, or (2) loss of interest or pleasure in activities. Moreover, the symptoms must cause either clinically significant levels of distress or impairment in at least one important area of functioning, such as social or occupational functioning, and must not be due directly to the use of drugs or medications, to a medical condition, or be accounted for by another psychological disorder.* Further, the episode must not represent a normal grief reaction to the death of a loved one—that is, **bereavement**.

1. Depressed mood during most of the day, nearly every day. Can be irritable mood in children or adolescents.
2. Greatly reduced sense of pleasure or interest in all or almost all activities, nearly every day for most of the day.
3. A significant loss or gain of weight (more than 5% of body weight in a month) without any attempt to diet, or an increase or decrease in appetite.
4. Daily (or nearly daily) insomnia or hypersomnia (oversleeping).
5. Excessive agitation or slowing down of movement responses nearly every day.
6. Feelings of fatigue or loss of energy nearly every day.
7. Feelings of worthlessness or misplaced or excessive or inappropriate guilt nearly every day.
8. Reduced ability to concentrate or think clearly or make decisions nearly every day.
9. Recurrent thoughts of death or suicide without a specific plan, or occurrence of a suicidal attempt or specific plan for committing suicide.

*The *DSM* includes separate diagnostic categories for mood disorders due to medical conditions or use of substances such as drugs of abuse.
Source. Adapted from the *DSM-IV-TR* (APA, 2000).

bereavement The experience of grief suffering following the death of a loved one.

Truth OR Fiction? REVISITED

Most people who experience a major depressive episode never have another one.

FALSE. Most people who experience a major depressive episode have recurrences.

The following case illustrates the range of features connected with major depressive disorder:

A Case of Major Depressive Disorder

A 38-year-old female clerical worker has suffered from recurrent bouts of depression since she was about 13 years of age. Most recently, she has been troubled by crying spells at work, sometimes occurring so suddenly she wouldn't have enough time to run to the ladies room to hide her tears from others. She has difficulty concentrating at work and feels a lack of enjoyment from work she used to enjoy. She harbors severe pessimistic and angry feelings, which have been more severe lately since she has been recently putting on weight and has been neglectful in taking care of her diabetes. She feels guilty that she may be slowly killing herself by not taking better care of her health. She sometimes feels that she deserves to be dead. She has been bothered by excessive sleepiness for the past year and a half, and her driving license has been suspended due to an incident the previous month in which she fell asleep while driving, causing her car to hit a telephone pole. She wakes up most days feeling groggy and just "out of it," and remains sleepy throughout the day. She has never had a steady boyfriend, and lives quietly at home with her mother, with no close friends outside of her family. During the interview, she cried frequently and answered questions in a low monotone, staring downward continuously.

—*Adapted from Spitzer et al., 1989, pp. 59–62*

Major depressive episodes may remit in a matter of months or last for a year or more (APA, 2000; USDHHS, 1999a). Some people experience a single episode with a full return to previous levels of functioning. However, the great majority of people with major depression, perhaps as many as 85%, have repeated occurrences (Mueller et al., 1999). The average person with major depression can expect to have four episodes during his or her lifetime (Judd, 1997). Relapses tend to be more frequent in people who continue to have some leftover depressive symptoms following a first depressive episode (Judd et al., 2000b). Given a pattern of repeated occurrences of major depressive episodes and prolonged symptomatology, many professionals have come to view major depression as a chronic, indeed lifelong disorder. On the positive side, the longer the period of recovery from major depression, the lower the risk of eventual relapse (Solomon et al., 2000).

Major depression versus bereavement. Major depression is distinguished from a normal grief reaction to the death of a loved one, which is termed bereavement. Major depression may occur in people whose bereavement becomes prolonged or seriously interferes with normal functioning.

Risk Factors in Major Depression Factors that place people at increased risk of developing major depression include age (initial onset is more common among younger adults than older adults); socioeconomic status (people lower down the socioeconomic ladder are at greater risk than those who are better off); and marital status (people who are separated or divorced have higher rates than married or never-married people).

Women are nearly twice as likely as men to develop major depression (APA, 2000; Blazer et al., 1994; Kessler et al., 1994) (see Figure 8.1). The difference in relative risk between males and females begins in early adolescence and persists through at least the mid-50s (Barefoot et al., 2001; Kessler et al., 1993). Although hormonal or other biologically linked gender differences may be involved, a panel convened by the American Psychological Association (APA) attributed the gender difference largely to the greater amount of stress that women encounter in contemporary life (Goleman, 1990b; McGrath et al., 1990). The panel concluded that women are more likely than men to encounter such stressful life factors as physical and sexual abuse, poverty, single parenthood, and sexism. Despite the gender difference in prevalence, the course of major depression is similar for both genders: Men and women with the disorder do not differ significantly in the likelihood of having recurrences, the frequency of recurrences, the severity or duration of recurrences, or the time to first recurrence (Eaton et al., 1997).

Differences in coping styles may also help explain women's greater proneness toward depression. Regardless of whether the factors that precipitate depression are biological, psychological, or social, one's coping responses may exacerbate or reduce the severity and duration of depressive episodes. Nolen-Hoeksema and her colleagues (1991; Nolen-Hoeksema, Morrow, & Fredrickson, 1993) propose that men are more likely to distract themselves when they are depressed, whereas women are more likely to amplify depression by ruminating about their feelings and their possible causes. Women may be more likely to sit at home when they are depressed and think about how they feel or try to understand the reasons they feel the way they do, whereas men may try to distract themselves by doing something they enjoy, such as going to a favorite hangout to get their mind off their feelings. On the other hand, men often turn to alcohol as a form of self-medication, which can lead to another set of psychological and social problems (Nolen-Hoeksema et al., 1993). Rumination is not limited to women, however. Both men and women who ruminate more following the loss of loved ones or when feeling down or sad are more likely to become depressed and to suffer longer and more severe depression than those who ruminate less (Just & Alloy, 1997; Nolen-Hoeksema, 2000).

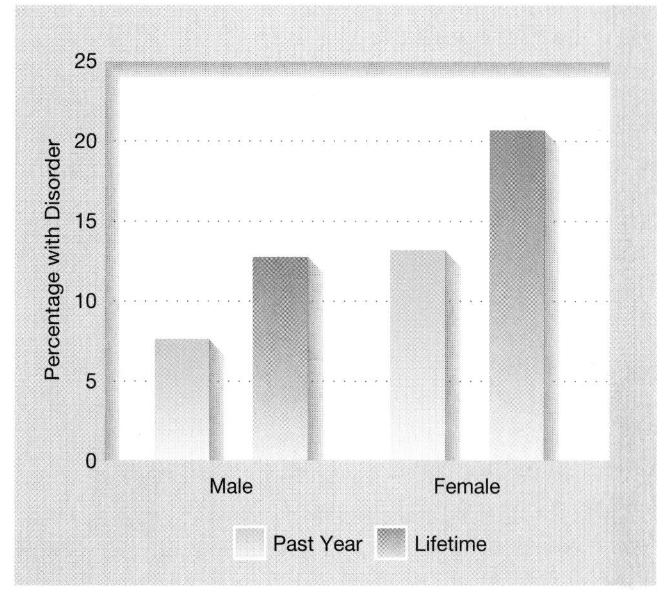

FIGURE 8.1 Prevalence of major depressive episodes by gender. Major depressive episodes affect about twice as many women as men.

Source. Kessler et al. (1994) National Comorbidity Survey (Kessler et al., 1994).

Although the gender gap in depression continues, it appears to be narrowing as more men are coming forward seeking help for depression. The male ego also seems to be battered by assaults from corporate downsizing and growing financial insecurity. Although long viewed by men as a sign of personal weakness, the stigma associated with depression shows signs of lessening, although not disappearing (*NBC Nightly News,* 1996).

Major depression typically develops in young adulthood, with an average age of onset in the mid-20s (APA, 2000). However, the disorder may affect even young children, although the risks are very low through age 14 (Lewinsohn et al., 1986). A multinational study of nine countries[1] showed that the rates of major depression have been rising in the United States and elsewhere (Cross-National Collaborative Group, 1992). In some countries, young people born after 1955 stood about three times greater likelihood of suffering a major depression than did their grandparents when they were the same age (Goleman, 1992b). The greatest increases were found in Florence, Italy; the least in Christchurch, New Zealand. In all countries, rates for depression were higher among women than men.

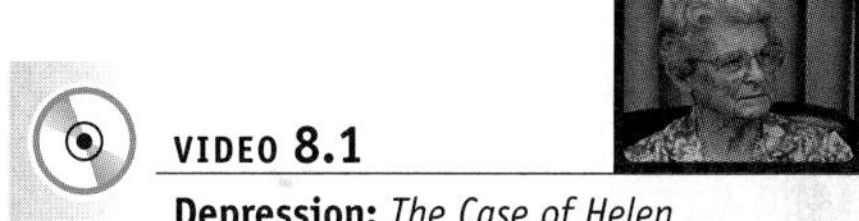

VIDEO 8.1

Depression: *The Case of Helen*

No one knows why depression has been on the rise in many cultures, but speculation focuses on social and environmental changes, such as increasing fragmentation of families due to relocations, exposure to wars and internal conflicts, and increased rates of violent crimes, as well as possible exposure to toxins or infectious agents in the environment that might affect mental as well as physical health (Cross-National Collaborative Group, 1992). One example is the dramatic increase in depression that occurred in the period 1950 to 1960 in Beirut, Lebanon. This was a period of chaotic political and demographic changes in the country. Depression dropped sharply in the following period, 1960 to 1970, a time of relative prosperity and stability in the country, but increased again between 1970 and 1980 during a time of social upheaval and internal warfare.

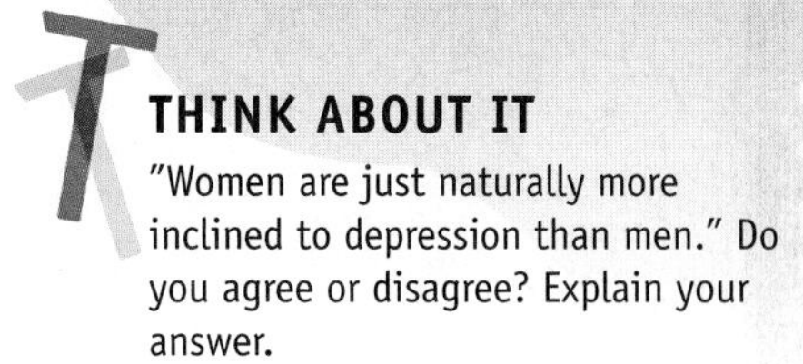

THINK ABOUT IT

"Women are just naturally more inclined to depression than men." Do you agree or disagree? Explain your answer.

Seasonal Affective Disorder Are you glum on gloomy days? Is your temper short during the brief days of winter? Are you dismal during the dark of long winter nights? Are you feeling up when the long sunny days of spring and summer return?

Many people report that their moods do vary with the weather. For some people, the changing of the seasons from summer into fall and winter leads to a type of depression called *seasonal affective (mood) disorder*—SAD.[2] The features of SAD include fatigue, excessive sleep,

Web Link 8.1

Facts About Depression

[1]United States, Canada, Puerto Rico, Italy, France, Germany, Lebanon, Taiwan, and New Zealand

[2]Seasonal affective disorder is not classified as a diagnostic category in its own right in the *DSM-IV* but is designated as a specifier of mood disorders in which major depressive episodes occur. For example, major depressive disorder that occurs seasonally would be given a diagnosis of major depressive disorder with seasonal pattern.

postpartum depression (PPD) Persistent and severe mood changes that occur after childbirth.

Light therapy. Exposure to bright artificial light for a few hours a day during the fall and winter months can often bring relief from seasonal affective disorder.

Truth OR Fiction? REVISITED

The bleak light of winter casts some people into a diagnosable state of depression.

TRUE. The changing of the seasons does lead to a depressive disorder in some people.

craving for carbohydrates, and weight gain. SAD tends to lift with the early buds of spring. It affects women more often than men and is most common among young adults.

Although the causes of SAD remain unknown, one possibility is that seasonal changes in light may alter the body's underlying biological rhythms that regulate such processes as body temperature and sleep-wake cycles (Lee et al., 1998). Another possibility is that some parts of the central nervous system may have deficiencies in transmission of the mood-regulating neurotransmitter serotonin during the winter months (Schwartz et al., 1997). Whatever the underlying cause, a trial of intense light therapy, called *phototherapy*, often helps relieve depression. Phototherapy typically consists of exposure to several hours of bright artificial light a day (e.g., Terman et al., 2001). The artificial light apparently supplements the meager sunlight the person otherwise receives. Patients can generally carry out some of their daily activities (for example, eating, reading, writing) during their phototherapy sessions. Improvement typically occurs within several days of phototherapy, but treatment is apparently required throughout the course of the winter season. Light directed at the eyes tends to be more successful than light directed at the skin (Sato, 1997).

Postpartum Depression Many, perhaps even most, new mothers experience mood changes, periods of tearfulness, and irritability following the birth of a child. These mood changes are commonly called the "maternity blues," "postpartum blues," or "baby blues." They usually last for a couple of days and are believed to be a normal response to hormonal changes that attend childbirth. Given these turbulent hormonal shifts, it would be "abnormal" for most women *not* to experience some changes in feeling states shortly following childbirth.

Some mothers, however, undergo severe mood changes that may persist for months or even a year or more. These problems in mood are referred to as **postpartum depression (PPD)**. *Postpartum* derives from the Latin roots *post*, meaning "after," and *papere*, meaning "to bring forth." PPD is often accompanied by disturbances in appetite and sleep, low self-esteem, and difficulties in maintaining concentration or attention. Between 8% and 15% of mothers experience a diagnosable postpartum depressive disorder of at least moderate severity (Campbell & Cohn, 1991; Gitlin & Pasnau, 1989). Postpartum depression is not unique to the United States; evidence from a study in an urban area in Portugal reported a similar prevalence rate (13%) (Augusto et al., 1996).

Women and depression. Women are more likely to suffer from major depression than men. A panel convened by the APA attributed the higher rates of depression among women to factors such as unhappy marriages, physical and sexual abuse, impoverishment, single parenthood, sexism, hormonal changes, childbirth, and excessive caregiving burdens. APA panel member Bonnie Strickland expressed surprise that even more women were not clinically depressed, since they are treated as second-class citizens.

Questionnaire

Are You Depressed?

This test, offered by the organizers of the National Depression Screening Day, can help you assess whether you are suffering from a depression. It is not intended for you to diagnose yourself, but rather to raise your awareness of concerns you may want to discuss with a professional.

	YES	NO
1. I feel downhearted, blue, and sad.	___	___
2. I don't enjoy the things that I used to.	___	___
3. I feel that others would be better off if I were dead.	___	___
4. I feel that I am not useful or needed.	___	___
5. I notice that I am losing weight.	___	___
6. I have trouble sleeping through the night.	___	___
7. I am restless and can't keep still.	___	___
8. My mind isn't as clear as it used to be.	___	___
9. I get tired for no reason.	___	___
10. I feel hopeless about the future.	___	___

Rating your responses: If you agree with at least five of the statements, including either item 1 or 2, and if you have had these complaints for at least 2 weeks, professional help is strongly recommended. If you answered "yes" to statement 3, seek consultation with a professional immediately. If you don't know whom to turn to, contact your college counseling center, neighborhood mental health center, or health provider.

Source. Adapted from J. E. Brody, "Myriad masks hide an epidemic of depression," *The New York Times*, September 30, 1992, p. C12.

Postpartum depression is considered a form of major depression in which the onset of the depressive episode begins within 4 weeks after childbirth (APA, 2000). Investigators find that postpartum depression typically is less severe than other forms of major depression and lifts relatively sooner than most (Whiffen & Gotlib, 1993). Yet some suicides are linked to postpartum depression (McQuiston, 1997). Although PPD may involve chemical or hormonal imbalances brought on by childbirth, factors associated with an increased risk include stress, single or first-time motherhood, financial problems, a troubled marriage, social isolation, lack of support from partners and family members, a history of depression, or having an unwanted, sick, or temperamentally difficult infant (Forman et al., 2000; Ritter et al., 2000; Swendsen & Mazure, 2000). Having PPD also increases the risk that the woman will suffer future depressive episodes (Philipps & O'Hara, 1991).

Postpartum depression is not limited to our culture. Recent reports find high rates of PPD among South African women (Cooper et al., 1999) and Chinese women from Hong Kong (D. T. S. Lee et al., 2001). In the South African sample, a lack of psychological and financial support from the baby's father was associated with an increased risk of the disorder in this sample, mirroring findings with U.S. samples.

Dysthymic Disorder

Major depressive disorder is severe and marked by a relatively abrupt change from one's preexisting state. A milder form of depression seems to follow a chronic course of development that often begins in childhood or adolescence (Klein et al., 2000a, 2000b). Earlier diagnostic formulations of this type of chronic sadness were labeled "depressive neurosis" or "depressive personality" (Brody, 1995a). It was so labeled in an effort to account for several features that are traditionally connected with neurosis, such as early childhood origins, a chronic course, and generally mild levels of severity. The *DSM* labels this form of depression **dysthymic disorder,** or *dysthymia*, which derives from Greek roots *dys*-, meaning "bad" or "hard" and *thymos*, meaning "spirit."

WWW **Web Link 8.2** Facts About Dysthymic Disorder

Persons with dysthymic disorder do feel "bad spirited" or "down in the dumps" most of the time, but they are not so severely depressed as those with major depressive disorder. Whereas major depressive disorder tends to be severe and time limited, dysthymic disorder

dysthymic disorder A mild but chronic type of depressive disorder.

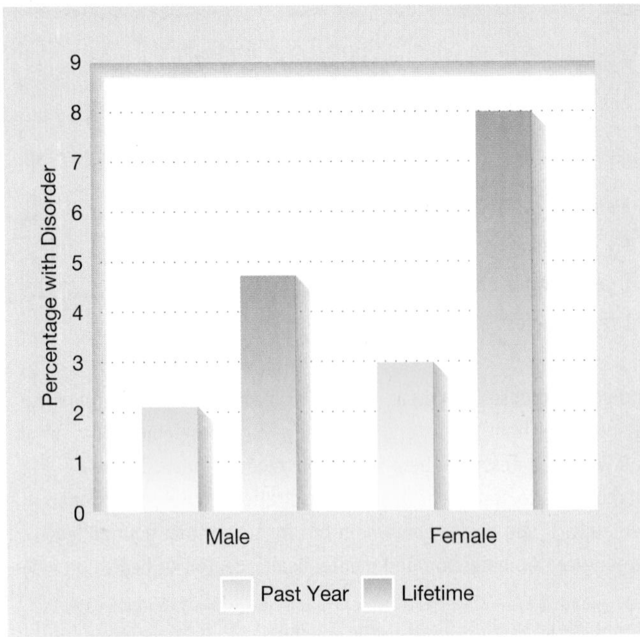

FIGURE 8.2 Prevalence of dysthymic disorder by gender. Like major depression, dysthymic disorder occurs in about twice as many women as men.

Source. Kessler et al. (1994) National Comorbidity Survey (Kessler et al., 1994).

is relatively mild and nagging, typically lasting for years (Klein et al., 2000b). Feelings of depression and social difficulties continue even after the person makes an apparent recovery (USDHHS, 1999a). The risk of relapse is also quite high (Klein et al., 2000a).

Dysthymia affects about 6% of the general population at some point in their lifetimes (APA, 2000). Like major depressive disorder, dysthymic disorder is more common in women than men (see Figure 8.2).

In dysthymic disorder, complaints of depression may become such a fixture of people's lives that they seem to be intertwined with their personality structures. The persistence of complaints may lead others to perceive them as whining and complaining (Akiskal, 1983). Although dysthymic disorder is less severe than major depressive disorder, persistent depressed mood and low self-esteem can affect the person's occupational and social functioning, as we see in the following case:

A Case of Dysthymic Disorder

The woman, a 28-year-old junior executive, complained of chronic feelings of depression since the age of 16 or 17. Despite doing well in college, she brooded about how other people were "genuinely intelligent." She felt she could never pursue a man she might be interested in dating because she felt inferior and intimidated. While she had extensive therapy through college and graduate school, she could never recall a time during those years when she did not feel somewhat depressed. She got married shortly after college graduation to the man she was dating at the time, although she didn't think that he was anything "special." She just felt she needed to have a husband for companionship and he was available. But they soon began to quarrel and she's lately begun to feel that marrying him was a mistake. She has had difficulties at work, turning in "slipshod" work and never seeking anything more than what was basically required of her and showing little initiative. While she dreams of acquiring status and money, she doesn't expect that she or her husband will rise in their professions because they lack "connections." Her social life is dominated by her husband's friends and their spouses and she doesn't think that other women would find her interesting or impressive. She lacks interest in life in general and expresses dissatisfaction with all facets of her life—her marriage, her job, her social life.

—Adapted from Spitzer et al., 1994, pp. 110–112

■

Some people are affected by both dysthymic disorder and major depression at the same time. The term **double depression** applies to those who have a major depressive episode superimposed upon a longer-standing dysthymic disorder (Keller, Hirschfeld, & Hanks, 1997). People suffering from double depression generally have more severe depressive episodes than do people with major depression alone (Klein et al., 2000b). Recent evidence suggests that virtually all people with dysthymia eventually develop double depression (Klein et al., 2000a).

We have noted that major depressive disorder and dysthymic disorder are depressive disorders in the sense that the disturbance of mood is only in one direction—down. Yet people with mood disorders may have fluctuations in mood in both directions that exceed the usual ups and downs of everyday life. These types of disorders are called bipolar disorders. Here we focus on the two types of these mood-swing disorders: (1) bipolar disorder, and (2) cyclothymic disorder.

double depression A diagnosis of both major depressive disorder and dysthymic disorder.

Bipolar Disorder

People with **bipolar disorder** ride an emotional roller coaster, swinging from the heights of elation to the depths of depression without external cause. The first episode may be either manic or depressive. Manic episodes, typically lasting from a few weeks to several months, are generally shorter in duration and end more abruptly than major depressive episodes. Some people with recurring bipolar disorder attempt suicide "on the way down" from the manic phase. They report that they would do nearly anything to escape the depths of depression they know lie ahead.

The *DSM* distinguishes between two general types of bipolar disorder, *bipolar I disorder* and *bipolar II disorder* (APA, 2000). In bipolar I disorder, the person experiences at least one full manic episode. In many cases, the person experiences mood swings between elation and depression with intervening periods of normal mood. Some cases present with no evidence of major depressive episodes, but it is assumed that such episodes may either develop in the future or have been overlooked in the past. In a few cases, called the mixed type, both a manic episode and a major depressive episode occur simultaneously.

Bipolar II disorder is associated with a milder form of mania. In bipolar II disorder, the person has experienced one or more major depressive episodes and at least one hypomanic episode. But the person has never had a full-blown manic episode. Whether bipolar I and bipolar II disorders represent qualitatively different disorders or different points along a continuum of severity of bipolar disorder remains to be determined.

Bipolar disorder is relatively uncommon, with reported lifetime prevalence rates from community surveys ranging from 0.4% to 1.6% for bipolar I disorder and about 0.5% for bipolar II disorder (APA, 2000; USDHHS, 1999a). Bipolar disorder typically develops around age 20 in both men and women. Only about 1 in 3 people with bipolar disorder receive any treatment (Goleman, 1994c). Sadly, about 1 in 5 of the people who go untreated eventually commit suicide (Hilts, 1994).

Unlike major depression, rates of bipolar I disorder appear about equal in men and women. In men, however, the onset of bipolar I disorder typically begins with a manic episode, whereas with women, it usually begins with a major depressive episode. The underlying reason for this gender difference remains unknown. Bipolar II disorder appears to be more common in women (APA, 2000).

Sometimes cases involve periods of "rapid cycling" in which the individual experiences two or more full cycles of mania and depression within a year without any intervening normal periods. Rapid cycling is relatively uncommon, but occurs more often among women than men (Leibenluft, 1996). It is usually limited to a year or less, but is associated with poorer social and job functioning (Coryell, Endicott, & Keller, 1992b) and a higher rate of relapse (Keller et al., 1993).

Manic Episode **Manic episodes,** or periods of mania, typically begin abruptly, gathering force within days. During a manic episode, the person experiences a sudden elevation or expansion of mood and feels unusually cheerful, euphoric, or optimistic. The person seems to have boundless energy and is extremely sociable, although perhaps to the point of becoming overly demanding and overbearing toward others. Other people recognize the sudden shift in mood to be excessive in the light of the person's circumstances. It is one thing to feel elated if one has just won the state lottery. It is another to feel euphoric because it's Wednesday.

People in a manic episode or phase are excited and may strike others as silly, by carrying jokes too far, for example. They tend to show poor judgment and to become argumentative, sometimes going so far as destroying property. Roommates may find them abrasive and avoid them. They tend to speak very rapidly (with **pressured speech**). Their thoughts and speech may jump from topic to topic (in a **rapid flight of ideas**). Others find it difficult to get a word in edgewise. They may also become extremely generous and make large charitable contributions they can ill afford or give away costly possessions. They may not be able to sit still or sleep restfully. They almost always show decreased need for sleep. They tend to awaken early yet feel well rested and

bipolar disorder A disorder characterized by mood swings between states of extreme elation and severe depression.

manic episode A period of unrealistically heightened euphoria, extreme restlessness, and excessive activity characterized by disorganized behavior and impaired judgment.

pressured speech An outpouring of speech in which words seem to surge urgently for expression.

rapid flight of ideas A characteristic of manic behavior involving rapid speech and changes of topics.

Truth OR Fiction? REVISITED

For no apparent cause, some people experience dramatic mood swings from the depths of depression to the heights of elation.

TRUE. Some people with bipolar disorder do ride an emotional roller coaster between periods of extreme elation and periods of extreme depression, without external cause.

Patty Duke The actress Patty Duke, who won an academy award for her performance as Helen Keller in the movie *The Miracle Worker,* was diagnosed with bipolar disorder as a young adult.

full of energy. They sometimes go for days without sleep and without feeling tired. Although they may have abundant stores of energy, they seem unable to organize their efforts constructively. Their elation impairs their ability to work and to maintain normal relationships.

Curiously, many observers have noted connections between mood disorders and creativity (Jamison, 1993; McDermott, 2001; Richards, 1994). Many distinguished writers, artists, and composers seemed to have suffered from mood disorders, especially bipolar disorder, including such luminaries as the artists Michelangelo and Vincent Van Gogh, the composers William Schumann and Peter Tchaikovsky, the novelists Virginia Woolf and Ernest Hemingway, and the poets Alfred Lord Tennyson, Emily Dickinson, Walt Whitman, and Sylvia Plath. Perhaps some creative people are able to channel the seemingly boundless energy and rapid stream of thoughts associated with manic periods to enhance their productivity and ability to express themselves in novel ways.

Web Link 8.3
National Depressive and Manic-Depressive Association

People in a manic episode generally experience an inflated sense of self-esteem that may range from extreme self-confidence to wholesale delusions of grandeur. They may feel capable of solving the world's problems or of composing symphonies, despite a lack of any special knowledge or talent. They may spout off about matters on which they know little, such as how to eliminate world hunger or create a new world order. It soon becomes clear that they are disorganized and incapable of completing their projects. They become highly distractible. Their attention is easily diverted by irrelevant stimuli like the sounds of a ticking clock or of people talking in the next room. They tend to take on multiple tasks, more than they can handle. They may suddenly quit their jobs to enroll in law school, wait tables at night, organize charity drives on weekends, and work on the great American novel in their "spare time." They tend to exercise poor judgment and fail to weigh the consequences of their actions. They may get into trouble as a result of lavish spending, reckless driving, or sexual escapades. In severe cases, they may experience disorders of thinking similar to those of people with schizophrenia. They may experience hallucinations or become grossly delusional, believing, for example, that they have a special relationship with God.

VIDEO 8.2
Bipolar Disorder: *The Case of Craig*

The following case provides a firsthand account of a manic episode. The early stages are dominated by euphoria, boundless energy, and an inflated sense of self. As mania intensifies, the individual may become confused:

A Case of Bipolar Disorder

When I start going into a high, I no longer feel like an ordinary housewife. Instead I feel organized and accomplished and I begin to feel I am my most creative self. I can write poetry easily. I can compose melodies without effort. I can paint. My mind feels facile and absorbs everything. I have countless ideas about improving the conditions of mentally retarded children, of how a hospital for these children should be run, what they should have around them to keep them happy and calm and unafraid. I see myself as being able to accomplish a great deal for the good of people. I have countless ideas about how the environment problem could inspire a crusade for the health and betterment of everyone. I feel able to accomplish a great deal for the good of my family and others. I feel pleasure, a sense of euphoria or elation. I want it to last forever. I don't seem to need much sleep. I've lost weight and feel healthy and I like myself. I've just bought six new dresses, in fact, and they look quite good on me. I feel sexy and men stare at me. Maybe I'll have an affair, or perhaps several. I feel capable of speaking and doing good in politics. I would like to help people with problems similar to mine so they won't feel hopeless.

It's wonderful when you feel like this. . . . The feeling of exhilaration—the high mood—makes me feel light and full of the joy of living. However, when I go beyond this stage, I become manic, and the creativeness becomes so magnified I begin to see things in my mind that aren't real. For instance, one night I created an entire movie, complete with cast, that I still think would be terrific. I saw the people as clearly as if watching them in real life. I also experienced complete terror, as if it were actually happening, when I knew that an assassination scene was about to take place. I cowered under the covers and became a complete shaking wreck. As you know, I went into a manic psychosis at that point. My screams awakened my husband, who tried to reassure me that we were in our bedroom and everything was the same. There was nothing to be afraid of. Nevertheless, I was admitted to the hospital the next day.

—*From Fieve, 1975, pp. 12–18*

Cyclothymic Disorder

Cyclothymia is derived from the Greek *kyklos*, which means "circle," and *thymos* ("spirit"). The notion of a circular-moving spirit is an apt description because this disorder involves a chronic cyclical pattern of mood disturbance characterized by mild mood swings of at least 2 years (1 year for children and adolescents). **Cyclothymic disorder** usually begins in late adolescence or early adulthood and persists for years. Few, if any, periods of normal mood last for more than a month or two. Neither the periods of elevated or depressed mood are severe enough to warrant a diagnosis of bipolar disorder, however. Estimates from community studies indicate lifetime prevalence rates for cyclothymic disorder of between 0.4% to 1% (4 to 10 people in 1,000), with men and women about equally likely to be affected (APA, 2000).

The periods of elevated mood are called hypomanic episodes, from the Greek prefix *hypo-*, meaning "under" or "less than." *Hypo*manic episodes are less severe than manic episodes and are not accompanied by the severe social or occupational problems associated with full-blown manic episodes. During hypomanic episodes, people may have an inflated sense of self-esteem, feel unusually charged with energy and alert, and may be more restless and irritable than usual. They may be able to work long hours with little fatigue or need for sleep. Their projects may be left unfinished when their moods reverse, however. Then they enter a mildly depressed mood state and find it difficult to summon the energy or interest to persevere. They feel lethargic and depressed, but not to the extent typical of a major depressive episode.

Social relationships may become strained by shifting moods, and work may suffer. Social invitations, eagerly sought during hypomanic periods, may be declined during depressed periods. Phone calls may not be returned as the mood slumps. Sexual interest waxes and wanes with the person's moods.

The boundaries between bipolar disorder and cyclothymic disorder are not yet clearly established. Some forms of cyclothymic disorder may represent a mild, early type of bipolar disorder. Approximately 33% of people with cyclothymic disorder eventually develop bipolar disorder, a figure that is about 33 times higher than the general population (USDHHS, 1999a). We presently lack the ability to distinguish persons with cyclothymia who are likely to eventually develop bipolar disorder (Howland & Thase, 1993). The following case presents an example of the mild mood swings that typify cyclothymic disorder:

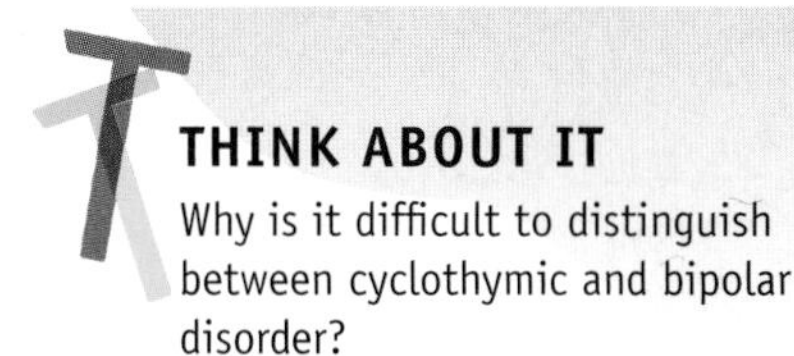
THINK ABOUT IT
Why is it difficult to distinguish between cyclothymic and bipolar disorder?

A Case of Cyclothymic Disorder

The man, a 29-year-old car salesman, reports that since the age of 14 he has experienced alternating periods of "good times and bad times." During his "bad" periods, which generally last between 4 and 7 days, he sleeps excessively and feels a lack of confidence, energy, and motivation, as if he were "just vegetating." Then his moods abruptly shift for a period of three or four days, usually upon awakening in the morning, and he feels

cyclothymic disorder A mood disorder characterized by a chronic pattern of mild mood swings that is not sufficiently severe to be classified as bipolar disorder.

aflush with confidence and sharpened mental ability. During these "good periods" he engages in promiscuous sex and uses alcohol, in part to enhance his good feelings and in part to help him sleep at night. The good periods may last upwards of 7–10 days at times, before shifting back into the "bad" periods, generally following a hostile or irritable outburst.

—Adapted from Spitzer et al., 1994, pp. 155–157

Quiz **8.1**
Types of Mood Disorders

Theoretical Perspectives on Mood Disorders

Mood disorders involve a complex interaction of biological and psychosocial influences (Cui & Vaillant, 1997). Though a full understanding of the causes of mood disorders presently lies beyond our grasp, we have begun to identify many of the important contributors to mood disorders, and in particular depression, the type of mood disorder studied most extensively by investigators. Here we begin with relationships between stress and the mood disorders and then consider the psychological and biological perspectives on mood disorders.

Stress and Mood Disorders

Stressful life events such as the loss of a loved one, the breakup of a romantic relationship, prolonged unemployment, physical illness, marital or relationship problems, economic hardship, pressure at work, or racism and discrimination increase the risks of developing a mood disorder or experiencing a recurrence of a mood disorder, especially major depression (Greenberger et al., 2000; Kendler, Thornton, & Gardner, 2000; Monroe et al., 2001). In one research example, investigators found that in about four of five cases, major depression was preceded by stressful life events (Mazure, 1998). People are also more likely to become depressed when they hold themselves responsible for undesirable events, such as school problems, financial difficulties, unwanted pregnancy, interpersonal problems, and problems with the law (Hammen & de Mayo, 1982).

Yet the relationship between stress and depression may cut both ways: Stressful life events may contribute to depression, and depressive symptoms in themselves may be stressful or lead to additional sources of stress, such as divorce or loss of employment (Cui & Vaillant, 1997; Daley et al., 1997). When you're depressed, for example, you may find it more difficult to keep up with your work at school or on the job, which can lead to more stress as your work backs up. The closer the stressful event taps the person's core concerns (failing at work or school, for instance), the more likely it is to precipitate a relapse in people who have a history of depression (Segal et al., 1992). Traumatic stressful events may play important roles in the cycling of bipolar disorder, although perhaps not in the onset of the disorder (Hammen & Gitlin, 1997; Miklowitz & Alloy, 1999).

Though stress is often implicated in depression, not everyone who encounters stress becomes depressed. Factors such as coping skills, genetic endowment, and availability of social support contribute to the likelihood of depression in the face of stressful events (USDHHS, 1999a). The development of depression may also be influenced by prior abuse or trauma. Consistent with the diathesis-stress model, researchers find that young women are more likely to develop depression in the face of stressful life events if they possessed a diathesis in the form of exposure to childhood adversities such as family violence or parental mental disorders or alcoholism (Hammen, Henry, & Daley, 2000). Moreover, physical or sexual abuse in childhood can disrupt the development of early attachment bonds to parents, setting the stage for later relationship problems and emotional disorders involving depression and anxiety (USDHHS, 1999a).

A strong marital relationship may provide a source of support during times of stress. Not surprisingly, people who are divorced or separated have higher rates of

Stress and depression. Depression is strongly associated with major stressors, such as prolonged unemployment and economic hardship. However, whether people consider themselves responsible for the hardship affects their likelihood of becoming depressed. In addition, it may be difficult to determine whether a person becomes depressed over losing a job or loses a job because of suffering from depression.

depression and suicide attempts than married people (Weissman et al., 1991). The availability of social support is also associated with quicker recoveries from episodes of both major depression and bipolar disorder (Johnson et al., 1999; Moos, Cronkite, & Moos, 1998).

People with major depression often lack skills needed to solve interpersonal problems with friends, coworkers, or supervisors (Marx, Williams, & Claridge, 1992). But those who take a more active approach to solving their interpersonal problems tend to have better clinical outcomes than depressed people who assume a more passive style of coping (Sherbourne, Hays, & Wells, 1995).

Psychodynamic Theories

The classic psychodynamic theory of depression of Freud (1917/1957) and his followers (e.g., Abraham, 1916/1948) holds that depression represents anger directed inward rather than against significant others. Anger may become directed against the self following either the actual or threatened loss of these important others.

Freud believed that **mourning,** or normal bereavement, is a healthy process by which one eventually comes to separate oneself psychologically from a person who is lost through death, separation, divorce, or other reason. Pathological mourning, however, does not promote healthy separation. Rather, it fosters lingering depression. Pathological mourning is likely to occur in people who hold powerful **ambivalent** feelings—a combination of positive (love) and negative (anger, hostility) feelings—toward the person who has departed or whose departure is feared. Freud theorized that when people lose, or even if they fear losing, an important figure about whom they feel ambivalent, their feelings of anger toward the other person turn to rage. Yet rage triggers guilt, which in turn prevents the person from venting anger directly at the lost person (called an "object").

mourning Normal feelings of grief following a loss.

ambivalent Holding conflicting feelings toward another person or a goal.

Social support as a buffer against depression. Social support appears to buffer the effects of stress and may reduce the risk of depression. People who lack important relationships and who rarely join in social activities are more likely to suffer from depression.

To preserve a psychological connection to the lost object, people *introject*, or bring inward, a mental representation of the object. They thus incorporate the other person into the self. Now anger is turned inward, against the part of the self that represents the inward representation of the lost person. This produces self-hatred, which in turn leads to depression.

From the psychodynamic viewpoint, bipolar disorder represents shifting dominance of the individual's personality between the ego and superego. In the depressive phase, the superego is dominant, producing exaggerated notions of wrongdoing and flooding the individual with feelings of guilt and worthlessness. After a time, the ego rebounds and asserts supremacy, producing feelings of elation and self-confidence that characterize the manic phase. The excessive display of ego eventually triggers a return of guilt, once again plunging the individual into depression.

While also emphasizing the importance of loss, more recent psychodynamic models focus more on issues relating to the individual's sense of self-worth or self-esteem. One model, called the *self-focusing model*, considers how people allocate their attentional processes after a loss (death of a loved one, a personal failure, etc.) (Pyszczynski & Greenberg, 1987). According to this model, depression-prone people experience a period of intense self-examination (self-focusing) following a major loss or disappointment. They become preoccupied with thoughts about the lost object (loved one) or important goal and remain unable to surrender hope of somehow regaining it.

Consider a person who must cope with the termination of a failed romantic relationship. It may be clear to all concerned that the relationship is beyond hope of revival. The self-focusing model proposes, however, that the depression-prone individual persists in focusing attention on restoring the relationship, rather than recognizing the futility of the effort and getting on with life. Moreover, the lost partner was a source of emotional support and someone upon whom the depression-prone individual had relied to maintain feelings of self-esteem. Following the loss, the depression-prone individual feels stripped of hope and optimism because these positive feelings had depended on the other person, now lost. The loss of self-esteem and feelings of security, not of the relationship per se, precipitates depression. If depression-prone people peg their self-worth to a specific occupational goal, such as success in a modeling career, failure triggers self-focusing and consequent depression. Only by surrendering the object or lost goal and fostering alternate sources of identity and self-worth can the cycle be broken.

Research Evidence Psychodynamic theorists focus on the role of loss in depression. Research does show that the losses of significant others (through death or divorce, for example) are often associated with the onset of depression (Paykel, 1982). Such losses may also lead to other psychological disorders, however. There is yet a lack of research to support Freud's view that repressed anger toward the departed loved one is turned inward in depression.

Research supporting the self-focusing model is mixed. On one hand, people who are depressed have been shown to engage in more self-focusing following failure experiences than do nondepressed people, and in relatively lower levels of self-focusing following successes (Pyszczynski & Greenberg, 1985, 1986). Also supporting the model is evidence from a laboratory study that diverting attention away from the self can reduce depressed affect in depressed subjects (Nix et al., 1995). On the other hand, self-focused attention is linked to disorders other than depression, including anxiety disorders, alcoholism, mania, and schizophrenia (Ingram, 1991). The general linkage between self-focused attention and psychopathology may limit the model's value as an explanation of depression.

Humanistic Theories

From the humanistic framework, people become depressed when they cannot imbue their existence with meaning and make authentic choices that lead to self-fulfillment. The world is then a drab place. People's search for meaning gives color and substance to their lives. Guilt may arise when people believe they have not lived up to their potentials. Humanistic psychologists challenge us to take a long hard look at our lives. Are they worthwhile and enriching? Or are they drab and routine? If the latter, it may be we have frustrated our needs for self-actualization. We may be settling, coasting through life. Settling can give rise to a sense of dreariness that becomes expressed in depressive behavior—lethargy, sullen mood, and withdrawal.

Like psychodynamic theorists, humanistic theorists also focus on the loss of self-esteem that can arise when people lose friends or family members, or suffer occupational setbacks or losses. We tend to connect our personal identity and sense of self-worth with our social roles as parents, spouses, students, or workers. When these role identities are lost, through the death of a spouse, the departure of children to college, or loss of a job, our sense of purpose and self-worth can be shattered. Depression is a frequent consequence of such losses. It is especially likely when we base our self-esteem on our occupational role or success. The loss of a job, a demotion, or a failure to achieve a promotion are common precipitants of depression, especially when we are reared to value ourselves on the basis of occupational success.

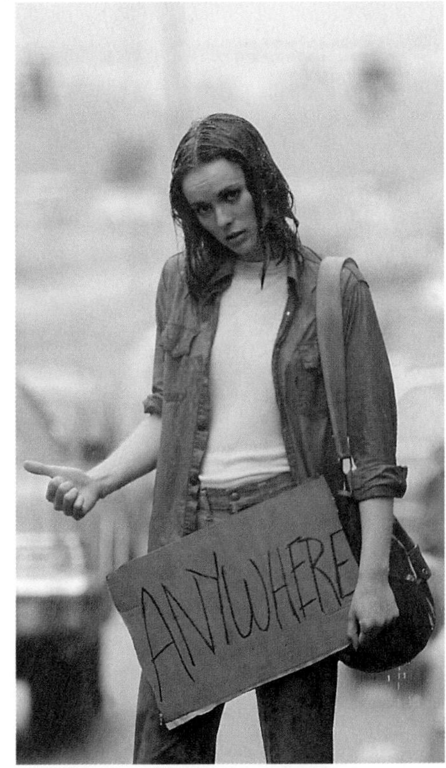

What happens when we lose our sense of direction? According to the humanistic-existential perspective, depression may result from the inability to find meaning and purpose in one's life.

Learning Theories

Whereas the psychodynamic perspectives focus on inner, often unconscious, determinants of mood disorders, learning theorists dwell more on situational factors, such as the loss of positive reinforcement. We perform best when levels of reinforcement are commensurate with our efforts. Changes in the frequency or effectiveness of reinforcement can shift the balance so that life becomes unrewarding.

Reinforcement and Depression Learning theorist Peter Lewinsohn (1974) proposed that depression results from an imbalance between behavioral output and reinforcement input from the environment. A lack of reinforcement for one's efforts can sap motivation and induce feelings of depression. A vicious cycle may ensue: Inactivity and social withdrawal deplete opportunities for reinforcement; lesser reinforcement exacerbates withdrawal. The low rate of activity typical of depression may also be a source of secondary gain or secondary reinforcement. Family members and other people may rally around people suffering from depression and release them from their responsibilities. Rather than help people who are struggling with depression regain normal levels of productive behavior, sympathy may thus backfire and maintain depressed behavior.

Reduction in reinforcement levels can occur for many reasons. A person who is recuperating at home from a serious illness or injury may find little that is reinforcing to do. Social reinforcement may plummet when people close to us, who were suppliers of reinforcement, die or leave us. People who suffer social losses are more likely to become depressed when they lack the social skills to form new relationships. Some first-year college students are homesick and depressed because they lack the skills to form rewarding new relationships. Widows and widowers may be at a loss as to how to ask someone for a date or start a new relationship.

Changes in life circumstances may also alter the balance of effort and reinforcement. A prolonged layoff may reduce financial reinforcements, which may in turn force painful cutbacks in lifestyle. A disability or an extended illness may also impair one's ability to ensure a steady flow of reinforcements.

Lewinsohn's model is supported by research findings that connect depression to a low level of positive reinforcement. In early work, Lewinsohn and Libet (1972) noted a correspondence between depressed moods and lower rates of participation in potentially reinforcing activities. People with depressive disorders were also found to report fewer pleasant activities than nondepressed people (MacPhillamy & Lewinsohn, 1974). However, it is conceivable that depression precedes rather than follows a reduction in reinforcement (J. M. Williams, 1984). In other words, depression may lead people to withdraw from socially reinforcing activities. Regardless of the root causes of depression, a behavioral treatment approach that encourages depressed people to increase their levels of pleasant activities and provides them with the skills to do so is often helpful in alleviating depression (DeRubeis & Crits-Christoph, 1998; Jacobson et al., 1996).

Interactional Theory The interactions between depressed persons and other people may help explain the former group's shortfall in positive reinforcement. Interactional theory, developed by psychologist James Coyne (1976), proposes that the adjustment to living

Strained relationships. Depressed people may encounter rejection in long-term relationships as the result of the stressful demands they place on others.

with a depressed person can become so stressful that the partner or family member becomes progressively less reinforcing toward the depressed person.

Interactional theory is based on the concept of reciprocal interaction. People's behavior influences, and in turn, is influenced by, the behavior of other. The theory holds that depression-prone people react to stress by demanding greater reassurance and social support. At first people who become depressed may succeed in garnering support. Over time, however, their demands and behavior begin to elicit anger or annoyance. Although loved ones may keep their negative feelings to themselves, so as not to further upset the depressed person, these feelings may surface in subtle ways that spell rejection. Depressed people may react to rejection with deeper depression and greater demands, triggering a vicious cycle of further rejection and more profound depression. They may also feel guilty about distressing their family members, which can exacerbate negative feelings about themselves.

Evidence shows that people who become depressed tend to encounter rejection in long-term relationships (Marcus & Nardone, 1992). Family members may find it stressful to adjust to the behavior of the person who is depressed, especially to such behaviors as withdrawal, lethargy, despair, and constant requests for reassurance. A recent study finds that people whose spouses are being treated for depression tend to report higher than average levels of emotional distress (Benazon, 2000).

All in all, research evidence generally supports Coyne's belief that people who suffer from depression elicit rejection from others, but there remains a lack of evidence to show that this rejection is mediated by negative emotions (anger and annoyance) that the depressed person induces in others (Segrin & Dillard, 1992). Rather, a growing body of literature suggests that depressed people may lack effective social skills, which may account for the fact that others often reject them (Segrin & Abramson, 1994). They tend to be unresponsive, uninvolved, and even impolite when they interact with others. For example, they tend to gaze very little at the other person, to take an excessive amount of time to respond, to show very little approval or validation of the other person, and to dwell on their problems and negative feelings (Segrin & Abramson, 1994). They even dwell on negative feelings when interacting with strangers. In effect, they turn other people off, setting the stage for rejection.

Whether social skills deficits are a cause or a symptom of depression remains to be determined. Whatever the case, impaired social behavior likely may play an important role in determining the persistence or recurrence of depression. As we shall see, some psychological approaches to treating depression (e.g., interpersonal psychotherapy and Lewinsohn's social skills training approach—discussed later) focus on helping people with depression better understand and overcome their interpersonal problems. This may help, in turn, alleviate depression or perhaps prevent future recurrences.

Cognitive Theories

Cognitive theorists relate the origin and maintenance of depression to the ways in which people see themselves and the world around them.

Aaron Beck's Cognitive Theory One of the most influential cognitive theorists, psychiatrist Aaron Beck (Beck, 1976; Beck et al., 1979), relates the development of depression to the adoption early in life of a negatively biased or distorted way of thinking—the **cognitive triad of depression** (see Table 8.4). The cognitive triad includes negative beliefs about oneself (e.g., "I'm no good"), the environment or the world at large (e.g., "This school is awful"), and the future (e.g., "Nothing will ever turn out right for me"). Cognitive theory holds that people who adopt this negative way of thinking are at greater risk of becoming depressed in the face of stressful or disappointing life experiences, such as getting a poor grade or losing a job.

Beck views these negative concepts of the self and the world as mental templates or *cognitive schemes* that are adopted in childhood on the basis of early learning experiences. Children may find that nothing they do is good enough to please their parents or teachers. As a result, they may come to regard themselves as basically incompetent and to perceive their future prospects as dim. These beliefs may sensitize them later in life to interpret any

cognitive triad of depression The view that depression derives from adopting negative views of oneself, the environment or world at large, and the future.

TABLE 8.4 The Cognitive Triad of Depression

Negative View of Oneself	Perceiving oneself as worthless, deficient, inadequate, unlovable, and as lacking the skills necessary to achieve happiness.
Negative View of the Environment	Perceiving the environment as imposing excessive demands and/or presenting obstacles that are impossible to overcome, leading continually to failure and loss.
Negative View of the Future	Perceiving the future as hopeless and believing that one is powerless to change things for the better. One expects of the future only continuing failure and unrelenting misery and hardship.

Note. According to Aaron Beck, depression-prone people adopt a habitual style of negative thinking—the so-called cognitive triad of depression.
Source. Adapted from Beck & Young, 1985; Beck et al., 1979.

failure or disappointment as a reflection of something basically wrong or inadequate about themselves. Minor disappointments and personal shortcomings become "blown out of proportion." Even a minor disappointment becomes a crushing blow or a total defeat, which can lead to depression.

The tendency to magnify the importance of minor failures is an example of an error in thinking that Beck labels a *cognitive distortion.* He believes cognitive distortions set the stage for depression in the face of personal losses or negative life events. Psychiatrist David Burns (1980) enumerated a number of the cognitive distortions associated with depression:

1. *All-or-Nothing Thinking.* Seeing events in black and white, as either all good or all bad. For example, one may perceive a relationship that ended in disappointment as a totally negative experience, despite any positive feelings or experiences that may have occurred along the way. Perfectionism is an example of all-or-nothing thinking. Perfectionists judge any outcome other than perfect success to be complete failure. They may consider a grade of B or even A− to be tantamount to an F. They may feel like abject failures if they fall a few dollars short of their sales quotas or receive a very fine (but less than perfect) performance evaluation. Perfectionism is connected with an increased vulnerability to depression as well as poor outcomes in treatment, whether the treatment involves antidepressant medication or psychological approaches such as cognitive therapy or interpersonal psychotherapy (Blatt et al., 1998; Minarik & Ahrens, 1996).
2. *Overgeneralization.* Believing that if a negative event occurs, it is likely to occur again in similar situations in the future. One may come to interpret a single negative event as foreshadowing an endless series of negative events. For example, receiving a letter of rejection from a potential employer leads one to assume that all other job applications will similarly be rejected.
3. *Mental Filter.* Focusing only on negative details of events, thereby rejecting the positive features of one's experiences. Like a droplet of ink that spreads to discolor an entire beaker of water, focusing only on a single negative detail can darken one's vision of reality. Beck called this cognitive distortion **selective abstraction,** meaning the individual selectively abstracts the negative details from events and ignores their positive features. One thus bases one's self-esteem on perceived weaknesses and failures, rather than on positive features, or on a balance of accomplishments and shortcomings. For example, a person receives a job evaluation that contains positive and negative comments but focuses only on the negative.

selective abstraction A cognitive distortion involving the tendency to focus only on the negative parts of experiences or events.

4. *Disqualifying the Positive.* This refers to the tendency to snatch defeat from the jaws of victory by neutralizing or denying your accomplishments. An example is dismissal of congratulations for a job well done by thinking and saying, "Oh, it's no big deal. Anyone could have done it." By contrast, taking credit where credit is due may help people overcome depression by increasing their belief they can make changes that will lead to a positive future (Needles & Abramson, 1990).
5. *Jumping to Conclusions.* Forming a negative interpretation of events, despite a lack of evidence. Two examples of this style of thinking are "mind reading" and "the fortune teller error." In *mind reading,* you arbitrarily jump to the conclusion that others don't like or respect you, as in interpreting a friend's not calling for a while as a rejection. The *fortune teller error* involves the prediction that something bad is always about to happen to oneself. The person believes the prediction of calamity is factually based even though there is an absence of evidence to support it. For example, the person concludes that a passing tightness in the chest *must* be a sign of heart disease, discounting the possibility of more benign causes.
6. *Magnification and Minimization.* Magnification, or *catastrophizing,* refers to the tendency to make mountains out of molehills—to exaggerate the importance of negative events, personal flaws, fears, or mistakes. Minimization is the mirror image, a type of cognitive distortion in which one minimizes or underestimates one's good points.
7. *Emotional Reasoning.* Basing reasoning on emotions—thinking, for example, "If I feel guilty, it must be because I've done something really wrong." One interprets feelings and events based on emotions rather than on fair consideration of evidence.
8. *Should Statements.* Creating personal imperatives or self-commandments—"shoulds" or "musts." For example, "I *should* always get my first serve in!" or, "I *must* make Chris like me!" By creating unrealistic expectations, **musterbation**—the label given this form of thinking by Albert Ellis—can lead one to become depressed when one falls short.
9. *Labeling and Mislabeling.* Explaining behavior by attaching negative labels to oneself and others. You may explain a poor grade on a test by thinking you were "lazy" or "stupid" rather than simply unprepared for the specific exam or, perhaps, ill. Labeling other people as "stupid" or "insensitive" can engender hostility toward them. Mislabeling involves the use of labels that are emotionally charged and inaccurate, such as calling yourself a "pig" because of a minor deviation from your usual diet.
10. *Personalization.* This refers to the tendency to assume you are responsible for other people's problems and behavior. You may assume your partner or spouse is crying because of something you have done (or not done) rather than recognizing that other causes may be involved.

Consider the errors in thinking illustrated in the following case example:

Errors in Thinking in a Case of Depression

Christie was a 33-year-old real estate sales agent who suffered from frequent episodes of depression. Whenever a deal fell through, she would blame herself: "If only I had worked harder . . . negotiated better . . . talked more persuasively . . . the deal would have been done." After several successive disappointments, each one followed by self-recriminations, she felt like quitting altogether. Her thinking became increasingly dominated by negative thoughts, which further depressed her mood and lowered her self-esteem: "I'm a loser . . . I'll never succeed . . . It's all my fault . . . I'm no good and I'm never going to succeed at anything."

Christie's thinking included cognitive errors such as the following: (1) personalization *(believing herself to be the sole cause of negative events);*

musterbation A rigid thought pattern characterized by the tendency to impose excessive, unrealistic demands on oneself or personal imperatives.

automatic thoughts Thoughts that seem to pop into one's mind.

cognitive-specificity hypothesis The belief that different emotional disorders are linked to particular kinds of automatic thoughts.

(2) labeling and mislabeling *(labeling herself to be a loser); (3)* overgeneralization *(predicting a dismal future on the basis of a present disappointment); and (4)* mental filter *(judging her personality entirely on the basis of her disappointments). In therapy, Christie was helped to think more realistically about events and not to jump to conclusions that she was automatically at fault whenever a deal fell through, or to judge her whole personality on the basis of disappointments or perceived flaws in herself. In place of this self-defeating style of thinking, she began to think more realistically when disappointments occurred, like telling herself, "Okay, I'm disappointed. I'm frustrated. I feel lousy. So what? It doesn't mean I'll never succeed. Let me discover what went wrong and try to correct it the next time. I have to look ahead, not dwell on disappointments in the past."*

—From the Authors' Files

Distorted thinking tends to be experienced as automatic, as if the thoughts had just popped into one's head. These **automatic thoughts** are likely to be accepted as statements of fact rather than as opinions or habitual ways of interpreting events.

Beck and his colleagues formulated a **cognitive-specificity hypothesis,** which proposes that different disorders, anxiety disorders and depressive disorders in particular, are characterized by different types of automatic thoughts. The results of one study (Beck et al., 1987) showed some interesting differences in the types of automatic thoughts people with depressive and anxiety disorders reported (see Table 8.5). People with diagnosable depression more often reported thoughts concerning themes of loss, self-deprecation, and pessimism. People with anxiety disorders more often reported thoughts concerning physical danger and other threats.

Research Evidence on Cognitions and Depression Supporting Beck's model is evidence linking distorted cognitions and negative thinking to depressive symptoms and clinical depression (e.g., Clark, Cook, & Snow, 1998; Stader & Hokanson, 1998). People who are depressed also tend to hold more pessimistic views of the future and are more critical of

TABLE 8.5 Automatic Thoughts Associated with Depression and Anxiety

Common Automatic Thoughts Associated with Depression	Common Automatic Thoughts Associated with Anxiety
1. I'm worthless.	**1.** What if I get sick and become an invalid?
2. I'm not worthy of other people's attention or affection.	**2.** I am going to be injured.
3. I'll never be as good as other people are.	**3.** What if no one reaches me in time to help?
4. I'm a social failure.	**4.** I might be trapped.
5. I don't deserve to be loved.	**5.** I am not a healthy person.
6. People don't respect me anymore.	**6.** I'm going to have an accident.
7. I will never overcome my problems.	**7.** Something will happen that will ruin my appearance.
8. I've lost the only friends I've had.	**8.** I am going to have a heart attack.
9. Life isn't worth living.	**9.** Something awful is going to happen.
10. I'm worse off than they are.	**10.** Something will happen to someone I care about.
11. There's no one left to help me.	**11.** I'm losing my mind.
12. No one cares whether I live or die.	
13. Nothing ever works out for me anymore.	
14. I have become physically unattractive.	

Source. Adapted from Beck et al., 1987.

A Closer Look

On Positive Illusions and Mental Health: Is It Adaptive to See Things As They Truly Are?

A common assumption exists that there is an objective reality "out there" and that the ability to perceive reality accurately is a basic feature of psychological adjustment. Individuals who perceive the world for what it is may be better able to adapt to their physical and social environments and avoid harm (Colvin & Block, 1994). Yet might it be more adaptive under some circumstances to hold certain positive biases or optimistic illusions about oneself and one's place in the world? Although we usually equate mental health with good reality testing, and mental illness with distorted perceptions and cognitions, the negative perceptions held by many depressed people may not be distorted at all but rather may be quite realistic. Perhaps it is the rest of us who tend to see the world through "rose-colored lenses" that cast too rosy a tint on our perceptions of our abilities and our likelihood of success. Normal human thought, investigators find, is characterized by mild distortions or illusions, such as inflated beliefs about ourselves, exaggerated perceptions about our ability to control events, and unrealistic optimism (Taylor et al., 2000). Mentally healthy people may manage to keep themselves out of the dumps by maintaining these illusions despite evidence to the contrary. In effect, we may need to maintain some positive illusions to maintain our spirits in coping with the "ups and downs" of life (Goode, 2001e). They may also help us keep our chins up and maintain expectancies of success in the future rather than accept negative outcomes as somehow fated for us.

Researchers devised a research project to test the hypothesis that depressed people view their own abilities to control events more realistically than nondepressed people. In essence, they created a laboratory situation in which subjects had no control over a laboratory task so they could assess which subjects nevertheless managed to maintain an *illusion of control*. The results showed that subjects who maintained an illusion of control showed less evidence of depression following exposure to stressful experiences afterward (Alloy & Clements, 1992). Perhaps the tendency to think we are in charge of our destinies, even incorrectly, reduces our susceptibility to depression. We should caution that more evidence is needed to support the link between positive illusions and psychological adjustment. Let us also note that holding extremely positive illusions, such as delusions of grandeur, is clearly associated with psychological problems such as schizophrenia and mania (Taylor & Brown, 1994).

Truth OR Fiction? REVISITED

In some ways, many "mentally healthy" people see things *less* realistically than people who are depressed.

TRUE. Researchers find that mentally healthy people may in some ways view the world less realistically than people who are depressed. Many people with depression, however, may be more accurate in their assessment of the extent to which they actually can exercise control over events.

Depressed people may see the proverbial glass not only as half empty but as both half empty and half full, as compared to nondepressed people, who may see the glass only as hall full. In a recent study, researchers found that depressed women attended equally to positive and negative stimuli in a laboratory task (the stimuli were either positive, neutral, or negative words presented on a computer display). Nondepressed women showed a positive bias by attending more to the positive or neutral stimuli (McCabe & Gotlib, 1995). This suggests that depressed individuals are not necessarily biased toward perceiving only the negative. Rather, what seems to set them apart from nondepressed people is their failure to maintain a positive bias.

themselves and others (Glara et al., 1993). Other findings indicate that dysfunctional attitudes (above a certain threshold) increase vulnerability to depression in the face of negative life events (Lewinsohn, Joiner, & Rohde, 2001).

The relationship between negative thinking and depression may depend more on the balance between negative and positive thoughts than on the presence of negative thoughts alone. Research using a thought-counting method showed that people who functioned well psychologically experienced both positive and negative thoughts, but the positive thoughts occurred one and a half to two times as often as the negative thoughts (R. M. Schwartz, 1986). People with a mild level of psychological dysfunction, by contrast, produced about equal numbers of positive and negative thoughts. Thinking positive thoughts may serve as a kind of buffer or shock absorber in helping people cope with negative life events without becoming depressed (Bruch, 1997; Lightsey, 1994a, 1994b).

learned helplessness A behavior pattern characterized by passivity and perceptions of lack of control.

Research also supports Beck's view that people who are depressed tend to magnify their shortcomings. In one study, college students were given a test that supposedly measured the presence of a personality trait and were asked to indicate the value of the trait to them. Students who weren't depressed tended to inflate the value of the trait when they were told they possessed a great deal of it. Students who were depressed, however, exaggerated the importance of the trait when they were informed they had done poorly on the test (Wenzlaff & Grozier, 1988). Whereas students who were depressed accentuated the negative, other students were more self-enhancing; they emphasized the importance of a quality they believed themselves to have.

All in all, there is broad research support for many aspects of the theory, including Beck's concept of the cognitive triad of depression and his view that people with depression think more negatively than nondepressed people about themselves, the future, and the world in general (Haaga, Dyck, & Ernst, 1991). Although dysfunctional cognitions (negative, distorted, or pessimistic thoughts) are more common among people who are depressed, the causal pathways remain unclear. We can't yet say whether dysfunctional or negative thinking causes depression or is merely a feature of depression. Thus the central theme of cognitive theory, that negative, distorted thoughts are causally related to depression, remains to be confirmed (Cole et al., 1998; Stader & Hokanson, 1998).

THINK ABOUT IT

Which of the cognitive distortions, if any, that are listed in the text characterize your way of thinking about disappointing experiences in your life? What were the effects of these thought patterns on your mood? How did they affect your feelings about yourself? How might you change these ways of thinking in the future?

Perhaps the causal linkages go both ways. Our thoughts may affect our moods and our moods may affect our thoughts (Kwon & Oei, 1994). Think in terms of a vicious cycle. People who feel depressed may begin thinking in more negative, distorted ways. The more negative and distorted their thinking becomes, the more depressed they feel, and the more depressed they feel, the more dysfunctional their thinking becomes. Alternatively, dysfunctional thinking may come first in the cycle, perhaps in response to a disappointing life experience, which then leads to a downcast mood. This in turn may accentuate negative thinking, and so on. We are still faced with the old "chicken or the egg" dilemma of determining which comes first in the causal sequence, distorted thinking or depression. Future research may help tease out these causal pathways. Even if it should become clear that distorted cognitions play no direct causal role in the initial onset of depression, the reciprocal interaction between thoughts and moods may play a role in maintaining depression and in determining the likelihood of recurrence (Kwon & Oei, 1994). We know, for example, that people who recover from depression but continue to hold distorted cognitions tend to be at greater risk of recurrence of depression (Rush & Weissenburger, 1994). Fortunately, evidence shows that dysfunctional attitudes tend to decrease with effective treatment for depression (Fava et al., 1994).

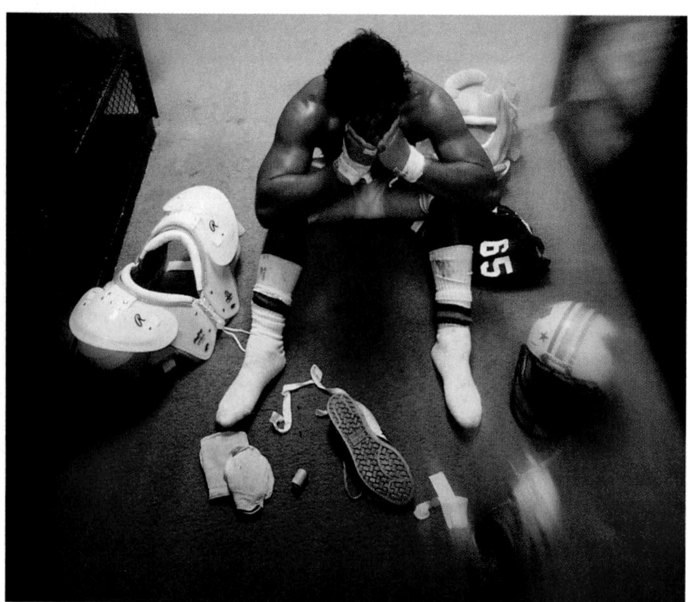

How could he have missed that tackle? This football player missed a crucial tackle and is rehashing it. He is putting himself down and telling himself that there is nothing he can do to improve his performance. Cognitive theorists believe that a person's self-defeating or distorted interpretations of negative events can set the stage for depression.

Learned Helplessness (Attributional) Theory The **learned helplessness** model proposes that people may become depressed because they learn to view themselves as helpless to control the reinforcements in their environments—or to change their lives for the better. The originator of the learned helplessness concept, Martin Seligman (1973, 1975), suggests that people learn to perceive themselves as helpless because of their experiences. The learned helplessness model therefore straddles the behavioral and the cognitive: Situational factors foster attitudes that lead to depression.

Seligman and his colleagues based the learned helplessness model on early laboratory studies of animals. In these studies, dogs exposed to an inescapable electric shock showed the "learned helplessness effect" by failing to learn to escape when the shock was later made escapable (Overmier & Seligman, 1967; Seligman & Maier, 1967). Exposure to uncontrollable forces apparently taught the animals they were helpless to change their situation. Animals who developed learned helplessness showed behaviors that were similar to those of people with depression, including lethargy, lack of motivation, and difficulty acquiring new skills (Maier & Seligman, 1976).

Seligman (1975, 1991) proposed that some forms of depression in humans might result from exposure to apparently uncontrollable situations. Such experiences can instill the expectation that future reinforcements will also be beyond the individual's control. A cruel vicious cycle may come into play in many cases of depression. A few failures may produce feelings of helplessness and expectations of further failure. Perhaps you know people who have failed certain subjects, such as mathematics. They may come to believe themselves incapable of succeeding in math. They may thus decide that studying for the quantitative section of the Graduate Record Exam is a waste of time. They then perform poorly, completing the self-fulfilling prophecy by confirming their expectations, which further intensifies feelings of helplessness, leading to lowered expectations, and so on, in a vicious cycle.

Although it stimulated much interest, Seligman's model failed to account for the low self-esteem typical of people who are depressed. Nor did it explain variations in the persistence of depression. Seligman and his colleagues (Abramson, Seligman, & Teasdale, 1978) offered a reformulation of the theory to meet such shortcomings. The revised theory held that perception of lack of control over reinforcement alone did not explain the persistence and severity of depression. It was also necessary to consider cognitive factors, especially the ways in which people explain their failures and disappointments to themselves.

Seligman and his colleagues recast helplessness theory in terms of the social psychology concept of **attributional style.** An attributional style is a personal style of explanation. When disappointments or failures occur, we may explain them in various characteristic ways. We may blame ourselves (an **internal attribution**), or we may blame the circumstances we face (an **external attribution**). We may see bad experiences as typical events (a **stable attribution**) or as isolated events (an **unstable attribution**). We may see them as evidence of broader problems (a **global attribution**) or as evidence of precise and limited shortcomings (a **specific attribution**). The revised helplessness theory—called the reformulated helplessness theory—holds that people who explain the causes of negative events (like failure in work, school, or romantic relationships) according to these three types of attributions are most vulnerable to depression:

1. Internal factors, or beliefs that failures reflect their personal inadequacies, rather than external factors, or beliefs that failures are caused by environmental factors;
2. Global factors, or beliefs that failures reflect sweeping flaws in personality rather than specific factors, or beliefs that failures reflect limited areas of functioning; and
3. Stable factors, or beliefs that failures reflect fixed personality factors rather than unstable factors, or beliefs that the factors leading to failures are changeable.

Let us illustrate these attributional styles with the example of a college student who goes on a disastrous date. Afterward he shakes his head in wonder and tries to make sense of his experience. An internal attribution for the calamity would involve self-blame, as in "I really messed it up." An external attribution would place the blame elsewhere, as in "Some couples just don't hit it off," or "She must have been in a bad mood." A stable attribution would suggest a problem that cannot be changed, as in "It's my personality." An unstable attribution, on the other hand, would suggest a transient condition, as in "It was probably the head cold." A global attribution for failure magnifies the extent of the problem, as in "I really have no idea what I'm doing when I'm with people." A specific attribution, in contrast, chops the problem down to size, as in "My problem is how to make small talk to get a relationship going."

The revised theory holds that each attributional dimension makes a specific contribution to feelings of helplessness. Internal attributions for negative events are linked to lower self-esteem. Stable attributions help explain the persistence—or, in medical terms, the chronicity—of helplessness cognitions. Global attributions are associated with the generality or pervasiveness of feelings of helplessness following negative events. Attributional style should be distinguished from negative thinking (Gotlib et al., 1993). You may think

attributional style A personal style for explaining cause-and-effect relationships between events.

internal attribution A belief that the cause of an event involved factors within oneself.

external attribution A belief that the cause of an event involved factors outside oneself.

stable attribution A belief that the cause of an event involved stable, rather than changeable, factors.

unstable attribution A belief that the cause of an event involved changeable, rather than stable, factors.

global attribution A belief that the cause of an event involved generalized, rather than specific, factors.

specific attribution A belief that the cause of an event involved specific, rather than generalized, factors.

Is it me? According to reformulated helplessness theory, the kinds of attributions we make concerning negative events can make us more or less vulnerable to depression. Attributing the breakup of a relationship to internalizing ("It's me"), globalizing ("I'm totally worthless"), and stabilizing ("Things are always going to turn out badly for me") causes can lead to depression.

negatively (pessimistically) or positively (optimistically), but still hold yourself to blame for your perceived failures.

Research generally supports the reformulated helplessness (attributional) model. As the model would predict, depressed people are generally more likely than others to attribute the causes of failures to internal, stable, and global factors (e.g., Seligman et al., 1988; Sweeney, Anderson, & Bailey, 1986). A depressive attributional style also predicts responsiveness to antidepressant medication in depressed patients (Levitan, Rector, & Bagby, 1998). Also supporting the model are findings that negative attributional styles and dysfunctional attitudes predict higher lifetime rates of major depression (Alloy et al., 2000). Attributional style may have a stronger relationship to depression in people who tend to think more about the causes of events, however (Haaga, 1995).

Biological Factors

Evidence has accumulated pointing to the important role of biological factors, especially genetics and neurotransmitter functioning, in the development of mood disorders. Recent investigations are examining the biological roots of depression at the neurotransmitter level as well as the genetic, molecular, and even cellular level.

Genetic Factors A growing body of knowledge implicates genetic factors in mood disorders. We know that mood disorders, including major depression and especially bipolar disorder, tend to run in families (Klein et al., 2001; USDHHS, 1999a). Families, however, share environmental similarities as well as genes. Family members may share blue eyes (an inherited attribute) but also a common religion (a cultural attribute). Yet evidence pointing to a genetic basis for mood disorders comes from studies showing that the closer the genetic relationship one shares with a person with a major mood disorder (major depression or bipolar disorder), the greater the likelihood that one will also suffer from a major mood disorder (e.g., Vincent et al., 1999).

Twin studies and adoptee studies provide additional evidence of a genetic contribution. A higher concordance (agreement) rate among monozygotic (MZ) twins than dizygotic (DZ) twins for a given disorder is taken as supportive evidence of genetic factors. Both types of twins share common environments, but MZ twins share 100% of their genes as compared to 50% for DZ twins. The concordance rate for major mood disorders (unipolar and bipolar disorders) between MZ twins ranges from 45% to 70%, which is more than double the rate between DZ twins (Kendler et al., 1992a, 1993b). This provides strong support for a genetic component, but is short of the 100% concordance we would expect if genetics were solely responsible for these disorders. Adoptee studies, which might provide corroborating evidence of genetic factors in mood disorders, are sparse.

All in all, researchers believe that heredity plays an important role in major depression (Kendler & Prescott, 1999; Sullivan, Neale, & Kendler, 2000). However, genetics isn't the only determinant of major depression, nor is it necessarily the most important determinant. Environmental factors, such as exposure to stressful life events, appear to play at least as great a role—if not a greater role—than genetics (Kendler & Prescott, 1999). It appears that major depression is a complex disorder that is caused by a combination of genetic and environmental factors (Sullivan, Neale, & Kendler, 2000).

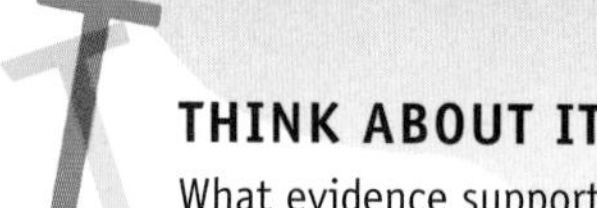

THINK ABOUT IT

What evidence supports a genetic contribution to mood disorders? Is genetics solely responsible? Why or why not?

Genetic factors may play a greater role in explaining bipolar disorder than unipolar depression (major depressive disorder) (Krehbiel, 2000). As for dysthymic disorder, data from twin studies indicates that the disorder may be relatively less influenced by genetic factors than either major depression or bipolar disorder (Torgersen, 1986).

Biochemical Factors and Brain Abnormalities in Depression Early research on the biological underpinnings of depression focused on deficits in neurotransmitter levels in the brain. Neurotransmitters were first suspected of playing a role in depression back in the 1950s. Findings were reported then that hypertensive patients who were taking the drug *reserpine* often became depressed. Reserpine depletes the supplies of various neurotransmitters in the brain, including norepinephrine and serotonin (USDHHS, 1999a). Then came the discovery that drugs that increase the brain levels of neurotransmitters such as norepinephrine and serotonin can relieve depression. These drugs, called antidepressants, include *tricyclics*, such as imipramine (trade name Tofranil) and amitriptyline (trade name Elavil); *monoamine oxidase* (MAO) inhibitors, such as phenelzine (trade name Nardil); and *selective serotonin-reuptake inhibitors* (SSRIs), such as fluoxetine (trade name Prozac) and sertraline (trade name Zoloft). But later research cast doubt on the belief that depression is caused simply by a lack of particular neurotransmitters in the brain. For example, while antidepressants increase the availability of neurotransmitters in the brain within hours, depressed patients do not usually begin to show a response to treatment for several weeks (Nierenberg et al., 2000). Therefore, it is unlikely that these drugs work by simply boosting levels of neurotransmitters in the brain (Duman, Heninger, & Nestler, 1997).

More complex views of the role of neurotransmitters in depression are evolving (Cravchik & Goldman, 2000). A widely held view today is that depression involves irregularities in (1) the numbers of receptors on receiving neurons where neurotransmitters dock (having either too many or too few); or (2) in the sensitivity of receptors to particular neurotransmitters (Yatham et al., 2000). Antidepressants may work by affecting either the number or sensitivity of receptors. Deficiencies of certain neurotransmitters may also play a role (Lambert et al., 2000). Complicating matters further, there are several different types of receptors for each neurotransmitter. There may also be many subtypes for each type (USDHHS, 1999a). The actions of particular antidepressants may be specific to certain types or subtypes of receptors.

Another avenue of research focuses on possible abnormalities in the *prefrontal cortex*, the area of the frontal lobes lying in front of the motor areas. Investigators find evidence of lower metabolic activity and size of the prefrontal cortex in clinically depressed people as compared to healthy controls (e.g., Damasio, 1997). The prefrontal cortex is involved in regulating neurotransmitters believed to be involved in mood disorders, including serotonin and norepinephrine, so it is not surprising that evidence points to irregularities in this region of the brain. Other investigators find that MRI scans of the brains of people with bipolar disorder show evidence of structural abnormalities in parts of the brain involved in regulating mood states (Strakowski et al., 1999).

We also find evidence of other biochemical pathways to depression, including roles for proteins that help nerve cells communicate (Kramer et al., 1998) and endocrine system activity involving the thyroid gland, the adrenal glands, and in women, fluctuations of the female sex hormones estrogen and progesterone (Marangell et al., 1997; Seeman, 1997; Stahl, 2001).

Tying It Together

Depression and other mood disorders involve interplay of multiple factors. Consistent with the *diathesis-stress model*, depression may reflect an interaction of biological factors (such as genetic factors, neurotransmitter irregularities, or brain abnormalities), psychological factors (such as cognitive distortions or learned helplessness), and social and environmental stressors (such as divorce or loss of a job).

Let us consider a possible causal pathway based on the diathesis-stress model (see Figure 8.3). Stressful life events, such as prolonged unemployment or a divorce, may have a depressing effect by reducing neurotransmitter activity in the brain. These biochemical effects may be more likely to occur or be more pronounced in people with a certain genetic predisposition or *diathesis* for depression. However, a depressive disorder may not develop, or may develop in a milder form, in people with more effective coping resources for handling stressful situations. For example, people who

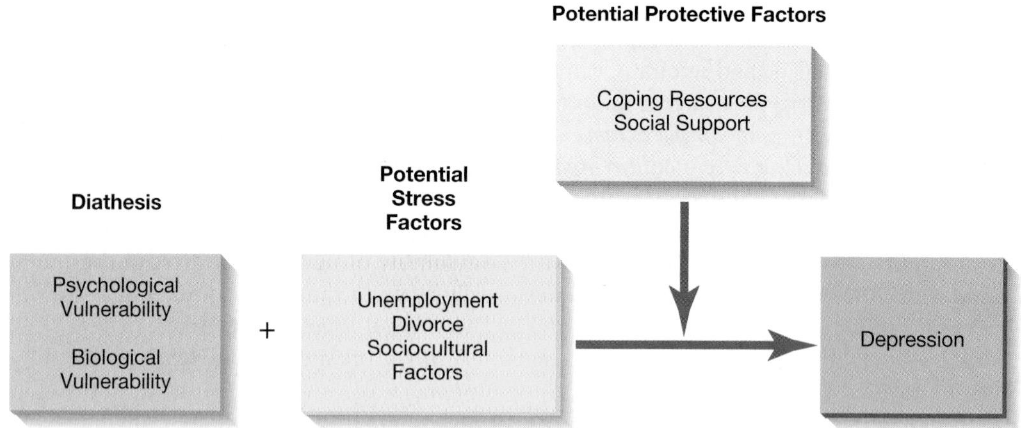

FIGURE 8.3 Diathesis-stress model of depression.

receive emotional support from others may be better able to withstand the effects of stress than those who have to go it alone. So too for people who make active coping efforts to meet the challenges they face in life.

Sociocultural factors may be major sources of stress that influence the development of mood disorders (Ostler et al., 2001). These factors include poverty; overcrowding; exposure to racism, sexism, and prejudice; violence in the home or community; unequal stressful burdens placed on women; and family disintegration. These factors may figure prominently in either precipitating mood disorders or accounting for their recurrence. Other sources of stress include negative life events such as the loss of a job, the development of a serious illness, the breakup of a romantic relationship, and the loss of a loved one.

THINK ABOUT IT

Jonathan becomes clinically depressed after losing his job and his girlfriend. Based on your review of the different theoretical perspectives on depression, explain how these losses may have figured in Jonathan's depression.

The diathesis for depression may take the form of a psychological vulnerability involving a depressive thinking style, one characterized by tendencies to exaggerate the consequences of negative events, to heap blame on oneself, and to perceive oneself as helpless to effect positive change. This cognitive diathesis may increase the risk of depression in the face of negative life events. These cognitive influences may also interact with a genetically based diathesis to further increase the risk of depression following stressful life events. Then too, the availability of social support from others may help bolster a person's resistance to stress during difficult times. People with more effective social skills may be better able to garner and maintain social reinforcement from others and thus be better able to resist depression than people lacking social skills. But biochemical changes in the brain might make it more difficult for the person to cope effectively and bounce back from stressful life events. Lingering biochemical changes and feelings of depression may exacerbate feelings of helplessness, compound the effects of the initial stressor, and so on.

Gender-related differences in coping styles may also come into play. Men and women may respond differently to feelings of depression. According to Nolen-Hoeksema, women are more likely to ruminate when facing emotional problems, and men are more likely to seek refuge in a bottle. These or other differences in coping styles may propel women into longer and more severe bouts of depression while setting the stage for the development of alcohol-related problems in men. As you can see, a complex web of contributing factors may be involved in the development of mood disorders.

Quiz 8.2
Theoretical Perspectives on Mood Disorders

Treatment of Mood Disorders

Just as theoretical perspectives suggest that many factors may be involved in the development of mood disorders, there are various approaches to treatment that derive from psychological and biological models. Here we focus on several of the leading contemporary approaches.

Psychodynamic Approaches

Traditional psychoanalysis aims to help people who become depressed understand their ambivalent feelings toward important people (objects) in their lives they have lost or whose loss was threatened. By working through feelings of anger toward these lost objects, they can turn anger outward—through verbal expression of feelings, for example—rather than leave it to fester and turn inward.

Traditional psychoanalysis can take years to uncover and deal with unconscious conflicts. Modern psychoanalytic approaches also focus on unconscious conflicts, but they are more direct, relatively brief, and focus on present as well as past conflicted relationships. A recent study supported the efficacy of structured, short-term dynamic therapy (Luborsky et al., 1996). Eclectic psychodynamic therapists may use behavioral methods to help clients acquire the social skills needed to develop a broader social network.

Interpersonal psychotherapy (IPT). IPT is usually a brief, psychodynamically oriented therapy that focuses on issues in the person's current interpersonal relationships. Like traditional psychodynamic approaches, IPT assumes that early life experiences are key issues in adjustment, but IPT focuses on the present—the here and now.

Newer models of psychotherapy for depression have emerged from the interpersonal school of psychodynamic therapy derived initially from the work of Harry Stack Sullivan (see Chapter 2) and other neo-Freudians, such as Karen Horney. One contemporary example is **interpersonal psychotherapy (IPT)** (Klerman et al., 1984). IPT is a brief form of therapy (usually no more than 9 to 12 months) that focuses on the client's current interpersonal relationships. The developers of ITP believe that depression occurs within an interpersonal context and that relationship issues need to be emphasized in treatment. IPT has been shown to be an effective treatment for major depression and shows promise in treating other psychological disorders, including dysthymic disorder and bulimia (DeRubeis & Crits-Christoph, 1998; Leichsenring, 2001).

Although IPT shares some features with traditional psychodynamic approaches (principally the belief that early life experiences and persistent personality features are important issues in psychological adjustment), it differs from traditional psychodynamic therapy by focusing primarily on clients' current relationships rather than on helping them acquire insight into unconscious internal conflicts of childhood origins. Although unconscious factors and early childhood experiences are recognized, therapy focuses on the present—the here and now.

IPT helps clients deal with unresolved or delayed grief reactions following the death of a loved one as well as role conflicts in present relationships (Weissman & Markowitz, 1994). The therapist helps clients express grief and come to terms with their loss while assisting them in developing new activities and relationships to help renew their lives. The therapist also helps clients identify areas of conflict in their present relationships, understand the issues that underlie them, and consider ways of resolving them. If the problems in a relationship are beyond repair, the therapist helps the client consider ways of ending it and establishing new relationships. In the case of Sal D., a 31-year-old TV repairman's assistant, depression was associated with marital conflict:

Interpersonal Psychotherapy in a Case of Depression

Sal began to explore his marital problems in the fifth therapy session, becoming tearful as he recounted his difficulty expressing his feelings to his wife because of feelings of being "numb." He felt that he had been "holding on" to his feelings, which was causing him to become estranged from his wife. The next session zeroed in on the similarities between himself and his father, in particular how he was distancing himself from his wife in a similar way to how his father had kept a distance from him. By session 7, a turning point had been reached. Sal expressed how he and his wife had become

interpersonal psychotherapy (IPT) A brief form of psychodynamic therapy that focuses on the client's current interpersonal relationships.

"emotional" and closer to one another during the previous week and how he was able to talk more openly about his feelings, and how he and his wife had been able to make a joint decision concerning a financial matter that had been worrying them for some time. When later he was laid off from his job, he sought his wife's opinion, rather than picking a fight with her as a way of thrusting his job problems on her. To his surprise he found that his wife responded positively—not "violently" as he had expected—to times when he expressed his feelings. In his last therapy session (session 12), Sal expressed how therapy had led to a "reawakening" within himself with respect to the feelings he had been keeping to himself—an openness that he hoped to create in his relationship with his wife.

—*Adapted from Klerman et al., 1984, pp. 111–113*

Behavioral Approaches

Behavioral treatment approaches presume that depressive behaviors are learned and can be unlearned. Behavior therapists aim to directly modify behaviors rather than to foster awareness of possible unconscious causes of these behaviors. Behavior therapy has been shown to produce substantial benefits in treating depression both in adults and adolescents (Craighead, Craighead, & Ilardi, 1998).

One illustrative behavioral program was developed by Lewinsohn and his colleagues (Lewinsohn et al., 1996). It consists of a 12-session, 8-week group therapy program organized as a course—the *Coping With Depression (CWD) Course.* The course helps clients acquire relaxation skills, increase pleasant activities, and build social skills that enable them to obtain social reinforcement. For example, students learn how to accept rather than deny compliments and how to ask friends to join them in activities to raise the frequency and quality of their social interactions. Participants are taught to generate a self-change plan, to think more constructively, and to develop a lifetime plan for maintaining treatment gains and preventing recurrent depression. The therapist is considered a teacher; the client, a student; the session, a class. Each participant is treated as a responsible adult who is capable of learning. The structure involves lectures, activities, and homework and each session follows a structured lesson plan. Depressed adolescents who received the CWD treatment showed lower rates of depression and increased activity levels compared to control subjects (Lewinsohn et al., 1996).

Cognitive Approaches

Cognitive theorists believe that distorted thinking plays a key role in the development of depression. Aaron Beck and his colleagues have developed a multicomponent treatment approach, called **cognitive therapy**, which focuses on helping people with depression learn to recognize and change their dysfunctional thinking patterns. Depressed people tend to focus on how they are feeling rather than on the thoughts that may underlie their feeling states. That is, they usually pay more attention to how bad they feel than to the thoughts that may trigger or maintain their depressed moods.

Cognitive therapy, like behavior therapy, involves a relatively brief therapy format, frequently 14 to 16 weekly sessions (Butler & Beck, 1995). Therapists use a combination of behavioral and cognitive techniques to help clients identify and change dysfunctional thoughts and develop more adaptive behaviors. For example, they assist clients in connecting thought patterns to negative moods by having them monitor the automatic negative thoughts they experience throughout the day by means of a thought diary or daily record. They note when and where negative thoughts occur and how they feel at the time. Once these disruptive thoughts are identified, the therapist helps the client challenge their validity and replace them with more adaptive thoughts. The following case example shows how a cognitive therapist works with a client to dispute the validity of

cognitive therapy Aaron Beck's form of therapy that helps clients recognize and correct distorted patterns of thinking.

thoughts that reflect the cognitive distortion called *selective abstraction* (the tendency to judge oneself entirely on the basis of specific weaknesses or flaws in character). The client judged herself to be totally lacking in self-control because she ate a single piece of candy while she was on a diet.

A Case Vignette of Cognitive Therapy

CLIENT: *I don't have any self-control at all.*

THERAPIST: *On what basis do you say that?*

C: *Somebody offered me candy and I couldn't refuse it.*

T: *Were you eating candy every day?*

C: *No, I just ate it this once.*

T: *Did you do anything constructive during the past week to adhere to your diet?*

C: *Well, I didn't give in to the temptation to buy candy every time I saw it at the store. . . . Also, I did not eat any candy except that one time when it was offered to me and I felt I couldn't refuse it.*

T: *If you counted up the number of times you controlled yourself versus the number of times you gave in, what ratio would you get?*

C: *About 100 to 1.*

T: *So if you controlled yourself 100 times and did not control yourself just once, would that be a sign that you are weak through and through?*

C: *I guess not—not through and through (smiles).*

—Adapted from Beck et al., 1979, p. 68

■

Ample evidence supports the effectiveness of cognitive therapy in treating major depression and reducing risks of recurrent episodes (DeRubeis et al., 1999, 2001; Jarrett et al., 2001; Leichsenring, 2001). The benefits achieved from cognitive therapy appear to be at least equal to those achieved from antidepressant medication in treating depression (DeRubeis et al., 1999, 2001; Jarrett et al., 1999). However, it remains an open question as to whether a combination approach of antidepressant medication and psychotherapy works better than either approach alone (Murray, 2000b). However, a combination of antidepressant medications and psychotherapy appears to be more effective than psychotherapy alone in treating the more severe, recurrent forms of depression (USDHHS, 1999a). Cognitive therapy, or cognitive behavioral therapy, also appears to produce about the same level of benefit as interpersonal psychotherapy, the most widely studied form of brief, psychodynamically oriented therapy for depression (Elkin et al., 1989; Leichsenring, 2001; Shapiro et al., 1995).

We have little research yet on the psychological treatment of dysthymia, although techniques used in treating major depression, such as cognitive therapy and interpersonal psychotherapy, have shown some promising results (Thase et al., 1997). Large-scale investigations of the effects of psychological treatments for bipolar disorder are underway. Early studies suggest that psychosocial treatments, such as cognitive-behavioral therapy and forms of interpersonal therapy and family therapy, may be effective adjuncts to drug therapy in the treatment of bipolar disorder (Lam et al., 2000; Otto, 2001).

Cognitive theorists suggest that cognitive errors can lead to depression if they are left to rummage around unchallenged in the individual's mind. Cognitive therapists help clients to recognize cognitive distortions and replace them with more rational alternative thoughts.

Table 8.6 shows some common examples of automatic thoughts, the types of cognitive distortions they represent, and some rational alternative responses.

TABLE 8.6 Cognitive Distortions and Rational Responses

Automatic Thought	Kind of Cognitive Distortion	Rational Response
I'm all alone in the world.	All-or-Nothing Thinking	It may feel like I'm all alone, but there are some people who care about me.
Nothing will ever work out for me.	Overgeneralization	No one can look into the future. Concentrate on the present.
My looks are hopeless.	Magnification	I may not be perfect looking, but I'm far from hopeless.
I'm falling apart. I can't handle this.	Magnification	Sometimes I just feel overwhelmed. But I've handled things like this before. Just take it a step at a time and I'll be okay.
I guess I'm just a born loser.	Labeling and Mislabeling	Nobody is destined to be a loser. Stop talking yourself down.
I've only lost 8 pounds on this diet. I should just forget it. I can't succeed.	Negative Focusing/Minimization/Disqualifying the Positive/Jumping to Conclusions/All-or-Nothing Thinking	Eight pounds is a good start. I didn't gain all this weight overnight, and I have to expect that it will take time to lose it.
I know things must really be bad for me to feel this awful.	Emotional Reasoning	Feeling something doesn't make it so. If I'm not seeing things clearly, my emotions will be distorted too.
I know I'm going to flunk this course.	Fortune Teller Error	Give me a break! Just focus on getting through this course, not on jumping to negative conclusions.
I know John's problems are really my fault.	Personalization	Stop blaming yourself for everyone else's problems. There are many reasons why John's problems have nothing to do with me.
Someone my age should be doing better than I am.	Should Statements	Stop comparing yourself to others. All anyone can be expected to do is their best. What good does it do to compare myself to others? It only leads me to get down on myself rather than get motivated.
I just don't have the brains for college.	Labeling and Mislabeling	Stop calling yourself names like "stupid." I can accomplish a lot more than I give myself credit for.
Everything is my fault.	Personalization	There you go again. Stop playing this game of pointing blame at yourself. There's enough blame to go around. Better yet, forget placing blame and try to think through how to solve this problem.
It would be awful if Sue turns me down.	Magnification	It might be upsetting, but it needn't be awful unless I make it so.
If people really knew me, they would hate me.	Mind Reader	What evidence is there for that? More people who get to know me like me than don't like me.
If something doesn't get better soon, I'll go crazy.	Jumping to Conclusions/Magnification	I've dealt with these problems this long without falling apart. I just have to hang in there. Things are not as bad as they seem.
I can't believe I have another pimple on my face. This is going to ruin my whole weekend.	Mental Filter	Take it easy. A pimple is not the end of the world. It doesn't have to spoil my whole weekend. Other people get pimples and seem to have a good time.

Biological Approaches

The most common biological approaches to treating mood disorders involve the use of antidepressant drugs and electroconvulsive therapy for depression and lithium carbonate for bipolar disorder.

Antidepressant Drugs Drugs used to treat depression include several classes of antidepressants: tricyclic antidepressants (TCAs), monoamine oxidase (MAO) inhibitors, and selective serotonin-reuptake inhibitors (SSRIs). All of these drugs increase brain levels and, perhaps, the actions of neurotransmitters. The increased availability of key neurotransmitters in the synaptic cleft may alter the sensitivity of postsynaptic neurons to these chemical messengers. Yet no one is exactly sure how antidepressants work in relieving depression (Januzzi & DeSanctis, 1999; Kupfer, 1999). Antidepressants tend to have a delayed effect, typically requiring several weeks of treatment before a therapeutic benefit is achieved. SSRIs not only lift mood, but in many cases also eliminate delusions that may accompany severe depression (Zanardi et al., 1996). Antidepressant medication is clearly effective in helping relieve major depression in many cases (Kupfer, 1999; Thase et al., 2000; USDHHS, 1999a). Antidepressants are also helpful in treating dysthymia (Hellerstein et al., 2000).

The different classes of antidepressants increase the availability of neurotransmitters, but in different ways (see Figure 8.4). The tricyclics, which include *imipramine* (trade name Tofranil), *amitriptyline* (Elavil), *desipramine* (Norpramin), and *doxepin* (Sinequan), are so named because of their three-ringed molecular structure. They increase levels in the brain of the neurotransmitters norepinephrine and serotonin by interfering with the reuptake (reabsorption by the transmitting cell) of these chemical messengers. The SSRIs (*fluoxextine*, trade name Prozac, is one) work in a similar fashion but have more specific effects on raising the levels of serotonin in the brain. The MAO inhibitors increase the availability of neurotransmitters by inhibiting the action of monoamine oxidase, an enzyme that normally breaks down or degrades neurotransmitters in the synaptic cleft. MAO inhibitors are used less widely than other antidepressants because of potentially serious interactions with certain foods and alcoholic beverages.

Though antidepressants increase the availability of neurotransmitters at the synaptic level in the brain, we still have much to learn about how they work to relieve depression. Most probably, the antidepressant effects of these drugs involve multiple therapeutic actions on more than one neurotransmitter system (Feighner, 1999; USDHHS, 1999a).

The potential side effects of tricyclics and MAO inhibitors include dry mouth, psychomotor retardation, constipation, blurred vision, and, less frequently, urinary retention, paralytic ileus (a paralysis of the intestines, which impairs the passage of intestinal contents), confusion, delirium, and cardiovascular complications, such as reduced blood

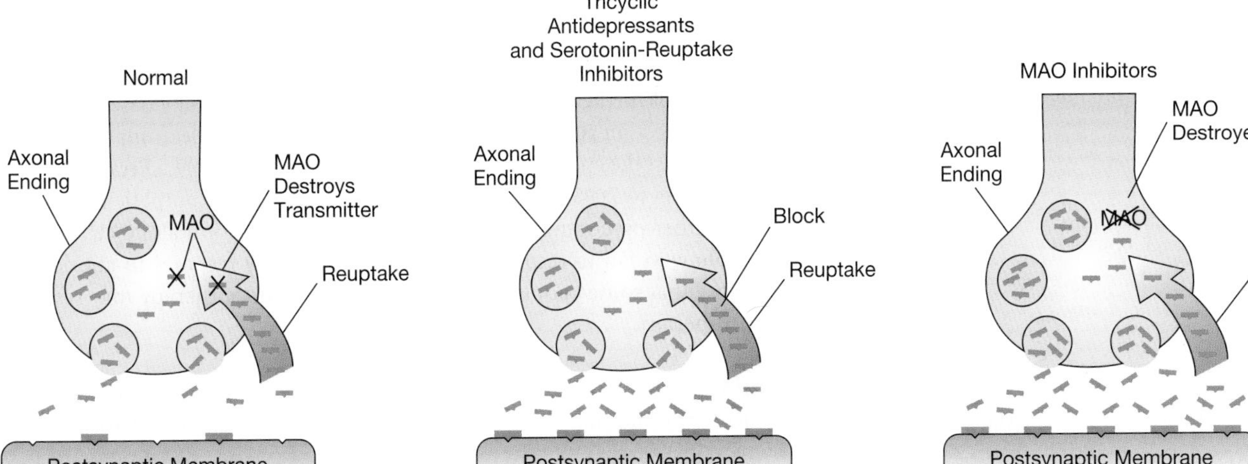

FIGURE 8.4 The actions of various types of antidepressants at the synapse.
Tricyclic antidepressants and serotonin-reuptake inhibitors both increase the availability of neurotransmitters by preventing their reuptake by the presynaptic neuron. Tricyclic antidepressants impede the reuptake of both norepinephrine and serotonin. MAO inhibitors work by inhibiting the action of monoamine oxidase, an enzyme that normally breaks down neurotransmitters in the synaptic cleft.

A Closer Look

St. John's Wort—A Natural "Prozac"?

Might a humble herb be a remedy for depression? The herb, called St. John's wort, or *Hypericum perforatum*, has been used for centuries to help heal wounds. Now, people are using it to relieve depression (Andrews, 1997). Nowhere is it more popular than Germany, where high-strength versions of the herb became the most widely used antidepressant on the market, outselling Prozac, its nearest competitor, by a margin of 4 to 1. Early small-scale studies in Europe provided preliminary support for the benefits of St. John's wort, with few reported side effects, in treating mild to moderate depression (Carey, 1998). The herb appears to increase the levels of serotonin in the brain by interfering with its reabsorption, the same mechanism believed to account for Prozac's benefits. Although people seeking help for depression may be attracted to the idea of using a natural product such as St. John's wort, more definitive studies are needed to establish its safety and effectiveness (Siegel, 2001). Hopes were lowered by the results of a 2001 study showing that St. John's wort worked no better than a placebo in treating major depression (Shelton et al., 2001; "St. John's Wort," 2001a). Whether or not it might help alleviate less severe forms of depression remains undetermined ("St. John's Wort," 2001b).

Truth OR Fiction? REVISITED

The most widely used remedy for depression in Germany is not a drug, but an herb.

TRUE. The most widely used remedy for depression in Germany is an herb, St. John's wort. Clinical trials are underway to evaluate its effectiveness.

pressure. Tricyclics are also highly toxic, which raises the prospect of suicidal overdoses if the drugs are used without close supervision.

The SSRIs such as Prozac and Zoloft are about equal in effectiveness to the older generation of tricyclics (Kupfer, 1999; McGrath et al., 2000; Noonan, 2000). Yet because they hold two major advantages, they have largely replaced the earlier generation of TCAs. The first advantage is that they are less toxic and so are less dangerous in overdose. Secondly, they have fewer of the common side effects (such as dry mouth, constipation, and weight gain) associated with the tricyclics and MAO inhibitors. Still, Prozac and other SSRIs may produce side effects such as upset stomach, headaches, agitation, insomnia, lack of sexual drive, and delayed orgasm ("Antidepressants Linked," 2000; Michelson et al., 2000).

One issue we need to address in discussing drug therapy is the high rate of relapse following discontinuation of medication (Kocsis et al., 1996). Psychologically based therapies may provide greater protection against relapse, presumably because the learning that occurs during therapy carries past the end of active treatment (Butler & Beck, 1995; Persaud, 2000).

Overall, about 50% to 70% of depressed patients treated on an outpatient basis respond favorably to either psychotherapy or antidepressant medication alone (USDHHS, 1999a). Some people who fail to respond to psychotherapy may respond to antidepressants. Recognize, too, that some people who fail to respond to drug therapy may respond favorably to psychotherapy.

Truth OR Fiction? REVISITED

The ancient Greeks and Romans used a chemical to curb turbulent mood swings that is still used today.

TRUE. The ancient Greeks and Romans did use a chemical substance to control mood swings that is still widely used today. It is called lithium.

Drug Treatments for Bipolar Disorder The drug lithium carbonate, a powdered form of the metallic element lithium, is the most widely used and recommended treatment for bipolar disorder. It could be said that the ancient Greeks and Romans were among the first to use lithium as a form of chemotherapy. They prescribed mineral water that contained lithium for people with turbulent mood swings.

Lithium is effective in stabilizing moods in people with bipolar disorder and reducing recurrent episodes of mania and depression (Baldessarini & Tondo, 2000; Grof & Alda, 2000). Yet lithium is generally more effective in treating manic than depressive symptoms (Sachs et al., 1994). People with bipolar disorder may need to use lithium indefinitely to control their mood swings, just as diabetics use insulin continuously to control their illness. Lithium is given orally in the form of a natural mineral salt, lithium carbonate. Despite more than 40 years of use as a therapeutic drug, we still can't say how lithium works (Price & Heninger, 1994).

Yet lithium treatment is no panacea. At least 30% to 40% of patients with mania either fail to respond to the drug or cannot tolerate it (Dubovsky, 2000; Duffy et al., 1998). Among responders, about 6 in 10 eventually relapse (Goleman, 1994c).

Lithium treatment must be closely monitored because of potential toxic effects and other side effects. Lithium can produce a mild impairment in memory, "the kind of thing that might make a productive person stop taking it," as one expert put it (Goleman, 1994c). The drug can also lead to weight gain, lethargy, and grogginess, and to a general slowing down of motor functioning. It can also produce gastrointestinal distress and lead to liver problems over the long term. For a number of reasons, many patients discontinue using lithium or fail to take it reliably (Johnson & McFarland, 1996).

Anticonvulsant drugs used in the treatment of epilepsy, such as *carbamazepine* (brand name Tegretol) and *divalproex* (brand name Depakote), are also used to stabilize moods and relieve manic symptoms in people with bipolar disorder (Alao & Dewan, 2001; Baldessarini, Tohen, & Tondo, 2000; Bowden et al., 2000). Anticonvulsant drugs may be of benefit in treating people with bipolar disorder who either do not respond to lithium or cannot tolerate the drug because of side effects. Anticonvulsant drugs usually cause fewer or less severe side effects than lithium. However, some patients have only a partial response to lithium or anticonvulsant drugs and some fail to respond at all. Thus there remains the need for alternative treatments or drug strategies to be developed, perhaps involving a combination of these or other drugs (Bowden et al., 2000; Tohen et al., 2000).

Electroconvulsive Therapy **Electroconvulsive therapy (ECT),** more commonly called *shock therapy,* continues to evoke controversy. The idea of passing an electric current through someone's brain may seem barbaric. Yet ECT is a generally safe and effective treatment for severe depression, and it can help relieve depression in many cases in which alternative treatments have failed.

WWW **Web Link 8.4**
Task Force Report on ECT

ECT involves the administration of an electrical current to the head. A current of between 70 to 130 volts is used to induce a convulsion that is similar to a grand mal epileptic seizure. ECT is usually administered in a series of 6 to 12 treatments distributed in a series of three per week, over a period of several weeks (USDHHS, 1999a). The patient is put to sleep with a brief-acting general anesthetic and given a muscle relaxant to avoid wild convulsions that might result in injury. As a result, spasms may be barely perceptible to onlookers. The patient awakens soon after the procedure and generally remembers nothing. Although ECT had earlier been used in the treatment of a wide variety of psychological disorders, including schizophrenia and bipolar disorder, the American Psychiatric Association recommended in 1990 that ECT be used only to treat major depressive disorder in people who do not respond to antidepressant medication.

ECT leads to significant improvement in about 50% to 60% of people with major depression who have failed to respond to antidepressant medication (Prudic et al., 1996; Sackeim, Prudic, & Devanand, 1990). ECT also results in shorter and less costly hospitalizations for major depression (Olfson et al., 1998). Although it often produces dramatic relief of symptoms of depression, no one knows exactly how ECT works. ECT produces such mammoth chemical and electrical changes in the body that it is difficult to pinpoint the mechanism of therapeutic action. One possibility is that ECT works by normalizing brain levels of certain neurotransmitters. Although ECT can be an effective short-term treatment of severe depression, it too is no panacea. Depression often returns at some later point, even among people who continue to be treated with antidepressant medication (Sackeim et al., 1994).

ECT may be administered to either both sides of the head (*bilateral ECT*) or to only one side of the head (*unilateral ECT*). Unilateral ECT is applied to the nondominant hemisphere of the brain, which, for most people, is the right side. High-dosage

electroconvulsive therapy (ECT) A method of treating severe depression by means of administering an electric current to the head.

unilateral ECT appears to be as effective as bilateral ECT, but produces less severe and persistent memory impairment (Lisanby et al., 2000; Sackeim et al., 2000; Sackeim & Vaughn McCall, 2001).

There is an understandable concern among patients, relatives, and professionals themselves about the possible risks of ECT, especially concerning memory loss for events occurring around the time of treatment (Weiner, 2000). As noted in Chapter 4, another nagging problem with ECT is a high rate of relapse following treatment (Sackeim, 2001; Sackeim et al., 2001). All in all, many professionals view ECT as a treatment of last resort, to be considered only after other treatment approaches have been tried and failed.

THINK ABOUT IT

If you were to become clinically depressed, which course of treatment would you prefer—medication, psychotherapy, or a combination? Explain.

Clinical Practice Guidelines for Depression

In summing up, let us note the recommendations of a government-sponsored expert panel for the treatment of depression. The guidelines were based on evidence from controlled studies showing the following treatments to be effective in treating depression (Depression Guideline Panel, 1993b):

- Antidepressant medication (tricyclics or selective serotonin-reuptake inhibitors)
- Three specific forms of psychotherapy: cognitive therapy, behavior therapy, and interpersonal psychotherapy
- A combination of one of the recommended forms of psychotherapy and antidepressant medication
- Other specified forms of treatment, including ECT and phototherapy for seasonal depression

Quiz 8.3
Treatment of Mood Disorders

Suicide

Suicidal thoughts are common enough. Under great stress, many, if not most, people have considered suicide. A recent nationally representative survey found that 13% of U.S. adults reported having experienced suicidal thoughts and 4.6% reported making a suicide attempt (Kessler, Borges, & Walters, 1999). More than half (54%) of one sample of 694 first-year college students reported they had contemplated suicide on at least one occasion (Meehan et al., 1991). In a large sample of adolescents in Oregon, nearly 1 in 5 (19%) reported having suicidal thoughts at some point in their lives (Lewinsohn, Rohde, & Seeley, 1996). It is fortunate that most people who have suicidal thoughts do not act on them. Among the first-year college students who had considered suicide, fewer than 1 in 5 had attempted suicide (Meehan et al., 1991). Still, each year in the United States some 500,000 people are treated in hospital emergency rooms for attempted suicide and more than 30,000 "succeed" in taking their lives (Foxhall, 2001a; National Strategy for Suicide Prevention, 2001). Suicide exacts a heavy toll on the nation, as you can see in statistics reported in a recent report from the U.S. Surgeon General (see Table 8.7).

Suicidal behavior is not a psychological disorder in itself. But it is often a feature or symptom of an underlying psychological disorder, usually a mood disorder, which is the reason we discuss it in this chapter. The federal government estimates that about 60% of people who commit suicide have suffered from a mood disorder (National Strategy for Suicide Prevention, 2001).

Who Commits Suicide?

Suicide is the third leading cause of death among 15 to 24 year olds in the United States, following unintentional injuries and homicide. The suicide rate among adolescents and younger adults nearly tripled in the period 1952 to 1995 (Centers for Disease Control, 2001c). Yet suicide rates increase with age and are highest among adults age 65 and older,

TABLE 8.7 U.S. Surgeon General's Report on Suicide: Cost to the Nation

- Every 17 minutes another life is lost to suicide. Every day, 86 Americans take their own life and over 1,500 attempt suicide.
- Suicide is now the eighth leading cause of death in Americans.
- For every two victims of homicide in the United States, there are three deaths from suicide.
- There are now twice as many deaths due to suicide than due to HIV/AIDS.
- Between 1952 and 1995, the incidence of suicide among adolescents and young adults nearly tripled.
- In the month prior to their suicide, 75% of elderly persons had visited a physician.
- Over half of all suicides occur in adult men, ages 25 to 65.
- Many who make suicide attempts never seek professional care immediately after the attempt.
- Males are four times more likely to die from suicide than are females.
- More teenagers and young adults die from suicide than from cancer, heart disease, AIDS, birth defects, stroke, pneumonia and influenza, and chronic lung disease, combined.
- Suicide takes the lives of more than 30,000 Americans every year.

Source. Center for Mental Health Services, 2001.

especially older White males (USDHHS, 1999a; National Strategy for Suicide Prevention, 2001; Pearson & Brown, 2000; see Figure 8.5).

Despite life-extending advances in medical care, some older adults may find the quality of their lives is less than satisfactory. With longer life, older people are more susceptible to diseases such as cancer and Alzheimer's, which can leave them with feelings of helplessness and hopelessness that, in turn, can give rise to suicidal thinking. Many older adults also suffer a mounting accumulation of losses of friends and loved ones as time progresses, leading to social isolation. These losses, as well as the loss of good health and of a responsible role in the community, may wear down the will to live. Not surprisingly, the highest suicide rates in older men are among those who are widowed or lead socially isolated lives. Society's increased acceptance of suicide in older people may also play a part. Whatever the causes, suicide has become an increased risk for elderly people. Perhaps society should focus its attention as much on the quality of life that is afforded our elderly and not simply on providing them the medical care that helps make longer life possible.

More women attempt suicide, but more men "succeed" (National Strategy for Suicide Prevention, 2001; USDHHS, 1999a). More males succeed, in large part, because they tend to choose quicker acting and more lethal means, such as handguns. Gender differences in suicide risk may mask the underlying factors. The common finding that men are more likely to take their own lives may be due to the fact that men are also more likely to have a history of alcohol and drug abuse and less likely to have children in the home. When these two factors were taken into account in a recent study, gender differences in suicide risk disappeared (Young et al., 1994).

Overall, Whites are about twice as likely as Blacks to commit suicide. Suicide rates among adolescents are highest among White males, but they are rising at a faster rate among Black males (CDC, 2001c). Native Americans are at much greater than average risk

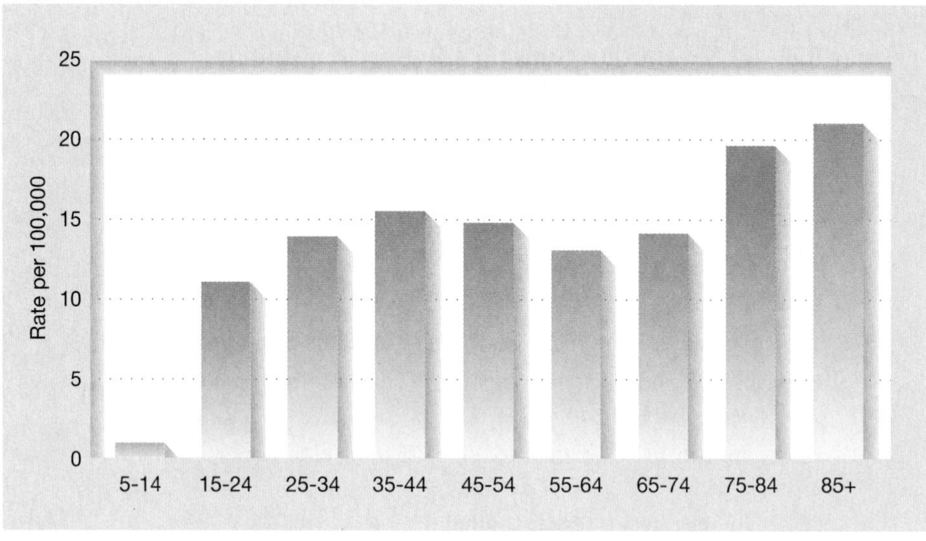

FIGURE 8.5
Suicide rates according to age.
Although adolescent suicides may be more highly publicized, adults, especially older adults, have higher suicide rates.

Source. Murphy, S. L. (2000). Deaths: Final data for 1998. *National vital statistics report,* 48 (11). Hyattsville, MD: National Center for Health Statistics. DHHS Publication No. (PHS) 20000-1120.

of suicide attempts and completed suicides. Overall, Native Americans (American Indians and Alaskan Natives) have a suicide rate that is 50% higher than other groups (Stinson, 2001). As noted in Chapter 2, male Native American adolescents and young adults have the highest suicide rates in the nation (USDHHS, 1999a).

Hopelessness and exposure to others who have attempted or completed suicide may contribute to the increased risk of suicide among Native American youth. Native American youth at greatest risk tend to be reared in communities that are largely isolated from the benefits of U.S. society at large. They perceive themselves as having relatively few opportunities to gain the skills necessary to join the work force in the larger society and are also relatively more prone to substance abuse, including alcohol abuse. Knowledge that peers have attempted or completed suicide renders suicide a highly visible escape from psychological pain.

Why Do People Commit Suicide?

To many lay observers, suicide seems so extreme an act that they believe only "insane" people (meaning people who are out of touch with reality) would commit suicide. However, suicidal thinking does not necessarily imply loss of touch with reality, deep-seated unconscious conflict, or a personality disorder. Having thoughts about suicide generally reflects a narrowing of the range of options people think are available to them to deal with their problems (Rotheram-Borus et al., 1990). That is, they are discouraged by their problems and see no other way out.

The risk of suicide is much greater among people with severe mood disorders, such as major depression and bipolar disorder (Bostwick & Pankratz, 2000). Major depression accounts for about 20% to 35% of suicide deaths in the United States (Angst, Angst, & Stassen, 1999). As many as one in five people with bipolar disorder eventually commit suicide (Cowan & Kandel, 2001). Experts believe that greater efforts toward diagnosing and treating mood disorders may result in lower suicide rates (Isacsson, 2000). Attempted or completed suicide is also linked to other psychological disorders, such as alcoholism and drug dependence, schizophrenia, panic disorder, antisocial personality disorder, posttraumatic stress disorders, borderline personality disorder, and a family history of suicide (e.g.,

Heikkinen et al., 1997; Hufford, 2001; Kotler et al., 2001; Roy, 2000). Suicide is also the leading cause of premature death among people with schizophrenia (Fenton et al., 1997). More than half the suicide attempters in a recent study had two or more psychological disorders (Beautrais et al., 1996).

Not all suicides are connected with psychological disorders. Some people suffering from painful and hopeless physical illness seek to escape further suffering by taking their own lives. These suicides are sometimes labeled "rational suicides" in the belief that they are based on a rational decision that life is no longer worth living in the light of continual suffering. However, in perhaps many of these cases the person's judgment and reasoning ability may be colored by an underlying and potentially treatable psychological disorder, such as depression. Other suicides are motivated by deep-seated religious or political convictions, as in the case of people who sacrifice themselves in acts of protest against their governments. A yet more horrific example is that of terrorists who kill others as well as themselves in the belief that their acts will be rewarded in an afterlife.

Suicide attempts often occur in response to highly stressful life events, especially "exit events" such as the death of a spouse, close friend, or relative; divorce or separation; a family member's leaving home; or the loss of a close friend. People who consider suicide in times of stress may lack problem-solving skills and be less able to find alternative ways of coping with the stressors they face. Underscoring the psychological impact of severe stress, researchers find suicides to be more common among survivors of natural disasters, especially severe floods (Krug et al., 1998).

Theoretical Perspectives on Suicide

The classic psychodynamic model views depression as the turning inward of anger against the internal representation of a lost love object. Suicide then represents inward-directed anger that turns murderous. Suicidal people, then, do not seek to destroy themselves. Instead, they seek to vent their rage against the internalized representation of the love object. In so doing, they destroy themselves as well, of course. In his later writings, Freud speculated that suicide may be motivated by the "death instinct," a tendency to return to the tension-free state that preceded birth. Existential and humanistic theorists relate suicide to the perception that life is meaningless and hopeless. Suicidal people report they find life duller, emptier, and more boring than nonsuicidal people (Mehrabian & Weinstein, 1985).

In the nineteenth century, social thinker Emile Durkheim (1958) noted that people who experienced **anomie**—who feel lost, without identity, rootless—are more likely to commit suicide. Sociocultural theorists likewise believe that alienation in today's society may play a role in suicide. In our modern, mobile society, people frequently move hundreds or thousands of miles to schools and jobs. Executives and their families may be relocated every 2 years or so. Military personnel and their families may be shifted about yet more rapidly. Many people are thereby socially isolated or cut off from their support groups. Moreover, city dwellers tend to limit or discourage informal social contacts because of crowding, overstimulation, and fear of crime. It is thus understandable that many people find few sources of support in times of crisis. In some cases, the availability of family support may not be helpful. Family members may be perceived as part of the problem, not part of the solution.

Learning theorists focus largely on the lack of problem-solving skills for handling significant life stress. According to Shneidman (1985), suicide attempters wish to escape unbearable psychological pain and may perceive no other way out. People who threaten or attempt suicide may also receive sympathy and support from loved ones and others, perhaps making future—and more lethal—attempts more likely. This is not to suggest that suicide attempts or gestures should be ignored. It is not the case that people who threaten suicide are merely seeking attention. Although those who have threatened suicide may not carry out the act, their threats should be taken seriously. People who commit suicide often tell others of their intentions or leave clues beforehand. Moreover, many people make

Truth OR Fiction? REVISITED

People who threaten suicide are basically attention seekers.

FALSE. Although people who threaten suicide may not carry out the act, their threats should be taken seriously. Most people who do commit suicide had told others of their intentions or had left clues.

anomie A feeling of rootlessness.

aborted suicide attempts in which they stop just before inflicting harm on themselves before they go on to make actual suicide attempts (Barber et al., 1998).

Social-cognitive theorists suggest that suicide may be motivated by positive expectancies and by approving attitudes toward the legitimacy of suicide (D. Stein et al., 1998). People who kill themselves may expect that they will be missed or eulogized after death, or that survivors will feel guilty for mistreating them. Suicidal psychiatric patients hold more positive expectancies concerning suicide than do nonsuicidal psychiatric samples. They more often expressed the belief that suicide would solve their problems, for example (Linehan et al., 1987). Suicide may represent a desperate attempt to deal with one's problems in one fell swoop rather than piecemeal.

Social-cognitive theorists also focus on the potential modeling effects of observing suicidal behavior in others, especially among teenagers who feel overwhelmed by academic and social stressors. A *social contagion*, or spreading of suicide in a community, may occur in the wake of suicides that receive widespread publicity. Teenagers, who seem to be especially vulnerable to these modeling effects, may even romanticize the suicidal act as one of heroic courage. The incidence of suicide among teenagers sometimes rises markedly in the period following news reports about suicide (Kessler et al., 1990). In the Oregon study, suicidal behavior of a friend was a risk factor in suicide attempts among adolescents (Lewinsohn, Rohde, & Seeley, 1996). Copycat suicides may be more likely to occur when reports of suicides are sensationalized such that other teenagers expect their demises to have broad impacts on their communities (Kessler et al., 1990).

Biological factors are also implicated in suicide. Reduced serotonin activity is found in many people who attempt or commit suicide (Ghanshyam et al., 1995; Mann & Malone, 1997). Since reduced availability of serotonin is linked to depression, the relationship with suicide is not surprising. Yet serotonin acts to curb or inhibit nervous system activity, so perhaps decreased serotonin activity leads to a *disinhibition* or release of impulsive behavior that takes the form of a suicidal act in vulnerable individuals. Suicide also tends to run in families, which hints of genetic factors. Evidence from a recent twin study showed that among nine twin pairs in which both twins committed suicide, seven were MZ twins and two were DZ twins (Roy et al., 1991). All in all, about one suicide attempter in four has a family member who has committed suicide (Sorensen & Rutter, 1991).

The presence of psychological disorders among other family members may be connected with suicide (Sorensen & Rutter, 1991). But what are the causal connections? Do people who attempt suicide inherit vulnerabilities to disorders that are connected with suicide? Does the family atmosphere subject its members to feelings of hopelessness? Does the suicide of one family member give others the idea of doing the same thing? Does one suicide create the impression that other family members are destined to kill themselves? These are all questions researchers need to address.

THINK ABOUT IT

What factors are related to suicide and suicide prevention? Did your reading of the text change your ideas about how you might deal with a suicidal threat by a friend or loved one? If so, how?

Suicide is connected with a complex web of factors, and its prediction is no simpler. Yet it is clear that many suicides could be prevented if people with suicidal feelings would receive treatment for disorders underlying suicidal behavior, including depression, schizophrenia, and alcohol and substance abuse ("Many Suicides Could Be Prevented," 1998). We also need strategies that emphasize the maintenance of hope during times of severe stress (Malone et al., 2000).

Predicting Suicide

"I don't believe it. I just saw him last week and he looked fine."
"She sat here just the other day, laughing with the rest of us. How were we to know what was going on inside her?"
"I knew he was depressed, but I never thought he'd do something like this. I didn't have a clue."
"Why didn't she just call me?"

A Closer Look

Suicide Prevention

Imagine yourself having an intimate conversation with a close campus friend, Chris. You know that things have not been good. Chris's grandfather died 6 weeks ago, and the two were very close. Chris's grades have been going downhill, and Chris's romantic relationship also seems to be coming apart at the seams. Still, you are unprepared when Chris says very deliberately, "I just can't take it anymore. Life is just too painful. I don't feel like I want to live anymore. I've decided that the only thing I can do is to kill myself."

When somebody discloses that he or she is contemplating suicide, you may feel bewildered and frightened, as if a great burden has been placed on your shoulders. It has. If someone confides suicidal thoughts to you, your goal should be to persuade him or her to see a professional, or to get the advice of a professional yourself as soon as you can. But if the suicidal person declines to talk to another person and you sense you can't break away for such a conference, there are some things you can do then and there:

Teen suicide. Suicidal teenagers may see no other way of handling their life problems. The availability of counseling and support services may help prevent suicide by assisting troubled teens in learning alternate ways of reducing stress and resolving conflicts with others.

1. *Draw the person out.* Shneidman advises framing questions like, "What's going on?" "Where do you hurt?" "What would you like to see happen?" (1985, p. 11). Such questions may prompt people to verbalize thwarted psychological needs and offer some relief. They also grant you the time to appraise the risk and contemplate your next move.
2. *Be sympathetic.* Show that you fathom how troubled the person is. Don't say something like, "You're just being silly. You don't really mean it."
3. *Suggest that means other than suicide can be discovered to work out the person's problems,* even if they are not apparent at the time. Shneidman (1985) notes that suicidal people can usually see only two solutions to their predicaments—either suicide or some kind of magical resolution. Professionals try to broaden the available alternatives of people who are suicidal.
4. *Inquire as to how the person expects to commit suicide.* People with explicit methods who also possess the means (for example, a gun or drugs) are at greater risk. Ask if you may hold on to the gun, drugs, or whatever, for a while. Sometimes the person agrees.
5. *Propose that the person accompany you to consult a professional right now.* Many campuses have hot lines that you or the suicidal individual can call. Many towns and cities have such hot lines and they can be called anonymously. Other possibilities include the emergency room of a general hospital, a campus health center or counseling center, or the campus or local police. If you are unable to maintain contact with the suicidal person, get professional assistance as soon as you separate.
6. *Don't say something like "You're talking crazy."* Such comments are degrading and injurious to the individual's self-esteem. Don't press the suicidal person to contact specific people, such as parents or a spouse. Conflict with them may have given rise to the suicidal thoughts.

Above all, keep in mind that your primary goal is to confer with a helping professional. Don't go it alone any longer than you have to.

Friends and family members often respond to news of a suicide with disbelief or guilt that they failed to pick up signs of the impending act. Yet even trained professionals find it difficult to predict who is likely to commit suicide.

Web Link 8.5
Surgeon General's Call to Action to Prevent Suicide

Evidence points to the pivotal role of hopelessness in predicting suicidal thinking and suicide attempts (Brown et al., 2000; Malone et al., 2000). In one study, psychiatric

Quiz 8.4
Suicide

Quiz 8.5
Chapter Exam

Research Update
Chapter 8

outpatients with hopelessness scores above a certain cutoff were 11 times more likely to commit suicide than those with scores below the cutoff (Beck et al., 1990). But *when* does hopelessness lead to suicide?

People who commit suicide tend to signal their intentions, often quite explicitly, such as by telling others about their suicidal thoughts (Denneby et al., 1996). Some attempt to cloak their intentions. Behavioral clues may still reveal suicidal intent, however. Edwin Shneidman, a leading researcher on suicide, found that 90% of the people who committed suicide had left clear clues, such as disposing of their possessions (Gelman, 1994). People contemplating suicide may also suddenly try to sort out their affairs, as in drafting a will or buying a cemetery plot. They may purchase guns despite lack of prior interest in firearms. When troubled people decide to commit suicide, they may seem to be suddenly at peace; they feel relieved of having to contend with life problems. This sudden calm may be misinterpreted as a sign of hope.

The prediction of suicide is not an exact science, even for experienced professionals. Many observable factors, such as hopelessness, do seem to be connected with suicide, but we cannot predict *when* a hopeless person will attempt suicide, if at all.

Overview of Mood Disorders

TYPES OF MOOD DISORDERS

	Description	Features
Major Depressive Disorder	Episodes of severe depression	• A range of features may be present, from downcast mood to appetite and sleep disturbance, to lack of interest and motivation • Seasonal affective disorder and postpartum depression are subtypes of major depression
Dysthymic Disorder	Long-standing mild depression	• Feeling "down in the dumps" most of the time, but not as severely depressed as people with major depressive disorder • Double depression is characterized by major depressive episodes occurring during the course of dysthymia
Bipolar Disorder	Mood swings between elation and depression	• The two general subtypes are bipolar I disorder and bipolar II disorder • In rapid cycling, mania and major depression alternate without intervening periods of normal mood
Cyclothymia	Milder mood swings than bipolar disorder	• Chronic, cyclical pattern of shifting mood states from hypomanic episodes to states of mild depression • Frequent periods of depressed mood or loss of interest or pleasure in activities, but not at the level of severity of a major depressive episode

CAUSAL FACTORS Multiple causes are involved, interacting with each other in complex ways

Biological Factors	• Genetic predispositions • Disturbed neurotransmitter functioning • Abnormalities in parts of the brain regulating mood states • Possible endocrine system involvement in mood states
Social-Environmental Factors	• Stressful life events, such as the loss of a loved one or prolonged unemployment
Behavioral Factors	• Lack of reinforcement • Negative interactions with others, leading to rejection
Emotional and Cognitive Factors	• In classic psychoanalytic theory, anger turned inward • Emotional difficulties coping with the loss of significant others • Lack of meaning or purpose in life • Negatively biased or distorted ways of thinking, or a depressive attributional style

TREATMENT APPROACHES Treatment may include one or more therapeutic approaches

Biomedical Treatment	• Antidepressant drugs (tricyclics, MAO inhibitors, SSRIs) to control depressive symptoms by influencing the availability of neurotransmitters in the brain • Lithium or anticonvulsant drugs to stabilize moods in bipolar patients • Electroconvulsive therapy (ECT) in severe cases of depression • Phototherapy for seasonal affective disorder
Cognitive-Behavioral Therapy	• To help clients correct distorted ways of thinking, develop more effective coping responses, and increase levels of positive reinforcement
Interpersonal Therapy	• To resolve interpersonal problems and lingering grief reactions

Summing Up

Types of Mood Disorders

What are mood disorders? Mood disorders are disturbances in mood that are unusually prolonged or severe and serious enough to impair daily functioning.

What are the major types of mood disorders? There are various kinds of mood disorders, including depressive (unipolar) disorders, such as major depressive disorder and dysthymic disorder, and disorders involving mood swings, such as bipolar disorder and cyclothymic disorder.

What is major depressive disorder? In major depression, people experience a profound change in mood that impairs their ability to function. There are many associated features of major depressive disorder, including downcast mood; changes in appetite; difficulty sleeping; reduced sense of pleasure in formerly enjoyable activities; feelings of fatigue or loss of energy; sense of worthlessness; excessive or misplaced guilt; difficulties concentrating, thinking clearly, or making decisions; repeated thoughts of death or suicide; attempts at suicide; and even psychotic behaviors (hallucinations and delusions).

What is dysthymic disorder? Dysthymic disorder is a form of chronic depression that is milder than major depressive disorder but may nevertheless be associated with impaired functioning in social and occupational roles.

What is bipolar disorder? In bipolar disorder, people experience fluctuating mood states that interfere with the ability to function. Bipolar I disorder is identified by one or more manic episodes. Bipolar II is characterized by the occurrence of at least one major depressive episode and one hypomanic episode, but without any full-blown manic episodes.

What are the features of a manic episode? Manic episodes are characterized by sudden elevation or expansion of mood and sense of self-importance, feelings of almost boundless energy, hyperactivity, and extreme sociability, which often takes a demanding and overbearing form. People in manic episodes tend to exhibit pressured or rapid speech, rapid "flight of ideas," and decreased need for sleep.

What is cyclothymic disorder? Cyclothymic disorder is a type of bipolar disorder characterized by a chronic pattern of mild mood swings that sometimes progresses to bipolar disorder.

Theoretical Perspectives on Mood Disorders

How is stress related to mood disorders? Exposure to life stress in associated with an increased risk of development and recurrence of mood disorders, especially major depression. Yet some people are more resilient in the face of stress, perhaps because of psychosocial factors such as social support and coping styles.

How do psychodynamic theorists conceptualize mood disorders? In classic psychodynamic theory, depression is viewed in terms of inward-directed anger. People who hold strongly ambivalent feelings toward people they have lost, or whose loss is threatened, may direct unresolved anger toward the inward representations of these people that they have incorporated or introjected within themselves, producing self-loathing and depression. Bipolar disorder is understood within psychodynamic theory in terms of the shifting balances between the ego and superego. More recent psychodynamic models, such as the self-focusing model, incorporate both psychodynamic and cognitive aspects in explaining depression in terms of the continued pursuit of lost love objects or goals that it would be more adaptive to surrender.

How do humanistic theorists view depression? Theorists working within the humanistic framework view depression as reflecting a lack of meaning and authenticity in a person's life.

How do learning theorists view depression? Learning perspectives focus on situational factors in explaining depression, such as changes in the level of reinforcement. When reinforcement is reduced, the person may feel unmotivated and depressed, which can occasion inactivity and further reduce opportunities for reinforcement. Coyne's interactional theory focuses on the negative family interactions that can lead the family members of people with depression to become less reinforcing toward them.

What are two major cognitive models of depression? Beck's cognitive model focuses on the role of negative or distorted thinking in depression. Depression-prone people hold negative beliefs toward themselves, the environment, and the future. This cognitive triad of depression leads to specific errors in thinking, or cognitive distortions, in response to negative events, which, in turn, lead to depression.

The learned helplessness model is based on the belief that people may become depressed when they come to view themselves as helpless to control the reinforcements in their environment or to change their lives for the better. A reformulated version of the theory held that the ways in which a people explain events—their attributions—determine their proneness toward depression in the face of negative events. The combination of internal, global, and stable attributions for negative events renders one most vulnerable to depression.

What role do biological factors play in mood disorders? Genetics appears to play a role in mood disorders, especially in explaining major depressive disorder and bipolar disorder. Imbalances in the neurotransmitter activity in the brain appear to be involved in depression and mania. The diathesis-stress model is used as an explanatory framework to illustrate how biological or psychological diatheses may interact with stress in the development of depression.

Treatment of Mood Disorders

What approaches to treatment are represented by each of the major theoretical perspectives? Psychodynamic treatment of depression has traditionally focused on helping the depressed person uncover and work through ambivalent feelings toward the lost object, thereby lessening the anger directed inward. Modern psychodynamic approaches tend to be more direct and briefer and focus more on developing adaptive means of achieving self-worth and resolving interpersonal conflicts. Learning theory approaches have focused on helping people with depression increase the frequency of reinforcement in their lives through such means as increasing the rates of pleasant activities in which they participate and assisting them in developing more effective social skills to increase their ability to obtain social reinforcements from others. Cognitive therapists focus on helping the person identify and correct distorted or dysfunctional thoughts and learn more adaptive behaviors. Biological approaches have focused on the use of antidepressant drugs and other biological treatments, such as electroconvulsive therapy (ECT). Antidepressant drugs may help normalize neurotransmitter functioning in the brain. Bipolar disorder is commonly treated with lithium.

Suicide

What factors are linked to suicide? Mood disorders are often linked to suicide. Although women are more likely to attempt suicide, more men actually succeed, probably because they select more lethal means. The elderly—not the young—are more likely to commit suicide, and the rate of suicide among the elderly appears to be increasing. People who attempt suicide are often depressed, but they are generally in touch with reality. They may, however, lack effective problem-solving skills and see no other way of dealing with life stress than suicide. A sense of hopelessness also figures prominently in suicides.

What are the major theoretical approaches to understanding suicide? These draw upon the classic psychodynamic model of anger turned inward; Durkeim's theory of social alienation; and learning, social-cognitive, and biologically based perspectives.

Why should you never ignore a person's threat to commit suicide? Although certainly not all people who threaten suicide go on to commit the act, many do. People who commit suicide often signal their intentions, such as by telling others about their suicidal thoughts.

CHAPTER NINE

Personality Disorders

Paul Klee
Beware of Red, 1940

Truth OR Fiction?

- Warning signs of personality disorders may appear in early childhood. (p. 267)
- People with schizoid personalities may have deeper feelings for animals than they do for people. (p. 269)
- People with psychopathic personalities inevitably run afoul of the law. (p. 272)
- Many notable figures in history, from Lawrence of Arabia to Adolf Hitler and even Marilyn Monroe, have been depicted as borderline personalities. (p. 275)
- Some people with dependent personality disorder have so much difficulty making independent decisions that they allow their parents to decide whom they will marry. (p. 281)
- It may be difficult to draw a clear line between normal variations in behavior and personality disorders. (p. 283)
- The diagnosis of some personality disorders may reflect sexist biases. (p. 283)
- Despite a veneer of self-importance, people with narcissistic personalities may harbor deep feelings of insecurity. (p. 285)
- People with antisocial personalities tend to remain unduly calm in the face of impending pain. (p. 290)

All of us have particular styles of behavior and ways of relating to others. Some of us are orderly, others sloppy. Some of us prefer solitary pursuits; others are more social. Some of us are followers; others are leaders. Some of us seem immune to rejection by others, whereas others avoid social initiatives for fear of getting shot down. When behavior patterns become so inflexible or maladaptive that they cause significant personal distress or impair people's social or occupational functioning, they may be diagnosed as personality disorders.

Chapter Outline

Types of Personality Disorders

> In most of us by the age of thirty, the character has set like plaster, and will never soften again.
>
> —William James

Personality disorders are excessively rigid patterns of behavior or ways of relating to others. Their rigidity prevents people from adjusting to external demands; thus the patterns ultimately become self-defeating. The disordered personality traits become evident by adolescence or early adulthood and continue through much of adult life, becoming so deeply ingrained that they are highly resistant to change. The warning signs of personality disorders may be detected during childhood, even in the troubled behavior of preschoolers. Children with psychological disorders or problem behaviors in childhood, such as conduct disorder, depression, anxiety, and immaturity, are at greater than average risk of later developing personality disorders (Bernstein et al., 1996; Kasen et al., 2001). Personality disorders appear to be quite common; a recent community survey of adults in Oslo, Norway, found that 13.4% of community residents showed evidence of one or more personality disorders (Torgersen, Kringlen, & Cramer, 2001).

Despite the self-defeating consequences of their behavior, people with personality disorders do not generally perceive a need to change. Using psychodynamic terms, the *DSM* notes that people with personality disorders tend to perceive their traits as **ego syntonic**—as natural parts of themselves. Consequently, people with personality disorders are more likely to be brought to the attention of mental health professionals by others than to seek services themselves. In contrast, people with anxiety disorders (Chapter 6) or mood disorders (Chapter 8) tend to view their disturbed behaviors as **ego dystonic.** They do not see their behaviors as parts of their self-identities and are thus more likely to seek help to relieve the distress caused by them.

The *DSM* groups clinical syndromes on Axis I and personality disorders on Axis II. Both clinical syndromes and personality disorders may thus be diagnosed in clients whose behavior meets the criteria for both classes of disorders. A person may have an Axis I mood disorder, for example, such as major depression, and also show the more enduring characteristics associated with an Axis II personality disorder.

The *DSM* groups personality disorders into three clusters:

Cluster A: People who are perceived as odd or eccentric. This cluster includes paranoid, schizoid, and schizotypal personality disorders.

Cluster B: People whose behavior is overly dramatic, emotional, or erratic. This grouping consists of antisocial, borderline, histrionic, and narcissistic personality disorders.

Cluster C: People who often appear anxious or fearful. This cluster includes avoidant, dependent, and obsessive-compulsive personality disorders.

Truth OR Fiction? REVISITED

Warning signs of personality disorders may appear in early childhood.

TRUE. Warning signs of personality disorder may be found in problem behaviors observed in young children, even preschoolers.

WWW **Web Link 9.1** Fact Sheet on Personality Disorders

Personality Disorders Characterized by Odd or Eccentric Behavior

This group of personality disorders includes paranoid, schizoid, and schizotypal disorders. People with these disorders often have difficulty relating to others, or they may show little or no interest in developing social relationships.

personality disorders Excessively rigid behavior patterns, or ways of relating to others, that ultimately become self-defeating.

ego syntonic Referring to behaviors or feelings that are perceived as natural parts of the self.

ego dystonic Referring to behaviors or feelings that are perceived to be alien to one's self-identity.

THINK ABOUT IT

Which behaviors or ways of relating to others do you view as intrinsic parts of yourself, or ego syntonic? Do you believe they are changeable? Why or why not?

Paranoid Personality Disorder The defining trait of the **paranoid personality disorder** is pervasive suspiciousness—the tendency to interpret other people's behavior as deliberately threatening or demeaning. People with the disorder are excessively mistrustful of others, and their relationships suffer for it. Though they may be suspicious of coworkers and supervisors, they can generally maintain employment.

The following case illustrates the unwarranted suspicion and reluctance to confide in others that typifies people with paranoid personalities:

A Case of Paranoid Personality Disorder

An 85-year-old retired businessman was interviewed by a social worker to determine the health care needs for himself and his infirm wife. The man had no history of treatment for a mental disorder. He appeared to be in good health and mentally alert. He and his wife had been married for 60 years, and it appeared that his wife was the only person he'd ever really trusted. He had always been suspicious of others. He would not reveal personal information to anyone but his wife, believing that others were out to take advantage of him. He had refused offers of help from other acquaintances because he suspected their motives. When called on the telephone, he would refuse to give out his name until he determined the nature of the caller's business. He'd always involved himself in "useful work" to occupy his time, even during the 20 years of his retirement. He spent a good deal of time monitoring his investments and had altercations with his stockbroker when errors on his monthly statement prompted suspicion that his broker was attempting to cover up fraudulent transactions.

—*Adapted from Spitzer et al., 1994, pp. 211–213*

■

Web Link 9.2 WWW
Specific Types of Personality Disorders

People who have paranoid personalities tend to be overly sensitive to criticism, whether real or imagined. They take offense at the smallest slight. They are readily angered and hold grudges when they think they have been mistreated. They are unlikely to confide in others because they believe that personal information may be used against them. They question the sincerity and trustworthiness of friends and associates. A smile or a glance may be viewed with suspicion. As a result, they have few friends and intimate relationships. When they do form an intimate relationship, they may suspect infidelity, although there is no evidence to back up their suspicions. They tend to remain hypervigilant, as if they must be on the lookout against harm. They deny blame for misdeeds, even when warranted, and are perceived by others as cold, aloof, scheming, devious, and humorless. They tend to be argumentative and may launch repeated lawsuits against those who they believe have mistreated them.

Clinicians need to weigh cultural and sociopolitical factors when arriving at a diagnosis of paranoid personality disorder. They may find members of immigrant or ethnic minority groups, political refugees, or people from other cultures to be guarded or defensive in their behavior. This behavior may reflect unfamiliarity with the language, customs, or rules and regulations of the majority culture; it may also reflect a cultural mistrust arising from a history of neglect or oppression against the individual's cultural or ethnic group. Such behavior should not be confused with paranoid personality disorder.

Although the suspicions of people with paranoid personality disorder are exaggerated and unwarranted, there is an absence of the outright paranoid delusions that characterize the thought patterns of people with paranoid schizophrenia (for example, believing the FBI is out to get them). People who have paranoid personalities are unlikely to seek treatment for themselves; they see others as causing their problems. The reported prevalence of paranoid personality disorder in the general population ranges from 0.5% to 2.5% (APA, 2000). The disorder is diagnosed in people receiving mental health treatment more often in men than women.

paranoid personality disorder A personality disorder characterized by suspiciousness of others' motives, but not to the point of delusion.

Schizoid Personality Disorder Social isolation is the cardinal feature of **schizoid personality disorder.** Often described as a loner or an eccentric, the person with a schizoid personality lacks interest in social relationships. The emotions of persons with schizoid personalities appear shallow or blunted, but not to the degree found in schizophrenia (see Chapter 13). People with this disorder seem rarely, if ever, to experience strong anger, joy, or sadness. They look distant and aloof. Their faces tend to show no emotional expression, and they rarely exchange social smiles or nods. They seem indifferent to criticism or praise and appear to be wrapped up in abstract ideas rather than in thoughts about people. Although they prefer to remain distant from others, they maintain better contact with reality than people with schizophrenia do. The prevalence of the disorder in the general population remains unknown.

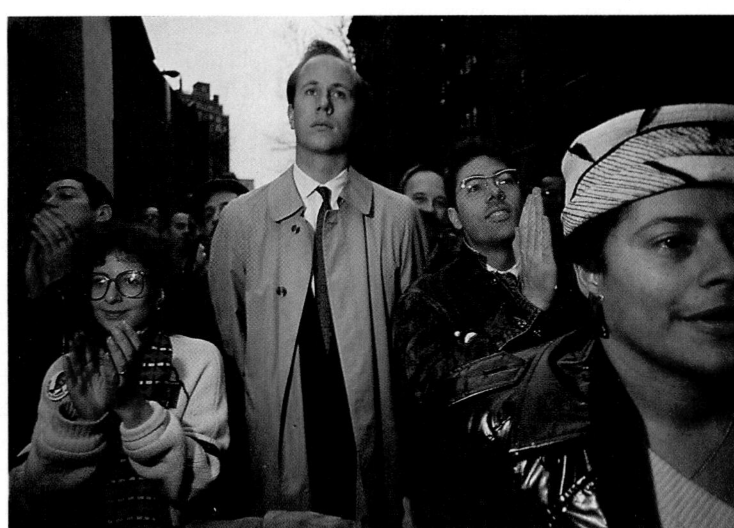

Schizoid personality. It is normal to be reserved about displaying one's feelings, especially when one is among strangers. But people with schizoid personalities rarely express emotions and are distant and aloof. Yet the emotions of people with schizoid personalities are not as shallow or blunted as they are in people with schizophrenia.

The schizoid personality pattern is usually recognized by early adulthood. Men with this disorder rarely date or marry. Women with the disorder are more likely to accept romantic advances passively and marry, but they seldom initiate relationships or develop strong attachments to their partners.

Akhtar (1987) claims that there may be discrepancies between outer appearances and the inner lives of people with schizoid personalities. Although they may appear to have little appetite for sex, for example, they may harbor voyeuristic wishes and become attracted to pornography. Akhtar also suggests that the distance and social aloofness of people with schizoid personalities may be somewhat superficial. They may also harbor exquisite sensitivity, deep curiosities about people, and wishes for love that they cannot express. In some cases, sensitivity is expressed in deep feelings for animals rather than people:

A Case of Schizoid Personality Disorder

John, a 50-year-old retired police officer, sought treatment a few weeks after his dog was hit by a car and died. Since the dog's death, John has felt sad and tired. He has had difficulty concentrating and sleeping. He lives alone and prefers to be by himself, limiting his contacts with others to a passing "Hello" or "How are you?" He feels that social conversation is a waste of time and feels awkward when others try to initiate a friendship. Though he avidly reads newspapers and keeps abreast of current events, he has no real interest in people. He works as a security guard and is described by his coworkers as a "loner" and a "cold fish." The only relationship he had was with his dog, with which he felt he could exchange more sensitive and loving feelings than he could share with people. At Christmas, he would "exchange gifts" with his dog, buying presents for the dog and wrapping a bottle of Scotch for himself as a gift from the animal. The only event that ever saddened him was the loss of his dog. In contrast, the loss of his parents failed to evoke an emotional response. He considers himself to be different from other people and is bewildered by the displays of emotionality that he sees in others.

—*Adapted from Spitzer et al., 1989, pp. 249–250*

■

Truth OR Fiction? REVISITED

People with schizoid personalities may have deeper feelings for animals than they do for people.

TRUE. People with a schizoid personality may show little or no interest in people but develop strong feelings for animals.

Schizotypal Personality Disorder **Schizotypal personality disorder** usually becomes evident by early adulthood. The diagnosis applies to people who have difficulties forming close relationships and whose behavior, mannerisms, and thought patterns are peculiar or odd, but not disturbed enough to merit a diagnosis of schizophrenia. They may be especially anxious in social situations, even when interacting with familiar people. Their social

schizoid personality disorder A personality disorder characterized by persistent lack of interest in social relationships, flattened affect, and social withdrawal.

schizotypal personality disorder A personality disorder characterized by eccentricities of thought and behavior, but without clearly psychotic features.

ideas of reference A form of delusional thinking in which a person reads personal meaning into the behavior of others or external events.

anxieties seem to be associated with paranoid thinking (e.g., fears that others mean them harm) rather than with concerns about being rejected or evaluated negatively by others (APA, 2000).

Schizotypal personality disorder may be slightly more common in males than in females and is believed to affect about 3% of the general population (APA, 2000). Clinicians need to be careful not to label as schizotypal certain behavior patterns that reflect culturally determined beliefs or religious rituals, such as beliefs in voodoo and other magical beliefs.

The eccentricity associated with the schizoid personality is limited to a lack of interest in social relationships. Schizotypal personality disorder refers to a wider range of odd behaviors, beliefs, and perceptions. Persons with the disorder may experience unusual perceptions or illusions, such as feeling the presence of a deceased family member in the room. They realize, however, that the person is not actually there. They may become unduly suspicious of others or paranoid in their thinking. They may develop **ideas of reference,** such as the belief that other people are talking about them. They may engage in "magical thinking," such as believing they possess a "sixth sense" (i.e., can foretell the future) or that others can sense their feelings. They may attach unusual meanings to words. Their own speech may be vague or unusually abstract, but it is not incoherent or filled with the loose associations that characterize schizophrenia. They may appear unkempt, display unusual mannerisms, and engage in unusual behaviors, such as talking to themselves in the presence of others. Their faces may register little emotion. Like people with schizoid personalities, they may fail to exchange smiles with, or nod at, others. Or they may appear silly and smile and laugh at the wrong times. They tend to be socially withdrawn and aloof, with few if any close friends or confidants. They seem to be especially anxious around unfamiliar people. We can see evidence of the social aloofness and illusions that are often associated with schizotypal personality disorder in this case:

A Case of Schizotypal Personality Disorder

Jonathan, a 27-year-old auto mechanic, had few friends and preferred science fiction novels to socializing with other people. He seldom joined in conversations. At times, he seemed to be lost in his thoughts, and his coworkers would have to whistle to get his attention when he was working on a car. He often showed a "queer" expression on his face. Perhaps the most unusual feature of his behavior was his reported intermittent experience of "feeling" his deceased mother standing nearby. These illusions were reassuring to him, and he looked forward to their occurrence. Jonathan realized they were not real. He never tried to reach out to touch the apparition, knowing it would disappear as soon as he drew closer. It was enough, he said, to feel her presence.

—From the Authors' Files

THINK ABOUT IT

Distinguish between schizoid and schizotypal personality disorder by giving examples of some ways in which a person with each disorder might behave.

Despite the *DSM*'s grouping of "schizotypal" behaviors with personality disorders, the schizotypal behavior pattern may fall within a spectrum of schizophrenia-related disorders that also includes paranoid and schizoid personality disorders, as well as schizoaffective disorder (discussed in Chapter 13) and schizophrenia itself. Schizotypal personality disorder may actually share a common genetic basis with schizophrenia (Kendler & Walsh, 1995). Biological relatives of people with schizotypal personality disorder are more likely than relatives of people with non-schizophrenia-related personality disorders (for example, histrionic, borderline, or narcissistic disorders) to be diagnosed as suffering from schizophrenia or a related disorder (Siever et al., 1990).

Let us note, however, that schizotypal personality disorder tends to follow a chronic course, and relatively few people diagnosed with the disorder go on to develop schizophrenia or other psychotic disorders (APA, 2000). Perhaps the emergence of schizophrenia in persons with this shared genetic predisposition is determined by such factors as stressful early family relationships.

Personality Disorders Characterized by Dramatic, Emotional, or Erratic Behavior

This cluster of personality disorders includes the antisocial, borderline, histrionic, and narcissistic types. The behavior patterns of these types are excessive, unpredictable, or self-centered. People with these disorders have difficulty forming and maintaining relationships.

antisocial personality disorder
A personality disorder characterized by antisocial and irresponsible behavior and lack of remorse for misdeeds.

Antisocial Personality Disorder People with **antisocial personality disorder** persistently violate the rights of others and often break the law. They disregard social norms and conventions, are impulsive, and fail to live up to interpersonal and vocational commitments. Yet they often show a superficial charm and are at least average in intelligence (Cleckley, 1976). Perhaps the features that are most striking about them are their low levels of anxiety in threatening situations and lack of guilt or remorse following wrongdoing. Punishment seems to have little if any effect on their behavior. Although parents and others have usually punished them for their misdeeds, they persist in leading irresponsible and impulsive lives.

Antisocial personality. Serial killer Ted Bundy, shown here shortly before his execution, killed without feeling or remorse but also displayed some of the superficial charm seen in some people with antisocial personality disorder.

Although women are more likely than men to develop anxiety and depressive disorders, men are more likely than women to receive diagnoses of antisocial personality disorder (Robins, Locke, & Reiger, 1991). The prevalence rates for the disorder in community samples range from about 3% to 6% in men and about 1% in women (APA, 2000; Kessler et al., 1994; see Figure 9.1). For the diagnosis of antisocial personality disorder to be applied, the person must be at least 18 years of age. The alternative diagnosis of conduct disorder is used with younger people (see Chapter 14). Many children with conduct disorders do not continue to show antisocial behavior as adults.

We once used terms like *psychopath* and *sociopath* to refer to the type of people who today are classified as having antisocial personalities, people whose behavior is amoral and asocial, impulsive, and lacking in remorse and shame. Some clinicians continue to use these terms interchangeably with *antisocial personality.* The roots of the word *psychopath* focus on the idea that there is something amiss (pathological) in the individual's psychological functioning. The roots of *sociopathy* center on the person's social deviance.

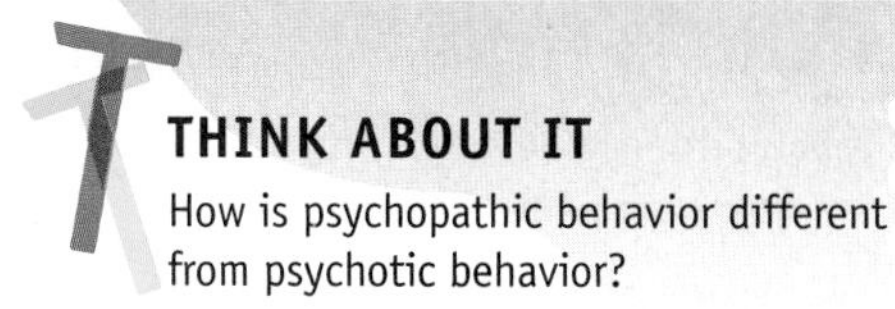

THINK ABOUT IT
How is psychopathic behavior different from psychotic behavior?

The pattern of behavior that characterizes antisocial personality disorder begins in childhood or adolescence and extends into adulthood. However, the antisocial and criminal behavior associated with the disorder tends to decline with age, and may actually disappear by the time the person reaches the age of 40. Not so for the underlying personality traits associated with the disorder—traits such as egocentricity; manipulativeness; lack of empathy, guilt, or remorse; and callousness toward others. These appear to be relatively stable even with increasing age (Harpur & Hare, 1994).

Much of our attention in this chapter focuses on antisocial personality disorder. Historically it is the personality disorder that has been most extensively studied by scholars and researchers.

Sociocultural Factors and Antisocial Personality Disorder Antisocial personality disorder cuts across all racial and ethnic groups. Researchers find no evidence of ethnic or racial differences in the rates of the disorder (Robins, Tipp, & Przybeck, 1991). The disorder is more common, however, among people in lower socioeconomic groups. One explanation is that people with antisocial personality disorder may drift downward occupationally, perhaps because their antisocial behavior makes it difficult for them to hold steady jobs or progress upward. It is possible too that people from lower socioeconomic levels are more likely to have been reared by parents who themselves modeled antisocial behavior. However, it is also possible that the diagnosis is misapplied to people living in hard-pressed communities who may engage in seemingly antisocial behaviors as a type of survival strategy (APA, 2000).

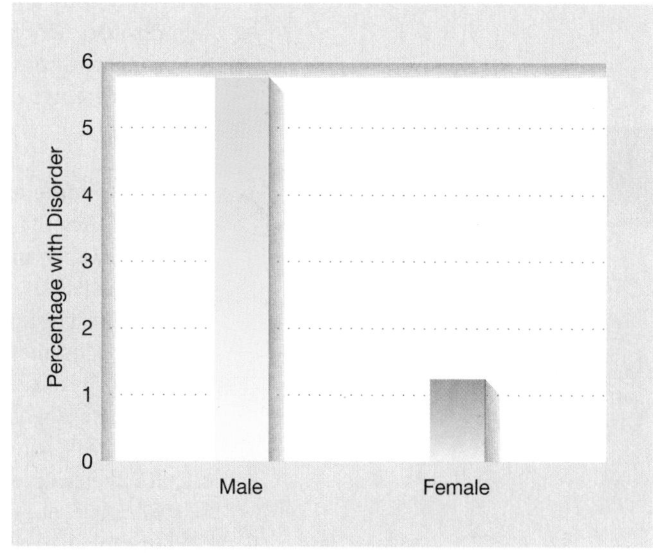

FIGURE 9.1 Lifetime prevalences of antisocial personality disorder by gender.
Antisocial personality disorder is more than five times as common among men than women. However, the disorder has been rising more rapidly among women in recent years.
Source. National Comorbidity Survey (Kessler et al, 1994).

Criminality or antisocial personality disorder? It is likely that many prison inmates could be diagnosed with antisocial personality disorder; however, people may become criminals or delinquents not because of a disordered personality but because they were raised in environments or exposed to subcultures that both encouraged and rewarded criminal behavior.

Antisocial Behavior and Criminality We may tend to think of antisocial behavior as synonymous with criminal behavior. Although a strong relationship does exist between the two, not all criminals show signs of psychopathy and not all people with psychopathic personalities become criminals (Lilienfeld & Andrews, 1996). Many are law-abiding and quite successful in their chosen occupations. Yet they possess a personality style characterized by a callous disregard of the interests and feelings of others.

Investigators have begun to view psychopathic personality as composed of two somewhat independent dimensions. The first is a personality dimension. It consists of such traits as superficial charm, selfishness, lack of empathy, callous and remorseless use of others, and disregard for their feelings and welfare. This type of psychopathic personality applies to people who have these kinds of psychopathic traits but don't become lawbreakers.

The second dimension is considered a behavioral dimension. It is characterized by the adoption of a generally unstable and antisocial lifestyle, including frequent problems with the law, poor employment history, and unstable relationships (Brown & Forth, 1997; Cooke & Michie, 1997). These two dimensions are not entirely separate; many psychopathic individuals show evidence of both sets of traits.

Truth OR Fiction? REVISITED

People with psychopathic personalities inevitably run afoul of the law.

FALSE. Not all criminals show signs of psychopathy and not all people with psychopathic personalities become criminals.

We should also note that people may become criminals or delinquents not because of a disordered personality but because they were reared in environments or subcultures that encouraged and rewarded criminal behavior. Although the behavior of criminals is deviant to society at large, it may be normal by the standards of their subcultures. We should also recognize that lack of remorse, which is a cardinal feature of antisocial personality disorder, does not characterize all criminals. Some criminals regret their crimes, and evidence of remorse is considered when a sentence is passed.

Only about half of prison inmates could be diagnosed with antisocial personality disorder (Robins et al., 1991). Conversely, fewer than half of the people with antisocial personality disorder run afoul of the law (Robins et al., 1991). Many fewer still fit (thankfully!) the stereotype of the psychopathic killer popularized in such films as *The Silence of the Lambs.*

VIDEO **9.1**

Antisocial Personality Disorder:
The Case of Paul

Profile of the Antisocial Personality Hervey Cleckley (1941) showed that the characteristics that define the psychopathic (antisocial) personality—self-centeredness, irresponsibility, impulsivity, and insensitivity to the needs of others—exist not only among criminals but also among many respected members of the community, including doctors, lawyers, politicians, and business executives.

Common features of people with antisocial personality disorder include failure to conform to social norms, irresponsibility, aimlessness and lack of long-term goals or plans, impulsive behavior, outright lawlessness, violence, chronic unemployment, marital problems, lack of remorse or empathy, substance abuse, a history of alcoholism, and a disregard for the truth and for the feelings and needs of others (Patrick, Cuthbert, & Lang, 1994; Robins et al., 1991). Irresponsibility may be seen in a personal history dotted by repeated, unexplained absences from work, abandonment of jobs without having other job opportunities to fall back on, or long stretches of unemployment despite available job opportunities. Irresponsibility extends to financial matters, where there may be repeated failure to repay debts, to pay child support, or to meet other financial responsibilities to one's family and dependents. The diagnostic features of antisocial personality disorder, as defined in the *DSM,* are shown in Table 9.1.

The following case represents a number of antisocial characteristics:

A Case of Antisocial Behavior

The 19-year-old male is brought by ambulance to the hospital emergency room in a state of cocaine intoxication. He's wearing a T-shirt with the imprint "Twisted Sister" on the front, and he sports a punk-style haircut. His mother is called and sounds groggy and confused on the phone; the doctors must coax her to come to the hospital. She later tells the doctors that her son has arrests for shoplifting and for driving while intoxicated. She suspects that he takes drugs, although she has no direct evidence. She believes that he is performing fairly well at school and has been a star member of the basketball team.

It turns out that her son has been lying to her. In actuality, he never completed high school and never played on the basketball team. A day later, his head cleared, the patient tells his doctors, almost boastfully, that his drug and alcohol use started at the age of 13, and that by the time he was 17, he was regularly using a variety of psychoactive substances, including alcohol, speed, marijuana, and cocaine. Lately, however, he has preferred cocaine. He and his friends frequently participate in drug and alcohol binges. At times they each drink a case of beer in a day along with downing other drugs. He steals car radios from parked cars and money from his mother to support his drug habit, which he justifies by adopting a (partial) "Robin Hood" attitude—that is, taking money only from people who have lots of it.

—Adapted from Spitzer et al., 1994, pp. 81–83

borderline personality disorder (BPD) A personality disorder characterized by abrupt shifts in mood, lack of a coherent sense of self, and unpredictable, impulsive behavior.

Although this case is suggestive of antisocial personality disorder, the diagnosis was maintained as provisional because the interviewer could not determine that the deviant behavior (lying, stealing, skipping school) began before the age of 15.

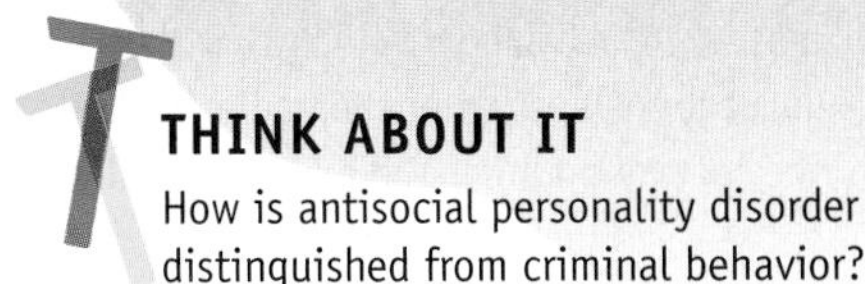

THINK ABOUT IT
How is antisocial personality disorder distinguished from criminal behavior?

Borderline Personality Disorder **Borderline personality disorder (BPD)** is characterized by a range of behavioral, emotional, and personality features (Sanislow, Grilo, & McGlashan, 2000). At the core is a pervasive pattern of instability in relationships, self-image, and mood, and a lack of control over impulses. People with borderline personality disorder

TABLE 9.1 Diagnostic Features of Antisocial Personality Disorder

(a) The person is at least 18 years old.
(b) There is evidence of a conduct disorder prior to the age of 15, as shown by such behavior patterns as truancy, running away, initiating physical fights, use of weapons, forcing someone into sexual activities, physical cruelty to people or animals, deliberate destruction of property or fire setting, lying, stealing, or mugging.
(c) Since the age of 15, there has been general indifference to and violation of the rights of other people, as shown by several of the following:
  (1) Lack of conformity to social norms and legal codes, as shown by law-breaking behavior that may or may not result in arrest, such as destruction of property, engaging in unlawful occupations, stealing, or harassing others.
  (2) Aggressive and highly irritable style of relating to others, as shown by repeated physical fights and assaults with others, possibly involving abuse of one's spouse or children.
  (3) Consistent irresponsibility, as shown by failure to maintain employment due to chronic absences, lateness, abandonment of job opportunities or extended periods of unemployment despite available work; and/or by failure to honor financial obligations, such as failing to maintain child support or defaulting on debts; and/or by lack of a sustained monogamous relationship.
  (4) Failure to plan ahead or impulsivity, as shown by traveling around without prearranged employment or clear goals.
  (5) Disregard for the truth, evidenced by repeated lying, conning others, or use of aliases for personal gain or pleasure.
  (6) Recklessness with regard to personal safety or the safety of other people, as shown by driving while intoxicated or repeated speeding.
  (7) Lack of remorse for misdeeds, as shown by indifference to the harm done to others, and/or by rationalizing that harm.

Source. Adapted from the *DSM-IV-TR* (APA, 2000).

A Closer Look

Did Samson Have Antisocial Personality Disorder?

Did Samson, the biblical figure who fought the Philistines but lost his great strength when his long hair was shorn by the wily Delilah, have antisocial personality disorder? We noted in Chapter 1 how case studies have been reported on deceased individuals, even historical figures who died several hundred years ago. But applying modern diagnostic criteria to individuals from biblical times requires quite a leap of faith. Yet four psychiatrists writing in a respected professional journal argue that Samson's behavior nearly 3,000 years ago clearly meets many of the criteria for antisocial personality disorder (Altschuler et al., 2001). The biblical account of his behavior shows him to have broken the law, to have lied repeatedly, to have acted impulsively and without regard for the safety of himself and others, to have initiated many physical fights, and to have lacked remorse for his actions—patterns of behavior that today would likely be recognized as signs of antisocial personality disorder (Goode, 2001b). As an example of his reckless disregard of his own safety, the authors point to Samson's revelation to Delilah that the secret of his strength lay in his uncut locks, even after she had made three unsuccessful attempts to pry the secret from him. The biblical account also relates that 3,000 Israelites—Samson's own people!—captured him and turned him over the Philistines. This indicates that his brutal behavior was not viewed as acceptable conduct in the time in which he lived. However, it cannot be determined based on the biblical record whether Samson showed evidence of conduct disorder prior to the age of 15, as the diagnostic criteria for antisocial personality disorder require. On a broader level, the authors argue that modern diagnostic concepts may help us form better understandings of behavior of historical figures that have long seemed puzzling and in need of explanation.

Samson: A case of antisocial personality disorder? The actor Victor Mature portrayed Samson in the 1949 film *Samson and Delilah.* Recently, several psychiatrists have claimed that the biblical record of Samson's behavior indicates that he met many of the criteria for antisocial personality disorder.

tend to be uncertain about their personal identities—their values, goals, careers, and perhaps even their sexual orientations. This instability in self-image or personal identity leaves them with nagging feelings of emptiness and boredom. They cannot tolerate being alone and will make desperate attempts to avoid feelings of abandonment (Gunderson, 1996). Fear of abandonment renders them clinging and demanding in their social relationships, but their clinging often pushes away the people on whom they depend. Signs of rejection may enrage them, straining their relationships further. Their feelings toward others are consequently intense and shifting. They alternate between extremes of adulation (when their needs are met) and loathing (when they feel scorned). They tend to view other people as all good or all bad, shifting abruptly from one extreme to the other. As a result, they may flit from partner to partner in a series of brief and stormy relationships. People they had idealized are treated with contempt when relationships end or when they feel the other person fails to meet their needs (Gunderson & Singer, 1986).

Many notable figures have been described as having personality features associated with borderline personality disorder, including Marilyn Monroe, Lawrence of Arabia, Adolf Hitler, and the philosopher Sören Kierkegaard (Sass, 1982). Some theorists believe we live in highly fragmented and alienating times that tend to create the problems in forming cohesive identities and stable relationships that characterize people with borderline personalities (Sass, 1982). "Living on the edge," or border, can be seen as a metaphor for an unstable society. Borderline personality disorder is believed to occur in about 2% of the general population (APA, 2000). Although it is diagnosed more often (about 75% of the time) in women, gender differences in prevalence rates for BPD in the general population remain undetermined.

The term *borderline personality* was originally used to refer to individuals whose behavior appeared on the border between neuroses and psychoses. People with borderline personality disorder generally maintain better contact with reality than people with psychoses, although they may show transient psychotic behaviors during times of stress. Generally speaking, they seem to be more severely impaired than most people with neuroses but not as dysfunctional as those with psychotic disorders.

Instability of moods is a central characteristic of borderline personality disorder (Sanislow et al., 2000). Moods run the gamut from anger and irritability to depression and anxiety, with each lasting from a few hours to a few days. People with BPD have difficulty controlling anger and are prone to fights or smashing things. They often act on impulse, such as eloping with someone they have just met. This impulsive and unpredictable behavior is often self-destructive, involving such behaviors as self-mutilation and suicidal gestures and actual attempts (e.g., Sanislow et al., 2000). It may also involve spending sprees, gambling, drug abuse, engaging in unsafe sexual activity, reckless driving, binge eating, or shoplifting. Impulsive acts of self-mutilation may involve such acts as scratching the wrists or burning cigarettes on the arms, as exemplified in the following dialogue:

CLIENT: I've got such repressed anger in me; what happens is . . . I can't *feel* it; I get anxiety attacks. I get very nervous, smoke too many cigarettes. So what happens to me is I tend to *explode.* Into tears or hurting myself or whatever . . . because I don't know how to contend with all those mixed up feelings.

INTERVIEWER: What was the more recent example of such an "explosion"?

CLIENT: I was alone at home a few months ago; I was frightened! I was trying to get in touch with my boyfriend and I couldn't . . . He was nowhere to be found. All my friends seemed to be busy that night and I had no one to talk to . . . I just got more and more nervous and more and more agitated. Finally, *bang!* . . . I took out a cigarette and lit it and stuck it into my forearm. I don't know why I did it because I didn't really care for him all that much. I guess I felt I had to do something dramatic . . ."

—*Adapted from Stone, 1980, p. 400*

Self-mutilation is sometimes carried out as an expression of anger or a means of manipulating others. Such acts may be intended to counteract self-reported feelings of "numbness," particularly in times of stress. Not surprisingly, frequent self-mutilation among people with BPD is associated with an increased risk of suicidal thinking (Dulit et al., 1994).

Individuals with BPD tend to have very troubled relationships with their families of origin and with others. They often have histories of traumatic experiences in childhood, such as parental losses or separations, abuse, neglect, or witnessing violence (Liotti et al., 2000). They tend to view their relationships as rife with hostility and to perceive others as rejecting and abandoning (Benjamin & Wonderlich, 1994). They also tend to be difficult to work with in psychotherapy, demanding a great deal of support from therapists, calling them at all hours or acting suicidally to elicit support, or dropping out of therapy prematurely. Their feelings toward therapists, as toward other people, undergo rapid alterations between idealization and outrage. These abrupt shifts in feelings are interpreted by psychoanalysts as signs of "splitting," or inability to reconcile the positive and negative aspects of one's experience of oneself and others.

From the modern psychodynamic perspective, borderline individuals cannot synthesize positive and negative elements of personality into complete wholes. They therefore fail to achieve fixed self-identities or images of others. Rather than viewing important figures in their lives as sometimes loving and as sometimes rejecting, they shift back and forth between

Truth OR Fiction? REVISITED

Many notable figures in history, from Lawrence of Arabia to Adolf Hitler and even Marilyn Monroe, have been depicted as borderline personalities.

TRUE. Many notable public figures have been described as having personality features associated with borderline personality disorder.

WWW **Web Link 9.3**
BPD Sanctuary

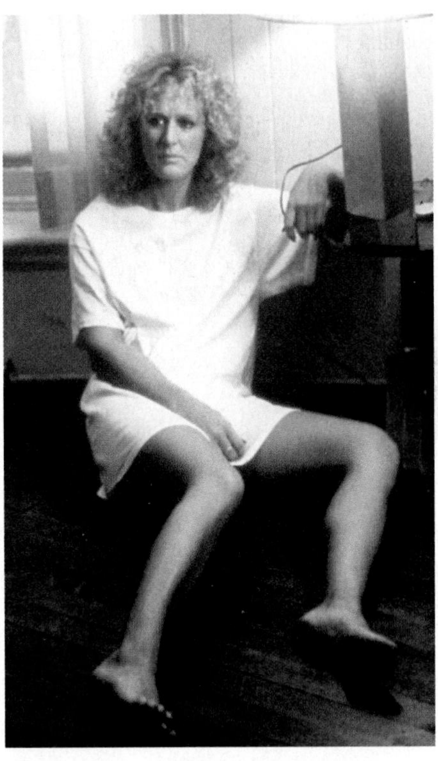

Borderline personality. In the movie *Fatal Attraction,* the actress Glenn Close played a character who exhibited many of the characteristics associated with borderline personality disorder, including impulsivity, extreme mood swings, and unstable relationships.

Over the top? Not all people who dress outrageously or flamboyantly have histrionic personalities. What other personality features characterize people with histrionic personality disorder?

viewing them as all good or all bad, between idealization and abhorrence. The psychoanalyst Otto Kernberg, a leading authority on borderline personality, tells of a woman in her 30s whose attitude toward him vacillated in such a way. According to Kernberg, the woman would respond to him in one session as the most wonderful therapist and feel that all her problems were solved. But several sessions later she would turn against him and accuse him of being unfeeling and manipulative, become very dissatisfied with the treatment she was receiving, and threaten to drop out and never come back (Sass, 1982). Borderline personality disorder remains in many ways a perplexing and frustrating problem.

Histrionic Personality Disorder **Histrionic personality disorder** involves excessive emotionality and an overwhelming need to be the center of attention. The term is derived from the Latin *histrio,* which means "actor." People with histrionic personality disorder tend to be dramatic and emotional, but their emotions seem shallow, exaggerated, and volatile. The disorder was formerly called *hysterical personality.* The following case example illustrates the excessively dramatic behaviors that are typical of someone with histrionic personality disorder:

A Case of Histrionic Personality Disorder

Marcella was a 36-year-old, attractive, but overly made up woman who was dressed in tight pants and high heels. Her hair was in a bird's nest of the type that had been popular when she was a teenager. Her social life seemed to bounce from relationship to relationship, from crisis to crisis. Marcella sought help from the psychologist at this time because her 17-year-old daughter, Nancy, had just been hospitalized for cutting her wrists. Nancy lived with Marcella and Marcella's current boyfriend, Morris, and there were constant arguments in the apartment. Marcella recounted the disputes that took place with high drama, waving her hands, clanging the bangles that hung from her bracelets, and then clutching her breast. It was difficult having Nancy live at home because Nancy had expensive tastes, was "always looking for attention," and flirted with Morris as a way of "flaunting her youth." Marcella saw herself as a doting mother and denied any possibility that she was in competition with her daughter.

Marcella came for a handful of sessions, during which she basically ventilated her feelings and was encouraged to make decisions that might lead to a reduction of some of the pressures on her and her daughter. At the end of each session she said, "I feel so much better" and thanked the psychologist profusely. At termination of "therapy," she took the psychologist's hand and squeezed it endearingly. "Thank you so much, Doctor," she said and made her exit.

—From the Authors' Files

■

The supplanting of *hysterical* with *histrionic* and the associated exchange of the roots *hystera* (meaning "uterus") and *histrio* allow professionals to distance themselves from the notion that the disorder is intricately bound up with being female. The disorder is diagnosed more frequently in women than men (Hartung & Widiger, 1998), although some studies using structured interview methods find similar rates of occurrence among men and women (APA, 2000). Whether the gender discrepancy in clinical practice reflects true differences in the underlying rates of the disorder, or diagnostic biases or other unseen factors, remains something of an open question (Corbitt & Widiger, 1995).

Despite a long-standing belief among clinicians that histrionic personality is closely related to conversion disorder (see Chapter 7), research has not borne out this connection (Kellner, 1992). People with conversion disorder are actually more likely to show features of dependent personality disorder than histrionic personality disorder.

histrionic personality disorder A personality disorder characterized by excessive need for attention, praise, reassurance, and approval.

People with histrionic personalities may become unusually upset by news of a sad event and cancel plans for the evening, inconveniencing their friends. They may exude exaggerated delight when they meet someone or become enraged when someone fails to notice their new hairstyle. They may faint at the sight of blood or blush at a slight faux pas. They tend to demand that others meet their needs for attention and play the victim when others fall short. If they feel a touch of fever, they may insist that others drop everything to rush them to the doctor. They tend to be self-centered and intolerant of delays of gratification; they want what they want when they want it. They grow quickly restless with routine and crave novelty and stimulation. They are drawn to fads. Others may see them as putting on airs or playacting, although they may evince a certain charm. They may enter a room with a flourish and embellish their experiences with flair. When pressed for details, however, they fail to color in the specifics of their tales. They tend to be flirtatious and seductive but are too wrapped up in themselves to develop intimate relationships or have deep feelings toward others. As a result, their associations tend be stormy and ultimately ungratifying. They tend to use their physical appearance as a means of drawing attention to themselves. Men with the disorder may act and dress in an overly "macho" manner to draw attention to themselves; women may choose very frilly, feminine clothing. Glitter supercedes substance.

People with histrionic personalities may be attracted to professions like modeling or acting, where they can hog the spotlight. Despite outward successes, they may lack self-esteem and strive to impress others to boost their self-worth. If they suffer setbacks or lose their place in the limelight, depressing inner doubts may emerge.

Narkissos. According to one version of the Greek myth, *Narkissos* fell in love with his reflection in a spring. Because of his excessive self-love, the gods transformed him into a flower—the narcissus.

Narcissistic Personality Disorder *Narkissos* was a handsome youth who, according to Greek myth, fell in love with his reflection in a spring. Because of his excessive self-love, in one version of the myth, he was transformed by the gods into the flower we know as the narcissus.

Persons with **narcissistic personality disorder** have an inflated or grandiose sense of themselves and an extreme need for admiration. They brag about their accomplishments and expect others to shower them with praise. They expect others to notice their special qualities, even when their accomplishments are ordinary, and they enjoy basking in the light of adulation. They are self-absorbed and lack empathy for others. Although they share certain features with histrionic personalities, such as demanding to be the center of attention, they have a much more inflated view of themselves and are less melodramatic than people with histrionic personality disorder. The label of borderline personality disorder (BPD) is sometimes applied to them, but people with narcissistic personality disorder are generally better able to organize their thoughts and actions. They tend to be more successful in their careers and are better able to rise to positions of status and power. Their relationships also tend to be more stable than those of people with BPD.

Narcissistic personality disorder is found among less than 1% of people in the general population (APA, 2000). Although more than half of the people diagnosed with the disorder are men, we cannot say whether there is an underlying gender difference in prevalence rates in the general population. A certain degree of narcissism may represent a healthful adjustment to insecurity, a shield from criticism and failure, or a motive for achievement (Goleman, 1988b). Excessive narcissistic qualities can become unhealthful, especially when the cravings for adulation are insatiable. Table 9.2 compares "normal" self-interest with self-defeating extremes of narcissism. Up to a point, self-interest fosters success and happiness. In more extreme cases, as with narcissism, it can compromise relationships and careers.

People with narcissistic personalities tend to be preoccupied with fantasies of success and power, ideal love, or recognition for brilliance or beauty. They, like people with

narcissistic personality disorder
A personality disorder characterized by adoption of an inflated self-image and demands for attention and admiration.

TABLE 9.2 Features of Normal Self-Interest Compared with Self-Defeating Narcissism

Normal Self-Interest	Self-Defeating Narcissism
Appreciating acclaim, but not requiring it in order to maintain self-esteem.	Craving adoration insatiably; requiring acclaim in order to feel momentarily good about oneself.
Being temporarily wounded by criticism.	Being inflamed or crushed by criticism and brooding about it extensively.
Feeling unhappy but not worthless following failure.	Having enduring feelings of mortification and worthlessness triggered by failure.
Feeling "special" or uncommonly talented in some way.	Feeling incomparably better than other people, and insisting upon acknowledgment of that preeminence.
Feeling good about oneself, even when other people are being critical.	Needing constant support from other people in order to maintain one's feelings of well-being.
Being reasonably accepting of life's setbacks, even though they can be painful and temporarily destabilizing.	Responding to life's wounds with depression or fury.
Maintaining self-esteem in the face of disapproval or denigration.	Responding to disapproval or denigration with loss of self-esteem.
Maintaining emotional equilibrium despite lack of special treatment.	Feeling entitled to special treatment and becoming terribly upset when one is treated in an ordinary manner.
Being empathic and caring about the feelings of others.	Being insensitive to other people's needs and feelings; exploiting others until they become fed up.

Source. Based on Goleman, 1988b, p. C1.

histrionic personalities, may gravitate toward careers in which they can receive adulation, such as modeling, acting, or politics. Although they tend to exaggerate their accomplishments and abilities, many people with narcissistic personalities are quite successful in their occupations. But they envy those who achieve even greater success. Insatiable ambition may prompt them to devote themselves tirelessly to work. They are driven to succeed, not so much for money as for the adulation that attends success.

Interpersonal relationships are invariably strained by the demands that people with narcissistic personality impose on others and by their lack of empathy with, and concern for, other people. They seek the company of flatterers and are often superficially charming and friendly and able to draw people to them. But their interest in people is one-sided: They seek people who will serve their interests and nourish their sense of self-importance (Goleman, 1988b). They have feelings of entitlement that lead them to exploit others. They treat sex partners as devices for their own pleasure or to brace their self-esteem, as in the case of Bill:

THINK ABOUT IT

We have referred to characters in movies in our illustrations of personality disorders characterized by emotional and erratic behavior. You can probably think of other examples of characters who manifest the traits of these disorders. Why are these characters so frequently encountered in entertainment media? What is the attraction for dramatists, actors, and viewers?

A Case of Narcissistic Personality Disorder

Most people agreed that Bill, a 35-year-old investment banker, had a certain charm. He was bright, articulate, and attractive. He possessed a keen sense of humor that drew people to him at social gatherings. He would always position himself in the middle of the room, where he could be the center of attention. The topics of conversation invariably focused on his "deals," the "rich and famous" people he had met, and his outmaneuvering of opponents. His next project was always bigger and more daring than the last. Bill loved an audience. His face would light up when others responded to him with praise or admiration for his business successes, which were always inflated beyond their true measure. But when the conversation shifted to other people, he would lose interest and excuse himself to make a drink or to call his answering machine. When hosting a party, he would urge guests to stay late and feel hurt if they had to leave early; he showed no sensitivity to, or awareness of, the needs of his friends.

The few friends he had maintained over the years had come to accept Bill on his own terms. They recognized that he needed to have his ego fed or that he would become cool and detached.

Bill had also had a series of romantic relationships with women who were willing to play the adoring admirer and make the sacrifices that he demanded—for a time. But they inevitably tired of the one-sided relationship or grew frustrated by Bill's inability to make a commitment or feel deeply toward them. Lacking empathy, Bill was unable to recognize other people's feelings and needs. His demands for constant attention from willing admirers did not derive from selfishness, but from a need to ward off underlying feelings of inadequacy and diminished self-esteem. It was sad, his friends thought, that Bill needed so much attention and adulation from others and that his many achievements were never enough to calm his inner doubts.

—From the Authors' Files

A person with a narcissistic personality? People with narcissistic personalities are often preoccupied with fantasies of success and power, ideal love, or recognition for their brilliance or beauty. They may pursue careers that provide opportunities for public recognition and adulation, such as acting, modeling, or politics. They may become deeply wounded by any hint that they are not as special as they believe themselves to be.

Personality Disorders Characterized by Anxious or Fearful Behavior

This cluster of personality disorders includes the avoidant, dependent, and obsessive-compulsive types. Although the features of these disorders differ, they share a component of fear or anxiety.

Avoidant Personality Disorder Persons with **avoidant personality disorder** are so terrified of rejection and criticism that they are generally unwilling to enter relationships without ardent reassurances of acceptance. As a result, they may have few close relationships outside their immediate families. They also tend to avoid group occupational or recreational activities for fear of rejection. They prefer to lunch alone at their desks. They shun company picnics and parties, unless they are perfectly sure of acceptance. Avoidant personality disorder, which appears to be equally common in men and women, is believed to affect between 0.5% and 1.0% of the general population (APA, 2000).

avoidant personality disorder A personality disorder characterized by avoidance of social relationships due to fears of rejection.

Unlike people with schizoid qualities, with whom they share the feature of social withdrawal, individuals with avoidant personalities have interest in, and feelings of warmth toward, other people. However, fear of rejection prevents them from striving to meet their needs for affection and acceptance. In social situations, they tend to hug the walls and avoid conversing with others. They fear public embarrassment, the thought that others might see them blush, cry, or act nervously. They tend to stick to their routines and exaggerate the risks or effort involved in trying new things. They may refuse to attend a party that is an hour away on the pretext that the late drive home would be too taxing. Consider the following case example:

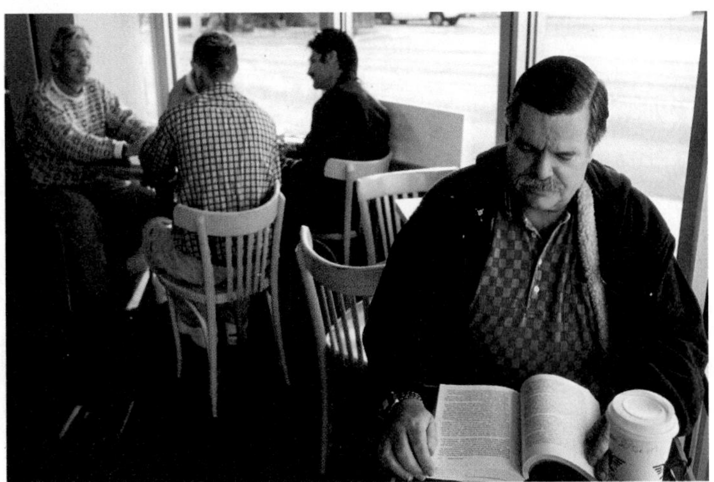

A person with an avoidant personality? People with avoidant personalities often keep to themselves because of fear of rejection.

A Case of Avoidant Personality Disorder

Harold, a 24-year-old accounting clerk, had dated but a few women, and he had met them through family introductions. He never felt confident enough to approach a woman on his own. Perhaps it was his shyness that first attracted Stacy. Stacy, a 22-year-old secretary, worked alongside Harold and asked him if he would like to get together sometime after work. At first Harold declined, claiming some excuse, but when Stacy asked again a week later, Harold agreed, thinking she must really like him if she were willing to pursue him. The relationship developed quickly, and soon they were dating virtually every night. The relationship was strained, however. Harold interpreted any slight hesitation in her voice as a lack of interest. He repeatedly requested reassurance that she cared about him, and he evaluated every word and gesture for evidence of her feelings. If Stacy said that she could not see him because of fatigue or illness, he assumed she was rejecting him and sought further reassurance. After several months, Stacy decided she could no longer accept Harold's nagging, and the relationship ended. Harold assumed that Stacy had never truly cared for him.

—From the Authors' Files

There is a good deal of overlap between avoidant personality disorder and social phobia, particularly with a severe subtype of social phobia that involves a generalized pattern of social phobia (excessive, irrational fear of most social situations) (Turner, Beidel, & Townsley, 1992; Widiger, 1992). Although research evidence shows that many cases of generalized social phobia occur in the absence of avoidant personality disorder (Holt, Heimberg, & Hope, 1992), relatively fewer cases of avoidant personality occur in the absence of generalized social phobia (Widiger, 1992). Thus avoidant personality disorder may represent a more severe form of social phobia (Hoffman et al., 1995). Still, the scientific jury is out on the question of whether avoidant personality disorder should be considered a severe form of generalized social phobia or a distinct diagnostic category as it is presently classified.

Dependent Personality Disorder **Dependent personality disorder** describes people who have an excessive need to be taken care of by others. This leads them to be overly submissive and clinging in their relationships and extremely fearful of separation. People with this disorder find it very difficult to do things on their own. They seek advice in making even the smallest decision. Children or adolescents with the problem may look to their parents to select their clothes, diets, schools or colleges, even their friends. Adults with the disorder allow others to make important decisions for them. Sometimes they are so dependent on others for making decisions that they allow their parents to determine whom they will marry, as in the case of Matthew:

A Case of Dependent Personality Disorder

Matthew, a 34-year-old single accountant who lives with his mother, sought treatment when his relationship with his girlfriend came to an end. His mother had objected to marriage because his girlfriend was of a different religion, and—because "blood is thicker than water"—Matthew acceded to his mother's wishes and ended the relationship. Yet he is angry with himself and at his mother because he feels that she is too possessive to ever grant him permission to get married. He describes his mother as a domineering woman who "wears the pants" in the family and is accustomed to having things her way. Matthew alternates between resenting his mother and thinking that perhaps she knows what's best for him.

Matthew's position at work is several levels below what would be expected of someone of his talent and educational level. Several times he has declined promotions in

dependent personality disorder
A personality disorder characterized by difficulty making independent decisions and overly dependent behavior.

order to avoid increased responsibilities that would require him to supervise others and make independent decisions. He has maintained close relationships with two friends since early childhood and has lunch with one of them on every working day. On days his friend calls in sick, Matthew feels lost. Matthew has lived his whole life at home, except for one year away at college. He returned home because of homesickness.

—*Adapted from Spitzer et al., 1994, pp. 179–180*

After marriage, people with dependent personality disorder may rely on their spouses to make decisions such as where they should live, which neighbors they should cultivate, how they should discipline the children, what jobs they should take, how they should budget money, and where they should vacation. Like Matthew, individuals with dependent personality disorder avoid positions of responsibility. They turn down challenges and promotions and work beneath their potential. They tend to be overly sensitive to criticism and are preoccupied with fears of rejection and abandonment. They may be devastated by the end of a close relationship or by the prospect of living on their own. Because of fear of rejection, they often subordinate their wants and needs to those of others. They may agree with outlandish statements about themselves and do degrading things in order to please others.

Although dependent personality disorder is diagnosed more frequently in women (APA, 2000; Bornstein, 1997), it is not clear that there is any underlying difference in the prevalence of the disorder between men and women (Corbitt & Widiger, 1995). The diagnosis is often applied to women who, for fear of abandonment, tolerate husbands who openly cheat on them, abuse them, or gamble away the family's resources. Underlying feelings of inadequacy and helplessness discourage them from taking effective action. In a vicious cycle, their passivity encourages further abuse, leading them to feel yet more inadequate and helpless. The diagnosis of women with this pattern is controversial and may be seen as unfairly "blaming the victim," because women in our society are often socialized to dependent roles. A panel convened by the American Psychological Association noted that women also encounter greater stress than men in contemporary life (Goleman, 1990b). Moreover, since women generally encounter greater social pressures to be passive, demure, or deferential than men, dependent behaviors in women may reflect cultural influences rather than an underlying personality disorder.

Dependent personality disorder has been linked to other psychological disorders, including major depression, bipolar disorder, and social phobia, and to physical problems, such as hypertension, cancer, and gastrointestinal disorders like ulcers and colitis (Bornstein, 1999; Loranger, 1996; Reich, 1996). There also appears to be a link between dependent personality and what psychodynamic theorists refer to as "oral" behavior problems, such as smoking, eating disorders, and alcoholism (Bornstein, 1993, 1999). Psychodynamic writers trace dependent behaviors to the utter dependence of the newborn baby and the baby's seeking of nourishment through oral means (suckling). From infancy, they suggest, people associate the provision of food with love. Food may come to symbolize love, and persons with dependent personalities may overeat to ingest love symbolically. Research shows that people with dependent personalities are more reliant on others for support and guidance than is the average person (Greenberg & Bornstein, 1988a). People with dependent personalities often attribute their problems to physical rather than emotional causes and seek support and advice from medical experts rather than psychologists or counselors (Greenberg & Bornstein, 1988b).

Obsessive-Compulsive Personality Disorder The defining features of **obsessive-compulsive personality disorder** involve an excessive degree of orderliness, perfectionism, rigidity, difficulty coping with ambiguity, difficulties expressing feelings, and meticulousness in work habits. About 1% of people in community samples are diagnosed with the disorder (APA, 2000). The disorder is about twice as common in men than women. Unlike obsessive-compulsive anxiety disorder, people with obsessive-compulsive

Truth OR Fiction? REVISITED

Some people with dependent personality disorder have so much difficulty making independent decisions that they allow their parents to decide whom they will marry.

TRUE. People with dependent personality disorder may be so dependent on others for making decisions that they allow their parents to determine whom they will marry.

obsessive-compulsive personality disorder A personality disorder characterized by rigid ways of relating to others, perfectionistic tendencies, lack of spontaneity, and excessive attention to detail.

"A place for everything, and everything in its place"? People with obsessive-compulsive personalities may have invented this maxim. Many such people have excessive needs for orderliness in their environment, as suggested in this Laurie Simmons photograph, *Red Library #2*.

personality disorder do not necessarily experience outright obsessions or compulsions. If they do, both diagnoses may be deemed appropriate.

Persons with obsessive-compulsive personality disorder are so preoccupied with the need for perfection that they cannot complete things in a timely fashion. Their efforts inevitably fall short of their expectations, and they force themselves to redo their work. Or they may ruminate about how to prioritize their assignments and never seem to get started working. They focus on details that others perceive as trivial. As the saying goes, they often fail to see the forest for the trees. Their rigidity impairs their social relationships; they insist on doing things their way rather than compromising. Their zeal for work keeps them from participating in, or enjoying, social and leisure activities. They tend to be stingy with money. They find it difficult to make decisions and postpone or avoid them for fear of making the wrong choice. They tend to be overly rigid in issues of morality and ethics because of inflexibility in personality rather than deeply held convictions. They tend to be overly formal in relationships and find it difficult to express feelings. It is hard for them to relax and enjoy pleasant activities; they worry about the costs of such diversions. Consider the following case example:

A Case of Obsessive-Compulsive Personality Disorder

Jerry, a 34-year-old systems analyst, was perfectionistic, overly concerned with details, and rigid in his behavior. Jerry was married to Marcia, a graphics artist. He insisted on scheduling their free time hour by hour and became unnerved when they deviated from his agenda. He would circle a parking lot repeatedly in search of just the right parking spot to ensure that another car would not scrape his car. He refused to have the apartment painted for over a year because he couldn't decide on the color. He had arranged all the books in their bookshelf alphabetically and insisted that every book be placed in its proper position.

Jerry never seemed to be able to relax. Even on vacation, he was bothered by thoughts of work that he had left behind and by fears that he might lose his job. He couldn't understand how people could lie on the beach and let all their worries evaporate in the summer air. Something can always go wrong, he figured, so how can people let themselves go?

—From the Authors' Files

Problems with the Classification of Personality Disorders

Questions remain about the reliability and validity of the diagnostic categories for personality disorders (Farmer, 2000). There may be too much overlap among the diagnoses to justify so many different categories. Agreement between raters on personality disorder diagnoses remains modest at best (Coolidge & Segal, 1998). The classification system also seems to blur the distinctions between normal and abnormal variations in personality. Some categories of personality disorder, moreover, may be based on sexist presumptions. Finally, there is concern that the diagnoses may confuse labels with explanations.

Undetermined Reliability and Validity The present *DSM* system sought to remove ambiguities in the diagnostic criteria of personality disorders by providing descriptive criteria that more tightly define particular disorders. The reliability and validity of the definitions used in the *DSM-IV* remains to be fully tested, however.

Problems Distinguishing Axis I from Axis II Disorders Some reviewers question whether Axis II personality disorders can be reliably differentiated from Axis I clinical syndromes such as anxiety or mood disorders (Farmer, 2000; Livesley et al., 1994). For example, clinicians may have difficulty distinguishing between obsessive-compulsive disorder and obsessive-compulsive personality disorder. Clinical syndromes are believed to be variable over time, whereas personality disorders are held to be generally more enduring patterns of disturbance. Yet evidence indicates that features of personality disorders may vary over time with changes in circumstances. On the other hand, some Axis I clinical syndromes (dysthymia, for example) follow a more or less chronic course.

Overlap Among Disorders There is also a high degree of overlap among the personality disorders (Westen & Shedler, 1999). Overlap undermines the *DSM*'s conceptual clarity or purity by increasing the number of cases that seem to fit two or more diagnostic categories (Livesley, 1985). Although some personality disorders have distinct features, many appear to share common traits, such as problems in romantic relationships (Daley, Burge, & Hammen, 2000). Moreover, the same person may have traits suggestive of dependent personality disorder (inability to make decisions or initiate activities independently) and of avoidant personality disorder (extreme social anxiety and heightened sensitivity to criticism). Overall, about two in three people with personality disorders meet diagnostic criteria for more than one type of personality disorder (Widiger, 1991). The high degree of overlap suggests that the personality disorders included in the *DSM* system may not be sufficiently distinct from one another (Westen & Schedler, 1999). Some so-called disorders may thus represent different aspects of the same disorder, not separate diagnostic categories.

Difficulty in Distinguishing Between Variations in Normal Behavior and in Abnormal Behavior Another problem with the diagnosis of personality disorders is that they involve traits which, in lesser degrees, describe the behavior of most normal individuals. Feeling suspicious now and then does not mean you have a paranoid personality disorder. The tendency to exaggerate your own importance does not mean you are narcissistic. You may avoid social interactions for fear of embarrassment or rejection without having an avoidant personality disorder, and you may be especially conscientious in your work without having an obsessive-compulsive personality disorder. Because the defining attributes of these disorders are commonly occurring personality traits, clinicians should only apply these diagnostic labels when the patterns are so pervasive that they interfere with the individual's functioning or cause significant personal distress. Yet it can be difficult to know where to draw the line between normal variations in behavior and personality disorders. We continue to lack data to guide us more precisely in determining the point at which a trait becomes sufficiently inflexible or maladaptive to justify a personality disorder diagnosis (Widiger & Costa, 1994).

Truth OR Fiction? REVISITED

It may be difficult to draw a clear line between normal variations in behavior and personality disorders.

TRUE. The boundaries between normal variations in behavior and personality disorders can be blurry.

Sexist Biases The construction of certain personality disorders may have sexist underpinnings. For example, diagnostic criteria for personality disorders label stereotypical feminine behaviors as pathological with greater frequency than is the case with stereotypical masculine behaviors. The concept of the histrionic personality, for example, seems a caricature of the traditional stereotype of the feminine personality: flighty, emotional, shallow, seductive, attention seeking. But if the feminine stereotype corresponds to a mental disorder, shouldn't we also have a diagnostic category that reflects the masculine stereotype of the "macho male"? It may be possible to show that overly masculinized traits are associated with significant distress or impairment in social or occupational functioning in certain males: Highly masculinized males often get into fights and experience difficulties working

Truth OR Fiction? REVISITED

The diagnosis of some personality disorders may reflect sexist biases.

TRUE. The concepts of histrionic and dependent personality disorders may indeed be sexist. It could be argued that the descriptions of these disorders are parodies of the traditional feminine gender-role stereotype.

Are there sexist biases in the conception of personality disorders? The concept of the histrionic personality disorder seems to be a caricature of the highly stereotyped feminine personality. Why, then, is there not also something akin to a macho male personality disorder, which caricatures the highly stereotyped masculine personality?

for female bosses. There is no personality disorder that corresponds to the "macho male" stereotype, however.

The diagnosis of dependent personality disorder may also unfairly stigmatize women who are socialized into dependent roles as having a "mental disorder." Women may be at greater risk of receiving diagnoses of histrionic or dependent personality disorders because clinicians perceive these patterns as more common among women or because women are more likely than men to be socialized into these behavior patterns.

Clinicians may also be biased in favor of perceiving women as having histrionic personality disorder and men as having antisocial personality disorder, even when they do not differ in symptomatology (Garb, 1997). Clinicians may also have a gender bias when it comes to diagnosing borderline personality disorder. In one study, researchers presented a hypothetical case example to a sample of 311 psychologists, social workers, and psychiatrists (Becker & Lamb, 1994). Half of the sample was presented with a case identified as a female; the other half read the identical case, except that it was identified as male. Clinicians more often diagnosed the case identified as female as having borderline personality disorder.

THINK ABOUT IT

Are some personality disorders more likely to be diagnosed in men or in women because of societal expectations rather than because of real underlying pathology? Have you ever assumed, for example, that women are "just dependent or hysterical" or that men are "just narcissists or antisocial"? What kinds of problems do these underlying assumptions pose for clinicians and researchers?

Confusing Labels with Explanations It may seem obvious that we should not confuse diagnostic labels with explanations, but in practice the distinction is sometimes clouded. If we confuse labeling with explanation, we may fall into the trap of circular reasoning. What is wrong, for example, with the logic of the following statements?

1. John's behavior is antisocial.
2. Therefore, John has an antisocial personality disorder.
3. John's behavior is antisocial because he has an antisocial personality disorder.

The statements are circular in reasoning because they (1) use behavior to make a diagnosis, and then (2) use the diagnosis as an explanation for the behavior. We may be guilty of circular reasoning in our everyday speech. Consider: "John never gets his work in on time; therefore, he is lazy. John doesn't get his work in because he's lazy." The label may be acceptable and useful in everyday conversation, but it lacks scientific rigor. For a construct such as *laziness* to have scientific rigor, we need to understand the causes of laziness and the factors that help maintain it. We should not confuse the label we attach to behavior with the cause of the behavior.

Moreover, labeling people with disturbing behavior as personality disordered tends to overlook the social and environmental contexts in which the behavior occurs. We need to attend to the impact of specific traumatic life events, which may occur with a greater range or intensity among members of one gender or cultural group, as important factors underlying patterns of maladaptive behavior. The conceptual underpinnings of the personality disorders lack such a perspective. Moreover, conceptualizations of personality disorders fail to account for the social inequalities in society and the differences in power between the genders or between dominant and minority cultures that may give rise to the types of problems identified as personality disorders. For example, Brown (1992) and Walker (1988) document the significant prevalence of a history of childhood physical and sexual abuse among women diagnosed with personality disorders. The ways in which people cope with abuse may come to be viewed as flaws in their character rather than as reflections of the dysfunctional societal factors that underlie abusive relationships.

THINK ABOUT IT

What are the major points of controversy concerning the classification of personality disorders? Explain the problems that result from using labels as explanations of behavior? Have you ever been "labeled" in this way? What kinds of real-life problems can this cause?

All in all, personality disorders are convenient labels for identifying common patterns of ineffective and ultimately self-defeating behavior, but labels do not explain their causes. Still, the development of an accurate descriptive system is an important step toward scientific explanation. The establishment of reliable diagnostic categories sets the stage for valid research into causation and treatment.

Quiz 9.1
Types of Personality Disorders

Theoretical Perspectives

In this section we consider theoretical perspectives on the personality disorders. Many of the theoretical accounts of disturbed personality derive from the psychodynamic model. We thus begin with a review of traditional and modern psychodynamic models.

Psychodynamic Perspectives

Traditional Freudian theory focused on problems arising from the Oedipus complex as the foundation for many abnormal behaviors, including personality disorders. Freud believed that children normally resolve the Oedipus complex by forsaking incestuous wishes for the parent of the opposite gender and identifying with the parent of the same gender. As a result, they incorporate the parent's moral principles in the form of a personality structure called the superego. Many factors may interfere with appropriate identification, however, such as having a weak or absent father or an antisocial parent. These factors may sidetrack the normal developmental process, preventing children from developing the moral constraints that prevent antisocial behavior and the feelings of guilt or remorse that normally follow behavior that is hurtful to others. Freud's account of moral development focused mainly on the development of males. He has been criticized for failing to account for the moral development of females.

More recent psychodynamic theories have generally focused on the earlier, pre-Oedipal period of about 18 months to 3 years, during which infants are theorized to begin to develop their identities as separate from those of their parents. These recent advances in psychodynamic theory focus on the development of the sense of self in explaining such disorders as narcissistic and borderline personality disorders.

Hans Kohut One of the principal shapers of modern psychodynamic concepts is Hans Kohut, whose views are labeled **self psychology.** Kohut focused much of his attention on the development of the narcissistic personality.

self psychology A theory that describes processes that normally lead to achievement of a cohesive sense of self.

Kohut (1966) believed that people with narcissistic personalities might mount a facade of self-importance to cover up deep feelings of inadequacy. The narcissist's self-esteem is like a reservoir that needs to be constantly replenished lest it run dry. A steady stream of praise and attention prevents the narcissist from withering with insecurity. A sense of grandiosity helps people with a narcissistic personality mask their underlying feelings of worthlessness. Failures or disappointments threaten to expose these feelings and drive the person into a state of depression. As a defense against despair, the person attempts to diminish the importance of disappointments or failures. People with narcissistic personalities may become enraged by others whom they perceive have failed to protect them from disappointment or have declined to shower them with reassurance, praise, and admiration. They may become infuriated by even the slightest criticism, no matter how well intentioned. They may mask feelings of rage and humiliation by adopting a facade of cool indifference. They can make difficult psychotherapy clients because they may become enraged when therapists puncture their inflated self-images to help them develop more realistic self-concepts.

Truth OR Fiction? REVISITED

Despite a veneer of self-importance, people with narcissistic personalities may harbor deep feelings of insecurity.

TRUE. Theorists such as Hans Kohut believe that people with narcissistic personalities might mount a facade of self-importance to cover up deep feelings of inadequacy.

Kohut believed that early childhood is characterized by a normal stage of "healthful narcissism." Infants feel powerful, as though the world revolves around them. Infants also normally perceive their parents as idealized towers of strength and wish to be one with them and to share their power (Edmundson, 2001; Strozier, 2001). Empathic parents reflect their children's inflated perceptions by making them feel that anything is possible and by nourishing their self-esteem (e.g., telling them how terrific and precious they are). Even empathic parents are critical from time to time, however, and puncture their children's grandiose sense of self. Or they fail to measure up to their children's idealized views of them. Gradually, unrealistic expectations dissolve and are replaced by more realistic appraisals. This process of childhood narcissism that eventually gives way to more realistic appraisals of self and others is perfectly normal. Earlier grandiose self-images form the basis for assertiveness later in childhood and set the stage for ambitious striving in adulthood. In adolescence, childhood idealization is transformed into realistic admiration for parents, teachers, and friends. In adulthood, these ideas develop into a set of internal ideals, values, and goals.

splitting An inability to reconcile the positive and negative aspects of the self and others, resulting in sudden shifts between positive and negative feelings.

symbiotic The state of oneness that normally exists between mother and infant.

separation-individuation The process by which an infant develops a separate identity from that of the mother.

Lack of parental empathy and support, however, sets the stage for pathological narcissism in adulthood. Children who are not prized by their parents may fail to develop a sturdy sense of self-esteem. They may be unable to tolerate even slight blows to their self-worth. They develop damaged self-concepts and feel incapable of being loved and admired because of perceived inadequacies or flaws. Pathological narcissism involves the construction of a grandiose facade of self-perfection that is merely a shell to cloak perceived inadequacies. The facade always remains on the brink of crumbling, however, and it must be continually shored up by a constant flow of reassurance that one is special and unique. This leaves the person vulnerable to painful blows to self-esteem following failure to achieve social or occupational goals. So needy of constant approval, the person with a narcissistic personality may fly into a rage when he or she feels slighted in any way.

Kohut's approach to therapy provides clients who have a narcissistic personality with an initial opportunity to express their grandiose self-images and to idealize the therapist. Over time, however, the therapist helps them explore the childhood roots of their narcissism and gently points out imperfections in both client and therapist to encourage clients to form more realistic images of the self and others.

Otto Kernberg Modern psychodynamic views of the borderline personality also trace the disorder to difficulties in the development of the self in early childhood. Otto Kernberg (1975), a leading psychodynamic theorist, views borderline personality in terms of a pre-Oedipal failure to develop a sense of constancy and unity in one's image of the self and others. Kernberg proposes that childhood failure to synthesize these contradictory images of good and bad results in a failure to develop a consistent self-image and in tendencies toward **splitting**—shifting back and forth between viewing oneself and other people as "all good" or "all bad."

In Kernberg's view, parents, even excellent parents, invariably fail to meet all their children's needs. Infants therefore face the early developmental challenge of reconciling images of the nurturing, comforting "good mother" with those of the withholding, frustrating "bad mother." Failure to reconcile these opposing images into a realistic, unified, and stable parental image may fixate children in the pre-Oedipal period. As adults, then, they may retain these rapidly shifting attitudes toward their therapists and others.

Separation-individuation. According to the influential psychodynamic theorist Margaret Mahler, young children undergo a process of separation-individuation by which they learn to differentiate their own identities from their mothers. She believed that a failure to successfully master this developmental challenge may lead to the development of a borderline personality.

Margaret Mahler Margaret Mahler, another influential modern psychodynamic theorist, explained borderline personality disorder in terms of childhood separation from the mother figure. Mahler and her colleagues (Mahler & Kaplan, 1977; Mahler, Pine, & Bergman, 1975) believed that during the first year infants develop a **symbiotic** attachment to their mothers. *Symbiosis* is a biological term derived from Greek roots meaning "to live together" and describes life patterns in which two species lead interdependent lives. In psychology, symbiosis is likened to a state of oneness in which the child's identity is fused with the mother's. Normally, children gradually differentiate their own identities or senses of self from their mothers. The process is called **separation-individuation.** Separation is the developing of a separate psychological and biological identity from the mother. Individuation involves recognizing the personal characteristics that define one's self-identity. Separation-individuation may be a stormy process. Children may vacillate between seeking greater independence and moving closer to, or "shadowing," the mother, which is seen as a wish for reunion. The mother may disrupt normal separation-individuation by refusing to let go of the child or by too quickly pushing the child toward independence. The tendencies of people with borderline personalities to react to others with ambivalence, to alternate between love and hate, are suggestive to Mahler of earlier ambivalences during the separation-individuation process. Borderline personality disorder may arise from the failure to master this developmental challenge.

All in all, psychodynamic theory provides a rich theoretical mine for the understanding of the development of several personality disorders. But some critics contend that theories of disorders such as borderline personality disorder and narcissistic personality disorder are based largely on inferences drawn from behavior and retrospective accounts of adults rather than on observations of children (Sass, 1982). Mahler's theory has been challenged by evidence that even infants show a certain degree of psychological differentiation from others (Klein, 1981). We may also question whether direct comparisons should be made between normal childhood experiences and abnormal behaviors in adulthood. For example, the ambivalences that characterize the adult borderline personality may bear only a superficial relationship, if any, to children's vacillations between closeness and separation with maternal figures during separation-individuation.

Later we underscore the links between abuse in childhood and later development of personality disorders. These linkages suggest that failure to form close-bonding relationships with parental caretakers in childhood plays a critical role in developing many of the maladaptive personality patterns classified as personality disorders.

THINK ABOUT IT

What further research is needed to support psychodynamic theories about the development of personality disorders?

Learning Perspectives

Learning theorists tend to focus more on the acquisition of behaviors than on the notion of enduring personality traits. Similarly, they think more in terms of maladaptive behaviors than of disorders of "personality" or "personality traits." Trait theorists believe that personality traits steer behavior, providing a framework for consistent behavior in diverse situations. Many critics (e.g., Mischel, 1993), however, argue that behavior is actually less consistent across situations than trait theorists would suggest. Behavior may depend more on situational demands than on inherent traits. For example, we may describe a person as lazy and unmotivated. But is this person always lazy and unmotivated? Aren't there some situations in which the person may be energetic and ambitious? What differences in these situations may explain differences in behavior? Learning theorists are interested in defining the learning histories and situational factors that give rise to maladaptive behaviors and the reinforcers that maintain them.

WWW Web Link 9.4
Personality Disorders Foundation

Learning theorists suggest it is in childhood that many important experiences occur that shape the development of the maladaptive habits of relating to others that constitute personality disorders. For example, children who are regularly discouraged from speaking their minds or exploring their environments may develop a dependent personality behavior pattern. Obsessive-compulsive personality disorder may be connected with excessive parental discipline or overcontrol in childhood. Theodore Millon (1981) suggests that children whose behavior is rigidly controlled and punished by parents, even for slight transgressions, may develop inflexible, perfectionistic standards. As these children mature, they may strive to develop themselves in an area in which they excel, such as schoolwork or athletics, as a way of avoiding parental criticism or punishment. But overattention to a single area of development may prevent them from becoming well rounded. They may thus squelch spontaneity and avoid new challenges or risks. They may also place perfectionistic demands on themselves, so as to avoid any risk of punishment or rebuke, and develop other behaviors associated with the obsessive-compulsive personality pattern.

Millon suggests that histrionic personality disorder may be rooted in childhood experiences in which social reinforcers, such as parental attention, are connected to the child's appearance and willingness to perform for others, especially in cases where reinforcers are dispensed inconsistently. Inconsistent attention teaches children not to take approval for granted and to strive for it continually. People with histrionic personalities may also have identified with parents who are dramatic, emotional, and attention-seeking. Extreme sibling rivalry would further heighten motivation to perform for attention from others.

Social-cognitive theories emphasize the role of reinforcement in explaining the origins of antisocial behaviors. Ullmann and Krasner (1975) proposed, for example, that people with antisocial personalities might have failed to learn to respond to other people as potential reinforcers. Most children learn to treat others as reinforcing agents because others reinforce them with praise when they behave appropriately and punish them for misbehavior. Reinforcement and punishment provide feedback (information about social expectations) that helps children modify their behavior to maximize the chances of future rewards and minimize the risks of future punishment. As a consequence, children become socialized. They become sensitive to the demands of powerful others, usually parents and teachers, and learn to regulate their behavior accordingly. They thus adapt to social expectations. They learn what to do and what to say, how to dress and how to act to obtain social reinforcement or approval from others.

What are the origins of antisocial personality disorder? Are youth who develop antisocial personalities largely "unsocialized" because their early learning experiences lack the consistency and predictability that help other children connect their behavior with rewards and punishments? Or are they very "socialized," but socialized to imitate the behavior of other antisocial youth? To what extent does criminal behavior or membership in gangs overlap with antisocial personality disorder?

People with antisocial personalities, by contrast, may not have become socialized in this way because their early learning experiences lacked the consistency and predictability that helped other children connect their behavior with rewards and punishments. Perhaps they were sometimes rewarded for doing the "right thing," but just as often not. They may have borne the brunt of harsh physical punishments that depended more on parental whims than on their own conduct. As adults, they may not place much value on what other people expect because there was no clear connection between their own behavior and reinforcement in childhood. They may have learned as children that there was little they could do to prevent punishment and so perhaps lost the motivation to try. Although Ullmann and Krasner's views may account for some features of antisocial personality disorder, they may not adequately address the development of the "charming" type of antisocial personality; people in this group are skillful at reading the social cues of others and in using them for personal advantage.

Social-cognitive theorist Albert Bandura (1973, 1986) has studied the role of observational learning in aggressive behavior, which is one of the common components of antisocial behavior. He and his colleagues (e.g., Bandura, Ross, & Ross, 1963) have shown that children acquire skills, including aggressive skills, by observing the behavior of others. Exposure to aggression may come from watching violent television programs or in observing parents who act violently toward each other. Bandura does not believe children and adults display aggressive behaviors in a mechanical way, however. Rather, people usually do not imitate aggressive behavior unless they are provoked and believe they are more likely to be rewarded than punished for it. When models get their way with others by acting aggressively, children may be more likely to imitate them. Children may also acquire antisocial behaviors such as cheating, bullying, or lying by direct reinforcement if they find that those behaviors help them avoid blame or manipulate others.

Social-cognitive psychologists have also shown that the ways in which people with personality disorders interpret their social experiences influence their behavior. Antisocial adolescents, for example, tend to incorrectly interpret other people's behavior as threatening (Dodge, 1985). Often, perhaps because of their family and community experiences, they presume that others intend them ill when they do not. In a promising cognitive therapy method based on such findings, called **problem-solving therapy,** antisocial adolescent boys have been encouraged to reconceptualize their social interactions as problems to be solved rather than as threats to their "manhood" (Lochman, 1992). They then generate nonviolent solutions to social confrontations and, like scientists, test the most promising ones. In the section on biological perspectives we also see that the antisocial personality's failure to profit from punishment is connected with a cognitive factor: the *meaning* of the aversive stimulus.

All in all, learning approaches to personality disorders, like the psychodynamic approaches, have their limitations. They are grounded in theory rather than in observations of family interactions that presage the development of personality disorders. Research is needed to determine whether childhood experiences proposed by psychodynamic and learning theorists actually lead to the development of particular personality disorders as hypothesized.

Family Perspectives

Many theorists have argued that disturbances in family relationships underlie the development of personality disorders. Consistent with psychodynamic formulations, researchers find that people with borderline personality disorder (BPD) *remember* their parents as having been more controlling and less caring than do reference subjects with other psychological disorders (Zweig-Frank & Paris, 1991). When people with BPD recall their earliest memories, they are more likely than other people to paint significant others as malevolent or evil. They portray their parents and others close to them as having been likely to injure them deliberately or to fail to help them escape injuries by others (Nigg et al., 1992).

A number of researchers have linked a history of physical or sexual abuse or neglect in childhood to the development of personality disorders, including BPD, in adulthood (e.g., Johnson et al., 1999; Trull, 2001; Wilkinson-Ryan & Westen, 2000). Perhaps the "splitting" observed in people with the disorder is a function of having learned to cope with unpredictable and harsh behavior from parental figures or other caregivers.

problem-solving therapy A form of therapy that focuses on helping people develop more effective problem-solving skills.

Again consistent with psychodynamic theory, family factors such as parental overprotection and authoritarianism have been implicated in the development of dependent personality traits that may hamper the development of independent behavior (Bornstein, 1992). Extreme fears of abandonment may also be involved, perhaps resulting from a failure to develop secure bonds with parental attachment figures in childhood due to parental neglect, rejection, or death. Subsequently, a chronic fear of being abandoned by other people with whom one has close relationships may develop, leading to the clinginess that typifies dependent personality disorder. Theorists also suggest that obsessive-compulsive personality disorder may emerge within a strongly moralistic and rigid family environment, which does not permit even minor deviations from expected roles or behavior (e.g., Oldham, 1994).

As with BPD, researchers find that childhood abuse or neglect is a risk factor in the development of antisocial personality disorder (APD) in adulthood (Luntz & Widom, 1994). In a view that straddles the psychodynamic and learning theories, the McCords (McCord & McCord, 1964) focus on the role of parental rejection or neglect in the development of APD. They suggest that children normally learn to associate parental approval with conformity to parental practices and values, and disapproval with disobedience. When tempted to transgress, children feel anxious for fear of losing parental love. Anxiety serves as a signal that encourages the child to inhibit antisocial behavior. Eventually, the child identifies with parents and internalizes these social controls in the form of a conscience. When parents do not show love for their children, this identification does not occur. Children do not fear loss of love because they have never had it. The anxiety that might have served to restrain antisocial and criminal behavior is absent.

Children who are rejected or neglected by their parents may not develop warm feelings of attachment to others. They may lack the ability to empathize with the feelings and needs of others, developing instead an attitude of indifference toward others. Or perhaps they retain a wish to develop loving relationships but lack the ability to experience genuine feelings.

Although family factors may be implicated in some cases of antisocial personality disorder, many neglected children do not later show antisocial or other abnormal behaviors. We are left to develop other explanations to predict which deprived children will develop antisocial personalities or other abnormal behaviors, and which will not.

THINK ABOUT IT

What features do the psychodynamic and family perspectives on personality disorders have in common? How do they differ?

Biological Perspectives

Little is known about biological factors in most personality disorders. Although many theorists see personality disorders as the expression of maladaptive personality traits, the potential biological facets of such traits remain for the most part unknown.

Genetic Factors We have little direct evidence of genetic transmission of personality disorders (Carey & DiLalla, 1994). We do have suggestive evidence of genetic factors, based in part on findings that the first-degree biological relatives (parents and siblings) of people with certain personality disorders, especially antisocial, schizotypal, and borderline types, are more likely to be diagnosed with these disorders than are members of the general population (APA, 2000; Battaglia et al., 1995; Nigg & Goldsmith, 1994).

Studies of familial transmission are limited because family members share common environments as well as genes. Hence researchers have turned to twin and adoptee studies to tease out genetic and environmental effects. Evidence from twin studies suggests that the dimensions of personality associated with particular personality disorders may have an inherited component (Livesley et al., 1993). Researchers examined the genetic contribution to 18 dimensions that underlie various personality disorders, including callousness, identity problems, anxiousness, insecure attachment, narcissism, social avoidance, self-harm, and oppositionality (negativity) (Livesley et al., 1993). Genetic influences were suggested by findings of greater correlations on a given trait among identical (monozygotic, or MZ) twins than among fraternal (dizygotic, or DZ) twins. A statistical measure of heritability, reflecting the percentage of variability on a given trait that is accounted for by genetics, was computed for each personality dimension. The results showed that 12 of the 18 dimensions had heritabilities in the 40% to 60% range, indicating a substantial genetic contribution to these characteristics. The highest heritabilities were for narcissism (64%) and identity problems (59%) and the lowest for conduct problems (0%) and submissiveness (25%). Bear in mind that these twins were selected from the general population, not from a sample of people with diagnosed personality disorders. Therefore, the results may not be generalizable to people with diagnosable disorders. Still, the findings suggest that genetics plays a role in varying degrees to the development of traits that underlie personality disorders. Certainly not all people possessing these traits develop personality disorders. It is possible, however, that people with a genetic predisposition for these traits may be more vulnerable to developing personality disorders if they encounter certain environmental influences, such as being reared in a dysfunctional family.

Evidence from adoption studies suggests that both genetic and environmental factors affect the risk of developing antisocial personality disorder. Adopted-away children of biological parents with antisocial personality disorder stand a greater than average risk of developing the disorder themselves, but their risk level also depends on the family environment of their adoptive families (APA, 2000). Providing further evidence of the role of heredity, investigators find striking similarities among identical twins who were reared apart on antisocial or psychopathic personality traits (DiLalla et al., 1996).

Lack of Emotional Responsiveness According to a leading theorist, Hervey Cleckley (1976), people with antisocial personalities can maintain their composure in stressful situations that would induce anxiety in most people. Lack of anxiety in response to threatening situations may help explain the failure of punishment to induce antisocial people to relinquish antisocial behavior. For most of us, the fear of getting caught and being punished are sufficient to inhibit antisocial impulses. People with antisocial personalities, however, often fail to inhibit behavior that has led to punishment in the past (Arnett, Smith, & Newman, 1997). They may not learn to inhibit antisocial or aggressive behavior because they experience little if any fear or anticipatory anxiety about being caught and punished.

In an early classic study, Lykken (1957) showed that prison inmates with antisocial personalities performed more poorly than normal controls on a shock-avoidance task, although their general learning ability did not differ from that of normals. The shock-avoidance task involved learning responses to avoid getting a mild electric shock. Lykken reasoned that the inmates who had antisocial personalities were less able to learn avoidance responses because they experienced unusually low levels of anxiety in anticipation of the shock.

Schachter and Latané (1964) found that prisoners with antisocial personalities performed significantly better on the Lykken avoidance-learning task when they were administered epinephrine (adrenaline), a hormone that increases heart rate and other indices of arousal of the autonomic nervous system. Their performance apparently improved because the epinephrine had heightened their anticipatory anxiety. Other researchers (Chesno & Kilmann, 1975) showed similar results after raising antisocial subjects' levels of autonomic arousal through bursts of aversive noise rather than injections of epinephrine.

Cognitive theorists can point to research showing that the effects of aversive stimuli on people with antisocial personality disorder may depend on the *meaning* or *value* of the stimuli. Anticipation of aversive stimulation in the form of electric shock may not foster avoidance learning in persons with antisocial personalities, but the threat of punishment in the form of loss of money may do so. In another classic study, Schmauk (1970) had subjects perform a maze-learning task under three different forms of punishment for incorrect responses: electric shock, loss of money (losing a quarter for each error from an initial "bankroll" of 40 quarters), and social disapproval (the experimenter said "Wrong" following each incorrect response). Under the shock and social punishment conditions, people with antisocial personalities performed more poorly than normal controls. They outperformed normal controls, however, when they were threatened with forfeiture of money. Though people with antisocial personalities may not be deterred from misconduct by the threat of physical punishment, they may be keenly sensitive to the loss of money. Perhaps they learn better from their mistakes when the cost is meaningful to them.

Truth OR Fiction? REVISITED

People with antisocial personalities tend to remain unduly calm in the face of impending pain.

TRUE. People with antisocial personalities tend to show little anxiety in anticipation of impending pain. This lack of emotional responsivity may help explain why the threat of punishment seems to have so little effect on deterring their antisocial behavior.

When people get anxious, their palms tend to sweat. This skin response, called the *galvanic skin response* (GSR), is a sign of activation of the sympathetic branch of the autonomic nervous system (ANS). In an early study, Hare (1965) showed that people with antisocial personalities had lower GSR levels when they were expecting painful stimuli than did normal controls. Apparently, the people with antisocial personalities experienced little anxiety in anticipation of impending pain.

Hare's findings of a weaker GSR response in people with antisocial personalities has been replicated a number of times (e.g., Arnett, 1997; Patrick, Cuthbert, & Lang, 1994). Other research generally supports the view that people with antisocial personalities are generally less aroused than others, both at times of rest and in situations in which they are faced with stress (Fowles, 1993). This lack of emotional responsivity may help explain why the threat of punishment seems to have so little effect on deterring their antisocial behavior. It is conceivable that the autonomic nervous system (ANS) of people with antisocial personalities is underresponsive to threatening stimuli.

The Craving-for-Stimulation Model Other investigators have attempted to explain the antisocial personality's lack of emotional response in terms of the levels of stimulation necessary to maintain an **optimum level of arousal.** Our optimum levels of arousal are the degrees of arousal at which we feel best and function most efficiently.

Psychopathic individuals appear to have exaggerated cravings for stimulation (Arnett et al., 1997). Perhaps they require a higher-than-normal threshold of stimulation to maintain an optimum state of arousal (Quay, 1965). That is, they may need more stimulation than other people to maintain interest or function normally.

A need for higher levels of stimulation may explain why people with psychopathic traits tend to become bored more easily than other people and more often gravitate to more stimulating but potentially dangerous activities, like the use of intoxicants such as drugs or alcohol, motorcycling, skydiving, high-stakes gambling, or sexual adventures. A higher-than-normal threshold for stimulation would not directly cause antisocial or criminal behavior; after all, part of the "right stuff" of the nation's respected astronauts includes sensation seeking. However, the threat of boredom and inability to tolerate monotony may influence some sensation seekers to drift into crime or reckless behavior (R. J. Smith, 1978).

Brain Abnormalities Studies utilizing sophisticated brain-imaging techniques link antisocial personality disorder to abnormalities in the prefrontal cortex of the frontal lobes (Damasio, 2000; Raine et al., 2000). The prefrontal cortex is the part of the brain responsible for inhibiting impulsive behavior, weighing the consequences of our actions, solving problems, and planning for the future (Angier, 2000a; Duncan et al., 2000). Brain abnormalities could help account for many features of APD, including lack of conscience, failure to inhibit impulsive behavior, low arousal states, poor problem-solving efforts, and failure to think about the consequences of one's behavior before acting (Raine et al., 2000). Nevertheless, the question of just how many people with APD are affected by underlying brain abnormalities remains to be determined.

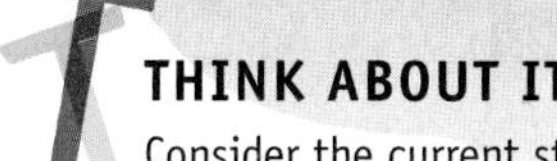

THINK ABOUT IT

Consider the current state of knowledge about biological causes of antisocial personality disorder. What social-policy issues does this information raise, if it is confirmed by further research?

Sociocultural Perspectives

The sociocultural perspective leads us to examine the social conditions that may contribute to the development of the behavior patterns identified as personality disorders. Because antisocial personality disorder is reported most frequently among people from lower socioeconomic classes, we might examine the role that the kinds of stressors encountered by disadvantaged families play in developing these behavior patterns. Many inner city neighborhoods are beset by social problems such as alcohol and drug abuse, teenage pregnancy, and disorganized and disintegrating families. These stressors are associated with an increased likelihood of child abuse and neglect, which may in turn contribute to lower self-esteem and breed feelings of anger and resentment in children. Neglect and abuse may become translated into the lack of empathy and a callous disregard for the welfare of others that are associated with antisocial personalities.

optimum level of arousal The level of arousal associated with peak performance and optimal feelings of well-being.

Questionnaire

The Sensation-Seeking Scale

Do you crave stimulation or seek sensation? Are you satisfied by reading or in watching television, or must you ride the big wave or bounce your motorbike over desert dunes? Zuckerman (1980) finds four factors related to sensation seeking: (1) pursuit of thrill and adventure, (2) disinhibition (that is, proclivity to express impulses), (3) pursuit of experience, and (4) susceptibility to boredom. Although some sensation seekers get involved with drugs or encounter trouble with the law, many are law abiding and limit their sensation seeking to sanctioned activities. Thus sensation seeking should not be interpreted as criminal or antisocial in itself.

Zuckerman developed several sensation-seeking scales that assess the levels of stimulation people seek to feel at their best and function efficiently. A brief form of one of them follows. To assess your own sensation-seeking tendencies, pick the choice, A or B, that best depicts you. Then compare your responses to those in the key at the end of the chapter.

1. A. I would like a job that requires a lot of traveling.
B. I would prefer a job in one location.

2. A. I am invigorated by a brisk, cold day.
B. I can't wait to get indoors on a cold day.

3. A. I get bored seeing the same old faces.
B. I like the comfortable familiarity of everyday friends.

4. A. I would prefer living in an ideal society in which everyone is safe, secure, and happy.
B. I would have preferred living in the unsettled days of our history.

5. A. I sometimes like to do things that are a little frightening.
B. A sensible person avoids activities that are dangerous.

6. A. I would not like to be hypnotized.
B. I would like to have the experience of being hypnotized.

7. A. The most important goal in life is to live it to the fullest and experience as much as possible.
B. The most important goal in life is to find peace and happiness.

8. A. I would like to try parachute jumping.
B. I would never want to try jumping out of a plane, with or without a parachute.

Sensation! Is there a connection between sensation seeking and antisocial personality disorder? Not all people who crave excitement have antisocial personalities. Yet people with antisocial personalities may have an excessive need for stimulation that makes them more likely to engage in antisocial or reckless behavior.

9. A. I enter cold water gradually, giving myself time to get used to it.
B. I like to dive or jump right into the ocean or a cold pool.

10. A. When I go on a vacation, I prefer the change of camping out.
B. When I go on a vacation, I prefer the comfort of a good room and bed.

11. A. I prefer people who are emotionally expressive even if they are a bit unstable.
B. I prefer people who are calm and even-tempered.

12. A. A good painting should shock or jolt the senses.
B. A good painting should give one a feeling of peace and security.

13. A. People who ride motorcycles must have some kind of unconscious need to hurt themselves.
B. I would like to drive or ride a motorcycle.

Source. From Zuckerman, M. Sensation seeking. In H. London & J. Exner (Eds.), *Dimensions of personality*. New York: John Wiley & Sons.

Children reared in poverty are also more likely to be exposed to deviant role models, such as neighborhood drug dealers. Maladjustment in school may lead to alienation and frustration with the larger society, leading to antisocial behavior (Siegel, 1992). Addressing the problem of antisocial personality may involve attempts at a societal level to redress social injustice and improve social conditions.

Little information is available about the rates of personality disorders in other cultures. One initiative in this direction involved a joint program sponsored by the World Health Organization (WHO) and the Alcohol, Drug Abuse, and Mental Health Administration (ADAMHA) of the U.S. government. The goal of the program was to develop and standardize diagnostic instruments that could be used to arrive at psychiatric diagnoses worldwide. One result of this effort was the development of the International Personality Disorder Examination (IPDE), a semistructured interview protocol for diagnosing personality disorders (Loranger et al., 1994). The IPDE was pilot-tested by psychiatrists and clinical psychologists in 11 different countries (India, Switzerland, the Netherlands, Great Britain, Luxembourg, Germany, Kenya, Norway, Japan, Austria, and the United States). The interview protocol had reasonably good reliability for diagnosing personality disorders among the different languages and cultures that were sampled. Although more research is needed to determine the rates of particular personality disorders in other countries, investigators found the borderline and avoidant types to be the most frequently diagnosed. Perhaps the characteristics associated with these personality disorders reflect some dimensions of personality disturbance that are commonly encountered throughout the world.

THINK ABOUT IT

Have you known anyone who you believe might fit the profile of the antisocial personality? What factors do you believe may have shaped this particular individual's personality development? How did the individual's personality affect his or her relationships with others?

Quiz 9.2
Theoretical Perspectives

Treatment of Personality Disorders

We began the chapter with a quote from the eminent psychologist William James, who suggested that people's personalities seem to be "set in plaster" by a certain age. His view may seem to be especially applicable to many people with personality disorders, who are typically highly resistant to change.

People with personality disorders usually see their behaviors, even maladaptive, self-defeating behaviors, as natural parts of themselves. Although they may be unhappy and distressed, they are unlikely to perceive their own behavior as causative. Like Marcella, whom we described as showing features of a histrionic personality disorder, they may condemn others for their problems and believe others, not they, need to change. Thus they usually do not seek help on their own. Or they begrudgingly acquiesce to treatment at the urging of others but drop out or fail to cooperate with the therapist. Or they may go for help when they feel overwhelmed by anxiety or depression and terminate treatment as soon as they find some relief rather than probe more deeply for the underlying causes of their problems. People with personality disorders also tend to respond more poorly to treatment of problems like depression than do others, perhaps because of the negative influence of their maladaptive behavioral patterns (Shea, Widiger, & Klein, 1992).

Psychodynamic Approaches

Psychodynamic approaches are often used to help people diagnosed with personality disorders become more aware of the roots of their self-defeating behavior patterns and learn more adaptive ways of relating to others. Progress in therapy may be hampered by difficulties in working therapeutically with people with personality disorders, especially clients with borderline and narcissistic personality disorders. Psychodynamic therapists often report that people with borderline personality disorder tend to have turbulent relationships with them, sometimes idealizing them, sometimes denouncing them as uncaring.

WWW **Web Link 9.5**
Online Screening for Personality Disorders (NYU School of Medicine)

Despite problems in treating people with personality disorders in psychotherapy, some promising results have been reported using psychodynamically oriented therapies (e.g., Bateman & Fonagy, 2001). One example involves a brief, structured form of psychodynamic therapy pioneered at New York's Beth Israel Medical Center (Winston et al., 1991). There, researchers reported that a relatively brief form of therapy that averaged 40 weeks of treatment resulted in significant improvement both in symptom complaints and the social adjustment of people with personality disorders (Winston et al., 1994). The treatment emphasized interpersonal behavior and used a more active, confrontational style in addressing the client's defenses than is the case in traditional psychoanalysis.

THINK ABOUT IT

What factors make it difficult to treat people with personality disorders? If you were a therapist, how might you attempt to overcome these difficulties?

THINK ABOUT IT

Have you known individuals whose personality traits or styles caused continuing difficulties in their relationships with others? Did their personalities relate to any one of the types of personality disorders discussed in this chapter? Did the person ever seek help from a mental health professional? If so, what was the outcome? If not, why not?

Behavioral Approaches

Behavior therapists see their task as changing clients' behaviors rather than their personality structures. Many behavioral theorists do not think in terms of clients' "personalities" at all, but rather in terms of acquired maladaptive behaviors that are maintained by reinforcement contingencies. Behavior therapists therefore focus on attempting to replace maladaptive behaviors with adaptive behaviors through techniques such as extinction, modeling, and reinforcement. If clients are taught behaviors likely to be reinforced by other people, the new behaviors may well be maintained.

Despite difficulties in treating borderline personality disorder (BPD), two groups of therapists headed by Aaron Beck (e.g., Arntz, 1994; Beck, Freeman, & Associates, 1990) and Marsha Linehan (Linehan, 1993; Linehan et al., 1991, 1994) report promising results using cognitive-behavioral techniques. Beck's approach focuses on helping the individual correct cognitive distortions that underlie tendencies to see oneself and others as either all good or all bad. Linehan's technique, called *dialectical behavior therapy* (DBT), combines behavior therapy and supportive psychotherapy. Behavioral techniques are used to help clients develop more effective social skills and problem-solving skills, which can help improve their relationships with others and ability to cope with negative events. Because people with BPD tend to be overly sensitive to even the slightest cues of rejection, therapists provide continuing acceptance and support, even when clients push the limits by becoming manipulative or overly demanding. While early results in using DBT are promising, researchers recognize that more research support is needed to support its efficacy in treating this challenging disorder (Scheel, 2000; Swenson, 2000; Turner, 2000).

Some antisocial adolescents have been placed, often by court order, in residential and foster-care programs that contain numerous behavioral treatment components. These residential programs have concrete rules and clear rewards for obeying them. At Achievement Place, for example, which was founded in Kansas in the 1960s and has been reproduced elsewhere, prosocial behaviors such as completing homework are systematically reinforced; antisocial behaviors, such as using profanity, are extinguished (Kirigin & Wolf, 1998). Some residential programs rely on *token economies,* in which prosocial behaviors are rewarded with tokens such as plastic chips that can be exchanged for privileges. Although participants in such programs may show improved behavior, it remains unclear whether such programs reduce the risk that adolescent antisocial behavior will continue into adulthood.

Biological Approaches

Drug therapy does not directly treat personality disorders. Antidepressants or antianxiety drugs are sometimes used to treat the emotional distress that individuals with personality disorders may encounter, however. Drugs do not alter the long-standing patterns of maladaptive behavior that may give rise to distress. However, a study indicates that the antidepressant Prozac can reduce aggressive behavior and irritability in impulsive and aggressive individuals with personality disorders (Coccaro & Kavoussi, 1997). Researchers suspect that impulsive and aggressive behavior may be related to serotonin deficiencies. Prozac and similar drugs act to increase the availability of serotonin in the synaptic connections in the brain.

Quiz 9.3
Treatment of Personality Disorders

Much remains to be learned about working with people who have personality disorders. The major challenges involve recruiting people who do not see themselves as being disordered into treatment and prompting them to develop insight into their self-defeating or injurious behaviors. Current efforts to help such people are too often reminiscent of the old couplet:

Quiz 9.4
Chapter Exam

> He that complies against his will,
> Is of his own opinion still.
>
> —Samuel Butler, *Hudibras*

In this chapter we have considered a number of problems in which people act out on maladaptive impulses yet fail to see how their behaviors are disrupting their lives. In the next chapter we explore other maladaptive behaviors that are frequently connected with lack of self-insight: behaviors involving substance abuse.

Research Update
Chapter 9

Overview of Personality Disorders

TYPES OF PERSONALITY DISORDERS

Personality Disorders Characterized by Odd or Eccentric Behavior

Paranoid Personality Disorder	• Pervasive suspiciousness of the motives of others but without outright paranoid delusions
Schizoid Personality Disorder	• Social aloofness and shallow or blunted emotions
Schizotypal Personality Disorder	• Persistent difficulty forming close social relationships and odd or peculiar beliefs and behaviors without clear psychotic features

Personality Disorders Characterized by Dramatic, Emotional, or Erratic Behavior

Antisocial Personality Disorder	• Chronic antisocial behavior, callous treatment of others, irresponsible behavior, and lack of remorse for wrongdoing
Borderline Personality Disorder	• Tumultuous moods and stormy relationships with others, unstable self-image, and lack of impulse control
Histrionic Personality Disorder	• Overly dramatic and emotional behavior; demands to be the center of attention; excessive needs for reassurance, praise, and approval
Narcissistic Personality Disorder	• Grandiose sense of self; extreme needs for admiration

Personality Disorders Characterized by Anxious or Fearful Behavior

Avoidant Personality Disorder	• Chronic pattern of avoiding social relationships due to fears of rejection
Dependent Personality Disorder	• Excessive dependence on others and difficulty making independent decisions
Obsessive-Compulsive Personality Disorder	• Excessive needs for orderliness and perfectionism, excessive attention to detail, rigid ways of relating to others

THEORETICAL PERSPECTIVES

Psychodynamic Perspectives

- To Kohut, the failure to replace childhood narcissism with more realistic appraisals of self and others underlies the development of narcissistic personality
- To Kernberg, the failure in early childhood to develop a cohesive sense of self and others leads to the development of borderline personality
- To Mahler, the failure to master the developmental challenge of separation-individuation early in life underlies the development of borderline personality

Learning Perspectives

- Behavioral features of personality disorders relate to learning experiences in childhood, including observational learning of deviant or aggressive behavior
- Lack of opportunity in childhood to learn explorative or independent behaviors may lead to dependent personality traits
- Excessive parental discipline or overcontrol may lead to obsessive-compulsive personality traits
- Inconsistent attention and reinforcement for attention-getting behaviors may lead to histrionic personality traits
- Lack of predictable and consistent reinforcement for socially approved behavior may lead to antisocial personality traits

Family Perspectives

- For antisocial personality disorder, parental rejection or neglect may lead to a failure to internalize parental values and failure to develop empathy
- Parental overprotection and authoritarianism may lead to the development of dependent personality traits

Sociocultural Perspectives

- Social or economic disadvantage and exposure to deviant role models may lead to the failure to develop properly socialized behaviors
- Physical or sexual abuse may underlie the development of borderline personality traits

Biological Perspectives

- Possible genetic influences on personality traits underlying personality disorders
- Possible inherited component of antisocial personality disorder
- For antisocial personality disorder, possible lack of emotional responsiveness in threatening situations
- For antisocial personality disorder, possible needs for higher levels of stimulation to maintain optimum levels of arousal
- For antisocial personality disorder, reduced activity in brain centers controlling impulsive behavior

▶

Overview of Personality Disorders (continued)

TREATMENT APPROACHES	Despite difficulties working therapeutically with individuals with personality disorders, promising results are emerging based on psychodynamic and cognitive-behavioral approaches
Drug Therapy	• Antidepressants or antianxiety drugs may be used to control symptoms but do not alter underlying patterns of behavior
Cognitive-Behavioral Therapy	• To help foster more adaptive behavior, to develop more effective social skills and problem-solving skills, and to replace faulty thinking with rational alternatives
Psychodynamic Therapy	• To help people understand the childhood roots of their problems and learn more effective ways of relating to others

Summing Up

Types of Personality Disorders

What are personality disorders? Personality disorders are maladaptive or rigid behavior patterns or personality traits associated with states of personal distress that impair the person's ability to function in social or occupational roles. People with personality disorders do not generally recognize a need to change themselves.

What are the classes of personality disorders within the DSM system? The *DSM* classifies personality disorders on Axis II and categorizes them according to the following clusters of characteristics: odd or eccentric behavior; dramatic, emotional, or erratic behavior; or anxious or fearful behavior.

What are the features associated with personality disorders characterized by odd or eccentric behavior? People with paranoid personality disorder are unduly suspicious and mistrustful of others, to the point that their relationships suffer. But they do not hold the more flagrant paranoid delusions typical of schizophrenia. Schizoid personality disorder describes people who have little if any interest in social relationships, show a restricted range of emotional expression, and appear distant and aloof. People with schizotypal personalities appear odd or eccentric in their thoughts, mannerisms, and behavior, but not to the degree found in schizophrenia.

What are the features associated with personality disorders characterized by dramatic, emotional, or erratic behavior? Antisocial personality disorder describes people who persistently engage in behavior that violates social norms and the rights of others and who tend to show no remorse for their misdeeds. Borderline personality disorder is defined in terms of instability in self-image, relationships, and mood. People with borderline personality disorder often engage in impulsive acts, which are frequently self-destructive. People with histrionic personality disorder tend to be highly dramatic and emotional in their behavior, whereas people diagnosed with narcissistic personality disorder have an inflated or grandiose sense of self, and like those with histrionic personalities, they demand to be the center of attention.

What are the features associated with personality disorders characterized by anxious or fearful behavior? Avoidant personality disorder describes people who are so terrified of rejection and criticism that they are generally unwilling to enter relationships without unusually strong reassurances of acceptance. People with dependent personality disorder are overly dependent on others and have extreme difficulty acting independently or making even the smallest decisions on their own. People with obsessive-compulsive personality disorder have various traits such as orderliness, perfectionism, rigidity, and overattention to detail, but are without the true obsessions and compulsions associated with obsessive-compulsive (anxiety) disorder.

What are some problems associated with the classification of personality disorders? Various controversies and problems attend the classification of personality disorders, including lack of demonstrated reliability and validity, too much overlap among the categories, difficulty in distinguishing between variations in normal behavior and abnormal behavior, underlying sexist biases in certain categories, and confusion of labels with explanations.

Theoretical Perspectives

How do traditional Freudian concepts of disturbed personality development compare with more recent psychodynamic approaches? Earlier Freudian theory focused on unresolved Oedipal conflicts in explaining normal and abnormal personality development. More recent psychodynamic theorists have focused on the pre-Oedipal period in explaining the development of such personality disorders as narcissistic and borderline personality.

How do learning theorists view personality disorders? Learning theorists view personality disorders in terms of maladaptive patterns of behavior rather than personality traits. Learning theorists seek to identify the early learning experiences and present reinforcement patterns that may explain the development and maintenance of personality disorders.

What is the role of family relationships in personality disorders? Many theorists argue that disturbed family relationships play formative roles in the development of personality disorders. For example, theorists have connected antisocial personality to parental rejection or neglect and parental modeling of antisocial behavior.

How do encoding strategies of antisocial adolescents differ from those of their peers? Antisocial adolescents are more likely to interpret social cues as provocations or intentions of ill will. This cognitive bias may lead them to be confrontative in their relationships with peers.

What roles might biological factors play in antisocial personality disorder? Research suggests that people with antisocial personalities may lack emotional responsiveness to physically threatening stimuli and have reduced levels of autonomic reactivity. People with antisocial personalities may also require higher levels of stimulation to maintain optimal levels of arousal.

What role do sociocultural factors play in the development of personality disorders? The effects of poverty, urban blight, and drug abuse can lead to family disorganization and disintegration, making it less likely that children will receive the nurturance and support they need to develop more socially adaptive behavior patterns. Sociocultural theorists believe that such factors may underlie the development of personality disorders, especially antisocial personality disorder.

Treatment of Personality Disorders

How do therapists approach the treatment of personality disorders? Therapists from different schools of therapy try to assist people with personality disorders to gain better awareness of their self-defeating behavior patterns and learn more adaptive ways of relating to others. Despite difficulties in working therapeutically with people with personality disorders, promising results have emerged from the use of relatively short-term psychodynamic therapy and cognitive-behavioral treatment approaches.

Key for Sensation-Seeking Scale

Because this is an abbreviated version of a questionnaire, no norms are applicable. However, answers that agree with the following key are suggestive of sensation seeking:

1. A
2. A
3. A
4. B
5. A
6. B
7. A
8. A
9. B
10. A
11. A
12. A
13. B

CHAPTER TEN

Substance Abuse and Dependence

Christopher Richard Wynne Nevinson
A Bursting Shell, 1915

Truth OR Fiction?

- More deaths are caused by illicit drugs, especially cocaine and heroin, than by legally available drugs. (p. 299)
- You cannot become psychologically dependent on a drug without first being physically addicted to it. (p. 303)
- Alcohol "goes to women's heads" more rapidly than to men's. (p. 308)
- Alcohol use at any level increases the risk of heart attacks. (p. 312)
- Heroin was developed during the search for a nonaddictive drug that would relieve pain as effectively as morphine. (p. 314)
- Coca-Cola originally contained cocaine. (p. 314)
- Habitual smoking is just a bad habit, not a physical addiction. (p. 317)
- Breast cancer is the leading cause of cancer deaths among U.S. women. (p. 317)
- Being able to "hold your liquor" better than most people helps prevent the development of problem drinking. (p. 322)
- A widely used treatment for addiction to heroin is based on the substitution of another addictive drug. (p. 331)

Our society is flooded with **psychoactive** substances, or drugs, that alter the mood and twist perceptions—substances that lift you up, calm you down, and turn you upside down. Many young people start using these substances because of peer pressure or because their parents and other authority figures tell them not to.

The old standby alcohol is the most popular drug on campus—whether the campus is a high school or college (Johnston, Bachman, & O'Malley, 1992). In fact, nearly 9 of 10 college students report using alcohol within the past year, as compared to 1 student in 3 who reports using any illicit drug. Under certain conditions, the use of substances that affect mood and behavior is normal enough, at least as gauged by statistical frequency and social standards. It is normal to start the day with caffeine in the form of coffee or tea, to take wine or coffee with meals, to meet friends for a drink after work, and to end the day with a nightcap. Many of us take prescription drugs that calm us down or ease our pain. Flooding the bloodstream with nicotine by means of smoking is normal in the sense that about 1 in 4 Americans smoke. Some psychoactive substances are illegal and are used illicitly, such as cocaine, marijuana, and heroin. Others are available by prescription, such as minor tranquilizers and amphetamines. Still others are available without prescription or over the counter, such as tobacco (which contains nicotine, a mild stimulant) and alcohol (which is a depressant). Ironically, the most widely and easily accessible substances—tobacco and alcohol—cause more deaths through sickness and accidents than all illicit drugs combined.

After falling steadily during the 1980s, use of illicit drugs, such as marijuana, among adolescents rose sharply during the early 1990s, before beginning to decline by the end of the decade (Machan, 2000; "Teen Drug Use," 2000). Still, nearly 1 in 10 (9%) 12- to 17-year-olds reports using illicit drugs, such as marijuana and heroin, during the past month (Stout, 2000). By the time young people get to their senior year in high school, about half have used an illicit drug (Johnston, O'Malley, & Bachman, 1996). Two in 5 have tried marijuana. Among college students, about half have smoked marijuana at least once. After declining for a number of years, cocaine abuse began creeping upwards in many U.S. cities during the late 1990s (LeDuff, 2000; Mathias, 2000).

Table 10.1 shows data on reported drug use compiled from a continuing government survey of young people in the United States. The results shown here focus on college students. Respondents are asked whether they have ever used a substance or used it during the past 30 days. Note the recent increase in cocaine use, after a period of steady decline. The use of marijuana also climbed sharply during the 1990s, partially reversing an earlier decline. The most dramatic increase, however, occurred with the use of ecstasy (MDMA).

The most widely used drugs on campus (excepting caffeine) remain alcohol and nicotine (in the form of cigarette smoking). Cigarette smoking is rising not only among college students but also among high school students. By 1999, nearly 3 of 10 high school students (28%) were lighting up (Centers for Disease Control, 2000). For every person who smokes marijuana for the first time each year, there are about 250 people who start smoking cigarettes (Stout, 2000).

Binge drinking has emerged as a major problem on college campuses. *Binge drinking* is usually defined as having five or more drinks (for men) or four or more drinks (for women) on one occasion. Nearly half of the nation's college students report they have engaged in binge drinking during the past 2 weeks (McGinn, 2000). The president of Pennsylvania State University claimed that binge drinking was a bigger problem on college campuses than the use of illicit drugs ("College Binge Drinking Worries," 1999). Nationally, alcohol dependence has been on the rise, due largely to increases in adolescent alcohol abuse leading eventually to outright dependence (Hill et al., 2000; Nelson, Heath, & Kessler, 1998).

Classification of Substance-Related Disorders

The *DSM-IV* classifies substance-related disorders into two major categories: substance use disorders and substance-induced disorders. **Substance use disorders** involve maladaptive use of psychoactive substances. These types of disorders include substance abuse and

Chapter Outline

Truth OR Fiction? REVISITED

More deaths are caused by illicit drugs, especially cocaine and heroin, than by legally available drugs.

FALSE. Two legally available substances, alcohol and tobacco, cause more deaths.

WWW Web Link 10.1
Binge Drinking

psychoactive Referring to chemical substances that have psychological effects.

substance use disorders Disorders that involve maladaptive use of psychoactive substances (e.g., substance dependence).

TABLE 10.1 Trends in Drug Use Among College Students During Lifetime and During Last 30 Days (in percentages)

Drug	Used . . .	1984	1988	1992	1996	2000
Marijuana	Ever	59.0	51.3	44.1	45.1	51.2
	Last 30 days	23.0	16.3	14.6	17.5	20.0
Inhalants	Ever	10.4	12.6	14.2	11.4	12.9
	Last 30 days	0.7	1.3	1.1	0.8	0.9
Hallucinogens (includes LSD)	Ever	12.9	10.2	12.0	12.6	14.4
	Last 30 days	1.8	1.7	2.3	1.9	1.4
Cocaine (includes crack)	Ever	21.7	15.8	7.9	5.0	9.1
	Last 30 days	7.6	4.2	1.0	0.8	1.4
MDMA ("ecstasy")	Ever	NA	NA	2.9	4.3	13.1
	Last 30 days	NA	NA	0.4	0.7	2.5
Heroin	Ever	0.5	0.3	0.5	0.7	1.7
	Last 30 days	0.0	0.1	0.0	0.0	0.2
Stimulants (other than cocaine	Ever	27.8	17.7	10.5	9.5	12.3
and crystal meth)	Last 30 days	5.5	1.8	1.1	0.9	2.9
Barbiturates	Ever	6.4	3.6	3.8	4.6	6.9
	Last 30 days	0.7	0.5	0.7	0.8	1.1
Alcohol	Ever	94.2	94.9	91.8	88.4	86.6
	Last 30 days	71.1	77.0	71.4	67.0	67.4
Cigarettes	Ever	NA	NA	NA	NA	NA
	Last 30 days	21.5	22.6	23.5	27.9	28.2

Source. Johnston, L. D., O'Malley, P. M., & Bachman, J. G. *National survey results on drug use* from *The Monitoring the Future Study* 1975–2000. Volume II., *College students and young adults, ages 19–40.* U.S. Department of Health and Human Services, Public Health Service, National Institutes of Health: National Institute on Drug Abuse, 2001, Tables 9-1 (p. 213) and 9-3 (p. 215).

substance-induced disorders Disorders that can be induced by using psychoactive substances, such as intoxication.

intoxication A state of drunkenness.

substance abuse The continued used of a psychoactive drug despite the knowledge that it is causing a social, occupational, psychological, or physical problem.

substance dependence Impaired control over the use of a psychoactive substance; often characterized by physiological dependence.

tolerance Physical habituation to a drug such that with frequent use, higher doses are needed to achieve the same effects.

substance dependence. **Substance-induced disorders** are those that can be induced by using psychoactive substances, such as intoxication, withdrawal syndromes, mood disorders, delirium, dementia, amnesia, psychotic disorders, anxiety disorders, sexual dysfunctions, and sleep disorders. Different substances have different effects, so some of these disorders apply to one, to a few, or to nearly all substances.

Substance **intoxication** refers to a state of drunkenness or "being high." These effects largely reflect the chemical actions of the psychoactive substances. The particular features of intoxication depend on which drug is ingested, the dose, the user's biological reactivity, and—to some degree—the user's expectations. Signs of intoxication often include confusion, belligerence, impaired judgment, inattention, and impaired motor and spatial skills. Extreme intoxication from use of alcohol, cocaine, opioids, and PCP can even result in death (yes, you can die from alcohol overdoses), either because of the substance's biochemical effects or because of behavior patterns—such as suicide—that are connected with psychological pain or impaired judgment brought on by use of the drug.

VIDEO 10.1
Substance Abuse:
Therapist Jean Obert

Substance Abuse and Dependence

Where does substance use end and abuse begin? According to the *DSM,* substance abuse involves a pattern of recurrent use that leads to damaging consequences. Damaging consequences may involve failure to meet one's major role responsibilities (e.g., as student, worker, or parent), putting oneself in situations where substance use is physically dangerous (e.g., mixing driving and substance use), encountering repeated problems with the law arising from substance use (e.g., multiple arrests for substance-related behavior), or having recurring social or interpersonal problems because of substance use (e.g., repeatedly getting into fights when drinking).

Two of the many faces of alcohol use—and abuse. Alcohol is our most widely used—and abused—drug. Many people use alcohol to celebrate achievements and happy occasions, as in the photograph on the left. Unfortunately, like the man in the photograph on the right, some people use alcohol to drown their sorrows, which may only compound their problems. Where exactly does substance use end and abuse begin? According to the *DSM,* use becomes abuse when it leads to damaging consequences.

When people repeatedly miss school or work because they are drunk or "sleeping it off," their behavior may fit the definition of **substance abuse.** A single incident of excessive drinking at a friend's wedding would not qualify. Nor would regular consumption of low to moderate amounts of alcohol be considered abusive so long as it is not connected with any impairment in functioning. Neither the amount nor the type of drug ingested, nor whether or not the drug is illicit, is the key to defining substance abuse according to the *DSM.* Rather, the determining feature of substance abuse is whether a pattern of drug-using behavior becomes repeatedly linked to damaging consequences.

withdrawal syndrome A characteristic cluster of symptoms following the sudden reduction or cessation of use of a psychoactive substance after physiological dependence has developed.

tachycardia Abnormally rapid heartbeat.

Substance abuse may continue for a long period of time or progress to **substance dependence,** a more severe type of substance use disorder in which abuse is associated with physiological signs of dependence (tolerance or withdrawal) *or* compulsive use of a substance. People who become compulsive users lack control over their drug use. They may be aware of how their drug use is disrupting their lives or damaging their health, but feel helpless or powerless to stop using drugs, even though they may want to. By the time they become dependent on a given drug, they've given over much of their lives to obtaining and using it. The diagnostic features associated with substance dependence are listed in Table 10.2.

Repeated use of a substance may alter the body's physiological reactions, leading to the development of tolerance or a physical withdrawal syndrome (see Table 10.2). **Tolerance** is a state of physical habituation to a drug such that with frequent use, higher doses are needed to achieve the same effect. A **withdrawal syndrome** (also called an *abstinence syndrome*) involves a characteristic cluster of withdrawal symptoms that occur when a dependent person abruptly stops using a particular substance following a period of heavy, prolonged use. People who experience a withdrawal syndrome often return to using the substance in order to relieve the discomfort associated with withdrawal, which serves to maintain the addictive pattern. Withdrawal symptoms vary with the particular type of drug. With alcohol dependence, typical withdrawal symptoms include dryness in the mouth, nausea or vomiting, weakness, **tachycardia,** anxiety and depression, headaches, insomnia, elevated blood pressure, and fleeting hallucinations.

Throes of withdrawal. Withdrawal symptoms are characteristic of the abstinence syndrome that develops when a person who is physiologically dependent on a drug abruptly suspends use of the drug.

delirium tremens A withdrawal syndrome that occurs following sudden decrease of drinking in people with chronic alcoholism.

delirium A state of mental confusion, disorientation, and extreme difficulty focusing attention.

disorientation A state of mental confusion and lack of awareness of time, place, or the identity of oneself or others.

TABLE 10.2 Diagnostic Features of Substance Dependence

Substance dependence is defined as a maladaptive pattern of use that results in significant impairment or distress, as shown by the following features occurring within the same year:

1. Tolerance for the substance, as shown by either
   (a) the need for increased amounts of the substance to achieve the desired effect or intoxication, or
   (b) marked reduction in the effects of continuing to ingest the same amounts.
2. Withdrawal symptoms, as shown by either
   (a) the withdrawal syndrome that is considered characteristic for the substance, or
   (b) the taking of the same substance (or a closely related substance, as when methadone is substituted for heroin) to relieve or to prevent withdrawal symptoms.
3. Taking larger amounts of the substance, or for longer periods of time than the individual intended (e.g., person had desired to take only one drink, but after taking the first, continues drinking until severely intoxicated).
4. Persistent desire to cut down or control intake of substance or lack of success in trying to exercise self-control.
5. Spending a good deal of time in activities directed toward obtaining the substance (e.g., visiting several physicians to obtain prescriptions or engaging in theft), in actually ingesting the substance, or in recovering from its use. In severe cases, the individual's daily life revolves around substance use.
6. The individual has reduced or given up important social, occupational, or recreational activities due to substance use (e.g., person withdraws from family events in order to indulge in drug use).
7. Substance use is continued despite evidence of persistent or recurrent psychological or physical problems either caused or exacerbated by its use (e.g., repeated arrests for driving while intoxicated).

Note. Not all of these features need be present for a diagnosis to be made.
Source. Adapted from the *DSM-IV-TR* (APA, 2000).

In some cases of chronic alcoholism, withdrawal produces a state of **delirium tremens,** or DTs. The DTs are usually limited to chronic, heavy users of alcohol who dramatically lower their intake of alcohol after many years of heavy drinking. The DTs involve intense autonomic hyperactivity (profuse sweating and tachycardia) and **delirium**—a state of mental confusion characterized by incoherent speech, **disorientation,** and extreme restlessness. Terrifying hallucinations—frequently of creepy, crawling animals—may also be present.

Substances that may lead to withdrawal syndromes include, in addition to alcohol, opioids, cocaine, amphetamines, sedatives and barbiturates, nicotine, and antianxiety agents (minor tranquilizers). Marijuana and hallucinogens like LSD are not recognized as producing a withdrawal syndrome, because of a lack of evidence that abrupt withdrawal from these substances reliably produces clinically significant withdrawal effects (APA, 2000).

In the *DSM* system, substance dependence is often, but not always, associated with the development of physiological dependence (Langenbucher et al., 2000). In some cases it involves a pattern of compulsive use without physiological dependence. For example, people may become compulsive users of marijuana, especially when they come to rely on the drug to help them cope with the stresses of daily life. Yet they may not require larger amounts of the substance to get "high" or experience distressing withdrawal symptoms when they cease using it. In most cases, however, substance dependence and physiological features of dependence occur together. Despite the fact that the *DSM* considers substance abuse and dependence to be distinct diagnostic categories, the borderline between the two is not always clear.

THINK ABOUT IT

What is the basis for determining when drug use becomes abuse or dependence? Have you or someone you know crossed the line between use and abuse?

An estimated 15 million people in the United States suffer from substance dependence (Cowan & Kandel, 2001). Alcohol dependence alone affects about 1 in 7 (14%) U.S. adults (Anthony, Warner, & Kessler, 1994; Warner et al., 1995). About 1 in 4 U.S. adults suffer from nicotine dependence resulting from regular use of tobacco products, most usually cigarettes (Breslau et al., 2001). About 1 in 13 (7.5%) have developed a dependence on an illicit drug, inhalant, or nonprescription use of tranquilizers or other psychiatric (psychotropic) drugs. Figure 10.1 gives the lifetime prevalence of drug dependence for various types of drugs.

People may abuse or become dependent on more than one psychoactive substance at the same time. People who abuse or become dependent on heroin, for instance, may also abuse or become dependent on other drugs, such as alcohol, cocaine, or stimulants—either simultaneously or successively. People who engage in these patterns of polydrug abuse face increased risk of harmful overdoses. Moreover, the "successful" treatment of one form of abuse may not affect, or in some cases may even exacerbate, abuse of other drugs.

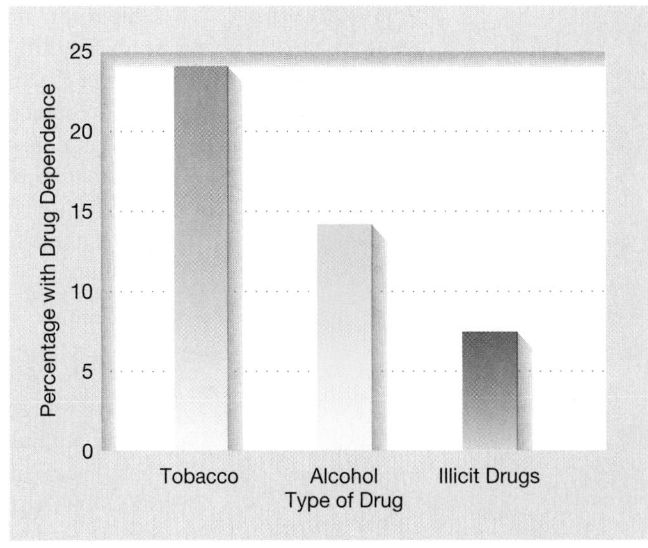

FIGURE 10.1 Lifetime prevalence of drug dependence by type of drug.
One in 4 adults in the United States suffers from tobacco dependence at some point. About 1 in 7 experiences alcohol dependence and about 1 in 13 develops a drug dependence on an illicit drug.
Source. National Comorbidity Survey (Anthony, Warner, & Kessler, 1994).

Addiction, Physiological Dependence, and Psychological Dependence

The *DSM* uses the terms *substance abuse* and *substance dependence* to classify people whose use of these substances impairs their functioning. It does not use the term *addiction* to describe these problems. Yet the concept of addiction is widespread among professionals and laypeople alike. But what is meant by addiction?

People define addiction in different ways. For our purposes, we define **addiction** as the habitual or compulsive use of a drug accompanied by evidence of physiological dependence. **Physiological dependence** means that one's body has changed as a result of the regular use of a psychoactive drug in such a way that it depends on a steady supply of the substance. The major signs of physiological dependence involve the development of tolerance and/or an abstinence syndrome. **Psychological dependence** involves the compulsive use of a drug to meet a psychological need, such as relying on a drug to cope with stress. As we noted earlier, you can become psychologically dependent on a drug without developing a physiological dependence or addiction.

On the other hand, people may become physiologically dependent on a drug but not become a compulsive user or psychologically dependent. For example, people recuperating from surgery are often given narcotics derived from opium as painkillers. Some may develop signs of physiological dependence, such as tolerance and a withdrawal syndrome, but not become habitual users or show a lack of control over the use of these drugs.

In recent years, the concept of addiction has also extended beyond the abuse of chemical substances to apply to many habitual forms of maladaptive behavior, such as pathological gambling. In the vernacular, we hear of people being addicted to love or shopping or to almost anything. The concept of addiction has even been extended to excessive use of the Internet (Griffiths, 1999; Young, 1999). Research evidence is accumulating that suggests that excessive Internet use carries similar risks as gambling (Jamison, 2000). Investigators find that it can lead to social isolation, depression, and problems at work or school. Many heavy Internet users show evidence of compulsive use, especially those who are drawn to the anonymous social contacts available through Internet chat rooms. They may feel a building up of tension prior to engaging in Internet activity, which is followed by a sudden relief when they log on. Many heavy users sacrifice sleep or forego family or work responsibilities to satisfy their compulsive needs. Internet chat rooms may also pose dangers of spreading sexually transmitted disease. Investigators found that seeking sexual partners through the Internet was a common practice among a

Truth OR Fiction? REVISITED

You cannot become psychologically dependent on a drug without first being physically addicted to it.

FALSE. You can become psychologically dependent on a drug without developing a physiological dependence.

addiction Impaired control over the use of a chemical substance, accompanied by physiological dependence.

physiological dependence A condition in which the drug user's body comes to depend on a steady supply of the substance.

psychological dependence Compulsive use of a substance to meet a psychological need.

Internet addiction? Compulsive, dependent use of the Internet and other forms of compulsive behavior, such as compulsive gambling or shopping, have been likened to forms of nonchemical addiction. Whether we label such maladaptive patterns of behavior as "addictions" depends on how we define the concept of addiction.

sample of people seeking HIV testing and counseling (McFarlane et al., 2000). In another study, an outbreak of syphilis was linked to a group of people who had met through an Internet chat room (Klausner et al., 2000).

Whether compulsive gambling or shopping or excessive Internet use constitute addictions depends on how we define our terms. For our purposes, we limit the term *addiction* to compulsive use of substances that produce physiological dependence. Although these other behaviors may be forms of compulsive behavior, we are reluctant to call them addictions because they do not involve physiological dependence on a chemical substance.

Racial/Ethnic Differences in Substance Dependence

Despite the popular stereotype that drug dependence is more frequent among ethnic minorities, this belief is not supported by evidence. The National Comorbidity Survey (NCS) shows that drug dependence is actually less common among African Americans than non-Hispanic White Americans and no more common among Hispanic Americans than among non-Hispanic White Americans (Anthony et al., 1994). Moreover, African American adolescents are much less likely than non-Hispanic White adolescents to develop substance abuse or dependence problems (Kilpatrick et al., 2000). In a later section we shall examine evidence on racial/ethnic group differences in alcohol use and abuse.

Pathways to Drug Dependence

Although the progression to substance dependence varies from person to person, some common pathways can be described according to the following stages (Weiss & Mirin, 1987):

1. *Experimentation.* During the stage of experimentation, or occasional use, the drug temporarily makes users feel good, even euphoric. Users feel in control and believe they can stop at any time.
2. *Routine use.* During the next stage, a period of routine use, people begin to structure their lives around the pursuit and use of drugs. Denial plays a major role at this stage, as users mask the negative consequences of their behavior to themselves and others. Values change. What had formerly been important, such as family and work, comes to matter less than the drugs.

   The following clinical interview illustrates how denial can mask reality. This 48-year-old executive was brought for a consultation by his wife. She complained his once-successful business was jeopardized by his erratic behavior, he was grouchy and moody, and he had spent $7,000 in the previous month on cocaine.

   CLINICIAN: Have you missed many days at work recently?
   EXECUTIVE: Yes, but I can afford to, since I own the business. Nobody checks up on me.
   CLINICIAN: It sounds like that's precisely the problem. When you don't go to work, the company stays open, but it doesn't do very well.
   EXECUTIVE: My employees are well trained. They can run the company without me.
   CLINICIAN: But that's not happening.
   EXECUTIVE: Then there's something wrong with them. I'll have to look into it.
   CLINICIAN: It sounds as if there's something wrong with you, but you don't want to look into it.
   EXECUTIVE: Now you're on my case. I don't know why you listen to everything my wife says.
   CLINICIAN: How many days of work did you miss in the last two months?
   EXECUTIVE: A couple.
   CLINICIAN: Are you saying that you missed only two days of work?

THINK ABOUT IT

Do you or someone you know show evidence of any nonchemical forms of addiction, such as compulsive shopping, gambling, or sexual behavior? How is this behavior affecting your (his or her) life? What can you (he or she) do about overcoming it?

EXECUTIVE: Maybe a few.
CLINICIAN: Only three or four days?
EXECUTIVE: Maybe a little more.
CLINICIAN: Ten? Fifteen?
EXECUTIVE: Fifteen.
CLINICIAN: All because of cocaine?
EXECUTIVE: No.
CLINICIAN: How many were because of cocaine?
EXECUTIVE: Less than fifteen.
CLINICIAN: Fourteen? Thirteen?
EXECUTIVE: Maybe thirteen.
CLINICIAN: So you missed thirteen days of work in the last two months because of cocaine. That's almost two days a week.
EXECUTIVE: That sounds like a lot but it's no big deal. Like I say, the company can run itself.
CLINICIAN: How long have you been using cocaine?
EXECUTIVE: About three years.
CLINICIAN: Did you ever use drugs or alcohol before that in any kind of quantity?
EXECUTIVE: No.
CLINICIAN: Then let's think back five years. Five years ago, if you had imagined yourself missing over a third of your workdays because of a drug, and if you had imagined yourself spending the equivalent of $84,000 a year on that same drug, and if you saw your once-successful business collapsing all around you, wouldn't you have thought that that was indicative of a pretty serious problem?
EXECUTIVE: Yes, I would have.
CLINICIAN: So what's different now?
EXECUTIVE: I guess I just don't want to think about it.

—From Weiss & Mirin, 1987, pp. 79–80

As routine drug use continues, problems mount. Users devote more resources to their drugs. Family bank accounts are ravaged, "temporary" loans are sought from friends and family for trumped-up reasons, and family heirlooms and jewelry are sold to pawnbrokers for a fraction of their value. Lying and manipulation become a way of life to cover up the drug use. The husband sells the TV set to a pawnbroker and forces the front door open to make it look like a burglary. The wife claims to have been robbed at knifepoint to explain the disappearance of a gold chain or engagement ring. Family relationships become strained as the mask of denial shatters and the consequences of drug abuse become apparent: days lost from work, unexplained absences from home, rapid mood shifts, depletion of family finances, failure to pay bills, stealing from family members, and missing family gatherings or children's birthday parties.

VIDEO 10.2
Substance Abuse: *Therapist Louise Roberts*

3. *Addiction or dependence.* Routine use becomes addiction or dependence when users feel powerless to resist drugs, either because they want to experience their effects or to avoid the consequences of withdrawal. Little or nothing else matters at this stage, as seen in the case of a 41-year-old architect, who related the following conversation with his wife:

A Case of Cocaine Dependence

She had just caught me with cocaine again after I had managed to convince her that I hadn't used in over a month. Of course I had been tooting (snorting) almost every day, but I had managed to cover my tracks a little better than usual. So she said to me that I was going to have to make a choice—either cocaine or her. Before she finished the

THINK ABOUT IT

Agree or disagree with the following statement and support your answer: "I have control over any drugs I am using."

Quiz **10.1**
Classification of Substance-Related Disorders

sentence, I knew what was coming, so I told her to think carefully about what she was going to say. It was clear to me that there wasn't a choice. I love my wife, but I'm not going to choose anything *over cocaine. It's sick, but that's what things have come to. Nothing and nobody comes before my coke.*

—*From Weiss & Mirin, 1987, p. 55*

Now let us examine the effects of different types of drugs of abuse and the consequences associated with their use and abuse.

Drugs of Abuse

Drugs of abuse are generally classified within three major groupings: (1) depressants, such as alcohol and opioids; (2) stimulants, such as amphetamines and cocaine; and (3) hallucinogens.

Depressants

A **depressant** is a drug that slows down or curbs the activity of the central nervous system. It reduces feelings of tension and anxiety, causes our movements to become sluggish, and impairs our cognitive processes. In high doses, depressants can arrest vital functions and cause death. The most widely used depressant, alcohol, can lead to death when taken in large amounts because of its depressant effects on respiration (breathing). Other effects are specific to the particular kind of depressant. For example, some depressants, such as heroin, produce a "rush" of pleasure. Here let us consider several of the major types of depressants.

Alcohol You may not have thought of alcohol as a drug, perhaps because it is so popular, or perhaps because it is ingested by drinking rather than by smoking or injection. But alcoholic beverages—such as wine, beer, and hard liquor—contain a depressant called *ethyl alcohol* (or *ethanol*). The concentration of the drug varies with the type of beverage (wine and beer have less pure alcohol per ounce than distilled spirits such as rye, gin, or vodka). Alcohol is classified as a depressant drug because it has biochemical effects similar to those of a class of minor tranquilizers, the benzodiazepines, which includes the well-known drugs *diazepam* (Valium) and *chlordiazepoxide* (Librium). We can think of alcohol as a type of over-the-counter tranquilizer.

Alcohol is used in many ways. It is our mealtime relaxant, our party social facilitator, our bedtime sedative. We observe holy days, laud our achievements, and express joyful wishes with alcohol. Adolescents assert their maturity with alcohol. Pediatricians have swabbed the painful gums of teething babies with alcohol. Alcohol even deals the deathblow to germs on surface wounds and is the active ingredient in antiseptic mouthwashes.

Most American adults drink alcohol at least occasionally. Most people who drink do so in moderation, but many develop significant problems with alcohol use (Garbutt et al., 1999; Miller & Brown, 1997). Alcohol is the most widely abused substance in the United States and worldwide. Alcohol dependence affects an estimated 14 million Americans, about 14% of the adult population. Alcohol abuse without dependence is believed to affect about 9% of adult Americans (Kessler et al., 1994). Many lay and professional people use the term ***alcoholism*** to refer to alcohol dependence. Though definitions of alcoholism vary, we use the term to refer to a physical dependence on, or addiction to, alcohol that is characterized by impaired control over the use of the drug.

The personal and social costs of alcoholism exceed those of all illicit drugs combined. Alcohol abuse is connected with lower productivity, loss of jobs, and downward movement in socioeconomic status. Estimates are that perhaps 30% to 40% of homeless people in the United States suffer from alcoholism (McCarty et al., 1991). About one in three suicides in this country and about the same proportion of deaths due to unintentional injury (such as from motor vehicle accidents) are believed to be alcohol-related (Hingson et al., 2000).

depressant A drug that lowers the level of activity of the central nervous system.

alcoholism An alcohol dependence disorder or addiction that results in serious personal, social, occupational, or health problems.

Questionnaire

Are You Hooked?

Are you dependent on alcohol? If you shake and shiver and undergo the tortures of the darned (our editor insisted on changing this word to maintain the decorum of a textbook) when you go without a drink for a while, the answer is clear enough. Sometimes the clues are more subtle, however.

The following questions, adapted from the National Council on Alcoholism's self-test, can shed some light on the question. Simply place a check mark in the yes or no column for each item. Then check the key at the end of the chapter.

	YES	NO
1. Do you sometimes go on drinking binges?	___	___
2. Do you tend to keep away from your family or friends when you are drinking?	___	___
3. Do you become irritated when your family or friends talk about your drinking?	___	___
4. Do you feel guilty now and then about your drinking?	___	___
5. Do you often regret the things you have said or done when you have been drinking?	___	___
6. Do you find that you fail to keep the promises you make about controlling or cutting down on your drinking?	___	___
7. Do you eat irregularly or not at all when you are drinking?	___	___
8. Do you feel low after drinking?	___	___
9. Do you sometimes miss work or appointments because of drinking?	___	___
10. Do you use more and more to get drunk or high?	___	___

Source. Adapted from *Newsweek*, February 20, 1989, p. 52.

More teenagers die from alcohol-related motor vehicle accidents than from any other cause (National Highway Traffic Safety Administration, 1988). All told, an estimated 100,000 people in the United States die from alcohol-related causes each year, mostly from alcohol-related motor vehicle crashes and diseases (Kalb, 2001b; Wood, Vinson, & Sher, 2001).

Alcohol, not cocaine or other drugs, is the drug of choice among young people today and the leading drug of abuse (Johnston et al., 1992). Drinking has become so integrated into college life that it has become essentially normative, as much a part of the college experience as attending a weekend football or basketball game. Despite the popular image of the person who develops alcoholism as a skid-row drunk, only a small minority of people with alcoholism fit the stereotype. The great majority of people with alcoholism are the type of people you're likely to see every day—your neighbors, coworkers, friends, and members of your own family. They are found in all walks of life and from every social and economic class. Many have families, hold good jobs, and live fairly comfortably. Yet alcoholism can have just as devastating an effect on the well-to-do as the indigent, leading to wrecked careers and marriages, to motor vehicle and other accidents, and to severe, life-threatening physical disorders, as well as exacting an enormous emotional toll.

WWW **Web Link 10.2**
How to Cut Down on Your Drinking

No one drinking pattern is exclusively associated with alcoholism. Some people with alcoholism drink heavily every day; others binge only on weekends. Others can abstain for lengthy periods of time but periodically "go off the wagon" and engage in episodes of binge drinking that may last for weeks or months.

WWW **Web Link 10.3**
Self-Screening for Alcoholism

Risk Factors for Alcoholism Investigators have identified a number of factors that place people at increased risk for developing alcoholism and alcohol-related problems. These include the following:

1. *Gender.* Men are more than twice as likely as women (20% vs. 8%, respectively) to develop alcohol dependence disorder (Grant, 1997). One possible reason for this gender difference is sociocultural; perhaps tighter cultural constraints are placed on women. Yet it may also be that alcohol hits women harder, and not only because women usually weigh less than men. Alcohol seems to "go to women's heads" more rapidly than men's. This is apparently because women metabolize less alcohol in the

Truth OR Fiction? REVISITED

Alcohol "goes to women's heads" more rapidly than to men's.

TRUE. Women metabolize less alcohol in the stomach than do men, which means that ounce for ounce women drinkers absorb more alcohol into their bloodstreams than do their male counterparts.

stomach than men do. Why is this? It appears that women have less of an enzyme that metabolizes alcohol in the stomach than men do (Lieber, 1990). Consequently, ounce for ounce women absorb more alcohol into their bloodstreams than do men. As a result, they are likely to become inebriated on less alcohol than men.

2. *Age.* The great majority of cases of alcohol dependence develop in young adulthood, typically before age 40 (Langenbucher & Chung, 1995). Although alcohol use disorders tend to develop somewhat later in women than in men, women who develop these problems experience similar health, social, and occupational problems by middle age as their male counterparts.
3. *Antisocial personality disorder.* Antisocial behavior in adolescence or adulthood increases the risk of later alcoholism. On the other hand, many people with alcoholism showed no antisocial tendencies in adolescence, and many antisocial adolescents do not abuse alcohol or other drugs as adults (Nathan, 1988).
4. *Family history.* The best predictor of problem drinking in adulthood appears to be a family history of alcohol abuse. Family members who drink may act as models ("set a poor example"). Moreover, the biological relatives of people with alcohol dependence may also inherit a predisposition that makes them more likely to develop problems with alcohol.
5. *Sociodemographic factors.* A lifetime history of alcohol dependence is more common among people of lower income and educational levels and among people living alone (Anthony et al., 1994).

Ethnicity and Alcohol Use and Abuse Higher rates of alcohol use and alcoholism are found to vary among American ethnic and racial groups (Beauvais, 1998; Lex, 1987; Moncher, Holden, & Trimble, 1990; Schinke, 1999). Jewish Americans, for example, have relatively low rates of alcoholism (Yeung & Greenwald, 1992), perhaps because Jews tend to expose children to the ritual use of wine within a religious context and to impose strong cultural restraints on excessive and underage drinking. Asian Americans also tend to drink less heavily than most other groups (Schinke, 1999). They too also place strong cultural constraints on excessive drinking. But a biological factor may also be involved. Asian Americans are more likely than other groups to show a flushing response to alcohol. Flushing is characterized by redness and feelings of warmth on the face, and, at higher doses, nausea, heart palpitations, dizziness, and headaches (Ellickson, Hays, & Bell, 1992). Genes that control the metabolism of alcohol are responsible for the flushing response (Begley, 2001b). Since people like to avoid these unpleasant experiences, flushing may serve as a natural defense against alcoholism by curbing excessive alcohol intake.

Women and alcohol. Women are less likely to develop alcoholism, in part because of greater cultural constraints on excessive drinking by women and perhaps because women absorb more pure alcohol into the bloodstream than men; thus women become more affected by the alcohol they consume than men who drink the same amount.

Hispanic American men and non-Hispanic White men have similar rates of alcohol consumption and alcohol-related physical problems (Caetano, 1987; Kessler et al., 1994). Hispanic American women, however, are much less likely to use alcohol and to develop alcohol use disorders than non-Hispanic White women. Why this difference? An important factor may be cultural expectations. Traditional Hispanic American cultures place severe restrictions on the use of alcohol by women, especially heavy drinking. However, with increasing acculturation, Hispanic American women in the United States apparently are becoming more similar to Euro-American women with respect to

Alcohol and ethnic diversity. The damaging effects of alcohol abuse appear to be taking the heaviest toll on African Americans and Native Americans. The prevalence of alcohol-related cirrhosis of the liver is nearly twice as high among African Americans than among White Americans, even though African Americans are less likely to develop alcohol abuse or dependence disorders. Jewish Americans have relatively low incidences of alcohol-related problems, perhaps because they tend to expose children to the ritual use of wine in childhood and impose strong cultural restraints on excessive drinking. Asian Americans tend to drink less heavily than most other Americans, in part because of cultural constraints and possibly because they have less biological tolerance of alcohol, as shown by a greater flushing response to alcohol.

alcohol use and abuse. Drinking rates among more acculturated Asian American groups are also comparable to those of the general population (Schinke, 1999).

Alcohol abuse is taking a heavy toll on African Americans. The prevalence of *cirrhosis* a degenerative, potentially fatal liver disease, is nearly twice as high in African Americans as in non-Hispanic White Americans. African Americans are also much more likely to develop alcohol-related coronary heart disease and oral and throat cancers (Rogan, 1986). Yet African Americans are much less likely than non-Hispanic White Americans to develop alcohol abuse or dependence (Anthony et al., 1994; Grant et al., 1994). Why, then, do African Americans suffer more from alcohol-related problems?

Socioeconomic factors may help explain these differences. African Americans are more likely to encounter the stresses of unemployment and economic hardship, and stress may compound the damage to the body caused by heavy alcohol consumption. African Americans also tend to have poorer access to medical services and may be less likely to receive early treatment for the medical problems caused by alcohol abuse.

American Indians are perhaps the American ethnic group that suffers most from alcohol-related problems. Though rates of drinking vary from tribe to tribe, the American Indian population has very high rates of problem drinking and alcohol-related consequences, such as cirrhosis of the liver, fetal abnormalities, and automobile and other accident fatalities (Beauvais, 1998; Rabasca, 2000a; Schinke, 1999).

Many Indian people believe the loss of their culture is largely responsible for their high rates of drinking-related problems (Beauvais, 1998). The disruption of traditional Indian culture caused by the appropriation of Indian lands and by attempts by European American society to sever Native Americans from their cultural traditions while denying them full access to the dominant culture resulted in severe cultural and social disorganization that may account for their high rates of depression and substance abuse (Kahn, 1982). Beset by such problems, Native American adults are also prone to child abuse and neglect. Abuse and neglect contribute to feelings of hopelessness and depression among adolescents, who then seek to escape their feelings through alcohol and other drugs (Berlin, 1987).

Conceptions of Alcoholism: Disease, Moral Defect, or Behavior Pattern? According to the medical perspective, alcoholism is a disease. E. M. Jellinek (1960), a leading proponent of the disease model, believed that alcoholism is a permanent, irreversible condition. Jellinek believed that once a person with alcoholism takes a drink, the biochemical effects of the drug on the brain create an irresistible physical craving for more. Jellinek's ideas have contributed to the view that, "Once an alcoholic, always an alcoholic." Alcoholics Anonymous (AA), which has adopted Jellinek's concepts, views people who suffer from alcoholism as either drinking or "recovering." In other words, alcoholism is never cured. Jellinek's concepts have also supported the idea that "just one drink" will cause the person with alcoholism to "fall off the wagon." In this view, the sole path to recovery is abstinence.

Yet not all professionals regard alcoholism as a disease in the medical sense. To some, the term is used as a label to describe a harmful pattern of alcohol ingestion and related behaviors. In this view, the "just-one-drink" hypothesis is not a biochemical inevitability. It is, instead, a common self-fulfilling prophecy, as we see later in the chapter.

Psychological Effects of Alcohol The effects of alcohol or other drugs vary from person to person. By and large they reflect the interaction of (1) the physiological effects of the substances, and (2) our interpretations of those effects. What do most people expect from alcohol? People frequently hold stereotypical expectations that alcohol will reduce states of tension, enhance pleasurable experiences, wash away their worries, and enhance their social skills. But what *does* alcohol actually do?

At a physiological level, alcohol, like the benzodiazepines (a family of antianxiety drugs; see Chapter 6), appears to heighten the sensitivity of GABA receptor sites. Because GABA is an inhibitory neurotransmitter, increasing the action of GABA reduces overall nervous system activity, producing feelings of relaxation. As people drink, their senses become clouded, and balance and coordination suffer. Still higher doses act on the parts of the brain that regulate involuntary vital functions, such as heart rate, respiration rate, and body temperature.

People may do many things when drinking that they would not do when sober, in part because of expectations concerning the drug, in part because of the drug's effects on the brain. For example, they may become more flirtatious or sexually aggressive or say or do things they later regret. Their behavior may reflect their expectation that alcohol has liberating effects and provides an external excuse for questionable behavior. Later, they can claim, "It was the alcohol, not me." The drug may impair the brain's ability to curb impulsive, risk-taking, or violent behavior (Curtin et al., 2001) (discussed further in Chapter 16), perhaps by interfering with information-processing functions. Although alcohol may make them feel more relaxed and self-confident, it may prevent them from exercising good judgment, which can lead them to make choices they would ordinarily reject, such as engaging in risky sex (Gordon & Carey, 1996). One of the lures of alcohol is that it induces short-term feelings of euphoria and elation that can drown self-doubts and self-criticism. Alcohol may also make people less capable of perceiving the unfortunate consequences of their behavior.

Alcohol in increasing amounts can dampen sexual arousal or excitement and impair our ability to perform sexually. As an intoxicant, alcohol also hampers coordination and motor ability, and slurs speech. These effects help explain why alcohol use is implicated in nearly 50% of the nation's fatal auto accidents, about 25% of fatal falls, and 30% to 50% of fatal fires and drownings (see Miller & Brown, 1997; Ravenholt, 1984). Figure 10.2 shows the relationship between alcohol dosage and impaired driving.

Alcohol and driving. Nearly half of the nation's fatal motor vehicle accidents, and about 25% of fatal falls, involve the use of alcohol.

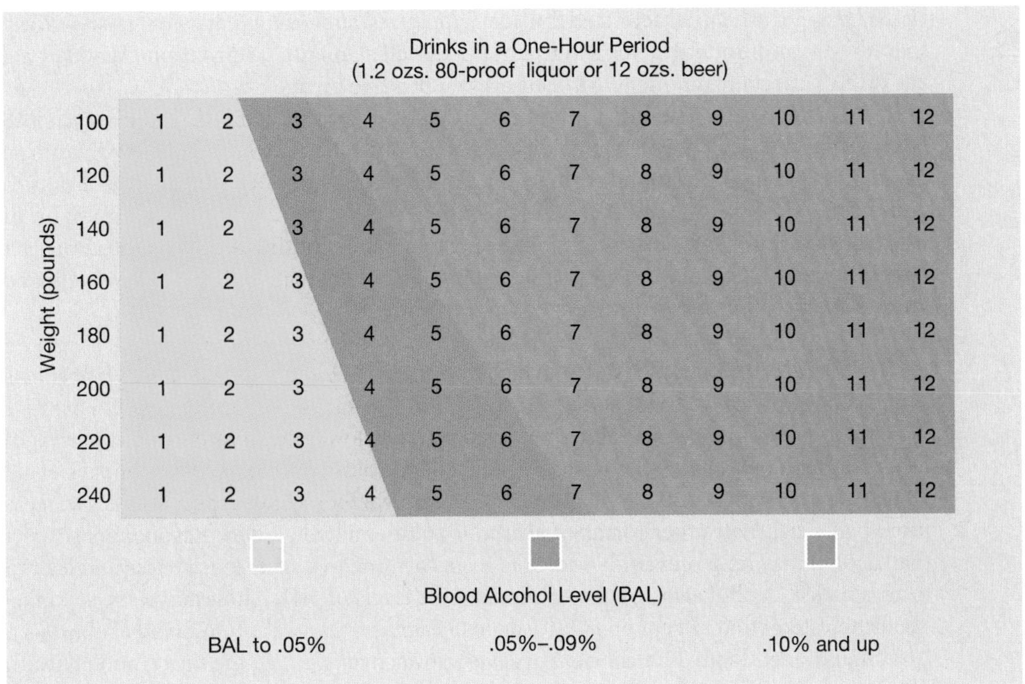

FIGURE 10.2 Alcohol intake and blood alcohol level.
Using this chart, you can estimate your blood alcohol level (BAL) as a function of your alcohol intake. For example, if you weigh 180 pounds and consume six beers in a 1-hour period, your BAL would be over .10%—a level beyond the 0.08% legal limit for driving in many states. But any amount of drinking may impair driving ability.

Source. Adapted from National Highway Traffic Safety Administration.

Physical Health and Alcohol Chronic, heavy alcohol use affects virtually every organ and body system, either directly or indirectly. Heavy alcohol use is linked to a higher risk of some forms of cancer, including cancer of the throat, esophagus, larynx, stomach, colon, liver, and possibly the bowels and breasts (e.g., Fuchs et al., 1995; Reichman, 1994). Heavy drinking is also linked to a wide range of other serious health concerns, including coronary heart disease, neurological disorders, and other forms of liver disease (Gordis, 1999). Two of the major forms of alcohol-related liver disease are *alcoholic hepatitis,* a serious and potentially life-threatening inflammation of the liver, and *cirrhosis of the liver,* a potentially fatal disease in which healthy liver cells are replaced with scar tissue.

Habitual drinkers tend to be malnourished, which can put them at risk of complications arising from nutritional deficiencies. Chronic drinking is thus associated with such nutritionally linked disorders such as cirrhosis of the liver (linked to protein deficiency) and **alcohol-induced persisting amnestic disorder** (connected with vitamin B deficiency). This condition, also known as *Korsakoff's syndrome,* is characterized by glaring confusion, disorientation, and memory loss for recent events (see Chapter 15).

All told, about 100,000 deaths annually in the United States result from various alcohol-related diseases and motor vehicle and other accidents (Potter, 1997). After tobacco, alcohol is the second leading cause of premature death in our society. Men who drink heavily stand nearly twice the risk of dying before the age of 65 as men who abstain; women who drink heavily are more than three times as likely to die before age 65 as are women who abstain ("NIAAA Report," 1990).

Mothers who drink during pregnancy place their fetuses at risk for infant mortality, birth defects, central nervous system dysfunctions, and later academic problems. Children whose mothers drink during pregnancy may develop fetal alcohol syndrome

THINK ABOUT IT

Do you use alcohol? How does it affect you physically and mentally? Have you drunk alcohol and driven? How do you feel about that? Have you ever done anything under the influence of alcohol that you later regretted? What? Why?

alcohol-induced persisting amnestic disorder A form of brain damage associated with chronic thiamine deficiency and alcoholism; characterized by memory loss, disorientation, and confabulations.

Truth OR Fiction? REVISITED

Alcohol use at any level increases the risk of heart attacks.

FALSE. Findings from recent studies show that moderate intake of alcohol is associated with a lower risk of heart attacks and lower death rates.

THINK ABOUT IT

Do you think it is wise to use alcohol in moderation to reduce the risk of cardiovascular disease? Why or why not?

barbiturates Types of depressants that are used to reduce anxiety or to induce sleep but that are highly addictive.

sedatives Types of depressants that reduce states of tension and restlessness and induce sleep.

(FAS), a syndrome characterized by facial features such as a flattened nose, widely spaced eyes, and underdeveloped upper jaw, as well as mental retardation (Wood et al., 2001). FAS affects from one to three of every 1,000 live births.

We don't know whether a minimum amount of alcohol is needed to produce FAS (Wood et al., 2001). Though the risk is greater among women who drank heavily during pregnancy, FAS has been found among children of mothers who drank as little as 2 ounces of alcohol a day during the first trimester (Astley et al., 1992). Although the question of whether there is any safe dose of alcohol during pregnancy continues to be debated, the fact remains that FAS is an entirely preventable birth defect. The safest course for women who know or suspect they are pregnant is not to drink. Period.

Moderate Drinking: Is There a Health Benefit? Despite this list of adverse effects associated with heavy drinking, evidence shows that moderate use of alcohol (1 to 2 drinks per day) is linked to lower risks of heart attacks, lower death rates, and lower risk of heart failure in older adults (Abramson et al., 2001; Goldberg et al., 2001; Gronbaek et al., 2000; "New Research," 2000). It remains unclear whether wine, especially red wine, is more beneficial than other forms of alcohol (Goldberg et al., 2001). Researchers suspect that alcohol may help prevent blood clots from forming that can clog arteries and lead to heart attacks. Alcohol also appears to increase the levels of HDL cholesterol, the so-called good cholesterol that sweeps away fatty deposits along artery walls (Goldberg et al., 2001). Although moderate use of alcohol may have a protective effect on the heart, public health officials caution that promoting the possible health benefits of alcohol may backfire by increasing the risks of alcohol abuse and dependence (Brody, 1994d). Health promotion efforts might be better directed toward finding safer ways of achieving the health benefits associated with moderate drinking than by encouraging alcohol consumption, such as by quitting smoking, lowering dietary fat and cholesterol, and exercising more regularly (Gaziano, 1993).

Barbiturates Estimates indicate that about 1% of the adult population meet criteria for a substance abuse or dependence disorder involving the use of barbiturates, sleep medication (hypnotics), or antianxiety agents at some point in their lives (Anthony & Helzer, 1991). **Barbiturates** such as amobarbital, pentobarbital, phenobarbital, and secobarbital are depressants, or **sedatives.** These drugs have several medical uses, including alleviation of anxiety and tension, anesthetizing of pain, and treatment of epilepsy and high blood pressure. Barbiturate use quickly leads to psychological dependence and physiological dependence in the form of both tolerance and development of a withdrawal syndrome.

Barbiturates are also popular street drugs because they are relaxing and produce a mild state of euphoria, or "high." High doses of barbiturates, like alcohol, produce drowsiness, slurred speech, motor impairment, irritability, and poor judgment—a particularly deadly combination of effects when their use is combined with operation of a motor vehicle. The effects of barbiturates last from 3 to 6 hours.

Because of synergistic effects, a mixture of barbiturates and alcohol is about four times as powerful as either drug used by itself. A combination of barbiturates and alcohol is implicated in the deaths of the entertainers Marilyn Monroe and Judy Garland. Even such widely used antianxiety drugs as Valium and Librium, which have a wide margin of safety when used alone, can be dangerous and lead to overdoses when their use is combined with alcohol (APA, 2000).

Physiologically dependent people need to be withdrawn carefully, and only under medical supervision, from sedatives, barbiturates, and antianxiety agents. Abrupt withdrawal can produce states of delirium that may involve visual, tactile, or auditory hallucinations and disturbances in thinking processes and consciousness. The longer the period of use and the higher the doses used, the greater the risk of severe withdrawal effects. Epileptic (grand mal) seizures and even death may occur if the individual undergoes untreated, abrupt withdrawal.

Opioids. **Opioids** are **narcotics,** a term used for addictive drugs that have pain-relieving and sleep-inducing properties. Opioids include both naturally occurring opiates (morphine, heroin, codeine) derived from the juice of the poppy plant and synthetic drugs (Demerol, Percodan, Darvon) manufactured in the laboratory to have opiate-like effects. The ancient Sumerians named the poppy plant *opium,* meaning "plant of joy."

Opioids produce a rush or intense feelings of pleasure, which is the primary reason for their popularity as street drugs. They also dull awareness of one's personal problems, which is attractive to people seeking a mental escape from stress.

The major medical application of opioids—natural or synthetic—is the relief of pain, or **analgesia.** Medical use of opioids, however, is carefully regulated because overdoses can lead to coma and even death. Yet some prescription opioids, especially the drug OxyContin, become drugs of abuse when they are used illicitly as street drugs (Tough, 2001). Street use of opioids is associated with many fatal overdoses and accidents. In a number of American cities, young men are more likely to die of a heroin overdose than in an automobile accident (Alter, 2001).

Estimates are that about 0.7% of the adult population (7 people in 1,000) currently have or have had an opiate abuse or dependence disorder (Anthony & Helzer, 1991). Once dependence sets in, it usually follows a chronic course, although brief periods of abstinence are frequent (APA, 2000).

Opiates become drugs of abuse because they produce euphoric states of pleasure, or a "rush." Their pleasurable effects derive from their ability to directly stimulate the brain's pleasure circuits—the same brain networks responsible for feelings of sexual pleasure or pleasure from eating a satisfying meal (Begley, 2001b).

Two revealing discoveries, made in the 1970s, showed that the brain produces chemicals of its own that have opiate-like effects. One was that neurons in the brain had receptor sites into which opiates fit—like a key in a lock. The second was that the human body produces its own opiate-like substances that dock at the same receptor sites as opiates do. These natural substances are labeled **endorphins,** which is short for "endogenous morphine"—that is, morphine coming from within. Endorphins play important roles in regulating natural states of pleasure and pain. Opioids mimic the actions of endorphins by docking at receptor sites intended for them, which in turn stimulates the brain centers that produce pleasurable sensations.

The withdrawal syndrome associated with opioids can be severe. It begins within 4 to 6 hours of the last dose. Flulike symptoms are accompanied by anxiety, feelings of restlessness, irritability, and cravings for the drug. Within a few days, symptoms progress to rapid pulse, high blood pressure, cramps, tremors, hot and cold flashes, fever, vomiting, insomnia, and diarrhea, among other symptoms. Although these symptoms can be uncomfortable, they are usually not devastating, especially when other drugs are prescribed to relieve them. Moreover, unlike withdrawal from barbiturates, the withdrawal syndrome rarely results in death.

Morphine **Morphine**—which receives its name from Morpheus, the Greek god of dreams—was introduced at about the time of the United States Civil War. Morphine, a powerful opium derivative, was used liberally to deaden pain from wounds. Physiological dependence on morphine became known as the "soldier's disease." There was little stigma attached to dependence until morphine became a restricted substance.

Heroin **Heroin,** the most widely used opiate, is a powerful depressant that can create a euphoric rush. Users of heroin claim that it is so pleasurable it can eradicate any thought of food or sex. Heroin was developed in 1875 during a search for a drug that would relieve

opioids Natural or synthetic drugs with strong addictive properties; natural opioids, referred to as opiates, are derived from the opium poppy.

narcotics Drugs that are used for pain relief and treatment of insomnia but that have strong addictive potential.

analgesia Relief from pain without loss of consciousness.

endorphins Natural substances that function as neurotransmitters in the brain and are similar in their effects to morphine.

morphine A strongly addictive narcotic derived from the opium poppy that relieves pain and induces feelings of well-being.

heroin A narcotic derived from morphine that has strong addictive properties.

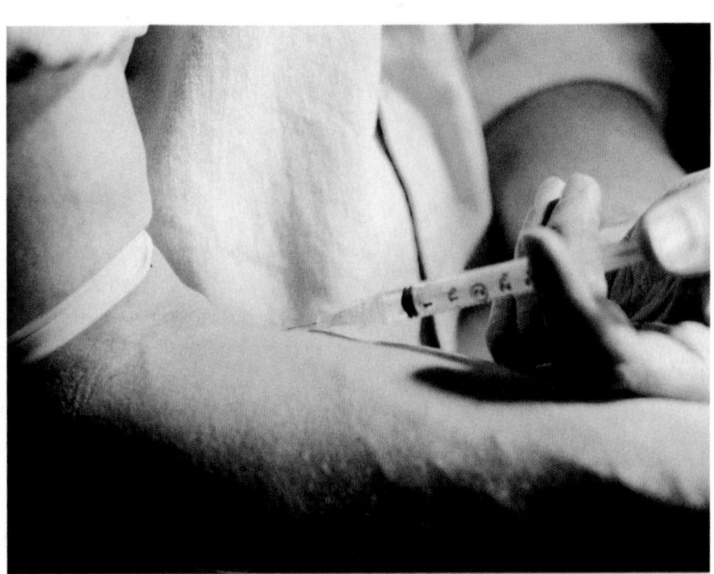

Shooting up. Heroin users often inject the substance directly into their veins. Heroin is a powerful depressant that provides a euphoric rush. Users often claim that heroin is so pleasurable that it obliterates any thought of food or sex.

Truth OR Fiction? REVISITED

Heroin was developed during the search for a nonaddictive drug that would relieve pain as effectively as morphine.

TRUE. Heroin was developed as researchers sought a drug that would relieve pain as effectively as morphine without causing its level of physical addiction.

pain as effectively as morphine, but without causing addiction. Chemist Heinrich Dreser transformed morphine into a new and stronger miracle drug, heroin, by means of a minor chemical change. He believed, erroneously, that heroin did not create physiological dependence.

Heroin is usually injected either directly beneath the skin (skin popping) or into a vein (mainlining). The positive effects are immediate. There is a powerful rush that lasts from 5 to 15 minutes and a state of satisfaction, euphoria, and well-being that lasts from 3 to 5 hours. In this state, all positive drives seem satisfied. All negative feelings of guilt, tension, and anxiety disappear. With prolonged usage, addiction can develop. Many physiologically dependent people support their habits through dealing (selling heroin), prostitution, or selling stolen goods. Heroin is a depressant, however, and its chemical effects do not directly stimulate criminal or aggressive behavior.

Stimulants

Stimulants such as amphetamines and cocaine are psychoactive substances that increase the activity of the nervous system. Effects vary somewhat from drug to drug, but some stimulants contribute to feelings of euphoria and self-confidence. Stimulants such as amphetamines, cocaine, and even caffeine (the stimulant found in coffee) increase the availability in the brain of the neurotransmitters norepinephrine and dopamine. High levels of these neurotransmitters therefore remain available in the synaptic gaps between neurons, maintaining high levels of nervous system activity and states of high arousal.

Amphetamines The **amphetamines** are a class of synthetic stimulants. Street names for stimulants include speed, uppers, bennies (for *amphetamine sulfate*; trade name Benzedrine), "meth" (for *methamphetamine*; trade name Methedrine), and dexies (for *dextroamphetamine;* trade name Dexedrine).

Amphetamines are used in high doses for their euphoric rush. They are often taken in pill form, or smoked in a relatively pure form called "ice" or "crystal meth." The most potent form of amphetamine, liquid methamphetamine, is injected directly into the veins and produces an intense and immediate rush. Some users inject methamphetamine for days on end to maintain an extended high. Eventually such highs come to an end. People who have been on extended highs sometimes "crash" and fall into a deep sleep or depression. Some people commit suicide on the way down. High doses can cause restlessness, irritability, hallucinations, paranoid delusions, loss of appetite, and insomnia.

More than 1 million people in the United States use "meth," almost three times as many as use heroin (Bonné, 2001). Physiological dependence can develop from using amphetamines, leading to an abstinence syndrome characterized most often by depression and fatigue, as well as by unpleasant, vivid dreams, insomnia or hypersomnia (excessive sleeping), increased appetite, and either a slowing down of motor behavior or agitation (APA, 2000). Psychological dependence is seen most often in people who use amphetamines as a way of coping with stress or depression.

Methamphetamine abuse can cause brain damage, producing deficits in learning and memory in addition to other effects (Blakeslee, 2001; Ernst et al., 2000; Volkow et al., 2001; Zickler, 2000). Violent behavior may also occur, especially when the drug is smoked or injected intravenously (APA, 2000). The hallucinations and delusions of **amphetamine psychosis** mimic the features of paranoid schizophrenia, which has encouraged researchers to study the chemical changes induced by amphetamines as possible causes of schizophrenia.

Truth OR Fiction? REVISITED

Coca-Cola originally contained cocaine.

TRUE. The original formula for Coca-Cola contained an extract of cocaine.

Ecstasy The drug *ecstasy,* or MDMA (3,4-methylenedioxymethamphetamine) is a designer drug, a chemical knockoff similar in chemical structure to amphetamine (Braun, 2001). It produces mild euphoria and hallucinations and has attracted a growing user base among young people, especially on college campuses and in clubs and "raves" in many cities (Hernandez, 2000; Mathias, 2000; "Dip in Youth Killing," 2000). Ecstasy is fast becoming the nation's most popular illicit drug (Kuhn & Wilson, 2001). Use by high school seniors nearly doubled during the latter half of the 1990s (Butterfield, 2001). The drug can produce adverse psychological effects, including depression, anxiety, insomnia, and even paranoia and

amphetamines Types of stimulants, such as Benzedrine or Dexedrine.

amphetamine psychosis A psychotic state induced by ingestion of amphetamines.

psychosis. The drug may also impair cognitive functioning, including learning ability and attention and may have long-lasting effects on memory (Gouzoulis-Mayfrank et al., 2000; Reneman, et al., 2001). The drug may also deplete levels of serotonin in the brain, a neurotransmitter linked to regulation of mood and appetite. This may explain why users of the drug can experience feelings of depression when they go off the drug ("Ecstasy Use," 2000). Physical side effects include higher heart rate and blood pressure, a tense or chattering jaw, and body warmth and/or chills (Braun, 2001). The drug can be lethal when taken in high doses (Kuhn & Wilson, 2001).

Ecstasy. Recreational use of the drug ecstasy has become popular in many clubs catering to young people. Yet even occasional use of the drug may affect cognitive functioning, such as learning, memory, and attention. High doses can be lethal.

Cocaine It might surprise you to learn that the original formula for Coca-Cola contained an extract of **cocaine.** In 1906, however, the company withdrew cocaine from its secret formula. The beverage was originally described as a "brain tonic and intellectual beverage," in part because of its cocaine content. Cocaine is a natural stimulant extracted from the leaves of the coca plant—the plant from which the soft drink obtained its name. Coca-Cola is still flavored with an extract from the coca plant, one that is not known to be psychoactive.

It was long believed that cocaine was not physically addicting. However, evidence supports the addictive properties of the drug in producing a tolerance effect and an identifiable withdrawal syndrome, which is characterized by depressed mood and disturbances in sleep and appetite (APA, 2000). Intense cravings for the drug and loss of ability to experience pleasure may also be present. Withdrawal symptoms are usually brief in duration and may involve a "crash," or period of intense depression and exhaustion following abrupt withdrawal.

Cocaine is usually snorted in powder form or smoked in the form of **crack,** a hardened form of cocaine that may contain more than 75% pure cocaine. Crack "rocks"—so called because they look like small white pebbles—are available in small ready-to-smoke amounts and considered to be the most habit-forming street drug available. Crack produces a prompt and potent rush that wears off in a few minutes. The rush from snorting is milder and takes a while to develop, but it tends to linger longer than the rush of crack.

Freebasing also intensifies the effects of cocaine. Cocaine in powder form is heated with ether, freeing the psychoactive chemical base of the drug, and then smoked. Ether, however, is highly flammable.

Next to marijuana, cocaine is the most widely used illicit drug in the United States. Nearly 3% (2.7%) of adults in the United States in the 15- to 54-year age range have a history of cocaine dependence (Anthony et al., 1994). Cocaine abuse is characterized by periodic binges lasting perhaps 12 to 36 hours, which are then followed by 2 to 5 days of abstinence, during which time the abuser may experience cravings that prompt another binge (Gawin et al., 1989). According to one estimate, between 10% and 15% of people who try snorting cocaine eventually develop cocaine abuse or dependence (Gawin, 1991).

The cocaine epidemic may have peaked in some respects (see Table 10.1 showing declining use among college students). Although the numbers of casual users of cocaine have greatly declined, there has been no corresponding reduction in the numbers of hard-core users.

Effects of Cocaine Like heroin, cocaine directly stimulates the brain's reward or pleasure circuits (Volkow et al., 1997). It also produces a sudden rise in blood pressure, constricts blood vessels (with associated reduction of the oxygen supply to the heart), and accelerates the heart rate. Overdoses can produce restlessness, insomnia, headaches, nausea, convulsions, tremors, hallucinations, delusions, and even sudden death. Such a death usually results from respiratory or cardiovascular collapse. Although intravenous use of cocaine carries the greatest risk of a lethal overdose, other forms of use can also cause fatal overdoses. Table 10.3 summarizes a number of the health risks of cocaine use.

cocaine A stimulant derived from the leaves of the coca plant.

crack The hardened, smokable form of cocaine.

freebasing A method of ingesting cocaine by heating it with ether to separate its most potent component (its "free base") and then smoking the extract.

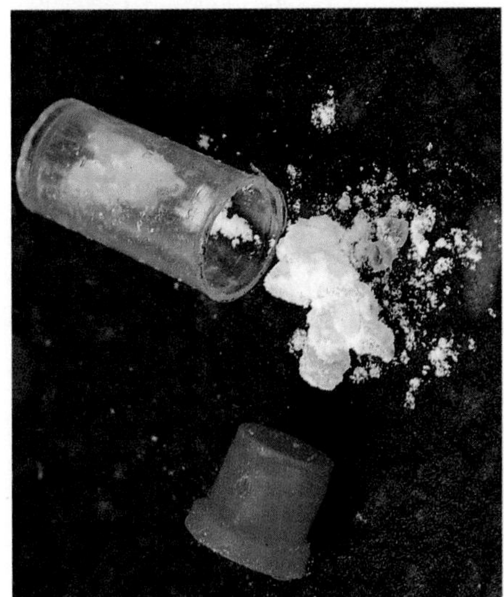

Crack. Crack "rocks" resemble small, white pebbles. Crack produces a powerful, prompt rush when smoked. Small, ready-to-smoke doses are available at prices that have made them affordable to adolescents.

TABLE 10.3 Health Risks of Cocaine Use

Physical Effects and Risks	
Effects	**Risks**
Increased heart rate	Accelerated heart rate may give rise to heart irregularities that can be fatal, such as ventricular tachycardia (extremely rapid contractions) or ventricular fibrillation (irregular, weakened contractions).
Increased blood pressure	Rapid or large changes in blood pressure may place too much stress on a weak-walled blood vessel in the brain, which can cause it to burst, producing cerebral hemorrhage or stroke.
Increased body temperature	Can be dangerous to some individuals.
Possible grand mal seizures (epileptic convulsions)	Some grand mal seizures are fatal, particularly when they occur in rapid succession or while driving a car.
Respiratory effects	Overdoses can produce gasping or shallow, irregular breathing that can lead to respiratory arrest.
Dangerous effects in special populations	Various special populations are at greater risk from cocaine use or overdose. People with coronary heart disease have died because their heart muscles were taxed beyond the capacity of their arteries to supply oxygen.
Medical Complications of Cocaine Use	
Nasal Problems	When cocaine is administered intranasally (snorted), it constricts the blood vessels serving the nose, decreasing the supply of oxygen to these tissues, leading to irritation and inflammation of the mucous membranes, ulcers in the nostrils, frequent nosebleeds, and chronic sneezing and nasal congestion. Chronic use may lead to tissue death of the nasal septum, the part of the nose that separates the nostrils, requiring plastic surgery.
Lung Problems	Freebase smoking may lead to serious lung problems within 3 months of initial use.
Malnutrition	Cocaine suppresses the appetite so that weight loss, malnutrition, and vitamin deficiencies may accompany regular use.
Seizures	Grand mal seizures, typical of epileptics, may occur due to irregularities in the electrical activity of the brain. Repeated use may lower the seizure threshold, described as a type of "kindling" effect.
Sexual Problems	Despite the popular belief that cocaine is an aphrodisiac, frequent use can lead to sexual dysfunctions, such as impotence and failure to ejaculate among males, and decreased sexual interest in both sexes. Although some people report initial increased sexual pleasure with cocaine use, they may become dependent on cocaine for sexual arousal or lose the ability to enjoy sex for extended periods following long-term use.
Other Effects	Cocaine use may increase the risk of miscarriage among pregnant women. Sharing of infected needles is associated with transmission of hepatitis, endocarditis (infection of the heart valve), and HIV. Repeated injections often lead to skin infections as bacteria are introduced into the deeper levels of the skin.

Source. Adapted from Weiss & Mirin (1987).

Repeated use and high-dose use of cocaine can lead to depression and anxiety (Weiss & Mirin, 1987). Depression may be severe enough to prompt suicidal behavior. Both initial and routine users report episodes of "crashing" (feelings of depression after a binge), although crashing is more common among long-term high-dose users. Psychotic behaviors, which can be induced by cocaine use as well as by use of amphetamines, tend to become more severe with continued use. Cocaine psychosis is usually preceded by a period of heightened suspiciousness, depressed mood, compulsive behavior, fault finding, irritability, and increasing paranoia (Weiss & Mirin, 1987). The psychosis may also include intense visual and auditory hallucinations and delusions of persecution.

Nicotine Habitual smoking is not merely a bad habit: It is also a form of physical addiction to a stimulant drug, nicotine, found in tobacco products including cigarettes, cigars, and smokeless tobacco (Kessler et al., 1997b). Smoking (or other tobacco uses) is the means of administering the drug to the body.

More than 400,000 lives in the United States are lost each year from smoking-related causes, mostly from lung cancer, cardiovascular disease, and chronic obstructive lung disease (Fried et al., 1998). This figure is nearly eight times the number that dies from motor vehicle accidents and about equal to the population of a city the size of Atlanta, Georgia. Smoking is implicated in 1 in 3 cancer deaths, including more than 100,000 deaths due to lung cancer (Boyle, 1993). Smokers overall stand twice the risk of dying from cancer as nonsmokers; among heavy smokers, the risk is four times as great (Bartecchi, MacKenzie, & Schrier, 1994).

The World Health Organization estimates that 1 billion people worldwide smoke, and more than 3 million die each year from smoking-related causes. Smoking is expected to become the world's leading cause of death by the year 2020 ("Smoking," 1996). Smoking may also pose risks to one's mental health. A recent study reported that cigarette smoking among adolescents may increase the risk of anxiety disorders in late adolescence and early adulthood (J. G. Johnson et al., 2000).

Largely because of health concerns, the percentage of Americans who smoke declined from 42% in 1966 to about 25% today (Miller & Brown, 1997). On the other hand, we are losing the battle against teenage smoking, which increased sharply during the early to late 1990s (Feder, 1996; Stolberg, 1998b). The greatest increase in teenage cigarette smoking was among African Americans, up by 80%.

Lung cancer, which in 90% of cases is caused by smoking, has now surpassed breast cancer as the leading killer of women. Although quitting smoking clearly has health benefits, it unfortunately does not reduce the risks to normal (nonsmoking) levels. The lesson is clear: If you don't smoke, don't start; but if you do smoke, quit.

Ethnic differences in smoking rates are shown in Figure 10.3. With the exception of Native Americans (American Indian/Alaskan Native), women in each ethnic group are less likely to smoke than their male counterparts. Figure 10.4 shows relationships between

Truth OR Fiction? REVISITED

Habitual smoking is just a bad habit, not a physical addiction.

FALSE. Habitual smoking involves physical addiction to nicotine, the stimulant drug found in tobacco.

WWW **Web Link 10.4**
Check Your Smoking IQ

Truth OR Fiction? REVISITED

Breast cancer is the leading cause of cancer deaths among U.S. women.

FALSE. Lung cancer has surpassed breast cancer as the leading cancer killer among women. It is also the leading cancer killer among men. Cigarette smoking is the culprit in the great majority of cases.

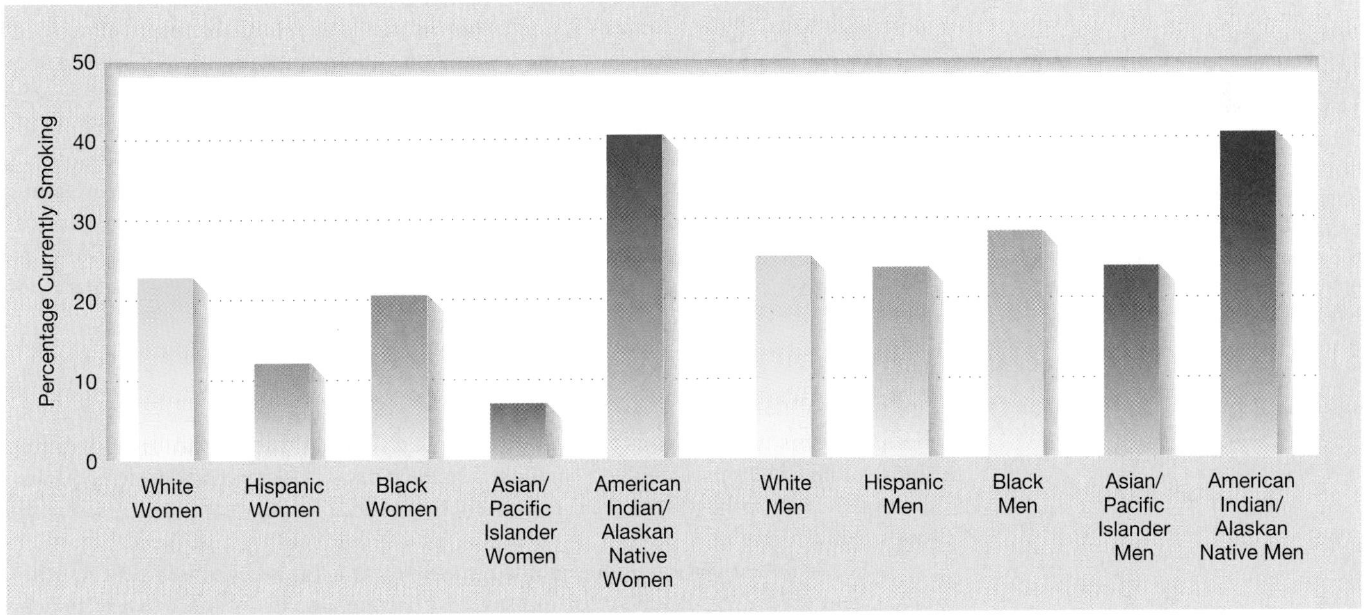

FIGURE 10.3 Gender and racial breakdown of cigarette smokers among U.S. adults. Smoking rates are highest for Native American men and women. Women in each ethnic group (with the exception of Native Americans) are less likely to smoke than their male counterparts. The rates for non-Hispanic White American men and women are comparable.

Source. Centers for Disease Control (2001b).

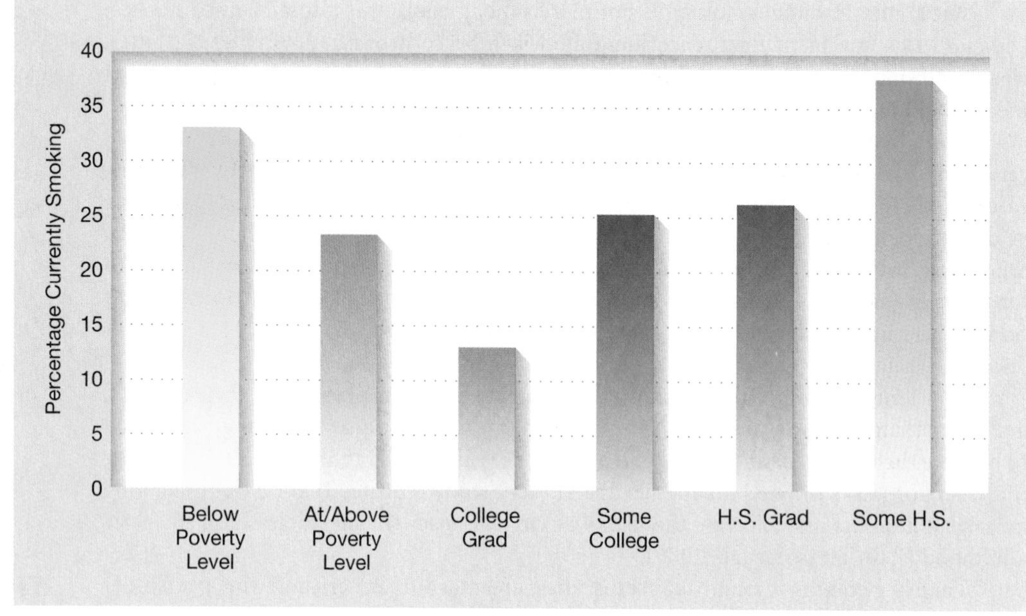

FIGURE 10.4 Cigarette smokers in the United States in relation to poverty status and level of education. Smoking is becoming increasingly more prevalent among the poorer and less educated members of society.

Source. Centers for Disease Control (2001b).

smoking rates and income and educational levels. Note how smoking is disproportionately represented among the poorer and less educated segments of the population.

Nicotine is delivered to the body through the use of tobacco products. As a stimulant it increases alertness but can also give rise to cold, clammy skin, nausea and vomiting, dizziness and faintness, and diarrhea—all of which account for the discomforts of novice smokers. Nicotine also stimulates the release of epinephrine, a hormone that generates a rush of autonomic activity, including rapid heartbeat and release of stores of sugar into the blood. Nicotine quells the appetite and provides a sort of psychological "kick." Nicotine also leads to the release of endorphins, the opiate-like hormones produced in the brain. This may account for the pleasurable feelings associated with tobacco use.

Habitual use of nicotine leads to a physiological dependence on the drug (Lichtenstein & Glasgow, 1992). Nicotine dependence is associated with both tolerance (intake rises to a level of a pack or two a day before leveling off) and a characteristic withdrawal syndrome. The withdrawal syndrome for nicotine includes such features as lack of energy, depressed mood, irritability, frustration, nervousness, impaired concentration, lightheadedness and dizziness, drowsiness, headaches, fatigue, irregular bowels, insomnia, cramps, lowered heart rate, heart palpitations, increased appetite, weight gain, sweating, tremors, and craving for cigarettes (APA, 2000; Klesges et al., 1997). It is nicotine dependence, not cigarette smoking per se, that is classifiable as a mental disorder in the *DSM* system. The great majority of regular smokers (80% to 90%) meet diagnostic criteria for nicotine dependence (APA, 2000).

THINK ABOUT IT

Speculate on the reasons that smoking is more prevalent among poorer and less educated segments of the U.S. population and on why smoking is on the rise among teenagers. What kinds of public health messages might be effective at reaching these at-risk groups?

Hallucinogens

Hallucinogens, also known as *psychedelics,* are a class of drugs that produce sensory distortions or hallucinations, including major alterations in color perception and hearing. Hallucinogens may also have additional effects, such as relaxation and euphoria, or, in some cases, panic.

The hallucinogens include such drugs as lysergic acid diethylamide (LSD), psilocybin, and mescaline. Psychoactive substances that are similar in effect to psychedelic drugs are marijuana (cannabis) and phencyclidine (PCP). Mescaline is derived from the peyote cactus and has been used for centuries by Native Americans in the Southwest, Mexico, and Central America in religious ceremonies, as has psilocybin, which is derived from certain mushrooms. LSD, PCP, and marijuana are more commonly used in the United States.

hallucinogens Substances that cause hallucinations.

Although tolerance to hallucinogens may develop, we lack evidence of a consistent or clinically significant withdrawal syndrome associated with their use (APA, 2000). Cravings following withdrawal may occur, however.

LSD LSD is the acronym for **lysergic acid diethylamide,** a synthetic hallucinogenic drug. In addition to the vivid parade of colors and visual distortions produced by LSD, users have claimed it "expands consciousness" and opens new worlds—as if they were looking into some reality beyond the usual reality. Sometimes they believe they have achieved great insights during the LSD "trip," but when it wears off they usually cannot follow through or even summon up these discoveries.

The effects of LSD are unpredictable and depend on the amount taken as well as the user's expectations, personality, mood, and surroundings (USDHHS, 1992). The user's prior experiences with the drug may also play a role, as users who have learned to handle the effects of the drug through past experience may be better prepared than new users.

Some users have unpleasant experiences with the drug, or "bad trips." Feelings of intense fear or panic may occur (USDHHS, 1992). Users may fear losing control or sanity. Some experience terrifying fears of death. Fatal accidents have sometimes occurred during LSD trips. **Flashbacks,** typically involving a reexperiencing of some of the perceptual distortions of the "trip," may occur days, weeks, or even years afterward. Flashbacks tend to occur suddenly and often without warning. Perceptual distortions may involve geometric forms, flashes of color, intensified colors, afterimages, or appearances of halos around objects, among others (APA, 2000). They may stem from chemical changes in the brain caused by the prior use of the drug. Triggers for flashbacks include entry into darkened environments, use of various drugs, anxiety or fatigue states, or stress (APA, 2000). Psychological factors, such as underlying personality problems, may also be involved in explaining why some users experience flashbacks. In some cases, a flashback may involve an imagined reenactment of the LSD experience.

lysergic acid diethylamide (LSD) A type of hallucinogen.

flashbacks The experience of sensory distortions or hallucinations occurring after use of LSD or other hallucinogenic drugs.

marijuana A hallucinogenic drug derived from the leaves and stems of the plant *Cannabis sativa.*

delta-9-tetrahydrocannabinol (THC) The active ingredient in marijuana.

hashish A drug derived from the resin of the plant *Cannabis sativa.*

Phencyclidine (PCP) Phencyclidine, or PCP—which is referred to as "angel dust" on the streets—was developed as an anesthetic in the 1950s but was discontinued as such when the hallucinatory side effects of the drug were discovered. A smokable form of PCP became popular as a street drug in the 1970s. By the mid-1980s, more than one in five young people in the 18 to 25 age range had used PCP (USDHHS, 1986b). However, its popularity has since waned, largely because of its unpredictable effects.

The effects of PCP, like most drugs, are dose related. In addition to causing hallucinations, PCP accelerates the heart rate and blood pressure and causes sweating, flushing, and numbness. PCP is classified as a *deliriant*—a drug capable of producing states of delirium. It also has dissociating effects, causing users to feel as if there is some sort of invisible barrier or wall between themselves and their environments. Dissociation can be experienced as pleasant, engrossing, or frightening, depending on the user's expectations, mood, setting, and so on. Overdoses can give rise to drowsiness and a blank stare, convulsions, and, now and then, coma; paranoia and aggressive behavior; and tragic accidents resulting from perceptual distortion or impaired judgment during states of intoxication.

Marijuana **Marijuana** is derived from the *Cannabis sativa* plant. Marijuana sometimes produces mild hallucinations, so it is regarded as a minor hallucinogen. The psychoactive substance in marijuana is **delta-9-tetrahydrocannabinol,** or THC. THC is found in branches and leaves of the plant but is highly concentrated in the resin of the female plant. **Hashish,** or "hash," is also derived from the resin. Although it is more potent than marijuana, hashish has similar effects.

WWW Web Link 10.5
Marijuana Facts

Use of marijuana exploded throughout the so-called swinging 1960s and the 1970s, but the drug then lost some (but not all) of its cachet. Still, marijuana remains our most widely used illegal drug, although its prevalence doesn't compare with alcohol's. Approximately 33% of people in the United States age 12 or older, nearly 70 million people, have tried marijuana at least once in their lives, and 5% are current users (USDHHS, 1993).

Cannabis (or marijuana) dependence is the most common form of illicit-drug dependence in the United States, affecting an estimated 4.2% of the adult population at some point in their lives (Anthony et al., 1994). Males are more likely than females to develop a cannabis use disorder (either abuse or dependence), and the rates of these disorders is greatest among young people age 18 to 30 (APA, 2000).

Low doses of the drug can produce relaxing feelings similar to drinking alcohol. Some users report that at low doses the drug makes them feel more comfortable in social gatherings. Higher doses, however, often lead users to withdraw into themselves. Some users believe the drug increases their capacity for self-insight or creative thinking, although the insights achieved under its influence may not seem so insightful once the drug's effects have passed. People may turn to marijuana, as to other drugs, to help them cope with life problems or to help them function when they are under stress. Strongly intoxicated people perceive time as passing more slowly. A song of a few minutes may seem to last an hour. There is increased awareness of bodily sensations, such as heartbeat. Smokers also report that strong intoxication heightens sexual sensations. Visual hallucinations may occur.

Strong intoxication can cause smokers to become disoriented. If their moods are euphoric, disorientation may be construed as harmony with the universe. Yet some smokers find strong intoxication disturbing. An accelerated heart rate and sharpened awareness of bodily sensations cause some smokers to fear their hearts will "run away" with them. Some smokers are frightened by disorientation and fear they will not "come back." High levels of intoxication now and then induce nausea and vomiting.

Cannabis dependence is associated more with patterns of compulsive use or psychological dependence than with physiological dependence. Although tolerance to many of the drug's effects may occur with chronic use, some users report reverse tolerance, or *sensitization.* A withdrawal syndrome has not been reliably demonstrated (APA, 2000). However, new research with animals points to some disturbing similarities between marijuana and addictive drugs like heroin and cocaine (Wickelgren, 1997). In one study researchers found that withdrawal from marijuana activated the same brain circuits involved in withdrawal from opioids, alcohol, and cocaine (Rodríguez de Fonseca et al., 1997). These brain circuits are also involved in producing feelings of anxiety when an animal or person is under stress (Blakeslee, 1997a). In another study, researchers determined that marijuana activated the same reward circuits in the brain as heroin (Tanda, Pontien, & Chiara, 1997). Although these studies were conducted with animals, researchers believe the underlying biological mechanisms may apply to humans as well (Blakeslee, 1997a).

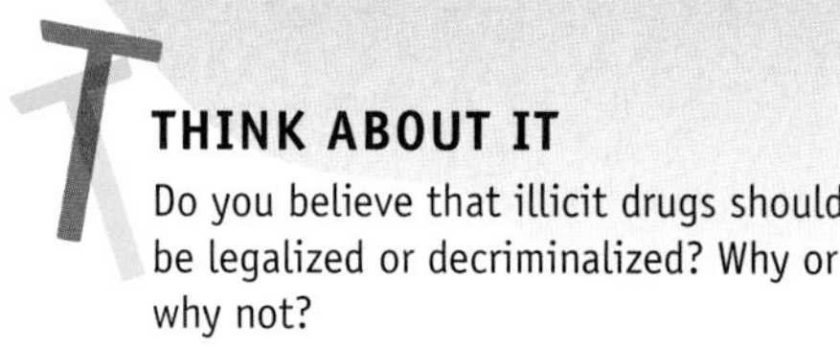

THINK ABOUT IT

Do you believe that illicit drugs should be legalized or decriminalized? Why or why not?

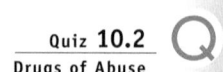

College students who are heavy users of marijuana show evidence of intellectual impairment, including diminished ability in tasks requiring attention, abstraction, and mental flexibility (Pope & Yurgelun-Todd, 1996). However, it is unclear whether these deficits are due to the drug or to characteristics of people who become heavy users (Lee, 1996). We do know that marijuana impairs perception and motor coordination and thus makes driving and the operation of other machines dangerous. It also impairs short-term memory and retards learning ability. Although it induces positive mood changes in many users, some people report anxiety and confusion; there are also occasional reports of psychotic reactions. Marijuana elevates heart rate and blood pressure and is linked to an increased risk of heart attacks in people with heart disease ("Another Worry," 2000). Finally, marijuana smoke contains carcinogenic hydrocarbons, so chronic users risk lung cancer and other respiratory diseases.

Quiz **10.2** Q
Drugs of Abuse

Theoretical Perspectives

People begin using psychoactive substances for various reasons. Some adolescents may start using drugs because of peer pressure or because they believe drugs make them seem more sophisticated or grown up. Some use drugs as a way of rebelling against their parents or society at large. Regardless of why people get started with drugs, they continue to use them because drugs produce pleasurable effects or because they find it difficult to stop. Most adolescents drink alcohol to "get high," not to establish that they are adults. Many people smoke cigarettes for the pleasure they provide. Others smoke to help them relax when they are tense and, paradoxically, to give them a kick or a lift when they are tired. Yet many would like to quit but find it difficult to break their addiction.

People who are anxious about their jobs or social lives may be drawn to the calming effects of alcohol, marijuana (in certain doses), tranquilizers, and sedatives. People with low self-confidence and self-esteem may be drawn to the ego-bolstering effects of amphetamines and cocaine. Many poor young people attempt to escape the poverty, anguish, and tedium of inner city life through using heroin and similar drugs. More well-to-do adolescents may rely on drugs to manage the transition from dependence to independence and major life changes concerning jobs, college, and lifestyles.

In the next sections we consider several major theoretical perspectives on substance abuse and dependence.

Biological Perspectives

We are beginning to learn more about the biological underpinnings of addiction. Much of the recent research has focused on neurotransmitters, especially dopamine, and on the role of genetic factors.

WWW **Web Link 10.6**
The Brain's Response to Drugs

Neurotransmitters Drugs such as nicotine, alcohol, amphetamines, heroin, cocaine, and even marijuana produce pleasurable effects by increasing the concentration of dopamine in the brain's pleasure or reward circuits—the network of neurons responsible for the pleasurable feelings we experience from sexual stimulation, or winning a sporting event, or even eating a scrumptious dessert (Begley, 2001b; O'Brien & McLellan, 1997; Volkow et al., 1997). The feelings of pleasure from using these drugs may range from mild happiness to euphoria.

But what of long-term use? Investigators suspect that chronic drug use reduces the numbers of receptors on the receiving neurons where dopamine docks (Begley, 2001b). It may also reduce the brain's ability to produce dopamine on its own (Blakeslee, 1997a). Consequently, the ability to derive pleasure from activities of everyday life, such as having a good meal, attending an enjoyable movie, and the like, wanes. The chronic drug user comes to rely on drugs to produce feelings of pleasure that the brain may no longer be able to produce on its own, and to avert depression, anxiety, and other disturbing feelings. Without drugs, life may not seem to be worth living. These changes in the dopamine system may explain the intense cravings and anxiety that accompany drug withdrawal and the difficulty people with chemical dependencies have maintaining abstinence.

The biochemical bases of drug use and abuse are complex and appear to involve other neurotransmitters besides dopamine. For example, researchers suspect that serotonin may also activate the brain's pleasure or reward circuits in response to cocaine, alcohol, and other drug use (Begley, 2001b; Rocha et al., 1998).

We also know that a group of neurotransmitters called endorphins have pain-blocking properties similar to those of opioids such as heroin. Endorphins and opiates dock at the same receptor sites in the brain. Normally, the brain produces a certain level of endorphins that maintains a sort of psychological steady state of comfort and potential to experience pleasure. However, when the body becomes habituated to a supply of opioids, it may stop producing endorphins. This makes the user dependent on opiates for feelings of comfort, relief from pain, and feelings of pleasure. When the habitual user stops using heroin or other opiates, feelings of discomfort and little aches and pains may be magnified until the body resumes adequate production of endorphins. This discomfort may account, at least in part, for the unpleasant withdrawal symptoms that opiate addicts experience. However, this model remains speculative, and more research is needed to document direct relationships between endorphin production and withdrawal symptoms.

Genetic Factors Evidence links genetic factors to various forms of substance use and abuse, including alcoholism, opiate addiction, and even cigarette smoking (Kendler Thornton, & Pederson, 2000; McLellan et al., 2000; Nurnberger et al., 2001; Wall et al., 2001). We focus our discussion on alcohol dependence, because this has been the area of the greatest research interest.

Alcoholism tends to run in families (APA, 2000; Wood et al., 2001). The closer the genetic relationship, the greater the risk. Familial patterns provide only suggestive evidence of genetic factors, because families share common environment as well as common genes. More definitive evidence comes from twin and adoptee studies.

Monozygotic (MZ) twins have identical genes, whereas fraternal or dizygotic (DZ) twins share only half of their genes. If genetic factors are involved, we would expect MZ twins to have higher concordance (agreement) rates for alcoholism than DZ twins. Evidence of higher concordance rates for alcoholism is found among MZ twins than DZ twins, although the results are more consistent for male samples than female samples (Wood et al., 2001).

A limitation of twin studies is that MZ twins may share more environmental as well as genetic similarity. That is, they may be treated more alike than DZ twins. However, evidence also shows that male adoptees whose biological parents suffered from alcoholism have an increased risk of developing alcoholism themselves, even if they are raised in nondrinking homes (Gordis, 1995; Schuckit, 1987). Among women, however, the rate of alcoholism in adopted-away daughters of parents with alcoholism is only slightly higher than that for adopted-away daughters of nonalcoholics, thus casting doubt on a strong genetic linkage to alcoholism in women (Svikis, Velez, & Pickens, 1994). All in all, genetic factors are believed to play a moderate role in male alcoholism and a modest role in female alcoholism (McGue, 1993). Other evidence points to a genetic contribution in other forms of substance abuse, including opioid, marijuana, cocaine, and nicotine dependence (Lerman et al., 1999; Sabol et al., 1999; Tsuang et al., 1998).

If alcoholism or other forms of substance abuse and dependence are influenced by genetic factors, what is it that is inherited? Some clues have begun to emerge. Researchers have linked alcoholism, nicotine dependence, and opioid addiction to genes involved in determining the structure of dopamine receptors in the brain (Kotler, 1997). We've mentioned that dopamine is involved in regulating states of pleasure, which leads researchers to suspect that genetic factors enhance feelings of pleasure derived from alcohol, which in turn may increase cravings for the drug (Altman, 1990a). In all likelihood there is no one "alcoholism gene" but rather a set of genes that interact with each other and with environmental factors to increase the risk of alcoholism (Devor, 1994). Other evidence suggests that a genetic vulnerability to alcoholism may involve a combination of factors, such as reaping greater pleasure from alcohol and a capacity for greater biological tolerance for the drug (Pihl, Peterson, & Finn, 1990; Pollock, 1992).

Other research has shown that men who have immediate biological relatives (parents or siblings) with a history of alcoholism tend to metabolize alcohol more rapidly than do men without a history of alcoholism in their immediate families (Schuckit & Rayes, 1979). People who metabolize alcohol relatively quickly can tolerate larger doses and are less likely to develop upset stomachs, dizziness, and headaches when they drink. Unfortunately, a lower sensitivity to the unpleasant effects of alcohol may make it difficult to know when to say when. Thus people who are better able to "hold their liquor" may be at greater risk of developing drinking problems. They may need to rely on other cues, such as counting their drinks, to learn to limit their drinking. People whose bodies more readily "put the brakes" on excess drinking may be less likely to develop problems in moderating their drinking than those with better tolerance.

Truth OR Fiction? REVISITED

Being able to "hold your liquor" better than most people helps prevent the development of problem drinking.

FALSE. Being able to "hold you liquor" may encourage you to drink more, which may set the stage for the development of problem drinking.

Other research suggests that men with a family history of alcoholism may be genetically predisposed to be unusually tense or nervous because of deficiencies of certain neurotransmitters in the brain (Goleman, 1990a, 1992b). Perhaps they turn to alcohol to help them relax.

Whatever the role of heredity in alcoholism, there is ample "room" for other factors, which is highlighted by the finding that at least one-third of people with alcoholism have no family history of the disorder (Schuckit, 1983). Most researchers today believe that alcoholism and other forms of substance dependence involve the actions of multiple genes together with social, cultural, and psychological factors (Devor, 1994; Dick et al., 2001).

Learning Perspectives

Learning theorists propose that substance-related behaviors are largely learned and can, in principle, be unlearned. They focus on the roles of operant and classical conditioning and observational learning. Substance abuse problems are not regarded as symptoms of diseases but rather as problem habits. Although learning theorists do not deny that genetic or biological factors may be involved in the genesis of substance abuse problems, they place a greater emphasis on the role of learning in the development and maintenance of these problem behaviors (McCrady, 1993, 1994). They also recognize that people who suffer from depression or anxiety may turn to alcohol as a way of relieving these troubling emotional states, however briefly. Evidence shows that emotional stress, such as anxiety or depression, often sets the stage for the development of substance abuse (Dixit & Crum, 2000; McGue, Slutske, & Iaono, 1999).

Drug use may become habitual because of the pleasure or positive reinforcement, or temporary relief from negative emotions like anxiety and depression, that drugs can produce. With drugs like cocaine, which appear capable of directly stimulating pleasure mechanisms in the brain, the positive reinforcement is direct and powerful.

Operant Conditioning People may initially use a drug because of social influence, trial and error, or social observation. In the case of alcohol, they learn that the drug can produce reinforcing effects, such as feelings of euphoria, and reductions in states of anxiety and tension. Alcohol may also release behavioral inhibitions. Alcohol can thus be reinforcing when it is used to combat depression (by producing euphoric feelings, even if short-lived), to combat tension (by functioning as a tranquilizer), or to help people sidestep moral conflicts (for example, by dulling awareness of moral prohibitions against sexual behavior or aggression). Social reinforcers are also made available by substance abuse, such as the approval of drug-abusing companions and, in the cases of alcohol and stimulants, the (temporary) overcoming of social shyness.

Alcohol and Tension Reduction Learning theorists have long maintained that one of the primary reinforcers for using alcohol is relief from states of tension or unpleasant states of arousal (Hussong et al., 2001; Wood et al., 2001). The *tension-reduction theory* proposes that the more often one drinks to reduce tension or anxiety, the stronger or more habitual the habit becomes. Viewed in this way, alcohol use can be likened to a form of self-medication, a way of easing psychological pain, at least temporarily, as in the following case example:

A Case of Self-Medication

"I use them (the pills and alcohol) to take away the hurt I feel inside." Joceyln, a 36-year-old mother of two, was physically abused by her husband, Phil. "I have no self-esteem. I just don't feel I can do anything," she told her therapist. Joceyln had escaped from an abusive family background by getting married at age 17, hoping that it would offer her a better life. The first few years of marriage were free of abuse but things changed when Phil lost his job and began to drink heavily. By then, Jocelyn had two young children and felt trapped. She blamed herself for her unhappy family life, for Phil's drinking, for her son's learning disability. "The only thing I can do is drink or do pills. At least then I don't have to think about things for awhile."

—*From the Authors' Files*

Self-medication? People who turn to other drugs or alcohol as a form of self-medication for anxiety or depression may only compound their problems by developing a substance use disorder.

Drugs, including nicotine from cigarette smoking, may be used as a form of self-medication for depression (Breslau et al., 1998). Stimulants like nicotine temporarily elevate the mood, whereas depressants like alcohol quell anxiety. Although nicotine, alcohol, and other drugs may temporarily alleviate emotional distress, they cannot resolve underlying personal or emotional problems. Rather than learning to resolve these problems, people who use drugs as forms of self-medication often find themselves facing additional substance-related problems.

Negative Reinforcement and Withdrawal Once people become physiologically dependent, *negative reinforcement* comes into play in maintaining the drug habit. In other words, people may resume using drugs to gain relief from unpleasant withdrawal symptoms. In operant conditioning terms, the resumption of drug use is negatively reinforced by relief from unpleasant withdrawal symptoms. For example, the addicted smoker who quits cold turkey may shortly return to smoking to fend off the discomfort of withdrawal. Smokers who are able to quit and maintain abstinence are occasionally bothered by urges to smoke but have learned to manage them.

The Conditioning Model of Cravings Classical conditioning may help explain drug cravings experienced by people with drug dependency. Drug cravings may have a biological basis, reflecting a bodily need to restore levels of the addictive substance. But they also come to be triggered by environmental cues associated with prior use of the substance (Kilts et al., 2001). These drug-related cues, such as the sight or aroma of an alcoholic beverage or the sight of a needle and syringe, may become conditioned stimuli that elicit a conditioned response in the form of strong desires or cravings for the drug (Drummond & Glautier, 1994). For example, socializing with certain companions ("drinking buddies") or even passing a liquor store may elicit conditioned cravings for alcohol. In recent research, alcoholic subjects showed distinctive changes in brain activity in areas of the brain that regulate emotion, attention, and appetitive behavior when they were shown pictures of alcoholic beverages (George et al., 2001). Social drinkers, by comparison, did not show this pattern of brain activation.

Sensations of anxiety or depression that were paired with the use of alcohol or drugs may also elicit cravings. The following case illustrates conditioned cravings to environmental cues:

A Case of Conditioned Drug Cravings

A 29-year-old man was hospitalized for the treatment of heroin addiction. After four weeks of treatment, he returned to his former job, which required him to ride the subway past the stop at which he had previously bought his drugs. Each day, when the subway doors opened at this location, [he] experienced enormous craving for heroin, accompanied by tearing, a runny nose, abdominal cramps, and gooseflesh. After the doors closed, his symptoms disappeared, and he went on to work.

—From Weiss & Mirin, 1987, p. 71

Similarly, some people are primarily "stimulus smokers." They reach for a cigarette in the presence of smoking-related stimuli, such as seeing someone smoke or smelling smoke. Smoking becomes a strongly conditioned habit because it is paired repeatedly with many situational cues—watching TV, finishing dinner, driving in the car, studying, drinking or socializing with friends, sex, and, for some, using the bathroom.

The conditioning model of craving is strengthened by research showing that people with alcoholism tend to salivate more than others at the sight and smell of alcohol (Monti et al., 1987). In Pavlov's classic experiment, a salivation response was conditioned in dogs by repeatedly pairing the sound of a bell (a neutral or conditioned stimulus) with the presentation of food powder (an unconditioned stimulus). Salivation among people who develop alcoholism can also be viewed as a conditioned response to alcohol-related cues. Whereas salivating to a bell may be harmless, salivating at a bottle of Scotch, or at a picture of a bottle in a magazine ad, can throw the person who suffers from alcoholism and is trying to remain abstinent into a tailspin. People with drinking problems who show the greatest salivary response to alcohol cues may be at highest risk of relapse. They may also profit from treatments designed to extinguish their responses to alcohol-related cues.

One such treatment, called *cue exposure training,* holds promise in the treatment of alcohol dependence and other forms of addictive behavior (Drummond & Glautier, 1994). In cue exposure treatment, the person is repeatedly seated in front of the drug or alcohol-related cues, such as open alcoholic beverages, while prevented from using the drug. This pairing of the cue (alcohol bottle) with nonreinforcement (by dint of preventing drinking) may lead to extinction of the conditioned craving. Cue exposure treatment may be combined with coping skills training to help people with substance abuse problems learn to cope with drug use urges without resorting to drug use (Monti et al., 1994). It has also been used to help nondependent problem drinkers learn to stop drinking after two or three drinks (Sitharthan et al., 1997).

Observational Learning The role of modeling or observational learning may at least partly explain the increased risk of substance abuse problems in adolescents in families with a history of a substance abuse or dependence (Kilpatrick et al., 2000). For example, parents who model inappropriate or excessive drinking may set the stage for alcohol use and abuse in their children. In one study, teens who said their fathers drank more than two drinks a day had about a 75% greater risk of developing substance abuse problems than did teens with fathers who were described as light drinkers or abstainers ("Teens Who Have Problems," 1999). Researchers also find that young men from families with a history of alcoholism were more strongly affected by exposure to others who modeled excessive drinking than were men without familial alcoholism (Chipperfield & Vogel-Sprott, 1988). Perhaps their parents had modeled excessive drinking and they had learned to regulate their own intake by observing the drinking behavior of others. When their drinking companions drink to excess, they may be more likely to follow their lead.

Cognitive Perspectives

Evidence supports the role of various cognitive factors in substance abuse and dependence, including expectancies and beliefs.

Outcome Expectancies, and Substance Abuse The beliefs and expectancies you hold concerning the effects of alcohol and other drugs clearly influence your decision to use them or not. People who hold positive expectancies about the effects of a drug are not only more likely to use the drug (Schafer & Brown, 1991) but also more likely to use larger quantities of the drug (Baldwin, Oei, & Young, 1994). One of the key factors in predicting alcohol use and misuse in adolescents is the degree to which their friends hold positive attitudes toward alcohol use (Scheier, Botvin, & Baker, 1997; Wood et al., 2001). Similarly, fifth and seventh graders who hold more positive impressions of smokers (for example, seeing them as cool, independent, or good looking) were more likely than their peers to become smokers by the time they reach the ninth grade (Dinh et al., 1995). Smoking prevention programs may need to focus on changing the image that young people hold of smokers long before they ever light up a cigarette themselves. Positive alcohol expectancies also appear in children even before drinking begins.

Among the most widely held positive expectancies concerning alcohol are that it reduces tension, helps divert attention from one's problems, heightens pleasure, and lessens anxiety in social situations and makes one more socially adept. The belief that alcohol helps make a person more socially adept (more relaxed, outgoing, assertive, and carefree in social interactions) appears to be an especially important factor in prompting drinking in adolescents and college students (Burke & Stephens, 1999; Smith et al., 1995).

Self-Efficacy Expectancies Part of the appeal of substances such as alcohol lies in their ability to enhance self-efficacy expectancies (beliefs in our ability to accomplish tasks) either directly (by enhancing feelings of energy, power, and well-being) or indirectly (by reducing stressful states of arousal, such as anxiety) (G. T. Wilson, 1987). Cocaine also enhances self-efficacy expectancies, an outcome sought in particular by performance-conscious athletes. People may therefore come to rely on substances in challenging situations where they doubt their abilities. Alcohol can also help protect one's sense of self-efficacy by shunting criticism for socially unacceptable behavior from the self to the alcohol. People who "screw up" while drinking can maintain their self-esteem by attributing their shortcomings to the alcohol.

Does One "Slip" Cause People with Substance Abuse or Dependence to Go on Binges? Perhaps What You Believe Is What You Get According to the disease model of alcoholism, abstainers who binge after just one drink do so largely for biochemical reasons. Experimental research, however, suggests that cognitive factors may be more important. In fact, the one-drink hypothesis may be explained by the drinker's expectancies rather than by the biochemical properties of alcohol.

Studies of the one-drink hypothesis, like many other studies on alcohol, are made possible by the fact that the taste of vodka can be cloaked by tonic water. In a classic study by Marlatt and his colleagues (1973), subjects were led to believe they were participating in a taste test. Alcohol-dependent subjects and social drinkers who were informed they were sampling an alcoholic beverage (vodka) drank significantly more than counterparts who were informed they were sampling a nonalcoholic beverage. The expectations of the alcohol-dependent subjects and the social drinkers alike were the crucial factors that predicted the amount consumed (see Figure 10.5). *The actual content of the beverages was immaterial.*

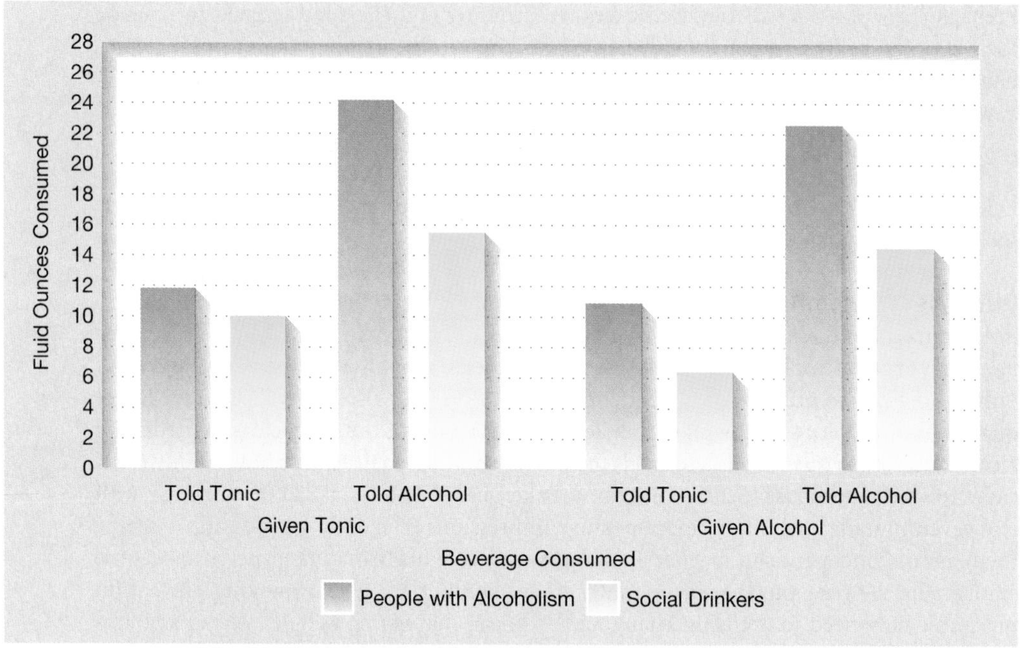

FIGURE 10.5 Must people who develop alcoholism fall off the wagon if they have one drink? It is widely believed that people who develop alcoholism will lose control if they have just one drink. Will they? If so, why? Laboratory research by Marlatt and his colleagues suggests that the tendency of people who suffer from alcoholism to drink to excess following a first drink may be the result of a self-fulfilling prophecy rather than a craving. Like the dieter who eats a piece of chocolate, people who develop alcoholism may assume that they have lost control because they have fallen off the wagon and then go on a binge. This figure shows that people with alcoholism who participated in the Marlatt study drank more when they were led to believe that the beverage contained alcohol, regardless of its actual content. It remains unclear, however, whether binge drinking by people with alcohol-related problems in real-life settings can be explained as a self-fulfilling prophecy.

Source. Adapted from Marlatt et al. (1973).

Marlatt (1978) explained the one-drink effect as a self-fulfilling prophecy. If people with alcohol-related problems believe that just one drink will cause a loss of control, they perceive the outcome as predetermined when they drink. Their drinking—even taking one drink—may thus escalate into a binge. When individuals who were formerly physiologically dependent on alcohol share this belief—which is propounded by many groups, including AA—they may interpret "just one drink" as "falling off the wagon." Marlatt's point is that the "mechanism" of falling off the wagon due to the consumption of one drink is cognitive, reflecting one's expectations about the effects of the drink, and not physiological. This expectation is an example of what Aaron Beck refers to as *absolutist thinking.* When we insist on seeing the world in black and white rather than shades of gray, we may interpret one bite of dessert as proof we are off our diets, or one cigarette as proof we are hooked again. Rather than telling ourselves, "Okay, I goofed, but that's it. I don't have to have more," we encode our lapses as catastrophes and transform them into relapses. Still, alcohol-dependent people who believe they may go on a drinking binge if they have just one drink are well advised to abstain rather than place themselves in a situation they feel they may not be able to manage.

Psychodynamic Perspectives

According to traditional psychodynamic theory, alcoholism reflects certain features of what is termed an *oral-dependent personality.* Alcoholism is, by definition, an oral behavior pattern. Psychodynamic theory also associates excessive alcohol use with other oral traits, such as dependence and depression, and traces the origins of these traits to fixation in the oral stage of psychosexual development. Excessive drinking in adulthood symbolizes an individual's efforts to attain oral gratification.

Psychodynamic theorists also view smoking as an oral fixation, although they have not been able to predict who will or will not smoke. Sigmund Freud smoked upward of 20 cigars a day despite several vain attempts to desist. Although he contracted oral cancer and had to have his jaw replaced, he would still not surrender his "oral fixation." He eventually succumbed to cancer of the mouth in 1939 at the age of 83, after years of agony.

Research support for these psychodynamic concepts is mixed. Although people who develop alcoholism often show dependent traits, it is unclear whether dependence contributes to or stems from problem drinking. Chronic drinking, for example, is connected with loss of employment and downward movement in social status, both of which would render drinkers more reliant on others for support. Moreover, an empirical connection between dependence and alcoholism does not establish that alcoholism represents an oral fixation that can be traced to early development.

Then too, many—but certainly not all—people who suffer from alcoholism have antisocial personalities characterized by independence seeking as expressed through rebelliousness and rejection of social and legal codes (Graham & Strenger, 1988). All in all, there doesn't appear to be a single alcoholic personality (Wood et al., 2001).

Peer pressure. Peer pressure is a major influence on alcohol and drug use among adolescents.

Sociocultural Perspectives

Drinking is determined, in part, by where we live, whom we worship with, and the social or cultural norms that regulate our behavior. Cultural attitudes can encourage or discourage problem drinking. As we have already

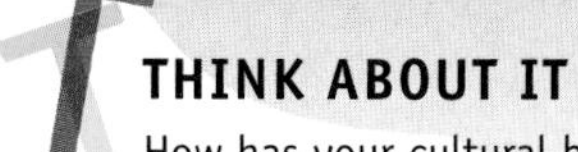

THINK ABOUT IT

How has your cultural background affected your attitudes and your use of drugs such as alcohol and tobacco?

seen, rates of alcohol abuse vary across ethnic and religious groups. Let us note some other sociocultural factors. Church attendance, for example, is generally connected with abstinence from alcohol. Perhaps people who are more willing to engage in culturally sanctioned activities, such as churchgoing, are also more likely to adopt culturally sanctioned prohibitions against excessive drinking. Rates of alcohol use also vary across cultures. For example, alcohol use is greater in Germany than in the United States, apparently because of a cultural tradition that makes the consumption of alcohol, especially beer, normative within German society (Cockerham, Kunz, & Lueschen, 1989).

Drug use by peers and peer pressure to use drugs are important influences in determining alcohol and drug use among adolescents ("Peers Sway," 2001; Simons-Morton et al., 2001; Wills & Cleary, 1999). Kids who start drinking before age 15 stand a fivefold higher risk of developing alcohol dependence in adulthood than do teens who began drinking at a later age (Kluger, 2001). Yet studies of Hispanic and African American adolescents show that support from family members can reduce the negative influence of drug-using peers on the adolescent's use of tobacco and other drugs (Farrell & White, 1998; Frauenglass et al., 1997).

Tying It Together

Substance abuse and dependence are complex patterns of behavior that involve an interplay of biological, psychological, and environmental factors. Genetic factors and the early home environment may give rise to predispositions (diatheses) to abuse and dependence. In adolescence and adulthood, positive expectations concerning drug use, together with social pressures and a lack of cultural constraints, affect drug use decisions and tendencies toward abuse. When physiological dependence occurs, people may use a substance to avoid withdrawal symptoms.

Genetic factors may create an inborn tolerance for certain drugs, such as alcohol, which can make it difficult to regulate usage, to know "when to say when." Some individuals may have genetic tendencies that lead them to become unusually tense or anxious. Perhaps they turn to alcohol or other drugs to quell their nervousness. Genetic predispositions may interact with environmental factors that increase the potential for drug abuse and dependence—factors such as pressure from peers to use drugs, parental modeling of excessive drinking or drug use, and family disruption that results in a lack of effective guidance or support. Cognitive factors, especially positive drug expectancies (e.g., beliefs that using drugs will enhance one's social skills or sexual prowess), may also raise the potential for alcohol or drug problems.

Sociocultural factors need to be taken into account in this matrix of factors, such as the availability of alcohol and other drugs, presence or absence of cultural constraints that might curb excessive or underage drinking, the glamorizing of drug use in popular media, and inborn tendencies (such as among Asians) to flush more readily following alcohol intake.

Learning factors also play important roles. Drug use may be *positively* reinforced by the pleasurable effects associated with the use of the drug (mediated perhaps by release of dopamine in the brain or by activation of endorphin receptors). It may also be *negatively* reinforced by the reduction of states of tension and anxiety that depressant drugs such as alcohol, heroin, and tranquilizers can produce. In a sad but ironic twist, people who become dependent on drugs may continue to use them solely because of the relief from withdrawal symptoms and cravings they encounter when they go without the drug.

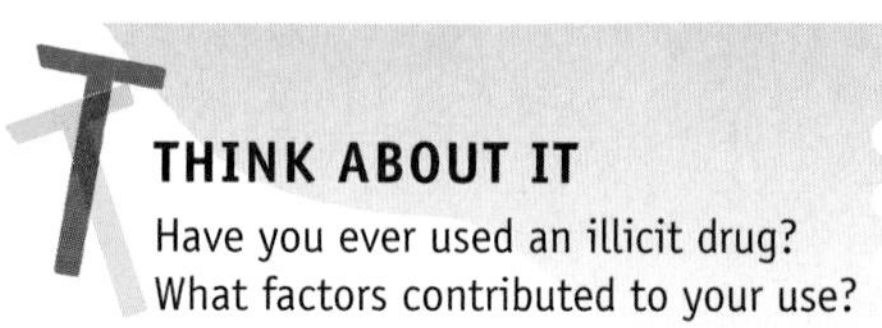

THINK ABOUT IT

Have you ever used an illicit drug? What factors contributed to your use?

Problems of substance abuse and dependence are best approached by investigating the distinctive constellation of factors that apply to each individual case. No single model or set of factors will explain each case, which is why we need to understand each individual's unique characteristics and personal history.

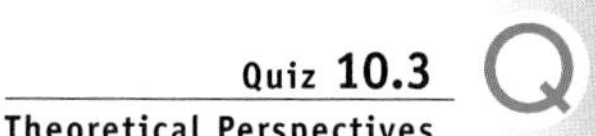

Quiz **10.3**
Theoretical Perspectives

Treatment of Substance Abuse and Dependence

There have been and remain a vast array of nonprofessional, biological, and psychological approaches to substance abuse and dependence. However, treatment has often been a frustrating endeavor. In many, perhaps most, cases, people with drug dependencies really do not want to discontinue the substances they are abusing. Most people who abuse cocaine, for example, like most abusers of alcohol and other drugs, do not seek treatment on their own. Those who do not seek treatment tend to be heavy abusers who deny the negative impact of cocaine on their lives and dwell within a social milieu that fails to encourage them to get help. When people do come for treatment, helping them through a withdrawal syndrome is usually straightforward enough, as we shall see. However, helping them pursue a life devoid of their preferred substances is more problematic. Moreover, treatment takes place in a setting—such as the therapist's office, a support group, a residential center, or a hospital—in which abstinence is valued and encouraged. Then the individual returns to the work, family, or street settings in which abuse and dependence were instigated and maintained. The problem of returning to abuse and dependence following treatment—that is, of *relapse*—can thus be more troublesome than the problems involved in initial treatment.

Another complication is that many people with substance abuse problems also have psychological disorders, and vice versa (McCrady & Langenbucher, 1996; Miller & Brown, 1997). Most clinics and treatment programs focus on the drug or alcohol problem, or the other psychological disorders, rather than treating all these problems simultaneously, however. This narrow focus results in poorer treatment outcomes, including more frequent rehospitalizations among those with these *dual diagnoses.* It has been estimated that 20% to 70% of people who have other psychological disorders—and 50% to 70% of the young adults with other psychological disorders—merit a dual diagnosis that includes substance abuse (Polcin, 1992).

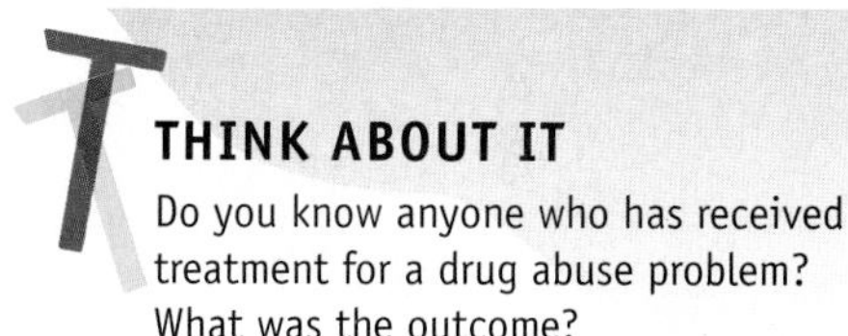

THINK ABOUT IT

Do you know anyone who has received treatment for a drug abuse problem? What was the outcome?

Biological Approaches

An increasing range of biological approaches is used in treating problems of substance abuse and dependence. For people with chemical dependencies, biological treatment typically begins with **detoxification**—that is, helping them through withdrawal from addictive substances.

Detoxification Detoxification is often carried out in a hospital setting to provide the support needed to help the person withdraw safely from the addictive substance. In the case of addiction to alcohol or barbiturates, hospitalization allows medical personnel to monitor the development of potentially dangerous withdrawal symptoms, such as convulsions. Antianxiety drugs, such as the benzodiazepines Librium and Valium, may help block more severe withdrawal symptoms such as seizures and delirium tremens (Mayo-Smith, 1997). Behavioral treatment using monetary rewards for abstinent behavior (judged by clean urine samples) may help improve outcomes during detoxification from opioids (Bickel et al., 1997). Detoxification to alcohol takes about a week. Detoxification is an important step toward staying clean, but it is only a start. Approximately half of all drug abusers relapse within a year of detoxification (Cowley, 2001a). Continuing support and use of structured forms of therapy, such as behavioral counseling and possible use of therapeutic drugs, may hopefully increase the chances of long-term success.

A number of therapeutic drugs are used in treating people with chemical dependencies, and more chemical compounds are in the testing stage (Kranzler, 2000). Here we survey some of the major drugs in use today.

Disulfiram The drug *disulfiram* (brand name Antabuse) discourages alcohol consumption because the combination of the two produces a violent response consisting of nausea, headache, heart palpitations, and vomiting (Kalb, 2001b). In some extreme cases, combining

detoxification The process of ridding the system of alcohol or other drugs under supervised conditions.

methadone An artificial narcotic that is used to help people who are addicted to heroin to abstain from it without a withdrawal syndrome.

disulfiram and alcohol can produce such a dramatic drop in blood pressure that the individual goes into shock and may even die. Although disulfiram has been used widely in alcoholism treatment, its effectiveness is limited because many patients who want to continue drinking simply stop using the drug. Others stop taking the drug because they believe that they can maintain abstinence without it. Unfortunately, many return to uncontrolled drinking. Another drawback is that the drug has toxic effects in people with liver disease, a frequent ailment of people who suffer from alcoholism. Little evidence supports the efficacy of the drug in the long run (Garbutt et al., 1999; Schuckit, 1996).

Antidepressants Antidepressants show promise in reducing cravings for cocaine following withdrawal. These drugs stimulate neural processes that regulate feelings of pleasure derived in everyday experiences. If pleasure can be more readily derived from non-drug-related activities, cocaine users may be less likely to return to using cocaine to induce pleasurable feelings. However, antidepressants have not yet produced consistent results in reducing relapse rates for cocaine dependence, so it is best to withhold judgment concerning their efficacy (O'Brien, 1996).

Web Link 10.7 WWW
A Guide to Quitting Smoking

Deficiencies of the neurotransmitter serotonin may play a role in producing desires or cravings for alcohol (Anton, 1994). Research is underway on whether alcohol cravings and consumption can be curbed by using serotonin-reuptake inhibitors (Prozac is one) and other drugs that help normalize serotonin activity in the brain (B. A. Johnson et al., 2000b; Kranzler, 2000). The actions of another neurotransmitter, dopamine, may account for the pleasurable or euphoric effects of alcohol. Drugs that mimic dopamine may be helpful in blocking the pleasurably reinforcing effects of alcohol.

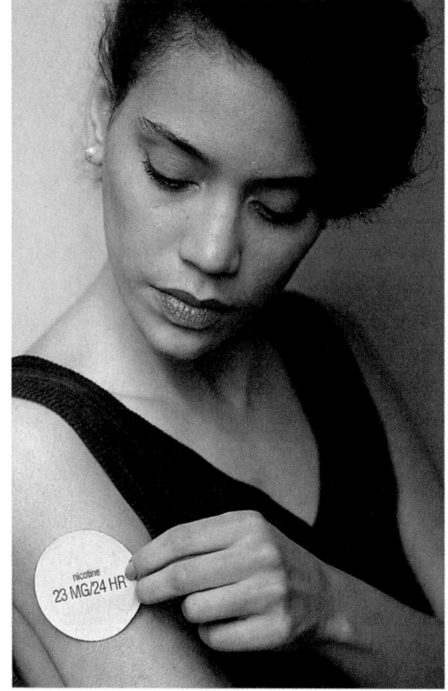

Is the path to abstinence from smoking skin deep? Forms of nicotine replacement therapy—such as nicotine transdermal (skin) patches and chewing gum that contains nicotine—allow people to continue to take in nicotine when they quit smoking. Though nicotine replacement therapy is more effective than a placebo in helping people quit smoking, it does not address the behavioral components of addiction to nicotine, such as the habit of smoking while drinking alcohol. For this reason, nicotine replacement therapy may be more effective if it is combined with behavior therapy that focuses on changing smoking habits.

Nicotine Replacement Therapy Most regular smokers, perhaps the great majority, are nicotine dependent. The use of nicotine replacements in the form of prescription gum (brand name Nicorette), transdermal (skin) patches, and a recently approved nasal spray may help smokers avoid the unpleasant withdrawal symptoms and cravings for cigarettes that may occur following smoking cessation (Tiffany, Cox, & Elash, 2000). After quitting smoking, ex-smokers can gradually wean themselves from the nicotine replacement.

Evidence shows that nicotine chewing gum and the nicotine patch are effective aids in quitting smoking (e.g., O'Brien & McKay, 1998; Skaar et al., 1997). The jury is still out on nicotine nasal sprays. Though nicotine replacement may help quell the physiological components of withdrawal, they have no effect on the behavioral patterns of addiction, such as the habit of smoking while drinking alcohol or socializing. As a result, nicotine replacement may be ineffective in promoting long-term changes unless it is combined with behavioral therapy that focuses on changing smoking habits ("Last Draw for Smokers," 1996).

In 1997, the government approved the use of the first non-nicotine-based antismoking drug, an antidepressant called *bupropion* (trade name Zyban). The drug has been shown to be more effective than placebo in helping smokers quit (Hurt et al., 1997). It is the first drug that works on reducing cravings for nicotine, in much the same way that other antidepressants are being used to reduce cocaine cravings.

Methadone Maintenance Programs **Methadone** is a synthetic opiate that has been used for more than 30 years in treating heroin addiction ("Beyond Methadone," 2000). It blunts cravings for heroin and helps curb the unpleasant symptoms that accompany withdrawal (P. G. O'Connor, 2000; Sees, 2000). Because methadone in normal doses does not produce a high or leave the user feeling drugged, it can help heroin addicts hold jobs and get their lives back on track (Cowley, 2001a; Fiellin et al., 2001; R. E. Johnson et al., 2000). However, like other opioids, methadone is highly addictive. For this reason, people treated with methadone can be conceptualized as swapping dependence on one drug for dependence on another. Yet because most methadone programs are publicly financed, they relieve people who are addicted to heroin of the need to engage in criminal activity to support their dependence on methadone.

Approximately 120,000 people in the United States participate in methadone programs. Although methadone can be taken indefinitely, individuals may be weaned from it without returning to using heroin. Although methadone treatment produces clear benefits in improved daily functioning, not everyone succeeds with methadone, even with counseling. Some addicts turn to other drugs such as cocaine to get high or return to using heroin. Others drop out of methadone programs (Goode, 2001c).

Recently, another type of synthetic narcotic, *buprenorphine,* has been brought to market. Many treatment providers prefer buprenorphine to methadone because it produces less of a sedative effective and can be taken in pill form only three times a week, whereas methadone is given in liquid form on a daily basis (O'Connor, 2001a). For maximum effectiveness, methadone or buprenorphine treatment should be combined with psychological counseling and psychosocial rehabilitation (P. G. O'Connor, 2000; E. O'Connor, 2001a; Rounsaville & Kosten, 2000).

Truth OR Fiction? REVISITED

A widely used treatment for addiction to heroin is based on the substitution of another addictive drug.

TRUE. Methadone, a synthetic narcotic, is widely used in treating heroin addiction.

Naloxone and Naltrexone **Naloxone** and **naltrexone** are sister drugs that block the high produced by heroin and other opioids. By blocking the opioid's effects, they may be useful in helping addicts avoid relapsing following opiate withdrawal (Anton et al., 2001; Dettmer et al., 2001).

Naltrexone (brand name ReVia) blocks the high from alcohol as well as from opiates. In a double-blind placebo-control study, naltrexone in combination with behavioral treatment cut the relapse rates in people treated for alcoholism by more than half (Volpicelli et al., 1994). Naltrexone doesn't prevent the person from taking a drink, but seems to blunt cravings for the drug (Kalb, 2001b). By blocking the pleasure produced by alcohol, the drug can help break the vicious cycle in which one drink creates a desire for another, leading to episodes of binge drinking.

A nagging problem with drugs such as naltrexone, naloxone, disulfiram, and methadone is that people with substance abuse problems may simply stop using them and return to their substance-abusing behavior. Nor do such drugs provide alternative sources of positive reinforcement that can replace the pleasurable states produced by drugs of abuse. Drugs such as these are only effective in the context of a broader treatment program, consisting of psychological counseling and other treatment components, such as job training, and stress management training—treatments designed to provide people with substance abuse problems the skills they need to embark on a life in the mainstream culture (Miller & Brown, 1997).

THINK ABOUT IT

What do you think of the concept of using methadone, a narcotic drug, to treat addiction to another narcotic drug, heroin? What are the advantages and disadvantages of this approach? Do you believe the government should support methadone maintenance programs? Why or why not?

naloxone A drug that prevents users from becoming high if they take heroin.

naltrexone A drug related to naloxone that blocks the high from alcohol as well as from opiates.

Culturally Sensitive Treatment of Alcoholism

Members of ethnic minority groups may resist traditional treatment approaches because they feel excluded from full participation in society. Native American women, for example, tend to respond less favorably to traditional alcoholism counseling than White women (Rogan, 1986). Hurlburt and Gade (1984) attribute this difference to the resistance of Native American women to "White man's" authority. They suggest that the early stages of intervention might be more successful in overcoming this resistance if treatment was provided by Native American counselors.

The use of counselors from the client's own ethnic group is an example of a culturally sensitive treatment approach. Culturally sensitive programs also address all facets of the human being, including racial and cultural identity, that nurture ethnic pride and help people resist the temptation to cope with stress through chemicals (Rogan, 1986). Culturally sensitive treatment approaches have been extended to other forms of drug dependence, including programs for smoking cessation (Nevid & Javier, 1997; Nevid, Javier, & Moulton, 1996).

Treatment providers may also be more successful if they recognize and incorporate indigenous forms of healing into the treatment process. For example, spirituality is an important aspect of traditional Native American culture, and spiritualists have played important roles as natural healers. Seeking the assistance of a spiritualist may help improve the counseling relationship.

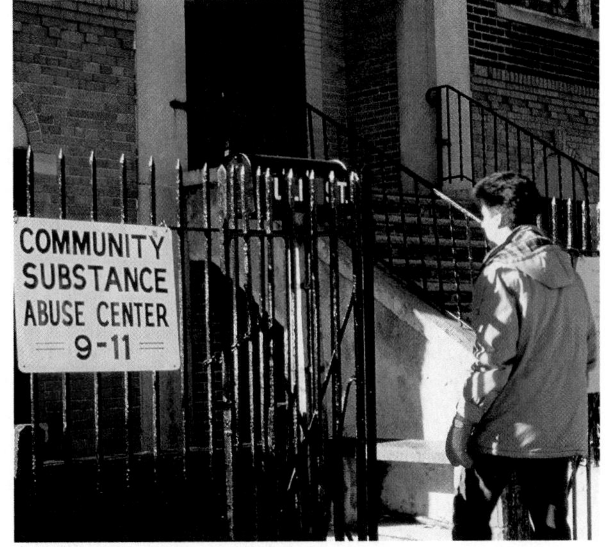

Culturally sensitive treatment. Culturally sensitive therapy or treatment addresses all aspects of the person, including ethnic factors and the nurturance of pride in one's cultural identity. Ethnic pride may help people resist the temptation to cope with stress through alcohol and other substances.

Web Link **10.8**
Alcoholics Anonymous (AA) World Services

Web Link **10.9**
National Association for Children of Alcoholics

Likewise, given the importance of the church in African American and Hispanic American culture, counselors working with people with alcohol use disorders from these groups may be more successful when they draw on clergy and church members as resources in the treatment process.

Nonprofessional Support Groups

Despite the complexity of the factors contributing to substance abuse and dependence, these problems are frequently handled by laypeople or nonprofessionals. Such people often have or had the same problems themselves. For example, self-help group meetings are sponsored by organizations such as Alcoholics Anonymous, Narcotics Anonymous, and Cocaine Anonymous. These groups promote abstinence and provide members an opportunity to discuss their feelings and experiences in a supportive group setting. More experienced group members (sponsors) support newer members during periods of crisis or potential relapse. The meetings are sustained by nominal voluntary contributions.

The most widely used nonprofessional program, Alcoholics Anonymous (AA), is based on the belief that alcoholism is a disease, not a sin. AA assumes that people who suffer from alcoholism are never cured, regardless of how long they abstain from alcohol or how well they control their drinking. Instead of being "cured," people who suffer from alcoholism are seen as "recovering." It is also assumed that people who suffer from alcoholism cannot control their drinking and need help to stop drinking. AA has more than 50,000 chapters in North America. AA is so deeply embedded in the consciousness of helping professionals that many of them automatically refer newly detoxified people to AA as the follow-up agency. About half of AA members have problems with illicit drugs as well as alcohol.

The AA experience is in part spiritual, in part group supportive, in part cognitive. AA follows a 12-step approach that focuses on accepting one's powerlessness over alcohol and turning one's will and life over to a higher power (Cowley, 2001a). This spiritual component may be helpful to some participants but not to others, who prefer not to appeal for divine support. (Other lay organizations, such as Rational Recovery, do not adopt a spiritual approach.) The later steps focus on examining one's character flaws, admitting one's wrongdoings, being open to a higher power (or God) for help to overcome one's character defects, making amends to others, and, in step 12, bringing the AA message to other people suffering from alcoholism (McCrady, 1994). Prayer and or meditation are urged upon members to help them get in touch with their higher power. The meetings themselves provide group support. So does the buddy, or sponsoring, system, which encourages members to call each other for support when they feel tempted to drink.

AA claims to have a high success rate, one in the neighborhood of 75% (Wallace, 1985). Critics note that percentages this high are based on personal testimonies rather than on careful surveys or experiments. Moreover, such estimates include only persons who attend meetings for extended periods. The dropout rate is high, with as many as 70% of members dropping out within 10 meetings, according to one survey. Even AA estimates that 50% of members drop out after 3 months ("Treatment of Alcoholism—Part II," 1996). It has been difficult to conduct controlled studies because AA does not keep records of its members in order to protect their anonymity and also because of an inability to conduct randomized clinical trials in AA settings (McCaul & Furst, 1994). On the other hand, the greater a person's involvement with AA, researchers find, the better the outcome in terms of days not drinking or using drugs (Morgenstern et al., 1997). We cannot say for certain whether regular participation in AA is responsible for better outcomes, or whether personal motivation may account for both greater AA participation and change in substance-using behavior. In all likelihood success is due to both factors. Nor can we say who is likely to succeed in AA and who is not.

Al-Anon, begun in 1951, is a spinoff of AA that supports the families and friends of people suffering from alcoholism. There are some 26,000 Al-Anon groups nationwide (Desmond, 1987). Another spin-off of AA, Alateen, provides support to children whose parents have alcoholism, helping them see they are not to blame for their parents' drinking and are thus undeserving of the guilt they may feel.

Al-Anon An organization that sponsors support groups for family members of people with alcoholism.

Residential Approaches

A residential approach to treatment involves a stay in a hospital or therapeutic residence. Hospitalization may be recommended when substance abusers cannot exercise self-control in their usual environments, or cannot tolerate withdrawal symptoms, and when their behavior is self-destructive or dangerous to others. Outpatient treatment is less costly and often indicated when withdrawal symptoms are less severe, clients are committed to changing their behavior, and environmental support systems, such as families, strive to help clients make the transition to a drug-free lifestyle. The great majority (nearly 90%) of people treated for alcoholism are treated on an outpatient basis (McCaul & Furst, 1994).

Most inpatient programs use an extended 28-day detoxification, or drying-out, period. Clients are helped through withdrawal symptoms in a few days. Then the emphasis shifts to counseling about the destructive effects of alcohol and combating distorted ideas or rationalizations. Consistent with the disease model, the goal of abstinence is urged.

Despite their popularity, researchers find that most people with alcohol use disorders do not require hospitalization (some certainly do). Studies comparing outpatient and inpatient programs reveal no overall difference in relapse rates (Miller & Hester, 1986). However, medical insurance may not cover outpatient treatment, which may encourage many people who may benefit from outpatient treatment to admit themselves for inpatient treatment.

A number of residential therapeutic communities are also in use. Some of them have part- or full-time professional staffs. Others are run entirely by laypeople. Residents are expected to remain free of drugs and take responsibility for their actions. They are often confronted about their excuses for failing to take responsibility for themselves and about their denial of the damage being done by their drug abuse. They share their life experiences to help one another develop productive ways of handling stress.

As with AA, we lack evidence from controlled studies demonstrating the efficacy of residential treatment programs. Also like AA, therapeutic communities have high numbers of early dropouts. Moreover, many former members of residential treatment programs who remain substance free during their time in residence relapse upon returning to the world outside. A recent study suggests that a day treatment therapeutic community may be as effective as a residential treatment facility (Guydish et al., 1998).

Psychodynamic Approaches

Psychoanalysts view substance abuse and dependence as symptomatic of conflicts that are rooted in childhood experiences. Focusing on substance abuse or dependence per se is seen to offer, at most, a superficial type of therapy. It is assumed that if the underlying conflicts are resolved, abusive behavior will also subside as more mature forms of gratification are sought. Traditional psychoanalysts also assume that programs directed solely at abusive behavior will be of limited benefit because they fail to address the underlying psychological causes of abuse. Although there are many reports of successful psychodynamic case studies of people with substance abuse problems, there is a dearth of controlled and replicable research studies. The effectiveness of psychodynamic methods for treating substance abuse and dependence thus remains unsubstantiated.

Behavioral Approaches

The use of behavior therapy or behavior modification in treating substance abuse and dependence focuses on modifying abusive and dependent behavior patterns. The issue to many behaviorally oriented therapists is not whether substance abuse and dependence are diseases but whether abusers can learn to change their behavior when they are faced with temptation.

Self-Control Strategies Self-control training focuses on helping abusers develop skills they can use to change their abusive behavior. Behavior therapists focus on three components of substance abuse:

1. The *antecedent* cues or stimuli (A's) that prompt or trigger abuse,
2. The abusive *behaviors* (B's) themselves, and
3. The reinforcing or punishing *consequences* (C's) that maintain or discourage abuse.

Table 10.5 shows the kinds of strategies used to modify the "ABC's" of substance abuse.

Aversive Conditioning In **aversive conditioning,** painful or aversive stimuli are paired with substance abuse or abuse-related stimuli to make abuse less appealing. In the case of problem drinking, tastes of different alcoholic beverages are usually paired with chemically induced nausea and vomiting or with electric shock (G. T. Wilson, 1991). As a consequence, alcohol may come to elicit an aversive conditioned response, such as fear or nausea that inhibits drinking. Relief from aversive responses then negatively reinforces avoidance of alcohol.

Social Skills Training Social skills training helps people develop effective interpersonal responses in social situations that prompt substance abuse. Assertiveness training, for example, may be used to teach people with alcohol-related problems how to fend off social pressures to drink. Behavioral marital therapy seeks to improve marital communication and a couple's problem-solving skills to relieve marital stresses that can trigger abuse. Couples may learn how to use written behavioral contracts. One such contract might stipulate that the person with a substance abuse problem agrees to abstain from drinking or to take Antabuse, and his or her spouse agrees to refrain from comments about past drinking and the probability of future lapses. The available evidence supports the utility of social skills training and behavioral marital therapy approaches in treating alcoholism (Finney & Monahan, 1996; O'Farrell et al., 1996).

Relapse-Prevention Training

The word **relapse** derives from Latin roots meaning "to slide back." From 50% to 90% of people who are successfully treated for substance abuse problems eventually relapse (Leary, 1996b). Because of the prevalence of relapse, behaviorally oriented therapists have devised a number of methods referred to as **relapse-prevention training.** Such training helps people with substance abuse problems cope with high-risk situations and to prevent *lapses,* (slips)—from becoming full-blown relapses (Marlatt & Gordon, 1985). High-risk situations include negative mood states, such as depression, anger, or anxiety; interpersonal conflict (e.g., marital problems or conflicts with employers); and socially conducive situations such as "the guys getting together." Participants learn to cope with these situations, for example, by learning self-relaxation skills to counter anxiety and learning to resist social pressures to resume use of the substance. Trainees are also taught to avoid practices that might prompt a relapse, such as keeping alcohol on hand for friends.

Although it contains many behavioral strategies, relapse-prevention training is a cognitive-behavioral technique in that it also focuses on the person's *interpretations* of any lapses or slips that may occur, such as smoking a first cigarette or taking a first drink following quitting. Clients are taught how to avoid the so-called **abstinence violation effect (AVE)**—the tendency to overreact to a lapse—by learning to reorient their thinking about lapses and slips. People who have a slip may be more likely to relapse if they attribute their slip to personal weakness, and experience shame and guilt, than if they attribute the slip to an external or transient event (Curry, Marlatt, & Gordon, 1987). For example, consider a skater who slips on the ice (Marlatt & Gordon, 1985). Whether or not the skater gets back up and continues to perform depends largely on whether the skater sees the slip as an isolated and correctable event or as a sign of complete failure. Evidence shows that the best predictor of progression from a first to a second lapse among ex-smokers was the feeling of giving up after the first lapse (Shiffman et al., 1996). But those who responded to a first lapse by using coping strategies were more likely to succeed in averting a subsequent lapse on the same day.

Participants in relapse-prevention training programs are encouraged to view lapses as temporary setbacks that provide opportunities to learn what kinds of situations lead to temptation and how they can avoid or cope with such situations. If they can learn to think,

aversive conditioning A behavior therapy technique in which a maladaptive response is paired with exposure to an aversive stimulus to develop a conditioned aversion.

relapse A recurrence of a problem behavior or disorder.

relapse-prevention training A cognitive-behavioral technique involving the use of behavioral and cognitive strategies for resisting temptations and preventing relapses.

abstinence violation effect (AVE) The tendency to overreact to a minor lapse with feelings of guilt and resignation that may trigger a relapse.

TABLE 10.5 Self-Control Strategies for Modifying the "ABC's" of Substance Abuse

1. Controlling the A's (Antecedents) of Substance Abuse

People who abuse or become dependent on psychoactive substances become conditioned to a wide range of external (environmental) and internal stimuli (bodily states). They may begin to break these stimulus-response connections by:

- Removing drinking and smoking paraphernalia from the home—all alcoholic beverages, beer mugs, carafes, ashtrays, matches, cigarette packs, lighters, etc.
- Restricting the stimulus environment in which drinking or smoking is permitted. Use the substance only in a stimulus-deprived area of their homes, such as the garage, bathroom, or basement. All other stimuli that might be connected to using the substance are removed—there is no TV, reading materials, radio, or telephone. In this way, substance abuse becomes detached from many controlling stimuli.
- Not socializing with others with substance abuse problems, by avoiding situations linked to abuse—bars, the street, bowling alleys, etc.
- Frequenting substance-free environments—lectures or concerts, a gym, museums, evening classes; and by socializing with nonabusers, sitting in nonsmoking cars of trains, eating in restaurants without liquor licenses.
- Managing the internal triggers for abuse. This can be done by practicing self-relaxation or meditation and not taking the substance when tense; by expressing angry feelings by writing them down or self-assertion, not by taking the substance; by seeking counseling for prolonged feelings of depression, not alcohol, pills, or cigarettes.

2. Controlling the B's (Behaviors) of Substance Abuse

People can prevent and interrupt substance abuse by:

- Using response prevention—breaking abusive habits by physically preventing them from occurring or making them more difficult (e.g., by not bringing alcohol home or cigarettes to the office).
- Using competing responses when tempted; by being prepared to handle substance-related situations with appropriate ammunition—mints, sugarless chewing gum, etc; by taking a bath or shower, walking the dog, walking around the block, taking a drive, calling a friend, spending time in a substance-free environment, practicing meditation or relaxation, or exercising when tempted, rather than using the substance.
- Making abuse more laborious—buying one can of beer at a time; storing matches, ashtrays, and cigarettes far apart; wrapping cigarettes in foil to make smoking more cumbersome; pausing for 10 minutes when struck by the urge to drink, smoke, or use another substance and asking oneself, "Do I really need *this* one?"

3. Controlling the C's (Consequences) of Substance Abuse

Substance abuse has immediate positive consequences such as pleasure, relief from anxiety and withdrawal symptoms, and stimulation. People can counter these intrinsic rewards and alter the balance of power in favor of nonabuse by:

- Rewarding themselves for nonabuse and punishing themselves for abuse.
- Switching to brands of beer and cigarettes they don't like.
- Setting gradual substance-reduction schedules and rewarding themselves for sticking to them.
- Punishing themselves for failing to meet substance-reduction goals. People with substance abuse problems can assess themselves, say, 10 cents for each slip and donate the cash to an unpalatable cause, such as a brother-in-law's birthday present.
- Rehearsing motivating thoughts or self-statements—such as writing reasons for quitting smoking on index cards. For example:

  Each day I don't smoke adds another day to my life.
  Quitting smoking will help me breathe deeply again.
  Foods will smell and taste better when I quit smoking.
  Think how much money I'll save by not smoking.
  Think how much cleaner my teeth and fingers will be by not smoking.
  I'll be proud to tell others that I kicked the habit.
  My lungs will become clearer each and every day I don't smoke.

Smokers can carry a list of 20 to 25 such statements and read several of them at various times throughout the day. They can become parts of one's daily routine, a constant reminder of one's goals.

A Closer Look

The Controlled Drinking Controversy

The disease model of alcoholism contends that people who suffer from the disease who have just one drink will lose control and go on a binge. Some professionals, however, like Linda and Mark Sobell (1973, 1984), have argued that behavior modification self-control techniques can teach many people with alcohol abuse or dependence to engage in **controlled drinking**—to have a drink or two without necessarily falling off the wagon.

The contention that people who develop alcoholism can learn to drink moderately remains controversial. The proponents of the disease model of alcoholism, who have wielded considerable political strength, stand strongly opposed to attempts to teach controlled social drinking.

Investigators have found that controlled social drinking may be a reasonable treatment goal for younger people with problem drinking who are less alcohol dependent but are headed on the road toward chronic alcoholism (e.g., Adamson & Sellman, 2001; Larimer et al., 1993 Miller & Muñoz, 1983; Sanchez-Craig & Wilkinson, 1986/1987). Evidence supporting controlled social drinking programs for people with chronic alcoholism, however, remains lacking. Yet interest in controlled drinking programs has waned, largely because of strong opposition from professionals and lay organizations committed to the abstinence model.

Controlled drinking programs may be best suited for younger persons with early-stage alcoholism or problem drinking, for those who reject goals of total abstinence or who have failed in programs requiring abstinence, and for those who do not show severe withdrawal symptoms (Marlatt et al., 1993; Rosenberg, 1993). Researchers also find that women tend to do better than men in controlled drinking programs (Marlatt et al., 1993).

Controlled drinking programs may actually represent a pathway to abstinence for people who would not otherwise enter abstinence-only treatment programs (Marlatt et al., 1993). That is, treatment in a controlled drinking program may be the first step toward giving up drinking completely. A large percentage (about one in four in one study; Miller et al., 1993) enters with the goal of achieving controlled drinking but become abstinent by the end of treatment. On the other hand, controlled social drinking may not be appropriate for people with established alcohol dependence and those who are taking medications that interact with alcohol or have other medical risks that might be aggravated by alcohol ("Treatment of Alcoholism—Part II," 1996).

"Okay, I had a slip, but that doesn't mean all is lost unless I believe it is," they are less likely to catastrophize lapses and subsequently relapse.

All in all, efforts to treat people with substance abuse and dependence problems have been mixed at best. Many abusers really do not want to discontinue use of these substances, although they would prefer, if possible, to avoid their negative consequences. The more effective substance abuse treatments programs involve multiple treatment approaches that match the needs of substance abusers and the range of problems they often encounter, including co-occurring (comorbid) psychiatric problems like depression (Brown et al., 1997; Kessler et al., 1997b; Rychtarik et al., 2000). *Comorbidity* (co-occurrence) of substance use disorders and other psychological disorders has become the rule in treatment facilities rather than the exception (Brems & Johnson, 1997). Substance abusers who have comorbid disorders or more severe psychological problems typically fare more poorly in treatment for their drug or alcohol problems (Simpson et al., 1999). For people with alcoholism and other substance abuse problems, a number of different therapies, including 12-step and cognitive-behavioral approaches, seem to work well if they are well delivered (Miller & Brown, 1997; Project MATCH Research Group, 1997).

The major problem is that as many as 80% of people in the United States with alcohol use disorders have no contact whatsoever with alcohol treatment programs or self-help organizations (Institute of Medicine, 1990). Clearly more needs to be done to help people whose use of alcohol and other drugs puts them at risk.

In the case of inner city youth who have become trapped within a milieu of street drugs and hopelessness, the availability of culturally sensitive drug counseling and job training opportunities would be of considerable benefit in helping them assume more productive social roles. The challenge is clear: to develop cost-effective ways of helping people recognize the negative effects of substances and forgo the powerful and immediate reinforcements they provide.

THINK ABOUT IT

Many teenagers today have parents who themselves smoked marijuana or used other drugs when they were younger. If you were one of those parents, what would you tell your kids about drugs?

controlled drinking An approach to treating problem drinkers whose goal is moderate social drinking rather than abstinence.

Quiz 10.4 Treatment of Substance Abuse and Dependence

Quiz 10.5 Chapter Exam

Research Update Chapter 10

Overview of Substance-Related Disorders

TYPES OF SUBSTANCE-RELATED DISORDERS

Substance Use Disorders	Maladaptive use of a psychoactive substance	• **Substance Abuse Disorder:** Pattern of drug-using behavior leading to negative consequences, such as repeatedly losing time from work or aggravating an underlying physical problem • **Substance Dependence Disorder:** A more severe form of substance use disorder, it is associated with physiological dependence or compulsive use of a substance
Substance-Induced Disorders	Physiological or psychological disorders induced by the use of a psychoactive substance	• Intoxication • Drug withdrawal syndromes • Mood disorders • Delirium • Dementia • Amnesia • Psychotic disorders

CAUSAL FACTORS Multiple factors interact in leading to problems of substance abuse and dependence

Biological Factors	• Pleasurable effects and addictive properties of drugs may depend on their effects on neurotransmitter systems in the brain • Genetic factors may create predispositions for substance-related disorders
Psychosocial Factors	• Positive reinforcement (pleasure inducing) and negative reinforcement (relief from states of tension or anxiety and avoidance or escape from unpleasant withdrawal symptoms) contribute to initiation and maintenance of drug use • Modeling of excessive drinking by family members and friends • Cravings may be conditioned responses to cues associated with prior drug use • In psychodynamic theory, alcohol and drug abuse represent forms of oral fixation and are linked to dependent personality traits
Cognitive Factors	• Positive outcome expectancies linked to drug use • Effects of drugs in boosting self-efficacy expectations • "Falling off the wagon" as a self-fulfilling prophecy
Sociocultural Factors	• Peer pressure from drug-using peers • Exposure to deviant subcultures (e.g., gang culture) in which drug use is commonplace or encouraged

TREATMENT APPROACHES Intensive, multicomponent treatment approaches generally work best

Biological Approaches	• Detoxification to help substance abusers withdraw safely from addictive drugs • Use of drugs that cause extreme nausea when combined with alcohol (Antabuse) • Use of antidepressants to control drug cravings • Use of chemical substitutes, such as nicotine replacement in place of cigarettes, or methadone in place of heroin • Use of drugs that block the high produced by opioids or alcohol (naloxone and naltrexone)
Behavioral Approaches	• To break drug-abusing patterns of behavior and strengthen more adaptive behaviors
Psychodynamic Approaches	• To help individuals with substance abuse problems identify and resolve underlying psychological conflicts
Other Treatment Approaches	• **Residential treatment approaches** and **nonprofessional support groups,** such as AA, to help individuals regain control over their lives and maintain abstinence in the community • **Relapse prevention training** to help individuals learn to resist drug temptations, to cope effectively with high-risk situations, and to prevent lapses from becoming relapses

Summing Up

Classification of Substance-Related Disorders

How does the* DSM *distinguish between substance abuse disorders and substance dependence disorders? According to the *DSM*, substance abuse disorders involve a pattern of recurrent use of a substance that repeatedly leads to damaging consequences. Substance dependence disorders involves impaired control over use of a substance and often include features of physiological dependence on the substance, as manifest by the development of tolerance or an abstinence syndrome.

***What do we mean by the terms* addiction *and* psychological dependence?** Although different people use the term *addiction* differently, it is used here to refer to the habitual or compulsive use of a substance combined with the development of physiological dependence. Psychological dependence involves compulsive use of a substance, with or without the development of physiological dependence.

Drugs of Abuse

What are depressants? Depressants are drugs that depress or slow down nervous system activity. They include alcohol, sedatives and minor tranquilizers, and opioids. Their effects include intoxication, impaired coordination, slurred speech, and impaired intellectual functioning. Chronic alcohol abuse is linked to alcohol-induced persisting amnestic disorder (Korsakoff's syndrome), cirrhosis of the liver, fetal alcohol syndrome, and other physical health problems. Barbiturates are depressants or sedatives that have been used medically for relief of anxiety and short-term insomnia, among other uses. Opioids such as morphine and heroin are derived from the opium poppy. Others are synthesized. Used medically for relief of pain, they are strongly addictive.

What are stimulants? Stimulants increase the activity of the nervous system. Amphetamines and cocaine are stimulants that increase the availability of neurotransmitters in the brain, leading to heightened states of arousal and pleasurable feelings. High doses can produce psychotic reactions that mimic features of paranoid schizophrenia. Habitual cocaine use can lead to a variety of health problems, and an overdose can cause sudden death. Repeated use of nicotine, a mild stimulant found in cigarette smoking, leads to physiological dependence.

What are hallucinogens? Hallucinogens are drugs that distort sensory perceptions and can induce hallucinations. They include LSD, psilocybin, and mescaline. Other drugs with similar effects are cannabis (marijuana) and phencyclidine (PCP). There is little evidence that these drugs induce physiological dependence, although psychological dependence may occur.

Theoretical Perspectives

How do the major theoretical perspectives view the causes of substance abuse and dependence? The biological perspective focuses on uncovering the biological pathways that may explain mechanisms of physiological dependence. The biological perspective spawns the disease model, which posits that alcoholism and other forms of substance dependence are disease processes. Learning perspectives view substance abuse disorders as learned patterns of behavior, with roles for classical and operant conditioning and observational learning. Cognitive perspectives focus on roles of attitudes, beliefs, and expectancies in accounting for substance use and abuse. Sociocultural perspectives emphasize the cultural, group, and social factors that underlie drug use patterns, including the role of peer pressure in determining adolescent drug use. Psychodynamic theorists view problems of substance abuse, such as excessive drinking and habitual smoking, as signs of an oral fixation.

Treatment

What treatments approaches are used to help people overcome problems of substance abuse and dependence? Biological approaches to substance abuse disorders include detoxification; the use of drugs such as disulfiram, methadone, naloxone, naltrexone, and antidepressants; and nicotine replacement therapy. Residential treatment approaches include hospitals and therapeutic residences. Nonprofessional support groups, such as Alcoholics Anonymous, promote abstinence within a supportive group setting.

Psychodynamic therapists focus on uncovering the inner conflicts originating in childhood that they believe lie at the root of substance abuse problems. Behavior therapists focus on helping people with substance-related problems change problem behaviors through such techniques as self-control training, aversive conditioning, and skills training approaches. Regardless of the initial success of a treatment technique, relapse remains a pressing problem in treating people with substance abuse problems. Relapse-prevention training employs cognitive-behavioral techniques to help recovering substance abusers cope with high-risk situations and to prevent lapses from becoming relapses by helping participants interpret lapses in less damaging ways.

Key for "Are You Hooked?" Questionnaire

Any yes answer suggests you may be dependent on alcohol. If you have answered any of these questions in the affirmative, we suggest you seriously examine what your drinking means to you.

CHAPTER ELEVEN

Eating Disorders, Obesity, and Sleep Disorders

Pablo Picasso
Girl Before a Mirror

Truth OR Fiction?

- Though others see them as but "skin and bones," young women with anorexia nervosa still see themselves as too fat. (p. 343)
- Dieting represents an abnormal eating pattern among American women. (p. 346)
- Bulimic women induce vomiting only after binges. (p. 347)
- Drugs used to treat depression may also help curb bulimic binges. (p. 349)
- The excess calories consumed by Americans each day could feed a country of 80 million people. (p. 353)
- Obesity is one of the most common psychological disorders in the United States. (p. 353)
- When you lose weight, your body starts putting the brakes on the rate at which it burns calories. (p. 355)
- Obese people lose fat cells when they diet. (p. 355)
- Most dieters eventually gain back the weight they lose. (p. 358)
- Many people suffer from sleep attacks in which they suddenly fall asleep without any warning. (p. 362)
- Some people literally gasp for breath hundreds of times during sleep without realizing it. (p. 363)

Jessica was a 20-year-old communications major when she consulted a psychologist for the first time. For years she had kept a secret from everyone, including her fiancé, Ken. She and Ken were planning to get married in 3 months. She had decided that it was time to finally confront the problem. She told the psychologist she didn't want to bring the problem into the marriage with her, that it wouldn't be fair to Ken. She said, "I don't want him to have to deal with this. I want to stop this before the marriage. I have to stop bingeing and throwing up." Jessica went on to describe her problem: "I go on binges and then throw it all up. It makes me feel like I am in control, but really I'm not." To conceal her secret, she would lock herself in the bathroom, run the water in the sink to mask the sounds, and induce vomiting. She would then clean up after herself and spray an air deodorant to mask any telltale odors. "The only one who suspects," she said with embarrassment, "is my dentist. He said my teeth are beginning to decay from stomach acid."

Jessica had *bulimia nervosa,* an eating disorder characterized by recurrent cycles of bingeing and purging. Eating disorders like *bulimia nervosa* and *anorexia nervosa* often affect young people of high school or college age, especially young women. Although rates of diagnosable eating disorders in college students are not as high as you might think, chances are you have known people with anorexia or bulimia or with disturbed eating patterns that fall within a spectrum of disturbed eating behaviors, such as repeated binge eating or excessive dieting. You probably also know people who suffer from obesity, a major health problem for increasing numbers of Americans. Other psychological disorders that commonly affect young adults are sleep disorders. The most common form of sleep disorder, chronic insomnia, affects many young people who are making their way in the world and tend to bring their worries and concerns to bed with them.

Chapter Outline

Eating Disorders

In a nation of plenty, some people literally starve themselves—sometimes to death. They are obsessed with their weight and desire to achieve an exaggerated image of thinness. Others engage in repeated cycles in which they binge on food and then attempt to purge their excess eating, such as by inducing vomiting. These dysfunctional patterns are, respectively, the two major types of eating disorders, **anorexia nervosa** and **bulimia nervosa. Eating disorders** are characterized by disturbed patterns of eating and maladaptive ways of controlling body weight. Like many other psychological disorders, anorexia and bulimia are often accompanied by other forms of psychopathology, including depression, anxiety disorders, and substance abuse disorders.

VIDEO 11.1
Eating Disorders:
Nutritionist Alise Thresh

Anorexia nervosa and bulimia nervosa were once considered very rare, but they are becoming increasingly common in the United States and other developed countries. The great majority of cases occur among women, especially young women. Although these disorders may develop in middle or even late adulthood, they typically begin during adolescence or early adulthood when the pressures to be thin are the strongest (Beck, Casper, & Andersen, 1996). As these social pressures have increased, so too have rates of eating disorders. Approximately 0.5% (1 in 200) females in our society develop anorexia nervosa (APA, 2000). The prevalence rate of bulimia nervosa among women is estimated to range between 1% and 3% (APA, 2000). A much larger percentage of young women show anorexic or bulimic behaviors, but not to the point that they would warrant a diagnosis of an eating disorder. Studies of college women indicate that perhaps 1 in 2 have binged and purged at least once (Fairburn & Wilson, 1993). Rates of anorexia and bulimia among males are about one-tenth those among females (APA, 2000).

anorexia nervosa An eating disorder characterized by maintenance of an abnormally low body weight, distortions of body image, intense fears of gaining weight, and, in females, amenorrhea.

bulimia nervosa An eating disorder characterized by recurrent binge eating followed by self-induced purging, accompanied by overconcern with body weight and shape.

eating disorder A psychological disorder characterized by disturbed patterns of eating and maladaptive ways of controlling body weight.

Anorexia Nervosa

The Case of Karen

Karen was the 22-year-old daughter of a renowned English professor. She had begun her college career full of promise at the age of 17, but two years ago, after "social problems" occurred, she had returned to live at home and taken progressively lighter course loads at

Web Link 11.1 WWW
Facts About Eating Disorders

a local college. Karen had never been overweight, but about a year ago her mother noticed that she seemed to be gradually "turning into a skeleton."

Karen spent literally hours every day shopping at the supermarket, butcher, and bakeries conjuring up gourmet treats for her parents and younger siblings. Arguments over her lifestyle and eating habits had divided the family into two camps. The camp led by her father called for patience; that headed by her mother demanded confrontation. Her mother feared that Karen's father would "protect her right into her grave" and wanted Karen placed in residential treatment "for her own good." The parents finally compromised on an outpatient evaluation.

At an even 5 feet, Karen looked like a prepubescent 11-year-old. Her nose and cheekbones protruded crisply. Her lips were full, but the redness of the lipstick was unnatural, as if too much paint had been dabbed on a corpse for the funeral. Karen weighed only 78 pounds, but she had dressed in a stylish silk blouse, scarf, and baggy pants so that not one inch of her body was revealed.

Karen vehemently denied that she had a problem. Her figure was "just about where I want it to be" and she engaged in aerobic exercise daily. A deal was struck in which outpatient treatment would be tried as long as Karen lost no more weight and showed steady gains back to at least 90 pounds. Treatment included a day hospital with group therapy and two meals a day. But word came back that Karen was artfully toying with her food—cutting it up, sort of licking it, and moving it about her plate—rather than eating it. After 3 weeks Karen had lost another pound. At that point, her parents were able to persuade her to enter a residential treatment program, where her eating behavior could be more carefully monitored.

—From the Authors' Files

At risk? Anorexia is most common among young women involved in ballet and modeling, in which a great emphasis is put upon achieving an ultrathin body image.

Anorexia derives from the Greek roots *an-*, meaning "without," and *orexis*, meaning "a desire for." *Anorexia* thus means "without desire for [food]," which is something of a misnomer, because loss of appetite among people with anorexia nervosa is rare. However, they may be repelled by food and refuse to eat more than is absolutely necessary to maintain a minimal weight for their ages and heights. Often, they starve themselves to the point where they become dangerously emaciated. By and large, anorexia nervosa develops in early to late adolescence, between the ages of 12 and 18, although earlier and later onsets are sometimes found.

The clinical features listed in Table 11.1 are used to diagnose anorexia nervosa. Although reduced body weight is the most obvious sign, the most prominent clinical feature is an intense fear of obesity. One common pattern of anorexia begins after menarche when the girl notices added weight and insists it must come off. The addition of body fat is normal in adolescent females: in an evolutionary sense, fat is added in preparation for childbearing and nursing (Angier, 1999). But anorexic women seek to rid their bodies of any additional weight and so turn to extreme dieting and, often, excessive exercise. These efforts continue unabated after the initial weight-loss goal is achieved, however—even after their families and others express concern. Another common pattern occurs among young women when they leave home to attend college and encounter difficulties ad-

TABLE 11.1 Diagnostic Features of Anorexia Nervosa

A. Refusal to maintain weight at or above the minimal normal weight for one's age and height; for example, a weight more than 15% below normal.
B. Strong fear of putting on weight or becoming fat, despite being thin.
C. A distorted body image in which one's body—or part of one's body—is perceived as fat, although others perceive the person as thin.
D. In case of females who have had menarche, absence of three or more consecutive menstrual periods.

Source. Adapted from the *DSM-IV-TR* (APA, 2000).

justing to the demands of college life and independent living. Anorexia is also more common among young women involved in ballet or modeling in which there is often a strong emphasis on maintaining an unrealistically thin body shape.

Although anorexia in women is far more common than in men, an increasing number of young men are presenting with anorexia. Many are involved in sporting activities, such as wrestling, in which they have experienced pressure to maintain a lower weight classification.

Adolescent girls and women with anorexia almost always deny that they are losing too much weight or wasting away. They may argue that their ability to engage in stressful exercise demonstrates their fitness. Women with eating disorders are more likely than normal women to view themselves as heavier than they are (Horne, Van Vactor, & Emerson, 1991). Others may see them as nothing but "skin and bones," but anorexic women have a distorted body image and may still see themselves as too fat. Although they literally starve themselves, they may spend much of the day thinking and talking about food, and even preparing elaborate meals for others (Rock & Curran-Celentano, 1996).

How do I see myself? A distorted body image is a key component of eating disorders.

Subtypes of Anorexia There are two general subtypes of anorexia, a *binge-eating/purging type* and a *restrictive type.* The first type is characterized by frequent episodes of binge eating and purging; the second type is not. Although repeated cycles of binge eating and purging occur in bulimia, bulimic individuals do not reduce their weight to anorexic levels. The distinction between the subtypes of anorexia is supported by differences in personality patterns. Individuals with the eating/purging type tend to have problems relating to impulse control, which in addition to binge-eating episodes may involve substance abuse or stealing (Garner, 1993). They tend to alternate between periods of rigid control and impulsive behavior. Those with the restrictive type tend to rigidly, even obsessively, control their diet and appearance.

VIDEO 11.2
Anorexia:
The Case of Tamora

Truth OR Fiction? REVISITED

Though others see them as but "skin and bones," young women with anorexia nervosa still see themselves as too fat.

TRUE. Others may see them as nothing but "skin and bones," but anorexic women have a distorted body image and may still see themselves as too fat.

Medical Complications of Anorexia Anorexia can lead to serious medical complications that in extreme cases can be fatal. Losses of as much as 35% of body weight may occur and anemia may develop. Females suffering from anorexia are also likely to encounter dermatological problems such as dry, cracking skin; fine, downy hair; even a yellowish discoloration that may persist for years after weight is regained. Cardiovascular complications include heart irregularities, hypotension (low blood pressure), and associated dizziness upon standing, sometimes causing blackouts. Decreased food ingestion can cause gastrointestinal problems such as constipation, abdominal pain, and obstruction or paralysis of the bowels or intestines. Menstrual irregularities are common, and **amenorrhea** (absence or suppression of menstruation) is part of the clinical definition of anorexia in females. Muscular weakness and abnormal growth of bones may occur, causing loss of height and **osteoporosis.**

The death rate from anorexia is estimated at 5% to 8% over a 10-year period, with most deaths due to suicide or medical complications associated with severe weight loss (Goleman, 1995g).

amenorrhea Absence of menstruation.

osteoporosis A physical disorder caused by calcium deficiency and characterized by brittle bones.

Bingeing. People with bulimia nervosa may cram thousands of calories during a single binge and then attempt to purge what they have consumed by forcing themselves to vomit.

Bulimia Nervosa

The Case of Nicole

Nicole has only opened her eyes, but already she wishes it was time for bed. She dreads going through the day, which threatens to turn out like so many other days of her recent past. Each morning she wonders, is this the day that she will be able to get by without being obsessed by thoughts of food? Or will she "blow it again" and spend the day gorging herself? Today is the day she will get off to a new start, she promises herself. Today she will begin to live like a normal person. Yet she is not convinced that it is really up to her.

Nicole starts the day with eggs and toast. Then she goes to work on cookies; doughnuts; bagels smothered with butter, cream cheese, and jelly; granola; candy bars; and bowls of cereal and milk—all within 45 minutes. Then she cannot take in any more food and turns her attention to purging what she has eaten. She goes to the bathroom, ties back her hair, turns on the shower to mask any noise she will make, drinks a glass of water, and makes herself vomit. Afterward she vows, "Starting tomorrow, I'm going to change." But she suspects that tomorrow may be just another chapter of the same story.

—*Adapted from Boskind-White & White, 1983, p. 29*

TABLE 11.2 Diagnostic Features of Bulimia Nervosa

A. Recurrent episodes of binge eating (gorging) as shown by both:
   (1) Eating an unusually high quantity of food during a 2-hour period, and
   (2) Sense of loss of control over food intake during the episode.
B. Regular inappropriate behavior to prevent weight gain such as self-induced vomiting, abuse of laxatives, diuretics or enemas, or by fasting or excessive exercise.
C. A minimum average of two episodes a week of binge eating and inappropriate compensatory behavior to prevent weight gain over a period of at least 3 months.
D. Persistent overconcern with the shape and weight of one's body.

Source. Adapted from the *DSM-IV* (APA, 2000).

Nicole suffers from bulimia nervosa. *Bulimia* derives from the Greek roots *bous,* meaning "ox" or "cow," and *limos,* meaning "hunger." The unpretty picture inspired by the origin of the term is one of continuous eating, like a cow chewing its cud. Bulimia nervosa is an eating disorder characterized by recurrent episodes of gorging on large quantities of food, followed by use of inappropriate ways to prevent weight gain. These may include purging by means of self-induced vomiting; use of laxatives, diuretics, or enemas; or fasting or engaging in excessive exercise (see Table 11.2). A woman with bulimia may use two or more strategies for purging, such as vomiting and laxatives (Tobin, Johnson, & Dennis, 1992). Although people with anorexia are extremely thin, bulimic individuals are usually of normal weight. However, they have an excessive concern about their shapes and weight.

Bulimic individuals typically gag themselves to induce vomiting. Most attempt to conceal their behavior. Fear of gaining weight is a constant factor. Although an overconcern with body shape and weight is a cardinal feature of bulimia and anorexia, bulimic individuals do not pursue the extreme thinness characteristic of anorexia. Their ideal weights are similar to those of women who do not suffer from eating disorders.

The binge itself usually occurs in secret, most commonly at home during unstructured afternoon or evening hours (Drewnowski, 1997; Guertin, 1999). A binge typically lasts from 30 to 60 minutes and involves consumption of forbidden foods that are generally sweet and rich in fat. Binge eaters typically feel they lack control over their bingeing and may consume 5,000 to 10,000 calories at a sitting. One young woman described eating everything available in the refrigerator, even to the point of scooping out margarine from its container with her finger. The episode continues until the binger is spent or exhausted, suffers painful stomach distension, induces vomiting, or runs out of food. Drowsiness, guilt, and depression usually ensue, but bingeing is initially pleasant because of release from dietary constraints.

The average age for onset of bulimia is the late teens, when concerns about dieting and dissatisfaction with bodily shape or weight are at their height. Bulimia nervosa typically affects (non-Hispanic) White women in late adolescence or early adulthood (APA,

2000). Despite the widespread belief that eating disorders, especially anorexia nervosa, are more common among affluent people, the available evidence shows no strong linkage between socioeconomic status and eating disorders (Wakeling, 1996). Beliefs that eating disorders are associated with high socioeconomic status may reflect the tendency for affluent patients to obtain treatment. Alternatively, it may be that the social pressures on young women to strive to achieve an ultrathin ideal have now generalized across all socioeconomic levels.

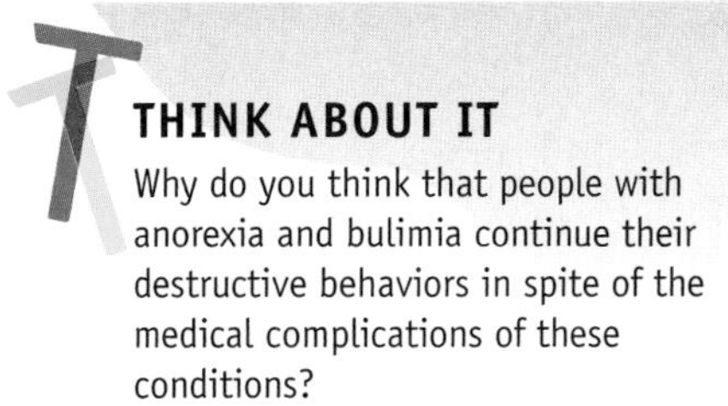

THINK ABOUT IT

Why do you think that people with anorexia and bulimia continue their destructive behaviors in spite of the medical complications of these conditions?

Medical Complications of Bulimia Bulimia is also associated with many medical complications. Many of these stem from repeated vomiting. There may be irritations of the skin around the mouth due to frequent contact with stomach acid, blockage of salivary ducts, decay of tooth enamel, and dental cavities. The acid from the vomit may damage taste receptors on the palate, making the person less sensitive to the taste of vomit with repeated purgings (Rodin et al., 1990). Decreased sensitivity to the aversive taste of vomit may play a role in maintaining the purging behavior. Cycles of bingeing and vomiting may cause abdominal pain, hiatal hernia, and other abdominal complaints. Stress on the pancreas may produce pancreatitis (inflammation), which is a medical emergency. Disturbed menstrual function is found in as many as 50% of normal weight women with bulimia (Weltzin et al., 1994). Excessive use of laxatives may cause bloody diarrhea and laxative dependency, so the person cannot have normal bowel movements without laxatives. In extreme cases, the bowel can lose its reflexive eliminatory response to pressure from waste material. Bingeing on large quantities of salty food may cause convulsions and swelling. Repeated vomiting or abuse of laxatives can lead to potassium deficiency, producing muscular weakness, cardiac irregularities, even sudden death—especially when diuretics are used. As with anorexia, menstruation may come to a halt.

VIDEO **11.3**

Bulimia: *The Case of Ann*

Causes of Anorexia and Bulimia

Like other psychological disorders, anorexia and bulimia involve a complex interplay of factors. But most significant of all are the social pressures felt by young women that lead them to base their self-worth on their physical appearance, especially their weight.

Sociocultural Factors Sociocultural theorists point to societal pressures and expectations placed on young women in our society as contributing to the development of eating disorders (Bemporad, 1996; Stice, 1994). The pressure to achieve an unrealistic standard of thinness, combined with the importance attached to appearance in defining the female role in our society, can lead young women to become dissatisfied with their bodies (Stice, 2001). Even in children as young as eight, girls express more dissatisfaction with their bodies than do boys (Ricciardelli & McCabe, 2001). Body dissatisfaction in young women may lead to excessive dieting and to the development of disturbed eating behaviors. The idealization of thinness in women can be illustrated in the changes in the body mass index (BMI) of winners of the Miss America pageant (Rubinstein & Caballero, 2000) (see Figure 11.1). Body mass index is a measure of height-adjusted weight.

The pressure to be thin falls most heavily on women. This pressure is so prevalent that dieting has become the normative pattern of eating among young

To be like Barbie. The Barbie doll has long represented a symbol of the buxom but thin feminine form that has become idealized in our culture. If women were to be proportioned like the classic Barbie doll, they would resemble the woman in the photograph on the right. To achieve this idealized form, the average woman would need to grow nearly a foot in height, reduce her waist by 5 inches, and add 4 inches to her bustline. What message do you think the Barbie-doll figure conveys to young girls?

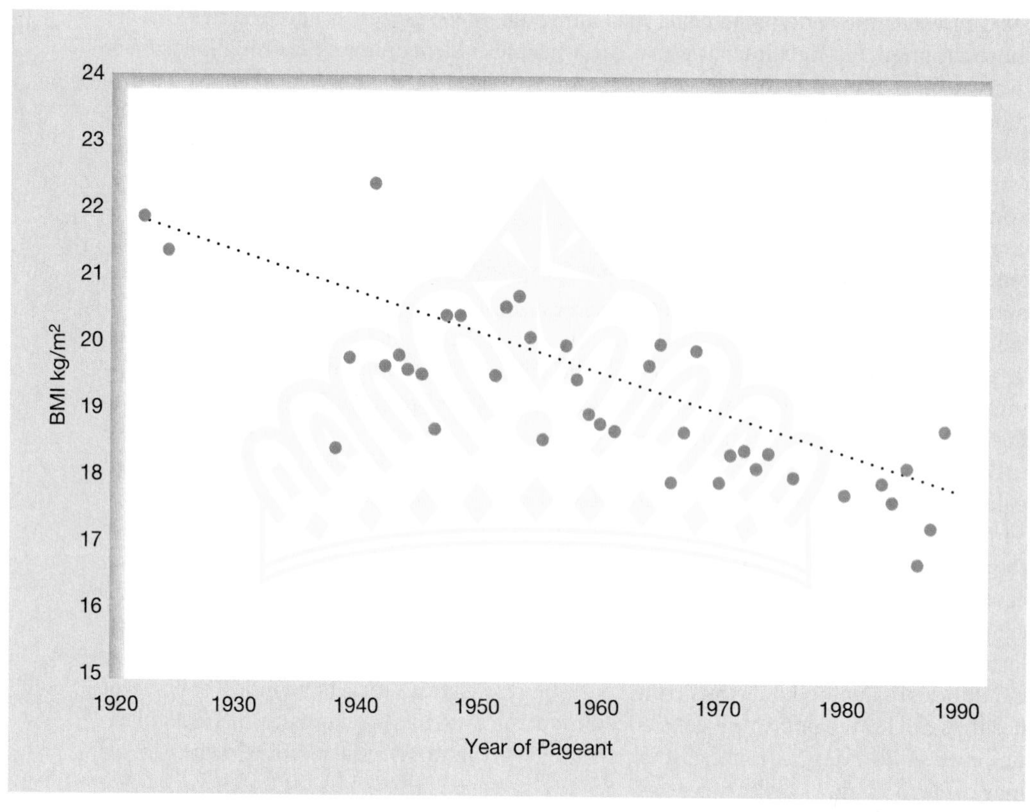

FIGURE 11.1 Thinner and Thinner.
Note the downward trend in the body mass index levels (BMIs) of Miss America contest winners over time. What might these data suggest about changes in society's view of the ideal female form?
Source. Rubinstein & Caballero (2000).

Truth OR Fiction? REVISITED

Dieting represents an abnormal eating pattern among American women.

FALSE. Dieting is so pervasive that it has essentially become a normative pattern of eating among American women.

American women. Four of five young women in the United States have gone on a diet by the time they reach their 18th birthdays. In actuality, the gender gap in obesity is quite small—27% of women versus 24% of men (see Figure 11.2). Moreover, gender differences in obesity do not arise until midlife.

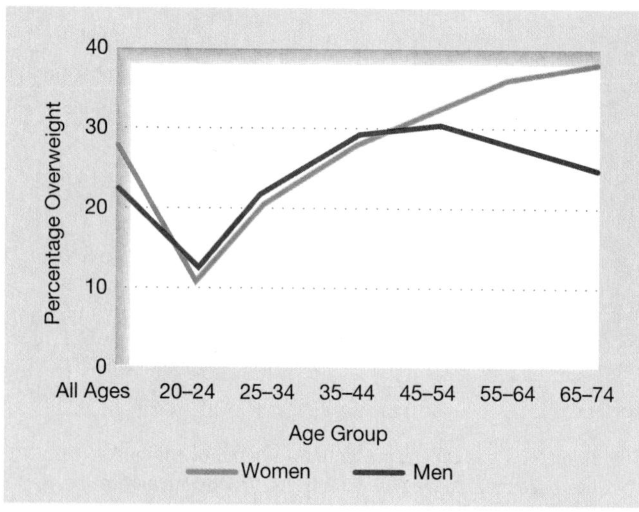

FIGURE 11.2 Prevalence of overweight by gender and age.
Source. National Heart, Lung, and Blood Institute (NHLBI), National Institutes of Health. (1993, March). *Data fact sheet: Obesity and cardiovascular disease.* Bethesda, MD: Author.

In support of the sociocultural model, evidence shows that eating disorders are less common, even rare, in non-Western countries (Stice, 1994; Wakeling, 1996). Even in Western cultures, eating disorders are more prevalent in the weight-obsessed United States than in other Western countries for which data is available, such as Greece and Spain, or in the most technologically advanced nation in the Far East, Japan (Stice, 1994). Rates of disordered eating behaviors and eating disorders also vary in the United States among ethnic groups, with higher rates in Euro American adolescents than in African American and other ethnic minority adolescents (Leon et al., 1995; Stice, 1994). One likely reason for this discrepancy is that body image and body dissatisfaction are less closely tied to body weight among minority women (Angier, 2000b). Yet disordered eating behaviors that may give rise to eating disorders are more common among African American women who identify more closely with the dominant White culture (Abrams, Allen, & Gray, 1993). Disturbed eating behaviors may also be more common among Native American adolescents than is commonly believed (Smith & Krejci, 1991). Investigators also caution that body dissatisfaction may be more prevalent among Hispanic and Asian girls than is generally recognized and may set the stage for distorted eating behaviors in these groups (Robinson et al., 1996). There are also signs that eating disorders may increase in the future in developing countries (Grange, Telch, & Tibbs, 1998).

Psychosocial Factors Although cultural pressures to conform to an ultrathin female ideal play a major role in eating disorders, the great majority of young women exposed to these pressures do not develop eating disorders. Other factors must be involved. One likely factor with bulimia at least is a history of rigid dieting (Patton et al., 1999; Rock & Curran-Celentano, 1996). Women with bulimia typically engage in extreme dieting characterized by very strict rules about what they can eat, how much they can eat, and how often they can eat (Drewnowski et al., 1994). Not surprisingly, they tend to spend more time thinking about their weight than do nonbulimic women (Zotter & Crowther, 1991).

Bulimic women tend to have been slightly overweight preceding the development of bulimia, and the initiation of the binge-purge cycle usually follows a period of strict dieting to lose weight. In a typical scenario, the rigid dietary controls fail, which prompts initial bingeing. This sets in motion a chain reaction in which bingeing leads to fear of weight gain, which prompts self-induced vomiting or excessive exercise to reduce any added weight. Some bulimic women become so concerned about possible weight gain that they resort to vomiting after every meal (Lowe, Golaves, & Murphy-Eberenz, 1998). Purging is negatively reinforced because it produces relief, or at least partial relief, from anxiety over gaining weight.

Truth OR Fiction? REVISITED

Bulimic women induce vomiting only after binges.

FALSE. Some bulimic women induce vomiting after every meal.

Body dissatisfaction is another important factor in eating disorders (Heatherton et al., 1997). Body dissatisfaction may lead to maladaptive attempts—through self-starvation and purging—to attain a desired body weight or shape. Bulimic and anorexic women tend to be extremely concerned about their body weight and shape (Fairburn et al., 1997). Even many normal weight children express concerns about their weight ("Nutrition, Obesity and Perception," 2001).

Cognitive factors are also involved. Young women with anorexia often have perfectionistic attitudes and high achievement strivings (Halmi et al., 2000). They may get down on themselves whenever they fail to meet their impossibly high standards. Their extreme dieting may give them a sense of control and independence that they may feel they lack in other aspects of their lives (Shafran & Mansell, 2001). Bulimic women tend to be both perfectionistic and dichotomous ("black or white") in their thinking patterns (Fairburn et al., 1997). Thus, they expect themselves to adhere perfectly to their rigid dietary rules and judge themselves as complete failures when they deviate even slightly. They also judge themselves harshly for episodes of binge eating and purging. These cognitive factors influence each other, as illustrated in Figure 11.3. In addition, women with bulimic tendencies

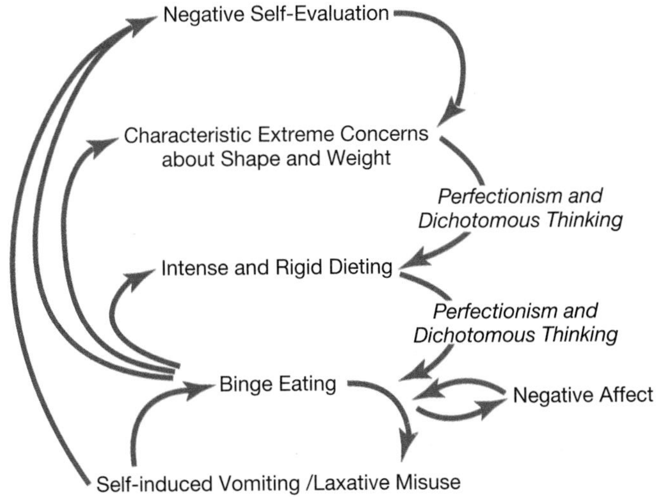

FIGURE 11.3 A cognitive model of bulimia nervosa.

Source. Fairburn, C. G. (1997). Eating disorders. In D. M. Clark & C. G. Fairburn, Eds., *Science and practice of cognitive behaviour therapy.* Oxford: Oxford University Press. Reprinted with permission of Oxford University Press.

tend to have a dysfunctional type of cognitive style that may lead to exaggerated beliefs about the negative consequences of gaining weight (Poulakis & Wertheim, 1993).

Psychodynamically oriented writers believe that girls with anorexia may have difficulties separating from their families and consolidating separate, individuated identities (Bruch, 1973; Minuchin, Rosman, & Baker, 1978). Perhaps anorexia represents the girl's unconscious effort to remain a prepubescent child. By maintaining the veneer of childhood, pubescent girls may avoid dealing with such adult issues as increased independence and separation from their families, sexual maturation, and assumption of personal responsibilities.

Some learning theorists view anorexia as a type of weight phobia. Excessive, irrational fears of putting on weight may reflect the tendencies in our culture to idealize the slender female form. Learning theorists also suggest that purging may be a type of compulsive ritual reinforced by the reduction of the fear of gaining weight that follows a binge-eating episode, just as compulsive hand washing in people with obsessive-compulsive may be reinforced by relief from disorder anxiety induced by obsessive thoughts.

A number of investigators have linked bulimia to problems in interpersonal relationships. Bulimic women tend to be shy and to have few if any close friends (Fairburn et al., 1997). One study of 21 college women with bulimia and a matched control group of 21 nonbulimic women found that those with bulimia had more social problems. They believed that less social support was available to them and reported more social conflict, especially with family members (Grissett & Norvell, 1992). They also rated themselves, and were judged by others, as less socially skillful than the control group. Although causal links between a lack of social skills and eating disorders remain to be elaborated, it is possible that enhancing the social skills of women with bulimia may increase the quality of their relationships and perhaps reduce their tendencies to use food in maladaptive ways.

Young women with bulimia also tend to have more emotional problems and lower self-esteem than other dieters (Fairburn et al., 1997). Bulimia often occurs together with many kinds of psychological disorders, including alcoholism, major depression, and anxiety disorders such as panic disorder, phobia, and generalized anxiety disorder (Kendler et al., 1991). Perhaps some forms of binge eating involve attempts at coping with emotional distress (Sherwood et al., 2000). Unfortunately, cycles of bingeing and purging serve to exacerbate emotional problems rather than relieve them. Bulimic women are also more likely than other women to have experienced childhood sexual and physical abuse (Kent & Waller, 2000; Wonderlich et al., 1997). Bulimia may develop in some cases as an ineffective means of coping with abuse.

What we have gained from this research is the understanding that bulimia often develops within a context of extreme, rigid dieting overlaying interpersonal, emotional, and cognitive factors.

Family Factors Eating disorders frequently develop against a backdrop of family conflicts (Fairburn et al., 1997; Wonderlich et al., 1997). Some theorists focus on the brutal effect of self-starvation on parents. They suggest that some adolescents use refusal to eat to punish their parents for feelings of loneliness and alienation they experience in the home. One related study compared the mothers of adolescent girls with eating disorders to the mothers of other girls. Mothers of the adolescents with eating disorders were more likely to be unhappy about their families' functioning, to have their own problems with eating and dieting, to believe their daughters ought to lose weight, and to regard their daughters as unattractive (Pike & Rodin, 1991). Is binge eating, as suggested by Humphrey (1986), a metaphoric effort to gain the nurturance and comfort that the mother is denying her daughter? Does purging represent the symbolic upheaval of negative feelings toward the family?

Families of young women with eating disorders tend to be more often conflicted, less cohesive and nurturing, yet more overprotective and critical than those of reference groups (Fairburn et al., 1997). The parents seem less capable of promoting independence in their

Questionnaire

The Fear of Fat Scale

The fear of becoming fat is a prime factor underlying eating disorders like anorexia and bulimia. The Goldfarb Fear of Fat scale (Goldfarb, Dykens, & Gerrard, 1985) measures the degree to which people fear becoming fat. The scale may help identify individuals at risk of developing eating disorders. It differentiates between anorexic and normal women, and between bulimic and nonbulimic women. Dieters, however, also score higher on the scale than nondieters.

To complete the Fear of Fat scale, read each of the following statements and write in the number that best represents your own feelings and beliefs. Then check the key at the end of the chapter.

1 = very untrue
2 = somewhat untrue
3 = somewhat true
4 = very true

____ **1.** My biggest fear is of becoming fat.

____ **2.** I am afraid to gain even a little weight.

____ **3.** I believe there is a real risk that I will become overweight someday.

____ **4.** I don't understand how overweight people can live with themselves.

____ **5.** Becoming fat would be the worst thing that could happen to me.

____ **6.** If I stopped concentrating on controlling my weight, chances are I would become very fat.

____ **7.** There is nothing that I can do to make the thought of gaining weight less painful and frightening.

____ **8.** I feel like all my energy goes into controlling my weight.

____ **9.** If I can eat even a little, I may lose control and not stop eating.

____ **10.** Staying hungry is the only way I can guard against losing control and becoming fat.

Source. Goldfarb, L. A., Dykens, E. M., & Gerrard, M. (1985). The Goldfarb Fear of Fat scale. *Journal of Personality Assessment, 49,* 329–332. Reprinted with permission.

daughters. Conflicts with parents over issues of autonomy are often implicated in the development of both anorexia nervosa and bulimia (Ratti, Humphrey, & Lyons, 1996). Yet it remains uncertain whether these family patterns contribute to the initiation of eating disorders or whether eating disorders go on to disrupt family life. The truth probably lies in an interaction between the two.

From the **systems perspective,** families are systems that regulate themselves in ways that minimize the open expression of conflict and reduce the immediate need for overt change. Within this perspective, girls who develop anorexia may be seen as helping maintain the shaky balances and harmonies found in dysfunctional families by displacing attention from family conflicts and marital tensions onto themselves (Minuchin et al., 1978). The girl may become the *identified patient,* although the family unit is actually dysfunctional.

Regardless of the factors that initiate eating disorders, social reinforcers may maintain them. Children with eating disorders may quickly become the focus of attention of their families, and receive attention from their parents that might otherwise be lacking.

Biological Factors Scientists suspect that abnormalities in brain mechanisms controlling hunger and satiety are involved in bulimia, most probably involving the brain chemical serotonin (Goode, 2000a). Serotonin plays a key role in regulating mood and appetite, especially appetite for carbohydrates. Low levels of the chemical, or lack of sensitivity of serotonin receptors in the brain, may prompt binge-eating episodes, especially carbohydrate bingeing (Levitan et al., 1997). This line of thinking is buttressed by evidence that antidepressants, such as Prozac, which increases serotonin activity, can decrease binge-eating episodes in bulimic women (Jimerson et al., 1997). We also know that many women with eating disorders are depressed or have a history of depression, and imbalances of serotonin are implicated in depressive disorders.

THINK ABOUT IT

How do sociocultural and psychosocial issues interact to cause eating disorders? How might we as a society change the social influences that lead many young women to develop disordered eating habits?

Truth OR Fiction? REVISITED

Drugs used to treat depression may also help curb bulimic binges.

TRUE. Antidepressants have been shown to be helpful in curbing binge eating among bulimic women.

systems perspective The view that problems reflect the systems (family, social, school, ecological, etc.) in which they occur.

Evidence also points to a role for genetic factors in eating disorders (Wade et al., 2000). Eating disorders also tend to run in families, which is suggestive of a genetic component (Goode, 2000a). Stronger evidence comes from a study of more than 2,000 female twins, which showed a much higher concordance rate for bulimia, 23% versus 9%, among monozygotic (MZ) twins than dizygotic (DZ) twins (Kendler et al., 1991). A greater concordance for anorexia was also found among MZ than DZ twins, 50% versus 5% (Holland, Sicotte, & Treasure, 1988). Nonetheless, genetic factors cannot fully account for the development of eating disorders. Perhaps in the manner of the diathesis-stress model, a genetic predisposition involving a dysfunction of neurotransmitter activity interacts with family, social, cultural, and environmental pressures in leading to the development of eating disorders (Strober & Humphrey, 1987).

Treatment of Anorexia Nervosa and Bulimia Nervosa

Web Link 11.2 WWW
Prevention of Eating Disorders

Eating disorders are difficult to treat. People with anorexia may be hospitalized, especially when weight loss is severe or body weight is falling rapidly. In the hospital they are usually placed on a closely monitored refeeding regimen. Behavior therapy is also commonly used, with rewards made contingent on adherence to the refeeding protocol (Rock & Curran-Celentano, 1996). Commonly used reinforcers include ward privileges and social opportunities.

Psychodynamic therapy is sometimes combined with behavior therapy to probe for deeper psychological conflicts. Family therapy may also be employed to help resolve underlying family conflicts. Behavior therapy has been shown to be effective in promoting weight gain of anorexic patients during hospitalization (Johnson, Tsoh, & Varnado, 1996). Individual or family therapy following hospitalization has also shown favorable long-term benefits (Eisler et al., 1997). Hospitalization may be helpful in breaking the binge-purge cycle in bulimia, but appears to be necessary only where eating behaviors are clearly out of control and outpatient treatment has failed, or where there is evidence of severe medical complications, suicidal thoughts or attempts, or substance abuse.

Cognitive-behavioral therapy (CBT) is useful in helping bulimic individuals challenge self-defeating thoughts and beliefs, such as unrealistic, perfectionistic expectations regarding dieting and body weight. Another common dysfunctional thinking pattern is dichotomous (all-or-nothing) thinking that predisposes them to purge when they slip even a little from their rigid diets. CBT also challenges tendencies to overemphasize appearance in determining self-worth. To eliminate self-induced vomiting, therapists may use the behavioral technique of *exposure with response prevention* that was developed for treatment of people with obsessive-compulsive disorder. In this technique, the bulimic patient is exposed to eating forbidden foods while the therapist stands by to prevent vomiting until the urge to purge passes. Bulimic individuals thus learn to tolerate violations of their dietary rules without resorting to purging. Cognitive-behavioral therapy (CBT) has been shown to reduce binge-eating and purging episodes in people with bulimia (Agras et al., 2000a; Anderson & Maloney, 2001; Goode, 2000a; Tuschen-Caffier, Pook, & Frank, 2001). Another form of psychotherapy, interpersonal therapy (IPT), has also been used effectively in treating bulimia. Interpersonal therapy focuses on resolving interpersonal problems in the belief that more effective interpersonal functioning will lead to healthier food habits and attitudes. Although IPT did not produce as good results as CBT in a recent trial (Agras et al., 2000b), it may be used as an alternative treatment in cases where CBT proves unsuccessful (Wilson & Fairburn, 1998).

Antidepressant drugs have also shown therapeutic benefits in treating bulimia (Goode, 2000a; Wilson & Fairburn, 1998). They are believed to work by decreasing the urge to binge through normalizing serotonin—the brain chemical involved in regulating appetite. Antidepressant medication, especially Prozac, also appears promising in treating anorexia, perhaps because it helps relieve underlying depression (Johnson, Tsoh, & Varnado, 1996).

binge-eating disorder (BED) A disorder characterized by recurrent eating binges without purging; classified as a potential disorder requiring further study.

A review of the available evidence suggests that cognitive-behavioral therapy is more effective than antidepressant medication in treating bulimia nervosa and carries a lower rate of relapse (Johnson, Tsoh, & Varnado, 1996; Wilson & Fairburn, 1998). CBT may be considered a treatment of first choice for bulimia, followed by the use of antidepressant medication if psychological treatment is not successful (Compas et al., 1997). In a recent study, antidepressant medication showed therapeutic benefits in treating patients who had failed to respond to cognitive-behavioral treatment (Walsh et al., 2000). Studies examining whether a combination CBT/medication treatment approach is more effective than either treatment component alone have thus far produced inconsistent results (Johnson, Tsoh, & Varnado, 1996; Walsh et al., 1997).

Although progress has been made in treating eating disorders, there is considerable room for improvement. Even in CBT, about half of treated patients show continued evidence of bulimic behavior (Compas et al., 1997; Wilson & Fairburn, 1998). Eating disorders can be tenacious and enduring problems, especially when excessive fears of body weight and distortions in body image continue beyond active treatment. A recent study reported that 10 years after an initial presentation with bulimia, approximately 30% of women still showed recurrent binge-eating or purging behaviors (Keel et al., 1999). Recovery from anorexia also tends to be a long process. A recent study of 88 German patients with anorexia showed that 50% of the patients did not recover sooner than 6 years after their first hospitalization (Herzog, Schellberg, & Deter, 1997).

Difficulties in treating eating disorders only buttresses the need for effective prevention programs that target young women at risk. Recently, investigators showed positive results from an Internet-based intervention that focused on changing disordered eating behaviors and attitudes and reducing body dissatisfaction in a sample of college women (Celio et al., 2000). The intervention had a significant impact on reducing risk factors for eating disorders (i.e., disordered attitudes and behaviors) as compared to a waiting list control condition.

THINK ABOUT IT

Do you know anyone who has suffered from an eating disorder? What factors discussed in the text might help you to better understand the causes of the person's problem?

Binge-Eating Disorder

People with **binge-eating disorder (BED)** have recurrent eating binges but do not purge themselves of the excess food afterwards. Binge-eating disorder is classified in the *DSM* manual as a potential disorder requiring further study. We presently know too little about the characteristics of people with BED to warrant its inclusion as an official diagnostic category. The criteria used for diagnosing the disorder also need further evaluation. Presently, persons who qualify for the diagnosis show evidence of bingeing at least 2 days a week for a period of 3 months (Stotland, 2000). During a binge, they continue to eat despite feeling uncomfortably full. They are embarrassed to be seen during a binge and feel guilty afterwards.

The available evidence indicates that unlike bulimia, BED is more commonly found among obese individuals (Spitzer et al., 1992). It is believed to affect about 2% of the population (Goode, 2000a). BED is frequently associated with depression and with a history of unsuccessful attempts at losing excess weight and keeping it off. People with BED tend to be older than those with anorexia or bulimia (Arnow, Kenardy, & Agras, 1992). Like other eating disorders, it is found more frequently among women.

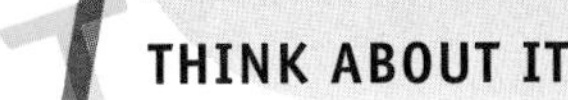

THINK ABOUT IT

How does binge-eating disorder differ from anorexia and bulimia? In what ways is it similar to substance abuse?

People with BED are often described as "compulsive overeaters." During a binge they feel a loss of control over their eating. BED may fall within a broader domain of compulsive behaviors characterized by impaired control over maladaptive behaviors, such as pathological gambling and substance abuse disorders. A history of dieting may play a role in some cases of BED, although it appears to be a less important factor in BED than in bulimia (Howard & Porzelius, 1999).

Cognitive-behavioral techniques have shown positive effects in treating binge-eating disorder (Wilson & Fairburn, 1998). Antidepressants, especially antidepressants of the SSRI family, may also reduce the frequency of binge-eating episodes by helping regulate serotonin levels in the brain (McElroy et al., 2000; Stotland, 2000).

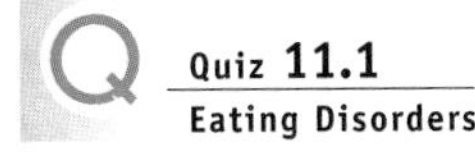

Quiz **11.1**
Eating Disorders

Overview of Eating Disorders

TYPES OF EATING DISORDERS

	Description	Features
Anorexia Nervosa	Self-starvation, resulting in a minimal weight for one's age and height or dangerously unhealthy weight	• Strong fears of gaining weight or becoming fat • Distorted self-image (perceiving oneself as fat despite extreme thinness) • Two general subtypes: binge-eating/purging type and restrictive type • Potentially serious, even fatal medical complications • Typically affects young, Euro-American women
Bulimia Nervosa	Recurrent episodes of binge eating followed by purging	• Weight is usually maintained within a normal range • Overconcern about body shape and weight • Binge-purge episodes may result in serious medical complications • Typically affects young, Euro-American women
Binge-Eating Disorder (a proposed disorder requiring further study)	Recurrent binge eating without compensatory purging	• Individuals with BED frequently are described as compulsive overeaters • Typically affects obese women who are older than those affected by anorexia or bulimia

CAUSAL FACTORS An interplay of multiple factors are at work in eating disorders

Sociocultural Factors	• Excessive pressures on young women to adhere to unrealistic standards of thinness
Psychological Factors	• Rigid or highly restrictive dieting may set the stage for loss of control following dietary transgressions, resulting in bulimic binges • Body dissatisfaction may prompt unhealthy ways of achieving desired body weight • Perceived lack of control over other aspects of life apart from dieting • Difficulties separating from one's family and establishing an individuated identity • Psychological needs for perfectionism and tendencies to think in dichotomous or black-and-white terms
Family Factors	• Families of eating disorder patients are often characterized by conflict, lack of cohesion and nurturing, and failure to foster independence and autonomy in their daughters • From a family systems perspective, a daughter's eating disorder may serve to maintain a shaky balance within a dysfunctional family by diverting attention away from family or marital problems
Biological Factors	• Possible imbalances in neurotransmitter systems in the brain regulating mood and appetite • Possible genetic influences

TREATMENT APPROACHES Often difficult to treat but a range of therapeutic approaches is available

Biomedical Treatment	• Hospitalization may be needed to help anorexic patients restore a healthy body weight or bulimic patients break binge-purge cycles in cases where outpatient therapy has failed • Antidepressant medication may be used to regulate appetite by altering brain chemistry or to relieve underlying depression
Psychotherapy	• Psychodynamic therapy aims at exploring and resolving underlying psychological conflicts
Cognitive-Behavioral Therapy	• To help individuals with an eating disorder challenge self-defeating thoughts and beliefs and develop healthier eating habits and thinking patterns • Behavior modification helps hospitalized anorexic patients regain weight by means of linking desired rewards to appropriate eating behaviors • Exposure with response prevention helps bulimic individuals tolerate eating forbidden foods without bingeing and purging
Family Therapy	• May be used to resolve family conflicts and improve communication among family members

Obesity: A National Epidemic

obesity A condition of excess body fat; generally defined by a BMI above 30.

Obesity has become a national epidemic. Consider some statistics:

- More Americans are overweight today than at any time since the government started tracking obesity in the 1960s. According to recent estimates, 61% of American adults are overweight and more than a quarter (26%) are obese ("CDC Says," 2000). The prevalence of obesity in the United States skyrocketed by about 50% during the 1990s (Mokdad et al., 1999, 2000).
- Obesity in children and adolescents is also on the upswing. About one in three children are overweight or at risk of becoming overweight (Cowley, 2000a). Obese children stand a better than 75% chance of becoming obese adults (Begley, 2000).
- Americans consume 815 billion calories of food each day, which is 200 billion calories more than necessary to maintain their weight at moderate levels of activity (C. D. Jenkins, 1988). The extra calories are enough to sustain a country of 80 million people.
- Some 65 million Americans diet every year and use nearly 30,000 approaches (Blumenthal, 1988). As many as one in four Americans are on a diet on any given day (French & Jeffery, 1994).
- More than 90% of dieters fail to keep pounds off permanently (Wilson, 1994). Whether they drop 15 pounds or 50, most dieters put almost all the pounds back on within a year (Brody, 1992a).

Truth OR Fiction? REVISITED

The excess calories consumed by Americans each day could feed a country of 80 million people.

TRUE. The excess calories consumed daily by Americans could feed a country of 80 million people. We, however, are paying the penalty of excess caloric intake in the form of obesity.

Despite all the money and effort spent on weight-loss products and programs, our collective waistlines are getting larger—a result, government health experts believe, of consuming too many calories and exercising too little ("CDC Says," 2000). The increased use of laborsaving devices (driving more, walking less) eventually translates into added inches to our waistlines (Bouchard, 1997). We are fast becoming a nation of couch potatoes and cyberslugs who are glued to the TV or computer screen ("*Did someone say supersize it?*").

Obesity is classified as a chronic medical disorder, not a psychological disorder (Atkinson, 1997). It is also a major risk factor in such chronic, potentially life-threatening diseases as heart disease, diabetes, and some forms of cancer (Devlin, Yanovski, & Wilson, 2000; "Diabetes," 2001; Michaud et al., 2001; Pinel, Assanand, & Lehman, 2000; "Third of Some Cancers," 2001). Experts estimate that 300,000 people in the United States die prematurely because of obesity-linked diseases (Mokdad et al., 2000; Must et al., 1999). Though obesity is a medical condition, it involves psychological factors in both its development and treatment. This reminds us of the complex interrelationships between the mind and body.

Truth OR Fiction? REVISITED

Obesity is one of the most common psychological disorders in the United States.

FALSE. Obesity is a medical disorder, not a psychological disorder.

Are You Obese?

The most widely used standard for determining obesity is the body mass index, or BMI. The BMI takes into account a person's body weight and height. It is calculated by dividing body weight (in kilograms) by the square of the person's height (in meters).

The National Institutes of Health has set a level of 25 as the cutoff for determining whether a person is overweight (see Figure 11.4) (Brody, 1998; Shapiro, 1998). This level is associated with a weight level about 20% above the recommended weight for a person's age and height. People in the range of 25 to 27 are considered slightly overweight. People with a BMI greater than 30 are considered obese.

Hazardous waist. Obesity is indeed a hazard to health and longevity.

What Causes Obesity?

What causes obesity? Body weight varies as a function of energy balance. When caloric intake exceeds energy output, the excess calories are stored in the body in the form of fat, leading to obesity (Esparza et al., 2000). The key to preventing obesity is to bring energy expenditure in line with energy (caloric) intake. Unfortunately, this is easier said than done.

The Body Mass Index

Federal health authorities are using this index to determine who is overweight. Under new guidelines, a body mass of **25** or more is considered overweight. To use the table, find the appropriate height in the left-hand column. Move across to a given weight. The number at the top of the column is the BMI at that height and weight. Pounds have been rounded off.

	BMI																
	19	20	21	22	23	24	25	26	27	28	29	30	31	32	33	34	35
HEIGHT (inches)	Body Weight (pounds)																
58	91	96	100	105	110	115	119	124	129	134	138	143	148	153	158	162	167
59	94	99	104	109	114	119	124	128	133	138	143	148	153	158	163	168	173
60	97	102	107	112	118	123	128	133	138	143	148	153	158	163	168	174	179
61	100	106	111	116	122	127	132	137	143	148	153	158	164	169	174	180	185
62	104	109	115	120	126	131	136	142	147	153	158	164	169	175	180	186	191
63	107	113	118	124	130	135	141	146	152	158	163	169	175	180	186	191	197
64	110	116	122	128	134	140	145	151	157	163	169	174	180	186	192	197	204
65	114	120	126	132	138	144	150	156	162	168	174	180	186	192	198	204	210
66	118	124	130	136	142	148	155	161	167	173	179	186	192	198	204	210	216
67	121	127	134	140	146	153	159	166	172	178	185	191	198	204	211	217	223
68	125	131	138	144	151	158	164	171	177	184	190	197	203	210	216	223	230
69	128	135	142	149	155	162	169	176	182	189	196	203	209	216	223	230	236
70	132	139	146	153	160	167	174	181	188	195	202	209	216	222	229	236	243
71	136	143	150	157	165	172	179	186	193	200	208	215	222	229	236	243	250
72	140	147	154	162	169	177	184	191	199	206	213	221	228	235	242	250	258
73	144	151	159	166	174	182	189	197	204	212	219	227	235	242	250	257	265
74	148	155	163	171	179	186	194	202	210	218	225	233	241	249	256	264	272
75	152	160	168	176	184	192	200	208	216	224	232	240	248	256	264	272	279
76	156	164	172	180	189	197	205	213	221	230	238	246	254	263	271	279	287

FIGURE 11.4 The body mass index.

Source. Adapted from G.A. Bray & D.S. Gray (1998). Obesity. Part I-Pathogenesis. *Western Journal of Medicine, 149,* 429–441. Reprinted from *Clinical guidelines on the identification, evaluation, and treatment of overweight and obesity in adults.* National Heart, Lung, and Blood Institute, 1998, Bethesda, MD.

Research suggests that a number of factors contribute to the imbalance between energy intake and expenditure that underlies obesity, including genetics, metabolic factors, lifestyle factors, and psychological factors.

Genetic Factors We know that obesity clearly runs in families (Bouchard, 1997). It was once assumed that obese parents encouraged their children to become heavy by setting poor examples. A study of Scandinavian adoptees strongly suggests a key role for heredity, however (Stunkard et al., 1986). It revealed that children's weight is more closely related to the weight of their biological parents than to that of their adoptive parents. Perhaps the strongest evidence for the role of genetics comes from a study of identical twins that showed that regard-

Web Link **11.3**
Guidelines on Overweight and Obesity WWW

less of whether the twins were reared together or apart, they wound up weighing virtually the same when they became adults (Stunkard et al., 1990). Consistent with a genetic explanation, fraternal twins varied much more in body weight (corrected for height) than did MZ twins.

Most experts in the field believe that genetics plays an important role in determining the risk of obesity (Devlin et al., 2000). But genetics doesn't tell the whole story. Obesity experts recognize that both genetics and environmental factors (diet and exercise patterns) contribute to obesity (Wing & Polley, 2001).

Metabolic Factors When we lose weight, especially significant amounts of weight, the body reacts as if it were starving. It responds to falling weight by slowing the **metabolic rate,** the rate at which it burns calories (Kolata, 1995c; Leibel, Rosenbaum, & Hirsch, 1995). This makes it difficult to continue losing more weight or even maintain the weight loss. Some theorists believe that mechanisms in the brain control the body's metabolism to keep body weight around a genetically determined **set point** (Keesey, 1980). You may be able to offset this metabolic adjustment by following a more vigorous exercise regimen. Vigorous exercise burns calories directly and may increase the metabolic rate by replacing fat tissue with muscle, especially if the exercise program involves weight-bearing activity. Also, ounce for ounce, muscle burns more calories than fat. Before starting an exercise regimen, check with your physician to determine which types of activity are best suited to your overall health condition.

Fat Cells The efforts of heavy people to keep a slender profile may be sabotaged by cells within their own bodies termed **fat cells.** No, fat cells are not cells that are fat. They are cells that store fat. Fat cells comprise fatty tissue in the body (also called *adipose tissue*). Obese people have more fat cells (Brownell & Wadden, 1992) than people who are not obese. Severely obese people may have some 200 billion fat cells, as compared to 25 or 30 billion in normal weight individuals. Why does this matter? As time passes after eating, the blood sugar level declines, drawing out fat from these cells to supply more nourishment to the body. The hypothalamus in the brain detects the depletion of fat in these cells. The hypothalamus then signals the cerebral cortex, triggering the hunger drive, which motivates eating and thereby replenishes the fat cells.

People with more fatty tissue send more signals of fat depletion to the brain than people who are equal in weight but who have fewer fat cells. As a result, they feel food deprived sooner. Sad to say, dieters do not expel fat cells; instead, they shrink them. Many dieters who are markedly obese, even successful dieters, thus complain they are constantly hungry as they struggle to maintain normal weights. People with high levels of adipose tissue are doubly beset, because fatty tissue metabolizes food more slowly than muscle.

How is the number of fat cells in our bodies determined? Unfortunately, heredity seems to play a role. Excessive food intake in early childhood may also have an influence, however (Brownell & Wadden, 1992).

Lifestyle Factors Obese people are typically less physically active than people of normal weight (Brownell & Wadden, 1992; "The Sedentary Society," 1996). Although correlational evidence is insufficient to establish causality, it is reasonable to assume that inactivity and overweight may interact with each other. In other words, inactivity may lead to weight gain, and weight gain may in turn lead people to become less active.

Other lifestyle factors, such as adopting a high-fat diet and eating larger portions, also contribute to obesity (Wing & Polley, 2001). Exposure to a constant bombardment of food-related cues in television commercials, print advertising, and the like, may also play a part. Even though evidence does not show obese people to be more sensitive to food cues than normal weight individuals, overresponsivity to these cues can lead to inappropriate food consumption in people of any weight class.

metabolic rate The rate at which energy is used in the body.

set point A value, such as body weight, that the body's regulatory mechanisms attempt to maintain.

fat cells Body cells specialized to store fat.

Truth OR Fiction? REVISITED

When you lose weight, your body starts putting the brakes on the rate at which it burns calories.

TRUE. Unfortunately, the body reacts to falling weight by slowing the metabolic rate, the rate at which it burns calories. This makes it difficult to continue losing weight or even maintain the weight loss.

Truth OR Fiction? REVISITED

Obese people lose fat cells when they diet.

FALSE. So far as we know, people do not lose fat cells as they lose weight.

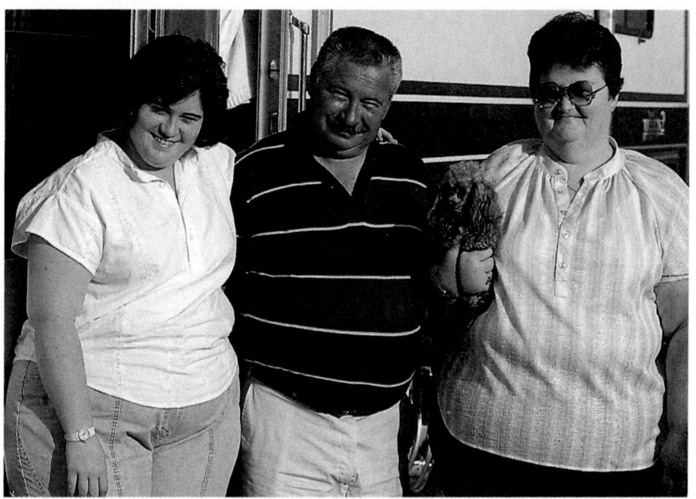

All in the family? Obesity tends to run in families. The question is, *Why?*

Psychological Factors According to psychodynamic theory, eating is the cardinal oral activity. Psychodynamic theorists believe that people who were fixated in the oral stage by conflicts concerning dependence and independence are likely to regress in times of stress to excessive oral activities such as overeating. Other psychological factors connected with overeating and obesity include low self-esteem, lack of self-efficacy expectancies, family conflicts, and negative emotions. Although the connections between these factors and obesity affect both men and women, women most frequently seek treatment for obesity, largely because of the greater pressures they experience to adhere to social expectations of thinness. Consider the cases of Joan and Terry:

The Case of Joan

Joan was trapped in the yo-yo syndrome, repeatedly dropping 20 pounds, then regaining it. Whenever Joan got stuck at a certain weight plateau, or started to regain weight, her self-esteem plummeted. She'd hear herself muttering, "Who am I kidding? I'm not worthy of being thin. I should just accept being fat."

An incident with her mother revealed how her negative thinking was often triggered. Joan had lost 24 pounds from an original weight of 174 and was beginning to feel good about herself. Most other people reinforced her by complimenting her on her weight loss. She called her mother, who lived in another state, to share the good news. Instead of jumping on the bandwagon, her mother cautioned her not to expect too much from her success. After all, her mother pointed out, she had been repeatedly disappointed in the past. The message came through loud and clear: Don't get your hopes up because you will only be more disappointed in the end. Joan's mother may have only meant to protect Joan from eventual disappointment, but the message she conveyed reinforced the negative view that Joan held of herself: You're a loser. Don't expect too much of yourself. Accept your reality. Don't try to change. You're a hopeless case.

As soon as she hung up the receiver, Joan rushed to the pantry. Without hesitation, she devoured three packages of chocolate chip cookies in a frenzied binge on the stairway. The next day she told her psychologist that the binge had reactivated childhood memories of bingeing on Oreo cookies while hiding in the stairwell.

The Case of Terry

For years Terry's husband had scrutinized every morsel she consumed. "Haven't you had enough?" he would ask derisively. The more he harped on her weight, the more anger she felt, although she did not express her feelings directly. The criticism did not apply only to her weight. She also heard "You're not smart enough . . . Why don't you take better care of yourself? . . . How come you're not sexy?" After years of assault on her self-esteem, Terry petitioned for divorce, convinced the single life could be no worse than her marriage.

While separated and awaiting the final divorce decree, Terry felt free to be herself for the first time in her adult life. However, she had not expected the effect freedom would have on her weight. She ballooned from 155 pounds to 186 pounds within a few months. She identified leftover resentment from her marriage as the driving factor. "There's no one to make me diet anymore," she said. Her overeating was like saying, "See, I can eat if I want to." Without her husband, she could express her anger and outrage toward him by eating to excess. Unfortunately, her mode of expressing anger was self-defeating. Terry's lingering resentment encouraged her to act spitefully rather than constructively.

—From the Authors' Files

■

Ethnic and Socioeconomic Differences in Obesity

Obesity does not affect ethnic/racial groups in our society in equal proportions. It is more prevalent among people of color, especially among women of color (see Figure 11.5) (National Institutes of Health, 1998, 1999). The question is, why?

Socioeconomic Factors Obesity is more prevalent among poorer people (Stunkard & Sørensen, 1993). Since people of color in our society are as a group lower in socioeconomic status than (non-Hispanic) White Americans, we should not be surprised that rates of obesity are higher among people of color, at least among women of color.

Why are people on the lower rungs of the socioeconomic ladder at greater risk of obesity? For one thing, more affluent people have greater access to information about nutrition and health. They are more likely to take health education courses. They have greater access to health care providers. Poorer people also exercise less regularly than more affluent people do. The fitness boom has been largely limited to more affluent people. They have the time and income to participate in organized fitness programs. Many poor people in the inner city also turn to food as a way of coping with the stresses of poverty, discrimination, crowding, and crime.

Acculturation Though acculturation may help immigrant people adapt more successfully to their new culture, it can become a double-edged sword if it involves adoption of unhealthful dietary practices of the host culture. Consider that Japanese American men living in California and Hawaii eat a higher-fat diet than Japanese men do. Not surprisingly, the prevalence of obesity is two to three times higher among Japanese American men than among men living in Japan (Curb & Marcus, 1991).

Acculturation may also contribute to high rates of obesity among Native Americans, who are more likely than White Americans to have diseases linked to obesity, such as cardiovascular disease and diabetes (Broussard et al., 1991). A study of several hundred Cree and Ojibwa Indians in Canada found that nearly 90% of the women in the 45- to 54-year-old age group were obese (Young & Sevenhuyser, 1989). The adoption of a high-fat Western-style

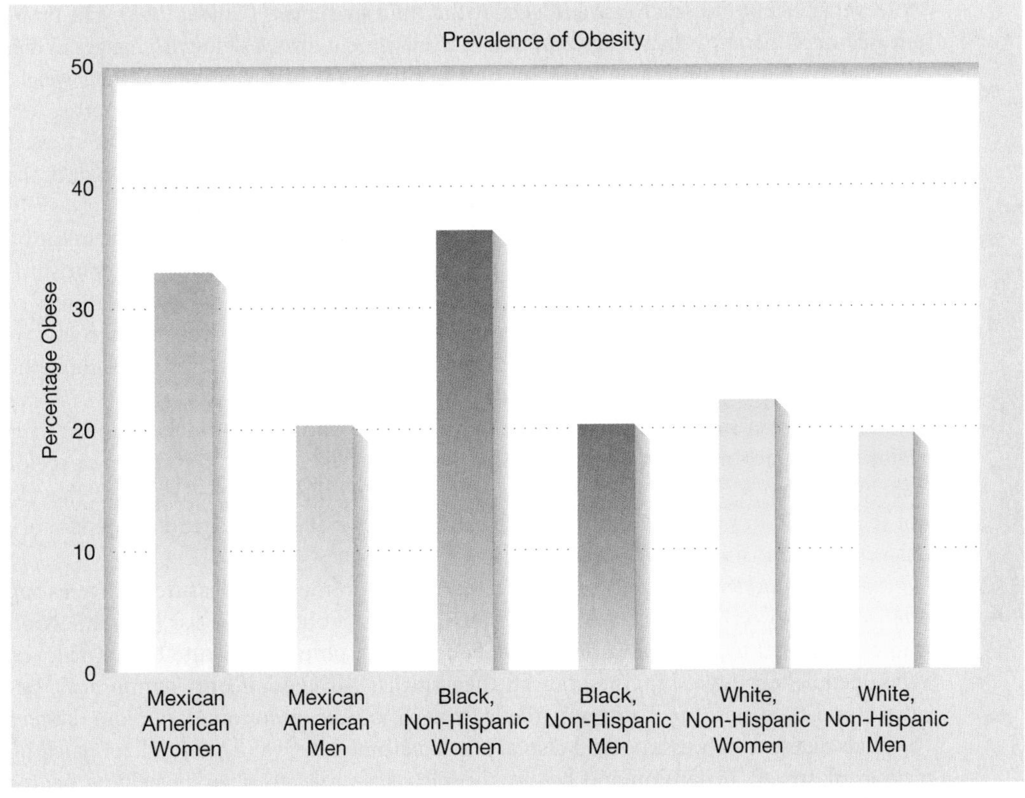

FIGURE 11.5 Rates of obesity (age 20 or higher).
This figure shows the rates of obesity among U.S. adults in relation to race and ethnicity.

Source. National Institutes of Health, National Heart, Lung, and Blood Institute. (1998). *Clinical guidelines on the identification, evaluation, and treatment of overweight and obesity in adults.* Bethesda, MD: Author.

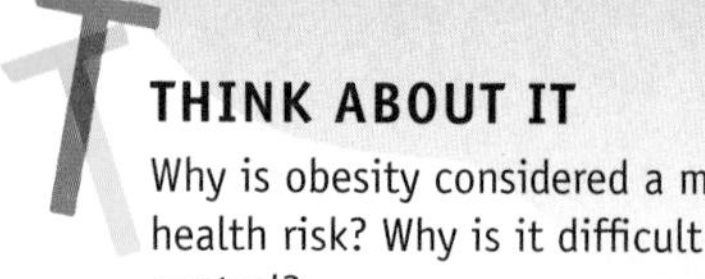

THINK ABOUT IT

Why is obesity considered a major health risk? Why is it difficult to control?

diet, the destruction of physically demanding native industries, and chronic unemployment combined with low levels of physical activity are cited as factors contributing to obesity among Native Americans in the United States and Canada.

Metabolic Factors Biological factors, such as genetic differences in metabolic rates, may also be involved. Researchers report that Black women had lower resting metabolic rates (rate at which calories or food energy is burned while at rest) than did a sample of White women (Brody, 1997a).

Whatever the underlying reasons accounting for racial/ethnic differences in obesity may be, evidence shows that these differences may be narrowing. Though rates of obesity rose for men and women in all racial/ethnic groups during the 1980s, they rose most sharply among Whites (Burros, 1994; McMurtrie, 1994).

What Can Be Done? Preventing obesity calls for strategies that apply regardless of ethnicity or income level, such as cutting back on fat intake and sugar consumption and adopting regular exercise habits. But certain health initiatives need to be specifically targeted toward the needs of the socially and economically disadvantaged groups, including the following (Jeffery, 1991):

- Increased access to health education;
- Requirements for health education curricula in all public schools;
- Guarantees of universal access to treatment of obesity;
- Increased access to healthful foods and recreational opportunities.

Truth OR Fiction? REVISITED

Most dieters eventually gain back the weight they lose.

TRUE. Most dieters regain the weight they've lost, and then some.

Facing the Challenge of Obesity

The goal of weight-loss programs should be to help people who are overweight achieve and maintain a healthy weight, not a cosmetically desirable weight that is the product of a thinness-obsessed society. Yet, for people who are overweight or obese, various treatment alternatives are available to help them reduce excess weight. Some people turn to diet drugs, but unfortunately the track record for these drugs in the long run is not very encouraging (Winter, 2000). Moreover, they may cause adverse and possibly harmful side effects (Chase, 1998). The problem with drug therapy is that long-term weight management involves lifestyle changes in diet and exercise. Even people whose heredity may work against them can control their weight within some broad limits through adopting a sensible diet, increasing activity and exercise levels, and changing problem eating habits (Brownell & Wadden, 1992; Foster et al., 1997; Ross et al., 2000). A modest weight loss on the order of 10% or 15% of body weight can reduce the health risks associated with obesity (Lane, 1994).

Behavior modification programs focus on helping individuals change problem eating habits by altering the "ABC's" of eating. The A's are the *antecedents* of eating—cues or stimuli that trigger eating. These include environmental stimuli such as the sight and aromas of food, internal stimuli such as sensations of hunger, and emotional states such as anxiety, fatigue, anger, depression, and boredom. Controlling the A's of eating involves redesigning the environment so people are not continuously bombarded by food-related cues.

The B's refer to eating *behaviors* themselves. People who eat too quickly prevent their brains from "catching up" to their stomachs, because it takes about 15 minutes or so for feelings of satiety to register in the brain after food reaches the stomach. Therefore, they do not realize that they are full, and they keep eating. The B's of eating extend to preparatory behaviors such as shopping, food storage habits, and so on.

The C's are the *consequences* of overeating. The immediate pleasure of overeating often overshadows the long-term negative consequences of obesity and risk to health. Food is also connected to other reward systems. Food can become a substitute friend or lover. When people feel depressed, they can lift their spirits with food, if only temporarily. Because food activates the parasympathetic branch of the autonomic nervous system (through digestive processes), food also acts as a natural sedative or tranquilizer, quelling feelings of anxiety or tension and helping people relax or get to sleep. In helping people

Taking it off and keeping it off. Health experts recognize that losing excess weight and keeping if off requires a lifelong commitment to adopting a sensible diet and engaging in regular exercise.

cope with the C's of eating, behavior therapists make the long-term benefits of sensible eating more immediate. Methods commonly used to address the ABCs of eating are shown in Table 11.3.

Behavior modification leads to modest weight losses that are generally well maintained through at least a year after treatment (Brownell & Wadden, 1992). However, long-term outcomes reveal a dimmer picture (Devlin et al., 2000). As many as 90% to 95% of people who lose weight by dieting or behavior modification return to their baseline weight levels within 5 years (Kolata, 2000a; Wilson, 1994). The problem is that changes in diet, eating habits, and exercise patterns that led to initial weight loss are not carried over into changes in lifestyles that would promote long-term maintenance of weight loss. The lesson here is clear-cut: People seeking to lose excess weight and keep it off need to make a lifelong commitment to following a sensible, calorie-sparing diet and engaging in regular exercise (Brody, 2000; Fogelholm et al., 2000; Kolata, 2000b; Wing & Polley, 2001).

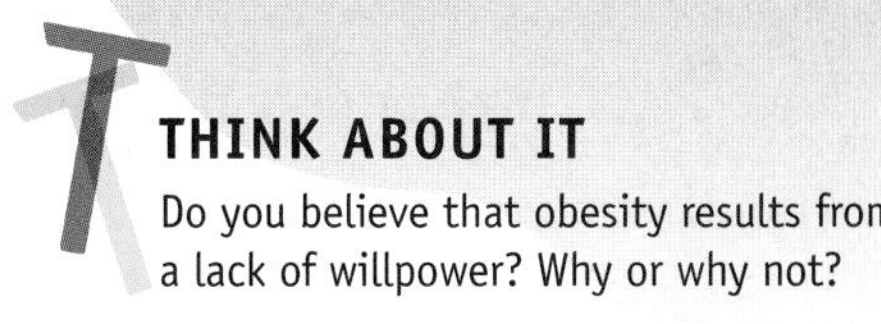

THINK ABOUT IT

Do you believe that obesity results from a lack of willpower? Why or why not?

Quiz **11.2**
Obesity

Sleep Disorders

Sleep is a biological function that remains in many ways a mystery. We know that sleep is restorative and that most of us need at least 7 or more hours of sleep a night to function at our best. Yet we cannot identify the specific biochemical changes occurring during sleep that account for its restorative function. We also know that many of us are troubled by sleep problems, although the causes of some of these problems remain obscure. Sleep problems of sufficient severity and frequency that they lead to significant personal distress or impaired functioning in social, occupational, or other roles are classified in the *DSM* system as **sleep disorders.**

WWW Web Link **11.4**
What Is Your Sleep IQ?

sleep disorders Persistent or recurrent sleep-related problems that cause distress or impaired functioning.

polysomnographic (PSG) recording Simultaneous measurement of multiple physiological responses during sleep or attempted sleep.

dyssomnias Sleep disorders involving disturbances in the amount, quality, or timing of sleep.

Highly specialized diagnostic facilities, called sleep disorders centers, have been established throughout the United States and Canada to provide a more comprehensive assessment of sleep problems than is possible in the typical office setting. People with sleep disorders typically spend a few nights at a sleep center, where they are wired to devices that track their physiological responses during sleep or attempted sleep—brain waves, heart and respiration rates, and so on. This form of assessment is termed **polysomnographic (PSG) recording,** because it involves simultaneous measurement of diverse physiological response patterns, including brain waves, eye movements, muscle movements, and respiration. Information obtained from physiological monitoring of sleep patterns is combined with that obtained from medical and psychological evaluations, subjective reports of sleep disturbance, and sleep diaries (i.e., daily logs compiled by the problem sleeper that track the length of time between retiring to bed and falling asleep, number of hours slept, nightly awakenings, daytime naps, and so on). Multidisciplinary teams of physicians and psychologists sift through this information to arrive at a diagnosis and suggest treatment approaches to address the presenting problem.

The *DSM* groups sleep disorders within two major categories: *dyssomnias* and *parasomnias.*

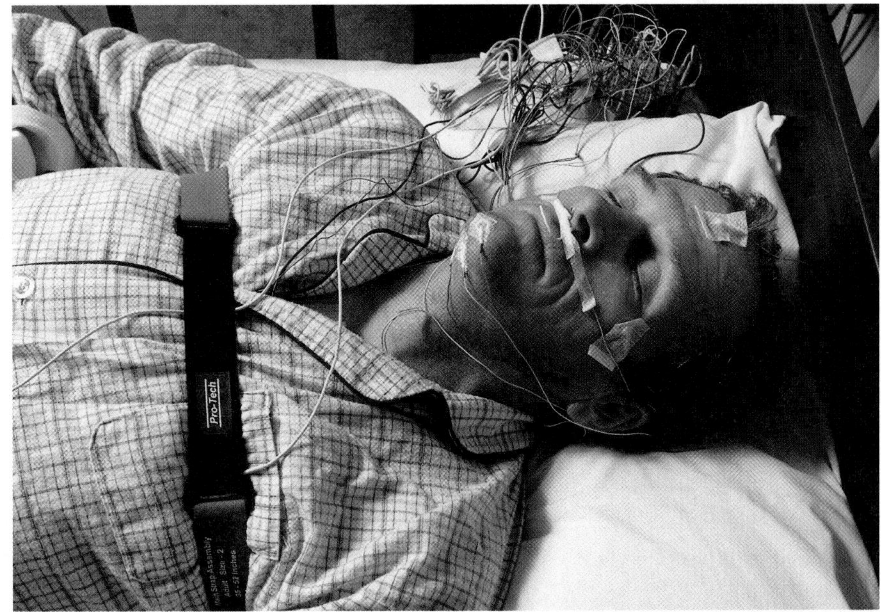

Sleep center. People with sleep disorders are often evaluated in sleep centers, where their physiological responses can be monitored as they sleep.

Dyssomnias

Dyssomnias are sleep disorders that are characterized by disturbances in the amount, quality, or timing of sleep. There are five specific types of

TABLE **11.3 Behavioral Techniques of Modifying the ABC's of Eating to Foster Weight Loss**

Changing the A's of Overeating	
Change the Environmental A's	Avoid settings that trigger overeating. (Eat at The Celery Stalk, not The Chocolate Gourmet.) Don't leave tempting treats around the house. Serve food on smaller plates. Use a lunch plate rather than a dinner plate. Don't leave seconds on the table. Serve preplanned portions. Do not leave open casseroles on the table. Immediately freeze leftovers. Don't keep them warm on the stove. Avoid the kitchen as much as possible. Disconnect eating from other stimuli, such as watching television, talking on the telephone, or reading. Establish food-free zones in your home. Imagine there is a barrier at the entrance to your bedroom that prevents the passage of food.
Control the Internal A's	Don't bury disturbing feelings in a box of cookies or a carton of mocha delight ice cream. Relabel feelings of hunger as signals that you're burning calories. Say to yourself, "It's okay to feel hungry. It doesn't mean I'm going to die or pass out. Each minute I delay eating, more calories are burned."
Changing the B's of Overeating	
Slow Down the Pace of Eating	Put down utensils between bites. Take smaller bites. Chew thoroughly. Savor each bite. Don't wolf bites down to make room for the next. Take a break during the meal. Put down your utensils and converse with your family or guests for a few minutes. (Give your rising blood sugar level a chance to signal your brain.) When you resume eating, ask yourself whether you need to finish every bite. Leave something over to be thrown away or enjoyed later.
Modify Shopping Behavior	Shop from a list. Don't browse through the supermarket. Shop quickly. Don't make shopping the high point of your day. Treat the supermarket like enemy territory. Avoid the aisles containing junk food and snacks. If you must walk down these aisles, put on mental blinders and look straight ahead. Never shop when hungry. Shop after meals, not before.
Practice Competing Responses	Substitute nonfood activities for food-related activities. When tempted to overeat, leave the house, take a bath, walk the dog, call a friend, or walk around the block. Substitute low-calorie foods for high-calorie foods. Keep lettuce, celery, or carrots in the middle of the refrigerator so they are available when you want a snack. Fill spare time with non-food-related activities: volunteer at the local hospital, play golf or tennis, join exercise groups, read in the library (rather than the kitchen), take long walks.
Break the Chain	Stretch the overeating chain. Before allowing yourself to snack, wait 10 minutes. Next time wait 15 minutes, and so on. Break the eating chain at its weakest link. It's easier to interrupt the eating chain by taking a route home that bypasses the bakery than to exercise self-control when you're placing your order.
Changing the C's of Overeating	
Reward Yourself for Meeting Calorie/Diet Goals	One pound of body weight is equivalent to 3,500 calories. To lose 1 pound per week, you need to cut 3,500 calories per week, or 500 calories per day, from your typical calorie intake level, assuming your weight has been stable. Reward yourself for meeting weekly calorie goals. Reward yourself with gifts you would not otherwise purchase for yourself, such as a special gift for yourself, like a cashmere sweater or tickets to a show. Repeat the reward program from week to week. If during some weeks you miss your calorie goals, don't lose heart. Get back on track next week.
Use Self-Punishment	Charge yourself for deviating from your diet. Send one dollar to a political candidate you despise, or to a hated cause, each time the chocolate cake wins.

dyssomnias: primary insomnia, primary hypersomnia, narcolepsy, breathing-related sleep disorder, and circadian rhythm sleep disorder.

Insomnia The term **insomnia** derives from the Latin *in-*, meaning "not" or "without," and, of course, *somnus,* meaning "sleep." Occasional bouts of insomnia, especially during times of stress, are not abnormal. But persistent insomnia characterized by recurrent difficulty getting to sleep or remaining asleep is an abnormal behavior pattern. (Pallesen et al., 2001). About one in three adult Americans experience chronic or persistent insomnia in any given year (Gillin, 1991). Chronic insomnia lasting a month or longer is often a sign of an underlying physical problem or a psychological disorder such as depression. If the underlying problem is treated successfully, chances are that normal sleep patterns will be restored. Chronic insomnia that cannot be accounted for by another psychological or physical disorder, or by the effects of drugs or medications, is classified as a sleep disorder called *primary insomnia.*

WWW Web Link 11.5
Facts About Insomnia

An estimated 14 million people in the United States, most of them over the age of 40, have primary insomnia (Nagourney, 2001b). People with primary insomnia have persistent difficulty falling asleep, remaining asleep, or achieving restorative sleep (sleep that leaves the person feeling refreshed and alert) for a period of a month or longer. Young people with primary insomnia usually complain it takes too long to get to sleep. Older people with insomnia are more likely to complain of waking frequently during the night, or of waking too early in the morning.

Primary insomnia leads to daytime fatigue and causes significant levels of personal distress or difficulties performing usual social, occupational, student, or other roles. Not surprisingly, there is a high rate of comorbidity (co-occurrence) between insomnia and other psychological problems, especially anxiety and depression (Breslau et al., 1996; Morin & Ware, 1996). Although the prevalence of primary insomnia is unknown, it is considered the most common form of sleep disturbance.

Psychological factors play a prominent role in primary insomnia. People troubled by primary insomnia tend to bring their anxieties and worries to bed with them, which raises their bodily arousal to a level that prevents natural sleep. Then they worry about not getting enough sleep, which only compounds their sleep difficulties. They may try to force themselves to sleep, which tends to backfire by creating more anxiety and tension, making sleep even less likely to occur. Sleep cannot be forced. Trying to make yourself fall asleep is likely to backfire. We can only set the scene for sleep by retiring when we are tired and relaxed and allowing sleep to occur naturally.

Hypersomnia The word *hypersomnia* is derived from the Greek *hyper,* meaning "over" or "more than normal," and the Latin *somnus,* meaning "sleep." Primary **hypersomnia** involves a pattern of excessive sleepiness during the day that continues for a month or longer. The excessive sleepiness (sometimes referred to as "sleep drunkenness") may take the form of difficulty awakening following a prolonged sleep period (typically 8 to 12 hours of sleep). Or there may be a pattern of daytime sleep episodes, occurring virtually every day, in the form of intended or unintended napping (such as inadvertently falling asleep while watching TV). Despite the fact that daytime naps often last an hour or more, the person does not feel refreshed upon awakening. The disorder is considered primary because it cannot be accounted for by inadequate amounts of sleep during the night due to insomnia or other factors (such as loud neighbors keeping the person up), by another psychological or physical disorder, or by drug or medication use.

Although many of us feel sleepy during the day from time to time, and may even drift off occasionally while reading or watching TV, the person with primary hypersomnia has more persistent and severe periods of sleepiness that typically lead to difficulties in daily functioning, such as missing important meetings because of difficulty awakening. Although the prevalence of the disorder is unknown, surveys of the general population show complaints relating to daytime sleepiness affecting between 0.5% to 5% of the adult population (APA, 2000).

insomnia Difficulties falling asleep, remaining asleep, or achieving restorative sleep.

hypersomnia A pattern of excessive sleepiness during the day.

narcolepsy A sleep disorder characterized by sudden, irresistible episodes of sleep.

cataplexy A brief, sudden loss of muscle control.

REM sleep The stage of sleep associated with dreaming and characterized by rapid eye movements under the closed eyelids.

breathing-related sleep disorder A sleep disorder in which sleep is repeatedly disrupted by difficulty with breathing normally.

apnea Temporary cessation of breathing.

Narcolepsy The word **narcolepsy** derives from the Greek *narke,* meaning "stupor" and *lepsis,* meaning "an attack." People with narcolepsy experience sleep attacks in which they suddenly fall asleep without any warning at various times during the day. They remain asleep for a period of about 15 minutes on the average. The person can be in the midst of a conversation at one moment and slump to the floor fast asleep a moment later. The diagnosis is made when sleep attacks occur daily for a period of 3 months or longer and are combined with the presence of one or both of the following conditions: (1) **cataplexy** (a sudden loss of muscular control); and (2) intrusions of **REM sleep** in the transitional state between wakefulness and sleep (APA, 2000). REM, or rapid eye movement, sleep is the stage of sleep associated with dreaming. It is so named because the sleeper's eyes tend to dart about rapidly under the closed lids. Narcoleptic attacks are associated with an almost immediate transition into REM sleep from a state of wakefulness. In normal sleep, REM typically follows several stages of non-REM sleep.

Cataplexy typically follows a strong emotional reaction such as joy or anger. It can range from a mild weakness in the legs to a complete loss of muscle control that results in the person suddenly collapsing (Dahl, 1992). People with narcolepsy may also experience *sleep paralysis,* a temporary state following awakening in which the person feels incapable of moving or talking. The person may also report frightening hallucinations, called *hypnagogic hallucinations,* which occur just before the onset of sleep and tend to involve visual, auditory, tactile, and kinesthetic (body movement) sensations.

Narcolepsy affects men and women equally and is a relatively uncommon disorder, affecting an estimated 0.02% (2 in 10,000) to 0.16% (16 in 10,000) people within the general adult population (APA, 1994). Unlike hypersomnia in which daytime sleep episodes follow a period of increasing sleepiness, narcoleptic attacks occur abruptly and are experienced as refreshing upon awakening. The attacks can be dangerous and frightening, especially if they occur when the person is driving or using heavy equipment or sharp implements. About 2 of 3 people with narcolepsy have fallen asleep while driving, and 4 of 5 have fallen asleep on the job (Aldrich, 1992). Household accidents resulting from falls are also common (Cohen, Ferrans, & Eshler, 1992). Not surprisingly, the disorder is associated with a lower quality of life in terms of general health and daily functioning (Ferrans, Cohen, & Smith, 1992). The cause or causes of narcolepsy remain unknown, but suspicion focuses on the loss of brain cells in the hypothalamus that produce a sleep-regulating chemical (Bazell, 2000; Mignot & Thorsby, 2001; Peyron et al., 2000; Thannickal et al., 2000).

Web Link 11.6 WWW
Facts About Narcolepsy

Truth OR Fiction? REVISITED

Many people suffer from sleep attacks in which they suddenly fall asleep without any warning.

FALSE. Sleep attacks are relatively uncommon. They are characteristic of a disorder called narcolepsy that affects between 2 and 16 persons in 10,000.

Breathing-Related Sleep Disorder People with a **breathing-related sleep disorder** experience repeated disruptions of sleep due to respiratory problems (APA, 2000). These frequent disruptions of sleep result in insomnia or excessive daytime sleepiness.

The subtypes of the disorder are distinguished in terms of the underlying causes of the breathing problem. The most common type is *obstructive sleep apnea,* which involves repeated episodes of either complete or partial obstruction of breathing during sleep (Zwilich, 2000). The disorder affects as many as 18 million Americans (Smith, 2001a). The word **apnea** derives from the Greek prefix *a-,* meaning "not, without," and *pneuma,* meaning "breath." The breathing difficulty results from the blockage of airflow in the upper airways, which is often caused by a structural defect, such as an overly thick palate or enlarged tonsils or adenoids. In cases of complete obstruction, the sleeper may literally stop breathing for periods of from 15 to 90 seconds as many as 500 times during the night! When these lapses of breathing occur, the sleeper may suddenly sit up, gasp for air, take a few deep breaths, and fall back asleep without awakening or realizing that breathing was interrupted.

Web Link 11.7 WWW
Facts About Sleep Apnea

Although a biological reflex kicks in to force a gasping breath after these brief interruptions of breathing, the frequent disruptions of normal sleep resulting from apneas can leave people feeling sleepy the following day, making it difficult for them to function effectively. Obstructive sleep apnea is a relatively common problem, with estimates indicating that the disorder affects approximately 1% to 10% of the adult population and perhaps an even higher percentage of older adults (APA, 2000). The disorder is most common in middle-aged men. Although men with apnea outnumber women by about a

2:1 ratio, the disorder often goes undiagnosed in women (Young et al., 1996). It is also more common among people who are obese, apparently because of a narrowing of the upper airways due to an enlargement of soft tissue (APA, 2000).

Not surprisingly, people who have sleep apnea report a poorer quality of life than unaffected people (Gall, Isaac, & Kryger, 1993). Sleep apnea is also a health concern because of its association with an increased risk of hypertension, a major risk factor for heart attacks and strokes (Nieto et al., 2000; Peppard et al., 2000; Zwilich, 2000).

Sleep apnea. Loud snoring may be a sign of obstructive sleep apnea, a breathing-related sleep disorder in which the person may temporarily stop breathing as many as 500 times during a night's sleep. Loud snoring, described by bed partners as reaching levels of industrial noise pollution, may alternate with momentary silences when breathing is suspended.

Circadian Rhythm Sleep Disorder Most bodily functions follow a cycle or an internal rhythm—called a circadian rhythm—that lasts about 24 hours. Even when people are relieved of scheduled activities and work duties and placed in environments that screen the time of day, they usually follow relatively normal sleep-wake schedules.

In **circadian rhythm sleep disorder,** this rhythm becomes grossly disturbed because of a mismatch between the sleep schedule demands imposed on the person and the person's internal sleep-wake cycle. The disruption in normal sleep patterns caused by the mismatch can lead to insomnia or hypersomnia. Like other sleep disorders, the mismatch must be persistent and severe enough to cause significant levels of distress or impair one's ability to function in social, occupational, or other roles. The jet lag that can accompany travel between time zones does not qualify because it is usually transient. However, frequent changes of time zones and frequent changes of work shifts (as encountered, for example, by nursing personnel) can induce more persistent or recurrent problems adjusting sleep patterns to scheduling demands, resulting in a circadian rhythm sleep disorder. Treatment may involve a program of making gradual adjustments in the sleep schedule to allow the person's circadian system to become aligned with changes in the sleep-wake schedule (Dahl, 1992).

Truth OR Fiction? REVISITED

Some people literally gasp for breath hundreds of times during sleep without realizing it.

TRUE. People with sleep apnea may gasp for breath hundreds of times during the night without realizing it.

Parasomnias

The **parasomnias** involve abnormal behaviors or physiological events taking place during sleep or at the threshold between wakefulness and sleep. Among the more common parasomnias are nightmare disorder, sleep terror disorder, and sleepwalking disorder.

Nightmare Disorder **Nightmare disorder** involves recurrent awakenings from sleep because of frightening dreams (nightmares). The nightmares typically involve lengthy story-like dreams that include threats of imminent physical danger to the individual, such as being chased, attacked, or injured. The person usually recalls the nightmare vividly upon awakening. Although alertness is regained quickly after awakening, anxiety and fear may linger and prevent a return to sleep. Perhaps half the adult population occasionally experiences nightmares, although the percentages of people having the intense, recurrent nightmares that produce the kind of emotional distress or difficulties in functioning that would lead to a diagnosis of nightmare disorder remains unknown.

Nightmares are often associated with traumatic experiences and are generally more frequent when the individual is under stress. Supporting the general link between trauma and nightmares, researchers report that the incidence of nightmares was greater among survivors of the 1989 San Francisco earthquake in the weeks following the quake than among comparison groups (Wood et al., 1992). An increased frequency of nightmares was also observed among children who were exposed to the 1994 Los Angeles earthquake (Kolbert, 1994).

circadian rhythm sleep disorder A sleep disorder characterized by a mismatch between the body's normal sleep-wake cycle and the demands of the environment.

parasomnias Sleep disorders involving abnormal behaviors or physiological events that occur during sleep or while falling asleep.

nightmare disorder A sleep disorder characterized by recurrent awakenings due to frightening nightmares.

sleep terror disorder A sleep disorder characterized by recurrent episodes of sleep terror resulting in abrupt awakenings.

sleepwalking disorder A sleep disorder involving repeated episodes of sleepwalking.

Nightmares generally occur during REM sleep. REM periods tend to become longer and the dreams occurring during REM more intense in the latter half of sleep, so nightmares usually occur late at night or toward morning. Although nightmares may contain great motor activity, as in fleeing from an assailant, dreamers show little muscle activity. The same biological processes that activate dreams—including nightmares—inhibit body movement, causing a type of paralysis. This is indeed fortunate, as it prevents the dreamer from jumping out of bed and running into a dresser or a wall in the attempt to elude the pursuing assailants from the dream.

Sleep Terror Disorder It typically begins with a loud, piercing cry or scream in the night. Even the most soundly asleep parent will be summoned to the child's bedroom as if shot from a cannon. The child (most cases involve children) may be sitting up, appearing frightened and showing signs of extreme arousal—profuse sweating with rapid heartbeat and respiration. The child may start talking incoherently or thrash about wildly, but remain asleep. If the child awakens fully, he or she may not recognize the parent or may attempt to push the parent away. After a few minutes the child falls back into a deep sleep and upon awakening in the morning remembers nothing of the experience. These terrifying attacks, called *sleep terrors*, are more intense than ordinary nightmares. Unlike nightmares, sleep terrors tend to occur during the first third of nightly sleep and during deep, non-REM sleep (Dahl, 1992).

A **sleep terror disorder** involves repeated episodes of sleep terrors resulting in abrupt awakenings that begin with a panicky scream (APA, 2000). If awakening occurs during a sleep terror episode, the person will usually appear confused and disoriented for a few minutes. The person may feel a vague sense of terror and be able to report some fragmentary dream images, but not the sort of detailed dreams typical of nightmares. Most of the time the person falls back asleep and remembers nothing of the experience the following morning.

Sleep terror disorder in children is typically outgrown during adolescence. More boys than girls are affected by the disorder, but among adults the gender ratio is about even. In adults, the disorder tends to follow a chronic course during which the frequency and intensity of the episodes waxes and wanes over time. Prevalence data on the disorder are lacking, but episodes of sleep terror are estimated to occur in 1% to 6% of children and in less than 1% of adults (APA, 2000). The cause of sleep terror disorder remains a mystery.

Sleepwalking Disorder **Sleepwalking disorder** involves repeated episodes in which the sleeper arises from bed and walks about the house while remaining fully asleep. Because these episodes tend to occur during the deeper stages of sleep in which there is an absence of dreaming, sleepwalking episodes do not seem to involve the enactment of a dream. In sleepwalking disorder, the occurrence of repeated episodes of sleepwalking are of sufficient severity to cause significant levels of personal distress or impaired functioning. Sleepwalking disorder is most common in children, affecting between 1% and 5% of children according to some estimates (APA, 2000). Between 10% and 30% of children are believed to have had at least one episode of sleepwalking. The prevalence of the disorder among adults is unknown, as are its causes. However, perhaps as many as 7% of adults have experienced occasional sleepwalking episodes (APA, 2000). The causes of sleepwalking remain obscure, although both genetic and environmental factors are believed to be involved (Hublin et al., 1997).

Although sleepwalkers typically avoid walking into things, accidents occasionally happen. Sleepwalkers tend to have a blank stare on their faces during these episodes. They are generally unresponsive to others and difficult to awaken. When they do awaken the following morning, they typically have little if any recall of the experience. If they are awakened during the episode, they may be disoriented or confused for a few minutes (as is the case with sleep terrors), but full alertness is soon restored. There is no basis to the belief that it is harmful to sleepwalkers to awaken them during episodes. Isolated incidents of violent behavior have been associated with sleepwalking, but these are rare occurrences and may well involve other forms of psychopathology.

THINK ABOUT IT

What are the distinguishing features of the major types of sleep disorders? Why are they categorized as psychological problem in the *DSM*?

Treatment of Sleep Disorders

The most common method for treating sleep disorders in the United States is the use of sleep medications called **hypnotics.** However, because of problems associated with these drugs, nonpharmacological treatment approaches, principally cognitive-behavioral therapy, have come to the fore.

Biological Approaches Antianxiety drugs are often used to treat insomnia, including a class of minor tranquilizers called benzodiazepines (for example, Valium, Librium, and Ativan) (Pallesen et al., 2001). (These drugs are also widely used in the treatment of anxiety disorders, as we saw in Chapter 6.) Another widely used drug, *zolpidem* (trade name Ambien), appears to be about as effective as the benzodiazepines but may produce fewer side effects and possibly fewer withdrawal effects (Kupfer & Reynolds, 1997). Nonetheless, all these drugs can produce chemical dependence if used regularly over time.

When used for the short-term treatment of insomnia, antianxiety drugs are generally effective in reducing the time it takes to get to sleep, increasing total length of sleep and reducing nightly awakenings (Nowell et al., 1998). They work by reducing arousal and inducing feelings of calmness, thereby making the person more receptive to sleep.

A number of problems are associated with using drugs to combat insomnia (Kryger, Roth, & Dement, 2000). Sleep-inducing drugs tend to suppress REM sleep, which may interfere with some of the restorative functions of sleep. They can also lead to a carryover or "hangover" the following day, which is associated with daytime sleepiness and reduced performance. Rebound insomnia can also follow discontinuation of the drug, causing worse insomnia than was originally the case. Rebound insomnia may be lessened, however, by tapering off the drug rather than abruptly discontinuing it. These drugs quickly lose their effectiveness at a given dosage level, so progressively larger doses must be used to achieve the same effect. High doses can be dangerous, especially if they are mixed with alcoholic beverages at bedtime. Regular use can also lead to physical dependence (addiction). Once dependence is established, withdrawal symptoms following cessation of use may occur, including agitation, tremors, nausea, headaches, and in severe cases, delusions or hallucinations.

Users can also become *psychologically* dependent on sleeping pills. That is, they can develop a psychological need for the medication and assume they will not be able to get to sleep without them. Because worry about going without drugs heightens bodily arousal, such self-doubts are likely to become self-fulfilling prophecies. Moreover, users may attribute their success in falling asleep to the pill and not to themselves, which strengthens reliance on the drugs and makes it harder to forgo using them.

WWW **Web Link 11.8**
National Sleep Foundation

Not surprisingly, there is little evidence of long-term benefits of drug therapy after withdrawal (Morin & Wooten, 1996). Relying on sleeping pills does nothing to resolve the underlying cause of the problem or help the person learn more effective ways of coping with it. If hypnotic drugs like benzodiazepines are to be prescribed at all for sleep problems, they should only be used for a brief time (a few weeks at most) and at the lowest possible dose (Dement, 1992; Kupfer & Reynolds, 1997). The aim should be to provide a temporary respite so the clinician can help the client find effective ways of handling the sources of stress and anxiety that contribute to insomnia.

Minor tranquilizers of the benzodiazepine family and tricyclic antidepressants are also used to treat the deep-sleep disorders—sleep terrors and sleepwalking. They seem to have a beneficial effect by decreasing the length of deep sleep and reducing partial arousals between sleep stages (Dahl, 1992). Use of sleep medications for these disorders, like primary insomnia, also incurs the risk of physiological and psychological dependence and thus should be used only in severe cases and only as a temporary means of "breaking the cycle." Other psychoactive drugs, such as stimulants, are sometimes used to help maintain wakefulness in people with narcolepsy and to combat daytime sleepiness in people with hypersomnia. Daily naps of 10 to 60 minutes, and coping support from mental health professionals or self-help groups, may also be of help to people with narcolepsy (Aldrich, 1992). Sleep apnea is sometimes treated with drugs that act on brain

Does going to bed mean going to sleep? People who use their beds for many other activities, including eating, reading, and watching television, may find that lying in bed loses its value as a cue for sleeping. Behavior therapists use stimulus control techniques to help people with insomnia create a stimulus environment associated with sleeping.

centers that stimulate breathing. Surgery may also be used to widen the upper airways. Mechanical devices may help maintain breathing during sleep, such as a nose mask that exerts pressure to keep the upper airway passages open or a battery-powered device that continuously blows air through the nose to prevent the airways from collapsing (e.g., Ballester et al., 1999).

Psychological Approaches Psychological approaches have by and large been limited to treatment of primary insomnia. Overall, cognitive-behavioral treatment approaches have produced substantial benefits in treating chronic insomnia, as measured by both substantial reductions in the time it takes to get to asleep and wakefulness during the night, as well as improved ratings of sleep quality (Currie et al., 2000; Edinger et al., 2001; Espie et al., 2001). In one recent study, two of three treatment participants were able to fall asleep within 30 minutes of retiring (Espie, Inglis, & Harvey, 2001). Sleep experts believe that CBT is just as effective as sleep medication in treating insomnia in the short-term and more effective in the long-term (Smith, 2001a).

Cognitive-behavioral techniques are short term in emphasis and focus on directly lowering states of physiological arousal, modifying maladaptive sleeping habits, and changing dysfunctional thoughts. Cognitive-behavioral therapists typically use a combination of techniques, including stimulus control, establishment of a regular sleep-wake cycle, relaxation training, and rational restructuring. Stimulus control involves changing the stimulus environment associated with sleeping. Under normal conditions, we learn to associate stimuli relating to lying down in bed with sleeping, so that exposure to these stimuli comes to induce feelings of sleepiness. But when people use their beds for many other activities—such as eating, reading, and watching television—the bed may lose its association with sleepiness. Moreover, the longer the person with insomnia lies in bed tossing and turning, the more the bed becomes associated with cues related to anxiety and frustration. Stimulus control techniques attempt to strengthen the connection between the bed and sleep by restricting as much as possible the activities spent in bed to sleeping. Typically, the person is instructed to limit the time spent in bed trying to fall sleep to 10 or 20 minutes at a time. If sleep does not occur within the designated time, the person is instructed to leave the bed and go to another room to restore a relaxed frame of mind before returning to bed, such as by sitting quietly, reading, watching TV, or practicing relaxation exercises. Moreover, the person is encouraged to establish a regular sleep-wake cycle by adopting more consistent sleeping and waking times (Riedel & Lichstein, 2001). Relaxation techniques, such as the Jacobson progressive relaxation approach, may also be practiced before bedtime to help reduce the level of physiological arousal. Based on a meta-analysis of outcome studies, it does not appear that the combination of stimulus control and relaxation training produces any larger benefit than either approach alone (Murtagh & Greenwood, 1995).

Rational restructuring involves substituting rational alternatives for self-defeating, maladaptive thoughts or beliefs (see accompanying "Closer Look" section for examples). The belief that failing to get a good night's sleep will lead to unfortunate, even disastrous, consequences the next day reduces the chances of falling asleep because it raises the level of anxiety and can lead the person to try unsuccessfully to force sleep to happen. Most of us do reasonably well if we lose sleep or even miss a night of sleep, even though we might like more.

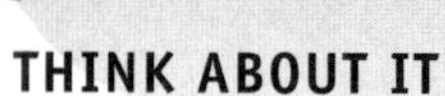

THINK ABOUT IT

What are the drawbacks of relying on sleep medications to combat chronic insomnia?

THINK ABOUT IT

Do your sleep habits help or hinder your sleeping patterns? Explain.

Quiz **11.3**
Sleep Disorders

Quiz **11.4**
Chapter Exam

Research Update
Chapter 11

A Closer Look

To Sleep, Perchance to Dream

Many of us have difficulty from time to time falling asleep or remaining asleep. Although sleep is a natural function and cannot be forced, we can develop more adaptive sleep habits that help us become more receptive to sleep. However, if insomnia or other sleep-related problems persist or become associated with difficulties functioning during the day, it would be worthwhile to have the problem checked out with a professional. Here are some techniques to help you acquire more adaptive sleep habits:

1. Retire to bed only when you feel sleepy.
2. Limit as much as possible your activities in bed to sleeping. Avoid watching TV or reading in bed.
3. If after 10 to 20 minutes of lying in bed you are unable to fall asleep, get out of bed, leave the bedroom, and put yourself in a relaxed mood by reading, listening to calming music, or practicing self-relaxation.
4. Establish a regular routine. Sleeping late to make up for lost sleep can throw off your body clock. Set your alarm for the same time each morning and get up, regardless of how many hours you have slept.
5. Avoid naps during the daytime. You'll feel less sleepy at bedtime if you catch *z's* during the afternoon.
6. Avoid ruminating in bed. Don't focus on solving your problems or organizing the rest of your life as you're attempting to sleep. Tell yourself that you'll think about tomorrow, tomorrow. Help yourself enter a more sleepful frame of mind by engaging in a mental fantasy or mind trip, or just let all thoughts slip away from consciousness. If an important idea comes to you, don't rehearse it in your mind. Jot it down on a handy pad so you won't lose it. But if thoughts persist, get up and follow them elsewhere.
7. Put yourself in a relaxed frame of mind before sleep. Some people unwind before bed by reading; others prefer watching TV or just resting quietly. Do whatever you find most relaxing. You may find it helpful to incorporate within your regular bedtime routine the techniques for lowering your level of arousal discussed earlier in this chapter, such as meditation or progressive relaxation.
8. Establish a regular daytime exercise schedule. Regular exercise during the day (not directly before bedtime) can help induce sleepiness upon retiring.
9. Avoid use of caffeinated beverages, such as coffee and tea, in the evening or late afternoon. Also, avoid drinking alcoholic beverages. Alcohol can interfere with normal sleep patterns (reduced total sleep, REM sleep, and sleep efficiency) even when consumed 6 hours before bedtime (Landolt et al., 1996).
10. Practice rational restructuring. Substitute rational alternatives for self-defeating thoughts. Here are some examples:

Self-Defeating Thoughts	Rational Alternatives
"I must fall asleep right now or I'll be a wreck tomorrow."	"I may feel tired, but I've been able to get by with little sleep before. I can make up for it tomorrow by getting to bed early."
"What's the matter with me that I can't seem to fall sleep?"	"Stop blaming yourself. You can't control sleep. Just let whatever happens, happen."
"If I don't get to sleep right now, I won't be able to concentrate tomorrow on the exam (conference, meeting, etc.)."	"My concentration may be off a bit, but I'm not going to fall apart. There's no point blowing things out of proportion. I might as well get up for a while and watch a little TV rather than lie here ruminating."

Overview of Sleep Disorders

TYPES OF SLEEP DISORDERS	Sleep disorders are often evaluated in specialized sleep centers where multiple physiological responses during sleep can be measured simultaneously in the form of polysomnographic (PSG) recording	
	Description	**Subtypes/Features**
Dyssomnias	Disturbances in the amount, quality, or timing of sleep	• **Insomnia:** Difficulty falling asleep, remaining asleep, or getting enough restful sleep • **Hypersomnia:** Excessive daytime sleepiness • **Narcolepsy:** Sudden attacks of sleep during the day • **Breathing-Related Sleep Disorder:** Sleep repeatedly interrupted due to difficulties breathing • **Circadian Rhythm Sleep Disorder:** Disruption of the internal sleep-wake cycle due to time changes in sleep patterns
Parasomnias	Disturbances occurring either during sleep or at the threshold between sleep and wakefulness	• **Nightmare Disorder:** Repeated awakenings due to nightmares • **Sleep Terror Disorder:** Repeated experiences of sleep terrors resulting in abrupt awakenings • **Sleepwalking Disorder:** Repeated episodes of sleepwalking

CAUSAL FACTORS	Many causes remain unspecified, but biological and psychosocial factors are prominent contributors
Biological Factors	• Underlying physical problems (in insomnia, apnea, and narcolepsy) • Possible genetic defects disrupting brain mechanisms controlling sleep (in narcolepsy) • Drug use interfering with normal sleep
Psychological Factors	• Psychological factors, such as anxiety or depression, that interfere with getting to sleep or remaining asleep • Frequent time shifting of sleep and waking times (in circadian rhythm sleep disorder) • Exposure to trauma (in nightmare disorder)

TREATMENT APPROACHES	Sleep medication may offer short-term relief for insomnia, but cognitive-behavioral therapy helps people change unhealthy sleep habits
Drug Therapy	• May be used for short-term relief of insomnia and to treat deep-sleep disorders (sleep terrors and sleepwalking), narcolepsy, and sleep apnea
Biomedical Treatment	• Surgery or mechanical devices may be used to open airways in apnea patients
Cognitive-Behavioral Therapy	• May be used to change maladaptive sleep habits and dysfunctional thoughts or beliefs about sleep

Summing Up

Eating Disorders

What are the major types of eating disorders? Two major types of eating disorders are included in the *DSM*: anorexia nervosa and bulimia nervosa. Anorexia nervosa involves maintenance of weight more than 15% below normal levels, intense fears of becoming overweight, distorted body image, and in females, amenorrhea. Bulimia nervosa involves preoccupation with weight control and body shape, repeated binges, and regular purging to keep weight down, which is characterized by self-starvation and failure to maintain normal body weight. Another type of eating disorder, binge-eating disorder (BED), is presently classified as a potential disorder requiring further study.

What factors are implicated in the development of eating disorders? Eating disorders typically begin in adolescence and affect many more females than males. Anorexia and bulimia are linked to preoccupations with weight control and maladaptive ways of trying to keep weight down. Many other factors are implicated in their development, including social pressures on young women to adhere to unrealistic standards of thinness, issues of control, underlying psychological problems, and conflict within the family, especially over issues of autonomy. People with BED tend to be older than those with anorexia or bulimia and to suffer from obesity.

What are the major forms of treatment for eating disorders? Severe cases of anorexia are often treated in an inpatient setting where a refeeding regimen can be closely monitored. Behavior modification and other psychological interventions, including psychotherapy and family therapy, may also be helpful. Most cases of bulimia are treated on an outpatient basis, with evidence supporting the therapeutic benefits of cognitive-behavioral therapy, interpersonal therapy, and antidepressant medication. Cognitive-behavioral therapy and antidepressant medication have shown positive effects in treating binge-eating disorder.

Obesity

Is obesity a psychological disorder? No, obesity is classified as a chronic disease, not a psychological disorder. It is a major risk factor linked to many serious chronic diseases, including heart disease and diabetes. Rates of obesity in the United States have been rising. The causes of obesity include genetic factors, metabolic factors, fat cells, lifestyle factors, and psychological factors.

Why is obesity difficult to treat? Quickie diets and diet pills don't work because long-term success in losing weight and keeping it off depends on making lasting changes in eating habits and exercise patterns.

Sleep Disorders

What are the major types of sleep disorders? Sleep disorders are classified in two major categories, dyssomnias and parasomnias. Dyssomnias are disturbances in the amount, quality, or timing of sleep. They include five specific types: primary insomnia, primary hypersomnia, narcolepsy, breathing-related sleep disorder, and circadian rhythm sleep disorder. Parasomnias are disturbed behaviors or abnormal physiological responses occurring either during sleep or at the threshold between wakefulness and sleep. Parasomnias include three major types: nightmare disorder, sleep terror disorder, and sleepwalking disorder.

What are the major forms of treatment for sleep disorders? The most common form of treatment of sleep disorders involves the use of antianxiety drugs. However, use of these drugs should be time limited because of the potential for psychological and/or physical dependence, among other problems associated with their use. Cognitive-behavioral interventions have produced substantial benefits in helping people with chronic insomnia.

Norms for the Fear of Fat Scale

Comparative scores are available for women only. You may compare your own score on the Fear of Fat scale to those obtained by the following groups:

Group	N	Mean
Nondieting college women (women satisfied with their weight)	49	17.30
General female college population	73	18.33
College women who are dissatisfied with their weight and have been on three or more diets during the past year	40	23.90
Bulimic college women (actively bingeing and purging)	32	30.00
Anorexic women in treatment	7	35.00

Source. Goldfarb, L.A., Dykens, E.M., & Gerrard, M. (1985). The Goldfarb Fear of Fat scale. *Journal of Personality Assessment, 49,* 329–332.

Keep the following in mind as you interpret your score:

1. The Goldfarb samples are quite small.
2. A score at a certain level does not place you in that group; it merely means that you report an equivalent fear of fat. In other words, a score of 33.00 does not indicate you have bulimia or anorexia. It means that your self-reported fear of fat approximates those reported by bulimic and anorexic women in the Goldfarb study.

CHAPTER TWELVE

Gender Identity Disorder, Paraphilias, and Sexual Dysfunctions

Bharati Chaudhuri
Invisible Tension, 1992

Truth OR Fiction?

- Despite changes in societal attitudes toward homosexuality, it is still classified as a mental disorder within the DSM system. (p. 372)
- Gay males and lesbians have a gender identity of the opposite sex. (p. 373)
- Wearing revealing bathing suits is a form of exhibitionism, according to clinical criteria. (p. 378)
- Becoming aroused while watching your partner disrobe or viewing an explicit movie falls within the clinical definition of voyeurism. (p. 380)
- Some people cannot become sexually aroused without pain or humiliation. (p. 381)
- Orgasm is a reflex. (p. 386)
- Premature ejaculation affects a relatively small proportion of men. (p. 387)
- The male sex hormone testosterone is produced by the body in both women and men. (p. 389)
- Though it is used mostly by men, Viagra can also help women overcome sexual dysfunctions. (p. 397)

Off the fog-bound shore of Ireland lies the isle of Inis Beag.[1] From the air, it is a verdant jewel, warm and enticing. From the ground, the perspective is different.

For example, the inhabitants of Inis Beag believe that normal women do not have orgasms and those who do must be deviant (Messenger, 1971). Premarital sex is virtually unknown. Women participate in sexual relations in order to conceive children and pacify their husbands' lustful urges. They need not be concerned about being called on for frequent performances, because the men of Inis Beag believe, groundlessly, that sex saps their strength. Relations on Inis Beag take place in the dark—literally and figuratively, and with nightclothes on. Consistent with local standards of masculinity, the man ejaculates as quickly as he can. Then he rolls over and goes to sleep, without concern for his partner's satisfaction. Women do not complain, however, as they are reared to believe it is abnormal for them to experience sexual pleasure.

If Inis Beag is not your cup of tea, perhaps the ambience of Mangaia will strike you as more congenial. Mangaia is a Polynesian pearl. Languidly, Mangaia lifts out of the azure waters of the Pacific. Inis Beag and Mangaia are on opposite sides of the world—literally and figuratively.

From childhood, Mangaian children are expected to explore their sexuality through masturbation (Marshall, 1971). Mangaian teenagers are encouraged by their elders to engage in sexual relations. They will be found on hidden beaches or beneath the sheltering fronds of palms, industriously practicing skills acquired from their elders. Mangaian women usually reach orgasm numerous times before their partners do. Young men vie to see who is more skillful in helping their partners attain multiple orgasms.

The inhabitants of Mangaia and Inis Beag have like anatomic features, and the same hormones pulse through their bodies. Their attitudes and cultural values about what is normal and abnormal differ vastly, however. Their attitudes affect their sexual behavior and the enjoyment they attain—or do not attain—from sex. In sex, as in other areas of behavior, the lines between the normal and the abnormal are not always drawn precisely. Sex, like eating, is a natural function. Yet this natural function has been profoundly affected by custom, folklore, superstition, and cultural, religious, and moral beliefs.

Even in the United States today, we find attitudes as diverse as those on Inis Beag and Mangaia. Some people feel guilty about any form of sexual activity and thus reap little if any pleasure from sex. Others, who see themselves as the children of the sexual revolution of the 1960s and 1970s, may worry about whether they have become free enough or skillful enough in their sexual activity.

Chapter Outline

Normal and Abnormal in Sexual Behavior

In the realm of sexual behavior, our conceptions of what is normal and what is not are clearly influenced by sociocultural factors. Various patterns of sexual behavior that might be considered abnormal in Inis Beag, such as masturbation, premarital intercourse, and oral-genital sex, are normal in American society from the standpoint of statistical frequency. For example, a recent national survey, based on a representative sample of 3,432 males and females between the ages of 18 and 59, found that 63% of the adult men and 42% of the adult women surveyed reported that they had masturbated during the previous year (Laumann et al., 1994). It is likely that many more practiced masturbation but were hesitant to admit so to interviewers.

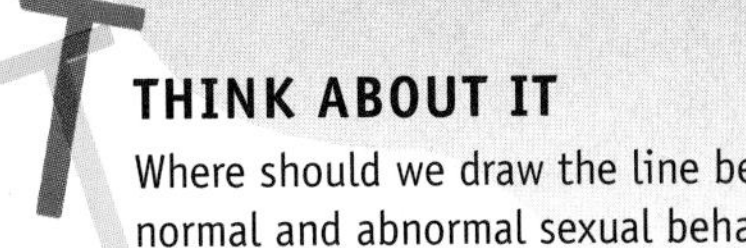

THINK ABOUT IT

Where should we draw the line between normal and abnormal sexual behavior? What criteria should we use?

Attitudes toward **homosexuality** vary widely from culture to culture and from time to time. Studies of tribal societies show societal attitudes ranging from condemnation to tolerance and acceptability (Ford & Beach, 1951). In our society, homosexuality was once considered a form of mental illness, but in 1973 the American Psychiatric Association decided to remove homosexuality from its listing of mental disorders. Though homosexuality

homosexuality A sexual orientation characterized by erotic interest in, and development of romantic relationships with, members of one's own gender.

[1] Inis Beag is actually a pseudonym for an Irish folk community.

What is normal and what is abnormal in the realm of sexual and sexually related behavior patterns? The cultural context must be considered in defining what is normal and what is abnormal. The people in these photographs—and the ways in which they cloak or expose their bodies—would be quite out of place in one another's societies.

Truth OR Fiction? REVISITED

Despite changes in societal attitudes toward homosexuality, it is still classified as a mental disorder within the *DSM* system.

FALSE. Homosexuality is no longer classified as a mental disorder.

Quiz **12.1**
Normal and Abnormal in Sexual Behavior

is no longer considered a mental disorder, lesbians and gay males continue to be targets of extreme hostility, fear, and prejudice (see nearby *Closer Look* feature).[2]

Sexual behavior may be considered abnormal if it is self-defeating, deviates from social norms, harms others, causes personal distress, or interferes with one's ability to function. The disorders we feature in this chapter—gender identity disorder, paraphilias, and sexual dysfunctions—meet one or more of the criteria of abnormality. In exploring these disorders, we touch on questions that probe the boundaries between abnormality and normality. Are some instances of voyeurism or exhibitionism normal and others abnormal? When is it considered abnormal to have difficulty becoming sexually aroused or reaching orgasm?

Gender Identity Disorder

VIDEO **12.1**
Gender Identity Disorder:
The Case of Denise

Our **gender identity** is our sense of being male or female. Gender identity is normally based on anatomic gender. In the normal run of things, our gender identity is consistent with our anatomic gender. In **gender identity disorder,** however, there is a conflict between one's anatomic gender and one's gender identity.

Gender identity disorder may begin in childhood. Children with the disorder find their anatomic gender to be a source of persistent and intense distress. The diagnosis is not used simply to label "tomboyish" girls and "sissyish" boys. It is applied to children who persistently repudiate their anatomic traits (girls might insist on urinating standing up or assert they do not want to grow breasts; boys may find their penis and testes revolting) or who are preoccupied with clothing or activities that are stereotypic of the other gender (see Table 12.1).

The diagnosis of gender identity disorder (formerly called *transsexualism*) applies to both children and adults who perceive themselves psychologically as members of the opposite gender and who show persistent discomfort with their anatomic gender.

Though the overall rate of gender identity disorder is not known, the disorder is believed to occur about five times more often in boys than girls (Zucker & Green, 1992). The disorder takes many paths. It can come to an end or abate markedly by adolescence, with

gender identity One's psychological sense of being female or male.

gender identity disorder A disorder in which the individual believes that her or his anatomical gender is inconsistent with her or his gender identity.

[2]In keeping with the suggestions of the American Psychological Association's (1991) Committee on Lesbian and Gay Concerns, we refer to *gay males* and *lesbians* rather than *homosexuals.* As noted by the Committee, there are several problems with the label *homosexual:* One, because it has been historically associated with concepts of deviance and mental illness, it may perpetuate negative stereotypes of gay men and lesbians. Two, the term is often used to refer to men only, thus rendering lesbians invisible. Third, it is often ambiguous in meaning—that is, does it refer to sexual behavior or sexual orientation?

the child becoming more accepting of her or his gender identity. Or it may persist into adolescence or adulthood, leading to a transsexual identity (Cohen-Kettenis et al., 2001). Or the child may also develop a gay male or lesbian sexual orientation at about the time of adolescence (Zucker & Green, 1992).

Many transexual adults undergo gender reassignment surgery. In these procedures, surgeons attempt to construct external genitalia that are close as possible to those of the opposite gender. People who undergo these operations can engage in sexual activity, even achieve orgasm, yet they are incapable of conceiving or bearing children because they lack the internal reproductive organs of their reconstructed gender. Investigators generally find favorable psychological outcomes following gender reassignment surgery (e.g., Cohen-Kettenis & van Goozen, 1997), especially when safeguards are taken to restrict surgical treatment to the most appropriate candidates. In one recent study, 14 male-to-female and 5 female-to-male patients were found to be functioning well socially and psychologically postoperatively, with none expressing regrets about the procedure (Cohen-Kettenis & van Goozen, 1997). In other research, none of 20 adolescent patients who underwent gender reassignment surgery later expressed regrets about their decision (Smith et al., 2001).

Web Link 12.1
Transgender Forum

Truth OR Fiction? REVISITED

Gay males and lesbians have a gender identity of the opposite sex.

FALSE. Gender identity should not be confused with sexual orientation. Gay males and lesbians have erotic interest in members of their own gender, but their gender identity is consistent with their anatomic sex.

Men seeking gender reassignment outnumber women applicants by perhaps 3 or 4 to 1, but outcomes are generally more favorable for female-to-male cases. One reason may be society's greater acceptance of women who desire to live as men. Another reason appears to be that females with gender identity disorder are generally better adjusted than their male counterparts before surgery (Kockott & Fahrner, 1988). Male-to-female patients whose surgery left no telltale signs (such as scarring of the breasts or leftover erectile tissue) were found to be better adjusted than those whose surgery was less successful in allowing them to "pass" as female (Ross & Need, 1989).

Gender identity should not be confused with sexual orientation. Gay males and lesbians have erotic interests in members of their own gender, but their gender identity (sense

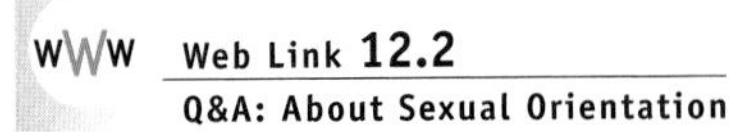
Web Link 12.2
Q&A: About Sexual Orientation

TABLE 12.1 Clinical Features of Gender Identity Disorder

(a) A strong, persistent identification with the other gender.

At least four of the following features are required to make the diagnosis in children:

(1) Repeated expression of the desire to be a member of the other gender (or expression of the belief that the child does belong to the other gender)

(2) Preference for wearing clothing stereotypical of members of the other gender

(3) Presence of persistent fantasies about being a member of the other gender, or assumption of parts played by members of the other gender in make-believe play

(4) Desire to participate in leisure activities and games considered stereotypical of the other gender

(5) Strong preference for playmates that belong to the other gender (at ages when children typically prefer playmates of their own gender).

Adolescents and adults typically express the wish to be of the other gender, frequently "pass" as a member of the other gender, and wish to live as a member of the other gender, or believe that their emotions and behavior typify the other gender.

(b) A strong, persistent sense of discomfort with one's anatomic gender or with the behaviors that typify the gender role of that gender.

In children, these features are commonly present: Boys state that their external genitals are repugnant or that it would be better not to have them, show aversion to "masculine" toys, games, and rough-and-tumble play. Girls prefer not to urinate while sitting, express the wishes not to grow breasts or to menstruate, or show an aversion to "feminine" clothing.

Adolescents and adults typically state that they were born the wrong gender and express the wish for medical intervention (e.g., hormone treatments or surgery) to rid them of their own sex characteristics and simulate the characteristics of the other gender.

(c) There is no "intersex condition," such as ambiguous sexual anatomy, that might give rise to such feelings.

(d) The features cause serious distress or impair key areas of occupational, social, or other functioning.

Source. Adapted from the *DSM-IV-TR* (APA, 2000).

A Closer Look

Homophobia: Social Prejudice or Personal Psychopathology?

Conceptions of abnormal behavior are societally constructed beliefs as to what behaviors are deemed normal or abnormal within a given culture and at a particular time. Mental health professionals base judgments of abnormality on evidence showing that a particular pattern of behavior either causes personal distress or interferes significantly with a person's ability to function in meeting social and occupational roles. Yet evidence does not support the view that lesbians, gay males, and bisexuals are any more psychologically disturbed than comparison heterosexual groups, despite the ostracism, prejudice, and discrimination they face in society (Coleman, 1987; Reiss, 1980).

George Weinberg (1972) coined the term **homophobia** to describe the persistent, irrational fear of lesbians and gay men. Fear and anxiety about lesbians and gay men or lesbian/gay sexual orientations has been deemed irrational because such fears are usually based on beliefs that are of questionable validity or have been overwhelmingly disputed. Examples of false beliefs about lesbians and gay men are that gay men are more likely to become child molesters than heterosexual men, that lesbians and gay men wish to be members of the other gender, that they do not make good parents, that their children will become lesbian or gay, that they are sexually promiscuous or indiscriminate in their sexual attractions, that their relationships are transitory and focused only on sexual interactions, and that they are responsible for the AIDS epidemic (Jenny, Roesler, & Poyer, 1994; Peplau, 1991). Many people who hold these beliefs feel justified in engaging in prejudice against lesbians and gay men that may range from personal rudeness or hostility to vandalism, harassment and even violent physical attacks (Freiberg, 1995; Katz, 1995).

As these attacks are commonplace in many parts of the United States, they create a climate of terror for lesbians and gay men and make their lives more difficult. The fact that these beliefs persist among many otherwise intelligent people, despite evidence to the contrary, and that they are often connected to expressions of violence that harms others makes their understanding an important social issue. It also raises important questions for behavioral scientists about what purpose these beliefs serve, what they mean to those who hold them, and how to alter them.

Many theoreticians use the term **heterosexism** (Herek, 1996) to describe a broader cultural ideology and resulting pattern of institutional discrimination against lesbians and gay men. They suggest that the antigay feelings of individuals are only a part of a broader institutional pattern that is embedded in our culture and is based on the cultural assumption that reproductive sexuality is the only outcome of psychosexual development that is psychologically healthy and morally correct. They go on to suggest that antigay sentiments are rewarded in our society more than they are punished and are not considered to be abnormal or pathological.

Still, others believe that homophobia represents a form of clinical pathology in that it is an irrational belief or belief system that persists in the face of evidence to the contrary. Marvin Kantor (1998) views homophobia as an emotional disorder in which the false beliefs about lesbians and gay men and the anxiety associated with them represents a symptom that is similar to that of a paranoid delusion. In this view, homophobic beliefs may represent people's underlying fears or anxieties about their own latent homosexual attractions or strivings or insecurity about their own masculinity or femininity.

Gay bashing. In a heinous case that brought gay bashing into the national spotlight in 1998, a gay 21-year-old University of Wyoming student, Matthew Shepard, was pistol-whipped, tied to a fencepost, and left to die by a group of assailants.

Violent physical and verbal attacks against lesbians and gay men are referred to as gay bashing and are considered a hate crime. According to Kantor, many people who feel the need to bash lesbians and gay men and/or those who actually do so are really attempting to reassure themselves that they do not have such feelings or attractions. By punishing those who express these forbidden wishes, the gay basher may be seeking to prove to himself and others that he is not one of them.

Research on people who are homophobic suggests that they tend to have rigid personalities and are intolerant of anything that deviates from their personal view of appropriate behavior (Kantor, 1998). Other research suggests that they are people who have not, to their knowledge, had direct contact with lesbians or gay men (Herek, 1996). Their beliefs are selectively drawn from a larger culture that has historically put forth in the media demeaning images of lesbians and gay men by depicting them as dangerous, perverted, depressed, or so much the focus of comic relief that they would not be taken seriously. Ignorance about lesbians and gay men, maintained by their invisibility, fuels homophobic attitudes.

Because of the harm that homophobia and heterosexism does to lesbians and gay men and because of its adverse effects on their mental health, these phenomena are worthy of our serious attention and understanding. Do heterosexism and homophobia constitute some form of social or personal pathology, or both?

of being male or female) is consistent with their anatomic sex. They do not desire to become members of the opposite gender or despise their own genitalia, as we may find in people with gender identity disorder.

Unlike a gay male or lesbian sexual orientation, gender identity disorder is very rare. People with gender identity disorder who are sexually attracted to members of their own anatomic gender are unlikely to consider themselves gay males or lesbians, however. Nature's gender assignment is a mistake in their eyes. From their perspective, they are trapped in the body of the wrong gender.

homophobia Hatred and fear of lesbians and gay males.

heterosexism The culturally based belief system that holds that only reproductive sexuality is psychologically healthy and morally correct.

paraphilias Sexual disorders in which the person experiences recurrent sexual urges and fantasies involving nonhuman objects, inappropriate or nonconsenting partners, or painful or humiliating situations.

Theoretical Perspectives

No one knows what causes gender identity disorder (Money, 1994). Psychodynamic theorists point to extremely close mother–son relationships, empty relationships between the mothers and fathers, and fathers who were absent or detached (Stoller, 1969). These family factors may foster strong identification with the mother in young males, leading to a reversal of expected gender roles and identity. Girls with weak, ineffectual mothers and strong, masculine fathers may overly identify with their fathers and develop a psychological sense of themselves as "little men."

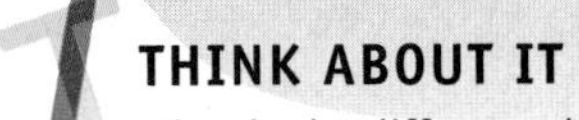

THINK ABOUT IT
What is the difference between gender identity disorder and a gay male or lesbian sexual orientation?

Learning theorists similarly point to father absence in the case of boys—to the unavailability of a strong male role model. Socialization patterns might have affected children who were reared by parents who had wanted children of the other gender and who strongly encouraged cross-gender dressing and patterns of play.

Nonetheless, the great majority of people with the types of family histories described by psychodynamic and learning theorists do not develop gender identity disorder. Perhaps these family factors play a role in combination with a biological predisposition. We know that people with gender identity disorder often showed cross-gender preferences in toys, games, and clothing very early in childhood. If there are critical early learning experiences in gender identity disorder, they may occur very early in life. Prenatal hormonal imbalances may also be involved. Perhaps the brain is "masculinized" or "feminized" by sex hormones during certain stages of prenatal development. The brain could become differentiated as to gender identity in one direction while the genitals develop in the other direction (Money, 1987). All in all, researchers suspect that gender identity disorders may develop as the result of an interaction in utero between the developing brain and the release of sex hormones (Zhou et al., 1995). Yet speculations about the origins of gender identity disorder remain unsubstantiated by hard evidence. But whatever the biological contributions to gender identity turn out to be, are people who are different by virtue of their gender identities necessarily suffering from a disease or disorder? The answer may depend on how the society or culture in which they live regards them, as we explore further in the accompanying "Closer Look" feature.

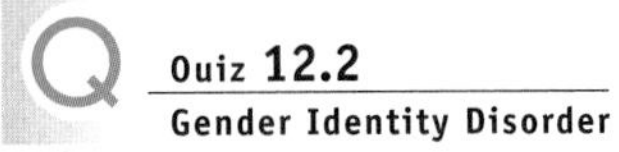

Quiz **12.2**
Gender Identity Disorder

Paraphilias

The word *paraphilia* was coined from the Greek roots *para,* meaning "to the side of," and *philos,* meaning "loving." In the **paraphilias,** people show sexual arousal ("loving") in response to atypical stimuli ("to the side of" normally arousing stimuli). According to the *DSM-IV,* paraphilias involve recurrent, powerful sexual urges and fantasies lasting 6 months or longer that center on (1) nonhuman objects such as underwear, shoes, leather, or silk, (2) humiliation or experience of pain in oneself or one's partner, or (3) children and other persons who do not or cannot grant consent. Although acting out on paraphilic urges is not required for a diagnosis (the person might be distressed by the urges but not act on them), people with paraphilias often engage in overt paraphilic behaviors such as exhibitionism and voyeurism.

THINK ABOUT IT
Do you think that anyone requesting a sex change operation should be entitled to receive one? If not, what criteria should be applied?

Some persons who receive the diagnosis can function sexually in the absence of paraphilic stimuli or fantasies. Others resort to paraphilic stimuli under stress. Still others cannot become sexually aroused unless these stimuli are used, in actuality or in fantasy. For

A Closer Look

Gender Identity Disorder: A Disorder or Culture-Specific Creation?

Though our concept of gender, or maleness and femaleness, reflects the biological division of the sexes, it is also a social construction. It is a culturally based concept for designating the roles, attributes, responsibilities, privileges, and traits that a given culture assigns to men and women. Gender roles embody the behaviors a specific culture considers appropriate and fundamental to being a man or a woman. Gender identity, however, is a psychological construct that reflects an individual's psychological sense of their own gender, of who and what they are. Most of the assumptions we make about gender identity are based on the construction of gender as a dichotomous, mutually exclusive category—either male or female. We also assume that people's concept of their own gender should be consistent with their biological (anatomic) sex. People whose gender is at odds with their biological sex may be classified as having gender identity disorder. We assume people should be satisfied with the gender assigned by nature. As gender is also a social construction—a concept society creates and uses to assign people to different categories and statuses—it is also embedded in a socio-political context in which a person's gender not only defines them personally but also designates their place in a social hierarchy that determines their access to gender-based privileges. For example, men in our society have traditionally been accorded greater access than women to high prestige occupations in business, medicine, law, and engineering. Though gender roles today are less rigid than they were a few generations ago, differences in social expectations continue. Certainly more women are working today, but by and large they earn less than men and are still expected to shoulder the lion's share of household and child-care responsibilities.

Just as traditional gender roles have been challenged, so too have traditional concepts of gender identity been challenged by people who are dissatisfied with their biological sex and/or wish to adopt the gender roles of the other sex. The assumptions we make about the fixed dichotomy of gender into maleness and femaleness are further challenged by cross-cultural studies that reveal the presence of cultures that have a socially sanctioned identity for persons who are regarded as neither male or female. In these cultures, individuals who adopt the gender roles and identity of the other biological sex are not considered pathological or undesirable. They are simply regarded as having a third or other gender.

Into the late 1800s, the Plains Indians and many other Western tribes responded with understanding when a young member of the tribe crossed traditional gender roles (Wade & Tavris, 1994). Crossing traditional gender roles was not merely allowed but was accorded a respectable status in these societies. More than half of the surviving native languages have words that describe people who were neither male nor female but something else. Their presence and their roles are depicted in ways that indicated they were socially acceptable and even desirable members of the tribe. Native tribal members believed that human beings contained male and female elements. In many tribes, "two spirit" was the term used for persons who were believed to embody a higher level of integration of their male and female spirit. Sometimes a *two spirit* person was a biological male who took on the tribe's female gender roles but was not considered a male or a female (Tafoya, 1996). On the other hand, a female could take on the role and behaviors associated with tribal males. She could be initiated into puberty as a male and could adopt male roles and activities thereafter, including marrying a female.

Native Americans are not the only cultural group that has categories for people who do not fit contemporary U.S. gender dichotomies. These societies highlight social contexts and cultures in which dichotomous categories of gender were not presumed to be the norm and in which inconsistencies between biological sex, gender identities, and gender roles were not considered a psychological problem. Given the variability of gender roles and identities that we observe across cultures, we may question the validity of the concept of what has been described as a "disorder."

Califia (1997) and others (Israel & Tarver, 1997; K. K. Wilson, 1997) argue that although people who are dissatisfied with their biological sex or who wish to adopt the gender role that is inconsistent with their biological sex are perhaps atypical, their conditions should neither constitute nor function as a marker of psychopathology. Rather, their dysphoria can be attributed to living in a society that insists that people fit into either of two arbitrarily designated categories and subjects them to ill treatment if they do not. Their plight may be compared to that of lesbians and gay men. We understand that the distress that lesbians and gay men often experience in reference to their sexual orientation is a function of the hostility and abuse they receive because of it. Their distress is not an inevitable consequence of their sexual orientation, but rather an appropriate response to the negative treatment they receive.

It could be argued that dissatisfaction with one's biological sex or an inconsistency between one's biological sex and desired gender role or identity is not a problem unless you live in a society that says it is and is intolerant of it. If the society you live in were intolerant of you for being different, it is understandable you would become distressed by it. The ill treatment that people with *transgender* identities (identities that cross traditional gender lines) in our society receive can be a significant source of personal distress (Reid & Whitehead, 1992). Looked at in this way, our concept of gender identity disorder may be understood as a social construction that reflects our culture's definition of gender and treatment of people who are different, rather than a diseased or disordered condition residing within the person. In other cultures in which concepts of mutually exclusive gender categories do not apply, the disorder is not held to exist. Just as beliefs about sexual orientation have changed over time, perhaps greater tolerance and a greater appreciation for the diversity of gender expression in human beings will lead us to conceptualize gender identity with greater flexibility.

some individuals, the paraphilia is their exclusive means of attaining sexual gratification. With the exceptions of sexual masochism and some isolated cases of other disorders, paraphilias are almost never diagnosed in women (Seligman & Hardenburg, 2000). Even with masochism, it is estimated that men receiving the diagnosis outnumber women by a ratio of 20 to 1 (APA, 2000).

Some paraphilias are relatively harmless and victimless. Among these are fetishism and transvestic fetishism. Others, such as exhibitionism and pedophilia, have unwilling victims. A most harmful paraphilia is sexual sadism when acted out with a nonconsenting partner. Voyeurism falls somewhere in between, because the "victim" does not typically know he or she is being watched.

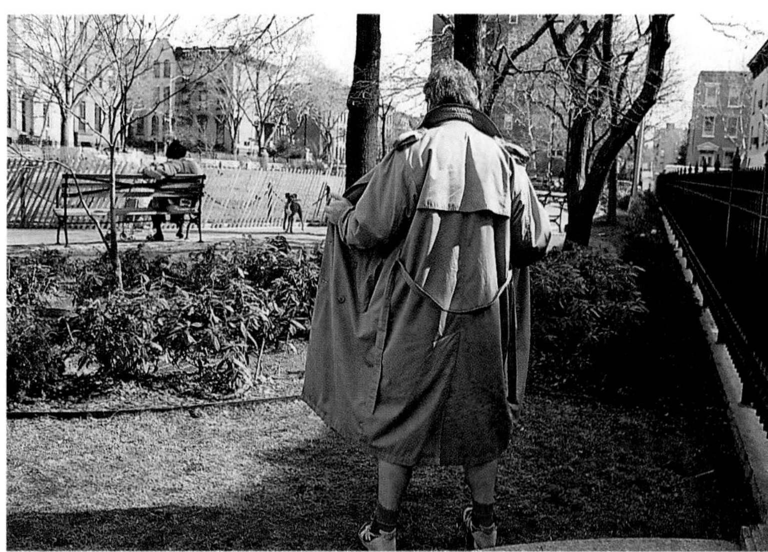

Exhibitionism. Exhibitionism is a type of paraphilia that characterizes people who seek sexual arousal or gratification through exposing themselves to unsuspecting victims. People with this disorder are usually not interested in actual sexual contact with their victims.

Exhibitionism

Exhibitionism involves recurrent, powerful urges to expose one's genitals to an unsuspecting stranger in order to surprise, shock, or sexually arouse the victim. The person may masturbate while fantasizing about or actually exposing himself (almost all cases involve men). The victims are almost always females.

The person diagnosed with exhibitionism is typically not interested in actual sexual contact with the victim and therefore not usually dangerous. Nevertheless, victims may believe themselves in great danger and may be traumatized by the act. Victims are probably best advised to show no reaction to people who expose themselves but to just continue on their way, if possible. It would be unwise to insult the person who exposed himself, lest it provoke a violent reaction. Nor do we recommend an exaggerated show of shock or fear; it tends to reinforce the person for the act of exposing himself.

Some researchers view exhibitionism as a means of indirectly expressing hostility toward women, perhaps because of perceptions of having been wronged by women in the past or of not being noticed or taken seriously by them (Geer, Heiman, & Leitenberg, 1984). Men with this disorder tend to be shy, dependent, and lacking in social and sexual skills, even socially inhibited (Dwyer, 1988). Some doubt their masculinity and harbor feelings of inferiority. Their victims' revulsion or fear boosts their sense of mastery of the situation and heightens their sexual arousal. Consider the following case example of exhibitionism:

A Case of Exhibitionism

Michael was a 26-year-old, handsome, boyish-looking married male with a 3-year-old daughter. He had spent about one-quarter of his life in reform schools and in prison. As an adolescent, he had been a fire-setter. As a young adult, he had begun to expose himself. He came to the clinic without his wife's knowledge because he was exposing himself more and more often—up to three times a day—and he was afraid that he would eventually be arrested and thrown into prison again.

Michael said he liked sex with his wife, but it wasn't as exciting as exposing himself. He couldn't prevent his exhibitionism, especially now, when he was between jobs and worried about where the family's next month's rent was coming from. He loved his daughter more than anything and couldn't stand the thought of being separated from her.

Michael's method of operation was as follows: He would look for slender adolescent females, usually near the junior high school and the senior high school. He would take his penis out of his pants and play with it while he drove up to a girl or a small group of girls. He would lower the car window, continuing to play with himself, and ask them for

exhibitionism A paraphilia in which one is sexually aroused by exposing one's genitals to a stranger.

fetishism A paraphilia in which one uses an inanimate object or body part as a focus of sexual interest and arousal.

transvestic fetishism A paraphilia in heterosexual males characterized by recurrent sexual urges involving dressing in female clothing.

directions. Sometimes the girls didn't see his penis. That was okay. Sometimes they saw it and didn't react. That was okay, too. When they saw it and became flustered and afraid, that was best of all. He would start to masturbate harder, and now and then he managed to ejaculate before the girls had departed.

Michael's history was unsettled. His father had left home before he was born, and his mother had drunk heavily. He was in and out of foster homes throughout his childhood, "all over" the capital district area of New York State. Before he was 10 years old, he was involved in sexual activities with neighborhood boys. Now and then, the boys also forced neighborhood girls into petting, and Michael had mixed feelings when the girls got upset. He felt bad for them, but he also enjoyed it. A couple of times girls seemed horrified at the sight of his penis, and it made him "really feel like a man. To see that look, you know, with a girl, not a woman, but a girl—a slender girl, that's what I'm after."

—From the Authors' Files

Truth OR Fiction? REVISITED

Wearing revealing bathing suits is a form of exhibitionism, according to clinical criteria.

FALSE. Wearing revealing bathing suits is not a form of exhibitionism in the clinical sense of the term. Virtually all people diagnosed with the disorder are men, and they are motivated by the wish to shock and dismay unsuspecting observers, not to show off the attractiveness of their bodies.

Wearing revealing bathing suits is not a form of exhibitionism in the clinical sense of the term. Virtually all people diagnosed with the disorder are men, and they are motivated by the wish to shock and dismay unsuspecting observers, not to show off the attractiveness of their bodies. Nor do professional strippers typically meet the clinical criteria for exhibitionism. Although they may seek to show off the attractiveness of their bodies, they are generally not motivated by the desire to expose themselves to unsuspecting strangers in order to arouse them or shock them. The chief motive of the stripteaser, of course, may simply be to earn a living.

Fetishism

The French word *fétiche* is thought to derive from the Portuguese *feitico,* referring to a "magic charm." In this case, the "magic" lies in the object's ability to arouse sexually. The chief feature of **fetishism** is recurrent, powerful sexual urges and arousing fantasies involving inanimate objects, such as an article of clothing (bras, panties, hosiery, boots, shoes, leather, silk, and the like). It is normal for men to like the sight, feel, and smell of their lovers' undergarments. Men with fetishism, however, may prefer the object to the person and may not be able to become sexually aroused without it. They often experience sexual gratification by masturbating while fondling the object, rubbing it, or smelling it; or by having their partners wear it during sexual activity.

The origins of fetishism may be traced to early childhood in many cases. Most individuals with a rubber fetish in one research sample were able to recall first experiencing a fetishistic attraction to rubber sometime between the age of 4 and 10 (Gosselin & Wilson, 1980).

Origins of fetishism? The conditioning model of the origins of fetishism suggests that men who develop fetishisms involving women's undergarments may have had experiences in childhood in which sexual arousal was repeatedly paired with exposure to their mother's undergarments. The developing fetish may have been strengthened by eroticizing the meaning of these stimuli by incorporating them within erotic fantasies or masturbatory activity.

Transvestic Fetishism

The chief feature of **transvestic fetishism** is recurrent, powerful urges and related fantasies involving cross-dressing for purposes of sexual arousal. Other people with fetishisms can be satisfied by handling objects such as women's clothing while they masturbate; people with transvestic fetishism want to wear them. They may wear full feminine attire and makeup or favor one particular article of clothing, such as women's stockings. Transvestic fetishism is reported only among heterosexual men. Typically, the man cross-dresses in private and imagines himself to be a woman who he is stroking as he masturbates. Some frequent transvestite clubs or become involved in a transvestic subculture.

Gay men may cross-dress to attract other men or because it is fashionable to masquerade as women in some social circles, not because they are sexually aroused by cross-dressing. Males with gender identity disorder cross-dress because of gender discomfort associated with wearing men's clothing. Because cross-dressing among gay men and men with gender identity disorder is performed for other reasons than sexual arousal or gratification, it is not considered a form of transvestic fetishism. Nor are female impersonators who cross-dress for theatrical purposes considered to have a form of transvestism. For reasons such as these, the diagnosis is limited to heterosexuals.

Most men with transvestism are married and engage in sexual activity with their wives, but they seek additional sexual gratification through dressing as women, as in the following case example:

WWW **Web Link 12.3**
National Council of Sexual Addiction and Compulsivity

A Case of Transvestic Fetishism

Archie was a 55-year-old plumber who had been cross-dressing for many years. There was a time when he would go out in public as a woman, but as his prominence in the community grew, he became more afraid of being discovered in public. His wife Myrna knew of his "peccadillo," especially since he borrowed many of her clothes, and she also encouraged him to stay at home, offering to help him with his "weirdness." For many years, his paraphilia had been restricted to the home. The couple came to the clinic at the urging of the wife. Myrna described how Archie had imposed his will on her for 20 years. Archie would wear her undergarments and masturbate while she told him how disgusting he was. (The couple also regularly engaged in "normal" sexual intercourse, which Myrna enjoyed.) The cross-dressing situation had come to a head because a teenaged daughter had almost walked into the couple's bedroom while they were acting out Archie's fantasies.

With Myrna out of the consulting room, Archie explained how he grew up in a family with several older sisters. He described how underwear had been perpetually hanging all around the one bathroom to dry. As an adolescent, Archie experimented with rubbing against articles of underwear, then with trying them on. On one occasion a sister walked in while he was modeling panties before the mirror. She told him he was a "dredge to society" and he straightaway experienced unparalleled sexual excitement. He masturbated when she left the room, and his orgasm was the strongest of his young life.

Archie did not think that there was anything wrong with wearing women's undergarments and masturbating. He was not about to give it up, regardless of whether his marriage was destroyed as a result. Myrna's main concern was finally separating herself from Archie's "sickness." She didn't care what he did anymore, so long as he did it by himself. "Enough is enough," she said.

That was the compromise the couple worked out in marital therapy. Archie would engage in his fantasies by himself. He would choose times when Myrna was not at home, and she would not be informed of his activities. He would also be very, very careful to choose times when the children would not be around.

Six months later the couple were together and content. Archie had replaced Myrna's input into his fantasies with transvestic-sadomasochistic magazines. Myrna said, "I see no evil, hear no evil, smell no evil." They continued to have sexual intercourse. After a while, Myrna even forgot to check to see which underwear had been used.

—From the Authors' Files

Voyeurism

The chief feature of **voyeurism** is either acting on or being strongly distressed by recurrent, powerful sexual urges and related fantasies involving watching unsuspecting people, generally strangers, who are undressed, disrobing, or engaging in sexual activity. The purpose of

voyeurism A paraphilia characterized by recurrent sexual urges involving watching unsuspecting others in sexual situations.

Truth OR Fiction? REVISITED

Becoming aroused while watching your partner disrobe or viewing an explicit movie falls within the clinical definition of voyeurism.

FALSE. Becoming sexually aroused while watching your partner disrobe is not voyeurism in the clinical sense of the term. Voyeurism involves watching unsuspecting persons disrobing or engaging in sexual activities.

watching, or "peeping," is to attain sexual excitement. The person who engages in voyeurism does not typically seek sexual activity with the person or persons being observed.

Are the acts of watching one's partner disrobe or viewing sexually explicit films forms of voyeurism? The answer is no. The people who are observed know they are being observed by their partners or will be observed by film audiences. Voyeuristic acts involve watching unsuspecting persons disrobing or engaging in sexual activities. Note that feelings of sexual arousal while watching our partners undress or observing sex scenes in R- and X-rated films fall within the normal spectrum of human sexuality.

During voyeuristic acts, the person usually masturbates while watching or while fantasizing about watching. Peeping may be the person's exclusive sexual outlet. Some people engage in voyeuristic acts in which they place themselves in risky situations. The prospect of being found out or injured apparently heightens the excitement.

Frotteurism

The French word *frottage* refers to the artistic technique of making a drawing by rubbing against a raised object. The chief feature of the paraphilia of **frotteurism** is recurrent, powerful sexual urges and related fantasies involving rubbing against or touching a nonconsenting person. Frotteurism or "mashing" generally occurs in crowded places, such as subway cars, buses, or elevators. It is the rubbing or touching, not the coercive aspect of the act, that is sexually arousing to the man. He may imagine himself enjoying an exclusive, affectionate sexual relationship with the victim. Because the physical contact is brief and furtive, people who commit frotteuristic acts stand only a small chance of being caught by authorities. Even the victims may not realize at the time what has happened or register much protest (Spitzer et al., 1989). In the following case example, a man victimized about 1,000 women over a period of years but was arrested only twice:

A Case of Frotteurism

A 45-year-old man was seen by a psychiatrist following his second arrest for rubbing against a woman in the subway. He would select as his target a woman in her 20s as she entered the subway station. He would then position himself behind her on the platform and wait for the train to arrive. He would then follow her into the subway car and when the doors closed would begin bumping against her buttocks, while fantasizing that they were enjoying having intercourse in a loving and consensual manner. About half of the time he would reach orgasm. He would then continue on his way to work. Sometimes when he hadn't reached orgasm, he would change trains and seek another victim. While he felt guilty for a time after each episode, he would soon become preoccupied with thoughts about his next encounter. He never gave any thought to the feelings his victims might have about what he had done to them. While he was married to the same woman for 25 years, he appears to be rather socially inept and unassertive, especially with women.

—Adapted from Spitzer et al., 1994, pp. 164–165; reprinted from Nevid, Fichner-Rathus, & Rathus, 1995, p. 570

frotteurism A paraphilia characterized by recurrent sexual urges involving bumping or rubbing against nonconsenting others for sexual gratification.

pedophilia A paraphilia involving recurrent, powerful sexual urges and related fantasies involving sexual activity with prepubescent children.

Pedophilia

Pedophilia derives from the Greek *paidos,* meaning "child." The chief feature of pedophilia is recurrent, powerful sexual urges and related fantasies involving sexual activity with prepubescent children (typically 13 years old or younger). Molestation of children may or may not occur. To be diagnosed with pedophilia, the person must be at least 16 years of age and at least 5 years older than the child or children toward whom the person is sexually attracted or has victimized. In some cases of pedophilia, the person is attracted only to children. In other cases, the person is attracted to adults as well.

Although some persons with pedophilia restrict their pedophilic activity to looking at or undressing children, others engage in exhibitionism, kissing, fondling, oral sex, and anal intercourse and, in the case of girls, vaginal intercourse (Knudsen, 1991). Not being worldly wise, children are often taken advantage of by molesters who inform them they are "educating" them, "showing them something," or doing something they will "like." Some men with pedophilia limit their sexual activity with children to incestuous relations with family members; others only molest children outside the family. Not all child molesters have pedophilia, however. The clinical definition of pedophilia is brought to bear only when sexual attraction to children is recurrent and persistent. Some molesters engage in these acts or experience pedophilic urges only occasionally or during times of opportunity.

Despite the stereotype, most cases of pedophilia do not involve "dirty old men" who hang around schoolyards in raincoats. Men with this disorder (virtually all cases involve men) are usually (otherwise) law-abiding, respected citizens in their 30s or 40s. Most are married or divorced and have children of their own. They are usually well acquainted with their victims, who are typically either relatives or friends of the family. Many cases of pedophilia are not isolated incidents. They may be a series of acts that begin when children are very young and continue for many years until they are discovered or the relationship is broken off (Finkelhor et al., 1990).

The origins of pedophilia are complex and varied. Some cases fit the stereotype of the weak, shy, socially inept, and isolated man who is threatened by mature relationships and turns to children for sexual gratification because children are less critical and demanding (Ames & Houston, 1990). In other cases, it may be that childhood sexual experiences with other children were so enjoyable that the man, as an adult, is attempting to recapture the excitement of earlier years. Or perhaps in some cases of pedophilia, men who were sexually abused in childhood by adults may now be reversing the situation in an effort to establish feelings of mastery. Men whose pedophilic acts involve incestuous relationships with their own children tend to fall at one extreme or the other on the dominance spectrum, either being very dominant or passive (Ames & Houston, 1990).

THINK ABOUT IT

Do you believe exhibitionists, voyeurs, and pedophiles should be punished, treated, or both? Do you think that people with these different types of paraphilias should be treated differently by society? Explain.

Sexual Masochism

Sexual masochism derives its name from the Austrian novelist Leopold Ritter von Sacher-Masoch (1836–1895), who wrote stories and novels about men who sought sexual gratification from women inflicting pain on them, often in the form of flagellation (being beaten or whipped). Sexual masochism involves strong, recurrent urges and fantasies relating to sexual acts that involve being humiliated, bound, flogged, or made to suffer in other ways. The urges are either acted on or cause significant personal distress. In some cases of sexual masochism, the person cannot attain sexual gratification in the absence of pain or humiliation.

Truth OR Fiction? REVISITED

Some people cannot become sexually aroused without pain or humiliation.

TRUE. Some people with sexual masochism cannot become sexually aroused unless they are subjected to pain or humiliation by others.

In some cases, sexual masochism involves binding or mutilating oneself during masturbation or sexual fantasies. In others, a partner is engaged to restrain (bondage), blindfold (sensory bondage), paddle, or whip the person. Some partners are prostitutes; others are consensual partners who are asked to perform the sadistic role. In some cases, the person may desire, for purposes of sexual gratification, to be urinated or defecated upon or subjected to verbal abuse.

A most dangerous expression of masochism is **hypoxyphilia,** in which participants are sexually aroused by being deprived of oxygen—for example by using a noose, plastic bag, chemical, or pressure on the chest during a sexual act, such as masturbation. The oxygen deprivation is usually accompanied by fantasies of asphyxiating or being asphyxiated by a lover. People who engage in this activity generally discontinue it before they lose consciousness, but occasional deaths due to suffocation have resulted from miscalculations (Blanchard & Hucker, 1991).

sexual masochism A paraphilia characterized by recurrent, powerful sexual urges and fantasies involving receiving humiliation or pain.

hypoxyphilia A form of sexual masochism in which a person seeks sexual gratification by being deprived of oxygen.

sexual sadism A paraphilia characterized by recurrent, powerful sexual urges and fantasies involving inflicting humiliation or pain.

sadomasochism Sexual activities involving the attainment or gratification by means of inflicting and receiving pain and humiliation

Sexual Sadism

Sexual sadism is named after the infamous Marquis de Sade, the 18th-century Frenchman who wrote stories about the pleasures of achieving sexual gratification by inflicting pain or humiliation on others. Sexual sadism is the flip side of sexual masochism. It involves recurrent, powerful urges and related fantasies to engage in acts in which the person is sexually aroused by inflicting physical suffering or humiliation on another person. People with this paraphilia either act out their fantasies or are disturbed by them. They may recruit consenting partners, who may be lovers or wives with a masochistic streak, or prostitutes. Still others stalk and assault nonconsenting victims and become aroused by inflicting pain or suffering on their victims. Sadistic rapists fall into this last group. Most rapists, however, do not seek to become sexually aroused by inflicting pain on their victims; they may even lose sexual interest when they see their victims in pain.

Many people have occasional sadistic or masochistic fantasies or engage in sex play involving simulated or mild forms of **sadomasochism** with their partners. Sadomasochism describes a mutually gratifying sexual interaction involving both sadistic and masochistic acts. Simulation may take the form of using a feather brush to strike one's partner, so that no actual pain is administered. People who engage in sadomasochism frequently switch roles during their encounters or from one encounter to another. The clinical diagnosis of sexual masochism or sadism is not usually brought to bear unless such people become distressed by their behavior or fantasies, or act them out in ways that are harmful to themselves or others.

Other Paraphilias

There are many other paraphilias. These include making obscene phone calls ("telephone scatologia"), necrophilia (sexual urges or fantasies involving sexual contact with corpses), partialism (sole focus on part of the body), zoophilia (sexual urges or fantasies involving sexual contact with animals), and sexual arousal associated with feces (coprophilia), enemas (klismaphilia), and urine (urophilia).

Theoretical Perspectives

Psychodynamic theorists see many paraphilias as defenses against leftover castration anxiety from the Oedipal period. The thought of the penis disappearing within the vagina is unconsciously equated with castration. The man who develops a paraphilia may avoid this threat of castration anxiety by displacing sexual arousal into safer activities—for example, undergarments, children, or watching others. By sequestering his penis under women's clothes, the man with transvestic fetishism engages in a symbolic act of denial that women do not have penises, which eases castration anxiety by unconsciously providing evidence of women's (and his own) safety. The shock and dismay shown by the victim of a man who exposes himself provides unconscious reassurance that he does, after all, have a penis. Sadism involves an unconscious identification with the man's father—the "aggressor" of his Oedipal fantasies—and relieves anxiety by giving him the opportunity to enact the role of the castrator. Some psychoanalytic theorists see masochism as a way of coping with conflicting feelings about sex. Basically, the man feels guilty about sex, but is able to enjoy it so long as he is being punished for it. Others view masochism as the redirection inward of aggressive impulses originally aimed at the powerful, threatening father. Like the child who is relieved when his inevitable punishment is over, the man gladly accepts humiliation and punishment in place of castration. These views remain speculative and controversial. We lack any direct evidence that men with paraphilias are handicapped by unresolved castration anxiety.

Learning theorists explain paraphilias in terms of conditioning and observational learning. Some object or activity becomes inadvertently associated with sexual arousal. The object or activity then gains the capacity to elicit sexual arousal. For example, a boy who glimpses his mother's stockings on the towel rack while he is masturbating may go on to develop a fetish for stockings (Breslow, 1989). Orgasm in the presence of the object reinforces the erotic connection, especially when it occurs repeatedly. Yet if fetishes were acquired by mechanical association, we might expect people to develop fetishes to stimuli that are inadvertently and repeatedly connected with sexual activity, such as bedsheets, pillows, even ceilings. But they do not. The *meaning* of the stimulus plays a primary role. The development of fetishes may depend on eroticizing certain types of stimuli (such as women's undergarments) by incorporating them within sexual fantasies and masturbation rituals.

Fetishes can often be traced to early childhood. Consider the development of rubber fetishes. Reinisch (1990) speculates that the earliest awareness of sexual arousal or response (such as erection) may have been connected with rubber pants or diapers such that an association was made between the two, setting the stage for the development of the fetish.

Like other patterns of abnormal behavior, paraphilias may involve multiple biological, psychological, and sociocultural factors. Money and Lamacz (1990) hypothesize a multifactorial model that traces the development of paraphilias to childhood. They suggest that childhood experiences etch a pattern, or "lovemap," which can be likened to a software program in the brain that determines the kinds of stimuli and behaviors that come to arouse people sexually. In the case of paraphilias, lovemaps become "vandalized" by early traumatic experiences, such as incest, physical abuse, neglect, or excessively harsh antisexual child rearing. Yet not all children who undergo such experiences develop paraphilias. Nor do all people with paraphilias have such traumatic experiences. Perhaps some children are more vulnerable to developing distorted lovemaps than others. The precise nature of such vulnerability remains to be defined.

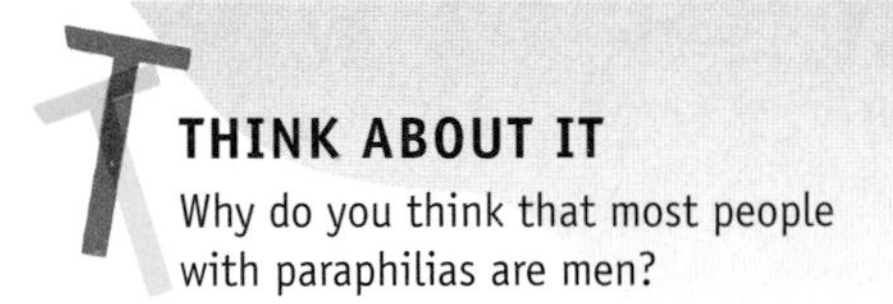

THINK ABOUT IT

Why do you think that most people with paraphilias are men?

Treatment of Paraphilias

People with paraphilia don't typically seek treatment on their own. They usually receive treatment in prison after they have been convicted of a sexual offense. Or they may be referred to a treatment provider by the courts. Under these circumstances, it is not surprising that sex offenders are often resistant or recalcitrant to treatment. Therapists recognize that treatment may be futile when clients lack the motivation to change their behavior. Nonetheless, evidence suggests that some forms of treatment, principally behavior therapy and cognitive behavioral therapy, may be helpful to sex offenders who seek to change their behavior.

One behavioral technique used in treating paraphilias is aversive conditioning. The goal of treatment is to induce a negative emotional response to inappropriate stimuli or fantasies. In this technique, a stimulus that elicits sexual arousal (for example, panties) is paired repeatedly with an aversive stimulus (for example, electric shock) in the hope that the stimulus will acquire aversive properties. A basic limitation of aversive conditioning is that it does not help the individual acquire more adaptive behaviors in place of maladaptive response patterns. This may explain why researchers find that a broad-based, cognitive-behavioral program for treating exhibitionism that emphasized the development of adaptive thoughts, the building of social skills, and the development of stress management skills was more effective than an alternative program based on aversion therapy (Marshall, Eccles, & Barbaree, 1991). *Covert sensitization* is a variation of aversive conditioning in which the pairing of an aversive stimulus and the problem behavior occurs in imagination.

Maletzky (1991, 1998) reported on the success rates of the largest treatment program study to date, based on more than 7,000 cases of rapists and sex offenders with paraphilias. Treatment incorporated a variety of behavioral techniques, including aversive conditioning and nonaversive methods, to help individuals acquire more adaptive behaviors. Though success rates exceeding 80% were reported, the criteria for success depended on part on self-reports of an absence of deviant sexual interests or behavior. As we know, self-reports may be biased, especially in offender groups. Secondly, lacking a control group, we cannot discount the possibility that other factors, such as fears of legal consequences, influenced the outcome.

Some promising results are also reported in using the antidepressant Prozac in treating voyeurism and fetishism (Lorefice, 1991; Perilstein, Lipper, & Friedman, 1991). Why Prozac? Prozac has been used effectively in treating obsessive-compulsive disorder (see Chapter 6). Researchers speculate that paraphilias may fall within an obsessive-compulsive spectrum. Many people with paraphilias report feeling compelled to carry out paraphilic acts, in much the same way that people with obsessive-compulsive disorder feel driven to perform compulsive acts. Paraphilias also tend to have an obsessional quality. The person experiences intrusive, repetitive urges to engage in paraphilic acts or thoughts that relate to the paraphilic object or situation. However, Maletzky (1998) cautions that these drugs may act to reduce sexual drives, rather than specifically target deviant sexual fantasies.

Quiz 12.3 Paraphilias

Overview of Paraphilias

MAJOR TYPES OF PARAPHILIAS	Atypical or deviant patterns of sexual gratification; excepting masochism, these disorders occur almost exclusively among males
Exhibitionism	Sexual gratification from exposing one's genitals in public
Voyeurism	Sexual gratification from observing unsuspecting others who are naked, undressing, or engaging in sexual arousal
Sexual masochism	Sexual gratification associated with the receipt of humiliation or pain
Fetishism	Sexual attraction to inanimate objects or particular body parts
Frotteurism	Sexual gratification associated with acts of bumping or rubbing against nonconsenting strangers
Sexual sadism	Sexual gratification associated with inflicting humiliation or pain on others
Transvestic fetishism	Sexual gratification associated with cross-dressing
Pedophilia	Sexual attraction to children
CAUSAL FACTORS	**Multiple causes may be involved**
Learning Perspective	• Atypical stimuli become conditioned stimuli for sexual arousal as the result of prior pairing with sexual activity • Atypical stimuli may become eroticized by incorporating them within erotic and masturbatory fantasies
Psychodynamic Perspective	• Unresolved castration anxiety from childhood leads to sexual arousal being displaced onto safer objects or activities
Multifactorial Perspective	• Sexual or physical abuse in childhood may corrupt normal sexual arousal patterns
TREATMENT APPROACHES	**Results remain questionable**
Biomedical Treatment	• Antidepressants to help individuals control deviant sexual urges or reduce sexual drives
Cognitive-Behavioral Therapy	• Including aversive conditioning (pairing deviant stimuli with aversive stimuli), covert sensitization (pairing the undesirable behavior with an aversive stimulus in imagination), and nonaversive methods that help individuals acquire more adaptive behaviors

Sexual Dysfunctions

sexual dysfunctions Persistent problems with sexual interest, arousal, or response.

Sexual dysfunctions involve problems with sexual interest, arousal, or response. Sexual dysfunctions are widespread in our society, affecting 43% of women and 31% of men, according to a recent national survey (Laumann, Paik, & Rosen, 1999). They are often significant sources of distress to the affected person and his or her partner. There are various types of sexual dysfunctions, but they tend to share some common features, as outlined in Table 12.2. Table 12.3 shows the estimated rates of several major types of sexual dysfunction based on recent community samples.

Some cases of sexual dysfunction have existed throughout the individual's lifetime, and are thus labeled *lifelong dysfunctions.* In the case of an *acquired dysfunction,* the problem begins following a period (or at least one occurrence) of normal functioning. In the case of a *situational dysfunction,* the problem occurs in some situations (for example, with one's spouse), but not in others (for example, with a lover or when masturbating), or at some times but not others. In the case of a *generalized dysfunction,* the problem occurs in all situations and at all times the individual engages in sexual activity.

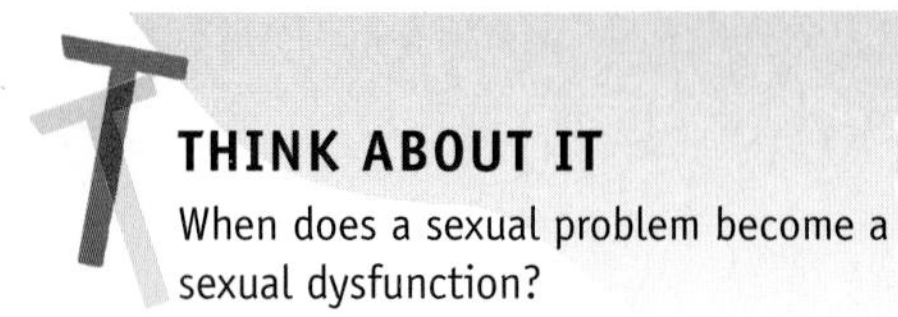

THINK ABOUT IT

When does a sexual problem become a sexual dysfunction?

To provide perspective on the sexual dysfunctions, we first describe normal patterns of sexual response. Then we explore the various types of sexual dysfunctions and the methods used to treat them.

TABLE 12.2 Common Features of Sexual Dysfunctions

Feature	Description
Fear of failure	Fears relating to failure to achieve or maintain erection or failure to reach orgasm.
Assumption of a spectator role rather than a performer role	Monitoring and evaluating your body's reactions during sex.
Diminished self-esteem	Thinking less of yourself for failure to meet your standard of normality.
Emotional effects	Guilt, shame, frustration, depression, anxiety.
Avoidance behavior	Avoiding sexual contacts for fear of failure to perform adequately; making excuses to your partner.

Source. From Nevid, Fichner-Rathus, & Rathus (1995), p. 445. Reprinted with permission.

TABLE 12.3 Estimated Prevalence of Various Current Sexual Dysfunctions (percentage of respondents reporting any problem)

Premature ejaculation	36–38
Erectile dysfunction	4–9
Male orgasmic disorder	4–10
Female orgasmic disorder	5–10

Source. Adapted from Spector, I. M., & Carey, M. P. (1990). Incidence and prevalence of the sexual dysfunctions: A critical review of the empirical evidence. *Archives of Sexual Behavior, 19,* 389–408.

When a source of pleasure becomes a source of anxiety. Sexual dysfunctions can be a source of intense personal distress and lead to friction between partners. Problems in communication can give rise to or exacerbate sexual dysfunctions.

The Sexual Response Cycle

Sexual dysfunctions interfere with the initiation or completion of the sexual response cycle. Much of our understanding of the sexual response cycle is based on the pioneering work of sex researchers William Masters and Virginia Johnson. Elaborating on their work and others, such as the sex therapist Helen Singer Kaplan, the *DSM* describes the sexual response cycle in terms of four distinct phases:

1. *Appetitive Phase.* This phase involves sexual fantasies and the desire to engage in sexual activity. The occurrence of sexual fantasies and desires are quite normal; the question is, "How much (or how little) sexual interest is normal?"
2. *Excitement Phase.* This phase involves the physical changes and feelings of pleasure that occur during the process of sexual arousal. In response to sexual stimulation, the heart rate, respiration rate, and blood pressure increase. Sexual excitement involves two essential sexual reflexes—erection in the man and vaginal lubrication ("wetness") in the woman. In men, erection occurs as blood vessels in chambers of loose tissue within the penis dilate to permit an increased blood flow to expand the tissues. In women, the breasts swell and the nipples become erect. Blood engorges the genitals, causing the clitoris to expand. The vagina lengthens and dilates, and lubrication appears as the engorgement of the blood vessels in the vagina forces moisture through capillary membranes.
3. *Orgasm Phase.* In both men and women, the building up of sexual tension reaches a peak and is released through involuntary rhythmic contractions of the pelvic muscles that are accompanied by feelings of pleasure. Orgasm, like erection and lubrication, is a reflex. In men, the contraction of the pelvic muscles forces semen to be expelled through the tip of the penis during ejaculation. In women, the pelvic muscles surrounding the outer third of the vagina contract reflexively. In men and women, the first contractions are strongest and spaced at 0.8-second intervals (five contractions in 4 seconds). Subsequent contractions are weaker and spread farther apart.

   People cannot will or force an orgasm. Nor can they will or force other sexual reflexes, such as erection and vaginal lubrication. We can only set the stage for these sexual responses and let them happen. Setting the stage for orgasm involves receiving adequate sexual stimulation and having an accepting attitude toward sexual pleasure. But trying to force an orgasm is likely to prevent it from happening.
4. *Resolution Phase.* Relaxation and a sense of well-being occur. During this phase, men are physiologically incapable of achieving erection and orgasm for a period of time. Women, however, may be able to maintain a high level of sexual excitement with continued stimulation, and experience multiple orgasms in swift succession. During the sexual revolution of the 1960s and 1970s, awareness of this capacity for multiple orgasm caused some women to think they ought not to be satisfied with just one orgasm. This is the flip side of the old saw that sexual enjoyment is appropriate for men only. In sex, as in other areas of life, oughts and shoulds are often arbitrary demands that elicit feelings of anxiety and inadequacy.

Truth OR Fiction? REVISITED

Orgasm is a reflex.

TRUE. Orgasm is a reflex. People cannot will or force an orgasm. Nor can they will or force other sexual reflexes, such as erection and vaginal lubrication. However, they can set the stage for these sexual responses and let them happen naturally.

Types of Sexual Dysfunctions

The *DSM-IV* groups most sexual dysfunctions within the following categories:

1. Sexual desire disorders
2. Sexual arousal disorders

3. Orgasm disorders
4. Sexual pain disorders

The first three categories correspond to the first three phases of the sexual response cycle.

Sexual Desire Disorders Sexual desire disorders involve disturbances in sexual appetite or an aversion to genital sexual activity. People with **hypoactive sexual desire disorder** have an absence or lack of sexual interest or desire. Typically there is either a complete or virtual absence of sexual fantasies. However, clinicians have not reached any universally agreed-upon criteria for determining the level of sexual desire that is considered normal (J. G. Beck, 1995). Individual clinicians must weigh various factors in reaching a diagnosis in cases of low sexual desire, such as the client's lifestyle (for example, in parents contending with the demands of infants or young children, lack of sexual energy or interest is to be expected), sociocultural factors (for example, culturally restrictive attitudes may restrain sexual desire or interest), the quality of the relationship between the client and her or his partner (declining sexual interest or activity may reflect relationship problems rather than diminished drive), and the client's age (desire normally declines but does not disappear with increasing age). Couples usually seek help when one or both partners recognize that the level of sexual activity in the relationship is deficient or has waned to the point that little desire or interest remains. Sometimes the lack of desire is limited to one partner. In other cases, both partners may feel sexual urges, but anger and conflict concerning other issues inhibit sexual interaction. Giving lie to the myth that men are always ready for sex, the numbers of men presenting with hypoactive sexual desire disorder appears to be on the rise (Letourneau & O'Donohue, 1993).

People with **sexual aversion disorder** have a strong aversion to genital sexual contact and avoid all or nearly all genital contact with a partner. They may, however, desire and enjoy affectionate contact or nongenital sexual contact. Their disgust with any form of genital contact may stem from childhood sexual abuse, rape, or other traumatic experiences. In other cases, deep-seated feelings of sexual guilt or shame may impair sexual response. In men, the diagnosis is often connected with a history of erectile failure (Spark, 1991). Such men may associate sexual opportunities with failure and shame. Their partners may also develop aversions to sexual contact because their sexual contacts have been so frustrating or emotionally painful.

Sexual Arousal Disorders Disorders of sexual arousal involve an inability to achieve or maintain the physiological responses involved in sexual arousal or excitement—vaginal lubrication in the woman or penile erection in the man—that are needed to allow completion of sexual activity.

In women, sexual arousal is characterized by lubrication of the vaginal walls that makes entry by the penis possible. In men, sexual arousal is characterized by erection. Almost all women now and then have difficulty becoming or remaining lubricated. Almost all men have occasional difficulty attaining or maintaining an erection through intercourse. The diagnoses of **female sexual arousal disorder** and **male erectile disorder** (also called *sexual impotence* or *erectile dysfunction*) are reserved for persistent or recurrent problems in becoming genitally aroused.

Orgasm Disorders Orgasm or sexual climax is an involuntary reflex that results in rhythmic contractions of the pelvic muscles and is usually accompanied by feelings of intense pleasure. In men, these contractions are accompanied by expulsion of semen. There are three specific types of orgasm disorders: **female orgasmic disorder, male orgasmic disorder,** and **premature ejaculation.**

Orgasmic disorders are seen as a pattern of difficulty reaching orgasm, or inability to reach orgasm following a normal level of sexual interest and arousal. The clinician needs to make a judgment about whether there is an "adequate" amount and type of stimulation to achieve an orgasmic response. A broad range of normal variation in sexual response needs to be considered. Many women, for example, require direct clitoral stimulation (by means

hypoactive sexual desire disorder Persistent or recurring lack of sexual interest or sexual fantasies.

sexual aversion disorder A type of sexual dysfunction characterized by aversion to and avoidance of genital sexual contact.

female sexual arousal disorder A sexual dysfunction in women involving difficulty becoming sexually aroused or lack of sexual excitement or pleasure during sexual activity.

male orgasmic disorder A sexual dysfunction in males involving difficulty achieving orgasm following a normal pattern of sexual interest and excitement.

female orgasmic disorder A sexual dysfunction in women involving difficulty reaching orgasm or inability to reach orgasm following a normal level of sexual interest and arousal.

male erectile disorder A sexual dysfunction in men characterized by difficulty achieving or maintaining erection during sexual activity.

premature ejaculation A sexual dysfunction in men characterized by ejaculation following minimal sexual stimulation.

Truth OR Fiction? REVISITED

Premature ejaculation affects a relatively small proportion of men.

FALSE. Premature ejaculation is extremely common, affecting about one in three men.

Web Link 12.4
Online Sexual Disorders Screening for Women (NYU School of Medicine)

Web Link 12.5
Online Sexual Disorders Screening for Men (NYU School of Medicine)

dyspareunia Persistent or recurrent pain experienced during or following sexual intercourse.

vaginismus A sexual dysfunction characterized by persistent or recurring contraction of the muscles surrounding the vaginal opening, making intercourse difficult or impossible.

of stimulation by her own hand or her partner's) in order to achieve orgasm during vaginal intercourse. This should not be considered abnormal, since the clitoris, not the vagina, is the woman's most erotically sensitive organ.

In men, a pattern of difficulty achieving orgasm following a normal pattern of sexual interest and excitement is termed *male orgasmic disorder.* This disorder is relatively rare and has received very little attention in the clinical literature (Dekker, 1993; Rosen & Leiblum, 1995). Men with this problem can usually reach orgasm through masturbation but not through intercourse. Because of its infrequency, there are only a few isolated case studies on the problem (Rathus, 1978).

Premature ejaculation refers to a pattern of ejaculating with minimal sexual stimulation. It can occur prior to, upon, or shortly after penetration, but before the man desires it. Note the subjective elements. In making the diagnosis, the clinician weighs the man's age, the novelty of the partner, and the frequency of sexual activity. Occasional experiences of rapid ejaculation, such as when the man is with a new partner, has had infrequent sexual contacts, or is very highly aroused, fall within the normal spectrum. More persistent patterns of premature ejaculation would occasion a diagnosis of the disorder. About one in three men experience premature ejaculation (Spector & Carey, 1990).

Sexual Pain Disorders In **dyspareunia,** sexual intercourse is associated with recurrent pain in the genital region. The pain cannot be explained fully by an underlying medical condition and so is believed to have a psychological component. However, many, perhaps even most, cases of pain during intercourse are traceable to an underlying medical condition, such as insufficient lubrication or a urinary tract infection. The *DSM* classifies these cases under a different diagnostic label, "Sexual Dysfunction Due to Medical Condition."

Vaginismus involves an involuntary spasm of the muscles surrounding the vagina when vaginal penetration is attempted, making sexual intercourse painful or impossible.

Theoretical Perspectives

Like most psychological disorders, sexual dysfunctions reflect the interplay of biological, psychological, and other factors.

Biological Perspectives Many cases of sexual dysfunction stem from underlying biological factors or from a combination of biological and psychological factors (Carey, Wincze & Meisler, 1998). Deficient testosterone production and thyroid overactivity or underactivity are among the many biological conditions that can lead to impaired sexual desire (Kresin, 1993). Medical conditions can also impair sexual arousal in both men and women. Diabetes, for instance, is the most common organic cause of erectile dysfunction, with estimates indicating that half of diabetic men eventually suffer some degree of erectile dysfunction (Thomas & LoPiccolo, 1994). Diabetes may also impair sexual response in women, with decreased vaginal lubrication being the most common consequence.

Biological factors may play a prominent role in as many as 70% to 80% of cases of erectile dysfunction (Brody, 1995b). Other biological factors that can impair sexual desire, arousal, and orgasm include nerve-damaging conditions such as multiple sclerosis; lung disorders; kidney disease; circulatory problems; damage caused by sexually transmitted diseases; and side effects of various drugs (Brody, 1995b; Segraves, 1988; Spark, 1991). Yet, even in cases of sexual dysfunction that are traced to physical causes, emotional problems such as anxiety and depression and marital conflict can compound the problem.

The male sex hormone testosterone plays a pivotal role in energizing sexual desire and sexual activity in both men and women (Tuiten et al., 2000; Yates, 2000). Both men and women produce testosterone in their bodies, although women produce smaller amounts. Men with deficient production of testosterone may lose sexual interest and the capacity for erections (Kresin, 1993; Spark, 1991). The adrenal glands and ovaries are the sites where testosterone is produced in women. Women who have these organs surgically removed because of invasive disease no longer produce testosterone and may gradually lose sexual interest and the capacity for sexual response. Recent evidence shows that replacement testosterone can improve sexual functioning in such cases (Shifren et al., 2000). We should recognize, however, that most men and women with sexual dysfunctions have normal hormone levels (Spark, 1991).

Psychodynamic Perspectives Psychodynamic hypotheses generally revolve around presumed conflicts of the phallic stage (Fenichel, 1945). Mature genital sexuality is believed to require successful resolution of the Oedipus and Electra complexes. Men with sexual dysfunctions are presumed to suffer from unconscious castration anxiety. Sexual intercourse elicits an unconscious fear of retaliation by the father, rendering the vagina unsafe. Erectile dysfunction "saves" the man from having to enter the vagina. Premature ejaculation allows him to "escape" rapidly and may also represent unconscious hatred of women (Kaplan, 1974). Orgasmic disorder prevents him from completing the act and unconsciously minimizes his guilt and fear. Rapid ejaculation serves the unconscious purpose of expressing hatred through soiling the woman and denying her sexual pleasure.

In women, unresolved penis envy engenders hostility toward men. The woman who remains fixated in the phallic stage punishes her partner for having a penis by not permitting the organ to bring her pleasure, as in female sexual arousal disorder. The clamping down of the vaginal muscles in vaginismus may express an unconscious wish to castrate her partner (Kaplan, 1974). In orgasmic disorder, she has failed to overcome penis envy and to develop mature sexuality, which involves transferring erotic feelings from the clitoris to the vagina. She thus prevents orgasm from occurring through intercourse. It is difficult to test the validity of the psychoanalytic concepts because they involve unconscious conflicts, like castration anxiety and penis envy, which cannot be scientifically observed. Evidence for these views relies on case studies that involve interpretation of patients' histories. Case study accounts are open to rival interpretations, however. We can say with certainty, however, that despite the traditional psychoanalytic conception, clitoral stimulation remains a key part of the woman's erotic response as she matures and is not a sign of an immature fixation.

Learning Perspectives Learning theorists focus on the role of conditioned anxiety in the development of sexual dysfunctions. The occurrence of physically or psychologically painful experiences associated with sexual activity may cause a person to respond to sexual encounters with anxiety that is strong enough to counteract sexual pleasure and performance. A history of sexual abuse or rape plays a role in many cases in women of sexual arousal disorder, sexual aversion disorder, orgasmic disorder, and vaginismus. People who have been sexually traumatized earlier in life may find it difficult to respond sexually when they develop intimate relationships. They may be flooded with feelings of helplessness, unresolved anger, or misplaced guilt, or experience flashbacks of the abusive experiences when they engage in sexual relations with their partners, preventing them from becoming sexually aroused or achieving orgasm (see Chapter 16 for a discussion of the psychological effects of rape).

Sexual fulfillment is also based on learning sexual skills. Sexual skills or competencies, like other types of skills, are acquired through opportunities for new learning. We learn about how our bodies and our partners respond sexually in various ways, including trial and error with our partners, by learning about our own sexual response through self-exploration (as in masturbation), by reading about sexual techniques, and perhaps by talking to others or viewing sex films or videotapes. Yet children who are raised to feel guilty or anxious about sex may have lacked such opportunities to develop sexual knowledge and skills. Consequently, they may respond to sexual opportunities with feelings of anxiety and shame rather than arousal and pleasure.

Cognitive Perspectives Albert Ellis (1977b) points out that irrational beliefs and attitudes may contribute to sexual dysfunctions. Consider the irrational beliefs that we must have the approval at all times of everyone who is important to us and that we must be thoroughly competent at everything we do. If we cannot accept the occasional disappointment of others, we may catastrophize the significance of a single frustrating sexual episode. If we insist that every sexual experience be perfect, we set the stage for inevitable failure.

Truth OR Fiction? REVISITED

The male sex hormone testosterone is produced by the body in both women and men.

TRUE. Both men and women produce testosterone in their bodies, although women produce smaller amounts than men.

Evidence also points to a role for personal attributions, or perceived causes of events. Attributing the cause for erectile difficulty to oneself rather than to the situation may play an important role in determining future sexual functioning (Weisberg et al., 2001).

The prominent sex therapist Helen Singer Kaplan (1974) noted problems that can occur with our ability to regulate our levels of sexual arousal. Men who ejaculate prematurely, for example, may have difficulty gauging their level of sexual arousal. As a consequence, they may not be able to temporarily suspend stimulation in time to delay ejaculation.

Most men respond to sexual arousal with positive emotions, such as joy and warmth. But for men with sexual dysfunctions, sexual arousal becomes disconnected from positive emotions (Rowland, Cooper, & Slob, 1996). Psychologist David Barlow (1986) proposed that anxiety may have inhibiting or arousing effects on sexual response depending on the man's thought processes (see Figure 12.1). For men with sexual dysfunctions, anxiety has inhibiting effects. Perhaps because they expect to fail in sexual encounters, their thoughts are focused on anticipated feelings of shame and embarrassment rather than on erotic stimuli. Concerns about failing increase autonomic arousal or anxiety, which leads them to focus even more attention on the consequences of failure, which in turn leads to dysfunctional performance. Failure experiences in turn lead to avoidance of sexual encounters because these situations have become encoded as opportunities for repeated failure, frustration, and self-defeat. Functional men, by contrast, expect to succeed and focus their attention on erotic stimuli, not on fears of failure. Their erotic attentional focus increases autonomic arousal or anxiety, but not to the point that it interferes with their sexual response. Mild anxiety may actually enhance their sexual arousal. By focusing on erotic cues, functional men become more aroused, successfully engage in sexual activity, and heighten their expectations of future successful performance—all leading to increased approach tendencies.

The cognitive model formulated by Barlow highlights the role of interfering cognitions in sexual dysfunctions. Interfering cognitions include expectancies of failure that are evoked by performance demands. They heighten anxiety to the point of impairing sexual performance. In a vicious cycle, the more people focus on these interfering cognitions, the more difficult it will be for them to perform sexually—and the more likely they will be to focus on interfering cognitions in the future. Although the model was derived from research on men, Barlow believes it may also help explain sexual dysfunctions in women.

Problems in Relationships As the saying goes, "It takes two to tango." Sexual relations are usually no better than other facets of relationships or marriages (Perlman & Abramson, 1982). Couples who harbor resentments toward one another may choose the sexual arena for combat. Communication problems, moreover, are linked to general marital dissatisfaction. Couples who find it difficult to communicate their sexual desires may lack the means to help their partners become more effective lovers.

The following case illustrates how sexual arousal disorder may be connected with problems in the relationship:

The Case of Paul and Petula

After living together for six months, Paul and Petula are contemplating marriage. But a problem has brought them to a sex therapy clinic. As Petula puts it, "For the last two months he hasn't been able to keep his erection after he enters me." Paul is 26 years old, a lawyer; Petula, 24, is a buyer for a large department store. They both grew up in middle-class, suburban families, were introduced through mutual friends and began having intercourse, without difficulty, a few months into their relationship. At Petula's urging, Paul moved into her apartment, although he wasn't sure he was ready for such a step. A week later he began to have difficulty maintaining his erection during intercourse, although he felt strong desire for his partner. When his erection waned, he would try again, but would lose his desire and be unable to achieve another erection. After a few times like this, Petula would become so angry that she

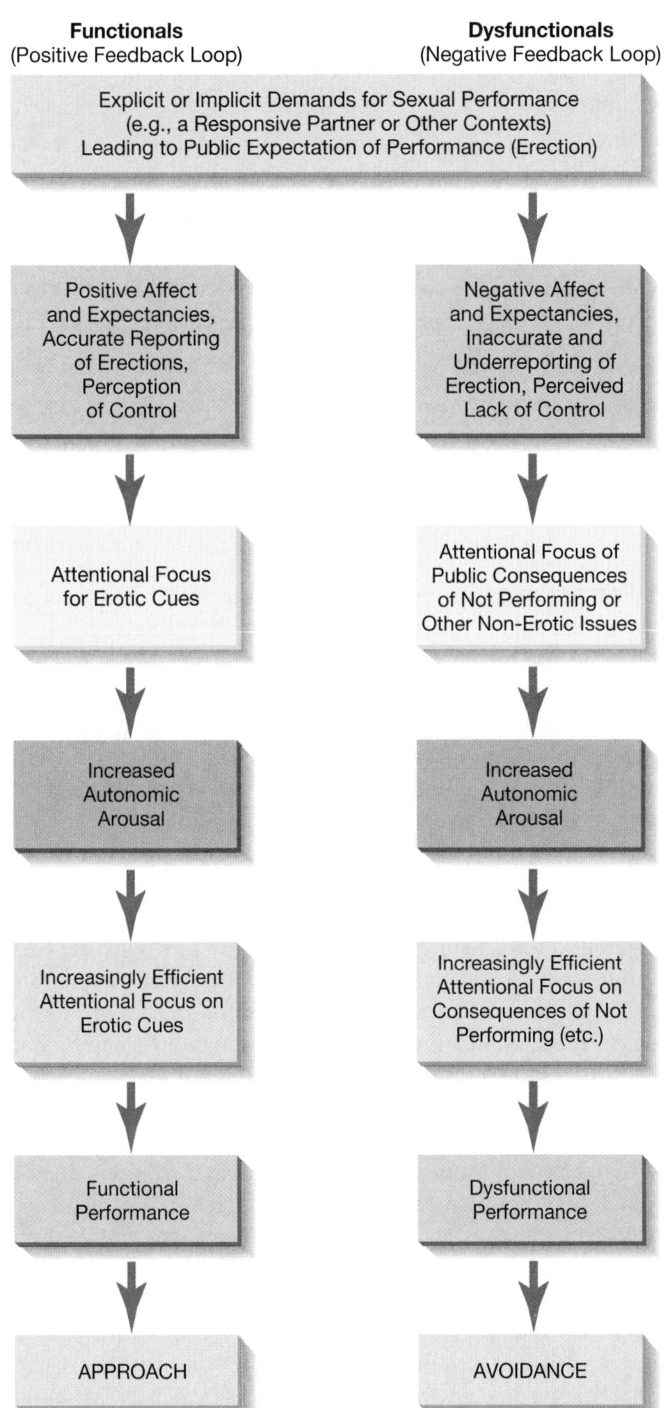

FIGURE 12.1 Barlow's model of erectile dysfunction.
In this model, past experience with erectile dysfunction leads men to expect that they will fail again. They consequently focus on anticipated feelings of shame and embarrassment when they engage in sexual relations rather than on erotic stimuli. These concerns heighten their anxiety, impairing their performance and distracting them from erotic cues. Functional men, by contrast, expect to succeed and focus more of their attention on erotic stimuli, which heightens their sexual response. Though they too may experience anxiety, it is not severe enough to distract them from erotic cues or impair their performance.

Source. Barlow, D. H. (1986), Causes of sexual dysfunction: The role of anxiety and cognitive interference. *Journal of Consulting and Clinical Psychology, 54*, p. 146.

began striking Paul in the chest and screaming at him. Paul, who at 200 pounds weighed more than twice as much as Petula, would just walk away, which angered Petula even more. It became clear that sex was not the only trouble spot in their relationship. Petula complained that Paul preferred to spend time with his friends and go to baseball games than to spend time with her. When together at home, he would become absorbed in watching sports events on television, and showed no interest in activities she enjoyed—attending the theater, visiting museums, etc. Since there was no evidence that the sexual difficulty was due to either organic problems or depression, a diagnosis of male erectile disorder was given. Neither Paul nor Petula was willing to discuss their nonsexual problems with a therapist. While the sexual problem was treated successfully with a form of sex therapy modeled after techniques developed by Masters and Johnson, and the couple later married, Paul's ambivalences continued, even well into their marriage, and there were future recurrences of sexual problems as well.

—Adapted from Spitzer et al., 1994, pp. 198–200

Sociocultural Perspectives Around the turn of the 20th century, an Englishwoman was quoted as saying she would "close her eyes and think of England" when her husband approached her for sexual relations. This old-fashioned stereotype suggests how sexual pleasure was once considered exclusively a male preserve—that sex, for women, was primarily a duty. Mothers usually informed their daughters of the conjugal duties before the wedding, and girls encoded sex as just one of the ways in which women serviced the needs of others. Women who harbor such stereotypical attitudes toward female sexuality may be unlikely to become aware of their sexual potentials. In addition, sexual anxieties may transform negative expectations into self-fulfilling prophecies. Sexual dysfunctions in men, too, may be linked to severely restricted sociocultural beliefs and sexual taboos.

Modern psychodynamic theorists recognize that anger and other negative feelings that women may hold toward men can lead to sexual dysfunctions. Yet they believe that these negative emotions stem from sociocultural factors rather than from penis envy. Women in our society are often socialized to sacrifice for and submit to their husbands, which may engender rebellion that finds expression through sexual dysfunctions.

Javier (1993) notes, for example, the idealization within many Hispanic cultures of the *marianismo* stereotype, which derives its name from the Virgin Mary. From this sociocultural perspective, the ideal virtuous woman "suffers in silence" as she submerges her needs and desires to those of her husband and children. She is the provider of joy, even in the face of her own pain or frustration. It is not difficult to imagine that some women who adopt these stereotypical expectations may find it difficult to assert their own needs for sexual gratification or may express resistance to this cultural ideal by becoming sexually unresponsive.

Sociocultural factors play an important role in erectile dysfunction as well. Investigators find a greater incidence of erectile dysfunction in cultures with more restrictive sexual attitudes toward premarital sex among females, toward sex in marriage, and toward extramarital sex (Welch & Kartub, 1978). Men in these cultures may be prone to develop sexual anxiety or guilt that may interfere with sexual performance.

In India, cultural beliefs that link the loss of semen to a draining of the man's life energy underlie the development of dhat syndrome, which involves an excessive fear of semen loss (see Chapter 7). Men with this condition sometimes develop erectile dysfunction because their fears about the risks of wasting precious seminal fluid interfere with their ability to perform sexually (Singh, 1985).

Psychological Factors Various psychological factors such as depression, anxiety, guilt, and low self-esteem can impair sexual interest or performance. One principal culprit is **performance anxiety,** a type of anxiety that involves an excessive concern about whether we will be able to perform successfully. People troubled by performance anxiety become

performance anxiety Fear relating to the threat of failure to perform adequately.

spectators during sex rather than performers. Their attention is focused on how their bodies are responding (or not responding) to sexual stimulation and on their concerns about the negative consequences of failing to perform adequately, rather than absorbing themselves in their erotic experiences. Men with performance anxiety may have difficulty achieving or maintaining an erection or may ejaculate prematurely; women may fail to become adequately aroused or have difficulty achieving orgasm. A vicious cycle may ensue in which each failure experience instills deeper doubts, which leads to more anxiety during sexual encounters, which occasions repeated failure, and so on.

In Western cultures, the connection between a man's sexual performance and his sense of manhood is deeply ingrained. The man who repeatedly fails to perform sexually may suffer a loss of self-esteem, become depressed, or feel he is no longer a man (Carey, Wincze, & Meisler, 1998). He may see himself as a total failure, despite other accomplishments in life. Sexual opportunities are construed as tests of his manhood, and he may respond to them by trying to will (force) an erection. Willing an erection may backfire, because erection is a reflex that cannot be forced. With so much of his self-esteem riding on the line whenever he makes love, it is little wonder that anxiety about the quality of his performance—performance anxiety—may mount to a point that it inhibits erection. The erectile reflex is controlled by the parasympathetic branch of the autonomic nervous system. Activation of the sympathetic nervous system, which occurs when we are anxious, can block parasympathetic control, preventing the erectile reflex from occurring. Ejaculation, in contrast, is under sympathetic nervous system control, so heightened levels of arousal, as in the case of performance anxiety, can trigger premature ejaculation.

Eugene, a client who suffered from erectile dysfunction, described his feelings of sexual inadequacy this way:

The Case of Eugene

I always felt inferior, like I was on probation, having to prove myself. I felt like I was up against the wall. You can't imagine how embarrassing this was. It's like you walk out in front of an audience that you think is a nudist convention and it turns out to be a tuxedo convention.

—From the Authors' Files

■

Another man, Timothy, described how performance anxiety led him to prepare for sexual relations as though he were psyching himself up for a big game:

The Case of Timothy

At work I have control over what I do. With sex, you don't have control over your sex organ. I know that my mind can control what my hands do. But the same is not true of my penis. I had begun to view sex as a basketball game. I used to play in college. When I would prepare for a game, I'd always be thinking, "Who was I guarding that night?" I'd try to psych myself up, sketching out in my mind how to play this guy, thinking through all possible moves and plays. I began to do the same thing with sex. If I were dating someone, I'd be thinking the whole evening about what might happen in bed. I'd always be preparing for the outcome. I'd sketch out in my mind how I was going to touch her, what I'd ask her to do. But all the time, right through dinner or the movies, I'd be worrying that I wouldn't get it up. I kept picturing her face and how disappointed she'd be. By the time we did go to bed, I was paralyzed with anxiety.

—From the Authors' Files

■

Women, too, may equate their self-esteem with their ability to reach frequent and intense orgasms. Yet when men and women try to bear down to will arousal or lubrication, or to force an orgasm, they may find that the harder they try, the more these responses elude them. Thirty or 40 years ago the pressures concerning sex often revolved around the issue "Should I or shouldn't I?" Today, however, the pressures for both men and women are often based more on achieving performance goals relating to proficiency at reaching orgasm and satisfying one's partner's sexual needs.

Sex Therapy

Until the groundbreaking research of sex researchers William Masters and Virginia Johnson in the 1960s, there was no effective treatment for most sexual dysfunctions. Psychoanalytic forms of therapy approached sexual dysfunctions indirectly, for example. It was assumed that sexual dysfunctions represented underlying conflicts, and the dysfunctions might abate if the underlying conflicts—the presumed causes of the dysfunctions—were resolved through psychoanalysis. A lack of evidence that psychoanalytic approaches reversed sexual dysfunctions led clinicians and researchers to develop other approaches that focus more directly on the sexual problems themselves.

Contemporary sex therapists assume that sexual dysfunctions can be treated by directly modifying the couple's sexual interactions and patterns of communication. Pioneered by Masters and Johnson, sex therapy employs a variety of relatively brief, cognitive-behavioral techniques that center on enhancing self-efficacy expectancies, improving a couple's ability to communicate, fostering sexual competencies (sexual knowledge and skills), and reducing performance anxiety. Therapists may also work with couples to help them iron out problems in the relationship that may impede sexual functioning. When feasible, both sex partners are involved in therapy. In some cases, however, individual therapy may be preferable, as we shall see.

THINK ABOUT IT

Can you think of examples in your own life in which you have been hampered by performance anxiety of one kind or another? What did you do about it?

Significant changes have occurred in the treatment of sexual dysfunctions in the past 20 years. There is greater emphasis now on the role of biological or organic factors in the development of sexual problems and greater use of medical treatments, such as the use of the drug *Viagra* in treating male erectile dysfunction (Rosen & Leiblum, 1995). But even men whose erectile problems can be traced to physical causes can benefit from sex therapy along with medical intervention (Carey, Wincze, & Meisler, 1998).

Let us briefly survey some of the more common sex therapy techniques for particular types of disorders.

Masters and Johnson. Sex therapists William Masters and Virginia Johnson.

Sexual Desire Disorders Sex therapists may try to help people with low sexual desire kindle their sexual appetite through the use of self-stimulation (masturbation) exercises together with erotic fantasies. Or in working with couples, the therapist might prescribe mutual pleasuring exercises the couple could perform at home or encourage them to expand their sexual repertoire in order to add novelty and excitement to their sex life. When a lack of sexual desire is connected with depression, the treatment would probably focus on relieving the underlying depression in the hope that sexual interest would rebound when the depression lifts. When problems of low sexual desire or sexual aversion appear to be rooted in deep-seated causes, sex therapist Helen Singer Kaplan (1987) recommended the use of insight-oriented approaches to help uncover and resolve underlying issues. Some cases of hypoactive sexual desire involve hormonal deficiencies, especially lack of testosterone (Rabkin, Wagner, & Rabkin, 2000). Testosterone replacement is effective only in the relatively few cases in which production of the hormone is truly deficient (Spark, 1991). A lack of sexual desire may also reflect relationship problems that may need to be addressed through couples therapy. Couples therapy might also be used when sexual aversion develops from problems in the relationship (Gold & Gold, 1993). In other cases of sexual aversion, a program of mutual pleasuring, beginning with partner stimulation in nongenital areas and gradually progressing to genital stimulation, may help desensitize fears about sexual contact.

Disorders of Arousal Women who have difficulty becoming sexually aroused and men with erectile problems are first educated to the fact that they need not "do" anything to become aroused. As long as their problems are psychological, not organic, they need only experience sexual stimulation under relaxed, nonpressured conditions, so that disruptive cognitions and anxiety do not inhibit reflexive responses.

Masters and Johnson have the couple counter performance anxiety by engaging in **sensate focus exercises.** These are nondemand sexual contacts—sensuous exercises that do not demand sexual arousal in the form of vaginal lubrication or erection. Partners begin by massaging one another without touching the genitals. The partners learn to "pleasure" each other and to "be pleasured" by means of following and giving verbal instructions and by guiding each other's hands. The method fosters both communication and sexual skills and countermands anxiety because there is no demand for sexual arousal. After several sessions, direct massage of the genitals is included in the pleasuring exercise. Even when obvious signs of sexual excitement are produced (lubrication or erection), the couple does not straightaway engage in intercourse, because intercourse might create performance demands. After excitement is achieved consistently, the couple engages in a relaxed sequence of other sexual activities, culminating eventually in intercourse. A number of similar sex therapy methods were employed in the following case:

Viagra. Viagra, the first drug approved to treat erectile dysfunction, became the fastest selling new drug in history after its introduction in 1998. Former senator and presidential candidate Bob Dole became a spokesperson for the drug and has appeared in many advertisements for it.

The Case of Victor

Victor P., a 44-year-old concert violinist, was eager to show the therapist reviews of his concert tour. A solo violinist with a distinguished orchestra, Victor's life revolved around practice, performances, and reviews. He dazzled audiences with his technique and the energy of his performance. As a concert musician, Victor had exquisite control over his body, especially his hands. Yet he could not control his erectile response in the same way. Since his divorce seven years earlier, Victor had been troubled by recurrent episodes of erectile failure. Time and time again he had become involved in a new relationship, only to find himself unable to perform sexually. Fearing repetition, he would sever the relationship. He was unable to face an audience of only one. For a while he dated casually, but then he met Michelle.

Michelle was a writer who loved music. They were a perfect match because Victor, the musician, loved literature. Michelle, a 35-year-old divorcée, was exciting, earthy, sensual, and accepting. The couple soon grew inseparable. He would practice while she would write—poetry mostly, but also short magazine pieces. Unlike some women Victor met who did not know Bach from Bartok, Michelle held her own in conversations with Victor's friends and fellow musicians over late night dinner at Sardi's. They kept their own apartments; Victor needed his own space and solitude for practice.

In the nine months of their relationship, Victor was unable to perform on the stage that mattered most to him—his canopied bed. It was just so frustrating, he said. "I would become erect and then just as I approach her to penetrate, pow! It collapses on me." Victor's history of nocturnal erections and erections during light petting suggested that he was basically suffering from performance anxiety. He was bearing down to force an erection, much as he might try to learn the fingering of a difficult violin piece. Each night became a command performance in which Victor served as his own severest critic. Victor became a spectator to his own performance, a role that Masters and Johnson refer to as ***self-spectatoring.*** *Rather than focus on his partner, his attention was riveted on the size of his penis. As noted by the late great pianist Vladimir Horowitz, the worst thing a pianist can do is watch his fingers. Perhaps the worst thing a man with erectile problems can do is watch his penis.*

To break the vicious cycle of anxiety, erectile failure, and more anxiety, Victor and Michelle followed a sex therapy program (Rathus & Nevid, 1977) modeled after the Masters-and-Johnson-type treatment. The aim was to restore the pleasure of sexual activity, unfettered by anxiety. The couple was initially instructed to abstain from

sensate focus exercises Mutual pleasuring activities focused on the partners taking turns giving and receiving physical pleasure.

self-spectatoring The tendency to observe one's behavior as if one were a spectator.

attempts at intercourse to free Victor from any pressure to perform. The couple progressed through a series of steps:

1. *Relaxing together in the nude without any touching, such as when reading or watching TV together.*
2. *Sensate focus exercises.*
3. *Genital stimulation of each other manually or orally to orgasm.*
4. *Nondemand intercourse (intercourse performed without any pressure on the man to satisfy his partner). The man may afterward help his partner achieve orgasm by using manual or oral stimulation.*
5. *Resumption of vigorous intercourse (intercourse involving more vigorous thrusting and use of alternative positions and techniques that focus on mutual satisfaction). The couple is instructed not to catastrophize occasional problems that may arise.*

The therapy program helped Victor overcome his erectile disorder. Victor was freed of the need to prove himself by achieving erection on command. He surrendered his post as critic. Once the spotlight was off the bed, he became a participant and not a spectator.

—*From the Authors' Files*

■

Disorders of Orgasm Women with orgasmic disorder often harbor underlying beliefs that sex is dirty or sinful. They may have been taught not to touch themselves. They are often anxious about sex and have not learned, through trial and error, what kinds of sexual stimulation will arouse them and help them reach orgasm. Treatment in these cases includes modification of negative attitudes toward sex. When orgasmic disorder reflects the woman's feelings about or relationship with her partner, treatment requires working through these feelings or enhancing the relationship.

In either case, Masters and Johnson work with the couple and first use sensate focus exercises to lessen performance anxiety, open channels of communication, and help the couple acquire pleasuring skills. Then during genital massage and, later, during intercourse, the woman directs her partner to use caresses and techniques that stimulate her. By taking charge, the woman becomes psychologically freed from the stereotype of the passive, submissive female role.

Many researchers find that a program of directed masturbation is most effective in helping *preorgasmic* women—women who have never achieved orgasm through any means (Baucom et al., 1998; LoPiccolo & Stock, 1986). Masturbation provides a chance to learn about one's own body and give oneself pleasure without reliance on a partner or need to attend to a partner's needs. Directed masturbation programs educate women about their sexual anatomy and encourage them to experiment with self-caresses in the privacy of their own homes. Women proceed at their own pace and are encouraged to incorporate sexual fantasies and imagery during self-stimulation exercises designed to heighten their level of sexual arousal. They gradually learn to bring themselves to orgasm, sometimes with the help of an electric vibrator. Once women can masturbate to orgasm, additional couples-oriented treatment can facilitate but does not guarantee transference to orgasm with a partner.

Although scant attention in the scientific literature has been focused on male orgasmic disorder, the standard treatment, barring any underlying organic problem, focuses on increasing sexual stimulation and reducing performance anxiety (LoPiccolo, 1990; LoPiccolo & Stock, 1986).

The most widely used approach to treating premature ejaculation, called the *stop-start* or *stop-and-go* technique, was introduced in 1956 by the urologist James Semans. The man and his partner suspend sexual activity just when he is about to ejaculate and then resume stimulation when his sensations subside. Repeated practice enables him to regulate ejaculation by sensitizing him to the cues that precede the ejaculatory reflex (making him more aware of his "point of no return," the point at which the ejaculatory reflex is triggered).

Vaginismus and Dyspareunia Vaginismus is a conditioned reflex involving the involuntary constriction of the vaginal opening. It involves a psychologically based fear of penetration, rather than a physical defect or disorder (LoPiccolo & Stock, 1986). Treatment for vaginismus involves a combination of relaxation techniques and the use of vaginal dilators to gradually desensitize the vaginal musculature. The woman herself regulates the insertion of dilators (plastic rods) of increasing diameter, always proceeding at her own pace to avoid any discomfort (LoPiccolo & Stock, 1986). The method is generally successful as long as it is unhurried. Because many women with vaginismus and dyspareunia have histories of rape or sexual abuse, psychotherapy may be part of the treatment program in order to deal with the psychological consequences of traumatic experiences.

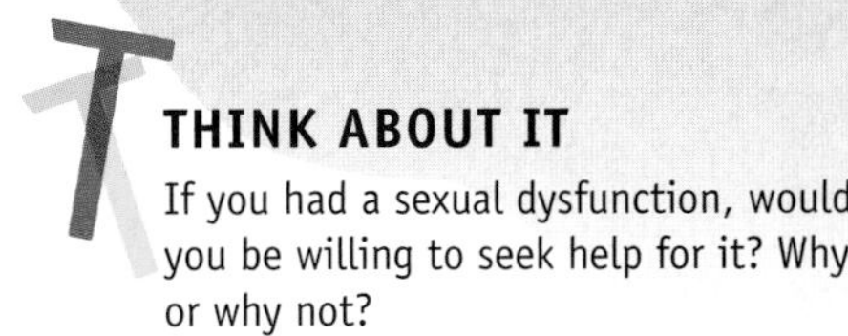

THINK ABOUT IT

If you had a sexual dysfunction, would you be willing to seek help for it? Why or why not?

Evaluation of Sex Therapy Success rates for sex therapy have been more impressive for some disorders than for others. High levels of success are reported in treating vaginismus in women and premature ejaculation in men (J. G. Beck, 1993; O'Donohue, Letourneau, & Geer, 1993). Reported success rates in treating vaginismus have ranged as high as 80% (Hawton & Catalan, 1990) to 100% (Masters & Johnson, 1970). Success rates in treating premature ejaculation with the stop-start procedure as high as 95% have been reported, but relapse rates tend to be high (Segraves & Althof, 1998). Success rates in treating erectile dysfunction with sex therapy techniques are more variable (Rosen, 1996), and we still lack methodologically sound studies needed to support the effectiveness of these techniques (O'Donohue et al., 1999). Outcomes of treatment for male orgasmic disorder also vary and are often disappointing (Dekker, 1993).

Although some progress has been made in treating sexual desire disorders, we would benefit from new treatment techniques because present techniques often fail to resolve the problem (J. G. Beck, 1995; Hawton, 1991). Better results are generally reported from directed masturbation programs for preorgasmic women, with success rates (percentage of women achieving orgasm) reported in a range of 70% to 90% (Rosen & Leiblum, 1995). However, much lower rates are reported when measured in terms of percentages of women reporting orgasm during sexual intercourse with their partners (Segraves & Althof, 1998). Some researchers believe that the final determination of the effectiveness of directed masturbation as a treatment alternative remains to be made (O'Donohue, Dopke, & Swingen, 1997).

Truth OR Fiction? REVISITED

Though it is used mostly by men, Viagra can also help women overcome sexual dysfunctions.

FALSE. We lack evidence that Viagra can help women overcome sexual dysfunctions.

Biological Treatments of Male Sexual Dysfunction

Biological treatments of erectile disorder have included a wide range of techniques, but are now focused primarily on the use of drugs to either induce erections or delay ejaculation. The most widely known example is *Viagra,* a drug that expands blood vessels in the penis, which has the effect of increasing blood flow to the penis that causes it to become erect (Goldstein et al., 1998; Kolata, 1998). Taken about an hour before sexual relations, the drug has helped 70% to 80% of erectile dysfunction patients achieve erections. Viagra became the fastest selling new drug in history soon after its release. Unfortunately, Viagra has failed to help women with sexual dysfunctions ("Viagra Fails to Help Women," 2000).

Hormone treatments may be helpful to men with abnormally low levels of male sex hormones but not those whose hormone levels are within normal limits (Spark, 1991). Because hormone treatments can have side effects, such as liver damage, they should not be undertaken lightly.

Vascular surgery may be effective in rare cases in which blockage in the blood vessels prevents blood from swelling the penis, or in which the penis is structurally defective (LoPiccolo & Stock, 1986).

Recent evidence indicates that SSRI-type antidepressant drugs may also help delay ejaculation in men with premature ejaculation (Kim & Seo, 1998; Segraves & Althof, 1998). These drugs affect the availability of neurotransmitters that may play a role in the brain's regulation of the ejaculatory reflex.

All in all, the success rates reported for treating sexual dysfunctions through psychological or biological approaches are quite encouraging, especially when we remember that only a few generations ago there were no effective treatments available.

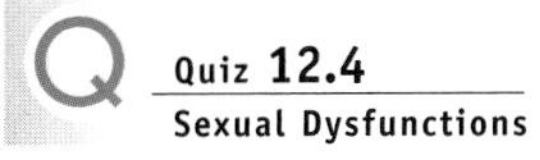

Quiz 12.4
Sexual Dysfunctions

Quiz 12.5
Chapter Exam

Research Update
Chapter 12

Overview of Sexual Dysfunctions

TYPES OF SEXUAL DYSFUNCTIONS	**Problems with sexual interest, arousal, or response that may exist throughout the person's lifetime (lifelong dysfunction) or be acquired at some point after a period of normal functioning (acquired dysfunction)**
Sexual Desire Disorders	• **Hypoactive sexual desire disorder:** Lack of sexual interest or desire • **Sexual aversion disorder:** Aversion to, and avoidance of, genital sexual contact
Sexual Arousal Disorders	• **Female sexual arousal disorder:** Difficulty becoming aroused or maintaining sexual arousal or excitement during sexual activity • **Male erectile disorder:** Difficulty achieving or maintaining erection during sexual activity
Orgasm Disorders	• **Female orgasmic disorder:** Difficulty achieving orgasm • **Male orgasmic disorder:** Difficulty achieving orgasm
Sexual Pain Disorders	• **Dyspareunia:** Pain during or following sexual intercourse not explainable by an underlying medical condition • **Vaginismus:** Involuntary contraction of the vaginal musculature, making penile penetration painful or impossible
CAUSAL FACTORS	
Biological Factors	• Disease or deficient sex hormone production may disrupt sexual desire, arousal, or response
Psychodynamic Factors	• Psychodynamic theorists speculate that unconscious conflicts dating from childhood may lie at the root of problems with sexual arousal or response
Psychosocial Factors	• Performance anxiety arising from excessive concerns about one's ability to perform sexually • History of sexual trauma or abuse • Lack of opportunity to acquire sexual skills • Exposure to negative attitudes and beliefs toward sexuality, especially female sexuality
Cognitive Factors	• Adoption of irrational beliefs, such as the belief that one should be perfectly competent at all times, may engender performance anxiety • In premature ejaculation, failure to gauge rising levels of sexual tension preceding ejaculation • Interfering cognitions, such as fears of failure, may impair normal sexual response
Relationship Factors	• Relationship problems and failure to communicate sexual needs
TREATMENT APPROACHES	**Most cases of sexual dysfunction can be treated successfully**
Biomedical Treatment	• Primarily involves use of drugs to treat erectile dysfunction or premature ejaculation
Cognitive-Behavioral Therapy	• Sex therapy—brief, cognitive-behavioral techniques that help individuals and couples develop more satisfying sexual relations and reduce performance anxiety

Summing Up

Normal and Abnormal in Sexual Behavior

Where do we draw the line between normal and abnormal sexual behavior? We may label sexual behavior as abnormal when it deviates from societal norms, or when it is self-defeating, harms others, causes, personal distress, or interferes with one's ability to function. Yet we should recognize that sexual behavior that is considered normal in one culture may be deemed abnormal in another.

Gender Identity Disorder

What is gender identity disorder? People with gender identity disorder find their anatomic gender to be a source of persistent and intense distress. People with the disorder may seek to change their sex organs to resemble those of the opposite gender, and many undergo gender reassignment surgery to accomplish this purpose.

How is gender identity disorder different than sexual orientation? Gender identity disorder involves a mismatch between one's psychological sense of being male or female and one's anatomic sex. Sexual orientation relates to the direction of one's sexual attraction—toward members of one's own gender or the opposite gender. Unlike people with gender identity disorder, people with a gay male or lesbian sexual orientation have a gender identity consistent with their anatomic gender.

Paraphilias

What are paraphilias? Paraphilias are sexual deviations involving patterns of arousal to stimuli such as nonhuman objects (for example, shoes or clothes), humiliation or the experience of pain in oneself or one's partner, or children.

What are the major types of paraphilia? Paraphilias include exhibitionism, fetishism, transvestic fetishism, voyeurism, frotteurism, pedophilia, sexual masochism, and sexual sadism. Although some paraphilias are essentially harmless (such as fetishism), others, such as pedophilia and sexual sadism, often harm nonconsenting victims.

What causes are implicated in paraphilias and what makes these disorders so difficult to treat? Paraphilias may be caused by the interaction of biological, psychological, and social factors. Efforts to treat paraphilias are compromised by the fact that most people with these disorders do not wish to change.

Sexual Dysfunctions

What are the major types of sexual dysfunctions? Sexual dysfunctions include sexual desire disorders (hypoactive sexual desire disorder and sexual aversion disorder), sexual arousal disorders (female sexual arousal disorder and male erectile disorder), orgasm disorders (female and male orgasmic disorders, and premature ejaculation), and sexual pain disorders (dyspareunia and vaginismus).

What causes sexual dysfunctions? Sexual dysfunctions can stem from biological factors (such as disease or the effects of alcohol and other drugs), psychological factors (such as performance anxiety, unresolved conflicts, or lack of sexual competencies), and sociocultural factors (such as sexually restrictive cultural learning).

What are the major goals of sex therapy? Sex therapists help people overcome sexual dysfunctions by enhancing self-efficacy expectancies, teaching sexual competencies, improving sexual communication, and reducing performance anxiety.

What biologically based treatments are available to help treat male sexual dysfunctions? These include hormone treatments, vascular surgery, and most commonly, the use of drugs to help induce erections (Viagra) or delay ejaculation (antidepressants).

CHAPTER THIRTEEN

Schizophrenia and Other Psychotic Disorders

Gino Severini
The Head

Truth OR Fiction?

- Individuals may show all the signs of schizophrenia for several months but still not be diagnosed with the disorder. (p. 404)
- Some people are deluded that they are loved by a famous person. (p. 407)
- The syndrome we identify as schizophrenia is experienced in virtually the same way in every culture that has been available for study. (p. 409)
- Visual hallucinations ("seeing things") are the most common type of hallucination in people with schizophrenia. (p. 413)
- Auditory hallucinations may be a form of inner speech. (p. 415)
- Some people with schizophrenia sustain unusual, uncomfortable positions for hours and will not respond to questions or communicate during these periods. (p. 417)
- A 54-year-old hospitalized woman diagnosed with schizophrenia was conditioned to cling to a broom by being given cigarettes as reinforcers. (p. 419)
- Living in a family that is hostile, critical, and unsupportive can increase the risk of relapse in people with schizophrenia. (p. 426)
- If you have two parents with schizophrenia, it's nearly certain that you will develop schizophrenia yourself. (p. 429)
- Drugs developed in the past few years not only treat schizophrenia, but also can cure it in many cases. (p. 431)

Schizophrenia is perhaps the most puzzling and disabling clinical syndrome. It is the psychological disorder that best corresponds to popular conceptions of madness or lunacy. It often elicits fear, misunderstanding, and condemnation rather than sympathy and concern. Schizophrenia strikes at the heart of the person, stripping the mind of the intimate connections between thoughts and emotions and filling it with distorted perceptions, false ideas, and illogical conceptions, as in the following case example:

Chapter Outline

Angela's "Hellsmen"

Angela, 19, was brought to the emergency room by her boyfriend Jaime because she had cut her wrists. When she was questioned, her attention wandered. She seemed transfixed by creatures in the air, or by something she might be hearing. It seemed as though she had an invisible earphone.

Angela explained that she had slit her wrists at the command of the "hellsmen." Then she became terrified. Later she related that the hellsmen had cautioned her not to disclose their existence. Angela had been fearful that the hellsmen would punish her for her indiscretion.

Jaime related that Angela and he had been living together for nearly a year. They had initially shared a modest apartment in town. But Angela did not like being around other people and persuaded Jaime to rent a cottage in the country. There Angela spent much of her days making fantastic sketches of goblins and monsters. She occasionally became agitated and behaved as though invisible beings were issuing directions. Her words would begin to become jumbled.

Jaime would try to persuade her to go for help, but she would resist. Then, about nine months ago, the wrist-cutting began. Jaime believed that he had made the bungalow secure by removing all knives and blades. But Angela always found a sharp object.

Then he would bring Angela to the hospital against her protests. Stitches would be put in, she would be held under observation for a while, and she would be medicated. She would recount that she cut her wrists because the hellsmen had informed her that she was bad and had to die. After a few days in the hospital, she would disavow hearing the hellsmen and insist on discharge.

Jaime would take her home. The pattern would repeat itself.

—*From the Authors' Files*

■

Schizophrenia touches every facet of the affected person's life. Acute episodes of schizophrenia are characterized by delusions, hallucinations, illogical thinking, incoherent speech, and bizarre behavior. Between acute episodes, people with schizophrenia may still be unable to think clearly and may lack an appropriate emotional response to the people and events in their lives. They may speak in a flat tone and show little if any facial expressiveness (Mandal, Pandey, & Prasad, 1998). Although researchers are immersed in probing the psychological and biological foundations of schizophrenia, the disorder remains in many ways a mystery. Schizophrenia is not the only type of psychotic disorder in which the person experiences a break with reality. In this chapter we also consider other psychotic disorders, including brief psychotic disorder, schizophreniform disorder, schizoaffective disorder, and delusional disorder.

History of the Concept of Schizophrenia

Although various forms of "madness" have afflicted people throughout the course of history, no one knows how long the behavior pattern we now label schizophrenia existed before it was first described as a medical syndrome by Emil Kraepelin in 1893. Modern

schizophrenia An enduring psychotic disorder that involves disturbed behavior, thinking, emotions, and perceptions.

dementia praecox The term given by Kraepelin to the disorder we now call schizophrenia.

Emil Kraepelin.

Eugen Bleuler.

conceptualizations of schizophrenia have been largely shaped by the contributions of Kraepelin, Eugen Bleuler, and Kurt Schneider.

Emil Kraepelin

Kraepelin (1856–1926), one of the fathers of modern psychiatry, called the disorder **dementia praecox.** The term derived from the Latin *dementis,* meaning "out" (*de-*) of one's "mind" (*mens*), and the roots that form the word *precocious,* meaning "before" one's level of "maturity." *Dementia praecox* thus refers to premature impairment of mental abilities. Kraepelin believed that dementia praecox was a disease process caused by specific, although unknown, pathology in the body.

Kraepelin wrote that dementia praecox involved the "loss of the inner unity of thought, feeling, and acting." The syndrome begins early in life, and the course of deterioration eventually results in complete "disintegration of the personality" (Kraepelin, 1909–1913, Vol. 2, p. 943). Kraepelin's description of dementia praecox includes behavior patterns such as delusions, hallucinations, and odd motor behaviors—the behavior patterns that typically characterize the disorder today.

Eugen Bleuler

In 1911, the Swiss psychiatrist Eugen Bleuler (1857–1939) renamed dementia praecox *schizophrenia,* from the Greek *schistos,* meaning "cut" or "split," and *phren,* meaning "brain." In doing so, Bleuler focused on the major characteristic of the syndrome, the splitting of the brain functions that give rise to cognition, feelings or affective responses, and behavior. A person with schizophrenia, for example, might giggle inappropriately when discussing an upsetting event, or might show no emotional expressiveness in the face of tragedy.

Although the Greek roots of *schizophrenia* mean "split brain," schizophrenia should not be confused with dissociative identity disorder (formerly multiple personality disorder), which is frequently referred to as "split personality" by laypeople. People with dissociative identity disorder (see Chapter 7) exhibit two or more alter personalities, but the alter personalities typically show better integrated cognitive, affective, and behavioral functioning than is the case in schizophrenia. In schizophrenia, the splitting cleaves cognition, affect, and behavior. There may thus be little agreement between the thoughts and the emotions, or between the individual's perceptions of reality and what is truly happening.

Although Bleuler accepted Kraepelin's description of the symptoms of schizophrenia, he did not accept Kraepelin's views that schizophrenia necessarily begins early in life and inevitably follows a deteriorating course. Bleuler proposed that schizophrenia follows a more variable course. In some cases, acute episodes occur intermittently. In others, there might be limited improvement rather than inevitable deterioration.

Bleuler believed that schizophrenia could be recognized on the basis of four primary features or symptoms. Today, we refer to them as the **four A's**:

1. *Associations.* **Associations,** or relationships among thoughts, become disturbed. We now call this type of disturbance "thought disorder" or "looseness of associations." Looseness of associations means ideas are strung together with little or no relationship among them; nor does the speaker appear to be aware of the lack of connectedness. The person's speech appears to others to be rambling and confused.
2. *Affect.* **Affect,** or emotional response, becomes flattened or inappropriate. The individual may show a lack of response to upsetting events, or burst into laughter upon hearing that a family member or friend has died.
3. *Ambivalence.* People with schizophrenia hold ambivalent or conflicting feelings toward others, such as loving and hating them at the same time.
4. *Autism.* **Autism** is a term that describes withdrawal into a private fantasy world that is not bound by principles of logic.

In Bleuler's view, hallucinations and delusions represent "secondary symptoms," symptoms that accompany the primary symptoms but do not define the disorder. In more recent years, however, other theorists such as Kurt Schneider (1957) have proposed that hallucinations and delusions are key, or primary, features of schizophrenia. Bleuler was strongly influenced by psychodynamic theory. He came to believe that the content of hallucinations and delusions could be explained by the attempt to replace the external world with a world of fantasy.

Bleuler's contributions led to the adoption of a broader definition of schizophrenia and brought the diagnostic category into more common use. Bleuler's ideas were especially influential in the United States. U.S.-trained professionals began to use the diagnosis more freely than their European counterparts, who were more influenced by Kraepelin's narrower definition of the disorder. The diagnosis of schizophrenia was broadened in the United States even beyond Bleuler's criteria to include people who showed combined features of schizophrenia and mood disorders. These cases are now generally classified separately from schizophrenia under the category of schizoaffective disorder. (We will cover this disorder later in this chapter.)

Kurt Schneider

Another influential developer of modern concepts of schizophrenia was the German psychiatrist Kurt Schneider (1887–1967). Schneider believed Bleuler's criteria (his "four A's") were too vague for diagnostic purposes and that they failed to adequately distinguish schizophrenia from other disorders. Schneider's (1957) most notable contribution was to discriminate between the features of schizophrenia that he believed are central to diagnosis, which he termed **first-rank symptoms,** and so-called **second-rank symptoms,** which he believed are found not only in schizophrenia, but also in other psychoses

four A's The primary characteristics of schizophrenia: loose ***A**ssociations,* blunted or inappropriate ***A**ffect,* ***A**mbivalence,* and ***A**utism.*

associations Relationships among thoughts.

affect Emotional responsiveness.

autism Withdrawal into a private fantasy world.

first-rank symptoms The primary features of schizophrenia, such as hallucinations and delusions.

second-rank symptoms Symptoms associated with schizophrenia that also occur in other mental disorders.

THINK ABOUT IT

Trace the development of the concept of schizophrenia in your own words. What were the areas of agreement and disagreement? Why is the disorder difficult to define?

Hallucinations. According to Kurt Schneider, hallucinations and delusions are numbered among the first-rank symptoms of schizophrenia—that is, the symptoms that are central to the diagnosis. So-called second-rank symptoms are found in other disorders as well. Schneider considered confusion and disturbances in mood to be second-rank symptoms.

TABLE 13.1 Major Clinical Features of Schizophrenia

A. Two or more of the following must be present for a significant portion of time over the course of a 1-month period:
   (1) delusions
   (2) hallucinations
   (3) speech that is either incoherent or characterized by marked loosening of associations
   (4) disorganized or catatonic behavior
   (5) negative features (e.g., flattened affect)

B. Functioning in such areas as social relations, work, or self-care during the course of the disorder is markedly below the level achieved prior to the onset of the disorder. If the onset develops during childhood or adolescence, there is a failure to achieve the expected level of social development.

C. Signs of the disorder have occurred continuously for a period of at least 6 months. This 6-month period must include an active phase lasting at least a month in which psychotic symptoms (listed in A), which are characteristic of schizophrenia, occur.

D. The disorder cannot be attributed to the effects of a substance (e.g., substance abuse or prescribed medication) or to a general medical condition.

Source. Adapted from the *DSM-IV-TR* (APA, 2000).

Truth OR Fiction? REVISITED

Individuals may show all the signs of schizophrenia for several months but still not be diagnosed with the disorder.

TRUE. A diagnosis of schizophrenia requires that signs of the disorder be present for at least 6 months.

and in some nonpsychotic disorders, such as personality disorders. In Schneider's view, if first-rank symptoms are present and cannot be accounted for by organic factors, a diagnosis of schizophrenia is justified. Hallucinations and delusions are prominent first-rank symptoms. Disturbances in mood and confused thinking are considered second-rank symptoms. Although Schneider's ranking of disturbed behaviors helped distinguish schizophrenia from other disorders, we now know that first-rank symptoms are sometimes found among people with other disorders, especially bipolar disorder. Although first-rank symptoms are clearly associated with schizophrenia, they are not unique to it.

Contemporary Diagnostic Practices

Today, the contributions of Kraepelin, Bleuler, and Schneider are expressed in modified form in the present *DSM* diagnostic system. However, the diagnostic code for schizophrenia is not limited, as Kraepelin had proposed, to cases where there is a course of progressive deterioration. Today, we recognize that schizophrenia does not necessarily follow a persistent downhill course (USDHHS, 1999a). As many as one-half to two-thirds of schizophrenia patients improve significantly over time, and some patients even recover fully (USDHHS, 1999a).

The present diagnostic code is tighter than earlier conceptualizations. It separates into other diagnostic categories cases in which there are disturbances of mood combined with psychotic behavior and those involving schizophrenic-like thinking but without overt psychotic behavior (schizotypal personality disorder). The *DSM-IV* criteria for schizophrenia also require that psychotic behaviors be present at some point during the course of the disorder and that signs of the disorder be present for at least 6 months. People with briefer forms of psychosis are placed in diagnostic categories that may be connected with more favorable outcomes. Table 13.1 describes the major clinical criteria for schizophrenia.

Quiz 13.1
History of the Concept of Schizophrenia

Other Forms of Psychosis

Although we tend to link psychotic behavior with schizophrenia, the *DSM* recognizes several different types of psychotic disorders.

Brief Psychotic Disorder

Some brief psychotic episodes do not progress to schizophrenia. The *DSM-IV* category of **brief psychotic disorder** applies to a psychotic disorder that lasts from a day to a month and is characterized by at least one of the following features: delusions, hallucinations, disorganized speech, or disorganized or catatonic behavior. Eventually there is a full return to the individual's prior level of functioning. Brief psychotic disorder is often linked to a significant stressor or stressors, such as the loss of a loved one or exposure to brutal traumas in wartime. Some cases in women involve a postpartum onset that begins within the first month after childbirth.

brief psychotic disorder A psychotic disorder lasting from a day to a month that often follows exposure to a major stressor.

schizophreniform disorder A psychotic disorder lasting less than six months in duration, with features that resemble schizophrenia.

delusional disorder A type of psychosis characterized by persistent delusions, often of a paranoid nature, that do not have the bizarre quality of the type found in paranoid schizophrenia.

Schizophreniform Disorder

Schizophreniform disorder consists of abnormal behaviors identical to those in schizophrenia that have persisted for at least 1 month but less than 6 months. They thus do not yet justify the diagnosis of schizophrenia. Although some cases have good outcomes, in others the disorder persists beyond 6 months and may be reclassified as schizophrenia or perhaps another form of psychotic disorder, such as schizoaffective disorder. Questions remain about the validity of the diagnosis, however (Strakowski, 1994). It may be more appropriate to diagnose people who show psychotic features of recent origin with a classification such as *psychotic disorder of an unspecified type* until additional information clearly indicates the specific type of disorder involved.

WWW Web Link **13.1**
Facts About Delusional Disorder

Delusional Disorder

Many of us, perhaps even most of us, feel suspicious of other people's motives at times. We may feel others have it in for us or believe others are talking about us behind our backs. For most of us, however, paranoid thinking does not take the form of outright delusions. The diagnosis of **delusional disorder** applies to people who hold persistent, clearly delusional beliefs, often involving paranoid themes. Delusional disorder is uncommon, affecting an estimated 5 to 10 people in 10,000 during their lifetimes (APA, 2000).

In delusional disorders, the delusional beliefs concern events that may possibly occur, such as the infidelity of a spouse, persecution by others, or attracting the love of a famous person. The apparent plausibility of these beliefs may lead others to take them seriously and check them out before concluding they are unfounded. Apart from the delusion, the individual's behavior does not show evidence of obviously bizarre or odd behavior, as we see in the following case example:

Is someone out to get you? People with delusional disorder often weave paranoid fantasies in their minds such that they confuse with reality.

Mr. Polsen's Hit Men

Mr. Polsen, a married 42-year-old postal worker, was brought to the hospital by his wife because he had been insisting that there was a contract out on his life. Mr. Polsen told the doctors that the problem had started some four months ago when he was accused by his supervisor of tampering with a package, an offense that could have cost him his job. When he was exonerated at a formal hearing, his supervisor was "furious" and felt publicly humiliated, according to Mr. Polsen. Shortly afterwards, Mr. Polsen reported, his co-workers began avoiding him, turning away from him when he walked by, as if they didn't want to see him. He then began to think that they were talking about him behind his back, although he could never clearly make out what they were saying. He gradually became convinced that his co-workers were avoiding him because his boss had put a contract on his life. Things remained about the same for two months, when Mr. Polsen began to notice several large white cars cruising up and down the street where he lived. This frightened him and he became convinced there were hit men in these cars. He then refused to leave his home without an escort and would run home in panic when he saw one of these cars approaching. Other than the reports of his belief that his life was in danger, his thinking and behavior appeared entirely normal on interview. He denied experiencing hallucinations and showed no other signs of psychotic behavior, except for the queer beliefs about his life being in danger. The diagnosis of Delusional Disorder, Persecutory type seemed the most appropriate, since there was no evidence that a contract had been taken on his life (hence, a persecutory delusion) and there was an absence of other clear signs of psychosis that might support a diagnosis of a schizophrenic disorder.

—*Adapted from Spitzer et al., 1994, pp. 177–179*

Mr. Polsen's delusional belief that "hit teams" were pursuing him was treated with antipsychotic medication in the hospital setting and faded in about 3 weeks. His belief that he had been the subject of an attempted "hit" stuck in his mind, however. A month following admission, he stated, "I guess my boss has called off the contract. He couldn't get away with it now without publicity" (Spitzer et al., 1994, p. 179).

Although delusions frequently occur in schizophrenia, delusional disorder is believed to be distinct from schizophrenia. Persons with delusional disorder do not exhibit the confused or jumbled thinking characteristic of schizophrenia. Hallucinations, when they occur, are not as prominent. Delusions in schizophrenia are embedded within a larger array of disturbed thoughts, perceptions, and behavior. In delusional disorders, the delusion itself may be the only clear sign of abnormality.

Delusional disorder should also be distinguished from another disorder in which paranoid thinking is present—paranoid personality disorder (see Chapter 9). People with paranoid personality disorder may hold exaggerated or unwarranted suspicions of others, but not the outright delusions that are found among people with delusional disorders or paranoid schizophrenia. People with paranoid personality disorder may believe that they were passed over for a promotion because their boss had it in for them, but they would not maintain the unfounded belief that their boss had put a contract on their life.

Various types of delusional disorder are described in Table 13.2. Delusional disorders are relatively uncommon. Once a delusion is established, it may persevere, although the individual's concern about it may wax and wane over the years. In other cases, the delusion

TABLE 13.2 Types of Delusional Disorder

Type	Description
Erotomanic Type	Delusional beliefs that someone else, usually someone of higher social status, such as movie star or political figure, is in love with you; also called *erotomania*.
Grandiose Type	Inflated beliefs about your worth, importance, power, knowledge, or identity, or beliefs that you hold a special relationship to a deity or a famous person. Cult leaders who believe they have special mystical powers of enlightenment may have delusional disorders of this type.
Jealous Type	Delusions of jealousy in which the person may become convinced, without due cause, that his or her lover is unfaithful. The delusional person may misinterpret certain clues as signs of unfaithfulness, such as spots on the bedsheets.
Persecutory Type	The most common type of delusional disorder, persecutory delusions involve themes of being conspired against, followed, cheated, spied upon, poisoned or drugged, or otherwise maligned or mistreated. Persons with such delusions may repeatedly institute court actions, or even commit acts of violence, against those who they perceived are responsible for their mistreatment.
Somatic Type	Delusions involving physical defects, disease, or disorder. Persons with these delusions may believe that foul odors are emanating from their bodies, or that internal parasites are eating away at them, or that certain parts of their body are unusually disfigured or ugly, or not functioning properly despite evidence to the contrary.
Mixed Type	Delusions typify more than one of the other types; no single theme predominates.

Source. Adapted from *DSM-IV-TR* (APA, 2000).

A Closer Look

The Love Delusion

Erotomania, or the love delusion, is a delusional disorder in which the individual believes he or she is loved by someone else, usually someone famous or of high social status. In reality, the individual has only a passing or nonexistent relationship with the alleged lover (R. L. Goldstein, 1986). Although the love delusion was once thought to be predominantly a female disorder, recent reports suggest it may not be a rarity among men. It has been suggested, for example, that John Hinckley Jr., who attempted to assassinate then-president Ronald Reagan reportedly to impress actress Jodie Foster, could be considered a case of erotomania (Stone, 1984). Although women with erotomania may have a potential for violence when their attentions are rebuffed, men with this condition appear more likely to threaten or commit acts of violence in the pursuit of the objects of their unrequited desires (Goldstein, 1986). Antipsychotic medications may reduce the intensity of the delusion but do not appear to eliminate it (Kelly, Kennedy, & Shanley, 2000; Segal, 1989). Nor is there evidence that psychotherapy helps people with erotomania. The prognosis is thus bleak, and people with erotomania may harass their love objects for many years. Mental health professionals also need to be aware of the potential for violence in the management of people who possess these delusions of love (Mullen, 2000; Segal, 1989). The following cases provide some examples of the love delusion:

Three Cases of Erotomania

Mr. A., a 35-year-old man, was described as a "love-struck" suitor of a daughter of a former President of the United States. He was arrested for repeatedly harassing the woman in an attempt to win her love, although they were actually perfect strangers. Refusing to adhere to the judge's warnings to stop pestering the woman, he placed numerous phone calls to her from prison and was later transferred to a psychiatric facility, still declaring they were very much in love.

Mr. B. was arrested for breaching a court order to stop pestering a famous pop singer. A 44-year-old farmer, Mr. B. had followed his love interest across the country, constantly bombarding her with romantic overtures. He was committed to a psychiatric hospital, but maintained the belief that she'd always wait for him.

Then there was Mr. C., a 32-year-old businessman, who believed a well-known woman lawyer had fallen in love with him following a casual meeting. He constantly called and sent flowers and letters, declaring his love. While she repeatedly rejected his advances and eventually filed criminal charges for harassment, he felt that she was only testing his love by placing obstacles in his path. He abandoned his wife and business and his functioning declined. When the woman continued to reject him, he began sending her threatening letters and was committed to a psychiatric facility.

—Adapted from Goldstein, 1986, p. 802

Truth OR Fiction? REVISITED

Some people are deluded that they are loved by a famous person.

TRUE. Some people do suffer from the delusion that they are loved by a famous person. They are said to have a delusional disorder, erotomanic type.

may disappear entirely for periods of time and then recur. Sometimes the disorder permanently disappears.

Schizophrenia-Spectrum Disorders

Some people have persistent patterns of unusual thinking or emotional responses that seem to lie within the broader spectrum of schizophrenic problems, but may not fit the stringent definition of schizophrenia. The schizophrenia spectrum includes related disorders that vary in severity from milder personality disorders (schizoid, paranoid, and schizotypal types) to schizophrenia itself.

Also classified within the schizophrenia spectrum is **schizoaffective disorder,** which is characterized by a "mixed bag" of symptoms including psychotic features such as hallucinations and delusions, together with major disturbances of mood, such as mania or

WWW **Web Link 13.2**
More About Schizoaffective Disorder

erotomania A delusional disorder characterized by the belief that one is loved by someone of high social status.

schizoaffective disorder A type of psychotic disorder in which individuals experience both severe mood disturbance and features associated with schizophrenia.

THINK ABOUT IT

What is schizophrenia? How is it diagnosed? How is schizophrenia distinguished from other disorders within the schizophrenic spectrum?

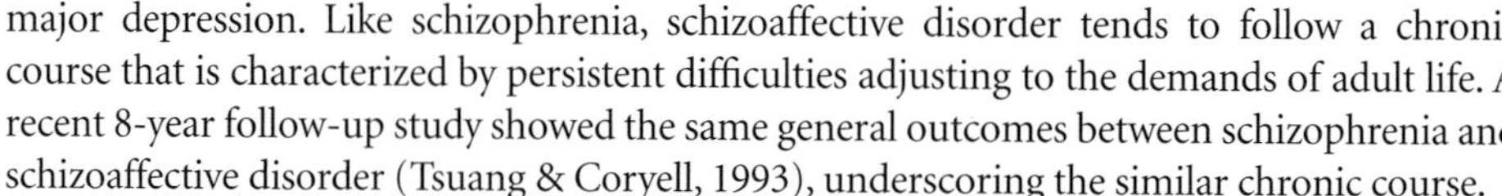

major depression. Like schizophrenia, schizoaffective disorder tends to follow a chronic course that is characterized by persistent difficulties adjusting to the demands of adult life. A recent 8-year follow-up study showed the same general outcomes between schizophrenia and schizoaffective disorder (Tsuang & Coryell, 1993), underscoring the similar chronic course.

The distinction between schizophrenia and schizophrenia-spectrum disorders may be more a matter of degree than of kind. Differences in genetic vulnerability or environmental stress may lead to the development of milder or more severe forms of a common schizophrenic-type disorder (Andreasen, 1987a).

Research shows that some schizophrenia-spectrum disorders appear to share a common genetic link (Begley, 1998; Fanous et al., 2001). Consistent with a common genetic basis, researchers find a greater than average incidence of schizoaffective disorders among the relatives of people with schizophrenia and a greater than average incidence of schizophrenia among the relatives of people with schizoaffective disorder (Kendler, Gruenberg, & Tsuang, 1985; Maj et al., 1991). Evidence also shows familial linkages between symptomatology in schizophrenia patients and schizotypal personality disorder symptomatology in their relatives (Fanous et al., 2001).

Quiz **13.2**
Other Forms of Psychosis

Schizophrenia

Schizophrenia typically develops in late adolescence or early adulthood, at the very time that people are making their way from the family into the outside world (Cowan & Kandel, 2001; Harrop & Trower, 2001). People who develop schizophrenia become increasingly disengaged from society. They fail to function in the expected roles of student, worker, or spouse, and their families and communities grow intolerant of their deviant behavior. The disorder typically develops in the late teens or early 20s, a time at which the brain is reaching full maturation. In about three of four cases, the first signs of schizophrenia appear by the age of 25 (Keith, Regier, & Rae, 1991).

In some cases, the onset of the disorder is acute. It occurs suddenly, within a few weeks or months. The individual may have been well adjusted and shown few if any signs of behavioral disturbance. Then a rapid transformation in personality and behavior leads to an acute psychotic episode.

In most cases, there is a slower, more gradual decline in functioning. It may take years before psychotic behaviors emerge, although early signs of deterioration may be observed. This period of deterioration is called the **prodromal phase.** It is characterized by waning interest in social activities and increasing difficulty in meeting the responsibilities of daily living. At first, such people seem to take less care of their appearance. They fail to bathe regularly or they wear the same clothes repeatedly. Over time, their behavior may become increasingly odd or eccentric. There are lapses in job performance or schoolwork. Their speech may become increasingly vague and rambling. At first these changes in personality may be so gradual that they raise little concern among friends and families. They may be attributed to "a phase" that the person is passing through. But as behavior becomes more bizarre—such as hoarding food, collecting garbage, or talking to oneself on the street—the acute phase of the disorder begins. Frankly psychotic symptoms develop, such as wild hallucinations, delusions, and increasingly bizarre behavior.

Following acute episodes, people who develop schizophrenia may enter the **residual phase,** in which their behavior returns to the level that was characteristic of the prodromal phase. Although flagrant psychotic behaviors may be absent during the residual phase, the person may continue to be impaired by a deep sense of apathy, by difficulties in thinking or speaking clearly, and by the harboring of unusual ideas, such as beliefs in telepathy or clairvoyance. Such patterns of behavior make it difficult to meet expected social roles as wage earners, marital partners, or students. Full return to normal behavior is uncommon but does occur in some cases. More commonly, a chronic pattern develops, which is characterized by occasional acutely psychotic episodes and continued cognitive, emotional, and motivational impairment between episodes (Wiersma et al., 1998; USDHHS, 1999a).

prodromal phase In schizophrenia, the period of decline in functioning that precedes the first acute psychotic episode.

residual phase In schizophrenia, the phase that follows an acute phase, characterized by a return to the level of functioning of the prodromal phase.

Prevalence of Schizophrenia

About 1% of the adult population in the United States is affected by schizophrenia, more than 2 million people in total (APA, 2000; Cowan & Kandel, 2001). According to the results of the World Health Organization (WHO) multinational study reported in Chapter 1, the rate of schizophrenia appears to be similar in both developed and developing cultures (Jablensky et al., 1992). The World Health Organization estimates that about 24 million people worldwide suffer from schizophrenia (Olson, 2001). Nearly 1 million people in the United States receive treatment for schizophrenia each year, with about a third of these requiring hospitalization (Grady, 1997a). The costs for treating schizophrenia are estimated at $30 billion annually and account for 75% of all expenditures in the United States for mental health treatment (Cowan & Kandel, 2001; "Schizophrenia Update—Part I," 1995).

Men tend to have a slightly higher risk of developing schizophrenia (APA, 2000). Women tend to develop the disorder somewhat later than men do, with the age of onset occurring most commonly between age 25 and the mid-30s in women and between age 18 and 25 in men (APA, 2000). Women also tend to achieve a higher level of functioning before the onset of the disorder and to have a less severe course of the disorder than do men (Häfner et al., 1998; USDHHS, 1999a).

Though the occurrence of schizophrenia appears to be universal across cultures, the course of the disorder and its symptoms may vary from culture to culture (Thakker & Ward, 1998). For example, visual hallucinations appear to be more common in some non-Western cultures (Ndetei & Singh, 1983; Ndetei & Vadher, 1984). In a study conducted in an English hospital in Kenya, researchers found that people of African, Asian, or Jamaican background with schizophrenia were about twice as likely to experience visual hallucinations as those of European background (Ndetei & Vadher, 1984).

Truth OR Fiction? REVISITED

The syndrome we identify as schizophrenia is experienced in virtually the same way in every culture that has been available for study.

FALSE. Both the course of schizophrenia and its features vary among cultures.

THINK ABOUT IT

Do the prevalence data for schizophrenia suggest that it is or is not a biologically based disorder? Explain your answer.

Major Features of Schizophrenia

Schizophrenia is a pervasive disorder that affects a wide range of psychological processes involving cognition, affect, and behavior (Arango, Kirkpatrick, & Buchanan, 2000). People with schizophrenia show a marked decline in occupational and social functioning. They may have difficulty holding a conversation, forming friendships, holding a job, or taking care of their personal hygiene. Yet no one behavior pattern is unique to schizophrenia, nor is any one behavior pattern invariably present among people with schizophrenia. People with schizophrenia may exhibit delusions, problems with associative thinking, and hallucinations at one time or another, but not necessarily all at once. There are also different kinds or types of schizophrenia, characterized by different behavior patterns.

Men with schizophrenia appear to differ from women with the disorder in several ways. They tend to show an earlier age of onset, have a poorer level of adjustment prior to showing signs of the disorder, and have more cognitive impairment, behavioral deficits, and a poorer response to drug therapy than do women with schizophrenia (Gorwood et al., 1995; Ragland et al., 1999). These differences have led researchers to speculate that men and women may tend to develop different forms of schizophrenia. Perhaps schizophrenia affects different areas of the brain in men and women, which may explain differences in the form or features of the disorder between the genders.

Here let us consider how schizophrenia affects thinking, speech, attentional and perceptual processes, emotional processes, and voluntary behavior.

WWW **Web Link 13.3**
Facts About Schizophrenia

VIDEO 13.1
Schizophrenia:
The Case of Georgiana

Disturbances of Thought and Speech Schizophrenia is characterized by disturbances in thinking and in the expression of thoughts through coherent, meaningful speech. Disturbances in thinking may be found in both the content and form of thought.

Disturbances in the Content of Thought The most prominent disturbance in the content of thought involves *delusions*, or false beliefs that remain fixed in the person's mind despite their illogical bases and lack of evidence to support them. They tend to remain unshakable even in the face of disconfirming evidence. Delusions may take many forms.

Some of the most common are:

- *Delusions of persecution* (e.g., "The CIA is out to get me")
- Delusions of reference ("People on the bus are talking about me," or "People on TV are making fun of me")
- *Delusions of being controlled* (believing that one's thoughts, feelings, impulses, or actions are controlled by external forces, such as agents of the devil)
- *Delusions of grandeur* (believing oneself to be Jesus or believing one is on a special mission, or having grand but illogical plans for saving the world)

People with delusions of persecution may think they are being pursued by the Mafia, FBI, CIA, or some other group. A woman we treated who had delusions of reference believed that television news correspondents were broadcasting coded information about her. A man with delusions of this type expressed the belief that his neighbors had bugged the walls of his house. Other delusions include beliefs that one has committed unpardonable sins, is rotting away from a horrible disease, or that the world or oneself does not really exist.

The Hospital at the North Pole

Though people with schizophrenia may feel hounded by demons or earthly conspiracies, Mario's delusions had a messianic quality. "I need to get out of here," he said to his psychiatrist. "Why do you need to leave?" the psychiatrist asked. Mario responded, "My hospital. I need to get back to my hospital." "Which hospital?" he was asked. "I have this hospital. It's all white and we find cures for everything wrong with people." Mario was asked where his hospital was located. "It's all the way up at the North Pole," he responded. His psychiatrist asked, "But how do you get there?" Mario responded, "I just get there. I don't know how. I just get there. I have to do my work. When will you let me go so I can help the people?"

—From the Authors' Files

Other commonly occurring delusions include *thought broadcasting* (believing one's thoughts are somehow transmitted to the external world so that others can overhear them), *thought insertion* (believing one's thoughts have been planted in one's mind by an external source), and *thought withdrawal* (believing that thoughts have been removed from one's mind). Mellor (1970) offers the following examples of thought broadcasting, thought insertion, and thought withdrawal:

- *Thought Broadcasting*: A 21-year-old student reported, "As I think, my thoughts leave my head on a type of mental ticker-tape. Everyone around has only to pass the tape through their mind and they know my thoughts." (p. 17)
- *Thought Insertion*: A 29-year-old housewife reported that when she looks out of the window, she thinks, "The garden looks nice and the grass looks cool, but the thoughts of [a man's name] come into my mind. There are no other thoughts there, only his. . . . He treats my mind like a screen and flashes his thoughts on it like you flash a picture." (p. 17)
- *Thought Withdrawal*: A 22-year-old woman experienced the following: "I am thinking about my mother, and suddenly my thoughts are sucked out of my mind by a phrenological vacuum extractor, and there is nothing in my mind, it is empty." (pp. 16–17)

Disturbances in the Form of Thought Unless we are engaged in daydreaming or purposefully letting our thoughts wander, our thoughts tend to be tightly knit together. The connections (or associations) between our thoughts tend to be logical and coherent.

People with schizophrenia tend to think in a disorganized, illogical fashion, however. In schizophrenia, the form or structure of thought processes as well as their content is often disturbed. Clinicians label this type of disturbance a **thought disorder.**

Thought disorder is recognized by the breakdown in the organization, processing, and control of thoughts. Looseness of associations, which we now regard as a cardinal sign of thought disorder, was one of Bleuler's four A's. The speech pattern of people with schizophrenia is often disorganized or jumbled, with parts of words combined incoherently or words strung together to make meaningless rhymes. Their speech may jump from one topic to another, but show little interconnectivity between the ideas or thoughts that are expressed. People with thought disorder are usually unaware that their thoughts and behavior appear abnormal. In severe cases their speech may become completely incoherent or incomprehensible.

Another common sign of thought disorder is poverty of speech (that is, speech that is coherent but so slow, limited in production, or vague that little informational value is conveyed). Less commonly occurring signs include **neologisms** (words made up by the speaker that have little or no meaning to others), **perseveration** (inappropriate but persistent repetition of the same words or train of thought), **clanging** (stringing together of words or sounds on the basis of rhyming, such as, "I know who I am but I don't know Sam"), and **blocking** (involuntary, abrupt interruption of speech or thought). Disconnected speech is more common and more severe among younger patients, while poverty of speech is found more often and is more severe among older patients (Harvey et al., 1997).

Many but not all people with schizophrenia show evidence of thought disorder. Some appear to think and speak coherently but have disordered content of thought, as seen by the presence of delusions. Nor is disordered thought unique to schizophrenia; it has even been found in milder form among people without psychological disorders (Andreasen & Grove, 1986), especially when they are tired or under stress. Disordered thought is also found among other diagnostic groups, such as persons with mania. Thought disorders in people experiencing a manic episode tend to be short-lived and reversible, however. In those with schizophrenia, thought disorder tends to be more persistent or recurrent.

Thought disorder occurs most often during acute episodes, but may linger into residual phases. Thought disorders that persist beyond acute episodes are connected with poorer prognoses, perhaps because lingering thought disorders reflect more severe disorders (Marengo & Harrow, 1987).

thought disorder A disturbance in thinking characterized by the breakdown of logical associations between thoughts.

neologisms New words.

perseveration The persistent repetition of the same thought or response.

clanging The tendency to string words together because they rhyme or sound alike.

blocking An involuntary interruption of speech.

Attentional Deficiencies To read this book you must screen out background noises and other environmental stimuli. Attention, the ability to focus on relevant stimuli, is basic to learning and thinking. Kraepelin and Bleuler suggested that schizophrenia involves a breakdown in the processes of attention. People with schizophrenia appear to have difficulty filtering out irrelevant distracting stimuli, a deficit that makes it nearly impossible to focus their attention and organize their thoughts (Asarnow et al., 1991). They may become easily distracted because of a brain abnormality that makes it difficult for them to allocate their attention to relevant tasks and filter out unessential information (Braff, 1993). Scientists have discovered a genetic defect tied to a brain abnormality that may explain this filtering deficit (Grady, 1997a). The mother of a son who had schizophrenia described her son's difficulties in filtering out extraneous sounds:

> His hearing is different when he's ill. One of the first things we notice when he's deteriorating is his heightened sense of hearing. He cannot filter out anything. He hears each and every sound around him with equal intensity. He hears the sounds from the street, in the yard, and in the house, and they are all much louder than normal. (Anonymous, 1985, p. 1; as cited in Freedman et al., 1987, p. 670)

WWW Web Link 13.4
The Schizophrenia Home Page

People with schizophrenia also appear to be *hypervigilant,* or acutely sensitive to extraneous sounds, especially during the early stages of the disorder. During acute episodes, they may become flooded by these stimuli, overwhelming their ability to make sense of their environments. Through measuring the brain's involuntary brain wave responses to auditory stimuli, researchers find that the brains of people with schizophrenia are less able

Filtering out extraneous stimuli. You probably have little difficulty filtering out unimportant stimuli, such as street sounds. But people with schizophrenia may be distracted by irrelevant stimuli and be unable to filter them out. Consequently, they may have difficulty focusing their attention and organizing their thoughts.

than those of other people to inhibit or screen out responses to distracting sounds (Braff, 1993; Freedman et al., 1987).

Investigators suspect that attentional deficits associated with schizophrenia are inherited to a certain extent (Finkelstein et al., 1997; Grady, 1997a). Although the underlying mechanism is not entirely clear, attentional deficits may be related to dysfunction in the subcortical parts of the brain that regulate attention to external stimuli, such as the basal ganglia (Cornblatt & Kelip, 1994). Scientists suspect there may be a "gating" mechanism in the brain responsible for filtering extraneous stimuli, much like closing a gate in a road can stem the flow of traffic (Freedman et al., 1987).

Links between attentional deficits and schizophrenia are supported by various studies focusing on the psychophysiological aspects of attention (Carter et al., 1997). Let us briefly review some of this research.

Eye Movement Dysfunction About one in three chronic schizophrenia patients show evidence of eye movement dysfunctions (Ross, 2000). *Eye movement dysfunction* (also called *eye tracking dysfunction*) involves abnormal movements of the eyes as they track a target that moves across the field of vision. Rather than steadily tracking the target, the eyes fall back and then catch up in a kind of jerky movement. Eye movement dysfunctions appear to involve a defect in the brain's involuntary attentional processes responsible for visual attention.

Eye movement dysfunctions are common among people with schizophrenia and among their first-degree relatives (parents and siblings), which suggests it might be a genetically transmitted trait, or *marker,* that is associated with genes involved in the development of schizophrenia (Holzman et al., 1997; Ross, 2000).

The role of eye movement dysfunction as a biological marker for schizophrenia is clouded, however, because it is not unique to schizophrenia; many people with bipolar disorder show similar types of dysfunction (Sweeney et al., 1994). Research is needed to identify markers that are more specific to schizophrenia. We should also note that not all people with schizophrenia or their family members show eye movement dysfunctions. This suggests there may be different underlying genetic pathways associated with schizophrenia.

Deficiencies in Event-Related Potentials Researchers have also studied brain wave patterns, called event-related potentials, or ERPs, that occur in response to external stimuli. ERPs can be broken down into components that emerge at various intervals following the presentation of a stimulus such as a flash of light or an auditory tone. Early components (brain wave patterns occurring within the first 250 milliseconds [ms], or one quarter of a second, of exposure to a stimulus) may be involved in registering the stimulus in the brain. Later components, such as the P300 component (a brain wave pattern that typically occurs about 300 ms, or three-tenths of a second, after a stimulus) may be involved in focusing attention on the stimulus.

People with schizophrenia often have early ERP components (less than 250 ms) of greater than expected magnitude in response to touch (Holzman, 1987). This pattern of brain wave activity suggests that abnormally high levels of sensory information are reaching higher brain centers in people with schizophrenia, producing a condition called *sensory overload.* This may help explain the difficulty that people with schizophrenia have in filtering out distracting stimuli. We also have evidence of lower than expected levels of P300 brain wave patterns in response to auditory tones (e.g., Salisbury et al., 1998). This evidence points to attentional deficits that may at least partly explain why people with schizophrenia have difficulty extracting meaningful information from stimuli (lights, sounds, touch) that impinge upon them. Studies of ERPs are thus consistent with the view that people with schizophrenia may be flooded with high levels of sensory information but have greater difficulty extracting useful information from it. As a result, they may be confused and find it difficult to filter out irrelevant stimuli such as extraneous noises. Although ERP research is promising, the meaning of ERP abnormalities in schizophrenia is not fully resolved (Tracy, Josiassen, & Bellack, 1995).

In sum, several lines of evidence point to underlying physiological deficits in the ability to attend to relevant stimuli and ignore distracting stimuli among people with schizophrenia (O'Leary et al., 1996). Although the search continues for biological markers for schizophrenia, no definitive biological pattern unique to schizophrenia has yet been found. Recent evidence indicates that training in attention skills may help reduce attentional deficits in schizophrenia patients (Medalia et al., 1998).

Perceptual Disturbances

Voices, Devils, and Angels

Every so often during the interview, Sally would look over her right shoulder in the direction of the office door and smile gently. When asked why she kept looking at the door, she said that the voices were talking about the two of us just outside the door and she wanted to hear what they were saying. "Why the smile?" Sally was asked. "They were saying funny things," she replied, "like maybe you thought I was cute or something."

Tom was flailing his arms wildly in the hall of the psychiatric unit. Sweat seemed to pour from his brow, and his eyes darted about with agitation. He was subdued and injected with haloperidol (brand name Haldol) to reduce his agitation. When he was about to be injected he started shouting, "Father, forgive them for they know not . . . forgive them . . . father . . ." His words became jumbled. Later, after he had calmed down, he reported that the ward attendants had looked to him like devils or evil angels. They were red and burning, and steam issued from their mouths.

—*From the Authors' Files*

Hallucinations, the most common form of perceptual disturbance in schizophrenia, are images perceived in the absence of external stimulation. They are difficult to distinguish from reality. For Sally, the voices coming from outside the consulting room were real enough, even though no one was there. Hallucinations may involve any of the senses. Auditory hallucinations ("hearing voices") are most common. Tactile hallucinations (such as tingling, electrical, or burning sensations) and somatic hallucinations (such as feeling like snakes are crawling inside one's belly) are also common. Visual hallucinations (seeing things that are not there), gustatory hallucinations (tasting things that are not present), and olfactory hallucinations (sensing odors that are not present) are rarer.

Truth OR Fiction? REVISITED

Visual hallucinations ("seeing things") are the most common type of hallucinations in people with schizophrenia.

FALSE. Auditory, not visual, hallucinations are the most common type of hallucinations among people with schizophrenia.

Auditory hallucinations occur in about 70% of cases of schizophrenia (Cleghorn et al., 1992). In auditory hallucinations, the voices may be experienced as female or male and as originating inside or outside one's head (Asaad & Shapiro, 1986). Hallucinators may hear voices conversing about them in the third person, debating their virtues or faults.

Some people with schizophrenia experience *command hallucinations,* voices that instruct them to perform certain acts, such as harming themselves or others (Rogers et al., 1990). Angela, for example, was instructed by the "hellsmen" to commit suicide. People with schizophrenia who experience command hallucinations are often hospitalized for fear they may harm themselves or others. There is a good reason for this. A recent study found that four of five people with command hallucinations reported obeying them, with nearly half reporting they had obeyed commands to harm themselves during the past month (Kasper, Rogers, & Adams, 1996). Yet command hallucinations often go undetected by professionals, because patients deny them or are unwilling to discuss them.

Hallucinations are not unique to schizophrenia. People with major depression and mania sometimes experience hallucinations. Nor are hallucinations invariably a sign of psychopathology. Cross-cultural evidence shows they are common and socially valued in some developing countries (Bentall, 1990). Even in developed countries like the United States, about 5% of respondents in nonpatient samples report experiencing hallucinations during the preceding year, mostly auditory hallucinations (Honig et al., 1998). Hallucinations in people without psychiatric conditions are often triggered by unusually low levels of sensory stimulation (lying in the dark in a soundproof room for extended time) or low levels of arousal (Teunisse et al., 1996). Unlike psychotic individuals, these people realize that their hallucinations are not real and feel in control of them.

People who are free of psychological disorders sometimes experience hallucinations during the course of a religious experience or ritual (Asaad & Shapiro, 1986). Participants in such experiences may report fleeting trancelike states with visions or other perceptual aberrations. All of us hallucinate nightly, if we consider dreams to be a form of hallucination (perceptual experience in the absence of external stimuli).

Hallucinations may also occur in response to hallucinogenic drugs, such as LSD. Hallucinations may also occur during grief reactions, when images of the deceased may appear, and in other stressful conditions. In most cases, grief-induced hallucinations can be differentiated from psychotic hallucinations in that the individual can distinguish them from reality. Bentall (1990) views the hallucinations of psychiatric patients as involving the lack of ability to distinguish between real and imaginary events. They tend to confuse real and imaginary (hallucinatory) events, that is.

Drug-induced hallucinations tend to be visual and often involve abstract shapes such as circles or stars, or flashes of light. Schizophrenic hallucinations, in contrast, tend to be more fully formed and complex. Hallucinations (for example, of bugs crawling on one's skin) may also occur during delirium tremens (the DTs), which often occur as part of the withdrawal syndrome for chronic alcoholism. Hallucinations may also occur as side effects of medications or in neurological disorders, such as Parkinson's disease.

The causes of psychotic hallucinations remain unknown, but speculations abound (Asaad & Shapiro, 1986). Disturbances in brain chemistry are suspected as playing a causal role. The neurotransmitter dopamine has been implicated, because antipsychotic drugs that block dopamine activity also tend to reduce hallucinations. Conversely, drugs that lead to increased production of dopamine tend to induce hallucinations. Because hallucinations resemble dreamlike states, it is also possible that hallucinations are connected to a failure of brain mechanisms that normally prevent dream images from intruding on waking experiences.

Hallucinations may also represent a type of subvocal inner speech (Cleghorn et al., 1992). Many of us talk to ourselves from time to time, although we usually keep our mutterings beneath our breaths (subvocal) and recognize the voice as our own. Might auditory hallucinations that occur among people with schizophrenia be projections of their own internal voices, or self-speech, onto external sources? In one experiment, 14 of 18 hallucinators who suffered from schizophrenia reported that the voices disappeared when they engaged in a procedure that prevented them from talking to themselves under their breath (Bick & Kinsbourne, 1987). Similar results were obtained for 18 of 21 normal subjects who reportedly experienced hallucinations in response to hypnotic suggestions.

Researchers find that brain activity in Broca's area, a part of the brain involved in controlling speech, was greater in men with schizophrenia when they were hearing voices than at a later time when they were no longer hallucinating (McGuire, Shah, & Murray, 1993). This same area is known to become active when people engage in inner speech (Paulesu, Frith, & Frackowisk, 1993). Researchers also find evidence of similar electrical activity in the auditory cortex of the brain during auditory hallucinations and in response to hearing real sounds (Tiihonen et al., 1992). This evidence supports the view that auditory hallucinations may be a form of inner speech (silent self-talk) that for some unknown rea-

A painting by a man with schizophrenia. This picture was painted by a young man who reported monsters—apparent hallucinations—like the one pictured here crawling on the floor. He also reported that the chairs next to his bed had turned into devils.

son is attributed to external sources rather than to one's own thoughts (Ford et al., 2001; Hoffman et al., 1999). This line of research has led to treatment applications in which behavior therapists attempt to teach hallucinators to reattribute their voices to themselves (Bentall, Haddock, & Slade, 1994). Hallucinators are also trained to recognize the situational cues associated with their hallucinations. For example,

> . . . one patient . . . recognized that her voices tended to become worse following family arguments. She became aware that the content of her voices reflected the things that she was feeling and thinking about her family but that she was unable to express. Specific targets and goals were then set to allow her to address these difficulties with her family, and techniques such as rehearsal, problem solving and cognitive restructuring were employed to help her work towards these goals. (Bentall, Haddock, & Slade, 1994, p. 58)

Although cognitive-behavioral approaches to treating hallucinations are still in their infancy, the early results are promising (e.g., Bentall, Haddock, & Slade, 1994; Wiersma et al., 2001). But let us note that even if theories linking subvocal speech to auditory hallucinations stand up to further scientific inquiry, they cannot account for hallucinations in other sensory modalities, such as visual, tactile, or olfactory hallucinations.

The brain mechanisms responsible for hallucinations are likely to involve a number of interconnected systems. One intriguing possibility is that defects in deeper brain structures may lead the brain to create its own reality. This alternative reality goes unchecked because the higher thinking centers in the brain, located in the frontal lobes of the cerebral cortex, fail to perform a reality check on these images to determine whether they are real, imagined, or hallucinated (Begley, 1995). Consequently, people may misattribute their own internally generated voices to outside sources. As we'll see later, evidence from other brain-imaging studies points to abnormalities in the frontal lobes in people with schizophrenia.

Truth OR Fiction? REVISITED

Auditory hallucinations may be a form of inner speech.

TRUE. Recent research suggests that auditory hallucinations may be a form of inner speech, which, for unknown reasons, becomes misattributed to external sources.

Emotional Disturbances Disturbances of affect or emotional response in schizophrenia are typified by blunted affect—also called *flat affect*—and by inappropriate affect. Flat affect is inferred from the absence of emotional expression in the face and voice. People with schizophrenia may speak in a monotone and maintain an expressionless face, or "mask." They may not experience a normal range of emotional response to people and events. Or their emotional responses may be inappropriate, such as giggling at bad news.

It is not fully clear, however, whether emotional blunting in people with schizophrenia is a disturbance in their ability to express emotions, to report the presence of emotions, or to actually experience emotions (Berenbaum & Oltmanns, 1990). Recent laboratory-based evidence shows that schizophrenia patients experience more intense negative emotions, but less intense positive emotions, than controls (Myin-Germeys, Delespaul, & deVries, 2000). In other words, schizophrenia patients may experience strong emotions (especially negative emotions), even if their experiences are not communicated to the world outside through their facial expressions or behavior. People with schizophrenia may lack the capacity to express their emotions outwardly (Kring & Neale, 1996).

Other Types of Impairment People who suffer from schizophrenia may become confused about their personal identities—the cluster of attributes and characteristics that define themselves as individuals and give meaning and direction to their lives. They may fail to recognize themselves as unique individuals and be unclear as to how much of what they experience are parts of themselves. In psychodynamic terms, this phenomenon is sometimes referred to as loss of *ego boundaries.* They may also have difficulty adopting a third-party perspective and fail to perceive their own behavior and verbalizations as socially inappropriate in a given situation because they are unable to see things from another person's point of view (Carini & Nevid, 1992). They also have difficulty recognizing or perceiving emotions in others (Penn et al., 2000).

Disturbances of volition are most often seen in the residual or chronic state. These disturbances are characterized by loss of initiative to pursue goal-directed activities. People with schizophrenia may be unable to carry out plans and may lack interest or drive. Apparent ambivalence toward choosing courses of action may block goal-directed activities.

stupor A state of relative or complete unconsciousness in which a person is not aware of, or responsive to, the environment.

disorganized type The subtype of schizophrenia characterized by disorganized behavior, bizarre delusions, and vivid hallucinations.

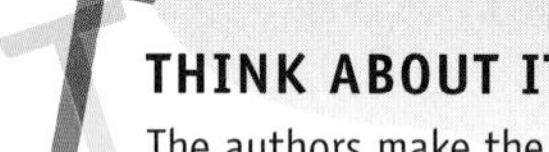

THINK ABOUT IT

The authors make the statement that schizophrenia is perhaps the most disabling type of psychological or mental disorder. What makes it so?

A young man diagnosed with disorganized schizophrenia. One of the characteristic features of disorganized schizophrenia is grossly inappropriate affect, as shown by this patient who continually giggles and laughs for no apparent reason.

People with schizophrenia may show highly excited or wild behavior, or slow to a state of **stupor.** They may exhibit odd gestures and bizarre facial expressions, or become unresponsive and curtail spontaneous movement. In extreme cases, as in catatonic schizophrenia, the person may seem unaware of the environment or maintain a rigid posture. Or the person may move about in an excited but seemingly purposeless manner.

People with schizophrenia also tend to show significant impairment in their interpersonal relationships. They tend to withdraw from social interactions and become absorbed in private thoughts and fantasies. Or they cling so desperately to others that they make them uncomfortable. They may become so dominated by their own fantasies that they essentially lose touch with the outside world. They also tend to have been introverted and peculiar even before the appearance of psychotic behavior (Berenbaum & Fujita, 1994). These early signs may be associated with a vulnerability to schizophrenia, at least in people with a genetic risk of developing the disorder.

Subtypes of Schizophrenia

The belief that there are different forms or types of schizophrenia traces back to Kraepelin, who listed three types of schizophrenia: paranoid, catatonic, and hebephrenic (now called disorganized type). The *DSM-IV* lists three specific types of schizophrenia: *disorganized, catatonic,* and *paranoid.* People with schizophrenia who display active psychotic features, such as hallucinations, delusions, incoherent speech, or confused or disorganized behavior, but who do not meet the specifications of the other types are considered to be of an *undifferentiated type.* Others who have no prominent psychotic features at the time of evaluation but have some residual features (for example, social withdrawal, peculiar behavior, blunted or inappropriate affect, strange beliefs or thoughts) would be classified as having a *residual type* of schizophrenia.

Here let us consider the specific types of schizophrenia recognized by the *DSM* system.

Disorganized Type The **disorganized type** of schizophrenia is associated with such features as confused behavior, incoherent speech, vivid and frequent hallucinations, flattened or inappropriate affect, and disorganized delusions that often involve sexual or religious themes. Social impairment is frequent among people with disorganized schizophrenia. They also display silliness and giddiness of mood, giggling and talking nonsensically. They often neglect their appearance and hygiene and lose control of their bladders and bowels. Consider the following case example:

A Case of Schizophrenia, Disorganized Type

A 40-year-old man who looks more like 30 is brought to the hospital by his mother, who reports that she is afraid of him. It is his twelfth hospitalization. He is dressed in a tattered overcoat, baseball cap, and bedroom slippers, and sports several medals around his neck. His affect ranges from anger (hurling obscenities at his mother) to giggling. He speaks with a childlike quality and walks with exaggerated hip movements and seems to measure each step very carefully. Since stopping his medication about a month ago, his mother reports, he had been hearing voices and looking and acting more bizarrely. He tells the interviewer he has been "eating wires" and lighting fires. His speech is generally incoherent and frequently falls into rhyme and clanging associations. His history reveals a series of hospitalizations since the age of 16. Between hospitalizations, he lives with his mother, who is now elderly, and often disappears for months at a time, but is eventually picked up by the police for wandering in the streets.

—Adapted from Spitzer et al., 1994, pp. 189–190

■

Catatonic Type The **catatonic type** is a subtype of schizophrenia characterized by markedly impaired motor behavior and a slowing down of activity that progresses to a stupor but may switch abruptly into an agitated phase. People with catatonic schizophrenia may show unusual mannerisms or grimacing, or maintain bizarre, apparently strenuous postures for hours, even though their limbs become stiff or swollen. A striking but less common feature is **waxy flexibility,** which involves the adoption of a fixed posture into which they have been positioned by others. They will not respond to questions or comments during these periods, which can last for hours. Later they may report they heard what others were saying at the time, however.

A Case of Schizophrenia, Catatonic Type

A 24-year-old man had been brooding about his life. He professed that he did not feel well, but could not explain his bad feelings. While hospitalized, he initially sought contact with people but a few days later was found in a statuesque position, his legs contorted in an awkward-looking position. He refused to talk to anyone and acted as if he couldn't see or hear anything. His face was an expressionless mask. A few days later, he began to talk, but in an echolalic or mimicking way. For example, he would respond to the question, "What is your name?" by saying, "What is your name?" He could not care for his needs and required to be fed by spoon.

—*Adapted from Arieti, 1974, p. 40*

A person diagnosed with catatonic schizophrenia. People with catatonic schizophrenia may remain in unusual, difficult positions for hours, even though their limbs become stiff or swollen. They may seem oblivious to their environment, even to people who are talking about them. Yet they may later say that they heard what was being said. Periods of stupor may alternate with periods of agitation.

Although catatonia is associated with schizophrenia, it may also occur in other physical and psychological disorders, including brain disorders, states of drug intoxication, metabolic disorders, and mood disorders ("What Is Catatonia?," 1995).

Paranoid Type The **paranoid type** of schizophrenia is characterized by preoccupations with one or more delusions or with the presence of frequent auditory hallucinations (APA, 2000). The behavior and speech of someone with paranoid schizophrenia does not show the marked disorganization typical of the disorganized type, nor is there a prominent display of flattened or inappropriate affect or catatonic behavior. Their delusions often involve themes of grandeur, persecution, or jealousy. They may believe, for example, that their spouse or lover is unfaithful despite a lack of evidence. They may also become highly agitated, confused, and fearful.

A Case of Schizophrenia, Paranoid Type

The 25-year-old woman was visibly frightened. She was shaking badly and had the look of someone who feared that she might be attacked at any moment. The night before she had been found cowering in a corner of the local bus station, mumbling incoherently, to herself having arrived in town minutes earlier on a bus from Philadelphia. The station manager had called the police, who took her to the hospital. She told the interviewer that she had to escape Philadelphia because the Mafia was closing in on her. She was a schoolteacher, she explained, at least until the voices started bothering her. The voices would tell her she was bad and had to be punished. Sometimes the voices were in her head, sometimes they spoke to her through the electrical wires in her apartment. The voices told her how someone

Truth OR Fiction? REVISITED

Some people with schizophrenia sustain unusual, uncomfortable positions for hours and will not respond to questions or communicate during these periods.

TRUE. People with the catatonic subtype of schizophrenia often maintain unusual and apparently uncomfortable positions and will fail to move or respond during these periods, which can last for hours.

catatonic type The subtype of schizophrenia characterized by gross disturbances in motor activity, such as catatonic stupor.

waxy flexibility A less common feature of catatonic schizophrenia, in which a person's limbs are moved into a certain posture or position by others and which the person then rigidly maintains.

paranoid type The subtype of schizophrenia characterized by hallucinations and systematized delusions, commonly involving themes of persecution.

from the Mafia would come to kill her. She felt that one of her neighbors, a shy man who lived down the hall, was in league with the Mafia. She felt the only hope she had was to escape. To go somewhere, anywhere. So she hopped on the first bus leaving town, heading nowhere in particular, except away from home.

—From the Authors' Files

On the run? People with paranoid schizophrenia hold systematized delusions that commonly involve themes of persecution and grandeur. They usually do not show the degree of confusion, disorganization, or disturbed motor behavior seen in people with catatonic or disorganized schizophrenia. Unless they are discussing the areas in which they are delusional, their thought processes can appear to be relatively intact.

Type I versus Type II Schizophrenia Some investigators have gone beyond the *DSM* topology in proposing other ways of typing schizophrenia. One alternative typology distinguishes between two basic types of schizophrenia, Type 1 and Type II (Crow, 1980a, 1980b, 1980c). Type I schizophrenia is characterized by the more flagrant symptoms, called **positive symptoms,** such as hallucinations, delusions, and looseness of associations, as well as by an abrupt onset, preserved intellectual ability, and a more favorable response to antipsychotic medication (Penn, 1998). Type II schizophrenia corresponds to a pattern consisting largely of the deficit or **negative symptoms** of schizophrenia. These involve a loss or reduction of normal functions, as shown by such features as lack of emotional expression, low or absent levels of motivation, loss of pleasure in activities, social withdrawal, and poverty of speech, as well as by a more gradual onset, intellectual impairment, and poorer response to antipsychotic drugs (USDHHS, 1999a).

One intriguing possibility is that Type I and Type II schizophrenia represent different pathological processes. The Type I pattern may involve a defect in the inhibitory (blocking) mechanisms in the brain that would normally control excessive or distorted behaviors. Underlying this malfunction may be a disturbance in the supply or regulation of dopamine in the brain, because antipsychotic drugs that regulate dopamine function generally have a favorable impact on positive symptoms. The negative symptoms associated with Type II schizophrenia represent the more enduring or persistent characteristics of schizophrenia. The Type II pattern is associated with poorer functioning before the person developed schizophrenia, or **premorbid functioning,** and with a more progressive decline in functioning leading to enduring disability (Earnst & Kring, 1997; McGlashan & Fenton, 1992). One possibility is that the negative symptoms associated with Type II schizophrenia are caused by structural damage in the brain.

The Type I–Type II distinction remains controversial, as evidence does not clearly support the existence of two distinct behavior patterns in schizophrenia. Some investigators (e.g., Kay, 1990) find that only a minority of people with schizophrenia can be classified as exhibiting either predominantly positive or negative symptoms. Positive and negative symptoms may not define distinct subtypes of schizophrenia but rather separate dimensions that can coexist in the same individual.

Perhaps, as recent research suggests, a three-dimensional model is most appropriate for grouping schizophrenic symptoms (Arango et al., 2000; USDHHS, 1999a). One dimension, a *psychotic dimension,* consists of delusional thinking and hallucinations. A *negative dimension* comprises negative symptoms, such as flat affect and poverty of speech and thought. The third dimension, labeled a *disorganized dimension,* includes inappropriate affect and thought disorder (disordered thought and speech). Although schizophrenic symptoms seem to cluster into these three dimensions, there is considerable overlap among the dimensions. Thus, while the three-factor model may have clinical value in representing particular clusters of symptoms, these dimensions do not appear to represent distinct subtypes of schizophrenia.

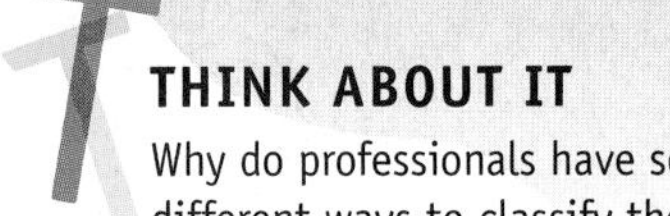

THINK ABOUT IT

Why do professionals have several different ways to classify the types of schizophrenia?

positive symptoms Flagrant symptoms of schizophrenia, such as hallucinations, delusions, bizarre behavior, and thought disorder.

negative symptoms Behavioral deficiencies associated with schizophrenia, such as social skills deficits, social withdrawal, flattened affect, poverty of speech and thought, psychomotor retardation, and failure to experience pleasure.

premorbid functioning The level of functioning before the person developed schizophrenia.

Theoretical Perspectives on Schizophrenia

The understanding of schizophrenia has been approached from each of the major theoretical perspectives. Though the underlying causes of schizophrenia remain elusive, they are presumed to involve biological abnormalities in combination with psychosocial and environmental influences (USDHHS, 1999a).

Psychodynamic Perspectives According to the psychodynamic perspective, schizophrenia represents the overwhelming of the ego by primitive sexual or aggressive drives or impulses arising from the id. These impulses threaten the ego and give rise to intense intrapsychic conflict. Under such a threat, the person regresses to an early period in the oral stage, referred to as *primary narcissism.* In this period, the infant has not yet learned that the world and it are distinct entities. Because the ego mediates the relationship between the self and the outer world, this breakdown in ego functioning accounts for the detachment from reality that is typical of schizophrenia. Input from the id causes fantasies to become mistaken for reality, giving rise to hallucinations and delusions. Primitive impulses may also carry more weight than social norms and be expressed in bizarre, socially inappropriate behavior.

Freud's followers, such as Erik Erikson and Harry Stack Sullivan, placed more emphasis on interpersonal than intrapsychic factors. Sullivan (1962), for example, who devoted much of his life's work to schizophrenia, emphasized the importance of impaired mother–child relationships, arguing that they can set the stage for gradual withdrawal from other people. In early childhood, anxious and hostile interactions between the child and parent lead the child to take refuge in a private fantasy world. A vicious cycle ensues: The more the child withdraws, the less opportunity there is to develop a sense of trust in others and the social skills necessary to establish intimacy. Then the weak bonds between the child and others prompt social anxiety and further withdrawal. This cycle continues until young adulthood. Then, faced with increasing demands at school or work and in intimate relationships, the person becomes overwhelmed with anxiety and withdraws completely into a world of fantasy.

Critics of Freud's views point out that schizophrenic behavior and infantile behavior are not much alike, so that schizophrenia cannot be explained by regression. Critics of Freud and modern psychodynamic theorists note that psychodynamic explanations are *post hoc,* or retrospective. Early child–adult relationships are recalled from the vantage point of adulthood rather than observed longitudinally. Psychoanalysts have not been able to demonstrate that hypothesized early childhood experiences or family patterns lead to schizophrenia.

Learning Perspectives Although learning theory may not account for schizophrenia, the principles of conditioning and observational learning may play a role in the development of some forms of schizophrenic behavior. From this perspective, people may learn to "emit" schizophrenic behaviors when these are more likely to be reinforced than normal behaviors.

Support for this view is found in operant conditioning studies in which bizarre behavior is shaped by reinforcement. Experiments involving individuals with schizophrenia show, for example, that reinforcement affects the frequency of bizarre versus normal verbalizations and that hospital patients can be shaped into performing odd behaviors. In a classic case example, Haughton and Ayllon (1965) conditioned a 54-year-old woman with chronic schizophrenia to cling to a broom. A staff member gave her the broom to hold and, when she did, another staff member gave her a cigarette. This pattern was repeated several times. Soon the woman could not be parted from the broom. But the fact that reinforcement can influence people to engage in peculiar behavior does not demonstrate that bizarre behaviors characterizing schizophrenia are learned behaviors determined by reinforcement.

Social-cognitive theorists suggest that modeling of schizophrenic behavior can occur within the mental hospital. In that setting, patients may begin to model themselves after fellow patients who act strangely. Hospital staff may inadvertently reinforce schizophrenic behavior by paying more attention to those patients who exhibit more bizarre behavior.

Truth OR Fiction? REVISITED

A 54-year-old hospitalized woman diagnosed with schizophrenia was conditioned to cling to a broom by being given cigarettes as reinforcers.

TRUE. A 54-year-old hospitalized woman with schizophrenia was in fact conditioned to cling to a broom by being given cigarettes as reinforcers. The issue is the extent to which principles of learning can account for the bizarre behavior patterns shown by people with schizophrenia.

Withdrawing into oneself? Harry Stack Sullivan and some other psychodynamic theorists see individuals with schizophrenia as withdrawing into private fantasy worlds, largely because of severely disturbed relationships with their mothers.

This understanding is consistent with the observation that schoolchildren who disrupt the class garner more attention from their teachers than well-behaved children do.

Perhaps some forms of schizophrenic behavior can be explained by the principles of modeling and reinforcement. However, many people come to display schizophrenic behavior patterns without prior exposure to other people with schizophrenia. In fact, the onset of schizophrenic behavior patterns is more likely to lead to hospitalization than to result from hospitalization.

Biological Perspectives Although we still have much to learn about the biological underpinnings of schizophrenia, most investigators today recognize that biological factors play a determining role.

Genetic Factors We now have compelling evidence that schizophrenia is strongly influenced by genetic factors (Charney, Nestler, & Bunney, 1999; Gottesman, 2001; USDHHS, 1999a). One source of evidence of genetic factors is based on familial studies. Schizophrenia, like many other disorders, tends to run in families (Erlenmeyer-Kimling et al., 1997). Cross-cultural evidence from studies in such countries as Sweden, Iceland, Ireland, as well as the United States, shows an increased risk of schizophrenia in people who have biological relatives with the disorder (Erlenmeyer-Kimling et al., 1997; Kendler & Diehl, 1993). Overall, first-degree relatives of people with schizophrenia (parents or siblings) have about a tenfold greater risk of developing schizophrenia than do members of the general population (APA, 2000; Kendler & Diehl, 1993).

The Genain quadruplets. Schizophrenia is more likely to affect individuals who have family members with the disorder. Here we see a photo of the Genain quadruplets, each of whom developed schizophrenia.

Further supporting a genetic linkage, the closer the genetic relationship between people diagnosed with schizophrenia and their family members, the greater the likelihood (or *concordance rate*) of schizophrenia in their relatives (Gottesman, 2001). Figure 13.1 shows the pooled results of European studies on family incidence of schizophrenia conducted from 1920 to 1987. However, the fact that families share common environments as well as common genes requires that we dig deeper to examine the genetic underpinnings of schizophrenia.

More support for a genetic contribution to schizophrenia is found in twin studies, which show concordance rates for the disorder among identical or monozygotic (MZ) twins of about 48% on the average, which is more than twice the rate found among fraternal or dizygotic (DZ) twins (about 17%) (Gottesman, 1991; Plomin et al., 1994). A twins study in Norway found an even greater spread: 48% concordance in MZ twins versus 3.6% in DZ twins (Onstad et al., 1991).

We should be cautious, however, not to overinterpret the results of twin studies. MZ twins not only share 100% genetic similarity, but they may also be treated more alike than DZ twins. Thus environmental factors may play a role in explaining the higher concordance rates found among MZ twins. To help sort out environmental from genetic factors, investigators have turned to adoption studies in which high-risk (HR) children (children of

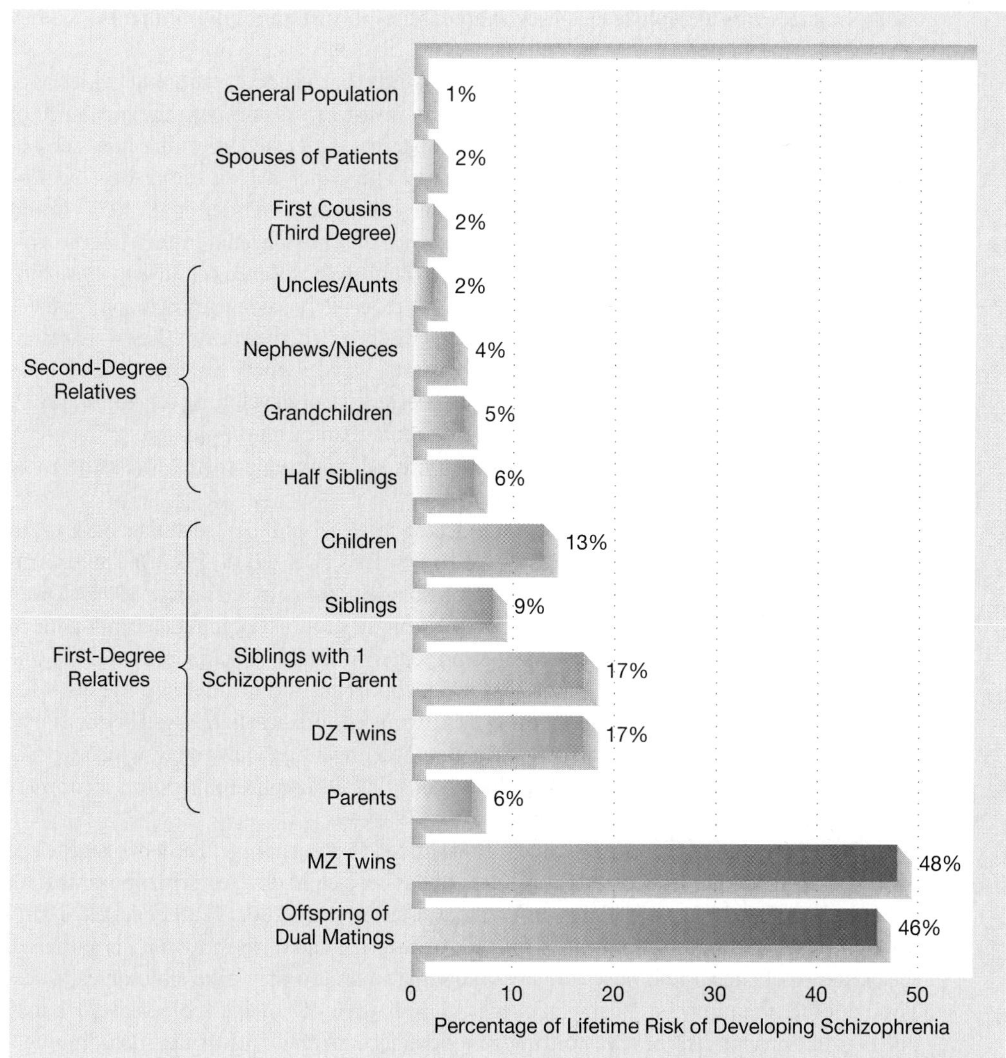

FIGURE 13.1 The familial risk of schizophrenia.
Generally speaking, the more closely one is related to people who have developed schizophrenia, the greater the risk of developing schizophrenia for oneself. Monozygotic (MZ) twins, whose genetic heritages overlap fully, are much more likely than dizygotic (DZ) twins, whose genes overlap by 50%, to be concordant for schizophrenia.
Source. Adapted from Gottesman et al. (1987).

one or more biological parents with schizophrenia) were adopted away shortly after birth and reared apart from their biological parents.

WWW **Web Link 13.5**
Genetics and Schizophrenia

Adoption studies provide the strongest evidence to date for a genetic contribution to schizophrenia. In perhaps the best known example, researchers in Denmark examined official registers and found 39 HR adoptees who had been reared apart from their biological mothers who had schizophrenia (Rosenthal et al., 1968, 1975). Three of the 39 HR adoptees (8%) were diagnosed with schizophrenia, as compared to 0% of a reference group of 47 adoptees whose biological parents had no psychiatric history.

Other investigators have approached the question of heredity in schizophrenia from the opposite direction. U.S. researcher Seymour Kety and Danish colleagues (Kety et al., 1975, 1978) used official records to find 33 index cases of children in Copenhagen, Denmark, who had been adopted early in life and were later diagnosed with schizophrenia. They compared the rates of diagnosed schizophrenia in the biological and adoptive relatives of the index cases with those of the relatives of a matched reference group that consisted of adoptees with no psychiatric history. The results strongly supported the genetic explanation. The incidence of diagnosed schizophrenia was greater among the biological relatives of the adoptees who had schizophrenia than among the biological relatives of the control adoptees. Adoptive relatives of both the index cases and control cases showed similar, *low* rates of schizophrenia. Similar results were found in later research that extended the scope of the investigation to the rest of Denmark (Kety et al., 1994). It thus appears that

cross-fostering study A method of determining heritability of a trait or disorder by examining differences in prevalence among adoptees reared by either adoptive or biological parents who possessed the trait or disorder in question.

dopamine theory The theory that proposes that schizophrenia involves overactivity of dopamine receptors in the brain.

neuroleptics A group of antipsychotic drugs (the "major tranquilizers") used in the treatment of schizophrenia, such as the phenothiazines (Thorazine, Mellaril, etc.).

THINK ABOUT IT

Summarize briefly the evidence that supports a genetic component to the development of schizophrenia. How strong is the genetic component? What other factors might be involved? Is the evidence that having a father over age 50 is a risk factor an argument for or against a genetic component?

family linkages in schizophrenia follow shared genes, not shared environments (Kinney et al., 1997; Moldin, 1994).

Still another approach, the **cross-fostering study,** has yielded additional evidence of genetic factors in schizophrenia. In this approach, investigators compare the incidence of schizophrenia among children whose biological parents either had or didn't have schizophrenia and who were reared by adoptive parents who either had or didn't have schizophrenia. Another Danish study by Wender and his colleagues (Wender et al., 1974) found the incidence of schizophrenia related to the presence of schizophrenia in the children's biological parents, but not in their adoptive parents. High-risk children (children whose biological parents had schizophrenia) were almost twice as likely to develop schizophrenia as those of nonschizophrenic biological parents, regardless of whether or not they were reared by a parent with schizophrenia. It is also notable that adoptees whose biological parents did not suffer from schizophrenia were placed at no greater risk of developing schizophrenia by being reared by an adoptive parent with schizophrenia than by a nonschizophrenic parent. In sum, a genetic relationship with a person with schizophrenia seems to be the most prominent risk factor for developing the disorder.

Though genetic factors are clearly implicated in schizophrenia, scientists have yet to discover any of the particular genes that may be involved (USDHHS, 1999a). The present understanding is that multiple genes are responsible (Buchsbaum & Hazlett, 1998; Cowan & Kandel, 2001). In sum, the evidence to date strongly supports a genetic component in schizophrenia. However, genetics alone does not determine risk of schizophrenia. For one thing, people may carry a high genetic risk of schizophrenia and not develop the disorder. For another, the rate of concordance among MZ twins, as noted earlier, is well below 100%, even though identical twins carry identical genes. The prevailing view today, which we discuss later, is the diathesis-stress model, which holds that schizophrenia involves a complex interplay of genetic and environmental factors.

Some cases of schizophrenia appear to be linked to paternal age. Fathers over the age of 50 in one study had three times the chances of having a child develop schizophrenia than did fathers under the age of 25 (Lewin, 2001; Malaspina et al., 2001; "Father's Age," 2001). Though scientists remain skeptical of the link to paternal age until more data is gathered, the existence of such a link raises the possibility that men too may have a biological clock. Though older men may be capable of siring children, their advancing biological clock may put them at greater risk of transmitting genetic defects to their offspring, including perhaps schizophrenia.

Biochemical Factors Contemporary biological investigations of schizophrenia have focused on the role of the neurotransmitter dopamine. The **dopamine theory** posits that schizophrenia involves an overreactivity of dopamine receptors in the brain—the receptor sites on postsynaptic neurons into which molecules of dopamine lock (Haber & Fudge, 1997).

The major source of evidence for the dopamine model is found in the effects of antipsychotic drugs called major tranquilizers or **neuroleptics.** The most widely used neuroleptics belong to a class of drugs called *phenothiazines,* which includes such drugs as Thorazine, Mellaril, and Prolixin. Neuroleptic drugs block dopamine receptors, thereby reducing the level of dopamine activity (Kane, 1996). As a consequence, neuroleptics inhibit excessive transmission of neural impulses that may give rise to schizophrenic behavior.

Another source of evidence supporting the role of dopamine in schizophrenia is based on the actions of amphetamines, a class of stimulant drugs. These drugs increase the concentration of dopamine in the synaptic cleft by blocking its reuptake by presynaptic neurons. When given in large doses to normal people, these drugs can lead to abnormal behavior states that mimic paranoid schizophrenia.

Overall, evidence points to irregularities in schizophrenia patients in the neural pathways in the brain that utilize dopamine (Meador-Woodruff et al., 1997). The specific nature of this abnormality remains unclear. We can't yet say whether the abnormality involves overreactivity of particular dopamine pathways or more complex interactions among dopamine systems. One possibility is that overreactivity of dopamine receptors may be involved in producing more flagrant behavior patterns (positive symptoms) but not the negative symptoms or deficits associated with schizophrenia. Decreased, rather than increased, dopamine reactivity may be connected with some of the negative symptoms of schizophrenia (Earnst & Kring, 1997; Okubo et al., 1997). We should also note that other neurotransmitters, such as norepinephrine, serotonin, and GABA, also appear to be involved in schizophrenia (Busatto et al., 1997).

Viral Infections Might schizophrenia be caused by a slow-acting virus that attacks the developing brain of a fetus or newborn child? Prenatal rubella (German measles), a viral infection, is a cause of later mental retardation. Could another virus give rise to schizophrenia?

Viral infections are most prevalent in the winter months. The viral theory could account for findings of an excess number of people who later develop schizophrenia being born in the winter (Mortensen et al., 1999; Tam & Sewell, 1995). However, we have yet to find an identified viral agent we can link to schizophrenia. Thus we must consider the viral theory of schizophrenia to be intriguing but inconclusive. Even if a viral basis for schizophrenia were discovered, it would probably account for but a small fraction of cases.

Brain Abnormalities Despite the widespread belief among professionals that schizophrenia is a brain disease, we are still asking the question, "Where is the pathology?" (Stevens, 1997). Researchers are trying to answer this question by using modern brain-imaging techniques, including PET scans, EEGs, CT scans, and MRIs, that probe the inner workings of the brain. Evidence from brain-imaging studies point to various abnormalities in the brains of many people with schizophrenia (e.g., Gur et al., 1998; Ettinger et al., 2001).

The clearest finding of structural damage in the brain is evidence of enlarged brain ventricles (hollow spaces in the brain) occurring in perhaps three of four schizophrenia patients (Coursey, Alford, & Safarjan, 1997) (see Figure 13.2). Enlarged ventricles are signs of loss of brain tissue (cell loss). Investigators also find that brains of schizophrenia patients are about 5% smaller, on average, in total volume than those of normal individuals, with the greatest volume reductions in the cerebral cortex (Cowan & Kandel, 2001).

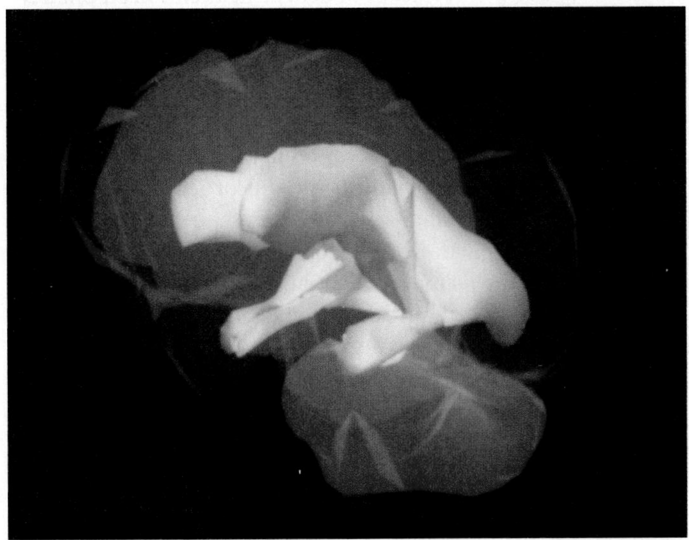

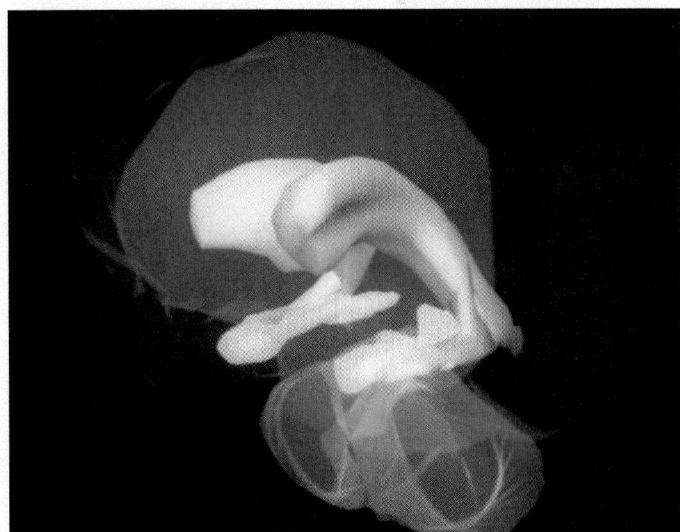

FIGURE 13.2 Structural changes in the brain of a person with schizophrenia as compared with that of a normal subject.
The magnetic resonance imaging (MRI) of the brain of a person with schizophrenia (left) shows a relatively shrunken hippocampus (yellow) and relatively enlarged, fluid-filled ventricles (gray) when compared to the structures of the normal subject (right). The MRI was conducted by schizophrenia researcher Nancy C. Andreasen.

Source: Gershon, E. S., & Rieder, R. O. (1992). Major disorders of mind and brain. *Scientific American, 267*(No. 3), p. 128.

Bear in mind, however, that not all people with schizophrenia show any evidence of enlarged ventricles or other signs of structural damage to brain tissue. This finding leads researchers to suspect that there may be several forms of schizophrenia that have different causal processes. One form may involve a degenerative loss of brain tissue (Knoll et al., 1998). Scientists suspect that some cases of schizophrenia may be associated with a progressive loss of brain tissue or perhaps the failure of the brain to have developed normally in the first place. They suspect that brain abnormalities may be the result of prenatal viral infections, inadequate fetal nutrition, genetic defects, or birth traumas or complications (McGlashan & Hoffman, 2000; McNeil, Cantor-Graae, & Weinberger, 2000; Rosso et al., 2000; Wahlbeck et al., 2001). Still, not all people with schizophrenia show signs of brain damage or abnormal brain development. Some forms of schizophrenia may involve reduced brain growth or degenerative loss of brain tissue early in life whereas others may not (Baaré et al., 2001; Knoll et al., 1998).

A considerable body of evidence points to abnormalities in the prefrontal cortex in schizophrenia patients (Glantz & Lewis, 2000; Gur et al., 2000a; Sanfilipo et al., 2000). Some of this evidence shows reduced brain wave activity in the prefrontal cortex of many schizophrenia patients (Kim et al., 2000; Ragland et al., 2001) (see Figure 13.3). Other evidence shows a loss of brain tissue in the prefrontal cortex in some schizophrenia patients (Fannon et al., 2000; Karp et al., 2001; Mathalno et al.. 2001; Staal et al., 2000).

The prefrontal cortex is the thinking and organizing center of the brain. It is the area of the frontal lobes of the cerebral cortex that lies in front of the motor cortex (the part of the brain that controls voluntary body movements). The prefrontal cortex is involved in controlling many cognitive and emotional functions, the kinds of functions that are often impaired in people with schizophrenia. The prefrontal cortex serves as a kind of mental clipboard for holding information needed to guide organized behavior (Casanova, 1997). Prefrontal abnormalities may explain why people with schizophrenia have difficulty organizing their thoughts and behavior and performing higher level cognitive tasks, such as formulating concepts, prioritizing information, and formulating goals and plans (Barch et al., 2001; Bertolino et al., 2000; Callicott et al., 2000). The prefrontal cortex is also involved in regulating attention, so findings of prefrontal abnormalities coincide with research evidence of deficits in attention among people with schizophrenia. These are intriguing findings that may provide clues as to the biological bases of schizophrenia.

We have yet other evidence of structural differences in the brains of schizophrenia patients in subcortical regions of the brain—structures in the brain lying beneath the cortex that are involved in regulating emotions, attention, and memory formation (e.g., Byrne et al., 2001; Ettinger et al., 2001; Gur et al., 2000b; Wright et al., 2000). Disturbances in brain physiology in subcortical regions, perhaps involving neurotransmitter imbalances or faulty connections (wiring) between neurons, may contribute to disturbances in thinking, attention, and emotions associated with schizophrenia.

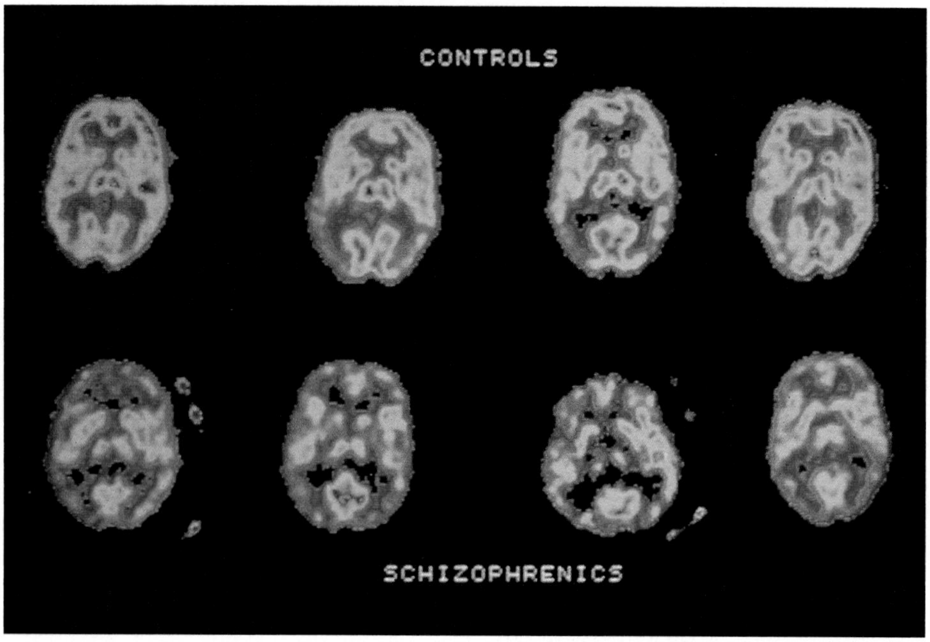

THINK ABOUT IT

What have we learned about the biological bases of schizophrenia? What don't we know? What might you say to critics who claim that schizophrenia is not a disease because no one has yet found any specific disease process in the brain that accounts for it?

FIGURE 13.3 PET scans of people with schizophrenia versus normals.
Positron emission tomography (PET scan) evidence of the metabolic processes of the brain show relatively less metabolic activity (indicated by less yellow and red) in the frontal lobes of the brains of people with schizophrenia. PET scans of the brains of four normal people are shown in the top row, and PET scans of the brains of four people with schizophrenia are shown below.

Family Theories Disturbed family relationships have long been regarded as playing a role in the development and course of schizophrenia (Miklowitz, 1994). Early family theories of schizophrenia focused on the role of a "pathogenic" family member, such as the **schizophrenogenic mother** (Fromm-Reichmann, 1948, 1950). In what some feminists view as historic psychiatric sexism, the schizophrenogenic mother was described as cold, aloof, overprotective, and domineering. She was characterized as stripping her children of self-esteem, stifling their independence, and forcing them into dependency on her. Children reared by such mothers were believed to be at special risk for developing schizophrenia if their fathers were passive and failed to counteract the pathogenic influences of the mother. Despite extensive research, however, mothers of people who develop schizophrenia do not fit the stereotypical picture of the schizophrenogenic mother (Hirsch & Leff, 1975).

schizophrenogenic mother A since-discarded concept of a cold but overprotective mother who, it was believed, was capable of causing schizophrenia in her children.

double-bind communications A communication pattern involving contradictory or mixed messages without acknowledging the inherent conflict.

In the 1950s, family theorists began to focus on the role of disturbed communications in the family. One of the more prominent theories, put forth by Gregory Bateson and his colleagues (1956), was that **double-bind communications** contribute to the development of schizophrenia. A double-bind communication transmits two mutually incompatible messages. In a double-bind communication with a child, a mother might freeze up when the child approaches her and then scold the child for keeping a distance. Whatever the child does, she or he is wrong. With repeated exposure to such double binds, the child's thinking may become disorganized and chaotic. The double-binding mother prevents discussion of her inconsistencies, because she cannot admit to herself that she is unable to tolerate closeness. Note this vignette:

A Case Example of Double-Bind Communication

A young man who had fairly well recovered from an acute schizophrenic episode was visited in the hospital by his mother. He was glad to see her and impulsively put his arm around her shoulders whereupon she stiffened. He withdrew his arm and she asked, "Don't you love me anymore?" He then blushed and she said, "Dear, you must not be so easily embarrassed and afraid of your feelings." The patient was able to stay with her only a few minutes more and following her departure he assaulted an aide.

—From Bateson et al., 1956, p. 251

■

Perhaps double-bind communications serve as a source of family stress that increases the risk of schizophrenia in genetically vulnerable individuals. In more recent years, investigators have broadened the investigation of family factors in schizophrenia by viewing the family in terms of a system of relationships among the members rather than singling out mother–child or father–child interactions. Research has begun to identify stressful factors in the family that may interact with a genetic vulnerability in leading to the development of schizophrenia. Two principal sources of family stress that have been studied are patterns of deviant communications and negative emotional expression in the family.

Communication Deviance Communication deviance (CD) is a pattern of unclear, vague, disruptive, or fragmented communication that is often found among parents and family members of schizophrenia patients. CD is characterized by speech that is hard to follow and from which it is difficult to extract any shared meaning (Wahlberg et al., 2001). High CD parents also have difficulty focusing on what their children are saying (Miklowitz, 1994). They tend to verbally attack their children rather than offer constructive criticism and may subject them to double-bind communications. They also tend to interrupt the child with intrusive, negative comments. They are prone to telling the child what she or he "really" thinks rather than allowing the child to formulate her or his own thoughts and feelings. Parents of people with schizophrenia show higher levels of communication deviance than parents of people without schizophrenia (Miklowitz, 1994).

Communication deviance may be one of the stress-related factors that increase the risk of development of schizophrenia in genetically vulnerable individuals (Goldstein, 1987).

Then too, the causal pathway may work in the opposite direction. Perhaps communication deviance is a parental reaction to the behavior of disturbed children. Parents may learn to use odd language as a way of coping with children who continually interrupt and confront them. Or perhaps parents and children share genetic traits that become expressed as disturbed communications and increased vulnerability toward schizophrenia, without there being a casual link between the two.

Expressed Emotion Another measure of disturbed family communications is called expressed emotion (EE). EE involves the tendency of family members to be hostile, critical, and unsupportive of their schizophrenic family members. People with schizophrenia whose families are high in EE tend to show poorer adjustment and have higher rates of relapse following release from the hospital than those with more supportive families (Cutting & Docherty, 2000; King & Dixon, 1999). Assessment of personality functioning shows high EE relatives to possess less empathy, tolerance, and flexibility than low EE relatives (Hooley & Hiller, 2000). High EE relatives also tend to believe that schizophrenia patients can exercise greater control over their behavior than do low EE relatives of the same patients (Weisman et al., 2000). Expressed emotion in relatives is also associated with a greater risk of relapse from other disorders, such as major depression and eating disorders (Butzlaff & Hooley, 1998).

Truth OR Fiction? REVISITED

Living in a family that is hostile, critical, and unsupportive can increase the risk of relapse in people with schizophrenia.

TRUE. Schizophrenia patients in families with hostile, critical, and unsupportive family members who are at greater risk of relapse than those from more supportive families.

Low EE families may actually serve to protect, or buffer, the family member with schizophrenia from the adverse impact of outside stressors and help prevent recurrent episodes (see Figure 13.4). Yet family interactions are a two-way street. Family members and patients influence each other and are influenced in turn. Disruptive behaviors by the schizophrenic family member frustrate other members of the family, prompting them to respond to the person in a less supportive and more critical and hostile way. This in turn can exacerbate the schizophrenia patient's disruptive behavior (Bellack & Mueser, 1993).

Relationships between expressed emotion and rates of recurrence of schizophrenia have been drawn largely from research with non-Hispanic White samples in England and the United States (Karno et al., 1987). Because family patterns are often influenced by cultural factors, researchers have begun to explore whether the EE construct has value in predicting recurrence of schizophrenia in other cultures. Some cross-cultural support for the prognostic value of the construct comes from a study of low-income, relatively unacculturated Mexican American family members of people with schizophrenia (Karno et al., 1987). Paralleling findings with the non-Hispanic White British and American families, high levels of EE (that is, critical, hostile, and emotionally overinvolved attitudes and behaviors) among key family members in the Mexican American sample were associated with an increased risk of relapse among schizophrenia patients living with their families after hospitalization. Evidence from samples of both Mexican American families and Anglo American families with high levels of expressed emotion show them to be more likely than low EE families to view the psychotic behavior of a schizophrenic family member to lie within the person's control (Weisman et al., 1993, 1998). The anger and criticism of high EE family members may stem from the perception that patients can and should exert greater control over their aberrant behavior.

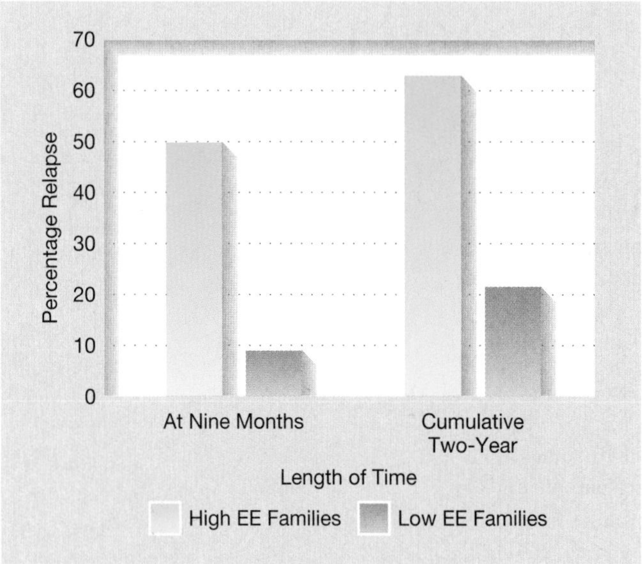

FIGURE 13.4 Relapse rates of people with schizophrenia in high and low EE families.

People with schizophrenia whose families are high in expressed emotion (EE) are at greater risk of relapse than those whose families are low in EE. Whereas low-EE families may help protect the family member with schizophrenia from environmental stressors, high-EE families may impose additional stress.

Source. Adapted from Leff & Vaughn (1981).

We also have evidence for cultural variations in expressed emotion. Much lower prevalences of high EE behaviors, as compared with the (non-Hispanic) White British and American families, were found among a sample of Mexican American families and families from northern India who had family members with schizophrenia (Wig et al., 1987). The (non-Hispanic) White families thus tended to be more openly critical of the members of their families who had schizophrenia than were family members from these other cultural backgrounds. The specific components of EE may also vary across cultures (Martins, de Lemos, & Bebbington, 1992).

Other researchers find that the extended family structure often found in traditional cultures may offer an emotional and financial buffer against the hardships imposed by the behavioral excesses and deficiencies of people with schizophrenia. In Western cultures, these burdens are more likely to be borne by the nuclear family (Lefley, 1990). Such differences remind us of the need to take cultural factors into account when examining relationships between family factors and schizophrenia.

Families of people with schizophrenia tend to have little if any preparation or training for coping with the stressful demands of caring for them (Winefield & Harvey, 1994). Rather than focusing so much attention on the negative influence of high EE family members, perhaps we should seek to help family members learn more constructive ways of relating to one another. Evidence shows that families can be helped to reduce their level of expressed emotion (Penn & Mueser, 1996).

Family Factors in Schizophrenia: Causes or Sources of Stress? No evidence supports the belief that family factors, such as negative family interactions, lead to schizophrenia in children who do not have a genetic vulnerability. What then is the role of family factors in schizophrenia? Within the diathesis-stress model, disturbed patterns of emotional interaction and communication in the family represent a source of potential stress that may increase the risks of developing schizophrenia among people with a genetic predisposition for the disorder. Perhaps these increased risks could be minimized or eliminated if families are taught to handle stress and to be less critical and more supportive of the members of their families with schizophrenia. Counseling programs that help family members of people with chronic schizophrenia learn to express their feelings without attacking or criticizing the person with schizophrenia may prevent family conflicts that damage the person's adjustment. The family member with schizophrenia may also benefit from efforts to reduce the level of contact with relatives who fail to respond to family interventions.

How families conceptualize mental disorders also has a bearing on how they relate to relatives who suffer from them. For example, the term *schizophrenia* is connected with a stigma in our society and with the expectation that the disorder is enduring (Jenkins & Karno, 1992). In contrast, to many Mexican Americans, a person with schizophrenia is perceived as suffering from *nervios* ("nerves"), a cultural label attached to a wide range of troubled behaviors, including anxiety, schizophrenia, and depression, and one that carries less stigmata and more positive expectations than the label of schizophrenia (Jenkins, 1988; Jenkins & Karno, 1992). Researchers believe the label *nervios* may have the effect of destigmatizing the person with schizophrenia:

> Since severe cases of *nervios* are not considered blameworthy or under an individual's control, the person who suffers its effects is deserving of sympathy, support, and special treatment. Moreover, severe cases of *nervios* are potentially curable. It is interesting to note that Mexican-descent relatives do not adopt another possible cultural label for craziness, *loco.* As a *loco,* the individual would be much more severely stigmatized and considered to be out of control with little chance for recovery. . . .
>
> Defining the problem as *nervios,* a common condition that in its milder forms afflicts nearly everyone, provides them a way of identifying with and minimizing the problem by claiming that the ill relative is "just like me, only more so." (Jenkins & Karno, 1992, pp. 17–18)

Family members may respond differently to relatives who have schizophrenia if they ascribe aspects of their behavior to a temporary or curable condition, which they believe can be altered by willpower, than if they believe the behavior is caused by a permanent brain abnormality. The degree to which relatives perceive family members with schizophrenia as having control over their disorders may be a critical factor in how they respond to them. Families may cope better with a family member with schizophrenia if they take a balanced view, believing on the one hand that people with schizophrenia can maintain some control

THINK ABOUT IT

Have you known anyone diagnosed with schizophrenia? What information do you have about the person's family history, family relationships, and stressful life events that might shed light on the development of the disorder?

over their behavior, while allowing that some of their odd or disruptive behavior is a product of their underlying disorder (Weisman et al., 1993). It remains to be seen whether these different ways in which family members conceptualize schizophrenia are connected with differences in the rates of recurrence of the disorder among affected family members.

Tying It Together: The Diathesis-Stress Model

In 1962, psychologist Paul Meehl proposed an integrative model that led to the development of the diathesis-stress model. Meehl suggested that certain people possess a genetic predisposition to schizophrenia that is expressed behaviorally only if they are reared in stressful environments (Meehl, 1962, 1972).

Later, Zubin and Spring (1977) formulated the diathesis-stress model, which views schizophrenia in terms of the interaction or combination of a *diathesis,* in the form of a genetic predisposition to develop the disorder, with environmental stress that exceeds the individual's stress threshold or coping resources. Environmental stressors may include psychological factors, such as family conflict, child abuse, emotional deprivation, or loss of supportive figures, as well as physical environmental influences, such as early brain trauma or injury. On the other hand, if environmental stress remains below the person's stress threshold, schizophrenia may never develop, even in persons at genetic risk (see Figure 13.5).

But what is the biological basis for the diathesis? No one has yet been able to find any specific brain abnormality present in all individuals who receive a schizophrenia diagnosis (Powchik et al., 1998; Stevens, 1997). Perhaps it shouldn't surprise us that a "one-size-fits-all" model doesn't apply. Schizophrenia is a complex disorder characterized by different subtypes and symptom complexes. There may be different causal processes in the brain explaining different forms of schizophrenia. What we now call *schizophrenia* may turn out to be more than one disorder (Buchanan & Carpenter, 1997; Knoll et al., 1998).

We noted two possible causal processes, one involving structural damage to brain tissue, the other involving disturbed neurotransmitter functioning. These factors, or perhaps a combination of the two, result in disrupted brain circuits involving the prefrontal cortex and its connections to lower brain regions responsible for organizing our thoughts, perceptions, emotions, and attentional processes. These neural networks are involved in processing information efficiently and turning it into meaningful thoughts and behavior. Defect in this circuitry may be involved in explaining the more flagrant, positive features of schizophrenia such as hallucinations, delusions, and thought disorder.

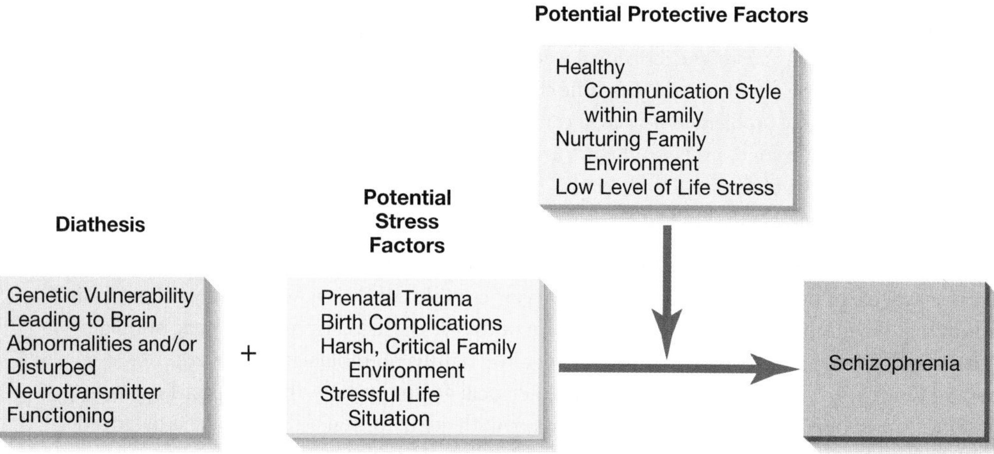

FIGURE 13.5 Diathesis-stress model of schizophrenia.

Research Evidence Supporting the Diathesis-Stress Model Several lines of evidence support the diathesis-stress model. One source of support is the fact that schizophrenia tends to develop in late adolescence or early adulthood, around the time that young people typically face the increased stress associated with developmental challenges relating to establishing independence and finding a role in life. Other evidence shows that psychosocial stress, such as harping criticisms from family members, worsens symptoms in people with schizophrenia, increasing risks of relapse (King & Dixon, 1995). However, the question of whether stress directly triggers the initial onset of schizophrenia in genetically vulnerable individuals still remains open to debate (Walker & Diforio, 1997).

Other sources of stress that may contribute to the development of schizophrenia in genetically vulnerable individuals involve sociocultural factors associated with poverty, such as overcrowding, poor diet and sanitation, impoverished housing, and inadequate health care (Kety, 1980). Other support for the diathesis-stress model comes from longitudinal studies of high-risk (HR) children who are at increased genetic risk of developing the disorder by virtue of having one or both parents with schizophrenia. Longitudinal studies of HR children (offspring of parents with schizophrenia) support the central tenet of the diathesis-stress model that heredity interacts with environmental influences in determining vulnerability to schizophrenia. Longitudinal studies track individuals over extended periods of time. Ideally they begin before the emergence of the disorder or behavior pattern in question and follow its course. In this way, investigators may identify early characteristics that predict the later development of a particular disorder, such as schizophrenia. These studies require a commitment of many years and substantial cost. Because schizophrenia occurs in only about 1% of the general population, researchers have focused on HR children because they are more likely to develop the disorder. Children with one parent with schizophrenia have about a 10% to 25% chance of developing schizophrenia, and those with two parents have about a 45% risk (Erlenmeyer-Kimling et al., 1997; Gottesman, 1991). Still, even children with two biological parents with schizophrenia stand a slightly better than even chance of not developing the disorder themselves.

Truth OR Fiction? REVISITED

If you have two parents with schizophrenia, it's nearly certain that you will develop schizophrenia yourself.

FALSE. Children of parents who both have schizophrenia have slightly less than a 50% chance (about 45%) of developing the disorder themselves.

The best known longitudinal study of HR children was undertaken by Sarnoff Mednick and his colleagues in Denmark. In 1962, the Mednick group identified 207 HR children (children whose mothers had schizophrenia) and 104 reference subjects who were matched for such factors as gender, social class, age, and education but whose mothers did not have schizophrenia (Mednick, Parnas, & Schulsinger, 1987). The children from both groups ranged in age from 10 to 20 years, with a mean of 15 years. None showed signs of disturbance when first interviewed.

Five years later, at an average age of 20, the children were reexamined. By then 20 of the HR children were found to have demonstrated abnormal behavior, although not necessarily a schizophrenic episode (Mednick & Schulsinger, 1968). The children who showed abnormal behavior, referred to as the HR "sick" group, were then compared with a matched group of 20 HR children from the original sample who remained well functioning (an HR "well" group) and a matched group of 20 low-risk (LR) subjects. It turned out that the mothers of the HR "well" offspring had experienced easier pregnancies and deliveries than those of the HR "sick" group or the LR group. Seventy percent of the mothers of the HR "sick" children had serious complications during pregnancy or delivery. Perhaps, consistent with the diathesis-stress model, complications during pregnancy, childbirth, or shortly after birth cause brain damage (a stress factor) that in combination with a genetic vulnerability leads to severe mental disorders in later life. Finnish researchers also find links between fetal and postnatal abnormalities and the development of schizophrenia in adulthood (Jones et al., 1998). The low rate of complications during pregnancy and birth in the HR "well" group in the Danish study suggests that normal pregnancies and births may actually help protect HR children from developing abnormal behavior patterns (Mednick et al., 1987).

Evaluation of these same HR subjects in the late 1980s, when they averaged 42 years of age and had passed through the period of greatest risk for development of schizophrenia, showed a significantly higher percentage of schizophrenia in the HR group than the LR comparison group, 16% versus 2%, respectively (Parnas et al., 1993).

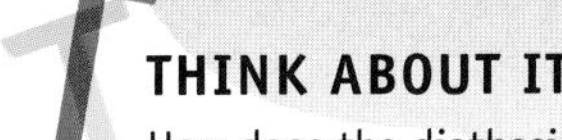

THINK ABOUT IT

How does the diathesis-stress model try to account for the development of schizophrenia? What evidence supports the model?

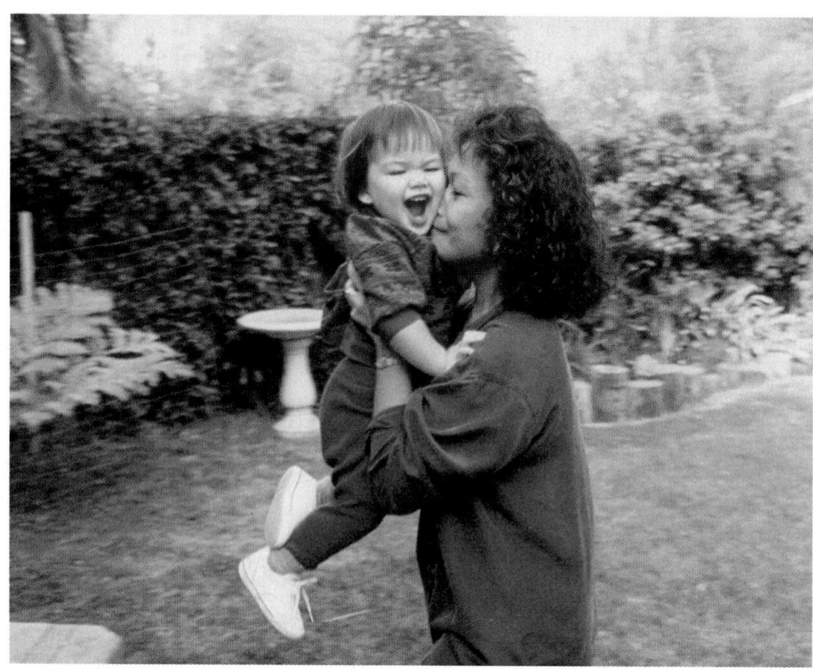

Protective factors in high-risk children. A supportive and nurturing environment may reduce the likelihood of developing schizophrenia among high-risk children.

Certain environmental factors, such as good parenting, may actually have a protective role in preventing the development of the disorder in people at increased genetic risk. In support of the role of early environmental influences, Mednick and his colleagues found that HR children who developed schizophrenia had poorer relationships with their parents than did HR children who did not go on to develop the disorder (Mednick et al., 1987). The presence of childhood behavior problems may also be a marker for the later development of schizophrenia-related disorders in HR children (Amminger et al., 1999).

Treatment Approaches

There is no cure for schizophrenia. Treatment is generally multifaceted, incorporating pharmacological, psychological, and rehabilitative approaches. Most people treated for schizophrenia in organized mental health settings receive some form of antipsychotic medication, which is intended to control the more flagrant behavior patterns, such as hallucinations and delusions, and decrease the risk of recurrent episodes.

Biological Approaches The advent in the 1950s of antipsychotic drugs—also referred to as major tranquilizers or neuroleptics—revolutionized the treatment of schizophrenia and provided the impetus for large-scale releases of mental patients to the community (deinstitutionalization). Antipsychotic medication helped control the more flagrant behavior patterns of schizophrenia and reduced the need for long-term hospitalization when taken on a maintenance or continuing basis after an acute episode (Kane, 1996; Sheitman et al., 1998). Yet for many patients with chronic schizophrenia, entering a hospital is like going through a revolving door. That is, they are repeatedly admitted and discharged within a relatively brief time frame. Many are simply discharged to the streets once they are stabilized on medication and receive little if any follow-up care or available housing. This often leads to a pattern of chronic homelessness punctuated by brief stays in the hospital. Only a small proportion of people with schizophrenia who are discharged from long-term care facilities are successfully reintegrated into the community (Bellack & Mueser, 1990).

Web Link 13.6 WWW
Treatment Guidelines for Schizophrenia

Commonly used antipsychotic drugs include the phenothiazines *chlorpromazine* (Thorazine), *thioridazine* (Mellaril), *trifluoperazine* (Stelazine), and *fluphenazine* (Prolixin). *Haloperidol* (Haldol), which is chemically distinct from the phenothiazines, produces similar effects.

Though we can't say with certainty how these drugs work, it appears they derive their therapeutic effect from blocking dopamine receptors in the brain. This reduces dopamine activity, which seems to quell the more flagrant signs of schizophrenia, such as hallucinations and delusions. The effectiveness of antipsychotic drugs has been repeatedly demonstrated in double-blind, placebo-controlled studies (Kane, 1996). Yet a substantial minority of people with schizophrenia receive little benefit from traditional neuroleptics, and no clear-cut factors determine who will best respond (Kane & Marder, 1993).

The major risk of long-term treatment with neuroleptic drugs (possibly excluding clozapine) is a potentially disabling side effect called **tardive dyskinesia (TD).** TD is an involuntary movement disorder that can affect any body part (Hansen, Casey, & Hoffman, 1997). It is irreversible in many cases, even when the neuroleptic medication is withdrawn. It occurs most often in patients who are treated with neuroleptics for 6 months or longer. TD can take different forms, the most common of which is frequent eye blinking. Common signs of the disorder include involuntary chewing and eye movements, lip smacking

tardive dyskinesia (TD) A disorder characterized by involuntary movements of the face, mouth, neck, trunk, or extremities and caused by long-term use of antipsychotic medication.

and puckering, facial grimacing, and involuntary movements of the limbs and trunk. In some cases, the movement disorder is so severe that patients have difficulty breathing, talking, or eating. Overall, about one in four people receiving long-term treatment with neuroleptics eventually develops TD (Jeste & Caligiuri, 1993).

TD is most common among older people and among women (Hansen et al., 1997). Unfortunately, we lack a safe and effective treatment for TD (Egan, Apud, & Wyatt, 1997; Sheitman et al., 1998). Although TD tends to improve gradually or stabilize over a period of years, many people with TD remain persistently and severely disabled.

The risk of this potentially disabling side effect requires physicians to weigh carefully the risks and benefits of long-term treatment with these drugs. Investigators have altered drug regimens in the attempt to reduce the risk of TD, such as by stopping medication in stable outpatients and starting it again when early symptoms reappear. However, intermittent medication schedules are associated with a twofold increase in the risk of relapse and have not been shown to lower the risk of TD (Kane, 1996).

A new generation of drugs, commonly referred to as atypical antipsychotic drugs (*clozapine, risperidone,* and *olanzapine* are examples), are as effective as the older drugs but carry fewer neurological side effects than conventional antipsychotics (Arranz et al., 2000; Kapur & Remington, 2000). They may help schizophrenia patients who have failed to respond to more conventional antipsychotics (Essock et al., 2000; Geddes et al., 2000; Wahlbeck et al., 1999). They also hold promise of providing more effective treatment of negative symptoms than do conventional antipsychotics.

One of these atypical antipsychotics, clozapine, is the only known drug that carries a minimal risk of TD (Conley & Buchanan, 1997; Kane, 1996). However, other side effects limit its use, especially the risk of a potentially lethal disorder in which the body produces inadequate supplies of white blood cells. Because of this risk, patients receiving the drug must receive frequent blood monitoring (USDHHS, 1999a). There are also promising findings that olanzapine may reduce the risk of TD relative to conventional antipsychotics (Tollefson et al., 1997), but more research is needed to reach a more definitive conclusion (Egan et al., 1997).

Antipsychotic drugs help control the more flagrant or bizarre features of schizophrenia, but are not a cure. People with chronic schizophrenia typically receive maintenance doses of antipsychotic drugs once their flagrant symptoms abate. Continued medication reduces the rate of relapse but is no panacea. Many patients, perhaps 15% to 20% per year, relapse even if they are maintained on continued medication (Kane, 1996). Yet estimates are that 75% of patients with schizophrenia who have been in remission for a year or more will relapse within 12 to 18 months if their medication is withdrawn (Kane, 1996). Still, not all people with schizophrenia require antipsychotic medication to maintain themselves in the community. Unfortunately, no one can yet predict which patients can manage effectively without continued medication.

Medication alone is insufficient to meet the multifaceted needs of people with schizophrenia. Drug therapy needs to be supplemented with psychoeducational programs that help schizophrenia patients develop better social skills and adjust to demands of community living. A wide array of treatment components are needed within a comprehensive model of care, including such elements as antipsychotic medication, medical care, family therapy, social skills training, crisis intervention, rehabilitation services, and housing and other social services (Penn & Mueser, 1996; USDHHS, 1999a). Treatment programs also need to ensure a continuity of care between the hospital and the community.

Sociocultural Factors in Treatment Investigators find that response to psychiatric medications and dosage levels varies with patient ethnicity (USDHHS, 1999a). Asians and Hispanics, for example, may require lower doses of neuroleptics than do Caucasians. Asians also tend to experience more side effects from the same dosage.

Ethnicity may also play a role in the family's involvement in the treatment process. In a study of 26 Asian Americans and 26 non-Hispanic White Americans with schizophrenia, family members of the Asian American patients were more frequently involved in the treatment program (Lin et al., 1991). For example, the Asian American

Truth OR Fiction? REVISITED

Drugs developed in the past few years not only treat schizophrenia, but also can cure it in many cases.

FALSE. Antipsychotic drugs help control the symptoms of schizophrenia but cannot cure the disorder.

THINK ABOUT IT

What are the relative risks and benefits of antipsychotic medication? Why is medication alone not sufficient to treat schizophrenia? Do you believe that people with schizophrenia should be treated indefinitely with antipsychotic drugs? Why or why not?

patients were more likely to be accompanied to their medication evaluation sessions by family members. The authors believe that the greater family involvement among Asian Americans represents the relatively stronger sense of family responsibility in Asian cultures. Non-Hispanic White Americans are more likely to emphasize individualism and self-responsibility.

Maintaining connections between the person with schizophrenia and the family and larger community is part of the cultural tradition in many Asian cultures, as well as in other parts of the world, such as Africa. The seriously mentally ill of China, for instance, retain strong supportive links to their families and workplaces, which helps increase their chances of being reintegrated into community life (Liberman, 1994). In traditional healing centers for the treatment of schizophrenia in Africa, the strong support that patients receive from the family and community members, together with a community centered lifestyle, are important elements of successful care (Peltzer & Machleidt, 1992).

Web Link 13.7 WWW
When Someone Has Schizophrenia

There is clear value in working with the family in treating schizophrenia in Asian Americans, as well as in other groups. Researchers find that failure to include the family often compromises the value of therapy for Asian Americans and causes many of them to drop out of therapy prematurely (Lin et al., 1978). Researchers in a hospital in Great Britain reported that living with family members was among the factors that might explain the lower relapse and rehospitalization rates found among Asian people with schizophrenia as compared to White or Afro-Caribbean people (Birchwood et al., 1992). Family interactions are not necessarily harmonious or conducive of better outcomes, however, as research on expressed emotion makes clear. Neglect or rejection of the person with schizophrenia within the family may play an important role in premature treatment termination and poorer outcomes.

Psychodynamic Therapy Freud did not believe that traditional psychoanalysis was well suited to the treatment of schizophrenia. The withdrawal into a fantasy world that typifies schizophrenia prevents the individual with schizophrenia from forming a meaningful relationship with the psychoanalyst. The techniques of classical psychoanalysis, Freud wrote, must "be replaced by others; and we do not know yet whether we shall succeed in finding a substitute" (as cited in Arieti, 1974, p. 532).

Other psychoanalysts, such as Harry Stack Sullivan and Frieda Fromm-Reichmann, adapted psychoanalytic techniques specifically for the treatment of schizophrenia. However, research has failed to demonstrate the effectiveness of psychoanalytic or psychodynamic therapy for schizophrenia. In the light of negative findings, some critics have argued that further research on the use of psychodynamic therapies for treating schizophrenia is not warranted (e.g., Klerman, 1984). However, promising results are reported for a form of individual psychotherapy called *personal therapy* that is grounded in the diathesis-stress model. Personal therapy helps patients cope more effectively with stress and helps them build social skills, such as learning how to deal with criticism from others. Preliminary evidence suggests that personal therapy may reduce relapse rates and improve social functioning, at least among schizophrenia patients living with their families (Bustillo et al., 2001; Hogarty et al., 1997a, 1997b).

Learning-Based Therapies Although few behavior therapists believe that faulty learning causes schizophrenia, learning-based interventions have been shown to be effective in modifying schizophrenic behavior and assisting people with the disorder to develop more adaptive behaviors that can help them adjust more effectively to living in the community. Therapy methods include techniques such as (1) selective reinforcement of behavior (like providing attention for appropriate behavior and extinguishing bizarre verbalizations through withdrawal of attention); (2) the token economy, in which individuals on inpatient units are rewarded for appropriate behavior with tokens, such as plastic chips, that can be exchanged for tangible reinforcers such as desirable goods or privileges; and (3) social skills training, in which clients are taught conversational skills and other appropriate social behaviors through coaching, modeling, behavior rehearsal, and feedback.

Promising results have emerged from studies that apply intensive learning-based approaches in hospital settings. A classic study by Paul and Lentz (1977) showed that a psychosocial treatment program based on a token-economy system improved adaptive behavior in the hospital, decreased need for medication, and lengthened community tenure following release in relation to a traditional, custodial-type treatment condition and a milieu approach that emphasized patient participation in decision making.

Overall, token economies have proven to be more effective than intensive milieu treatment and traditional custodial treatment in improving social functioning and reducing psychotic behavior (Glynn & Mueser, 1992; Mueser & Liberman, 1995). However, the many prerequisites may limit the applicability of this approach. Such programs require strong administrative support, skilled treatment leaders, extensive staff training, and continuous quality control (Glynn & Mueser, 1986).

Recently, investigators have shown promising results in using cognitive-behavioral approaches in reducing or even eliminating hallucinations or delusions in patients with schizophrenia (Bouchard et al., 1996; Bustillo et al., 2001). More research is needed to demonstrate the clinical utility of using CBT to treat psychotic symptoms in general clinical practice.

Social skills training (SST) involves programs that help individuals acquire a range of social and vocational skills. People with schizophrenia are often deficient in basic social skills involving assertiveness, interviewing skills, and general conversational skills, skills that may be needed to adjust successfully to community living. Controlled studies have shown that SST improves social skills and adaptive functioning of schizophrenia patients in the community (Hunter, Bedell, & Corrigan, 1997; Penn, 1998). The effectiveness of SST in enhancing patients' social skills is not limited to our culture; it was recently demonstrated in a sample of Chinese schizophrenia patients in Hong Kong (Tsang, 2001). However, SST has not yet shown clear benefits in reducing either relapse rate or improving employment status (Bustillo et al., 2001).

Although different approaches to skills training have been developed, the basic model uses role-playing exercises within a group format. Participants practice skills such as starting or maintaining conversations with new acquaintances and receive feedback and reinforcement from the therapist and other group members. The first step might be a dry run in which the participant role-plays the targeted behavior, such as asking strangers for bus directions. The therapist and other group members then praise the effort and provide constructive feedback. Role-playing is augmented by techniques such as modeling (observation of the therapist or other group members enacting the desired behavior), direct instruction (specific directions for enacting the desired behavior), shaping (reinforcement for successive approximations to the target behavior), and coaching (therapist use of verbal or nonverbal prompts to elicit a particular desired behavior in the role play). Participants are given homework assignments to practice the behaviors in the settings in which they live, such as on the hospital ward or in the community. The aim is to enhance generalization or transfer of training to other settings. Training sessions may also be run in stores, restaurants, schools, and other real-life settings.

Psychosocial Rehabilitation People with schizophrenia typically have difficulties functioning in social and occupational roles. These problems limit their ability to adjust to community life even in the absence of overt psychotic behavior.

A number of self-help clubs (commonly called clubhouses) and more structured psychosocial rehabilitation centers have sprung up to help people with schizophrenia find a place in society. Many centers were launched by nonprofessionals or by people with schizophrenia themselves, largely because mental health agencies often failed to provide comparable services (Anthony & Liberman, 1986). The clubhouse movement began in 1948 with the founding of Fountain House by a group of formerly hospitalized people with schizophrenia (Foderaro, 1994). There are now more than 200 clubhouses modeled after Fountain House across the country, and some 50 or more in other countries including Sweden, Japan, and Australia. Although no one actually lives in the clubhouse, it serves as a kind of

THINK ABOUT IT
What psychosocial interventions are useful in the treatment of schizophrenia? What limitations are there in current treatment models?

self-contained community that provides members with social support and help in finding educational opportunities and paid employment.

Multiservice rehabilitation centers typically offer housing as well as job and educational opportunities. These centers often make use of skills training approaches to help clients learn how to handle money, resolve disputes with family members, develop friendships, take buses, cook their own meals, shop, and so on.

Family Intervention Programs Family conflicts and negative family interactions can heap stress on family members with schizophrenia, increasing the risk of recurrent episodes (Marsh & Johnson, 1997). Researchers and clinicians have worked with families of people with schizophrenia to help them cope with the burdens of care and assist them in developing more cooperative, less confrontational ways of relating to others. The specific components of family interventions vary from program to program, but they usually share some common features, such as a focus on the practical aspects of everyday living, educating family members about schizophrenia, teaching them how to relate in a less hostile way to family members with schizophrenia, improving communication in the family, and fostering effective problem-solving and coping skills for handling family problems and disputes. Evidence shows that structured family intervention programs can reduce friction in the family, improve social functioning in schizophrenia patients, and even reduce relapse rates (Bustillo et al., 2001; Mueser et al., 2001; Penn & Mueser, 1996). However, the benefits appear to be modest, and questions remain about whether relapses are prevented or merely delayed.

Quiz 13.3
Schizophrenia

In sum, no single treatment approach meets all the needs of people with schizophrenia. The conceptualization of schizophrenia as a lifelong disability underscores the need for long-term treatment interventions that incorporate antipsychotic medication, family therapy, supportive or cognitive-behavioral forms of therapy, vocational training, and provision of decent housing and other social support services (Bustillo et al., 2001; Huxley, Rendall, & Sederer, 2000; Sensky et al., 2000; Tarrier et al., 2000). These interventions should be coordinated and integrated within a comprehensive model of treatment to be most effective in helping the individual achieve maximal social adjustment (Coursey et al., 1997). Treatment services are also more likely to improve client functioning in certain areas, such as improving work or independent living, when they are specifically targeted toward those areas (Brekke et al., 1997). This model may consist of drug therapy, hospitalization as needed, inpatient learning-based programs, family intervention programs, skills training programs, social self-help clubs, and structured rehabilitation programs.

Quiz 13.4
Chapter Exam

Research Update
Chapter 13

Overview of Schizophrenia

CLINICAL FEATURES OF SCHIZOPHRENIA

Disturbed Thought Processes	• Delusions (fixed false ideas) and thought disorder (disorganized thinking and incoherent speech)
Attentional Deficiencies	• Difficulty attending to relevant stimuli and screening out irrelevant stimuli
Perceptual Disturbances	• Hallucinations (sensory perceptions in the absence of external stimulation)
Emotional Disturbances	• Flat (blunted) or inappropriate emotions
Other Impairments	• Confusion about personal identity, lack of volition, excitable behavior or states of stupor, odd gestures or bizarre facial expressions, and impaired ability to relate to others

MAJOR SUBTYPES OF SCHIZOPHRENIA*

Disorganized Type	• Confused and bizarre behavior, incoherent speech, vivid hallucinations, flat or inappropriate affect, and disorganized delusions
Catatonic Type	• Gross disturbances in motor activity in which behavior may slow to a stupor but abruptly shift to a highly agitated state
Paranoid Type	• Delusions (typically of themes of grandeur, persecution, or jealousy) and frequent auditory hallucinations

*Variations of schizophrenia distinguished in terms of specific subtypes. Another way of subtyping schizophrenia is based on distinguishing between Type I schizophrenia, characterized by more flagrant symptoms (positive symptoms), and Type II schizophrenia, characterized by deficit symptoms (negative symptoms).

CAUSAL FACTORS*

Biological Factors	• Strong evidence of a major genetic contribution • Irregularities in neurotransmitter systems in the brain, especially in brain circuits that utilize the neurotransmitter dopamine • Underlying brain abnormalities in many cases, such as structural damage or deterioration of brain tissue or disturbed brain circuitry in parts of the brain regulating cognitive and emotional functioning • Possible role of viral infections affecting the developing brain prenatally or during early life
Psychosocial Factors	• Stressful experiences may contribute to the development of schizophrenia in genetically vulnerable individuals.

*The specific causes remain unknown, but most researchers believe they reflect an interaction of genetic and stress-related factors, as represented by the diathesis-stress model.

TREATMENT APPROACHES

A comprehensive treatment approach incorporating biomedical, psychosocial, and family interventions is recommended.

Biomedical Treatment	• Antipsychotic drugs are used to control psychotic symptoms.
Psychosocial Treatment	• Learning-based approaches, such as the token economy system and social skills training, can help schizophrenia patients develop more adaptive behaviors.
Psychosocial Rehabilitation	• Self-help clubs and structured residential programs can help schizophrenia patients adjust to community living.
Family Intervention Programs	• Family interventions are used to improve communication in the family and reduce levels of family conflict and stress.

Summing Up

History of the Concept of Schizophrenia

What is schizophrenia and how prevalent is it? Schizophrenia is a chronic psychotic disorder characterized by acute episodes involving a break with reality, as manifest by such features as delusions, hallucinations, illogical thinking, incoherent speech, and bizarre behavior. Residual deficits in cognitive, emotional, and social areas of functioning persist between acute episodes. Schizophrenia is believed to affect about 1% of the population.

What key historical figures in psychiatry influenced our conceptions of schizophrenia? Emil Kraepelin was the first to describe the syndrome we identify as schizophrenia. He labeled the disorder *dementia praecox* and believed it was a disease that develops early in life and follows a progressively deteriorating course. Eugen Bleuler renamed the disorder *schizophrenia* and believed its course is more variable. He also distinguished between primary symptoms (the four A's) and secondary symptoms. Kurt Schneider distinguished between first-rank symptoms that define the disorder and second-rank symptoms that occur in schizophrenia and other disorders.

Other Forms of Psychosis

What are other forms of psychotic disorders besides schizophrenia? These include brief psychotic disorder (a psychotic disorder lasting less than a week that may be reactive to a significant stressor), schizophreniform disorder (symptoms identical to those of schizophrenia that lasting for a month to less than 6 months), delusional disorder (denoted by delusions that are apparently plausible and less bizarre than those in schizophrenia), and schizoaffective disorder (combination of psychotic symptoms and significant mood disturbance).

What are schizophrenia-spectrum disorders? The term encompasses the schizophrenic-type disorders that range in severity from milder personality disorders, such as schizotypal and schizoid types, to frankly psychotic disorders, such as schizophrenia itself and schizoaffective disorder.

Schizophrenia

What are the major phases of schizophrenia? Schizophrenia usually develops in late adolescence or early adulthood. Its onset may be abrupt or gradual. Gradual onset is preceded by a prodromal phase, a period of gradual deterioration that precedes the onset of acute symptoms. Acute episodes, which may occur periodically throughout life, are typified by clear psychotic symptoms, such as hallucinations and delusions. Between acute episodes the disorder is characterized by a residual phase in which the person's level of functioning is similar to that which was present during the prodromal phase.

What are the most prominent features of schizophrenia? Among the more prominent features of schizophrenia are disorders in the content of thought (delusions) and the form of thought (thoughtdisorder), as well as the presence of often severe perceptual distortions (hallucinations) and emotional disturbances (flattened or inappropriate affect). There are also dysfunctions in brain processes regulating attention to the external world.

What are the specific subtypes of schizophrenia? The disorganized type is associated with grossly disorganized behavior and thought processes. The catatonic type is associated with grossly impaired motor behaviors, such as maintenance of fixed postures and muteness for long periods. The paranoid type is characterized by paranoid delusions and frequent auditory hallucinations. The undifferentiated type is a catchall category applying to cases in which schizophrenic episodes don't clearly fit the other types. The residual type applies to individuals with schizophrenia who do not have prominent psychotic behaviors at the time of evaluation. Researchers have also distinguished between two general types of schizophrenia: Type I, characterized by positive symptomatology, more abrupt onset, better response to antipsychotic medication, and better preserved intellectual ability, and Type II, characterized by negative symptomatology, more gradual onset, poorer response to antipsychotic medication, and greater cognitive impairment.

How is schizophrenia conceptualized within traditional psychodynamic theory and learning perspectives? In the traditional psychodynamic model, schizophrenia represents a regression to a psychological state corresponding to early infancy in which the proddings of the id produce bizarre, socially deviant behavior and give rise to hallucinations and delusions. Learning theorists propose that some form of schizophrenic behavior may result from lack of social reinforcement, which leads to gradual detachment from the social environment and increased attention to an inner world of fantasy. Modeling and selective reinforcement of bizarre behavior may explain some schizophrenic behaviors in the hospital setting.

What do we know about the biological bases of schizophrenia? Compelling evidence for a strong genetic component in schizophrenia comes from studies of family patterns of schizophrenia, twin studies, and adoption studies. The mode of genetic transmission remains unknown. Most researchers believe the neurotransmitter dopamine plays a role in schizophrenia, especially in the more flagrant features of the disorder. Viral factors may also be involved, but definite proof of viral involvement is lacking. Evidence of brain dysfunctions and structural damage in schizophrenia is accumulating, but researchers are uncertain about causal pathways.

How is schizophrenia conceptualized within the diathesis-stress model? The diathesis-stress model posits that schizophrenia results from an interaction of a genetic predisposition (the diathesis) and environmental stressors (e.g., family conflict, child abuse, emotional deprivation, loss of supportive figures, early brain trauma).

How are family factors related to the development and course of schizophrenia? Family factors such as communication deviance and expressed emotion may act as sources of stress that increase the risk of development or recurrence of schizophrenia among people with a genetic predisposition.

How does the treatment of schizophrenia involve a multifaceted approach? Contemporary treatment approaches tend to be multifaceted, incorporating pharmacological and psychosocial approaches. Antipsychotic medication is not a cure but tends to stem the more flagrant aspects of the disorder and to reduce the need for hospitalization and the risk of recurrent episodes.

What types of psychosocial interventions have shown promising results? These are principally learning-based approaches, such as token economy systems and social skills training. They help increase adaptive behavior of schizophrenia patients. Psychosocial-rehabilitation approaches help people with schizophrenia adapt more successfully to occupational and social roles in the community. Family intervention programs help families cope with the burdens of care, communicate more clearly, and learn more helpful ways of relating to the patient.

CHAPTER FOURTEEN

Abnormal Behavior in Childhood and Adolescence

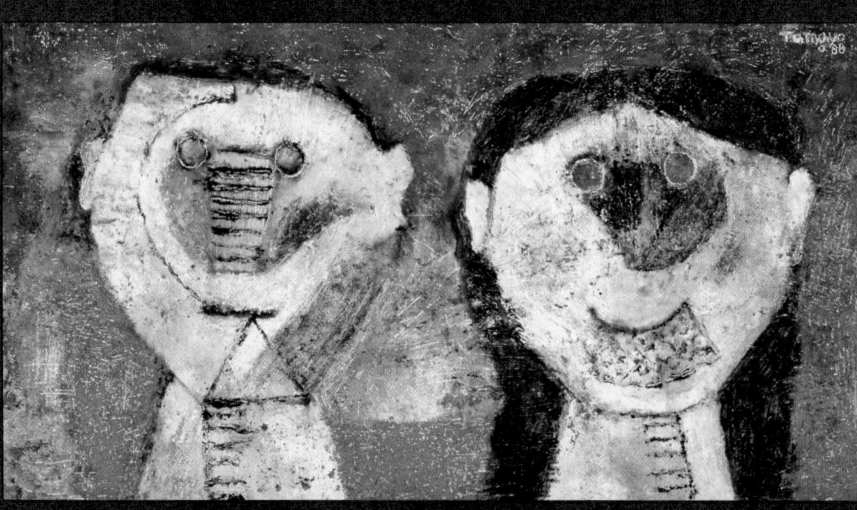

Rufino Tamayo
Dos Caras

Truth OR Fiction?

- Many behavior patterns considered normal for children would be considered abnormal in adults. (p. 439)
- Therapists have used Puerto Rican folktales to help Puerto Rican children adjust to the demands of living in mainstream U.S. society. (p. 440)
- Some people can recall verbatim every story they read in a newspaper. (p. 453)
- Maternal smoking during pregnancy may put children at increased risk of attention-deficit hyperactivity disorder (ADHD). (p. 461)
- Children who are hyperactive are often given depressants to help calm them down. (p. 461)
- Major depression rarely occurs before adulthood. (p. 466)
- Some children refuse to go to school because they believe that terrible things may happen to their parents while they are away. (p. 467)
- Problems of persistent bed-wetting in childhood generally persist into adolescence. (p. 472)

Psychological problems in childhood and adolescence often have a special poignancy. They affect children at ages when they have little capacity to cope. Some of these problems, such as autism and mental retardation, prevent children from fulfilling their developmental potentials. Some psychological problems in childhood and adolescents mirror the types of problems found in adults—problems such as mood disorders and anxiety disorders. In some cases, the problems are unique to childhood, such as separation anxiety; in others, such as ADHD or attention-deficit hyperactivity disorder, the problem manifests itself differently in childhood than in adulthood.

Chapter Outline

Normal and Abnormal Behavior in Childhood and Adolescence

To determine what is normal and abnormal among children and adolescents, we consider, in addition to the criteria outlined in Chapter 1, the child's age and cultural background (USDHHS, 1999a). Many problems are first identified when the child enters school. They may have existed earlier but been tolerated, or unrecognized as problematic, in the home. Sometimes the stress of starting school contributes to their onset. Keep in mind, however, that what is socially acceptable at one age, such as intense fear of strangers at about 9 months, may be socially unacceptable at more advanced ages. Many behavior patterns that would be considered abnormal among adults—such as intense fear of strangers and lack of bladder control—are perfectly normal for children at certain ages.

Cultural Beliefs About What Is Normal and Abnormal

Cultural beliefs help determine whether people view behavior as normal or abnormal. People who base judgments of normality only on standards derived from their own cultures risk being ethnocentric when they view the behavior of people in other cultures as abnormal (Kennedy, Scheirer, & Rogers, 1984). The problem is of special concern regarding child psychopathology. Because children rarely label their own behavior as abnormal, definitions of normality depend largely on how a child's behavior is filtered through the lenses by which parents in a particular culture view that behavior. Cultures may vary with respect to the types of behaviors they classify as unacceptable or abnormal as well as the threshold for labeling child behaviors as deviant or socially unacceptable. Researchers find that parents in different cultures do judge the unusualness of behavior from different perspectives (Lambert et al., 1992).

For example, researchers posed the question, "When a child has psychological problems, what determines whether adults will consider the problem serious or whether they will seek professional help?" (Weisz et al., 1988, p. 601). To explore this question, researchers presented vignettes to Thai and American parents, teachers, and clinical psychologists. The vignettes depicted two children, one with problems characterized by "overcontrol" (for example, shyness and fears) and one with problems characterized by undercontrol (for example, disobedience and fighting). The Thai parents rated *both* sets of problems as less serious and worrisome than American parents (see Figure 14.1), and as more likely to improve without treatment as time passed. Such an interpretation is embedded within traditional Thai-Buddhist beliefs and values, which tolerate broad variations in children's behavior. They assume that change is inevitable and that children's behavior will eventually change for the better. Differences between cultural groups were greater for parents and teachers than for psychologists, which suggests that professional training in a common scientific tradition might offset cultural differences.

Truth OR Fiction? REVISITED

Many behavior patterns considered normal for children would be considered abnormal in adults.

TRUE. Many behavior patterns that would be considered abnormal among adults—such as intense fear of strangers and lack of bladder control—are perfectly normal for children at certain ages.

Culturally Sensitive Therapy

Psychotherapy with children has been approached from various perspectives and differs in important respects from therapy with adults. Children may not have the verbal skills to express their feelings through speech or the ability to sit in a chair through a therapy

play therapy A form of psychodynamic therapy in which play activities and objects are used to help children symbolically enact conflicts or express underlying feelings.

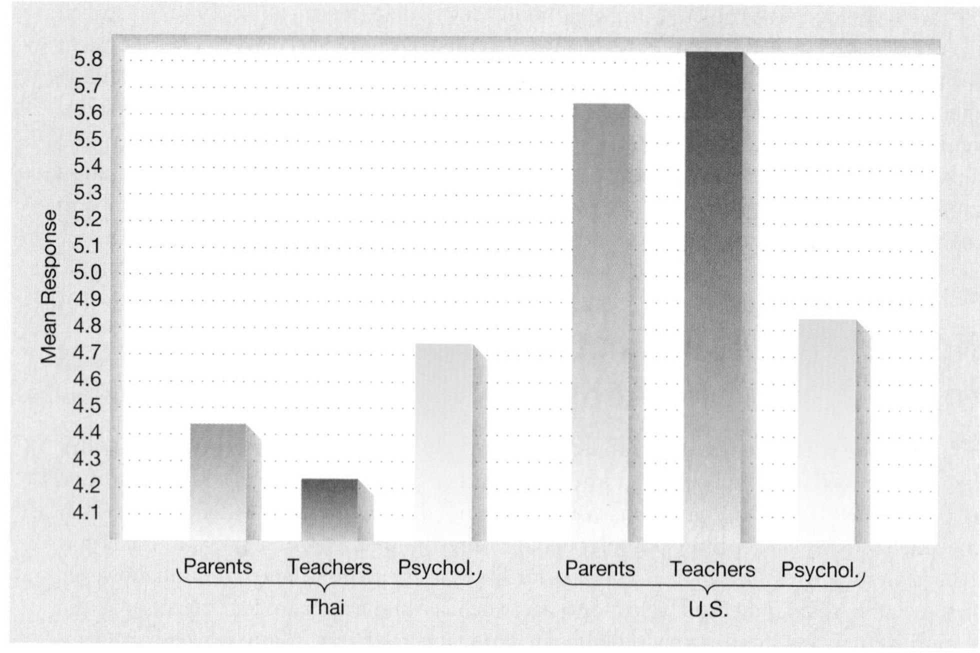

FIGURE 14.1 Ratings by Thai and U.S. parents of the seriousness of children's behavioral problems.
Researchers presented vignettes of children with problems characterized by overcontrol (for example, shyness and fears) and undercontrol (for example, disobedience and fighting) to Thai and American parents. The Thai parents rated both sets of problems as less serious and worrisome than American parents, and as more likely to improve without treatment. The Thai parents apparently assume that people change and that their children's behavior will eventually change for the better.

Source. Weisz, J. R., et al., (1988). Thai and American perspectives on over- and undercontrolled child behavior problem: Exploring the threshold model among parents, teachers, and psychologists. *Journal of Consulting and Clinical Psychology, 56,* 601–609. Copyright © 2001 by the American Psychological Association. Reprinted with permission.

Truth OR Fiction? REVISITED

Therapists have used Puerto Rican folktales to help Puerto Rican children adjust to the demands of living in mainstream U.S. society.

TRUE. Therapists have used adapted Puerto Rican folktales (*cuentos*) to help Puerto Rican children adjust to the demands of living in mainstream U.S. society.

session. Therapy methods must be tailored to the level of the child's cognitive, physical, social, and emotional development. For example, psychodynamic therapists have developed techniques of **play therapy** in which children enact family conflicts symbolically through their play activities, such as by playacting with dolls or puppets. Or they might be given drawing materials and asked to draw pictures, in the belief that their drawings will reflect their underlying feelings.

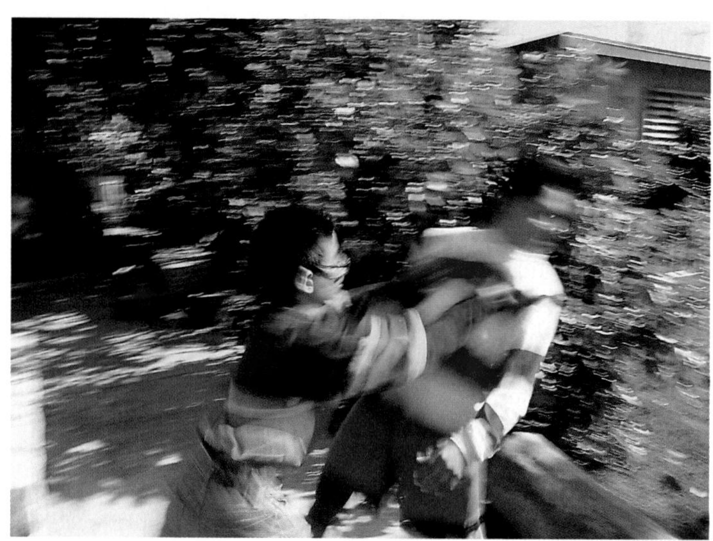

How serious is this problem? Thai parents might judge the behavior shown by these children to be less serious than American parents would. Thai-Buddhist values tolerate broad variations in children's behavior and assume that it will change for the better.

Research suggests that culturally sensitive therapies, ones specifically tailored to the cultural backgrounds and needs of children from diverse cultural groups, are important in establishing effective therapeutic relationships with children. Costantino and his colleagues (1986), for example, adapted traditional Puerto Rican folktales, or *cuentos,* as modeling examples in treating Puerto Rican children with behavior problems. The *cuentos* featured child protagonists who served as models for adaptive behavior. The stories were read aloud by the therapists and the children's mothers, and were followed by group discussion of the behavior and feelings of the main character and the moral of the story. The final element in such sessions was role playing, in which children were given the opportunity to imitate the adaptive behavior exemplified by the main character in of the story.

The value of culturally sensitive therapies is not limited to the treatment of Puerto Rican children, of course. Evidence of their value in working with other cultural groups is mounting. For example, researchers find that ethnic-specific treatment programs for Asian American children appear to be more effective in reducing early treatment dropouts and increasing functioning at termination as compared to mainstream outpatient mental health centers (Yeh, Takeuchi, & Sue, 1994).

Play therapy. In play therapy, children may enact scenes with dolls or puppets that symbolically represent conflicts occurring within their own families.

Prevalence of Mental Health Problems in Children and Adolescents

Just how common are mental health problems among America's children and adolescents? According to a recent report from the U.S. Surgeon General, 1 in 10 children suffer from a mental disorder severe enough to impair their development ("A Children's Mental Illness 'Crisis'," 2001). More American children suffer from mental disorders than from diabetes, AIDS, and leukemia combined (Chamberlin, 2001). Yet 60% to 80% of children with mental health disorders fail to get the treatment they need (Goldberg, 2001). Children who have *internalized* problems, such as anxiety and depression, are at higher risk of going untreated than are children with *externalized* problems (problems involving acting out or aggressive behavior) that tend to be disruptive or annoying to others.

Boys are at greater risk for developing many childhood problems, ranging from autism to hyperactivity to elimination disorders. Problems of anxiety and depression also affect boys more often than girls. In adolescence, however, anxiety and mood disorders become more common among girls and remain so throughout adulthood (USDHHS, 1999a).

Let us now examine the types of psychological disorders that may affect children and adolescents. We will examine the features of these disorders, their causes, and the treatments used to help children who suffer from them.

Quiz **14.1**
Normal and Abnormal Behavior in Childhood and Adolescence

WWW Web Link **14.1**
Teenage Brain: A Work in Progress

Pervasive Developmental Disorders

Children with **pervasive developmental disorders (PDDs)** show markedly impaired behavior or functioning in multiple areas of development. These disorders generally become evident in the first few years of life and are often associated with mental retardation. They were generally classified as forms of psychoses in early editions of the *DSM*. They were thought to reflect childhood forms of adult psychoses like schizophrenia because they share features such as social and emotional impairment, oddities of communication, and stereotyped motor behaviors. Research has shown that they are distinct from schizophrenia and other psychoses, however. Only very rarely in these children is there evidence of the prominent hallucinations or delusions that would justify a diagnosis of schizophrenia.

The major type of pervasive developmental disorder, which is our focus here, is autistic disorder (autism). **Asperger's disorder,** a milder form of pervasive developmental disorder, is characterized by poor social interactions and stereotyped behavior. However, in contrast to autism, Asperger's disorder does not involve significant language or cognitive deficits (APA, 2000; Szatmari et al., 2000). Other, less common types of pervasive developmental disorder include **Rett's disorder,** a disorder reported only in females; and **childhood disintegrative disorder,** a rare condition that appears to be more common among males.

Though the rate of PDDs remains unclear, a recent community study of preschool children in England showed that 0.6% of the children (6 in 1,000) met criteria for one or another PDD, most commonly autism (Chakrabarti & Fombonne, 2001). The diagnostic features of autistic disorder are found in Table 14.1. Table 14.2 lists the diagnostic features of other types of pervasive developmental disorders.

pervasive developmental disorders (PDDs) A class of developmental disorders characterized by significantly impaired behavior or functioning in multiple areas of development.

Asperger's disorder A pervasive developmental disorder characterized by social deficits and stereotyped behavior but without the significant language or cognitive delays associated with autism.

Rett's disorder A pervasive development disorder characterized by a range of physical, behavioral, motor, and cognitive abnormalities that begin after a few months of apparently normal development.

childhood disintegrative disorder A pervasive developmental disorder involving loss of previously acquired skills and abnormal functioning following a period of apparently normal development during the first two years of life.

TABLE 14.1 Diagnostic Features of Autistic Disorder

A. Diagnosis requires a combination of features from the following groups. Not all of the features from each group need be present for a diagnosis to be made.

Impaired Social Interactions

1. Impairment in the nonverbal behaviors such as facial expressiveness, posture, gestures, and eye contact that normally regulate social interaction
2. Does not develop age-appropriate peer relationships
3. Failure to express pleasure in the happiness of other people
4. Does not show social or emotional reciprocity (give and take)

Impaired Communication

1. Delay in development of spoken language (nor is there an effort to compensate for this lack through gestures)
2. When speech development is adequate, there is nevertheless lack of ability to initiate or sustain conversation
3. Shows abnormalities in form or content of speech (e.g., stereotyped or repetitive speech, as in echolalia; idiosyncratic use of words; speaking about the self in the second or third person—using "you" or "he" to mean "I")
4. Does not show spontaneous social or imaginative (make-believe) play

Restricted, repetitive, and stereotyped behavior patterns

1. Shows restricted range of interests
2. Insists on routines (e.g., always uses same route to go from one place to another)
3. Shows stereotyped movements (e.g., hand flicking, head banging, rocking, spinning)
4. Shows preoccupation with parts of objects (e.g., repetitive spinning of wheels of toy car) or unusual attachments to objects (e.g., carrying a piece of string)

B. Onset occurs prior to the age of 3 through display of abnormal functioning in at least one of the following: social behavior, communication, or imaginative play.

Source. Adapted from the *DSM-IV-TR* (APA, 2000).

Autism

A Case of Autism

Peter nursed eagerly, sat and walked at the expected ages. Yet some of his behavior made us vaguely uneasy. He never put anything in his mouth. Not his fingers nor his toys—nothing. . . .

More troubling was the fact that Peter didn't look at us, or smile, and wouldn't play the games that seemed as much a part of babyhood as diapers. He rarely laughed, and when he did, it was at things that didn't seem funny to us. He didn't cuddle, but sat upright in my lap, even when I rocked him. But children differ and we were content to let Peter be himself. We thought it hilarious when my brother, visiting us when Peter was 8 months old, observed that "That kid has no social instincts, whatsoever." Although Peter was a first child, he was not isolated. I frequently put him in his playpen in front of the house, where the schoolchildren stopped to play with him as they passed. He ignored them, too.

It was Kitty, a personality kid, born two years later, whose responsiveness emphasized the degree of Peter's difference. When I went into her room for the late feeding, her little head bobbed up and she greeted me with a smile that reached from her head to her toes. And the realization of that difference chilled me more than the wintry bedroom.

Peter's babbling had not turned into speech by the time he was 3. His play was solitary and repetitious. He tore paper into long thin strips, bushel baskets of it every day. He spun the lids from my canning jars and became upset if we tried to divert him. Only rarely could I catch his eye, and then saw his focus change from me to the reflection in my glasses. . . .

[Peter's] adventures into our suburban neighborhood had been unhappy. He had disregarded the universal rule that sand is to be kept in sandboxes, and the children themselves had punished him. He walked around a sad and solitary figure, always carrying a toy airplane, a toy he never played with. At that time, I had not heard the word that was to dominate our lives, to hover over every conversation, to sit through every meal beside us. That word was autism.

—Adapted from Eberhardy, 1967

autism A pervasive developmental disorder characterized by failure to relate to others, lack of speech, disturbed motor behaviors, intellectual impairment, and demands for sameness in the environment.

Autism, or *autistic disorder,* is one of the severest disorders of childhood. It is a chronic, lifelong condition. Children with autism, like Peter, seem utterly alone in the world, despite parental efforts to bridge the gulf that divides them.

TABLE 14.2 Diagnostic Features of Other Pervasive Developmental Disorders

Disorder	Diagnostic Features
Asperger's disorder	• Markedly impaired social interactions (e.g., failure to maintain eye contact or to develop age-appropriate peer relationships, or failure to seek out others to share enjoyable activities or interests) • Development of narrow, repetitive, and stereotyped behaviors, interests, and activities (e.g., twisting hands or fingers; rigidly adhering to fixed routines or rituals that lack any clear purpose; fascination with train schedules) • No clinically significant delay in language or cognitive development or in development of self-help skills or adaptive behaviors apart from social interactions
Rett's disorder	After apparently normal development during the first few months of life, the following abnormalities develop: • Slowing of head growth • Deteriorating motor skills (loss of purposeful hand skills) • Development of stereotyped hand movements, typically resembling hand-wringing or hand washing • Development of poorly coordinated gait or movement of whole body • Loss of social interest • Severe deficits in language development • Typically associated with profound or severe mental retardation
Childhood disintegrative disorder	After apparently normal development for at least the first 2 years of life: • Significant loss of previously acquired skills in such areas as understanding or using language, social or adaptive functioning, bowel or bladder control, play, or motor skills • Abnormal functioning as shown by impaired social interactions or communication, and development of narrow, stereotyped, and repetitive behaviors, interests, or activities

Source. Adapted from the *DSM-IV-TR* (APA, 2000).

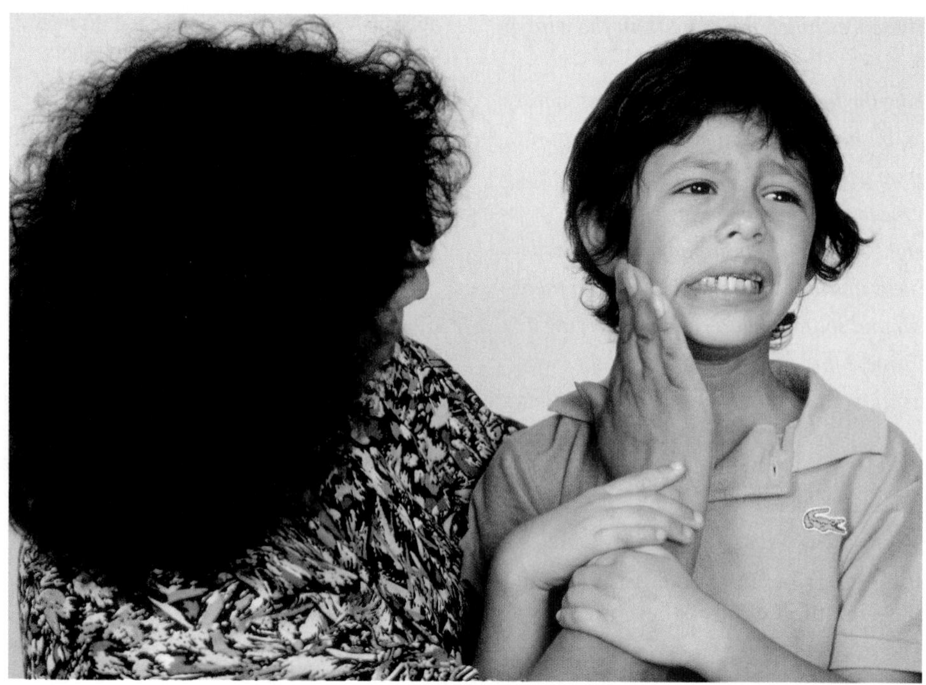

Autism. Autism, one of the most severe childhood disorders, is characterized by pervasive deficits in the ability to relate to and communicate with others, and by a restricted range of activities and interests. Children with autistic disorder lack the ability to relate to others and seem to live in their own private worlds.

The word *autism* derives from the Greek *autos,* meaning "self." The term was first used in 1906 by the Swiss psychiatrist Eugen Bleuler to refer to a peculiar style of thinking among people with schizophrenia (autism is one of Bleuler's "four As"). Autistic thinking is the tendency to view oneself as the center of the universe, to believe that external events somehow refer to oneself. In 1943, another psychiatrist, Leo Kanner, applied the diagnosis "early infantile autism" to a group of disturbed children who seemed unable to relate to others, as if they lived in their own private worlds. Unlike children suffering from mental retardation, these children seemed to shut out any input from the outside world, creating a kind of "autistic aloneness" (Kanner, 1943).

Medical authorities today believe that autism is more common than was previously believed, affecting an estimated 2 to 20 persons in 10,000 in the general population (APA, 2000; Fox, 2000). The disorder, which occurs mostly among boys, generally becomes evident in toddlers between 18 and 30 months of age (Rapin, 1997). However, it is not until about age 6 that the average child is first diagnosed with the disorder (Fox, 2000). Delays in diagnosis can be detrimental, as children with autism generally do better the earlier they are diagnosed and treated (Fox, 2000).

Web Link 14.2 WWW
Facts About Autism

Autism seems always to have been with Peter. In the following case example, the disorder apparently developed between the ages of 12 and 24 months:

The Case of Eric

"People used to say to me they hoped they [would have] a baby just like mine," Sarah said of Eric, 3 years old at the time. As an infant, Eric smiled endearingly, laughed, and hugged. He uttered a dozen words by his first birthday. By 16 months he had memorized the alphabet and could read some signs. "People were very impressed," Sarah said.

Gradually, things changed, but it took months for Sarah to realize that Eric had a problem. At the age of 2, other members of Eric's play group bubbled with conversation. Eric had abandoned words completely. Instead, Eric combined letters and numbers in idiosyncratic ways, such as "B–T–2–4–6–Z–3."

Eric grew increasingly withdrawn. His diet was essentially self-limited to peanut butter and jelly sandwiches. He spent hour after hour arranging letters and numbers on a magnetic board. But the "symptom" that distressed Sarah most was impossible to measure: when she gazed into Eric's eyes, she no longer saw a "sparkle."

—Adapted from Martin, 1989

■

Children with autism are often described by their parents as having been "good babies" early in infancy. This generally means they were not demanding. As they develop, however, they begin to reject physical affection, such as cuddling, hugging, and kissing. Their speech development begins to fall behind the norm. Although Eric did quite well through his first 16 months, there are often signs of social detachment beginning as early as the first year of life, such as failure to look at other people's faces (Osterling & Dawson, 1994). The clinical features of the disorder appear prior to 3 years of age (APA, 2000). Autism is four to five times more common among males than females (APA, 2000).

Features of Autism Perhaps the most poignant feature of autism is the child's utter aloneness (see Table 14.1). Other features include language and communication problems and ritualistic or stereotyped behavior. The child may also be mute, or if some language skills are present, they may be characterized by peculiar usage, as in echolalia (parroting back what the child has heard in a high-pitched monotone); pronoun reversals (using "you" or "he" instead of "I"); use of words that have meaning only to those who have intimate knowledge of the child; and tendencies to raise the voice at the end of sentences, as if asking a question. Nonverbal communication may also be impaired or absent. For example, children with autistic disorder may not engage in eye contact or display facial expressions. They are also slow to respond to adults who try to grab their attention, if they attend at all (Leekam & López, 2000). Although they may be unresponsive to others, researchers find they are capable of displaying strong emotions, especially strong negative emotions such as anger, sadness, and fear (Capps et al., 1993; Kasari et al., 1993).

A primary feature of autism is repeated purposeless stereotyped movements—interminably twirling, flapping the hands, or rocking back and forth with the arms around the knees. Some children with autism mutilate themselves, even as they cry out in pain. They may bang their heads, slap their faces, bite their hands and shoulders, or pull out their hair. They may also throw sudden tantrums or panics. Another feature of autism is aversion to environmental changes—a feature termed "preservation of sameness." When familiar objects are moved even slightly from their usual places, children with autism may throw tantrums or cry continually until their placement is restored. Like Eric, children with autistic disorder may insist on eating the same food every day.

Children with autism are bound by ritual. The teacher of a 5-year-old girl with autistic disorder learned to greet her every morning by saying, "Good morning, Lily, I am very, very glad to see you" (Diamond, Baldwin, & Diamond, 1963). Although Lily would not respond to the greeting, she would shriek if the teacher omitted even one of the *verys*.

Children who develop autism appear to have failed to develop a differentiated self-concept, a sense of themselves as distinct individuals. Despite their unusual behavior, they are often quite attractive and often have an "intelligent look" about them. However, as measured by scores on standardized tests, their intellectual development tends to lag below the norm. Three of four show evidence of mental retardation (Rapin, 1997). Even those who function at an average level of intelligence show deficits in activities requiring the ability to symbolize, such as recognizing emotions, engaging in symbolic play, and solving problems conceptually. They also display difficulty in attending to tasks that involve interacting with other people. The relationship between autism and intelligence is clouded, however, by difficulties in administering standardized IQ tests to these children. Testing requires cooperation, a skill that is dramatically lacking in children with autism. At best, we can only estimate their intellectual ability.

Theoretical Perspectives The causes of autism remain unknown, but are presumed to involve underlying brain abnormalities. Early, discredited views of autism viewed the child's aloofness as a reaction to parents who were cold and detached—"emotional refrigerators" who lacked the ability to establish warm relationships with their children. Research failed to support the assumption—so devastating to many parents—that they are in fact frosty and remote (Hoffmann & Prior, 1982). Of course there is truth to the notion that children with autism and their parents do not relate to one another very well, but causal connections are clouded. Rather than rejecting their children and thus fostering autism, parents may grow somewhat aloof because their efforts to relate to their children repeatedly meet with failure. Aloofness then becomes a result of autism, not a cause.

Psychologist O. Ivar Lovaas and his colleagues (1979) offered a cognitive-learning perspective on autism. They suggest that children with autism have perceptual deficits that limit them to processing only one stimulus at a time. As a result, they are slow to learn by

VIDEO 14.1
Autism: *Dr. Kathy Pratt*

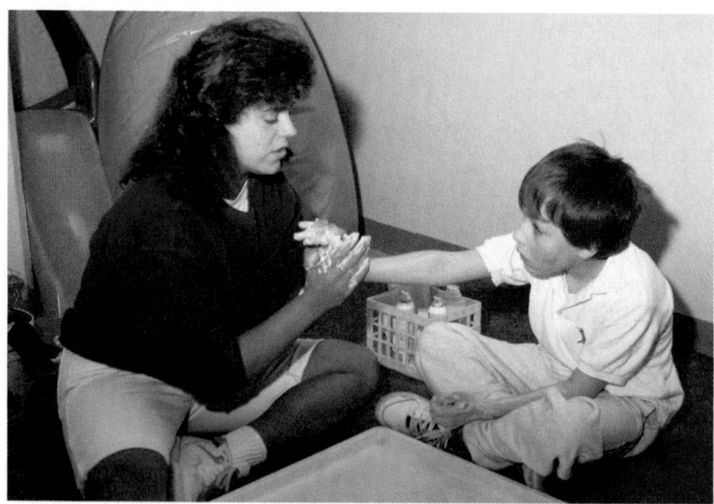

Establishing contact. One of the principal therapeutic tasks in working with children with autism is the establishment of interpersonal contact. Behavior therapists use reinforcers to increase adaptive social behaviors, such as paying attention to the therapist and playing with other children. Behavior therapists may also use punishments to suppress self-mutilative behavior.

means of classical conditioning (association of stimuli). From the learning theory perspective, children become attached to their primary caregivers because they are associated with primary reinforcers such as food and hugging. Children with autism, however, attend either to the food or to the cuddling and do not connect it with the parent.

Cognitive theorists have focused on the kinds of cognitive deficits shown by children with autism and the possible relationships among them. Children with autism appear to have difficulty integrating information from various senses (Rutter, 1983). At times they seem unduly sensitive to stimulation. At other times they become so insensitive that an observer might wonder whether they are deaf. Perceptual and cognitive deficits seem to diminish their capacity to make use of information—to comprehend and apply social rules.

But what is the basis of these perceptual and cognitive deficits? The many impairments associated with autism, including mental retardation, language deficits, bizarre motor behavior, even seizures, suggest an underlying neurological basis involving some form of brain damage or neurochemical imbalance in the brain (Perry et al., 2001; Stokstad, 2001). Evidence from MRI and PET scan studies show abnormalities in the brains of boys and men with autistic disorder, including enlarged ventricles indicative of a loss of brain cells (Haznedar et al., 2000; Piven et al., 1997). But researchers have yet to pinpoint any particular brain disturbance that could account for autism (Rapin, 1997; Zilbovicius et al., 1995). Perhaps autism stems from multiple causes involving more than one type of brain abnormality (Ritvo & Ritvo, 1992). Scientists suspect that the underlying causes of the disorder may involve defective genes or prenatal exposure to toxic agents (O'Connor, 2001b; Stokstad, 2001). Still, the underlying causes of autism remain a mystery.

Treatment Although there is no cure for autism, 30 years of research have supported the efficacy of intensive behavioral treatment programs that apply principles of learning to reduce disturbed behaviors and improve learning and communication skills in autistic children (USDHHS, 1999a). No other treatment approach has yielded comparable results (Gill, 2001). Behavioral approaches are based largely on operant conditioning methods in which rewards and punishments are systematically applied to increase the child's ability to attend to others, to play with other children, to develop academic skills, and to eliminate self-mutilative behavior.

Because children who suffer from autism show behavioral deficits, a central focus of behavior modification is the development of new behavior. New behaviors are maintained by reinforcements, so it is important to teach these children, who often respond to people as they would to a piece of furniture, to accept people as reinforcers. People can be established as reinforcers by pairing praise with such primary reinforcers as food. Then social reinforcement (praise) and primary reinforcers (food) can be used to shape and model toileting behaviors, speech, and social play. The involvement of families and residential treatment personnel in these behavioral programs prompts the maintenance and generalization of behavioral changes (Romanczyk, 1986).

Techniques based on extinction (withholding reinforcement following a response) are sometimes used to eliminate self-mutilative behaviors such as head banging. However, extinction often fails to eliminate the behavior. The problem seems to be that many repetitive behavior patterns—such as rocking and self-injurious behaviors—are maintained by internal reinforcements such as increased stimulation. Therefore, withdrawal of social reinforcers may have little if any effect. Aversive stimulation such as spanking and, in extreme cases, electric shock, may be used in cases when more benign approaches prove ineffective. Brief bursts of mild but painful electrical stimulation can eliminate self-mutilation within a minute of application (Lovaas, 1977). Using electric shock with children raises moral, ethical, and legal concerns, of course. Lovaas has countered that failure to eliminate self-injurious behavior places the child at greater risk of physical harm and denies children the opportunity to participate in other kinds of therapy. The use of aversive stimulation should be combined with positive reinforcement for acceptable alternate behaviors.

The most effective behavioral treatment programs are highly intensive and structured, offering a great deal of individual instruction (Rapin, 1997). The classic example is the program developed by psychologist O. Ivar Lovaas of UCLA. In a classic study by Lovaas (1987), autistic children received more than 40 hours of one-to-one behavior modification each week for at least 2 years. Significant intellectual and educational gains were reported for 9 of the 19 children (47%) in the program. The children who improved achieved normal IQ scores and were able to succeed in the first grade. Only 2% of a control group that did not receive the intensive treatment achieved similar gains. Treatment gains were well maintained at the time of a follow-up when the children were 11 years old (McEachin, Smith, & Lovaas, 1993).

Although intensive behavioral programs may produce impressive gains, longer term follow-ups remain to be reported. We should also note that some children make great progress and others do not (Smith, 1999). Children who are better functioning at the start of treatment typically gain the most.

Biological approaches have had only limited impact in the treatment of autism. This may be changing. One line of research has shown that drugs that enhance serotonin activity, such as SSRIs, can reduce repetitive thoughts and behavior and aggression and lead to some improvement in social relatedness and language use in adults with autism (McDougle et al., 1996). The effects of these drugs on children with autism remain to be seen. Other research has focused on drugs normally used to treat schizophrenia, such as Haldol, which blocks dopamine activity. Several controlled studies show Haldol to be helpful in many cases in reducing social withdrawal and repetitive motor behavior (such as rocking behavior), aggression, hyperactivity, and self-injurious behavior (McBride et al., 1996). We have not seen drugs lead to consistent improvement in the cognitive and language development in children with autism, however.

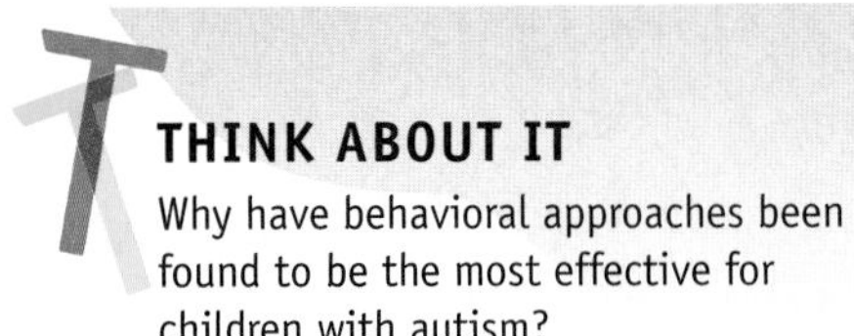

THINK ABOUT IT

Why have behavioral approaches been found to be the most effective for children with autism?

Autistic traits generally continue into adulthood to one degree or another. Yet some autistic children do go on to achieve college degrees and are able to function independently (Rapin, 1997). Others need continuing treatment throughout their lives, even institutionalized care. Even the highest functioning adults with the disorder manifest deficient social and communication skills and a highly limited range of interests and activities (APA, 2000).

Quiz **14.2**
Pervasive Developmental Disorders

Mental Retardation

About 1% of the general population is affected by **mental retardation,** a broad-ranging delay in the development of cognitive and social functioning (APA, 2000). The course of mental retardation is variable. Many children with mental retardation improve over time, especially if they receive support, guidance, and enriched educational opportunities. Those who are reared in impoverished environments may fail to improve or may deteriorate further in relation to other children.

Mental retardation is diagnosed by a combination of three criteria: (1) low scores on formal intelligence tests (an IQ score of approximately 70 or below); (2) evidence of impaired functioning in performing life tasks expected of someone of the same age in a given cultural setting; and (3) development of the disorder before the age of 18 (APA, 2000; Robinson, Zigler, & Gallagher, 2001).

The *DSM* classifies mental retardation according to level of severity, as shown in Table 14.3. Most children with mental retardation (about 85%) fall into the mildly retarded range. These children are generally capable of meeting basic academic demands, such as learning to read simple passages. As adults they are generally capable of independent functioning, although they may require some guidance and support. Table 14.4 provides a description of the deficits and abilities associated with various degrees of mental retardation.

mental retardation A generalized delay or impairment in the development of intellectual and adaptive abilities.

Down syndrome A condition caused by the presence of an extra chromosome on the 21st pair and characterized by mental retardation and various physical anomalies.

TABLE 14.3 Levels of Mental Retardation

Degree of Severity	Approximate IQ Range	Percentage of People with Mental Retardation within the Range
Mild mental retardation	50–55 to approximately 70	Approximately 85%
Moderate mental retardation	35–40 to 50–55	10
Severe mental retardation	20–25 to 35–40	3–4
Profound mental retardation	Below 20 or 25	1–2

Source. Adapted from the *DSM-IV-TR* (APA, 2000).

Causes of Retardation

The causes of mental retardation may be primarily biological in nature, or primarily psychosocial, or a combination of both (APA, 2000). Biological causes include chromosomal and genetic disorders, infectious diseases, and maternal alcohol use during pregnancy. However, more than half of the cases of mental retardation remain unexplained, with most of these falling in the mild range of severity (Flint et al., 1995). These unexplained cases might involve cultural or familial causes, such as being raised in an impoverished home environment. Or perhaps they involve an interaction of psychosocial and genetic factors, the nature of which remains poorly understood (Thaper et al., 1994).

Web Link 14.3
American Association on Mental Retardation WWW

Down Syndrome and Other Chromosomal Abnormalities The most common chromosomal abnormality resulting in mental retardation is **Down syndrome**

TABLE 14.4 Levels of Retardation, Typical Ranges of IQ Scores, and Types of Adaptive Behaviors Shown

Approximate IQ Score Range	Preschool Age 0–5 Maturation and Development	School Age 6–21 Training and Education	Adult 21 and Over Social and Vocational Adequacy
Mild 50–70	Often not noticed as retarded by casual observer, but is slower to walk, feed self, and talk than most children.	Can acquire practical skills and useful reading and arithmetic to a 3rd to 6th grade level with special education. Can be guided toward social conformity.	Can usually achieve social and vocational skills adequate to self-maintenance; may need occasional guidance and support when under unusual social or economic stress.
Moderate 35–49	Noticeable delays in motor development, especially in speech; responds to training in various self-help activities.	Can learn simple communication, elementary health and safety habits, and simple manual skills; does not progress in functional reading or arithmetic.	Can perform simple tasks under sheltered conditions; participates in simple recreation; travels alone in familiar places; usually incapable of self-maintenance.
Severe 20–34	Marked delay in motor development; little or no communication skill; may respond to training in elementary self-help—e.g., self-feeding.	Usually walks, barring specific disability; has some understanding of speech and some response; can profit from systematic habit training.	Can conform to daily routines and repetitive activities; needs continuing direction and supervision in protective environment.
Profound Below 20	Gross retardation; minimal capacity for functioning in sensorimotor areas; needs nursing care.	Obvious delays in all areas of development; shows basic emotional responses; may respond to skillful training in use of legs, hands, and jaws; needs close supervision.	May walk, may need nursing care, may have primitive speech; will usually benefit from regular physical activity; incapable of self-maintenance.

Source. From *Essentials of psychology* (6 ed.) by S.A. Rathus (1996). Copyright © 2001. Reprinted with permission of Brooks/Cole, an imprint of the Wadsworth Group, a division of Thomson Learning. FAX 800-730-2215.

(formerly called Down's syndrome), which is characterized by an extra or third chromosome on the 21st pair of chromosomes, resulting in 47 chromosomes rather than the normal complement of 46 (Wade, 2000). Down syndrome occurs in about 1 in 800 births. It usually occurs when the 21st pair of chromosomes in either the egg or the sperm fails to divide normally, resulting in an extra chromosome. Chromosomal abnormalities become more likely as parents age, so expectant couples in their mid-30s or older often undergo prenatal genetic tests to detect Down syndrome and genetic abnormalities. Down syndrome can be traced to a defect in the mother's chromosomes in about 95% of cases (Antonarakas et al., 1991), with the remainder attributable to defects in the father's sperm.

fragile X syndrome An inherited form of mental retardation caused by a mutated gene on the X chromosome.

People with Down syndrome are recognizable by certain physical features, such as a round face, broad, flat nose, and small, downward-sloping folds of skin at the inside corners of the eyes that gives the impression of slanted eyes. A protruding tongue, small, squarish hands and short fingers, a curved fifth finger, and disproportionately small arms and legs in relation to their bodies also characterize children with Down syndrome. Nearly all of these children have mental retardation, and many suffer from physical problems, such as malformations of the heart and respiratory difficulties. Sadly, most die by middle age. In their later years, they tend to suffer memory losses and experience childish emotions that represent a form of senility.

Children with Down syndrome suffer various deficits in learning and development. They tend to be uncoordinated and to lack proper muscle tone, which makes it difficult for them to carry out physical tasks and engage in play activities like other children. Down syndrome children suffer memory deficits, especially for information presented verbally, which makes it difficult for them to learn in school. They also have difficulty following instructions from teachers and expressing their thoughts or needs clearly in speech. Despite their disabilities, most can learn to read, write, and perform simple arithmetic, if they receive appropriate schooling and the right encouragement.

Although less common than Down syndrome, chromosomal abnormalities on the sex chromosome may also result in mental retardation, such as in Klinefelter's syndrome and Turner's syndrome. Klinefelter's syndrome, which only occurs among males, is characterized by the presence of an extra X chromosome, resulting in an XXY chromosomal pattern rather than the XY pattern that men normally have. Estimates of the prevalence of Klinefelter's syndrome range from 1 in 500 to 1 in 1,000 male births (Brody, 1993c). Men with this XXY pattern fail to develop appropriate secondary sex characteristics, resulting in small, underdeveloped testes, low sperm production, enlarged breasts, poor muscular development, and infertility. Mild retardation or learning disabilities frequently occur among these men. Men with Klinefelter's syndrome often don't discover they have the condition until they undergo tests for infertility.

Found only among females is Turner's syndrome, which is characterized by the presence of a single X sex chromosome instead of the normal two. Although such girls develop normal external genitals, their ovaries remain poorly developed, producing reduced amounts of estrogen. As women, they tend to be shorter than average and infertile. They also tend to show evidence of mild retardation, especially in skills relating to math and science.

Striving to achieve. Most children with Down syndrome can learn basic academic skills if they are afforded opportunities to learn and provided with the right encouragement.

Fragile X Syndrome and Other Genetic Abnormalities Fragile X syndrome is the most common type of inherited (genetic) mental retardation (Kwon et al., 2001). It is the second most common form of retardation overall, after Down syndrome (Plomin et al., 1994). The disorder is believed to be caused by a mutated gene on the X chromosome (Hagerman, 1996). The defective gene is located in an area of the chromosome that appears fragile, hence the name **fragile X syndrome.** Fragile X syndrome causes mental retardation in every 1,000 to 1,500 males and (generally less severe) mental handicaps in every 2,000 to 2,500 females (Angier, 1991b; Rousseau et al., 1991). The effects of fragile X syndrome range from mild learning disabilities to retardation so profound that those affected can hardly speak or function.

Females normally have two X chromosomes, whereas males have only one. For females, having two X chromosomes seems to provide some protection against the disorder

phenylketonuria (PKU) A genetic disorder that prevents the metabolization of phenylpyruvic acid, leading to mental retardation unless the diet is strictly controlled.

cytomegalovirus A source of infection that, in pregnant women, carries a risk of mental retardation to the unborn child.

if the defective gene turns up on one of the two chromosomes (Angier, 1991b). This may explain why the disorder usually has more profound effects on males than on females. Yet the mutation does not always manifest itself. Many males and females carry the fragile X mutation but show no clinical evidence of it. Yet they can pass along the syndrome to their offspring.

A genetic test can detect the presence of the mutation by direct DNA analysis (Rousseau et al., 1991) and may be of help to prospective parents in genetic counseling. Prenatal testing of the fetus is also possible (Sutherland et al., 1991). Although there is no treatment for fragile X syndrome, identifying the defective gene is the first step toward understanding how the protein produced by the gene functions to produce the disability, which may lead to the development of treatments in the future (Angier, 1991b).

Phenylketonuria (PKU) is a genetic disorder that occurs in 1 in 10,000 births (Plomin et al., 1994). It is caused by a recessive gene that prevents the child from metabolizing the amino acid phenylalanine, which is found in many foods. Consequently, phenylalanine and its derivative, phenylpyruvic acid, accumulate in the body, causing damage to the central nervous system that results in mental retardation and emotional disturbance. The presence of PKU can be detected among newborns by analyzing blood or urine samples. Although there is no cure for PKU, children with the disorder may suffer less damage or develop normally if they are placed on a diet low in phenylalanine soon after birth (Brody, 1990). Such children receive protein supplements that compensate for their nutritional loss.

Today, various prenatal tests can detect the presence of chromosomal abnormalities and genetic disorders. In *amniocentesis,* which is usually conducted about 14 to 15 weeks following conception, a sample of amniotic fluid is drawn with a syringe from the amniotic sac that contains the fetus. Cells from the fetus can then be separated from the fluid, allowed to grow in a culture, and examined for abnormalities, including Down syndrome. Blood tests are used to detect carriers of other disorders.

Prenatal Factors Some cases of mental retardation are caused by maternal infections or substance abuse during pregnancy. Rubella (German measles) in the mother, for example, can be passed along to the unborn child, causing brain damage that results in retardation, and it may play a role in autism. Although the mother may experience mild symptoms or none at all, the effects on the fetus can be tragic. Other maternal diseases that may cause retardation in the child include syphilis, **cytomegalovirus,** and genital herpes.

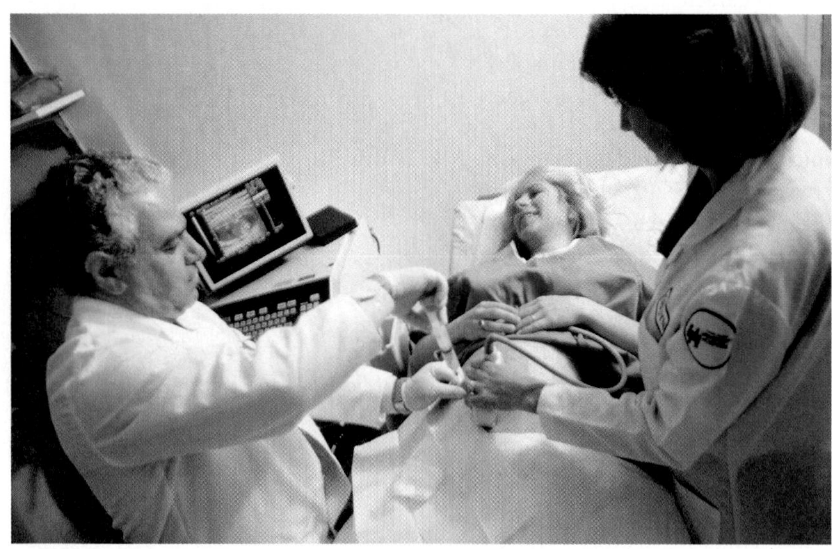

Amniocentesis. In amniocentesis, a physician extracts a sample of amniotic fluid to test for biochemical and chromosomal abnormalities. Here the physician uses ultrasound to determine the location of the fetus to help prevent accidental injury to it while placing the syringe in the mother's abdomen.

Widespread programs that immunize women against rubella before pregnancy and tests for syphilis during pregnancy have reduced the risk of transmission of these infections to children. Most children who contract genital herpes from their mothers do so during delivery by coming into contact with the herpes simplex virus in the birth canal. Caesarean sections (C-sections) reduce the risk of the baby's coming into contact with the virus during childbirth.

Drugs that the mother ingests during pregnancy may pass through the placenta to the child. Some can cause severe birth deformities and mental retardation. Children whose mothers drink alcohol during pregnancy are often born with fetal alcohol syndrome (described in Chapter 10). FAS is among the most prominent causes of mental retardation. Maternal smoking during pregnancy has also been linked to the development of attention-deficit hyperactivity disorder in children (Milberger et al., 1996).

Birth complications, such as oxygen deprivation or head injuries, place children at increased risk for neurological disorders, including mental retardation. Prematurity also places children at risk of retardation and other developmental problems. Brain infections, such as encephalitis and meningitis, or traumas during infancy and early childhood can cause mental retardation and other health problems. Children who ingest toxins, such as paint chips containing lead, may also suffer brain damage that produces mental retardation.

Environmental hazards. Children who are exposed to environmental hazards, such as paint chips containing lead-based paint, may suffer brain damage that can lead to mental retardation.

Cultural-Familial Causes Most cases of mental retardation fall in the mild range of severity. In most of these cases, there is no apparent biological cause or distinguishing physical feature that sets the child apart from other children. Psychosocial factors, such as an impoverished home or social environment that is intellectually unstimulating, or parental neglect or abuse, may play a causal or contributing role in the development of mental retardation in such children. Supporting a family linkage is evidence from a study in Atlanta in which mothers who failed to finish high school were four times more likely than better educated mothers to have children with mild retardation (Drews et al., 1995).

These cases are considered **cultural-familial retardation.** Children in impoverished families may lack toys, books, or opportunities to interact with adults in intellectually stimulating ways. Consequently, they may fail to develop appropriate language skills or become unmotivated to learn the skills that are valued in contemporary society. Economic burdens, such as the need to hold multiple jobs, may prevent their parents from spending time reading to them, talking to them at length, and exposing them to creative play or trips to museums and parks. They may spend most of their days glued to the TV set. The parents, most of whom were also reared in poverty, may lack the reading or communication skills to help shape the development of these skills in their children. A vicious cycle of poverty and impoverished intellectual development may be repeated from generation to generation.

Children with this form of retardation may respond dramatically when provided with enriched learning experiences, especially at the earlier ages. Social programs like Head Start, for example, have helped children at risk of cultural-familial retardation to function within the normal range of ability (e.g., Barnett & Escobar, 1990).

Intervention

The services that children with mental retardation require to meet the developmental challenges they face depend in part on the level of severity and type of retardation (Dykens & Hodapp, 1997; Snell, 1997). With appropriate training, children with mild retardation may approach a sixth grade level of competence. They can acquire vocational skills that allow them to support themselves minimally through meaningful work. Many such children can be mainstreamed in regular classes. At the other extreme, children with severe or profound mental retardation may need institutional care or placement in a residential care facility in the community, such as a group home. Placement in an institution is often based on the need to control destructive or aggressive behavior, not because of severity of the individual's intellectual impairment. Consider the case of a child with moderate retardation:

A Case of Moderate Mental Retardation

The mother pleaded with the emergency room physician to admit her 15-year-old son, claiming that she couldn't take it anymore. Her son, a Down syndrome patient with an IQ of 45, had alternated since the age of 8 between living in institutions and at home. Each visiting day he pleaded with his mother to take him home, and after about a year at each placement, she would bring him home but find herself unable to control his behavior. During temper tantrums, he would break dishes and destroy furniture and had

cultural-familial retardation A mild form of mental retardation that is influenced by impoverishment of the home environment.

Imparting skills. In 1975, Congress enacted legislation that requires public schools to provide children with disabilities with educational programs that meet their individual needs.

recently become physically assaultive toward his mother, hitting her on the arm and shoulder during a recent scuffle when she attempted to stop him from repeatedly banging a broom on the floor.

—*Adapted from Spitzer et al., 1989, pp. 338–340*

Controversy remains concerning whether children with mental retardation should be mainstreamed in regular classes or placed in special education classes. Although some children with mild retardation may achieve better when they are mainstreamed, others may not do so well in regular classes. They may find these classes overwhelming and withdraw from their schoolmates. There has also been a trend toward deinstitutionalization of people with more severe mental retardation, a policy shift motivated in large part by public outrage over the appalling conditions that existed in many institutions serving this population. The Developmentally Disabled Assistance and Bill of Rights Act, which Congress passed in 1975, provided that persons with mental retardation have the right to receive appropriate treatment in the least restrictive treatment setting. Nationwide, the population of institutions for people with mental retardation shrunk by nearly two-thirds from the 1970s to the 1990s.

People with mental retardation who are capable of functioning in the community have the right to receive less restrictive care than is provided in large institutions. Many are capable of living outside the institution and have been placed in supervised group homes. Residents typically share household responsibilities and are encouraged to participate in meaningful daily activities, such as training programs or sheltered workshops. Others live with their families and attend structured day programs. Adults with mild retardation often work in outside jobs and live in their own apartments or share apartments with other persons with mild retardation. Although the large-scale dumping of mental patients in the community from psychiatric institutions resulted in massive social problems and swelled the ranks of America's homeless population, deinstitutionalization of people with mental retardation has largely been a success story that has been achieved with rare dignity (Winerip, 1991).

Children and adults with mental retardation may need psychological counseling to help them adjust to life in the community. Many have difficulty making friends and may become socially isolated. Problems with self-esteem are also common, especially because people who have mental retardation are often demeaned and ridiculed. Supportive counseling may be supplemented with behavioral techniques that help them acquire skills in areas such as personal hygiene, work, and social relationships. More structured behavioral approaches can be used to teach persons with more severe retardation such basic hygienic behaviors as toothbrushing, self-dressing, and hair combing.

Other behavioral treatment techniques include social skills training, which focuses on increasing the individual's ability to relate effectively to others, and anger management training to help individuals develop more effective ways of handling conflicts without aggressive acting out (Huang & Cuvo, 1997; Rose, 1996).

Children with mental retardation stand perhaps a three to four times greater-than-normal chance of developing other psychological disorders, such as attention-deficit hyperactivity disorder (ADHD), depression, or anxiety disorders (Borthwick-Duffy, 1994). As many as three of four boys with fragile X syndrome, for example, develop ADHD (Matson & Sevin, 1994). Mental health professionals have been slow to recognize the prevalence of mental health problems among people with mental retardation, perhaps because of a long-held conceptual distinction between emotional impairment on the one hand and intellectual deficits on the other (Nezu, 1994). Many professionals even assumed (wrongly) that people with mental retardation were somehow immune from psychological problems or that they lacked the necessary verbal ability to benefit from psychotherapy (Bütz, Bowlling, & Bliss, 2000; Nezu, 1994). Given these commonly held beliefs, it is perhaps not surprising that many of the psychological problems of people with mental retardation have gone unrecognized and untreated (Reiss & Valenti-Hein, 1994). However, evidence shows that people with mental retardation can benefit from psychotherapy (Bütz et al., 2000).

Quiz 14.3 Q
Mental Retardation

A Closer Look

Savant Syndrome

Got a minute? Try the following:

1. Without referring to a calendar, calculate the day of the week that March 15, 2079, will fall on.
2. List the prime numbers between 1 and 1 billion. (Hint: the list starts 1, 2, 3, 5, 7, 11, 13, 17 . . .)
3. Repeat verbatim the newspaper stories you read over coffee this morning.
4. Sing accurately every note played by the first violin in Beethoven's Ninth Symphony.

These tasks are impossible for all but a very few. Ironically, people who are most likely to be able to accomplish these feats suffer from autism, mental retardation, or both. Such a person is commonly called an *idiot savant*. The term *savant* is derived from the French *savoir,* meaning "to know." The label *savant syndrome* is preferable to the pejorative term *idiot savant* to refer to someone with severe mental deficiencies who possesses some remarkable mental abilities. The prevalence of the savant syndrome among people with mental retardation is estimated at about .06%, or about 1 case in 2,000 (Hill, 1977). The emergence of the savant syndrome is also closely linked to infantile autism (Miller, 1999). Most people with savant syndrome, like most people with autism, are male (Treffert, 1988). Among a sample of 5,400 people with autism, 531 (9.8%) were reported by parents to have the savant syndrome (Rimland, 1978). Because they want to think of their children as special, however, parents might overreport the incidence of the savant syndrome.

Several hundred people with the savant syndrome have been described in this century. They are reported to have shown remarkable but circumscribed mental skills, such as calendar calculating, rare musical talent, even accomplished poetry (Dowker, Hermelin, & Pring, 1996)—all of which stand in contrast to their limited general intellectual abilities. People with the savant syndrome also have outstanding memories. Just as we learn about health by studying illness, we may be able to learn more about normal mechanisms of memory by studying people in whom memory stands apart from other aspects of mental functioning (e.g., Kelly, Macaruso, & Sokol, 1997).

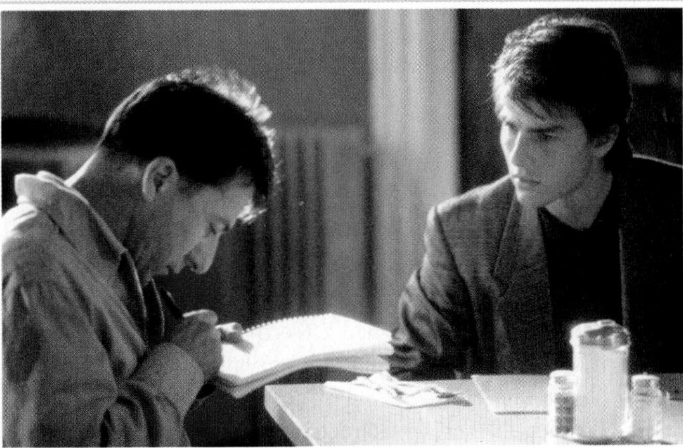

Savant syndrome. Dustin Hoffman (left) won the Best Actor Oscar for his portrayal in the film *Rainman* of a man with autism who showed a remarkable capacity for numerical calculation. Tom Cruise (right) played his brother. Hoffman was able to capture the sense of emotional detachment and isolation of his character.

The savant syndrome phenomenon occurs more frequently in males by a ratio of about 6 to 1. The special skills of people with the savant syndrome tend to appear out of the blue and may disappear as suddenly. Some people with the syndrome engage in lightning calculations. A 19th-century enslaved person in Virginia, Thomas Fuller, "was able to calculate the number of seconds in 70 years, 17 days, and 12 hours in a minute and one half, taking into account the 17 leap years that would have occurred in the period" (S. C. Smith, 1983). There are also cases of persons with the syndrome who were blind but could play back any musical piece, no matter how complex, or repeat long passages of foreign languages without losing a syllable. Some people with the syndrome make exact estimates of elapsed time. One could reportedly repeat verbatim the contents of a newspaper he had just heard; another could repeat backward what he had just read (Tradgold, 1914, cited in Treffert, 1988).

Truth OR Fiction? REVISITED

Some people can recall verbatim every story they read in a newspaper.

TRUE. Ability to recall news stories verbatim is found in some individuals with the savant syndrome.

Various theories have been presented to explain the savant syndrome (Treffert, 1988). Some believe that children with the savant syndrome have unusually well-developed memories that allow them to record and scan vast amounts of information. It has been suggested that people with the savant syndrome may inherit two sets of hereditary factors, one for retardation and the other for special abilities. Perhaps it is coincidental that their special abilities and their mental handicaps were inherited in common. Other theorists suggest that the left and right hemispheres of their cerebral cortexes are organized in an unusual way. This latter belief is supported by research suggesting that the special abilities they possess often involve skills associated with right hemisphere functioning. Still other theorists suggest they learn special skills to compensate for their lack of more general skills, perhaps as a means of coping with their environment, or perhaps as a means of earning social reinforcements. Perhaps their skills in concrete functions, like calculation, compensate for their lack of abstract thinking ability. Linguists like Noam Chomsky (1965) theorize that people are neurologically "prewired" to grasp the deep structure that underlies all human languages. Perhaps, as the neurologist Oliver Sacks speculates (1985b), the ▶

brain circuits of some people with the savant syndrome are wired with a "deep arithmetic"—an innate structure for perceiving mathematical relationships that is analogous to the prewiring that allows people to perceive and produce language.

Recent research has pointed to possible gender-linked left hemisphere damage occurring prenatally or congenitally. Compensatory right hemisphere development might then take place, establishing specialized brain circuitry that processes concrete and narrowly defined kinds of information (Treffert, 1988). An environment that reinforces savant abilities and provides opportunities for practice and concentration would give further impetus to the development of these unusual abilities. Still the savant syndrome remains a mystery.

Learning Disorders

Nelson Rockefeller served as governor of New York State and as vice president of the United States. He was brilliant and well educated. However, despite the best of tutors, he always had trouble reading. Rockefeller suffered from **dyslexia,** a term derived from the Greek roots *dys-*, meaning "bad," and *lexikon,* meaning "of words." Dyslexia is the most common type of **learning disorder** (also called a *learning disability*) (Shaywitz, 1998). It accounts for perhaps 80% of cases of learning disability and describes individuals who have trouble reading despite possessing at least average intelligence (Miller-Medzon, 2000). Mental retardation involves a general delay in intellectual development. People with learning disorders, by contrast, may be intelligent, even gifted, but show poor development in reading, math, or writing skills to a point that it impairs their school performance or daily functioning. Today, about one in eight children (about 12%) is placed in a program for the learning disabled, and the percentage of children participating in such programs is growing (Levine, 2000).

Learning disorders tend to be chronic disorders that continue to affect development into adulthood. Children with learning disorders tend to perform poorly in school. They are often viewed as failures by their teachers and their families. It is not surprising that most of them develop low expectations and problems with self-esteem.

THINK ABOUT IT

Do you think children with mental retardation should be mainstreamed within regular classes? Why or why not?

Types of Learning Disorders

Types of learning disorders include *mathematics disorder, disorder of written expression,* and *reading disorder.*

Mathematics Disorder *Mathematics disorder* describes children with deficiencies in arithmetic skills. They may have problems understanding basic mathematical terms or operations, such as addition or subtraction; decoding mathematical symbols (+, =, etc.); or learning multiplication tables. The problem may become apparent as early as the first grade (age 6) but is not generally recognized until about the second or third grade.

Web Link 14.4 WWW
Facts About Learning Disabilities

Disorder of Written Expression *Disorder of written expression* refers to children with grossly deficient writing skills. The deficiency may be characterized by errors in spelling, grammar, or punctuation, or by difficulty in composing sentences and paragraphs. Severe writing difficulties generally become apparent by age 7 (second grade), although milder cases may not be recognized until the age of 10 (fifth grade) or later.

Reading Disorder Reading disorder—*dyslexia*—characterizes children who have poorly developed skills in recognizing words and comprehending written text. Dyslexia is estimated to affect about 4% of school-age children (APA, 2000). Children with dyslexia may read slowly with difficulty, and distort, omit, or substitute words when reading aloud. They have trouble decoding letters and letter combinations and translating them into the appropriate sounds (Miller-Medzon, 2000). They may also misperceive letters as upside down (for example, confusing *w* for *m*) or in reversed images (*b* for *d*). Dyslexia is usually apparent by the age of 7, coinciding with the second grade, although it is sometimes recognized in 6-year-olds. Children

dyslexia A learning disorder characterized by impaired reading ability.

learning disorder A deficiency in a specific learning ability in the context of normal intelligence and exposure to learning opportunities.

and adolescents with dyslexia tend to be more prone than their peers to depression, to have lower self-worth and feelings of competence in their academic work, and to have signs of attention-deficit hyperactivity disorder (Boetsch, Green, & Pennington, 1996).

More boys are diagnosed with reading disorder than girls, but this difference may have more to do with a bias toward identifying the disorder in males rather than an underlying gender difference in the rate of the disorder (APA, 2000). Boys with dyslexia are more likely than girls to show disruptive behavior in class and so are more likely to be referred for evaluation. Carefully conducted studies find similar rates of the disorder in boys and girls (APA, 2000; Shaywitz, 1998).

Rates of dyslexia vary with respect to native language. Rates of dyslexia are high in English-speaking and French-speaking countries, where the language contains a large number of ways of spelling words containing the same meaningful sounds (e.g., the same "o" sound in the words "toe" and "tow") than in Italy, where the language has a smaller ratio of sounds to letter combinations ("Dyslexia," 2001; Paulesu et al., 2001).

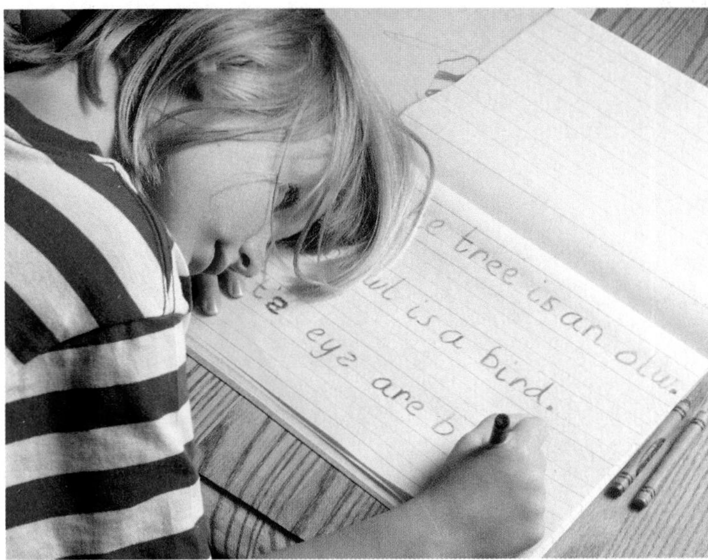

Dyslexia. Children with dyslexia have difficulty decoding words. Note the reversal of the letters *w* and *l* in the word *owl* in this picture of a girl with dyslexia completing a writing exercise.

Theoretical Perspectives

Hypotheses of the origins of learning disorders tend to focus on cognitive-perceptual problems and possible underlying neurological factors. Many children with learning disorders have problems with visual or auditory perception. They may lack the capacity to copy words or to discriminate geometric shapes. Other children have short attention spans or show hyperactivity, which may also be suggestive of an underlying brain abnormality.

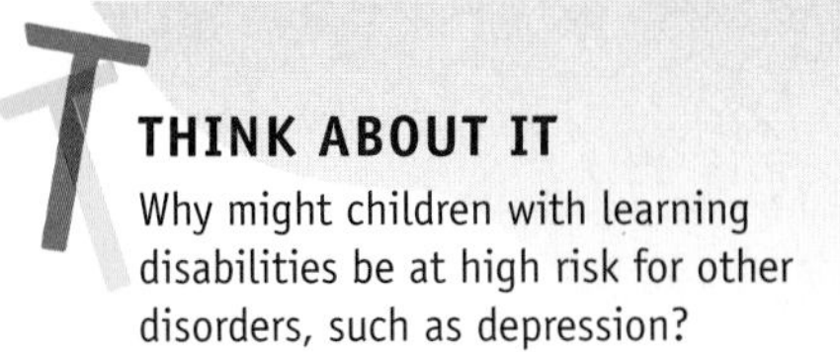

THINK ABOUT IT

Why might children with learning disabilities be at high risk for other disorders, such as depression?

Much of the research on learning disorders has focused on dyslexia. Although no one can say with certainty what causes dyslexia, mounting evidence points to underlying deficits in how the brain processes visual and auditory information (e.g., Azar, 2000b; Miller-Medzon, 2000; Murray, 2000a). Evidence of impaired visual processing in the brains of people with dyslexia points to a possible defect in a visual relay station in the brain through which visual information flows from the eye to the visual cortex for processing (Livingstone et al., 1991). As a result, the brains of people with dyslexia may not be able to decipher a rapid succession of visual stimuli needed to decode letters and words. Words may thus become blurry, fuse together, or seem to jump off the page—problems reported by people with dyslexia.

Additional evidence from PET studies shows lower levels of activity in dyslexic people in parts of the brain involved in language processing and reading (Helmuth, 2001; Paulesu et al., 2001). These differences in brain function are observed in dyslexic subjects from England, France, and Italy, which suggest that the disorder has a common biological basis despite the differences in the rates of the disorder across these countries.

Sensory pathways for hearing and even touch are also implicated in dyslexia. Some forms of dyslexia may be traceable to abnormalities in brain circuits responsible for processing the rapid flow of sounds (Blakeslee, 1994b). This flaw in brain circuitry may make it difficult to understand rapidly occurring speech sounds, such as the sounds corresponding to the letters *b* and *p* in syllables like *ba* and *pa.* Problems in discerning the differences between many basic speech sounds can make it difficult for people with dyslexia to learn to speak correctly and later, perhaps, to learn to read. They continue to have problems distinguishing in rapid speech between words like *boy* and *toy* or *pet* and *bet.* If defects in brain circuitry responsible for relaying and processing sensory data are involved in learning disorders, as the evidence now suggests, it may lead the way to the development of specialized treatment programs to help children adjust to their sensory capabilities.

Taken together, evidence points to a neurological basis to the cognitive problems in dyslexia (Paulesu et al., 2001). The underlying dysfunction may have a genetic basis, as evidence connects genetic factors with a greater risk of dyslexia (Shaywitz, 1998; Nagourney, 2001a). As we can see from Figure 14.2, people with dyslexic parents are at increased risk of

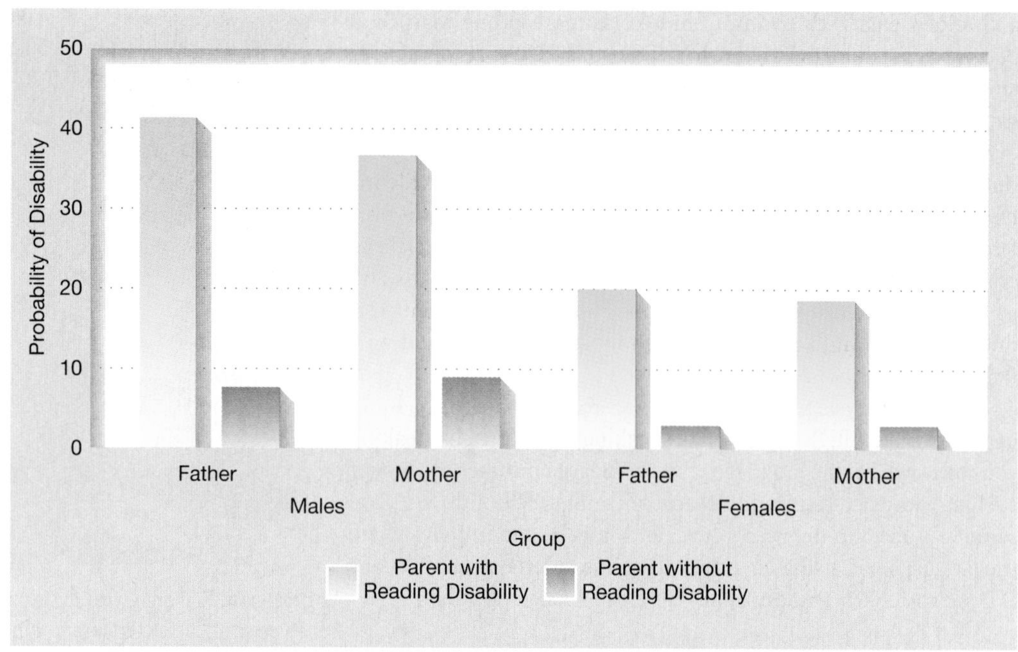

FIGURE 14.2 Familial risk of developmental reading disorder (dyslexia).
Boys are at greater risk than girls of developing dyslexia, and children of both genders whose parents with dyslexia are at relatively greater risk. Although these data are consistent with a genetic explanation of the etiology of dyslexia, it is also possible that parents with dyslexia do not provide their children with the types of stimulation such as books and reading bedtime stories that foster reading skills.

Source. Adapted from Vogler et al. (1985).

having the disorder themselves (Vogler, DeFries, & Decker, 1985). Moreover, higher rates of concordance (agreement) for dyslexia are found between identical (MZ) than fraternal (DZ) twins, 70% versus 40%, respectively (Plomin et al., 1994). Suspicion has focused on the role that particular genes may play in causing subtle defects in brain circuitry involved in reading.

Intervention

Interventions for learning disorders have generally been approached from the following perspectives (Lyon & Moats, 1988):

1. *The psychoeducational model.* Psychoeducational approaches emphasize children's strengths and preferences, rather than attempt to correct assumed underlying deficiencies. For example, a child who retains auditory information better than visual information might be taught verbally, for example, using tape recordings rather than written materials.
2. *The behavioral model.* The behavioral model assumes that academic learning is built on a hierarchy of basic skills, or "enabling behaviors." To read effectively, one must first learn to recognize letters, then attach sounds to letters, then combine letters and sounds into words, and so on. The child's learning competencies are assessed to determine where deficiencies lie in the hierarchy of skills. An individualized program of instruction and reinforcement helps the child acquire the skills needed to perform academic tasks.
3. *The medical model.* This model assumes that learning disorders are symptoms of biologically based deficiencies in cognitive processing. Proponents suggest that remediation should be directed at the underlying pathology rather than the learning disability. If the child has a visual defect that makes it difficult to follow a line of text, treatment should aim to remediate the visual deficit, perhaps through visual-tracking exercises. Improvement in reading ability would be expected to follow.

4. *The neuropsychological model.* This approach borrows from the psychoeducational and medical models. It assumes that learning disorders reflect underlying biologically based deficits in processing information (medical model). It also assumes that educational programs should be adapted to take into account these underlying deficits and individualized to the child's needs (Levine, 2000).
5. *The linguistic model.* The linguistic approach focuses on children's basic language deficiencies, such as failing to recognize how sounds and words are strung together to create meaning, which can give rise to problems in reading, spelling, and finding the words to express themselves. Adherents to this model teach language skills sequentially, helping the student grasp the structure and use of words (Shaywitz, 1998; Wagner & Torgesen, 1987).
6. *The cognitive model.* This model focuses on how children organize their thoughts when they learn academic material. Within this perspective, children are helped to learn by (1) recognizing the nature of the learning task, (2) applying effective problem-solving strategies to complete tasks, and (3) monitoring the success of their strategies. Children with arithmetic problems might be guided to break down a math problem into its component tasks, think through the steps necessary to complete each task, and evaluate their performance at each step to judge how to proceed. Children progress through a systematic approach to problem solving that can be applied to diverse academic tasks.

communication disorders A class of psychological disorders characterized by difficulties in understanding or using language.

Evaluation of Treatment Approaches The medical model is currently limited by lack of evidence that underlying deficiencies are correctable or that such improvements foster academic skills (Hinshaw, 1992; Lyon & Moats, 1988). There is also a lack of evidence for the psychoeducational approach (Brady, 1986; Lyon & Moats, 1988). Although the neuropsychological approach has not yet been fully tested, interventions focused on changing the child's learning strategies in order to circumvent apparent underlying neuropsychological deficits have thus far failed to demonstrate significant gains in children with severe forms of learning disability (Hinshaw, 1992). The interventions showing the most promising results to date are those that provide direct instruction in the academic skills in which the child is deficient, such as oral and written language skills (Hinshaw, 1992). The behavioral model has also shown some promising results in improving the performance of children who are deficient in reading and arithmetic skills (Koorland, 1986). Whether the gains from behavioral training generalize to classroom performance beyond the training setting remains to be seen. The linguistic approach has received some support, but not enough to advocate widespread use in treating children with reading and spelling deficiencies (Lyon & Moats, 1988). The cognitive model, too, has received some support, but many children with learning disorders have not developed enough basic knowledge in their problem areas to use it to think through problems.

Many children who have learning disorders are placed in special education programs or classes. Yet programs for learning-disabled children vary widely in quality and we still lack firm evidence demonstrating their long-term effectiveness (Hinshaw, 1992; Wingert & Kantrowitz, 1997).

THINK ABOUT IT

Do you think people with learning disorders should be given special consideration when given standardized tests, like the Scholastic Aptitude Test (SAT), such as having extra time? Why or why not?

Quiz **14.4**
Learning Disorders

Communication Disorders

Communication disorders involve difficulties in understanding or using language. The categories of communication disorders include *expressive language disorder, mixed receptive/expressive language disorder, phonological disorder,* and *stuttering.* Each of these disorders interferes with academic or occupational functioning or ability to communicate socially. Table 14.5 lists the *DSM-IV* classification of learning disorders and communication disorders.

Expressive language disorder involves impairment in the use of spoken language, such as slow vocabulary development, errors in tense, difficulties recalling words, and problems producing sentences of appropriate length and complexity for the individual's age. Affected children may also have a phonological (articulation) disorder, compounding their speech problems.

TABLE 14.5 *DSM-IV* Classification of Learning Disorders and Communication Disorders

Learning Disorders	Reading Disorder Mathematics Disorder Disorder of Written Expression
Communication Disorders	Expressive Language Disorder Mixed Receptive/Expressive Language Disorder Phonological Disorder Stuttering

Source. Adapted from the *DSM-IV-TR* (APA, 2000).

Mixed receptive/expressive language disorder refers to children who have difficulties both understanding and producing speech. There may be difficulty understanding words or sentences. In some cases, children have difficulty understanding certain word types (such as words expressing differences in quantity—*large, big,* or *huge*), spatial terms (such as *near* or *far*), or sentence types (such as sentences that begin with the word *unlike*). Other cases are marked by difficulties understanding simple words or sentences.

Phonological disorder involves difficulties in articulating the sounds of speech in the absence of defects in the oral speech mechanism or neurological impairment. Children with the disorder may omit, substitute, or mispronounce certain sounds—especially *ch, f, l, r, sh,* and *th* sounds, which are usually articulated properly by the time children reach the early school years. It may sound as if they are uttering "baby talk." In more severe cases, there are problems articulating sounds usually mastered during the preschool years: *b, m, t, d, n,* and *h.* Speech therapy is often helpful, and milder cases often resolve themselves by the age of 8.

Stuttering involves disturbances in the ability to speak fluently with appropriate timing of speech sounds. The lack of normal fluency must be inappropriate for the person's age in order to justify the diagnosis. Stuttering usually begins between 2 and 7 years of age and affects about 1 child in 100 before puberty (APA, 2000). The disorder is characterized by one or more of the following characteristics: (1) repetitions of sounds and syllables; (2) prolongations of certain sounds; (3) interjections of inappropriate sounds; (4) broken words, such as pauses occurring within a spoken word; (5) blocking of speech; (6) circumlocutions (substitutions of alternative words to avoid problematic words); (7) displaying an excess of physical tension when emitting words; and (8) repetitions of monosyllabic whole words (for example, "I-I-I-I am glad to meet you") (APA, 2000). Stuttering occurs predominantly among males by a ratio of about 3 to 1. Stuttering remits in upward of 80% of children, typically before age 16. As many as 60% of cases show remission without any treatment. Stuttering is believed to involve an interaction of genetic and environmental factors (Felsenfeld, 1996; Yairi, Ambrose; & Cox, 1996). Underlying social anxiety or social phobias may be involved in some cases, at least among adults with stuttering problems (De-Carle & Pato, 1996; Schneier, Wexler, & Liebowitz, 1997). Treatment of communication disorders is generally approached with speech therapy and with psychological counseling for social anxiety or other emotional problems.

Quiz 14.5 Communication Disorders

Attention-Deficit and Disruptive Behavior Disorders

The category of *attention-deficit and disruptive behavior disorders* refers to a diverse range of problem behaviors, including *attention-deficit hyperactivity disorder* (ADHD), *conduct disorder* (CD), and *oppositional defiant disorder* (ODD). These disorders are socially disruptive and usually more upsetting to other people than to the children who receive these diagnoses. Although there are differences among these disorders, the rate of co-morbidity (co-occurrence) among these disorders is very high (Jensen, Martin, & Cantwell, 1997).

attention-deficit hyperactivity disorder (ADHD) A behavior disorder characterized by excessive motor activity and inability to focus one's attention.

hyperactivity An abnormal behavior pattern characterized by difficulty in maintaining attention and extreme restlessness.

Attention-Deficit Hyperactivity Disorder

Many parents believe that their children are not attentive toward them—that they run around on whim and do things in their own way. Some inattention, especially in early childhood, is normal enough. In **attention-deficit hyperactivity disorder (ADHD),** however, children display impulsivity, inattention, and **hyperactivity** that are considered inappropriate to their developmental levels.

ADHD is divided into three subtypes: a predominantly inattentive type, a predominantly hyperactive or impulsive type, or a combination type characterized by high levels of both inattention and hyperactivity-impulsivity (APA, 2000). The disorder is usually first diagnosed during elementary school, when problems with attention or hyperactivity-impulsivity make it difficult for the child to adjust to school. Although signs of hyperactivity are often observed earlier, many overactive toddlers do not go on to develop ADHD.

ADHD is the most commonly diagnosed psychological disorder in children today (Bradley & Golden, 2001). The disorder is estimated to affect between 3% and 7% of school-age children, or some two million American youngsters in total (Shute, Locy, & Pasternak, 2000; Wingert, 2000; APA, 2000). ADHD is diagnosed two to nine times more often in boys than girls (APA, 2000). Although inattention appears to be the basic problem, associated problems include inability to sit still for more than a few moments, bullying, temper tantrums, stubbornness, and failure to respond to punishment (see Table 14.6).

Activity and restlessness impair the ability of children with ADHD to function in school. They seem incapable of sitting still. They fidget and squirm in their seats, butt into

TABLE 14.6 Diagnostic Features of Attention-Deficit Hyperactivity Disorder (ADHD)

Kind of Problem	Specific Behavior Pattern
Lack of attention	Fails to attend to details or makes careless errors in schoolwork, etc.
	Has difficulty sustaining attention in schoolwork or play
	Doesn't appear to pay attention to what is being said
	Fails to follow through on instructions or to finish work
	Has trouble organizing work and other activities
	Avoids work or activities that require sustained attention
	Loses work tools (e.g., pencils, books, assignments, toys)
	Becomes readily distracted
	Is forgetful in daily activities
Hyperactivity	Fidgets with hands or feet or squirms in his or her seat
	Leaves seat in situations such as the classroom in which remaining seated is required
	Is constantly running around or climbing on things
	Has difficulty playing quietly
Impulsivity	Frequently "calls out" in class
	Fails to wait his/her turn in line, games, etc.

To receive a diagnosis of ADHD, the disorder must begin by the age of 7; must have significantly impaired academic, social, or occupational functioning; and must be characterized by a designated number of clinical features shown in this table occurring over a 6-month period in at least two settings such as at school, at home, or at work.

Source. Adapted from the *DSM-IV-TR* (APA, 2000).

other children's games, have outbursts of temper, and may engage in dangerous behavior, such as running into the street without looking. All in all, they can drive parents and teachers to despair.

Where does "normal" age-appropriate overactivity end and hyperactivity begin? Assessment of the degree of hyperactive behavior is crucial, because many normal children are called "hyper" from time to time. Some critics of the ADHD diagnosis argue that it merely labels children who are difficult to control as mentally disordered or sick. Most children, especially boys, are highly active during the early school years. Proponents of the diagnosis counter that there is a difference in quality between normal overactivity and ADHD. Normally overactive children are usually goal directed and can exert voluntary control over their behavior. But children with ADHD appear hyperactive without reason and do not seem to be able to conform their behavior to the demands of teachers and parents. Put another way: Most children can sit still and concentrate for a while when they want to do so; children who are hyperactive seemingly cannot.

Web Link 14.5 WWW
Diagnosis and Treatment of Attention Deficit Hyperactivity Disorder

Although children with ADHD tend to be of average or above-average intelligence, they often underachieve in school. They are frequently disruptive in the classroom and tend to get into fights (especially the boys). They may fail to follow or remember instructions or complete assignments. They are more likely to have learning disabilities, to repeat grades, and to be placed in special education classes (Faraone et al., 1993). They also show a higher frequency of physical injuries and hospital admissions than their peers ("Children with Hyperactivity," 2001; Leibson et al., 2001). They also stand a greater risk of mood disorders, anxiety disorders, and problems getting along with family members (Biederman et al., 1996a, b). Compared to their peers, ADHD boys tend to lack empathy, or awareness of other people's feelings (Braaten & Rosén, 2000). Not surprisingly, children with ADHD tend to be unpopular with their classmates. The disorder often persists into adolescence and adulthood. Although ADHD symptoms tend to decline as children age, the disorder often persists in milder form into adolescence and adulthood (Biederman, Mick, & Faraone, 2000; Faraone et al., 2000; Wender et al., 2000). Children with ADHD are more likely than their peers to go on to become delinquents, to be suspended from school, and to require continued interventions during adolescence (Lambert et al., 1987) (see Figure 14.3).

VIDEO 14.2
ADHD: *Dr Raun Melmed*

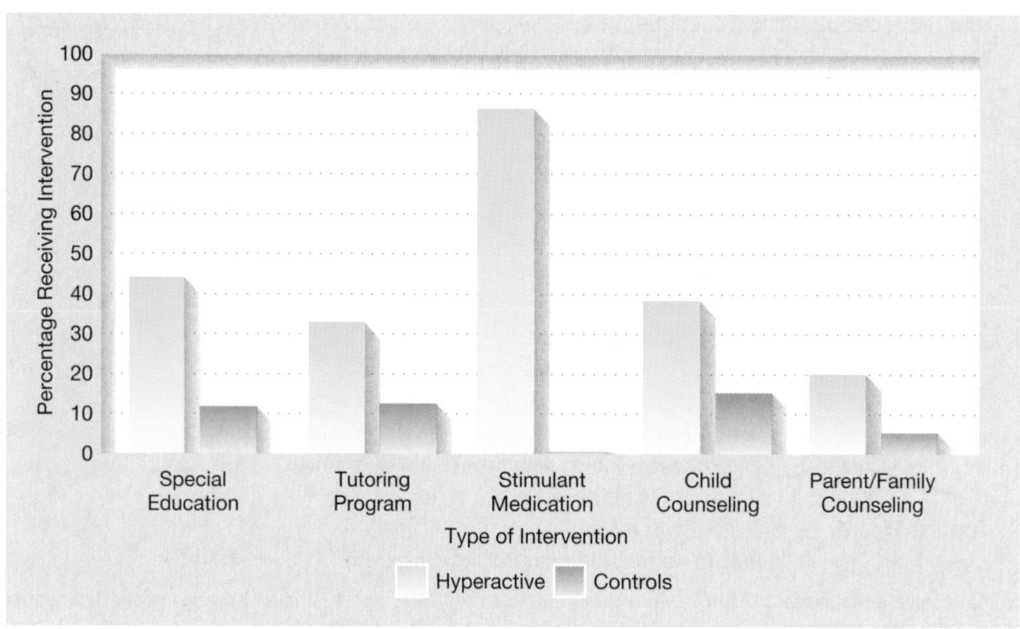

FIGURE 14.3 Interventions received by adolescents who were earlier diagnosed as hyperactive. Adolescents who were diagnosed as hyperactive during childhood were more likely than other adolescents to receive the kinds of interventions shown in this figure.

Source. Adapted from Lambert et al. (1987).

Theoretical Perspectives Although the causes of ADHD are not known, both biological and environmental influences appear to be involved. Investigators believe that genetic factors make a substantial genetic contribution to ADHD (Bradley & Golden, 2001; Faraone et al., 2001). Evidence of a genetic contribution comes largely from studies showing higher concordance rates for ADHD among monozygotic (MZ) twins than DZ (dizygotic) twins (Sherman et al., 1997). Yet environmental factors and the gene-environment interaction also play important roles (Bradley & Golden, 2001). For example, ADHD is much more common in children whose mothers smoked during pregnancy than in other children (Milberger et al., 1996). Maternal smoking may cause defects in the brain during prenatal development. Investigators continue to track down other possible environmental factors, such as a high level of family conflict, emotional stress during pregnancy, and poor parenting skills in handling children's misbehavior.

Investigators are also seeking to track down the parts of the brain that may be affected in children with ADHD. One prominent view today is that ADHD involves a genetically determined pattern of underactivity of the frontal lobes of the cerebral cortex, the parts of the brain responsible for inhibiting impulses and maintaining self-control (Barkley, 1997, 2001). Neuropsychological testing, EEG studies, and MRI studies point to subtle brain abnormalities in children and adolescents with ADHD in areas of the brain involved in regulating attention, arousal, control of motor (movement) behavior, and communication between the left and right hemispheres (e.g., Castellanos et al., 2001; Murray, 2000; Semrud-Clikeman et al., 2000). Yet different subtypes of ADHD may reflect dysfunctions in different regions of the brain (Bradley & Golden, 2001). Irregularities in nerve pathways in the brain that use the neurotransmitter serotonin may also help explain the impulsive and hyperactive components of ADHD (Quist & Kennedy, 2001).

Attention-deficit hyperactivity disorder (ADHD). ADHD is more common in boys than girls and is characterized by attentional difficulties, restlessness, impulsivity, excessive motor behavior (continuous running around or climbing), and temper tantrums.

Treatment It seems odd that drugs used to help many ADHD children calm down and attend better in school belong to a class of stimulants that includes *Ritalin* (methylphenidate), *Cylert* (pemoline), and other long-acting stimulants that can be taken in once-a-day doses (Rugino & Copley, 2001). Stimulant drugs have the paradoxical effect of calming down children with ADHD and increasing their attention spans. Although the use of stimulant medication is not without its critics, it is clear that these drugs can help many children with ADHD calm down and concentrate better on tasks and schoolwork, perhaps for the first times in their lives (Greenhill, 1998). These drugs not only improve attention in ADHD children but also reduce impulsivity, overactivity, and disruptive, annoying, or aggressive behavior (Gillberg et al., 1997; Hinshaw, 1992). Stimulant medication appears to be safe and effective when carefully monitored and successful in helping about three of four children with ADHD (Barkley, 1997). Improvements are noted at home as well as in school. The normal (voluntary) high activity levels shown in physical education classes and on weekends are not disrupted, however.

Truth OR Fiction? REVISITED

Maternal smoking during pregnancy may put children at increased risk of attention-deficit hyperactivity disorder (ADHD).

TRUE. Maternal smoking during pregnancy is associated with an increased risk of ADHD in children.

Use of stimulant medication increased dramatically during the 1990s (Gibbs, 1998; Zito et al., 2000). We do not know what accounts for the seemingly paradoxical effects of stimulants in calming children with ADHD, although it is suspected that these drugs work on neurotransmitter systems in the brain. These drugs heighten dopamine activity in the frontal lobes of the brain, the area regulating attention and control of impulsive behavior. Thus, the drugs may help ADHD children focus their attention and avoid acting out impulsively (Casey et al., 1997).

Although stimulant medication can help reduce restlessness and increase attention in school, these gains may not translate into improved academic performance ("Attention Deficit Disorder—Part II," 1995; Rutter, 1997). However, a recent study showed that combining stimulant medication and behavior modification succeeded in boosting academic performance in teenagers with ADHD, including performance on such measures as quiz scores and daily assignments (Carpenter, 2001a; Evans, Pelham, & Smith, 2001). One problem with stimulant medication, as with many other uses of psychotropic drugs, is a high rate of relapse once the child stops taking the medication (Greenhill, 1998). Also, the range of effectiveness is limited, as in the following case example:

Truth OR Fiction? REVISITED

Children who are hyperactive are often given depressants to help calm them down.

FALSE. Children with ADHD are often given stimulant drugs like Ritalin, not depressants. These stimulants have a paradoxical effect of calming them down and increasing their attention spans.

Eddie Hardly Ever Sits Still

Nine-year-old Eddie is a problem in class. His teacher complains that he is so restless and fidgety that the rest of the class cannot concentrate on their work. He hardly ever sits still. He is in constant motion, roaming the classroom, talking to other children while they are working. He has been suspended repeatedly for outrageous behavior, most recently swinging from a fluorescent light fixture and unable to get himself down. His mother reports that Eddie has been a problem since he was a toddler. By the age of 3 he had become unbearably restless and demanding. He has never needed much sleep and always awakened before anyone else in the family, making his way downstairs and wrecking things in the living room and kitchen. Once, at the age of 4, he unlocked the front door and wandered into traffic, but was rescued by a passerby.

Psychological testing shows Eddie to be average in academic ability, but to have a "virtually nonexistent" attention span. He shows no interest in television or in games or toys that require some concentration. He is unpopular with peers and prefers to ride his bike alone or to play with his dog. He has become disobedient at home and at school and has stolen small amounts of money from his parents and classmates.

Eddie has been treated with methylphenidate (Ritalin), but it was discontinued because it had no effect on his disobedience and stealing. However, it did seem to reduce his restlessness and increase his attention span at school.

Adapted from Spitzer et al., 1989, pp. 315–317

Then there's the matter of side effects. Although short-term side effects (e.g., loss of appetite or insomnia) usually subside within a few weeks or may be eliminated by lowering the dose, stimulant drugs may lead to other effects, including a slowdown of physical growth (Wingert, 2000). Fortunately, children taking stimulant medication eventually catch up to their peers in physical stature (Gittelman-Klein & Mannuzza, 1990; Gorman, 1998).

With so many children on Ritalin and similar drugs, critics claim we are too ready to seek a "quick fix" for problem behavior in children rather than examine other factors contributing to the child's problem, such as problems in the family (Gibbs, 1998). As one pediatrician put it, "It takes time for parents and teachers to sit down and talk to kids. . . . It takes less time to get a child a pill" (Hancock, 1996, p. 52). Whatever the benefits of stimulant medication, medication alone typically fails to bring the social and academic behavior of children with ADHD into a normal range (Hinshaw, 1992). Drugs cannot teach new skills. So attention has focused on whether a combination of stimulant medication and behavioral or cognitive-behavioral techniques can produce greater benefits than either approach alone. Cognitive-behavioral treatment of ADHD combines behavior modification, typically based on the use of reinforcement (for example, a teacher praising the child with ADHD for sitting quietly) and cognitive modification (for example, training the child to silently talk himself or herself through the steps involved in solving challenging academic problems). Thus far evidence is mixed on the value of combining cognitive-behavioral therapy with medication (Braswell & Kendall, 2001; Hinshaw, Klein, & Abikoff, 1998).

THINK ABOUT IT

What are the risks and benefits of using stimulant medication, like Ritalin, in treating ADHD in children? If you had a child with ADHD, would you consider using these drugs? Why or why not?

Conduct Disorder

Although it also involves disruptive behavior, **conduct disorder** differs in important ways from ADHD. Whereas children with ADHD seem literally incapable of controlling their behavior, children with conduct disorder purposefully engage in patterns of antisocial behavior that violate social norms and the rights of others. Whereas children with ADHD throw temper tantrums, children diagnosed as conduct disordered are intentionally aggressive and cruel. Like antisocial adults, many conduct-disordered children are callous and apparently do not experience guilt or remorse for their misdeeds. They may steal or de-

conduct disorder A psychological disorder in childhood and adolescence characterized by disruptive, antisocial behavior.

stroy property. In adolescence they may commit rape, armed robbery, even homicide. They may cheat in school—when they bother to attend—and lie to cover their tracks. They frequently engage in substance abuse and sexual activity.

oppositional defiant disorder (ODD) A psychological disorder in childhood and adolescence characterized by excessive oppositionality or tendencies to refuse requests from parents and others.

Rates of conduct disorder in the general population range from less than 1% to more than 10%, depending on the particular study (APA, 2000). Conduct disorders are much more common among boys than girls, and the disorder typically takes a somewhat different form in boys than girls. In boys, conduct disorder is more likely to be manifested by stealing, fighting, vandalism, or disciplinary problems at school, whereas in girls it is more likely to involve lying, truancy, running away, substance use, and prostitution. Children with conduct disorder often present with other disorders or problem behaviors, including ADHD, social withdrawal, and major depression (Lambert et al., 2001).

Conduct disorder is typically a chronic or persistent disorder (Lahey et al., 1995). Longitudinal studies show that elementary school children with conduct disorder are more likely than other children to engage in delinquent acts as early adolescents (Tremblay et al., 1992). Antisocial behavior in the form of delinquent acts (stealing, truancy, vandalism, fighting or threatening others, and so on) during early adolescence (ages 14 to 15) has also been found to predict alcohol and substance abuse in late adolescence, especially among boys (Boyle et al., 1992). Another form of conduct disorder may involve a cluster of personality traits that have different origins than antisocial behavior (Wootton et al., 1997). These personality traits include callousness (uncaring, mean, cruel) and an unemotional way of relating to others (Barry et al., 2000).

Oppositional Defiant Disorder

Debate continues among professionals over the issue of whether conduct disorder (CD) and **oppositional defiant disorder (ODD)** are separate disorders or variations of a common disruptive behavior disorder (Rey, 1993). Or perhaps ODD is a precursor or milder form of conduct disorder (Abikoff & Klein, 1992; Biederman et al., 1996a). At present, the two disorders are conceptualized as related but separate. ODD is more closely related to nondelinquent (negativistic) conduct disturbance, and conduct disorder involves more outright delinquent behavior in the form of truancy, stealing, lying, and aggression (Rey, 1993). Yet oppositional defiant disorder, which typically develops earlier than CD, may lead to the development of antisocial behavior and conduct disorder at later ages (Loeber, Lahey, & Thomas, 1991).

Oppositional defiant disorder (ODD). ODD is characterized by negativistic and oppositional behavior in response to directives from parents, teachers, or other authority figures. Children with ODD may act spitefully or vindictively toward others, but do not typically show the cruelty, aggressivity, and delinquent behavior associated with conduct disorder. Yet questions remain about whether the two disorders are truly distinct or are variations of a common underlying disorder involving disruptive behavior patterns.

Children with ODD tend to be negativistic or oppositional. They are defiant of authority, which is exhibited by their tendency to argue with parents and teachers and refuse to follow requests or directives from adults. They may deliberately annoy other people, become easily angered or lose their temper, become touchy or easily annoyed, blame others for their mistakes or misbehavior, feel resentful toward others, or act in spiteful or vindictive ways toward others (Angold & Costello, 1996; APA, 2000). The disorder typically begins before age 8 and develops gradually over a period of months or years. It typically starts in the home environment but may extend to other settings, such as school.

ODD is one of the most common diagnoses among children (Doll, 1996). Studies show that among children diagnosed with a psychological disorder, about one in three are judged to meet criteria for ODD (Rey, 1993). A recent view of epidemiological studies estimated the prevalence of ODD among children in the general community at about 6% (Rey, 1993). ODD is more common overall among boys than girls. However, this overall effect masks a gender shift over age. Among children 12 years of age or younger, ODD appears to be more than twice as common among boys. Yet among adolescents, a higher prevalence is reported in girls (Rey, 1993). By contrast, most studies find conduct disorder to be more common in boys than girls across all age groups.

Theoretical Perspectives on ODD and CD The causal factors in ODD remain obscure. Some theorists believe that oppositionality is an expression of an underlying child temperament described as the "difficult-child" type (Rey, 1993). Others believe that unresolved parent-child conflicts or overly strict parental control may lie at the root of the disorder. Psychodynamic theorists look at ODD as a sign of fixation at the anal stage of psychosexual development, when conflicts between the parent and child may emerge over toilet training. Leftover conflicts may later become expressed in the form of rebelliousness against parental wishes (Egan, 1991). Learning theorists view oppositional behaviors as arising from parental use of inappropriate reinforcement strategies. In this view, parents may inappropriately reinforce oppositional behavior by "giving in" to the child's demands whenever the child refuses to comply with the parent's wishes, which can become a pattern.

Family factors are also implicated in the development of conduct disorder. Some forms of conduct disorder appear to be linked to ineffective parenting styles, such as failure to provide positive reinforcement for appropriate behavior and use of harsh and inconsistent discipline for misbehavior. Families of children with CD tend to be characterized by negative, coercive interactions (Dadds et al., 1992). Children with CD are often very demanding and noncompliant in relating to their parents and other family members. Family members often reciprocate by using negative behaviors, such as threatening or yelling at the child or using physical means of coercion. Parental aggression against children with conduct behavior problems is common, including pushing, grabbing, slapping, spanking, hitting, or kicking (Jourile et al., 1997). Parents of children with oppositional defiant disorders or severe conduct disorder display high rates of antisocial personality disorder and substance abuse (Frick et al., 1992). It's not too much of a stretch to speculate that parental modeling of antisocial behaviors can lead to antisocial conduct in children.

Conduct disorders often occur in a context of parental distress, such as marital conflict. Another factor is maternal depression. Depressed mothers tend to display poor parenting behaviors—such as vague and interrupted commands—that may foster disruptive behavior in their children (Forehand et al., 1988). Mothers of children with conduct disorders are also more likely than other mothers to be inconsistent in their use of discipline and less able to supervise their children's behavior (Frick et al., 1992). Maternal smoking during pregnancy has also been linked to a greater likelihood of conduct disorder in sons (Wakschlag et al., 1997). Perhaps maternal smoking affects the developing fetus in ways that lead to conduct problems, or perhaps there are other characteristics of mothers who smoke, such as ineffective parenting skills, that set the stage for childhood behavior problems.

Some investigations focus on the ways in which children with disruptive behavior disorders process information. For example, children who are overly aggressive in their behavior tend to be biased in their processing of social information: They may assume that others intend them ill when they do not (Lochman, 1992). They usually blame others for the scrapes they get into. They believe that they are misperceived and treated unfairly. They may believe that aggression leads to favorable results (Dodge et al., 1997). They are also less able than their peers to generate alternative, nonviolent, responses to social conflicts (Lochman & Dodge, 1994).

Genetic factors may interact with family or other psychosocial factors in the development of conduct disorder (APA, 2000; O'Connor et al., 1998; Slutske et al., 1998). Genetic factors may also be involved in the development of oppositional defiant disorder.

Treatment The treatment of conduct disorders remains a challenge. Although there is no established pharmacological treatment approach, a recent study indicates that Ritalin may be effective in reducing antisocial behavior in CD children and adolescents (Klein et al., 1997). Traditional psychotherapy has not generally been shown to help disruptive children change their behavior. Placing children with conduct disorders in residential treatment programs that establish explicit rules and clear rewards for obeying them may offer greater promise (e.g., Henggeler et al., 1986). Such programs usually rely on operant conditioning procedures that involve systematic use of rewards and punishments.

Many children with conduct disorders, especially boys, display aggressive behavior and have problems controlling their anger. Many can benefit from programs designed to help them learn anger coping skills that they can use to handle conflict situations without resorting to violent behavior. Cognitive-behavioral therapy has been used to teach boys who engage in antisocial and aggressive behavior to reconceptualize social provocations as problems to be solved rather than as challenges to their manhood that must be answered with violence. They have been trained to use calming self-talk to inhibit impulsive behavior and control anger whenever they experience social taunts or provocations and to generate and try out nonviolent solutions to social conflicts (Lochman & Lenhart, 1993). Other programs present child models on videotape demonstrating skills of anger control. The results of these programs appear promising (Kazdin & Weisz, 1998; Webster-Stratton & Hammond, 1997). Sometimes the disruptive child's parents are brought into the treatment process (Kazdin, Siegel, & Bass, 1992).

Henggeler and his colleagues (1986) have developed a "family-ecological" approach based on Urie Bronfenbrenner's (1979) ecological theory. Like Bronfenbrenner, Henggeler sees children as embedded within various social systems—family, school, criminal justice, community, and so on. He focuses on how juvenile offenders affect and are affected by the systems with which they interact. The techniques themselves are not unique. Rather, the family-ecological approach tries to change children's relationships with multiple systems to end disruptive interactions. This multiple systems or *multisystemic therapy (MST) approach* has shown promising results in the treatment of juvenile offenders in terms of reducing the frequency of subsequent arrests in comparison with youths who received typical youth services from a county youth services department (Henggeler, Melton, & Smith, 1992; Henggeler et al., 1997; Kazdin, 1998).

The following example illustrates the involvement of the parents in the behavioral treatment of a case of oppositional defiant disorder:

time out A behavioral technique in which a person who behaves in an undesirable way is removed from a reinforcing environment and placed in an unreinforcing environment for a short time.

A Case of Oppositional Defiant Disorder

Billy was a 7-year-old second grader referred by his parents. The family was relocated frequently because the father was in the navy. Billy usually behaved when his father was taking care of him, but he was noncompliant with his mother and yelled at her when she gave him instructions. His mother was incurring great stress in the effort to control Billy, especially when her husband was at sea.

Billy had become a problem at home and in school during the first grade. He ignored and violated rules in both settings. Billy failed to carry out his chores and frequently yelled at and hit his younger brother. When he acted up, his parents would restrict him to his room or the yard, take away privileges and toys, and spank him. But all of these measures were used inconsistently. He also played on the railroad tracks near his home and twice the police had brought him home after he had thrown rocks at cars.

A home observation showed that Billy's mother often gave him inappropriate commands. She interacted with him as little as possible and showed no verbal praise, physical closeness, smiles, or positive facial expressions or gestures. She paid attention to him only when he misbehaved. When Billy was noncompliant, she would yell back at him and then try to catch him to force him to comply. Billy would then laugh and run from her.

Billy's parents were informed that the child's behavior was a product of inappropriate cueing techniques (poor directions), a lack of reinforcement for appropriate behavior, and lack of consistent sanctions for misbehavior. They were taught the appropriate use of reinforcement, punishment, and ***time out.*** *The parents then charted Billy's problem behaviors to gain a clearer idea of what triggered and maintained them. They were shown how to reinforce acceptable behavior and use time out as a contingent punishment for misbehavior. Billy's mother was also taught relaxation training to help desensitize her to Billy's disruptions. Biofeedback was used to enhance the relaxation response.*

During a 15-day baseline period, Billy behaved in a noncompliant manner about four times per day. When treatment was begun, Billy showed an immediate drop to about one instance of noncompliance every two days. Follow-up data showed that instances of noncompliance were maintained at a bearable level of about one per day. Fewer behavioral problems in school were also reported, even though they had not been addressed directly.

—*Adapted from Kaplan, 1986, pp. 227–230*

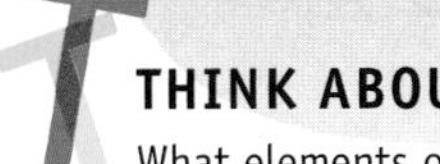

THINK ABOUT IT

What elements of Billy's treatment do you believe contributed to its effectiveness? Would this type of approach be likely to be as effective with a conduct disorder?

Quiz 14.6
Attention-Deficit and Disruptive Behavior Disorders

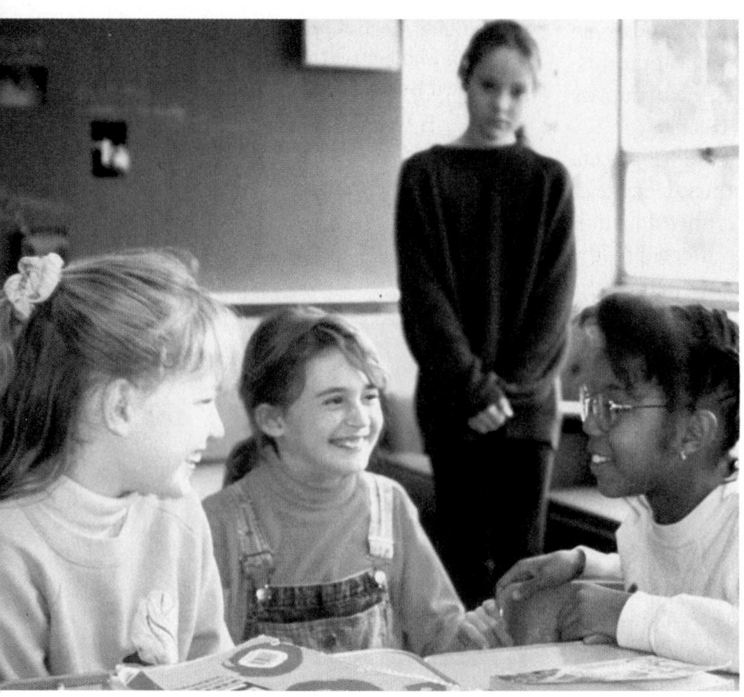

Social avoidance. Socially avoidant children tend to be excessively shy and withdrawn and have difficulty interacting with other children.

Anxiety and Depression

Anxieties and fears are a normal feature of childhood, just as they are a normal feature of adult life. Childhood fears—of the dark or of small animals—are commonplace and are usually outgrown naturally. Anxiety is considered abnormal, however, when it is excessive and interferes with normal academic or social functioning or becomes troubling or persistent. Children, like adults, may suffer from different types of diagnosable anxiety disorders, including specific phobias, social phobias, generalized anxiety disorder (GAD), PTSD, and mood disorders, including major depression and bipolar disorder. Although these disorders may develop at any age, we will consider further a type of disorder that typically develops during early childhood: separation anxiety disorder.

Children may also show the more general pattern of avoidance of social interactions that characterizes avoidant personality disorder. Although children who are socially avoidant or have social anxiety disorder (also called social phobia) may have warm relationships with family members, they tend to be shy and withdrawn around others. Their avoidance of people outside the family interferes with their development of social relationships with their peers. Such problems tend to develop after normal fear of strangers fades, at age 2½ or later. Their distress at being around other children at school can also affect their academic progress. Having a social anxiety disorder during adolescence or young adult also increases the likelihood of later developing a depressive disorder (Stein et al., 2001).

Truth OR Fiction? REVISITED

Major depression rarely occurs before adulthood.

FALSE. Major depression is actually quite common in later childhood and adolescence.

We may think of childhood as the happiest time of life. Most children are protected by their parents and are unencumbered by adult responsibilities. From the perspective of aging adults, their bodies seem made of rubber and free of aches. They have apparently boundless energy. Despite the stereotype of a happy childhood, clinical depression is common among children and adolescents. Estimates indicate that perhaps 8% to 9% of children in the 10- to 13-year age range experience major depression during any given 1-year period (Goleman, 1994a). Although rare, major depression has even been found among preschoolers. Although there is no discernible gender difference in the risk of depression in childhood, a prominent gender differences appears after the age of 15, with adolescent girls becoming about twice as likely to become depressed as adolescent boys (Hankin et al., 1998; Lewinsohn, Rohde, & Seeley, 1994).

Web Link 14.6 WWW
Depression in Children and Adolescents

Separation Anxiety Disorder

It is normal for young children to show anxiety when they are separated from their caregivers. Mary Ainsworth (1989), who has chronicled the development of attachment behaviors, notes that separation anxiety is a normal feature of the child-caregiver relationship and begins during the first year. The sense of security normally provided by bonds of attachment apparently encourages children to explore their environments and become progressively independent of their caregivers (Bowlby, 1988).

Separation anxiety disorder is diagnosed when separation anxiety is persistent and excessive or inappropriate for the child's developmental level. That is, 3-year-olds ought to be able to attend preschool without nausea and vomiting brought on by anxiety. Six-year-olds ought to be able to attend first grade without persistent dread that something awful will happen to themselves or their parents. Children with this disorder tend to cling to their parents and follow them around the house. They may voice concerns about death and dying and insist that someone stay with them while they are falling asleep. Other features of the disorder include nightmares, stomachaches, nausea and vomiting when separation is anticipated (as on school days), pleading with parents not to leave, or throwing tantrums when parents are about to depart. Children may refuse to attend school for fear that something will happen to their

separation anxiety disorder A childhood disorder characterized by extreme fear of separation from parents or other caretakers.

parents while they are away. The disorder affects about 4% of children and young adolescents and occurs more frequently, according to community-based studies, among females (APA, 2000). The disorder may persist into adulthood, leading to an exaggerated concern about the well-being of one's children and spouse and difficulty tolerating any separation from them.

In previous years, separation anxiety disorder was usually referred to as *school phobia.* Separation anxiety disorder may occur at preschool ages, however. Today, most cases in which younger children refuse to attend school are viewed as forms of separation anxiety. In adolescence, however, refusal to attend school is also frequently connected with academic and social concerns, in which case the label of separation anxiety disorder would not apply.

The development of separation anxiety disorder frequently follows a stressful life event, such as illness, the death of a relative or pet, or a change of schools or homes. Alison's problems followed the death of her grandmother:

Alison's Fear of Death

Alison's grandmother died when Alison was 7 years old. Her parents decided to permit her request to view her grandmother in the open coffin. Alison took a tentative glance from her father's arms across the room, then asked to be taken out of the room. Her 5-year-old sister took a leisurely close-up look, with no apparent distress.

Alison had been concerned about death for two or three years by this time, but her grandmother's passing brought on a new flurry of questions: "Will I die?", "Does everybody die?", and so on. Her parents tried to reassure her by saying, "Grandma was very, very old, and she also had a heart condition. You are very young and in perfect health. You have many, many years before you have to start thinking about death."

Alison also could not be alone in any room in her house. She pulled one of her parents or her sister along with her everywhere she went. She also reported nightmares about her grandmother and, within a couple of days, insisted on sleeping in the same room with her parents. Fortunately, Alison's fears did not extend to school. Her teacher reported that Alison spent some time talking about her grandmother, but her academic performance was apparently unimpaired.

Alison's parents decided to allow Alison time to "get over" the loss. Alison gradually talked less and less about death, and by the time 3 months had passed, she was able to go into any room in her house by herself. She wanted to continue to sleep in her parents' bedroom, however. So her parents "made a deal" with her. They would put off the return to her own bedroom until the school year had ended (a month away), if Alison would agree to return to her own bed at that time. As a further incentive, a parent would remain with her until she fell asleep for the first month. Alison overcame the anxiety problem in this fashion with no additional delays.

—*From the Authors' Files*

Perspectives on Anxiety Disorders in Childhood

Theoretical understandings of excessive anxiety in children to some degree parallel explanations of anxiety disorders in adults. Psychoanalytic theorists argue that childhood anxieties and fears, like their adult counterparts, symbolize unconscious conflicts. Cognitive theorists focus on the role of cognitive biases underlying anxiety reactions. In support of the cognitive model, investigators find that highly anxious children show cognitive biases in processing information, such as interpreting ambiguous situations as threatening, expecting negative outcomes, doubting their ability to deal with problem situations, and engaging in negative self-talk (e.g., Bögels & Zigerman, 2000; Weems et al., 2001). Expecting the worst,

Truth OR Fiction? REVISITED

Some children refuse to go to school because they believe that terrible things may happen to their parents while they are away.

TRUE. Some children with separation anxiety disorder do refuse to go to school because they believe that terrible things may happen to their parents while they are away.

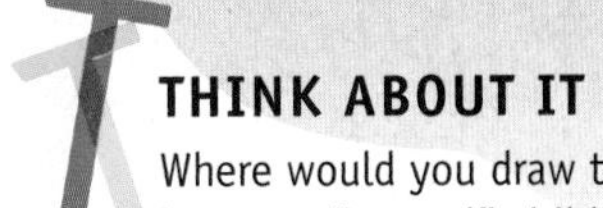

THINK ABOUT IT

Where would you draw the line between "normal" childhood fears and specific phobias?

Separation anxiety. In separation anxiety disorder, a child shows persistent anxiety when separated from her or his parents that is inconsistent with her or his developmental level. Such children tend to cling to their parents and resist even brief separations.

combined with having low self-confidence, encourages avoidance of feared activities—with friends, in school, and elsewhere. Negative expectations also heighten feelings of anxiety to the point where they may impair performance in the classroom or the athletic field.

Learning theorists suggest that the occurrence of generalized anxiety may touch on broad themes, such as fears of rejection or failure that carry across situations. Underlying fears of rejection or self-perceptions of inadequacy may generalize to most areas of social interaction and achievement. Genetics may also play a role in separation anxiety and other anxiety disorders (Coyle, 2001).

Whatever the causes, overanxious children may profit from the anxiety-control techniques we discussed in Chapter 6, such as gradual exposure to phobic stimuli and relaxation training. Cognitive techniques such as replacing anxious self-talk with coping self-talk may also be helpful. Cognitive-behavioral approaches have produced impressive results in treating childhood anxiety disorders (Barrett et al., 2001; Beidel, Turner, & Morris, 2000; Braswell & Kendall, 2001). Treatment with the drug fluvoxamine (Luvox), a selective serotonin-reuptake inhibitor, also shows good therapeutic effects in treating children and adolescents with various types of anxiety disorders (Coyle, 2001; Riddle et al., 2001; Walkup et al., 2001).

Depression in Childhood and Adolescence

Children and adolescents may suffer from diagnosable mood disorders, including bipolar disorder and major depression. Like depressed adults, depressed children and adolescents typically have feelings of hopelessness, more distorted thinking patterns and tendencies to blame themselves for negative events, and lower self-esteem, self-confidence, and perceptions of competence than their nondepressed peers (Lewinsohn et al., 1994; Kovacs, 1996). They often report episodes of sadness and crying, feelings of apathy, as well as insomnia, fatigue, and poor appetite. They may also experience suicidal thoughts or even attempt suicide. Yet depression in children may be associated with some distinctive features as well, such as refusal to attend school, fears of parents' dying, and clinging to parents. Depression may also be masked by behaviors that do not appear directly related to depression. Conduct disorders, academic problems, physical complaints, and even hyperactivity may stem, now and then, from unrecognized depression. Among adolescents, aggressive and sexual acting out may also be signs of underlying depression.

Is this child too young to be depressed? Although we tend to think of childhood as the happiest and most carefree time of life, depression is actually quite common among children and adolescents. Depressed children may report feelings of sadness and lack of interest in previously enjoyable activities. Many, however, do not report or are not aware of feelings of depression, even though they may look depressed to observers. Depression may also be masked by other problems, such as conduct or school-related problems, physical complaints, and overactivity.

One thing we should recognize is that depressed children or adolescents may fail to label what they are feeling as depression. They may not report feeling sad even though they appear sad to others and may be tearful (Goleman, 1994a). Part of the problem is cognitive-developmental. Children are not usually capable of recognizing internal feeling states until about the age of 7. They may not be able to identify negative feeling states in themselves, including depression, until adolescence (Larson et al., 1990). Even adolescents may not recognize what they are experiencing as depression.

The average length of a major depressive episode in childhood or adolescence is about 11 months, but an individual episode may last for as long as 18 months in some cases (Goleman, 1994a). Moderate levels of depression, however, may persist for years, severely impacting school performance and social functioning (Nolen-Hoeksema & Girgus, 1994). Adolescent depression is associated with an increased risk of future major depressive episodes and suicide attempts in adulthood (Weissman, 1999). About three of four children who become depressed from age 8 to 13 have a recurrence later in life (Goleman, 1994a).

Depressed children may also lack various skills, including academic, athletic, and social skills (Seroczynski, Cole, & Maxwell, 1997). They may find it hard to concentrate in school and may suffer from impaired memory, making it difficult for them to keep their grades up (Goleman, 1994a). They often keep their feelings to themselves, which may prevent their parents from recognizing the problem and seeking help for them. Negative feelings may also be expressed in the form of anger, sullenness, or impatience, leading to conflicts with parents that in turn can accentuate and prolong depression in the child.

Childhood depression rarely occurs by itself. Depressed children typically experience other psychological disorders, especially anxiety disorders and conduct or oppositional defiant disorders (Hammen & Compas, 1994). Eating disorders are also common among depressed adolescents, at least among females (Rohde, Lewinsohn, & Seeley, 1991). Overall, childhood depression increases the chances that a child will develop another psychological disorder by at least 20-fold (Angold & Costello, 1993). A sizeable percentage of depressed adolescents (between 20% and 40%) later develop bipolar disorder (USDHHS, 1999a).

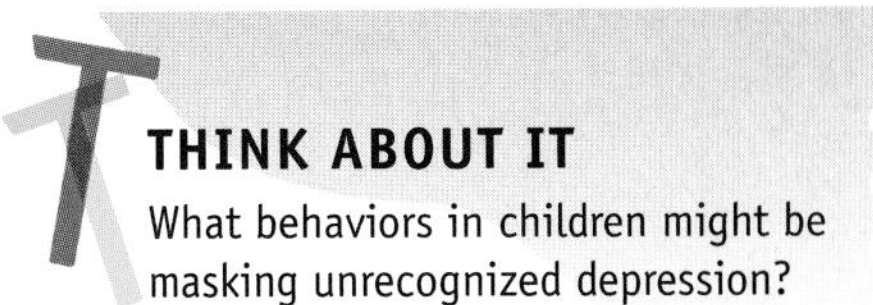

THINK ABOUT IT

What behaviors in children might be masking unrecognized depression?

Correlates and Treatment of Depression in Childhood and Adolescence

Depression and suicidal behavior in childhood are frequently related to family problems and conflicts. Children who are exposed to stressful life events affecting the family, such as parental conflict or unemployment, stand an increased risk of depression, especially younger children (Nolen-Hoeksema, Girgus, & Seligman, 1992). Stressful life events and a lack of social support from friends and family also figure into the profile of adolescents who become depressed (Lewinsohn et al., 1994). Depression in adolescents may be triggered by such stressful life events as conflicts with parents and dissatisfaction with school grades. In girls, disturbed eating behaviors and body dissatisfaction after puberty often predict which girls will go on to develop major depression during adolescence (Stice et al., 2000). Interestingly, the relationship between loss of a parent in childhood and later depression during childhood or adolescence is not a consistent finding; some studies show a connection whereas others do not (Lewinsohn et al., 1994).

As children mature and their cognitive abilities increase, however, cognitive factors, such as attributional styles, appear to play a stronger role in the development of depression. Older children (sixth and seventh graders) who adopt a more helpless or pessimistic explanatory style (attributing negative events to internal, stable, and global causes, and attributing positive events to external, unstable, and specific causes) are more likely than children with a more optimistic explanatory style to develop depression (Nolen-Hoeksema et al., 1992). Researchers also find that adolescents who are depressed tend to hold more dysfunctional attitudes and to adopt a more helpless explanatory style than do their nondepressed peers (Lewinsohn et al., 1994). Like their adult counterparts, children and adolescents with depression tend to adopt a cognitive style that is characterized by negative attitudes toward themselves and the future (Garber, Weiss, & Shanley, 1993). All in all, the distorted cognitions of depressed children include the following:

1. Expecting the worst (pessimism)
2. Catastrophizing the consequences of negative events
3. Assuming personal responsibility for negative outcomes, even when it is unwarranted
4. Selectively attending to the negative aspects of events

Although there are links between cognitive factors and depression, it remains to be determined whether children become depressed because of a depressive mindset or whether depression causes changes in thinking patterns (USDHHS, 1999a). Genetic factors also appear to play a role in explaining depressive symptoms, at least among adolescents (O'Connor et al., 1998). The role of genetics in childhood depression requires further study, however (Kovacs et al., 1997).

Adolescent girls face a greater risk of depression than do adolescent boys, perhaps because they typically face more social challenges than boys—challenges such as pressures to narrow their interests and pursue feminine-typed activities (Nolen-Hoeksema & Girgus, 1994). Girls who adopt a more passive, ruminative style of coping may be at greatest risk of becoming depressed. Gender differences in adolescent depression may have a cultural component. A recent study reported that gender differences in depressive symptoms were greater among U.S. adolescents than among Chinese adolescents (Greenberger et al., 2000). The reasons for these cross-cultural differences remain to be explored.

Accumulating evidence supports the effectiveness of cognitive-behavioral therapy (CBT) in treating depressed children and adolescents (Berman et al., 2000; Braswell & Kendall, 2001; Lewinsohn & Clarke, 1999). Although individual approaches vary, CBT usually involves a coping skills model in which children or adolescents receive social skills training (e.g., learning how to start a conversation or make friends) to increase the likelihood of obtaining social reinforcement (Kazdin & Weisz, 1998). CBT typically also includes training in problem-solving skills and ways of increasing the frequency of rewarding activities and countering depressive styles of thinking. In addition, family therapy may be useful in helping families resolve underlying conflicts and reorganize their relationships in ways that members can become more supportive of each other.

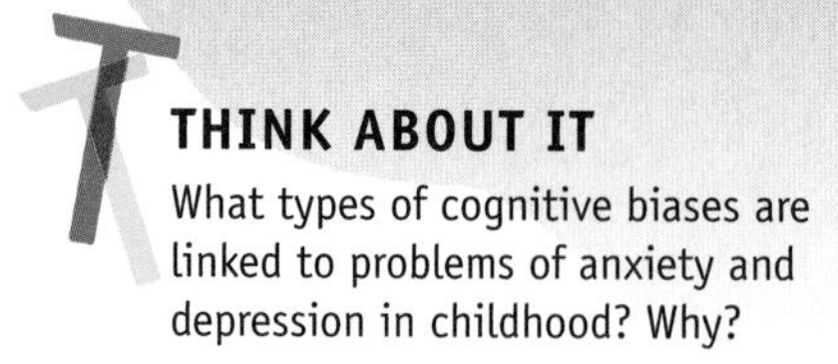

THINK ABOUT IT

What types of cognitive biases are linked to problems of anxiety and depression in childhood? Why?

We should not assume that because antidepressants are effective in treating depression in adults they will work as well, or be as safe, when used with children (Bitiello & Jensen, 1997). However, SSRI-type antidepressants, such as Prozac, have shown promise in treating depression in children and adolescents (USDHHS, 1999a). Lithium is also used with generally favorable results in treating children and adolescents with bipolar disorder (USDHHS, 1999a).

Suicide Among Children and Adolescents Suicide is relatively uncommon among younger children and adolescents, with statistics showing less than 1 case per 100,000 persons among younger children and fewer than 2 cases per 100,000 among 10- to 14-year-olds (Brody, 1992b; "Report: Adolescent Suicide," 1995). Despite the low frequency, the rate among the 10- to 14-year-olds more than doubled from 1980 to the early 1990s, although it fell slightly for the population under age 25 on the whole (Gelman, 1994). The suicide rate for adolescents in the 15 to 19 age range is considerably higher, some 8 per 100,000 persons (USDHHS, 1991b) and has risen sharply since the mid-20th century. More ready access to guns, together with a rising incidence of family disruption and disintegration, may help explain the rise in adolescent suicides. These official statistics only account for reported suicide; some apparent accidental deaths, such as those due to falling from a window, may be suicides as well.

Despite the commonly held view that children and adolescents who talk about suicide are only venting their feelings, young people who do intend to kill themselves may very well talk about it beforehand (Brody, 1992b). In fact, those who discuss their plans are the ones most likely to carry them out. Moreover, children and adolescents who have survived suicide attempts are most likely to try it again (Brody, 1992b). Unfortunately, parents tend not to take their children's suicidal talk seriously. They often refuse treatment for their children, or terminate treatment prematurely.

Several factors are associated with an increased risk of suicide among children and adolescents (Levy, Jurkovic, & Spirito, 1995; Lewinsohn et al., 1994, 2001; Neiger, 1988):

1. *Gender.* Girls, like women, are three times more likely than boys to attempt suicide. Boys, like men, are more likely to succeed, however, perhaps because boys, like men, are more apt to use lethal means, such as guns. The presence of a loaded handgun in the house turns out to be the greatest risk factor for completed suicide among children, even those as young as 5 (Brody, 1992b).
2. *Age.* Young people in late adolescence or early adulthood (ages 15 to 24) are at greater risk than younger adolescents.
3. *Geography.* Adolescents in less populated areas are more likely to commit suicide. Adolescents in the rural western regions of the United States have the highest suicide rate.
4. *Race.* The suicide rates for African American, Asian American, and Hispanic American youth are about 30% to 60% lower than that of non-Hispanic White youth (USDHHS, 1991a). Among young adults age 15 to 24, Native Americans commit suicide nearly twice as often as do European Americans. Although European American teens are more likely to commit suicide than African American teens, rates of suicide are rising fastest among African American teenage males (Shaffer, Gould, & Hicks, 1994).
5. *Depression and hopelessness.* Major depression with features of hopelessness and low self-esteem are major risk factors for suicide among adolescents, as it is among adults (USDHHS, 1999a).
6. *Previous suicidal behavior.* One quarter of adolescents who attempt suicide are repeaters. More than 80% of adolescents who take their lives have talked about it before doing so. Suicidal teenagers may carry lethal weapons, talk about death, make

suicide plans, or engage in risky or dangerous behavior. A family history of suicide also increases risk of teenage suicide (Mann et al., 1996).

7. *Family problems.* Family problems are present among about 75% of adolescent suicide attempters. The problems include family instability and conflict, physical or sexual abuse, loss of a parent due to death or separation, and poor parent-child communication (Asarnow, Carlson, & Guthrie, 1987; Wagner, 1997).
8. *Stressful life events.* Many suicides among young people are directly preceded by stressful or traumatic events, such as breaking up with a girlfriend or boyfriend, having an unwanted pregnancy, getting arrested, having problems at school, moving to a new school, or having to take an important test.
9. *Substance abuse.* Addiction in the adolescent's family, or by the adolescent, is a factor.
10. *Social contagion.* Adolescent suicides sometimes occur in clusters, especially when a suicide or a group of suicides receives widespread publicity (Kessler et al., 1990; USDHHS, 1999a). Adolescents may romanticize suicide as a heroic act of defiance. There are often suicides or attempts among the siblings, friends, parents, or adult relatives of suicidal adolescents. Adolescent suicides may occur in bunches in a community, especially when adolescents are subjected to mounting academic pressures, such as competing for admission to college. Perhaps the suicide of a family member or schoolmate renders suicide a more "real" option for managing stress or punishing others. Perhaps the other person's suicide gives the adolescent the impression that he or she is "doomed" to commit suicide. Note the case of Pam:

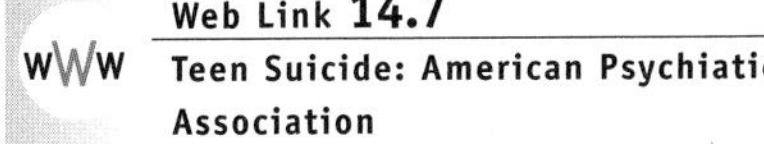

Web Link 14.7
Teen Suicide: American Psychiatic Association

Pam, Kim, and Now Brian

Pam was an exceptionally attractive 17-year-old who was hospitalized after cutting her wrists. "Before we moved to [an upper-middle-class town in Westchester County]," she told the psychologist, "I was the brightest girl in the class. Teachers loved me. If we had had a yearbook, I'd have been the most likely to succeed. Then we moved, and suddenly I was hit with it. Everybody was bright, or tried to be. Suddenly I was just another ordinary student planning to go to college.

"Teachers were good to me, but I was no longer special, and that hurt. Then we all applied to college. Do you know that 90 percent of the kids in the high school go on to college? I mean four-year colleges? And we all knew—or suspected—that the good schools had quotas on kids from here. I mean you can't have 30 kids from our senior class going to Yale or Princeton or Wellesley, can you? You're better off applying from Utah.

"Then Kim got her early-acceptance rejection from Brown. Kim was number one in the class. Nobody could believe it. Her father'd gone to Brown and Kim had almost 1500 SATs. Kim was out of commission for a few days—I mean she didn't come to school or anything—and then, boom, she was gone. She offed herself, kaput, no more, the end. Then Brian was rejected from Cornell. A few days later, he was gone, too. And I'm like, 'These kids were better than me.' I mean their grades and their SATs were higher than mine, and I was going to apply to Brown and Cornell. I'm like, 'What chance do I have? Why bother?' "

—From the Authors' Files

THINK ABOUT IT

Considering the risk factors presented in the text, do you know anyone who is at high risk? What could be done to lower the risk of that person committing suicide?

You can see how catastrophizing cognitions can play a role in such tragic cases. Consistent with the literature on suicide among adults, suicidal children make little use of active problem-solving strategies in handling stressful situations. They may see no other way out of their perceived failures or stresses. As with adults, one approach to working with suicidal children involves helping them challenge distorted thinking and generate alternate strategies for handling the problems and stressors they face.

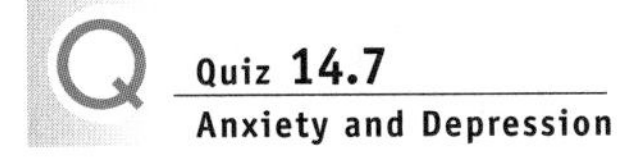

Quiz 14.7
Anxiety and Depression

TABLE 14.7 Diagnostic Features of Enuresis

The child repeatedly wets bedding or clothes (whether intentional or involuntary).

The child's chronological age is at least 5 (or child is at an equivalent developmental level).

The behavior occurs at least twice a week for 3 months or causes significant impairment in functioning or distress.

The disorder does not have an organic basis.

Source. Adapted from the *DSM-IV-TR* (APA, 2000).

Truth OR Fiction? REVISITED

Problems of persistent bed-wetting in childhood generally persist into adolescence.

FALSE. Most cases of enuresis usually resolve themselves by adolescence at the latest.

enuresis Failure to control urination after one has reached the "normal" age for attaining such control.

Elimination Disorders

Fetuses and newborn children eliminate waste products reflexively. As children develop, they are trained to inhibit the natural reflexes that govern urination and bowel movements. In the classic *Patterns of Child Rearing,* Robert Sears and his colleagues (1957) reported that American children were toilet trained, on the average, at 18 months. However, nighttime bladder accidents occurred frequently until about 24 months. Today, most children in the United States achieve bladder control between the ages of 2 and 3. Many continue to have nighttime accidents for another year or so, however. Enuresis and encopresis are disorders involving problems with elimination that are not due to organic causes.

Enuresis

Enuresis derives from the Greek roots *en-*, meaning "in," and *ouron,* meaning "urine." **Enuresis** is failure to control urination after one has reached the "normal" age for attaining such control. Conceptions of what age is normal for achieving control can vary among clinicians. *DSM-IV* standards are shown in Table 14.7. Enuresis, like so many other developmental disorders, is more common among boys. Enuresis is estimated to affect 7% of boys and 3% of girls by age 5. The disorder usually resolves itself by adolescence if not earlier, although in about 1% of cases the problem continues into adulthood (APA, 2000).

Enuresis may occur during nighttime sleep only, during waking hours only, or both during nighttime sleep and waking hours. Nighttime-only enuresis is the most common type, and accidents occurring during sleep are referred to as *bed-wetting.* Achieving bladder control at night is more difficult than achieving daytime control. When asleep at night, children must learn to wake up when they feel the pressure of a full bladder and then go to the bathroom to relieve themselves. The younger the "trained" child is, the more likely she or he is to wet the bed at night. It is perfectly normal for children who have acquired daytime control over their bladders to have nighttime accidents for a year or more. Bed-wetting usually occurs during the deepest stage of sleep and may reflect immaturity of the nervous system. The diagnosis of enuresis applies in cases of repeated bed-wetting or daytime wetting of clothes by children of at least 5 years of age.

Theoretical Perspectives Numerous psychological explanations of enuresis have been advanced. Psychodynamic explanations suggest that enuresis may represent the expression of hostility toward children's parents because of harsh toilet training. It may represent regression in response to the birth of a sibling or some other stressor or life change, such as starting school or suffering the death of a parent or relative. Learning theorists point out that enuresis occurs most commonly in children whose parents attempted to train them early. Early failures may have connected anxiety with efforts to control the bladder. Conditioned anxiety, then, induces rather than curbs urination.

Evidence from a 1995 Danish study strongly suggests that *primary enuresis,* the most prevalent form of the disorder, which characterizes children who have persistent bed-wetting and have never established urinary control, is genetically transmitted (Eiberg, Berendt, & Mohr, 1995; Goleman, 1995e). We don't yet understand the genetic mechanism accounting for the transmission of the disorder, but one possibility is that it may involve genes that regulate the rate of development of motor control over eliminatory reflexes by the cerebral cortex. Although genetic factors appear to be involved in the transmission of primary enuresis, it is likely that environmental and behavioral factors also come into play in determining the development and course of the disorder. The other type of enuresis, *secondary enuresis,* characterizes children who develop the problem after having established urinary control and is associated with occasional bed-wetting. Genetic factors are apparently not involved in this type of enuresis (Goleman, 1995e).

Treatment Enuresis usually resolves itself as children mature. Behavioral methods have been shown to be helpful when enuresis endures or causes parents or children great distress, however. Such methods condition children to wake up when their bladders are full. One reasonably dependable example is Mowrer's bell-and-pad method.

The problem in bed-wetting is that children with enuresis continue to sleep despite bladder tension that awakens most other children. As a consequence, they reflexively urinate in bed. Psychologist O. Hobart Mowrer pioneered the bell-and-pad method, in which a special pad is placed beneath the sleeping child. When the pad is wet, an electrical circuit closes, causing a bell to ring and the sleeping child to waken. After several repetitions, most children learn to wake up in response to bladder tension—*before* they wet the pad. The technique is usually explained through principles of classical conditioning. In the bell-and-pad method, tension in children's bladder is paired repeatedly with a stimulus (a bell) that wakes them up. The bladder tension (conditioned stimulus, or CS) comes to elicit the same response (waking up—the conditioned response, or CR) that is elicited by the bell (the unconditioned stimulus, or US). Variations of the bell-and-pad method have also been used successfully with adults who have enuresis (van Son, Mulder, & Londen, 1990).

Both psychological treatment, usually involving the Mowrer urine alarm technique or some variation, or drug therapy, are often helpful in treating enuresis. The drug fluvoxamine, an SSRI-type antidepressant, works on brain systems that control urination (Kano & Arisaka, 2000; Horrigan, & Barnhill, 2000). The available evidence points to better results with psychological treatment, however (Houts, Berman, & Abramson, 1994).

encopresis Lack of control over bowel movements that is not caused by an organic problem in a child who is at least 4 years old.

Encopresis

Encopresis derives from the Greek roots *en-* and *kopros,* meaning "feces." **Encopresis** is lack of control over bowel movements that is not caused by an organic problem. The child must have a chronological age of at least 4, or in children with developmental delays, a mental age of at least 4 years (APA, 2000). About 1% of 5-year-olds have encopresis. Soiling, like enuresis, is more common among boys. Encopresis is rare among adolescents in their middle teens, except among teens who are profoundly or severely retarded. Soiling may be voluntary or involuntary and is not caused by an organic problem, except in cases in which constipation is involved (APA, 2000). Among the possible predisposing factors are inconsistent or incomplete toilet training and psychosocial stressors, such as the birth of a sibling or beginning school.

Soiling, unlike enuresis, is more likely to happen during the day than at night. It can thus be keenly embarrassing to the child. Classmates often avoid or ridicule soilers. Because feces have a strong odor, teachers may find it hard to act as though nothing of consequence has happened. Parents, too, are eventually galled by recurrent soiling and may increase their demands for self-control and employ powerful punishments for failure. Because of all this, the child may start to hide soiled underwear. Such children may distance themselves from classmates, or feign sickness to stay at home. Their levels of anxiety concerning soiling increase. Because anxiety (arousal of the sympathetic branch of the autonomic nervous system) promotes bowel movements, control may become yet more elusive.

THINK ABOUT IT

Do you know any children who were treated for a psychological disorder? What treatments did they receive? What were the outcomes?

When soiling is involuntary, it is often associated with constipation, impaction, or retention that results in subsequent overflow. Constipation may be related to psychological factors, such as fears associated with defecating in a particular place or with a more general pattern of negativistic or oppositional behavior. Or constipation may be related to physiological factors, such as complications from an illness or from medication. Much less frequently, encopresis is deliberate or intentional.

Quiz **14.8**
Elimination Disorders

Soiling often appears to follow harsh punishment of an accident or two, particularly in children who are already highly stressed or anxious. Harsh punishment may rivet children's attention on soiling. They may then ruminate about soiling, raising their level of anxiety so that self-control is impaired.

Quiz **14.9**
Chapter Exam

Operant conditioning methods may be helpful in dealing with soiling. These employ rewards (by praise and other means) for successful attempts at self-control and mild punishments for continued accidents (for example, gentle reminders to attend more closely to bowel tension and having the child clean her or his own underwear). When encopresis persists, thorough medical and psychological evaluation is recommended to determine possible causes and appropriate treatments.

Research Update
Chapter 14

Overview of Disorders of Childhood and Adolescence

TYPES OF DISORDERS

	Description	Major Types/Levels of Severity	Features
Pervasive Developmental Disorder	Marked impairment in multiple areas of development	• Autistic Disorder • Asperger's Disorder	• **Autism:** Major deficits in relating to others, impaired language and cognitive functioning, and restricted range of activities and interests • **Asperger's Disorder:** Poor social interactions and stereotyped behavior but without the significant language or cognitive deficits of autism
Mental Retardation	A broad-based delay in the development of cognitive and social functioning	• Deficits vary with level of severity from mild to profound	• Diagnosed on the basis of low IQ score and poor adaptive functioning
Learning Disorders	Deficiencies in specific learning abilities in the context of at least average intelligence and exposure to learning opportunities	• Mathematics Disorder • Disorder of Written Expression • Reading Disorder (Dyslexia)	• **Mathematics Disorder:** Difficulty understanding basic mathematical operations • **Disorder of Written Expression:** Grossly deficient writing skills • **Reading Disorder:** Difficulty recognizing words and comprehending written text
Communication Disorders	Difficulties in understanding or using language	• Expressive Language Disorder • Mixed Receptive/ Expressive Language Disorder • Phonological Disorder • Stuttering	• **Expressive Language Disorder:** Difficulty using spoken language • **Mixed Receptive/Expressive Disorder:** Difficulty understanding and producing speech • **Phonological Disorder:** Difficulty articulating the sounds of speech • **Stuttering:** Difficulty speaking fluently without interruption
Attention-Deficit and Disruptive Behavior Disorders	Patterns of disturbed behavior that are generally disruptive to others and to adaptable social functioning	• Attention-Deficit Hyperactivity Disorder (ADHD) • Conduct Disorder (CD) • Oppositional Defiant Disorder (ODD)	• **ADHD:** Problems of impulsivity, inattention, and hyperactivity • **CD:** Antisocial behavior that violates social norms and the rights of others • **ODD:** Pattern of noncompliant, negativistic, or oppositional behavior
Anxiety and Mood Disorders	Emotional disorders affecting children and adolescents	• Separation Anxiety Disorder • Specific Phobia • Social Phobia • Generalized Anxiety Disorder • Major Depression • Bipolar Disorder	• Anxiety and depression often have similar features in children as in adults, but some differences exist • Children may suffer from school phobia as a form of separation anxiety • Depressed children may fail to label their feelings as depression or may show behaviors that mask depression, such as conduct problems and physical complaints
Elimination Disorders	Persistent problems with controlling urination or defecation that cannot be explained by organic causes	• Enuresis (lack of control over urination) • Encopresis (lack of control over defecation)	• **Enuresis:** Nighttime-only enuresis (bed-wetting) is the most common type • **Encopresis:** Occurs most often during daytime hours

CAUSAL FACTORS	TREATMENT APPROACHES
• **Autism:** Causes unknown, but are presumed to involve underlying brain abnormalities, possibly resulting from genetic defects or prenatal exposure to toxic agents	• **Autism:** Intensive, long-term behavioral treatment to improve adaptive behavior and communication skills
• Chromosomal abnormalities, such as Down syndrome • Genetic abnormalities, such as fragile X syndrome • Prenatal infections or maternal substance abuse • Cultural-familial causes	• Psychoeducational interventions to foster development of academic skills and adaptive behaviors; institutional care may be needed in more severe cases
• Causes unclear, but may involve abnormalities in brain circuits for processing visual and auditory information (in dyslexia) • Genetic factors are also implicated in dyslexia	• Interventions based on one or more of the following theoretical models: —Psychoeducational model —Behavioral model —Medical model —Neuropsychological model —Linguistic model —Cognitive model
• **Stuttering:** Causes unclear, but may involve a combination of genetic and environmental influences	• Speech therapy and possible additional psychological counseling for social anxiety associated with speech impairment
• Family factors, such as unresolved parent-child conflict; negative marital conflict; and coercive parent-child interactions • Poor parenting behaviors, such as lack of reinforcement for appropriate behavior • Possible genetic component and subtle brain abnormalities associated with ADHD	• **ADHD:** Drug therapy (Ritalin or other stimulant drug), cognitive-behavioral therapy to help develop more appropriate behaviors and attentional skills • Parent training to assist parents in use of more appropriate reinforcement • **CD:** Residential treatment programs, anger management programs, and more broadly based multisystemic therapy to help develop more appropriate social behaviors
• Cognitive factors, such as dysfunctional thinking patterns, are observed in depressed and anxious children as well as adults • Stressful life events, family conflicts and problems, and lack of social support • Genetic factors may also play a role, especially in adolescent depression	• Cognitive-behavioral therapy to help anxious and depressed children develop healthier thinking patterns and coping skills • SSRI-type antidepressants, such as Prozac, may be helpful, but more research examining their effectiveness is needed
• **Enuresis:** Psychological conflicts, conditioned anxiety, and biological factors (genetic) may be involved • **Encopresis:** Causes may involve a combination of such factors as constipation, inconsistent or incomplete toilet training, psychosocial stressors, and anxiety or other psychological factors	• **Enuresis:** Some variation of the bell-and-pad method is commonly used • **Encopresis:** Operant conditioning methods may be employed (rewards for successful efforts at self-control, mild punishment for continued accidents)

Summing Up

Normal and Abnormal Behavior in Childhood and Adolescence

What factors do we need to consider in distinguishing normal and abnormal behavior in childhood and adolescence? In addition to the criteria described in Chapter 1, we need to take into account the child's age and cultural background.

Pervasive Developmental Disorders

What are pervasive developmental disorders? Pervasive developmental disorders involve marked deficiencies in multiple areas of development. They develop within the first years of life and are often associated with mental retardation. Autistic disorder is the most prominent type of pervasive developmental disorder.

What are the clinical features of autism? Children with autism shun affectionate behavior, engage in stereotyped behavior, attempt to preserve sameness, and tend to have peculiar speech habits such as echolalia, pronoun reversals, and idiosyncratic speech. The causes of autism remain unknown, but gains in academic and social functioning have been obtained through the use of intensive behavior therapy.

Mental Retardation

What is mental retardation and how is it assessed? Mental retardation is a general delay in the development of intellectual and adaptive abilities. It is assessed by intelligence tests and measures of functional ability. Most cases fall in the mildly retarded range.

What are the causes of mental retardation? Mental retardation is caused by chromosomal abnormalities, such as Down syndrome; genetic disorders, such as fragile X syndrome and phenylketonuria; prenatal factors, such as maternal diseases and alcohol use; and familial/cultural factors associated with intellectually impoverished home environments.

Learning Disorders

What are learning disorders? Learning disorders (also called learning disabilities) are specific deficits in the development of arithmetic, writing, or reading skills.

What are the causes of learning disorders and approaches to treatment? The causes remain under study but most probably involve underlying brain dysfunctions that make it difficult to process or decode visual and auditory information. Intervention focuses mainly on attempts to remediate specific skill deficits.

Communication Disorders

What are communications disorders? These disorders are characterized by impaired understanding or use of language. The specific types of communications disorders include expressive language disorder, mixed receptive/expressive language disorder, phonological disorder, and stuttering.

Attention-Deficit and Disruptive Behavior Disorders

What are attention-deficit and disruptive behavior disorders? This category includes attention-deficit hyperactivity disorder, conduct disorder, and oppositional defiant disorder. ADHD is characterized by impulsivity, inattention, and hyperactivity. Children with conduct disorders intentionally engage in antisocial behavior. Children with ODD show negativistic or oppositional behavior but not outright delinquent or antisocial behavior characteristic of conduct disorder. However, ODD may lead to the development of conduct disorder.

How are these disorders treated? Stimulant medication is generally effective in reducing hyperactivity, but it has not led to general academic gains. Behavior therapy may help ADHD children adapt better to school. Behavior therapy may also be helpful in modifying behaviors of children with conduct disorders and oppositional defiant disorder.

Anxiety and Depression

What types of anxiety disorders affect children? Anxiety disorders that occur commonly among children and adolescents include specific phobias, social phobia, and generalized anxiety disorder. Children may also show separation anxiety disorder, which involves excessive anxiety at times when they are separated from their parents. Cognitive biases, such as expecting negative outcomes, negative self-talk, and interpreting ambiguous situations as threatening, figure prominently in anxiety disorders in children and adolescents, as they often do among adults.

What are the distinguishing features of depression in childhood and adolescence? Depressed children, especially younger children, may not report or be aware of feeling depressed. Depression may also be masked by seemingly unrelated behaviors, such as conduct disorders. Depressed children also tend to show cognitive biases associated with depression in adulthood, such as adoption of a pessimistic explanatory style and distorted thinking. Although rare, suicide in children does occur and threats should be taken seriously. Risk factors for adolescent suicide include gender, age, geography, race, depression, past suicidal behavior, strained family relationships, stress, substance abuse, and social contagion.

Elimination Disorders

What are elimination disorders? These are problems of impaired control over urination (enuresis) and bowel movements (encopresis) that cannot be accounted for by organic causes. Both disorders are more common in boys.

What is the bell-and-pad method for treating enuresis? The bell-and-pad method conditions children with enuresis to respond to bladder tension.

CHAPTER FIFTEEN

ognitive Disorders and Disorders Related to Aging

Paul Klee
Allegorische Figurine

Truth OR Fiction?

- A man with a brain tumor patted the heads of fire hydrants and parking meters in the belief that they were children. (p. 480)
- The most often identified cause of delirium is ingestion of toxic mushrooms. (p. 482)
- After a motorcycle accident, a medical student failed to recognize the woman he had married a few weeks earlier. (p. 484)
- Dementia is a normal part of the aging process. (p. 485)
- People who occasionally become forgetful in middle age are probably suffering from the early stages of Alzheimer's disease. (p. 488)
- A common pain reliever found in most people's medicine cabinets has been found to reduce the risk of Alzheimer's disease when taken regularly. (p. 493)
- A famous folksinger and songwriter was misdiagnosed with alcoholism and spent several years in mental hospitals until the correct diagnosis was made. (p. 496)
- A form of dementia is linked to mad-cow disease. (p. 497)

In *The Man Who Mistook His Wife for a Hat,* neurologist Oliver Sacks (1985a) tells of Dr. P., a distinguished musician and teacher who had lost the ability to recognize objects visually. For example, Dr. P. failed to recognize the faces of his students at the music school. When a student spoke, however, Dr. P. immediately recognized his or her voice. Not only did the professor fail to discriminate faces visually, but sometimes he perceived faces where none existed. He patted the heads of fire hydrants and parking meters, which he took to be children. He warmly addressed the rounded knobs on furniture. These peculiarities were generally dismissed as jokes and laughed off by Dr. P. and his colleagues. After all, Dr. P. was well known for his oddball humor and jests. But Dr. P.'s music remained as accomplished as ever, his general health seemed fine, and so these misperceptions seemed little to be concerned about.

Not until 3 years later did Dr. P. seek a neurological evaluation. His ophthalmologist had found that although Dr. P.'s eyes were healthy, he had problems interpreting visual stimulation. So he made the referral to Dr. Sacks, a neurologist. When Sacks engaged Dr. P. in conversation, Dr. P.'s eyes fixated oddly on miscellaneous features of Dr. Sack's face—his nose, then his right ear, then his chin, sensing parts of his face but apparently not connecting them in a meaningful pattern. When Dr. P. sought to put on his shoe after a physical examination, he confused his foot with the shoe. When preparing to leave, Dr. P. looked around for his hat, and then . . .

> [Dr. P.] reached out his hand, and took hold of his wife's head, tried to lift it off, to put it on. He had apparently mistaken his wife for a hat! His wife looked as if she was used to such things. (Sacks, 1985a, p. 10)

Dr. P.'s peculiar behavior may seem amusing to some, but his loss of visual perception was tragic. Although Dr. P. could identify abstract forms and shapes—a cube, for example—he no longer recognized the faces of his family, or his own. Some features of particular faces would strike a chord of recognition. For example, he could recognize a picture of Einstein from the distinctive hair and mustache, and a picture of his own brother from the square jaw and big teeth. But he was responding to isolated features, not grasping the facial patterns as wholes.

Sacks recounts a final test:

> It was still a cold day, in early spring, and I had thrown my coat and gloves on the sofa.
> "What is this?" I asked, holding up a glove.
> "May I examine it?" he asked, and, taking it from me, he proceeded to examine it as he had examined the geometrical shapes.
> "A continuous surface," he announced at last, "infolded on itself. It appears to have"—he hesitated—"five outpouchings, if this is the word."
> "Yes," I said cautiously. "You have given me a description. Now tell me what it is."
> "A container of some sort?"
> "Yes," I said, "and what would it contain?"
> "It would contain its contents!" said Dr. P., with a laugh. "There are many possibilities. It could be a change-purse, for example, for coins of five sizes. It could . . ."
> I interrupted the blarney flow. "Does it not look familiar? Do you think it might contain, might fit, a part of your body?"
> No light of recognition dawned on his face.
> No child would have the power to see and speak of "a continuous surface . . . infolded on itself," but any child, any infant, would immediately know a glove as a glove, see it as familiar, as going with a hand. Dr. P. didn't. He saw nothing as familiar. Visually, he was lost in a world of lifeless abstractions. (Sacks, 1985a, p. 13)

Later, we might add, Dr. P. accidentally put the glove on his hand, exclaiming, "My God, it's a glove!" (Sacks, 1985a, p. 13). His brain immediately seized the pattern of **tactile** information, although his visual brain centers were powerless to provide a confirmatory interpretation. Dr. P., that is, showed lack of visual knowledge—a symptom referred to as visual **agnosia,** derived from Greek roots meaning "without knowledge." Still, Dr. P.'s musical abilities and verbal skills remained intact. He was able to function, to dress himself, take a

Chapter Outline

tactile Pertaining to the sense of touch.

agnosia A disturbance of sensory perception, usually affecting visual perception.

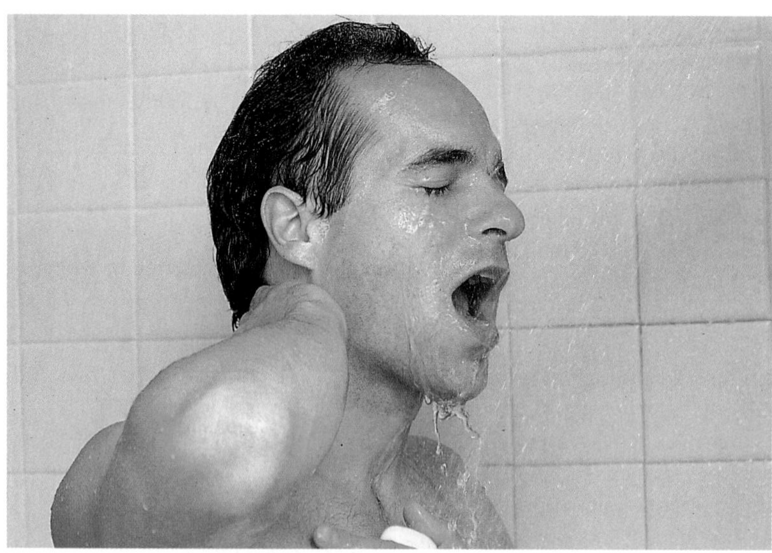

Does this man's singing help him coordinate his actions? In a celebrated case study, Dr. Oliver Sacks discussed the case of "Dr. P.," who was discovered to be suffering from a brain tumor that impaired his ability to interpret visual cues. Yet he could continue to eat meals and wash and dress himself so long as he could sing to himself.

shower, and eat his meals by singing various songs to himself—for example, eating songs and dressing songs—that helped him coordinate his actions. However, if his dressing song was interrupted while he was dressing himself, he would lose his train of thought and be unable to recognize not only the clothes his wife had laid out but also his own body. When the music stopped, so did his ability to make sense of the world. Sacks later learned that Dr. P. had a massive tumor in the area of the brain that processes visual information. Dr. P. was apparently unaware of his deficits, having filled his visually empty world with music in order to function and imbue his life with meaning and purpose.

Dr. P.'s case is unusual in the peculiarity of his symptoms, but it illustrates the universal dependence of psychological functioning on an intact brain. The case also shows how some people adjust—sometimes so gradually that the changes are all but imperceptible—to developing physical or organic problems. Dr. P.'s visual problems might have been relatively more debilitating in a person who was less talented or who had less social support to draw on. In this chapter, we focus on cognitive disorders, which are psychological disorders that arise from injuries or diseases that affect the brain.

Truth OR Fiction? REVISITED

A man with a brain tumor patted the heads of fire hydrants and parking meters in the belief that they were children.

TRUE. A man with a brain tumor patted the heads of fire hydrants and parking meters in the belief that they were children. The tumor caused dysfunction in the parts of his brain that processed visual information.

Cognitive Disorders

Cognitive disorders involve disturbances in thinking or memory that represent a marked change from the individual's prior level of functioning (APA, 2000). Cognitive disorders are not psychologically based; they are caused by physical or medical conditions, or drug use or withdrawal, that affect the functioning of the brain. In some cases the specific cause of the cognitive disorder can be identified; in others, it cannot be pinpointed. Although these disorders are biologically based, we can see in the case of Dr. P. that psychological and environmental factors play key roles in determining the impact and range of disabling symptoms as well as the individual's ability to cope with deterioration of cognitive and physical abilities.

Our ability to perform cognitive functions—to think, reason, and store and recall information—is dependent on the functioning of the brain. Cognitive disorders arise when the brain is either damaged or impaired in its ability to function due to injury, illness, exposure to toxins, or use or abuse of psychoactive drugs. When brain damage results from an injury or stroke, deterioration in intellectual, social, and occupational functioning can be rapid and severe. In the case of a progressive form of deterioration, such as Alzheimer's disease (discussed in the next section of this chapter), the decline is more gradual but leads eventually to a state of virtual helplessness. The extent and location of brain damage largely determine the range and severity of impairment. By and large, the more widespread the damage, the greater and more extensive the impairment in functioning. The location of the damage is also critical because many brain structures or regions perform specialized functions. Damage to the temporal lobe, for example, is associated with defects in memory and attention, whereas damage to the occipital lobe may result in visual-spatial deficits, such as Dr. P.'s loss of ability to recognize familiar faces.

Web Link 15.1 WWW
Brain Disorders Network

People who suffer from cognitive disorders may become completely dependent on others to meet basic needs in feeding, toileting, and grooming. In other cases, although some assistance in meeting the demands of daily living may be required, people are able to function at a level that permits them to live semi-independently.

There are three major types of cognitive disorders: delirium, dementia, and amnestic disorders (see Table 15.1).

cognitive disorders Mental disorders characterized by impaired cognitive abilities and daily functioning in which biological causation is either known or presumed.

TABLE 15.1 Major Types of Delirium, Dementia, and Amnestic Disorder

Delirium	Delirium Due to a General Medical Condition
	Substance Intoxication Delirium
	Substance Withdrawal Delirium
Dementia	Dementia of the Alzheimer's Type
	Vascular Dementia
	Dementia Due to HIV Disease
	Dementia Due to Head Trauma
	Dementia Due to Parkinson's Disease
	Dementia Due to Huntington's Disease
	Dementia Due to Pick's Disease
	Dementia Due to Creutzfeldt-Jakob Disease
	Dementia Due to Other General Medical Conditions (e.g., hypothyroidism or brain tumor)
	Substance-Induced Persisting Dementia
Amnestic disorder	Amnestic Disorder Due to a General Medical Condition
	Substance-Induced Persisting Amnestic Disorder

Source. Adapted from *DSM–IV-TR* (APA, 2000).

THINK ABOUT IT

What are cognitive disorders? What role does social support play in determining the adjustment of people affected by cognitive disorders?

Delirium

Delirium derives from the Latin roots *de-*, meaning "from," and *lira,* meaning "line" or "furrow." It means straying from the line, or the norm, in perception, cognition, and behavior. **Delirium** involves a state of extreme mental confusion in which people have difficulty concentrating and speaking clearly and coherently (see Table 15.2). People suffering from delirium may find it difficult to tune out irrelevant stimuli or shift their attention to new tasks. They may speak excitedly, but their speech carries little if any meaning. Disorientation as to time (not knowing the current date, day of the week, or time) and place (not knowing where you are) is common. Disorientation to person (the identities of oneself and others) is not. People in a state of delirium may experience terrifying hallucinations, especially visual hallucinations. Disturbances in perceptions often occur, such as misinterpretations of sensory stimuli (for example, confusing an alarm clock for a fire bell) or illusions (for instance, feeling as if the bed has an electrical charge passing through it). There can be a dramatic slowing down of movement into a state resembling catatonia. There may be

TABLE 15.2 Features of Delirium

Domain	**Level of Severity**		
	Mild	**Moderate**	**Severe**
Emotion	Apprehension	Fear	Panic
Cognition and perception	Confusion, racing thoughts	Disorientation, delusions	Meaningless mumbling, vivid hallucinations
Behavior	Tremors	Muscle spasms	Seizures
Autonomic activity	Abnormally fast heartbeat (tachycardia)	Perspiration	Fever

Source. Adapted from Freemon (1981), p. 82.

delirium A state of mental confusion, disorientation, and inability to focus attention.

dementia Profound deterioration of mental functioning, involving impaired memory, thinking, judgment, and language use.

senile dementias Forms of dementia that begin after age 65.

presenile dementias Forms of dementia that begin at or before age 65.

amnestic disorders Disturbances of memory associated with inability to learn new material or recall past events.

rapid fluctuations between restlessness and stupor. Restlessness is characterized by insomnia, agitated, aimless movements, even bolting out of bed or striking out at nonexistent objects. This may alternate with periods in which victims have to struggle to stay awake.

Delirium can result from a variety of medical conditions (Lichtenberg & Duffy, 2000). These include head trauma; metabolic disorders, such as hypoglycemia (low blood sugar); fluid or electrolyte imbalances; seizure disorders (epilepsy); deficiencies of the B vitamin thiamine; brain lesions; and various diseases that affect the functioning of the central nervous system, including Parkinson's disease, Alzheimer's disease, viral encephalitis (a type of brain infection), liver disease, and kidney disease (APA, 2000). Delirium may also occur due to exposure to toxic substances (such as eating certain poisonous mushrooms), as a side effect of using certain medications, or during states of drug or alcohol intoxication. Delirium may also result from abrupt cessation of use of psychoactive substances, especially alcohol, usually after periods of chronic, heavy use. Abrupt withdrawal from psychoactive drugs, especially alcohol, is the most common cause of delirium (Freemon, 1981). People with chronic alcoholism who abruptly stop drinking may experience a form of delirium called *delirium tremens* or DTs. During an acute episode of the DTs, the person may be terrorized by wild and frightening hallucinations, such as "bugs crawling down walls" or on the skin. The DTs can last for a week or more and are best treated in a hospital setting, where the patient can be carefully monitored and the symptoms treated with mild tranquilizers and environmental support. Although there are many known causes of delirium, in many cases the specific cause cannot be identified.

Truth OR Fiction? REVISITED

The most often identified cause of delirium is ingestion of toxic mushrooms.

FALSE. Though ingestion of certain types of poisonous mushrooms may result in delirium, the most frequently identified cause of delirium is abrupt withdrawal from alcohol or other drugs.

Whatever the cause, delirium involves a generalized disturbance of the brain's metabolic processes and imbalances in the levels of neurotransmitters. As a result, the ability to process information is impaired and confusion reigns. The abilities to think and speak clearly, to interpret sensory stimuli accurately, and to attend to the environment decline. Delirium may occur abruptly, as in cases resulting from seizures or head injuries, or gradually over hours or days, as in cases involving infections, fever, or metabolic disorders. During the course of delirium, the person's mental state will often fluctuate between periods of clarity ("lucid intervals"), which are most common in the morning, and periods of confusion and disorientation. Delirium is generally worse in the dark and following sleepless nights.

Unlike dementia, in which there is a steady deterioration of mental ability, states of delirium often clear up spontaneously when the underlying organic or drug-related cause is resolved. The course of delirium is relatively brief, usually lasting about a week, but rarely longer than a month. However, if the underlying cause persists or leads to further deterioration, delirium may progress to coma or death.

Dementia

Dementia involves a profound deterioration in mental functioning characterized by severe problems with memory and by one or more of the cognitive deficits listed in Table 15.3 (APA, 2000). There are more than 70 known causes of dementia, including brain diseases such as Alzheimer's disease and Pick's disease, and infections or disorders that affect the functioning of the brain, such as meningitis, HIV infection, and encephalitis. In some cases, the dementia can be halted or reversed, especially when it is caused by certain types of tumors and treatable infections, or when it results from depression or substance abuse. Most dementias are progressive and irreversible, however, including the most common form, dementia of the Alzheimer's type (Kasl-Godley & Gatz, 2000). Alzheimer's disease accounts for more than half of the cases of dementia.

THINK ABOUT IT

Do you expect people to become senile as they age? If so, what is the basis of your opinion?

Dementias usually occur in people over the age of 80. Those that begin after age 65 are called late-onset or **senile dementias.** Those that begin at age 65 or earlier are termed early-onset or **presenile dementias.**

Web Link 15.2 WWW
Dementia: What Are the Warning Signs?

Amnestic Disorders

Amnestic disorders (commonly called amnesias) are characterized by a dramatic decline in memory functioning that is not connected with states of delirium or dementia. Amnesia

TABLE 15.3 Cognitive Deficits in Dementia

Cognitive Deficit	Definition	Description
Aphasia	Impaired ability to comprehend and/or produce speech	There are several types of aphasia. In sensory or receptive aphasia, people have difficulty understanding written or spoken language, but retain the ability to express themselves through speech. In motor aphasia, the ability to express thoughts through speech is impaired, but the person can understand spoken language. A person with a motor aphasia may not be able to summon up the names of familiar objects or may scramble the normal order of words.
Apraxia	Impaired ability to perform purposeful movements despite an absence of any defect in motor functioning.	There may be an inability to tie a shoelace or button a shirt, although the person can describe how these activities should be performed and despite the fact that there is nothing wrong with the person's arm or hand. The person may have difficulty pantomiming the use of an object (e.g., combing one's hair).
Agnosia	Inability to recognize objects despite an intact sensory system.	Agnosias may be limited to specific sensory channels. A person with a visual agnosia may not be able to identify a fork when shown a picture of the object, although he or she has an intact visual system and may be able to identify the object if allowed to touch it and manipulate it by hand. Auditory agnosia is marked by impairment in the ability to recognize sounds; in tactile agnosia, people are unable to identify objects (such as coins or keys) by holding them or touching them.
Disturbance in executive functioning	Deficits in planning, organizing, or sequencing activities or in engaging in abstract thinking.	An office manager who formerly handled budgets and scheduling loses the ability to manage the flow of work in the office or adapt to new demands. An English teacher loses the ability to extract meaning from a poem or story.

Source. Adapted from *DSM-IV-TR* (APA, 2000).

involves an inability to learn new information (deficits in short-term memory) or to recall previously accessible information or past events from one's life (deficits in long-term memory). Problems with short-term memory may be revealed by an inability to remember the names of, or to recognize, people whom the person met 5 or 10 minutes earlier. Immediate memory, as measured by ability to repeat back a series of numbers, seems to be unimpaired in states of amnesia. The number series is unlikely to be recalled later, however, no matter how often it is rehearsed.

Unlike the memory disorders of dissociative amnesia and dissociative fugue discussed in Chapter 7, amnestic disorder results from a physical cause. Amnestic disorders frequently follow a traumatic event, such as a blow to the head, an electric shock, or an operation. A head injury may prevent people from remembering events that occurred shortly before the accident. The victim of an automobile accident or a football player who is knocked unconscious may be unable to remember events that occurred within several minutes of the injury. The automobile accident victim may not remember anything that transpired after getting into the car. The football player who is rendered amnestic from a blow to the head during the game may not remember anything after leaving the locker room. In some cases, memories for the remote past are retained but recent memories are

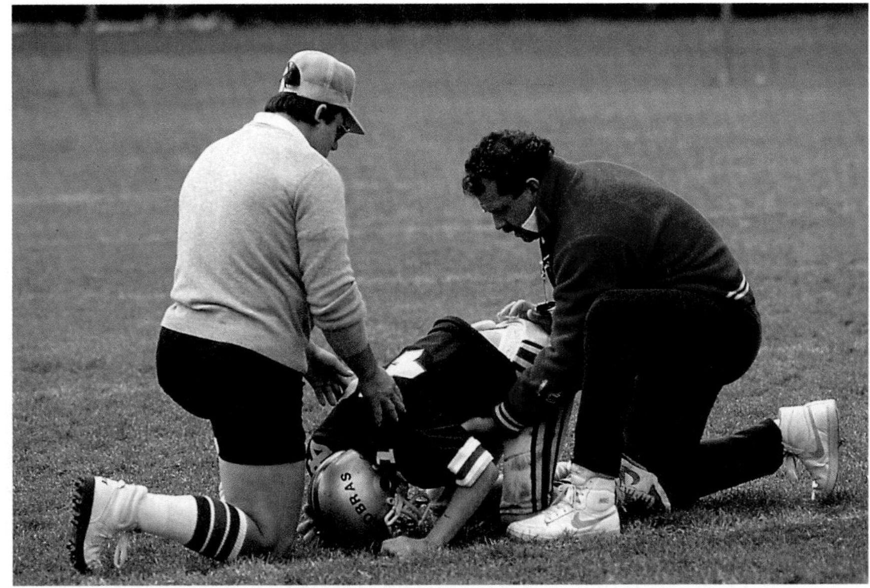

Amnestic disorder. An amnestic disorder syndrome can follow a traumatic injury such as a blow to the head. This football player may not be able to recall the events that happened just prior to his being tackled, nor the collision itself.

lost. People with amnesia may be more likely to remember events from their childhood than last evening's dinner, for example. Consider the following case:

Who Is She?

A medical student was rushed to the hospital after he was thrown from a motorcycle. His parents were with him in his hospital room when he awakened. As his parents were explaining what had happened to him, the door suddenly flew open and his flustered wife, whom he had married a few weeks earlier, rushed into the room, leaped onto his bed, and began to caress him and expressed her great relief that he was not seriously injured. After several minutes of expressing her love and reassurance, his wife departed and the flustered student looked at his mother and asked: "Who is she?"

—*Adapted from Freemon, 1981, p. 96*

Truth OR Fiction? REVISITED

After a motorcycle accident, a medical student failed to recognize the woman he had married a few weeks earlier.

TRUE. After a motorcycle accident, a medical student failed to recognize the woman he had married a few weeks earlier.

The medical student's long-term memory loss included memories dating not only to the accident but also further back to the time before he was married or had met his wife. Like most victims of posttraumatic amnesia, the medical student recovered his memory fully.

People with an amnestic disorder may experience disorientation, more commonly involving disorientation to place (not knowing where one is at the time) and time (not knowing the day, month, and year) than disorientation as to self (not knowing one's own name). They may also lack insight into their memory loss and attempt to deny or mask their memory deficits even when evidence of their impairment is presented to them (APA, 2000). They may also attempt to fill the gaps in their memories with imaginary events. Or they may admit they have a memory problem but appear apathetic about it, showing a kind of emotional blandness.

Although people with amnestic disorder may suffer profound memory losses, their general intelligence tends to remain within a normal range. Memory loss in pure amnesia may thus be distinguished from that occurring in progressive dementias like Alzheimer's disease, in which memory and intellectual functioning both deteriorate. Early detection and diagnosis of the causes of memory problems are vital to many sufferers, because 20% to 30% of them are correctable (Cohen, 1986).

Other causes of amnesia include brain surgery; **hypoxia,** or sudden loss of oxygen to the brain; brain infection or disease; **infarction,** or blockage of the blood vessels supplying the brain; and chronic, heavy use of certain psychoactive substances, most commonly alcohol.

Alcohol-Induced Persisting Amnestic Disorder (Korsakoff's Syndrome) A common cause of amnestic disorder is thiamine deficiency linked to chronic abuse of alcohol. Alcohol abusers tend to take poor care of their nutritional needs and may not follow a diet rich enough in vitamin B_1, or thiamine. Thiamine deficiencies may produce an irreversible form of memory loss called *alcohol-induced persisting amnestic disorder,* which is more commonly known as **Korsakoff's syndrome.** The word *persisting* is used because the memory deficits persist even years after the person stops drinking (APA, 2000). Korsakoff's syndrome is not limited to people with chronic alcoholism, however. It has been reported in other groups who experience thiamine deficiencies during times of deprivation, such as among prisoners of war.

People with Korsakoff's syndrome have major gaps in their memory of past experiences (Phaf, Geurts, & Eling, 2000). Their memory deficits are believed to result from the loss of brain tissue due to bleeding in the brain. Despite their memory losses, patients with Korsakoff's syndrome may retain their general level of intelligence. They are often described as being superficially friendly but lacking in insight, unable to discriminate between actual events and wild stories they invent to fill the gaps in their memories. They sometimes become grossly disoriented and confused and require custodial care.

Korsakoff's syndrome often follows an acute attack of **Wernicke's disease,** another brain disorder caused by thiamine deficiency. Wernicke's disease is characterized by confu-

hypoxia Decreased supply of oxygen to the brain or other organs.

infarction The development of an infarct, or area of dead or dying tissue, resulting from the blocking of blood vessels normally supplying the tissue.

Korsakoff's syndrome A syndrome associated with chronic alcoholism that is characterized by memory loss and disorientation (also called *alcohol-induced persisting amnestic disorder*).

Wernicke's disease A brain disorder, associated with chronic alcoholism, characterized by confusion, disorientation, and difficulty maintaining balance while walking.

sion and disorientation, difficulty maintaining balance while walking (**ataxia**), and paralysis of the muscles that control eye movements. These symptoms may pass, but the person is often left with Korsakoff's syndrome and enduring memory impairment. If, however, Wernicke's disease is treated promptly with major doses of vitamin B_1, Korsakoff's syndrome may not develop. Once Korsakoff's syndrome has set in, it is usually permanent, although slight improvement is possible with treatment.

ataxia Loss of muscle coordination.

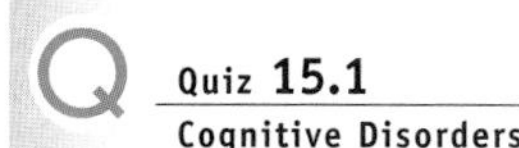

Psychological Disorders Related to Aging

Many physical changes occur with aging. Changes in calcium metabolism cause the bones to grow brittle and heighten the risk of breaks from falls. The skin grows less elastic, creating wrinkles and folds. The senses become less keen, so older people see and hear less acutely. Older people need more time (called *reaction time*) to respond to stimuli, whether they are driving or taking intelligence tests. For example, older drivers require more time to react to traffic signals and other cars. The immune system functions less effectively with increasing age, so people become more vulnerable to illness as they age. The skin becomes less elastic and cuts more easily. The sense of hearing declines, as does the elasticity of the lenses of the eyes, which makes it more difficult to focus on close objects or fine print.

Cognitive changes occur as well as we age. It is normal for people in later life to experience some decline in memory functioning and general cognitive ability, as measured by tests of intelligence, or IQ tests. The decline is sharpest on timed items, such as the performance scales of the Wechsler Adult Intelligence Scale. Although some declines in cognitive ability (reading comprehension, spatial ability as in map reading, or basic mathematical reasoning) in later life is common, it is not universal. Studies show that 20% to 30% of people in their 80s perform about as well on intelligence tests as those in their 30s and 40s (Goleman, 1994d). Some abilities, such as vocabulary and accumulated store of knowledge, hold up rather well in later life. However, people typically experience some reduction in memory as they age, especially memory for names or recent events. But apart from the occasional social embarrassment resulting from forgetting a person's name, cognitive declines experienced by most people as they age do not significantly interfere with their ability to meet their social or occupational responsibilities. Declines in cognitive functioning may also be offset to a certain extent by increased knowledge and experience.

Truth OR Fiction? REVISITED

Dementia is a normal part of the aging process.

FALSE. Dementia is not a normal part of aging. It is a sign of a degenerative brain disease.

The important point here is that dementia, or senility, is not the result of a normal process of aging (USDHHS, 1999a). It is a sign of degenerative brain disease. Screening and testing on neurological and neuropsychological tests can help distinguish dementias from normal aging processes. Generally speaking, the decline in intellectual functioning in dementia is more rapid and severe.

Let us now turn to consider relationships between various psychological disorders and aging, beginning with anxiety disorders.

Anxiety Disorders and Aging

Though anxiety disorders may develop at any point in life, they tend to be less prevalent among older adults than their younger counterparts (Flint, 1994). Still, anxiety disorders are the most common type of mental disorder affecting older adults and are about twice as common as mood disorders, such as depression ("Anxiety," 2000). Approximately 1 in 10 adults over the age of 55 suffers from a diagnosable anxiety disorder (USDHHS, 1999a). Older women are more likely to be affected by anxiety disorders than older men, by a ratio of two to one (Stanley & Beck, 2000). The most frequently occurring anxiety disorders among older adults are generalized anxiety disorder (GAD) and phobic disorders. Panic disorder is rare. Most cases of agoraphobia affecting older adults tend to be of recent origin and may involve the loss of social support systems due to the death of a spouse or close friends. Then again, some older individuals who are frail may have realistic fears of falling on the street and may be misdiagnosed as agoraphobic if they refuse to leave the house alone. Generalized anxiety disorder may arise from the perception that one lacks control

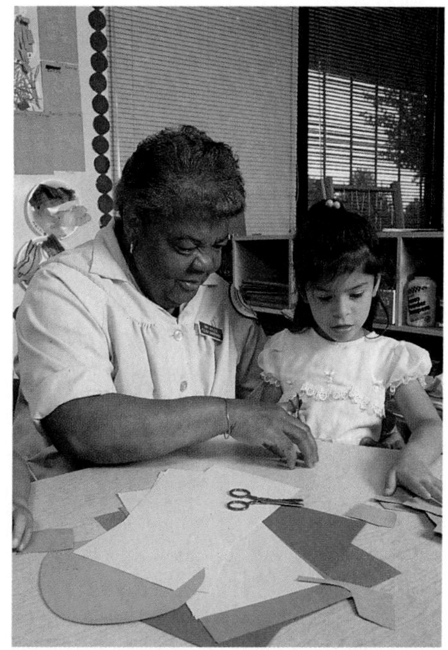

What changes take place as we age? How do they affect our moods? Although some declines in cognitive and physical functioning are connected with aging, older adults who remain active and engage in rewarding activities can be highly satisfied with their lives.

Agoraphobia? Or a need for support? Some older adults may refuse to venture away from home on their own because of realistic fears of falling in the street. They may be in need of social support, not therapy.

over one's life, which may develop in later life as the person contends with infirmity, loss of friends and loved ones, and lessened economic opportunities. Mild tranquilizers, such as the benzodiazepines (Valium is one), are commonly used to quell anxiety in older adults. Yet psychological interventions, such as cognitive-behavior therapy, may be an appropriate alternative to the use of psychiatric drugs (Stanley & Beck, 2000).

Depression and Aging

Though the risks of major depression also decline with age, depression is a major problem faced by many older adults (Karel & Hinrichsen, 2000; Unützer et al., 1997). In some cases, depression is a continuation of a lifelong pattern; in other cases, it first arises in later life. Between 8% and 20% of older adults experience some symptoms of depression (USDHHS, 1999a), with perhaps about 3% of them suffering from major depressive disorder (Rimer, 1999; Steffens et al., 2000). Rates of depression are higher still among residents of nursing homes. Though fewer older adults suffer from major depression than do younger adults, suicide is more frequent among older adults, especially older males (Knight & Satre, 1999; USDHHS, 1999a).

Depression in later life is also associated with a faster rate of physical decline and a higher mortality rate (Brenda et al., 1998; Zubenko et al., 1997). Depression may be linked to a higher mortality rate because of coexisting medical conditions or perhaps because of a lack of compliance with taking necessary medications.

Older people of color have many of the same concerns as other older adults, such as reduced opportunities for social participation and loneliness, but they are also likely to have encountered social stresses such as discrimination and poverty, and among immigrant groups, acculturative stress and English language deficiency. A history of discrimination and prejudice may make it difficult for people of color to trust therapists who are not of their own racial or ethnic group (Freed, 1992). The importance of acculturative stress was underscored in a study of Mexican American older adults that showed those who were minimally acculturated to U.S. society to have greater rates of depression than either highly acculturated or bicultural individuals (Zamanian et al., 1992).

Web Link 15.3 WWW
Age Page: Depression

Depressive disorders occur commonly in people suffering from various brain disorders, several of which, like Alzheimer's disease and stroke, disproportionately affect older people (Teri & Wagner, 1992). Researchers estimate that depressive disorders occur in as many as half of stroke victims and about a third to a half of people affected by Alzheimer's disease or Parkinson's disease (e.g., Chemerinski et al., 2001; Heun et al., 2001; Lyketsos et al., 2000). In the case of Parkinson's disease, depression may be not only a reaction to coping with the disease, but also result from neurobiological changes in the brain that are caused by the disease (Rao, Huber, & Bornstein, 1992).

The availability of social support appears to buffer the effects of stress, bereavement, and illness, thereby reducing the risk of depression. Social support is especially important to older people who are challenged because of physical disability. However, coping with a depressed spouse can take its toll, leading to an increased risk of depression in the caretaking spouse (Tower & Kasl, 1996).

On the other hand, participation in volunteer organizations and religious institutions is associated with a lower risk of depression among older people (Palinkas, Wingard, & Barrett-Connor, 1990). These forms of social participation may provide not only a sense of meaning and purpose but also a needed social outlet.

Older people may be especially vulnerable to depression because of the stress of coping with life changes associated with the so-called golden years—retirement; physical illness or incapacitation; placement in a residential facility or nursing

Elder stress. Older people of color may experience similar stresses as other older adults, but they are also more likely to have been exposed to additional social stresses, such as discrimination and poverty.

home; the deaths of a spouse, siblings, lifetime friends, and acquaintances; or the need to care for a spouse whose health is declining. Retirement, whether voluntary or forced, may sap the sense of meaningfulness in life and lead to a loss of role identity. Deaths of relatives and friends induce grief and remind older people of their own advanced age as well as reducing the availability of social support. Older adults may feel incapable of forming new friendships or finding new goals in life.

Evidence suggests that the chronic strain of coping with a family member with dementia can lead to depression in the caregiver, even in the absence of any prior vulnerability to depression. Nearly half of Alzheimer's caregivers become depressed (Small et al., 1997).

Despite the prevalence of depression in older people, physicians often fail to recognize it or to treat it appropriately (Rimer, 1999). In one study of more than 500 elderly people in Ontario who committed suicide, nearly 9 of 10 were found to have gone untreated (Duckworth & McBride, 1996). Health care providers may be less likely to recognize depression among older people than in middle-aged or young people because they tend to focus more on the older patient's physical complaints or because depression in older people is often masked by physical complaints and sleeping problems.

Most older people with memory deficits do not suffer from Alzheimer's disease. They are more likely to have memory losses due to depression or other factors like chronic alcohol use or the effects of small strokes (Bäckman & Forsell, 1994). The good news is that the memory impairment lapses that can accompany depression in many older adults often lift when the underlying depression is resolved.

Evidence indicates that treatments for depression that are effective for younger people, such as antidepressant medication, cognitive-behavioral therapy, and interpersonal psychotherapy, as well as ECT, are also effective in treating geriatric depression (Bondareff et al., 2000; Karel & Hinrichsen, 2000; J. W. Williams et al., 2000; Zeiss & Breckenridge, 1997). In fact, older adults benefit as much, although perhaps more slowly, from pharmacological and psychological interventions as midlife or younger adults (Reynolds et al., 1996; Scogin & McElreath, 1994). These finding should help put to rest the belief that psychotherapy is not appropriate for older people. We lack evidence, however, showing whether any particular form of psychotherapy is clearly superior in treating depression in older people.

THINK ABOUT IT

Why do you suppose that depression is so common among older people? To what extent might it be related to the diminished role expectations placed on older people in our society? In what ways might society provide more meaningful social roles for people as they age?

Sleep Problems and Aging

Sleep problems, especially insomnia, are common among older people (Lichstein et al., 2001). Insomnia in late adulthood is actually more prevalent than depression (Morgan, 1996). People are likely to experience sleep problems as they age, which may to a certain extent reflect age-related changes in sleep physiology, such as tendencies to wake up earlier in the morning (Martin, Shochat, & Ancoli-Israel, 2000). However, sleep problems may be a feature of other psychological disorders, such as depression, dementia, and anxiety disorders, as well as medical illness (Lamberg, 2000). Psychosocial factors, such as loneliness and the related difficulty of sleeping alone after loss of a spouse, may also be involved. Dysfunctional cognitions, such as excessive concerns about the negative consequences of losing sleep and perceptions of hopelessness and helplessness about controlling sleep, may play a role in perpetuating insomnia in older people (Morin et al., 1993a).

Mild tranquilizers are often used in treating late-life insomnia. However, problems such as dependence and withdrawal symptoms caution against their long-term use. Fortunately, researchers find that behavioral approaches, similar to those described in Chapter 11, are effective in treating insomnia in later life (Lichstein, Wilson, & Johnson, 2000; Martin, Shochat, & Ancoli-Israel, 2000). Moreover, older adults are as capable of benefiting from the treatment as younger adults.

A study of sleep apnea (temporary cessation of breathing during sleep) in a geriatric population showed that between 25% and 42% of the people studied had five or more apneas per hour of sleep (Ancoli-Israel et al., 1991). Apnea may involve more than a sleep problem; it is also linked to an increased risk of dementia and of cardiovascular disorders (Strollo & Rogers, 1996).

Alzheimer's disease (AD) A progressive brain disease characterized by gradual loss of memory and intellectual functioning, personality changes, and eventual loss of ability to care for oneself.

Dementia of the Alzheimer's Type

Alzheimer's disease (AD) is a degenerative brain disease that leads to a progressive and irreversible form of dementia, characterized by loss of memory and other cognitive functions. As noted, it accounts for more than half of the cases of dementia in the general population (see Figure 15.1) (Selkoe, 1992; Tune, 1998). Alzheimer's disease affects about 4 million Americans and is the fourth leading cause of death among adults (Cowan & Kandel, 2001).

Although AD is strongly connected with aging, it is a disease and not a consequence of normal aging (Butler, 2001). Alzheimer's affects about 1 in 10 Americans over the age of 65 and about half of those over the age of 85 (Lemonick & Park, 2001). The risk of AD nearly doubles in each 5-year period after age 60 (USDHHS, 1999a). Women are at higher risk of developing the disease than are men, though this may be a consequence of women tending to live longer. As the U.S. population on the whole continues to age, the numbers of Americans with AD is expected to nearly quadruple during the first half of the century,

Truth OR Fiction? REVISITED

People who occasionally become forgetful in middle age are probably suffering from the early stages of Alzheimer's disease.

FALSE. Occasional memory loss or forgetfulness in middle to later adulthood is a normal consequence of the aging process and not a sign of the early stages of Alzheimer's disease.

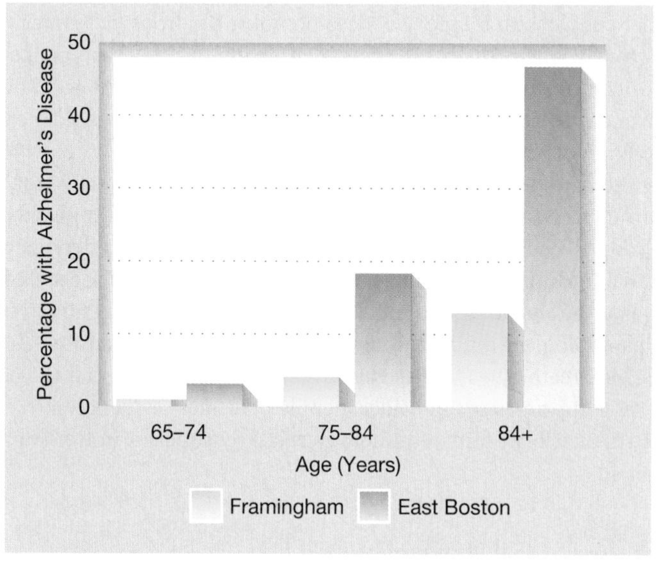

(A)

Distribution, by Probable Cause*

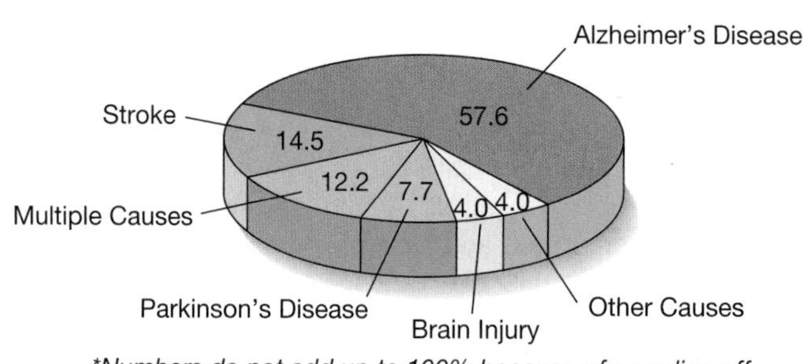

Numbers do not add up to 100% because of rounding off.

(B)

Some Treatable Causes of Dementia

Medications	Certain Tumors or Infections of the Brain
Emotional Depression	Blood Clots Pressing on the Brain
Vitamin B_{12} Deficiency	Metabolic Imbalances (Including Thyroid, Kidney or Liver Disorders)
Chronic Alcoholism	

(C)

FIGURE 15.1 Prevalence of Alzheimer's disease among older population of two communities in Massachusetts and distribution of probable causes of dementia. Studies of East Boston and Framingham in Massachusetts show that the prevalence of Alzheimer's disease increases with age among older people. The figures reported for Framingham are lower than those reported in East Boston, most probably because the Framingham study employed narrower criteria in diagnosing the disease. Note in part B that Alzheimer's disease is considered the probable cause in most cases of dementia (57.6%).

rising to nearly 14 million by the year 2000 (Kawas & Brookmeyer, 2001; Lemonick & Park, 2001). Economic costs of AD are estimated at more than $90 billion a year in the United States (Cowan & Kandel, 2001).

The dementia associated with AD involves a progressive deterioration of mental abilities involving memory, language, and problem solving. Occasional memory loss or forgetfulness in middle life (e.g., forgetting where one put one's glasses) is a normal consequence of the aging process and not a sign of the early stages of Alzheimer's disease (Bazell, 2000). People in later life (and some of us not quite that advanced in years) complain of not remembering names as well as they used to, or of forgetting names that were once well known to them. Although mild forgetfulness may concern people, it need not impair their social or occupational functioning.

Suspicions of AD are raised when cognitive impairment is more severe and pervasive, affecting the individual's ability to meet the ordinary responsibilities of daily work and social roles. Over the course of the illness, people with AD may get lost in parking lots or in stores, or even in their own homes (Kolata, 1994b). The wife of an AD patient describes how AD has affected her husband: "With no cure, Alzheimer's robs the person of who he is. It is painful to see Richard walk around the car several times because he can't find the door" (Morrow, 1998a, p. D4). Agitation, wandering behavior, depression, and aggressive behavior become common as the disease progresses (Chen et al., 1999; Slone & Gleason, 1999).

People with AD may become confused or delusional in their thinking and may sense that their mental ability is slipping away but not understand why. Bewilderment and fear may lead to paranoid delusions or beliefs that their loved ones have betrayed them, robbed them, or don't care about them. They may forget the names of their loved ones or fail to recognize them. They may even forget their own names.

Psychotic features such as delusions and hallucinations were found in about one in three people with AD in a recent study (Jeste et al., 1992). The appearance of psychotic symptoms appears to be connected with greater cognitive impairment and more rapid deterioration. People with Alzheimer's disease are frequently depressed or suicidal, but their doctors may overlook the danger signs or disregard them (Teri & Wagner, 1992).

Alzheimer's disease was first described in 1907 by the German physician Alois Alzheimer (1864–1915). During an autopsy of a 56-year-old woman who had suffered from severe dementia, he found two brain abnormalities now regarded as signs of the disease: plaques (portions of degenerative brain tissue) and neurofibrillary tangles (twisted bundles of nerve cells) (Näslund et al., 2000) (see Figure 15.2). The darkly shaded areas in the photo to the right in Figure 15.2 show the diminished brain activity associated with Alzheimer's disease.

Alzheimer's disease. Alzheimer's disease (AD) has struck a number of notable people, including former President Ronald Reagan, here shown with his wife, Nancy, at his first public appearance after being diagnosed with AD.

WWW **Web Link 15.4**
Facts About Alzheimer's Disease

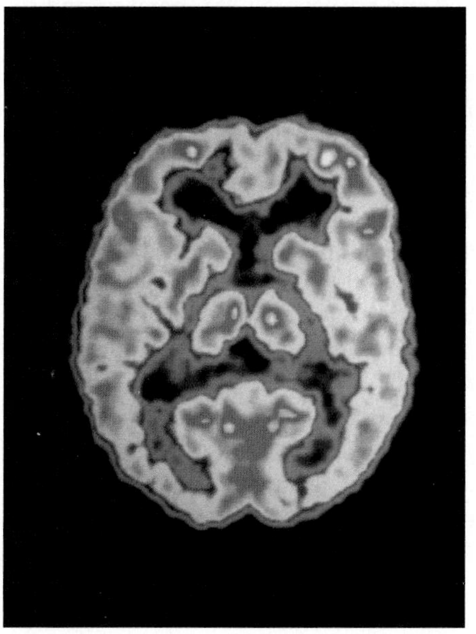

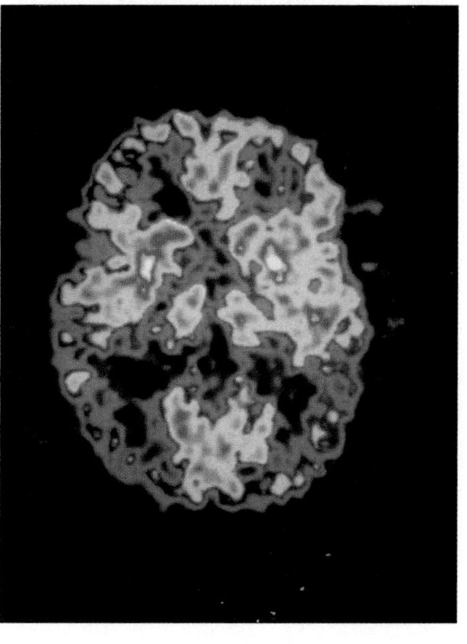

FIGURE 15.2 PET scans of brains from a healthy aged adult (left) and a patient with Alzheimer's disease (right).
The darkly shaded areas in the photo to the right suggest how the neurological changes associated with Alzheimer's disease (as marked by the excess of dark shading) impair brain activity.

Source. Friedland, R. P., Case Western Reserve University, courtesy of *Clinical Neuroimaging.* Copyright © 1988 by John Wiley and Sons, Inc.

Alzheimer's disease. Alzheimer's disease can devastate patients' families. Spouses usually provide the bulk of daily care. This man has been caring for his wife for several years, and he believes that his hugs and kisses sometimes prompt his wife to murmur his name.

Diagnosis There is no clear-cut diagnostic test for AD. The diagnosis of AD is generally based on a process of exclusion and given only when other possible causes of dementia are eliminated. Other medical and psychological conditions may mimic AD, such as severe depression resulting in memory loss and impaired cognitive functioning. Consequently, misdiagnoses may occur, especially in the early stages of the disease. A confirmatory diagnosis of AD can be made only upon inspection of brain tissue by biopsy or autopsy. However, biopsy is rarely performed because of the risk of hemorrhaging or infection, and autopsy, of course, occurs too late to help the patient.

Features of Alzheimer's Disease Alzheimer's disease develops gradually but progresses steadily, leading to dementia by about 3 years following onset (Cooke, 1994; Skoog, 2000). Earlier age of onset of AD appears to be associated with poorer cognitive functioning, even when the duration of the illness is taken into account. When AD strikes earlier in life, it may involve a more severe form of the disease.

The early stages of the disease are marked by limited memory problems and subtle personality changes ("New Use of Brain Scan," 2000). People may at first have trouble managing their finances; remembering recent events or basic information such as telephone numbers, area codes, zip codes, and the names of their grandchildren; and performing numerical computations (Reisberg et al., 1986). A business executive who once managed millions of dollars may become unable to perform simple arithmetic. There may be subtle personality changes, such as signs of withdrawal in people who had been outgoing or irritability in people who had been gentle. In these early stages, people with AD generally appear neat and well groomed and are generally cooperative and socially appropriate.

As AD progresses to a level of moderate severity, assistance may be required in managing everyday tasks. People with AD in the moderately severe range may be unable to select clothes for the season or the occasion. They may be unable to recall their addresses or names of family members. When they drive, they begin making mistakes, such as failing to stop at stop signs or accelerating when they should be braking.

Families who helplessly watch their loved ones slowly deteriorate have been described as attending a "funeral that never ends" (Aronson, 1988). Living with a person with advanced AD may seem like living with a stranger, so profound are the changes in the person's personality and behavior. The difficulties imposed on the families by the symptoms of advanced AD, such as wandering away, aggressiveness, destructiveness, incontinence, screaming, and remaining awake at night all contribute to the level of stress imposed on caregivers (Gurland & Cross, 1986).

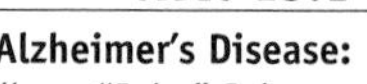

VIDEO 15.1

Alzheimer's Disease:
The Case of Wilburn "John" Johnson

Some people with AD are not aware of their deficits. Others deny them. At first they may attribute their problems to other causes, such as stress or fatigue. Denial may protect people with AD in the early or mild stages of the disease from recognition that their intellectual abilities are in decline (Reisberg et al., 1986). Or the recognition that one's mental abilities are slipping away may lead to depression.

Cognitive impairment becomes more severe as the disease progresses. At the moderately severe level, people encounter difficulties in various aspects of personal functioning, such as toileting and bathing themselves. There are large gaps in their memories for recent events. They may not be able to recall their complete addresses but may remember parts of them. Or they may forget the name of the president but be able to recall his last name if given the first name. They may fail to recognize familiar people or forget their names. They often make mistakes in recognizing themselves in mirrors. Memory for remote events is also affected. They are generally unable to recall the names of their schools, parents, or birthplaces. They may no longer be able to speak in full sentences. Verbal responses may be limited to a few words.

Movement and coordination functions deteriorate further. People with AD at the moderately severe level may begin walking in shorter, slower steps. They may no longer be able to sign their names, even when assisted by others. They may have difficulty handling a knife and fork. Agitation becomes a prominent feature at this stage, and victims may act out in response to the threat of having to contend with an environment that no longer seems controllable. They may pace or fidget, or display aggressive behavior such as yelling, throwing, or hitting. Patients may wander off because of restlessness and be unable to find their way back.

People with advanced AD may start talking to themselves or experience visual hallucinations or paranoid delusions. They may believe someone is attempting to harm them or is stealing their possessions, or that their spouses are unfaithful to them. They may believe their spouses are actually other people.

At the most severe stage, cognitive functions decline to the point where people become essentially helpless. They may lose the ability to speak or control body movement (Fuchs, 2001). They become incontinent; are unable to communicate, walk, or even sit up; and require assistance in toileting and feeding. In the end state, seizures, coma, and death result.

The major features of Alzheimer's disease, such as memory loss, disorientation, and behavior problems, are illustrated by the following case:

THINK ABOUT IT

What is Alzheimer's disease? What have we learned about the biological underpinnings of the disease? What don't we know?

A Case of Dementia of the Alzheimer's Type

A 65-year-old draftsman began to have problems remembering important details at work; at home he began to have difficulty keeping his financial records up-to-date and remembering to pay bills on time. His intellectual abilities progressively declined, forcing him eventually to retire from his job. Behavioral problems began to appear at home, as he grew increasingly stubborn and even verbally and physically abusive toward others when he felt thwarted.

On neurological examination, he displayed disorientation as to place and time, believing that the consultation room was his place of employment and that the year was "1960 or something," when it was actually 1982. He had difficulty with even simple memory tests, failing to remember any of six objects shown to him ten minutes earlier, not recalling the names of his parents or siblings, or the name of the president of the United States. His speech was vague and filled with meaningless phrases. He couldn't perform simple arithmetical computations, but he could interpret proverbs correctly.

Shortly following the neurological consultation, the man was placed in a hospital since his family was no longer able to control his increasingly disruptive behavior. In the hospital, his mental abilities continued to decline, while his aggressive behavior was largely controlled by major tranquilizers (antipsychotic drugs). He was diagnosed as suffering from a primary degenerative dementia of the Alzheimer type. He died at age 74, some 8 years following the onset of his symptoms.

—Adapted from Spitzer et al., 1989, pp. 131–132

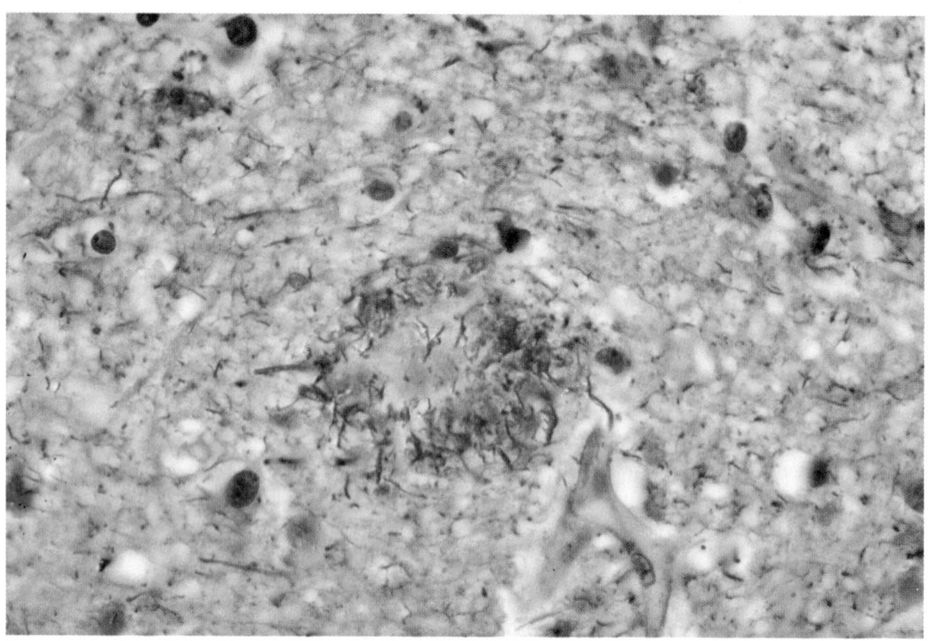

Plaques linked to Alzheimer's disease. In Alzheimer's disease, nerve tissue in the brain degenerates, forming steel-wool-like clumps or plaques composed of beta-amyloid protein fragments.

Causal Factors We don't know either what causes AD or how to prevent it or cure it. AD may involve multiple causes, so we may need to look for different causal pathways. We do know that plaques, the steel-wool-like clumps that form in the brains of people with Alzheimer's disease, are composed of a material called *beta amyloid,* which consists of fibrous protein fragments (Iwata et al., 2001; Skoog, 2000). For some unknown reason, possibly involving a genetic mutation, these fragments break off from a larger protein during metabolism and cluster together in strings that attract remnants of other nerve cells, forming plaques. These plaques may be responsible for destroying adjacent brain tissue, leading to the death of brain cells across a large area of the brain, which in turn may lead to the memory loss, confusion, and other symptoms of the disease.

Genetics plays an important role in AD (Ertekin-Taner et al., 2000; Myers et al., 2000; USDHHS, 1999a). Late-onset AD is linked to genes that regulate production of a protein called *apolipoprotein E* (APOe) (Voelker, 2000). This protein transports cholesterol through the bloodstream. People possessing defects in these genes may begin to show signs of brain deterioration even before memory symptoms appear ("New Use of Brain Scan," 2000). Their brains also appear to work harder when performing a memory task, which suggests that the brain is trying to compensate for underlying abnormalities (Bookheimer et al., 2000). Other genes may also be involved in AD, including genes involved in cell metabolism (the process of converting food into energy), and in the production of beta amyloid (Bertram et al., 2000; Ertekin-Taner et al., 2000; Li, 2000).

THINK ABOUT IT

What treatments are available for Alzheimer's disease? What are their limitations?

Treatment AD patients show reduced levels of the neurotransmitter acetylcholine (ACh), which may result from death of brain cells in an area of the brain (the nucleus basalis of Meynert) that manufactures the chemical. Presently several drugs are available that produce modest results in slowing the cognitive decline in AD ("Agency Approves," 2001; Skoog, 2000; USDHHS, 1999a). These drugs all work by inhibiting the breakdown of ACh, which increases the availability of the chemical in the brain. These drugs may slow the progression of the disease but are far from a cure. Psychosocial interventions, such as memory training programs, may also help AD patients make optimal use of their remaining abilities (Kasl-Godley & Gatz, 2000).

There are some promising leads on the prevention front. Preliminary evidence shows that people who regularly take anti-inflammatory drugs, such as the widely used pain reliever ibuprofen (brand names *Advil, Nuprin,* or *Motrin*), are less likely than others to develop AD (Hager & Peyser, 1997; Veld et al., 2001). Ibuprofen may reduce the risk of AD by curbing brain

Boosting memory. Memory training programs are not a cure, but they may help Alzheimer's patients make the best use of their remaining abilities.

Questionnaire

Examining Your Attitudes Toward Aging

What are your assumptions about late adulthood? Do you see older people as basically different from the young in their behavior patterns and their outlooks, or just as a few years more mature?

To evaluate the accuracy of your attitudes toward aging, mark each of the following items true (T) or false (F). Then turn to the answer key at the end of the chapter.

	TRUE	FALSE
1. By age 60 most couples have lost their capacity for satisfying sexual relations.	___	___
2. Older people cannot wait to retire.	___	___
3. With advancing age people become more externally oriented, less concerned with the self.	___	___
4. As individuals age, they become less able to adapt satisfactorily to a changing environment.	___	___
5. General satisfaction with life tends to decrease as people become older.	___	___
6. As people age they tend to become more homogeneous—that is, all old people tend to be alike in many ways.	___	___
7. For the older person, having a stable intimate relationship is no longer highly important.	___	___
8. The aged are susceptible to a wider variety of psychological disorders than young and middle-aged adults.	___	___
9. Most older people are depressed much of the time.	___	___
10. Church attendance increases with age.	___	___
11. The occupational performance of the older worker istypically less effective than that of the younger adult.	___	___
12. Most older people are just not able to learn new skills.	___	___
13. Compared to younger persons, older people tend to think more about the past than the present or the future.	___	___
14. Most people in later life are unable to live independently and reside in nursing-home-like institutions.	___	___

Source. Adapted from Rathus, S. A., & Nevid, J. S. (1995)3. *Adjustment and growth: The challenges of life.* (6th ed.), p. 440. Reprinted with permission of Brooks/Cole, an imprint of Wadsworth Group, a division of Thomson Learning. FAX 800-730-2215.

inflammation associated with the disease (M. F. Weiner, 1996). But further research is needed to demonstrate conclusively whether anti-inflammatory drugs do in fact reduce the risks of AD (Breitner & Zandi, 2001; Kolata, 2001b).

Researchers report initial success with an experimental vaccine in reducing or preventing buildup of plaques in the brains of laboratory mice ("Alzheimer's Vaccine," 2001). Although tests remain to be done with humans, the vaccine offers hope that we may one day be able to prevent Alzheimer's disease or stem the further progression of the disease in people already affected.

Vascular Dementia

The brain, like other living tissues, depends on the bloodstream to supply it with oxygen and glucose and to carry away its metabolic wastes. A stroke, also called a **cerebrovascular accident (CVA),** occurs when part of the brain becomes damaged because of a disruption in its blood supply, usually as the result of a blood clot that becomes lodged in an artery that services the brain and obstructs circulation. The areas of the brain affected may be damaged or destroyed, leaving the victim with disabilities in motor, speech, and cognitive functions. Death may also occur. Vascular dementia (formerly called *multi-infarct dementia*) is a form of dementia that results from repeated strokes. (An *infarct* refers to the death of tissue caused by insufficient blood supply.) Vascular dementia, the second most common form of dementia, most often affects people in later life but at somewhat earlier ages than dementia due to Alzheimer's disease. It accounts for about 20% of cases of dementia ("Update on Alzheimer's Disease," Part I, 1995) and appears to be more common in men

Truth OR Fiction? REVISITED

A common pain reliever found in most people's medicine cabinets has been found to reduce the risk of Alzheimer's disease when taken regularly.

TRUE. Recent evidence suggests that regular use of the pain reliever ibuprofen may reduce the risk of AD. However, physicians caution against routine use of the drug for this purpose because of the risks of side effects.

cerebrovascular accident (CVA) Damage to part of the brain because of a disruption in its blood supply, usually as the result of a blood clot.

Pick's disease A form of dementia, similar to Alzheimer's disease, but distinguished by specific abnormalities (Pick's bodies) in nerve cells and absence of neurofibrillary tangles and plaques.

Parkinson's disease A progressive disease of the basal ganglia characterized by muscle tremor and shakiness, rigidity, difficulty walking, poor control of fine motor movements, lack of facial muscle tonus, and, in some cases, cognitive impairment.

than women (APA, 2000). Unlike AD, heredity does not appear to play a major role in vascular dementia (Bergem et al., 1997).

Single strokes may produce gross impairments in specific functions, such as aphasia, but single strokes do not typically cause the more generalized cognitive declines that characterize dementia. Vascular dementia generally results from multiple strokes that occur at different times and that have cumulative effects on a wide range of mental abilities.

Features of Vascular Dementia The symptoms of vascular dementia are similar to those of dementia of the Alzheimer's type, including impaired memory and language ability, agitation and emotional instability, and loss of ability to care for one's own basic needs. However, AD is characterized by an insidious onset and a gradual decline of mental functioning, whereas vascular dementia typically occurs abruptly and follows a stepwise course of deterioration involving a pattern of rapid declines in cognitive functioning that are believed to reflect the effects of additional strokes (Brinkman et al., 1986; Kasl-Godley & Gatz, 2000). Some cognitive functions in people with vascular dementia may remain relatively intact in the early course of the disorder, leading to a pattern of patchy deterioration in which islands of mental competence remain while other abilities suffer gross impairment, depending on the particular areas of the brain that have been damaged by multiple strokes.

Quiz 15.2
Psychological Disorders Related to Aging

Dementias Due to General Medical Conditions

We have examined relationships between aging and psychological disorders such as dementia and depression. Next we consider a number of physical disorders that affect psychological functioning in various ways.

Dementia Due to Pick's Disease

Pick's disease causes a progressive dementia that is symptomatically similar to AD. Symptoms include memory loss and social inappropriateness, such as a loss of modesty or the display of flagrant sexual behavior. Diagnosis is confirmed only upon autopsy by the *absence* of the neurofibrillary tangles and plaques that are found in AD and by the presence of other abnormal structures—Pick's bodies—in nerve cells. Pick's disease is believed to account for perhaps 5% of dementias. Unlike AD, it becomes evident most often between the ages of 50 and 60, although it can occur at later ages (APA, 2000). The risk declines with advancing age after 70. Men are more likely than women to suffer from Pick's disease.

Pick's disease appears to run in families, and a genetic component is suspected in its etiology (Brun, 1996; Hutton, 2001). It has been estimated that members of the immediate family of victims of Pick's disease have an overall risk of 17% of contracting the disease by age 75 (Heston, White, & Mastri, 1987).

Battling Parkinson's disease. Actor Michael J. Fox has been waging a personal battle against Parkinson's disease and has brought national attention to the need to fund research efforts to develop more effective treatments for this degenerative brain disease.

Dementia Due to Parkinson's Disease

Dementia occurs in approximately 20% to 60% of people with **Parkinson's disease** (APA, 2000), a slowly progressing neurological disease that affects between one half million and one million people in the United States, including the former heavyweight champion Muhammad Ali and the actor Michael J. Fox (Cowan & Kandel, 2001; Cowley, 2000b). The disease affects men and women about equally and most often strikes between the ages of 50 and 69.

Parkinson's disease is characterized by uncontrollable shaking or tremors, rigidity, disturbances in posture (leaning forward), and lack of control over body movements. People with Parkinson's disease may be able to exercise control over their shaking or tremors, but only briefly. Some cannot walk at all. Others walk laboriously, in a crouch. Some execute voluntary body movements with difficulty, have poor control over fine motor movements, such as finger control, and have sluggish reflexes. They may look expressionless, as if they are wearing masks, a symptom that apparently reflects the degeneration of brain tissue that controls facial muscles. It is particularly difficult for patients to engage in sequences of complex movements, such as those required to sign their names. People with Parkinson's disease may be unable to coordinate two movements at the same time, as seen in this description of a Parkinson's patient who had difficulty walking and reaching for his wallet at the same time:

THINK ABOUT IT

How might your underlying attitudes toward older people affect the ways in which you relate to them? How might societal attitudes toward older people affect the allocation of funds for social services and for medical research?

Motor Impairment in a Case of Parkinson's Disease

A 58-year-old man was walking across the hotel lobby in order to pay his bill. He reached into his inside jacket pocket for his wallet. He stopped walking instantly as he did so and stood immobile in the lobby in front of strangers. He became aware of his suspended locomotion and resumed his stroll to the cashier; however, his hand remained rooted in his inside pocket, as though he were carrying a weapon he might display once he arrived at the cashier.

—*Adapted from Knight, Godfrey, & Shelton, 1988*

Despite the severity of motor disability, cognitive functions seem to remain intact during the early stages of the disease. Dementia is more common in the later stages of the disease or among those with more severe forms of the disease (APA, 2000). The form of dementia associated with Parkinson's disease typically involves a slowing down of thinking processes, impaired ability to think abstractly or plan or organize a series of actions, and difficulty retrieving memories. Overall, the cognitive impairments associated with Parkinson's disease tend to be more subtle than those associated with Alzheimer's disease (Knight et al., 1988). People with Parkinson's disease often become socially withdrawn and are at greater-than-average risk for depression (Cummings, 1992). Depression may be due to difficulty in coping with the disease or to the biochemical changes that are part and parcel of the disease (Rao et al., 1992).

Parkinson's disease is caused by the destruction or impairment of dopamine-producing nerve cells in an area of the brain called the *substantia nigra* ("black substance") that is involved in regulating body movement (Health Resources, 2000; Kolata, 2001a). The causes of the disease remain unknown, but genetic factors appear to be involved in at least some forms of the disease (Blakeslee, 2000c; Olson, 2000; Sveinbjornsdottir et al., 2000).

Whatever the underlying cause, the symptoms of the disease—the uncontrollable tremors, shaking, rigid muscles, and difficulty walking—are tied to deficiencies in the amount of dopamine in the brain. The drug L-dopa increases levels of dopamine in the brain and brought hope to Parkinson's patients when it was first used in treating the disease in the 1970s. L-dopa is converted in the brain into dopamine.

L-dopa helps control the symptoms of the disease and slows its progression, but it does not cure it (Parkinson Study Group, 2000). About 80% of people with Parkinson's disease show significant improvement in their tremors and motor symptoms following treatment with L-dopa. After a few years, however, L-dopa begins to lose its effectiveness, and the disease continues to progress. Several other drugs are in the experimental stage, offering hope for further advances in treatment. Another source of hope comes from experimental use of electrical stimulation of deep-brain structures (Deep-Brain Stimulation for Parkinson's Disease Study Group, 2001) and from genetic studies that may one day lead to an effective gene therapy for the disease (Kordower et al., 2000; Olson, 2000).

WWW **Web Link 15.5**
National Parkinson Foundation, Inc.

Dementia Due to Huntington's Disease

Huntington's disease, also known as *Huntington's chorea,* was first recognized by the neurologist George Huntington in 1872. Huntington's disease involves a progressive deterioration of the basal ganglia, especially of the caudate nucleus and the putamen, which primarily affects neurons that produce ACh and GABA.

Huntington's disease An inherited degenerative disease that is characterized by jerking and twisting movements, paranoia, and mental deterioration.

Woody Guthrie. The folksinger Woody Guthrie died from Huntington's disease in 1967, after 22 years of battling the disease.

The most prominent physical symptoms of the disease are involuntary, jerky movements of the face (grimaces), neck, limbs, and trunk—in contrast to the poverty of movement that typifies Parkinson's disease. These twitches are termed *choreiform,* which derives from the Greek *choreia,* meaning "dance." Unstable moods, alternating with states of apathy, anxiety, and depression, are common in the early stages of the disease. As the disease progresses, paranoia may develop and people may become suicidally depressed. Difficulties retrieving memories in the early course of the disease may develop into dementia as the disease progresses. Eventually, there is loss of control of bodily functions, leading to death occurring within about 15 years after onset of the disease.

Huntington's disease, which afflicts about 1 in 10,000 people, typically begins in the prime of adulthood, between the ages of 30 and 45 ("Researchers Gain Insight," 2001). Men and women are equally likely to develop the disease (APA, 2000). One of the victims of the disease was the folksinger Woody Guthrie, who gave us the beloved song "This Land Is Your Land," among many others. He died of Huntington's disease in 1967, after 22 years of battling the malady. Because of the odd, jerky movements associated with the disease, Guthrie, like many other Huntington's victims, was misdiagnosed as suffering from alcoholism. He spent several years in a number of mental hospitals before the correct diagnosis was made.

Huntington's disease is caused by a genetic defect on a single defective gene (Cowan & Kandel, 2001; Nucifora Jr. et al., 2001; Tamminga, 1997). It is transmitted genetically from either parent to children of either gender. People who have a parent with Huntington's disease stand a 50% chance of inheriting the gene. People who inherit the gene eventually contract the disease.

Until recently, children of Huntington's disease victims had to wait until the symptoms developed—usually in midlife—to learn whether they had inherited the disease. A genetic test has been developed that can detect carriers of the defective gene, those who will eventually develop the disease should they live long enough. Eventually, perhaps, genetic engineering may provide a means of modifying the defective gene or its effects. Because researchers have not yet developed ways to cure or control Huntington's disease, some potential carriers, like folksinger Arlo Guthrie, son of Woody Guthrie, preferred not knowing whether they inherited the gene. Meanwhile, research continues. One promising development in experimental work with monkeys involves the use of brain implants that release a substance that protects the kind of brain cells killed off by Huntington's disease ("Tests Suggest," 1997). We don't yet know whether the use of such implants could be used to treat the disease in humans.

Web Link 15.6
The Huntington's Disease Society of America
WWW

Truth OR Fiction? REVISITED

A famous folksinger and songwriter was misdiagnosed with alcoholism and spent several years in mental hospitals until the correct diagnosis was made.

TRUE. The folksinger and songwriter was Woody Guthrie, whose Huntington's disease went misdiagnosed for years.

THINK ABOUT IT

The availability of a genetic test for Huntington's disease means that a child of someone with Huntington's can find out whether he or she carries the gene that will eventually lead to the disease. If you faced this situation, do you think you would want to know whether you carried the gene? Why or why not?

Dementia Due to HIV Disease

Human immunodeficiency virus (HIV), the virus that causes AIDS, can invade the central nervous system causing a cognitive disorder—dementia due to HIV disease. The most typical signs of dementia due to HIV disease include forgetfulness and impaired concentration and problem-solving ability (APA, 2000). Common behavioral features of the dementia are apathy and social withdrawal. As AIDS progresses, the dementia grows more severe, taking the form of delusions, disorientation, further impairments in memory and thinking processes, and perhaps even delirium. In its later stages, the dementia may resemble the profound deficiencies found among people with advanced Alzheimer's disease.

Dementia is rare in persons with HIV who have not yet developed full-blown AIDS. Yet one in four people with AIDS develops some form of cognitive impairment that may progress to dementia (Center for Mental Health Services, 1994). Signs of intellectual impairment short of full-blown dementia may also occur earlier than the onset of AIDS (Baldeweg et al., 1997). People with HIV who show early signs of intellectual impairment appear to be at greater risk of early death from AIDS ("Cognitive Impairment," 1996; Wilkie et al., 1998).

Dementia Due to Creutzfeldt-Jakob Disease

Creutzfeldt-Jakob disease is a rare and fatal brain disease (Cowley, 2001b). It is characterized by the formation of small cavities in the brain that resemble the holes in a sponge. Dementia is a common feature of the disease. The disease typically affects people in the 40- to 60-year-old age range, although it may develop in adults at any age (APA, 2000). There are no treatments for the disease and death usually results within months of onset of symptoms. In about 5% to 15% of cases there is evidence of familial transmission, which suggests that a genetic component may be involved in determining susceptibility to the disease. The human form of mad-cow disease, a fatal illness spread by eating infected beef, is a variant of Creutzfeldt-Jakob disease (Cowan & Kandel, 2001; "How Mad-Cow Disease Jumped to Humans," 2001; McNeil, 2001).

Heading toward brain damage? Multiple blows to the head may lead to a progressive form of dementia.

Dementia Due to Head Trauma

Head trauma can injure the brain. Jarring, banging, or cutting brain tissues, usually because of accident or assault, are causes of such injuries. Progressive dementia due to head trauma is more likely to result from multiple head traumas (as in the case of boxers who receive multiple blows to the head during their careers) than to a single blow or head trauma (APA, 2000). Yet even a single head trauma can have psychological effects, and if severe enough, can lead to physical disability or death. Specific changes in personality following traumatic injury to the brain vary with the site and extent of the injury, among other factors (Prigatano, 1992). Damage to the frontal lobe, for example, is associated with a range of emotional changes involving alterations of mood and personality.

Neurosyphilis

General paresis (from the Greek *parienai,* meaning "to relax") is a form of dementia—or "relaxation" of the brain in its most negative connotation—that results from neurosyphilis, a form of syphilis in which the disease organism, in a late stage of infection, directly attacks the brain and central nervous system. General paresis is of historical significance to abnormal psychology. The 19th-century discovery of the connection between this form of dementia and a concrete physical illness, syphilis, strengthened the medical model and held out the promise that organic causes would eventually be found for other abnormal behavior patterns. Syphilis is a sexually transmitted disease caused by the bacterium *Treponema pallidum.*

General paresis is associated with physical symptoms such as tremors, slurred speech, impaired motor coordination, and, eventually, paralysis—all of which are suggestive of relaxed control over the body. Psychological signs include shifts in mood states, blunted emotional responsiveness, and irritability; delusions; changes in personal habits, such as suspension of personal grooming and hygiene; and progressive intellectual deterioration, including impairments of memory, judgment, and comprehension. Some people with general paresis grow euphoric and entertain delusions of grandiosity. Others become lethargic and depressed. Eventually, people with general paresis lapse into a state of apathy and confusion, characterized by the inability to care for themselves or to speak intelligibly. Death eventually ensues, either because of renewed infection or because of the damage caused by the existing infection.

Late-stage syphilis once accounted for 10% to 30% of admissions to psychiatric hospitals. However, advances in detection and the development of antibiotics that cure the infection have sharply reduced the incidence of late-stage syphilis and the development of general paresis. The effectiveness of treatment depends on when antibiotics are introduced and the extent of central nervous system damage. In cases where extensive tissue damage has been done, antibiotics can stem the infection and prevent further damage, thereby producing some improvement in intellectual performance. They cannot restore people to their original levels of functioning, however.

Truth OR Fiction? REVISITED

A form of dementia is linked to mad-cow disease.

TRUE. A form of dementia is caused by the human form of mad-cow disease.

General paresis A form of dementia resulting from neurosyphilis.

Quiz 15.3
Dementias Due to General Medical Conditions

Quiz 15.4
Chapter Exam

Research Update
Chapter 15

Overview of Cognitive Disorders

TYPES OF COGNITIVE DISORDERS

	Description	Features
Delirium	States of extreme mental confusion interfering with concentration and ability to speak coherently	• Difficulty filtering out irrelevant stimuli or shifting attention • Excited speech that conveys little meaning • Disorientation as to time and place • Frightening hallucinations or other perceptual distortions • Motor behavior may slow to a stupor or fluctuate between states of restlessness and stupor • Mental states may fluctuate between lucid intervals and periods of confusion
Dementia	Profound deterioration of mental functioning, including memory	• Most forms are irreversible and progressive, such as dementia of the Alzheimer's type • Associated with specific cognitive deficits, such as aphasia, apraxia, agnosia, and disturbance in executive functioning • Types of dementias are grouped by age of onset into senile dementias (beginning after age 65) and presenile dementias (beginning at age 65 or earlier)
Amnestic Disorder	Profound deficit in memory not associated with delirium or dementia	• May affect short-term memory and/or long-term memory • Person may be disoriented, especially as to place and time

CAUSAL FACTORS

Delirium

- Medical conditions, such as head trauma, metabolic disorders, low blood sugar, fluid or electrolyte imbalances, epilepsy, vitamin B deficiencies, and brain lesions
- Brain diseases, such as Parkinson's disease and Alzheimer's disease
- Abrupt withdrawal from alcohol in cases of chronic alcoholism (called delirium tremens, or DTs)

Dementia

- Brain diseases, such as Alzheimer's disease, Pick's disease, Parkinson's disease, Huntington's disease, HIV disease, and Creutzfeldt-Jakob disease
- Neurosyphilis
- Multiple strokes (vascular dementia)
- Brain tumors
- Head trauma
- Brain infections such as meningitis and encephalitis
- Causes of Alzheimer's disease remain unknown, but evidence points to a genetic contribution

Amnestic Disorder

- Physical causes such as a blow to the head
- Complications from brain surgery
- Brain infection
- Blockage of blood supply to the brain
- Heavy use of certain psychoactive substances (as in Korsakoff's syndrome)

TREATMENT APPROACHES

Delirium

- May clear up spontaneously or when the underlying medical condition is treated successfully
- Monitoring in hospital setting may be needed, especially for the DTs

Dementia

- Available treatments for dementia of the Alzheimer's type are limited to drugs that may slow the progression of the disease but are not a cure

Amnestic Disorder

- Memory may return spontaneously or with effective treatment of underlying conditions

Summing Up

Cognitive Disorders

What are cognitive disorders? Cognitive disorders involve disturbances of thinking or memory that represent a marked decline in intellectual or memory functioning. They are caused by physical or medical conditions, or drug use or withdrawal, affecting the functioning of the brain.

What is delirium? Delirium is a state of mental confusion characterized by symptoms such as impaired attention, disorientation, disorganized thinking and rambling speech, reduced level of consciousness, and perceptual disturbances. Delirium is most commonly caused by alcohol withdrawal, as in the form of delirium tremens (DTs).

What is dementia? Dementia involves cognitive deterioration or impairment, as evidenced by memory deficits, impaired judgment, personality changes, and disorders of higher cognitive functions such as problem-solving ability and abstract thinking.

Is dementia a normal part of aging? No, dementia is not a normal consequence of aging but a sign of a degenerative brain disorder. There are various causes of dementias, including Alzheimer's disease and Pick's disease, and brain infections or disorders.

What are amnestic disorders? Amnestic disorders involve deficits in short-term or long-term memory. The most common cause of amnestic syndrome is alcohol-induced persisting amnestic disorder, or Korsakoff's syndrome, which involves a thiamine deficiency typically associated with patterns of chronic alcohol abuse.

Psychological Disorders Related to Aging

What types of psychological problems affect people in later life? Generalized anxiety disorder and phobic disorders are the most commonly occurring anxiety disorders among older people. Depression to varying degrees is common among people in later life and may be associated with memory deficits that may lift as the depression clears. Dementia of the Alzheimer's type and vascular dementia primarily affect people in later life. They involve irreversible and progressive memory impairment. Certain sleep disorders, such as insomnia and sleep apnea, are also common among older people.

What are Alzheimer's disease and vascular dementia? Alzheimer's disease (AD) is a progressive brain disease characterized by progressive loss of memory and cognitive ability, as well as deterioration in personality functioning and self-care skills. There is no cure or effective treatment for AD. Research into its causes has focused on genetic factors and imbalances in neurotransmitters, especially acetylcholine. Vascular dementia results from multiple strokes (blood clots that block the supply of blood to parts of the brain, damaging or destroying brain tissue).

Dementias Due to General Medical Conditions

What other general medical conditions can lead to dementia? Various other medical conditions can lead to dementia, including Pick's disease, Parkinson's disease, Huntington's disease, Creutzfeldt-Jakob disease, HIV disease, head trauma, and neurosyphilis.

Scoring Key for Attitudes Toward Aging Scale

1. False. Most healthy couples continue to engage in satisfying sexual activities into their 70s and 80s.
2. False. This is too general a statement. Those who find their work satisfying are less desirous of retiring.
3. False. In late adulthood, we tend to become more concerned with internal matters—our physical functioning and our emotions.
4. False. Adaptability remains reasonably stable throughout adulthood.
5. False. Age itself is not linked to noticeable declines in life satisfaction. Of course, we may respond negatively to disease and losses, such as the death of a spouse.
6. False. Although we can predict some general trends for older adults, we can also do so for younger adults. Older adults, like their younger counterparts, are heterogeneous in personality and behavior patterns.
7. False. Older adults with stable intimate relationships are more satisfied.
8. False. We are susceptible to a wide variety of psychological disorders at all ages.
9. False. Only a minority are depressed.
10. False. Actually, church attendance declines, but not verbally expressed religious beliefs.
11. False. Although reaction time may increase and general learning ability may undergo a slight decline, older adults usually have little or no difficulty at familiar work tasks. In most jobs, experience and motivation are more important than age.
12. False. Learning may just take a bit longer.
13. False. Older adults do not direct a higher proportion of thoughts toward the past than do younger people. Regardless of our age, we may spend more time daydreaming at any age if we have more time on our hands.
14. False. Fewer than 10% of older adults require some form of institutional care.

CHAPTER SIXTEEN

Violence and Abuse

Rufino Tamayo
Title Unknown (Abstract)

Truth OR Fiction?

- Despite all the talk about crime, the United States has a relatively low homicide rate as compared with other industrialized countries. (p. 501)
- A male teen in the United States stands a greater chance of dying from a gunshot wound than from all natural causes of death combined. (p. 501)
- Though alcohol use is linked to aggression, it is rarely involved in homicides. (p. 510)
- Women who remain with men who abuse them suffer from a form of masochism. (p. 517)
- The most common form of child maltreatment is physical abuse. (p. 519)
- Women are more likely to be raped by men they know than by strangers. (p. 522)
- Child molesters typically use physical force to compel children into performing sexual acts. (p. 526)
- In some cases, sexual relations between clients and therapists are therapeutically justified. (p. 533)

There was a time not so long ago when people would leave their doors unlocked, walk through city parks long into the night, and open their front doors to strangers without hesitation. Today, fear of violent crime and the threat of terrorist attacks increasingly govern our behavior. We live behind triple-locked doors, equip our homes and cars with the most sophisticated security systems, and endure long waits at security checkpoints at the nation's airports. Some of us rarely if ever go out at night, even in our own neighborhoods. Yet we are much more likely to be attacked or killed by people we know than by strangers—by our spouses, family members, friends or acquaintances, or people we date.

By international standards, the United States stands out as a violent culture. Homicide is the second leading cause of death, after accidents, among young people in the 15 to 24 age range. It is the leading cause of death among Black males in this age group (Coontz & Franklin, 1997). The murder rate in the United States is nearly 10 times what it is in Japan; the robbery rate is nearly 150 times greater (Kristof, 1995). The homicide rate in the United States is especially high (see Figure 16.1), exceeding by a wide margin the homicide rates of other developed nations (United Nations, 1998).

Although other factors are clearly involved in the high rate of homicides in the United States, a contributing factor is easy access to firearms. About three of four homicides of young people result from the use of firearms, as compared to fewer than one in four in other developed nations (Kristof, 1995). A national survey showed that one in five teenagers in the United States carries a weapon ("One in Five," 1998). More male teens in the United States die from firearms than from all natural causes combined (M. L. Rosenberg, 1993). It is too easy for many young people in the United States to acquire guns, and an alarming number of them carry them (O'Donnell, 1995).

Although the homicide rate in the United States remains high, there is good news. The murder rate overall, and homicides of young persons involving the use of firearms, began dropping in the early 1990s (Fingerhut, Ingram, & Feldman, 1998). Moreover, violent crime overall declined steadily during the 1990s, the longest period of decline in 25 years (Butterfield, 1997a, 1997c).

In this chapter we focus on violent and abusive behavior. We will see that violence and abuse take many forms, from outright physical aggression to sexual harassment. We will also see that the search for the origins of violent and abusive behavior is best approached from a multifactorial model that takes into account sociocultural, psychological, and biological factors.

Chapter Outline

Violence and Abnormal Behavior

In Chapter 1 we introduced you to the idea that abnormal behavior can be defined in a number of ways, including unusualness, social deviance or unacceptability, faulty perceptions or interpretations of reality, severe personal distress, maladaptiveness, and dangerousness. Given these definitions, is violent behavior abnormal?

The answer largely depends on the context in which the behavior occurs. The behavior of a professional football player or hockey player may indeed be violent, but it is not what we would usually consider to be abnormal, except perhaps if it exceeds the boundaries of what would be considered sporting. (In hockey, it may be difficult to know where to draw the line.) Prizefighters earn their livings in a sport that is by its nature violent. Even though the physical pounding a prizefighter endures in the ring may lead to the development of physical and psychological disorders (see Chapter 15), prizefighting is not classified as abnormal behavior.

Warfare or combat is also inherently violent and certainly dangerous to self and others. Yet we don't typically consider the violent behavior of soldiers who fight wars or the general officers who direct them as abnormal. Perhaps we might consider the behavior of the nations themselves that wage war or their leaders as abnormal. But those who risk their lives and commit violent acts against soldiers of other nations are generally seen as loyal or patriotic, perhaps even heroic, not abnormal.

Where, then, do we draw the line between violent behavior that is deemed normal and that which is deemed abnormal? We adopt the following standard. We consider violent

Truth OR Fiction? REVISITED

Despite all the talk about crime, the United States has a relatively low homicide rate as compared with other industrialized countries.

FALSE. The United States leads the industrialized world in homicide rates.

Truth OR Fiction? REVISITED

A male teen in the United States stands a greater chance of dying from a gunshot wound than from all natural causes of death combined.

TRUE. A male teenager is more likely to die from a gunshot wound than from all natural causes combined.

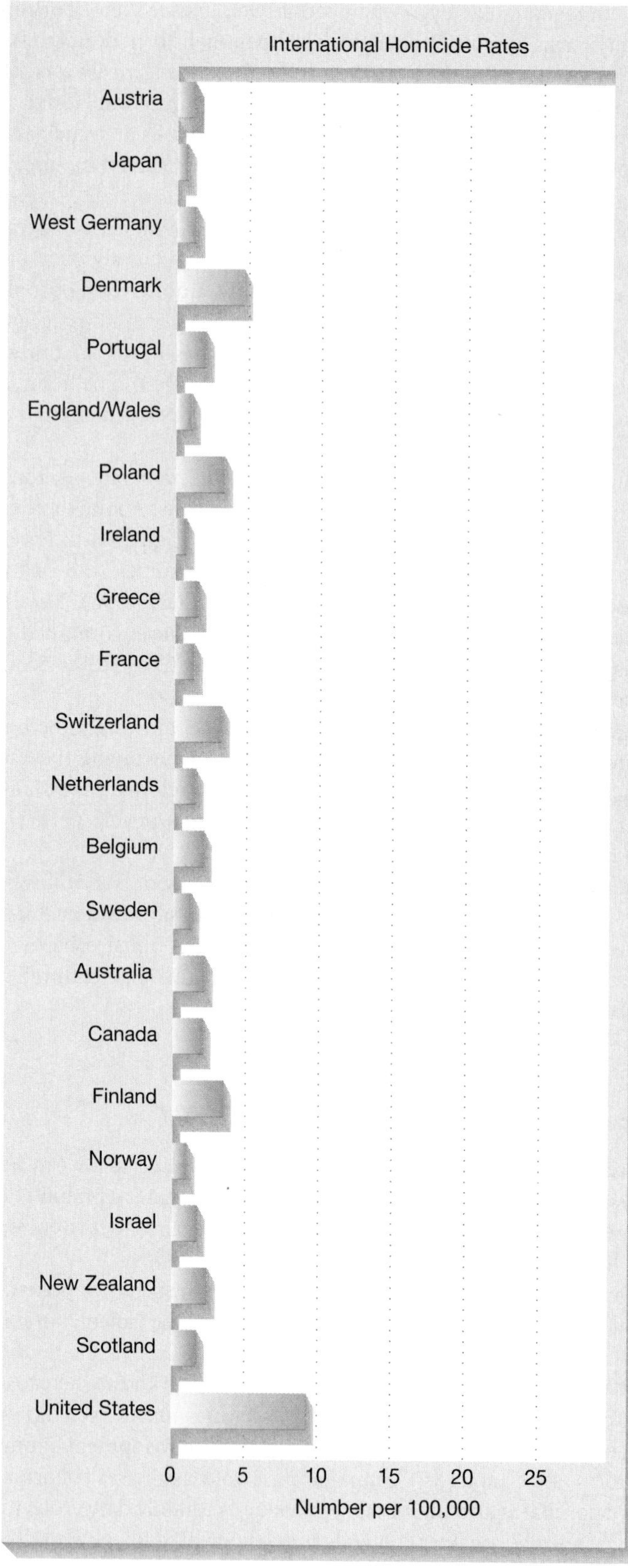

FIGURE 16.1 Homicide rates: U.S. versus other developed countries.
The homicide rate in the United States far exceeds the rates in other developed countries. The easy availability of firearms is clearly a contributing factor to the high rate of homicides in the United States.

Source. United Nations. (1998). *Demographic yearbook.* New York: Author.

behavior to be abnormal if it (1) occurs outside a socially sanctioned context, and (2) is either self-defeating or dangerous (harmful to oneself or others). The football player who slams an opponent into the ground is rewarded with a multimillion-dollar contract; the one who slams his wife against the wall commits a violent act that is abnormal as well as criminal. The spouse abuser's behavior is clearly dangerous (harmful to others and perhaps to self), socially unacceptable, and also maladaptive or self-defeating, because it can lead to arrest, marital dissolution, and retaliation, as well as failing to resolve the underlying conflict. It may also be associated with severe personal distress.

America the violent. The homicide rate in the United States is 10 times what it is in Japan. Young men in the United States are more likely to die as the result of homicide than young men in any of the other developed nations listed in Figure 16.1. These empty shoes of young gunshot victims are a poignant reminder of the many needless deaths caused by homicides.

Violence and Psychological Disorders

A common perception exists that people with psychological disorders are especially prone to violence. Some people with psychological disorders do commit violent acts, just as assuredly as do some people without diagnosable psychological disorders. The great majority of people with psychological disorders, however, are nonviolent (Lamberg, 1998). Substance or alcohol abuse, and a history of criminal behavior are much more strongly tied to violent crimes than are mental disorders (Bonta, Law, & Hanson, 1998).

On the other hand, evidence points to an increased risk of violence associated with some mental disorders, such as schizophrenia, especially during times of active hallucinations and delusions (Hodgkins et al., 1998; Link & Stueve, 1998). Most of the violent acts committed by former patients occur within their networks of families and friends (Link & Stueve, 1998; Steadman et al., 1998). Factors such as substance abuse and poor adherence to medication in people with severe mental disorders (schizophrenia or bipolar disorder) are also linked to a higher risk of violent behavior (e.g., Steadman et al., 1998; Swartz et al., 1998; Tiihonen et al., 1997).

As many as one in two people with schizophrenia have either an alcohol or illicit drug dependence disorder (Miller & Brown, 1997; Ziedonis & Trudeau, 1997). People with these dual diagnoses are labeled *MICAs* (mentally ill chemical abusers). They sometimes engage in violent behavior when they stop taking their psychiatric medication and return to using alcohol or drugs. Alcohol and other drugs may contribute to violence by impairing behavioral controls over impulses. We should also note that violent behavior is sometimes a feature of antisocial personality disorder (see Chapter 9) and conduct disorder in children and adolescents (see Chapter 14). All told, however, only a small proportion of aggressive or violent behavior on our streets and in our homes can be attributed to diagnosable psychological disorders. We need to consider other explanations derived from biological, social-cognitive, and sociocultural perspectives to provide a fuller accounting of violent and abusive behavior.

Quiz **16.1**
Violence and Abnormal Behavior

Explaining Human Aggression

Are human beings basically aggressive by nature? Or is aggression learned behavior? There is no shortage of opinions among observers of the human condition as to the origins of human aggression. We have yet to come to any generally accepted theory of violence and aggression. Here, let us consider how several of the major theoretical perspectives have approached the problem. We also offer our own speculations on what these perspectives may teach us about our capacity to aggress and do violence against others.

Biological Perspectives

The animal world is not a peaceable kingdom. Animals in their natural habitat face a constant struggle to survive as either prey or predator. The classic biological view of

instinct A fixed, inborn pattern of behavior that is specific to members of a particular species.

sociobiology Biological perspective that explains psychological traits as behavioral tendencies that increased our ancestors' chances of survival and were passed down genetically.

aggression holds that it is a product of **instinct.** Instincts are fixed, inborn patterns of behavior that are specific to members of a particular species. We say that birds instinctively build nests and salmon instinctively return upstream to spawn. Classic instinct theory holds that predators possess an aggressive instinct that has a survival function. A predator is more likely to survive and be around long enough to pass along its genetic inheritance to its offspring if it responds "instinctively" by attacking its prey. Within species, aggression usually occurs among males and typically involves a way of establishing dominance or access to mates. The more aggressive males may thus have greater access to the most fertile females and be more likely to pass along their genetic inheritance to future generations, presumably including genes controlling aggressive behavior, than would their more placid brethren.

To what extent might human aggression be instinctual? An early proponent of the belief that human aggression is a product of instinct was Sigmund Freud. As he looked upon the destruction and devastation that human beings had wrought in the Great War (as World War I was called at the time), Freud came to believe there must be an underlying instinct that accounts for human aggression, which he dubbed the "death instinct." The "death instinct" was basically self-destructive in its aim, having as its ultimate purpose the return to the tension-free state that preceded birth. The death instinct can give rise to self-destructive behaviors, including suicide. Sometimes it is turned against others in the form of outward aggression, violence, and warfare. In Freud's view, instinctual impulses such as aggression and sex strive for expression and must be released in some form. The ego is responsible for finding socially appropriate outlets for channeling these impulses. Thus, aggressive impulses may be "vented" by participating in rough-and-tumble sports such as football or hockey or by seeking to "trounce" your opponent in tennis or bridge. They may also be expressed vicariously, such as by watching a violent movie or sporting event. The venting of aggressive impulses, which is termed *catharsis,* is believed to act as something of a safety valve, an acceptable way of "letting off steam." Outward aggression arises when the ego is unable to contain these destructive impulses, which can occur if the ego is weak or overtaxed.

Does aggression have a genetic component? Sociobiologists believe that aggressive tendencies that might have helped ancestral humans to survive might be part of our genetic inheritance.

Sociobiological Views Today, most scholars reject the instinct theory of human behavior in large part because it fails to account for the diversity that exists among humans and the importance of learning and culture in determining behavior. A newer biological perspective, called **sociobiology,** has emerged. Sociobiologists do not explain human aggression on the basis of instinct. Rather, they believe that we inherit behavioral tendencies or dispositions, including aggressive tendencies, that increased the chances of survival of our early ancestors and were passed down along the genetic highway all the way to us (Gaulin & McBurney, 2001; Goode, 2000b; Thornhill & Palmer, 2000). They believe that we inherit not only physical traits such as hair color or height, but also behavioral traits such as aggression, even though such traits may no longer be adaptive in modern civilization.

The period of time since the dawning of modern civilization is but a brief moment in our history as a species, a mere blink of the eye. Sociobiologists believe that aggressiveness may be a behavioral trait or disposition that had survival value to our earliest ancestors who eked out a bare-bones existence in groups of hunter-gatherers ordered along distinct gender roles in which men hunted and killed animals and women collected edible roots and shrubs. Sociobiologists see contemporary evidence showing that boys and men tend to be more aggressive than girls or women as consistent with this evolutionary perspective (Knight, Fabes, & Higgins, 1996). They may also see attraction to violence in contemporary media and in video games as a by-product of our aggressive inheritance.

Sociobiological or evolutionary viewpoints remain controversial. Critics claim that humans are not marionettes whose strings are pulled by invisible genetic masters. They argue that culture, learning, and personal choice are more important determinants of our behavior than genetics (Eagly & Wood, 1991). On the other hand, contemporary evolutionary models argue for an interactionist approach that posits that biological factors interact with social and environmental factors in creating conditions that lead to aggressive behavior (Buss & Shackelford, 1997). They believe there is ample room for both biology and culture in explaining human behavior. Here, let us consider evidence that may uncover the biological underpinnings of aggressive behavior.

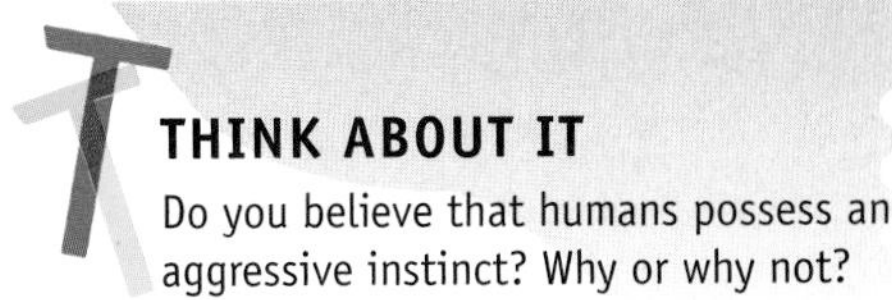

THINK ABOUT IT

Do you believe that humans possess an aggressive instinct? Why or why not?

Neurobiological Bases of Aggression Investigations of brain mechanisms in aggression have focused attention on the role of the hypothalamus. If we electrically stimulate certain parts of the hypothalamus of other animals such as rats and monkeys, we can elicit attack behavior or other stereotypical violent responses. This leads us to believe that the hypothalamus may act as a control center in regulating aggressive behavior, at least in other animals. However, the human brain is much more complex than that of other species and our behavior depends more on learning than on neural reflexes. Whatever brain mechanisms may be involved in regulating aggression in humans may be subject to the overriding influences of culture and learning.

Contemporary neurobiological research on aggression has focused largely on the role of neurotransmitters, especially serotonin and the male sex hormone testosterone (e.g., Virkkunen & Linnoila, 1993; Virkkunen et al., 1994).

Serotonin acts as an inhibitory neurotransmitter in some parts of the brain, especially the *limbic system,* a part of the brain involved in regulating primitive drives such as hunger, thirst, and aggression. The limbic system also plays key roles in learning, memory, and the regulation of emotions. Researchers suspect that serotonin helps put the brakes on primitive behaviors, including acts of impulsive aggression (Cowley & Underwood, 1998). In one study, men who had committed crimes involving acts of impulsive violence were found to have abnormally low levels of serotonin in their brains, whereas men who had committed violent criminal acts involving premeditation and planning had normal levels of serotonin (Toufexis, 1993). Serotonin activity has also been implicated in aggressive behavior in young boys (Pine et al., 1997). Dysfunctions in serotonin activity may also have implications for explaining the impulsive, aggressive behavior observed in people with certain personality disorders, such as borderline personality disorder (Siever & Trestman, 1993).

In animal research, investigators find that if you destroy certain parts of the brain involved in serotonin production, the level of aggressive behavior increases (Siever & Trestman, 1993). This can be observed in increased mouse-killing behavior in rats. Other research shows that more dominant male monkeys have higher levels of serotonin than lower-ranking males. A leading investigator, John Mann of Columbia University, calls serotonin a "behavioral seat belt" because of its role as a restraining mechanism in holding back impulsive behavior, including sexual and aggressive impulses (Bjork et al., 2000; Cowley & Underwood, 1998; Davidson, Putnam, & Larson, 2000). Yet other researchers are taking more of a "wait and see" attitude. They believe it is premature to make any definitive statements about the role of serotonin in human aggression (Berman, Tracy, & Coccaro, 1997).

Testosterone is also implicated in aggression, in part because men tend to be more aggressive than women (Buss & Kenrick, 1998; Segell, 2000). Although testosterone is produced in both men and women, the levels in men are much higher. Researchers find that teenage boys with elevated levels of testosterone are more likely to respond aggressively to provocations than their peers (Olweus, 1987). Others report that violent criminals have unusually high levels of testosterone (Virkkunen & Linnoila, 1993). High testosterone levels are also found among female prisoners who are high in aggressive dominance (Dabbs & Hargrove, 1997).

WWW **Web Link 16.1**
Youth Violence: A Report of the Surgeon General

Researchers also find cross-cultural evidence supporting a link between testosterone and aggressive behavior in adult men. A study of men from the !Kung San tribe of Namibia, Africa, showed that men with higher levels of testosterone were more violent than those with lower levels (Christiansen & Winkler, 1992). Because these studies are correlational in nature, we cannot conclude that testosterone plays a causal role in aggression. Yet it gives us reason to believe that high levels of testosterone may be a link in a chain leading to aggressive behavior. The picture is clouded, however, by evidence, admittedly preliminary in nature, linking lowered levels of testosterone to aggressive behavior in men ("Testosterone Wimping Out?," 1995). Although more research is needed on the links between testosterone and aggression in men, it is possible that either excesses or deficits of the hormone may be involved in mediating aggressive behavior in men.

More evidence of biological factors in human aggression comes from genetic studies of criminality and violent behavior. Are propensities toward violent behavior inherited? Evidence from both twin studies and adoptee studies points to a genetic contribution to criminality (Carey, 1992). However, not all criminals are violent; some commit property offenses. The heritability of violent criminality per se is less clear and remains to be more fully explored in future research.

Social-Cognitive Perspectives

Social-cognitive theorists such as Albert Bandura (1973, 1986) propose that aggression is learned behavior acquired in the same way as other behaviors. The roles of modeling and reinforcement are highlighted in the learning of aggressive behavior. Children may learn to imitate violent behavior they observe at home, in the schoolyard, or on television or in other media. If they are then reinforced for acting aggressively, such as by getting their way with their peers or earning peer approval or respect, their tendency to aggress may become stronger over time.

A child's exposure to aggressive models may begin in the home in the form of witnessing spousal violence or suffering violence firsthand in the form of physical abuse or punishment meted out by parents. One lesson the child may draw from such experiences is that violence in the context of interpersonal relationships is an acceptable way of getting others to do what you want them to do, or of punishing them when they fail to comply with your requests. Evidence shows that aggressive or violent children often come from homes where parents and other family members model aggressive behavior ("Risk Factors," 2000).

Childhood exposure to crime and violence is widespread, especially among urban youth. A study of urban high school students showed that 93% reported having witnessed a violent act and 44% had been victimized themselves (Berman et al., 1997). Not surprisingly, young people exposed to violence, or who have been victims themselves, are at greater risk of mental health problems, including PTSD, depression, anxiety, as well as trouble with teachers, than are their nonvictimized peers (Boney-McCoy & Finkelhor, 1996; Freeman, Shaffer, & Smith, 1996).

Psychologist David Lykken (1993) points to a lack of proper socialization as a root cause of violence in our society. He argues that the inability or unwillingness of parents, especially single parents, to socialize their children—to teach them right from wrong—leads children to become overly aggressive.

Social-cognitive theorists also incorporate roles for expectancies and competencies in explaining aggressive behavior. People who expect that aggressive behavior will produce positive outcomes are more likely to act in kind. People who lack competencies for solving interpersonal problems without resorting to aggression may be more likely to act aggressively in conflict situations.

Social-cognitive theorists also focus on the ways in which people interpret confrontational or conflict situations (Berkowitz, 1994). When people see other people's motives as hostile, they are more likely to act aggressively than when they form a more benign interpretation of the other's behavior. Evidence shows that young people who have problems with aggression tend to distort other people's motives. They tend to assume other people intend them harm when they do not (Crick & Dodge, 1994). In a similar way, men who commit date rape may misread the woman's expressed wishes, believing that the woman is merely playing "hard to get" when she resists his sexual overtures. Cognitive theorists also recognize that people are more likely to act aggressively when they magnify the importance of a perceived insult (e.g., Lochman & Dodge, 1994).

Web Link **16.2**
Raising Children to Resist Violence: What You Can Do WWW

Social-cognitive theorists argue against the catharsis view of aggression. The expression of aggression, even in controlled situations such as a sporting event, may not reduce the potential for aggression but actually increase it by providing additional opportunities for reinforcement.

Effects of Prior Abuse and Victimization Violence may also beget violence from one generation to another. Many people who engage in abusive or violent behavior were themselves abused as children. In one recent study, dangerously violent adolescents had higher levels of exposure to violence and victimization than nonviolent matched controls (Flannery, Singer, & Wester, 2001). In another study, teenage boys who were physically abusive toward their dating partners were more likely than nonviolent boys to have been maltreated during childhood (Wolfe et al., 2001).

Physically abused children often begin to show violent behavior early in childhood (Davis & Boster, 1992). Neglected children may be at even greater risk of becoming violent later in life. However, the pathway between child abuse (which includes physical abuse, neglect, and maltreatment) and later violent or abusive behavior is neither direct nor certain. Most studies show that the majority of abused children do not become delinquents or violent offenders (Widom, 1989a, 1989b). Child abuse may lead to other outcomes, such as withdrawal or self-destructive behavior, rather than outward aggression. We need to expand our knowledge of how other factors, such as exposure to violence in media, interact with child abuse to increase the potential for later violent behavior.

The intergenerational transmission of violent behavior may involve a modeling effect. If children are abused, or if they observe one parent battering the other, they may learn that violent behavior is an acceptable means of dealing with interpersonal conflicts. Children who are exposed to violence in the home or were abused themselves may fail to establish secure, loving attachments with their parents and a sense of empathy and respect for the feelings of others, which may set the stage for wanton acts of violent cruelty toward others. It is also conceivable that there is a genetic component underlying the intergenerational transmission of violent behavior.

Effects of Media Violence Another modeling effect is the influence of exposure to violence in the media, especially violence on television. It is estimated that the average child, who watches 2 to 4 hours of TV daily, will have seen on the TV screen some 8,000 murders, and 100,000 other acts of violence, by the time he or she completes elementary school (Eron, 1993). But does exposure to media violence contribute to violent behavior?

Relationships between media exposure to violence and aggressive behavior in children are complex and may cut both ways. More aggressive children may be more inclined to watch violent programming (DeAngelis, 1993). Yet most experts believe that exposure to violent media contributes to aggressive and violent behavior in children and adolescents ("Health Groups," 2000; Huesmann & Miller, 1994). Classic experiments by psychologist Albert Bandura and his colleagues showed that children imitated aggressive behavior they observed on television, even when the models were cartoon characters (e.g., Bandura, Ross, & Ross, 1963). Figure 16.2 illustrates how children imitate aggressive behavior of an adult model that was shown hitting a toy "Bobo" doll. In other laboratory-based studies, both children and adults are found to act more aggressively when exposed to violence on television or other media (DeAngelis, 1993). Evidence also points to increased aggressive behavior in boys and men following exposure to violent video games (Anderson & Dill, 2000).

Exposure to violence in the media may contribute to aggressive behavior in several ways (Eron, 1993; "Health Groups," 2000; Huesmann & Miller, 1994). It may prime or kindle aggressive thoughts or impulses. Repeated exposure to media violence may lessen inhibitions for using violence to resolve conflicts, especially when characters on TV or in the movies are rewarded for acting violently or are shown "getting away with it." Repeated exposure to media violence may also have an emotional numbing or habituating effect, leading viewers to become desensitized to acts of real violence and perhaps less sensitive to the fate of victims. Having said this, evidence suggests that exposure to violence on TV is clearly not as powerful a predictor of aggressive behavior in children as violence observed in the home, the schools, or the community (Gunter & McAleer, 1990).

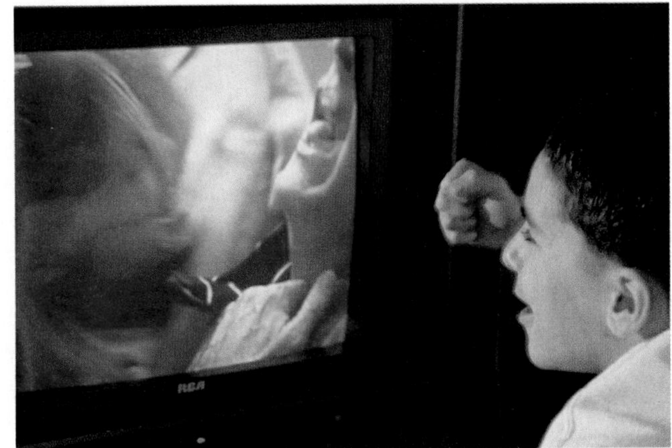

What's Johnny learning? Research evidence supports the view that exposure to violent media contributes to aggressive behavior in children. What mechanisms may account for these effects?

FIGURE 16.2 A classic experiment in the imitation of aggressive models. Research by Albert Bandura and his colleagues has shown that children frequently imitate the aggressive behavior that they observe. In the top row, an adult model strikes a clown doll. The lower rows show a boy and a girl imitating the aggressive behavior.

THINK ABOUT IT

Given the evidence provided in the text, do you think that a biological or a socio-cognitive explanation for violence is more persuasive? Explain your answer.

Parents may take a more active role in limiting their children's exposure to violent programming. Television shows carry labels indicating whether violent behavior is present, so parents can screen what their children watch. In addition, new television sets are equipped with V-chips that permit parents to block violent programming (Rutenberg, 2001). Parents can also impress upon children the differences between violence in the media and real violence. They can inform them that media violence is staged and is not real, that television actors, unlike real victims, can simply wipe away fake blood. They can help children understand how the media sensationalizes or glamorizes violent behavior and that most people in real life deal with conflict situations through peaceful means.

Sociocultural Perspectives

Web Link 16.3 WWW
Children and Television Violence

From the sociocultural perspective, violent behavior is rooted in underlying social causes, many of which go hand in hand, such as poverty, lack of opportunity, family breakdown, and exposure to deviant role models. Social stressors such as prolonged unemployment also play a role. Data drawn from the Epidemiologic Catchment Area (ECA) study showed that people who were nonviolent and employed when first interviewed but were later laid off were nearly six times more likely to engage in violent behavior than others who remained employed (Catalano et al., 1993). In countries such as Japan that are characterized by social cohesiveness, strong family ties, and a relatively balanced distribution of wealth, violent crime is but a small fraction of what it is in the more fragmented U.S. society (Kristof, 1995). Children in the United States who grow up in economically disadvantaged, inner city neighborhoods show higher levels of aggression than do less disadvantaged children (Guerra et al., 1995). In helping to explain the poverty-aggression connection, let us note that poorer children are generally exposed to greater levels of life stress, including stress associated with exposure to neighborhood violence. They also tend to be more accepting of aggressive behavior (Guerra et al., 1995). Children who are exposed to more life stress and those who adopt more accepting beliefs concerning aggression tend to behave more aggressively than do other children.

Sociocultural roots of violence. According to the sociocultural perspective, violent behavior is rooted in social ills, such as poverty, family breakdown, and social decay, that give rise to deviant subcultures, such as gangs.

Another factor linking poverty and violent behavior is the gang subculture. In America's poorer neighborhoods, a subculture of violence has sprung up, organized around deviant peer groups or gangs, in which protecting "turf" and proving one's manhood by taking up the gun or the knife has become the social norm. To young people who feel alienated from society and who hold little or no hope for their future, the sense of belongingness and acceptance that comes from joining a gang can be a tempting lure that may be difficult to resist, especially for those youngsters from disintegrated or disrupted families that are unable to provide them with support and moral guidance.

Sociocultural theorists also examine the role of violence as a social influence tactic. Violence, or the threat of violence, may be viewed as a form of coercion used to get people to comply with one's wishes, whether it involves strong-arm tactics of organized crime "enforcers" or abusive spouses who demand that their partners accede to their demands.

Sociocultural perspectives on violence also consider how cultural values and methods of child rearing have a way of breeding violence. In other cultures, such as in Thailand and Jamaica, aggression in children is actively discouraged and politeness and deference is fostered (Tharp, 1991). By contrast, our culture socializes children to be competitive and independent. Cultural differences may explain findings that children in Thailand and Jamaica are more likely than U.S. children to be "overcontrolled" and to complain of sleeping problems, fears, and physical problems, whereas children in the United States are more likely to be "undercontrolled" and perceived by others as argumentative, disobedient, and belligerent (Tharp, 1991).

Homicide Rates and Ethnicity Homicide is more prevalent among racial and ethnic minority groups (Council of Economic Advisors, 1998). We shouldn't be surprised by these differences, as homicide and other violent crime is more common among people at lower socioeconomic levels (Parker & Pruitt, 2000). Traditionally disadvantaged minorities in the United States are disproportionately represented at the lower rungs of the economic ladder.

Although African Americans constitute about 13% of the population, about 50% of the murder victims in the United States are African American (U.S. Department of Justice, 1994). Homicides committed by or against young African Americans are especially high. The chances of African American men in the 15 to 34 age range becoming victims of homicide are about nine times greater than the chances of non-Hispanic White American men in the same age range. The proportion of homicides committed by young men is more than seven times greater among African Americans (85.6 per 100,000) than non-Hispanic White Americans (11.2 per 100,000). The odds of becoming a victim of homicide are almost as great among young Hispanic American men as young African American men (Tardiff et al., 1994). In the great majority of homicides in the United States, the victim and the perpetrator were of the same race and knew each other. Blacks tend to kill Blacks; Whites tend to kill Whites.

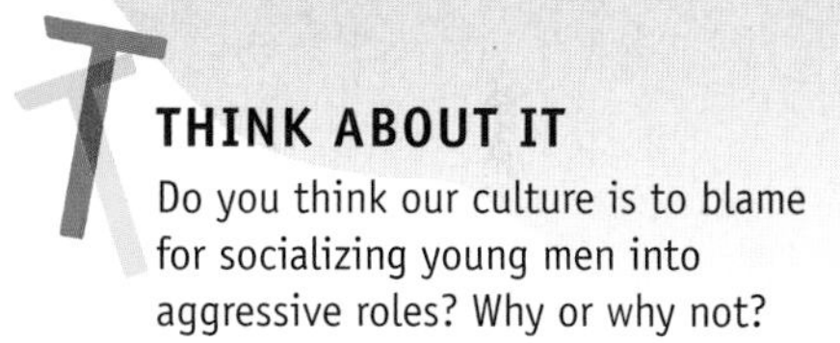

THINK ABOUT IT

Do you think our culture is to blame for socializing young men into aggressive roles? Why or why not?

We need to consider other factors in explaining ethnic differences in homicide rates besides socioeconomic status. Another second factor is the proliferation of firearms, especially in poorer, predominantly minority communities. Yet another factor is illicit drug use, which is often linked to street crime, especially in poorer communities.

Alcohol and Aggression

Alcohol use is implicated in nearly 40% of violent crimes, including more than 60% of homicides, at least 25% of serious assaults, and more than 25% of rapes (Collins & Messerschmidt, 1993; Gordis, 1999; Martin, 1992). The U.S. Justice Department estimates that more than one of three adult offenders had been drinking just preceding the commission of their crimes (Cable News Network, 1998).

A cross-cultural study of violent crime data from 11 countries in different parts of the world showed that in nearly two of three violent crimes overall, the perpetrator had been drinking at the time the crime was committed (Murdoch, Pihl, & Ross, 1990). The risks of homicide, suicide, and violent death are greater among alcohol and illicit drug users than nonusers (Rivara et al., 1997). Even people living with alcohol or drug users who

Truth OR Fiction? REVISITED

Though alcohol use is linked to aggression, it is rarely involved in homicides.

FALSE. Alcohol use is implicated in more than 60% of homicides and plays a role in many cases of other violent crimes.

don't drink or use drugs themselves are at greater risk of being killed than are people living in drug-free households (Wren, 1997a).

Although linkages between alcohol and aggressive behavior are correlational in nature, a growing body of experimental findings points to alcohol playing a causal role in both verbal and physical aggression (e.g., Giancola & Zeichner, 1997; Ito, Miller, & Pollock, 1996). Several factors may be involved in explaining alcohol's effects. For one thing, alcohol has certain cognitive effects, such as impairing decision-making ability. We may be less able to weigh the consequences of our actions when we have been drinking, which in some circumstances may lead us to act aggressively with little if any regard for the consequences that may result. Alcohol use may also loosen inhibitions or restraints (a process called *disinhibition*) that normally curtail impulsive behavior, including acts of impulsive violence. Alcohol also has a relaxing effect and may make the person less sensitive to anxiety-arousing cues relating to potential punishment that might ordinarily serve to inhibit aggressive behavior. Perhaps, too, people may act aggressively when drinking because of their expectations about the effects of alcohol, rather than because of its biochemical properties per se. Alcohol and other drugs may also make it more difficult for people to perceive the motives of others accurately, leading them to perceive a malevolent intent in other's behavior that can trigger a violent response.

Not everyone who drinks becomes aggressive, however. Relationships between violent behavior and alcohol or other drug use are complex and may be moderated by a number of factors, including the dosage level and the user's biological sensitivity to the drug's effects, the user's relationship to the victim, the setting of the encounter, as well as other situational, individual, and sociocultural factors. For example, some people become more violent when they drink than do others, and violent behavior tends to occur more often when people drink in some situations, such as sporting events, than in others. Differences in the levels or actions of the neurotransmitter serotonin may also be involved. Researchers have linked tendencies to become violent under the influence of alcohol to low levels of serotonin in the brain, which suggests a deficiency of serotonin may lower the threshold for violent behavior following alcohol use (Virkkunen & Linnoila, 1993). All in all, it appears that the interplay of alcohol and violent behavior involves complex relationships between the chemical effects of alcohol on the brain and environmental cues, which under certain circumstances may lead to violent behavior.

Emotional Factors in Violent Behavior

Emotional factors, especially frustration and anger, often figure prominently in aggressive behavior. Frustration is the emotional state associated with the thwarting or blocking of one's attempt to achieve a goal. According to the classic *frustration-aggression* hypothesis, frustration always produces aggression, and aggression is always a consequence of frustration. Consider the following example. Let's say you attend a movie but are unable to enjoy it because someone sitting in front of you is constantly talking and says you should be the one to move if it bothers you so much. You may feel frustrated because your goal of enjoying the movie is thwarted, but will you attack the person you hold responsible? Perhaps, but perhaps not. We've come to recognize that although frustration often plays a role in aggression, it may lead to other responses than aggression. In the preceding example, you might leave the theater or complain to the manager rather than instigating a fight. Moreover, aggression often has other causes than frustration. For example, aggression may involve a response to a direct provocation or retaliation for perceived wrongdoing. Aggression in response to frustration is more likely when it induces anger and when the person blames the other person for being responsible for the situation.

Anger is often a catalyst or instigator of violent or aggressive behavior. A husband strikes his wife in a "fit of anger" when he feels frustrated that dinner is not waiting for him when he returns from work. The child abuser lashes out in anger when the child fails to comply quickly enough with his or her demands. In the schoolyard, a slight provocation is blown out of proportion, eliciting an angry response that quickly escalates into a physical confrontation.

Problems with controlling anger figure prominently in personality disorders, especially borderline and antisocial personality disorders. People with borderline personality disorder often show a lack of control over their anger, having frequent temper outbursts or displays of impulsive, aggressive behavior directed at themselves or others. People with anti-social personalities may channel feelings of resentment and anger against family or society in general into violent or aggressive behavior. But anger management is not only a problem for people with personality disorders. In Chapter 5, we noted how problems related to anger and hostility are implicated as risk factors in cardiovascular disorders. We saw in Chapter 14 how children and adolescents with conduct disorders often have problems controlling their anger and need to learn anger coping skills, such as calming self-talk, to handle conflict situations without resorting to violent behavior. Cognitive-behavior therapists have made important inroads in helping people with anger management problems, as we discuss further in the nearby "A Closer Look" section.

Tying It Together

Human aggression is a complex behavior that cannot be explained by any single cause. Although some aggressive displays in other animals may be a product of "instinct," instinct theories fail to account for the variation in aggressive behavior in humans. A terrorist bombing, a mugging on a street corner, organized warfare, and a husband slamming his wife against the wall are all considered forms of aggression, but the motives that underlie them reflect different political, social, and psychological factors.

Although we cannot reduce human aggression to the level of instinct, increasing evidence points to roles for biological factors in human aggression. Biology may have more to teach us about the more impulsive forms of violent behavior, such as explosive acts of rage, than the more calculated forms of violence, such as reprisals for perceived wrongdoing or premeditated, violent crimes. Biological factors may also play a more direct role in lowering the threshold for violence in people who use alcohol or other drugs.

We also need to consider roles for culture and learning. Violent behavior is practically unknown in some cultures, but all too common in others, including unfortunately our own. Children are exposed to cultural attitudes that legitimize certain forms of violence. Young boys learn from peer influences and from television and movie role models that conflicts are settled with fists or weapons rather than words. Cognitive factors such as ways of interpreting provocations, expectations that violence will lead to positive outcomes, and tendencies to practice angering self-statements in conflict situations help account for individual differences in aggressive behavior.

Consider one possible causal pathway involving multiple factors leading to impulsive violence. Young people who are exposed to aggressive role models in the home and the community may learn that violent behavior is an acceptable way to respond to conflict, perhaps even the expected way. They may lack resources for handling stress or channeling anger in more constructive ways. We can further speculate that a biological predisposition for impulsive, violent behavior, perhaps mediated by a serotonin imbalance, may further increase the potential for violence in people exposed to these learning experiences. Alcohol use may also enter the mix of factors that raise the potential for violence, perhaps even to a greater extent in people with these biological vulnerabilities.

What can be done to prevent violence? One illustrative example comes from the state of Washington, where school children are exposed to nonaggressive ways of handling problem situations. The curriculum focuses on the following skills:

1. *Empathy training,* which helps children identify their own feelings and become more aware of other children's feelings;
2. *Anger management training,* in which children are taught coping skills to control anger;
3. *Impulse control training,* in which students learn problem-solving skills to handle problem situations.

The program has had some success in reducing physically aggressive behavior and fostering more appropriate social behavior. What can each of us do in our own lives to counter violence?

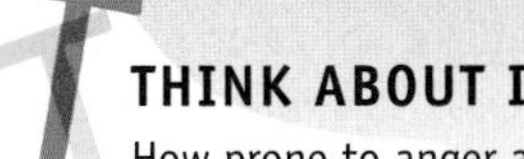

THINK ABOUT IT

How prone to anger are you? To violence? What factors make it more or less likely that you will act out aggressively or violently in a given situation?

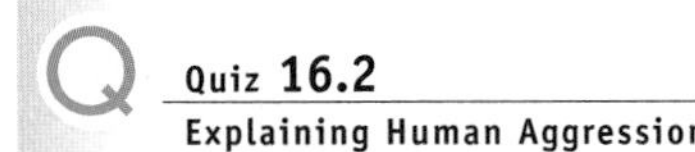

Quiz **16.2**
Explaining Human Aggression

A Closer Look

Anger Management

Cognitive-behavioral therapists assist people with anger control problems by helping them identify and correct anger-inducing thoughts they experience in situations in which they are provoked by others. Clients are taught to scrutinize the fleeting thoughts they have in confrontative situations and to recognize the cognitive distortions underlying these thoughts, such as tendencies to personalize a stranger's rudeness as a personal affront and demanding that others live up to their expectations.

Psychologist Ray Novaco (1974, 1977) developed a treatment program for anger control based on stress inoculation therapy, a type of cognitive-behavioral treatment that helps people manage anticipated stressors by developing new coping skills, such as self-relaxation, and by countering disruptive thoughts that occur in confrontative situations (e.g., "Who does he think he is? I'll show him!") with more adaptive self-statements (e.g., "Relax, don't get steamed up. The guy's just a jerk."). Like a vaccine that inoculates you against a virus by exposing you to an inert variant of the microbe, stress inoculation therapy exposes participants to a managed "dose" of the stressor by having them imagine themselves keeping their "cool" in confrontative situations by practicing coping responses and adaptive self-statements. Other cognitive-behavioral approaches, such as problem-solving therapy, are also used to help people with anger management problems generate alternative, nonviolent solutions to conflict situations. A key facet in these treatment programs is helping people rethink provocations as problems to be solved rather than as threats demanding an aggressive response.

Anger management training programs have been used with a number of different groups, including adults with anger control problems and violent youth. Treatment programs for violent youth often require a broader, multifaceted approach that involves cognitive, behavioral, and emotional components (Davis & Boster, 1993; Deffenbacher et al., 1996). The cognitive component helps participants rethink minor provocations so they don't flare out of control and to use calming self-talk in angering situations. The behavioral component may include training in relaxation skills to help participants calm themselves down in angering situations and in social problem-solving skills, which helps them develop alternative ways of coping with confrontative situations. The emotional component may involve the opportunity to receive emotional support within the context of supportive group therapy sessions.

Anger and aggression. Anger is often an instigator of aggressive behavior. What techniques do cognitive-behavioral therapists use to help people control anger and manage frustrating or confrontative situations without resorting to violence?

Coping with Frustrating or Confrontational Situations

What we tell ourselves about the motives that underlie other people's behavior can increase our arousal and prompt an aggressive response. Let us suggest a few coping responses that may help you tone down your response to frustrating or confrontational situations:

1. *Attend to your internal states of arousal.* When you feel yourself getting "hot under the collar," tell yourself, "Stop and think." Take stock of any angering thoughts you may have.
2. *Pause for a moment to consider the evidence.* Are you taking the situation too personally? Are you overreacting to it? Are you jumping to conclusions about the other person's motives? Are there other ways of viewing the person's behavior, other than as a personal affront?
3. *Practice adaptive self-statements,* such as "I can deal with this. Easy does it."
4. *Practice a competing response to anger, such as calming mental imagery, or meditative or self-relaxation exercises.* Or disrupt an anger response by taking a walk around the block, watching TV, or reading. Or, to paraphrase Mark Twain, count to 10 when you're feeling angry. If that doesn't work, count to 100. The time-honored

Domestic Violence

Web Link **16.4**
Controlling Anger—Before It Controls You WWW

More than 2 million women in the United States are severely beaten by their husbands each year (Koss et al., 1994). About one in eight husbands has committed an act of spousal violence (Holtzworth-Munroe, 1995). More than 1,000 women annually in the United States

technique of counting to 10 interrupts the tendency to respond impulsively to provocations. It gives you extra time to collect your thoughts, and even more importantly, to diffuse any self-angering thoughts with some rational counters: "Hey, calm down. This is not worth getting bent out of shape. Stay cool."

5. *Counter anger with empathy.* Rather than saying, "What a despicable person he is to act that way," think, "Maybe he's having a rough day," or, "She's just jealous of me and is acting out like a child." Or, "He must be a very unhappy person to act that way." Or, "She may have reasons—or thinks she has reasons—for acting this way. Anyway, that's her problem, not mine. Don't take it so personally. Better to focus on solving this problem rather than getting steamed."
6. *Think through alternative, nonviolent solutions to the problem or situation and formulate a plan of action.*
7. *Give yourself a mental pat on the back for coping assertively, not aggressively, in the situation.*

Table 16.1 offers examples of angering self-statements and rational alternatives for some common confrontational situations.

TABLE 16.1 Anger Management: Substituting Rational Responses for Angering Self-Statements

Situation	Angering Self-Statement	Rational Alternative
A provocateur says, "So what are you going to do about it?"	"That jerk. Who does he think he is? I'll teach him a lesson he won't forget!"	"He must really have a problem if he acts this way. But that's his problem. I don't have to act at his level."
You get caught in a monster traffic jam.	"Why does this always happen to me? I can't stand this."	"This may be inconvenient, but it's not the end of the world. Don't blow it out of proportion. Everyone gets caught in traffic every now and then. Just relax and listen to some music."
You're waiting behind someone in the checkout line at the supermarket who has to cash a check. It seems like it's taking hours.	"He (she) has some nerve holding up the line. It's so unfair for someone to make other people wait. I'd like to tell him off!"	"It will only take a few minutes. People have a right to cash their checks in the market. Just relax and read a magazine off the rack while you wait."
You're cruising looking for a parking spot when suddenly another car cuts you off and seizes a vacant parking space.	"No one should be allowed to treat me like this. I'd like to punch him out."	"Don't expect other people to always be considerate of your interests. Stop personalizing things." "Relax, there's no sense going to war over this."
Your spouse or partner comes home several hours later than expected, without calling ahead.	"It's so unfair. I can't let him (her) treat me like this."	"Make it fair. Explain how you feel without putting him (her) down."
You're watching a movie at the theater and someone sitting next to you is talking throughout the picture.	"Don't they have any regard for other people's rights? I'm so angry with these people I could tear their heads off."	"Even if they're inconsiderate it doesn't mean I have to get angry about it or ruin my enjoyment of the movie. If they don't quiet down when I ask them, I'll just change my seat or call the manager."
A person insults you or treats you disrespectfully.	"I just can't walk away from this like nothing happened."	"Of course you can. When did anger ever settle anything? There are better ways of handling this than getting steamed."

die as the result of beatings by partners in intimate relationships (Rennison, 2001). Women stand a greater chance of being attacked, raped, injured, or killed by their current or former male partners than by other types of assailants (Koss et al., 1994). Yet it may surprise you to learn that evidence from community samples in the United States and other countries, such as New Zealand, show that women are as likely as men are, if not more likely, to

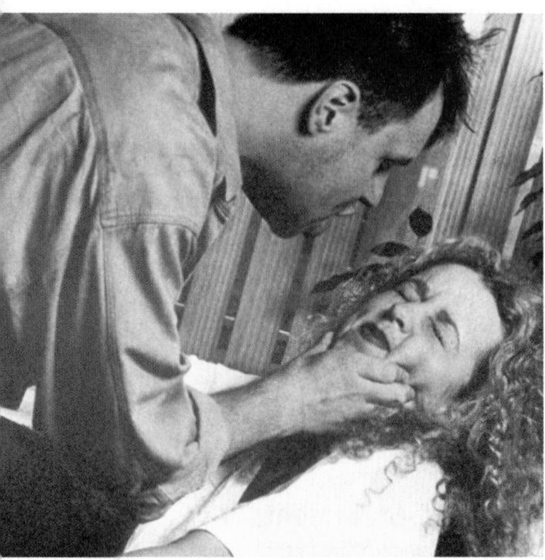

Spouse abuse. About one in four couples in the United States report incidences of domestic violence. Though both spouses may abuse one another, women suffer a much greater incidence of severe abuse at the hands of their partners and resulting physical injury and psychological effects.

commit acts of violence against their partner (Magdol et al., 1997; Margolin & Burman, 1993). In about half of the couples in which partner abuse occurs, both partners engage in acts of physical abuse against one another. However, the sheer frequency of battering does not take into account differences in the severity of the abuse. Women are much more likely than men to suffer severe abuse and to sustain physical injuries, including severe injuries such as broken bones and damage to internal organs (O'Leary, 1995). For this reason, our focus is on the male batterer.

Different motivations may be involved in accounting for partner violence in men and women. One hypothesis gaining interest is that men tend to attack while women tend to react. That is, male violence toward women may stem from factors that threaten their traditional position of dominance in relationships, such as unemployment and drug abuse. Violence perpetrated by women may arise from the stress of coping with an abusive partner (Magdol et al., 1997).

Although domestic violence cuts across all strata of our society, it is most commonly reported among people of lower socioeconomic levels. This may reflect a greater level of stress experienced by people who are struggling financially. It may also reflect a reporting bias, a tendency for upper income groups to use personal physicians who may be less inclined to report incidents of domestic violence than health care providers in public facilities who tend to serve less affluent groups (Margolin & Burman, 1993). Some evidence suggests that it is income disparity between husband and wife, with the wife earning more than the husband, that contributes to wife abuse, not poverty per se (McCloskey, 1996).

Ethnic or racial group differences appear unrelated to the risk of domestic violence when we take into account socioeconomic level and other sociodemographic factors (Margolin & Burman, 1993). The lack of opportunity and alienation from the larger society experienced by many economically disadvantaged people, including many ethnic minority group members, increases the potential for mental health problems and violence and abuse within domestic relationships (Nelson et al., 1992).

THINK ABOUT IT

Relate what you learned about causes of violence in the previous section to the information just presented about psychological patterns of spouse abusers. Which causes seem most important?

Psychological Characteristics of Male Batterers

Although there is no one psychological profile of male batterers, they tend to show higher levels of anger, hostility, impulsivity, verbal aggression, early problem behavior, and antisocial and borderline personality traits than nonbatterers (e.g., Holtzworth-Munroe, Rehman, & Herron, 2000; Murphy et al., 2001). Male batterers also tend to externalize or minimize blame for their actions and feel inadequate or dissatisfied with themselves (Flournoy & Wilson, 1991). Other characteristics associated with increased risk of battering include youthful age, lower income and occupational and educational status, high levels of stress, problem behaviors during adolescence, lack of assertive self-expression, exposure to parental violence in childhood, being the victim of physical abuse by one's mother during teenage years, and alcohol use, especially heavy use (e.g., Feldman, 1997; Magdol et al., 1998).

Patterns of Abuse

Battering typically occurs within a larger pattern of abuse involving physical abuse of children and sexual abuse of spouses. Battering often starts before marriage vows are taken, and although it may involve milder forms of aggression at first, such as pushing and grabbing, it may escalate if nothing is done to stop it (Holtzworth-Munroe, 1995).

Relationship factors, especially marital conflict, also play a role in initiating and maintaining spousal abuse. Moreover, as compared to nonviolent couples, physically aggressive couples show poorer problem-solving skills (Anglin & Holtzworth-Munroe, 1997). They also focus less of their efforts on solving their problems while showing more negative interactions in which angry responses from one spouse beget angry responses from the other (Margolin & Burman, 1993).

Spousal violence often follows an event that serves as a trigger for the abuser to lose control (Ryan, 1993). The triggering event may involve criticism or rejection from the spouse or incidents that lead the man to feel trapped, insecure, or threatened (Bitler, Linnoila, & George, 1994). The use of alcohol or other drugs further raises the risk that such events will lead to a battering episode. An analysis of some 62 episodes of domestic violence showed that about 90% of the assailants reported using alcohol or other drugs on the day of the assault (Brookoff et al., 1997).

Male batterers often have low self-esteem and a sense of personal inadequacy (Murphy, Meyer, & O'Leary, 1994). They may become excessively dependent on their wives for emotional support and feel threatened if they perceive their partners becoming more independent or developing separate interests from their own. Spousal violence may represent an inappropriate way of responding to this emotional threat. Abusers also tend to have poor problem-solving skills in handling conflict situations with their spouses (Else et al., 1993), which may explain (although not condone) why they turn to the use of physical force when a triggering event occurs.

Sociocultural Viewpoints

Writing from a sociocultural perspective, feminist theorists consider domestic violence to be a product of the differential power relationships that exist between men and women in our society. Men are socialized into dominant roles in which they expect women to be subordinate to their wishes (Koss et al., 1994). Cross-cultural evidence shows that the husband's need for control is an underlying contributor to wife abuse (Wilson & Daly, 1996). Men also learn that aggressive displays of masculine power are socially sanctioned and are even glorified in some settings, such as on the athletic field. These role expectations, together with a willingness to accept interpersonal violence as an appropriate means of resolving differences, creates a context for spousal abuse in situations in which the man's sense of control is threatened when he perceives his partner as failing to meet his needs or respect his wishes. Men who batter may also have less power in their relationships and may attempt to make up for their lack of power by using physical force.

Feminist theorists further point out that domestic violence exists because our society condones it (Gardiner, 1992). The man who beats his wife may be taken aside and "talked to" by a police officer rather than arrested on the spot. Even if he is arrested and convicted, his punishment is likely to be less severe (sometimes just a "slap on the wrist") than if he had assaulted a stranger.

Effects of Domestic Violence

In addition to the risk of physical injury, domestic violence can lead to posttraumatic stress disorder (PTSD) and other psychological effects, especially depression and low self-esteem (Cascardi et al., 1995; O'Leary, 1995; Shalev, Yehuda, & McFarlane, 2000; Watson et al., 1997). In one study, more than three fourths of battered women presented with evidence of diagnosable PTSD (Watson et al., 1997). Typically, the more severe the abuse, the greater the level of psychological distress. Battering may also be a contributing factor to alcohol or substance use disorders. Domestic violence may also lead the abused woman to flee the abusive situation, which in the absence of other resources may lead to homelessness. For the male abuser, battering may lead to a severing of family ties that may also ultimately lead to homelessness.

Exposure to domestic violence also takes an emotional toll on the children (Graham-Bermann, & Edleson, 2001; Rossman, 2001). More than 3 million children in the United States witness violence between their parents each year (DeAngelis, 1995b). In a study of 62 incidents of domestic violence in Memphis, Tennessee, children witnessed 85% of the assaults (Brookoff et al., 1997). Witnessing interparental violence has direct negative effects on the child's emotional health and behavior, leading in many cases to depression and behavior problems. Even nonphysical forms of parental aggression, such as insults, threats, or kicking furniture, is associated with more behavioral and emotional problems in children (Jouriles et al., 1996). Parental modeling of spouse abuse may also set the stage for perpetuating an intergenerational pattern of abuse from generation to

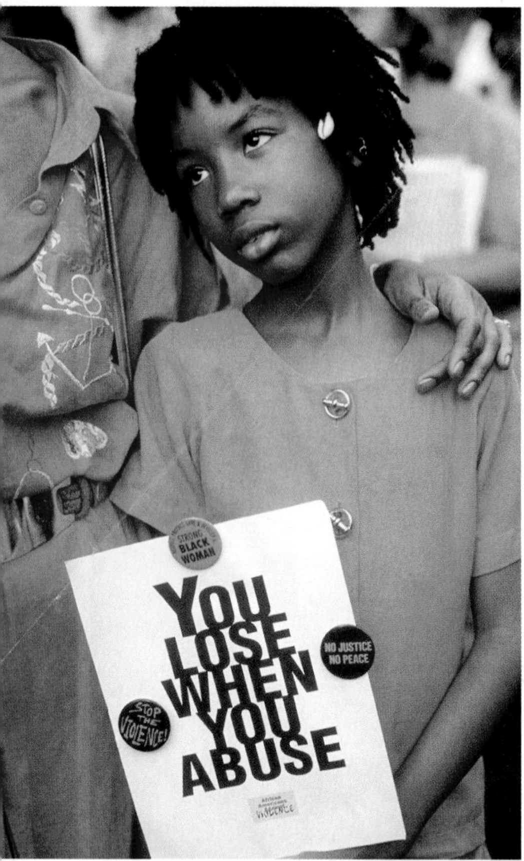

What about the children? Children too are affected by spousal abuse. Children exposed to interparental abuse may become depressed or develop behavioral problems. Childhood exposure may also set the stage for the perpetuation of spousal abuse from one generation to another. Many children are also beaten by the abusers.

generation. Clearly, it sets a poor example by modeling the use of violence as an acceptable means of resolving interpersonal conflicts. Many abusers, as well as many battered spouses, also have histories of being emotionally or physically abused during childhood. Moreover, domestic violence often occurs in a context in which children are physically abused by the offender as well (O'Leary et al., 2000). The greater the frequency of domestic violence, the more likely that children too will suffer physical abuse at the hands of the violent father or mother (Ross, 1996).

Why Don't Battered Women Just Leave?

About half of the abused women who seek professional assistance return to their abusive husbands or partners (Strube, 1988). Why do women remain in abusive relationships or take back abusive partners after a separation? Are they driven by unconscious, masochistic needs to punish themselves?

Professionals who work with abused women argue that women in abusive relationships are better understood as trauma survivors than as masochists (Strube, 1988). Battered women have much in common with other trauma survivors, such as former political hostages and victims of kidnapping. Leonore Walker (1979) coined the term *battered woman syndrome* to describe the traumatizing effects of battering, which includes feelings of helplessness and impaired coping ability that can make it difficult for the battered woman to leave the abuser and establish a new life on her own. Compounding their problems, many battered wives lack the economic means to establish an independent household of their own for themselves and their children. Consequently, many fear becoming destitute if they leave the batterer.

Spousal violence may exact a greater emotional toll than other forms of trauma because the agent of abuse is someone whom the woman trusted and loved and with whom she may need to continue a relationship, especially if children are involved. Betrayal of trust may intensify the woman's emotional reactions and further impair her ability to cope.

A review of the literature on the effects of spousal violence identifies several psychological factors that may affect the battered woman's ability to cope effectively (Follingstad, Neckerman, & Vormbrock, 1988):

1. *Shattering the myth of personal invulnerability.* Many people maintain an illusion that they are somehow immune to the traumas or tragic events that befall others, such as violent crime, crippling traffic accidents, or fatal cancers. This illusion, which helps maintain a sense of personal security, may be shattered in women exposed to spousal abuse, leaving them feeling vulnerable. If a woman is brutalized by the man in whom she placed her trust, she may come to think that other bad things will also befall her. The shield of invulnerability cracks. Her sense of safety and security in her own home that most of us take for granted may be destroyed.
2. *Reduced problem-solving ability.* Trauma may hamper the woman's ability to weigh alternative solutions to her problems. Survivors may become so preoccupied with the immediate problem of preventing recurrent beatings that they cannot focus on ways of making the larger life changes that will extricate them and their children from the marriages. Women may even come to believe that they cannot take control of events and that submission is the only realistic way of preventing further abuse. Some abused women sink to a state of despair in which they essentially "give up" trying and decide to return to the abuser rather than seek alternatives (Newman, 1993).
3. *Stress-related reactions.* Battered women may experience a form of posttraumatic stress disorder that can further impair their coping ability. Like soldiers who have had traumatic combat experiences, battered women may reexperience beatings in the form of nightmares, flashbacks, and intrusive images of abuse. They may become numbed to their environment, have sleep disturbances, feel anxious in the presence of cues or reminders of beatings, and avoid situations or stimuli connected with beatings. Anxiety and feelings of dread may lead the battered woman to withdraw from the outside world and cut off support from other people. Feelings of pessimism,

anger, guilt, and depression are also common, along with suicidal thoughts and attempts. Problems with alcohol or drug abuse may also develop, further impairing the woman's ability to function effectively.

4. *Thought conversion.* Hostages and kidnap victims who are held for a lengthy period of time sometimes experience a conversion in attitudes and come to regard their captors with positive or sympathetic feelings, at the same time developing negative feelings toward potential rescuers. In a similar way, battered women who are subjected to chronic abuse may come to view their tormenters in more sympathetic terms.
5. *Finding meaning in the abuse.* According to existential psychiatrist Viktor Frankl (1959), people have a basic psychological need to find meaning in their experiences, even in brutal abuse. Some battered women may try to find meaning in the abuse, or even justify it, by using rationalization. Rationalizations (e.g., "He didn't really mean it . . . it was really the alcohol . . . it was really my fault.") as well as the use of denial (e.g., "It wasn't really that bad.") tend to perpetuate abusive relationships.
6. *Learned helplessness.* Domestic violence can lead to feelings of helplessness, which can sap the woman's motivation to seek to overcome the trauma and make it less likely she will seek help to extricate herself from the abusive relationship (Walker, 1979; Wilson et al., 1992). Helplessness may develop as the abused woman finds that her repeated attempts to make changes in the relationship fail or when her requests for external help are met with frustrating "run-arounds" or lack of concern from criminal justice or mental health personnel. She may come to believe that nothing she can do will prevent the battering or extricate herself from the situation.
7. *Difficulties handling troubling emotions.* Battered women may have difficulty handling their anger toward the abuser, perhaps because they have learned, from his example, that anger is a dangerous or uncontrollable emotion (Carmen, Rieker, & Mills, 1984). Unexpressed anger may become redirected inward in the form of self-blame and self-loathing, which, in turn, can lead to feelings of resignation and depression and to self-destructive behaviors ranging from substance abuse to suicide attempts. Depression may further hamper the woman's ability to change her life.

We also need to take into account the cultural expectations placed on women. Women in many cultures are expected to adhere to a self-sacrificial ideal that a woman's role is to sacrifice her needs for the sake of her children and family. If she is abused, she may see her role as "suffering in silence" lest her response to the situation threaten family stability. Yet she may face a conflict between two conflicting moral values: (1) that "good women provide a strong family base for their children," and (2) that "people should not harm those they love" (Pilowsky, 1993). The turning point to taking effective action may come when she asserts as a stronger moral imperative the belief that self-respect and self-care are the moral rights of every woman.

In sum, battered women are trauma survivors, not masochists. Labeling them as masochists has the unfortunate effect of blaming the victim for the abuse. When we see battered women as survivors, rather than masochists, we supplant blaming them with an effort to understand the psychological factors that may lead them to feel trapped in their relationships. Such understanding may facilitate the development of programs to help abused women make hard decisions and counteract tendencies toward rationalization, self-hatred, and denial. Viewing the woman who suffers abuse as a survivor also shifts blame away from her and onto the batterer where it belongs.

Truth OR Fiction? REVISITED

Women who remain with men who abuse them suffer from a form of masochism.

FALSE. Battered women are trauma survivors, not masochists. Labeling them as masochists has the unfortunate effect of blaming the victim for the abuse.

Treatment of Batterers and Abused Partners

Before treatment can begin, the battering must stop. Only then can any therapeutic attempt at healing or reconciliation begin. Couples therapy or family therapy may be useful in treating couples and families with a history of domestic violence (O'Leary, 1995). In some cases, each spouse receives individual therapy along with marital or couples therapy. The therapist may help the couple understand rage as an expression of a sense of inner powerlessness and assist them to better understand each other's emotional pain and learn more productive ways of handling anger and resolving conflicts without resorting to violence (Mones & Panitz, 1994). In some cases, the relationship cannot be saved and the couple may need help in coping with the consequences of divorce.

Group therapy for male batterers may allow batterers to feel secure enough to express their inner feelings and be confronted by other group members if they avoid taking responsibility for their abusive behavior. Support groups for battered women are also available in many communities, often led by women who were formerly abused themselves. They provide mutual support and help battered women recognize the cycle of violence, develop escape strategies, weigh alternatives to marriage, enhance self-esteem, and decrease self-blame.

Abusers may be mandated by the courts to receive treatment as an alternative to incarceration. Unfortunately, evidence casts doubt on whether court-mandated treatment reduces future acts of domestic violence as compared to the deterrent effects of traditional criminal justice penalties (Rosenfeld, 1992). Another problem is that many abusers simply stop attending therapy, despite a court order requiring their attendance.

Quiz **16.3**
Domestic Violence

Child Abuse

Each year about one million children in the United States are identified as victims of child abuse (Eckenrode et al., 2000; Golden, 2000). More than 1,000 children in the United States die each year as the result of abuse or neglect. As horrific as these numbers are, they greatly understate the problem, as most incidents of child maltreatment are never publicly identified.

Child abuse cuts across all ethnic, racial, and national boundaries. A random sample of households in Ontario, Canada, showed that 31% of the adult men and 21% of the adult women reported a history of physical abuse in childhood (MacMillan et al., 1997). Although reliable statistics in less developed countries are often hard to come by, a survey in the African nation of Nigeria revealed a large percentage of children and adolescents reported having seen an abused or neglected child and believed that child abuse and neglect was widespread in their country (Ebigbo, 1993). Sad to say, child abuse may be universal, although particular rates of child abuse may vary across countries.

The child's own parents are the perpetrators of abuse in the great majority of cases. Although the mother is identified as the abuser in about 60% of cases of physical abuse (DeAngelis, 1995b), it's known that mothers assume a disproportionate share of child-care responsibilities and constitute the great majority of single-parent heads of households.

Child abuse does not arise out of a vacuum, but may be seen as progressing through a series of increasingly coercive behaviors by the abusive parent. Cognitive factors, such as blaming the child for the abuse (e.g., "If he didn't want to get hit, he should not have left his clothes lying around"), serve to justify the abuse and may contribute to the progression to more abusive behaviors. Abusive parents tend to see their children's misbehavior as intentional, even when it is not. They also rely more heavily on physical punishment and less on reasoning as a means of controlling their children than do nonabusive parents (Belsky, 1993). They often lack appropriate parenting and problem-solving skills for dealing with child behavior problems and have a low tolerance for demands made by children (Milner, 1993; Pogge, 1992). Yet, it does not appear that parents who physically abuse their children are more likely than nonabusers to have diagnosable psychological disorders (Pogge, 1992). Nor do they have any clearly identifiable psychological traits that set them apart.

Modeling of excessive use of physical punishment by parents may help explain intergenerational transmission of child abuse. When children are abused, or when they observe their parents resorting to violence against one another when stressed or angry, they may come to view violence as an acceptable response for handling conflict and child disobedience.

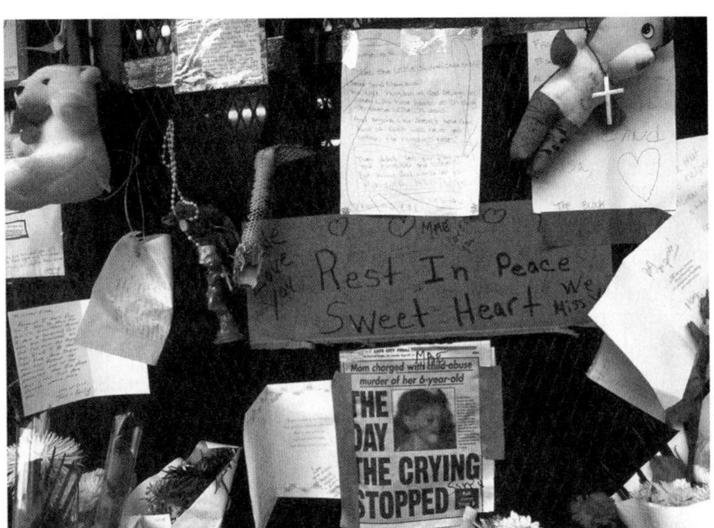

Child abuse. About 1 million children in the U.S. are identified as victims of child abuse each year. More than 1,000 die each year as the result of abuse or neglect at the hands of their parents or caretakers. Here we see a memorial organized by community residents in New York City after the death of a 6-year-old victim of child abuse.

TABLE 16.2 Types of Child Abuse

Type of Abuse	Definition
Physical abuse	Nonaccidental physical injury of a child caused by a parent or caretaker. The injuries may range from superficial bruises to broken bones, burns, and serious internal injuries and may result in death in some cases.
Physical neglect	Failing to provide children with, or withholding from them, adequate food, shelter, clothing, hygiene, medical care, or supervision needed to promote their growth and development.
Sexual abuse	The sexual exploitation of children involving acts ranging from nontouching offenses, such as exhibitionism, to genital fondling, sexual intercourse, or involving them in the production of pornography.
Emotional maltreatment	The use of constant harsh criticism of the child involving the use of verbally abusive language, or emotional neglect, which is characterized by the withholding of physical and emotional contact needed to promote normal emotional development and, in some extreme cases, physical development.

Source. Adapted from Alpert & Green (1992), pp. 228–229.

Types of Child Abuse

Child abuse includes several types of physical, sexual, and emotional maltreatment or neglect. A widely used set of definitions of the different types of maltreatment was developed by the New York State Federation on Child Abuse and Neglect (New York City Board of Education, 1984) (see Table 16.2). Despite these definitions, it can be difficult to determine where to draw the line between "acceptable" spanking or hitting and child abuse. Neglect is the most common form of abuse, representing nearly half (49%) of substantiated cases (Daro & Wiese, 1995). Physical abuse accounts for 21% of cases, sexual abuse for 11%, emotional maltreatment for 3%, and other forms for 16% (see Figure 16.3).

Mandatory reporting laws in all 50 states require mental health professionals, including psychologists, social workers, and marriage and family counselors, to notify child protective officials if they come to know of child abuse or have reasonable suspicions that abuse is occurring. Despite mandatory reporting laws, a great many cases of child abuse go unreported.

Truth OR Fiction? REVISITED

The most common form of child maltreatment is physical abuse.

FALSE. Neglect, not physical abuse, is the most common form of child abuse.

Risk Factors in Child Abuse

A number of parental factors are associated with an increased risk of child abuse, including stress, witnessing family violence in one's family of origin, being abused during one's own childhood, failure to develop an appropriate attachment to one's children, poor anger

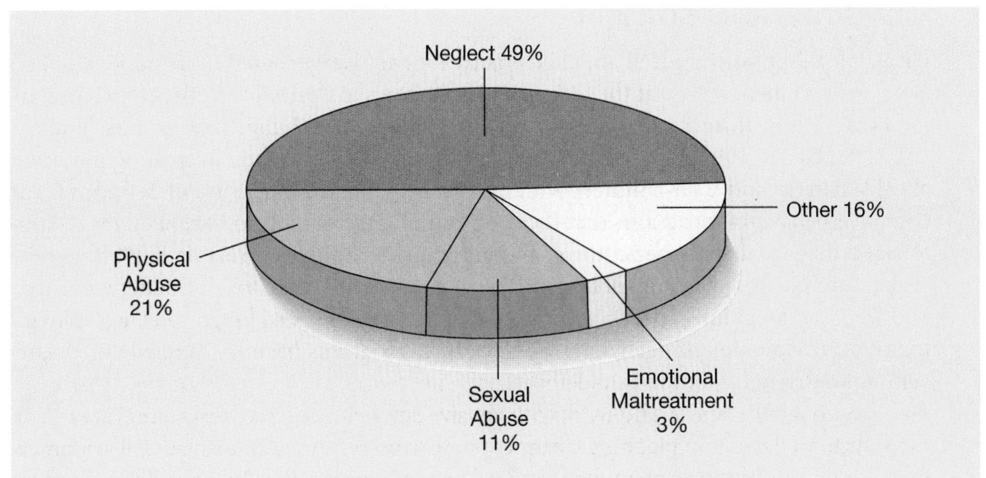

FIGURE 16.3 Substantiated cases of child maltreatment by type.
This figure shows the breakdown of substantiated cases of child maltreatment reported by child protective agencies in 36 states in the United States for 1993 and 1994. Child neglect was the most frequent type of child maltreatment, followed by physical abuse.

Source. National Committee to Prevent Child Abuse (NCPCA); Daro & Wiese (1995).

management skills, alcohol or substance abuse, holding rigid rules concerning child rearing, and acceptance of violence as a means of resolving conflicts and controlling children's behavior (Belsky, 1993; Pogge, 1992). Moreover, teenage parents, undereducated parents, and single parents are more likely than other parents to physically abuse their children (Christmas, Wodarski, & Smokowski, 1996; DeAngelis, 1995b). Investigators find no discernible difference in the risk of child abuse across ethnic groups or socioeconomic levels of parents when the parent's level of stress is taken into account (Pogge, 1992).

Stress, a major risk factor in child abuse in its own right, also underlies other risk factors such as poverty and single parenthood. Sources of parental stress include unemployment or job-related problems, medical problems, financial pressures, marital conflict, and living in an unstable and unsafe environment.

The stress of adapting to changing family structures is another risk factor in child maltreatment. A great many families in our society have disintegrated or are in the process of coming apart at the seams. For some parents, stressful demands of changes in family structure lead to feelings of powerlessness and lower self-esteem that, in turn, may lead to child neglect or to irrational, impulsive acts of physical abuse of children.

Effects of Child Abuse

The physical injuries suffered by physically abused children are disturbing and often tragic, ranging from welts and bruises to broken bones and massive internal injuries, which sometimes result in death. The emotional wounds of abuse and neglect may run even deeper and be longer lasting. Abused or neglected children often have difficulties forming healthy peer relationships and healthy attachments to others. They may lack the capacity for empathy or fail to develop a sense of conscience or concern about the welfare of others. They may act out in ways that mirror the cruelty they've experienced in their lives, such as by torturing or killing animals, setting fires, or aggressing against smaller, more vulnerable children.

Other common psychological effects of neglect and abuse include lowered self-esteem, depression, immature behaviors such as bed-wetting or thumb-sucking, suicide attempts and suicidal thinking, poor school performance, behavior problems, and failure to venture beyond the home to explore the outside world (Golden, 2000; Saywitz et al., 2000; Shonk & Cicchetti, 2001; Wolfe et al., 2001). Child abuse is also associated with an increased risk in later life of such psychological disorders as bulimia, dissociative identity disorder (multiple personality), PTSD, depression, substance abuse, and borderline personality disorder (e.g., Lipman, MacMillan, & Boyle, 2001; McCauley et al., 1997). In adulthood, abused or maltreated children are more likely to engage in criminal behavior, to be unemployed or to hold lower paying jobs, to have completed fewer years of education, and to have higher suicide rates as compared to nonabused controls (Widom, 1991).

Child Abuse Treatment

Given the scope of the problem, child maltreatment clearly represents a national emergency. Government efforts at the federal and state levels to deal with the problem have thus far been largely unsuccessful, despite the expenditures of billions of dollars annually (Alpert & Green, 1992). Nonetheless, some progress has been made in treating physically abusive parents and their children, with most gains reported in the use of behavioral and cognitive-behavioral programs that focus on training the parents in various skills relating to stress management, anger control, and parenting techniques (Wekerle & Wolfe, 1993). Parent training programs aim at helping the abusive parents learn to cope better with stress and improve their interactions with their children (DeAngelis, 1995b). Yet we still lack long-term follow-ups to determine whether such programs have lasting benefits in preventing recurrent episodes of child abuse.

Although therapists who work with abusive families recognize the importance of ethnic and cultural issues in planning treatment programs, research exploring the development of culturally sensitive treatment interventions remains lacking. We also know little about the effects of child abuse treatment programs that target the children themselves, in part because virtually all programs are geared toward treating the abusive parents (DeAngelis, 1995a). Also lacking are studies specifically targeting child neglect as distinguished from physical or emotional abuse, despite evidence that neglect can have even more damaging consequences than physical abuse (DeAngelis, 1995b). In fact, nearly half of the child fatalities due to maltreatment are the result of neglect, not physical abuse (Daro & Wiese, 1995).

Preventing Child Abuse

Much of the focus in preventing child abuse has centered on training new and expectant parents, and especially teen parents, in parenting skills. Although skills training approaches have been shown to increase parenting knowledge and to enhance child-rearing skills, we still await evidence from large well-controlled studies that such programs succeed in reducing the rates of child abuse in at-risk families (Wekerle & Wolfe, 1993). A major limitation in the research to date is that virtually all participants in these parenting programs are women; the needs of men who are potentially at risk of becoming abusive parents have yet to be addressed on any widespread basis.

Because stress associated with poverty is a major risk factor in child abuse, significant progress in the battle against child abuse and neglect may come only when we deter young people from having children until they are financially and psychologically prepared to raise children (Belsky, 1993; Christmas, Wodarski, & Smokowski, 1996). Yet child abuse is not limited to young parents or the poor. Finding solutions to the problems of child abuse will likely involve ways of helping people acquire the parenting and communication skills they need to be caring, effective parents, as well as adhering to a national policy of zero tolerance for child abuse. Parenting programs may need to be integrated within the high school curriculum, as driving education programs are today.

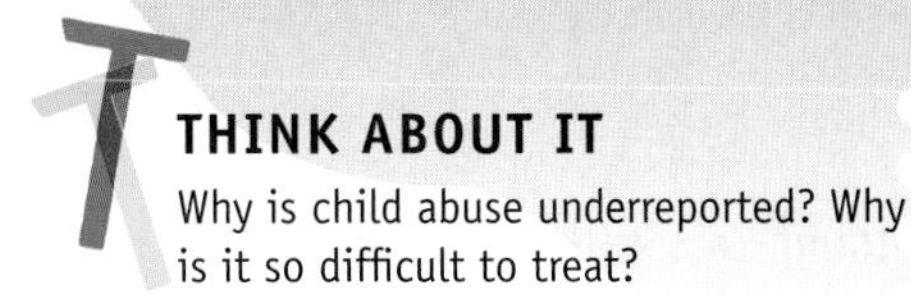

THINK ABOUT IT

Why is child abuse underreported? Why is it so difficult to treat?

Quiz **16.4**
Child Abuse

Sexual Aggression

We use the term *sexual aggression* to include both acts of outright sexual violence, such as rape or sexual assault, as well as sexual harassment, in which one person aggresses against another sexually by subjecting the other person to unwanted sexual overtures, demands, or lewd comments. Sexual molestation of children is also by definition a form of sexual aggression, even if no direct force is used to obtain sexual favors, because children by virtue of their developmental level are deemed incapable of providing informed consent.

Rape

Although rape is not a diagnosable mental disorder, it certainly meets several of the criteria used to define abnormal behavior. It is socially unacceptable, violates social norms, and is grievously harmful to its victims. Rape may also be associated with some clinical syndromes, especially some forms of sexual sadism.

Forcible rape is the use of force, violence, or threats of violence to coerce someone into sexual intercourse. **Statutory rape** is sexual intercourse with a person who is unable to give consent, either because of being under the age of consent or because of mental disability, even though the person may cooperate with the rapist.

Incidence of Rape The federal government reports about 153,000 completed rapes and about 67,000 attempted rapes are committed each year in the United States (U.S. Bureau of Justice Statistics, 1999). Yet most rapes go unreported to police. The prevalence of rapes reported to authorities in the United States is 20 times greater than the rate in Japan and 13 times greater than the rate in Great Britain ("Women Under Assault," 1990).

There is some good news to report, however. The numbers of reported rapes began declining during the mid-1990s. Unreported rapes also appear to be on the decline. Whether this decline represents an ebbing of sexual crimes against women remains to be seen.

forcible rape Forced sexual intercourse with a nonconsenting person.

statutory rape Sexual intercourse with a minor, even with the minor's cooperation.

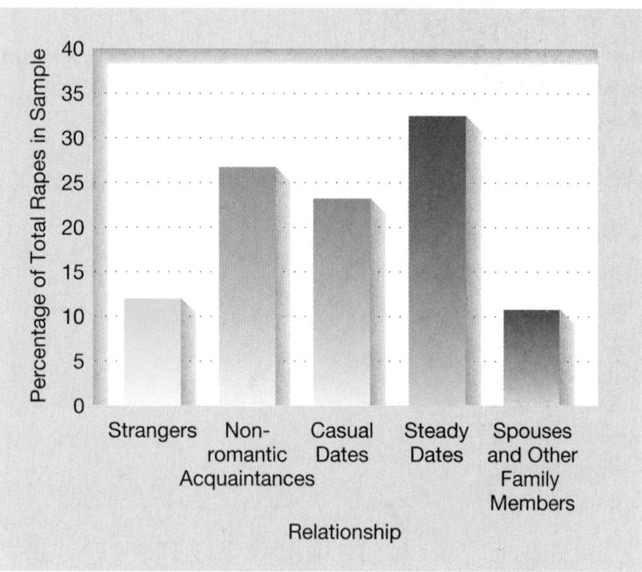

FIGURE 16.4 Relative percentages of stranger rapes and acquaintance rapes.
According to a large-scale survey of more than 3,000 college women at 32 colleges, the great majority of rapes of college women were committed by men whom the women were acquainted with, including dates, nonromantic acquaintances, and family members.
Source. Adapted from Koss (1988).

Based on a compilation of official records and large-scale surveys of victims, researchers estimate that between 14% and 25% of U.S. women have been raped or will be raped at some point during their lifetimes (Brener et al., 1999; Calhoun & Atkeson, 1991; Polaschek, Ward, & Hudson, 1997). In about four of five cases of rape overall the woman was acquainted with the assailant (Gibbs, 1991), though typically they were not intimate acquaintances or lovers. Figure 16.4 shows the relationship patterns of rapist and victim based on a large, national survey of college women (Koss, Gidycz, & Wisniewski, 1987). This survey revealed a disturbingly high percentage of the women reporting they had experienced rape (15.4%) or attempted rape (12.1%) (Koss et al., 1987). In nearly 90% of the rapes in the college sample, the woman was acquainted with the assailant (Koss, 1988). In any given year, about 3% of college women in the United States suffer a rape or attempted rape (Fisher, Cullen, & Turner, 2000).

Any woman is at risk of rape, including women and girls of all ages, races, and socioeconomic levels.[1] However, younger women are most at risk, especially adolescent girls. In a recent survey of 9th- through 12th-grade female students in Massachusetts, about one in five students reported either being physically or sexually abused by a dating partner (Silverman et al., 2001). More than half of the reported rapes in the United States are committed against girls under the age of 18; about one in six rapes are committed against girls under the age of 12 ("U. S. Finds," 1994). The younger the woman, the more likely she is to be acquainted with the assailant.

Web Link 16.5 WWW
Rape Fact Sheet

Truth OR Fiction? REVISITED

Women are more likely to be raped by men they know than by strangers.

TRUE. Most rapes are not committed by strangers, but by men with whom the women were acquainted.

Theoretical Perspectives on Rape There is no single kind of rape or rapist. Rape has more to do with violent impulses and issues of power and control than sexual gratification. Rapists are no more likely than other offenders to have the kinds of psychological disorders coded on Axis I of the *DSM-IV* (Polaschek et al., 1997). Some rapists have such feelings of shyness and inadequacy that they report being unable to find willing partners; others are basically antisocial and tend to act out on their impulses regardless of the cost to the victim. Many of these antisocial rapists have long records as violent offenders (Siegel, 1992). Many were sexually abused themselves as children (Dhawan & Marshall, 1996). For still other rapists, violent cues appear to enhance sexual arousal, so they are motivated to combine sex with aggression (Barbaree & Marshall, 1991). College men who report having used force or threat of force to gain sexual favors showed greater penile arousal in a laboratory study in response to depictions of rape scenes than did noncoercive men (Lohr, Adams, & Davis, 1997). Some rapists who were abused as children may humiliate women as a way of expressing anger and power over women, and of taking revenge.

On the basis of clinical experience with more than 1,000 rapists, Groth and Hobson (1983) hypothesize the existence of three basic kinds of rape: anger rape, power rape, and sadistic rape. The anger rape is a savage, unpremeditated attack triggered by feelings of hatred and resentment. Anger rapists often use more force than necessary to gain compliance to take revenge for humiliations they have suffered—or believe they have suffered—at the hands of women. The power rapist is basically motivated by the desire to control another person. Groth and Hobson (1983) suggest that power rapists use rape to try "to resolve disturbing doubts about [their] masculine identity and worth, [or] to combat deep-seated feelings of insecurity and vulnerability" (p. 165). Sadistic rapes frequently employ torture and bondage, merging sex and aggression. Sadistic rapists are most likely to mutilate their victims.

Sociocultural factors also need to be considered. Although some rapists show evidence of psychopathology on psychological tests, especially psychopathic traits, many do not (Brown & Forth, 1997; Herkov et al., 1996). The very normality of many rapists on psychological instruments suggests that socialization factors play an important role. Some

[1]Although male rapes do occur, the great majority of cases involve women who are raped.

sociocultural theorists argue that our culture actually breeds rapists by socializing men into sexually dominant and aggressive roles associated with stereotypical concepts of masculinity (e.g., Hall & Barongan, 1997). For example, consider the aggressive competitiveness expressed by this male college student in his views about dating relationships:

> A man is supposed to view a date with a woman as a premeditated scheme for getting the most sex out of her. Everything he does, he judges in terms of one criterion—"getting laid." He's supposed to constantly pressure her to see how far he can get. She is his adversary, his opponent in a battle, and he begins to view her as a prize, an object, not a person. While she's dreaming about love, he's thinking about how to conquer her. (Powell, 1991, p. 55)

Evidence shows that sexually coercive college men tend to have an excessively masculinized personality orientation, adhere more strictly to traditional gender roles, view women as adversaries in the "mating game," and hold more accepting attitudes toward the use of violence against women than do noncoercive men (e.g., O'Donohue, McKay, & Schewe, 1996; Polaschek et al., 1997). They are also more prone to blame rape survivors than rapists and to become more sexually aroused by portrayals of rape than are men who hold less rigid stereotypes. Sociocultural influences also reinforce themes that may underlie rape, such as the cultural belief that a masculine man is expected to be sexually assertive and overcome a woman's resistance until she "melts" in his arms (Stock, 1991).

Date rape is a form of acquaintance rape. According to recent surveys of high school girls in Massachusetts and Minnesota, between 1 in 10 and 1 in 5 report a history of being physically or sexually assaulted by a dating partner (Goode, 2001f; Stenson, 2001b). College men on dates frequently perceive their dates' protests as part of an adversarial sex game. One male undergraduate said, "Hell, no" when asked whether a date had consented to sex. He added, "But she didn't say no, so she must have wanted it, too. It's the way it works" (Celis, 1991). Consider the comments of Jim, a man who raped a woman he had just met at a party:

A Case of Date Rape

She looked really hot, wearing a sexy dress that showed off her great body. We started talking right away. I knew that she liked me by the way she kept smiling and touching my arm while she was speaking. She seemed pretty relaxed so I asked her back to my place for a drink . . . When she said yes, I knew that I was going to be lucky!

When we got to my place, we sat on the bed kissing. At first, everything was great. Then, when I started to lay her down on the bed, she started twisting and saying she didn't want to. Most women don't like to appear too easy, so I knew that she was just going through the motions. When she stopped struggling, I knew that she would have to throw in some tears before we did it.

She was still very upset afterwards, and I just don't understand it! If she didn't want to have sex, why did she come back to the room with me? You could tell by the way she dressed and acted that she was no virgin, so why she had to put up such a big struggle I don't know.

—From Trenton State College, 1991

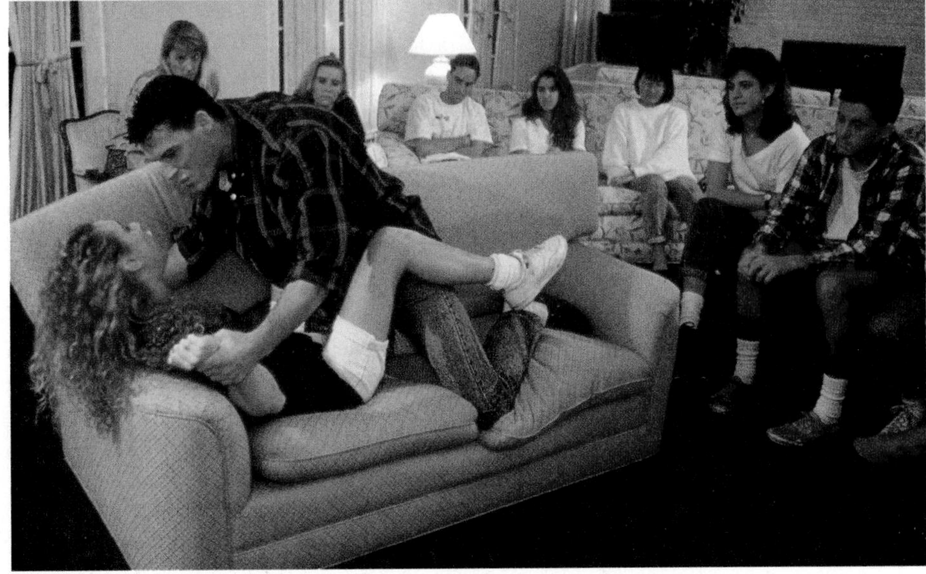

Date rape. Many rape awareness workshops, like the one depicted here, have been held on college campuses in order to combat the problem of date rape on campus.

Rape crisis counseling. Rape crisis centers help rape survivors cope with the trauma of rape. They provide them with emotional support and help them obtain medical, psychological, and legal services.

Let us rebut these beliefs. Accepting a date is not the equivalent of consenting to intercourse. Accompanying a man to his room or apartment is not the equivalent of consenting to intercourse. Kissing and petting are not the equivalent of consenting to intercourse. When a woman fails to consent or says no, the man must take no for an answer.

Effects of Rape Women who are raped suffer more than the rape itself. They report loss of appetite, headaches, irritability, anxiety and depression, and menstrual irregularity in the wake of a rape. Some become sullen, withdrawn, and mistrustful. In some cases, women show an unrealistic composure, which often gives way to venting of feelings later on. Because of society's tendency to blame the victim for the assault, some survivors also have misplaced feelings of guilt, shame, and self-blame. Survivors often develop sexual dysfunctions such as lack of sexual desire and difficulty becoming sexually aroused. Survivors show signs of posttraumatic stress disorder (PTSD), including intrusive memories of the rape, nightmares, emotional numbing, and heightened autonomic arousal (Gilboa-Schechtman & Foa, 2001; Nishith, Mechanic, & Resick, 2000). Psychological problems experienced by rape survivors, such as depression and anxiety, often continue for years after the assault (Sadler et al., 2000).

Web Link 16.6 WWW
FYI: Acquaintance Rape

Treatment of Rape Survivors Treatment of rape survivors is often a two-phase process that first assists women in coping with the immediate aftermath of rape and then helps them with their long-term adjustment. Crisis intervention provides women with emotional support and information to help them see to their immediate needs as well as to help them develop strategies for coping with the trauma. Longer-term treatment may be designed to help rape survivors cope with undeserved feelings of guilt and shame, lingering feelings of anxiety and depression, and the interpersonal and sexual problems they may develop with the men in their lives. Unfortunately, most rape survivors do not seek help from mental health professionals, rape crisis centers, or rape survivor assistance programs (Kimerling & Calhoun, 1994). The cultural stigma associated with seeking help for mental health problems may discourage a fuller utilization of psychological services.

THINK ABOUT IT
Do you believe a woman who flirts with men in a bar is looking for sex, whether or not she says "no"? Why or why not?

THINK ABOUT IT
Did you ever attempt to coerce someone into sexual activity? Did you feel justified at the time? How do you feel about it now?

Child Sexual Abuse

Few crimes are as heinous as sexual abuse of children. The consequences of sexual abuse can be severe and long lasting, leading to emotional problems and difficulties developing intimate relationships that last long into the future. Estimates suggest that at least 20% or 25% of women have a history of childhood sexual abuse (Bradley & Follingstad, 2001). As shocking as these figures may be, surveys may undershoot the mark, as people may not be willing to report abuse. Although girls are more likely than boys to be sexually abused, estimates are that boys constitute perhaps one-fifth to one-third of sexually abused children (DeAngelis, 1995b; Finkelhor, 1990).

Sexual abuse, like other forms of violence and abuse, cuts across all socioeconomic and family background characteristics (Finkelhor, 1993). Neither social class nor ethnicity appears to be associated with relative risk. However, children from less cohesive or disintegrating families appear more likely to be victimized than children from intact families. The average age at which children are first sexually abused ranges between 7 and 10 for boys and 6 and 12 for girls (Knudsen, 1991).

Patterns of Child Sexual Abuse Sexual abuse of children includes a range of sexual acts such as fondling, kissing, exhibitionism, touching of genitals, oral sex, anal intercourse, and

A Closer Look

Rape Prevention

Given the incidence of rape, it is important to be aware of strategies that may prevent it. By listing strategies for rape prevention, we do not mean to imply that rape survivors are somehow responsible for falling prey to an attack. The responsibility for any act of sexual violence lies with the perpetrator, not with the person who is assaulted, and perhaps with society for fostering attitudes that underlie sexual violence. On a societal level, we need to do a better job of socializing young men to acquire prosocial and respectful attitudes toward women. Exposure to feminist, egalitarian, and multicultural education may help promote more respectful attitudes in young men (Hall & Barongan, 1997; Kershner, 1996).

Preventing Stranger Rape

- Establish signals and plans with other women in the building or neighborhood.
- List first initials only in the phone directory and on the mailbox.
- Use dead-bolt locks.
- Lock windows and install iron grids on first-floor windows.
- Keep doorways and entries well lit.
- Have keys handy for the car or the front door.
- Do not walk by yourself after dark.
- Avoid deserted areas.
- Do not allow strange men into the apartment or the house without checking their credentials.
- Keep the car door locked and the windows up.
- Check out the backseat of the car before getting in.
- Don't live in a risky building.
- Don't give rides to hitchhikers (that includes women hitchhikers).
- Don't converse with strange men on the street.
- Shout "Fire!" not "Rape!" People flock to fires but circumvent scenes of violence.

Preventing Date Rape

- Avoid getting into secluded situations until you know your date very well.
- Be wary when a date attempts to control you in any way, such as frightening you by driving rapidly or taking you some place you would rather not go.
- Stay sober. We often do things we would not otherwise do—including sexual activity with people we might otherwise reject—when we have had too many drinks. Be aware of your limits.
- Be very assertive and clear concerning your sexual intentions. Some rapists, particularly date rapists, tend to misinterpret women's wishes. If their dates begin to implore them to stop during kissing or petting, they construe pleading as "female game playing." So if kissing or petting is leading where you don't want it to go, speak up.
- When dating a person for the first time, try to date in a group.
- Encourage your college or university to offer educational programs about date rape. The University of Washington, for example, offers students lectures and seminars on date rape and provides women with escorts to get home. Many universities require all incoming students to attend orientation sessions on rape prevention.
- Talk to your date about his attitudes toward women. If you get the feeling that he believes that men are in a war with women, or that women try to "play games" with men, you may be better off dating someone else.

Sources. Adapted from Boston Women's Health Book Collective, 1984; Rathus & Fichner-Rathus, 1994.

among girls, vaginal intercourse (Knudsen, 1991). Because children are not deemed capable of giving voluntary consent, any sexual act between an adult and a child is considered a form of sexual abuse, even if there is no force or physical threat used or the child gives consent.

Sexual abuse of children is more likely to be committed by family members than by strangers, but girls are even more likely than boys to be abused by a family member or acquaintance (Faller, 1989). Boys are more often threatened or suffer physical injuries during a sexual assault than is the case with girls (Knudsen, 1991). A study at the University of New Mexico found that while the frequency of childhood sexual abuse (about 30% overall) was similar between Hispanic and non-Hispanic White college women, Hispanic women reported that perpetrators were more likely to be extended family members and less likely to be members of the immediate family or non-family members than did non-Hispanic White women (Arroyo, Simpson, & Aragon, 1997).

It may surprise you to learn that the abuser seldom uses physical force. Most of the time the abuser is able to use manipulation, deception, or threat of force to obtain the

Truth OR Fiction? REVISITED

Child molesters typically use physical force to compel children into performing sexual acts.

FALSE. In most cases, the molester is able to use manipulation, deception, or threat of force to obtain the child's compliance rather than direct force.

child's compliance rather than direct force. Children are not worldly wise and are typically submissive to adult authority. They may be easily misled or manipulated by an unscrupulous adult, especially when the abuser is someone with whom the child has had a trusting relationship. Threat of force may be used if guile and deception fail to achieve compliance. Although most abused children suffer one incident of abuse, in some cases a pattern of abuse occurs that continues for a period of months or even years. Children who are abused by family members are most likely to suffer repeated incidents of abuse.

The most common type of abuse of girls or boys involves genital fondling (Knudsen, 1991). In one female sample, intercourse occurred in only 4% of cases, as compared to genital fondling that occurred in 38% and exhibitionism that occurred in 20% (Knudsen, 1991). However, in cases of sexual abuse of girls by family members, a common pattern involves a gradual progression of abuse that begins with affectionate fondling during the preschool years, progresses to oral sex or mutual masturbation during middle childhood, and then to vaginal or anal intercourse in preadolescence or adolescence.

Characteristics of Abusers The great majority of abusers of both boys and girls are men. Authorities in the field estimate that men perpetrate perhaps 95% of cases of abuse involving young girls and 80% involving young boys (Finkelhor & Russell, 1984). But not all sex offenders are men, as the case of a Seattle woman, Mary Kay LeTourneau, illustrates:

The Case of Mary Kay LeTourneau

She wore red coveralls—the uniform of the female prisoner. She looked exhausted and disheveled as she stood for sentencing. She is Mary Kay LeTourneau, a 35-year-old former grade school teacher in suburban Seattle. She is the woman who had sexual relations, and a baby, with a former student—a 13-year-old boy. LeTourneau wept as her lawyer pleaded for mercy and made no comment when the judge offered her a chance to speak. She was sentenced to seven years and five months in prison and denounced by the judge who had been lenient some months earlier, in a case that has raised questions as to whether female sex offenders can be held to different standards than men. In November of 1997, she had pleaded guilty to raping the boy and promised King County Superior Court Judge Linda Lau that she would have no further contact with him.

"I give you my word," LeTourneau had said. "It will not happen again." LeTourneau is the mother of four children by her ex-husband. She had no previous criminal record and was given a suspended prison sentence and ordered to undergo treatment. But in February of 1998, after hearing testimony that LeTourneau had again seen the boy, and possibly planned to flee with him, Judge Lau revoked the suspension and sent LeTourneau to prison. Prosecutors, state legislators and professionals who treat sex offenders note that LeTourneau had escaped an earlier prison sentence largely because she was attractive and presented herself as being in love with an emotionally mature boy.

The victim? The boy repeatedly protested to the media that he was not a victim, that he was in love with his former teacher and deeply saddened by her conviction. LeTourneau had claimed that the boy is emotionally mature beyond his years. But people who treat sex offenders say that they often justify sex with underage people by saying that the victim is emotionally mature.

The saga apparently continues. In April 2000 the boy, 16 at the time, filed law suits against the city of Des Moines, Washington, and his local school district for emotional suffering, lost income and the cost of raising his two children, who were in the custody of his mother ("Boy to Sue," 2000).

—Adapted from Rathus, Nevid, & Fichner-Rathus, 2002

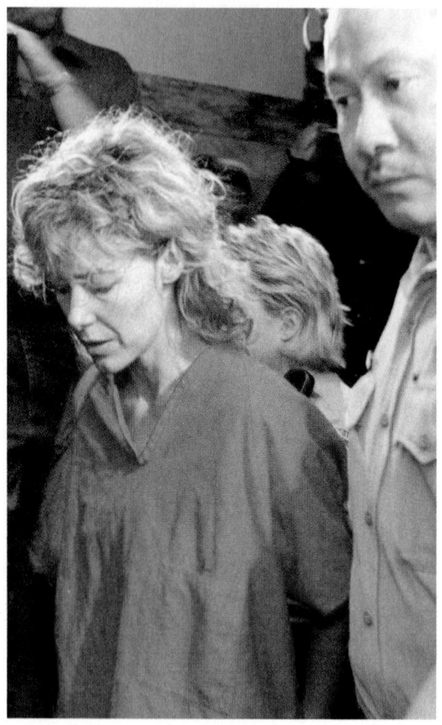

Former teacher sent to prison for child rape. Mary Kay LeTourneau is shown here being led away from the courtroom after the judge revoked her suspended sentence and ordered her to be sent to prison for having continued contact with the boy whom she was convicted of raping.

Although most abusers are adult males, some are adolescent males, many of whom were abused themselves as young boys and may be imitating their own victimization (Muster, 1992). Gay males and lesbians account for only a small percentage of abusers of either boys or girls.

Despite the popular stereotype, most abusers do not fit the profile of the proverbial stranger hanging around the schoolyard. Most cases of child sexual abuse, perhaps as many as 75% to 80%, involve assailants who have some kind of relationship with the child or the child's family, typically a relative or step-relative, a family friend, or a neighbor (Zielbauer, 2000). In many cases (estimates range from 10% to 50% of cases), the molester is a family member, typically a father, uncle, or stepfather.

Family members who discover that a child has been abused are less likely to report the abuse to authorities when the offender is a family member, perhaps because to do so might shame the family and out of concern that they might be held accountable for failing to protect the child. In a Boston study, none of the parents whose children were abused by family members notified the authorities, as compared to about one in four parents whose children were abused by acquaintances and about three of four whose children were abused by strangers (Finkelhor, 1984).

VIDEO 16.1

Child Sexual Abuse: *The Case of Karen*

In an early study, Gebhard and his colleagues (1965) found that many fathers who had sexually abused their daughters were religiously devout, fundamentalistic, and moralistic. Such men, when sexually frustrated, may be less likely to find extramarital and extrafamilial sexual outlets. The father may be under stress but fails to find adequate emotional and sexual support from his wife. He may turn to a daughter as a wife surrogate. The girl is often mature enough to have assumed household chores and may become, in her father's eyes, the "woman of the house." The wife may be surprised by revelations of incest, despite obvious clues and even the daughter's repeated complaints. Sometimes the wife seems involved in a tacit conspiracy to allow the abuse to continue to preserve the family or because of fear of the abuser.

Banning (1989) offers a sociocultural explanation of why a disproportionate number of molesters are men. Banning argues that men are socialized in our culture to seek younger and weaker partners they can more easily dominate. In the extreme, this pattern can lead to sexual interest in young girls, who because of their age can be more easily dominated than adult women. Yet child molestation may also be motivated by pedophilia, an unusual pattern of sexual arousal characterized by a sexual interest in children, sometimes to the exclusion of more appropriate (adult) stimuli (see Chapter 12). A cycle of sexual abuse and other forms of sexual violence is a common finding, with children who suffer sexual abuse engaging in sexual abuse against their own children or committing other types of sexual offenses as adults (McCloskey & Bailey, 2000; Romano & De-Luca, 1996).

THINK ABOUT IT

Is there a "typical" child abuser? How can a parent best protect a child from possible abuse?

Effects of Child Sexual Abuse The effects of child sexual abuse are variable and no one pattern applies in all cases (Price et al., 2001). Though some sexually abused children may show no clear psychological effects, most exhibit some types of psychological problems, most commonly anxiety, depression, aggressive behavior, poor self-esteem, eating disorders, premature sexual behavior or promiscuity, suicidal thinking, and substance abuse (e.g., Kisiel & Lyons, 2001; Meston & Heiman, 2000; Saywitz et al., 2000). Regressive behavior in the form of thumb sucking, or recurrences of childhood fears, such as fear of the dark or of strangers, are common. Psychological problems may continue into adulthood in the form of PTSD, anxiety, depression, substance abuse and relationship problems (Bradley & Follingstad, 2001; Kendler et al., 2000b; Read et al., 2001). Late adolescence and early adulthood are particularly difficult times for survivors of child sexual abuse because unresolved feelings of anger and guilt and a deep sense of mistrust can prevent the expected development of intimate relationships (Jackson et al., 1990). Although much of the research on effects of childhood sexual abuse has focused on female survivors, a significant proportion of male survivors also suffer adverse psychological effects into adulthood (Dhaliwal et al., 1996). It is a myth to believe that sexual abuse has little effect on boys.

Some child survivors retreat into a personal fantasy world or refuse to leave the house. We noted in Chapter 7 how many cases of dissociative identity disorder (multiple

So there really was a monster in her bedroom.

For many kids, there's a real reason to be afraid of the dark.

Last year in Indiana, there were 6,912 substantiated cases of sexual abuse. The trauma can be devastating for the child and for the family. So listen closely to the children around you.

If you hear something you don't want to believe, perhaps you should. For helpful information on child abuse prevention, contact the LaPorte County Child Abuse Prevention Council, 7451 Johnson Road, Michigan City, IN 46360. (219) 874-0007

LaPorte County Child Abuse Prevention Council

So there really was a monster in her bedroom. Not all monsters are imagined. Some, like incest perpetrators, are family members.

personality) have been linked to a history in childhood of retreating into fantasy to cope with sexual abuse. Childhood sexual abuse is also linked to the later development of borderline personality disorder (Murray, 1993; Weaver & Clum, 1995).

Although the effects of child sexual abuse are more similar than not between boys and girls (e.g., both genders tend to experience fears and sleep disturbances), there are some important differences. The clearest gender difference is that boys tend to develop "externalized" behavior problems such as excessive aggressive behavior, whereas girls tend to experience "internalized" problems, such as depression (Finkelhor et al., 1990).

Although long-term effects of child sexual abuse are common, they appear to be greater among survivors who were abused by their fathers or stepfathers, who were abused at an earlier age, who were subjected to penetration, who suffered more prolonged and severe abuse, or who were forced to submit or threatened with physical force (DeAngelis, 1995b). Adult survivors of childhood sexual abuse who blame themselves for the abuse tend to have more psychological problems than those who blame the perpetrator (Feinauer & Stuart, 1996). The use of cognitive coping skills in adulthood, such as disclosing and discussing the abuse but not dwelling on it, appears to differentiate well-adjusted and poorly adjusted college women who suffered sexual abuse as children (Himelein & McElrath, 1996).

When the offender is a father or other family member, the effects of abuse are amplified by the deep feeling of the betrayal of trust by the offender as well as by their mothers or other family members whom they perceive as having failed to protect them. They may have felt powerless to control their bodies or their lives and may find it difficult to ever develop a trusting relationship in adulthood.

Researchers have begun to examine ethnic group differences in the effects of child sexual abuse. In one example, investigators at a child sex abuse clinic compared samples of Asian American, African American, Hispanic American, and non-Hispanic White American child survivors (Rao, DiClemente, & Ponton, 1992). The results showed that Asian American children were more likely to develop suicidal thoughts and less likely to overtly display anger or engage in sexual acting out than were children from the other groups.

Gail Wyatt (1990) investigated childhood sexual abuse in a sample of 126 African American women and 122 Euro-American (non-Hispanic White) women in Los Angeles County. The subjects were matched on such variables as marital status, number of children, and education. The operational definition of childhood sexual abuse was broad, including acts such as exhibitionism, fondling, oral sex, and sexual intercourse. The prevalence of abuse was similar in both groups. Nearly one woman in two had suffered at least one incident of abuse, and nearly 40% of these incidences had gone unreported to authorities. The African American women were somewhat less likely to report abuse to their immediate family members or to the police. However, they were nearly twice as likely to have informed extended family members than were the Euro-American women, a finding that underscores the importance of the extended family among African Americans. African American women were more likely than Euro-American women (35% versus 22%) to avoid reporting abuse for fear of repercussions. The African American women may have felt more vulnerable to the financial adversity that their families would have had to endure if the abuser—often their mothers' boyfriends or their stepfathers on whom the family was financially dependent—had been forced to leave the home. Euro-American women more often expressed fear of being blamed for the abuse as a reason for nonreporting (36%) than did African American women (23%). Other investigators report that Mexican American women, raised in a culture in which women are often expected to "suffer in silence," are also less likely than Euro-American women to report rape and sexual abuse (Lira, Koss, & Russo, 1999).

Web Link **16.7** WWW
FYI: Child Sexual Abuse

The psychological impact of abuse in the Gail Wyatt study was similar for both groups. Women from both groups were likely to have felt violated and to harbor feelings of fear, disgust, and anger. Sexual problems in adulthood were also common in both groups. However, the African American women were more prone to report that the abusive experiences from childhood affected their general attitudes toward men and led them to avoid men who in some ways reminded them of their abusers.

Treating Survivors of Child Sexual Abuse Because most cases of child sexual abuse go unreported, survivors of abuse may not receive psychotherapy for overcoming trauma-related feelings of anger and (misplaced) guilt until adulthood. Sex therapy may help adult survivors overcome sexual dysfunctions and fears. Group therapy may help them face their feelings in a supportive setting with people who have undergone similar trauma.

When sexual abuse is uncovered in childhood, a multicomponent treatment approach may be recommended, including individual therapy for the child survivor, joint therapy with the child and nonoffending parent, and family therapy (Celano et al., 1996; Cohen & Mannarino, 1997). Therapy with the child survivor typically focuses on providing support, addressing issues of betrayal and powerlessness, and helping the child see that he or she is not to blame.

Survivors of child sexual abuse. Most cases of child sexual abuse go unreported. This means that survivors of abuse may not receive any psychological counseling to help them deal with their traumatic experiences until adulthood, if then.

Treating the Offenders Convicted rapists and child molesters are criminals and typically sentenced to prison as a form of punishment, not treatment. They may receive psychological treatment during their incarceration in the hope that it will help prepare them for their eventual release and deter them from committing future offenses when they do reenter society. Treatment programs are more likely to be successful when they address the long-term adjustment of the sex offender and do not expect a permanent "cure" to result from a single round of prison-based treatment sessions (Hanson, Steffy, & Gauthier, 1993). The most widely used form of treatment for incarcerated sex offenders is group therapy, predicated on the belief that although sex offenders may trick counselors, they cannot so readily fool each another (Kaplan, 1993).

Group therapy may be supplemented by cognitive-behavior therapy techniques such as covert sensitization, which we discussed in Chapter 12. Also coming into greater use is *empathy training,* which attempts to increase the offender's sensitivity toward his victim by having him write about his crime from what he imagines would be the victim's perspective. The fact remains, however, that the great majority of incarcerated sex offenders receive little or nothing in the way of psychological treatment in prison (Goleman, 1992a).

A biologically based treatment involves the use of antiandrogen (testosterone-reducing) drugs such as Depo-Provera. These drugs lower the sex drive, which may help offenders control their urges to offend, at least so long as they continue using the drug (Roesler & Witztum, 2000). The effects of antiandrogens are reversible when the drugs are discontinued. However, problems with compliance with taking the drugs consistently present a major obstacle to their widespread use. Moreover, taking antiandrogens does not help rapists resolve the psychological factors underlying their sexual attacks, particularly resentment and anger toward women and needs for dominance and power. Nor does it assist men with pedophilia to learn to respond to more adaptive erotic cues, rather than to images of children. Nor do drugs provide offenders with opportunities to develop social skills they need to develop consensual sexual relationships. Despite these drawbacks, the use of antiandrogens appears to be about equal in effectiveness to cognitive-behavioral treatments in preventing recidivism of sex offenders, with both treatment approaches producing medium-sized treatment effects in reducing recidivism in relation to an absence of treatment (Hall, 1995). Antiandrogens may also be helpful when they are used in conjunction with psychological counseling (Leary, 1998).

THINK ABOUT IT

Do you know anyone who was sexually abused during childhood? What were the effects? Are there any lingering effects? Was the abuser reported to authorities? Why or why not?

Overall, efficacy studies of treatment of sex offenders show a modest benefit in reducing recidivism rates associated with comprehensive cognitive-behavioral treatment, or psychological treatment combined with antiandrogen therapy (Hall, 1995; Polaschek et al., 1997). Generally speaking, better outcomes are achieved with child molesters and exhibitionists than with rapists. Not surprisingly, recidivism rates are higher among sex offenders who fail to complete treatment and who show deviant sexual interest, such as penile arousal to child pornography (Hanson & Bussiére, 1998). More extreme measures, such as surgical castration (removal of the testes), have been used in some European countries but not as yet in the United States. The effects of surgical castration on recidivism are not clear.

One treatment alternative that remains largely unexplored is targeting men who are sexually attracted to children or who have a predisposition to rape *before* they commit abusive acts. Perhaps early intervention can prevent sexual violence from occurring.

THINK ABOUT IT

Do you believe we should simply punish sex offenders, or should we be concerned about "treating" them? Explain your opinion.

Unfortunately, most men who would likely be good candidates for early intervention do not come forward for treatment until they are convicted of a sexual crime.

Preventing Child Sexual Abuse Sexual abuse prevention includes community efforts, such as enactment and enforcement of laws requiring teachers and helping professionals to report cases of suspected child sexual or physical abuse to child protective services. Many states have enacted laws that require convicted sex offenders to register with the local police. These laws, called "Megan's laws," after a 7-year-old New Jersey girl who was killed by a neighbor with a history of sexual assault, are intended to let members of the community know of the presence of sex offenders (Weber, 1996).

At the individual level, school-based sexual abuse prevention programs reach about two of three children in the United States (Goleman, 1993c). Although these programs vary in their content, most teach children to avoid contact with strangers and to distinguish between acceptable touching (a member of their family embracing them affectionately or patting them on the head) and "bad" or unacceptable touching. However, evidence attesting to the efficacy of child sexual abuse prevention programs remains limited.

Children are more likely to report incidents of sexual abuse if they are reassured that they will be believed and not be blamed, that their parents will continue to love them, and that they and their families will be protected from the abuser. Ensuring the credibility of children's reports of sexual abuse remains a thorny issue, however, as children are easily suggestible and can be led to believe that abuse occurred even when it did not, especially if investigators direct the inquiry in ways that plant ideas in the children's minds. Recall too (see Chapter 7) the controversy that has swirled around the issue of recovered memories of childhood sexual or physical abuse that the individual only becomes aware of at some point in adulthood, often during therapy or hypnosis. The judicial system is faced with the daunting challenge of distinguishing between true and faulty memories, especially when there is an absence of independent verification of abuse.

THINK ABOUT IT

What do you believe society should do to reduce the prevalence of rape and sexual abuse?

Sexual Harassment

Sexual harassment is a form of sexual coercion in which one person subjects another to unwanted sexual comments, overtures, gestures, physical contact, or direct demands for sexual favors as a condition of employment, retention, or advancement (see Table 16.3). Women tend to perceive a wider ranger of behaviors as harassing than do men (Rotundo, Nguyen, & Sackett, 2001). For example, women are more likely to perceive behaviors involving expression of derogatory attitudes, dating pressures, and physical sexual contact (e.g., fondling or kissing) as forms of harassment than do men. Men and women are in stronger agreement in classifying such extreme behaviors as rape, requests

TABLE 16.3 Examples of Sexual Harassment

Verbal harassment or abuse
Subtle pressure for sexual activity
Remarks about a person's clothing, body, or sexual activities
Leering or ogling at a person's body
Unwelcome touching, patting, or pinching
Brushing against a person's body
Demands for sexual favors accompanied by implied or overt threats concerning one's job or student status
Physical assault

Source. Adapted from Powell (1991).

sexual harassment Speech, gestures, demands, or physical contact of a sexual nature that is unwelcomed by the person to whom it is directed.

for sexual involvement as a condition of employment or promotion, and unwanted pressure or requests for sexual involvement as forms of sexual harassment.

Sexual harassment can occur in many settings, including the workplace, school, or therapist's consulting room. The great majority of cases of sexual harassment involve men harassing women. Sexual harassment in the workplace is considered a form of sex discrimination, and employers can be held accountable if sexual harassment creates a hostile or abusive work environment or interferes with an employee's work performance. A 1998 decision by the U.S. Supreme Court held that persons can bring a sexual harassment action even if they did not suffer any career setback, such as a dismissal or loss of a promotion, as the result of the harassment (Greenhouse, 1998). Employers can be accountable if they either *knew* that harassment was occurring or *should have known* and failed to promptly correct the situation (McKinney & Maroules, 1991).

Sexual harassment? Sexual harassment is behavior of a sexual nature that is unwelcome by the recipient. Sexual harassment on the job creates a work environment that is hostile or abusive to the person on the receiving end. It may also interfere with the person's ability to perform his or her job. Would you say that the man's behavior here could be interpreted as harassment? What would you do if you were sexually harassed?

Although sexual harassment may involve many motives, it typically has more to do with the abuse of power than with sexual motivation (Goleman, 1991). Harassers usually hold a dominant position in relation to the person who is harassed and abuse their authority by taking advantage of the other person's vulnerability. Resentment and hostility directed at women who venture beyond the traditional gender boundaries and enter traditional male occupations may be expressed in the form of sexual taunts and overtures by men as a way of "keeping women in their place" (Fitzgerald, 1993a).

Although laws prohibit sexual harassment, and people subjected to harassment can sue to have the harassment stopped or can obtain monetary awards for emotional damages they suffer, relatively few formal complaints are filed. Why? For one thing, harassment may be difficult to prove because it usually occurs without corroborating witnesses or evidence. Like survivors of rape, people who are harassed may fear that they will not be believed or will suffer retaliation by the harasser, lose their jobs, or have their reputations soured in the industries in which they work.

What should you do if you are sexually harassed? The accompanying "A Closer Look" section discusses some options for you to consider.

Prevalence of Sexual Harassment Sexual harassment is the most frequently occurring form of sexual victimization in the United States (Fitzgerald, 1993a). Overall, estimates indicate that at least 33% of women, and 15% of men, encounter sexual harassment in a work or academic setting (Choi et al., 1998; Stawar, 1999). Sexual harassment even extends downward to teenage workers. A recent survey of 16- and 17-year-old girls with jobs found that nearly half (49%) said they had been victims of sexual harassment (Goodstein & Connelly, 1998). Moreover, a recent national survey of high school students showed that about 80% of the boys and girls reported being sexually harassed by their peers (Smith, 2001b).

Sexual harassment directed at women is especially common in worksites that are traditional male preserves, such as the construction site, the shipyard, and the firehouse (Fitzgerald, 1993a). Despite the increased attention to the problem of sexual harassment in the military, studies actually show a decline in the prevalence of harassment in these settings (Seppa, 1997). Cross-cultural research, though limited, shows high frequencies of sexual harassment in other developed countries that have been studied; in Japan, about 70% of women report incidents of harassment; in Europe, about 50% (Castro, 1992).

A Closer Look

How to Resist Sexual Harassment

What would you do if an employer or a professor sexually harassed you? How would you handle it? Would you try to ignore it and hope it would stop? What actions might you take? We offer some suggestions, adapted from Powell (1991), that might be helpful. Recognize, however, that responsibility for sexual harassment always lies with the perpetrator and the organization that permits sexual harassment to take place, not with the person subjected to the harassment.

1. *Convey a professional attitude.* Harassment may be stopped cold by responding to the harasser with a businesslike, professional attitude. If a harassing professor suggests that you come back after school to review your term paper so the two of you will be undisturbed, set limits assertively. Tell the professor that you'd feel more comfortable discussing the paper during regular office hours. The harasser should quickly get the message that you wish to maintain a strictly professional relationship. If the harasser persists, do not blame yourself. You are only responsible for your own actions. When the harasser persists, a more direct response may be appropriate: "Professor Jones, I'd like to keep our relationship on a purely professional basis, okay?"
2. *Avoid being alone with the harasser.* If you are being harassed by your professor but need some advice about preparing your term paper, approach him or her after class when other students are milling about, not privately during office hours. Or bring a friend to wait outside the office while you consult the professor.
3. *Maintain a record.* Keep a record of all incidents of harassment as documentation in the event you decide to lodge an official complaint. The record should include the following: (1) where the incident took place; (2) the date and time; (3) what happened, including the exact words that were used, if you can recall them; (4) how you felt; and (5) the names of any witnesses. Some people who have been subjected to sexual harassment have carried a hidden tape recorder during contacts with the harasser. Such recordings may not be admissible in a court of law, but they are persuasive in organizational grievance procedures. A hidden tape recorder may be illegal in your state, however. It is thus advisable to check the law.
4. *Talk with the harasser.* It may be uncomfortable to address the issue directly with a harasser, but doing so puts the offender on notice that you are aware of the harassment and want it to stop. It may be helpful to frame your approach in terms of a description of the specific offending actions (e.g., "When we were alone in the office, you repeatedly attempted to touch me or brush up against me"); your feelings about the offending behavior ("It made me feel like my privacy was being violated. I'm very upset about this and haven't been sleeping well"); and what you would like the offender to do ("So I'd like you to agree never to attempt to touch me again, okay?"). Having a talk with the harasser may stop the harassment. If the harasser denies the accusations, it may be necessary to take further action.
5. *Write a letter to the harasser.* Set down on paper a record of the offending behavior, and put the harasser on notice that the harassment must stop. Your letter might (1) describe what happened ("Several times you have made sexist comments about my body"); (2) describe how you feel ("It made me feel like a sexual object when you talked to me that way"); and (3) describe what you would like the harasser to do ("I want you to stop making sexist comments to me").
6. *Seek support.* Support from people you trust can help you through the often-trying process of resisting sexual harassment. Talking with others allows you to express your feelings and receive emotional support, encouragement, and advice. In addition, it may strengthen your case if you have the opportunity to identify and talk with other people who have been harassed by the offender.
7. *Consider filing a complaint.* Companies and organizations, such as universities and colleges, are required by law to respond reasonably to complaints of sexual harassment. In large organizations, a designated official (sometimes an ombudsman, affirmative action officer, or sexual harassment adviser) is usually charged to handle such complaints. Set up an appointment with this official to discuss your experiences. Ask about the grievance procedures in the organization and your right to confidentiality. Have available a record of the dates of the incidents, what happened, how you felt about it, and so on.

   The two major government agencies that handle charges of sexual harassment are the Equal Employment Opportunity Commission (look under the government section of your phone book for the telephone number of the nearest office) and your state Human Rights Commission (listed in your phone book under state or municipal government). These agencies may offer advice on how you can protect your legal rights and proceed with a formal complaint.
8. *Consider legal remedies.* Sexual harassment is illegal and actionable. If you are considering legal action, consult an attorney familiar with this area of law. You may be entitled to back pay (if you were fired for reasons arising from the sexual harassment), job reinstatement, and punitive damages.

Source. Adapted from Nevid et al., 1995. Reprinted with permission.

Sexual harassment on college campuses is also common. A survey of 2,000 college women showed that about half had been subjected to sexual harassment from professors, most commonly in the form of crude or degrading remarks (Fitzgerald et al., 1988). Nearly 1 in 3 reported unwanted sexual attention and 1 in 10 reported unwanted sexual contact, including fondling and outright sexual assaults. A compilation of results from various studies showed that 7% to 27% of men also reported some exposure to sexual harassment on campus (McKinney & Maroules, 1991).

Sexual harassment may also occur between doctor and patient, or between therapist and client. Professionals may use their power and influence to pressure patients or clients into having sexual relations. In some cases, harassment is disguised by unscrupulous therapists who make it seem that sexual contact would be therapeutically beneficial. Clients tend to perceive therapists as experts whose suggestions carry great authority. Thus, they may be vulnerable to exploitation by therapists who abuse the trust that clients place in their hands. Let's be absolutely clear here: There is no therapeutic justification for a therapist to have sex with a client. The ethical codes of psychologists and other mental health professionals specifically prohibit any kind of sexual contact between therapists and clients. There is no therapeutic justification for sexual relations between client and therapist. Any therapist who makes a sexual overture toward a client, or tries to persuade a client to have sexual relations, is acting unethically. Professionals too may be sexually harassed. Surveys show that more than one-third of female doctors (Frank, Brogan, & Schiffman, 1998) and more than one-half of female psychologists (deMayo, 1997) report they have been sexually harassed by patients.

Effects of Sexual Harassment Sexual harassment, like other forms of sexual abuse, can have damaging psychological effects, including anxiety and lowered self-esteem (e.g., Koss et al., 1994). Women subjected to sexual taunts or outright demands for sexual favors in the workplace may feel forced to resign. Women attending college who are unable to stop sexual harassment by professors may drop courses, switch majors, or even transfer to other schools, often at great personal sacrifice (Fitzgerald, 1993a, 1993b). Forms of sexual harassment such as verbal taunts, sexual teasing, and staring may even have similar effects as sexual abuse, leading to disturbances in body image and eating behavior (Weiner & Thompson, 1997).

The effects of sexual harassment are compounded by the attitude held by many in our society that people who complain of harassment or rape are somehow to blame for their difficulties (Powell, 1991). People who bring harassment complaints may be perceived as exaggerating the incident or taking things too seriously. Women, especially, are subjected to stereotypical expectations that they are to be nice and demure—to be passive and never to "make a scene." A woman who seeks to protect her rights by filing a complaint may be labeled as a "troublemaker." "Women are damned if they assert themselves and victimized if they don't" (Powell, 1991, p. 114).

We focused in this chapter on interpersonal aggression, acts in which one person does violence to another. Yet, throughout history, human aggression in the form of organized conflict and warfare has claimed far more lives than individual aggression and has left widespread devastation in its wake. With the terrorist attack on America on September 11, 2001, the horrific effects of organized violence came to our shores. As we look ahead in this new millenium, we might wonder whether psychologists and other social scientists might use their skills in conflict resolution to help people mediate conflicts that give rise to war and acts of terrorism. Can psychologists succeed where leaders of government, kings and queens, philosophers, and great historic figures have failed? Certainly the causes of organized warfare are complex, which belies any attempt at finding a simple solution. Yet, perhaps psychologists can use their unique perspectives to help people learn nonaggressive ways of resolving conflicts. For example, the cognitive perspective might give us insights into ways in which leaders of nations interpret situations that might set the stage for a warlike response to international conflicts. Efforts toward developing a "peace psychology" are still in their infancy (see Feshbach, 1994). We cannot say whether such efforts will succeed, but we can say it would be a far greater shame not to try.

Truth OR Fiction? REVISITED

In some cases, sexual relations between clients and therapists are therapeutically justified.

FALSE. There is no therapeutic justification for sexual relations between a client and therapist. It is unethical conduct on the part of the therapist.

THINK ABOUT IT

Have you ever been subjected to sexual harassment? What was the outcome? Would you do anything differently if it were to happen again?

THINK ABOUT IT

Suppose you were to be subject to sexual harassment and confided in a friend who tells you, "Forget it. It's no big deal. It happens to everyone sooner or later." How would you respond?

Quiz **16.5**
Sexual Aggression

Quiz **16.6**
Chapter Exam

Research Update
Chapter 16

Summing Up

Violence and Abnormal Behavior

Is violent or aggressive behavior abnormal? Applying the criteria discussed in Chapter 1, we may consider violent or aggressive behavior to be abnormal when it is not socially sanctioned and is either self-defeating or results in harm to self or others.

What is the relationship between violent behavior and psychological disorders? Most people with psychological disorders are nonviolent. However, evidence shows an increased rate of violent behavior is associated with some abnormal behavior patterns, such as acute episodes of schizophrenia, or when severe mental disorders are compounded by problems of alcohol or substance abuse or poor adherence to medication.

Explaining Human Aggression

What are the major biological perspectives on human aggression? The instinct model of human aggression does not hold favor today largely because of criticism that it fails to account for diversity in aggressive behavior in humans and the importance of learning and culture. Contemporary biological models focus on the neurobiological underpinnings of aggression, including brain mechanisms involved in regulating aggressive behavior, the role of neurotransmitters, especially serotonin, and the role of the male sex hormone testosterone.

How do social-cognitive theorists view human aggression? Social-cognitive theorists view aggressive behavior as learned behavior that is influenced by modeling (exposure to aggressive models), expectancies (expectations of positive outcomes of aggressive behavior), competencies (skills needed to handle conflict situations through nonviolent means), and interpretation of social cues in confrontative situations (e.g., misreading other people's motives or intentions).

How do sociocultural theorists view human aggression? Sociocultural theorists believe that violent behavior is rooted in underlying social causes, such as poverty, lack of opportunity, family breakdown, exposure to deviant role models, and social stressors such as unemployment. They also focus on the role of violence as a social influence tactic.

How might alcohol use contribute to aggressive behavior? Alcohol use may make it difficult to weigh the consequences of one's behavior, loosen inhibitions, and decrease sensitivity to anxiety-arousing cues that might otherwise serve to inhibit aggressive impulses.

How are emotional factors implicated in aggression? Emotional states such as frustration and anger may act as catalysts for aggressive behavior, especially when the person blames someone else for being responsible for the frustrating or angering situation.

What general conclusion might we reach about the underlying causes of human aggression? Most theorists recognize that human aggression is a complex form of behavior that cannot be explained by any single cause. We may come to a better understanding of human aggression by adopting a multiple causation model that takes into account the contributions and interactions of biological, psychological, and sociocultural factors.

Domestic Violence

What factors are linked to domestic violence? Relationship problems, alcohol use, and feelings of inadequacy and low self-esteem on the part of the batterer often figure prominently in battering relationships. Spousal battering typically occurs within a larger pattern of abuse that may also involve child abuse and sexual aggression against the spouse. Battering incidents often follow a triggering event that leads the abuser to lose control.

How do feminist theorists view domestic violence? Feminist theorists believe that domestic violence arises from the differential power relationships that exist between men and women in our society. They point out that domestic violence occurs within a larger social context in which men are socialized into dominant roles while women are expected to be subordinate to their interests.

What are the effects of domestic violence? The effects range from physical injury to PTSD and other emotional problems, especially depression and self-esteem. Women who remain in abusive relationships should be viewed as survivors of trauma, not as masochists.

Child Abuse

What are the different forms of child abuse and maltreatment? There are several different forms of child abuse and maltreatment, including neglect, physical abuse, sexual abuse, and emotional maltreatment. Neglect is the most frequently occurring form of abuse.

What factors are linked to child abuse? Stress, poor parenting and anger management skills, and a history of being physically abused during childhood or witnessing family violence are implicated among various risk factors for predicting which parents are at increased risk of abusing their children.

What are the effects of child abuse? The effects of child abuse range from physical injuries, even death, to emotional consequences, such as difficulties forming healthy attachments, low self-esteem, suicidal thinking, depression, and failure to explore the outside world, among other problems. The emotional and behavioral consequences of child abuse and neglect often extend into adulthood.

Sexual Aggression

What factors underlie rape? Rape has more to do with violent impulses and issues of power than with pursuit of sexual gratification. The desires to dominate women or express hatred toward them are prominent motives for rape. Although some rapists show clear evidence of underlying psychopathology, many do not. Sociocultural theorists argue that cultural attitudes and practices may socialize young men into sexually aggressive or dominant roles.

What effects does rape have on survivors? Although the effects are variable, rape survivors often suffer a range of immediate and long-term psychological effects.

What is child sexual abuse? Any sexual act involving children is a form of sexual abuse, even if no direct force or threat of force is used, or the child consents. The great majority of assailants have had some prior relationship with the child or child's family.

What are the effects of child sexual abuse? Although the effects vary, survivors typically experience a range of emotional and behavioral problems in childhood and adolescence. Survivors may also have difficulty forming or maintaining intimate relationships in adulthood.

What is sexual harassment? Sexual harassment involves unwelcome sexual comments, overtures, gestures, physical contact, or direct demands for sexual favors as a condition of employment, retention, or advancement. Sexual harassment may occur in many places, including the workplace, school, or doctor's or therapist's office. Like other forms of abuse, sexual harassment can have damaging psychological consequences.

What factors underlie sexual harassment? Sexual harassment typically has more to do with the abuse of power than sexual motivation. It is often used by men as a tactic to "keep women in their place," especially in situations that have been traditional male preserves.

CHAPTER SEVENTEEN

[Ab]normal Psychology and Society

P. Filonov
Man in the World

Truth OR Fiction?

- People can be committed to psychiatric facilities because of eccentric behavior. (p. 538)
- Psychologists and other mental health professionals can accurately predict the dangerousness of people they evaluate. (p. 540)
- Therapists are not obligated to breach client confidentiality even to warn intended victims of threats of violence made against them by their clients. (p. 542)
- A ruling in a legal case in Alabama established that patients in mental hospitals in the state may be required to perform general housekeeping duties in the facility. (p. 545)
- An attempt to assassinate the president of the United States was seen by millions of television viewers, but the would-be assassin was found not guilty by a court of law. (p. 547)
- The insanity defense is used in a large number of trials, usually successfully. (p. 548)
- People who are found not guilty of a crime by reason of insanity may remain confined to a mental hospital indefinitely—for many years longer than they would have been sentenced to prison, if they had been found guilty. (p. 550)
- It is possible for a defendant to be held to be competent to stand trial but still be judged not guilty of a crime by reason of insanity. (p. 552)

Chapter Outline

Larry Hogue—the "wild man" of West 96th Street. Hogue, a homeless veteran of the Vietnam War who dwells in the alleyways and doorways of Manhattan's Upper West Side. Hogue, a middle-aged man who goes barefoot in winter, eats from garbage cans, and mutters to himself (Dugger, 1992). Hogue, who reportedly stalked a teacher and threatened to cook and eat her fawn-colored Akita. Hogue, who reportedly becomes violent when he smokes crack and was once arrested for pushing a schoolgirl in front of a school bus (Shapiro, 1992). (Miraculously, she escaped injury.) Hogue, who had been shuttled in and out of state psychiatric hospitals and prisons more than 40 times (Wickenhaver, 1992). Hogue, for whom the criminal justice systems and mental health systems are nothing but revolving doors. Hogue, whom many regard as the living embodiment of the cracks in our mental health, criminal justice, and social services systems.

Typically, Hogue would improve during a brief hospital stay and be released, only to return to using crack instead of his psychiatric medication. His behavior would then deteriorate (Dugger, 1994).

What does society do about Larry Hogue? What does society do about Joyce Brown?

Joyce Brown? Joyce Brown was a middle-aged woman who also lived on the streets of New York. At one time she slept above a hot air vent on the sidewalk on the Upper East Side of Manhattan, in the midst of some of the most expensive real estate in the world. Sometimes she was observed defecating in her clothes or on the sidewalk. She hurled insults at strangers and refused to go to a shelter, preferring to live in the streets, despite the obvious dangers of potential attack and the risks of exposure to the elements.

In New York, a program was begun to provide outreach services to homeless people in need of psychological treatment. Teams of specialists, each consisting of a nurse, a social worker, and a psychiatrist, were charged with the task of identifying and monitoring behaviorally disordered homeless people, helping them obtain services, bringing them soup and sandwiches, and taking them to the hospital if they were deemed to represent an immediate threat to themselves or others under the authority provided by state laws governing psychiatric commitment, even if it was against the individual's will.

Brown had been picked up and brought against her will to a city hospital for evaluation, where she was diagnosed with paranoid schizophrenia and judged to be in need of treatment. She resisted treatment and claimed that she had a right to live her life as she saw fit, even if it offended other people. As long as she committed no crime, what right did society have to deprive her of her liberty? Yes, she admitted, she had defecated in the streets. But there were no public rest rooms available, and establishments such as restaurants had refused her access.

Brown sued for her release, and while her case meandered through the courts, she remained in the hospital, although her doctors were prevented from medicating her against her will. Because she refused medication, the doctors released her, claiming there was little they could do for her. Although we've lost track of Joyce Brown, Larry Hogue turned up a few years later in his old haunts on the Upper West Side of Manhattan, where he was seen panhandling. Local residents expressed fears that he would return to terrorizing them (Holloway, 1998).

The cases of Larry Hogue and Joyce Brown touch on the more general issue of how to balance the rights of the individual with the rights of society. Do people, for example, have the right to live on the streets under unsanitary conditions? There are those who argue that a just and humane society has the right and responsibility to care for people who are perceived incapable of protecting their own best interests, even if "care" means involuntarily committing them to a psychiatric institution. Do people who are obviously mentally disturbed have the right to refuse treatment? Do psychiatric institutions have the right to inject them with antipsychotic and other drugs against their will? Should mental patients with a history of disruptive or violent behavior be hospitalized indefinitely or permitted to live in supervised residences in the community once their conditions are stabilized? When severely disturbed people break the law, should society respond to them with the criminal justice system or with the mental health system?

"The Wild Man of West 96th Street." Larry Hogue, the so-called "Wild Man of West 96th Street" in New York City, has become a symbol of the cracks in the mental health, criminal justice, and social services systems.

THINK ABOUT IT

Do you believe that schizophrenia patients who wander city streets mumbling to themselves and living in cardboard boxes should be hospitalized against their will? Why or why not?

In this chapter we consider psychiatric commitment and other issues that arise from society's response to abnormal behavior, such as the rights of patients in institutions, the use of the insanity defense in criminal cases, and the responsibility of professionals to warn individuals who may be placed at risk by the dangerous behavior of their clients.

Psychiatric Commitment and Patient's Rights

Legal placement of people in psychiatric institutions against their will is called **civil,** or psychiatric, **commitment.** Through civil commitment, individuals deemed to be mentally disordered and to be a threat to themselves or others may be involuntarily confined to psychiatric institutions to provide them with treatment and help ensure their own safety and that of others. Civil commitment should be distinguished from *legal* or *criminal commitment,* in which an individual who has been acquitted of a crime by reason of insanity is placed in a psychiatric institution for treatment. In **legal commitment,** a criminal's unlawful act is judged by a court of law to result from a mental disorder or defect, and the person is committed to a psychiatric hospital where treatment can be provided rather than incarcerated in a prison.

Civil commitment should also be distinguished from *voluntary hospitalization,* in which an individual voluntarily seeks treatment in a psychiatric institution, and can, with adequate notice, leave the institution when she or he so desires. Even in such cases, however, when the hospital staff believes that a voluntary patient presents a threat to her or his own welfare or to others, they may petition the court to change the patient's legal status from voluntary to involuntary.

Civil commitment in a psychiatric hospital usually requires that a petition be filed by a relative or a professional. Psychiatric examiners may be empowered by the court to evaluate the person in a timely fashion, after which a judge hears psychiatric testimony and decides whether or not to commit the individual. In the event of commitment, the law usually requires periodic legal review and recertification of the patient's involuntary status. The legal process is intended to ensure that people are not indefinitely "warehoused" in psychiatric hospitals. Hospital staff must demonstrate the need for continued inpatient treatment.

Truth OR Fiction? REVISITED

People can be committed to psychiatric facilities because of eccentric behavior.

FALSE. People cannot be committed because they are eccentric. The U.S. Supreme Court has determined that people must be judged mentally ill and present a clear and present danger to themselves or others to be psychiatrically committed.

Legal safeguards are usually in place to protect people's civil rights in commitment proceedings. Defendants have the right to due process and to be assisted by an attorney, for example. But when individuals are deemed to present a clear and imminent threat to themselves or others, the court may order immediate hospitalization until a more formal commitment hearing can be held. Such emergency powers are usually limited to a specific period, usually 72 hours. During this time a formal commitment petition must be filed with the court, or the individual has a right to be discharged.

Standards for psychiatric commitment have been tightened over the past generation, and the rights of individuals who are subject to commitment proceedings are more strictly protected. In the past, psychiatric abuses were more commonplace. People were often committed without clear evidence that they posed a threat. Not until 1979, in fact, did the U.S. Supreme Court rule, in *Addington* v. *Texas,* that in order for individuals to be hospitalized involuntarily, they must be judged both to be "mentally ill" and to present a clear and present danger to themselves or others. Thus people cannot be committed because of their eccentricity.

Web Link 17.1 WWW
Facts About Psychiatric Hospitalization

Few would argue that contemporary tightening of civil commitment laws provides greater protection of the rights of the individual. Even so, some critics of the psychiatric system have called for the complete abolition of psychiatric commitment on the grounds that commitment deprives the individual of liberty in the name of therapy, and that such a loss of liberty cannot be justified in a free society. Perhaps the most vocal and persistent critic of the civil commitment statutes is psychiatrist Thomas Szasz (Szasz, 1970). Szasz argued that the label of *mental illness* is a societal invention that transforms social deviance into medical illness. In Szasz's view, people should not be deprived of their liberty because their behavior is perceived to be socially deviant or disruptive. According to Szasz, people who violate the law should be prosecuted for criminal behavior, not confined to a psychiatric hospital.

civil commitment The legal process of placing a person in a mental institution, even against his or her will.

legal commitment The legal process of confining a person found not guilty by reason of insanity in a mental institution.

Although psychiatric commitment may prevent some individuals from acting violently, it does violence to many more by depriving them of liberty:

> The mental patient, we say, *may be* dangerous: he may harm himself or someone else. But we, society, *are* dangerous: we rob him of his good name and of his liberty, and subject him to tortures called "treatments." (Szasz, 1970, p. 279)

Szasz's strident opposition to institutional psychiatry and his condemnation of psychiatric commitment focused attention on abuses in the mental health system. Szasz also persuaded many professionals to question the legal, ethical, and moral bases of coercive psychiatric treatment in the forms of involuntary hospitalization and forced medication. Many caring and concerned professionals draw the line at abolishing psychiatric commitment, however. They argue that people may not be acting in their considered best interests when they threaten suicide or harm to others, or when their behavior becomes so disorganized that they cannot meet their basic needs.

THINK ABOUT IT
Do you believe we should abolish psychiatric commitment? Why or why not?

Predicting Dangerousness

To be psychiatrically committed, people must be judged to be dangerous to themselves or others. Professionals are thus responsible for making accurate predictions of dangerousness to determine whether people should be involuntarily hospitalized or maintained involuntarily in the hospital. But how accurate are professionals in predicting dangerousness? Do professionals have special skills or clinical wisdom that renders their predictions accurate, or are their predictions no more accurate than those of laypeople?

Unfortunately, psychologists and other mental health professionals who rely on their clinical judgments are not very accurate when it comes to predicting dangerousness of the people they treat. Mental health professionals tend to overpredict dangerousness—that is, to label many individuals as dangerous when they are not (Monahan, 1981). Clinicians tend to err on the side of caution in overpredicting the potential for dangerous behavior, perhaps because they believe that failure to predict violence may have more serious consequences than overprediction. Overprediction of dangerousness does deprive many people of liberty on the basis of fears that turn out to be groundless. According to Szasz and other critics of the practice of psychiatric commitment, the commitment of the many to prevent the violence of the few is a form of preventive detention that violates the basic principles on which the United States was founded.

The leading professional organizations, the American Psychological Association (1978) and the American Psychiatric Association (1998), have both gone on record as stating that neither psychologists nor psychiatrists, respectively, can reliably predict violent behavior. As a leading authority in the field, John Monahan of the University of Virginia, put

Should she be committed to a psychiatric institution? People must be judged as dangerous in order to be psychiatrically hospitalized against their wills. This photograph of emergency workers pulling a woman away from a ledge after she threatened to jump leaves little doubt about the dangerousness of her behavior. But professionals have not demonstrated that they can reliably predict future dangerousness.

Truth OR Fiction? REVISITED

Psychologists and other mental health professionals can accurately predict the dangerousness of people they evaluate.

FALSE. Psychologists and other mental health professionals who rely on their clinical judgments are not very accurate when it comes to predicting the dangerousness of the people they treat. The best predictor of future violence is a history of past violence.

the issue, "When it comes to predicting violence, our crystal balls are terribly cloudy" (Rosenthal, 1993, p. A1).

The accuracy of clinician predictions of dangerousness is significantly better than predictions based on chance alone, but is still often inaccurate (Kaplan, 2000). Clinician predictions are generally also less accurate than predictions based on evidence of past violent behavior (Gardner et al., 1996; Mossman, 1994). Basically, clinicians do not possess any special knowledge or ability for predicting violence beyond that of the average person. In fact, a layperson supplied with information concerning an individual's past violent behavior may be more accurate in predicting the individual's potential for future violence than the clinician who bases a prediction solely on information obtained from a clinical interview (Mossman, 1994). Unfortunately, although past violent behavior may be the best predictor of future violence, hospital staff may not be permitted access to criminal records or may lack the time or resources to track down these records (Rosenthal, 1993). The prediction problem has been cited by some as grounds for the abandonment of dangerousness as a criterion for civil commitment.

Although their crystal balls may be cloudy, mental health professionals who work in institutional settings continue to be called on to make these predictions—deciding whom to commit and whom to discharge based largely on how they appraise the potential for violence. Clinicians may be more successful in predicting violence by basing predictions on a composite of factors, including evidence of past violent behavior, than on any single factor (Shaffer, Waters, & Adams, 1994). Not surprisingly, the accuracy of clinician predictions of violence tends to be greater when clinicians agree with one another than when they disagree (McNiel, Lam, & Binder, 2000). Accuracy is also improved when clinicians make short-term predictions of dangerousness, such as predictions of imminent violence (Binder, 1999). Another variable that may increase predictability is substance abuse. The potential for violence is heightened in people with serious psychiatric disorders when they use alcohol, crack, or other drugs (Rosenthal, 1993; Tardiff et al., 1997). It is also heightened among schizophrenia patients who experience command hallucinations—voices commanding them to harm themselves or others (McNiel, Lam, & Binder, 2000).

Various factors may lead to inaccurate predictions of dangerousness, including the following.

The *Post Hoc* Problem Recognizing violent tendencies after a violent incident occurs (*post hoc*) is easier than predicting it (*ad hoc*). It is often said that hindsight is 20/20. Like Monday morning quarterbacking, it is easier to piece together fragments of people's prior behaviors as evidence of their violent tendencies *after* they have committed acts of violence. Predicting a violent act before the fact is a more difficult task, however.

The Problem in Leaping from the General to the Specific Generalized perceptions of violent tendencies may not predict specific acts of violence. Most people who have "general tendencies" toward violence may never act on them. Nor is classification within a diagnostic category associated with aggressive or dangerous behavior, such as antisocial personality disorder, a sufficient basis for predicting specific violent acts in individuals (Bloom & Rogers, 1987).

Problems in Defining Dangerousness One difficulty in assessing the predictability of dangerousness is the lack of agreement in defining the criteria for labeling behavior as violent or dangerous. There is no universal agreement on the definition of violence or dangerousness. Most people would agree that crimes such as murder, rape, and assault are acts of violence. There is less agreement, even among authorities, for labeling other acts—for example, driving recklessly, harshly criticizing one's spouse or children, destroying property, selling drugs, shoving into people at a tavern, or stealing cars—as violent or dangerous. Consider, also, the "dangerous behavior" of business owners and corporate executives who produce and market cigarettes despite widespread knowledge of the death and disease caused by these substances. Clearly, the determination of which behaviors are regarded as dangerous involves moral and political judgments within a given social context (Monahan, 1981).

Base-Rate Problems The prediction of dangerousness is complicated by the fact that violent acts such as murder, assault, or suicide are infrequent or rare events at the individual level within the general population, even if newspaper headlines sensationalize them regularly. Other rare events—such as earthquakes—are also difficult to predict with any degree of certainty concerning when or where they will strike.

The relative difficulty of making predictions of infrequent or rare events is known as the *base-rate problem.* Consider as an example the problem of suicide prediction. If the suicide rate in a given year has a low base rate of about 1% of a clinical population, the likelihood of accurately predicting that any given person in this population will commit suicide is not very favorable. You would be correct 99% of the time if you predicted that any given individual in this population would *not* commit suicide in a given year. But to predict the nonoccurrence of suicide in every case would mean you would fail to predict the relatively few cases in which suicide does occur, even though virtually all of your predictions would likely be correct. Yet predicting the one likely case of suicide among each 100 people in the population is likely to be tricky. You are likely to be wrong more often than not if you made predictions of suicide in a given year in only 3 cases out of 100 (even if 1 of the 3 did commit suicide).

When clinicians make predictions, they weigh the relative risks of incorrectly failing to predict the occurrence of a behavior (a *false negative*) against the consequences of incorrectly predicting it (a *false positive*). Clinicians often err on the side of caution by overpredicting dangerousness. From the clinician's perspective, erring on the side of caution might seem like a no-lose situation. Yet many people committed to an institution under such circumstances are denied their liberty when they would not actually have acted violently against themselves or others.

Is he suicidal? One reason clinicians have difficulty predicting violent behaviors such as suicide or murder is that they are relatively infrequent acts. Clinicians often err on the side of caution and so have a tendency toward overpredicting dangerousness.

The Unlikelihood of Disclosure of Direct Threats of Violence How likely is it that truly dangerous people will disclose their intentions to a health professional who is evaluating them or to their own therapist? The client in therapy is not likely to inform a therapist of a clear threat, such as "I'm going to kill _____ next Wednesday morning." Threats are more likely to be vague and nonspecific, as in "I'm so sick of _____; I could kill her," or "I swear he's driving me to murder." In such cases, therapists must infer dangerousness from hostile gestures and veiled threats. Vague, indirect threats of violence are less reliable indicators of dangerousness than specific, direct threats.

The Difficulty of Predicting Behavior in the Community from Behavior in the Hospital Mental health professionals fall well short of the mark when making long-term predictions of dangerousness (Buchanan, 1999). They are often wrong when making predictions of dangerousness of hospitalized patients following their release from the hospital. One reason is that they often base their predictions on patients' behavior in the hospital. But violent or dangerous behavior may be situation specific. A model patient who is able to adapt to a structured environment like that of a psychiatric hospital may be unable to cope with pressures of independent communal life. Clinicians are generally more accurate when their predictions are based on the patient's past behavior in the community, such as a history of violent incidents, rather than on behavior in the hospital setting (Klassen & O'Connor, 1988).

As we explore in the nearby "A Closer Look" section, the problem of predicting dangerousness also arises when therapists need to evaluate the seriousness of threats made by their patients against others.

THINK ABOUT IT

If you were called on to evaluate whether an individual posed a danger to himself or herself or others, on what criteria would you base your judgment?

WWW **Web Link 17.2**
Confidentiality Issues

Patients' Rights

We have considered society's right to hospitalize involuntarily people who are judged to be mentally ill and to pose a threat to themselves or others. What happens following commitment, however? Do involuntarily committed patients have the right to receive or demand treatment? Or can society just warehouse them in psychiatric facilities indefinitely without treating them? Consider the opposite side of the coin as well: May people who are involuntarily committed refuse treatment? Such issues—which have been brought into public light

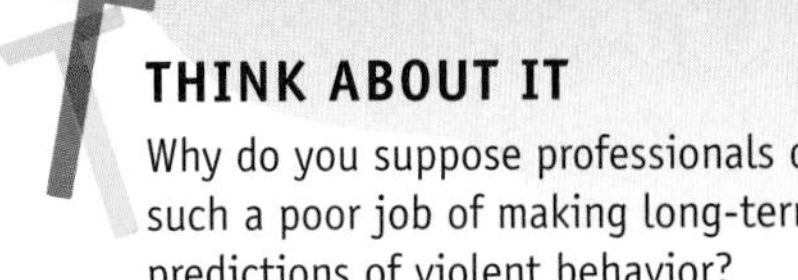

THINK ABOUT IT

Why do you suppose professionals do such a poor job of making long-term predictions of violent behavior?

A Closer Look

The Duty to Warn

One of the most difficult dilemmas a therapist may face is whether to disclose confidential information that may protect third parties from harm. Part of the difficulty lies in determining whether the client has made a bona fide threat against another person. The other part of the dilemma is that information a client discloses in psychotherapy is generally protected as privileged communication, which carries a right to confidentiality. But this right is not absolute. Courts in some states have determined that a therapist is obliged to breach confidentiality under certain conditions, such as when there is clear and compelling evidence that an individual poses a serious threat to others.

Truth OR Fiction? REVISITED

Therapists are not obligated to breach client confidentiality even to warn intended victims of threats of violence made against them by their clients.

FALSE. Therapists are obligated under some state laws to breach client confidentiality in order to warn people when threats of violence are made against them by their clients.

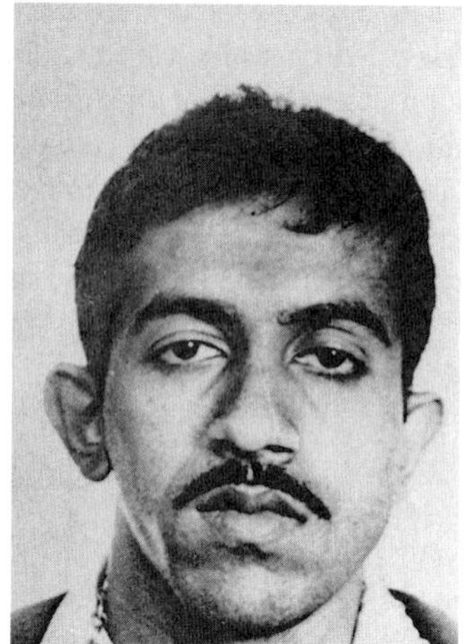

Tatiana Tarasoff and Prosenjit Poddar. Poddar, Tatiana's killer, was a rejected suitor who had made threats against her to his therapist at a university health center. Poddar was subsequently convicted of voluntary manslaughter in her death. A suit brought by Tatiana's parents against the university led to a landmark court ruling that established an obligation on therapists to warn third parties of threats made against them by their clients.

A California case, *Tarasoff* v. *the Regents of the University of California*, established the legal basis for the therapist's **duty to warn.** In 1969, a graduate student at the University of California at Berkeley, Prosenjit Poddar, a native of India, became depressed when his romantic overtures toward a young woman, Tatiana Tarasoff, were rebuffed. Poddar entered psychotherapy with a psychologist at a student health facility, during the course of which he informed the psychologist that he intended to kill Tatiana when she returned from her summer vacation. The psychologist, concerned about Poddar's potential for violence, first consulted with his colleagues and then notified the campus police. He informed them that Poddar was dangerous and recommended he be taken to a facility for psychiatric treatment.

Poddar was subsequently interviewed by the campus police. They believed he was rational and released him after he promised to keep his distance from Tatiana. Poddar then terminated treatment with the psychologist, and shortly afterward killed Tatiana. He shot her with a pellet gun when she refused to allow him entry to her home and then repeatedly stabbed her as she fled into the street. Poddar was found guilty of the lesser sentence of voluntary manslaughter rather than murder based on testimony of three psychiatrists that Poddar suffered from diminished mental capacity and paranoid schizophrenia. Under California law, his diminished capacity prevented the finding of malice that was necessary for conviction on a charge of first- or second-degree murder. Following a prison term, Poddar returned to India, where he reportedly made a new life for himself (Schwitzgebel & Schwitzgebel, 1980).

Tatiana's parents, however, sued the university. They claimed that the university health center had failed in its responsibility to warn Tatiana of the threat made against her by Poddar. The Supreme Court of

by landmark court cases—fall under the umbrella of *patients' rights.* Generally speaking, the history of abuses in the mental health system, as highlighted in such popular books and movies as *One Flew Over the Cuckoo's Nest,* have led to a tightening of standards of care and adoption of legal guarantees to protect patients' rights. The legal status of some issues, such as the right to treatment, remains unsettled, however.

duty to warn The therapist's obligation to warn third parties of threats made against them by clients.

Right to Treatment We might assume that mental health institutions that accept people for treatment would provide them with treatment. Not until the 1972 landmark federal court case of *Wyatt* v. *Stickney,* however, did a federal court establish a minimum standard of care to

the State of California agreed with the parents. They ruled that a therapist who has reason to believe that a client poses a serious threat to another person is obligated to warn the potential victim. This obligation is not met by notifying police. This ruling imposed on therapists a duty-to-warn obligation when their clients show the potential for violence by making threats against others.

The ruling recognized that the rights of the intended victim outweigh the rights of confidentiality. Under *Tarasoff,* the therapist does not merely have a *right* to breach confidentiality and warn potential victims of danger, but is *obligated* by law to divulge such confidences to the victim.

The duty-to-warn provision poses ethical and practical dilemmas for psychologists and other psychotherapists. Psychotherapists, in states that apply the *Tarasoff* obligation, have a duty to assess the potential violence of their clients, even though professionals are not generally able to predict dangerousness with a high degree of accuracy. Under *Tarasoff,* then, therapists may actually feel obliged to protect their personal interests and those of others by breaching confidentiality on the mere suspicion that their clients harbor violent intentions toward third parties. Because there are very few cases in which clients' threats are carried out, the *Tarasoff* ruling may deny many clients their rights to confidentiality in order to prevent such rare instances. Although some clinicians may "overreact" to *Tarasoff* and breach confidentiality without sufficient cause, it can be argued that the interests of the few potential victims outweigh the interests of the many who may suffer a loss of confidentiality.

Although therapists apparently possess no special ability to predict dangerousness, the *Tarasoff* ruling obliges them to judge whether or not their clients' disclosures indicate a clear intent to harm others. In the *Tarasoff* case, the threat was obvious enough to prompt the therapist to breach confidentiality by requesting the help of campus police. In most cases, however, threats are not so clear-cut. There remains a lack of clear criteria for determining whether or not a therapist "should have known" that a client was dangerous before a violent act occurs (Fulero, 1988). In the absence of guidelines that specify the criteria therapists should use to fulfill their duty to warn, they must rely on their best subjective judgments.

Although the intent of the *Tarasoff* decision was to protect potential victims, it may inadvertently increase the risks of violence when applied to clinical practice (Stone, 1976). For example,

1. *Clients may be less willing to confide in their therapists.* Under the obligations imposed on therapists by *Tarasoff,* clients may be less willing to confide violent urges to their therapists, making it more difficult for therapists to help them diffuse these feelings before they are acted on.
2. *Potentially violent people may be less likely to enter therapy.* People with violent tendencies may be less willing to enter therapy for fear that disclosures made to a therapist may be revealed.
3. *Therapists may be less likely to probe violent tendencies for fear of legal complications.* To protect themselves and their careers, therapists may avoid asking clients questions concerning potential violence in the belief that they are legally protected if they remain ignorant of them (Wise, 1978). Therapists might also avoid accepting patients for treatment who are believed to have violent tendencies.

It is unclear whether *Tarasoff* has protected lives or endangered lives. It is clear, however, that *Tarasoff* has raised concerns for clinicians who are trying to meet their legal responsibilities under *Tarasoff* and their clinical responsibilities to their clients.

A survey of psychiatric residents in San Francisco showed that nearly half had issued Tarasoff-type warnings (Binder & McNiel, 1996). In most cases, the intended victim had already been aware of the threat. Most of the patients were told by their therapists that warnings had been issued. Though issuing a warning produced no clear effects on the therapeutic relationship in most cases, negative effects were reported in some cases.

The *Tarasoff* case was brought in California, and the decision applied only to that state (DeBell & Jones, 1997b). Other states vary in their statutes that apply in duty-to-warn cases (Schaffer, 2000). In most states, therapists are granted discretion in deciding whether or not to notify a threatened person or law enforcement officials when threats are made by their clients in the therapeutic context (USDHHS, 1999a). Therapists must be aware of the statutes and legal precedents that exist in the particular states in which they practice. Therapists must also not lose sight of the primary therapeutic responsibility to their clients when legal issues arise. They must balance the obligation to meet their responsibilities under duty-to-warn provisions with the need to help their clients resolve the feelings of rage and anger that give rise to violent threats.

be provided by hospitals. The case was a class action suit against Stickney, the commissioner of mental health for the State of Alabama, brought on behalf of Ricky Wyatt, a mentally retarded young man, and other patients at a state hospital and school in Tuscaloosa.

The federal district court in Alabama held both that the hospital had failed to provide treatment to Wyatt and others and that living conditions at the hospital were inadequate and dehumanizing. The court described the hospital dormitories as "barnlike structures" that afforded no privacy to the residents. The bathrooms had no partitions between stalls, the patients were outfitted with shoddy clothes, the wards were filthy and crowded, the kitchens were unsanitary, and the food was substandard. In addition, the staff

THINK ABOUT IT

Do you believe therapists should be obligated to breach confidentiality in cases in which their patients make threats against others? Why or why not? What concerns have therapists raised about the duty to warn? Do you believe their concerns are warranted?

What are the rights of mental patients? Popular books and films such as *One Flew Over the Cuckoo's Nest* starring Jack Nicholson have highlighted many of the abuses of mental hospitals. In recent years, a tightening of standards of care and the adoption of legal safeguards have led to better protection of the rights of patients in mental hospitals.

Web Link 17.3 WWW
National Alliance for the Mentally Ill

was inadequate in numbers and poorly trained. The case of *Wyatt* v. *Stickney* established certain patient rights, including the right not to be required to perform work that is performed for the sake of maintaining the facility. The court held that mental hospitals must, at a minimum, provide the following:

1. A humane psychological and physical environment,
2. Qualified staff in numbers sufficient to administer adequate treatment, and
3. Individualized treatment plans (*Wyatt* v. *Stickney*, 334 Supp., p. 1343, 1972).

The court established that the state was obliged to provide adequate treatment for people who were involuntarily confined to psychiatric hospitals. The court further ruled that to commit people to hospitals for treatment involuntarily, and then not to provide treatment, violated their rights to due process under the law.

A listing of some of the rights granted institutionalized patients under the court's ruling is shown in Table 17.1. Although the ruling of the court was limited to Alabama, many other states have followed suit and revised their mental hospital standards to ensure that involuntarily committed patients are not denied basic rights.

Other court cases have further clarified patients' rights.

O'Connor* v. *Donaldson The 1975 case of Kenneth Donaldson is another landmark in patients' rights. Donaldson, a former patient at a state hospital in Florida, sued two hospital doctors on the grounds that he had been involuntarily confined without receiving treatment for 14 years, despite the fact that he posed no serious threat to himself or others. Donaldson had been originally committed on the basis of a petition filed by his father, who had perceived him as delusional. Despite the fact that Donaldson received no treatment during his confinement and was denied grounds privileges and occupational training, his repeated requests for discharge were denied by the hospital staff. He was finally released when he threatened to sue the hospital. Once discharged, Donaldson did sue his doctors and was awarded damages of $38,500 from O'Connor, the superintendent of the hospital. The case was eventually argued before the U.S. Supreme Court.

TABLE 17.1 Partial Listing of the Patient's Bill of Rights Under *Wyatt* v. *Stickney*

1. Patients have rights to privacy and to be treated with dignity.
2. Patients shall be treated under the least restrictive conditions that can be provided to meet the purposes that commitment was intended to serve.
3. Patients shall have rights to visitation and telephone privileges unless special restrictions apply.
4. Patients have the right to refuse excessive or unnecessary medication. In addition, medication may not be used as a form of punishment.
5. Patients shall not be kept in restraints or isolation except in emergency conditions in which their behavior is likely to pose a threat to themselves or others and less restrictive restraints are not feasible.
6. Patients shall not be subject to experimental research unless their rights to informed consent are protected.
7. Patients have the right to refuse potentially hazardous or unusual treatments, such as lobotomy, electroconvulsive shock, or aversive behavioral treatments.
8. Unless it is dangerous or inappropriate to the treatment program, patients shall have the right to wear their own clothing and keep possessions.
9. Patients have rights to regular exercise and to opportunities to spend time outdoors.
10. Patients have rights to suitable opportunities to interact with the opposite gender.
11. Patients have rights to humane and decent living conditions.
12. No more than six patients shall be housed in a room and screen or curtains must be provided to afford a sense of privacy.
13. No more than eight patients shall share one toilet facility, with separate stalls provided for privacy.
14. Patients have a right to nutritionally balanced diets.
15. Patients shall not be required to perform work that is performed for the sake of maintenance of the facility.

Court testimony established that although the hospital staff had not perceived Donaldson to be dangerous during his hospitalization, they had refused to release him. The hospital doctors argued that continued hospitalization had been necessary because they had believed Donaldson was unlikely to adapt successfully to community living. The doctors had prescribed antipsychotic medications as a course of treatment, but Donaldson had refused to take them because of his Christian Science beliefs. As a result, he received only custodial care.

The Supreme Court held that "mental illness [alone] cannot justify a State's locking a person up against his will and keeping him indefinitely in simple custodial confinement. There is still no constitutional basis for confining such persons involuntarily if they are dangerous to no one and can live safely in freedom" (p. 2493). The ruling addressed patients who are not considered dangerous. It is not yet clear whether the same constitutional rights would be applied to committed patients who are judged to be dangerous.

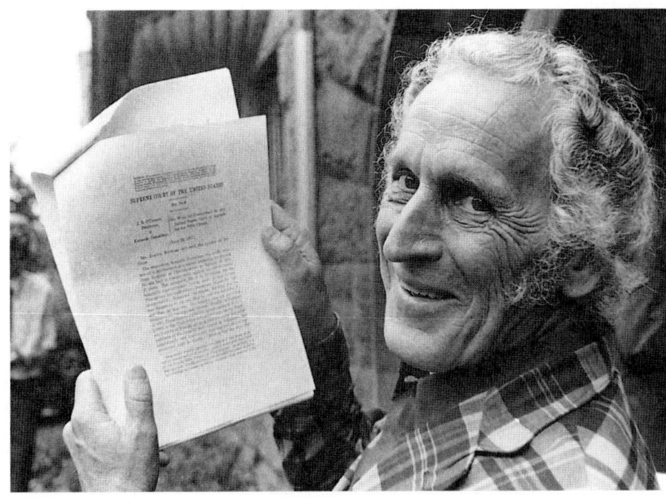

Kenneth Donaldson. Donaldson points to the U.S. Supreme Court decision that ruled that people who are considered mentally ill but not dangerous cannot be confined against their will if they can be maintained safely in the community.

In its ruling in *O'Connor* v. *Donaldson,* the Supreme Court did not deal with the larger issue of the rights of patients to receive treatment. The ruling does not directly obligate state institutions to treat involuntarily committed, nondangerous people because the institutions may elect to release them instead.

The Supreme Court did touch on the larger issue of society's rights to protect itself from individuals who are perceived as offensive. In delivering the opinion of the Court, Justice Potter Stewart wrote,

> May the State fence in the harmless mentally ill solely to save its citizens from exposure to those whose ways are different? One might as well ask if the State, to avoid public uneasiness, could incarcerate all who are physically unattractive or socially eccentric. Mere public intolerance or animosity cannot constitutionally justify the deprivation of a person's physical liberty. (*O'Connor* v. *Donaldson,* 95 S. Ct. 2486, 1975)

WWW **Web Link 17.4**
Mental Health Bill of Rights

Truth OR Fiction? REVISITED

A ruling in a legal case in Alabama established that patients in mental hospitals in the state may be required to perform general housekeeping duties in the facility.

FALSE. The Alabama case of *Wyatt* v. *Stickney* established certain patient rights, including the right not to be required to perform work that is performed for the sake of maintaining the facility.

Youngberg* v. *Romeo In a 1982 case, *Youngberg* v. *Romeo,* the U.S. Supreme Court more directly addressed the issue of the patient's right to treatment. Even so, it seemed to retreat somewhat from the patients' rights standards established in *Wyatt* v. *Stickney.* Nicholas Romeo, a 33-year-old man with profound retardation who was unable to talk or care for himself, had been institutionalized in a state hospital and school in Pennsylvania. While in the state facility he had a history of injuring himself through his violent behavior and was often kept in restraints. The case was brought by the patient's mother, who alleged that the hospital was negligent in not preventing his injuries and in routinely using physical restraints for prolonged periods while not providing adequate treatment.

The Supreme Court ruled that involuntarily committed patients, like Nicholas, have a right to be confined in less restrictive conditions, such as being freed from physical restraints whenever it is reasonable to do so. The Supreme Court ruling also included a limited recognition of the committed patient's right to treatment. The Court held that institutionalized patients have a right to minimally adequate training to help them function free of physical restraints, but only to the extent that such training can be provided in reasonable safety. The determination of reasonableness, the Court held, should be made on the basis of the judgment of the qualified professionals in the facility. The federal courts should not interfere with the internal operations of the facility, the Court held, because "there's no reason to think judges or juries are better qualified than appropriate professionals in making such decisions" (p. 2462). The courts should only second-guess the judgments of qualified professionals, the Supreme Court held, when such judgments are determined to depart from professional standards of practice. But the Supreme Court did not address the broader issues of the rights of committed patients to receive training that might eventually enable them to function independently outside the hospital.

Quiz 17.1
Psychiatric Commitment and Patient's Rights

The related issue of whether people with severe psychological disorders who reside in the community have a constitutional right to receive mental health services (and whether states are obliged to provide these services) continues to be argued in the courts at both the state and federal levels (Perlin, 1994).

THINK ABOUT IT
Under what circumstances do you think someone should be confined to a mental hospital? What rights should patients have under such circumstances?

THINK ABOUT IT
What legal rights do individuals have regarding obtaining or refusing treatment?

Right to Refuse Treatment Consider the following scenario. A person, John Citizen, is involuntarily committed to a mental hospital for treatment. The hospital staff determines that John suffers from a psychotic disorder, paranoid schizophrenia, and should be treated with antipsychotic medication. John, however, decides not to comply with treatment. He claims that the hospital has no right to treat him against his will. The hospital staff seeks a court order to mandate treatment, arguing it makes little sense to commit people involuntarily unless the hospital is empowered to treat them as the staff deems fit.

Does an involuntary patient, like John, have the right to refuse treatment? If so, does this right conflict with states' rights to commit people to mental institutions to receive treatment for their disorders? One might also wonder whether people who are judged in need of involuntary hospitalization are competent to make decisions about which treatments are in their best interests.

The rights of committed patients to refuse psychotropic medications was tested in the 1979 case *Rogers* v. *Okin,* in which a Massachusetts federal district court imposed an injunction on a Boston state hospital that prohibited the forced medication of patients except in emergency situations. The court ruled that committed patients could not be forcibly medicated, except in the case of emergency—for example, when patients' behaviors pose a significant threat to themselves or others. The court recognized that a patient may be unwise to refuse medication, but it held that a patient with or without a mental disorder has the right to exercise bad judgment so long as the effects of the "error" do not impose "a danger of physical harm to himself, fellow patients, or hospital staff" (p. 1367).

Despite the concerns some mental health professionals raised that patients who refused medications would be "rotting with their rights on" (Gutheil, 1980), legal protections ensuring patients' rights to refuse psychiatric treatments do not appear to have had seriously damaging or disruptive effects on mental health services or on the people receiving these services (Kapp, 1994).

Although statutes and regulations vary from state to state, cases in which hospitalized patients refuse medications are often first brought before an independent review panel. If the panel rules against the patient, the case may then be brought before a judge, who makes the final decision about whether or not the patient is to be forcibly medicated. In actual practice, the number of refusals of medication is generally low, about 10% overall. Furthermore, between 70% and 90% of refusals that reach the review process are eventually overridden. Hospitalized patients in a recent study who refused medication tended to be more assaultive, were more likely to require seclusion and restraint, and had longer hospitalizations than did compliant patients (Kasper et al., 1997). However, the refusal episodes tended to be brief, about 3 days on the average, and all initial refusers were eventually treated.

Our discussion of legal issues and abnormal behavior now turns to the controversy concerning the insanity defense.

Not guilty by reason of insanity. In 1981, John Hinckley Jr. attempted to assassinate President Ronald Reagan outside the Washington Hilton Hotel. Here Reagan winces as he was shot in the upper left side. Hinckley was later found not guilty by reason of insanity. The public outrage over the Hinckley verdict led to a reexamination of the insanity plea in many states.

The Insanity Defense

As President Ronald Reagan stepped out of the Washington Hilton on March 31, 1981, gunshots rang out. Secret Service agents formed a human shield around the president as another agent shoved him into a waiting limousine, which then sped away to a hospital. At first, the president did not know he had been wounded. He said later that it sounded like firecrackers. Agents seized the gunman, John Hinckley, a 25-year-old drifter. Not only the president had been wounded. James Brady, his press secretary, was hit by a stray bullet that shattered his spine, leaving him partially paralyzed. A Secret Service agent was also shot.

Hinckley had left a letter in his hotel room revealing his hope that his assassination of the president would impress a young actress, Jodie Foster. Hinckley had never met Foster but had a crush on her.

There was never any question at the trial that Hinckley had fired the wounding bullets, but the prosecutor was burdened to demonstrate beyond a reasonable doubt that at the time of the assassination attempt, Hinckley had the capacity to control his behavior and appreciate its wrongfulness. The defense presented testimony that portrayed Hinckley as an incompetent schizophrenic who suffered under the delusion that he would achieve a "magic union" with Foster as a result of killing the president. The prosecutor portrayed Hinckley as making a conscious and willful choice to kill the president. The prosecution further argued that whatever mental disorder Hinckley might have had did not prevent him from controlling his behavior.

The jury sided with the defense. A man who was seen by millions of TV viewers attempting to assassinate President Reagan was found "not guilty by reason of insanity." The verdict led to a public outcry across the country, with many calling for the abolition of the insanity defense. Public opinion polls taken several days after the Hinckley verdict was returned showed the public had little confidence in the psychiatric testimony that had been offered at the trial (Slater & Hans, 1984). One objection focused on the fact that once the defense presented evidence to support a plea of insanity, the federal prosecutor had the responsibility of proving *beyond a reasonable doubt* that the defendant was sane. It can be difficult enough to demonstrate that someone is sane or insane in the present, so imagine the problems that attend proving someone was sane at the time a criminal act was committed.

In the aftermath of the Hinckley verdict, the federal government and many states changed their statutes to shift the burden of proof to the defense to prove insanity (Ogloff, Roberts, & Roesch, 1993). Even the American Psychiatric Association went on record as stating that psychiatric expert witnesses should not be called on to render opinions about whether defendants can control their behavior. In the opinion of the psychiatric association, these are not medical judgments that psychiatrists are trained to provide.

Perceptions of the use of the **insanity defense** tend to stray far from the facts. The public grossly overestimates the number of cases in which the insanity defense is used and how often it succeeds. The public estimates that the insanity defense is used in about 1 in 3 felony cases, but in actuality it is used in fewer than 1% of cases (Silver, Cirincione, & Steadman, 1994; Steadman et al., 1993). A study in Baltimore showed that the insanity defense plea was used in a minuscule 1/100th of 1% of indictments brought to trial (Janofsky et al., 1996). Whereas the public believes that 44% of defendants who claim insanity are acquitted, the actual figure (based on a review of the records in eight states) was 26% (Silver et al., 1994; Steadman et al., 1993). In actual practice, out of 1,000 felony cases, there are fewer than 10 in which the insanity defense is used, and only about 2 or 3 in which it is used successfully. Thus despite public perceptions to the contrary, the use of the insanity defense is actually rather rare, and the rate of acquittals is rarer still ("Insanity," 1992).

The public also overestimates the proportion of defendants acquitted on the basis of insanity who are set free rather than confined to mental health institutions and underestimates the length of hospitalization of those who are confined (Silver et al., 1994). The facts show that people found not guilty on the basis of insanity are often confined to mental hospitals for longer periods of time than they would have served in prison (Lymburner & Roesch, 1999). The net result is that although changes in the insanity defense, or its abolition, may prevent some few flagrant cases of abuse, they are not likely to afford the public much broader protection.

In the wake of the Hinckley acquittal, a number of states adopted a new type of verdict, the "guilty-but-mentally-ill" (GBMI) verdict (Boudouris, 2000; Maeder, 1985). The

Jodie Foster. John Hinckley reportedly attempted to assassinate President Reagan in order to impress Ms. Foster, whom he had seen in the film *Taxi Driver* but never met. In the film, Robert DeNiro, who portrayed a person with paranoid schizophrenia, rescued Foster's character from a life of prostitution. Hinckley's attorneys claimed that their client experienced similar rescue fantasies.

Truth OR Fiction? REVISITED

An attempt to assassinate the president of the United States was seen by millions of television viewers, but the would-be assassin was found not guilty by a court of law.

TRUE. A man who was seen by millions of TV viewers attempting to assassinate President Reagan was found "not guilty by reason of insanity" by a court of law.

insanity defense A legal defense in which a defendant in a criminal case pleads innocent on the basis of insanity.

Truth OR Fiction? REVISITED

The insanity defense is used in a large number of trials, usually successfully.

FALSE. The insanity defense is rarely used in felony cases and the rate of acquittals based on the defense is even rarer.

GBMI verdict offers juries the option of finding a defendant both guilty and mentally ill, if they determine that the defendant is mentally ill but that the mental illness did not cause the defendant to commit the crime. The GBMI verdict provides that people so convicted may be imprisoned but also receive treatment while in prison. The GBMI verdict has now been adopted in many states as a supplement to, not a replacement of, the insanity defense (Ogloff et al., 1993).

The GBMI verdict is something of an in-between determination, in that the defendant is neither acquitted nor found guilty in the traditional sense of the term. The verdict has sparked a great deal of controversy. Although GBMI was intended to reduce the number of NGRI (not guilty by reason of insanity) verdicts, this outcome has apparently not occurred. All in all, the GBMI verdict appears to be a social experiment that has not yet proved its usefulness and is seen by some as merely a means of stigmatizing defendants who are found guilty as also being mentally ill (Maeder, 1985).

In a celebrated case, John E. du Pont, an heir to the du Pont family fortune, was found guilty but mentally ill following trial for the 1996 murder of Olympic gold medalist wrestler David Schultz ("Du Pont Heir Found Guilty," 1997). Du Pont was subsequently remanded to a state psychiatric hospital for treatment rather than to prison. At trial, it was reported that du Pont exhibited bizarre behavior and suffered from delusions of being spied on by Nazis and of having his body inhabited by bugs.

Although the public outrage over the Hinckley and other celebrated insanity verdicts has led to a reexamination of the insanity defense, society has long held to the doctrine of free will as a basis for determining responsibility for wrongdoing. The doctrine of free will, as applied to criminal responsibility, requires that people can be held guilty of a crime only if they are judged to have been in control of their actions at the time of commission of the crime. Not only must it be determined by a court of law that a defendant had committed a crime beyond a reasonable doubt, but the issue of the individual's state of mind must be considered as well in determining guilt. The court must thus rule not only on whether a crime was committed but also on whether an individual is held morally responsible and deserving of punishment. The insanity defense is based on the belief that when a criminal act derives from a distorted state of mind, and not from the exercise of free will, the individual should not be punished but rather treated for the underlying mental disorder. The insanity defense has a long legal history.

Guilty but mentally ill? John du Pont, an heir to the du Pont family fortune, was tried for the 1996 murder of Olympic gold medalist wrestler David Schultz. Du Pont was found guilty but mentally ill, a verdict that remains controversial in part because it is seen by some people as merely a way of stigmatizing defendants who are found guilty as also being mentally ill.

Legal Bases of the Insanity Defense

We can find in modern law three major court rulings that bear on the insanity defense. The first involved a case in Ohio in 1834 in which it was ruled that people could not be held responsible if they are compelled to commit criminal actions because of impulses they are unable to resist.

The second major legal test of the insanity defense is referred to as the M'Naghten rule, based on a case in England in 1843 of a Scotsman, Daniel M'Naghten, who had intended to assassinate the prime minister of England, Sir Robert Peel. Instead, he killed Peel's secretary, whom he had mistaken for the prime minister. M'Naghten claimed that the voice of God had commanded him to kill Peel. The English court acquitted M'Naghten on the basis of insanity, finding that the defendant had been " ... labouring under such a defect of reason, from disease of the mind, as not to know the nature and quality of the act he was doing; or, if he did know it, that he did not know he was doing what was wrong." The M'Naghten rule holds that people do not bear criminal responsibility if, by reason of a mental disease or defect, they either have no knowledge of their actions or are unable to tell right from wrong.

To find the third major case that helped lay the foundation for the modern insanity defense we must jump more than 100 years to 1954 and the case of *Durham* v. *United States.* In this case, the presiding judge, David Bazelon, held that the "accused [person] is not criminally responsible if his unlawful act was the product of mental disease or mental defect" (pp. 874–875). Under the Durham rule, juries were expected to decide not only whether the accused suffered from a mental disease or defect but also whether this mental condition was causally connected to the criminal act. The court recognized that criminal intent is a precondition of criminal responsibility:

> The legal and moral traditions of the western world require that those who, of their own free will and with evil intent . . . commit acts which violate the law, shall be criminally responsible for those acts. Our traditions also require that where such acts stem from and are the product of a mental disease or defect . . . moral blame shall not attach, and hence there will not be criminal responsibility. (*Durham* v. *United States,* 214 F2d 862, D.C. circ. 1954)

The intent of the Durham rule was to reject as outmoded the two earlier standards of legal insanity, the irresistible impulse rule and the "right–wrong" principle under the M'Naghten rule. Judge Bazelon argued that the "right–wrong test" was outmoded because the concept of "mental disease" is broader than the ability to recognize right from wrong. The legal basis of insanity should thus not be judged on just one feature of a mental disorder, such as deficient reasoning ability. The irresistible impulse test was denied because the court recognized that in certain cases, criminal acts arising from "mental disease or defect" might occur in a cool and calculating manner rather than in the manner of a sudden, irresistible impulse.

The Durham rule, however, has proved to be unworkable for several reasons, such as a lack of precise definitions of such terms as *mental disease* or *mental defect* (Maeder, 1985). Courts were confused, for example, about issues such as whether a personality disorder (e.g., antisocial personality disorder) constituted a "disease." It also proved difficult for juries to draw conclusions about whether an individual's "mental disease" was causally connected to the criminal act. Without clear or precise definitions of terms, juries came to rely increasingly on expert psychiatric testimony, often serving as little more than rubber stamps as their verdicts endorsed the testimony of expert witnesses (Maeder, 1985).

By 1972, the Durham rule had failed and was replaced in many jurisdictions by a set of legal guidelines formulated by the American Law Institute (ALI) to define the legal basis of insanity, or some variation of these standards (Ogloff et al., 1993). These guidelines, which essentially combine the M'Naghten principle with the irresistible impulse principle, include the following provisions:

1. A person is not responsible for criminal conduct if at the time of such conduct as a result of mental disease or defect he lacks substantial capacity either to appreciate the criminality (wrongfulness) of his conduct or to conform his conduct to the requirements of law.
2. . . . the terms "mental disease or defect" do not include an abnormality manifested only by repeated criminal or otherwise antisocial conduct. (American Law Institute, 1962, p. 66)

The first guideline incorporates aspects of the M'Naghten test (being unable to appreciate right from wrong) and the irresistible impulse test (being unable to conform one's behavior to the requirements of law) of insanity. The second guideline asserts that repeated criminal behavior (such as a pattern of drug dealing) is not sufficient in itself to establish a mental disease or defect that might relieve the individual of criminal responsibility. Although many legal authorities believe the ALI guidelines are an improvement over earlier tests, questions remain as to whether a jury composed of ordinary citizens can be expected to make complex judgments about the defendant's state of mind, even on the basis of expert testimony. Under the ALI guidelines, juries must determine whether defendants lack substantial capacity to be aware of, or capable of, conforming their behavior to the law. By adding the term *substantial capacity* to the legal test, the ALI guidelines may also broaden the legal basis of the insanity defense because this may imply that defendants need not be completely incapable of controlling their criminal actions in order to meet the legal test of not guilty by reason of insanity.

Under our system of justice, juries must struggle with the complex question of determining criminal responsibility, not merely criminal actions. But what of those individuals who successfully plead not guilty by reason of insanity? Should they be committed to a

THINK ABOUT IT

Describe how the test for legal insanity has changed over time. How does this reflect changing ideas about mental illness and abnormal behavior?

mental institution for a fixed sentence, as they might have been had they been incarcerated in a penal institution? Or should their commitments be of an indeterminate term and their release depend on their mental status? The legal basis for answering such questions was decided in the following case of a man, Michael Jones, whose acquittal by reason of insanity resulted in involuntary commitment to a mental hospital for a period that turned out to be seven times longer than the maximum prison sentence for the crime.

Determining the Length of Criminal Commitment

The issue of determinate versus indeterminate commitment was addressed in the case of Michael Jones (*Jones* v. *United States*), who was arrested in 1975 and charged with petty larceny for attempting to steal a jacket from a Washington, D.C., department store. Jones was first committed to a public mental hospital, St. Elizabeth's Hospital (which, incidentally, is the hospital where John Hinckley remains committed as of this writing). Jones was diagnosed by a hospital psychologist as suffering from paranoid schizophrenia and was kept hospitalized until he was judged competent to stand trial, about 6 months later. Jones offered a plea of not guilty by reason of insanity, which the court accepted without challenge, remanding him to St. Elizabeth's. Despite the fact that Jones's crime carried a maximum sentence of 1 year in prison, Jones's repeated attempts to obtain release were denied in subsequent court hearings.

Truth OR Fiction? REVISITED

People who are found not guilty of a crime by reason of insanity may remain confined to a mental hospital indefinitely—for many years longer than they would have been sentenced to prison, if they had been found guilty.

TRUE. People who are found not guilty of a crime by reason of insanity may remain confined to a mental hospital indefinitely.

The U.S. Supreme Court eventually heard his appeal *7 years* after he was hospitalized. The Supreme Court reached its decision in 1983. It ruled against Jones's appeal and affirmed the decision of the lower courts that he was to remain in the hospital. The Supreme Court thereby established a principle that individuals who are acquitted by reason of insanity "constitute a special class that should be treated differently" than civilly committed individuals. They may be committed for an indefinite period to a mental institution under criteria that require a less stringent level of proof of dangerousness than would ordinarily be applied in cases of civil commitment. Thus people found not guilty by reason of insanity may remain confined to a mental hospital for many years longer than they would have been sentenced to prison had they been found guilty.

Among other things, the Supreme Court ruling in *Jones* v. *United States* provides that the usual and customary sentences that the law provides for particular crimes have no bearing on criminal commitment. In the words of the Court,

> different considerations underlie commitment of an insanity acquittee. As he was not convicted, he may not be punished. His confinement rests on his continuing illness and dangerousness. There simply is no necessary correlation between severity of the offense and length of time necessary for recovery. (*Jones* v. *United States,* 103 S.Ct. 3043, 1983)

The ruling held that a person who is criminally committed may be confined "to a mental institution until such time as he has regained his sanity or is no longer a danger to society" (p. 3053). As in the case of Michael Jones, people who are acquitted on the basis of insanity may remain confined in a mental institution for longer periods of time than they would have been sentenced to prison. It is also possible, however, that such persons could be released earlier than they might have been released from prison, if their "mental condition" is found to have improved. A person acquitted of a serious crime by virtue of the insanity defense may even be released within a few weeks or months, although, practically speaking, it is doubtful that improvement would be established so quickly. Public outrage over a speedy release, especially for a major crime, might also prevent rapid release.

The indeterminateness of legal or criminal commitment raises various questions. Is it reasonable to deny people like Michael Jones their liberty for an indefinite and possibly lifelong term for a relatively minor crime, such as petty larceny? On the other hand, is justice served by acquitting perpetrators of heinous crimes by reason of insanity and then allowing them the opportunity for an early release if they are deemed by professionals to be able to rejoin society?

The Supreme Court's ruling in *Jones* v. *United States* seems to imply that we must separate the notion of legal sentencing from that of legal or criminal commitment. The former, in which sentences are scaled according to the seriousness of the crime, rests on the principle that the punishment should fit the crime. In legal or criminal commitment, however, persons acquitted of their crimes by reason of insanity are guiltless in the eyes of the law. They must be treated, not punished, until such time that their mental status has improved to the point that permits them to safely reenter society.

THINK ABOUT IT

Explain in your own words why there are different standards for legal sentencing and for legal commitment.

Perspectives on the Insanity Defense

The insanity defense places special burdens on juries. In assessing criminal responsibility, the jury is expected to determine not only that a crime was committed by the accused, but also the defendant's state of mind at the time of commission of the crime. In rejecting the Durham decision, courts have relieved psychiatrists and other expert witnesses from bearing the burden of responsibility for determining whether or not the defendant's behavior is a product of a "mental disease or defect." But is it reasonable to assume that juries of people from all walks of life are better able to assess defendants' states of mind than mental health professionals? In particular, how can a jury evaluate the testimony of expert conflicting witnesses? The task imposed on the jury is made even more difficult by the mandate to decide whether or not the defendant was mentally incapacitated *at the time of the crime.* The defendant's courtroom behavior may bear little resemblance to his or her behavior during the crime.

Thomas Szasz and others who deny the existence of mental illness itself have raised another challenge to the insanity defense. If mental illness does not exist, then the insanity defense becomes groundless. Szasz argues that the insanity defense is ultimately degrading because it strips people of personal responsibility for their behavior. People who break laws are criminals, Szasz argues, and should be prosecuted and sentenced accordingly. Acquittal of defendants by reason of insanity treats them as nonpersons, as unfortunates who are not deemed to possess the basic human qualities of free choice, self-determination, and personal responsibility. We are each responsible for our behavior, Szasz contends, and we should each be held accountable for our misdeeds.

Szasz argues that the insanity defense has historically been invoked in crimes that were particularly heinous or perpetrated against persons of high social rank. When persons of low social rank commit crimes against persons of higher status, Szasz argues, the effect of the insanity defense is to direct attention away from the social ills that may have motivated the crime. Despite Szasz's contention, however, the insanity defense is invoked in many cases of less shocking crimes or in cases involving persons from similar social classes.

How, then, are we to evaluate the insanity defense? To abolish the insanity defense in all forms would be to reverse hundreds of years of a legal tradition that has recognized that people are not to be held responsible for their criminal behavior when their ability to control themselves is impaired by a mental disorder or defect.

Consider a hypothetical example. John Citizen commits a crime, say a heinous crime like murder, while acting on a delusional belief that the victim was intent on assassinating him. The accused claims that voices from his TV set informed him of the identity of the assailant and commanded him to kill the assailant to save himself and other potential victims. Cases like this are thankfully rare. Few mentally disturbed persons, even few people with psychotic features, commit violent crimes, and even fewer commit murder.

In reaching a judgment on the insanity plea, we need to consider whether we believe the law should allow special standards to apply in cases such as our hypothetical case, or whether one standard of criminal responsibility should apply to all. If we assert the legitimacy of the insanity defense in some cases, we still need a standard of insanity that can be interpreted and applied by juries of ordinary citizens. The furor over the Hinckley verdict suggests that issues concerning the insanity plea remain unsettled.

THINK ABOUT IT

Do you believe the insanity verdict should be abolished? Should it be replaced with another type of verdict, such as the guilty-but-mentally-ill verdict? Why or why not?

Competency to Stand Trial

There is a basic rule of law that those who stand accused of crimes must be able to understand the charges and proceedings brought against them and be able to participate in their own defense. The concept of *competency to stand trial* should not be confused with the legal

Truth OR Fiction? REVISITED

It is possible for a defendant to be held to be competent to stand trial but still be judged not guilty of a crime by reason of insanity.

TRUE. A defendant can be held competent to stand trial but still be judged not guilty of a crime by reason of insanity.

THINK ABOUT IT

What is the difference between the concept of competency to stand trial and the legal defense of insanity?

Quiz **17.2**
The Insanity Defense

Quiz **17.3**
Chapter Exam

defense of insanity. A defendant can be held competent to stand trial but still be judged not guilty of a crime by reason of insanity. A clearly delusional person, for example, may understand the court proceedings and be able to confer with defense counsel, but still be acquitted by reason of insanity. On the other hand, a person may be incapable of standing trial at a particular time, but be tried and acquitted or convicted at a later time when competency is restored.

Far more people are confined to mental institutions on the basis of a determination that they lack the mental competence to stand trial than on the basis of the insanity verdict (Roesch et al., 1999). There may be 45 people committed under the mental competency to stand trial criteria for every one committed following a verdict of not guilty by reason of insanity (Steadman, 1979).

People declared incompetent to stand trial are generally confined to a mental institution until they are deemed competent or a determination is made that they are unlikely to ever regain competency. Abuses may occur, however, if the accused are kept incarcerated for indefinite periods awaiting trial. In earlier years, it was not uncommon for people who were declared incompetent to have their trials delayed for months or years until they were judged ready to stand trial, if indeed they were ever judged to be competent. In 1972, however, the U.S. Supreme Court ruled, in the case of *Jackson* v. *Indiana,* that a person could not be kept in a mental hospital awaiting trial longer than it would take to determine whether treatment was likely to restore competency. If it did not seem the person would ever become competent, even with treatment, the individual would have to be either released or committed under the procedures for civil commitment.

A 1992 ruling by the U.S. Supreme Court, in the case of *Medina* v. *California,* held that the burden of proof for determining competency to stand trial lies with the defendant, not the state (Greenhouse, 1992). This decision may speed the trials of many people who commit crimes and whose competency is in question.

We opened this book by noting that despite the public impression that abnormal behavior affects only the few, it actually affects nearly every one of us in one way or another. Let us close by suggesting that if we all work together to foster research into the causes, treatment, and prevention of abnormal behavior, perhaps we can meet the multifaceted challenges that abnormal behavior poses to so many of us and to our society at large.

Summing Up

Psychiatric Commitment and Patient's Rights

What is the difference between civil commitment and legal commitment? The legal process by which people are placed in psychiatric institutions against their will is called civil or psychiatric commitment. Civil commitment is intended to provide treatment to people who are deemed to suffer from mental disorders and to pose a threat to themselves or others. Legal or criminal commitment, by comparison, involves the placement of a person in a psychiatric institution for treatment who has been acquitted of a crime by reason of insanity. In voluntary hospitalization, people voluntarily seek treatment in a psychiatric facility and can leave of their own accord unless a court rules otherwise.

How successful are mental health professionals in predicting the dangerousness of patients they evaluate? Although people must be judged dangerous to be placed involuntarily in a psychiatric facility, mental health professionals have not demonstrated any special ability to predict dangerousness. Factors that may account for the failure to predict dangerousness include (1) recognizing violent tendencies post hoc is easier than predicting it; (2) generalized perceptions of violent tendencies may not predict specific acts of violence; (3) lack of agreement in defining violence or dangerousness; (4) base-rate problems; (5) unlikelihood of direct threats of violence; and (6) predictions based on hospital behavior may not generalize to community settings.

What is meant by the duty to warn? Although information disclosed by a client to a therapist generally carries a right to confidentiality, the California *Tarasoff* ruling held that therapists have a duty or obligation to warn third parties of threats made against them by their clients.

What landmark court cases dealt with the rights of patients in psychiatric facilities? In *Wyatt* v. *Stickney,* a court in Alabama imposed a minimum standard of care. In *O'Connor* v. *Donaldson,* the U.S. Supreme Court ruled that nondangerous mentally ill people could not be held in psychiatric facilities against their will if such people could be maintained safely in the community. In *Youngberg* v. *Romeo,* the Supreme Court ruled that involuntarily confined patients have a right to less restrictive types of treatment and to receive training to help them function. Court rulings, such as that of *Rogers* v. *Okin* in Massachusetts, have established that patients have a right to refuse medication, except in case of emergency.

The Insanity Defense

What are the legal bases of the insanity defense? Three court cases established legal precedents for the insanity defense. In 1834, a court in Ohio applied a principle of irresistible impulse as the basis of an insanity defense. The M'Naghten rule, based on a case in England in 1843, treated the failure to appreciate the wrongfulness of one's action as the basis of legal insanity. The Durham rule was based on a case in the United States in 1954 in which it was held that persons did not bear criminal responsibility if their criminal behavior was the product of "mental disease or mental defect." Another set of standards developed by the American Law Institute that combines the M'Naghten and irresistible impulse principles has been adopted in some form in 22 states. People who are criminally committed may be hospitalized for an indefinite period of time, with their eventual release dependent on a determination of their mental status.

What is meant by the legal concept of competency to stand trial? People who are accused of crimes but are incapable of understanding the charges against them or assisting in their own defense can be found incompetent to stand trial and remanded to a psychiatric facility. In the case of *Jackson* v. *Indiana,* the U.S. Supreme Court placed restrictions on the length of time a person judged incompetent to stand trial could be held in a psychiatric facility.

GLOSSARY

A

Abnormal psychology. The branch of psychology that deals with the description, causes, and treatment of abnormal behavior patterns.

Abstinence violation effect (AVE). The tendency to overreact to a minor lapse with feelings of guilt and resignation that may trigger a relapse.

Acculturation. The process of adapting to a new culture.

Acquired immunodeficiency syndrome (AIDS). An immunological disease caused by HIV.

Acrophobia. Excessive, irrational fear of heights.

Acute stress disorder (ASD). A traumatic stress reaction occurring in the days and weeks following exposure to a traumatic event.

Addiction. Impaired control over the use of a chemical substance, accompanied by physiological dependence.

Adjustment disorder. A maladaptive reaction to an identified stressor, which is characterized by impaired functioning or signs of emotional distress that exceed what would normally be expected.

Adoptee studies. Studies that compare the traits and behavior patterns of adopted children to those of their biological parents and their adoptive parents.

Affect. Emotional responsiveness.

Agnosia. A disturbance of sensory perception, usually affecting visual perception.

Agoraphobic. Relating to excessive, irrational fear of open places.

Al-Anon. An organization that sponsors support groups for family members of people with alcoholism.

Alarm reaction. The first stage of the GAS, characterized by heightened sympathetic activity.

Alcoholism. An alcohol dependence disorder or addiction that results in serious personal, social, occupational, or health problems.

Alcohol-induced persisting amnestic disorder. A form of brain damage associated with chronic thiamine deficiency and alcoholism; characterized by memory loss, disorientation, and confabulation.

Alzheimer's disease (AD). A progressive brain disease characterized by gradual loss of memory and intellectual functioning, personality changes, and eventual loss of ability to care for oneself.

Ambivalent. Holding conflicting feelings toward another person or a goal.

Amenorrhea. Absence of menstruation.

Amnestic disorders. Disturbances of memory associated with inability to learn new material or recall past events.

Amphetamine psychosis. A psychotic state induced by ingestion of amphetamines.

Amphetamines. Types of stimulants, such as Benzedrine or Dexedrine.

Amygdala. Limbic system structure involved in processing threatening stimuli.

Anal expulsive. A loosening of restraint, as seen in extreme sloppiness or messiness.

Anal fixation. Attachment to objects and behaviors characteristic of the anal stage.

Anal retentive. Excessive need for self-control and orderliness.

Anal stage. The psychosexual stage during toddlerhood in which pleasure is sought primarily through anal activities.

Analgesia. Relief from pain without loss of consciousness.

Analogue measure. A measure taken in a contrived situation meant to simulate a real-life situation.

Analogue. Something that resembles something else.

Analytical psychology. Jung's theory, emphasizing the collective unconscious, archetypes, and the self as the unifying force of personality.

Anomie. A feeling of rootlessness.

Anorexia nervosa. An eating disorder characterized by maintenance of an abnormally low body weight, distortions of body image, intense fears of gaining weight, and, in females, amenorrhea.

Antianxiety drugs. Drugs that combat anxiety and reduce states of muscle tension.

Antibodies. Substances produced by white blood cells that identify and target antigens for destruction.

Antidepressants. Drugs used to treat depression.

Antigens. Substances that trigger an immune response.

Antipsychotic drugs. Drugs used to treat schizophrenia or other psychotic disorders.

Antisocial personality disorder. A personality disorder characterized by antisocial and irresponsible behavior and lack of remorse for misdeeds.

Anxiety. An emotional state characterized by physiological arousal, unpleasant feelings of tension, and a sense of apprehension or foreboding.

Anxiety disorder. A type of mental disorder whose most prominent feature is anxiety.

Anxiety sensitivity. A fear of anxiety and anxiety-related symptoms.

Apnea. Temporary cessation of breathing.

Archetypes. Primitive images or concepts that reside in the collective unconscious.

Arteriosclerosis. A disease involving thickening and hardening of the arteries.

Asperger's disorder. A pervasive developmental disorder characterized by social deficits and stereotyped behavior but without the significant language or cognitive delays associated with autism.

Associations. Relationships among thoughts.

Ataxia. Loss of muscle coordination.

Atherosclerosis. The buildup of fatty deposits along artery walls that leads to the formation of artery-clogging plaque.

Attention-deficit hyperactivity disorder (ADHD). A behavior disorder characterized by excessive motor activity and inability to focus one's attention.

Attributional style. A personal style for explaining cause-and-effect relationships between events.

Autism. A pervasive developmental disorder characterized by failure to relate to others, lack of speech, disturbed motor behaviors, intellectual impairment, and demands for sameness in the environment.

Autism. Withdrawal into a private fantasy world.

Automatic thoughts. Thoughts that seem to pop into one's mind.

Autonomic nervous system. The division of the peripheral nervous system that regulates the activities of the glands and involuntary functions.

Aversive conditioning. A behavior therapy technique in which a maladaptive response is paired with exposure to an aversive stimulus to develop a conditioned aversion.

Avoidant personality disorder. A personality disorder characterized by avoidance of social relationships due to fears of rejection.

Axon. The long, thin part of a neuron along which nerve impulses travel.

B

Barbiturates. Types of depressants that are used to reduce anxiety or to induce sleep but that are highly addictive.

Basal ganglia. An assemblage of neurons located between the thalamus and cerebrum, involved in coordinating motor (movement) processes.

Baseline. The rate at which a behavior occurs before treatment.

Behavior therapy. The therapeutic application of learning-based techniques.

Behavioral assessment. The approach to clinical assessment that focuses on the objective recording and description of the problem behavior.

Behavioral interview. Clinical interview that focuses on relating the problem behavior to antecedent stimuli and reinforcement consequences.

Behavioral rating scale. A scale used to record the frequency of occurrence of target behaviors.

Behaviorism. The school of psychology that defines psychology as the study of observable behavior.

Benzodiazepines. The class of antianxiety drugs that includes Valium and Xanax.

Bereavement. The experience of grief suffering following the death of a loved one.

Binge-eating disorder (BED). A disorder characterized by recurrent eating binges without purging; classified as a potential disorder requiring further study.

Biofeedback training (BFT). A method of feeding back to the individual information about bodily functions so that the person can gain some degree of control over these functions.

Bipolar. Characterized by opposite ends of a dimension or continuum, as in bipolar disorder.

Bipolar disorder. A disorder characterized by mood swings between states of extreme elation and severe depression.

Blind. A state of being unaware of whether one has received an experimental treatment.

Blocking. An involuntary interruption of speech.

Borderline personality disorder (BPD). A personality disorder characterized by abrupt shifts in mood, lack of a coherent sense of self, and unpredictable, impulsive behavior.

Brain electrical activity mapping (BEAM). Imaging technique involving computer analysis of data from multiple electrodes to reveal areas of the brain with relatively high or low levels of activity.

Breathing-related sleep disorder. A sleep disorder in which sleep is repeatedly disrupted by difficulty with breathing normally.

Brief psychotic disorder. A psychotic disorder lasting from a day to a month that often follows exposure to a major stressor.

Bulimia nervosa. An eating disorder characterized by recurrent binge eating followed by self-induced purging, accompanied by overconcern with body weight and shape.

C

Cardiovascular disease. A disease or disorder of the cardiovascular system, such as coronary heart disease or hypertension.

Case study. A carefully drawn biography based on clinical interviews, observations, and psychological tests.

Castration anxiety. The young boy's unconscious fear that he will be castrated as punishment for his incestuous desire for his mother.

Cataplexy. A brief, sudden loss of muscle control.

Catastrophize. To exaggerate the negative consequences of events.

Catatonic type. The subtype of schizophrenia characterized by gross disturbances in motor activity, such as catatonic stupor.

Catecholamines. A group of substances that includes neurotransmitters (dopamine and norepinephrine) and hormones (epinephrine and norepinephrine).

Catharsis. The discharge of states of tension associated with repression of threatening impulses or material; the purging or free expression of feelings.

Central nervous system. The brain and spinal cord.

Cerebellum. A structure in the hindbrain involved in coordination and balance.

Cerebral cortex. The wrinkled surface area of the cerebrum, it is responsible for processing sensory stimuli and controlling higher mental functions, such as thinking and use of language.

Cerebrovascular accident (CVA). Damage to part of the brain because of a disruption in its blood supply, usually as the result of a blood clot.

Cerebrum. The large mass of the forebrain, consisting of the two cerebral hemispheres.

Childhood disintegrative disorder. A pervasive developmental disorder involving loss of previously acquired skills and abnormal functioning following a period of apparently normal development during the first two years of life.

Choleric. Having or showing bad temper.

Chromosomes. The structures found in the nuclei of cells that carry the units of heredity, or genes.

Circadian rhythm sleep disorder. A sleep disorder characterized by a mismatch between the body's normal sleep-wake cycle and the demands of the environment.

Civil commitment. The legal process of placing a person in a mental institution, even against his or her will.

Clanging. The tendency to string words together because they rhyme or sound alike.

Claustrophobia. Excessive, irrational fear of small, enclosed spaces.

Client-centered therapy. Another term for *person-centered therapy.*

Closed-ended questions. Questionnaire or test items with a limited range of response options.

Cocaine. A stimulant derived from the leaves of the coca plant.

Cognitive disorders. Mental disorders characterized by impaired cognitive abilities and daily functioning in which biological causation is either known or presumed.

Cognitive restructuring. A cognitive therapy method that involves replacing irrational thoughts with rational alternatives.

Cognitive therapy. Aaron Beck's form of therapy that helps clients recognize and correct distorted patterns of thinking.

Cognitive triad of depression. The view that depression derives from adopting negative views of oneself, the environment or world at large, and the future.

Cognitive-behavioral therapy (CBT). A learning-based approach to therapy incorporating cognitive and behavioral techniques.

Cognitive-specificity hypothesis. The belief that different emotional disorders are linked to particular kinds of automatic thoughts.

Collective unconscious. The storehouse of archetypes and racial memories.

Comatose. In a coma, a state of deep, prolonged unconsciousness.

Communication disorders. A class of psychological disorders characterized by difficulties in understanding or using language.

Compulsion. A repetitive or ritualistic behavior that the person feels compelled to perform.

Compulsion to utter. The urge to verbally express repressed material.

Computed tomography (CT scan). Computer-enhanced imaging of the internal structures of the brain by passing a narrow X-ray beam through the head.

Concordance. Agreement.

Concurrent validity. A type of test validity based on the statistical relationship between the test and a criterion measure taken at the same time.

Conditional positive regard. Valuing other people on the basis of whether their behavior meets one's approval.

Conditioned response. In classical conditioning, a learned response to a previously neutral stimulus.

Conditioned stimulus. A previously neutral stimulus that evokes a conditioned response after repeated pairings with an unconditioned stimulus that had previously evoked that response.

Conditions of worth. Standards by which one judges the worth or value of oneself or others.

Conduct disorder. A psychological disorder in childhood and adolescence characterized by disruptive, antisocial behavior.

Confidentiality. Protection of the identity of participants by keeping records secure and not disclosing their identities.

Congruence. The fit between one's thoughts, behaviors, and feelings.

Conscious. To Freud, the part of the mind that corresponds to our present awareness.

Construct validity. The degree to which treatment effects can be accounted for by the theoretical mechanisms (constructs) represented in the independent variables; the degree to which a test measures the hypothetical construct that it purports to measure.

Content validity. The degree to which the content of a test covers a representative sample of the content it is designed to measure.

Contrasted groups approach. A method of determining concurrent validity that uses a test's ability to differentiate between members of two or more comparison groups.

Control subjects. In an experiment, subjects who do not receive the experimental treatment.

Controlled drinking. An approach to treating problem drinkers that has as its goal to moderate social drinking rather than abstinence.

Conversion disorder. A type of somatoform disorder characterized by loss or impairment of physical function in the absence of any apparent organic cause.

Corpus callosum. A thick bundle of fibers that connects the two cerebral hemispheres.

Correlation. A relationship or association between variables.

Countertransference. The transfer of feelings or attitudes that the analyst holds toward other persons onto the client.

Couples therapy. Therapy that focuses on resolving conflicts in distressed couples.

Crack. The hardened, smokable form of cocaine.

Creative self. To Adler, the self-aware part of the personality that strives to achieve its potential.

Criterion validity. The degree to which a test correlates with an independent, external criterion or standard.

Critical thinking. Adoption of a questioning attitude and careful scrutiny of claims and arguments in the light of evidence.

Cross-fostering study. A method of determining heritability of a trait or disorder by examining differences in prevalence among adoptees reared by either adoptive or biological parents who possessed the trait or disorder in question.

Cultural-familial retardation. A mild form of mental retardation that is influenced by impoverishment of the home environment.

Culture-bound syndrome. A pattern of abnormal behavior that is found within only one or a few cultures.

Cyclothymic disorder. A mood disorder characterized by a chronic pattern of mild mood swings that is not sufficiently severe to be classified as bipolar disorder.

Cytomegalovirus. A source of infection that, in pregnant women, carries a risk of mental retardation to the unborn child.

D

Debriefed. To be fully informed about an experiment after it takes place.

Defense mechanisms. The reality-distorting strategies used by the ego to shield the self from awareness of anxiety-provoking materials.

Deinstitutionalization. The policy of shifting care for patients with severe or chronic mental health problems from inpatient facilities to community-based facilities; the practice of discharging hospitalized mental patients into the community and of reducing the need for new admissions through alternative treatment approaches.

Delirium. A state of mental confusion, disorientation, and extreme difficulty focusing attention.

Delirium tremens. A withdrawal syndrome that occurs following sudden decrease of drinking in people with chronic alcoholism.

Delta-9-tetrahydrocannabinol (THC). The active ingredient in marijuana.

Delusion. A firmly held but inaccurate belief that persists despite evidence that it has no basis in reality.

Delusional disorder. A type of psychosis characterized by persistent delusions, often of a paranoid nature, that do not have the bizarre quality of the type often found in paranoid schizophrenia.

Dementia. Deterioration of mental functioning, involving impaired memory, thinking, judgment, and language use.

Dementia praecox. The term given by Kraepelin to the disorder now called schizophrenia.

Demonological model. The model that explains abnormal behavior in terms of supernatural forces.

Demonology. The idea that abnormal behavior is caused by supernatural forces.

Dendrites. The rootlike structures at the ends of neurons that receive nerve impulses from other neurons.

Dependent personality disorder. A personality disorder characterized by difficulty making independent decisions and overly dependent behavior.

Dependent variables. Outcomes of an experiment believed to be dependent on the effects of an independent variable.

Depersonalization disorder. A disorder characterized by persistent or recurrent episodes of depersonalization.

Depersonalization. Feelings of unreality or detachment from one's self or one's body.

Depressant. A drug that lowers the level of activity of the central nervous system.

Derealization. Loss of the sense of reality of one's surroundings, experienced in terms of strange changes in the environment or in the passage of time.

Description. The representation of observations without making interpretations or drawing inferences.

Detoxification. The process of ridding the system of alcohol or other drugs under supervised conditions.

Deviation IQ. An intelligence quotient obtained by determining the deviation between the person's score and the norm (mean).

Dhat syndrome. A culture-bound somatoform disorder, found primarily among Asian Indian males, characterized by excessive fears over the loss of seminal fluid.

Diathesis. A vulnerability or predisposition to a particular disorder.

Diathesis-stress model. A model that posits that abnormal behavior problems involve the interaction of a vulnerability or predisposition and stressful life events or experiences.

Disorganized type. The subtype of schizophrenia characterized by disorganized behavior, bizarre delusions, and vivid hallucinations.

Disorientation. A state of mental confusion and lack of awareness of time, place, or the identity of oneself or others.

Displacement. A defense mechanism in which one transfers sexual or aggressive impulses toward less threatening or safer objects or persons.

Displacing. Transferring impulses toward threatening or unacceptable objects onto more acceptable or safer objects.

Dissociative amnesia. A dissociative disorder in which a person experiences memory loss without any identifiable organic cause.

Dissociative disorder. Any of a group of disorders characterized by a disruption, or dissociation, of the functions of identity, memory, or consciousness.

Dissociative fugue. A dissociative disorder in which a person suddenly flees from his or her life situation, travels to a new location, assumes a new identity, and has amnesia for personal material.

Dissociative identity disorder. A dissociative disorder in which a person has two or more distinct, or alter, personalities.

Distress. A state of physical or emotional pain or suffering.

Dizygotic (DZ) twins. Twins that develop from separate fertilized eggs.

Dopamine theory. The theory that proposes that schizophrenia involves overactivity of dopamine receptors in the brain.

Double depression. A diagnosis of both major depressive disorder and dysthymic disorder.

Double-bind communications. A communication pattern involving contradictory or mixed messages without acknowledging the inherent conflict.

Down syndrome. A condition caused by the presence of an extra chromosome on the 21st pair and characterized by mental retardation and various physical anomalies.

Downward drift hypothesis. The theory that explains the linkage between low socioeconomic status and behavior problems by suggesting that problem behaviors lead people to drift downward in social status.

Duty to warn. The therapist's obligation to warn third parties of threats made against them by clients.

Dyslexia. A learning disorder characterized by impaired reading ability.

Dyspareunia. Persistent or recurrent pain experienced during or following sexual intercourse.

Dyssomnias. Sleep disorders involving disturbances in the amount, quality, or timing of sleep.

Dysthymic disorder. A mild but chronic type of depressive disorder.

E

Eating disorder. A psychological disorder characterized by disturbed patterns of eating and maladaptive ways of controlling body weight.

Eclectic therapy. An approach to psychotherapy that incorporates principles or techniques from various systems or theories.

Ego. The psychic structure that corresponds to the concept of the self, governed by the reality principle and characterized by the ability to tolerate frustration.

Ego analysts. Psychodynamically oriented therapists who are influenced by ego psychology.

Ego dystonic. Referring to behaviors or feelings that are perceived to be alien to one's self-identity.

Ego ideal. The set of higher social values and moral ideals embodied in the superego.

Ego identity. The achievement of a firm sense of personal identity.

Ego psychology. Modern psychodynamic approach that focuses more on the conscious strivings of the ego than on the hypothesized unconscious functions of the id.

Ego syntonic. Referring to behaviors or feelings that are perceived as natural parts of the self.

Electra complex. The conflict that occurs during the phallic stage of development, in which the young girl desires her father and perceives her mother as a rival.

Electroconvulsive therapy (ECT). A method of treating severe depression by administering electrical shock to the head.

Electroencephalograph (EEG). An instrument for measuring the electrical activity of the brain.

Electromyograph (EMG). An instrument for measuring muscle tension.

Emotion-focused coping. A coping style that attempts to minimize emotional responsiveness rather than deal with the stressor directly.

Empathy. The ability to understand someone's experiences and feelings from that person's point of view.

Encopresis. Lack of control over bowel movements that is not caused by an organic problem in a child who is at least 4 years old.

Endocrine system. The system of ductless glands that secrete hormones directly into the bloodstream.

Endorphins. Natural substances that function as neurotransmitters in the brain and are similar in their effects to morphine.

Enuresis. Failure to control urination after one has reached the "normal" age for attaining such control.

Epidemiological method. A method of research that involves tracking the rates of occurrence of a disorder among different groups.

Erogenous zone. A part of the body that is sensitive to sexual stimulation.

Eros. Freud's concept of the basic life instinct that seeks to preserve and perpetuate life.

Erotomania. A delusional disorder characterized by the belief that one is loved by someone of high social status.

Etiological. Relating to cause or origin.

Exhaustion stage. The third stage of the GAS, characterized by lowered resistance, increased parasympathetic activity, and possible physical deterioration.

Exhibitionism. A paraphilia in which one is sexually aroused by exposing one's genitals to a stranger.

Exorcism. A ritual intended to expel demons from a person believed to be possessed.

Expectancies. Beliefs about expected outcomes.

Experimental method. A scientific method that aims to discover cause-and-effect relationships by manipulating independent variables and observing the effects on the dependent variables.

Experimental subjects. In an experiment, subjects who receive the experimental treatment.

External attribution. A belief that the cause of an event involved factors outside oneself.

External validity. The degree to which experimental results can be generalized to other settings and conditions.

F

Face validity. The degree to which the content of a test bears an apparent relationship to the constructs it is designed to measure.

Factitious disorder. A disorder characterized by intentional fabrication of psychological or physical symptoms for no apparent gain.

False negative. An incorrect appraisal that a person is free of a disorder, when in fact he or she has the disorder.

False positive. An incorrect appraisal that a person has a particular disorder.

Family therapy. Therapy in which the family, not the individual, is the unit of treatment.

Fat cells. Body cells specialized to store fat.

Fear-stimulus hierarchy. An ordered series of increasingly fearful stimuli.

Female orgasmic disorder. A sexual dysfunction in women involving difficulty reaching orgasm or inability to reach orgasm following a normal level of sexual interest and arousal.

Female sexual arousal disorder. A sexual dysfunction in women involving difficulty becoming sexually aroused or lack of sexual excitement or pleasure during sexual activity.

Fetishism. A paraphilia in which one uses an inanimate object or body part as a focus of sexual interest and arousal.

Fight-or-flight reaction. The inborn tendency to respond to a threat by either fighting or fleeing.

First-rank symptoms. The primary features of schizophrenia, such as hallucinations and delusions.

Fixation. Arrested development in the form of attachment to objects of an earlier developmental stage.

Flashbacks. The experience of sensory distortions or hallucinations occurring after use of LSD or other hallucinogenic drugs.

Flooding. A form of exposure therapy in which subjects are exposed to high levels of fear-inducing stimuli.

Forced-choice formats. Test questions that require respondents to select from among a limited number of choices.

Forcible rape. Forced sexual intercourse with a nonconsenting person.

Four A's. The primary characteristics of schizophrenia: loose Associations, blunted or inappropriate Affect, Ambivalence, and Autism.

Fragile X syndrome. An inherited form of mental retardation caused by a mutated gene on the X chromosome.

Free association. The method of verbalizing thoughts as they occur without a conscious attempt to edit or censure them.

Freebasing. A method of ingesting cocaine by heating it with ether to separate its most potent component (its "free base") and then smoking the extract.

Frotteurism. A paraphilia characterized by recurrent sexual urges involving bumping or rubbing against nonconsenting others for sexual gratification.

Functional analysis. Analysis of behavior in terms of antecedent stimuli and reinforcement consequences.

Functional magnetic resonance imaging (fMRI). Type of MRI used to identify parts of the brain that become active when people engage in particular tasks.

G

Galvanic skin response (GSR). A measure of the change in electrical activity of the skin that accompanies sympathetic nervous system arousal.

Gamma-aminobutyric acid (GABA). An inhibitory neurotransmitter believed to play a role in anxiety.

Gender identity disorder. A disorder in which the individual believes that her or his anatomical gender is inconsistent with her or his gender identity.

Gender identity. One's psychological sense of being female or male.

Gender roles. The characteristic ways in which males and females are expected to behave in a given culture.

General adaptation syndrome (GAS). The body's three-stage response to states of prolonged or intense stress.

Generalized anxiety disorder (GAD). A type of anxiety disorder characterized by general feelings of dread and foreboding and heightened states of bodily arousal.

Genes. The units, found on chromosomes, that carry heredity.

Genetics. The science of heredity.

Genital stage. The final stage of psychosexual development, characterized by expression of libido through sexual intercourse with an adult of the opposite sex.

Genotype. The set of traits specified by an individual's genetic code.

Genuineness. The ability to recognize and express one's true feelings.

Global attribution. A belief that the cause of an event involved generalized, rather than specific, factors.

Gradual exposure. A behavior therapy technique for overcoming fears through direct exposure to increasingly fearful stimuli.

Group therapy. Therapy method in which a group of clients meet together with a therapist.

H

Halfway houses Supervised community residences that provide a bridge between institutional facilities and independent community living.

Hallucination. A perception that occurs in the absence of an external stimulus and that is confused with reality.

Hallucinogens. Substances that cause hallucinations.

Hashish. A drug derived from the resin of the plant *Cannabis sativa.*

Health psychologist. A psychologist who studies the relationships between psychological factors and physical illness.

Heroin. A narcotic derived from morphine that has strong addictive properties.

Heterosexism. The culturally based belief system that holds that only reproductive sexuality is psychologically healthy and morally correct.

Histrionic personality disorder. A personality disorder characterized by excessive need for attention, praise, reassurance, and approval.

Homophobia. Hatred and fear of lesbians and gay males.

Homosexuality. A sexual orientation characterized by erotic interest in, and development of romantic relationships with, members of one's own gender.

Hormones. Substances secreted by endocrine glands that regulate body functions and promote growth and development.

Human immunodeficiency virus (HIV). The virus that causes AIDS.

Humors. According to the ancient Hippocratic belief system, the vital bodily fluids (phlegm, black bile, blood, yellow bile).

Huntington's disease. An inherited degenerative disease that is characterized by jerking and twisting movements, psychotic behavior, and mental deterioration.

Hyperactivity. An abnormal behavior pattern characterized by difficulty in maintaining attention and extreme restlessness.

Hypersomnia. A pattern of excessive sleepiness during the day.

Hyperventilation. A pattern of overly rapid breathing associated with states of anxiety.

Hypnosis. A trancelike state, induced by suggestion, in which the individual responds to the commands of the hypnotist.

Hypnotics. Drugs, including anesthetics and sedatives, that induce partial or complete unconsciousness and are used to treat sleep disorders.

Hypoactive sexual desire disorder. Persistent or recurring lack of sexual interest or sexual fantasies.

Hypomanic. Referring to a mild state of mania, or elation.

Hypothalamus. A structure in the forebrain involved in regulating body temperature, emotion, and motivation.

Hypothesis. An assumption that is tested through experimentation.

Hypothyroidism. A physical condition, caused by deficiency of the hormone thyroxin, characterized by sluggishness and lowered metabolism.

Hypoxia. Decreased supply of oxygen to the brain or other organs.

Hypoxyphilia. A paraphilia in which a person seeks sexual gratification by being deprived of oxygen.

I

Id. The unconscious psychic structure, present at birth, that contains primitive instincts and is regulated by the pleasure principle.

Ideas of persecution. A form of delusional thinking characterized by false beliefs that one is being persecuted or victimized by others.

Ideas of reference. A form of delusional thinking in which a person reads personal meaning into the behavior of others or external events.

Identification. The process of incorporating the personality or behavior of others.

Immune system. The body's system of defense against disease.

Incidence. The number of new cases of a disorder that occurs within a specific period of time.

Independent variables. Factors that are manipulated in experiments.

Individual psychology. Adler's psychodynamic theory.

Individual response specificity. The belief that people respond to the same stressor in different ways.

Infarction. The development of an infarct, or area of dead or dying tissue, resulting from the blocking of blood vessels normally supplying the tissue.

Inference. A conclusion that is drawn from data.

Inferiority complex. The feelings of inferiority that Adler believed to be a central source of motivation.

Informed consent. The principle that subjects should receive enough information about an experiment beforehand to decide freely whether to participate.

Insanity defense. A legal defense in which a defendant in a criminal case pleads innocent on the basis of insanity.

Insight. The attainment of awareness and understanding of one's true motives and feelings.

Insomnia. Difficulties falling asleep, remaining asleep, or achieving restorative sleep.

Instinct. A fixed, inborn pattern of behavior that is specific to members of a particular species.

Intelligence. (1) The capacity to understand the world and respond to its challenges; (2) the trait measured by intelligence tests.

Intelligence quotient (IQ). A measure of intelligence based on scores on an intelligence test; the ratio between a respondent's mental age and actual age.

Internal attribution. A belief that the cause of an event involved factors within oneself.

Internal consistency. Cohesiveness or interrelationships of items on a test or scales.

Internal locus of control. Perception of one's ability to control reinforcements or affect outcomes.

Internal validity. The degree to which manipulation of the independent variables can be causally related to changes in the dependent variables.

Interpersonal psychotherapy (IPT). A brief form of psychodynamic therapy that focuses on the client's current interpersonal relationships.

Interrater reliability. Consistency of or agreement between raters.

Intoxication. A state of drunkenness.

Introject. To unconsciously incorporate features of another person's personality into one's own ego structure.

Involuntary. Automatic or without conscious direction.

K

Knobs. The swollen endings of axon terminals.

Koro syndrome. A culture-bound somatoform disorder, found primarily in China, in which people fear that their genitals are shrinking.

Korsakoff's syndrome. A syndrome associated with chronic alcoholism that is characterized by memory loss and disorientation (also called alcohol persisting amnestic disorder).

L

La belle indifférence A French expression ("beautiful indifference" in English) describing the lack of concern over one's symptoms displayed by some people with conversion disorder.

Latency stage. The psychosexual stage in middle childhood characterized by repression of sexual impulses.

Latent content. The underlying or symbolic content of dreams.

Learned helplessness. A behavior pattern characterized by passivity and perceptions of lack of control.

Learning disorder. A deficiency in a specific learning ability in the context of normal intelligence and exposure to learning opportunities.

Legal commitment. The legal process of confining a person found not guilty by reason of insanity in a mental institution.

Leukocytes. White blood cells.

Libido. The energy of Eros; sexual drive or energy.

Limbic system. A group of forebrain structures involved in learning, memory, and basic drives.

Longitudinal study. A research study in which subjects are followed over time.

Lysergic acid diethylamide (LSD). A type of hallucinogen.

M

Magnetic resonance imaging (MRI). A computer-generated image of the brain formed by measuring the signals emitted when the head is placed in a strong magnetic field.

Maintaining ethnic identity. Some studies point to psychological benefits in immigrant groups that adapt to the host culture while maintaining ethnic identity.

Major depressive disorder. A severe mood disorder characterized by major depressive episodes.

Male erectile disorder. A sexual dysfunction in men characterized by difficulty achieving or maintaining erection during sexual activity.

Male orgasmic disorder. A sexual dysfunction in males involving difficulty achieving orgasm following a normal pattern of sexual interest and excitement.

Malingering. Faking illness.

Managed care systems. Health care delivery systems that impose limits on the number of treatment sessions they will approve for payment and the fees they will allow for reimbursement.

Manic episode. A period of unrealistically heightened euphoria, extreme restlessness, and excessive activity characterized by disorganized behavior and impaired judgment.

Manic. Relating to mania, as in the manic phase of bipolar disorder.

Manifest content. The reported content or apparent meaning of dreams.

Mantra. A word or phrase that is repeated to induce a state of relaxation and narrowing of consciousness.

Marijuana. A hallucinogenic drug derived from the leaves and stems of the plant *Cannabis sativa.*

Medical model. A biological perspective in which abnormal behavior is viewed as symptomatic of underlying illness.

Medulla. An area of the hindbrain involved in regulation of heartbeat and respiration.

Melancholia. A state of severe depression.

Mental age. The age equivalent that corresponds to a person's level of intelligence as measured by the Stanford-Binet Intelligence Scale.

Mental retardation. A generalized delay or impairment in the development of intellectual and adaptive abilities.

Mental status examination. A structured clinical assessment to determine various aspects of the client's mental functioning.

Meta-analysis. A statistical technique for combining the results of different studies into an overall average.

Metabolic rate. The rate at which energy is used in the body.

Methadone. An artificial narcotic that is used to help people who are addicted to heroin to abstain from it without a withdrawal syndrome.

Modeling. Learning by observing and imitating the behavior of others; a behavior therapy technique for helping an individual acquire a new behavior by means of having a therapist or another individual demonstrate a target behavior that is then imitated by the client

Monoamine oxidase (MAO) inhibitors. A group of antidepressant drugs that increase the availability of neurotransmitters in the brain by inhibiting the actions of an enzyme that breaks down neurotransmitters.

Monozygotic (MZ) twins. Twins that develop from the same fertilized egg and therefore share identical genes.

Mood disorder. A type of disorder characterized by disturbances of mood.

Mood. The pervasive quality of an individual's emotional experience.

Moral principle. The principle that governs the superego to set and enforce moral standards.

Moral therapy. A 19th-century treatment approach that emphasized treating hospitalized patients with care and understanding.

Morphine. A strongly addictive narcotic derived from the opium poppy that relieves pain and induces feelings of well-being.

Mourning. Normal feelings of grief following a loss.

Munchausen syndrome. A type of factitious disorder characterized by the feigning of medical symptoms.

Musterbation. A rigid thought pattern characterized by the tendency to impose excessive, unrealistic demands on oneself or personal imperatives.

Myocardial infarction. A breakdown of heart tissue due to an obstruction in the blood vessels that supply blood to the heart.

N

Naloxone. A drug that prevents users from becoming high if they take heroin.

Naltrexone. A drug related to naloxone that blocks the high from alcohol as well as from opiates.

Narcissistic personality disorder. A personality disorder characterized by adoption of an inflated self-image and demands for attention and admiration.

Narcolepsy. A sleep disorder characterized by sudden, irresistible episodes of sleep.

Narcotics. Drugs that are used for pain relief and treatment of insomnia but that have strong addictive potential.

Naturalistic-observation method. A research method in which subjects' behavior is observed and measured in their natural environments.

Negative correlation. A statistical relationship between two variables such that increases in one are associated with decreases in the other.

Negative reinforcers. Reinforcers that on removal increase the frequency of behavior.

Negative symptoms. Behavioral deficiencies associated with schizophrenia, such as social skills deficits, social withdrawal, flattened affect, poverty of speech and thought, psychomotor retardation, and failure to experience pleasure.

Neo-Freudians. Theorists, such as Jung, Adler, Horney, and Sullivan, who, in comparison with Freud, placed greater emphasis on the importance of cultural and social influences and lesser emphasis on sexual impulses.

Neologisms. New words.

Neuroleptics. A group of antipsychotic drugs (the "major tranquilizers") used in the treatment of schizophrenia, such as the phenothiazines (Thorazine, Mellaril, etc.).

Neurons. Nerve cells.

Neuropsychological. Pertaining to the relationships between the brain and behavior.

Neurosis. A nonpsychotic form of disturbed behavior characterized by problems involving anxiety.

Neuroticism. A trait that involves characteristics such as anxious behavior, apprehension about the future, and avoidance behavior.

Neurotransmitters. Chemical substances that transmit messages from one neuron to another.

Nightmare disorder. A sleep disorder characterized by recurrent awakenings due to frightening nightmares.

Nonspecific factors. Factors not specific to any one form of psychotherapy, such as therapist attention and support, and the engendering of positive expectancies of change.

O

Obesity. A condition of excessive body fat; generally defined by a body mass index (BMI) above 30.

Object relations. The person's relationships to the internalized representations, or "objects," of others' personalities that have been introjected within the person's ego structure.

Object-relations theory. The psychodynamic viewpoint that focuses on the influences of internalized representations of the personalities of parents and other strong attachment figures (called "objects").

Objective tests. Tests that allow a limited range of response options and can therefore be scored objectively.

Obsession. A recurrring thought or image that the individual cannot control.

Obsessive-compulsive personality disorder. A personality disorder characterized by rigid ways of relating to others, perfectionistic tendencies, lack of spontaneity, and excessive attention to detail.

Oedipus complex. The conflict that occurs during the phallic stage of development, in which the young boy desires his mother and perceives his father as a rival.

Open-ended questions. Questions that provide an unlimited range of response options.

Opioids. Natural or synthetic drugs with strong addictive properties; natural opioids, referred to as opiates, are derived from the opium poppy.

Oppositional defiant disorder (ODD). A psychological disorder in childhood and adolescence characterized by excessive oppositionality, or tendencies to refuse requests from parents and others.

Optimum level of arousal. The level of arousal associated with peak performance and optimal feelings of well-being.

Oral stage. The psychosexual stage during infancy in which pleasure is sought primarily through oral activities.

Osteoporosis. A physical disorder caused by calcium deficiency and characterized by brittle bones.

P

Panic disorder. A type of anxiety disorder characterized by repeated episodes of intense anxiety or panic.

Paranoid personality disorder. A personality disorder characterized by suspiciousness of others' motives, but not to the point of delusion.

Paranoid. Referring to irrational suspicions.

Paranoid type. The subtype of schizophrenia characterized by hallucinations and systematized delusions, commonly involving themes of persecution.

Paraphilias. Sexual disorders in which the person experiences recurrent sexual urges and fantasies involving nonhuman objects, inappropriate or nonconsenting partners, or painful or humiliating situations.

Parasomnias. Sleep disorders involving abnormal behaviors or physiological events that occur during sleep or while falling asleep.

Parasympathetic. Pertaining to the division of the autonomic nervous system whose activity reduces states of arousal and regulates bodily processes that replenish energy reserves.

Parkinson's disease. A progressive disease of the basal ganglia characterized by muscle tremor and shakiness, rigidity, difficulty walking, poor control of fine motor movements, lack of facial muscle tonus, and, in some cases, cognitive impairment.

Pathogens. Disease-causing organisms.

Pedophilia. A paraphilia involving recurrent, powerful sexual urges and related fantasies involving sexual activity with prepubescent children.

Performance anxiety. Fear relating to the threat of failure to perform adequately.

Peripheral nervous system. The somatic and autonomic nervous systems.

Perseveration. The persistent repetition of the same thought or response.

Person-centered therapy. The establishment of a warm, accepting therapeutic relationship that frees clients to engage in self-exploration and achieve self-acceptance.

Personality disorders. Excessively rigid behavior patterns, or ways of relating to others, that ultimately become self-defeating.

Pervasive developmental disorders. A class of developmental disorders characterized by significantly impaired behavior or functioning in multiple areas of development.

Phallic stage. A psychosexual stage in early childhood during which pleasure is sought primarily through the phallic region and the child develops incestuous desires for the parent of the opposite sex.

Phenothiazines. A group of antipsychotic drugs ("major tranquilizers") used to treat schizophrenia.

Phenotype. An individual's actual or expressed traits.

Phenylketonuria (PKU). A genetic disorder that prevents the metabolization of phenylpyruvic acid, leading to mental retardation unless the diet is strictly controlled.

Phlegmatic. Slow and solid.

Phrenologist. Someone who studies the bumps on a person's head to determine the person's underlying traits.

Physiological dependence. A condition in which the drug user's body comes to depend on a steady supply of the substance.

Pick's disease. A form of dementia, similar to Alzheimer's disease, but distinguished by specific abnormalities (Pick's bodies) in nerve cells and absence of neurofibrillary tangles and plaques.

Placebo. An inert medication or bogus treatment that is intended to control for expectancy effects.

Play therapy. A form of psychodynamic therapy in which play activities and objects are used to help children symbolically enact conflicts or express underlying feelings.

Pleasure principle. The governing principle of the id, involving demands for immediate gratification of needs.

Polysomnographic (PSG) recording. Simultaneous measurement of multiple physiological responses during sleep or attempted sleep.

Pons. A structure in the hindbrain involved in respiration.

Population. A total group of people, other organisms, or events.

Positive correlation. A statistical relationship between two variables such that increases in one are associated with increases in the other.

Positive reinforcers. Reinforcers that, introduced, increase when the frequency of behavior.

Positive symptoms. Flagrant symptoms of schizophrenia, such as hallucinations, delusions, bizarre behavior, and thought disorder.

Positron emission tomography (PET scan). An imaging technique that forms a computer-generated image by tracing the amount of glucose used in various regions of the brain.

Possession. A superstitious belief in which abnormal behavior is taken as a sign that the person is possessed by demons or the devil.

Postpartum depression (PPD). Persistent and severe mood changes that occur after childbirth.

Posttraumatic stress disorder (PTSD). A prolonged maladaptive reaction to a traumatic event.

Powerful drive for superiority. Complex of feelings that motivates us to achieve prominence and social dominance.

Preconscious. To Freud, the part of the mind whose contents lie outside of present awareness but can be brought into awareness by focusing attention.

Predictive validity. The degree to which a test score is predictive of some future behavior or outcome.

Prefrontal lobotomy. A form of psychosurgery, no longer in use, in which certain neural pathways in the brain are severed in order to control disturbed behavior.

Pregenital. Referring to characteristics typical of psychosexual stages that precede the genital stage.

Premature ejaculation. A sexual dysfunction in men characterized by ejaculation following minimal sexual stimulation.

Premorbid functioning. The level of functioning before the person developed schizophrenia.

Prepared conditioning. The belief that people are genetically prepared to acquire fear responses to certain stimuli, such as snakes or large animals.

Presenile dementias. Forms of dementia that begin before age 65.

Pressured speech. An outpouring of speech in which words seem to surge urgently for expression.

Prevalence. The overall number of cases of a disorder in a population within a specific period of time.

Primary gains. Relief from underlying anxiety gained through the development of neurotic symptoms.

Primary prevention. Efforts designed to prevent problems from arising.

Primary process thinking. In infancy, the mental process by which the id seeks gratification by imagining that it possesses what it desires; thinking that is illogical or magical.

Primary reinforcers. Reinforcers that fulfill basic needs, such as water, food, warmth, and relief from pain.

Proband. The case first diagnosed of a given disorder.

Problem-focused coping. A coping style that attempts to confront the stressor directly.

Problem-solving therapy. A form of therapy that focuses on helping people develop more effective problem-solving skills.

Prodromal phase. In schizophrenia, the period of decline in functioning that precedes the first acute psychotic episode.

Projection. A defense mechanism in which one's own sexual or aggressive impulses are attributed to another person.

Projective tests. Psychological tests that present ambiguous stimuli onto which the examinee is thought to project his or her personality and unconscious motives.

Psychiatrist. A physician who specializes in the diagnosis and treatment of emotional disorders.

Psychic. Relating to mental phenomena.

Psychoactive. Referring to chemical substances that have psychological effects.

Psychoanalysis. The method of psychotherapy developed by Sigmund Freud.

Psychoanalytic theory. The theoretical model of personality developed by Sigmund Freud; also called psychoanalysis.

Psychodynamic model. The theoretical model of Freud and his followers, in which abnormal behavior is viewed as the product of clashing forces within the personality.

Psychodynamic therapy. Therapy that helps individuals gain insight into, and resolve, unconscious conflicts.

Psychological dependence. Compulsive use of a substance to meet a psychological need.

Psychological disorders. Abnormal behavior patterns that involve a disturbance of psychological functioning or behavior.

Psychological hardiness. A cluster of stress-buffering traits characterized by commitment, challenge, and control.

Psychologist. A person with advanced graduate training in psychology.

Psychometric approach. An assessment method that relies on psychological tests to identify and measure the traits that compose an individual's personality.

Psychoneuroimmunology. The study of relationships between psychological factors and immunological functioning.

Psychopharmacology. The field of study that examines the effects of therapeutic or psychiatric drugs.

Psychosexual. Pertaining to Freud's stages of development, in which libido becomes expressed through different erogenous zones during different stages.

Psychosis. A severe form of disturbed behavior characterized by impaired ability to interpret reality and difficulty meeting the demands of daily life.

Psychosomatic. Pertaining to a physical disorder in which psychological factors play a causal or contributing role.

Psychotherapy. A structured form of treatment derived from a psychological framework which consists of one or more verbal interactions or treatment sessions between a client and a therapist.

Punishments. Unpleasant stimuli that reduce the frequency of the behaviors they follow.

R

Random sample. A sample that is drawn in such a way that every member of a population has an equal chance of being included.

Rapid flight of ideas. A characteristic of manic behavior involving rapid speech and changes of topics.

Rational-emotive behavior therapy (REBT). A therapeutic approach that focuses on helping clients replace irrational, maladaptive beliefs with alternative, more adaptive beliefs.

Reactivity. The tendency for the behavior being observed to be influenced by the way in which it is measured.

Reality principle. The governing principle of the ego, which involves considerations of social acceptability and practicality.

Reality testing. The ability to perceive the world accurately and to distinguish between reality and fantasy.

Rebound anxiety. The experiencing of strong anxiety following withdrawal from a tranquilizer.

Receptor site. A part of a dendrite on a receiving neuron that is structured to receive a neurotransmitter.

Reinforcement. A stimulus or event that increases the frequency of the response that it follows.

Relapse. A recurrence of a problem behavior or disorder.

Relapse-prevention training. A cognitive-behavioral technique involving the use of behavioral and cognitive strategies for resisting temptations and preventing relapses.

Reliable. In psychological assessment, the consistency of a measure or diagnostic instrument or system.

REM sleep. The stage of sleep associated with dreaming and characterized by rapid eye movements under the closed eyelids.

Repression. A defense mechanism involving the unconscious ejection of anxiety-provoking ideas, images, or impulses.

Residual phase. In schizophrenia, the phase that follows an acute phase, characterized by a return to the level of functioning of the prodromal phase.

Resistance stage. The second stage of the GAS, involving the body's attempt to withstand prolonged stress and preserve resources.

Resistance. The blocking of thoughts or feelings that would evoke anxiety if they were consciously experienced.

Reticular activating system. Brain structure involved in processes of attention, sleep, and arousal.

Rett's disorder. A pervasive development disorder characterized by a range of physical, behavioral, motor, and cognitive abnormalities that begin after a few months of apparently normal development.

Reversal design. An experimental design that consists of repeated measurement of a subject's behavior through a sequence of alternating baseline and treatment phases.

Reward. A pleasant stimulus or event that increases the frequency of the response that it follows.

Role diffusion. A state of confusion, aimlessness, and heightened susceptibility to the suggestions of others, associated with failure to acquire a firm sense of identity during adolescence.

S

Sadomasochism. Sexual activities involving the attainment or gratification by means of inflicting and receiving pain and humiliation.

Sample. Part of a population.

Sanguine. Having a cheerful disposition.

Sanism. The negative stereotyping of people who are identified as mentally ill.

Schizoaffective disorder. A type of psychotic disorder in which individuals experience both severe mood disturbance and features associated with schizophrenia.

Schizoid personality disorder. A personality disorder characterized by persistent lack of interest in social relationships, flattened affect, and social withdrawal.

Schizophrenia. An enduring psychotic disorder that involves disturbed behavior, thinking, emotions and perceptions.

Schizophreniform disorder. A psychotic disorder lasting less than six months in duration with features that resemble schizophrenia.

Schizophrenogenic mother. The since-discarded concept of a cold but overprotective mother who it was believed was capable of causing schizophrenia in her children.

Schizotypal personality disorder. A personality disorder characterized by eccentricities of thought and behavior, but without clearly psychotic features.

Scientific method. A method of conducting scientific research in which theories or assumptions are examined in the light of evidence.

Second-rank symptoms. Symptoms associated with schizophrenia that also occur in other mental disorders.

Secondary gains. Side benefits associated with neurotic or other disorders, such as expressions of sympathy, increased attention, and release from responsibilities.

Secondary prevention. Efforts to ameliorate existing problems at an early stage.

Secondary process thinking. The reality-based thinking processes and problem-solving activities of the ego.

Secondary reinforcers. Stimuli that gain reinforcement value through their association with established reinforcers, such as money and social approval.

Sedatives. Types of depressants that reduce states of tension and restlessness and induce sleep.

Selection factor. A type of bias in which differences between experimental and control groups result from differences in the subjects placed in the groups, not from the independent variable.

Selective abstraction. A cognitive distortion involving the tendency to focus only on the negative parts of experiences or events.

Selective serotonin-reuptake inhibitors (SSRIs). A group of antidepressant drugs that increase the availability of serotonin in the brain by interfering with its reuptake by the transmitting neuron.

Self psychology. A theory that describes processes that normally lead to achievement of a cohesive sense of self.

Self-actualization. In humanistic psychology, the tendency to strive to become all that one is capable of being. The motive that drives one to reach one's full potential and express one's unique capabilities.

Self-monitoring. The process of observing or recording one's own behaviors, thoughts, or emotions.

Self-report personality test. A structured personality test in which individuals give information about themselves by responding to items that require a limited type of response, such as "yes-no" or "agree-disagree."

Self-spectatoring. The tendency to observe one's behavior as if one were a spectator.

Semi-structured interview. Interview in which the clinician follows a general outline of questions designed to gather essential information but is free to ask them in any order and to branch off in other directions.

Senile dementias. Forms of dementia that begin after age 65.

Sensate focus exercises. Mutual pleasuring activities focused on the partners taking turns giving and receiving physical pleasure.

Sensitivity. The ability of a diagnostic instrument to correctly identify people who have the disorder the test is intended to detect.

Separation anxiety disorder. A childhood disorder characterized by extreme fear of separation from parents or other caretakers.

Separation-individuation. The process by which an infant develops a separate identity from that of the mother.

Set point. A value, such as body weight, that the body's regulatory mechanisms attempt to maintain.

Sexual aversion disorder. A type of sexual dysfunction characterized by aversion to and avoidance of genital sexual contact.

Sexual dysfunctions. Persistent problems with sexual interest, arousal, or response.

Sexual harassment. Speech, gestures, demands, or physical contact of a sexual nature that is unwelcomed by the person to whom such actions are directed.

Sexual masochism. A paraphilia characterized by recurrent, powerful sexual urges and fantasies involving receiving humiliation or pain.

Sexual sadism. A paraphilia characterized by recurrent, powerful sexual urges and fantasies involving inflicting humiliation or pain.

Significant. In statistics, a magnitude of difference that is taken as indicating meaningful differences between groups.

Single-case experimental design. A type of case study in which the subject is used as his or her own control.

Sleep disorders. Persistent or recurrent sleep-related problems that cause distress or impaired functioning.

Sleep terror disorder. A sleep disorder characterized by recurrent episodes of sleep terror resulting in abrupt awakenings.

Sleepwalking disorder. A sleep disorder involving repeated episodes of sleepwalking.

Social phobia. Excessive fear of social interactions or situations.

Social-cognitive theory. A learning-based theory that emphasizes observational learning and incorporates roles for both situational and cognitive variables in determining behavior.

Sociobiology. Biological perspective that explains psychological traits as behavioral tendencies that increased our ancestors' chances of survival and were passed down genetically.

Soma. A cell body.

Somatic nervous system. The division of the peripheral nervous system that relays information from the sense organs to the brain and transmits messages from the brain to the skeletal muscles.

Somatization disorder. A type of somatoform disorder involving recurrent multiple complaints that cannot be explained by any physical cause.

Somatoform disorders. A group of disorders characterized by complaints of physical problems or symptoms that cannot be explained by physical causes.

Specific attribution. A belief that the cause of an event involved specific, rather than generalized, factors.

Specific phobia. A persistent and excessive fear of a specific object or situation.

Specificity. The ability of a diagnostic instrument to avoid classifying people as having a characteristic or disorder when they truly do not have the characteristic or disorder.

Splitting. An inability to reconcile the positive and negative aspects of the self and others, resulting in sudden shifts between positive and negative feelings.

Stable attribution. A belief that the cause of an event involved stable, rather than changeable, factors.

Standard scores. Scores that indicate the relative standing of raw scores in relation to the distribution of normative scores.

Statutory rape. Sexual intercourse with a minor, even with the minor's consent.

Steroids. A group of hormones that includes testosterone, estrogen, progesterone, and corticosteroids.

Stress. A demand made on an organism to adapt or adjust.

Stressor. A source of stress.

Stroke. Blocking of a blood vessel that supplies the brain due to a blood clot.

Structural hypothesis. The belief that the clashing forces within the personality can be divided into three structures: id, ego, and superego.

Structured interview. Interview that follows a preset series of questions in a particular order.

Stupor. A state of relative or complete unconsciousness in which a person is not aware of or responsive to the environment.

Substance abuse. The continued used of a psychoactive drug despite the knowledge that it is causing a social, occupational, psychological, or physical problem.

Substance dependence. Impaired control over the use of a psychoactive substance; often characterized by physiological dependence.

Substance use disorders. Disorders that involve maladaptive use of psychoactive substances, such as substance abuse and substance dependence.

Substance-induced disorders. Disorders that can be induced by using psychoactive substances, such as intoxication.

Superego. The psychic structure that incorporates the values of the parents and important others and that is governed by the moral principle; consists of two parts, the conscience and the ego ideal.

Survey method. A research method in which large samples of people are questioned by means of a survey instrument.

Symbiotic. The state of oneness that normally exists between mother and infant.

Sympathetic. Pertaining to the division of the autonomic nervous system whose activity leads to heightened states of arousal.

Synapse. The junction between the terminal knob of one neuron and the dendrite or soma of another through which nerve impulses pass.

Syndromes. Clusters of symptoms that are characteristic of particular disorders.

Systematic desensitization. A behavior therapy technique for overcoming phobias by means of exposure to progressively more fearful stimuli while one remains deeply relaxed.

Systems perspective. The view that problems reflect the systems (family, social, school ecological, etc.) in which they occur.

T

Tachycardia. Abnormally rapid heartbeat.

Tactile. Pertaining to the sense of touch.

Taijin-kyofu-sho. A psychiatric syndrome, found in Japan, involving excessive fear of offending or embarrassing others.

Tardive dyskinesia (TD). A disorder characterized by involuntary movements of the face, mouth, neck, trunk, or extremities and caused by long-term used of antipsychotic medication.

Temporal stability. The consistency of test responses over time, as measured by test-retest reliability.

Terminals. The small branching structures at the tips of axons.

Test-retest reliability. A method of measuring the reliability of a test by means of comparing the scores of the same subjects on different occasions.

Thalamus. A structure in the forebrain involved in relaying sensory information to the cortex and in processes related to sleep and attention.

Theory. A formulation of the relationships underlying observed events.

Thermistor. A device for registering body temperature.

Thought disorder. A disturbance in thinking characterized by the breakdown of logical associations between thoughts.

Time out. A behavioral technique in which a person who behaves in an undesirable way is removed from a reinforcing environment and placed in an unreinforcing environment for a short time.

Token economy. Behavioral treatment program in which a controlled environment is constructed such that people are reinforced for desired behaviors by receiving tokens that may be exchanged for desired rewards.

Tolerance. Physical habituation to use of a drug; habituation to a drug such that, with frequent use, higher doses are needed to achieve the same effects.

Transcendental meditation (TM). A form of meditation that focuses on repeating a mantra to induce a meditative state.

Transference relationship. The client's transfer onto the analyst of feelings or attitudes the client holds toward important figures in his or her life.

Transvestic fetishism. A paraphilia in heterosexual males characterized by recurrent sexual urges involving dressing in female clothing.

Trephination. A harsh, prehistoric practice of cutting a hole in a person's skull, possibly in an attempt to release demons.

Tricyclics. A group of antidepressant drugs that increase the activity of norepinephrine and serotonin by interfering with the reuptake of these neurotransmitters.

Two-factor model. A theoretical model that accounts for the development of phobic reactions on the basis of classical and operant conditioning.

Type A behavior pattern (TABP). A behavior pattern characterized by a sense of time urgency, competitiveness, and hostility.

U

Unconditional positive regard. The expression of unconditional acceptance of another person's basic worth as a person.

Unconditioned response. An unlearned response.

Unconditioned stimulus. A stimulus that elicits an unlearned response.

Unconscious. To Freud, the part of the mind that lies outside the range of ordinary awareness and that contains instinctual urges.

Unipolar. Pertaining to a single pole, or direction.

Unobtrusive. Not interfering or conspicuous.

Unstable attribution. A belief that the cause of an event involved changeable, rather than stable, factors.

Unstructured interview. Interview in which the clinician adopts his or her own style of questioning rather than following any standard format.

V

Vaginismus. A sexual dysfunction characterized by persistent or recurring contraction of the muscles surrounding the vaginal opening, making intercourse difficult or impossible.

Validity scales. Groups of test items that are used to detect whether the results of a test are valid.

Validity. The degree to which a test or diagnostic system measures the traits or constructs it purports to measure.

Variables. Conditions that are measured (dependent variables) or manipulated (independent variables) in experiments.

Voyeurism. A paraphilia characterized by recurrent sexual urges involving watching unsuspecting others in sexual situations.

W

Waxy flexibility. A feature of catatonic schizophrenia in which a person's limbs are moved into a certain posture or position, which the person then rigidly maintains.

Weaning. The process of accustoming a child to eat solid food.

Wernicke's disease. A brain disorder, associated with chronic alcoholism, characterized by confusion, disorientation, and difficulty maintaining balance while walking.

Withdrawal syndrome. A characteristic cluster of symptoms following the sudden reduction or cessation of use of a psychoactive substance after physiological dependence has developed.

Worldview. The prevailing view of the times (English translation of German term *Weltanschaung*).

REFERENCES

A

Achenbach, T. M., & Edelbrock, C. S. (1979). The Child Behavior Profile: I. Boys aged 12–16 and girls aged 6–11 and 12–16. *Journal of Consulting and Clinical Psychology, 47*, 223–233.

A children's mental illness 'crisis': Report: 1 in 10 children suffers enough to impair development. (2001, January 3). *Associated Press Web Posting*. Retrieved January 5, 2001, from http://www.msnbc.com/news/510934.asp.

Acklin, M. W., et al. (2000). Interobserver agreement, intraobserver reliability, and the Rorschach comprehensive system. *Journal of Personality Assessment, 74*, 15–47.

Adamson, S. J., & Sellman, J. D. (2001). Drinking goal selection and treatment outcome in out-patients with mild-moderate alcohol dependence. *Drug and Alcohol Review, 20*, 351–359.

Adler, J. (1999, June 14). Stress. *Newsweek*, pp. 58–63.

Affleck, G., Tennen, H., Croog, S., & Levine, S. (1987). Causal attribution, perceived benefits, and morbidity after a heart attack: An 8-year study. *Journal of Consulting and Clinical Psychology, 55*, 29–35.

Agency approves fourth Alzheimer's treatment. (2001, March 1). *CNN Web Posting*. Retrieved March 8, 2001, from http://www.cnn.com/2001/HEALTH/conditions/03/01/alzheimers.drug.ap/index.html.

Agras, W. S., et al. (2000a). Outcome predictors for the cognitive behavior treatment of bulimia nervosa: Data from a multisite study. *American Journal of Psychiatry, 157*, 1302–1308.

Agras, W. S., et al. (2000b). A multicenter comparison of cognitive-behavioral therapy and interpersonal psychotherapy for bulimia nervosa. *Archives of General Psychiatry, 57*, 459–466.

Ainsworth, M. D. S. (1989). Attachments beyond infancy. *American Psychologist, 44*, 709–716.

Akhtar, A. (1987). Schizoid personality disorder: A synthesis of developmental, dynamic, and descriptive features. *American Journal of Psychotherapy, 41*, 499–517.

Akhtar, S. (1988). Four culture-bound psychiatric syndromes in India. *The International Journal of Social Psychiatry, 34*, 70–74.

Akiskal, H. S. (1983). Dysthymic disorder: Psychopathology of proposed chronic depressive subtypes. *American Journal of Psychiatry, 140*, 11–20.

Alao, A. O., & Dewan, M. J. (2001). Evaluating the tolerability of the newer mood stabilizers. *Journal of Nervous & Mental Disease, 189*, 60–63.

Aldarondo, E. (1996). Risk marker analysis of the cessation and persistence of wife assault. *Journal of Consulting and Clinical Psychology, 64*, 1010–1019.

Aldrich, M. S. (1992). Narcolepsy. *Neurology, 42* (7, Suppl. 6), 34–43.

Allderidge, P. (1979). Hospitals, madhouses and asylums: Cycles in the care of the insane. *British Journal of Psychiatry, 134*, 1476–1478.

Alloy, L. B., & Clements, C. M. (1992). Illusion of control: Invulnerability to negative affect and depressive symptoms after laboratory and natural stressors. *Journal of Abnormal Psychology, 101*, 2234–2245.

Alloy, L. B., et al. (2000). The Temple-Wisconsin cognitive vulnerability to depression project: Lifetime history of Axis I psychopathology in individuals at high and low cognitive risk for depression. *Journal of Abnormal Psychology, 109*, 403–418.

Alpert, J. L., & Green, D. (1992). Child abuse and neglect: Perspectives on a national emergency. *Journal of Social Distress and the Homeless, 1*, 223–236.

Alter, J. (2001, February 12). The war on addiction. *Newsweek*, pp. 36–39.

Altman, L. K. (1990a, April 18). Scientists see a link between alcoholism and a specific gene. *The New York Times*, pp. A1, A18.

Altman, L. K. (1994a, February 22). Stomach microbe offers clues to cancer as well as ulcers. *The New York Times*, p. C3.

Altman, L. K. (2001, January 30). The AIDS questions that linger. *The New York Times*, pp. F1, F6.

Altschuler, E. L., et al. (2001). Did Samson have antisocial personality disorder? *Archives of General Psychiatry, 58*, 202.

Alzheimer's vaccine passes key test. (2001, July 23). *CNN Web Posting*. Retrieved July 25, 2001, from http://www.cnn.com/2001/HEALTH/07/23/alzheimers.vaccine/index.html.

American Association of Mental Retardation (AAMR). (1992). *Mental retardation: Definition, classification, and systems of supports* (9th ed.). Washington, DC: Author.

American Law Institute. (1962). Model penal code: Proposed official draft. Philadelphia: Author.

American Psychiatric Association. (1991). The APA task force report on benzodiazepine dependence, toxicity, and abuse [Editorial]. *American Journal of Psychiatry, 148*, 151–152.

American Psychiatric Association. (1994). *DSM-IV: Diagnostic and statistical manual of mental disorders* (4th ed.). Washington, DC: Author.

American Psychiatric Association. (1998). *Fact sheet: Violence and mental illness*. Washington, DC: Author.

American Psychiatric Association. (2000). *DSM-IV-TR: Diagnostic and statistical manual of mental disorders* (4th ed., Text Revision). Washington, DC: Author.

American Psychological Association. (1978). Report of the Task Force on the Role of Psychology in the Criminal Justice System. *American Psychologist, 33*, 1099–1113.

American Psychological Association. (1992). Rules and Procedures. *American Psychologist, 47*, 1612–1628.

American Psychological Association. (1992). *Big world, small screen: The role of television in American society*. Washington, DC: Author.

Amering, M., & Katschnig, H. (1990). Panic attacks and panic disorder in cross-cultural perspective. *Psychiatric Annals, 20*, 511–516.

Ames, M. A., & Houston, D. A. (1990). Legal, social, and biological definitions of pedophilia. *Archives of Sexual Behavior, 19*, 333–342.

Amminger, G. P., et al. (1999). Relationship between childhood behavioral disturbance and later schizophrenia in the New York High-Risk Project. *American Journal of Psychiatry, 156*, 525–530.

Ancoli-Israel, S., et al. (1991). Dementia in institutionalized elderly: Relation to sleep apnea. *Journal of the American Geriatrics Society, 39*, 258–263.

Andersen, B. L. (1992). Psychological interventions for cancer patients to enhance the quality of life. *Journal of Consulting and Clinical Psychology, 60*, 552–568.

Andersen, B. L., & Golden-Kreutz, D. M. (2001). Cancer. In D. W. Johnston & M. Johnston (Eds.), *Health psychology, Vol. 8. Comprehensive clinical psychology* (pp. 217–236). Amsterdam, Netherlands: Elsevier Science Publishers.

Andersen, B. L., Golden-Kreutz, D. M., & DiLillo, V. (2001). Cancer. In A. Baum, T. A. Revenson, & J. E. Singer (Eds.), *Handbook of health psychology* (pp. 709–726). Mahwah, NJ: Lawrence Erlbaum Associates.

Anderson, C. A., & Dill, K. E. (2000). Video games and aggressive thoughts, feelings, and behavior in the laboratory and in life. *Journal of Personality and Social Psychology, 78*, 772–790.

Anderson, D. Q., & Maloney, K. C. (2001). The efficacy of cognitive-behavioral therapy on the core symptoms of bulimia nervosa. *Clinical Psychology Review, 21*, 971–988.

Anderson, E. M., & Lambert, M. J. (1995). Short-term dynamically oriented psychotherapy: A review and meta-analysis. *Clinical Psychology Review, 15*, 503–514.

Anderson, E. M., & Lambert, M. J. (2001). A survival analysis of clinically significant change in outpatient psychotherapy. *Journal of Clinical Psychology, 57*, 875–888.

Anderson, L. P. (1991). Acculturative stress: A theory of relevance to Black Americans. *Clinical Psychology Review, 11*, 685–702.

Andreasen, N. C. (1987a). The diagnosis of schizophrenia. *Schizophrenia Bulletin, 13*, 1–8.

Andreasen, N. C. (1999). Understanding the causes of schizophrenia. *New England Journal of Medicine, 340*, 645–647.

Andreasen, N. C., & Grove, W. M. (1986). Thought, language, and communication in schizophrenia: Diagnosis and prognosis. *Schizophrenia Bulletin, 12*, 348–359.

Andreasen, N. C., et al. (1997). Hypofrontality in schizophrenia: Distributed dysfunctional circuits in neuroleptic-naive patients. *The Lancet, 349*, 1730–1734.

Andrews, B., et al. (2000). Predicting PTSD symptoms in victims of violent crime: The role of shame, anger, and childhood abuse. *Journal of Abnormal Psychology, 109*, 69–73.

Andrews, E. L. (1997, September 9). In Germany, humble herb is a rival to Prozac. *The New York Times*, pp. C1, C7.

Angier, N. (1991a, May 30). Gene causing common type of retardation is discovered. *The New York Times*, pp. A1, B11.

Angier, N. (1991b, August 4). Kids who can't sit still. *The New York Times*, Section 4A, pp. 30–33.

Angier, N. (1999). *Woman: An intimate geography*. Boston: Houghton Mifflin.

Angier, N. (2000a, July 21). Study finds region of brain may be key problem solver. *The New York Times*, pp. C1, C4.

Angier, N. (2000b, November 7). Who is fat? It depends on culture. *The New York Times*, pp. F1–F2.

Anglin, K., & Holtzworth-Munroe, A. (1997). Comparing the responses of maritally violent and nonviolent spouses to problematic marital and nonmarital situations: Are the skills deficits of physically aggressive husbands and wives global? *Journal of Family Psychology, 11*, 301–313.

Angold, A., & Costello, E. J. (1993). Depressive comorbidity in children and adolescents: Empirical, theoretical, and methodological issues. *American Journal of Psychiatry, 150*, 1779–1791.

Angold, A., & Costello, J. (1996). Toward establishing an empirical basis for the diagnosis of oppositional defiant disorder. *Journal of the American Academy of Child and Adolescent Psychiatry, 35*, 1205–1212.

Angst, J., Angst, F., & Stassen, H. H. (1999). Suicide risk in patients with major depressive disorder. *The Journal of Clinical Psychiatry, 60*, 57–62.

Antonarakas, S. E., et al. (1991). Prenatal origin of the extra chromosomein trisomy 21 as indicated by analysis of DNA polymorphisms. *The New England Journal of Medicine, 324*, 872–876.

Anonymous. (1985, June 13). Schizophrenia—A mother's agony over her son's pain. *Chicago Tribune*, Section 5, pp. 1–3.

Another worry for aging baby boomers—pot and their hearts. (2000, March 3). *CNN Web Posting*. Retrieved March 13, 2000, from www.canoe.ca/Health0003/03_pot.html.

Ansell, B. J. (2001, January 9). Fearing one fate, women ignore a killer. *The New York Times*, p. F8.

Anthony, J. C., & Helzer, J. E. (1991). Syndromes of drug abuse and dependence. In L. N. Robins & D. A. Regier (Eds.), *Psychiatric disorders in America: The Epidemiologic Catchment Area Study* (pp. 116–154). New York: The Free Press.

Anthony, J. C., Warner, L. A., & Kessler, R. C. (1994). Comparative epidemiology of dependence on tobacco, alcohol, controlled substances, and inhalants: Basic findings from the National Comorbidity Survey. *Experimental and Clinical Psychopharmacology, 2*, 244–268.

Anthony, W. A., Cohen, M., & Kennard, W. (1990). Understanding the current facts and principles of mental health systems planning. *American Psychologist, 45*, 1249–1256.

Anthony, W. A., & Liberman, R. P. (1986). The practice of psychiatric rehabilitation: Historical, conceptual, and research base. *Schizophrenia Bulletin, 12*, 542–559.

Antidepressants linked to sexual side effects. (2000, February 7). *CNN Web Posting*. Retrieved February 8, 2000, from www.cnn.com/2000/HEALTH/02/07/antidepressant.sex.wmd/.

Anton, R. F. (1994). Medications for treating alcoholism. *Alcohol Health and Research World, 18*, 265–271.

Anton, R. F., et al. (2001). Posttreatment results of combining naltrexone with cognitive-behavior therapy for the treatment of alcoholism. *Journal of Clinical Psychopharmacology, 21*, 72–77.

Antoni, M. H., Levine, J., Tischer, P., Green, C., & Millon, T. (1986). Refining personality assessments by combining MCMI high-point profiles and MMPI codes: IV. MMPI 89/98. *Journal of Personality Assessment, 50*, 65–72.

Anxiety: Most common mental health problem. (2000, February 2). *CNN Web Posting*. Retrieved February 4, 2000, from www.cnn.com/2000/HEALTH/02/02/mental.health.wmd/.

Arango, C., Kirkpatrick, B., &. Buchanan, R. W. (2000). Neurological signs and the heterogeneity of schizophrenia. *American Journal of Psychiatry, 157*, 566–572.

Arieti, S. (1974). *Interpretation of schizophrenia* (2nd ed.). New York: Basic Books.

Arndt, J., & Greenberg, J. (1996). Fantastic accounts can take many forms: False memory construction? Yes. Escape from self? We don't think so. *Psychological Inquiry, 7*, 127–132.

Arnett, P. A. (1997). Autonomic responsivity in psychopaths: A critical review and theoretical proposal. *Clinical Psychology Review, 17*, 903–936.

Arnett, P. A., Smith, S. S., & Newman, J. P. (1997). Approach and avoidance motivation in psychopathic criminal offenders during passive avoidance. *Journal of Personality and Social Psychology, 72*, 1413–1428.

Arnow, B., Kenardy, J., & Agras, W. S. (1992). Binge eating among the obese: A descriptive study. *Journal of Behavioral Medicine, 15*, 155–170.

Arntz, A. (1994). Treatment of borderline personality disorder: A challenge for cognitive-behavioral therapy. *Behaviour Research and Therapy, 32*, 419–430.

Aronson, M. K. (1988). Patients and families: Impact and long-term-management implications. In M. K. Aronson (Ed.), *Understanding Alzheimer's disease* (pp. 74–78). New York: Charles Scribners & Sons.

Arranz, M. J., et al. (2000). Pharmacogenetic prediction of clozapine response. *Lancet, 355*, 1615–1616.

Arroyo, J. A., Simpson, T. L., & Aragon, A. S. (1997). Childhood sexual abuse among Hispanic and non-Hispanic White college women. *Hispanic Journal of Behavioral Sciences, 19*, 57–68.

Asaad, G., & Shapiro, B. (1986). Hallucinations: Theoretical and clinical overview. *American Journal of Psychiatry, 143*, 1088–1097.

Asarnow, J. R., Carlson, G. A., & Guthrie, D. (1987). Coping strategies, self-perceptions, hopelessness, and perceived family environments in depressed and suicidal children. *Journal of Consulting and Clinical Psychology, 55*, 361–366.

Asarnow, R. F., et al. (1991). Span of apprehension in schizophrenia. In J. Zubin, S. Steinhauer, & J. Gruzelier (Eds.), *Handbook of schizophrenia: Vol. 5. Neuropsychology, psychophysiology, and information-processing* (pp. 353–370). Amsterdam: Elsevier Science.

Asthma affects 15 million U.S. adults. (2001, August 16). *MSNBC Web Posting, Associated Press*. Retrieved August 16, 2001, from http://www.msnbc.com/news/615104.asp?0dm=H16MH.

Astley, S. J., et al. (1992). Analysis of facial shape in children gestationally exposed to marijuana, alcohol, and/or cocaine. *Pediatrics, 89*, 67–77.

Atkinson, R. L. (1997). Use of drugs in the treatment of obesity. *Annual Review of Nutrition, 17*, 383–403.

Attention Deficit Disorder—Part II. (1995, May). *The Harvard Mental Health Letter, 11*(11), 1–3.

Augusto, A., et al. (1996). Post-natal depression in an urban area of Portugal: Comparison of childbearing women and matched controls. *Psychological Medicine, 26*, 135–141.

Azar, B. (2000a). The debate over child care isn't over yet. *Monitor on Psychology, 31*, pp. 32–34.

Azar, B. (2000b). Brain studies point to inefficient connections in people with dyslexia. *Monitor on Psychology, 31*, pp. 32–34.

Azrin, N. H., & Peterson, A. L. (1989). Reduction of an eye tick by controlled blinking. *Behavior Therapy, 20*, 467–473.

B

Baaré, W. F. C., et al. (2001). Volumes of brain structures in twins discordant for schizophrenia. *Archives of General Psychiatry, 58*, 33–40.

Bäckman, L., & Forsell, Y. (1994). Episodic memory functioning in a community-based sample of old adults with major depression: Utilization of cognitive support. *Journal of Abnormal Psychology, 103*, 361–370.

Baer, L., et al. (1995). Cingulotomy for intractable obsessive-compulsive disorder: Prospective long-term follow-up of 18 patients. *Archives of General Psychiatry, 52*, 384–392.

Bagby, R. M., Nicholson, R., & Buis, T. (1998). Effectiveness of the MMPI-2 validity indicators. *Journal of Personality Assessment*, 70, 405–415.

Bailar, J. C., III. (2001). The powerful placebo and the wizard of Oz. *The New England Journal of Medicine, 344*, 1630–1632.

Baker, R. C., & Kirschenbaum, D. S. (1993). Self-monitoring may be necessary for successful weight control. *Behavior Therapy, 24*, 377–394.

Baldessarini, R. J., Tohen, M., & Tondo, L. (2000). Maintenance treatment in bipolar disorder. *Archives of General Psychiatry, 57*, 490–492.

Baldessarini, R. J., & Tondo, L. (2000). Does lithium treatment still work? Evidence of stable responses over three decades. *Archives of General Psychiatry, 57*, 187–190.

Baldeweg, T., et al. (1997). Neurophysiological changes associated with psychiatric symptoms in HIV-infected individuals without AIDS. *Biological Psychiatry, 41*, 474–487.

Baldwin, A. R., Oei, T. P., & Young, R. (1994). To drink or not to drink: The differential role of alcohol expectancies and drinking refusal self-efficacy in quantity and frequency of alcohol consumption. *Cognitive Therapy & Research, 17*, 511–530.

Ballester, E., et al. (1999). Evidence of the effectiveness of continuous positive airway pressure in the treatment of sleep apnea/hypopnea syndrome. *American Journal of Respiratory and Critical Care Medicine, 159*, 495–501.

Banaji, M. R., & Kihlstrom, J. F. (1996). The ordinary nature of alien abduction memories. *Psychological Inquiry, 7*, 132–135.

Bandura, A. (1973). *Aggression: A social learning analysis*. Englewood Cliffs, NJ: Prentice-Hall.

Bandura, A. (1982). Self-efficacy mechanism in human agency. *American Psychologist, 37*, 122–147.

Bandura, A. (1986). *Social foundations of thought and action: A social-cognitive theory*. Englewood Cliffs, NJ: Prentice-Hall.

Bandura, A. (2001). Social cognitive theory: An agentic perspective. *Annual Review of Psychology, 52*, 1–26.

Bandura, A., Barr-Taylor, C., Williams, S. L., Mefford, I. N., & Barchas, J. D. (1985). Catecholamine secretion as a function of perceived coping self-efficacy. *Journal of Consulting and Clinical Psychology, 53*, 406–414.

Bandura, A., Jeffery, R. W., & Wright, C. L. (1974). Efficacy of participant modeling as a function of response induction aids. *Journal of Abnormal Psychology, 83*, 56–64.

Bandura, A., Ross, S. A., & Ross, D. (1963). Imitation of film-mediated aggressive models. *Journal of Abnormal and Social Psychology, 66*, 3–11.

Banning, A. (1989). Mother-son incest: Confronting a prejudice. *Child Abuse and Neglect, 13*, 563–570.

Barbaree, H. E., & Marshall, W. L. (1991). The role of male sexual arousal in rape: Six models. *Journal of Consulting and Clinical Psychology, 59*, 621–631.

Barber, J., et al. (2000). Alliance predicts patients' outcome beyond in-treatment change in symptoms. *Journal of Consulting and Clinical Psychology, 68*, 1027–1032.

Barber, M. E., et al. (1998). Aborted suicide attempts: A new classification of suicidal behavior. *American Journal of Psychiatry, 155*, 385–389.

Barch, D. M., et al. (2001). Selective deficits in prefrontal cortex function in medication-naive patients with schizophrenia. *Archives of General Psychiatry, 58*, 280–288.

Barefoot, J. C., et al. (2001). A longitudinal study of gender differences in depressive symptoms from age 50 to 80. *Psychology and Aging, 16*, 342–345.

Barkham, M., et al. (1996). Dose-effect relations in time-limited psychotherapy for depression. *Journal of Consulting and Clinical Psychology, 64*, 927–935.

Barkley, R. A. (1997). *ADHD and the nature of self-control*. New York: Guilford Press.

Barkley, R. A. (2001). Executive function and ADHD: A reply. *Journal of the American Academy of Child & Adolescent Psychiatry, 40*, 501–502.

Barkley, R. A., et al. (1976). Evaluation of a token system for juvenile delinquents in a residential setting. *Journal of Behavior Therapy and Experimental Psychiatry, 7*, 227–230.

Barksy, A. J., et al. (2001). Hypochondriacal patients' appraisal of health and physical risks. *American Journal of Psychiatry, 158*, 788–794.

Barlow, D. H. (1986). Causes of sexual dysfunction: The role of anxiety and cognitive interference. *Journal of Abnormal Psychology, 54*, 140–148.

Barlow, D. H., Esler, J. L., & Vitali, A. E. (1998). Psychosocial treatments for panic disorders, phobias, and generalized anxiety disorder. In P. E. Nathan & J. M. Gorman (Eds.), *A guide to treatments that work* (pp. 288–318). New York: Oxford University Press.

Barlow, D. H., Gorman, J. M., Shear, M. K., & Woods, S. W. (2000). Cognitive-behavioral therapy, imipramine, or their combination for panic disorder: A randomized controlled trial. *Journal of the American Medical Association, 283*, 2529–2536.

Barnett, W. S., & Escobar, C. M. (1990). Economic costs and benefits of early intervention. In S. J. Meisels & J. P. Shonkoff (Eds.), *Handbook of early childhood intervention*. New York: Cambridge University Press.

Barrett, P. M., et al. (2001). Cognitive-behavioral treatment of anxiety disorders in children: Long-term (6-year) follow-up. *Journal of Consulting and Clinical Psychology, 69*, 135–141.

Barry, C. T., et al. (2000). The importance of callous–unemotional traits for extending the concept of psychopathy to children. *Journal of Abnormal Psychology, 109*, 335–340.

Barsky, A. J., Wyshak, G., & Klerman, G. L. (1992). Psychiatric comorbidity in *DSM-III-R* hypochondriasis. *Archives of General Psychiatry, 49*, 101–108.

Barsky, A. J., et al. (1994). Histories of childhood trauma in adult hypochondriacal patients. *American Journal of Psychiatry, 151*, 397–401.

Barsky, A. J., et al. (1998). A prospective 4- to 5-year study of *DSM-III-R* hypochondriasis. *Archives of General Psychiatry, 55*, 737–744.

Bartecchi, C. E., MacKenzie, T. D., & Schrier, R. W. (1994). The human costs of tobacco use (First of two parts). *New England Journal of Medicine, 330*, 907–912.

Basch, M. F. (1980). *Doing psychotherapy*. New York: Basic Books.

Basoglu, M., et al. (1997). Double-blindness procedures, rater blindness, and ratings of outcome. *Archives of General Psychiatry, 54*, 744–748.

Bateman, A. B., & Fonagy, P. (2001). Treatment of borderline personality disorder with psychoanalytically oriented partial hospitalization: An 18-month follow-up. *American Journal of Psychiatry, 158*, 36–42.

Bateson, G. D., Jackson, D., Haley, J., & Weakland, J. (1956). Toward a theory of schizophrenia. *Behavioral Science, 1*, 251–264.

Battaglia, M., et al. (1995). A family study of schizotypal disorder. *Schizophrenia Bulletin, 21*, 33–45.

Baucom, D. H., et al. (1998). Empirically supported couple and family interventions for marital distress and adult mental health problems. *Journal of Consulting and Clinical Psychology, 66*, 53–88.

Bazell, R. (2000, August 29). Cause of narcolepsy pinpointed. *MSNBC Web Posting*. Retrieved August 29, 2000, from http://www.msnbc.com/news/452884.asp.

Beautrais, A. L., et al. (1996). Prevalence and comorbidity of mental disorders in persons making serious suicide attempts: A case-control study. *American Journal of Psychiatry, 153*, 1009–1014.

Beauvais, F. (1998). American Indians and alcohol. *Alcohol Health and Research World*, 22, 253–259.

Bebbington, P. (1993). Transcultural aspects of affective disorders. *International Review of Psychiatry, 5*, 145–156.

Bechtoldt, H., Norcross, J. C., Wyckoff, L. A., Pokrywa, M. L., & Campbell, L. F. (2001). Theoretical orientations and employment settings of clinical and counseling psychologists: A comparative study. *The Clinical Psychologist, 54*(1), 3–6.

Beck, A. T. (1976). *Cognitive therapy and the emotional disorders*. New York: International Universities Press.

Beck, A. T., Brown, G., Steer, R. A., Eidelson, J. I., & Riskind, J. H. (1987). Differentiating anxiety and depression: A test of the cognitive content-specificity hypothesis. *Journal of Abnormal Psychology, 96*, 179–183.

Beck, A. T., & Clark, D. A. (1997). An information processing model of anxiety: Automatic and strategic processes. *Behaviour Research and Therapy, 35*, 49–58.

Beck, A. T., Freeman, A., & Associates. (1990). *Cognitive therapy of personality disorders*. New York: Guilford Press.

Beck, A. T., Rush, A. J., Shaw, B. F., & Emery, G. (1979). *Cognitive therapy of depression*. New York: Guilford Press.

Beck, A. T., Ward, C. H., Mendelson, M., Mock, J., & Erbaugh, J. (1961). An inventory for measuring depression. *Archives of General Psychiatry, 4*, 561–571.

Beck, A. T., & Young, J. E. (1985). Depression. In D. H. Barlow (Ed.), *Clinical handbook of psychological disorders* (pp. 206–244). New York: Guilford Press.

Beck, A. T., et al. (1990). Relationship between hopelessness and ultimate suicide: A replication with psychiatric outpatients. *American Journal of Psychiatry, 147*, 190–195.

Beck, D., Casper, R., & Andersen, A. (1996). Truly late onset of eating disorders: A study of 11 cases averaging 60 years of age at presentation. *International Journal of Eating Disorders, 20*, 389–395.

Beck, J. G. (1993). Vaginismus. In W. O'Donohue & J. H. Geer (Eds.), *Handbook of sexual dysfunctions: Assessment and treatment* (pp. 381–397). Boston: Allyn & Bacon.

Beck, J. G. (1995). Hypoactive sexual desire disorder: An overview. *Journal of Consulting and Clinical Psychology, 63*, 919–927.

Becker, D., & Lamb, S. (1994). Sex bias in the diagnosis of borderline personality disorder and posttraumatic stress disorder. *Professional Psychology: Research and Practice, 25*, 55–61.

Begley, S. (1995, November 20). Lights of madness. *Newsweek*, pp. 76–77.

Begley, S. (1998, January 26). Is everybody crazy? *Newsweek*, pp. 48–56.

Begley, S. (2000, July 3). What families should do. *Newsweek*, pp. 44–47.

Begley, S. (2001a) Fall. AIDS at 20. *Newsweek*, pp. 35–37.

Begley, S. (2001b), June 11. How it all starts inside your brain. *Newsweek*, pp. 40–42.

Beidel, D. C., Turner, S. M., & Morris, T. L. (2000). Behavioral treatment of childhood social phobia. *Journal of Consulting and Clinical Psychology, 68*, 1072–1080.

Beitman, B. D., Goldfried, M. R., & Norcross, J. C. (1989). The movement toward integrating the psychotherapies: An overview. *American Journal of Psychiatry, 146*, 138–147.

Bellack, A. S., & Mueser, K. T. (1990). Schizophrenia. In A. S. Bellack, M. Hersen, & A. E. Kazdin (Eds.), *International handbook of behavior modification and therapy* (2nd ed., pp. 353–370). New York: Plenum Press.

Bellack, A. S., & Mueser, K. T. (1993). Psychosocial treatment for schizophrenia. *Schizophrenia Bulletin, 19*, 317–336.

Belsky, J. (1993). Etiology of child maltreatment: A developmental-ecological analysis. *Psychological Bulletin, 114*, 413–434.

Bemporad, J. R. (1996). Self-starvation through the ages: Reflections on the pre-history of anorexia nervosa. *International Journal of Eating Disorders, 19*, 217–237.

Benazon, N. R. (2000). Predicting negative spousal attitudes toward depressed persons: A test of Coyne's interpersonal model. *Journal of Abnormal Psychology, 109*, 500–554.

Bender, L. (1938). A visual motor gestalt test and its clinical use. *Research Monograph of the American Orthopsychiatric Association, 3, XI*, 176.

Benet, S. (1974). *Abkhasians: The long living people of the Caucasus*. New York: Holt, Rinehart & Winston.

Benjamin, L., & Wonderlich, S. A. (1994). Social perceptions and borderline personality disorder: The relation to mood disorders. *Journal of Abnormal Psychology, 103*, 610–624.

Bennett, D. (1985). Rogers: More intuition in therapy. *APA Monitor, 16*, p. 3.

Benson, H. (1975). *The relaxation response*. New York: Morrow.

Benson, H., Manzetta, B. R., & Rosner, B. (1973). Decreased systolic blood pressure in hypertensive subjects who practiced meditation. *Journal of Clinical Investigation, 52*, 8.

Bentall, R. P. (1990). The illusion of reality: A review and integration of psychological research on hallucinations. *Psychological Bulletin, 107*, 82–95.

Bentall, R. P., Haddock, G., & Slade, P. (1994). Cognitive behavior therapy for persistent auditory hallucinations: From theory to therapy. *Behavior Therapy, 25*, 51–66.

Berenbaum, H., & Fujita, F. (1994). Schizophrenia and personality: Exploring the boundaries and connections between vulnerability and outcome. *Journal of Abnormal Psychology, 103*, 148–158.

Berenbaum, H., & Oltmanns, T. F. (1990). Emotional experience and expression in schizophrenia and depression. *Journal of Abnormal Psychology, 101*, 37–44.

Bergem, A. L. M., et al. (1997). Heredity in late-onset Alzheimer's disease and vascular dementia. *Archives of General Psychiatry, 54*, 264–270.

Bergner, R. M. (1997). What is psychopathology? And so what? *Clinical Psychology: Science and Practice, 4*, 235–248.

Berkowitz, L. (1994). Is something missing? Some observations prompted by the cognitive-neoassociationist view of anger and emotional aggression. In L. R. Huesmann (Ed.), *Aggressive behavior: Current perspectives*. New York: Plenum Press.

Berlin, I. N. (1987). Effects of changing Native American cultures on child development. *Journal of Community Psychology, 15*, 299–306.

Berman, M. E., Tracy, J. I., & Coccaro, E. F. (1997). The serotonin hypothesis of aggression revisited. *Clinical Psychology Review, 17*, 651–665.

Berman, S. L., et al. (1997). The impact of exposure to crime and violence on urban youth. *American Journal of Orthopsychiatry, 66*, 329–336.

Berman, S. L., et al. (2000). Predictors of outcome in exposure-based cognitive and behavioral treatments for phobic and anxiety disorders in children. *Behavior Therapy, 31*, 713–731.

Bernstein, A. S. (1987). Orienting response research in schizophrenia: Where we have come and where we might go. *Schizophrenia Bulletin, 13*, 623–641.

Bernstein, A. S., et al. (1988). Schizophrenia is associated with altered orienting activity: Depression with electrodermal (cholinergic?) deficit and normal orienting response. *Journal of Abnormal Psychology, 97*, 3–12.

Bernstein, D. P., et al. (1996). Childhood antecedents of adolescent personality disorders. *American Journal of Psychiatry, 153*, 907–913.

Bernstein, E. M., & Putnam, F. W. (1986). Development, reliability, and validity of a dissociation scale. *The Journal of Nervous and Mental Disease, 174*, 727–735.

Bernstein, R. L., & Gaw, A. C. (1990). Koro: Proposed classification for *DSM-IV*. *American Journal of Psychiatry, 147*, 1670–1674.

Bertolino, A., et al. (2000). Specific relationship between prefrontal neuronal-acetyl aspartate and activation of the working memory cortical network in schizophrenia. *American Journal of Psychiatry, 157*, 16–25.

Bertram, L., et al. (2000). Evidence for genetic linkage of Alzheimer's disease to chromosome 10q. *Science, 290*, 2302–2303.

Beutler, L. E. (1995). Common factors and specific effects. *Clinical Psychology: Science and Practice, 2*, 79–82.

Beutler, L. E., Harwood, T. M., & Caldwell, R. (2001). Cognitive-behavioral therapy and psychotherapy integration. In K. S. Dobson (Ed.), *Handbook of cognitive-behavioral therapies* (2nd ed., pp. 138–170). New York: Guilford Press.

Beyond methadone: More options for treating heroin addiction. (2000, November 1). *MSNBC Web Posting*. Retrieved November 7, 2000, from http://www.msnbc.com/news/484097.asp.

Bick, P. A., & Kinsbourne, M. (1987). Auditory hallucinations and subvocal speech in schizophrenic patients. *American Journal of Psychiatry, 144*, 222–225.

Bickel, W. K., et al. (1997). Effects of adding behavioral treatment to opioid detoxification with buprenorphine. *Journal of Consulting and Clinical Psychology, 65*, 803–810.

Biederman, J., Mick, E., & Faraone, S. V. (2000). Age-dependent decline of symptoms of attention-deficit hyperactivity disorder: Impact of remission definition and symptom type. *American Journal of Psychiatry, 157*, 816–818.

Biederman, J., et al. (1996a). Is childhood oppositional defiant disorder a precursor to adolescent conduct disorder? Findings from a four-year follow-up study of children with ADHD. *Journal of the American Academy of Child and Adolescent Psychiatry, 35*, 1193–1204.

Biederman, J. S., et al. (1996b). A prospective 4-year follow-up study of attention-deficit hyperactivity and related disorders. *Archives of General Psychiatry, 53*, 437–446.

Bigler, E. D., & Ehrenfurth, J. W. (1981). The continued inappropriate singular use of the Bender Visual Motor Gestalt Test. *Professional Psychology, 12*, 562–569.

Binder, J. L., & Strupp, H. H. (1997). Negative process: A recurrently discovered and underestimated facet of therapeutic process and outcome in the individual psychotherapy of adults. *Clinical Psychology: Science and Practice, 4*, 121–139.

Binder, R. L. (1999). Are the mentally ill dangerous? *Journal of the American Academy of Psychiatry and the Law, 27*, 189–201.

Binder, R. L., & McNiel, D. E. (1996). Application of the Tarasoff ruling and its effect on the victim and the therapeutic relationship. *Psychiatric Services, 47*, 1212–1215.

Biofeedback applications as central or adjunctive treatment. (1997, April). *Clinician's Research Digest, 15*(4), 5.

Biran, M. (1988). Cognitive and exposure treatment for agoraphobia: Re-examination of the outcome research. *Journal of Cognitive Psychotherapy: An International Quarterly, 2*, 165–178.

Birchwood, M., et al. (1992). The influence of ethnicity and family structure on relapse in first-episode schizophrenia: A comparison of Asian, Afro-Caribbean, and White patients. *British Journal of Psychiatry, 161*, 783–790.

Birnbaum, M. H., Martin, H., & Thomann, K. (1996). Visual function in multiple personality disorder. *Journal of the American Optometric Association, 67*, 327–334.

Bitiello, B., & Jensen, P. S. (1997). Medication development, testing in children and adolescents. *Archives of General Psychiatry, 54*, 871–876.

Bitler, D. A., Linnoila, M., & George, D. T. (1994). Psychosocial and diagnostic characteristics of individual initiating domestic violence. *Journal of Nervous and Mental Disease, 182*, 583–585.

Bjork, J. M., et al. (2000). Differential behavioral effects of plasma tryptophan depletion and loading in aggressive and nonaggressive men. *Neuropsychopharmacology, 22*, 357–359.

Blackman, S. J. (1996). Has drug culture become an inevitable part of youth culture? A critical assessment of drug education. *Educational Review, 48*, 131–142.

Blackwood, H. R., et al. (1996). A locus for bipolar affective disorder on chromosome 4p. *Nature Genetics, 12*, 427–430.

Blais, M. A., et al. (2001). Predicting *DSM-IV* cluster B personality disorder criteria from MMPI-2 and Rorschach data: A test of incremental validity. *Journal of Personality Assessment, 76*, 150–168.

Blakeslee, S. (1994b, August 16). New clue to cause of dyslexia seen in mishearing of fast sounds. *The New York Times*, pp. C1, C10.

Blakeslee, S. (1997a, June 27). Brain studies tie marijuana to other drugs. *The New York Times*, p. A16.

Blakeslee, S. (1997b, December 16). Suicide rate higher in 3 gambling cities, study says. *The New York Times*, p. A16.

Blakeslee, S. (2000, November 5). Pesticide found to produce Parkinson's symptoms in rats. *The New York Times*, p. A38.

Blakeslee, S. (2001, March 6). Drug's effect on brain is extensive, study finds. *The New York Times*, p. F3.

Blanchard, E. B., & Diamond, S. (1996). Psychological treatment of benign headache disorders. *Professional Psychology, 27*, 541–547.

Blanchard, E. B., et al. (1990). A controlled evaluation of thermal biofeedback and thermal feedback combined with cognitive therapy in the treatment of vascular headache. *Journal of Consulting and Clinical Psychology, 58*, 216–224.

Blanchard, R., & Hucker, S. J. (1991). Age, transvestism, bondage, and concurrent paraphilic activities in 117 fatal cases of autoerotic asphyxia. *British Journal of Psychiatry, 159*, 371–377.

Blankstein, K. R., & Segal, Z. V. (2001). Cognitive assessment: Issues and methods. In K. S. Dobson (Ed.), *Handbook of cognitive-behavioral therapies* (2nd ed., pp. 40–85). New York: Guilford Press.

Blatt, S. J., et al. (1998). When and how perfectionism impedes the brief treatment of depression: Further analyses of the National Institute of Mental Health Treatment of Depression Collaborative Research Program. *Journal of Consulting and Clinical Psychology, 66*, 423–428.

Blazer, D. G., et al. (1994). The prevalence and distribution of major depression in a National Comorbidity Survey. *American Journal of Psychiatry, 151*, 979–986.

Bliss, E. L., & Jeppsen, E. A. (1985). Prevalence of multiple personality among inpatients and outpatients. *American Journal of Psychiatry, 142*, 250–251.

Bloom, B. L. (1992). Computer-assisted psychological intervention: A review and commentary. *Clinical Psychology Review, 12*, 169–197.

Bloom, J. D., & Rogers, J. L. (1987). The legal basis of forensic psychiatry: Statutorily mandated psychiatric diagnosis. *American Journal of Psychiatry, 144*, 847–853.

Blumenthal, D. (1988, October 9). Dieting reassessed. *The New York Times Magazine, Part 2: The Good Health Magazine*, pp. 24–25, 53–54.

Blumenthal, R., & Endicott, J. (1997). Barriers to seeking treatment for major depression. *Depression and Anxiety, 4*, 273–278.

Boetsch, E. A., Green, P. A., & Pennington, B. F. (1996). Psychosocial correlates of dyslexia across the life span. *Development and Psychopathology, 8*, 539–562.

Bögels, S. M., & Zigerman, D. (2000). Dysfunctional cognitions in children with social phobia, separation anxiety disorders, and generalized anxiety disorder. *Journal of Abnormal Child Psychology, 28*, 205–211.

Bolton, P. (2001). Cross-cultural validity and reliability testing of a standard psychiatric assessment instrument without a gold standard. *Journal of Nervous & Mental Disease, 189*, 238–242.

Bondareff, W., et al. (2000). Comparison of sertraline and nortriptyline in the treatment of major depressive disorder in late life. *American Journal of Psychiatry, 157*, 745–750.

Boney-McCoy, S., & Finkelhor, D. (1996). Is youth victimization related to trauma symptoms and depression after controlling for prior symptoms and family relationships? A longitudinal, prospective study. *Journal of Consulting and Clinical Psychology, 64*, 1406–1416.

Bonné, J. (2001, February 6). Meth's deadly buzz. *MSNBC.com Special Report*. Retrieved February 8, 2001, from http://www.msnbc.com/news/510835.asp?bt=nm&btu=http://www.msnbc.com/tools/newstools/d/news_menu.asp&cp1=1.

Bonta, J., Law, M., & Hanson, K. (1998). The prediction of criminal and violent recidivism among mentally disordered offenders: A meta-analysis. *Psychological Bulletin, 123*, 123–142.

Bookheimer, S. Y., et al. (2000). Patterns of brain activation in people at risk for Alzheimer's disease. *The New England Journal of Medicine, 343*, 450–456.

Boren, T., Faulk, P., Roth, K. A., Larson, G., & Normark, S. (1993). Attachment of *Helicobacter pylori* to human gastric epithelium mediated by blood group antigens. *Science, 262*, 1892–1895.

Bornstein, M. R., Bellack, A. S., & Hersen, M. (1977). Social-skills training for unassertive children: A multiple-baseline analysis. *Journal of Applied Behavior Analysis, 10*, 183–195.

Bornstein, R. F. (1992). The dependent personality: Developmental, social, and clinical perspectives. *Psychological Bulletin, 112*, 3–23.

Bornstein, R. F. (1993). *The dependent personality*. New York: The Guilford Press.

Bornstein, R. F. (1997). Dependent personality disorder in the *DSM-IV* and beyond. *Clinical Psychology: Science and Practice, 4*, 175–187.

Bornstein, R. F. (1999). Criterion validity of objective and projective dependency tests: A meta-analytic assessment of behavioral prediction. *Psychological Assessment, 11*, 48–57.

Bornstein, R. F. (1999). Dependent and histrionic personality disorders. In T. Millon et al. (Eds.), *Oxford textbook of psychopathology. Oxford textbooks in clinical psychology, Vol. 4* (pp. 535–554). New York: Oxford University Press.

Borthwick-Duffy, S. A. (1994). Epidemiology and prevalence of psychopathology in individuals with dual diagnoses. *Journal of Consulting and Clinical Psychology, 62*, 17–27.

Boskind-White, M., & White, W. C. (1983). *Bulimarexia: The binge-purge cycle*. New York: W. W. Norton.

Boston Women's Health Book Collective. (1984). *The new our bodies, ourselves*. New York: Simon & Schuster.

Bostwick, J. M., & Pankratz, V. S. (2000). Affective disorders and suicide risk: A reexamination. *American Journal of Psychiatry*, 157, 1925–1932.

Botelho, R. J., & Richmond, R. (1996). Secondary prevention of excessive alcohol use: Assessing the prospects of implementation. *Family Practice, 13*, 182–193.

Bouchard, C. (1997). Obesity in adulthood—The importance of childhood and parental obesity. *The New England Journal of Medicine, 337*, 926–927.

Bouchard, C., Rhéaume, J., & Ladouceur, R. (1999). Responsibility and perfectionism in OCD: An experimental study. *Behaviour Research & Therapy, 37*, 239–248.

Bouchard, S., et al. (1996). Cognitive restructuring in the treatment of psychotic symptoms in schizophrenia: A critical analysis. *Behavior Therapy, 27*, 257–277.

Boudouris, J. (2000). The insanity defense in Polk County, Iowa. *American Journal of Forensic Psychology, 18*, 41–79.

Bouton, M. E., Mineka, S., & Barlow, D. H. (2001). A modern learning theory perspective on the etiology of panic disorder. *Psychological Review, 108*, 4–32.

Bouwer, C., & Stein, D. J. (1997). Association of panic disorder with a history of traumatic suffocation. *American Journal of Psychiatry, 154*, 1566–1570.

Bowden, C. L., et al. (2000). A randomized, placebo-controlled 12-month trial of divalproex and lithium in treatment of outpatients with bipolar I disorder. *Archives of General Psychiatry, 57*, 481–489.

Bowers, T. G., & Clum, G. A. (1988). Relative contribution of specific and nonspecific treatment effects: Meta-analysis of placebo-controlled behavior therapy research. *Psychological Bulletin, 103*, 315–323.

Bowlby, J. (1988). *A secure base*. New York: Basic Books.

Boy to sue city over tryst with teacher. (2000, April 14). *Reuters News Agency*.

Boyd-Franklin, N. (1989). *Black families in therapy: A multisystems approach*. New York: Guilford Press.

Boyle, M. H., et al. (1992). Predicting substance use in late adolescence: Results from the Ontario Child Health Study Follow-up. *American Journal of Psychiatry, 149*, 761–767.

Boyle, P. (1993). The hazards of passive—and active—smoking. *New England Journal of Medicine, 328*, 1708–1709.

Braaten, E. B., & Rosén, L. E. (2000). Self-regulation of affect in attention deficit–hyperactivity disorder (ADHD) and non-ADHD boys: Differences in empathic responding. *Journal of Consulting and Clinical Psychology, 68*, 313–321.

Braddock, D. (1992). Community mental health and mental retardation services in the United States: A comparative study of resource allocation. *American Journal of Psychiatry, 149*, 175–183.

Bradley, J. D. D., & Golden, C. J. (2001). Biological contributions to the presentation and understanding of attention-deficit/hyperactivity disorder: A review. *Clinical Psychology Review, 21*, 907–929.

Bradley, R. G., & Follingstad, D. R. (2001). Utilizing disclosure in the treatment of the sequelae of childhood sexual abuse. A theoretical and empirical review. *Clinical Psychology Review, 21*, 1–32.

Brady, K., et al. (2000). Efficacy and safety of sertraline treatment of posttraumatic stress disorder: A randomized controlled trial. *Journal of the American Medical Association, 283*, 1837–1844.

Brady, S. (1986). Short-term memory, phonological processing, and reading ability. *Annals of Dyslexia, 36*, 138–153.

Braff, D. L. (1993). Information processing and attention dysfunction in schizophrenia. *Schizophrenia Bulletin, 19*, 233–259.

Braswell, L., & Kendall, P. C. (2001). Cognitive-behavioral therapy with youth. In K. S. Dobson (Ed.), *Handbook of cognitive-behavioral therapies* (2nd ed., pp. 246–294). New York: The Guilford Press.

Braun, B. G. (Ed.). (1986). *Treatment of multiple personality disorder*. Washington, DC: American Psychiatric Press.

Braun, S. (2001, Spring). Seeking insight by prescription. *Cerebrum*, pp. 10–21.

Breaux, C., Matsuoka, J. K., & Ryujin, D. H. (1995, August). *National utilization of mental health services by Asian/Pacific Islanders*. Paper presented at the meeting of the American Psychological Association, New York, NY.

Breitner, J. C. S., & Zandi, P. P. (2001). Do nonsteroidal anti-inflammatory drugs reduce the risk of Alzheimer's disease? *New England Journal of Medicine, 345*, 1567–1568.

Brekke, J. S., et al. (1997). The impact of service characteristics on functional outcomes from community support programs for persons with schizophrenia: A growth curve analysis. *Journal of Consulting and Clinical Psychology, 65*, 464–475.

Bremmer, J. D., et al. (1996). Chronic PTSD in Vietnam combat veterans: Course of illness and substance abuse. *American Journal of Psychiatry, 153*, 369–375.

Brems, C., & Johnson, M. E. (1997). Clinical implications of the co-occurrence of substance use and other psychiatric disorders. *Professional Psychology: Research and Practice, 28*, 437–447.

Brenda, W. J. H., et al. (1998). Depressive symptoms and physical decline in community-dwelling older persons. *Journal of the American Medical Association, 279*, 1720–1726.

Brener, N. D., McMahon, P. M., Warren, C. W., & Douglas, K. A. (1999). Forced sexual intercourse and associated health-risk behaviors among female college students in the United States. *Journal of Consulting and Clinical Psychology, 67*, 252–259.

Breslow, N. (1989). Sources of confusion in the study and treatment of sadomasochism. *Journal of Social Behavior and Personality, 4*, 263–274.

Breslau, N., et al. (1996). Sleep disturbance and psychiatric disorders: A longitudinal epidemiological study of young adults. *Biological Psychiatry, 39*, 411–418.

Breslau, N., et al. (1997a). Sex differences in posttraumatic stress disorder. *Archives of General Psychiatry, 54*, 1044–1048.

Breslau, N., et al. (1997b). Psychiatric sequelae of posttraumatic stress disorder in women. *Archives of General Psychiatry, 54*, 81–87.

Breslau, N., et al. (1998). Major depression and stages of smoking: A longitudinal investigation. *Archives of General Psychiatry, 55*, 161–166.

Breslau, N., et al., (2001). Nicotine dependence in the United States: Prevalence, trends, and smoking persistence. *Archives of General Psychiatry, 58*, 810–816.

Brewin, C. R., Andrews, B., & Valentine, J. D. (2000). Meta-analysis of risk factors for posttraumatic stress disorder in trauma-exposed adults. *Journal of Consulting and Clinical Psychology, 68*, 748–766.

Brinkman, S. D., Largen, J. W., Jr., Cushman, L., Braun, P. R., & Block, R. (1986). Clinical validators: Alzheimer's disease and multi-infarct dementia. In L. W. Poon (Ed.), *Handbook for clinical memory assessment of older adults* (pp. 307–313). Washington, DC: American Psychological Association.

Brody, J. E. (1988a, May 5). Sifting fact from myth in the face of asthma's growing threat to American children. *The New York Times*, p. B19.

Brody, J. E. (1990, June 7). A search to bar retardation in a new generation. *The New York Times*, p. B9.

Brody, J. E. (1992a, May 15). Study finds liquid diet works (but not for the 50% who quit). *The New York Times*, p. B7.

Brody, J. E. (1992b, June 16). Suicide myths cloud efforts to save children. *The New York Times*, pp. C1, C3.

Brody, J. E. (1992c, September 30). Myriad masks hide an epidemic of depression. *The New York Times*, p. C12.

Brody, J. E. (1993c, December 15). Living with a common genetic abnormality. *The New York Times*, p. C17.

Brody, J. E. (1994b, February 9). Depression in the elderly: Old notions hinder help. *The New York Times*, p. C13.

Brody, J. E. (1994d, December 28). Wine for the heart: Over all, risks may outweigh benefits. *The New York Times*, p. C10.

Brody, J. E. (1995a, January 18). Dysthymia: Help for chronic sadness. *The New York Times*, p. C8.

Brody, J. E. (1995b, August 2). With more help available for impotence, few men seek it. *The New York Times*, p. C9.

Brody, J. E. (1996a, August 7). Relaxation method may aid health. *The New York Times*, p. C10.

Brody, J. E. (1996b, November 14). Decline seen in death rates from cancer as a whole. *The New York Times*, p. A21.

Brody, J. E. (1996c, November 20). Controlling anger is good medicine for the heart. *The New York Times*, p. C15.

Brody, J. E. (1997a, March 26). Race and weight. *The New York Times*, p. C8.

Brody, J. E. (1998, June 16). Gaining weight on sugar-free, fat-free diets. *The New York Times*, p. F7.

Brody, J. E. (2000, October 17). One-two punch for losing pounds: Exercise and careful diet. *The New York Times*, p. F6.

Bronfrenbrenner, U. (1979). *The ecology of human development: Experiments by nature and design*. Cambridge, MA: Harvard University Press.

Brookoff, D., et al. (1997). Characteristics of participants in domestic violence: Assessment at the scene of domestic assault. *Journal of the American Medical Association, 277*, 1369–1373.

Broussard, B. A., et al. (1991). Prevalence of obesity in American Indians and Alaska Natives. *American Journal of Clinical Nutrition, 53* (6 Suppl.), 1535S–1542S.

Brown, D. R., Ahmed, F., Gary, L. E., & Milburn, N. G. (1995). Major depression in a community sample of African Americans. *American Journal of Psychiatry*, 373–378.

Brown, G. K., Beck, A. T., Steer, R. A., & Grisham, J. R. (2000). Risk factors for suicide in psychiatric outpatients: A 20-year prospective study. *Journal of Consulting and Clinical Psychology, 68*, 371–377.

Brown, L. S. (1992). A feminist critique of the personality disorders. In L. Brown & M. Balou (Eds.), *Personality and psychopathology: Feminist reappraisals* (pp. 206–228). New York: Guilford Press.

Brown, L. S. (1997, November). Recovered memories of abuse: Research and clinical update. *Clinician's Research Digest, Supplemental Bulletin, 17*, 1–2.

Brown, R. A., et al. (1997). Cognitive-behavioral treatment for depression in alcoholism. *Journal of Consulting and Clinical Psychology, 65*, 715–726.

Brown, S. L., & Forth, A. E. (1997). Psychopathy and sexual assault: Static risk factors, emotional precursors, and rapist subtypes. *Journal of Consulting and Clinical Psychology, 65*, 848–857.

Brown, T. A., et al. (2001). Reliability of *DSM–IV* anxiety and mood disorders: Implications for the classification of emotional disorders. *Journal of Abnormal Psychology, 110*, 49–58.

Brownell, K. D., & Wadden, T. A. (1992). Etiology and treatment of obesity: Understanding a serious, prevalent, and refractory disorder. *Journal of Consulting and Clinical Psychology, 60*, 505–517.

Bruce, T. J. (1996). Predictors of alprazolam discontinuation with and without cognitive behavior therapy for panic disorder: A reply. *American Journal of Psychiatry, 153*, 1109–1110.

Bruch, H. (1973). *Eating disorders: Obesity, anorexia and the person within*. New York: Basic Books.

Bruch, M. A. (1997). Positive thoughts or cognitive balance as a moderator of the negative life events–dysphoria relationship: A reexamination. *Cognitive Therapy and Research, 21*, 25–38.

Brun, A. (1996). Frontal lobe degeneration of non-Alzheimer type. *Acta Neurologica Scandinavica Supplementum, 168*, 28–30.

Bryant, R. A. (2001). Posttraumatic stress disorder and traumatic brain injury: Can they co-exist *Clinical Psychology Review, 21*, 931–948.

Buchanan, A. (1999). Risk and dangerousness. *Psychological Medicine, 29*, 465–473.

Buchanan, R. W., & Carpenter, W. T., Jr. (1997). The neuroanatomies of schizophrenia. *Schizophrenia Bulletin, 23*, 367–372.

Buchsbaum, M. S., & Hazlett, E. A. (1998). Positron emission tomography studies of abnormal glucose metabolism in schizophrenia. *Schizophrenia Bulletin, 24*, 343–364.

Bullman, T. A., & Kang, H. K. (1994). Posttraumatic stress disorder and the risk of traumatic deaths among Vietnam veterans. *Journal of Nervous and Mental Disease, 182*, 604–610.

Buriel, R., Calzada, S., & Vazquez, R. (1982). The relationship of traditional Mexican American culture to adjustment and delinquency among three generations of Mexican American male adolescents. *Hispanic Journal of Behavioral Sciences, 4*, 41–55.

Burke, R. S., & Stephens, R. S. (1999). Social anxiety and drinking in college students: A social cognitive theory analysis. *Clinical Psychology Review, 19*, 513–530.

Burnam, M. A., Hough, R. L., Karno, M., Escobar, J. I., & Telles, C. A. (1987). Acculturation and lifetime prevalence of psychiatric disorders among Mexican Americans in Los Angeles. *Journal of Health and Social Behavior, 28*, 89–102.

Burns, D. D. (1980). *Feeling good: The new mood therapy*. New York: Morris.

Burns, D. D., & Beck, A. T. (1978). Modification of mood disorders. In J. P. Foreyt & D. P. Rathjen (Eds.), *Cognitive behavior therapy: Research and application* (pp. 109–134). New York: Plenum Press.

Burns, D. D., & Nolen-Hoeksema, S. (1992). Therapeutic empathy and recovery from depression in cognitive-behavioral therapy: A structural equation model. *Journal of Consulting and Clinical Psychology, 60*, 441–449.

Burros, M. (1994, July 17). Despite awareness of risks, more in U.S. are getting fat. *The New York Times*, pp. A1, A8.

Burton, N., & Lane, R. C. (2001). The relational treatment of dissociative identity disorder. *Clinical Psychology Review, 21*, 301–320.

Busatto, G. F., et al. (1997). Correlation between reduced in vivo benzodiazepine receptor binding and severity of psychotic symptoms in schizophrenia. *American Journal of Psychiatry, 154*, 56–63.

Buss, D. M., & Kenrick, D. T. (1998). Evolutionary social psychology. In D. T. Gilbert, S. T. Fiske, & G. Lindzey (Eds.), *The handbook of social psychology* (4th ed., Vol. 2, pp. 982–1026). Boston, MA: McGraw-Hill, Inc.

Buss, D. M., & Shackelford, T. K. (1997). Human aggression in evolutionary psychological perspective. *Clinical Psychology Review, 17*, 605–619.

Bustillo, J. R., et al. (2001). The psychosocial treatment of schizophrenia: An update. *American Journal of Psychiatry, 158*, 163–175.

Butler, A. C., & Beck, A. T. (1995, Summer). Cognitive therapy for depression. *The Clinical Psychologist, 48*, 3–5.

Butler, G. (1989). Issues in the application of cognitive and behavioral strategies to the treatment of social phobia. *Clinical Psychology Review, 9*, 91–106.

Butler, L. D., et al. (1996). Hypnotizability and traumatic experience: A diathesis-stress model of dissociative symptomatology. *American Journal of Psychiatry, 153*(Suppl.), 42–63.

Butler, R. N. (2001, Fall/Winter). The myth of old age. *Newsweek Special Issue*, p. 33.

Butterfield, F. (1997a, January 5). Serious crime decreased for fifth year in a row. *The New York Times*, p. A10.

Butterfield, F. (1997c, June 2). Homicides plunge 11 percent in U.S., F.B.I. report says. *The New York Times*, pp. A1, B10.

Butterfield, F. (2001). Violence rises as club drug spreads out into the streets. *The New York Times*, pp. A1, A14.

Bütz, M. R., Bowlling, J. B., & Bliss, C. A. (2000). Psychotherapy with the mentally retarded: A review of the literature and the implications. *Professional Psychology: Research and Practice, 31*, 42–47.

Butzlaff, R. L., & Hooley, J. M. (1998). Expressed emotion and psychiatric relapse. *Archives of General Psychiatry, 55*, 547–552.

Byne, W., et al. (2001). Magnetic resonance imaging of the thalamic mediodorsal nucleus and pulvinar in schizophrenia and schizotypal personality disorder. *Archives of General Psychiatry, 58*, 133–140.

C

Cable News Network. (1998a, April 6). Alcohol remains large factor in violent crime. *CNN Interactive* [Online].

Cable News Network. (1998b, May 20). New studies show a genetic link to the complex disease of alcoholism, *Cable News Network Web Posting*. Retrieved May 23, 1998, from www.cnn.com/HEALTH/9805/20/genetic.alcoholism/.

Caetano, R. (1987). Acculturation and drinking patterns among U.S. Hispanics. *British Journal of Addiction, 82*, 789–799.

Calhoon, S. K. (1996). Confirmatory factor analysis of the Dysfunctional Attitude Scale in a student sample. *Cognitive Therapy and Research, 20*, 81–91.

Calhoun, K. S., & Atkeson, B. M. (1991). *Treatment of rape victims: Facilitating social adjustment*. New York: Pergamon Press.

Calhoun, P. S., et al. (2000). Drug use and validity of substance use self-reports in veterans seeking help for posttraumatic stress disorder. *Journal of Consulting and Clinical Psychology, 68*, 923–927.

Califia, P. (1997). *Sex changes: The politics of transgenderism*. San Francisco: Cleis.

Callicott, J. H., et al. (2000). Selective relationship between prefrontal n-acetylaspartate measures and negative symptoms in schizophrenia. *American Journal of Psychiatry, 157*, 1646–1651.

Cameron, N. (1963). *Personality development and psychopathology: A dynamic approach*. Boston: Houghton Mifflin.

Campbell, S. B., & Cohn, J. F. (1991). Prevalence and correlates of postpartum depression in first-time mothers. *Journal of Abnormal Psychology, 100*, 594–599.

Can stress make you sick? (1998). *Harvard Health Letter, 23*(6), pp. 1–3.

Cancer rates inch down, mostly for men. (1999, April 20). *Cable News Network*. Retrieved April 23, 1999, from http://www.cnn.com.

Capps, L., et al. (1993). Parental perception of emotional expressiveness in children with autism. *Journal of Consulting and Clinical Psychology, 61*, 475–484.

Cardno, A. G., et al. (1999). Heritability estimates for psychotic disorders in the Maudsley twin psychosis series. *Archives of General Psychiatry, 56*, 162–168.

Cardozo, B. L., et al. (2000). Mental health, social functioning, and attitudes of Kosovar Albanians following the war in Kosovo. *Journal of the American Medical Association, 284*, 569–577.

Carey, B. (1998, January/February). The sunshine supplement. *Health*, pp. 52–55.

Carey, G. (1992). Twin imitation for antisocial behavior: Implications for genetic and family environment research. *Journal of Abnormal Psychology, 101*, 18–25.

Carey, G., & DiLalla, D. L. (1994). Personality and psychopathology: Genetic perspectives. *Journal of Abnormal Psychology, 103*, 32–43.

Carey, M. P., Wincze, J. P., & Meisler, A. W. (1998). Sexual dysfunction: Male erectile disorder. In D. H. Barlow (Ed.), *Clinical handbook for psychological disorders* (pp. 442–480). New York: Guilford Publication.

Carini, M. A., & Nevid, J. S. (1992). Social appropriateness and impaired perspective in schizophrenia. *Journal of Clinical Psychology, 48*, 170–177.

Carlin, A. S., Hoffman, H. G., & Weghorst, S. (1997). Virtual reality and tactile augmentation in the treatment of spider phobia. *Behaviour Research and Therapy, 35*, 1153–1158.

Carmen, E. H., Rieker, P. P., & Mills, T. (1984). Victims of violence and psychiatric illness. *American Journal of Psychiatry, 141*, 378–383.

Carney, R. M., Freedland, K. E., & Jaffe, A. S. (2001). Depression as a risk factor for coronary heart disease mortality. *Archives of General Psychiatry, 58*.

Carpenter, S. (2000, September). Psychologists tackle neuroimaging at APA-sponsored Advanced Training Institute. *Monitor on Psychology*, pp. 42–43.

Carpenter, S. (2001a, May). Stimulants boost achievement in ADHD teens. *Monitor on Psychology*, pp. 26–27.

Carpenter, S. (2001b, September). A new reason for keeping a diary. *Monitor on Psychology*, pp. 68–70.

Carter, C. S., et al. (1997). Anterior cingulate gyrus dysfunction and selective attention deficits in schizophrenia: H2O PET study during single-trial Stroop task performance. *American Journal of Psychiatry, 154*, 1670–1675.

Carter, M. M., et al. (1995). Effects of a safe person on induced distress following a biological challenge in panic disorder with agoraphobia. *Journal of Abnormal Psychology, 104*, 156–163.

Carver, C. S., & Gaines, J. G. (1987). Optimism, pessimism, and postpartum depression. *Cognitive Therapy & Research, 11*, 449–462.

Casanova, M. R. (1997). Functional and anatomical aspects of prefrontal pathology in schizophrenia. *Schizophrenia Bulletin, 23*, 517–519.

Cascardi, M., et al. (1995). Characteristics of women physically abused by their spouses and who seek treatment regarding marital conflict. *Journal of Consulting and Clinical Psychology, 63*, 616–623.

Casey, B. J., et al. (1997). Implication of right frontostriatal circuitry in response inhibition and attention-deficit hyperactivity disorder. *Journal of the American Academy of Child and Adolescent Psychiatry, 36*, 374–383.

Castellanos, F. X., et al. (2001). Quantitative brain magnetic resonance imaging in girls with attention-deficit/hyperactivity disorder. *Archives of General Psychiatry, 58*, 289–295.

Castro, J. (1992, January 20). Sexual harassment: A guide. *Time Magazine*, p. 37.

Catalano, R., et al. (1993). Using ECA survey data to examine the effect of job layoffs on violent behavior. *Hospital and Community Psychiatry, 44*, 874–879.

Catz, S. L., & Kelly, J. A. (2001). Living with HIV disease. In A. Baum, T. A. Revenson, & J. E. Singer (Eds.), *Handbook of health psychology* (pp. 841–850). Mahwah, NJ: Erlbaum.

CDC says 61 percent of U.S. adults overweight. (2000, December 15). *CNN Web Posting*. Retrieved December 19, 2000, from http://www.cnn.com/2000/HEALTH/diet.fitness/12/15/fat.america.ap/index.html.

Celano, M., et al. (1996). Treatment of traumagenic beliefs among sexually abused girls and their mothers: An evaluation study. *Journal of Abnormal Child Psychology, 24*, 1–17.

Celio, A. A., et al. (2000). Reducing risk factors for eating disorders: comparison of an Internet- and a classroom-delivered psychoeducational program. *Journal of Consulting and Clinical Psychology, 68*, 650–657.

Celis, W. (1991, January 2). Students trying to draw line between sex and an assault. *The New York Times*, pp. 1, B8.

Center for Mental Health Services (CMHS). (1994). *Mental health statistics*. Office of Consumer, Family and Public Information, Center for Mental Health Services, U.S. Department of Health and Human Services. Rockville, MD: Author.

Center for Mental Health Services (2001, May). *National Strategy for Suicide Prevention: Goals and objectives for action: Summary. A joint effort of SAMHS, CDC, NIH, and HRSA*. Washington, DC: Author.

Centers for Disease Control (CDC). (2000). Tobacco use among middle and high school students—United States, 1999. *Morbidity and Mortality Weekly Report, 49*, 49–53.

Centers for Disease Control (CDC). (2001a). Self-reported asthma prevalence among adults: United States, 2000. *Morbidity and Mortality Weekly Report, 50*, 682–686.

Centers for Disease Control (CDC). (2001b, October 12). Cigarette smoking among adults—United States, 1999. *Morbidity and Mortality Weekly Report, 50*(40). Retrieved November 15, 2001, from http://www.cdc.gov/tobacco/research_data/adults_prev/mm5040.htm.

Centers for Disease Control (CDC). (2001c). *Suicide in the United States*. National Center for Injury Prevention and Control, Centers for Disease Control. Atlanta: Author.

Chadda, R. K., & Ahuja, N. (1990). Dhat syndrome: A sex neurosis of the Indian subcontinent. *British Journal of Psychiatry, 156*, 577–579.

Chakrabarti, S., & Fombonne, E. (2001). Pervasive developmental disorders in preschool children. *Journal of the American Medical Association, 285*, 3093–3099.

Chamberlin, J. (2001, July/August). Putting a face on child mental illness. *Monitor on Psychology*, pp. 28–29.

Chambless, D. L., &. Ollendick, T. H. (2001). Empirically supported psychological interventions: Controversies and evidence. *Annual Review of Psychology, 52*, 685–716.

Chambless, D. L., et al. (1998, Winter). Update on empirically validated therapies, II. *The Clinical Psychologist, 51*, 3–16.

Chan, D. W. (1991). The Beck Depression Inventory: What difference does the Chinese version make? *Psychological Assessment, 3*, 616–622.

Chang, S. C. (1984). Review of I. Yamashita "Taijin-kyofu." *Transcultural Psychiatric Research Review, 21*, 283–288.

Charney, D. S., Nestler, E. J., & Bunney, B. S. (1999). *Neurobiology of mental illness*. New York: Oxford University Press.

Chase, M. (1998, March 2). A new diet drug hits the marketplace, with potential risks. *The Wall Street Journal*, p. B1.

Chemerinski, E., et al. (2001). The specificity of depressive symptoms in patients with Alzheimer's Disease. *American Journal of Psychiatry, 158*, 68–72.

Chemtob, C. M., et al. (1997). Cognitive-behavioral treatment for severe anger in posttraumatic stress disorder. *Journal of Consulting and Clinical Psychology, 65*, 184–189.

Chen, J., et al. (2001). Racial differences in the use of cardiac catheterization after acute myocardial infarction. *The New England Journal of Medicine, 344*, 1443–1449.

Chen, P., Ganguli, M., Mulsant, B. H., & DeKosky, S. T. (1999). The temporal relationship between depressive symptoms and dementia: A community-based prospective study. *Archives of General Psychiatry, 56*, 261–266.

Chermack, S. T., & Giancola, P. R. (1997). The relation between alcohol and aggression: An integrated biopsychosocial conceptualization. *Clinical Psychology Review, 17*, 621–649.

Chesno, F. A., & Kilmann, P. R. (1975). Effects of stimulation intensity on sociopathic avoidance learning. *Journal of Abnormal Psychology, 84*, 144–151.

Cheung, F. (1991). The use of mental health services by ethnic minorities. In H. F. Myers et al. (Eds.), *Ethnic minority perspectives on clinical training and services in psychology* (pp. 23–31). Washington, DC: American Psychological Association.

Cheung, F. M., & Ho, R. M. (1997). Standardization of the Chinese MMPI-A in Hong Kong: A preliminary study. *Psychological Assessment, 9*, 499–502.

Cheung, F., Song, W., & Butcher, J. N. (1991). An infrequency scale for the Chinese MMPI. *Psychological Assessment, 3*, 648–653.

Children with hyperactivity disorder prone to injury. (2001, January 2). *CNN Web Posting*. Retrieved January 4, 2001, from http://www.cnn.com/2001/HEALTH/children/01/02/bc.health.hyperactive.reut/index.html.

Ching, J. W. J., et al. (1995). Perceptions of family values and roles among Japanese Americans: Clinical considerations. *American Journal of Orthopsychiatry, 65*, 216–224.

Chipperfield, B., & Vogel-Sprott, M. (1988). Family history of problem drinking among young male social drinkers: Modeling effects on alcohol consumption. *Journal of Abnormal Psychology, 97*, 423–428.

Chobanian, A. V. (2001). Control of hypertension: An important national priority. *The New England Journal of Medicine, 345*, 534–535.

Choi, K. H., Binson, D., Adelson, M., & Catania, J. A. (1998). Sexual harassment, sexual coercion, and HIV risk among U.S. adults 18–49 years. *AIDS and Behavior, 2(1)*, 33–40.

Chowdhury, A. N. (1996). The definition and classification of Koro. *Culture, Medicine and Psychiatry, 20*, 41–65.

Christiansen, B. A., & Goldman, M. S. (1983). Alcohol-related expectancies versus demographic/background variables in the prediction of adolescent drinking. *Journal of Consulting and Clinical Psychology, 52*, 249–257.

Christiansen, K., & Winkler, E. M. (1992). Hormonal, anthropometrical, and behavioral correlates of physical aggression in Kung San men of Namibia. *Aggressive Behavior, 18*, 271–280.

Christmas, A. L., Wodarski, J. S., & Smokowski, P. R. (1996). Risk factors for physical child abuse: A practice theoretical paradigm. *Family Therapy, 23*, 233–248.

Chu, J. A., et al. (1999). Memories of childhood abuse: Dissociation, amnesia, and corroboration. *American Journal of Psychiatry, 156*, 749–755.

Ciesla, J. A., & Roberts, J. E. (2001). Meta-analysis of the relationship between HIV infection and risk for depressive disorders. *American Journal of Psychiatry, 158*, 725–730.

Clark, D. A., Cook, A., & Snow, D. (1998). Depressive symptom differences in hospitalized, medically ill, depressed psychiatric inpatients and nonmedical controls. *Journal of Abnormal Psychology, 107*, 38–48.

Clark, D. M. (1986). A cognitive approach to panic. *Behaviour Research and Therapy, 24*, 461–470.

Clark, D. M., et al. (1997). Misinterpretation of body sensations in panic disorder. *Journal of Consulting and Clinical Psychology, 65*, 203–213.

Clark, S. E., & Loftus, E. F. (1996). The construction of space alien abduction memories. *Psychological Inquiry, 7*, 140–143.

Clay, R. A. (2001a, January). To the heart of the matter. *Monitor on Psychology*, pp. 42–45.

Clay, R. A. (2001b, January). Bringing psychology to cardiac care. *Monitor on Psychology*, pp. 46–49.

Cleckley, H. (1976). *The mask of sanity* (5th ed.). St. Louis: Mosby.

Cleghorn, J. M., et al. (1992). Toward a brain map of auditory hallucinations. *American Journal of Psychiatry, 149*, 1062–1069.

Clemence, A., & Handler, L. O. (2001). Psychological assessment on internship: A survey of training directors and their expectations for students. *Journal of Personality Assessment, 76*, 18–47.

Clemetson, L. (2000, November 6). The new victims of hate. *Newsweek*, p. 61.

Coccaro, E. F., & Kavoussi, R. J. (1997). Fluoxetine and impulsive aggressive behavior in personality-disordered subjects. *Archives of General Psychiatry, 54*, 1081–1088.

Cockerham, W. C., Kunz, G., & Lueschen, G. (1989). Alcohol use and psychological distress: A comparison of Americans and West Germans. *The International Journal of the Addictions, 24*, 951–961.

Cognitive impairment linked to early death in HIV-infected patients. (1996, March 29). *Reuters News Service*.

Cohen, A. C., et al. (1993). Factors determining the decision to institutionalize dementing individuals: A prospective study. *The Gerontologist, 22*, 714–720.

Cohen, D. (1986). Psychopathological perspectives: Differential diagnosis of Alzheimer's disease and related disorders. In L. W. Poon (Ed.), *Handbook for clinical memory assessment of older adults* (pp. 81–88). Washington, DC: American Psychological Association.

Cohen, F. L., Ferrans, C. E., & Eshler, B. (1992). Reported accidents in narcolepsy. *Loss, Grief and Care, 5*, 71–80.

Cohen, J. A., & Mannarino, A. P. (1997). A treatment study for sexually abused preschool children: Outcome during a one-year follow-up. *Journal of the American Academy of Child and Adolescent Psychiatry, 36*, 1228–1235.

Cohen, S., Tyrrell, D. A. J., & Smith, A. P. (1991). Psychological stress and susceptibility to the common cold. *The New England Journal of Medicine, 325*, 606–612.

Cohen, S., et al. (1998). Types of stressors that increase susceptibility to the common cold in healthy adults. *Health Psychology, 17*, 214–223.

Cohen-Kettensi, P. T. (2001). Gender identity disorder in *DSM? Journal of American Academy of Child & Adolescent Psychiatry, 40*, 391.

Cohen-Kettenis, P. T., & van Goozen, S. H. M. (1997). Sex reassignment of adolescent transsexuals: A follow-up study. *Journal of the American Academy of Child and Adolescent Psychiatry, 36*, 263–271.

Cole, D. A., et al. (1998). A longitudinal look at the relation between depression and anxiety in children and adolescents. *Journal of Consulting and Clinical Psychology, 66*, 451–460.

Coleman, D., & Baker, F. M. (1994). Misdiagnosis of schizophrenia in older, Black veterans. *Journal of Nervous and Mental Disease, 182*, 527–528.

Coleman, E. (1987). Bisexuality: Challenging our understanding of sexual orientation. *Sexuality and Medicine, 1*, 225–242.

College binge drinking worries (1999, August 27). *Associated Press, NewsReal, Inc.*

Collins, J. J., & Messerschmidt, P. M. (1993). Epidemiology of alcohol-related violence. *Alcohol Health and Research World, 17*, 93–100.

Collins, P. H. (1990). *Black feminist thought: Knowledge, consciousness, and the politics of empowerment*. Boston: Unwin Hyman.

Colvin, C. R., & Block, J. (1994). Do positive illusions foster mental health? An examination of the Taylor and Brown formulation. *Psychological Bulletin, 16*, 3–20.

Comas-Diaz, L., & Griffith, E. (1988). Introduction: On culture and psychotherapeutic care. In L. Comas-Diaz & E. Griffith (Eds.), *Clinical guidelines in cross-cultural mental health*. New York: Wiley.

Combs, B. J., Hales, D. R., & Williams, B. K. (1980). *An invitation to health*. Menlo Park, CA: Benjamin/Cummings.

Comings, D. E. (1997). Genetic aspects of childhood behavioral disorders. *Child Psychiatry and Human Development, 27*, 139–150.

Compas, B. E., et al. (1998). Sampling of empirically supported psychological treatments from health psychology: Smoking, chronic pain, cancer, and bulimia nervosa. *Journal of Consulting and Clinical Psychology, 66*, 89–112.

Cone, J. D. (1999). Introduction to the special section on self-monitoring: A major assessment method in clinical psychology. *Psychological Assessment, 11*, 411–414.

Conley, R. R., & Buchanan, R. W. (1997). Evaluation of treatment-resistant schizophrenia. *Schizophrenia Bulletin, 23*, 663–674.

Connors, G. J., et al. (1997). The therapeutic alliance and its relationship to alcoholism treatment participation and outcome. *Journal of Consulting and Clinical Psychology, 65*, 588–598.

Cooke, D. J., & Michie, C. (1997). An item response theory analysis of the Hare Psychopathy Checklist—Revised. *Psychological Assessment, 9*, 3–14.

Cooke, R. (1994, November 8). Memory's foe: Progress seen in battle vs. brain killer. *New York Newsday*, p. A16.

Cookson, W. O. C. M., & Moffatt, M. R. (1997). Asthma—An epidemic in the absence of infection? *Science, 275*, 41–42.

Coolidge, F. L., & Segal, D. L. (1998). Evolution of personality disorder diagnosis in the *Diagnostic and Statistical Manual of Mental Disorders*. *Clinical Psychology Review, 18*, 585–599.

Cooney, N. L., et al. (1997). Alcohol cue reactivity, negative-mood reactivity, and relapse in treated alcoholic men. *Journal of Abnormal Psychology, 106*, 243–250.

Coons, P. M. (1986). Treatment progress in 20 patients with multiple personality disorder. *Journal of Nervous and Mental Disease, 174*, 715–721.

Coons, P. M., Bowman, E. S., & Pellow, T. A. (1989). Post-traumatic aspects of the treatment of victims of sexual abuse and incest. *Psychiatric Clinics of North America, 12*, 325–327.

Coontz, S., & Franklin, D. (1997, October 28). When the marriage penalty is marriage. *The New York Times*, p. A23.

Cooper, P. J., et al. (1999). Post-partum depression and the mother-infant relationship in a South African peri-urban settlement. *British Journal of Psychiary, 175*, 554–558.

Corbitt, E. M., & Widiger, T. A. (1995). Sex differences among the personality disorders: An exploration of the data. *Clinical Psychology: Science and Practice, 2*, 225–238.

Cordes, C. (1985). Common threads found in suicide. *APA Monitor, 16* (10), 11.

Cordova, M. J., et al. (1995). Frequency and correlates of posttraumatic-stress-disorder-like symptoms after treatment for breast cancer. *Journal of Consulting and Clinical Psychology, 63,* 981–986.

Cornblatt, B. A., & Kilep, J. G. (1994). Impaired attention, genetics, and the pathophysiology of schizophrenia. *Schizophrenia Bulletin, 20,* 31–46.

Coronado, S. F., & Peake, T. H. (1992). Culturally sensitive therapy: Sensitive principles. *Journal of College Student Psychotherapy, 7,* 63–72.

Cororve, M. B., & Gleaves, D. H. (2001). Body dysmorphic disorder: A review of conceptualizations, assessment, and treatment strategies. *Clinical Psychology Review, 21,* 949–970.

Coryell, W., Endicott, J., & Keller, M. (1992b). Rapidly cycling affective disorder: Demographics, diagnosis, family history, and course. *Archives of General Psychiatry, 49,* 126–131.

Coryell, W., et al. (1996). Importance of psychotic features to long-term course in major depressive disorder. *American Journal of Psychiatry, 153,* 483–489.

Costa, P. T., Jr., & McCrae, R. R. (1985). Hypochondriasis, neuroticism, and aging: When are somatic complaints unfounded? *American Psychologist, 40,* 19–28.

Costantino, G., et al. (1986). Cuento therapy: A culturally sensitive modality for Puerto Rican children. *Journal of Consulting and Clinical Psychology, 54,* 639–645.

Council of Economic Advisers for the President's Initiative on Race. (1998). *Changing America: Indicators of social and economic well-being by race and Hispanic origin.* Washington, DC: U.S. Government Printing Office.

Coursey, R. D., Alford, J., & Safarjan, B. (1997). Significant advances in understanding and treating serious mental illness. *Professional Psychology: Research and Practice, 28,* 205–216.

Cowan, W. M., & Kandel, E. R. (2001). Prospects for neurology and psychiatry. *Journal of the American Medical Association, 285,* 594–600.

Cowley, G. (2000a, July 3). Generation XXL. *Newsweek,* pp. 40–44.

Cowley, G. (2000b, May 22). The new war on Parkinson's. *Newsweek,* pp. 52–58.

Cowley, G. (2001a, February 12). New ways to stay clean. *Newsweek,* pp. 45–47.

Cowley, G. (2001b, March 12). Cannibals to cows: The path of a deadly disease. *Newsweek,* pp. 53–61.

Cowley, G. (2001c, June 11). Can he find a cure? *Newsweek,* pp. 39–41.

Cowley, G., & Underwood, A. (1998, January 5). A little help from serotonin. *Newsweek,* pp. 78–81.

Cox, W. M., & Klinger, E. (1988). A motivational model of alcohol use. *Journal of Abnormal Psychology, 97,* 168–180.

Coyle, J. T. (2001). Drug treatment of anxiety disorders in children. *The New England Journal of Medicine, 344,* 1326–1327.

Coyne, J. C. (1976). Toward an interactional description of depression. *Psychiatry, 39,* 14–27.

Craighead, W. E., Craighead, L. W., & Ilardi, S. S. (1998). Psychosocial treatments for major depressive disorder. In P. E. Nathan & J. M. Gorman (Eds.), *A guide to treatments that work* (pp. 226–239). New York: Oxford University Press.

Craske, M. G., Brown, T. A., & Barlow, D. H. (1991). Behavioral treatment of panic disorder: A two-year follow-up. *Behavior Therapy, 22,* 289–304.

Cravchik, A., & Goldman, D. (2000). Neurochemical individuality: Genetic diversity among human dopamine and serotonin receptors and transporters. *Archives of General Psychiatry, 57,* 1105–1114.

Crick, N. R., & Dodge, K. A. (1994). A review and reformulation of social information-processing mechanisms in children's social adjustment. *Psychological Bulletin, 115,* 74–101.

Crits-Christoph, P. (1992). The efficacy of brief dynamic psychotherapy: A meta-analysis. *American Journal of Psychiatry, 149,* 151–158.

Cross-National Collaborative Group. (1992). The changing rate of major depression: Cross-national comparisons. *Journal of the American Medical Association, 268,* 3098–3105.

Crow, T. J. (1980a). Molecular pathology of schizophrenia: More than one disease process? *British Medical Journal, 280,* 66–68.

Crow, T. J. (1980b). Positive and negative schizophrenic symptoms and the role of dopamine. *British Journal of Psychiatry, 137,* 383–386.

Crow, T. J. (1980c). Positive and negative schizophrenic symptoms and the role of dopamine: A debate. *British Journal of Psychiatry, 137,* 379–383.

Cuéllar, I., & Roberts, R. E. (1997). Relations of depression, acculturation, and socioeconomic status in a Latino sample. *Hispanic Journal of Behavioral Sciences, 19,* 230–238.

Cui, X-J., & Vaillant, G. E. (1997). Does depression generate negative life events? *Journal of Nervous and Mental Disease, 185,* 145–150.

Cummings, J. L. (1992). Depression and Parkinson's disease: A review. *American Journal of Psychiatry, 149,* 443–454.

Curb, J. D., & Marcus, E. B. (1991). Body fat and obesity in Japanese-Americans. *American Journal of Clinical Nutrition, 53,* 1552S–1555S.

Currie, S. R., et al. (2000). Cognitive-behavioral treatment of insomnia secondary to chronic pain. *Journal of Consulting and Clinical Psychology, 68,* 407–416.

Curry, S., Marlatt, G. A., Gordon, J. R. (1987). Abstinence violation effect: Validation of an attributional construct with smoking cessation. *Journal of Consulting and Clinical Psychology, 55,* 145–149.

Cutrona, C. E., et al. (2000). Direct and moderating effects of community context on the psychological well-being of African American women. *Journal of Personality and Social Psychology, 79,* 1088–1101.

Cutting, L. P., & Docherty, N. M. (2000). Schizophrenia outpatients' perceptions of their parents: Is expressed emotion a factor? *Journal of Abnormal Psychology, 109,* 266–272.

D

Dabbs, J. M., & Hargrove, M. F. (1997). Age, testosterone, and behavior among female prison inmates. *Psychosomatic Medicine, 59,* 477–480.

Dadds, M. R., et al. (1992). Childhood depression and conduct disorder: II. An analysis of family interaction patterns in the home. *Journal of Abnormal Psychology, 101,* 505–513.

Dahl, R. E. (1992). The pharmacologic treatment of sleep disorders. *Psychiatric Clinics of North America, 15,* 161–178.

Daley, S. E., Burge, D., & Hammen, C. (2000). Borderline personality disorder symptoms as predictors of 4-year romantic relationship dysfunction in young women: Addressing issues of specificity. *Journal of Abnormal Psychology, 109,* 451–460.

Daley, S. E., et al. (1997). Predictors of the generation of episodic stress: A longitudinal study of late adolescent women. *Journal of Abnormal Psychology, 106,* 251–259.

Damasio, A. R. (1997). Towards a neuropathology of emotion and mood. *Nature, 386,* 769–770.

Damasio, R. (2000). A neural basis for sociopathy. *Archives of General Psychiatry, 57,* 128–129.

Daro, D., & Wiese, D. (1995). *Current trends in child abuse reporting and fatalities: NCPCA's 1994 annual fifty state survey.* Chicago, IL: National Committee to Prevent Child Abuse.

Davidson, J. R. T., et al. (2001). Multicenter, double-blind comparison of sertraline and placebo in the treatment of posttraumatic stress disorder. *Archives of General Psychiatry, 58,* 485–492.

Davidson, P. R., & Parker, K. C. H. (2001). Eye movement desensitization and reprocessing (EMDR): A meta-analysis. *Journal of Consulting and Clinical Psychology, 69,* 305–316.

Davidson, R. J. (2000). Affective style, psychopathology, and resilience: Brain mechanisms and plasticity. *American Psychologist, 55,* 1196–1214.

Davidson, R. J., Putnam, K. M., & Larson, C. L. (2000). Dysfunction in the neural circuitry of emotion regulation—A possible prelude to violence. *Science, 289,* 591–594.

Davis, D. L., & Boster, L. H. (1992). Cognitive-behavioral-expressive interventions with aggressive and resistant youths. *Child Welfare, 71,* 557–573.

Davis, D. L., & Boster, L. H. (1993). Cognitive-behavioral-expressive interventions with aggressive and resistant youth. *Residential-Treatment for Children and Youth, 10,* 55–68.

Daw, J. (2001, April). Survey uncovers communication breakdown in the treatment of depression. *Monitor on Psychology,* p. 69.

DeAngelis, T. (1993, August). It's back: TV violence, concern for kid viewers. *APA Monitor, 24*(8), p. 16.

DeAngelis, T. (1994b, November). Ethnic-minority issues recognized in DSM-IV. *APA Monitor,* p. 36.

DeAngelis, T. (1995a, April). New threat associated with child abuse. *APA Monitor, 26*(4), pp. 1, 38.

DeAngelis, T. (1995b, April). Research documents trauma of abuse. *APA Monitor, 26*(4), p. 34.

DeBell, C., & Jones, R. D. (1997a). As good as it seems? A review of EMDR experimental research. *Professional Psychology: Research and Practice, 28,* 153–163.

DeBell, C., & Jones, R. D. (1997b). Privileged communication at last? An overview of *Jaffee v. Redmond. Professional Psychology: Research and Practice, 28,* 559–566.

De-Carle, A. J., & Pato, M. T. (1996). Social phobia and stuttering. *American Journal of Psychiatry, 153,* 1367–1368.

Deep-Brain Stimulation for Parkinson's Disease Study Group. (2001). Deep-brain stimulation of the subthalamic nucleus or the pars interna of the globus pallidus in Parkinson's disease. *New England Journal of Medicine, 345,* 956–963.

Deffenbacher, J. L., et al. (2000). An application of Beck's cognitive therapy to general anger reduction. *Cognitive Therapy & Research, 24,* 689–697.

Dekker, J. (1993). Inhibited male orgasm. In W. O'Donohue & J. H. Geer (Eds.), *Handbook of sexual dysfunctions: Assessment and treatment* (pp. 279–301). Boston: Allyn & Bacon.

De La Cancela, V., & Guzman, L. P. (1991). Latino mental health service needs: Implications for training psychologists. In H. F. Myers et al. (Eds.), *Ethnic minority perspectives on clinical training and services in psychology* (pp. 59–64). Washington, DC: American Psychological Association.

Delahanty, D. L., & Baum, A. (2001). Stress and breast cancer. In A. Baum, T. A. Revenson, & J. E. Singer (Eds.), *Handbook of health psychology* (pp. 747–756). Mahwah, NJ: Erlbaum.

deMayo, R. A. (1997). Patient sexual behavior and sexual harassment: A national survey of female psychologists. *Professional Psychology: Research and Practice, 28,* 58–62.

Dement, W. C. (1992). The proper use of sleeping pills in the primary care setting. *Journal of Clinical Psychiatry, 53*(12, Suppl.), 50–56.

Denneby, J. A., et al. (1996). Case-control study of suicide by discharged psychiatry patients. *British Medical Journal, 312,* 1580.

Denollet, J., et al. (1996). Personality as independent predictor of long-term mortality in patients with coronary heart disease. *Lancet, 347,* 417–421.

Depression Guideline Panel. (1993a). *Depression in primary care: Vol. 1. Detection and diagnosis.* Clinical Practice Guideline No. 5. Rockville, MD: U.S. Department of Health and Human Services, Public Health Service, Agency for Health Care Policy and Research (AHCPR Pub. No. 93–0550).

Depression Guideline Panel. (1993b). *Depression in primary care: Vol. 2. Treatment of major depression.* Clinical Practice Guideline No. 5. Rockville, MD: U.S. Department of Health and Human Services, Public Health Service, Agency for Health Care Policy and Research (AHCPR Pub. No. 93–0551).

DeRubeis, R. J., & Crits-Christoph, P. (1998). Empirically supported individual and group psychological treatments for adult mental disorders. *Journal of Consulting and Clinical Psychology, 66*, 37–52.

DeRubeis, R. J., Tang, T. Z., & Beck, A. T. (2001). Cognitive therapy. In K. S. Dobson (Ed.), *Handbook of cognitive-behavioral therapies* (2nd ed., pp. 349–392). New York: Guilford Press.

DeRubeis, R. J., et al. (1999). Medications versus cognitive behavior therapy for severely depressed outpatients: Meta-analysis of four randomized comparisons. *American Journal of Psychiatry, 156*, 1007–1013.

De Silva, P. (1993). Post-traumatic stress disorder: Cross-cultural aspects. *International Review of Psychiatry, 5*, 217–229.

Desmond, E. W. (1987, November). Out in the open: Changing attitudes and new research give fresh hope to alcoholics. *Time Magazine*, pp. 80–90.

Dettmer, K., et al. (2001). Take home naloxone and the prevention of deaths from opiate overdose: Two pilot schemes. *British Medical Journal, 322*, 895–896.

Devan, G. S. (1987). Koro and schizophrenia in Singapore. *British Journal of Psychiatry, 150*, 106–107.

DeVeaugh-Geiss J. (1994). Pharmacologic therapy of obsessive compulsive disorder. *Advances in Pharmacology, 30*, 35–52.

Devlin, M. J., Yanovski, S. Z., & Wilson, G. T. (2000). Obesity: What mental health professionals need to know. *American Journal of Psychiatry, 157*, 854–866.

Devor, E. J. (1994). A developmental-genetic model of alcoholism: Implications for genetic research. *Journal of Consulting and Clinical Psychology, 62*, 1108–1115.

Dhaliwal, G. K., et al. (1996). Adult male survivors of childhood sexual abuse: Prevalence, sexual abuse characteristics, and long-term effects. *Clinical Psychology Review, 16*, 619–639.

Dhawan, S., & Marshall, W. L. (1996). Sexual abuse histories of sexual offenders. *Sexual Abuse Journal of Research and Treatment, 8*, 7–15.

Diabetes as looming epidemic. (2001, January 30). *The New York Times*, p. F8.

Diamond, S., Baldwin, R., & Diamond, R. (1963). *Inhibition and choice.* New York: Harper & Row.

Dick, D. M., Rose, R. J., Viken, R. J., Kaprio, J., & Koskenvuo, M. (2001). Exploring gene-environment interactions: Socioregional moderation of alcohol use. *Journal of Abnormal Psychology, 110*, 625–632.

DiLalla, D. L., Carey, G., Gottesman, I. I., & Bouchard, T. J., Jr. (1996). Heritability of MMPI personality indicators of psychopathology in twins reared apart. *Journal of Abnormal Psychology, 105*, 491–499.

Dinh, K. T., et al. (1995). Children's perceptions of smokers and nonsmokers: A longitudinal study. *Health Psychology, 14*, 32–40.

Dip in youth killing, but not in youth drug use. (2000, December 15). *The New York Times*, p. A 28.

Dixit, A. R., & Crum, R. M. (2000). Prospective study of depression and the risk of heavy alcohol use in women. *American Journal of Psychiatry, 157*, 801–807.

Dixon, L., et al. (1997). Assertive community treatment and medication compliance in the homeless mentally ill. *American Journal of Psychiatry, 154*, 1302–1304.

Dobson, K. S., & Dozois, D. J. A. (2001). Historical and philosophical bases of the cognitive-behavioral therapies. In K. S. Dobson (Ed.), *Handbook of cognitive-behavioral therapies* (2nd ed., pp. 3–40). New York: Guilford Press.

Dodge, K. A. (1985). Attributional bias in aggressive children. *Advances in Cognitive Behavioral Research and Therapy, 4*, 73–110.

Dodge, K. A., et al. (1997). Reactive and proactive aggression in school children and psychiatrically impaired chronically assaultive youth. *Journal of Abnormal Psychology, 106*, 37–51.

Doleys, D. M. (1977). Behavioral treatments for nocturnal enuresis in children: A review of the literature. *Psychological Bulletin, 8*, 30–54.

Doll, B. (1996). Prevalence of psychiatric disorders in children and youth: An agenda for advocacy by school psychology. *School Psychology Quarterly, 11*, 20–46.

Donker, F. J. S. (2000). Cardiac rehabilitation: A review of current developments. *Clinical Psychology Review, 20*, 923–943.

Dorahy, M. J. (2001). Dissociative identity disorder and memory dysfunction: The current state of experimental research and its future directions. *Clinical Psychology Review, 21*, 771–795.

Doubt cast on power of placebos: Study finds sham treatment less effective than once thought. (2001, May 23). *MSNBC Web Posting.* Retrieved May 24, 2001, from http://www.msnbc.com/news/577411.asp.

Dougall, A. L., & Baum, A. (2001). Stress, health, and illness. In A. Baum, T. A. Revenson, & J. E. Singer (Eds.), *Handbook of health psychology* (pp. 339–348). Mahwah, NJ: Erlbaum.

Dowker, A., Hermelin, B., & Pring, L. (1996). A savant poet. *Psychological Medicine, 26*, 913–924.

Drake, R. E., et al. (1991). Housing instability and homelessness among rural schizophrenic patients. *American Journal of Psychiatry, 148*, 211–215.

Drewnowski, A. (1997). Taste preferences and food intake. *Annual Review of Nutrition, 17*, 237–253.

Drewnowski, A., et al. (1994). Eating pathology and *DSM-III-R* bulimia nervosa: A continuum of behavior. *American Journal of Psychiatry, 151*, 1217–1219.

Drews, C. D., et al. (1995). Variation in the influence of selected sociodemographic risk factors for mental retardation. *American Journal of Public Health, 85*, 329–334.

Drummond, D. C., & Glautier, S. (1994). A controlled trial of cue exposure treatment in alcohol dependence. *Journal of Consulting and Clinical Psychology, 62*, 809–817.

Druss, B. G., Rosenheck, R. A., & Sledge, W. H. (2000). Health and disability costs of depressive illness in a major U.S. corporation. *American Journal of Psychiatry, 157*, 1274–1278.

Dryden, W. (1984). *Rational-emotive therapy: Fundamentals and innovations.* London: Croom Helm.

Dryden, W., & Ellis, A. (2001). Rational emotive behavior therapy. In K. S. Dobson (Ed.), *Handbook of cognitive-behavioral therapies* (2nd ed., pp. 295–348). New York: Guilford Press.

Du Pont heir found guilty of murder but mentally ill. (1997, February 26). *The New York Times*, p. A10.

Dubovsky, S. (2000, September). Lithium: The oldest specific psychotropic medication. *Journal Watch for Psychiatry*, pp. 73, 76.

Duckworth, G., & McBride, H. (1996). Suicide in old age: A tragedy of neglect. *Canadian Journal of Psychiatry, 41*, 217–222.

Duffy, A., et al. (1998). Psychiatric symptoms and syndromes among adolescent children of parents with lithium-responsive or lithium-nonresponsive bipolar disorder. *American Journal of Psychiatry, 155*, 431–433.

Duffy, F. H. (1994). The role of quantified electroencephalography in psychological research. In G. Dawson & K. W. Fischer (Eds.), *Human behavior and the developing brain* (pp. 93–132). New York: Guilford Press.

Dugger, C. W. (1992, September 3). Threat only when on crack, homeless man foils system. *The New York Times*, pp. A1, B4.

Dugger, C. W. (1994, July 15). Larry Hogue is arrested in Westchester. *The New York Times*, pp. B1, B2.

Dugger, C. W. (1995, January 23). Slipping through cracks and out the door. *The New York Times*, pp. B1, B2.

Dulit, R. A., et al. (1994). Clinical correlates of self-mutilation in borderline personality disorder. *American Journal of Psychiatry, 151*, 1305–1311.

Duman, R. S., Heninger, G. R., & Nestler, E. J. (1997). A molecular and cellular theory of depression. *Archives of General Psychiatry, 54*, 597–606.

Dumas, J. E., Serketich, W. J., & LaFreniere, P. J. (1995). "Balance of power": A transactional analysis of control in mother-child dyads involving socially competent, aggressive, and anxious children. *Journal of Abnormal Psychology, 104*, 104–113.

Duncan, J., et al. (2000). A neural basis for general intelligence. *Science, 289*, 457–460.

Duncan, R. D., et al. (1996). Childhood physical assault as a risk factor for PTSD, depression, and substance abuse: Findings from a national survey. *American Journal of Orthopsychiatry, 66*, 437–447.

Durham v. United States, 214 F. 2d 862 (DC Circ 1954).

Durkheim, E. (1958). *Suicide.* (J. A. Spaulding & G. Simpson, Trans.). New York: Free Press. (Original work published 1897).

Dwyer, M. (1988). Exhibitionism/voyeurism. *Journal of Social Work and Human Sexuality, 7*, 101–112.

Dykens, E. M., & Hodapp, R. M. (1997). Treatment issues in genetic mental retardation syndromes. *Professional Psychology: Research and Practice, 28*, 263–270.

Dyslexia: The interaction of culture and biology. (2001, March 15). *CNN Web Posting.* Retrieved March 16, 2001, from http://www.cnn.com/2001/fyi/teachers.ednews/03/15/dyslexia.reading.ap.

E

Eagly, A. H., & Wood, W. (1991). Explaining sex differences in social behavior: A meta-analytic perspective. *Personality and Social Psychology Bulletin, 17*, 306–315.

Earnst, K .S., & Kring, A. M. (1997). Construct validity of negative symptoms: An empirical and conceptual review. *Clinical Psychology Review, 17*, 167–189.

Eaton, W. W., Dryman, A., & Weissman, M. M. (1991). Panic and phobia. In L. N. Robins & D. A. Regier (Eds.), *Psychiatric disorders in America: The Epidemiologic Catchment Area Study* (pp. 155–179). New York: Free Press.

Eaton, W. W., et al. (1994). Panic and panic disorder in the United States. *American Journal of Psychiatry, 151*, 413–420.

Eaton, W. W., et al. (1997). Natural history of diagnostic interview schedule/*DSM-IV* major depression: The Baltimore Epidemiologic Catchment Area follow-up. *Archives of General Psychiatry, 54*, 993–999.

Eberhardy, F. (1967). The view from "the couch." *Journal of Child Psychological Psychiatry, 8*, 257–263.

Ebigbo, P. O. (1993). Situation analysis of child abuse and neglect in Nigeria. *Journal of Psychology in Africa, 1*, 159–178.

Eckenrode, J., et al. (2000). Preventing child abuse and neglect with a program of nurse home visitation: The limiting effects of domestic violence. *Journal of the American Medical Association, 284*, 1385–1391.

Eckhardt, C. I., Barbour, K. A., & Stuart, G. L. (1997). Anger and hostility in maritally violent men: Conceptual distinctions, measurement issues, and literature review. *Clinical Psychology Review, 17*, 333–358.

Ecstasy use depletes brain chemical, study finds. (2000, July 25). *CNN Web Posting.* Retrieved July 26, 2000, from http://www.cnn.com/2000/HEALTH/07/25/ecstasy.brain.reut/index.html.

Edelson, E. (1998, March 9). Migraines come into focus. *Newsday*, p. C7.

Edinger, J. D., et al. (2001). Cognitive behavioral therapy for treatment of chronic primary insomnia: A randomized controlled trial. *Journal of the American Medical Association, 285*, 1856–1864.

Edman, J. L., & Johnson, R. C. (1999). Filipino American and Caucasian American beliefs about the causes and treatment of mental problems. *Cultural Diversity and Ethnic Minority Psychology, 5*, 380–386.

Edmundson, M. (2001, June 3). I'm O.K, and then some. *The New York Times Book Review*, p. 33.

Edwards, E. D., & Egbert-Edwards, M. (1990). American Indian adolescents: Combating problems of substance use and abuse through a community model. In A. R. Stiffman & L. E. Davis (Eds.), *Ethnic issues in adolescent mental health* (pp. 285–302). Newbury Park, CA: Sage Publications.

Egan, J. (1991). Oppositional defiant disorder. In J. M. Wiener (Ed.), *Textbook of child and adolescent psychiatry*. Washington, DC: American Psychiatric Press.

Egan, M. F., Apud, J., & Wyatt, R. J. (1997). Treatment of tardive dyskinesia. *Schizophrenia Bulletin, 23*, 583–609.

Ehlers, A. (1995). A 1-year prospective study of panic attacks: Clinical course and factors associated with maintenance. *Journal of Abnormal Psychology, 104*, 164–172.

Ehlers, A., Mayou, R. A., & Bryant, B. (1998). Psychological predictors of chronic posttraumatic stress disorder after motor vehicle accidents. *Journal of Abnormal Psychology, 107*, 508–519.

Eiberg, H., Berendt, I., & Mohr, J. (1995). Assignment of dominant inherited nocturnal euresis (ENUR1) to chromosome 13q. *Nature Genetics, 10*, 354–356.

Eichstedt, J. A., & Arnold, S. L. (2001). Childhood-onset obsessive-compulsive disorder. A tic-related subtype of OCD? *Clinical Psychology Review, 21*, 137–157.

Eisenbruch, M. (1992). Toward a culturally sensitive DSM: Cultural bereavement in Cambodian refugees and the traditional healer as taxonomist. *Journal of Nervous and Mental Disease, 180*, 8–10.

Eisler, I., et al. (1997). Family and individual therapy in anorexia nervosa: A 5-year follow-up. *Archives of General Psychiatry, 54*, 1025–1030.

Elkin, I., et al. (1989). National Institute of Mental Health treatment of depression collaborative research program: General effectiveness of treatments. *Archives of General Psychiatry, 46*, 971–982.

Ellason, J. W., & Ross, C. A. (1997). Two-year follow-up of inpatients with dissociative identity disorder. *American Journal of Psychiatry, 154*, 832–839.

Ellickson, P. L., Hays, R. D., & Bell, R. M. (1992). Stepping through the drug use sequence: Longitudinal scalogram analysis of initiation and regular use. *Journal of Abnormal Psychology, 101*, 441–451.

Elliott, A. J., et al. (1998). Randomized, placebo-controlled trial of paroxetine versus imipramine in depressed HIV-positive outpatients. *American Journal of Psychiatry, 155*, 367–372.

Elliott, D. M. (1997). Traumatic events: Prevalence and delayed recall in the general population. *Journal of Consulting and Clinical Psychology*, 811–820.

Ellis, A. (1977a). *Anger: How to live with and without it.* Secaucus, NJ: Citadel Press.

Ellis, A. (1977b). The basic clinical theory of rational-emotive therapy. In A. Ellis & R. Grieger (Eds.), *Handbook of rational-emotive therapy*. New York: Springer.

Ellis, A. (1993). Reflections on rational-emotive therapy. *Journal of Consulting and Clinical Psychology, 61*, 199–201.

Ellis, A. (1997). Using rational emotive behavior therapy techniques to cope with disability. *Professional Psychology: Research and Practice, 28*, 17–22.

Ellis, A. (2001, January). "Intellectual" and "emotional" insight revisited. *NYS Psychologist, 13*, 2–6.

Ellis, A., & Dryden, W. (1987). *The practice of rational emotional therapy*. New York: Springer.

Ellis, A., Young, J., & Lockwood, G. (1989). Cognitive therapy and rational-emotive therapy: A dialogue. *Journal of Cognitive Psychotherapy, 1*, 205–256.

Else, L., et al. (1993). Personality characteristics of men who physically abuse women. *Hospital and Community Psychiatry, 44*, 54–58.

Engdahl, B., et al. (1997). Posttraumatic stress disorder in a community group of former prisoners of war: A normative response to severe trauma. *American Journal of Psychiatry, 154*, 1576–1581.

Epping-Jordan, J. E., Compas, B. E., & Howell, D. C. (1994). Predictors of cancer progression in young adult men and women: Avoidance, intrusive thoughts, and psychological symptoms. *Health Psychology, 13*, 539–547.

Epping-Jordan, J. E., et al. (1999). Psychological adjustment in breast cancer: Processes of emotional distress. *Health Psychology, 18*, 315–326.

Erlenmeyer-Kimling, L., et al. (1997). The New York high-risk project: Prevalence and comorbidity of Axis I disorders in offspring of schizophrenic parents at 25-year follow-up. *Archives of General Psychiatry, 54*, 1096–1102.

Ernst, T., et al. (2000). Evidence for long-term neurotoxicity associated with methamphetamine abuse: A 1HMRS study. *Neurology, 54*, 1344–1349.

Eron, L. D. (1993, August). Cited in DeAngelis, T. (1993b). It's back: TV violence, concern for kid viewers. *APA Monitor, 24*(8), p. 16.

Ertekin-Taner, N., et al. (2000). Linkage of plasma A-42 to a quantitative locus on chromosome 10 in late-onset Alzheimer's disease pedigrees. *Science, 290*, 2303–2304.

Escobar, J. I. (1998). Immigration and mental health: Why are immigrants better off? *Archives of General Psychiatry, 55*, 781–782.

Escobar, J. I., Hoyos Nervi, C., & Gara, M. (2000). Immigration and mental health: Mexican-Americans in the United States. *Harvard Review of Psychiatry, 8*, 64–72.

Escobar, J. I., & Vega, W. A. (2000). Commentary: Mental health and immigration's AAAs: Where are we and where do we go from here? *Journal of Nervous & Mental Disease, 188*, 736–740.

Esparza, J., et al. (2000). Daily energy expenditure in Mexican and USA Pima Indians: Low physical activity as a possible cause of obesity. *International Journal of Obesity and Related Metabolic Disorders, 24*, 55–59.

Espie, C. A., Inglis, S. J., & Harvey, L. (2001). Predicting clinically significant response to cognitive behavior therapy for chronic insomnia in general medical practice: Analyses of outcome data at 12 months posttreatment. *Journal of Consulting and Clinical Psychology, 69*, 58–66.

Espie, C. A., et al. (2001). The clinical effectiveness of cognitive behaviour therapy for chronic insomnia: Implementation and evaluation of a sleep clinic in general medical practice. *Behaviour Research and Therapy, 39*, 45–60.

Essock, S. M., et al. (2000). Cost-effectiveness of clozapine compared with conventional antipsychotic medication for patients in state hospitals. *Archives of General Psychiatry, 57*, 987–994.

Esterling, B. A., L'Abate, L., Murray, E. J., & Pennebaker, J. W. (1999). Empirical foundations for writing in prevention and psychotherapy: Mental and physical health outcomes. *Clinical Psychology Review, 19*, 79–96.

Ettinger, U., et al. (2001). Magnetic resonance imaging of the thalamus in first-episode psychosis. *American Journal of Psychiatry, 158*, 116–118.

Evans, S. W., et al. (2001). Dose-response effects of methylphenidate on ecologically valid measures of academic performance and classroom behavior in adolescents with ADHD. *Experimental and Clinical Psychopharmacology, 9*, 163–175.

Exner, J. E. (1991). *The Rorschach: A comprehensive system: Vol. 2. Interpretation*. New York: Wiley.

Exner, J. E. (1993). *The Rorschach: A comprehensive system: Vol. 1. Basic foundations* (3rd ed.). New York: Wiley.

Extinguishing Alzheimer's. (1998, June 23). *The New York Times*, p. F7.

F

Fabian, J. L. (1991). "Koro: Proposed classification for *DSM-IV*": Comment. *American Journal of Psychiatry, 148*, 1766.

Fabrega, H. (1990). Hispanic mental health research: A case for cultural psychiatry. *Journal of Behavioral Sciences, 12*, 339–365.

Fabrega, H., Jr. (1992). Diagnosis interminable: Toward a culturally sensitive *DSM-IV*. *Journal of Nervous and Mental Disease, 180*, 5–7.

Fairburn, C. G., & Wilson, G. T. (Eds.). (1993). *Binge eating: Nature, assessment, and treatment*. New York: Guilford Press.

Fairburn, C. G., et al. (1997). Risk factors for bulimia nervosa: A community-based case-control study. *Archives of General Psychiatry, 54*, 509–517.

Faller, K. C. (1989). Characteristics of a clinical sample of sexually abused children: How boy and girl victims differ. *Child Abuse and Neglect, 13*, 281–291.

Fallon, B. A., et al. (1993). Fluoxetine for hypochondriacal patients without major depression. *Journal of Clinical Psychopharmacology, 13*, 438–441.

Falsetti, S., & Resnick, H. S. (2000). Treatment of PTSD using cognitive and cognitive behavioral therapies. *Journal of Cognitive Psychotherapy, 14*, 261–285.

Fannon, D., et al. (2000). Features of structural brain abnormality detected in first-episode psychosis. *American Journal of Psychiatry, 157*, 1829–1834.

Fanous, A., et al. (2001). Relationship between positive and negative symptoms of schizophrenia and schizotypal symptoms in nonpsychotic relatives. *Archives of General Psychiatry, 58*, 669–673.

Faraone, S. V., Kremen, W. S., & Tsuang, M. T. (1990). Genetic transmission of major affective disorders: Quantitative models and linkage analyses. *Psychological Bulletin, 108*, 109–127.

Faraone, S. V., et al. (1993). Intellectual performance and school failure in children with attention deficit hyperactivity disorders and their siblings. *Journal of Abnormal Psychology, 102*, 616–623.

Faraone, S. V., et al. (2000). Assessing symptoms of attention deficit hyperactivity disorder in children and adults: Which is more valid? *Journal of Consulting and Clinical Psychology, 68*, 830–842.

Faraone, S. V., et al. (2001). Meta-analysis of the association between the 7-repeat allele of the dopamine d4 receptor gene and attention deficit hyperactivity disorder. *American Journal of Psychiatry*, 158, 1052–1057.

Farber, B. A., Brink, D. C., & Raskin, P. M. (1996). *The psychotherapy of Carl Rogers: Cases and commentary* (pp. 74–75). New York: The Guilford Press.

Farberman, R. K. (1997). Public attitudes about psychologists and mental health care: Research to guide the American Psychological Association Public Education Campaign. *Professional Psychology: Research and Practice, 28*, 128–136.

Farmer, R. F. (2000). Issues in the assessment and conceptualization of personality disorders. *Clinical Psychology Review, 20*, 823–851.

Farr, C. B. (1994). Benjamin Rush and American psychiatry. *American Journal of Psychiatry, 151*(Suppl.), 65–73.

Farrell, A. D., Camplair, P. S., & McCullough, L. (1987). Identification of target complaints by computer interview: Evaluation of the Computerized Assessment System for Psychotherapy Evaluation and Research. *Journal of Consulting and Clinical Psychology, 55*, 691–700.

Farrell, A. D., & White, K. S. (1998). Peer influences and drug use among urban adolescents: Family structure and parent/adolescent relationship as protective factors. *Journal of Consulting and Clinical Psychology, 66,* 248–258.

Father's age tied to schizophrenia. (2001, April 12). *MSNBC News Service, Web Posting.* Retrieved April 13, 2001, from http://www.msnbc.com/news/558493.asp.

Fava, M., et al. (1994). Dysfunctional attitudes in major depression: Changes with pharmacotherapy. *Journal of Nervous and Mental Disease, 182,* 45–49.

Fawzy, F. I., & Fawzy, N. W. (1994). A structured psychoeducational intervention for cancer patients. *General Hospital Psychiatry, 16,* 149–192.

Feder, B. J. (1996, May 24). Increase in teen-age smoking sharpest among black males. *The New York Times,* p. A20.

Feinauer, L. L., & Stuart, D. A. (1996). Blame and resilience in women sexually abused as children. *American Journal of Family Therapy, 24,* 31–40.

Feldman, C. M. (1997). Childhood precursors of adult interpartner violence. *Clinical Psychology: Science and Practice, 4,* 307–333.

Felsenfeld, S. (1996). Progress and needs in the genetics of stuttering. *Journal of Fluency Disorders, 21,* 77–103.

Fenichel, O. (1945). *The psychoanalytic theory of neurosis.* New York: Norton.

Fenton, W. S., et al. (1997). Symptoms, subtype, and suicidality in patients with schizophrenia spectrum disorders. *American Journal of Psychiatry, 154,* 199–204.

Ferguson-Peters, M. (1985). Racial socialization of young Black children. In H. & J. L. McAdoo (Eds.), *Black children* (pp. 159–173). Beverly Hills, CA: Sage Publications.

Ferketich, A. K., et al. (2000). Depression as an antecedent to heart disease among women and men in the NHANES I Study. *Archives of Internal Medicine, 160,* 1261–1268.

Ferrans, C. E., Cohen, F. L., & Smith, K. M. (1992). The quality of life of persons with narcolepsy. *Loss, Grief and Care, 5,* 23–32.

Feshbach, S. (1994). Nationalism, patriotism, and aggression: A clarification of functional differences. In L. R. Huesmann (Ed.), *Aggressive behavior: Current perspectives* (pp. 275–291). New York: Plenum Press.

Fiellin, D. A., et al. (2001). Methadone maintenance in primary care: A randomized controlled trial. *Journal of the American Medical Association, 286,* 1724–1731.

Fieve, R. R. (1975). *Moodswings: The third revolution in psychiatry.* New York: Morrow.

Fiez, J. A. (2001). Bridging the gap between neuroimaging and neuropsychology: Using working memory as a case-study. *Journal of Clinical & Experimental Neuropsychology, 23,* 19–31.

Fingerhut, L. A., Ingram, D. D., & Feldman, J. J. (1998). Homicide rates among U.S. teenagers and young adults: Differences by mechanism, level of urbanization, race, and sex, 1987 through 1995. *Journal of the American Medical Association, 280,* 423–427.

Finkelhor, D. (1984). *Child sexual abuse: Theory and research.* New York: Free Press.

Finkelhor, D. (1990). Early and long-term effects of child sexual abuse: An update. *Professional Psychology: Research and Practice, 21,* 325–330.

Finkelhor, D. (1993). Epidemiological factors in the identification of child abuse. Special issue: Clinical recognition of sexually abused children. *Child Abuse and Neglect, 17,* 67–70.

Finkelhor, D., & Russell, D. (1984). Women as perpetrators: Review of the evidence. In D. Finkelhor (Ed.), *Child sexual abuse: Theory and research.* New York: Free Press.

Finkelhor, D., et al. (1990). Sexual abuse in a national survey of adult men and women: Prevalence, characteristics, and risk factors. *Child Abuse and Neglect, 14,* 19–28.

Finkelstein, J. R. J., et al. (1997). Attentional dysfunctions in neuroleptic-naive and neuroleptic-withdrawn schizophrenic patients and their siblings. *Journal of Abnormal Psychology, 106,* 203–212.

Finney, J. W., & Monahan, S. C. (1996). The cost-effectiveness of treatment for alcoholism: A second approximation. *Journal of Studies on Alcohol, 57,* 229–243.

Fishbain, D. A. (1991). "Koro: Proposed classification for *DSM-IV*": Comment. *American Journal of Psychiatry, 148,* 1765–1766.

Fisher, B. S., Cullen, F. T., & Turner, M. (2000). *The sexual victimization of college women.* Washington, DC: U.S. Bureau of Justice Statistics.

Fisman, S., & Takhar, J. (1996). "Fear of alien abduction": Reply. *Journal of the American Academy of Child and Adolescent Psychiatry, 35,* 556–557.

Fitzgerald, L. F. (1993a). Sexual harassment: Violence against women in the workplace. *American Psychologist, 48,* 1070–1076.

Fitzgerald, L. F. (1993b). *Sexual harassment in higher education: Concepts and issues.* Washington, DC: National Education Association.

Fitzgerald, L. F., et al. (1988). The incidence and dimensions of sexual harassment in academia and the workplace. *Journal of Vocational Behavior, 32,* 152–175.

Flannery, D. J., Singer, M. I., & Wester, K. (2001). Violence exposure, psychological trauma, and suicide risk in a community sample of dangerously violent adolescents. *Journal of American Academy of Child & Adolescent Psychiatry, 40,* 435–442.

Flint, A. J. (1994). Epidemiology and comorbidity of anxiety disorders in the elderly. *American Journal of Psychiatry, 151,* 640–649.

Flint, J., et al. (1995). The detection of subtelomeric chromosomal rearrangements in idiopathic mental retardation. *Nature Genetics, 9,* 132–140.

Flournoy, P. S., & Wilson, G. L. (1991). Assessment of MMPI profiles of male batterers. *Violence and Victims, 6,* 309–320.

Foa, E. B. (1990, August/September). Obsessive-compulsive disorder. In *American Psychiatric Association, DSM-IV Update.* Washington, DC: American Psychiatric Association.

Foa, E. B. (1996). The efficacy of behavioral therapy with obsessive-compulsives. *The Clinical Psychologist, 49,* 19–21.

Foa, E. B., & Kozak, M. J. (1995). *DSM-IV* field trial: Obsessive-compulsive disorder. *American Journal of Psychiatry, 152,* 90–96.

Foa, E. B., et al. (1999). A comparison of exposure therapy, stress inoculation training, and their combination for reducing posttraumatic stress disorder in female assault victims. *Journal of Consulting and Clinical Psychology, 67,* 194–200.

Foderaro, L. W. (1994, November 8). "Clubhouse" helps mentally ill find the way back. *The New York Times,* p. B1.

Fogelholm, M., Kukkonen-Harjual, K., Nenonen, A., & Pasenen, M. (2000). Effects of walking training on weight maintenance after a very-low-energy diet in premenopausal obese women: A randomized controlled trial. *Archives of Internal Medicine, 160,* 2177–2184.

Follette, W. C., & Houts, A. C. (1996). Models of scientific progress and the role of theory and taxonomy development: A case study of the *DSM. Journal of Consulting and Clinical Psychology, 64,* 1120–1132.

Follingstad, D. R., Neckerman, A. P., & Vormbrock, J. (1988). Reactions to victimization and coping strategies of battered women: The ties that bind. *Clinical Psychology Review, 8,* 373–390.

Ford, C. S., & Beach, F. A. (1951). *Patterns of sexual behavior.* New York: Harper & Row.

Ford, J. M., et al. (2001). Cortical responsiveness during inner speech in schizophrenia: An event-related potential study. *American Journal of Psychiatry, 158,* 1914–1916.

Forehand, R., Brody, G., Slotkin, J., Fauber, R., McCombs, A., & Long, N. (1988). Young adolescent and maternal depression: Assessment, interrelations, and family predictors. *Journal of Consulting and Clinical Psychology, 56,* 422–426.

Forman, D. N., et al. (2000). Postpartum depression: Identification of women at risk. *British Journal of Obstetrics and Gynaecology, 107,* 1210–1217.

Foster, G. D., et al. (1997). What is a reasonable weight loss? Patients' expectations and evaluations of obesity treatment outcomes. *Journal of Consulting and Clinical Psychology, 65,* 79–85.

Foster, S. L., & Cone, J. D. (1986). Design and use of direct observation procedures. In A. R. Ciminiero, K. S. Calhoun, & H. E. Adams (Eds.), *Handbook of behavioral assessment* (2nd ed., pp. 253–324). New York: Wiley.

Fowles, D. C. (1993). Electrodermal activity and antisocial behavior: Empirical findings and theoretical issues. In J. C. Roy et al. (Eds.), *Psychological theories of drinking and alcoholism* (pp. 181–226). New York: Guilford Press.

Fox, J. A., & Zawitz, M. W. (2000). *Homicide trends in the United States: 1998 Update.* Washington, DC: U.S. Department of Justice, Bureau of Justice Statistics.

Fox, M. (2000, August 21). Autism checks urged for all babies. *Reuters Limited, MSNBC Web Posting.* Retrieved August 23, 2000, from http://www. msnbc. com/news/449244.asp.

Fox, M. (2001, January 29). Strong family-heart disease tie found. *Reuters Web Posting.* Retrieved January 31, 2001, from http://www.msnbc.com/news/523336.asp.

Foxhall, K. (2000a, October). Platform for a long-term push. *Monitor on Psychology,* p. 30.

Foxhall, K. (2000b, October). Dispatches from the prescription privileges fronts. *Monitor on Psychology,* pp. 30–31.

Foxhall, K. (2001a, January). Suicide by profession: Lots of confusion, inconclusive data. *Monitor on Psychology,* p. 19.

Foxhall, K. (2001b, March). Study finds marital stress can triple women's risk of recurrent coronary event. *Monitor on Psychology, 32,* p. 14.

Foy, D. W., Resnick, H. S., Sipprele, R. C., & Carroll, E. M. (1987). Premilitary, military, and postmilitary factors in the development of combat-related posttraumatic stress disorder. *The Behavior Therapist, 10,* 3–9.

Frackiewicz, E. J., Sramek, J. J., Herrera, J. M., & Cutler, N. R. (1999). Review of neuroleptic dosage in different ethnic groups. In J. M. Herrera et al. (Eds.), *Cross cultural psychiatry* (pp. 107–130). Chichester, England: Wiley.

Frank, E., Brogan, D., & Schiffman, M. (1998). Harassment among U.S. women physicians. *Archives of Internal Medicine, 158,* 352–358.

Frankl, V. E. (1959). *Man's search for meaning.* Boston: Beacon Press.

Franklin, M. E., et al. (2000). Effectiveness of exposure and ritual prevention for obsessive-compulsive disorder: Randomized compared with nonrandomized samples. *Journal of Consulting and Clinical Psychology, 68,* 594–602.

Fraser, J. S. (1996). All that glitters is not always gold: Medical offset effects and managed behavioral health care. *Professional Psychology: Research & Practice, 27,* 335–344.

Frauenglass, S., et al. (1997). Family support decreases influence of deviant peers on Hispanic adolescent's substance use. *Journal of Clinical Child Psychology, 26,* 15–23.

Freed, A. O. (1992). Discussion: Minority elderly. *Journal of Geriatric Psychiatry, 25,* 105–111.

Freedman, R., et al. (1987). Neurobiological studies of sensory gating in schizophrenia. *Schizophrenia Bulletin, 13,* 669–678.

Freeman, L. N., Shaffer, D., & Smith, H. (1996). The neglected victims of homicide: The needs of young siblings of murder victims. *American Journal of Orthopsychiatry, 66,* 337–345.

Freemon, F. R. (1981). *Organic mental disease.* Jamaica, NY: Spectrum.

Freiberg, P. (1995, June). Psychologists examine attacks on homosexuals. *APA Monitor, 26*(6), pp. 30–31.

French, S. A., & Jeffery, R. W. (1994). Consequences of dieting to lose weight: Effects on physical and mental health. *Health Psychology, 13,* 195–212.

Freud, S. (1957). Mourning and melancholia (1917). In J. Rickman (Ed.), *A general selection from the works of Sigmund Freud.* Garden City, NY: Doubleday.

Freud, S. (1959a). Analysis of a phobia in a 5-year-old boy. In A. J. Strachey (Ed. & Trans.), *Collected papers. Vol. 3.* New York: Basic Books. (Original work published 1909)

Freud, S. (1964). New introductory lectures. In *Standard edition of the complete psychological works of Sigmund Freud (Vol. 22).* London: Hogarth. (Original work published in 1933)

Freund, K., & Blanchard, R. (1986). The concept of courtship disorder. *Journal of Sex and Marital Therapy, 12,* 79–92.

Frick, P. J., et al. (1992). Familial risk factors to oppositional defiant disorder and conduct disorder: Parental psychopathology and maternal parenting. *Journal of Consulting and Clinical Psychology, 60,* 49–55.

Fried, L. P., et al. (1998). Risk factors for 5 year mortality in older adults: The Cardiovascular Health Study. *Journal of the American Medical Association, 279,* 585–592.

Friedman, M., & Rosenman, R. H. (1974). *Type A behavior and your heart.* New York: Harper & Row.

Friedman, M., & Ulmer, D. (1984). *Treating Type A behavior and your heart.* New York: Fawcett Crest.

Friedman, M., et al. (1986). Alteration of type A behavior and its effect on cardiac recurrences in postmyocardial infarction patients: Summary results of the recurrent coronary prevention project. *American Heart Journal, 112,* 653–665.

Fromm-Reichmann, F. (1948). Notes on the development of treatment of schizophrenics by psychoanalytic psychotherapy. *Psychiatry, 11,* 263–273.

Fromm-Reichmann, F. (1950). *Principles of intensive psychotherapy.* Chicago: University of Chicago Press.

Frueh, B. C., et al. (1996). Trauma management therapy: A preliminary evaluation of a multicomponent behavioral treatment for chronic combat-related PTSD. *Behaviour Research and Therapy, 34,* 533–543.

Fuchs, C. S., et al. (1995). Alcohol consumption and mortality among women. *The New England Journal of Medicine, 332,* 1245–1250.

Fuchs, M. (2001, June 9). For Alzheimer's patients, some solace, if not hope. *The New York Times,* pp. B1, B6.

Fulero, S. M. (1988). Tarasoff: 10 years later. *Professional Psychology: Research and Practice, 19,* 184–190.

G

Gabbard, G. O., et al. (1997). The economic impact of psychotherapy: A review. *American Journal of Psychiatry, 154,* 147–155.

Galassi, J. P. (1988). Four cognitive-behavioral approaches: Additional considerations. *The Counseling Psychologist, 16*(1), 102–105.

Gall, R., Isaac, L., & Kryger, M. (1993). Quality of life in mild obstructive sleep apnea. *Sleep, 16*(Suppl.), S59–S61.

Gallant, J. E. (2000). Strategies for long-term success in the treatment of HIV infection. *Journal of the American Medical Association, 283,* 1329–1334.

Ganellen, R. J. (1996). Comparing the diagnostic efficiency of the MMPI, MCMI-II, and Rorschach: A review. *Journal of Personality Assessment, 67,* 219–243.

Garb, H. N. (1997). Race bias, social class bias, and gender bias in clinical judgment. *Clinical Psychology: Science and Practice, 4,* 99–120.

Garb, H. N. (2000). Computers will become increasingly important for psychological assessment: Not that there's anything wrong with that! *Psychological Assessment, 12,* 31–39.

Garber, J., Weiss, B., & Shanley, N. (1993). Cognitions, depressive symptoms, and development in adolescents. *Journal of Abnormal Psychology, 102,* 47–57.

Garbutt, J. C., et al. (1999). Pharmacological treatment of alcohol dependence. *Journal of the American Medical Association, 281,* 1318–1325.

Garcia, M., & Marks, G. (1989). Depressive symptomatology among Mexican-American adults: An examination with the CES-D scale. *Psychiatry Research, 27,* 137–148.

Gardiner, S. (1992). Out of harm's way: Intervention with children in shelters. Special issue: Feminist perspectives in child and youth care practices. *Journal of Child and Youth Care, 7,* 41–48.

Gardner, W., et al. (1996). Clinical versus actuarial predictions of violence in patients with mental illnesses. *Journal of Consulting and Clinical Psychology, 64,* 602–609.

Garfield, S. L. (1994). Eclecticism and integration in psychotherapy: Developments and issues. *Clinical Psychology: Science and Practice, 1,* 123–137.

Garner, D. M. (1993). Binge eating in anorexia nervosa. In C. G. Fairburn & G. T. Wilson (Eds.), *Binge eating: Nature, assessment, and treatment* (pp. 50–76). New York: Guilford.

Gatchel, R. J. (2001). Biofeedback and self-regulation of physiological activity: A major adjunctive treatment modality in health psychology. In A. Baum, T. A. Revenson, & J. E. Singer (Eds.), *Handbook of health psychology* (pp. 95–104). Mahwah, NJ: Erlbaum.

Gaulin, S. J. C., & McBurney, D. H. (2001). *Psychology: An evolutionary approach.* Upper Saddle River, NJ: Prentice-Hall.

Gauthier, J. G., Ivers, H., & Carrier, S. (1996). Nonpharmacological approaches in the management of recurrent headache disorders and their comparison and combination with pharmacotherapy. *Clinical Psychology Review, 16,* 543–571.

Gawin, F. H. (1991). Cocaine addiction: Psychology and neurophysiology. *Science, 251,* 1581.

Gawin, F. H., et al. (1989). Desipramine facilitation of initial cocaine abstinence. *Archives of General Psychiatry, 46,* 117–121.

Gaziano, J. M. (1993). Moderate alcohol intake, increased levels of high-density lipoprotein and its subfractions, and decreased risk of myocardial infarction. *New England Journal of Medicine, 329,* 1829–1834.

Ge, X., et al. (1994). Parents' stressful life events and adolescent depressed mood. *Journal of Health and Social Behavior, 35,* 28–44.

Gebhard, P. H., Gagnon, J. H., Pomeroy, W. B., & Christenson, C. V. (1965). *Sex offenders: An analysis of types.* New York: Harper & Row.

Geddes, J., et al. (2000). Atypical antipsychotics in the treatment of schizophrenia: Systematic overview and meta-regression analysis. *British Medical Journal, 321,* 1371–1376.

Geer, J., Heiman, J., & Leitenberg, H. (1984). *Human sexuality.* Englewood Cliffs, NJ: Prentice-Hall.

Gelman, D. (1994, April 18). The mystery of suicide. *Newsweek,* pp. 44–49.

George, M. S., et al. (2001). Activation of prefrontal cortex and anterior thalamus in alcoholic subjects on exposure to alcohol-specific cues. *Archives of General* Psychiatry, *58,* 345–352.

Gershon, E. S., & Rieder, R. O. (1992). Major disorders of mind and brain. *Scientific American, 267*(3), 126–133.

Gershuny, B. S., & Thayer, J. F. (1999). Relations among psychological trauma, dissociative phenomena, and trauma-related distress: A review and integration. *Clinical Psychology Review, 19,* 631–657.

Ghanshyam, N., et al. (1995). Platelet serotonin-2A receptors: A potential biological marker for suicidal behavior. *American Journal of Psychiatry, 152,* 850–855.

Giancola, P. R., & Zeichner, A. (1997). The biphasic effects of alcohol on human physical aggression. *Journal of Abnormal Psychology, 106,* 598–607.

Gibbs, N. (1991, June 3). When is it rape? *Time Magazine,* pp. 48–54.

Gibbs, N. (1998, November 30). The age of Ritalin. *Time,* pp. 84–94.

Gidron, Y., & Davidson, K. (1996). Development and preliminary testing of a brief intervention for modifying CHD-predictive hostility components. *Journal of Behavioral Medicine, 19,* 203–220.

Gidron, Y., Davidson, K., & Bata, I. (1999). The short-term effects of a hostility-reduction intervention on male coronary heart disease patients. *Health Psychology, 18,* 416–420.

Giembycz, M. A., & O'Connor, B. J. (2000). *Asthma: Epidemiology, anti-inflammatory therapy and future trends.* Boston: Birkhauser.

Gil, K. M., et al. (1990). The relationship of negative thoughts to pain and psychological distress. *Behavior Therapy, 21,* 349–362.

Gilbert, S. (1997a, January 22). Lag seen in aid for depression. *The New York Times,* p. C9.

Gilbert, S. (1997b, June 25). Social ties reduce risk of a cold. *The New York Times,* p. C11.

Gilboa-Schechtman, E., & Foa, E. B. (2001). Patterns of recovery from trauma: The use of intraindividual analysis. *Journal of Abnormal Psychology,* 110, 392–400.

Gill, A. R. (2001). Interventions for autism. In reply. *Journal of the American Medical Association, 286,* 670–671. [Letter]

Gillberg, C., et al. (1997). Long-term stimulant treatment of children with attention-deficit hyperactivity disorder symptoms: A randomized, double-blind, placebo-controlled trial. *Archives of General Psychiatry, 54,* 857–864.

Gillin, J. C. (1991). The long and the short of sleeping pills. *The New England Journal of Medicine, 324,* 1735–1736.

Gilroy, L., et al. (2000). Controlled comparison of computer-raided vicarious exposure versus live exposure in the treatment of spider phobia. *Behavior Therapy, 31,* 733–744.

Gitlin, M. J., & Pasnau, R. O. (1989). Psychiatric syndromes linked to reproductive function in women: A review of current knowledge. *American Journal of Psychiatry, 146,* 1413–1422.

Gittelman-Klein, R., & Mannuzza, S. (1990). Hyperactive boys almost grown up. *Archives of General Psychiatry, 45,* 1131–1134.

Glantz, K., et al. (1996). Virtual reality (VR) for psychotherapy: From the physical to the social environment. *Psychotherapy, 33,* 464–473.

Glantz, L. A., &. Lewis, D. A. (2000). Decreased dendritic spine density on prefrontal cortical pyramidal neurons in schizophrenia. *Archives of General Psychiatry, 57,* 65–73.

Glara, M. A., et al. (1993). Perceptions of self and other in major depression. *Journal of Abnormal Psychology, 102,* 93–100.

Glaser, R., et al. (1987). Stress-related immune suppression: Health implications. *Brain, Behavior, and Immunity, 1,* 7–20.

Glaser, R., Kiecolt-Glaser, J. K., Speicher, C. E., & Holliday, J. E. (1985). Stress, loneliness, and changes in herpes virus latency. *Journal of Behavioral Medicine, 8,* 249–260.

Glass, R. M. (2000). Panic disorder—It's real and it's treatable. *Journal of the American Medical Association, 283,* 2573–2574. [Editorial]

Glass, R. M. (2001). Electroconvulsive therapy: Time to bring it out of the shadows. *Journal of the American Medical Association, 285,* 1346–1348. [Editorial]

Gleaves, D. H. (1996). The sociocognitive model of dissociative identity disorder: A reexamination of the evidence. *Psychological Bulletin, 120*, 42–59.

Glynn, S., & Mueser, K. T. (1986). Social learning for chronic mental inpatients. *Schizophrenia Bulletin, 12*, 648–668.

Glynn, S., & Mueser, K. T. (1992). Social learning. In R. P. Liberman (Ed.), *Handbook of psychiatric rehabilitation* (pp. 127–152). New York: Macmillan.

Goddard, A. W., et al. (2001). Reductions in occipital cortex GABA levels in panic disorder detected with 1h-magnetic spectroscopy. *Archives of General Psychiatry, 58*, 556–561.

Goenjian, A. K., et al. (2001). Posttraumatic stress and depressive reactions among Nicaraguan adolescents after Hurricane Mitch. *American Journal of Psychiatry, 158*, 788–794.

Goetz, K. L., & Price, T. R. P. (1994). The case of koro: Treatment response and implications for diagnostic classification. *Journal of Nervous and Mental Disease, 182*, 590–591.

Goff, D, C., & Summs, C. A. (1993). Has multiple personality disorder remained consistent over time? A comparison of past and recent cases. *Journal of Nervous and Mental Disease, 181*, 595–600.

Goisman, R. M., et al. (1994). Panic, agoraphobia, and panic disorder with agoraphobia: Data from a multicenter anxiety disorders study. *Journal of Nervous and Mental Disease, 182*, 72–79.

Gold, S. R., & Gold, R. G. (1993). Sexual aversions: A hidden disorder. In W. O'Donohue & J. H. Geer (Eds.), *Handbook of sexual dysfunctions: Assessment and treatment* (pp. 83–102). Boston: Allyn & Bacon.

Goldberg, C. (2001, July 9). Children trapped by mental illness. *The New York Times*, pp. A1, A11.

Goldberg, I. J., et al. (2001). Wine and your heart: A science advisory for healthcare professionals from the Nutrition Committee, Council on Epidemiology and Prevention, and Council on Cardiovascular Nursing of the American Heart Association. *Circulation, 103*, 472–475.

Golden, C. J., Hammeke, T. A., & Purisch, A. D. (1980). *The Luria-Nebraska Neuropsychological Battery: Manual*. Los Angeles: Western Psychological Services.

Golden, O. (2000). The federal response to child abuse and neglect. *American Psychologist, 55*, 1050–1053.

Goldfarb, L. A., Dynens, E. M., & Gerrard, M. (1985). The Goldfarb fear of fat scale. *Journal of Personality Assessment, 49*, 329–332.

Goldstein, A. J., et al. (2000). EMDR for panic disorder with agoraphobia: Comparison with waiting list and credible attention-placebo control conditions. *Journal of Consulting and Clinical Psychology, 68*, 947–956.

Goldstein, I., et al. (1998). Oral sildenafil in the treatment of erectile dysfunction. *The New England Journal of Medicine, 338*, 1397–1404.

Goldstein, M. J. (1987). The UCLA high-risk project. *Schizophrenia Bulletin, 13*, 505–514.

Goldstein, R. L. (1986). Erotomania. *American Journal of Psychiatry, 143*, 802.

Goleman, D. (1988b, November 1). Narcissism looming larger as root of personality woes. *The New York Times*, pp. C1, C16.

Goleman, D. (1990a, June 26). Scientists pinpoint brain irregularities in drug addicts. *The New York Times*, pp. C1, C7.

Goleman, D. (1990b, December 6). Women's depression is higher. *The New York Times*.

Goleman, D. (1991, October 22). Sexual harassment: It's about power, not lust. *The New York Times*, pp. C1, C12.

Goleman, D. (1992a, April 14). Therapies offer hope for sex offenders. *The New York Times*, pp. C1, C11.

Goleman, D. (1992b, October 14). Study ties genes to drinking in women as much as in men. *The New York Times*, p. C14.

Goleman, D. (1993c, October 6). Abuse-prevention efforts aid children. *The New York Times*, p. C13.

Goleman, D. (1993e, December 7). Stress and isolation tied to a reduced life span. *The New York Times*, p. C5.

Goleman, D. (1994a, January 11). Childhood depression may herald adult ills. *The New York Times*, pp. C1, C10.

Goleman, D. (1994c, April 19). Revamping psychiatrists' bible. *The New York Times*, pp. C1, C11.

Goleman, D. (1994d, April 26). Mental decline in aging need not be inevitable. *The New York Times*, p. C1, C10.

Goleman, D. (1995b, May 2). Biologists find site of working memory. *The New York Times*, pp. C 1, C9.

Goleman, D. (1995e, July 1). A genetic clue to bed-wetting is located. *The New York Times*, p. A8.

Goleman, D. (1995g, October 4). Eating disorder rates surprise experts. *The New York Times*, p. C11.

Goode, E. (2000a, October 24). Watching volunteers, experts seek clues to eating disorders. *The New York Times*, pp. F1, F6.

Goode, E. (2000b, March 14). Human nature: Born or made? *The New York Times*, pp. F1, F9.

Goode, E. (2001a, February 20). What's in an inkblot? Some say, not much. *The New York Times*, pp. F1, F4.

Goode, E. (2001b, February 20). Samson diagnosis: Antisocial personality disorder, with muscles. *The New York Times*, p. F7.

Goode, E. (2001c, May 22). For users of heroin, decades of despair. *The New York Times*, p. F5.

Goode, E. (2001d, August 27). Disparities seen in mental care for minorities. *The New York Times*, pp. A1, A12.

Goode, E. (2001e, April 10). "A new improved me": Now appearing everywhere. *The New York Times*, p. F7.

Goode, E. (2001f, August 1). Study says 20% of girls reported abuse by a date. *The New York Times*, p. A10.

Goodstein, L., & Connelly, M. (1998, April 30). Teen-age poll finds a turn to the traditional. *The New York Times*, p. A20.

Gordis, E. (1995). The National Institute on Alcohol Abuse and Alcoholism. *Alcohol Health & Research World, 19*, 5–11.

Gordis, E. (1999, May 5). *What we know: Conceptual advances in alcohol research*. National Institute on Alcohol Abuse and Alcoholism (NIAAA). Retrieved July 13, from http://www.niaaa.nih.gov/about/conceptual.htm.

Gordon, C. M., & Carey, M. P. (1996). Alcohol's effects on requisites for sexual risk reduction in men: An initial experimental investigation. *Health Psychology, 15*, 56–60.

Gormally, J., Sipps, G., Raphael, R., Edwin, D., & Varvil-Weld, D. (1981). The relationship between maladaptive cognitions and social anxiety. *Journal of Consulting and Clinical Psychology, 49*, 300–301.

Gorman, C. (1998, November 30). How does it work? *Time*, p. 92.

Gorman, J. M., Kent, J. M., Sullivan, G. M., & Coplan, J. D. (2000). Neuroanatomical hypothesis of panic disorder, revised. *American Journal of Psychiatry, 57*, 493–505.

Gorman, J. M., et al. (2001). Physiological changes during carbon dioxide inhalation in patients with panic disorder, major depression, and premenstrual dysphoric disorder: Evidence for a central fear mechanism. *Archives of General Psychiatry, 58*, 125–131.

Gorwood, P., et al. (1995). Gender and age at onset in schizophrenia: Impact of family history. *American Journal of Psychiatry, 152*, 208–212.

Gosselin, C., & Wilson, G. (1980). *Sexual variations*. New York: Simon & Schuster.

Gotlib, I. H., et al. (1993). Negative cognitions and attributional style in depressed adolescents: An examination of stability and specificity. *Journal of Abnormal Psychology, 102*, 607–615.

Gottesman, I. I. (1991). *Schizophrenia genetics: The origins of madness*. New York: Freeman.

Gottesman, I. I., McGuffin, P., & Farmer, A. E. (1987). Clinical genetics as clues to the "real" genetics of schizophrenia. *Schizophrenia Bulletin, 13*, 23–47.

Gould, R., Miller, B. L., Goldberg, M. A., & Benson, D. F. (1986). The validity of hysterical signs and symptoms. *The Journal of Nervous and Mental Disease, 174*, 593–597.

Gould, R. A., et al. (1997). Cognitive-behavioral and pharmacological treatment for social phobia: A meta-analysis. *Clinical Psychology: Science and Practice, 4*, 291–306.

Gouzoulis-Mayfrank, E., et al. (2000). Impaired cognitive performance in drug free users of recreational ecstasy (MDMA). *Journal of Neurology, Neurosurgery, & Psychiatry, 68*, 719–725.

Grady, D. (1997a, January 21). Brain-tied gene defect may explain why schizophrenics hear voices. *The New York Times*, pp. C1, C3.

Graham, J. R. (2000). *MMPI-2: Assessing personality and psychopathology*. New York: Oxford University Press.

Graham, J. R., & Strenger, V. E. (1988). MMPI characteristics of alcoholics: A review. *Journal of Consulting and Clinical Psychology, 56*, 197–205.

Graham-Bermann, S. A., & Edleson, J. L. (Eds.). (2001). *Domestic violence in the lives of children: The future of research, intervention, and social policy* (pp. 35–65). Washington, DC: American Psychological Association.

Grange, D., Telch, C. F., & Tibbs, J. (1998). Eating attitudes and behaviors in 1,435 South African caucasian and non-caucasian college students. *American Journal of Psychiatry, 155*, 250–254.

Grant, B. F. (1997). Prevalence and correlates of alcohol use and *DSM-IV* alcohol dependence in the United States: Results of the National Longitudinal Alcohol Epidemiologic Survey. *Journal of Studies on Alcohol, 58*, 464–473.

Grant, B. F., et al. (1994). Prevalence of *DSM-IV* alcohol abuse and dependence: United States, 1992. *Alcohol Health & Research World, 18*, 243–248.

Gray-Little, B., & Hafdahl, A. R. (2000). Factors influencing racial comparisons of self-esteem: A quantitative review. *Psychological Bulletin*, 126, 26–54.

Greenberg, R. P., & Bornstein, R. F. (1988a). The dependent personality: I. Risk for physical disorders. *Journal of Personality Disorders, 2*, 126–135.

Greenberg, R. P., & Bornstein, R. F. (1988b). The dependent personality: II. Risk for psychological disorders. *Journal of Personality Disorders, 2*, 136–143.

Greenberg, R. P., Bornstein, R. F., Greenberg, M. D., & Fisher, S. (1992). A meta-analysis of antidepressant outcome under "blinder" conditions. *Journal of Consulting and Clinical Psychology, 60*, 664–669.

Greenberg, R. P., et al. (1994). A meta-analysis of fluoxetine outcome in the treatment of depression. *Journal of Nervous and Mental Disease, 182*, 547–551.

Greenberger, E., Chen, C., &. Tally, S. R. (2000). Family, peer, and individual correlates of depressive symptomatology among U.S. and Chinese adolescents. *Journal of Consulting and Clinical Psychology, 68*, 209–219.

Greene, B. (2000). Gender, sex and culture: Gender and culture. In A. Kazdin (Ed.), *Encyclopedia of psychology*. Washington, DC: American Psychological Association Press.

Greene, B. A. (1985). Considerations in the treatment of Black patients by White therapists. *Psychotherapy, 22*, 389–393.

Greene, B. A. (1986). When the therapist is White and the patient is Black: Considerations for psychotherapy in the feminist heterosexual and lesbian communities. *Women & Therapy, 5*, 41–65.

Greene, B. A. (1990). Sturdy bridges: The role of African American mothers in the socialization of African American children. *Women & Therapy, 10*, 205–225.

Greene, B. A. (1992b). Black feminist psychotherapy. In E. Wright (Ed.), *Psychoanalysis and feminism: A critical dictionary* (pp. 34–35). Oxford, U.K.: Basil Blackwell.

Greene, B. A. (1992c). Still here: A perspective on psychotherapy with African American women. In J. Chrisler & D. Howard (Eds.), *New directions in feminist psychology*. New York: Springer.

Greene, B. A. (1993a, Spring). Psychotherapy with African American women: The integration of feminist and psychodynamic approaches. *Journal of Training and Practice in Professional Psychology, 7*, 49–66.

Greene, B. A. (1993b). African American women. In L. Comas-Diaz & B. Greene (Eds.), *Women of color and mental health*. New York: Guilford Press.

Greenhill, L. L. (1998). Childhood attention deficit hyperactivity disorder: Pharmacological treatments. In P. E. Nathan & J. M. Gorman (Eds.), *A guide to treatments that work* (pp. 42–64). New York: Oxford University Press.

Greenhouse, C. J. (1998). Tuning to a key of gladness. *Law and Society Review, 32*, 5–21.

Greenhouse, L. (1992, June 23). Defendants must prove incompetency. *The New York Times*, p. A17.

Greenwald, R. (1996). The information gap in the EMDR controversy. *Professional Psychology: Research & Practice, 27*, 67–72.

Griffith, J. (1983). Relationship between acculturation and psychological impairment in adult Mexican-Americans. *Hispanic Journal of Behavioral Sciences, 5*, 431–459.

Griffiths, K. M., & Christensen, H. (2000). Quality of web-based information on treatment of depression: Cross sectional survey. *British Medical Journal, 16*, 1511–1515.

Griffiths, M. (1999). Internet addiction: Fact or fiction? *Psychologist, 12*, 246–250.

Grissett, N. I., & Norvell, N. K. (1992). Perceived social support, social skills and quality of relationships in bulimic women. *Journal of Consulting and Clinical Psychology, 60*, 293–299.

Grissom, R. J. (1996). The magical number .7 + -.2: Meta-meta-analysis of the probability of superior outcome in comparisons involving therapy, placebo, and control. *Journal of Consulting and Clinical Psychology, 64*, 973–982.

Grob, G. N. (1983). *Mental illness and American society, 1875–1940*. Princeton, NJ: Princeton University Press.

Grob, G. N. (1994). *The mad among us: A history of the care of America's mentally ill*. New York: Free Press.

Grof, P., & Alda, M. (2000). Discrepancies in the efficacy of lithium. *Archives of General Psychiatry, 57*, 191.

Gronbaek, M., et al. (2000). Type of alcohol consumed and mortality from all causes, coronary hear disease, and cancer. *Annals of Internal Medicine, 133*, 411–419.

Groth, A., & Hobson, W. (1983). The dynamics of sexual assault. In L. Schlesinger & E. Revitch (Eds.), *Sexual dynamics of antisocial behavior*. Springfield, IL: Thomas.

Guarnaccia, P. J., Angel, R., & Worobey, J. L. (1991). The impact of marital status and employment status on depressive affect for Hispanic Americans. *Journal of Community Psychology, 19*, 136–149.

Guarnaccia, P. J., & Rodriguez, O. (1996). Concepts of culture and their role in the development of culturally competent mental health services. *Hispanic Journal of Behavioral Sciences, 18*, 419–443.

Guerra, N. G., et al. (1995). Stressful events and individual beliefs as correlates of economic disadvantage and aggression among urban children. *Journal of Consulting and Clinical Psychology, 63*, 518–528.

Guertin, T. L. (1999). Eating behavior of bulimics, self-identified binge eaters, and noon-eating disordered individuals: What differentiates these populations? *Clinical Psychology Review, 19*, 1–24.

Gullette, E. C. D., et al. (1997). Effects of mental stress on myocardial ischemia during daily life. *Journal of the American Medical Association, 277*, 1521–1526.

Gunderson, J. G. (1996). The borderline patient's intolerance of aloneness: Insecure attachments and therapist availability. *American Journal of Psychiatry, 153*, 752–758.

Gunderson, J. G., & Phillips, K. A. (1991). A current view of the interface between borderline personality disorder and depression. *American Journal of Psychiatry, 148*, 967–975.

Gunderson, J. G., & Singer, M. T. (1986). Defining borderline patients: An overview. In M. H. Stone (Ed.), *Essential papers on borderline disorders* (pp. 453–474). New York: New York University Press.

Gunter, B., & McAleer, J. L. (1990). *Children and television: The one eyed monster?* Florence, KY: Taylor and Francis/Routledge.

Gur, R. E., & Pearlson, G. D. (1993). Neuroimaging in schizophrenia research. *Schizophrenia Bulletin, 19*, 337–353.

Gur, R. E., et al. (1998). A follow-up magnetic resonance imaging study of schizophrenia: Relationship of neuroanatomical changes to clinical and neurobehavioral measures. *Archives of General Psychiatry, 55*, 145–152.

Gur, R. E., et al. (2000a). Reduced dorsal and orbital prefrontal gray matter volumes in schizophrenia. *Archives of General Psychiatry, 57*, 761–768.

Gur, R. E., et al. (2000b). Temporolimbic volume reductions in schizophrenia. *Archives of General Psychiatry, 57*, 769–775.

Guralnik, O., Schmeidler, J., & Simeon, D. (2000). Feeling unreal: Cognitive processes in depersonalization. *American Journal of Psychiatry, 157*, 103–109.

Gurland, B. J., & Cross, P. S. (1986). Public health perspectives on clinical memory testing of Alzheimer's disease and related disorders. In L. W. Poon (Ed.), *Handbook for clinical memory assessment of older adults* (pp. 11–20). Washington, DC: American Psychological Association.

Gutheil, T. G. (1980). In search of true freedom: Drug refusal, involuntary medication, and "rotting with your rights on." *American Journal of Psychiatry, 137*, 327–328.

Guthrie, P. C., & Mobley, B. D. (1994). A comparison of the differential diagnostic efficiency of three personality disorder inventories. *Journal of Clinical Psychology, 50*, 656–665.

Gutierrez, P. M., & Silk, K. R. (1998). Prescription privileges for psychologists: A review of the psychological literature. *Professional Psychology: Research and Practice, 29*, 213–222.

Guydish, J., et al. (1998). Drug abuse day treatment: A randomized clinical trial comparing day and residential treatment programs. *Journal of Consulting and Clinical Psychology, 66*, 280–289.

Guze, S. B. (1993). Genetics of Briquet's syndrome and somatization disorder: A review of family, adoption, and twin studies. *Annals of Clinical Psychiatry, 5*, 225–230.

H

Haaga, D. A. F. (1995). Metatraits and cognitive assessment: Application to attributional style and depressive symptoms. *Cognitive Therapy and Research, 19*, 121–142.

Haaga, D. A. F., Dyck, M. J., & Ernst, D. (1991). Empirical status of cognitive theory of depression. *Psychological Bulletin, 110*, 215–236.

Haber, S. N., & Fudge, J. L. (1997). The interface between dopamine neurons and the amygdala: Implications for schizophrenia. *Schizophrenia Bulletin, 23*, 471–482.

Häfner, H., et al. (1998). Causes and consequences of the gender difference in age at onset of schizophrenia. *Schizophrenia Bulletin, 24*, 99–113.

Hager, M., & Peyser, M. (1997, March 24). Battling Alzheimer's. *Newsweek*, p. 66.

Hagerman, R. J. (1996). Fragile X syndrome. *Child and Adolescent Psychiatric Clinics of North America, 5*, 895–911.

Hall, G. C. (1995). Sexual offender recidivism revisited: A meta-analysis of recent treatment studies. *Journal of Consulting and Clinical Psychology, 63*, 802–809.

Hall, G. C., & Barongan, C. (1997). Prevention of sexual aggression: Sociocultural risk and protective factors. *American Psychologist, 52*, 5–14.

Hall, R. L. (1996). Escaping the self or escaping the anomaly? *Psychological Inquiry, 7*, 143–148.

Halmi, K. A., et al. (2000). Perfectionism in anorexia nervosa: Variation by clinical subtype, obsessionality, and pathological eating behavior. *American Journal of Psychiatry, 157*, 1799–1805.

Hammen, C., & Compas, B. E. (1994). Unmasking unmasked depression in children and adolescents: The problem of comorbidity. *Clinical Psychology Review, 14*, 585–603.

Hammen, C., & de Mayo, R. (1982). Cognitive correlates of teacher stress and depressive symptoms: Implications for attributional models of depression. *Journal of Abnormal Psychology, 91*, 96–101.

Hammen, C., & Gitlin, M. (1997). Stress reactivity in bipolar patients and its relation to prior history of disorder. *American Journal of Psychiatry, 154*, 856–857.

Hammen, C., Henry, R., & Daley, S. E. (2000). Depression and sensitization to stressors among young women as a function of childhood adversity. *Journal of Consulting and Clinical Psychology, 68*, 782–787.

Hancock, L. (1996, March 18). Mother's little helper. *Newsweek*, pp. 51–56.

Hankin, B. L., et al. (1998). Development of depression from preadolescence to young adulthood: Emerging gender differences in a 10-year longitudinal study. *Journal of Abnormal Psychology, 107*, 128–140.

Hansen, T. E., Casey, D. E., & Hoffman, W. F. (1997). Neuroleptic intolerance. *Schizophrenia Bulletin, 23*, 567–582.

Hanson, R. K., & Bussiere, M. T. (1998). Predicting relapse: A meta-analysis of sexual offender recidivism studies. *Journal of Consulting and Clinical Psychology, 66*, 348–362.

Hanson, R. K., Steffy, R. A., & Gauthier, R. (1993). Long-term recidivism of child molesters. *Journal of Consulting and Clinical Psychology, 61*, 646–652.

Hare, R. D. (1965). Temporal gradient of fear arousal in psychopaths. *Journal of Abnormal Psychology, 70*, 442–445.

Hare, R. D. (1986). Criminal psychopaths. In J. C. Yuille (Ed.), *Police selection and training: The role of psychology* (pp. 187–206). Dordrecht, Netherlands: Martinos Nijhoff.

Hare, R. D., Hart, S. D., & Harpur, T. J. (1991). Psychopathy and the *DSM-IV* criteria for antisocial personality disorder. *Journal of Abnormal Psychology, 100*, 391–398.

Harpur, T. J., & Hare, R. D. (1994). Assessment of psychopathy as a function of age. *Journal of Abnormal Psychology, 103*, 604–609.

Harrop, C., & Trower, P. (2001). Why does schizophrenia develop at late adolescence? *Clinical Psychology Review, 21*, 241–266.

Hartung, C. M., & Widiger, T. A. (1998). Gender differences in the diagnosis of mental disorders: Conclusions and controversies of the *DSM-IV*. *Psychological Bulletin, 123*, 260–278.

Harvey, A. G., & Bryant, R. A. (1999). The relationship between acute stress disorder and posttraumatic stress disorder: A 2-year prospective evaluation. *Journal of Consulting and Clinical Psychology, 67*, 985–988.

Harvey, A. G., & Bryant, R. A. (2000). Two-year prospective evaluation of the relationship between acute stress disorder and posttraumatic stress disorder following mild traumatic brain injury. *American Journal of Psychiatry, 157*, 629–631.

Harvey, P. D., et al. (1997). Age-related differences in formal thought disorder in chronically hospitalized schizophrenic patients: A cross-sectional study. *American Journal of Psychiatry, 154*, 205–210.

Haughton, E., & Ayllon, T. (1965). Production and elimination of symptomatic behavior. In L. P. Ullmann & L. Krasner (Eds.), *Case studies in behavior modification*. New York: Holt, Rinehart and Winston.

Havassy, B. E., Hall, S. M., & Wasserman, D. A. (1991). Social support and relapse: Commonalities among alcoholics, opiate users, and cigarette smokers. *Addictive Behaviors, 16*, 235–246.

Hawkrigg, J. J. (1975). Agoraphobia. *Nursing Times, 71*, 1280–1282.

Hawton, K. (1991). Sex therapy. Special issue: The changing face in behavioral psychotherapy. *Behavioral Psychotherapy, 19*, 131–136.

Hawton, K., & Catalan, J. (1990). Sex therapy for vaginismus: Characteristics of couples and treatment outcomes. *Sexual and Marital Therapy, 5*, 39–48.

Haznedar, M. M., et al. (2000). Limbic circuitry in patients with autism spectrum disorders studied with positron emission tomography and magnetic resonance imaging. *American Journal of Psychiatry, 157*, 1994–2001.

Headache coping strategies depend on the cause. (2000, August 14). *CNN Web Posting*. Retrieved August 20, 2000, from http://www.cnn.com/2000/HEALTH/08/14/headache.redux/index.html.

Health groups directly link media to child violence. (2000, July 26). *CNN Web Posting*. Retrieved July 27, 2000, from http://www.cnn.com/2000/HEALTH/children/07/26/children.violence.ap/.

Health Resources: Neurosurgery: //On Call®. (2000, March 16). *Parkinson's disease*. American Association of Neurological Surgeons/Congress of Neurological Surgeons. Retrieved April 4, 2000, from http://www.neurosurgery.org/health/patient/detail.asp?DisorderID=46.

Heatherton, T. F., et al. (1997). A 10-year longitudinal study of body weight, dieting, and eating disorder symptoms. *Journal of Abnormal Psychology, 106*, 117–125.

Heidrich, S. M., Forsthoff, C. A., & Ward, S. E. (1994). Psychological adjustment in adults with cancer: The self as mediator. *Health Psychology, 13*, 346–353.

Heikkinen, M. E., et al. (1997). Psychosocial factors and completed suicide in personality disorders. *Acta Psychiatrica Scandinavica, 95*, 49–57.

Hellerstein, D. J., et al. (2000). Double-blind comparison of sertraline, imipramine, and placebo in the treatment of dysthymia: Effects on personality. *American Journal of* Psychiatry, *157*, 1445–1452.

Helmuth, L. (2001). Commentary: Dyslexia: Same brains, different languages. *Science, 291*, 2064.

Helzer, J. E., Burnam, A., & McEvoy, L. T. (1991). Alcohol abuse and dependence. In L. N. Robins & D. A. Regier (Eds.), *Psychiatric disorders in America: The Epidemiologic Catchment Area Study* (pp. 81–115). New York: Free Press.

Henggeler, S. W., Melton, G. B., & Smith, L. A. (1992). Family preservation using multisystemic therapy: An effective alternative to incarcerating serious juvenile offenders. *Journal of Consulting and Clinical Psychology, 60*, 953–961.

Henggeler, S. W., et al. (1986). Multisystemic treatment of juvenile offenders: Effects on adolescent behavior and family interaction. *Developmental Psychology, 22*, 132–141.

Henggeler, S. W., et al. (1997). Multisystemic therapy with violent and chronic juvenile offenders and their families: The role of treatment fidelity in successful dissemination. *Journal of Consulting and Clinical Psychology, 65*, 821–833.

Henkin, W. A. (1985). Toward counseling the Japanese in America: A cross-cultural primer. *Journal of Counseling and Development, 63*, 500–503.

Herbert, J. D., et al. (2000). Science and pseudoscience in the development of eye movement desensitization and reprocessing. *Clinical Psychology Review, 20*, 945–972.

Herek, G. M. (1996). Heterosexism and homophobia. In R. P. Cabaj & T. S. Stein (Eds.), *Textbook of homosexuality and mental health* (pp. 101–113). Washington, DC: American Psychiatric Association Press.

Herkov, M. J., et al. (1996). MMPI differences among adolescent inpatients, rapists, sodomists, and sexual abusers. *Journal of Personality Assessment, 66*, 81–90.

Hernandez, R. (2000, August 2). In new drug battle, use of ecstasy among young soars. *The New York Times*, p. A21.

Hertz, M. R. (1986). Rorschach bound: A 50-year memoir. *Journal of Personality Assessment, 50*, 396–416.

Herzog, W., Schellberg, D., & Deter, H. C. (1997). First recovery in anorexia nervosa patients in the long-term course: A discrete-time survival analysis. *Journal of Consulting and Clinical Psychology, 65*, 169–177.

Heston, L. L., White, J. A., & Mastri, A. R. (1987). Pick's disease: Clinical genetics and natural history. *Archives of General Psychiatry, 44*, 409–411.

Hettema, J. M., Neale, M. C., & Kendler, K. S. (2001). A review and meta-analysis of the genetic epidemiology of anxiety disorders. *American Journal of Psychiatry, 158*, 1568–1578.

Heun, R., et al. (2001). A family study of Alzheimer disease and early- and late-onset depression in elderly patients. *Archives of General Psychiatry, 58*, 190–196.

Hewitt, P. L., Flett, G. L., & Ediger, E. (1996). Perfectionism and depression: Longitudinal assessment of a specific vulnerability hypothesis. *Journal of Abnormal Psychology, 105*, 276–280.

Hilchey, T. (1994, November 11). High anxiety raises risk of heart failure in men, study finds. *The New York Times*, p. A17.

Hill, A. L. (1977). Idiot savants: Rate of incidence. *Perceptual and Motor Skills, 44*, 161–162.

Hill, J. O., & Peters, J. C. (1998). Environmental contributions to the obesity epidemic. *Science, 280*, 1371–1374.

Hill, K. G., et al. (2000). Early adult outcomes of adolescent binge drinking: Person- and variable-centered analyses of binge drinking trajectories. *Alcohol: Clinical Experimental Research, 24*, 892–901.

Hill, S. Y. (1980). Introduction: The biological consequences. In *Alcoholism and alcohol abuse among women: Research issues*. Rockville, MD: National Institute on Alcohol Abuse and Alcoholism.

Hilts, P. J. (1991, October 9). Report is critical of mental clinics. *The New York Times*, p. L25.

Himelein, M. J., & McElrath, J. V. (1996). Resilient child sexual abuse survivors: Cognitive coping and illusion. *Child Abuse and Neglect, 20*, 747–758.

Hingson, R. W., et al. (2000). Age of drinking onset and unintentional injury involvement after drinking. *Journal of the American Medical Association, 284*, 1527–1533.

Hinshaw, S. P. (1992). Academic underachievement, attention deficits, and aggression: Comorbidity and implications for intervention. *Journal of Consulting and Clinical Psychology, 60*, 893–903.

Hinshaw, S. P., Klein, R. G., & Abikoff, H. (1998). Childhood attention deficit hyperactive disorder: Nonpharmacological and combination treatments. In P. E. Nathan & J. M. Gorman (Eds.), *A guide to treatments that work* (pp. 26–41). New York: Oxford University Press.

Hirsch, S. R., & Leff, J. P. (1975). *Abnormalities in parents of schizophrenics*. Oxford, U.K.: Oxford University Press.

Hirschfield, R. M. A., et al. (1997). The National Depressive and Manic-Depressive Association consensus statement on the undertreatment of depression. *Journal of the American Medical Association, 277*, 333–340.

Hodgins, S., et al. (1998). In reply. *Archives of General Psychiatry, 55*, 87–88.

Hoffman, R. E., et al. (1999). Selective speech perception alterations in schizophrenic patients reporting hallucinated "voices." *American Journal of Psychiatry, 156*, 393–399.

Hoffman, S. G. (2000a). Treatment of social phobia: Potential mediators and moderators. *Clinical Psychology: Science and Practice, 7*(1), 3–16.

Hoffman, S. G. (2000b). Self-focused attention before and after treatment of social phobia. *Behavior Research and Therapy, 38*, 717–725.

Hoffman, S. G., et al. (1995). Psychophysiological differences between subgroups of social phobia. *Journal of Abnormal Psychology, 104*, 224–231.

Hoffman, W., & Prior, M. (1982). Neuropsychological dimensions of autism in children: A test of the hemispheric dysfunction hypothesis. *Journal of Clinical Neuropsychology, 4*, 27–42.

Hogarty, G. E., Schroeder, N. R., Ulrich, R., Mussare, N., Peregino, F., & Herron, E. (1979b). Fluphenazine and social therapy in the aftercare of schizophrenic patients. *Archives of General Psychiatry, 36*, 1283–1294.

Hogarty, G. E., et al. (1997a). Three-year trials of personal therapy among schizophrenic patients living with or independent of family, II: Effects on adjustment of patients. *American Journal of Psychiatry, 154*, 1514–1524.

Holland, A. J., Sicotte, N., & Treasure, J. (1988). Anorexia nervosa: Evidence of a genetic basis. *Journal of Psychosomatic Research, 32*, 561–571.

Hollander, E., et al. (1992). Serotonergic function in obsessive-compulsive disorder: Behavioral and neuroendocrine responses to oral m-chlorophenylpiperazine and fenfluramine in patients and health volunteers. *Archives of General Psychiatry, 49*, 21–28.

Hollingshead, A. B., & Redlich, F. C. (1958). *Social class and mental illness: A community study*. New York: Wiley.

Hollon, S. D., Evans, M. D., & DeRubeis, R. J. (1990). Cognitive mediation of relapse prevention following treatment for depression: Implications of differential risk. In R. E. Ingram (Ed.), *Contemporary psychological approaches to depression: Theory, research, and treatment* (pp. 117–136). New York: Plenum Press.

Hollon, S. D., & Kendall, P. C. (1980). Cognitive self-statements in depression: Development of an automatic thoughts questionnaire. *Cognitive Therapy and Research, 4*, 383–395.

Holloway, L. (1998, February 12). A mental patient skips care and West Siders worry again. *The New York Times*, p. B5.

Holmes, D. S., Solomon, S., Cappo, B. M., & Greenberg, J. L. (1983). Effects of transcendental meditation versus resting on physiological and subjective arousal. *Journal of Personality and Social Psychology, 44*, 1244–1252.

Holmes, G. R., Offen, L., & Waller, G. (1997). See no evil, hear no evil, speak no evil: Why do relatively few male victims of childhood sexual abuse receive help for abuse-related issues in adulthood? *Clinical Psychology Review, 17*, 69–88.

Holroyd, K. A., et al. (2001). Management of chronic tension-type headache with tricyclic antidepressant medication, stress management therapy, and their combination: A randomized controlled trial. *Journal of the American Medical Association, 285*, 2208–2215.

Holt, C. S., Heimberg, R. G., & Hope, D. A. (1992). Avoidant personality disorder and the generalized subtype of social phobia. *Journal of Abnormal Psychology, 101*, 318–325.

Holtzworth-Munroe, A. (1995). Marital violence. *The Harvard Mental Health Letter, 12*, pp. 4–6.

Holtzworth-Munroe, A., Rehman, U., & Herron, K. (2000). General and spouse-specific anger and hostility in subtypes of maritally violent men and nonviolent men. *Behavior Therapy, 31*, 603–630.

Holzman, P. S. (1987). Recent studies of psychophysiology in schizophrenia. *Schizophrenia Bulletin, 13*, 49–75.

Holzman, P. S., et al. (1997). Smooth pursuit eye tracking in twins: A critical commentary. *Archives of General Psychiatry, 54*, 429–431.

Honig, A., et al. (1998). Auditory hallucinations: A comparison between patients and nonpatients. *Journal of Nervous and Mental Disease, 186*, 646–651.

Hooley, J. M., & Hiller, J. B. (2000). Personality and expressed emotion. *Journal of Abnormal Psychology, 109*, 40–44.

Horne, L. R., Van Vactor, J. C., & Emerson, S. (1991). Disturbed body image in patients with eating disorders. *American Journal of Psychiatry, 148*, 211–215.

Horrigan, J. P., & Barnhill, J. L. J. (2000). "Fluvoxamine and enuresis:" Comment. *Journal of the American Academy of Child & Adolescent Psychiatry, 39*, 1465–1466.

Houts, A. C., Berman, J. S., & Abramson, H. (1994). Effectiveness of psychological and pharmacological treatments for nocturnal enuresis. *Journal of Consulting and Clinical Psychology, 62*, 373–745.

How mad-cow disease jumped to humans. (2001, February). *Tufts University Health & Nutrition Letter*, p. 18.

Howard, C. E., & Porzelius, L. K. (1999). The role of dieting in binge eating disorder: Etiology and treatment implications. *Clinical Psychology Review, 19*, 25–44.

Howard, K. I., Kopta, S. M., Krause, M. S., & Orlinksy, D. E. (1986). The dose-effect relationship in psychotherapy. *American Psychologist, 41*, 159–164.

Howland, R. H., & Thase, M. E. (1993). A comprehensive review of cyclothymic disorder. *Journal of Nervous and Mental Disease, 18*, 485–493.

Hrobjartsson, A., & Gotzsche, P. C. (2001). Is the placebo powerless? An analysis of clinical trials comparing placebo with no treatment. *The New England Journal of Medicine, 344*, 1594–1602.

Huang, L. H. (1994). An integrative approach to clinical assessment and intervention with Asian-American adolescents. *Journal of Clinical Child Psychology, 23*, 21–31.

Huang, W., & Cuvo, A. J. (1997). Social skills training for adults with mental retardation in job-related settings. *Behavior Modification, 21*, 3–44.

Hublin, C., et al. (1997). Prevalence and genetics of sleepwalking: A population-based twin study. *Neurology, 48*, 177–181.

Hudziak, J. J., et al. (1996). Clinical study of the relation of borderline personality disorder to Briquet's syndrome (hysteria), somatization disorder, antisocial personality disorder, and substance abuse disorders. *American Journal of Psychiatry, 153*, 1598–1606.

Huesmann, L. R., & Miller, L. S. (1994). Long-term effects of repeated exposure to media violence in childhood. In L. R. Huesmann (Ed.), *Aggressive behavior: Current perspectives*. New York: Plenum Press.

Hufford, M. R. (2001). Alcohol and suicidal behavior. *Clinical Psychology Review, 21*, 797–811.

Humphrey, L. L. (1986). Family dynamics in bulimia. In S. C. Feinstein et al. (Eds.), *Adolescent psychiatry*. Chicago: University of Chicago Press.

Hunsley, J., & Bailey, J. H. (1999). The clinical utility of the Rorschach: Unfulfilled promises and an uncertain future. *Psychological Assessment*, 11, 266–277.

Hunter, R. H., Bedell, J. R., & Corrigan, P. W. (1997). Current approaches to assessment and treatment of persons with serious mental illness. *Professional Psychology: Research & Practice, 28*, 217–228.

Hurlburt, G., & Gade, E. (1984). Personality differences between Native American and Caucasian women alcoholics: Implications for alcoholism counseling. *White Cloud Journal, 3*, 35–39.

Hurt, R. D., et al. (1997). A comparison of sustained-release bupropion and placebo for smoking cessation. *The New England Journal of Medicine, 337*, 1195–1202.

Hussong, A. M., et al. (2001). Specifying the relations between affect and heavy alcohol use among young adults. *Journal of Abnormal Psychology, 110*, 449–461.

Hutton, M. (2001). Missense and splice site mutations in tau associated with FTDP-17: Multiple pathogenic mechanisms. *Neurology, 56*(Suppl. 4), S21–S25.

Huxley, N. A., Rendall, M., & Sederer, L. (2000). Psychosocial treatments in schizophrenia: A review of the past 20 years. *Journal of Nervous & Mental Disease, 188*, 187–201.

Hyman, D. J., & Pavlik, V. N. (2001). Characteristics of patients with uncontrolled hypertension in the United States. *The New England Journal of Medicine, 345, 479–486.*

I

Ickovics, J. R., et al. (2001). Mortality, CD4 cell count decline, and depressive symptoms among HIV-seropositive women: Longitudinal analysis from the HIV Epidemiology Research Study. *Journal of the American Medical Association, 285*, 1466–1474.

Ilardi, S. S., & Craighead, W. E. (1994). The role of nonspecific factors in cognitive-behavior therapy for depression. *Clinical Psychology: Science and Practice, 1*, 138–156.

in 't Veld, B. A., et al. (2001). Nonsteroidal antiinflammatory drugs and the risk of Alzheimer's disease. *New England Journal of Medicine, 345*, 1515–1521.

Ingersoll, S. L, & Patton, S. O. (1991). *Treating perpetrators of sexual abuse*. Lexington, MA: Lexington Books.

Ingraham, L. J., et al. (1995). Twenty-five year followup of the Israeli High-Risk Study: Current and lifetime psychopathology. *Schizophrenia Bulletin, 21*, 183–192.

Ingram, R. E. (1991). Tilting at windmills: A response to Pyszczynski, Greenberg, Hamilton, and Nix. *Psychological Bulletin, 110*, 544–550.

Ingram, R. E., Miranda, J., & Segal, Z. V. (1998). *Cognitive vulnerability to depression*. New York: Guilford Press.

Ingram, R. E., & Siegle, G. J. (2001). Cognition and clinical science: From revolution to evolution. In K. S. Dobson (Ed.), *Handbook of cognitive-behavioral therapies* (2nd ed., pp. 111–137). New York: Guilford Press.

Insanity: A defense of last resort. (1992, February 3). *Newsweek*, p. 49

Institute of Medicine. (1990). *Broadening the base of treatment for alcohol problems*. Washington, DC: National Academy Press.

Ioannidis, J. P. A., & Karassa, P. (2001). Comparison of evidence of treatment effects in randomized and nonrandomized studies. *Journal of the American Medical Association, 286*, 821–830.

Ioannidis, J. P. A., et al. (2001). Comparison of evidence of treatment effects in randomized and nonrandomized studies. *Journal of the American Medical Association, 286*, 821–830.

Iribarren, C., et al. (2000). Association of hostility with coronary artery calcification in young adults: The CARDIA Study. *Journal of the American Medical Association, 283*, 546–551.

Irle, E., et al. (1998). Obsessive-compulsive disorder and ventromedial frontal lesions: Clinical and neuropsychological findings. *American Journal of Psychiatry, 155*, 255–263.

Ironson, G., et al. (1997). Posttraumatic stress symptoms, intrusive thoughts, loss, and immune function after hurricane Andrew. *Psychosomatic Medicine, 59*, 128–141.

Isacsson, G. (2000). Suicide prevention—A medical breakthrough? *Acta Psychiatrica Scandinavica, 102*, 113–117.

Israel, G. E., & Tarver II, D. E. (1997). *Transgender care: Recommended guidelines, practical information, and personal accounts*. Philadelphia: Temple University Press.

Ito, T. A., Miller, N., & Pollock, V. E. (1996). Alcohol and aggression: A meta-analysis on the moderating effects of inhibitory cues, triggering events, and self-focused attention. *Psychological Bulletin, 120*, 60–82.

Iwata, N., et al. (2001). Metabolic regulation of brain abeta by neprilysin. *Science, 292*, 1550–1552.

J

Jablensky, A., et al. (1992). Schizophrenia: Manifestations, incidence and course in different cultures: A World Health Organization ten-country study. *Psychological Medicine, 20*(Monograph Suppl.), 1–97.

Jackson, J. L. (1999). Psychometric considerations in self-monitoring assessment. *Psychological Assessment, 11*, 439–447.

Jackson, J., et al. (1990). Young adult women who report childhood intrafamilial sexual abuse: Subsequent adjustment. *Archives of Sexual Behavior, 19*, 211–221.

Jacobs, M. K., et al. (2001). A Comparison of computer-based versus traditional individual psychotherapy. *Professional Psychology: Research and Practice, 32*, 92–96.

Jacobs, W., Newman, G. H., & Burns, J. C. (2001). The Homeless Assessment Program: A service-training model for providing disability evaluations for homeless, mentally ill individuals. *Professional Psychology: Research and Practice, 32*, 319–323.

Jacobson, N. S., Wilson, L., & Tupper, C. (1988). The clinical significance of treatment gains resulting from exposure-based interventions for agoraphobia: A reanalysis of outcome data. *Behavior Therapy, 19*, 539–554.

Jacobson, N. S., et al. (1996). A component analysis of cognitive-behavioral treatment for depression. *Journal of Consulting and Clinical Psychology, 64*, 295–304.

Jamison, B. (2000, June 13). Obsessive Internet use poses risk of isolation, depression, researchers say. WebMD.com. Retrieved June 29, 2000, from http://www.webmd.com.

Jamison, K. R. (1993). *Touched with fire*. New York: Free Press.

Janofsky, J. S., et al. (1996). Insanity defense pleas in Baltimore City: An analysis of outcome. *American Journal of Psychiatry, 153*, 1464–1468.

Januzzi, J., & DeSanctis, R. (1999). Looking to the brain to save the heart. *Cerebrum, 1*, 31–43.

Jarrett, R. B., et al. (1999). Treatment of atypical depression with cognitive therapy or phenelzine. *Archives of General Psychiatry, 56*, 431–437.

Jarrett, R. B., et al. (2001). Preventing recurrent depression using cognitive therapy with and without a continuation phase: A randomized clinical trial. *Archives of General Psychiatry, 58*, 381–388.

Javier, R. A. (1993). Cited in Rathus, S. A. (1993). *Psychology* (5th ed.). Fort Worth, TX: Harcourt Brace Jovanovich.

Jaycox, L. H ., Reivich, K. J., Gillham, J., & Seligman, M. E. P. (1994). Prevention of depressive symptoms in school children. *Behaviour Research and Therapy, 32*, 801–816.

Jeffery, R. W. (1991). Population perspectives on the prevention and treatment of obesity in minority populations. *American Journal of Clinical Nutrition, 53*(6 Suppl.), 1621A–1624S.

Jellinek, E. M. (1960). *The disease concept of alcoholism*. New Haven, CT: College and University Press.

Jemmott, J. B., et al. (1983, June 25). Academic stress, power motivation, and decrease in secretion rate of salivary secretory immunoglobin A. *Lancet*, 1400–1402.

Jenike, M. A., et al. (1997). Placebo-controlled trial of fluoxetine and phenelzine for obsessive-compulsive disorder. *American Journal of Psychiatry, 154*, 1261–1264.

Jenkins, C. D. (1988). Epidemiology of cardiovascular diseases. *Journal of Consulting and Clinical Psychology, 56*, 324–332.

Jenkins, J. H. (1988). Ethnopsychiatric interpretations of schizophrenic illness: The problem of nervios within Mexican-American families. *Culture, Medicine, and Psychiatry, 12*, 301–329.

Jenkins, J. H., & Karno, M. (1992). The meaning of expressed emotion: Theoretical issues raised by cross-cultural research. *American Journal of Psychiatry, 149*, 9–21.

Jenny, C., Roesler, T. A., & Poyer, K. L. (1994). Are children at risk for sexual abuse by homosexuals? *Pediatrics, 94,* 41–44.

Jensen, P. S., Martin, D., & Cantwell, D. P. (1997). Comorbidity in ADHD: Implications for research practice, and *DSM-V. Journal of the American Academy of Child and Adolescent Psychiatry, 36,* 1065–1079.

Jerome, L. W., et al. (2000). The coming of age in telecommunications in psychological research and practice. *American Psychologist, 55,* 507–421.

Jeste, D. V., & Caligiui, M. P. (1993). Tardive dyskinesia. *Schizophrenia Bulletin, 19,* 303–315.

Jeste, D. V., Lindamer, L. A., Evans, J., & Lacro, J. P. (1996). Relationship of ethnicity and gender to schizophrenia and pharmacology of neuroleptics. *Psychopharmacology Bulletin, 32,* 243–251.

Jeste, D. V., et al. (1992). Cognitive deficits of patients with Alzheimer's disease with and without delusions. *American Journal of Psychiatry, 149,* 184–188.

Jian, W., et al. (1996). Mental stress-induced myocardial ischemia and cardiac events. *Journal of the American Medical Association, 275,* 1651–1656.

Jimerson, D. C., et al. (1997). Decreased serotonin function in bulimia nervosa. *Archives of General Psychiatry, 54,* 529–534.

Johnson, B. A., et al. (2000). Ondansetron for reduction of drinking among biologically predisposed alcoholic patients: A randomized controlled trial. *Journal of the American Medical Association, 284,* 963–971.

Johnson, J. G., et al. (1999). A longitudinal investigation of social causation and social selection processes involved in the association between socioeconomic status and psychiatric disorders. *Journal of Abnormal Psychology, 108,* 490–499.

Johnson, J. G., et al. (2000). Association between cigarette smoking and anxiety disorders during adolescence and early adulthood. *Journal of the American Medical Association, 284,* 2348–2351.

Johnson, R. E., et al. (2000). A comparison of levomethadyl acetate, buprenorphine, and methadone for opioid dependence. *The New England Journal of Medicine, 343,* 1290–1297.

Johnson, R. J., & McFarland, B. H. (1996). Lithium use and discontinuation in a health maintenance organization. *American Journal of Psychiatry, 153,* 993–1000.

Johnson, S., et al. (1999). Social support and the course of bipolar disorder. *Journal of Abnormal Psychology, 108,* 558–566.

Johnson, W. G., Tsoh, J. Y., & Varnado, P. J. (1996). Eating disorders: Efficacy of pharmacological and psychological interventions. *Clinical Psychology Review, 16,* 457–478.

Johnston, L. D., Bachman, J. G., & O'Malley, P. M. (1992, January 25). *Monitoring the future: A continuing study of the lifestyles and values of youth.* Ann Arbor, MI: The University of Michigan News and Information Services.

Johnston, L. D., O'Malley, P. M., and Bachman, J. G. (1996). *National Survey Results on Drug Use from the Monitoring the Future Study, 1975–1995. Volume I. Secondary School Students.* Washington, DC: U.S. Department of Health and Human Services, Public Health Service, National Institutes of Health: National Institute on Drug Abuse.

Joiner, T. E., Alfano, M. S., & Metalsky, G. I. (1992). When depression breeds contempt: Reassurance seeking, self-esteem, and rejection of depressed college students by their roommates. *Journal of Abnormal Psychology, 101,* 165–173.

Jones v. United States, 103 S. Ct. 3043 (1983).

Jones, E. (1953). *The life and work of Sigmund Freud.* New York: Basic Books.

Jones, P. B., et al. (1998). Schizophrenia as a long-term outcome of pregnancy, delivery, and perinatal complications: A 28-year follow-up of the 1966 North Finland general population birth cohort. *American Journal of Psychiatry, 155,* 355–364.

Jouriles, E. N., et al. (1996). Physical violence and other forms of marital aggression: Links with children's behavior problems. *Journal of Family Psychology, 10,* 223–234.

Jouriles, E. N., et al. (1997). Psychometric properties of family members' reports of parental physical aggression toward clinic-referred children. *Journal of Consulting and Clinical Psychology, 65,* 309–318.

Judd, L. J. (1997). The clinical course of unipolar major depressive disorders. *Archives of General Psychiatry, 54,* 989–991.

Judd, L. L., et al. (2000a). Psychosocial disability during the long-term course of unipolar major depressive disorder. *Archives of General Psychiatry, 57,* 375–380.

Judd, L. L., et al. (2000b). Does incomplete recovery from first lifetime major depressive episode herald a chronic course of illness? *American Journal of Psychiatry, 157,* 1509–1511.

Just, N., Abramson, L. Y., & Alloy, L. B. (2001). Remitted depression studies as tests of the cognitive vulnerability hypotheses of depression onset. A critique and conceptual analysis. *Clinical Psychology Review, 21,* 63–83.

Just, N., & Alloy, L. B. (1997). The response styles theory of depression: Tests and an extension of the theory. *Journal of Abnormal Psychology, 106,* 221–229.

K

Kahn, M. W. (1982). Cultural clash and psychopathology in three aboriginal cultures. *Academic Psychology Bulletin, 4,* 553–561.

Kalb, C. (2001a, January 22). Seeing a virtual shrink. *Newsweek,* pp. 34–37.

Kalb, C. (2001b). Can this pill stop you from hitting the bottle? *Newsweek,* pp. 46–48.

Kalichman, S. C. (2000). HIV transmission risk behaviors of men and women living with HIV-AIDS: Prevalence, predictors, and emerging clinical intervention. *Clinical Psychology: Science and Practice, 7,* 32–47.

Kammeyer, K. C. W. (1990). *Marriage and family: A foundation for personal decisions* (2nd ed.). Boston: Allyn & Bacon.

Kammeyer, K. C. W., Ritzer, G., & Yetman, N. R. (1990). *Sociology: Experiencing changing societies.* Boston: Allyn & Bacon.

Kane, J. M. (1996). Drug therapy: Schizophrenia. *The New England Journal of Medicine, 334,* 34–41.

Kane, J. M., & Marder, S. R. (1993). Psychopharmalogic treatment of schizophrenia. *Schizophrenia Bulletin, 19,* 287–302.

Kanner, A. D., Coyne, J. C., Schaefer, C., & Lazarus, R. S. (1981). Comparison of two modes of stress measurement: Daily hassles and uplifts versus major life events. *Journal of Behavioral Medicine, 4,* 1–39.

Kanner, L. (1943). Autistic disturbances of affective content. *Nervous Child, 2,* 217–240.

Kano, K., & Arisaka, O. (2000). Fluvoxamine and enuresis. *Journal of the American Academy of Child & Adolescent Psychiatry, 39,* 1464–1465.

Kantor, M. (1998). *Homophobia: Description, development and dynamics of gay bashing.* Westport, CT: Praeger.

Kaplan, D. (1993, January 18). The incorrigibles. *Newsweek,* pp. 48–50.

Kaplan, H. S. (1974). *The new sex therapy: Active treatment of sexual dysfunctions.* New York: Brunner/Mazel.

Kaplan, H. S. (1987). *Sexual aversion, sexual phobias, and panic disorder.* New York: Brunner/Mazel.

Kaplan, R. M. (2000). Two pathways to prevention. *American Psychologist, 55,* 382–396.

Kaplan, S. J. (1986). *The private practice of behavior therapy: A guide for behavioral practitioners.* New York: Plenum Press.

Kapp, M. B. (1994). Treatment and refusal rights in mental health: Therapeutic justice and clinical accommodation. *American Journal of Orthopsychiatry, 64,* 223–234.

Kapur, S., & Remington, G. (2000). Atypical antipsychotics: Patients value the lower incidence of extrapyramidal side effects. *British Medical Journal, 321,* 1360–1361.

Karel, J. J., & Hinrichsen, G. (2000). Treatment of depression in late life: Psychotherapeutic interventions. *Clinical Psychology Review, 20,* 707–729.

Karel, M. J. (1997). Aging and depression: Vulnerability and stress across adulthood. *Clinical Psychology Review, 17,* 847–879.

Karlen, N. (1995, May 29). Greetings from Minnesober. *The New York Times Magazine,* pp. 32–35.

Karno, M., et al. (1987). Expressed emotions and schizophrenic outcome among Mexican-American Families. *Journal of Nervous and Mental Disease, 175,* 143–151.

Karp, B. I., et al. (2001). Abnormal neurologic maturation in adolescents with early-onset schizophrenia. *American Journal of Psychiatry, 158,* 118–122.

Kasari, D., et al. (1993). Affective development and communication in children with autism. In A. P. Kaiser & D. B. Gray (Eds.), *Enhancing children's communication: Research foundation for intervention* (pp. 201–222). New York: Brookes.

Kasen, S., et al. (2001). Childhood depression and adult personality disorder: Alternative pathways of continuity. *Archives of General Psychiatry, 58,* 231–236

Kasl-Godley, J., & Gatz, M. (2000). Psychosocial interventions for individuals with dementia: An integration of theory, therapy, and a clinical understanding of dementia. *Clinical Psychology Review, 20,* 755–782.

Kasper, J. A., et al. (1997). Prospective study of patients' refusal of antipsychotic medication under a physician discretion review procedure. *American Journal of Psychiatry, 154,* 483–489.

Kasper, M. E., Rogers, R., & Adams, P. A. (1996). Dangerousness and command hallucinations: An investigation of psychotic inpatients. *Bulletin of the American Academy of Psychiatry and the Law, 24,* 219–224.

Katz, J. N. (1995). *The invention of heterosexuality.* New York: Dutton.

Kawachi, I., et al. (1994). Prospective study of phobic anxiety and risk of coronary heart disease in men. *Circulation, 89,* 1992–1997.

Kawas, C. H., & Brookmeyer, R. (2001). Aging and the Public Health Effects of Dementia. *The New England Journal of Medicine, 344,* 1160–1161. [Editorial]

Kay, S. R. (1990). Significance of the positive-negative distinction in schizophrenia. *Schizophrenia Bulletin, 16,* 635–652.

Kazdin, A. E. (1992). *Research design in clinical psychology* (2nd ed.). Boston: Allyn & Bacon.

Kazdin, A. E. (1998). Psychosocial treatments for conduct disorder in children. In P. E. Nathan & J. M. Gorman (Eds.), *A guide to treatments that work* (pp. 65–89). New York: Oxford University Press.

Kazdin, A. E., Siegel, T. D., & Bass, D. (1992). Cognitive problem-solving skills training and parent management training in the treatment of antisocial behavior in children. *Journal of Consulting and Clinical Psychology,* 733–747.

Kazdin, A. E., & Weisz, J. R. (1998). Identifying and developing empirically supported child and adolescent treatments. *Journal of Consulting and Clinical Psychology, 66,* 19–36.

Keane, T. M. (1998). Psychological and behavioral treatments for post-traumatic stress disorder. In P. E. Nathan & J. M. Gorman (Eds.), *A guide to treatments that work* (pp. 398–407). New York: Oxford University Press.

Kearins, J. M. (1981). Visual spatial memory in Australian aboriginal children in desert regions. *Cognitive Psychology, 13,* 434–460.

Keel, P. K., et al. (1999). Long-term outcome of bulimia nervosa. *Archives of General Psychiatry, 56,* 63–69.

Keesey, R. E. (1980). A set-point analysis of the regulation of body weight. In A. J. Stunkard (Ed.), *Obesity.* Philadelphia: Saunders.

Keith, S. J., Regier, D. A., & Rae, D. S. (1991). Schizophrenic disorders. In L. N. Robins & D. A. Regier (Eds.), *Psychiatric disorders in America: The Epidemiologic Catchment Area Study* (pp. 33–52). New York: Free Press.

Keller, M. B., Hirschfeld, R. M. A., & Hanks, D. L. (1997). Double depression: A distinctive subtype of unipolar depression. *Journal of Affective Disorders, 45*, 65–73.

Keller, M. B., et al. (1993). Bipolar I: A five-year prospective follow-up. *Journal of Nervous and Mental Disease, 181*, 238–245.

Kellner, R. (1992). Diagnosis and treatment of hypochondriacal syndromes. *Psychosomatics, 33*, 278–289.

Kelly, J. A., Brasfield, T. L., & St. Lawrence, J. S. (1991). Predictors of vulnerability to AIDS risk behavior relapse. *Journal of Consulting and Clinical Psychology, 59*, 163–166.

Kelly, J. A., et al. (1995). Factors predicting continued high-risk behavior among gay men in small cities: Psychological, behavioral, and demographic characteristics related to unsafe sex. *Journal of Consulting and Clinical Psychology, 63*, 101–107.

Kelly, J. A., et al. (1998). Implications of HIV treatment advances for behavioral research on AIDS: Protease inhibitors and new challenges in HIV secondary prevention. *Health Psychology, 17*, 310–319.

Kelly, S. J., Macaruso, P., & Sokol, S. M. (1997). Mental calculation in an autistic savant: A case study. *Journal of Clinical and Experimental Neuropsychology, 19*, 172–184.

Kendell, R. E. (1983). Hysteria. In G. F. M. Russell & L. A. Hersov (Eds.), *Handbook of psychiatry (Vol. 4). The neuroses and personality disorders* (pp. 232–246). Cambridge: Cambridge University Press.

Kendler, K. (1992). Cited in Goleman, D. (1992, October 14). Study ties genes to drinking in women as much as in men. *The New York Times*, p. C14.

Kendler, K. S. (1994). Twin studies of psychiatric illness: Current status and future directions. *Archives of General Psychiatry, 50*, 905–918.

Kendler, K. S. (2001). A psychiatric dialogue on the mind-body problem. *American Journal of Psychiatry, 158*, 989–1000.

Kendler, K. S., & Diehl, S. R. (1993). The genetics of schizophrenia: A current, genetic-epidemiologic perspective. *Schizophrenia Bulletin, 19*, 261–295.

Kendler, K. S., & Gardner, C. O. (1998). Boundaries of major depression: An evaluation of *DSM-IV* criteria. *American Journal of Psychiatry, 155*, 172–177.

Kendler, K., S., Gardner, C. O., & Prescott, C. A. (2000). Corrections to 2 prior articles. *Archives of General Psychiatry, 57*, 94–95. [Letter]

Kendler, K. S., Gruenberg, A. M., & Tsuang, M.T. (1985). Psychiatric illness in first-degree relatives of schizophrenic and surgical control patients, a family study using *DSM-III* criteria. *Archives of General Psychiatry, 42*, 770–779.

Kendler, K. S., Myers, J. M., & Neale, M. C. (2000). A multidimensional twin study of mental health in women. *American Journal of Psychiatry, 157*, 521–527.

Kendler, K. S., & Prescott, C. A. (1999). A population-based twin study of lifetime major depression in men and women. *Archives of General Psychiatry, 56*, 39–44.

Kendler, K. S., & Thornton, L. M., & Gardner, C. O. (2000). Stressful life events and previous episodes in the etiology of major depression in women: An evaluation of the "kindling" hypothesis. *American Journal of Psychiatry, 157*, 1243–1251.

Kendler, K. S., Thornton, L. M., & Pedersen, N. L. (2000). Tobacco consumption in Swedish twins reared apart and reared together. *Archives of General Psychiatry, 57*, 886–892.

Kendler, K. S., & Walsh, D. (1995). Schizotypal personality disorder in parents and the risk for schizophrenia in siblings. *Schizophrenia Bulletin, 21*, 47–52.

Kendler, K. S., et al. (1991). The genetic epidemiology of bulimia nervosa. *American Journal of Psychiatry, 148*, 1627–1637.

Kendler, K. S. et al. (1992a). A population-based twin study of major depression in women: The impact of varying definitions of illness. *Archives of General Psychiatry, 49*, 257–266.

Kendler, K. S., et al. (1992c). The genetic epidemiology of phobias in women: The interrelationship of agoraphobia, social phobia, situational phobia, and simple phobia. *Archives of General Psychiatry, 49*, 273–281.

Kendler, K. S., et al. (1993a). A pilot Swedish twin study of affective illness, including hospital- and population-ascertained subsamples. *Archives of General Psychiatry, 50*, 699–706.

Kendler, K. S., et al. (1993b). The lifetime history of major depression in women: Reliability of diagnosis and heritability. *Archives of General Psychiatry, 50*, 863–870.

Kendler, K .S., et al. (1997). Resemblance of psychotic symptoms and syndromes in affected sibling pairs from the Irish study of high-density schizophrenia families: Evidence for possible etiologic heterogeneity. *American Journal of Psychiatry, 154*, 191–198.

Kendler, K. S., et al. (2000a). Clinical features of schizophrenia and linkage to chromosomes 5q, 6p, 8p, and 10p in the Irish study of high-density schizophrenia families. *American Journal of Psychiatry, 157*, 402–408.

Kendler, K. S., et al. (2000b). Childhood sexual abuse and adult psychiatric and substance use disorders in women: An epidemiological and co-twin control analysis. *Archives of General Psychiatry, 57*(10), 953–959.

Kendler, K. S., et al. (2001). The genetic epidemiology of irrational fears and phobias in men. *Archives of General Psychiatry, 58*, 257–265.

Kennedy, S., Scheirer, J., & Rogers, A. (1984). The price of success: Our monocultural science. *American Psychologist, 390*, 966–967.

Kent, A., & Waller, G. (2000). Childhood emotional abuse and eating psychopathology. *Clinical Psychology Review, 20*, 887–903.

Kent, J. M., et al. (2001). Specificity of panic response to CO2 inhalation in panic disorder: A comparison with major depression and premenstrual dysphoric disorder. *American Journal of Psychiatry, 158*, 58–67.

Kernberg, O. F. (1975). *Borderline conditions and pathological narcissism*. New York: Jason Aronson.

Kershner, R. (1996). Adolescent attitudes about rape. *Adolescence, 31*, 29–33.

Kessler, R. C. (1994). The National Comorbidity Survey: Preliminary results and future directions. *International Journal of Methods in Psychiatric Research, 4*, 114.1–114.13.

Kessler, R. C., Borges, G., & Walters, E. E. (1999). Prevalence of and risk factors for lifetime suicide attempts in the National Comorbidity Survey. *Archives of General Psychiatry, 56*, 617–626.

Kessler, R. C., et al. (1990). Clustering of teenage suicides after television news stories about suicides: A reconsideration. *American Journal of Psychiatry, 145*, 1379–1383.

Kessler, R. C., et al. (1993). Sex and depression in the National Comorbidity Survey I: Lifetime prevalence, chronicity and recurrence. *Journal of Affective Disorders, 29*, 85–96.

Kessler, R. C., et al. (1994). Lifetime and 12-month prevalence of *DSM-III-R* psychiatric disorders in the United States: Results from the National Comorbidity Survey. *Archives of General Psychiatry, 51*, 8–19.

Kessler, R. C., et al. (1995). Posttraumatic stress disorder in the National Comorbidity Survey. *Archives of General Psychiatry, 52*, 1048–1060.

Kessler, R. C., et al. (1997a). Differences in the use of psychiatric outpatient services between the United States and Ontario. *The New England Journal of Medicine, 336*, 551–557.

Kessler, R. C., et al. (1997b). Lifetime co-occurrence of *DSM-III-R* alcohol abuse and dependence with other psychiatric disorders in the National Comorbidity Survey. *Archives of General Psychiatry, 54*, 313–321.

Kety, S. S. (1980). The syndrome of schizophrenia: Unresolved questions and opportunities for research. *British Journal of Psychiatry, 136*, 421–436.

Kety, S. S., Rosenthal, D., Wender, P. H., Schulsinger, F., & Jacobsen, B. (1975). Mental illness in the biological and adoptive families of adoptive individuals who have become schizophrenic: A preliminary report based on psychiatric interviews. In R. R. Fieve, D. Rosenthal, & H. Brill (Eds.), *Genetic research in psychiatry*. Baltimore: The Johns Hopkins University Press.

Kety, S. S., Rosenthal, D., Wender, P. H., Schulsinger, F., & Jacobsen, B. (1978). The biological and adoptive families of adopted individuals who become schizophrenic. In C. Wynne, R. L. Cromwell, & S. Mathysse (Eds.), *The nature of schizophrenia* (pp. 25–37). New York: Wiley.

Kety, S., et al. (1994). Mental illness in the biological and adoptive relatives of schizophrenic adoptees: Replication of the Copenhagen study in the rest of Denmark. *Archives of General Psychiatry, 51*, 442–455.

Keyes, D. (1982). *The minds of Billy Milligan*. New York: Bantam Books.

Kiecolt-Glaser, J. K., & Glaser, R. (1992). Psychoneuroimmunology: Can psychological interventions modulate immunity? *Journal of Consulting and Clinical Psychology, 60*, 569–575.

Kiecolt-Glaser, J. K., Speicher, C. E., Holliday, J. E., & Glaser, R. (1984). Stress and the transformation of lymphocytes in Epstein-Barr virus. *Journal of Behavioral Medicine, 7*, 1–12.

Kiecolt-Glaser, J. K., et al. (1987b). Marital quality, marital disruption, and immune function. *Psychosomatic Medicine, 49*, 13–34.

Kiecolt-Glaser, J. K., et al. (1988). Marital discord and immunity in males. *Psychosomatic Medicine, 50*, 213–229.

Kiecolt-Glaser, J., et al. (1995). Slowing of wound healing by psychological stress. *Lancet, 346*, 1194–1196.

Kiesler, C. A., & Sibulkin, A. E. (1987). *Mental hospitalization: Myths and facts about a national crisis*. Newbury Park, CA: Sage.

Kiesler, D. J. (1999). *Beyond the disease model of mental disorders*. Westport, CT: Praeger Publishers.

Kilpatrick, D. G., et al. (2000). Risk factors for adolescent substance abuse and dependence: Data from a national sample. *Journal of Consulting and Clinical Psychology, 68*, 19–30.

Kilts, C. D., et al. (2001). Neural activity related to drug craving in cocaine addiction. *Archives of General Psychiatry, 58*, 334–341.

Kim, J. J., et al. (2000). Regional neural dysfunctions in chronic schizophrenia studied with positron emission tomography. *American Journal of* Psychiatry, *157*, 549–559.

Kim, S. C., & Seo, K. K. (1998). Efficacy and safety of fluoxetine, sertraline, and clomipramine in patients with prompter ejaculation: A double-blind, placebo controlled study. *Journal of Urology, 159*, 425–427.

Kimerling, R., & Calhoun, K. S. (1994). Somatic symptoms, social support, and treatment seeking among sexual assault victims. *Journal of Consulting and Clinical Psychology, 62*, 333–340.

King, S., & Dixon, M. J. (1995). Expressed emotion, family dynamics, and symptom severity in a predictive model of social adjustment for schizophrenic young adults. *Schizophrenia Research, 14*, 121–132.

King, S., & Dixon, M. J. (1999). Expressed emotion and relapse in young schizophrenia outpatients. *Schizophrenia Bulletin, 25*, 377–386.

Kinney, D. K., et al. (1997). Thought disorder in schizophrenic and control adoptees and their relatives. *Archives of General Psychiatry, 54*, 475–479.

Kirigin, K. A., & Wolf, M. M. (1998). Application of the teaching-family model to children and adolescents with conduct disorder. In V. B. Van Hasselt & M. Hersen (Eds.), *Handbook of psychological treatment protocols for children and adolescents. The LEA series in personality and clinical psychology* (pp. 359–380). Mahwah, NJ: Erlbaum.

Kirmayer, L. J., Robbins, J. M., & Paris, J. (1994). Somatoform disorders: Personality and the social matrix of somatic distress. *Journal of Abnormal Psychology, 103*, 125–136.

Kisiel, C. L., & Lyons, J. S. (2001). Dissociation as a mediator of psychopathology among sexually abused children and adolescents. *American Journal of Psychiatry, 158*, 1034–1039.

Kissel, R. C., Whitman, T. L., & Reid, D. H. (1983). An institutional staff training and self-management program for developing multiple self-care skills in severely-profoundly retarded individuals. *Journal of Applied Behavior Analysis, 16*, 395–415.

Klassen, D., & O'Connor, W. A. (1988). Predicting violence in schizophrenic and non-schizophrenic patients: A prospective study. *Journal of Community Psychology, 16*, 217–227.

Klausner, J. D., et al. (2000). Tracing a syphilis outbreak through cyberspace. *The Journal of the American Medical Association, 284*, 447–449.

Klein, D. F. (1994). "Klein's suffocation theory of panic": Reply. *Archives of General Psychiatry, 51*, 506.

Klein, D. N., Lewinsohn, P. M., Seeley, J. R., & Rohde, P. (2001). A family study of major depressive disorder in a community sample of adolescents. *Archives of General Psychiatry, 58*, 13–20.

Klein, D. N., Schwartz, J. E., Rose, S., & Leader, J. B. (2000a). Five-year course and outcome of dysthymic disorder: A prospective, naturalistic follow-up study. *American Journal of Psychiatry, 157*, 931–939.

Klein, D. N., Taylor, E. B., Dickstein, S., & Harding, K. (1988). Primary early-onset dysthymia: Comparison with primary nonbipolar nonchronic major depression on demographic, clinical, familial, personality, and socioenvironmental characteristics and short-term outcome. *Journal of Abnormal Psychology, 97*, 387–398.

Klein, D. N., et al. (2000b). Comparison of *DSM-III-R* chronic major depression and major depression superimposed on dysthymia (double depression): Validity of the distinction. *Journal of Abnormal Psychology, 109*, 419–427.

Klein, M. (1981). On Mahler's autistic and symbiotic phases: An exposition and evaluation. *Psychoanalysis and Contemporary Thought, 4*, 69–105.

Klein, R. G., et al. (1997). Clinical efficacy of methylphenidate in conduct disorder with and without attention deficit hyperactivity disorder. *Archives of General Psychiatry, 4*, 1073–1080.

Kleinman, A. (1987). Anthropology and psychiatry: The role of culture in cross-cultural research on illness. *British Journal of Psychiatry, 151*, 447–454.

Klerman, G. L. (1984). Ideology & science in the individual psychotherapy of schizophrenia. *Schizophrenia Bulletin, 10*, 608–612.

Klerman, G. L., Weissman, M. M., Rounsaville, B. J., & Chevron, E. S. (1984). *Interpersonal psychotherapy of depression*. New York: Basic Books.

Klesges, R. C., et al. (1997). How much weight gain occurs following smoking cessation? A comparison of weight gain using both continuous and point prevalence abstinence. *Journal of Consulting and Clinical Psychology, 65*, 286–291.

Klonoff, E. A., & Landrine, H. (1997). *Preventing misdiagnosis of women: A guide to physical disorders that have psychiatric symptoms*. Thousand Oaks, CA: Sage.

Kluft, R. P. (1986). Three high functioning multiples. *Journal of Nervous and Mental Disease, 174*, 722–726.

Kluft, R. P. (1988). The dissociative disorders. In J. Talbott, R. Hales, & S. Yudofsky (Eds.), *Textbook of psychiatry*. Washington, DC: American Psychiatric Press.

Kluger, J. (2001, June 18). How to manage teen drinking (the smart way). *Time*, pp. 42–44.

Knight, B. G., & Satre, D. D. (1999). Cognitive behavioral psychotherapy with older adults. *Clinical Psychology: Science and Practice, 6*, 188–203.

Knight, G. P., Fabes, R. A., & Higgins, D A. (1996). Concerns about drawing causal inferences from meta-analyses: An example in the study of gender differences in aggression. *Psychological Bulletin, 119*, 410–421.

Knight, R. G., Godfrey, H. P. D., & Shelton, E. J. (1988). The psychological deficits associated with Parkinson's disease. *Clinical Psychology Review, 8*, 391–410.

Knoll, J. L., IV., et al. (1998). Heterogeneity of the psychoses: Is there a neurodegenerative psychosis? *Schizophrenia Bulletin, 24*, 365–379.

Knudsen, D. D. (1991). Child sexual coercion. In E. Grauerholz & M. A. Koralewski (Eds.), *Sexual coercion: A sourcebook on its nature, causes, and prevention* (pp. 17–28). Lexington, MA: Lexington Books.

Kobak, K. A., et al. (1996). Computer-administered clinical rating scales: A review. *Psychopharmacology, 127*, 291–301.

Kobak, K. A., et al. (1997). A computer-administered telephone interview to identify mental disorders. *Journal of the American Medical Association, 278*, 905–910.

Kobasa, S. C. (1979). Stressful life events, personality, and health: An inquiry into hardiness. *Journal of Personality and Social Psychology, 37*, 1–11.

Kobasa, S. C., Maddi, S. R., & Kahn, S. (1982). Hardiness and health: A prospective study. *Journal of Personality and Social Psychology, 42*, 168–177.

Kockott, G., & Fahrner, E. (1988). Male-to-female and female-to-male transsexuals: A comparison. *Archives of Sexual Behavior, 17*, 539–545.

Kocsis, J. H., et al. (1996). Maintenance therapy for chronic depression: A controlled clinical trial of desipramine. *Archives of General Psychiatry, 53*, 769–774.

Kogan, A. O., & Guilford, P. M. (1998). Side effects of short-term 10,000-lux light therapy. *American Journal of Psychiatry, 155*, 293–294.

Kogon, M. M., et al. (1997). Effects of medical and psychotherapeutic treatment on the survival of women with metastatic breast carcinoma. *Cancer, 80*, 225–230.

Kohut, H. (1966). Forms and transformations of narcissism. *Journal of the American Psychoanalytic Association, 14*, 243–272.

Kolata, G. (1994a, February 3). Sweeteners-hyperactivity link is discounted. *The New York Times*, p. A19.

Kolata, G. (1994b, November 11). A simpler test for Alzheimer's is reported. *The New York Times*, p. A20.

Kolata, G. (1995b, February 9). Landmark in Alzheimer research: Breeding mice with the disease. *The New York Times*, p. A20.

Kolata, G. (1995c, March 9). Metabolism found to adjust for a body's natural weight. *The New York Times*, pp. A1, A22.

Kolata, G. (1998, March 28). U.S. approves sale of impotence pill; huge market seen. *The New York Times*, pp. A1, A8.

Kolata, G. (2000a, October 18). Days off are not allowed, experts argue. *The New York Times*, pp. A1, A20.

Kolata, G. (2000b, October 17). How the body knows when to gain or lose. *The New York Times*, pp. F1, F8.

Kolata, G. (2001a, March 8). Parkinson's research is set back by failure of fetal cell implants. *The New York Times*, pp. A1, A16.

Kolata, G. (2001b, November 22). Hints of an Alzheimer's aid in anti-inflammatory drugs. *The New York Times*, p. A24.

Kolbert, E. (1994, January 21). Demons replace dolls and bicycles in world of children of the quake. *The New York Times*, p. A19.

Kolko, D. J., & Rickard-Figueroa, J. L. (1985). Effects of video games on the adverse corollaries of chemotherapy in pediatric oncology patients: A single-case analysis. *Journal of Consulting and Clinical Psychology, 53*, 223–228.

Koorland, M. A. (1986). Applied behavior analysis and the correction of learning disabilities. In J. K. Torgesen & B. Y. L. Wong (Eds.), *Psychological and educational perspectives on learning disabilities* (pp. 297–328). Orlando, FL: Academic Press.

Kordower, J. H., et al. (2000). Neurodegeneration prevented by lentiviral vector delivery of GDNF in primate models of Parkinson's disease. *Science, 290*, 767–773.

Koren, D., Arnon, I., & Klein, E. (1999). Acute stress response and posttraumatic stress disorder in traffic accident victims: A one-year prospective, follow-up study. *American Journal of Psychiatry, 156*, 367–373.

Korotitsch, W. J., &. Nelson-Gray, R. O. (1999). An overview of self-monitoring research in assessment and treatment. *Psychological Assessment, 11*, 415–425.

Koss, M. P. (1988). Stranger and acquaintance rape: Are there differences in the victim's experience? *Psychology of Women Quarterly, 12*, 1–24.

Koss, M. P., Gidycz, C. A., & Wisniewski, N. (1987). The scope of rape: Incidence and prevalence of sexual aggression and victimization in a national sample of higher education students. *Journal of Consulting and Clinical Psychology, 55*, 162–170.

Koss, M. P., et al. (1994). *No safe haven: Male violence against women at home, at work, and in the community*. Washington, DC: American Psychological Association.

Kotler, M. (1997). Excess dopamine D4 receptor (D4DR) exon III seven repeat allele in opioid-dependent subjects. *Molecular Psychiatry, 2*, 251–254.

Kotler, M., et al. (2001). Anger, impulsivity, social support, and suicide risk in patients with posttraumatic stress disorder. *Journal of Nervous & Mental Disease, 189*, 162–167.

Kovacs, M. (1996). Presentation and course of major depressive disorder during childhood and later years of the life span. *Journal of the American Academy of Children and Adolescent Psychiatry, 35*, 705–715.

Kovacs, M., et al. (1997). A controlled family history study of childhood-onset depressive disorder. *Archives of General Psychiatry, 54*, 613–623.

Kraepelin, E. (1909–1913). *Psychiatrie* (8th ed.). Leipzig: J. A. Barth.

Kramer, M. S., et al. (1998, September 11). Distinct mechanism for antidepressant activity by blockade of central substance P receptors. *Science*, pp. 1640–1645.

Krantz, D. S., Contrada, R. J., Hills, D. R., & Friedler, E. (1988). Environmental stress and biobehavioral antecedents of coronary heart disease. *Journal of Consulting and Clinical Psychology, 56*, 333–341.

Kranzler, H. R. (2000). Medications for alcohol dependence: New vistas. *Journal of the American Medical Association, 284*, 1016–1017. [Editorial]

Kranzler, H. R., et al. (1996). Comorbid psychiatric diagnosis predicts three year outcomes in alcoholics: A posttreatment natural history study. *Journal of Studies in Alcohol, 57*, 619–626.

Krehbiel, K. (2000, October). Diagnosis and treatment of bipolar disorder. *Monitor on Psychology*, p. 22.

Kresin, D. (1993) Medical aspects of inhibited sexual desire disorder. In W. O'Donohue & J. H. Geer (Eds.), *Handbook of sexual dysfunctions: Assessment and treatment* (pp. 15–52). Boston: Allyn & Bacon.

Kring, A. M., & Neale, J. M. (1996). Do schizophrenic patients show a disjunctive relationship among expressive, experiential, and psychophysiological components of emotion? *Journal of Abnormal Psychology, 105*, 249–257.

Kristof, N. D. (1995, May 14). Japanese say no to crime: Tough methods, at a price. *The New York Times*, pp. A1, A8.

Krug, E., et al. (1998). Suicide after natural disasters. *The New England Journal of Medicine, 338*, 373–378.

Kryger, M. H., Roth, T., & Dement, W. C. (Eds.). (2000). *Principles and practice of sleep medicine* (3rd ed.). Philadelphia: W. B. Saunders.

Kubisyzn, T. W., et al. (2000). Empirical support for psychological assessment in clinical health care settings. *Professional Psychology: Research and Practice, 31*, 119–130.

Kuhn, C. M., & Wilson, W. A. (2001, Spring). Our dangerous love affair with ecstasy. *Cerebrum*, pp. 22–33.

Kuiper, N. A., & Martin, R. A. (1993). Humor and self-concept. *Humor International Journal of Humor Research, 6*, 251–270.

Kupfer, D. J. (1999). Research in affective disorders comes of age. *American Journal of Psychiatry, 156*, 165–167. [Editorial]

Kupfer, D. J, & Reynolds, C. F. (1997). Current concepts: Management of insomnia. *The New England Journal of Medicine, 336*, 341–346.

Kupfersmid, J. (1995). Does the Oedipus complex exist? *Psychotherapy, 32*, 535–547.

Kutchins, H., & Kirk, S. A. (1995, May). *DSM-IV*: Does bigger and newer mean better? *The Harvard Mental Health Letter, 11*(11), 4–6.

Kwon, H., et al. (2001). Functional neuroanatomy of visuospatial working memory in Fragile X syndrome: Relation to behavioral and molecular measures. *American Journal of Psychiatry, 158*, 1040–1051.

Kwon, S., & Oei, T. P. S. (1994). The roles of two levels of cognitions in the development, maintenance, and treatment of depression. *Clinical Psychology Review, 14*, 331–358.

L

Ladouceur, R., et al. (2000). Efficacy of a cognitive–behavioral treatment for generalized anxiety disorder: Evaluation in a controlled clinical trial. *Journal of Consulting and Clinical Psychology, 68*, 957–964.

Lahey, B. B., et al. (1995). Four-year longitudinal study of conduct disorder in boys: Patterns and predictors of persistence. *Journal of Abnormal Psychology, 104*, 83–93.

Lam, D. H., et al. (2000). Cognitive therapy for bipolar illness: A pilot study of relapse prevention. *Cognitive Therapy & Research, 24*, 503–520.

Lamb, H. R., & Lamb, D. M. (1990). Factors contributing to homelessness among the chronically and severely mentally ill. *Hospital and Community Psychiatry, 41*, 301–305.

Lamberg, L. (1998). Mental illness and violent acts: Protecting the patient and the public. *Journal of the American Medical Association, 280*, 407–408.

Lamberg, L. (2000). Sleep disorders, often unrecognized, complicate many physical illnesses. *Journal of the American Medical Association, 284*, 2173–2175.

Lambert, E. W., et al. (2001). Looking for the disorder in conduct disorder. *Journal of Abnormal Psychology*, 110, 110–123.

Lambert, G., et al. (2000). Reduced brain norepinephrine and dopamine release in treatment-refractory depressive illness: Evidence in support of the catecholamine hypothesis of mood disorders. *Archives of General Psychiatry, 57*, 787–793.

Lambert, M. C., et al. (1992). Jamaican and American adult perspectives on child psychopathology: Further exploration of the threshold model. *Journal of Consulting and Clinical Psychology, 60*, 146–149.

Lambert, M. J., & Bergin, A. E. (1994). The effectiveness of psychotherapy. In A. E. Bergin & S. L. Garfield (Eds.), *Handbook of psychotherapy and behavior change* (4th ed., pp. 72–113). New York: Wiley.

Lambert, M. J., & Okiishi, J. C. (1997). The effects of the individual psychotherapist and implications for future research. *Clinical Psychology: Science and Practice, 4*, 66–75.

Lambert, N. M., Hartsough, C. S., Sassone, D., & Sandoval, J. (1987). Persistence of hyperactivity symptoms from childhood to adolescence and associated outcomes. *American Journal of Orthopsychiatry, 57*, 22–32.

Landerman, L. R., et al. (1994). The relationship between insurance coverage and psychiatric disorder in predicting use of mental health services. *American Journal of Psychiatry, 151*, 1785–1790.

Landolt, H. P., et al. (1996). Late-afternoon ethanol intake affects nocturnal sleep and the sleep EEG in middle-aged men. *Journal of Clinical Psychopharmacology, 16*, 428–436.

Lane, E. (1994, December 6). Losing weight isn't enough. *New York Newsday*, p. A6.

Lang, P. J. (1968). Fear reduction and fear behavior: Problems in treating a construct. In J. M. Schlein (Ed.), *Research in psychotherapy, Vol. III* (pp. 90–102). Washington, DC: American Psychological Association.

Lang, P.J., & Lazovik, A. D. (1963). Experimental desensitization of phobia. *Journal of Abnormal and Social Psychology, 66*, 519–525.

Langenbucher, J. W., & Chung, T. (1995). Onset and staging of *DSM-IV* alcohol dependence using mean age and survival-hazard methods. *Journal of Abnormal Psychology, 104*, 346–354.

Langenbucher, J., et al. (2000). Toward the *DSM–V*: The withdrawal-gate model versus the *DSM-IV* in the diagnosis of alcohol abuse and dependence. *Journal of Consulting and Clinical Psychology, 68*, 799–809.

Lara, M. E., Leader, J., & Klein, D. N. (1997). The association between social support and course of depression: Is it confounded with personality? *Journal of Abnormal Psychology, 106*, 478–482.

Larimer, M. E., Marlatt, G. A., Baer, J. S., Quigley, L. A., Blume, A. W., & Hawkins, E. H. (1998). Harm reduction for alcohol problems: Expanding access to and acceptability of prevention and treatment services. In G. Marlatt (Ed.), *Harm reduction: Pragmatic strategies for managing high-risk behaviors* (pp. 69–121). New York: Guilford Press.

Larson, R. W., Raffaelli, M., Richards, M. H., Ham, M., & Jewell, L. (1990). Ecology of depression in late childhood and early adolescence: A profile of daily states and activities. *Journal of Abnormal Psychology, 99*, 92–102.

Last draw for smokers. (1996, October). *UC Berkeley Wellness Letter*, 13, 2–3.

Lauerman, C. (2000, November 7). Psychological counseling is now just a computer click away. Retrieved November 21, 2000, from http://www.psycport.com/news/2000/11/07/Knigt/3822-0076-MEDE-THERAPY.TB.html.

Laughter may be best medicine for heart disease. (2000, November 15) *CNN Web Posting*. Retrieved November 18, 2000, from http://www.cnn.com/2000/HEALTH/11/15/heart.laughter.reut/index.html.

Laumann, E. O., Gagnon, J. H., Michael, R. T., & Michaels, S. (1994). *The social organization of sexuality: Sexual practices in the United States*. Chicago: University of Chicago Press.

Laumann, E. O., Paik, A., & Rosen, R. C. (1999). Sexual dysfunction in the United States: Prevalence and predictors. *Journal of the American Medical Association, 281*, 537–544.

Lawson, D. M. (1983). Alcoholism. In M. Hersen (Ed.), *Outpatient behavior therapy: A clinical guide* (pp. 143–172). New York: Grune & Stratton.

Lawson, W. B. (1986). Racial and ethnic factors in psychiatric research. *Hospital and Community Psychiatry, 37*, 50–54.

Lawson, W. B. (1996). Clinical issues in the pharmacotherapy of African-Americans. *Psychopharmacology Bulletin, 32*, 275–281.

Lawson, W. B. (1999). Psychiatric diagnosis of African Americans. In J. M. Herrera, W. B. Lawson, & J. J. Sramek (Eds.), *Cross cultural psychiatry* (pp. 99–104). Chichester, England: Wiley.

Lazarus, A. A. (1992). Multimodal therapy: Technical eclecticism with minimal integration. In J. C. Norcross & M. R. Goldfried (Eds.), *Handbook of psychotherapy integration* (pp. 231–263). New York: Basic Books.

Lazarus, R. S., & Folkman, S. (1984). *Stress, appraisal, and coping*. New York: Springer.

Leary, W. E. (1996b, December 18). Responses of alcoholics to therapies seem similar. *The New York Times*, p. A17.

Leary, W. E. (1998, Feburary 12). New therapy offers promise in treatment of pedophiles. *The New York Times*, p. A11.

Leavitt, F., & Labott, S. M. (1997). Criterion-related validity of Rorschach analogues of dissociation. *Psychological Assessment, 9*, 244–249.

LeDuff, C. (2000, August 21). Cocaine quietly reclaims its hold as good times return. *The New York Times*, pp. B1, B2.

Lee, C. C., & Richardson, B. L. (1991). *Multicultural issues in counseling: New approaches to diversity*. Alexandria, VA: AACD.

Lee, D. T. S., et al. (2001). A psychiatric epidemiological study of postpartum Chinese women. *American Journal of Psychiatry, 158*, 220–226.

Lee, I-M., et al. (2001). Physical activity and coronary heart disease in women: Is "no pain, no gain" passé? *Journal of the American Medical Association, 285*, 1447–1454.

Lee, T. M. C., et al. (1998). Seasonal affective disorder. *Clinical Psychology: Science and Practice, 5*, 275–290.

Leedham, B., et al. (1995). Positive expectations predict health after heart transplantation. *Health Psychology, 14*, 74–79.

Leekam, S. R, & López, B. (2000). Attention and joint attention in preschool children with autism. *Developmental Psychology, 36*, 261–273.

Lefcourt, H. M., & Martin, R. A. (1986). *Humor and life stress: Antidote to adversity*. New York: Springer-Verlag.

Leff, J., & Vaughn, C. (1981). The role of maintenance therapy and relatives' expressed emotion in relapse of schizophrenia: A two-year follow-up. *British Journal of Psychiatry, 139*, 102–104.

Lefley, H. P. (1990). Culture and chronic mental illness. *Hospital and Community Psychiatry, 41*, 277–286.

Lehrer, M., et al. (1994). Relaxation and music therapies for asthma among patients prestabilized on asthma medication. *Journal of Behavioral Medicine, 17*, 1–24.

Lehrer, P. M., Sargunaraj, D., & Hochron, S. (1992). Psychological approaches to the treatment of asthma. *Journal of Consulting and Clinical Psychology, 60*, 639–643.

Leibel, R. L., Rosenbaum, M., & Hirsch, J. (1995). Changes in energy expenditure resulting from altered body weight. *New England Journal of Medicine, 332*, 621–628.

Leibenluft, E. (1996). Women with bipolar illness: Clinical and research issues. *American Journal of Psychiatry, 153*, 163–173.

Leibson, C. L., et al. (2001). Use and costs of medical care for children and adolescents with and without attention-deficit/hyperactivity disorder. *Journal of the American Medical Association, 285*, 60–66.

Leichsenring, F. (2001). Comparative effects of short-term psychodynamic psychotherapy and cognitive-behavioral therapy in depression: A meta-analytic approach. *Clinical Psychology Review, 21*, 401–419.

Lemonick, M. D., & Park, A. (2001, May 14). Alzheimer's: The Nun study. *Time*, pp. 54–64.

Leocani, L., et al. (2001). Abnormal pattern of cortical activation associated with voluntary movement in obsessive-compulsive disorder: An EEG study. *American Journal of Psychiatry, 158*, 140–142.

Leon, G. R., et al. (1995). Prospective analysis of personality and behavioral vulnerabilities and gender influences in the later development of disordered eating. *Journal of Abnormal Psychology, 104*, 140–149.

Lerman, C., et al. (1999). Evidence suggesting the role of specific genetic factors in cigarette smoking. *Health Psychology, 18*, 14–20.

Lesch, K. P., et al. (1996). Association of anxiety-related traits with a polymorphism in the serotonin transporter gene regulatory region. *Science, 274*, 1527–1531.

Leserman, J., et al. (2000). Impact of stressful life events, depression, social support, coping, and cortisol on progression to AIDS. *American Journal of Psychiatry, 157,* 1221–1228.

Leshner, A. I. (1999). Science is revolutionizing our view of addiction and what to do about it *American Journal of Psychiatry, 156,* 1–3.

Lesser, I. (1992, December). *Ethnic differences in response to psychotropic drugs.* Paper presented at a symposium, Anxiety Disorders in African Americans, presented by the State University of New York Health Science Center at Brooklyn, Brooklyn, NY.

Letourneau, E., & O'Donohue, W. (1993). Sexual desire disorders. In W. O'Donohue & J. H. Geer (Eds.), *Handbook of sexual dysfunctions: Assessment and treatment* (pp. 53–81). Boston: Allyn & Bacon.

Levenson, J. L., & Bemis, C (1991). The role of psychological factors in cancer onset and progression. *Psychosomatics, 32,* 124–132.

Levenstein, S., et al. (1999). Stress and peptic ulcer disease. *Journal of the American Medical Association, 281,* 10–11.

Levine, A. (2000, December 22). Tomorrow's education, made to measure. *The New York Times,* p. A 33.

Levitan, R. D., Rector, N. A., & Bagby, R. M. (1998). Negative attributional style in seasonal and nonseasonal depression. *American Journal of Psychiatry, 155,* 428–430.

Levitan, R. D., et al. (1997). Hormonal and subjective responses to intravenous metachlorophenylpiperazine in bulimia nervosa. *Archives of General Psychiatry, 54,* 521–527.

Levy, S. R., Jurkovic, G. L., & Spirito, A. (1995). A multisystems analysis of adolescent suicide attempters. *Journal of Abnormal Child Psychology, 23,* 221–234.

Lewin, T. (2001, April 15). Ask not for whom the clock ticks. *The New York Times Week in Review,* p. 4.

Lewinsohn, P. M. (1974). A behavioral approach to depression. In R. J. Friedman & M. M. Katz (Eds.), *The psychology of depression: Contemporary theory and research.* Washington, DC: Winston-Wiley.

Lewinsohn, P. M., & Clarke, G. N. (1999). Psychosocial treatments for adolescent depression. *Clinical Psychology Review, 19,* 329–342.

Lewinsohn, P. M., Clarke, G. N., Rhode, P., Hops, H., & Seely, J. (1996). A course in coping: A cognitive-behavioral approach to the treatment of adolescent depression. In D. Hibbs & P. S. Jensen (Eds.), *Psychosocial treatments for child and adolescent disorders: Empirically based strategies for clinical practice* (pp. 109–135). Washington, DC: American Psychological Association.

Lewinsohn, P. M., Duncan, E. M., Stanton, A. K., & Hautzinger, M. (1986). Age at first onset for nonpolar depression. *Journal of Abnormal Psychology, 95,* 378–383.

Lewinsohn, P. M., Joiner, T. E., & Rohde, P. (2001). Evaluation of cognitive diathesis-stress models in predicting major depressive disorder in adolescents. *Journal of Abnormal Psychology, 110,* 203–215.

Lewinsohn, P. M., & Libet, J. M. (1972). Pleasant events, activity schedules and depression. *Journal of Abnormal Psychology, 79,* 291–295.

Lewinsohn, P. M., Rohde, P., & Seeley, J. R. (1994). Psychosocial risk factors for future adolescent suicide attempts. *Journal of Consulting and Clinical Psychology, 62,* 297–305.

Lewinsohn, P. M., Rohde, P., & Seeley, J. R. (1996). Adolescent suicidal ideation and attempts: Prevalence, risk factors, and clinical implications. *Clinical Psychology: Science and Practice, 3,* 25–46.

Lewinsohn, P. M., Teri, L., & Wasserman, D. (1983). Depression. In M. Hersen (Ed.), *Outpatient behavior therapy: A practical guide* (pp. 81–108). New York: Grune & Stratton.

Lewinsohn, P. M., et al. (1994). Adolescent psychopathology: II. Psychosocial risk factors for depression. *Journal of Abnormal Psychology, 103,* 302–315.

Lewinsohn, P. M., et al. (1996). A course in coping: A cognitive-behavioral approach to the treatment of adolescent depression. In D. Hibbs & P. S. Jensen (Eds.), *Psychosocial treatments for child and adolescent disorders: Empirically based strategies for clinical practice* (pp. 109–135). Washington, DC: American Psychological Association.

Lewinsohn, P. M., et al. (2001). Gender differences in suicide attempts from adolescence to young adulthood. *Journal of American Academy of Child & Adolescent Psychiatry, 40,* 427–434.

Lewis, D. O., et al. (1997). Objective documentation of child abuse and dissociation in 12 murderers with dissociative identity disorder. *American Journal of Psychiatry, 154,* 1703–1710.

Lewis, R. J., Dlugokinski, E. L., Caputo, L. M., & Griffin, R. B. (1988). Children at risk for emotional disorders: Risk and resource dimensions. *Clinical Psychology Review, 8,* 417–440.

Lewis-Hall, F. (1992, December). *Overview of DSM-III-R: Focus on panic disorder and obsessive-compulsive disorder.* Paper presented at a symposium, Anxiety Disorders in African Americans, presented by the State University of New York Health Science Center at Brooklyn, Brooklyn, NY.

Lex, B. W. (1987). Review of alcohol problems in ethnic minority groups. *Journal of Consulting and Clinical Psychology, 55,* 293–300.

Ley, R. (1997). The Ondine curse, false suffocation alarms, trait-state suffocation fear, and dyspnea-suffocation fear in panic attacks. *Archives of General Psychiatry, 54,* 677.

Li, Y. M., et al. (2000). Photoactivated-secretase inhibitors directed to the active site covalently label presenilin, 1. *Nature, 405,* 689–693.

Liberman, R. P. (1994). Treatment and rehabilitation of the seriously mentally ill in China: Impressions of a society in transition. *American Journal of Orthopsychiatry, 64,* 68–77.

Lichstein, K. L., Wilson, N. M., & Johnson, C. T. (2000). Psychological treatment of secondary insomnia. *Psychology and Aging, 15,* 232–240.

Lichstein, K. L., et al. (2001). Primary versus secondary insomnia in older adults: Subjective sleep and daytime functioning.*Psychology and Aging, 16,* 264–271.

Lichtenberg, P. A., & Duffy, M. (2000). Psychological assessment and psychotherapy in long-term care. *Clinical Psychology: Science and Practice,* 317–328.

Lichtenstein, E., & Glasgow, R. E. (1992). Smoking cessation: What have we learned over the past decade? *Journal of Consulting and Clinical Psychology, 60,* 518–527.

Lieber, C. S. (1990, January 14). Cited in "Barroom biology: How alcohol goes to a woman's head." *The New York Times,* p. E24.

Liebowitz, M. R., et al. (2000). Social phobia or social anxiety disorder: What's in a name? *Archives of General Psychiatry,* 57, 191–192.

Lightsey, O. W., Jr. (1994a). Positive automatic cognitions as moderators of the negative life event-dysphoria relationship. *Cognitive Therapy and Research, 18,* 353–365.

Lightsey, O. W., Jr. (1994b). "Thinking positive" as a stress buffer: The role of positive automatic cognitions in depression and happiness. *Journal of Counseling Psychology, 41,* 325–334.

Lilienfeld, S. O. (1997). The relation of anxiety sensitivity to higher and lower order personality dimensions: Implications for the etiology of panic attacks. *Journal of Abnormal Psychology, 106,* 539–544.

Lilienfeld, S. O., & Andrews, B. P. (1996). Identifying noncriminal psychopaths. *Journal of Personality Assessment, 66,* 488–524.

Lilienfeld, S. O., & Marino, L. (1995). Mental disorder as a Roschian concept: A critique of Wakefield's "harmful dysfunction" analysis. *Journal of Abnormal Psychology, 104,* 411–420.

Lilienfeld, S. O., Wood, J. M, & Garb, H. N. (2000). The scientific status of projective techniques. *Psychological Science in the Public Interest, 1,* 27–66.

Lin, K., et al. (1991). Ethnicity and family involvement in the treatment of schizophrenic patients. *Journal of Nervous & Mental Disease, 179,* 631–633.

Lin, T. Y., et al. (1978). Ethnicity and patterns of help-seeking. *Culture, Medicine, and Psychiatry, 2,* 3–14.

Lindsey, K. P., & Paul, G. L. (1989). Involuntary commitments to public mental institutions: Issues involving the overrepresentation of blacks and assessment of relevant functioning. *Psychological Bulletin, 106,* 171–183.

Linehan, M. M. (1993). *Cognitive-behavioral treatment of borderline personality disorder.* New York: Guilford Press.

Linehan, M. M., Camper, P., Chiles, J. A., Strosahl, K., & Shearin, E. (1987). Interpersonal problem solving and parasuicide. *Cognitive Therapy and Research, 11,* 1–12.

Linehan, M., et al. (1991). Cognitive-behavioral treatment of chronically parasuicidal borderline patients. *Archives of General Psychiatry, 48,* 1060–1064.

Linehan, M. M., et al. (1994). Interpersonal outcome of cognitive behavioral treatment for chronically suicidal borderline patients. *American Journal of Psychiatry, 151,* 1771–1776.

Link, B. G., & Stueve, A. (1998). New evidence on the violence risk posed by people with mental illness. *Archives of General Psychiatry, 55,* 403–404.

Liotti, G., et al. (2000). Predictive factors for borderline personality disorder. *Acta Psychiatrica Scandinavica, 102,* 282–289.

Lipman, E. L., MacMillan, H. L., & Boyle, M. H. (2001). Childhood abuse and psychiatric disorders among single and married mothers. *American Journal of Psychiatry, 158,* 73–77.

Lipsey, M. W., & Wilson, D. B. (1993). The efficacy of psychology, educational, and behavioral treatment: Confirmation from meta-analysis. *American Psychologist, 48,* 1181–1209.

Lipsey, M. W., & Wilson, D. B. (1995). Reply to comments on Lispey and Wilson (1993). *American Psychologist, 50,* 113–115.

Lipton, R. B., et al. (1998). Efficacy and safety of acetaminophen, aspirin, and caffeine in alleviating migraine headache pain. *Archives of Neurology, 55,* 210–217.

Lipton R. B, et al. (2000a). Stratified care vs. step care strategies for migraine: The Disability in Strategies of Care (DISC) Study: A randomized trial. *Journal of the American Medical Association, 284,* 2599–2605.

Lipton, R. B., et al. (2000b). Migraine, quality of life, and depression. *Neurology, 55,* 629–635.

Lira, L. R., Koss, M. P., & Russo, N. F. (1999). Mexican American women's definitions of rape and sexual abuse. *Hispanic Journal of Behavioral Sciences, 21*(3), 236–265.

Lisanby, S. H., et al. (2000). The effects of electroconvulsive therapy on memory of autobiographical and public events. *Archives of General Psychiatry, 57,* 581–590.

Litz, B. T. (1992). Emotional numbing in combat-related post-traumatic stress disorder: A critical review and reformulation. *Clinical Psychology Review, 12,* 417–432.

Livesley, W. J. (1985). The classification of personality disorder, II: The problem of criteria. *Canadian Journal of Psychiatry, 30,* 359–362.

Livesley, W. J., et al. (1993). Genetic and environmental contributions to dimensions of personality disorder. *American Journal of Psychiatry, 150,* 1826–1831.

Livesley, W. J., et al. (1994). Categorical distinctions in the study of personality disorder: Implications for classification. *Journal of Abnormal Psychology, 103,* 6–17.

Livingstone, M., et al. (1991). Physiological and anatomical evidence for a magnocellular defect in developmental dyslexia. *Proceedings of the National Academy of Sciences, 88,* 7943–7947.

Lobel, M., et al. (2000). The impact of prenatal maternal stress and optimistic disposition on birth outcomes in medically high-risk women. *Health Psychology, 19,* 544–553.

Lochman, J. E. (1992). Cognitive-behavioral intervention with aggressive boys: Three-year follow-up and preventive effects. *Journal of Consulting and Clinical Psychology, 60,* 426–432

Lochman, J. E., & Dodge, K. A. (1994). Social-cognitive processes of severely violent, moderately aggressive, and nonaggressive boys. *Journal of Consulting and Clinical Psychology, 62,* 366–374.

Lochman, J. E., & Lenhart, L. (1993). Anger coping intervention for aggressive children: Conceptual models and outcome effects. *Clinical Psychology Review, 13,* 785–805.

Loeber, R., Lahey, B. B., & Thomas, C. (1991). Diagnostic conundrum of oppositional defiant disorder and its comorbid conditions: Effects of age and gender. *Journal of Consulting and Clinical Psychology, 59,* 379–390.

Loewenstein, R. J. (1991). Psychogenic amnesia and psychogenic fugue: A comprehensive review. *Annual Review of Psychiatry, 10,* 223–247.

Loftus, E. F. (1993). The reality of repressed memories. *American Psychologist, 48,* 518–537.

Loftus, E. F. (1996). The myth of repressed memory and the realities of science. *Clinical Psychology: Science and Practice, 3,* 356–365.

Loftus, E. F. (1997). Creating childhood memories. *Applied Cognitive Psychology, 11,* S75–S86.

Lohman, J. J. H. M. (2001). Treatment strategies for migraine headache. *Journal of the American Medical Association, 285,* 1014.

Lohr, B. A., Adams, H. E., & Davis, J. M. (1997). Sexual arousal to erotic and aggressive stimuli in sexually coercive and noncoercive men. *Journal of Consulting and Clinical Psychology, 106,* 230–242.

LoPiccolo, J. (1990). Sexual dysfunction. In A. S. Bellack, M. Hersen, & A. E. Kazdin (Eds.), *International handbook of behavior modification therapy* (2nd ed., pp. 557–564). New York: Plenum Press.

LoPiccolo, J., & Stock, W. E. (1986). Treatment of sexual dysfunction. *Journal of Consulting and Clinical Psychology, 54,* 158–167.

Loranger, A. W. (1996). Dependant personality disorder: Age, sex, and Axis I comorbidity. *Journal of Nervous and Mental Disease, 184,* 17–21.

Loranger, A. W., et al. (1994). The international personality disorder examination: The World Health Organization/Alcohol, Drug, Abuse and Mental Health Administration International Pilot Study of Personality Disorders. *Archives of General Psychiatry, 51,* 215–224.

Lorefice, L. S. (1991). Fluoxetine treatment of a fetish. *Journal of Clinical Psychiatry, 52,* 41.

Lovaas, O. I. (1977). *The autistic child: Language development through behavior modification.* New York: Halstead Press.

Lovaas, O. I. (1987). Behavioral treatment and normal educational and intellectual functioning in young autistic children. *Journal of Consulting and Clinical Psychology, 55,* 3–9.

Lovaas, O. I., Koegel, R. L., & Schreibman, L. (1979). Stimulus overselectivity in autism: A review of the research. *Psychological Bulletin, 86,* 1236–1254.

Lowe, M. R., Gleaves, D. H., Murphy-Eberenz, K. P. (1998). On the relation of dieting and bingeing in bulimia nervosa. *Journal of Abnormal Psychology, 107,* 263–271.

Lubin, B., Larsen, R. M., Matarazzo, J. D., & Seever, M. (1985). Psychological test usage patterns in five professional settings. *American Psychologist, 40,* 857–861.

Luborsky, I., et al. (1996). Factors in outcomes of short-term dynamic psychotherapy for chronic vs. non-chronic major depression. *Journal of Psychotherapy: Practice and Research, 5,* 152–159.

Luborsky, L., et al. (1988). *Who will benefit from psychotherapy? Predicting therapeutic outcomes.* New York: Basic Books.

Luntz, B. K., & Widom, C. S. (1994). Antisocial personality disorder in abused and neglected children grown up. *American Journal of Psychiatry, 151,* 670–674.

Lurigio, A. J., & Lewis, D. A. (1989). Worlds that fail: A longitudinal study of urban mental patients. *Journal of Social Issues, 45,* 79–90.

Lutgendorf, S. K., et al. (1997). Cognitive-behavioral stress management decreases dysphoric mood and herpes simplex virus-type 2 antibody titer in symptomatic HIV-seropositive gay men. *Journal of Consulting and Clinical Psychology, 65,* 31–43.

Lyketsos C. G., et al. (2000). Randomized, placebo-controlled, double-blind clinical trial of sertraline in the treatment of depression complicating Alzheimer's disease: Initial results from the Depression in Alzheimer's Disease Study. *American Journal of Psychiatry, 157,* 1686–1689.

Lykken, D. T. (1957). A study of anxiety in the sociopathic personality. *Journal of Abnormal and Social Psychology, 55,* 6–10.

Lykken, D. T. (1993). Predicting violence in the violent society. *Applied and Preventive Psychology, 2,* 13–20.

Lymburner, J. A., & Roech, R. (1999). The insanity defense: Five years of research (1993–1997). *International Journal of Law and Psychiatry, 22,* 213–240.

Lyon, F. R., & Moats, L. C. (1988). Critical issues in the instruction of the learning disabled. *Journal of Consulting and Clinical Psychology, 56,* 830–835.

M

Machan, D. (2000, December). Forget the champagne. *Forbes,* pp. 118–120.

MacMillan, H. L., et al. (1997). Prevalence of child physical and sexual abuse in the community: Results from the Ontario health supplement. *Journal of the American Medical Association, 278,* 131–135.

MacPhillamy, D. J., & Lewinsohn, P. M. (1974). Depression as a function of levels of desired and obtained pleasure. *Journal of Abnormal Psychology, 83,* 651–657.

Maddi, S. R., & Kobasa, S. C. (1984). *The hardy executive: Health under stress.* Homewood, IL: Dow Jones-Irwin.

Maeder, T. (1985). *Crime and madness: The origins and evolution of the insanity defense.* New York: Harper & Row.

Magdol, L., et al. (1997). Gender differences in partner violence in a birth cohort of 21-year olds: Bridging the gap between clinical and epidemiological approaches. *Journal of Consulting and Clinical Psychology, 65,* 68–78.

Magdol, L., et al. (1998). Developmental antecedents of partner abuse: A prospective longitudinal study. *Journal of Abnormal Psychology, 107,* 375–389.

Maher, W. B., & Maher, B. A. (1985). Psychopathology: I. From ancient times to the eighteenth century. In G. A. Kimble & K. Schlesinger (Eds.), *Topics in the history of psychology* (Vol. 2). Hillsdale, NJ: Erlbaum.

Mahler, M., & Kaplan, L. (1977). Developmental aspects in the assessment of narcissistic and so-called borderline personalities. In P. Hartocollis (Ed.), *Borderline personality disorders: The concept, the syndrome, the patient* (pp. 71–85). New York: International Universities Press.

Mahler, M. S., Pine, F., & Bergman, A. (1975). The borderline syndrome: The role of the mother in the genesis and psychic structure of the borderline personality. *International Journal of Psychoanalysis, 56,* 163–177.

Maier, S. F., & Seligman, M. E. P. (1976). Learned helplessness: Theory and evidence. *Journal of Experimental Psychology (General), 105,* 3–46.

Maier, S. F., Watkins, L. R., & Fleshner, M. (1994). Psychoneuroimmunology; The interface between behavior, brain, and immunity. *American Psychologist, 49,* 1004–1017.

Maier, T. (1995, February 21). Drug hailed as a "magic bullet" has skeptics. *Newsday,* p. B23.

Maj, M., et al. (1991). A family study of *DSM–III–R* schizoaffective disorder, depressive type, compared with schizophrenia and psychotic and nonpsychotic major depression. *American Journal of Psychiatry, 148,* 612–616.

Malaspina, D., et al. (2001). Advancing paternal age and the risk of schizophrenia. *Archives of General Psychiatry, 58,* 361–367.

Maldonado, J. R., Butler, L. D., & Spiegel, D. (1998). Treatments for dissociative disorders. In P. E. Nathan & J. M. Gorman (Eds.), *A guide to treatments that work* (pp. 423–446). New York: Oxford University Press.

Maletsky, B. M. (1980). Self-referred vs. court-referred sexually deviant patients: Success with assisted covert sensitization. *Behavior Therapy, 11,* 306–314.

Maletzky, B. M. (1991). *Treating the sexual offender.* Newbury Park, CA: Sage.

Maletzky, B. M. (1998). The paraphilias: Research and treatment. In P. E. Nathan & J. M. Gorman (Eds.), *A guide to treatments that work* (pp. 472–500). New York: Oxford University Press.

Malmo, R. B., & Shagass, C. (1949). Physiological study of symptom mechanism in psychiatric patients under stress. *Psychosomatic Medicine, 11,* 25–29.

Malone, K. M., et al. (2000). Protective factors against suicidal acts in major depression: Reasons for living. *American Journal of Psychiatry, 157,* 1084–1088.

Mandal, M. K., Pandey, R., & Prasad, A. B. (1998). Facial expression of emotions and schizophrenia: A review. *Schizophrenia Bulletin, 24,* 399–412.

Mann, J. J., & Malone, K. M. (1997). Cerebrospinal fluid amines and higher-lethality suicide attempts in depressed inpatients. *Biological Psychiatry, 41,* 162–171.

Mann, J. J., et al. (1996). Postmortem studies of suicide victims. In S. J. Watson (Ed.), *Biology of schizophrenia and affective disease* (pp. 179–221). Washington, DC: American Psychiatric Press.

Many suicides could be prevented if people would watch for signs. (1998, June 16). *St. Louis Post-Dispatch,* p. D2.

Marangell, L. B., et al. (1997). Inverse relationship of peripheral thyrotropin-stimulating hormone levels to brain activity in mood disorders. *American Journal of Psychiatry, 154,* 224–230.

Marcus, D. K., & Nardone, M. E. (1992). Depression and interpersonal rejection. *Clinical Psychology Review, 12,* 433–449.

Marengo, J., & Harrow, M. (1987). Schizophrenic thought disorder at follow-up: A persistent or episodic course? *Archives of General Psychiatry, 44,* 651–659.

Margolin, G., & Burman, B. (1993). Wife abuse versus marital violence: Different terminologies, explanations, and solutions. *Clinical Psychology Review, 13,* 59–73.

Mark, D. H. (1998). Editor's Note. *Journal of the American Medical Association, 279,* 151.

Markowitz, J. C., et al. (1998). Treatment of depressive symptoms in human immunodeficiency virus-positive patients. *Archives of General Psychiatry, 55,* 452–457.

Markovitz, J. H., et al. (1993). Psychological predictors of hypertension in the Framingham Study: Is there tension in hypertension? *Journal of the American Medical Association, 270,* 2439–2443.

Marks, I., et al. (1998a). Computer-aided treatments of mental health problems. *Clinical Psychology: Science and Practice, 5,* 151–170.

Marks, I., et al. (1998b). Treatment of posttraumatic stress disorder by exposure and/or cognitive restructuring: A controlled study. *Archives of General Psychiatry, 55,* 317–325.

Marks, M., & De Silva, P. (1994). The "match/mismatch" mode of fear: Empirical status and clinical implications. *Behaviour Research and Therapy, 32*, 759–770.

Marlatt, G. A. (1978). Craving for alcohol, loss of control, and relapse: A cognitive-behavioral analysis. In P. E. Nathan, G. A. Marlatt, & T. Loberg (Eds.), *Alcoholism: New directions in behavioral research and treatment* (pp. 271–314). New York: Plenum Press.

Marlatt, G. A., Demming, B., & Reid, J. B. (1973). Loss of control drinking in alcoholics: An experimental analogue. *Journal of Abnormal Psychology, 81*, 233–241.

Marlatt, G. A., & Gordon, J. R. (1985). *Relapse prevention: Maintenance strategies in the treatment of addictive behaviors*. New York: Guilford Press.

Marlatt, G. A., et al. (1993). Harm reduction for alcohol problems: Moving beyond the controlled drinking controversy. *Behavior Therapy, 24*, 461–504.

Marlatt, G. A., et al. (1998). Screening and brief intervention for high-risk college student drinkers: Results from a 2-year follow-up assessment. *Journal of Consulting and Clinical Psychology, 66*, 604–615.

Marsh, D. T., & Johnson, D. L. (1997). The family experience of mental illness: Implications for intervention. *Professional Psychology: Research & Practice, 28*, 229–237.

Marshall, D. (1971). Sexual behavior on Mangaia. In D. Marshall & R. Suggs (Eds.), *Human sexual behavior: Variations in the ethnographic spectrum*. Englewood Cliffs, NJ: Prentice-Hall.

Marshall, W. L., Eccles, A., & Barbaree, H. E. (1991). The treatment of exhibitionists: A focus on sexual deviance versus cognitive and relationship features. *Behaviour Research and Therapy, 29*, 129–135.

Martin, D. (1989, January 25). Autism: Illness that can steal a child's sparkle. *The New York Times*, p. B1.

Martin, D. J., Garske, J. P., & Davis, M. K. (2000). Relation of the therapeutic alliance with outcome and other variables: A meta-analytic review. *Journal of Consulting and Clinical Psychology, 68*, 438–450.

Martin, J., Shochat, T., & Ancoli-Israel, S. (2000). Assessment and treatment of sleep disturbances in older adults. *Clinical Psychology Review, 20*, 783–805.

Martin, P. R., & Seneviratne, H. M. (1997). Effects of food deprivation and a stressor on head pain. *Health Psychology, 16*, 310–318.

Martin, R. A., & Lefcourt, H. M. (1983). Sense of humor as a moderator of the relation between stressors and moods. *Journal of Personality and Social Psychology, 45*, 1313–1324.

Martin, R. A., et al. (1993). Humor, coping with stress, self-concept, and psychological well-being. *Humor International Journal of Humor Research, 6*, 89–104.

Martin, S. E. (1992). The epidemiology of alcohol-related interpersonal violence. *Alcohol Health and Research World, 16*, 230–237.

Martins, C., de Lemos, A. I., & Bebbington, P. E. (1992). A Portuguese/Brazilian study of expressed emotion. *Social Psychiatry and Psychiatric Epidemiology, 27*, 22–27.

Marx, E. M., Williams, J. M. G., & Claridge, G. C. (1992). Depression and social problem solving. *Journal of Abnormal Psychology, 101*, 78–86.

Mason, M. (1994, September). Why ulcers run in families. *Health*, pp. 44, 48.

Masters, W. H., & Johnson, V. E. (1970). *Human sexual inadequacy*. Boston: Little, Brown.

Mathalno, D. H., et al. (2001). Progressive brain volume changes and the clinical course of schizophrenia in men: A longitudinal magnetic resonance imaging study. *Archives of General Psychiatry, 58*, 148–157.

Mathias, R. (2000). Cocaine, marijuana, and heroin abuse up, methamphetamine abuse down. *NIDA Notes, 15*(3), 4–5.

Matson, J. L., & Sevin, J. A. (1994). Theories of dual diagnosis in mental retardation. *Journal of Consulting and Clinical Psychology, 62*, 6–16.

Mayo-Smith, M. F. (1997). Pharmacological management of alcohol withdrawal. A meta-analysis and evidence-based practice guideline. American Society of Addiction Medicine Working Group on Pharmacological Management of Alcohol Withdrawal. *Journal of the American Medical Association, 278*, 144–151.

Mays, V. M. (1985). The Black American and psychotherapy: The dilemma. *Psychotherapy, 22*, 379–388.

Mazure, C. M. (1998). Life stressors as risk factors in depression. *Clinical Psychology: Science and Practice, 5*, 291–313.

McBride, P. A., Anderson, G. M., & Shapiro, T. (1996). Autism research: Bringing together approaches to pull apart the disorder. *Archives of General Psychiatry, 53*, 980–983.

McCabe, S. B., & Gotlib, I H. (1995). Selective attention and clinical depression: Performance on a deployment-of-attention task. *Journal of Abnormal Psychology, 104*, 241–245.

McCarty, D., et al. (1991). Alcoholism, drug abuse, and the homeless. *American Psychologist, 46*, 1139–1148.

McCaul, M. E, & Furst, J. (1994). Alcoholism treatment in the United States. *Alcohol Health & Research World, 18*, 253–260.

McCauley, J., et al. (1997). Clinical characteristics of women with a history of childhood abuse: Unhealed wounds. *Journal of the American Medical Association, 277*, 1362–1368.

McClelland, D. C., Alexander, C., & Marks, E. (1982). The need for power, stress, immune functions, and illness among male prisoners. *Journal of Abnormal Psychology, 91*, 61–70.

McCloskey, L. A. (1996). Socioeconomic and coercive power within the family. *Gender and Society, 10*, 449–463.

McCloskey, L. A., & Bailey, J. A. (2000). The intergenerational transmission of risk for child sexual abuse. *Journal of Interpersonal Violence, 15*, 1019–1035.

McConaghy, N. (1990). Sexual deviation. In A. S. Bellack, M. Hersen, & A. E. Kazdin (Eds.), *International handbook of behavior modification and therapy* (2nd ed., pp. 565–580). New York: Plenum Press.

McCord, J. (1983). A 40-year perspective on effects of child abuse and neglect. *Child Abuse and Neglect, 7*, 265–270.

McCord, W., & McCord, J. (1964). *The psychopath: An essay on the criminal mind*. New York: D. Van Nostrand.

McCrady, B. S. (1993). Alcoholism. In D. H. Barlow (Ed.), *Clinical handbook of psychological disorders* (2nd ed., pp. 362–393). New York: Guilford Press.

McCrady, B. S. (1994). Alcoholics Anonymous and behavior therapy: Can habits be treated as diseases? Can diseases be treated as habits? *Journal of Consulting and Clinical Psychology, 62*, 1159–1166.

McCrady, B. S., & Langenbucher, J. W. (1996). Alcohol treatment and health care system reform. *Archives of General Psychiatry, 53*, 737–746.

McDermott, J. F. (2001). Emily Dickinson revisited: A study of periodicity in her work. *American Journal of Psychiatry, 158*, 686–690.

McDermut, W., Miller, I. W., & Brown, R. A. (2001). The efficacy of group psychotherapy for depression: A meta-analysis and review of the empirical research. *Clinical Psychology: Science and Practice, 8*, 98–116.

McDougle, C. J., et al. (1996). A double-blind, placebo-controlled study of fluvoxamine in adults with autistic disorder. *Archives of General Psychiatry, 53*, 1001–1008.

McEachin, J. J., Smith, T., & Lovaas, O. I. (1993). Long-term outcome for children with autism who received early intensive behavioral treatment. *American Journal on Mental Retardation, 97*, 359–372.

McElroy, S., et al. (2000). Placebo-controlled trial of sertraline in the treatment of binge eating disorder. *American Journal of Psychiatry, 157*, 1004–1006.

McFarlane, M., Bull, S. S., & Rietmeijer, C. A. (2000). The Internet as a newly emerging risk environment for sexually transmitted diseases. *Journal of the American Medical Association, 284*, 443–446.

McGinn, D. (2000, November 21). Scouting a dry campus. *Newsweek*, pp. 83–84.

McGinn, L. K., & Sanderson, W. C. (2001). What allows cognitive behavioral therapy to be brief: Overview, efficacy, and crucial factors facilitating brief treatment. *Clinical Psychology: Science and Practice, 8*, 23–37.

McGlashan, T. H., & Fenton, W. S. (1992). The positive-negative distinction in schizophrenia: Review of natural history validators. *Archives of General Psychiatry, 49*, 63–72.

McGlashan, T. H., & Hoffman, R. E. (2000). Schizophrenia as a disorder of developmentally reduced synaptic connectivity. *Archives of General Psychiatry, 57*, 637–648.

McGovern, P. G., et al. (1996). Recent trends in acute coronary heart disease. *The New England Journal of Medicine, 334*, 884–890.

McGovern, T. F. (1991). Ethical Considerations. In L. R. Dippel & T. J. Hutton (Eds.), *Caring for the Alzheimer patient: A practical guide* (2nd ed., pp. 169–177). Amherst, NY: Prometheus Books.

McGrath, E., Keita, G. P., Strickland, B. R., & Russo, N. F. (1990). *Women and depression: Risk factors and treatment issues*. Washington DC: American Psychological Association.

McGrath, P. J., et al. (2000). A placebo-controlled study of fluoxetine versus imipramine in the acute treatment of atypical depression. *American Journal of Psychiatry, 157*, 344–350.

McGue, M. (1993). From proteins to cognitions: The behavioral genetics of alcoholism. In R. Plomin & G. E. McClearn (Eds.), *Nature, nurture & psychology* (pp. 245–268). Washington, DC: American Psychological Association.

McGue, M., Slutske, W., & Iaono, W. G. (1999). Personality and substance use disorders: II. Alcoholism versus drug use disorders. *Journal of Consulting and Clinical Psychology, 67*, 394–404.

McGuire, P. K., Shah, G. M. S., & Murray, R. M. (1993). Increased blood flow in Broca's area during auditory hallucinations in schizophrenia. *The Lancet, 342*, 703–706.

McIntosh, H. (1998, November). Autism is likely to be linked to several genes. *APA Monitor, 29*(11), p. 13.

McKenna, M. C., et al. (1999). Psychosocial factors and the development of breast cancer: A meta-analysis. *Health Psychology, 18*, 520–531.

McKinney, K., & Maroules, N. (1991). Sexual harassment. In E. Grauerholz & M. A. Koralewski (Eds.), *Sexual coercion: A sourcebook on its nature, causes, and prevention* (pp. 29–44). Lexington, MA: Lexington Books.

McLean, P. D., et al. (2001). Cognitive versus behavior therapy in the group treatment of obsessive-compulsive disorder. *Journal of Consulting and Clinical Psychology, 69*, 205–214.

McLellan, A. T., et al. (1994). Similarity of outcome predictors across opiate, cocaine, and alcohol treatments: Role of treatment services. *Journal of Consulting and Clinical Psychology, 62*, 1141–1158.

McLellan, A. T., et al. (2000). Drug dependence, a chronic medical illness: Implications for treatment, insurance, and outcomes evaluation. *Journal of the American Medical Association, 284*, 1689–1695.

McMurtrie, B. (1994, July 19). Overweight fatten ranks. *New York Newsday*, p. A26.

McNally, R. (1987). Preparedness and phobias: A review. *Psychological Bulletin, 101*, 283–303.

McNally, R. J., Cassiday, K. L., & Calamari, J. E. (1990). Taijin-kyofu-sho in a Black American woman: Behavioral treatment of a "culture-bound" anxiety disorder. *Journal of Anxiety Disorders, 4*, 83–87.

McNally, R. J., & Eke, M. (1996). Anxiety sensitivity, suffocation fear, and breath-holding duration as predictors of response to carbon dioxide challenge. *Journal of Abnormal Psychology, 105*, 146–149.

McNally, R. J., et al. (1995). Clinical versus nonclinical panic: A test of suffocation false alarm theory. *Behaviour Research & Therapy, 33*, 127–131.

McNeil, D. G, Jr. (2001, February 4). Epidemic errors. *The New York Times Week in Review*, pp. 1, 5.

McNeil, T. F., Cantor-Graae, E., & Weinberger, D. R. (2000). Relationship of obstetric complications and differences in size of brain structures in monozygotic twin pairs discordant for schizophrenia. *American Journal of Psychiatry, 157*, 203–212.

McNiel, D. E., Lam, J. N., & Binder, R. L. (2000). Relevance of interrater agreement to violence risk assessment. *Journal of Consulting and Clinical Psychology, 68*, 6, 1111–1115.

McNiel, D. E., et al. (2000). The relationship between command hallucinations and violence. *Psychiatric Services, 51*, 1288–1292.

McNulty, J. L., et al. (1997). Comparative validity of MMPI-2 scores of African American and Caucasian mental health center clients. *Psychological Assessment, 9*, 464–470.

McQuiston, J. T. (1997, February 5). New mother on Long Island suffering from depression is found, apparently a suicide. *The New York Times*, p. B5.

Meacham, J. (2000, September 18). The new face of race. *Newsweek*, pp. 38–41.

Mead, M. (1935). *Sex and temperament in three primitive societies*. New York: Morrow.

Meador-Woodruff, J. H., et al. (1997). Dopamine receptor transcript expression in striatum and prefrontal and occipital cortex: Focal abnormalities in orbitofrontal cortex in schizophrenia. *Archives of General Psychiatry, 54*, 1089–1095.

Medalia, A., et al. (1998). Effectiveness of attention training in schizophrenia. *Schizophrenia Bulletin, 24*, 147–152.

Mednick, S. A., Parnas, J., & Schulsinger, F. (1987). The Copenhagen High-Risk project, 1962–86. *Schizophrenia Bulletin, 13*, 485–495.

Mednick, S. A., & Schulsinger, F. (1968). Some premorbid characteristics related to breakdown in children with schizophrenic mothers. In D. Rosenthal & S. S. Kety (Eds.), *The transmission of schizophrenia* (pp. 267–291). New York: Pergamon Press.

Meehan, P. J., et al. (1991). Attempted suicide among young adults: Progress toward a meaningful estimate of prevalence. *American Journal of Psychiatry, 149*, 41–44.

Meehl, P. E. (1962). Schizotaxia, schizotypy, schizophrenia. *American Psychologist, 17*, 827–838.

Meehl, P. E. (1972). A critical afterword. In I. I. Gottesman & J. Shields (Eds.), *Schizophrenia and genetics: A twin study vantage point* (pp. 367–415). New York: Academic Press.

Mehrabian, A., & Weinstein, L. (1985). Temperament characteristics of suicide attempters. *Journal of Consulting and Clinical Psychology, 53*, 544–546.

Meichenbaum, D. (1993). Changing conceptions of cognitive behavior modification: Retrospect and prospect. *Journal of Consulting and Clinical Psychology, 61*, 202–204.

Meichenbaum, D., & Deffenbacher, J. L. (1988). Stress inoculation training. *The Counseling Psychologist, 16*(1), 69–90.

Melani, D. (2001, January 17). Emotions can pull trigger on heart attack. *Evansville Courier & Press, Scripps Howard News Service*. Retrieved January 19, 2001, from http://www.psycport.com/news/2001/01/17/eng-courier press_features/eng-courierpress_features_134435_74_ 9803814571351.html.

Melchert, T. P. (1996). Childhood memory and a history of different forms of abuse. *Professional Psychology: Research and Practice, 27*, 438–446.

Mellor, C. S. (1970). First rank symptoms of schizophrenia. *British Journal of Psychiatry, 177*, 15–23.

Mental health problems cost North America and EU $120 billion. (2000, October 10). United Press International. *PsycPort News Story*, American Psychological Association. Retrieved October 28, 2000, from http://www.psycport.com/news/2000/10/10/up/0000-0148-switzerland-mentalhea.html.

Merckelbach, H., Arntz, A., & de Jong, P. (1991). Conditioning experiences in spider phobics. *Behaviour Research and Therapy, 29*, 301–304.

Merckelbach, H., et al. (1996). The etiology of specific phobias: A review. *Clinical Psychology Review, 16*, 337–361.

Messenger, J. (1971). Sex and repression in an Irish folk community. In D. Marshall & R. Suggs (Eds.), *Human sexual behavior: Variations in the ethnographic spectrum*. Englewood Cliffs, NJ: Prentice-Hall.

Messer, S. B. (2001a). Empirically supported treatments: What's a nonbehaviorist to do? In B. D. Slife & R. N. Williams (Eds.), *Critical issues in psychotherapy: Translating new ideas into practice* (pp. 3–19). Thousand Oaks, CA: Sage.

Messer, S. B. (2001b). What makes brief psychodynamic therapy time efficient? *Clinical Psychology: Science and Practice, 8*, 5–22.

Meston, C. M., & Heiman, J. R. (2000). Sexual abuse and sexual function: An examination of sexually relevant cognitive processes. *Journal of Consulting and Clinical Psychology, 68*(3), 399–406.

Meyer, G. J. (1997). Assessing reliability: Critical corrections for a critical examination of the Rorschach comprehensive system. *Psychological Assessment, 9*, 480–489.

Meyer, G. J. (2000). Incremental validity of the Rorschach Prognostic Rating Scale over the MMPI Ego Strength Scale and IQ. *Journal of Personality Assessment, 74*, 365–370.

Meyer, G. J., et al. (2001). Psychological testing and psychological assessment: A review of evidence and issues. *American Psychologist, 56*, 128–165.

Meyer, T. J., & Mark, M. M. (1995). Effects of psychosocial interventions with adult cancer patients: A meta-analysis of randomized experiments. *Health Psychology, 14*, 101–108.

Michaud, D. S., et al. (2001). Physical activity, obesity, height, and the risk of pancreatic cancer. *Journal of the American Medical Association, 286*, 921–929.

Michaud, E. (2000, October). Women's secret terror. *Prevention*, pp. 118–127.

Michelson, D., Bancroft, J., Targum, S., Kim, Y., & Tepner, R. (2000). Female sexual dysfunction associated with antidepressant administration: A randomized, placebo-controlled study of pharmacologic intervention. *American Journal of Psychiatry, 157*, 239–243.

Mignot, E., & Thorsby, E. (2001). Narcolepsy and the HLA System. *The New England Journal of Medicine, 344*, 692.

Miklowitz, D. J. (1994). Family risk indicators in schizophrenia. *Schizophrenia Bulletin, 20*, 137–149.

Miklowitz, D. J., & Alloy, L. B. (1999). Psychosocial factors in the course and treatment of bipolar disorder: Introduction to the special section. *Journal of Abnormal Psychology, 108*, 555–557.

Milberger, S., et al. (1996). Is maternal smoking during pregnancy a risk factor for attention deficit hyperactivity disorder in children? *American Journal of Psychiatry, 153*, 1138–1142.

Milberger, S., et al. (1997). Pregnancy, delivery, and infancy complications and attention deficit hyperactivity disorder: Issues of gene-environment interaction. *Biological Psychiatry, 41*, 65–75.

Miller, A. (2000, Fall/Winter). Growing up in the new family. *Newsweek Special Issue*, pp. 80–84.

Miller, E. (1987). Hysteria: Its nature and explanation. *British Journal of Clinical Psychology, 26*, 163–173.

Miller, G. E., & Cohen, S. (2001). Psychological interventions and the immune system: A meta-analytic review and critique. *Health Psychology, 20*, 47–63.

Miller, L. K. (1999). The savant syndrome: Intellectual impairment and exceptional skills. *Psychological Bulletin, 125*, 31–46.

Miller, S. D., et al. (1991). Optical differences in multiple personality disorder: A second look. *Journal of Nervous & Mental Disease, 179*, 132–135.

Miller, T. Q., et al. (1991). Reasons for the trend toward null findings in research on Type A behavior. *Psychological Bulletin, 110*, 469–485.

Miller, W. R., & Brown, S. A., (1997). Why psychologists should treat alcohol and drug problems. *American Psychologist, 52*, 1269–1279.

Miller, W. R., & Hester, R. K. (1986). Inpatient alcoholism treatment: Who benefits? *American Psychologist, 41*, 794–805.

Miller, W. R., Leckman, A. L., Delaney, H. D., & Tinkcom, M. (1993). Long-term follow-up of behavioral self-control training. *Journal of Studies on Alcohol, 53*, 249–261.

Miller, W. R., & Muñoz, R. F. (1983). *How to control your drinking* (2nd ed.). Albuquerque: University of New Mexico Press.

Miller-Medzon, K. (2000, August 20). Early dyslexia detection leads to normal learning. *Boston Herald*, pp. 1, 11.

Millon, T. (1981). *Disorders of personality DSM-III: Axis II*. New York: Wiley.

Millon, T. (1982). *Millon Clinical Multiaxial Inventory manual* (3rd ed.). Minneapolis: National Computer Systems.

Milner, J. S. (1993). Social information processing and physical child abuse. *Clinical Psychology Review, 13*, 275–294.

Minarik, M. L., & Ahrens, A. H. (1996). Relations of eating and symptoms of depression and anxiety to the dimensions of perfectionism among undergraduate women. *Cognitive Research & Therapy, 20*, 155–169.

Mineka, S. (1991, August). Paper presented to the annual meeting of the American Psychological Association, San Francisco. (Cited in Turkington, C. [1991]). Evolutionary memories may have phobia role. *APA Monitor, 22*(11), 14.

Minuchin, S., Rosman, B. L., & Baker, L. (1978). *Psychosomatic Families: Anorexia nervosa in context*. Cambridge, MA: Harvard University Press.

Mischel, W. (1993). *Introduction to personality* (5th ed.). Forth Worth, TX: Harcourt Brace Jovanovich.

Mitka, M. (2000). Psychiatrists help survivors in the Balkans. *Journal of the American Medical Association, 283*, 1277–1278.

Modestin, J. (1992). Multiple personality disorder in Switzerland. *American Journal of Psychiatry, 149*, 88–92.

Mokdad, A. H., Serdula, M. K., Dietz, W. H., Bowman, B. A., Marks, J. S., & Kaplan, J. P. (1999). The spread of the obesity epidemic in the United States, 1991–1998. *Journal of the American Medical Association, 282*, 1519–1522.

Mokdad, A. H., et al. (2000). The continuing epidemic of obesity in the United States. *Journal of the American Medical Association, 284*, 1650–1651. [Research Letters]

Mokuau, N. (1990). The impoverishment of native Hawaiians and the social work challenge. *Health and Social Work, 15*, 235–242.

Moldin, S. O. (1994). Indicators of liability to schizophrenia: Perspectives from genetic epidemiology. *Schizophrenia Bulletin, 20*, 169–184.

Monahan, J. (1981). *A clinical prediction of violent behavior*. DHHS Publication, Adm. 81–921. Rockville, MD: National Institutes of Mental Health.

Monahan, J. (1992). Mental disorder and violent behavior: Perceptions and evidence. *American Psychologist, 47*, 511–521.

Moncher, M. S., Holden, G. W., & Trimble, J. E. (1990). Substance abuse among Native-American youth. *Journal of Consulting and Clinical Psychology, 58*, 408–415.

Mones, A. G., & Panitz, P. E. (1994). Marital violence: An integrated systems approach. *Journal of Social Distress and the Homeless, 3*, 39–51.

Money, J. (1987). Sin, sickness, or status? Homosexual gender identity and psychoneuroendocrinology. *American Psychologist, 42*, 384–399.

Money, J. (1994). The concept of gender identity disorder in childhood and adolescence after 39 years. *Journal of Sex and Marital Therapy, 20*, 163–177.

Money, J., & Lamacz, M. (1990). *Vandalized lovemaps.* Buffalo, NY: Prometheus Books.

Monroe, S. M., et al. (1999). Life events and depression in adolescence: Relationship loss as a prospective risk factor for first onset of major depressive disorder. *Journal of Abnormal Psychology, 108*, 606–614.

Monroe, S. M., et al. (2001). Life stress and the symptoms of major depression. *Journal of Nervous & Mental Disease, 189*, 168–175

Monti, P. M., et al. (1987). Reactivity of alcoholics and nonalcoholics to drinking cues. *Journal of Abnormal Psychology, 96*, 122–126.

Monti, P. M., et al. (1994). Cue exposure with coping skills treatment for male alcoholics: A preliminary investigation. *Journal of Consulting and Clinical Psychology, 61*, 1011–1019.

Moos, R. H., Cronkite, R. C., & Moos, B. S. (1998). Family and extrafamily resources and the 10-year course of treated depression. *Journal of Abnormal Psychology, 107*, 450–460.

Moos, R. H., McCoy, L., & Moos, B. S. (2000). Global Assessment of Functioning (GAF) ratings: Determinants and roles as predictors of one-year treatment outcomes. *Journal of Clinical Psychology, 56*, 449–461.

Moran, M. G. (1991). Psychological factors affecting pulmonary and rheumatologic diseases: A review. *Psychosomatics, 32*, 14–23.

Morgan, D. L., & Morgan, R. K. (2001). Single-participant research design: Bringing science to managed care. *American Psychologist, 56*, 119–127.

Morgan, K. (1996). Mental health factors in late-life insomnia. *Reviews in Clinical Gerontology, 6*, 75–83.

Morgenstern, J., et al. (1997). The comorbidity of alcoholism and personality disorders in a clinical population: Prevalence rates and relation to alcohol typology variables. *Journal of Abnormal Psychology, 106*, 74–84.

Morin, C. M., & Ware, J. C. (1996). Sleep and psychopathology. *Applied and Preventive Psychology, 5*, 211–224.

Morin, C. M., & Wooten, V. (1996). Psychological and pharmacological approaches to treating insomnia: Critical issues in assessing their separate and combined effects. *Clinical Psychology Review, 16*, 521–542.

Morin, C. M., et al. (1993a). Dysfunctional beliefs and attitudes about sleep among older adults with and without insomnia complaints. *Psychology and Aging, 8*, 463–467.

Morin, C. M., et al. (1993b). Cognitive-behavior therapy for late-life insomnia. *Journal of Consulting and Clinical Psychology, 61*, 137–146.

Morin, C. M., et al. (1999). Behavioral and pharmacological therapies for late-life insomnia: A randomized controlled trial. *Journal of the American Medical Association, 281*, 991–999.

Morrison, J. (1989). Childhood sexual histories of women with somatization disorder. *American Journal of Psychiatry, 146*, 239–241.

Morrow, D. J. (1998a, March 5). Stumble on the road to market. *The New York Times*, p. D1.

Mortensen, P. B., et al. (1999). Effects of family history and place and season of birth on the risk of schizophrenia. *New England Journal of Medicine, 340*, 603–608.

Mossman, D. (1994). Assessing predictions of violence: Being accurate about accuracy. *Journal of Consulting and Clinical Psychology, 62*, 783–792.

Mowrer, O. H. (1948). Learning theory and the neurotic paradox. *American Journal of Orthopsychiatry, 18*, 571–610.

Mueller, T. I., et al. (1999). Recurrence after recovery from major depressive disorder during 15 years of observational follow-up. *The American Journal of Psychiatry, 156*, 1000–1006.

Mueser, K. T., & Liberman, R. P. (1995). Behavior therapy in practice. In B. Bongar & L. E. Beutler (Eds.), *Comprehensive textbook of psychotherapy: Theory and practice* (pp. 84–110). New York: Oxford.

Mueser, K. T., et al. (2001). Family treatment and medication dosage reduction in schizophrenia: Effects on patient social functioning, family attitudes, and burden. *Journal of Consulting and Clinical Psychology, 69*, 3–12.

Muñoz, R. F. , Mrazek, P. J., & Haggerty, R. J. (1996). Institute of Medicine Report on Prevention of Mental Disorders: Summary and commentary. *American Psychologist, 51*, 1116–1121.

Murdoch, D., Pihl, R. O., & Ross, D. (1990). Alcohol and crimes of violence: Present issues. *International Journal of the Addictions, 25*, 1065–1081.

Murphy, C. M., Meyer, S. L., & O'Leary, K. D. (1994). Dependency characteristics of partner assaultive men. *Journal of Abnormal Psychology, 103*, 729–735.

Murphy, C. M., et al. (2001). Correlates of intimate partner violence among male alcoholic patients. *Journal of Consulting and Clinical Psychology, 69*, 528–540.

Murray, B. (2000a). From brain scan to lesson plan. *Monitor on Psychology, 31*, pp. 22–28.

Murray, B. (2000b, July/August). Psychology seeks to replicate groundbreaking research on the success of drug/psychotherapy treatment. *Monitor on Psychology*, p. 13.

Murray, H. A. (1943). *Thematic Apperception Test: Pictures and manual.* Cambridge, MA: Harvard University Press.

Murray, J. B. (1993). Relationship of childhood sexual abuse to borderline personality disorder, posttraumatic stress disorder, and multiple personality disorder. *Journal of Psychology, 127*, 657–676.

Murstein, B. I., & Mathes, S. (1996). Projection on projective techniques & pathology: The problem that is not being addressed. *Journal of Personality Assessment, 66*, 337–349.

Murtagh, D. R. R., & Greenwood, K. M. (1995). Identifying effective psychological treatments for insomnia: A meta-analysis. *Journal of Consulting and Clinical Psychology, 63*, 79–89.

Must, A., et al. (1999). The disease burden associated with overweight and obesity. *Journal of the American Medical Association, 282*, 1523–1529.

Muster, N. J. (1992). Treating the adolescent victim-turned-offender. *Adolescence, 27*, 441–450.

Mutler, A. (2000, August 3). One-fourth of Kosovo's population suffering mental anguish in the aftermath of war. *Associated Press Web Listing.* Copyrighted by Associated Press. Retrieved August 3, 2000, from http://psycport.com/news/2000/08/03/wstm-/2537-2706-Kosovo-StillatWar.html

Myers, A., et al. (2000). Susceptibility locus for Alzheimer's disease on chromosome 10. *Science, 290*, 2304–2305.

Myin-Germeys, I., Delespaul, P. A. E. G., & deVries, M. W. (2000). Schizophrenia patients are more emotionally active than is assumed based on their behavior. *Schizophrenia Bulletin, 26*, 847–853.

N

Nagourney, E. (2001a, April 10). Geography of dyslexia is explored. *The New York Times*, p. F7.

Nagourney, E. (2001b, April 24). A good night's sleep, without the pills. *The New York Times*, p. F8.

Näslund, J., et al. (2000). Correlation between elevated levels of amyloid-peptide in the brain and cognitive decline. *Journal of the American Medical Association, 283*, 1571–1577.

Nathan, P. E. (1988). The addictive personality is the behavior of the addict. *Journal of Consulting and Clinical Psychology, 56*, 183–188.

Nathan, P. E. (1994). *DSM-IV*: Empirical, accessible, not yet ideal. *Journal of Clinical Psychology, 50*, 103–110.

Nathan, P. E., Stuart, S. P., & Dolan, S. L. (2000). Research on psychotherapy efficacy and effectiveness: Between Scylla and Charybdis? *Psychological Bulletin, 126*, 964–981.

National Center for Health Statistics. (1996b). News releases and fact sheets. *Highlights of a new report from the National Center for Health Statistics (NCHS), Monitoring health care in America: Quarterly fact sheet.* Washington, DC: US Department of Health, Education, and Welfare.

National Highway Traffic Safety Administration. (1988). *Fatal accident reporting system: 1987.* Washington, DC: U.S. Department of Transportation.

National Institute on Alcohol Abuse and Alcoholism. (1990). *7th Special Report to Congress on Alcohol and Health.* Rockville, MD: Author.

National Institutes of Health, National Heart, Lung, and Blood Institute. (1998). *Clinical guidelines on the identification, evaluation, and treatment of overweight and obesity in adults*: Bethesda, MD: Author.

National Institutes of Health, Office of Research on Women's Health, Office of the Director. (1999). *Women of color health data book: Adolescents to seniors.* NIH Publication 99-4247. Bethesda, MD: Author.

National Strategy for Suicide Prevention. (2001, May). *Goals and Objectives for Action: Summary.* A joint effort of SAMHSA, CDC, NIH, and HRSA. The Center for Mental Health Services. Rockville, MD: Author.

NBC Nightly News. (1996, November 11). National Broadcasting Company.

Ndetei, D. M., & Singh, A. (1983). Hallucinations in Kenyan schizophrenic patients. *Acta Psychiatrica Scandinavica, 67*, 144–147.

Ndetei, D. M., & Vadher, A. (1984). A comparative cross-cultural study of the frequencies of hallucination in schizophrenia. *Acta Psychiatrica Scandinavica, 70*, 545–549.

Neal, A. M., & Turner, S. M. (1991). Anxiety disorders research with African Americans: Current status. *Psychological Bulletin, 109*, 400–410.

Needles, D. J., & Abramson, L. Y. (1990). Positive life events, attributional style, and hopefulness: Testing a model of recovery from depression. *Journal of Abnormal Psychology, 99*, 156–165.

Negy, C., & Snyder, D. K. (1997). Ethnicity and acculturation: Assessing Mexican American couples' relationships using the marital satisfaction inventory—revised. *Psychological Assessment, 9*, 414–421.

Neiger, B. L. (1988). Adolescent suicide: Character traits of high-risk teenagers. *Adolescence, 23*, 469–475.

Neighbors, H. (1992, December). *The help seeking behavior of black Americans: A summary of the National Survey of Black Americans.* Paper presented at a symposium, Anxiety Disorders in African Americans, presented by the State University of New York Health Science Center at Brooklyn, Brooklyn, NY.

Nelson, C. B., Heath, A. C., & Kessler, R. C. (1998). Temporal progression of alcohol dependence symptoms in the U.S. household population: Results from the National Comorbidity Survey. *Journal of Consulting and Clinical Psychology, 66*, 474–483.

Nelson, S. H., et. al. (1992). An overview of mental health services for American Indians and Alaska natives in the 1990s. *Hospital and Community Psychiatry, 43*, 257–261.

Nemiah, J. C. (1978). Psychoneurotic disorders. In A. M. Nicholi (Ed.), *Harvard guide to modern psychiatry.* Cambridge, MA: Harvard University Press.

Nestadt, G., Samuels, J., Riddle, M., Bienvienu, O. J. III, Liang, K. Y., LaBuda, M., et al. (2000). A family study of obsessive-compulsive disorder. *Archives of General Psychiatry, 57,* 358–363.

Nestadt, G., et al. (2000). A family study of obsessive-compulsive disorder. *Archives of General Psychiatry, 57,* 358–363.

Neugebauer, R. (1979). Medieval and early modern theories of mental illness. *Archives of General Psychiatry, 36,* 477–484.

Nevid, J. S., Fichner-Rathus, L., & Rathus, S. A. (1995). *Human sexuality in a world of diversity* (2nd ed.). Boston: Allyn & Bacon.

Nevid, J. S., & Javier, R. A. (1997). Preliminary investigation of a culturally-specific smoking cessation intervention for Hispanic smokers. *American Journal of Health Promotion, 11,* 198–207.

Nevid, J. S., Javier, R. A., & Moulton, J. (1996). Factors predicting participant attrition in a community-based culturally-specific smoking cessation program for Hispanic smokers. *Health Psychology, 15,* 226–229.

Neville, H. A., et al. (1996). The impact of multicultural training on white racial identity attitudes and therapy competencies. *Professional Psychology: Research & Practice, 27,* 83–89.

New research could open doors to better migraine treatment. (2000, June 13). *CNN Web Posting.* Retrieved June 15, 2000, from http://www.cnn.com/2000/HEALTH/06/13/migraine/?related.

New research supports health benefits of red wine. (2000, July 3). *Cable News Network.* Retrieved July 5, 2000, from http://www.cnn.com/2000/HEALTH/diet.fitness/07/03/french.paradox/?related.

New use of brain scan may yield delays in Alzheimer's symptoms. (2000, May 15). *CNN Web Posting.* Retrieved May 16, 2000, from http://www.cnn.com/2000/HEALTH/aging/05/15/alzheimers.diagnosis/.

New York City Board of Education. (1984). *Child abuse and neglect prevention training manual: Working together to make a difference.* New York: Office of the Chief Executive for Instruction, Office of Student Progress, and Guidance Services Unit.

Newman, K. D. (1993). Giving up: Shelter experiences of battered women. *Public Health Nursing, 10,* 108–113.

Newman, L. S., & Baumeister, R. F. (1996). Toward an explanation of the UFO abduction phenomenon: Hypnotic elaboration, extraterrestrial sadomasochism, and spurious memories. *Psychological Inquiry, 7,* 99–126.

Nezu, A. M. (1994). Introduction to special section: Mental retardation and mental illness. *Journal of Consulting and Clinical Psychology, 62,* 4–5.

NIAAA report links drinking and early death. (1990, October). *The Addiction Letter, 6,* p. 5.

Niccols, G. A. (1994). Fetal alcohol syndrome: Implications for psychologists. *Clinical Psychology Review, 14,* 91–111.

Nicholson, R. A., et al. (1997). Utility of MMPI-2 indicators of response distortion: Receiver operating characteristic analysis. *Psychological Assessment, 9,* 471–479.

Nickerson, K. J., Helms, J. E., & Terrell, F. (1994). Cultural mistrust, opinions about mental illness, and Black students' attitudes toward seeking psychological help from White counselors. *Journal of Counseling Psychology, 41,* 378–385.

Nierenberg, A. A., et al. (2000). Timing of onset of antidepressant response with fluoxetine treatment. *American Journal of Psychiatry, 157,* 1429–1435.

Nieto, F. J., et al. (2000). Association of sleep-disordered breathing, sleep apnea, and hypertension in a large community-based study. *Journal of the American Medical Association, 283,* 1829–1836.

Nigg, J. T., & Goldsmith, H. H. (1994). Genetics of personality disorders: Perspectives from personality and psychopathology research. *Psychological Bulletin, 115,* 346–380.

Nigg, J. T., et al. (1992). Malevolent object representations in borderline personality disorder and major depression. *Journal of Abnormal Psychology, 101,* 61–67.

Niles, M. A., et al. (1995). Predictors of initial smoking cessation and relapse through the first 2 years of the Lung Health Study. *Journal of Consulting and Clinical Psychology, 63,* 60–69.

Nishith, P., Mechanic, M. B., & Resick, P. A. (2000). Prior interpersonal trauma: The contribution to current PTSD symptoms in female rape victims. *Journal of Abnormal Psychology, 109,* 20–25.

Nix, G., Watson, C., Pyszczynski, T., & Greenberg, J. (1995). Reducing depressive affect through external focus of attention. *Journal of Social and Clinical Psychology, 14,* 36–52.

Nolen-Hoeksema, S. (1991). Responses to depression and their effects on the duration of depressive episodes. *Journal of Abnormal Psychology, 100,* 569–582.

Nolen-Hoeksema, S. (2000). The role of rumination in depressive disorders and mixed anxiety/depressive symptoms. *Journal of Abnormal Psychology, 109,* 504–511.

Nolen-Hoeksema, S., & Girgus, J. S. (1994). The emergence of gender differences in depression during adolescence. *Psychological Bulletin, 115,* 424–443.

Nolen-Hoeksema, S., Girgus, J. S., & Seligman, M. E. P. (1992). Predictors and consequences of childhood depressive symptoms: A 5-year longitudinal study. *Journal of Abnormal Psychology, 101,* 405–422.

Nolen-Hoeksema, S., Morrow, J., & Fredrickson, B. L. (1993). Response styles and the duration of episodes of depressed mood. *Journal of Abnormal Psychology, 102,* 20–28.

Noonan, D. (2000, September 25). Why drugs cost so much. *Newsweek,* pp. 22–30.

North, C. S., et al. (1999). Psychiatric disorders among survivors of the Oklahoma City bombing. *Journal of the American Medical Association, 282,* 755–762.

Novaco, R. W. (1974). *A treatment program for the management of anger through cognitive and relaxation control.* Doctoral Dissertation, Indiana University.

Novaco, R. W. (1977). A stress inoculation approach to anger management in the training of law enforcement officers. *American Journal of Community Psychology, 5,* 327–346.

Nowell, P. D., et al. (1998). Effective treatments for selected sleep disorders. In P. E. Nathan & J. M. Gorman (Eds.), *A guide to treatments that work* (pp. 531–543). New York: Oxford University Press.

Noyes, R., et al. (1993). The validity of *DSM-III-R* hypochondriasis. *Archives of General Psychiatry, 50,* 961–970.

Nucifora Jr., F. C., et al. (2001). Interference by Huntington and atrophin-1 with cbp-mediated transcription leading to cellular toxicity. *Science, 291,* 2423–2428.

Nurnberger, J. I., et al. (2001). Evidence for a locus on chromosome 1 that influences vulnerability to alcoholism and affective disorder. *American Journal of Psychiatry, 158,* 718–724.

Nutrition, obesity and perception. (2001, January 9). *CNN Web Posting.* Retrieved January 11, 2001, from http://www.cnn.com/2001/HEALTH/children/01/09/overweight.kids/index.html.

O

O'Brien, C. P. (1996). Recent developments in the pharmacotherapy of substance abuse. *Journal of Consulting and Clinical Psychology, 64,* 677–686.

O'Brien, C. P., & McKay, J. (1998). Psychopharmacological treatments of substance use disorders. In P. E. Nathan & J. M. Gorman (Eds.), *A guide to treatments that work* (pp. 127–155). New York: Oxford University Press.

O'Brien, C. P., & McLellan, A. T. (1997). Addiction medicine. *Journal of the American Medical Association, 277,* 1840–1841.

O'Connor v. Donaldson, 95 S. Ct. 2486 (1975).

O'Connor, E. (2001a, January). Law sanctions new treatment for heroin addiction—and recommends psychological counseling. *Monitor on Psychology,* p. 18.

O'Connor, E. (2001b, February). Researchers pinpoint potential cause of autism. *Monitor on Psychology,* p. 13.

O'Connor, P. G. (2000). Treating opioid dependence—new data and new opportunities. *The New England Journal of Medicine, 343,* 1332–1334. [Editorial]

O'Connor, T. G., et al. (1998). Co-occurrence of depressive symptoms and antisocial behavior in adolescence: A common genetic liability. *Journal of Abnormal Psychology, 107,* 27–37.

O'Donnell, Clifford R. (1995). Firearm deaths among children and youth. *American Psychologist, 50,* 771–776.

O'Donohue, W., Dopke, C. A., & Swingen, D. N. (1997). Psychotherapy for female sexual dysfunction: A review. *Clinical Psychology Review, 17,* 537–566.

O'Donohue, W., Letourneau, E., & Geer, J. H. (1993). Premature ejaculation. In W. O'Donohue & J. H. Geer (Eds.), *Handbook of sexual dysfunctions: Assessment and treatment* (pp. 303–333). Boston: Allyn & Bacon.

O'Donohue, W., McKay, J. S., & Schewe, P. A. (1996). Rape: The roles of outcome expectancies and hypermasculinity. *Sexual Abuse Journal of Research and Treatment, 8,* 133–141.

O'Donohue, W. T., et al. (1999). Psychotherapy for male sexual dysfunction: A review. *Clinical Psychology Review, 19,* 591–630.

O'Farrell, T. J., et al. (1996). Cost-benefit and cost-effectiveness analyses of behavioral marital therapy as an addition to outpatient alcoholism treatment. *Journal of Substance Abuse, 8,* 145–166.

O'Leary, A. (1990). Stress, emotion, and human immune functions. *Psychological Bulletin, 108,* 382–383.

O'Leary, D. S., et al. (1996). Auditory attentional deficits in patients with schizophrenia: A positron emission tomography study. *Archives of General Psychiatry, 53,* 633–641.

O'Leary, K. D. (1995, July). Assessment and treatment of partner abuse. *Clinician's Research Digest, Supplemental Bulletin 12,* pp. 1–2.

O'Leary, K., et al. (2000). Co-occurrence of partner and parent aggression: Research and treatment implications. *Behavior Therapy, 31,* 631–648.

Oei, T. P. S., & Shuttlewood, G. J. (1996). Specified and nonspecific factors in psychotherapy: A case of cognitive therapy for depression. *Clinical Psychology Review, 16,* 83–103.

Oetting, E. R., Beauvais, F., & Edwards, R. (1988). Alcohol and Indian youth: Social and psychological correlates and prevention. *Journal of Drug Issues, 18,* 87–102.

Ogloff, J. R. P., Roberts, C. F., & Roesch, R. (1993). The insanity defense: Legal standards and clinical assessment. *Applied & Preventive Psychology, 2,* 163–178.

Öhman, A., & Mineka, S. (2001). Fears, phobias, and preparedness: Toward an evolved module of fear and fear learning. *Psychological Review, 108,* 483–522.

Okubo, Y., et al. (1997). Decreased prefrontal dopamine D1 receptors in schizophrenia revealed by PET. *Nature, 385,* 634–636.

Oldham, J. M. (1994). Personality disorders: Current perspectives. *Journal of the American Medical Association, 272,* 213–220.

Olesen, J. (1994). Understanding the biologic basis of migraine. *New England Journal of Medicine, 331,* 1713–1714.

Olfson, M., et al. (1998). Use of ECT for the inpatient treatment of recurrent major depression. *American Journal of Psychiatry, 155,* 22–29.

Olfson, M., et al. (2000). Barriers to the treatment of social anxiety. *American Journal of Psychiatry, 157,* 542–548.

Olson, E. (2001, October 7). Countries lag in treating mental illness, W. H. O. says. *The New York Times*, p. A24.

Olson, L. (2000). Combating Parkinson's disease—Step three. *Science, 290*, 721–724.

Olweus, D. (1987). Testosterone and adrenaline. In S. K. Mednick et al. (Eds.), *The causes of crime: New biological approaches* (pp. 263–282). Cambridge, UK: Cambridge University Press.

One in five teen-agers is armed, a survey finds. (1998). *The New York Times*, p. A19.

Onstad, S., Skre, I., Torgensen, S., & Kringlen, E. (1991). Twin concordance for *DSM-III-R* schizophrenia. *Acta Psychiatrica Scandinavica, 83*, 395–401.

Ormel, J., et al. (2001). The interplay and etiological continuity of neuroticism, difficulties, and life events in the etiology of major and subsyndromal, first and recurrent depressive episodes in later life. *American Journal of Psychiatry, 158*, 885–891.

Orne, M. T., et al. (1996). "Memories" of anomalous and traumatic autobiographical experiences: Validation and consolidation of fantasy through hypnosis. *Psychological Inquiry, 7*, 168–172.

Orsillo, S. M., et al. (1996). Current and lifetime psychiatric disorders among veterans with war zone-related posttraumatic stress disorder. *Journal of Nervous & Mental Disease, 184*, 307–313.

Ortega, A. N., et al. (2000). Acculturation and the lifetime risk of psychiatric and substance use disorders among Hispanics. *Journal of Nervous & Mental Disease, 188*, 728–735.

Orth-Gomér, K., et al. (2000). Marital stress worsens prognosis in women with coronary heart disease: The Stockholm Female Coronary Risk Study. *Journal of the American Medical Association, 284*, 3008–3014.

Osborne, L. (2001, May 6). Regional disturbances. *The New York Times*, p. A17.

Öst, L. (1987). Age of onset in different phobias. *Journal of Abnormal Psychology, 96*, 223–229.

Öst, L. (1992). Blood and injection phobia: Background and cognitive, physiological, and behavioral variables. *Journal of Abnormal Psychology, 101*, 68–74.

Osterling, J., & Dawson, G. (1994). Early recognition of children with autism: A study of first birthday home videotapes. *Journal of Autism and Developmental Disorder, 24*, 247–257.

Ostler, K., et al. (2001). Influence of socio-economic deprivation on the prevalence and outcome of depression in primary care. *British Medical Journal, 178*, 12–17.

Otto, M. W. (2001, February). A pilot study of CBT for bipolar disorders. *Journal Watch Psychiatry, 7*, 9.

Otto, M. W., Pollack, M. H., & Maki, K. M. (2000). Empirically supported treatments for panic disorder: Costs, benefits, and stepped care. *Journal of Consulting and Clinical Psychology, 68*, 556–563.

Ouellette, S. C., & DiPlacido, J (2001). Personality's role in the protection and enhancements of health: Where the research has been, where it is stuck, how it might move. In A. Baum, T. A. Revenson, & J. E. Singer (Eds.), *Handbook of health psychology* (pp 3–318). Mahwah, NJ: Lawrence Erlbaum Associates, Inc.

Ouimette, P. C., Finney, J. W., & Moos, R. H. (1997). Twelve-step and cognitive-behavioral treatment for substance abuse: A comparison of treatment effectiveness. *Journal of Consulting and Clinical Psychology, 65*, 230–240.

Overholser, J. C. (2000). Cognitive-behavioral treatment of panic disorder. *Psychotherapy: Theory, Research, Practice, Training, 37*, 247–256.

Overmier, J. B. L., & Seligman, M. E. P. (1967). Effect of inescapable shock upon subsequent escape and avoidance learning. *Journal of Comparative and Physiological Psychology, 63*, 28–33.

P

Palinkas, L. A., Wingard, D. L., & Barrett-Connor, E. (1990). The biocultural context of social networks and depression among the elderly. *Social Science and Medicine, 4*, 441–447.

Pallesen, S., et al. (2001). Clinical assessment and treatment of insomnia. *Professional Psychology: Research and Practice, 32*, 115–124.

Parker, G., Gladstone, G., & Chee, K. T. (2001). Depression in the planet's largest ethnic group: The Chinese. *American Journal of Psychiatry, 158*, 857–864.

Parker, K. F., & Pruitt, M. V. (2000). Poverty, poverty concentration, and homicide. *Social Science Quarterly, 81*, 555–570.

Parkinson Study Group. (2000). Pramipexole vs. levodopa as initial treatment for Parkinson disease: A randomized controlled trial. *Journal of the American Medical Association, 284*, 1931–1938.

Parnas, J., et al. (1993). Lifetime *DSM-III-R* diagnostic outcomes in the offspring of schizophrenic mothers: Results from the Copenhagen High-Risk Study. *Archives of General Psychiatry, 50*, 707–714.

Patrick, C. J., Cuthbert, B. N., & Lang, P. J. (1994). Emotion in the criminal psychopath: Fear image processing. *Journal of Abnormal Psychology, 103*, 523–534.

Patton, G. C., et al. (1999). Onset of adolescent eating disorders: Population based cohort study over 3 years. *British Medical Journal, 318*, 765–768.

Paul, G. L., & Lentz, R. J. (1977). *Psychosocial treatment of chronic mental patients: Milieu versus social-learning programs*. Cambridge, MA: Harvard University Press.

Paulesu, E., Frith, C. D., & Frackowisk, R. S. J. (1993). The neural correlates of the verbal component of working memory. *Nature, 362*, 342–344.

Paulesu, E., et al. (2001). Dyslexia: Cultural diversity and biological unity. *Science, 291*, 2165–2167.

Pauli, P., et al. (1997). Behavioral and neurophysiological evidence for altered processing of anxiety-related words in panic disorder. *Journal of Abnormal Psychology, 106*, 213–220.

Paykel, E. S. (1982). Life events and early environments. In E. S. Paykel (Ed.), *Handbook of affective disorders*. New York: Guilford Press.

Pearson, J. L., & Brown, G. K. (2000). Suicide prevention in late life: Directions for science and practice. *Clinical Psychology Review, 20*, 685–705.

Peers sway a child's interest in smoking, drinking as early as 6th grade. (2001, January 23). *CNN Web Posting*. Retrieved January 24, 2001, from http://www.cnn.com/2001/HEALTH/01/23/teen.drinking/index.html.

Peltzer, K., & Machleidt, W. (1992). A traditional (African) approach towards the therapy of schizophrenia and its comparison with Western models. *Therapeutic Communities International Journal for Therapeutic and Supportive Organizations, 13*, 229–242.

Pendery, M. L., Maltzman, I. M., & West, L. J. (1982). Controlled drinking by alcoholics? New findings and a reevaluation of a major affirmative study. *Science, 217*, 169–174.

Pengilly, J. W., & Dowd, E. T. (2000). Hardiness and social support as moderators of stress. *Journal of Clinical Psychology, 56*, 813–820.

Penn, D. L. (1998, June). Assessment and treatment of social dysfunction in schizophrenia. *Clinician's Research Digest, Supplemental Bulletin 18*.

Penn, D. L., & Mueser, K. T. (1996). Research update on the psychosocial treatment of schizophrenia. *American Journal of Psychiatry, 153*, 607–617.

Penn, D. L., et al. (2000). Emotion recognition in schizophrenia: Further investigation of generalized versus specific deficit models. *Journal of Abnormal Psychology, 109*, 555–558.

Penninx, B. W., et al. (2000). The protective effect of emotional vitality on adverse health outcomes in disabled older women. *Journal of the American Geriatrics Society, 48*, 1359–1366.

Penninx, B. W., et al. (2001). Depression and cardiac mortality: Results from a community-based longitudinal study. *Archives of General Psychiatry, 58*, 221–227.

Petrie, K. J., Booth, R. J., & Pennebaker, J. W. (1998). The immunological effects of thought suppression. *Journal of Personality and Social Psychology, 75*, 1264–1272.

Peplau, L. A. (1991). Lesbian and gay relationships. In J. Gonsiorek & J. Weinrich (Eds.), *Homosexuality: research implications for public policy* (pp. 177–196). Newbury Park, CA: Sage.

Peppard, P. E., Young, T., Palta, M., & Skatrud, J. (2000). Prospective study of the association between sleep-disordered breathing and hypertension. *The New England Journal of Medicine, 342*, 1378–1384.

Perilstein, R. D., Lipper, S., & Friedman, L. J. (1991). Three cases of paraphilias responsive to fluoxetine treatment. *Journal of Clinical Psychiatry, 52*, 169–170.

Perlin, M. L. (1994). Law and the delivery of mental health services in the community. *American Journal of Orthopsychiatry, 64*, 194–208.

Perlman, J. D., & Abramson, P. R. (1982). Sexual satisfaction among married & cohabitating individuals. *Journal of Consulting and Clinical Psychology, 50*, 458–460.

Perlman, L. M. (2001). Nonspecific, unintended, and serendipitous effects in psychotherapy. *Professional Psychology: Research and Practice, 32*, 283–288.

Perry, E. K., et al. (2001). Cholinergic activity in autism: Abnormalities in the cerebral cortex and basal forebrain. *American Journal of Psychiatry, 158*, 1058–1066.

Persaud, R. (2000). Recurrent depression and stressful life events: In reply. *Archives of General Psychiatry, 57*, 617. [Letter]

Peterson, E. D., et al. (1997). Racial variation in the use of coronary-revascularization procedures—Are the differences real? Do they matter? *The New England Journal of Medicine, 336*, 480–486.

Pettingale, K. W. (1985). Towards a psychobiological model of cancer: Biological considerations. Special issue: Cancer and the mind. *Social Science and Medicine, 20*, 779–787.

Peyron, C., et al. (2000). A mutation in a case of early onset narcolepsy and a generalized absence of hypocretin peptides in human nacroleptic brains. *Nat Med, 6*, 991–997.

Phaf, R. H., Geurts, H., & Eling, P. A. T. M. (2000). Word frequency and word stem completion in Korsakoff patients. *Journal of Clinical & Experimental Neuropsychology, 22*, 817–829.

Philipps, L. H., & O'Hara, M. W. (1991). Prospective study of postpartum depression: 4½-year follow-up of women and children. *Journal of Abnormal Psychology, 100*, 151–155.

Phinney, J. (1989). Stages of ethnic identity in minority group adolescents. *Journal of Early Adolescence, 9*, 34–49.

Phinney, J., & Alipuria, L. (1990). Ethnic identity in older adolescents from four ethnic groups. *Journal of Adolescence, 13*, 171–183.

Phinney, J., Lochner, B., & Murphy, R. (1990). Ethnic identity development and psychological adjustment in adolescence. In A. Stiffman & L. Davis (Eds.), *Ethnic issues in adolescent mental health*. Newbury Park. CA: Sage.

Pianta, R. C., & Egeland, B. (1994). Relation between depressive symptoms and stressful life events in a sample of disadvantaged mothers. *Journal of Consulting and Clinical Psychology, 62*, 1229–1234.

Pihl, R. O., Peterson, J., & Finn, P. (1990). Inherited predispostion to alcoholism: Characteristics of sons of male alcoholics. *Journal of Abnormal Psychology, 99*, 291–301.

Pike, K. M., & Rodin, J. (1991). Mothers, daughters, and disordered eating. *Journal of Abnormal Psychology, 101*, 198–204.

Pilowsky, J. E. (1993). The courage to leave: An exploration of Spanish-speaking women victims of spousal abuse. *Canadian Journal of Community Mental Health, 12*, 15–29.

Pinderhughes, E. (1989). *Understanding race, ethnicity and power: Keys to efficacy in clinical practice.* New York: Free Press.

Pine, D. S., et al. (1997). Neuroendocrine response to fenfluramine challenge in boys: Associations with aggressive behavior and adverse rearing. *Archives of General Psychiatry, 54*, 839–846.

Pinel, J. P. J., Assanand, S., & Lehman, D. R. (2000). Hunger, eating, and ill health. *American Psychologist, 55*, 1105–1116.

Piven, J., et al. (1997). An MRI study of the corpus callosum in autism. *American Journal of Psychiatry, 154*, 1051–1056.

Plomin, R., DeFries, J., McClearn, G. E., & Rutter, M. (1997). *Behavioral genetics* (3rd ed.). New York: Freeman.

Plomin, R., Owen, M. J., & McGuffin, P. (1994). The genetic basis of complex human behaviors. *Science, 264*, 1733–1739.

Pogge, D. L. (1992). Risk factors in child abuse and neglect. *Journal of Social Distress and the Homeless, 1*, 237–248.

Polaschek, D. L. L., Ward, T., & Hudon, S. M. (1997). Rape and rapists: Theory and treatment. *Clinical Psychology Review, 17*, 117–144.

Polcin, D. L. (1992). Issues in the treatment of dual diagnosis clients who have chronic mental illness. *Professional Psychology: Research and Practice, 23*, 30–37.

Pollock, V. E. (1992). Meta-analysis of subjective sensitivity to alcohol in sons of alcoholics. *American Journal of Psychiatry, 149*, 1534–1538.

Pope, H. G., Jr., & Yurgelun-Todd, D. (1996). The residual cognitive effects of heavy marijuana use in college students. *Journal of the American Medical Association, 275*, 521–527.

Potter, J. D. (1997). Hazards and benefits of alcohol. *The New England Journal of Medicine, 337*, 1763–1764.

Poulakis, Z., & Wertheim, E. H. (1993). Relationships among dysfunctional cognitions, depressive symptoms, and bulimic tendencies. *Cognitive Therapy and Research, 17*, 549–559.

Powchik, P., et al. (1998). Postmortem studies in schizophrenia. *Schizophrenia Bulletin, 24*, 325–341.

Powell, E. (1991). *Talking back to sexual pressure.* Minneapolis: CompCare Publishers.

Price, J. L., Hilsenroth, M. J., Petretic-Jackson, P. A., & Bonge, D. (2001). A review of individual psychotherapy outcomes for adult survivors of childhood sexual abuse. *Clinical Psychology Review, 21*, 1095–1121.

Price, L. H., & Heninger, G. R. (1994). Lithium in the treatment of mood disorders. *New England Journal of Medicine, 331*, 591–598.

Prichard, J. C. (1835). *Treatise on insanity.* London: Gilbert & Piper.

Prigatano, G. P. (1992). Personality disturbances associated with traumatic brain injury. *Journal of Consulting and Clinical Psychology, 60*, 360–368.

Prigerson, H. G., et al. (2001). Combat trauma: Trauma with highest risk of delayed onset and unresolved posttraumatic stress disorder symptoms, unemployment, and abuse among men. *Journal of Nervous & Mental Disease, 189*, 99–108.

Project MATCH Research Group. (1997). Matching alcoholism treatments to client heterogeneity: Project MATCH posttreatment drinking outcomes. *Journal of Studies on Alcohol, 58*, 7–29.

Prudic, J., et al. (1996). Resistance to antidepressant medications and short-term clinical response to ETC. *American Journal of Psychiatry, 153*, 985–992.

Pumariega, A. J. (1986). Acculturation and eating attitudes in adolescent girls: A comparative correlational study. *Journal of the American Academy of Child Psychiatry, 25*, 276–279.

Purdum, T. S. (2001, March 30). Non-Hispanic whites a minority, California census figures show. *The New York Times*, pp. A1, A18.

Putnam, F. W., & Carlson, E. B. (1994). "Screening for multiple personality disorder with the Dissociative Experiences Scale": A reply. *American Journal of Psychiatry, 151*, 1249–1250.

Putnam, F. W., Guroff, J. J., Silberman, E. K., Barban, L., & Post, R. M. (1986). The clinical phenomenology of multiple personality disorder: Review of 100 recent cases. *Journal of Clinical Psychiatry, 47*, 285–293.

Pyszczynski, T., & Greenberg, G. (1985). Depression and preference for self-focusing stimuli after success and failure. *Journal of Personality and Social Psychology, 49*, 1066–1075.

Pyszczynski, T., & Greenberg, G. (1986). Evidence for a depressive self-focusing style. *Journal of Research in Personality, 20*, 95–106.

Pyszczynski, T., & Greenberg, J. (1987). Self-regulatory perseveration and the depressive self-focusing style: A self-awareness theory of reactive depression. *Psychological Bulletin, 102*, 122–138.

Q

Quay, H. C. (1965). Psychopathic personality as pathological stimulation seeking. *American Journal of Psychiatry, 122*, 180–183.

Quist, J., & Kennedy, J. L. (2001). Genetics of childhood disorders: XXIII. ADHD, Part 7: The serotonin system. *Journal of American Academy of Child & Adolescent Psychiatry, 40*, 253–256.

R

Rabasca, L. (2000a, March). Listening instead of preaching. *Monitor on Psychology, 31*, pp. 50–51.

Rabasca, L. (2000b, July/August). Therapy that starts online but aims to continue in the psychologist's office. *Monitor on Psychology*, p. 15.

Rabkin, J. G., Wagner, G. J., & Rabkin R. (2000). A double-blind, placebo-controlled trial of testosterone therapy for HIV-positive men with hypogonadal symptoms. *Archives of General Psychiatry, 57*, 141–147.

Rachman, S. (2000). Joseph Wolpe (1915–1997). *American Psychologist, 55*, 431.

Rachman, S., & Bichard, S. (1988). The overprediction of fear. *Clinical Psychology Review, 8*, 303–312.

Rachman, S. J. (1994). Overprediction of fear: A review. *Behaviour Research and Therapy, 32*, 683–690.

Ragland, J. D., et al. (1999). Neuropsychological laterality indices of schizophrenia: Interactions with gender. *Schizophrenia Bulletin, 25*, 79–89.

Ragland, J. D., et al. (2001). Effect of schizophrenia on frontotemporal activity during word encoding and recognition: A PET cerebral blood flow study. *American Journal of* Psychiatry, *158*, 1114–1125.

Räikkönen. K., et al. (1999). Effects of hostility on ambulatory blood pressure and mood during daily living in healthy adults. *Health Psychology, 18*, 44–53.

Raine, A., et al. (2000). Reduced prefrontal gray matter volume and reduced autonomic activity in antisocial personality disorder. *Archives of General Psychiatry, 57*, 119–127.

Rao, K., DiClemente, R. J., & Ponton, L. E. (1992). Child sexual abuse of Asians compared with other populations. *Journal of the American Academy of Child and Adolescent Psychiatry, 31*, 880–886.

Rao, S. M., Huber, S. J., & Bornstein, R. A. (1992). Emotional changes with multiple sclerosis and Parkinson's disease. *Journal of Consulting and Clinical Psychology, 60*, 369–378.

Rapee, R. M. (1987). The psychological treatment of panic attacks: Theoretical conceptualization and review of evidence. *Clinical Psychology Review, 7*, 427–438.

Rapee, R. M. (1991). Generalized anxiety disorder: A review of clinical features and theoretical concepts. *Clinical Psychology Review, 11*, 419–440.

Rapin, I. (1997). Autism. *The New England Journal of Medicine, 337*, 97–104.

Rathus, S. A. (1978). Treatment of recalcitrant ejaculatory incompetence. *Behavior Therapy, 9*, 962.

Rathus, S. A. (1999). *Psychology* (7th ed.). Fort Worth: Harcourt Brace College Publishers.

Rathus, S. A., & Fichner-Rathus, L. (1994). *Making the most of college* (2nd ed.). Englewood Cliffs, NJ: Prentice-Hall.

Rathus, S. A., & Nevid, J. S. (1977). *Behavior therapy*. Garden City, NY: Doubleday.

Rathus, S. A., Nevid, J. S., & Fichner-Rathus, L. (2002). *Human sexuality in a world of diversity* (5th ed.). Boston: Allyn & Bacon.

Ratti, L. A., Humphrey, L. L., & Lyons, J. S. (1996). Structural analysis of families with a polydrug-dependent, bulimic, or normal adolescent daughter. *Journal of Consulting & Clinical Psychology, 64*, 1255–1262.

Rauch, S. L., & Jenike, M. A. (1998). Pharmacological treatment of obsessive complusive disorder. In P. E. Nathan & J. M. Gorman (Eds.), *A guide to treatments that work* (pp. 358–376). New York: Oxford University Press.

Rauschenberger, S. L., & Lynn, S. J. (1995). Fantasy proneness, *DSM-III-R* Axis I psychopathology, and dissociation. *Journal of Abnormal Psychology, 104*, 373–380.

Ravenholt, R. (1984). Addiction mortality in the United States. 1980: Tobacco, alcohol, and other substances. *Population and Development Review, 10*, 697–724.

Read, J. et al. (2001). Assessing suicidality in adults: Integrating childhood trauma as a major risk factor. *Professional Psychology: Research and Practice, 32*, 367–372.

Ready, T. (2000, June 7). Meditation apparently good for the heart as well as the mind. *CNN Web Posting.* Retrieved June 8, 2000, from http://www.cnn.com/2000/HEALTH/06/07/minding.heart.wmd/.

Redd, W. H. (1995). Behavioral research in cancer as a model for health psychology. *Health Psychology, 14*, 99–100.

Redd, W. H., & Jacobsen, P. (2001). Behavioral intervention in comprehensive cancer care. In A. Baum, T. A. Revenson, & J. E. Singer (Eds.), *Handbook of health psychology* (pp. 757–776). Mahwah, NJ: Erlbaum.

Reed, G. M., McLaughlin, C. J., & Milholland, K. (2000). Ten interdisciplinary principles for professional practice in telehealth: Implications for psychology. *Professional Psychology: Research and Practice, 31*, 170–178.

Regehr, C., Hill, J., & Glancy, G. D. (2000). Individual predictors of traumatic reactions in firefighters. *Journal of Nervous & Mental Disease, 188*, 333–339.

Reich, J. (1996). The morbidity of *DSM-III-R* dependent personality disorder. *Journal of Nervous & Mental Disease, 184*, 22–26.

Reich, J., & Noyes, R. (1986). Letters to the Editor: Differentiating schizoid and avoidant personality disorders. *American Journal of Psychiatry, 143*, 1061–1063.

Reichman, M. E. (1994). Alcohol and breast cancer. *Alcohol Health and Research World, 18*, 182–183.

Reid, B. V., & Whitehead, T. L. (1992). Introduction. In T. L. Whitehead & B. V. Reid (Eds.), *Gender constructs and social issues* (pp. 1–9). Chicago: University of Illinois.

Reinisch, J. M. (1990). *The Kinsey Institute new report on sex: What you must know to be sexually literate.* New York: St. Martin's Press.

Reisberg, B., Ferris, S. H., DeLeon, M. J., & Crook, T. (1982). The Global Deterioration Scale for Assessment of Primary Degenerative Dementia. *American Journal of Psychiatry, 139*, 1136–1139.

Reisberg, B., et al. (1986). Assessment of presenting symptoms. In L. W. Poon (Ed.), *Handbook for clinical memory assessment of older adults* (pp. 108–128). Washington, DC: American Psychological Association.

Reisner, A. D. (1994). Multiple personality disorder diagnosis: A house of cards? *American Journal of Psychiatry, 151*, 629.

Reisner, A. D. (1996). Repressed memories: True and false. *Psychological Record, 46*, 563–579.

Reiss, B. F. (1980). Psychological tests in homosexuality. In J. Marmor (Ed.), *Homosexual behavior* (pp. 296–311). New York: Basic Books.

Reiss, S., & Valenti-Hein, D. (1994). Development of a psychopathology rating scale for children with mental retardation. *Journal of Consulting and Clinical Psychology, 62*, 28–33.

Reneman, L., et al. (2001). Cortical serotonin transporter density and verbal memory in individuals who stopped using 3,4-methylenedioxymethamphetamine (MDMA or "Ecstasy"): Preliminary findings. *Archives of General Psychiatry, 58*, 901–906.

Renfrey, G. S. (1992). Cognitive-behavior therapy and the Native American client. *Behavior Therapy, 23*, 321–340.

Rennison, C. M. (2001). *Intimate partner violence and age of victim, 1993 to 1999.* Washington, DC: U.S. Department of Justice, Bureau of Justice Statistics.

Report: Adolescent suicide rates rise. (1995, April 21). *New York Newsday*, p. A54.

Researchers gain insight into Huntington's. (2001, March 22). *CNN Web Posting.* Retrieved March 24, 2001, from http://www.cnn.com/2001/HEALTH/conditions/03/22/huntingtons.gene/index.html.

Rey, J. M. (1993). Oppositional defiant disorder. *American Journal of Psychiatry, 150*, 1769–1778.

Reynolds, C. F., III, et al. (1996). Treatment outcome in recurrent major depression: A post hoc comparison of elderly ("Young Old") and midlife patients. *American Journal of Psychiatry, 153*, 1288–1292.

Ribisl, K. M., et al. (2000). English language use as a risk factor for smoking initiation among Hispanic and Asian American Adolescents: Evidence for mediation by tobacco-related beliefs and social norms. *Health Psychology, 19*, 403–410.

Ricciardelli, L. A., & McCabe, M. P. (2001). Children's body image concerns and eating disturbance: A review of the literature. *Clinical Psychology Review, 21*, 325–344.

Richards, J. C., Edgar, L. V., & Gibbon, P. (1996). Cardiac acuity in panic disorder. *Cognitive Therapy and Research, 20*, 361–376.

Richards, R. (1994). Creativity and bipolar mood swings: Why the association? In M. P. Shaw & M. A. Runco (Eds.), *Creativity and affect* (pp. 44–72). Norwood: Ablex.

Riddle, M. A., et al. (2001). Fluvoxamine for children and adolescents with obsessive-compulsive disorder: A randomized controlled, multicenter trial. *Journal of the American Academy of Child and Adolescent Psychiatry, 40*, 222–229.

Ridley, C. R. (1984). Clinical treatment of the nondisclosing Black client: A therapeutic paradox. *American Psychologist, 39*, 1234–1244.

Riedel, B. R. W., & Lichstein, K. L. (2001). Strategies for evaluating adherence to sleep restriction treatment for insomnia. *Behaviour Research and Therapy, 39*, 201–212.

Riether, A. M., & Stoudemire, A. (1988). Psychogenic fugue states: A review. *Southern Medical Journal, 81*, 568–571.

Riley, V. (1981). Psychoneuroendocrine influences on immunocompetence and neoplasia. *Science, 212*, 1100–1109.

Rimer, S. (1999, September 5). Gaps seen in treating depression in elderly. *The New York Times*, pp. 1, 18.

Rimland, B. (1978). The savant capabilities of autistic children and their cognitive implications. In G. Serban (Ed.), *Cognitive defects in the development of mental illness.* New York: Brunner-Mazel.

Risk factors help ID violence-prone youths, psychiatrists say. (2000, May 16). *American Heart Association Web Posting.* Retrieved May 19, 2000, from http://www.medicus.clk/AspNyheder/.

Ritter, C., et al. (2000). Stress, psychosocial resources, and depressive symptomatology during pregnancy in low-income, inner-city women. *Health Psychology, 19*, 576–585.

Ritvo, E. R., & Ritvo, R. (1992). "The UCLA-University of Utah Epidemiologic Survey of Autism: The etiologic role of rare diseases": Reply. *American Journal of Psychiatry, 149*, 146–147.

Rivara, F. P., et al. (1997). Alcohol and illicit drug abuse and the risk of violent death in the home. *Journal of the American Medical Association, 278*, 569–575.

Robins, L. N., Locke, B. Z., & Reiger, D. A. (1991). An overview of psychiatric disorders in America. In L. N. Robins & D. A. Regier (Eds.), *Psychiatric disorders in America: The Epidemiologic Catchment Area Study* (pp. 328–366). New York: Free Press.

Robins, L. N., Tipp, J., & Przybeck, T. (1991). Antisocial personality. In L. N. Robins & D. A. Regier (Eds.), *Psychiatric disorders in America: The Epidemiologic Catchment Area Study* (pp. 258–290). New York: Free Press.

Robinson, N. M., Zigler, E., & Gallagher, J. J. (2001). Two tails of the normal curve: Similarities and differences in the study of mental retardation and giftedness. *American Psychologist, 55*, 1413–1424.

Robinson, T., et al. (1996). Ethnicity and body dissatisfaction: Are Hispanic and Asian girls at increased risk of eating disorders. *Journal of Adolescent Health, 19*, 384–393.

Rocha, B. A., et al. (1998). Increased vulnerability to cocaine in mice lacking the serotonin-1 B receptor. [Letter]. *Nature, 393*, 175.

Rock, C. L., & Curran-Celentano. J. (1996). Nutritional management of eating disorders. *The Psychiatric Clinics of North America, 19*, 701–713.

Rodin, J., Bartoshuk, L., Peterson, C., & Schank, D. (1990). Bulimia and taste: Possible interactions. *Journal of Abnormal Psychology, 99*, 32–39.

Rodríguez de Fonseca, F., et al. (1997). Activation of corticotropin-releasing factor in the limbic system during cannabinoid withdrawal. *Science, 276*, 2050–2054.

Rodriguez, G. (2001, February 11). *The New York Times Week in Review*, Section 4, pp. 1, 4.

Rodriguez, N., et al. (1997). Posttraumatic stress disorder in adult female survivors of childhood sexual abuse: A comparison study. *Journal of Consulting and Clinical Psychology, 65*, 53–59.

Roesch, R., Zapf, P. A., Golding, S. L., & Skeem, J. L. (1999). Defining and assessing competence to stand trial. In A. K. Hess, I. B. Weiner, et al. (Eds.), *The handbook of forensic psychology* (2nd ed., pp. 327–349). New York: Wiley.

Roesler, A., & Witztum, E. (2000). Pharmacotherapy of paraphilias in the next millennium. *Behavioral Sciences & the Law, 18*(1), 43–56.

Rogan, A. (1986, Fall). Recovery from alcoholism: Issues for black and Native American alcoholics. *Alcohol Health and Research World, 10*, 42–44.

Rogers, C. R. (1951). *Client-centered therapy.* Boston: Houghton Mifflin.

Rogers, R., et al. (1990). The clinical presentation of command hallucinations in a forensic population. *American Journal of Psychiatry, 147*, 1304–1307.

Rogler, L. H., Cortes, D. E., & Malgady, R. G. (1991). Acculturation and mental health status among Hispanics: Convergence and new directions for research. *American Psychologist, 46*, 584–597.

Rohde, P., Lewinsohn, P. M., & Seeley, J. R. (1991). Comorbidity of unipolar depression: II. Comorbidity with other mental disorders in adolescents and adults. *Journal of Abnormal Psychology, 101*, 214–222.

Romanczyk, R. G. (1986). Some thoughts on future trends in the education of individuals with autism. *The Behavior Therapist, 8*, 162–164.

Romano, E., & De-Luca, R. V. (1996). Characteristics of perpetrators with histories of sexual abuse. *International Journal of Offender Therapy and Comparative Criminology, 40*, 147–156.

Rose, J. (1996). Anger management: A group treatment program for people with mental retardation. *Journal of Developmental and Physical Disabilities, 8*, 133–149.

Rosen, J. C. (1996). Body dysmorphic disorder: Assessment and treatment. In J. K. Thompson (Ed.), *Body image, eating disorders, and obesity* (pp. 149–170). Washington, DC: American Psychological Association.

Rosen, R. C. (1996). Erectile dysfunction: The medicalization of male sexuality. *Clinical Psychology Review, 16*, 497–519.

Rosen, R. C., & Leiblum, S. R. (1995). Treatment of sexual disorders in the 1990s: An integrated approach. *Journal of Consulting and Clinical Psychology, 63*, 877–890.

Rosenberg, H. (1993). Prediction of controlled drinking by alcoholics and problem drinkers. *Psychological Bulletin, 113*, 129–130.

Rosenberg, M. L. (1993). Promoting safety and nonviolent conflict resolution in adolescence. In S. G. Millstein, A. C. Petersen, & E. O. Nightingale (Eds.), *Promoting adolescent health: Third symposium on research opportunities in adolescence.* New York: Carnegie Council on Adolescent Development.

Rosenfeld, B. D. (1992). Court-ordered treatment of spouse abuse. *Clinical Psychology Review, 12*, 205–226.

Rosenheck, R. (2000). Cost-effectiveness of services for mentally ill homeless people: The application of research to policy and practice. *American Journal of Psychiatry, 157*, 1563–1570.

Rosenthal, D., et al. (1968). Schizophrenics' offspring reared in adoptive homes. In D. Rosenthal & S. S. Kety (Eds.), *The transmission of schizophrenia.* Oxford: Pergamon Press.

Rosenthal, D., et al. (1975). Parent-child relationships and psychopathological disorder in the child. *Archives of General Psychiatry, 32*, 466–476.

Rosenthal, E. (1993, April 9). Who will turn violent? Hospitals have to guess. *The New York Times*, pp. A1, C12.

Ross, C. A., Norton, G. R., & Wozney, K. (1989). Multiple personality disorder: An analysis of 236 cases. *Canadian Journal of Psychiatry, 34*, 413–418.

Ross, C. A., et al. (1990). Structured interview data on 102 cases of multiple personality disorder from four centers. *American Journal of Psychiatry, 147*, 596–601.

Ross, C. A., et al. (1991). The frequency of multiple personality disorder among psychiatric inpatients. *American Journal of Psychiatry, 148*, 1717–1720.

Ross, D. E. (2000). The deficit syndrome and eye tracking disorder may reflect a distinct subtype within the syndrome of schizophrenia. *Schizophrenia Bulletin, 26*, 855–866.

Ross, M., & Need, J. (1989). Effects of adequacy of gender reassignment surgery on psychological adjustment: A follow-up of fourteen male-to-female patients. *Archives of Sexual Behavior, 18*, 145–153.

Ross, R., Dagnone, D., Jones, P. J., Smith, H., Paddags, A., & Hudson, R. (2000). Reduction in obesity and related comorbid conditions after diet-induced weight loss or exercise-induced weight loss in men: A randomized, controlled trial. *Annals of Internal Medicine, 133*, 92–103.

Ross, S. M. (1996). Risk of physical abuse to children of spouse-abusing parents. *Child Abuse and Neglect, 20*, 589–598.

Rossman, B. B. R. (2001). Longer term effects of children's exposure to domestic violence. In S. A. Graham-Bermann & J. L. Edleson (Eds.), *Domestic violence in the lives of children: The future of research, intervention, and social policy* (pp. 35–65). Washington, DC: American Psychological Association.

Rosso, I. M., et al. (2000). Obstetric risk factors for early-onset schizophrenia in a Finnish birth cohort. *American Journal of Psychiatry, 157*, 816–818.

Rothbaum, B. O. (1996). Virtual reality exposure therapy in the treatment of fear of flying: A case report. *Behaviour Research and Therapy, 34*, 477–481.

Rothbaum, B. O., et al. (1995). Effectiveness of computer-generated (virtual reality) graded exposure in the treatment of acrophobia. *American Journal of Psychiatry, 152*, 626–628.

Rotheram-Borus, M. J., Trautman, P. D., Dopkins, S. C., & Shrout, P. E. (1990). Cognitive style and pleasant activities among female adolescent suicide attempters. *Journal of Consulting and Clinical Psychology, 58*, 554–561.

Rotter, J. B. (1966). Generalized expectancies for internal vs. external control of reinforcement. *Psychological Monographs, 1*, 210–609.

Rotter, J. B. (1990). Internal versus external control of reinforcement: A case history of a variable. *American Psychologist, 45*, 489–493.

Rotundo, M., Nguyen, D-H., & Sackett, P. R. (2001). A meta-analytic review of gender differences in perceptions of sexual harassment. *Journal of Applied Psychology, 86*, 914–922.

Rounsaville, B. J., & Kosten, T. R. (2000). Treatment for opioid dependence: Quality and access. *Journal of the American Medical Association, 283*, 1337–1339. [Editorial]

Rousseau, F., et al. (1991). Direct diagnosis by DNA analysis of the fragile X syndrome of mental retardation. *New England Journal of Medicine, 325*, 1673–1681.

Rowland, D. L., Cooper, S. E., & Slob, A. K. (1996). Genital and psychoaffective response to erotic stimulation in sexually functional and dysfunctional men. *Journal of Abnormal Psychology, 105*, 194–203.

Roy, A., et al. (1991). Suicide in twins. *Archives of General Psychiatry, 48*, 29–32.

Roy, E. (2000). Relation of family history of suicide to suicide attempts in alcoholics. *American Journal of Psychiatry, 157*, 2050–2051.

Roy-Byrne, P. P., & Cowley, D. S. (1998). Pharmacological treatment of panic, generalized anxiety, and phobic disorders. In P. E. Nathan & J. M. Gorman (Eds.), *A guide to treatments that work* (pp. 319–338). New York: Oxford University Press.

Rubin, L. J. (1996). Childhood sexual abuse: False accusations of "false memory"? *Professional Psychology: Research and Practice, 27*, 447–451.

Rubinstein, S., & Caballero, B. (2000). Is Miss America an undernourished role model? *Journal of the American Medical Association, 283*, 1569.

Rugino, T., & Copley, T. C. (2001). Effects of modafinil in children with attention-deficit/hyperactivity disorder: An open-label study. *Journal of American Academy of Child & Adolescent Psychiatry, 40*, 230–235.

Ruscio, A. M., Borkovec, T. D., & Ruscio, J. (2001). A taxometric investigation of the latent structure of worry. *Journal of Abnormal Psychology, 110*, 413–422.

Rush, A. J., & Weissenburger, J. E. (1994). Do thinking patterns predict depressive symptoms? *Cognitive Therapy and Research, 10*, 225–236.

Rutenberg, J. (2001, July 25). Survey shows few parents use TV V-Chip to limit children's viewing. *The New York Times*, pp. E1, E7.

Rutter, M. (1983). Cognitive deficits in the pathogenesis of autism. *Journal of Child Psychology & Psychiatry, 24*, 513–531.

Rutter, M. (1997). Implications of genetic research for child psychiatry. *Canadian Journal of Psychiatry, 42*, 569–576.

Ryan, G. (1993). Working with perpetrators of sexual abuse and domestic violence. *Pastoral Psychology, 41*, 303–319.

Rychtarik, R. G., et al. (2000). Treatment settings for persons with alcoholism: Evidence for matching clients to inpatient versus outpatient care. *Journal of Consulting and Clinical Psychology, 68*, 277–289.

Ryder, A. G., Alden, L. E., & Paulhus, D. L. (2000). Is acculturation unidimensional or bidimensional? A head-to-head comparison in the prediction of personality, self-identity, and adjustment. *Journal of Personality and Social Psychology, 79*(1), 49–65.

S

Sabol, S. Z.., et al. (1999). A genetic association for cigarette smoking behavior. *Health Psychology, 18*, 7–13.

Sachdev, P., & Hay, P. (1996). Site and size of lesion and psychosurgical outcome in obsessive-compulsive disorder: A magnetic resonance imaging study. *Biological Psychiatry, 39*, 739–742.

Sachs, G. S., Lafer, B., Truman, C., Noeth, M. & Thibault, A. B. (1994). Lithium monotherapy: Miracle, myth and misunderstanding. *Psychiatric Annals, 24*, 299–306.

Sachs, S. (2001, March 11). Redefining minority. *The New York Time, Week in Review*, Section 4, pp. 1, 4.

Sack, W. H., Clarke, G. N., & Seeley, J. (1996). Multiple forms of stress in Cambodian adolescent refugees. *Child Development, 67*, 107–116.

Sackheim, H. A., Prudic, J., & Devanand, D. P. (1990). Treatment of medication-resistant depression with electroconvulsive therapy. In A. Tasman, et al. (Eds.), *Review of psychiatry, Vol. 9*. Washington, DC: American Psychiatric Press.

Sackeim, H. A., & Vaughn McCall, W. (2001). In reply. *Archives of General Psychiatry, 58*, 608–609. [Letter]

Sackheim, H. A., et al. (1994). Effects of stimulus intensity and electrode placement on the efficacy and cognitive effects of electroconvulsive therapy. *New England Journal of Medicine, 328*, 839–846.

Sackeim, H. A., et al. (2000). A prospective, randomized, double-blind comparison of bilateral and right unilateral electroconvulsive therapy at different stimulus intensities. *Archives of General Psychiatry, 57*, 425–434.

Sackeim, H. A., et al. (2001). Continuation pharmacotherapy in the prevention of relapse following electroconvulsive therapy. *Journal of the American Medical Association, 285*, 1299–1307.

Sacks, O. (1985a). *The man who mistook his wife for a hat and other clinical tales*. New York: Summit.

Sadler, A. G., Booth, B. M., Nielson, D., & Doebbeling, B. N. (2000). Health-related consequences of physical and sexual violence: Women in the military. *Obstetrics & Gynecology, 96*(3), 473–480.

Safran, J. D., & Messer, S. B. (1997). Psychotherapy integration: A postmodern critique. *Clinical Psychology: Science and Practice, 4*, 140–152.

Salgado de Snyder, V. N. (1987). Factors associated with acculturative stress and depressive symptomatology among married Mexican immigrant women. *Psychology of Women Quarterly, 11*, 475–488.

Salgado de Snyder, V. N., Cervantes, R. C., & Padilla, A. M. (1990). Gender and ethnic differences in psychosocial stress and generalized distress among Hispanics. *Sex Roles, 22*, 441–453.

Salisbury, D. F., et al. (1998). First-episode schizophrenic psychosis differs from first-episode affective psychosis and controls in P300 amplitude over left temporal lobe. *Archives of General Psychiatry, 55*, 173–180.

Salkovskis, P. M., & Clark, D. M. (1993). Panic disorder and hypochondriasis. Special issue: Panic, cognitions and sensations. *Advances in Behaviour Research and Therapy, 15*, 23–48.

Salokangas, R. K. R., & Saarinen, S. (1998). Deinstitutionalization and schizophrenia in Finland: I. Discharged patients and their care. *Schizophrenia Bulletin, 24*, 457–467.

Sammons, M. T., & Brown, A. B. (1997). The Department of Defense psychopharmacology demonstration project: An evolving program for postdoctoral education in psychology. *Professional Psychology: Research and Practice, 28*, 107–112.

Sanchez, E. G., & Mohl, P. C. (1992). Psychotherapy with Mexican-American patients. *American Journal of Psychiatry, 149*, 626–630.

Sanchez-Craig, M., Annis, H. M., Bornet, A. R., & MacDonald, K. R. (1984). Random assignment to abstinence or controlled drinking: Evaluation of a cognitive-behavioral program for problem drinkers. *Journal of Consulting and Clinical Psychology, 52*, 390–403.

Sanchez-Craig, M., & Wilkinson, D. A. (1986/1987). Treating problem drinkers who are not severely dependent on alcohol. *Drugs and Society, 1*, 39–67.

Sanders, B., & Green, J. A. (1995). The factor structure of the Dissociative Experiences Scale in college students. *Dissociation Progress in the Dissociative Disorders, 7*, 23–27.

Sanderson, W. C., & Barlow, D. H. (1990). A description of patients diagnosed with *DSM-III-R* generalized anxiety disorder. *Journal of Nervous & Mental Disease, 178*, 588–591.

Sanderson, W. C., & Rego, S. A. (2000). Empirically supported treatment for panic disorder: Research, theory, and application of cognitive behavioral therapy. *Journal of Cognitive Psychotherapy, 14*, 219–244.

Sanfilipo, M., et al. (2000). Volumetric measure of the frontal and temporal lobe regions in schizophrenia: Relationship to negative symptoms. *Archives of General Psychiatry, 57*, 471–480.

Sanislow, C. A., Grilo, C. M., & McGlashan, T. H. (2000). Factor analysis of the *DSM-III-R* borderline personality disorder criteria in psychiatric inpatients. *American Journal of Psychiatry, 157*, 1629–1633.

Sar, V., et al. (1996). Structured interview data on 35 cases of dissociative identity disorder in Turkey. *American Journal of Psychiatry, 153*, 1329–1333.

Sass, L. (1982, August 22). The borderline personality. *The New York Times Magazine*, pp. 12–15, 66–67.

Satcher, D. (2000). Mental health: A report of the Surgeon General—Executive summary. *Professional Psychology: Research and Practice, 31*, 5–13.

Satir, V. (1967). *Conjoint family therapy* (rev. ed.). Palo Alto, CA: Science and Behavior Books.

Sato, T. (1997). Seasonal affective disorder and phototherapy: A critical review. *Professional Psychology: Research and Practice, 28*, 164–169.

Saywitz, K. J., Mannarino, A. P., Berliner, L., & Cohen, J. A. (2000). Treatment for sexually abused children and adolescents. *American Psychologist, 55*, 1040–1049.

Schachter, S., & Latané, B. (1964). Crime, cognition, and the autonomic nervous system. In D. Levine (Ed.), *Nebraska symposium on motivation* (Vol. 12, pp. 221–273). Lincoln: University of Nebraska Press.

Schacter, D. L. (1999). The seven sins of memory: Insights from psychology and cognitive neuroscience. *American Psychologist, 54*, 182–203.

Schafer, D. W. (1986). Recognizing multiple personality patients. *American Journal of Psychotherapy, 40*, 500–510.

Schafer, J., & Brown, S. (1991). Marijuana and cocaine effect expectancies and drug use patterns. *Journal of Consulting and Clinical Psychology, 59*, 558–565.

Schaffer, S. J. (2000, November/December). New NY Tarasoff/confidentiality ruling. *NYSPA Notebook*, p. 5.

Scheel, K. R. (2000). The empirical basis of dialectical behavior therapy: Summary, critique, and implications. *Clinical Psychology: Science and Practice, 7*, 68–86.

Scheflin, A. W., & Brown, D. (1996). Repressed memory or dissociative amnesia: What the science says. *Journal of Psychiatry and Law, 24*, 143–188.

Scheier, L. M., Botvin, G. J., & Baker, E. (1997). Risk and protective factors as predictors of adolescent alcohol involvement and transitions in alcohol use: A prospective analysis. *Journal of Studies on Alcohol, 58*, 652–767.

Scheier, M. F., & Carver, C. S. (1985). Optimism, coping, and health: Assessment and implications of generalized outcome expectancies. *Health Psychology, 4*, 219–247.

Scheier, M. F., & Carver, C. S. (1992). Effects of optimisim on psychological and physical well-being: Theoretical overview and empirical update. Special issue: Cognitive perspectives in health psychology. *Cognitive Therapy and Research, 16*, 201–228.

Scheier, M. F., et al. (1999). Optimism and rehospitalization after coronary artery bypass graft surgery. *Archives of Internal Medicine, 159*, 829–935.

Schepis, M. R., Reid, D. H., & Fitzgerald, J. R. (1987). Group instruction with profoundly retarded persons: Acquisition, generalization, and maintenance of a remunerative work skill. *Journal of Applied Behavior Analysis, 20*, 97–105.

Schizophrenia Update—Part I (1995, June). *The Harvard Mental Health Letter, 11*, 1–4.

Schmauk, F. J. (1970). Punishment, arousal, and avoidance learning in sociopaths. *Journal of Abnormal Psychology, 76*, 325–335.

Schmidt, N. B., Lerew, D. R., & Jackson, R. J. (1997). The role of anxiety sensitivity in the pathogenesis of panic: Prospective evaluation of spontaneous panic attacks during acute stress. *Journal of Abnormal Psychology, 106*, 355–364.

Schmidt, N. B., Trakowski, J. H., & Staab, J. P. (1997). Extinction of panicogenic effects of a 35% CO_2 challenge in patients with panic disorder. *Journal of Abnormal Psychology, 106*, 630–638.

Schmidt, N. B., et al. (2000). Dismantling cognitive-behavioral treatment for panic disorder: Questioning the utility of breathing retraining. *Journal of Consulting and Clinical Psychology, 68*, 417–424.

Schmitt, E. (2001a, April 1). U.S. now more diverse, ethnically and racially. *The New York Times*, p. A20.

Schmitt, E. (2001b, March 13). For 7 million people in census, one race category isn't enough. *The New York Times*, pp. A1, A14.

Schmitt, E. (2001c, April 1). U.S. now more diverse, ethnically and racially. *The New York Times*, p. A20.

Schneider, K. (1957). Primäre und sekundäre Symptome bei der Schizophrenia. *Fortschritte der Neurologie Psychiatrie, 25*, 487–490.

Schneiderman, N., Antoni, M. H., Saab, P. G., & Ironson, G. (2001). Health psychology: Psychosocial and biobehavioral aspects of chronic disease management. *Annual Review of Psychology, 52*, 555–580.

Schneier, F. R., Wexler, K. B., & Liebowitz, M. R. (1997). Social phobia and stuttering. *American Journal of Psychiatry, 154*, 131.

Schnurr, P. P., Ford, J. D., & Friedman, M. J. (2000). Predictors and outcomes of posttraumatic stress disorder in World War II veterans exposed to mustard gas. *Journal of Consulting and Clinical Psychology, 68*, 258–268.

Schnyder, U., et al. (2001). Incidence and prediction of posttraumatic stress disorder symptoms in severely injured accident victims. *American Journal of Psychiatry, 158*, 594–599.

Schoenman, T. J. (1984). The mentally ill witch in text books of abnormal psychology: Current status and implications of a fallacy. *Professional Psychiatry, 15*, 299–314.

Schteingart, J. S., et al. (1995). Homeless and child functioning in the context of risk and protective factors moderating child outcomes. *Journal of Clinical Child Psychology, 24*, 320–331.

Schuckit, M. A. (1983). Subjective responses to alcohol in sons of alcoholics and control. *Archives of General Psychiatry, 41*, 879–884.

Schuckit, M. A. (1987). Biological vulnerability to alcoholism. *Journal of Consulting and Clinical Psychology, 55*, 301–309.

Schuckit, M. A. (1996). Recent developments in the pharmacotherapy of alcohol dependence. *Journal of Consulting and Clinical Psychology, 64*, 669–676.

Schuckit, M. A., & Rayes, U. (1979). Ethanol ingestion: Differences in blood acetaldehyde concentrations in relatives of alcoholics. *Science, 203*, 54–55.

Schuckit, M. A., et al. (1999). Clinical implications for four drugs of the *DSM-IV* distinction between substance dependence with and without a physiological component. *American Journal of Psychiatry, 156*, 41–49.

Schwartz, B. S., et al. (1998). Epidemiology of tension-type headache. *Journal of the American Medical Association, 279*, 381–383.

Schwartz, J. M. (1998). Neuroanatomical aspects of cognitive-behavior therapy response in obsessive-compulsive disorder. *British Journal of Psychiatry, 173*, 38–44.

Schwartz, R. H., Voth, E. A., Sheridan, M. J. (1997). Marijuana to prevent nausea and vomiting in cancer patients: A survey of clinical oncologists. *Southern Medical Journal, 90*, 167–172.

Schwartz, R. M. (1986). The internal dialogue: On the asymmetry between positive and negative thoughts. *Cognitive Therapy & Research, 10*, 591–605.

Schwitzgebel, R. L., & Schwitzgebel, R. K. (1980). *Law and psychological practice*. New York: Wiley.

Scogin, F., & McElreath, L. K. (1994). Efficacy of psychosocial treatments for geriatric depression: A quantitative review. *Journal of Consulting and Clinical Psychology, 62*, 69–74.

Scroppo, J. C., Drob, S. L., Weinberger, J. L., & Eagle, P. (1998). Identifying dissociative identity disorder: A self-report and projective study. *Journal of Abnormal Psychology, 107*, 272–284.

Sears, R. R., Maccoby, E. E., & Levin, H. (1957). *Patterns of child rearing*. New York: Harper & Row.

The Sedentary Society. (1996, August). *Harvard Heart Letter, 6*, 3–4.

See, L. (1999, November). My face doesn't match my race. *Self*, pp. 60–61.

Seeman, M. V. (1997). Psychopathology in women and men: Focus on female hormones. *American Journal of Psychiatry, 154*, 1641–1647.

Sees, K. L., et al. (2000). Methadone maintenance vs. 180-day psychosocially enriched detoxification for treatment of opioid dependence. *Journal of the American Medical Association, 283*, 1303–1310.

Segal, J. H. (1989). Erotomania revisited: From Kraepelin to *DSM-III-R*. *American Journal of Psychiatry, 146*, 1261–1266.

Segal, S. P., Bola, J. R., & Watson, M. A. (1996). Race, quality of care, and antipsychotic prescribing practices in psychiatric emergency services. *Psychiatric Services, 47*, 282–286.

Segal, Z.. V., et al. (1992). Cognitive and life stress predictors of relapse in remitted unipolar depressed patients: Test of the congruency hypothesis. *Journal of Abnormal Psychology, 101*, 26–36.

Segell, M. (2000, October 24). Testosterone's not so bad after all. *MSNBC Web Posting*. Retrieved October 28, 2000, from http://www.msnbc.com/news/480175.asp.

Segerstrom, S. C., et al. (1998). Optimism is associated with mood, coping, and immune change in response to stress. *Journal of Personality and Social Psychology, 74*, 1646–1655.

Segraves, R. (1988). Drugs and desire. In S. Leiblum & R. Rosen (Eds.), *Sexual desire disorders*. New York: Guilford Press.

Segraves, R. T., & Althof, S. (1998). Psychotherapy and pharmacotherapy of sexual dysfunctions. In P. E. Nathan & J. M. Gorman (Eds.), *A guide to treatments that work* (pp. 447–471). New York: Oxford University Press.

Segrin, C., & Abramson, L. Y. (1994). Negative reactions to depressive behaviors: A communication theories analysis. *Journal of Abnormal Psychology, 103*, 655–668.

Segrin, C., & Dillard, J. P. (1992). The international theory of depression: A meta-analysis of the research literature. *Journal of Social and Clinical Psychology, 11*, 43–70.

Seligman, L., & Hardenburg, S. A. (2000). Assessment and treatment of paraphilias. *Journal of Counseling & Development, 78*, 107–113.

Seligman, M. E. P. (1973). Fall into helplessness. *Psychology Today, 7*, 43–48.

Seligman, M. E. P. (1975). *Helplessness: On depression, development, and death*. San Francisco: Freeman.

Seligman, M. E. P. (1991). *Learned optimism*. New York: Knopf.

Seligman, M. E. P. (1998, August). *Prevention of depression and positive psychology*. Paper presented at the meeting of the American Psychological Association, San Francisco.

Seligman, M. E. P., & Maier, S. F. (1967). Failure to escape traumatic shock. *Journal of Experimental Psychology, 74*, 1–9.

Seligman, M. E. P., & Rosenhan, D. L. (1984). *Abnormal psychology*. New York: W. W. Norton.

Seligman, M. E. P., et al. (1988). Explanatory style change during cognitive therapy for unipolar depression. *Journal of Abnormal Psychology, 97*, 13–18.

Selkoe, D. J. (1992). Aging brain, aging mind. *Scientific American, 267*(3), 134–142.

Selvin, B. W. (1993, June 1). Transsexuals are coming to terms with themselves and society. *New York Newsday*, pp. 55, 58, 59.

Selye, H. (1976). *The stress of life* (Rev. ed.) New York: McGraw-Hill.

Semrud-Clikeman, M., et al. (2000). Using MRI to examine brain-behavior relationships in males with attention deficit disorder with hyperactivity. *Journal of the American Academy of Child & Adolescent Psychiatry, 39*, 477–484.

Sensky, T., et al. (2000). A randomized controlled trial of cognitive-behavioral therapy for persistent symptoms in schizophrenia resistant to medication. *Archives of General Psychiatry, 57*, 165–172.

Seppa, N. (1997, June). Children's TV remains steeped in violence. *APA Monitor, 28*(6), p. 36.

Seroczynki, A. D., Cole, D. A., & Maxwell, S. E. (1997). Cumulative and compensatory effects of competence and incompetence on depressive symptoms in children. *Journal of Abnormal Psychology, 106*, 586–597.

Shadish, W. R., et al. (2000). The effects of psychological therapies under clinically representative conditions: A meta-analysis. *Psychological Bulletin, 126*, 512–529.

Shadish, W. R., Jr., Lurigio, A. J., & Lewis, D. A. (1989). After deinstitutionalization: The present and future of mental health long-term care policy. *Journal of Social Issues, 45*, 1–15.

Shaffer, C. E., Jr., Waters, W. F., & Adams, S. G., Jr. (1994). Dangerousness: Assessing the risk of violent behavior. *Journal of Consulting and Clinical Psychology, 62*, 1064–1068.

Shaffer, D., Gould, M., & Hicks, R. C. (1994). Worsening suicide rate in black teenagers. *American Journal of Psychiatry, 151*, 1810–1812.

Shafran, R., & Mansell, W. (2001). Perfectionism and psychopathology: A review of research and treatment. *Clinical Psychology Review, 21*, 879–906.

Shalev, A. Y., Yehuda, R., & McFarlane, A. C. (Eds.). (2000). *International handbook of human response to trauma*. New York: Kluwer Academic/Plenum Publishers.

Shapiro, D. A., et al. (1995). Effects of treatment duration and severity of depression on the maintenance of gains after cognitive-behavioral and psychodynamic interpersonal psychotherapy. *Journal of Consulting and Clinical Psychology, 63*, 378–387.

Shapiro, E. (1992, August 22). Fear returns to sidewalks of West 96th Street. *The New York Times*, pp. B3–B4.

Shapiro, F. (1995). *Eye movement desensitization and reprocessing: Basic principles, protocols, and procedures*. New York: Guilford Press.

Shapiro, L. 1998, (June 15). Fat, fatter: But who's counting? *Newsweek*, p. 55.

Sharkansky, E. J., King, D. W., King, L. A., & Wolfe, J. (2000). Coping with Gulf War combat stress: Mediating and moderating effects. *Journal of Abnormal Psychology, 109*, 188–197.

Sharp, T. J., & Harvey, A. G. (2001). Chronic pain and posttraumatic stress disorder: Mutual maintenance? *Clinical Psychology Review, 21*, 857–877.

Shaw, R., Cohen, F., Doyle, B., & Palesky, J. (1985). The impact of denial and repressive style on information gain and rehabilitation outcomes in myocardial infarction patients. *Psychosomatic Medicine, 47*, 262–273.

Shaywitz, S. E. (1998). Dyslexia. *The New England Journal of Medicine, 338*, 307–312.

Shea, M. T., Widiger, T. A., & Klein, M. H. (1992). Comorbidity of personality disorders and depression: Implications for treatment. *Journal of Consulting and Clinical Psychology, 60*, 857–868.

Sheitman, B. B., et al. (1998). Pharmacological treatments of schizophrenia. In P. E. Nathan & J. M. Gorman (Eds.), *A guide to treatments that work* (pp. 167–189). New York: Oxford University Press.

Shelton, R. C., et al. (2001). Effectiveness of St John's Wort in major depression: A randomized controlled trial. *Journal of the American Medical Association, 285*, 1978–1986.

Sherbourne, C. D., Hays, R. D., & Wells, K. B. (1995). Personal and psychosocial risk factors for physical and mental health outcomes and course of depression among depressed patients. *Journal of Consulting and Clinical Psychology, 63*, 345–355.

Sherbourne, C. D., et al. (2000). Impact of psychiatric conditions on health-related quality of life in persons with HIV infection. *American Journal of Psychiatry, 157*, 248–254.

Sherman, D. K., et al. (1997). Twin concordance for attention deficit hyperactivity disorder: A comparison of teachers' and mothers' reports. *American Journal of Psychiatry, 154*, 532–535.

Sherwood, N. E., et al. (2000). The perceived function of eating for bulimic, subclinical bulimic, and non-eating disordered women. *Behavior Therapy, 31*, 777–793.

Sheung-Tak, C. (1996). A critical review of Chinese koro. *Culture, Medicine and Psychiatry, 20*, 67–82.

Shiffman, S., et al. (1996). Progression from a smoking lapse to relapse: Prediction from abstinence violation effects, nicotine dependence, and lapse characteristics. *Journal of Consulting and Clinical Psychology, 64*, 993–1002.

Shifren, J. L., et al. (2000). Transdermal testosterone treatment in women with impaired sexual function after oophorectomy. *The New England Journal of Medicine, 343*, 682–688.

Shneidman, E. S. (1985). *Definition of suicide*. New York: Wiley.

Shnek, Z. M., et al. (2001). Psychological factors and depressive symptoms in ischemic heart disease. *Health Psychology*, 141–145.

Shonk, S. M., & Cicchetti, D. (2001). Maltreatment, competency deficits, and risk for academic and behavioral maladjustment. *Developmental Psychology, 37*, 3–17.

Shopper, M. (1996). Fear of alien abduction. *Journal of the American Academy of Child and Adolescent Psychiatry, 35*, 555–556.

Shute, N., Locy, T., & Pasternak, D. (2000, March 6). The perils of pills. *U.S. News & World Report*, pp. 44–50.

Shweder, R. (1985). Cross-cultural study of emotions. In A. Kleinman & B. Good (Eds.), *Culture and depression*. Berkeley: University of California Press.

Siegel, J. (2001, January 18). St. John's Wort: Nature's antidepressant? *HealthGate Web Posting. NBCI.com*. Retrieved January 28, 2001, from http://healthgate.nbci.com/getcontent.asp?siteid=nbci&docid=/healthy/alternative/2000/johnwort/index

Siegel, L. J. (1992). *Criminology* (4th ed.). St. Paul, MN: West.

Siever, L., & Trestman, R. L. (1993). The serotonin system and aggressive personality disorder. *International Clinical Psychopharmacology, 8*(Suppl. 2), 33–39.

Siever, L. J., et al. (1990). Increased morbid risk for schizophrenia-related disorders in relatives of schizotypal personality disordered patients. *Archives of General Psychiatry, 47*, 634–640.

Silberstein, R. B., et al. (1998). Functional brain electrical activity mapping in boys with attention-deficit/hyperactivity disorder. *Archives of General Psychiatry, 55*, 1105–1112.

Silberstein, S. D., et al. (2000). Rizaptriptan in the treatment of menstrual migraine. *Obstetrics & Gynecology, 96*, 237–242.

Silva, R. R., et al. (2000). Stress and vulnerability to posttraumatic stress disorder in children and adolescents. *American Journal of Psychiatry, 157*, 1229–1235.

Silver, E., Cirincione, C., & Steadman, H. J. (1994). Demythologizing inaccurate perceptions of the insanity defense. *Law and Human Behavior, 18*, 63–70.

Silverman, J. G., et al. (2001). Dating violence against adolescent girls and associated substance use, unhealthy weight control, sexual risk behavior, pregnancy, and suicidality. *Journal of the American Medical Association, 286*, 572–579.

Simeon, D., et al. (1997). Feeling unreal: 30 Cases of *DSM-III-R* depersonalization disorder. *American Journal of Psychiatry, 154*, 1107–1113.

Simeon, D., et al. (2000). Feeling unreal: A PET study of depersonalization disorder. *American Journal of Psychiatry, 157*, 1782–1788.

Simeon, D., et al. (2001). The role of childhood interpersonal trauma in depersonalization disorder. *American Journal of Psychiatry, 158*, 1027–1033.

Simon, G. E. (1998). Management of somatoform and factitious disorders. In P. E. Nathan & J. M. Gorman (Eds.), *A guide to treatments that work* (pp. 408–422). New York: Oxford University Press.

Simons-Morton, B., et al. (2001). Peer and parent influences on smoking and drinking among early adolescents. *Health Education & Behavior, 28*, 95–107.

Simpson, D. D., et al. (1999). A national evaluation of treatment outcomes for cocaine dependence. *Archives of General Psychiatry, 57*, 507–514.

Singh, G. (1985). Dhat syndrome revisisted. *Indian Journal of Psychiatry, 27*, 119–122.

Sitharthan, T., et al. (1997). Cue exposure in moderation drinking: A comparison with cognitive-behavior therapy. *Journal of Consulting and Clinical Psychology, 65*, 878–882.

Skaar, K. L., et al. (1997). Smoking cessation 1: An overview of research. *Behavioral Medicine, 23*, 5–13.

Skinner, B. F. (1938). *The behavior of organisms: An experimental analysis*. New York: Appleton.

Skoog, G., & Skoog, I. (1999). A 40-year follow-up of patients with obsessive-compulsive disorder. *Archives of General Psychiatry, 56*, 121–127.

Skoog, I. (2000). Detection of preclinical Alzheimer's disease. *The New England Journal of Medicine, 343*, 502–503.

Slater, D., & Hans, V. P. (1984). Public opinion of forensic psychiatry following the Hinckley verdict. *American Journal of Psychiatry, 141*, 675–679.

Sleek, S. (1994, January). Many methods employed to breach autism's walls. *APA Monitor*, pp. 30–31.

Slone, D. G., & Gleason, C. E. (1999). Behavior management planning for problem behaviors in dementia: A practical model. *Professional Psychology: Research and Practice, 30*, 27–36.

Slutske, W. S., et al. (1998). Common genetic risk factors for conduct disorder and alcohol dependence. *Journal of Abnormal Psychology, 107*, 363–374.

Small, J. G., et al. (1997). Quetiapine in patients with schizophrenia: A high- and low-dose double-blind comparison with placebo. *Archives of General Psychiatry, 54*, 549–557.

Smith, D. (1982). Trends in counseling and psychotherapy. *American Psychologist, 37*, 802–809.

Smith, D. (2001a, October). Sleep psychologists in demand. *Monitor on Psychology*, pp. 36–38.

Smith, D. (2001b, September). Harassment in the hallways. *Monitor on Psychology*, pp. 38–40.

Smith, G. R. (1994). The course of somatization and its effects on utilization of health care resources. *Psychosomatics, 35*, 263–267.

Smith, G. T., et al. (1995). Expectancy for social faciliation from drinking: The divergent paths of high-expectancy and low-expectancy adolescents. *Journal of Abnormal Psychology, 104*, 32–40.

Smith, J. E., & Krejci, J. (1991). Minorities join the majority: Eating disturbances among Hispanic and Native American youth. *International Journal of Eating Disorders, 10*, 179–186.

Smith, M. L., & Glass, G. V. (1977). Meta-analysis of psychotherapy otucome studies. *American Psychologist, 32*, 752–760.

Smith, M. L., Glass, G. V., & Miller, T. I. (1980). *The benefits of psychotherapy*. Baltimore, MD: Johns Hopkins University Press.

Smith, R. E., Smoll, F. L., & Ptacek, J. T. (1990). Conjunctive moderator variables in vulnerability and reliency research: Life stress, social support and coping skills, and adolescent sport injuries. *Journal of Personality and Social Psychology, 58*, 360–370.

Smith, S. C. (1983). *The great mental calculators*. New York: Columbia University Press.

Smith, T. (1999). Outcome of early intervention for children with autism. *Clinical Psychology: Science and Practice, 6*, 33–49.

Smith, T. W., Snyder, C. R., & Perkins, S. C. (1983). The self-serving function of hypochondriacal complaints: Physical symptoms as self-handicapping strategies. *Journal of Personality and Social Psychology, 44*, 787–797.

Smith, Y. L. S., et al. (2001). Adolescents with gender identity disorder who were accepted or rejected for sex reassignment surgery: A prospective follow-up study. *Journal of American Academy of Child & Adolescent Psychiatry, 40*, 472–481.

Smoking will be world's biggest killer. (1996, September 17). *Newsday*, p. A21.

Smyth, J. M. (1998). Written emotional expression: Effect sizes, outcome types, and moderating variables. *Journal of Consulting and Clinical Psychology, 66*, 174–184.

Smyth, J. M., et al. (1999). Effects of writing about stressful experiences on symptom reduction in patients with asthma or rheumatoid arthritis. *Journal of the American Medical Association, 281*, 1304–1309.

Smyth, J. M., & Pennebaker, J. W. (2001). What are the health effects of disclosure? In A. Baum, T. A. Revenson, & J. E. Singer (Eds.), *Handbook of health psychology* (pp. 339–348). Mahwah, NJ: Erlbaum.

Snell, M. E. (1997). Teaching children and young adults with mental retardation in school programs: Current research. *Behaviour Change, 14*, 73–105.

Sobell, L. C., et al. (1990). Behavior therapy. In A. S. Bellack & M. Hersen (Eds.), *Comparative treatment of adult disorders* (pp. 479–505). New York: Wiley.

Sobell, M. B., & Sobell, L. C. (1973). Alcoholics treated by individualized behavior therapy: One year treatment outcome. *Behaviour Research & Therapy, 11*, 599–618.

Sobell, M. B., & Sobell, L. C. (1976). Second year treatment outcome of alcoholics treated by individualized behavior therapy: Results. *Behaviour Research and Therapy, 14*, 195–215.

Sobell, M. B., & Sobell, L. C. (1984). The aftermath of heresy: A response to Pendery et al.'s critique of "Individualized behavior therapy for alcoholics." *Behaviour Research and Therapy, 22*, 413–440.

Solomon, D. A., et al. (1997). Recovery from major depression: A 10-year prospective follow-up across multiple episodes. *Archives of General Psychiatry, 54,* 1001–1006.

Solomon, D. A., et al. (2000). Multiple recurrences of major depressive disorder. *American Journal of Psychiatry, 157,* 229–233.

Sorenson, S. B., & Rutter, C. M. (1991). Transgenerational patterns of suicide attempt. *Journal of Consulting and Clinical Psychology, 59,* 861–866.

Southwick, S. M., et al. (1997). Noradrenergic and serotonergic function in posttraumatic stress disorder. *Archives of General Psychiatry, 54,* 749–758.

Spanos, N. P. (1978). Witchcraft in histories of psychiatry: A critical analysis and an alternative conceptualization. *Psychological Bulletin, 85,* 417–439.

Spanos, N. P. (1994). Multiple identity enactments and multiple personality disorder: A sociocognitive perspective. *Psychological Bulletin, 116,* 143–165.

Spanos, N. P., Weekes, J. R., & Bertrand, L. D. (1985). Multiple personality: A social psychological perspective. *Journal of Abnormal Psychology, 94,* 362–376.

Spark, R. F. (1991). *Male sexual health: A couple's guide.* Mount Vernon, NY: Consumer Reports Books.

Spechler, S. J., Fischbach, L., & Feldman, M. (2000). Clinical aspects of genetic variability in helicobacter pylori. *Journal of the American Medical Association, 283,* 1264–1266.

Spector, I. P., & Carey, M. P. (1990). Incidence and prevalence of the sexual dysfunctions: A critical review of the empirical literature. *Archives of Sexual Behavior, 19,* 389–408.

Spencer, D. J. (1983). Psychiatric dilemmas in Australian aborigines. *International Journal of Social Psychiatry, 29*(3), 208–214.

Spiegel, D., et al. (1989, October 14). Effect of psychosocial treatment on survival of patients with metastatic breast cancer. *Lancet,* pp. 888–891.

Spiegel, D. A., & Bruce, T. J. (1997). Benzodiazepines and exposure-based cognitive behavior therapies for panic disorder: Conclusions from combined treatment trials. *American Journal of Psychiatry, 154,* 773–781.

Spitzer, R. L., Gibbon, M., Skodol, A. E., Williams, J. B. W., & First, M. B. (1989). *DSM-III-R casebook.* Washington, DC: American Psychiatric Press.

Spitzer, R. L., et al. (1992). Binge eating disorder: A multisite field trial of the diagnostic criteria. *International Journal of Eating Disorders, 11,* 191–203.

Spitzer, R. L., et al. (1994). *DSM-IV case book* (4th ed.). Washington, DC: American Psychiatric Press.

Squier, L. H., & Domhoff, G. W. (1998). The presentation of dreaming and dreams in introductory psychology textbooks: A critical examination with suggestions for textbook authors and course instructors. *Dreaming, 8,* 149–168.

St. John's wort doesn't help with major depression. (2001a, June). *Tufts University Health & Nutrition Letter,* p. 2.

St. John's wort ineffective in severe depression. (2001b, April 17). *CNN Web Posting.* Retrieved April 19, 2001, from http://www.cnn.com/2001/HEALTH/conditions/04/17/st.johns.wort/index.html.

St. Lawrence, J. S., et al. (1995a). Comparison of education versus behavioral skills training interventions in lowering sexual HIV-risk behavior of substance-dependent adolescents. *Journal of Consulting and Clinical Psychology, 63,* 154–157.

St. Lawrence, J. S., et al. (1995b). Cognitive-behavioral intervention to reduce African American adolescents' risk for HIV infection. *Journal of Consulting and Clinical Psychology, 63,* 221–237.

Staal, W. G., et al. (2000). Structural brain abnormalities in patients with schizophrenia and their healthy siblings. *American Journal of Psychiatry, 157,* 416–421.

Stader, S. R., & Hokanson, J. E. (1998). Psychosocial antecedents of depressive symptoms: An evaluation using daily experiences methodology. *Journal of Abnormal Psychology, 107,* 17–26.

Stahl, S. M. (2001). Sex and psychopharmacology: Is natural estrogen a psychotropic drug in women? *Archives of General Psychiatry, 58,* 537–538. [Letter]

Stamler, J., et al. (1999). Low risk-factor profile and long-term cardiovascular and noncardiovascular mortality and life expectancy: Findings for 5 large cohorts of young adult and middle-aged men and women. *Journal of the American Medical Association, 282,* 2012–2018.

Stamm, B. H., & Perednia, D. A. (2000). Evaluating psychosocial aspects of telemedicine and telehealth systems. *Professional Psychology: Research and Practice, 31,* 184–189.

Stanley, M. A., & Beck, J. G. (2000). Anxiety disorders. *Clinical Psychology Review, 20,* 731–754.

Stanley, M. A., & Turner, S. M. (1995). Current status of pharmacological and behavioral treatment of obsessive-compulsive disorder. *Behavior Therapy, 26,* 163–186.

Stawar, T. (1999). A model for sexual harassment behavior. *Forensic Examiner, 8*(9–10), 30–34.

Steadman, H. J. (1979). *Beating a rap: Defendants found incompetent to stand trial.* Chicago: University of Chicago Press.

Steadman, H. J., et al. (1993). *Before and after Hinckley: Evaluating insanity defense reform.* New York: Guilford Press.

Steadman, H. J., et al. (1998). Violence by people discharged from acute psychiatric inpatient facilities and by others in the same neighborhoods. *Archives of General Psychiatry, 55,* 393–401.

Steer, R. A., et al. (1994). Psychometric properties of the Cognition Checklist with psychiatric outpatients and university students. *Psychological Assessment, 6,* 67–70.

Stefanek, M. E., Ollendick, T. H., Baldock, W. P., Francis, G., & Yaeger, N. J. (1987). Self-statements in aggressive, withdrawn, and popular children. *Cognitive Therapy and Research, 11,* 229–239.

Steffens, D. C., et al. (2000). Prevalence of depression and its treatment in an elderly population: The Cache County Study. *Archives of General Psychiatry, 57,* 601–607.

Stein, D., et al. (1998). The association between attitudes toward suicide and suicidal ideation in adolescents. *Acta Psychiatrica Scandinavia, 97,* 195–201.

Stein, M. B., Baird, A., & Walker, J. R. (1996). Social phobia in adults with stuttering. *American Journal of Psychiatry, 153,* 278–280.

Stein, M. B., Jan, K. L., & Livesley, W. J. (1999). Heritability of anxiety sensitivity: A twin study. *American Journal of Psychiatry, 156,* 246–251.

Stein, M. B., & Kean, Y. M. (2000). Disability and quality of life in social phobia: Epidemiologic findings. *American Journal of Psychiatry, 157,* 1606–1613.

Stein, M. B., Torgrud, L. J., & Walker, J. R. (2000). Social phobia symptoms, subtypes, and severity: Findings from a community survey. *Archives of General Psychiatry, 57,* 1046–1052.

Stein, M. B., Walker, J. R., & Forde, D. R. (1996). Public-speaking fears in a community sample: Prevalence, impact on functioning, and diagnostic classification. *Archives of General Psychiatry, 53,* 169–174.

Stein, M. B., et al. (2001). Social anxiety disorder and the risk of depression: A prospective community study of adolescents and young adults. *Archives of General Psychiatry, 58,* 251–256.

Steinberg, M. (1991). The spectrum of depersonalization: Assessment and treatment. *Annual Review of Psychiatry, 10,* 223–247.

Steketee, G., & Foa, E. B. (1985). Obsessive-compulsive disorder. In D. H. Barlow (Ed.), *Clinical handbook of psychological disorders* (pp. 69–144). New York: Guilford Press.

Stemberger, R. R., et al. (1995). Social phobia: An analysis of possible developmental factors. *Journal of Abnormal Psychology, 194,* 526–531.

Stenson, J. (2001a, August 26). Burden of mental illness in America falls on minorities. *MSNBC Web Posting.* Retrieved August 26, 2001, from http://www.msnbc.com/news/619545.asp.

Stenson, J. (2001b, August 26). Many teens abused by dating partners. *MSNBC Web Posting.* Retrieved August 27, 2001, from http://www.msnbc.com/news/619696.asp.

Stern, E., & Silbersweig, D. A. (2001). Advances in functional neuroimaging methodology for the study of brain system underlying human neuropsychological function and dysfunction. *Journal of Clinical & Experimental Neuropsychology, 23,* 3–18.

Sternberg, E. M. (2000). *The balance within: The science connecting health and emotions.* New York: Freeman.

Steven, J. E. (1995, January 30). Virtual therapy. *The Boston Globe,* pp. 25, 29.

Stevens, J. R. (1997). Anatomy of schizophrenia revisited. *Schizophrenia Bulletin, 23,* 373–383.

Stevens, S. E., Hynan, M. T., & Allen, M. (2000). A meta-analysis of common factor and specific treatment effects across the outcome domains of the phase model of psychotherapy. *Clinical Psychology: Science and Practice, 7,* 273–290.

Stewart, M. W., et al. (1994). Differential relationships between stress and disease activity for immunologically distinct subgroups of people with rheumatoid arthritis. *Journal of Abnormal Psychology, 1103,* 251–258.

Stice, E. (1994). Review of the evidence for a sociocultural model of bulimia nervosa and an exploration of the mechanisms of action. *Clinical Psychology Review, 14,* 633–661.

Stice, E. (2001). A prospective test of the dual-pathway model of bulimic pathology: Mediating effects of dieting and negative affect. *Journal of Abnormal Psychology, 110,* 124–135.

Stice, E., et al. (2000). Body-image and eating disturbances predict onset of depression among female adolescents: A longitudinal study. *Journal of Abnormal Psychology, 109,* 438–444.

Stock, W. E. (1991). Feminist explanations: Male power, hostility, and sexual coercion. In E. Grauerholz & M. A. Koralewski (Eds.), *Sexual coercion: A sourcebook on its nature, causes, and prevention* (pp. 61–73). Lexington, MA: Lexington Books.

Stokstad, E. (2001). New hints into the biological basis of autism. *Science, 294,* pp. 34–37.

Stolberg, S. G. (1998a, March 13). New cancer cases decreasing in U.S. as deaths do, too. *The New York Times,* p. A1, A14.

Stolberg, S. G. (1998b). Rise in smoking by young blacks erodes a success story. *The New York Times,* p. A24.

Stolberg, S. G. (2001, May 10). Blacks found on short end of heart attack procedure. *The New York Times,* p. A20.

Stoller, R. J. (1969). Parental influences in male transsexualism. In R. Green & J. Money (Eds.), *Transsexualism and sex reassignment.* Baltimore: Johns Hopkins University Press.

Stone, A. (1976). The Tarasoff decisions: Suing psychotherapists to safeguard society. *Harvard Law Review, 90,* 358–378.

Stone, A. (1984). *Law, psychiatry and morality.* Washington, DC: American Psychiatric Press.

Stone, A. A., et al. (1994). Daily events are associated with a secretory immune response to an oral antigen in men. *Health Psychology, 13,* 440–446.

Stone, A. A. et al. (2000b). Structured writing about stressful events: Exploring potential psychological mediators of positive health effects. *Health Psychology, 19,* 619–624.

Stone, M. H. (1980). *The borderline syndromes: Constitution, personality, and adaptation.* New York: McGraw-Hill.

Stotland, N. L. (2000, July 26). Sertraline: An effective treatment for binge-eating disorder? *Journal Watch Women's Health,* pp. 1, 4.

Stout, D. (2000, September 1). Use of illegal drugs is down among young, survey finds. *The New York Times*, p. A18.

Stover, E. S., et al. (1996). Perspectives from the National Institute of Mental Health: Preventing or living with AIDS. *Annals of Behavioral Medicine, 18*, 58–60.

Strakowski, S. M. (1994). Diagnostic validity of schizophreniform disorder. *American Journal of Psychiatry, 15*, 815–824.

Strakowski, S. M., et al. (1999). Brain magnetic resonance imaging of structural abnormalities in bipolar disorder. *Archives of General Psychiatry, 56*, 254–260.

Strassberg, D. S. (1997). A cross-national validity study of four MMPI-2 content scales. *Journal of Personality Assessment, 69*, 596–606.

Stricker, G., & Gold, J. R. (1999). The Rorschach: Toward a nomothetically based, idiographically applicable configurational model. *Psychological Assessment, 11*, 240–250.

Stricker, G., & Gold, J. R. (2001, January). An introduction to psychotherapy integration. *NYS Psychologist, 13*, 7–12.

Strober, M., & Humphrey, L. L. (1987). Familial contributions to the etiology and course of anorexia nervosa and bulimia. *Journal of Consulting and Clinical Psychology, 55*, 654–659.

Strollo, P. J., & Rogers, R. M. (1996). Obstructive sleep apnea. *The New England Journal of Medicine, 334*, 99–104.

Strozier, C. B. (2001). *Heinz Kohut: The making of a psychoanalyst*. New York: Farrar, Straus & Giroux.

Strube, M. J. (1988). The decision to leave an abusive relationship: Empirical evidence and theoretical issues. *Psychological Bulletin, 104*, 236–250.

Strupp, H. H. (1992). The future of psychodynamic psychotherapy. *Psychotherapy, 29*, 21–27.

Stuart, G. L., et al. (2000). Effectiveness of an empirically based treatment for panic disorder delivered in a service clinic setting: 1-year follow-up. *Journal of Consulting and Clinical Psychology, 68*, 506–512.

Stunkard, A. J., & Sørensen, T. I. A. (1993). Obesity and socioeconomic status—A complex relation. *New England Journal of Medicine, 329*, 1036–1037.

Stunkard, A. J., et al. (1986). An adoption study of human obesity. *New England Journal of Medicine, 314*, 193, 198.

Stunkard, A. J., et al. (1990). A separated twin study of the body mass index. *New England Journal of Medicine, 322*, 1483–1487.

Stuss, D. T., Gow, C. A., & Hetherington, C. R. (1992). "No longer Gage." Frontal lobe dysfunction and emotional changes. *Journal of Consulting and Clinical Psychology, 60*, 349–359.

Sue, S., et al. (1995). Psychopathology among Asian Americans: A model minority? *Cultural Diversity and Mental Health, 1*, 39–51.

Suinn, R. M. (2001). The terrible twos—Anger and anxiety: Hazardous to your health. *American Psychologist, 56*, 27–36.

Sullivan, H. S. (1962). *Schizophrenia as a human process*. New York: Norton.

Sullivan, P. F., Neale, M. C., & Kendler, K. S. (2000). Genetic epidemiology of major depression: Review and meta-analysis. *American Journal of Psychiatry, 157*, 1552–1562.

Sulloway, F. J. (1983). *Freud: Biologist of the mind*. New York: Basic Books.

Suls, J., Wan, C. K., & Blanchard, E. B. (1994). A multilevel data-analytic approach for evaluation of relationships between daily life stressors and symptomatology: Patients with irritable bowel syndrome. *Health Psychology, 13*, 103–113.

Sutherland, G. R., et al. (1991). Prenatal diagnosis of fragile X syndrome by direct detection of the unstable DNA sequence. *New England Journal of Medicine, 325*, 1720–1722.

Sutker, P. B., et al. (1995). War zone stress, personal resources, and PTSD in Persian Gulf War returnees. *Journal of Abnormal Psychology, 104*, 444–452.

Sveinbjornsdottir, S., et al. (2000). Familial aggregation of Parkinson's disease in Iceland. *New England Journal of Medicine, 343*, 1765–1770.

Svikis, D. S., Velez, M. L., & Pickens, R. W. (1994). Genetic aspects of alcohol use and alcoholism in women. *Alcohol Health & Research World, 18*, 192–196.

Swartz, M., et al. (1991). Somatization disorder. In L. N. Robins & D. A. Regier (Eds.), *Psychiatric disorders in America: The Epidemiologic Catchment Area Study* (pp. 220–257). New York: Free Press.

Swartz, M. S., et al. (1998). Violence and severe mental illness: The effects of substance abuse and nonadherence to medication. *American Journal of Psychiatry, 155*, 226–231.

Sweeney, J. A., et al. (1994). Eye tracking dysfunction in schizophrenia: Characterization of component eye movement abnormalities, diagnostic specificity, and the role of attention. *Journal of Abnormal Psychology, 103*, 222–230.

Sweeney, P. D., Anderson, K., Bailey, S. (1986). Attributional style in depression: A meta-analytic review. *Journal of Personality and Social Psychology, 50*, 974–991.

Swendsen, J. D., & Mazure, C. M. (2000). Life stress as a risk factor for postpartum depression: Current research and methodological issues. *Clinical Psychology: Science and Practice, 7*, 17–31.

Swenson, C. R. (2000). How can we account for DBT's widespread popularity? *Clinical Psychology: Science and Practice*, 87–91.

Szanto, K., et al. (1996). Suicide in elderly depressed patients. Is active vs. passive suicidal ideation a clinically valid distinction? *American Journal of Geriatric Psychiatry, 4*, 197–207.

Szasz, T. S. (1961). *The myth of mental illness*. New York: Harper & Row.

Szasz, T. S. (1970). *Ideology and insanity: Essays on the psychiatric dehumanization of man*. New York: Doubleday Anchor.

Szasz, T. S. (2000). Second commentary on "Aristotle's function argument." *Philosophy, Psychiatry, & Psychology, 7*(1), 3–16.

Szatmari, P., et al. (2000). Two-year outcome of preschool children with autism or Asperger's syndrome. *American Journal of Psychiatry, 157*, 1980–1987.

Szymanski, S., Kane, J. M., & Lieberman, J. A. (1991). A selective review of biological markers in schizophrenia. *Schizophrenia Bulletin, 17*, 99–111.

T

Tafoya, T. N. (1996). Native two-spirit people. In R. P. Cabaj & T. S. Stein (Eds.), *Textbook of homosexuality and mental health* (pp. 603–617). Washington, DC: American Psychiatric Press.

Takahashi, C. (2001, April 8). Selling to Gen Y: A far cry from Betty Crocker. *The New York Times Week in Review*, p. 3.

Tamminga, C. A. (1997). Clinical genetics, II. *American Journal of Psychiatry, 154*, 1046.

Tanda, G., Pontien, & Chiara (1997). Cannabinoid and heroin activation of mesolimbic dopamine transmission by a common opioid receptor mechanism. *Science, 276*, 2048–2050.

Tarasoff v. Regents of the University of California, 131 Cal Rptr. 14, 551 P. 2d 344 (1976).

Tardiff, K., et al. (1994). Homicide in New York City: Cocaine use and firearms. *Journal of the American Medical Association, 272*, 43–46.

Tardiff, K., et al. (1997). Violence by patients admitted to a private psychiatric hospital. *American Journal of Psychiatry, 154*, 88–93.

Tarrier, N., et al. (1999). A randomized trial of cognitive therapy and imaginal exposure in the treatment of chronic posttraumatic stress disorder. *Journal of Consulting and Clinical Psychology, 67*, 13–18.

Tarrier, N., et al. (2000). Two-year follow-up of cognitive-behavioral therapy and supportive counseling in the treatment of persistent symptoms in chronic schizophrenia. *Journal of Consulting and Clinical Psychology, 68*, 917–922.

Task Force on Promotion and Dissemination of Psychological Procedures. (1995). Training in and dissemination of empirically validated psychological treatments: Report and recommendations. *The Clinical Psychologist, 48*(1), 3–24.

Tate, D. F., Wing, R. R., & Winett, R. A. (2001). Internet technology to deliver a behavioral weight loss program. *Journal of the American Medical Association, 285*, 1172–1177.

Taylor, S. (1995). Assessment of obsessions and compulsions: Reliability, validity, and sensitivity to treatment effects. *Clinical Psychology Review, 15*, 261–296.

Taylor, S., & Rachman, S. J. (1994). Klein's suffocation theory of panic. *Archives of General Psychiatry, 51*, 505–506.

Taylor, S., et al. (2001). Posttraumatic stress disorder arising after road traffic collisions: Patterns of response to cognitive–behavior therapy. *Journal of Consulting and Clinical Psychology, 69*, 541–551.

Taylor, S. E., & Brown, J. D. (1994). Positive illusions and well-being revisited: Separating fact from fiction. *Psychological Bulletin, 116*, 21–27.

Taylor, S. E., et al. (2000). Psychological resources, positive illusions, and health. *American Psychologist, 55*, 99–109.

Teen drug use continues to decline. (2000, August 31). *MSNBC Web Posting*. Retrieved September 1, 2000, from http://www.msnbc.com/news/453784.asp.

Teri, L., & Wagner, A. (1992). Alzheimer's disease and depression. *Journal of Consulting and Clinical Psychology, 60*, 379–391.

Terman, J. S., et al. (2001). Circadian time of morning light administration and therapeutic response in winter depression. *Archives of General Psychiatry, 58*, 69–75.

Testosterone wimping out? (1995, July). *Newsweek*, p. 61.

Tests suggest possible protection from Huntington's symptoms. (1997, March 27). *The New York Times*, p. A 20.

Teunisse, R. J., et al. (1996). Visual hallucinations in psychologically normal people: Charles Bonnnet's syndrome. *Lancet, 347*, 794–797.

Thakker, J., & Ward, T. (1998). Culture and classification: The cross-cultural application of the *DSM-IV*. *Clinical Psychology Review, 18*, 501–529.

Thannickal, T. C., et al. (2000). Reduced number of hypocretin neurons in human narcolepsy. *Neuron, 17*, 469–474.

Thaper, A., et al. (1994). The genetics of mental retardation. *British Journal of Psychiatry, 164*, 747–758.

Tharp, R. G. (1991). Cultural diversity and treatment of children. *Journal of Consulting and Clinical Psychology, 59*, 799–812.

Thase, M. E., et al. (1997). Treatment of major depression with psychotherapy or psychotherapy-pharmacotherapy combinations. *Archives of General Psychiatry, 54*, 1009–1015.

Thase, M. E., et al. (2000). Treatment of men with major depression: A comparison of sequential cohorts treated with either cognitive-behavioral therapy or newer generation antidepressants. *Journal of Clinical Psychiatry, 61*, 466–472.

Theorell, T. (1992). Critical life changes: A review of research. *Psychotherapy and Psychosomatics, 57*, 108–117.

Third of some cancers tied to weight: WHO examines impact of obesity, sedentary lifestyle. (2001, April 5). *MSNBC Web Posting*. Retrieved April 7, 2001, from http://www.msnbc.com/news/555510.asp.

Thoits, P. A. (1983). Dimensions of life events as influences upon the genesis of psychological distress and associated conditions: An evaluation and synthesis of the literature. In H. B. Kaplan (Ed.), *Psychosocial stress: Trends in theory and research*. New York: Academic Press.

Thomas, A. M., & LoPiccolo, J. (1994). Sexual functioning in persons with diabetics: Issues in research, treatment, and education. *Clinical Psychology Review, 14*, 61–86.

Thornhill, R., & Palmer, C. T. (2000). *A natural history of rape*. Cambridge, MA: MIT Press.

Tienari, P. (1991). Interaction between genetic vulnerability and family environment: The Finnish adoptive family study of schizophrenia. *Acta Psychiatrica Scandinavica, 84*, 460–465.

Tienari, P. (1992). Implications of adoption studies on schizophrenia. *British Journal of Psychiatry, 161*(18, Suppl.), 52–58.

Tienari, P., et al. (1987). Genetic and psychosocial factors in schizophrenia: The Finnish Adoptive Family Study. *Schizophrenia Bulletin, 13*, 477–484.

Tienari, P., et al. (1990). Adopted-away offspring of schizophrenics and controls: The Finnish adoptive family study of schizophrenia. In L. Robins & M. Rutter (Eds.), *Straight and devious pathways from childhood to adulthood*. New York: Cambridge University Press.

Tiffany, S. T., Cox, L. S., & Elash, C. A. (2000). Effects of transdermal nicotine patches on abstinence-induced and cue-elicited craving in cigarette smokers. *Journal of Consulting and Clinical Psychology, 68*, 233–240.

Tiihonen, J., et al. (1992). Modified activity of the human auditory cortex during auditory hallucinations. *American Journal of Psychiatry, 149*, 255–257.

Tiihonen, J., et al. (1997). Specific major mental disorders and criminality: A 26-year prospective study of the 1996 Northern Finland Birth Cohort. *American Journal of Psychiatry, 154*, 840–845.

Tillfors, M., Furmark, T., Ekselius, L., & Fredrikson, M. (2001). Social phobia and avoidant personality disorder as related to parental history of social anxiety: A general population study. *Behaviour Research and Therapy, 39*, 289–298.

Timbrook, R. E., & Graham, J. R. (1994). Ethnic differences on the MMPI-2? *Psychological Assessment, 6*, 212–217.

Time capsule. (2000, November). *Monitor on Psychology*, p. 10.

Timpson, J., et al. (1988). Depression in a Native Canadian in Northwestern Ontario: Sadness, grief or spiritual illness? *Canada's Mental Health, 36*(2–3), 5–8.

Tobin, D. L., Johnson, C. L., & Dennis, A. B. (1992). Divergent forms of purging behavior in bulimia nervosa patients. *International Journal of Eating Disorders, 11*, 17–24.

Tohen, M., et al. (2000). Efficacy of olanzapine in acute bipolar mania: A double-blind, placebo-controlled study. *Archives of General Psychiatry, 57*, 841–849.

Tollefson, G. D., et al. (1997a). Olanzapine versus haloperidol in the treatment of schizophrenia and schizoaffective and schzophreniform disorders: Results of an international collaborative trial. *American Journal of Psychiatry, 154*, 448–456.

Tollefson, G. D., et al. (1997b). Blind, controlled, long-term study of the comparative incidence of treatment-emergent tardive dyskinesia with olanzapine or haloperidol. *American Journal of Psychiatry, 154*, 1248–1254.

Tolomiczenko, G. S., Sota, T., & Goering, P. N. (2000). Personality assessment of homeless adults as a tool for service planning. *Journal of Personality Disorders, 14*, 152–161.

Torgersen, S. (1986). Genetic factors in moderately severe and mild affective disorder. *Archives of General Psychiatry, 43*, 222–226.

Torgersen, S., Kringlen, E., & Cramer, V. (2001). The prevalence of personality disorders in a community sample. *Archives of General Psychiatry, 58*, 590–596.

Toufexis, A. (1993, April 19). Seeking the roots of violence. *Time Magazine*, pp. 52–53.

Tough, P. (2001, July 29). The alchemy of OxyContin. *The New York Times Magazine*, pp. 32–37, 52, 62–63.

Tower, R. B., & Kasl, S. V. (1996). Depressive symptoms across older spouses: Longitudinal influences. *Psychology and Aging, 11*, 683–697.

Tracy, J. I., Josiassen, R. C., & Bellack, A. S. (1995). Neuropsychology of dual diagnosis: Understanding the combined effects of schizophrenia and substance use disorders. *Clinical Psychology Review, 15*, 67–98.

Tradgold, A. F. (1914). *Mental deficiency*. New York: Wainwood.

Trask, P. C., & Sigmon, S. T. (1997). Munchausen syndrome: A review and new conceptualization. *Clinical Psychology: Science and Practice, 4*, 346–358.

Traven, N. D., et al. (1995). Coronary heart disease mortality and sudden death: Trends and patterns in 35- to 44-year-old white males, 1970–1990. *American Journal of Epidemiology, 142*, 45–52.

Treatment of alcoholism—Part II. (1996, September) *The Harvard Mental Health Letter, 13*, 1–5.

Treffert, D. A. (1988). The idiot savant: A review of the syndrome. *American Journal of Psychiatry, 145*, 563–572.

Tremblay, R. E., et al. (1992). Early disruptive behavior, poor school achievement, delinquent behavior, and delinquent personality: Longitudinal analyses. *Journal of Consulting and Clinical Psychology, 60*, 64–72.

Trenton State College. (1991, Spring). *Sexual Assault Victim Education and Support Unit (SAVES-U) Newsletter*.

Trimble, J. E. (1991). The mental health service and training needs of American Indians. In H. F. Myers et al. (Eds.), *Ethnic minority perspectives on clinical training and services in psychology* (pp. 43–48). Washington, DC: American Psychological Association.

Trull, T. J. (2001). Structural relations between borderline personality disorder features and putative etiological correlates. *Journal of Abnormal Psychology, 110*, 471–481.

Tsang, H. W.-H. (2001). Applying social skills training the context of vocational rehabilitation for people with schizophrenia. *Journal of Nervous & Mental Disease, 189*, 90–98.

Tseng, W., et al. (1992). Koro epidemics in Guangdong, China: A questionnaire survey. *Journal of Nervous & Mental Disease, 180*, 117–123.

Tsuang, D., & Coryell, W. (1993). An 8-year follow-up of patients with *DSM-III-R* psychotic depression, schizoaffective disorder, and schizophrenia. *American Journal of Psychiatry, 150*, 1182–1188.

Tsuang, M. T., et al. (1998). Co-occurrence of abuse of different drugs in men: The role of drug-specific and shared vulnerabilities. *Archives of General Psychiatry, 55*, 967–972.

Tuiten, A., et al. (2000). Time course of effects of testosterone administration on sexual arousal in women. *Archives of General Psychiatry, 57*, 149–153.

Tune, L. (1998). Treatments for dementia. In P. E. Nathan & J. M. Gorman (Eds.), *A guide to treatments that work* (pp. 90–126). New York: Oxford Press.

Turner, R. M. (2000). Understanding dialectical behavior therapy. *Clinical Psychology: Science and Practice, 7*, 95–98.

Turner, S. M., & Beidel, D. C. (1989). Social phobia: Clinical syndrome, diagnosis, and comorbidity. *Clinical Psychology Review, 9*, 3–18.

Turner, S. M., Beidel, D. C., & Jacob, R. G. (1994). Social phobia: A comparison of behavior therapy and atenolol. *Journal of Consulting and Clinical Psychology, 62*, 350–358.

Turner, S. M., Beidel, D. C., & Townsley, R. M. (1992). Social phobia: A comparison of specific and generalized subtypes and avoidant personality disorder. *Journal of Abnormal Psychology, 101*, 326–331.

Turner, S. M., McCann, B. S., Beidel, D. C., & Mezzich, J. E. (1986b). *DSM-III* classification of the anxiety disorders: A psychometric study. *Journal of Abnormal Psychology, 95*, 168–172.

Turovksy, J., & Barlow, D. H. (1995, Summer). Albany Panic Control Treatment (PCT) for panic disorder and agoraphobia. *The Clinical Psychologist, 48*(3), 5–6.

Tuschen-Caffier, B., Pook, M., & Frank, M. (2001). Evaluation of manual-based cognitive-behavioral therapy for bulimia nervosa in a service setting. *Behaviour Research and Therapy, 39*, 299–308.

Tutty, L. M. (1992). The ability of elementary school children to learn child sexual abuse prevention concepts. *Child Abuse and Neglect, 16*, 369–384.

U

U.S. Bureau of Justice Statistics. (1999). *National Crime Victimization Survey*. Washington, DC: Author.

U.S. Department of Health and Human Services (USDHHS). (1986b). *NIDA capsules: Heroin*. No. 11. Rockville, MD: U.S. Department of Health and Human Services, Public Health Service, Alcohol, Drug Abuse, and Mental Health Administration, National Institute on Drug Abuse.

U.S. Department of Health and Human Services (USDHHS). (1991b). *Vital statistics of the United States 1988*. (Vol. 2. Part A. Mortality.) Washington, DC: U.S. Government Printing Office. (DHHS Pub. No. PHS 91-1101).

U.S. Department of Health and Human Services (USDHHS). (1992). *NIDA capsules: LSD (lysergic acid diethylamide)* No. 39. Rockville, MD: U.S. Department of Health and Human Services, Public Health Service, Alcohol, Drug Abuse, and Mental Health Administration, National Institute on Drug Abuse, National Institute on Drug Abuse.

U.S. Department of Health and Human Services (USDHHS). (1993). *National household survey on drug abuse: Highlights 1991*. (DHHS Publication No. (SMA) 93-1979). Washington, DC: U.S. Government Printing Office.

U.S. Department of Health and Human Services (USDHHS). (1999a). *Mental health: A report of the Surgeon General*. Rockville, MD: U.S. Department of Health and Human Services, Substance Abuse and Mental Health Services Administration, Center for Mental Health Services, National Institutes of Health, National Institute of Mental Health.

U.S. Department of Health and Human Services (USDHHS). (1999b). *Mental health: A report of the Surgeon General—Executive summary*. Rockville, MD: U.S. Department of Health and Human Services, Substance Abuse and Mental Health Services Administration, Center for Mental Health Services, National Institutes of Health, National Institute of Mental Health.

U.S. Department of Health and Human Services (USDHHS). (2001). *Mental health: Culture, race, and ethnicity: A supplement to mental health: A report of the Surgeon General—Executive summary*. Rockville, MD: U.S. Department of Health and Human Services, Substance Abuse and Mental Health Services Administration, Center for Mental Health Services, National Institutes of Health, National Institute of Mental Health.

U.S. Department of Justice. (1994). *Crime in the United States: 1993. Uniform crime reports*. Washington, DC: Author.

U.S. finds heavy toll of rapes on young. (1994, June 23). *The New York Times*, p. A12.

Ullmann, L. P., & Krasner, L. (1975). *A psychological approach to abnormal behavior* (2nd ed.). Englewood Cliffs, NJ: Prentice-Hall.

Unger, J. B., Cruz, T. B., & Rohrbach, L. A. (2000). English language use as a risk factor for smoking initiation among Hispanic and Asian American adolescents: Evidence for mediation by tobacco-related beliefs and social norms. *Health Psychology, 19,* 403–410.

Unützer, J., et al. (1997). Depressive symptoms and the cost of health services in HMO patients aged 65 years and older: A 4-year prospective study. *Journal of the American Medical Association, 277,* 1618–1623.

Update on Alzheimer's Disease. Part I. (1995, February). *Harvard Mental Health Letter, 11*(6), 1–5.

Ursano, R. J., et al. (1999). Acute and chronic posttraumatic stress disorder in motor vehicle accident victims. *American Journal of Psychiatry, 156,* 589–595.

V

Van Ameringen, M. A., et al. (2001). Sertraline treatment of generalized social phobia: A 20-week, double-blind, placebo-controlled study. *American Journal of Psychiatry, 158,* 275–281.

Van Balkom, A. J. L. M., et al. (1997). A meta-analysis of the treatment of panic disorder with or without agoraphobia: A comparison of psychopharmacological, cognitive-behavioral, and combination treatments. *Journal of Nervous & Mental Disease, 185,* 510–516.

Van Ommeren, M., et al. (2001). Psychiatric disorders among tortured Bhutanese refugees in Nepal. *Archives of General Psychiatry, 58,* 475–482.

Van Praag, H. M. (1988). Editorial: Biological psychiatry. *Journal of Nervous and Mental Disease, 176,* 195–199.

Van Son, M. J. M., Mulder, G., & Londen, A. V. (1990). The effectiveness of dry bed training for nocturnal enuresis in adults. *Behaviour Research and Therapy, 28,* 347–349.

Vega, W. A., et al. (1998). Lifetime prevalence of *DSM-III-R* psychiatric disorders among urban and rural Mexican Americans in California. *Archives of General Psychiatry, 55,* 771–778.

Viagra fails to help women. (2000, May 22). MSNBC staff and wire reports. *MSNBC Web Posting.* Retrieved July 8, 2000, from http://www.msnbc.com.

Viglione, D. J. (1999). A review of recent research addressing the utility of the Rorschach. *Psychological Assessment, 11,* 251–265.

Vincent, J. B., et al. (1999). Genetic association analysis of serotonin system genes in bipolar affective disorder. *American Journal of Psychiatry, 156,* 136–138.

Virkkunen, M., & Linnoila, M. (1993). Brain serotonin, Type II alcoholism and impulsive violence. *Journal of Studies on Alcohol* (Suppl. 11), 163–169.

Virkkunen, M., et al. (1994). CSF biochemistries, glucose metabolism, and diurnal activity rhythms in alcoholic, violent offenders, fire setters, and healthy volunteers. *Archives of General Psychiatry, 51,* 20–27.

Visser, S., & Bouman, T. K. (2001). The treatment of hypochondriasis: Exposure plus response prevention vs. cognitive therapy. *Behaviour Research and Therapy, 39,* 423–442.

Voelker, R. (2000). Recessive Alzheimer gene? *Journal of the American Medical Association, 284,* 1777.

Vogler, G. P., DeFries, J. C., & Decker, S. N. (1985). Family history as an indicator of risk for reading disability. *Journal of Learning Disabilities, 18,* 419–421.

Volkow, N. D., et al. (1997). Relationship between subjective effects of cocaine and dopamine transporter occupancy. *Nature, 386,* 827–830.

Volkow, N. D., et al. (2001). Association of dopamine transporter reduction with psychomotor impairment in methamphetamine abusers. *American Journal of Psychiatry, 158,* 377–382.

Volpicelli, J. R., et al. (1994). Naltrexone and the treatment of alcohol dependence. *Alcohol Health & Research World, 18,* 272–278.

W

Wade, C., & Tavris, C. (1994). The longest war: Gender and culture. In W. J. Lonner & R. S. Malpass (Eds.), *Psychology and culture* (pp. 121–126). Boston: Allyn & Bacon.

Wade, N. (2000, May 9). Scientists decode Down syndrome chromosome. *The New York Times,* pp. C1, C3.

Wade, N. (2001a, February 11). Genome analysis shows humans survive on low number of genes. *The New York Times,* pp. 1, 42.

Wade, N. (2001b, February 11). Long-held beliefs are challenged by new human genome analysis: Genome analysis shows humans survive on low number of genes. *The New York Times,* p. A20.

Wade, T. D., et al. (2000). Anorexia nervosa and major depression: Shared genetic and environmental risk factors. *American Journal of Psychiatry, 157,* 469–471.

Wagner, B. M. (1997). Family risk factors for child and adolescent suicidal behavior. *Psychological Bulletin, 121,* 246–298.

Wagner, R. K., & Torgesen, J. K. (1987). The nature of phonological processing and its causal role in the acquisition of reading skills. *Psychological Bulletin, 101,* 192–212.

Wahlbeck, K., et al. (1999). Evidence of clozapine's effectiveness in schizophrenia: A systematic review and meta-analysis of randomized trials. *American Journal of Psychiatry, 156,* 990–999.

Wahlbeck, K. et al. (2001). Association of schizophrenia with low maternal body mass index, small size at birth, and thinness during childhood. *Archives of General Psychiatry, 58,* 48–52.

Wahlberg, K. E., et al. (2001). Long-term stability of communication deviance. *Journal of Abnormal Psychology, 110,* 443–448.

Wakefield, H., & Underwager, R. (1996). Commentary on Kenneth Pope's review. *Clinical Psychology: Science and Practice, 3,* 366–371.

Wakefield, J. C. (1992a). The concept of mental disorder: On the boundary between biological facts and social values. *American Psychologist, 47,* 373–388.

Wakefield, J. C. (1992b). Disorder as harmful dysfunction: A conceptual critique of *DSM-III-R's* definition of mental disorder. *Psychological Review, 99,* 232–247.

Wakefield, J. C. (1997). Normal inability versus pathological disability: Why Ossorio's definition of mental disorder is not sufficient. *Clinical Psychology: Science and Practice, 4,* 249–258.

Wakefield, J. C. (2001). Evolutionary history versus current causal role in the definition of disorder: Reply to McNally. *Behaviour Research and Therapy, 39,* 347–366.

Wakeling, A. (1996). Epidemiology of anorexia nervosa. *Psychiatry Research, 62,* 3–9.

Wakschlag, L. S., et al. (1997). Maternal smoking during pregnancy and the risk of conduct disorder in boys. *Archives of General Psychiatry, 54,* 670–676.

Walker, E. F., & Diforio, D. (1997). Schizophrenia: A neural diathesis-stress model. *Psychological Review, 104,* 667–685.

Walker, L. E. (1979). *The battered woman.* New York: Harper & Row.

Walker, L. E. (1988). The battered woman syndrome. In G. T. Hotaling, D. Finkelhor, J. T. Kirkpatrick, & M. A. Straus (Eds.), *Family abuse and its consequences: New directions in research* (pp. 139–148). Newbury Park, CA: Sage.

Walkup, J. T., et al. (2001). Fluvoxamine for the treatment of anxiety disorders in children and adolescents. The Research Unit on Pediatric Psychopharmacology Anxiety Study Group. *The New England Journal of Medicine, 344,* 1279–1285.

Wall, T. L., et al. (2001). A genetic association with the development of alcohol and other substance use behavior in Asian Americans. *Journal of Abnormal Psychology, 110,* 173–178.

Wallace, J. (1985). The alcoholism controversy. *American Psychologist, 40,* 372–373.

Waller, N. G., & Ross, C. A. (1997). The prevalence of biometric structure of pathological dissociation in the general population: Taxometric and behavior genetic findings. *Journal of Abnormal Psychology, 106,* 499–510.

Walsh, B. T., et al. (1997). Medication and psychotherapy in the treatment of bulimia nervosa. *American Journal of Psychiatry, 154,* 523–531.

Walsh, B. T., et al. (2000). Fluoxetine for bulimia nervosa following poor response to psychotherapy. *American Journal of Psychiatry, 157,* 1332–1334.

Wampold, B. E., et al. (1997a). A meta-analysis of outcome studies comparing bona fide psychotherapies: Empirically, "All must have prizes." *Psychological Bulletin, 122,* 203–215.

Wampold, B. E., et al. (1997b). The flat earth as a metaphor for the evidence for uniform efficacy of bona fide psychotherapies: Reply to Crits-Christoph (1997) and Howard et al. (1997). *Psychological Bulletin, 122,* 226–230.

Wang, X., et al. (2000). Longitudinal study of earthquake-related PTSD in a randomly selected community sample in North China. *American Journal of Psychiatry, 157,* 1260–1266.

Warheit, G. J., Vega, W. A., Auth, J., & Meinhardt, K. (1985). Psychiatric symptoms and dysfunctions among Anglos and Mexican Americans: An epidemiological study. In J. R. Greenley (Ed.), *Research in community and mental health* (pp. 3–32). London: JAI Press.

Warner, L. A., et al. (1995). Prevalence and correlates of drug use and dependence in the United States. *Archives of General Psychiatry, 52,* 219–229.

Waterman, J., et al. (1986). Challenges for the future. In K. MacFarlane, et al. (Eds.), *Sexual abuse of young children: Evaluation and treatment* (pp. 315–332). New York: Guilford Press.

Watson, C. G., et al. (1997). Lifetime prevalences of nine common psychiatric/personality disorders in female domestic abuse survivors. *Journal of Nervous & Mental Disease, 185,* 645–647.

Weaver, T. L., & Clum, G. A. (1995). Psychological distress associated with interpersonal violence: A meta-analysis. *Clinical Psychology Review, 15,* 115–140.

Weber, B. (1996, July 6). First arrests in New York under sex-offender law. *The New York Times,* p. B24.

Webster-Stratton, C., & Hammond, M. (1997). Treating children with early-onset conduct problems: A comparison of child and parent training interventions. *Journal of Consulting and Clinical Psychology, 65,* 93–109.

Wechsler, D. (1975). Intelligence defined and undefined: A relativistic appraisal. *American Psychologist, 30,* 135–139.

Weems, C. F., Berman, S. L., Silverman, W. K., & Saavedra, L. M. (2001). Cognitive errors in youth with anxiety disorders: The linkages between negative cognitive errors and anxious symptoms. *Cognitive Therapy & Research, 25,* 559–575.

Weinberg, G. (1972). *Society and the healthy homosexual.* New York: St. Martin's Press.

Weine, S. M., et al. (2000). Profiling the trauma related symptoms of Bosnian refugees who have not sought mental health services. *Journal of Nervous & Mental Disease, 188,* 416–421.

Weiner, I. B. (1994). The Rorschach inkblot method (RIM) is not a test: Implications for theory and practice. *Journal of Personality Assessment, 62,* 498–504.

Weiner, K. E., & Thompson, J. K. (1997). Overt and covert sexual abuse: Relationship to body image and eating disturbance. *International Journal of Eating Disorders, 22,* 273–284.

Weiner, M. F. (1996, July). What new treatments for Alzheimer's disease are being explored? *The Harvard Mental Health Letter, 13*(1), 8.

Weiner, R. D. (2000). Retrograde amnesia with electroconvulsive therapy characteristics and implications. *Archives of General Psychiatry, 57*, 591–592.

Weisberg, R. B., et al. (2001). Causal attributions and male sexual arousal: The impact of attributions for a bogus erectile difficulty on sexual arousal, cognitions, and affect. *Journal of Abnormal Psychology, 110*, 324–334.

Weisman, A., et al. (1993). An attributional analysis of expressed emotion in Mexican-American families with schizophrenia. *Journal of Abnormal Psychology, 102*, 601–606.

Weisman, A. G., et al. (1998). Expressed emotion, attributions, and schizophrenia symptom dimensions. *Journal of Abnormal Psychology, 107*, 355–359.

Weisman, A. G., et al. (2000). Controllability perceptions and reactions to symptoms of schizophrenia: A within-family comparison of relatives with high and low expressed emotion. *Journal of Abnormal Psychology, 109*, 167–171.

Weiss, R. D., & Mirin, S. M. (1987). *Cocaine*. Washington, DC: American Psychiatric Press.

Weissman, A. N., & Beck, A. T. (1978, November). *Development and validation of the Dysfunctional Attitudes Scale: A preliminary investigation*. Paper presented at the meeting of the American Educational Research Association, Toronto, Canada.

Weissman, M. M. (1999). Depressed adolescents grown up. *Journal of the American Medical Association, 281*, 1707–1713.

Weissman, M. M., & Markowtiz, J. C. (1994). Interpersonal psychotherapy: Current status. *Archives of General Psychiatry, 51*, 599–606.

Weissman, M. M., et al. (1989). Suicidal ideation and suicide attempts in panic disorder and attacks. *The New England Journal of Medicine, 321*, 1209–1214.

Weissman, M. M., et al. (1991). Affective disorders. In L. N. Robins & D. A. Regier (Eds.), *Psychiatric disorders in America: The Epidemiologic Catchment Area Study* (pp. 53–80). New York: Free Press.

Weisz, J. R., Pilkonis, P. A., Woody, S. R., & Follette, W. C. (2000). Stressing the (other) three Rs in the search for empirically supported treatments: Review procedures, research quality, relevance to practice and the public interest. *Clinical Psychology: Science and Practice, 7*, 243–258.

Weisz, J. R., et al. (1988). Thai and American perspectives on over-and undercontrolled child behavior problems: Exploring the threshold model among parents, teachers, and psychologists. *Journal of Consulting and Clinical Psychology, 56*, 601–609.

Wekerle, C., & Wolfe, D. A. (1993). Prevention and child physical abuse and neglect: Promising new directions. *Clinical Psychology Review, 13*, 501–540.

Welch, M. R., & Kartub, P. (1978). Socio-cultural correlates of incidence of impotence: A cross-cultural study. *Journal of Sex Research, 14*, 218–230.

Welkowitz, L. A., et al. (1999). Instructional set and physiological response to CO_2 inhalation. *American Journal of Psychiatry, 156*, 745–748.

Wells, K. B., et al. (2000). Impact of disseminating quality improvement programs for depression in managed primary care: A randomized controlled trial. *Journal of the American Medical Association*, 283, 212–220.

Weltzin, T. E., et al. (1994). Prediction of reproductive status in women with bulimia nervosa by past high weight. *American Journal of Psychiatry, 151*, 136–138.

Wender, P. H., Rosenthal, D., Kety, S. S., Schulsinger, F., & Welner, J. (1974). Cross-fostering: A research strategy for clarifying the role of genetic and experiential factors in the etiology of schizophrenia. *Archives of General Psychiatry, 30*, 121–128.

Wender, P. H., et al. (2000). ADHD in adults. *Journal of the American Academy of Child & Adolescent Psychiatry, 39*, 543.

Wenzlaff, R. M., & Grozier, S. A. (1988). Depression and the magnification of failure. *Journal of Abnormal Psychology, 97*, 90–93.

Wertlieb, D., Weigel, C., & Feldstein, M. (1987). Stress, social support, and behavior symptoms in middle childhood. *Journal of Clinical Child Psychology, 16*, 204–211.

West, M. A. (1985). Meditation and somatic arousal reduction. *American Psychologist, 40*, 717–719.

Westen, D. (1998). The scientific legacy of Sigmund Freud: Toward a psychodynamically informed psychological science. *Psychological Bulletin, 124*, 333–371.

Westen, D., & Shedler, J. (1999). Revising and assessing axis II, Part II: Toward an empirically based and clinically useful classification of personality disorders. *The American Journal of Psychiatry, 156*, 273–285.

Wetzler, S., & Marlowe, D. B. (1993). The diagnosis and assessment of depression, mania, and psychosis by self-report. *Journal of Personality Assessment, 60*, 1–31.

Wexler, B. E., et al. (2001). Functional magnetic resonance imaging of cocaine craving. *American Journal of Psychiatry, 158*, 86–95.

What is catatonia? (1995, February). *Harvard Mental Health Letter, 11*(8), p. 8.

What is PTSD? (1996). *American Journal of Psychiatry, 154*, 143–145. [Editorial]

Whiffen, V. E., & Gotlib, I. H. (1993). Comparison of postpartum and nonpostpartum depression: Clinical presentation, psychiatric history, and psychosocial functioning. *Journal of Consulting and Clinical Psychology, 61*, 485–493.

Wickelgren, I. (1997, June 27). Marijuana: Harder than thought? *Science, 276*, 1967.

Wickenhaver, J. (1992, September 8). After the "Wild Man": Can an insane system be cured? *Manhattan Spirit*, pp. 13, 28.

Wickizer, T. M., Lessler, D., & Travis, K .M. (1996). Controlling inpatient psychiatric utilization through managed care. *American Journal of Psychiatry, 153*, 339–345.

Widiger, T. A. (1991). *DSM-IV* reviews of the personality disorders: Introduction to special series. *Journal of Personality Disorder, 5*, 122–134.

Widiger, T. A. (1992). Generalized social phobia versus avoidant personality disorder: A commentary on three studies. *Journal of Abnormal Psychology, 101*, 340–343.

Widiger, T. A., & Clark, L. A. (2000). Toward *DSM-V* and the classification of psychopathology. *Psychological Bulletin, 126*, 946–963.

Widiger, T. A., & Costa, P. T., Jr. (1994). Personality and personality disorders. *Journal of Abnormal Psychology, 103*, 78–91.

Widom, C. S. (1989a). Child abuse, neglect, and adult behavior: Research design and findings on criminality, violence, and child abuse. *American Journal of Orthopsychiatry, 59*, 355–367.

Widom, C. S. (1989b). Does violence beget violence? A critical examination of the literature. *Psychological Bulletin, 106*, 3–28.

Widom, C. S. (1991). Childhood victimization: Risk factor for delinquency. In M. E. Colten & S. Gore (Eds.), *Adolescent stress: Causes and consequences* (pp. 201–221). New York: DeGruyter.

Wiersma, D., et al. (1998). Natural course of schizophrenic disorders: A 15-year follow-up of a Dutch incidence cohort. *Schizophrenia Bulletin, 24*, 75–85.

Wiersma, D., et al. (2001). Cognitive behaviour therapy with coping training for persistent auditory hallucinations in schizophrenia: A naturalistic follow-up study of the durability of effects. *Acta Psychiatrica Scandinavica, 103*, 393–399.

Wig, N. N., et al. (1987). Distribution of expressed emotion components among relatives of schizophrenic patients in Aarhus and Chandigarh. *British Journal of Psychiatry, 151*, 160–165.

Wilbur, C. B. (1986). Psychoanalysis and multiple personality disorder. In B. G. Braun (Ed.), *Treatment of multiple personality disorder*. Washington, DC: American Psychiatric Press.

Wilkie, F. L., et al. (1998). Mild cognitive impairment and risk of mortality in HIV-1 infection. *Journal of Neuropsychiatry and Clinical Neuroscience, 10*, 125–132.

Wilkinson, D. J. C., et al. (1998). Sympathetic activity in patients with panic disorder at rest, under laboratory mental stress, and during panic attacks. *Archives of General Psychiatry, 55*, 511–520.

Wilkinson-Ryan, T., & Westen, D. (2000). Identity disturbance in borderline personality disorders: An empirical investigation. *American Journal of Psychiatry, 157*, 528–541.

Williams, J. B., et al. (1992). The Structured Clinical Interview for DSM-III—(SCID). II: Multisite test-retest reliability. *Archives of General Psychiatry, 49*, 630–636.

Williams, J. E., et al. (2000). Anger proneness predicts coronary heart disease risk: Prospective analysis from the Atherosclerosis Risk in Communities (ARIC) study. *Circulation, 101*, 2034–2039.

Williams, J. M. (1984). *The psychological treatment of depression: A guide to the theory and practice of cognitive-behavior therapy*. New York: Free Press.

Williams, J. W., et al. (2000). Treatment of dysthymia and minor depression in primary care: A randomized controlled trial in older adults. *Journal of the American Medical Association, 284*, 1519–1526.

Williams, K., E., Chambless, D. L., & Ahrens, A. (1997). Are emotions frightening? An extension of the fear of fear construct. *Behaviour Research and Therapy, 35*, 239–248.

Williams, P. G., Wiebe, D. J., & Smith, T. W. (1992). Coping processes as mediators of the relationship between hardiness and health. *Journal of Behavioral Medicine, 15*, 237–255.

Williams, R. B., et al. (1997). Psychosocial correlates of job strain in a sample of working women. *Archives of General Psychiatry, 54*, 543–548.

Wills, T. A., & Cleary, S. D. (1999). Peer and adolescent substance use among 6th–9th graders: Latent growth analyses of influence versus selection mechanisms. *Health Psychology, 18*, 453–463.

Wills, T. A., & Filer Fegan, M. (2001). Social networks and social support. In A. Baum, T. A. Revenson, & J. E. Singer (Eds.), *Handbook of health psychology* (pp 3–18). Makwah, NJ: Lawrence Erlbaum Associates, Inc.

Wilson, G. T. (1987). Chemical aversion conditioning treatment for alcoholism: A re-analysis. *Behaviour Research and Therapy, 25*, 503–516.

Wilson, G. T. (1991). Chemical aversion conditioning in the treatment of alcoholism: Further comments: *Behaviour Research & Therapy, 29*, 415–419.

Wilson, G. T. (1994). Behavioral treatment of childhood obesity: Theoretical and practical implications. *Health Psychology, 13*, 371–372.

Wilson, G. T. (1997). Behavior therapy at century close. *Behavior Therapy, 28*, 449–457.

Wilson, G. T., & Fairburn, C. G. (1998). Treatment for eating disorders. In P. E. Nathan & J. M. Gorman (Eds.), *A guide to treatments that work* (pp. 501–530). New York: Oxford University Press.

Wilson, K., et al. (1992). Levels of learned helplessness in abused women. *Women and Therapy, 13*, 53–67.

Wilson, K. K. (1997, April). *The disparate classification of gender and sexual orientation in American psychiatry. Psychiatry On-Line*. Retrieved April 28, 1997, from http://www.priory.com/psych/disparat.htm.

Wilson, M. I., & Daly, M. (1996). Male sexual proprietariness and violence against wives. *Current Directions in Psychological Science, 5*, 2–7.

Wilson, S. A., Becker, L. A., & Tinker, R. H. (1997). Fifteen-month follow-up of eye movement desensitization and reprocessing (EMDR) treatment for posttraumatic stress disorder and psychological trauma. *Journal of Consulting and Clinical Psychology, 65*, 1047–1056.

Winefield, H. R., & Harvey, E. J. (1994). Needs of family caregivers in chronic schizophrenia. *Schizophrenia Bulletin, 20*, 557–566.

Winerip, M. (1991, December 18). Soldier in battle for the retarded. *The New York Times*, pp. B1, B6.

Winerip, M. (1999, May 23). Bedlam on the streets. *The New York Times*, pp. 42–49.

Wing, R. R., & Polley, B. A. (2001). Obesity. In A. Baum, T. A. Revenson, & J. E. Singer (Eds), *Handbook of health psychology* (pp. 263–279). Mahwah, NJ: Erlbaum.

Wingert, P. (2000, December 4). No more "afternoon nasties." *Newsweek*, p. 59.

Wingert, P., & Kantrowitz, B. (1997, October 27). Why Andy couldn't read. *Newsweek*, pp. 54–64.

Winkleby, M. A., Kramer, H. C., Ahn, D. K., & Varady, A. N. (1998). Ethnic and socioeconomic differences in cardiovascular disease risk factors. *Journal of the American Medical Association, 280*, 356–362.

Winston, A., et al. (1991). Brief psychotherapy of personality disorders. *Journal of Nervous & Mental Disease, 179*, 188–193.

Winston, A., et al. (1994). Short-term psychotherapy of personality disorders. *American Journal of Psychiatry, 51*, 190–194.

Winter, G. (2000, October 29). Fraudulent marketers capitalize on demand for sweat-free diets. *The New York Times*, pp. A1, A26.

Wise, T. P. (1978). Where the public peril begins: A survey of psychotherapists to determine the effects of Tarasoff. *Stanford Law Review, 135*, 165–190.

Wittchen, H., et al. (1994). *DSM-III-R* generalized anxiety disorder in the National Comorbidity Survey. *Archives of General Psychiatry, 51*, 355–363.

Wolfe, D. A., et al. (2001). Child maltreatment: Risk of adjustment problems and dating violence in adolescence. *Journal of American Academy of Child & Adolescent Psychiatry, 40*, 282–289.

Wolpe, J. (1958). *Psychotherapy by reciprocal inhibition*. Stanford, CA: Stanford University Press.

Wolpe, J., & Lazarus, A. A. (1966). *Behavior therapy techniques*. New York: Pergamon Press.

Wolpe, J., & Rachman, S. (1960). Psychoanalytic "evidence": A critique based on Freud's case of Little Hans. *Journal of Nervous and Mental Disease, 131*, 135–147.

Women under assault. (1990, July 16). *Newsweek*, p. 23.

Wonderlich, S. A., et al. (1997). Relationship of childhood sexual abuse and eating disorders. *Journal of the American Academy of Child and Adolescent Psychiatry, 36*, 1107–1115.

Wong, J. L., & Whitaker, D. J. (1993). Depressive mood states and their cognitive and personality correlates in college students: They improve over time. *Journal of Clinical Psychology, 49*, 615–621.

Wood, J. M., Nezworski, M. T., & Stejskal, W. J. (1996). The comprehensive system for the Rorschach: A critical examination. *Psychological Science, 7*, 3–10.

Wood, J. M., Nezworski, M. T., & Stejskal, W. J. (1997). The reliability of the comprehensive system for the Rorschach: A comment on Meyer. *Psychological Assessment, 9*, 490–494.

Wood, J. M., et al. (1992). Effects of 1989 San Francisco earthquake on frequency and content of nightmares. *Journal of Abnormal Psychology, 101*, 219–234.

Wood, M. D., Vinson, D. C., & Sher, K. J. (2001). Alcohol use and misuse. In A. Baum, T. A. Revenson, & J. E. Singer (Eds.), *Handbook of health psychology* (pp. 280–320). Mahwah, NJ: Erlbaum.

Woodward, A. M., Dwinell, A. D., & Arons, B. S. (1992). Barriers to mental health care for Hispanic Americans: A literature review and discussion. *Journal of Mental Health Administration, 19*, 224–236.

Wootton, J. M., et al. (1997). Ineffective parenting and childhood conduct problems: The moderating role of callous-unemotional traits. *Journal of Consulting and Clinical Psychology, 65*, 301–308.

Wren, C. S. (1997a, August 20). Saying "no" to drugs but dying in violence. *The New York Times*, p. A16.

Wren, C. S. (2001, June 26). Powell, at U.N., asks war on AIDS. *The New York Times*, pp. A1, A4.

Wright, I. C., et al. (2000). Meta-analysis of regional brain volumes in schizophrenia. *American Journal of* Psychiatry, *157*, 16–25.

Wulfert, E., Greenway, D. E., & Dougher, M. J. (1996). A logical functional analysis of reinforcement-based disorders: Alcoholism and pedophilia. *Journal of Consulting and Clinical Psychology, 64*, 1140–1151.

Wyatt v. Stickney, 334 F. Supp. 1341 (1972).

Wyatt, G. E. (1990). The aftermath of child sexual abuse of African American and White American women: The victim's experience. *Journal of Family Violence, 5*, 61–81.

Y

Yairi, E., Ambrose, N., & Cox, N. (1996). Genetics of stuttering: A critical review. *Journal of Speech and Hearing Research, 39*, 771–784.

Yates, W. R. (2000). Testosterone in psychiatry: Risks and benefits. *Journal of the American Medical Association, 57*, 155–156.

Yatham, L. M., et al. (2000). Brain serotonin2 receptors in major depression: A positron emission tomography study. *Archives of General Psychiatry, 57*, 850–858.

Yeh, M., Takeuchi, D. T., & Sue, S. (1994). Asian-American children treated in the mental health system: A comparison of parallel and mainstream outpatient service centers. *Journal of Clinical Child Psychology, 23*, 5–12.

Yeung, P. P., & Greenwald, S. (1992). Jewish Americans and mental health: Results of the NIMH Epidemiologic Catchment Area Study. *Social Psychiatry and Psychiatric Epidemiology, 27*, 292–297.

Young, A. S., et al. (2001). The quality of care for depressive and anxiety disorders in the United States. *Archives of General Psychiatry, 58*, 55–61.

Young, K. S. (1999). Evaluation and treatment of Internet addiction. In L. VanderCreek & T. L. Jackson (Eds.), *Innovations in clinical practice: A source book* (Vol. 17, pp. 19–31). Sarasota, FL: Professional Resource Press/Professional Resource Exchange.

Young, M. A., et al. (1994). Interactions of risk factors in predicting suicide. *American Journal of Psychiatry, 51*, 434–435.

Young, T., et al. (1996). The gender bias in sleep apnea diagnosis. *Archives of Internal Medicine, 156*, 2445–2451.

Young, T. J., & French, L. A. (1996). Suicide and homicide rates among U.S. Indian health service areas: The income inequality hypothesis. *Social Behavior and Personality, 24*, 365–366.

Young, T. K., & Sevenhuyser, G. (1989). Obesity in northern Canadian Indians: Patterns, determinants, and consequences. *American Journal of Clinical Nutrition, 49*, 786–793.

Youngberg v. Romeo, 102 S. Ct. 2452, 2463 (1982).

Z

Zahn-Waxler, C., et al. (1996). Behavior problems in 5-year-old monozygotic and dizygotic twins: Genetic and environmental influences, patterns of regulation, and internalization of control. *Development and Psychopathology, 8*, 103–122.

Zalewski, C., & Archer, R. P. (1991). Assessment of borderline personality disorder: A review of MMPI and Rorschach findings. *Journal of Nervous & Mental Disease, 179*, 338–345.

Zamanian, K., et al. (1992). Acculturation and depression in Mexican-American elderly. *Gerontologist, 11*, 109–121.

Zanardi, R., et al. (1996). Double-blind controlled trial of sertraline versus paroxetine in the treatment of delusional depression. *American Journal of Psychiatry, 153*, 1631–1633.

Zane, N., & Sue, S. (1991). Culturally responsive mental health services for Asian Americans: Treatment and training issues. In H. F. Myers et al. (Eds.), *Ethnic minority perspectives on clinical training and services in psychology* (pp. 49–58). Washington, DC: American Psychological Association.

Zeiss, A. M., & Breckenridge, J. S. (1997). Treatment of late life depression: A response to the NIH Consensus Conference. *Behavior Therapy, 28*, 3–21.

Zhou, J-N., et al. (1995). A sex difference in the human brain and its relation to transsexuality. *Nature, 378*, 68–70.

Zickler, P. (2000). Brain imaging studies show long-term damage from methamphetamine abuse. *NIDA Notes, 15*(3), 11, 13.

Ziedonis, D. M., & Trudeau, K. (1997). Motivation to quit using substances among individuals with schizophrenia: Implications for a motivation-based treatment model. *Schizophrenia Bulletin, 23*, 229–238.

Zielbauer, P. (2000, May 22). Sex offender listings on Web set off debate. *The New York Times online*. Retrived May 24, 2000, from http://www.nytimes.com

Zilbovicius, M., et al. (1995). Delayed maturation of the frontal cortex in childhood autism. *American Journal of Psychiatry, 152*, 248–252.

Zito, J. M., Safer, D. J., dosReis, S., Gardner, J. F., Boles, M., & Lynch, F. (2000). Trends in the prescribing of psychotropic medications to preschoolers. *Journal of the American Medical Association, 283*, 1025–1030.

Zlotnick, C., Bruce, S. E., Shea, M. T., & Keller, M. B. (2001). Delayed posttraumatic stress disorder (PTSD) and predictors of first onset of PTSD in patients with anxiety disorders. *Journal of Nervous & Mental Disease, 189*, 404–406.

Zoellner, L. A., Craske, M., G., & Rapee, R. M. (1996). Stability of catastrophic cognitions in panic disorder. *Behaviour Research and Therapy, 34*, 399–402.

Zoellner, L. A., Foa, E. B., Brigidi, B. D., & Przeworski, A. (2000). Are trauma victims susceptible to "false memories"? *Journal of Abnormal Psychology, 109*, 517–524.

Zola, S. M. (1999). Memory, amnesia, and the issue of recovered memory: Neurobiological aspects. *Clinical Psychology Review, 19*, 915–932.

Zorumski, C. F., & Isenberg, K. E. (1991). Insights into the structure and function of GABA-benzodiazepine receptors: Ion channels and psychiatry. *American Journal of Psychiatry, 148*, 162–172.

Zotter, D. L., & Crowther, J. H. (1991). The role of cognitions in bulimia nervosa. *Cognitive Therapy & Research, 15*, 413–426.

Zubenko, G. S., et al. (1997). Mortality of elderly patients with psychiatric disorders. *American Journal of Psychiatry, 154*, 1360–1368.

Zubin, J., & Spring, B. (1977). Vulnerability—New view of schizophrenia. *Journal of Abnormal Psychology, 86*, 103–126.

Zucker, K. J., & Green, R. (1992). Psychosexual disorders in children and adolescents. *Journal of Child Psychology and Psychiatry, 33*, 107–151.

Zuckerman, M. (1980). Sensation seeking. In H. London & J. Exner (Eds.), *Dimensions of personality*. New York: Wiley.

Zvolensky, M. J., & Eifert, G. H. (2001). A review of psychological factors/processes affecting anxious responding during voluntary hyperventilation and inhalations of carbon dioxide-enriched air. *Clinical Psychology Review, 21*, 375–400.

Zvolensky, M. J., et al. (2001). Assessment of anxiety sensitivity in young American Indians and Alaska Natives. *BTR, Behaviour Research and Therapy, 39*, 477–493.

Zweig-Frank, H., & Paris, J. (1991). Parents' emotional neglect and overprotection according to the recollections of patients with borderline personality disorder. *American Journal of Psychiatry, 148*, 648–651.

Zwilich, C. W. (2000). Is untreated sleep apnea a contributing factor for chronic hypertension? *Journal of the American Medical Association*, 283, 000. [Editorial]

PHOTO CREDITS

Chapter 1 Page 2 Wassily Kandinsky, (1866–1944) Russian, "Sweet Pink"/Sotheby's, London/SuperStock/© 2003 Artist Rights Society (ARS), New York/ADAGP, Paris; p. 5 (top) SuperStock, Inc.; p. 5 (left) Bob Daemmrich/Stock Boston; p. 5 (right) Steve Goldberg; p. 6 Catherine Karnow/Woodfin Camp & Associates; p. 9 Bierwert/American Museum of Natural History; p. 10 The Granger Collection; p. 11 The Granger Collection; p. 12 Courtesy of the Library of Congress; p. 13 CORBIS; p. 15 CORBIS; p. 16 (left) The Granger Collection; p. 16 (right) The Granger Collection; p. 21 Irven De Vore/Anthro-Photo File.

Chapter 2 Page 32 Alejandro Xul Solar, (1925), Latin American "Patria B"/Christie's Images/SuperStock/Museo Xul Solar Fundacion Pan Klub; p. 39 Ron Chapple/Getty Images Inc.; p. 41 (top) AP/Wide World Photos; p. 41 (bottom) Lew Merrim/Photo Researchers, Inc.; p. 42 Frank Siteman/Stock Boston; p. 44 Lucas Films/Picture Desk, Inc./Kobal Collection; p. 45 (left) Culver Pictures Inc.; p. 45 (center) Library of Congress; p. 45 (right) PhotoEdit; p. 48 (top) The Granger Collection; p. 48 (bottom) B.F. Skinner Foundation; p. 51 (left) Nancy Sheehan/Index Stock Imagery, Inc.; p. 51 (right) Laura Dwight Photography; p. 52 (left) Corbis/Rogers Ressmeyer/CORBIS; p. 52 (right) Ann Kaplan/Corbis/CORBIS; p. 52 (bottom) Tracey Saar; p. 53 Gabor Demjen/Stock Boston; p. 55 (left) Albert Ellis Institute; p. 55 (right) Aaron T. Beck, M.D.; p. 56 Hugh Rogers.

Chapter 3 Page 63 Dubuffet, Jean. "Ontogenese", (1975), © Copyright ARS, NY. Musee des Beaux-Arts Andre Malraux, Le Havre, France; p. 66 Greenlar/The Image Works; p. 68 K. Kai/The Image Works; p. 73 The Image Works; p. 74 Spencer Grant/PhotoEdit; p. 76 Pictor, New York; p. 83 Pearson Education/PH College; p. 84 Reprinted by permission of the publishers from Henry A. Murray, THEMATIC APPERCEPTION TEST, Cambridge, Mass.: Harvard University Press, © 1943 by the President and Fellows of Harvard College, © 1971 by Henry A. Murray; p. 84 (Video 3.1) Penn State; p. 90 Susan Rosenberg/Photo Researchers, Inc.; p. 92 David York/Medichrome/The Stock Shop, Inc.; p. 93 (top) Priest, Chris/Photo Researchers, Inc.; p. 93 (bottom) Brookhaven National Laboratory; p. 94 (top) Visuals Unlimited; p. 94 (left and right) Magnetic Resonance Imaging of the Brain in Schizophrenia, pg. 35–44, 1990; 47:35/Archive of General Psychiatry/American Medical Association.

Chapter 4 Page 96 Paul Klee (1979–1940), Swiss, "Variation II", Bauhaus Archive/ET Archive, London/SuperStock/© 2003 Artist Rights Society (ARS), New York/VG Bild-Kunst, Bonn; p. 100 Zigy Kaluzny/Getty Images, Inc.; p. 101 (top) Corbis/Sygma; p. 101 (bottom) Albert Rocarols/Getty Images Inc.; p. 102 Grantpix/Photo Researchers, Inc.; p. 104 Susan Rosenberg/Photo Researchers, Inc.; p. 105 (Video 4.1) Penn State; p. 111 Will Hart; p. 112 D. Young-Wolff/PhotoEdit; p. 113 Zigy Kaluzny/Getty Images Inc.; p. 115 From the *Wall Street Journal* – permission, by Cartoon Features Syndicate; p. 116 Rhoda Sydney; p. 121 James Wilson/Woodfin Camp Associates; p. 123 Bisson Bernard/Corbis/Sygma; p. 124 Michael Newman/PhotoEdit.

Chapter 5 Page 130 Diana Ong, "Medicine Man", B. (1940), SuperStock, Inc.; p. 133 Phyllis Picardi/Stock Boston; p. 134 (top) Biology Media/Science Source/Photo Researchers, Inc.; p. 134 (bottom) Marcus Townsend/CORBIS; p. 136 (left) Esbin/Anderson/Omni-Photo Communications, Inc.; p. 136 (right) Bob Daemmrich/Stock Boston; p. 138 A. Ramey/PhotoEdit; p. 139 Lawrence Migdale/Stock Boston; p. 14 Charles Gupton/Stock Boston; p. 143 Jeff Greenberg/David R. Frazier Photolibrary, Inc.; p. 146 Joel Gordon Photography; p. 148 (top) William Johnson/Stock Boston; p. 148 (bottom) Zigy Kaluzny/Getty Images Inc.; p. 153 Andy Levin/Photo Researchers, Inc.; p. 154 Alon Reininger/Woodfin Camp & Associates.

Chapter 6 Page 158 © Miriam Schapiro, "Free Fall", (1985), paper and acrylic on canvas, 90" x 64". Collection: Mr. and Mrs. Lee Sachs, Steinbaum Krauss Gallery, NYC.; p. 162 Walter Smith/CORBIS; p. 162 (Video 6.1) Pearson-Patients as Educators; p. 166 (left) Geri Engberg/Geri Engberg Photography; p. 166 (top) SuperStock, Inc.; p. 166 (bottom) David E. Dempster/Pearson Education/PH College; p. 169 Tony Stone Images; p. 170 Picture Desk, Inc./Kobal Collection; p. 170 (Video 6.2) Pearson-Patients as Educators; p. 172 (left) Allan Tannenbaum/The Image Works; p. 172 (right) Charles Piatiau/CORBIS; p. 174 Getty Images Inc./085652-001; p. 175 M.H. Sharp/Photo Researchers, Inc.; p. 185 Laima E. Druskis/Pearson Education/PH College; p. 188 Georgia Tech Communications; p. 189 Lori Grinker/Contact Press Images, Inc.

Chapter 7 Page 194 Naoki Okamoto. "Untitled". Living-Asian. B. (1990), SuperStock, Inc.; p. 195 AP/Wide World Photos; p. 197 (left) Film Still Archive/The Museum of Modern Art; p. 197 (center) Film Still Archive/The Museum of Modern Art; p. 197 (right) Film Still Archive/The Museum of Modern Art; p. 197 (Video 7.1) Penn State; p. 199 (Video 7.2) CDA; p. 204 Werner H. Muller/Peter Arnold, Inc.; p. 207 Frank Siteman/Stock Boston; p. 211 Matthew Borkoski/Stock Boston; p. 213 Karp, Ken/Omni-Photo Communications, Inc.; p. 214 Margo Granitsas/The Image Works; p. 216 Pictor, New York; p. 217 UPI/CORBIS.

Chapter 8 Page 224 Alexej von Jawlensky, "Evening", (1929), ARS/Christie's Images/CORBIS; p. 226 (Video 8.1) Pearson-Patients as Educators; p. 227 Joel Gordon Photography; p. 228 Nate Guidry/Impact Visuals Photos and Graphics, Inc.; p. 230 (top) Dan McCoy/Rainbow; p. 230 (bottom) Ted Mahieu/Corbis/Stock Market; p. 233 Jonathan Exley/Getty Images, Inc.; p. 233 (Video 8.2) Pearson-Patients as Educators; p. 237 Lisa Quinones/Black Star; p. 238 Barry Yee/Liaison Agency, Inc.; p. 239 Carroll Seghers/Photo Researchers, Inc.; p. 240 Patrick Ramsey/International Stock Photography Ltd.; p. 246 SuperStock, Inc.; p. 247 Sybil Shackman; p. 251 Michal Heron/Ms. Michal Heron; p. 263 Joel Gordon/Joel Gordon Photography.

Chapter 9 Page 266 Klee Paul, "Beware of Red", (1940), Oil on canvas. Private Collection, Milan, Italy © Photograph by Erich Lessing. © 1998 Artists Rights Society (ARS), New York/VG Bild-Kunst, Bonn./Art Resource, N.Y.; p. 269 Savino/The Image Works; p. 271 Mark Foley/AP/Wide World Photos; p. 272 Greg Smith/Corbis/SABA Press Photos, Inc.; p. 272 (Video 9.1) Pearson-Patients as Educators; p. 274 Picture Desk, Inc./Kobal Collection; p. 275 Picture Desk, Inc./Kobal Collection; p. 276 Douglas Kirkland/CORBIS; p. 277 CORBIS; p. 279 (top) Joel Gordon Photography; p. 279 (bottom) David R. Frazier Photolibrary, Inc.; p. 282 Laima E. Druskis/Pearson Education/PH College; p. 284 Bob Daemmrich/Stock Boston; p. 286 Bob Daemmrich/Stock Boston; p. 288 Richard Renaldi; p. 292 Andy Belcher/ImageState.

Chapter 10 Page 298 Christopher Richard Wynne Nevinson, "A Bursting Shell", (1915), London, Tate Gallery. B. 1889—d. 1946. 928 x 724 x 76 mm./ Art Resource, N.Y.; p. 300 (Video 10.1) CDA; p. 301 (left) Walter Bibikow/ Getty Images Inc.; p. 301 (right) Lawrence Migdale/Getty Images Inc.; p. 301 (bottom) MGM-Pathe/Kobal Collection; p. 304 Mario Beauregard/ CORBIS; p. 305 (Video 10.2) CDA; p. 308 Ken Fisher/Getty Images Inc.; p. 309 (left) D. Wells/The Image Works; p. 309 (right) Paul Chesley/Getty Images Inc.; p. 310 AP/Wide World Photos; p. 313 Don Farrall/Getty Images, Inc./PhotoDisc, Inc.; p. 315 (top) Peter Matthews/Black Star; p. 315 (bottom) Joel Gordon Photography; p. 323 Robert Essel/Corbis/Stock Market; p. 327 Richard Hutchings/PhotoEdit; p. 330 Joel Gordon Photography; p. 331 Jeff Greenberg/Lonely Planet Images/Photo 20-20.

Chapter 11 Page 340 Pablo Picasso, (1881–1973), Spanish, "Girl Before a Mirror"/Superstock/© 2003 Estate of Pablo Picasso/Artists Rights Society (ARS), New York. SuperStock, Inc.; p. 341 (Video 11.1) CDA; p. 342 Sisse Brimberg/Woodfin Camp & Associates; p. 343 Tony Freeman/PhotoEdit; p. 343 (Video 11.2) CDA; p. 344 Michael Weisbrot/The Image Works; p. 345 Jill Greenberg/Jill Greenberg Studio, Inc.; p. 345 (Video 11.3) CDA; p. 353 Tony Freeman/PhotoEdit; p. 355 Joel Gordon Photography; p. 358 Alex Wong/Getty Images, Inc.; p. 359 Russell D. Curtis/Photo Researchers, Inc.; p. 363 Pictor, New York; p. 366 Ogust/The Image Works.

Chapter 12 Page 370 Bharati Chaudhuri. "Invisible Tension", (1992), (1951)/SuperStock, Inc.; p. 372 (left) George Holton/Photo Researchers, Inc.; p. 372 (right) Lee Snyder/Photo Researchers, Inc.; p. 372 (Video 12.1) Pearson-Patients as Educators; p. 374 Corbis/Sygma; p. 377 Randy Matusow; p. 378 Fredrik D. Bodin; p. 386 Roy Morsch/Corbis/Stock Market; p. 394 Ira Wyman/Corbis/Sygma; p. 395 AP/Wide World Photos.

Chapter 13 Page 400 Gino Severini, (1883–1966), Italian, "The Head", Pushkin Museum of Fine Arts, Moscow, Russia/SuperStock/© 2003 Artist Rights Society (ARS), New York/ADAGP, Paris/SuperStock Inc.; p. 402 (left) The Granger Collection; p. 402 (right) CORBIS; p. 403 R. Flynt/The Image Works; p. 405 Pictor, New York; p. 409 (Video 13.1) Pearson-Patients as Educators; p. 412 Jonathan Nourok/PhotoEdit; p. 414 "Soulful Feelings", © Artwork by Danny Gayder, a participating artist of VSA arts, website *http://www.vsarts.org.;* p. 416 Bob Benyas Photography; p. 417 Grunnitus/Photo Researchers, Inc.; p. 418 Mitchell Funk/Getty Images Inc.; p. 419 CORBIS; p. 420 Monte S. Buchsbaum, M.D., Mount Sinai School of Medicine, New York, N.Y.; p. 423 (left) Nancy C. Andreasen, M.D./University of Iowa Hospitals & Clinics; p. 423 (right) Nancy C. Andreasen, M.D./University of Iowa Hospitals & Clinics; p. 424 Monte S. Buchsbaum, M.D./Mount Sinai School of Medicine, New York, N.Y.; p. 430 Amy Etra/PhotoEdit.

Chapter 14 Page 438 Rufino Tamayo (1899–1991), "Dos Caras," Schalkwijk/Art Resource; p. 440 Tony Freeman/PhotoEdit; p. 441 Michael Newman/PhotoEdit; p. 444 Erika Stone; p. 445 (Video 14.1) CDA; p. 446 George Goodwin; p. 449 Richard Hutchings/Photo Researchers, Inc.; p. 450 Yoav Levy/Phototake NYC; p. 451 Tony Freeman/PhotoEdit; p. 452 Paul Conklin/PhotoEdit; p. 453 Stephen Vaughan/Globe Photos, Inc.; p. 455 Will & Deni McIntyre/Photo Researchers, Inc.; p. 460 (Video 14.2) CDA; p. 461 Jose Azel/Aurora & Quanta Productions; p. 463 Pascal Quittemelle/Stock Boston; p. 466 Jeff Greenberg/PhotoEdit; p. 467 David Young-Wolff/PhotoEdit; p. 468 Jean Claude LeJeune/Stock Boston.

Chapter 15 Page 478 Paul Klee, (1879–1940), Swiss, "Allegorische Figurine", Christie's Images/SuperStock/© 2003 Artist Rights Society (ARS), New York/VG Bild-Kunst, Bonn; p. 480 Frank Siteman/Stock Boston; p. 483 Jeff Persons/Stock Boston; p. 485 Steve Weber/Getty Images Inc.; p. 486 (top) Spencer Grant/Stock Boston; p. 486 (bottom) Robert Ginn/PhotoEdit; p. 489 (top) Mark J. Terrill/AP/Wide World Photos; p. 489 (left) Robert P. Friedland, MD, Case Western University; p. 489 (right) Robert P. Friedland, MD, Case Western University; p. 490 AP/Wide World Photos; p. 490 (Video 15.1) Films for Humanities; p. 492 (top) Cecil Fox/Science Source/Photo Researchers, Inc.; p. 492 (bottom) Karen Kasmauski/Woodfin Camp & Associates; p. 494 AP/Wide World Photos; p. 496 Robin Carson/Library of Congress; p. 497 AP/Wide World Photos.

Chapter 16 Page 500 Rufino Tamayo, (1899–1991), "Title Unknown" (Abstract)/ Superstock/MUSEUM RUFINO TAMAYO; p. 503 Wally McNamee/Corbis/Sygma; p. 504 Jerry Wechter/eStock Photography LLC; p. 507 Peter Byron/Photo Researchers, Inc; p. 508 Albert Bandura; p. 509 Harvey Finkle; p. 512 Freda Leinwand Photography; p. 514 Esbin-Anderson/ The Image Works; p. 516 Les Stone/Corbis/Sygma; p. 518 Stephen Ferry/ Liaison Agency Inc.; p. 523 Steve McCurry/Magnum Photos, Inc.; p. 524 R. Sidney/The Image Works; p. 526 Anthony Bolante/Reuters/Getty Images, Inc.; p. 526 (Video 16.1) CDA; p. 528 LaPorte County Child Abuse Prevention Council; p. 529 Doug Plummer; p. 531 John Coletti/Index Stock Imagery, Inc.

Chapter 17 Page 536 P. Filonov, (1883–1944), "Man in the World", Russian State Museum, St. Petersburg, Russia/SuperStock; p. 537 AP/Wide World Photos; p. 539 AP/Wide World Photos; p. 541 Custom Medical Stock Photo, Inc.; p. 542 (left) AP/Wide World Photos; p. 542 (right) AP/Wide World Photos; p. 544 Picture Desk, Inc./Kobal Collection; p. 545 AP/Wide World Photos; p. 546 AP/Wide World Photos; p. 547 AP/Wide World Photos; p. 548 AP/Wide World Photos.

AUTHOR INDEX

SUBJECT INDEX